BROADCASTING & CABLE

YEARBOOK 2008

Published by
ProQuest LLC
630 Central Avenue
New Providence, NJ 07974 USA

Matt Dunie, President

For sales inquiries Telephone: 908-286-1090, Toll-free: 1-888-BOWKER2 (1-888-269-5372); Fax: 908-219-0182
E-mail address: info@bowker.com; URL: http://www.bowker.com

International Standard Book Number
ISBN 13: 978-1-60030-110-0

International Standard Serial Number
0000-1511

Library of Congress Control Number
71-649524

Printed and Bound in the United States of America

BROADCASTING & CABLE

YEARBOOK 2008

BROADCASTING & CABLE YEARBOOK 2008
was prepared by ProQuest's Serials Editorial Department in collaboration
with R.R. Bowker's Data Services and Information Technology Departments

Product Development
Yvette Diven, Director, Product Management, Serials

Editorial
Laurie Kaplan, Director, Serials
Nancy Bucenec, Managing Editor
Valerie Mahon, Managing Editor
Joseph A. Esser, Associate Editor
Jennifer Williams and Carolyn Hamilton, Assistant Editors

Data Services
Doreen Gravesande, Senior Director, Production
Ralph Coviello, Manager, Manufacturing Services
Myriam Nunez, Project Manager, Content Integrity
Lorena Soriano, Project Manager, Production
Gunther Stegman, Project Manager, Production

Information Technology Group
Dina Dvinyanova and Steve Gorski, Programmer Analysts

Computer Operations Group
John Nesselt, UNIX Administrator

Table of Contents

Index to Sections

D

E

F

N

O

P

R

S

T

U

V

W

Index to Advertisers

Glossary of Terms Used in *Broadcasting & Cable Yearbook*

AM—Amplitude modulation. Also referring to audio service broadcast over 535 khz-1705 khz.

Analog—A continuous electrical signal that carries information in the form of variable physical values, such as amplitude or frequency modulation.

Basic cable service—Package of programming on cable systems eligible for regulation by local franchising authorities under 1992 Cable Act, including all local broadcast signals and PEG (public, educational and government) access channels.

Cable television—System that transmits original programming, and programming of broadcast television stations, to consumers over wired network.

CC—Closed captioning. Method of transmitting textual information over television channel's vertical blanking interval; transmissions are deciphered with decoders; decoded transmissions appear as text superimposed over television image.

Clear channel—AM radio station allowed to dominate its frequency with up to 50 kw of power; their signals are generally protected for distance of up to 750 miles at night.

Closed circuit—The method of transmission of programs or other material that limits its target audience to a specific group rather than the general public.

Coaxial cable—Cable with several common axis lines under protective sheath used for television signal transmissions.

Common carrier—Telecommunication company that provides communications transmission services to the public.

DAB—Digital audio broadcasting. Modulations for sending digital rather than analog audio signals by either terrestrial or satellite transmitter with audio response up to compact disc quality (20 khz).

DBS—Direct broadcast satellite. High powered satellite authorized to broadcast direct to homes.

Digital—A discontinuous electrical signal that carries information in binary fashion. Data is represented by a specific sequence of off-on electrical pulses.

Directional antenna—An antenna that directs most of its signal strength in a specific direction rather than at equal strength in all directions. Used chiefly in AM radio operation.

Downlink—Earth station used to receive signals from satellites.

Earth station—Equipment used for transmitting or receiving satellite communications.

EDTV—Enhanced-definition television. Proposed intermediate systems for evolution to full HDTV, usually including slightly improved resolution and sound, with a wider (16:9) aspect ratio.

Effective competition—Market status under which cable TV systems are exempt from regulation of basic tier rates by local franchising authorities, as defined in 1992 Cable Act. To claim effective competition, a cable system must compete with at least one other multichannel provider that is available to at least 50%

of an area's households and is subscribed to by more than 15%

of the households.

Encryption—System for scrambling signals to prevent unauthorized reception.

ENG—Electronic news gathering.

ETV—Educational television.

Fiber-optic cable—Wires made of glass fiber used to transmit video, audio, voice or data providing vastly wider bandwidth than standard coaxial cable.

Field—Half of the video information in the frame of a video picture. The NTSC system displays 59.94 fields per second.

FM—Frequency modulation. Also referring to audio service broadcast over 88 mhz-108 mhz.

Footprint—Area on earth within which a satellite's signal can be received.

Frame—A full video picture. The NTSC system displays 29.97 525-line frames per second.

Frequency—The number of cycles a signal is transmitted per second, measured in hertz.

Geostationary orbit—Orbit 22,300 miles above earth's equator where satellites circle earth at same rate earth rotates.

ghz—Gigahertz. One billion hertz (cycles) per second.

HDTV—High-definition television.

Headend—Facility in cable system from which all signals originate. (Local and distant television stations, and satellite programming, are picked up and amplified for retransmission through system.)

Hertz—A measurement of frequency. One cycle per second equals one hertz (hz).

Independent television—Television stations that are not affiliated with networks and that do not use the networks as a primary source of their programming.

Information services—Broad term used to describe full range of audio, video and data transmission services that can be transmitted over the air or by cable.

Interactive—Allowing two-way data flow.

Interlaced scanning—Television transmission technique in which each frame is divided into two fields. NTSC system interleaves odd-numbered lines with even-numbered lines at a transmission rate of 59.94 fields per second.

ITFS—Instructional Television Fixed Service.

khz—Kilohertz. One thousand hertz (cycles) per second.

LED—Light emitting diode. Type of semiconductor that lights up when activated by voltage.

LO—Local origination channel.

MDS—Multipoint distribution service.

mhz—Megahertz. One million hertz (cycles) per second.

Microwave—Frequencies above 1,000 mhz.

MSO—Multiple cable systems operator.

Must carry—Legal requirement that cable operators carry local broadcast signals. Cable systems with 12 or fewer channels must carry at least three broadcast signals; systems with 12 or more channels must carry up to one-third of their capacity; systems with 300 or fewer subscribers are exempt. The 1992 Cable Act requires broadcast station to waive must-carry rights if it chooses to negotiate retransmission compensation (see "Retransmission consent").

NTSC-National Television System Committee. Committee that recommended current American standard color television.

PCM—Pulse code modulation. Conversion of voice signals into digital code.

PPV—Pay-per-view.

Progressive scanning—TV system where video frames are transmitted sequentially, unlike interlaced scanning in which frames are divided into two fields.

PSA—Public service announcement.

PTV—Public television.

Public radio—Radio stations and networks that are operated on a noncommercial basis.

Public television-Television stations and networks that operate as noncommercial ventures.

RCC—Radio common carrier. Common carriers whose major businesses include radio paging and mobile telephone services.

Retransmission consent—Local TV broadcasters' right to negotiate a carriage fee with local cable operators, as provided in 1992 Cable Act.

SCA—Subsidiary communications authorizations. Authorizations granted to FM broadcasters for using subcarriers on their channels for other communications services.

Shortwave—Transmissions on frequencies of 6-25 mhz.

SHF—Super high frequency.

Signal-to-noise ratio—The ratio between the strength of an electronically produced signal to interfering noises in the same bandwidth.

SMATV—Satellite master antenna television.

STV—Subscription television.

Superstation—Local television station whose signal is retransmitted via satellite to cable systems beyond reach of over-the-air signal.

Teletext—A one-way electronic publishing service that can be transmitted over the vertical blanking interval of a standard television signal or the full channel of a television station or cable television system. The major use today is for closed-captioning.

Translator—Broadcast station that rebroadcasts signals of other stations without originating its own programming.

Transponder—Satellite transmitter/receiver that picks up signals transmitted from earth, translates them into new frequencies and amplifies them before retransmitting them back to ground.

UHF—Ultra high frequency band (300 mhz-3,000 mhz), which includes TV channels 14-83.

Uplink—Earth station used for transmitting to satellite.

VHF—Very high frequencies (30 mhz-300 mhz), which include TV channels 2-13 and FM radio.

Videotext—Two-way interactive service that uses either two-way cable or telephone lines to connect a central computer to a television screen.

List of Abbreviations Used in *Broadcasting & Cable Yearbook*

***** . . . noncommercial
a . . . annual
A&E . . . Arts & Entertainment
actg . . . acting
admin . . . administrative
adv . . . advertising
affil . . . affiliate
affrs . . . affairs
AFRTS . . . Armed Forces Radio and TV Service
alt . . . alternate
ant . . . antenna
AOR . . . album-oriented rock
AP . . . Associated Press
assn . . . association
assoc . . . associate
asst . . . assistant
atty . . . attorney
aur . . . aural
aux . . . auxiliary
bcst . . . broadcast
bcstg . . . broadcasting
bcstr . . . broadcaster
bd . . . board
BET . . . Black Entertainment Television
bi-m . . . every two months
bk rev . . . book reviews
bldg . . . building
bor . . . borough
btfl . . . beautiful
C-SPAN . . . Cable Satellite Public Affairs Network
CATV . . . community antenna television
CBC . . . Canadian Broadcasting Corp.
CEO . . . chief executive officer
ch . . . channel
CH . . . critical hours
chg . . . charge
CHR . . . contemporary hit radio
chmn . . . chairman
circ . . . circulation
coml . . . commercial
contemp . . . contemporary
COO . . . chief operating officer
coord . . . coordinator
CP . . . construction permit
CRTC . . . Canadian Radio-television and Telecommunications Commission
C&W . . . country & western
D . . . day
d . . . daily
DA . . . directional antenna
dance rev . . . dance reviews
DBS . . . direct broadcast satellite
dev . . . development
dir . . . director
div . . . diverse
DMA . . . Designated Market Area
dups . . . duplicates
Eds . . . editors
Ed Bd . . . Editorial Board
educ . . . educational
engr . . . engineer
engrg . . . engineering
EPG . . . Electronic Program Guide
ERP . . . effective radiated power
ESPN . . . Entertainment & Sports Programming Network
ETV . . . educational television
exec . . . executive
FCC . . . Federal Communications Commission
film rev . . . film reviews
fortn . . . fortnightly
Fr . . . French
g . . . ground
gen . . . general
Ger . . . German
govt . . . government
HAAT . . . height above average terrain
HBO . . . Home Box Office
horiz . . . horizontal polarization
hqtrs . . . headquarters
ind . . . independent
info . . . information
instal . . . installation
ISBN . . . International Standard Book Number
ISSN . . . International Standard Serial Number
illus . . . illustrations
irreg . . . irregular
It . . . Italian
khz . . . kilohertz
kw . . . kilowatts
loc . . . local
LPTV . . . low power television
LS . . . local sunset
lstng . . . listening
lw . . . long wave
m . . . meters
MDS . . . Multipoint Distribution Service
mdse . . . merchandising
mfg . . . manufacturing
mgng . . . managing
mgr . . . manager
mgmt . . . management
mhz . . . megahertz
mi . . . miles
mktg . . . marketing
MMDS . . . Multichannel Multipoint Distribution Service
mo . . . month
mod . . . modification
MOR . . . middle of the road
MSO . . . multiple system operator
mthy . . . monthly
MTV . . . Music Television
mus . . . music
music rev . . . music reviews
mw . . . medium wave
N . . . night
na . . . not available
NAB . . . National Association of Broadcasters
natl . . . national
net . . . network
NPR . . . National Public Radio
nwspr . . . newspaper
off . . . officer
opns . . . operations
per . . . personnel
play rev . . . play reviews (theatre reviews)
Pol . . . Polish
pop . . . population
PR . . . public relations
pres . . . president
PRI . . . Public Radio International
progmg . . . programming
progsv . . . progressive
prom . . . promotion
PSA . . . presunrise authority, public service announcement
ptnr . . . partner
pub affrs . . . public affairs
publ . . . publicity
q . . . quarterly
quad . . . quadraphonic
record rev . . . record reviews
rel . . . relations
relg . . . religion
rep . . . representative
RFE . . . Radio Free Europe
rgn . . . region
rgnl . . . regional
RL . . . Radio Liberty
rsch . . . research
s-a . . . twice annually
s-m . . . twice monthly
s-w . . . twice weekly
sec . . . secretary
sep . . . separate
sh . . . shares
SH . . . specified hours
sls . . . sales
SMATV . . . satellite master antenna television
Sp . . . Spanish
sr . . . senior
ST . . . shares time
stn . . . station
sub . . . subscriber

supt .... superintendent
supvr .... supervisor
svcs .... services
sw .... short wave
t .... terrain
tech .... technical
tele rev .... television reviews
3/m .... three times a month
3/y .... three times a year
TNN .... The Nashville Network
traf .... traffic
trans .... translators
treas .... treasurer
twp .... township
TWX .... Teletypewriter Exchange
U .... unlimited
UHF .... ultra high frequency
UPI .... United Press International
UPN .... United Paramount Network
var .... variety
vert .... vertical polarization
VHF .... very high frequency
video rev .... video reviews
vis .... visual
VOA .... Voice of America
vp .... vice president
w .... watts
wkly .... weekly

Section A
Industry Overview

Television Markets Ranked by Number of TV Homes

Listed below are the Nielsen Media Research Designated Market Areas (DMAs) ranked by the number of television households. The estimates are from September 2006.

Rank	Designated Market Area	TV Households
1	New York	7,366,950
2	Los Angeles	5,611,110
3	Chicago	3,455,020
4	Philadelphia	2,941,450
5	San Francisco-Oakland-San Jose	2,383,570
6	Dallas-Fort Worth	2,378,660
7	Boston (Manchester)	2,372,030
8	Washington, DC (Hagerstown)	2,272,120
9	Atlanta	2,205,510
10	Houston	1,982,120
11	Detroit	1,938,320
12	Tampa-St. Petersburg, Sarasota	1,755,750
13	Phoenix	1,725,000
14	Seattle-Tacoma	1,724,450
15	Minneapolis-St. Paul	1,678,430
16	Miami-Fort Lauderdale	1,538,620
17	Cleveland	1,537,500
18	Denver	1,431,910
19	Orlando-Daytona Beach-Melbourne	1,395,830
20	Sacramento-Stockton-Modesto	1,368,680
21	St. Louis	1,228,980
22	Pittsburgh	1,163,150
23	Portland, OR	1,117,990
24	Baltimore	1,097,290
25	Indianapolis	1,060,550
26	Charlotte	1,045,240
27	San Diego	1,030,020
28	Hartford & New Haven	1,014,630
29	Raleigh-Durham (Fayetteville)	1,006,330
30	Nashville	944,100
31	Kansas City	913,280
32	Columbus, OH	898,030
33	Cincinnati	886,910
34	Milwaukee	882,990
35	Salt Lake City	839,170
36	Greenville-Spartanburg-Asheville-Anderson	826,290
37	San Antonio	774,470
38	West Palm Beach-Fort Pierce	772,140
39	Grand Rapids-Kalamazoo-Battle Creek	734,670
40	Birmingham (Anniston, Tuscaloosa)	723,210
41	Harrisburg-Lancaster-Lebanon-York	713,960
42	Norfolk-Portsmouth-Newport News	712,790
43	Las Vegas	671,630
44	Memphis	664,290
45	Albuquerque-Santa Fe	662,380
46	Oklahoma City	662,380
47	Greensboro-High Point-Winston Salem	660,570
48	Louisville	648,190
49	Buffalo	639,990
50	Jacksonville, Brunswick	639,110
51	Providence-New Bedford	633,950
52	Austin	602,340
53	Wilkes Barre-Scranton	590,170

Rank	Designated Market Area	TV Households
54	New Orleans	566,960
55	Fresno-Visalia	557,380
56	Albany-Schenectady-Troy	554,970
57	Little Rock-Pine Bluff	539,900
58	Dayton	531,120
59	Mobile-Pensacola (Fort Walton Beach)	524,200
60	Knoxville	523,010
61	Richmond-Petersburg	517,800
62	Tulsa	513,090
63	Lexington	483,520
64	Fort Myers-Naples	479,130
65	Charleston-Huntington	477,040
66	Flint-Saginaw-Bay City	474,430
67	Wichita-Hutchinson Plus	445,860
68	Roanoke-Lynchburg	445,840
69	Green Bay-Appleton	434,760
70	Tucson (Sierra Vista)	433,310
71	Toledo	425,820
72	Honolulu	419,160
73	Des Moines-Ames	417,900
74	Portland-Auburn	409,180
75	Omaha	403,560
76	Springfield-Holyoke	402,310
77	Spokane	395,490
78	Rochester, NY	392,630
79	Syracuse	386,940
80	Paducah-Cape Girardeau-Harrisburg-Mt. Vernon	384,510
81	Shreveport	381,200
82	Champaign & Springfield-Decatur	378,150
83	Columbia, SC	377,940
84	Huntsville-Decatur, Florence	375,270
85	Madison	369,220
86	Chattanooga	347,380
87	Jackson, MS	343,550
88	South Bend-Elkhart	334,370
89	Cedar Rapids-Waterloo & Dubuque	333,270
90	Burlington-Plattsburgh	327,480
91	Harlingen-Weslaco-Brownsville-McAllen	327,070
92	Tri-Cities, TN-VA	326,560
93	Baton Rouge	322,540
94	Colorado Springs-Pueblo	316,630
95	Waco-Temple-Bryan	311,690
96	Davenport-Rock Island-Moline	308,360
97	Savannah	298,130
98	Johnstown-Altoona	294,160
99	El Paso	293,700
100	Charleston, SC	290,110
101	Evansville	289,730
102	Fort Smith-Fayetteville-Springdale-Rogers	280,510
103	Youngstown	276,550
104	Lincoln & Hastings-Kearney	275,970
105	Myrtle Beach-Florence	272,340
106	Fort Wayne	271,550
107	Greenville-New Bern-Washington	270,420
108	Tallahassee-Thomasville	266,210
109	Springfield, MO	264,480
110	Reno	261,250
111	Tyler-Longview (Lufkin & Nacogdoches)	258,860

Rank	Designated Market Area	TV Households
112	Lansing	256,190
113	Traverse City-Cadillac	248,680
114	Augusta	247,450
115	Sioux Falls (Mitchell)	247,000
116	Peoria-Bloomington	243,280
117	Montgomery (Selma)	241,130
118	Boise	238,990
119	Fargo-Valley City	235,320
120	Eugene	231,710
121	Macon	230,180
122	Santa Barbara-Santa Maria-San Luis Obispo	227,700
123	Lafayette, LA	225,650
124	Monterey-Salinas	218,390
125	Yakima-Pasco-Richland-Kennewick	213,780
126	Bakersfield	210,960
127	La Crosse-Eau Claire	209,870
128	Columbus, GA	207,180
129	Corpus Christi	194,160
130	Chico-Redding	193,590
131	Amarillo	190,590
132	Columbus-Tupelo-West Point	187,150
133	Rockford	184,560
134	Wausau-Rhinelander	180,640
135	Monroe-El Dorado	178,200
136	Wilmington	174,170
137	Duluth-Superior	171,780
138	Topeka	171,310
139	Columbia-Jefferson City	170,260
140	Beaumont-Port Arthur	167,090
141	Medford-Klamath Falls	164,780
142	Erie	157,860
143	Sioux City	156,480
144	Joplin-Pittsburg	154,640
145	Albany, GA	153,190
146	Wichita Falls & Lawton	152,380
147	Lubbock	151,610
148	Salisbury	150,790
149	Palm Springs	149,880
150	Bluefield-Beckley-Oak Hill	145,550
151	Terre Haute	144,880
152	Bangor	143,170
153	Rochester-Mason City-Austin	143,090
154	Anchorage	142,230
155	Wheeling-Steubenville	140,950
156	Panama City	140,790
157	Binghamton	138,220
158	Minot-Bismarck-Dickinson	135,550
159	Odessa-Midland	135,270
160	Biloxi-Gulfport	134,320
161	Sherman, TX-Ada, OK	124,330
162	Gainesville	119,590
163	Idaho Falls-Pocatello	116,560
164	Abilene-Sweetwater	114,210
165	Hattiesburg-Laurel	111,580
166	Clarksburg-Weston	109,020
167	Yuma-El Centro	107,360
168	Missoula	106,250
169	Utica	106,080

Rank	Designated Market Area	TV Households
170	Billings	103,710
171	Quincy-Hannibal-Keokuk	103,690
172	Dothan	99,410
173	Elmira	96,690
174	Jackson, TN	95,070
175	Lake Charles	94,840
176	Watertown	94,050
177	Rapid City	93,870
178	Marquette	89,670
179	Alexandria, LA	89,600
180	Jonesboro	89,500
181	Harrisonburg	87,630
182	Charlottesville	83,850
183	Bowling Green	76,910
184	Greenwood-Greenville	76,830
185	Meridian	74,440
186	Grand Junction-Montrose	69,560
187	Laredo	65,790
188	Lafayette, IN	64,680
189	Parkersburg	63,850
190	Great Falls	63,510
191	Twin Falls	61,160
192	Butte-Bozeman	60,560
193	Eureka	59,360
194	Bend, OR	57,790
195	Cheyenne-Scottsbluff	54,030
196	Lima	53,180
197	San Angelo	52,930
198	Casper-Riverton	52,400
199	Ottumwa-Kirksville	51,470
200	Mankato	51,090
201	St. Joseph	45,840
202	Fairbanks	33,240
203	Zanesville	33,090
204	Presque Isle	31,170
205	Victoria	30,450
206	Helena	25,970
207	Juneau	23,910
208	Alpena	17,600
209	North Platte	15,480
210	Glendive	3,980

Top 25 Cable/Satellite Operators

Ranked by Basic Subscribers*

Rank	Company	Subscribers
1	Comcast Cable Comm.	24,236.0
2	DirecTV	16,188.0
3	Time Warner Cable	13,448.0
4	EchoStar	13,415.0
5	Cox Communications	5,451.0
6	Charter Communications	5,415.4
7	Cablevision Systems	3,139.0
8	Bright House Networks (e)	2,321.0
9	Suddenlink Communications (&, e)	1,422.2
10	Mediacom LLC (&)	1,362.0
11	Insight Communications	1,344.0
12	CableOne (&)	703.2
13	WideOpenWest (e)	363.4
14	RCN Corp. (&)	354.0
15	Bresnan (e)	296.6
16	Service Electric (e)	290.4
17	Atlantic Broadband	288.7
18	Armstrong Group of Co.	232.6
19	Midcontinent Communications (&)	200.9
20	Pencor Services (e)	184.4
21	Knology Holdings	181.0
22	Millennium Digital Media (e)	158.5
23	Buckeye CableSystem	146.8
24	Northland Communications	144.6
25	General Communication	140.2

* As of March 2007.
Unless otherwise noted, counts include owned and managed subscribers.
(&) Counts include recent sale or acquisition.
(e) Estimate.

Major Cable Systems/Clusters

Ranked by Basic Subscribers

Rank	Company	Subscribers
1	Cablevision Systems Corp., Greater New York Area, NY	3,127,000
2	Time Warner Cable, Ohio	2,400,000
3	Time Warner Cable, Texas	2,100,000
4	Time Warner Cable, Carolinas	2,100,000
5	Time Warner Cable, Los Angeles	2,000,000
6	Comcast, Philadelphia, PA	1,800,000
7	Time Warner Cable, New York (excluding NY City)	1,800,000
8	Comcast, Boston, MA	1,600,000
9	Comcast, Chicago, IL	1,600,000
10	Time Warner Cable, New York City	1,600,000
11	Comcast, San Francisco Bay Area, CA	1,500,000
12	Bright House Networks., Tampa Bay, FL	1,052,030
13	Comcast, Seattle, WA	1,000,000
14	Cox Communications, AZ	944,460
15	Comcast, Washington, DC	900,000
16	Bright House Networks., Central FL	800,835
17	Comcast, Atlanta, GA	800,000
18	Comcast, Detroit, MI	800,000
19	Comcast, Central PA	721,000
20	Mediacom, South Central	703,092
21	Comcast, Denver, CO	700,000
22	Comcast, Houston, TX*	700,000
23	Comcast, Miami, FL	700,000
24	Comcast, New York	700,000
25	Comcast, Pittsburgh, PA	700,000
26	Mediacom, North Central	676,908
27	Charter Comm., Central States	633,600
28	Comcast, West Palm Beach Region, FL	612,000
29	Comcast, Baltimore, MD	600,000
30	Time Warner Cable, Wisconsin	570,198
31	Cox Communications, San Diego, CA	543,606
32	Charter Comm., South Carolina	532,400
33	Comcast, Portland, OR	500,000
34	Comcast, Sacramento, CA	500,000
35	Comcast, St. Paul/Minneapolis, MN	500,000
36	Cox Communications, New England	460,752
37	Cox Communications, Oklahoma	450,900
38	Charter Comm., Tennessee	431,300
39	Cox Communications, Las Vegas	425,957
40	Cox Communications, Hampton Roads, VA	419,325
41	Time Warner Cable, Hawaii	401,146
42	Time Warner Cable, California (excluding L.A.)	400,000
43	Charter Comm., Alabama	343,500
44	Charter Comm., Minnesota/Nebraska	341,000
45	Comcast, Nashville, TN	332,000
46	Comcast, Albuquerque/Santa Fe, NM; Arizona	311,000
47	Cox Communications, Kansas	310,199
48	Comcast, Michigan Market (Grand Rapids & Lansing)	310,000
49	Time Warner Cable, Kansas City, MO	300,017
50	Comcast, Indianapolis, IN	300,000

Rank	Company	Subscribers
51	Charter Comm., North Wisconsin	291,400
52	Charter Comm., Georgia	289,800
53	Time Warner Cable, New England	287,000
54	Charter Comm., L.A. Metro	286,500
55	Insight Comm., Louisville, KY	285,654
56	Comcast, Jacksonville, FL	283,000
57	Cox Communications, Orange County, CA	277,000
58	Charter Comm., Northwest	274,700
59	Cox Communications, Northern Virginia	264,730
60	Comcast, Richmond, VA	258,000
61	Charter Comm., New England	256,100
62	Comcast, Salt Lake City, UT	253,000
63	Comcast, Chico/Yuba/Fresno,Stockton/Modesto, CA	250,000
64	Charter Comm., Southern Wisconsin	232,700
65	Comcast, Memphis, TN	226,641
66	Cox Communications, Omaha, NE	205,115
67	Charter Comm., Texas	201,800
68	Charter Comm., Northern Michigan	200,800
69	Comcast, Tampa/Sarasota, FL	200,000
70	Charter Comm., Eastern Michigan	198,800
71	Charter Comm., West Michigan	193,100
72	Charter Comm., Central California	184,100
73	Cox Communications, New Orleans, LA	181,876
74	Comcast, Knoxville, TN	177,632
75	Cox Communications, Gulf Coast/Florida	169,852
76	Charter Comm., Louisiana	163,100
77	Comcast, Salisbury, MD	150,000
78	Comcast, Chattanooga, TN	142,899
79	Charter Comm., Nevada	140,800
80	Atlantic Broadband, Western PA	138,774
81	Charter Comm., Inland Empire	137,800
82	Buckeye CableSystems, Toledo, OH	127,613
83	Insight Comm., Peoria, IL	126,530
84	Insight Comm., Northeast Indiana	119,749
85	Insight Comm., Springfield, IL	118,841
86	Comcast, Orlando, FL	100,000
	Total	**51,449,630**

All data as of December 31, 2006 except as noted.

* As of January 1, 2007.

U.S. Sales of Television Receivers 1983-2007

Data compiled by Consumer Electronics Association

	Analog Color TV		Digital TV		LCD TV		TV/VCR/DVD Combinations		Projection TV	
Year	**Units**	**Dollars**	**Units**	**Dollars**	**Units**	**Dollars**	**Units**	**Dollars**	**Units**	**Dollars**
1983	11,179	3,443	n/a	n/a	n/a	n/a	n/a	n/a	n/a	n/a
1984	13,092	3,875	n/a	n/a	n/a	n/a	n/a	n/a	n/a	n/a
1985	13,993	4,114	n/a	n/a	n/a	n/a	n/a	n/a	266	488
1986	15,399	4,481	n/a	n/a	n/a	n/a	n/a	n/a	304	530
1987	16,805	4,890	n/a	n/a	n/a	n/a	n/a	n/a	293	527
1988	17,768	4,691	n/a	n/a	n/a	n/a	n/a	n/a	302	529
1989	19,557	5,359	n/a	n/a	n/a	n/a	n/a	n/a	265	478
1990	18,453	5,148	n/a	n/a	n/a	n/a	424	178	351	626
1991	17,951	5,134	n/a	n/a	n/a	n/a	662	265	380	683
1992	21,056	6,591	n/a	n/a	n/a	n/a	936	375	404	714
1993	23,005	7,316	n/a	n/a	n/a	n/a	1,629	599	465	841
1994	24,715	7,225	n/a	n/a	n/a	n/a	2,017	710	636	1,117
1995	23,231	6,798	n/a	n/a	n/a	n/a	2,205	723	820	1,417
1996	22,384	6,492	n/a	n/a	n/a	n/a	2,199	697	887	1,426
1997	21,293	6,036	n/a	n/a	n/a	n/a	2,311	684	917	1,361
1998	22,204	6,122	14	43	n/a	n/a	3,147	832	1,070	1,577
1999	23,218	6,199	121	295	n/a	n/a	4,148	1,014	1,232	1,632
2000	24,175	6,503	648	1,426	832	107	4,964	968	1,216	1,481
2001	21,167	5,130	1,460	2,648	845	101	4,630	790	933	1,060
2002	22,469	5,782	4,145	6,383	935	246	4,870	733	681	733
2003	20,791	4,756	5,532	8,692	1,253	664	4,373	778	276	293
2004	19,934	3,526	8.002	12,300	1,842	1,574	3,643	867	97	85
2005	16,935	2,790	11,369	15,563	4,077	3,258	3,348	650	20	15
2006 (e)	8,753	1,070	23,924	23,661	10,500	8,480	2,022	348	5	3
2007 (p)	1,520	127	29,204	26,301	15,831	12,224	875	144	-	-

Dollar figure in millions and represent factory sales price multiplied by unit sales to dealers.

Unit figures include distributor sales and factory direct sales to dealers. All unit figures are in thousands (add 000).

n/a=not available

(e)=estimate

(p)=projection

Television Sets in Use

	In Home (000)	Avg. Sets Per HH
1970	81,040	1.39
1971	85,290	1.42
1972	89,770	1.45
1973	95,330	1.47
1974	100,020	1.51
1975	105,460	1.54
1976	108,890	1.56
1977	113,440	1.59
1978	118,630	1.63
1979	124,570	1.67
1980	128,190	1.68
1981	132,260	1.70
1982	142,460	1.75
1983	148,910	1.79
1984	149,180	1.78
1985	155,410	1.83
1986	157,500	1.83
1987	162,750	1.86
1988	168,260	1.90
1989	175,580	1.94
1990	193,320	2.10
1991	193,200	2.08
1992*	192,480	2.09
1993	200,565	2.15
1994	211,443	2.24
1995	217,067	2.28
1996	222,753	2.32
1997	228,740	2.36
1998	235,010	2.40
1999	240,320	2.42
2000	244,990	2.43
2001	248,160	2.43
2002	254,360	2.41
2003	260,230	2.44
2004	268,260	2.47
2005	287,000	2.62
2006	301,380	2.73
2007	310,840	2.79

1970-79, as of September of prior year; 1980 to date as of January of calendar year; excludes Alaska and Hawaii prior to 1989.

* Reflects adjustments to conform to the 1990 census.

Source: Television Bureau of Advertising, Nielsen Media Research.

53 Years of Station Transactions

Dollar volume of transactions (number of stations changing hands)

YEAR	RADIO ONLY*	GROUPS*	TV ONLY	TOTAL
1954	$10,224,047 (187)	$26,213,323 (18)	$23,906,760 (27)	$60,344,130
1955	27,333,104 (242)	22,351,602 (11)	23,394,660 (29)	$73,079,366
1956	32,563,378 (316)	65,212,055 (24)	17,830,395 (21)	$115,605,828
1957	48,207,470 (357)	47,490,884 (28)	28,489,206 (38)	$124,187,560
1958	49,868,123 (407)	60,872,618 (17)	16,796,285 (23)	$127,537,026
1959	65,544,653 (436)	42,724,727 (15)	15,227,201 (21)	$123,496,581
1960	51,763,285 (345)	24,648,400 (10)	22,930,225 (21)	$99,341,910
1961	55,532,516 (282)	42,103,708 (13)	31,167,943 (24)	$128,804,167
1962	59,912,520 (306)	18,822,745 (8)	23,007,638 (16)	$101,742,903
1963	43,457,584 (305)	25,045,726 (3)	36,799,768 (16)	$105,303,078
1964	52,296,480 (430)	67,185,762 (20)	86,274,494 (36)	$205,756,736
1965	55,933,300 (389)	49,756,993 (15)	29,433,473 (32)	$135,123,766
1966	76,633,762 (367)	28,510,500 (11)	30,574,054 (31)	$135,718,316
1967	59,670,053 (316)	32,086,297 (9)	80,316,223 (30)	$172,072,573
1968	71,310,709 (316)	47,556,634 (9)	33,588,069 (20)	$152,455,412
1969	108,866,538 (343)	35,037,000 (5)	87,794,032 (32)	$231,697,570
1970	86,292,899 (268)	1,038,465 (3)	87,454,078 (19)	$174,785,442
1971	125,501,514 (270)	750,000 (2)	267,296,410 (27)	$393,547,924
1972	114,424,673 (239)	0 (0)	156,905,864 (37)	$271,330,537
1973	160,933,557 (352)	2,812,444 (4)	66,635,144 (25)	$230,381,145
1974	168,998,012 (369)	19,800,000 (5)	118,983,462 (24)	$307,781,474
1975	131,065,860 (363)	0 (0)	128,420,101 (22)	$259,485,961
1976	180,663,820 (413)	1,800,000 (3)	108,459,657 (32)	$290,923,477
1977	161,236,169 (344)	0 (0)	128,635,435 (25)	$289,871,604
1978	331,557,239 (586)	30,450,000 (5)	289,721,159 (51)	$651,728,398
1979	335,597,000 (546)	463,500,000 (52)	317,581,000 (47)	$1,116,678,000
1980	339,634,000 (424)	27,000,000 (3)	534,150,000 (35)	$900,784,000
1981	447,838,060 (625)	78,400,000 (6)	227,950,000 (24)	$754,188,060
1982	470,722,833 (597)	0 (0)	527,675,411 (30)	$998,398,244
1983	621,077,876 (669)	332,000,000 (10)	1,902,701,830 (61)	$2,855,779,706
1984	977,024,266 (782)	234,500,000 (2)	1,252,023,787 (82)	$2,463,548,053
1985	1,414,816,073 (1,558)	962,450,000 (218)	3,290,995,000 (99)	$5,668,261,073
1986	1,490,131,426 (959)	1,993,021,955 (192)	2,709,516,490 (128)	$6,192,669,871
1987	1,236,355,748 (775)	4,610,965,000 (132)	1,661,832,724 (59)	$7,509,153,472
1988	1,841,630,156 (845)	1,326,250,000 (106)	1,779,958,042 (70)	$4,947,838,198
1989	1,148,524,765 (663)	533,599,078 (40)	1,541,055,033 (84)	$3,223,178,876
1990	868,636,700 (1,045)	411,037,150 (60)	696,952,350 (75)	$1,976,626,200
1991	534,694,500 (793)	206,995,500 (61)	273,365,000 (38)	$1,015,055,000
1992	603,192,980 (667)	318,176,050 (24)	124,004,000 (41)	$1,045,373,030
1993	815,450,000 (633)	756,722,000 (NA)	1,728,711,000 (101)	$3,300,883,000
1994	970,400,000 (494)	1,800,000,000 (154)	2,200,000,000 (89)	$4,970,400,000
1995	792,440,000 (524)	2,790,000,000 (213)	4,740,000,000 (112)	$8,322,440,000
1996	2,840,820,000 (671)	12,034,000,000 (345)	10,488,000,000 (99)	$25,362,820,000
1997	2,461,570,000 (630)	14,580,000,000 (329)	6,400,000,000 (108)	$23,441,570,000
1998	1,596,210,000 (589)	14,080,000,000 (271)	7,120,000,000 (90)	$22,796,210,000
1999	1,718,000,000 (382)	26,880,000,000 (196)	4,720,000,000 (86)	$33,318,000,000
2000**	24,900,000,000 (1,794)	0 (0)	8,800,000,000 (154)	$33,700,000,000
2001**	3,800,000,000 (1,000)	0 (0)	4,900,000,000 (108)	$8,700,000,000
2002**	5,594,141,000 (836)	0 (0)	2,529,039,000 (249)	$8,123,180,000
2003**	2,400,000,000 (950)	0 (0)	520,000,000 (97)	$2,920,000,000
2004**	1,897,422,000 (901)	0 (0)	871,923,000 (66)	$2,769,345,000
2005**	2,791,531,000 (895)	0 (0)	2,842,439,000 (86)	$5,633,970,000
2006**	22,871,247,000 (2,101)	0 (0)	18,127,686,000 (180)	$40,998,933,000
TOTAL	**$90,108,898,648**	**$85,110,886,616**	**$94,767,600,403**	**$269,987,385,667**

Note: Dollar volume figures represent total considerations reported for all transactions with exception of minority interest transfers in which control of stations did not change hands and stations sold as part of larger company transactions. Although all states have been approved by the FCC, they may not necessarily have reached final closing. Prior to 1978, combined AM-FM facilities were counted as one station in computing total number of stations traded. Now AM-FM combinations are counted as two stations.

*Starting in 1993, the Radio only column includes only stand alone AM and FM deals and the Groups column contains AM-FM combos and all other multiple station deals. In previous years the AM-FM combos were included under Radio only.

**Figures for 2000 to 2006 courtesy of BIA Financial Network.

Record of Television Station Growth Since Television Began

	TV Authorized	On Air
Jan. 1, 1946*	9	6
Jan. 1, 1947*	52	
Jan. 1, 1948*	73	17
Jan. 1, 1949	124	50
Jan. 1, 1950	111	97
Jan. 1, 1951	109	107
Jan. 1, 1952	108	108
Jan. 1, 1953*	273	129
Jan. 1, 1954	567	356
Jan. 1, 1955	576	439 [1]
Jan. 1, 1956	590	482 [2]
Jan. 1, 1957	631	511
Jan. 1, 1958	657	544 [3]
Jan. 1, 1959	666	562 [4]
Jan. 1, 1960	673	573 [5]
Jan. 1, 1961	634	583
Jan. 1, 1962	654	563
Jan. 1, 1963	662	579
Jan. 1, 1964	661	582
Jan. 1, 1965	676	586
Jan. 1, 1966	702	596
Jan. 1, 1967	769	623
Jan. 1, 1968	818	644
Jan. 1, 1969	834	672
Jan. 1, 1970	1,038	872
Jan. 1, 1971	1,025	892
Jan. 1, 1972	1,004	905
Jan. 1, 1973	1,001	922
Jan. 1, 1974	1,002	938
Jan. 1, 1975	1,010	952
Jan. 1, 1976	1,030	962
Jan. 1, 1977	1,029	984
Jan. 1, 1978	1,045	986
Jan. 1, 1979	1,059	992
Jan. 1, 1980	1,094	1,013
Jan. 1, 1981	1,143	1,019
Jan. 1, 1982	1,168	1,020
Jan. 1, 1983	1,276	1,090
Jan. 1, 1984	1,318	1,149
Jan. 1, 1985	1,505	1,194
Oct. 30, 1986	1,493	1,220
Oct. 31, 1987	1,558	1,285
Jan. 1, 1988	1,615	1,342
Jan. 1, 1989	1,683	1,395
Jan. 1, 1990	1,684	1,436
Jan. 1, 1991	1,690	1,469
Jan. 1, 1992	1,688	1,488
Jan. 1, 1993	1,688	1,505
Jan. 1, 1994		1,518
Jan. 1, 1995		1,520
Jan. 1, 1996		1,544
Jan. 1, 1997		1,554
Jan. 1, 1998		1,564
Jan. 1, 1999		1,589
Jan. 1, 2000		1,616
Jan. 1, 2001		1,663
Jan. 1, 2002		1,686
Jan. 1, 2003		1,719
Jan. 1, 2004		1,733
Jan. 1, 2005		1,748
Jan. 1, 2006		1,750
Jan. 1, 2007		1,756

*Comparable figures for all services not available at this date.

[1] Includes stations with Special Temporary Authorizations (STAs), which either had not started operations as of this date, had started but had gone dark, or had received authorizations but turned them back with or without operating.

[2] Includes 2 licenses that had suspended operation and 37 stations with STAs in same category as footnote 1.

[3] Includes 7 licenses that had suspended operation and 40 stations with STAs in same category as footnote 1.

[4] Includes 6 licenses that had suspended operation and 38 stations with STAs in same category as footnote 1.

[5] Includes 10 licenses that had suspended operation and 38 stations with STAs in same category as footnote 1.

Top 20 Cable Networks

Primetime Mon-Sun 8PM to 11PM

	Rating	Total Homes (000)
USA	1.7	1933
DSNY	1.6	1821
TNT	1.4	1596
FOXNC	1.1	1249
TBSC	1.1	1209
LIF	1.0	1153
AEN	1.0	1135
TOON(1)	1.0	1099
NAN(1)	0.9	1058
DISC	0.9	968
FX	0.9	964
ESPN	0.8	946
COURT	0.8	916
HALL	0.8	867
SPIKE	0.8	860
HGTV	0.8	855
HIST	0.8	836
SCIFI	0.7	821
AMC	0.7	793
CMDY	0.7	791

	Total day Rating	Total Homes (000)
NICK	1.5	1666
DSNY	1.0	1132
NAN	0.9	1058
TNT	0.9	998
USA	0.9	987
ADSM	0.9	955
TOON	0.8	890
LIF	0.7	770
TBSC	0.7	744
FOXNC	0.6	718
AEN	0.6	702
ESPN	0.5	578
DISC	0.5	561
COURT	0.5	545
HALL	0.5	544
FX	0.5	529
HIST	0.5	525
HGTV	0.4	499
MTV	0.4	499
SPIKE	0.4	495

2007 Year-To-Date 1/1/2007-05/13/2007.
Total Programming Day=M - Sun 6AM - 6AM or Individual Network's Total Programming Day and Sun 6AM - 3AM for Most Current Week
(1) Cable Network did not telecast during the entire daypart.
Coverage area ratings are within each cable network's universe.
Total U.S. ratings and household projections are based on 111.4 million TV homes.
This report includes only those cable networks who supply program names to the industry.

Source: Nielsen Media Research. Reprinted with permission.

Top 100 Television Programs

Full Season: 9/18/06-5/23/07

Rank	Program	Network	% U.S. Households
1	SUPER BOWL XLI (6:27P)	CBS	42.6
2	AFC CHAMPIONSHIP (6:44P)	CBS	26.4
3	ACADEMY AWARDS(S)	ABC	23.6
4	TOSTITOS BCS NATL CHAMP(S)	FOX	17.4
5	AMERICAN IDOL-WEDNESDAY	FOX	17.3
6	AMERICAN IDOL-TUESDAY	FOX	16.8
7	ROAD TO OSCARS 2007(S)	ABC	16.7
8	NBC NFL PLAYOFF GAME 2(S)	NBC	16.4
9	FOX NFC PLAYOFF-SAT(S)	FOX	16.3
10	AMERICAN IDOL THU SP-3/8(S)	FOX	16.0
11	AMERICAN IDOL THU SP-3/1(S)	FOX	15.5
12	CRIMINAL MINDS-SUPER BOWL(S)	CBS	14.9
13	SMARTER THAN 5TH TUE-2/27(S)	FOX	14.6
14	AMERICAN IDOL THU SP-2/22(S)	FOX	14.1
15	SMARTER THAN 5TH WED-2/28(S)	FOX	13.4
16	DANCING WITH THE STARS	ABC	13.3
17	GOLDEN GLOBE AWARDS(S)	NBC	13.2
18	DANCING W/THE STARS-MON	ABC	12.7
19	DANCING W/STARS RESULTS	ABC	12.7
20	CSI	CBS	12.2
21	CBS NCAA BSKBL CHAMPSHIPS(S)	CBS	12.2
22	GRAMMY AWARDS(S)	CBS	12.1
23	GREY'S ANATOMY-THU 9PM	ABC	12.1
24	DANCING W/STARS RESULT-TU	ABC	11.8
25	DANCING WITH-STARS-RECAP(S)	ABC	11.7
26	FOX WORLD SERIES GAME 2(S)	FOX	11.6
27	HOUSE	FOX	11.1
28	DESPERATE HOUSEWIVES	ABC	10.8
29	FOX MLB NLCS: GM 7(S)	FOX	10.8
30	CSI: MIAMI	CBS	10.7
31	NBC SUNDAY NIGHT FOOTBALL	NBC	10.5
32	GREY'S ANATOMY SP-11/23(S)	ABC	10.5
33	FOX WORLD SERIES GAME 4(S)	FOX	10.4
34	FOX WORLD SERIES GAME 5(S)	FOX	10.3
35	OPRAH WINFREY OSCAR SPEC(S)	ABC	10.3
36	FOX WORLD SERIES GAME 3(S)	FOX	10.2
37	CMA AWARDS(S)	ABC	9.9
38	CSI - THANKSGIVING(S)	CBS	9.9
39	HOUSE ELECTION-11/7(S)	FOX	9.7
40	NCIS - ELECT NT SP(S)	CBS	9.5
41	BARBARA WALTERS SP-2/25(S)	ABC	9.5
42	FOX WORLD SERIES GM2-PRE(S)	FOX	9.4
43	WITHOUT A TRACE	CBS	9.4
44	BOB BARKER: 50 YEARS(S)	CBS	9.3
45	GREY'S ANATOMY SP-9/21(S)	ABC	9.3
46	SURVIVOR:COOK ISL. FINALE(S)	CBS	9.3
47	ALLSTATE SUGAR BOWL(S)	FOX	9.3
48	HALLMARK HALL OF FAME(S)	CBS	9.3
49	DEAL OR NO DEAL-MON	NBC	9.2
50	SURVIVOR: COOK ISLANDS	CBS	9.1
51	TWO AND A HALF MEN	CBS	9.1
52	WITHOUT A TRACE-SPECIAL(S)	CBS	9.0
53	NCIS	CBS	9.0
54	DEAL OR NO DEAL-MON 12/25(S)	NBC	9.0

Rank	Program	Network	% U.S. Households
55	CSI 10PM-SPECIAL(S)	CBS	8.9
56	COLD CASE	CBS	8.9
57	CBS NCAA BSKBL CHAMP SA-2(S)	CBS	8.9
58	CSI: NY	CBS	8.9
59	PRICE IS RIGHT-MILLN 5/16(S)	CBS	8.9
60	20/20 SP EDITION-9/27(S)	ABC	8.9
61	CRIMINAL MINDS	CBS	8.8
62	24 PRVW SP-1/14 8P(S)	FOX	8.8
63	BARBARA WALTERS SP-11/16(S)	ABC	8.7
64	SHARK	CBS	8.7
65	60 MINUTES	CBS	8.7
66	BOSTON LEGAL SP-11/26(S)	ABC	8.6
67	HALLMARK HALL OF FAME(S)	CBS	8.6
68	CSI - THU 8P SPECIAL(S)	CBS	8.4
69	ALLSTATE BCS SELECT SHOW(S)	FOX	8.4
70	SURVIVOR: FIJI	CBS	8.4
71	COLD CASE 10PM SPECIAL(S)	CBS	8.4
72	TOSTITOS FIESTA BOWL(S)	FOX	8.4
73	CBS TUESDAY MOVIE SPECIAL(S)	CBS	8.3
74	TWO AND A HALF MEN 930PM(S)	CBS	8.3
75	60 MINUTES SPECIAL(S)	CBS	8.3
76	NFL REGULAR SEASON L	ESPN	8.2
77	LOST	ABC	8.2
78	SHOW ME THE MONEY-PREVIEW(S)	ABC	8.1
79	HEROES	NBC	8.0
80	FOX MLB NLCS GAME 6(S)	FOX	8.0
81	FOX WORLD SERIES GAME 1(S)	FOX	8.0
82	SURVIVOR:COOK ISL REUNION(S)	CBS	7.9
83	CSI: MIAMI THURSDAY-SP(S)	CBS	7.9
84	LAW AND ORDER:SVU	NBC	7.9
85	E.R. 11/23(S)	NBC	7.9
86	CSI: MIAMI 1/1(S)	CBS	7.8
87	RULES OF ENGAGEMENT	CBS	7.8
88	CSI - 8PM SPECIAL(S)	CBS	7.8
89	CSI: MIAMI - SPCL(S)	CBS	7.8
90	SURVIVOR: FIJI FINALE(S)	CBS	7.8
91	EXTREME MAKEOVER:HOME ED.	ABC	7.7
92	SHARK - THANKSGIVING(S)	CBS	7.7
93	24 SP-MON 2/12 8P(S)	FOX	7.7
94	SURVIVOR: FIJI - WED SPCL(S)	CBS	7.7
95	COLD CASE 8PM SPECIAL(S)	CBS	7.6
96	TWO AND A HALF MEN 1/1(S)	CBS	7.6
97	WONDERFUL WRLD-DISNEY-SPL(S)	ABC	7.6
98	PEOPLE'S CHOICE AWARDS(S)	CBS	7.6
99	CSI THU 8P-SPECIAL(S)	CBS	7.6
100	IDENTITY 12/18(S)	NBC	7.6

Top Ten Cable Programs

Rank	Program	Network	% U.S. Households
76	NFL REGULAR SEASON L	ESPN	8.2
350	MLB DIVISIONAL SERIES L	ESPN	4.4
368	2007 NBA ALLSTAR GAME(S)	TNT	4.2
370	SPONGEBOB THE MOVIE	NICK	4.2
436	TNT ORIGINAL(S)	TNT	3.6
447	S KIDS CHOICE 07	NICK	3.5
485	NBA ALLSTAR TIP OFF(S)	TNT	3.3
487	PLANET EARTH 1Q07(S)	DISC	3.3
499	PLANET EARTH 2Q07(S)	DISC	3.2
504	WWE ENTERTAINMENT	USA	3.2

Note: Shows identified by date or as specials (s) were one-time programs, programs that aired outside the regular time slot, or episodes that extended beyond regularly scheduled time periods.

Source: Television Bureau of Advertising, based on data from Nielsen Galaxy Explorer. A household rating is the percentage of the 111.4 million homes in the U.S. with TV sets.

Television Advertising Shares

	Network*	Spot	Local	Synd.*	Cable	Total
1970	$1,658	$1,234	$704	-	-	$3,596
1971	1,593	1,145	796	-	-	3,534
1972	1,804	1,318	969	-	-	4,091
1973	1,968	1,377	1,115	-	-	4,460
1974	2,145	1,497	1,212	-	-	4,854
1975	2,306	1,623	1,334	-	-	5,263
1976	2,857	2,154	1,710	-	-	6,721
1977	3,460	2,204	1,948	-	-	7,612
1978	3,975	2,607	2,373	-	-	8,955
1979	4,599	2,873	2,682	-	-	10,154
1980	5,130	3,269	2,967	50	72	11,488
1981	5,540	3,746	3,368	75	160	12,889
1982	6,144	4,364	3,765	150	290	14,713
1983	6,955	4,827	4,345	300	452	16,879
1984	8,318	5,488	5,084	420	733	20,043
1985	8,060	6,004	5,714	520	989	21,287
1986	8,342	6,570	6,514	600	1,173	23,199
1987	8,500	6,846	6,833	762	1,321	24,262
1988	9,172	7,147	7,270	901	1,641	26,131
1989	9,110	7,354	7,612	1,288	2,095	27,459
1990	9,863	7,788	7,856	1,109	2,631	29,247
1991	9,533	7,110	7,565	1,253	3,145	28,606
1992	10,249	7,551	8,079	1,370	3,830	31,079
1993	10,209	7,800	8,435	1,576	4,451	32,471
1994	10,942	8,993	9,464	1,734	5,209	36,342
1995	11,600	9,119	9,985	2,016	6,166	38,886
1996	13,081	9,803	10,944	2,218	7,778	43,824
1997	13,020	9,999	11,436	2,438	8,750	45,643
1998	13,736	10,659	12,169	2,609	10,340	49,513
1999	13,961	10,500	12,680	2,870	12,570	52,581
2000	15,888	12,264	13,542	3,108	15,455	60,257
2001	14,300	9,223	12,256	3,102	15,736	54,617
2002	15,000	10,920	13,114	3,034	16,297	58,365
2003	15,030	9,948	13,520	3,434	18,814	60,746
2004	16,713	11,370	14,507	3,674	21,527	67,791
2005	16,128	10,040	14,260	3,865	23,654	67,947
2006	16,676	11,626	14,887	3,691	25,025	71,905

all figures in millions
*Fox is included in syndication prior to 1990; it is included in network starting in 1990. UPN and WB added to syndication in 1995.

Source: Television Bureau of Advertising and Universal McCann.

Top 25 TV Advertisers

Jan. 1, 2006-Dec. 31, 2006

Rank	Parent Company	Jan - Dec $
1	PROCTER & GAMBLE CO	$2,692,177,807
2	GENERAL MOTORS CORP	$1,737,556,731
3	AT&T INC	$1,399,528,564
4	DAIMLERCHRYSLER AG	$1,323,972,581
5	FORD MOTOR CO	$1,306,644,059
6	TIME WARNER INC	$1,151,161,249
7	JOHNSON & JOHNSON	$1,006,862,298
8	TOYOTA MOTOR CORP	$953,558,219
9	VERIZON COMMUNICATIONS INC	$939,016,203
10	HONDA MOTOR CO LTD	$876,436,970
11	WALT DISNEY CO	$875,430,474
12	YUM! BRANDS INC	$857,247,771
13	NISSAN MOTOR CO LTD	$848,331,720
14	SONY CORP	$812,592,604
15	GLAXOSMITHKLINE PLC	$779,166,229
16	SPRINT NEXTEL CORP	$745,472,736
17	PEPSICO INC	$735,118,200
18	MCDONALDS CORP	$718,214,302
19	GENERAL MOTORS CORP-DA	$709,852,606
20	KRAFT FOODS INC	$692,526,719
21	GENERAL ELECTRIC CO	$649,164,735
22	BERKSHIRE HATHAWAY INC	$625,266,303
23	FORD MOTOR CO-DA	$617,447,044
24	TOYOTA MOTOR CORP-DA	$606,000,718
25	US GOVERNMENT	$570,941,416
	Total	**$24,229,688,258**

Source: Nielsen Monitor-Plus.

Top 25 TV Advertising Categories

Jan. 1, 2006-Dec. 31, 2006

Product Category	Jan - Dec $
Autos-Factory	$8,481,279,208
RESTAURANT-QUICK SVC	$3,736,183,335
Pharmaceutical	$3,716,756,426
MOTION PICTURE	$3,092,416,267
TELEPH SVCS-WIRELESS	$2,621,042,471
Autos-Dealer Association	$2,395,395,600
STORE-DEPT	$2,208,454,069
CREDIT CARD SVCS	$1,466,173,135
POLITICAL CAMPAIGN	$1,314,810,805
DIR RESP PROD	$1,242,108,224
RESTAURANT	$1,210,345,898
AUTO INSURANCE	$1,051,899,838
Autos-Dealership	$1,016,772,568
RECORDINGS-VIDEO	$981,029,116
BEER	$958,409,169
STORE-FURNITURE	$946,562,387
FINANCIAL-INVESTMENT SVCS	$784,474,741
INTERNET SVC PROVIDER	$686,445,227
CEREAL	$661,158,352
BANK SVCS	$603,911,853
STORE-HOME IMPROVEMENT	$602,549,834
TELECOMM SVCS & SYS	$563,753,593
PROFESSIONAL ORGN	$561,060,443
WEBSITE-AUTO INSURANCE	$540,016,695
STORE-ELECTRONICS	$510,875,420
Total	**$41,953,884,674**

Source: Nielsen Monitor-Plus.

A Brief History of Broadcasting and Cable

By Mark K. Miller, freelance writer and former managing editor, *Broadcasting & Cable* magazine

We could start our history of the electronic media in many places. We chose November 2, 1920. On that day, in Pittsburgh, Westinghouse Electric and Manufacturing Co.'s KDKA, generally acknowledged as the first licensed commercial radio station in the United States, broadcast the results of the Harding-Cox presidential elections. The broadcast demonstrated that radio was more than a novelty, that it could have real impact on our culture and our politics. The broadcast touched off a revolution that continues to this day. To give you a clear sense of the history, our history takes it decade by decade, assigning a theme to each.

1920s: Radio Begins Finding Its Way

Although we begin with that KDKA broadcast in 1920, there had been a lot of radio activity prior to that. Inventors and hobbyists had been filling the airwaves with signals, often on a very haphazard basis, for years. And government experiments during World War I contributed much to the growing volume of radio knowledge.

On August 20, 1920, WWJ(AM) Detroit, owned by the Detroit News, began what it claimed to be the first regular broadcasting schedule when it inaugurated daily broadcasts.

In 1922 the superheterodyne circuit as a broadcast receiver is demonstrated by its inventor, Edwin H. Armstrong, and will prove to become the industry standard. By May, there are 80 licensed radio stations in the United States; by year's end the number has grown to 569.

The rest of the decade is filled with examples of radio's surging growth and the unfettered vision of inventors around the world. In 1923 Dr. Vladimir K. Zworykin files for a U.S. patent for an all-electronic television system.

Stations begin to link themselves into "chains" or networks via telephone lines and numerous experiments test the feasibility of transmitting short-wave signals across the oceans, from airplanes and ships.

The 1924 Republican convention in Cleveland and the Democratic convention in New York are broadcast over networks, and in 1925 President Calvin Coolidge's inaugural ceremony is broadcast by 24 stations in a transcontinental network. The government was taking notice of the burgeoning industry; in 1926 the Federal Radio Commission is created in response to the chaos caused by the explosive growth of broadcasting.

But it was the launch of the National Broadcasting Co. on November 15, 1926, that marked the beginning of the network system of broadcasting that exists to this day. NBC was a joint venture of radio equipment manufacturers RCA (30 percent), General Electric (50 percent), and Westinghouse (20 percent). The initial broadcast was carried by 25 stations ranging from the East Coast to St. Louis and Kansas City, Missouri, and was estimated to have been heard by almost half of the nation's five million homes equipped with radios. In 1928 NBC would establish a permanent coast-to-coast radio network.

NBC's entry was followed in 1927 by that of the Columbia Broadcasting System, which debuted with a basic network of 16 stations. Two years later, William S. Paley, 27, purchases a controlling interest and is elected president.

There is also continuing activity on the television front. In 1927 Philo T. Farnsworth applies for a patent on his image dissector television camera tube and in 1929 Russian inventor Vladimir Zworykin demonstrates his kinescope, or cathode ray television receiver, before a meeting of the Institute of Radio Engineers.

1930s: The Rise of Radio Entertainment

It was in the 1930s that the "American Plan" of advertising-supported radio flourished, making it extremely unlikely that proponents (and there were many) of the "European Plan" of a government-operated medium would prevail. The attractiveness of free, high-quality entertainment during the Great Depression resulted in larger and larger audiences for the networks, exactly what companies needed to advertise their products.

As listening skyrocketed (about 12 million U.S. homes had a radio in 1930, by 1940 the figure was 28.5 million and car radios were becoming standard equipment), so too did the fortunes of the major networks: NBC's Red and Blue, CBS and Mutual (plus numerous regional nets). With radio personalities such as Jack Benny, Charles Correll, and Freeman Gosden (Amos 'n' Andy), Eddie Cantor, Burns and Allen, and Major Bowes attracting larger and larger audiences, advertisers wanted more precise listening figures. So in 1936 A.C. Nielsen Co. proposed its Audimeter, which would be attached to a sampling of radio sets and measure audience size.

Work continued on television. By 1937 there were 17 experimental TV stations operating and President Franklin D. Roosevelt was seen on TV when he opened the 1939 New York World's Fair.

1940s: The Rise of Radio News

As Europe was engulfed in war and the United States appeared headed toward the conflict, Americans turned to their radios to stay informed. During 1940, the networks' typical weekly schedules contained 56 quarter hours during the day as opposed to only 33 in evening programming in 1939 and none during daytime hours. CBS had Edward R. Murrow in London and his team of correspondents across Europe, while NBC had Fred Bate, Max Jordan, William Kierker, and many others. When the United States was attacked by Japan on December 7, 1941, network news reporting preempted regular programming. And the audience for President Roosevelt's broadcast to the nation on December 9, the day after war was declared, attracted the largest audience to that time, about 90 million. By the end of the week, all the networks and most stations were operating around the clock. For the first time, a war was heard by the people back home.

In addition to war reporting, coverage of domestic news was on the rise. With a tremendous amount of public interest in the 1944 presidential campaign, the networks canceled all commercial programs that would have interfered with their coverage of the Republican and Democratic conventions and used more than 300 reporters, technicians, and officials at each.

When the war ended and the era of the atomic bomb and the cold war began, the networks put their news departments to work on radio documentaries. CBS established a documentary unit that later resulted in a separate series, Ed Murrow and Fred Friendly's Hear It Now.

1950s: The Rise of Television Entertainment

While television had been in development since the 1920s and there had been experimental broadcasts since the 1930s, it wasn't until after World War II that the networks were able to concentrate on developing programming, manufacturers were able to return to making sets, and the public was able to afford them. (At the beginning of 1952 about 19 million U.S. homes had a TV set, and by the end of the decade that number was about 46.5 million.) The post-war economy was booming and the country's optimism was reflected in many of those early TV shows. Some of radio's stars made the transition to TV, as did many of the radio show formats. Soap operas (*The Guiding Light* on CBS), comedies (*Life of Riley* on NBC), westerns (*Gunsmoke* on CBS), dramas (*Kraft Television Theater* on NBC and ABC), variety (*Toast of the Town with Ed Sullivan* on CBS), and quiz shows (*Twenty One* on NBC) emerged as early favorites. And the TV syndication business was born when Frederic W. Ziv began selling shows such as 1951's *Bold Venture* with Humphrey Bogart and Lauren Bacall to local and regional advertisers and stations.

At the beginning of the decade RCA and CBS were in a battle: the companies each wanted FCC approval of a system for color TV. RCA eventually prevailed in 1953 because programs broadcast in its "compatible color" could still be watched on existing black and white sets. Another innovation that was to change television programming dramatically was unveiled in 1956 when Ampex Corp. demonstrated its videotape recorder at the National Association of Radio and Television Broadcasters convention and received $4 million in orders.

1960s: The Rise of Television News

Television journalism came of age in the 1960s. As the decade began, the FCC suspended its equal-time requirement for presidential and vice presidential candidates (following the suggestion of CBS President Frank Stanton), paving the way for the four "Great Debates" between Vice President Richard Nixon and Senator John F. Kennedy. The first debate, broadcast from Chicago on September 26, was seen by 75 million viewers, a record at that time. It changed the course of political campaigning and, quite probably, the course of the election. In 1961, the newly elected President Kennedy, recognizing the influence of the medium, allowed television to cover his press conferences. In 1963 the networks expanded their evening newscasts from 15 minutes to a half-hour. The nation was stunned when President Kennedy was assassinated in November 1963 and it's been said that television news came of age with its coverage. For four days following, the networks suspended normal programming and commercials with NBC-TV on the air for more than 71 hours, CBS-TV for 55, and ABC-TV for 60. The network coverage cost an unprecedented $32 million and CBS research showed that 93 percent of U.S. homes watched coverage of JFK's burial and that the average set was in use for more than 13 consecutive hours.

Americans were transfixed by coverage of the space race, beginning with Alan Shepard's suborbital flight in 1961 and culminating in live pictures of the moon landing in 1969. Americans were also presented with almost nightly images not so uplifting, with coverage of the civil rights movement and the Vietnam war (the networks established news bureaus in Saigon in 1965) becoming almost nightly subjects. The decade ended on two disappointing notes for broadcasters. Shortly after the election of Richard Nixon in 1968, Vice President Spiro Agnew began a series of speeches attacking the media and accusing the press of bias against the administration. And in 1969 the Supreme Court's decision in the Red Lion case upheld the FCC's fairness doctrine and personal attack rules, saying they "enhance rather than abridge the freedoms of speech and press protected by the First Amendment."

1970s: The Rise of FM and Satellites

While the technique of broadcasting using frequency modulation was patented by Edwin Armstrong in 1933, the first station built in 1939, and an FM band allocated by the FCC in 1940, growth of the new radio service was very slow. Delayed by World War II and a suspicion by many AM station owners that the new service would offer unwanted competition, it wasn't until the late 1960s that large numbers of consumers began buying FM receivers. Another handicap was beginning to be resolved in 1971 when car manufacturers began to include FM-equipped radios as standard equipment in about 20 percent of the new models (that number would rise to 50 percent within five years).

To differentiate FM stations from those on the AM band, and to take advantage of FM's higher fidelity, programmers developed new formats. One of the most innovative was the "underground" or "progressive" sound introduced by Tom Donahue at KMPX(FM) San Francisco in the late 1960s that played album tracks not heard on the tightly formatted top 40 AM stations. But "underground" wasn't the only sound to be heard on FM or even the predominant one. There was also the very aboveground sound of carefully researched syndicated formats. Coupled with the increased use of automation equipment, services like Drake-Chenault's "Solid Gold" and Bonneville's easy listening formats turned many money-losing FM stations into profit centers. This success was translated into value as multimillion-dollar prices for FM stations became common. And for the first time industry observers began predicting that FM would overtake AM as the band for music, with AM becoming primarily a news and information medium.

The phenomenon of the 1970s was the development of international broadcasting live via satellites. RCA inaugurated the nation's first domestic satellite communications service in 1974, using a Canadian satellite, later launching its Satcom series of birds. But the breakthrough came in 1975 when Home Box Office, Time Inc.'s pay cable subsidiary, announced plans to extend its service from the Northeast to nationwide via satellite.

The next year, 1976, Ted Turner, owner of two TV stations, begins using the satellite to distribute the signal of his Atlanta UHF station, WTCG, to cable systems across the country, dubbing it the "superstation." Buoyed by his success, in 1978 Turner announced plans to sell his other station, WRET-TV Charlotte, North Carolina, and use the money to start CNN, a 24-hour cable news service to be distributed by satellite.

Later in 1978, the FCC moved to enhance the competitive environment of satellite-distributed TV superstations by endorsing an "open entry" policy for the resale carriers that wished to feed local stations to cable television systems. Then in April 1979, a former Nixon administration staffer and cable trade reporter, Brian Lamb, persuaded a critical mass of the cable industry to support the Cable Satellite Public Affairs Network (C-SPAN). The new service provided satellite gavel-to-gavel television coverage of the House of Representatives proceedings.

1980s: The Rise of Cable Television

Building on the innovation of satellite delivery pioneered by HBO, the cable industry began to be viewed as more than just a relay service for TV stations. Now it could offer alternative channels. (The strategy paid off in increased demand: in 1980 cable had 20 percent penetration of U.S. TV households and by the end of the decade it was seen in 60 percent.) Ted Turner launched his Cable News Network on June 1, 1980, sending 24-hour-a-day news to 172 cable systems from its Atlanta studios.

Innovations and new cable channels appeared in almost every year of the decade. Most pay cable channels begin to scramble their satellite signals and in 1998 AT&T demonstrates laser-modulated fiber optics offering wider bandwidth and greater signal quality than coaxial cable.

1990s: The Rise of Mega-Media Companies

The modern media era really began in 1989 with the announcement that Time Inc. and Warner Communications Inc. had agreed to swap stock and merge into what would be the largest media and entertainment company in the world. The media landscape would only continue to change, and at an accelerating pace. A deregulatory wind was blowing through Washington in the 1990s.

In 1992 the FCC raised the cap on the number of radio stations a company could own from 12 AM and 12 FM to 18 of each and also permitted two of each service to be co-located in the same large market. A flurry of duopoly deals followed. The next year, the commission gave the big three TV networks a conditional OK to enter the lucrative network rerun business, when it lifts its financial interest and syndication rules (full repeal comes in 1995).

The year 1994 sees the radio caps raised again, to 20 AM and 20 FM. The FCC also said that it was time to acknowledge the dramatic changes in the video marketplace with equally dramatic deregulation of its TV ownership policies. The commission proposed new rules that would allow broadcasters to own as many stations as they want as long as they remain within the cap on total national audience reach. At the same time, the commission proposed raising that cap of 25 percent of the nation's TV households by 5 percent every three years to a maximum of 50 percent.

In another mega-merger, Viacom buys Paramount Communications. This is quickly followed in 1995 by announcements of Disney's $18.5 billion purchase of ABC, Time Warner's purchase of Turner Broadcasting System in an $8 billion stock swap, Westinghouse's $5.4 billion purchase of CBS, and Comcast's $1.6 billion purchase of the Scripps cable holdings. The big four TV networks get more competition in 1995 as The WB and United Paramount Network debut.

More deregulation appears with the Telecommunications Act of 1996, which eliminated cable rate regulation and the bar to telephone company-cable competition, resulting in AT&T and other phone companies offering packages that included cable, telephone, and Internet services and cable companies offering phone service. (This resulted in AT&T becoming the country's largest cable operator in 1999 when it bought TCI for $50 billion.) The 1996 act also eliminated the cap on radio station ownership and companies wasted no time in expanding their portfolios both through acquisitions and mergers (including Westinghouse/CBS-Infinity, which merged in 1996 in a $4.9 billion deal).

In 1997, TV group owner Bud Paxson began the seventh broadcast TV network with his PAX TV. Radio deals continued to proliferate, with a huge upsurge in station sales in 1998 and the mega-merger of Clear Channel's $6.35 billion purchase of Jacor.

The decade ends in a flurry of activity in 1999 with the FCC allowing ownership of two TV and up to six radio stations in top markets. AT&T bought Media One, Viacom spent $36 billion for CBS and the two largest radio groups, Clear Channel and AMFM, combined, leaving the merged Clear Channel owning 830 stations.

2000s: The Rise of Digital

Digital technology began to make a major impact. Cable used it to increase significantly the number of channels it offered to justify higher monthly fees and to meet the competition from channel-rich satellite technology. In contrast to cable, TV broadcasting has been slow to put digital technology to work. With digital, TV broadcasters could offer multiple channels of conventional TV, HDTV, or a little bit of both. But digital TV broadcasting has been mired in standards disputes and the resulting chicken-and-egg problem of receivers and programming: consumers are slow to replace their analog sets if there is little digital programming to watch and producers are hesitant to invest in new production equipment if few viewers are watching. But the transition continues.

The first digital TV station, WRAL-DT Raleigh, North Carolina, went on the air in 1996 and the FCC made digital channel assignments to all analog stations in 1997. Today, most stations are broadcasting two signals, one analog and one digital. But in February 2009, all must end the analog transmission and broadcast digitally only. Recognizing that millions of analog TV sets will still be in use in 2009, the government is setting up a program to help some analog set owners buy analog-to-digital converters.

The Internet became a TV medium, with the quality and quantity of video available over the Web rapidly increasing throughout the first half of the decade. A major milestone was reached in October 2005 when Disney announced it would make some of its most popular ABC primetime shows available to users of the new Apple Video iPods. The price of each video download: $1.99. The announcement prompted the other big media companies to begin "repurposing" primetime programming on the net. By the summer of 2006, all recognized that the Web would eventually become a TV medium at least the equal of broadcasting, cable, and satellite.

The first push for digital audio (other than that delivered by DBS or digital cable) is satellite-delivered radio. Two players, XM Satellite Radio and Sirius Satellite Radio, offer nationwide subscription service to receivers in cars and homes. Terrestrial radio is just beginning to make the digital transition. In-band, on-channel (IBOC) technology developed by iBiquity Digital Corp. allows radio stations to overlay their analog service with a digital one. As of August 2005, approximately 900 stations had been licensed by iBiquity to broadcast digitally. JVC, Kenwood, and Panasonic are among the companies offering consumer IBOC receivers.

A Chronology of the Electronic Media

From Isaac Newton to Janet Jackson: A Chronology of the Electronic Media

By Mark K. Miller, freelance writer and former managing editor, *Broadcasting & Cable* magazine

1666

Sir Isaac Newton performs basic experiments on the spectrum.

1794

Allessandro Volta of Italy invents the voltaic cell, a primitive battery.

1827

George Ohm of Germany shows the relationship between resistance, amperage and voltage. Sir Charles Wheatstone of England invents an acoustic device to amplify sounds that he calls a "microphone."

1844

Samuel F.B. Morse tests the first telegraph with "What hath God wrought?" message sent on link between Washington and Baltimore.

1858

First trans-Atlantic cable completed. President James Buchanan and Queen Victoria exchange greetings.

1867

James Clerk Maxwell of Scotland develops the electromagnetic theory.

1875

George R. Carey of Boston proposes a system that would transmit and receive moving visual images electrically.

1876

Alexander Graham Bell invents the telephone.

1877

Thomas A. Edison applies for a patent on a "phonograph or talking machine."

1878

Sir William Cooke of England passes high voltage through a wire in a sealed glass tube, causing a pinkish glow—evidence of cathode rays. It's the first step toward the development of the vacuum tubes.

1884

Paul Nipkow of Germany patents a mechanical, rotating facsimile scanning disk.

1886

Heinrich Hertz of Germany proves that electromagnetic waves can be transmitted through space at the speed of light and can be reflected and refracted.

1895

Wilheim Conrad Roentgen of Germany discovers X-rays.

Guglielmo Marconi sends and receives his first wireless signals across his father's estate at Bologna, Italy.

1896

Marconi applies for British patent for wireless telegraphy. He receives an American patent a year later.

1899

Marconi flashes the first wireless signals across the English Channel.

1900

Constantin Perskyi (France) coins the word television at the International Electricity Congress, part of the 1900 Paris Exhibition.

1901

Marconi at Newfoundland, Canada, receives the first trans-Atlantic signal, the letter "S," transmitted from Poldhu, England.

1906

Dr. Lee de Forest invents the audion, a three-element vacuum tube, having a filament, plate and grid, which leads to the amplification of radio signals.

1910

Enrico Caruso and Emmy Destinn, singing backstage at the Metropolitan Opera House in New York, broadcast through the De Forest radiophone and are heard by an operator on the SS Avon at sea and by wireless amateurs in Connecticut.

United States approves an act requiring certain passenger ships to carry wireless equipment and operators.

1912

The *Titanic* disaster proves the value of wireless at sea; 705 lives saved. Jack Phillips and Harold Bride are the ship's wireless operators.

1920

On August 20, 8MK (later, WWJ) in Detroit, owned by the Detroit News, starts what is later claimed to be regular broadcasting.

The Westinghouse Co.'s KDKA(AM) Pittsburgh broadcasts the Harding-Cox election on returns November 2 as the country's first licensed commercial radio station.

1921

The Dempsey-Carpentier fight is broadcast from Boyle's Thirty Acres in Jersey City through a temporarily installed transmitter at Hoboken, New Jersey. Major J. Andrew White was the announcer. This event gave radio a tremendous boost.

1922

The superheterodyne circuit is demonstrated by its inventor, Edwin H. Armstrong. It dramatically improves AM radio reception.

WEAF(AM) New York broadcasts what is claimed to be the first commercially sponsored program on September 7. The advertiser is the Queensborough Corp., a real estate organization.

WOI(AM) Ames, Iowa, goes on air as the country's first licensed educational station.

1923

Dr. Vladimir K. Zworykin files for a U.S. patent for an all-electronic television system.

A "chain" broadcast features a telephone tie-up between WEAF(AM) New York and WNAC(AM) Boston.

1924

The Republican convention in Cleveland and the Democratic convention in New York are broadcast over networks.

1925

President Calvin Coolidge's inaugural ceremony is broadcast by 24 stations in a transcontinental network.

1926

President Coolidge signs the Dill-White Radio Bill creating the Federal Radio Commission and ending the chaos on the radio dial caused by the wild growth of broadcasting.

National Broadcasting Co. is organized on November 1 with WEAF(AM) and WJZ(AM) in New York as key stations and Merlin Hall Aylesworth as president. Headquarters are at 711 Fifth Ave., New York.

1927

The Columbia Broadcasting System goes on the air with a basic network of 16 stations. Major J. Andrew White is president.

Philo T. Farnsworth applies for a patent on his image dissector television camera tube.

1928

NBC establishes a permanent coast-to-coast radio network.

1929

William S. Paley, 27, is elected president of the Columbia Broadcasting System.

Vladimir Zworykin demonstrates his kinescope or cathode ray television receiver before a meeting of the Institute of Radio Engineers on November 19.

1930

Experimental TV station W2XBS is opened by National Broadcasting Co. in New York.

1931

Experimental television station W2XAB is opened by Columbia Broadcasting System in New York.

The first issue of *Broadcasting* magazine appears on October 15.

The National Association of Broadcasters reports that more than half of the nation's radio stations are operating without a profit.

1932

CBS, NBC, and New York area stations, notably WOR(AM), go into round-the-clock operations to cover the Lindbergh kidnapping, radio's biggest spot-news reporting job to date.

NBC lifts its ban on recorded programs for its owned-and-operated stations, but continues to bar them from network use.

NBC withdraws prohibitions against price mentions

on the air during daytime hours; two months later, both NBC and CBS allow price mentions at nighttime as well.

1933

Associated Press members vote to ban network broadcasts of AP news and to restrict local broadcasts to bulletins to stipulated times with air credit to member newspapers.

The American Newspaper Publishers Association declares radio program schedules are advertising and should be published only if paid for.

CBS assigns publicity director Paul White to organize a nationwide staff to collect news for network broadcast. General Mills agrees to sponsor twice-daily newscasts.

1934

Congress passes the Communication Act, which, among other things, replaces the Federal Radio Commission with the Federal Communications Commission.

1935

RCA announces that it is taking television out of the laboratory for a $1 million field-test program.

1936

A year of TV demonstrations begins in June with the Don Lee Broadcasting System's first public exhibition of cathode ray television in the U.S., using a system developed by Don Lee TV director Harry Lubcke. One month later, RCA demonstrates its system of TV with transmissions from the Empire State Building, and Philco follows with a seven-mile transmission in August.

FM (frequency modulation) broadcasting, a new radio system invented by Major Edwin H. Armstrong, is described at an FCC hearing as static-free, free from fading and cross-talk, having uniformity day and night in all seasons and greater fidelity of reproduction.

A.C. Nielsen, revealing his firm's acquisition of the MIT-developed "Audimeter," proposes a metered tuning method of measuring radio audience size.

1937

WLS(AM) Chicago recording team of Herb Morrison, announcer, and Charles Nehlsen, engineer, on a routine assignment at Lakehurst, New Jersey, records an on-the-spot account of the explosion of the German dirigible Hindenburg. NBC breaks its rigid rule against recordings to put it on the network.

1938

Broadcasting publishes the first facsimile newspaper in a demonstration at the National Association of Broadcasters convention.

1939

After 15 years of litigation, the patent for iconoscope-kinescope tubes, the basis for electronic television, is granted to Dr. Vladimir Zworykin.

A telecast of the opening ceremonies of the New York World's Fair marks the start of a regular daily television schedule by RCA-NBC in New York.

The first baseball game ever televised—Princeton vs. Columbia—appears on NBC.

1940

The FCC authorizes commercial operation of FM, but puts TV back into the laboratory until the industry reaches an agreement on technical standards.

CBS demonstrates a system of color TV developed by its chief TV engineer, Dr. Peter Goldmark.

1941

Bulova Watch Co., Sun Oil Co., Lever Bros. and Procter & Gamble sign as sponsors of the first commercial telecasts on July 1 over NBC's WNBT(TV) New York (until then W2XBS).

President Roosevelt's broadcast to the nation on December 9, the day after war is declared, has the largest audience in radio history—about 90 million listeners.

1942

The Advertising Council is organized by advertisers, agencies, and media to put the talents and techniques of advertising at the disposal of the government to inspire and instruct the public concerning the war effort.

1943

Edward J. Noble buys the Blue Network from RCA for $8 million in cash. RCA had two networks, NBC Red and NBC Blue.

1944

With the FCC approval of the transfer of owned stations, the Blue Network assumes the name of its holding company, the American Broadcasting Co.

1945

Pooled coverage of the Nazi surrender in May brings the American people full details of the end of the war in Europe. Peace heralds a communications boom: Not only will programming restrictions end, but new station construction, frozen for the duration, will proceed at an explosive pace soon after V-J Day in August.

1946

A telecast of the Louis-Conn heavyweight title fight, sponsored by Gillette Safety Razor Co. on a four-city hookup, reaches an estimated 100,000 viewers and convinces skeptics that television is here to stay.

RCA demonstrates its all-electronic system of color TV.

Bristol-Myers is the first advertiser to sponsor a television network program—Geographically Speaking—which debuted October 27 on NBC TV's two-station network.

1947

Radio comedian Fred Allen uses a gag, which NBC had ruled out, about network vice presidents, and is cut off the air while he tells it. The story is front-page news across the country as the sponsor's ad agency demands a rebate for 35 seconds of dead air.

1948

Texaco puts an old-style vaudeville show on NBC TV; the hour-long series stars Milton Berle.

1949

The Academy of Television Arts & Sciences presents the first Emmy Awards at ceremonies televised by KTSL(TV) Los Angeles.

1950

General Foods drops actress Jean Muir, who denies any communist affiliations or sympathies, from the cast of *The Aldrich Family* (NBC TV) after protests against her appearance by "a number of groups." The Joint Committee Against Communism claims credit for her removal, announcing a drive to "cleanse" radio and television of pro-communist actors, directors, and writers.

The FCC approves CBS's color TV system, effective November 20. The network promises 20 hours of color programming a week within two months. TV set manufacturers are divided, however, over whether to make sets, since the CBS system is incompatible with black-and-white broadcasts. In the meantime, RCA continues work on its color system.

1951

Witness Frank Costello's hands provide TV's picture of the week as he refuses to expose his face to cameras covering New York hearings on organized crime of the Senate Crime Investigation Committee, chaired by Senator Estes Kefauver (D-Tenn.)

Sixteen advertisers sponsor the first commercial color telecast, an hour-long program on a five-station East Coast CBS TV hook-up.

Bing Crosby Enterprises announces the development of a system for recording video and audio programs on magnetic tape. The pictures shown at demonstrations are described as "hazy" but "viewable." A year later the images are described as improved "more than 20-fold."

1952

By rushing equipment across the country, from Bridgeport, Connecticut, to Portland, Oregon, KPTV(TV) Portland goes on the air as the first commercial UHF TV station.

1953

With the end of daylight-saving time, CBS TV and NBC TV inaugurate "hot kinescope" systems to put programs on the air on the West Coast at the same clock hour as in the East.

RCA demonstrates black-and-white and color TV programs recorded on magnetic tape. RCA-NBC Board Chairman David Sarnoff says two years of finishing touches are needed before the system is ready for market.

The FCC approves RCA's compatible (with black-and-white transmission) color TV standards. System supplants the incompatible CBS system.

1954

CBS President Frank Stanton broadcasts the first network editorial, urging that radio and TV be allowed to cover congressional hearings.

1955

A contract between the DuMont TV network and Jackie Gleason Enterprises calls for Gleason's *The Honeymooners* to be done as a filmed program for CBS TV on Saturday nights.

1956

Ampex Corp. unveils the first practical videotape recorder at the National Association of Radio and Television Broadcasters convention in Chicago. The company takes in $4 million in orders.

1957

Videotape recorders are seen as the solution to the TV networks' daylight-saving time problems.

1958

Subliminal TV messages are put under the spotlight at hearings in Los Angeles and Washington.

The BBDO ad agency converts live commercials to videotape.

1959

Sixty-eight TV stations defy the broadcasters' code of conduct by refusing to drop Preparation H commercials.

The quiz show scandal climaxes when famed *Twenty-One* prizewinner Charles Van Doren admits to a House committee that he had been provided with answers and strategies in advance. The sad ending to the quiz show era prompts cancellation of big-prize shows and vows by NBC and CBS to end deceptive practices.

1960

A satellite sends weather reports back from a 400-mile-high orbit.

RKO-Zenith plans a $10 million test of an on-air pay TV system in Hartford, Connecticut.

Sam Goldwyn offers a package of movies to television.

The last daytime serial on network radio ends.

The opening Kennedy-Nixon debate attracts the largest TV audience to date.

1961

FCC Chairman Newton Minnow shakes up the National Association of Broadcasters convention with his assessment of TV programming: Although it occasionally shines with programs like *Twilight Zone* and *CBS Reports*, it is, more than anything, from sign-on to sign-off "a vast wasteland."

Off-network shows become popular as syndicated fare.

The Ampex "electronic editor" permits inserts and additions to be made in videotape without physical splices.

ABC TV engineers develop a process for the immediate playback of videotape recordings in slow motion.

1962

John Glenn's orbital space flight is seen by 135 million TV viewers.

Telstar, AT&T's orbiting satellite, provides a glamorous debut for global television.

1963

Astronaut Gordon Cooper sends back the first TV pictures from space.

All radio and TV network commercials and entertainment programming are canceled following the assassination of President Kennedy. In the same week, the first trans-Pacific broadcast via satellite previews live TV coverage of the 1964 Olympics in Tokyo.

1964

The government and the tobacco companies each ponder their next move after the surgeon general's report links cigarette smoking and lung cancer. Within weeks, American Tobacco drops sports broadcasts, radio stations begin to ban cigarette ads and CBS TV orders a de-emphasis of cigarette use on programs.

1965

Early Bird, the first commercial communications satellite, goes into stationary orbit, opening trans-Atlantic circuits for TV use.

1966

Fred W. Friendly quits as president of CBS News when his new boss, John Schneider, CBS group vice president for broadcasting, cancels coverage of a Senate hearing on the Vietnam War and runs a rerun of *I Love Lucy* instead.

Network TV viewers see live close-up pictures of the moon—sent back by Surveyor I—as they come into the Jet Propulsion Laboratory.

1967

ABC Radio introduces a radical plan: four networks instead of one, each tailored to suit different station formats.

President Johnson signs the Public Broadcasting Act into law, establishing the Corporation for Public Broadcasting, federal funding mechanism.

1968

The Children's Television Workshop is created by the Ford Foundation, the Carnegie Corp., and the Office of Education to develop a 26-week series of hour-long color programs for preschool children. *Sesame Street* is the result.

The U.S. Supreme Court gives the FCC jurisdiction over all cable TV systems.

Pictures taken inside Apollo 7 in flight and sent back to Earth revive public interest in the space program.

NBC TV earns the life-long ire of sports fans when it cuts off the end of a Jets-Raiders game to air its made-for-TV movie *Heidi*. Viewers miss the Raiders' two-touchdowns-in-nine-seconds defeat of the Jets.

1969

The Corporation for Public Broadcasting plans the creation of the Public Broadcasting Service to distribute programming to noncommercial TV stations.

In the same week that ABC-TV announces its $8 million *Monday Night Football* deal (games to begin in 1970), Apollo 10 sends back the first color TV pictures of the moon and of Earth from the moon.

The world watches live coverage of Neil Armstrong's walk on the moon.

1970

House and Senate conferees agree on legislation to outlaw cigarette advertising on radio and TV, but change the bill's effective date from January 1, 1971, to January 2, so commercials can appear on New Year's Day football telecasts.

The FCC rules that TV stations in the top 50 markets cannot accept more than three hours of network programming between 7 and 11 p.m., and bars them from domestic syndication and from acquiring subsidiary rights in independently produced programs.

1971

National Public Radio debuts with a 90-station interconnected lineup.

1972

Judge Benjamin Hooks of Memphis, Tennessee, is nominated to the FCC. He becomes the first black to serve on a federal regulatory agency.

Home Box Office Inc., New York, is formed as a subsidiary of Sterling Communications to provide pay-cable TV systems with live and film programming.

1973

Western Union becomes the first company to receive federal permission to launch a commercial communications satellite in the U.S.

Broadcast media around the world open their coverage of the Senate select committee's investigation of the Watergate scandal.

1974

RCA inaugurates the nation's first domestic satellite communications service, using a Canadian satellite.

More than 110 million viewers watch President Nixon announce his resignation.

1975

Home Box Office, Time Inc.'s pay cable subsidiary, announces that it will inaugurate a satellite delivery network in the fall.

1976

Ampex Corp. and CBS develop the electronic still-store system, which uses a digital recording technique to store 1,500 frames in random mode, each accessible in 100 milliseconds.

Cable network launches include Showtime and Univision.

1977

ABC's eight-day telecast of the miniseries *Roots* becomes the most watched program in television history, with ratings in the mid-40s and shares in the mid-60s. Eighty million people watch at least some part of the final episode.

Sony unveils its Betamax videocassette in August and later the same month RCA introduces its SelectaVision home videotape recorder.

1978

The U.S. Supreme Court upholds the FCC in the "seven dirty words" case involving Pacifica's WBAI(FM) New York. The ruling says the FCC may regulate and punish for the broadcasting of "indecent material."

1979

Ampex demonstrates its digital videotape recorder at the Society of Motion Picture and Television Engineers conference in San Francisco in February. Sony unveils its version two months later.

Cable network launches include C-SPAN, ESPN, The Movie Channel, and Nickelodeon.

1980

"Who Shot J.R.?" episode of *Dallas* garners the highest rating for any program in modern TV history, with a 53.3 rating and a 76 share.

Cable network launches include Cable News Network, Black Entertainment Television, the Learning Channel, Bravo, and USA Network.

1981

With five ENG cameras rolling, the shooting of President Reagan becomes history's most heavily covered assassination attempt.

The first U.S. demonstration of high-definition television (HDTV) takes place at the annual convention of the Society of Motion Picture and Television Engineers. The Japanese Broadcasting Corp.'s (NHK) 1,125-line analog system draws raves from engineers and filmmakers.

Cable network launches include MTV: Music Television and the Eternal Word Television Network.

1982

Having reached a settlement with the Justice Department to divest itself of its 23 local telephone companies, communications giant AT&T hopes to lead the country into the "information age." The National Cable Television Association, Congress, and the FCC wonder what the agreement has wrought.

Cable network launches include the Weather Channel and the Playboy Channel.

1983

Reagan appointee Mark Fowler, chairman of the FCC, tells a common carrier conference that the U.S. is heading toward a regulation-free telecommunications marketplace.

In February, the two-and-a-half-hour final episode of CBS's *M*A*S*H* is the most watched program in TV history, garnering a 60.3 rating and a 77 share.

Cable network launches include the Disney Channel and Country Music Television.

1984

The U.S. Supreme Court rules that home videotaping is legal.

Congress passes the Cable Telecommunications Act of 1984, landmark legislation deregulating cable. Law accelerates the growth of cable.

Cable network launches include the Arts & Entertainment Network (A&E), American Movie Classics, and Lifetime.

1985

Ted Turner makes inquiries at the FCC about a possible takeover of CBS. Later, in March, media company Capital Cities Communications purchases ABC for $3.5 billion. Turner's efforts to acquire CBS fail by the end of July, when a federal judge approves the network's stock buyback plan.

The Advanced Television Services Committee (ATSC) votes in favor of the NHK HDTV standard: 1,125 lines, 60 fields, 2:1 interlace, 5.33:3 ratio. This standard is put forward by the U.S. to the International Radio Consultative Committee (CCIR) for consideration as the international standard. The CCIR adopts the recommendation later in the year.

Having lost his bid to buy CBS, Ted Turner makes a $1.5 billion offer for MGM/UA.

Cable network launches include The Discovery Channel, Home Shopping Network, and VH-1.

1986

MGM and Color Systems Technology sign an agreement for the conversion of 100 of the studio's black-and-white films to color.

Cable network launches include C-SPAN2 and QVC.

1987

Fox Broadcasting Co. introduces its primetime lineup with 108 affiliates in its bid to become the fourth major U.S. commercial television network.

The National Association of Broadcasters and the Association for Maximum Service Television broadcast HDTV over standard TV channels during public demonstrations in Washington.

President Reagan vetoes legislation to write the fairness doctrine into law. The doctrine required broadcast stations to allow opposing views of issues, but critics claimed that it discouraged open debate.

Cable network launches include Movietime (renamed E! Entertainment Television in 1990), The Travel Channel, and Telemundo.

1988

The FCC adopts preliminary ground rules for HDTV. It tentatively decides to require HDTV broadcasts to be compatible with NTSC sets and says it will not make additional spectrum available outside the VHF and UHF bands for HDTV because there is enough already available to accommodate the service.

Cable network launches include Turner Network Television.

1989

Time Inc. and Warner Communications agree to swap stock and merge into what will be world's largest media and entertainment company.

1990

Digital audio broadcasting is demonstrated at the National Association of Broadcasters convention and is heralded as the HDTV of radio.

General Instrument revolutionizes the development of high-definition television by proposing an all-digital system. The video compression system also has implications for satellite transmissions.

Cable network launches include CNBC and The Inspiration Network (INSP).

1991

The U.S. air attack on Iraq begins January 16 with dramatic live coverage from network reporters in Baghdad. CNN is the lone network to maintain contact with its Baghdad reporters through the night.

Free to move around Moscow and ready to commit resources to coverage, television and radio provide gripping details of the short-lived Soviet coup and the collapse of communism in the Soviet Union. During his detention in the Crimea, Soviet President Mikhail Gorbachev keeps track of events by listening to the BBC, Voice of America, and Radio Liberty.

Cable network launches include Court TV, Comedy Central, and Encore.

1992

In March, the Supreme Court let stand an appeals court ruling that struck down the FCC's around-the-clock ban on broadcast indecency as unconstitutional and requiring the commission to establish a safe harbor—a part of the day when few children are tuning in and during which radio and TV stations may broadcast without fear of FCC sanctions for indecency.

General Instrument and MIT show the first over-the-air digital HDTV transmission to Washington lawmakers and regulators. The 12-minute transmission of 1,050-line video was broadcast by noncommercial WETA-TV Washington.

The FCC raises the limit on radio stations a single company may own from 12 AM and 12 FM to 30 of each, then backpedals and lowers the caps to 18 each, with no more than two AMs and two FMs in large markets and three stations—only two in the same service—in small markets.

Fox expands its programming lineup to seven nights a week, ending its status as a "weblet" and becoming the fourth full-fledged commercial TV network in the U.S.

The FCC unanimously approves allowing broadcast TV networks to purchase cable systems that serve no more than 10 percent of U.S. homes and up to 50 percent of a particular market's homes.

The FCC tells TV broadcasters they will have five years to begin broadcasting in HDTV once the agency adopts a standard and makes channels available.

Cable network launches include The Cartoon Network and the Sci Fi Channel.

1993

Warner Bros. announces it will launch a fifth broadcast TV network in 1994.

The FCC expands the AM band's upper limit from 1605 kHz to 1705 kHz.

General Instrument, Zenith, AT&T, and the ATRC join forces as the "Grand Alliance" to develop a single HDTV system. Later in the year, the Grand Alliance announces its support of the emerging MPEG-2 digital compression HD system: six-channel, CD-quality Dolby AC-3 music system; 1,920-pixel by 1,080-line interlaced scanning picture; and progressive scanning.

Paramount Communications begins talks with TV stations about forming a fifth broadcast TV network.

Southwestern Bell and Cox Cable form a $4.9-billion partnership.

Cable network launches include ESPN2 and the Television Food Network.

1994

Two companies, Hubbard's United States Satellite Broadcasting and Hughes's DirecTV, begin direct broadcast satellite transmissions to 18-inch home dish antennas from a shared satellite.

Paramount and Viacom merge in a deal worth $9.2 billion, forming the world's most powerful entertainment company. Viacom's Sumner Redstone becomes the new company's chairman. Later in the year, Viacom adds Blockbuster Entertainment to its portfolio.

Cable network launches include FX, Home & Garden TV, the International Film Channel, Starz!, Trio, the Game Show Network, and Turner Classic Movies.

1995

Seagram pays $7 billion for the 80 percent of Hollywood studio MCA Inc. owned by Matsushita Electric Industrial Co. Seagram is controlled by the Bronfman family and is headed by President/CEO Edgar Bronfman Jr.

The Megamedia Age begins when, in the same week, Walt Disney Co. announces it is buying Capital Cities/ABC for $18.5 billion and then Westinghouse Electric Co. releases word of its purchase of CBS Inc. for $5.4 billion.

Time Warner and Turner Broadcasting System agree to merge in an $8 billion stock swap deal.

Live television coverage of the verdict in the O.J. Simpson murder trial sets viewing records when 150 million people watch the jury return a "not guilty" verdict.

Microsoft buys 50 percent stake in NBC's cable channel *America's Talking* for $250 million. AT's talk format will be dropped and the network will become a news operation after being rechristened MSNBC.

The FCC repeals its Prime Time Access and Fin-Syn rules. These rules restricted the major broadcast networks from owning interest in their own primetime programming.

Cable network launches include CNN/fn, The Golf Channel, Great American Country, the History Channel, and the Outdoor Life Network.

1996

Congress passes—and President Clinton signs—the Telecommunications Act of 1996, the first major overhaul of telecommunication legislation since 1934. Its key provisions include: replacing the 12-station TV ownership limit with a national home coverage cap of 35 percent; eliminating the national ownership limits on radio stations and allowing one company to own different numbers of stations locally, depending on the market size; requiring TV sets sold in the U.S. to be equipped with a V-chip to enable blocking of channels based on encoded ratings; deregulating cable rates.

Westinghouse/CBS buys Infinity Broadcasting for $4.9 billion, creating the country's largest radio station group in terms of earnings. The deal results in Westinghouse/CBS owning 83 radio stations in 15 markets.

The FCC releases its first list of proposed digital TV channel assignments for all U.S. analog television stations.

In July, WRAL-HD Raleigh, North Carolina, begins HDTV transmission on channel 32 under an experimental FCC license, making it the first HDTV station to broadcast in the U.S.

The Washington-based Model HDTV Station Project demonstrates live, over-the-air digital TV transmission and reception. A few months later, it bounces digital signals off a satellite and displays them on a receiver.

Cable network launches include Animal Planet, Fox News Channel, MSNBC, the Sundance Channel, and TVLand.

1997

After several starts and stops, the TV industry unveils content-based V-chip ratings to mixed reviews. Recalcitrant NBC maintains it will not implement the new ratings.

Paxson Communications chief Bud Paxson announces plans to launch a new television network, Pax Net, using his 73 owned UHF stations as a base and airing family friendly off-network programming.

ABC Television Network President Preston Padden and Sinclair Broadcasting President David Smith say broadcasters ought to consider using DTV channels for broadcasting multiple channels of conventional TV rather than a single channel of HDTV.

Hearst Corp. (8 TVs) and Argyle Television (6 TVs) join their TV stations and create a new company, Hearst-Argyle Television Inc., that is valued at $1.8 billion.

The FCC gives TV broadcasters a second channel for the delivery of HDTV and other digital services and said that all network affiliates in the top 10 markets have 24 months to start broadcasting a digital signal; those in markets 11-30 have 30 months; all other commercial stations have five years. Noncommercial broadcasters have six.

DTV service provider EchoStar plans to launch two satellites that will give it the ability to provide local broadcast TV signals to about 43 percent of the U.S.

Cable network launches include WE.

1998

The National Association of Broadcasters agrees to support plans by satellite TV providers to retransmit local TV station signals into their markets as long as the satellite services carry all a market's signals.

At 2:17 p.m. on February 27, WFAA-TV Dallas broadcast what it claims is the first non-experimental HDTV signal (in 1080i, 16:9 format). The broadcast began with a half-hour of taped HD programming, followed by a live simulcast of the station's NTSC programming that was upconverted to HDTV. The next month, Sinclair Broadcasting becomes the first TV group owner to broadcast multiple digital channels.

AT&T pays $50 billion for cable system giant Tele-Communications Inc.

Paxson Communications launches its broadcast television network, now called Pax TV, with a lineup of 90 stations covering about 75 percent of U.S. TV homes.

Radio group owner Clear Channel Communications purchases competitor Jacor Communications for $4.4 billion. The deal gives Clear Channel 453 stations in 101 markets. The year's other big deals include: Chancellor Media's purchase of Capstar Broadcasting for $3.9 billion; Hearst-Argyle Television's purchase of Pulitzer Broadcasting for $1.85 billion; Chancellor's purchase of LIN Television for $1.5 billion; and Sinclair Broadcast Group's purchase of Sullivan Broadcasting for $1 billion.

CBS is the first broadcast TV network to air a live HDTV sports event with its Nov. 8 telecast of the New York Jets-Buffalo Bills NFL game. It is carried by CBS stations in New York; Philadelphia; Washington; Cincinnati; Charlotte, North Carolina; Raleigh, North Carolina; and Columbus, Ohio.

Hughes Electronics Corp., parent of DBS provider DirecTV, announces deal to buy rival U.S. Satellite Broadcasting from Hubbard Broadcasting for $1.3 billion. The DBS business now has three providers: DirecTV, EchoStar, and Primestar.

Cable network launches include BBC America, the Biography Channel, Cinemax, Tech TV, and Toon Disney.

1999

Hughes Electronics Corp., parent of DBS provider DirecTV, buys rival Primestar for $1.1 billion plus stock. The DBS business now has two providers: DirecTV and EchoStar.

Paxson Broadcasting sells its 30 percent interest in The Travel Channel to the cable channel's 70 percent owner, Discovery Channel.

MSO Comcast offers $58 billion for MediaOne Group's cable systems. AT&T then comes in with a $69 billion offer that has AT&T swapping and selling Comcast systems with 2 million subscribers for roughly $9 billion. In return, Comcast agrees to withdraw its $58 billion offer.

CBS pays $2.5 billion for syndication giant King World Productions, whose properties include the hit shows *Oprah, Wheel of Fortune,* and *Jeopardy!*

FCC votes to allow a broadcaster to own two TV stations in a market under certain conditions and liberalizes its radio/TV cross-ownership restrictions. A flood of station deals follow.

Viacom Inc. buys CBS Corp. for $36 billion, merging Viacom's Paramount Station Group, UPN network, cable networks, and other properties, with those of CBS.

Clear Channel Communications pays $23.5 billion in stock and assumption of debt for the 443 radio stations of AMFM Inc., the country's largest radio broadcaster. Clear Channel will have to divest about 100 stations to comply with FCC and Justice Department regulations. Those spinoffs will bring Clear Channel $4.3 billion.

Legislation takes affect allowing satellite delivery of local television stations in their markets, increasing DBS providers' ability to compete with cable.

2000

America Online Inc. and Time Warner merge in a deal worth $181 billion. The merged company, AOL Time Warner, combines the company that serves the largest number of Internet users with the largest producer of TV shows and movies and cable programming, plus cable systems passing 20 percent of U.S. homes.

Tribune Co. buys Times Mirror Co. for $6.5 billion, acquiring seven daily newspapers and various magazines. The deal will give Tribune co-ownership of TV stations and major daily newspapers in the top three markets and the assets to sell packages of multimedia advertising to clients on national, regional, and local levels.

Harry Pappas, head of Pappas Television, the country's largest privately held TV station group, announces plans to launch Azteca America, the third U.S. Hispanic television network (Univision and Telemundo are the others) in 2001.

Cable network launches include Oxygen.

2001

FCC approves the $5.4-billion sale of Chris-Craft Broadcasting's ten TV stations to Fox Television.

DBS operator EchoStar Communications engineers a $26-billion bid for competitor DirecTV, owned by GM's Hughes Corp. The move follows attempts by Rupert Murdoch's News Corp. to acquire DirecTV. But regulatory reviews keep the deal in limbo.

XM Satellite Radio begins broadcasting a nationwide radio service of 200 channels from two satellites—"Rock" and "Roll"—in orbit above the equator. The Washington-based company charges subscribers $9.95 a month for the service. A rival, New York-based Sirius Satellite Radio, plans to launch a similar service later in the year.

The September 11 terrorist attacks on New York and Washington result in around-the-clock news coverage, dropping commercials. It's estimated that the networks lost $200 million-$300 million in the first four days of coverage. Four FM and nine New York TV stations whose antennas were on top of the World Trade Center are knocked off the air and several stations lost employees who had been manning the transmitters in Tower 1. Across the country, broadcasters raised money and arranged blood drives. The fall TV season is delayed, late-night talk/comedy shows are put on hiatus, the Emmy Awards are postponed, and several industry gatherings are canceled.

NBC buys Telemundo, the No. 2 U.S. Spanish-language TV network, for $2.7 billion.

Comcast negotiates $72 billion merger with rival cable operator AT&T Broadband, topping bids by AOL Time Warner and Cox Communications.

Cable network launches include ABC Family, Hallmark Channel, and National Geographic Television.

2002

Sirius Satellite Radio launches its satellite-delivered subscription radio service in four markets in February, then rolls out nationally in July. Sirius follows XM Satellite Radio to become the second U.S. satellite radio programmer.

Prompted by lawsuits from Fox, Viacom, NBC, and Time Warner, a three-judge panel of the federal appeals court in Washington refuses to uphold an FCC rule limiting a TV station group owner's audience reach to 35 percent of U.S. TV households and strikes down a rule barring a cable system from owning TV stations in its market. The court orders the FCC to rewrite or justify the ownership limit rule.

Tom Brokaw of NBC News announces he will step down as evening news anchor after the 2004 presidential election, to be succeeded by NBC's Brian Williams. Brokaw will then focus on in-depth reporting projects.

The Securities and Exchange Commission begins a formal investigation into the accounting practices of cable MSO Adelphia Communications. Five of Adelphia's top executives—including founder John Rigas and his two sons, Michael and Tim—are arrested on fraud charges, alleging that the family used the company as a "personal piggy bank," financing various personal transactions, including $3.1 billion in loans for stock and family businesses.

The FCC mandates that all TV sets must be equipped with digital tuners by 2007 and proposes strong copy-protection measures intended to prevent widespread copying and streaming of content over the Internet.

Lifestyle diva Martha Stewart, whose media empire included TV, magazines, and books, is investigated by the Justice Department for allegedly lying to federal authorities looking into insider trading involving Stewart's sale of ImClone Systems stock the day before it became public that the Food and Drug Administration had denied the company's application to market a new cancer drug.

In October, both the FCC and the Department of Justice reject DBS operator EchoStar Communications' proposed $26-billion purchase of competitor DirecTV, and a revised agreement fails to sway either agency. In December, EchoStar withdrew its merger request from the FCC. Rupert Murdoch's News Corp., whose previous bid for DirecTV had been rebuffed, puts together a new deal.

2003

Rupert Murdoch's News Corp. receives FCC and Justice Department approval of its deal to acquire 34 percent of DBS operator DirecTV's parent company Hughes Electronics for $6.6 billion in cash and stock.

New York City's Metropolitan Television Alliance agrees to place a new broadcast tower for New York-area television stations on top of the Freedom Tower, a 1,776-foot office tower that will be built on the site of the World Trade Center, where the stations' towers were located prior to 9/11. The MTVA comprises all the city's major TV broadcasters. After the terrorist attacks, most of the stations operated from backup facilities atop the Empire State Building. Ground is expected to be broken on the Freedom Tower in the summer of 2004, and broadcasters should begin operating from the tower by 2008.

The FCC releases new media ownership rules in response to a federal appeals court ruling in 2002. Among the changes: raising the national coverage cap for TV groups from 35 percent to 45 percent; allowing ownership of two TV stations (duopoly) in markets with five or more commercial stations; allowing ownership of three TV stations (triopoly) in markets with at least 18 stations; newspaper-TV cross-ownership is permitted in markets with at least four TV stations; radio-TV cross-ownership now include newspapers in the formula—owners in markets with nine or more TV stations face no cross-ownership restrictions per se but are limited by individual radio and TV limits applicable to specific markets. TV-duopoly owners would not be permitted to own newspapers in markets with fewer than nine TV stations. In markets with three or fewer TV stations, no cross-ownership of TV, radio, or newspapers is permitted. In markets with four to eight TV stations, an owner may form one of the following combos: (1) A daily newspaper, one TV station, up to one-half the number of radio stations permitted to one owner in that market. (2) A daily newspaper, the total number of radio stations permitted to one owner there, no TV stations. (3) Two TV stations and the total number of radio stations permitted there. Congress quickly reacts with legislation introduced by Rep. John Dingell (D-Mich.), which would restore the 35 percent cap. Other critics of the new rules challenge them in federal court.

Liberty Media pays $7.9 billion for Comcast's 56 percent stake in home shopping giant QVC. With 2002 sales of $4.4 billion, QVC is not just the largest shopping network, it's the second-largest television network of any kind.

A panel of federal appeals court judges in Philadelphia agrees with public advocacy groups and imposes a stay of the FCC's new broadcast-ownership rules scheduled to take effect on September 4. The stay will remain in effect until lawsuits to overturn the new rules are settled. The Philadelphia court then decides to retain the case attacking the new FCC broadcast-ownership limits rather than granting broadcast networks' pleas to transfer it to a court in Washington.

The Bush White House brokered a surprise compromise over media deregulation by agreeing to permanently set the national TV station ownership cap at 39 percent of U.S. television households. That percentage allows Fox and Viacom to retain all their stations. Wielding a threat to veto a catch-all spending bill over a provision that would roll the limit back to 35 percent, aides to President Bush persuaded Senate Appropriations Committee Chairman Ted Stevens (R-Alaska) to back down from the tighter limit. Stevens's action came less than a week after he had persuaded reluctant House leadership to go along with the old level. The compromise splits the difference between the 45 percent limit set by the FCC in June and the previous 35 percent level that rank-and-file lawmakers on both sides of Capitol Hill had been pushing to reinstate. The agreement is part of a spending bill that funds the FCC and many other agencies in fiscal 2004.

After a 36-year run, the California Cable Telecommunications Association's annual Western Cable Show makes its curtain call in December, citing consolidation in the cable industry and economic pressure.

Cable network launches include Spike TV.

2004

NBC gets Federal Trade Commission approval for its $14 billion purchase of Vivendi Universal Entertainment, its last regulatory hurdle. The FCC was not required to review the deal because it involved no station licenses. Among other things, NBC acquires USA Network and the Sci Fi Network. The new entity will be called NBC Universal.

Congress and the FCC react swiftly to the "wardrobe malfunction" that bared Janet Jackson's breast during the MTV-produced half-time entertainment in CBS TV's Super Bowl broadcast. Congress passes legislation that dramatically increases the limits on FCC fines for indecency violations.

Congress and the FCC take the first steps toward punishing stations that air "excessively" violent shows. Under orders from leaders of the House Commerce Committee, FCC Chairman Michael Powell by the end of the year will start investigating whether the commission should restrict onscreen violence. Cable can't count on immunity either. Growing ranks of lawmakers say cable must do more to make sure that children aren't exposed to potentially traumatizing content.

A panel of federal appeals court judges in Philadelphia concludes that the FCC wasn't justified in its June

2003 decision relaxing ownership restrictions in the newspaper, television, and radio industries. The rules, which were blocked from taking effect in September 2003, have been sent back to the FCC for a rewrite. A frustrated FCC Chairman Michael Powell criticized the decision, claiming that it created a "clouded and confused state of media law" and makes it nearly impossible for his agency to design standards for ownership limits.

Cable network launches include TV One.

2005

George W. Bush, on January 20, becomes the first president to have his inauguration covered in HDTV. ABC News deploys 36 HD cameras and four HD production vehicles throughout the parade route to give viewers an unparalleled view of American history.

President Bush chooses FCC commissioner Kevin Martin to be chairman of the agency.

In a King Solomon-like answer to critics that Viacom has become too big to grow, Chairman Sumner Redstone proposes cleaving it in half. The resulting companies would be Viacom and CBS Corp.

Longtime ABC World News Tonight anchor Peter Jennings, 67, died August 7 at his home in Manhattan, four months after being diagnosed with lung cancer.

Following Hurricane Katrina, local TV broadcasters and cable operators in the Gulf Coast area say rebuilding their stations and plants could take several months.

The Disney-ABC Television Group announces that three ABC shows, Desperate Housewives, Lost, and Night Stalker will be available for purchase from the Apple iTunes store for $1.99 an episode. The announcement prompts the other big media companies to begin "repurposing" primetime programming on the Internet. It's soon clear that the Web is the next big TV medium.

2006

In January, PBS dips into the ranks of its member stations and selects Paula Kerger of WNET New York to succeed Pat Mitchell as president of the noncommercial "network."

After battling to be the broadcasting fifth network for 11 years and mostly lackluster years, WB and UPN stun the broadcasting industry in January by deciding to merger into The CW. To fill the vacuum created by the loss of one network, Fox creates My Network Television, a mini network built around telenovelas, a popular Spanish TV format. Both CW and MNT debut in September.

Two years after the Janet Jackson "wardrobe malfunction" at the Super Bowl, broadcasters are still feeling the fallout. In March, the FCC issues another round of fines topped by $3.6 million against CBS affiliates for airing an episode of Without a Trace. A few months later, Congress increases ten-fold the base indecency fine to $325,000 per incident.

Ending a year of speculation, CBS announces in March the hiring of Katie Couric, the popular co-host of NBC's Today Show, to anchor the CBS Evening News. With new set and features, she begins her reign as anchor on September 5. Longtime anchor Dan Rather resigned from the job in March 2005 after botching a 60 Minutes story critical of President Bush's military record. CBS News's Washington Bureau Chief Bob Schieffer anchored the news during the Rather-Couric interregnum.

2007

On January 29 ION Media Networks Inc. changed the name of its TV network from "I" to ION Television.

In 2007 The Sopranos ended an eight-year run on HBO. There was much speculation about the final moments of the finale when the show faded to black.

After years of acquiring stations, on April 20 Clear Channel Communications Inc. entered into an agreement to sell its Television Group. And, as of June 30, the company had entered into definitive agreements to sell 389 radio stations in 77 markets.

The FCC and the Rules of Broadcasting

The following is the FCC's own overview of the laws and regulations that govern TV and radio stations. It is edited and printed here with the permission of the FCC. The agency invites those with questions about its rules and how it operates to visit its web site (www.fcc.gov) or to call 1-888-CALLFCC (1-888-225-5322).

THE FCC AND ITS REGULATORY AUTHORITY

The Communications Act

The FCC was created by Congress in the Communications Act of 1934 for the purpose, in part, of "regulating interstate and foreign commerce in communication by wire and radio so as to make available, so far as possible, to all the people of the United States a rapid, efficient, Nation-wide, and worldwide wire and radio communications service.." The Communications Act authorizes the FCC to "make such regulations not inconsistent with law as it may deem necessary to prevent interference between stations and to carry out the provisions of [the] Act.

How the FCC Adopts Regulations

Like most other federal agencies, the FCC cannot adopt regulations without first notifying and seeking comment from the public. The agency releases a document called a Notice of Proposed Rulemaking, where it explains the specific regulations being proposed and set a deadline for public comment. After receiving the comments, the FCC has several options: (1) adopt the proposed rules; (2) adopt a modified version of the proposed rules; (3) ask for public comment on additional issues relating to the proposals; or (4) end the rulemaking proceeding without adopting any rules at all. The FCC also establishes broadcast regulatory policies through individual cases that it decides.

The FCC and the Media Bureau

The FCC has five commissioners who are appointed by the President and confirmed by the Senate. Under the commissioners are various operating bureaus, one of which is the Media Bureau. The Media Bureau has day-to-day responsibility for developing, recommending and administering the rules governing radio and television stations. These rules are in Title 47 of the Code of Federal Regulations ("CFR"), Parts 73 and 74. The rules of practice and procedure are in Part 1 of Title 47.

FCC REGULATION OF BROADCAST TV AND RADIO

The FCC allocates new stations based both on the relative needs of communities for additional broadcast outlets and on engineering standards that prevent interference between stations. Whenever it looks at a broadcast station application/Mwhether to build, modify, renew or sell/Mit must determine if granting it would serve the public interest. The FCC expects stations to be aware of the important problems or issues in their communities and to foster public understanding by presenting some programs and/or announcements about local issues. However, broadcasters/Mnot the FCC or any other government agency/Mare responsible for selecting all the material they air. The Communications Act prohibits the FCC from censoring broadcast matter and, therefore, its role in overseeing the content of programming is limited. It is authorized to fine a station or revoke its license if it has, among other things, aired obscene language, broadcast indecent language when children are likely to be in the audience, broadcast some types of lottery information, or solicited money under false pretenses.

THE LICENSING OF TV AND RADIO STATIONS

Commercial and Noncommercial-Educational Stations

The FCC licenses radio and TV stations to be either commercial or noncommercial-educational. Commercial stations generally support themselves by advertising. In contrast, noncommercial-educational stations (including public stations) generally support themselves by contributions from listeners and viewers, and they may also receive government funding. Noncommercial-educational stations may also receive contributions from for-profit entities, and they may acknowledge such contributions or underwriting donations with announcements naming and generally describing the entity. However, noncommercial-educational stations may not broadcast promotional announcements or commercials on behalf of for-profit entities.

Applications to Build New Stations; Length of the License Period

To build a new TV or radio station, a citizen must first apply to the FCC for a construction permit. The applicant must demonstrate that he is qualified to construct and operate as proposed in the application. After the applicant has built the station, he must file a license application, where he certifies that he has constructed the station consistently with the construction permit. The FCC licenses radio and TV stations for a period of up to eight years. Before the FCC renews a station's license, it must first determine whether it has served the public interest. In addition, to have its license renewed, a station must certify that: (1) it has sent us certain specified reports that we require; (2) its ownership is consistent with Section 310(b) of the Communications Act, which restricts interests held by foreign governments and non-citizens; (3) there has not been a judgment against it by a court or administrative body under federal, state, or local law; and (4) it has placed certain specified material in its public inspection file (see below).

Employment Discrimination and Equal Employment Opportunity (EEO)

The FCC requires all radio and TV stations to afford equal opportunity in employment. We also prohibit employment discrimination on the basis of race, color, religion, national origin, or sex.

Public Participation in Licensing Process

Renewal Applications. Citizens can file a formal protest against a station by filing a formal petition to deny its renewal application, or by sending us an informal objection to the application. The citizen must file a petition to deny the application by the end of the first day of the last full calendar month of the expiring license term. (For example, if the license expires on December 31, the petition must be filed by the end of the day on December 1.) Before a citizen files a petition to deny an application, he should check the FCC's rules and policies to make sure that the petition complies with the procedural requirements. Before their licenses expire, stations have to broadcast announcements giving the date the license will expire, the date on which a renewal application must be filed, and the date by which formal petitions against it must be filed. A citizen can file an informal objection at any point until the FCC grant or deny the application.

Other Types of Applications. Citizens may also participate formally in the application process when a station is sold (technically called an assignment of the license), undergoes a major stock transfer (technically called a transfer of control), or proposes major construction. The station owner is required to run a series of advertisements in the closest local newspaper when it files these types of applications. Later, the FCC will also run a Public Notice (all FCC Public Notices are placed on our Internet home page at www.fcc.gov) and open a 30-day period during which you may file petitions to deny these applications. As with renewal applications, you can also file an informal objection at any point until we either grant or deny the application.

BROADCAST PROGRAMMING

The FCC and Freedom of Speech

The First Amendment and federal law generally prohibit us from censoring broadcast material and from interfering with freedom of expression in broadcasting. Individual radio and TV stations are responsible for selecting everything they broadcast and for determining how they can best serve their communities. Stations are responsible for choosing their entertainment programming, as well as their programs concerning local issues, news, public affairs, religion, sports events, and other subjects. They also decide how their programs (including call-in shows) will be conducted and whether to edit or reschedule material for broadcasting. The FCC does not substitute its judgment for that of the station, and it does not advise stations on artistic standards, format, grammar, or the quality of their programming. This also applies to a station's commercials, with the exception of commercials for political candidates during an election (see below).

Access to Station Facilities

Stations are not required to broadcast everything that is offered or suggested to them. Except as required by the Communications Act, stations have no obligation to have any particular person participate in a broadcast or to present that person's remarks. Further, no federal law or rule requires stations to broadcast "public service announcements" of any kind. The FCC generally does not require stations to keep the material they broadcast.

Station Identification

Stations must make identification announcements when they sign on and off for the day. They must also make the announcements hourly, as close to the hour as possible, at a natural programming break. TV stations may make these announcements on-screen or by voice only. Official station identification includes the station's call letters followed by the community or communities specified in its license as the station's location. Between the call letters and its community, the station may insert the name of the licensee, the station's channel number, and/or its frequency. However, we do not allow any other insertion.

Broadcast Journalism

Under the First Amendment and the Communications Act, the FCC cannot tell stations how to select material for news programs, and we cannot prohibit the broadcasting of an opinion on any subject. We also do not review anyone's qualifications to gather, edit, announce, or comment on the news; these decisions are the station's responsibility.

Broadcasts by Candidates for Public Office (Equal Time)

When a qualified candidate for public office has been permitted to use a station, the law requires the station to "afford equal opportunities to all other such candidates for that office." The Act also states that the station "shall have no power of censorship over the material broadcast" by the candidate. The FCC exempts the following: (1) An appearance by a legally qualified candidate on a bona fide newscast, interview or documentary (if the appearance of the candidate is incidental to the presentation of the subject covered by the documentary); or (2) on-the-spot coverage of bona fide news events, including political conventions and related incidental activities.

Children's Television Programming

Throughout its license term, every TV station must serve the educational and informational needs of children both through its overall programming. It must also broadcast programming that is specifically designed to serve those needs. The FCC considers programming to be educational and informational if it furthers the educational and informational needs of children 16 years old and under (this includes their intellectual/cognitive or social/emotional needs). A program is considered to be "specifically designed to serve educational and information needs of children" if (1) that it its principal purpose; (2) it is aired between the hours of 7:00 a.m. and 10:00 p.m.; (3) it is a regularly scheduled weekly program; and (4) it is at least 30 minutes in length. Commercial TV stations must identify programs specifically designed to educate and inform children at the beginning of the program, in a form left to their discretion, and must provide information identifying such programs to publishers of program guides. Additionally, in TV programs aimed at children 12 and under, advertising may not exceed 10.5 minutes an hour on weekends and 12 minutes an hour on weekdays.

Criticism, Ridicule, and Humor Concerning Individuals, Groups, and Institutions

The First Amendment's guarantee of freedom of speech protects programming that stereotypes or otherwise offends people with regard to their religion, race, national background, gender, or other characteristics. It also protects broadcasts that criticize or ridicule established customs and institutions, including the government and its officials.

Clear and Present Danger

The First Amendment protects advocacy of using force or of violating the law. However, the Supreme Court has said that the government may curtail speech if it is both: (1) intended to incite or produce dangerous activity; and (2) likely to succeed in achieving that result. Even where this "clear and present danger" test is met, the FCC believes that any review that might lead to a curtailment of speech should be performed by the appropriate criminal law enforcement authorities, and not by the FCC.

Obscenity and Indecency

Federal law prohibits the broadcasting of obscene programming and regulates the broadcasting of "indecent" language and images. Obscene speech is not protected by the First Amendment and cannot be broadcast at any time. To be obscene, material must have all three of the following characteristics: (1) an average person, applying contemporary community standards, must find that the material, as a whole, appeals to the prurient interest; (2) the material must depict or describe, in a patently offensive way, sexual conduct specifically defined by applicable law; and (3) the material, taken as a whole, must lack serious literary, artistic, political, or scientific value. Indecent speech is protected by the First Amendment and cannot be outlawed. However, the courts have upheld Congress's prohibition of the broadcast of indecent speech during times of the day when there is a reasonable risk that children may be in the audience. The FCC has decided that those times fall between 6:00 a.m. and 10:00 p.m. In other words, stations may only broadcast indecent material between 10 p.m. and 6 a.m. Indecent speech is defined as "language or material that, in context, depicts or describes, in terms patently offensive as measured by contemporary community standards for the broadcast medium, sexual or excretory organs or activities." Profanity that does not fall under one of the above two categories is fully protected by the First Amendment and cannot be regulated.

Violent Programming

Neither the law nor FCC rules regulate violent programming. However, the law requires TV sets with screens 13 inches or larger to be equipped with V-chip technology, which allows parents to program their TV sets to block display of TV programming that carries a certain rating. The rating system, voluntarily created by the television industry, tags programming that contains sexual, violent, or other indecent material programmers believe may be harmful to children.

Station-Conducted Contests

Stations that broadcast or advertise information about a contest that they conduct must fully and accurately disclose the material terms of the contest, and they must conduct the contest substantially as announced or advertised. Contest descriptions may not be false, misleading, or deceptive with respect to any material term. Material terms include the factors that define the operation of the contest and affect participation.

Broadcast Hoaxes

Broadcasting false information concerning a crime or a catastrophe violates the FCC's rules if: (1) the station knew the information was false; (2) broadcasting the false information directly caused substantial public harm; (3) and it was foreseeable that broadcasting the false information would cause substantial public harm. In this context, a "crime" is an act or omission that makes the offender subject to criminal punishment by law, and a "catastrophe" is a disaster or imminent disaster involving violent or sudden events affecting the public. "Public harm" must begin immediately; it must cause direct and actual damage to property or to the health or safety of the general public, or diversion of law enforcement or other public health and safety authorities from their duties.

Lotteries

The law prohibits broadcasting any advertisement for a lottery or any information concerning a lottery. A lottery is any game, contest, or promotion that contains the elements of prize, chance, and "consideration" (a legal term that means an act or promise that is made to induce someone into an agreement). There are a number of exceptions to this prohibition. Some of the exceptions are: (1) lotteries conducted by a state acting under the authority of state law, where the advertisement or information is broadcast by a radio or TV station licensed to a location in that state or in any other state that conducts such a lottery; (2) gaming conducted by an Indian Tribe under the Indian Gaming Regulatory Act; (3) lotteries authorized or not otherwise prohibited by the state in which they are conducted, and which are conducted by a not-for-profit organization or a governmental organization; and (4) lotteries conducted as a promotional activity by commercial organizations that are clearly occasional and ancillary to the primary business of that organization, as long as the lotteries are authorized or not otherwise prohibited by the state in which they are conducted.

Soliciting Funds

No federal law prohibits broadcast requests for funds for legal purposes (including appeals by stations for contributions to meet their operating expenses) if the money or other valuable things contributed are used for the announced purposes. It is up to an individual station to decide whether to permit fund solicitations. Fraud by wire, radio or television is prohibited by federal law and may lead to FCC sanctions, as well as to criminal prosecution by the U.S. Department of Justice.

Broadcasting Telephone Conversations

Before recording a telephone conversation for broadcast, or broadcasting a telephone conversation live, a station must inform any party to the call of its intention to broadcast the conversation. However, this does not apply to conversations whose broadcast can reasonably be presumed (for example, telephone calls to programs where the station customarily broadcasts the calls).

BROADCASTING AND ADVERTISING

Business Practices, Advertising Rates, and Profits

Except with respect to political advertisements, the FCC does not regulate a station's advertising rates or its profits. Rates charged for broadcast time are matters for negotiation between sponsors and stations. Further, except for certain classes of political advertisements, stations are free to accept or reject any advertising.

Sponsorship Identification

Sponsorship identification or disclosure must accompany any material that is broadcast in exchange for money, service, or anything else of value paid to a station, either directly or indirectly. This announcement must clearly say that the time was purchased and by whom. In the case of advertisements for commercial products or services, it is sufficient to announce the sponsor's corporate or trade name, or the name of the sponsor's product when it is clear that the mention of the product constitutes a sponsorship identification.

Underwriting Announcements on Noncommercial-Educational Stations

Noncommercial educational stations may acknowledge contributions over the air, but they may not promote the goods and services of for-profit donors or underwriters. Acceptable "enhanced underwriting" acknowledgements may include (1) logograms and slogans that identify but do not promote; (2) location information; (3) value-neutral descriptions of a product line or service; and (4) brand names, trade names, and product service listings. However, such acknowledgements may not interrupt a noncommercial station's regular programming.

Amount of Advertising

Except with respect to children's television programming, no law or regulation limits the amount of commercial matter that a station may broadcast. In TV programs aimed at children 12 and under, advertising may not exceed 10.5 minutes an hour on weekends and 12 minutes an hour on weekdays.

Loud Commercials

In surveys and technical studies of broadcast advertising, the FCC has found that loudness is a judgment that varies with each listener and is influenced by many factors (such as an announcement's content and style). We have also found no evidence that stations deliberately raise audio and modulation levels to emphasize commercial messages. Broadcast licensees have primary responsibility for the adoption of equipment and procedures to avoid objectionably loud commercials. Citizens should address any complaint about such messages to the station. They should identify each message by the sponsor or product's name and by the date and time of the broadcast.

False or Misleading Advertising

The Federal Trade Commission has primary responsibility for determining whether an advertisement is false or deceptive and for taking action against the sponsor. Also, the Food and Drug Administration has primary responsibility for the safety of food and drug products. Citizens should contact these agencies regarding advertisements that they believe may be false or misleading.

Offensive Advertising

Unless a broadcast advertisement is found to be in violation of a specific law or regulation, the government cannot take action against it. If a citizen thinks that an advertisement is offensive because of the kind of item advertised, the scheduling of the announcement, or the way the message is presented, then he should address his complaint directly to the stations and networks involved. This will help them become better informed about audience opinion.

Tobacco and Alcohol

The law prohibits advertising for cigarettes, little cigars, smokeless tobacco, or chewing tobacco on radio, TV, or any other medium of electronic communication under the FCC's jurisdiction. The law does not ban the advertising of smoking accessories, cigars, pipes, pipe tobacco, or cigarette-making machines. Congress has not enacted any law prohibiting broadcast advertising for any kind of alcoholic beverage. Also, the FCC does not have a rule or policy regulating advertisements for alcoholic beverages. Most broadcasters have voluntarily abstained from advertising hard liquor, although they accept beer and wine advertising.

Subliminal Programming

The FCC sometimes receives complaints regarding the alleged use of subliminal techniques in radio and TV programming. Subliminal programming is designed to be perceived on a subconscious level only. Regardless of whether it is effective, the use of subliminal perception is inconsistent with a station's obligation to serve the public interest because the broadcast is intended to be deceptive. However, it is not specifically prohibited.

INTERFERENCE

Blanketing Interference

Some people who are close to a radio station's transmitting antenna may experience impaired reception of other stations. This is called "blanketing" interference. The FCC requires the station causing the interference to resolve most interference complaints received within the first year of operation at no cost to the person complaining. However, stations are not required to resolve interference complaints based on malfunctioning or mistuned receivers, improperly installed antenna systems, or the use of high gain antennas or antenna booster amplifiers. Mobile receivers and non-radio frequency (RF) devices such as tape recorders or CD players are also excluded. Stations are not financially responsible for resolving interference complaints located outside the blanketing contour.

THE LOCAL PUBLIC INSPECTION FILE

Requirement to Maintain a Public Inspection File

The FCC requires all TV and radio stations and applicants for new stations to maintain a file available for public inspection containing documents relevant to the station's operation. The public inspection file generally must be maintained at the station's main studio. The station must make its public inspection file available at its main studio at any time during regular business hours. A station that chooses to maintain all or part of its public file in a computer database must provide a computer terminal to those who wish to review the file. Stations must keep the following materials in their public inspection file:

The License. Stations must keep a copy of their current FCC license in the public file, together with any material documenting FCC-approved modifications to the license. The license reflects the station's technical parameters (authorized frequency, call letters, operating power, transmitter location, etc.), as well as any special conditions imposed by the FCC on the station's operation. The license also indicates when it was issued and when it will expire.

Applications and Related Materials. The public file must contain copies of all applications that are pending before either the FCC or the courts. These include applications to sell the station or to modify its facilities (for example, to increase power, change the antenna system, or change the transmitter location); copies of any construction or sales application whose grant required the FCC to waive our rules; applications that required the FCC to waive its rules; renewal applications granted for less than a full license term (the FCC grants short-term renewals when it is concerned about the station's performance over the previous term).

Citizen Agreements. Stations must keep a copy of any written agreements they make with local viewers or listeners. These "citizen agreements" deal with programming, employment, or other issues of community concern. The station must keep these agreements in the public file for as long as they are in effect.

Contour Maps. The public file must contain copies of any service contour maps or other information submitted with any application filed with the FCC that reflects the station's service area and/or its main studio and transmitter location. These documents must stay in the file for as long as they remain accurate.

Material Relating to an FCC Investigation or a Complaint. The station must keep this material until the FCC notifies it that the material may be discarded. Since the FCC is not involved in disputes regarding matters unrelated to the Communications Act or our rules, stations do not have to keep material relating to such matters in the public file.

Ownership Reports and Related Material. The public file must contain a copy of the most recent, complete Ownership Report filed for the station. This report has the names of the owners of the station and their ownership interests, lists any contracts related to the station that are required to be filed with the FCC, and identifies any interest held by the station licensee in other broadcast stations.

List of Contracts Filed with the FCC. Stations have to keep either a copy of all the contracts that they have to file with the FCC, or an up-to-date list identifying all such contracts. If the station keeps a list and you ask to see copies of the actual contracts, the station must give them to you within seven days. Such contracts include network affiliation contracts; contracts relating to ownership or control of the licensee or permittee or its stock. Examples include articles of incorporation, bylaws, agreements providing for the assignment of a license or permit or affecting stock ownership or voting rights (stock options, pledges, or proxies), and mortgage or loan agreements that restrict the licensee or permittee's freedom of operation; management consultant agreements with independent contractors, and station management contracts that provide for a percentage of profits or sharing of losses.

Political File. Stations must keep a file containing records of all requests for broadcast time made by or for a candidate for public office. The file must identify how the station responded to such requests and (if the request was granted) the charges made, a schedule of the time purchased, the times the spots actually aired, the rates charged, and the classes of time purchased. The file must also reflect any free time provided to a candidate. The station must keep the political records for two years after the spot airs. You can find the political broadcasting rules elsewhere in this manual.

Letters and E-Mail from the Public. Commercial stations must keep written comments and suggestions received from the public regarding their operation for at least three years. Noncommercial stations are not subject to this requirement.

Issues/Programs List. Every three months, all stations must prepare and place in their file a list of programs that have provided their most significant treatment of community issues during the preceding three months. The list must briefly describe both the issue and the programming where the issue was discussed. The stations must keep these lists for the entire license term.

Children's Television Programming Reports. The Children's Television Act of 1990 and our rules require all TV stations to air programming that serves the educational and informational needs of children 16 and under, including programming that is specifically designed to serve such needs. In addition, commercial TV stations must make and retain Children's Television Programming Reports identifying the educational and informational programming for children aired by the station. (Noncommercial stations are not required to prepare these reports.) The report must include the name of the person at the station responsible for collecting comments on the station's compliance with the law. The station has to prepare these reports each calendar quarter, and it must place them in the public file separately from the file's other material. Stations must keep the reports for the remainder of their license terms. You can also view each station's reports on our web site at http://www.fcc.gov/mb/policy/kidstv.html.

Records Regarding Children's Programming Commercial Limits. The Children's Television Act of 1990 and our rules limit the type and amount of advertising that may be aired in TV programming directed to children 12 and under. On weekends, commercial television stations may air no more than 10.5 minutes of commercials per hour during children's programming, and no more than 12 minutes on weekdays. Stations must keep records that substantiate compliance with these limits.

Radio Time Brokerage Agreements. A time brokerage agreement is a type of contract that generally involves a station's sale of discrete blocks of air time to a broker, who then supplies the programming to fill that time and sells the commercial spot announcements to support the programming. Commercial radio stations must keep a copy of every agreement involving: (1) time brokerage of that station; or (2) time brokerage by any other station owned by the same licensee.

List of Donors. Noncommercial TV and radio stations must keep a list of donors supporting specific programs for two years after the program airs.

Local Public Notice Announcements. When someone files an application to build a new station or to renew, sell, or modify an existing station, the FCC often requires the applicant to make a series of local announcements to inform the public of the application's existence and nature.

Must-Carry or Retransmission Consent Election. There are two ways that a broadcast TV station can choose to be carried on a cable TV system: "must-carry" and "retransmission consent." All TV stations are generally entitled to be carried on cable television systems in their local markets. A station that chooses to exercise this must-carry right receives no compensation from the cable system. Instead of exercising their must-carry rights, TV stations may choose to receive compensation from a cable system in return for granting permission to the cable system to carry the station. This option is available only to commercial TV stations. Every three years, commercial TV stations must decide whether their relationship with each local cable system will be governed by must-carry or by retransmission consent agreements. Each commercial station must keep a copy of its decision in the public file for the three-year period to which it pertains. Noncommercial stations are not entitled to compensation in return for carriage on a cable system, but they may request mandatory carriage on the system. A noncommercial station making this request must keep a copy of the request in the public file for the duration of the period to which it applies.

Section B

Broadcast Television

TV Group Ownership

A

ABC Inc., 77 W. 66th St., New York, NY, 10023-6298. Phone: (212) 456-7777. Web Site: www.abc.com. Ownership: ABC Enterprises Inc., 100%. Note: ABC Enterprises Inc. is 100% owned by Disney Enterprises Inc. Disney Enterprises Inc. is 100% owned by The Walt Disney Co.

Stns: 10 TV. WLS, Chicago; WJRT-TV, Flint-Saginaw-Bay City, MI; KFSN-TV, Fresno-Visalia, CA; KTRK, Houston; KABC, Los Angeles; WABC-TV, New York; WPVI, Philadelphia; WTVD, Raleigh-Durham (Fayetteville), NC; KGO, San Francisco-Oakland-San Jose; WTVG, Toledo, OH.

Stns: 42 AM. 3 FM. KDIS-FM Little Rock, AR; KMIK Tempe, AZ; KSPN(AM) Los Angeles, CA; KMKY Oakland, CA; KDIS(AM) Pasadena, CA; KIID(AM) Sacramento, CA; KDDZ Arvada, CO; WDZK Bloomfield, CT; WBWL Jacksonville, FL; WMYM(AM) Miami, FL; WDYZ(AM) Orlando, FL; WMNE(AM) Riviera Beach, FL; WWMI Saint Petersburg, FL; WDWD Atlanta, GA; WSDZ Belleville, IL; WMVP Chicago, IL; WRDZ La Grange, IL; WRDZ-FM Plainfield, IN; KQAM Wichita, KS; WDRD(AM) Newburg, KY; WBYU New Orleans, LA; WMKI(AM) Boston, MA; WFDF(AM) Farmington Hills, MI; KDIZ Golden Valley, MN; KPHN Kansas City, MO; WGFY Charlotte, NC; WWJZ Mount Holly, NJ; KALY Los Ranchos de Albuquerque, NM; WDDY(AM) Albany, NY; WEPN(AM) New York, NY; WWMK Cleveland, OH; KMUS(AM) Sperry, OK; KDZR(AM) Lake Oswego, OR; WEAE Pittsburgh, PA; WDDZ(AM) Pawtucket, RI; KESN(FM) Allen, TX; KMIC(AM) Houston, TX; KMKI Plano, TX; KRDY(AM) San Antonio, TX; KWDZ(AM) Salt Lake City, UT; WDZY Colonial Heights, VA; WRJR(AM) Portsmouth, VA; WHKT Portsmouth, VA; KKDZ Seattle, WA; WKSH(AM) Sussex, WI.

Robert A. Iger, pres; Phillip J. Meek, pres; Lawrence J. Pollock, chmn owned TV stns.

ACME Communications Inc., 2101 E. Fourth St., Suite 202A, Santa Ana, CA, 92705. Phone: (714) 245-9499. Fax: (714) 245-9494. E-mail: t.allen@acmecomm.com Web Site: www.acmecommunications.com. Ownership: Alta Cpmmunications; Seaport Capital.

Stns: 8 TV. KASY, Albuquerque-Santa Fe, NM; KWBQ, Albuquerque-Santa Fe, NM; KRWB-TV, Albuquerque-Santa Fe, NM; WBUI, Champaign & Springfield-Decatur, IL; WBDT, Dayton, OH; WIWB, Green Bay-Appleton, WI; WBXX, Knoxville, TN; WBUW, Madison, WI.

Jamie Kellner, chmn/CEO; Doug Gealy, pres/COO; Tom Allen, exec VP & CFO.

Access.1 Communications Corp., 11 Penn Plaza, 16th Fl., New York, NY, 10001. Phone: (212) 714-1000. Fax: (212) 714-1563. Ownership: Sydney L. Small, 54.12%; Black Enterprise/Greenwich Street Capital Partners, 19.47%; MESBIC Ventures Inc., 5.54%; Chesley Maddox-Dorsey, 2.84%; and Adriane Gaines, 1.85%.

Stns: 1 TV. WMGM-TV, Philadelphia.

Stns: 7 AM. 13 FM. KSYR(FM) Benton, LA; KDKS-FM Blanchard, LA; KBTT(FM) Haughton, LA; KLKL(FM) Minden, LA; KOKA Shreveport, LA; WMGM(FM) Atlantic City, NJ; WGYM(AM) Hammonton, NJ; WTKU-FM Ocean City, NJ; WJSE-FM Petersburg, NJ; WTAA(AM) Pleasantville, NJ; WOND(AM) Pleasantville, NJ; WWRL New York, NY; KOYE(FM) Frankston, TX; KOOI-FM Jacksonville, TX; KYKX-FM Longview, TX; KFRO Longview, TX; KCUL Marshall, TX; KCUL-FM Marshall, TX; KTAL-FM Texarkana, TX; KKUS(FM) Tyler, TX.

Sydney L. Small, chmn/CEO.

Allbritton Communications Co., 1000 Wilson Blvd., Suite 2700, Arlington, VA, 22209. Phone: (703) 647-8700. Fax: (703) 647-8707. Web Site: www.allbritton.com.

Stns: 8 TV. WCFT, Birmingham (Anniston, Tuscaloosa), AL; WJSU, Birmingham (Anniston, Tuscaloosa), AL; WCIV, Charleston, SC; WHTM, Harrisburg-Lancaster-Lebanon-York, PA; KATV, Little Rock-Pine Bluff, AR; WSET-TV, Roanoke-Lynchburg, VA; KTUL, Tulsa, OK; WJLA-TV, Washington, DC (Hagerstown, MD).

Allbritton Communications, through affiliated company, publishes the *Enfield* (CT) *Press, The Longmeadow* (MA) *News, Westfield* (MA) *Evening News* & *The Penny Saver,* Westfield, MA.

Also owns Cable-NewsChannel 8, Arlington, VA. All 100% owned.

Frederick Ryan, pres; Robert L. Allbritton, CEO.

AsianMedia Group LLP, 1990 S. Bundy Dr., c/o KSCI, Suite 850, Los Angeles, CA, 90025. Phone: (310) 478-1818. Fax: (310) 479-8118. E-mail: info@la18.tv Web Site: www.la18.tv. Ownership: Leonard Green & Partners, LLP.

Stns: 2 TV. KIKU-TV, Honolulu, HI; KSCI, Los Angeles.

Peter Mathes, chmn/CEO; John Chang, CFO.

B

Bahakel Communications, Box 32488, Charlotte, NC, 28232. Phone: (704) 372-4434. Fax: (704) 335-9904. Ownership: The Cy N. Bahakel Trust Dated January 12, 2005, 100%.

Stns: 6 TV. WCCU, Champaign & Springfield-Decatur, IL; WRSP-TV, Champaign & Springfield-Decatur, IL; WCCB, Charlotte, NC; WABG-TV, Greenwood-Greenville, MS; WAKA, Montgomery-Selma, AL; WFXB, Myrtle Beach-Florence, SC.

Stns: 4 AM. 5 FM. KILO-FM Colorado Springs, CO; KYZX(FM) Pueblo West, CO; KOKZ-FM Waterloo, IA; KWLO(AM) Waterloo, IA; KXEL Waterloo, IA; KFMW-FM Waterloo, IA; WABG(AM) Greenwood, MS; WDEF-FM Chattanooga, TN; WDOD Chattanooga, TN.

Cy N. Bahakel, pres; Beverly Poston, exec VP/COO; Stephen Bahakel, Sr VP radio div; Russell Schwartz, Sr VP business affrs/gen counsel; Bill Napier, VP eng/tech; Anna Rufty, VP Hum Res.

Banks Broadcasting Inc., 1124 Merrill St., Winnetka, IL, 60093. Phone: (847) 446-9995. Fax: (847) 446-9997. Ownership: LIN Television; Banc of America; 21st Century Group.

Stns: 1 TV. KNIN, Boise, ID.

Lyle Banks, pres/CEO.

Barrington Broadcasting Group, LLC., 2500 W. Higgins Rd., Suite 155, Hoffman Estates, IL, 60169-7275. Phone: (847) 884-1877. Fax: (847) 755-3045. E-mail: info@barringtontv.com Web Site: www.barringtontv.com. Ownership: Pilot Group LP, 100% of votes.

Stns: 18 TV. WFXL, Albany, GA; KVIH, Amarillo, TX; KVII-TV, Amarillo, TX; KXRM, Colorado Springs-Pueblo, CO; WACH, Columbia, SC; KRCG, Columbia-Jefferson City, MO; WEYI-TV, Flint-Saginaw-Bay City, MI; WBSF, Flint-Saginaw-Bay City, MI; KGBT-TV, Harlingen-Weslaco -Brownsville-McAllen, TX; WLUC, Marquette, MI; WPDE, Myrtle Beach-Florence, SC; KTVO, Ottumwa, IA-Kirksville, MO; WHOI, Peoria-Bloomington, IL; KHQA, Quincy, IL-Hannibal, MO-Keokuk, IA; WSTM, Syracuse, NY; WNWO, Toledo, OH; WPBN, Traverse City-Cadillac, MI; WTOM, Traverse City-Cadillac, MI.

Paul M. McNicol, sr VP; K. James Yager, CEO; Chris Cornelius, pres/COO; Warren Spector, CFO; Paul McNicol, sec; Mary Flodin, sr VP; Keith Bland, sr VP.

Beach TV Properties Inc., Box 9556, Panama City Beach, FL, 32417. Phone: (850) 234-2773.

Stns: 2 TV. WAWD, Mobile, AL-Pensacola (Ft. Walton Beach), FL; WPCT, Panama City, FL.

Byron J. Colley, pres.

Bela LLC, 7500 N.W. 72 Ave., Medley, FL, 33166. Phone: (305) 530-1322. Ownership: Star Studios LLC, 51%; Cranston LLC, 44%; and Leibowitz Family Broadcasting LLC, 5%.

Stns: 2 TV. KMOH, Phoenix (Prescott), AZ; KBEH, Santa Barbara-Santa Maria-San Luis Obispo, CA.

Robert Behar, pres; Matthew L. Leibowitz, VP.

Belo Corp, (Television Group). 400 S. Record St., Dallas, TX, 75202. Phone: (214) 977-6600. Fax: (214) 977-6603. Web Site: www.belo.com. Ownership: Belo Corp.

Stns: 19 TV. KVUE, Austin, TX; KTVB, Boise, ID; WCNC, Charlotte, NC; WFAA-TV, Dallas-Ft. Worth; KHOU-TV, Houston; WHAS, Louisville, KY; WWL, New Orleans, LA; WVEC, Norfolk-Portsmouth-Newport News, VA; KTVK, Phoenix (Prescott), AZ; KASW, Phoenix (Prescott), AZ; KGW, Portland, OR; KENS, San Antonio, TX; KING, Seattle-Tacoma, WA; KONG, Seattle-Tacoma, WA; KREM, Spokane, WA; KSKN, Spokane, WA; KMOV, St. Louis, MO; KMSB, Tucson (Sierra Vista), AZ; KTTU, Tucson (Sierra Vista), AZ.

The publishing division of Belo Corp., publishes the following dailies: *The Dallas* (TX) *Morning News* and *The Press-Enterprise,* Riverside, CA; *Denton Record Chronicle,* Denton, TX; *The Providence Journal,* Providence, RI; *Arlington Morning News,* Arlington, VA. Other interests: News cable channels (100% owned): Northwest Cable News, Texas Cable News, 24/7 Newschannel (distribution in ID from Boise, ID). News cable channel partnerships in the following DMAs: Phoenix, AZ; New Orleans; and Norfolk-Portsmouth-Newport News, VA.

Robert W. Decherd, chmn.

Block Communications Inc., 6450 Monroe St., Sylvania, OH, 43560. Phone: (419) 724-6448. Fax: (419) 724-6167. Web Site: www.blockcommunications.com. Ownership: Estate of William Block, Allan Block, John R. Block.

Stns: 4 TV. KTRV-TV, Boise, ID; WLIO, Lima, OH; WDRB, Louisville, KY; WMYO, Louisville, KY.

Block Communications Inc. publishes the *Toledo* OH *Blade* & *Pittsburgh* (PA) *Post-Gazette.*

Allan Block, chmn; Gary J. Blair, exec VP; Jodi Miehls, treas; David Huey, pres.

BlueStone Television LLC, 8415 E. 21st St. N., Suite 120, Wichita, KS, 67206. Phone: (316) 315-0076. Fax: (316) 315-0345. E-mail: info@bluestonetv.com Web Site: www.bluestonetv.com. Ownership: Providence Equity Partners IV L.P., 95.8% votes, 99.5% total assets. Note: Group also owns KTXE-LP San Angelo, TX.

Stns: 7 TV. KTXS-TV, Abilene-Sweetwater, TX; KTVM, Butte-Bozeman, MT; KRCR, Chico-Redding, CA; KAEF, Eureka, CA; KCFW, Missoula, MT; KECI-TV, Missoula, MT; WCYB-TV, Tri-Cities, TN-VA.

Sandy DiPasquale, pres/CEO; John Grossi, VP; K.J. Lager, controller.

Bonneville International Corporation, Broadcast House, Box 1160, Salt Lake City, UT, 84110-1160. Phone: (801) 575-7500. Fax: (801) 575-7521. Web Site: www.bonnint.com. Ownership: Deseret Management Corp. Deseret Management Corp. owns *The Deseret Morning News,* a Salt Lake City, UT, daily.

Stns: 1 TV. KSL, Salt Lake City, UT.

Stns: 9 AM. 19 FM. KTAR-FM Glendale, AZ; KPKX(FM) Phoenix, AZ; KMVP Phoenix, AZ; KTAR(AM) Phoenix, AZ; KBWF(FM) San Francisco, CA; KDFC-FM San Francisco, CA; KOIT San Francisco, CA; KOIT-FM San Francisco, CA; WTWP(AM) Washington, DC; WTOP-FM Washington, DC; WDRV(FM) Chicago, IL; WILV(FM) Chicago, IL; WMVN(FM) East St. Louis, IL; WARH(FM) Granite City, IL; WTMX-FM Skokie, IL; WWDV(FM) Zion, IL; WIL-FM Saint Louis, MO; WIL(AM) Saint Louis, MO; KSL-FM Midvale, UT; KRSP-FM Salt Lake City, UT; KSFI-FM Salt Lake City, UT; KSL Salt Lake City, UT; KUTR(AM) Taylorsville, UT; WTWP-FM Warrenton, VA; WTLP(FM) Braddock Heights, MD; WTWT(AM) Frederick, MD; WFED(AM) Silver Spring, MD; WPRS-FM Waldorf, MD.

Bruce T. Reese, pres/CEO; Robert A. Johnson, exec VP & COO.

C

CBS Television Stations Group, 524 W. 57th St., 3rd Fl., New York, NY, 10019. Phone: (212) 975-4321. Web Site: cbslocal.com. Ownership: Viacom Inc., 100%.

Stns: 37 TV. WUPA, Atlanta; KEYE, Austin, TX; WJZ, Baltimore, MD; WSBK, Boston (Manchester, NH); WBZ, Boston (Manchester, NH); WBBM, Chicago; KTVT, Dallas-Ft. Worth; KTXA, Dallas-Ft. Worth; KCNC, Denver, CO; WKBD, Detroit; WWJ-TV, Detroit; WFRV, Green Bay-Appleton, WI; KCAL, Los Angeles; KCBS, Los Angeles; WJMN, Marquette, MI; WFOR-TV, Miami-Ft. Lauderdale, FL; WBFS, Miami-Ft. Lauderdale, FL; WCCO, Minneapolis-St. Paul, MN; KCCO, Minneapolis-St. Paul, MN; KCCW, Minneapolis-St. Paul, MN; WUPL, New Orleans, LA; WCBS-TV, New York; WGNT, Norfolk-Portsmouth-Newport News, VA; KYW, Philadelphia; WPSG, Philadelphia; KDKA, Pittsburgh, PA; WPCW, Pittsburgh, PA; WLWC, Providence, RI-New Bedford, MA; KMAX, Sacramento-Stockton-Modesto, CA; KOVR, Sacramento-Stockton-Modesto, CA; KUSG, Salt Lake City, UT; KUTV, Salt Lake City, UT; KBCW, San Francisco-Oakland-San Jose; KPIX, San Francisco-Oakland-San Jose; KSTW, Seattle-Tacoma, WA; WTOG, Tampa-St. Petersburg (Sarasota), FL; WTVX, West Palm Beach-Ft. Pierce, FL.

Tom Kane, pres/CEO; Anton Guitano, exec VP.

CHUM Ltd., 1331 Yonge St., Toronto, ON, M4T 1Y1. Canada. Phone: (416) 925-6666. Fax: (416) 926-1380. Web Site: www.chumlimited.com.

Stns: 15 TV. CKVR, Barrie, ON; CKX, Brandon, MB; CKAL, Calgary, AB; CKEM, Edmonton, AB; CJAL, Edmonton, AB; CKX-1, Foxwarren, MB; CFPL-TV, London, ON; CHRO-TV-43, Ottawa, ON; CHRO, Pembroke, ON; CHMI-TV, Portage la Prarie, MB; CITY-TV, Toronto, ON; CKVU, Vancouver, BC; CIVI-TV, Victoria, BC; CHWI-TV, Windsor, ON; CKNX, Wingham, ON.

Stns: 14 AM. 16 FM. CKCE-FM Calgary, AB; CHBN-FM Edmonton, AB; CKST Vancouver, BC; CFUN Vancouver,

BC; CHQM-FM Vancouver, BC; CFAX Victoria, BC; CHBE-FM Victoria, BC; CFWM-FM Winnipeg, MB; CFRW(AM) Winnipeg, MB; CJCH Halifax, NS; CJPT-FM Brockville, ON; CFJR-FM Brockville, ON; CKLC Kingston, ON; CFLY-FM Kingston, ON; CFCA-FM Kitchener, ON; CKKW Kitchener, ON; CKLY-FM Lindsay (city of Kawartha Lakes), ON; CHST-FM London, ON; CKKL-FM Ottawa, ON; CJMJ-FM Ottawa, ON; CFGO Ottawa, ON; CFRA Ottawa, ON; CKPT Peterborough, ON; CKQM-FM Peterborough, ON; CHUM Toronto, ON; CIDR-FM Windsor, ON; CIMX-FM Windsor, ON; CKLW Windsor, ON; CKWW Windsor, ON; CKGM Montreal, PQ.

Stephen Tapp, exec VP; Peter Miller, VP; David Kirkwood, exec VP; Sarah Crawford, VP; Mary Powers, VP; Denise Cooper, VP; Alan Mayne, CFO; Paul Ski, pres & CHUM Radio.

CTV Inc., Box 9, Station O, Scarborough, ON, M4A 2M9. Canada. Phone: (416) 332-5000. Fax: (416) 332-5283. Web Site: www.ctv.ca. Ownership: Bell Globemedia Inc., 100%.

Stns: 40 TV. CKYB, Brandon, MB; CJCH-6, Caledonia, NS; CFCN, Calgary, AB; CKCD, Campbellton, NB; CJCH-1, Canning, NS; CKCW-1, Charlottetown, PE; CKCK-1, Colgate, SK; CJOH-8, Cornwall, ON; CJOH-6, Deseronto, ON; CFRN, Edmonton, AB; CICI-1, Elliot Lake, ON; CKMC-1, Golden Prairie, SK; CJCH, Halifax, NS; CITO-2, Kearns, ON; CKCO, Kitchener, ON; CFCN-5, Lethbridge, AB; CKBQ, Melfort, SK; CKCW, Moncton, NB; CFCF, Montreal, PQ; CKNY, North Bay, ON; CJOH-TV, Ottawa, ON; CIPA, Prince Albert, SK; CKCK, Regina, SK; CKCW-TV-2, Saint Edward, PE; CKLT, Saint John, NB; CKCO-3, Sarnia, ON; CFQC, Saskatoon, SK; CHBX, Sault Ste. Marie, ON; CICI, Sudbury, ON; CKMC, Swift Current, SK; CJCB, Sydney, NS; CITO, Timmins, ON; CFTO, Toronto, ON; CKAM, Upsalquitch Lake, NB; CIVT, Vancouver, BC; CIEW, Warmley, SK; CKCO-2, Wiarton, ON; CKCK-2, Willow Bunch, SK; CKY, Winnipeg, MB; CICC, Yorkton, SK.

Ivan Fecan, CEO; Rick Brace, pres.

California Oregon Broadcasting Inc., Box 1489, Medford, OR, 97501. Phone: (541) 779-5555. Fax: (541) 779-1151. E-mail: cobiadmin@kobi5.com Ownership: Patricia C. Smullin and Carol Anne Smullin Brown. Other interests: Cable TV: Crestview Cable TV (systems in Oregon).

Stns: 3 TV. KLSR, Eugene, OR; KOBI, Medford-Klamath Falls, OR; KOTI, Medford-Klamath Falls, OR.

Patricia C. Smullin, owner.

CanWest Global Communications Corp., 201 Portage Ave., 31st Fl., Winnipeg, MB, R3B 3L7. Canada. Phone: (204) 956-2025. Fax: (204) 947-9841. E-mail: bleslie@canwest.com Web Site: www.canwestglobal.com. Ownership: Asper Family 85% voting shares, 45% of equity.

Stns: 16 TV. CICT, Calgary, AB; CHEK-5, Campbell River, BC; CHCA-TV-1, Coronation, AB; CITV, Edmonton, AB; CIHF, Halifax, NS; CHCH, Hamilton, ON; CHBC, Kelowna, BC; CISA-TV, Lethbridge, AB; CJNT, Montreal, PQ; CKMI, Quebec City, PQ; CHCA-TV, Red Deer, AB; CFRE, Regina, SK; CFSK, Saskatoon, SK; CHAN, Vancouver, BC; CHEK, Victoria, BC; CKND-TV, Winnipeg, MB.

Stns: 1 FM. CHAL-FM Halifax, NS.

Leonard Asper, pres/CEO; David Asper, exec VP.

Capital Community Broadcasting Inc., 360 Egan Dr., Juneau, AK, 99801-1748. Phone: (907) 586-1670. Fax: (907) 586-3612.

Stns: 1 TV. KTOO-TV, Juneau, AK.

Stns: 3 FM. KRNN(FM) Juneau, AK; KXLL(FM) Juneau, AK; KTOO-FM Juneau, AK.

Capitol Broadcasting Co. Inc., Box 12000, Raleigh, NC, 27605. Phone: (919) 821-8555. Fax: (919) 821-8733. Web Site: www.cbc-raleigh.com. Ownership: Capitol Holding Co. Inc.

Stns: 4 TV. WMYT-TV, Charlotte, NC; WJZY, Charlotte, NC; WRAL-TV, Raleigh-Durham (Fayetteville), NC; WRAZ, Raleigh-Durham (Fayetteville), NC.

Stns: 2 FM. WCMC-FM Creedmoor, NC; WRAL(FM) Raleigh, NC.

James F. Goodmon, pres/CEO; Vicke S. Murray, sec; Daniel P. McGrath, VO/CFO; Michael D. Hill, VP/gen counsel; James R. Hefner, III, VP & tv.

Cascade Broadcasting Group L.L.C., 60 E. Sir Francis Drake Blvd., Suite 300, Larkspur, CA, 94939. Phone: (415) 925-6500.

Stns: 2 TV. WBKI-TV, Louisville, KY; KWBA, Tucson (Sierra Vista), AZ.

Chambers Communications Corp., Box 7009, Eugene, OR, 97401. Phone: (541) 485-5611. Fax: (541) 342-1568. Web Site: www.cmc.net/chambers. E-mail: kezi@kezi.com Ownership: Carolyn S. Chambers. Other interests: Oregon cable TV.

Stns: 4 TV. KOHD, Bend, OR; KEZI, Eugene, OR; KDKF, Medford-Klamath Falls, OR; KDRV, Medford-Klamath Falls, OR.

Scott Chambers, pres; Carolyn Chambers, CEO.

Christian Faith Broadcasting Inc., 3809 Maple Ave., Castalia, OH, 44824. Phone: (419) 684-5311. Fax: (419) 684-5378. E-mail: wggn@lrbcg.com

Stns: 2 TV. WGGN, Cleveland-Akron (Canton), OH; WLLA, Grand Rapids-Kalamazoo-Battle Creek, MI.

Stns: 3 FM. WJKW-FM Athens, OH; WGGN-FM Castalia, OH; WLRD(FM) Willard, OH.

Shelby Gillam, pres; Rusty Yost, VP.

Christian Television Corporation Inc., 6922 142nd Ave. N., Largo, FL, 33771. Phone: (727) 535-5622. Fax: (727) 531-2497. Web Site: www.ctnonline.com. Ownership: Robert D'Andrea, 25% of votes; Virginia Oliver, 25% of votes; Jimmy Smith, 25% of votes; and Wayne Wetzel, 25% of votes.

Stns: 2 TV. KFXB, Cedar Rapids-Waterloo-Iowa City & Dubuque, IA; WCLF, Tampa-St. Petersburg (Sarasota), FL.

Robert D'Andrea, pres.

Citadel Communications Co. LTD., (Coronet Communications, Capital Communications, Citadel Comm. LLC.). 99 Pondfield Rd., Bronxville, NY, 10708. Phone: (914) 793-3400. Fax: (914) 793-3693. E-mail: citnyltd@aol.com Ownership: (Coronet Communications, Capital Communications, Citadel Comm. LLC.)

Stns: 4 TV. WHBF-TV, Davenport, IA-Rock Island-Moline, IL; WOI, Des Moines-Ames, IA; KLKN, Lincoln & Hastings-Kearney, NE; KCAU, Sioux City, IA.

Philip J. Lombardo, pres.

Clear Channel Communications Inc., 200 E. Basse Rd., San Antonio, TX, 78209. Phone: (210) 822-2828. Fax: (210) 822-2299. E-mail: markpmays@clearchannel.com Web Site: www.clearchannel.com. Ownership: Thomas O. Hicks, 6.5%; L. Lowry Mays, 5.2%. Publicly traded company with the majority of its shares owned by the investing public.

Stns: 35 TV. WXXA-TV, Albany-Schenectady-Troy, NY; KGET-TV, Bakersfield, CA; WIVT, Binghamton, NY; WKRC-TV, Cincinnati, OH; WETM, Elmira (Corning), NY; KMTR, Eugene, OR; KTCW, Eugene, OR; KMCB, Eugene, OR; KTVF, Fairbanks, AK; KGPE, Fresno-Visalia, CA; WHP-TV, Harrisburg-Lancaster-Lebanon-York, PA; WJKT, Jackson, TN; WTEV-TV, Jacksonville, FL; WAWS, Jacksonville, FL; KLRT, Little Rock-Pine Bluff, AR; KASN, Little Rock-Pine Bluff, AR; WPTY, Memphis, TN; WLMT, Memphis, TN; WPMI-TV, Mobile, AL-Pensacola (Ft. Walton Beach), FL; WJTC, Mobile, AL-Pensacola (Ft. Walton Beach), FL; KION-TV, Monterey-Salinas, CA; WHAM-TV, Rochester, NY; KTVX, Salt Lake City, UT; KUCW, Salt Lake City, UT; WOAI-TV, San Antonio, TX; KFTY, San Francisco-Oakland-San Jose; KCOY, Santa Barbara-Santa Maria-San Luis Obispo, CA; KVOS, Seattle-Tacoma, WA; WSYR-TV, Syracuse, NY; KMYT-TV, Tulsa, OK; KOKI, Tulsa, OK; WWTI, Watertown, NY; KSAS-TV, Wichita-Hutchinson Plus, KS; KAAS, Wichita-Hutchinson Plus, KS; KOCW, Wichita-Hutchinson Plus, KS.

Stns:304 AM. 684 FM. KASH-FM Anchorage, AK; KBFX(FM) Anchorage, AK; KENI Anchorage, AK; KGOT-FM Anchorage, AK; KTZN Anchorage, AK; KYMG-FM Anchorage, AK; KFBX(AM) Fairbanks, AK; KIAK-FM Fairbanks, AK; KKED-FM Fairbanks, AK; KAKQ-FM Fairbanks, AK; WSTH-FM Alexander City, AL; WMJJ-FM Birmingham, AL; WRTR(FM) Brookwood, AL; WZBQ-FM Carrollton, AL; WHOS Decatur, AL; WDRM-FM Decatur, AL; WTXT-FM Fayette, AL; WAGH-FM Fort Mitchell, AL; WAAX Gadsden, AL; WGMZ-FM Glencoe, AL; WTAK-FM Hartselle, AL; WENN(FM) Hoover, AL; WBHP Huntsville, AL; WDXB(FM) Jasper, AL; WHLW(FM) Luverne, AL; WWMG(FM) Millbrook, AL; WKSJ-FM Mobile, AL; WNTM(AM) Mobile, AL; WMXC-FM Mobile, AL; WRKH-FM Mobile, AL; WHAL(AM) Phenix City, AL; WGSY-FM Phenix City, AL; WBFA-FM Smiths, AL; WZHT-FM Troy, AL; WQEN(FM) Trussville, AL; WACT Tuscaloosa, AL; WQRV(FM) Tuscumbia, AL; KHKN(FM) Benton, AR; KMJX-FM Conway, AR; KKIX-FM Fayetteville, AR; KEZA-FM Fayetteville, AR; KWHN(AM) Fort Smith, AR; KMAG-FM Fort Smith, AR; KWHF(FM) Harrisburg, AR; KDJE(FM) Jacksonville, AR; KNEA(AM) Jonesboro, AR; KIYS(FM) Jonesboro, AR; KFIN(FM) Jonesboro, AR; KSSN-FM Little Rock, AR; KMXF-FM Lowell, AR; KHLR(FM) Maumelle, AR; KFXR-FM Chinle, AZ; KTZR-FM Green Valley, AZ; KOHT-FM Marana, AZ; KZZP-FM Mesa, AZ; KYOT-FM Phoenix, AZ; KOY Phoenix, AZ; KNIX-FM Phoenix, AZ; KMXP-FM Phoenix, AZ; KGME(AM) Phoenix, AZ; KESZ-FM Phoenix, AZ; KXEW South Tucson, AZ; KWFM(AM) Tucson, AZ; KWMT-FM Tucson, AZ; KRQQ-FM Tucson, AZ; KNST Tucson, AZ; KTTI-FM Yuma, AZ; KQSR(FM) Yuma, AZ; KZXY-FM Apple Valley, CA; KIXW Apple Valley, CA; KHYL-FM Auburn, CA; KBFP(AM) Bakersfield, CA; KBKO-FM Bakersfield, CA; KHTY(AM) Bakersfield, CA; KUSS(FM) Carlsbad, CA; KBFP-FM Delano, CA; KDFO-FM Delano, CA; KRDU Dinuba, CA; KHTS-FM El Cajon, CA; KALZ(FM) Fowler, CA; KCBL Fresno, CA; KHGE(FM) Fresno, CA; KATJ-FM George, CA; KRAB-FM Green Acres, CA; KURQ(FM) Grover Beach, CA; KRZR-FM Hanford, CA; KAVL Lancaster, CA; KSMY(FM) Lompoc, CA; KBIG-FM Los Angeles, CA; KFI Los Angeles, CA; KYSR-FM Los Angeles, CA; KTLK(AM) Los Angeles, CA; KOST-FM Los Angeles, CA; KIIS-FM Los Angeles, CA; KHHT(FM) Los Angeles, CA; KLAC Los Angeles, CA; KSTT-FM Los Osos-Baywood Park, CA; KIXA-FM Lucerne Valley, CA; KMRQ(FM) Manteca, CA; KTOM-FM Marina, CA; KJSN-FM Modesto, CA; KFIV Modesto, CA; KVVS(FM) Mojave, CA; KTPI(AM) Mojave, CA; KKGN(AM) Oakland, CA; KNEW Oakland, CA; KOCN-FM Pacific Grove, CA; KOSO-FM Patterson, CA; KSTE Rancho Cordova, CA; KGGI-FM Riverside, CA; KDIF Riverside, CA; KOSS-FM Rosamond, CA; KGBY-FM Sacramento, CA; KFBK Sacramento, CA; KDON-FM Salinas, CA; KION(AM) Salinas, CA; KPRC-FM Salinas, CA; KKDD San Bernardino, CA; KTDD(AM) San Bernardino, CA; KGB-FM San Diego, CA; KIOZ-FM San Diego, CA; KMYI(FM) San Diego, CA; KOGO(AM) San Diego, CA; KLSD(AM) San Diego, CA; KMEL-FM San Francisco, CA; KIOI-FM San Francisco, CA; KISQ-FM San Francisco, CA; KKSF-FM San Francisco, CA; KYLD-FM San Francisco, CA; KUFX-FM San Jose, CA; KSJO-FM San Jose, CA; KSLY-FM San Luis Obispo, CA; KVEC San Luis Obispo, CA; KXFM-FM Santa Maria, CA; KSNI-FM Santa Maria, CA; KSMA Santa Maria, CA; KQOD-FM Stockton, CA; KWSX(AM) Stockton, CA; KCNL(FM) Sunnyvale, CA; KTPI-FM Tehachapi, CA; KMYT(FM) Temecula, CA; KTMQ(FM) Temecula, CA; KBOS-FM Tulare, CA; KFSO-FM Visalia, CA; KEZL(AM) Visalia, CA; KRSX-FM Yermo, CA; KBCO-FM Boulder, CO; KBPI(FM) Denver, CO; KRFX-FM Denver, CO; KPTT(FM) Denver, CO; KHOW Denver, CO; KOA Denver, CO; KIIX(AM) Fort Collins, CO; KIBT(FM) Fountain, CO; KSME(FM) Greeley, CO; KGHF Pueblo, CO; KCSJ Pueblo, CO; KCCY(FM) Pueblo, CO; KDZA-FM Pueblo, CO; KVUU-FM Pueblo, CO; KPHT(FM) Rocky Ford, CO; KKZN(AM) Thornton, CO; KCOL(AM) Wellington, CO; KTCL(FM) Wheat Ridge, CO; KKLI-FM Widefield, CO; WPKX-FM Enfield, CT; WKCI-FM Hamden, CT; WKSS-FM Hartford, CT; WHCN-FM Hartford, CT; WELI New Haven, CT; WAVZ New Haven, CT; WPHH(FM) Waterbury, CT; WWYZ-FM Waterbury, CT; WWDC-FM Washington, DC; WMZQ-FM Washington, DC; WASH-FM Washington, DC; WBIG-FM Washington, DC; WWRC(AM) Washington, DC; WIHT(FM) Washington, DC; WTEM Washington, DC; WROO(FM) Callahan, FL; WBTP(FM) Clearwater, FL; WXTB-FM Clearwater, FL; WMMV Cocoa, FL; WJRR-FM Cocoa Beach, FL; WTKS-FM Cocoa Beach, FL; WSRZ-FM Coral Cove, FL; WTZB(FM) Englewood, FL; WHYI-FM Fort Lauderdale, FL; WBGG-FM Fort Lauderdale, FL; WMIB(FM) Fort Lauderdale, FL; WOLZ-FM Fort Myers, FL; WLDI-FM Fort Pierce, FL; WKGR-FM Fort Pierce, FL; WSYR-FM Gifford, FL; WJBT-FM Green Cove Springs, FL; WFUS(FM) Gulfport, FL; WOLL-FM Hobe Sound, FL; WFXJ(AM) Jacksonville, FL; WPLA(FM) Jacksonville, FL; WQIK-FM Jacksonville, FL; WAIL(FM) Key West, FL; WKEY-FM Key West, FL; WEOW(FM) Key West, FL; WCKT(FM) Lehigh Acres, FL; WMMB Melbourne, FL; WEBZ(FM) Mexico Beach, FL; WINZ(AM) Miami, FL; WIOD Miami, FL; WLVE-FM Miami Beach, FL; WMGE(FM) Miami Beach, FL; WFLA-FM Midway, FL; WMGF(FM) Mount Dora, FL; WBTT(FM) Naples Park, FL; WFKS(FM) Neptune Beach, FL; WQTM(AM) Orlando, FL; WRUM(FM) Orlando, FL; WPAP-FM Panama City, FL; WDIZ Panama City, FL; WFBX(FM) Parker, FL; WTKX-FM Pensacola, FL; WYCL-FM Pensacola, FL; WFLF(AM) Pine Hills, FL; WFKZ(FM) Plantation Key, FL; WCTH(FM) Plantation Key, FL; WZJZ(FM) Port Charlotte, FL; WPBH(FM) Port St. Joe, FL; WCCF(AM) Punta Gorda, FL; WXSR-FM Quincy, FL; WZZR(FM) Riviera Beach, FL; WCTQ(FM) Sarasota, FL; WSDV(AM) Sarasota, FL; WCVU(FM) Solana, FL; WAVW(FM) Stuart, FL; WTNT-FM Tallahassee, FL; WMTX(FM) Tampa, FL; WHNZ(AM) Tampa, FL; WFLA(AM) Tampa, FL; WXXL-FM Tavares, FL; WKEZ-FM Tavenier, FL; WLTQ-FM Venice, FL; WDDV(AM) Venice, FL; WZTA(AM) Vero Beach, FL; WCZR(FM) Vero Beach, FL; WQOL-FM Vero Beach, FL; WRLX-FM West Palm Beach, FL; WJNO(AM) West Palm Beach, FL; WBZT(AM) West Palm Beach, FL; WJYZ Albany, GA; WJIZ-FM Albany, GA; WKLS(FM) Atlanta, GA; WGST Atlanta, GA; WUBL(FM) Atlanta, GA; WIBL(FM) Augusta, GA; WBBQ-FM Augusta, GA; WEKL(FM) Augusta, GA; WRAK-FM Bainbridge, GA; WBZY(FM) Bowdon, GA; WSOL-FM Brunswick, GA; WWVA-FM Canton, GA; WDAK Columbus, GA; WSHE(AM) Columbus, GA; WVRK-FM Columbus, GA; WMRZ(FM)

Dawson, GA; WVVM(AM) Dry Branch, GA; WQBZ-FM Fort Valley, GA; WIBB-FM Fort Valley, GA; WPCH(FM) Gray, GA; WMGP(FM) Hogansville, GA; WLCG Macon, GA; WPRW-FM Martinez, GA; WCOH Newnan, GA; WLTM(FM) Peachtree City, GA; WRXR-FM Rossville, GA; WTKS(AM) Savannah, GA; WSOK Savannah, GA; WAEV-FM Savannah, GA; WTLY(FM) Thomasville, GA; WOBB-FM Tifton, GA; WRBV-FM Warner Robins, GA; WEBL(FM) Warner Robins, GA; KSSK(AM) Honolulu, HI; KDNN(FM) Honolulu, HI; KHVH Honolulu, HI; KHBZ(AM) Honolulu, HI; KUCD(FM) Pearl City, HI; KSSK-FM Waipahu, HI; KASI Ames, IA; KCCQ(FM) Ames, IA; KKSY(FM) Anamosa, IA; KPTL(FM) Ankeny, IA; KGRS-FM Burlington, IA; KBUR Burlington, IA; KMJM(AM) Cedar Rapids, IA; WMT Cedar Rapids, IA; WMT-FM Cedar Rapids, IA; KCHA Charles City, IA; KCHA-FM Charles City, IA; KLKK(FM) Clear Lake, IA; KMXG(FM) Clinton, IA; KCQQ(FM) Davenport, IA; WOC Davenport, IA; WLLR-FM Davenport, IA; WHO(AM) Des Moines, IA; KXNO(AM) Des Moines, IA; KDRB(FM) Des Moines, IA; KKDM-FM Des Moines, IA; KKEZ-FM Fort Dodge, IA; KWMT Fort Dodge, IA; KBKB Fort Madison, IA; KBKB-FM Fort Madison, IA; KXKT(FM) Glenwood, IA; KXIC Iowa City, IA; KKRQ-FM Iowa City, IA; KXFT(FM) Manson, IA; KIAI(FM) Mason City, IA; KGLO(AM) Mason City, IA; KCZE(FM) New Hampton, IA; KSMA-FM Osage, IA; KWSL Sioux City, IA; KSEZ-FM Sioux City, IA; KMNS Sioux City, IA; KGLI-FM Sioux City, IA; KLLP-FM Chubbuck, ID; KID Idaho Falls, ID; KID-FM Idaho Falls, ID; KPKY-FM Pocatello, ID; KWIK Pocatello, ID; KCDA(FM) Post Falls, ID; KEZJ-FM Twin Falls, ID; KLIX Twin Falls, ID; KATZ-FM Alton, IL; WVON(AM) Berwyn, IL; WLIT-FM Chicago, IL; WGRB(AM) Chicago, IL; WGCI-FM Chicago, IL; WNUA-FM Chicago, IL; WKSC-FM Chicago, IL; KMJM-FM Columbia, IL; KUUL(FM) East Moline, IL; WVZA-FM Herrin, IL; WDDD Johnston City, IL; WDDD-FM Marion, IL; WFXN(AM) Moline, IL; WTAO-FM Murphysboro, IL; WVAZ-FM Oak Park, IL; WQUL-FM West Frankfort, IL; WFRX West Frankfort, IL; WTFX-FM Clarksville, IN; WRZX-FM Indianapolis, IN; WFBQ-FM Indianapolis, IN; WNDE Indianapolis, IN; WQMF-FM Jeffersonville, IN; WZKF(FM) Salem, IN; KZCH(FM) Derby, KS; KZSN(FM) Hutchinson, KS; KRBB-FM Wichita, KS; KTHR(FM) Wichita, KS; WSFE(AM) Burnside, KY; WSEK(FM) Burnside, KY; WKED-FM Frankfort, KY; WKYW-FM Frankfort, KY; WFKY Frankfort, KY; WXRA(AM) Georgetown, KY; WBUL-FM Lexington, KY; WMXL-FM Lexington, KY; WLKT-FM Lexington-Fayette, KY; WKJK Louisville, KY; WAMZ(FM) Louisville, KY; WHAS Louisville, KY; WKRD(AM) Louisville, KY; WLUE(FM) Louisville, KY; WMKJ(FM) Mt. Sterling, KY; WUBT(FM) Russellville, KY; WCND Shelbyville, KY; WKRD-FM Shelbyville, KY; WSFC Somerset, KY; WKEQ(FM) Somerset, KY; WLLK-FM Somerset, KY; WKQQ-FM Winchester, KY; WJBO Baton Rouge, LA; WYNK-FM Baton Rouge, LA; KRVE-FM Brusly, LA; WSKR Denham Springs, LA; KYRK(FM) Houma, LA; WRNO-FM New Orleans, LA; WQUE-FM New Orleans, LA; WODT New Orleans, LA; WYLD New Orleans, LA; WYLD-FM New Orleans, LA; WNOE-FM New Orleans, LA; WRNX-FM Amherst, MA; WJMN-FM Boston, MA; WXKS Everett, MA; WKOX Framingham, MA; WHYN Springfield, MA; WHYN-FM Springfield, MA; WSNE-FM Taunton, MA; WNNZ Westfield, MA; WTAG Worcester, MA; WSRS-FM Worcester, MA; WKCG-FM Augusta, ME; WWBX-FM Bangor, ME; WABI Bangor, ME; WLKE-FM Bar Harbor, ME; WBFB-FM Belfast, ME; WCME-FM Boothbay Harbor, ME; WQSS-FM Camden, ME; WGUY(FM) Dexter, ME; WKSQ-FM Ellsworth, ME; WFAU Gardiner, ME; WVOM-FM Howland, ME; WIGY-FM Madison, ME; WRKD Rockland, ME; WFZX(FM) Searsport, ME; WTOS-FM Skowhegan, ME; WUBB-FM York Center, ME; WTKA Ann Arbor, MI; WWWW-FM Ann Arbor, MI; WBCK Battle Creek, MI; WBFN(AM) Battle Creek, MI; WDTW(AM) Dearborn, MI; WNIC-FM Dearborn, MI; WMXD-FM Detroit, MI; WDTW-FM Detroit, MI; WDFN Detroit, MI; WJLB-FM Detroit, MI; WKQI-FM Detroit, MI; WBFX(FM) Grand Rapids, MI; WBCT-FM Grand Rapids, MI; WTKG Grand Rapids, MI; WOOD Grand Rapids, MI; WOOD-FM Grand Rapids, MI; WMAX-FM Holland, MI; WRCC(FM) Marshall, MI; WKBZ(AM) Muskegon, MI; WMUS(FM) Muskegon, MI; WSNX-FM Muskegon, MI; WSHZ(FM) Muskegon, MI; WMRR-FM Muskegon Heights, MI; WLBY(AM) Saline, MI; KQQL-FM Anoka, MN; KNFX Austin, MN; KQHT-FM Crookston, MN; KLDJ-FM Duluth, MN; KKCB-FM Duluth, MN; WEBC Duluth, MN; KMFX-FM Lake City, MN; KTCZ-FM Minneapolis, MN; KTLK-FM Minneapolis, MN; KFAN Minneapolis, MN; KFXN Minneapolis, MN; KBMX(FM) Proctor, MN; KDWB-FM Richfield, MN; KWEB(AM) Rochester, MN; KRCH(FM) Rochester, MN; KEEY-FM Saint Paul, MN; KSNR-FM Thief River Falls, MN; KMFX Wabasha, MN; KSWF(FM) Aurora, MO; KTOZ-FM Pleasant Hope, MO; KSLZ(FM) Saint Louis, MO; KSD(FM) Saint Louis, MO; KLOU(FM) Saint Louis, MO; KATZ(AM) Saint Louis, MO; KIGL(FM) Seligman, MO; KXUS-FM Springfield, MO; KGMY Springfield, MO; WESE-FM Baldwyn, MS; WMJY-FM Biloxi, MS; WWKZ(FM) Columbus, MS; WJKX-FM Ellisville, MS; WBVV(FM) Guntown, MS; WFOR Hattiesburg, MS; WUSW-FM Hattiesburg, MS; WHER-FM Heidelberg, MS; WHAL-FM Horn Lake, MS; WJDX Jackson, MS; WHLH(FM) Jackson, MS; WMSI-FM Jackson, MS; WZRX Jackson, MS; WQJQ-FM Kosciusko, MS; WNSL-FM Laurel, MS; WEEZ Laurel, MS; WJDQ(FM) Marion, MS; WYHL(AM) Meridian, MS; WMSO(FM) Meridian, MS; WBUV(FM) Moss Point, MS; WWZD-FM New Albany, MS; WHTU(FM) Newton, MS; WQYZ-FM Ocean Springs, MS; WKNN-FM Pascagoula, MS; WZLD(FM) Petal, MS; WKMQ(AM) Tupelo, MS; WTUP Tupelo, MS; WZKS-FM Union, MS; WSTZ-FM Vicksburg, MS; KISN(FM) Belgrade, MT; KKBR-FM Billings, MT; KBBB(FM) Billings, MT; KCTR-FM Billings, MT; KBUL Billings, MT; KMMS(AM) Bozeman, MT; KMMS-FM Bozeman, MT; KZMY(FM) Bozeman, MT; KLCY(AM) East Missoula, MT; KLYQ Hamilton, MT; KBAZ(FM) Hamilton, MT; KMHK-FM Hardin, MT; KPRK Livingston, MT; KXLB(FM) Livingston, MT; KYSS-FM Missoula, MT; KGVO Missoula, MT; KZIN-FM Shelby, MT; KSEN Shelby, MT; KENR(FM) Superior, MT; WWNC Asheville, NC; WKSL(FM) Burlington, NC; WMKS(FM) Clemmons, NC; WDCG-FM Durham, NC; WGBT(FM) Eden, NC; WPEK(AM) Fairview, NC; WQNQ(FM) Fletcher, NC; WMYI(FM) Hendersonville, NC; WLYT-FM Hickory, NC; WMAG(FM) High Point, NC; WVBZ(FM) High Point, NC; WRFX-FM Kannapolis, NC; WCDG(FM) Moyock, NC; WRVA-FM Rocky Mount, NC; WEND-FM Salisbury, NC; WIBT(FM) Shelby, NC; WKKT-FM Statesville, NC; WMXF(AM) Waynesville, NC; WRDU-FM Wilson, NC; WTQR-FM Winston-Salem, NC; KFYR Bismarck, ND; KBMR Bismarck, ND; KSSS-FM Bismarck, ND; KXMR Bismarck, ND; KQDY(FM) Bismarck, ND; KLTC Dickinson, ND; KZRX-FM Dickinson, ND; KCAD(FM) Dickinson, ND; KKXL Grand Forks, ND; KJKJ-FM Grand Forks, ND; KIZZ-FM Minot, ND; KMXA-FM Minot, ND; KYYX-FM Minot, ND; KZPR-FM Minot, ND; KRRZ Minot, ND; KCJB Minot, ND; KTGL-FM Beatrice, NE; KHUS(FM) Bennington, NE; KIBZ(FM) Crete, NE; KLMY(FM) Lincoln, NE; KMCX-FM Ogallala, NE; KOGA Ogallala, NE; KOGA-FM Ogallala, NE; KQBW(FM) Omaha, NE; KFAB Omaha, NE; KGOR-FM Omaha, NE; KZKX-FM Seward, NE; KSFT-FM South Sioux City, NE; WGIP Exeter, NH; WERZ-FM Exeter, NH; WGXL-FM Hanover, NH; WTSL Hanover, NH; WXXK-FM Lebanon, NH; WGIR Manchester, NH; WGIR-FM Manchester, NH; WVRR-FM Newport, NH; WHEB-FM Portsmouth, NH; WQSO-FM Rochester, NH; WHCY-FM Blairstown, NJ; WSUS-FM Franklin, NJ; WHTZ-FM Newark, NJ; WNNJ Newton, NJ; WNNJ-FM Newton, NJ; KABQ Albuquerque, NM; KBQI(FM) Albuquerque, NM; KZRR-FM Albuquerque, NM; KPEK-FM Albuquerque, NM; KCQL Aztec, NM; KKFG-FM Bloomfield, NM; KTEG(FM) Bosque Farms, NM; KSYU-FM Corrales, NM; KTRA-FM Farmington, NM; KDAG(FM) Farmington, NM; KGLX(FM) Gallup, NM; KFMQ-FM Gallup, NM; KAZX(FM) Kirtland, NM; KABQ-FM Santa Fe, NM; KXTC-FM Thoreau, NM; KWNR-FM Henderson, NV; KSNE-FM Las Vegas, NV; KPLV(FM) Las Vegas, NV; KWID(FM) Las Vegas, NV; WPYX-FM Albany, NY; WHRL-FM Albany, NY; WPHR-FM Auburn, NY; WKKF(FM) Ballston Spa, NY; WINR Binghamton, NY; WVOR(FM) Canandaigua, NY; WCTW-FM Catskill, NY; WKGB-FM Conklin, NY; WWDG(FM) DeRuyter, NY; WALK(AM) East Patchogue, NY; WRWC(FM) Ellenville, NY; WELG(AM) Ellenville, NY; WENE(AM) Endicott, NY; WMRV-FM Endicott, NY; WBBI(FM) Endwell, NY; WCPV-FM Essex, NY; WBBS-FM Fulton, NY; WRWD-FM Highland, NY; WFXF(FM) Honeoye Falls, NY; WHUC Hudson, NY; WZCR(FM) Hudson, NY; WKGS-FM Irondequoit, NY; WKTU(FM) Lake Success, NY; WIXT(AM) Little Falls, NY; WSKU(FM) Little Falls, NY; WBWZ-FM New Paltz, NY; WAXQ-FM New York, NY; WLTW-FM New York, NY; WWPR-FM New York, NY; WEAV Plattsburgh, NY; WVTK(FM) Port Henry, NY; WKIP Poughkeepsie, NY; WPKF(FM) Poughkeepsie, NY; WRNQ-FM Poughkeepsie, NY; WOKR(FM) Remsen, NY; WADR Remsen, NY; WDVI(FM) Rochester, NY; WHTK Rochester, NY; WHAM Rochester, NY; WRNY Rome, NY; WUMX(FM) Rome, NY; WTRY-FM Rotterdam, NY; WRVE-FM Schenectady, NY; WGY Schenectady, NY; WCRR(FM) South Bristol Township, NY; WHEN Syracuse, NY; WWHT-FM Syracuse, NY; WYYY-FM Syracuse, NY; WSYR Syracuse, NY; WOFX(AM) Troy, NY; WUTQ Utica, NY; WOUR-FM Utica, NY; WMXW-FM Vestal, NY; WSKS(FM) Whitesboro, NY; WXZO(FM) Willsboro, NY; WHLO Akron, OH; WARF(AM) Akron, OH; WNCO Ashland, OH; WYBL(FM) Ashtabula, OH; WFUN Ashtabula, OH; WREO-FM Ashtabula, OH; WXEG-FM Beavercreek, OH; WNUS-FM Belpre, OH; WKDD(FM) Canton, OH; WLZT(FM) Chillicothe, OH; WKKJ(FM) Chillicothe, OH; WBEX Chillicothe, OH; WCHI Chillicothe, OH; WSAI(AM) Cincinnati, OH; WKRC Cincinnati, OH; WLW Cincinnati, OH; WVMX-FM Cincinnati, OH; WCKY(AM) Cincinnati, OH; WTAM Cleveland, OH; WMMS-FM Cleveland, OH; WMVX-FM Cleveland, OH; WMJI-FM Cleveland, OH; WGAR-FM Cleveland, OH; WMJK(FM) Clyde, OH; WBVB-FM Coal Grove, OH; WCOL-FM Columbus, OH; WYTS(AM) Columbus, OH; WNCI-FM Columbus, OH; WTVN(AM) Columbus, OH; WLWD(FM) Columbus Grove, OH; WMMX-FM Dayton, OH; WTUE-FM Dayton, OH; WONE Dayton, OH; WONW Defiance, OH; WZOM-FM Defiance, OH; WDFM-FM Defiance, OH; WZOO-FM Edgewood, OH; WDKF(FM) Englewood, OH; WZRX-FM Fort Shawnee, OH; WXXR(FM) Fredericktown, OH; WFXN-FM Galion, OH; WDSJ(FM) Greenville, OH; WBWR(FM) Hilliard, OH; WSRW Hillsboro, OH; WSRW-FM Hillsboro, OH; WBKS(FM) Ironton, OH; WIRO Ironton, OH; WLQT-FM Kettering, OH; WIMA Lima, OH; WIMT(FM) Lima, OH; WXXF(FM) Loudonville, OH; WMAN Mansfield, OH; WLTP(AM) Marietta, OH; WRVB-FM Marietta, OH; WMRN Marion, OH; WDIF-FM Marion, OH; WKFS-FM Milford, OH; WMVO(AM) Mount Vernon, OH; WQIO-FM Mount Vernon, OH; WNDH-FM Napoleon, OH; WBBG(FM) Niles, OH; WPFX-FM North Baltimore, OH; WHOF(FM) North Canton, OH; WFXJ-FM North Kingsville, OH; WBUK(FM) Ottawa, OH; WMLX-FM Saint Mary's, OH; WCPZ(FM) Sandusky, OH; WVKF(FM) Shadyside, OH; WSWR-FM Shelby, OH; WIZE Springfield, OH; WTTF Tiffin, OH; WVKS-FM Toledo, OH; WRVF(FM) Toledo, OH; WSPD Toledo, OH; WIOT(FM) Toledo, OH; WCWA Toledo, OH; WYNT-FM Upper Sandusky, OH; WCHO(AM) Washington Court House, OH; WKBN Youngstown, OH; WNIO(AM) Youngstown, OH; KTBT(FM) Broken Arrow, OK; KIZS(FM) Collinsville, OK; KXXY-FM Oklahoma City, OK; KHBZ-FM Oklahoma City, OK; KTOK(AM) Oklahoma City, OK; KTST-FM Oklahoma City, OK; KQLL-FM Owasso, OK; KZBB-FM Poteau, OK; KKBD(FM) Sallisaw, OK; KMOD-FM Tulsa, OK; KTBZ(AM) Tulsa, OK; KAKC(AM) Tulsa, OK; KKCW-FM Beaverton, OR; KPOJ(AM) Portland, OR; KEX Portland, OR; WAEB Allentown, PA; WSAN(AM) Allentown, PA; WZZO-FM Bethlehem, PA; WTKT(AM) Harrisburg, PA; WKBO Harrisburg, PA; WHP Harrisburg, PA; WRBT-FM Harrisburg, PA; WRKK Hughesville, PA; WLAN Lancaster, PA; WLAN-FM Lancaster, PA; WVRT(FM) Mill Hall, PA; WRFF(FM) Philadelphia, PA; WIOQ-FM Philadelphia, PA; WISX(FM) Philadelphia, PA; WUBA(AM) Philadelphia, PA; WDAS-FM Philadelphia, PA; WUSL-FM Philadelphia, PA; WKST-FM Pittsburgh, PA; WDVE-FM Pittsburgh, PA; WPGB(FM) Pittsburgh, PA; WBGG(AM) Pittsburgh, PA; WXDX-FM Pittsburgh, PA; WRFY-FM Reading, PA; WRAW(AM) Reading, PA; WBYL(FM) Salladasburg, PA; WBLJ-FM Shamokin, PA; WAKZ(FM) Sharpsville, PA; WRAK Williamsport, PA; WHJJ Providence, RI; WHJY-FM Providence, RI; WWBB-FM Providence, RI; WKSP(FM) Aiken, SC; WYKZ(FM) Beaufort, SC; WLTY-FM Cayce, SC; WEZL-FM Charleston, SC; WLTQ(AM) Charleston, SC; WALC-FM Charleston, SC; WCOS Columbia, SC; WVOC Columbia, SC; WNOK(FM) Columbia, SC; WSCC-FM Goose Creek, SC; WLFJ(AM) Greenville, SC; WESC-FM Greenville, SC; WGVL Greenville, SC; WLVH-FM Hardeeville, SC; WBZT-FM Mauldin, SC; WRFQ-FM Mt. Pleasant, SC; WYNF(AM) North Augusta, SC; WXLY-FM North Charleston, SC; WXBT(FM) West Columbia, SC; WUSY-FM Cleveland, TN; WPTN Cookeville, TN; WGSQ-FM Cookeville, TN; WGIC-FM Cookeville, TN; WHUB Cookeville, TN; WRVW-FM Lebanon, TN; WKZP(FM) McMinnville, TN; WAKI McMinnville, TN; WBMC McMinnville, TN; WDIA Memphis, TN; WEGR(FM) Memphis, TN; WHRK-FM Memphis, TN; KJMS(FM) Memphis, TN; WREC(AM) Memphis, TN; WSIX-FM Nashville, TN; WLAC Nashville, TN; WLND(FM) Signal Mountain, TN; WSMT Sparta, TN; WRKK-FM Sparta, TN; WTZX Sparta, TN; WTRZ(FM) Spencer, TN; KASE-FM Austin, TX; KPEZ-FM Austin, TX; KVET-FM Austin, TX; KYKR-FM Beaumont, TX; KLVI Beaumont, TX; KVNS(AM) Brownsville, TX; KTEX-FM Brownsville, TX; KKYS-FM Bryan, TX; KNFX-FM Bryan, TX; KMXR-FM Corpus Christi, TX; KRYS-FM Corpus Christi, TX; KUNO Corpus Christi, TX; KZPS-FM Dallas, TX; KFXR(AM) Dallas, TX; KDMX(FM) Dallas, TX; KHKS-FM Denton, TX; KRPT(FM) Devine, TX; KBFM(FM) Edinburg, TX; KTSM(AM) El Paso, TX; KHEY(AM) El Paso, TX; KPRR-FM El Paso, TX; KEGL-FM Fort Worth, TX; KDGE(FM) Fort Worth-Dallas, TX; KHFI-FM Georgetown, TX; KCOL-FM Groves, TX; KBRQ(FM) Hillsboro, TX; KBME Houston, TX; KTRH Houston, TX; KPRC Houston, TX; KODA-FM Houston, TX; KHMX-FM Houston, TX; KTBZ-FM Houston, TX; KKRW-FM Houston, TX; KIIZ-FM Killeen, TX; KAGG(FM) Madisonville, TX; KHKZ(FM) Mercedes, TX; KQXX-FM Mission, TX; KLFX-FM Nolanville, TX; KIOC-FM Orange, TX; KKMY-FM Orange, TX; KSAB-FM Robstown, TX; KFMK-FM Round Rock, TX; KAJA(FM) San Antonio, TX; WOAI San Antonio, TX; KQXT-FM San Antonio, TX; KTKR San Antonio, TX; KXXM-FM San Antonio, TX; KNCN-FM Sinton, TX; KWTX Waco, TX; KWTX-FM Waco, TX; WACO-FM Waco, TX; KBGO(FM) Waco, TX; KJMY(FM) Bountiful, UT; KXRV(FM) Centerville, UT; KODJ-FM Salt Lake City, UT; KNRS Salt Lake City, UT; KZHT(FM) Salt Lake City, UT; KOSY-FM Spanish Fork, UT; WYYD-FM

Amherst, VA; WSNZ(FM) Appomattox, VA; WKAV Charlottesville, VA; WCHV Charlottesville, VA; WCJZ(FM) Charlottesville, VA; WSUH(FM) Crozet, VA; WACL-FM Elkton, VA; WFQX-FM Front Royal, VA; WKCY Harrisonburg, VA; WJJX-FM Lynchburg, VA; WROV-FM Martinsville, VA; WOWI-FM Norfolk, VA; WKUS(FM) Norfolk, VA; WBTJ(FM) Richmond, VA; WRXL-FM Richmond, VA; WRVQ-FM Richmond, VA; WRVA Richmond, VA; WRNL Richmond, VA; WTVR-FM Richmond, VA; WZBL(FM) Roanoke, VA; WHTE-FM Ruckersville, VA; WSNV(FM) Salem, VA; WKDW Staunton, VA; WCYK-FM Staunton, VA; WSVO-FM Staunton, VA; WKSI-FM Stephens City, VA; WJJS-FM Vinton, VA; WKCI(AM) Waynesboro, VA; WTFX(AM) Winchester, VA; WJCD(FM) Windsor, VA; WAZR-FM Woodstock, VA; WEZF-FM Burlington, VT; WCVR-FM Randolph, VT; WTSJ(AM) Randolph, VT; WTSM(FM) Springfield, VT; WMXR-FM Woodstock, VT; KNBQ(FM) Centralia, WA; KELA Centralia-Chehalis, WA; KMNT(FM) Chehalis, WA; KFNK(FM) Eatonville, WA; KQSN(FM) Naches, WA; KIXZ-FM Opportunity, WA; KOLW(FM) Othello, WA; KEYW-FM Pasco, WA; KFLD Pasco, WA; KUBE-FM Seattle, WA; KJR Seattle, WA; KJR-FM Seattle, WA; KKZX-FM Spokane, WA; KPTQ(AM) Spokane, WA; KQNT(AM) Spokane, WA; KHHO Tacoma, WA; KDBL(FM) Toppenish, WA; KIJZ(FM) Vancouver, WA; KXRX-FM Walla Walla, WA; KIT Yakima, WA; KUTI(AM) Yakima, WA; WISM-FM Altoona, WI; WQRB-FM Bloomer, WI; WATQ(FM) Chetek, WI; WBIZ Eau Claire, WI; WIBA Madison, WI; WTSO Madison, WI; WMEQ(AM) Menomonie, WI; WQBW(FM) Milwaukee, WI; WISN Milwaukee, WI; WRIT-FM Milwaukee, WI; WOKY Milwaukee, WI; WKKV-FM Racine, WI; WMAD(FM) Sauk City, WI; WXXM(FM) Sun Prairie, WI; WMIL-FM Waukesha, WI; WMRE Charles Town, WV; WVHU(AM) Huntington, WV; WTCR-FM Huntington, WV; WTCR Kenova, WV; WAMX-FM Milton, WV; WZZW Milton, WV; WHNK(AM) Parkersburg, WV; WDMX-FM Vienna, WV; WEGW-FM Wheeling, WV; WBBD Wheeling, WV; WKWK-FM Wheeling, WV; WWVA(AM) Wheeling, WV; KIGN(FM) Burns, WY; KKTL Casper, WY; KWYY-FM Casper, WY; KTRS-FM Casper, WY; KTWO Casper, WY; KLEN-FM Cheyenne, WY; KQMY(FM) Cheyenne, WY; KOWB Laramie, WY; KCGY(FM) Laramie, WY; KRVK(FM) Midwest, WY; KGAB Orchard Valley, WY; WSMJ(FM) Baltimore, MD; WPOC-FM Baltimore, MD; WCAO Baltimore, MD; WTNT(AM) Bethesda, MD; WFMD(AM) Frederick, MD; WFRE(FM) Frederick, MD; WWFG-FM Ocean City, MD; WDKZ(FM) Salisbury, MD; WTGM Salisbury, MD; WSBY-FM Salisbury, MD; WOSC-FM Bethany Beach, DE; WDOV Dover, DE; WLBW-FM Fenwick Island, DE; WDSD-FM Smyrna, DE; WWTX(AM) Wilmington, DE; WILM Wilmington, DE

L. Lowry Mays, chmn; Mark P. Mays, CEO; Randall T. Mays, pres; Kenneth E. Wyker, sr VP; Herbert W. Hill Sr., sr VP; Don Perry, exec VP.

Cocola Broadcasting Companies LLC, 706 W. Herndon Ave., Fresno, CA, 93650. Phone: (559) 435-7000. Fax: (559) 435-3201. E-mail: info@cocolatv.com Web Site: www.cocolatv.com. Ownership: Gary M. Cocola, owner

Stns: 3 TV. KKJB, Boise, ID; KGMC, Fresno-Visalia, CA; KBBC-TV, Los Angeles.

Gary M. Cocola, pres/CEO.

Cogeco Radio-Television Inc., 612 St. Jacques, Suite 100, Montreal, PQ, H3C 5R1. Canada. Phone: (514) 390-6035. Fax: (514) 390-6070. Ownership: Cogeco Inc., 100%.

Stns: 1 TV. CFAP, Quebec City, PQ.

Stns: 4 FM. CJMF-FM Quebec, PQ; CJEC-FM Quebec, PQ; CFGE-FM Sherbrooke, PQ; CJEB-FM Trois Rivieres, PQ.

Rene Guimond, pres/CEO; Luc Doyon, VP; Therese David, VP; Guy Meunier, sls VP; Jacques Boiteau, gen mgr; Geoffrey O. Brow, gen mgr.

Communications Corp. of America, Box 53708, Lafayette, LA, 70505-3708. Phone: (337) 237-1142. Fax: (337) 237-1373. Ownership: Apollo Capital Management II & T. Galloway, jointly.

Stns: 10 TV. WGMB, Baton Rouge, LA; KTSM-TV, El Paso (Las Cruces, NM), TX; WEVV-TV, Evansville, IN; KVEO, Harlingen-Weslaco-Brownsville-McAllen, TX; KADN, Lafayette, LA; KPEJ, Odessa-Midland, TX; KMSS, Shreveport, LA; KETK, Tyler-Longview (Lufkin & Nacogdoches), TX; KWKT, Waco-Temple-Bryan, TX; KYLE, Waco-Temple-Bryan, TX.

Thomas R. Galloway, chmn; Wayne Elmore, pres/CEO.

Cordillera Communications Inc., 600 E. Superior St., Suite 203, Duluth, MN, 55802. Phone: (218) 625-3045. Fax: (218) 625-3047. Web Site: www.cordillera.tv. Ownership: Evening Post Publishing Co., 100%.

Stns: 11 TV. KTVQ, Billings, MT; KXLF-TV, Butte-Bozeman, MT; KBZK, Butte-Bozeman, MT; KOAA, Colorado Springs-Pueblo, CO; KRIS-TV, Corpus Christi, TX; KRTV, Great Falls, MT; KATC, Lafayette, LA; WLEX, Lexington, KY; KPAX-TV, Missoula, MT; KSBY, Santa Barbara-Santa Maria-San Luis Obispo, CA; KVOA, Tucson (Sierra Vista), AZ.

Terrance Hurley, pres; Lamont Wallis, VP; Andrew Suk, VP tech & engrg.

Cornerstone TeleVision Inc., 1 Signal Hill Dr., Wall, PA, 15148-1499. Phone: (412) 824-3930. Fax: (412) 824-5442. E-mail: info@ctvn.org Web Site: www.ctvn.org. Ownership: Nonprofit.

Stns: 2 TV. WKBS, Johnstown-Altoona, PA; WPCB, Pittsburgh, PA.

Ron Hembree, pres.

Corus Entertainment Inc., 630 3rd Ave. S.W., Suite 105, Calgary, AB, T2P 4L4. Canada. Phone: (403) 444-4244. Fax: (403) 444-4242. Web Site: www.corusent.com. Ownership: J.R. Shaw controls an aggregate of 80% of the voting rights.

Stns: 3 TV. CKWS, Kingston, ON; CHEX-TV-2, Oshawa, ON; CHEX, Peterborough, ON.

Stns: 18 AM. 25 FM. CFGQ-FM Calgary, AB; CKRY-FM Calgary, AB; CKNG-FM Edmonton, AB; CISN-FM Edmonton, AB; CHED Edmonton, AB; CFMI-FM New Westminster, BC; CKNW New Westminster, BC; CHMJ(AM) Vancouver, BC; CJOB Winnipeg, MB; CJZZ-FM Winnipeg, MB; CIQB-FM Barrie, ON; CHAY-FM Barrie, ON; CJXY-FM Burlington, ON; CJDV-FM Cambridge, ON; CKCB-FM Collingwood, ON; CJSS-FM Cornwall, ON; CFLG-FM Cornwall, ON; CJUL(AM) Cornwall, ON; CJOY Guelph, ON; CHML Hamilton, ON; CFFX(AM) Kingston, ON; CFMK-FM Kingston, ON; CKBT-FM Kitchener-Waterloo, ON; CFPL London, ON; CILQ-FM North York, ON; CKRU Peterborough, ON; CFHK-FM St. Thomas, ON; CFNY-FM Toronto, ON; CFMJ(AM) Toronto, ON; CKDK-FM Woodstock, ON; CJRC-FM Gatineau, PQ; CFOM-FM Levis, PQ; CHMP-FM Longueuil, PQ; CFEL-FM Montmagny, PQ; CKAC Montreal, PQ; CINW(AM) Montreal, PQ; CHRC Quebec, PQ; CKRS(AM) Saguenay, PQ; CIME-FM Saint Jerome, PQ; CHLT Sherbrooke, PQ; CHLT-FM Sherbrooke, PQ; CHLN Trois Rivieres, PQ; CINF(AM) Verdun, PQ.

John M. Cassaday, pres.

Cox Radio Inc., 6205 Peachtree Dunwoody Rd., Atlanta, GA, 30328. Phone: (678) 645-0000. E-mail: cxr.info@cox.com Web Site: coxradio.com. Ownership: Cox Enterprises Inc., 100%. Note: Cox Enterprises Inc. also owns 100% of Cox Television (see listing in section B under TV Group Ownership).

Stns: 14 AM. 65 FM. WZZK-FM Birmingham, AL; WAGG(AM) Birmingham, AL; WPSB(AM) Birmingham, AL; WNCB(FM) Gardendale, AL; WBPT(FM) Homewood, AL; WBHJ(FM) Midfield, AL; WBHK-FM Warrior, AL; WEZN-FM Bridgeport, CT; WPLR-FM New Haven, CT; WNLK(AM) Norwalk, CT; WFOX(FM) Norwalk, CT; WSTC(AM) Stamford, CT; WCTZ(FM) Stamford, CT; WFYV-FM Atlantic Beach, FL; WHQT-FM Coral Gables, FL; WCFB-FM Daytona Beach, FL; WSUN-FM Holiday, FL; WJGL(FM) Jacksonville, FL; WMXQ-FM Jacksonville, FL; WOKV Jacksonville, FL; WAPE-FM Jacksonville, FL; WPYO(FM) Maitland, FL; WHDR(FM) Miami, FL; WFLC-FM Miami, FL; WEDR-FM Miami, FL; WDUV-FM New Port Richey, FL; WHTQ-FM Orlando, FL; WDBO Orlando, FL; WMMO-FM Orlando, FL; WOKV-FM Ponte Vedra Beach, FL; WXGL(FM) Saint Petersburg, FL; WPOI(FM) Saint Petersburg, FL; WHPT(FM) Sarasota, FL; WWRM(FM) Tampa, FL; WSB Atlanta, GA; WSB-FM Atlanta, GA; WBTS(FM) Doraville, GA; WSRV(FM) Gainesville, GA; WALR-FM La Grange, GA; KCCN-FM Honolulu, HI; KRTR(AM) Honolulu, HI; KINE-FM Honolulu, HI; KRTR-FM Kailua, HI; KPHW(FM) Kaneohe, HI; KKNE(AM) Waipahu, HI; WSFR(FM) Corydon, IN; WVEZ(FM) Louisville, KY; WPTI(FM) Louisville, KY; WRKA-FM Saint Matthews, KY; WBAB(FM) Babylon, NY; WGBB Freeport, NY; WBLI-FM Patchogue, NY; WHFM-FM Southampton, NY; WHIO Dayton, OH; WHKO-FM Dayton, OH; WHIO-FM Piqua, OH; WZLR(FM) Xenia, OH; KKCM(FM) Sand Springs, OK; KWEN-FM Tulsa, OK; KJSR-FM Tulsa, OK; KRAV-FM Tulsa, OK; KRMG Tulsa, OK; WJMZ-FM Anderson, SC; WHZT(FM) Seneca, SC; KTHT(FM) Cleveland, TX; KHPT(FM) Conroe, TX; KONO-FM Helotes, TX; KHTC(FM) Lake Jackson, TX; KKBQ-FM Pasadena, TX; KCYY(FM) San Antonio, TX; KKYX San Antonio, TX; KONO San Antonio, TX; KISS-FM San Antonio, TX; KSMG(FM) Seguin, TX; KPWT(FM) Terrell Hills, TX; WDYL-FM Chester, VA; WKHK-FM Colonial Heights, VA; WKLR-FM Fort Lee, VA; WMXB-FM Richmond, VA.

Cox Enterprises Inc. owns the following daily newspapers: The (Grand Junction, CO) *Daily Sentinel; Palm Beach* (FL) *Daily News* and *The Palm Beach* (FL) *Post; The Atlanta* (GA) *Journal & Constitution; The Daily Advance* (Elizabeth City), *The* (Greenville) *Daily Reflector* and the *Rocky Mount Telegram*, all NC; *Dayton Daily News* and the *Springfield News-Sun*, both OH; *Austin American-Statesman, Longview News-Journal, The Lufkin Daily News, News Messenger* (Marshall), *The* (Nacogdoches) *Daily Sentinel* and the *Waco Tribune-Herald*, all TX. Cox also owns weekly newspapers and shoppers in CO, FL, NC, OH, and TX.

Robert F. Neil, pres/CEO; Marc. W. Morgan, COO; Neil O. Johnston, CFO; Richard A. Reis, group VP.

Cox Television, Box 105357, Atlanta, GA, 30348-5357. Phone: (678) 645-0000. Fax: (678) 645-5250. Web Site: www.coxenterprises.com. Ownership: Cox Enterprises Inc., 100%. Note: Cox Enterprises Inc. also owns 100% of Cox Radio Inc. (see listing in section D under Radio Group Ownership).

Stns: 14 TV. WSB-TV, Atlanta; WSOC-TV, Charlotte, NC; WAXN-TV, Charlotte, NC; WHIO, Dayton, OH; KFOX, El Paso (Las Cruces, NM), TX; WJAC, Johnstown-Altoona, PA; WFTV, Orlando-Daytona Beach-Melbourne, FL; WRDQ, Orlando-Daytona Beach-Melbourne, FL; WPXI, Pittsburgh, PA; KRXI-TV, Reno, NV; KTVU, San Francisco-Oakland-San Jose; KICU, San Francisco-Oakland-San Jose; KIRO, Seattle-Tacoma, WA; WTOV-TV, Wheeling, WV-Steubenville, OH.

Andrew S. Fisher, pres.

Cunningham Broadcasting Corporation, 2000 W. 41st St., Baltimore, MD, 21211. Phone: (410) 662-9688. Fax: (410) 662-0816.

Stns: 5 TV. WNUV, Baltimore, MD; WTAT, Charleston, SC; WTTE, Columbus, OH; WRGT-TV, Dayton, OH; WMYA-TV, Greenville-Spartanburg, SC-Asheville, NC-Anderson, SC.

Robert Simmons, CEO.

The Curators of the University of Missouri, (Business Services Division). University of Missouri, 316 University Hall, Columbia, MO, 65211. Phone: (573) 882-2388. Fax: (573) 882-0010. Web Site: www.umsystem.edu. Ownership: (Business Services Division).

Stns: 1 TV. KOMU, Columbia-Jefferson City, MO.

Stns: 5 FM. KBIA(FM) Columbia, MO; KCUR-FM Kansas City, MO; KMNR-FM Rolla, MO; KMST(FM) Rolla, MO; KWMU-FM Saint Louis, MO.

Michael Dunn, gen mgr; Martin Siddall, gen mgr.

D

Dispatch Broadcast Group, 770 Twin Rivers Dr., Columbus, OH, 43215. Phone: (614) 460-3700. Fax: (614) 460-2809. Web Site: www.10tv.com. Ownership: Dispatch Printing Company

Stns: 2 TV. WBNS, Columbus, OH; WTHR, Indianapolis, IN.

Stns: 1 AM. 1 FM. WBNS Columbus, OH; WBNS-FM Columbus, OH.

Owns *The Columbus* (OH) *Dispatch, This Week* & *Ohio Magazine.*

Tamara J. Clapsaddle, controller; Michael J. Fiorile, pres.

Diversified Communications, 121 Free St., Box 7437, Portland, ME, 04112-7437. Phone: (207) 842-5400. Fax: (207) 842-5405. Web Site: www.divbusiness.com. Ownership: Horace A. Hildreth Jr., Josephine H. Detmer. See Cross-Ownership, Sect. A. Cable TV: New England Cablevision Inc.

Stns: 2 TV. WABI-TV, Bangor, ME; WCJB-TV, Gainesville, FL.

David H. Lowell, pres.

R.H. Drewry Group, Box 708, Lawton, OK, 73502. Phone: (580) 353-0820. Phone: (580) 355-7000. Fax: (580) 357-3811. Ownership: R.H. Drewry owns 69% of KSWO-TV. KFDA-TV is a joint venture owned by Lawton Cablevision (50%), KSWD-TV (45%), KSWO(AM) (2 1/2%) and KRHD-AM-FM (2 1/2%). KWAB(TV) and KWES-TV are owned by KSWO Television Inc. (50%) and Lawton Cablevision Inc. (50%). KXXV(TV) is owned by Centrex Television L.P. Cable TV.

Stns: 5 TV. KFDA, Amarillo, TX; KWAB, Odessa-Midland, TX; KWES, Odessa-Midland, TX; KXXV, Waco-Temple-Bryan, TX; KSWO, Wichita Falls, TX & Lawton, OK.

Robert H. Drewry, pres; Larry Patton, VP.

Duhamel Broadcasting Enterprises, Box 1760, Rapid City, SD, 57709. Phone: (605) 342-2000. Fax: (605) 342-7305. Web Site: www.kotatv.com. Ownership: William F. Duhamel, 63%; Peter A. and Lois G. Duhamel, 37%.

Stns: 4 TV. KDUH, Cheyenne, WY-Scottsbluff, NE; KHSD-TV, Rapid City, SD; KOTA-TV, Rapid City, SD; KSGW-TV, Rapid City, SD.

Stns: 1 AM. 1 FM. KOTA Rapid City, SD; KDDX(FM) Spearfish, SD.

William F. Duhamel, pres.

E

Eagle Creek Broadcasting LLC, 2111 University Park Dr., Suite 650, Okemos, MI, 48864. Phone: (517) 347-4141. Phone: (517) 347-4675. E-mail: bradybw1@comcast.net Ownership: Alta Communications VIII L.P., 86.21% votes, 21.29% total assets; Brian W. Brady, 7.36% votes, 4.89% total assets.

Stns: 2 TV. KZTV, Corpus Christi, TX; KVTV, Laredo, TX.

Brian Brady, pres/CEO.

Emmis Communications Corp., 3500 W. Olive Ave., Suite 1450, Burbank, CA, 91436. Phone: (818) 238-9154. Fax: (818) 238-9158. Web Site: www.emmis.com. Ownership: Jeffrey H. Smulyan, approximately 61% votes.

Stns: 1 TV. WVUE, New Orleans, LA.

Stns: 2 AM. 20 FM. KPWR-FM Los Angeles, CA; KMVN(FM) Los Angeles, CA; WKQX-FM Chicago, IL; WLUP-FM Chicago, IL; WNOU(FM) Indianapolis, IN; WIBC(AM) Indianapolis, IN; WYXB(FM) Indianapolis, IN; WLHK(FM) Shelbyville, IN; WTHI-FM Terre Haute, IN; WWVR-FM West Terre Haute, IN; KSHE(FM) Crestwood, MO; KFTK(FM) Florissant, MO; KIHT-FM Saint Louis, MO; KPNT-FM Sainte Genevieve, MO; WRKS(FM) New York, NY; WQHT-FM New York, NY; WQCD(FM) New York, NY; KLBJ Austin, TX; KGSR-FM Bastrop, TX; KROX-FM Buda, TX; KDHT(FM) Cedar Park, TX; KBPA(FM) San Marcos, TX.

The publishing unit of Emmis Communications publishes seven magazines: *Allanta Magazine, Cincinnati Magazine, Los Angeles Magazine, Wildlife Journal, Indianapolis Monthly* and *Texas Monthly*.

Rick Cummings, pres.

Entravision Communications Corp., 2425 Olympic Blvd., Suite 6000W, Santa Monica, CA, 90404. Phone: (310) 447-3872. Fax: (310) 447-3899. E-mail: kthompson@entravision.com Web Site: www.entravision.com. Ownership: Walter F. Ulloa, Philip W. Wilkinson, Paul Zevnik.

Stns: 18 TV. KLUZ, Albuquerque-Santa Fe, NM; WUNI, Boston (Manchester, NH); KVSN, Colorado Springs-Pueblo, CO; KORO, Corpus Christi, TX; KCEC, Denver, CO; KINT, El Paso (Las Cruces, NM), TX; KTFN, El Paso (Las Cruces, NM), TX; KNVO, Harlingen-Weslaco -Brownsville-McAllen, TX; WUVN, Hartford & New Haven, CT; KLDO, Laredo, TX; KINC, Las Vegas, NV; KSMS, Monterey-Salinas, CA; KUPB, Odessa-Midland, TX; WVEN-TV, Orlando-Daytona Beach-Melbourne, FL; KPMR, Santa Barbara-Santa Maria-San Luis Obispo, CA; WVEA-TV, Tampa-St. Petersburg (Sarasota), FL; WJAL, Washington, DC (Hagerstown, MD); KVYE, Yuma, AZ-El Centro, CA.

Stns: 12 AM. 37 FM. KVVA-FM Apache Junction, AZ; KMIA(AM) Black Canyon City, AZ; KDVA(FM) Buckeye, AZ; KLNZ-FM Glendale, AZ; KZLZ-FM Kearny, AZ; KRRN(FM) Kingman, AZ; KSSE(FM) Arcadia, CA; KSEH(FM) Brawley, CA; KCVR-FM Columbia, CA; KXSE(FM) Davis, CA; KWST(AM) El Centro, CA; KSSD(FM) Fallbrook, CA; KLOK-FM Greenfield, CA; KMXX-FM Imperial, CA; KCVR Lodi, CA; KRCX-FM Marysville, CA; KDLE(FM) Newport Beach, CA; KTSE-FM Patterson, CA; KLYY(FM) Riverside, CA; KBMB(FM) Sacramento, CA; KDLD(FM) Santa Monica, CA; KSES-FM Seaside, CA; KNTY(FM) Shingle Springs, CA; KMBX(AM) Soledad, CA; KLOB-FM Thousand Palms, CA; KMIX-FM Tracy, CA; KSSC(FM) Ventura, CA; KPVW(FM) Aspen, CO; KMXA Aurora, CO; KJMN-FM Castle Rock, CO; KXPK-FM Evergreen, CO; WLQY Hollywood, FL; KRZY Albuquerque, NM; KRZY-FM Santa Fe, NM; KQRT(FM) Las Vegas, NV; KRNV-FM Reno, NV; KKPS-FM Brownsville, TX; KVLY-FM Edinburg, TX; KSVE El Paso, TX; KOFX-FM El Paso, TX; KINT-FM El Paso, TX; KHRO(AM) El Paso, TX; KYSE(FM) El Paso, TX; KFRQ-FM Harlingen, TX; KGOL Humble, TX; KBZO Lubbock, TX; KZPL(FM) Port Isabel, TX; KAIQ(FM) Wolfforth, TX; WACA Wheaton, MD.

Walter F. Ulloa, chmn/CEO; Philip Wilkinson, pres/COO; Larry Safir, exec VP.

Equity Media Holdings Corp., 1 Shackleford Dr., Suite 400, Little Rock, AR, 72211. Phone: (501) 219-2400. Fax: (501) 221-1101.

Stns: 24 TV. WNGS, Buffalo, NY; KBTZ, Butte-Bozeman, MT; KWWF, Cedar Rapids-Waterloo-Iowa City & Dubuque, IA; KDEV, Cheyenne, WY-Scottsbluff, NE; KTUW, Cheyenne, WY-Scottsbluff, NE; KTVC, Eugene, OR; KLMN, Great Falls, MT; KWBF, Little Rock-Pine Bluff, AR; KYPX, Little Rock-Pine Bluff, AR; WMQF, Marquette, MI; KMMF, Missoula, MT; KEYU, Oklahoma City, OK; KUOK, Oklahoma City, OK; WPXS, Paducah, KY-Cape Girardeau, MO-Harrisburg-Mount Vernon, IL; WBIF, Panama City, FL; KEGS, Reno, NV; KBNY, Salt Lake City, UT; KBCJ, Salt Lake City, UT; KUTF, Salt Lake City, UT; KUTH, Salt Lake City, UT; KQUP, Spokane, WA; KWBM, Springfield, MO; KPBI, Springfield, MO; WNYI, Syracuse, NY.

Larry Morton, pres; Greg Fess, VP; Max Hooper, VP; James Hearnsberger, VP; Lori Withrow, sec; Emilia Chastain, treas.

F

Family Stations Inc., 290 Hegenberger Rd., Oakland, CA, 94621. Phone: (510) 568-6200. Fax: (510) 568-6190. Ownership: Nonprofit corporation.

Stns: 1 TV. WFME, New York.

Stns: 12 AM. 57 FM. WBFR-FM Birmingham, AL; KEAF(FM) Fort Smith, AR; KPHF-FM Phoenix, AZ; KFRB-FM Bakersfield, CA; KHAP(FM) Chico, CA; KFRJ(FM) China Lake, CA; KFRP(FM) Coalinga, CA; KECR El Cajon, CA; KFNO-FM Fresno, CA; KXBC(FM) Garberville, CA; KEFR-FM Le Grand, CA; KFRN Long Beach, CA; KEBR Rocklin, CA; KEAR-FM Sacramento, CA; KEAR(AM) San Francisco, CA; KHFR(FM) Santa Maria, CA; KFRS-FM Soledad, CA; KPRA-FM Ukiah, CA; KFRY(FM) Pueblo, CO; WCTF Vernon, CT; WMFL-FM Florida City, FL; WFTI-FM Saint Petersburg, FL; WWFR(FM) Stuart, FL; WFRP(FM) Americus, GA; WFRC-FM Columbus, GA; KDFR(FM) Des Moines, IA; KEGR(FM) Fort Dodge, IA; KYFR Shenandoah, IA; WJCH-FM Joliet, IL; WQLZ(FM) Taylorville, IL; KPOR-FM Emporia, KS; WOFR(FM) Schoolcraft, MI; KFRT(FM) Butte, MT; KFRD(FM) Butte, MT; KFRW(FM) Great Falls, MT; KBFR(FM) Bismarck, ND; WKDN-FM Camden, NJ; WFME-FM Newark, NJ; KXFR(FM) Socorro, NM; WFBF-FM Buffalo, NY; WFRH-FM Kingston, NY; WFRS-FM Smithtown, NY; WFRW-FM Webster, NY; WCUE Cuyahoga Falls, OH; WOTL-FM Toledo, OH; WYTN-FM Youngstown, OH; KYOR(FM) Newport, OR; KPFR(FM) Pine Grove, OR; KQFE-FM Springfield, OR; WUFR(FM) Bedford, PA; WEFR-FM Erie, PA; WFRJ-FM Johnstown, PA; WXFR(FM) State College, PA; WFCH-FM Charleston, SC; KKAA(AM) Aberdeen, SD; KQFR(FM) Rapid City, SD; KQKD(AM) Redfield, SD; KIFR(FM) Alice, TX; KEDR(FM) Bay City, TX; KTXB-FM Beaumont, TX; KUFR-FM Salt Lake City, UT; KARR Kirkland, WA; KJVH-FM Longview, WA; WWJA(FM) Janesville, WI; WMWK-FM Milwaukee, WI; WJJO-FM Watertown, WI; WFSI-FM Annapolis, MD; WBGR Baltimore, MD; WBMD Baltimore, MD.

Harold Camping, pres.

Fisher Communications Inc., 100 4th Ave. N., Suite 440, Suite 1525, Seattle, WA, 98109. Phone: (206) 404-7000. Fax: (206) 404-7050. Web Site: www.fsci.com.

Stns: 12 TV. KBCI-TV, Boise, ID; KCBY, Eugene, OR; KPIC, Eugene, OR; KVAL, Eugene, OR; KIDK-TV, Idaho Falls-Pocatello, ID; KATU, Portland, OR; KUNP, Portland, OR; KUNS-TV, Seattle-Tacoma, WA; KOMO, Seattle-Tacoma, WA; KLEW-TV, Spokane, WA; KIMA-TV, Yakima-Pasco -Richland-Kennewick, WA; KEPR, Yakima-Pasco -Richland-Kennewick, WA.

Stns: 4 AM. 4 FM. KIKF(FM) Cascade, MT; KXGF Great Falls, MT; KINX(FM) Great Falls, MT; KQDI Great Falls, MT; KQDI-FM Great Falls, MT; KVI(AM) Seattle, WA; KOMO(AM) Seattle, WA; KPLZ(FM) Seattle, WA.

Collen Brown, pres/CEO; Sheri Leonard, asst.

Fort Myers Broadcasting Co., 2824 Palm Beach Blvd., Fort Myers, FL, 33916. Phone: (239) 334-1111. Fax: (239) 334-0744. E-mail: manaager@winktv.com Web Site: winktv.com. Ownership: Brian A. McBride.

Stns: 1 TV. WINK-TV, Ft. Myers-Naples, FL.

Stns: 3 AM. 2 FM. WINK(AM) Fort Myers, FL; WINK-FM Fort Myers, FL; WNPL(AM) Golden Gate, FL; WPTK(AM) Pine Island Center, FL; WTLQ-FM Punta Rassa, FL.

Brian McBride, pres/CEO; Gary Gardner, VP/gen mgr.

Forum Communications Co., Box 2020, Fargo, ND, 58107. Phone: (701) 235-7311. Fax: (701) 241-5406. Web Site: www.in-forum.com.

Stns: 4 TV. WDAY-TV, Fargo-Valley City, ND; WDAZ-TV, Fargo-Valley City, ND; KBMY, Minot-Bismarck-Dickinson, ND; KMCY, Minot-Bismarck-Dickinson, ND.

Stns: 1 AM. 1 FM. WZUU-FM Allegan, MI; WDAY Fargo, ND.

Forum Communications Co. owns the *Alexandria* (MN) *Echo Press; The Pioneer,* Bemidj, MN; *Detroit Lakes* (MN) *Tribune; The Becker County Record,* Detroit Lakes, MN; *Park Rapids* (MN) *Enterprise; The Wadena* (MN) *Pioneer Journal; West Central Daily Tribune,* Willmar, MN; *The Daily Globe;* Worthington, MN; *The Daily Republic,* Mitchell SD; *The Dickinson Press,* Dickinson, ND & *The* (ND) *Forum.*

William C. Marcil, pres.

Fox Television Stations Inc., 1999 S. Bundy Dr., Los Angeles, CA, 90025-5235. Phone: (310) 584-2000. Web Site: www.newscorp.com. Ownership: Fox Entertainment Group Inc., 85.2% voting interest. Note: Fox Entertainment Group Inc. is a wholly-owned subsidiary of News Corp.

Stns: 37 TV. WAGA, Atlanta; KTBC, Austin, TX; WUTB, Baltimore, MD; WBRC, Birmingham (Anniston, Tuscaloosa), AL; WFXT, Boston (Manchester, NH); WPWR-TV, Chicago; WFLD, Chicago; WJW, Cleveland-Akron (Canton), OH; KDFI, Dallas-Ft. Worth; KDFW, Dallas-Ft. Worth; KDVR, Denver, CO; KFCT, Denver, CO; WJBK, Detroit; WOGX, Gainesville, FL; WGHP, Greensboro-High Point-Winston Salem, NC; KRIV, Houston; KTXH, Houston; WDAF-TV, Kansas City, MO; KTTV, Los Angeles; KCOP, Los Angeles; WHBQ, Memphis, TN; WITI, Milwaukee, WI; KFTC, Minneapolis-St. Paul, MN; KMSP, Minneapolis-St. Paul, MN; WFTC, Minneapolis-St. Paul, MN; WNYW, New York; WWOR, New York; WRBW, Orlando-Daytona Beach-Melbourne, FL; WOFL, Orlando-Daytona Beach-Melbourne, FL; WTXF, Philadelphia; KUTP, Phoenix (Prescott), AZ; KSAZ-TV, Phoenix (Prescott), AZ; KSTU, Salt Lake City, UT; KTVI, St. Louis, MO; WTVT, Tampa-St. Petersburg (Sarasota), FL; WTTG, Washington, DC (Hagerstown, MD); WDCA, Washington, DC (Hagerstown, MD).

Kevin Hale, gen mgr; Jack Abernethy, CEO; Roger Ailes, chmn; Dennis Swanson, pres.

Freedom Communications Inc., Broadcast Division, Box 19549, Irvine, CA, 92623-9549. Phone: (949) 253-2315. Fax: (949) 798-3527. Web Site: www.freedom.com. Ownership: Freedom Communications Holdings Inc., 100%.

Stns: 9 TV. WCWN, Albany-Schenectady-Troy, NY; WRGB, Albany-Schenectady-Troy, NY; KFDM-TV, Beaumont-Port Arthur, TX; WTVC, Chattanooga, TN; WWMT, Grand Rapids-Kalamazoo-Battle Creek, MI; WLAJ, Lansing, MI; KTVL, Medford-Klamath Falls, OR; WLNE, Providence, RI-New Bedford, MA; WPEC, West Palm Beach-Ft. Pierce, FL.

Freedom Communications Inc., publishes 28 daily & 37 wkly nwsprs in 12 states.

Doreen Wade, pres.

G

Gannett Broadcasting, (Division of Gannett Co. Inc.). 7950 Jones Branch Dr., Mclean, VA, 22107. Phone: (703) 854-6760. Fax: (703) 854-2005. Web Site: www.gannett.com. Ownership: (Division of Gannett Co. Inc.)

Stns: 22 TV. WATL, Atlanta; WXIA, Atlanta; WLBZ, Bangor, ME; WGRZ, Buffalo, NY; WKYC, Cleveland-Akron (Canton), OH; WLTX, Columbia, SC; KUSA-TV, Denver, CO; WZZM, Grand Rapids-Kalamazoo-Battle Creek, MI; WFMY-TV, Greensboro-High Point-Winston Salem, NC; WJXX, Jacksonville, FL; WTLV, Jacksonville, FL; WBIR, Knoxville, TN; KTHV, Little Rock-Pine Bluff, AR; WMAZ-TV, Macon, GA; KARE, Minneapolis-St. Paul, MN; KNAZ, Phoenix (Prescott), AZ; KPNX, Phoenix (Prescott), AZ; WCSH, Portland-Auburn, ME; KXTV, Sacramento -Stockton-Modesto, CA; KSDK, St. Louis, MO; WTSP, Tampa-St. Petersburg (Sarasota), FL; WUSA, Washington, DC (Hagerstown, MD).

Gannett owns 101 daily newspapers, including the national newspaper *USA Today*, and non-daily newspapers throughout the country.

Dave Lougee, pres.

Glenwood Communications Corp., 222 Commerce St., Kingsport, TN, 37660. Phone: (423) 246-9578. Fax: (423) 246-6261. E-mail: golz@wkpttv.com Web Site: www.wkpttv.com. Ownership: William M. Boyd; Hugh N. Boyd Trust.

Stns: 1 TV. WKPT-TV, Tri-Cities, TN-VA.

Stns: 4 AM. 2 FM. WOPI Bristol, TN; WKTP Jonesborough, TN; WKPT Kingsport, TN; WTFM-FM Kingsport, TN; WMEV Marion, VA; WMEV-FM Marion, VA.

George E. DeVault Jr., pres.

Global BC, (A Division of Global Communications Ltd.). 7850 Enterprise St., Burnaby, BC, V5A 1V7. Canada. Phone: (604) 420-2288. Fax: (604) 422-6427. Web Site: www.canada.com. Ownership: CanWest Global Communications Corp., 100% (see listing).

Stns: 6 TV. CHAN-2, Bowen Island, BC; CHAN-5, Brackendale, BC; CHAN-1, Chilliwack, BC; CHKM-TV, Kamloops, BC; CHAN-3, Squamish, BC; CHAN-7, Whistler, BC.

Brett Manlove, VP sls; Roy Gardner, gen mgr; Fatbir Nijjar, VP finance; John O'Connor, VP.

Global Television, (a division of CanWest MediaWorks Inc.). 81 Barber Greene Rd., Don Mills, ON, M3C 2A2. Canada. Phone: (416) 446-5311. Fax: (416) 446-5447. Web Site: www.globaltv.com. Ownership: CanWest Global Communications Corp., 100% (see listing).

Stns: 9 TV. CIII-2, Bancroft, ON; CIII-7, Midland, ON; CIII-29, Oil Springs, ON; CIII-6, Ottawa, ON; CIII-4, Owen Sound, ON; CIII, Paris, ON; CIII-27, Peterborough, ON; CIII-22, Stevenson, ON; CIII-41, Toronto, ON.

W.K. Hunt, VP/gen mgr; Kathleen Dore, pres.

Granite Broadcasting Corp., 767 Third Ave., 34th Fl., New York, NY, 10017. Phone: (212) 826-2530. Fax: (212) 826-2858. Web Site: www.granitetv.com.

Stns: 10 TV. WBNG, Binghamton, NY; WKBW, Buffalo, NY; WMYD, Detroit; KBJR-TV, Duluth, MN-Superior, WI; KRII, Duluth, MN-Superior, WI; KSEE, Fresno-Visalia, CA; WISE-TV, Ft. Wayne, IN; WEEK, Peoria-Bloomington, IL; KBWB, San Francisco-Oakland-San Jose; WTVH, Syracuse, NY.

W. Don Cornwell, chmn/CEO; Larry Willis, CFO; Stuart Beck, pres; Ellen McClain, CEO; John Deushane, COO.

Grant Communications Inc., 915 Middle River Dr., Suite 409, Fort Lauderdale, FL, 33304. Phone: (954) 568-2000. Fax: (954) 568-2015.

Stns: 7 TV. KGCW-TV, Davenport, IA-Rock Island-Moline, IL; KLJB-TV, Davenport, IA-Rock Island-Moline, IL; WZDX, Huntsville-Decatur (Florence), AL; WLAX, La Crosse-Eau Claire, WI; WEUX, La Crosse-Eau Claire, WI; WFXR-TV, Roanoke-Lynchburg, VA; WWCW, Roanoke-Lynchburg, VA.

Mark Ryan, CFO.

Gray Television Inc., Box 1867, Albany, GA, 31702-1867. Phone: (229) 888-9390. Fax: (229) 888-9374. Web Site: www.graytvinc.com. E-mail: cindy.holden@gcslink.com Ownership: Bull Run Corp., Datasouth Computer Corp. and affiliated companies. Other interests: Porta Phone Paging Inc. and Lynqx.

Stns: 34 TV. WRDW-TV, Augusta, GA; WBKO-TV, Bowling Green, KY; WSAZ-TV, Charleston-Huntington, WV; WCAV, Charlottesville, VA; KKTV, Colorado Springs-Pueblo, CO; WTVY, Dothan, AL; KKCO, Grand Junction-Montrose, CO; WITN-TV, Greenville-New Bern-Washington, NC; WHSV-TV, Harrisonburg, VA; WVLT-TV, Knoxville, TN; WEAU-TV, La Crosse-Eau Claire, WI; WILX, Lansing, MI; WKYT-TV, Lexington, KY; WYMT-TV, Lexington, KY; KGIN, Lincoln & Hastings-Kearney, NE; KOLN, Lincoln & Hastings-Kearney, NE; WMTV, Madison, WI; WTOK-TV, Meridian, MS; WOWT, Omaha, NE; WJHG-TV, Panama City, FL; WTAP-TV, Parkersburg, WV; KOLO, Reno, NV; WIFR, Rockford, IL; KXII, Sherman, TX-Ada, OK; WNDU, South Bend-Elkhart, IN; WCTV, Tallahassee, FL-Thomasville, GA; WSWG, Tallahassee, FL-Thomasville, GA; WIBW-TV, Topeka, KS; KWTX-TV, Waco-Temple-Bryan, TX; KBTX-TV, Waco-Temple-Bryan, TX; WSAW-TV, Wausau-Rhinelander, WI; KAKE, Wichita-Hutchinson Plus, KS; KUPK-TV, Wichita-Hutchinson Plus, KS; KLBY, Wichita-Hutchinson Plus, KS.

Gray Communiations publishes, through Albany Herald Publishing Co., *The Albany* (GA) *Herald* & publishes, through The Rockdale Citizen Publishing Co., *The Rockdale Citizen*, Conyers, GA & *The Gwinnett Daily Post*, Lawrenceville, GA. Gray Communications also publishes *The Goshen News*, Goshen, IN.

James Ryan, VP/CFO; J. Mack Robinson, chmn/CEO; Robert A. Beizer, VP law & dev; Wayne Martin, rgnl VP TV; Robert S. Prather Jr., pres/COO.

Griffin Communications L.L.C., 7401 N. Kelley Ave., Oklahoma City, OK, 73111. Phone: (405) 843-6641. Fax: (405) 841-9135. Web Site: www.griffincommunications.net.

Stns: 3 TV. KWTV, Oklahoma City, OK; KOTV, Tulsa, OK; KQCW, Tulsa, OK.

David Griffin, pres; Steve Foerster, VP; Dick Dutton, VP; Joyce Reed, VP; Kathy Haney, VP; Ted Strickland, CFO.

Groupe TVA Inc., Tele-4/CFCM-TV, 1000 Myrand Ave., Ste.-Foy, PQ, G1V 2W3. Canada. Phone: (418) 688-9330. Fax: (418) 681-4239. Ownership: Quebecor Media Inc., 99.91%.

Stns: 8 TV. CJPM, Chicoutimi, PQ; CFTM, Montreal, PQ; CFCM, Quebec City, PQ; CFER, Rimouski, PQ; CFER-TV-2, Sept-Iles, PQ; CHLT, Sherbrooke, PQ; CKXT-TV, Toronto, ON; CHEM, Trois-Rivieres, PQ.

Serge Gouin, pres.

H

HITV License Subsidiary Inc., 1100 Wilson Blvd., Suite 3000, Arlington, VA, 22209. Phone: (703) 247-7500. Fax: (703) 247-7505. Web Site: www.mcgcapital.com. Ownership: HITV Operating Co. Inc., 100% of votes.

Stns: 3 TV. KGMD, Hilo, HI; KGMB, Honolulu, HI; KGMV, Wailuku, HI.

Michael McHugh, pres.

Hearst-Argyle Television Inc., 300 W. 57th St., 39th Fl., New York, NY, 10019. Phone: (212) 887-6800. Fax: (212) 887-6855. Web Site: www.hearstargyle.com. Ownership: The Hearst Corp., 65%.

Stns: 35 TV. KOAT, Albuquerque-Santa Fe, NM; KOCT, Albuquerque-Santa Fe, NM; KOVT, Albuquerque-Santa Fe, NM; WBAL, Baltimore, MD; WCVB-TV, Boston (Manchester, NH); WMUR, Boston (Manchester, NH); WNNE-TV, Burlington, VT-Plattsburgh, NY; WPTZ, Burlington, VT-Plattsburgh, NY; WLWT, Cincinnati, OH; KCCI, Des Moines-Ames, IA; KHBS, Ft. Smith-Fayetteville -Springdale-Rogers, AR; KHOG, Ft. Smith-Fayetteville -Springdale-Rogers, AR; WXII, Greensboro-High Point-Winston Salem, NC; WYFF, Greenville-Spartanburg, SC-Asheville, NC-Anderson, SC; WGAL, Harrisburg -Lancaster-Lebanon-York, PA; KHVO, Hilo, HI; KITV, Honolulu, HI; WAPT, Jackson, MS; KMBC-TV, Kansas City, MO; KCWE, Kansas City, MO; WLKY, Louisville, KY; WISN, Milwaukee, WI; KSBW, Monterey-Salinas, CA; WDSU, New Orleans, LA; KOCO-TV, Oklahoma City, OK; KETV, Omaha, NE; WESH, Orlando-Daytona Beach-Melbourne, FL; WKCF, Orlando-Daytona Beach-Melbourne, FL; WTAE-TV, Pittsburgh, PA; WMTW-TV, Portland-Auburn, ME; KQCA, Sacramento-Stockton-Modesto, CA; KCRA, Sacramento -Stockton-Modesto, CA; WMOR, Tampa-St. Petersburg (Sarasota), FL; KMAU, Wailuku, HI; WPBF, West Palm Beach-Ft. Pierce, FL.

KHOG-TV Fayetteville and KHBS(TV) Fort Smith, both AR; KCRA-DT Sacramento, KCRA-TV Sacramento and KSBW(TV) Salinas, all CA; WESH(TV) and WESH-DT Daytona Beach, FL; KHVO(TV) Hilo, KHVO-DT Hilo, KITV(TV) Honolulu, KITV-DT Honolulu, KMAU(TV) Wailuku and KMAU-DT Wailuku, all HI; KCCI(TV) Des Moines, IA; WLKY-TV Louisville, KY; WDSU(TV) New Orleans, LA; WBAL-DT and WBAL-TV Baltimore, MD; WCVB-DT and WCVB-TV Boston, MA; WAPT(TV) Jackson, MS; KMBC-TV Kansas City, MO; KETV(TV) Omaha, NE; KOAT-TV Albuquerque, KOCT(TV) Carlsbad, KOFT(TV) Farmington and KOVT(TV) Silver City, all NM; WPTZ(TV) North Pole, NY; WXII(AM) Kernersville and WXII-TV Winston-Salem, both NC; WLWT(TV) and WLWT-DT Cincinnati, OH; KOCO-TV Oklahoma City, OK; WGAL(TV) Lancaster, WTAE-DT Pittsburgh and WTAE-TV Pittsburgh, all PA; WYFF(TV) Greenville, SC; WNNE-TV Hartford, VT; WISN-DT and WISN-TV Milwaukee, WI. Stns managed by Hearst-Argyle Television, owned by The Hearst Corp.: WWWB(TV) Lakeland and WPBF(TV) Tequesta, both FL.

David Barrett, pres/CEO.

Heritage Broadcasting Co of MI, Box 627, Cadillac, MI, 49601. Phone: (231) 775-3478. Fax: (231) 775-3671. Web Site: www.9and10news.com. Ownership: Heritage Broadcasting Group Inc., 100%.

Stns: 2 TV. WWTV, Traverse City-Cadillac, MI; WWUP, Traverse City-Cadillac, MI.

Mario F. Lacobelli, pres; William E. Kring, VP/gen mgr.

Hoak Media Corporation, 500 Crescent Ct., Suite 220, Dallas, TX, 75201. Phone: (972) 960-4848. Fax: (972) 960-4899. Web Site: www.hoakmedia.com.

Stns: 14 TV. KREG, Denver, CO; KVLY, Fargo-Valley City, ND; KREX, Grand Junction-Montrose, CO; KREY, Grand Junction-Montrose, CO; KHAS, Lincoln & Hastings-Kearney, NE; KMOT, Minot-Bismarck-Dickinson, ND; KFYR, Minot-Bismarck-Dickinson, ND; KQCD, Minot-Bismarck-Dickinson, ND; KUMV, Minot-Bismarck-Dickinson, ND; KNOP-TV, North Platte, NE; KPRY, Sioux Falls (Mitchell), SD; KABY, Sioux Falls (Mitchell), SD; KSFY-TV, Sioux Falls (Mitchell), SD; KAUZ, Wichita Falls, TX & Lawton, OK.

Eric Van den Branden, pres.

Hubbard Broadcasting Inc., 3415 University Ave., St. Paul, MN, 55114. Phone: (651) 646-5555. Fax: (651) 642-4103. E-mail: jmahoney@hbi.com

Stns: 13 TV. WNYT, Albany-Schenectady-Troy, NY; KOBG-TV, Albuquerque-Santa Fe, NM; KOB, Albuquerque-Santa Fe, NM; KOBF, Albuquerque-Santa Fe, NM; KOBR, Albuquerque-Santa Fe, NM; WDIO, Duluth, MN-Superior, WI; WIRT, Duluth, MN-Superior, WI; KRWF, Minneapolis-St. Paul, MN; KSAX, Minneapolis-St. Paul, MN; KSTP, Minneapolis-St. Paul, MN; KSTC-TV, Minneapolis-St. Paul, MN; KAAL, Rochester, MN-Mason City, IA-Austin, MN; WHEC, Rochester, NY.

Stns: 2 AM. 2 FM. WFMP(FM) Coon Rapids, MN; KSTP(AM) Saint Paul, MN; KSTP-FM Saint Paul, MN; WIXK(AM) New Richmond, WI.

Stanley S. Hubbard, chmn/pres/CEO; Stanley E. Hubbard II, VP; Virginia H. Morris, VP; Robert W. Hubbard, VP; Julia D. Coyte, VP; Gerald D. Deeney, sr VP/treas/CFO; Harold C. Crump, VP; C. Thomas Newberry, VP; Linda S. Tremere, VP; Sue J. Cook, VP; Edward J. Aiken, VP; Kari Rominski, sec; Gary R. Macomber, asst sec.

I

ION Media Networks Inc., 601 Clearwater Park Rd., West Palm Beach, FL, 33401-6233. Phone: (561) 659-4122. Fax: (561) 655-7246. Fax: (561) 659-4252. E-mail: sethgrossman@paxson.com Web Site: www.ionmedianetworks.com. Ownership: Publicly traded company on the Amex ticker (ION). Note: ION Media Networks also owns Infomall TV Network, three state radio networks (Alabama Radio Network, Florida Radio Network, Tennessee Radio Network), three rgnl sports networks (University of Florida Sports Network, University of Miami Sports Network, Penn State Sports Network), outdoor adv.

Stns: 57 TV. WYPX, Albany-Schenectady-Troy, NY; WPXA, Atlanta; WPXH, Birmingham (Anniston, Tuscaloosa), AL; WPXG, Boston (Manchester, NH); WBPX, Boston (Manchester, NH); WPXJ, Buffalo, NY; KPXR, Cedar Rapids-Waterloo-Iowa City & Dubuque, IA; WLPX, Charleston-Huntington, WV; WCPX, Chicago; WVPX, Cleveland-Akron (Canton), OH; KPXD, Dallas-Ft. Worth; KPXC, Denver, CO; KFPX, Des Moines-Ames, IA; WPXD, Detroit; WZPX, Grand Rapids-Kalamazoo-Battle Creek, MI; WGPX, Greensboro-High Point-Winston Salem, NC; WPXU-TV, Greenville-New Bern-Washington, NC; WEPX, Greenville-New Bern-Washington, NC; WHPX, Hartford & New Haven, CT; KPXO-TV, Honolulu, HI; KPXB, Houston; WIPX, Indianapolis, IN; WPXC-TV, Jacksonville, FL; KPXE, Kansas City, MO; WPXK, Knoxville, TN; WUPX-TV, Lexington, KY; KPXN, Los Angeles; WPXM, Miami-Ft. Lauderdale, FL; WPXE, Milwaukee, WI; KPXM, Minneapolis-St. Paul, MN; WNPX, Nashville, TN; WPXN, New York; WPXV, Norfolk-Portsmouth-Newport News, VA; KWWT, Odessa-Midland, TX; KOPX, Oklahoma City, OK; WOPX, Orlando-Daytona Beach-Melbourne, FL; WPPX, Philadelphia; KPPX, Phoenix (Prescott), AZ; KPXG, Portland, OR; WDPX, Providence, RI-New Bedford, MA; WPXQ, Providence, RI-New Bedford, MA; WRPX, Raleigh-Durham (Fayetteville), NC; WFPX, Raleigh-Durham (Fayetteville), NC; WPXR, Roanoke-Lynchburg, VA; KSPX, Sacramento -Stockton-Modesto, CA; KUPX, Salt Lake City, UT; KPXL, San Antonio, TX; KKPX, San Francisco-Oakland-San Jose; KWPX, Seattle-Tacoma, WA; KGPX, Spokane, WA; WXPX, Tampa-St. Petersburg (Sarasota), FL; KTPX, Tulsa, OK; WWPX, Washington, DC (Hagerstown, MD); WPXW, Washington, DC (Hagerstown, MD); WTPX, Wausau-Rhinelander, WI; WPXP, West Palm Beach-Ft. Pierce, FL; WQPX, Wilkes Barre-Scranton, PA.

R. Brandon Burgess, CEO.

International Broadcasting Corp., 1554 Bori St., San Juan, PR, 00927-6113. Phone: (787) 274-1800. Fax: (787) 281-9758. Ownership: Pedro Roman Collazo, 100%. Note: Pedro Roman Collazo, as an individual, owns WVOZ(AM) San Juan, PR.

Stns: 3 TV. WVEO, Aguadilla, PR; WVOZ, Ponce, PR; WTCV, San Juan, PR.

Stns: 7 AM. 1 FM. WRSJ(AM) Bayamon, PR; WGIT(AM) Canovanas, PR; WVOZ-FM Carolina, PR; WIBS Guayama, PR; WXRF Guayama, PR; WTIL Mayaguez, PR; WEKO(AM) Morovis, PR; WCHQ(AM) Quebradillas, PR.

Pedro Roman Callazo, pres; Margarita Nazario, gen mgr.

J

Journal Communications Inc., 333 W. State St., Milwaukee, WI, 53203. Phone: (414) 224-2616. Fax: (414) 224-2469. Web Site: www.jc.com.

Stns: 9 TV. KIVI, Boise, ID; WFTX, Ft. Myers-Naples, FL; WGBA, Green Bay-Appleton, WI; WSYM, Lansing, MI; KTNV, Las Vegas, NV; WTMJ, Milwaukee, WI; KMTV, Omaha, NE; KMIR-TV, Palm Springs, CA; KGUN, Tucson (Sierra Vista), AZ.

Stns: 8 AM. 22 FM. KGMG-FM Oracle, AZ; KMXZ-FM Tucson, AZ; KQTH(FM) Tucson, AZ; KJOT(FM) Boise, ID; KGEM(AM) Boise, ID; KCID(AM) Caldwell, ID; KTHI(FM) Caldwell, ID; KRVB(FM) Nampa, ID; KQXR-FM Payette, ID; KYQQ(FM) Arkansas City, KS; KFXJ(FM) Augusta, KS; KFTI-FM Newton, KS; KICT-FM Wichita, KS; KFTI(AM) Wichita, KS; KSGF-FM Ash Grove, MO; KZRQ-FM Mount Vernon, MO; KSPW(FM) Sparta, MO; KSGF(AM) Springfield, MO; KSRZ-FM Omaha, NE; KKCD-FM Omaha, NE; KEZO-FM Omaha, NE; KXSP(AM) Omaha, NE; KQCH(FM) Omaha, NE; KXBL(FM) Henryetta, OK; KFAQ(AM) Tulsa, OK; WMYU(FM) Karns, TN; WKHT(FM) Knoxville, TN; WQBB Powell, TN; WWST(FM) Sevierville, TN; WTMJ Milwaukee, WI.

Journal Communications Inc., publisher of the morning *Milwaukee* (WI) *Journal Sentinel*, owns 100% of Journal Broadcast Corp.

Douglas G. Kiel, pres.

K

KB Prime Media L.L.C., 1320 Lafayette Rd., Gladwyne, PA, 19035. Phone: (610) 526-2927. Fax: (610) 526-0679. E-mail: guyonturner@compuserve.com Ownership: W.W. Keen Butcher, 80%; Guyon W. Turner, 20%.

Stns: 2 TV. WTLF, Tallahassee, FL-Thomasville, GA; WSWB, Wilkes Barre-Scranton, PA.

W.W. Keen Butcher, member/CEO; Guyon W. Turner, pres/member.

KEVN Inc., Box 677, Rapid City, SD, 57709. Phone: (605) 394-7777. Fax: (605) 348-9128. E-mail: news@blackhillsfox.com Web Site: www.blackhillsfox.com.

Stns: 2 TV. KEVN-TV, Rapid City, SD; KIVV-TV, Rapid City, SD.

Cindy McNeill, VP/gen mgr; Robert Slocum, CFO; Bill Reyner, pres; Kathy Silk, rgnl sls mgr.

KHQ Inc., Box 600, Spokane, WA, 99210-4102. Phone: (509) 448-6000. Fax: (509) 448-3231. Ownership: Cowles Publishing Company, 100%.

Stns: 3 TV. KHQ, Spokane, WA; KNDO, Yakima-Pasco -Richland-Kennewick, WA; KNDU, Yakima-Pasco -Richland-Kennewick, WA.

Cowles Publishing owns *Spokesman-Review*, Spokane, WA.

Lon C. Lee, pres.

KM Communications Inc., 3654 Jarvis Ave., Skokie, IL, 60076. Phone: (847) 674-0864. Fax: (847) 674-9188. Web Site: www.kmcommunications.com.

Stns: 2 TV. KWKB, Cedar Rapids-Waterloo-Iowa City & Dubuque, IA; KEJB, Monroe, LA-El Dorado, AR.

Stns: 2 AM. 9 FM. WPNG(FM) Pearson, GA; KTKB-FM Hagatna, GU; KTKB(AM) Tamuning, GU; KQMG Independence, IA; KQMG-FM Independence, IA; WLCN(FM) Atlanta, IL; WMKB(FM) Earlville, IL; KBWM(FM) Breckenridge, TX; KKEV(FM) Centerville, TX; KBDK(FM) Leakey, TX; KHMR(FM) Lovelady, TX.

Myoung Hwa Bae, pres; Kevin J. Bae, VP/gen mgr.

L

LIN Television Corporation, 4 Richmond Sq., Providence, RI, 02906. Phone: (401) 454-2880. Fax: (401) 454-2817. E-mail: deborah.jacobson@lintv.com Web Site: www.lintv.com. Ownership: Hicks, Muse, Tate & Furst 47%.

Stns: 34 TV. KBIM, Albuquerque-Santa Fe, NM; KREZ, Albuquerque-Santa Fe, NM; KRQE, Albuquerque-Santa Fe, NM; KXAM, Austin, TX; KXAN-TV, Austin, TX; WIVB-TV, Buffalo, NY; WNLO, Buffalo, NY; WAND, Champaign & Springfield-Decatur, IL; WWHO, Columbus, OH; WDTN, Dayton, OH; WANE-TV, Ft. Wayne, IN; WOOD-TV, Grand Rapids-Kalamazoo-Battle Creek, MI; WOTV, Grand Rapids-Kalamazoo-Battle Creek, MI; WLUK-TV, Green Bay-Appleton, WI; WCTX, Hartford & New Haven, CT; WTNH, Hartford & New Haven, CT; WNDY, Indianapolis, IN; WISH-TV, Indianapolis, IN; WLFI, Lafayette, IN; WNJX, Mayaguez, PR; WBPG, Mobile, AL-Pensacola (Ft. Walton Beach), FL; WALA-TV, Mobile, AL-Pensacola (Ft. Walton Beach), FL; WAVY-TV, Norfolk-Portsmouth-Newport News, VA; WVBT, Norfolk-Portsmouth-Newport News, VA; WTIN, Ponce, PR; WKPV, Ponce, PR; WPRI, Providence, RI-New Bedford, MA; WJPX, San Juan, PR; WAPA, San Juan, PR; WJWN, San Sebastian, PR; WWLP, Springfield-Holyoke, MA; WTHI, Terre Haute, IN; WUPW, Toledo, OH; WIRS, Yauco, PR.

Vincent L. Sadusky, CEO; Bart Catalane, CFO.

Lake Superior Community Broadcasting Corp., 1390 Bagley St., Alpena, MI, 49707. Phone: (989) 356-3434. Ownership: Stephen A. Marks, 100%.

Stns: 2 TV. WBKP, Marquette, MI; WBUP, Marquette, MI.

Landmark Communications Inc., (Landmark Broadcast Division.). 150 W. Brambleton Ave., Norfolk, VA, 23510. Phone: (757) 446-2000. Fax: (757) 446-2179. Web Site: www.landmarkcommunications.com. Ownership: (Landmark Broadcast Division.)

Stns: 2 TV. KLAS, Las Vegas, NV; WTVF, Nashville, TN.

Landmark Communications Inc. publishes the following daily newspapers: *Citrus County Crhonicle*, Crystal River, FL; *News-Enterprise*, Elizabethtown, KY; *The Carroll County Times*, Westminster, MD; *Los Alamos Monitor*, Los Alamos, NM; *News & Record*, Greensboro, NC; *The Virginian-Pilot*, Norfolk, VA; *Roanoke Times*, Roanoke, VA. Landmark Community Newspapers, Shelbyville, KY, publishes four community dailies, four tri-wklys, nine semi-wklys, 29 wklys, 39 shoppers & free newspapers, & 38 special-interest publications. Landmark Communications owns 49.9% of Capital-Gazette Communications Inc., publisher of *The Capital* (a daily newspaper in Annapolis, MD), *The Maryland Gazette* (a twice-weekly newspaper in Glen Burnie, MD), *Washingtonian Magazine* & weekly newspapers in Bowie & Crofton, MD.

Frank Batten Jr., chmn; Decker Anstrom, pres.

Le Sea Broadcasting, Box 12, South Bend, IN, 46624. Phone: (574) 291-8200. Fax: (574) 291-9043. E-mail: leseabroadcasting@lesea.com Web Site: www.lesea.com.

Stns: 8 TV. KWHD, Denver, CO; KWHH, Hilo, HI; KWHE, Honolulu, HI; WHMB-TV, Indianapolis, IN; WHNO, New Orleans, LA; WHME, South Bend-Elkhart, IN; KWHB, Tulsa, OK; KWHM, Wailuku, HI.

Stns: 1 AM. 3 FM. WHPZ-FM Bremen, IN; WHME-FM South Bend, IN; WHPD(FM) Dowagiac, MI; WDOW Dowagiac, MI.

Peter Sumrall, pres/CEO.

Liberman Broadcasting Inc., 1845 Empire Ave., Burbank, CA, 91504. Phone: (818) 729-5300. Fax: (818) 729-5678. E-mail: LBIinfo@lbimedia.com Web Site: www.lbimedia.com. Ownership: Lenard D. Liberman, 47.5-49% votes, 40-42.5% equity; Jose Liberman 2003 Annuity Trust, 23.75-24.5% votes, 20-21.25% equity; Esther Liberman 2003 Annuity Trust, 23.75-24.5% votes, 20-21.25% equity; public shareholders of Liberman Broadcasting Inc., 2-5% votes, 15-20% equity.

Stns: 3 TV. KMPX, Dallas-Ft. Worth; KZJL, Houston; KRCA, Los Angeles.

Stns: 6 AM. 14 FM. KEBN(FM) Garden Grove, CA; KBUE(FM) Long Beach, CA; KHJ(AM) Los Angeles, CA; KBUA(FM) San Fernando, CA; KWIZ-FM Santa Ana, CA; KTCY(FM) Azle, TX; KXGJ-FM Bay City, TX; KQQK(FM) Beaumont, TX; KBOC(FM) Bridgeport, TX; KJOJ Conroe, TX; KIOX-FM El Campo, TX; KJOJ-FM Freeport, TX; KEYH Houston, TX; KQUE Houston, TX; KNOR(FM) Krum, TX; KZZA(FM) Muenster, TX; KZMP-FM Pilot Point, TX; KTJM-FM Port Arthur, TX; KSEV Tomball, TX; KZMP(AM) University Park, TX.

Lenard Liberman, pres; Brett Zane, CEO.

Lincoln Financial Media, 100 N. Greene St., Greensboro, NC, 27420. Phone: (336) 691-3000. Fax: (336) 691-3222. Web Site: www.lincolnfinancialmedia.com. Ownership: Lincoln National Corp., 100%.

Stns: 3 TV. WCSC, Charleston, SC; WBTV, Charlotte, NC; WWBT, Richmond-Petersburg, VA.

Stns: 6 AM. 12 FM. KSOQ-FM Escondido, CA; KIFM-FM San Diego, CA; KBZT(FM) San Diego, CA; KSON San Diego, CA; KSON-FM San Diego, CA; KYGO-FM Denver, CO; KKFN Denver, CO; KQKS-FM Lakewood, CO; KEPN(AM) Lakewood, CO; KJCD(FM) Longmont, CO; WLYF(FM) Miami, FL; WMXJ-FM Pompano Beach, FL; WAXY South Miami, FL; WQXI Atlanta, GA; WSTR-FM Smyrna, GA; WBT Charlotte, NC; WLNK(FM) Charlotte, NC; WBT-FM Chester, SC.

Ed Hull, pres & Lincoln Financial Sports; John Shreves, pres & Lincoln Financial TV; Don Benson, pres & Lincoln Financial Radio.

Local TV LLC, 1717 Dixie Hwy., Suite 650, Ft. Wright, KY, 41011. Phone: (859) 448-2700. Fax: (859) 331-6014. Web Site: www.localtvllc.com. Ownership: Oak Hill Capital Partners II LP, 64.87%; Local TV B-Corp A Inc., 23.29%; Local TV B-Corp B Inc., 8.97%; Oak Hill Capital Management Partners II LP, 1.70%; Benjamin L. Homel (aka Randy Michaels), less than 2%; and Robert L. Lawrence, less than 1%.

Stns: 9 TV. WQAD, Davenport, IA-Rock Island-Moline, IL; WHO-TV, Des Moines-Ames, IA; KFSM-TV, Ft. Smith-Fayetteville-Springdale-Rogers, AR; WHNT, Huntsville-Decatur (Florence), AL; WREG, Memphis, TN; WTKR, Norfolk-Portsmouth-Newport News, VA; KAUT-TV, Oklahoma City, OK; KFOR, Oklahoma City, OK; WNEP, Wilkes Barre-Scranton, PA.

Randy Michaels, CEO; Robert L. Lawrence, pres/COO; Pam Taylor, CFO.

M

MAX Media L.L.C., 900 Laskin Rd., Virginia Beach, VA, 23451. Phone: (757) 437-9800. Fax: (757) 437-0034. Web Site: www.maxmediallc.com. Ownership: MBG-GG LLC, 42.0345%; MBG Quad-C Investors I Inc., 41.4124%; Aardvarks Also LLC, 6.1967%; Colonnade Max Investors Inc., 4.8671%; Quad-C Max Investors Inc., 4.6799%; MBG Quad-C Investors II Inc., 0.6221%; and Quad-C Max Investors II Inc., 0.1872%.

Stns: 10 TV. WMEI, Arecibo, PR; KULR-TV, Billings, MT; WNKY, Bowling Green, KY; KWYB, Butte-Bozeman, MT; WVIF, Christiansted, VI; KTMF, Missoula, MT; WPFO, Portland-Auburn, ME; WGTQ, Traverse City-Cadillac, MI; WGTU, Traverse City-Cadillac, MI; KYTX, Tyler-Longview (Lufkin & Nacogdoches), TX.

Stns: 12 AM. 23 FM. KVLD(FM) Atkins, AR; KCAB Dardanelle, AR; KVOM Morrilton, AR; KVOM-FM Morrilton, AR; KWKK-FM Russellville, AR; WCIL Carbondale, IL; WUEZ(FM) Carterville, IL; WXLT(FM) Christopher, IL; WOOZ-FM Harrisburg, IL; WJPF Herrin, IL; KZIM Cape Girardeau, MO; KEZS-FM Cape Girardeau, MO; KGIR Cape Girardeau, MO; KCGQ-FM Gordonville, MO; KLSC(FM) Malden, MO; KMAL(AM) Malden, MO; KWOC Poplar Bluff, MO; KJEZ-FM Poplar Bluff, MO; KKLR-FM Poplar Bluff, MO; KGKS-FM Scott City, MO; KSIM(AM) Sikeston, MO; WQDK-FM Ahoskie, NC; WGAI(AM) Elizabeth City, NC; WCMS-FM Hatteras, NC; WCXL(FM) Kill Devil Hills, NC; WFYY(FM) Bloomsburg, PA; WYGL-FM Elizabethville, PA; WWBE-FM Mifflinburg, PA; WLGL-FM Riverside, PA; WYGL Selinsgrove, PA; WCMS(AM) Newport News, VA; WGH-FM Newport News, VA; WXMM(FM) Norfolk, VA; WVBW(FM) Suffolk, VA; WXEZ-FM Yorktown, VA.

John A. Trinder, pres.

Malara Broadcast Group Inc., 5880 Midnight Pass Rd., Suite 701, Siesta Key, FL, 34242. Phone: (941) 312-0214. Ownership: TCM Media Associates LLC, 100%.

Stns: 2 TV. KDLH, Duluth, MN-Superior, WI; WPTA, Ft. Wayne, IN.

Manship Stations, Box 2906, Baton Rouge, LA, 70821. Phone: (225) 387-2222. Fax: (225) 336-2246. Web Site: www.2theadvocate.com.

Stns: 2 TV. WBRZ, Baton Rouge, LA; KRGV, Harlingen-Weslaco-Brownsville-McAllen, TX.

Also owns Baton Rouge *Morning Advocate* & Saturday & Sunday *Advocate*.

Richard F. Manship, pres.

Mark III Media Inc., 2312 Sagewood, Casper, WY, 82601. Phone: (307) 235-3962. Ownership: Julie Jaffe, 35%; Jennifer Lechter, 35%; and Mark R. Nalbone, 30%.

Stns: 3 TV. KGWC, Casper-Riverton, WY; KGWL, Casper-Riverton, WY; KGWR, Salt Lake City, UT.

Julie Jaffe, pres.

McGraw-Hill Broadcasting Co., c/o KGTV(TV), 4600 Air Way, San Diego, CA, 92102. Phone: (619) 237-6212. Fax: (619) 262-2275. E-mail: equinn@kgtv.com Web Site: www.mcgraw-hill.com. Ownership: The McGraw-Hill Companies, 100%.

Stns: 4 TV. KERO, Bakersfield, CA; KMGH, Denver, CO; WRTV, Indianapolis, IN; KGTV, San Diego, CA.

McGraw-Hill Inc., owner of the McGraw-Hill Broadcasting Co., publishes *Business Week* magazine & various trade publications.

Edward J. Quinn, pres; Tim Boling, engrg dir.

McKinnon Broadcasting Co., 5002 S. Padre Island Dr., Corpus Christi, TX, 78411. Phone: (361) 986-8300. Fax: (361) 986-8411. Ownership: Michael McKinnon.

Stns: 3 TV. KBMT, Beaumont-Port Arthur, TX; KIII, Corpus Christi, TX; KUSI, San Diego, CA.

Michael McKinnon, pres/CEO.

Media General Broadcast Group, 111 N. 4th St., Richmond, VA, 23219. Phone: (804) 775-4600. Fax: (804) 775-4601. Web Site: www.mgbg.com. Ownership: Media General Inc.

Stns: 23 TV. KALB-TV, Alexandria, LA; WJBF, Augusta, GA; WVTM, Birmingham (Anniston, Tuscaloosa), AL; WCBD-TV, Charleston, SC; WRBL, Columbus, GA; WCMH, Columbus, OH; WNCT-TV, Greenville-New Bern-Washington, NC; WNEG, Greenville-Spartanburg, SC-Asheville, NC-Anderson, SC; WYCW, Greenville-Spartanburg, SC-Asheville, NC-Anderson, SC; WSPA-TV, Greenville-Spartanburg, SC-Asheville, NC-Anderson, SC; WHLT, Hattiesburg-Laurel, MS; WJTV, Jackson, MS; WCWJ, Jacksonville, FL; WTVQ-TV, Lexington, KY; WKRG, Mobile, AL-Pensacola (Ft. Walton Beach), FL; WBTW, Myrtle Beach-Florence, SC; WMBB, Panama City, FL; WJAR, Providence, RI-New Bedford, MA; WNCN, Raleigh-Durham (Fayetteville), NC; WSLS, Roanoke-Lynchburg, VA; WSAV-TV, Savannah, GA; WFLA-TV, Tampa-St. Petersburg (Sarasota), FL; WJHL, Tri-Cities, TN-VA.

The Dothan Eagle, Opelika-Auburn News, The Enterprise Ledger, all AL; *The Denver Post* (20% ownership), CO; *The Tampa Tribune,* (Sebring) *Highlands Today,* (Brooksville) *Hernando Today, Jackson County Floridan,* FL; *Winston-Salem Journal,* (Concord & Kannapolis) *Independent Tribune, Hickory Daily Record, Statesville Record & Landmark, The* (Morganton) *News Herald, The Reidsville Review, The* (Eden) *Daily News, The* (Marion) *McDowell News,* NC; *The* (Florence) *Morning News,* SC; *Richmond Times-Dispatch, Bristol Herald Courier, The* (Lynchburg) *News & Advance, The* (Charlottesville) *Daily Progress, Potomac* (Woodbridge) *News, Danville Register & Bee, The* (Waynesboro) *News Virginian, Manassas Journal Messenger, Culpeper Star-Exponent, Virginia Business (monthly magazine), VA. Media General also owns nearly 100 weeklies and other periodicals and Media General News Service, DC.*

James A. Zimmerman, pres; Edward H. Deichman Jr., VP finance; Richard W. Roberts, VP; Ardell Hill, VP; Peter McCampbell, VP/dir sls; Daniel Bradley, VP; Steve Gleason, progmg VP; Catherine Gugerty, dir mktg; James Conschafter, VP; Tom Conway, VP; Paul Gaulke, mktg VP.

Meredith Broadcasting Group, Meredith Corp., 1716 Locust St., Des Moines, IA, 50309-3023. Phone: (515) 284-2159. Fax: (515) 284-2514. Web Site: www.meredith.com.

Ownership: Meredith Broadcasting is an operating group of Meredith Corp., Des Moines, IA.

Stns: 12 TV. WGCL-TV, Atlanta; WFLI, Chattanooga, TN; WNEM-TV, Flint-Saginaw-Bay City, MI; WHNS, Greenville-Spartanburg, SC-Asheville, NC-Anderson, SC; WFSB, Hartford & New Haven, CT; KSMO-TV, Kansas City, MO; KCTV, Kansas City, MO; KVVU, Las Vegas, NV; WSMV, Nashville, TN; KPHO-TV, Phoenix (Prescott), AZ; KPTV, Portland, OR; KPDX, Portland, OR.

Stns: 1 AM. WNEM(AM) Bridgeport, MI.

The publishing group includes:

Magazines: *American Baby, American Patchwork & Quilting, Better Homes & Gardens, Country Home, Country Home Country Gardens, Creative Home, Decorating, Do It Yourself, Garden, Deck, and Landscape, Garden Shed, Ladies' Home Journal, Midwest Living, MORE, Renovation Style, Successful Farming, Traditional Home,* and *Wood,* along with more than 170 special interest titles.

Paul Karpowic, pres; Douglas Lowe, exec VP.

Mission Broadcasting Inc., 7650 Chippewa Rd., Suite 305, Brecksville, OH, 44141. Phone: (440) 526-2227. Fax: (330) 336-8454. Fax: (440) 546-1903. E-mail: dpthatcher@sbcglobal.net

Stns: 13 TV. KRBC, Abilene-Sweetwater, TX; KCIT, Amarillo, TX; KHMT, Billings, MT; WFXP, Erie, PA; KODE, Joplin, MO-Pittsburg, KS; KAMC, Lubbock, TX; WTVO, Rockford, IL; KSAN-TV, San Angelo, TX; KOLR, Springfield, MO; WFXW-TV, Terre Haute, IN; WUTR, Utica, NY; KJTL, Wichita Falls, TX & Lawton, OK; WYOU, Wilkes Barre-Scranton, PA.

David Smith, pres; Nancie Smith, VP; Dennis Thatcher, COO.

Montecito Broadcast Group LLC, 559 San Ysidro Rd., Suite I, Montecito, CA, 93108. Phone: (805) 969-9278. Fax: (805) 969-2399. Web Site: sjlhost.com. Ownership: Blackstone ECC Communications Partners L.P., 69.50144%; Blackstone ECC Capital Partners IV L.P., 23.17084%; Blackstone Family Communications Partnership I L.P., 4.43626%.

Stns: 9 TV. KHAW-TV, Hilo, HI; KHON-TV, Honolulu, HI; KOIN, Portland, OR; KSNT-TV, Topeka, KS; KAII-TV, Wailuku, HI; KSNW, Wichita-Hutchinson Plus, KS; KSNC, Wichita-Hutchinson Plus, KS; KSNG, Wichita-Hutchinson Plus, KS; KSNK, Wichita-Hutchinson Plus, KS.

George D. Lillly, CEO.

Morris Multimedia Inc., 27 Abercorn St., Savannah, GA, 31401. Phone: (912) 233-1281. Fax: (912) 238-2059. Web Site: www.morrismultimedia.com. Ownership: Charles H. Morris.

Stns: 4 TV. WXXV, Biloxi-Gulfport, MS; WDEF, Chattanooga, TN; WCBI-TV, Columbus-Tupelo-West Point, MS; WMGT-TV, Macon, GA.

Daily newspapers, weekly newspapers.

Charles H. Morris, pres.

Mountain Broadcasting Corp., 99 Clinton Rd., West Caldwell, NJ, 07006. Phone: (973) 852-0300. Fax: (973) 808-5516. Ownership: Sun Young Joo, 66% votes, 35.2% total assets; John H. Joo, 14% votes, 6.3% total assets; Victor C. Joo, 14% votes, 5.6% total assets; Sun Hoo Joo, 6% votes, 2.9% total assets; and Hansen Lau, 5.7% total assets.

Stns: 1 TV. WMBC-TV, New York.

Stns: 3 AM. WPWA Chester, PA; WBTK(AM) Richmond, VA; WWGB Indian Head, MD.

Sun Young Joo, pres.

Multicultural Television Broadcasting LLC, 449 Broadway, New York, NY, 10013. Phone: (212) 431-4300. Ownership: Arthur Liu, 51%; and Yvonne Liu, 49%.

Stns: 5 TV. WMFP, Boston (Manchester, NH); WOAC, Cleveland-Akron (Canton), OH; WSAH, New York; WRAY, Raleigh-Durham (Fayetteville), NC; KCNS, San Francisco-Oakland-San Jose.

Morgan Murphy Stations (Evening Telegram Co), Box 44965, Madison, WI, 53744-4965. Phone: (608) 271-4321. Fax: (608) 271-6111. E-mail: talkback@wisctv.com Web Site: www.channel3000.com. Ownership: Evening Telegram Co. owns 100% of KVEW(TV), KXLY-AM-FM-TV, KXLY-DT and KAPP(TV). Evening Telegram Co. owns 84.4% of Television Wisconsin Inc., with an additional 15.2% of the stn held by Evening Telegram stockholders.

Stns: 5 TV. WKBT, La Crosse-Eau Claire, WI; WISC-TV, Madison, WI; KXLY-TV, Spokane, WA; KAPP, Yakima-Pasco -Richland-Kennewick, WA; KVEW, Yakima-Pasco -Richland-Kennewick, WA.

Stns: 4 AM. 3 FM. KXLX(AM) Airway Heights, WA; KXLY Spokane, WA; KZZU-FM Spokane, WA; KEZE-FM Spokane, WA; WGLR Lancaster, WI; WPVL Platteville, WI; WPVL-FM Platteville, WI.

The Evening Telegram principals own *Madison Magazine,* Madison, WI.

Elizabeth Murphy Burns, pres; George Nelson, exec VP; David Sanks, exec VP; Steve Herling, exec VP; Darrell Blue, VP/gen mgr; Scott Chorski, VP/gen mgr.

N

NBC Universal Television Stations, 30 Rockefeller Plaza, New York, NY, 10112. Phone: (212) 664-4444. Fax: (212) 664-5830. Web Site: www.nbcuni.com. Ownership: General Electric Co. owns 100% of NBC Telemundo Holding Co., which is the single controlling shareholder of NBC Telemundo Inc. NBC Telemundo Inc. in turn owns 100% of NBC Telemundo License Co. and NBC License Co.

Stns: 10 TV. WMAQ, Chicago; KXTX, Dallas-Ft. Worth; WVIT, Hartford & New Haven, CT; KNBC, Los Angeles; WTVJ, Miami-Ft. Lauderdale, FL; WNBC, New York; WCAU, Philadelphia; KNSD, San Diego, CA; KNTV, San Francisco-Oakland-San Jose; WRC, Washington, DC (Hagerstown, MD).

John Wallace, pres.

Neuhoff Family L.P., 1501 N. Washington, Danville, IL, 61832. Phone: (217) 442-1700. Phone: (217) 787-9200. Fax: (217) 431-1489. E-mail: mhulvey@cooketech.net Web Site: www.wdnlfm.com. Ownership: Neuhoff Corp., North Palm Beach, FL, 100% of votes.

Stns: 1 TV. KMVT, Twin Falls, ID.

Stns: 2 AM. 5 FM. WDAN(AM) Danville, IL; WDNL(FM) Danville, IL; WRHK-FM Danville, IL; WXAJ(FM) Hillsboro, IL; WFMB(AM) Springfield, IL; WFMB-FM Springfield, IL; WCVS-FM Virden, IL.

Mike Hulvey, gen mgr; Geoff Neuhoff, pres.

New Vision Television LLC, 3500 Lenox Rd., Suite 640, Atlanta, GA, 30326. Phone: (404) 995-4711. Fax: (404) 995-4712. Web Site: www.newvisiontv.com. Ownership: HBK NV LLC, 97.5%; Jason Elkin, 2%; and John Heinen, 0.5%.

Stns: 2 TV. WIAT, Birmingham (Anniston, Tuscaloosa), AL; KIMT, Rochester, MN-Mason City, IA-Austin, MN.

Jason Elkin, chmn/CEO; John Heinen, pres/COO.

NewCap Inc., 745 Windmill Rd., Dartmouth, NS, B3B1C2. Canada. Phone: (902) 468-7557. Fax: (902) 468-7558. Web Site: www.ncc.ca. Ownership: H.R. Steele, Blavin & Company.

Stns: 2 TV. CITL, Lloydminster, AB; CKSA, Lloydminster, AB.

Stns: 23 AM. 33 FM. CKBA Athabasca, AB; CJPR-FM Blairmore, AB; CJEG-FM Bonnyville, AB; CIXF-FM Brooks, AB; CIBQ Brooks, AB; CIQX-FM Calgary, AB; CFUL-FM Calgary, AB; CFCW-FM Camrose, AB; CFCW Camrose, AB; CKDQ Drumheller, AB; CKRA-FM Edmonton, AB; CIRK-FM Edmonton, AB; CFXE-FM Edson, AB; CJXK-FM Grand Centre (Cold Lake), AB; CKVH High Prairie, AB; CFXH-FM Hinton, AB; CKSA-FM Lloydminster, AB; CKGY-FM Red Deer, AB; CIZZ-FM Red Deer, AB; CHLW(AM) Saint Paul, AB; CHSL-FM Slave Lake, AB; CKSQ(AM) Stettler, AB; CKKY Wainwright, AB; CKWY-FM Wainwright, AB; CFOK(AM) Westlock, AB; CKJR Wetaskiwin, AB; CFXW-FM Whitecourt, AB; CHNK-FM Winnipeg, MB; CKJS Winnipeg, MB; CFRK-FM Fredericton, NB; CKIM Baie Verte, NF; CHVO Carbonear, NF; CFLC-FM Churchill Falls, NF; CKVO Clarenville, NF; CKXX-FM Corner Brook, NF; CFCB Corner Brook, NF; CKGA Gander, NF; CKXD-FM Gander, NF; CFLN Goose Bay, NF; CKCM Grand Falls, NF; CKXG-FM Grand Falls-Windsor, NF; CHCM Marystown, NF; CFNW Port au Choix, NF; CFCV-FM Saint Andrews, NF; VOCM(AM) Saint John's, NF; CKIX-FM Saint John's, NF; CFSX Stephenville, NF; CFLW Wabush, NF; CKUL-FM Halifax, NS; CIHT-FM Ottawa, ON; CILV-FM Ottawa, ON; CHNO-FM Sudbury, ON; CJUK-FM Thunder Bay, ON; CKTG-FM Thunder Bay, ON; CHTN-FM Charlottetown, PE; CKQK-FM Charlottetown, PE.

H.R. Steele, chmn; Scott Weatherby, CEO; R.G. Steele, pres/CEO.

Newfoundland Broadcasting Co., (NTV & OZ Networks). Box 2020, St. John's, NF, A1C 5S2. Canada. Phone: (709) 722-5015. Fax: (709) 726-5107. E-mail: ozfm@ozfm.com Web Site: www.ntv.ca. Ownership: Geoffrey W. Stirling, 89.95%; G. Scott Stirling, 10%; and others, 0.05%.

Stns: 6 TV. CJOM, Argentia, NF; CJWB, Bonavista, NF; CJWN, Corner Brook, NF; CJOX-1, Grand Bank, NF; CJCN, Grand Falls, NF; CJSV, Stephenville, NF.

Stns: 8 FM. CJOZ-FM Bonavista Bay, NF; CJKK-FM Clarenville, NF; CKOZ-FM Corner Brook, NF; CIOZ-FM Marystown, NF; CHOS-FM Rattling Brook, NF; CKSS-FM Red Rocks, NF; CHOZ-FM Saint John's, NF; CIOS-FM Stephenville, NF.

Scott G. Stirling, pres/CEO; Doug Neal, engrg dir.

News-Press & Gazette Co., Box 29, St. Joseph, MO, 64502. Phone: (816) 271-8500. Fax: (816) 271-8695. Ownership: David R. Bradley Jr., Henry H. Bradley, Lyle E. Leimkuhler. Cable TV: NPG Cable of Arizona.

Stns: 6 TV. KTVZ, Bend, OR; KRDO-TV, Colorado Springs-Pueblo, CO; KVIA-TV, El Paso (Las Cruces, NM), TX; KJCT, Grand Junction-Montrose, CO; KIFI, Idaho Falls-Pocatello, ID; KESQ-TV, Palm Springs, CA.

Stns: 2 AM. 1 FM. KESQ Indio, CA; KUNA-FM La Quinta, CA; KRDO(AM) Colorado Springs, CO.

News-Press & Gazette Co. publishes the *St. Joseph News-Press,* St. Joseph, MO.

John Kueneke, pres.

Newsweb Corp., 1645 W. Fullerton Ave., Chicago, IL, 60614. Phone: (773) 975-0401. Fax: (773) 975-1301. Ownership: Fred Eychaner, 100%.

Stns: 2 TV. KTVD, Denver, CO; KUPN, Denver, CO.

Stns: 5 AM. 4 FM. WKIE-FM Arlington Heights, IL; WSBC(AM) Chicago, IL; WCFJ Chicago Heights, IL; WCPT(AM) Crystal Lake, IL; WDEK-FM De Kalb, IL; WKIF-FM Kankakee, IL; WRZA(FM) Park Forest, IL; WAIT(AM) Willow Springs, IL; WNDZ Portage, IN.

Fred Eychaner, CEO; Charley Gross, COO.

Nexstar Broadcasting Group Inc., 909 Lake Carolyn Pkwy., Suite 1450, Irving, TX, 75039. Phone: (972) 373-8800. Fax: (972) 373-8888. Web Site: www.nexstar.tv.

Stns: 31 TV. KTAB, Abilene-Sweetwater, TX; KAMR, Amarillo, TX; KBTV, Beaumont-Port Arthur, TX; KSVI, Billings, MT; WCFN-TV, Champaign & Springfield-Decatur, IL; WCIA-TV, Champaign & Springfield-Decatur, IL; WDHN, Dothan, AL; WJET-TV, Erie, PA; WTVW, Evansville, IN; KFTA-TV, Ft. Smith-Fayetteville-Springdale-Rogers, AR; KNWA-TV, Ft. Smith-Fayetteville-Springdale-Rogers, AR; WFFT-TV, Ft. Wayne, IN; WLYH-TV, Harrisburg -Lancaster-Lebanon-York, PA; WTAJ-TV, Johnstown-Altoona, PA; KSNF, Joplin, MO-Pittsburg, KS; KARK, Little Rock-Pine Bluff, AR; KLBK, Lubbock, TX; KARD, Monroe, LA-El Dorado, AR; KMID, Odessa-Midland, TX; WMBD, Peoria-Bloomington, IL; WROC-TV, Rochester, NY; WQRF, Rockford, IL; KLST, San Angelo, TX; KTAL, Shreveport, LA; KSFX-TV, Springfield, MO; KQTV, St. Joseph, MO; WTWO-TV, Terre Haute, IN; WFXV, Utica, NY; WHAG-TV, Washington, DC (Hagerstown, MD); KFDX, Wichita Falls, TX & Lawton, OK; WBRE, Wilkes Barre-Scranton, PA.

Perry Sook, chmn/pres/CEO; Matt Devine, CFO; Duane Lammers, execVP/COO.

Northwest Broadcasting Inc., 2111 University Park Dr., Suite 650, Okemos, MI, 48864. Phone: (517) 347-4141. Fax: (517) 347-4675. E-mail: bradybw1@comcast.net Ownership: LPTV: WBPN-LP Morris, NY; and KCYU-LP Yakima, WA.

Stns: 4 TV. WICZ, Binghamton, NY; KMVU, Medford-Klamath Falls, OR; KAYU-TV, Spokane, WA; KFFX-TV, Yakima-Pasco-Richland-Kennewick, WA.

Brian Brady, pres/CEO.

P

Pappas Telecasting Companies, 500 S. Chinowth Rd., Visalia, CA, 93277. Phone: (559) 733-7800. Fax: (559) 733-7878. Web Site: www.pappastv.com. Ownership: Harry J. Pappas.

Stns: 19 TV. WLGA, Columbus, GA; KDMI, Des Moines-Ames, IA; KCWI-TV, Des Moines-Ames, IA; KDBC-TV, El Paso (Las Cruces, NM), TX; KMPH-TV, Fresno-Visalia, CA; KFRE-TV, Fresno-Visalia, CA; WWAZ-TV, Green Bay-Appleton, WI; WCWG, Greensboro-High Point-Winston Salem, NC; KAZH, Houston; KWNB, Lincoln & Hastings-Kearney, NE; KHGI, Lincoln & Hastings-Kearney, NE; KAZA-TV, Los Angeles; KPTM, Omaha, NE; KREN, Reno, NV; KTNC, San Francisco-Oakland-San Jose; KUNO-TV, San Francisco-Oakland-San Jose; KPTH, Sioux City, IA; KCWK, Yakima-Pasco-Richland-Kennewick, WA; KSWT, Yuma, AZ-El Centro, CA.

Stns: 2 AM. KMPH(AM) Modesto, CA; KTRB(AM) San Francisco, CA.

Harry J. Pappas, chmn/CEO; Dennis J. Davis, pres/COO; Bruce M. Yeager, exec VP/CFO.

Parker Broadcasting Inc., 5341 Tate Ave., Plano, TX, 75093. Phone: (214) 704-7559. Ownership: Barry J.C. Parker, 100%.

Stns: 2 TV. KXJB, Fargo-Valley City, ND; KFQX, Grand Junction-Montrose, CO.

The Jim Pattison Broadcast Group, 460 Pemberton Terrace, Kamloops, BC, V2C 1T5. Canada. Phone: (250) 372-3322. Fax: (250) 374-0445. Web Site: www.jpbroadcast.com. Ownership: Jim Pattison Group.

Stns: 4 TV. CFJC-TV, Kamloops, BC; CHAT, Medicine Hat, AB; CHAT-1, Pivot, AB; CKPG, Prince George, BC.

Stns: 2 AM. 26 FM. CIBW-FM Drayton Valley, AB; CJXX-FM Grande Prairie, AB; CHLB-FM Lethbridge, AB; CFMY-FM Medicine Hat, AB; CHAT-FM Medicine Hat, AB; CHUB-FM Red Deer, AB; CFDV-FM Red Deer, AB; CHBW-FM Rocky Mountain House, AB; CJBZ-FM Taber, AB; CKLR-FM Courtenay, BC; CHBZ-FM Cranbrook, BC; CHDR-FM Cranbrook, BC; CJDR-FM Fernie, BC; CKBZ-FM Kamloops, BC; CIFM-FM Kamloops, BC; CKLZ-FM Kelowna, BC; CKOV Kelowna, BC; CKWV-FM Nanaimo, BC; CHWF-FM Nanaimo, BC; CIBH-FM Parksville, BC; CHPQ-FM Parksville, BC; CJAV-FM Port Alberni, BC; CKDV-FM Prince George, BC; CKKN-FM Prince George, BC; CJJR-FM Vancouver, BC; CKBD Vancouver, BC; CKKQ-FM Victoria, BC; CJZN-FM Victoria, BC.

Rick Arnish, pres; Bill Dinicol, VP Finance; Bruce Davis, VP sls; Loretta Lewis, admin asst.

Pegasus Broadcast Television Inc., 225 City Line Ave., Suite 200, Bala Cynwyd, PA, 19004. Phone: (610) 934-7000. Fax: (610) 934-7072. Web Site: www.pgtv.com. Ownership: Marshall W. Pagon, 100%.

Stns: 5 TV. WDSI-TV, Chattanooga, TN; WPXT, Portland-Auburn, ME; WTLH, Tallahassee, FL-Thomasville, GA; WQMY, Wilkes Barre-Scranton, PA; WOLF, Wilkes Barre-Scranton, PA.

Howard E. Verlin, exec VP; Marshall Pagon, chmn/CEO; Denise Rolfe, VP.

Piedmont Television Holdings LLC, 7621 Little Ave., Suite 506, Charlotte, NC, 28226. Phone: (704) 341-0944. Fax: (704) 341-0945. Web Site: www.piedmonttv.com. Ownership: Richard L. Gorman, Kathy R. Gorman.

Stns: 7 TV. KTBY, Anchorage, AK; WYDO, Greenville-New Bern-Washington, NC; WFXI, Greenville-New Bern-Washington, NC; WGXA, Macon, GA; KTVE, Monroe, LA-El Dorado, AR; WJCL, Savannah, GA; WKBN, Youngstown, OH.

Paul Brissette, pres/CEO; Bill Fielder, CFO.

Pollack Broadcasting Co., 5500 Poplar Ave. #1, Memphis, TN, 38119. Phone: (901) 685-3993. Fax: (901) 685-3995. E-mail: wpollack@midsouth.rr.com Ownership: William H. Pollack, 100%. Note: Group also owns KWCE-LP Alexandria, LA.

Stns: 2 TV. KLAX, Alexandria, LA; KIEM, Eureka, CA.

Stns: 3 AM. 3 FM. KBOA-FM Piggott, AR; KCRV Caruthersville, MO; KCRV-FM Caruthersville, MO; KBOA Kennett, MO; KTMO(FM) New Madrid, MO; KMIS Portageville, MO.

William H. Pollack, pres.

Post-Newsweek Stations Inc., 550 W. Lafayette, Detroit, MI, 48226. Phone: (313) 223-2260. Fax: (313) 223-2263. Ownership: Post-Newsweek Stations is a subsidiary of the publicly traded Washington Post Co.

Stns: 6 TV. WDIV, Detroit; KPRC-TV, Houston; WJXT, Jacksonville, FL; WPLG, Miami-Ft. Lauderdale, FL; WKMG, Orlando-Daytona Beach-Melbourne, FL; KSAT, San Antonio, TX.

The Washington Post Co. publishes the *Washington* DC *Post*, the *Everett* (WA) *Herald, Newsweek* magazine, *Newsweek International* (New York, NY), *Newsweek Japan* & *Newsweek Korea*.

Prime Cities Broadcasting Inc., Box 4026, 3130 E. Broadway Ave, Bismarck, ND, 58502-4026. Phone: (701) 355-0026. Fax: (701) 250-7244. E-mail: kndx@fox26.tv Web Site: www.fox26.tv. Ownership: John Tupper, Bruce Fox.

Stns: 2 TV. KNDX, Minot-Bismarck-Dickinson, ND; KXND, Minot-Bismarck-Dickinson, ND.

John B. Tupper, pres/CEO; Gary O'Halloran, gen mgr.

Q

Quincy Newspapers Inc., 130 S. Fifth St., Quincy, IL, 62301. Phone: (217) 223-5100. Fax: (217) 223-5019. Web Site: www.qni.biz.

Stns: 12 TV. WVVA, Bluefield-Beckley-Oak Hill, WV; KWWL, Cedar Rapids-Waterloo-Iowa City & Dubuque, IA; WQOW-TV, La Crosse-Eau Claire, WI; WXOW-TV, La Crosse-Eau Claire, WI; WKOW, Madison, WI; WGEM-TV, Quincy, IL-Hannibal, MO-Keokuk, IA; KTTC, Rochester, MN-Mason City, IA-Austin, MN; WREX-TV, Rockford, IL; KTIV, Sioux City, IA; WSJV, South Bend-Elkhart, IN; WAOW-TV, Wausau-Rhinelander, WI; WYOW, Wausau-Rhinelander, WI.

Stns: 1 AM. 1 FM. WGEM Quincy, IL; WGEM-FM Quincy, IL.

Quincy Newspapers Inc. owns the *Quincy* (IL) *Herald-Whig*, and the *New Jersey Herald*, Newton, NJ.

Thomas A. Oakley, pres.

R

RNC MEDIA Inc., 380 Murdoch, Rowyn-Norando, PQ, J9X 1G5. Canada. Phone: (514) 866-8686. Fax: (514) 866-8056. Web Site: www.radionord.com.

Stns: 5 TV. CKRN-3, Bearn-Fabre, PQ; CFGS-TV, Gatineau, PQ; CHOT-TV, Gatineau, PQ; CKRN-TV, Rouyn-Noranda, PQ; CFVS, Val d'Or, PQ.

Stns: 9 FM. CHPR-FM Hawkesbury, ON; CKNU-FM Donnacona, PQ; CHLX-FM Gatineau, PQ; CFTX-FM Gatineau, PQ; CJLA-FM Lachute, PQ; CKLX-FM Montreal, PQ; CHOI-FM Quebec, PQ; CHOA-FM Rouyn-Noranda, PQ; CHGO-FM Val d'Or, PQ.

Pierre R. Brosseau, pres.

Ramar Communications II Ltd., Box 3757, Lubbock, TX, 79452. Phone: (806) 745-3434. Fax: (806) 748-1949. Web Site: www.ramarcom.com. E-mail: bmoran@ramarcom.com Ownership: Ray Moran, 51%; Brad Moran, 49%.

Lubbock, TX 79423, 9800 University Ave.

Stns: 4 TV. KTLL-TV, Albuquerque-Santa Fe, NM; KUPT, Albuquerque-Santa Fe, NM; KTEL-TV, Albuquerque-Santa Fe, NM; KJTV-TV, Lubbock, TX.

Stns: 1 AM. 3 FM. KLZK(FM) Brownfield, TX; KJTV(AM) Lubbock, TX; KXTQ-FM Lubbock, TX; KSTQ-FM Plainview, TX.

Ray Moran, chmn; Brad Moran, pres.

Raycom Media Inc., 201 Monroe St., RSA Tower, 20th Fl, Montgomery, AL, 36104. Phone: (334) 206-1400. Fax: (334) 206-1555. Web Site: www.raycommedia.com. Ownership: Raycom Media Inc. Other interests: Raycom Sports, New York, NY; Charlotte, NC; Ft. Lauderdale, FL; Nashville, TN; and Chicago, IL.

Stns: 38 TV. WALB, Albany, GA; KASA-TV, Albuquerque-Santa Fe, NM; WAFB, Baton Rouge, LA; WLOX, Biloxi-Gulfport, MS; WXIX, Cincinnati, OH; WUAB, Cleveland-Akron (Canton), OH; WOIO, Cleveland-Akron (Canton), OH; WIS, Columbia, SC; WTVM, Columbus, GA; WDFX, Dothan, AL; WFIE-TV, Evansville, IN; WDAM-TV, Hattiesburg-Laurel, MS; KHBC, Hilo, HI; KHNL, Honolulu, HI; KFVE, Honolulu, HI; WAFF, Huntsville-Decatur (Florence), AL; WLBT, Jackson, MS; KAIT, Jonesboro, AR; WTNZ, Knoxville, TN; KPLC, Lake Charles, LA; WAVE, Louisville, KY; KCBD, Lubbock, TX; WMC, Memphis, TN; WSFA, Montgomery-Selma, AL; WMBF-TV, Myrtle Beach-Florence, SC; KFVS-TV, Paducah, KY-Cape Girardeau, MO-Harrisburg-Mount Vernon, IL; WPGX, Panama City, FL; WTVR, Richmond-Petersburg, VA; WTOC, Savannah, GA; KSLA, Shreveport, LA; WTOL, Toledo, OH; KOLD, Tucson (Sierra Vista), AZ; KTRE, Tyler-Longview (Lufkin & Nacogdoches), TX; KLTV, Tyler-Longview (Lufkin & Nacogdoches), TX; KOGG, Wailuku, HI; WFLX, West Palm Beach-Ft. Pierce, FL; WECT, Wilmington, NC; WWAY, Wilmington, NC.

Paul McTear, pres/CEO.

Red River Broadcast Co. L.L.C., Box 9115, Fargo, ND, 58106. Phone: (701) 277-1515. Fax: (701) 277-1830. Ownership: Curtis Squire Inc., 100%. Myron Kunin owns 100% of Curtis Squire Inc.

Stns: 7 TV. KQDS-TV, Duluth, MN-Superior, WI; KVRR, Fargo-Valley City, ND; KBRR, Fargo-Valley City, ND; KJRR, Fargo-Valley City, ND; KNRR, Fargo-Valley City, ND; KDLT-TV, Sioux Falls (Mitchell), SD; KDLV-TV, Sioux Falls (Mitchell), SD.

Ro Grignon, pres; Kathy M. Lau, VP/gen mgr.

Reiten Television Inc., Box 1686, Minot, ND, 58702-1686. Phone: (701) 852-2104. Fax: (701) 838-9360. Web Site: www.kxmc.com.

Stns: 4 TV. KXMA, Minot-Bismarck-Dickinson, ND; KXMB, Minot-Bismarck-Dickinson, ND; KXMC, Minot-Bismarck-Dickinson, ND; KXMD, Minot-Bismarck-Dickinson, ND.

David Reiten, pres.

Roberts Broadcasting Co., 1408 N. Kingshighway Blvd., St. Louis, MO, 63113. Phone: (314) 367-4600. Fax: (314) 367-0174. Web Site: www.upn46stl.com. Ownership: St. Louis/Denver LLC.

Stns: 5 TV. WZRB, Columbia, SC; KTFD-TV, Denver, CO; WAZE-TV, Evansville, IN; WRBJ, Jackson, MS; WRBU, St. Louis, MO.

Michael Roberts, CEO; Steven C. Roberts, pres.

Rockfleet Broadcasting Inc., 575 Madison Ave., 10th Fl., New York, NY, 10022. Phone: (212) 605-0401. Phone: (631) 204-0830. Fax: (212) 605-0402. Fax: (631) 204-0832. Ownership: Rockfleet Holdings, 100%.

Stns: 4 TV. WVII, Bangor, ME; WFQX-TV, Traverse City-Cadillac, MI; WFUP, Traverse City-Cadillac, MI; WJFW-TV, Wausau-Rhinelander, WI.

R. Joseph Fuchs, pres/CEO.

Rogers Broadcasting Ltd., 777 Jarvis St., Toronto, ON, M4Y 3B7. Canada. Phone: (416) 935-8200. Web Site: www.rogers.com. Ownership: Rogers Media Inc., 100%. Note: Rogers Media Inc. is 100% owned by Rogers Communications Inc.

Stns: 4 TV. CHNU-TV, Fraser Valley, BC; CJMT-TV, Toronto, ON; CFMT, Toronto, ON; CIIT-TV, Winnipeg, MB.

Stns: 5 AM. 35 FM. CFFR Calgary, AB; CKIS-FM Calgary, AB; CHMN-FM Canmore, AB; CKER-FM Edmonton, AB; CHDI-FM Edmonton, AB; CKYX-FM Fort McMurray, AB; CJOK-FM Fort McMurray, AB; CFGP-FM Grande Prairie, AB; CFRV-FM Lethbridge, AB; CJRX-FM Lethbridge, AB; CKQC-FM Abbotsford, BC; CKGO-FM-1 Boston Bar, BC; CKCL-FM Chilliwack, BC; CKSR-FM Chilliwack, BC; CFSR-FM Hope, BC; CISP-FM Pemberton, BC; CKKS-FM Sechelt, BC; CKWX Vancouver, BC; CKIZ-FM Vernon, BC; CHTT-FM Victoria, BC; CIOC-FM Victoria, BC; CISW-FM Whistler, BC; CITI-FM Winnipeg, MB; CKY-FM Winnipeg, MB; CKNI-FM Moncton, NB; CHNI-FM Saint John, NB; CJNI-FM Halifax, NS; CKAT North Bay, ON; CHUR-FM North Bay, ON; CICX-FM Orillia, ON; CHAS-FM Sault Ste. Marie, ON; CJQM-FM Sault Ste. Marie, ON; CKBY-FM Smiths Falls, ON; CIGM(AM) Sudbury, ON; CJMX-FM Sudbury, ON; CJRQ-FM Sudbury, ON; CKGB-FM Timmins, ON; CJQQ-FM Timmins, ON; CJCL Toronto, ON; CJAQ-FM Toronto, ON.

Rael Merson, pres.

S

Saga Communications Inc., 73 Kercheval Ave., Suite 201, Grosse Pointe Farms, MI, 48236. Phone: (313) 886-7070. Fax: (313) 886-7150. E-mail: chapsburg@sagacom.com Web Site: www.sagacommunications.com. Ownership: Edward K. Christian, 56.5% of the voting stock. Other Interests: Illinois Radio Network, Michigan Radio Network, Michigan Farm Radio Network.

Stns: 3 TV. WXVT, Greenwood-Greenville, MS; KOAM, Joplin, MO-Pittsburg, KS; KAVU, Victoria, TX.

Stns: 27 AM. 54 FM. KEGI(FM) Jonesboro, AR; KDXY(FM) Lake City, AR; KJBX(FM) Trumann, AR; KLTI-FM Ames, IA; KRNT Des Moines, IA; KSTZ(FM) Des Moines, IA; KIOA(FM) Des Moines, IA; KAZR(FM) Pella, IA; KICD Spencer, IA; KICD-FM Spencer, IA; KLLT(FM) Spencer, IA; WLRW-FM Champaign, IL; WIXY-FM Champaign, IL; WXTT(FM) Danville, IL; WYMG(FM) Jacksonville, IL; WABZ(FM) Sherman, IL; WQQL(FM) Springfield, IL; WTAX(AM) Springfield, IL; WDBR(FM) Springfield, IL; WCFF(FM) Urbana, IL; WCVQ-FM Fort Campbell, KY; WJQI(AM) Fort Campbell, KY; WVVR(FM) Hopkinsville, KY; WZZP(FM) Hopkinsville, KY; WEGI(FM) Oak Grove, KY; WHNP(AM) East Longmeadow, MA; WHMQ(AM) Greenfield, MA; WPVQ(FM) Greenfield, MA; WHMP Northampton, MA; WLZX(FM) Northampton, MA; WAQY-FM Springfield, MA; WRSI(FM) Turners Falls, MA; WSNI(FM) Winchendon, MA; WVAE(AM) Biddeford, ME; WBAE Portland, ME; WPOR(FM) Portland, ME; WZAN Portland, ME; WMGX(FM) Portland, ME; WGAN(AM) Portland, ME; WYNZ-FM Westbrook, ME; WOXL-FM Biltmore Forest, NC; WYSE(AM) Canton, NC; WTMT(FM) Weaverville, NC; WMLL(FM) Bedford, NH; WKNE(FM) Keene, NH; WZBK(AM) Keene, NH; WKBK(AM) Keene, NH; WZID(FM) Manchester, NH; WFEA Manchester, NH; WINQ(FM) Winchester, NH; WYXL-FM Ithaca, NY; WQNY-FM Ithaca, NY; WNYY(AM) Ithaca, NY; WHCU Ithaca, NY; WBCO(AM) Bucyrus, OH; WQEL-FM Bucyrus, OH; WSNY(FM) Columbus, OH; WODB(FM) Delaware, OH; WJZA-FM Lancaster, OH; WJZK(FM) Richwood, OH; KMIT-FM Mitchell, SD; KLQT(FM) Wessington Springs, SD; WKFN(AM) Clarksville, TN; WINA Charlottesville, VA; WQMZ(FM) Charlottesville, VA; WVAX(AM) Charlottesville, VA; WWWV-FM Charlottesville, VA; WCNR(FM) Keswick, VA; WJOI Norfolk, VA; WAFX-FM Suffolk, VA; WKVT Brattleboro, VT; WKVT-FM Brattleboro, VT; WRSY(FM) Marlboro, VT; KGMI Bellingham, WA; KPUG Bellingham, WA; KBAI(AM) Bellingham, WA; WJZX(FM) Brookfield, WI; WJMR-FM Menomonee Falls, WI; WHQG(FM) Milwaukee, WI; WJYI Milwaukee, WI; WKLH-FM Milwaukee, WI.

Edward K. Christian, pres/CEO; Marcia Lobaito, VP business affrs; Sam Bush, CFO; Warren Lada Sr., VP opns.

SagamoreHill Broadcasting LLC, 3825 Inverness Way, Augusta, GA, 30901. Phone: (706) 855-9506. Ownership: Duff Ackerman & Goodrich QP Fund II L.P., 74.777%; Broadcast Media Group LLC, 12.35%; Duff Ackerman & Goodrich II L.P., 7.0635%; DAG GP Fund II LLC, 4.2514%; and Louis Wall, 0.77%.

Stns: 5 TV. KGWN, Cheyenne, WY-Scottsbluff, NE; KSTF, Cheyenne, WY-Scottsbluff, NE; KGNS, Laredo, TX;

WBMM, Montgomery-Selma, AL; WNCF, Montgomery-Selma, AL.

SagamoreHill Midwest LLC, 3825 Inverness Way, Augusta, GA, 30907. Phone: (706) 855-8506. Ownership: Louis Wall, 100%.
Stns: 2 TV. WWMB, Myrtle Beach-Florence, SC; KXLT, Rochester, MN-Mason City, IA-Austin, MN.

Sage Broadcasting Corp., 406 S. Irving, San Angelo, TX, 76903. Phone: (325) 655-6006. Fax: (325) 655-8461. Ownership: Suzanne S. Brown, 32%; Sherry S. Hawk, 32%; Paris R. Schindler, 32%; Anne Marie Carter, 3%; and Timothy R. Brown, 0.33%.
Stns: 2 TV. KXVA, Abilene-Sweetwater, TX; KIDY, San Angelo, TX.

Sainte Partners II L.P., Box 4159, Modesto, CA, 95352-4159. Phone: (209) 523-0777. Fax: (209) 523-0839. E-mail: csmith@sainte.tv Ownership: Chester Smith, gen ptnr, & Naomi Smith, gen ptnr; & other limited ptnrs.
Stns: 2 TV. KCVU, Chico-Redding, CA; KBVU, Eureka, CA.

Schurz Communications Inc., 225 W. Colfax Ave., South Bend, IN, 46626. Phone: (219) 287-1001. Fax: (219) 287-2257. E-mail: mburdick@schurz.com Web Site: www.schurz.com. Ownership: Franklin D. Schurz Jr., James M. Schurz, Scott C. Schurz and Mary Schurz, trustees.
Stns: 9 TV. WAGT, Augusta, GA; WDBJ, Roanoke-Lynchburg, VA; WSBT, South Bend-Elkhart, IN; KYTV, Springfield, MO; KBSD, Wichita-Hutchinson Plus, KS; KBSH, Wichita-Hutchinson Plus, KS; KBSL, Wichita-Hutchinson Plus, KS; KWCH, Wichita-Hutchinson Plus, KS; KSCW, Wichita-Hutchinson Plus, KS.
Stns: 4 AM. 8 FM. WASK-FM Battle Ground, IN; WXXB(FM) Delphi, IN; WASK Lafayette, IN; WKOA-FM Lafayette, IN; WSBT South Bend, IN; WNSN-FM South Bend, IN; KFXS-FM Rapid City, SD; KKLS Rapid City, SD; KKMK-FM Rapid City, SD; KOUT-FM Rapid City, SD; KRCS-FM Sturgis, SD; KBHB Sturgis, SD.
Schurz Communications publishes the following nwsprs: *Imperial Valley Press, Southside Times-Beech, Times; Bedford Times-Mail, Bloomington Herald-Times* , & *South Bend Tribune, Martinsville Reporter, Danville Advocate-Messenger, The Herald Mail Co.; Daily American,* Somerset, PA.
Marcia K. Burdick, sr VP bcstg; Franklin D. Schurz Jr., chmn; Todd F. Schurz, pres.

Scripps Howard Broadcasting Co., Box 5380, 312 Walnut St., 28th Fl., Cincinnati, OH, 45201. Phone: (513) 977-3000. Fax: (513) 977-3728. Web Site: www.scripps.com. Ownership: The E.W. Scripps Co.
Stns: 10 TV. WMAR, Baltimore, MD; WCPO, Cincinnati, OH; WEWS, Cleveland-Akron (Canton), OH; WXYZ, Detroit; KMCI, Kansas City, MO; KSHB-TV, Kansas City, MO; KNXV, Phoenix (Prescott), AZ; WFTS, Tampa-St. Petersburg (Sarasota), FL; KJRH, Tulsa, OK; WPTV, West Palm Beach-Ft. Pierce, FL.
Newspapers include: *Abilene Reporter-News; The Albuquerque Tribune; Anderson Independent-Mail; Birmingham Post-Herald; The Cincinnati Post; The Commercial Appeal* (Memphis); *Corpus Christi Caller-Times; Daily Camera* (Boulder); *Evansville Courier & Press; The Gleaner* (Henderson); *The Knoxville News-Sentinel; Naples Daily News; Redding Record Searchlight; Rocky Mountain News* (Denver); *San Angelo Standard-Times; The Stuart News; The Sun* (Bremerton); *The Tribune* (Ft. Pierce); *Ventura County Star; Vero Beach Press Journal; Wichita Falls Times Record News.*
William B. Peterson, sr VP, TV station group & The E.W. Scripps Co.

Sinclair Broadcast Group Inc., 10706 Beaver Dam Rd., Hunt Valley, MD, 21030. Phone: (410) 568-1500. Fax: (410) 568-1533. E-mail: ir@sbgnet.com Web Site: www.sbgi.net. Ownership: Smith brothers (major shareholders).
Stns: 49 TV. WBFF, Baltimore, MD; WABM, Birmingham (Anniston, Tuscaloosa), AL; WLDM, Birmingham (Anniston, Tuscaloosa), AL; WNYO-TV, Buffalo, NY; WUTV, Buffalo, NY; KGAN-TV, Cedar Rapids-Waterloo-Iowa City & Dubuque, IA; WICD-TV, Champaign & Springfield-Decatur, IL; WICS-TV, Champaign & Springfield-Decatur, IL; WMMP, Charleston, SC; WCHS, Charleston-Huntington, WV; WVAH, Charleston-Huntington, WV; WSTR-TV, Cincinnati, OH; WSYX, Columbus, OH; WKEF, Dayton, OH; KDSM-TV, Des Moines-Ames, IA; WSMH, Flint-Saginaw-Bay City, MI; WMYV, Greensboro-High Point-Winston Salem, NC; WXLV, Greensboro-High Point-Winston Salem, NC; WLOS, Greenville-Spartanburg, SC-Asheville, NC-Anderson, SC; KVCW, Las Vegas, NV; KVMY, Las Vegas, NV; WDKY-TV, Lexington, KY; WMSN, Madison, WI; WCGV, Milwaukee, WI; WVTV, Milwaukee, WI; WUCW, Minneapolis-St. Paul, MN; WEAR, Mobile, AL-Pensacola (Ft. Walton Beach), FL; WFGX, Mobile, AL-Pensacola (Ft. Walton Beach), FL; WUXP-TV, Nashville, TN; WZTV, Nashville, TN; WTVZ-TV, Norfolk-Portsmouth-Newport News, VA; KOCB, Oklahoma City, OK; KOKH, Oklahoma City, OK; KBSI, Paducah, KY-Cape Girardeau, MO-Harrisburg-Mount Vernon, IL; WYZZ, Peoria-Bloomington, IL; WPGH, Pittsburgh, PA; WPMY, Pittsburgh, PA; WGME, Portland-Auburn, ME; WLFL, Raleigh-Durham (Fayetteville), NC; WRDC, Raleigh-Durham (Fayetteville), NC; WRLH-TV, Richmond-Petersburg, VA; WUHF, Rochester, NY; KMYS, San Antonio, TX; KABB, San Antonio, TX; WGGB, Springfield-Holyoke, MA; KDNL, St. Louis, MO; WSYT, Syracuse, NY; WTWC-TV, Tallahassee, FL-Thomasville, GA; WTTA, Tampa-St. Petersburg (Sarasota), FL.
David D. Smith, pres/CEO; J.Duncan Smith, sec; Frederick Smith, treas; David B. Amy, exec VP.

Smith Media License Holdings LLC, 1215 Cole St., St. Louis, MO, 63106. Phone: (314) 853-7736. Ownership: Smith Media LLC, 100%.
Stns: 6 TV. KIMO, Anchorage, AK; WFFF, Burlington, VT-Plattsburgh, NY; KATN, Fairbanks, AK; KJUD, Juneau, AK; KEYT, Santa Barbara-Santa Maria-San Luis Obispo, CA; WKTV, Utica, NY.

South Central Communications Corp., Box 3848, Evansville, IN, 47736. Phone: (812) 463-7950. Fax: (812) 463-7915. Web Site: www.southcentralcommunications.net. Ownership: John D. Engelbrecht, 80%, J.P. Engelbrecht, 20%.
Stns: 1 TV. WMAK, Knoxville, TN.
Stns: 1 AM. 11 FM. WEJK(FM) Boonville, IN; WLFW(FM) Chandler, IN; WEOA Evansville, IN; WIKY-FM Evansville, IN; WABX-FM Evansville, IN; WSTO-FM Owensboro, KY; WIMZ-FM Knoxville, TN; WJXB-FM Knoxville, TN; WQJK(FM) Maryville, TN; WCJK(FM) Murfreesboro, TN; WJXA-FM Nashville, TN; WRJK(FM) Norris, TN.
John D. Engelbrecht, pres; J.P. Engelbrecht, VP.

Southeastern Media Holdings Inc., 3500 Colonnade Pkwy., Suite 600, Birmingham, AL, 35243. Phone: (205) 298-7100. Fax: (205) 298-7104. Ownership: Community Newspaper Holdings Inc., 100%. Web site: www.cnhi.com
Stns: 4 TV. WFXG, Augusta, GA; WXTX, Columbus, GA; WUPV, Richmond-Petersburg, VA; WSFX-TV, Wilmington, NC.
Community Newspaper Holdings Inc. is the parent company for daily, wkly and semiweekly newspapers published in more than 200 communities throughout the U.S.
Michael E. Reed, pres/CEO.

Southern Broadcast Corp. of Sarasota, 1477 10th St., Sarasota, FL, 34236. Phone: (941) 923-8840. Fax: (941) 924-3971. Ownership: Calkins Media Inc., 100% of total assets.
Stns: 3 TV. WAAY-TV, Huntsville-Decatur (Florence), AL; WTXL, Tallahassee, FL-Thomasville, GA; WWSB, Tampa-St. Petersburg (Sarasota), FL.

Spanish Broadcasting System Inc., 2601 South Bayshore Dr., PH 2, Coconut Grove, FL, 33133. Phone: (305) 441-6901. Fax: (305) 446-5148. Web Site: www.spanishbroadcasting.com. Ownership: Raul Alarcon Sr., Raul Alarcon Jr., Jose Grimalt.
Stns: 1 TV. WSBS-TV, Miami-Ft. Lauderdale, FL.
Stns: 20 FM. KLAX-FM East Los Angeles, CA; KXOL-FM Los Angeles, CA; KRZZ(FM) San Francisco, CA; WRMA-FM Fort Lauderdale, FL; WCMQ-FM Hialeah, FL; WXDJ-FM North Miami Beach, FL; WLEY-FM Aurora, IL; WPAT-FM Paterson, NJ; WSKQ-FM New York, NY; WODA(FM) Bayamon, PR; WCMA-FM Fajardo, PR; WMEG-FM Guayama, PR; WZET(FM) Hormigueros, PR; WNOD(FM) Mayaguez, PR; WIOB-FM Mayaguez, PR; WIOC-FM Ponce, PR; WZMT-FM Ponce, PR; WEGM(FM) San German, PR; WZNT-FM San Juan, PR; WIOA(FM) San Juan, PR.
Raul Alarcon Sr., chmn; Raul Alarcon Jr., pres/CEO; Jose Grimalt, exec VP.

Standard Broadcasting Corp., 2 St. Clair Ave. W., Suite 1100, Toronto, ON, M4V 1L6. Canada. Phone: (416) 960-9911. Fax: (416) 323-6828. Ownership: Slaight Communications Inc., 100%.
Stns: 2 TV. CJDC, Dawson Creek, BC; CFTK, Terrace, BC.
Stns: 22 AM. 33 FM. CJAY-FM Calgary, AB; CKMX Calgary, AB; CIBK-FM Calgary, AB; CFBR-FM Edmonton, AB; CFRN Edmonton, AB; CFMG-FM Saint Albert, AB; CFKC Creston, BC; CJDC Dawson Creek, BC; CKRX-FM Fort Nelson, BC; CKNL-FM Fort St. John, BC; CHRX-FM Fort St. John, BC; CKGR Golden, BC; CKIR Invermere, BC; CKFR(AM) Kelowna, BC; CHSU-FM Kelowna, BC; CILK-FM Kelowna, BC; CKTK-FM Kitimat, BC; CKKC-FM Nelson, BC; CKZX-FM New Denver, BC; CJOR Osoyoos, BC; CKOR Penticton, BC; CJMG-FM Penticton, BC; CHTK Prince Rupert, BC; CIOR Princeton, BC; CKCR Revelstoke, BC; CISL Richmond, BC; CKXR-FM Salmon Arm, BC; CHOR Summerland, BC; CFTK Terrace, BC; CJFW-FM Terrace, BC; CJAT-FM Trail, BC; CKZZ-FM Vancouver, BC; CICF-FM Vernon, BC; CKXA-FM Brandon, MB; CKX-FM Brandon, MB; CFQX-FM Selkirk, MB; CKMM-FM Winnipeg, MB; CKOC Hamilton, ON; CHAM Hamilton, ON; CKLH-FM Hamilton, ON; CJBK(AM) London, ON; CJBX-FM London, ON; CIQM-FM London, ON; CKSL London, ON; CKQB-FM Ottawa, ON; CHVR-FM Pembroke, ON; CHRE-FM Saint Catharines, ON; CHTZ-FM Saint Catharines, ON; CKTB Saint Catharines, ON; CFRB Toronto, ON; CJEZ-FM Toronto, ON; CFMX-FM Toronto, ON; CJFM-FM Montreal, PQ; CJAD Montreal, PQ; CHOM-FM Montreal, PQ.
Gary Slaight, pres/CEO.

Sunbeam Television Corp., 1401 79th St. Causeway, Miami, FL, 33141. Phone: (305) 751-6692. Fax: (305) 757-2266. E-mail: 7news@wsvn.com Web Site: www.wsvn.com. Ownership: Edmund N. Ansin, 83.09% votes, 80.15% assets; James L. Ansin, 6.3% votes, 7.06% assets; Andrew L. Ansin, 6.3% votes, 7.06% assets; and Andrew L. Ansin, James L. Ansin and Stephanie L. Ansin (as trustees for Stephanie L. Ansin), 4.31% votes, 5.73% assets.
Stns: 3 TV. WHDH, Boston (Manchester, NH); WLVI, Boston (Manchester, NH); WSVN, Miami-Ft. Lauderdale, FL.
Edmund N. Ansin, pres; Deisy Bermudez, progmg dir.

Sunbelt Communications Co., c/o KVBC(TV), 1500 Foremaster Ln., Las Vegas, NV, 89101. Phone: (702) 642-3333. Fax: (702) 657-3423. E-mail: ch3@kvbc.com Web Site: www.kvbc.com. Ownership: James E. Rogers.
Stns: 12 TV. KCWY, Casper-Riverton, WY; KBAO, Great Falls, MT; KBBJ, Great Falls, MT; KTVH, Helena, MT; KJWY, Idaho Falls-Pocatello, ID; KPVI, Idaho Falls-Pocatello, ID; KSWY, Rapid City, SD; KRNV, Reno, NV; KVNV, Salt Lake City, UT; KENV, Salt Lake City, UT; KXTF, Twin Falls, ID; KYMA, Yuma, AZ-El Centro, CA.
Ralph Toddre, pres.

Surtsey Media LLC, 73 Kercheval Ave., Suite 100, Grosse Pointe Farms, MI, 48236. Phone: (313) 884-7878. Ownership: Dana C. Raymant, 100%.
Stns: 2 TV. KFJX-TV, Joplin, MO-Pittsburg, KS; KVCT, Victoria, TX.

T

Sarkes Tarzian Inc., Box 62, Bloomington, IN, 47402. Phone: (812) 332-7251. Fax: (812) 331-4575. Ownership: Tom Tarzian; Mary Tarzian estate.
Stns: 2 TV. WRCB-TV, Chattanooga, TN; KTVN, Reno, NV.
Stns: 1 AM. 3 FM. WTTS-FM Bloomington, IN; WGCL Bloomington, IN; WLDE-FM Fort Wayne, IN; WAJI(FM) Fort Wayne, IN.
Tom Tarzian, opns VP; Bob Davis, CFO; Geoff Vargo, pres radio; Valerie Comey, gen councel.

Tele Inter-Rives Ltee., 298 Boulevard Theriault, Riviere-du-Loup, PQ, G5R 4C2. Canada. Phone: (418) 867-1341. Fax: (418) 867-4710. Web Site: www.cimt.ca. Ownership: 101885 Canada Ltee., 54.37%; Groupe TVA Inc., 44.66% (see listing); and Marc Simard, 0.97%.
Stns: 4 TV. CHAU, Carleton, PQ; CIMT, Riviere-du-Loup, PQ; CKRT, Riviere-du-Loup, PQ; CFTF-TV, Riviere-du-Loup, PQ.
Marc Simard, pres.

Telemundo Television Stations, 2290 W. 8th Ave., Hialeah, FL, 33010. Phone: (305) 884-8200. Fax: (305) 889-7950. Web Site: www.telemundo.com. Ownership: General Electric Co., 100%. See also NBC Universal Television Stations (see listing).
Stns: 14 TV. WNEU, Boston (Manchester, NH); WSNS, Chicago; KDEN, Denver, CO; KNSO, Fresno-Visalia, CA; KTMD, Houston; KVEA, Los Angeles; KWHY, Los Angeles; WSCV, Miami-Ft. Lauderdale, FL; WNJU, New York; KTAZ, Phoenix (Prescott), AZ; KVDA, San Antonio, TX; KSTS, San Francisco-Oakland-San Jose; WKAQ-TV, San Juan, PR; KHRR, Tucson (Sierra Vista), AZ.
Vincent Sadusky, CFO/treas; Juan Antunez, VP gen counsel/dev. VP/finance; Ibra Morales, pres.

Tele-Quebec, 1000 rue Fullum, Montreal, PQ, H2K 3L7. Canada. Phone: (514) 521-2424. Fax: (514) 873-4413. E-mail: info@telequebec.qc.ca Web Site: www.telequebec.tv. Ownership: La Societe de radio-television du Quebec is a para-governmental organization. Its mandate is to manage an educ TV net throughout the province of Quebec.
Stns: 11 TV. CIVF, Baie-Trinite, PQ; CIVP, Chapeau, PQ; CIVV, Chicoutimi, PQ; CIVO-TV, Gatineau, PQ; CIVM, Montreal, PQ; CIVQ, Quebec City, PQ; CIVB, Rimouski, PQ; CIVA, Rouyn, PQ; CIVG, Sept-Iles, PQ; CIVS,

Sherbrooke, PQ; CIVC, Trois-Rivieres, PQ.

Paul Beaugrand Champagne, pres/dir gen; Mario Clement, dir mgr programming; Line Simoneau, dir mgr admin/fianances/human resources; Denis Belisle, dir sec gen; Cecile Bellemare, dir dev projects; Jacques Legace, dir dev institutional; Danielle Beaudry, dir production.

Tri-State Christian Television, Box 1010, Marion, IL, 62959. Phone: (618) 997-9333. Fax: (618) 997-1859. Web Site: www.tct.tv. Ownership: Nonprofit corporation. LPTV: WDWO-CA Detroit, MI; and WDYR-CA Dyersburg, TN.

Stns: 6 TV. WNYB, Buffalo, NY; WAQP, Flint-Saginaw-Bay City, MI; WINM, Ft. Wayne, IN; WTLJ, Grand Rapids-Kalamazoo-Battle Creek, MI; WLXI, Greensboro-High Point-Winston Salem, NC; WTCT, Paducah, KY-Cape Girardeau, MO-Harrisburg-Mount Vernon, IL.

Garth W. Coonce, pres; Shane Chaney, CFO.

Tribune Broadcasting Co., 435 N. Michigan Ave., Suite 1800, Chicago, IL, 60611. Phone: (312) 222-3333. Fax: (312) 329-0611. Web Site: www.tribune.com. Ownership: Robert R. McCormick Tribune Foundation, 13.3%; The Chandler Trusts, 11.6%; Vanguard Fiduciary Trust Co., 6.9%.

Stns: 24 TV. WGN, Chicago; KDAF, Dallas-Ft. Worth; KWGN, Denver, CO; WXMI, Grand Rapids-Kalamazoo-Battle Creek, MI; WPMT, Harrisburg-Lancaster-Lebanon-York, PA; WTIC, Hartford & New Haven, CT; WTXX, Hartford & New Haven, CT; KHCW, Houston; WTTK, Indianapolis, IN; WTTV, Indianapolis, IN; WXIN, Indianapolis, IN; KTLA, Los Angeles; WSFL-TV, Miami-Ft. Lauderdale, FL; WGNO, New Orleans, LA; WNOL, New Orleans, LA; WPIX, New York; WPHL, Philadelphia; KRCW-TV, Portland, OR; KTXL, Sacramento-Stockton-Modesto, CA; KSWB, San Diego, CA; KCPQ, Seattle-Tacoma, WA; KMYQ, Seattle-Tacoma, WA; KPLR, St. Louis, MO; WDCW, Washington, DC (Hagerstown, MD).

Stns: 1 AM. WGN(AM) Chicago, IL.

John Reardon, pres/CEO; John Vitanovec, exec VP.

Trinity Broadcasting Network, 2442 Michelle Dr., Tustin, CA, 92780. Phone: (714) 832-2950. Fax: (714) 730-0657. E-mail: comments@tbn.org. Web Site: www.tbn.org. Ownership: Nonprofit corporation.

Stns: 24 TV. KNAT, Albuquerque-Santa Fe, NM; WHSG-TV, Atlanta; WTJP-TV, Birmingham (Anniston, Tuscaloosa), AL; WELF-TV, Chattanooga, TN; WWTO, Chicago; WDLI-TV, Cleveland-Akron (Canton), OH; KDTX, Dallas-Ft. Worth; WKOI-TV, Dayton, OH; KAAH-TV, Honolulu, HI; WCLJ-TV, Indianapolis, IN; KTBN, Los Angeles; WBUY-TV, Memphis, TN; WHFT-TV, Miami-Ft. Lauderdale, FL; WMPV, Mobile, AL-Pensacola (Ft. Walton Beach), FL; WMCF, Montgomery-Selma, AL; WPGD-TV, Nashville, TN; WTBY-TV, New York; KTBO, Oklahoma City, OK; WHLV-TV, Orlando-Daytona Beach-Melbourne, FL; WGTW-TV, Philadelphia; KPAZ, Phoenix (Prescott), AZ; KTBW, Seattle-Tacoma, WA; KTAJ-TV, St. Joseph, MO; KDOR-TV, Tulsa, OK.

Paul F. Crouch, pres; Rod Henke, VP sls; Ben Miller, VP engrg; Janice Crouch, VP progmg.

Tyler Media Broadcasting Corp., 5101 S. Shields Blvd., Oklahoma City, OK, 73129. Phone: (405) 616-5500. Fax: (405) 616-5505. Web Site: www.kkng.com. Ownership: Ty A. Tyler, Tony J. Tyler and Tony J. Tyler 2000 Irrevocable Trust, Tony J. Tyler, trustee.

Stns: 1 TV. KTUZ-TV, Oklahoma City, OK.

Stns: 2 AM. 3 FM. KOJK(FM) Blanchard, OK; KOCY(AM) Del City, OK; KKNG-FM Newcastle, OK; KTUZ-FM Okarche, OK; KTLR(AM) Oklahoma City, OK.

Skip Stow, market mgr; Robert De Negri, CFO.

U

United Communications Corp., 5800 7th Ave., Kenosha, WI, 53140. Phone: (262) 657-1000. Fax: (262) 657-6226. E-mail: hbrown@kenoshanews.com Web Site: www.kenoshanews.com. Ownership: Howard J. Brown, Lucy Brown Minn, Sarah Brown Russ, Amy Brown Tuchler. Note: Group also owns LPTV stn WNYF-CA Watertown, NY.

Stns: 2 TV. KEYC-TV, Mankato, MN; WWNY, Watertown, NY.

Other media: Dailies: *Kenosha News,* Kenosha, WI; *Sun Chronicle,* Attleboro, MA; *Public Opinion,* Watertown, SD. Weeklies: *Zion-Benton News,* Zion, IL; *Foxboro Reporter,* Foxboro, MA; *Lake Geneva Regional News,* Lake Geneva, WI. Shoppers: *Bulletin,* Kenosha, WI; *News-Bargaineer,* Zion, IL; *Coteau Shopper,* Watertown, SD.

Howard J. Brown, pres; Kenneth Dowdell, VP; Ronald Montemurro, VP.

Univision Communications Inc., 5999 Center Dr., Los Angeles, CA, 90045. Phone: (310) 216-3434. Fax: (310) 556-3568. Web Site: www.univision.net/corp/en/overview.jsp. Ownership: Broadcasting Media Partners Inc.

Stns: 38 TV. KTFQ-TV, Albuquerque-Santa Fe, NM; WUVG-TV, Atlanta; KNIC-TV, Austin, TX; KUVI-TV, Bakersfield, CA; WUTF-TV, Boston (Manchester, NH); WLII, Caguas, PR; WXFT-TV, Chicago; WGBO, Chicago; WQHS, Cleveland-Akron (Canton), OH; KUVN-TV, Dallas-Ft. Worth; KSTR, Dallas-Ft. Worth; KFTV, Fresno-Visalia, CA; KTFF-TV, Fresno-Visalia, CA; KXLN, Houston; KFTH-TV, Houston; KMEX, Los Angeles; KFTR-TV, Los Angeles; WAMI, Miami-Ft. Lauderdale, FL; WLTV, Miami-Ft. Lauderdale, FL; WFTY-TV, New York; WFUT-TV, New York; WXTV, New York; WOTF-TV, Orlando-Daytona Beach-Melbourne, FL; WUVP-TV, Philadelphia; KFPH-TV, Phoenix (Prescott), AZ; KTVW, Phoenix (Prescott), AZ; WSUR, Ponce, PR; WUVC-TV, Raleigh-Durham (Fayetteville), NC; KUVS-TV, Sacramento -Stockton-Modesto, CA; KTFK-TV, Sacramento -Stockton-Modesto, CA; KWEX, San Antonio, TX; KFSF-TV, San Francisco-Oakland-San Jose; KDTV, San Francisco-Oakland-San Jose; WFTT-TV, Tampa-St. Petersburg (Sarasota), FL; KFTU-TV, Tucson (Sierra Vista), AZ; KUVE-TV, Tucson (Sierra Vista), AZ; KAKW-TV, Waco-Temple-Bryan, TX; WFDC-TV, Washington, DC (Hagerstown, MD).

Ray Rodriguez, pres/COO.

V

VCY America Inc., 3434 W. Kilbourn Ave., Milwaukee, WI, 53208. Phone: (414) 935-3000. Fax: (414) 935-3015. E-mail: vcy@vcyamerica.org Web Site: www.vcyamerica.org.

Stns: 1 TV. WVCY, Milwaukee, WI.

Stns: 1 AM. 14 FM. KVCY-FM Fort Scott, KS; KCVS(FM) Salina, KS; WVCN(FM) Baraga, MI; WVCM(FM) Iron Mountain, MI; WJIC-FM Zanesville, OH; KVCF(FM) Freeman, SD; KVCX-FM Gregory, SD; KVFL(FM) Pierre, SD; WVCF-FM Eau Claire, WI; WVFL(FM) Fond du Lac, WI; WVCY-FM Milwaukee, WI; WVCY Oshkosh, WI; WVCX(FM) Tomah, WI; WEGZ(FM) Washburn, WI; WVRN(FM) Wittenberg, WI.

Vic Eliason, VP/gen mgr; Jim Schneider, progmg dir.

The Victory Television Network, Box 22007, Little Rock, AR, 72221-2007. Phone: (501) 223-2525. Fax: (501) 221-3837. E-mail: jim.grant@vtntv.com Web Site: www.vtntv.com. Ownership: Agape Church Inc.

Stns: 3 TV. KVTJ, Jonesboro, AR; KVTN, Little Rock-Pine Bluff, AR; KVTH, Little Rock-Pine Bluff, AR.

Jim Grant, gen mgr; Pastor Happy Caldwell, pres.

W

WTVA Inc., Box 350, Tupelo, MS, 38802. Phone: (662) 842-7620. Fax: (662) 844-7061. Web Site: www.wtva.com. Ownership: WTVA Inc. ownership: Mary Jane Spain, 51%; Margaret Spain, 40%. Note: estate of Frank K. Spain owns 100% of WMDN Inc., licensee of WMDN(TV) Meridian, MS.

Stns: 3 TV. WTVA, Columbus-Tupelo-West Point, MS; WMDN, Meridian, MS; KTFL, Phoenix (Prescott), AZ.

Waterman Broadcasting Corp., Box 7578, Fort Myers, FL, 33911-7578. Phone: (239) 939-2020. Fax: (239) 939-7903. Web Site: www.water.net. Ownership: Bernard Waterman, Edith Waterman.

Stns: 2 TV. WVIR-TV, Charlottesville, VA; WBBH, Ft. Myers-Naples, FL.

Bernard Waterman, pres; Steve Pontius, exec VP; Joe Ernest, VP.

Weigel Broadcasting Co., 26 N. Halsted St., Chicago, IL, 60661. Phone: (312) 705-2600. Fax: (312) 705-2656. Web Site: www.wciu.com. Ownership: Weigel Broadcasting Co., limited ptnr; Madison Halsted LLC, gen ptnr.

Stns: 2 TV. WCIU-TV, Chicago; WDJT, Milwaukee, WI.

Norman Shapiro, pres; Howard Shapiro, chmn; Neal Savin, exec VP.

West Virginia Media Holdings LLC, Box 11848, Charleston, WV, 25339-1848. Phone: (304) 720-6527. Fax: (304) 345-7280. Ownership: West Virginia Medio Partners, LP.

Stns: 4 TV. WVNS-TV, Bluefield-Beckley-Oak Hill, WV; WOWK, Charleston-Huntington, WV; WBOY, Clarksburg-Weston, WV; WTRF-TV, Wheeling, WV-Steubenville, OH.

Bray Cary, pres/CEO; Marty Becker, chmn.

Mel Wheeler Inc., 5009 S. Hulen, Suite 101, Fort Worth, TX, 76132-1989. Phone: (817) 294-7644. Fax: (817) 294-8519. Ownership: Estate of Mel Wheeler, 68%; Clark Wheeler, 10.2%; Leonard Wheeler, 11.1%; Steve Wheeler, 10.6%.

Stns: 2 TV. KPOB, Paducah, KY-Cape Girardeau, MO-Harrisburg-Mount Vernon, IL; WSIL-TV, Paducah, KY-Cape Girardeau, MO-Harrisburg-Mount Vernon, IL.

Stns: 2 AM. 4 FM. WVBE-FM Lynchburg, VA; WXLK-FM Roanoke, VA; WVBE(AM) Roanoke, VA; WSLQ-FM Roanoke, VA; WSLC-FM Roanoke, VA; WFIR(AM) Roanoke, VA.

Leonard Wheeler, pres; Clark Wheeler, VP; Gretchen Cummings, sec/treas.

White Knight Holdings Inc., Box 3058, Lafayette, LA, 70502. Phone: (337) 237-9965. Fax: (337) 235-5872.

Stns: 4 TV. WVLA, Baton Rouge, LA; WNTZ, Jackson, MS; KSHV, Shreveport, LA; KFXK, Tyler-Longview (Lufkin & Nacogdoches), TX.

Sheldon Galloway, pres.

Wilderness Communications LLC, 3501 Northwest Evangeline Thruway, Carencro, LA, 70520. Phone: (337) 896-1600. Fax: (337) 896-2695. Ownership: Chatelain Group.

Stns: 2 TV. KBCA, Alexandria, LA; KLWB, Lafayette, LA.

Withers Broadcasting Co., Box 1508, Mount Vernon, IL, 62864. Phone: (618) 242-3500. Fax: (618) 242-4444. Ownership: W. Russell Withers Jr., 100%.

Stns: 2 TV. WDTV, Clarksburg-Weston, WV; WDHS, Marquette, MI.

Stns: 10 AM. 14 FM. KOKX Keokuk, IA; KRNQ(FM) Keokuk, IA; WKIB(FM) Anna, IL; WRUL-FM Carmi, IL; WROY Carmi, IL; WCEZ(FM) Carthage, IL; WILY Centralia, IL; WEBQ-FM Eldorado, IL; WISH-FM Galatia, IL; WEBQ Harrisburg, IL; WMOK Metropolis, IL; WZZT-FM Morrison, IL; WYNG(FM) Mount Carmel, IL; WMIX Mount Vernon, IL; WMIX-FM Mount Vernon, IL; WSSQ-FM Sterling, IL; WSDR Sterling, IL; WZZL-FM Reidland, KY; WGKY-FM Wickliffe, KY; KGMO(FM) Cape Girardeau, MO; KAPE Cape Girardeau, MO; KUGT Jackson, MO; KRHW Sikeston, MO; KBXB(FM) Sikeston, MO.

W. Russell Withers Jr., pres.

Woods Communications Corp., One WCOV Ave., Montgomery, AL, 36111. Phone: (334) 288-7020. Fax: (334) 288-5414. Web Site: www.wcov.com. Ownership: David D. Woods, 100%.

Stns: 2 TV. KLCW-TV, Lubbock, TX; WCOV, Montgomery-Selma, AL.

David Woods, pres/CEO.

Wooster Republican Printing Co., (dba Dix Communications). 212 E. Liberty St., Wooster, OH, 44691. Phone: (330) 264-3511. Fax: (330) 263-5013. Web Site: www.dixcom.com. Ownership: (dba Dix Communications).

Stns: 1 TV. KFBB, Great Falls, MT.

Stns: 3 AM. 6 FM. WNDT(FM) Alachua, FL; WOGK(FM) Ocala, FL; WNDD-FM Silver Springs, FL; WKVX Wooster, OH; WQKT(FM) Wooster, OH; WTBO Cumberland, MD; WKGO-FM Cumberland, MD; WFRB Frostburg, MD; WFRB-FM Frostburg, MD.

Wooster Republican Printing Co. publishes *The Daily Record*, Wooster, OH.

Robert C. Dix, TV div chmn; G. Charles Dix, VP; Dale E. Gerber, CFO.

Word Broadcasting Network Inc., Box 19229, Louisville, KY, 40259. Phone: (502) 964-3304. Fax: (502) 966-9692. Web Site: www.wbna21.com. Ownership: Robert W. Rodgers, 20%; Gregory A. Holt, 20%; Melissa Fraser, 20%; Cleddie Kieth, 20%; and Margaret A. Rodgers, 20%.

Stns: 1 TV. WBNA, Louisville, KY.

Stns: 3 AM. WYMM(AM) Jacksonville, FL; WVHI Evansville, IN; WYRM(AM) Norfolk, VA.

Bob Rogers, pres; Greg Holt, VP.

Y

Young Broadcasting Inc., 599 Lexington Ave., 47th Fl., New York, NY, 10022. Phone: (212) 754-7070. Fax: (212) 758-1229. Web Site: www.youngbroadcasting.com. Ownership: Vincent J. Young, Gabelli Asset Management Inc., New South Capital Management Inc.

Stns: 14 TV. WCDC, Albany-Schenectady-Troy, NY; WTEN, Albany-Schenectady-Troy, NY; KWQC, Davenport, IA-Rock Island-Moline, IL; WBAY, Green Bay-Appleton, WI; WATE, Knoxville, TN; KLFY, Lafayette, LA; WLNS, Lansing, MI; WKRN, Nashville, TN; KCLO, Rapid City, SD; WRIC-TV, Richmond-Petersburg, VA; KRON-TV, San Francisco-Oakland-San Jose; KDLO, Sioux Falls (Mitchell), SD; KELO-TV, Sioux Falls (Mitchell), SD; KPLO, Sioux Falls (Mitchell), SD.

Vincent Young, chmn; James Morgan, exec VP & CFO; Deborah McDermott, pres.

Key to Television Listings

Television listings include TV stations in the United States, its territories and Canada. All collected data for these listings include information current to summer 2007. To use the television key, see boldface numbers and corresponding explanations.

(1) WOF-TV—(2) Analog channel: 17. Digital Channel: 53. Analog hrs: 24 2,200 kw vis, 20 kw aur, ant 500t/300g. TL: N36 49 21 W108 47 32 (CP: Ant 750t/550g) **(3)**On air date: Apr 13, 1952. **(4)** Box 100, Dothan, AL 36301. Phone: (909) 555-1000. FAX: (909) 999-9999. Web Site: www.wof.tv. **(5)** Licensee: WOF Broadcasting Co. **(6)** Group owner: Acme Stations (acq 7-20-69; $2 million; **(6a)** FTR 7-29-69). **(7)** Population served: 230,000 **(8)** Natl. Network: CBS. **(9)** Natl. Rep: Jones, Tri-State. Washington Atty: Goltz & Stick. **(10)** News staff: 3; News: 10 hrs wkly. **(11) Key Personnel:**

Jud Jones . . . . .pres & gen mgr
D. Spark . . . . .chief engr

(1) Station call letters as assigned by the Federal Communications Commission (FCC) or Canadian Radio-television and Telecommunications Commission (CRTC).

(2) Analog channel and, where applicable, digital channel, hours of operation (analog and digital), power, antenna, location and construction permit. WOF-TV operates with 2,200 kilowatts (effective radiated power) visual and 20 kilowatts aural. Its antenna is 500 feet above average terrain and 300 feet above ground. N36 49 21 W108 47 32ö refers to the geographical coordinates (latitude and longitude) of the transmitter location. WOF-TV holds a construction permit for an antenna height change to 750 feet above average terrain, 550 feet above ground.

(3) Date station first went on the air (regardless of subsequent ownership changes).

(4) Address and zip code, telephone and fax number, web site and e-mail address.

(5) Licensee name.

(6) Ownership and date of acquisition (if not original owner). If a station has been sold, any available sale information is listed following the acquisition date. WOF-TV is owned by Acme Stations.

(6a) FTR date refers to Broadcasting & Cable magazine's weekly "For the Record" column that appeared in the magazine until June 8, 1998, where station sales were recorded as received from the FCC.

(7) Population served refers to the station's potential market.

(8) Network programming. WOF-TV's national network is CBS.

(9) Representatives and Washington attorney. Sales representatives are listed with the national rep first, then regional.

(10) Number of staff providing local news and number of local news aired weekly.

(11) Key personnel.

An asterisk (*) preceding station call letters indicates noncommercial stations.

Directory of TV Stations in the United States

Alabama

Anniston

see Birmingham (Anniston, Tuscaloosa), AL market

Birmingham (Anniston, Tuscaloosa), AL (DMA 40)

WABM— Analog channel: 68. Digital channel: 36. Analog hrs: 24 1,442 kw vis, 144 kw aur. ant 1,029t TL: N33 27 57 W86 47 45 (CP: 5,000 kw vis, ant 1,030t. TL: N33 27 37 W86 51 07) On air date: January 1986. 651 Beacon Pkwy. W., Suite 105, Birmingham, AL, 35209. Phone: (205) 943-2168. Fax: (205) 290-2114. Web Site: www.wabm68.com. Licensee: Birmingham (WABM-TV) Licensee Inc. **Group owner:** Glencairn Ltd. (acq 2-1-2002). Population served: 522,420 Natl. Network: MyNetworkTV.

Key Personnel:
Steve Marks . . . . . CFO
Scott Campbell . . . . . gen mgr
Charlie Slaight . . . . . gen sls mgr & adv dir
Lucrecia Rubio . . . . . progmg dir & engrg dir
Peggy Johnson . . . . . news dir
John Batspm . . . . . chief of engrg

***WBIQ—** Analog channel: 10. Digital channel: 53. Analog hrs: 24 Digital hrs: 24 316 kw vis, 31.6 kw aur. 1,325t/1,042g TL: N33 29 19 W86 47 58 On air date: Apr 28, 1955. 2112 11th Ave. S., Suite 400, Birmingham, AL, 35205. Phone: (205) 328-8756. Fax: (205) 251-2192. Web Site: www.aptv.org. Licensee: Alabama ETV Commission. Population served: 1,400,000 Natl. Network: PBS. Washington Atty: Dow, Lohnes & Albertson, PLLC.

Key Personnel:
Allan Pizzato . . . . . CEO
Charles Grantham . . . . . COO
Pauline Howland . . . . . CFO
John Brady . . . . . dev VP & dev dir

WBRC— Analog channel: 6. Digital channel: 50. Analog hrs: 24 100 kw vis, 10 kw aur. ant 1,377t/1,010g TL: N33 29 19 W86 47 58 (CP: Ant 1,086t) On air date: July 1, 1949. Box 6, Birmingham, AL, 35201. 1720 Valley View Dr., Mooresville, AL 35209. Phone: (205) 322-6666. Fax: (205) 583-4386. Web Site: www.wbrc.com. Licensee: Fox Television Stations Inc. Group owner: (group owner; (acq 7-21-95). Population served: 1,000,000 Natl. Network: Fox. Natl. Rep: TeleRep. News staff: 53; News: 21 hrs wkly.

Key Personnel:
Mike McClain . . . . . VP
Dennis Leonard . . . . . gen mgr
Roy Gardner . . . . . opns mgr
Mike Lewis . . . . . sls VP & gen sls mgr
Sonya Ridderhoff . . . . . natl sls mgr
Wayne Farr . . . . . progmg dir
Jerry Thorn . . . . . chief of engrg

WCFT-TV— Analog channel: 33.1,225 kw vis, 203 kw aur. 540t/442g TL: N33 10 27 W87 29 09 (CP: 4,741 kw vis, ant 2,066t) On air date: Oct 27, 1965. 800 Concourse Pkwy., Suite 200, Birmingham, AL, 35244. 4000 37th St. E., Tuscaloosa, AL 35405. Phone: (205) 403-3340. Fax: (205) 403-3329. Web Site: www.abc3340.com. Licensee: TV Alabama Inc. Group owner: Allbritton Communications Co. (acq 1996; $20 million). Population served: 84,000 Natl. Network: ABC. Washington Atty: Hogan & Hartson.

Key Personnel:
Mike Murphy . . . . . gen mgr
Gary Watkins . . . . . opns dir

***WCIQ—** Analog channel: 7. Digital channel: 56. Analog hrs: 24 Digital hrs: 24 316 kw vis, 31.6 kw aur. 2,000t/537g TL: N33 29 07 W85 48 33 On air date: Jan 7, 1955. 2112 11th Ave. S., Suite 400, Birmingham, AL, 35205. Phone: (205) 328-8756. Fax: (205) 251-2192. Web Site: www.aptv.org. Licensee: Alabama ETV Commission. Natl. Network: PBS. Washington Atty: Dow, Lohnes & Albertson, PLLC.

Key Personnel:
Allan Pizzato . . . . . CEO
Charles Grantham . . . . . COO
Pauline Howland . . . . . CFO & gen mgr
Polly Anderson . . . . . dev VP

WDBB— Analog channel: 17. Digital channel: 18. Analog hrs: 24 2,240 kw vis. ant 2,214t/1,966g TL: N33 28 51 W87 24 03 On air date: Oct 1, 1984. 651 Beacon Pkwy. W., Suite 105, Birmingham, AL, 35209. Phone: (205) 943-2168. Fax: (205) 290-2114. Web Site: www.wtto21.com. Licensee: WDBB-TV Inc. Ownership: Cecil Heftel; H. Carl Parmer; D&C L.L.C. (acq 1-19-95; $1.5 million). Population served: 250,000 Natl. Rep: Adam Young. Washington Atty: Fletcher, Heald & Hildreth. Wire Svc: NOAA Weather Wire Svc: Weather Wire News staff: 20; News: 15 hrs wkly.

Key Personnel:
Scott Campbell . . . . . gen mgr
Steve Marks . . . . . CEO & opns mgr
Amy Hughes . . . . . sls dir & rgnl sls mgr
Lucrecia Rubio . . . . . mktg dir & progmg dir
Mary Ann Huie . . . . . adv dir
Peggy Johnson . . . . . progmg dir & news dir
John Batson . . . . . engrg mgr & chief of engrg

WIAT— Analog channel: 42. Digital channel: 30. Analog hrs: 24 Digital hrs: 24 2,163 kw vis, 216 kw aur. ant 1,382t/1,134g TL: N33 29 02 W86 48 21 (CP: 5,000 kw vis) On air date: Oct 17, 1965. 2075 Goldencrest Dr., Birmingham, AL, 35209. Phone: (205) 322-4200. Fax: (205) 320-2710. Web Site: www.wiat.com. Licensee: NVT Birmingham Licensee LLC. Group owner: Media General Broadcast Group (acq 10-6-2006; $35 million with KIMT(TV) Mason City, IA). Population served: 700,000 Natl. Network: CBS. Natl. Rep: MMT. Washington Atty: Dow, Lohnes & Albertson. News: 6 hrs wkly.

Key Personnel:
Bill Ballard . . . . . pres, VP & gen mgr
Greg Butler . . . . . opns dir & opns mgr
Kathy Pozgar . . . . . gen sls mgr
Bill Payer . . . . . news dir

WJSU-TV— Analog channel: 40. Digital channel: 9. Analog hrs: 24 5,000 kw vis. ant 1,299t/538g TL: N33 36 24 W86 25 03 On air date: Oct 26, 1969. 800 Concourse Pkwy., Suite 200, Birmingham, AL, 35244. 1330 Noble St., Suite 40 Radio Bldg., Mooresville, AL 36202. Phone: (205) 403-3340. Fax: (205) 403-3329. Web Site: www.abc3340.com. Licensee: TV Alabama Inc. Group owner: Allbritton Communications Co. (acq 1-24-2000). Population served: 91,000 Natl. Network: ABC. Washington Atty: Haley, Bader & Potts. News staff: 16; News: 9 hrs wkly.

Key Personnel:
Mike Murphy . . . . . gen mgr
Gary Watkins . . . . . opns mgr

WLDM— Analog channel: 23.890 kw vis. ant 872t/669g TL: N33 03 15 W87 32 57 On air date: 2000. 651 Beacon Pkwy. W., Suite 105, Birmingham, AL, 35209. Phone: (205) 943-2168. Fax: (205) 290-2114. Licensee: The Board of Trustees of the University of Alabama. (acq 11-30-2004; donation).

Key Personnel:
Scott Campbell . . . . . gen mgr
Ron Snyder . . . . . natl sls mgr
Lucrecia Romeo . . . . . progmg dir
Peggy Johnson . . . . . news dir
John Batson . . . . . chief of engrg

WPXH— Analog channel: 44. Digital channel: 45. Analog hrs: 24 1,750 kw vis, 175 kw aur. 1,000t/500g TL: N33 57 11 W86 13 00 (CP: 5,000 kw vis, ant 1,115t. TL: N33 53 27 W86 28 18) On air date: Apr 26, 1986. 2085 Goldencrest Dr., Birmingham, AL, 35209. Phone: (205) 870-4404. Fax: (205) 870-0744. Licensee: Paxson Communications License Co. L.L.C. Group owner: Paxson Communications Corp. Population served: 1,000,000 Natl. Network: i Network. Washington Atty: Fletcher, Heald & Hildreth. News: 8 hrs wkly.

WTJP-TV— Analog channel: 60. Digital channel: 26. Analog hrs: 24 5,000 kw vis, 500 kw aur. ant 1,139t TL: N33 48 53 W86 26 55 On air date: July 22, 1986. 313 Rosedale Ave., Gadsden, AL, 35901-5361. Phone: (256) 546-8860. Fax: (256) 543-8623. Web Site: www.tbn.org. Licensee: Trinity Christian Center of Santa Ana Inc. dba Trinity Broadcasting Network. Group owner: Trinity Broadcasting Network (acq 5-8-2000). Population served: 1,600,000

Key Personnel:
Paul F. Crouch . . . . . CEO & pres
Terry Hickey . . . . . exec VP
Gary Hodges . . . . . gen mgr & gen sls mgr
Curtiss Kemp . . . . . chief of engrg

WTTO— Analog channel: 21. Digital channel: 28.1,042 kw vis, 104.2 kw aur. ant 1,342t/1,058g TL: N33 30 42 W86 48 24 (CP: 5,000 kw vis, ant 1,340t) On air date: Apr 21, 1982. 651 Beacon Pkwy. W., Suite 105, Huntsville, AL, 35209. Phone: (205) 943-2168. Fax: (205) 290-2114 / (205) 250-6788. Web Site: www.wtto21.com. Licensee: WTTO Licensee LLC. (acq 12-21-90). Natl. Network: CW. Natl. Rep: Millennium Sales & Marketing. Washington Atty: Arter & Hadden.

Key Personnel:
Chris Hummel . . . . . CEO & CFO
Scott Campbell . . . . . gen mgr
Amy Hughes . . . . . rgnl sls mgr & prom dir
Lucrecia Rubio . . . . . adv dir & progmg dir
Peggy Johnson . . . . . news dir
John Batson . . . . . engrg dir & chief of engrg

WVTM-TV— Analog channel: 13. Digital channel: 52.316 kw vis, 47.4 kw aur. ant 1,340t/1,073g TL: N33 29 26 W86 47 48 On air date: May 1949. 1732 Valley View Dr., Birmingham, AL, 35209. Phone: (205) 558-7300 (news) / (205) 933-1313. Fax: (205) 933-7516 (sales). E-mail: newscomments@nbc13.com Web Site: www.nbc13.com. Licensee: Media General Communications Inc. Group owner: NBC TV Stations Division (acq 6-26-2006; grpsl). Population served: 2,500,000 Natl. Network: NBC. Natl. Rep: Harrington, Righter & Parsons. News staff: 70; News: 24 hrs wkly.

Key Personnel:
Clark Dumornay . . . . . opns mgr
Joe Tracy . . . . . sls VP
Ed Moran . . . . . natl sls mgr
Mike Sherry . . . . . mktg mgr & prom dir
Yvette Miley . . . . . progmg dir
Yvette M. Miley . . . . . news dir
Terese Messick . . . . . pub affrs dir
Chuck Blackwood . . . . . engrg dir

Decatur

see Huntsville-Decatur (Florence), AL market

Dothan, AL (DMA 172)

WDFX-TV— Analog channel: 34. Analog hrs: 21 1,120 kw vis. 466t TL: N31 12 30 W85 36 51 On air date: Feb 23, 1991. 2221 Ross Clark Cir., Dothan, AL, 36301. Phone: (334) 794-3434. Fax: (334) 794-0034. Web Site: www.wdfxfox34.com. Licensee: Raycom America License Subsidiary LLC. Group owner: Raycom Media Inc. (acq 10-14-2003; grpsl). Population served: 104,000 Natl. Network: Fox. Washington Atty: Borsari & Paxson.

Key Personnel:
Eric Steffens . . . . . gen mgr
Melinda Chaney . . . . . gen mgr & rgnl sls mgr
Jennifer Otto . . . . . mktg mgr & prom dir
Wes Roten . . . . . chief of engrg

WDHN— Analog channel: 18. Analog hrs: 24 1,080.4 kw vis, 108 kw aur. 730t/796g TL: N31 14 30 W85 18 48 On air date: Aug 7, 1970. Box 6237, Dothan, AL, 36302. 5274 E. Hwy. 52, Webb, AL 36302. Phone: (334) 793-1818. Fax: (334) 793-2623. Web Site: www.wdhn.com. Licensee: Nexstar Broadcasting Inc. Group owner: Nexstar Broadcasting Group Inc. (acq 8-1-2003; $40 million with KARK-TV Little Rock, AR). Population served: 277,000 Natl. Network: ABC. Washington Atty: Fletcher, Heald & Hildreth. News staff: 8; News: 7 hrs wkly.

Key Personnel:
Mike Smith . . . . . VP, gen mgr & progmg dir
Janie Hinson . . . . . gen sls mgr
Yolanda Everett . . . . . prom dir
Mike Quinn . . . . . news dir
Edna Darrow . . . . . pub affrs dir
Neal Riddle . . . . . chief of engrg

WTVY— Analog channel: 4. Digital channel: 36. Analog hrs: 24 Digital hrs: 24 Note: CBS is on WTVY(TV) ch 4, CW and MyNetworkTV are on WTVY-DT ch 36. 100 kw vis, 20 kw aur. 1,670t/1,909g TL: N30 55 10 W85 44 28 On air date: Feb 12, 1955. Box 1089, Dothan, AL, 36302. 285 N. Foster St., Dothan, AL 36303. Phone: (334) 792-3195. Fax: (334) 793-3947. Web Site: www.wtvynews4.com. Licensee: Gray Television Licensee Inc. Group owner: Gray Television Inc. (acq 8-29-2002; grpsl). Population served: 230,000 Natl. Network: CBS, CW, MyNetworkTV Digital Network: Note: CBS is on WTVY(TV) ch 4, CW and MyNetworkTV are on WTVY-DT ch 36. Natl. Rep: Continental Television Sales. Wire Svc: AP News staff: 20; News: 16 hrs wkly.

Key Personnel:
Patrick Dalbey . . . . . VP & gen mgr
Richard Morgan . . . . . gen sls mgr
Millicent Smith . . . . . natl sls mgr & prom dir
Judy Calhoun . . . . . rgnl sls mgr
Mike Doherty . . . . . prom mgr

Katie McManus . . . . . news dir
Wade Thomaston . . . . . chief of engrg

Florence

see Huntsville-Decatur (Florence), AL market

Huntsville-Decatur (Florence), AL (DMA 84)

WAAY-TV— Analog channel: 31. Digital channel: 32.1,255 kw vis, 125 kw aur. ant 1,790t/999g TL: N34 44 15 W86 32 02 On air date: Aug 1, 1959. 1000 Monte Sano Blvd., Huntsville, AL, 35801. Phone: (256) 533-3131. Fax: (256) 533-6616. Web Site: www.waaytv.com. Licensee: WAAY-TV License LLC. Group owner: Piedmont Television Holdings LLC (acq 1-31-2007; $41.645 million). Population served: 866,000 Natl. Network: ABC. Washington Atty: Cohn & Marks.
Key Personnel:
Ray Depa . . . . . VP & gen mgr
Ben Boles . . . . . opns mgr & prom mgr
Chris Kidd. . . . . gen sls mgr
Dave Keller. . . . . progmg dir
Al Carl . . . . . news dir
Jim Bowman . . . . . chief of engrg

WAFF— Analog channel: 48. Digital channel: 49. Analog hrs: 24 Digital hrs: 24 1,170 kw vis, 234 kw aur. 1,900t/1,526g TL: N34 42 39 W86 32 07 On air date: July 4, 1954. 1414 N. Memorial Pkwy., Huntsville, AL, 35801. Phone: (256) 533-4848. Fax: (256) 533-1337. E-mail: webmaster@waff.com. Web Site: www.waff.com. Licensee: Raycom America License Subsidiary LLC. Group owner: Raycom Media Inc. (acq 3-16-97; grpsl). Population served: 879,000 Natl. Network: NBC. Natl. Rep: Harrington, Righter & Parsons. Washington Atty: Covington & Burling. Wire Svc: AP News staff: 43; News: 26 hrs wkly.
Key Personnel:
Lee Meredith . . . . . VP & gen mgr
Dale Stafford . . . . . gen sls mgr
Susan Craft . . . . . rgnl sls mgr
Becky Nichols . . . . . mktg mgr
Tracey Gallien . . . . . news dir
J.T. Harriman. . . . . engrg dir

***WFIQ—** Analog channel: 36. Digital channel: 22. Analog hrs: 24 Digital hrs: 24 851 kw vis. ant 725t/515g TL: N34 34 40 W87 46 54 On air date: Aug 16, 1967. 2112 11th Ave. S., Suite 400, Birmingham, AL, 35205. Phone: (205) 328-8756. Fax: (205) 251-2192. Web Site: www.aptv.org. Licensee: Alabama ETV Commission. Natl. Network: PBS. Washington Atty: Dow, Lohnes & Albertson, PLLC.
Key Personnel:
Allan Pizzato . . . . . CEO
Charles Grantham . . . . . CEO & COO
Pauline Howland . . . . . CFO
John Brady . . . . . dev VP

WHDF— Analog channel: 15. Digital channel: 14. Analog hrs: 24 Digital hrs: 24 2,510 kw vis. ant 1,414t/1,315g TL: N35 00 09 W87 08 09 On air date: Oct 29, 1957. 200 Andrew Jackson Way, Huntsville, AL, 35801. 840 Cypress Mill Rd., Florence, AL 35630. Phone: (256) 767-1515 /1550. Fax: (256) 764-7750. Web Site: www.thevalleyscw.tv. Licensee: Huntsville TV L.L.C. Ownership: James L. Lockwood Jr., 100% Population served: 500,000 Natl. Network: CW. Natl. Rep: Blair Television.
Key Personnel:
Shanda Love . . . . . CEO & gen sls mgr
Louann Thomson . . . . . gen mgr
Brian Capaldo . . . . . stn mgr
Tim Rovere . . . . . chief of engrg

***WHIQ—** Analog channel: 25. Digital channel: 24. Analog hrs: 24 Digital hrs: 24 1,230 kw vis. ant 1,155t/312g TL: N34 44 14 W86 31 46 On air date: November 1965. 2112 11th Ave. S., Suite 400, Birmingham, AL, 35205. Phone: (205) 328-8756. Fax: (205) 251-2192. Web Site: www.aptv.org. Licensee: Alabama ETV Commission. Population served: 259,550 Natl. Network: PBS. Washington Atty: Dow, Lohnes & Albertson, PLLC.
Key Personnel:
Allan Pizzato . . . . . CEO
Charles Grantham . . . . . COO
Pauline Howland . . . . . CFO
John Brady . . . . . dev VP

WHNT-TV— Analog channel: 19. Digital channel: 59. Analog hrs: 24 1,279 kw vis, 254 kw aur. ant 1,750t/944g TL: N34 44 19 W86 31 56 On air date: Nov 28, 1963. 200 Holmes Ave., Huntsville, AL, 35801. Box 19, Huntsville, AL 35801. Phone: (256) 533-1919. Fax: (256) 533-4503. Fax: (256) 536-9468 (news). E-mail: feedback@whnt19.com Web Site: www.whnt.com. Licensee: Local TV Alabama License LLC. Group owner: The New York Times Co. (acq 5-7-2007; grpsl). Population served: 251,100 Natl. Network: CBS. Washington Atty: Koteen & Naftalin.
Key Personnel:
Tharon Honeycutt . . . . . gen mgr
Robert Alverson. . . . . opns mgr & progmg dir
Stan Pylant . . . . . sls VP
Heather Carlton . . . . . rgnl sls mgr
Holy Griggs. . . . . mktg dir
Kevin Osgood . . . . . news dir
Steve King. . . . . engrg dir

WYLE— Analog channel: 26. Digital channel: 20. Analog hrs: 24 690 kw vis, 69 kw aur. 756t TL: N34 34 38 W87 46 57 On air date: Apr 19, 1986. Box 850, Sheffield, AL, 35660. 700 W. Montgomery Ave., Sheffield, AL 35660. Phone: (256) 381-2600. Fax: (256) 383-3157. E-mail: etccom@bellsouth.net Licensee: ETC Communications Inc. Ownership: Les White, 100%. (acq 6-93). Population served: 480,000 Washington Atty: Irwin, Campbell & Tannenwood. News staff: 9; News: 15 hrs wkly.
Key Personnel:
Karen Snead . . . . . gen mgr & progmg dir
Bud Hayle. . . . . gen sls mgr
Les White . . . . . CEO & film buyer
Donna Kemp . . . . . engrg mgr

WZDX— Analog channel: 54. Digital channel: 41.Note: Fox is on WZDX(TV) ch 54, MyNetworkTV is on WZDX-DT ch 41. 2,400 kw vis, 240 kw aur. ant 1,699t/906g TL: N34 44 12 W86 31 59 On air date: Apr 14, 1985. Box 3889, Huntsville, AL, 35810. Phone: (256) 533-5454. Fax: (256) 533-5315. Web Site: www.fox54.com. Licensee: Huntsville Television Acquisition Licensing LLC. Group owner: Grant Communications (acq 4-90; $6.1 million). Natl. Network: Fox, MyNetworkTV Digital Network: Note: Fox is on WZDX(TV) ch 54, MyNetworkTV is on WZDX-DT ch 41.
Key Personnel:
Kevin Tucker . . . . . stn mgr & gen sls mgr
Thomas Grant . . . . . natl sls mgr
Frank White. . . . . prom mgr
Linda Jones . . . . . progmg dir
Rose Anna Martincak . . . . . pub affrs dir
Harry Wilkins. . . . . chief of engrg

Louisville

see Columbus, GA market

Mobile, AL-Pensacola (Ft. Walton Beach), FL (DMA 59)

WALA-TV— Analog channel: 10. Digital channel: 9. Analog hrs: 24 316 kw vis, 47 kw aur. ant 1,246t/1,200g TL: N30 41 17 W87 47 54 On air date: Jan 14, 1953. 1501 Satchel Paige Dr., Mobile, AL, 36606. Phone: (251) 434-1010. Fax: (251) 434-1073 / 1061. E-mail: fox10@wala.emmis.com Web Site: www.fox10tv.com. Licensee: LIN of Alabama LLC. Group owner: Emmis Communications Corp. (acq 11-30-2005; grpsl). Population served: 1,509,000 Natl. Network: Fox. Natl. Rep: TeleRep. Washington Atty: Fisher, Wayland, Cooper, Leader & Zaragoza. News staff: 45; News: 23 hrs.
Key Personnel:
Matt Pumo . . . . . gen sls mgr
Mike Kelly. . . . . natl sls mgr
Kristen Mosley. . . . . mktg mgr & prom mgr
Bob Cashen . . . . . news dir
Roland Fields . . . . . chief of engrg

WAWD— Analog channel: 58. Digital channel: 49.490 kw vis. ant 161t TL: N30 23 35 W86 29 41 On air date: Aug 1, 1998. 8317 Front Beach Rd., Suite 23, Panama City, FL, 32407. Phone: (850) 234-2773. Fax: (850) 234-1179. Web Site: www.tripsmarter.com. Licensee: Beach TV Properties Inc. Group owner: Beach TV Properties Inc. (acq 10-29-99; $175,000). Washington Atty: Baraff, Koerner, Olender & Hochberg.
Key Personnel:
Robin Quinlan . . . . . gen mgr
Mike Hartzog . . . . . stn mgr

WBPG— Analog channel: 55.1510 kw vis. ant 1,010t TL: N30 36 37 W87 36 26 On air date: Sept 1, 2001. 1501 Satchel Paige Dr., Mobile, AL, 36606. Phone: (251) 434-1010. Fax: (251) 434-1073 / 1061. Licensee: LIN of Alabama LLC. Group owner: Emmis Communications Corp. (acq 7-7-2006; grpsl). Population served: 1,200,000 Natl. Network: CW. Natl. Rep: TeleRep.
Key Personnel:
Carey Golden . . . . . gen sls mgr
Mike Kelly. . . . . natl sls mgr
Kristen Mosley . . . . . mktg mgr
Roland Fields . . . . . chief of engrg

WEAR-TV—(Pensacola, FL) Analog channel: 3. Analog hrs: 24 100 kw vis, 20 kw aur. ant 1,886t/1,847g TL: N30 36 45 W87 38 43 On air date: Jan 13, 1954. Box 12278, Pensacola, FL, 32581. 4990 Mobile Hgwy, Pensacola, FL 32506. Phone: (850) 456-3333. Fax: (850) 455-0159. E-mail: comments@wear.sbjnet.com Web Site: www.weartv.com. Licensee: WEAR Licensee L.L.C. Group owner: Sinclair Broadcast Group Inc. (acq 10-8-97). Population served: 440,800 Natl. Network: ABC. Natl. Rep: Millennium Sales & Marketing. Washington Atty: Shaw Pittman LLP. News staff: 30; News: 14 hrs wkly.
Key Personnel:
Carl Leahy . . . . . pres
Peter Neuman . . . . . news dir
David Brown . . . . . chief of engrg

***WEIQ—** Analog channel: 42. Digital channel: 41. Analog hrs: 24 Digital hrs: 24 1,170 kw vis, 117 kw aur. 600t/545g TL: N30 39 33 W87 53 33 On air date: Nov 6, 1964. 2112 11th Ave. S., Suite 400, Birmingham, AL, 35205. Phone: (205) 328-8756. Fax: (205) 251-2192. Web Site: www.aptv.org. Licensee: Alabama ETV Commission. Natl. Network: PBS. Washington Atty: Dow, Lohnes & Albertson, PLLC.
Key Personnel:
Allan Pizzato . . . . . CEO
Charles Grantham . . . . . CEO & COO
Pauline Howland . . . . . CFO
Kathie Martin . . . . . dev VP

WFBD— Analog channel: 48. Digital channel: 48.5,000 kw vis. ant 454t/452g TL: N30 30 52 W86 13 12 Not on air, target date: unknown: 118 S. Bellevue, Suite 222, Memphis, TN, 38104. Phone: (901) 516-8970. Permittee: George S. Flinn Jr. Ownership: George S. Flinn Jr., 100%.

WFGX— Analog channel: 35. Digital channel: 25. Analog hrs: 24 635 kw vis, 63.6 kw aur. ant 280t/250g TL: N30 26 36 W86 35 56 On air date: Apr 7, 1987. Box 12278, Pensacola, FL, 32581. 4990 Mobile Hgwy., Fort Walton Beach, FL 32506. Phone: (850) 456-3333. Fax: (850) 453-4335. E-mail: wfgx@wfgxtv.com Web Site: www.wfgxtv.com. Licensee: WFGX Licensee LLC. Group owner: Sinclair Broadcast Group Inc. (acq 3-31-2004; $520,000). Population served: 216,300 Natl. Network: MyNetworkTV. Washington Atty: Shaw Pittman LLP.
Key Personnel:
David D. Smith . . . . . pres
Carl Leahy . . . . . gen mgr
Joe Smith . . . . . opns mgr

WHBR— Analog channel: 33. Analog hrs: 24 5,000 kw vis, 500 kw aur. 1,365t/1,330g TL: N30 37 35 W87 38 50 (CP: 3,500 kw vis) On air date: Jan 27, 1986. Box 2633, Pensacola, FL, 32513. 6500 Pensacola Blvd., Pensacola, FL 32505. Phone: (850) 473-8633. Fax: (850) 473-8671. E-mail: dmayo@whbr.org Web Site: www.whbr.org. Licensee: Christian Television of Pensacola/Mobile Inc. Ownership: David C. Gibbs III, Wayne Wetzel , Bill Anderson and Ginny Oliver. (acq 12-16-97). Washington Atty: Gammon & Grange.
Key Personnel:
Bob D'Andrea . . . . . pres
Wayne Wetzel. . . . . VP
David Mayo . . . . . gen mgr

WJTC— Analog channel: 44. Digital channel: 45. Analog hrs: 24 Digital hrs: 24 3,289 kw vis, 328.9 kw aur. 1,493t TL: N30 35 18 W87 33 16 On air date: December 1984. 661 Azalea Rd., Mobile, AL, 36609. Phone: (251) 602-1544. Fax: (251) 602-1547. E-mail: wjtc@wjtc.com Web Site: www.utv44.com. Licensee: Clear Channel Broadcasting Licenses Inc. Group owner: Clear Channel Communications Inc. (acq 3-21-2001). Population served: 1,229,000 Natl. Rep: Millennium Sales & Marketing. Washington Atty: Wiley, Rein & Fielding.
Key Personnel:
Donita Todd . . . . . gen mgr
Tim Woodard . . . . . opns mgr
Shea Grandquest. . . . . gen sls mgr
Nona Simmons . . . . . prom mgr & prom

WKRG-TV— Analog channel: 5. Digital channel: 27.100 kw vis, 20 kw aur. ant 1,906t/1,879g TL: N30 41 20 W87 49 49 On air date: Sept 5, 1955. 555 Broadcast Dr., Mobile, AL, 36606. Phone: (251) 479-5555. Fax: (251) 473-8130. Fax: TWX: 810-741-4263. Web Site: www.krg.com. Licensee: Media General Communications Inc. Group owner: Media General Broadcast Group (acq 3-27-2000; grpsl). Population served: 425,700 Natl. Network: CBS. Washington Atty: Wiley, Rein & Fielding.
Key Personnel:
Joe Goleniowski . . . . . pres, VP & gen mgr
Warren Fiihr . . . . . gen sls mgr
Robin Delaney . . . . . mktg dir
Darrel Taylor. . . . . progmg mgr
Dan Cates . . . . . news dir
Jim Richard . . . . . chief of engrg

WMPV-TV— Analog channel: 21. Analog hrs: 24 4,336 kw vis, 433.6 kw aur. 1,400t TL: N30 35 18 W87 33 16 On air date: Dec 19, 1985.

1668 S. Beltline Hwy., Mobile, AL, 36693. Phone: (251) 661-2101. Fax: (251) 661-7121. Web Site: www.tbn.org. Licensee: Trinity Broadcasting Network. Group owner: (group owner; (acq 5-8-2000; grpsl). Washington Atty: Fisher, Wayland, Cooper, Leader & Zaragoza.
Key Personnel:
Linda Dixon . . . gen mgr
Heather McCollum . . . progmg dir
LaTroynnda Cunningham . . . pub affrs dir
Alvin Goins . . . chief of engrg

WPAN— Analog channel: 53.3088 kw vis, 309 kw aur. 720t/749g TL: N30 24 09 W86 59 35 On air date: Feb 14, 1984. Box 18126, Pensacola, FL, 32523. 2105 W. Gregory St., Pensacola, FL 32523. Phone: (850) 433-1766. Fax: (850) 433-1641. Licensee: Franklin Media Inc. Ownership: John L. Franklin, 20%; Delores A. Franklin, 20%; Joseph C. Denison, 20%; Robert Gatlin, 20%; Glyn Lowery, 20% (acq 5-23-88). Washington Atty: Pepper & Corazzini.

WPMI-TV— Analog channel: 15. Digital channel: 47. Analog hrs: 24 5,000 kw vis, 500 kw aur. ant 1,847t/1,847g TL: N30 36 40 W87 36 27 On air date: Mar 12, 1982. 661 Azalea Rd., Mobile, AL, 36609-1515. Phone: (251) 602-1500. Fax: (251) 602-1547. Web Site: www.wpmi.com. Licensee: Clear Channel Broadcasting Licenses Inc. Group owner: Clear Channel Communications Inc. Population served: 1,229,000 Natl. Network: NBC. Natl. Rep: Millennium Sales & Marketing. Washington Atty: Wiley, Rein & Fielding. News staff: 50; News: 5 hrs wkly.
Key Personnel:
Shea Grandquest . . . gen mgr & gen sls mgr
Tim Woodward . . . opns mgr
Robert Herron . . . natl sls mgr
Ric Phillips . . . rgnl sls mgr
Jean Stanley . . . prom mgr
Kelly Barher . . . progmg dir
Joe Raia . . . news dir
Tim Reid . . . chief of engrg

***WSRE**— Analog channel: 23. Digital channel: 31.3,020 kw vis, 610 kw aur. 487t/466g TL: N30 26 36 W87 14 03 On air date: Sept 11, 1967. Bldg. 23, 1000 College Blvd., Pensacola, FL, 32504-8998. Phone: (850) 484-1200. Fax: (850) 484-1255. E-mail: rolandphillips@wsre.pbs.org Web Site: www.wsre.org. Licensee: District Board of Trustees of Pensacola Junior College. (acq 8-31-71). Population served: 251,000 Natl. Network: PBS.

Montgomery-Selma, AL
(DMA 117)

***WAIQ**— Analog channel: 26. Digital channel: 27. Analog hrs: 24 Digital hrs: 24 1,420 kw vis, 142 kw aur. 600t/525g TL: N32 22 52 W86 17 30 On air date: Dec 18, 1962. 2112 11th Ave. S., Suite 400, Birmingham, AL, 35205. Phone: (205) 328-8756. Fax: (205) 251-2192. Web Site: www.aptv.org. Licensee: Alabama ETV Commission. Natl. Network: PBS. Washington Atty: Dow, Lohnes & Albertson, PLLC.
Key Personnel:
Allan Pizzato . . . CEO
Charles Grantham . . . COO
Pauline Howland . . . pres & CFO
John Brady . . . gen mgr & dev VP

WAKA— Analog channel: 8. Analog hrs: 24 316 kw vis, 63.5 kw aur. 1,760t/1,757g TL: N32 08 58 W86 46 48 On air date: Mar 17, 1960. Box 230667, Montgomery, AL, 36123. 3020 East Blvd., Montgomery, AL 36123. Phone: (334) 271-8888. Fax: (334) 272-6444. Web Site: www.waka.com. Licensee: Alabama Broadcasting Partners. Group owner: Bahakel Communications (acq 8-85). Population served: 600,000 Natl. Network: CBS. Wire Svc: U.S. Weather Service News staff: 22; News: 9 hrs wkly.
Key Personnel:
Jim Caruthers . . . gen mgr
Steffanie Patterson . . . gen sls mgr & rgnl sls mgr
Mark Smith . . . progmg dir
Rob Martin . . . news dir
Thomas Mayberry . . . chief of engrg

WBIH— Analog channel: 29. Digital channel: 29.3,900 kw vis. ant 1,338t/1,171g TL: N32 32 27 W86 50 33 On air date: 2002. 225 N. Memorial, Suite 222, Prattville, AL, 36067. Phone: (334) 491-2900. Fax: (334) 491-2929. Licensee: Flinn Broadcasting Corp.

WBMM— Analog channel: 22. Digital channel: 24.2,820 kw vis. ant 1,118t TL: N32 04 05 W85 56 41 On air date: June 1, 2002. 3251 Harrison Rd., Montgomery, AL, 36109. Phone: (334) 270-3200. Fax: (334) 271-6348. Web Site: www.cwmontgomery.com. Licensee: SagamoreHill Broadcasting of Alabama LLC. Group owner: Equity Broadcasting Corp. (acq 7-26-2006; $2 million). Natl. Network: CW.

WCOV-TV— Analog channel: 20. Analog hrs: 24 2,667 kw vis, 266.7 kw aur. 2,043t TL: N31 58 32 W86 09 46 On air date: Apr 23, 1953. c/o WCOV-TV, One WCOV Ave., Montgomery, AL, 36111. Box 250045, Montgomery, AL 36125. Phone: (334) 288-7020. Fax: (334) 288-5414. E-mail: mail@wcov.com Web Site: www.wcov.com. Licensee: Woods Communications Corp. Group owner: (group owner; (acq 12-1-85; $4 million; FTR: 6-10-85). Population served: 484,987 Natl. Network: Fox. Natl. Rep: Millennium Sales & Marketing. Washington Atty: Kenkel, Barnard & Edmundson.

***WDIQ**— Analog channel: 2. Digital channel: 11. Analog hrs: 24 100 kw vis, 10 kw aur. 695t/566g TL: N31 33 16 W86 23 32 On air date: Aug 8, 1956. 2112 11th Ave. S., Suite 400, Birmingham, AL, 35205. Phone: (205) 328-8756. Fax: (205) 251-2192. Web Site: www.aptv.org. Licensee: Alabama ETV Commission. Population served: 400,000 Natl. Network: PBS. Washington Atty: Dow, Lohnes & Albertson, PLLC.
Key Personnel:
Allan Pizzato . . . CEO
Charles Grantham . . . CEO & COO
Pauline Howland . . . CFO
John Brady . . . dev VP

***WIIQ**— Analog channel: 41. Digital channel: 19. Analog hrs: 24 Digital hrs: 24 447 kw vis, 44.7 kw aur. 1,082t/999g TL: N32 22 01 W87 52 03 On air date: Sept 13, 1971. 2112 11th Ave. S., Suite 400, Birmingham, AL, 35205-2884. Phone: (205) 328-8756. Fax: (205) 251-2192. Web Site: www.aptv.org. Licensee: Alabama ETV Commission. Natl. Network: PBS. Washington Atty: Dow, Lohnes & Albertson, PLLC.
Key Personnel:
Allan Pizzato . . . CEO
Charles Grantham . . . COO
Pauline Howland . . . CFO
John Brady . . . dev VP

WMCF-TV— Analog channel: 45. Digital channel: 46. Analog hrs: 24 600 kw vis, 60 kw aur. ant 1,010t/1,169g TL: N32 24 11 W86 11 48 (CP: 619.4 kw vis) On air date: Oct 12, 1985. 300 Mendel Pkwy. W., Montgomery, AL, 36117. Phone: (334) 272-0045. Fax: (334) 277-6635. Licensee: Trinity Broadcasting Network. Group owner: (group owner; (acq 5-8-2000; grpsl). Washington Atty: Baraff, Keorner, Olender & Hochberg.
Key Personnel:
P. Crouch . . . pres
Aaron Motley . . . gen mgr, opns mgr & progmg mgr
Linda Bell . . . pub affrs dir
Larry Dean . . . chief of engrg

WNCF— Analog channel: 32. Digital channel: 51. Analog hrs: 24 Digital hrs: 24 3,000 kw vis, 545 kw aur. ant 1,788t/1,771g TL: N32 08 30 W86 44 42 On air date: Mar 12, 1964. 3251 Harrison Rd., Montgomery, AL, 36109. Phone: (334) 270-3200. Fax: (334) 271-6348. E-mail: gsingleton@wncftv.com Web Site: www.wncftv.com. Licensee: Channel 32 Montgomery L.L.C. (acq 1999; $8 million). Population served: 667,000 Natl. Network: ABC. Natl. Rep: Blair Television. Washington Atty: Wiley, Rein & Fielding. News staff: 2; News: one hr wkly.
Key Personnel:
Jesse Grear . . . gen mgr
Katy Hodges . . . gen sls mgr
Laura Balentine . . . rgnl sls mgr & prom dir
Lois Crenshaw . . . progmg dir
Mack Paulk . . . chief of engrg

WRJM-TV— Analog channel: 67. Digital channel: 48. Analog hrs: 24 2,820 kw vis. ant 1066t/913g TL: N32 03 37 W85 57 02 On air date: Dec. 5, 2000. Josie Park Broadcasting Inc., 285 E. Broad St., Ozark, AL, 36360. 315 S. Three Notch St., Troy, AL 36081. Phone: (334) 670-6766. Fax: (334) 670-6717. E-mail: wrjm67@troycable.net Web Site: www.wrjm.com. Licensee: Josie Park Broadcasting Inc. Ownership: H. Jack Misell, 67%; Walter P. Lunsford, 33% Population served: 611,750 Natl. Network: MyNetworkTV. Washington Atty: Borsari and Assoc, PLC.
Key Personnel:
Jack Misell . . . CEO & gen mgr
Walter P. Lunsford . . . VP
Vincent Hodges . . . stn mgr & prom mgr
Boyd Mizell . . . opns mgr
Buddy Johnson . . . natl sls mgr
Sonny Strassburger . . . rgnl sls mgr
Don Hess . . . progmg dir
Jenny Dykes . . . pub affrs dir
Dan Mizell . . . engrg VP

WSFA— Analog channel: 12. Analog hrs: 24 316 kw vis, 63.2 kw aur. 2,000t/1,935g TL: N31 58 32 W86 09 46 On air date: Dec 25, 1954. 12 E. Delano Ave., Montgomery, AL, 36105. Phone: (334) 288-1212. Fax: (334) 613-8301. Fax: (334) 613-8303. Web Site: www.wsfa.com. Licensee: Libco Inc. Group owner: Liberty Corp. (acq 1-13-2006; grpsl). Population served: 210,600 Natl. Network: NBC. Natl. Rep: Harrington, Righter & Parsons. Washington Atty: Dow, Lohnes & Albertson.
Key Personnel:
Hoyt Andres . . . gen mgr & stn mgr
Mark Wilder . . . opns dir
James Belton . . . natl sls mgr
Lewis Fryer . . . rgnl sls mgr
Edith Parten . . . mktg dir
Denise Vickers . . . news dir
Craig Young . . . engrg mgr
Ken Thayer . . . chief of engrg

Opelika

see Columbus, GA market

Selma

see Montgomery-Selma, AL market

Tuscaloosa

see Birmingham (Anniston, Tuscaloosa), AL market

Alaska

Anchorage, AK
(DMA 154)

***KAKM**— Analog channel: 7. Analog hrs: 24 162 kw vis, 16.2 kw aur. 780t/808g TL: N61 25 22 W149 52 20 On air date: May 7, 1975. 3877 University Dr., Anchorage, AK, 99508. Phone: (907) 563-7070. Fax: (907) 273-9192. E-mail: questions@kakm.org Web Site: www.kakm.org. Licensee: Alaska Public Telecommunications Inc. Population served: 250,000 Natl. Network: PBS. Washington Atty: Dow, Lohnes & Albertson.
Key Personnel:
Paul Stankovich . . . gen mgr
Will Peterson . . . dev dir & dev mgr

KDMD— Analog channel: 33. Analog hrs: 24 5,000 kw vis, 500 kw aur. 98t TL: N61 09 57 W149 54 01 On air date: February 1990. 1310 E. 66th Ave., Anchorage, AK, 99518-1915. Phone: (907) 562-5363. Fax: (907) 562-5346. E-mail: stationmail@kdmd.tv Web Site: www.kdmd.tv. Licensee: Ketchikan TV LLC. (acq 6-19-2002). Population served: 300,000
Key Personnel:
David Drucker . . . CEO
Bill Vanderpoel . . . pres
Andy Tierney . . . gen mgr & sls VP
Don Nelson . . . chief of engrg

KIMO— Analog channel: 13. Digital channel: 12. Analog hrs: 24 316 kw vis, 31.6 kw aur. ant 781t TL: N61 25 22 W149 52 20 On air date: Oct 31, 1967. 2700 E. Tudor Rd., Anchorage, AK, 99507. Phone: (907) 561-1313. Fax: (907) 561-1377. E-mail: info@aksuperstation.com Web Site: www.aksuperstation.com. Licensee: Smith Media License Holdings LLC. Group owner: Smith Broadcasting Group Inc. (acq 11-8-2004; grpsl). Population served: 250,000 Natl. Network: ABC, CW. Natl. Rep: Continental Television Sales.
Key Personnel:
Sean Bradley . . . VP & gen mgr
Chris Munroe . . . sls dir
Shawn McCalip . . . prom mgr
Terri Bradley . . . progmg dir
Ty Hardt . . . news dir
George Heacock . . . chief of engrg

KTBY— Analog channel: 4. Analog hrs: 24 38.9 kw vis, 7.8 kw aur. 180t TL: N61 13 11 W149 53 24 On air date: Dec 2, 1983. 440 E. Benson Blvd., Anchorage, AK, 99503. Phone: (907) 274-0404. Fax: (907) 264-5180. Licensee: Piedmont Television of Anchorage License LLC. Group owner: Piedmont Television Holdings LLC Population served: 290,000 Natl. Network: Fox. Washington Atty: Cohn & Marks.
Key Personnel:
Sean Bradley . . . gen mgr
Jeff Glaser . . . gen sls mgr
Terri Bradley . . . progmg dir

KTUU-TV— Analog channel: 2. Digital channel: 10. Analog hrs: 24 100 kw vis, 10 kw aur. ant 721t/715g TL: N61 25 22 W149 52 20 On air date: December 1953. Tudor Park, 701 E. Tudor Rd., Suite 220, Delta, AK, 99503. Phone: (907) 762-9202. Fax: (907) 561-0882. Fax: (907) 563-3318. E-mail: ktuu@ktuu.com Web Site: www.ktuu.com. Licensee: Channel 2 Broadcasting Co. Ownership: Residential and Z&L Trust. (acq 3-9-2001). Population served: 208,100 Natl. Network: NBC. News staff: 40; News: 12 hrs wkly.

Key Personnel:

Greg D. Zaser . . . . pres
Al Bramstedt Jr. . . . . gen mgr
Trent McNelly . . . . opns mgr
Andy MacLeod . . . . gen sls mgr
Nancy Johnson. natl sls mgr, mktg dir, prom mgr, progmg mgr & film buyer
Dianna Rowedder . . . . adv dir & adv mgr
John Tracy . . . . news dir
Barry Sowinski . . . . pub affrs dir
Leland Verschueren . . . . chief of engrg

KTVA— Analog channel: 11. Analog hrs: 24 45 kw vis, 5 kw aur. 300t/392g TL: N61 11 33 W149 54 01 On air date: Dec 11, 1953. 1007 W. 32nd Ave., Anchorage, AK, 99503. Phone: (907) 273-3192. Fax: (907) 273-3189. E-mail: 11news@ktva.com Web Site: www.ktva.com. Licensee: Alaska Broadcasting Company Inc. Ownership: MediaNews Group Inc. (acq 5-25-2000; grpsl). Population served: 306,000 Natl. Network: CBS. Rgnl. Rep: Rgnl rep: Art Moore Washington Atty: Wilkinson, Barker, Knauer & Quinn. News staff: 16; News: 7 hrs wkly.

Key Personnel:

Jerry Bever . . . . gen mgr & stn mgr
Bush Houston . . . . opns mgr
Laurie Bruce . . . . gen sls mgr
Cyd Terhune . . . . progmg dir
Staci Chil . . . . news dir
Tom Lambert . . . . chief of engrg

KYES-TV— Analog channel: 5. Digital channel: 6. Analog hrs: 24 5.9 kw vis, 50 kw aur. ant 820t/160g TL: N61 20 10 W149 30 49 On air date: November 1989. 3700 Woodland Dr., Suite 800, Anchorage, AK, 99517. Phone: (907) 248-5937. Fax: (907) 339-3889. Web Site: www.yes.com. Licensee: Fireweed Communications LLC. Ownership: Jeremy Lansman, 51%; Carol Schatz, 49% (acq 12-11-91; $100 & assumption of debt; FTR: 1-6-92) Population served: 254,479 Natl. Network: MyNetworkTV. Washington Atty: Benjamin Perez. News: one hr wkly.

Key Personnel:

Jeremy Lansman . . . . pres & chief of engrg
Carol Schatz . . . . gen mgr & progmg mgr
Roy Nederbrock . . . . chief of opns
Lori Erickson . . . . . . gen sls mgr, natl sls mgr & rgnl sls mgr
Maryann Spinella . . . . prom mgr

Bethel

***KYUK-TV—** Analog channel: 4.4.68 kw vis, 933 w aur. ant 206t/253g TL: N60 47 33 W161 46 22 On air date: August 1973. Pouch 468, Bethel, AK, 99559. Phone: (907) 543-3131. Fax: (907) 543-3130. Web Site: www.kyuk.org. Licensee: Bethel Broadcasting Inc. Population served: 12,000 Natl. Network: PBS, ABC, CBS. Washington Atty: Wilkinson, Barker, Knauer & Quinn. Wire Svc: DAC News staff: 5; News: 3 hrs wkly.

Key Personnel:

Joan Hamilton . . . . chmn
Ron Daugherty . . . . gen mgr
Jose Seibert . . . . progmg dir & engrg dir
Rebroadcasts KUAC-TV Fairbanks 100%.

Fairbanks, AK (DMA 202)

KATN— Analog channel: 2. Digital channel: 18.28.2 kw vis, 5.5 kw aur. ant 200t/151g TL: N64 50 42 W147 42 52 On air date: Mar 1, 1955. 516 2nd Ave., Suite 400, Fairbanks, AK, 99701. Phone: (907) 452-2125. Fax: (907) 456-8225. E-mail: info@aksuperstation.com Web Site: www.aksuperstation.com. Licensee: Smith Media License Holdings LLC. Group owner: Smith Broadcasting Group Inc. (acq 11-8-2004; grpsl). Population served: 84,800 Natl. Network: ABC. News: 6p - 11p wkly.

Key Personnel:

Sean Bradley . . . . VP & gen mgr
Mike Hammer . . . . opns VP
Jeff Glaser . . . . sls VP
Rita Corwin . . . . mktg dir & prom dir
Terri Bradley . . . . progmg dir
Ty Hardt . . . . news dir
Gerilynne Buonocore . . . . pub affrs dir
George Heacock . . . . engrg dir
Rebroadcasts KIMO-TV Anchorage 99%.

KFXF— Analog channel: 7. Analog hrs: 20 7.8 kw vis. ant 879t TL: N64 55 20 W147 42 55 On air date: Feb 27, 1995. 3650 Bradock St., Fairbanks, AK, 99701. Phone: (907) 452-3697. Fax: (907) 456-3428. Web Site: www.TVTV.com. Licensee: Tanana Valley Television Co. Ownership: Bill St. Pierre 60%, Mike Young, Dave Wike. Population served: 80,000 Natl. Network: Fox. Washington Atty: Baker & Hostetler. News staff: 4; News: 10 hr wkly.

Key Personnel:

Christine Fry . . . . opns mgr & progmg dir
John Hoff . . . . gen mgr & gen sls mgr
Darryl Lewis . . . . news dir
Dave Sala . . . . chief of engrg

KJNP-TV— Analog channel: 4. Digital channel: 20. Analog hrs: 24 Digital hrs: 24 18.66 kw vis, 2.8 kw aur. 1,619t/191g TL: N64 52 44 W148 03 10 On air date: Dec 7, 1981. Box 56359, 2501 Mission Rd., North Pole, AK, 99705-1359. Phone: (907) 488-2216. Fax: (907) 488-5246. E-mail: kjnp@mosquitonet.com Web Site: www.mosquitonet.com/~kjnp. Licensee: Evangelistic Alaska Missionary Fellowship. Washington Atty: Fletcher, Heald & Hildreth.

Key Personnel:

Genevieve Nelson . . . . CEO
Yvonne Carriker . . . . pres
Richard T. Olson . . . . VP
Julie Beaver . . . . stn mgr

KTVF— Analog channel: 11. Digital channel: 26.50 kw vis, 5 kw aur. 50t/168g TL: N64 50 36 W147 42 48 On air date: Feb 17, 1955. 3528 International Way, Fairbanks, AK, 99701. Phone: (907) 458-1800. Fax: (907) 458-1820. Web Site: www.webcenter11.com. Licensee: Ackerley Broadcasting Operations LLC. Group owner: Clear Channel Communications Inc. (acq 6-14-2002; grpsl). Population served: 85,000 Natl. Network: NBC. Natl. Rep: Adam Young. Washington Atty: Wilkinson, Barker, Knauer & Quinn. News staff: 6; News: 5 hrs wkly.

Key Personnel:

Bill Wright . . . . gen mgr
Richard Port . . . . stn mgr & opns mgr
Deedee Caciari . . . . gen sls mgr
Celia Vissers . . . . progmg dir
Bob Miller . . . . news dir
William Tanner . . . . chief of engrg

***KUAC-TV—** Analog channel: 9. Digital channel: 24.46.8 kw vis, 8.33 kw aur. ant 500t/151g TL: N64 54 42 W147 46 38 On air date: Dec 22, 1971. Box 755620, Univ. of Alaska-Fairbanks, 312 Tanana Dr., Fairbanks, AK, 99775-5620. Phone: (907) 474-7491. Fax: (907) 474-5064. E-mail: comments@kuac.org Web Site: www.kuac.org. Licensee: University of Alaska. Population served: 70,000 Natl. Network: PBS.

Key Personnel:

Greg Petrowich . . . . CEO & sls dir
Claudia Clark . . . . gen mgr & progmg dir
Gregg Petrowich . . . . gen mgr
Jeremy Cate . . . . opns mgr
Gretchen Gordon . . . . dev dir
Tammy Tragis-McCook . . . . mktg mgr
Joseph Forgue . . . . chief of engrg
Anne Biberman . . . . sls
Keith Martin . . . . engr

Juneau, AK (DMA 207)

KJUD— Analog channel: 8. Digital channel: 11. Analog hrs: 24 Note: ABC is on KJUD(TV) ch 8, CW is on KJUD-DT ch 11. 239 w vis, 47 w aur. ant 1,160t/69g TL: N58 18 06 W134 26 29 On air date: Feb 19, 1956. 175 S. Franklin St., Senate Bldg., Juneau, AK, 99801. Phone: (907) 561-1313 / (907) 586-3145. Fax: (907) 561-1377 / (907) 463-3041. E-mail: info@aksuperstation.com Web Site: www.aksuperstation.com. Licensee: Smith Media License Holdings LLC. Group owner: Smith Broadcasting Group Inc. (acq 11-8-2004; grpsl). Population served: 59,825 Natl. Network: ABC, CW Digital Network: Note: ABC is on KJUD(TV) ch 8, CW is on KJUD-DT ch 11. Washington Atty: Kaye, Scholer, Fierman, Hays & Handler. News staff: 9; News: 5 hrs wkly.

Key Personnel:

John Bradley . . . . gen mgr
Jeff Glaser . . . . gen sls mgr
Terri Bradley . . . . progmg dir
Ty Hardt . . . . news dir
Gerilynne Buonocore . . . . pub affrs dir
George Heacock . . . . chief of engrg

KTNL-TV— Analog channel: 13. Analog hrs: 24 199 w vis, 30 w aur. ant -782t/155g TL: N57 03 27 W135 20 02 (CP: 2.25 kw vis, ant -843t. TL: N57 03 02 W135 20 03) On air date: Sept 1, 1966. 520 Lake St., Sitka, AK, 99835. Phone: (907) 747-5749. Fax: (907) 747-8440. E-mail: stationmail@ktnl.tv Web Site: www.ktnl.tv. Licensee: Ketchikan TV LLC. (acq 6-19-2002). Population served: 52,000 Natl. Network: CBS. Washington Atty: Wilkinson, Barker, Knauer L.L.P.

Key Personnel:

David Drucker . . . . CEO
Bill Vanderpoel . . . . VP
Charlene Nelson . . . . gen mgr
Garrett Leighton . . . . opns mgr
Amanda McMellon . . . . sls dir

***KTOO-TV—** Analog channel: 3. Digital channel: 6. Analog hrs: 24 Digital hrs: 24 2.45 kw vis, 490 w aur. -1,016t/259g TL: N58 18 04 W134 25 21 On air date: Oct 1, 1978. 360 Egan Dr., Juneau, AK, 99801. Phone: (907) 586-1670. Fax: (907) 586-3612. E-mail: ktoo@ktoo.org Web Site: www.ktoo.org. Licensee: Capital Community Broadcasting Inc. Population served: 60,000 Natl. Network: PBS. Washington Atty: Schwartz, Woods & Miller. News staff: one; News: one hr wkly.

Key Personnel:

Bill Legere . . . . pres & gen mgr
Jim Mahan . . . . stn mgr
Cheryl Levitt . . . . dev dir
William Judy . . . . engrg dir
Rebroadcasts KUAC-TV Fairbanks 95%.

Ketchikan

KUBD— Analog channel: 4.960 w vis. 571t TL: N55 20 59 W131 40 12 On air date: Feb 11, 2000. 516 Stedman St., Ketchikan, AK, 99901. Phone: (907) 225-4613. Fax: (907) 247-5365. Licensee: Ketchikan TV LLC. Ownership: David M. Drucker, 100% (acq 6-19-2002).

Arizona

Phoenix (Prescott), AZ (DMA 13)

***KAET—** Analog channel: 8. Digital channel: 29. Analog hrs: 24 316 kw vis, 47.9 kw aur. ant 1,756t/346g TL: N33 20 00 W112 03 49 On air date: Jan 30, 1961. Box 871405, Tempe, AZ, 85287. Stauffer Hall B-Wing, Arizona State Univ., Tempe, AZ 85287. Phone: (480) 965-8888. Fax: (480) 965-1000. Web Site: www.kaet.asu.edu. Licensee: Arizona Board of Regents. Population served: 3,300,000 Natl. Network: PBS. Washington Atty: Covington & Burling. Foreign lang progmg: SP 1 News staff: 7; News: 3 hrs wkly.

Key Personnel:

Greg Giczi . . . . gen mgr
Beth Vershure . . . . stn mgr
John Martinez . . . . opns mgr
Kelly McCullough . . . . . dev dir, dev mgr, mktg dir & mktg mgr
John Menzies . . . . prom mgr
Michael Philipsen . . . . news dir
Joseph Manning . . . . engrg mgr

KASW— Analog channel: 61. Digital channel: 49.2,510 kw vis. ant 1,774t TL: N33 20 01 W112 03 44 On air date: September 1995. c/o TV Stn KTVK, 5555 N. 7th Ave., Phoenix, AZ, 85013. Phone: (602) 207-3333. Fax: (602) 207-3477. Web Site: www.azfamily.com. Licensee: KASW-TV Inc. Group owner: Belo Corp., Broadcast Division (acq 1-24-2000). Natl. Network: CW.

Key Personnel:

Dean Apostalides . . . . gen mgr
Skip Cass . . . . stn mgr
Rick Soltesz . . . . gen sls mgr
Brock Kruzie . . . . natl sls mgr
Scott Rein . . . . rgnl sls mgr
Mark Demopoulos . . . . progmg dir
Mike Stone . . . . chief of engrg

KAZT-TV— Analog channel: 7. Analog hrs: 24 8.79 kw vis, 1.76 kw aur. 2,814t/120g TL: N34 41 15 W112 07 01 On air date: Sept 5, 1982. 4343 E. Camelback Rd., Suite 130, Phoenix, AZ, 85018. 3211 Tower Rd, Prescott, AZ 86305. Phone: (602) 224-0027. Fax: (602) 224-2214. E-mail: rbergamo@kaz.tv Web Site: www.kaz.tv. Licensee: KAZT L.L.C. Ownership: Londen Media Group L.L.C., 99%; and Ron Bergamo, 1% (acq 4-1-2002; $7.336 million). Population served: 4,500,000 Natl. Rep: Petry Television Inc. Washington Atty: Shaw Pittman.

Key Personnel:

Ron Bergamo . . . . gen mgr
Richard Howe . . . . stn mgr
Michael Hagerty . . . . prom dir, progmg dir & progmg mgr

KCFG— Analog channel: 9. Analog hrs: 24 1 kw vis. ant 1,948t TL: N35 14 26 W111 35 48 On air date: 2001. 2616 North Steves Blvd., Flagstaff, AZ, 86004. Phone: (928) 526-5234. Fax: (928) 526-1172. Web Site: www.kcfg.net. Licensee: KM Television of Flagstaff L.L.C.

***KDTP—** Analog channel: 11.1.6 kw vis. ant 177t/174g TL: N34 55 05 W110 08 25 On air date: 2001. Box 612066, Dallas, TX, 75261.

Phone: (602) 207-3939. Web Site: www.daystar.com. Licensee: Community Television Educators Inc.

KFPH-TV— Analog channel: 13. Digital channel: 27. Analog hrs: 24 316 kw vis. ant 1,555t/239g TL: N34 58 05 W111 30 29 On air date: 1991. 2158 N. 4th St., Flagstaff, AZ, 86004. Phone: (928) 527-1300. Fax: (928) 527-1394. Web Site: www.univision.com. Licensee: TeleFutura Partnership of Flagstaff. Group owner: Univision Communications Inc. (acq 10-2-2001; $19.113 million plus assumption of liabilities with KFTU-TV Douglas). Natl. Network: TeleFutura (Spanish). Foreign lang progmg: SP 168

KMOH-TV— Analog channel: 6. Digital channel: 19.100 kw vis, 10 kw aur. ant 1,920t TL: N35 01 57 W114 21 56 On air date: Feb 22, 1988. 950 Flynn Rd., Camarillo, CA, 93012. Phone: (805) 388-0081. Fax: (305) 863-5701. Licensee: Phoenix 6 TV LLC. Group owner: (group owner; (acq 12-8-2004). Population served: 138,000 Foreign lang progmg: SP 168
Satellite of KBEH(TV) Oxnard, CA.

KNAZ-TV— Analog channel: 2. Digital channel: 22. Analog hrs: 24 100 kw vis, 5 kw aur. ant 1,597t/284g TL: N34 58 06 W111 30 28 On air date: May 2, 1970. 2201 N. Vickey St., Flagstaff, AZ, 86004. Box 3360, Flagstaff, AZ 86004. Phone: (928) 526-2232. Fax: (928) 526-8110. E-mail: 2news@knaztv2.com Licensee: Multimedia Holdings Corp. Group owner: Gannett Broadcasting (acq 1997; $6.25 million with KMOH-TV Kingman). Population served: 150,000 Natl. Network: NBC. Washington Atty: Dow, Lohnes & Albertson.
Key Personnel:
Jerome Parra . . . gen mgr
Scott Jones . . . opns mgr & sls dir
Stan Pierce . . . gen sls mgr
Marge Divine . . . progmg dir
Kim Smith . . . news dir
Jon Koger . . . chief of engrg

KNXV-TV— Analog channel: 15. Digital channel: 56. Analog hrs: 24 631 kw vis, 63.1 kw aur. ant 1,710t/282g TL: N33 20 00 W112 03 46 On air date: Sept 9, 1979. 515 N. 44th St., Phoenix, AZ, 85008. Phone: (602) 273-1500. Fax: (602) 685-3000. E-mail: news15@abc15.com Web Site: www.abc15.com. Licensee: Scripps Howard Broadcasting Co. Group owner: (group owner, see Cross-Ownership; (acq 1-9-85; $26.6 million). Population served: 2,877,000 Natl. Network: ABC. Natl. Rep: Eagle Television Sales. Washington Atty: Baker & Hostetler.
Key Personnel:
Janice Todd . . . gen mgr
Ryan Steward . . . opns mgr, engrg mgr & chief of engrg
Kimberly Steele . . . natl sls mgr
Jim Hart . . . prom mgr
Amy Wilson . . . progmg mgr
Bob Sullivan . . . news dir
Colleen Reid . . . pub affrs dir
Will Bruner . . . chief of engrg

***KPAZ-TV—** Analog channel: 21. Digital channel: 20. Digital hrs: 24 1,282 kw vis, 247 kw aur. 2,143t/178g TL: N33 20 03 W112 03 42 On air date: Sept 16, 1967. 3551 E. McDowell Rd., Phoenix, AZ, 85008. Phone: (602) 273-1477. Fax: (602) 267-9427. Licensee: Trinity Broadcasting of Arizona Inc. Group owner: Trinity Broadcasting Network (acq 1977). Washington Atty: Joseph E. Dunne III.
Key Personnel:
Oralena Valero . . . stn mgr
Gary Nichols . . . chief of engrg

KPHO-TV— Analog channel: 5. Digital channel: 17. Analog hrs: 24 Digital hrs: 24 100 kw vis, 10 kw aur. 1,768t/387g TL: N33 20 02 W112 03 40 On air date: Dec 4, 1949. 4016 N. Black Canyon Hwy., Phoenix, AZ, 85017. Phone: (602) 264-1000. Fax: (602) 650-5510. Fax: (602) 650-5545. E-mail: cbs5news@kpho.com Web Site: www.kpho.com. Licensee: Meredith Corp. Group owner: Meredith Broadcasting Group, Meredith Corp., see Cross-Ownership (acq 6-25-52; grpsl; FTR: 6-30-52). Population served: 1,600,000 Natl. Network: CBS. Natl. Rep: Harrington, Righter & Parsons. Washington Atty: Dow Lohnes. Wire Svc: Weather Wire News staff: 60; News: 30.5 hrs wkly.
Key Personnel:
Steve Hammel . . . gen mgr
Mitch Nye . . . gen sls mgr
Seth Parker . . . progmg VP & progmg dir
Tom Bell . . . news dir & engrg dir

KPNX—Mesa, Analog channel: 12. Digital channel: 36. Analog hrs: 24 316 kw vis, 46.8 kw aur. 1,780t/350g TL: N33 20 00 W112 03 48 On air date: Apr 23, 1953. Box 711, Phoenix, AZ, 85004. 1101 N. Central Ave., Phoenix, AZ 85001. Phone: (602) 257-1212. Fax: (602) 261-6135. Fax: (602) 257-6619 (news). E-mail: webmaster@12news.com Licensee: Multimedia Holdings Corp. Group owner: Gannett Broadcasting (acq 6-7-79; grpsl; FTR: 6-11-79). Population served: 2,989,000 Natl. Network: NBC. News staff: 70.
Key Personnel:
John Misner . . . gen mgr
Dan Mayasich . . . gen sls mgr

KPPX— Analog channel: 51.4,875 kw vis, 487 kw aur. 1,749t/354g TL: N33 20 03 W112 03 38 On air date: Feb 15, 1999. 1101 N. Central Ave., Phoenix, AZ, 85004. Phone: (602) 808-0729. Fax: (602) 808-8864. Web Site: www.ionline.tv. Licensee: America 51 L.P. Group owner: Paxson Communications Corp. (acq 1-2-01; $6.6 million for 51%). Natl. Network: i Network. Washington Atty: Skadden, Arps, Slate, Meagher & Flom.

KSAZ-TV— Analog channel: 10. Digital channel: 31. Analog hrs: 24 316 kw vis, 47 kw aur. 1,700t/264g TL: N33 20 03 W112 03 43 (CP: Ant 1,829t) On air date: Oct 24, 1953. 511 W. Adams St., Phoenix, AZ, 85003. Phone: (602) 257-1234. Fax: (602) 262-0177 (602) 262-0181. Fax: (602) 262-0456 (sales). Licensee: KSAZ License Inc. Group owner: Fox Television Stations Inc. (acq 11-96; grpsl). Population served: 4,000,000 Natl. Network: Fox. Natl. Rep: Fox Stations Sales. Washington Atty: Koteen & Naftalin. News staff: 96; News: 38 hrs wkly.
Key Personnel:
Patrick Nevin . . . gen mgr
Paul Austill . . . opns mgr
Mellynda Hartel. . . . natl sls mgr
David Saline . . . progmg dir
Doug Bannard . . . news dir
Jim Kauffman . . . chief of engrg

KTAZ— Analog channel: 39. Digital channel: 39. Analog hrs: 24 1.6 kw vis. ant 177t/174g TL: N34 55 05 W110 08 25 On air date: July 4, 2000. 222 Navajo Blvd., Holbrook, AZ, 85040. Phone: (928) 524-1652. Fax: (928) 524-6459. Licensee: NBC Telemundo License Co. Group owner: Telemundo Group Inc. (acq 9-26-2002; $7.5 million with KPHZ-LP Phoenix and KPSW-LP Phoenix). Natl. Network: Telemundo (Spanish). Washington Atty: Wiley, Rein, Fielding. Foreign lang progmg: SP 168
Key Personnel:
Araceli De Leon. . . . gen mgr
Phillip Williams . . . stn mgr
David Carr. . . . chief of engrg

KTFL— Analog channel: 4.63 kw vis. ant 1,597t/256g TL: N34 58 04 W111 30 30 Box 350, Tupelo, MS, 38802. 1359 Rd. 681, Tupelo, MS 38802. Phone: (662) 842-7620. Fax: (662) 844-7061. Licensee: WTVA Inc. Group owner: (group owner). Washington Atty: Garvey, Schubert & Barer.
Key Personnel:
Mark Ledbetter . . . gen mgr
Larry Harris . . . gen sls mgr
Josh Ward . . . prom dir
Ed Bishop . . . progmg dir
Robert Davidson . . . news dir
Wendell Robbinson . . . chief of engrg

KTVK— Analog channel: 3. Digital channel: 24. Analog hrs: 24 Digital hrs: 24 100 kw vis, 15.1 kw aur. ant 1,670t/231g TL: N33 20 01 W112 03 45 (CP: Ant 1,778t) On air date: Feb 28, 1955. 5555 N. 7th Ave., Phoenix, AZ, 85013. Phone: (602) 207-3333. E-mail: feedback@azfamily.com Web Site: www.azfamily.com. Licensee: KTVK Inc. Group owner: Belo Corp., Broadcast Division (acq 9-3-99; $315 million cash including 50% of Arizona News Channel). Population served: 1,720,000 Natl. Rep: TeleRep. Wire Svc: AP News staff: 100+; News: 48 hrs wkly.
Key Personnel:
Mark Higgins . . . pres & gen mgr
Jamie Aitken . . . stn mgr
Marie McGlynn . . . sls VP
Teri Lane. . . . mktg dir
Sandy Breland . . . news dir
Jim Cole . . . engrg dir

KTVW-TV— Analog channel: 33.2,290 kw vis, 229 kw aur. 1,710t/282g TL: N33 20 00 W112 03 46 (CP: Ant 1,673t) On air date: Sept 2, 1979. 6006 South 30th St., Phoenix, AZ, 85042. Phone: (602) 243-3333. Fax: (602) 276-8658. Licensee: KTVW License Partnership G.P. Group owner: Univision Communications Inc. (acq 5-17-89; $23 million; FTR: 6-5-89). Population served: 500,000 Natl. Network: Univision (Spanish). Foreign lang progmg: SP 168
Key Personnel:
Jose Luis Padilla . . . gen mgr
Carlos Flys. . . . opns VP
Andrew Deschapelles . . . rgnl sls mgr
Laura De La Mata . . . rgnl sls mgr
Javier Ramis . . . mktg dir & prom dir
Virginia Luna . . . progmg dir
Marco Flores . . . news dir
Tom Foy . . . chief of engrg

KUTP— Analog channel: 45. Digital channel: 26. Analog hrs: 24 2,750 kw vis, 275 kw aur. 1,792t/381g TL: N33 20 01 W112 03 32 On air date: Dec 23, 1985. 511 W. Adam St., Phoenix, AZ, 85003. Phone: (602) 257-1234. Fax: (602) 262-0177/ 5123. Web Site: www.kutp.com. Licensee: Fox Television Stations Inc. Group owner: (group owner; (acq 7-31-2001; grpsl). Population served: 1,122,800 Natl. Network: MyNetworkTV. Washington Atty: Wilmer, Cutler & Pickering.
Key Personnel:
Patrick Nevin . . . gen mgr
Jim Kauffman . . . opns dir, opns mgr & chief of engrg
David Saline . . . gen sls mgr & progmg dir
Mellynda Hartel . . . gen sls mgr
Doug Bannard. . . . progmg dir & news dir

Prescott

see Phoenix (Prescott), AZ market

Sierra Vista

see Tucson (Sierra Vista), AZ market

Tucson (Sierra Vista), AZ (DMA 70)

KFTU-TV— Analog channel: 3. Analog hrs: 24 100 kw vis. ant 30t/180g TL: N31 22 08 W109 31 45 On air date: 2001. 1111 G Ave., Douglas, AZ, 85607. Phone: (520) 805-1773. Fax: (520) 805-1768. Licensee: Univision Partnership of Douglas. Group owner: Univision Communications Inc. (acq 10-2-2001; $19.113 million plus assumption of liabilities with KFPH-TV Flagstaff). Foreign lang progmg: SP 168
Key Personnel:
Jose Luis Padilla . . . gen mgr
Carlos Flys . . . opns mgr
Alfonso Romero . . . rgnl sls mgr
Javier Ramis . . . prom dir & adv dir
Salvador Ocano. . . . progmg dir
Marco Flores . . . news dir
Tom Foy . . . chief of engrg

KGUN— Analog channel: 9. Digital channel: 35. Analog hrs: 24 110 kw vis, 21.94 kw aur. ant 3,739t/220g TL: N32 24 53 W110 42 58 On air date: June 3, 1956. 7280 E. Rosewood St., Tucson, AZ, 85710. Phone: (520) 722-5486. Fax: (520) 733-7099. Fax: (520) 733-7070. Web Site: www.kgun9.com. Licensee: Journal Broadcast Corp. Group owner: Emmis Communications Corp. (acq 12-5-2005; grpsl). Population served: 1,050,000 Natl. Network: ABC. Natl. Rep: MMT. Washington Atty: Reed Smith LLP. Wire Svc: NOAA Weather News staff: 47; News: 22 hrs wkly.
Key Personnel:
Kelly Donnell . . . opns dir, opns mgr & prom mgr
Adam Johnston . . . gen sls mgr
Kara Quintela . . . natl sls mgr
Thor Wasbotten . . . news dir
Stephen Somerville . . . chief of engrg

KHRR— Analog channel: 40. Digital channel: 42. Analog hrs: 24 1,550 kw vis, 155 kw aur. ant 2,030t/184g TL: N32 14 55 W111 06 57 On air date: Jan 1, 1985. 5151 E. Broadway, Suite 600, Tucson, AZ, 85711. Phone: (520) 322-6888. Fax: (520) 319-9148. Web Site: www.telemundo.com. Licensee: NBC Telemundo License Co. Group owner: Telemundo Group Inc. (acq 1-1-2003; $20 million with KDRX-CA Phoenix). Population served: 333,650 Natl. Network: Telemundo (Spanish). Foreign lang progmg: SP 168 News staff: 7; News: 5 hrs wkly.
Key Personnel:
Araceli De Leon. . . . gen mgr
Lupita Celaya . . . opns dir & progmg dir
Martha Muniz . . . prom dir & prom mgr
Sergio Pedroza. . . . news dir & pub affrs dir

KMSB-TV— Analog channel: 11. Digital channel: 25. Analog hrs: 8:30 AM-5:30 PM 316 kw vis, 31.6 kw aur. ant 1,662t/200g TL: N31 42 18 W110 55 26 On air date: Feb 1, 1967. 1855 N. 6th Ave., Tucson, AZ, 85705-5061. Phone: (520) 770-1123. Fax: (520) 629-7185. Web Site: www.kmsb.com. Licensee: Belo TV Inc. Group owner: Belo Corp., Broadcast Division (acq 2-28-97; grpsl). Natl. Network: Fox. Natl. Rep: TeleRep. Washington Atty: Wiley, Rein & Fielding.
Key Personnel:
Tod A. Smith . . . VP & gen mgr
Lou Medran . . . opns dir
Claudia Montgomery . . . gen sls mgr
Jim Ferreira . . . rgnl sls mgr
Betsy Green . . . mktg dir
Harry West . . . prom dir, progmg dir & film buyer
Bob Lee . . . pub affrs dir
Roy Mitchell . . . chief of engrg

KOLD-TV— Analog channel: 13. Digital channel: 32. Analog hrs: 24 Digital hrs: 24 302 kw vis, 3 kw aur. ant 2,040t/187g TL: N32 14 56 W110 06 58 On air date: Jan 13, 1953. 7831 N. Business Park Dr.,

Tucson, AZ, 85743. Phone: (520) 744-1313. Fax: (520) 744-5233. Web Site: www.kold.com. Licensee: KOLD License Subsidiary LLC. Group owner: Raycom Media Inc. (acq 9-12-96). Population served: 1,100,000 Natl. Network: CBS. Natl. Rep: Harrington, Righter & Parsons. Washington Atty: Covington & Burling. Wire Svc: NWS (National Weather Service) Wire Svc: AP News: 27 hrs wkly.

Key Personnel:

Jim Arnold . . . VP & gen mgr
Bob Gaff . . . opns mgr
Adam Weyne . . . gen sls mgr
Bob Duffy . . . rgnl sls mgr
Lec Coble . . . mktg dir
Michelle Germano . . . news dir
Stewart Roman . . . chief of engrg

KTTU-TV— Analog channel: 18. Digital channel: 19. Analog hrs: 8:30 AM-5:30 PM 2510 kw vis, 251 kw aur. ant 1,970t/200g TL: N32 14 55 W111 06 57 On air date: Dec 31, 1984. 1855 N. 6th Ave., Tucson, AZ, 85705. Phone: (520) 624-0180. Fax: (520) 629-7185. Web Site: www.kttu.com. Licensee: KTTU-TV Inc. Group owner: Belo Corp., Broadcast Division (acq 2-28-2002; $18 million). Population served: 750,000 Natl. Network: MyNetworkTV. Natl. Rep: TeleRep. Washington Atty: Hogan & Hartson. Foreign lang progmg: SP 1

Key Personnel:

Lou Medran . . . pres & opns mgr
Jim Watson . . . gen mgr & natl sls mgr
Tod A. Smith . . . VP & gen mgr
Michael Hornfeck . . . rgnl sls mgr
Bob Richardson . . . progmg dir & news dir
Roy Mitchell . . . chief of engrg

***KUAS-TV**— Analog channel: 27.151 kw vis, 15.1 kw aur. 570t/15g TL: N32 12 53 W111 00 21 (CP: 30.2 kw vis) On air date: January 1986. Box 210067, University of Arizona, Carthage, AZ, 85721-0067. Phone: (520) 621-5828. Fax: (520) 621-4122 (news). Web Site: www.kuat.org. Licensee: Arizona Board of Regents, University of Arizona. Natl. Network: PBS. Washington Atty: Dow, Lohnes & Albertson. Foreign lang progmg: SP 1

Key Personnel:

Jack Parris . . . gen mgr
Rudy Casillas . . . progmg dir & progmg mgr
Rebecca Kunsberg . . . film buyer
Hector Gonzalez . . . news dir
John Anderson . . . chief of engrg

***KUAT-TV**— Analog channel: 6. Digital channel: 30. Analog hrs: 6 AM-1 AM Digital hrs: 24 35.5 kw vis, 3.5 kw aur. 3,630t/196g TL: N32 24 55 W110 42 54 On air date: Mar 8, 1959. Box 210067, University of Arizona, Tucson, AZ, 85721-0067. Phone: (520) 621-5828. Fax: (520) 621-4122 (news). Web Site: www.kuat.org. Licensee: Arizona Board of Regents, University of Arizona. Population served: 800,000 Natl. Network: PBS. Washington Atty: Dow, Lohnes & Albertson. Foreign lang progmg: SP 7

Key Personnel:

Jack Gibson . . . gen mgr
Michael Serres . . . prom mgr
Rudy Casillas . . . progmg dir & film buyer
Peter Michaels . . . news dir
David Ross . . . chief of engrg

KUVE-TV— Analog channel: 46.1,679 kw vis. ant 3,592t/174g TL: N32 24 54 W110 42 56 On air date: 2002. 2301 N. Forbes Blvd., Suite 103, Tucson, AZ, 85745. Phone: (520) 204-1245. Fax: (520) 204-1247. Licensee: Univision Television Group Inc. Group owner: Univision Communications Inc. (acq 9-24-2003; $12.3 million). Natl. Network: Univision (Spanish).

KVOA— Analog channel: 4. Digital channel: 23. Analog hrs: 24 Digital hrs: 24 35 kw vis, 18 kw aur. 3,610t/223g TL: N32 12 53 W111 00 20 On air date: Sept 15, 1953. Box 5188, Tucson, AZ, 85703-0188. 209 W. Elm St., Tucson, AZ 85705-6538. Phone: (520) 792-2270. Fax: (520) 620-1309. Web Site: www.kvoa.com. Licensee: KVOA Communications Inc. Group owner: Cordillera Communications Inc. (acq 12-31-93; $13.25 million; FTR: 11-15-93). Population served: 1,061,000 Natl. Network: NBC. Natl. Rep: Millennium Sales & Marketing. Washington Atty: Dow, Lohnes & Albertson. News: 22 hrs wkly.

Key Personnel:

Gary R. Nielsen . . . pres & gen mgr
Dave Kerrigan . . . opns mgr
Yvette Perez . . . mktg dir
Kathleen Choal . . . news dir

KWBA— Analog channel: 58. Digital channel: 44.5,000 kw vis, 500 kw aur. ant 1,086t TL: N31 45 33 W110 48 02 On air date: Jan 1, 1999. 3055 N. Campbell Ave., Suite 113, Tucson, AZ, 85719. Phone: (520) 889-5800. Fax: (520) 889-5855. E-mail: soundoff@kwba.com Web Site: www.kwba.com. Licensee: Tucson Communications L.L.C. Group owner: Cascade Broadcasting Group L.L.C. (acq 12-20-01; grpsl). Natl. Network: CW. Washington Atty: Shaw Pittman.

Key Personnel:

Greg Kunz . . . CEO
Carol La Fever . . . COO
Tom Hettle . . . CFO
Andrew Stewart . . . VP & gen mgr
Doug McClure . . . VP & sls VP
Jackie Anderson . . . VP & natl sls mgr
Jay Clifford . . . rgnl sls mgr
Gene Steinberg . . . mktg VP & progmg VP
Ken Cummings . . . prom dir & progmg mgr
Alicia Knighton . . . prom mgr
Mac Powas . . . chief of engrg

Yuma, AZ-El Centro, CA

(DMA 167)

KAJB— Analog channel: 54.5,000 kw vis. ant 1,404t/308g TL: N33 03 02 W114 49 38 On air date: 2002. 1803 N. Imperial Ave., El Centro, CA, 92243. Phone: (760) 482-7777. Fax: (760) 482-0099. Licensee: Calipatria Broadcasting Com. L.L.C. Ownership: Kenneth D. Pollin. (acq 12-16-97; $30,000). Foreign lang progmg: SP 168

Key Personnel:

Eric Chavez . . . gen mgr
Albert Valdez . . . stn mgr

KECY-TV— Analog channel: 9. Digital channel: 48. Analog hrs: 24 Note: Fox is on KECY-TV ch 9, ABC is on KECY-DT ch 48. 316 kw vis, 31.6 kw aur. ant 1,720t/460g TL: N33 03 19 W114 49 39 On air date: Dec 11, 1968. 1965 S. 4th Ave., Suite B, Yuma, AZ, 85364. Phone: (928) 539-9990. Fax: (928) 343-0218. Licensee: Pacific Media Corp. Ownership: Robinson O. Everett, 50%; and Estate of Kathrine Everett, 50% (acq 5-97). Population served: 250,000 Natl. Network: Fox, ABC Digital Network: Note: Fox is on KECY-TV ch 9, ABC is on KECY-DT ch 48. Natl. Rep: Millennium Sales & Marketing. Washington Atty: Baraff, Koerner, Olender & Hochberg.

Key Personnel:

Deborah Weekes . . . gen mgr
Darin Coragata . . . gen sls mgr
Deborah Weeks . . . natl sls mgr
Jesus Corona . . . prom mgr
Adriana Sanchez . . . progmg mgr

KSWT— Analog channel: 13. Digital channel: 16. Analog hrs: 5:45 AM-3 AM Note: CBS is on KSWT(TV) ch 13, CW is on KSWT-DT ch 16. 316 kw vis, 31.6 kw aur. 1,700t/203g TL: N33 03 17 W114 49 34 On air date: Dec 1, 1963. 1301 S. 3rd Ave., Yuma, AZ, 85364. Phone: (928) 782-5113. Fax: (928) 783-0866. E-mail: kswt@adelphia.net Web Site: www.kswt.com. Licensee: Pappas Arizona License LLC. Group owner: Pappas Telecasting Companies (acq 9-8-2000; $5.375 million). Population served: 303,000 Natl. Network: CBS, CW Digital Network: Note: CBS is on KSWT(TV) ch 13, CW is on KSWT-DT ch 16. Washington Atty: Paul, Hastings, Janofsky & Walker. News staff: 8; News: 5 hrs wkly.

KVYE— Analog channel: 7. Analog hrs: 24 316 kw vis. 895t TL: N33 03 21 W115 49 44 On air date: July 1996. 1803 N. Imperial Ave., El Centro, CA, 92243. Phone: (760) 482-7777. Fax: (760) 482-0099. Web Site: www.entravision.com. Licensee: Entravision Holdings L.L.C. Group owner: Entravision Communications Co. L.L.C. (acq 2-19-98; $500,000 for CP). Population served: 147,000 Natl. Network: Univision (Spanish). Washington Atty: Thompson, Hine & Flory L. Foreign lang progmg: SP 24 News staff: 8; News: 5 hrs wkly.

Key Personnel:

Walter Ulloa . . . CEO
Philip Wilkinson . . . CFO
Albert Valdez . . . gen mgr & opns mgr
Eric Chavez . . . gen mgr

KYMA— Analog channel: 11. Analog hrs: 24 316 kw vis, 31.6 kw aur. 518t/1,617g TL: N33 03 10 W114 49 40 On air date: January 1988. 1385 S. Pacific Ave., Yuma, AZ, 85365-1725. Phone: (928) 782-1111. Fax: (928) 782-5401. Web Site: www.kyma.com. Licensee: Yuma Broadcasting Co. Group owner: Sunbelt Communications Co. (acq 6-6-89; $60,000; FTR: 6-26-89). Population served: 86,900 Natl. Network: NBC. Washington Atty: Dow, Lohnes & Albertson. News staff: 24; News: 12 hrs wkly.

Key Personnel:

Paul Heebink . . . gen mgr
Barbara Monroy . . . progmg dir
Luis Cruz . . . news dir
Robbie Decorse . . . chief of engrg

Arkansas

El Dorado

see Monroe, LA-El Dorado, AR market

Eureka Springs

see Springfield, MO market

Fayetteville

see Ft. Smith-Fayetteville-Springdale-Rogers, AR market

Ft. Smith-Fayetteville -Springdale-Rogers, AR

(DMA 102)

***KAFT**— Analog channel: 13.316 kw vis, 31.6 kw aur. 1,660t/1,138g TL: N35 48 53 W94 01 41 On air date: Sept 18, 1976. Box 1250, Conway, AR, 72033. 350 S. Donaghey, Conway, AR 72034. Phone: (501) 450-1727. Phone: (501) 682-2386. Fax: (501) 682-4122. E-mail: info@aetn.org Web Site: www.aetn.org. Licensee: Arkansas Educational Television Commission. Population served: 657,000 Natl. Network: PBS. Rgnl. Network: SECA. Washington Atty: Dow, Lohnes & Albertson.

Key Personnel:

Allen Weatherly . . . gen mgr
Tony Brooks . . . stn mgr
Robert Bland . . . opns dir
Mona Dixon . . . dev dir

KFSM-TV— Analog channel: 5. Digital channel: 18. Analog hrs: 24 Digital hrs: 24 100 kw vis, 12.7 kw aur. 1,086t/1,173g TL: N35 30 43 W94 21 38 On air date: Dec 3, 1956. Box 369, 318 N. 13th St., Fort Smith, AR, 72902. Phone: (479) 783-3131. Fax: (479) 783-3295. Web Site: www.5newsonline.com. Licensee: Local TV Arkansas License LLC. Group owner: The New York Times Co. (acq 5-7-2007; grpsl). Population served: 883,500 Natl. Network: CBS. Washington Atty: Covington & Burling. News staff: 36; News: 28 hrs wkly.

Key Personnel:

Debby Etzkorn . . . pres & progmg dir
Robert L. Lawrence . . . pres
Van Comer . . . gen mgr
Mark LaCrue . . . gen sls mgr
Rose Smith . . . prom mgr
Dale Cox . . . news dir
Larry Duncan . . . chief of engrg

KFTA-TV— Analog channel: 24. Digital channel: 27. Analog hrs: 24 2,510 kw vis, 251 kw aur. ant 1,040t/499g TL: N35 42 37 W94 08 15 On air date: Nov 12, 1978. 15 South Block Ave., Suite 101, Fayetteville, AR, 72701. Phone: (479) 571-5100. Fax: (479) 571-8914. E-mail: news@knwa.com Web Site: www.myfox24.com. Licensee: Nexstar Broadcasting Inc. (acq 12-21-2004; $10 million with KNWA-TV Rogers). Population served: 690,000 Natl. Network: Fox. Natl. Rep: TeleRep. Washington Atty: Holland & Knight. News staff: 45; News: 20.5 hrs wkly.

Key Personnel:

Blake Russell . . . VP & gen mgr
Cheryl Gwym . . . opns mgr

KHBS— Analog channel: 40.3,160 kw vis, 316 kw aur. 2,000t/500g TL: N35 04 16 W94 40 46 On air date: July 28, 1971. 2415 N. Albert Pike, Fort Smith, AR, 72904-5698. Phone: (479) 783-4040. Fax: (479) 785-5375. E-mail: comments@thehometownchannel.com Web Site: www.thehometownchannel.com. Licensee: KHBS Hearst-Argyle Television Inc. Group owner: Hearst-Argyle Television Inc. (acq 7-16-97; grpsl). Population served: 153,700 Natl. Network: ABC. Washington Atty: Wiley, Rein & Fielding. News staff: 30; News: 15 hrs wkly.

KHOG-TV— Analog channel: 29. Analog hrs: 24 (Su-F) 1410 kw vis, 150 kw aur. 890t/556g TL: N36 00 57 W94 04 59 On air date: December 1977. 2415 N. Albert Pike, Fort Smith, AR, 72904. Phone: (479) 783-4040 / (479) 521-1010. Fax: (479) 785-5375 / (479) 479-9124. E-mail: comments@thehometownchannel.com Web Site: www.thehometownchannel.com. Licensee: KHBS Hearst-Argyle Television Inc. Group owner: Hearst-Argyle Television Inc. (acq 7-16-97; grpsl). Population served: 250,000 Natl. Network: ABC. Washington Atty: Wiley, Rein & Fielding.

KNWA-TV— Analog channel: 51. Digital channel: 50. Analog hrs: 22 182 kw vis. ant 469t/184g TL: N36 12 15 W94 06 05 (CP: 151 kw vis, ant 875t/478g. TL: N36 24 47.7 W93 57 16.7) On air date: Aug 23, 1989. 15 South Block St., Suite 101, Fayetteville, AR, 72701. Phone: (479) 571-5100. Fax: (479) 571-8914. Web Site: www.knwa.com. Licensee: Nexstar Broadcasting Inc. (acq 12-21-2004; $10 million with KFTA-TV Fort Smith). Population served: 107,830 Natl. Network: NBC. Natl. Rep: Blair Television. Washington Atty: Drinker, Biddle & Reath. News staff: 27; News: 17 hrs wkly.

Key Personnel:
Blake Russell . . . . VP & gen mgr
Mike Vaughn. . . . . gen sls mgr
Chad Beckham. . . . . natl sls mgr
Rob Heverling . . . . news dir
Lisa Kelsey . . . . sls
Satellite of KFTA-TV Fort Smith.

KWOG— Analog channel: 57. Digital channel: 39. Analog hrs: 24 Digital hrs: 24 171.68 kw vis. ant 384t TL: N36 11 07 W94 17 49 On air date: Dec 11, 1995. Box 612066, Dallas, TX, 75261-2066. Phone: (817) 571-1229. Fax: (817) 571-8962. Web Site: ww2.daystar.com. Licensee: Word of God Fellowship Inc. Ownership: Corrine Lamb, 20%; Jimmie F. Lamb, 20%; John T. Calender, 20%; Joni T. Lamb, 20%; and Marcus D. Lamb, 20% (acq 7-7-2006; $1.5 million).

Key Personnel:
Harvey Rogers . . . . chief of engrg
Marcus Lamb . . . . pres & chief of engrg

Harrison

see Springfield, MO market

Jonesboro, AR (DMA 180)

KAIT— Analog channel: 8. Digital channel: 9. Analog hrs: 24 316 kw vis, 47.9 kw aur. ant 1,750t/1,799g TL: N35 53 17 W90 56 09 On air date: July 15, 1963. Box 790, Jonesboro, AR, 72403-0790. 472 Country Rd. 766, Jonesboro, AZ 72401. Phone: (870) 931-8888. Fax: (870) 933-8058 (news). Fax: (870) 931-1371(sales). Web Site: www.kait8.com. Licensee: KAIT License Subsidiary LLC. Group owner: Liberty Corp. (acq 1-13-2006; grpsl). Population served: 704,000 Natl. Network: ABC. Natl. Rep: Harrington, Righter & Parsons. Washington Atty: Covington & Burling. Wire Svc: Weather Wire Wire Svc: AP News staff: 32; News: 17 hrs wkly.

Key Personnel:
Tim Ingram . . . . VP & gen mgr
Ronnie Weston . . . . opns dir
Stephanie Duckworth . . . . sls dir & gen sls mgr
Ralph Caudill . . . . natl sls mgr
Jeremy Shirley . . . . mktg dir & pub affrs dir
Randy Parrott. . . . . news dir
Gerald Erickson. . . . . chief of engrg

***KTEJ**— Analog channel: 19.1,230 kw vis, 123 kw aur. 1,020t/969g TL: N35 54 14 W90 46 14 On air date: May 1, 1976. Box 1250, Conway, AR, 72033. 350 S. Donaghey, Conway, AR 72034. Phone: (501) 450-1727. Phone: (501) 682-2386. Fax: (501) 682-4122. Web Site: www.aetn.org. Licensee: Arkansas Educational Television Commission. Population served: 505,000 Natl. Network: PBS. Rgnl. Network: SECA. Washington Atty: Dow, Lohnes & Albertson.

Key Personnel:
Allen Weatherly . . . . gen mgr
Tony Brooks . . . . stn mgr
Robert Bland . . . . opns dir
Mona Dixon . . . . dev dir

KVTJ— Analog channel: 48. Analog hrs: 24 1,000 kw vis. 1,023t/1,791g TL: N35 53 17 W90 56 09 On air date: June 6, 1998. 701 Napa Valley Dr., Little Rock, AR, 72211. Phone: (501) 223-2525. Fax: (501) 221-3837. E-mail: jim.grant@vtntv.com Web Site: www.vtntv.com. Licensee: Agape Church Inc. Group owner: The Victory Television Network (acq 1995).

Little Rock-Pine Bluff, AR (DMA 57)

KARK-TV— Analog channel: 4. Digital channel: 32. Analog hrs: 24 100 kw vis, 20 kw aur. 1,650t/1,175g TL: N34 47 57 W92 29 59 On air date: Apr 15, 1954. 1401 W. Capitol Ave., Suite 104, Little Rock, AR, 72201. Phone: (501) 340-4444. Fax: (501) 376-1852. Web Site: www.kark.com. Licensee: Nexstar Broadcasting Inc. Group owner: Nexstar Broadcasting Group Inc. (acq 8-1-2003; $40 million with WDHN(TV) Dothan, AL). Population served: 1,177,000 Natl. Network: NBC. Natl. Rep: Petry Television Inc. Wire Svc: Photofax News staff: 40; News: 22 hrs wkly.

Key Personnel:
Rich Rogala . . . . VP & gen mgr
Craig Castrellon . . . . gen sls mgr
Cindy Rochelle . . . . natl sls mgr
Ed Tudor . . . . prom dir
Mary Mobbs . . . . progmg dir
Bill Addington . . . . chief of engrg

KASN— Analog channel: 38. Digital channel: 39.5,000 kw vis, 500 kw aur. ant 2,008t/1,910g TL: N34 26 31 W92 13 03 On air date: June 17, 1986. 10800 Colonel Glenn, Little Rock, AR, 72204. Phone: (501) 225-0038 / 0016. Fax: (501) 225-0428. Web Site: www.upn38tv.com. Licensee: Clear Channel Broadcasting Licenses Inc. Group owner: (group owner; (acq 12-24-91; $14,299,652; FTR: 2-3-92). Population served: 481,000 Natl. Network: CW.

Key Personnel:
Chuck Spohn . . . . gen mgr
Vickie McRae . . . . gen sls mgr
Holly Rose . . . . rgnl sls mgr
Jim Hays. . . . . mktg dir
Miranda Morris . . . . progmg dir
Michael Fabac . . . . news dir
Alan Finne. chief of engrg

KATV— Analog channel: 7.316 kw vis, 36.5 kw aur. 2,272t/2,000g TL: N34 28 23 W92 12 11 On air date: Dec 18, 1953. Box 77, Little Rock, AR, 72203. Phone: (501) 324-7777. Fax: (501) 324-7899. Web Site: www.katv.com. Licensee: KATV L.L.C. Group owner: Allbritton Communications Co. (acq 2-14-83; grpsl). Population served: 169,700 Natl. Network: ABC. Washington Atty: Hogan & Hartson.

Key Personnel:
Dale Nicholson . . . . pres & gen mgr
Mark Rose . . . . gen sls mgr
Richard Farrester . . . . progmg dir
Randy Dixon . . . . news dir
Fred Anderson . . . . chief of engrg

***KEMV**— Analog channel: 6.100 kw vis, 10 kw aur. 1,390t/995g TL: N35 48 47 W92 17 24 On air date: Nov 11, 1980. Box 1250, Conway, AR, 72033. 350 S. Donaghey, Conway, AR 72034. Phone: (501) 450-1727. Phone: (501) 682-2386. Fax: (501) 682-4122. Web Site: www.aetn.org. Licensee: Arkansas Educational Television Commission. Population served: 655,000 Natl. Network: PBS. Rgnl. Network: SECA. Washington Atty: Dow, Lohnes & Albertson.

Key Personnel:
Allen Weatherly . . . . gen mgr
Tony Brooks . . . . stn mgr
Robert Bland . . . . opns dir
Mona Dixon . . . . dev dir & dev mgr

***KETG**— Analog channel: 9.316 kw vis, 31.6 kw aur. 1,070t/1,110g TL: N33 54 26 W93 06 46 On air date: Oct 2, 1976. Box 1250, Conway, AR, 72033. 350 S. Donaghey St., Conway, AR 72034. Phone: (501) 450-1727. Phone: (501) 682-2386. Fax: (501) 682-4122. Web Site: www.aetn.org. Licensee: Arkansas Educational Television Commission. Population served: 379,000 Natl. Network: PBS. Rgnl. Network: SECA. Washington Atty: Dow, Lohnes & Albertson.

Key Personnel:
Allen Weatherly . . . . gen mgr
Tony Brooks . . . . stn mgr
Robert Bland . . . . opns dir
Mona Dixon . . . . dev dir

***KETS**— Analog channel: 2. Digital channel: 5.100 kw vis, 10 kw aur. 1,780t/2,000g TL: N34 28 23 W92 12 11 On air date: Dec 4, 1966. Box 1250, Conway, AR, 72033. 350 S. Donaghey, Conway, AR 72034. Phone: (501) 450-1727. Phone: (501) 682-2386. Fax: (501) 682-4122. Web Site: www.aetn.org. Licensee: Arkansas Eucational Television Commission. Population served: 1,523,000 Natl. Network: PBS. Rgnl. Network: SECA. Washington Atty: Dow, Lohnes & Albertson.

Key Personnel:
Allen Weatherly . . . . gen mgr
Tony Brooks . . . . stn mgr
Robert Bland . . . . opns dir
Mona Dixon . . . . dev dir

***KKAP**— Analog channel: 36. Analog hrs: 24 2,570 kw vis. ant 1,135t TL: N34 47 56 W92 29 44 On air date: 2001. 3901 Hwy 121, Bedford, TX, 76034. Phone: (817) 571-1229. Fax: (817) 571-7458. Web Site: www.daystart.com. Licensee: Educational Broadcasting Corp. Ownership: Dr. Tracy Harris, 14%; Eric S. Erickson, 14%; Gregory Fess, 14%; Joni T. Lamb, 14%; Larry Morton, 14%; Marcus Lamb, 14%; and Max Hooper, 14% (acq 7-6-2001; $1 million).

***KLEP**— Analog channel: 17.14.9 kw vis, 1.49 kw aur. 530t/499g TL: N35 43 25 W91 26 40 On air date: Jan 1, 1985. 1502 N. Hill St., Newark, AR, 72562. Phone: (870) 799-8969. Phone: (870) 799-8691. Fax: (870) 799-8647. Licensee: Newark Public School System.

KLRT— Analog channel: 16. Analog hrs: 24 5,000 kw vis, 500 kw aur. 1,772t/1,266g TL: N34 47 57 W92 29 52 On air date: June 26, 1983. 10800 Colonel Glenn, Little Rock, AR, 72204. Phone: (501) 225-0016. Fax: (501) 225-0428. Web Site: www.fox16.com. Licensee: Clear Channel Radio Licenses Inc. Group owner: Clear Channel Communications Inc. (acq 6-19-91; $6.6 million; FTR: 7-8-91). Natl. Network: Fox. Washington Atty: Crowell & Moring.

Key Personnel:
Chuck Spohn . . . . gen mgr
Vicki McRae . . . . gen sls mgr
Jim Hays. . . . . mktg dir
Logan Wilcoxson . . . . prom dir
Miranda Morris . . . . progmg dir
Michael Fabac . . . . news dir & pub affrs dir
Alan Finne. chief of engrg

KTHV— Analog channel: 11. Analog hrs: 24 Digital hrs: 24 316 kw vis, 38 kw aur. 1,709t TL: N34 47 57 W92 29 59 On air date: Nov 27, 1955. Box 269, Little Rock, AR, 72203. 720 Izard St., Little Rock, AR 72201. Phone: (501) 376-1111. Fax: (501) 376-3324. E-mail: 11listens@todaysthu.com Web Site: www.todaysthv.com. Licensee: Arkansas Television Co. Group owner: Gannett Broadcasting (acq 11-30-94; $27 million; FTR: 1-16-95). Population served: 152,483 Natl. Network: CBS. Natl. Rep: Blair Television. Washington Atty: Wiley, Rein & Fielding. News staff: 55; News: 20 hrs wkly.

Key Personnel:
Larry Audas. . . . . chmn, pres & gen mgr
Alison Fletcher . . . . . opns mgr, chief of opns & chief of engrg
Leslie Heizman. . . . . gen sls mgr
Chad Kelley . . . . natl sls mgr
Joanne Canelli . . . . rgnl sls mgr
David Craft . . . . mktg VP, mktg dir, prom VP & prom dir
Bobbie Rawlins . . . . progmg dir
Mark Raines . . . . news dir
Theba Lolley . . . . pub affrs dir

KVTH— Analog channel: 26. Analog hrs: 24 245 kw vis, 24.5 kw aur. 941t TL: N32 22 20 W93 02 47 (CP: 5,000 kw vis, ant 902t. TL: N34 22 17 W93 02 16) On air date: 1991. 701 Napa Valley Dr., Little Rock, AR, 72211. Phone: (501) 223-2525. Fax: (501) 221-3837. E-mail: jimgrant@vtntv.com Web Site: www.vtntv.com. Licensee: Agape Church. Group owner: The Victory Television Network (acq 1994).
Satellite of KVTN(TV) Pine Bluff.

KVTN— Analog channel: 25. Digital channel: 24. Analog hrs: 24 Digital hrs: 24 4,370 kw vis, 43.7 kw aur. ant 594t/594g TL: N34 31 55 W92 02 41 On air date: Dec 1, 1988. 701 Napa Valley Dr., Little Rock, AR, 72211. Phone: (501) 223-2525. Fax: (501) 221-3837. E-mail: jim.grant@vtntv.com Web Site: www.vtntv.com. Licensee: Agape Church Inc. Group owner: The Victory Television Network (acq 6-87; $41,000; FTR: 5-1-87). Washington Atty: John Fiorini.

Key Personnel:
Jim Grant . . . . gen mgr
Andrea Qualls . . . . prom dir
Kim Worden . . . . progmg dir
Ron Brown . . . . engrg dir

KWBF— Analog channel: 42. Digital channel: 44. Analog hrs: 24 5,000 kw vis. ant 512t TL: N34 52 28 W92 00 35 On air date: Dec 1, 1997. 1 Shackleford Dr., Suite 400, Little Rock, AR, 72211. Phone: (501) 219-2400. Fax: (501) 219-1210. Licensee: River City Broadcasting Inc. Group owner: Equity Broadcasting Corp. (acq 7-28-00; $7.5 million). Natl. Network: MyNetworkTV.

Key Personnel:
Angie Hughes . . . . gen mgr & opns mgr
Steve Soldinger . . . . gen mgr
Nathan Stamp . . . . progmg dir
Doug Krile. . . . . news dir & pub affrs dir
Paul Brandenburg. . . . . chief of engrg

KYPX— Analog channel: 49. Digital channel: 49.3,020 kw vis. ant 574t TL: N33 16 19 W92 42 12 On air date: 1997. 1 Shackelford Dr., Little Rock, AR, 72211. Phone: (501) 219-2400. Fax: (501) 604-8004. Licensee: Arkansas 49 Inc. Group owner: Equity Broadcasting Corp. (acq 9-15-99).

Pine Bluff

see Little Rock-Pine Bluff, AR market

Rogers

see Ft. Smith-Fayetteville-Springdale-Rogers, AR market

Springdale

see Ft. Smith-Fayetteville-Springdale-Rogers, AR market

California

Bakersfield, CA
(DMA 126)

KBAK-TV— Analog channel: 29. Digital channel: 33. Analog hrs: 24 Digital hrs: 24 1,700 kw vis, 340 kw aur. 3,730t/230g TL: N35 27 11 W118 35 25 On air date: Aug 23, 1953. Box 2929, Bakersfield, CA, 93303. 1901 Westwind Dr., Bakersfield, CA 93301. Phone: (661) 327-7955. Fax: (661) 327-5603. Web Site: eyeoutforyou.com. Licensee: Westwind Communications LLC. (acq 12-7-95). Population served: 566,000 Natl. Network: CBS. Natl. Rep: Continental Television Sales. Washington Atty: Brooks, Pierce, McLendon, Humphrey & Leonard. Wire Svc: AP News staff: 40; News: 30.5 hrs wkly.
Key Personnel:
Peter Desnoes . . . chmn
Pete Capra . . . opns mgr & chief of engrg
Cindi Dias. . . . gen sls mgr
Tracy Peoples. . . . prom dir & engrg dir
Wayne Lansche. . . . CEO, pres, gen mgr & prom mgr
Nancy Clarke. . . . progmg dir & progmg mgr
Meaghan St.Pierre. . . . news dir

KERO-TV— Analog channel: 23. Analog hrs: 24 1,760 kw vis, 64.6 kw aur. 3,700t/183g TL: N35 27 14 W118 35 37 On air date: Sept 26, 1953. 321 21st St., Bakersfield, CA, 93301. Phone: (661) 637-2323. Fax: (661) 322-1701. Web Site: www.thebakersfieldchannel.com. Licensee: McGraw-Hill Broadcasting Co. Group owner: (group owner; (acq 3-8-72; grpsl; FTR: 3-13-72). Population served: 2,034,000 Natl. Network: ABC. Natl. Rep: Harrington, Righter & Parsons. Washington Atty: Holland & Knight.
Key Personnel:
Craig Jahelka . . . VP, gen mgr, progmg VP & progmg dir
Steve McEvoy . . . gen sls mgr
Steve Taylor. . . . prom mgr
Todd Karli . . . news dir
Tom Wimberly. . . . engrg mgr & chief of engrg

KGET-TV— Analog channel: 17. Digital channel: 25.5,000 kw vis, 500 kw aur. ant 1,400t/288g TL: N35 26 20 W118 44 23 On air date: Nov 8, 1959. 2120 L St., Bakersfield, CA, 93301. Phone: (661) 283-1700. Fax: (661) 283-1794. Web Site: www.kget.com. Licensee: Ackerley Broadcasting Operations LLC. Group owner: Clear Channel Communications Inc. (acq 6-14-2002; grpsl). Population served: 523,000 Natl. Network: NBC. News staff: 33; News: 27 hrs wkly.
Key Personnel:
Teri Riley . . . VP & gen mgr
Kristi Spitzer . . . sls dir
Jim Tripeny . . . prom mgr
Shirley Sanford . . . progmg dir
John Pilios . . . news dir
Kathleen McNeil . . . pub affrs dir
Tom Ballew . . . chief of engrg

KUVI-TV— Analog channel: 45. Digital channel: 55. Analog hrs: 24 5,000 kw vis, 500 kw aur. ant 1,325t/144g TL: N36 26 20 W118 44 24 On air date: Dec 18, 1988. 5801 Truxtun Ave., Bakersfield, CA, 93309. Phone: (661) 324-0045. Fax: (661) 334-2693. Web Site: www.kuvi45.com. Licensee: KUVI License Partnership G.P. Group owner: Univision Communications Inc. (acq 2-4-98; $14,010,800). Population served: 508,000 Natl. Network: MyNetworkTV. Washington Atty: Shaw Pittman.
Key Personnel:
Denise Snanoudt . . . natl sls mgr
Maritere Alvarez-Jackson. . . . rgnl sls mgr
Teresa Ford . . . gen mgr, opns mgr, progmg dir & film buyer
Maria Herrandez . . . pub affrs dir
Ken Richter . . . chief of engrg

Calipatria

see Yuma, AZ-El Centro, CA market

Chico-Redding, CA
(DMA 130)

KCVU— Analog channel: 30. Digital channel: 20. Analog hrs: 24 Digital hrs: 24 2,510 kw vis, 252 kw aur. 2,447t/250g TL: N39 57 45 W121 42 40 On air date: November 1990. 300 Main St., Chico, CA, 95928-5438. Phone: (530) 893-1234. Fax: (530) 899-5475. E-mail: info@fox30.com Web Site: www.fox30.com. Licensee: Sainte Partners II L.P. Group owner: (group owner) Population served: 445,000 Natl. Network: Fox. Natl. Rep: Millennium Sales & Marketing. Washington Atty: Fletcher, Heald & Hildreth. News: 0 hrs wkly.
Key Personnel:
Doug Holroyd. . . . gen mgr
Bert Westhoff . . . natl sls mgr
Glenn Taylor. . . . rgnl sls mgr
Paula Murphy . . . progmg mgr
Ken Rice . . . chief of engrg
Betsy Brewer . . . prom

KHSL-TV— Analog channel: 12. Digital channel: 43. Analog hrs: 24 Note: CBS is on KHSL-TV ch 12, CW is on KHSL-DT ch 43. 316 kw vis, 38 kw aur. ant 1,300t/287g TL: N39 57 30 W121 42 48 On air date: Aug 29, 1953. 3460 Silverbell Rd., Chico, CA, 95973. Phone: (530) 342-0141. Fax: (530) 342-4905. E-mail: khsltv@khsltv.com Web Site: www.khsltv.com. Licensee: Catamount Broadcasting of Chico-Redding Inc. Ownership: Catamount Holdings LLC, 100% votes Group owner: Catamount Broadcast Group (acq 7-30-98; $10 million). Population served: 480,000 Natl. Network: CBS, CW Digital Network: Note: CBS is on KHSL-TV ch 12, CW is on KHSL-DT ch 43. Natl. Rep: Continental Television Sales. Washington Atty: Haley, Bader & Potts. News staff: 34; News: 20 hrs wkly.
Key Personnel:
John Stall . . . gen mgr
Marnie McDonald. . . . gen sls mgr
Morgan Schmidt . . . prom dir & prom mgr
Shannon Bomar. . . . progmg dir
Trisha Coder . . . news dir
Dave Sien. . . . chief of engrg

***KIXE-TV—** Analog channel: 9. Digital channel: 18. Analog hrs: 18 Digital hrs: 18 115 kw vis, 12 kw aur. 3,590t/99g TL: N40 36 09 W122 39 01 On air date: Oct 5, 1964. 603 N Market St., Redding, CA, 96003-3609. Phone: (530) 243-5493. Fax: (530) 243-7443. E-mail: channel9@kixe.org Web Site: www.kixe.org. Licensee: Northern California Educational TV Association Inc. Population served: 500,000 Natl. Network: PBS. Washington Atty: Schwartz, Woods & Miller. News: one hr wkly.
Key Personnel:
Myron A. Tisdel . . . pres & gen mgr
Renee Cooper . . . CFO
Mike Lampella . . . opns dir
Anne Kerns . . . dev dir
Fred Gaines . . . rgnl sls mgr
Rob Keenan . . . progmg dir & progmg mgr
Sue Maxey . . . chief of engrg

KNVN— Analog channel: 24. Analog hrs: 24 5,000 kw vis, 600 kw aur. 1,849t/997g TL: N40 15 31 W122 05 20 On air date: Sept 24, 1985. 3460 Silverbell Rd., Chico, CA, 95973. Phone: (530) 894-6397. Fax: (530) 342-2405. E-mail: news@knvn.com Web Site: www.knvn.com. Licensee: Chico License L.L.C. (acq 6-2000; $9.2 million). Natl. Network: NBC. Washington Atty: Leventhal, Senter & Lerman.
Key Personnel:
John Stall . . . gen mgr
Scott Howard. . . . news dir

KRCR-TV— Analog channel: 7. Digital channel: 34.115 kw vis, 22.4 kw aur. ant 3,620t/126g TL: N40 36 10 W122 39 00 On air date: Aug 1, 1956. Box 992217, 755 Auditorium Dr., Redding, CA, 96001. Phone: (530) 243-7777. Fax: (530) 243-0217. E-mail: info@krcrtv.com Web Site: www.krcrtv.com. Licensee: BlueStone License Holdings Inc. Group owner: Lamco Communcations Inc. (acq 6-15-2004; grpsl). Population served: 180,000 Natl. Network: ABC. Natl. Rep: Petry Television Inc. News staff: 21; News: 16 hrs wkly.
Key Personnel:
Penny Loll . . . progmg dir
Jennifer Scarbrough . . . news dir
Lance Cratty . . . chief of engrg

El Centro

see Yuma, AZ-El Centro, CA market

Eureka, CA
(DMA 193)

KAEF— Analog channel: 23.141 kw vis, 14 kw aur. 1,672t TL: N40 43 36 W123 58 18 (CP: 195 kw vis, 19 kw aur) On air date: Aug 1, 1987. 540 E St., Eureka, CA, 95501. Phone: (707) 444-2323. Fax: (707) 445-9451. E-mail: kaeftv@kadf.com Licensee: BlueStone License Holdings Inc. Group owner: Lamco Communications Inc. (acq 6-15-2004; grpsl). Natl. Network: ABC. Washington Atty: Koteen & Naftalin. News staff: 2; News: 2 hrs wkly.

KBVU— Analog channel: 29. Analog hrs: 24 66.1 kw vis. 1,096t TL: N40 49 32 W124 00 05 On air date: July 20, 1994. 730 7th St., Suite 201, Eureka, CA, 95501. Phone: (707) 442-2999. Fax: (707) 441-0111. Web Site: www.eurekatelevision.tv. Licensee: Sainte Partners II L.P. Group owner: (group owner; acq 7-24-97). Natl. Network: Fox. Natl. Rep: Millennium Sales & Marketing. Washington Atty: Womble, Carlyle, Sandrige & Rice.
Key Personnel:
Chester Smith . . . CEO
Don Smullin . . . gen mgr

***KEET—** Analog channel: 13. Digital channel: 11. Analog hrs: 24 Digital hrs: 24 182 kw vis. ant 1,804t/420g TL: N40 43 38 W123 58 16 (CP: ant 1,803t/419g. TL: N40 43 38.9 W123 58 17) On air date: Apr 14, 1969. Box 13, Eureka, CA, 95502. 7246 Humboldt Hill Rd., Eureka, CA 95503. Phone: (707) 445-0813. Fax: (707) 445-8977. E-mail: letters@keet.pbs.org Web Site: www.keet.com. Licensee: Redwood Empire Pub TV Inc. Population served: 130,000 Natl. Network: PBS.
Key Personnel:
Ronald L. Schoenherr . . . CEO & pres
Seth Frankel . . . opns dir
Karen Barnes. . . . dev dir & progmg dir
Claire Reynolds . . . prom dir
Joel Householter . . . chief of engrg

KIEM-TV— Analog channel: 3.100 kw vis, 10 kw aur. 1,650t/249g TL: N40 43 52 W123 57 06 On air date: Oct 25, 1953. 5650 S. Broadway, Eureka, CA, 95503. Phone: (707) 443-3123. Fax: (707) 442-6084. Web Site: www.kiem-tv.com. Licensee: Pollack/Belz Broadcasting Co. L.L.C. Group owner: Pollack Broadcasting Co. (acq 5-1-96; $3 million). Population served: 139,400 Natl. Network: NBC.
Key Personnel:
Robert Browning . . . gen mgr
Phil Wright . . . opns mgr & prom mgr
Hank Ingham . . . gen sls mgr
Shawna Brisco . . . progmg dir
Bob Brown. . . . news dir
Bob Mottaz . . . chief of engrg

KVIQ— Analog channel: 6. Digital channel: 17.100 kw vis, 10.5 kw aur. ant 1,740t/377g TL: N40 43 36 W123 28 18 On air date: Apr 1, 1958. 730 7th St., Suite 201, Eureka, CA, 95501. Phone: (707) 443-3061. Fax: (707) 443-4435. Web Site: www.kviq.com. Licensee: Raul Broadcasting Co. of Eureka Inc. Ownership: Raul Palazuelos, 100% Group owner: Clear Channel Communications Inc. (acq 4-25-2005; $2 million). Population served: 24,337 Natl. Network: CBS.
Key Personnel:
John Burgess. . . . gen mgr
Penny King . . . stn mgr & gen sls mgr
Rick St. Charles. . . . prom dir
Lauren Faucett . . . progmg dir
Deve Silurrbrand . . . news dir
Jim Mixou . . . chief of engrg

Fresno-Visalia, CA
(DMA 55)

KAIL— Analog channel: 53. Digital channel: 7. Analog hrs: 24 Digital hrs: 24 2,510 kw vis, 251 kw aur. ant 1,906t/140g TL: N37 04 23 W119 15 52 On air date: Dec 18, 1961. 1590 Alluvial Ave., Clovis, CA, 93611. Phone: (559) 299-9753. Fax: (559) 299-1523. Web Site: www.kail.tv. Licensee: Trans-America Broadcasting Corp. Ownership: Albert J. Williams, 79.9%; Jack M. Reeder, 20.1%. (acq 12-23-66; $236,500; FTR: 12-26-66). Population served: 1,700,000 Natl. Network: MyNetworkTV. Washington Atty: Miller & Fields. News staff: 3; News: 5 hrs wkly.
Key Personnel:
Albert J. Williams . . . pres
Charles Williams . . . gen mgr
Mike Nicassio. . . . opns mgr & chief of opns
Dave Hetrick. . . . gen sls mgr
Robert Jenkins . . . progmg dir
Terrence Kendrials . . . prom VP & prom

KFRE-TV—Sanger, Analog channel: 59. Digital channel: 36. Analog hrs: 24 4,287 kw vis. ant 2,102t/246g TL: N37 04 37 W119 26 01 On air date: July 17, 1985. 5111 E. McKinley Ave., Fresno, CA, 93727. Phone: (559) 435-5900/(559) 252-5900. Fax: (559) 255-0275. Web Site: www.kfre.com. Licensee: KFRE(TV) License LLC. Group owner: Pappas Telecasting Companies (acq 12-22-2003; $25 million). Population served: 1,583,500 Natl. Network: CW. Washington Atty: Cohn & Marks. News staff: 5; News: 2 hrs wkly.
Key Personnel:
Charles Pfaff . . . stn mgr
Ken Felder . . . stn mgr & gen sls mgr
Mark Hodorowski . . . mktg dir, progmg dir & progmg mgr

KFSN-TV— Analog channel: 30. Digital channel: 9. Analog hrs: 24 Digital hrs: 24 316 kw vis, 31.6 kw aur. 4,750t/271g TL: N37 04 38 W119 26 00 On air date: May 10, 1956. 1777 G St., Fresno, CA, 93706-1688. Phone: (559) 442-1170. Fax: (559) 233-5844 (sls). Fax: (559) 266-5024 (news). Web Site: www.abc30.com. Licensee: KFSN Television LLC. Group owner: (group owner). Population served: 527,770 Natl. Network: ABC. Natl. Rep: ABC National Television Sales. Washington Atty: ABC Legal. News staff: 50; News: 30 hrs wkly.

Key Personnel:
Bob A. Hall . . . . pres & gen mgr
Charlene Ciavaglia . . . . progmg mgr
Joel Davis . . . . news dir
Beth Marney. . . . . pub affrs dir
Ron Neil . . . . rgnl sls mgr & engrg dir

KFTV— Analog channel: 21. Digital channel: 20. Analog hrs: 24 Digital hrs: 24 5,000 kw vis, 605 kw aur. ant 1,984t/216g TL: N37 04 22 W119 25 50 On air date: July 1972. 3239 W. Ashlan Ave., Fresno, CA, 93722. Phone: (559) 222-2121. Fax: (559) 222-2890. Fax: (559) 222-0917. Web Site: www.univision.net. Licensee: KFTV L.P., G.P. Group owner: Univision Communications Inc. (acq 8-87). Population served: 637,000 Natl. Network: Univision (Spanish). Washington Atty: Shaw Pittman LLP. Foreign lang progmg: SP 168 News staff: 58; News: 8 hrs wkly.
Key Personnel:
Maria L. Gutierrez . . . . gen mgr
Brett Covish . . . . opns mgr & gen sls mgr
Ken Holden . . . . chief of engrg

KGMC—Clovis, Analog channel: 43. Digital channel: 44. Analog hrs: 24 Digital hrs: 24 1,025 kw vis, 102 kw aur. 2,001t TL: N36 44 45 W119 16 57 On air date: Sept 11, 1992. 706 W. Herndon Ave., Fresno, CA, 93650. Phone: (559) 432-4300. Phone: (559) 435-7000. Fax: (559) 435-3201. E-mail: info@cocolatv.com Web Site: www.cocolatv.com. Licensee: Gary M. Cocola. Group owner: Cocola Broadcasting Companies (acq 11-19-92). Population served: 1,400,000 Washington Atty: Dow, Lohnes & Albertson.
Key Personnel:
Gary M. Cocola . . . . CEO & pres
Nick Giotto . . . . opns dir
Kevin Mosesian . . . . natl sls mgr
Nick Giotto . . . . rgnl sls mgr
Joe Verdugo . . . . adv dir
Nori Zahari. progmg VP
Al Kinney. engrg mgr

KGPE— Analog channel: 47. Digital channel: 34. Analog hrs: 24 2,500 kw vis. ant 1,958t/226g TL: N37 04 14 W119 25 31 On air date: Oct 1, 1953. 4880 N. First St., Fresno, CA, 93720. Phone: (559) 222-2411. Fax: (559) 222-5593. E-mail: programming@cbs47.tv Web Site: www.cbs47.tv. Licensee: Ackerley Broadcasting Operations, LLC/dba Clear Channel Television-KGPE-TV. Group owner: Clear Channel Communications Inc. (acq 6-14-2002; grpsl). Population served: 1,453,000 Natl. Network: CBS. Natl. Rep: Millennium Sales & Marketing. Washington Atty: Wiley, Rein & Fielding. News staff: 50; News: 28hrs wkly.
Key Personnel:
Steve Stendlove. . . . . exec VP
TB A . . . . chief of opns & gen sls mgr
David King . . . . natl sls mgr
Tom Long . . . . sls dir & rgnl sls mgr
Jim Holland . . . . news dir

KMPH-TV—Visalia, Analog channel: 26. Digital channel: 28.2,950 kw vis, 442 kw aur. ant 2,730t/252g TL: N36 17 12 W118 50 20 (CP: Ant 2,571t. TL: N36 40 02 W118 52 42) On air date: Oct 11, 1971. 5111 E. McKinley Ave., Fresno, CA, 93727-2033. Phone: (559) 255-2600. Fax: (559) 255-0275. E-mail: viewercomments@kmph.com Web Site: www.kmph.com. Licensee: KMPH(TV) License LLC. Group owner: Pappas Telecasting Companies (acq 6-1-78; $3,105,550). Population served: 1,432,000 Natl. Network: Fox. Natl. Rep: TeleRep. Washington Atty: Fletcher, Heald & Hildreth. News staff: 21; News: 7 hrs wkly.
Key Personnel:
Harry J. Pappas . . . . pres
LeBon Abercrombie . . . . exec VP
John Carpenter . . . . VP
Charlie Pfaff . . . . gen mgr
Ken Felder . . . . gen sls mgr
Mark Hodorowski. . . . . prom mgr
Debbie Sweeney . . . . progmg dir
Roger Gadley. news dir
Jim Boston . . . . chief of engrg

KNSO— Analog channel: 51. Digital channel: 5.5,000 kw vis. ant 1,830t/174g TL: N37 04 19 W119 25 49 On air date: Mar 22, 1996. 30 River Park Pl. W., Suite 200, Fresno, CA, 93720. Phone: (559) 252-5101. Web Site: www.kcso33.com. Licensee: NBC Telemundo License Co. Group owner: Telemundo Group Inc. (acq 4-30-2003; $33 million). Natl. Network: Telemundo (Spanish). Foreign lang progmg: SP 110

***KNXT**— Analog channel: 49. Digital channel: 50. Analog hrs: 24 Digital hrs: 50 2,140 kw vis, 214 kw aur. 2,739t TL: N36 17 14 W118 50 17 On air date: Nov 2, 1986. 1550 N. Fresno St., Fresno, CA, 93703. Phone: (559) 488-7440. Fax: (559) 488-7444. E-mail: knxt49@hotmail.com Web Site: www.dioceseoffresno.org. Licensee: Board of Directors Diocese of Fresno Education Corp. Population served: 1,600,000 Foreign lang progmg: SP 7

Key Personnel:
Bishop John T. Steinback. . . . . pres
Colin Dougherty . . . . gen mgr

KSEE— Analog channel: 24. Digital channel: 38. Analog hrs: 24 Digital hrs: 24 1,600 kw vis, 320 kw aur. 2,350t/321g TL: N36 44 45 W119 16 53 On air date: June 1, 1953. 5035 E. McKinley Ave., Fresno, CA, 93727-1964. Phone: (559) 454-2424. Fax: (559) 454-2487. Web Site: www.ksee24.com. Licensee: Ksee License, Inc. Group owner: (group owner; (acq 12-93; $32 million with WTVH(TV) Syracuse, NY; FTR: 8-30-93). Population served: 165,972 Natl. Network: NBC. Natl. Rep: Harrington, Righter & Parsons. Washington Atty: Akin, Gump, Strauss, Hauer & Feld.
Key Personnel:
Todd McWilliams . . . . pres & gen mgr
George Hillis. . . . . chief of opns, progmg mgr & chief of engrg
Jan Katzenberger . . . . gen sls mgr & prom dir
Kathleen Goble. . . . . natl sls mgr
Chris Zanghi. . . . . rgnl sls mgr
Julie Akins . . . . news dir

KTFF-TV— Analog channel: 61. Digital channel: 48. Analog hrs: 24 2,510 kw vis, 251 kw aur. 1,443t TL: N36 17 14 W118 50 17 On air date: May 6, 1992. 3239 W. Ashlan Ave., Fresno, CA, 93722-4402. Phone: (559) 439-6100. Fax: (559) 439-5950. Web Site: www.telefutura.com. Licensee: TeleFutura Fresno LLC. Group owner: Univision Communications Inc. (acq 2-7-2003; $35 million). Natl. Network: TeleFutura (Spanish).
Key Personnel:
Maria L. Gutierrez . . . . gen mgr
Brett Covish . . . . opns mgr
Jose Elgorriga . . . . gen sls mgr
Darrell Jennings . . . . rgnl sls mgr
Samuel Belilty . . . . progmg dir & news dir
Ken Holden . . . . chief of engrg

***KVPT**— Analog channel: 18. Analog hrs: 6 AM-11:59 PM 562 kw vis, 112 kw aur. ant 2,220t/245g TL: N36 44 45 W119 16 52 On air date: Apr 10, 1977. 1544 Van Ness Ave., Fresno, CA, 93721. Phone: (559) 266-1800. Fax: (559) 650-1880. E-mail: web@kvpt.org Web Site: www.kvpt.org. Licensee: Valley Public Television Inc. (acq 11-1-87). Population served: 2,500,000 Natl. Network: PBS. Washington Atty: Fletcher, Heald & Hildreth. Foreign lang progmg: SP 6 News staff: 35.
Key Personnel:
Paula Castadio . . . . CEO & mktg dir
Douglas E. Noll . . . . chmn
Phyllis Brotherton. CFO
Jim Page . . . . opns mgr & mktg dir
Eva Torres . . . . dev dir & sls dir
Jerry Lee. . . . . progmg VP & progmg dir
Rodger Hixon . . . . chief of engrg

Los Angeles (DMA 2)

KABC-TV— Analog channel: 7. Digital channel: 53. Analog hrs: 24 159 kw vis, 31.7 kw aur. ant 2,970t/234g TL: N34 13 37 W118 03 58 (CP: 141.25 kw vis, 28.9 kw aur, ant 3,213t) On air date: Sept 16, 1949. 500 Circle Seven Dr., Glendale, CA, 91201. Phone: (818) 863-7777. Fax: (818) 863-7080. E-mail: abc7@abc.com Web Site: www.abc7.com. Licensee: ABC Inc. Group owner: (group owner; (acq 1-6-86; grpsl; FTR: 7-15-85). Population served: 15,077,000 Natl. Network: ABC. Wire Svc: UPI

KAZA-TV— Analog channel: 54.5,000 kw vis. ant 1,220t TL: N33 21 00 W118 21 05 On air date: July 28, 2001. 500 S. Chinowth Rd., Visalia, CA, 93277. Phone: (559) 733-7800. Phone: (818) 241-5400. Fax: (559) 733-7878. Web Site: pappastv.com. Licensee: Pappas Southern California License LLC. Group owner: Pappas Telecasting Companies. Foreign lang progmg: SP 168
Key Personnel:
Harry J. Pappas . . . . CEO
Eduardo Urbiola. . . . . gen mgr
Fernando Acosta . . . . gen mgr & opns mgr
Alberto Ezquerro . . . . gen sls mgr
Ramon Delgado. . . . . progmg dir
Oscar Salcedo . . . . news dir
Joe Berardi . . . . chief of engrg

KBBC-TV—Analog channel: 20.4.2 kw vis. ant 3,031t/30g TL: N37 24 43 W118 11 06 Not on air, target date: unknown: Cocola Broadcasting Companies, 706 W. Herndon Ave., Fresno, CA, 93650-1033. Phone: (559) 435-7000. Fax: (559) 435-3201. Web Site: www.cocolatv.com. Permittee: Bellagio Broadcasting LLC.

KCAL— Analog channel: 9. Digital channel: 43. Analog hrs: 24 141 kw vis, 28.2 kw aur. 3,184t/464g TL: N34 13 38 W118 04 00 On air date: Oct 6, 1948. 6121 Sunset Blvd., Hollywood, CA, 90028. Phone: (323) 467-9999/(323) 460-3000. Fax: (323) 464-2526. E-mail: kcalnews@cbs.com Web Site: www.kcal.com. Licensee: Viacom Television Stations Group of Los Angeles LLC. Group owner: Viacom Television Stations Group (acq 5-3-2002; $650 million). Population served: 5,006,380 Natl. Network: CBS. Natl. Rep: Adam Young. Wire Svc: Conus Wire Svc: World Television News Service

KCBS-TV— Analog channel: 2. Digital channel: 60. Analog hrs: 24 36.3 kw vis, 7.26 kw aur. 3,632t/974g TL: N34 13 57 W118 04 18 On air date: May 6, 1948. 6121 Sunset Blvd., Los Angeles, CA, 90028. Phone: (323) 460-3000. Fax: (323) 460-3733. Web Site: www.cbs2.com. Licensee: CBS Inc. Group owner: CBS (acq 12-27-50; $3.6 million; FTR: 1-1-51). Population served: 14,400,000 Natl. Network: CBS. News staff: 104; News: 26 hrs wkly.
Key Personnel:
Paul Latham. . . . . CFO
Don Corsini . . . . gen mgr
Melanie Steensland . . . . opns dir
Garen Vandebeek . . . . prom dir
Virginia Hunt . . . . progmg dir
Nancy Bauer Gonzalez . . . . news dir
Craig Harrison . . . . engrg dir

***KCET**— Analog channel: 28. Digital channel: 59. Analog hrs: 19 2,455 kw vis, 245.5 kw aur. ant 3,038t/330g TL: N34 13 26 W118 03 44 On air date: Sept 28, 1964. 4401 Sunset Blvd., Los Angeles, CA, 90027. Phone: (323) 666-6500. Fax: (323) 953-5523. Web Site: www.kcet.org. Licensee: Community TV of Southern California. Population served: 4,600,000 Natl. Network: PBS. Rgnl. Network: Eastern Educ., Pacific Mtn. Washington Atty: Arent, Fox, Kintner, Plotkin & Kahn.
Key Personnel:
Al Jerome. . . . . CEO, pres & gen mgr
Debbie Hinton . . . . CFO
Roger Terracina . . . . opns mgr & dev VP

KCOP— Analog channel: 13. Digital channel: 66.161 kw vis, 32.4 kw aur. ant 2,972t/187g TL: N34 13 29 W118 03 48 On air date: Sept 17, 1948. 1999 S. Bundy Dr., Los Angeles, CA, 90025. Phone: (310) 584-2000. Fax: (310) 584-2024. Web Site: www.upn13.com. Licensee: Fox Television Stations Inc. Group owner: (group owner; (acq 7-31-2001; grpsl). Population served: 149,000 Natl. Network: MyNetworkTV. Washington Atty: Wilmer, Cutler & Pickering.

KDOC-TV— Analog channel: 56. Digital channel: 32. Analog hrs: 24 2,390 kw vis. ant 3,024t/249g TL: N34 13 36 W118 03 58 (CP: 2,340 kw vis, ant 3,113t/321g) On air date: Oct 1, 1982. 18021 Cowan, Irvine, CA, 92614-6023. Phone: (949) 442-9800. Fax: (949) 261-5956/(949) 221-4171. E-mail: mmerker@kdoctv.net Web Site: www.kdoctv.net. Licensee: Ellis Communications KDOC Licensee LLC. (acq 2006; $149.5 million). Washington Atty: Cohn & Marks.
Key Personnel:
Pat Boone. . . . . pres
Calvin Brack . . . . VP
John Davis. . . . . gen mgr
Tom Jimenez . . . . gen sls mgr
Dale Foshee. rgnl sls mgr
Michelle Merker. . . . . mktg dir, prom VP & pub affrs dir
John Atkinson . . . . progmg mgr
Roger Knipp . . . . chief of engrg

KFTR-TV—Ontario, Analog channel: 46. Analog hrs: 24 2,450 kw vis, 372 kw aur. 3,040t/323g TL: N34 13 37 W118 03 58 On air date: Apr 21, 1984. 5999 Center Dr., Los Angeles, CA, 90045. Phone: (310) 348-3411. Fax: (310) 348-4849. Licensee: Univision Partnership of Southern California. Group owner: Univision Communications Inc. (acq 5-21-2001; grpsl). Natl. Network: TeleFutura (Spanish). Washington Atty: Wiley, Rein & Fielding. Foreign lang progmg: SP 168
Key Personnel:
Jorge Delgado . . . . gen mgr
Mark Dante . . . . gen sls mgr
Luis De La Parra. . . . . mktg dir & prom dir
Amy Rico . . . . progmg mgr
Chris Homer . . . . chief of engrg

KHIZ— Analog channel: 64. Digital channel: 44. Analog hrs: 24 Digital hrs: 24 3,160 kw vis, 627 kw aur. ant 1,699t/203g TL: N34 36 34 W117 17 11 On air date: 1987. Box 1468, Victorville, CA, 92393-1468. 15605 Village Dr., Victorville, CA 92394. Phone: (760) 241-6464. Fax: (760) 241-0056. Web Site: www.khiztv.com. Licensee: Sunbelt Television Inc. Ownership: Initial Broadcasting of California LLC, 93%; TVPlus LLC, 7% (acq 11-14-2005). Population served: 5,000,000 Washington Atty: Wilkinson, Barker & Knauer. News staff: 10; News: 5 hrs wkly.
Key Personnel:
Peter White . . . . pres
Garrett Law . . . . gen mgr
Stella Montoya . . . . progmg dir

KJLA— Analog channel: 57. Digital channel: 49. Analog hrs: 24 5,000 kw vis. ant 1,738t/397g TL: N34 18 10 W119 13 41 On air date: October 1990. 2323 Corinth Ave., West Los Angeles, CA, 90064. Phone: (310) 943-5288. Fax: (310) 943-5299. E-mail: kjlainfo@kjla.com

Web Site: www.kjla.com. Licensee: KJLA LLC. Ownership: LATV LLC (acq 11-14-94; FTR: 1-2-95). Population served: 13,000,000 Washington Atty: Thompson Hine & Flory. Foreign lang progmg: SP 20

Key Personnel:

Walter Ulloa . . . . . pres
Ed Safa . . . . . CFO
Daniel Crowe . . . . . exec VP
Francis Wilkinson . . . . . exec VP & gen mgr
Mike Seros . . . . . opns dir
Richard Deanda . . . . . dev dir & rgnl sls mgr

***KLCS—** Analog channel: 58. Analog hrs: 6 AM-midnight 550 kw vis, 110 kw aur. 3,050t/180g TL: N34 13 26 W118 03 45 On air date: Nov 5, 1973. 1061 W. Temple St., Los Angeles, CA, 90012. Phone: (213) 625-6958/(213) 241-4000. Fax: (213) 481-1019. E-mail: info@klcs.org Web Site: www.klcs.org. Licensee: Los Angeles Unified School District. Ownership: Los Angeles Unified School Dist. Population served: 15,000,000 Natl. Network: PBS. Washington Atty: Cohn & Marks. News: 5 hrs wkly.

Key Personnel:

Dr. Janalyn W. Glymph . . . . . gen mgr & stn mgr
Myles Jang . . . . . dev dir
Sabrina Thomas . . . . . opns dir, progmg dir & progmg mgr

KMEX-TV— Analog channel: 34. Digital channel: 35. Analog hrs: 24 1,950 kw vis, 195 kw aur. ant 2,940t/170g TL: N34 13 35 W118 03 56 On air date: Sept 30, 1962. 5999 Center Dr., Los Angeles, CA, 90045. Phone: (310) 216-3434. Fax: (310) 348-3459. Web Site: www.univision.com (keyword: Los Angeles). Licensee: KMEX License Partnership G.P. Group owner: Univision Communications Inc. Population served: 281,606 Natl. Network: Univision (Spanish). Wire Svc: UPI Foreign lang progmg: SP 168 News: 17 hrs wkly.

Key Personnel:

A. Jerrold Perenchio . . . . . CEO & chmn
Jose Delgado . . . . . pres
George Blank . . . . . CFO
Jorge Delgado . . . . . gen mgr & opns mgr
Mark Dante . . . . . gen sls mgr
Luis De la Parra . . . . . mktg dir
Antoinette Gill . . . . . progmg dir
Jorge Mattey . . . . . news dir
Christina Sanchez-Camino . . . . . pub affrs dir
Chris Homer . . . . . engrg mgr & chief of engrg

KNBC— Analog channel: 4. Digital channel: 36.44.7 kw vis, 7.76 kw aur. 3,200t/496g TL: N34 13 32 W118 03 52 On air date: 1998. 3000 W. Alameda Ave., Burbank, CA, 91523. Phone: (818) 840-4444. Fax: (818) 840-3003. Web Site: www.nbc4.tv. Licensee: NBC Telemundo License Co. Group owner: NBC TV Stations Division (acq 6-5-86). Natl. Network: NBC. Natl. Rep: NBC TV Stations Sales. Wire Svc: Reuters Wire Svc: UPI

Key Personnel:

Robert Long . . . . . VP
Linda Sullivan . . . . . gen mgr
Mike McCarthy . . . . . stn mgr & sls VP
Robert L. Long . . . . . news dir

***KOCE-TV—** Analog channel: 50. Digital channel: 48. Analog hrs: 24 Digital hrs: 24 2,354 kw vis. ant 3,113t/321g TL: N34 13 35 W118 03 57 On air date: November 1972. Box 2476, 15751 Gothard St., Huntington Beach, CA, 92647. Phone: (714) 895-5623. Fax: (714) 895-0852. E-mail: koce@cccd.edu Web Site: www.koce.org. Licensee: KOCE-TV Foundation. (acq 11-1-2004; $25.5 million). Population served: 8,220,000 Natl. Network: PBS. Washington Atty: Vorys, Sater, Seymour & Pease. News: 3 hrs wkly.

Key Personnel:

Robert Brown . . . . . chmn
Michael Taylor . . . . . VP & news dir
Mel Rogers . . . . . gen mgr
Nancy Weed . . . . . stn mgr
Bette Kain . . . . . dev dir
Judith Schaefer . . . . . prom dir & pub affrs dir
Patricia Petric . . . . . progmg dir
Roger Yoakum . . . . . chief of engrg

KPXN— Analog channel: 30. Digital channel: 38.3,800 kw vis, 251 kw aur. ant 2,345t TL: N34 11 15 W117 41 58 On air date: Jan 7, 1994. 3000 W. Alameda Ave., Suite 32622, Burbank, CA, 91523. Phone: (818) 840-4444. Fax: (818) 840-2129. Web Site: www.ionline.tv. Licensee: Paxson Los Angeles License Inc. Group owner: Paxson Communications Corp. (acq 3-22-95; $18 million; FTR: 6-19-95). Population served: 13,000,000 Natl. Network: i Network.

Key Personnel:

Alisha Wofford . . . . . VP
Paula Madison . . . . . gen mgr & chief of engrg
Rob Word . . . . . stn mgr & progmg dir
Mark Douglas . . . . . gen sls mgr
Steve Vinke . . . . . chief of engrg

KRCA— Analog channel: 62. Digital channel: 68.2,630 kw vis. ant 2,936t/138g TL: N34 12 50 W118 03 40 On air date: Dec 17, 1988. 1813 Victory Pl., Burbank, CA, 91504. Phone: (818) 563-5722. Fax: (818) 972-2694. E-mail: info@lbimedia.com Licensee: KRCA License LLC. Group owner: Liberman Broadcasting Inc. (acq 6-18-90). Foreign lang progmg: SP 52

Key Personnel:

Michael Sheron . . . . . gen sls mgr
Winter Horton . . . . . gen mgr & progmg dir
Chris Buchanan . . . . . chief of engrg

KSCI— Analog channel: 18. Digital channel: 61. Analog hrs: 24 2,583 kw vis. ant 2,949t/164g TL: N34 12 47.8 W118 03 41 On air date: June 30, 1977. 1990 S. Bundy Dr., Suite 850, Los Angeles, CA, 90025. Phone: (310) 478-1818. Fax: (310) 479-8118. E-mail: info@kscitv.com Web Site: www.la18.tv. Licensee: KSLS Inc. Group owner: Asian Media Group (acq 1-18-01; $165 million cash for 69.4%). Population served: 6,000,000 Washington Atty: Wilkinson, Barker, Knauer LLP. News staff: 12; News: 12 hrs wkly.

Key Personnel:

Peter Mathes . . . . . chmn
Brian Reed . . . . . gen mgr

KTBN-TV—Santa Ana, Analog channel: 40. Digital channel: 23.631 kw vis, 126 kw aur. ant 2,890t/202g TL: N34 13 32 W118 03 44 On air date: Jan 5, 1967. Box A, Santa Ana, CA, 92711. 2442 Michelle Dr., Tustin, CA 92780. Phone: (714) 832-2950. Fax: (714) 665-2191. Web Site: www.tbn.org. Licensee: Trinity Broadcasting Network. Group owner: (group owner; (acq 8-2-74; $1,266,400; FTR: 8-19-74). Washington Atty: Joseph E. Dunne III.

KTLA— Analog channel: 5. Digital channel: 31. Analog hrs: 24 44.7 kw vis, 6.7 kw aur. 6,176t/473g TL: N34 13 36 W118 03 56 (CP: Ant 3,201t) On air date: Jan 22, 1947. 5800 Sunset Blvd., Los Angeles, CA, 90028. Phone: (323) 460-5500. Fax: (323) 460-5405. Web Site: www.ktla.com. Licensee: KTLA Inc. Group owner: Tribune Broadcasting Co. (acq 12-23-85; $510 million). Population served: 14,443,000 Natl. Network: CW. Natl. Rep: TeleRep. Washington Atty: Sidley & Austin. Wire Svc: Reuters Wire Svc: NWS (National Weather Service) Foreign lang progmg: SP 22 News: 22 hrs wkly.

Key Personnel:

Vinnie Malcolm . . . . . VP, gen mgr & stn mgr
Gordon Peppars . . . . . gen sls mgr
Jymm Adams . . . . . prom dir
Jeff Wald . . . . . news dir
Ray Gonzales . . . . . pub affrs dir
Chris Neuman . . . . . engrg dir
Dave Cox . . . . . chief of engrg

KTTV— Analog channel: 11. Digital channel: 65. Analog hrs: 24 166 kw vis, 20 kw aur. 2,940t/237g TL: N34 13 29 W118 03 47 On air date: Jan 1, 1949. 1999 S. Bundy Dr., Los Angeles, CA, 90025. Phone: (310) 584-2000. Fax: (310) 584-2024. Web Site: www.fox11la.com. Licensee: Fox Television Stations Inc. Group owner: (group owner; (acq 11-14-86; grpsl). Population served: 14,254,000 Natl. Network: Fox. News: 25 hrs wkly.

Key Personnel:

Jack Abernathy . . . . . CEO
Kevin Hale . . . . . gen mgr

***KVCR-TV—** Analog channel: 24. Digital channel: 26.1,318 kw vis, 131.8 kw aur. 3,166t/215g TL: N30 57 57 W117 17 05 On air date: Sept 11, 1962. 701 S. Mt. Vernon Ave., San Bernardino, CA, 92410. Phone: (909) 384-4444. Fax: (909) 885-2116. E-mail: info@kvcr.pbs.org Web Site: www.kvcr.org. Licensee: San Bernardino Community College District. Population served: 17,000,000 Natl. Network: PBS. Rgnl. Network: Pacific.

Key Personnel:

Larry R. Ciecalone . . . . . gen mgr
Al Gondos . . . . . opns dir
Kenn Couch . . . . . dev dir
Lillian Vasquez . . . . . prom dir
Don Leiffer . . . . . progmg dir
Thomas Guptill . . . . . chief of engrg

KVEA—Corona, Analog channel: 52.2,630 kw vis, 263 kw aur. 2,890t/200g TL: N34 13 27 W118 03 45 (CP: 2,570 kw vis, ant 2,939t) On air date: June 29, 1966. 3000 W. Alameda, Burbank, CA, 91523. Phone: (818) 260-5700. Fax: (818) 260-5222. Web Site: www.kvea.com. Licensee: NBC Telemundo License Co. Group owner: Telemundo Group Inc. (acq 4-12-2002; grpsl). Population served: 711,000 Natl. Network: Telemundo (Spanish). Foreign lang progmg: SP 168

KVMD— Analog channel: 31. Digital channel: 23.12 kw vis. ant 295t TL: N34 09 15 W116 11 50 6448 Hallee Rd., Suite 3, Joshua Tree, CA, 92252. Phone: (760) 366-9881. Fax: (760) 366-1342. Web Site: www.kumd-tv.com. Licensee: KVMD Licensee Co. LLC. Ownership: Ronald L. Ulloa, 100% (acq 4-5-2001; $900,000).

Key Personnel:

Larry Peterson . . . . . gen mgr, stn mgr, opns mgr & gen sls mgr
Ken Brown . . . . . chief of engrg

KWHY-TV— Analog channel: 22. Digital channel: 42. Analog hrs: 20 2,630 kw vis, 257 kw aur. 2,916t/182g TL: N34 13 36 W118 03 59 On air date: Mar 25, 1963. 3400 W. Olive Ave., Suite 600, Burbank, CA, 91505-5539. Phone: (818) 260-5822. Fax: (818) 260-5805. Licensee: NBC Telemundo License Co. Group owner: Telemundo Group Inc. (acq 4-12-2002; grpsl). Population served: 5,000,000 Washington Atty: Cohn & Marks. Wire Svc: Reuters Foreign lang progmg: SP 87 News staff: 6; News: 12 hrs wkly.

KXLA— Analog channel: 44. Digital channel: 51.2,340 kw vis. ant 3,113t/325g TL: N34 13 35 W118 03 58 On air date: 2001. 2323 Corinth Ave., Los Angeles, CA, 90064. Phone: (310) 478-0055. Fax: (310) 478-8070. Web Site: www.kxlatv.com. Licensee: Rancho Palos Verdes Broadcasters Inc. Ownership: RPVB Lender Inc. (acq 8-27-01; up to $40 million for stock with KXLA-DT Rancho Palos Verdes).

Key Personnel:

Ron Ulloa . . . . . pres, gen mgr & progmg dir
Ken Brown . . . . . chief of engrg

Modesto

see Sacramento-Stockton-Modesto, CA market

Monterey-Salinas, CA (DMA 124)

KCBA— Analog channel: 35. Analog hrs: 24 2,328 kw vis, 283 kw aur. 2,414t/355g TL: N36 45 19 W121 30 05 On air date: Nov 1, 1981. 1550 Moffett St., Salinas, CA, 93905. Phone: (831) 422-3500. Fax: (831) 754-1120. Web Site: www.iknowcentralcoast.com/ www.kcba.com. Licensee: Seal Rock Broadcasters LLC. Ownership: George V. Kristie and Lance W. Anderson, mgng members (acq 1-5-00; $11 million). Natl. Network: Fox. Washington Atty: Rubin, Winston, Diercks, Harris & Cooke. News: 10 hrs wkly.

Key Personnel:

Mark Faylor . . . . . VP & gen mgr
Chris Chidlaw . . . . . gen mgr & stn mgr
Camilla Boolootian . . . . . gen sls mgr & prom mgr
Eric Casalla . . . . . progmg dir
Monica Escobedo . . . . . film buyer
Denise Clodjeaux . . . . . news dir
Adam Perez . . . . . engrg dir & chief of engrg

KION-TV—Monterey, Analog channel: 46.1,350 kw vis, 135 kw aur. 2,530t/222g TL: N36 32 05 W121 37 14 On air date: Feb 2, 1969. 1550 Moffett St., Salinas, CA, 93905. Phone: (831) 784-1702/(831) 422-3500. Fax: (831) 784-6395. Web Site: www.kion46.com. Licensee: Ackerley Broadcasting Operations LLC. Group owner: Clear Channel Communications Inc. (acq 6-14-2002; grpsl). Population served: 516,000 Natl. Network: CBS. Washington Atty: Wiley, Rein & Fielding. Wire Svc: UPI News staff: 25; News: 8 hrs wkly.

***KQET—** Analog channel: 25. Digital channel: 58.52.5 kw vis. ant 2,198t TL: N36 45 23 W121 30 05 (CP: 501 kw vis, ant 2,214t) On air date: November 1989. c/o KTEH, 1585 Schallenburger Rd., San Jose, CA, 95110-1301. Phone: (408) 795-5400. Fax: (408) 995-5446. Web Site: www.kteh.org. Licensee: Northern California Public Broadcasting Inc. (acq 10-1-2006). Natl. Network: PBS.

Key Personnel:

Tom Fanella . . . . . CEO
Judy Armstrong . . . . . gen mgr & dev dir

KSBW— Analog channel: 8. Digital channel: 8. Analog hrs: 24 Digital hrs: 24 158 kw vis, 15.8 kw aur. 2,940t/1,552g TL: N37 03 30 W121 46 33 On air date: Sept 11, 1953. 238 John St., Salinas, CA, 93901. Phone: (831) 758-8888. Fax: (831) 424-3750. Web Site: www.theksbwchannel.com. Licensee: Hearst-Argyle Stations Inc. Group owner: Hearst-Argyle Television Inc. (acq 6-1-98). Population served: 740,000 Natl. Network: NBC. Natl. Rep: Eagle Television Sales. Washington Atty: Brooks, Pierce, McLendon, Humphrey & Leonard. News staff: 31; News: 31 hrs wkly.

Key Personnel:

Joseph W. Heston . . . . . pres & gen mgr
Jose Camacho . . . . . opns mgr, mktg mgr & engrg mgr
Wendy Hillan . . . . . gen sls mgr & natl sls mgr
Bill Mushrush . . . . . mktg dir, prom dir & adv dir
Britt Govea . . . . . mktg mgr
Karen Pren . . . . . progmg mgr
Lawton Dodd . . . . . news dir
Theresa Wright . . . . . pub affrs dir
Don Engelhardt . . . . . engrg dir

KSMS-TV—Monterey, Analog channel: 67.1,260 kw vis. 2,299t TL: N36 45 23 W121 30 05 On air date: Sept 1, 1986. 67 Garden Ct., Monterey, CA, 93940. Phone: (831) 373-6767. Fax: (831) 373-6700. Web Site: www.entravision.com. Licensee: Entravision Holdings L.L.C. Group owner: Entravision Communications Co. L.L.C. (acq 4-25-97). Natl. Network: Univision (Spanish). Foreign lang progmg: SP 168

Key Personnel:
Philip Wilkinson . . . . pres & VP
Aaron Scoby . . . . gen mgr
Jeanie Harrison . . . . gen sls mgr

Oakland

see San Francisco-Oakland-San Jose market

Palm Springs, CA (DMA 149)

KESQ-TV— Analog channel: 42. Digital channel: 52. Analog hrs: 24 Digital hrs: 24 Note: ABC is on KESQ-TV ch 42, CW is on KESQ-DT ch 52. 316 kw vis. ant 745t/72g TL: N33 51 58 W116 26 02 On air date: Oct 5, 1968. 42-650 Melanie Pl., Palm Desert, CA, 92211. Phone: (760) 773-0342. Fax: (760) 773-5107. Web Site: www.kesq.com. Licensee: Gulf-California Broadcast Co. (acq 4-24-96; $19.4 million). Population served: 329,000 Natl. Network: ABC, CW Digital Network: Note: ABC is on KESQ-TV ch 42, CW is on KESQ-DT ch 52. Natl. Rep: Continental Television Sales. Washington Atty: Smithwick & Belendiuk. Wire Svc: News 1 Wire Svc: AP News staff: 40; News: 17 hrs wkly.
Key Personnel:
Bob Allen . . . . pres, exec VP & gen mgr
Todd Graham . . . . opns mgr
Barry Gorfine. . . . . sls dir & gen sls mgr
Ken Spalding. . . . . progmg dir & progmg mgr
Tony Ballew . . . . news dir
Dave Swartz . . . . engrg dir & chief of engrg

KMIR-TV— Analog channel: 36. Digital channel: 46. Analog hrs: 24 Digital hrs: 24 490 kw vis. 679t/123g TL: N33 52 00 W116 25 56 On air date: Oct 26, 1968. 72-920 Parkview Dr., Palm Desert, CA, 92260. Phone: (760) 568-3636/(760) 340-1623. Fax: (760) 568-1176. E-mail: news@kmir6.com Web Site: www.kmir6.com. Licensee: Journal Broadcast Corp. Group owner: Journal Broadcast Group Inc. (acq 6-11-99; $28.1 million). Population served: 318,000 Natl. Network: NBC. Washington Atty: Koteen & Naftalin. News staff: 30; News: 24 hrs wkly.
Key Personnel:
Dianne Downey . . . . VP & gen mgr
Frank Keller . . . . opns mgr
Tony Billett . . . . gen sls mgr
Wendy Degnan. . . . . natl sls mgr
Scott Johnson . . . . rgnl sls mgr & adv dir
Mayra Mancilla. . . . . progmg mgr
Russ Kilgore . . . . news dir
Tim Balint . . . . chief of engrg

Redding

see Chico-Redding, CA market

Sacramento-Stockton-Modesto, CA (DMA 20)

***KBSV**— Analog channel: 23.15.1 kw vis. ant 154t TL: N37 35 21 W120 57 23 Not on air, target date: unknown: Box 4116, Modesto, CA, 95352. Phone: (209) 538-9801. Fax: (209) 538-2795. E-mail: kssv@aol.com Web Site: www.betnahrain.org/kbsv. Licensee: Bet-Nahrain.
Key Personnel:
Dr. Sargon Dadesho. pres
Shemiran Daniel . . . . VP & gen mgr

KCRA-TV— Analog channel: 3. Digital channel: 35. Analog hrs: 24 100 kw vis, 10 kw aur. ant 1,938t/2,000g TL: N38 15 52 W121 29 22 On air date: Sept 3, 1955. 3 Television Cir., Sacramento, CA, 95814-0794. Phone: (916) 446-3333. Fax: (916) 441-4050 (news). Web Site: www.thekcrachannel.com. Licensee: Hearst-Argyle Stations Inc. Group owner: Hearst-Argyle Television Inc. (acq 2-18-99). Natl. Network: NBC. Natl. Rep: Petry Television Inc. Washington Atty: Koteen & Naftalin. News: 55 hrs wkly.

KMAX-TV— Analog channel: 31. Digital channel: 21.5,000 kw vis, 500 kw aur. ant 1,830t/2,000g TL: N38 15 52 W121 29 22 On air date: Oct 5, 1974. 2713 Kovr Dr., West Sacramento, CA, 95605-1600. Phone: (916) 374-1313. Fax: (916) 374-1459. Web Site: www.cw31.com. Licensee: Sacramento Television Stations Inc. Group owner: Viacom Television Stations Group (acq 3-24-98; $100 million). Population served: 254,413 Natl. Network: CW. Foreign lang progmg: SP 33
Key Personnel:
Bruno Cohen . . . . VP & gen mgr
Kevin Walsh. gen sls mgr
Gavin Joe. . . . . natl sls mgr
Drew Fowler. . . . . prom mgr
Rita Gazitano . . . . progmg mgr
Brent Baader . . . . news dir
Bob Hess . . . . chief of engrg

KOVR—Stockton, Analog channel: 13. Digital channel: 25. Analog hrs: 24 Digital hrs: 24 316 kw vis, 47.4 kw aur. ant 2,001t/2,011g TL: N38 14 24 W121 30 03 On air date: Sept 5, 1954. 2713 KOVR Dr., West Sacramento, CA, 95605. Phone: (916) 374-1313. Fax: (916) 374-1459. Fax: (916) 374-1304. Web Site: www.cbs13.com. Licensee: Sacramento Television Stations Inc. Group owner: Sinclair Broadcast Group Inc. (acq 4-29-2005; $285 million). Population served: 2,250,000 Natl. Network: CBS. News staff: 120; News: 21 hrs wkly.
Key Personnel:
Bruno Cohen . . . . gen mgr
Kevin Walch . . . . gen sls mgr
Rita Gazitano . . . . progmg dir & progmg mgr
Denise Dituri . . . . pub affrs dir
Bob Hess . . . . engrg dir & chief of engrg

KQCA—Stockton, Analog channel: 58. Digital channel: 46.5,000 kw vis, 500 kw aur. ant 1,761t/1,769g TL: N38 15 54 W121 29 24 On air date: Apr 13, 1986. 58 Television Cir., Sacramento, CA, 95814-0794. Phone: (916) 446-3333. Fax: (916) 554-4658. Web Site: www.my58.com. Licensee: Hearst-Argyle Stations Inc. Group owner: Hearst-Argyle Television Inc. (acq 1-24-2000; less than $1 million). Population served: 2,100,000 Natl. Network: MyNetworkTV. Washington Atty: Skadden, Arps, Slate, Meagher & Flom.
Key Personnel:
Elliot Troshinsky. gen mgr
Jerry Brehm . . . . sls dir & natl sls mgr
Patrick Donnelly . . . . gen sls mgr
Jim Caselli . . . . mktg dir & progmg dir
Gene Robinson . . . . prom mgr
Dan Weiser . . . . news dir
Stefan Hadl . . . . pub affrs dir & engrg dir

KSPX— Analog channel: 29. Digital channel: 48. Analog hrs: 24 5,000 kw vis, 500 kw aur. ant 1,296t/1,300g TL: N38 37 49 W120 51 20 (CP: 5,000 kw vis, ant 1,053t) On air date: Aug 27, 1990. 3352 Mather Field Rd., Rancho Cordova, CA, 95670. Phone: (916) 368-2929. Fax: (916) 368-0225. Licensee: Paxson Sacramento License Inc. Group owner: Paxson Communications Corp. (acq 5-25-2000; $17.725 million). Natl. Network: i Network. Washington Atty: Wiley, Rein & Fielding. News staff: 75; News: 7 hrs wkly.
Key Personnel:
Jim Eaton. . . . . gen sls mgr
Lee Roberts . . . . gen mgr & pub affrs dir
Frank Ernandes. . . . . engrg dir & chief of engrg

KTFK-TV— Analog channel: 64. Analog hrs: 24 1,950 kw vis, 195 kw aur. ant 2,980t/90g TL: N37 53 35 W121 53 58 On air date: July 11, 1988. 1710 Arden Way, Sacramento, CA, 95815. Phone: (916) 927-1900. Licensee: TeleFutura Sacramento LLC. Group owner: Univision Communications Inc. (acq 12-1-2003; $65 million). Population served: 8,941,512 Natl. Network: TeleFutura (Spanish). Foreign lang progmg: SP 168
Key Personnel:
Diego Ruiz . . . . gen mgr & stn mgr
Steve Stuck . . . . gen sls mgr

KTXL— Analog channel: 40. Analog hrs: 8:30 AM-5:30 PM 5,000 kw vis, 1,000 kw aur. ant 1,962t/1,968g TL: N38 16 18 W121 30 18 On air date: Oct 26, 1968. 4655 Fruitridge Rd., Sacramento, CA, 95820-5299. Phone: (916) 454-4422. Fax: (916) 739-1079. Web Site: www.ktxl.com. Licensee: Channel 40 Inc. Group owner: Tribune Broadcasting Co. (acq 3-25-97; grpsl). Population served: 8,245,000 Natl. Network: Fox. Natl. Rep: TeleRep. News: 7 hrs wkly.
Key Personnel:
Audrey L. Farrington . . . . VP & gen mgr
Bill Gee . . . . opns mgr, progmg dir & progmg mgr
Mike Armstrong . . . . gen sls mgr
Eric Byers. . . . . rgnl sls mgr
Pam Schoen . . . . prom dir
Steve Kraycik. . . . . news dir
Elyse Dietrich . . . . pub affrs dir
Jack Davis. chief of engrg

KUVS-TV—Modesto, Analog channel: 19. Digital channel: 18.5,000 kw vis, 560 kw aur. 1,877t/250g TL: N38 14 20 W121 28 52 On air date: Aug 26, 1966. 1710 Arden Way, Sacramento, CA, 95815. 1150 9th St., Suite 1505, Modesto, CA 95354. Phone: (916) 927-1900. Fax: (916) 614-1902. Web Site: www.univision.com. Licensee: KUVS License Partnership G.P. Group owner: Univision Communications Inc. (acq 3-6-97; $40 million). Population served: 3,480,000 Natl. Network: Univision (Spanish). Washington Atty: Shaw Pittman. Foreign lang progmg: SP 168 News staff: 30; News: 12 hrs wkly.

***KVIE**— Analog channel: 6. Digital channel: 53. Analog hrs: 24 100 kw vis, 10 kw aur. ant 1,804t/1,811g TL: N38 16 18 W121 30 22 On air date: Feb 23, 1959. 2595 Capitol Oaks Dr., Sacramento, CA, 95833. Phone: (916) 929-5843. Fax: (916) 929-7215. E-mail: publicinfo@kvie.org Web Site: www.kvie.org. Licensee: KVIE Inc. Population served: 4,200,000 Natl. Network: PBS. Washington Atty: Dow, Lohnes & Albertson.
Key Personnel:
David Hosley . . . . pres & gen mgr
David Lowe . . . . mktg VP
Jan Tilmon. . . . . progmg VP
Michael Wall . . . . chief of engrg

KXTV— Analog channel: 10. Digital channel: 61. Analog hrs: 24 Digital hrs: 24 316 kw vis, 5.13 kw aur. 1,953t/1,960g TL: N38 14 24 W121 30 03 On air date: Mar 20, 1955. 400 Broadway, Sacramento, CA, 95818-2041. Phone: (916) 441-2345. Fax: (916) 321-3384. Web Site: www.news10.net. Licensee: KXTV Inc. Group owner: Gannett Broadcasting (acq 1999; swap with KVUE-TV Austin, TX). Population served: 2,200,000 Natl. Network: ABC. Natl. Rep: Blair Television. Washington Atty: Wiley, Rein & Fielding.
Key Personnel:
Russell Postell . . . . pres & gen mgr
Kelly Bradley . . . . gen sls mgr
Dustin Snyder . . . . natl sls mgr & rgnl sls mgr
Ron Comings. . . . . news dir
Rod Robinson . . . . chief of engrg

Salinas

see Monterey-Salinas, CA market

San Diego, CA (DMA 27)

KFMB-TV— Analog channel: 8. Digital channel: 55. Analog hrs: 7 AM-1 AM 316 kw vis, 63.2 kw aur. ant 745t/249g TL: N32 50 17 W117 14 57 On air date: May 16, 1949. 7677 Engineer Rd., San Diego, CA, 92111. Box 85888, San Diego, CA 92186. Phone: (858) 571-8888. Fax: (858) 495-9363. Web Site: www.kfmb.com. Licensee: Midwest Television Inc. Ownership: Elisabeth Meyer Kimmel, 51% of class B voting stock; August C. Meyer Jr., 49% of class B voting stock Group owner: (group owner; (acq 4-17-2007; with KFMB-AM-FM San Diego). Population served: 710,000 Natl. Network: CBS. Natl. Rep: TeleRep. Washington Atty: Covington & Burling. Wire Svc: UPI
Key Personnel:
Ed Trimble . . . . pres & gen mgr
Rich Lochmann . . . . opns mgr & engrg mgr
John Marquiss . . . . sls dir
Sheri Kowalke . . . . rgnl sls mgr

KGTV— Analog channel: 10. Digital channel: 25. Analog hrs: 24 316 kw vis, 31.6 kw aur. ant 745t/223g TL: N32 50 20 W117 14 56 On air date: Sept 13, 1953. Box 85347, San Diego, CA, 92186. Phone: (619) 237-1010. Fax: (619) 262-1302. Web Site: www.10news.com. Licensee: McGraw-Hill Broadcasting Co. Group owner: (group owner; (acq 6-1-72; grpsl; FTR: 3-13-72). Population served: 735,000 Natl. Network: ABC. Natl. Rep: Harrington, Righter & Parsons. Washington Atty: Dow, Lohnes & Albertson, PLLC. News staff: 80; News: 35 hrs wkly.
Key Personnel:
Derek Dalton . . . . gen mgr
Mike Biltucci . . . . opns dir & opns mgr
Ken Rycyzn . . . . gen sls mgr
Gary Brown . . . . news dir
Pam Smith . . . . engrg mgr
Ron Eden . . . . chief of engrg

KNSD— Analog channel: 39. Digital channel: 40.2,510 kw vis. and 1,893t/154g TL: N32 41 48 W116 56 06 On air date: Nov 14, 1965. 225 Broadway, San Diego, CA, 92101. Phone: (619) 231-3939. Fax: (619) 578-0225. E-mail: feedback@nbcsandiego.com Web Site: www.nbcsandiego.com. Licensee: Station Venture Operations LP. Group owner: NBC TV Stations Division (acq 3-2-98; with KXAS-TV Fort Worth, TX). Natl. Network: NBC. Washington Atty: Pepper & Corazzini. Foreign lang progmg: SP 3
Key Personnel:
Phylliss Schwartz . . . . gen mgr
Randy Mickler . . . . opns mgr

***KPBS**— Analog channel: 15. Digital channel: 30. Analog hrs: 22 3,310 kw vis, 302 kw aur. ant 1,876t/182g TL: N32 41 53 W116 56 03 On air date: June 25, 1967. 5200 Campanile Dr., Hickory, CA, 92182-5400. Phone: (619) 594-1515. Fax: (619) 594-3812. E-mail: letters@kpbs.org Web Site: www.kpbs.org. Licensee: Board of Trustees, California State University for San Diego State University. Population served: 2,000,000 Natl. Network: PBS. Washington Atty: Bryan Cave.
Key Personnel:
Doug Myrland. . . . . gen mgr
Keith York . . . . progmg dir
Michael Marcotte . . . . news dir

KSWB-TV— Analog channel: 69. Digital channel: 19. Analog hrs: 24 (W-S) 4,790 kw vis, 479 kw aur. 1,950t/151g TL: N32 41 47 W116 56 07 On air date: Oct 1, 1984. 7191 Engineer Rd., San Diego, CA, 92111. Phone: (858) 492-9269. Fax: (858) 268-0401. Web Site: www.kswbtv.com. Licensee: KSWB Inc. Group owner: Tribune Broadcasting Co. (acq 1996; $70.5 million). Population served: 943,500 Natl. Network: CW. Natl. Rep: TeleRep. Washington Atty: Sidley & Austin. News: 14 hrs wkly.

KUSI-TV— Analog channel: 51. Digital channel: 18. Analog hrs: 24 2,820 kw vis, 288 kw aur. ant 1,916t/177g TL: N32 41 49 W116 56 05 On air date: Sept 13, 1982. Box 719051, San Diego, CA, 92171. 4575 Viewridge Ave., San Diego, CA 92171. Phone: (858) 571-5151. Fax: (858) 505-5050. E-mail: flaherty@kusi.com Web Site: www.kusi.com. Licensee: Channel 51 of San Diego Inc. Group owner: McKinnon Broadcasting Co. (acq 6-29-90; FTR: 4-30-90). Washington Atty: Cohn & Marks. News staff: 50; News: 25 hrs wkly.
Key Personnel:
Mike McKinnon . . . . pres, gen mgr & opns VP
Craig Hume . . . . news dir
Richard Large . . . . engrg dir & chief of engrg

XETV—(Tijuana, MEX) Analog channel: 6. Analog hrs: 24 100 kw vis, 50 kw aur. 1,000t/550g On air date: Jan 29, 1953. 8253 Ronson Rd., San Diego, CA, 92111. Phone: (858) 279-6666. Fax: (858) 268-9388. E-mail: comments@fox6.com Web Site: www.fox6.com. Licensee: Radio-Television SA. Ownership: Grupo Televisa, 100%. Population served: 920,570 Natl. Network: Fox. Washington Atty: Leventhal, Senter & Lerman. News: 20 hrs wkly.
Key Personnel:
Rodrigo Salazar . . . . CFO
Richard Doutre Jones . . . . VP & gen mgr
Bob Anderson . . . . opns dir & opns mgr
Chuck Dunning. . . . . gen sls mgr
Harry Melkerson . . . . natl sls mgr
Lynda DiLorenzo . . . . rgnl sls mgr
Scott Dillon . . . . rgnl sls mgr
Judy Albrecht. . . . . prom dir & prom mgr
Deirdre Bianchi . . . . progmg dir & progmg mgr
Raphael Ahlgren . . . . pub affrs dir
Gary Stigall . . . . chief of engrg

XEWT-TV—(Tijuana, MEX) Analog channel: 12. Analog hrs: 20 325 kw vis, 32.5 kw aur. 1,000t/200g TL: N32 30 06 W117 02 23 On air date: July 12, 1960. Box 434537, San Diego, CA, 92143. 637 Third Ave., Suite B, Chulavista, CA 91910. Phone: (800) TELEV 12. Phone: (619) 585-9398. Fax: (619) 585-9463. Web Site: www.televisa.com. Licensee: Televisora de Calimex, SA. Ownership: Televisa, S.A. Population served: 2,200,000 Washington Atty: Leventhal, Senter & Lerman. Foreign lang progmg: SP 168 News staff: 50; News: 11 hrs wkly.
Key Personnel:
Ricardo Azcarraga . . . . gen mgr
Lourdes Numez . . . . opns mgr

San Francisco-Oakland-San Jose (DMA 5)

KBCW— Analog channel: 44. Digital channel: 45. Analog hrs: 24 5,000 kw vis, 500 kw aur. 1,610t/977g TL: N37 45 19 W122 27 06 On air date: Jan 2, 1968. 855 Battery St., San Francisco, CA, 94111-1509. Phone: (415) 765-8144. Fax: (415) 765-8844. Web Site: www.kbcwtv.com / www.cwbayarea.com. Licensee: San Francisco Television Station KBCW Inc. Group owner: Viacom Television Stations Group (acq 11-6-2001; swap with WDCA(TV) Washington, DC; and KTXH(TV) Houston, TX). Population served: 6,373,000 Natl. Network: CW. Washington Atty: Hogan & Hartson.
Key Personnel:
Ron Lonsinotti . . . . pres & gen mgr
Steve Poitras . . . . VP
Steve Poitres. . . . . VP & stn mgr
Rosemary Roach . . . . pub affrs dir

KBWB— Analog channel: 20. Digital channel: 19.Note: Azteca America is on KBWB-DT ch 19. 3,470 kw vis, 347 kw aur. ant 1,548t/820g TL: N37 45 19 W122 27 06 On air date: Apr 1, 1968. 2500 Marin St., San Francisco, CA, 94124. Phone: (415) 821-2020. Fax: (415) 821-1518. Web Site: www.yourtv20.com. Licensee: KBWB License Inc. Group owner: Granite Broadcasting Corp. (acq 7-20-98; $173.75 million). Population served: 1,200,000 Natl. Rep: MMT.
Key Personnel:
Bob Anderson . . . . pres, CFO, CFO & gen mgr
Dennis McNamara . . . . stn mgr & sls VP
Dave Figura . . . . opns mgr & mktg dir
Steve Jones. . . . . gen sls mgr & natl sls mgr
Jennifer King . . . . prom mgr
Michele Ball . . . . progmg dir
Frank Brucks. . . . . chief of engrg

KCNS— Analog channel: 38. Digital channel: 39. Analog hrs: 24 Digital hrs: 24 5,000 kw vis, 500 kw aur. ant 1,443t/726g TL: N37 45 20 W122 27 05 On air date: Jan 3, 1986. 1550 Bryant St., Suite 740, San Francisco, CA, 94103. Phone: (415) 863-3800. Fax: (415) 863-3998. E-mail: kcnstv@pacbell.net Web Site: www.kcnstv.com. Licensee: MTB San Francisco Licensee LLC. Group owner: Scripps Howard Broadcasting Co. (acq 2006; grpsl). Population served: 7,150,000

***KCSM-TV—** Analog channel: 43. Analog hrs: 24 536 kw vis. ant 1,404t/672g TL: N37 45 19 W122 27 06 On air date: Oct 12, 1964. 1700 W. Hillsdale Blvd., San Mateo, CA, 94402. Phone: (650) 574-6586. Fax: (650) 524-6975. Web Site: www.kcsm.org. Licensee: San Mateo County Community College District. Population served: 487,000 Natl. Network: PBS. Washington Atty: Tierney & Swift. Foreign lang progmg: SP 8
Key Personnel:
Marilyn Lawrence . . . . gen mgr
Alisa Clancy . . . . opns mgr
Shelly Rogers . . . . dev dir
Michelle Muller . . . . engrg dir

KDTV— Analog channel: 14.3,980 kw vis, 398 kw aur. 2,509t/439g TL: N37 29 57 W121 52 16 (CP: 3,390 kw vis, ant 2,509t. TL: N37 29 57 W121 52 16) On air date: Aug 13, 1975. 50 Fremont St., 41st Fl., Elizabeth City, CA, 94105. Phone: (415) 538-8000. Fax: (415) 538-8053. Web Site: www.univison.com. Licensee: KDTV L.P. Group owner: Univision Communications Inc. Population served: 1,023,300 Natl. Network: Univision (Spanish). Washington Atty: Fisher, Wayland, Cooper, Leader & Zaragoza. Foreign lang progmg: SP 168 News staff: 15; News: 5 hrs wkly.
Key Personnel:
Jim VanTassell . . . . VP & opns mgr
Marcela Medina . . . . gen mgr
Ernie Rizzuti. . . . . gen sls mgr
Maria Rodriquez. . . . . progmg dir
Sandra Thomas . . . . news dir
Mike Roberts. . . . . chief of engrg

KFSF-TV— Analog channel: 66. Digital channel: 34. Analog hrs: 24 3,470 kw vis, 346 kw aur. ant 1,528t/797g TL: N37 45 19 W122 27 16 On air date: Nov 25, 1986. 50 Fremont St., 4th Fl., San Francisco, CA, 94105. Phone: (415) 538-6466. Fax: (415) 538-8053. Licensee: Univision Spanish Media Inc. Group owner: Univision Communications Inc. (acq 12-18-2001; $39 million). Population served: 5,500,000 Natl. Network: TeleFutura (Spanish). Foreign lang progmg: SP 168
Key Personnel:
Jim VanTassell. . . . . CFO & opns mgr
Marcela Medina . . . . gen mgr
Ernie Rizzuto . . . . stn mgr & gen sls mgr
Maria S. Rodriguez . . . . dev VP & progmg dir
Sandra Thomas . . . . news dir
Mike Roberts. . . . . chief of engrg

KFTY— Analog channel: 50.302 kw vis, 60.4 kw aur. 3,080t/172g TL: N38 40 10 W122 37 52 On air date: May 1, 1981. 533 Mendocino Ave., Santa Rosa, CA, 95401. Phone: (707) 526-5050. Fax: (707) 526-7429. Web Site: www.kfty.com. Licensee: Ackerley Broadcasting Operations LLC. Group owner: Clear Channel Communications Inc. (acq 6-14-2002; grpsl). Population served: 1,500,000 Washington Atty: Rubin, Winston, Diercks, Harris & Cooke. Wire Svc: NWS (National Weather Service) Wire Svc: Bay City News Service News staff: 14; News: 7 hrs wkly.
Key Personnel:
John Burgess. VP, gen mgr, rgnl sls mgr, mktg VP, mktg dir, prom VP, prom dir, progmg VP, progmg dir & news dir
Richard Starkey . . . . opns mgr & chief of opns
Rob Rector . . . . gen sls mgr
Eric Casella. . . . . progmg dir & news dir
Brad Thompson. . . . . chief of engrg

KGO-TV— Analog channel: 7. Digital channel: 24.316 kw vis, 63.2 kw aur. ant 1,670t/977g TL: N37 45 20 W122 27 05 On air date: May 5, 1949. 900 Front St., San Francisco, CA, 94111-1450. Phone: (415) 954-7777. Fax: (415) 956-6402. Web Site: www.abc7news.com. Licensee: KGO-TV Inc. Group owner: ABC Inc. (acq 6-27-86; grpsl; FTR: 7-15-85). Population served: 5,094,700 Natl. Network: ABC.

KICU-TV— Analog channel: 36. Analog hrs: 24 4,098 kw vis, 409.8 kw aur. 2,250t/600g TL: N37 29 17 W121 51 59 On air date: Oct 3, 1967. 2102 Commerce Dr., San Jose, CA, 95131-1804. Phone: (408) 953-3636. Web Site: www.kicu.com. Licensee: KTVU Partnership. Group owner: Cox Broadcasting (acq 2-18-00; $130 million). Population served: 5,824,520 Washington Atty: Leventhal, Senter & Lerman. News staff: 3; News: .5 hr wkly.
Key Personnel:
Tom Raponi . . . . gen mgr
Robert Martinez . . . . sls VP & gen sls mgr
Carolyn Chang . . . . progmg dir & progmg mgr
Chuck Pracna . . . . chief of engrg

KKPX— Analog channel: 65. Analog hrs: 24 3,060 kw vis, 1,179 kw aur. 2,667t/223g TL: N37 06 41 W121 50 30 On air date: Nov 12, 1986. 848 Battery St., San Francisco, CA, 94111. Phone: (415) 276-1400. Fax: (415) 276-1401. Web Site: www.pax.tv. Licensee: Paxson San Jose License Inc. Group owner: Paxson Communications Corp. (acq 3-22-95; $5 million; FTR: 6-19-95). Natl. Network: i Network. Washington Atty: Joseph E. Dunne III.
Key Personnel:
Carol Denham . . . . gen mgr
Bob Getsla . . . . chief of engrg

***KMTP-TV—** Analog channel: 32. Analog hrs: 24 1,334 kw vis, 267 kw aur. 1,610t/885g TL: N37 45 20 W122 27 05 On air date: Aug 31, 1991. 1010 Cooperation Way, Palo Alto, CA, 94303. Phone: (415) 777-3232. Fax: (415) 552-3209. E-mail: kmtpgm@pacbell.net Web Site: www.kmtp.org. Licensee: Minority Television Project. Population served: 2,000,000 Natl. Network: PBS.
Key Personnel:
Arlene Stevens. . . . . opns dir, progmg dir & news dir
Booker T. Wade Jr. . . . . gen mgr & dev dir

KNTV— Analog channel: 11. Digital channel: 12. Analog hrs: 24 80 kw vis, 8 kw aur. 2,770t/291g TL: N37 06 40 W121 50 34 On air date: Sept 12, 1955. 2450 N. 1st St., San Jose, CA, 95131. Phone: (408) 286-1111. Fax: (408) 422-4425. Web Site: www.nbc11.com. Licensee: NBC Telemundo License Co. Group owner: NBC TV Stations Division (acq 4-30-2002; $230 million). Population served: 700,000 Natl. Network: NBC. Natl. Rep: Harrington, Righter & Parsons. Washington Atty: Akin, Gump, Strauss, Hauer & Feld. News: 20 hrs wkly.
Key Personnel:
Rich Cerussi . . . . gen mgr
Caroline Chang . . . . progmg dir
Jim Sanders . . . . news dir

KPIX-TV— Analog channel: 5. Digital channel: 29.100 kw vis, 10 kw aur. 1,660t/980g TL: N37 45 20 W122 27 05 On air date: Dec 22, 1948. 855 Battery St., San Francisco, CA, 94111-1597. Phone: (415) 362-5550. Fax: (415) 765-8844. Web Site: www.cbs5.com. Licensee: CBS Broadcasting Inc. Group owner: Viacom Television Stations Group (acq 5—4-2000; grpsl). Population served: 2,253,220 Natl. Network: CBS. Natl. Rep: CBS TV Stations National Sales. Washington Atty: Wilkes, Artis, Hedrick & Lane. Wire Svc: Reuters News staff: 87; News: 20 hrs wkly.
Key Personnel:
Ron Longinotti. . . . . pres, VP & gen mgr
Dee Joyce . . . . mktg dir
Tom Spitz . . . . progmg dir
Dan Rosenheim. . . . . news dir
Rosemary Roach . . . . pub affrs dir
Mike Englehaupt . . . . chief of engrg

***KQED—** Analog channel: 9. Digital channel: 30. Analog hrs: 24 316 kw vis, 37 kw aur. 1,670t/980g TL: N37 45 17 W122 27 06 On air date: June 10, 1954. 2601 Mariposa St., San Francisco, CA, 94110-1426. Phone: (415) 864-2000. Fax: (415) 553-2241. E-mail: (name)@kqed.org Web Site: www.kqed.org. Licensee: Northern California Public Broadcasting Inc. Population served: 6,200,000 Natl. Network: PBS. Washington Atty: Arnold & Porter LLP. News staff: 8; News: 1/2 hr wkly.
Key Personnel:
Jeff Clarke . . . . CEO & pres
Michael Isip. . . . . stn mgr
Traci Eckels. . . . . dev VP
Donald Derheim . . . . sls VP

***KRCB—** Analog channel: 22. Digital channel: 23. Analog hrs: 10 AM-11 PM 68.823 kw vis, 6.882 kw aur. 2,034t TL: N38 25 07 W122 40 33 On air date: Dec 2, 1984. 5850 Labath Ave., Rohnert Park, CA, 94928. Phone: (707) 584-2000. Fax: (707) 585-1363. E-mail: viewer@krcb.org Web Site: www.krcb.org. Licensee: Rural California Broadcasting Corp. Population served: 2,300,000 Natl. Network: PBS. Foreign lang progmg: SP 3
Key Personnel:
Nancy Dobbs . . . . CEO, pres & gen mgr
Stan Marvin . . . . progmg dir

KRON-TV— Analog channel: 4. Digital channel: 57. Analog hrs: 24 100 kw vis, 15.1 kw aur. 1,680t/977g TL: N37 45 19 W122 27 06 On air date: Nov 15, 1949. 1001 Van Ness Ave., San Francisco, CA, 94109. Phone: (415) 441-4444. Fax: (415) 561-8142. Web Site: www.kron.com. Licensee: Young Broadcasting of San Francisco Inc. Group owner: Young Broadcasting Inc. (acq 6-26-2000; $823 million). Population served: 6,700,000 Natl. Network: MyNetworkTV. Natl. Rep: Adam Young. Washington Atty: Brooks, Pierce. Wire Svc: Conus News: 42 hrs wkly.
Key Personnel:
Mark Antonitis . . . . pres & gen mgr
Mary Kennedy. . . . . stn mgr
Mark Mano. . . . . opns dir & opns mgr
Sarah Squiers . . . . dev dir & rgnl sls mgr
Karen Orofino . . . . sls VP & gen sls mgr

Ben Holland . . . . . natl sls mgr
Jeffrey Weinstock . . . . . mktg dir, prom dir & adv dir
Pat Patton . . . . . progmg VP
Stacy Owen . . . . . news dir
Javier Valeucia . . . . . pub affrs dir
Craig Porter . . . . . chief of engrg

KSTS— Analog channel: 48. Digital channel: 49.2,510 kw vis. ant 2,257t/420g TL: N37 29 57 W121 52 16 On air date: May 31, 1981. 2450 N 1st St., San Jose, CA, 95131-1002. Phone: (408) 435-8848 / (408) 944-4848. Fax: (408) 433-5921. Web Site: www.ksts.com. Licensee: NBC Telemundo License Co. Group owner: Telemundo Group Inc. (acq 4-12-2002; grpsl). Population served: 1,700,000 Natl. Network: Telemundo (Spanish). Washington Atty: Hogan & Hartson. Wire Svc: Reuters Wire Svc: UPI Foreign lang progmg: SP 100 News staff: 10; News: 3 hrs wkly.

***KTEH—** Analog channel: 54. Digital channel: 50.661 kw vis, 132 kw aur. 1,922t/137g TL: N37 29 07 W121 51 57 On air date: October 1964. 1585 Schallenburger Rd., San Jose, CA, 95110-1301. Phone: (408) 795-5400. Fax: (408) 995-5446. Web Site: www.kteh.org. Licensee: Northern California Public Broadcasting Inc. (acq 10-1-2006). Population served: 453,000 Natl. Network: PBS. Rgnl. Network: Pacific. Washington Atty: Schwartz, Woods & Miller. Foreign lang progmg: SP 3
Key Personnel:
Thomas Fanella . . . . . CEO
Judy Armstrong . . . . . pres & dev dir

KTLN-TV— Analog channel: 68. Analog hrs: 24 1,100 kw vis, 110 kw aur. ant 1,319t/59g TL: N38 09 00 W122 35 31 On air date: Aug 31, 1998. 400 Tamal Plaza, Suite 428, Corte Madera, CA, 94925. Phone: (415) 924-7500. Fax: (415) 924-0264. E-mail: ktln@tln.com Web Site: www.ktln.tv. Licensee: Christian Communications of Chicagoland Inc. (acq 12-10-98; $500,000). Population served: 7,000,000
Key Personnel:
Jerry K. Rose . . . . . CEO & pres
James Nichols . . . . . CFO
Debra Fraser . . . . . gen mgr
Brian Avery . . . . . stn mgr

KTNC-TV— Analog channel: 42. Digital channel: 63. Analog hrs: 24 1,290 kw vis. ant 2,808t TL: N37 53 34 W121 53 53 On air date: June 19, 1983. 1700 Montgomery St., Suite 400, San Francisco, CA, 94111. Phone: (415) 398-4242. Fax: (415) 352-1800. E-mail: rpineda@ktnc.com Web Site: www.ktnc.com. Licensee: KTNC License LLC. Group owner: Pappas Telecasting Companies (acq 10-29-97). Population served: 10,000,000 Washington Atty: Fletcher, Heald & Hildreth. Foreign lang progmg: SP 168
Key Personnel:
Dennis Davis . . . . . CEO
Harry Pappas . . . . . chmn & pres
Fernando Acosta . . . . . gen mgr
LeBon Abercrombie . . . . . dev dir
Roberto Pineda . . . . . gen sls mgr

KTSF— Analog channel: 26. Digital channel: 27. Analog hrs: 24 Digital hrs: 24 2,510 kw vis, 500 kw aur. ant 1,380t/259g TL: N37 41 12 W122 26 03 On air date: Sept 4, 1976. 100 Valley Dr., Brisbane, CA, 94005-1350. Phone: (415) 468-2626. Fax: (415) 467-7559. E-mail: admin@ktsftv.com Web Site: www.ktsf.com. Licensee: Lincoln Broadcasting Co., a California L.P. Population served: 6,000,000 Washington Atty: Law Office of Michael D. Berg. Wire Svc: Bay City News Service Wire Svc: AP Wire Svc: CNN News staff: 40; News: 24 hrs wkly.
Key Personnel:
Lillian L. Howell . . . . . chmn
Lincoln C. Howell . . . . . pres
Michael Sherman . . . . . gen mgr
Mike Fusaro . . . . . chief of opns & engrg dir
Lisa Yokota . . . . . mktg dir
Victor Marino . . . . . progmg dir
Rose Shirinian . . . . . news dir

KTVU—Oakland, Analog channel: 2. Digital channel: 56. Analog hrs: 24 100 kw vis, 20 kw aur. 1,811t/980g TL: N37 45 20 W122 27 05 On air date: Mar 3, 1958. Box 22222, Oakland, CA, 94623. Phone: (510) 834-1212. Fax: (510) 272-9957. Web Site: www.ktvu.com. Licensee: KTVU Partnership. Group owner: Cox Enterprises (acq 10-16-63; $12.36 million; FTR: 10-21-63). Population served: 2,253,000 Natl. Network: Fox. Natl. Rep: TeleRep. Washington Atty: Dow, Lohnes & Albertson. Wire Svc: Reuters Wire Svc: NWS (National Weather Service)
Key Personnel:
Jeff Block . . . . . VP, gen mgr & stn mgr
Tom Raponi . . . . . gen mgr & sls dir
Greg Bilte . . . . . gen sls mgr
Dan Haass . . . . . natl sls mgr
Phil Adams . . . . . natl sls mgr
Caroline Chang . . . . . progmg mgr
Ed Chapuis . . . . . news dir

Rosy Chu . . . . . pub affrs dir
Don Thompson . . . . . engrg mgr
Ken Manley . . . . . chief of engrg

KUNO-TV— Analog channel: 8. Digital channel: 15.225 kw vis, 22.5 kw aur. ant 2,446t/186g TL: N39 41 38 W123 34 43 On air date: Feb 1, 1990. 500 S. Chinowth Rd., Visalia, CA, 93277. Phone: (707) 964-8888. Fax: (707) 964-8150. Licensee: Concord License LLC. Group owner: Pappas Telecasting Companies (acq 6-12-97; $1.75 million). Natl. Rep: Blair Television.

San Jose

see San Francisco-Oakland-San Jose market

San Luis Obispo

see Santa Barbara-Santa Maria-San Luis Obispo, CA market

Santa Barbara-Santa Maria-San Luis Obispo, CA (DMA 122)

KBEH— Analog channel: 63. Digital channel: 24. Analog hrs: 24 Digital hrs: 24 1,782 kw vis, 513 kw aur. 1,801t/335g TL: N34 19 49 W119 01 24 On air date: Aug 17, 1985. 950 Flynn Rd., Camarillo, CA, 93012. Phone: (805) 388-0081. Fax: (305) 863-5701. Web Site: www.canal63.com. Licensee: Bela TV LLC. Group owner: Bela LLC (acq 3-5-2004). Washington Atty: Wiley, Rein & Fielding. Foreign lang progmg: SP 168
Key Personnel:
Lou Bardfield . . . . . gen mgr & chief of engrg
Mara Rankin . . . . . gen sls mgr

KCOY-TV— Analog channel: 12. Digital channel: 19.115 kw vis, 22.9 kw aur. 1,940t/140g TL: N34 54 37 W120 11 08 On air date: Mar 16, 1964. 1211 W. McCoy Ln., Santa Maria, CA, 93455. Phone: (805) 925-1200. Fax: (805) 349-2740. E-mail: daveulrickson@clearchannel.com Web Site: www.kcoy.com. Licensee: Ackerley Broadcasting Operations LLC. Group owner: Clear Channel Communications Inc. (acq 2002; grpsl). Population served: 462,000 Natl. Network: CBS. Natl. Rep: Continental Television Sales. Washington Atty: Covington & Burling. News staff: 42; News: 28 hrs wkly.
Key Personnel:
Jeffery MacDougall . . . . . VP & gen mgr
Larry Barnes . . . . . sls dir & gen sls mgr
Tracy Reiner . . . . . rgnl sls mgr
Laurie Pipan . . . . . prom mgr
Shirley Stanford . . . . . progmg mgr
Jimmy Sprague . . . . . chief of engrg

KEYT-TV— Analog channel: 3. Digital channel: 27. Analog hrs: 24 hrs 50 kw vis, 5.9 kw aur. ant 3,010t/210g TL: N34 31 32 W119 57 28 On air date: July 24, 1953. 730 Miramonte Dr., Santa Barbara, CA, 93109. Phone: (805) 882-3933. Fax: (805) 882-3934. E-mail: keyt@aol.com Web Site: www.keyt.com. Licensee: Smith Media License Holdings LLC. Group owner: Smith Broadcasting Group Inc. (acq 11-8-2004; grpsl). Population served: 900,000 Natl. Network: ABC. Natl. Rep: Blair Television. Washington Atty: Hogan & Hartson. Wire Svc: AP Wire Svc: CNN News staff: 25; News: 21 hrs wkly.
Key Personnel:
Bob Grissom . . . . . VP, gen mgr & natl sls mgr
Dave Fete . . . . . opns mgr
Jim Valice . . . . . gen sls mgr
Jeff Martin . . . . . prom mgr & pub affrs dir
Renee Foley . . . . . progmg dir
Paul Vercammen . . . . . news dir
Dave Williams . . . . . chief of engrg

KPMR— Analog channel: 38. Digital channel: 21.2,690 kw vis. ant 2,877t TL: N34 31 32 W119 57 28 On air date: April 1, 2001. Entravision Communications Corp., 2425 Olympic Blvd., Suite 6000W, Santa Monica, CA, 90404. Phone: (805) 685-3800 / (310) 447-3870. Fax: (805) 685-6892. Web Site: www.entravision.com. Licensee: Entravision Holdings LLC. Group owner: Entravision Communications Corp. (acq 11-2-2000; $4.75 million). Natl. Network: Univision (Spanish). Foreign lang progmg: SP 168
Key Personnel:
Chris Roman . . . . . gen mgr
Michael Scanlon . . . . . gen sls mgr
Andres Angulo . . . . . news dir

KSBY— Analog channel: 6. Digital channel: 15. Analog hrs: 24 Note: NBC is on KSBY(TV) ch 6, CW is on KSBY-DT ch 15. 100 kw vis, 12 kw aur. ant 2,250t/452g TL: N35 21 37 W120 39 17 On air date: May 1953. 1772 Calle Joaquin, San Luis Obispo, CA, 93405. Phone: (805) 541-6666. Fax: (805) 541-5142. E-mail: ksby@ksby.com Web Site: www.ksby.com. Licensee: KSBY Communications Inc. Group owner: New Vision Group LLC (acq 2-18-2005; $67.75 million). Population served: 494,800 Natl. Network: NBC, CW Digital Network: Note: NBC is on KSBY(TV) ch 6, CW is on KSBY-DT ch 15. Natl. Rep: TeleRep. Washington Atty: Latham & Watkins.
Key Personnel:
Wade O'Hagen . . . . . CFO
Tim Perry . . . . . gen mgr & gen sls mgr
Carl Edge . . . . . opns dir, progmg dir & progmg mgr
Madeline Palaszeuski . . . . . prom dir & prom mgr
Madeline Palaszewski . . . . . pub affrs dir
Aaron Klohs . . . . . chief of engrg

KTAS— Analog channel: 33. Digital channel: 34. Analog hrs: 24 Digital hrs: 24 60.3 kw vis. ant 1,443t/75g TL: N35 21 38 W120 39 21 On air date: 1990. Box 172, Santa Maria, CA, 93456. 330 W. Carmen Ln., Santa Maria, CA 93458. Phone: (805) 928-7700. Fax: (805) 928-8606. E-mail: ktastv@fix.net Licensee: Raul and Consuelo Palazuelos. (acq 7-3-97). Natl. Network: Telemundo (Spanish). Washington Atty: Wiley, Rein & Fielding, LLC. Foreign lang progmg: SP 168 News staff: 6; News: 3 hrs wkly.
Key Personnel:
Sandy Keefer . . gen mgr, opns mgr, chief of opns & gen sls mgr
Roger Hernandez . . . . . news dir
Telemundo.

Santa Maria

see Santa Barbara-Santa Maria-San Luis Obispo, CA market

Stockton

see Sacramento-Stockton-Modesto, CA market

Visalia

see Fresno-Visalia, CA market

Yreka City

see Medford-Klamath Falls, OR market

Colorado

Colorado Springs-Pueblo, CO (DMA 94)

KKTV— Analog channel: 11. Digital channel: 10. Analog hrs: 24 Digital hrs: 24 Note: CBS is on KKTV(TV) ch 11, MyNetworkTV is on KKTV-DT ch 10. 234 kw vis, 46.8 kw aur. ant 2,380t/351g TL: N38 44 41 W104 51 41 On air date: Dec 7, 1952. Box 2110, Colorado Springs, CO, 80901. 3100 N. Nevada Ave., Colorado Springs, CO 80901. Phone: (719) 634-2844. Fax: (719) 632-0808. Fax: (719) 442-6981. Web Site: www.kktv.com. Licensee: WEAU Licensee Corp. Group owner: Gray Television Inc. (acq 10-25-02; grpsl). Population served: 700,000 Natl. Network: CBS, MyNetworkTV Digital Network: Note: CBS is on KKTV(TV) ch 11, MyNetworkTV is on KKTV-DT ch 10. Natl. Rep: Continental Television Sales. News staff: 42; News: 27 hrs wkly.
Key Personnel:
Robert Prather . . . . . pres
Charles Peterson . . . . . VP & gen mgr
Emily Edwards . . . . . opns mgr
Marion Houghton . . . . . gen sls mgr
Michelle Hughes . . . . . mktg dir & prom mgr
Charles Hogetvedt . . . . . progmg dir & progmg mgr
Nick Matesi . . . . . news dir
John Burrell . . . . . chief of engrg

KOAA-TV—Pueblo, Analog channel: 5.100 kw vis, 10 kw aur. 1,310t/977g TL: N38 22 25 W104 33 27 On air date: June 13, 1953. Box 195, 2200 7th Ave., Pueblo, CO, 81002-0195. 530 Communications Cir., Colorado Springs, CO 81003. Phone: (719) 544-5781. Phone: (719) 228-6277. Fax: (719) 228-6265. Web Site: www.koaa.com. Licensee: Sangre De Cristo Communications Inc. Group owner: Evening Post Publishing Co. (acq 8-6-76; $4.5 million; FTR: 8-30-76). Population served: 500,000 Natl. Network: NBC. Natl. Rep: Harrington, Righter & Parsons. Washington Atty: Dow, Lohnes & Albertson. News staff: 23; News: 7 hrs wkly.
Key Personnel:
David Whitaker . . . . . pres & gen mgr
Tom Wright . . . . . gen sls mgr & natl sls mgr

Ron Eccher . . . . progmg dir & film buyer
Cindy Aubrey . . . . news dir
Patricia Cone . . . . pub affrs dir
Quentin Henry . . . . chief of engrg

KRDO-TV— Analog channel: 13. Digital channel: 24. Analog hrs: 24 282 kw vis, 29 kw aur. ant 2,080t/100g TL: N38 44 41 W104 51 38 On air date: Sept 21, 1953. 399 S. 8th St., Colorado Springs, CO, 80905. Phone: (719) 632-1515. Fax: (719) 475-0815. Web Site: www.krdo.com. Licensee: Pikes Peak Television Inc. Group owner: (group owner) (acq 6-26-2006; $45 million with KJCT(TV) Grand Junction). Population served: 761,000 Natl. Network: ABC. Natl. Rep: Airtime TV. Rgnl. Rep: Rgnl rep: Blair Television
Key Personnel:
David Bradley Jr. . . . . pres
Neil Klockziem . . . . gen mgr

***KTSC—** Analog channel: 8. Analog hrs: 24 316 kw vis, 63.2 kw aur. 1,224t/972g TL: N38 22 25 W104 33 27 (CP: 233 kw vis, ant 2,386t. TL: N38 44 41 W104 51 38) On air date: Feb 3, 1971. 2200 Bonforte Blvd., Pueblo, CO, 81001-4901. Phone: (719) 543-8800. Fax: (719) 549-2208. E-mail: ktsc@rmpbs.org Web Site: www.rmpbs.org. Licensee: Rocky Mountain Public Broadcasting Network Inc. (acq 1999; $2.375 million). Population served: 219,930 Natl. Network: PBS.
Key Personnel:
Tom Scheel. . . . . chmn & dev dir
James Morgese . . . . pres
Wynona Sullivan . . . . stn mgr
Tiffany Q. Tyson. . . . . mktg dir
Donna Sanford . . . . progmg dir
Ian Hartley. chief of engrg

KVSN— Analog channel: 48.950 kw vis. ant 2,280t/271g TL: N38 44 42 W104 51 37 Not on air, target date: unknown: 777 Grant St., Suite 500, Denver, CO, 80203. Phone: (303) 832-0050. Fax: (303) 832-3410. Web Site: www.entravision.com. Permittee: Entravision Holdings LLC.
Key Personnel:
Walter F. Ulloa . . . . CEO
Mario Carrera. . . . . gen mgr

KXRM-TV— Analog channel: 21. Digital channel: 22. Analog hrs: 24 1,054 kw vis, 105.4 kw aur. ant 2,085t/125g TL: N38 44 40 W104 51 37 (CP: 1,700 kw vis, ant 2,152t) On air date: Dec 24, 1984. 560 Wooten Rd., Colorado Springs, CO, 80915. Phone: (719) 596-2100. Fax: (719) 591-4180. E-mail: info@kxrm.com Web Site: www.yourfavoritestation.com. Licensee: Barrington Colorado Springs License LLC. Group owner: Raycom Media Inc. (acq 8-11-2006; grpsl). Natl. Network: Fox. Natl. Rep: TeleRep. Washington Atty: Covington & Durling. Wire Svc: CNN News: 3.5 hrs wkly.
Key Personnel:
K. James Yager . . . . CEO
Steve Dant . . . . VP & gen mgr
Donna D'Amico . . . . gen sls mgr & natl sls mgr
Dan Corken . . . . rgnl sls mgr
Susan Corbin . . . . mktg mgr
Patti Clements . . . . progmg dir
Larry W. Douglas. . . . . film buyer
Joe Duckett . . . . chief of engrg

Denver, CO (DMA 18)

***KBDI-TV—** Broomfield, Analog channel: 12. Digital channel: 38. Analog hrs: 24 229 kw vis, 22.9 kw aur. ant 2,420t/66g TL: N39 40 55 W105 29 49 On air date: Feb 22, 1980. 2900 Welton St., Denver, CO, 80205. Phone: (303) 296-1212. Fax: (303) 296-6650. Web Site: www.kbdi.org. Licensee: Colorado Public Television Inc. Population served: 500,000 Natl. Network: PBS. Washington Atty: Mintz, Levin, Cohn, Ferris, Glovsky & Popeo.
Key Personnel:
Dr. Willard Rowland . . . . CEO, chmn & pres
Kim Johnson . . . . opns VP
Darrow Hodges . . . . dev VP & adv VP
Kirby McClure. . . . . progmg dir

KCEC— Analog channel: 50. Digital channel: 51.2,498 kw vis. 764t TL: N39 43 47 W105 07 16 On air date: Oct 19, 1990. 777 Grant St., 5th Fl., Monroe, CO, 80203. Phone: (303) 832-0050. Fax: (303) 832-3410. Licensee: Entravision Holdings L.L.C. Group owner: Entravision Communications Co. L.L.C. (acq 4-25-97). Population served: 624,000 Natl. Network: Univision (Spanish). Foreign lang progmg: SP 168
Key Personnel:
Walter Ulloa . . . . pres
Mario M. Carrera . . . . gen mgr
Don Daboub. . . . . gen sls mgr
Mark Goodrich . . . . natl sls mgr
Kathy Berumen . . . . prom dir
Erma Atencio . . . . progmg mgr
Rodolfo Cardenas . . . . news dir

Jamie Moreno . . . . pub affrs dir
Carl Cutforth . . . . engrg mgr

KCNC-TV— Analog channel: 4. Digital channel: 35. Analog hrs: 24 100 kw vis, 35.1 kw aur. 1,480t/833g TL: N39 43 48 W105 14 02 On air date: Dec 24, 1953. 1044 Lincoln St., Denver, CO, 80203. Phone: (303) 861-4444. Fax: (303) 830-6537. Web Site: news4colorado.com. Licensee: CBS Television Stations Inc. Group owner: Viacom Television Stations Group (acq 9-10-95; grpsl). Population served: 2,642,000 Natl. Network: CBS. Natl. Rep: CBS TV Stations National Sales. Wire Svc: Conus Wire Svc: PR Newswire Wire Svc: Medialink
Key Personnel:
Walt DeHaven . . . . gen mgr
David Layne. . . . . opns mgr
David Rash . . . . sls dir
Kevin Dorsey. . . . . sls dir & gen sls mgr
Roxanne Marati . . . . mktg dir
Wendy Holmes . . . . progmg dir

KDEN— Analog channel: 25. Digital channel: 29.5,000 kw vis. ant 1,066t/964g TL: N40 05 47 W104 54 04 On air date: 1999. 1120 Lincoln St., Suite 800, Denver, CO, 80203-2137. Phone: (303) 832-0402 (Denver). Fax: (303) 832-0777. Licensee: NBC Telemundo License Co. (acq 7-13-2006; $42 million). Natl. Network: Telemundo (Spanish). Washington Atty: NBC Legal. Foreign lang progmg: SP 168 News staff: one; News: one hr wkly.
Key Personnel:
Don Brown . . . . CEO
Clara Rivas . . . . VP & gen mgr

KDVR— Analog channel: 31. Digital channel: 32. Analog hrs: 24 5,000 kw vis, 500 kw aur. 1,038t/440g TL: N39 43 45 W105 14 12 On air date: Aug 10, 1983. 100 E. Speer Blvd., Denver, CO, 80203. Phone: (303) 595-3131. Phone: (888) 595-3131. Fax: (303) 566-2931. Fax: (303) 566-7631. Web Site: www.kdvr.com. Licensee: Fox Television Stations Inc. Group owner: (group owner; (acq 1995; $70 million). Natl. Network: Fox.
Key Personnel:
Bill Schneider. . . . . gen mgr
Ray Dowdle . . . . gen sls mgr
John Hirsch . . . . progmg dir & film buyer
Bill Dallman . . . . news dir

KFCT— Analog channel: 22. Analog hrs: 24 1,860 kw vis. ant 840t/705g TL: N40 38 23 W104 49 05 On air date: October 1994. c/o TV Stn KDVR, 100 E. Speer Blvd., Denver, CO, 80203. Phone: (303) 595-3131. Fax: (303) 566-2931. Web Site: www.kdvr.com. Licensee: Fox Television Stations Inc. Group owner: (group owner; (acq 1995). Natl. Network: Fox.
Key Personnel:
Ray Dowdle . . . . sr VP & gen sls mgr
William Schneider . . . . gen mgr
Catherine Andrey . . . . natl sls mgr
Sheryl Personett . . . . rgnl sls mgr
Clyde Becker . . . . prom VP & prom mgr
John Hirsch . . . . progmg mgr
Bill Dallman . . . . news dir
Jon Takayama . . . . pub affrs dir
Skip Erickson . . . . engrg VP & engrg mgr
Satellite of KDVR(TV) Denver.

KFNR— Analog channel: 11. Analog hrs: 24 1.66 kw vis, 166 w aur. 230t/449g TL: N41 46 15 W107 14 25 On air date: Apr 16, 1986. 1856 Skyview Dr., Casper, WY, 82601. Phone: (307) 577-5923/5924. Fax: (307) 577-5928. E-mail: klwy@coffey.com Licensee: First National Broadcasting Corp. Natl. Network: Fox. Washington Atty: Irwin, Campbell & Tannenwald.
Key Personnel:
Mark Nalbone. . . . . gen mgr
Terry Lane . . . . opns mgr & progmg dir
Tina Nalbone . . . . gen sls mgr & rgnl sls mgr
Joe Lownden . . . . prom dir
Dave Ericson. . . . . chief of engrg

***KMAS-TV—** Analog channel: 24. Digital channel: 10. Analog hrs: 24 5,500 kw vis, 500 kw aur. ant 515t/98g TL: N40 27 43 W106 51 02 On air date: May 1988. 1089 Bannock St., Denver, CO, 80204. Phone: (303) 892-6666. Fax: (303) 620-5600. Web Site: www.krma.org. Licensee: Rocky Mountain Public Broadcasting Network Inc. Group owner: Telemundo Group Inc. (acq 2-2-2007). Population served: 15,000 Natl. Network: PBS.
Rebroadcasts KRMA-TV Denver 100%.

KMGH-TV— Analog channel: 7.316 kw vis, 50 kw aur. ant 1,017t/249g TL: N39 43 46 W105 14 12 On air date: Nov 1, 1953. 123 Speer Blvd., Denver, CO, 80203. Phone: (303) 832-7777. Fax: (303) 832-0119. Web Site: www.thedenverchannel.com. Licensee: McGraw-Hill Broadcasting Co. Inc. Group owner: McGraw-Hill Broadcasting Co. (acq 6-1-72; grpsl; FTR: 3-13-72). Population served: 4,662,800 Natl. Network: ABC. Natl. Rep: Harrington, Righter & Parsons.
Key Personnel:
Darrell K. Brown . . . . gen mgr
Barry Edmond . . . . opns mgr
John Curry . . . . gen sls mgr
Laura Horgis. natl sls mgr

KPXC-TV— Analog channel: 59. Digital channel: 43. Analog hrs: 24 5,000 kw vis, 500 kw aur. ant 1,168t/85g TL: N39 40 24 W105 13 03 On air date: Sept 10, 1987. 3001 S. Jamaica Ct., Suite 200, Aurora, CO, 80014. Phone: (303) 751-5959. Fax: (303) 751-5993. Web Site: www.ionline.tv/stations/list.cfm. Licensee: Paxson Denver License Inc. Group owner: Paxson Communications Corp. (acq 7-1-96; grpsl). Natl. Network: i Network. Washington Atty: Cole, Raywid & Braverman. News: 24 hrs wkly.
Key Personnel:
Bud Paxson . . . . pres
Christy Bradford . . . . rgnl sls mgr
Mark Cornetta . . . . gen sls mgr & progmg mgr
Brian Schauer . . . . chief of engrg

KREG-TV— Analog channel: 3. Digital channel: 23.67.6 kw vis, 6.76 kw aur. 2,530t/256g TL: N39 25 05 W107 22 01 On air date: Dec 15, 1983. Box 789, Grand Junction, CO, 81502. Phone: (970) 963-3333. Fax: (970) 242-0886. Web Site: www.kregtv.com. Licensee: Hoak Media of Colorado LLC. Group owner: Hoak Media Corporation (acq 10-10-2003; grpsl). Population served: 135,450 Natl. Network: CBS. Natl. Rep: Petry Television Inc. Washington Atty: Gardner, Carton & Douglas.

***KRMA-TV—** Analog channel: 6. Digital channel: 18. Analog hrs: 24 Digital hrs: 24 100 kw vis, 15.1 kw aur. 880t/213g TL: N39 43 48 W105 15 00 (CP: Ant 1,230t. TL: N39 43 48 W105 15 00) On air date: Jan 30, 1956. 1089 Bannock St., Denver, CO, 80204. Phone: (303) 892-6666. Fax: (303) 620-5600. Web Site: www.rmpbs.org. Licensee: Rocky Mountain Public Broadcasting Network Inc. Population served: 2,000,000 Natl. Network: PBS. Washington Atty: Dow & Lohnes. News: one hr wkly.
Key Personnel:
James N. Morgese . . . . pres & gen mgr
Suzanne Banning . . . . dev dir & dev mgr
Donna Sanford . . . . progmg dir
John Anderson . . . . engrg dir & chief of engrg

***KRMT—** Analog channel: 41. Digital channel: 40. Analog hrs: 24 2,240 kw vis. ant 1,128t/169g TL: N39 35 59 W105 12 35 On air date: Jan 4, 1994. 12014 W. 64th Ave., Arvada, CO, 80004. Phone: (303) 423-4141. Fax: (303) 424-0571. Web Site: www.daystar.com. Licensee: Word of God Fellowship Inc. (acq 5-29-97; $1.95 million). Population served: 2,000,000 Washington Atty: Hogan & Hartson. Foreign lang progmg: SP 3
Key Personnel:
Marcus D. Lamb . . . . CEO & pres
Joni Lamb . . . . exec VP
Janice Smith. . . . . natl sls mgr

KTFD-TV— Analog channel: 14. Digital channel: 15.2,400 kw vis. ant 1,151t/138g TL: N39 40 18 W105 13 12 On air date: March 1986. 777 Grant St., 5th Fl., Denver, CO, 80203. Phone: (303) 832-1414. Fax: (303) 832-3410. Licensee: Spanish Television of Denver Inc. Group owner: Roberts Broadcasting Co. (acq 2-27-2003). Natl. Network: TeleFutura (Spanish). Foreign lang progmg: SP 168
Key Personnel:
Mario Carrera. . . . . gen mgr
Chris Matthews . . . . opns mgr & natl sls mgr
Luis Caneda . . . . prom dir & progmg dir

KTVD— Analog channel: 20. Digital channel: 19. Analog hrs: 24 5,000 kw vis, 500 kw aur. ant 1,256t/243g TL: N39 40 18 W105 13 12 On air date: Dec 1, 1988. Box 3166, Englewood, CO, 80155. Phone: (303) 792-2020. Fax: (303) 790-4633. E-mail: feedback@upn20.tv Web Site: www.upn20.tv. Licensee: Twenver Broadcast Inc. Group owner: Newsweb Corp. Natl. Network: MyNetworkTV. Washington Atty: Fletcher, Heald & Hildreth.

KUPN— Analog channel: 3. Analog hrs: 21 60.6 kw vis, 6 kw aur. ant 760t/604g TL: N40 34 57 W103 01 56 On air date: Jan 1, 1964. c/o KTVD(TV), 11203 E. Peakview Ave., Centennial, CO, 80111. Phone: (303) 792-2020. Fax: (303) 790-4633. Licensee: Channel 20 TV Co. Group owner: Newsweb Corp. (acq 8-31-99; $240,000). Washington Atty: Covington & Burling. News: 17 hrs wkly.
Key Personnel:
Greg Armstrong . . . . gen mgr
Hilary Castine . . . . prom mgr
Rita McCoy . . . . pub affrs dir
Michael Dant. . . . . chief of engrg
Rebroadcasts KTVD(TV) Denver 100%.

KUSA-TV— Analog channel: 9. Digital channel: 16.316 kw vis, 45.3 kw aur. ant 918t/246g TL: N39 43 46 W105 14 08 On air date: Oct 12, 1952. 500 Speer Blvd., Denver, CO, 80203. Phone: (303) 871-9999. Fax: (303) 698-4719 (sales). E-mail: kusa@9news.com Web Site: www.9news.com. Licensee: Multimedia Cablevision Inc. Group owner: Gannett Broadcasting (Division of Gannett Co. Inc.) (acq 6-7-79; grpsl; FTR: 6-11-79). Population served: 3,535,000 Natl. Network: NBC. Washington Atty: Wiley, Rein & Fielding.
Key Personnel:
Mark Cornetta . . . pres & gen mgr
Patricia Wilson . . . gen sls mgr
Patti Dennis . . . news dir
Don Perez . . . engrg dir

KWGN-TV— Analog channel: 2. Digital channel: 34. Analog hrs: 24 Digital hrs: 24 100 kw vis, 20 kw aur. 1,050t/449g TL: N39 43 58 W105 14 08 On air date: July 18, 1952. 6160 S. Wabash Way, Greenwood Village, CO, 80111. Phone: (303) 740-2222. Fax: (303) 740-2847. Web Site: www.cw2.com. Licensee: KWGN Inc. Group owner: Tribune Broadcasting Co., see Cross-Ownership (acq 3-3-66; $3.5 million). Population served: 2,600,000 Natl. Network: CW. Natl. Rep: TeleRep. Washington Atty: Sidley & Austin. News: 29.5 hrs wkly.
Key Personnel:
Dennis O'Brien . . . CFO
Matt Mansi . . . sls dir & gen sls mgr
Natalie Grant . . . progmg dir
Carl Bilek . . . news dir
Beverly Martinez . . . pub affrs dir
Don Rooney . . . engrg dir

KWHD— Analog channel: 53. Digital channel: 46. Analog hrs: 24 Digital hrs: 24 5,000 kw vis, 1,000 kw aur. 713t TL: N39 25 58 W104 39 18 On air date: July 1, 1990. 12999 E. Adam Aircraft, Englewood, CO, 80112. Phone: (303) 799-8853. Fax: (303) 792-5303. Web Site: www.kwhdtv53.com. Licensee: LeSea Broadcasting. Group owner: (group owner) Population served: 1,578,660, Foreign lang progmg: SP 1
Key Personnel:
Pete Sumrall . . . CEO
Dan Smith . . . gen mgr

***KWYP-TV—** Analog channel: 8. Analog hrs: 24 37 kw vis. ant 1,043t/144g TL: N41 17 17 W105 26 42 On air date: December 2004. Wyoming Public Television, 2660 Peck Ave., Riverton, WY, 82501. Phone: (307) 856-6944. Fax: (307) 856-3893. Web Site: wyoptv.org. Licensee: Central Wyoming College. Natl. Network: PBS.
Key Personnel:
JoAnne McFarland . . . pres
Dan Schiedel . . . gen mgr
Rudy Calvert . . . progmg dir
Bob Spain . . . chief of engrg

Durango

see Albuquerque-Santa Fe, NM market

Grand Junction-Montrose, CO (DMA 186)

KFQX— Analog channel: 4. Digital channel: 15.10.7 kw vis. ant 1,384t TL: N39 03 56 W108 44 52 On air date: 2000. 345 Hillcrest Manor, Grand Junction, CO, 81501. Phone: (970) 242-5285. Fax: (970) 242-0886. Web Site: www.krextv.com. Licensee: Parker Broadcasting of Colorado LLC. (acq 12-27-2004). Natl. Network: Fox.
Key Personnel:
Dennis Adkins . . . gen mgr
Shawna Greiger . . . sls VP & mktg VP
Maranda Wolter . . . prom VP
Shelley Nelson . . . progmg VP
Keira Bresnalhan . . . news dir
Maranda Wolter . . . pub affrs dir
Don May . . . engrg VP

KJCT— Analog channel: 8.120.2 kw vis, 12 kw aur. 2,720t/141g TL: N39 02 55 W108 15 06 On air date: Oct 22, 1979. Box 3788, Grand Junction, CO, 81502. 8 Foresight Cir., Grand Junction, CO 81505. Phone: (970) 245-8880. Fax: (970) 245-8249. Web Site: www.kjct8.com. Licensee: Pikes Peak Television Inc. Group owner: (group owner) (acq 6-26-2006; $45 million with KRDO-TV Colorado Springs). Population served: 175,000 Natl. Network: ABC. Natl. Rep: Blair Television. Rgnl. Rep: Rgnl rep: Blair Television Washington Atty: Fletcher, Heald & Hildreth.
Key Personnel:
David Bradley Jr. . . . pres
Neil Klockziem . . . gen mgr

KKCO— Analog channel: 11. Digital channel: 12. Analog hrs: 24 Note: NBC is on KKCO(TV) ch 11, CW is on KKCO-DT ch 12. 155 kw vis. 1,407t TL: N39 04 00 W108 44 41 On air date: July 19, 1996. 2325 Interstate Ave., Grand Junction, CO, 81505. Phone: (970) 243-1111. Fax: (970) 243-1770. Web Site: www.nbc11news.com. E-mail: billv@nbc11news.com Licensee: Gray Television Licensee Inc. (acq 1-31-2005; $13.5 million with translator K50EZ Montrose). Population served: 200,000 Natl. Network: NBC, CW Digital Network: Note: NBC is on KKCO(TV) ch 11, CW is on KKCO-DT ch 12. Natl. Rep: Millennium Sales & Marketing. Washington Atty: Wood, Maines & Brown, Chartered. News staff: 35; News: 28 hrs wkly.
Key Personnel:
J. Mack Robinson . . . pres
Debbie Varecha . . . CFO
Paul Varecha . . . opns VP & prom dir
Sandy Moore . . . progmg dir
Peter Franklin . . . mus dir
Jean Reynolds . . . news dir
William Varecha. gen mgr, dev VP, sls VP, mktg VP, film buyer & engrg dir
Roger LaFrance . . . chief of engrg

KREX-TV— Analog channel: 5. Digital channel: 2. Analog hrs: 20 12.9 kw vis, 2.5 kw aur. 10t/343g TL: N39 05 15 W108 33 56 On air date: May 22, 1954. Box 789, 345 Hillcrest Manor, Grand Junction, CO, 81502. Phone: (970) 242-5000. Fax: (970) 242-0886. Web Site: www.krextv.com. Licensee: Hoak Media of Colorado LLC. Group owner: Hoak Media Corporation (acq 11-12-2003; grpsl). Population served: 139,000 Natl. Network: CBS. Natl. Rep: Petry Television Inc. News staff: 15; News: 30 hrs wkly.
Key Personnel:
Dennis Adkins . . . gen mgr
Shelly Nelson . . . opns mgr
*Satellite of KREY-TV Montrose.

KREY-TV— Analog channel: 10.6.17 kw vis, 1.36 kw aur. 79t/113g TL: N38 31 02 W107 51 12 On air date: Aug 26, 1956. 614 N. First, Montrose, CO, 81401. Phone: (970) 249-9601. Fax: (970) 249-9610. E-mail: kreytv@gwe.net Licensee: Hoak Media of Colorado LLC. Group owner: Hoak Media Corporation (acq 10-10-2003; grpsl). Population served: 17,000 Natl. Network: CBS, NBC.

***KRMJ—** Analog channel: 18. Digital channel: 17. Analog hrs: 24 Digital hrs: 24 186 kw vis. 2,896t TL: N39 03 14 W108 15 13 On air date: January 1997. c/o KRMA-TV, 1089 Bannock St., Denver, CO, 80204. Phone: (970) 245-1818. Fax: (970) 255-2700. Web Site: www.rmpbs.org. Licensee: Rocky Mountain Public Broadcasting Network Inc. Natl. Network: PBS. Washington Atty: Dow, Lohnes & Albertson.
Key Personnel:
James Morgese . . . pres & gen mgr
Angie Salazar . . . stn mgr
Suzanne Banning . . . dev dir
Donna Sanford . . . progmg dir
John Anderson . . . engrg dir & chief of engrg
Rebroadcasts KRMA-TV Denver 99.9%.

Montrose

see Grand Junction-Montrose, CO market

Pueblo

see Colorado Springs-Pueblo, CO market

Connecticut

Bridgeport

see New York market

Hartford & New Haven, CT (DMA 28)

WCTX— Analog channel: 59. Digital channel: 39. Analog hrs: 24 5,000 kw vis, 500 kw aur. ant 1,029t TL: N41 25 23 W72 57 06 On air date: Apr 3, 1995. 8 Elm St., New Haven, CT, 06510. Phone: (203) 782-5900. Fax: (203) 782-5995. Licensee: WTNH Broadcasting Inc. Group owner: LIN Television Corporation (acq 3-2002). Population served: 1 m,ill,ion Natl. Network: MyNetworkTV. Natl. Rep: Petry Television Inc. News: 8.5 hrs wkly.
Key Personnel:
Jon Hitchcock . . . gen mgr
Roger Hess . . . sls dir
Karen Rorke . . . natl sls mgr & rgnl sls mgr
Sem Dietrich . . . natl sls mgr & rgnl sls mgr
Mary Lee Weber . . . mktg dir
Deanna Banas-Kluk . . . progmg dir
Kirk Varner . . . news dir
Francine DuVerger . . . engrg dir & chief of engrg

***WEDH—** Analog channel: 24. Analog hrs: 6 AM-midnight (M-S); 7:30 AM-midnight (Su) 692 kw vis, 69 kw aur. 860t/455g TL: N41 46 27 W72 48 20 On air date: Oct 1, 1962. 1049 Asylum Ave., Hartford, CT, 06105. Phone: (860) 278-5310. Fax: (860) 275-7500. Web Site: www.cptv.org. Licensee: Connecticut Public Broadcasting Inc. Population served: 837,990 Natl. Network: PBS. Rgnl. Network: Eastern Educ. Washington Atty: Schwartz, Woods & Miller.
Key Personnel:
Jerry Franklin . . . CEO, pres & gen mgr
Meg Sakellarides . . . CFO
Larry Rifkin . . . exec VP & progmg VP
Haig Papasian . . . opns VP
Joseph Zareski . . . opns dir
Dean Orton . . . dev dir & progmg dir

***WEDN—** Analog channel: 53. Digital channel: 45. Analog hrs: 6 AM-midnight (M-S); 7:30 AM-midnight (S 801.64 kw vis, 80.16 kw aur. 480t/476g TL: N41 31 11 W72 10 04 On air date: Mar 5, 1967. 1049 Asylum Ave., Hartford, CT, 06105. Phone: (860) 278-5310. Fax: (860) 275-7403. Web Site: www.cptv.org. Licensee: Connecticut Public Broadcasting. Population served: 400,000 Natl. Network: PBS. Rgnl. Network: Eastern Educ.
Key Personnel:
Jerry Franklin . . . CEO, pres & gen mgr
Meg Sakellarides . . . CFO
Larry Rifkin . . . exec VP
Haig Papasian . . . opns VP & opns dir
Christopher Flynn . . . dev VP

***WEDY—** Analog channel: 65.4.47 kw vis, 500 w aur. ant 270t/102g TL: N41 19 42 W72 54 25 On air date: November 1974. 1049 Asylum Ave., Hartford, CT, 06105. Phone: (860) 278-5310. Fax: (860) 275-7500. Web Site: www.cptv.org. Licensee: Connecticut Public Broadcasting Inc. Population served: 1,300,000 Natl. Network: PBS.
Key Personnel:
Jerry Franklin . . . CEO, pres & gen mgr
Meg Sakellarides . . . CFO
Larry Rifkin . . . exec VP & progmg VP
Haig Papasian . . . opns VP
Joseph Zareski . . . opns dir
Dean Orton . . . dev VP & dev dir
Rebroadcasts WEDH(TV) Hartford 100%.

WFSB— Analog channel: 3. Digital channel: 33. Analog hrs: 24 100 kw vis, 20 kw aur. 904t/518g TL: N41 46 30 W72 48 20 On air date: Sept 23, 1957. 3 Constitution Plaza, Hartford, CT, 06103-1821. Phone: (860) 728-3333. Fax: (860) 247-8940. Fax: (860) 728-0263 (News Room). Web Site: www.wfsb.com. Licensee: Meredith Corp. dba WFSB. Group owner: Meredith Broadcasting Group, Meredith Corp. (acq 9-4-97; $159 million). Population served: 3,888,000 Natl. Network: CBS. Natl. Rep: Harrington, Righter & Parsons. Washington Atty: Garvey, Schubert & Barer. News: 36 hrs wkly.
Key Personnel:
Klarn DePalma . . . gen mgr
John Ahearn . . . opns mgr
Bill Whittle . . . gen sls mgr
Stephanie Turner . . . mktg dir
Shelly Smith . . . prom mgr
Gary Brown . . . news dir
Victor Zarrillio . . . engrg dir

WHPX— Analog channel: 26. Digital channel: 34.2,792 kw vis, 279 kw aur. 1,251t TL: N41 25 05 W72 11 55 On air date: Sept 15, 1986. 3 Shaws Cove, Suite 226, New London, CT, 06320. Phone: (860) 444-2626. Fax: (860) 440-2601. Web Site: www.ionline.tv. Licensee: Paxson Hartford License Inc. Group owner: Paxson Communications Corp. (acq 2-25-00; grpsl). Natl. Network: i Network.

WTIC-TV— Analog channel: 61.5,000 kw vis, 1,000 kw aur. 1,692t/1,339g TL: N41 42 13 W72 49 57 On air date: Sept 17, 1984. One Corporate Ctr., Hartford, CT, 06103. Phone: (860) 527-6161. Fax: (860) 727-0158. Web Site: www.fox61.com. Licensee: 61 Licensee Inc. Group owner: Tribune Broadcasting Co. (acq 7-15-97; grpsl). Natl. Network: Fox. Natl. Rep: TeleRep. News staff: 31; News: 6 hrs wkly.

WTNH-TV— Analog channel: 8. Digital channel: 10.166 kw vis, 16.6 kw aur. 1,210t/909g TL: N41 25 23 W72 57 06 (CP: 175 kw vis, ant 863t) On air date: June 15, 1948. Box 1859, 8 Elm St., New Haven, CT, 06510. Phone: (203) 784-8888. Fax: (203) 789-2010. E-mail: wtnh@wtnh.com Web Site: www.wtnh.com. Licensee: LIN Television Inc. Group owner: LIN Television Corporation (acq 12-94; $120.17 million; FTR: 1-2-95). Population served: 1,000,000 Natl. Network: ABC. Natl. Rep: Petry Television Inc. Washington Atty: Lin Legal. News staff: 77; News: 32 hrs wkly.
Key Personnel:
Jon Hitchcock . . . gen mgr

Jamie Holowaty . . . . opns dir
Roger Hess. sls dir & sls dir
Roger Megroz . . . . natl sls mgr
Phil Jermain . . . . rgnl sls mgr
Mary Lee Weber . . . . mktg dir & pub affrs dir
Deanna Barns-Kluk . . . . progmg mgr
Kirk Varner . . . . news dir
Francine DuVerger . . . . engrg dir

WTXX— Analog channel: 20. Digital channel: 12.2,239 kw vis, 223.9 kw aur. ant 1,200t/1,013g TL: N41 31 04 W73 01 07 On air date: Sept 4, 1953. One Corporate Ctr., Hartford, CT, 06103. Phone: (860) 520-6573. Phone: (203) 758-3900. Fax: (860) 727-0158. E-mail: newsteam@foxx61.com Web Site: www.fox61.com. Licensee: WTXX Inc. (acq 8-6-2001). Population served: 1,000,000 Natl. Network: CW. Natl. Rep: MMT.

WUVN— Analog channel: 18. Analog hrs: 24 3,272 kw vis, 327.2 kw aur. 980t TL: N41 46 30 W72 48 04 On air date: Sept 25, 1954. 1 Constitution Plaza, Suite 7, Hartford, CT, 06103. Phone: (860) 278-1818. Fax: (860) 278-1811. Web Site: www.wuvntv.com. Licensee: Entravision Holdings LLC. Group owner: Entravision Communications Corp. (acq 1-4-01; $18 million). Natl. Network: Univision (Spanish). Washington Atty: Wiley, Rein & Fielding. News: 5 hrs wkly.
Key Personnel:
Robert Smith . . . . opns mgr
Ulysses Arrigoitia . . . . gen mgr & gen sls mgr
Rob Donner . . . . natl sls mgr
Meg Godin . . . . mktg mgr, prom mgr & adv mgr
Renee Barbour. . . . . progmg mgr
Dania Alexandrino . . . . mus dir
Sara Suarez . . . . news dir
Fran Vaccain. . . . . chief of engrg

WVIT—New Britain, Analog channel: 30. Analog hrs: 24 5,000 kw vis, 500 kw aur. 1,479t/1,051g TL: N41 42 00 W72 49 59 On air date: Feb 13, 1953. 1422 New Britain Ave., West Hartford, CT, 06110. Phone: (860) 521-3030. Fax: (860) 521-4860. Fax: (860) 521-3110. Licensee: NBC Telemundo License Co. Group owner: NBC TV Stations Division (acq 12-07-97; trade). Population served: 912,970 Natl. Network: NBC. Natl. Rep: NBC TV Stations Sales.
Key Personnel:
Dave Doebler. . . . . pres & gen mgr
David Bondanza . . . . opns dir & engrg dir
Bill Nandi . . . . opns mgr & chief of opns
Steve Smith . . . . sls VP
Eric Bloom . . . . natl sls mgr
Marcie Miller. . . . . mktg mgr
Maria Famicielli . . . . prom dir & adv dir
Ronni Attenello . . . . progmg dir
B.J. Finnell. . . . . news dir
LaVerne Jefferys . . . . pub affrs dir

New Haven

see Hartford & New Haven, CT market

Delaware

Seaford

see Salisbury, MD market

Wilmington

see Philadelphia market

District of Columbia

Washington, DC (Hagerstown, MD) (DMA 8)

WDCA— Analog channel: 20. Digital channel: 35. Analog hrs: 24 4,000 kw vis, 400 kw aur. ant 770t/809g TL: N38 57 49 W77 06 18 On air date: Apr 20, 1966. 5151 Wisconsin Ave. N.W., Truro, DC, 20016-4124. Phone: (202) 895-3050. Fax: (202) 895-3340. E-mail: upn20wdca@paramount.com Web Site: www.upn20wdca.com. Licensee: Fox Television Stations Inc. Group owner: (group owner; (acq 11-6-2001; with KTXH(TV) Houston, TX in swap for KBHK-TV San Francisco, CA). Population served: 5,000,000 Natl. Network: MyNetworkTV. Natl. Rep: Fox Stations Sales. Washington Atty: Leventhal, Senter & Lerman.

WDCW— Analog channel: 50. Digital channel: 51. Analog hrs: 24 Digital hrs: 24 4,168 kw vis. ant 828t/735g TL: N38 57 44 W77 01 36 On air date: November 1981. 2121 Wisconsin Ave. N.W., Suite 350, Washington, DC, 20007. Phone: (202) 965-5050. Fax: (202) 965-0050. Web Site: thecwdc.trb.com. Licensee: WBDC Broadcasting Inc. Group owner: Tribune Broadcasting Co. (acq 11-16-99). Population served: 3,000,000 Natl. Network: CW. Natl. Rep: Harrington, Righter & Parsons.
Key Personnel:
Dennis Fitzsimons . . . . CEO
Eric Meyrowitz . . . . VP & gen mgr
Chip Shenkan . . . . gen sls mgr
Brett Burke . . . . natl sls mgr
Jim Byrne . . . . mktg dir & prom mgr
John Handley . . . . chief of engrg

***WETA-TV—** Analog channel: 26. Digital channel: 27. Analog hrs: 6:30 AM-2:30 AM 2,254 kw vis. 756t TL: N38 57 49 W77 06 18 (CP: Ant 764t) On air date: Oct 2, 1961. 2775 S. Quincy St., Arlington, VA, 22206. Phone: (703) 998-2600. Fax: (703) 998-3401. Web Site: www.weta.org. Licensee: Greater Washington Educational Telecommunications Association Inc. Population served: 3,000,000 Natl. Network: PBS. Rgnl. Network: Eastern Educ. Washington Atty: Dow, Lohnes & Albertson.
Key Personnel:
Sharon Rockefeller . . . . pres
Joe Bruns . . . . CFO, exec VP & chief of opns
Karen Fritz . . . . gen mgr
Chad Davis. . . . . progmg dir

WFDC-TV— Analog channel: 14. Digital channel: 15.2,680 kw vis. ant 567t/407g TL: N38 56 24 W77 04 54 On air date: Aug 3, 1993. 101 Constitution Ave. N.W., # LL, Washington, DC, 20001. Phone: (301) 589-0030. Fax: (301) 495-9556. E-mail: rguernica@entravisiondc.com Web Site: www.entravision.com. Licensee: TeleFutura D.C. LLC. Group owner: Univision Communications Inc. (acq 6-1-2001; $30 million). Natl. Network: Univision (Spanish). Foreign lang progmg: SP 168
Key Personnel:
Rudy Guernica . . . . gen mgr
Ernesto Clavijo . . . . news dir
Fred Willard . . . . chief of engrg

***WFPT—** Analog channel: 62. Digital channel: 28.3,160 kw vis. ant 453t/476g TL: N39 17 53 W77 20 35 On air date: 1986. 11767 Owings Mills Blvd., Owings Mills, MD, 21117-1499. Phone: (410) 356-5600. Fax: (410) 581-6579. E-mail: comments@mpt.org Web Site: www.mpt.org. Licensee: Maryland Public Broadcasting Commission. Population served: 600,000 Natl. Network: PBS. Washington Atty: Schwartz, Woods & Miller.
Key Personnel:
Robert Shuman . . . . CEO & pres
Larry Unger . . . . CFO
Kirby Storms . . . . chief of engrg
Rebroadcasts WMPB(TV) Baltimore 100%.

WHAG-TV— Analog channel: 25. Digital channel: 55. Analog hrs: 24 Digital hrs: 24 1,352 kw vis, 135.2 kw aur. 1,230t/453g TL: N39 39 35 W77 57 57 On air date: Jan 3, 1970. 13 E. Washington St., Hagerstown, MD, 21740. Phone: (301) 797-4400. Fax: (301) 733-1735. Fax: (301) 745-4093. E-mail: hbreslin@ nbc25.com Web Site: www.your4state.com. Licensee: Nexstar Broadcasting Inc. Group owner: Nexstar Broadcasting Group Inc. (acq 12-31-2003; grpsl). Population served: 730,469 Natl. Network: NBC. Natl. Rep: Petry Television Inc. Washington Atty: Drinker Biddle & Reath L.L.P. News: 20 hrs wkly.
Key Personnel:
Hugh J. Breslin . . . . VP & gen mgr
Hugh Breslin . . . . gen sls mgr & progmg dir
Melissa Fountain . . . . prom mgr
Mark Kraham . . . . news dir
Michael Doty . . . . chief of engrg

***WHUT-TV—** Analog channel: 32. Digital channel: 33. Analog hrs: 18 500 kw vis, 158 kw aur. 700t/809g TL: N38 57 49 W77 06 18 On air date: Nov 17, 1980. 2222 4th St. N.W., Washington, DC, 20059. Phone: (202) 806-3200. Fax: (202) 806-3300. E-mail: j-lawson@howard.edu Web Site: www.whut.org. Licensee: Howard University. Population served: 2,200,000 Natl. Network: PBS. Washington Atty: Arnold & Porter. Wire Svc: Bloomberg Financial
Key Personnel:
Jennifer Lawson. . . . . gen mgr
Samuel Hyder. . . . . opns dir
Elizabeth Ventura . . . . dev dir

WJAL— Analog channel: 68. Digital channel: 16. Analog hrs: 24 4,000 kw vis, 400 kw aur. ant 1,335t/275g TL: N39 53 31 W77 58 02 On air date: May 5, 1987. Box 190, Chambersburg, PA, 17201-0190. 262 Swamp Fox Rd., Chambersburg, PA 17201. Phone: (717) 375-4000. Fax: (717) 375-4052. E-mail: adsales@wjal.com Web Site: www.wjal.com. Licensee: Entravision Holdings LLC. Group owner: Entravision Communications Corp. (acq 6-20-2001; $10.7 million including $400,000 bridge loan). Population served: 1,361,190 Washington Atty: Leventhal, Senter & Lerman. News: 5 hrs local news wkly.
Key Personnel:
Steve Ullom. . . . . stn mgr
C. Griffen . . . . engr

WJLA-TV— Analog channel: 7. Digital channel: 39. Analog hrs: 24 316 kw vis, 48 kw aur. 770t/640g TL: N38 57 01 W77 04 47 On air date: Oct 3, 1947. 1100 Wilson Blvd., 6th Fl., Arlington, VA, 22209. Phone: (703) 236-9552. Fax: (703) 236-2345. Web Site: www.wjla.com. Licensee: ACC Licensee Inc. Group owner: Allbritton Communications Co. (acq 1-76; grpsl). Population served: 5,517,000 Natl. Network: ABC. Washington Atty: Dow, Lohnes & Albertson, PLLC. News: 24 hrs wkly.

***WNVC—** Analog channel: 56. Digital channel: 57. Analog hrs: 24 1,230 kw vis, 123 kw aur. ant 1,049t/689g TL: N38 52 28 W77 13 24 (CP: 1,260 kw vis, 126 kw aur, ant 705t/345g) On air date: June 1, 1983. 8101A Lee Hwy., Falls Church, VA, 22042. Phone: (703) 770-7100. Web Site: WWW.mhznetworks.org. Licensee: Commonwealth Public Broadcasting Corp. (acq 6-4-2004). Population served: 4,000,000 Washington Atty: Wiley, Rein & Fielding.

***WNVT—** Analog channel: 53. Digital channel: 30. Analog hrs: 24 2,290 kw vis, 229 kw aur. ant 751t/651g TL: N38 37 42 W77 26 20 On air date: Mar 1, 1972. 8101A Lee Hwy., Falls Church, VA, 22042. Phone: (703) 770-7100. Web Site: www.mhznetworks.org. Licensee: Commonweath Public Broadcasting Corp. (acq 6-4-2004). Population served: 1,500,000 Washington Atty: Wiley, Rein & Fielding.

WPXW— Analog channel: 66. Digital channel: 43. Analog hrs: 24 5,000 kw vis, 500 kw aur. 560t/455g TL: N38 47 16 W77 19 49 (CP: 4,330 kw vis) On air date: Mar 26, 1978. 6199 Old Arrington Ln., Fairfax Stn., VA, 22039. Phone: (703) 503-7966. Fax: (703) 503-1225. Web Site: www.ionline.tv. Licensee: Paxson Washington License Inc. Group owner: Paxson Communications Corp. (acq 4-16-97; $30 million). Natl. Network: i Network. Washington Atty: Wilmer, Cutler & Pickering.

WRC-TV— Analog channel: 4. Digital channel: 48. Analog hrs: 24 100 kw vis, 15.1 kw aur. 778t/662g TL: N38 56 24 W77 04 54 On air date: June 27, 1947. 4001 Nebraska Ave. N.W., Washington, DC, 20016. Phone: (202) 885-4000. Fax: (202) 885-4104. Web Site: www.nbc4.com. Licensee: NBC Telemundo License Co. Group owner: NBC TV Stations Division. Population served: 4,000,000 Natl. Network: NBC. Natl. Rep: NBC TV Stations Sales. Wire Svc: UPI

WTTG— Analog channel: 5. Digital channel: 36.100 kw vis, 15 kw aur. 770t/705g TL: N38 57 21 W77 04 57 On air date: Jan 1, 1947. 5151 Wisconsin Ave. N.W., Washington, DC, 20016. Phone: (202) 244-5151. Fax: (202) 244-1745. Web Site: www.fox5dc.com. Licensee: Fox Television Stations Inc. Group owner: (group owner; (acq 3-86; grpsl). Population served: 4,000,000 Natl. Network: Fox. Natl. Rep: TeleRep. Washington Atty: Hogan & Hartson.

WUSA— Analog channel: 9. Digital channel: 34. Analog hrs: 24 316 kw vis, 31.6 kw aur. 780t/639g TL: N38 57 01 W77 04 47 On air date: Jan 16, 1949. 4100 Wisconsin Ave. N.W., Washington, DC, 20016. Phone: (202) 895-5999. Fax: (202) 364-6163. E-mail: 9news@wusatv9.com Web Site: www.wusatv9.com. Licensee: The Detroit News Inc. Group owner: Gannett Broadcasting (Division of Gannett Co. Inc.) (acq 2-18-86). Population served: 756,510 Natl. Network: CBS. Natl. Rep: Blair Television. Washington Atty: Reed Smith LLP. News: 39 hrs wkly.

***WVPY—** Analog channel: 42. Analog hrs: 24 141 kw vis. 1,305t/101g TL: N38 57 36 W78 19 52 On air date: Aug 22, 1996. c/o WVPT, 298 Port Republic Rd., Harrisonburg, VA, 22801. Phone: (540) 434-5391. Fax: (540) 434-7084. Web Site: www.wvpt.net. Licensee: Shenandoah Valley Educational TV Corp. Population served: 345,000 Natl. Network: PBS. Washington Atty: Covington & Burling.
Key Personnel:
Tony Mancari. . . . . exec VP & engrg VP
Richard Parker . . . . gen mgr
Wanda Zimmerman . . . . progmg dir
Rebroadcasts WVPT Staunton 100%.

***WWPB—** Analog channel: 31. Digital channel: 44. Analog hrs: 24 4,070 kw vis. ant 1,223t/390g TL: N39 39 04 W77 58 15 On air date: 1986. 11767 Owings Mills Blvd., Owings Mills, MD, 21117-1499. Phone: (410) 356-5600. Fax: (410) 581-6579. E-mail: comments@mpt.org

Web Site: www.mpt.org. Licensee: Maryland Public Broadcasting Commission. Population served: 600,000 Natl. Network: PBS. Washington Atty: Schwartz, Woods & Miller.
Key Personnel:
Robert Shuman . . . . CEO & pres
Larry Unger . . . . CFO
Kirby Storms . . . . chief of engrg
Rebroadcasts WMPB(TV) Baltimore 100%.

WWPX— Analog channel: 60. Digital channel: 12. Analog hrs: 19 2,040 kw vis. ant 984t/200g TL: N39 27 27 W78 03 52 On air date: Oct 1, 1991. 74 Swinging Bridge Rd., Martinsburg, WV, 25401. Phone: (304) 267-4950. Web Site: www.ionline.tv. Licensee: Paxson Martinsburg License Inc. Group owner: Paxson Communications Corp. (acq 6-1-2000). Natl. Network: i Network. Washington Atty: Cohn & Marks.

Florida

Daytona Beach

see Orlando-Daytona Beach-Melbourne, FL market

Destin

see Mobile, AL-Pensacola (Ft. Walton Beach), FL market

Fort Walton Beach

see Mobile, AL-Pensacola (Ft. Walton Beach), FL market

Ft. Lauderdale

see Miami-Ft. Lauderdale, FL market

Ft. Myers-Naples, FL (DMA 64)

WBBH-TV— Analog channel: 20. Digital channel: 15.5,000 kw vis, 500 kw aur. 1,482t/1,500g TL: N26 49 27 W81 45 51 On air date: Dec 19, 1968. 3719 Central Ave., Fort Myers, FL, 33901. Phone: (239) 939-2020. Fax: (239) 939-3244. E-mail: comments@nbc-2.com Web Site: www.nbc-2.com. Licensee: Waterman Broadcasting Corp. of Fla. Group owner: Waterman Broadcasting Corp. Natl. Network: NBC. Natl. Rep: Continental Television Sales. Washington Atty: Cohn & Marks. News staff: 80; News: 32 hrs wkly.
Key Personnel:
Bernard Waterman . . . . pres
Gerry Poppe. . . . . CFO
Steven Pontius . . . . exec VP & gen mgr
Bob Beville. . . . . sls dir

WFTX—Cape Coral, Analog channel: 36. Digital channel: 35. Analog hrs: 24 4,550 kw vis, 450 kw aur. ant 1,503t/1,450g TL: N26 47 43 W81 48 04 On air date: Oct 14, 1985. 621 S.W. Pine Island Rd., Cape Coral, FL, 33991. Phone: (239) 574-3636. Fax: (239) 574-2025. Web Site: www.fox4florida.com. Licensee: Journal Broadcast Corp. (acq 12-5-2005; grpsl). Population served: 413,000 Natl. Network: Fox. Washington Atty: Dow, Lohnes & Albertson.
Key Personnel:
Judy Kenney . . . . VP & gen mgr
Brent Struense . . . . mktg dir
Forrest Carr . . . . news dir

***WGCU**— Analog channel: 30.1,321 kw vis, 158 kw aur. 963t/992g TL: N26 48 54 W81 45 44 On air date: Aug 15, 1983. 10501 FGCU Blvd. S., West Palm Beach, FL, 33965. Phone: (239) 590-2300 / 7072. Fax: (239) 590-2310. Web Site: www.wgcu.org. Licensee: Board of Trustees, Florida Gulf Coast University. (acq 11-16-01). Natl. Network: PBS. Washington Atty: Cohn & Marks.
Key Personnel:
Kathleen Davey . . . . gen mgr & stn mgr
Michael Stepp . . . . opns mgr
Joseph Maggio . . . . chief of opns & progmg mgr
Amy Tardiff . . . . rgnl sls mgr & news dir

WINK-TV— Analog channel: 11. Digital channel: 9. Analog hrs: 24 Digital hrs: 24 316 kw vis, 31.6 kw aur. 1,478t/1,519g TL: N26 48 01 W81 45 48 On air date: Mar 18, 1954. 2824 Palm Beach Blvd., Fort Myers, FL, 33916. Phone: (239) 334-1111. Fax: (239) 334-0744. E-mail: webmaster@winktv.com Web Site: www.winktv.com. Licensee: Fort Myers Broadcasting Co. Group owner: (group owner) Population served: 1,061,000 Natl. Network: CBS. Natl. Rep: Blair Television. Washington Atty: Leibowitz & Associates. News staff: 55; News: 32 hrs wkly.
Key Personnel:
Brian A. McBride . . . . CEO & pres
Gary W. Gardner . . . . VP & gen mgr
Wayne Simons . . . . sls dir
Jesse Daniels . . . . natl sls mgr & rgnl sls mgr

WRXY-TV— Analog channel: 49. Digital channel: 33. Analog hrs: 24 5,000 kw vis, 500 kw aur. ant 800t/790.5g TL: N26 47 08 W81 47 41 On air date: Jan 29, 1995. Box 50490, Ft. Myers, FL, 33994-0490. 40000 Horseshoe Rd., Punta Gorda, FL 33982. Phone: (239) 543-7200. Fax: (239) 543-6800. E-mail: wrxy@wrxytv.com Licensee: West Coast Christian Television Inc. Ownership: David C. Gibbs III, Wayne Wetzel and Bill Anderson. (acq 12-16-97).
Key Personnel:
Paul Lodato . . . . gen mgr
Ken Griffith . . . . sls dir, mktg mgr & progmg mgr
Steven Speheger . . . . chief of opns, engrg dir & chief of engrg

WXCW— Analog channel: 46. Digital channel: 45. Analog hrs: 24 5,000 kw vis, 500 kw aur. ant 1,496t/1,496g TL: N26 47 08 W81 47 40 On air date: Oct 22, 1990. 3451 Bonita Bay Blvd., Suite 101, Bonita Springs, FL, 34134. Phone: (239) 498-4600. Fax: (239) 498-0146. Web Site: www.wb6tv.com. Licensee: Sun Broadcasting Inc. Ownership: Joseph C. Schwartzel, 100% Group owner: Acme Communications Inc. (acq 2-16-2007; $45 million). Natl. Network: CW. Washington Atty: Dickstein Shapiro Morin & Oshinsky L.L.P.
Key Personnel:
Bill Scaffide . . . . gen mgr
Jack Spiess . . . . chief of opns

WZVN-TV— Analog channel: 26. Digital channel: 41. Analog hrs: 24 5,000 kw vis, 500 kw aur. 1,206t/1,224g TL: N26 25 22 W81 37 49 On air date: Aug 21, 1974. 3719 Central Ave., Fort Myers, FL, 33901. Phone: (239) 939-2020. Fax: (239) 939-3244 (news). Fax: (239) 939-4801. Web Site: abc-7.com. Licensee: Montclair Communications Inc. Ownership: Lara Kunkler. (acq 10-10-1996; $21.3 million). Population served: 451,700 Natl. Network: ABC. Natl. Rep: Continental Television Sales. Washington Atty: Irwin, Campbell & Tannenwald. News staff: 85; News: 19.5 hrs wkly.
Key Personnel:
Lara Kunkler. . . . . pres
Laura Kunkler. gen mgr
Chris Rhodes . . . . opns mgr & mktg dir

Ft. Pierce

see West Palm Beach-Ft. Pierce, FL market

Ft. Walton Beach

see Mobile, AL-Pensacola (Ft. Walton Beach), FL market

Gainesville, FL (DMA 162)

WCJB-TV— Analog channel: 20. Digital channel: 16. Analog hrs: 24 Digital hrs: 24 2,818 kw vis, 282 kw aur. ant 1,049t/985g TL: N29 32 11 W82 24 00 On air date: Apr 7, 1971. 6220 N.W. 43rd St., Gainesville, FL, 32653. Phone: (352) 377-2020. Fax: (352) 373-6516. E-mail: tv20news@wcjb.com Web Site: www.wcjb.com. Licensee: Diversified Broadcasting Inc. Group owner: Diversified Communications (acq 12-1-76; FTR: 11-1-76). Population served: 370,710 Natl. Network: ABC, CW. Washington Atty: Irwin, Campbell & Tannenwald. News staff: 35; News: 17 hrs wkly.

WGFL— Analog channel: 53. Digital channel: 28. Analog hrs: 24 5,000 kw vis, 53 kw aur. 911t TL: N29 37 47 W82 34 24 (CP: 1,343 kw vis, ant 925t) On air date: Sept 20, 1997. 4190 N.W. 93 Ave., Gainesville, FL, 32653. Phone: (352) 375-5300. Fax: (352) 371-9353. Web Site: www.wgfl.com. Licensee: WGFL License Corp. Ownership: Pegasus Satellite Communications Inc., debtor-in-possession (acq 1-12-2004; $4.075 million with WYPN-CA Gainesville and WLCF-LP Lake City). Population served: 400,000 Natl. Network: CBS, MyNetworkTV.
Key Personnel:
Todd Senter . . . . gen mgr
Sue Edwards . . . . opns dir

WOGX— Analog channel: 51. Digital channel: 31. Analog hrs: 6 AM-2 AM 2,750 kw vis, 275 kw aur. ant 918t/855g TL: N29 21 32 W82 19 53 On air date: Nov 1, 1983. 1551 S.W. 37th Ave., Ocala, FL, 34474. 35 Skyline Dr., Lake Mary, FL 32746. Phone: (352) 873-6951. Phone: (407) 644-3535. Fax: (352) 237-5423. Web Site: www.wofl.com. Licensee: Fox Television Stations Inc. Group owner: (group owner; (acq 6-17-2002; with WOFL(TV) Orlando). Natl. Network: Fox. Natl. Rep: TeleRep. Washington Atty: Skadden, Arps, Slate, Meagher & Flom. News staff: 35; News: 7 hrs wkly.

***WUFT**— Analog channel: 5. Analog hrs: 21 100 kw vis, 20 kw aur. 860t/869g TL: N29 42 34 W82 23 40 On air date: Nov 10, 1958. 2200 Weimer Hall, Univ. of Florida, Gainesville, FL, 32611. Phone: (352) 392-5551. Fax: (352) 392-5731. E-mail: info@wuft.org Web Site: www.wuft.org. Licensee: Board of Trustees, University of Florida. Population served: 250,000 Natl. Network: PBS. Washington Atty: Schwartz, Woods & Miller. News: 3 hrs wkly.
Key Personnel:
Richard Lehner . . . . gen mgr
Titus Rush . . . . stn mgr
Brent Williams . . . . dev dir
Rob Carr . . . . chief of engrg

Jacksonville, FL (DMA 50)

WAWS— Analog channel: 30. Digital channel: 32.2,789 kw vis, 278.9 kw aur. ant 991t/1,030g TL: N30 16 53 W81 36 15 (CP: 5,000 kw vis, 508 kw aur, ant 991t/997g) On air date: Feb 15, 1981. 11700 Central Pkwy., Jacksonville, FL, 32224. Phone: (904) 642-3030. Fax: (904) 642-5665. E-mail: fox30news@ccjax.com Web Site: www.fox30online.com. Licensee: Clear Channel Radio Licenses Inc. Group owner: Clear Channel Communications Inc. (acq 8-5-92). Population served: 1,100,000 Natl. Network: Fox. Natl. Rep: Continental Television Sales. Washington Atty: Holland & Knight.

WCWJ— Analog channel: 17. Digital channel: 34. Analog hrs: 24 4,680 kw vis, 500 kw aur. ant 1,049t TL: N30 16 36 W81 33 47 On air date: Feb 19, 1966. Box 17000, Jacksonville, FL, 32245-7000. 9117 Hogan Rd., Jacksonville, FL 32216. Phone: (904) 641-1700. Fax: (904) 641-0306. Web Site: www.wjwb.com. Licensee: Media General Broadcasting Inc. Group owner: Media General Broadcast Group (acq 12-23-82; $18 million; FTR: 11-8-82). Population served: 1,000,000 Natl. Network: CW. Washington Atty: Dow, Lohnes & Albertson.
Key Personnel:
Mike Liff . . . . pres & gen mgr
George Birnbaum. . . . . opns mgr

***WJCT**— Analog channel: 7. Analog hrs: 18 316 kw vis, 31.6 kw aur. 910t/1,030g TL: N30 16 53 W81 34 15 On air date: Sept 10, 1958. 100 Festival Park Ave., Jacksonville, FL, 32202. Phone: (904) 353-7770. Fax: (904) 358-6331. E-mail: wjct@wjct.org Web Site: www.wjct.org. Licensee: WJCT Inc. Population served: 650,000 Natl. Network: PBS. Washington Atty: Schwartz, Woods & Miller.
Key Personnel:
Michael T. Boylan . . . . CEO & pres
Steven Wallace . . . . chmn
Jocelyn Enriquez . . . . CFO
Rick Johnson . . . . sr VP & opns VP
Jeri Cirillo. . . . . dev VP

***WJEB-TV**— Analog channel: 59. Digital channel: 44.3,311 kw vis. ant 948t TL: N30 16 34 W81 33 53 On air date: May 29, 1991. 3101 Emerson Expwy., Jacksonville, FL, 32277. Phone: (904) 399-8413. Fax: (904) 399-8423. E-mail: prayer@wjeb.org Web Site: www.wjeb.org. Licensee: Jacksonville Educators Broadcasting Inc. Natl. Network: PBS.
Key Personnel:
Collette D. Snowden . . . . gen mgr & stn mgr
Clayton Roney . . . . engrg mgr

WJXT— Analog channel: 4. Analog hrs: 24 100 kw vis, 20 kw aur. 930t/996g TL: N30 16 23 W81 33 13 (CP: Ant 436t) On air date: Sept 15, 1949. Box 5270, Jacksonville, FL, 32247. 4 Broadcast Pl., Jacksonville, FL 32207. Phone: (904) 399-4000. Fax: (904) 399-1828. E-mail: jaxnews@news4jax.com Web Site: www.news4jax.com. Licensee: Post-Newsweek Stations, Fla. Inc. Group owner: Post-Newsweek Stations Inc. (acq 1-28-53; grpsl; FTR: 2-2-53). Population served: 671,400 Natl. Rep: MMT. Washington Atty: Covington & Burling. News staff: 75; News: 51 hrs wkly.
Key Personnel:
Ann Sutton . . . . VP, stn mgr & natl sls mgr
Larry Blackerby. . . . . VP, gen mgr & gen mgr
Mo Ruddy-Baker. . . . . VP
Tina Schultz . . . . opns mgr
Wayne Reid . . . . gen sls mgr
Mike Guerrieri. mktg dir
Mo Ruddy . . . . news dir

WJXX— Analog channel: 25. Analog hrs: 24 5,000 kw vis. 659t/725g TL: N30 04 27 W81 48 23 On air date: Feb 9, 1997. 1070 East Adam

St., Jacksonville, FL, 32202. Phone: (904) 354-1212. Fax: (904) 353-3455. E-mail: news@firstcoastnews.com Web Site: www.firstcoastnews.com. Licensee: Gannett River States Publishing Corp. Group owner: Gannett Broadcasting (acq 3-15-00; $81 million). Natl. Network: ABC.
Key Personnel:
Ken Tonning . . . . . pres, gen mgr & opns dir
Glenn Sebold . . . . . progmg dir
Mike McCormick . . . . . news dir

WPXC-TV— Analog channel: 21. Digital channel: 24. Analog hrs: 24 5,000 kw vis. ant 1,371t TL: N31 08 22 W81 56 15 On air date: Apr 2, 1990. 7434 Blythe Island Hwy., Brunswick, GA, 31523. Phone: (912) 267-0021. Fax: (912) 261-9582. Web Site: www.ionline.tv. Licensee: Paxson Jax License Inc. Group owner: Paxson Communications Corp. (acq 12-6-2000; $3.07 million). Natl. Network: i Network. Washington Atty: Fleischman & Walsh.

WTEV-TV— Analog channel: 47. Digital channel: 19. Analog hrs: 24 5,000 kw vis, 500 kw aur. ant 980t/990g TL: N30 16 34 W81 33 58 On air date: Aug 1, 1980. 11700 Central Pkwy., Jacksonville, FL, 32224. Phone: (904) 642-3030. Fax: (904) 642-5665. E-mail: cbs47news@ccjax.com Web Site: cbs47.com. Licensee: Clear Channel Broadcasting Licenses Inc. Group owner: Clear Channel Communications Inc. (acq 2-13-2001; grpsl). Population served: 1,100,000 Natl. Network: CBS. Natl. Rep: Millennium Sales & Marketing. Washington Atty: Holland & Knight. News staff: 72; News: 22 hrs wkly.
Key Personnel:
Marc Hefner . . . . . gen sls mgr
Anne Baudeaux . . . . . mktg mgr
Doug Crall . . . . . news dir & chief of engrg

WTLV— Analog channel: 12. Analog hrs: 24 316 kw vis, 31.6 kw aur. 1,049t/999g TL: N30 16 23 W81 33 13 On air date: Sept 1, 1957. 1070 E. Adams St., Jacksonville, FL, 32202. Phone: (904) 354-1212. Fax: (904) 633-8899. E-mail: news@firstcoastnews.com Web Site: www.firstcoastnews.com. Licensee: Multimedia Cablevision Inc. Group owner: Gannett Broadcasting (Division of Gannett Co. Inc.) (acq 5-12-75; $11,401,217; FTR: 4-14-75). Population served: 1,265,000 Natl. Network: NBC. Natl. Rep: Blair Television. Washington Atty: Reed, Smith, Shaw & McClay. News staff: 54; News: 17 hrs wkly.
Key Personnel:
Ken Tonning . . . . . pres & gen mgr
Sam Folley . . . . . natl sls mgr

***WXGA-TV**— Analog channel: 8.316 kw vis, 47.9 kw aur. ant 1,030t/1,089g TL: N31 13 17 W82 34 24 On air date: Dec 4, 1961. 6433 TV-Tower Rd., Milwood, GA, 31552. Phone: (912) 338-5200. Web Site: www.gpb.org. Licensee: Georgia Public Telecommunications Commission. Natl. Network: PBS.

Melbourne

see Orlando-Daytona Beach-Melbourne, FL market

Miami-Ft. Lauderdale, FL (DMA 16)

WAMI-TV— Analog channel: 69. Digital channel: 47. Analog hrs: 24 5,000 kw vis, 500 kw aur. 866t/869g TL: N25 57 59 W80 12 33 On air date: Aug 10, 1988. 8550 N.W. 33rd St., Miami, FL, 33122. Phone: (305) 421-1900. Fax: (305) 463-9154. Web Site: www.univision.com. Licensee: TeleFutura Miami LLC. Group owner: Univision Communications Inc. (acq 6-6-01; grpsl). Natl. Network: TeleFutura (Spanish). Washington Atty: Hogan & Hartson. News staff: 35; News: 3 hrs wkly.
Key Personnel:
Luis Fernandez Rocha . . . . . VP
Luis Fernandez Rocha . . . . . gen mgr
Marilyn Hansen . . . . . gen sls mgr

WBFS-TV— Analog channel: 33. Digital channel: 32.5,000 kw vis, 500 kw aur. ant 924g TL: N25 57 59 W80 12 33 On air date: Dec 9, 1984. 8900 N.W. 18th Terr., Doral, FL, 33172. Phone: (305) 621-3333. Fax: (305) 628-3900. E-mail: upn33@wbfs.com Web Site: www.upn33.com. Licensee: Viacom Stations Group of Miami Inc. Group owner: Viacom Television Stations Group. Natl. Network: MyNetworkTV.

WFOR-TV— Analog channel: 4. Digital channel: 22. Analog hrs: 24 100 kw vis, 10 kw aur. ant 997t/997g TL: N25 58 07 W80 13 20 On air date: Mar 21, 1949. 8900 N.W. 18th Terr., Doral, FL, 33172. Phone: (305) 591-4444. Fax: (305) 639-4444. Web Site: www.cbs4news.com. Licensee: CBS Television Stations Inc. Group owner: Viacom Television Stations Group. Population served: 3,909,276 Natl. Network: CBS. Natl. Rep: CBS TV Stations National Sales. Wire Svc: NWS (National Weather Service) News staff: 90; News: 30 hrs wkly.

Key Personnel:
Shaun McDonald . . . . . pres & gen mgr
Michael Applebaum . . . . . gen mgr & gen sls mgr
Tom Cury . . . . . sls dir
Tracy Letize . . . . . progmg dir
Shannon High-Bassalik . . . . . news dir
Juan Andrea . . . . . engrg dir
Franklin Anderson . . . . . chief of engrg

WGEN-TV— Analog channel: 8. Digital channel: 12. Analog hrs: 24 2.65 kw vis. ant 179t/168g TL: N24 33 18 W81 48 05 On air date: Jan 1, 1995. 527 Southard St. W., Key West, FL, 33040-6871. Phone: (305) 293-4333. Fax: (305) 293-4007. E-mail: amonge@wegentv.com Web Site: www.wgent.tv.com. Licensee: Sonia Licensed Subsidiary LLC. Ownership: Community Property Trust under the De La Pena Family Trust of 11/26/01, 50%; and Husband's Separate Trust under the De La Pena Family Trust of 11/26/01, 50% (acq 4-16-2004; $2.75 million).

***WHFT-TV**— Analog channel: 45. Digital channel: 46. Analog hrs: 24 Digital hrs: 24 2,400 kw vis, 240 kw aur. 1,020t TL: N25 59 34 W80 10 27 (CP: 2,541 kw vis, ant 1,010t) On air date: Mar 17, 1975. 3324 Pembroke Rd., Pembroke Park, FL, 33021. Phone: (954) 962-1700. Fax: (954) 962-2817. Web Site: www.tbn.org. Licensee: Trinity Broadcasting of Florida Inc. Group owner: Trinity Broadcasting Network (acq 5-14-80; $10 million). Population served: 1,000,000 Washington Atty: Colby M. May. Foreign lang progmg: SP 0
Key Personnel:
Paul F. Crouch . . . . . pres
T. Hines . . . . . gen mgr

***WLRN-TV**— Analog channel: 17. Digital channel: 20. Analog hrs: 24 2,820 kw vis, 283 kw aur. ant 1,014t/1,010g TL: N25 57 30 W80 12 44 On air date: June 26, 1962. 172 N.E. 15th St., Miami, FL, 33135. Phone: (305) 995-1717. Fax: (305) 995-2299. E-mail: info@wlrn.org Web Site: www.wlrn.org. Licensee: The School Board of Miami-Dade County, FL. Population served: 4,200,000 Natl. Network: PBS. Rgnl. Network: SECA. Washington Atty: Leibowitz & Associates.
Key Personnel:
Karen Echols . . . . . CFO
John LaBonia . . . . . gen mgr
Bernadette Siy . . . . . stn mgr & dev mgr

WLTV— Analog channel: 23.4,470 kw vis. 974t TL: N25 58 07 W80 13 20 On air date: Nov 15, 1967. 9405 N.W. 41st St., Miami, FL, 33178. Phone: (305) 470-2323. Fax: (305) 471-3959. Web Site: www.univision.net. Licensee: WLTV L.P. Group owner: Univision Communications Inc. (acq 8-7-88). Population served: 4,600,000 Natl. Network: Univision (Spanish). Washington Atty: Wiley, Rein & Fielding. Foreign lang progmg: SP 168
Key Personnel:
Lois Fernandez-Rocha . . . . . VP & gen mgr
Raul Perez-Liste . . . . . opns dir

***WPBT**— Analog channel: 2. Analog hrs: 24 100 kw vis, 20 kw aur. 932t/923g TL: N25 57 30 W80 12 44 On air date: Aug 12, 1955. Box 610002, Miami, FL, 33261-0002. 14901 N.E. 20th Ave., Miami, FL 33261-0002. Phone: (305) 949-8321. Fax: (305) 944-4211. E-mail: channel2@channel2.org Web Site: www.channel2.org. Licensee: Community Television Foundation of South Florida Inc. Population served: 1,300,000 Natl. Network: PBS. Washington Atty: Wilmer, Cutler & Pickering.
Key Personnel:
Rick Schneider . . . . . CEO, pres & gen mgr
Dave Mullins . . . . . mktg VP
Jody Rafkind . . . . . prom mgr

WPLG— Analog channel: 10. Digital channel: 9. Analog hrs: 24 316 kw vis, 47.9 kw aur. 1,042t/1,046g TL: N25 57 59 W80 12 44 On air date: Nov 20, 1961. 3900 Biscayne Blvd., Miami, FL, 33137. Phone: (305) 576-1010. Fax: (305) 325-2381. Web Site: www.local10.com. Licensee: Post-Newsweek Stations, Fla. Inc. Group owner: Post-Newsweek Stations Inc. (acq 9-27-69; grpsl; FTR: 10-6-69). Population served: 3,274,000 Natl. Network: ABC. Natl. Rep: MMT. Washington Atty: Covington & Burling.
Key Personnel:
David Boylan . . . . . VP & gen mgr
Sharon Harrison . . . . . opns mgr
Mimi Del Ca . . . . . progmg mgr

WPXM— Analog channel: 35. Analog hrs: 24 3,240 kw vis, 324 kw aur. 335t TL: N24 41 05 W80 18 52 On air date: Oct 15, 1992. 9100 S. Dadeland Blvd., Suite 1804, Miami, FL, 33156. Phone: (954) 622-6835. Fax: (954) 622-6843. Web Site: www.ionline.tv. Licensee: Paxson Communications License Co. L.L.C. Group owner: Paxson Communications Corp. (acq 12-12-97). Natl. Network: i Network.

WSBS-TV— Analog channel: 22. Digital channel: 3. Analog hrs: 24 11.2 kw vis, 1.1 kw aur. ant 203t TL: N24 33 18 W81 48 07 On air date: June 1, 1993. 2601 S. Bayshore Dr., Suite 2020, Coconut Grove, FL, 33133. Phone: (305) 644-4800. Fax: (786) 470-1667. E-mail: info@mega.tv Web Site: www.mega.tv. Licensee: WDLP Licensing Inc. (acq 2-28-2006; $37.25 million with WSBS-CA Miami). Population served: 35,000 Foreign lang progmg: SP 168

WSCV— Analog channel: 51. Analog hrs: 24 5,000 kw vis. 827t TL: N25 57 59 W80 12 33 On air date: Dec 6, 1968. 15000 S.W. 27th St., Miramar, FL, 33027. Phone: (954) 622-6000. Fax: (954) 622-6107. E-mail: info@t51.com Web Site: www.t51.com. Licensee: NBC Telemundo License Co. Group owner: Telemundo Group Inc. (acq 4-12-2002; grpsl). Population served: 2,000,000 Natl. Network: Telemundo (Spanish). Foreign lang progmg: SP 168 News staff: 70; News: 3 hrs wkly.
Key Personnel:
Don Browne . . . . . CEO
Vince Sadusky . . . . . CFO
Michael Rodriguez . . . . . gen mgr
Jorge Carballo . . . . . gen sls mgr
Alan Brydger . . . . . mktg mgr
Maria Christina Barros . . . . . progmg dir

WSFL-TV— Analog channel: 39. Digital channel: 19. Analog hrs: 24 5,000 kw vis, 500 kw aur. ant 905t/905g TL: N25 58 07 W80 13 20 On air date: Oct 16, 1982. 2055 Lee St., Hollywood, FL, 33020. Phone: (954) 925-3939. Phone: (305) 949-3900. Fax: (954) 922-3965. Web Site: cwsfl.com. Licensee: Channel 39 Inc. Group owner: Tribune Broadcasting Co. (acq 7-15-97; grpsl). Population served: 3,247,000 Natl. Network: CW. Natl. Rep: TeleRep. News: 3.5 hrs wkly.
Key Personnel:
John Reardon . . . . . pres
Cam Trinh . . . . . CFO
Rich Engberg . . . . . VP & gen mgr
Mike Ward . . . . . gen sls mgr
Mark Drury . . . . . prom dir & prom mgr
Wendy Logsdon . . . . . progmg dir & progmg mgr

WSVN— Analog channel: 7. Digital channel: 8. Analog hrs: 24 316 kw vis, 30.2 kw aur. 950t/1,002g TL: N25 57 49 W80 12 44 On air date: July 29, 1956. 1401 79th St. Causeway, Miami, FL, 33141. Phone: (305) 751-6692. Fax: (305) 757-2266. Fax: TWX: (810) 848-6151. Web Site: www.wsvn.com. Licensee: Sunbeam Television Corp. (acq 10-4-67; FTR: 10-16-67). Population served: 333,859 Natl. Network: Fox. Natl. Rep: Harrington, Righter & Parsons. Washington Atty: Koteen & Naftalin.
Key Personnel:
Edmund N. Ansin . . . . . pres
Steven Cejas . . . . . exec VP
Robert W. Leider . . . . . gen mgr

WTVJ— Analog channel: 6. Analog hrs: 24 100 kw vis, 15.9 kw aur. 1,842t/1,841.5g TL: N25 32 24 W80 28 07 On air date: Sept 20, 1967. NBC 6 WTVD, 15000 S.W. 27th St., Miramar, FL, 33027. Phone: (954) 622-6000. Web Site: www.nbc6.net. Licensee: NBC Telemundo License Co. Group owner: NBC TV Stations Division (acq 1995). Population served: 3,594,000 Natl. Network: NBC. Natl. Rep: NBC TV Stations Sales. Wire Svc: NWS (National Weather Service)
Key Personnel:
Ardyth R. Diercks . . . . . pres, gen mgr, gen mgr & sls VP
Meg Green . . . . . pres & CFO

Naples

see Ft. Myers-Naples, FL market

Orlando-Daytona Beach-Melbourne, FL (DMA 19)

WACX-DT—Leesburg, Digital channel: 40. Analog hrs: 24 1,000 kw vis. ant 1,619t/1,591g TL: N28 35 11.6 W81 04 58.2 On air date: Mar 6, 1982. Box 608040, Orlando, FL, 32860. 285 W. Central Pkwy., Altamonte Springs, FL 32714. Phone: (407) 263-4040. E-mail: superchannel@superchannel.com Web Site: www.wacxtv.com. Licensee: Associated Christian Television System Inc. (acq 6-8-83; FTR: 7-4-83). Population served: 3,500,000 Washington Atty: Koerner & Olender. Foreign lang progmg: SP 0
Key Personnel:
Claud Bowers . . . . . CEO, pres, pres & gen mgr
Carol Gentry . . . . . gen sls mgr

***WBCC**— Analog channel: 68. Digital channel: 30. Analog hrs: 6 AM-midnight Digital hrs: 6 AM-midnight 6,000 kw vis. 613t TL: N28 18 26 W80 54 48 (CP: 2,844 kw vis, 284 kw aur, ant 941t) On air date: Jan 12, 1988. 1519 Clearlake Rd., Cocoa, FL, 32922. Phone: (321) 433-7111. Fax: (321) 433-7154. E-mail: wbcc@brevardcc.edu Web Site: www.wbcctv.org. Licensee: Brevard Community College. Population served: 1.5 m,ill,ion Natl. Network: PBS.

Key Personnel:
Dr. Tom Gamble . . . . pres
Joe Williams . . . . gen mgr
Phillip Wallace . . . . stn mgr

***WCEU—** Analog channel: 15. Digital channel: 33. Analog hrs: 20 Digital hrs: 20 708 kw vis, 201.4 kw aur. ant 577t/567g TL: N29 10 24 W81 09 24 On air date: Feb 1, 1988. Box 9245, Daytona Beach, FL, 32120-2811. 1200 W. International Speedway Blvd., Daytona Beach, FL 32114. Phone: (386) 506-4415. Fax: (386) 506-4427. E-mail: wceu@wceu-pbs.org Web Site: www.wceu.org. Licensee: Daytona Beach Community College. (acq 6-30-2002). Population served: 3,500,000 Natl. Network: PBS. Washington Atty: Fletcher, Heald & Hildreth.
Key Personnel:
Bruce E. Dunn . . . . gen mgr & gen sls mgr
Michael Dietz . . . . progmg dir
Bill Schwartz . . . . chief of engrg

WESH—Daytona Beach, Analog channel: 2. Digital channel: 11.100 kw vis, 10 kw aur. 1,650t/1,670g TL: N28 56 17 W81 18 58 On air date: June 11, 1956. 1021 N. Wymore Rd., Winter Park, FL, 32789. Phone: (407) 645-2222. Fax: (407) 539-7812. Fax: (407) 539-7948 (news). Web Site: www.wesh.com. Licensee: WESH-TV Broadcasting. Group owner: Hearst-Argyle Television Inc. (acq 1999; grpsl). Population served: 1,735,900 Natl. Network: NBC. Washington Atty: Brooks, Pierce, McLendon, Humphrey & Leonard.
Key Personnel:
Wlliam P. Bauman . . . . pres & VP
William Bauman . . . . gen mgr
Bob Fein . . . . stn mgr
Rick Scharf . . . . opns mgr
Rob Halpern . . . . gen sls mgr
Jessica A. Derle . . . . natl sls mgr
Linda Kitchens . . . . progmg dir
Barb Maushard . . . . news dir

WFTV— Analog channel: 9. Digital channel: 39. Analog hrs: 24 316 kw vis, 31.6 kw aur. ant 1,570t/1,543g TL: N28 36 08 W81 05 37 On air date: Feb 1, 1958. Box 999, Orlando, FL, 32802. 490 E. South St., Orlando, FL 32801-2841. Phone: (407) 841-9000. Fax: (407) 422-1887. Fax: (407) 481-2891 (news). Web Site: www.wftv.com. Licensee: WFTV-TV Holdings Inc. Group owner: Cox Communications Inc. (acq 8-85; $185 million). Population served: 2,400,000 Natl. Network: ABC. News staff: 80; News: 25 hrs wkly.
Key Personnel:
Shawn Bartelt . . . . VP & gen mgr
Chip Reif . . . . opns mgr
Lisa Hines . . . . gen sls mgr
Bob St. Charles . . . . mktg mgr
Bob Jordan . . . . news dir

WHLV-TV— Analog channel: 52. Digital channel: 53. Analog hrs: 24 4,680 kw vis, 468 kw aur. ant 934t/1,005g TL: N28 18 26 W80 54 48 On air date: Aug 16, 1982. Box C-11949, Santa Ana, CA, 92711. Phone: (407) 423-5200. Fax: (407) 423-8153. E-mail: ed@tv52.org Web Site: www.tv52.org. Licensee: Trinity Christian Center of Santa Ana Inc. Group owner: (group owner; (acq 9-19-2006; $50 million). Population served: 3,068,000 Washington Atty: Gammon & Grange.
Key Personnel:
Ken Mikesell . . . . gen mgr
Ed Griffis . . . . opns VP
Eileen Kelly . . . . progmg dir & progmg
Marshall Royalty . . . . chief of engrg

WKCF—Clermont, Analog channel: 18. Digital channel: 17. Analog hrs: 24 5,000 kw vis. ant 1,683t TL: N28 35 12 W81 04 58 On air date: December 1988. 1021 North Wymore Rd., Winter Park, FL, 32789. Phone: (407) 645-1818. Fax: (407) 647-4163. E-mail: wb18wkcf@wb18.com Web Site: www.wb18.com. Licensee: Orlando Hearst-Argyle Television Inc. Group owner: Emmis Communications Corp. (acq 7-7-2006; $217.5 million). Population served: 1,345,700 Natl. Network: CW. Natl. Rep: Harrington, Righter & Parsons.
Key Personnel:
Bill Bauman . . . . gen mgr & gen sls mgr
John Soapes . . . . gen mgr & gen sls mgr
Steve Rifkin . . . . opns dir & mktg dir
Joe Addalia . . . . engrg dir

WKMG-TV— Analog channel: 6. Analog hrs: 24 74.1 kw vis, 14.8 kw aur. 1,460t/1,484g TL: N28 36 08 W81 05 37 (CP: 100 kw vis, 20 kw aur, ant 1,840t) On air date: July 1, 1954. 4466 N. John Young Pkwy., Orlando, FL, 32804. Phone: (407) 291-6000. Fax: (407) 521-1204. Fax: (407) 298-2122 (news). Web Site: www.local6.com. Licensee: Post-Newsweek Stations Orlando Inc. Group owner: Post-Newsweek Stations Inc. (acq 9-4-97). Population served: 3,600,000 Natl. Network: CBS. Natl. Rep: MMT. Washington Atty: Covington & Burling. News staff: 80; News: 24 hrs wkly.
Key Personnel:
Alan Frank . . . . pres & gen sls mgr
Henry Maldonado . . . . VP & gen mgr

WLCB-TV— Analog channel: 45. Analog hrs: 24 1,200 kw vis. ant 462t/377g TL: N28 40 48 W81 49 16 On air date: December 2000. 31 Skyline Dr., Lake Mary, FL, 32746. Phone: (407) 423-5200. Fax: (407) 422-0120. E-mail: info@tv52.org Web Site: www.tv52.org. Licensee: Good Life Broadcasting Inc. Group owner: (group owner; (acq 11-19-2001; with WTGL-TV Cocoa). Population served: 2,000,000 Washington Atty: Leventhal, Senter & Lerman.

***WMFE-TV—** Analog channel: 24. Digital channel: 24. Analog hrs: 24 Digital hrs: 18 1,350 kw vis. ant 1,246t/1,220g TL: N28 36 08 W81 05 37 On air date: Mar 25, 1965. 11510 E. Colonial Dr., Orlando, FL, 32817-4699. Phone: (407) 273-2300. Fax: (407) 206-2791. Web Site: wmfe.org. Licensee: Community Communications Inc. Ownership: Community Licensee, 100%. Population served: 1,301,000 Natl. Network: PBS. Washington Atty: Schwartz, Woods & Miller.
Key Personnel:
Stephen McKenney Steck . . . . CEO
Joy Barrett Sebol . . . . chmn
Bethany Mott . . . . pres & mktg VP
Jose A. Fajardo . . . . pres
Aldo Vivona . . . . sr VP
Michael Crane . . . . progmg VP

WOFL— Analog channel: 35. Digital channel: 22. Analog hrs: 24 2,570 kw vis, 513 kw aur. 1,479t/1,486g TL: N28 36 17 W81 05 13 On air date: Oct 15, 1979. 35 Skyline Dr., Lake Mary, FL, 32746. Phone: (407) 644-3535. Fax: (407) 333-3535. Web Site: www.wofl.com. Licensee: Fox Television Stations Inc. Group owner: (group owner; (acq 6-17-2002; with WOGX(TV) Ocala). Population served: 1,500,000 Natl. Network: Fox.
Key Personnel:
Stan Knott . . . . gen mgr
Rick Snyder . . . . prom VP
Terry Walden . . . . progmg mgr

WOPX— Analog channel: 56. Digital channel: 48. Analog hrs: 24 5,000 kw vis. ant 1,548t/1,522g TL: N28 05 37 W81 07 28 On air date: June 1986. 7091 Grand Natl Dr., Suite 100, Orlando, FL, 32819. Phone: (407) 370-5600. Fax: (407) 363-1757. Web Site: www.ionline.tv. Licensee: Paxson Orlando License Inc. Group owner: Paxson Communications Corp. (acq 12-12-97; $13,161,274). Population served: 685,000 Natl. Network: i Network. Washington Atty: Dow, Lohnes & Albertson.

WOTF-TV— Analog channel: 43. Analog hrs: 24 4,170 kw vis, 854 kw aur. 1,049t/1,005g TL: N28 18 26 W80 54 48 On air date: July 5, 1982. 1021 North Wymore Rd., Winter Park, FL, 32789. Phone: (321) 254-4343. Fax: (321) 254-9343. Web Site: www.univision.com. Licensee: Univision of Melbourne Inc. Group owner: Univision Communications Inc. (acq 5-21-2001; grpsl). Washington Atty: Dow, Lohnes & Albertson. Foreign lang progmg: SP 168
Key Personnel:
Sylvia Willis . . . . VP
Bill Bauman . . . . gen mgr

WRBW— Analog channel: 65. Digital channel: 41. Analog hrs: 24 5,000 kw vis. 1,525t TL: N28 34 51 W81 04 32 On air date: June 6, 1994. 35 Skyline Dr., Lake Mary, FL, 32746. Phone: (407) 644-3535. Fax: (407) 741-5048. E-mail: wrbw@wrbw.com Web Site: www.wrbw.com. Licensee: Fox Television Stations Inc. Group owner: (group owner; (acq 7-31-2001; grpsl). Natl. Network: MyNetworkTV.
Key Personnel:
Stan Knott . . . . gen mgr
Terry Walden . . . . opns dir & progmg mgr

WRDQ— Analog channel: 27. Digital channel: 14. Analog hrs: 24 5,000 kw vis, 500 kw aur. ant 1,866t/1,840g TL: N28 16 44 W81 01 25 On air date: 2000. 490 E. South St., Orlando, FL, 32801. Phone: (407) 841-9000. Fax: (407) 422-1414. Web Site: www.wrdq.com. Licensee: WFTV-TV Holdings Inc. Group owner: Cox Communications Inc. (acq 2-1-2001).
Key Personnel:
Shawn Bartelt . . . . gen mgr
Mario Mendosa . . . . gen sls mgr
Bob Jordan . . . . news dir

WVEN-TV—Daytona Beach, Analog channel: 26. Analog hrs: 24 2,750 kw vis, 2,750 kw aur. 1,063t TL: N29 17 10 W81 29 37 On air date: October 1988. 523 Douglas Ave., Suite 100, Altamonte Springs, FL, 32714. Phone: (407) 774-2626. Fax: (407) 774-3384. Licensee: Entravision Holdings L.L.C. Group owner: Entravision Communications Corp. (acq 9-15-00; $22.55 million). Natl. Network: Univision (Spanish). Foreign lang progmg: SP 168

Panama City, FL (DMA 156)

WBIF— Analog channel: 51.1,970 kw vis. ant 833t/856g TL: N30 30 42 W85 29 17 On air date: 2002. Tiger Eye Broadcasting Corp., 3400 Lakeside Dr., Suite 500, Miramar, FL, 33027. Phone: (954) 431-3144. Fax: (954) 431-3591. Licensee: EBC Panama City Inc. Group owner: Equity Broadcasting Corp. (acq 3-23-2004; $1.2 million).
Key Personnel:
James Gallenger . . . . gen mgr
Tom Cury . . . . sls dir
Tracy Letize . . . . progmg dir

***WFSG—** Analog channel: 56. Digital channel: 38. Analog hrs: 5 AM-11:30 PM 1,147 kw vis, 48.6 kw aur. 509t/499g TL: N30 22 02 W85 55 29 On air date: July 11, 1988. Public TV Ctr., 1600 Red Barber Plaza, Tallahassee, FL, 32310. Phone: (850) 487-3170. Fax: (850) 487-3093. E-mail: mail@wfsu.org Web Site: www.wfsu.org. Licensee: Board of Regents of Florida. (acq 2-28-86). Population served: 100,000 Natl. Network: PBS. Washington Atty: Cohn & Marks.
Key Personnel:
Patrick Keating . . . . gen mgr
Charles Allen . . . . dev dir
Jannie Whitt . . . . prom dir & prom mgr
Mike Dunn . . . . progmg dir
David Lauther . . . . chief of engrg
Rebroadcasts WFSU-TV Tallahassee.

WJHG-TV— Analog channel: 7. Digital channel: 8. Analog hrs: 24 Note: NBC is on WJHG-TV ch 7, CW and MyNetworkTV are on WJHG-DT ch 8. 316 kw vis, 34 kw aur. ant 870t/887g TL: N30 26 00 W85 24 51 On air date: Dec 1, 1953. 8195 Front Beach Rd., Panama City Beach, FL, 32407. Phone: (850) 234-7777. Fax: (850) 233-6647. Web Site: www.wjhg.com. Licensee: WEAU Licensee Corp. Group owner: Gray Television Inc. (acq 6-29-60; $340,000; FTR: 7-4-60). Population served: 79,000 Natl. Network: NBC, CW, MyNetworkTV Digital Network: Note: NBC is on WJHG-TV ch 7, CW and MyNetworkTV are on WJHG-DT ch 8. Natl. Rep: Continental Television Sales. Washington Atty: Venable, Baetjer, Howard & Civiletti.
Key Personnel:
Jon McKee . . . . opns mgr
Tracy Connors . . . . gen mgr & gen sls mgr

WMBB— Analog channel: 13. Analog hrs: 24 316 kw vis, 63 kw aur. 1,549t/1,464g TL: N30 21 09 W85 23 26 On air date: Oct 3, 1973. 613 Harrison Ave., Panama City, FL, 32401-2623. Phone: (850) 769-2313. Fax: (850) 769-8231. E-mail: dstrong@wmbb.com Web Site: www.wmbb.com. Licensee: Media General Communications Inc. Group owner: Media General Broadcast Group (acq 3-27-2000; grpsl). Population served: 192,200 Natl. Network: ABC. Natl. Rep: Blair Television. Washington Atty: Covington & Burling. News staff: 24; News: 143 hrs wkly.
Key Personnel:
Bill Byrd . . . . gen mgr
Marc Morriston . . . . prom dir

WPCT— Analog channel: 46.126 kw vis. ant 194t/180g TL: N30 10 59 W85 46 42 On air date: 1997. Box 9556, Panama City Beach, FL, 32407. Phone: (850) 234-2773. Fax: (850) 234-1179. Web Site: www.tripsmarter.com. Licensee: Beach TV Properties Inc. Group owner: (group owner)
Key Personnel:
Jud Colley . . . . pres
Mike Hartzog . . . . gen mgr

WPGX— Analog channel: 28. Digital channel: 9. Analog hrs: 24 1,260 kw vis, 126 kw aur. ant 748t TL: N30 23 42 W85 32 02 On air date: May 21, 1988. Fox TV Ctr., 637 Luverne Ave., Panama City, FL, 32401. Box 208, Panama City, FL 32402. Phone: (850) 784-0028. Fax: (850) 784-1773. Licensee: WPGX License Subsidiary LLC. Group owner: Raycom Media Inc. (acq 12-15-2003; grpsl). Population served: 250,000 Natl. Network: Fox. Natl. Rep: Millennium Sales & Marketing. Washington Atty: Leventhal, Senter & Lerman.
Key Personnel:
David Cavileer . . . . VP & chief of engrg
Tim Cabrey . . . . gen mgr
Sue Stewart . . . . opns mgr & prom mgr

Pensacola

see Mobile, AL-Pensacola (Ft. Walton Beach), FL market

Sarasota

see Tampa-St. Petersburg (Sarasota), FL market

St. Petersburg

see Tampa-St. Petersburg (Sarasota), FL market

Tallahassee, FL-Thomasville, GA (DMA 108)

WCTV—(Thomasville, GA) Analog channel: 6. Digital channel: 46. Analog hrs: 24 Note: CBS is on WCTV(TV) ch 6, MyNetworkTV is on WCTV-DT ch 46. 97.5 kw vis, 19.5 kw aur. ant 2,031t/2,000g TL: N30 40 13 W83 56 26 On air date: Sept 15, 1955. 1801 Halstead Blvd., Tallahassee, FL, 32309. Phone: (850) 893-6666. Fax: (850) 893-5193. Web Site: www.wctv6.com. Licensee: Gray Television Licensee Inc. Group owner: Gray Television Inc. (acq 1996; $165 million with WVLT-TV Knoxville, TN). Natl. Network: CBS, MyNetworkTV Digital Network: Note: CBS is on WCTV(TV) ch 6, MyNetworkTV is on WCTV-DT ch 46. Natl. Rep: Continental Television Sales.
Key Personnel:
Nick Waller. . . . . pres & gen mgr
Chris Mossman . . . . sls dir & gen sls mgr
Ella Paris . . . . natl sls mgr
Heather Pryor . . . . rgnl sls mgr
Mike Smith. . . . . news dir

***WFSU-TV**— Analog channel: 11.316 kw vis, 31.6 kw aur. 777t/774g TL: N30 21 29 W84 36 39 On air date: Sept 20, 1960. 1600 Red Barber Plaza, Tallahassee, FL, 32310. Phone: (850) 487-3170. Fax: (850) 487-3093. E-mail: mail@wfsu.org Web Site: www.wfsu.org. Licensee: Florida Board of Regents & Florida State University. Population served: 170,000 Natl. Network: PBS.
Key Personnel:
Patrick Keating . . . . gen mgr
Charles Allen . . . . dev dir
Beckie Hamilton. progmg dir

WFXU— Analog channel: 57. Digital channel: 48. Analog hrs: 24 2,500 kw vis. 443t TL: N30 33 00 W83 00 46 On air date: July 1998. Box 949, Midway, FL, 32343. Phone: (850) 576-4990. Fax: (850) 576-0200. E-mail: fox49@fox49.com Web Site: www.fox49.com. Licensee: WFXU License Corp. Ownership: Pegasus Communications Corp., 100% (acq 3-29-2002; $250,914). Natl. Network: CW. Washington Atty: Shaw Pittman LLP.
Key Personnel:
Mark Pagon. . . . . CEO
Jack Paris . . . . sr VP
David Hinterschied. gen mgr
Tyrone Hayes. . . . . opns mgr & chief of opns
Tana Kenny . . . . gen sls mgr
Don Abel . . . . mktg mgr, progmg dir & progmg mgr

WSWG— Analog channel: 44. Digital channel: 43. Analog hrs: 24 Digital hrs: 24 Note: CBS is on WSWG(TV) ch 44, MyNetworkTV is on WSWG-DT ch 43. 1,700 kw vis, 257 kw aur. ant 920t/950g TL: N31 10 18 W83 21 57 (CP: 1,365 kw vis, ant 922t) On air date: Sept 1, 1995. Box 1987, Moultrie, GA, 31776. 107 2nd Ave. S.W., Moultrie, GA 31768. Phone: (229) 985-1340. Fax: (229) 985-7549. Web Site: wswgtv.com. Licensee: Gray Television Licensee Inc. (acq 11-10-2005; $3.75 million). Natl. Network: CBS, MyNetworkTV Digital Network: Note: CBS is on WSWG(TV) ch 44, MyNetworkTV is on WSWG-DT ch 43. Washington Atty: Robert Bizer.
Key Personnel:
Chris Mossman . . . . pres & gen sls mgr
Nick Waller. . . . . gen mgr
Jared Yost . . . . rgnl sls mgr
Jim Killinger . . . . chief of engrg

WTLF— Analog channel: 0. Digital channel: 24.24 kw vis. ant 128t/171g TL: N30 29 40 W84 25 03 On air date: 2004. Box 949, Midway, FL, 32343. 950 Commerce Blvd., Midway, FL 32343. Phone: (850) 576-4990. Fax: (850) 576-0200. Licensee: KB Prime Media L.L.C. Group owner: (group owner).
Key Personnel:
Guyon W. Turner . . . . pres
David Hinterschied. gen mgr

WTLH—(Bainbridge, GA) Analog channel: 49. Digital channel: 50. Analog hrs: 24 Note: Fox is on WTLH(TV) ch 49, CW is on WTLH-DT ch 50. 5,000 kw vis. ant 1,958t/1,961g TL: N30 40 51 W83 58 21 On air date: 1989. 950 Commerce Blvd., Box 949, Midway, FL, 32343. Phone: (850) 576-4990. Fax: (850) 576-0200. E-mail: fox49@fox49.com Web Site: www.fox49.com. Licensee: WTLH License Corp., debtor-in-possession. Group owner: Pegasus Broadcast Television Inc. (acq 1996; $5.595 million). Population served: 569,000 Natl. Network: Fox, CW Digital Network: Note: Fox is on WTLH(TV) ch 49, CW is on WTLH-DT ch 50. Natl. Rep: Petry Television Inc. Washington Atty: Shaw Pittman LLP.
Key Personnel:
Mark Pagon. . . . . CEO
Jack Paris . . . . exec VP & sr VP
David Hinterschied. gen mgr
Don Abel . . . . progmg dir
Mike Brown . . . . chief of engrg
Tyrone Hayes . . . . chief of opns & opns

WTWC-TV— Analog channel: 40. Analog hrs: 24 3,160 kw vis, 316 kw aur. 880t/821g TL: N30 35 11 W84 14 11 On air date: Apr 21, 1983. 8440 Deerlake Rd. S., Tallahassee, FL, 32312. Phone: (850) 893-4140. Fax: (850) 893-6974. Web Site: www.wtwc40.com. Licensee: WTWC Licensee L.L.C. Group owner: Sinclair Broadcast Group Inc. (acq 1999; grpsl). Natl. Network: NBC. Natl. Rep: Millennium Sales & Marketing. Washington Atty: Dow, Lohnes & Albertson. News staff: 30; News: 14 hrs wkly.
Key Personnel:
Bob W. Franklin. . . . . gen mgr
Mike Plumber . . . . stn mgr

WTXL-TV— Analog channel: 27. Digital channel: 22. Analog hrs: 24 Digital hrs: 24 3,000,000 watts. 1,700 TL: N30 34 27 W84 12 09 On air date: Sept 16, 1976. 8440 Deer Lake Rd., Tallahassee, FL, 32312. Phone: (850) 893-4140. Fax: (850) 668-1460. Web Site: www.wtxl.com. Licensee: Southern Broadcast Corp. of Sarasota. (acq 11-30-2005; $12 million). Population served: 268,000 Natl. Network: ABC. Washington Atty: Keck, Mahin & Cate. News staff: 26; News: 24 hrs wkly.
Key Personnel:
Gary Shorts . . . . CEO
Mike Plummer. . . . . stn mgr
Steve Rollison. . . . . gen mgr & news dir

Tampa-St. Petersburg (Sarasota), FL (DMA 12)

WCLF— Analog channel: 22. Digital channel: 21. Analog hrs: 24 5000 kw vis, 500 kw aur. ant 1,410t/1,538g TL: N28 11 04 W82 45 39 On air date: October 1979. Box 6922, Clearwater, FL, 33758. 6922 142nd Ave., Largo, FL 33771. Phone: (727) 535-5622. Fax: (727) 531-2497. Web Site: www.ctnonline.com. Licensee: Christian Television Corporation Inc. Group owner: (group owner; (acq 12-16-97). Population served: 10,000,000 Washington Atty: Gammon & Grange. Foreign lang progmg: SP 2

***WEDU**—Tampa, Analog channel: 3. Digital channel: 54. Analog hrs: 24 Digital hrs: 24 100 kw vis, 20 kw aur. 1,551t TL: N27 49 48 W82 15 59 On air date: Oct 27, 1958. Box 4033, Tampa, FL, 33677-4033. 1300 North Blvd., Tampa, FL 33607. Phone: (813) 254-9338. Fax: (813) 253-0826. Web Site: www.wedu.org. Licensee: Florida West Coast Pub Broadcasting Inc. Population served: 3,200,000 Natl. Network: PBS. Washington Atty: Schwartz, Woods & Miller.
Key Personnel:
Richard M. Lobo . . . . CEO & pres
Patrick Perkins. . . . . CFO
Frank Wolynski . . . . opns VP
Susanna Grady . . . . dev VP
Ellyne Lonergan . . . . progmg VP
Mike Seymour . . . . progmg

WFLA-TV—Tampa, Analog channel: 8. Digital channel: 7.316 kw vis, 31.6 kw aur. ant 1,545t/1,535g TL: N27 50 32 W82 15 46 On air date: Feb 14, 1955. Box 1410, Tampa, FL, 33601. 200 South Parker St., Tampa, FL 33608. Phone: (813) 228-8888. Fax: (813) 221-5787. Web Site: www.wfla.com. Licensee: Media General Communications Inc. Group owner: Media General Broadcast Group, see Cross-Ownership (acq 1965; $17.5 million). Population served: 3,332,000 Natl. Network: NBC. Natl. Rep: Harrington, Righter & Parsons. Washington Atty: Dow, Lohnes & Albertson. Wire Svc: UPI
Key Personnel:
Eric Land . . . . pres & gen mgr
S. Rick McEwen. . . . . opns dir

WFTS—Tampa, Analog channel: 28. Digital channel: 29. Analog hrs: 24 2.63 kw vis, 260 w aur. 1,546t/1,649g TL: N27 50 32 W82 15 46 On air date: Dec 14, 1981. 4045 N. Himes Ave., Tampa, FL, 33607. Phone: (813) 354-2828. Fax: (813) 878-2828. Web Site: www.abcactionnews.com. Licensee: Tampa Bay Television Inc. Group owner: Scripps Howard Broadcasting Co., see Cross-Ownership (acq 1-2-86; grpsl). Natl. Network: ABC. Natl. Rep: Eagle Television Sales. Washington Atty: Baker & Hostetler.
Key Personnel:
Bill Carey . . . . VP & gen mgr
Jack Winter . . . . opns mgr
Chris Raynor . . . . prom mgr

WFTT-TV—Tampa, Analog channel: 50. Digital channel: 47. Analog hrs: 24 Digital hrs: 24 4,200 kw vis, 420 kw aur. 1,600t/1,580g TL: N27 50 32 W82 15 46 On air date: Feb 1, 1988. 2610 W. Hillsborough Ave., Tampa, FL, 33614. Phone: (813) 872-6262. Fax: (813) 998-3600. Web Site: univision.com. Licensee: TeleFutura Tampa LLC. Group owner: Univision Communications Inc. (acq 5-21-2001; grpsl). Population served: 1,600,000 Natl. Network: TeleFutura (Spanish). Washington Atty: Wiley, Rein & Fielding. Foreign lang progmg: SP 168
Key Personnel:
Lilly Gonzalez. . . . . gen mgr
Nelson Castillo . . . . gen sls mgr
Steve Hess . . . . chief of engrg

WMOR-TV— Analog channel: 32. Digital channel: 19. Analog hrs: 24 5,000 kw vis, 500 kw aur. ant 1086t/1087g TL: N27 50 15 W81 56 53 On air date: Apr 24, 1986. 7201 E. Hillsborough Ave., Tampa, FL, 33610-4126. Phone: (813) 626-3232. Fax: (813) 622-7732. Web Site: www.moretv32.com. Licensee: WMOR-TV Company. Group owner: Hearst-Argyle Television Inc. (acq 1996; $25.5 million). Population served: 3,177,000 Natl. Rep: MMT. Washington Atty: Brooks, Pierce, McLendon, Humphrey & Leonard.
Key Personnel:
Ken Lucas . . . . VP & gen mgr
Roy Tym . . . . natl sls mgr
Bonita Elias . . . . rgnl sls mgr
Pete George . . . . prom mgr
Joseph Pauly . . . . progmg mgr

WTOG—Saint Petersburg, Analog channel: 44. Digital channel: 59. Analog hrs: 24 5,000 kw vis, 285 kw aur. ant 1,649t/1,507g TL: N27 49 48 W82 15 59 On air date: Nov 4, 1968. 365 105th Terr. N.E., Saint Petersburg, FL, 33716. Phone: (727) 576-4444. Fax: (727) 570-4458. Web Site: cw44.com. Licensee: CBS Operations Inc. Group owner: Viacom Television Stations Group (acq 9-19-96). Population served: 1,710,400 Natl. Network: CW. Natl. Rep: TeleRep. Wire Svc: NWS (National Weather Service)

WTSP—Saint Petersburg, Analog channel: 10. Digital channel: 24.316 kw vis, 31.6 kw aur. 1,549t/1,538g TL: N28 11 04 W82 45 39 On air date: July 17, 1965. 11450 Gandy Blvd., Saint Petersburg, FL, 33702. Phone: (727) 577-1010. Fax: (727) 578-7637. Web Site: www.wtsp.com. Licensee: Pacific and Southern Co. Group owner: Gannett Broadcasting (acq 12-31-96). Population served: 4,583,600 Natl. Network: CBS. Natl. Rep: Blair Television. Washington Atty: Wiley, Rein & Fielding.
Key Personnel:
Sam Rosenwasser. . . . . gen mgr
Lee Griffin. . . . . opns mgr
Pete Nikiel . . . . mktg mgr

WTTA—Saint Petersburg, Analog channel: 38. Digital channel: 57. Analog hrs: 24 5,000 kw vis, 500 kw aur. ant 1,867t/1,500g TL: N27 50 32 W82 15 46 On air date: June 21, 1991. 7622 Bald Cypress Pl., Tampa, FL, 33614. Phone: (813) 886-9882. Fax: (813) 880-8100 / 8154. E-mail: comments@wtta38.com Web Site: www.wtta38.com. Licensee: Bay Television Inc. Population served: 3,617,000 Natl. Network: MyNetworkTV. Washington Atty: Shaw Pittman.

WTVT—Tampa, Analog channel: 13. Digital channel: 12. Analog hrs: 24 316 kw vis, 47.4 kw aur. 1,416t/1,549g TL: N27 49 09 W82 14 26 On air date: April 1955. Box 31113, Tampa, FL, 33631-3113. 3213 W. Kennedy Blvd., Tampa, FL 33609. Phone: (813) 876-1313. Fax: (813) 871-3135. E-mail: news@wtvt.com Web Site: www.wtvt.com. Licensee: New World Communications of Tampa Inc. Group owner: Fox Television Stations Inc. (acq 11-96; grpsl). Population served: 2,935,000 Natl. Network: Fox. Washington Atty: Hogan & Hartson. Wire Svc: Conus News staff: 100; News: 46 hrs wkly.
Key Personnel:
Bob Linger . . . . exec VP & gen mgr
Jim Benedict . . . . opns dir

***WUSF-TV**—Tampa, Analog channel: 16. Analog hrs: 24 1,620 kw vis, 162 kw aur. 1,010t TL: N27 50 53 W82 15 48 On air date: Sept 12, 1966. Univ. of South Florida, 4202 Fowler Ave., Tampa, FL, 33620. Phone: (813) 974-4000. Fax: (813) 974-4806. E-mail: pholley@wusf.org Web Site: www.wusf.org. Licensee: University of South Florida. Population served: 3,000,000 Natl. Network: PBS. Washington Atty: Cohn & Marks.
Key Personnel:
Jo Ann Urofsky . . . . gen mgr
Pat Holly . . . . stn mgr
Jeff Hammel. . . . . opns mgr
Cathy Coccia . . . . dev dir
Susan Geiger. . . . . progmg dir

WVEA-TV— Analog channel: 62. Digital channel: 25. Analog hrs: 24 Digital hrs: 24 5,000 kw vis, 500 kw aur. 572t/570g TL: N27 06 01 W82 22 18 On air date: May 3, 1991. 2610 W. Hillsborough Ave., Tampa, FL, 33614-6132. Phone: (813) 872-6262. Fax: (813) 998-3600. Web Site: www.wvea.entravision.com. Licensee: Entravision Holdings L.L.C. Group owner: Entravision Communications Corp. (acq 1999; $17 million). Population served: 1,600,000 Natl. Network: Univision (Spanish). Washington Atty: Thompson Hine, LLP. Wire Svc: AP Foreign lang progmg: SP 168 News staff: 14; News: 3.5 hrs wkly.
Key Personnel:
Lilly Gonzalez. . . . . gen mgr

Nelson Castillo . . . . gen sls mgr
Pilar Ortiz . . . . news dir
Bill Mierisch . . . . chief of engrg

WWSB— Analog channel: 40. Digital channel: 52. Analog hrs: 24 2,871 kw vis, 431 kw aur. ant 771t/814g TL: N27 33 27 W82 21 59 On air date: Oct 23, 1971. 1477 10th St., Sarasota, FL, 34236. Phone: (941) 923-8840. Fax: (941) 924-3971/ (941) 923-8709. E-mail: generalmanager@wwsb.tv Web Site: www.wwsb.tv. Licensee: Southern Broadcast Corp. of Sarasota. (acq 3-26-86; $40,500). Population served: 157,310 Natl. Network: ABC. Washington Atty: Leibowitz & Associates. News staff: 45; News: 21 hrs wkly.
Key Personnel:
Gary Shorts . . . . CEO
J. Manuel Calvo. . . . . pres & gen mgr
Jason Wildanstein . . . . opns dir

WXPX— Analog channel: 66. Digital channel: 42.2,240 kw vis, 224 kw aur. ant 1,525t/1,496g TL: N27 24 30 W82 15 00 On air date: Aug 1, 1994. 4444 66th St N., Clearwater, FL, 33764-7204. Phone: (813) 314-5462. Fax: (813) 314-5464. Web Site: www.ionline.tv. Licensee: Paxson Communications License Co. L.L.C. Group owner: Paxson Communications Corp. (acq 12-15-97). Natl. Network: i Network.

West Palm Beach-Ft. Pierce, FL (DMA 38)

WFGC— Analog channel: 61. Digital channel: 49. Analog hrs: 24 5,000 kw vis, 500 kw aur. ant 410t TL: N26 45 47 W80 12 19 On air date: May 21, 1993. 1900 S. Congress Ave., Suite B, West Palm Beach, FL, 33406. Phone: (561) 642-3361. Fax: (561) 967-5961. E-mail: comments@wfgc.com Web Site: www.wfgc.com. Licensee: Christian TV of Palm Beach County Inc. Ownership: David C. Gibbs III, Wayne Wetzel and Bill Anderson. (acq 12-16-97). Washington Atty: Gammon & Grange.
Key Personnel:
Wayne Wetzel . . . . pres
Neville Chankersingh . . . . CFO
Mike Gonzales . . . . gen mgr
Chris Mavrois . . . . chief of engrg

WFLX— Analog channel: 29. Digital channel: 28. Analog hrs: 24 5,000 kw vis, 500 kw aur. ant 1,540t/1,533g TL: N26 34 37 W80 14 32 On air date: Aug 14, 1982. 4119 W. Blue Heron Blvd., West Palm Beach, FL, 33404. Phone: (561) 845-2929. Fax: (561) 863-1238. Web Site: www.wflx.com. Licensee: Raycom National Inc. Group owner: Raycom Media Inc. (acq 1998). Natl. Network: Fox. Natl. Rep: Harrington, Righter & Parsons.
Key Personnel:
John Spinola . . . . VP & gen mgr
John Heislman. . . . . gen sls mgr & natl sls mgr

WPBF— Analog channel: 25. Analog hrs: 24 5,000 kw vis, 500 kw aur. 1,529t/1,549g TL: N26 50 09 W80 05 44 On air date: Jan 1, 1989. 3970 RCA Blvd., Suite 7007, Palm Beach Gardens, FL, 33410. Phone: (561) 694-2525. Fax: (561) 624-1089. Web Site: www.wpbfnews.com. Licensee: WPBF-TV Co. Group owner: Hearst-Argyle Television Inc. (acq 8-1-97). Population served: 772,000 Natl. Network: ABC. Natl. Rep: Continental Television Sales. News staff: 48; News: 25 hrs wkly.
Key Personnel:
Victoria Regan . . . . VP & gen mgr
Caroline Scollard . . . . gen sls mgr

WPEC— Analog channel: 12. Digital channel: 13. Analog hrs: 24 Digital hrs: 24 316 kw vis, 56.2 kw aur. 980t/1,027g TL: N26 35 17 W80 12 28 On air date: Jan 1, 1955. Box 198512, West Palm Beach, FL, 33419-8512. 1100 Fairfield Dr., West Palm Beach, FL 33419-8512. Phone: (561) 844-1212. Fax: (561) 842-1212. Web Site: www.wpecnews12.com. Licensee: Freedom Broadcasting of Florida Licensee L.L.C. Group owner: Freedom Broadcasting Inc. (acq 2-1-96; $150 million). Population served: 1,400,000 Natl. Network: CBS. Natl. Rep: TeleRep. Washington Atty: Latham & Watkins. News: 24 hrs wkly.
Key Personnel:
Doreen Wade . . . . pres & VP
Diana Wilkin. . . . . VP & gen mgr
Donn Colee . . . . stn mgr & progmg mgr
Doug Wolfmueller. . . . . gen sls mgr
Mary Gregg . . . . natl sls mgr
Jim Posey. . . . . rgnl sls mgr
Steve Hunsicker. news dir
Keith Betts. . . . . engrg dir & chief of engrg

***WPPB-TV—** Analog channel: 63. Analog hrs: 24 5,000 kw vis. ant 1,000t/1,000g TL: N25 59 10 W80 11 36 On air date: 2001. Broward Education Communications Network, 6600 S.W. Nova Dr., Fort Lauderdale, FL, 33317. Phone: (754) 321-1000. Fax: (754) 321-1180. E-mail: feedback@becon.tv Web Site: www.becon.tv. Licensee: The School Board of Broward County, Florida. (acq 3-31-2000). Population served: 1,600,000

WPTV— Analog channel: 5. Analog hrs: 24 100 kw vis, 20 kw aur. 990t/1,031g TL: N26 35 20 W80 12 43 On air date: Aug 22, 1954. 100 Banyan Blvd., West Palm Beach, FL, 33401. Phone: (561) 655-5455. Fax: (561) 653-5719. E-mail: newstips@wptv.com Web Site: www.wptv.com. Licensee: Scripps Howard Broadcasting Co. Group owner: (group owner; (acq 12-27-61; $2 million; FTR: 12-25-61). Population served: 1,200,000 Natl. Network: NBC. Natl. Rep: Harrington, Righter & Parsons. Washington Atty: Baker & Hostetler. Wire Svc: Reuters News staff: 75; News: 27 hrs wkly.
Key Personnel:
Kenneth W. Lowe . . . . CEO
Joseph G. Ne Castro . . . . CFO
Brian Lawlor . . . . sr VP, VP & gen mgr

WPXP— Analog channel: 67. Analog hrs: 24 1,000 kw vis, 200 kw aur. 492t TL: N26 47 59 W80 04 33 On air date: 1998. 500 Australian Ave., Suite 200, 601 Clearwater Park Rd., West Palm Beach, FL, 33401. Phone: (561) 686-6767. Fax: (561) 682-3475. Web Site: www.ionline.tv. Licensee: Hispanic Broadcasting Inc. Ownership: Paxson Communications Corp., 90%; Betti Lidsky, 10%. Group owner: Paxson Communications Corp. (acq 1998). Natl. Network: i Network.

***WTCE-TV—** Analog channel: 21.2,285.6 kw vis, 457.12 kw aur. 973t/1,002g TL: N27 26 05 W80 21 42 On air date: May 1990. 3601 N. 25th St., Fort Pierce, FL, 34946. Phone: (772) 489-2701. Fax: (772) 489-6833. Web Site: www.wtcebellsouth.net. Licensee: Jacksonville Educators Broadcasting Inc. (acq 7-6-90; $630,089). Population served: 151,000

WTVX— Analog channel: 34. Digital channel: 50.5,000 kw vis, 500 kw aur. ant 1,492t/1,520g TL: N27 07 20 W80 23 21 On air date: Apr 5, 1966. 1700 Palm Beach Lakes Blvd. Ste 150, 4411 Beacon Cir., West Palm Beach, FL, 33401-2017. Phone: (561) 841-3434. Fax: (561) 848-9150. Web Site: www.wtvx.com. Licensee: Channel 34 Television Station LLC. Group owner: Viacom Television Stations Group (acq 11-13-2001). Population served: 898,000 Natl. Network: CW. Natl. Rep: TeleRep. Washington Atty: Haley, Bader & Potts.

***WXEL-TV—** Analog channel: 42. Digital channel: 27. Analog hrs: 24 Digital hrs: 24 2,140 kw vis, 2.14 kw aur. ant 1,443t/1,440g TL: N26 34 37 W80 14 32 On air date: July 7, 1982. Box 6607, West Palm Beach, FL, 33405-6607. 3401 South Congress Ave., Boynton Beach, FL 33426. Phone: (561) 737-8000. Fax: (561) 369-3067. E-mail: jcarr@wxel.org Web Site: www.wxel.org. Licensee: Barry Telecommunications Inc. Ownership: Barry University (acq 4-16-97). Population served: 2,000,000 Natl. Network: PBS. Washington Atty: Schwartz, Woods & Miller.
Key Personnel:
Jerry Carr. . . . . CEO
Bernard Henneberg . . . . CFO
Jerry Carr . . . . gen mgr
Fred Flaxman . . . . dev VP
Ross Cooper . . . . sls dir, mktg dir & mktg mgr
Lee Rowand . . . . prom dir

Georgia

Albany, GA (DMA 145)

***WABW-TV—** Analog channel: 14.5,000 kw vis, 500 kw aur. 1,240t/1,224g TL: N31 08 05 W84 06 16 On air date: Jan 1, 1967. 260 14th St. N.W., Atlanta, GA, 30318. Phone: (404) 685-2400. Fax: (404) 685-2591. E-mail: ask@gpb.org Web Site: www.gpb.org. Licensee: Georgia Public Telecommunications Commission. Natl. Network: PBS.

***WACS-TV—** Analog channel: 25. Analog hrs: 19 363 kw vis, 36.3 kw aur. 1,045t/1,096g TL: N31 56 15 W84 33 15 On air date: Mar 6, 1967. 260 14th St. N.W., Atlanta, GA, 30318. Phone: (404) 685-2400. Fax: (404) 685-2591. E-mail: ask@gpb.org Web Site: www.gpb.org. Licensee: Georgia Public Telecommunications Commission. Natl. Network: PBS. Washington Atty: Arent, Fox, Kintner, Plotkin & Kahn.

WALB— Analog channel: 10.316 kw vis, 43.6 kw aur. 964t/1,000g TL: N31 19 52 W83 51 44 On air date: Apr 7, 1954. Box 3130, 1709 Stuart Ave., Albany, GA, 31706. Phone: (229) 446-1010. Fax: (229) 446-4000. E-mail: walb@walb.com Web Site: www.walb.com. Licensee: Libco Inc. Group owner: Liberty Corp. (acq 1-13-2006; grpsl). Population served: 424,000 Natl. Network: NBC. Natl. Rep: Continental Television Sales. News staff: 5; News: 15 hrs wkly.
Key Personnel:
James Wilcox. . . . . pres & gen mgr
Dawn Hobby . . . . news dir

WFXL— Analog channel: 31. Digital channel: 12. Analog hrs: 24 1,580 kw vis, 150 kw aur. ant 990t/1,000g TL: N31 19 52 W83 51 43 On air date: Feb 14, 1982. Box 4050, Albany, GA, 31706. 1201 Stuart Ave., Albany, GA 31707. Phone: (229) 435-3100. Fax: (229) 903-8240. Web Site: www.wfxl.com. Licensee: Barrington Albany License LLC. Group owner: Raycom Media Inc. (acq 8-11-2006; grpsl). Population served: 385,000 Natl. Network: Fox. Natl. Rep: MMT. Washington Atty: Covington & Burling. News staff: 7; News: 4 hrs wkly.
Key Personnel:
Jenny Collins . . . . gen mgr
Deborah Owens . . . . stn mgr
Pat Coffman. opns mgr
Teri Underwood . . . . gen sls mgr
Terry Graham. news dir
Ken Clubb . . . . chief of engrg
Brandi Fickel . . . . mktg

WSST-TV— Analog channel: 55. Analog hrs: 24 Digital hrs: 24 100 kw vis. 410t TL: N31 53 35 W83 48 18 On air date: May 22, 1989. Box 917, 112 S. 7th St., Detroit, GA, 31015. Phone: (229) 273-0001. Fax: (229) 273-8894. E-mail: wsst@sowega.net Web Site: www.wsst.com. Licensee: Sunbelt-South Telecommunications Ltd. Ownership: William B. Goodson, 65%; Phillip A. Streetman, 35%. Population served: 600,000 Washington Atty: Law Offices of Scott Cinnamon.
Key Personnel:
Phillip Streetman . . . . sr VP, VP, stn mgr & news dir
Sara J. Howell . . . . gen sls mgr
Lee W. Wright . . . . progmg dir

Atlanta (DMA 9)

WAGA— Analog channel: 5. Digital channel: 27. Analog hrs: 24 100 kw vis, 20 kw aur. 1,076t/1,103g TL: N33 47 51 W84 20 02 On air date: April 1949. 1551 Briarcliff Rd. N.E., Atlanta, GA, 30306. Phone: (404) 875-5555. Fax: (404) 898-0238. Web Site: www.fox5atlanta.com. Licensee: Fox Television Stations Inc. Group owner: (group owner; (acq 11-96; grpsl). Population served: 4,000,000 Natl. Network: Fox. News: 38 hrs wkly.
Key Personnel:
Gene McHugh . . . . gen mgr
Neil Mazur . . . . opns VP & opns dir

***WATC—** Analog channel: 57. Digital channel: 41. Analog hrs: 24 Digital hrs: 24 398 kw vis, 39.8 kw aur. 423t TL: N33 48 40 W84 21 51 On air date: Apr 14, 1996. 1862 Enterprise Dr., Norcross, GA, 30093. Phone: (770) 300-9828. Fax: (770) 300-9838. Web Site: www.watc.tv. Licensee: Community Television Inc. (acq 6-3-93; $79,866; FTR: 6-28-93).
Key Personnel:
James Thompson . . . . pres & gen mgr
Joanne Thompson . . . . VP
Greg West . . . . prom dir & progmg dir

WATL— Analog channel: 36. Digital channel: 25.2,682 kw vis, 402 kw aur. 1,170t/1,174g TL: N33 48 27 W84 20 26 On air date: July 5, 1976. One Monroe Pl., Atlanta, GA, 30324. Phone: (404) 881-3600. Fax: (404) 881-3749. Web Site: www.myatltv.com. Licensee: Gannett Georgia L.P. Group owner: Tribune Broadcasting Co. (acq 8-7-2006; $180 million). Population served: 4,112,000 Natl. Network: MyNetworkTV. Natl. Rep: Blair Television. Washington Atty: Wiley, Rein & Fielding.
Key Personnel:
Bob Walker. . . . . gen mgr
Jack Walsh . . . . gen sls mgr
Steven Fredericks. . . . . natl sls mgr

WGCL-TV— Analog channel: 46. Digital channel: 19. Analog hrs: 24 Digital hrs: 24 2,333 kw vis, 233 kw aur. 1,089t/1,145g TL: N33 48 27 W84 20 26 On air date: June 6, 1971. Box 93524, Atlanta, GA, 30377. 425 14th St. NW, Atlanta, GA 30318. Phone: (404) 325-4646. Fax: (404) 327-3004. Fax: (404) 327-3003. E-mail: cbs46news@cbs46.com Web Site: www.cbs46.com. Licensee: Meredith Corp. Group owner: Meredith Broadcasting Group, Meredith Corp. (acq 2-22-99; $370 million swap with KCPQ(TV) Tacoma, WA). Population served: 5,725,787 Natl. Network: CBS. Natl. Rep: Harrington, Righter & Parsons. Washington Atty: Dow, Lohnes & Albertson, PLLC. News staff: 80; News: 14.5 hrs wkly.
Key Personnel:
Andy Alford . . . . VP & gen mgr
Joanna Hemleb . . . . gen sls mgr
Rick Erbach . . . . news dir

***WGTV—** Analog channel: 8. Analog hrs: 6 AM-midnight 316 kw vis, 56.2 kw aur. ant 1,132t/420g TL: N33 48 18 W84 08 40 On air date:

May 23, 1960. 260 14th St. N.W., Atlanta, GA, 30318. Phone: (404) 685-2400. Fax: (404) 685-2431. Web Site: www.gpb.org. Licensee: Georgia Public Telecommunications Commission. Ownership: Ga. Public TV Net. Population served: 1,500,000 Natl. Network: PBS. Washington Atty: Arent, Fox, Kintner, Plotkin & Kahn.

***WHSG-TV—** Analog channel: 63. Analog hrs: 24 5,000 kw vis. 1,191t TL: N33 44 22 W84 00 14 On air date: Feb 22, 1991. 1550 Agape Way, Decatur, GA, 30035. Phone: (404) 288-1156. Fax: (404) 288-5613. Web Site: www.tbn.org. Licensee: Trinity Broadcasting Network Inc. Group owner: (group owner; (acq 11-21-89). Population served: 90,000

***WPBA—** Analog channel: 30. Analog hrs: 24 1,380 kw vis. 1,096t TL: N33 45 35 W84 20 07 On air date: Feb 17, 1958. 740 Bismark Rd. N.E., Atlanta, GA, 30324-4102. Phone: (678) 686-0321. Fax: (678) 686-0356. Web Site: www.WPBA.org. Licensee: Board of Education of the City of Atlanta. Population served: 2,500,000 Natl. Network: PBS. Washington Atty: Schwartz, Woods & Miller. News: 1.5 hrs wkly.

WPXA— Analog channel: 14. Analog hrs: 24 3,890 kw vis. 2,021t/787g TL: N34 18 47 W84 38 55 On air date: Jan 15, 1988. 601 Clearwater Park Dr., West Palm Beach, FL, 33401. Phone: (404) 885-7646. Fax: (404) 881-9553. Web Site: www.ionline.tv. Licensee: TV-14 Inc. Group owner: Paxson Communications Corp. (acq 7-13-94; $9.5 million). Population served: 3,900,000 Natl. Network: i Network. Washington Atty: Dow, Lohnes & Albertson.

Key Personnel:

Lowell Paxson . . . . . CEO
James Bocock . . . . . pres
Scott Jolles. . . . . gen mgr

WSB-TV— Analog channel: 2. Digital channel: 39. Analog hrs: 24 Digital hrs: 24 100 kw vis, 20 kw aur. ant 1,037t/1,076g TL: N33 45 51 W84 21 42 On air date: Sept 29, 1948. 1601 W. Peachtree St. N.E., Atlanta, GA, 30309. Phone: (404) 897-7000. Fax: (404) 897-6246 (gen mgr). E-mail: talk2us@wsbtv.com Web Site: www.wsbtv.com. Licensee: Georgia Television Company. Group owner: Cox Broadcasting Natl. Network: ABC. Natl. Rep: TeleRep. Washington Atty: Dow, Lohnes & Albertson.

Key Personnel:

Bill Hoffman . . . . . VP & gen mgr
David Lamothe . . . . . opns dir
Jane Williams. . . . . sls dir
Steve Riley. . . . . mktg dir
Art Rogers . . . . . progmg dir
Marian Pittman . . . . . news dir
Jocelyn Dorsey . . . . . pub affrs dir
Gary Alexander. . . . . engrg dir

WTBS— Analog channel: 17.2,224 kw vis, 224 kw aur. 1,093t/1,042g TL: N33 46 57 W84 23 20 On air date: Sept 1, 1967. 1050 Techwood Dr. N.W., Atlanta, GA, 30318. Phone: (404) 827-1717. Fax: (404) 885-2148. Web Site: www.tbssuperstation.com. Licensee: Superstation Inc., div of Turner Broadcasting System. Ownership: R.E. Turner III, 32.03%; TCI, 24.43%; Time-Warner, 20.34%. (acq 1-70). Population served: 60,714,000 Natl. Rep: TBS.

WUPA— Analog channel: 69. Digital channel: 43. Analog hrs: 24 Digital hrs: 24 1,000 kw vis. ant 1,102t/1,056g TL: N33 44 40 W84 21 36 On air date: Aug 22, 1981. Phoenix Business Park, 2700 Northeast Expwy., Atlanta, GA, 30345. Phone: (404) 325-6969. Fax: (404) 633-4567. Web Site: www.cwatlantatv.com. Licensee: Viacom Stations Group of Atlanta Inc. Group owner: Viacom Television Stations Group (acq 5-4-2000; grpsl). Population served: 5,422,000 Natl. Network: CW. Washington Atty: Wiley, Rein & Fielding.

WUVG-TV— Analog channel: 34.1,258 kw vis, 125.8 kw aur. 1,351t/1,236g TL: N34 12 27 W83 47 38 (CP: 5,000 kw vis, ant 1,443t) On air date: April 1989. 3350 Peach Tree Rd., Suite 1250, Atlanta, GA, 33026. Phone: (404) 926-2300. Fax: (404) 926-2320. Web Site: www.univision.com. Licensee: Univision Partnership of Atlanta. Group owner: Univision Communications Inc. (acq 6-6-01; grpsl). Population served: 750,000 Natl. Network: Univision (Spanish). Washington Atty: William M. Barnard. Foreign lang progmg: SP 168

WXIA-TV— Analog channel: 11. Digital channel: 10.316 kw vis, 63.2 kw aur. 1,048t/1,040g TL: N33 45 24 W84 19 55 On air date: Sept 30, 1951. 1611 W. Peachtree St. N.E., Atlanta, GA, 30309. Phone: (404) 892-1611. Fax: (404) 881-0675. Web Site: www.11alive.com. Licensee: Gannett Georgia L.P. Group owner: Gannett Broadcasting, division of Gannett Co. Inc. (acq 6-7-79; grpsl; FTR: 6-11-79). Population served: 1,592,000 Natl. Network: NBC. Washington Atty: Reed Smith LLP.

Augusta, GA (DMA 114)

WAGT— Analog channel: 26. Digital channel: 30. Analog hrs: 24 Digital hrs: 24 65 kw vis, 6.5 kw aur. ant 1,590t/1,478g TL: N33 25 15 W81 50 19 On air date: Dec 24, 1968. 905 Broad St, Augusta, GA, 30901. Phone: (706) 826-0026. Fax: (706) 724-4028. Fax: (706) 724-7491. Web Site: www.nbcaugusta.com. Licensee: WAGT Television Inc. Group owner: Schurz Communications Inc. (acq 7-1-80; $5 million). Population served: 267,730 Natl. Network: NBC, CW. Natl. Rep: Petry Television Inc. Washington Atty: Hogan & Hartson. News staff: 28; News: 8 hrs wkly.

Key Personnel:

Dewayne Jones . . . . . gen sls mgr
Ed Everest . . . . . prom mgr
Scott Brady . . . . . news dir
Dave DeFrehn . . . . . chief of engrg

***WCES-TV—** Analog channel: 20. Analog hrs: 24 5,000 kw vis, 500 kw aur. ant 1,480t/1,465g TL: N33 15 33 W82 17 09 On air date: Sept 12, 1966. Box 525, Wrens, GA, 30833. Phone: (706) 547-0293. Fax: (800) 222-4788. E-mail: viewerservices@gpb.org Web Site: www.gpb.org. Licensee: Georgia Public Telecommunications Commission. Natl. Network: PBS.

***WEBA-TV—** Analog channel: 14. Digital channel: 33. Analog hrs: 24 661 kw vis. ant 800t/809g TL: N33 11 13 W81 23 54 On air date: Sept 5, 1967. 1101 George Rogers Blvd., Columbia, SC, 29201. Phone: (803) 737-3545. Fax: (803) 737-3495. E-mail: mail@myetv.org Web Site: www.myetv.org. Licensee: South Carolina ETV Commission. Natl. Network: PBS.

Key Personnel:

Maurice "Moss" Bresnahan . . . . . CEO, pres & sr VP
L.W. Griffin Jr. . . . . . engrg VP

WFXG— Analog channel: 54. Analog hrs: 24 4,517 kw vis, 451 kw aur. 1,120t/1,506g TL: N33 25 00 W81 50 60 (CP: 2,491 kw vis, ant 1,164t) On air date: May 23, 1991. Box 204540, Augusta, GA, 30917-4540. 3933 Washington Rd., Augusta, GA 30907. Phone: (706) 650-5400. Fax: (706) 650-8411. E-mail: gtomlinson@wfxg.com Web Site: www.wfxg.com. Licensee: Southeastern Media Holdings Inc. Group owner: (group owner; (acq 12-1-2003; $40 million with WXTX(TV) Columbus). Natl. Network: Fox. Washington Atty: Miller & Fields.

WJBF— Analog channel: 6. Digital channel: 42.100 kw vis, 20 kw aur. ant 1,624t/1,472g TL: N33 24 20 W81 50 01 On air date: Nov 23, 1953. Box 1404, Augusta, GA, 30903. 1001 Reynolds St., Augusta, GA 30901. Phone: (706) 722-6664. Fax: (706) 722-0022. Web Site: www.wjbf.com. Licensee: Media General Broadcasting of South Carolina Holdings Inc. Group owner: Media General Broadcast Group (acq 3-27-2000; grpsl). Population served: 613,000 Natl. Network: ABC. Washington Atty: Dow, Lohnes & Albertson, PLLC. News: 22 hrs wky.

Key Personnel:

Gene Kirkconnell . . . . . gen mgr
Bill Stewart . . . . . gen sls mgr
Charles Coleman . . . . . natl sls mgr
Scot Seabolt. . . . . rgnl sls mgr
Cil Frazier . . . . . mktg dir
Mary Jones . . . . . progmg dir & progmg
Mark Rosen . . . . . news dir
Cary Hale . . . . . chief of engrg

WRDW-TV— Analog channel: 12. Digital channel: 31. Analog hrs: 24 Digital hrs: 24 Note: CBS is on WRDW-TV ch 12, MyNetworkTV is on WRDW-DT ch 31. 316 kw vis, 30.2 kw aur. ant 1,590t/1,506g TL: N33 24 29 W81 50 36 On air date: Feb 14, 1954. Box 1212, Augusta, GA, 30903-1212. 1301 Georgia Ave., North Augusta, SC 29841. Phone: (803) 278-1212. Fax: (803) 279-8316. Web Site: www.wrdw.com. Licensee: Gray Television Licensee Inc. Group owner: Gray Television Inc. (acq 1-4-96; $34 million). Population served: 604,000 Natl. Network: CBS, MyNetworkTV Digital Network: Note: CBS is on WRDW-TV ch 12, MyNetworkTV is on WRDW-DT ch 31. Natl. Rep: Continental Television Sales. News staff: 35; News: 20 hrs wkly.

Key Personnel:

John Ray . . . . . pres & gen mgr
Estelle Parsley. . . . . opns dir & news dir
Joe Tonsing . . . . . gen sls mgr
Michael Oates . . . . . natl sls mgr & rgnl sls mgr
Edward Elser. . . . . chief of engrg

Bainbridge

see Tallahassee, FL-Thomasville, GA market

Brunswick

see Jacksonville, FL market

Chatsworth

see Chattanooga, TN market

Columbus, GA (DMA 128)

***WGIQ—** Analog channel: 43. Digital channel: 44. Analog hrs: 24 Digital hrs: 24 5,000 kw vis. 902t/715g TL: N31 43 05 W85 26 03 On air date: Sept 9, 1968. 2112 11th Ave. S., Suite 400, Birmingham, AL, 35205. Phone: (205) 328-8756. Fax: (205) 251-2192. Web Site: www.aptv.org. Licensee: Alabama ETV Commission. Natl. Network: PBS. Washington Atty: Dow, Lohnes & Albertson, PLLC.

Key Personnel:

Allan Pizzato . . . . . CEO
Charles Grantham . . . . . CEO & COO
Pauline Howland . . . . . CFO
John Brady . . . . . dev VP

***WJSP-TV—** Analog channel: 28. Analog hrs: 24 5,000 kw vis, 500 kw aur. ant 1,512t/1,102g TL: N32 51 08 W84 42 04 On air date: Aug 10, 1964. 609 Whitehouse Pkwy, Warm Springs, GA, 31830. 260 14th St, N.W., Atlanta, GA 30318. Phone: (706) 655-2145. Fax: (404) 685-2431. E-mail: ask@gpb.org Web Site: www.gpb.org. Licensee: Georgia Public Telecommunications Commission. Natl. Network: PBS.

WLGA— Analog channel: 66. Digital channel: 31. Analog hrs: 24 794.3 kw vis, 79.43 kw aur. ant 679t TL: N32 38 33 W85 14 13 On air date: May 16, 1982. 1800 Pepperell Pkwy., Opelika, AL, 36801. Phone: (334) 745-0066. Fax: (334) 749-5768. E-mail: mbrooks@pappastv.com Web Site: www.wlgatv.com. Licensee: Pappas Telecasting of Opelika L.P. (a Delaware limited partnership). Group owner: (group owner; (acq 1996; $1.6 million). Natl. Network: CW. Washington Atty: Paul, Hastings, Janofsky & Walker.

Key Personnel:

Harry J. Pappas . . . . . pres
Mike Brooks . . . . . gen mgr
Walter Dix. . . . . opns mgr

WLTZ— Analog channel: 38. Digital channel: 35. Analog hrs: 24 Digital hrs: 24 1,070 kw vis, 209 kw aur. 1,310t/1,319g TL: N32 27 29 W84 53 08 On air date: Oct 29, 1970. 6140 Buena Vista Rd., Columbus, GA, 31907. Box 12289, Columbus, GA 31917. Phone: (706) 561-3838. Fax: (706) 563-8467. Fax: (706) 561-3880 (sales). E-mail: wltz@wltz.com Web Site: www.wltz.com. Licensee: Lewis Broadcasting Corp. Ownership: J.C. Lewis Jr., 100%. (acq 7-1-81; $3.25 million). Population served: 189,600 Natl. Network: NBC. Natl. Rep: Blair Television. Washington Atty: Wiley, Rein & Fielding.

Key Personnel:

J. Curtis Lewis III . . . . . pres
Charles Izlar. . . . . CFO
Tom Breazeale. . . . . VP & gen mgr
Charles Collins . . . . . opns dir

WRBL— Analog channel: 3. Digital channel: 15. Analog hrs: 24 Digital hrs: 24 100 kw vis, 12 kw aur. ant 1,780t/1,749g TL: N30 19 25 W84 46 46 On air date: Nov 15, 1953. Box 270, Columbus, GA, 31902-0270. 1350 13th Ave., Columbus, GA 31902-0270. Phone: (706) 323-3333. Fax: (706) 327-6655. Fax: (706) 323-0841. Web Site: www.wrbl.com. Licensee: Media General Broadcasting of South Carolina Holdings Inc. Group owner: Media General Broadcast Group (acq 3-27-2000; grpsl). Population served: 769,000 Natl. Network: CBS. Natl. Rep: MMT. Washington Atty: Dow, Lohnes & Albertson.

WTVM— Analog channel: 9.284 kw vis, 52.5 kw aur. 1,650t/1,749g TL: N32 19 25 W84 46 46 On air date: Oct 6, 1953. Box 1848, Harrisonburg, GA, 31902. Phone: (706) 324-6471. Fax: (706) 322-7527. E-mail: newsleader@wtvm.com Web Site: www.wtvm.com. Licensee: Raycom America License Subsidiary LLC. Group owner: Raycom Media Inc. (acq 1996; grpsl). Population served: 496,000 Natl. Network: ABC. Natl. Rep: Harrington, Righter & Parsons. Washington Atty: Powell, Goldstein, Frazer & Murphy.

Key Personnel:

Lee Brantley. . . . . VP & gen mgr
Rick Moll . . . . . news dir

WXTX— Analog channel: 54. Analog hrs: 24 1,000 kw vis, 100 kw aur. 1,140t/1,121g TL: N32 27 49 W84 52 37 (CP: TL: N32 27 40 W84 52 43) On air date: June 17, 1983. Box 1848, Columbus, GA, 31902. Phone: (706) 324-6471. Fax: (706) 322-7527. E-mail: programming@wxtx.com Web Site: www.wxtx.com. Licensee: Southeastern Media Holdings Inc. Group owner: (group owner; (acq 12-1-2003; $40 million with WFXG(TV) Augusta). Natl. Network: Fox. Washington Atty: Fisher, Wayland, Cooper, Leader & Zaragoza.

Key Personnel:

Lee Brantley . . . . . gen mgr
Rick Moll . . . . . news dir

Dalton

see Chattanooga, TN market

Macon, GA
(DMA 121)

WGNM— Analog channel: 64. Digital channel: 45. Analog hrs: 24 52 kw vis, 5.2 kw aur. ant 608t/685g TL: N32 44 58 W83 33 35 (CP: 1,150 kw vis, ant 731t. TL: N32 45 51 W83 33 32) On air date: Nov 30, 1990. 178 Steven Dr., Macon, GA, 31210. Phone: (478) 474-8400. Fax: (478) 474-4777. E-mail: recet@wgnm.com Web Site: www.wgnm.com. Licensee: Christian Television Network Inc. Ownership: Jimmy Smith, 25%; Robert T. D'Andrea, 25%; Virginia Oliver, 25%; and Wayne Wetzel, 25% (acq 12-31-2003; $3 million). Population served: 210,000 Washington Atty: Allen & Harold. News: one hr wkly.
Key Personnel:
Robert D'Andrea . . . . . pres
Rip Kenley . . . . . gen mgr

WGXA— Analog channel: 24. Digital channel: 16.Note: Fox is on WGXA(TV) ch 24, MyNetworkTV is on WGXA-DT ch 16. 1,290 kw vis, 252 kw aur. ant 800t/898g TL: N32 45 08 W83 33 38 On air date: Apr 21, 1982. Box 340, 599 Martin Luther King Blvd., Macon, GA, 31201. Phone: (478) 745-2424. Fax: (478) 745-6057 (news). Web Site: www.fox24.com. Licensee: Piedmont Television of Macon License LLC. Group owner: GOCOM Communications (acq 1-27-2000). Natl. Network: Fox, MyNetworkTV Digital Network: Note: Fox is on WGXA(TV) ch 24, MyNetworkTV is on WGXA-DT ch 16. Washington Atty: Leibowitz & Spencer.

WMAZ-TV— Analog channel: 13. Digital channel: 4. Analog hrs: 24 Digital hrs: 24 316 kw vis, 62 kw aur. 780t/1,209g TL: N32 45 10 W83 33 32 On air date: Sept 27, 1953. Box 5008, Macon, GA, 31208. 1314 Gray Hwy., Macon, GA 31211. Phone: (478) 752-1313. Fax: (478) 752-1331. Web Site: www.13wmaz.com. Licensee: Gannett Georgia L.P. Group owner: Gannett Broadcasting (acq 12-4-95). Population served: 563,000 Natl. Network: CBS. Washington Atty: Wiley, Rein & Fielding. News staff: 39; News: 27 hrs wkly.
Key Personnel:
Dodie Cantrell . . . . . VP & gen mgr
Frank Shurling . . . . . gen sls mgr
Jeff Dudley . . . . . mktg dir & prom dir

WMGT-TV— Analog channel: 41. Digital channel: 40. Analog hrs: 24 760 kw vis, 154 kw aur. ant 893t/837g TL: N32 45 12 W83 33 46 On air date: Aug 26, 1968. Box 4328, Macon, GA, 31208-4328. 301 Poplar St., Macon, GA 31201. Phone: (478) 745-4141. Fax: (478) 742-2626. E-mail: info@wmgt.com Web Site: www.wmgt.com. Licensee: Morris Network Inc. Group owner: Morris Multi-Media (acq 11-30-78; $2.8 million; FTR: 12-18-78). Population served: 531,000 Natl. Network: NBC. Natl. Rep: Millennium Sales & Marketing. Washington Atty: McFadden, Evans & Sill.
Key Personnel:
Dean Hinson . . . . . pres
Carl Bruce . . . . . gen mgr
Debbie Wright . . . . . progmg dir
Mike Roberts . . . . . news dir

***WMUM-TV—** Analog channel: 29. Digital channel: 7.5,000 kw vis, 50 kw aur. ant 1,087t/1,168g TL: N32 28 11 W83 15 17 On air date: Jan 1, 1968. 243 Cary Salem Rd., Cochran, GA, 31014. Phone: (478) 934-3095. Licensee: Georgia Public Telecommunications Commission. Natl. Network: PBS.

WPGA-TV— Analog channel: 58. Analog hrs: 24 1,170 kw vis, 117 kw aur. 810t TL: N32 33 20 W83 44 14 On air date: March 1995. 1691 Forsyth St., Macon, GA, 31201. Phone: (478) 745-5858. Fax: (478) 745-5800. Web Site: www.wpga.tv. Licensee: Radio Perry Inc. Population served: 473,000 Natl. Network: ABC. Washington Atty: Brown, Nietert & Kaufman. News: 3 hrs wkly.
Key Personnel:
Lowell Register . . . . . pres
Debbie Hart . . . . . gen mgr
Len Register . . . . . opns mgr
Julie Register . . . . . prom mgr

Savannah, GA
(DMA 97)

WGSA— Analog channel: 34. Digital channel: 35. Analog hrs: 24 Digital hrs: 24 316 kw vis. ant 482t TL: N31 45 53 W82 13 38 On air date: May 1, 1992. 401 Mall Blvd., Suite 202A, Savannah, GA, 31406. Phone: (912) 692-8000. Fax: (912) 692-0400. Licensee: Southern TV Corp. Ownership: Dan L. and Betty Jo Johnson, 42.5%. (acq 6-3-98; $3.2 million). Natl. Network: CW. Washington Atty: Irwin, Campbell & Tannenwald.
Key Personnel:
Dan L. Johnson . . . . . CEO, chmn & pres
Charles E. Robb . . . . . CFO
Jo Johnson . . . . . exec VP
Fred Pierce . . . . . gen mgr

WJCL— Analog channel: 22. Digital channel: 23. Analog hrs: 24 3,830 kw vis, 383 kw aur. 1,430t/1,478g TL: N32 03 30 W81 20 20 On air date: July 18, 1970. 10001 Abercorn St., Savannah, GA, 31406. Phone: (912) 925-0022. Fax: (912) 921-2235. E-mail: comments@wjcl.com Web Site: www.abc22tv.com. Licensee: Piedmont Television of Savannah License LLC. Group owner: Piedmont Television Holdings LLC (acq 12-8-99; grpsl). Population served: 721,000 Natl. Network: ABC. Natl. Rep: Petry Television Inc. Washington Atty: Cohn & Marks. News staff: 14; News: 5 hrs wkly.
Key Personnel:
Paul Brissette . . . . . CEO
Dave Tillery . . . . . gen mgr
Dave German . . . . . opns mgr

***WJWJ-TV—** Analog channel: 16. Analog hrs: 6 AM-3 AM 851 kw vis, 169 kw aur. 1,279t/1,320g TL: N32 42 44 W80 40 49 On air date: Sept 19, 1976. Box 1165, Beaufort, SC, 29901. 925 Ribaut Rd., Beaufort, SC 29902. Phone: (843) 524-0808. Fax: (843) 524-1016. E-mail: wjwj@hargray.com Web Site: www.wjwj.org. Licensee: South Carolina ETV Commission. Population served: 245,000 Natl. Network: PBS.
Key Personnel:
Scott Johnson . . . . . opns mgr
Juan Singleton . . . . . news dir
Mike Milburn . . . . . chief of engrg

WSAV-TV— Analog channel: 3. Digital channel: 39.Note: NBC is on WSAV-TV ch 3, MyNetworkTV is on WSAV-DT ch 39. 100 kw vis, 20 kw aur. ant 1,476t/1,532g TL: N32 03 32 W81 17 57 On air date: Feb 1, 1956. 1430 E. Victory Dr., Savannah, GA, 31404. Phone: (912) 651-0300. Fax: (912) 651-0304. Web Site: www.wsav.com. Licensee: Media General Broadcasting Inc. Group owner: Media General Broadcast Group (acq 7-25-97; grpsl). Population served: 230,000 Natl. Network: NBC, MyNetworkTV Digital Network: Note: NBC is on WSAV-TV ch 3, MyNetworkTV is on WSAV-DT ch 39. Washington Atty: Dow, Lohnes & Albertson.
Key Personnel:
Jim Berman . . . . . VP
Brad Moses . . . . . gen mgr
Dave Stagnitto . . . . . progmg dir
Kevin Brennan . . . . . news dir

WTGS— Analog channel: 28. Analog hrs: 24 5,000 kw vis, 500 kw aur. 1,499t/1,527g TL: N32 02 48 W81 20 27 On air date: Sept 1, 1985. 10001 Abercorn St., Savannah, GA, 31406. Phone: (912) 925-2287. Fax: (912) 925-7026. E-mail: comments@wjcl.com Web Site: www.fox28tv.com. Licensee: Bluenose Broadcasting of Savannah L.L.C. Ownership: Stephen C. Brissette (acq 10-31-01). Natl. Network: Fox. Washington Atty: Goldberg, Godles, Wiener & Wright. News: 5 hrs wkly.
Key Personnel:
Dave Tillery . . . . . gen mgr
Stephen C. Brissette . . . . . pres & gen mgr
Dave German . . . . . opns mgr
Jennifer Burns . . . . . natl sls mgr
Kurt Hetager . . . . . prom mgr & pub affrs dir
Erik Schrader . . . . . news dir
Ed Youmans . . . . . chief of engrg

WTOC-TV— Analog channel: 11.316 kw vis, 31.6 kw aur. 1,470t/1,531g TL: N32 03 14 W81 21 01 On air date: Feb 14, 1954. Box 8086, Savannah, GA, 31412. 11 The News Place-Chatham Center, Savannah, GA 31405. Phone: (912) 234-1111. Fax: (912) 238-5133. Web Site: www.wtoc.com. Licensee: Raycom America License Subsidiary LLC. Group owner: Raycom Media Inc. (acq 4-15-97; grpsl). Population served: 558,300 Natl. Network: CBS. Natl. Rep: Harrington, Righter & Parsons. Washington Atty: Covinton & Burling.
Key Personnel:
William Cathcart . . . . . VP & gen mgr
Craig Harney . . . . . opns mgr

***WVAN-TV—** Analog channel: 9. Analog hrs: 18 316 kw vis, 34.7 kw aur. ant 1,050t/1,086g TL: N32 08 48 W81 37 05 On air date: Sept 16, 1963. 260 14th St. N.W., 86 Vandiver St., Atlanta, GA, 30318. Phone: (912) 653-4996. Fax: (404) 685-2591. E-mail: ask@gpb.org Web Site: www.gpb.org. Licensee: Georgia Public Telecommunications Commission. Population served: 205,900 Natl. Network: PBS.

Thomasville

see Tallahassee, FL-Thomasville, GA market

Toccoa

see Greenville-Spartanburg, SC-Asheville, NC-Anderson, SC market

Valdosta

see Tallahassee, FL-Thomasville, GA market

Waycross

see Jacksonville, FL market

Hawaii

Hilo

KGMD-TV— Analog channel: 9. Digital channel: 8.9.77 kw vis, 1.71 kw aur. ant -290t/285g TL: N19 43 00 W155 08 13 On air date: May 15, 1955. c/o KGMB, 1534 Kapiolani Blvd., Honolulu, HI, 96814. Phone: (808) 973-5462. Fax: (808) 941-8153. E-mail: kgmbnews@kgmb9.com Web Site: www.kgmb.com. Licensee: HITV License Subsidiary Inc. Group owner: Emmis Communications Corp. (acq 5-25-2007; grpsl). Natl. Network: CBS.

KHAW-TV— Analog channel: 11. Digital channel: 21.Note: Fox is on KHAW-TV ch 11, CW is on KHAW-DT ch 21. 2.09 kw vis, 275 w aur. ant -620t/139g TL: N19 43 56 W155 04 04 On air date: Nov 27, 1961. c/o KHON-TV, 88 Piikoi St., Honolulu, HI, 96814. Phone: (808) 591-2222. Fax: (808) 591-9085. E-mail: khan@khon.emmis.com Web Site: www.khon.com. Licensee: Montecito Hawaii License LLC. Group owner: Emmis Communications Corp. (acq 3-31-2006; grpsl). Natl. Network: Fox, CW Digital Network: Note: Fox is on KHAW-TV ch 11, CW is on KHAW-DT ch 21. Washington Atty: Gardner, Carton & Douglas. News staff: 45; News: 25 hrs wkly.
Key Personnel:
Rick Blangiardi . . . . . gen mgr
Rick Langardy . . . . . stn mgr
Satellite of KHON-TV Honolulu.

KHBC-TV— Analog channel: 2. Digital channel: 22. Analog hrs: 24 Digital hrs: 24 2.29 kw vis, 1.37 kw aur. ant -577t/132g TL: N19 43 51 W155 04 11 On air date: Aug 22, 1983. c/o KHNL, 150-B Puuhale Rd., Honolulu, HI, 96819. Phone: (808) 847-3246. Fax: (808) 845-3616. E-mail: news8@khnl.com Web Site: www.khnl.com. Licensee: KHNL/KFVE License Subsidiary LLC. Group owner: Raycom Media Inc. (acq 9-2-99; grpsl). Natl. Network: NBC. Washington Atty: Covington & Burling.
Rebroadcasts KHNL(TV) Honolulu 100%.

KHVO— Analog channel: 13. Digital channel: 18. Analog hrs: 24 Digital hrs: 24 100 kw vis, 20 kw aur. -823t/80g TL: N19 43 57 W155 04 04 On air date: May 15, 1960. 801 S. King St., Honolulu, HI, 96813. Phone: (808) 535-0400. Licensee: Hearst-Argyle Stations Inc. Group owner: Hearst-Argyle Stations Inc. (acq 7-16-97; grpsl). Natl. Network: ABC.
Satellite of KITV Honolulu.

KWHH— Analog channel: 14. Digital channel: 23. Analog hrs: 24 13.2 kw vis, 1.32 kw aur. ant -557t/145g TL: N19 43 51 W155 04 11 On air date: Oct 1, 1989. Century Square, 1188 Bishop St., Suite 502, Honolulu, HI, 96813. Phone: (808) 538-1414. Fax: (808) 526-0326. E-mail: kwhe@lesea.com Web Site: www.lesea.com. Licensee: Le Sea Broadcasting Corp. Group owner: (group owner; (acq 10-1-89; $8,277; FTR: 12-26-89). Washington Atty: Gardner, Carton & Douglas.
Key Personnel:
Peter Sumrall . . . . . pres
Anthony Hale . . . . . CFO
Tony Boquer . . . . . gen mgr
Rebroadcasts KWHE Honolulu 100%.

Honolulu, HI
(DMA 72)

***KAAH-TV—** Analog channel: 26. Analog hrs: 24 75.9 kw vis, 7.59 kw aur. 2,118t TL: N21 23 45 W158 05 58 On air date: Dec 23, 1982. 1152 Smith St., Honolulu, HI, 96817. Phone: (808) 521-5826. Fax: (808) 599-6238. E-mail: tbnkaahtv26@hotmail.com Web Site: www.tbn.org. Licensee: Trinity Christian Center of Santa Ana Inc. dba Trinity Broadcasting Network. Group owner: Trinity Broadcasting Network (acq 7-1-2000; grpsl). Population served: 1,250,000 Washington Atty: Colby May.
Key Personnel:
Paul F. Crouch . . . . . pres
Paul Crouch Jr. . . . . . exec VP
Jan Crouch . . . . . VP
Cheryl Witbeck . . . . . gen mgr
Satellite of KTBN-TV Los Angeles (Santa Ana), CA 95%.

***KALO**— Analog channel: 38. Analog hrs: 24 Digital hrs: 24 206 kw vis. ant 1,899t/46g TL: N21 23 33 W158 05 43 On air date: 2000. Box 8969, Honolulu, HI, 96830. Phone: (808) 591-8282. Fax: (808) 591-1250. Web Site: www.kalo-tv.com. Licensee: Pacifica Broadcasting Co. Population served: 1,400,000 Washington Atty: Fletcher, Heald & Hildreth, P.L.C., Harry F. Cole, Esq.
Christian programming.

KBFD— Analog channel: 32. Digital channel: 33. Analog hrs: 24 Digital hrs: 18 146 kw vis, 14.6 kw aur. 405t/428g TL: N21 18 49 W157 51 43 On air date: Mar 7, 1986. Century Sq., 1188 Bishop St., Honolulu, HI, 96813. Phone: (808) 521-8066. Fax: (808) 521-5233. Web Site: www.kbfd.com. Licensee: Allen Broadcasting Corp. Ownership: Kea Sung Chung, 8%; Ok Soon Chung, 10%; June Ho Chung, 52%; Yon Hee Chung 30%. Population served: 1,200,000 Washington Atty: Wilkinson, Barker, Knauer L.L.P. News staff: 4; News: 6 hrs wkly.
Key Personnel:
Kea Sung Chung. . . . . CEO & pres
June Ho Chung . . . . exec VP
Jeff Chung . . . . gen mgr

KFVE— Analog channel: 5. Digital channel: 23. Analog hrs: 24 Digital hrs: 24 95.5 kw vis, 19.8 kw aur. ant 161t/2,640g TL: N21 24 03 W158 06 10 On air date: Feb 7, 1988. 150 B Puuhale Rd., Honolulu, HI, 96819. Phone: (808) 847-3246. Fax: (808) 845-3616. Web Site: www.ksthehometeam.com. Licensee: KHNL/KFVE License Subsidiary LLC. Group owner: Raycom Media Inc. (acq 12-28-99). Natl. Network: MyNetworkTV. Washington Atty: Brown, Nietert & Kaufman. News: 7 hrs wkly.

KGMB— Analog channel: 9. Digital channel: 22.209 kw vis, 29.5 kw aur. ant -50t/436g TL: N21 17 46 W157 50 36 (CP: Ant 45t/495g) On air date: Dec 1, 1962. 1534 Kapiolani Blvd., Honolulu, HI, 96814. Phone: (808) 973-5462. Fax: (808) 973-9354. E-mail: kgmbnews@pixi.com Web Site: www.kgmb.com. Licensee: HITV License Subsidiary Inc. Group owner: Emmis Communications Corp. (acq 5-25-2007; grpsl). Population served: 1,150,000 Natl. Network: CBS. Natl. Rep: Harrington, Righter & Parsons.

***KHET**— Analog channel: 11. Analog hrs: 17 148 kw vis, 29.5 kw aur. -75t/431g TL: N21 17 46 W157 50 36 On air date: Apr 15, 1966. Box 11599, Harrisonburg, HI, 96822. 2350 Dole St., Harrisonburg, HI 96822. Phone: (808) 973-1000. Fax: (808) 973-1090. Web Site: www.pbshawaii.org. Licensee: Hawaii Public Broadcasting Authority. Population served: 1,100,000 Natl. Network: PBS. Washington Atty: Wilkes, Artis, Hedrick & Lane.

KHNL— Analog channel: 13. Digital channel: 35. Analog hrs: 24 Digital hrs: 24 316 kw vis, 46.8 kw aur. ant 20t TL: N21 17 46 W157 50 19 On air date: July 4, 1962. 150 B. Puuhale Rd., Honolulu, HI, 96819. Phone: (808) 847-3246. Fax: (808) 845-3616. E-mail: news8@khnl.com Web Site: www.khnl.com. Licensee: KHNL/KFVE License Subsidiary LLC. Group owner: Raycom Media Inc. (acq 9-2-99; grpsl). Population served: 412,000 Natl. Network: NBC. Natl. Rep: TeleRep. Washington Atty: Covington & Burling. Wire Svc: CNN Wire Svc: AP Wire Svc: NBC News: 23 hrs wkly.

KHON-TV— Analog channel: 2. Digital channel: 8. Analog hrs: 5 AM-4 AM Digital hrs: 5 am-4 am Note: Fox is on KHON-TV ch 2, CW is on KHON-DT ch 8. 100 kw vis, 20 kw aur. ant 59t/498g TL: N21 17 39 W157 50 18 On air date: Dec 15, 1952. 88 Piikoi St., Honolulu, HI, 96814. Phone: (808) 591-2222. Fax: (808) 591-9085. E-mail: khon@khon.emmis.com Web Site: www.khon.com. Licensee: Montecito Hawaii License LLC. Group owner: Emmis Communications Corp. (acq 3-31-2006; grpsl). Population served: 1,150,000 Natl. Network: Fox, CW Digital Network: Note: Fox is on KHON-TV ch 2, CW is on KHON-DT ch 8. Natl. Rep: MMT. News staff: 45; News: 25 hrs wkly.
Key Personnel:
Joe McNamara . . . . pres & gen mgr
William Spellman . . . . gen sls mgr

KIKU-TV— Analog channel: 20. Digital channel: 19. Analog hrs: 24 467 kw vis, 46.7 kw aur. ant 2,040t TL: N21 23 51 W158 06 01 On air date: Dec 30, 1983. 737 Bishop St., Suite 1430, Honolulu, HI, 96813. Phone: (808) 847-2021. Fax: (808) 841-3326. Web Site: www.kikutv.com. Licensee: KHLS Inc. Group owner: Asian Media Group (acq 1-18-2001; $165 million cash for 69.4%). Natl. Rep: Petry Television Inc. Washington Atty: Skadden, Arps, Slate, Meagher & Flom.

KITV— Analog channel: 4. Digital channel: 40. Analog hrs: 24 Digital hrs: 24 100 kw vis, 20 kw aur. 50t/495g TL: N21 17 37 W157 50 34 (CP: 56.9 kw vis, ant 5,458t) On air date: Apr 16, 1954. 801 S. King St., Honolulu, HI, 96813. Phone: (808) 535-0400. Fax: (808) 536-8777. Web Site: www.thehawaiichannel.com. Licensee: Hearst-Argyle Stations Inc. Group owner: Hearst-Argyle Television Inc. (acq 7-16-97; grpsl). Population served: 324,871 Natl. Network: ABC. Washington Atty: Brooks, Pierce, McLendon, Humphrey & Leonard. News staff: 40; News: 20 hrs wkly.

KKAI— Analog channel: 50. Analog hrs: 24 19 kw vis. ant 1,115t/107g TL: N21 19 23 W157 40 53 On air date: 2005. 875 Waimanu St., Suite 110, Honolulu, HI, 96813. Phone: (808) 922-5300. Fax: (808) 441-0092. E-mail: info@kkai.tv Web Site: www.kkai.tv. Licensee: Kailua Television LLC. Ownership: Kailua Television Partners. Washington Atty: Fletcher, Heald & Hildreth.

KLEI— Analog channel: 6. Digital channel: 25. Analog hrs: 24 Digital hrs: 24 52.5 kw vis, 6.7 kw aur. 2,910t TL: N19 42 56 W155 55 00 On air date: 1988. Box 8969, Honolulu, HI, 96830. Phone: (808) 262-2000. Fax: (808) 254-1313. E-mail: info@klei.com Web Site: klei-tv.com. Licensee: Aina'e Co. Ltd. (acq 9-14-94; FTR: 9-26-94). Natl. Network: i Network. Washington Atty: Harry F. Cole, Esq. of Fletcher, Heald & Hildreth.
CW.

KPXO-TV— Analog channel: 66. Digital channel: 41. Analog hrs: 24 95.5 kw vis. ant 2,073t TL: N21 19 49 W157 45 24 On air date: 1998. 875 Waimanu St., Suite 630, Honolulu, HI, 96813. Phone: (808) 591-1275. Fax: (808) 591-1409. Web Site: www.ionmedia.tv. Licensee: Paxson Hawaii License Inc. Group owner: Paxson Communications Corp. (acq 8-12-98; $6.9 million). Natl. Network: i Network.

KUPU— Analog channel: 56. Analog hrs: 24 19 kw vis. ant 1,115t/107g TL: N21 19 23 W157 40 53 On air date: 2004. Box 235770, Honolulu, HI, 96823. Phone: (808) 943-0007. Fax: (808) 440-1375. E-mail: info@oceaniachurch.org Web Site: www.kupu.tv. Licensee: Oceania Christian Church. Ownership: Oceania Christian Church, 100% of total assets (acq 9-28-2006; $5 million). News staff: 12; News: 14 hrs wkly.

***KWBN**— Analog channel: 44.302 kw vis. 1,902t TL: N21 23 45 W158 05 58 On air date: 2000. 3901 Hwy. 121, Bedford, TX, 76021. Phone: (817) 571-1229. Fax: (817) 571-7458. Web Site: www.daystar.com. Licensee: Ho'ona'auao Community Television Inc.

KWHE— Analog channel: 14. Digital channel: 31.75.9 kw vis. ant 26t TL: N21 18 49 W157 51 43 On air date: Mar 7, 1988. Century Square, 1188 Bishop St., Suite 502, Honolulu, HI, 96813. Phone: (808) 538-1414. Fax: (808) 526-0326. E-mail: kwhe@lesea.com Web Site: kwhe.com. Licensee: LeSea Broadcasting Corp. Group owner: (group owner; (acq 8-15-86; $825,000; FTR: 6-16-86). Washington Atty: Gardner, Carton & Douglas.
Key Personnel:
Peter Sumrall . . . . pres
Anthony Hale . . . . CFO
Tony Boquer . . . . gen mgr
Mauro Pena . . . . chief of engrg

Wailuku

KAII-TV— Analog channel: 7. Digital channel: 36. Analog hrs: 5 AM-4 AM Note: Fox is on KAII-TV ch 7, CW is on KAII-DT ch 36. 29.8 kw vis, 5.9 kw aur. ant 5,940t/75g TL: N20 42 41 W156 15 26 On air date: Nov 17, 1958. 88 Piikoi St., Honolulu, HI, 96814. Phone: (808) 591-2222. Fax: (808) 593-8479. E-mail: khon@khon.emmis.com Web Site: www.khon.com. Licensee: Montecito Hawaii License LLC. Group owner: Emmis Communications Corp. (acq 3-31-2006; grpsl). Natl. Network: Fox, CW Digital Network: Note: Fox is on KAII-TV ch 7, CW is on KAII-DT ch 36. Washington Atty: Gardner, Carton & Douglas.
Key Personnel:
Rick Blangiardi . . . . gen mgr
Rick Langardy. . . . . stn mgr
Satellite of KHON(TV) Honolulu 100%.

KGMV— Analog channel: 3. Digital channel: 24.14.1 kw vis, 2.69 kw aur. ant 5,950t/60g TL: N20 42 41 W156 15 35 On air date: Apr 24, 1955. c/o KGMB, 1534 Kapiolani Blvd., Honolulu, HI, 96814. Phone: (808) 973-5462. Fax: (808) 973-9354. E-mail: kmbnews@pixi.com Web Site: www.kgmb.com. Licensee: HITV License Subsidiary Inc. Group owner: Emmis Communications Corp. (acq 5-25-2007; grpsl). Population served: 8,280 Natl. Network: CBS.

KMAU— Analog channel: 12. Digital channel: 29. Analog hrs: 24 27.5 kw vis, 4.36 kw aur. 5,910t/70g TL: N20 42 43 W156 15 26 (CP: 56.9 kw vis, ant 5,458t. TL: N20 42 16 W156 16 35) On air date: Nov 28, 1955. 801 S. King St., Honolulu, HI, 96813. Phone: (808) 535-0400. Licensee: Hearst-Argyle Stations Inc. Group owner: Hearst-Argyle Television Inc. (acq 7-16-97; grpsl). Population served: 324,871 Natl. Network: ABC.
Satellite of KITV Honolulu.

***KMEB**— Analog channel: 10.30.9 kw vis, 6 kw aur. 5,940t/47g TL: N20 42 40 W156 15 34 On air date: Sept 22, 1966. Box 11599, Honolulu, HI, 96822. 2350 Dole St., Honolulu, HI 96822. Phone: (808) 973-1000. Fax: (808) 973-1090. Web Site: www.pbshawaii.org. Licensee: Hawaii Public Broadcasting Authority. Natl. Network: PBS.
Satellite of *KHET Honolulu.

KOGG— Analog channel: 15. Digital channel: 16. Analog hrs: 24 759 kw vis, 75.9 kw aur. ant 5,653t/75g TL: N20 42 34 W156 15 54 On air date: Aug 22, 1989. c/o KHNL, 150-B Puuhale Rd., Honolulu, HI, 96819. Phone: (808) 847-3246. Fax: (808) 845-3616. E-mail: news@khnl.com Web Site: www.khnl.com. Licensee: KHNL/KFVE License Subsidiary LLC. Group owner: Raycom Media Inc. (acq 9-2-99; grpsl). Natl. Network: NBC.
Satellite of KHNL(TV) Honolulu.

KWHM— Analog channel: 21.44.25 kw vis. -371t TL: N20 53 25 W156 30 22 On air date: 1993. Century Square, 1188 Bishop St., Suite 502, Honolulu, HI, 96813. Phone: (808) 538-1414. Fax: (808) 526-0326. Web Site: www.lesea.com. Licensee: Le Sea Broadcasting Corp. Group owner: (group owner) Washington Atty: Gardner, Carton & Douglas.
Key Personnel:
Peter Sumrall . . . . pres
Anthony Hale . . . . CFO
Tony Boquer . . . . gen mgr
Mauro Pena . . . . chief of engrg
Satellite of KWHE Honolulu.

Idaho

Boise, ID
(DMA 118)

***KAID-TV**— Analog channel: 4. Analog hrs: 24 57.2 kw vis, 5.7 kw aur. 2,474t/142g TL: N43 45 16 W116 05 56 On air date: Dec 31, 1971. 1455 N. Orchard St., Boise, ID, 83706. Phone: (208) 373-7220. Fax: (208) 373-7245. E-mail: idptv@idahoptv.org Web Site: www.idahoptv.org. Licensee: Idaho State Board of Education. Population served: 157,000 Natl. Network: PBS. Washington Atty: Fletcher, Heald & Hildreth. News staff: 5; News: 3 hrs wkly.
Key Personnel:
Peter Morrill . . . . gen mgr
Kim Philipps . . . . dev dir

KBCI-TV— Analog channel: 2. Digital channel: 28. Analog hrs: 24 Digital hrs: 24 65 kw vis, 7.0l kw aur. 2,550t/100g TL: N43 45 17 W116 05 53 On air date: Nov 26, 1953. 140 N. 16th St., Boise, ID, 83702. Phone: (208) 472-2222. Fax: (208) 472-2212. E-mail: comments@kbcitv.com Web Site: www.2news.tv.com. Licensee: Fisher Broadcasting - Idaho TV L.L.C. Group owner: Fisher Broadcasting Company (acq 7-1-99; grpsl). Population served: 604,000 Natl. Network: CBS. Washington Atty: Shaw Pittman. News staff: 37; News: 19 hrs wkly.
Key Personnel:
Colleen Brown . . . . CEO
Robert Thomas . . . . VP & gen mgr
Jeff Bishop . . . . opns mgr
Richard Brace . . . . prom dir & progmg dir
Yvonne Simons . . . . news dir
Walt Baker . . . . pub affrs dir

KIVI—Nampa, Analog channel: 6. Analog hrs: 24 60.3 kw vis, 12.0 kw aur. ant 2,660t/210g TL: N43 45 20 W116 05 55 On air date: Feb 1, 1974. 1866 E. Chisholm Dr., Nampa, ID, 83687. Phone: (208) 336-0500. Fax: (208) 381-6682. Web Site: www.kivitv.com. Licensee: Journal Broadcast Corp. Group owner: Journal Communications Inc. (acq 11-15-2001). Population served: 580,000 Natl. Network: ABC. News staff: 45; News: 19.5 hrs wkly.
Key Personnel:
Bob Rosenthal . . . . VP & gen mgr
Ken Richie . . . . gen sls mgr
Norma Petty. . . . . rgnl sls mgr
Jason Knose . . . . mktg dir
Scott Picken . . . . news dir
Jeff Hoffert. . . . . chief of engrg

KKJB— Analog channel: 39. Analog hrs: 24 1,295 kw vis. ant 1,752t/164g TL: N43 44 23 W116 08 15 On air date: 2005. Cocola Broadcasting Companies, 706 W. Herndon Ave., Fresno, CA, 93650. Phone: (208) 331-3900. Fax: (559) 435-3201. E-mail: info@cocolatv.com Web Site: www.cocolatv.com. Licensee: Boise Telecasters L.P. Group owner: Cocola Broadcasting Companies (acq 5-7-2004; $3 million for CP). Washington Atty: Dow, Lohnes & Albertson, LLPC.
Key Personnel:
Gary Cocola. . . . . CEO
Terry Dolph . . . . stn mgr
Nick Giotto . . . . natl sls mgr & rgnl sls mgr
Nori Zahari. . . . . progmg VP
Ralph Malhrich . . . . chief of engrg

KNIN-TV— Analog channel: 9. Digital channel: 10. Analog hrs: 24 162 kw vis. ant 2,690t/210g TL: N43 45 18 W116 05 52 On air date: 1993. 816 W. Bannock St., Suite 402, Boise, ID, 83702. Phone: (208) 331-0909. Fax: (208) 344-0119. Web Site: www.knin.com. Licensee: Banks-Boise Inc. Group owner: Banks Broadcasting Inc. (acq 5-7-2001). Natl. Network: CW. Natl. Rep: Blair Television. Washington Atty: Covington & Burling.
Key Personnel:
Lyle Banks . . . . pres
Larry Newton . . . . gen mgr & gen sls mgr

KTRV-TV— Analog channel: 12. Digital channel: 13. Analog hrs: 24 178 kw vis, 18.2 kw aur. ant 2,760t/220g TL: N43 45 18 W116 05 52 On air date: Oct 18, 1981. One 6th St. N., Nampa, ID, 83687. Phone: (208) 466-1200. Fax: (208) 467-6958. E-mail: comments@ktrv.com Web Site: www.fox12idaho.com. Licensee: Idaho Independent Television Inc. Group owner: Block Communications Inc. (acq 4-23-85; $4.9 million; FTR: 3-25-85). Population served: 500,000 Natl. Network: Fox. Washington Atty: Dow, Lohnes & Albertson. News staff: 12; News: 3 hrs wkly.
Key Personnel:
Rick Joseph . . . . pres & gen mgr
Ed Crampton . . . . opns mgr
Ken Hunter . . . . sls dir
C.J. Gish . . . . progmg mgr
Kelly Cross . . . . news dir
Daniel Paixao . . . . chief of engrg

KTVB— Analog channel: 7. Digital channel: 7. Analog hrs: 24 160 kw vis, 26.2 kw aur. 2,645t/226g TL: N43 45 16 W116 05 56 On air date: July 12, 1953. PO Box 7, Boise, ID, 83707. 5407 Fairview Ave, Boise, ID 83706. Phone: (208) 375-7277. Fax: (208) 378-1762. E-mail: info@ktvb.com Web Site: www.ktvb.com. Licensee: KTVB-TV Inc. Group owner: Belo Corp., Broadcast Division (acq 1997; grpsl). Population served: 450,000 Natl. Network: NBC. Natl. Rep: TeleRep. Washington Atty: Wiley, Rein & Fielding. News: 27 hrs wkly.
Key Personnel:
Douglas L. Armstrong . . . . pres
Douglas Armstrong . . . . gen mgr
Paul Budell . . . . opns mgr
Kristi Edmunds . . . . gen sls mgr
Tom Zito . . . . natl sls mgr
Mark Danielson . . . . news dir
Richard Strack . . . . chief of engrg
Brad Bond . . . . sls

Coeur d'Alene

see Spokane, WA market

Idaho Falls-Pocatello, ID (DMA 163)

KFXP— Analog channel: 31.2,140 kw vis. 1,466t TL: N42 55 15 W112 20 44 On air date: July 17, 1998. 902 E. Sherman St., Pocatello, ID, 83201. Phone: (208) 232-6666. Fax: (208) 232-6678. Web Site: www.kpvi.com. Licensee: Compass Communications of Idaho Inc. Natl. Network: Fox.
Key Personnel:
Bill Fouch . . . . gen mgr
Patrick Anderson . . . . gen sls mgr
Rockky Hansen . . . . prom dir
Brenda Baumgartner . . . . news dir
Robin Estopinal . . . . chief of engrg

KIDK-TV— Analog channel: 3. Analog hrs: 24 100 kw vis, 14.4 kw aur. 1,600t/200g TL: N43 29 51 W112 39 50 On air date: Dec 20, 1953. Box 1255 E. 17 th St., Idaho Falls, ID, 83404. 145 S. Arthur, Pocatello, ID 83204. Phone: (208) 522-5100. Fax: (208) 535-0946. E-mail: comments@kidk.com Web Site: www.kidk.com. Licensee: Fisher Broadcasting - S.E. Idaho TV L.L.C. Group owner: Fisher Broadcasting Company (acq 12-4-01 grpsl). Population served: 102,130 Natl. Network: CBS. Washington Atty: Shaw, Pittman. News staff: 20; News: 15 hrs wkly.

KIFI-TV— Analog channel: 8. Digital channel: 9. Analog hrs: 24 316 kw vis, 63.1 kw aur. 1,520t/180g TL: N43 30 02 W112 39 36 On air date: Jan 21, 1961. Box 2148, Idaho Falls, ID, 83403. 1915 N. Yellowstone Hwy, Idaho Falls, ID 83401. Phone: (208) 525-2520. Fax: (208) 522-1930. Fax: (208) 529-2443 (news). E-mail: localnews8@aol.com Web Site: www.localnews8.com. Licensee: NPG of Idaho Inc. Group owner: (group owner) (acq 6-15-2005; $12.5 million). Population served: 300,000 Natl. Network: ABC. Washington Atty: Shaw Pittman LLP. News staff: 15; News: 30 hrs wkly.
Key Personnel:
David Bradley Jr. . . . . pres
Tim Larson . . . . gen mgr
Monte Young . . . . gen sls mgr

***KISU-TV—** Analog channel: 10. Analog hrs: 24 122 kw vis, 12.2 kw aur. 1,527t/144g TL: N43 30 02 W112 39 36 On air date: July 7, 1971. Campus Box 8111, Pocatello, ID, 83209. 1455 N. Orchard St., Boise, ID 83706. Phone: (208) 282-2857. Fax: (208) 282-2848. E-mail: idptv@idahoptv.org Web Site: www.idahoptv.org. Licensee: Idaho State Board of Education. Natl. Network: PBS.
Key Personnel:
Peter Morrill . . . . gen mgr
Kim Neilsen . . . . stn mgr
Dave Turnmire . . . . chief of engrg

KJWY— Analog channel: 2. Analog hrs: 24 178 w vis, 17.8 w aur. 997t TL: N43 27 42 W110 45 10 On air date: 1991. Box 7454, Jackson, WY, 83002. Phone: (307) 733-2066. Fax: (307) 733-4834. Web Site: www.kjwy2.com. Licensee: Two Ocean Broadcasting Co. Group owner: Sunbelt Communications Co. (acq 11-95; grpsl). Population served: 25,000 Natl. Network: NBC. News staff: one; News: 5 hrs wkly.
Key Personnel:
James Rogers . . . . CEO
Ralph Toddre . . . . exec VP
Christel Rahme . . stn mgr, sls dir, adv dir, progmg dir & news dir
Robin Estopinal . . . . engrg dir

KPIF— Analog channel: 15. Analog hrs: 24 5,000 kw vis. ant 1,073t/262g TL: N42 51 50 W112 31 10 On air date: 2006. 5023 Rainbow Ln., Chubbuck, ID, 83202-7607. Phone: (208) 237-5743. Web Site: www.kpif.net. Licensee: Pocatello Channel 15 L.L.C. Ownership: Myoung Hwa Bae, 100% (acq 12-1-2003). Natl. Network: CW.

KPVI—Pocatello, Analog channel: 6. Analog hrs: 24 100 kw vis, 17.4 kw aur. 1,530t/619g TL: N42 55 15 W112 20 44 On air date: Apr 26, 1974. Box 667, Pocatello, ID, 83204. Phone: (208) 232-6666. Fax: (208) 233-6678. Web Site: www.kpvi.com. Licensee: Oregon Trail Broadcasting Co. Group owner: Sunbelt Communications Co. (acq 11-15-95). Population served: 309,000 Natl. Network: NBC. Natl. Rep: Blair Television. Wire Svc: AP News staff: 23; News: 17 hrs wkly. Rebroadcasts: KJWY Jackson Hole, WY.

Lewiston

see Spokane, WA market

Moscow

see Spokane, WA market

Pocatello

see Idaho Falls-Pocatello, ID market

Twin Falls, ID (DMA 191)

***KBGH—** Analog channel: 19.75.9 kw vis. 528t TL: N42 43 47 W114 24 52 On air date: 1997. Box 1238, Twin Falls, ID, 83303-1238. 315 Falls Ave., Twin Falls, ID 83301. Phone: (208) 732-6351. Fax: (208) 736-2137. Web Site: www.radio.boisestate.edu. Licensee: College of Southern Idaho. Natl. Network: PBS.

KIDA— Analog channel: 5.10.25 kw vis. ant 1,830t/59g TL: N43 38 36 W114 23 49 On air date: 2003. Blue Lakes Blvd. N., Suite 101, Twins Falls, ID, 83301. Phone: (954) 732-9539. E-mail: mturnerco@aol.com Licensee: Marcia T. Turner dba Turner Enterprises.

***KIPT-TV—** Analog channel: 13.22.4 kw vis. 528t/69g TL: N42 43 47 W114 24 52 On air date: Jan 18, 1992. c/o KAID, 1455 N. Orchard St., Boise, ID, 83706. Phone: (208) 373-7220. Fax: (208) 373-7245. E-mail: idptv@idahoptv.org Web Site: www.idahoptv.org. Licensee: State Board of Education, State of Idaho. Natl. Network: PBS. Washington Atty: Fletcher, Heald & Hildreth.
Key Personnel:
Peter Morrill . . . . gen mgr
Kim Philipps . . . . dev dir
Rebroadcasts KAID Boise 100%.

KMVT— Analog channel: 11. Digital channel: 16. Analog hrs: 24 Digital hrs: 20.5 316 kw vis, 31.6 kw aur. 1,190t/690g TL: N42 43 48 W114 24 52 On air date: May 30, 1955. 1100 Blue Lakes Blvd. N., Twin Falls, ID, 83301. Phone: (208) 733-1100. Fax: (208) 733-4649. Web Site: www.kmvt.com. Licensee: Neuhoff Family L.P. Group owner: (group owner; (acq 8-3-2004; $17.3 million). Population served: 160,100 Natl. Network: CBS. Natl. Rep: Continental Television Sales. Washington Atty: Cohn & Marks. News staff: 14; News: 10 hrs wkly.

Key Personnel:
Lee Wagner . . . . gen mgr & progmg dir
Fred Bartlett . . . . opns mgr
Lisa Collins . . . . gen sls mgr
Paul Johnson . . . . mktg mgr & prom mgr
Joe Martin . . . . news dir
Rodger Martin . . . . chief of engrg

KXTF— Analog channel: 35. Analog hrs: 24 100 kw vis. 538t TL: N42 43 42 W114 24 43 On air date: Jan 31, 1989. 1061 Blue Lakes Blvd. N., Twin Falls, ID, 83301. Phone: (208) 733-0035. Fax: (208) 733-0160. Web Site: www.kxtf.com. Licensee: Sunbelt Broadcasting Co. Group owner: Sunbelt Communications Co. Population served: 163,300 Natl. Network: Fox. Washington Atty: Hamel & Park. News: 2 hrs wkly.
Key Personnel:
Bill Fouch . . . . gen mgr
Patrick Anderson . . . . gen sls mgr
Rocky Hanson . . . . prom mgr
Brenda Baumgartner . . . . news dir
Robin Estopinal . . . . chief of engrg

Illinois

Bloomington

see Peoria-Bloomington, IL market

Carbondale

see Paducah, KY-Cape Girardeau, MO-Harrisburg-Mount Vernon, IL market

Champaign & Springfield-Decatur, IL (DMA 82)

WAND— Analog channel: 17. Digital channel: 18. Analog hrs: 24 5,000 kw vis, 1,000 kw aur. ant 1,290t/1,314g TL: N39 37 07 W88 49 55 On air date: Aug 16, 1953. 904 Southside Dr., Decatur, IL, 62521. Phone: (217) 424-2500. Fax: (217) 424-2583. Web Site: www.wandtv.com. Licensee: WAND Television Inc. Group owner: LIN Television Corporation (acq 2-27-95; FTR: 5-22-95). Population served: 1,205,000 Natl. Network: NBC. Washington Atty: Schwartz, Woods & Miller.
Key Personnel:
Mike Johnson . . . . gen mgr
Denise Daniels . . . . gen sls mgr
Tracey Cole . . . . stn mgr & progmg mgr
Jim Platzer . . . . news dir
Hal Campbell . . . . chief of engrg

WBUI— Analog channel: 23. Digital channel: 22. Analog hrs: 24 Digital hrs: 24 1,951 kw vis, 195.1 kw aur. ant 1,030t/1,351g TL: N39 56 56 W88 50 12 On air date: May 14, 1984. 2510 Pkwy Ct., Decatur, IL, 62526. Phone: (217) 428-2323. Fax: (217) 428-6455. E-mail: promotions @centralillinoiscw.com Web Site: www.centralillinoiscw.com. Licensee: Acme TV Licenses of Illinois L.L.C. Group owner: Acme Communications Inc. (acq 6-14-99; $13.3 million). Population served: 382,460 Natl. Network: CW. Natl. Rep: MMT.
Key Personnel:
Bill Snider . . . . gen mgr
Allen White . . . . gen sls mgr
Scott Washburn . . . . opns mgr & chief of engrg

WCCU— Analog channel: 27. Digital channel: 26. Analog hrs: 24 Digital hrs: 24 3,360 kw vis, 218 kw aur. ant 442t/829 TL: N40 18 42 W87 54 48 On air date: 1987. 119 W. Church St., Champaign, IL, 61820. Phone: (217) 403-9927. Fax: (217) 403-1007. Web Site: www.myfoxchampaign.com. Licensee: Springfield Broadcasting Partners. Group owner: Bahakel Communications (acq 7-20-92). Natl. Network: Fox.
Key Personnel:
Peter O'Brien . . . . gen mgr
Randy Stone . . . . gen sls mgr
Jeff Kaufmann . . . . prom dir
Jack Richardson . . . . chief of engrg
Rebroadcasts WRSP-TV Springfield 100%.

WCFN-TV— Analog channel: 49. Digital channel: 53. Analog hrs: 24 200 kw vis, 20 kw aur. ant 620t/655g TL: N39 47 27 W89 30 53 On air date: 1987. Box 20, 509 S. Neil St., Champaign, IL, 61820. Phone: (217) 356-8333. Fax: (217) 373-3680. E-mail: program@wcia.com Web Site: www.wcfn.tv. Licensee: Nexstar Finance Inc. Group owner:

Nexstar Broadcasting Group Inc. (acq 5-19-2000; grpsl). Natl. Network: MyNetworkTV. Natl. Rep: Blair Television. Washington Atty: Drinker-Biddle-Reath.
Key Personnel:
Russ Hamilton . . . . gen mgr
Don Osika . . . . sls dir
Linda Voorhees . . . . rgnl sls mgr
Peter Carlson . . . . prom mgr
Holly Kennedy . . . . progmg dir
Darren Martin . . . . chief of engrg

WCIA-TV— Analog channel: 3.100 kw vis, 20 kw aur. 940t/981g TL: N40 06 23 W88 26 59 On air date: Nov 14, 1953. Box 20, 509 S. Neil, Champaign, IL, 61824-0020. Phone: (217) 356-8333. Fax: (217) 373-3680. E-mail: webmaster@wcia.com Web Site: www.wcia.com. Licensee: Nexstar Finance Inc. Group owner: Nexstar Broadcasting Group Inc. (acq 5-19-2000; grpsl). Natl. Network: CBS. Washington Atty: Covington & Burling.
Key Personnel:
Russ Hamilton . . . . gen mgr
Don Osika . . . . sls dir
Linda Voorhees . . . . rgnl sls mgr
Peter Carlson . . . . prom mgr
Holly Kennedy . . . . progmg dir
Darren Martin . . . . chief of engrg

***WEIU-TV—** Analog channel: 51. Analog hrs: 24 48.5 kw vis, 4.85 kw aur. 234t/213g TL: N39 28 43 W88 10 21 On air date: July 1, 1986. 1521 Buzzard Hall, 600 Lincoln Ave., Charleston, IL, 61920. Radio & TV Ctr., Eastern Illinois Univ., Charleston, IL 61920. Phone: (217) 581-5956. Phone: (877) 727-9348. Fax: (217) 581-6650. E-mail: weiu@weiu.net Web Site: www.weiu.net. Licensee: Eastern Illinois University. Population served: 500,000 Natl. Network: PBS. Washington Atty: Cohn & Marks. News staff: 25; News: 3 hrs wkly.
Key Personnel:
Denis Roche . . . . gen mgr
Ke'an Rogers . . . . prom mgr
Linda Kingery . . . . progmg mgr
Kelly Runyon . . . . news dir
Kevin Armstrong . . . . chief of engrg

WICD-TV— Analog channel: 15. Digital channel: 41. Analog hrs: 24 358 kw vis, 35 kw aur. ant 1,300t/1,338g TL: N40 04 11 W87 54 45 On air date: Apr 24, 1959. 250 S. Country Fair Dr., Champaign, IL, 61821-2920. Phone: (217) 351-8500. Fax: (217) 351-6056. Web Site: www.wicd15.com. Licensee: WICD License L.L.C. Group owner: Sinclair Broadcast Group Inc. (acq 7-2-99; $81 million with WICS(TV) Springfield). Population served: 332,000 Natl. Network: ABC. Washington Atty: Wiley, Rein & Fielding. News staff: 20; News: 17 hrs wkly.
Key Personnel:
David Smith . . . . CEO
Gary Hackler . . . . VP & stn mgr
Tim Mathis . . . . pres & gen mgr
David Schroeder . . . . news dir

WICS-TV— Analog channel: 20. Digital channel: 42. Analog hrs: 24 Digital hrs: 24 676 kw vis, 67.6 kw aur. ant 1,430t/1,458g TL: N39 48 15 W89 27 40 On air date: Oct 30, 1953. Box 3920, Springfield, IL, 62703-1999. 2680 E. Cook St., Springfield, IL 62703-1999. Phone: (217) 753-5620. Fax: (217) 753-8177. E-mail: comments@wics.com Web Site: www.wics.com. Licensee: WICS Licensee L.L.C. Group owner: Sinclair Broadcast Group Inc. (acq 7-1-99). Population served: 336,000 Natl. Network: ABC. Washington Atty: Dow, Lohnes PLLC. News staff: 32; News: 22 hrs wkly.
Key Personnel:
Tim Mathis . . . . gen mgr
Jim Waldeck . . . . opns dir

***WILL-TV—** Analog channel: 12. Analog hrs: 24 316 kw vis, 63.1 kw aur. 990t/1,047g TL: N40 02 18 W88 40 10 On air date: Aug 1, 1955. Campbell Hall for Public Telecommunications, 300 N. Goodwin Ave., Urbana, IL, 61801-2316. Phone: (217) 333-1070. Fax: (217) 244-6386. E-mail: will-tv@uiuc.edu Web Site: www.will.uiuc.edu. Licensee: University of Illinois Board of Trustees. Population served: 1,000,000 Natl. Network: PBS. Washington Atty: Dow, Lohnes & Albertson.
Key Personnel:
Carl Caldwell . . . . stn mgr
David Thiel . . . . progmg dir & chief of engrg
Rick Finnie . . . . chief of engrg

WRSP-TV— Analog channel: 55. Digital channel: 44. Analog hrs: 24 Digital hrs: 24 2,000 kw vis, 200 kw aur. ant 1,442t/1,449g TL: N39 47 57 W89 26 46 On air date: June 1, 1979. 3003 Old Rochester Rd., Springfield, IL, 62703. Phone: (217) 523-8855. Fax: (217) 523-4410. Web Site: www.myfoxspringfield.com. Licensee: Springfield Independent Television Inc. Group owner: Bahakel Communications (acq 7-20-92). Population served: 384,100 Natl. Network: Fox.
Key Personnel:
Jeff Kaufmann . . . . pres & prom mgr
Peter O'Brien . . . . gen mgr
Randy Stone . . . . gen sls mgr
Jack Richardson . . . . chief of engrg

***WSEC—** Analog channel: 14. Digital channel: 15. Analog hrs: 6 AM-midnight Digital hrs: 6 AM-midnight 28.25 kw vis, 2.83 kw aur. 313t/339g TL: N39 44 08 W90 10 32 On air date: Aug 21, 1984. Box 6248, Springfield, IL, 62708. Phone: (217) 483-7887. Fax: (217) 483-1112. Web Site: www.wsec.tv. Licensee: West Central Illinois Educational Telecommunication Corp. Population served: 915,000 Natl. Network: PBS. Washington Atty: Dow, Lohnes & Albertson.
Key Personnel:
Jerold Gruebel . . . . CEO, pres & gen mgr
Richard Plotkin . . . . opns VP & engrg VP

Chicago (DMA 3)

WBBM-TV— Analog channel: 2. Digital channel: 3. Analog hrs: 24 Digital hrs: 24 35.4 kw vis, 7.08 kw aur. 1,350t/1,456g TL: N41 53 56 W87 37 23 On air date: August 1940. 630 N. McClurg Ct., Chicago, IL, 60611. Phone: (312) 202-2222. Fax: (312) 943-7193. Web Site: www.cbs2chicago.com. Licensee: CBS Broadcasting Inc. Group owner: Viacom Television Stations Group (acq 2-9-53; $6 million; FTR: 2-16-53). Population served: 8,000,000 Natl. Network: CBS. Natl. Rep: CBS TV Stations National Sales. Wire Svc: Reuters
Key Personnel:
Joseph Ahern . . . . gen mgr
Al Connor . . . . gen sls mgr
Will Sliger . . . . prom dir
Fran Preston . . . . progmg dir
Carol Fowler . . . . news dir
Tom Schnecke . . . . chief of engrg

WCIU-TV— Analog channel: 26. Digital channel: 27. Analog hrs: 24 Digital hrs: 24 5,000 kw vis, 500 kw aur. 1,555t/1,552g TL: N41 52 44 W87 38 10 On air date: Feb 6, 1964. 26 N. Halsted St., Chicago, IL, 60661. Phone: (312) 705-2600. Fax: (312) 705-2656. Web Site: www.wciu.com. Licensee: WCIU-TV L.P. Group owner: Weigel Broadcasting Co. Population served: 9,194,000 Washington Atty: Cohn & Marks.
Key Personnel:
Neil Sabin . . . . gen mgr
Fred Weintraub . . . . stn mgr
Brad Lesak . . . . gen sls mgr
Sean Long . . . . progmg dir
Dave Rosenberg . . . . news dir
Kyle Walker . . . . chief of engrg

WCPX— Analog channel: 38. Digital channel: 43.3,630 kw vis. ant 1,673t/1,667g TL: N41 52 44 W87 38 08 On air date: May 31, 1976. 333 S. Desplains St., Suite 101, Chicago, IL, 60661-8735. Phone: (312) 376-8520. Fax: (312) 575-8735. Web Site: www.ionline.tv. Licensee: Paxson Chicago License Inc. Group owner: Paxson Communications Corp. (acq 8-11-98; $120 million including all interest in KWOK(TV) Novato, CA and other telecasting progmg rights). Population served: 10,000,000 Natl. Network: i Network. Washington Atty: Dow, Lohnes & Albertson.

WFLD— Analog channel: 32. Digital channel: 31.5,000 kw vis, 500 kw aur. 1,415t/1,456g TL: N41 53 55 W87 37 23 On air date: Jan 6, 1966. 205 N. Michigan Ave., Chicago, IL, 60601. Phone: (312) 565-5532. Fax: (312) 565-5517. Web Site: www.myfoxchicago.com. Licensee: Fox Television Stations Inc. Group owner: (group owner; (acq 11-14-86; grpsl). Population served: 825,300 Natl. Network: Fox. Natl. Rep: Fox Stations Sales. News: 35.5 hrs wkly.

WGBO-TV— Analog channel: 66.5,000 kw vis, 500 kw aur. 1,296t/1,456g TL: N41 53 56 W87 37 23 On air date: Sept 17, 1981. 541 N. Fairbanks Ct., Suite 1100, Chicago, IL, 60611. Phone: (312) 670-1000. Fax: (312) 494-6492. Web Site: www.univision.com. Licensee: Combined Broadcasting of Chicago Inc. Group owner: Univision Communications Inc. (acq 2-27-95; FTR: 5-22-95). Population served: 200,000 Natl. Network: Univision (Spanish). Washington Atty: Shaw Pittman LLP. Foreign lang progmg: SP 168 News staff: 18.
Key Personnel:
Vincent Cordero . . . . sr VP & gen mgr
Heather Bostounes . . . . gen sls mgr
Francisco Garcia . . . . prom dir
Yolanda Lopez De Otero . . . . news dir
George Molnar . . . . chief of engrg

WGN-TV— Analog channel: 9. Digital channel: 19. Analog hrs: 24 Digital hrs: 24 110 kw vis, 22 kw aur. ant 1,360t/1,453g TL: N41 53 56 W87 37 23 On air date: Apr 5, 1948. 2501 W. Bradley Pl., Chicago, IL, 60618-4718. Phone: (773) 528-2311. Fax: (773) 528-6857 (main). Web Site: wgntv.com. Licensee: WGN Continental Broadcasting Co. Group owner: Tribune Broadcasting Co., see Cross-Ownership Population served: 92,000,000 Natl. Network: CW. Natl. Rep: TeleRep. Washington Atty: Sidley & Austin. Wire Svc: AP Wire Svc: City News Bureau News: 32 hrs wkly.
Key Personnel:
Tom Ehlmann . . . . VP & gen mgr
Marty Wilke . . . . stn mgr & sls dir
Zach Smith . . . . natl sls mgr
Joanne Stern . . . . prom dir
Greg Caputo . . . . news dir

WJYS-TV— Analog channel: 62. Digital channel: 36. Analog hrs: 24 Digital hrs: 24 5,000 kw vis, 300 kw aur. 741 TL: N41 33 10 W87 47 09 (CP: Ant 479t) On air date: Mar 2, 1991. 18600 S. Oak Park Ave., Tinley Park, IL, 60477. Phone: (708) 633-0001. Fax: (708) 633-0040. E-mail: cs@wjystv62.net Licensee: Jovon Broadcasting Corp. Population served: 7,100,000
Key Personnel:
Joseph Stroud . . . . gen mgr
Eric Ferguson . . . . chief of engrg

WLS-TV— Analog channel: 7. Digital channel: 52.55 kw vis, 11.2 kw aur. ant 1,688t/1,710 TL: N41 52 44 W87 38 10 On air date: Sept 17, 1948. 190 N. State St., Chicago, IL, 60601. Phone: (312) 750-7777. Fax: (312) 750-7015. Web Site: www.abc7chicago.com. Licensee: WLS Television Inc. Group owner: ABC Inc. (acq 6-27-86; grpsl; FTR: 7-15-85). Population served: 3,200,000 Natl. Network: ABC. Wire Svc: PR Newswire Wire Svc: Dow Jones Financial News Services Wire Svc: Sports Wire News staff: 151; News: 8 hrs wkly.
Key Personnel:
Emily L. Barr . . . . pres & gen mgr
Joseph Trimarco . . . . opns mgr
Ed Pearson . . . . gen sls mgr
Tom Hebel . . . . prom dir
Ellen Crawley . . . . progmg dir
Jennifer Graves . . . . news dir
Kal Hassan . . . . engrg dir

WMAQ-TV— Analog channel: 5. Digital channel: 29.40.1 kw vis, 8.0 kw aur. ant 1,320t/1,456g TL: N41 53 56 W87 37 23 On air date: January 1948. 454 N. Columbus Dr., Chicago, IL, 60611-5555. Phone: (312) 836-5555. Web Site: www.nbc5.com. Licensee: NBC Telemundo License Co. Group owner: NBC TV Stations Division (acq 6-5-86). Population served: 3,204,710 Natl. Network: NBC. Natl. Rep: NBC TV Stations Sales.
Key Personnel:
Larry Wert . . . . pres & gen mgr
Patrica Golden . . . . sls VP
Toni Falvo . . . . mktg dir & progmg dir
Jan Jaros . . . . engrg dir

WPWR-TV—(Gary, IN) Analog channel: 50. Digital channel: 51. Analog hrs: 24 5,000 kw vis, 600 kw aur. ant 1,620t/1,624g TL: N41 52 44 W87 38 10 On air date: Jan 18, 1987. 205 N. Michigan Ave., Chicago, IL, 60614. Phone: (312) 565-5532. Fax: (312) 565-5517. Web Site: www.my50chicago.com. Licensee: Fox Television Stations Inc. Group owner: (group owner; (acq 8-21-2002; $425 million). Population served: 6,000,000 Natl. Network: MyNetworkTV. Natl. Rep: Fox Stations Sales.
Key Personnel:
Patrick Mullen. . . . . gen mgr
John Baich . . . . chief of engrg

WSNS— Analog channel: 44.4,260 kw vis, 500 kw aur. 1,420t/1,456g TL: N41 53 56 W87 37 23 On air date: Apr 5, 1970. 454 N. Columbus Dr., Chicago, IL, 60611. Phone: (312) 836-3000. Fax: (312) 836-3034. Web Site: www.telemundochicago.com. Licensee: NBC Telemundo License Co. Group owner: Telemundo Group Inc. (acq 4-12-2002; grpsl). Population served: 1,800,000 Natl. Network: Telemundo (Spanish). Washington Atty: Cohn & Marks. Foreign lang progmg: SP 138 News staff: 12; News: 5 hrs wkly.

***WTTW-TV—** Analog channel: 11. Digital channel: 47.60.3 kw vis, 12 kw aur. 1,630t/1,710g TL: N41 52 44 W87 38 10 On air date: Sept 6, 1955. 5400 N. St. Louis Ave., Chicago, IL, 60625. Phone: (773) 583-5000. Fax: (773) 583-3046. Web Site: www.networkchicago.com. Licensee: Window to the World Communications Inc. Population served: 10,000,000 Natl. Network: PBS. Washington Atty: Schwartz, Woods & Miller. News staff: 15; News: 5 hrs wkly.
Key Personnel:
Daniel Schmidt. . . . . CEO & pres
Farrell Frentress. . . . . exec VP
Howard Fisher . . . . exec VP & gen sls mgr
Reese Marcusson. . . . . CFO & opns VP
Donna Davies . . . . dev VP
Dan Sales . . . . progmg VP & progmg dir

WWTO-TV— Analog channel: 35. Analog hrs: 24 117.5 kw vis, 11.7 kw aur. 1,900t/1,371g TL: N41 16 51 W88 56 13 On air date: Dec 1, 1986. 420 E. Stevenson Rd., Ottawa, IL, 61350. Phone: (815) 434-2700. Fax: (815) 434-2458. Web Site: www.tbn.org. Licensee: Trinity Broadcasting Network. Group owner: (group owner; (acq 7-1-2000; grpsl). Washington Atty: Joseph E. Dunne III.

Key Personnel:
Charlie Boyd . . . . pres & chief of engrg
Marlene Zepeda . . . . gen mgr & stn mgr

WXFT-TV— Analog channel: 60.5,000 kw vis, 500 kw aur. 1,600t/1,621g TL: N41 52 44 W87 38 10 On air date: Apr 20, 1982. 541 N. Fairbanks Ct., Suite 1100, Chicago, IL, 60611. Phone: (312) 670-1000. Fax: (312) 467-5821. Web Site: www.univision.com. Licensee: TeleFutura Chicago LLC. Group owner: Univision Communications Inc. (acq 5-21-2001; grpsl). Natl. Network: TeleFutura (Spanish). Washington Atty: Wiley, Rein & Fielding. Wire Svc: City News Bureau
Key Personnel:
Vincent Cordero . . . . gen mgr
Heather Bastounes . . . . gen sls mgr
Francisco Garcia . . . . prom dir
Yolanda Lopez De Otero . . . . news dir
George Molnar . . . . chief of engrg

***WYCC—** Analog channel: 20.2,421 kw vis, 242.1 kw aur. 1,239t/1,110g TL: N41 53 56 W87 37 23 On air date: Sept 20, 1965. 7500 S. Pulaski Rd., Chicago, IL, 60652. Phone: (773) 838-7878. Fax: (773) 581-2071. E-mail: comments@wycc.org Web Site: www.wycc.org. Licensee: College Dist. #508, County of Cook. (acq 11-3-81). Natl. Network: PBS. Washington Atty: Dow, Lohnes & Albertson. Foreign lang progmg: SP 5
Key Personnel:
Maria Moore . . . . gen mgr
Arthur Wood . . . . stn mgr
Larry Eskridge . . . . prom mgr & engrg dir
Cynthia Syperek . . . . progmg dir & progmg mgr

***WYIN—** Analog channel: 56. Digital channel: 17. Analog hrs: 24 Digital hrs: 24 1,353 kw vis, 1.3 kw aur. 1,003t/998g TL: N41 20 56 W87 24 02 (CP: Ch 17, N41 52 44 W87 38 10, 300 Kw) On air date: Nov 15, 1987. 8625 Indiana Pl., Merrillville, IN, 46410. Phone: (219) 756-5656. Fax: (219) 755-4312. E-mail: mail@lakeshoreptv.com Web Site: www.lakeshoreptv.com. Licensee: Northwest Indiana Public Broadcasting Inc. Population served: 8,737,442 Natl. Network: PBS. Washington Atty: Schwartz, Woods & Miller.

Decatur

see Champaign & Springfield-Decatur, IL market

East St. Louis

see St. Louis, MO market

Harrisburg

see Paducah, KY-Cape Girardeau, MO-Harrisburg-Mount Vernon, IL market

Marion

see Paducah, KY-Cape Girardeau, MO-Harrisburg-Mount Vernon, IL market

Moline

see Davenport, IA-Rock Island-Moline, IL market

Mount Vernon

see Paducah, KY-Cape Girardeau, MO-Harrisburg-Mount Vernon, IL market

Olney

see Terre Haute, IN market

Peoria-Bloomington, IL (DMA 116)

WAOE— Analog channel: 59. Digital channel: 39.331 kw vis. ant 584t TL: N40 43 26 W89 29 04 On air date: 2000. 2907 Springfield Rd., East Peoria, IL, 61611. Phone: (309) 674-5900. Fax: (309) 674-5959. Web Site: my59.tv. Licensee: Four Seasons Peoria LLC. Ownership: Venture Technologies Group LLC, 48.5%; Malibu Broadcasting LLC, 48.5%; and Paul H. Koplin, 3% (acq 9-15-99). Natl. Network: MyNetworkTV.
Key Personnel:
Mark DeSantis . . . . VP & gen mgr
Rose Bortolussi . . . . stn mgr
Pete Russell . . . . gen sls mgr
Tim Campbell . . . . prom mgr
Jim Garrott . . . . news dir

WEEK-TV— Analog channel: 25. Digital channel: 57. Analog hrs: 24 Digital hrs: 24 2,400 kw vis, 239 kw aur. 680t/604 TL: N40 37 48 W89 32 51 On air date: Feb 1, 1953. 2907 Springfield Rd., East Peoria, IL, 61611. Phone: (309) 698-2525. Fax: (309) 698-9663 (sales). Fax: (309) 698-3737 (news). E-mail: news25@week.com Web Site: www.week.com. Licensee: WEEK-TV License Inc. Group owner: Granite Broadcasting Corp. (acq 10-31-88; $33 million). Population served: 467,800 Natl. Network: NBC. Washington Atty: Akin, Gump, Strauss, Hauer & Feld. News staff: 26; News: 16 hrs wkly.
Key Personnel:
Mark DeSantis . . . . gen mgr
Dennis Riley . . . . chief of engrg

WHOI— Analog channel: 19. Digital channel: 40. Analog hrs: 24 Note: ABC is on WHOI(TV) ch 19, CW is on WHOI-DT ch 40. 2,240 kw vis, 224 kw aur. ant 636t/632g TL: N40 39 11 W89 35 14 On air date: Oct 20, 1953. 500 N. Stewart St., Creve Coeur, IL, 61610. Phone: (309) 698-1919. Fax: (309) 698-1910. Web Site: www.hoinews.com. Licensee: Barrington Broadcasting Peoria Corp. Group owner: Barrington Broadcasting Corp. (acq 4-30-2004; $23.5 million with KHQA-TV Hannibal, MO). Population served: 532,000 Natl. Network: ABC, CW Digital Network: Note: ABC is on WHOI(TV) ch 19, CW is on WHOI-DT ch 40. Natl. Rep: Harrington, Righter & Parsons. Washington Atty: Covington & Burling. News: 7 hrs wkly.
Key Personnel:
Leo T Henning . . . . gen mgr
Jon Skorburg . . . . stn mgr
Tom Stemmler . . . . opns mgr & prom mgr
Valerie Bricka . . . . rgnl sls mgr
Donna Thompson . . . . progmg dir
Jolie Alois . . . . news dir
Jim Malone . . . . chief of engrg

WMBD-TV— Analog channel: 31. Digital channel: 30. Analog hrs: 24 2,050 kw vis, 406 kw aur. ant 635t/548g TL: N40 38 07 W89 32 19 On air date: Jan 1, 1958. 3131 N. University, Peoria, IL, 61604. Phone: (309) 688-3131. Fax: (309) 686-8650. Web Site: www.wmbd.com. Licensee: Nexstar Finance Inc. Group owner: Nexstar Broadcasting Group Inc. (acq 1999). Population served: 547,760 Natl. Network: CBS. Washington Atty: Covington & Burling.
Key Personnel:
Coby Cooper . . . . VP & gen mgr
Barry Allentuck . . . . sls dir
Nancy Linebaugh . . . . natl sls mgr
Travis Herriford . . . . rgnl sls mgr
David Tomlianovich . . . . prom dir
Herman Marvel . . . . news dir

***WTVP—** Analog channel: 47. Digital channel: 46. Analog hrs: 24 1,410 kw vis, 251 kw aur. 710t/599g TL: N40 37 44 W89 34 12 On air date: June 23, 1971. 101 State St, Peoria, IL, 61602. Phone: (309) 677-4747. Fax: (309) 677-4730. E-mail: wtvpmail@wtvp.pbs.org Web Site: www.wtvp.org. Licensee: Illinois Valley Public Telecommunication Corp. Population served: 750,000 Natl. Network: PBS. Washington Atty: Dow, Lohnes PLLC.
Key Personnel:
Chet Tomczyk . . . . CEO, pres & gen mgr
Jon Cecil . . . . chmn & VP
Jackie Luebcke . . . . opns mgr
John Morris . . . . dev VP
Linda Miller . . . . progmg VP
David Schenk . . . . engrg VP

WYZZ-TV— Analog channel: 43.1,200 kw vis, 112 kw aur. 979t/1,006g TL: N40 38 45 W89 10 45 (CP: 5,000 kw vis, ant 965t) On air date: Oct 18, 1982. 3131 N. University, Peoria, IL, 61604. Phone: (309) 688-3131. Fax: (309) 686-8650. Web Site: www.22fox.com. Licensee: WYZZ Licensee Inc. Group owner: Sinclair Broadcast Group Inc. (acq 1996; $23 million). Natl. Network: Fox. Natl. Rep: Harrington, Righter & Parsons. Washington Atty: Shaw Pittman LLP.
Key Personnel:
Kevin Harlan . . . . gen mgr
Barry Allantuck . . . . sls dir
Nancy Linebaugh . . . . natl sls mgr
Curt Bolak . . . . rgnl sls mgr
Kirby Matthews . . . . prom dir
Chris Manson . . . . news dir
Herman Marvel . . . . chief of engrg

Quincy, IL-Hannibal, MO-Keokuk, IA (DMA 171)

KHQA-TV—(Hannibal, MO) Analog channel: 7. Digital channel: 29. Analog hrs: 24 Digital hrs: 24 316 kw vis. ant 889t/805g TL: N39 58 22 W91 19 54 On air date: Sept 23, 1953. 301 S. 36th St., Quincy, IL, 62301. Phone: (217) 222-6200. Fax: (217) 228-3164/(217) 222-5078. E-mail: khqa@khqa.com Web Site: www.khqa.com Licensee: Barrington Broadcasting Quincy Corp. Group owner: Barrington Broadcasting Corp. (acq 4-30-2004; $23.5 million with WHOI(TV) Peoria, IL). Population served: 41,100 Natl. Network: CBS. Natl. Rep: Continental Television Sales. Washington Atty: Covington & Burling. Wire Svc: CBS Wire Svc: FNS News staff: 12; News: 13 hrs wkly.
Key Personnel:
Robert B. Sherman . . . . CEO
Jon Van Ness . . . . gen mgr & stn mgr
Leo Henning . . . . gen mgr & prom dir
Mava Clingingsmith . . . . gen sls mgr & natl sls mgr
Cindy Johnson . . . . rgnl sls mgr
Jim Malone . . . . chief of engrg

WGEM-TV— Analog channel: 10. Digital channel: 54. Analog hrs: 24 Digital hrs: 24 Note: NBC is on WGEM-TV ch 10, CW and Fox are on WGEM-DT ch 54. 316 kw vis, 31.6 kw aur. 780t/673g TL: N39 57 03 W91 19 54 On air date: Sept 4, 1953. Box 80, 513 Hampshire, Quincy, IL, 62306. Phone: (217) 228-6600. Fax: (217) 228-6670. Web Site: wgem.com. Licensee: Quincy Broadcasting Co. Group owner: Quincy Newspapers Inc. Population served: 350,000 Natl. Network: NBC, CW, Fox Digital Network: Note: NBC is on WGEM-TV ch 10, CW and Fox are on WGEM-DT ch 54. Natl. Rep: Blair Television. Washington Atty: Wilkinson, Barker, Knauer & Quinn. Wire Svc: AP News staff: 25; News: 20 hrs wkly.
Key Personnel:
Thomas A. Oakley . . . . CEO
Ralph M. Oakley . . . . VP
Tom Allen . . . . gen mgr
Frank Forgey . . . . stn mgr

***WMEC—** Analog channel: 22. Digital channel: 21. Analog hrs: 6 AM-midnight Digital hrs: 6 AM-midnight 24.15 kw vis, 2.42 kw aur. 519t/535g TL: N40 25 40 W90 40 58 On air date: Oct 1, 1984. Box 6248, Springfield, IL, 62708. Phone: (217) 483-7887. Phone: (800) 232-3605. Fax: (217) 483-1112. Web Site: www.wmec.tv. Licensee: West Central Illinois Educational Telecommunications Corp. Population served: 915,000 Natl. Network: PBS. Washington Atty: Dow, Lohnes & Albertson.
Key Personnel:
Jerold Gruebel . . . . CEO, pres & gen mgr
Richard Plotkin . . . . opns VP & opns dir
Ed Strong . . . . adv dir

***WQEC—** Analog channel: 27. Digital channel: 34. Analog hrs: 6 AM-midnight Digital hrs: 6 AM-midnight 14.8 kw vis, 1.48 kw aur. 567t/495g TL: N39 58 44 W91 18 33 On air date: Mar 11, 1985. Box 6248, Springfield, IL, 62708. Phone: (217) 483-7887. Phone: (800) 232-3605. Fax: (217) 483-1112. Web Site: www.wqec.tv. Licensee: West Central Illinois Educational Telecommunications Corp. Population served: 915,000 Natl. Network: PBS. Washington Atty: Dow, Lohnes & Albertson.
Key Personnel:
Jerold Gruebel . . . . CEO, pres & gen mgr
Richard Plotkin . . . . opns VP & opns dir
Ed Strong . . . . adv dir

WTJR— Analog channel: 16. Digital channel: 32. Analog hrs: 24 179.2 kw vis, 31.5 kw aur. ant 1,025t/948g TL: N39 58 18 W91 19 42 On air date: Jan 1, 1986. 222 North 6th St., Quincy, IL, 62301. Phone: (217) 228-1616. E-mail: tv16@wtjr.org Web Site: www.wtjr.org. Licensee: Christian Television Network Inc. Ownership: Robert D'Andrea, 20% votes; Virginia Oliver, 20% votes; Jimmy Smith, 20% votes; Wayne Wetzel, 20% votes; Robert S. Young, 20% votes (acq 5-23-2006; $2.1 million). Population served: 350,000
Key Personnel:
Donette Douglas . . . . gen mgr & stn mgr
Jim Wilson . . . . engr

Rock Island

see Davenport, IA-Rock Island-Moline, IL market

Rockford, IL (DMA 133)

WIFR—Freeport, Analog channel: 23. Analog hrs: 24 676 kw vis, 85.2 kw aur. 720t/731g TL: N42 17 48 W89 10 15 On air date: Sept 12, 1965. Box 123, Rockford, IL, 61105. 2523 N. Meridian Rd., Rockford, IL 61101. Phone: (815) 987-5300. Fax: (815) 987-0981. E-mail: talkto23@wifr.com Web Site: www.wifr.com. Licensee: WEAU Licensee Corp. Group owner: Gray Television Inc. (acq 8-29-2002; grpsl). Population served: 187,500 Natl. Network: CBS. Natl. Rep: Continental Television Sales. Washington Atty: Covington & Burling. News staff: 19; News: 19 hrs wkly.
Key Personnel:
Greg Graber . . . . VP & gen mgr
Dave Smith . . . . opns mgr & news dir

Jeff Clark . . . . opns mgr
Tim Myers. . . . . gen sls mgr

WQRF-TV— Analog channel: 39. Digital channel: 42.525 kw vis, 5.25 kw aur. 700 TL: N42 17 26 W89 09 51 On air date: Nov 27, 1978. 1917 N. Meridian Rd., Rockford, IL 61101. Phone: (815) 963-5413. Fax: (815) 963-6113. Licensee: Nexstar Finance Inc. Group owner: Nexstar Broadcasting Group Inc. (acq 12-31-03; grpsl). Natl. Network: Fox. Washington Atty: Arter & Hadden.
Key Personnel:
Marshall Porter. . . . . VP & gen mgr
Eileen Boucek. . . . . stn mgr
Tina Mickelson . . . . sls dir
Sean Anderson . . . . prom mgr
Jose Cabezas . . . . progmg mgr
Kent Harrell . . . . news dir
Mike Real . . . . chief of engrg

WREX-TV— Analog channel: 13. Digital channel: 54. Analog hrs: 24 Digital hrs: 24 Note: NBC is on WREX-TV ch 13, CW is on WREX-DT ch 54. 316 kw vis, 39.8 kw aur. 710t/652g TL: N42 17 50 W89 14 24 On air date: Oct 1, 1953. Box 530, Rockford, IL, 61105. 10322 W. Auburn Rd., Rockford, IL 61103. Phone: (815) 335-2213. Fax: (815) 335-2055. E-mail: wrex@wrex.com Web Site: www.wrex.com. Licensee: WREX Television LLC. Group owner: Quincy Newspapers Inc., see Cross-Ownership (acq 5-22-2001; grpsl). Population served: 452,000 Natl. Network: NBC, CW Digital Network: Note: NBC is on WREX-TV ch 13, CW is on WREX-DT ch 54. Natl. Rep: Blair Television. Washington Atty: Wilkinson, Barker & Knauer.
Key Personnel:
John Chadwick. . . . . VP & gen mgr
Gerry Meinders . . . . prom dir & chief of engrg
Kim Arney. . . . . progmg mgr
Maggie Hradecky . . . . news dir

WTVO— Analog channel: 17. Digital channel: 16.Note: ABC is on WTVO(TV) ch 17, MyNetworkTV is on WTVO-DT ch 16. 646 kw vis. ant 666t/676g TL: N42 17 14 W89 10 15 On air date: May 3, 1953. Box 470, Rockford, IL, 61105. 1917 N. Meridian Rd., Rockford, IL 61101. Phone: (815) 963-5413. Fax: (815) 963-6113. Web Site: www.mystateline.com. Licensee: Mission Broadcasting Inc. Group owner: (group owner; (acq 1-4-2005; $20,750,000). Population served: 350,000 Natl. Network: ABC, MyNetworkTV Digital Network: Note: ABC is on WTVO(TV) ch 17, MyNetworkTV is on WTVO-DT ch 16. Washington Atty: Wiley, Rein & Fielding. News staff: 16; News: 7 hrs wkly.
Key Personnel:
Marshall Porter. . . . . VP & gen mgr
Eileen Boucek. . . . . stn mgr
Tina Mickelson . . . . sls dir
Sean Anderson . . . . prom mgr
Jose Cabezas . . . . progmg mgr
Kent Harrell . . . . news dir
Mike Real . . . . chief of engrg

Springfield

see Champaign & Springfield-Decatur, IL market

Indiana

Elkhart

see South Bend-Elkhart, IN market

Evansville, IN (DMA 101)

WAZE-TV— Analog channel: 19. Digital channel: 20. Analog hrs: 24 1,143 kw vis, 114.3 kw aur. ant 1,194t/1,053g TL: N37 24 46 W87 31 32 (CP: 2,676 kw vis, ant 790t) On air date: September 1997. 1277 N. St. Joseph Ave., Evansville, IN, 47720. Phone: (812) 425-1900. Fax: (812) 423-3405. E-mail: cw19@roberts-companies.com Web Site: www.cwaze.com. Licensee: Roberts Broadcasting Co. of Evansville, IN LLC. Group owner: (group owner; (acq 12-14-2006; $1 million). Population served: 350,000 Natl. Network: CW. Washington Atty: Fletcher, Heald & Hildreth.
Key Personnel:
Greg Pittman . . . . gen mgr, opns dir & gen sls mgr
Bryan Gibbs. . . . . prom mgr
Lisa Pillow . . . . pres & progmg dir
Jim McFarland . . . . chief of engrg

WEHT— Analog channel: 25.60 kw vis, 6.2 kw aur. 1,030t/988g TL: N37 51 56 W87 34 04 On air date: Sept 11, 1953. Box 25, Evansville, IN, 47701. 800 Marywood Dr., Henderson, KY 42420. Phone: (800) 879-8549. Fax: (270) 827-0561. E-mail: contactus@news25.us Web Site: www.news25.us. Licensee: Gilmore Broadcasting Corp. Ownership: National City Bank of Michigan/Illinois, directed by G. Lennon & M. Lemieux (acq 1-15-2003). Population served: 185,000 Natl. Network: ABC. Washington Atty: Wiley, Rein & Fielding.
Key Personnel:
Doug Padgett. . . . . gen mgr
Mike Riley . . . . stn mgr & gen sls mgr
Melisse Marks . . . . prom mgr
Genny Powers . . . . progmg mgr
Mark Glover . . . . news dir
Darren Gibson . . . . chief of engrg

WEVV-TV— Analog channel: 44. Digital channel: 45. Analog hrs: 24 Note: CBS is on WEVV(TV) ch 44, MyNetworkTV is on WEVV-DT ch 45. 1,250 kw vis, 125 kw aur. 1,000t/1,000g TL: N37 53 17 W87 32 37 On air date: Nov 17, 1983. 44 Main St., Evansville, IN, 47708-1450. Phone: (812) 464-4444. Fax: (812) 465-4559. Web Site: www.wevv.com. Licensee: Comcorp of Indiana License Corp. Group owner: Communications Corp. of America (acq 1999; $27.5 million). Natl. Network: CBS Digital Network: Note: CBS is on WEVV(TV) ch 44, MyNetworkTV is on WEVV-DT ch 45. Washington Atty: Leventhal, Senter & Lerman. News staff: 30; News: 9 hrs wkly.
Key Personnel:
Dan Robbins . . . . gen mgr
John Bennett . . . . opns mgr & chief of engrg
Sandy Eickhoff. . . . . gen sls mgr
Tim Black . . . . prom dir
Joanne Provenzano . . . . progmg dir

WFIE-TV— Analog channel: 14. Digital channel: 46. Analog hrs: 24 2,510 kw vis. ant 1,017t/905g TL: N37 53 14 W87 31 07 On air date: Nov 9, 1953. Box 1414, Evansville, IN, 47701. 1115 Mt. Auburn Rd., Evansville, IN 47720. Phone: (812) 426-1414. Fax: (812) 426-1945. E-mail: wfie@14wfie.com Web Site: www.14wfie.com. Licensee: Libco Inc. Group owner: Liberty Corp. (acq 1-13-2006; grpsl). Population served: 290,000 Natl. Network: NBC. Washington Atty: Dow, Lohnes & Albertson. News staff: 35.
Key Personnel:
Lucy Himstedt . . . . gen mgr
Laura Lovejoy . . . . gen sls mgr
Gil Mazur . . . . mktg dir
Bill Cummings . . . . news dir
Jim Sears . . . . chief of engrg
Kirk Williams . . . . progmg

***WKMA—** Analog channel: 35. Digital channel: 42.617 kw vis, 61.7 kw aur. 1,040t/998g TL: N37 11 25 W87 30 47 On air date: Sept 23, 1968. 600 Cooper Dr., Lexington, KY, 40502. Phone: (606) 258-7000. Fax: (606) 258-7399. Web Site: www.ket.org. Licensee: Kentucky Authority for Educational TV. Natl. Network: PBS.
Key Personnel:
Malcolm Wall . . . . chmn
Craig Cornwell . . . . opns mgr & progmg dir
Tim Bischoff . . . . dev VP & mktg dir
Robert Ball . . . . engrg dir

***WKOH—** Analog channel: 31. Digital channel: 30.692 kw vis, 55 kw aur. 460t/500g TL: N37 51 06 W87 19 43 (CP: 617 kw vis, 708 kw aur max) On air date: Mar 1, 1979. 600 Cooper Dr., Lexington, KY, 40502. Phone: (606) 258-7000. Fax: (606) 258-7399. Web Site: www.ket.org. Licensee: Kentucky Authority for Educational TV. Natl. Network: PBS. Rgnl. Network: SECA. Washington Atty: Kenkel, Barnard & Edmundson.
Key Personnel:
Malcolm Wall . . . . chmn
Craig Cornwell. . . . . opns mgr & progmg mgr
Tim Bischoff . . . . mktg dir

***WNIN-TV—** Analog channel: 9. Digital channel: 12. Analog hrs: 18 Digital hrs: 18 282 kw vis, 56.2 kw aur. 570t/570g TL: N38 01 27 W87 21 43 On air date: Mar 16, 1970. 405 Carpenter St., Evansville, IN, 47708. Phone: (812) 423-2973. Fax: (812) 428-7548. E-mail: wnin@wnin.org Web Site: www.wnin.org. Licensee: Tri-State Public Teleplex Inc. (acq 9-12-73). Population served: 138,764 Natl. Network: PBS. Washington Atty: Dow, Lohnes PLLC.
Key Personnel:
David Dial . . . . pres & gen mgr
Bonnie Rheinhardt . . . . . . opns VP, prom mgr & progmg VP
Tonya Wolf . . . . gen sls mgr
Don Hollingsworth . . . . mktg VP & chief of engrg

WTVW— Analog channel: 7. Digital channel: 28. Analog hrs: 24 316 kw vis, 63.2 kw aur. 1,013t/880g TL: N38 01 27 W87 21 43 On air date: Aug 26, 1956. 477 Carpenter St., Evansville, IN, 47708. Phone: (812) 424-7777. Fax: (812) 421-4040. Web Site: www.tristatehomepage.com. Licensee: Nexstar Broadcasting Inc. Group owner: Nexstar Broadcasting Group Inc. (acq 10-30-2003; grpsl). Population served: 686,500 Natl. Network: Fox. Natl. Rep: Blair Television. Washington Atty: Arter & Hadden. News: 22.5 hrs wkly.
Key Personnel:
Jeff Fischer . . . . gen sls mgr
Jay Hiett . . . . rgnl sls mgr
Mike Smith . . . . gen mgr & progmg mgr
Bob Walters . . . . news dir
Dan Jordan . . . . chief of engrg

Ft. Wayne, IN (DMA 106)

WANE-TV— Analog channel: 15. Analog hrs: 24 2,450 kw vis. ant 827t/804g TL: N41 05 38 W85 10 48 On air date: Sept 26, 1954. Box 1515, Fort Wayne, IN, 46801. 2915 W. State Blvd., Fort Wayne, IN 46808. Phone: (260) 424-1515. Fax: (260) 407-1607. Web Site: www.wane.com. Licensee: Indiana Broadcasting L.L.C. Group owner: LIN Television Corporation (acq 11-14-94; FTR: 12-12-94). Population served: 268,610 Natl. Network: CBS. News staff: 35; News: 22 hrs wkly.
Key Personnel:
Alan Riebe. . . . . gen mgr & natl sls mgr
Jim Riecken . . . . opns mgr
Tom Antisdel. . . . . rgnl sls mgr
Jerry Grider . . . . prom dir
Nancy Applegate . . . . progmg mgr
Ted Linn. . . . . news dir
Mark Johnson . . . . chief of engrg

WFFT-TV— Analog channel: 55. Digital channel: 36. Analog hrs: 24 600 kw vis, 38 kw aur. 780t/805g TL: N41 06 33 W85 11 44 On air date: Dec 21, 1977. 3707 Hillegas Rd., Fort Wayne, IN, 46808-1351. Box 8655, Fort Wayne, IN 46808. Phone: (260) 471-5555. Fax: (260) 484-4331. E-mail: fox55@wfft.com Web Site: www.wfft.com. Licensee: Nexstar Finance Inc. Group owner: Nexstar Broadcasting Group Inc. (acq 12-31-03; grpsl). Population served: 1,855,000 Natl. Network: Fox. Natl. Rep: Blair Television. Washington Atty: Drinker, Biddle & Reath.
Key Personnel:
Perry A. Sook . . . . CEO & pres
Matt Devine . . . . CFO
Tim Busch . . . . sr VP
Robert M. Blacher . . . . VP & gen mgr
Bill Ritchhart. . . . . gen sls mgr

***WFWA—** Analog channel: 39. Digital channel: 40. Analog hrs: 24 Digital hrs: 24 1,380 kw vis, 138 kw aur. ant 730t/744g TL: N41 06 13 W85 11 28 On air date: Dec 1, 1989. 2501 E. Coliseum Blvd., Fort Wayne, IN, 46805-1562. Phone: (260) 484-8839. Fax: (260) 482-3632. E-mail: info@wfwa.org Web Site: www.wfwa.org. Licensee: Fort Wayne Public Television. Population served: 274,300 Natl. Network: PBS. Washington Atty: Wiley Rein LLP.
Key Personnel:
Roger Rhodes. . . . . pres, gen mgr & stn mgr
Claudia Johnson . . . . opns VP & opns mgr
Toni Kayumi. . . . . gen sls mgr & mktg mgr
Matt Kyle . . . . chief of engrg & engr
Mark Ryan . . . . prom

WINM— Analog channel: 63. Analog hrs: 24 Digital hrs: 24 5,000 kw vis, 500 kw aur. 499g TL: N41 27 15 W84 48 10 (CP: 1,374 kw vis, ant 473t) On air date: Mar 10, 1983. Box 159, Butler, IN, 46721. Phone: (419) 298-3703. Fax: (419) 298-3707. E-mail: winm@tct.tv Licensee: Tri-State Christian TV. Group owner: (group owner; (acq 1-24-91; $400,000; FTR: 2-11-91).
Key Personnel:
Leo Vogt. . . . . gen mgr
Lee Gilbert . . . . chief of engrg

WISE-TV— Analog channel: 33. Digital channel: 19. Analog hrs: 24 Note: NBC is on WISE-TV ch 33, MyNetworkTV is on WISE-DT ch 19. 594 kw vis, 59 kw aur. ant 770t/793g TL: N41 05 40 W85 10 36 On air date: Nov 21, 1953. 3401 Butler Rd., Harrisonburg, IN, 46808. Phone: (260) 422-7474. Fax: (260) 422-7702. Web Site: www.indianasnewscenter.com. Licensee: WISE-TV License LLC. Group owner: New Vision Group LLC (acq 3-8-2005; $44.2 million). Population served: 222,480 Natl. Network: NBC, MyNetworkTV Digital Network: Note: NBC is on WISE-TV ch 33, MyNetworkTV is on WISE-DT ch 19. Washington Atty: Drinker, Biddle & Reath, LLP. News staff: 32; News: 14 hrs wkly.
Key Personnel:
Sherry Avara. . . . . rgnl sls mgr
Peter Neumann . . . . news dir
Bret Angel . . . . chief of engrg

WPTA— Analog channel: 21. Digital channel: 24. Analog hrs: 24 Digital hrs: 24 Note: ABC is on WPTA(TV) ch 21, CW is on WPTA-DT ch 24. 562 kw vis, 55 kw aur. ant 760t/770g TL: N41 06 08 W85 11

04 On air date: Sept 28, 1957. 3401 Butler Rd., Fort Wayne, IN, 46808. Phone: (260) 483-0584. Fax: (260) 483-2568. E-mail: news@indianasnewscenter.com Web Site: www.indianasnewscenter.com. Licensee: Malara Broadcast Group of Fort Wayne License LLC. Group owner: (group owner; (acq 12-8-2004; $45.9 million). Population served: 562,700 Natl. Network: ABC, CW Digital Network: Note: ABC is on WPTA(TV) ch 21, CW is on WPTA-DT ch 24. Washington Atty: Akin, Gump, Strauss, Hauer & Feld. News staff: 31; News: 31 hrs wkly.

Key Personnel:
Anthony Malara . . . pres
Jerry Giesler . . . pres & gen mgr
Dan Hoffman . . . gen mgr & sls dir
Doug Barrow . . . stn mgr
Tad Frank . . . prom dir
Peter Neumann . . . news dir

Gary

see Chicago market

Hammond

see Chicago market

Indianapolis, IN (DMA 25)

WCLJ-TV— Analog channel: 42. Digital channel: 56.5,000 kw vis, 500 kw aur. 1,039t TL: N39 24 12 W86 08 50 On air date: August 1987. 2528 U.S. 31 S., Greenwood, IN, 46143. Phone: (317) 535-5542. Fax: (317) 535-8584. Licensee: Trinity Broadcasting of Indiana Inc. Group owner: Trinity Broadcasting Network.

Key Personnel:
Mark Crouch . . . gen mgr
Ken Harl . . . chief of engrg

***WDTI**— Analog channel: 69. Digital channel: 44.9.77 kw vis. ant 548t/636g TL: N39 50 25 W86 10 34 On air date: April 1992. Indianapolis Community Television Inc., 3901 Hwy. 121 S., Bedford, TX, 76021. Phone: (817) 571-1229. Phone: (817) 571-7458. Web Site: www.daystar.com/schedules.htm. Licensee: Indianapolis Community Television Inc. Ownership: Dr. Alan Bullock, 14.3%; Joni T. Lamb, 14.3%; Marcus D. Lamb, 14.3%; Peter Keenan, 14.3%; Rob Price, 14.3%; and Vernon Piercey, 14.3% (acq 8-2-2004; $4 million). Population served: 750,000 Washington Atty: Koerner & Olender.

***WFYI-TV**— Analog channel: 20. Digital channel: 21. Analog hrs: 24 Digital hrs: 24 1,135 kw vis, 114 kw aur. 847t/867g TL: N39 53 59 W86 12 01 On air date: Oct 4, 1970. 1401 N. Meridian St., Indianapolis, IN, 46202. Phone: (317) 636-2020. Fax: (317) 633-7418. E-mail: sjensen@wfyi.org Web Site: www.wfyi.org. Licensee: Metropolitan Indianapolis Public Broadcasting. Population served: 792,500 Natl. Network: PBS.

Key Personnel:
Lloyd Wright . . . pres
Anthony Lorenz . . . CFO
Alan Cloe . . . exec VP & sr VP
Jeanelle Adamak . . . exec VP

WHMB-TV— Analog channel: 40. Digital channel: 16. Analog hrs: 24 Digital hrs: 24 2,090 kw vis, 209 kw aur. 991t/1,007g TL: N39 53 39 W86 12 19 On air date: Jan 25, 1971. Box 50450, Indianapolis, IN, 46250. 10511 Greenfield Ave., Noblesville, IN 46060. Phone: (317) 773-5050. Fax: (317) 776-4051. E-mail: kpasson@lesea.com Web Site: www.whmbtv.com. Licensee: LeSea Broadcasting of Indianapolis Inc. Group owner: (group owner; (acq 8-15-72; $354,618; FTR: 9-4-72). Population served: 2,700,000 Washington Atty: Gardner, Carton & Douglas. News: one hr wkly.

Key Personnel:
Pete Sumrall . . . CEO & VP
Tony Hale . . . pres & CFO
Keith Passon . . . gen mgr & gen sls mgr

***WIPB**— Analog channel: 49. Analog hrs: 24 676 kw vis, 67.6 kw aur. ant 510t/548g TL: N40 09 38 W85 22 42 On air date: May 8, 1953. Edmund F. Ball Bldg., Ball State Univ., Muncie, IN, 47306. Phone: (765) 285-1249. Fax: (765) 285-5548. E-mail: wipb@bsu.edu Web Site: www.bsu.edu/wipb. Licensee: Ball State University. (acq 10-31-71; $125,000; FTR: 12-6-71). Population served: 450,000 Natl. Network: PBS. Rgnl. Network: CEN. Washington Atty: Schwartz, Woods & Miller.

Key Personnel:
Alice Cheney . . . gen mgr
Bob Fairchild . . . chief of engrg

WIPX— Analog channel: 63. Digital channel: 27. Analog hrs: 24 2,000 kw vis, 200 kw aur. ant 1,053t/2,300g TL: N39 24 16 W86 08 37 On air date: Dec 27, 1988. 601 Clearwater Park Rd., West Palm Beach, FL, 33401. Phone: (317) 486-0633. Web Site: www.ionline.tv. Licensee: Paxson Indianapolis License Inc. Group owner: Paxson Communications Corp. (acq 2-18-2000; grpsl). Population served: 825,000 Natl. Network: i Network. Washington Atty: Fisher, Wayland, Cooper, Leader & Zaragoza.

WISH-TV— Analog channel: 8. Digital channel: 9.316 kw vis, 42.7 kw aur. 990t/997g TL: N39 45 39 W86 00 21 (CP: TL: N39 53 25 W86 12 20) On air date: July 1, 1954. 1950 N. Meridian St., Indianapolis, IN, 46202. Phone: (317) 923-8888. Fax: (317) 926-1144 (sales). E-mail: news@wishtv.com Web Site: www.wishtv.com. Licensee: Indiana Broadcasting L.L.C. Group owner: LIN Television Corporation (acq 11-14-94; FTR: 12-12-94). Population served: 1,339,500 Natl. Network: CBS. Washington Atty: Covington & Burling. Wire Svc: UPI

Key Personnel:
Jeff White . . . gen mgr
Julie Zoumbaris . . . gen sls mgr
Scott Hainey . . . prom dir
Lance Carwile . . . progmg mgr
Kevin Finch . . . news dir
Terry Van Bibber . . . chief of engrg

WNDY-TV— Analog channel: 23. Digital channel: 32. Analog hrs: 24 5,000 kw vis, 600 kw aur. ant 1,082t TL: N40 08 57 W85 56 15 On air date: Nov 1, 1987. 1950 N. Meridian St., Indianapolis, IN, 46202. Phone: (317) 823-8888. Fax: (317) 926-1144 (sales). E-mail: news@wndy.com Web Site: www.wndy.com. Licensee: Indiana Broadcasting LLC. Group owner: Viacom Television Stations Group (acq 3-31-2005; $85 million with WWHO(TV) Chillicothe, OH). Natl. Network: MyNetworkTV. Washington Atty: Dow, Lohnes PLLC.

Key Personnel:
Jeff White . . . gen mgr
Julie Zoumbaris . . . gen sls mgr
Scott Hainey . . . prom dir
Lance Carwile . . . progmg mgr
Kevin Finch . . . news dir
Terry VanBibber . . . chief of engrg

WRTV— Analog channel: 6. Digital channel: 25. Analog hrs: 24 Digital hrs: 24 100 kw vis, 20 kw aur. 990t/1,019g TL: N39 53 59 W86 12 02 On air date: May 30, 1949. 1330 N. Meridian St., Indianapolis, IN, 46202. Phone: (317) 635-9788. Phone: (317) 269-1440 (news). Fax: (317) 269-1400. E-mail: newstips@theindychannel.com Web Site: www.theindychannel.com. Licensee: McGraw-Hill Broadcasting Co. Inc. Group owner: McGraw-Hill Broadcasting Co. (acq 6-1-72). Population served: 1,100,000 Natl. Network: ABC. Natl. Rep: Harrington, Righter & Parsons. Washington Atty: Holland & Knight.

Key Personnel:
Don Lundy . . . gen mgr
Sally Kohn . . . gen sls mgr
Paul Montgomery . . . prom mgr & progmg mgr
Jason Heath . . . news dir
Brian Vetor . . . chief of engrg

WTHR— Analog channel: 13. Analog hrs: 24 316 kw vis, 63.2 kw aur. 980t/1,039g TL: N39 55 43 W86 10 55 On air date: Oct 30, 1957. Box 1313, Indianapolis, IN, 46206. 1000 N. Meridian St., Indianapolis, IN 46204. Phone: (317) 636-1313. Fax: (317) 636-3717. Fax: (317) 632-6720 (news). Web Site: www.wthr.com. Licensee: VideoIndiana Inc. Group owner: Dispatch Broadcast Group (acq 10-1-75; $17.65 million; FTR: 9-1-75). Population served: 1,077,400 Natl. Network: NBC. Washington Atty: Sidley & Austin.

Key Personnel:
Rich Pegram . . . pres & gen mgr
Tim Warner . . . gen sls mgr
Jeff Dutton . . . prom dir
Rod Porter . . . progmg dir
Roger Bishop . . . chief of engrg

***WTIU**— Analog channel: 30. Digital channel: 14. Analog hrs: 24 200 kw vis, 39.8 kw aur. 710t/647g TL: N39 08 32 W86 29 43 (CP: 832 kw, ant 708 t) On air date: March 1969. Radio-TV Bldg., Indiana Univ., 1229 E. 7th St., Bloomington, IN, 47405. Phone: (812) 855-5900. Phone: (812) 855-8000. Fax: (812) 855-0729. E-mail: wtiu@indiana.edu Web Site: www.wtiu.indiana.edu. Licensee: Trustees of Indiana University. Population served: 500,000 Natl. Network: PBS. Washington Atty: Crowell & Moring. News staff: 2; News: 3 hrs wkly.

Key Personnel:
Perry Metz . . . CEO & gen mgr
Phil Meyer . . . stn mgr
Brad Howard . . . opns mgr

WTTK— Analog channel: 29. Digital channel: 54. Analog hrs: 24 3,090 kw vis, 309 kw aur. ant 775t/774g TL: N40 20 20 W85 57 15 On air date: May 1988. 6910 Network Place, Indianapolis, IN, 46278. Phone: (800) 968-4434. Fax: (317) 687-6531, (317) 687-6534. Web Site: www.thecw4.com. Licensee: Tribune Denver Radio Inc. Group owner: Tribune Broadcasting Co. (acq 7-12-2002; $125 million with WTTV(TV) Bloomington). Population served: 900,000 Natl. Network: CW. Natl. Rep: TeleRep.

Key Personnel:
Jerry Martin . . . gen mgr
Tim McNamara . . . gen sls mgr
Kurt Tovey . . . prom dir
Harry Ford . . . progmg dir
Rich Kittlestved . . . engrg dir
Satellite of WTTV(TV) Bloomington.

WTTV—Bloomington, Analog channel: 4. Digital channel: 48. Analog hrs: 24 55 kw vis, 11 kw aur. ant 1,200t/1,170g TL: N39 24 26 W86 08 52 On air date: Nov 11, 1949. 6910 Network Pl., Indianapolis, IN, 46278. Phone: (317) 632-5900. Fax: (317) 687-6532. Web Site: www.thecw4.com. Licensee: Tribune Denver Radio Inc. Group owner: Tribune Broadcasting Co. (acq 7-12-2002; $125 million with WTTK(TV) Kokomo). Population served: 744,624 Natl. Network: CW. Natl. Rep: TeleRep. Washington Atty: Sidley Austin.

Key Personnel:
Jerry Martin . . . VP & gen mgr
Tim McNamara . . . gen sls mgr
Kurt Tovey . . . prom dir
Harry Ford . . . progmg dir
Rich Kttilstved . . . engrg dir

WXIN— Analog channel: 59. Digital channel: 45. Analog hrs: 24 2,090 kw vis, 209 kw aur. 990t/1,030g TL: N39 53 20 W86 12 07 On air date: Feb 1, 1984. 6910 Network Pl., Indianapolis, IN, 46278. Phone: (317) 632-5900. Fax: (317) 687-6532. Web Site: www.fox59.com. Licensee: Tribune Television Comp. Group owner: Tribune Broadcasting Co. (acq 7-15-97; grpsl). Population served: 963,320 Natl. Network: Fox. Natl. Rep: TeleRep. Washington Atty: Sidley Austin. News: 19 hrs wkly.

Key Personnel:
Jerry Martin . . . VP & gen mgr
Tim McNamara . . . sls dir
Kurt Tovey . . . prom dir
Harry Ford . . . progmg dir
Rich Kittilstved . . . engrg dir

Lafayette, IN (DMA 188)

WLFI-TV— Analog channel: 18. Digital channel: 11. Analog hrs: 24 1,490 kw vis, 298 kw aur. 778t/755 TL: N40 23 20 W86 36 46 On air date: June 15, 1953. 2605 Yeager Rd., West Lafayette, IN, 47906. Phone: (765) 463-1800. Fax: (765) 463-7979. Web Site: www.wlfi.com. Licensee: Primeland Television Inc. Group owner: LIN Television Corporation (acq 4-1-2000; in exchange for 67% of WAND(TV) Decatur, IL). Population served: 342,100 Natl. Network: CBS. Natl. Rep: Petry Television Inc. Wire Svc: UPI News staff: 30; News: 22 hrs wkly.

Key Personnel:
Tom Combs . . . stn mgr
Chris Hilgendorf . . . opns mgr
Jenny Olszewski . . . sls dir
Deb McMahan . . . prom mgr
Rick Thedwall . . . progmg mgr
Chris Morisse . . . news dir
Mark Brooks . . . chief of engrg

Richmond

see Dayton, OH market

Salem

see Louisville, KY market

South Bend-Elkhart, IN (DMA 88)

WHME-TV— Analog channel: 46. Analog hrs: 24 1,600 kw vis, 160 kw aur. 1,000t/982g TL: N41 35 43 W86 09 38 On air date: July 27, 1974. 61300 S. Ironwood Rd., South Bend, IN, 46614. Phone: (574) 291-8200. Fax: (574) 291-9043. Web Site: www.whme.com. Licensee: Lester Sumrall Evangelistic Association. Group owner: Le Sea Broadcasting (acq 6-10-77; $496,000; FTR: 7-27-77). Population served: 477,400 Washington Atty: Gardner, Carton & Douglas.

Key Personnel:
Peter Sumrall . . . gen mgr
Mike Swinehart . . . opns mgr
Anna Riblet . . . rgnl sls mgr
Wes Hylton . . . VP & chief of engrg

WNDU-TV— Analog channel: 16. Digital channel: 42. Analog hrs: 24 4,170 kw vis. ant 1,069t/1,007g TL: N41 36 20 W86 12 46 On air date: July 15, 1955. Box1616, South Bend, IN, 46634. 54516 State Rd. 933, SouthBend, IN 46634. Phone: (574) 631-1616. Fax: (574) 631-1600. E-mail: newscenter16@wndu.com Web Site: www.wndu.com. Licensee: Michiana Telecasting Corp. (acq 3-6-2006; $85 million). Population served: 294,530 Natl. Network: NBC.
Key Personnel:
CJ Beutein . . . . . pres & news dir
Matt Jaquint . . . . . gen mgr
Howard Voss . . . . . gen sls mgr
Michael Fowler . . . . . prom dir
George Molnar . . . . . chief of engrg

***WNIT-TV—** Analog channel: 34. Digital channel: 35. Analog hrs: 24 Digital hrs: 24 708 kw vis, 77 kw aur. 530t/500g TL: N41 36 59 W86 11 43 (CP: 691.8 kw vis, ant 264t) On air date: Feb 14, 1974. Box 3434, Elkhart, IN, 46515. 2300 Charger Blvd., Elkhart, IN 46514. Phone: (574) 674-5961. Phone: (574) 675-9648. Fax: (574) 262-8497. E-mail: wnit@wnit.org Web Site: www.wnit.org. Licensee: Michiana Public Broadcasting Corp. Population served: 478,000 Natl. Network: PBS. Washington Atty: Dow, Lohnes & Albertson.
Key Personnel:
Amy Cassidy . . . . . CFO
Mary Pruess . . . . . pres & gen mgr
Brian Hoover . . . . . opns dir
Jim Faucett . . . . . dev VP & dev dir

WSBT-TV— Analog channel: 22. Analog hrs: 24 4,790 kw vis, 479 kw aur. 1,070t/1,047g TL: N41 37 00 W86 13 01 On air date: Dec 21, 1952. 300 W. Jefferson Blvd., South Bend, IN, 46601. Phone: (574) 233-3141. Fax: (574) 288-6630. E-mail: wsbtnews@wsbt.com Web Site: www.wsbt.com. Licensee: WSBT Inc. Group owner: Schurz Communications Inc., see Cross-Ownership Population served: 771,300 Natl. Network: CBS. Natl. Rep: Harrington, Righter & Parsons. Washington Atty: Wilmer Hale. News staff: 40; News: 22 hrs wkly.
Key Personnel:
John Mann . . . . . pres & gen mgr
Bob Johnson . . . . . opns dir & progmg dir
Meg Sauer . . . . . opns dir & news dir
Scott Leiter . . . . . prom mgr

WSJV—Elkhart, Analog channel: 28. Analog hrs: 24 5,000 kw vis, 500 kw aur. ant 1,086t/1,045g TL: N41 36 58 W86 11 38 On air date: Mar 15, 1954. Box 28, South Bend, IN, 46624. Phone: (574) 679-9758. Fax: (574) 294-1267. E-mail: fox28@fox28.com Web Site: www.fox28.com. Licensee: WSJV Television Inc. Group owner: Quincy Newspapers Inc., see Cross-Ownership (acq 3-31-75; $3.2 million; FTR: 4-14-75). Population served: 774,000 Natl. Network: Fox. Washington Atty: Wilkinson, Barker, Knauer & Quinn. News staff: 22; News: 16 hrs wkly.
Key Personnel:
Thomas A. Oakley . . . . . pres
Stephen Morris . . . . . gen mgr

Terre Haute, IN
(DMA 151)

WFXW-TV— Analog channel: 38. Digital channel: 39. Analog hrs: 24 2,140 kw vis, 214 kw aur. ant 976t/1,004g TL: N39 13 58 W87 23 49 On air date: Apr 3, 1973. Box 299, Terre Haute, IN, 47808. 10849 US Hwy. 41, Terre Haute, IN 47850. Phone: (812) 696-2121. Fax: (812) 696-2755. Web Site: www.mywabashvalley.com. Licensee: Mission Broadcasting Inc. Group owner: (group owner). Population served: 70,286 Natl. Network: Fox. News: 2 hrs wkly.
Key Personnel:
Duane Lammers . . . . . gen mgr
Lois Mathis . . . . . stn mgr
Chris Collins . . . . . gen sls mgr
Chris O'Nea; . . . . . prom mgr
Tom McClanahan . . . . . gen mgr & news dir
Bruce Yowell . . . . . chief of engrg

WTHI-TV— Analog channel: 10. Digital channel: 24. Analog hrs: 24 Digital hrs: 24 316 kw vis, 31.6 kw aur. ant 960t/993g TL: N39 14 36 W87 23 07 On air date: July 22, 1954. PO Box 1486, Terre Haute, IN, 47808. 918 Ohio St., Terre Haute, IN 47807. Phone: (812) 232-9481. Fax: (812) 232-8953. Web Site: www.wthitv.com. Licensee: Indiana Broadcasting LLC. Group owner: Emmis Communications Corp. (acq 11-30-2005; grpsl). Population served: 164,800 Natl. Network: CBS. Natl. Rep: Petry Television Inc.

WTWO-TV— Analog channel: 2. Digital channel: 36. Analog hrs: 24 100 kw vis, 19.5 kw aur. 950t/999g TL: N39 14 33 W87 23 29 On air date: Sept 1, 1965. Box 299, Terre Haute, IN, 47808. 10849 N. U.S. Hwy. 41, Farmersburg, IN 47850. Phone: (812) 696-2121. Fax: (812) 696-2755. E-mail: station@wtwo.com Web Site: www.mywabashvalley.com. Licensee: Nexstar Broadcasting Inc. Group owner: Nexstar Broadcasting Inc. (acq 2-14-97; with KQTV(TV) Saint Joseph, MO). Population served: 162,320 Natl. Network: NBC. Washington Atty: Drinker, Biddle & Reath LLP. News staff: 25; News: 20 hrs wkly.
Key Personnel:
Duane Lammers . . . . . gen mgr
Richard Haddox . . . . . stn mgr
Derek Brown . . . . . gen sls mgr
Chris Collins . . . . . rgnl sls mgr
Chris O'Neal . . . . . prom mgr
Tom McClanahan . . . . . news dir
Bruce Yowell . . . . . chief of engrg

***WUSI-TV—** Analog channel: 16. Digital channel: 19. Analog hrs: 24 Digital hrs: 24 977 kw vis, 195 kw aur. ant 930t/976g TL: N38 50 18 W88 07 46 On air date: Aug 19, 1968. 1100 Lincoln Dr., Suite 1003, SUI Mailcode 6602, Carbondale, IL, 62901. 1003 Communications Bldg., Caarbondale, IL 62901-6602. Phone: (618) 453-4343. Fax: (618) 453-6186. Web Site: www.wsiu.org. Licensee: Board of Trustees, Southern Illinois University. Population served: 151,180 Natl. Network: PBS. Washington Atty: Cohn & Marks. Wire Svc: AP
Key Personnel:
Dr. Candis S. Isberner . . . . . gen mgr & stn mgr
Robert Henderson . . . . . opns
Trina Thomas . . . . . progmg
Jack Hammer . . . . . engr
Rebroadcasts WSIU-TV Carbondale 100%.

***WVUT—** Analog channel: 22. Digital channel: 52. Analog hrs: 7am-11:30 pm 1,150 kw vis, 115 kw aur. 570t/560g TL: N38 39 06 W87 28 37 On air date: Feb 15, 1968. Davis Hall, 1200 N. 2nd St., Vincennes, IN, 47591. Phone: (812) 888-4345. Fax: (812) 882-2237. E-mail: wvut@vinu.edu Web Site: www.vubroadcastinng.org. Licensee: Board of Trustees for the Vincennes Univ. (acq 9-16-76; FTR: 10-11-76). Population served: 250,000 Natl. Network: PBS. Washington Atty: Fletcher, Heald & Hildreth. Wire Svc: UPI News staff: 4; News: 5 hrs wkly.
Key Personnel:
Al Rerko . . . . . gen mgr
Jill Ballinger . . . . . opns mgr
Sharon Keifer . . . . . progmg dir

Iowa

Ames

see Des Moines-Ames, IA market

Cedar Rapids-Waterloo-Iowa City & Dubuque, IA
(DMA 89)

KCRG-TV— Analog channel: 9. Digital channel: 52.316 kw vis, 63.2 kw aur. ant 1,988t/1,926g TL: N42 18 59 W91 51 31 On air date: Oct 15, 1953. Box 816, Cedar Rapids, IA, 52406-0816. 501 2nd Ave. S.E., Cedar Rapids, IA 52401. Phone: (319) 395-9999. Fax: (319) 398-8378. Web Site: www.kcrg.com. Licensee: Cedar Rapids TV Co. Ownership: The Gazette Co., 100% Group owner: The Gazette Co. (acq 8-12-54; $101,500; FTR: 8-23-54). Population served: 323,380 Natl. Network: ABC. Washington Atty: Wiley, Rein & Fielding.
Key Personnel:
Joseph F. Hladky III . . . . . CEO & pres
John Phelan III . . . . . gen mgr

KFXA— Analog channel: 28. Digital channel: 27. Analog hrs: 24 4,470 kw vis. ant 1,473t/1,486g TL: N42 05 25 W92 05 13 On air date: August 1995. Box 3131, Cedar Rapids, IA, 52406-3131. Phone: (319) 393-2800. Fax: (319) 395-7028. E-mail: kfxa@kfxa.tv Web Site: www.kfxa.tv. Licensee: Second Generation of Iowa Ltd. Ownership: Tom Embrescia, Larry Blum. Natl. Network: Fox. News staff: 6; News: 13.5 hrs wkly.
Key Personnel:
Larry Blum . . . . . pres
Joe Denk . . . . . gen mgr
Greg Stuart . . . . . opns mgr

KFXB— Analog channel: 40. Digital channel: 43. Analog hrs: 22 646 kw vis, 64.6 kw aur. ant 841t TL: N42 31 05 W90 37 16 On air date: Sept 12, 1976. 744 Main St., Dubuque, IA, 52001. Phone: (563) 690-1704. Fax: (563) 557-9383. Web Site: www.kfxb.net. Licensee: Christian Television Network of Iowa Inc. Group owner: (group owner; (acq 8-2-2004). Population served: 90,000 Natl. Network: Fox. Natl. Rep: Millennium Sales & Marketing. News staff: 12; News: 5 hrs wkly.

KGAN-TV— Analog channel: 2. Analog hrs: 24 100 kw vis, 20 kw aur. ant 1,450t/1,355g TL: N42 17 39 W91 53 10 On air date: Sept 30, 1953. Box 3131, Cedar Rapids, IA, 52406. 600 Old Marion Rd. N.E., Cedar Rapids, IA 52406. Phone: (319) 395-9060. Fax: (319) 395-0987. E-mail: kgan@kgan.com Web Site: www.kgan.com. Licensee: KGAN Licensee L.L.C. Group owner: Sinclair Broadcast Group Inc. (acq 1999; grpsl). Population served: 117,040 Natl. Network: CBS. Natl. Rep: Millennium Sales & Marketing. Washington Atty: Shaw Pittman. Wire Svc: AP Wire Svc: CBS News staff: 35; News: 12 hrs wkly.
Key Personnel:
Michael Sullivan . . . . . gen mgr & stn mgr
Ruth Barnett . . . . . opns dir

***KIIN-TV—** Analog channel: 12. Digital channel: 45.316 kw vis, 31.6 kw aur. 1,440t/1,449g TL: N41 43 14 W91 20 29 On air date: Feb 8, 1970. Box 6450, Iowa Public TV, Johnston, IA, 50131-6450. 6450 Corporate Dr., Johnston, IA 50131. Phone: (515) 242-3100. E-mail: public_information@iptv.org Web Site: www.iptv.org. Licensee: Iowa Public Broadcasting Board. Natl. Network: PBS. Washington Atty: Dow, Lohnes PLLC.

KPXR— Analog channel: 48. Analog hrs: 24 5,000 kw vis, 500 kw aur. 466t TL: N42 04 51 W91 41 45 (CP: 2,920 kw vis, ant 1,060t) On air date: May 3, 1997. 1957 Blairs Ferry Rd. N.E., Cedar Rapids, IA, 52402-5819. Phone: (319) 378-1260. Fax: (319) 378-0076. Web Site: www.ionline.tv. Licensee: Paxson Communications License Co. L.L.C. Group owner: Paxson Communications Corp. (acq 7-15-97; $5 million). Population served: 442,400

***KRIN—** Analog channel: 32. Digital channel: 35.5,000 kw vis, 500 kw aur. ant 1,851t/1,795g TL: N42 18 59 W91 51 31 On air date: Dec 15, 1974. Box 6450, Iowa PublicTV, Johnston, IA, 50131-6450. 6450 Corporate Dr., Johnston, IA 50131-6450. Phone: (515) 242-3100. E-mail: public_information@iptv.org Web Site: www.iptv.org. Licensee: Iowa Public Broadcasting Board. Natl. Network: PBS. Washington Atty: Dow, Lohnes PLLC.

KWKB— Analog channel: 20. Digital channel: 25. Analog hrs: 24 5,000 kw vis. ant 1,446t/1,460g TL: N41 43 28 W91 21 07 On air date: 1999. 1547 Baker Ave., West Branch, IA, 52358. Phone: (319) 643-5952. Fax: (319) 643-3124. E-mail: wb20@kwkb.com Web Site: www.kwkb.com. Licensee: KM Television of Iowa L.L.C. Natl. Network: CW, MyNetworkTV.
Key Personnel:
Mark Robbins . . . . . gen mgr
Jeff Hoffman . . . . . opns mgr & chief of engrg
Jim Walker . . . . . rgnl sls mgr
Trish Wethington . . . . . progmg dir & progmg dir

KWWF— Analog channel: 22. Analog hrs: 24 500 kw vis. ant 92t/174g TL: N42 29 51 W92 20 07 On air date: April 4, 2003. 1 Shackleford Dr., Little Rock, AR, 72211. 501 Sycamore St., Suite 710, Waterloo, IA 50703. Phone: (319) 287-5841. Licensee: EBC Waterloo Inc. Group owner: Equity Broadcasting Corp. (acq 8-6-2004; $5 million with WNYI(TV) Ithaca, NY).

KWWL—Waterloo, Analog channel: 7. Digital channel: 55. Analog hrs: 24 Digital hrs: 24 316 kw vis, 27 kw aur. 1,980t/2,000g TL: N42 24 04 W91 50 43 On air date: November 1953. 500 E. 4th St., Huntsville, IA, 50703. Phone: (319) 291-1200. Fax: (319) 274-0466. E-mail: kwwl@kwwl.com Web Site: www.kwwl.com. Licensee: KWWL Television Inc. Group owner: Raycom Media Inc. (acq 7-1-2006; $63 million). Population served: 322,400 Natl. Network: NBC. Natl. Rep: Blair Television. Washington Atty: Covington & Burling. Wire Svc: AP News staff: 40; News: 22 hrs wkly.
Key Personnel:
Chris Hussey . . . . . mktg mgr
Don Morehead . . . . . mus dir
Jon Okerstrom . . . . . news dir
Kim Leer . . . . . opns

Council Bluffs

see Omaha, NE market

Davenport, IA-Rock Island-Moline, IL
(DMA 96)

KGCW-TV— Analog channel: 26. Digital channel: 41. Analog hrs: 24 54.3 kw vis, 5.43 kw aur. ant 315t TL: N40 49 25 W91 08 22 On air date: Jan 6, 1988. 937 E. 53rd St., Davenport, IA, 52807. Phone: (563) 386-1818. Fax: (563) 386-8543. Web Site: www.kgcwtv.com. Licensee: Burlington Television Acquisition Licensing LLC. (acq 1995; $400,000). Population served: 590,000 Natl. Network: CW. Natl. Rep: TeleRep. Washington Atty: Wilkinson, Barker, Knauer & Quinn.
Key Personnel:
Kathy DeBoeuf . . . . . stn mgr
Jaime Horowitz . . . . . gen sls mgr
Tony Wilkins . . . . . natl sls mgr & rgnl sls mgr
Tim Emmerson . . . . . prom mgr

John Bain. . . . . progmg mgr

KLJB-TV— Analog channel: 18. Digital channel: 49. Analog hrs: 24 Digital hrs: 24 3,000 kw vis, 300 kw aur. ant 1,010t/993g TL: N41 19 17 W90 22 47 (CP: Ant 989t/942g) On air date: July 28, 1985. 937 E. 53rd St., Suite D, Davenport, IA, 52807. Phone: (563) 386-1818. Fax: (563) 386-8543. Web Site: www.kljb.com. Licensee: Quad Cities Television Acquisition Licensing LLC. Population served: 737,000 Natl. Network: Fox. Natl. Rep: TeleRep. Washington Atty: Wilkinson, Barker, Knauer & Quinn. News: 3 hrs wkly.
Key Personnel:
Kathy DeBoeuf . . . . stn mgr & opns mgr
Jaime Horowitz. . . . . gen sls mgr
Tony Wilkins. . . . . natl sls mgr & rgnl sls mgr
Tim Emmerson. . . . . prom mgr
John Bain. progmg mgr

***KQCT—** Analog channel: 36.6.76 kw vis, 676 w aur. 213t TL: N41 31 58 W90 34 40 On air date: December 1991. 6600 34th Ave., Moline, IA, 61265. Phone: (309) 796-2424. Fax: (309) 796-2484. E-mail: wqpt@bhc.edu Web Site: www.wqpt.org. Licensee: Iowa Public Broadcasting Board. (acq 7-2-0202; $200,000).
Key Personnel:
Rick Best . . . . gen mgr
Lora Adams . . . . dev dir
Lora Adams . . . . mktg dir
Jerry Myers . . . . progmg mgr
Steve Ellis. . . . . chief of engrg
Satellite of *WQPT-TV Moline IL 100%.

KWQC-TV— Analog channel: 6. Digital channel: 56. Analog hrs: 24 Digital hrs: 24 100 kw vis, 15.1 kw aur. ant 940t/978g TL: N41 32 49 W90 28 35 On air date: Oct 31, 1949. 805 Brady St., Davenport, IA, 52803. Phone: (563) 383-7000. Fax: (563) 383-7129. Web Site: www.kwqc.com. Licensee: Young Broadcasting of Davenport Inc. Group owner: Young Broadcasting Inc. (acq 4-15-96; $55 million). Population served: 98,469 Natl. Network: NBC. Natl. Rep: Adam Young. Washington Atty: Wiley, rein & Fielding.
Key Personnel:
Cathie Whiteside. . . . . stn mgr
John Hegeman. . . . . opns mgr
Jeff Glass. . . . . rgnl sls mgr
Trish Tague. mktg dir
Jeff Bilyeu . . . . prom mgr
April Samp. news dir
Doug Bierman . . . . chief of engrg

WHBF-TV— Analog channel: 4. Analog hrs: 5 AM-2 AM 100 kw vis, 10 kw aur. 1,342t/1,383g TL: N41 32 49 W90 28 35 On air date: July 1, 1950. 231 18th St., Rock Island, IL, 61201. Phone: (309) 786-5441. Fax: (309) 788-4975. E-mail: sales@cbs4qc.com Web Site: www.cbs4qc.com. Licensee: Coronet Communications Co. Group owner: Citadel Communications Co. Ltd. (acq 3-16-87; grpsl; FTR: 11-17-86). Population served: 312,000 Natl. Network: CBS. Natl. Rep: Continental Television Sales. Washington Atty: Latham & Watkins. News staff: 22; News: 7 hrs wkly.
Key Personnel:
J.D. Walls. . . . . pres, opns dir & progmg
Martha Huggins . . . . CFO, VP & gen mgr
Todd Grady . . . . natl sls mgr
Patty Gilbert. . . . . prom mgr
Arthur Steadman . . . . news dir
Ron Schmidt. . . . . chief of engrg
Steve Garman. . . . . sls

WQAD-TV— Analog channel: 8. Analog hrs: 24 282 kw vis, 23.2 kw aur. 1,010t/1,066g TL: N41 18 44 W90 22 47 On air date: Aug 1, 1963. 3003 Park 16th St., Moline, IL, 61265. Phone: (309) 764-8888. Fax: (309) 764-5763. E-mail: wqad@wqad.com Web Site: www.wqad.com. Licensee: Local TV Illinois License LLC. Group owner: The New York Times Co. (see Cross-Ownership). (acq 5-7-2007; grpsl). Population served: 322,200 Natl. Network: ABC. News staff: 75; News: 22.5 hrs wkly.
Key Personnel:
Dale R. Woods . . . . pres & gen mgr
Trent Poindexter. . . . . VP & gen sls mgr
Leigh Geramanis . . . . news dir
Rick Serre. . . . . chief of engrg

***WQPT-TV—** Analog channel: 24. Digital channel: 23. Analog hrs: 18 Digital hrs: 18 148 kw vis, 14.8 kw aur. 320t/355g TL: N41 28 31 W90 26 50 (CP: 1,200 kw vis, ant 905t) On air date: Nov 3, 1983. 6600 34th Ave., Moline, IL, 61265. Phone: (309) 796-2424. Fax: (309) 796-2484. E-mail: bestr@bhc.edu Web Site: www.wqpt.org. Licensee: Black Hawk College. Population served: 500,000 Natl. Network: PBS. Washington Atty: Drinker, Biddle & Reath.
Key Personnel:
Rick Best . . . . gen mgr
Lora Adams . . . . dev dir & mktg dir
Jerry Myers . . . . progmg mgr
Steve Ellis . . . . chief of engrg

Des Moines-Ames, IA (DMA 73)

KCCI— Analog channel: 8. Digital channel: 31. Analog hrs: 24 Digital hrs: 24 316 kw vis, 31.6 kw aur. 1,953t/1,997g TL: N41 48 35 W93 37 16 On air date: July 31, 1955. 888 9th St., Des Moines, IA, 50309-1288. Phone: (515) 247-8888. Fax: (515) 244-0202. Fax: (515) 471-8910. E-mail: Web Site: www.kcci.com. Licensee: KCCI Television Inc. Group owner: Hearst-Argyle Television Inc. (acq 1999; grpsl). Natl. Network: CBS. Natl. Rep: Eagle Television Sales. Washington Atty: Brooks, Pierce, McLendon, Humprey & Leonard, LLP. News staff: 50; News: 30 hrs wkly.
Key Personnel:
Paul Fredericksen . . . . pres & gen mgr
Bob Day . . . . opns dir
Dave Porepp . . . . gen sls mgr
Anne Marie Caudron . . . . natl sls mgr
Nanci Elder. mktg dir
Dave Busiek . . . . news dir
Steve Houg . . . . chief of engrg

KCWI-TV— Analog channel: 23. Analog hrs: 24 5,000 kw vis. ant 2,011t TL: N41 49 47 W93 36 56 On air date: Jan 20, 2001. 2701 S.E. Convenience Blvd., Suite 1, Ankeny, IA, 50021. Phone: (515) 964-2323. Fax: (515) 965-6900. E-mail: yourstation@kcwi23.com Web Site: www.kcwi23.com. Licensee: KPWB License LLC. Group owner: Pappas Telecasting Companies. Natl. Network: CW.
Key Personnel:
Ted Stephens. gen mgr
Jim Cordero . . . . opns mgr

***KDIN-TV—** Analog channel: 11. Digital channel: 50.316 kw vis, 31.6 kw aur. 1,973t/2,000g TL: N41 48 33 W93 36 53 On air date: Apr 27, 1959. Box 6450, Iowa Public TV, Johnston, IA, 50131-6450. 6450 Corporate Dr., Johnston, IA 50131. Phone: (515) 242-3100. E-mail: public_information@iptv.org Web Site: www.iptv.org. Licensee: Iowa Public Broadcasting Board. Population served: 360,000 Natl. Network: PBS. Washington Atty: Dow, Lohnes PLLC. Wire Svc: NWS (National Weather Service)

KDMI— Digital channel: 56.1,000 kw vis. ant 1,942t/2,893g TL: N41 49 47 W93 36 56 On air date: 2006. 2701 S.E. Convenience Blvd., Suite 1, Ankeny, IA, 50021. Phone: (515) 964-2323. Fax: (515) 965-6900. Licensee: KDMI License LLC. (acq 12-15-2005; $1 million for CP). Natl. Network: CW, MyNetworkTV.

KDSM-TV— Analog channel: 17. Digital channel: 16. Analog hrs: 24 Digital hrs: 24 3,020 kw vis, 311 kw aur. 1,516t/1,503g TL: N41 48 01 W93 36 27 On air date: 1983. 4023 Fleur Dr., Des Moines, IA, 50321. Phone: (515) 287-1717. Fax: (515) 287-0064. E-mail: programming@kdsm17.com Web Site: www.kdsm.com. Licensee: KDSM Licensee L.L.C. Group owner: Sinclair Broadcast Group Inc. Population served: 249,413 Natl. Network: Fox. Natl. Rep: Millennium Sales & Marketing. Washington Atty: Dow, Lohnes & Albertson. News staff: 7; News: 4 hrs wkly.
Key Personnel:
Mike Wilson . . . . VP & gen mgr
Beth Grant. opns VP
Eric Johnson . . . . gen sls mgr
Carolyn Lawrence. . . . . natl sls mgr
Julie Quick-Alcorn . . . . progmg dir
Doug Hammond . . . . chief of engrg

KEFB— Analog channel: 34.87.1 kw vis. ant 492t/459g TL: N41 58 49 W93 44 23 On air date: 2005. Box 201, Huxley, IA, 50124-0201. Phone: (515) 597-3138. Licensee: Family Educational Broadcasting Inc.

KFPX— Analog channel: 39. Analog hrs: 24 4,440 kw vis. ant 505t TL: N41 49 05 W93 12 32 On air date: 1998. 4570 114th St., Urbandale, IA, 50322. Phone: (515) 331-3939. Fax: (515) 331-1312. Web Site: www.ionline.tv. Licensee: Paxson Des Moines License Inc. Group owner: Paxson Communications Corp. Natl. Network: i Network.
Key Personnel:
Doug Bognar . . . . rgnl sls mgr
Dave Ohmstede . . . . chief of engrg
Marsha Theis . . . . progmg

***KTIN—** Analog channel: 21. Digital channel: 25.1,580 kw vis, 158 kw aur. 1,160t/1,206g TL: N42 49 02 W94 24 40 On air date: Apr 8, 1977. Box 6450, Iowa Public TV, Johnston, IA, 50131-6450. 6450 Corporate Dr., Johnston, IA 50131. Phone: (515) 242-3100. E-mail: public_information@iptv.org Web Site: www.iptv.org. Licensee: Iowa Public Broadcasting Board. Natl. Network: PBS. Washington Atty: Dow, Lohnes PLLC.

WHO-TV— Analog channel: 13. Digital channel: 19. Analog hrs: 24 Digital hrs: 19 316 kw vis, 47.9 kw aur. ant 1,970t/2,000g TL: N41 48 33 W93 36 53 On air date: Apr 15, 1954. 1801 Grand Ave., Des Moines, IA, 50309. Phone: (515) 242-3500. Fax: (515) 242-3743. Fax: (515) 242-3796 (news). Web Site: www.whotv.com. Licensee: Local TV Iowa License LLC. Group owner: The New York Times Co. (acq 5-7-2007; grpsl). Natl. Network: NBC. Natl. Rep: Millennium Sales & Marketing. Washington Atty: Covington & Burling. News staff: 50; News: 30 hrs wkly.
Key Personnel:
Robert L. Lawrence . . . . pres
Rebecca Jess. . . . . VP
James Boyer . . . . gen mgr
Ross Reardon . . . . gen sls mgr
Tim Gardner. . . . . prom mgr
Rod Petersen. . . . . news dir
Brad Olk . . . . chief of engrg

WOI-TV—Ames, Analog channel: 5. Analog hrs: 5 AM-2 AM 100 kw vis, 20 kw aur. 1,850t/2,000g TL: N41 48 33 W93 36 53 On air date: Feb 21, 1950. 3903 Westown Pkwy., West Des Moines, IA, 50266. Phone: (515) 457-9645. Fax: (515) 457-1034. E-mail: info@myabc5.com Web Site: www.myabc5.com. Licensee: Capital Communications Co. Inc. Group owner: Citadel Communications Company Ltd., Coronet Communications Co. (acq 3-1-94; $12.7 million). Population served: 559,700 Natl. Network: ABC. Natl. Rep: Continental Television Sales. Washington Atty: Latham & Watkins. Wire Svc: AP News staff: 25; News: 15 hrs wkly.
Key Personnel:
Philip J. Lombardo. . . . . CEO & chmn
Ray Cole . . . . pres & gen mgr
Randy Shelton . . . . opns dir

Dubuque

see Cedar Rapids-Waterloo-Iowa City & Dubuque, IA market

Iowa City

see Cedar Rapids-Waterloo-Iowa City & Dubuque, IA market

Keokuk

see Quincy, IL-Hannibal, MO-Keokuk, IA market

Mason City

see Rochester, MN-Mason City, IA-Austin, MN market

Ottumwa, IA-Kirksville, MO (DMA 199)

KTVO—(Kirksville, MO) Analog channel: 3. Digital channel: 33. Analog hrs: 24 100 kw vis, 14.3 kw aur. ant 1,112t/1,050g TL: N40 31 47 W92 26 29 On air date: Nov 21, 1955. Box 949, Hwy. 63 N., Kirksville, MO, 63501. 15518 Hwy. 63 N., Kirksville, MO 63501. Phone: (660) 627-3333. Phone: (641) 682-3333. Fax: (660) 627-1885. Fax: (641) 682-1572. Web Site: www.ktvo.com. Licensee: Barrington Kirksville License LLC. Group owner: Raycom Media Inc. (acq 8-11-2006; grpsl). Population served: 44,500 Natl. Network: ABC. Natl. Rep: Harrington, Righter & Parsons. Washington Atty: Covington & Burling. News staff: 16; News: 14 hr wkly.
Key Personnel:
Crystal Amini-Rad . . . . gen mgr
Merle Snyder . . . . natl sls mgr
Melissa Billington. . . . . progmg mgr
Marlene Speas . . . . news dir
Teresa Johnson . . . . mktg dir & news dir
John Wise. . . . . chief of engrg

KYOU-TV— Analog channel: 15. Digital channel: 14. Analog hrs: 24 Digital hrs: 24 2090 kw vis, 209 kw aur. 1,200t TL: N41 11 42 W91 57 15 On air date: June 29, 1987. 820 W. 2nd St., Ottumwa, IA, 52501. Phone: (641) 684-5415. Fax: (641) 682-5173. E-mail: reception@kyoutv.com Web Site: www.kyoutv.com. Licensee: Ottumwa Media Holdings LLC. Ownership: Thomas B. Henson, 90%; and Macon B. Moye, 10% (acq 12-15-2003; $4 million). Natl. Network: Fox. Natl. Rep: MMT. Washington Atty: Covington & Burling.
Key Personnel:
Dianne Little . . . . gen mgr & gen sls mgr
Dave Cecil . . . . opns mgr
Phil Benjamin . . . . chief of engrg

Red Oak

see Omaha, NE market

Sioux City, IA (DMA 143)

KCAU-TV— Analog channel: 9. Analog hrs: 20-21 245 kw vis, 49 kw aur. 2,020t/2,000g TL: N42 35 12 W96 13 57 On air date: Mar 28, 1953. 625 Douglas St., Sioux City, IA, 51101. Phone: (712) 277-2345. Fax: (712) 277-3733. Web Site: www.kcautv.com. Licensee: Citadel Communications Co. Ltd. Group owner: Citadel Communications Co. Ltd., Coronet Communications Co. (acq 10-1-85; $15 million). Population served: 166,000 Natl. Network: ABC. Washington Atty: Latham & Watkins.
Key Personnel:
Roger Moody . . . VP, VP & gen mgr
Brent Nelson . . . progmg dir
Dan Ackerman . . . chief of engrg

KMEG— Analog channel: 14. Digital channel: 39. Analog hrs: 20 280 kw vis, 75.9 kw aur. ant 1,152t/1,000g TL: N42 30 53 W96 18 13 On air date: Sept 5, 1967. 100 Gold Cir., Dakota Dunes, SD, 57049. Phone: (712) 277-3554. Fax: (712) 277-4732. E-mail: kmegtv@kmeg.com Web Site: www.kmegtv.com. Licensee: Waitt Broadcasting Inc. Ownership: Waitt Media Inc., 100% of total assets Group owner: (group owner; (acq 6-23-98; $12.25 million). Population served: 289,800 Natl. Network: CBS. Natl. Rep: Harrington, Righter & Parsons. Washington Atty: Wilkinson, Barker, Knauer & Quinn.
Key Personnel:
Norman Waitt Jr. . . . CEO & pres
Steve Seline . . . chmn
Mike Delich . . . pres
John Schuele . . . CFO
Scott Eymer . . . gen mgr
Paul Miller . . . stn mgr
Mary Ann Johnson . . . opns mgr & gen sls mgr

KPTH— Analog channel: 44. Digital channel: 49. Analog hrs: 24 Fox is on KPTH(TV) ch 44, MYNetworkTV is on KPTH-DT ch 49. 5,000 kw vis. ant 1,948t TL: N42 35 16 W96 13 22 On air date: May 9, 1999. 100 Gold Cir., North Sioux City, SD, 57049. Phone: (402) 241-4400. Fax: (402) 241-4444/(402) 241-4046. E-mail: yourstation@kpth.com Web Site: www.kpth.com. Licensee: Pappas Telecasting of Sioux City L.P. (a DE limited partnership). Group owner: Pappas Telecasting Companies Population served: 280,000 Natl. Network: Fox, MyNetworkTV Digital Network: Fox is on KPTH(TV) ch 44, MYNetworkTV is on KPTH-DT ch 49. Natl. Rep: Harrington, Righter & Parsons.
Key Personnel:
Howard Shrier . . . CEO
Harry Pappas . . . chmn
Scott Eymer . . . gen mgr
Mary Johnson . . . gen sls mgr
Mary Ann Johnson . . . rgnl sls mgr & mktg mgr
Ed Bok . . . chief of engrg

***KSIN—** Analog channel: 27. Digital channel: 28.4,070 kw vis, 407 kw aur. 1,070 TL: N42 30 53 W96 18 13 On air date: Jan 4, 1975. Box 6450, Iowa Public TV, Johnston, IA, 50131-6450. 6450 Corporate Dr., Joohnston, IA 50131. Phone: (515) 242-3100. E-mail: public_information@iptv.org Web Site: www.iptv.org. Licensee: Iowa Public Broadcasting Board. Natl. Network: PBS. Washington Atty: Dow, Lohnes PLLC.

KTIV— Analog channel: 4. Digital channel: 41. Analog hrs: 24 Digital hrs: 24 Note: NBC is on KTIV(TV) ch 4, CW is on KTIV-DT ch 41. 100 kw vis, 20 kw aur. 1,920t/2,000g TL: N42 35 12 W96 13 57 On air date: Oct 9, 1954. 3135 Floyd Blvd., Sioux City, IA, 51108. Phone: (712) 239-4100. Fax: (712) 239-2621. E-mail: ktiv4@ktiv.com Web Site: www.ktiv.com. Licensee: KTIV Television Inc. Group owner: Quincy Newspapers Inc., see Cross-Ownership (acq 11-20-89). Population served: 156,950 Natl. Network: NBC, CW Digital Network: Note: NBC is on KTIV(TV) ch 4, CW is on KTIV-DT ch 41. Natl. Rep: Blair Television. Washington Atty: Wilkinson, Barker, Knauer & Quinn. News: 19 hrs wkly.
Key Personnel:
Jerry Watson . . . VP
Adrian Wisner . . . gen mgr & gen sls mgr
David Madsen . . . stn mgr
Bridget Breen . . . news dir

***KXNE-TV—** Analog channel: 19. Digital channel: 16. Analog hrs: 18 Digital hrs: 18 1,682.67 kw vis. 1,141t/1093g TL: N42 14 15 W97 16 41 (CP: 776 kw vis, ant 1,122t/1,149g) On air date: Nov 10, 1967. 1800 N. 33rd St., Lincoln, NE, 68503. Phone: (402) 472-3611. Fax: (402) 472-1785. E-mail: net1@unl.edu Web Site: www.netnebraska.org. Licensee: Nebraska Educational Telecommunications Commission. Natl. Network: PBS. Washington Atty: Dow, Lohnes & Albertson. News staff: 3; News: 30 min.wkly.
Key Personnel:
Rod Bates . . . gen mgr
Steven Graziano . . . prom mgr
Steven Graziano . . . adv mgr & progmg mgr
Satellite of *KUON-TV Lincoln.

Waterloo

see Cedar Rapids-Waterloo-Iowa City & Dubuque, IA market

Kansas

Hutchinson Plus

see Wichita-Hutchinson Plus, KS market

Lawrence

see Kansas City, MO market

Pittsburg

see Joplin, MO-Pittsburg, KS market

Topeka, KS (DMA 138)

KSNT-TV— Analog channel: 27. Digital channel: 28. Analog hrs: 24 Digital hrs: 24 912 kw vis, 138 kw aur. 1,050t/1,149g TL: N39 05 34 W95 47 04 On air date: Dec 28, 1967. Box 2700, Topeka, KS, 66601. 6835 N.W. Hwy. 24, Topeka, KS 66618. Phone: (785) 582-4000. Fax: (785) 582-5283. Fax: (785) 582-4783. E-mail: 27news@ksnt.com Web Site: www.ksnt.com. Licensee: Montecito Television License Corp. of Topeka. Group owner: Emmis Communications Corp. (acq 1-27-2006; grpsl). Population served: 294,000 Natl. Network: NBC, CW. Natl. Rep: Harrington, Righter & Parsons. Wire Svc: AP News staff: 30; News: 25 hrs wkly.
Key Personnel:
Matt Broxterman . . . VP & rgnl sls mgr
Jean Turnbough . . . gen mgr, gen sls mgr & natl sls mgr
Nate Hill . . . prom dir & news dir
Charlie Good . . . chief of engrg

KTKA-TV— Analog channel: 49. Digital channel: 48. Analog hrs: 24 2,690 kw vis. ant 1,486t/1,415g TL: N39 01 34 W95 55 01 On air date: June 19, 1983. Box 4949, Topeka, KS, 66604. 2121 S.W. Chelsea Dr., Topeka, KS 66614. Phone: (785) 273-4949. Fax: (785) 273-7811. E-mail: 49email@ktka.tv Web Site: www.ktka.tv. Licensee: Free State Communications LLC. Ownership: Orbiter LLC, 100% (acq 8-17-2005; $6.2 million). Natl. Network: ABC. Natl. Rep: Millennium Sales & Marketing. Washington Atty: Cohn & Marks. News staff: 21; News: 15 hrs wkly.
Key Personnel:
Ann Niccum . . . prom mgr
Angie Cox . . . progmg dir
Ike Walker . . . news dir
Kathy Mohn . . . gen mgr & engrg mgr

***KTWU—** Analog channel: 11. Digital channel: 23. Analog hrs: 24 Digital hrs: 24 316 kw vis, 31.6 kw aur. ant 991t/903g TL: N39 03 50 W95 45 49 On air date: Oct 21, 1965. 1700 S.W. College Ave, Topeka, KS, 66621. Phone: (785) 670-1111. Fax: (785) 670-1112. E-mail: ktwu-press@lists.washburn.edu Web Site: ktwu.washburn.edu. Licensee: Washburn University of Topeka. Population served: 151,000 Natl. Network: PBS.
Key Personnel:
Eugene Williams . . . gen mgr
Cindy Barry . . . dev dir
Kevin Goodman . . . mktg dir
Val VanDerSluis . . . progmg dir
Duane Loyd . . . chief of engrg

WIBW-TV— Analog channel: 13. Digital channel: 44. Analog hrs: 24 Digital hrs: 24 Note: CBS is on WIBW-TV ch 13, MyNetworkTV is on WIBW-DT ch 44. 204 kw vis, 40.7 kw aur. 1,380t/1,255 TL: N39 00 19 W96 02 58 On air date: Nov 15, 1953. 631 S.W. Commerce Pl., Topeka, KS, 66615. Phone: (785) 272-6397. Fax: (785) 272-0117. E-mail: 13news@wibw.com Web Site: www.wibw.com. Licensee: WEAU Licensee Corp. Group owner: Gray Television Inc. (acq 8-29-2002; grpsl). Population served: 850,000 Natl. Network: CBS, MyNetworkTV Digital Network: Note: CBS is on WIBW-TV ch 13, MyNetworkTV is on WIBW-DT ch 44. Natl. Rep: Continental Television Sales. Washington Atty: Covington & Burling. News staff: 26; News: 30.5 hrs wkly.
Key Personnel:
Jim Ogle . . . gen mgr
Mark Doan . . . opns dir & chief of engrg
Lisa Chapman . . . rgnl sls mgr
Sharon Cole . . . progmg mgr
Jon Janes . . . news dir

Wichita-Hutchinson Plus, KS (DMA 67)

KAAS-TV— Analog channel: 18. Digital channel: 17. Analog hrs: 24 238.3 kw vis, 23.83 kw aur. ant 663t/466g TL: N39 06 16 W97 36 30 On air date: April 1988. 316 N. West St., Wichita, KS, 67203. Phone: (316) 942-2424. Fax: (316) 942-8927. E-mail: programming@foxkansas.com Web Site: www.foxkansas.com. Licensee: Clear Channel Broadcasting Licenses Inc. Group owner: Clear Channel Communications Inc. (acq 8-2-90). Natl. Network: Fox. Washington Atty: Wiley, Rein & Fielding.
Key Personnel:
Kent Cornish . . . gen mgr
Jon Deeble . . . chief of opns
Jeff McClausland . . . sls dir & rgnl sls mgr
Jim Hanning . . . natl sls mgr
Shawn Wheat . . . prom mgr
Fatma Al-Tamim . . . progmg dir
Dave Caruso . . . chief of engrg
Satellite of KSAS-TV Wichita.

KAKE-TV— Analog channel: 10. Digital channel: 21. Analog hrs: 24 316 kw vis, 44.7 kw aur. 1,030t/1,079g TL: N37 46 54 W97 31 10 On air date: Oct 19, 1954. 1500 North West St., Wichita, KS, 67203. Phone: (316) 943-4221. Fax: (316) 943-5493. Web Site: www.kake.com. Licensee: Gray Television Licensee, Inc. Group owner: Gray Television Inc. (acq 8-29-2002; grpsl). Population served: 300,000 Natl. Network: ABC. Natl. Rep: Continental Television Sales. Washington Atty: Covington & Burling. News staff: 40; News: 16 hrs wkly.
Key Personnel:
Terry Cole . . . pres & gen mgr
Dave Grant . . . stn mgr & news dir
Patrick Myers . . . opns dir
Dan Wall . . . gen sls mgr
Bryan Frye . . . mktg dir & prom mgr

KBSD-TV— Analog channel: 6.100 kw vis, 10 kw aur. 720t/600g TL: N37 38 28 W100 20 40 On air date: July 24, 1957. 100 Airport Rd., Dodge City, KS, 67801. Phone: (620) 227-3121. Fax: (620) 225-1675. Web Site: www.kbsd6.com. Licensee: Sunflower Broadcasting Inc. Group owner: Media General Broadcast Group (acq 9-25-2006; grpsl). Population served: 80,000 Natl. Network: CBS.
Key Personnel:
Joan Barnett . . . gen mgr
Les Bach . . . stn mgr & chief of engrg
Tony Thompson . . . gen sls mgr
David Bell . . . prom dir
Laverne Goering . . . progmg dir
Michelle Gors . . . news dir
Rebroadcasts KWCH-(TV) Wichita 95%.

KBSH-TV— Analog channel: 7. Analog hrs: 24 316 kw vis, 33.6 kw aur. 710t/812g TL: N38 53 01 W99 20 15 On air date: Sept 1, 1958. 2300 Hall St., Hays, KS, 67601. Phone: (785) 625-5277. Fax: (785) 625-1161. Web Site: www.kbsh7.com. Licensee: Sunflower Broadcasting Inc. Group owner: Media General Broadcast Group (acq 9-25-2006; grpsl). Population served: 60,000 Natl. Network: CBS. Natl. Rep: Harrington, Righter & Parsons. Washington Atty: Dow, Lohnes & Albertson. News staff: 1; News: 24 hrs wkly.
Key Personnel:
Todd F. Schurz . . . pres
Joan Barrett . . . gen mgr
Brian McDonough . . . gen sls mgr
Ken Clifford . . . rgnl sls mgr
David Bell . . . mktg mgr
Laverne Goering . . . prom dir
Michelle Gors . . . progmg VP & news dir
Les Bach . . . chief of engrg
Rebroadcasts KWCH-TV Wichita 90%.

KBSL-TV— Analog channel: 10. Analog hrs: 24 316 kw vis, 56.2 kw aur. ant 990t/975 TL: N39 28 09 W101 33 20 On air date: Apr 28, 1959. Box 629, Goodland, KS, 67735. 3023 W. 31 St., Goodland, KS 67735. Phone: (785) 899-2321. Fax: (785) 899-3138. E-mail: kbsltv@eaglecom.net Web Site: www.kwch.com. Licensee: Sunflower Broadcasting Inc. Group owner: Media General Broadcast Group (acq 9-25-2006; grpsl). Natl. Network: CBS.
Key Personnel:
Joan Barrett . . . gen mgr
Brian McDonough . . . rgnl sls mgr

Dennis Massier . . . . chief of engrg
Satellite of KWCH-TV Hutchinson.

***KDCK—** Analog channel: 21. Digital channel: 21. Digital hrs: 18 190 kw vis. 250t TL: N37 49 30 W100 10 36 On air date: March 1998. Box 9, 604 Elm St., Bunker Hill, KS, 67626. Phone: (785) 483-6990. Fax: (785) 483-4605. E-mail: shptv@shptv.org Web Site: www.shptv.org. Licensee: Smoky Hills Public Television. Natl. Network: PBS. Washington Atty: Dow, Lohnes & Albertson.
Key Personnel:
Lawrence Holden . . . . CEO & gen mgr
Terry Cutler . . . . chief of opns & news dir
KOOD-TV.

KLBY— Analog channel: 4. Digital channel: 17. Analog hrs: 24 Digital hrs: 24 100 kw vis, 21 kw aur. ant 770t/3,420g TL: N39 15 09 W101 21 09 On air date: July 4, 1984. 2900 E. Schulman Ave., Garden City, KS, 67846. Phone: (620) 275-1560. Web Site: www.kake.com. Licensee: Gray Television Licensee Inc. Group owner: Gray Television Inc. (acq 8-29-2002; grpsl). Population served: 165,000 Natl. Network: ABC. Washington Atty: Covington & Burling. News staff: 2.
Satellite of KAKE-TV Wichita.

KMTW— Analog channel: 36. Digital channel: 35. Analog hrs: 24 Digital hrs: 24 3,467 kw vis. ant 1,063t TL: N37 56 23 W97 30 42 On air date: Jan 6, 2001. 316 N. West St., Wichita, KS, 67203. Phone: (316) 942-2424. Fax: (316) 942-8927. Web Site: www.mytvwichita.com. Licensee: Mercury Broadcasting Co. Inc. Ownership: Van H. Archer III, 100% Group owner: (group owner; (acq 7-1-2001). Population served: 718,125 Natl. Network: MyNetworkTV. Natl. Rep: Millennium Sales & Marketing. Washington Atty: Fletcher, Heald & Hildreth.
Key Personnel:
Kent Cornish . . . . gen mgr
Jeff Causland. sls dir
Jim Hanning . . . . natl sls mgr
Fatma Al-Tamim . . . . progmg dir
David Caruso . . . . chief of engrg

KOCW— Analog channel: 14. Analog hrs: 24 150 kw vis. ant 525t TL: N38 37 54 W98 50 52 On air date: 2001. 316 N. West St., Wichita, KS, 67203. Phone: (316) 942-2424. Fax: (316) 942-8927. E-mail: programming@foxkansas.com Web Site: www.foxkansas.com. Licensee: Clear Channel Broadcasting Licenses Inc. Group owner: Clear Channel Communications Inc. Natl. Network: Fox. Washington Atty: Wiley, Rein & Fielding.
Key Personnel:
Kent Cornish . . . . gen mgr
Jon Deeble. . . . . opns dir
Jeff McCousland . . . . sls dir
Jim Hanning . . . . natl sls mgr
Shawn Wheat. prom dir
Fatma Al-Tamim . . . . progmg dir
David Caruso . . . . engrg dir
Satellite of KSAS-TV Wichita 100%.

***KOOD—** Analog channel: 9. Digital channel: 16. Analog hrs: 18 Digital hrs: 18 316 kw vis, 31.6 kw aur. 2,959t/1,119g TL: N38 46 16 W98 44 17 On air date: Nov 10, 1982. Box 9, 604 Elm St., Bunker Hill, KS, 67626. Phone: (785) 483-6990. Fax: (785) 483-4605. Web Site: www.shptv.org. Licensee: Smoky Hills Public Television Corp. Natl. Network: PBS. Washington Atty: Dow, Lohnes & Albertson. Foreign lang progmg: SP 1
Key Personnel:
Jayne Heller . . . . CEO & dev mgr
Larry Holden . . . . gen mgr
Jane Habiger . . . . mktg dir
Mary-Pat Waymaster . . . . progmg dir
Terry Cutler . . . . chief of engrg
Rebroadcasts KOOD(TV), Hays, 100%.

***KPTS-TV—** Analog channel: 8. Digital channel: 29. Analog hrs: 24 229 kw vis, 22.9 kw aur. 800t/785g TL: N38 03 21 W97 46 35 On air date: Jan 7, 1970. 320 W. 21st St. N., Wichita, KS, 67203. Phone: (316) 838-3090. Fax: (316) 838-8586. E-mail: tv8@kpts.org Web Site: www.kpts.org. Licensee: Kansas Public Telecommunications Service Inc. (acq 1979). Population served: 120,250 Natl. Network: PBS. Washington Atty: Dow, Lohnes PLLC.
Key Personnel:
Don Checots . . . . gen mgr
Jesse Huxman. . . . . progmg mgr
Dave McClintock . . . . engrg dir

KSAS-TV— Analog channel: 24. Digital channel: 26. Analog hrs: 24 3,300 kw vis, 331 kw aur. ant 1,120t/1,165g TL: N37 46 40 W97 30 37 On air date: Aug 24, 1985. 316 N. West St., Wichita, KS, 67203. Phone: (316) 942-2424. Fax: (316) 942-8927. E-mail: programming@foxkansas.com Web Site: www.foxkansas.com. Licensee: Clear Channel Broadcasting Licenses Inc. Group owner: Clear Channel Communications Inc. (acq 8-5-92). Population served: 548,050 Natl. Network: Fox. Natl. Rep: Millennium Sales & Marketing. Washington Atty: Wiley, Rein & Fielding.
Key Personnel:
Kent Cornish . . . . gen mgr
Jon Deeble . . . . chief of opns
Jeff McClausland . . . . sls dir
Jim Hanning . . . . natl sls mgr
Shawn Wheat. prom dir
Fatma Al-Tamim. . . . . progmg dir
Dave Caruso . . . . chief of engrg

KSCW— Analog channel: 33. Digital channel: 31. Analog hrs: 24 2,300 kw vis. ant 1,076t/1,076g TL: N37 48 01 W97 31 29 On air date: 2000. 200 W. Douglas, 7th Fl., Wichita, KS, 67202. Phone: (316) 303-0700. Fax: (316) 303-0160 (sales; traffic). Fax: (316) 303-9807. E-mail: programming@kansascw.com Web Site: www.kansascw.com. Licensee: Sunflower Broadcasting Inc. Group owner: Banks Broadcasting Inc. (acq 7-20-2007; $6.8 million). Population served: 443,690 Natl. Network: CW.
Key Personnel:
Joan M. Barrett . . . . pres & gen mgr
Marty Heffner . . . . opns mgr, chief of opns & engrg VP
Marcus Wilkerson. . . . . gen sls mgr
Lisa Bryce . . . . progmg mgr
Shawn Hilferty . . . . mktg dir, prom dir & pub affrs dir

KSNC— Analog channel: 2. Digital channel: 22.100 kw vis, 17.8 kw aur. 970t/1,005g TL: N38 25 54 W98 46 18 On air date: Nov 28, 1954. 833 N. Main St., Wichita, KS, 67203. Phone: (316) 265-3333. Fax: (316) 292-1197. E-mail: ksnc@ksn.com Web Site: www.ksn.com. Licensee: Montecito Television License Corp. of Wichita. Group owner: Emmis Communications Corp. (acq 1-27-2006; grpsl). Population served: 25,000 Natl. Network: NBC. Natl. Rep: TeleRep.
Key Personnel:
Al Buck . . . . gen mgr
Dan Shurtz . . . . gen sls mgr & rgnl sls mgr
Drew Rhodes . . . . gen sls mgr
Gregg Cox . . . . prom mgr
Betty Erickson . . . . progmg mgr
Todd Spessard . . . . news dir
Warren Kunkle . . . . chief of engrg

KSNG— Analog channel: 11. Digital channel: 16.200 kw vis, 24.5 kw aur. 800t/837g TL: N37 46 40 W100 52 08 On air date: Nov 5, 1958. 833 N. Main St., Wichita, KS, 67203. Phone: (316) 265-3333. Fax: (316) 292-1197. Web Site: www.ksn.com. Licensee: Montecito Television License Corp. of Wichita. Group owner: Emmis Communications Corp. (acq 1-27-2006; grpsl). Population served: 165,000 Natl. Network: NBC. Natl. Rep: TeleRep. Washington Atty: Latham & Watkins.
Key Personnel:
Al Buck . . . . gen mgr
Drew Rhodes . . . . gen sls mgr
Dan Shurtz . . . . rgnl sls mgr
Gregg Cox . . . . prom mgr & pub affrs dir
Betty Erickson . . . . progmg mgr
Todd Spessard . . . . news dir
Warren Kunkle . . . . chief of engrg

KSNK— Analog channel: 8. Digital channel: 12. Analog hrs: 24 295 kw vis, 60 kw aur. 709t/676g TL: N39 49 48 W100 42 04 On air date: Nov 28, 1959. 833 N. Main St., Wichita, KS, 67203. Phone: (316) 265-3333. Fax: (316) 292-1197. Web Site: www.ksn.com. Licensee: Montecito Television License Corp. of Wichita. Group owner: Emmis Communications Corp. (acq 1-27-2006; grpsl). Population served: 682,080 Natl. Network: NBC. Natl. Rep: TeleRep. Washington Atty: Latham & Watkins.
Key Personnel:
Al Buch . . . . gen mgr
Drew Rhodes . . . . gen sls mgr
Dan Shurtz . . . . rgnl sls mgr
Gregg Cox . . . . prom mgr
Betty Erickson . . . . progmg dir & progmg mgr
Todd Spessaud . . . . news dir
Warren Kunkle . . . . chief of engrg

KSNW— Analog channel: 3. Digital channel: 45.100 kw vis, 20 kw aur. 1,000t/1,071g TL: N37 46 37 W97 31 01 On air date: Sept 1, 1955. 833 N. Main St., Worcester, KS, 67203. Phone: (316) 265-3333. Fax: (316) 292-1197. E-mail: news@ksn.com Web Site: www.ksn.com. Licensee: Montecito Television License Corp. of Wichita. Group owner: Emmis Communications Corp. (acq 1-27-2006; grpsl). Population served: 1,307,352 Natl. Network: NBC. Natl. Rep: TeleRep. Washington Atty: Wiley, Rein & Fielding.
Key Personnel:
Al Buch . . . . gen mgr
Drew Rhodes. . . . . gen sls mgr & news dir
Dan Shurtz . . . . rgnl sls mgr
Greg Cox. . . . . prom mgr
Betty Erickson . . . . progmg dir & progmg mgr
Todd Spessard . . . . news dir
Warren Kunkle . . . . chief of engrg

***KSWK—** Analog channel: 3. Digital channel: 8. Analog hrs: 18 100 kw vis, 20 kw aur. 561t/586g TL: N37 49 38 W101 06 35 On air date: Mar 15, 1989. Box 9, 604 Elm St., Bunker Hill, KS, 67626. Phone: (785) 483-6990. Fax: (785) 483-4605. Web Site: www.shptv.org. Licensee: Smoky Hills Public Television Corp. Washington Atty: Dow, Lohens PLLC.
Key Personnel:
Lawrence Holden . . . . CEO & gen mgr
Jayne Heller. . . . . dev dir
Mary-Pat Waymaster . . . . progmg dir
Terry Cutler . . . . chief of engrg

KUPK-TV— Analog channel: 13. Digital channel: 18. Analog hrs: 24 Digital hrs: 24 225 kw vis, 45 kw aur. ant 870t/881g TL: N37 39 01 W100 40 06 On air date: Nov 8, 1964. 2900 E. Schulman Ave., Garden City, KS, 67846-9064. Phone: (620) 275-1560. Web Site: www.kake.com. Licensee: Gray Television Licensee Inc. Group owner: Gray Television Inc. (acq 8-29-2002; grpsl). Population served: 435,000 Natl. Network: ABC. Washington Atty: Covington & Burling. News staff: 2; News: 7 hrs wkly.
Satellite of KAKE-TV Wichita.

KWCH-TV—Hutchinson, Analog channel: 12. Digital channel: 19. Analog hrs: 24 Digital hrs: 24 316 kw vis, 63.1 kw aur. ant 1,522t/1,504g TL: N38 03 40 W97 45 49 On air date: July 1, 1953. 2815 E. 37th St. N., Wichita, KS, 67219. Box 12, Wichita, KS 67201. Phone: (316) 838-1212. Fax: (316) 831-6198. Web Site: www.kwch.com. Licensee: Sunflower Broadcasting Inc. Group owner: Media General Broadcast Group (acq 9-25-2006; grpsl). Population served: 425,000 Natl. Network: CBS. Rgnl. Network: Kansas Net. Natl. Rep: Harrington, Righter & Parsons. Washington Atty: Dow, Lohnes & Albertson. Wire Svc: NWS (National Weather Service) Wire Svc: AP News staff: 42; News: 24 hrs wkly.
Key Personnel:
Todd F. Schurz . . . . pres
Marshall Morton . . . . CFO
Gary Hoipemer. . . . . sr VP & VP
Joan Barrett . . . . gen mgr
Brian McDonough. . . . . gen sls mgr
Tim Vanderzwaag. . . . . natl sls mgr
David Bell . . . . prom mgr & pub affrs dir
Laverne Goering . . . . progmg dir & film buyer
Michele Gors . . . . news dir
Les Bach . . . . chief of engrg

Kentucky

Ashland

see Charleston-Huntington, WV market

Bowling Green, KY (DMA 183)

WBKO-TV— Analog channel: 13. Digital channel: 33. Analog hrs: 24 Note: ABC is on WBKO(TV) ch 13, CW and Fox are on WBKO-DT ch 33. 316 kw vis, 31.6 kw aur. ant 741t/562g TL: N37 03 49 W86 26 07 On air date: June 3, 1962. Box 13000, Bowling Green, KY 42102-9800. 2727 Russellville Rd., Bowling Green, KY 42102-9800. Phone: (270) 781-1313. Fax: (270) 781-1814. Web Site: www.wbko.com. Licensee: WEAU Licensee Corp. Group owner: Gray Television Inc. (acq 8-29-2002; grpsl). Population served: 67,300 Natl. Network: ABC, CW, Fox Digital Network: Note: ABC is on WBKO(TV) ch 13, CW and Fox are on WBKO-DT ch 33. Natl. Rep: Continental Television Sales. Washington Atty: Covington & Burling. News staff: 22; News: 17 hrs wkly.
Key Personnel:
Brad Odil. . . . . sr VP, stn mgr, sls VP & gen sls mgr
Rick McCue . . . . VP & gen mgr
Tammy Martin . . . . . prom dir, prom dir, prom mgr & engrg dir
Barbara Powell. . . . . progmg mgr
Henry Chu . . . . news dir
Wilburn England . . . . chief of engrg

***WKGB-TV—** Analog channel: 53. Digital channel: 48.562 kw vis, 112 kw aur. 810t/618g TL: N37 05 22 W86 38 05 On air date: Sept 23, 1968. 600 Cooper Dr., Lexington, KY, 40502. Phone: (859) 258-7000. Fax: (859) 258-7399. Web Site: www.ket.org. Licensee: Kentucky Authority for Educational TV. Natl. Network: PBS.
Key Personnel:
Malcolm Wall . . . . opns dir
Tim Bischoff . . . . mktg dir
Craig Cornwell . . . . progmg dir

***WKYU-TV—** Analog channel: 24. Digital channel: 18. Analog hrs: 8 AM-12 PM (M-F); 8 AM-12 PM (S, Su) 400 kw vis, 20 kw aur. ant 648t/603g TL: N37 03 52 W86 26 07 On air date: Jan 17, 1989. Academic Complex 153, Western Kentucky Univ., Bowling Green, KY, 42101-1034. Phone: (270) 745-2400. Fax: (270) 745-2084. E-mail: wkyupbs@wkyu.edu Web Site: www.wkyu.org. Licensee: Western Kentucky University. Natl. Network: PBS. Washington Atty: Leventhal, Senter & Lerman. News staff: one; News: one hr wkly.

Key Personnel:
Gary Ransdell . . . . . pres
Jack Hanes . . . . . gen mgr
Linda Gerossky . . . . . stn mgr

WNKY— Analog channel: 40. Analog hrs: 24 Digital hrs: 24 631 kw vis, 63.1 kw aur. 561t TL: N37 02 10 W86 10 20 (CP: 776 kw vis, 77.6 kw aur, ant 800t) On air date: Dec 15, 1991. 325 Emmett Ave., Bowling Green, KY, 42101. Phone: (270) 781-2140. Fax: (270) 842-7140. E-mail: wnky@nbc40.tv Licensee: MMK License LLC. Group owner: MAX Media L.L.C. (acq 3-1-2003; $7 million). Natl. Network: NBC. Natl. Rep: Millennium Sales & Marketing. Washington Atty: Williams & Mullen, P.C.

Key Personnel:
Ed Groves . . . . . pres & gen mgr
Greg Fotos . . . . . gen sls mgr
Gerald Keith . . . . . prom dir
Ellen Grundy . . . . . progmg dir
Heather Davison . . . . . pub affrs dir

Covington

see Cincinnati, OH market

Harlan

see Knoxville, TN market

Lexington, KY (DMA 63)

WDKY-TV— Analog channel: 56. Digital channel: 4. Analog hrs: 24 Digital hrs: 24 5,000 kw vis, 500 kw aur. ant 1,154t/1,126g TL: N37 52 51 W84 19 16 On air date: Feb 10, 1986. Chevy Chase Plaza, 836 Euclid Ave., Suite 201, Lexington, KY, 40502. Phone: (859) 269-5656. Fax: (859) 269-3774. Web Site: www.wdky56.com. Licensee: WDKY Licensee L.L.C. Group owner: Sinclair Broadcast Group Inc. (acq 1996; $63 million with KOCB(TV) Oklahoma City, OK). Population served: 487,900 Natl. Network: Fox. Natl. Rep: Millennium Sales & Marketing. Washington Atty: Fisher, Wayland, Cooper, Leader & Zaragoza. News: 7 hrs wkly.

Key Personnel:
Marvin Bartlett . . . . . CEO & news dir
Michael Brickey . . . . . gen mgr
Kevin Neumann . . . . . gen sls mgr & natl sls mgr
Jeff Sleete . . . . . mktg VP & mktg dir
Rick White . . . . . natl sls mgr & progmg dir
Dave Koller . . . . . chief of engrg

***WKHA—** Analog channel: 35. Digital channel: 16.417 kw vis, 83.2 kw aur. 1,260t/608g TL: N37 11 34 W83 11 16 On air date: June 6, 1968. 600 Cooper Dr., Lexington, KY, 40502. Phone: (859) 258-7000. Fax: (859) 258-7399. Web Site: www.ket.org. Licensee: Kentucky Authority for Educational TV. Natl. Network: PBS.

Key Personnel:
Mike Brower . . . . . opns mgr
Robert Ball . . . . . dev VP
Tim Bischoff . . . . . mktg dir
Craig Cornwell . . . . . progmg dir & engrg dir

***WKLE—** Analog channel: 46. Digital channel: 42.1,050 kw vis, 105 kw aur. 870t/850g TL: N37 52 45 W84 19 33 On air date: Sept 23, 1968. 600 Cooper Dr., Lexington, KY, 40502. Phone: (859) 258-7000. Fax: (859) 258-7399. Web Site: www.ket.org. Licensee: Kentucky Authority for Educational TV. Population served: 3,500,000 Natl. Network: PBS. Rgnl. Network: SECA. Washington Atty: Kenkel, Barnard & Edmundson.

Key Personnel:
Craig Cornwell . . . . . opns mgr & progmg dir
Robert Ball . . . . . progmg dir

***WKMR—** Analog channel: 38. Digital channel: 15.575 kw vis, 115 kw aur. 960t/607g TL: N38 10 38 W83 24 18 On air date: Sept 23, 1968. 600 Cooper Dr., Lexington, KY, 40502. Phone: (859) 258-7000. Fax: (606) 258-7399. Web Site: www.ket.org. Licensee: Kentucky Authority for Educational TV. Natl. Network: PBS.

Key Personnel:
Craig Cornwell . . . . . opns mgr & progmg mgr
Tim Bischoff . . . . . mktg dir
Robert Ball . . . . . engrg dir

***WKSO-TV—** Analog channel: 29. Digital channel: 14.584 kw vis, 117 kw aur. 1,460t/1,000g TL: N37 10 00 W84 49 28 On air date: Sept 23, 1968. 600 Cooper Dr., Lexington, KY, 40502. Phone: (606) 258-7000. Fax: (606) 258-7399. Web Site: www.ket.org. Licensee: Kentucky Authority for Educational TV. Natl. Network: PBS.

Key Personnel:
Craig Cornwell . . . . . opns mgr & progmg mgr
Tim Bischoff . . . . . mktg dir
Robert Ball . . . . . engrg dir

WKYT-TV— Analog channel: 27. Digital channel: 13. Analog hrs: 24 Note: CBS is on WKYT-TV ch 27, CW is on WKYT-DT ch 13. 1,510 kw vis, 151 kw aur. 984t/992g TL: N38 02 22 W84 24 11 On air date: Sept 30, 1957. Box 55037, Lexington, KY, 40555-5037. 2851 Winchester Rd., Lexington, KY 40509. Phone: (859) 299-0411. Fax: (859) 299-5531. E-mail: wmartin@wkyt.com Web Site: www.wkyt.com. Licensee: Gray Television Licensee, Inc. Group owner: Gray Television Inc. (acq 1-21-76; FTR: 2-9-76). Population served: 871,000 Natl. Network: CBS, CW Digital Network: Note: CBS is on WKYT-TV ch 27, CW is on WKYT-DT ch 13. Natl. Rep: Harrington, Righter & Parsons. Washington Atty: Venable, Baetjer, Howard & Civiletti. News staff: 50; News: 43 hrs wkly.

Key Personnel:
Wayne Martin . . . . . pres & gen mgr
Michael D. Kanarek . . . . . opns VP
Chris Martin . . . . . sls VP
Barbara Howard . . . . . progmg dir
Bill Bryant . . . . . news dir

WLEX-TV— Analog channel: 18. Digital channel: 39. Analog hrs: 24 1,104 kw max vis, 110 kw aur. 640t/670g TL: N37 55 23 W84 09 14 On air date: Mar 15, 1955. Box 1457, Lexington, KY, 40588-1457. 1065 Russell Cave Rd., Lexington, KY 40505. Phone: (859) 259-1818. Fax: (859) 255-2418. Fax: TWX: 859-254-1272. E-mail: wlextv@wlextv.com Web Site: www.wlextv.com. Licensee: WLEX Communications L.L.C. Group owner: Cordillera Communications Inc. (acq 7-8-99; $99.1 million). Population served: 720,300 Natl. Network: NBC. Washington Atty: Dow, Lohnes PLLC.

Key Personnel:
Tim Gilbert . . . . . chmn, pres & gen mgr
Sandra Byron . . . . . CFO
Sean Franklin . . . . . opns mgr
Chris Fedele . . . . . sls dir
Mary West . . . . . gen sls mgr
Sandy Stevenson . . . . . natl sls mgr
Chip Alfred . . . . . prom dir & pub affrs dir
Teresa Cassidy . . . . . progmg dir & progmg
Bruce Carter . . . . . news dir
Tony Michalski . . . . . chief of engrg

WLJC-TV— Analog channel: 65. Digital channel: 7. Analog hrs: 24 92.75 kw vis, 9.275 kw aur. 665 TL: N37 36 23 W83 41 16 (CP: 73.45 kw vis, 7.345 kw aur, ant 646.2t) On air date: Oct 16, 1982. PO Box Y, 219 Radio Station Loop, Beattyville, KY, 41311. Phone: (606) 464-3600. Fax: (606) 464-5021. E-mail: wljc@wljc.com Web Site: www.wljc.com. Licensee: Hour of Harvest Inc. Natl. Rep: Rgnl Reps. Washington Atty: Fletcher, Heald & Hildreth.

Key Personnel:
Margaret Drake . . . . . pres
Jonathan Drake . . . . . gen mgr
Rachel Bogale . . . . . opns mgr
Kim Mitchell . . . . . gen sls mgr & progmg dir
Allan Mulford . . . . . chief of engrg

WTVQ-TV— Analog channel: 36. Digital channel: 40. Analog hrs: 24 Note: digital ch 36.2 (part of digital ch 40) broadcasts weather 24 hrs each day. 1,580 kw vis, 158 kw aur. 994t/1,000g TL: N38 02 03 W84 23 39 On air date: June 2, 1968. 6940 Man-O-War Blvd., Lexington, KY, 40509-8412. Phone: (859) 294-3636. Fax: (859) 293-5002. E-mail: programming@wtvq.com Web Site: www.wtvq.com. Licensee: Media General Communications Inc. Group owner: Media General Broadcast Group (acq 3-21-97; grpsl). Population served: 1,981,200 Natl. Network: ABC Digital Network: Note: digital ch 36.2 (part of digital ch 40) broadcasts weather 24 hrs each day. Natl. Rep: MMT. Washington Atty: Dow, Lohnes & Albertson. Wire Svc: UPI News staff: 40; News: 27 hrs wkly.

Key Personnel:
Mark Pimentel . . . . . VP & gen mgr
Mitch Bukata . . . . . mktg dir
Tai Takahashi . . . . . news dir

WUPX-TV— Analog channel: 67. Digital channel: 21.5,000 kw vis. ant 1,463t/1,310g TL: N37 54 26 W83 38 01 On air date: June 1998. 2166 McCausey Ridge Rd., Frenchburg, KY, 40322. Phone: (606) 784-7932. Fax: (606) 768-9278. Web Site: www.ionline.tv. Licensee: Paxson Lexington License Inc. Group owner: Paxson Communications Corp. (acq 4-27-2001; $8 million).

WYMT-TV— Analog channel: 57. Digital channel: 12. Analog hrs: 24 Digital hrs: 24 2,630 kw vis, 263 kw aur. 1,560t/1,029g TL: N37 11 38 W83 10 52 On air date: Oct 20, 1969. Box 1299, 199 Black Gold Blvd., Hazard, KY, 41702. Phone: (606) 436-5757. Fax: (606) 439-3760. Web Site: www.wymtnews.com. Licensee: Gray Television Licensee Inc. Group owner: Gray Television Inc. (acq 9-2-94). Population served: 216,670 Natl. Network: CBS. Natl. Rep: Harrington, Righter & Parsons. News staff: 15.

Key Personnel:
Ernestine Cornett . . . . . gen mgr
James Boggs . . . . . gen sls mgr
Edna Eldridge . . . . . prom dir, prom mgr & progmg dir
Neil Middleton . . . . . news dir
Phillip Hayes . . . . . adv mgr & chief of engrg

Louisville, KY (DMA 48)

WAVE— Analog channel: 3.100 kw vis, 10 kw aur. 1,820t/1,690g TL: N38 27 23 W85 25 28 On air date: Nov 24, 1948. Box 32970, Louisville, KY, 40232. 725 S. Floyd St., Louisville, KY 40203. Phone: (502) 585-2201. Fax: (502) 561-4115. Web Site: www.wave3.com. Licensee: Libco Inc. Group owner: Liberty Corp. (acq 1-13-2006; grpsl). Population served: 1,500,000 Natl. Network: NBC. Natl. Rep: Harrington, Righter & Parsons. Washington Atty: Covington & Burling.

Key Personnel:
Steve Langford . . . . . gen mgr
Nick Ulmer . . . . . gen sls mgr
Bob Mack . . . . . mktg dir
Dan Foos . . . . . progmg dir
Jim Sears . . . . . chief of engrg

WBKI-TV— Analog channel: 34. Digital channel: 19. Analog hrs: 24 5,000 kw vis, 595.74 kw aur. ant 1,269t/1,086g TL: N37 31 51 W85 26 45 On air date: Apr 6, 1983. 1601 Alliant Ave., Louisville, KY, 40299. Phone: (502) 809-3400. Fax: (502) 266-6262. E-mail: hr@wb34.com Web Site: www.cwlouisville.com. Licensee: Louisville Communications L.L.C. Group owner: Cascade Broadcasting Group L.L.C. (acq 6-9-2000). Natl. Network: CW. Washington Atty: Shaw Pittman.

Key Personnel:
Carol LaFever . . . . . CEO
Kim Grau . . . . . gen sls mgr & rgnl sls mgr
Dana Meredith . . . . . natl sls mgr
Myrna Jane Jaspan . . . . . progmg mgr
Mac Powas . . . . . engrg dir

WBNA— Analog channel: 21. Digital channel: 8. Analog hrs: 24 Digital hrs: 24 2,000 kw vis, 200 kw aur. 696t TL: N38 01 59 W85 45 16 On air date: Apr 2, 1986. 3701 Fern Valley Rd., Louisville, KY, 40219. Phone: (502) 964-2121. Fax: (502) 966-9692. Web Site: www.wbna-21.com. Licensee: Word Broadcasting Network Inc. Group owner: (group owner) Natl. Network: i Network. Washington Atty: Pepper & Corazzini.

Key Personnel:
Tom Fawbush . . . . . pres & gen mgr
Harry Monroe . . . . . chief of engrg

WDRB— Analog channel: 41. Digital channel: 49. Analog hrs: 24 Digital hrs: 24 5,000 kw vis, 500 kw aur. 1,283t/1,003g TL: N38 21 00 W85 50 57 On air date: Feb 28, 1971. 624 W. Muhammad Ali Blvd., Louisville, KY, 40203. Phone: (502) 584-6441. Fax: (502) 589-5559. Web Site: www.fox41.com. Licensee: Independence Television Co. Group owner: Block Communications Inc. (acq 3-84; $10 million; FTR: 1-2-84). Population served: 298,451 Natl. Network: Fox. Natl. Rep: TeleRep. Washington Atty: Dow, Lohnes & Albertson. News staff: 45; News: 35 hrs wkly.

Key Personnel:
Bill Lamb . . . . . pres & gen mgr
Harry Beam . . . . . opns mgr
Marti Hazel . . . . . gen sls mgr
James Reed . . . . . natl sls mgr
Steve Ballard . . . . . CFO & natl sls mgr
Barry Fulmer . . . . . news dir
Gary Schroder . . . . . chief of engrg

WHAS-TV— Analog channel: 11. Digital channel: 55.135 kw vis, 13.5 kw aur. 1,290t/973g TL: N38 21 23 W85 50 52 On air date: Mar 27, 1950. 520 W. Chestnut St., Louisville, KY, 40201. Phone: (502) 582-7711. Fax: (502) 582-7279. E-mail: whasprogramming@whas11.com Web Site: www.whas11.com. Licensee: Belo Kentucky Inc. Group owner: Belo Corp., Broadcast Division (acq 2-97; grpsl). Population served: 550,500 Natl. Network: ABC. Washington Atty: Covington & Burling.

Key Personnel:
Allan Cohen . . . . . gen mgr
Lori Morgan . . . . . gen sls mgr
Kirk Szesny . . . . . mktg dir
Joy Pritchett . . . . . progmg dir
Aaron Ramey . . . . . news dir
Neal Metersky . . . . . chief of engrg

WKMJ— Analog channel: 68. Digital channel: 38.1,170 kw vis, 230 kw aur. 835t/550g TL: N38 22 02 W85 49 53 On air date: Aug 31, 1970. 600 Cooper Dr., Lexington, KY, 40502. Phone: (859) 258-7000. Fax: (859) 258-7399. Web Site: www.ket.org. Licensee: Kentucky Authority for Educational TV. Natl. Network: PBS.
Key Personnel:
Craig Cornwell. . . . . opns mgr & progmg mgr
Mike Brower . . . . dev mgr
Tim Bischoff . . . . mktg dir & pub affrs dir
Robert Ball . . . . engrg dir

***WKPC-TV**— Analog channel: 15. Digital channel: 17.263 kw vis, 46.8 kw aur. 860t/549g TL: N38 22 02 W85 49 53 (CP: 525 kw vis, 589 kw aur max, ant 860t) On air date: Sept 5, 1958. 600 Cooper Dr., Lexington, KY, 40502. Phone: (859) 258-7000. Fax: (859) 258-7399. Web Site: www.ket.org. Licensee: Kentucky Authority for Educational Television. Population served: 519,000 Natl. Network: PBS. Washington Atty: Schwartz, Woods & Miller.
Key Personnel:
Craig Cornwell. . . . . opns mgr & progmg mgr
Mike Brower. . . . . dev VP
Tim Bischoff . . . . mktg dir & progmg VP
Robert Ball . . . . chief of engrg

***WKZT-TV**— Analog channel: 23. Digital channel: 43. Analog hrs: 24 575 kw vis, 115 kw aur. 650t/655g TL: N37 40 55 W85 50 32 On air date: Sept 23, 1968. 600 Cooper Dr., Lexington, KY, 40502. Phone: (859) 258-7000. Fax: (859) 258-7399. Web Site: www.ket.org. Licensee: Kentucky Authority for Educational TV. Natl. Network: PBS.
Key Personnel:
Mike Brower . . . . dev mgr
Tim Bischoff . . . . opns mgr & mktg dir
Craig Cornwell . . . . progmg mgr & engrg dir

WLKY-TV— Analog channel: 32. Digital channel: 26. Analog hrs: 24 Digital hrs: 24 4,300 kw vis, 430 kw aur. 1,260t/989g TL: N38 22 10 W85 50 02 On air date: Sept 18, 1961. Box 6205, 1918 Mellwood Ave., Louisville, KY, 40206. Phone: (502) 893-3671. Fax: (502) 897-2384. Web Site: www.wlky.com. Licensee: Hearst-Argyle Properties Inc. Group owner: Hearst-Argyle Television Inc. (acq 3-18-99; grpsl). Population served: 2,060,000 Natl. Network: CBS. Washington Atty: Brooks, Pierce, McLendon, Humphrey & Leonard. Wire Svc: UPI News: 37.5 hrs wkly.
Key Personnel:
David Levy . . . . opns mgr
Greg Baird . . . . gen sls mgr
Jim Carter . . . . pres, gen mgr & progmg mgr

WMYO— Analog channel: 58. Digital channel: 51. Analog hrs: 24 Digital hrs: 24 5,000 kw vis. ant 1,305t TL: N38 42 29 W86 05 57 On air date: Mar 15, 1994. 624 W. Muhammed Ali Blvd., Louisville, KY, 40203. Phone: (502) 584-6441. Fax: (502) 589-5559. Web Site: www.wmyo.com. Licensee: Independence Television Co. Group owner: Block Communications Inc. (acq 3-30-2001). Natl. Network: MyNetworkTV. Natl. Rep: TeleRep. Washington Atty: Dow, Lohnes & Albertson.
Key Personnel:
Bill Lamb . . . . pres & gen mgr
Steve Ballard . . . . CFO
Harry Beam . . . . opns mgr
Marti Hazel . . . . gen sls mgr
James Reed. natl sls mgr
Gary Schroder . . . . chief of engrg

Madisonville

see Evansville, IN market

Newport

see Cincinnati, OH market

Owensboro

see Evansville, IN market

Owenton

see Cincinnati, OH market

Paducah, KY-Cape Girardeau, MO-Harrisburg-Mount Vernon, IL (DMA 80)

KBSI— Analog channel: 23. Digital channel: 22. Analog hrs: 24 1,860 kw vis, 186 kw aur. 1,768t/1,524g TL: N37 24 23 W89 33 44 On air date: Sept 10, 1983. 806 Enterprise, Cape Girardeau, MO, 63703. Phone: (573) 334-1223. Fax: (573) 334-1208. Web Site: www.kbsi23.com. Licensee: KBSI Licensee L.P. Group owner: Sinclair Broadcast Group Inc. (acq 1998; grpsl). Population served: 850,000 Natl. Network: Fox. Natl. Rep: Millennium Sales & Marketing.
Key Personnel:
Tom Tipton. . . . . gen mgr
Rob Chronister . . . . opns dir
Jennifer Chronister . . . . gen sls mgr
Jean Graham . . . . natl sls mgr
Chuck Moffitt . . . . prom mgr
Alan Muster . . . . progmg dir
Chris Girard . . . . chief of engrg

KFVS-TV— Analog channel: 12. Digital channel: 57. Analog hrs: 24 316 kw vis, 63.2 kw aur. 2,001t/1,678g TL: N37 25 46 W89 30 14 On air date: Oct 3, 1954. PO Box 100, Cape Girardeau, MT, 63702. 310 Broadway, Cape Girardeau, MT 63702. Phone: (573) 335-1212. Fax: (573) 335-6303. E-mail: manager@kfvs12.com Web Site: www.kfvs12.com. Licensee: Raycom America License Subsidiary LLC. Group owner: Raycom Media Inc. (acq 4-97; grpsl). Population served: 910,000 Natl. Network: CBS. Natl. Rep: Harrington, Righter & Parsons. Washington Atty: Covington & Burling. News staff: 45; News: 28 hrs wkly.
Key Personnel:
Mike Smythe . . . . gen mgr
Mike Wunderlich. . . . . opns dir
Joe Trepasso . . . . gen sls mgr
Brad Zaruba. . . . . natl sls mgr
Karen Wade. . . . . rgnl sls mgr
Paul Keener . . . . mktg dir
Dan Timpe . . . . prom dir
Kathy Cowan. . . . . progmg dir & pub affrs dir
Mark Little . . . . news dir
Arnold Killian. . . . . chief of engrg

KPOB-TV— Analog channel: 15. Analog hrs: 24 389 kw vis, 38.9 kw aur. 600t/526g TL: N36 48 02 W90 27 03 On air date: Sept 15, 1967. 1416 Country Air Dr., Carterville, IL, 62918. Phone: (618) 985-2333. Fax: (618) 985-3709. Web Site: www.wsiltv.com. Licensee: Mel Wheeler Inc. Group owner: (group owner; (acq 5-12-83; $6.6 million; FTR: 6-6-83). Natl. Network: ABC. Washington Atty: Brooks, Pierce, McLendon, Humphrey & Leonard.
Key Personnel:
Steve Wheeler . . . . gen mgr
Harold McDaniel . . . . opns mgr
Pat Victoria . . . . chief of engrg
Satellite of WSIL-TV Harrisburg IL.

WDKA— Analog channel: 49. Digital channel: 50. Analog hrs: 24 2,610 kw vis, 275 kw aur. ant 1,079t/1,848g TL: N37 23 42 W88 56 23 On air date: June 1997. 806 Enterprise St., Cape Girardeau, MO, 63703. Phone: (573) 334-1223. Fax: (573) 334-1208. Web Site: www.wdka49.com. Licensee: WDKA Acquisition Corp. Natl. Network: MyNetworkTV.
Key Personnel:
Tom Tipton . . . . gen mgr
Rob Chronister . . . . opns dir
Jennifer Chronister . . . . gen sls mgr
Jean Graham . . . . natl sls mgr
Chuck Moffitt . . . . prom dir
Alan Muster . . . . progmg dir
Chris Girard . . . . chief of engrg

***WKMU**— Analog channel: 21. Digital channel: 36.575 kw vis, 115 kw aur. 660t/655g TL: N36 41 33 W88 32 10 On air date: Oct 9, 1968. 600 Cooper Dr., Lexington, KY, 40502. Phone: (859) 258-7000. Fax: (859) 258-7399. Web Site: www.ket.org. Licensee: Kentucky Authority for Educational TV. Natl. Network: PBS.
Key Personnel:
Craig Cornwell . . . . opns dir & progmg mgr
Tim Bischoff . . . . mktg dir
Robert Ball . . . . engrg dir

***WKPD**— Analog channel: 29. Digital channel: 41.676 kw vis, 67.6 kw aur. 709t TL: N36 57 42 W88 41 22 (CP: 145 kw vis, ant 499t. TL: N37 05 38 W88 40 19) On air date: May 31, 1971. 600 Cooper Dr., Lexington, KY, 40502. Phone: (859) 258-7000. Fax: (859) 258-7399. Web Site: www.ket.org. Licensee: Kentucky Authority for Educational TV. (acq 2-28-78). Natl. Network: PBS. Rgnl. Network: SECA. Washington Atty: Kenkel, Barnard & Edmundson.
Key Personnel:
Mike Brower. opns mgr
Tim Bischoff . . . . mktg dir
Craig Cornwell . . . . progmg mgr
Robert Ball . . . . engrg dir

WPSD-TV— Analog channel: 6. Digital channel: 32. Analog hrs: 24 100 kw vis, 13.8 kw aur. ant 1,581t/1,593g TL: N37 11 31 W88 58 53 On air date: May 28, 1957. Box 1197, Paducah, KY, 42002-1197. 100 Television Ln., Paducah, KY 42003. Phone: (270) 415-1900. Fax: (270) 415-2020. E-mail: bevans@wpsdtv.com Web Site: www.wpsdtv.com. Licensee: WPSD-TV LLC. Ownership: Paxton Media Group Inc., see Cross-Ownership. (acq 12-3-01). Population served: 333,000 Natl. Network: NBC. Washington Atty: Covington & Burling. News staff: 49; News: 23 hrs wkly.
Key Personnel:
Richard Paxton . . . . pres & gen mgr
Bill Evans . . . . opns VP
Mark Hall . . . . opns mgr
David Jernigan . . . . sls VP & progmg VP
Bob Crosno . . . . rgnl sls mgr
Cathy Crecelius . . . . prom mgr & pub affrs dir
Griff Potter. . . . . news dir
Joey Gill . . . . chief of engrg

WPXS— Analog channel: 13. Digital channel: 21. Analog hrs: 24 302 kw vis, 30.2 kw aur. ant 991t/441g TL: N38 32 39 W88 55 26 (CP: 316 kw vis, 31.6 kw aur) On air date: Mar 1, 1983. 4751 Cartter Rd., Kell, IL, 62853. Phone: (618) 822-6900. Fax: (618) 822-6526. E-mail: wpxs@mvn.net Licensee: EBC St. Louis Inc. Group owner: Equity Broadcasting Corp. (acq 4-26-2001; $17.75 million with KDUO(TV) Flagstaff, AZ). Population served: 506,000

WSIL-TV— Analog channel: 3. Digital channel: 34. Analog hrs: 24 Digital hrs: 24 100 kw vis, 20 kw aur. 1,120t/1,000g TL: N37 36 46 W88 52 20 On air date: December 1953. 1416 Country Aire Dr, Carterville, IL, 62918. Phone: (618) 985-2333. Fax: (618) 985-3709. Web Site: www.wsiltv.com. Licensee: WSIL TV Inc. Group owner: Mel Wheeler Inc. (acq 5-12-83; grpsl; FTR: 6-6-83). Population served: 927,000 Natl. Network: ABC. Natl. Rep: Continental Television Sales. Washington Atty: Brooks, Pierce, McLendon, Humprey & Leonard.

***WSIU-TV**— Analog channel: 8. Digital channel: 40. Analog hrs: 24 Digital hrs: 24 316 kw vis, 40.7 kw aur. ant 890t/861g TL: N38 06 11 W89 14 40 On air date: November 1961. 1003 Communications Bldg., 1100 Lincoln Dr., Carbondale, IL, 62901. Phone: (618) 453-4343. Fax: (618) 453-6186. Web Site: www.wsiu.org. Licensee: Board of Trustees of Southern Illinois University. Population served: 326,000 Natl. Network: PBS. Washington Atty: Cohn & Marks. Wire Svc: AP News staff: one; News: 2 hrs wkly.
Key Personnel:
Dr. Candis S. Isberner . . . . CEO & gen mgr
Delores Kerstein . . . . CFO
Robert Henderson . . . . opns dir
Renee Dillard. . . . . dev dir & mktg dir
Monica Tichenor. . . . . prom dir
Trina Thomas . . . . progmg
Jack Hammer . . . . engr
Rebroadcasts WUSI-TV Olney 99%.

WTCT— Analog channel: 27. Digital channel: 17. Analog hrs: 24 2,600 kw vis, 260 kw aur. 775t/500g TL: N37 33 26 W89 01 24 On air date: Aug 16, 1981. Box 698, 11717 Rt. 37 N., Marion, IL, 62959. Phone: (618) 997-4700. Fax: (618) 993-9778. Web Site: www.tct.tv. Licensee: Tri-State Christian TV. Group owner: (group owner; (acq 5-29-84; $1.2 million).
Key Personnel:
Fortune Brayfield . . . . gen mgr & stn mgr
Wes Hall . . . . chief of engrg

Pikeville

see Charleston-Huntington, WV market

Louisiana

Alexandria, LA (DMA 179)

KALB-TV— Analog channel: 5. Digital channel: 35. Analog hrs: 24 Note: NBC is on KALB-TV ch 5, CBS is on KALB-DT ch 35. 100 kw vis, 20 kw aur. 1,590t/1,586g TL: N31 02 15 W92 29 45 On air date: Sept 29, 1954. Box 951, Alexandria, LA, 71309. 605 Washington St., Alexandria, LA 70301. Phone: (318) 445-2456. Fax: (318) 442-7427. E-mail: news@kalb.com Web Site: www.kalb.com. Licensee: Media General Communications Inc. Group owner: Media General Broadcast Group (acq 3-21-97; grpsl). Population served: 504,400 Natl. Network:

NBC, CBS Digital Network: Note: NBC is on KALB-TV ch 5, CBS is on KALB-DT ch 35. Washington Atty: Dow, Lohnes & Albertson.

Key Personnel:

Les Golmon . . . . pres & gen mgr
Gary Coullard . . . . opns dir & chief of engrg
Tom Pears III . . . . gen sls mgr
Keith Holt . . . . mktg dir
Shannon Tassin . . . . progmg dir

KBCA— Analog channel: 41. Analog hrs: 24 5,000 kw vis. ant 993t/975g TL: N30 54 17 W92 37 28 On air date: June 1, 2005. Delta Media Corp., 3501 Northwest Evangeline Thruway, Carencro, LA, 70520. Phone: (337) 896-1600. Fax: (337) 896-2695. Web Site: www.cwtv41.com. Licensee: Wilderness Communications LLC. (acq 4-1-2006). Natl. Network: CW. Natl. Rep: Roslin Television Sales. Washington Atty: Fletcher, Heald & Hildreth.

Key Personnel:

Charles Chatelain. . . . . pres
Eddie Blanchard. gen mgr
Leila Dablan. gen sls mgr

KLAX-TV— Analog channel: 31. Digital channel: 32. Analog hrs: 24 1,309 kw vis, 131 kw aur. 1,092t/1,028g TL: N31 33 54 W92 32 59 On air date: Mar 3, 1983. Box 8818, 1811 England Dr., Alexandria, LA, 71303. Phone: (318) 473-0031. Fax: (318) 442-4646. Web Site: www.klax-tv.com. Licensee: Pollack-Belz Communication Co. Inc. (acq 6-3-88; $1.1 million). Natl. Network: ABC. Natl. Rep: Blair Television. Washington Atty: Wood, Maines & Brown. News staff: 4; News: 4 hrs wkly.

Key Personnel:

William H. Pollack. . . . . pres
David Carlson . . . . CFO
Ken Nolan . . . . gen mgr & gen sls mgr
Lisa Ballance . . . . rgnl sls mgr
Frances Yeager . . . . progmg dir
D. Herbert. chief of engrg

***KLPA-TV**— Analog channel: 25. Digital channel: 25. Analog hrs: 24 2,040 kw vis, 204 kw aur. 1,360t/1,329g TL: N31 33 56 W92 32 50 On air date: July 1, 1983. 7733 Perkins Rd., Baton Rouge, LA, 70810. Phone: (225) 767-5660. Phone: (800) 272-8161. Fax: (225) 767-4299. Web Site: www.lpb.org. Licensee: Louisiana Educational Television Authority. Population served: 1,000,000 Natl. Network: PBS. Washington Atty: Schwartz, Woods & Miller.

Key Personnel:

Beth Courtney . . . . CEO & pres
Bob Neese . . . . prom mgr
Jennifer Howze . . . . progmg dir
Randy Ward . . . . opns mgr & engrg dir

Baton Rouge, LA
(DMA 93)

WAFB— Analog channel: 9. Digital channel: 46. Analog hrs: 24 Digital hrs: 24 316 kw vis, 63 kw aur. 1,670t/1,729g TL: N30 21 58 W91 12 47 On air date: Apr 19, 1953. 844 Government St., Baton Rouge, LA, 70802. Phone: (225) 383-9999. Fax: (225) 379-7891. Fax: TWX: 510-993-3406. E-mail: news@wafb.com Web Site: www.wafb.com. Licensee: WAFB License Subsidiary LLC. Group owner: Raycom Media Inc. (acq 12-31-96; grpsl). Population served: 266,640 Natl. Network: CBS. Natl. Rep: Harrington, Righter & Parsons. Washington Atty: Covington & Burling.

Key Personnel:

Nick Simonette . . . . VP, gen mgr & progmg mgr
Vicki Kellum . . . . gen sls mgr
Ellen Salmon . . . . rgnl sls mgr
Andree Zamarlik . . . . prom mgr
Vicki Zimmerman . . . . news dir
Dale Russell . . . . chief of engrg

WBRZ— Analog channel: 2. Analog hrs: 24 100 kw vis, 10 kw aur. 1,689t TL: N30 17 48 W91 11 36 On air date: Apr 14, 1955. Box 2906, Baton Rouge, LA, 70821. 1650 Highland Rd., Baton Rouge, LA 70821. Phone: (225) 387-2222. Fax: (225) 336-2246. E-mail: news@wbrz.com Web Site: www.2theadvocate.com. Licensee: Louisiana Television Broadcasting LLC. Group owner: Manship Stations (acq 1958; $548,000). Population served: 723,000 Natl. Network: ABC. News staff: 47; News: 22 hrs wkly.

Key Personnel:

James "Rocky" Daboval. . . . . gen mgr
Jim Reardon . . . . sls dir
Steve Storey. gen sls mgr
Denise Akers . . . . mktg dir
Michelle Martone . . . . progmg dir & progmg mgr
Chuck Bark . . . . news dir
Clyde Pierce . . . . engrg dir & chief of engrg

WGMB— Analog channel: 44. Analog hrs: 24 3,871 kw vis. 1,164t TL: N30 19 35 W91 16 36 On air date: Aug 11, 1991. 10000 Perkins Rd, Baton Rouge, LA, 70810. Phone: (225) 769-0044. Fax: (225) 769-9462. Web Site: www.fox44.com. Licensee: Comcorp of Baton Rouge. Group owner: Communications Corp. of America (acq 2-13-95; FTR: 5-8-95). Population served: 710,500 Natl. Network: Fox. Washington Atty: Fletcher, Heald & Hildreth.

Key Personnel:

Phil Waterman . . . . CFO & gen mgr
Tom Poehler . . . . natl sls mgr
Lee Stolf . . . . rgnl sls mgr & adv dir
Meisie Pacris . . . . prom mgr
Destiny Kelley . . . . progmg dir
Karen Mire . . . . pub affrs dir
Cecil Connella . . . . chief of engrg

***WLPB-TV**— Analog channel: 27. Digital channel: 25. Analog hrs: 24 2,570 kw vis, 257 kw aur. 994t/1,030g TL: N30 22 22 W91 12 16 On air date: Sept 6, 1975. 7733 Perkins Rd., Baton Rouge, LA, 70810. Phone: (225) 767-5660. Phone: (800) 272-8161. Fax: (225) 767-4299. Web Site: www.lpb.org. Licensee: Louisiana Educational Television Authority. Population served: 1,000,000 Natl. Network: PBS. Washington Atty: Schwartz, Woods & Miller. News staff: 3; News: one hr wkly.

Key Personnel:

Beth Courtney . . . . CEO & pres
Jennifer Howze . . . . opns mgr & progmg dir
Bob Neese . . . . prom mgr
Randy Ward . . . . engrg dir

WVLA—Analog channel: 33. Digital channel: 34. Analog hrs: 22 5,000 kw vis, 1,000 kw aur. ant 1,750t TL: N30 19 35 W91 16 36 On air date: Oct 16, 1971. 10000 Perkins Rd., Baton Rouge, LA, 70810. Phone: (225) 766-3233. Fax: (225) 768-9200. Web Site: www.nbc33tv.com. Licensee: Knight Broadcasting of Baton Rouge. Group owner: White Knight Holdings Inc. (acq 1996; $23.975 million). Population served: 773,400 Natl. Network: NBC. Washington Atty: Pillsbury, Winthrop, Shaw & Pittman LLP. News staff: 3; News: 6 hrs wkly.

Key Personnel:

Phil Waterman . . . . gen mgr & stn mgr
Brooks Hogg . . . . sls dir
Tom Poehler . . . . natl sls mgr
Doreen Morgan . . . . mktg mgr
Scott Thomson. . . . . prom mgr
Suzanne Marva . . . . progmg dir
Jeff Hamburger . . . . news dir
Elaine Harrison . . . . pub affrs dir
Terry Freeman . . . . engrg VP
Cecil Connella . . . . chief of engrg

Lafayette, LA
(DMA 123)

KADN— Analog channel: 15. Digital channel: 16.2,630 kw vis, 231 kw aur. ant 1,181t/1,282g TL: N30 21 44 W92 12 53 On air date: Feb 28, 1980. 123 N. Easy St., Lafayette, LA, 70506. Phone: (337) 237-1500. Fax: (337) 237-2526. Web Site: www.kadn.com. Licensee: Comcorp of Louisiana License Corp. (acq 12-9-2004; $13,125,000). Population served: 650,000 Natl. Network: Fox. Washington Atty: Fletcher, Heald & Hildreth.

Key Personnel:

Tom Poehler. . . . . pres, gen mgr & natl sls mgr
Morgan Polito . . . . rgnl sls mgr
Katie Flash . . . . prom mgr
Vikki Chapman. . . . . progmg mgr
Tony Guillory. . . . . chief of engrg

KATC— Analog channel: 3. Analog hrs: U'nlimited 100 kw vis, 20 kw aur. 1,740t/1,793g TL: N30 02 19 W92 22 15 On air date: Sept 19, 1962. Box 63333, Lafayette, LA, 70596-3333. 1103 Eraste Landry Rd., Lafayette, LA 70596-3333. Phone: (337) 235-3333. Fax: (337) 235-9363. E-mail: webmaster@katctv.com Web Site: www.katc.com. Licensee: KATC Communications Inc. Group owner: Cordillera Communications Inc. (acq 1995; $24.5 million). Population served: 618,500 Natl. Network: ABC. Natl. Rep: Continental Television Sales. Washington Atty: Dow, Lohnes. Wire Svc: AP Wire Svc: CNN News staff: 45; News: 19.5 hrs wkly.

Key Personnel:

Andrew Shenkan . . . . gen mgr
Bonnie R. Will . . . . gen sls mgr
Arte Richard . . . . mktg dir & mktg mgr
James Warner . . . . news dir
Don Mouton . . . . chief of engrg

KLFY-TV— Analog channel: 10. Digital channel: 56. Analog hrs: 24 309 kw vis, 44.7 kw aur. 1,738t/1,761g TL: N30 19 18 W92 22 41 (CP: 295 kw vis) On air date: June 3, 1955. Box 90665, Lafayette, LA, 70509. Phone: (337) 981-4823. Fax: (337) 984-8323. Web Site: www.klfy.com. Licensee: Young Broadcasting of Louisiana Inc. Group owner: Young Broadcasting Inc. (acq 5-28-88; $51 million; FTR: 12-14-87). Population served: 205,190 Natl. Network: CBS. Natl. Rep: Adam Young. Washington Atty: Wiley, Rein & Fielding. News staff: 24; News: 14 hrs wkly.

Key Personnel:

Mike Barras . . . . gen mgr
Spencer Bienvenu . . . . gen sls mgr
Carolyn Chretien . . . . progmg dir
C.J. Hoyt . . . . news dir
Rodney Evans . . . . chief of engrg

***KLPB-TV**— Analog channel: 24. Digital channel: 24. Analog hrs: 24 2,140 kw vis, 214 kw aur. 1,190t/1,225g TL: N30 02 38 W92 22 14 On air date: May 2, 1981. 7733 Perkins Rd., Baton Rouge, LA, 70810. Phone: (225) 767-5660. Phone: (800) 272-8161. Fax: (225) 767-4299. Web Site: www.lpb.org. Licensee: Louisiana Educational Television Authority. Population served: 1,000,000 Natl. Network: PBS. Washington Atty: Schwartz, Woods & Miller.

Key Personnel:

Beth Courtney . . . . CEO & pres
Bob Neese . . . . prom mgr
Jennifer Howze . . . . progmg dir
Randy Ward . . . . chmn & chief of engrg

KLWB— Analog channel: 50. Analog hrs: 24 5,000 kw vis. ant 994t/998g TL: N30 20 32 W91 58 32 On air date: June 1, 2006. 3501 Northwest Evangeline Thruway, Carencro, LA, 70520. Phone: (337) 896-1600. Fax: (337) 896-2695. Web Site: www.cwtv50.com. Licensee: Wilderness Communications LLC. (acq 8-1-2006). Natl. Network: CW. Natl. Rep: Roslin Television Sales. Washington Atty: Fletcher, Heald & Hildreth.

Key Personnel:

Eddie Blanchard. gen mgr
Dave Pierce . . . . gen sls mgr
Connie Hanks. progmg dir

Lake Charles, LA
(DMA 175)

***KLTL-TV**— Analog channel: 18. Digital channel: 20. Analog hrs: 24 1,260 kw vis, 126 kw aur. ant 1,030t/1,058g TL: N30 23 59 W93 00 10 On air date: May 5, 1981. 7733 Perkins Rd., Baton Rouge, LA, 70810. Phone: (225) 767-5660. Phone: (800) 272-8161. Fax: (225) 767-4299. Web Site: www.lpb.org. Licensee: Louisiana Educational Television Authority. Population served: 1,000,000 Natl. Network: PBS. Rgnl. Network: SECA. Washington Atty: Schwartz, Woods & Miller.

Key Personnel:

Beth Courtney . . . . CEO & pres
Bob Neese. . . . . chmn & prom mgr
Jennifer Howze . . . . progmg dir
Randy Ward . . . . engrg dir

KPLC— Analog channel: 7. Analog hrs: 24 Digital hrs: 24 295 kw vis, 55 kw aur. 1,480t/1,519g TL: N30 23 43 W93 00 08 On air date: September 1954. Box 1490, Lake Charles, LA, 70602. 320 Division St., Lake Charles, LA 70602. Phone: (337) 439-9071. Fax: (337) 437-7600. E-mail: vbilbo@kplctv.com Web Site: www.kplctv.com. Licensee: Libco Inc. Group owner: Liberty Corp. (acq 1-13-2006; grpsl). Population served: 1,450,100 Natl. Network: NBC. Natl. Rep: Harrington, Righter & Parsons. Washington Atty: Covington & Burling LP. News staff: 28.

Key Personnel:

Jim Serra . . . . gen mgr
Diana Mayo . . . . opns mgr
John Ware . . . . gen sls mgr
Stephanie Cormeaux . . . . rgnl sls mgr
Robin Daugereau . . . . progmg dir
Scott Flannagan. news dir
John Scott. chief of engrg

KVHP— Analog channel: 29. Digital channel: 30. Analog hrs: 24 700 kw vis, 131 kw aur. 453t/404g TL: N30 11 50 W93 13 12 (CP: 2,507 kw vis, ant 1,292t) On air date: Dec 12, 1982. 129 W. Prien Lake Rd., Lake Charles, LA, 70601. Phone: (337) 474-1316. Fax: (337) 477-0715. E-mail: info@watchfox.com Web Site: www.watchfox.com. Licensee: National Communications Inc. (acq 10-3-96). Population served: 236,000 Natl. Network: Fox. Washington Atty: Baraff, Koerner, Olender & Hochberg. News staff: 20; News: 9 hrs wkly.

Key Personnel:

Madelyn Bonnot. . . . . gen mgr
Madelyn Bennet . . . . gen sls mgr
Mary Stevens . . . . natl sls mgr
Crystal Miller. . . . . prom dir & prom mgr
Kim Anderson. . . . . progmg dir
Mark Ewing . . . . chief of engrg

Monroe, LA-El Dorado, AR
(DMA 135)

KAQY— Analog channel: 11.316 kw vis. 1,771t/1,929g TL: N32 05 41 W92 10 39 On air date: Dec 10, 1998. Box 4309, Monroe, LA, 71211. 3100 Sterlington Rd., Monroe, LA 71203. Phone: (318) 325-3011. Fax:

(318) 327-7519. Web Site: www.abc-11.com. Licensee: Monroe Broadcasting Inc. Ownership: Charles H. Chatelain, 100% (acq 11-98). Natl. Network: ABC.

Key Personnel:
Joe Currie . . . gen mgr
Carolyn Clampit . . . stn mgr & gen sls mgr
Mike Halbrook . . . prom mgr
Doug Ginn . . . progmg mgr
Pat O'Brien . . . chief of engrg

KARD— Analog channel: 14. Digital channel: 36.5,000 kw vis, 500 kw aur. 2,049t/1,929g TL: N32 05 41 W92 10 39 On air date: Oct 6, 1974. 200 Pavilion Rd., West Monroe, LA, 71292. Phone: (318) 323-1972. Fax: (318) 322-0926. Web Site: www.kard.com. Licensee: Nexstar Finance Inc. Group owner: Nexstar Broadcasting Group Inc. (acq 12-31-03; grpsl). Natl. Network: Fox. Washington Atty: Arter & Hadden.

Key Personnel:
Mark Cummings . . . gen mgr
Chris Tingle . . . gen sls mgr
Esther Phillips . . . prom mgr
Irma Campbell . . . progmg mgr
Randall Kamma . . . news dir
Joey Guy . . . chief of engrg

KEJB— Analog channel: 43.5,000 kw vis. ant 1,738t/1,706g TL: N33 04 41 W92 13 41 On air date: October 2003. 1001 N. 11th St., Monroe, LA, 71201. Phone: (318) 322-4394. Fax: (318) 322-8732. Web Site: www.kejb.com. Licensee: KM Television of El Dorado L.L.C. Group owner: KM Communications Inc. Natl. Network: MyNetworkTV.

Key Personnel:
Wayne Gentry . . . gen mgr
Terri Egloff . . . stn mgr & rgnl sls mgr
Jeremy Tucker . . . opns dir
Vince Anderson . . . sls

***KETZ**— Analog channel: 0. Digital channel: 12.4,000 kw vis. ant 1,765t/1,738g TL: N33 04 41 W92 13 41 On air date: 2006. Box 1250, Conway, AR, 72033. Phone: (501) 682-2386. Fax: (501) 682-4122. Web Site: www.aetn.org. Licensee: Arkansas Educational Television Commission. Natl. Network: PBS.

***KLTM-TV**— Analog channel: 13. Digital channel: 13. Analog hrs: 24 316 kw vis, 31.6 kw aur. 1,777t/1,989g TL: N32 11 45 W92 04 10 On air date: Sept 8, 1976. 7733 Perkins Rd., Baton Rouge, LA, 70810. Phone: (225) 767-5660. Phone: (800) 272-8161. Fax: (225) 767-4299. Web Site: www.lpb.org. Licensee: Louisiana Educational Television Authority. Population served: 1,000,000 Natl. Network: PBS. Washington Atty: Schwartz, Woods & Miller.

Key Personnel:
Beth Courtney . . . CEO & pres
Bob Neese . . . prom mgr
Jennifer Howze . . . progmg dir
Randy Ward . . . prom mgr & engrg dir

KMCT-TV— Analog channel: 39. Digital channel: 38. Analog hrs: 24 560 kw vis, 56 kw aur. 498t/500g TL: N32 30 21 W92 08 54 On air date: Apr 7, 1986. 701 Parkwood Dr., West Monroe, LA, 71291-5435. Phone: (318) 322-1399. Fax: (318) 323-3783. E-mail: lamb@lambbroadcasting.org Web Site: www.lambbroadcasting.org. Licensee: Louisiana Christian Broadcasting Inc. Ownership: Lamb Broadcasting Inc., 100% Group owner: (group owner; (acq 7-13-2004). Washington Atty: Hardy, Chautin & Balkin. News: 6 hrs wkly.

Key Personnel:
Mike Reed . . . pres & gen mgr
David Thompson . . . chief of engrg

KNOE-TV— Analog channel: 8. Digital channel: 7. Analog hrs: 24 Digital hrs: 24 Note: CBS is on KNOE-TV ch 8, CW is on KNOE-DT ch 7. 316 kw vis, 31.6 kw aur. ant 1,930t/1,985g TL: N32 11 50 W92 04 14 On air date: Sept 27, 1953. Box 4067, Monroe, LA, 71211. 1400 Oliver Rd., Monroe, LA 71201. Phone: (318) 388-8888. Fax: (318) 388-0070. Fax: (318) 322-8774. E-mail: knoetv@knoe.com Web Site: www.knoe.com. Licensee: Noe Corp. LLC. Ownership: Betty S. Noe Grantor Retained Annuity Trust. Population served: 428,000 Natl. Network: CBS, CW Digital Network: Note: CBS is on KNOE-TV ch 8, CW is on KNOE-DT ch 7. Natl. Rep: Blair Television. Washington Atty: Cohn & Marks. Wire Svc: CBS Wire Svc: AP Wire Svc: CNN News staff: 28; News: 22 hrs wkly.

Key Personnel:
George Noe . . . pres
Roy Frostenson . . . gen mgr
Tom Cole . . . opns mgr
John Matherne . . . gen sls mgr
Taylor Henry . . . news dir
Jerry Harkins . . . chief of engrg

KTVE— Analog channel: 10. Digital channel: 27.316 kw vis, 63.1 kw aur. 2,027t/2,001g TL: N33 04 41 W92 13 31 On air date: Dec 3, 1955. 200 Pavillion Rd., West Monroe, LA, 71292. Phone: (318) 323-1972. Fax: (318) 322-9718. Web Site: www.nbc10news.net. Licensee: Piedmont Television of Monroe/El Dorado License LLC. Group owner: Piedmont Television Holdings LLC. Population served: 171,600 Natl. Network: NBC. Washington Atty: Cohn & Marks. News staff: 26; News: 13 hrs wkly.

Key Personnel:
Mark Cummings . . . gen mgr
Chris Tingle . . . gen sls mgr
Esther Phillips . . . prom mgr
Sharon Jones . . . progmg mgr
Randall Kamma . . . news dir
Joe Holland . . . chief of engrg

New Orleans, LA (DMA 54)

WDSU— Analog channel: 6. Analog hrs: 24 100 kw vis, 20 kw aur. 928t/928g TL: N29 56 59 W89 57 28 On air date: Dec 18, 1948. 846 Howard Ave., New Orleans, LA, 70113. Phone: (504) 679-0600. Fax: (504) 679-0745. E-mail: feedback6@wdsu.com Web Site: www.wdsu.com. Licensee: New Orleans Hearst-Argyle Television Inc. Group owner: Hearst-Argyle Television Inc. (acq 1999; grpsl). Population served: 675,076 Natl. Network: NBC. Natl. Rep: Eagle Television Sales. Washington Atty: Brooks, Pierce, McLendon, Humphrey & Leonard. Wire Svc: AP News staff: 60; News: 32 hrs. wkly.

Key Personnel:
Joel Vilmenay . . . pres & gen mgr
Wendy Walters . . . dev dir
Frank Raterman . . . gen sls mgr
Joseph Schiltz . . . prom dir
Johnathan Shelley . . . progmg dir & news dir
Joy Maurice . . . progmg dir & pub affrs dir
Chet Guillot . . . chief of engrg

WGNO— Analog channel: 26. Digital channel: 15. Analog hrs: 24 3,140 kw vis. ant 1,014t/1,015g TL: N29 58 57 W89 56 58 On air date: Oct 16, 1967. One Gallaria Blvd, Suite 850, Metairce, LA, 70001. Phone: (504) 525-3838. Fax: (504) 569-0908. E-mail: wgno-tv@tribune.com Web Site: www.abc26.com. Licensee: WGNO Inc. Group owner: Tribune Broadcasting Co. (acq 9-1-83; $21 million; FTR: 8-8-83). Population served: 1,231,000 Natl. Network: ABC. Natl. Rep: TeleRep. Washington Atty: Schnader, Harrison, Segal & Lewis.

Key Personnel:
Larry Delia . . . gen mgr
John Cruse . . . gen sls mgr
Bob Noonan . . . news dir
Steve Zanolini . . . chief of engrg

WHMM-DT— Analog channel: 0. Digital channel: 42.1,000 kw vis. ant 964t/964g TL: N29 58 41 W89 56 26 On air date: June 1, 2007. Box 50790, New Orleans, LA, 70150. Phone: (504) 913-1540. Fax: (504) 340-4737. Web Site: www.mayavision.tv. Licensee: Mayavision Inc. Ownership: Ernesto Schweikert III, 100% (acq 4-20-2006; $950,000). Natl. Network: Telemundo (Spanish). Foreign lang progmg: SP 168

WHNO— Analog channel: 20. Analog hrs: 24 5,000 kw vis, 500 kw aur. 905t TL: N29 55 11 W90 01 29 On air date: October 1994. 839 Saint Charles Ave. Suite 309, New Orleans, LA, 70130-3744. Phone: (504) 681-0210. Fax: (504) 681-0180. E-mail: whno@lesea.com Web Site: www.whno.com. Licensee: Le Sea Broadcasting Corp. Group owner: Le Sea Broadcasting News: 10 hrs wkly.

Key Personnel:
Dean Powery . . . gen mgr
Steve Warnecke . . . gen mgr & progmg mgr
Bob Lawrence . . . chief of engrg
Ivan Hinson . . . sls

***WLAE-TV**— Analog channel: 32.55 kw vis, 11 kw aur. 1,020t/1,045g TL: N29 58 57 W89 57 09 (CP: 2,290 kw vis, 229 kw aur) On air date: July 8, 1984. 3330 N. Causeway Blvd., Suite 345, Metairie, LA, 70002-3573. Phone: (504) 866-7411. Fax: (504) 840-9838. Web Site: www.pbs.org/wlae. Licensee: Educational Broadcasting Foundation Inc. Population served: 270,000 Natl. Network: PBS. Washington Atty: Marmet & McCombs. Foreign lang progmg: SP 3

WNOL-TV— Analog channel: 38.5,000 kw vis, 500 kw aur. 1,049t/1,049g TL: N29 58 41 W89 56 26 On air date: Mar 25, 1984. 1400 Poydras St., Suite 745, New Orleans, LA, 70112-5100. Phone: (504) 525-3838. Fax: (504) 569-0908. E-mail: wnoltv@tribune.com Web Site: www.neworleanscw38.com. Licensee: WGNO Inc. Group owner: Tribune Broadcasting Co. (acq 1-12-2000; approximately $95 million for remaining 67% with WATL(TV) Atlanta, GA). Population served: 1,000,000 Natl. Network: CW. Natl. Rep: MMT.

Key Personnel:
Larry Delia . . . pres & gen mgr
Bob Noonan . . . gen mgr & news dir
John Cruse . . . gen sls mgr
Steve Zanolini . . . chief of engrg

WPXL— Analog channel: 49. Digital channel: 50. Analog hrs: 24 5,000 kw vis, 500 kw aur. ant 945t/948g TL: N29 55 11 W90 01 29 On air date: Mar 19, 1989. 3900 Veterans Memorial Blvd., Suite 202, Metairie, LA, 70002. Phone: (504) 887-9795. Fax: (504) 887-1518. Web Site: www.ionline.tv. Licensee: Flinn Broadcasting Corp. (acq 4-30-93; $135,000; FTR: 5-24-93).

Key Personnel:
Ami Jenkins . . . opns mgr & progmg dir
Matt Pate . . . gen mgr & rgnl sls mgr
Ernie Harvey . . . chief of engrg

WUPL— Analog channel: 54. Digital channel: 24. Analog hrs: 24 Digital hrs: 24 4,376 kw vis, 437.6 kw aur. ant 658t On air date: June 1, 1995. 1024 N Ramport St, New Orleans, LA, 70116. Phone: (504) 529-4444. Web Site: wupltv.com. Licensee: CBS Radio Stations Inc. Group owner: Viacom Television Stations Group. Natl. Network: MyNetworkTV.

Key Personnel:
Bud Brown . . . gen mgr
Mike Zikmund . . . gen sls mgr
Christopher Merrifield . . . prom dir
Carol St.Martin . . . progmg mgr
Robert Gass . . . chief of engrg

WVUE— Analog channel: 8. Digital channel: 29. Analog hrs: 24 316 kw vis, 31.6 kw aur. ant 990t/1,046g TL: N29 57 14 W89 56 58 On air date: Feb 1, 1959. 1025 S. Jefferson Davis Pkwy., New Orleans, LA, 70125. Phone: (504) 486-6161. Fax: (504) 483-1101. E-mail: info@fox8live.com Web Site: www.fox8live.com. Licensee: Emmis Television License LLC. Group owner: Emmis Communications Corp. (acq 7-16-98; grpsl). Population served: 1,648,000 Natl. Network: Fox. Natl. Rep: Harrington, Righter & Parsons. Washington Atty: Brooks, Pierce, McLendon, Humphrey & Leonard. News staff: 54; News: 21 hrs wkly.

Key Personnel:
Vanessa Oubre . . . VP & gen mgr
Johnny Faith . . . gen sls mgr
Michelle Kehoe Ogden . . . natl sls mgr
Dee Dee Indovina . . . rgnl sls mgr
Mimi Strawn . . . news dir

WWL-TV— Analog channel: 4. Digital channel: 36. Analog hrs: 24 100 kw vis, 10 kw aur. 1,000t/1,049g TL: N29 54 23 W90 02 23 On air date: Sept 7, 1957. 1024 N. Rampart St., New Orleans, LA, 70116. Phone: (504) 529-4444. Fax: (504) 529-6483. Web Site: www.wwltv.com. Licensee: WWL-TV Inc. Group owner: Belo Corp. (acq 1994). Population served: 609,000 Natl. Network: CBS. Natl. Rep: TeleRep. Washington Atty: Holland & Knight.

Key Personnel:
Bud Brown . . . gen mgr
Mike Zikmund . . . gen sls mgr
Christopher Merrifield . . . prom dir & pub affrs dir
Carol St. Martin . . . progmg mgr
Robert Gass . . . chief of engrg

***WYES-TV**— Analog channel: 12. Digital channel: 11. Analog hrs: 24 Digital hrs: 24 316 kw vis, 31.6 kw aur. 1,010t/1,046g TL: N29 56 58 W29 57 14 On air date: Apr 1, 1957. Stn currently dark Box 24026, New Orleans, LA, 70184. Phone: (504) 486-5511. Web Site: www.wyes.org. Licensee: Greater New Orleans Educational TV Foundation. Population served: 675,000 Natl. Network: PBS. Washington Atty: Schwartz, Woods & Miller.

Shreveport, LA (DMA 81)

***KLTS-TV**— Analog channel: 24. Analog hrs: 24 1,620 kw vis, 162 kw aur. 1,070t/1,080g TL: N32 40 41 W93 55 35 On air date: Aug 9, 1978. 7733 Perkins Rd., Baton Rouge, LA, 70810. Phone: (225) 767-5660. Fax: (225) 767-4299. Web Site: www.lpb.org. Licensee: Louisiana Education Television Authority. Population served: 1,000,000 Natl. Network: PBS. Washington Atty: Schwartz, Woods & Miller.

Key Personnel:
Beth Courtney . . . CEO & pres
Bob Neese . . . prom mgr
Jennifer Howze . . . progmg dir
Randy Ward . . . engrg dir

KMSS-TV— Analog channel: 33. Digital channel: 34. Analog hrs: 24 4,570 kw vis, 457 kw aur. 1,813t/1,781g TL: N32 36 51 W93 48 59 On air date: Oct 6, 1985. 3519 Jewella Ave., Shreveport, LA, 71109. Phone: (318) 631-5677. Fax: (318) 631-4194. Web Site: www.kmsstv.com. Licensee: Comcorp of Texas License Corp. Group owner: Communications Corp. of America (acq 10-94). Natl. Network: Fox. Washington Atty: Fletcher, Heald & Hildreth, PLC.

Key Personnel:
Paula Hayward . . . gen mgr
Susan Newman . . . gen sls mgr
Mike Halbrook . . . prom mgr & pub affrs dir

Doug Ginn . . . . . progmg dir
Pat O'Brien . . . . . chief of engrg

KPXJ— Analog channel: 0. Digital channel: 21.3,020 kw vis. ant 469t TL: N32 44 40 W93 22 54 On air date: 1999. Box 4066, Shreveport, LA, 71104. 312 E. Kings Hwy, Shreveport, LA 71104. Phone: (318) 861-5800. Fax: (318) 219-4600. Web Site: www.kpxj21.com. Licensee: Minden Television Co. LLC. Ownership: Lauren Wray Ostendorff, 100% (acq 5-7-2004; $10 million). Natl. Network: CW. Washington Atty: Garvey, Schubert & Barer.
Key Personnel:
Lauren Wray Ostendorff . . . . . pres
George Sirven . . . . . gen mgr

KSHV— Analog channel: 45. Digital channel: 44.1,660 kw vis. 662t TL: N32 35 38 W93 51 39 (CP: 2,982 kw vis, ant 1,664t) On air date: Apr 15, 1994. 3519 Jewella Ave., Shreveport, LA, 71109. Phone: (318) 631-4545. Fax: (318) 621-9688. Web Site: www.kshv.com. Licensee: White Knight Broadcasting of Shreveport License Corp. Group owner: White Knight Holdings Inc. (acq 1995; $3.8 million). Population served: 410,000 Natl. Network: MyNetworkTV. Washington Atty: Pillsbury, Winthrop, Shaw Pittman, LLC.
Key Personnel:
Paula Hayward . . . . . gen mgr
Issac Turner . . . . . sls dir & progmg
Susan Newman . . . . . gen sls mgr & rgnl sls mgr
Mike Halbrook . . . . . mktg dir & prom mgr
Pat O'Brien . . . . . chief of engrg

KSLA-TV— Analog channel: 12. Digital channel: 0.316 kw vis, 40.7 kw aur. 1,800t/1,800g TL: N32 40 29 W93 55 59 On air date: Jan 1, 1954. 1812 Fairfield Ave., Shreveport, LA, 71101. Phone: (318) 222-1212. Fax: (318) 677-6703. E-mail: ksla@ksla.com Web Site: ksla.com. Licensee: KSLA License Subsidiary LLC. Group owner: Raycom Media Inc. (acq 9-1-96; grpsl). Population served: 461,600 Natl. Network: CBS. Natl. Rep: TeleRep. Wire Svc: UPI
Key Personnel:
James Smith . . . . . VP & gen mgr
Cindy Delaney . . . . . sls dir
Barbara Bennett. . . . . prom dir
Jayne Ruben . . . . . news dir
Ted Small . . . . . chief of engrg

KTAL-TV—(Texarkana, TX) Analog channel: 6.100 kw vis, 10 kw aur. 1,580t/1,552g TL: N32 54 12 W94 00 23 On air date: Aug 16, 1953. 3150 N. Market St., Shreveport, LA, 71107. Box 7428, Shreveport, LA 71107. Phone: (318) 629-6000. Fax: (318) 629-6001. Web Site: www.arklatexhomepage.com. Licensee: Nexstar Finance Inc. Group owner: Nexstar Broadcasting Group Inc. (acq 9-11-2000; $35.25 million). Natl. Network: NBC. Washington Atty: Covington & Burling.
Key Personnel:
Scott Thomas. . . . . gen mgr
Cheryl Olive . . . . . gen sls mgr
Jean Byrd . . . . . progmg dir
Andrew Pontz . . . . . news dir
Kevin Southernland . . . . . chief of engrg

KTBS-TV— Analog channel: 3. Analog hrs: 24 100 kw vis, 20 kw aur. 1,780t/1,800g TL: N32 41 08 W93 46 00 On air date: Sept 3, 1955. Box 44227, Shreveport, LA, 71134-4227. 312 E. Kings Hwy., Shreveport, LA 71104-3554. Phone: (318) 861-5800. Fax: (318) 219-4680. E-mail: ktbsnews@ktbs.com Web Site: www.ktbs.com. Licensee: KTBS Inc. Ownership: Helen H. Wray, Florence H. Wray, George D. Wray Jr. Population served: 750,000 Natl. Network: ABC. Washington Atty: Fletcher, Heald & Hildreth. Wire Svc: AP
Key Personnel:
Lauren Wray Ostendorff . . . . . pres
George Sirven . . . . . gen mgr
Linda Howard . . . . . rgnl sls mgr
Cheryl May . . . . . prom mgr
Randy Bain . . . . . news dir
Dale Cassidy. . . . . chief of engrg
Bernadette Collier . . . . . progmg

Maine

Auburn

see Portland-Auburn, ME market

Bangor, ME (DMA 152)

WABI-TV— Analog channel: 5. Digital channel: 19. Analog hrs: 24 Digital hrs: 24 Note: CBS is on WABI-TV ch 5, CW is on WABI-DT ch 19. 40 kw vis, 6 kw aur. ant 1,316t/490g TL: N44 42 13 W69 04 47 On air date: Jan 25, 1953. 35 Hildreth St., Bangor, ME, 04401. Phone: (207) 947-8321. Fax: (207) 941-9378. E-mail: wabi@wabi.tv Web Site: www.wabi.tv. Licensee: Community Broadcasting Service. Group owner: Diversified Communications (acq 10-7-53; $125,000; FTR: 10-12-53). Population served: 200,000 Natl. Network: CBS, CW Digital Network: Note: CBS is on WABI-TV ch 5, CW is on WABI-DT ch 19. Natl. Rep: Continental Television Sales. Washington Atty: Irwin, Campbell & Tannenwald. Wire Svc: AP News staff: 30; News: 25 hrs wkly.
Key Personnel:
Michael Young . . . . . VP & gen mgr
Tom Gass . . . . . sls dir
Steve Hiltz . . . . . progmg dir
Jim Morris . . . . . news dir
Dale Carter . . . . . chief of engrg
Keith Allen . . . . . opns

WLBZ— Analog channel: 2. Analog hrs: 24 51.3 kw vis, 10.2 kw aur. 630t/99g TL: N44 44 10 W68 40 17 On air date: Sept 12, 1954. 329 Mt. Hope Ave., Bangor, ME, 04401. Phone: (207) 942-4821. Fax: (207) 945-6816. Fax: (207) 942-2109 (news). Web Site: www.wlbz2.com. Licensee: Pacific and Southern Co. Inc. Group owner: Gannett Broadcasting (acq 1998; $110 million with WCSH(TV) Portland). Population served: 214,800 Natl. Network: NBC. Washington Atty: Wiley, Rein & Fielding. News staff: 22; News: 33 hrs wkly.
Key Personnel:
Judy Haran. . . . . gen mgr
Mark Parent . . . . . prom dir
Mike Marshall. . . . . progmg dir
Heather Seavey. . . . . news dir
Mike Curry . . . . . news dir
Dave Mundee . . . . . engrg dir & chief of engrg

***WMEB-TV—** Analog channel: 12. Digital channel: 9.299 kw vis, 30 kw aur. 990t/369g TL: N44 45 36 W68 33 59 On air date: Sept 23, 1963. 1450 Lisbon St., Lewiston, ME, 04240. Phone: (800) 884-1717. Phone: (207) 783-9101. Fax: (207) 783-5193. Fax: (207) 942-2857. E-mail: comments@mpbn.org Web Site: www.mpbn.org. Licensee: Maine Public Broadcasting Corp. (acq 6-23-92; FTR: 7-13-92). Population served: 151,000 Natl. Network: PBS. Rgnl. Network: Eastern Educ. Washington Atty: Dow, Lohnes PLLC. Foreign lang progmg: SP 1
Key Personnel:
Jim Dowe . . . . . gen mgr
Lou Morin. . . . . mktg mgr
Jeff Pierce. . . . . prom dir & prom mgr
Charles Beck . . . . . progmg mgr
Keith Shortall. . . . . news dir

***WMED-TV—** Analog channel: 13. Digital channel: 15.31.6 kw vis, 6.2 kw aur. 430t/190g TL: N45 01 44 W67 19 24 (CP: 100 kw vis, ant 439t) On air date: September 1965. 1450 Lisbon St., Lewiston, ME, 04240. 65 Texas Ave, Bangor, ME 04401. Phone: (800) 884-1717. Phone: (207) 783-9101. Fax: (207) 783-5193. Fax: (207) 942-2857. E-mail: comments@mpbn.net Web Site: www.mpbn.org. Licensee: Maine Public Broadcasting Corp. (acq 6-23-92; FTR: 7-13-92). Natl. Network: PBS. Washington Atty: Dow, Lohnes PLLC. Foreign lang progmg: SP 1
Key Personnel:
Jim Dowe . . . . . gen mgr
Lou Morin. . . . . mktg mgr
Jeff Pierce . . . . . prom mgr
Charles Beck . . . . . progmg mgr
Keith Shortall. . . . . news dir

WVII-TV— Analog channel: 7. Digital channel: 14.316 kw vis, 31.6 kw aur. ant 819t/137g TL: N44 45 35 W68 34 01 On air date: Oct 15, 1965. 371 Target Industrial Cir., Bangor, ME, 04401. Phone: (207) 945-6457. Fax: (207) 942-0511. E-mail: tv7news@wvii.com Web Site: www.wvii.com. Licensee: Bangor Communications LLC. Group owner: Rockfleet Broadcasting Inc. Population served: 429,000 Natl. Network: ABC, Fox. Washington Atty: Mullin, Rhyne, Emmons & Topel. News staff: 15; News: 6 hrs wkly.
Key Personnel:
Mike Palmer . . . . . VP
Mike Palmer . . . . . gen mgr
Keryn Smith. . . . . gen sls mgr & natl sls mgr
Gene Hardin . . . . . prom mgr & pub affrs dir
George Thomas. . . . . news dir
Mike Staples . . . . . chief of engrg

Portland-Auburn, ME (DMA 74)

***WCBB—** Analog channel: 10. Digital channel: 17.309 kw vis, 30.9 kw aur. 1,000t/641g TL: N44 09 16 W70 00 37 On air date: Nov 13, 1961. 1450 Lisbon St., Lewiston, ME, 04240. Phone: (800) 884-1717. Phone: (207) 783-9101. Fax: (207) 783-5193. Fax: (207) 942-2857. E-mail: comments@mpbn.net Web Site: www.mpbn.net. Licensee: Maine Public Broadcasting Corp. (acq 6-23-92; FTR: 7-13-92). Population served: 600,000 Natl. Network: PBS. Rgnl. Network: Eastern Educ. Washington Atty: Dow, Lohnes PLLC. Foreign lang progmg: SP 2
Key Personnel:
Jim Dowe . . . . . gen mgr
Lou Morin. . . . . mktg mgr
Jeff Pierce . . . . . prom mgr
Charles Becks . . . . . progmg mgr
Keith Shortall . . . . . news dir

WCSH— Analog channel: 6. Digital channel: 44.100 kw vis, 20 kw aur. ant 2,000t/1,292g TL: N43 51 32 W70 42 40 On air date: Dec 1, 1953. One Congress Sq., Portland, ME, 04101. Phone: (207) 828-6666. Fax: (207) 828-6620. E-mail: wcsh6@wcsh6.com Web Site: www.wcsh6.com. Licensee: Pacific and Southern Co. Inc. Group owner: Gannett Broadcasting (acq 1-98). Population served: 866,000 Natl. Network: NBC. Washington Atty: Wiley, Rein & Fielding.
Key Personnel:
Steve Thaxton . . . . . pres & gen mgr
Dave Abel. . . . . gen sls mgr
Mike Marshall. . . prom VP, prom mgr, progmg VP & progmg mgr
Mike Curry. . . . . news dir
Dave Mundee . . . . . chief of engrg

WGME-TV— Analog channel: 13. Digital channel: 38.295 kw vis, 29.5 kw aur. ant 1,609t/1,665g TL: N43 55 28 W70 29 28 On air date: May 16, 1954. 81 Northport Dr., Portland, ME, 04103. Phone: (207) 797-1313. Fax: (207) 878-3505. E-mail: tvmail@wgme.com Web Site: www.wgme.com. Licensee: WGME Licensee L.L.C. Group owner: Sinclair Broadcast Group Inc. (acq 5-3-99; grpsl). Population served: 340,000 Natl. Network: CBS. Washington Atty: Dow, Lohnes PLLC. News staff: 55; News: 25 hrs wkly.
Key Personnel:
Alisa Burris . . . . . pres & prom
Terry Cole . . . . . gen mgr
Don Barr . . . . . gen sls mgr
Robb Atkinson . . . . . rgnl sls mgr & news dir
Kate Reilly . . . . . progmg dir
Craig Clark . . . . . engrg dir
Gary Legters . . . . . opns

***WLED-TV—** Analog channel: 49. Digital channel: 48. Analog hrs: 24 93.3 kw vis, 9.33 kw aur. 1,280t/400g TL: N44 21 14 W71 44 23 On air date: Feb 7, 1968. 268 Mast Rd., Durham, NH, 03824-4601. Phone: (603) 868-1100. Fax: (603) 868-7552. E-mail: themailbox@nhptv.org Web Site: www.nhptv.org. Licensee: University of New Hampshire. Natl. Network: PBS. Washington Atty: Schwartz, Woods & Miller.
Key Personnel:
Peter A. Frid . . . . . gen mgr
Dennis Malloy . . . . . dev dir & sls dir
Jeff Moris. . . . . prom mgr
Brian Shepperd. . . . . engrg dir & chief of engrg
Rebroadcasts *WENH Durham 100%.

***WMEA-TV—** Analog channel: 26. Digital channel: 45.589 kw vis, 117 kw aur. 800t/550g TL: N43 25 00 W70 48 09 On air date: March 1975. 1450 Lisbon St, Lewiston, ME, 04240. 65 Texas Ave., Bangor, ME 04401. Phone: (207) 783-9101. Fax: (207) 942-2857. Fax: (207) 783-5193. E-mail: comments@mpbn.net Web Site: www.mpbn.net. Licensee: Maine Public Broadcastig Corp. (acq 6-23-92; FTR: 7-13-92). Population served: 500,000 Natl. Network: PBS. Washington Atty: Dow, Lohnes PLLC.
Key Personnel:
Christopher Amann . . . . . CFO
P. James Dowe . . . . . pres & gen mgr

WMTW-TV—Poland Spring, Analog channel: 8. Digital channel: 46. Analog hrs: 24 316 kw vis. ant 1,994t/1,630g TL: N43 50 44 W70 45 43 On air date: Aug 31, 1954. Box 8, Auburn, ME, 04210. 99 Danville Corner Rd, Auburn, ME 04210. Phone: (207) 782-1800. Phone: (207) 775-1800. Fax: (207) 783-7371. Fax: (207) 782-2165. E-mail: wmtw@wmtw.com Web Site: www.wmtw.com. Licensee: Hearst-Argyle Properties Inc. Group owner: Hearst-Argyle Television Inc. (acq 5-11-2004; $37.5 million). Natl. Network: ABC. Natl. Rep: Eagle Television Sales. Washington Atty: Brooks & Pierce. Wire Svc: AP News: 13.5 hrs wkly.
Key Personnel:
Ken Bauder . . . . . gen mgr
Gary Jensen. . . . . gen sls mgr
Gloria Shallcross . . . . . progmg mgr
George Matz . . . . . news dir

WPFO— Analog channel: 23. Analog hrs: 24 5,000 kw vis. 1,086t TL: N44 09 15 W70 00 37 On air date: Apr 24, 2003. 233 Oxford St., Suite 35, Portland, ME, 04101. Phone: (207) 828-0023. Fax: (207) 347-7330. Web Site: www.fox23me.com. Licensee: CMCG Portland License LLC. Group owner: MAX Media L.L.C. (acq 4-7-2003; $10 million with WVIF(TV) Christiansted, VI). Population served: 400,000 Natl. Network: Fox. Natl. Rep: TeleRep. Washington Atty: Williams Mullen & Garvey, Schubert & Barer. News staff: one; News: 18.5 hrs wkly.

Key Personnel:
Dirk Brinkerhoff . . . gen mgr
Tom McArthur . . . gen sls mgr
Teresa Pinney . . . prom mgr
Arnie Marzen . . . news dir
Dave Cox . . . chief of engrg

WPME— Analog channel: 35. Digital channel: 28. Analog hrs: 24 1,100 kw vis, 110 kw aur. ant 840t TL: N43 51 06 W70 19 40 (CP: 5,000 kw vis, ant 912t) On air date: Aug 1, 1997. 4 Ledgeview Dr., Westbrook, ME, 04092. Phone: (207) 774-0051. Fax: (207) 774-6849. E-mail: comments@ourmaine.com Web Site: www.ourmaine.com. Licensee: Pegasus Satellite Communications Inc., debtor-in-possession. Ownership: Pegasus Communications Corp., 100% Group owner: (group owner; (acq 2-10-2005; $3,775,523). Population served: 865,000 Natl. Network: MyNetworkTV. Washington Atty: Covington & Burling. News: 3.5 hrs wkly.
Key Personnel:
Douglas Finck . . . gen mgr & stn mgr
Dooouglas Finck. . . . gen sls mgr
Cory Culleton . . . rgnl sls mgr
Matt Maloney . . . progmg mgr
Roy Ouellette . . . chief of engrg

WPXT— Analog channel: 51. Digital channel: 43. Analog hrs: 24 3,035 kw vis, 303 kw aur. ant 1,000t/720g TL: N43 51 06 W70 19 40 On air date: 1986. 4 Ledgeview Dr., Westbrook, ME, 04092. Phone: (207) 774-0051. Fax: (207) 774-6849. E-mail: comments@ourmaine.com Web Site: www.ourmaine.com. Licensee: HMW Inc., debtor-in-possession. Group owner: Pegasus Broadcast Television Inc. (acq 5-17-96; $17.5 million). Population served: 865,000 Natl. Network: CW. Washington Atty: Covington & Burling. News staff: 25; News: 3.5 hrs wkly.
Key Personnel:
Douglas Finck . . . gen mgr & gen sls mgr
Cory Culleton . . . rgnl sls mgr
Matt Maloney . . . progmg mgr
Roy Ouellette . . . chief of engrg

Presque Isle, ME (DMA 204)

WAGM-TV— Analog channel: 8. Digital channel: 16.Note: CBS is on WAGM-TV ch 8, Fox is on WAGM-DT ch 16. 120.0 kw vis, 12.0 kw aur. 1,148t/142g TL: N46 43 04 W67 48 34 On air date: Oct 13, 1956. 12 Brewer Rd., Presque Isle, ME, 04769. Phone: (207) 764-4461. Fax: (207) 764-5329. E-mail: wagmtv@wagmtv.com Web Site: www.wagmtv.com. Licensee: NEPSK Inc. Ownership: Peter P. Kozloski, 100%. (acq 3-8-91; grpsl; FTR: 4-1-91). Population served: 89,000 Natl. Network: CBS, Fox Digital Network: Note: CBS is on WAGM-TV ch 8, Fox is on WAGM-DT ch 16. Washington Atty: Koteen & Naftalin. News staff: 12; News: 14 hrs wkly.
Key Personnel:
Gordon Wark . . . gen mgr
Linda Connolly. gen sls mgr, natl sls mgr, rgnl sls mgr, prom dir, prom mgr & progmg mgr
Jon Gulliver . . . news dir
Brett Lovley . . . chief of engrg

***WMEM-TV—** Analog channel: 10. Digital channel: 12.299 kw vis, 30 kw aur. 1,090t/158g TL: N46 33 05 W67 48 37 On air date: Feb 17, 1964. 1450 Lisbon St., Lewiston, ME, 04240-3514. 65 Texas Ave., Bangor, ME 04401. Phone: (800) 884-1717. Phone: (207) 783-9101. Fax: (207) 783-5193. Fax: (207) 942-2857. E-mail: comments@mpbn.net Web Site: www.mpbc.net. Licensee: Maine Public Broadcasting Network. (acq 6-23-92; FTR: 7-13-92). Population served: 151,000 Natl. Network: PBS. Washington Atty: Dow, Lohnes PLLC. Foreign lang progmg: SP 2

Maryland

Baltimore, MD (DMA 24)

WBAL-TV— Analog channel: 11. Digital channel: 59. Analog hrs: 24 316 kw vis, 31.6 kw aur. 1,000t/998g TL: N39 20 05 W76 39 03 On air date: Mar 11, 1948. 3800 Hooper Ave., Baltimore, MD, 21211. Phone: (410) 467-3000. Fax: (410) 338-6460. Web Site: www.wbaltv.com. Licensee: WBAL Hearst-Argyle Television Inc. Group owner: (group owner) Population served: 905,759 Natl. Network: NBC. Natl. Rep: Eagle Television Sales. Washington Atty: Brooks, Pierce, McLendon, Humphrey & Leonard. Wire Svc: UPI News staff: 63; News: 24 hrs wkly.
Key Personnel:
Jordan Wertlieb . . . gen mgr
Michelle Butt . . . news dir
Wanda Draper . . . progmg dir & pub affrs dir

WBFF— Analog channel: 45. Digital channel: 46.1,292 kw vis, 258 kw aur. 1,266t TL: N39 20 10 W76 38 59 On air date: Apr 11, 1971. 2000 W. 41st St., Baltimore, MD, 21211. Phone: (410) 467-4545. Fax: (410) 467-5090. Web Site: www.foxbaltimore.com. Licensee: Chesapeake Television Licensee L.L.C. Group owner: Sinclair Broadcast Group Inc. (acq 9-10-90; grpsl; FTR: 10-15-90). Population served: 905,759 Natl. Network: Fox. Natl. Rep: TeleRep. Washington Atty: Shaw Pittman.
Key Personnel:
William Fanshawe . . . gen mgr
Russell Lucas. . . . opns dir & opns mgr
Peter Paisley. . . . sls dir & gen sls mgr
Sharon Wylie . . . mktg dir & mktg mgr
Peter Ferraro . . . prom mgr
Mary Press. . . . progmg dir
Scott Livingston . . . news dir
David Hackney . . . engrg mgr

WJZ-TV— Analog channel: 13. Digital channel: 38.316 kw vis, 31.6 kw aur. 990t TL: N39 20 05 W76 39 03 On air date: Nov 2, 1948. 3725 Malden Ave., Baltimore, MD, 21211. Phone: (410) 466-0013. Fax: (410) 578-7502. E-mail: ness@wjz.com Web Site: www.wjz.com. Licensee: Viacom Inc. Group owner: Viacom Television Stations Group (acq 6-28-57; $4.4 million; FTR: 7-1-57). Population served: 935,759 Natl. Network: CBS. Washington Atty: Wilkes, Artis, Hedrick & Lane. Wire Svc: UPI
Key Personnel:
Jay B. Newman . . . VP & gen mgr
Vee Bennedetto . . . opns dir
Sara Scott . . . sls dir
Gail Bending . . . news dir
Susan Otradovec. . . . pub affrs dir
Rick Seaby . . . engrg dir
Michelle Dowd-Wood . . . progmg

WMAR-TV— Analog channel: 2. Digital channel: 52. Analog hrs: 24 100 kw vis, 11 kw aur. 1,000t/999g TL: N39 20 05 W76 39 03 On air date: Oct 27, 1947. 6400 York Rd., Baltimore, MD, 21212. Phone: (410) 377-2222. Fax: (410) 377-0493. E-mail: hooper@wmar.com Web Site: www.abc2news.com. Licensee: Scripps-Howard Broadcasting Co. Group owner: (group owner; (acq 1991; $125 million; FTR: 9-3-90). Population served: 905,759 Natl. Network: ABC. Natl. Rep: Harrington, Righter & Parsons. Washington Atty: Baker & Hostetler. Wire Svc: AP
Key Personnel:
Bill Hooper . . . VP & gen mgr
Shirley Pridgeon. . . . sls dir
Darlene Dorman. . . . progmg dir
David Silverstein . . . news dir
Paul Garnet . . . chief of engrg

***WMPB—** Analog channel: 67. Digital channel: 29. Analog hrs: 24 646 kw vis, 76.38 kw aur. ant 820t/672g TL: N39 27 01 W76 46 37 On air date: 1986. 11767 Owings Mills Blvd., Owings Mills, MD, 21117-1499. Phone: (410) 356-5600. Fax: (410) 581-6579. E-mail: comments@mpt.org Web Site: www.mpt.org. Licensee: Maryland Public Broadcasting Commission. Population served: 600,000 Natl. Network: PBS. Washington Atty: Schwartz, Woods & Miller.
Key Personnel:
Robert J. Shuman . . . CEO & pres
Larry D. Unger. . . . CFO
George Beneman . . . opns VP
Kirby Storms . . . chief of engrg

***WMPT—** Analog channel: 22. Digital channel: 42. Analog hrs: 24 5,000 kw vis. ant 895t/827g TL: N39 00 36 W76 36 33 On air date: 1986. 11767 Owings Mills Blvd., Owings Mills, MD, 21117-1499. Phone: (410) 356-5600. Fax: (410) 581-6579. E-mail: comments@mpt.org Web Site: www.mpt.org. Licensee: Maryland Public Broadcasting Commission. Population served: 3000000 Natl. Network: PBS. Washington Atty: Schwartz, Woods & Miller.
Key Personnel:
Robert Shuman . . . CEO & pres
Larry Unger . . . CFO
Kirby Storms. . . . chief of engrg
Rebroadcasts WMPB(TV) Baltimore 100%.

WNUV— Analog channel: 54. Digital channel: 15. Analog hrs: 24 5,000 kw vis, 500 kw aur. 1,148t/998g TL: N39 17 15 W76 45 38 On air date: July 1, 1982. 2000 W. 41st St., Baltimore, MD, 21211. Phone: (410) 467-8854. Fax: (410) 467-5093. Web Site: www.wbbaltimore.com. Licensee: Baltimore (WNUV-TV) Licensee Inc. Group owner: Cunningham Broadcasting Corporation (acq 1-9-2002). Natl. Network: CW. Washington Atty: Arter & Hadden.
Key Personnel:
William Fanshawe . . . gen mgr
Russell Lucas . . . opns mgr & opns mgr
Billy Robbins . . . gen sls mgr & rgnl sls mgr
Sharon Wylie . . . mktg mgr
Peter Ferraro . . . prom mgr
Mary Press. . . . progmg dir
David Hackney . . . chief of engrg

WUTB— Analog channel: 24. Digital channel: 41. Analog hrs: 24 1,170 kw vis, 117 kw aur. 1,069t/996g TL: N39 17 15 W76 45 38 On air date: December 1985. 4820 Seton Dr., Suite M-N, Baltimore, MD, 21215. Phone: (410) 358-2400. Fax: (410) 764-7232. E-mail: wttg-hr@foxtv.com Web Site: www.my24wutb.com. Licensee: Fox Television Stations Inc. Group owner: (group owner; (acq 7-31-2001; grpsl). Natl. Network: MyNetworkTV. Natl. Rep: Fox Stations Sales. Washington Atty: Law Offices of Hogan & Hartson.
Key Personnel:
Alan J. Sawyer. . . . VP & gen mgr
Brock Abernathy . . . natl sls mgr
Shirley Pridgeon . . . rgnl sls mgr
Dan Carlin. . . . progmg VP
Duane Myers. . . . chief of engrg

Frederick

see Washington, DC (Hagerstown, MD) market

Hagerstown

see Washington, DC (Hagerstown, MD) market

Oakland

see Pittsburgh, PA market

Salisbury, MD (DMA 148)

WBOC-TV— Analog channel: 16. Digital channel: 21. Analog hrs: 24 Note: CBS is on WBOC-TV ch 16, Fox is on WBOC-DT ch 21. 4,070 kw vis, 407 kw aur. 980t/1,003g TL: N38 30 16 W75 38 35 On air date: July 15, 1954. Box 2057, Salisbury, MD, 21802-2057. 1729 N. Salisbury Blvd., Salisbury, MD 21801. Phone: (410) 749-1111. Fax: (410) 749-2361. E-mail: wboc@wboc.com Web Site: www.wboc.com. Licensee: WBOC Inc. Ownership: Draper Holdings Business Trust. (acq 9-80; $8 million). Population served: 85,500 Natl. Network: CBS, Fox Digital Network: Note: CBS is on WBOC-TV ch 16, Fox is on WBOC-DT ch 21. Washington Atty: Covington & Burling. News staff: 40; News: 31 hrs wkly.
Key Personnel:
Rick Jordan . . . VP & gen mgr
David Speicher. . . . natl sls mgr
Bob Bachman . . . rgnl sls mgr
Steve Bach . . . rgnl sls mgr
Mary Borger. . . . mktg mgr & prom mgr
John Dearing. . . . news dir
Danny Panicella . . . chief of engrg

***WCPB—** Analog channel: 28. Digital channel: 56. Analog hrs: 24 2,190 kw vis. ant 515t/515g TL: N38 23 09 W75 35 33 On air date: 1986. 11767 Owings Mills Blvd., Owings Mills, MD, 21117-1499. Phone: (410) 356-5600. Fax: (410) 581-6579. E-mail: comments@mpt-org Web Site: www.mpt.org. Licensee: Maryland Public Broadcasting Commission. Population served: 600,000 Natl. Network: PBS. Washington Atty: Schwartz, Woods & Miller.
Key Personnel:
Robert Shuman . . . CEO & pres
Larry Unger . . . CFO
George Benaman II . . . VP
Kirby Storms . . . chief of engrg
Rebroadcasts WMPB(TV) Baltimore 100%.

***WDPB—** Analog channel: 64. Digital channel: 44. Analog hrs: 24 191 kw vis, 19.1 kw aur. ant 640t/657g TL: N38 39 15 W75 36 42 On air date: December 1982. The Linden Bldg., 625 Orange St., Wilmington, DE, 19801. Phone: (302) 888-1200. Fax: (302) 575-0346. E-mail: whyydbc@whyy.org Web Site: www.whyy.org. Licensee: WHYY Inc. (acq 2-28-86). Natl. Network: PBS. Washington Atty: Schwartz, Woods & Miller. News staff: 10; News: 3 hrs wkly.
Key Personnel:
Molly Dickinson Shephard. . . . CEO & chmn
David A. Othmer. . . . VP
William J. Marrazzo . . . pres & stn mgr

WMDT— Analog channel: 47. Digital channel: 53. Analog hrs: 20 Note: ABC is on WMDT(TV) ch 47, CW is on WMDT-DT ch 53. 2,190 kw vis, 219 kw aur. ant 997t/1,024g TL: N38 30 06 W75 44 09 On air date: Apr 11, 1980. Box 4009, Salisbury, MD, 21803-4009. 202 Downtown Plaza, Salisbury, MD 21801. Phone: (410) 742-4747. Fax: (410) 742-5767. E-mail: wmdt@wmdt.com Web Site: www.wmdt.com. Licensee: Delmarva Broadcast Service G.P. Ownership: Marion B. Brechner, 80%; Berl M. Brechner, 20%. (acq 1982). Population served: 248,969 Natl. Network: ABC, CW Digital Network: Note: ABC is on WMDT(TV) ch 47, CW is on WMDT-DT ch 53. Washington Atty: Cohn & Marks. News staff: 23; News: 12 hrs wkly.

Key Personnel:
Kathleen McLain . . . . gen mgr
Phil Bankert . . . . gen sls mgr
Michael Polk. . . . . progmg mgr
Dawn Mitchell. . . . . news dir
Bill Hoctor . . . . chief of engrg

Massachusetts

Adams

see Albany-Schenectady-Troy, NY market

Boston (Manchester, NH) (DMA 7)

WBPX— Analog channel: 68. Digital channel: 32. Analog hrs: 24 1,350 kw vis, 135 kw aur. 870t/885g TL: N42 20 50 W71 04 59 On air date: January 1979. 1120 Soldiers Field Rd., Boston, MA, 02134. Phone: (617) 787-6868. Fax: (617) 787-4114. Web Site: www.ionline.tv. Licensee: Paxson Boston-68 License Inc. Group owner: Paxson Communications Corp. (acq 5-2-2000; grpsl). Population served: 6,366,400 Natl. Network: i Network. Wire Svc: Reuters News: 20 hrs wkly.
Key Personnel:
Robert Gilbert. . . . . gen mgr & sls VP
Dianne McLaughlin . . . . opns mgr
William Spitzer. dev VP

WBZ-TV— Analog channel: 4. Digital channel: 30. Analog hrs: 24 60.3 kw vis, 9.75 kw aur. ant 1,160t/1,199g TL: N42 18 37 W71 14 14 On air date: June 9, 1948. 1170 Soldiers Field Rd., Boston, MA, 02134. Phone: (617) 787-7000. Fax: (617) 787-5969. E-mail: webmaster@wbztv.com Web Site: www.wbztv.com. Licensee: Viacom Inc. Group owner: Viacom Television Stations Group. Population served: 641,071 Natl. Network: CBS. Natl. Rep: CBS TV Stations National Sales. Washington Atty: Wilkes, Artis, Hedrick & Lane. Foreign lang progmg: SP 1 News staff: 81; News: 20 hrs wkly.
Key Personnel:
Ed Piette . . . . pres & gen mgr
Angie Kucharski . . . . VP & stn mgr
Jack Barry . . . . opns dir & engrg dir
Helen Wynyard. . . . . gen sls mgr
Wendy McMahon . . . . mktg dir
Christine Ferrara . . . . progmg mgr
Jennifer Street . . . . news dir

WCVB-TV— Analog channel: 5. Digital channel: 20. Analog hrs: 24 Digital hrs: 24 61.7 kw vis. ant 1,158t TL: N42 18 37 W71 14 14 On air date: Mar 19, 1972. 5 TV Pl., Needham, MA, 02494. Phone: (781) 449-0400. Fax: (781) 433-4490/(781) 433-4022. Web Site: www.thebostonchannel.com. Licensee: WCVB Hearst-Argyle Television Inc. Group owner: Hearst-Argyle Television Inc. (acq 7-16-97; grpsl). Population served: 5,874,000 Natl. Network: ABC. Natl. Rep: Eagle Television Sales. Washington Atty: Brooks, Pierce, McLendon, Humphrey & Leonard. Wire Svc: PR Newswire Foreign lang progmg: SP 1 News: 30 hrs wkly.
Key Personnel:
Bill Fine . . . . pres & gen mgr
Gloria Spence . . . . CFO
Elizabeth Cheng . . . . VP
Joseph Rebelo. opns mgr
Peter Hennessey . . . . gen sls mgr

***WEKW-TV**— Analog channel: 52. Digital channel: 49. Analog hrs: 24 Digital hrs: 24 95.5 kw vis, 9.55 kw aur. 1,080t/455g TL: N43 02 00 W72 22 04 On air date: May 21, 1968. 268 Mast Rd., Durham, NH, 03824-4601. Phone: (603) 868-1100. Fax: (603) 868-7552. E-mail: themailbox@nhptv.org Web Site: www.nhptv.org. Licensee: University of New Hampshire. Population served: 631,000 Natl. Network: PBS. Washington Atty: Schwartz, Woods & Miller.
Key Personnel:
Peter A. Frid . . . . gen mgr
Dennis Malloy . . . . sls dir
Jeff Morris . . . . adv mgr
Mercedes Sabio. . . . . progmg dir
Brian Shepperd. . . . . chief of engrg

***WENH-TV**— Analog channel: 11. Digital channel: 57. Analog hrs: 24 Digital hrs: 24 316 kw vis, 31.6 kw aur. 970t/390g TL: N43 10 33 W71 12 29 On air date: July 6, 1959. 268 Mast Rd., Durham, NH, 03824-4601. Phone: (603) 868-1100. Fax: (603) 868-7552. E-mail: themailbox@nhptv.org Web Site: www.nhptv.org. Licensee: University of New Hampshire. Population served: 631,000 Natl. Network: PBS. Rgnl. Network: Eastern Educ. Washington Atty: Schwartz, Woods & Miller. News: 2 hrs wkly.

WFXT— Analog channel: 25. Digital channel: 31. Analog hrs: 21 1,380 kw vis, 106 kw aur. 1,170t/1,101g TL: N42 18 12 W71 13 08 On air date: Oct 10, 1977. WFXT-TV 25 Fox Dr., Dedham, MA, 02027. Phone: (781) 467-2525. Fax: (781) 467-7213. Web Site: www.myfoxboston.com. Licensee: Fox Television Stations Inc. Group owner: (group owner; (acq 7-95; FTR: 3-6-95). Population served: 2,126,300 Natl. Network: Fox. News staff: 70; News: 7 hrs wkly.
Key Personnel:
Gregg Kelley . . . . VP & gen mgr
Chris Tzianabos . . . . gen sls mgr
Lisa Graham . . . . news dir
Steve Harrington . . . . engrg VP
Tricia Maloney. . . . . progmg

***WGBH-TV**— Analog channel: 2. Digital channel: 19. Analog hrs: 24 87.1 kw vis, 8.71 kw aur. 1,040t/1,199g TL: N42 13 37 W71 14 14 On air date: May 2, 1955. One Guest Ave, Boston, MA, 02135. Phone: (617) 300-5400. Fax: (617) 300-1026. E-mail: feedback@wgbh.org Web Site: www.wgbh.org. Licensee: WGBH Educational Foundation. Population served: 1,550,000 Natl. Network: PBS. Rgnl. Network: Eastern Educ. Washington Atty: Covington & Burling.
Key Personnel:
Jonathan C. Abbott . . . . COO & exec VP
Henry Becton Jr. . . . . pres
Marita Rivero . . . . VP & gen mgr

***WGBX-TV**— Analog channel: 44. Digital channel: 43.977 kw vis, 97.7 kw aur. 1,080t/1,119g TL: N42 18 37 W71 14 14 On air date: Sept 25, 1967. One Guest St, Boston, MA, 02135. Phone: (617) 300-5400. Fax: (617) 300-1013. E-mail: feedback@wgbh.org Web Site: www.wgbh.org. Licensee: WGBH Educational Foundation. Population served: 2,100,000 Natl. Network: PBS. Washington Atty: Covington & Burling.
Key Personnel:
Jonathan C. Abbott . . . . COO, exec VP & progmg dir
Henry P. Becton Jr. pres
Marita Rivero . . . . VP, gen mgr, gen mgr & prom VP

WHDH-TV— Analog channel: 7. Digital channel: 42.316 kw vis, 63.2 kw aur. 1,000t/1,069g TL: N42 18 40 W71 13 00 On air date: June 21, 1948. 7 Bulfinch Pl., Boston, MA, 02114. Phone: (617) 725-0777. Fax: (617) 248-5420. Web Site: www.whdh.com. Licensee: WHDH-TV Co. (acq 6-3-93; $204 million; FTR: 6-21-93). Population served: 5,330,400 Natl. Network: NBC. Natl. Rep: TeleRep. Washington Atty: Holland & Knight.
Key Personnel:
Randi Goldklank . . . . VP, VP & gen mgr
Chris Wayland . . . . gen sls mgr
J.T. Smith . . . . natl sls mgr
Marc Lehner. . . . . rgnl sls mgr
Joan McCready . . . . progmg mgr
Linda Miele . . . . news dir
Jim Shultis. . . . . engrg dir

WLVI-TV—Cambridge, Analog channel: 56. Digital channel: 41. Analog hrs: 24 2,240 kw vis, 166 kw aur. ant 1,186t/1,201g TL: N42 18 12 W71 13 08 On air date: Aug 31, 1953. 7 Bulfinch Pl, Boston, MA, 02114. Phone: (617) 725-0777. Fax: (617) 248-5420. Web Site: www.cw56.com. Licensee: WHDH-TV. Group owner: Tribune Broadcasting Co. (acq 11-15-2006; $113.7 million). Population served: 641,071 Natl. Network: CW. Natl. Rep: TeleRep. Washington Atty: Holland & Knight. News staff: 44; News: 7 hrs wkly.
Key Personnel:
Randi Goldklank . . . . VP & gen mgr
Robert P. Burns . . . . gen sls mgr
Heather Hazelton . . . . natl sls mgr
Joan McCready . . . . progmg dir
Linda Miele . . . . news dir
Jim Shultis. . . . . engrg dir & chief of engrg

WMFP— Analog channel: 62. Digital channel: 18. Analog hrs: 24 Digital hrs: 24 5,000 kw vis. 610t/669g TL: N42 21 29 W71 03 40 On air date: Oct 16, 1987. 1 Beacon St. 35th Fl, Boston, MA, 02108. Phone: (617) 720-1062. Licensee: MTB Boston Licensee LLC. Group owner: Scripps Howard Broadcasting Co. (acq 11-15-2006; grpsl). Washington Atty: Hopkins & Sutter.

WMUR-TV— Analog channel: 9. Digital channel: 59. Analog hrs: 24 Digital hrs: 24 282 kw vis, 33.5 kw aur. 1,030t/227g TL: N42 58 59 W71 35 19 On air date: Mar 9, 1954. 100 S. Commercial St., Manchester, NH, 03101. Phone: (603) 641-9000. Fax: (603) 641-9005 (admin). E-mail: storyideas@wmur.com Web Site: www.wmur.com. Licensee: Hearst-Argyle Properties Inc. Group owner: Hearst-Argyle Properties Inc. (acq 3-28-01; $185 million). Population served: 100,000 Natl. Network: ABC. Natl. Rep: Eagle Television Sales. Washington Atty: Brooks, Pierce, McLendon, Humphrey & Leonard. Wire Svc: AP News staff: 54; News: 29 hrs wkly.

WNEU— Analog channel: 60. Digital channel: 34. Analog hrs: 24 Digital hrs: 24 1,410 kw vis, 141 kw aur. ant 1,010t/138g TL: N42 59 02 W71 35 20 On air date: 1987. One Sundial Ave., Suite 501, Manchester, NH, 03103. Phone: (603) 647-6060. E-mail: wneu@comcast.net Licensee: NBC Telemundo License Co. Group owner: Telemundo Group Inc. (acq 10-22-2002; $26 million). Population served: 1,500,000 Natl. Network: NBC, Telemundo (Spanish). Washington Atty: Davis Wright Tremaine L.L.P. Foreign lang progmg: SP 168
Key Personnel:
Donna Sill . . . . stn mgr
David Raymond. . . . . chief of engrg

WPXG— Analog channel: 21. Analog hrs: 24 1,860 kw vis, 186 kw aur. 1,128t TL: N43 11 04 W71 19 12 (CP: Ant 1,050t) On air date: Apr 16, 1984. 1120 Soldiers Field Rd., Boston, MA, 02134. Phone: (617) 787-6868. Fax: (617) 787-4114. Web Site: www.ionline.tv. Licensee: Paxson Boston-68 License Inc. Group owner: Paxson Communications Inc. (acq 5-2-2000; grpsl).
Key Personnel:
Robert Gilbert . . . . . gen mgr, sls VP, mktg mgr & progmg mgr
Paul Strieby . . . . chief of engrg
Satellite of WBPX Boston.

WSBK-TV— Analog channel: 38. Digital channel: 39. Analog hrs: 24 3,160 kw vis, 316 kw aur. ant 1,161t/1,013g TL: N42 18 12 W71 13 08 On air date: Oct 12, 1964. 1170 Soldiers Field Rd., Boston, MA, 02134. Phone: (617) 787-7000. Fax: (617) 787-5969. E-mail: webmaster@tv38.com Web Site: tv38.com. Licensee: Viacom Inc. Group owner: Viacom Television Stations Group (acq 2-27-95; FTR: 5-22-95). Population served: 2,140,000 News: 4 hrs wkly.
Key Personnel:
Ed Piette. . . . . pres, gen mgr & opns dir
Angie Kucharski . . . . VP & stn mgr
Helen Wynyard. . . . . gen sls mgr
Wendy McMahon . . . . mktg dir
Christine Ferrara . . . . progmg mgr
Jennifer Street . . . . news dir
Jack Barry. chief of engrg

WUNI— Analog channel: 27. Digital channel: 29. Analog hrs: 24 Digital hrs: 24 1,150 kw vis, 245 kw aur. 1,528t/1,349g TL: N42 20 09 W71 42 57 On air date: Jan 2, 1970. 33 Fourth Ave., Needham, MA, 02494. Phone: (781) 433-2727. Fax: (781) 433-2750. Fax: (781) 433-2701. E-mail: feedback@wunitv.com Web Site: www.wunitv.com. Licensee: Entravision 27 L.L.C. Group owner: Entravision Communications Corp. (acq 1-4-01; $47.5 million). Population served: 3,500,000 Natl. Network: Univision (Spanish). Washington Atty: Thompson Hine LLP. Wire Svc: AP Foreign lang progmg: SP 168 News staff: 13; News: 3 hrs wkly.
Key Personnel:
Alexander von Lichtenberg . . . . gen mgr
Bob Kerrigan . . . . opns mgr
Scott McGavick . . . . gen sls mgr
Rob Donner . . . . natl sls mgr
Meg Godin . . . . mktg mgr & prom mgr
Sara Suarez . . . . news dir
Fran Vaccari . . . . chief of engrg

WUTF-TV— Analog channel: 66. Digital channel: 23. Analog hrs: 24 Digital hrs: 24 3,311 kw vis. ant 1,168t/1,227g TL: N42 23 02 W71 29 37 On air date: Feb 12, 1985. 71 Parmenter Rd., Hudson, MA, 01749. Phone: (978) 562-0660. Fax: (978) 562-1166. Web Site: univision.com. Licensee: Univision Partnership of Massachusetts. Group owner: Univision Communications Inc. (acq 5-21-2001; grpsl). Population served: 2,200,000 Natl. Network: TeleFutura (Spanish). Washington Atty: Shaw Pittman. Foreign lang progmg: SP 168
Key Personnel:
Rolo Duartas . . . . gen mgr
Richard A. Peper . . . . chief of engrg
Scott McGavick. . . . . gen sls mgr & chief of engrg

WWDP— Analog channel: 46. Digital channel: 52. Analog hrs: 24 501 kw vis. ant 351t/384g TL: N42 01 36 W71 03 35 On air date: Sept 15, 1996. 6740 Shady Oak Rd., Eden Prairie, MN, 55344. Phone: (952) 943-6000. Fax: (952) 943-6566. Web Site: www.shopnbc.com. Licensee: Norwell Television LLC. Ownership: ValueVision Media Acquisition Inc. (acq 4-1-2003).

***WYDN**— Analog channel: 48. Digital channel: 47.240 kw vis, 100 kw aur. ant 807t TL: N42 18 14 W71 53 51 On air date: 2000. Box 612066, Dallas, TX, 75261-2066. 99Asnebumskit Rd., Paxton, MA 01612. Phone: (817) 571-1229. Fax: (817) 571-7458. E-mail: comments@daystar.com Web Site: www.daystar.com. Licensee: Educational Public TV Corp.

WZMY-TV— Analog channel: 50. Digital channel: 35. Analog hrs: 24 Digital hrs: 24 4,790 kw vis, 479 kw aur. ant 699t TL: N42 44 07 W71 23 36 On air date: Sept 5, 1983. 11 A Street, Derry, NH, 03038. Phone: (603) 845-1000. Fax: (603) 434-8627. Web Site: www.mytvstation.tv. Licensee: ShootingStar Broadcasting of New England LLC. Ownership: ShootingStar Inc., 100% of votes; Alta ShootingStar Corp., approximately 98% of the nonvoting preferred membership units (acq 9-16-2004; $31 million). Population served: 5,200,000 Natl. Network: MyNetworkTV. Natl. Rep: Blair Television. Washington Atty: Leventhal, Senter & Lerman. Wire Svc: AP News: 5 hrs wkly.
Key Personnel:
Diane Sutter. . . . . CEO & pres
Alfredo Fonseca . . . . mktg dir
Gene Steinberg . . . . progmg mgr
Mike DeBolasi . . . . news dir

Holyoke

see Springfield-Holyoke, MA market

New Bedford

see Providence, RI-New Bedford, MA market

Pittsfield

see Albany-Schenectady-Troy, NY market

Springfield-Holyoke, MA (DMA 76)

***WGBY-TV—** Analog channel: 57. Analog hrs: 6:45 AM-3 AM 776 kw vis, 155 kw aur. 1,000t/141g TL: N42 14 30 W72 38 56 On air date: Sept 26, 1971. 44 Hampden St., Springfield, MA, 01103. Phone: (413) 781-2801. Fax: (413) 731-5093. E-mail: feedback@wgby.org Web Site: www.wgby.org. Licensee: WGBH Educational Foundation. Population served: 203,000 Natl. Network: PBS. Washington Atty: Covington & Burling.
Key Personnel:
Russell Peotter . . . . VP
Russell J. Peotter . . . . gen mgr
Charley Rose . . . . mktg dir
Lynn Roginski . . . . progmg dir & opns
Ray Miller . . . . chief of engrg

WGGB-TV— Analog channel: 40. Digital channel: 55. Analog hrs: 24 Digital hrs: 24 4,250 kw vis, 425 kw aur. 1,056t/167g TL: N42 14 30 W72 38 56 On air date: Apr 14, 1953. 1300 Liberty St., Springfield, MA, 01104. Phone: (413) 733-4040. Fax: (413) 781-1363. Web Site: www.wggb.com. Licensee: WGGB Licensee L.L.C. Group owner: Sinclair Broadcast Group Inc. (acq 1999; grpsl). Population served: 345,000 Natl. Network: ABC.
Key Personnel:
James McKeever. . . . . gen mgr & prom dir
Dean Davidson . . . . opns dir, engrg dir & chief of engrg
Patrick Berry . . . . opns dir & rgnl sls mgr
Carol Moran. . . . . progmg mgr
Kathy Tobin . . . . news dir

WWLP— Analog channel: 22. Digital channel: 11. Analog hrs: 24 Digital hrs: 24 4,170 kw vis, 417 kw aur. 877t/530g TL: N42 05 05 W72 42 14 On air date: Mar 17, 1953. Box 2210, Springfield, MA, 01102-2210. One Broadcast Ctr., Chieopee, MA 01013. Phone: (413) 377-2200. Fax: (413) 377-2261. Web Site: www.wwlp.com. Licensee: WWLP Broadcasting L.L.C. Group owner: LIN Television Corporation (acq 10-20-2000; about $128 million). Natl. Network: NBC. Natl. Rep: Blair Television. Washington Atty: Covington & Burling. Wire Svc: AP
Key Personnel:
William Pepin. . . . . gen mgr
John Baran . . . . stn mgr
Carl Miller . . . . sls dir
Lowell McLane . . . . natl sls mgr
Anna Giza . . . . prom dir
Michael Garreffi . . . . news dir
Dave Cote. . . . . chief of engrg

Vineyard Haven

see Providence, RI-New Bedford, MA market

Michigan

Alpena, MI (DMA 208)

WBKB-TV— Analog channel: 11. Analog hrs: 24 316 kw vis, 32.4 kw aur. 665t/500g TL: N44 42 25 W83 31 23 On air date: Sept 22, 1975. 1390 Bagley St., Alpena, MI, 49707. Phone: (989) 356-3434. Fax: (989) 356-4188. E-mail: wbkbtv@speednetllc.com Web Site: www.wbkbtv.com. Licensee: Thunder Bay Broadcasting Corp. Ownership: Stephen A. Marks, 88.56%. Population served: 76,000 Natl. Network: CBS. Natl. Rep: Millennium Sales & Marketing. Washington Atty: Cohn & Marks. Wire Svc: UPI News staff: 6; News: 5.5 hrs wkly.
Key Personnel:
Stephen A. Marks. pres
Mary Compton . . . . gen sls mgr & prom dir
Barb Bowen . . . . rgnl sls mgr
Mark Nowak . . . . chief of engrg

***WCML-TV—** Analog channel: 6. Digital channel: 57. Analog hrs: 6:45 AM-midnight 100 kw vis, 15.1 kw aur. ant 1,472t/1,349g TL: N45 08 17 W84 09 44 On air date: 1975. Central Michigan Univ., 1999 E. Campus Dr., Mt. Pleasant, MI, 48859. Phone: (989) 774-3105. Fax: (989) 774-4427. E-mail: schud1ra@cmich.edu Web Site: www.wcmu.org. Licensee: Central Michigan University. Natl. Network: PBS. Washington Atty: Dow, Lohnes & Albertson.
Key Personnel:
Ed Grant . . . . stn mgr
Linda Dielman . . . . progmg dir & progmg mgr
Randy Kapenga. . . . . engrg dir & chief of engrg
Satellite of WCMU-TV Mt. Pleasant.

Battle Creek

see Grand Rapids-Kalamazoo-Battle Creek, MI market

Bay City

see Flint-Saginaw-Bay City, MI market

Cadillac

see Traverse City-Cadillac, MI market

Detroit (DMA 11)

WADL— Analog channel: 38. Digital channel: 39.1,243 kw vis, 248 kw aur. 630t TL: N42 33 15 W82 53 15 On air date: May 20, 1989. 22590 15 Mile Rd., Clinton Twp, MI, 48035-2814. Phone: (586) 790-3838. Fax: (586) 790-3841. Web Site: www.wadldetroit.com. Licensee: Adell Broadcasting Corp. Ownership: Adell Broadcasting Corp. (acq 9-3-03). Population served: 1,800,000
Key Personnel:
Kevin Adell . . . . CEO & gen sls mgr
Lewis Gibbs . . . . pres
Nicole Mayo . . . . gen mgr
Jamie Harrington . . . . progmg mgr
Tom Ponsart . . . . chief of engrg

WDIV— Analog channel: 4. Digital channel: 45. Analog hrs: 24 100 kw vis, 10 kw aur. 1,004t TL: N42 28 58 W83 12 19 On air date: Mar 4, 1947. 550 W. Lafayette Blvd., Detroit, MI, 48226. Phone: (313) 222-0444. Phone: (313) 222-0454. Fax: (313) 222-0592. Web Site: www.clickondetroit.com. Licensee: Post-Newsweek Stations, Michigan Inc. Group owner: Post-Newsweek Stations Inc. (acq 6-24-78). Population served: 1,753,000 Natl. Network: NBC. Natl. Rep: MMT. Washington Atty: Covington & Burling.
Key Personnel:
Steve Wasserman . . . . VP & gen mgr
Ted Pearse . . . . sls VP
Neil Goldstein. news dir
Marcus Williams . . . . chief of engrg

WJBK— Analog channel: 2. Digital channel: 58. Analog hrs: 24 100 kw vis, 10 kw aur. 1,000t/1,057g TL: N42 27 38 W83 12 47 On air date: Oct 24, 1948. Box 2000, Southfield, MI, 48037. 16550 W. Nine Mile Rd., Southfield, MI 48075. Phone: (248) 557-2000. Fax: (248) 557-6343. E-mail: contact@myfoxdetroit.com Web Site: www.myfoxdetroit.com. Licensee: WJBK License Inc. Group owner: Fox Television Stations Inc. (acq 1-22-97; grpsl). Population served: 1,800,000 Natl. Network: Fox. Wire Svc: Reuters News staff: 140; News: 36 hrs wkly.
Key Personnel:
Jeff Murri . . . . VP & gen mgr
Sheila Bruce. . . . . opns mgr & sls VP
Ann Marie Carlton . . . . natl sls mgr
Terry D'Esposito. . . . . mktg dir
Keith Stironek. prom dir
Connie Davis. . . . . progmg dir & progmg mgr
Dana Hahn. . . . . news dir
Lee Thomas . . . . news dir
Katie Fehr . . . . pub affrs dir
Tim Redmond . . . . chief of engrg

WKBD— Analog channel: 50. Digital channel: 14. Analog hrs: 24 Digital hrs: 24 2,340 kw vis, 209 kw aur. ant 960t/1,053g TL: N42 29 01 W83 18 44 On air date: Jan 10, 1965. 26905 W. 11 Mile Rd., Southfield, MI, 48034. Phone: (248) 355-7000. Fax: (248) 359-7494. E-mail: shows@wkbdtv.com Web Site: www.cw50detroit.com. Licensee: Detroit Television Station WKBD Inc. Group owner: Viacom Television Stations Group (acq 9-1-93; $105 million; FTR: 9-13-93). Population served: 5,521,787 Natl. Network: CW. Natl. Rep: TeleRep.
Key Personnel:
Trey Fabacher . . . . VP & gen mgr
Michael Michell . . . . stn mgr
Patrick Donnelly . . . . gen sls mgr
Mike Shippey . . . . natl sls mgr
Pam Shecter . . . . prom mgr
Paul A. Prange . . . . progmg dir
Edward Foxworth. . . . . pub affrs dir
Chuck Davis . . . . engrg dir & chief of engrg

WMYD— Analog channel: 20. Digital channel: 21.1,200 kw vis, 120 kw aur. ant 961t/1,030g TL: N42 29 01 W83 18 44 (CP: 1,500 kw vis, ant 1,063t/1,055g. TL: N42 26 52.9 W83 10 23.3) On air date: Sept 15, 1968. 27777 Franklin Rd., Suite 1220, Southfield, MI, 48034. Phone: (248) 355-2020. Fax: (248) 355-0368. Web Site: www.tv20detroit.com. Licensee: WXON License Inc. Group owner: Granite Broadcasting Corp. (acq 1-31-97; $175 million). Population served: 4,000,000 Natl. Network: MyNetworkTV. Washington Atty: Akin, Gump, Strauss, Hauer & Feld.
Key Personnel:
Sarah Norat-Phillips . . . . gen mgr
David Bangura . . . . stn mgr & gen sls mgr
Carolyn Worford . . . . opns dir, opns mgr & progmg dir
Ken Frierson. . . . . natl sls mgr
Dan Riley . . . . chief of engrg

WPXD— Analog channel: 31. Digital channel: 33. Analog hrs: 24 1,230 kw vis, 217 kw aur. ant 1,080t/1,044g TL: N42 22 25 W84 04 10 On air date: Jan 12, 1981. 3975 Varsity Dr., Ann Arbor, MI, 48108. Phone: (734) 973-7900. Fax: (734) 973-7906. E-mail: helenskinner@ionmedia.tv Web Site: www.ionline.tv. Licensee: Paxson Communications License Co. L.L.C. Group owner: Paxson Communications Corp. (acq 12-11-97; $35 million including LPTV ch). Washington Atty: Verner, Liipfert, Bernhard, McPherson & Hand.
Key Personnel:
Helen Skinner . . . . opns mgr
Robert Thompson . . . . pres & chief of engrg

***WTVS—** Analog channel: 56. Digital channel: 43. Analog hrs: 24 Digital hrs: 24 2,200 kw vis, 200 kw aur. 961t/1,020g TL: N42 29 01 W83 18 44 On air date: Oct 3, 1955. Riley Broadcast Ctr, 1 Clover Ct, Wixom, MI, 48393. Phone: (313) 873-7200. Fax: (313) 876-8118. E-mail: email@dptv.org Web Site: www.detroitpublictv.org. Licensee: Detroit Educational Television Foundation. Population served: 4,500,000 Natl. Network: PBS. Rgnl. Network: CEN. Rgnl. Rep: Rgnl rep: Karole White Washington Atty: Schwartz, Woods & Miller.
Key Personnel:
Daniel Alpert . . . . COO, gen mgr & stn mgr
John Wenzel . . . . CFO
Tim Wilson . . . . mktg dir
Dave Devereaux . . . . prom VP
Daniel Gaitens . . . . progmg dir
John O'Donnell . . . . pub affrs dir
Helge Blucher . . . . engrg VP & chief of engrg

WWJ-TV— Analog channel: 62. Digital channel: 44. Analog hrs: 24 Digital hrs: 24 5000 kw vis, 500 kw aur. 1073t/1,059g TL: N42 26 52 W83 10 23 (CP: 5,000 kw vis, ant 1,073t. TL: N42 26 52 W83 10 23) On air date: September 1975. 26905 W. 11-Mile Rd., Southfield, MI, 48034. Phone: (248) 355-7000. Fax: (248) 359-7499. E-mail: shows@wwjtv.com Web Site: www.wwjtv.com. Licensee: CBS Broadcasting Inc. Group owner: Viacom Television Stations Group (acq 1995; $24 million). Population served: 4,854,707 Natl. Network: CBS. Natl. Rep: CBS TV Stations National Sales. Washington Atty: Hogan & Hartson.
Key Personnel:
Trey Fabacher . . . . VP & gen mgr
Michael Michell . . . . stn mgr
Stephen Danowski . . . . gen sls mgr
Darren Pich . . . . natl sls mgr
Scott Cote . . . . rgnl sls mgr
Pam Shecter . . . . prom dir

Paul Prange . . . . progmg dir
Edward Foxworth. . . . . pub affrs dir
Chuck Davis . . . . engrg dir & chief of engrg

WXYZ-TV— Analog channel: 7. Digital channel: 41. Analog hrs: 24 316 kw vis, 31.6 kw aur. 1,000t/1,073g TL: N42 28 15 W83 15 00 On air date: Oct 9, 1948. 20777 W.10 Mile Rd., Southfield, MI, 48037. Phone: (248) 827-7777. Fax: (248) 827-9444. E-mail: talkback@wxyz.com Web Site: www.detnow.com. Licensee: Channel 7 Detroit Inc. Group owner: Scripps Howard Broadcasting Co., see Cross-Ownership (acq 1-2-86; grpsl). Population served: 4,500,000 Natl. Network: ABC. Natl. Rep: Eagle Television Sales. Washington Atty: Baker & Hostetler. Wire Svc: Reuters
Key Personnel:
Grace Gilchrist. . . . . VP & gen mgr
Mike Murri . . . . gen sls mgr
Steve Kopicki . . . . natl sls mgr
Mike MacLean . . . . rgnl sls mgr
Marla Drutz. progmg dir
Andrea Parquet-Taylor . . . . news dir
Ray Thurber . . . . engrg dir & chief of engrg

Flint-Saginaw-Bay City, MI (DMA 66)

WAQP— Analog channel: 49. Digital channel: 48. Analog hrs: 24 Digital hrs: 24 1,000 kw vis, 100 kw aur. ant 1,023t/1,049g TL: N43 13 18 W84 03 14 On air date: Mar 26, 1985. 2865 Trautner Dr., Saginaw, MI, 48604-9483. Phone: (989) 249-5969. E-mail: waqp@tct.tv Licensee: TCT of Michigan Inc. Group owner: Tri-State Christian Television.
Key Personnel:
Garth W. Coonce . . . . pres
Shane Chaney. CFO
Tina Coonce . . . . sr VP
Michael Socier . . . . gen mgr
John W. Dady . . . . chief of engrg

WBSF— Analog channel: 46. Analog hrs: 24 1,600 kw vis. ant 1,004t/1,023g TL: N43 28 26.75 W83 50 44.65 On air date: Sept. 15, 2006. 2225 W. Willard Rd., Clio, MI, 48420-8847. Phone: (810) 687-1000. Phone: (989) 755-0525. Fax: (810) 687-4925. E-mail: mail@nbc25.net Web Site: www.cw46online.com. Licensee: Barrington Bay City License LLC. (acq 4-14-2005; $4.5 million for CP). Population served: 104,868 Natl. Network: CW. Natl. Rep: Harrington, Righter & Parsons. Washington Atty: Covington & Burling LLP. Wire Svc: AP
Key Personnel:
Matt Kreiner . . . . gen mgr
Jon Bengston . . . . opns dir & progmg dir
Becky Butcher . . . . gen sls mgr
Jeff Reinarz . . . . prom mgr
Don Shafer. news dir
TB D . . . . chief of engrg

***WCMU-TV**— Analog channel: 14. Analog hrs: 6 AM-midnight 200 kw vis, 40 kw aur. 520t/547g TL: N43 34 24 W84 46 21 On air date: Mar 29, 1967. Central Michigan Univ., 1999 E. Campus Dr., Mount Pleasant, MI, 48859. Phone: (989) 774-3105. Fax: (989) 774-4427. E-mail: schud1ra@cmich.edu Web Site: www.wcmu.org. Licensee: Central Michigan University. Population served: 151,000 Natl. Network: PBS. Washington Atty: Dow, Lohnes & Albertson.
Key Personnel:
Ed Grant . . . . gen mgr & stn mgr
Linda Dielman . . . . progmg dir & progmg mgr
Randy Kapenga. . . . . engrg dir & chief of engrg

***WDCP-TV**— Analog channel: 19. Digital channel: 18. Analog hrs: 24 1,290 kw vis, 129 kw aur. 459t/493g TL: N43 33 43 W83 58 54 On air date: Oct 12, 1964. 1961 Delta Rd., University Center, MI, 48710. Phone: (877) 472-7677. Fax: (989) 686-0155. E-mail: wdcq@delta.edu Web Site: www.delta.edu/broadcasting. Licensee: Delta College. Population served: 1,200,000 Natl. Network: PBS. Washington Atty: Cohn & Marks.
Key Personnel:
Jean Goodnow . . . . pres
Barry Baker . . . . gen mgr
Pam Clark . . . . stn mgr
Tom Garnett . . . . chief of engrg

***WDCQ-TV**— Analog channel: 35. Digital channel: 15. Analog hrs: 24 Digital hrs: 24 85.1 kw vis, 8.51 kw aur. 508t/469g TL: N43 41 26 W82 56 29 On air date: Dec 12, 1986. University Ctr., 1961 Delta Rd., University Center, MI, 48710. Phone: (877) 472-7677. Fax: (989) 686-0155. E-mail: wdcq@delta.edu Web Site: www.delta.edu/broadcasting. Licensee: Delta College. Population served: 1,200,000 Natl. Network: PBS. Washington Atty: Cohn & Marks.
Key Personnel:
Jean Goodnow . . . . pres
Barry Baker . . . . gen mgr
Pam Clark . . . . stn mgr
Tom Garnett . . . . chief of engrg

WEYI-TV—Saginaw, Analog channel: 25. Digital channel: 30. Analog hrs: 24 Digital hrs: 24 Note: NBC is on WEYI-TV ch 25, CW is on WEYI-DT ch 30. 2,040 kw vis, 203 kw aur. ant 1,296t/1,292g TL: N43 13 01 W83 43 17 On air date: Apr 5, 1953. 2225 W. Willard Rd., Clio, MI, 48420. Phone: (810) 687-1000. Phone: (989) 755-0525. Fax: (810) 687-4925. E-mail: mail@nbc25.net Web Site: www.nbc25online.com. Licensee: Barrington Broadcasting Flint Corp. Group owner: Barrington Broadcasting Corp. (acq 5-14-2004; $24 million). Population served: 104,868 Natl. Network: NBC Digital Network: Note: NBC is on WEYI-TV ch 25, CW is on WEYI-DT ch 30. Natl. Rep: Harrington, Righter & Parsons. Washington Atty: Covington & Burling LLP. Wire Svc: AP News: 19.5 hrs wkly.
Key Personnel:
Matt Kreiner . . . . gen mgr & opns mgr
Becky Butcher . . . . gen sls mgr
Jeff Reinarz . . . . prom mgr
Jon Bengston . . . . progmg dir & opns
Don Shafer. news dir
TB D . . . . chief of engrg

***WFUM**— Analog channel: 28. Digital channel: 52.2,400 kw vis, 240 kw aur. 967t/929g TL: N42 53 57 W83 27 42 On air date: Aug 23, 1980. Michigan-Television, Univ. of Michigan-Flint, 303 E. Kearsley St., Flint, MI, 48502. Phone: (810) 762-3028. Fax: (810) 233-6017. E-mail: information@michigantelevision.org Web Site: www.michigantelevision.org. Licensee: Board of Regents, University of Michigan. Population served: 3,485,000 Natl. Network: PBS. Washington Atty: Dow, Lohnes PLLC.
Key Personnel:
Jennifer White . . . . stn mgr
Wayne Henderson. . . . . chief of engrg

WJRT-TV— Analog channel: 12. Digital channel: 36. Analog hrs: 24 Digital hrs: 24 316 kw vis, 31.6 kw aur. 940t/999g TL: N43 13 48 W84 03 35 On air date: Oct 12, 1958. 2302 Lapeer Rd., Flint, MI, 48503. Phone: (810) 233-3130. Fax: (810) 257-2834. E-mail: wjrt@abc.com Web Site: www.abc12.com. Licensee: Flint License Subsidiary Corp., a wholly owned subsidiary of WJRT Inc. Group owner: ABC Inc. (acq 1995; $155 million with WTVG(TV) Toledo, OH). Population served: 828,350 Natl. Network: ABC. Natl. Rep: ABC National Television Sales. Wire Svc: ESSA Weather Service Wire Svc: AP News staff: 55; News: 34 hrs wkly.
Key Personnel:
Thomas Bryson . . . . pres & gen mgr
Daniel C. Aube. . . . . gen sls mgr
Cheri Foss . . . . natl sls mgr & rgnl sls mgr
Skip Orvis. . . . . natl sls mgr & engrg dir
Brock Rice . . . . rgnl sls mgr
Sara Jo Gallock . . . . mktg dir & progmg dir
James Bleicher . . . . news dir

WNEM-TV—Bay City, Analog channel: 5. Digital channel: 22.Note: CBS is on WNEM-TV ch 5, MyNetworkTV is on WNEM-DT ch 22. 100 kw vis, 20 kw aur. ant 1,029t/1,049g TL: N43 28 13 W83 50 35 On air date: Feb 16, 1954. Box 531, Saginaw, MI, 48606. 107 N. Franklin St., Saginaw, MI 48607. Phone: (989) 755-8191. Fax: (989) 758-2111. Fax: (989) 758-2112. E-mail: wnem@wnem.com Web Site: www.wnem.com. Licensee: Meredith Corp. Group owner: Meredith Broadcasting Group, Meredith Corp. (acq 4-16-69; $11.5 million; FTR: 4-21-69). Population served: 1,251,000 Natl. Network: CBS, MyNetworkTV Digital Network: Note: CBS is on WNEM-TV ch 5, MyNetworkTV is on WNEM-DT ch 22. Natl. Rep: TeleRep. Washington Atty: Haley, Bader & Potts.
Key Personnel:
Al Blinke . . . . VP & gen mgr
Carl Prutting. . . . . gen sls mgr
Bryan Resnik . . . . rgnl sls mgr
Karen Frey . . . . prom dir, progmg dir & news dir
Ian Rubin . . . . news dir
Mike Tamme . . . . chief of engrg

WSMH— Analog channel: 66. Analog hrs: 24 1,170 kw vis, 117 kw aur. 940t/989g TL: N43 13 18 W84 03 14 (CP: 5,000 kw vis, ant 941t) On air date: Dec 15,1984. G-3463 W. Pierson Rd., Flint, MI, 48504. Phone: (810) 785-8866. Fax: (810) 785-8963. Web Site: www.wsmh66.com. Licensee: WSMH Licensee L.L.C. Group owner: Sinclair Broadcast Group Inc. (acq 2-28-96; $33 million). Population served: 294,000 Natl. Network: Fox. Natl. Rep: Harrington, Righter & Parsons.
Key Personnel:
David Schwartz . . . . gen mgr
Chad Conklin . . . . gen sls mgr
Gale Garrison . . . . natl sls mgr
John Grover . . . . chief of engrg

Grand Rapids-Kalamazoo-Battle Creek, MI (DMA 39)

***WGVK**— Analog channel: 52. Analog hrs: 18 44.7 kw vis, 4.47 kw aur. 410t/295g TL: N42 18 24 W85 39 26 On air date: Oct 1, 1984. 301 W. Fulton St., Grand Rapids, MI, 49504-6492. Phone: (616) 331-6666. Fax: (616) 331-6625. E-mail: wgvu@gvsu.edu Web Site: www.wgvu.org. Licensee: Grand Valley State University. Population served: 180,000 Natl. Network: PBS. Washington Atty: Mark Van Bergh. News staff: 15; News: one hr wkly.
Key Personnel:
Michael T. Walenta. . . . . gen mgr
Pamela Holtz. . . . . dev mgr & prom mgr
Gary Hunt . . . . sls dir
Carrie Corbin . . . . progmg dir
Fred Martino . . . . news dir
Robert Lumbert. . . . . engrg dir
Rebroadcasts WGVU-TV Grand Rapids 100%.

***WGVU-TV**— Analog channel: 35. Digital channel: 11. Analog hrs: 20 1,000 kw vis, 100 kw aur. 857t/859g TL: N42 57 35 W85 53 45 On air date: Dec 17, 1972. 301 W. Fulton St., Grand Rapids, MI, 49504-6492. Phone: (616) 331-6666. Fax: (616) 331-6625. E-mail: wgvu@gvsu.edu Web Site: www.wgvu.org. Licensee: Board of Control, Grand Valley State University. Population served: 638,940 Natl. Network: PBS. Washington Atty: Mark Van Bergh.
Key Personnel:
Michael Walenta. . . . . gen mgr
Gary Hunt . . . . sls dir
Pamela Holtz. . . . . dev mgr & prom mgr
Carrie Corbin . . . . progmg mgr
Bob Lumbert. . . . . engrg dir
Rebroadcasts WGVK (TV) Kalamazoo 100%.

WLLA— Analog channel: 64. Digital channel: 45. Analog hrs: 24 2,510 kw vis, 65 kw aur. 1,046t/339g TL: N42 33 52 W85 27 31 On air date: June 30, 1987. Box 3157, Kalamazoo, MI, 49003. 7048 E. Kilgore Rd., Kalamazoo, MI 49003. Phone: (269) 345-6421. Fax: (269) 345-5665. E-mail: deloris@wlla.com Web Site: www.wlla.com. Licensee: Christian Faith Broadcasting Inc. Group owner: (group owner; (acq 1-13-86; $35,000; FTR: 12-9-85). Population served: 2,400,000

WOOD-TV— Analog channel: 8. Digital channel: 7. Analog hrs: 24 316 kw vis, 31.6 kw aur. 991t TL: N42 41 13 W85 30 35 On air date: Aug 15, 1949. Box B, Grand Rapids, MI, 49501. 120 College Ave. S.E., Grand Rapids, MI 49503. Phone: (616) 456-8888. Fax: (616) 771-9676. E-mail: woodtv@woodtv.com Web Site: www.woodtv.com. Licensee: Wood License Co. LLC. Group owner: LIN Television Corporation (acq 6-30-99). Population served: 1,871,000 Natl. Network: NBC. Washington Atty: Covington & Burling.
Key Personnel:
Diane Kniowski . . . . pres & gen mgr
Craig Cole . . . . opns mgr & progmg dir
Brent Denny. . . . . gen sls mgr & natl sls mgr
Molly Kelly . . . . mktg mgr & prom mgr
Ethan Beute . . . . adv mgr
Patti McGethgain . . . . news dir
Eva Cooper . . . . pub affrs dir
Ken Selvig. . . . . engrg dir & chief of engrg

WOTV— Analog channel: 41. Analog hrs: 24 5,000 kw vis, 200 kw aur. ant 1,079t/938g TL: N42 34 15 W85 28 07 On air date: July 24, 1971. Box B, Grand Rapids, MI, 49501. 120 College Ave. S.E., Grand Rapids, MI 49503. Phone: (269) 968-9341. Fax: (269) 660-1222. E-mail: wotv@wotv.com Web Site: www.wotv.com. Licensee: Wood License Co. LLC. Group owner: LIN Television Corporation (acq 12-6-2001; $2.25 million). Population served: 1,871,000 Natl. Network: ABC. Washington Atty: Covington & Burling. News staff: 25; News: 13 hrs wkly.
Key Personnel:
Diane Kniowski. . . . . . . pres, gen mgr, gen mgr & gen sls mgr
Ann Marie Young . . . . sls dir
Swaina Noble . . . . natl sls mgr
Molly Kelly . . . . mktg mgr
Ethan Beute. . . . . prom mgr
Craig Cole . . . . progmg dir
Patti McGethgain . . . . news dir
Dave Morse . . . . chief of engrg

WTLJ— Analog channel: 54. Digital channel: 24. Analog hrs: 24 4,395 kw vis, 440 kw aur. 1,000t/989g TL: N42 57 25 W85 54 07 On air date: Nov 1, 1986. 10290 48th Ave., Allendale, MI, 49401. Phone: (616) 895-4154. Fax: (616) 892-4401. E-mail: wtlj@tct.tv Web Site: www.tct.tv. Licensee: TCT of Michigan Inc. Group owner: Tri-State Christian Television (acq 1-15-92; $1.5 million; FTR: 2-10-92). Population served: 2,000,000
Key Personnel:
Vic Van Deventer. . . . . gen mgr & stn mgr
Vic VanDeventer . . . . natl sls mgr, mktg mgr & progmg dir
Frank Ayre. chief of engrg

WWMT— Analog channel: 3. Digital channel: 2. Analog hrs: 24 Digital hrs: 24 Note: CBS is on WWMT(TV) ch. 3, CW is on WWMT-DT ch. 2. 100 kw vis, 20 kw aur. 1,000t/1,130g TL: N42 37 56 W85 32 16 On air date: June 1, 1950. 590 W. Maple St., Kalamazoo, MI, 49008. Phone: (269) 388-3333. Fax: (269) 388-8228. E-mail:

newschannel3@wwmt.com Web Site: www.wwmt.com. Licensee: Freedom Broadcasting of Michigan Licensee L.L.C. Group owner: Freedom Communications Inc., Broadcast Division (acq 7-18-98; $170 million with WLAJ(TV) Lansing). Population served: 1,861,000 Natl. Network: CBS, CW Digital Network: Note: CBS is on WWMT(TV) ch. 3, CW is on WWMT-DT ch. 2. Natl. Rep: TeleRep. Washington Atty: Akin, Gump, Strauss, Hauer & Feld. News: 25 hrs wkly.

Key Personnel:
James Lutton . . . VP & gen mgr
James Wagner . . . gen sls mgr
Jeff Watts . . . prom dir
Susan Abraham . . . progmg dir
Kathy Younkin . . . news dir
Jim Steffey . . . chief of engrg

WXMI— Analog channel: 17. Digital channel: 19. Analog hrs: 21 1,300 kw vis, 130 kw aur. 802t/1,081g TL: N42 41 15 W85 31 57 On air date: March 1982. 3117 Plaza Dr. N.E., Grand Rapids, MI, 49525. Phone: (616) 364-8722. Fax: (616) 364-8506. E-mail: feedback@wxmi.com Web Site: www.wxmi.com. Licensee: Tribune Television Holdings Inc. Group owner: Tribune Broadcasting Co. (acq 6-5-98; grpsl). Population served: 722,900 Natl. Network: Fox. Natl. Rep: Harrington, Righter & Parsons. News staff: 30; News: 4 hrs wkly.

Key Personnel:
Patricia Hamilton . . . VP & gen mgr
Jeff Cartwright . . . sls dir & progmg dir
Tim Dye . . . sls dir & news dir
Pennie Westers . . . rgnl sls mgr
Travis Henkaline . . . prom dir
Mark Krause . . . progmg dir
Dale Scholten . . . chief of engrg

WZPX— Analog channel: 43. Digital channel: 44. Analog hrs: 24 5,000 kw vis, 500 kw aur. 1,058t TL: N42 40 45 W85 03 57 On air date: 1996. 2610 Horizon Dr. S.E., Suite E, Grand Rapids, MI, 49546. Phone: (616) 222-4343. Fax: (616) 493-2677. Web Site: www.ionline.tv. Licensee: Paxson Battle Creek License Inc. Group owner: Paxson Communications Corp. (acq 3-13-00; grpsl). Population served: 4,500,000 Natl. Network: i Network.

WZZM-TV— Analog channel: 13. Digital channel: 39. Analog hrs: 24 Digital hrs: 24 295 kw vis, 63 kw aur. ant 1,000t/991g TL: N43 18 34 W85 54 44 On air date: Nov 1, 1962. Box Z, Grand Rapids, MI, 49501. 645 Three Mile Rd., N.W., Grand Rapids, MI 49544. Phone: (616) 785-1313. Fax: (616) 785-1301. E-mail: management@wzzm13.com Web Site: www.wzzm13.com. Licensee: Combined Communications Corp. of Oklahoma Inc. Group owner: Gannett Broadcasting (acq 1-27-97; grpsl). Population served: 635,000 Natl. Network: ABC. Natl. Rep: Blair Television. News: 26 hrs wkly.

Key Personnel:
Janet Mason . . . pres & gen mgr
Chuck Mikowski . . . VP & prom dir
Kim Krause . . . gen sls mgr
Tim Geraghty . . . news dir
Catherine Behrendt . . . progmg

Kalamazoo

see Grand Rapids-Kalamazoo-Battle Creek, MI market

Lansing, MI (DMA 112)

WHTV— Analog channel: 18. Digital channel: 34. Analog hrs: 24 8.91 kw vis. ant 239t TL: N42 14 08 W84 24 00 On air date: 1999. 2820 E. Saginaw St., Lansing, MI, 48912-4240. Phone: (517) 372-9497. Fax: (517) 372-9499. E-mail: info@my18.tv Web Site: www.my18.tv. Licensee: Spartan-TV LLC. Population served: 250,000 Natl. Network: MyNetworkTV.

Key Personnel:
Lori Harper . . . stn mgr
Corey Cummings . . . chief of engrg

WILX-TV— Analog channel: 10. Digital channel: 57. Analog hrs: 24 Digital hrs: 24 309 kw vis, 61.7 kw aur. 970t/983g TL: N42 26 33 W84 34 21 On air date: Mar 15, 1959. 500 American Rd., Lansing, MI, 48911. Phone: (517) 393-0110. Fax: (517) 393-8555. E-mail: news@wilx.com Web Site: www.wilx.com. Licensee: Gray Television Licensee Inc. Group owner: Gray Television Inc. (acq 8-29-2002; grpsl). Population served: 592,400 Natl. Network: NBC. Natl. Rep: Continental Television Sales. News staff: 45.

Key Personnel:
Mike King . . . gen mgr
John O'Brien . . . gen sls mgr
Paul Crockett . . . rgnl sls mgr
Craig Tucker . . . prom mgr
Kevin Ragan . . . news dir
Gary King . . . chief of engrg

***WKAR-TV**— Analog channel: 23. Digital channel: 55.1,100 kw vis, 219 kw aur. 975t/1,038g TL: N42 42 08 W84 24 51 On air date: Jan 15, 1954. 283 Communication Arts Bldg., Michigan State Univ., East Lansing, MI, 48824-1212. Phone: (517) 432-9527. Fax: (517) 353-7124. E-mail: mail@wkar.org Web Site: www.wkar.org. Licensee: Michigan State University. Population served: 3,000,000 Natl. Network: PBS. Washington Atty: Schwartz, Woods & Miller.

Key Personnel:
De Anne Hamilton . . . gen mgr
Cindy Herfindahl . . . dev dir & film buyer
Jeanie Croope . . . prom dir & prom mgr
Kent Wieland . . . stn mgr & progmg dir
Gary Blievernicht . . . engrg dir

WLAJ— Analog channel: 53. Digital channel: 51. Analog hrs: 24 Note: ABC is on WLAJ(TV), CW is on WLAJ-DT ch 51. 1,660 kw vis, 166 kw aur. ant 976t/1,009g TL: N42 25 11 W84 31 26 (CP: 3,320 kw vis, 332 kw aur, ant 981t/1,014g) On air date: Oct 13, 1990. 5815 S. Pennsylvania Ave., Lansing, MI, 48911-5230. Phone: (517) 394-5300. Fax: (517) 887-0077. Web Site: www.wlaj.com. Licensee: WLAJ License Inc. Group owner: Freedom Communications Inc., Broadcast Division (acq 6-22-98; $170 million with WWMT(TV) Kalamazoo). Natl. Network: ABC, CW Digital Network: Note: ABC is on WLAJ(TV), CW is on WLAJ-DT ch 51. News staff: 10; News: 5 hrs wkly.

Key Personnel:
Jim Wareham . . . VP & gen mgr
Susan Angel . . . pres & rgnl sls mgr
Jim Fordyce . . . news dir
Mike Winsky . . . chief of engrg

WLNS-TV— Analog channel: 6. Digital channel: 59.100 kw vis, 20 kw aur. 1,000t/1,023g TL: N42 41 14 W84 22 35 On air date: May 1, 1950. Box 40226, Lansing, MI, 48901. 2820 E. Saginaw St., Lansing, MI 48912. Phone: (517) 372-8282. Fax: (517) 374-7610. E-mail: wlns@wlns.com Web Site: www.wlns.com. Licensee: Young Broadcasting of Lansing Inc. Group owner: Young Broadcasting Inc. (acq 9-15-86; $72 million; FTR: 4-14-86). Population served: 1,000,000 Natl. Network: CBS. Washington Atty: Wiley, Rein & Fielding.

Key Personnel:
Clay Koenig . . . gen mgr & gen sls mgr
Don Marmichael . . . gen mgr
Gene Shanahan . . . opns mgr
Teresa Morton . . . progmg dir
Phil Hendrix . . . news dir
Cory Cumming . . . chief of engrg

WSYM-TV— Analog channel: 47. Digital channel: 38. Analog hrs: 24 Digital hrs: 24 1,350 kw vis, 135 kw aur. 1,000t/1,036g TL: N42 28 03 W84 39 06 On air date: Dec 1, 1982. 600 W. St. Joseph St., Suite 47, Lansing, MI, 48933. Phone: (517) 484-7747. Fax: (517) 484-3144. E-mail: fox47news@fox47news.com Web Site: www.fox47news.com. Licensee: Journal Broadcast Corp. Group owner: Journal Broadcast Group Inc. (acq 11-9-85; FTR: 12-3-84). Natl. Network: Fox. Natl. Rep: Harrington, Righter & Parsons. Washington Atty: Crowell & Moring. News staff: 22; News: 10 hrs wkly.

Key Personnel:
Lyle Schulze . . . VP & gen mgr
Gary Baxter . . . gen sls mgr
Jami Anderson . . . natl sls mgr
Kip Bohne . . . mktg mgr & prom mgr

Marquette, MI (DMA 178)

WBKP— Analog channel: 5. Digital channel: 11. Analog hrs: 24 100 kw vis. ant 968t TL: N47 02 11 W88 41 43 On air date: Oct 30, 1996. 2025 U.S. 41 W., Marquette, MI, 49855. Phone: (906) 225-5700. Fax: (906) 225-5598. E-mail: abc510@wbkp.com Web Site: www.wbuptv.com. Licensee: Lake Superior Community Broadcasting Corp. Group owner: (group owner; (acq 1-15-2004; $500,000 with WBUP(TV) Ishpeming). Natl. Network: CW. News staff: 7; News: 10 hrs wkly.

Key Personnel:
Ken Lindeman . . . chmn & stn mgr
Bob McDonald . . . gen sls mgr
Randy Carlisle . . . prom mgr
Steve Marks . . . progmg mgr
Gerry Heyn . . . chief of engrg

WBUP— Analog channel: 10.133 kw vis. ant 344t/310g TL: N46 21 10 W87 51 15 On air date: Jan 30, 2003. 2025 US 41 W., Marquette, MI, 49855. Phone: (906) 225-5700. Fax: (906) 225-5598. E-mail: abc510@wbuptv.com Web Site: www.wbuptv.com. Licensee: Lake Superior Community Broadcasting Corp. Group owner: (group owner; (acq 1-15-2004; $500,000 with WBKP(TV) Calumet). Natl. Network: ABC. Washington Atty: Latham and Watkins.

Key Personnel:
Ken Lindeman . . . stn mgr
Bob McDonald . . . gen sls mgr
Randy Carlisle . . . prom mgr
Gerry Heyn . . . chief of engrg

WDHS— Analog channel: 8. Digital channel: 22. Analog hrs: 24 2 kw vis, 200 w aur. 508t TL: N45 49 14 W88 02 39 (CP: 30.2 kw, ant 623t) On air date: September 1986. 1500 W. B St., Iron Mountain, MI, 49801-8888. Phone: (906) 776-8888. Fax: (906) 776-8888. Licensee: W. Russell Withers Jr. Group owner: Withers Broadcasting Co. Population served: 200,000

WJMN-TV— Analog channel: 3. Analog hrs: 24 100 kw vis, 20 kw aur. 1,192t/1,048g TL: N48 06 04 W86 56 52 On air date: Oct 7, 1969. Box 19055, Green Bay, WI, 54307. Phone: (920) 437-5411. Phone: (906) 226-3023 (sales). Fax: (920) 437-5769 (news). Licensee: Liberty Media. Group owner: Viacom Television Stations Group. Population served: 300,000 Natl. Network: CBS. Natl. Rep: TeleRep.

Key Personnel:
R. Perry Kidder . . . pres, VP & gen mgr
Dale Mitchell . . . opns dir & engrg dir
Jackie Stewart . . . sls dir
Mike Smith . . . natl sls mgr
Kit Overlock . . . rgnl sls mgr
Kristen Kent . . . mktg mgr & pub affrs dir
Monica Zegers . . . prom dir & adv mgr
Jay Schabow . . . progmg mgr
H. Lee Hitter . . . news dir
Satellite of WFRV-TV Green Bay, WI.

WLUC-TV— Analog channel: 6. Digital channel: 35. Analog hrs: 21 Digital hrs: 21 100 kw vis, 20 kw aur. ant 978t/1,018g TL: N46 20 11 W87 50 55 On air date: Apr 29, 1956. 177 U.S. Hwy. 41 E., Negaunee, MI, 49866. Phone: (906) 475-4161. Phone: (906) 475-4141 (news). Fax: (906) 475-4824. Fax: (906) 475-5070 (news). E-mail: tv6@wluctv6.com Web Site: www.wluctv6.com. Licensee: Barrington Marquette License LLC. Group owner: Raycom Media Inc. (acq 8-11-2006; grpsl). Population served: 292,600 Natl. Network: NBC. Natl. Rep: Harrington, Righter & Parsons. Washington Atty: Covington & Burling. Wire Svc: AP News staff: 24; News: 16 hrs wkly.

Key Personnel:
Brad Van Sluyters . . . VP & gen mgr
Dan DiLoreto . . . natl sls mgr
Rob Jamros . . . rgnl sls mgr
Kim Parker . . . mktg mgr
Brian Cabell . . . news dir
Sonny Reschka . . . chief of engrg

WMQF— Analog channel: 19.5,000 kw vis. ant 646t/469g TL: N46 30 52 W87 28 37 On air date: 2003. Equity Broadcasting Corp., 1 Shackleford Dr., Suite 400, Little Rock, AR, 72211. Phone: (501) 219-2400. Fax: (501) 604-8004. Licensee: Marquette Broadcasting Inc. Group owner: Equity Broadcasting Corp. (acq 7-18-2001). Natl. Network: Fox.

***WNMU-TV**— Analog channel: 13. Digital channel: 33. Analog hrs: 18 316 kw vis, 63.1 kw aur. 1,090t/1,000g TL: N46 21 09 W87 51 32 On air date: Dec 28, 1972. Northern Michigan Univ., 1401 Presque Isle Ave., Marquette, MI, 49855-5301. Phone: (906) 227-9668. Fax: (906) 227-2905. E-mail: tv13@nmu.edu Web Site: www.nmu.edu/wnmutv. Licensee: Board of Control of Northern Michigan University. Population served: 250,000 Natl. Network: PBS. Washington Atty: Cohn & Marks. Wire Svc: UPI News: one hr wkly.

Key Personnel:
Eric Smith . . . gen mgr
Bruce Turner . . . stn mgr

Saginaw

see Flint-Saginaw-Bay City, MI market

Traverse City-Cadillac, MI (DMA 113)

***WCMV**— Analog channel: 27.274.2 kw vis, 27.42 kw aur. 587t/303g TL: N44 08 22 W85 20 28 On air date: Sept 7, 1984. Central Michigan Univ., 1999 E. Campus Dr., Mt. Pleasant, MI, 48859. Phone: (989) 774-3105. Fax: (989) 774-4427. E-mail: schud1r@cmich.edu Web Site: www.wcmu.org. Licensee: Central Michigan University. Natl. Network: PBS. Washington Atty: Dow, Lohnes & Albertson.

Key Personnel:
Edwards Grant . . . gen mgr & stn mgr
Linda Dielman . . . progmg dir & progmg mgr
Randy Kapenga . . . prom mgr & engrg dir
Rebroadcasts WCMU(TV) Mt. Pleasant 100%.

***WCMW**— Analog channel: 21. Digital channel: 58.224.4 kw vis, 22.44 kw aur. ant 340t/298g TL: N44 03 57 W86 19 58 On air date: Sept 7, 1984. Central Michigan Univ., 1999 E. Campus Dr., Mt. Pleasant, MI, 48859. Phone: (989) 774-3105. Fax: (989) 774-4427. E-mail: schud1ra@cmich.edu Web Site: www.wcmu.org. Licensee: Central Michigan University. Natl. Network: PBS. Washington Atty: Dow, Lohnes & Albertson.

Key Personnel:
Ed Grant . . . . . stn mgr
Linda Dielman . . . . . progmg dir & progmg mgr
Randy Kapenga. . . . . engrg dir & chief of engrg
Satellite of WCMU-TV, Mt. Pleasant.

WFQX-TV— Analog channel: 33. Analog hrs: 24 776 kw vis, 77.6 kw aur. 1,023t/650g TL: N44 08 53 W85 20 45 On air date: Oct 12, 1989. 7669 S. 45 Rd., Cadillac, MI, 49601. Phone: (231) 775-9813. Fax: (231) 775-1898. E-mail: info@fox33.com Web Site: www.fox33.com. Licensee: Rockfleet Broadcasting II L.L.C. Group owner: Rockfleet Broadcasting Inc. (acq 2-1-2000; with WFUP(TV) Vanderbilt). Population served: 188,120 Natl. Network: Fox. Washington Atty: Fletcher, Heald & Hildreth. News: 3 hrs wkly.
Key Personnel:
Bruce Pfeiffer . . . . . gen mgr & sls VP
Ginny Buzzell . . . . . prom VP & prom mgr
Greg Buzzell. . . . . opns VP, engrg VP & chief of engrg

WFUP— Analog channel: 45. Analog hrs: 24 851 kw vis, 85.1 kw aur. 1,063t/2,168g TL: N45 10 12 W84 45 04 On air date: Sept 24, 1992. 7669 S. 45 Rd., Cadillac, MI, 49601. Phone: (231) 775-9813. Fax: (231) 775-1898. E-mail: info@fox33.com Web Site: www.fox33.com. Licensee: Rockfleet Broadcasting II L.L.C. Group owner: Rockfleet Broadcasting Inc. (acq 2-1-00; with WFQX-TV Cadillac). Population served: 142,291 Natl. Network: Fox. Washington Atty: Fletcher, Heald & Hildreth. News staff: 5; News: 3 hrs wkly.
Key Personnel:
Bruce Pfeiffer . . . . . gen mgr, sls VP & natl sls mgr
Greg Buzzell . . . . . opns VP, opns mgr & engrg VP
Julia Horchner . . . . . opns mgr & progmg dir
Ginny Buzzell . . . . . prom VP & prom mgr
Quentin Parker . . . . . news dir & chief of engrg
Rebroadcasts WFQX(TV) Cadillac 100%.

WGTQ— Analog channel: 8. Digital channel: 9.316 kw vis, 163.6 kw aur. 978t/864g TL: N46 03 06 W84 06 40 On air date: Nov 3, 1976. 201 East Font St., Traverse City, MI, 49684. Phone: (231) 946-2900. Fax: (231) 946-1600. E-mail: wgtu@wgtu.com Web Site: www.wgtu.com. Licensee: MTC License LLC. Group owner: MAX Media L.L.C. (acq 8-13-03; $7.75 million with WGTU(TV) Traverse City). Natl. Network: ABC. Washington Atty: Latham and Watkins.
Key Personnel:
Jeff Cash . . . . . gen mgr
Betsy Bard . . . . . gen sls mgr
Craig Toomey . . . . . rgnl sls mgr
Lori Puckett . . . . . prom mgr
Lynne Bennett . . . . . progmg mgr
Ronald Stark. . . . . chief of engrg
Satellite of WGTU(TV) Traverse City, rebroadcast 100%.

WGTU— Analog channel: 29. Digital channel: 31. Analog hrs: 24 1,303 kw vis. ant 1,289t/1,216g TL: N44 44 53 W85 04 08 On air date: Aug 23, 1971. 201 E. Front St., Traverse City, MI, 49684. Phone: (231) 946-2900. Fax: (231) 946-1600. E-mail: wgtu@wgtu.com Web Site: www.wgtu.com. Licensee: MTC License LLC. Group owner: MAX Media L.L.C. Population served: 400,000 Natl. Network: ABC. Washington Atty: Latham & Watkins.
Key Personnel:
Jeff Cash . . . . . gen mgr
Betsy Bard . . . . . gen sls mgr
Craig Toomey . . . . . rgnl sls mgr
Lori Puckett . . . . . prom mgr
Lynne Bennett . . . . . progmg mgr
Ronald Stark. . . . . opns mgr & chief of engrg

WPBN-TV— Analog channel: 7. Digital channel: 50. Analog hrs: 24 316 kw vis, 63.2 kw aur. ant 1,348t/1,130g TL: N44 16 33 W85 42 49 On air date: Sept 13, 1954. Box 546, Traverse City, MI, 49685. 8518 M-72 West, Traverse City, MI 49686. Phone: (231) 947-7770. Fax: (231) 947-0354. Fax: (231) 947-1229. E-mail: tv7-4@tv7-4.com Web Site: www.tv7-4.com. Licensee: Barringtron Traverse City License LLC. Group owner: Raycom Media Inc. (acq 8-11-2006; grpsl). Population served: 300,000 Natl. Network: NBC. Natl. Rep: Harrington, Righter & Parsons. Washington Atty: Covington & Burling.
Key Personnel:
Julie A. Brinks . . . . . gen sls mgr & gen mgr
Kim St. Mary . . . . . mktg mgr
Mary Speck . . . . . progmg mgr
Doug DeYoung . . . . . news dir
Mike Miller. . . . . chief of engrg

WTOM-TV— Analog channel: 4. Digital channel: 35. Analog hrs: 24 100 kw vis, 20 kw aur. ant 620t/590g TL: N45 39 01 W84 20 37 On air date: May 16, 1959. Box 546, Traverse City, MI, 49685. 8518 M-72 West, Traverse City, MI 49686. Phone: (231) 947-7770. Fax: (231) 947-1229. Fax: (231) 947-0354. E-mail: tv7-4@tv7-4.com Web Site: www.tv7-4.com. Licensee: Barrington Traverse City License LLC. Group owner: Raycom Media Inc. (acq 8-11-2006; grpsl). Natl. Network: NBC. Washington Atty: Covington & Burling.
Key Personnel:
Julie A. Banks . . . . . gen mgr
Julie A. Brinks . . . . . gen sls mgr
Kim St. Mary . . . . . mktg mgr
Mary Speck . . . . . progmg mgr
Doug DeYoung . . . . . news dir
Mike Miller. . . . . chief of engrg
Satellite of WPBN-TV Traverse City.

WWTV— Analog channel: 9. Digital channel: 40. Analog hrs: 24 316 kw vis, 63.1 kw aur. 1,635t/1,295g TL: N44 08 12 W85 20 33 On air date: Dec 11, 1953. Box 627, Cadillac, MI, 49601. Phone: (231) 775-3478. Fax: (231) 775-3671. Web Site: www.9and10news.com. Licensee: Heritage Broadcasting Co. of Michigan. Ownership: . (acq 3-3-89; grpsl; FTR: 3-20-89). Population served: 300,000 Natl. Network: CBS. Washington Atty: Wamble Carlyle. Wire Svc: UPI
Key Personnel:
William Kring. . . . . CFO & gen mgr
John DeMarsh . . . . . sls VP & gen sls mgr
Tessia Klix . . . . . prom mgr
Sherri Magiera . . . . . progmg dir
Kevin Dunaway . . . . . news dir
Lowell Shore . . . . . chief of engrg
Satellite of WWUP-TV Sault Ste. Marie.

WWUP-TV— Analog channel: 10. Digital channel: 49.316 kw vis, 31.6 kw aur. ant 1,214t/1,092g TL: N46 03 36 W84 05 57 On air date: June 15, 1962. Box 627, Cadillac, MI, 49601. Phone: (231) 775-3478. Fax: (231) 775-3671. Web Site: www.9and10news.com. Licensee: Heritage Broadcasting Co. of Michigan. (acq 3-3-89; grpsl; FTR: 3-20-89). Natl. Network: CBS. Washington Atty: Wamble Carlyle.
Key Personnel:
William Kring . . . . . gen mgr
John DeMarsh . . . . . gen sls mgr
Tessia Klix . . . . . prom mgr
Sherri Magiera . . . . . progmg dir
Kevin Dunaway . . . . . news dir
Lowell Shore. . . . . chief of engrg

Minnesota

Austin

see Rochester, MN-Mason City, IA-Austin, MN market

Crookston

see Fargo-Valley City, ND market

Duluth, MN-Superior, WI (DMA 137)

KBJR-TV—(Superior, WI) Analog channel: 6. Digital channel: 19.Note: NBC is on KBJR-TV ch 6, MyNetworkTV is on KBJR-DT ch 19. 100 kw vis, 20 kw aur. ant 1,010t/804g TL: N46 47 21 W92 06 51 On air date: Mar 11, 1954. 246 S. Lake Ave., Duluth, MN, 55802-2304. Phone: (218) 720-9600. Fax: (218) 720-9699. E-mail: news6@kbjr.com Web Site: www.northlandnewscenter.com. Licensee: KBJR License, Inc. Group owner: Granite Broadcasting Corp. (acq 11-1-88; $12.8 million; FTR: 9-23-74). Population served: 325,000 Natl. Network: NBC, MyNetworkTV Digital Network: Note: NBC is on KBJR-TV ch 6, MyNetworkTV is on KBJR-DT ch 19. Washington Atty: Akin, Gump, Strauss, Hauer & Feld. News staff: 80; News: 10 hrs wkly.
Key Personnel:
Robert Wilmers . . . . . gen mgr
David Jensch . . . . . stn mgr

KCWV— Analog channel: 27.1,750 kw vis. ant 879t/695g TL: N46 47 15 W92 07 21 Not on air, target date: unknown: 275 Goodwyn, Memphis, TN, 38111. Phone: (901) 375-9324. Permittee: George S. Flinn III.

KDLH— Analog channel: 3. Digital channel: 33. Analog hrs: 5:30 AM-2 AM 100 kw vis, 20 kw aur. ant 990t/816g TL: N46 47 07 W92 07 15 On air date: Mar 14, 1954. 246 S. Lake Ave., Duluth, MN, 55802. Phone: (218) 733-0303. Fax: (218) 727-7515. Web Site: www.kdlh.com. Licensee: Malara Broadcast Group of Duluth Licensee LLC. Group owner: New Vision Group LLC (acq 3-14-2005; $10.8 million). Population served: 203,100 Natl. Network: CBS, CW. Natl. Rep: Harrington, Righter & Parsons. Washington Atty: Wolf Bloch. News staff: 17; News: 12 hrs wkly.
Key Personnel:
Anthony J Malara . . . . . pres
Kelli Latuska . . . . . gen mgr
Carl Keller. . . . . sls dir & natl sls mgr
Nate Stoltman . . . . . mktg dir, mktg mgr & prom mgr
Jeff Reinarz . . . . . adv dir
Barb Wentworth . . . . . progmg dir
Derrick Hinds . . . . . news dir
Larry Erickson . . . . . engrg dir

KQDS-TV— Analog channel: 21. Digital channel: 17. Analog hrs: 18 955 kw vis. 590t TL: N46 47 41 W92 07 05 (CP: 45.36 kw vis) On air date: November 1994. 2001 London Rd., Duluth, MN, 55812. Phone: (218) 728-1622. Fax: (218) 728-1557. E-mail: dhileman@kqdsfox21.tv Licensee: KQDS Acquisition Corp. Group owner: Red River Broadcast Co. LLC (acq 10-21-98; grpsl). Population served: 172,000 HH Natl. Network: Fox. Natl. Rep: Harrington, Righter & Parsons. News staff: 20; News: 2.5 hrs wkly.
Key Personnel:
Ro Grignon . . . . . pres
Kathy Lau. . . . . VP & opns dir
Dave Hileman . . . . . gen mgr, gen sls mgr & natl sls mgr
Julie Moravchik . . . . . news dir

KRII— Analog channel: 11.316 kw vis. ant 657t/607g TL: N47 51 39.3 W92 56 46 On air date: Nov 27, 2002. 246 South Lake Ave., Duluth, MN, 55802-2304. Phone: (218) 720-9600. Fax: (218) 720-9660. E-mail: news6@kbjr.com Web Site: www.news6.tv. Licensee: Channel 11 License Inc. Group owner: Granite Broadcasting Corp. (acq 5-9-2001; grpsl). Population served: 70,000 Natl. Network: NBC.
Key Personnel:
David Jensch . . . . . stn mgr
Vincent Nelson . . . . . sls dir
Chris Hussey . . . . . prom mgr
Barb Wentworth . . . . . progmg mgr
Derrick Hinds. . . . . news dir
Larry Erickson . . . . . engrg dir
Satellite of KBJR-TV Superior, WI.

WDIO-TV— Analog channel: 10. Digital channel: 43. Analog hrs: 24 316 kw vis, 105 kw aur. 987t/836g TL: N46 47 13 W92 07 17 On air date: Jan 24, 1966. Box 16897, Duluth, MN, 55816-0897. 10 Observation Rd., Duluth, MN 55811-3506. Phone: (218) 727-6864. Fax: (218) 727-4415. E-mail: news@wdio.com Web Site: www.wdio.com. Licensee: WDIO-TV L.L.C. Group owner: Hubbard Broadcasting Inc. (acq 12-87; grpsl). Population served: 176,000 Natl. Network: ABC. Washington Atty: Fletcher, Heald & Hildreth. News staff: 18; News: 9 hrs wkly.
Key Personnel:
George Couture. . . . . gen mgr
Deb Messer . . . . . sls dir
Jeff Laumdergan . . . . . natl sls mgr
Jeff Laumdergan . . . . . prom dir
Dave Poirer . . . . . progmg dir
Steve Goodspeed . . . . . news dir
Mike Hatlestad . . . . . chief of engrg

***WDSE-TV—** Analog channel: 8. Digital channel: 38.316 kw vis, 31.6 kw aur. ant 950t/788g TL: N46 47 31 W92 07 21 On air date: Sept 1, 1964. 632 Niagara Ct., Duluth, MN, 55811-3098. Phone: (218) 724-8567. Fax: (218) 724-4269. E-mail: email@wdse.org Web Site: www.wdse.org. Licensee: Duluth-Superior Area Educ TV Corp. Population served: 151,000 Natl. Network: PBS. Washington Atty: Arnold & Porter.
Key Personnel:
Allen Harmon. . . . . gen mgr
Cheryl Leeper . . . . . gen sls mgr
Beth Lyden . . . . . prom mgr
Ron Anderson . . . . . progmg mgr
Rex Greenwell . . . . . chief of engrg

WIRT— Analog channel: 13. Digital channel: 36.125 kw vis, 21.6 kw aur. 670t/476g TL: N47 22 52 W92 57 18 On air date: Sept 1, 1967. Box 16897, Duluth, MN, 55816-0897. 10 Observation Rd., Duluth, MN 55811-3506. Phone: (218) 727-6864. Fax: (218) 727-4415. E-mail: news@wdio.com Web Site: www.wdio.com. Licensee: WDIO-TV L.L.C. Group owner: Hubbard Broadcasting Inc. (acq 12-87; grpsl). Natl. Network: ABC.
Key Personnel:
George Couture . . . . . VP & gen mgr
Deb Messer . . . . . sls dir
Jeff Laumdergan . . . . . natl sls mgr & prom dir
Dave Poirer . . . . . progmg dir
Steve Goodspeed . . . . . news dir
Mike Hatlestad . . . . . chief of engrg
Satellite of WDIO-TV Duluth.

Mankato, MN (DMA 200)

KEYC-TV— Analog channel: 12. Digital channel: 38. Analog hrs: 5 AM-1 AM Digital hrs: 5 AM-1 AM Note: CBS is on KEYC-TV ch 12, Fox is on KEYC-DT ch 38. 316 kw vis, 63 kw aur. 1,045t/1,116g TL: N43

56 14 W94 24 41 On air date: Oct 5, 1960. Box 128, Mankato, MN, 56002. 1570 Lookout Dr., N. Mankato, MN 56003. Phone: (507) 625-7905. Fax: (507) 625-5745. E-mail: keyc@keyc.com Web Site: www.keyc.tv. Licensee: United Communications Corp. Group owner: (group owner; (acq 10-14-77; $5 million). Population served: 387,250 Natl. Network: CBS, Fox Digital Network: Note: CBS is on KEYC-TV ch 12, Fox is on KEYC-DT ch 38. Natl. Rep: Continental Television Sales. Washington Atty: Wood, Maines & Brown. Wire Svc: AP Wire Svc: CBS News staff: 12; News: 10 hrs wkly.

Key Personnel:
Dennis M. Wahlstrom . . . . VP, gen mgr & natl sls mgr
Sharon Freitag . . . . opns mgr
John Ginther. . . . . rgnl sls mgr
Jan Ellanson . . . . prom mgr, progmg dir & progmg mgr
Terry Rudenick . . . . chief of engrg

Minneapolis-St. Paul, MN (DMA 15)

KARE—Minneapolis, Analog channel: 11. Digital channel: 35.316 kw vis, 31.6 kw aur. 1,440t/1,375g TL: N45 03 44 W93 08 21 On air date: Sept 1, 1953. 8811 Olson Memorial Hwy., Minneapolis, MN, 55427. Phone: (763) 546-1111. Fax: (763) 546-8590. Web Site: www.kare11.com. Licensee: Multimedia Holdings Corp. Group owner: Gannett Broadcasting (division of Gannett Co. Inc.) (acq 4-13-83; $75 million; FTR: 5-7-83). Population served: 2,500,000 Natl. Network: NBC.

Key Personnel:
John Remes . . . . pres & gen mgr
Tom Lindner . . . . news dir
Jeff Phillips . . . . chief of engrg

***KAWB**— Analog channel: 22. Digital channel: 28. Analog hrs: 24 Digital hrs: 24 214 kw vis. 745t/677g TL: N46 25 21 W94 27 41 On air date: Mar 1, 1988. 1500 Birchmont Dr. NE #9, Bemidji, MN, 56601-2600. Phone: (218) 751-3407. Fax: (218) 751-3142. E-mail: viewerservices@lakelandptv.org Web Site: www.lakelandptv.org. Licensee: Northern Minnesota Public TV Inc. Ownership: Community Population served: 160,000 Natl. Network: PBS. Washington Atty: Dow, Lohnes PLLC. News: 2.5 hrs wkly.

Key Personnel:
Rollin Buck . . . . pres
Bill Sanford. . . . . gen mgr
Dan Hegstad . . . . stn mgr
Jess Skala . . . . opns mgr
Sharon Pugh . . . . dev dir & dev mgr

***KAWE**— Analog channel: 9. Digital channel: 18. Analog hrs: 24 Digital hrs: 24 316 kw vis, 31.6 kw aur. 1,080t/1,000g TL: N47 42 03 W94 29 15 On air date: June 1, 1980. 1500 Birchmont Dr. NE #9, Bemidji, MN, 56601-2699. Phone: (218) 751-3407. Fax: (218) 751-3142. E-mail: viewerservices@lakelandptv.org Web Site: www.lakelandptv.org. Licensee: Northern Minnesota Public TV Inc. Ownership: Community Population served: 165,000 Natl. Network: PBS. Washington Atty: Dow, Lohnes PLLC. News staff: 7; News: 2.5 hrs wkly.

Key Personnel:
Rollin Buck . . . . pres
Bill Sanford. . . . . gen mgr
Jess Skala . . . . opns mgr
Sharon Pugh . . . . dev dir & dev mgr

KCCO-TV— Analog channel: 7. Digital channel: 24.316 kw vis, 63.1 kw aur. 1,120t/1,133g TL: N45 41 03 W95 08 14 On air date: Oct 8, 1958. 90 S. 11th St., Minneapolis, MN, 55403. Phone: (612) 339-4444. Fax: (612) 330-2627. E-mail: wcconewstips@wcco.com Web Site: www.wcco.com. Licensee: CBS Broadcasting Inc. Group owner: Viacom Television Stations Group. Population served: 150,000 Natl. Network: CBS. Natl. Rep: TeleRep. Washington Atty: Rosenman & Colin. Wire Svc: NOAA Weather

Key Personnel:
Tom Bourassa . . . . gen sls mgr
Susan Adams Loyd . . . . . progmg dir, VP, gen mgr & stn mgr
Jeff Kiernan . . . . news dir
Gary Kroger . . . . engrg dir

KCCW-TV— Analog channel: 12. Digital channel: 20.316 kw vis, 63.1 kw aur. 930t/999g TL: N46 56 03 W94 27 25 On air date: Jan 1, 1964. 90 S. 11th St., Minneapolis, MN, 55403. Phone: (612) 339-4444. Fax: (612) 330-2627. Web Site: www.wcco.com. Licensee: CBS Broadcasting Inc. Group owner: Viacom Television Stations Group. Population served: 150,000 Natl. Network: CBS. Washington Atty: Rosenman & Colin.

Key Personnel:
Tom Bourassa . . . . gen sls mgr
Susan Adams Loyd . . . . . progmg dir, VP, gen mgr & stn mgr
Jeff Kiernam . . . . news dir
Gary Kroger . . . . engrg dir

KFTC— Analog channel: 26. Analog hrs: 24 5,000 kw vis. 354t TL: N47 22 18 W94 52 56 Not on air, target date: 2000: 11358 Viking Dr., Eden Prairie, MN, 55344. Phone: (952) 944-9999. Fax: (952) 942-0286. E-mail: fox9news@foxtv.com Web Site: www.my29tv.com. Licensee: Fox Television Stations Inc. Group owner: (group owner; (acq 9-21-2001; grpsl). Natl. Network: MyNetworkTV.

Key Personnel:
Carol Rueppel . . . . gen mgr
Bill Dallman . . . . news dir
Marc Majeres . . . . chief of engrg
Satellite of WFTC Minneapolis.

KMSP-TV—Minneapolis, Analog channel: 9. Digital channel: 26. Analog hrs: 24 316 kw vis, 31.6 kw aur. 1,427t/1,430g TL: N44 51 32 W93 25 09 On air date: Jan 9, 1955. 11358 Viking Dr., Eden Prairie, MN, 55344-7258. Phone: (952) 944-9999. Fax: (952) 942-0286. E-mail: fox9news@foxtv.com Web Site: www.myfox9.com. Licensee: Fox Television Stations Inc. Group owner: (group owner; (acq 7-31-2001; grpsl). Population served: 1,100,000 Natl. Network: Fox. Wire Svc: AP News staff: 64; News: 19.5 hrs wkly.

Key Personnel:
Carol Rueppel . . . . gen mgr
Bill Dallman . . . . news dir
Marc Majeres . . . . chief of engrg

KPXM— Analog channel: 41. Digital channel: 40. Analog hrs: 24 2,750 kw vis, 275 kw aur. 1,469t/1,498g TL: N45 23 00 W93 42 30 On air date: Nov 24, 1982. 22601 176th St., Big Lake, MN, 55309. Phone: (763) 263-8666. Fax: (763) 263-6600. Web Site: www.ionline.tv. Licensee: Paxson Communications of Minneapolis/41 Inc. Group owner: Paxson Communications Corp. (acq 10-1-96; $12 million). Population served: 1,100,000 Natl. Network: i Network. Washington Atty: Mullin, Rhyne, Emmons & Topel.

Key Personnel:
Sherry Black. . . . . opns mgr
Doug Bognar . . . . rgnl sls mgr
Joe Brunke . . . . chief of engrg

KRWF— Analog channel: 43. Digital channel: 27. Analog hrs: 24 Digital hrs: 24 1,230 kw vis, 123 kw aur. 548t/536g TL: N44 29 03 W95 29 27 On air date: Apr 14, 1987. Box 189, Alexandria, MN, 56308. 415 Fillmore St., Alexandria, MN 56308. Phone: (320) 763-5729. Fax: (320) 763-4627. E-mail: ksax@ksax.com Web Site: ksax.com. Licensee: KSAX-TV Inc. Group owner: Hubbard Broadcasting Inc. Natl. Network: ABC. Washington Atty: Holland & Knight.

Key Personnel:
Robert Hubbard . . . . gen mgr
Edward Smith . . . . stn mgr & gen sls mgr
Larry Eckblad . . . . chief of engrg

KSAX— Analog channel: 42. Digital channel: 36. Analog hrs: 24 Digital hrs: 24 2,770 kw vis, 277 kw aur. 1,176t/1,164g TL: N45 41 59 W95 10 36 On air date: Sept 15, 1987. Box 189, Worcester, MN, 56308. 415 Fillmore St., Alexandria, MN 56308. Phone: (320) 763-5729. Fax: (320) 763-4627. E-mail: ksax@ksax.com Web Site: www.ksax.com. Licensee: KSAX-TV Inc. Group owner: Hubbard Broadcasting Inc. Natl. Network: ABC. Natl. Rep: Petry Television Inc. Washington Atty: Holland & Knight. Wire Svc: AP

Key Personnel:
Robert Hubbard . . . . gen mgr
Edward Smith . . . . stn mgr & gen sls mgr
Larry Eckblad . . . . chief of engrg

KSTC-TV—Minneapolis, Analog channel: 45. Digital channel: 44. Analog hrs: 24 5,000 kw vis, 500 kw aur. ant 1,410t/1,315g TL: N45 03 45 W93 08 21 On air date: 1995. 3415 University Ave., St. Paul, MN, 55114-2099. Phone: (651) 645-4500. Fax: (651) 523-7320. Web Site: www.kstc45.com. Licensee: KSTC.TV LLC. Group owner: Hubbard Broadcasting Inc. (acq 4-24-2000). Population served: 3,100,000 Washington Atty: Holland & Knight LLP.

Key Personnel:
Susan Wenz . . . . gen mgr
Andy Stavast . . . . gen sls mgr
Joe Johnston . . . . mktg mgr
Michael E. Smith . . . . progmg dir & progmg mgr
Christopher Berg . . . . news dir
Dick Rice . . . . chief of engrg

KSTP-TV—Saint Paul, Analog channel: 5. Digital channel: 50. Analog hrs: 24 Digital hrs: 24 100 kw vis, 15.1 kw aur. 1,430t/1,375g TL: N45 03 45 W93 08 22 On air date: Apr 23, 1948. 3415 University Ave., Saint Paul, MN, 55114. Phone: (651) 646-5555. Fax: (651) 642-4172. Web Site: www.kstp.com. Licensee: KSTP-TV LLC. Group owner: Hubbard Broadcasting Inc. Natl. Network: ABC. Natl. Rep: Petry Television Inc. Washington Atty: Holland and Knight. Wire Svc: AP News: 27.5 hrs wkly.

Key Personnel:
Robert Hubbard. . . . . pres & gen mgr
Monica Doyle . . . . opns mgr
John McCormick . . . . gen sls mgr
Andrea Creech. . . . . mktg mgr
Michael Smith . . . . progmg dir & progmg mgr
Christopher Berg . . . . news dir
Dick Rice . . . . chief of engrg

***KTCA-TV**—Saint Paul, Analog channel: 2. Digital channel: 34. Analog hrs: 24 Digital hrs: 24 100 kw vis, 20 kw aur. 1,336t/1,372g TL: N45 03 30 W93 07 27 On air date: Sept 3, 1957. 172 E. 4th St., St. Paul, MN, 55101. Phone: (651) 222-1717. Fax: (651) 229-1282. Web Site: www.tpt.org. Licensee: Twin Cities Public TV Inc. Population served: 1,400,000 Natl. Network: PBS.

Key Personnel:
Jim Pagliarini . . . . CEO
Dan Thomas. . . . . COO & progmg dir
Jim Paliarini . . . . pres & exec VP
Stephen Usery . . . . mktg VP & news dir
Bruce Jacobs . . . . chief of engrg

***KTCI-TV**—Saint Paul, Analog channel: 17. Digital channel: 16. Analog hrs: 24 Digital hrs: 24 331 kw vis, 33.1 kw aur. ant 1,298t/1,471g TL: N45 03 29 W93 07 27 On air date: May 3, 1965. 172 E. 4th St., Saint Paul, MN, 55101. Phone: (651) 222-1717. Fax: (651) 229-1282. E-mail: viewerservices@tpt.org Web Site: www.tpt.org. Licensee: Twin Cities Public Television Inc. Population served: 1,400,000 Natl. Network: PBS.

Key Personnel:
Jim Pagliarini . . . . CEO & pres
Dan Thomas . . . . COO
Stephen Usery . . . . mktg VP
Bruce Jacobs . . . . chief of engrg

***KWCM-TV**— Analog channel: 10. Digital channel: 31. Analog hrs: 24 316 kw vis, 37.1 kw aur. 1,250t/1,274g TL: N45 10 03 W96 00 02 On air date: Feb 7, 1966. 120 W. Schlieman Ave., Appleton, MN, 56208-1351. Phone: (320) 289-2622. Fax: (320) 289-2634. E-mail: yourtv@pioneer.org Web Site: www.pioneer.org. Licensee: West Central Minnesota Educational TV Co. Population served: 750,000 Natl. Network: PBS. Washington Atty: Fletcher, Heald & Hildreth.

Key Personnel:
Les Heen . . . . pres & gen mgr
Jon Panzer. . . . . stn mgr, engrg VP & chief of engrg
Shirley Schwarz . . . . progmg dir

WCCO-TV—Minneapolis, Analog channel: 4. Digital channel: 32.100 kw vis, 10 kw aur. 1,430t/1,375g TL: N45 03 45 W93 08 21 On air date: July 1, 1949. 90 S. 11th St., Minneapolis, MN, 55403. Phone: (612) 339-4444. Fax: (612) 330-2627. Web Site: wcco.com. Licensee: CBS Broadcasting Inc. Group owner: Viacom Television Stations Group (acq 2-92; grpsl; FTR: 7-26-76). Population served: 434,400 Natl. Network: CBS. Natl. Rep: CBS TV Stations National Sales. Wire Svc: WU Wire Svc: Reuters News staff: 75; News: 23 hrs wkly.

Key Personnel:
Tom Bourassa . . . . gen sls mgr
Susan Adams Loyd . . . . progmg dir, VP & gen mgr
Jeff Kiernan . . . . news dir
Gary Kroger . . . . engrg dir

WFTC—Minneapolis, Analog channel: 29. Digital channel: 21. Analog hrs: 24 5,000 kw vis, 500 kw aur. 1,223t/1,205g TL: N45 03 30 W93 07 27 On air date: October 1982. 11358 Viking Dr., Eden Prairie, MN, 55344. Phone: (952) 944-9999. Fax: (952) 942-0286. E-mail: fox9news@foxtv.com Web Site: www.my29tv.com. Licensee: Fox Television Stations Inc. Group owner: (group owner; (acq 10-1-2001; grpsl). Natl. Network: MyNetworkTV. Wire Svc: AP News: 3.5 hrs wkly.

Key Personnel:
Carol Rueppel . . . . gen mgr
Bill Dallman . . . . news dir
Marc Majeres . . . . chief of engrg

***WHWC-TV**— Analog channel: 28. Digital channel: 27. Analog hrs: 24 Digital hrs: 24 1,100 kw vis, 54.5 kw aur. 1,151t/1197g TL: N45 02 47 W91 51 42 On air date: Nov 18, 1973. 3319 W. Beltline Hwy., Madison, WI, 53713. Phone: (608) 264-9600. Fax: (608) 264-9664. Web Site: www.ecb.org. Licensee: State of Wisconsin-Educational Communications Board. Natl. Network: PBS. Washington Atty: Dow, Lohnes & Albertson.

Key Personnel:
Mike Edgette . . . . opns dir
Jon Miskowski . . . . dev dir & chief of engrg
Michael Bridgeman . . . . prom mgr
Mary Clare Sorenson . . . . adv dir & progmg mgr
Kathy Bissen . . . . news dir
Dick Taugher. . . . . chief of engrg

WUCW—Minneapolis, Analog channel: 23. Digital channel: 22. Analog hrs: 24 4,570 kw vis, 457 kw aur. 1,150t/1,450g TL: N45 03 30 W93 07 27 On air date: Sept 22, 1982. 1640 Como Ave., St. Paul, MN, 55108. Phone: (651) 646-2300. Fax: (651) 646-1220. Web Site: www.thecwtc.com. Licensee: KLGT Licensee L.L.C. Group owner: Sinclair Broadcast Group Inc. (acq 3-16-98; $52.5 million). Population served: 1,700,000 Natl. Network: CW.

Key Personnel:
Joe Tracy . . . . gen mgr
Eric Lazar . . . . natl sls mgr
Tom Burke . . . . rgnl sls mgr
Cece Smith . . . . progmg mgr
Steve Lunde . . . . chief of engrg

Rochester, MN-Mason City, IA-Austin, MN
(DMA 153)

KAAL— Analog channel: 6. Digital channel: 33.100 kw vis, 10 kw aur. 1,049t/1,000g TL: N43 37 42 W93 09 12 On air date: Aug 17, 1953. Box 577, Austin, MN, 55912. 1701 10th Pl. N.E., Austin, MN 55912. Phone: (507) 437-6666. Fax: (507) 433-9560. Web Site: www.kaaltv.com. Licensee: KAAL-TV LLC. Group owner: Hubbard Broadcasting Inc. (acq 12-13-00; $9.5 million). Natl. Network: ABC. Washington Atty: Schwartz, Woods & Miller. News staff: 18; News: 11 hrs wkly.
Key Personnel:
David Harbert . . . . gen mgr & news dir
Ila Teskey . . . . gen sls mgr & rgnl sls mgr
Sheryl Barlon . . . . mktg dir & progmg dir
Dan Collado . . . . prom mgr
Janet Anderson . . . . prom mgr
Wendell Nelson . . . . chief of engrg
Jan Thompson . . . . progmg

KIMT— Analog channel: 3. Digital channel: 42. Analog hrs: 21 100 kw vis, 10 kw aur. 1,510t/1,525g TL: N43 22 20 W92 49 59 (CP: 5 kw aur, 97.7 kw vis, ant 1,550t/1,569g) On air date: May 15, 1954. 112 N. Pennsylvania Ave., Mason City, IA, 50401. Phone: (641) 423-2540. Fax: (641) 423-9309. E-mail: mail@kimt.com Web Site: www.kimt.com. Licensee: NVT Mason City Licensee LLC. Group owner: Media General Broadcast Group (acq 10-6-2006; $35 million with WIAT-TV Birmingham, AL). Population served: 343,000 Natl. Network: CBS. Natl. Rep: Harrington, Righter & Parsons. Washington Atty: Covington & Burling. News staff: 21; News: 19 hrs wkly.
Key Personnel:
Steve Martinson . . . . VP & gen mgr
Michael Fitzgerald . . . . gen sls mgr
Jerome Risting . . . . prom mgr & progmg dir
John Murray . . . . news dir
Steve Reiter . . . . chief of engrg
Wayne Kohlhaas . . . . rgnl sls mgr & sls

***KSMQ-TV—** Analog channel: 15. Digital channel: 20.1,215.24 kw vis, 121.8 kw aur. ant 380t/448g TL: N43 40 34 W93 00 09 On air date: Oct 17, 1972. 2000 8th Ave. N.W., Austin, MN, 55912. Phone: (507) 433-0678. Fax: (507) 433-0670. E-mail: ksmq@ksmq.org Web Site: www.ksmq.org. Licensee: Southern Minnesota Quality Broadcasting Inc. (acq 5-27-2005). Population served: 650,000 Natl. Network: PBS. Rgnl. Network: CEN. Washington Atty: Schwartz, Woods & Miller.
Key Personnel:
Sandra Session-Robertson . . . . CEO & pres
Suzi Stone . . . . progmg dir
Shawn Weitzel . . . . chief of engrg

KTTC— Analog channel: 10. Digital channel: 36. Analog hrs: 24 Note: NBC is on KTTC(TV) ch 10, CW is on KTTC-DT ch 36. 316 kw vis, 46.8 kw aur. ant 1,260t/1,314g TL: N43 34 15 W92 25 37 On air date: July 16, 1953. 6301 Bandel Rd. N.W., Rochester, MN, 55901. Phone: (507) 288-4444. Fax: (507) 288-6324. Fax: (507) 288-6278 (news). E-mail: kttc@kttc.com Web Site: www.kttc.com. Licensee: KTTC TV Inc. Group owner: Quincy Newspapers Inc. (acq 7-1-76; $4.25 million; FTR: 5-24-76). Population served: 500,000 Natl. Network: NBC, CW Digital Network: Note: NBC is on KTTC(TV) ch 10, CW is on KTTC-DT ch 36. Natl. Rep: Blair Television. Washington Atty: Wilkinson, Barker, Knauer & Quinn. Wire Svc: AP Wire Svc: CNN Wire Svc: NBC News staff: 15; News: 11 hrs wkly.
Key Personnel:
Jerry Watson . . . . gen mgr
Elizabeth Dahlen . . . . stn mgr & mktg dir
Dave Ferber . . . . natl sls mgr
Rita Duda . . . . prom mgr
Vickie Broughton . . . . progmg dir & progmg mgr
Tim Morgan . . . . engrg dir & chief of engrg

KXLT-TV— Analog channel: 47. Digital channel: 46. Analog hrs: 24 1510 kw vis. ant 1,125 t TL: N43 38 34 W92 31 35 On air date: Aug 1, 1987. 6301 Bandel Rd. N.W., Rochester, MN, 55901. Phone: (507) 252-4747. Fax: (507) 252-5050. E-mail: comments@fox47kxlt.com Web Site: www.fox47kxlt.com. Licensee: SagamoreHill of Minnesota Licenses LLC. (acq 3-31-2005; $2.05 million). Natl. Network: Fox. Washington Atty: Rosenman & colin. News: 7 hrs wkly.
Key Personnel:
Louis Wall . . . . gen mgr
Liz Dahlen . . . . stn mgr
Kristopher Lake . . . . gen sls mgr
Rita Duda . . . . prom dir & prom mgr
Samantha Bishop . . . . progmg mgr
Tim Morgan . . . . engrg dir & chief of engrg

***KYIN—** Analog channel: 24. Digital channel: 18.1,740 kw vis, 174 kw aur. 1,430t/1,565g TL: N43 22 25 W92 51 00 On air date: May 14, 1977. Box 6450, Iowa Public TV, Johnston, IA, 50131-6450. 6450 Corporate Dr., Johnston, IA 50131. Phone: (515) 242-3100. E-mail: public_information@iptv.org Web Site: www.iptv.org. Licensee: Iowa Public Broadcasting Board. Natl. Network: PBS. Washington Atty: Dow, Lohnes PLLC.

St. Paul

see Minneapolis-St. Paul, MN market

Thief River Falls

see Fargo-Valley City, ND market

Worthington

see Sioux Falls (Mitchell), SD market

Mississippi

Biloxi-Gulfport, MS
(DMA 160)

WLOX— Analog channel: 13. Digital channel: 39. Analog hrs: 24 316 kw vis, 57.5 kw aur. 1,340t/1,319g TL: N30 43 23 W89 05 28 On air date: Sept 15, 1962. 208 De Buys Rd., Biloxi, MS, 39531. Phone: (228) 896-1313. Fax: (228) 896-0749. E-mail: wlox@wlox.com Web Site: www.wlox.com. Licensee: Libco Inc. Group owner: Liberty Corp. (acq 1-13-2006; grpsl). Population served: 120,700 Natl. Network: ABC. Washington Atty: Covington & Burling. Wire Svc: AP News staff: 44; News: 18 hrs wkly.
Key Personnel:
Leon Long . . . . VP & gen mgr
Dave Vincent . . . . stn mgr
Roger Garrett . . . . opns mgr
Linda Sherman . . . . gen sls mgr
Don Moore . . . . rgnl sls mgr
Darlene Duffano . . . . progmg dir
David Vincent . . . . news dir
John Armstrong . . . . chief of engrg

***WMAH-TV—** Analog channel: 19. Digital channel: 16. Analog hrs: 24 Digital hrs: 24 1,480 kw vis, 148 kw aur. 1,558t/1,537g TL: N30 45 14 W88 56 44 (CP: 1,620 kw vis, ant 1,568t/1,547g) On air date: Jan 14, 1972. 3825 Ridgewood Rd., Jackson, MS, 39211. Phone: (601) 432-6565. Fax: (601) 432-6654 / (601) 432-6311. Web Site: www.mpbonline.org. Licensee: Mississippi Authority for Educational TV. Natl. Network: PBS. Washington Atty: Schwartz, Woods & Miller.
Key Personnel:
Marie Antoon . . . . exec VP & gen mgr
Bob Buie . . . . opns dir
teresa Collier . . . . news dir

WXXV-TV— Analog channel: 25. Digital channel: 48. Analog hrs: 24 2,240 kw vis, 224 kw aur. 1,780t/1,540g TL: N30 44 48 W89 03 30 On air date: Feb 14, 1987. Box 2500, Gulfport, MS, 39505. 14351 Hwy. 49 N., Gulfport, MS 39503. Phone: (228) 832-2525. Fax: (228) 832-4442. E-mail: info@wxxv25.com Web Site: www.wxxv25.com. Licensee: Morris Network of Mississippi Inc. Group owner: Morris Network Inc. (acq 5-22-97; $17.475 million). Natl. Network: Fox. Natl. Rep: Millennium Sales & Marketing. Washington Atty: Fletcher, Heald & Hildroth.
Key Personnel:
Dean Hinson . . . . pres
Bobby Edwards . . . . gen mgr
Deidre Pyron . . . . prom dir
Ray Luke . . . . opns dir, gen sls mgr & chief of engrg

Columbus-Tupelo-West Point, MS
(DMA 132)

WCBI-TV— Analog channel: 4. Digital channel: 35. Analog hrs: 24 Note: CBS is on WCBI-TV ch 4, CW and MyNetworkTV are on WCBI-DT ch 35. 100 kw vis, 10 kw aur. 1,996t/1,800g TL: N33 45 06 W88 52 40 On air date: July 13, 1956. 201 5th St. S., Columbus, MS, 39701. Phone: (662) 327-4444. Fax: (662) 329-1004. E-mail: comments@wcbi.com Web Site: www.wcbi.com. Licensee: WCBI-TV LLC. Group owner: Morris Multi-Media (acq 12-31-03; $20 million). Population served: 165,000 Natl. Network: CBS, CW, MyNetworkTV Digital Network: Note: CBS is on WCBI-TV ch 4, CW and MyNetworkTV are on WCBI-DT ch 35. Washington Atty: Fletcher, Heald & Hildreth. Wire Svc: Weather Wire
Key Personnel:
Bobby Berry . . . . gen mgr, progmg dir & progmg mgr
Derek Rogers . . . . sls dir & gen sls mgr
Susan Bell . . . . prom mgr
Russ Geller . . . . news dir
Gary Savage . . . . pub affrs dir & chief of engrg

WKDH— Analog channel: 45. Analog hrs: 20 1,500 kw vis. ant 1,610t/1,368g TL: N33 47 40 W89 05 16 On air date: June 2002. Box 1645, Tupelo, MS, 38802-1645. Phone: (662) 842-7620. Fax: (662) 842-6342. Fax: (662) 844-7061. Web Site: www.wkdh.com. Licensee: Southern Broadcasting Inc. Natl. Network: ABC. Washington Atty: Garvey, Schubert & Barer.
Key Personnel:
Walter Spain . . . . pres
Gerald Stanford . . . . chief of engrg

WLOV-TV— Analog channel: 27. Digital channel: 16. Analog hrs: 5:30 AM-1:30 AM Digital hrs: 5:30 AM-1:30 AM 1.80t/1,442g TL: N33 47 40 W89 05 16 On air date: May 29, 1983. Box 1732, Tupelo, MS, 38802. Phone: (662) 842-2227. Fax: (662) 844-7061. E-mail: manager@wlov.com Web Site: www.wlov.com. Licensee: Lingard Broadcasting Corp. Ownership: Jack Lingard, 100%. (acq 4-12-94). Natl. Network: Fox. Natl. Rep: Continental Television Sales. Washington Atty: Law Office of Robert E. Levine. Wire Svc: AP News: 2.5 hrs wkly.
Key Personnel:
Jennifer Dennington . . . . stn mgr
Ed Bishop . . . . progmg dir
Marty Davis . . . . chief of engrg

***WMAA—** Analog channel: 43.81 kw vis. ant 668t/658g TL: N33 50 31 W88 41 48 Not on air, target date: unknown: 3825 Ridgewood Rd., Jackson, MS, 39211. Phone: (601) 432-6565. Fax: (601) 432-6392. Web Site: www.mpbonline.org. Licensee: Mississippi Authority for Educational Television. Natl. Network: PBS.
Key Personnel:
Marie Antoon . . . . gen mgr
Ron Evans . . . . news dir
Lee Tapley . . . . pub affrs dir

***WMAB-TV—** Analog channel: 2. Digital channel: 10. Analog hrs: 24 Digital hrs: 24 100 kw vis, 10 kw aur. ant 1,250t/1,091g TL: N33 21 07 W89 08 56 On air date: July 4, 1971. 3825 Ridgewood Rd., Jackson, MS, 39211. Phone: (601) 432-6565. Fax: (601) 432-6654. Fax: (601) 432-6311. Web Site: www.mpbonline.org. Licensee: Mississippi Authority for Educational TV. Population served: 734,380 Natl. Network: PBS. Washington Atty: Schwartz, Woods & Miller. News staff: 4.
Key Personnel:
Marie Antoon . . . . gen mgr
Bob Buie . . . . opns dir
Randy Tinney . . . . prom dir
Teresa Collier . . . . news dir

***WMAE-TV—** Analog channel: 12. Digital channel: 55. Analog hrs: 24 Digital hrs: 24 100 kw vis, 8.91 kw aur. ant 741t/538g TL: N34 40 00 W88 45 05 On air date: Aug 11, 1974. 3825 Ridgewood Rd., Jackson, MS, 39211. Phone: (601) 432-6565. Fax: (601) 432-6654. Fax: (601) 432-6311. Web Site: www.mpbonline.org. Licensee: Mississippi Authority for Educational TV. Natl. Network: PBS. Washington Atty: Schwartz, Woods & Miller.
Key Personnel:
Marie Antoon . . . . gen mgr
Teresa Collier . . . . news dir

WTVA— Analog channel: 9. Digital channel: 8. Analog hrs: 24 Digital hrs: 24 316 kw vis, 31.6 kw aur. 1,781t/1,585g TL: N33 47 40 W89 05 16 On air date: Mar 18, 1957. Box 350, Tupelo, MS, 38802. 1359 Rd. 681, Tupelo, MS 38802. Phone: (662) 842-7620. Fax: (662) 844-7061. E-mail: manager@wtva.com Web Site: www.wtva.com. Licensee: WTVA Inc. Group owner: (group owner). Natl. Network: NBC. Washington Atty: Garvey, Schubert & Barer. Wire Svc: AP News: 17 hrs wkly.
Key Personnel:
Jane Spain . . . . pres
Mark Ledbetter . . . . gen mgr
Jon Ball . . . . opns mgr
Ed Bishop . . . . progmg dir

Greenville

see Greenwood-Greenville, MS market

Greenwood-Greenville, MS (DMA 184)

WABG-TV— Analog channel: 6. Digital channel: 32. Analog hrs: 24 Note: ABC is on WABG-TV ch 6, Fox is on WABG-DT ch 32. 100 kw vis, 10 kw aur. ant 2,000t/2,136g TL: N33 22 23 W90 32 31 On air date: Oct 20, 1959. Box 1243, 849 Washington Ave., Greenville, MS, 38701. Box 720, 2001 Garrard Ave., Greenwood, MS 38930. Phone: (662) 332-0949. Fax: (662) 344-1814. Web Site: www.wabg.com. Licensee: Mississippi Broadcasting Partners. Group owner: Bahakel Communications Natl. Network: ABC, Fox Digital Network: Note: ABC is on WABG-TV ch 6, Fox is on WABG-DT ch 32. Natl. Rep: Continental Television Sales. News staff: 16; News: 13 hrs wkly.
Key Personnel:
Sherry Nelson . . . . stn mgr
Jonas Oswalt . . . . rgnl sls mgr
Donnie Reid . . . . progmg dir & progmg mgr
Pam Chatman . . . . news dir
Larry Nixon . . . . chief of engrg

***WMAI—** Analog channel: 31.759 kw vis. ant 315t/312g TL: N33 44 01 W90 42 50 Not on air, target date: unknown: 3825 Ridgewood Rd., Jackson, MS, 39211. Phone: (601) 432-6565. Fax: (610) 432-6392. Web Site: www.mpbonline.org. Licensee: Mississippi Authority for Educational Television. Natl. Network: PBS.

***WMAO-TV—**Greenwood, Analog channel: 23. Digital channel: 25. Analog hrs: 24 Digital hrs: 24 537 kw vis, 53.7 kw aur. ant 1,040t/1,061g TL: N33 22 34 W90 32 31 On air date: Sept 15, 1972. 3825 Ridgewood Rd., Jackson, MS, 39211. Phone: (601) 432-6565. Fax: (601) 432-6654. Fax: (601) 432-6311. Web Site: www.mpbonline.org. Licensee: Mississippi Authority for Educational TV. Washington Atty: Schwartz, Woods & Miller.
Key Personnel:
Marie Antoon . . . . exec VP & gen mgr
Bob Buie. . . . . opns dir
Teresa Collier. . . . . dev dir & news dir

WXVT— Analog channel: 15. Digital channel: 17. Analog hrs: 24 Digital hrs: 24 2,746 kw vis, 549 kw aur. 887t/919g TL: N33 39 26 W90 42 18 On air date: Nov 7, 1980. 3015 E. Reed Rd., Greenville, MS, 38703. Phone: (662) 334-1500. Fax: (662) 378-8122. Web Site: www.wxvt.com. Licensee: Saga Broadcasting LLC. Group owner: Saga Communications Inc. (acq 7-1-99; $5.2 million). Population served: 137,000 Natl. Network: CBS.
Key Personnel:
Darren Lehrmann . . . . gen mgr
Larry Cazavan . . . . gen sls mgr
Carolyn Byars. . . . . progmg dir
Earl Phelps . . . . news dir
Paul Serio . . . . chief of engrg

Gulfport

see Biloxi-Gulfport, MS market

Hattiesburg-Laurel, MS (DMA 165)

WDAM-TV—Laurel, Analog channel: 7. Digital channel: 28. Analog hrs: 21 Digital hrs: 21 316 kw vis, 47 kw aur. 510t/575g TL: N31 27 12 W89 17 05 On air date: June 8, 1956. Box 16269, Hattiesburg, MS, 39404-6269. 2362 Hwy. 11 N., Moselle, MS 39459. Phone: (601) 544-4730. Fax: (601) 584-9302. E-mail: info@wdam.com Web Site: www.wdam.com. Licensee: WDAM License Subsidiary Inc. Group owner: Raycom Media Inc. (acq 9-24-96; grpsl). Population served: 244,500 Natl. Network: NBC. Natl. Rep: Harrington, Righter & Parsons. Washington Atty: Covington & Burling. News staff: 26; News: 20 hrs wkly.
Key Personnel:
Jim Cameron . . . . VP & gen mgr
Ted Palmer . . . . gen sls mgr
Wanda Morrison . . . . natl sls mgr
Pam McGovern . . . . mktg mgr & prom dir
Betty Young . . . . progmg dir
Randy Swan . . . . news dir
Jim Wilkinson . . . . engrg mgr & chief of engrg

WHLT— Analog channel: 22. Analog hrs: 24 1,200 kw vis, 120 kw aur. 800t/707g TL: N31 24 20 W89 14 13 On air date: Jan 12, 1987. 5912 Hwy. 49, The Cloverleaf Center, Suite A, Hattiesburg, MS, 39401. Phone: (601) 545-2077. Fax: (601) 545-3589. E-mail: wbabbidge@whlt.com Web Site: www.cbs22thehub.com. Licensee: Media General Broadcasting Inc. Group owner: Media General Broadcast Group (acq 7-25-97; grpsl). Natl. Network: CBS. Natl. Rep: MMT.
Key Personnel:
Robert Romine . . . . gen mgr
Wally Babbidge . . . . stn mgr & gen sls mgr
Gary Wolverton . . . . prom mgr
Jackie McDonald . . . . progmg dir
Gary Wright. . . . . engrg mgr & chief of engrg

Holly Springs

see Memphis, TN market

Jackson, MS (DMA 87)

WAPT— Analog channel: 16. Digital channel: 21. Analog hrs: 24 Digital hrs: 24 4,780 kw vis, 478 kw aur. 1,178t/1,072g TL: N32 16 39 W90 17 41 On air date: Oct 3, 1970. 7616 Channel 16 Way, Jackson, MS, 39209. Phone: (601) 922-1607. Fax: (601) 922-1663. Web Site: www.wapt.com. Licensee: WAPT Hearst-Argyle Television Inc. Group owner: Hearst-Argyle Televison Inc. (acq 7-16-97; grpsl). Population served: 280,000 Natl. Network: ABC. News staff: 26; News: 14 hrs wkly.
Key Personnel:
Stuart Kellogg . . . . gen mgr
Jeff Wolfe. . . . . gen sls mgr & rgnl sls mgr
Linda Bozone . . . . progmg mgr
Bruce Barkley. news dir
Tom Bondurant . . . . chief of engrg

WDBD— Analog channel: 40. Digital channel: 41. Analog hrs: 24 Digital hrs: 24 5,000 kw vis,. ant 1,961t/1,853g TL: N32 12 49 W90 22 56 On air date: Nov 30, 1984. One Great Pl., Jackson, MS, 39209. Phone: (601) 922-1234. Fax: (601) 922-0268. Web Site: www.gomiss.com. Licensee: Jackson Television L.L.C. Ownership: Sheldon H. Galloway, 100% voting control (acq 9-30-2003; $13.4 million with WXMS-LP Jackson). Natl. Network: Fox. Washington Atty: Fisher, Wayland, Cooper, Leader & Zaragoza.
Key Personnel:
Mike Dunlop . . . . gen mgr
Mike Garza . . . . sls dir
Loretta Nichols. . . . . prom mgr
Sylvia Walker . . . . progmg dir
Robert Flanagan . . . . chief of engrg

WJTV— Analog channel: 12. Digital channel: 52. Analog hrs: 24 Digital hrs: 24 316 kw vis, 63.1 kw aur. 1,630t/1,615g TL: N32 14 26 W90 24 15 On air date: Mar 15, 1954. 1820 TV Rd., Jackson, MS, 39204-4148. Phone: (601) 372-6311. Fax: (601) 372-8798. Web Site: www.wjtv.com. Licensee: Media General Broadcasting Inc. Group owner: Media General Broadcast Group (acq 7-25-97; grpsl). Population served: 477,300 Natl. Network: CBS. Washington Atty: Wiley, Rein & Fielding. News staff: 34; News: 21 hrs wkly.
Key Personnel:
Bob Romine . . . . gen mgr
Mark Chapman . . . . sls dir & gen sls mgr
Gary Wolverton . . . . mktg dir
David Bunger. prom dir
Jackie McDonald . . . . progmg mgr
Rick Russell . . . . news dir
Steve Schrader. . . . . chief of engrg

WLBT— Analog channel: 3. Digital channel: 9. Analog hrs: 24 Digital hrs: 24 95.7 kw vis, 19.1 kw aur. 2,419t/1,999g TL: N32 12 46 W90 22 54 On air date: Dec 28, 1953. 715 S. Jefferson St., Jackson, MS, 39201. Phone: (601) 948-3333. Fax: (601) 960-4412. E-mail: news@wlbt.com Web Site: www.wlbt.com. Licensee: Civco Inc. Group owner: Liberty Corp. (acq 1-13-2006; grpsl). Population served: 320,000 Natl. Network: NBC. Natl. Rep: Continental Television Sales. Washington Atty: Dow, Lohnes & Albertson. Wire Svc: AP
Key Personnel:
Dan Modisett . . . . gen mgr
Frankie Thomas . . . . gen sls mgr
Dennis Smith . . . . news dir
Curtis McKnight. . . . . chief of engrg

***WMAU-TV—** Analog channel: 17. Digital channel: 18. Analog hrs: 24 Digital hrs: 24 550 kw vis, 55 kw aur. 1,121t/1,066g TL: N31 22 19 W90 45 05 On air date: Jan 14, 1972. 3825 Ridgewood Rd., Jackson, MS, 39211. Phone: (601) 432-6565. Fax: (610) 432-6654 / (601) 432-6311. Web Site: www.mpbonline.org. Licensee: Mississippi Authority for Educational TV. Natl. Network: PBS. Washington Atty: Schwartz, Woods & Miller.
Key Personnel:
Marie Antoon . . . . gen mgr
Teresa Collier. . . . . news dir & engrg dir

***WMPN-TV—** Analog channel: 29. Digital channel: 20. Analog hrs: 24 Digital hrs: 24 912 kw vis, 91.2 kw aur. ant 1,958t/1,997g TL: N32 12 46 W90 22 54 On air date: Feb 1, 1970. 3825 Ridgewood Rd., Jackson, MS, 39211. Phone: (601) 432-6565. Fax: (601) 432-6654. Fax: (601) 432-6311. Web Site: www.mpbonline.org. Licensee: Mississippi Authority for Educational TV. Population served: 1,100,000 Natl. Network: PBS. Washington Atty: Schwartz, Woods & Miller.
Key Personnel:
Marie Antoon . . . . gen mgr
Bob Buie. . . . . opns dir
Teresa Collier. . . . . news dir

***WMYC—** Analog channel: 32.769 kw vis. ant 36t TL: N32 50 48 W90 23 18 Not on air, target date: unknown: 3825 Ridgewood Rd., Jackson, MS, 39211. Phone: (601) 432-6565. Fax: (601) 432-6311. Web Site: www.mpbonline.org. Licensee: Mississippi Authority for Educational Television. Natl. Network: PBS.

WNTZ— Analog channel: 48. Digital channel: 49. Analog hrs: 24 Digital hrs: 24 1,170 kw vis, 117 kw aur. ant 843t/848g TL: N31 30 33 W91 24 19 (CP: 2,765 kw vis, ant 1,240t) On air date: Nov 16, 1985. 1777 Jackson St., Alexandria, LA, 71301. Phone: (318) 443-4700. Fax: (318) 443-4899. Web Site: www.fox48tv.com. Licensee: White Knight Broadcasting of Natchez License Corp. Group owner: White Knight Holdings Inc. (acq 6-22-98). Natl. Network: Fox, MyNetworkTV. Natl. Rep: Millennium Sales & Marketing. Washington Atty: Shaw Pittman L.L.P.
Key Personnel:
Sharon Rachal . . . . gen mgr, gen mgr & gen sls mgr
Vikki Chapman. . . . . progmg mgr
Tony Guillory. . . . . chief of engrg

WRBJ— Analog channel: 34.1,400 kw vis. ant 1,229t/1,242g TL: N32 07 18 W89 32 52 On air date: January 2006. 745 N. State St., Jackson, MS, 39202. Phone: (601) 974-5700. Fax: (601) 974-5711. Web Site: cw34jackson.com. Licensee: Roberts Broadcasting of Jackson, MS, LLC. Group owner: Roberts Broadcasting Co. Natl. Network: CW. Natl. Rep: Harrington, Righter & Parsons. Washington Atty: Fletcher, Heald & Hildreth.
Key Personnel:
Robin Jackson . . . . natl sls mgr
Charles Flowers . . . . chief of engrg

WUFX— Analog channel: 35. Analog hrs: 24 5,000 kw vis. ant 830t/800g TL: N32 19 35 W90 37 03 On air date: 2004. One Great Place, Jackson, MS, 39209. Phone: (601) 922-1234. Fax: (601) 922-0268. Web Site: www.gomiss.com. Licensee: Mississippi Television LLC. Ownership: JW Mississippi LLC (acq 6-8-2007). Natl. Network: MyNetworkTV.
Key Personnel:
Mike Dunlop . . . . gen mgr
Mike Garza . . . . sls dir
Loretta Nichols. . . . . prom mgr
Sylvia Walker. . . . . progmg dir
Robert Flanagan . . . . chief of engrg

WWJX— Analog channel: 51.5,000 kw vis. ant 1,260t/1,200g TL: N32 14 26 W90 24 15 Not on air, target date: unknown: 188 S. Bellevue, Suite 222, Memphis, TN, 38104. Phone: (901) 516-8970. Permittee: George S. Flinn Jr. Ownership: George S. Flinn Jr., 100%

Laurel

see Hattiesburg-Laurel, MS market

Meridian, MS (DMA 185)

***WGBC—** Analog channel: 30. Digital channel: 31. Analog hrs: 24 Digital hrs: 24 89.1 kw vis, 8.91 kw aur. 610t/405g TL: N32 19 34 W88 41 12 (CP: 1,600 kw vis, ant 613t/408g) On air date: Sept 15, 1991. Box 2424, Meridian, MS, 39302. 1151 Crestview Cir., Meridian, MS 39301. Phone: (601) 485-3030. Fax: (601) 693-9889. Web Site: www.wgbctv.com. Licensee: Robert M. Ledbetter Jr. Ownership: Robert M. Ledbetter Jr., 100% (acq 4-21-2006; $750,000). Population served: 64,700 Natl. Network: NBC. Natl. Rep: Millennium Sales & Marketing. Washington Atty: Wiley, Rein & Fielding.

***WMAW-TV—** Analog channel: 14. Digital channel: 44. Analog hrs: 24 Digital hrs: 24 550 kw vis, 55 kw aur. ant 1,210t/1,069g TL: N32 08 18 W89 05 36 On air date: Jan 14, 1972. 3825 Ridgewood Rd., Jackson, MS, 39211. Phone: (601) 432-6565. Fax: (601) 432-6654. Fax: (601) 432-6311. Web Site: www.mpbonline.org. Licensee: Mississippi Authority for Educational TV. Population served: 499,869 Natl. Network: PBS. Washington Atty: Schwartz, Woods & Miller. News staff: 4.
Key Personnel:
Marie Antoon . . . . gen mgr
Teresa Collier. . . . . news dir

WMDN— Analog channel: 24. Digital channel: 26. Analog hrs: 24 Digital hrs: 24 724 kw vis, 72.4 kw aur. 662t/388g TL: N32 19 40 W88 41 31 (CP: 724 kw vis, ant 581t. TL: N32 18 43 W88 41 33) On air date: June 10, 1986. Box 2424, Meridian, MS, 39302. 1151Crestview Cir., Meridian, MS 39301. Phone: (601) 693-2424. Fax: (601) 693-7126. E-mail: administration@wmdn.net Web Site: www.wmdntv.com. Licensee: WMDN Inc. (acq 7-18-2006). Population served: 45,083 Natl. Network: CBS. Natl. Rep: Millennium Sales & Marketing. Washington Atty: Garvey, Schubert & Barer.

WTOK-TV— Analog channel: 11. Digital channel: 49.Note: ABC is on WTOK-TV ch 11, CW and Fox are on WTOK-DT ch 49. 316 kw vis, 47.9 kw aur. ant 536t/315g TL: N32 19 38 W88 41 28 On air date: Sept 27, 1953. Box 2988, Meridian, MS, 39302. 815 23rd Ave., Meridian, MS 39301. Phone: (601) 693-1441. Fax: (601) 483-3266. Web Site: www.wtok.com. Licensee: Gray Television Licensee Corp. Group owner: Gray Television Inc. (acq 8-29-2002; grpsl). Population served: 635,000 Natl. Network: ABC, CW, Fox Digital Network: Note: ABC is on WTOK-TV ch 11, CW and Fox are on WTOK-DT ch 49. Natl. Rep: Continental Television Sales. Washington Atty: Wiley, Rein & Fielding, LLP. Wire Svc: AP News staff: 15; News: 15 hrs wkly.
Key Personnel:
Tim Walker. . . . . gen mgr & opns mgr
Julie Walker. . . . . prom mgr
Matt Willis . . . . progmg dir
John Johnson . . . . news dir & pub affrs dir
Brad LeBrun . . . . chief of engrg

Oxford

see Memphis, TN market

Tupelo

see Columbus-Tupelo-West Point, MS market

West Point

see Columbus-Tupelo-West Point, MS market

Missouri

Cape Girardeau

see Paducah, KY-Cape Girardeau, MO-Harrisburg-Mount Vernon, IL market

Columbia-Jefferson City, MO (DMA 139)

KMIZ— Analog channel: 17. Digital channel: 22. Analog hrs: 24 1,580 kw vis, 400 kw aur. 1,141t/1,113g TL: N38 46 29 W92 33 22 On air date: Dec 5, 1971. 501 Business Loop 70 E., Columbia, MO, 65201. Phone: (573) 449-0917. Fax: (573) 875-7078. E-mail: info@kmiz.com Web Site: www.kmiz.com. Licensee: JW Broadcasting LLC. Ownership: Alta/JW Broadcasting Investor Corp., 52.38%; and DJ Broadcasting LLC, 47.62% (acq 10-21-03). Population served: 379,700 Natl. Network: ABC, MyNetworkTV, Fox. Natl. Rep: Petry Television Inc. Washington Atty: Covington & Burling. Wire Svc: AP News staff: 12; News: 19 hrs wkly.
Key Personnel:
Randy Wright . . . . VP & gen mgr
Mark Hotchkiss . . . . sls dir & gen sls mgr
Paul Robinson . . . . prom mgr
Cynthia Clark . . . . film buyer
Curtis Varns . . . . news dir
Rick Hartford . . . . chief of engrg

KNLJ— Analog channel: 25. Digital channel: 20. Analog hrs: 24 2,040 kw vis, 204 kw aur. 1,028t/945g TL: N38 42 16 W92 05 20 On air date: Mar 30, 1986. Box 2525, New Bloomfield, MO, 65603-2525. 9810 State Rd. AE, New Bloomfield, MO 65603. Phone: (573) 896-5105. Fax: (573) 896-4376. E-mail: traffic@knlj.tv Web Site: www.knlj.tv. Licensee: New Life Evangelistic Center Inc. Population served: 149,000 Washington Atty: John H. Midlen Jr.
Key Personnel:
Larry Rice . . . . gen mgr
Charles Hale . . . . gen sls mgr
James Shackleford. . . . . progmg dir
Shawn Baker. . . . . chief of engrg

KOMU-TV— Analog channel: 8. Digital channel: 36. Analog hrs: 24 Digital hrs: 24 316 kw vis, 31.6 kw aur. 790t/774g TL: N38 53 16 W92 15 48 On air date: Dec 21, 1953. 5550 Hwy. 63 S., Columbia, MO, 65201. Phone: (573) 882-8888. Fax: (573) 884-8888. Web Site: www.komu.com. Licensee: The Curators of the University of Missouri. Group owner: (group owner) Population served: 157,510 Natl. Network: NBC, CW. Natl. Rep: Millennium Sales & Marketing. Washington Atty: Shaw Pittman. News: 20 hrs wkly.
Key Personnel:
Martin Siddall . . . . gen mgr
Al Leitl . . . . gen sls mgr
Tom Dugan . . . . natl sls mgr
Matt Garrett . . . . prom dir
Stacey Woelfel . . . . news dir
Chris Swisher . . . . chief of engrg

KRCG— Analog channel: 13. Digital channel: 12. Analog hrs: 5 AM-2 AM 316 kw vis, 47.4 kw aur. ant 1,010t/929g TL: N38 41 28 W92 05 43 On air date: Feb 13, 1955. Box 659, Jefferson City, MO, 65102. 10188 Old Hwy. 54 N., New Bloomfield, MO 65063. Phone: (573) 896-5144. Fax: (573) 896-5193. E-mail: info@krcg.com Web Site: www.krcg.com. Licensee: Barrington Broadcasting Missouri Corp. Group owner: (group owner; (acq 12-27-2004; $38 million). Population served: 960,700 Natl. Network: CBS. Washington Atty: Pepper & Corazzini. News staff: 14; News: 14 hrs wkly.
Key Personnel:
Betsy Farris . . . . gen mgr
Lee Gordon. . . . . stn mgr, progmg dir & progmg mgr
Wendy Gustofson. . . . . natl sls mgr
Roger Hulett. . . . . rgnl sls mgr
Gregg Palermo . . . . prom mgr & news dir
K.J. Lambein. . . . . prom mgr & news dir
Jim Malone . . . . chief of engrg

Hannibal

see Quincy, IL-Hannibal, MO-Keokuk, IA market

Jefferson City

see Columbia-Jefferson City, MO market

Joplin, MO-Pittsburg, KS (DMA 144)

KFJX-TV— Analog channel: 14. Analog hrs: 24 hrs Digital hrs: 24 hrs. 5,000 kw vis. ant 1,112t/1,069g TL: N37 18 46 W94 48 59 On air date: 2003. Box 659, Pittsburg, KS, 66762-0659. Phone: (417) 782-1414. Fax: (417) 206-4081. E-mail: ddishman@fox14tv.com Web Site: www.fox14tv.com. Licensee: Surtsey Media LLC. Group owner: (group owner; (acq 3-7-2003). Population served: 350,000 Natl. Network: Fox. News staff: 4; News: 3 hrs wkly.

KOAM-TV— Analog channel: 7. Digital channel: 13. Analog hrs: 24 316 kw vis, 63.1 kw aur. 1,090t/1,159g TL: N37 13 15 W94 42 25 On air date: Dec 13, 1953. Box 659, Pittsburg, KS, 66762-0659. 2950 N.E. Hwy. 69, Pittsburg, KS 66762-0659. Phone: (417) 624-0233. Fax: (417) 624-3115, sls & admin. E-mail: email@koamtv.com Web Site: www.koamtv.com. Licensee: Saga Quad States Communications LLC. Group owner: Saga Communications Inc. (acq 10-12-94; $8.55 million). Population served: 343,000 Natl. Network: CBS. Natl. Rep: Continental Television Sales. News: 19 hrs wkly.
Key Personnel:
Danny Thomas . . . . pres, gen mgr & film buyer
Vance Lewis . . . . prom dir
Kristi Spencer. . . . . news dir
Larry White . . . . chief of engrg

KODE-TV— Analog channel: 12. Digital channel: 43. Analog hrs: 24 316 kw vis, 63.2 kw aur. 1,020t/999g TL: N37 04 36 W94 32 10 On air date: Sept 26, 1954. Box 46, Joplin, MO, 64802. 1928 W. 13th St., Joplin, MO 64802. Phone: (417) 623-7260. Fax: (417) 623-3736. Web Site: www.kode-tv.com. Licensee: Mission Broadcasting of Joplin Inc. Group owner: Mission Broadcasting Inc. (acq 2-27-2002; $6 million). Natl. Network: ABC. Washington Atty: Cohn. News staff: 18; News: 16 hrs wkly.
Key Personnel:
Shirley Morton . . . . gen mgr & stn mgr
Gary Hood . . . . gen sls mgr
Janice Rohman . . . . progmg dir
Larry Young . . . . news dir
Jeff Hadley . . . . chief of engrg

***KOZJ**— Analog channel: 26. Digital channel: 25.51.3 kw vis, 5.13 kw aur. 932t/849g TL: N37 04 36 W94 32 10 (CP: 832 kw vis, 8.3 kw aur) On air date: June 1, 1986. Box 1226, Joplin, MO, 64802. 403 S. Main St., Joplin, MO 64801. Phone: (417) 782-2226. Fax: (417) 782-7222. E-mail: mail@optv.org Web Site: www.optv.org. Licensee: Board of Governors of Southwest Missouri State University. (acq 4-25-2001; $1.3 million assumption of debt with KOZK(TV) Springfield). Natl. Network: PBS. Washington Atty: Dow, Lohns PLLC.
Key Personnel:
Brent Moore . . . . pres & chief of engrg
Tammy Wiley . . . . gen mgr
Norma Scott. . . . . dev dir
Tom Carter. . . . . progmg dir
Satellite of *KOZK Springfield.

KSNF— Analog channel: 16. Digital channel: 46.2,570 kw vis, 257 kw aur. 1,027t/1,013g TL: N37 04 33 W94 33 16 On air date: Sept 2, 1967. Box 1393, Joplin, MO, 64802. 1502 Cleveland, Joplin, MO 64801. Phone: (417) 781-2345. Fax: (417) 782-2417. Web Site: www.ksntv.com. Licensee: Nexstar Finance Inc. Group owner: Nexstar Broadcasting Group Inc. (acq 11-6-97; grpsl). Population served: 343,000 Natl. Network: NBC. Washington Atty: Arter & Hadden.
Key Personnel:
John Hoffman. . . . . gen mgr
Debra Palmer . . . . sls dir & gen sls mgr
Larry Young . . . . news dir
Jeff Hadley . . . . chief of engrg
Robin Richey . . . . progmg

Kansas City, MO (DMA 31)

***KCPT**— Analog channel: 19. Digital channel: 18. Analog hrs: 24 Digital hrs: 24 1,150 kw vis, 115 kw aur. 1,171t TL: N39 04 59 W94 28 49 On air date: Mar 29, 1961. 125 E. 31st St., Kansas City, MO, 64108. Phone: (816) 756-3580. Fax: (816) 931-2500. E-mail: kcpt@kcpt.org Web Site: www.kcpt.org. Licensee: Public TV 19 Inc. (acq 1-1-72; $22,226; FTR: 2-14-72). Population served: 1,300,000 Natl. Network: PBS. Washington Atty: Arter & Hadden.
Key Personnel:
Victor Hogstrom . . . . CEO & pres
William T. Reed . . . . gen mgr
Bob Hagg . . . . chief of engrg

KCTV— Analog channel: 5. Digital channel: 24.100 kw vis, 15.1 kw aur. 1,130t/1,042g TL: N39 04 15 W94 34 57 On air date: Sept 27, 1953. Box 5555, Kansas City, MO, 64109-0155. 4500 Shawnee Mission Pkwy., Fairway, KS 66205. Phone: (913) 677-5555. Fax: (913) 677-7109. E-mail: kctv5@kctv5.com Web Site: www.kctv5.com. Licensee: Meredith Corp. Group owner: Meredith Broadcasting Group, Meredith Corp., see Cross-Ownership (acq 10-1-53; $2 million). Natl. Network: CBS. Natl. Rep: TeleRep.
Key Personnel:
Kirk Black. . . . . VP & gen mgr
Dave Duncan . . . . gen sls mgr
Michael Cornette . . . . natl sls mgr
Jason Mullenix . . . . rgnl sls mgr
Pam Carder . . . . rgnl sls mgr
Beth Green. . . . . progmg dir
Tracy Brogden-Miller . . . . news dir
Tom Casey . . . . engrg dir & chief of engrg

KCWE— Analog channel: 29. Digital channel: 31. Analog hrs: 24 5,000 kw vis. ant 1,174t/1,089g TL: N39 05 01 W94 30 57 On air date: September 1996. 1049 Central, Kansas City, MO, 64105. Phone: (816) 221-9999. Fax: (816) 760-9149. Web Site: www.thekansascitychannel.com. Licensee: KCWE LMA Inc. (acq 8-15-2006; $10.96 million). Population served: 875,090 Natl. Network: CW. Natl. Rep: MMT.
Key Personnel:
Robert Liepold . . . . pres & gen mgr
Peggy Madigan . . . . sls dir
Jerry Agresti . . . . engrg dir

KMBC-TV— Analog channel: 9. Digital channel: 7. Analog hrs: 24 316 kw vis, 36.8 kw aur. 1,171t/1,124g TL: N39 05 01 W94 30 57 On air date: Aug 1, 1952. 1049 Central, Kansas City, MO, 64105. Phone: (816) 221-9999. Fax: (816) 760-9245. Web Site: www.thekansascitychannel.com. Licensee: KMBC Hearst-Argyle Television Inc. Group owner: Hearst-Argyle Television Inc. (acq 7-16-97; grpsl). Population served: 875,090 Natl. Network: ABC. Natl. Rep: Eagle Television Sales. Washington Atty: Brooks, Pierce, McLendon, Humphrey & Leonard. Wire Svc: AP Wire Svc: CNN Wire Svc: ABC News staff: 55; News: 28 hrs wkly.
Key Personnel:
Wayne Godsey. . . . . VP & gen mgr
Peggy Madigan. . . . . sls dir & gen sls mgr
Michael Sipes. . . . . news dir
Jerry Agresti . . . . engrg dir & chief of engrg

KMCI— Analog channel: 38. Digital channel: 36. Analog hrs: 24 Digital hrs: 24 5,000 kw vis, 494 kw aur. ant 1,069t/1,128g TL: N38 58 42 W94 32 01 On air date: February 1988. 4720 Oak St., Kansas City,

MO, 64112. Phone: (816) 753-4141. Fax: (816) 932-4122. E-mail: comments@38thespot.com Web Site: www.38thespot.com. Licensee: Scripps Howard Broadcasting Co. Group owner: (group owner; (acq 2-3-2000). Population served: 2,182,000 Natl. Rep: Harrington, Righter & Parsons.
Key Personnel:
Craig Allison . . . VP & gen mgr
Alan Fuchsman . . . gen sls mgr
Dana Boyd . . . progmg dir
Debbie Bush . . . news dir
Jay Nix . . . chief of engrg

***KMOS-TV—** Analog channel: 6. Digital channel: 15. Analog hrs: 18 100 kw vis, 25 kw aur. 772t/797g TL: N38 44 47 W93 16 30 On air date: Dec 22, 1979. University of Central Missouri, Wood 11, Warrensburg, MO, 64093. Phone: (660) 543-4155. Fax: (660) 543-8863. E-mail: kmos@kmos.org Web Site: www.kmos.org. Licensee: Central Missouri State University. (acq 6-6-78; $1,000). Population served: 1,500,000 Natl. Network: PBS. Washington Atty: Shaw Pittman.
Key Personnel:
Donald W. Peterson . . . gen mgr
Fred Hunt . . . opns mgr
Michael O'Keefe . . . progmg mgr
Dorothy McGrath . . . pub affrs dir
John Long . . . chief of engrg

KPXE— Analog channel: 50. Digital channel: 51. Analog hrs: 6 AM-1 AM 678 kw vis, 67.8 kw aur. 1,119t/1,164g TL: N39 01 19 W94 30 50 On air date: Dec 1, 1978. 4720 Oak St., Kansas City, MO, 64112. Phone: (816) 924-5050. Fax: (816) 931-1818. Web Site: www.ionline.tv. Licensee: Paxson Kansas City License Inc. Group owner: Paxson Communications Corp. (acq 3-3-97; $16.4 million). Population served: 750,000 Natl. Network: i Network. Washington Atty: Wiley, Rein & Fielding. Foreign lang progmg: SP 1
Key Personnel:
Frank Barajas . . . gen mgr
Alan Fuchsman . . . adv dir
Dave Campbell . . . chief of engrg

KSHB-TV— Analog channel: 41. Digital channel: 42. Analog hrs: 24 Digital hrs: 24 3,450 kw vis. ant 1,036t/1,097g TL: N38 58 42 W94 32 01 On air date: Sept 28, 1970. 4720 Oak St., Kansas City, MO, 64112. Phone: (816) 753-4141. Fax: (816) 932-4122. E-mail: programming @nbcactionnews.com Web Site: www.nbcactionnews.com. Licensee: Scripps Howard Broadcasting Co. Group owner: (group owner; (acq 10-28-77; FTR: 10-3-77). Population served: 2,182,000 Natl. Network: NBC. Natl. Rep: Harrington, Righter & Parsons. News staff: 75; News: 32 hrs wkly.
Key Personnel:
Craig Allison . . . VP, gen mgr & prom dir
John McKenna . . . gen sls mgr
Dana Boyd . . . progmg dir
Debbie Bush . . . news dir
Jay Nix . . . chief of engrg

KSMO-TV— Analog channel: 62. Digital channel: 47.1,795 kw vis, 179 kw aur. ant 1,119t/1,167g TL: N39 05 26 W94 28 18 On air date: Dec 7, 1983. 4500 Shawnee Mission Pkwy, Fairway, KS, 66205. Phone: (913) 621-6262. Fax: (913) 621-4703. E-mail: ksmo@myKSMOtv.com Web Site: www.myksmotv.com. Licensee: Meredith Corp. Group owner: Sinclair Broadcast Group Inc. (acq 9-29-2005; $26.8 million). Population served: 1,681,165 Natl. Network: MyNetworkTV. Natl. Rep: Millennium Sales & Marketing.
Key Personnel:
Kirk Black . . . VP & gen mgr
Darrin McDonald . . . gen sls mgr
Beth Green . . . progmg dir & progmg
Tom Casey . . . chief of engrg

WDAF-TV— Analog channel: 4. Digital channel: 34. Analog hrs: 24 Digital hrs: 24 100 kw vis, 10 kw aur. ant 1,130t/1,163g TL: N39 04 20 W94 35 45 On air date: Oct 16, 1949. 3030 Summit, Kansas City, MO, 64108. Phone: (816) 753-4567. Fax: (816) 931-3984. Web Site: www.myfoxkc.com. Licensee: WDAF License Inc. Group owner: Fox Television Stations Inc. (acq 1-23-97; grpsl). Population served: 1,562,000 Natl. Network: Fox. Natl. Rep: Fox Stations Sales. News staff: 115; News: 49 hrs wkly.
Key Personnel:
Cheryl McDonald . . . VP & gen mgr
Kelly Satalowich . . . sls VP
Matt Rankin . . . progmg dir
Jim Moore . . . engrg VP

Kirksville

see Ottumwa, IA-Kirksville, MO market

Poplar Bluff

see Paducah, KY-Cape Girardeau, MO-Harrisburg-Mount Vernon, IL market

Springfield, MO (DMA 109)

KOLR-TV— Analog channel: 10. Digital channel: 52. Analog hrs: 24 316 kw vis, 31.6 kw aur. 2,070t/1,887g TL: N37 13 08 W92 56 56 On air date: Mar 14, 1953. 2650 E. Division, Springfield, MO, 65803. Phone: (417) 862-1010. Fax: (417) 862-6439. E-mail: dwasson@kolr10.com Web Site: www.ozarksfirst.com. Licensee: Mission Broadcasting Inc. Group owner: (group owner; (acq 12-17-2003). Population served: 1,472,000 Natl. Network: CBS. News staff: 30; News: 22 hrs wkly.
Key Personnel:
Mark Gordon . . . gen mgr & natl sls mgr
Dean Wasson . . . stn mgr & progmg mgr
Dave Thomason . . . gen sls mgr & prom mgr
Dave Bowen . . . prom mgr & chief of engrg
Polly Van Doren-Orr . . . news dir
David Smith . . . pub affrs dir & engrg dir

***KOZK—** Analog channel: 21. Digital channel: 23. Analog hrs: 24 1,410 kw vis, 141 kw aur. 1,791t/2,001g TL: N37 13 08 W92 56 56 On air date: Jan 21, 1975. 901 S. National, Springfield, MO, 65897. Phone: (417) 836-3500. Fax: (417) 863-3569. E-mail: mail@optv.org Web Site: www.optv.org. Licensee: Board of Governors of Missouri State University. Population served: 650,000 Natl. Network: PBS. Washington Atty: Dow, Lohnes PLLC.
Key Personnel:
Brent Moore . . . gen mgr & chief of engrg
Tammy Wiley . . . gen mgr
Norma Scott . . . dev dir
Tom Carter . . . progmg dir

KPBI— Analog channel: 34. Analog hrs: 24 454 kw vis. ant 688t/351g TL: N36 26 30 W93 58 25 (CP: 3,236 kw vis, ant 697t/351g) On air date: 2000. 1 Shackleford Dr., Suite 400, Little Rock, AR, 72211. Phone: (501) 219-2400. Fax: (501) 221-7908. Licensee: TV 34 Inc. Group owner: Equity Broadcasting Corp. (acq 8-23-99).
Key Personnel:
Greg Fess . . . pres
Glenn Charlesworth . . . CFO
Max Hooper . . . exec VP
James Hearnsberger . . . VP
Debbie James . . . gen sls mgr
Terrill Weiss . . . stn mgr & gen sls mgr
Frank White . . . prom dir
Nathan Stamp . . . progmg dir
Doug Krile . . . news dir
Sheryl Lackey . . . pub affrs dir
Paul Brandenburg . . . engrg dir
Don Jones . . . engrg mgr

KRBK— Analog channel: 49.5,000 kw vis. ant 1,519t/1,375g TL: N37 49 10 W92 44 52 Not on air, target date: unknown: 1 S. Memorial Dr., 20th Fl., St. Louis, MO, 63102. Phone: (314) 345-1000. Permittee: Koplar Communications International Inc. Ownership: Edward J. Koplar, 100%

KSFX-TV— Analog channel: 27. Digital channel: 28. Analog hrs: 24 5,000 kw vis, 500 kw aur. ant 1,694t/1,569g TL: N37 13 08 W92 56 56 On air date: Sept 22, 1968. 2650 E. Division St., Springfield, MO, 65803. Phone: (417) 862-2727. Fax: (417) 831-4209. Web Site: www.ozarksfirst.com. Licensee: Nexstar Finance Inc. Group owner: Nexstar Broadcasting Group Inc. (acq 12-31-2003; grpsl). Population served: 890,000 Natl. Network: Fox. Washington Atty: Arter & Hadden.
Key Personnel:
Dave Thomason . . . VP & gen sls mgr
Mark Gordon . . . gen mgr
Dean Wasson . . . stn mgr
Dave Bowen . . . prom mgr
Nancy Bingaman . . . progmg mgr
Polly Van Doren-Orr . . . news dir
David Smith . . . engrg dir

KSPR— Analog channel: 33. Digital channel: 19. Analog hrs: 24 5,010 kw vis, 112 kw aur. 1,995t/1,816g TL: N37 13 08 W92 56 56 (CP: 1,000 kw aur) On air date: Mar 17, 1983. Box 6030, Springfield, MO, 65801-6030. 1359 St. Louis St., Springfield, MO 65802. Phone: (417) 831-1333. Fax: (417) 831-4125. Web Site: www.springfield33.com. Licensee: Perkin Media LLC. Ownership: William N. Perkin, 100% Group owner: Piedmont Television Holdings LLC (acq 2007; $20.629 million). Natl. Network: ABC. Washington Atty: Sciarrino and Associates PLLC. News staff: 21; News: 9 hrs wkly.
Key Personnel:
Brad Belote . . . news dir
Neal Evans . . . chief of engrg

KWBM— Analog channel: 31.5,000 kw vis. ant 1,110t/1,063g TL: N36 42 18 W93 03 45 On air date: 2002. 1736 E. Sunshine St., Suite 815, Springfield, MO, 65804. Phone: (417) 336-0031. Fax: (417) 336-3199. Licensee: EBC Harrison Inc. (acq 9-21-2004; $8,666,670). Natl. Network: MyNetworkTV.

KYTV— Analog channel: 3. Digital channel: 44. Analog hrs: 24 Note: NBC is on KYTV(TV) ch 3, CW is on KYTV-DT ch 44. 100 kw vis, 20 kw aur. ant 2,040t/2,000g TL: N37 10 11 W92 56 30 On air date: Oct 1, 1953. 999 W. Sunshine, Springfield, MT, 65807. Phone: (417) 268-3000. Fax: (417) 268-3100. E-mail: ky3@ky3.com Web Site: www.ky3.com. Licensee: KY-3 Inc. Group owner: Schurz Communications Inc. (acq 2-19-87; $50.8 million; FTR: 1-19-87). Natl. Network: NBC, CW Digital Network: Note: NBC is on KYTV(TV) ch 3, CW is on KYTV-DT ch 44. News staff: 40; News: 40 hrs wkly.
Key Personnel:
Mike Scott . . . gen mgr
Bryan Cochran . . . gen sls mgr
Jeff Benscoter . . . mktg dir & news dir
Dan McGrane . . . prom mgr
Trenna Underhill . . . progmg dir & progmg mgr
Tom McKleroy . . . chief of engrg

St. Joseph, MO (DMA 201)

KQTV— Analog channel: 2. Digital channel: 53. Analog hrs: 20 Digital hrs: 20 100 kw vis, 20 kw aur. 810t/750g TL: N39 46 12 W94 47 53 On air date: Sept 27, 1953. Box 8369, Saint Joseph, MO, 64508. 4000 & Faraon St., Saint Joseph, MO 64506. Phone: (816) 364-2222. Fax: (816) 364-3787. Fax: TWX: 910-777-7872. E-mail: kq2@kq2.com Web Site: www.kq2.com. Licensee: Nexstar Broadcasting Inc. Group owner: Nexstar Broadcasting Group Inc. (acq 2-14-97; with WTWO(TV) Terre Haute, IN). Population served: 256,000 Natl. Network: ABC. Natl. Rep: Blair Television, Petry Television Inc. Washington Atty: Drinker, Biddle & Reath. News staff: 18; News: 16 hrs wkly.
Key Personnel:
Heather Shearin . . . VP & gen mgr
Steve Cline . . . opns mgr
Doug Conrad . . . gen sls mgr
Jim Conlon . . . prom mgr
Jill Jensen . . . news dir
Bill Smith . . . chief of engrg

KTAJ-TV— Analog channel: 16. Digital channel: 21. Analog hrs: 24 Digital hrs: 24 5,000 kw vis, 500 kw aur. 1,071t/1,027g TL: N39 39 03 W94 40 11 On air date: Oct 6, 1986. 4402 A S. 40th St., Saint Joseph, MO, 64503. Phone: (816) 364-1616. Fax: (816) 364-6729. E-mail: ktaj@tbn.org Web Site: www.tbn.org. Licensee: Trinity Christian Center of Santa Ana Inc. dba Trinity Broadcasting Network. Group owner: Trinity Broadcasting Network (acq 5-8-2000; grpsl).
Key Personnel:
Paul Crouch . . . pres
Jan Crouch . . . VP
Julie A. Cluck. stn mgr, mktg mgr, prom mgr, adv mgr & progmg mgr
Andrae Hannon . . . pub affrs dir
Jeff Landers . . . engrg dir & chief of engrg

St. Louis, MO (DMA 21)

KDNL-TV— Analog channel: 30. Digital channel: 31. Analog hrs: 24 2,190 kw vis, 219 kw aur. ant 1,099t/1,123g TL: N38 34 50 W90 19 45 On air date: June 8, 1969. 1215 Cole St., Saint Louis, MO, 63106. Promotions/Tape Delivery, 1261 Dublin Rd., Columbus, OH 43215. Phone: (314) 436-3030. Fax: (314) 259-5504. Web Site: www.abcstlouis.com. Licensee: KDNL Licensee L.L.C. Group owner: Sinclair Broadcast Group Inc. (acq 5-30-96). Population served: 1,088,550 Natl. Network: ABC. Natl. Rep: Millennium Sales & Marketing. Washington Atty: Shaw, Pittman.
Key Personnel:
Tom Tipton . . . gen mgr
Jim Wright . . . opns dir & engrg dir
John Strassner . . . gen sls mgr
Andrea Schaffer . . . natl sls mgr
Mike Held . . . rgnl sls mgr
Sandra Habeck . . . progmg mgr

***KETC—** Analog channel: 9. Digital channel: 39. Analog hrs: 24 295 kw vis, 29.5 kw aur. 1,070t/1,073g TL: N38 28 56 W90 23 53 On air date: Sept 20, 1954. 3655 Olive St., Saint Louis, MO, 63108. Phone: (314) 512-9000. Fax: (314) 512-9005. Web Site: www.ketc.org. Licensee: St. Louis Regional Educational and Public Television Commission. Population served: 2,900,000 Natl. Network: PBS. Washington Atty: Dow, Lohnes & Albertson.
Key Personnel:
Dick Skalski . . . COO & sr VP
Richard Skalski . . . CFO
Chrys Marlow. . . . opns VP, opns dir, engrg VP & chief of engrg

Patrick Murphy . . . . prom VP
Patti Kistler. . . . . progmg VP & progmg dir

KMOV— Analog channel: 4. Digital channel: 56. Analog hrs: 24 100 kw vis, 15 kw aur. 1,097t/1,201g TL: N38 31 47 W90 17 58 On air date: July 8, 1954. One Memorial Dr., Saint Louis, MO, 63102. Phone: (314) 621-4444. Fax: (314) 444-3367. Fax: (314) 621-4755. E-mail: channel4@kmov.com Web Site: www.kmov.com. Licensee: KMOV-TV Inc. Group owner: Belo Corp., Broadcast Division (acq 6-02-97; grpsl). Population served: 3,500,000 Natl. Network: CBS. Natl. Rep: TeleRep. Washington Atty: Wiley, Rein & Fielding. News staff: 70; News: 29 hrs wkly.
Key Personnel:
Allan Cohen . . . . pres & gen mgr
Jim Rothschild . . . . opns dir & prom mgr
Robert Totsch . . . . gen sls mgr
Paul Conaty . . . . natl sls mgr
Liz Mullen. . . . . progmg mgr
Jenie Garner . . . . news dir
Walt Nichol . . . . engrg dir & chief of engrg

KNLC— Analog channel: 24. Digital channel: 14.3,090 kw vis, 309 kw aur. 1,000t TL: N38 21 40 W90 32 58 On air date: Sept 12, 1982. Box 924, Saint Louis, MO, 63188. 1411 Locust St., St. Louis, MO 63188. Phone: (314) 436-2424. Fax: (314) 436-2434. E-mail: judy@knlc.tv Web Site: www.knlc.tv. Licensee: New Life Evangelistic Center Inc. Population served: 1,200,000 Washington Atty: Midlen & Guillot.
Key Personnel:
Larry Rice . . . . pres, gen mgr & stn mgr
Ray Redlich . . . . VP
Judy Redlich . . . . sls dir
Victor Anderson . . . . progmg dir
Jim Barnes . . . . chief of engrg

KPLR-TV— Analog channel: 11. Digital channel: 26. Analog hrs: 24 316 kw vis, 32 kw aur. ant 1,010t/1,033g TL: N38 31 47 W90 17 58 On air date: Apr 28, 1959. 2250 Ball Dr., Saint Louis, MO, 63146. Phone: (314) 447-1111. Fax: (314) 447-6404. E-mail: administration4@tribune.com Web Site: www.cb11tv.com. Licensee: KPLR Inc. Group owner: Tribune Broadcasting Co. (acq 3-21-2003; $275 million with KWBP(TV) Salem, OR). Population served: 1,906,000 Natl. Network: CW. Natl. Rep: TeleRep. News: 4 hrs wkly.
Key Personnel:
Sheldon Ripson . . . . sr VP & news dir
Bill Lanesey . . . . gen mgr
Glen P. Callanan . . . . gen sls mgr
Suzi Schrappen . . . . prom dir
Gwen Moore . . . . progmg dir
Greg Boling . . . . chief of engrg

KSDK— Analog channel: 5. Digital channel: 35. Analog hrs: 24 100 kw vis, 20 kw aur. 1,090t/1,148g TL: N38 34 05 W90 19 55 On air date: Feb 8, 1947. 1000 Market St., Saint Louis, MO, 63101. Phone: (314) 421-5055. Fax: (314) 444-5164. E-mail: comments@ksdk.com Web Site: www.ksdk.com. Licensee: Multimedia KSDK Inc. Group owner: Gannett Broadcasting (acq 11-30-95; grpsl). Population served: 2,500,000 Natl. Network: NBC. Natl. Rep: Blair Television. Washington Atty: Wiley, Rein & Fielding.
Key Personnel:
Lynn Beall . . . . pres & gen mgr
Julie Heskett . . . . VP
Mike Meara . . . . gen sls mgr
Mike Shipley . . . . mktg mgr & news dir
Jeff Winget . . . . prom dir & news dir
Rebecca Rahm. . . . . progmg dir & progmg mgr
Dave Hummert . . . . chief of engrg

KTVI— Analog channel: 2. Digital channel: 43. Analog hrs: 24 100 kw vis, 20 kw aur. ant 1,089t/1,014g TL: N38 32 07 W90 22 23 On air date: Aug 10, 1953. 5915 Berthold Ave., Saint Louis, MO, 63110. Phone: (314) 647-2222. Fax: (314) 644-7419. Web Site: www.myfoxstl.com. Licensee: KTVI License Inc. Group owner: Fox Television Stations Inc. (acq 11-96; grpsl). Population served: 622,236 Natl. Network: Fox. Washington Atty: Pepper & Corazzini. News: 37.5 hrs wkly.
Key Personnel:
Spencer Koch . . . . gen mgr
Kurt Krueger. gen sls mgr
Cindy Rosen. natl sls mgr
Steve Mills . . . . rgnl sls mgr
Kathryn Hansen. prom dir
Elaine Claspill. . . . . progmg dir
Kingsley Smith . . . . news dir
Ernie Dachel. . . . . chief of engrg

WRBU— Analog channel: 46. Digital channel: 47. Analog hrs: 24 5,000 kw vis, 500 kw aur. 1,749t/849g TL: N38 23 18 W90 29 16 On air date: September 1989. 1408 N. Kingshighway Blvd., Suite 300, St. Louis, MO, 63113. Phone: (314) 256-4600. Fax: (314) 256-4655. E-mail: contest@upn46stl.com Web Site: www.my46stl.com. Licensee: Roberts Broadcasting Co. Group owner: (group owner). Population served: 2,200,000 Natl. Network: UPN. Natl. Rep: Harrington, Righter & Parsons. Washington Atty: Dow, Lohnes & Albertson. News staff: 4.
Key Personnel:
Gregg Filandrinos . . . . gen mgr
Bonni Burns . . . . sls dir
Monica Nettles-Johnson . . . . prom dir & progmg dir
Didier Baptist . . . . news dir
Chris Meisch . . . . chief of engrg

Montana

Billings, MT

(DMA 170)

KHMT— Analog channel: 4. Digital channel: 22. Analog hrs: 24 100 kw vis, 10 kw aur. 1,062t/628g TL: N45 44 29 W108 08 19 On air date: Aug 16, 1995. 445 S. 24th St. W., Billings, MT, 59102. Phone: (406) 652-4743. Fax: (406) 652-6963. Web Site: www.abc6.fox4.tv. Licensee: Mission Broadcasting Inc. Group owner: (group owner; (acq 12-30-2003). Population served: 150,000 Natl. Network: Fox. Washington Atty: Drinker, Biddle & Reath L.L.P.
Key Personnel:
Jerry Jones . . . . gen mgr & opns mgr
Sandra Zoldwski . . . . gen sls mgr & natl sls mgr
Patricia King . . . . progmg dir
Ron Walden . . . . chief of engrg

KSVI— Analog channel: 6. Digital channel: 18. Analog hrs: 24 100 kw vis, 10 kw aur. 817t TL: N45 48 26 W108 20 25 On air date: 1993. 445 S. 24th St. W., Billings, MT, 59102. Phone: (406) 652-4743. Fax: (406) 652-6963. Web Site: www.abc6fox4.tv. Licensee: Nexstar Broadcasting Inc. Group owner: Nexstar Broadcasting Group Inc. (acq 12-31-2003; grpsl). Population served: 150,000 Natl. Network: ABC. Washington Atty: Drinker, Biddle & Reath L.L.P.
Key Personnel:
Jerry Jones . . . . gen mgr
Sandra Zoldowski. . . . . gen sls mgr
Patricia King . . . . progmg dir
Ron Walden . . . . chief of engrg

KTVQ— Analog channel: 2. Digital channel: 10. Analog hrs: 24 Digital hrs: 24 Note: CBS is on KTVQ(TV) ch 2, CW is on KTVQ-DT ch 10. 100 kw vis, 10.2 kw aur. ant 670t/383g TL: N45 46 00 W108 27 27 On air date: Nov 9, 1953. 3203 3rd Ave. N., Billings, MT, 59101. Phone: (406) 252-5611. Fax: (406) 252-9938. E-mail: news@ktvq.com Web Site: www.ktvq.com. Licensee: Evening Post Publishing Co. Group owner: Cordillera Communications Inc. (acq 1994; $8.5 million). Population served: 87,500 Natl. Network: CBS, CW Digital Network: Note: CBS is on KTVQ(TV) ch 2, CW is on KTVQ-DT ch 10. Natl. Rep: Harrington, Righter & Parsons. Washington Atty: Dow, Lohnes & Albertson. Wire Svc: AP News staff: 21; News: 17 hrs wkly.
Key Personnel:
Monty Wallis . . . . gen mgr
John Webber . . . . progmg dir & chief of engrg
Jon Stepanek. . . . . news dir

KULR-TV— Analog channel: 8. Digital channel: 11. Analog hrs: 24 Digital hrs: 24 316 kw vis, 38.9 kw aur. 750t/531g TL: N45 45 35 W108 27 14 On air date: Mar 15, 1958. 2045 Overland Ave., Billings, MT, 59102. Phone: (406) 656-8000. Fax: (406) 652-8207. E-mail: generalmanager@kulr.com Web Site: www.kulr8.com. Licensee: MMM License II LLC. Group owner: MAX Media L.L.C. (acq 6-9-2004; $11 million). Population served: 242,000 Natl. Network: NBC. Wire Svc: AP News staff: 24; News: 20 hrs wkly.
Key Personnel:
John A. Trinder. . . . . pres
Bruce Cummings . . . . gen mgr
Rafael Archille . . . . gen sls mgr
Kris Aschim . . . . prom dir
Blaire Martin . . . . news dir
Rebroadcasts KYUS(TV) Miles City.

KYUS-TV— Analog channel: 3. Analog hrs: 24 10.4 kw vis, 1 kw aur. 102t/42g TL: N46 24 48 W105 51 04 On air date: 2003. Not on air, target date: unknown: Stn currently dark c/o KXGN-TV, 210 S. Douglas, Glendive, MT, 59330. Phone: (406) 377-3377. Fax: (406) 365-2181. E-mail: kxgnkdzn@midrivers.com Licensee: KYUS-TV Broadcasting Corp. Ownership: Stephen A. Marks Population served: 14,321 Natl. Network: NBC. Natl. Rep: Adam Young.
Key Personnel:
Stephen A. Marks. . . . . pres
Paul Sturlaugson . . . . gen mgr

Bozeman

see Butte-Bozeman, MT market

Butte-Bozeman, MT

(DMA 192)

KBTZ— Analog channel: 24.3,000 kw vis. ant 1,925t TL: N46 00 27 W112 26 30 On air date: Aug 1, 2002. Equity Broadcasting Corp., 1 Shackleford Dr., Suite 400, Little Rock, AR, 72211. 5115 US Hwy. 93 S., Missoula, MT 59804. Phone: (406) 542-8900. Fax: (501) 604-8004. Licensee: Montana License Sub, Inc. Group owner: Equity Broadcasting Corp. Natl. Network: Fox, MyNetworkTV.
Key Personnel:
Linda Gray. . . . . pres & gen mgr
Leslie Stoll . . . . gen sls mgr
Terry Cudderford . . . . rgnl sls mgr
Linda Julius . . . . progmg mgr
Mike Warner . . . . chief of engrg

KBZK— Analog channel: 7. Digital channel: 13. Analog hrs: 20 Digital hrs: 20 Note: CBS is on KBZK(TV) ch 7, CW is on KBZK-DT ch 13. 43.7 kw vis, 4.37 kw aur. ant 816t TL: N45 40 24 W110 52 02 (CP: Ant 1,122t) On air date: September 1987. Box 6040, Bozeman, MT, 59715. 1128 E. Main, Bozeman, MT 59715. Phone: (406) 586-3280. Fax: (406) 586-4135. E-mail: receptionist@kbzk.com Web Site: kbzk.com. Licensee: KCTZ Communications Inc. Group owner: Cordillera Communications Inc. (acq 12-93). Population served: 75,000 Natl. Network: CBS, CW Digital Network: Note: CBS is on KBZK(TV) ch 7, CW is on KBZK-DT ch 13. Washington Atty: Dow, Lohnes & Albertson. Wire Svc: AP News staff: 3.
Key Personnel:
Terry Hurley . . . . pres
Pat Cooney . . . . gen mgr
Tim Gazy . . . . stn mgr, gen sls mgr & mktg mgr
John Sherer . . . . news dir
Andy Suk . . . . engrg dir
Ron Schlosser . . . . chief of engrg

KTVM— Analog channel: 6. Digital channel: 33. Analog hrs: 24 Digital hrs: 24 100 kw vis, 10 kw aur. ant 1,940t/213g TL: N46 00 29 W112 26 30 On air date: May 12, 1970. 201 S. Wallace, Suite A5, Bozeman, MT, 59715. Phone: (406) 586-0296. Fax: (406) 586-0554. E-mail: news@ktvm.com Web Site: www.ktvm.com. Licensee: BlueStone License Holdings Inc. Group owner: Lamco Communications Inc. (acq 6-15-2004; grpsl). Population served: 59,300 Natl. Network: NBC. Natl. Rep: Continental Television Sales. Washington Atty: Hogan & Hartson, L.L.P.
Key Personnel:
Charlie Henrich . . . . gen mgr
Scott Bruce . . . . rgnl sls mgr
Jean Zosel . . . . prom dir & progmg dir
Charlie Cannaliato. . . . . chief of engrg

***KUSM—** Analog channel: 9. Digital channel: 8. Analog hrs: 24 Digital hrs: 24 44 kw vis. ant 817t/305g TL: N45 40 24 W110 52 02 On air date: Oct 1, 1984. Box 173340, Visual Communications, Bldg. 183, Montana State Univ., Bozeman, MT, 59717-3340. Phone: (406) 994-3437. Fax: (406) 994-6545. E-mail: kusm@montanapbs.org Web Site: www.montanapbs.org. Licensee: Montana State University. Population served: 46,000 Natl. Network: PBS.
Key Personnel:
Eric Hyyppa . . . . gen mgr
Lisa Titus . . . . dev dir
Amy Colson . . . . prom dir
Aaron Pruitt . . . . progmg dir
Dean Lawver. . . . . engrg dir & chief of engrg

KWYB— Analog channel: 18. Digital channel: 19. Analog hrs: 24 Digital hrs: 24 2,684 kw vis. 1,955t TL: N46 00 29 W112 26 30 On air date: September 1996. 505 W. Park, Butte, MT, 59701. Phone: (406) 782-7185. Fax: (406) 723-9269. Web Site: www.kwyb.com. Licensee: MMM License LLC. Group owner: MAX Media LLC (acq 2-5-2001; grpsl). Natl. Network: ABC. Washington Atty: Reddy, Begley & McCormick.
Key Personnel:
Linda Gray . . . . pres, pres & gen mgr
Leslie Stoll . . . . natl sls mgr
Linda Julius . . . . prom VP & progmg mgr
Mike Warner . . . . chief of engrg

KXLF-TV— Analog channel: 4. Digital channel: 4. Analog hrs: 24 Digital hrs: 24 Note: CBS is on KXLF-TV ch 4, CW is on KXLF-DT ch 5. 100 kw vis, 20 kw aur. ant 1,890t/202g TL: N46 00 27 W112 26 30 On air date: Aug 14, 1953. 1003 S. Montana, Butte, MT, 59701. Phone: (406) 496-8400. Fax: (406) 782-8906. E-mail: kxlf@kxlf.com Web Site: www.montanasnewsstation.com. Licensee: KXLF Communications Inc. Group owner: Cordillera Communications Inc. (acq 12-15-86; grpsl; FTR: 9-29-86). Population served: 134,400 Natl. Network: CBS,

CW Digital Network: Note: CBS is on KXLF-TV ch 4, CW is on KXLF-DT ch 5. Natl. Rep: Harrington, Righter & Parsons. Washington Atty: Dow, Lohnes & Albertson. Wire Svc: AP News staff: 12; News: 16 hrs wkly.
Key Personnel:
Pat Cooney . . . gen mgr, stn mgr & gen sls mgr
Mariya Peck . . . prom dir & prom mgr
Lynn Hopewell . . . progmg dir & progmg mgr
John Sherer . . . news dir
Ron Schlosser . . . chief of engrg

Glendive, MT
(DMA 210)

KXGN-TV— Analog channel: 5. Analog hrs: 24 14.8 kw vis, 2.9 kw aur. 500t/146g TL: N47 03 15 W104 40 45 On air date: Nov 1, 1957. 210 S. Douglas, Glendive, MT, 59330. Phone: (406) 377-3377. Fax: (406) 365-2181. E-mail: kxgnkdzn@midrivers.com Web Site: www.glendivebroadcasting.com. Licensee: Glendive Broadcasting Corp. Ownership: Stephen A. Marks. Group owner: Glendive Broadcasting Corp. Population served: 6,305 Natl. Network: CBS, NBC. Washington Atty: Davis Wright Tremaine LLP. Wire Svc: UPI
Key Personnel:
Stephen A. Marks . . . pres
Paul Sturlaugson . . . gen mgr, gen sls mgr, adv mgr & news dir
Lauri Harbig . . . progmg dir
Mike Huseby . . . chief of engrg

Great Falls, MT
(DMA 190)

KBAO— Analog channel: 13. Analog hrs: 22 5 kw vis. ant 1,929t TL: N47 10 46 W109 32 05 On air date: Dec 5, 2001. Box 6125, Helena, MT, 59604. Phone: (406) 457-1212. Fax: (406) 442-5106. Web Site: www.ktvh.com. Licensee: Beartooth Communications Co. Group owner: (group owner). Natl. Network: NBC.
Key Personnel:
Kathy Ernst . . . gen mgr
Casey Kysler-West . . . news dir
Mike Anderson . . . chief of engrg

KBBJ— Analog channel: 9. Analog hrs: 22 316 kw vis. 482t TL: N48 29 39 W109 42 48 On air date: Dec 5, 2001. Box 6125, Helena, MT, 59604. Phone: (406) 457-1212. Fax: (406) 442-5106. Web Site: www.ktvh.com. Licensee: Beartooth Communications Co. Group owner: (group owner). Natl. Network: NBC.
Key Personnel:
Kathy Ernst . . . gen mgr
Casey Kyler-West . . . news dir
Mike Anderson . . . chief of engrg

KFBB-TV— Analog channel: 5. Digital channel: 8. Analog hrs: 18 Digital hrs: 18 100 kw vis, 20 kw aur. 590t/540g TL: N47 32 08 W111 17 02 On air date: Mar 21, 1954. Box 1139, Great Falls, MT, 59403-1139. 3200 Old Havre Hwy., Black Eagle, MT 59414. Phone: (406) 453-4377. Fax: (406) 727-9703. E-mail: kfbb@kfbb.com Web Site: www.kfbb.com. Licensee: KFBB L.L.C. Group owner: Dix Communications (acq 7-1-82; $5.2 million; FTR: 5-17-82). Population served: 161,800 Natl. Network: ABC. Washington Atty: Baker & Hostetler. Wire Svc: Direct Line Weather Wire News staff: 6; News: 6 hrs wkly.
Key Personnel:
Danette Sukut . . . gen mgr, stn mgr & gen sls mgr
Julie Klesh . . . news dir
Roy Davis . . . chief of engrg
Linda Julius . . . progmg

KLMN— Analog channel: 26.5,000 kw vis. ant 574t TL: N47 32 23 W111 17 06 On air date: 2003. 118 6th St. S., Great Falls, MT, 59401. 5115 US Hwy. 93 S., Missoula, MT 59804. Phone: (406) 251-1360. Fax: (406) 251-1364. Licensee: Montana License Sub, Inc. Group owner: Equity Broadcasting Corp. Natl. Network: MyNetworkTV.

KRTV— Analog channel: 3. Digital channel: 7. Analog hrs: 20 Digital hrs: 20 Note: CBS is on KRTV(TV) ch 3, CW is on KRTV-DT ch 7. 100 kw vis, 10 kw aur. ant 590t/550g TL: N47 32 09 W111 17 02 On air date: Oct 5, 1958. Box 2989, Great Falls, MT, 59403. 3300 Old Havre Hwy., Black Eagle, MT 59414. Phone: (406) 791-5400. Fax: (406) 791-5479. E-mail: krtv@krtv.com Web Site: www.montanasnewsstation.com. Licensee: KRTV Communications Inc. Group owner: Cordillera Communications Inc. (6-86). Population served: 80,000 Natl. Network: CBS, CW Digital Network: Note: CBS is on KRTV(TV) ch 3, CW is on KRTV-DT ch 7. Natl. Rep: Harrington, Righter & Parsons. Washington Atty: Dow, Lohnes & Albertson. Wire Svc: AP Wire Svc: CNN News staff: 15; News: 15 hrs wkly.
Key Personnel:
Jon Saunders . . . gen mgr
Art Taft . . . prom mgr
Jerry Howard . . . news dir
Marlowe Rames . . . engrg dir
Roxie Rattray . . . progmg

KTGF— Analog channel: 16. Digital channel: 45. Analog hrs: 24 2,040 kw vis, 204 kw aur. ant 1,046t/830g TL: N47 36 26 W111 21 27 On air date: Sept 21, 1986. 118 6th St. S., Box 169, Great Falls, MT, 59405. Phone: (406) 761-8816. Fax: (406) 454-3484. E-mail: ktgf@ktgf.com Web Site: www.ktgf.com. Licensee: Destiny Licenses LLC. Ownership: Destiny Communications LLC, 100% Group owner: MAX Media LLC (acq 11-24-2004; $3 million with translator K47DP Lewistown). Natl. Network: Fox. Washington Atty: Garvey, Schubert & Barer.
Key Personnel:
Darnell Washington . . . CEO, pres & stn mgr
Linda Scutari . . . natl sls mgr
Jennifer Rimmel . . . progmg mgr

Helena, MT
(DMA 206)

KMTF— Analog channel: 10. Digital channel: 29. Analog hrs: 24 Digital hrs: 24 316 kw vis. ant 1,899t TL: N46 35 47 W112 17 47 On air date: 1998. 100 W. Lyndale Ave., Helena, MT, 59601. Phone: (406) 457-1010. Fax: (406) 457-2758. E-mail: cw10@surewest.net Web Site: www.cwhelena.com. Licensee: Rocky Mountain Broadcasting Co. Natl. Network: CW.

KTVH— Analog channel: 12. Digital channel: 14. Analog hrs: 24 Digital hrs: 24 105 kw vis, 10.5 kw aur. 2,250t/176g TL: N46 49 35 W111 42 33 On air date: Jan 1, 1958. Box 6125, Helena, MT, 59604. 100 W. Lyndale, Helena, MT 59601. Phone: (406) 457-1212. Fax: (406) 442-5106. Web Site: www.ktvh.com. Licensee: Beartooth Communications Co. Group owner: Sunbelt Communications Co. (acq 7-9-97). Population served: 140,000 Natl. Network: NBC. Washington Atty: Gerald S. Rourke. News staff: 8; News: 7 hrs wkly.
Key Personnel:
Kathy Ernst . . . gen mgr
Casey Kyler-West . . . news dir
Mike Anderson . . . chief of engrg

Missoula, MT
(DMA 168)

KCFW-TV— Analog channel: 9. Digital channel: 38. Analog hrs: 24 Digital hrs: 24 26.5 kw vis, 5.3 kw aur. ant 2,794t/240g TL: N48 00 48 W114 21 55 On air date: June 10, 1968. Box 857, Kalispell, MT, 59901. 401 First Ave. E., Kalispell, MT 59901. Phone: (406) 755-5239. Fax: (406) 752-8002. E-mail: news@kcfw.com Web Site: www.nbcmontana.com. Licensee: BlueStone License Holdings Inc. Group owner: Lamco Communications Inc. (acq 6-15-2004; grpsl). Population served: 79,600 Natl. Network: NBC. Washington Atty: Cohn & Marks. News staff: 5; News: 6 hrs wkly.
Key Personnel:
Charlie Henrich . . . gen mgr
Jacque Walawander . . . gen sls mgr
Jean Zosel . . . prom dir & progmg dir
Wade Muehlhof . . . stn mgr & news dir
Chris Neuhausen . . . chief of engrg

KECI-TV— Analog channel: 13. Digital channel: 40. Analog hrs: 24 Digital hrs: 24 302 kw vis, 30.2 kw aur. 2,001t/290g TL: N47 01 04 W114 00 47 On air date: July 1, 1954. 340 W. Main, Missoula, MT, 59802. Box 5268, Missoula, MT 59806-5268. Phone: (406) 721-2063. Fax: (406) 721-2083/(406) 549-6507. E-mail: news@keci.tv Web Site: www.nbcmontana.com. Licensee: BlueStone License Holdings Inc. Group owner: Lamco Communications Inc. (acq 6-15-2004; grpsl). Population served: 104,700 Natl. Network: NBC. Natl. Rep: Continental Television Sales. Washington Atty: Hogan & Hartson. News: 14 hrs wkly.
Key Personnel:
Charlie Henrich . . . gen mgr & natl sls mgr
Jacque Walawander . . . rgnl sls mgr
Jean Zosel . . . mktg dir, prom dir & progmg dir
Jim Harmon . . . news dir
Charlie Cannaliato . . . chief of engrg

KMMF— Analog channel: 17. Analog hrs: 24 3,020 kw vis. ant 1,689 ft TL: N46 48 30 W113 58 38 On air date: May 1, 2001. 5115 US Hwy. 93 S., Missoula, MT, 59804. Phone: (406) 542-8900. Fax: (406) 728-4800. Web Site: www.ktmf.com. Licensee: Montana License Sub, Inc. Group owner: Equity Broadcasting Corp. Natl. Network: Fox, MyNetworkTV. Washington Atty: Irwin, Caampbell & Tannenwald.
Key Personnel:
Linda Gray . . . pres & gen mgr
Leslie Stoll . . . gen sls mgr
Terry Cudderford . . . rgnl sls mgr
Linda Julius . . . progmg dir
Mike Warner . . . chief of engrg

KPAX-TV— Analog channel: 8. Digital channel: 8. Analog hrs: 24 Note: CBS is on KPAX-TV ch 8, CW is on KPAX-DT ch 7. 275 kw vis, 49 kw aur. 2,150t/284g TL: N47 01 06 W114 00 41 On air date: June 5, 1970. 1049 W. Central, Missoula, MT, 59801. Phone: (406) 542-4400. Fax: (406) 543-7111. E-mail: office@kpax.com Web Site: www.kpax.com. Licensee: KPAX-TV Communications Inc. Group owner: Cordillera Communications Inc. Population served: 207,200 Natl. Network: CBS, CW Digital Network: Note: CBS is on KPAX-TV ch 8, CW is on KPAX-DT ch 7. Washington Atty: Dow, Lohnes & Albertson. News: 15 hrs wkly.
Key Personnel:
Bob Hermes . . . gen mgr
Jim McLean . . . sls dir & gen sls mgr
James Rafferty . . . prom mgr
Tammy Engle . . . progmg dir
Joel Lundstad . . . news dir
Larry Arbaugh . . . chief of engrg

KTMF— Analog channel: 23. Digital channel: 36. Analog hrs: 24 1,820 kw vis. 2,054t TL: N47 01 10 W114 00 46 On air date: Nov 16, 1990. 5115 US Highway 93 S, Missoula, MT, 59804. Phone: (406) 542-8900. Fax: (406) 728-4800. E-mail: ktmf@ktmf.com Web Site: www.ktmf.com. Licensee: MMM License LLC. Group owner: MAX Media LLC (acq 2-5-2001; grpsl). Natl. Network: ABC. Washington Atty: Reddy, Begley & McCormick.
Key Personnel:
Linda Gray . . . pres & gen mgr
Leslie Stoll . . . gen sls mgr & natl sls mgr
Linda Julius . . . progmg dir & progmg mgr
Mike Warner . . . chief of engrg
Rebroadcasts KTMF (LP) Kalispell 90%.

***KUFM-TV—** Analog channel: 11. Digital channel: 27. Analog hrs: 24 125 kw vis. 2,116t/259g (CP: 3 kw vis, ant 2,070t) On air date: Jan 18, 1997. PARTV 180, 32 Campus Dr., Univ. of Montana, Missoula, MT, 59812. Phone: (406) 243-4101. Fax: (406) 243-3299. E-mail: kufm@montanapbs.org Web Site: www.montanapbs.org. Licensee: The University of Montana. Natl. Network: PBS.
Key Personnel:
William Marcus . . . gen mgr
Daniel Dauterive . . . opns dir
Rebroadcasts KUSM(TV) Bozeman 85%.

Nebraska

Alliance

see Rapid City, SD market

Hastings

see Lincoln & Hastings-Kearney, NE market

Kearney

see Lincoln & Hastings-Kearney, NE market

Lincoln & Hastings-Kearney, NE
(DMA 104)

KCWL-TV— Analog channel: 51. Analog hrs: 24 51.4 kw vis. ant 410t/490g TL: N40 51 10 W96 40 36 On air date: June 26, 2006. 707 N. 48th St., Suite B, Lincoln, NE, 68504. Phone: (402) 464-5694. Fax: (402) 466-1311. Web Site: www.kcwl51.com. Licensee: Lincoln Broadcasting LLC. Ownership: World Investments Inc. Natl. Network: CW. Washington Atty: Hogan & Hartson.
Key Personnel:
William Conley . . . pres
Deb Jacob . . . stn mgr

KGIN— Analog channel: 11. Digital channel: 32. Analog hrs: 24 Digital hrs: 24 Note: CBS is on KGIN(TV) ch 11, MyNetworkTV is on KGIN-DT ch 32. 316 kw vis, 55 kw aur. 1,010t/1,069g TL: N40 35 20 W98 48 10 On air date: Oct 1, 1961. Box 1069, Grand Island, NE, 68801. 123 N. Locust St., Grand Island, NE 68802. Phone: (308) 382-6100. Fax: (308) 382-3216. E-mail: kgin1011@hotmail.com Web Site: www.kolnkgin.com. Licensee: WEAU Licensee Corp. Group owner: Gray Television Inc. (acq 7-30-98; grpsl). Population served: 240,800 Natl. Network: CBS, MyNetworkTV Digital Network: Note: CBS is on KGIN(TV) ch 11, MyNetworkTV is on KGIN-DT ch 32. Natl. Rep: Continental Television Sales. Washington Atty: Pepper & Corazzini. Satellite of KOLN Lincoln.

KHAS-TV— Analog channel: 5. Digital channel: 21. Analog hrs: 6 AM-12:35 AM Digital hrs: 6AM-12:35AM 100 kw vis, 20 kw aur. 731t/768g TL: N40 39 06 W98 23 04 On air date: Jan 1, 1956. 6475 Osborne Dr. W., Hastings, NE, 68901. Phone: (402) 463-1321. Fax: (402) 463-6551. E-mail: khas@khastv.com Web Site: www.khastv.com. Licensee: Hoak Media of Nebraska License LLC. Group owner: (group owner; (acq 11-7-2005; with KNOP-TV North Platte). Population served: 600,000 Natl. Network: NBC. Washington Atty: Fletcher, Heald & Hildreth. Wire Svc: Skycom
Key Personnel:
Ulysses Carlini . . . . . gen mgr
Alan Uerling . . . . . opns mgr
Connie Caldwell . . . . . gen sls mgr
Connie Cardwell . . . . . rgnl sls mgr
Jackie Ackerman . . . . . prom mgr
Jackie Arkerman. . . . . progmg dir
Dennis Kellogg . . . . . news dir

KHGI-TV— Analog channel: 13. Digital channel: 36. Analog hrs: 24 Digital hrs: 24 316 kw vis, 31.6 kw aur. 1,110t/1,163g TL: N40 39 28 W98 52 04 On air date: Dec 24, 1953. Box 220, Kearney, NE, 68848-0220. 1078 25th Rd., Axtell, NE 68924. Phone: (308) 743-2494. Fax: (308) 743-2644. E-mail: comments@nebraska.tv Web Site: www.nebraska.tv. Licensee: Pappas Telecasting of Central Nebraska L.P. (DE limited partnership). Group owner: Pappas Telecasting Companies (acq 7-1-96; grpsl). Population served: 257,910 Natl. Network: ABC. Natl. Rep: Harrington, Righter & Parsons. News staff: 26; News: 17.5 hrs wkly.
Key Personnel:
Mark Baumert . . . . . CEO & news dir
Scott Swenson. . . . . pres & progmg mgr
Jerry Fuehrer . . . . . exec VP & chief of engrg
Janet Noll . . . . . gen mgr
Susan Christensen . . . . . gen sls mgr
Anita Wragge . . . . . prom mgr

***KHNE-TV**— Analog channel: 29. Digital channel: 28. Analog hrs: 18 Digital hrs: 18 605 kw vis, 60.5 kw aur. 1,220t/1,238g TL: N40 46 20 W98 05 21 On air date: Nov 17, 1968. 1800 N. 33rd St., Lincoln, NE, 68503. Phone: (402) 472-3611. Fax: (402) 472-1785. E-mail: net1@unl.edu Web Site: www.netnebraska.org. Licensee: Nebraska Educational Telecommunications Commission. Natl. Network: PBS. Washington Atty: Dow, Lohnes & Albertson. News staff: 3; News: 30 min. wkly. Satellite of *KUON-TV Lincoln.

KLKN— Analog channel: 8. Digital channel: 31. Analog hrs: 24 316 kw vis, 31.6 kw aur. 2,001t/1,974g TL: N41 32 28 W97 40 45 On air date: Dec 3, 1964. 3240 S. 10th St., Lincoln, NE, 68502. Phone: (402) 434-8000. Fax: (402) 436-2236. E-mail: 8@klkntv.com Web Site: www.klkntv.com. Licensee: Citadel Communications Co. L.L.C. Group owner: Citadel Communications Co. Ltd. (acq 11-15-86; FTR: 7-28-86). Population served: 655,000 Natl. Network: ABC. Natl. Rep: Millennium Sales & Marketing. Washington Atty: Latham & Watkins. Wire Svc: AP News staff: 21; News: 22 hrs wkly.
Key Personnel:
Roger Moody . . . . . gen mgr
Jeff Swanson . . . . . opns dir
Kay Wonderlich . . . . . natl sls mgr
Phil Maddern . . . . . rgnl sls mgr
Mark Haggar . . . . . news dir
Dan Ackerman . . . . . chief of engrg

***KLNE-TV**— Analog channel: 3. Digital channel: 26. Analog hrs: 18 Digital hrs: 18 100 kw vis, 10 kw aur. 1,062t/1,065g TL: N40 23 05 W99 27 30 On air date: Sept 6, 1965. 1800 N. 33rd St., Lincoln, NE, 68503. Phone: (402) 472-3611. Fax: (402) 472-1785. E-mail: net1@unl.edu Web Site: www.netnebraska.org. Licensee: Nebraska Educational Telecommunications Commission. Natl. Network: PBS. Washington Atty: Dow, Lohnes & Albertson.
Key Personnel:
Rod Bates . . . . . gen mgr
Steven Graziano . . . . . prom mgr & adv mgr
Steve Graziano . . . . . progmg mgr
Satellite of *KUON-TV Lincoln.

***KMNE-TV**— Analog channel: 7. Digital channel: 15. Analog hrs: 18 Digital hrs: 18 316 kw vis, 31.6 kw aur. 1,484t/1,524g TL: N42 20 05 W99 29 01 On air date: Sept 1, 1967. 1800 N. 33rd St., Lincoln, NE, 68503. Phone: (402) 472-3611. Fax: (402) 472-1785. E-mail: net1@unl.edu Web Site: www.netnebraska.org. Licensee: Nebraska Educational Telecommunications Commission. Natl. Network: PBS. Washington Atty: Dow, Lohnes & Albertson.
Key Personnel:
Rod Bates . . . . . gen mgr
Steven Graziano . . . . . prom mgr, adv mgr & progmg mgr
Satellite of *KUON-TV Lincoln.

KOLN— Analog channel: 10. Digital channel: 25. Analog hrs: 24 Digital hrs: 24 Note: CBS is on KOLN(TV) ch 10, MyNetworkTV is on KOLN-DT ch 25. 316 kw vis, 36.3 kw aur. ant 1,530t/1,500g TL: N40 48 08 W97 10 46 On air date: Feb 18, 1953. Box 30350, Lincoln, NE, 68503. 40th & W Sts., Lincoln, NE 68503. Phone: (402) 467-4321. Fax: (402) 467-9210. E-mail: info@kolnkgin.com Web Site: www.kolnkgin.com. Licensee: Gray Television Licensee Inc. Group owner: Gray Television Inc. (acq 7-30-98; grpsl). Population served: 240,800 Natl. Network: CBS, MyNetworkTV Digital Network: Note: CBS is on KOLN(TV) ch 10, MyNetworkTV is on KOLN-DT ch 25. Natl. Rep: Continental Television Sales. Washington Atty: Venable, Baetijer, Howard & Civiletti.
Key Personnel:
Frank Jonas . . . . . pres
Lisa Guill . . . . . gen mgr & opns mgr
Kris Ryan . . . . . gen sls mgr
Marty Winters . . . . . natl sls mgr
Troy Frankforter . . . . . prom mgr & progmg mgr
Randy Lube . . . . . news dir
Mindy Burback. . . . . pub affrs dir
Brent Haun . . . . . chief of engrg

KSNB-TV— Analog channel: 4. Digital channel: 34. Analog hrs: 24 Digital hrs: 24 100 kw vis, 12.6 kw aur. 1,131t/1,086g TL: N40 05 13 W97 55 13 (CP: 2,000t/1,959g) On air date: Oct 1, 1965. Box 220, Kearney, NE, 68848-0220. 1078 25 Rd., Axell, NE 68924. Phone: (308) 743-2494. Fax: (308) 743-2644. E-mail: comments@nebraska.tv Web Site: www.nebraska.tv. Licensee: Colins Broadcasting Co. (acq 2-17-99; $333,333). Population served: 257,910 Natl. Network: Fox. Natl. Rep: Harrington, Righter & Parsons. News: 13 hrs wkly.
Key Personnel:
Vince Barresi . . . . . gen mgr
Susan Christensen . . . . . gen sls mgr
Anita Wragge . . . . . prom mgr
Scott Swenson. . . . . progmg mgr
Mark Baumert . . . . . news dir
Jerry Fuehrer. . . . . chief of engrg
Satellite of KTVG Grand Island.

KTVG— Analog channel: 17. Digital channel: 19. Analog hrs: 24 Digital hrs: 24 3,890 kw vis, 21.9 kw aur. 610t/270g TL: N40 43 44 W98 34 13 On air date: December 1992. Box 220, Kearney, NE, 68848. 1078 25 Rd., Axtell, NE 68924. Phone: (308) 734-2794. Fax: (308) 743-2644. E-mail: comments@nebraska.tv Web Site: www.nebraska.tv. Licensee: Hill Broadcasting Inc. Natl. Network: Fox. Natl. Rep: Harrington, Righter & Parsons.
Key Personnel:
Vince Barresi . . . . . gen mgr
Susan Christensen . . . . . gen sls mgr
Anita Wragge . . . . . prom mgr
Scott Swenson. . . . . progmg mgr
Mark Baumert . . . . . news dir
Jerry Fuehrer. . . . . chief of engrg

***KUON-TV**— Analog channel: 12. Digital channel: 40. Analog hrs: 18 Digital hrs: 18 316 kw vis, 31.6 kw aur. 830t/879g TL: N41 08 18 W96 27 19 On air date: Nov 1, 1954. 1800 N. 33rd St., Lincoln, NE, 68503. Phone: (402) 472-3611. Fax: (402) 472-1785. E-mail: net1@unl.edu Web Site: www.netnebraska.org. Licensee: University of Nebraska. (acq 7-28-54; FTR: 8-2-54). Population served: 670,000 Natl. Network: PBS. Washington Atty: Dow, Lohnes & Albertson. News staff: 3; News: 30 min. wkly.
Key Personnel:
Rod Bates . . . . . gen mgr
Steven Graziano . . . . . prom mgr, adv mgr & progmg mgr

KWNB-TV— Analog channel: 6. Digital channel: 18. Analog hrs: 24 Digital hrs: 24 100 kw vis, 21.6 kw aur. 737t/586g TL: N40 37 29 W101 01 58 On air date: Feb 9, 1956. Box 220, Kearney, NE, 68848. 1078 25 Rd., Axtell, NE 68924. Phone: (308) 743-2794. Fax: (308) 743-2644. Web Site: www.nebraska.tv. Licensee: Pappas Telecasting of Central Nebraska L.P. (DE limited partnership). Group owner: Pappas Telecasting Companies (acq 7-1-96; grpsl). Population served: 257,910 Natl. Network: ABC. Natl. Rep: Harrington, Righter & Parsons. News staff: 26; News: 17.5 hrs wkly.
Key Personnel:
Vince Barresi . . . . . gen mgr
Susan Christensen . . . . . gen sls mgr
Anita Wragge . . . . . prom mgr
Scott Swensen . . . . . progmg dir
Mark Baumert . . . . . news dir
Jerry Fuehrer. . . . . chief of engrg
Rebroadcasts KHGI-TV, Kearney, 100%.

McCook

see Wichita-Hutchinson Plus, KS market

Merriman

see Sioux Falls (Mitchell), SD market

Norfolk

see Sioux City, IA market

North Platte, NE (DMA 209)

KNOP-TV— Analog channel: 2. Digital channel: 22. Analog hrs: 18 100 kw vis, 15 kw aur. 630t/608g TL: N41 12 13 W100 43 58 On air date: Dec 2, 1958. Box 749, North Platte, NE, 69103. N. Hwy. 83, North Platte, NE 69101. Phone: (308) 532-2222. Fax: (308) 532-9579. E-mail: lewysknop@knoptv.com Web Site: www.knop.msnbc.com. Licensee: Hoak Media of Nebraska License LLC. Group owner: (group owner; (acq 11-7-2005; with KHAS-TV Hastings). Population served: 14,500 Natl. Network: NBC. Natl. Rep: Blair Television. Washington Atty: Fletcher, Heald & Hildreth. Wire Svc: AP News staff: 10; News: 15 hrs wkly.
Key Personnel:
Lewys Carlini . . . . . gen mgr
Darlene Lyman. . . . . gen sls mgr
Gregg Hoover. . . . . prom dir
Jacques Harms . . . . . news dir
Mike McNeil . . . . . chief of engrg

***KPNE-TV**— Analog channel: 9. Digital channel: 16. Analog hrs: 18 Digital hrs: 18 316 kw vis, 31.6 kw aur. 1,020t/1,006g TL: N41 01 16 W101 09 10 On air date: Sept 12, 1966. 1800 N. 33rd St., Lincoln, NE, 68503. Phone: (402) 472-3611. Fax: (402) 472-1785. E-mail: net1@unl.edu Web Site: www.netnebraska.org. Licensee: Nebraska Educational Telecommunications Commission. Natl. Network: PBS. Washington Atty: Dow, Lohnes & Albertson.
Key Personnel:
Rod Bates . . . . . gen mgr
Steven Graziano . . . . . prom mgr, adv mgr & progmg mgr
Satellite of *KUON-TV Lincoln.

Omaha, NE (DMA 75)

***KBIN**— Analog channel: 32. Digital channel: 33.575 kw vis, 57.5 kw aur. ant 317t/163g TL: N41 15 14 W95 50 07 On air date: Sept 7, 1975. Box 6450, Iowa Public TV, Johnston, IA, 50131-6450. 6450 Corporate Dr., Johnston, IA 50131. Phone: (515) 242-3100. E-mail: public_information@iptv.org Web Site: www.iptv.org. Licensee: Iowa Public Broadcasting Board. Natl. Network: PBS. Washington Atty: Dow, Lohnes PLLC.

KETV— Analog channel: 7. Digital channel: 20. Analog hrs: 24 Digital hrs: 24 316 kw vis, 31.6 kw aur. ant 1,373t/1,330g TL: N41 18 32 W96 01 33 On air date: Sept 17, 1957. 2665 Douglas St., Omaha, NE, 68131-2699. Phone: (402) 345-7777. Fax: (402) 522-7755. Fax: (402) 522-7761. Web Site: www.ketv.com. Licensee: KETV Hearst-Argyle Television Inc. Group owner: Hearst-Argyle Television Inc. (acq 3-18-99; grpsl). Population served: 1,504,300 Natl. Network: ABC. Natl. Rep: Eagle Television Sales. News: 28 hrs wkly.
Key Personnel:
Sarah Smith . . . . . pres & gen mgr
Brian Sather. . . . . gen sls mgr
Rose Ann Shannon . . . . . news dir
Warren Behrens . . . . . chief of engrg
Linda Hood . . . . . progmg

***KHIN**— Analog channel: 36. Digital channel: 35.2,040 kw vis, 204 kw aur. ant 1,560t/1,503g TL: N41 20 40 W95 15 21 On air date: Sept 7, 1975. Box 6450, Iowa Public TV, Johnston, IA, 50131-6450. 6450 Corporate Dr., Johnston, IA 50131. Phone: (515) 242-3100. E-mail: public_information@iptv.org Web Site: www.iptv.org. Licensee: Iowa Public Broadcasting Board. Natl. Network: PBS. Washington Atty: Dow, Lohnes PLLC.

KMTV— Analog channel: 3. Digital channel: 45. Analog hrs: 24 Digital hrs: 24 100 kw vis, 20 kw aur. ant 1,371t/1,409g TL: N41 18 25 W96 01 37 On air date: Sept 1, 1949. 10714 Mockingbird Dr., Omaha, NE, 68127. Phone: (402) 592-3333. E-mail: feedback@action3news.com Web Site: www.action3news.com. Licensee: Journal Broadcast Corp. Group owner: Emmis Communications Corp. (acq 3-27-2007). Population served: 357,800 Natl. Network: CBS. News: 23 hrs wkly.
Key Personnel:
Steve Wexler . . . . . gen mgr
Eric Hanneman. . . . . gen sls mgr
Todd Long . . . . . mktg dir
Renee Rich . . . . . progmg dir
Ken Dudzik . . . . . news dir
Scott Krayenhagen . . . . . chief of engrg

KPTM— Analog channel: 42. Digital channel: 43. Analog hrs: 24 Digital hrs: 24 Note: Fox is on KPTM(TV) ch 42, MyNetworkTV is on

KPTM-DT ch 43. 5,000 kw vis, 500 kw aur. ant 1,558t/1,479g TL: N41 04 15 W96 13 30 (CP: 4,800 kw vis, ant 1,558t/1,484g. TL: N41 04 14 W96 13 33)) On air date: Apr 6, 1986. 4625 Farnam St., Omaha, NE, 68132. Phone: (402) 558-4200. Fax: (402) 554-4290. E-mail: contact42@kptm.com Web Site: www.kptm.com. Licensee: KPTM (TV) License LLC. Group owner: Pappas Telecasting Companies (acq 3-14-86). Natl. Network: Fox, MyNetworkTV Digital Network: Note: Fox is on KPTM(TV) ch 42, MyNetworkTV is on KPTM-DT ch 43. Natl. Rep: TeleRep. Washington Atty: Paul, Hastings, Janofsky & Walker. Wire Svc: FNS News staff: 20; News: 7 hrs wkly.
Key Personnel:
Randy Oswald . . . . exec VP & gen mgr
Chris McDade . . . . opns mgr
Jeff Miller . . . . gen sls mgr
Joe Radske . . . . gen sls mgr & news dir
Tim Moan . . . . rgnl sls mgr
Sam Lawson . . . . prom mgr
Darlene Goldsberry . . . . progmg mgr

KXVO— Analog channel: 15. Digital channel: 38. Analog hrs: 24 Digital hrs: 24 5,000 kw vis, 500 kw aur. ant 1,558t/1,464g TL: N41 04 15 W96 13 30 On air date: June 10, 1995. 4625 Farnam St., Omaha, NE, 68132. Phone: (402) 554-1500. Fax: (402) 554-4290. Web Site: www.kxvo.com. Licensee: Mitts Telecasting Co. Ownership: (LMA to Pappas Telecasting Companies) (acq 6-13-2000; $972,000). Natl. Network: CW. Natl. Rep: TeleRep. Washington Atty: Bryan Cave.
Key Personnel:
Randy Oswald . . . . exec VP & gen mgr
Jeff Miller . . . . gen sls mgr
Tim Moan . . . . rgnl sls mgr
Sam Lawson . . . . prom mgr
Darlene Goldsberry . . . . progmg mgr
Chris McDade . . . . opns

***KYNE-TV—** Analog channel: 26. Digital channel: 17. Analog hrs: 18 Digital hrs: 18 525 kw vis, 52.5 kw aur. 426t/396g TL: N41 15 28 W96 00 32 On air date: Oct 19, 1965. 1800 N. 33rd St., Lincoln, NE, 68503. Phone: (402) 472-3611. Fax: (402) 472-1785. E-mail: net1@unl.edu Web Site: www.netnebraska.org. Licensee: Nebraska Educational Telecommunications Commission. Population served: 500,000 Natl. Network: PBS. Washington Atty: Dow, Lohnes & Albertson. News staff: 3; News: 30 min. wkly.
Key Personnel:
Rod Bates . . . . gen mgr
Steven Graziano . . . . prom mgr, adv mgr & progmg mgr
Satellite of *KUON-TV Lincoln.

WOWT— Analog channel: 6. Digital channel: 22. Analog hrs: 24 Digital hrs: 24 100 kw vis, 20 kw aur. 1,371t/1,344g TL: N41 18 40 W96 01 37 On air date: Aug 29, 1949. 3501 Farnam St., Omaha, NE, 68131-3356. Phone: (402) 346-6666. Fax: (402) 233-7880. E-mail: sixonline@wowt.com Web Site: www.wowt.com. Licensee: Gray Television Licensee Inc. Group owner: Gray Television Inc. (acq 10-2002; grpsl). Population served: 347,328 Natl. Network: NBC. Natl. Rep: Continental Television Sales. Washington Atty: Fletcher, Heald & Hildreth. News staff: 52; News: 37 hrs wkly.
Key Personnel:
Frank Jonas . . . . gen mgr
Don Felton . . . . gen sls mgr
Vic Richards . . . . prom dir
Gail Backer . . . . progmg mgr
John Clark . . . . news dir
Rick Klutts . . . . chief of engrg

Scottsbluff

see Cheyenne, WY-Scottsbluff, NE market

Nevada

Elko

see Salt Lake City, UT market

Ely

see Salt Lake City, UT market

Las Vegas, NV (DMA 43)

KBLR— Analog channel: 39. Digital channel: 40. Analog hrs: 24 Digital hrs: 24 1,320 kw vis, 132 kw aur. ant 1,204t/223g TL: N36 00 31 W115 00 22 (CP: 2,800 kw vis) On air date: Apr 20, 1989. 73 Spectrum Blvd., Las Vegas, NV, 89101. Phone: (702) 258-0039. Fax: (702) 258-0556. Web Site: http://telemundo.yahoo.com. Licensee: Summit Media Limited Partnership. Ownership: Scott Gentry, Bruce F. Becker, William O'Connell, et al. (acq 1993; $1.5 million; FTR: 9-20-93). Population served: 1,300,000 Natl. Network: Telemundo (Spanish). Washington Atty: KMZ Rosenman. Foreign lang progmg: SP 168 News staff: 8; News: 2 hrs wkly.
Key Personnel:
Carlos Sanchez . . . . VP & gen mgr
Bill George . . . . natl sls mgr
Brenda Macias . . . . news dir

KINC— Analog channel: 15. Analog hrs: 24 5,000 kw vis. 1,260t TL: N36 00 32 W115 00 22 On air date: October 1995. 500 Pilot Rd., Suite D, Las Vegas, NV, 89119. Phone: (702) 434-0015. Fax: (702) 434-0527. Web Site: www.kinc.entravision.com. Licensee: Entravision Holdings L.L.C. Group owner: Entravision Communications Co. L.L.C. Population served: 1,625,000 Natl. Network: Univision (Spanish). Washington Atty: Thompson Hine L.L.P. Foreign lang progmg: SP 168 News: 5 hrs wkly.
Key Personnel:
Chris Roman . . . . gen mgr
J.R. Des Amours . . . . natl sls mgr
Karina Barcena . . . . prom dir & prom mgr

KLAS-TV— Analog channel: 8. Digital channel: 7. Analog hrs: 24 Digital hrs: 24 316 kw vis, 57.5 kw. 2,001t/272g TL: N35 56 44 W115 02 33 On air date: July 22, 1953. 3228 Channel 8 Dr., Las Vegas, NV, 89109. Phone: (702) 792-8888. Fax: (702) 734-7437. Web Site: www.lasvegasnow.com. Licensee: KLAS Inc., a Nevada Corp. Group owner: Landmark Communications Inc. (acq 7-1-78; $8 million). Population served: 315,000 Natl. Network: CBS. Washington Atty: Hogan & Hartson.
Key Personnel:
Emily Neilson . . . . pres & gen mgr
Linda Bonnici . . . . sls VP
Michael Watkins . . . . gen sls mgr
Doug Kramer . . . . chief of engrg

***KLVX—** Analog channel: 10. Digital channel: 11. Analog hrs: 24 Digital hrs: 24 295 kw vis, 29.5 kw aur. 1,220t/176g TL: N36 00 27 W115 00 24 On air date: Mar 25, 1968. 4210 Channel 10 Dr., Las Vegas, NV, 89119. Phone: (702) 799-1010. Fax: (702) 799-5586. Web Site: www.klvx.org. Licensee: Clark County School District Board of Trustees. Population served: 986,152 Natl. Network: PBS. Washington Atty: Wiley, Rein & Fielding.
Key Personnel:
Tom Axtell . . . . gen mgr
TB D . . . . dev mgr
Cyndy Robbins . . . . progmg dir

KMCC— Analog channel: 34. Digital channel: 32. Analog hrs: 24 Digital hrs: 24 416 kw vis. ant -207t/270g TL: N35 10 08 W114 38 09 On air date: 2005. 3100 S. Needles Hwy., Suite 1700, Laughlin, NV, 89029. Phone: (702) 298-2222. Fax: (702) 298-3495. Licensee: Mojave Broadcasting Co. Ownership: Suzanne E. Rogers, 48% of votes; Perry C. Rogers, 28% of votes; and Kimberly Rogers Cell, 24% of votes. Population served: 64,051 Natl. Rep: Blair Television. Washington Atty: Wiley, Rein & Fielding.

KTNV— Analog channel: 13. Digital channel: 12. Analog hrs: 24 Digital hrs: 24 316 kw vis, 31.6 kw aur. 2,001t/259g TL: N35 56 43 W115 02 32 On air date: May 4, 1956. 3355 S. Valley View Blvd., Las Vegas, NV, 89102. Phone: (702) 876-1313. Fax: (702) 871-1961. E-mail: desk@ktnv.com Web Site: www.ktnv.com. Licensee: Journal Broadcast Corp. Group owner: Journal Broadcast Group Inc. (acq 6-29-79). Population served: 1,564,000 Natl. Network: ABC. Natl. Rep: Petry Television Inc. Washington Atty: Crowell & Moring. News staff: 60; News: 22 hrs wkly.
Key Personnel:
Jim Prather . . . . exec VP, VP & gen mgr
Thom Poterfield . . . . gen sls mgr
Jim Koonce . . . . mktg mgr
Karin Movesian . . . . prom mgr & news dir
Marie Shea . . . . progmg dir

KVBC— Analog channel: 3. Digital channel: 2. Digital hrs: 24 100 kw vis, 10 kw aur. ant 1,263t/226g TL: N36 00 32 W115 00 19 On air date: Oct 1, 1979. 1500 Foremaster Ln., Las Vegas, NV, 89101. Phone: (702) 642-3333. Fax: (702) 657-3152 (news). E-mail: news3@kvbc.com Web Site: www.kvbc.com. Licensee: Valley Broadcasting Co. Ownership: James Rogers, 49.97%; Louis Wiener Jr. Estate, 30%; Janet Rogers, 12.53%. Population served: 1,101,000 Natl. Network: NBC. Washington Atty: Dow, Lohnes & Albertson. News staff: 55; News: 29 hrs wkly.
Key Personnel:
James E. Rogers . . . . CEO
Ralph Toddre . . . . pres
Lisa Howfield . . . . gen mgr
Joanne Nasby . . . . gen sls mgr & natl sls mgr
Dale Wyman . . . . prom dir
Dick Tuiniga . . . . news dir
Mark Guranik . . . . engrg dir & chief of engrg

KVCW— Analog channel: 33. Digital channel: 29. Analog hrs: 24 Digital hrs: 24 1,350 kw vis, 500 kw aur. ant 1,906t/80g TL: N35 56 44 W115 02 31 On air date: Aug 1, 1989. 3830 S. Jones Blvd., Las Vegas, NV, 89103. Phone: (702) 952-4600. Fax: (702) 873-1233. Web Site: www.thecwlasvegas.tv. Licensee: Channel 33 Inc. Group owner: Sinclair Broadcast Group Inc. (acq 2-23-2000; $33 million for stock). Population served: 1,566,000 Natl. Network: CW. Natl. Rep: Adam Young. Washington Atty: Fletcher, Heald & Hildreth.
Key Personnel:
Rob Weisbord . . . . gen mgr
Chris Cohen . . . . sls dir
Tom Gonzalez . . . . progmg dir
Mike brown . . . . chief of engrg

KVMY— Analog channel: 21. Digital channel: 22. Analog hrs: 24 Digital hrs: 24 400 kw vis, 40 kw aur. ant 1,160t/115g TL: N36 00 26 W115 00 24 (CP: 2,576 kw vis) On air date: July 31, 1984. 3830 S. Jones Blvd., Las Vegas, NV, 89103. Phone: (702) 382-2121. Fax: (702) 382-1351. Web Site: www.mylvtv.com. Licensee: KUPN Licensee L.L.C. Group owner: Sinclair Broadcast Group Inc. (acq 5-30-97; $87 million). Population served: 864,000 Natl. Network: MyNetworkTV. Washington Atty: Dow, Lohnes & Albertson.
Key Personnel:
David Smith . . . . CEO, pres & opns mgr
Rob Weisbord . . . . gen mgr
Chris Cohen . . . . sls dir
Mike Brown . . . . chief of engrg

KVVU-TV—Henderson, Analog channel: 5. Digital channel: 51. Analog hrs: 24 Digital hrs: 24 100 kw vis, 20 kw aur. 1,191t/140g TL: N36 00 26 W115 00 23 On air date: October 1967. 25 TV-5 Dr., Henderson, NV, 89014. Phone: (702) 435-5555. Fax: (702) 436-2507. Web Site: www.fox5vegas.com. Licensee: KVVU Broadcasting Corp. Group owner: Meredith Broadcasting Group, Meredith Corp., see Cross-Ownership (acq 5-85; $36 million). Population served: 1,246,200 Natl. Network: Fox. Natl. Rep: TeleRep. Washington Atty: Haley, Bader & Potts.
Key Personnel:
Holly Steuart . . . . gen mgr
Jill Saarela . . . . rgnl sls mgr

Reno, NV (DMA 110)

KAME-TV— Analog channel: 21. Digital channel: 20. Analog hrs: 24 Digital hrs: 24 631 kw vis, 63.1 kw aur. ant 620t/152g TL: N39 19 07 W119 47 51 On air date: Oct 1, 1983. 4920 Brookside Ct., Reno, NV, 89502. Phone: (775) 856-2121. Fax: (775) 856-2100. Web Site: www.foxreno.com. Licensee: Broadcast Development Corp. (acq 2-28-94). Population served: 500,000 Natl. Network: MyNetworkTV. Natl. Rep: TeleRep. Washington Atty: Bryan Cave.
Key Personnel:
Steve Cummings . . . . VP & gen mgr
Ray Stofer . . . . chief of opns & chief of engrg
Amy Chapman . . . . rgnl sls mgr

KEGS— Analog channel: 7.22.9 kw vis. ant 1,469t TL: N38 03 05 W117 13 30 On air date: 2002. Equity Broadcasting Corp., 1 Shackleford Dr., Suite 400, Little Rock, AR, 72211. Phone: (501) 219-2400. Fax: (501) 604-8004. Licensee: Nevada Channel 3 Inc.

***KNPB—** Analog channel: 5. Analog hrs: 24 5.01 kw vis, 1 kw aur. 459t/93g TL: N39 35 01 W119 47 52 On air date: October 1983. 1670 N. Virginia St., Reno, NV, 89503. Phone: (775) 784-4555. Fax: (775) 784-1438. E-mail: info@knpb.org Web Site: www.knpb.org. Licensee: Channel 5 Public Broadcasting Inc. Natl. Network: PBS. Washington Atty: Schwartz, Woods & Miller.
Key Personnel:
Pat Miller . . . . gen mgr
Barbara Harmon . . . . progmg mgr

KOLO-TV— Analog channel: 8. Analog hrs: 24 166 kw vis, 30.2 aur. 2,929t/119g TL: N39 18 49 W119 53 00 On air date: Sept 27, 1953. 4850 Ampere Dr., Reno, NV, 89502. Phone: (775) 858-8888. Fax: (775) 858-8855. Web Site: www.kolotv.com. Licensee: Gray Television Licensee Corp. Group owner: Gray Television Inc. (acq 12-10-2002; $41.5 million with K12IX Austin, K58AO Crystal Bay, K03DN Ely/McGill and K49CK Stead/Lawton, all NV). Population served: 432,400 Natl. Network: ABC. Natl. Rep: Millennium Sales & Marketing. Washington Atty: Haley, Bader & Potts. News: 22 hrs wkly.

KREN-TV— Analog channel: 27. Digital channel: 26. Analog hrs: 24 Digital hrs: 24 1,820 kw vis, 182 kw aur. ant 2,923t/139g TL: N39 18 47 W119 52 59 On air date: Nov 1, 1985. 5166 Meadowood Mall Cir., Reno, NV, 89502-6502. Phone: (775) 333-2727. Fax: (775) 327-6868. E-mail: renofrontdesk@kren.com Web Site: www.kren.com. Licensee:

Reno License LLC. Group owner: Pappas Telecasting Companies (acq 2-10-95; $3 million; FTR: 5-8-95). Natl. Network: CW. Natl. Rep: Harrington, Righter & Parsons. Washington Atty: Fletcher, Heald & Hildreth.

Key Personnel:
Harry Pappas . . . . . CEO
Leopoldo L. Ramos . . . . . gen mgr
Bill May. . . . . rgnl sls mgr
Leslie Sadley . . . . . prom dir
Debbie Sweeney . . . . . progmg VP
Leslie SAdley . . . . . pub affrs dir
James Ocon . . . . . chief of engrg

KRNV— Analog channel: 4. Digital channel: 7. Analog hrs: 24 Digital hrs: 24 17.4 kw vis, 3.4 kw aur. 420t/92g TL: N39 35 03 W119 48 06 On air date: Sept 30, 1962. Box 7160, Reno, NV, 89510. 1790 Vassar St., Reno, NV 89502. Phone: (775) 322-4444. Fax: (775) 785-1208. Fax: (775) 785-1206. E-mail: comments@krnv.com Web Site: www.krnv.com. Licensee: Sierra Broadcasting Co. Ownership: James E. Rogers. Group owner: Sunbelt Communications Co. (acq 9-13-89). Population served: 172,863 Natl. Network: NBC. Washington Atty: Gerald S. Rourke. News staff: 26; News: 14 hrs wkly.

Key Personnel:
Mary Beth Farrell . . . . . gen mgr & stn mgr
John Finkbohner . . . . . opns mgr
Mark Murakami. . . . . rgnl sls mgr
Jon Killoran . . . . . news dir

KRXI-TV— Analog channel: 11. Digital channel: 44. Analog hrs: 24 Digital hrs: 24 178 kw vis, 17.8 kw aur. 2,808t/207g TL: N39 35 25 W119 55 40 On air date: Dec 3, 1995. 4920 Brookside Ct., Reno, NV, 89502. Phone: (775) 856-1100. Fax: (775) 856-1101. Web Site: www.foxreno.com. Licensee: KTVU Partnership. Natl. Network: Fox. Natl. Rep: TeleRep. Washington Atty: Dow, Lohnes & Albertson.

Key Personnel:
Steve Cmmings . . . . . gen mgr
Steve Cummings . . . . . gen sls mgr
Mandy Anderson . . . . . rgnl sls mgr
Mike Arnold . . . . . chief of engrg

KTVN— Analog channel: 2. Digital channel: 13. Analog hrs: 24 Digital hrs: 24 89.1 kw vis, 8.9 kw aur. 2,152t/187g TL: N39 15 29 W119 42 37 On air date: June 4, 1967. Box 7220, Reno, NV, 89510. 4925 Energy Way, Reno, NV 89502. Phone: (775) 858-2222. Fax: (775) 861-4298. E-mail: ktvn@ktvn.com Web Site: www.ktvn.com. Licensee: Sarkes Tarzian Inc. Group owner: (group owner; (acq 8-13-80; $12.5 million). Population served: 497,000 Natl. Network: CBS. Washington Atty: Leventhal, Senter & Lerman. News staff: 33; News: 19.5 hrs wkly.

Key Personnel:
Tom Tarzian. . . . . chmn
Tom Tolar . . . . . pres
Bob Davis. . . . . CFO
Lawson Fox . . . . . gen mgr
John Richardson . . . . . gen sls mgr
Sharon Facque. . . . . natl sls mgr
Ann Burns . . . . . prom mgr & pub affrs dir
Pat Hall . . . . . progmg dir
Jason Pasco . . . . . news dir
Jack Antonio . . . . . chief of engrg

KWNV— Analog channel: 7. Analog hrs: 24 890 w vis. 2,132t TL: N41 00 41 W117 45 59 On air date: 1998. c/o KRNV, 1790 Vassar St., Reno, NV, 89502. Phone: (775) 322-4444. Fax: (775) 785-1208. Licensee: Sierra Broadcasting Co. Natl. Network: NBC. Rebroadcasts KRNV Reno 100%.

New Hampshire

Concord

see Boston (Manchester, NH) market

Derry

see Boston (Manchester, NH) market

Durham

see Boston (Manchester, NH) market

Keene

see Boston (Manchester, NH) market

Littleton

see Portland-Auburn, ME market

Manchester

see Boston (Manchester, NH) market

Merrimack

see Boston (Manchester, NH) market

New Jersey

Atlantic City

see Philadelphia market

Burlington

see Philadelphia market

Camden

see Philadelphia market

Linden

see New York market

Montclair

see New York market

New Brunswick

see New York market

Newark

see New York market

Newton

see New York market

Paterson

see New York market

Secaucus

see New York market

Trenton

see Philadelphia market

Vineland

see Philadelphia market

West Milford

see New York market

Wildwood/Atlantic City

see Philadelphia market

New Mexico

Albuquerque-Santa Fe, NM (DMA 45)

KASA-TV— Analog channel: 2. Digital channel: 27. Analog hrs: 24 Digital hrs: 24 28.2 kw vis, 2.82 kw aur. 1,968t/178g TL: N35 46 50 W106 31 35 On air date: May 8, 1981. 13 Broadcast Plaza S.W., Albuquerque, NM, 87104. Phone: (505) 243-2285. Fax: (505) 248-1464. E-mail: kasa@kasa.com Web Site: www.kasa.com. Licensee: KASALicense Subsidiary, LLC. Group owner: Raycom Media Inc. (acq 9-2-99; grpsl). Population served: 1,664,000 Natl. Network: Fox. Natl. Rep: TeleRep. Washington Atty: Covington & Burling. News: 7 hrs wkly.

Key Personnel:
Bill Anderson . . . . . gen mgr
Jim Giudecess . . . . . natl sls mgr
Frank Montoya . . . . . rgnl sls mgr
Parker Harms. . . . . rgnl sls mgr & mktg dir
Michelle Donaldson . . . . . news dir

KASY-TV— Analog channel: 50. Digital channel: 45. Analog hrs: 24 Digital hrs: 24 1,450 kw vis. ant 4,153t TL: N35 12 45 W106 26 56 On air date: Oct 6, 1995. 8341 Washington St. N.E., Albuquerque, NM, 87113. Phone: (505) 797-1919. Fax: (505) 938-4401. Web Site: www.my50.tv. Licensee: Acme Television Licenses of New Mexico L.L.C. Group owner: Acme Communications Inc. (acq 6-18-99; $25.4 million). Natl. Network: MyNetworkTV. Washington Atty: Leventhal, Senter & Lerman.

Key Personnel:
Stan Gill . . . . . gen mgr
Dan Marchese . . . . . gen sls mgr
Chris Iller . . . . . prom mgr
Larry Oliver . . . . . chief of engrg

***KAZQ**— Analog channel: 32. Digital channel: 17. Analog hrs: 24 Digital hrs: 24 263 kw vis. ant 4,090t/110g TL: N35 12 51 W106 27 01 On air date: Oct 12, 1987. 4501 Montgomery Blvd. N.E., Albuquerque, NM, 87109. Phone: (505) 884-8355. Fax: (505) 883-1229. E-mail: kazq32@kazq32.org Web Site: www.kazq32.org. Licensee: Alpha-Omega Broadcasting of Albuquerque Inc. Population served: 800,000 Washington Atty: Midlen Law Center. Foreign lang progmg: SP 2

Key Personnel:
Brenton Franks . . . . . gen mgr & opns VP
Jeffrey Helmers . . . . . gen mgr, dev dir & mktg mgr
Howard Holley . . . . . progmg dir

KBIM-TV— Analog channel: 10. Digital channel: 41. Analog hrs: 24 Digital hrs: 24 316 kw vis, 40.7 kw aur. ant 1,999t/1,839g TL: N33 03 20 W103 49 12 On air date: Feb 26, 1966. Box 910, Roswell, NM, 88201. 214 N. Main St., Roswell, NM 88201. Phone: (505) 622-2120. Fax: (505) 623-6606. Web Site: www.kbimtv.com. Licensee: LIN of New Mexico LLC. Group owner: Emmis Communications Corp. (acq 11-30-2005; grpsl). Population served: 300,000 Natl. Network: CBS. Washington Atty: Reed, Smith, Shaw & McClay. News staff: 11; News: 6 hrs wkly.

Key Personnel:
Gene Munsey . . . . . VP, gen mgr & gen sls mgr
Marcus Damberger . . . . . opns mgr
Rebroadcasts KRQE(TV) Albuquerque 90%.

KCHF— Analog channel: 11. Digital channel: 10. Analog hrs: 24 Digital hrs: 24 263 kw vis, 26.3 kw aur. 2,027t/272g TL: N35 47 15 W106 31 35 On air date: Jan 21, 1984. Box 4338, Albuquerque, NM, 87196. 27556 I 25 &. Frontage Rd., Santa Fe, NM 87508. Phone: (505) 345-1991 (radio). Phone: (505) 473-1111. Fax: (505) 345-5669. Web Site: www.kchf.com. Licensee: Son Broadcasting Inc. Population served: 660,000 Washington Atty: Gammon & Grange. Foreign lang progmg: SP 4

Key Personnel:
Belarmino R. Gonzalez . . . . . CEO, chmn & pres
Annette Garcia . . . . . gen mgr
Mary Kay Gonzales . . . . . progmg dir
Rob Ramseyer . . . . . chief of engrg

KLUZ-TV— Analog channel: 41. Digital channel: 42. Analog hrs: 24 Digital hrs: 24 1,200 kw vis. ant 4,120t/118g TL: N35 12 41 W106 26 56 On air date: September 1987. 2725 Broadbent Pkwy. N.E., Suite E, Albuquerque, NM, 87107. Phone: (505) 342-4141. Phone: (505) 344-5589. Fax: (505) 344-8714. Web Site: www.univision.com. Licensee: Entravision Holdings L.L.C. Group owner: Entravision Communications Co. L.L.C. (acq 3-21-99). Population served: 819,330 Natl. Network: Univision (Spanish). Foreign lang progmg: SP 168 News staff: 10; News: 1/2 hr wkly.

Key Personnel:
Walter Ulloa . . . . . CEO
Phillip Wilkinson . . . . . pres
John DiLorenzo . . . . . CFO
Margarita Wilder . . . . . gen mgr & gen sls mgr

Kambiz Victory . . . . chief of engrg

KNAT— Analog channel: 23. Digital channel: 24. Analog hrs: 24 Digital hrs: 24 1,200 kw vis, 120 kw aur. 4,130t/128g TL: N35 12 54 W106 27 02 On air date: Oct 17, 1975. 1510 Coors Rd. N.W., Albuquerque, NM, 87121. Phone: (505) 836-6585. Fax: (505) 831-8725. E-mail: cmansfield@tbn.org Web Site: www.tbn.org. Licensee: Trinity Broadcasting Network. Group owner: (group owner; (acq 5-8-2000; grpsl). Population served: 800,000 Natl. Network: NBC. Washington Atty: Joseph E. Dunne III. Foreign lang progmg: SP 3

***KNMD-TV—** Analog channel: 0. Digital channel: 9. Digital hrs: 24 200 w vis. ant 4,070t/46g TL: N35 12 45 W106 26 58 On air date: Sept 12, 2004. 1130 University Blvd. N.E., Albuquerque, NM, 87102. Phone: (505) 277-2121. Fax: (505) 277-2191. E-mail: viewer@knme.org Web Site: www.knmetv.org. Licensee: The Regents of the University of New Mexico. Population served: 857,000 Natl. Network: PBS. Washington Atty: Dow, Lohnes & Albertson, LLC.

Key Personnel:

Ted A. Garcia . . . . CEO & gen mgr
Jim Gale . . . . engrg dir

***KNME-TV—** Analog channel: 5. Digital channel: 35. Analog hrs: 24 Digital hrs: 24 26.9 kw vis, 5.6 kw aur. ant 4,228t/172g TL: N35 12 44 W106 26 57 On air date: May 5, 1958. 1130 University Blvd. N.E., Albuquerque, NM, 87102. Phone: (505) 277-2121. Fax: (505) 277-2191. Web Site: www.knmetv.org. Licensee: Regents of University of New Mexico and Board of Education, Albuquerque. Population served: 945,000 Natl. Network: PBS. Washington Atty: Dow, Lohnes & Albertson. News staff: 4; News: one hr wkly.

Key Personnel:

Ted A. Garcia . . . . CEO & gen mgr
Joanne Bachmann . . . . dev dir
Jim Gale . . . . engrg dir

KOAT-TV— Analog channel: 7. Digital channel: 21. Analog hrs: 24 Digital hrs: 24 87.1 kw vis, 17.4 kw aur. 4,240t/233g TL: N35 12 53 W106 27 01 On air date: Sept 28, 1953. Box 25982, Albuquerque, NM, 87125. 3801 Carlisle N.E., Albuquerque, NM 87107. Phone: (505) 884-7777. Fax: (505) 884-6282. E-mail: koatdesk@hearst.com Web Site: www.koat.com. Licensee: KOAT Hearst-Argyle Television Inc. Group owner: Hearst-Argyle Television Inc. (acq 3-18-99; grpsl). Population served: 543,751 Natl. Network: ABC. Washington Atty: Brooks, Pierce, McLendon, Humphrey & Leonard. Wire Svc: UPI

KOBF— Analog channel: 12. Analog hrs: 24 316 kw vis, 31.6 kw aur. 410t/209g TL: N36 41 43 W108 13 14 On air date: 1972. Box 1620, Farmington, NM, 87499. 825 W. Broadway, Farmington, NM 87401. Phone: (505) 326-1141. Fax: (505) 327-5196. Web Site: www.kob.com. Licensee: KOB-TV L.L.C. Group owner: Hubbard Broadcasting Inc. (acq 9-19-83; $2.35 million; FTR: 8-15-83). Population served: 109,000 Natl. Network: NBC. Washington Atty: Fletcher, Heald & Hildreth. News staff: 6; News: 4 hrs wkly.
Satelite of KOB-TV Albuquerque.

KOBG-TV— Analog channel: 6. Analog hrs: 24 Digital hrs: 24 6 kw vis. ant 1,647t/190g TL: N32 51 49 W108 14 27 On air date: 2001. Box 1351, Albuquerque, NM, 87103. Phone: (505) 243-4411. Fax: (505) 764-2522. E-mail: kobtv@kob.com Web Site: www.kob.com. Licensee: KOB-TV LLC. Natl. Network: NBC.
Satellite of KOB-TV Albuquerque.

KOBR— Analog channel: 8. Digital channel: 38. Analog hrs: 24 Digital hrs: 24 316 kw vis, 52.5 kw aur. ant 1,748t/1,542g TL: N33 22 31 W103 46 12 On air date: June 24, 1953. 124 E. 4th St., Roswell, NM, 88201. Phone: (505) 625-8888. Fax: (505) 625-8866. E-mail: kobtv@kob.com Web Site: www.kob.com. Licensee: Stanley S. Hubbard Revocable Trust. Group owner: Hubbard Broadcasting Inc. (acq 8-10-2001). Population served: 307,000 Natl. Network: NBC. Washington Atty: Fletcher, Heald & Hildreth. News staff: 4.

Key Personnel:

Stanley S. Hubbard . . . . pres
Charlie Blanco . . . . gen mgr
Dusty Deane . . . . prom mgr
Wayne Koontz . . . . chief of engrg
Rebroadcasts KOB-TV Albuquerque 90%.

KOB-TV— Analog channel: 4. Digital channel: 26. Analog hrs: 24 Digital hrs: 24 26.9 kw vis, 2.7 kw aur. ant 4,198t/178g TL: N35 12 42 W106 26 57 On air date: Nov 29, 1948. Box 1351, Albuquerque, NM, 87103. 4 Broadcast Plaza S.W., Albuquerque, NM 87104. Phone: (505) 243-4411. Fax: (505) 764-2522. E-mail: kobtv@kob.com Web Site: www.kob.com. Licensee: KOB-TV L.L.C. Group owner: Hubbard Broadcasting Inc. (acq 3-15-57; grpsl; FTR: 3-18-57). Population served: 568,700 Natl. Network: NBC. Washington Atty: Fletcher, Heald & Hildreth. Wire Svc: UPI News: 12 hrs wkly.

Key Personnel:

Mike Burgess . . . . VP
Susan Raybon . . . . gen mgr
Susan Connor. . . . . stn mgr & opns mgr
Jeff Finkel. . . . . gen sls mgr & natl sls mgr
Vince Gasparich. . . . . prom dir
Juanita Garay . . . . progmg mgr
Rhonda Aubrey . . . . news dir
Joan Lucas . . . . pub affrs dir
Sean Anker . . . . engrg dir & chief of engrg

KOCT— Analog channel: 6. Digital channel: 19.100 kw vis, 10 kw aur. 1,201t/1,048g TL: N32 47 39 W104 12 27 On air date: August 1959. Box 25982, Albuquerque, NM, 87125. 3801 Carlisle N.E., Albuquerque, NM 87107. Phone: (505) 884-7777. Fax: (505) 884-6282. E-mail: koatdesk@hearst.com Web Site: www.koat.com. Licensee: KOAT Hearst-Argyle Television Inc. Group owner: Hearst-Argyle Television Inc. (acq 3-18-99; grpsl). Natl. Network: ABC. Washington Atty: Verner, Liipfert, Bernhard, McPherson & Hand.
Satellite of KOAT-TV Albuquerque.

KOFT— Analog channel: 3. Digital channel: 8. Analog hrs: 24 Digital hrs: 24 100 kw vis. ant 413t/269g TL: N36 41 48 W108 10 39 Not on air, target date: unknown: Box 25982, Albuquerque, NM, 87125. 3801 Carlisle N.E., Albuquerque, NM 87107. Phone: (505) 884-7777. Fax: (505) 884-6282. E-mail: koatdesk@hearst.com Web Site: www.koat.com. Permittee: KOAT Hearst-Argyle Television Inc. Natl. Network: ABC. Washington Atty: Verner, Liipfert, Bernhard, McPherson & Hand.

KOVT— Analog channel: 10. Digital channel: 12. Analog hrs: 24 Digital hrs: 24 8.71 kw vis, 871 w aur. 1,591t/112g TL: N32 51 46 W108 14 28 On air date: Sept 9, 1987. Box 25982, c/o KOAT-TV, Albuquerque, NM, 87125. 3801 Carlisle N.E., Albuquerque, NM 87107. Phone: (505) 884-7777. Fax: (505) 884-6282. E-mail: koatdesk@hearst.com Web Site: www.koat.com. Licensee: KOAT Hearst-Argyle Television Inc. Group owner: Hearst-Argyle Television Inc. (acq 3-18-99; grpsl). Population served: 550,000 Natl. Network: ABC. Washington Atty: Verner, Liipfert, Bernhard, McPherson & Hand.
Satellite of KOAT-TV Albuquerque.

KREZ-TV— Analog channel: 6. Digital channel: 15.6.17 kw vis. ant 361t/141g TL: N37 15 46 W107 53 58 On air date: Sept 4, 1965. 158 Bodo Dr., Durango, CO 81303. Phone: (970) 259-6666. Fax: (970) 247-8472. Licensee: LIN of Colorado LLC. Group owner: Emmis Communications Corp. (acq 11-30-2005; grpsl). Population served: 66,000 Natl. Network: CBS, NBC.

***KRMU—** Analog channel: 20. Digital channel: 20. Digital hrs: 24 8.51 kw vis. ant 308t/46g TL: N37 15 45 W107 54 07 On air date: Jan 2005. 1089 Bannock St., Denver, CO, 80204. Phone: (303) 892-6666. Fax: (303) 620-5600. Web Site: www.rmpbs.org. Licensee: Rocky Mountain Public Broadcasting Network, Inc. Permittee: Rocky Mountain Public Broadcasting Network Inc. Natl. Network: PBS. Washington Atty: Dow, Lohnes & Albertson.

Key Personnel:

James N. Morgese . . . . pres & gen mgr
Suzanne Banning. dev dir
Donna Sanford . . . . progmg dir
John Anderson . . . . engrg dir
Rebroadcasts KRMJ-TV Grand Junction 100%.

KRPV— Analog channel: 27. Analog hrs: 24 871 kw vis, 67 kw aur. 377t/500g TL: N33 24 58 W104 33 59 On air date: Sept 15, 1986. Box 61000, Midland, TX, 79711. Box 967, 2606 S. Main, Roswell, NM 88203-0967. Phone: (800) 707-0420. Fax: (505) 622-3424. E-mail: info@ptcbglc.com Web Site: www.godslearningchannel.com. Licensee: Prime Time Christian Broadcasting. Population served: 272,826 Foreign lang progmg: SP 4

Key Personnel:

Al Cooper. . . . . CEO, pres & gen mgr
Tommy Cooper . . . . VP
Jeff Tveit. . . . . stn mgr & gen sls mgr

KRQE— Analog channel: 13. Digital channel: 16. Analog hrs: 24 Digital hrs: 24 89.1 kw vis, 9 kw aur. ant 4,178t/141g TL: N35 12 40 W106 26 57 On air date: Oct 3, 1953. 13 Broadcast Plaza S.W., Albuquerque, NM, 87104. Phone: (505) 243-2285. Fax: (505) 248-1464. E-mail: kresoundoff@krge.com Web Site: www.krqe.com. Licensee: LIN of New Mexico LLC. Group owner: Emmis Communications Corp. (acq 11-30-2005; grpsl). Population served: 1,710,000 Natl. Network: CBS. Natl. Rep: Harrington, Righter & Parsons. Washington Atty: Reed, Smith, Shaw & McClay. Wire Svc: CBS News staff: 65; News: 24 hrs wkly.

Key Personnel:

Bill Anderson . . . . gen mgr
Gina Galindo . . . . opns dir
Mary Lou Davis . . . . natl sls mgr
Frank Montoya. . . . . rgnl sls mgr
Parker Harms. . . . . mktg dir
Don Pierce . . . . progmg dir & progmg mgr
Michelle Donaldson . . . . news dir
Frank Lilley . . . . engrg dir

KRWB-TV— Analog channel: 24.5,000 kw vis. ant 420t/449g TL: N33 06 01 W104 15 15 On air date: 2004. 8341 Washington St. N.E., Albuquerque, NM, 87113. Phone: (505) 797-1919. Fax: (505) 938-4401. Licensee: Acme Television Licenses of New Mexico LLC. Group owner: ACME Communications Inc. (acq 1-7-2004).

Key Personnel:

Stan Gill. . . . . gen mgr
Larry Oliver . . . . chief of engrg

KTEL-TV— Analog channel: 25. Digital channel: 39. Analog hrs: 24 Digital hrs: 24 100 kw vis. ant 440t TL: N32 26 09 W104 11 14 On air date: 2001. Box 30068, Albuquerque, NM, 87190. 2400 Monroe St. N. E., Albuquerque, NM 87110. Phone: (505) 884-5353. Fax: (505) 889-8390. E-mail: gzavala@kteltv.com Licensee: Ramar Communications II Ltd. Group owner: (group owner; (acq 8-10-99; $10,000). Natl. Network: Telemundo (Spanish). Foreign lang progmg: SP 168

Key Personnel:

Ray Moran . . . . CEO
Brad Moran . . . . pres
Gabriel Zavala . . . . gen mgr

KTFQ-TV— Analog channel: 14. Analog hrs: 24 5,000 kw vis. ant 1,233t/1,046g TL: N35 24 44 W106 43 32 On air date: Apr 28, 1999. 2725 Broadbent Pkwy. N.E., Suite E, Albuquerque, NM, 87107. Phone: (505) 342-4141. Fax: (505) 344-8714. Web Site: www.univision.com. Licensee: TeleFutura Albuquerque LLC. Group owner: Univision Communications Inc. (acq 5-30-2003; $20 million). Natl. Network: TeleFutura (Spanish). Foreign lang progmg: SP 168

KTLL-TV— Analog channel: 33. Analog hrs: 24 166 kw vis. ant 400t/66g TL: N37 15 46 W107 53 45 On air date: 2002. Box 3757, Ramar Communications, Lubbock, TX, 79452. Phone: (806) 745-3434. Fax: (806) 748-1949. Web Site: www.fox34.com. Licensee: Ramar Communications II Ltd. Group owner: (group owner; (acq 1-24-2001). Natl. Network: Telemundo (Spanish). Foreign lang progmg: SP 168

Key Personnel:

Brad Moran . . . . pres & gen mgr
Gabriel Zavala . . . . gen mgr

KUPT— Analog channel: 29. Digital channel: 16. Analog hrs: 24 75.9 kw vis. ant 522t/502g TL: N32 43 28 W103 05 46 On air date: 1989. Box 3757, Lubbock, TX, 79452. 9800 University Ave., Lubbock, TX 77551-5556. Phone: (806) 745-3434. Fax: (806) 748-1949. Licensee: Ramar Communications II Ltd. Group owner: (group owner; (acq 6-4-97; $200,000). Population served: 30,000 Natl. Network: UPN. Washington Atty: Leventhal, Senter & Lerman.

Key Personnel:

Brad Moran . . . . pres
Chuck Hinez . . . . gen mgr
Scott Cawthron. opns mgr
Terri Holt. . . . . progmg dir

KWBQ— Analog channel: 19. Digital channel: 29. Analog hrs: 24 Digital hrs: 24 5,000 kw vis. ant 1,971t TL: N35 46 50 W106 31 35 On air date: December 1998. 8341 Washington St. N.E., Albuquerque, NM, 87113. Phone: (505) 797-1919. Fax: (505) 344-1145. Web Site: www.newmexicoscw.tv. Licensee: Acme TV Licenses of New Mexico L.L.C. Group owner: ACME Communications Inc. Natl. Network: CW.

Key Personnel:

Stan Gill. . . . . gen mgr
Dan Marchese . . . . gen sls mgr
Chris Iller . . . . prom mgr
Larry Oliver . . . . chief of engrg

Clovis

see Amarillo, TX market

Las Cruces

see El Paso (Las Cruces, NM), TX market

Portales

see Amarillo, TX market

Santa Fe

see Albuquerque-Santa Fe, NM market

New York

Albany-Schenectady-Troy, NY (DMA 56)

WCDC— Analog channel: 19. Digital channel: 36. Analog hrs: 24 Digital hrs: 24 538 kw vis, 53 kw aur. 3,688t/248g TL: N42 38 14 W73 10 07 On air date: Feb 5, 1954. 341 Northern Blvd., Albany, NY, 12204. Phone: (518) 436-4822. Fax: (518) 462-6065. E-mail: news@news10.com Web Site: www.wten.com. Licensee: Young Broadcasting of Albany Inc. Group owner: Young Broadcasting Inc. (acq 10-11-89; grpsl; FTR: 9-11-89). Natl. Network: ABC. Washington Atty: Wiley, Rein & Fielding. News staff: 37; News: 15 hrs wkly. Satellite of WTEN-TV Albany.

WCWN— Analog channel: 45. Digital channel: 43. Analog hrs: 24 Digital hrs: 24 2,950 kw vis. ant 1,109t TL: N42 37 37 W74 00 40 On air date: Sept 27, 1993. 1400 Balltown Rd., Schenectady, NY, 12309. Phone: (518) 346-6666. Fax: (518) 381-3770. Web Site: www.capitalregionscw.com. Licensee: Freedom Broadcasting of New York Licensee L.L.C. Group owner: Tribune Broadcasting Co. (acq 12-5-2006; $17 million). Natl. Network: CW.
Key Personnel:
Robert Furlong . . . . . VP & gen mgr
Fred Lass . . . . . chief of engrg
Nicole Parianos . . . . . mktg

***WMHT**— Analog channel: 17. Digital channel: 34. Analog hrs: 24 Digital hrs: 24 2,000 kw vis, 200 kw aur. ant 983t/271g TL: N42 38 13 W74 00 06 On air date: May 2, 1962. 4 Global View, Troy, NY, 12180. Phone: (518) 880-3400. Fax: (518) 880-3409. E-mail: email@wmht.org Web Site: www.wmht.org. Licensee: WMHT Educational Telecommunications. Population served: 600,000 Natl. Network: PBS. Rgnl. Network: Eastern Educ. Washington Atty: Schwartz, Woods & Miller. Foreign lang progmg: SP 1

WNYA— Analog channel: 51. Analog hrs: 24 1,580 kw vis. ant 1,000t/141g TL: N42 30 09 W73 18 58 On air date: 9/2003. 1400 Balltown Rd., Schenectady, MA, 12309. Phone: (518) 381-3751. Fax: (518) 381-3740. Web Site: www.mytv4albany.com. Licensee: Venture Technologies Group LLC. Ownership: Lawrence H. Rogow (acq 7-30-2003). Natl. Network: MyNetworkTV. Natl. Rep: TeleRep. Washington Atty: Wiley, Rein LLP.

WNYT— Analog channel: 13. Digital channel: 12.178 kw vis, 19.9 kw aur. 1,171t/737g TL: N42 47 08 W73 37 44 On air date: June 15, 1956. Box 4035, 715 N. Pearl St., Albany, NY, 12204. Phone: (518) 436-4791. Phone: (518) 207-4700. Web Site: www.wnyt.com. Licensee: WNYT-TV LLC. Group owner: Hubbard Broadcasting Inc. (acq 9-19-96). Population served: 1,245,000 Natl. Network: NBC. Natl. Rep: Petry Television Inc. News: 27 hrs wkly.
Key Personnel:
Stephen Baboulis . . . . . gen mgr
Tony McManus . . . . . gen sls mgr
Paul Lewis . . . . . news dir
Richard Klein . . . . . engrg dir
Maryann Ryan . . . . . progmg

WRGB—Schenectady, Analog channel: 6. Digital channel: 39. Analog hrs: 24 Digital hrs: 24 93.3 kw vis, 11 kw aur. 1,020t/314g TL: N42 38 13 W73 59 45 On air date: Jan 13, 1928. 1400 Balltown Rd., Schenectady, NY, 12309. Phone: (518) 346-6666. Fax: (518) 381-3707 (progm). Fax: (518) 381-3736 (gen mgr). Web Site: www.wrgb.com. Licensee: Freedom Broadcasting of New York Licensee L.L.C. Group owner: Freedom Communications Inc. (acq 3-4-86; FTR: 11-25-85). Population served: 1,280,600 Natl. Network: CBS. Natl. Rep: TeleRep. Washington Atty: Latham & Watkins. Wire Svc: AP News: 26.5 hrs wkly.
Key Personnel:
Scott Flanders . . . . . CEO
Doreen Wade . . . . . pres
Robert J. Furlong . . . . . VP, gen mgr & progmg mgr
Matt Sames . . . . . gen sls mgr
Robert Hewitt . . . . . natl sls mgr
Lisa Jackson . . . . . prom mgr
Beau Duffy . . . . . news dir
Fred Lass . . . . . engrg dir & chief of engrg

WTEN— Analog channel: 10. Digital channel: 26. Analog hrs: 24 Digital hrs: 24 316 kw vis, 31.6 kw aur. ant 1,000t/276g TL: N42 38 15 W73 59 54 On air date: Oct 14, 1953. 341 Northern Blvd., Albany, NY, 12204. Phone: (518) 436-4822. Fax: (518) 462-6065. E-mail: news@news10.com Web Site: www.wten.com. Licensee: Young Broadcasting of Albany Inc. Group owner: Young Broadcasting Inc. (acq 10-11-89; grpsl). Population served: 1,258,000 Natl. Network: ABC. Washington Atty: Wiley, Rein & Fielding. News staff: 50; News: 22 hrs wkly.

WXXA-TV— Analog channel: 23. Digital channel: 7. Analog hrs: 24 Digital hrs: 23 3,020 kw vis, 302 kw aur. ant 1,200t/465g TL: N42 37 01 W74 00 46 On air date: July 30, 1982. 28 Corporate Cir., Albany, NY, 12203. Phone: (518) 862-2323. Phone: (518) 862-0995. Fax: (518) 862-0865. Fax: (518) 862-0930. E-mail: news@fox23news.com Web Site: www.fox23news.com. Licensee: Clear Channel Broadcasting Inc. Group owner: Clear Channel Communications Inc. (acq 12-1-94; $25.5 million). Population served: 1,288,000 Natl. Network: Fox. Natl. Rep: Millennium Sales & Marketing. Washington Atty: Wiley, Rein & Fielding. News staff: 70; News: 23.5 hrs wkly.
Key Personnel:
Lowry Mays . . . . . CEO
Bill Moll . . . . . pres & exec VP
Steve Kimatian . . . . . exec VP
Jeffrey Whitson . . . . . gen mgr
Chuck Hunt . . . . . sls dir
Ardelle Hirsch . . . . . natl sls mgr & mktg
Paul Pelliccia . . . . . progmg dir
Gene Ross . . . . . news dir
Sargent Cathrall . . . . . chief of engrg

WYPX— Analog channel: 55. Digital channel: 50. Analog hrs: 24 Digital hrs: 24 5,000 kw vis, 500 kw aur. ant 722t/720g TL: N42 59 04 W74 10 56 On air date: Dec 14, 1987. 1 Charles Blvd, Guilderland, NY, 12084. Phone: (518) 464-0143. Fax: (518) 464-0633. Web Site: www.ionline.tv. Licensee: Channel 55 of Albany Inc. Group owner: Paxson Communications Corp. (acq 6-1-96; $2.5 million). Natl. Network: i Network. Natl. Rep: NBC TV Stations Sales. News: 5 hrs wkly.
Key Personnel:
Dean Goodman . . . . . CEO
Lowell "Bud" Paxson . . . . . chmn
Renee Osterlitz . . . . . stn mgr
Steve Appel . . . . . gen sls mgr
Claude Pine . . . . . pub affrs dir & chief of engrg

Binghamton, NY (DMA 157)

WBNG-TV— Analog channel: 12. Digital channel: 7. Analog hrs: 23 Digital hrs: 23 166 kw vis, 18.2 kw aur. ant 1,210t/785g TL: N42 02 33 W75 57 06 On air date: Dec 1, 1949. 560 Columbia Dr., Johnson City, NY, 13790. Phone: (607) 729-8812. Fax: (607) 797-6211. E-mail: wbng@wbngtv.com Web Site: www.wbng.com. Licensee: WBNG License Inc. Group owner: Television Station Group LLC (acq 7-26-2006; $45 million). Population served: 154,400 Natl. Network: CBS, CW. Natl. Rep: Continental Television Sales. Washington Atty: Latham & Watkins. Wire Svc: AP News: 24.5 hrs wkly.
Key Personnel:
Bob Krummunecker . . . . . gen sls mgr
Kate Garger . . . . . progmg dir
Greg Catlin . . . . . news dir
Mike Calkins . . . . . chief of engrg

WICZ-TV— Analog channel: 40. Digital channel: 8.468 kw vis, 46.8 kw aur. 1,230t/934g TL: N42 03 22 W75 56 39 On air date: Nov 1, 1957. 4600 Vestal Pkwy. E., Vestal, NY, 13850. Phone: (607) 770-4040. Fax: (607) 798-7950. E-mail: fox40@wicz.com Web Site: www.wicz.com. Licensee: Stainless Broadcasting L.P. Group owner: Northwest Broadcasting Inc. (acq 7-15-97; $16 million cash-out merger with KTVZ(TV) Bend, OR). Population served: 337,000 Natl. Network: Fox. Washington Atty: Leventhal, Senter & Lerman. News staff: 13; News: 2 hrs wkly.
Key Personnel:
Brian Brady . . . . . CEO
Bill Quarles . . . . . CFO
John Leet . . . . . gen mgr

WIVT— Analog channel: 34. Digital channel: 4. Analog hrs: 22 Digital hrs: 22 2,820 kw vis. ant 928t/528g TL: N42 03 39 W75 56 36 On air date: Nov 25, 1962. 203 Ingraham Hill Rd., Binghamton, NY, 13903. Phone: (607) 771-3434. Fax: (607) 723-1034. Web Site: www.NewsChannel34.com. Licensee: Central NY News Inc. Group owner: Clear Channel Communications Inc. (acq 6-14-2002; grpsl). Population served: 400,000 Natl. Network: ABC. Natl. Rep: Millennium Sales & Marketing.
Key Personnel:
Steve Kimatian . . . . . pres
John Birchall . . . . . VP & gen mgr
John King . . . . . opns VP, engrg VP & chief of engrg
Abiodun Sadik . . . . . chief of opns
Maura Burtis . . . . . natl sls mgr & rgnl sls mgr
Jim La Vasser . . . . . prom mgr
Vince Spacolia . . . . . progmg dir
Jim Ehmke . . . . . news dir

***WSKG-TV**— Analog channel: 46. Digital channel: 42. Analog hrs: 24 Digital hrs: 24 603 kw vis, 60.3 kw aur. ant 1,230t/927g TL: N42 03 22 W75 56 39 On air date: May 12, 1968. Box 3000, Binghamton, NY, 13902. 601 Gates Rd., Vestal, NY 13850. Phone: (607) 729-0100. Fax: (607) 729-7328. E-mail: wskg_mail@pbs.org Web Site: www.wskg.org. Licensee: WSKG Public Telecommunications Council. Natl. Network: PBS. Rgnl. Network: Eastern Educ. Washington Atty: Dow, Lohnes & Albertson.
Key Personnel:
Brian Sicora . . . . . CEO, pres & gen mgr
Nancy Christensen . . . . . opns dir & gen sls mgr

Buffalo, NY (DMA 49)

WGRZ-TV— Analog channel: 2. Digital channel: 33.100 kw vis, 20 kw aur. ant 941t/899g TL: N42 43 06 W73 22 48 On air date: Aug 14, 1954. 259 Delaware Ave., Buffalo, NY, 14202. Phone: (716) 849-2222. Fax: (716) 849-7602. Web Site: www.wgrz.com. Licensee: Multimedia Entertainment Inc. Group owner: Gannett Broadcasting (acq 1-27-97; grpsl). Population served: 1,325,500 Natl. Network: NBC. Wire Svc: Newsweek Wire Svc: CNBC Wire Svc: NBC Wire Svc: UPI

WIVB-TV— Analog channel: 4. Digital channel: 39. Analog hrs: 24 Digital hrs: 24 100 kw vis, 20 kw aur. 1,201t/1,060g TL: N42 39 33 W78 37 33 On air date: May 14, 1948. 2077 Elmwood Ave., Buffalo, NY, 14207. Phone: (716) 874-4410. Fax: (716) 879-4896. Web Site: www.wivb.com. Licensee: WIVB Broadcasting L.L.C. Group owner: LIN Television Corporation (acq 12-16-97). Population served: 1,200,00 Natl. Network: CBS. Washington Atty: Covington & Burling.
Key Personnel:
Dan Meyers . . . . . gen mgr, mktg mgr & adv mgr
Diane Breen . . . . . progmg dir
Dennis Majewicz . . . . . chief of engrg

WKBW-TV— Analog channel: 7. Digital channel: 38. Analog hrs: 24 Digital hrs: 24 100 kw vis, 18.2 kw aur. 1,420t/1,076g TL: N42 38 15 W78 37 12 On air date: Nov 30, 1958. 7 Broadcast Plaza, Buffalo, NY, 14202. Phone: (716) 845-6100. Fax: (716) 842-1855. Fax: TWX: 710-522-1846. Web Site: www.wkbw.com. Licensee: Granite Broadcasting Corp. Group owner: (group owner; (acq 1995; $13.42 million). Population served: 1,720,000 Natl. Network: ABC. Natl. Rep: TeleRep. Washington Atty: Akin, Gump, Strauss, Haver & Feld.
Key Personnel:
William Ransom . . . . . gen mgr
Mike Anger . . . . . chief of engrg

***WNED-TV**— Analog channel: 17. Digital channel: 43. Analog hrs: 24 Digital hrs: 24 2,510 kw vis, 251 kw aur. 1,082t/1,091g TL: N43 01 48 W78 55 15 On air date: Mar 30, 1959. Box 1263, Buffalo, NY, 14240. Horizons Plaza, 140 Lower Terr., Buffalo, NY 14202. Phone: (716) 845-7000. Fax: (716) 845-7036. Web Site: www.wned.org. Licensee: Western New York Public Broadcasting Association. Population served: 8,000,000 Natl. Network: PBS. Rgnl. Network: Eastern Educ. Washington Atty: Schwartz, Woods & Miller.
Key Personnel:
Donald Boswell . . . . . CEO & pres
Michael Trapper . . . . . CFO, gen sls mgr, mktg mgr & adv mgr
Richard Daly . . . . . sr VP
Ron Santora . . . . . progmg dir

WNGS— Analog channel: 67. Analog hrs: 24 34.7 kw vis, 3.47 kw aur. ant 469t/161g TL: N42 24 16 W78 39 55 On air date: 1997. 7 Broadcast Plaza, Buffalo, NY, 14202. 1 Shackleford Dr., Suite 400, Little Rock, AR 72211. Phone: (716) 942-3000. Phone: (501) 219-2400. Fax: (716) 942-3010. Fax: (501) 221-1101. Licensee: EBC Buffalo Inc. Group owner: Equity Broadcasting Corp. (acq 6-25-2004; $5 million). Population served: 500,000 Natl. Network: UPN.

WNLO— Analog channel: 23. Digital channel: 32. Analog hrs: 24 Digital hrs: 24 955 kw vis, 95.5 kw aur. ant 1,030t/1,039g TL: N43 01 48 W78 55 15 On air date: May 13, 1987. 2077 Elmwood Ave., Buffalo, NY, 14207. Phone: (716) 874-4410. Fax: (716) 879-4896. Web Site: www.wivb.com. Licensee: WIVB Broadcasting L.L.C. Group owner: LIN Television Corporation (acq 6-6-2001; $26.2 million). Population served: 1,200,000 Natl. Network: CW. Washington Atty: Covington & Burling.
Key Personnel:
Dan Meyers . . . . . mktg mgr & adv mgr
Diane Breen . . . . . progmg dir
Dennis Majewicz . . . . . chief of engrg

WNYB— Analog channel: 26. Digital channel: 42. Analog hrs: 24 Digital hrs: 24 5,000 kw vis, 500 kw aur. 1,519t/1,059g TL: N42 23 36 W79 13 44 On air date: Sept 24, 1988. 5775 Big Tree Rd., Orchard Park, NY, 14127. Phone: (716) 662-2659. Fax: (716) 667-2499. E-mail: wnyb@tct.tv Web Site: www.tct.tv. Licensee: Faith Broadcasting Network Inc. Group owner: Tri-State Christian Television (acq 5-7-2002).

WNYO-TV— Analog channel: 49. Digital channel: 34. Analog hrs: 24 Digital hrs: 24 4,900 kw vis, 414 kw aur. ant 1,233t/1,047g TL: N42 46 58 W78 27 28 On air date: Sept 2, 1987. 699 Hertel Ave., Suite 100, Buffalo, NY, 14207. Phone: (716) 447-3200. Fax: (716) 875-4919.

Web Site: www.mytvbuffalo.com. Licensee: New York Television Inc. Group owner: Sinclair Broadcast Group Inc. (acq 1-25-2002; $51.5 million for stock). Population served: 645,000 Natl. Network: MyNetworkTV.
Key Personnel:
Nick Maginni . . . . . gen mgr
Donald Stewart . . . . . opns dir & engrg dir
Jose Chapa . . . . . gen sls mgr
Mike Bice . . . . . prom mgr

WPXJ-TV— Analog channel: 51.708 kw vis. ant 1,456t TL: N42 53 42 W78 00 56 On air date: 2000. 601 Clearwater Park Rd., West Palm Beach, FL, 33401. 259 Delaware Ave., Buffalo, NY 14202. Phone: (716) 852-1818. Phone: (716) 849-1818. Fax: (716) 852-8288. Web Site: www.ionline.tv. Licensee: Paxson Buffalo License Inc. (acq 7-15-97; $3 million). Natl. Network: i Network.
Key Personnel:
Lou Friefeld. . . . . VP
Ken Beedle . . . . . gen mgr
Barb Lipka . . . . . opns mgr

WUTV— Analog channel: 29. Digital channel: 14. Analog hrs: 24 Digital hrs: 14 1,050 kw vis, 105 kw aur. 920t/959g TL: N43 01 32 W78 55 43 (CP: 3,980 kw vis, 398 kw aur) On air date: Dec 21, 1970. 699 Hertel Ave., Suite 100, Buffalo, NY, 14207. Phone: (716) 477-3200. Fax: (716) 875-4919. Web Site: www.wutv.com. Licensee: WUTV Licensee LLC. Group owner: Sinclair Broadcast Group Inc. (acq 12-10-01; grpsl). Population served: 645,000 Natl. Network: Fox. Natl. Rep: Harrington, Righter & Parsons. Washington Atty: Arter & Hadden.
Key Personnel:
Nick Magnini . . . . . gen mgr
Donald Stewart . . . . . opns dir & engrg dir
Jose Chapa . . . . . gen sls mgr
Mike Bice. . . . . prom mgr

Corning

see Elmira (Corning), NY market

Elmira (Corning), NY (DMA 173)

WENY-TV— Analog channel: 36. Digital channel: 55. Analog hrs: 24 Note: ABC is on WENY-TV ch 36, CW is on WENY-DT ch 55. 468 kw vis, 85.4 kw aur. ant 1,050t/840g TL: N42 06 20 W76 52 17 On air date: Nov 19, 1969. 474 Old Ithaca Rd., Horseheads, NY, 14845. Phone: (607) 739-3636. Fax: (607) 739-1418. E-mail: info@weny.com Web Site: www.weny.com. Licensee: Lilly Broadcasting L.L.C. Ownership: Lilly Broadcasting, LLC (acq 10-17-99; $4.8 million). Population served: 250,000 Natl. Network: ABC, CW Digital Network: Note: ABC is on WENY-TV ch 36, CW is on WENY-DT ch 55. Washington Atty: Cordon & Kelly. News: 10 hrs wkly.
Key Personnel:
Kevin Lilly . . . . . CEO & pres
Brian Lilly. . . . . exec VP & VP
Nick White . . . . . VP
Peter Veto . . . . . gen mgr, stn mgr & opns mgr
Bridgid Allinger. . . . . gen sls mgr
Patrick Reilly . . . . . prom dir
Scott Cook. . . . . news dir

WETM-TV— Analog channel: 18. Digital channel: 2. Analog hrs: 24 Digital hrs: 24 166 kw vis, 22.4 kw aur. ant 1,220t/843g TL: N42 06 20 W76 52 17 (CP: 603 kw vis, ant 1,233t) On air date: Sept 10, 1956. Box 1207, Elmira, NY, 14901. 101 E. Water St., Elmira, NY 14901. Phone: (607) 733-5518. Fax: (607) 734-1176. E-mail: info@wetmtv.com Web Site: www.wetmtv.com. Licensee: Central NY News Inc. Group owner: Smith Broadcasting Group Inc. (acq 10-1-2004; $13 million). Population served: 91,000 Natl. Network: NBC. Washington Atty: Hogan & Hartson. News staff: 21; News: 14 hrs wkly.
Key Personnel:
Randy Reid . . . . . gen mgr
Bob Cibulsky . . . . . gen sls mgr & rgnl sls mgr

***WSKA—** Analog channel: 30. Analog hrs: 24 813 kw vis. ant 794t/331g TL: N42 01 55 W76 47 02 On air date: July 2006. Not on air, target date: unknown: Box 3000, Binghamton, NY, 13902-3000. Phone: (607) 729-0100. Fax: (607) 729-7328. E-mail: wskg-mail@pbs.org Web Site: www.wskg.org. Permittee: WSKG Public Telecommunications Council. Natl. Network: PBS.
rebroadcast of WSKG(TV) Binghamton.

WYDC— Analog channel: 48. Digital channel: 50. Analog hrs: 24 Digital hrs: 24 136 kw vis. ant 423t TL: N42 02 29 W77 15 18 On air date: September 1994. 33 E. Market St., Corning, NY, 14830. Phone: (607) 937-5000. Fax: (607) 937-4019. E-mail: jmattison@wydctv.com Web Site: www.wydctv.com. Licensee: WYDC Inc. Ownership: Bill Christian, CEO (acq 11-19-97; $1.75 million). Population served: 94,000 Natl. Network: Fox. Washington Atty: Drinker Biddle & Reath LLP.

New York (DMA 1)

WABC-TV— Analog channel: 7. Digital channel: 45. Analog hrs: 24 Digital hrs: 24 64.6 kw vis, 6.5 kw aur. ant 1,611t/1,730g TL: N40 42 43 W74 00 49 On air date: Aug 10, 1948. 7 Lincoln Sq., New York, NY, 10023. Phone: (212) 456-7777. Fax: (212) 456-2290. Web Site: www.7online.com. Licensee: ABC Inc. Group owner: (group owner). Natl. Network: ABC. Natl. Rep: ABC National Television Sales.
Key Personnel:
Dave Davis . . . . . pres & gen mgr
Evelyn del Cerro . . . . . opns mgr
Scott Simensky. . . . . sls VP & gen sls mgr
Alyson Roznee . . . . . mktg VP & mktg mgr
Janine DiCarlo . . . . . mktg VP
Art Moore . . . . . progmg VP & progmg dir
Kenny Plotnik. . . . . news dir
Saundra Thomas . . . . . pub affrs dir
Bill Beam . . . . . engrg VP & engrg dir
Kurt Hanson . . . . . chief of engrg

WCBS-TV— Analog channel: 2. Digital channel: 56. Analog hrs: 24 Digital hrs: 24 21.4 kw vis, 4.07 kw aur. ant 1,581t/1,589g TL: N40 42 43 W74 00 49 On air date: July 1, 1941. 524 W. 57 St., New York, NY, 10019. Phone: (212) 975-4321. Fax: (212) 975-4677. E-mail: cbsnewyork@cbs.com Web Site: www.wcbstv.com. Licensee: CBS Broadcasting Inc. Group owner: Viacom Television Stations Group. Natl. Network: CBS. Natl. Rep: CBS TV Stations National Sales.
Key Personnel:
Peter Dunn. . . . . pres & gen mgr
Vincent McCarthy . . . . . natl sls mgr
David M. Friend. . . . . news dir

***WEDW—** Analog channel: 49. Digital channel: 52. Analog hrs: 24 1,950 kw vis. 728t TL: N41 16 43 W73 11 08 On air date: Dec 17, 1967. 1049 Asylum Ave., Hartford, CT, 06105. Phone: (860) 278-5310. Fax: (860) 275-7500. Web Site: www.cptv.org. Licensee: Connecticut Public Broadcasting. Population served: 2,500,000 Natl. Network: PBS.
Key Personnel:
Jerry Franklin . . . . . pres, gen mgr & engrg VP
Meg Sakellarides . . . . . CFO
Haig Papasian . . . . . opns VP
Joseph Zareski . . . . . opns dir & chief of engrg
Dean Orton . . . . . dev dir & news dir
Larry Rifkin . . . . . progmg VP

***WFME-TV—** Analog channel: 66. Digital channel: 29. Analog hrs: 24 Digital hrs: 24 24 kw vis. ant 711t TL: N74 12 03 W41 07 14 On air date: 1997. 289 Mt. Pleasant Ave., West Orange, NJ, 07052. Phone: (973) 736-3600. Fax: (973) 736-4832. Fax: (510) 562-1023. E-mail: wfme@wfme.net Web Site: www.familyradio.com. Licensee: Family Stations Inc. Group owner: (group owner).
Key Personnel:
Harold Camping. . . . . gen mgr
Charles Menut . . . . . stn mgr & rgnl sls mgr
Charles H. Menut . . . . . chief of engrg

WFTY-TV— Analog channel: 67. Digital channel: 23. Analog hrs: 24 2,630 kw vis, 263 kw aur. 720t/678g TL: N40 53 23 W72 57 13 On air date: November 1973. 3200 Expressway Dr. S., Islandia, NY, 11749. Phone: (631) 582-6700. Fax: (631) 582-8337. Web Site: www.univision.com. Licensee: Univision New York LLC. Group owner: Univision Communications Inc. (acq 5-21-2001; grpsl). Population served: 3,000,000 Washington Atty: Wiley, Rein & Fielding. Foreign lang progmg: SP News: 4 hrs wkly.
Key Personnel:
Cristina Schwarz . . . . . VP & gen mgr
David Marinace. . . . . chief of engrg

WFUT-TV— Analog channel: 68. Digital channel: 53. Analog hrs: 24 2,630 kw vis. 1,440t/1,430g TL: N40 44 54 W73 59 10 On air date: Sept 29, 1974. Univison 41, 500 Frank W. Burr Blvd., 6th Fl., Teaneck, NJ, 07666. Phone: (201) 287-4042. Fax: (201) 287-9422. Licensee: Univision Partnership of New Jersey. Group owner: Univision Communications Inc. (acq 5-21-2001; grpsl). Washington Atty: Shaw, Pittman. Foreign lang progmg: SP 168
Key Personnel:
Ramon Pineda . . . . . VP
Morris Marotta. . . . . stn mgr
John De Simon. . . . . sls VP
Norma Morato . . . . . news dir
Rebroadcasts WFTY, Smithtown, NY, 100%.

***WLIW—** Analog channel: 21. Digital channel: 22. Analog hrs: 24 Digital hrs: 24 3,160 kw vis, 316 kw aur. 400t/429g TL: N40 47 19 W73 27 09 On air date: Jan 6, 1969. Box 21, Plainview, NY, 11803. Phone: (516) 367-2100. Fax: (516) 692-7629. Web Site: www.wliw.org. Licensee: Educational Broadcasting Corp. (acq 1-31-2003). Population served: 2,000,000 Natl. Network: PBS. Washington Atty: Schwartz, Woods & Miller. Wire Svc: UPI
Key Personnel:
Neal Shapiro. . . . . pres
Terrel Cass. . . . . gen mgr

WLNY— Analog channel: 55. Digital channel: 57. Analog hrs: 24 Digital hrs: 24 425 kw vis. ant 635t/613g TL: N40 53 50 W72 54 56 On air date: Apr 28, 1985. 270 S. Service Rd., Suite 55, Melville, NY, 11747. Phone: (631) 777-8855. Fax: (631) 777-8180. E-mail: ny55@aol.com Web Site: www.wlnytv.com. Licensee: WLNY LP. Ownership: WLNY GP Inc., gen ptnr; WLNY LP Inc., limited ptnr. WLNY Holdings Inc. owns 100% of the voting stock. Population served: 4,200,000 Washington Atty: Cohn & Marks. News: 2 hrs wkly.
Key Personnel:
Marvin R. Chauvin . . . . . CEO
David Feinblatt . . . . . pres & gen mgr
Gerald Diorio. . . . . opns VP
Elliot Simmons . . . . . sls VP & rgnl sls mgr
Andy Starr . . . . . natl sls mgr
Richard Rose. . . . . news dir
Richard Mulliner . . . . . engrg mgr

WMBC-TV— Analog channel: 63. Digital channel: 18. Analog hrs: 24 Digital hrs: 24 2,190 kw vis, 109 kw aur. 731t TL: N41 00 36 W74 35 39 (CP: 2,188 kw vis) On air date: Apr 26, 1993. 99 Clinton Rd., West Caldwell, NJ, 07006. Phone: (973) 852-0300. Fax: (973) 808-5516. E-mail: info@wmbctv.com Web Site: www.wmbctv.com. Licensee: Mountain Broadcasting Corp. Population served: 13,000,000 Washington Atty: Fleischman & Walsh. News staff: 8; News: 5 hrs wkly.
Key Personnel:
Hansen Lau . . . . . pres & news dir
Victor C. Joo . . . . . gen mgr
Joon S. Joo . . . . . chief of engrg

WNBC— Analog channel: 4. Digital channel: 28. Analog hrs: 24 Digital hrs: 24 17.4 kw vis, 3.47 kw aur. 1,689t/1,728g TL: N40 42 43 W74 00 49 On air date: July 1, 1941. 30 Rockefeller Plaza, New York, NY, 10112. Phone: (212) 664-4444. Fax: (212) 664-2994 (news). Web Site: www.wnbc.com. Licensee: NBC Telemundo License Co. Group owner: NBC TV Station Division (acq 6-5-86; grpsl). Natl. Network: NBC.
Key Personnel:
Frank Comerford . . . . . pres & gen mgr
Karen Seminara . . . . . CFO
Dan Forman . . . . . sr VP, stn mgr & news dir
Mathew Braatz . . . . . opns dir
Mark Lund. . . . . sls VP
David Hyman. . . . . prom VP
Adele Rifkin . . . . . progmg dir

***WNET—**(Newark, NJ) Analog channel: 13. Digital channel: 61. Analog hrs: 24 Digital hrs: 24 60.3 kw vis, 5 kw aur. ant 1,640t/1,652g TL: N40 42 43 W74 00 49 On air date: Jan 2, 1948. 450 W. 33rd St., New York, NY, 10001-2605. Phone: (212) 560-1313. Fax: (212) 560-1314. Web Site: www.thirteen.org. Licensee: Educational Broadcasting Corp. (acq 1970). Natl. Network: PBS. Rgnl. Network: Eastern Educ. Washington Atty: Leventhal, Senter & Lerman. News: 5 hrs wkly.
Key Personnel:
Dr. William Baker. . . . . CEO & pres
Neal Shapiro. . . . . pres

***WNJB—** Analog channel: 58. Digital channel: 8. Analog hrs: 24 Digital hrs: 24 1,321 kw vis, 132 kw aur. 726t/401g TL: N40 37 17 W74 30 15 On air date: June 5, 1973. Box 777, Trenton, NJ, 08625-0777. 25 S. Stockton St., Trenton, NJ 08608-1832. Phone: (609) 777-5000. Fax: (609) 633-2920. E-mail: audience@njn.org Web Site: www.njn.net. Licensee: New Jersey Public Broadcasting Authority. Population served: 9,800,000 Natl. Network: PBS. Washington Atty: Schwartz, Woods & Miller. News: 2 hrs wkly.
Key Personnel:
Lisa Bair Miller . . . . . progmg dir
Janice Selinger. . . . . engrg dir
Satellite of *WNJT(TV) Trenton.

***WNJN—** Analog channel: 50. Digital channel: 51. Analog hrs: 24 Digital hrs: 24 1,225 kw vis. ant 764t/590g TL: N40 51 53 W74 12 03 On air date: June 5, 1973. Box 777, Trenton, NJ, 08625-0777. 25 S. Stockton St., Trenton, NJ 08608-1832. Phone: (609) 777-5000. Fax: (609) 633-2920. E-mail: audience@njn.org Web Site: www.njn.net. Licensee: New Jersey Public Broadcasting Authority. Population served: 3,000,000 Natl. Network: PBS. Washington Atty: Schwartz, Woods & Miller. News: 3 hrs wkly.
Satellite of *WNJT Trenton.

WNJU—(Linden, NJ) Analog channel: 47. Digital channel: 36. Analog hrs: 24 Digital hrs: 24 4,570 kw vis, 977 kw aur. ant 1,508t/1,730g TL: N40 42 43 W74 00 49 On air date: May 16, 1965. 2200 Fletcher Ave., 6th Floor, Fort Lee, NJ, 07024. Phone: (201) 969-4247. Fax: (201) 969-4120. Web Site: www.telemundo47.com. Licensee: NBC Telemundo License Co. Group owner: Telemundo Group Inc. (acq 4-12-2002; grpsl). Natl. Network: Telemundo (Spanish). Washington Atty: Hogan & Hartson. Foreign lang progmg: SP 168
Key Personnel:
Manuel Martinez. . . . . gen mgr
Lenny Stole . . . . . chief of engrg

***WNYE-TV**— Analog channel: 25. Digital channel: 24. Analog hrs: 24 Digital hrs: 24 2,450 kw vis, 245 kw aur. ant 1,296t/1,289g TL: N40 44 54 W73 59 10 On air date: Apr 3, 1967. One Centre St., 27th Fl, New York, NY, 10007. Phone: (212) 669-7400. Fax: (212) 669-8448. E-mail: tv@tv.nyc.gov Web Site: www.nyc.gov/tv. Licensee: New York City Dept. of Info Technology & Telecommunications. Population served: 18,000,000 Washington Atty: Arnold & Porter.
Key Personnel:
Arick Wierson. . . . . gen mgr
Trevor Scotland . . . . . progmg mgr
Chang Kim . . . . . chief of engrg

WNYW— Analog channel: 5. Digital channel: 44. Analog hrs: 24 Digital hrs: 24 17.4 kw vis, 1.74 kw aur. 1,689t/1,729g TL: N40 42 43 W74 00 49 On air date: May 2, 1944. 205 E. 67th St., New York, NY, 10021. Phone: (212) 452-5555. Fax: (212) 452-5750. Web Site: www.myfoxny.com. Licensee: Fox Television Stations Inc. Group owner: (group owner; (acq 11-14-86; grpsl). Population served: 7,900,000 Natl. Network: Fox.
Key Personnel:
Al Shjarback . . . . . opns VP & engrg VP
Scott Mathews . . . . . news dir
Audrey Pass. . . . . pub affrs dir
Edward Harris . . . . . engrg dir

WPIX— Analog channel: 11. Digital channel: 33. Analog hrs: 24 Digital hrs: 24 58.9 kw vis, 11.7 kw aur. 1,660t/1,760g TL: N40 42 43 W74 00 49 On air date: June 15, 1948. 220 E. 42nd St., New York, NY, 10017. Phone: (212) 949-1100. Fax: (212) 210-2591. Web Site: www.cw11.com. Licensee: WPIX Inc. Group owner: Tribune Broadcasting Co. Population served: 12,000,000 Natl. Network: CW. Natl. Rep: TeleRep. Washington Atty: Sidley & Austin. Wire Svc: UPI News: 19.5 hrs wkly.
Key Personnel:
Betty Ellen Berlamino . . . . . VP & gen mgr
Bob Marra . . . . . gen sls mgr
Karen Scott . . . . . news dir
Michael Gano . . . . . chief of engrg

WPXN-TV— Analog channel: 31. Digital channel: 30. Analog hrs: 24 Digital hrs: 24 55 kw vis, 5.5 kw aur. ant 1,543t/1,569g TL: N40 42 43 W74 00 49 On air date: Nov 1, 1962. 1330 Avenue of the Americas, 32nd Fl, New York, NY, 10019. Phone: (212) 757-3100. Fax: (212) 956-2661. Web Site: www.ionline.tv. Licensee: Paxson Communications License Co. L.L.C. Group owner: Paxson Communications Corp. (acq 3-4-98; $257.5 million). Population served: 1,025,000 Natl. Network: i Network.
Key Personnel:
Mildred Diaz. . . . . opns mgr
Jack Davidson . . . . . chief of engrg

WRNN-TV— Analog channel: 62. Digital channel: 48. Analog hrs: 24 Digital hrs: 24 5,000 kw vis, 500 kw aur. ant 1,939t/276g TL: N42 05 06 W74 06 00 On air date: Dec 15, 1985. 800 Westchester Ave., Suite S-640, Rye Brook, NY, 10573. Phone: (914) 417-2700. Fax: (914) 696-0279. E-mail: comments@rnntv.com Web Site: www.rnntv.com. Licensee: WRNN License Co. LLC. (acq 7-31-2001). Washington Atty: Baker & Hostetler. News staff: 9; News: 82 hrs wkly.

WSAH— Analog channel: 43. Digital channel: 42.2.5 kw vis, 2 kw aur. ant 620t/300g TL: N41 21 43 W73 06 48 On air date: Sept 28, 1987. 7 Wakely St., Seymour, CT, 06483. Phone: (203) 881-1153. Fax: (203) 881-1302. Licensee: MTB Bridgeport-NY Licensee LLC. Group owner: Scripps Howard Broadcasting Co. (acq 11-15-2006; grpsl). Washington Atty: Crowell & Moring.

WTBY-TV— Analog channel: 54. Digital channel: 27. Analog hrs: 24 Digital hrs: 24 5,000 kw vis, 500 kw aur. 852t/894g TL: N41 43 09 W73 59 47 On air date: Apr 19, 1981. 451 Fishkill Ave. #4, Beacon, NY, 12508-1247. Phone: (845) 896-4610. Fax: (845) 896-4614. E-mail: wtby@tbn.org Web Site: www.tbn.org. Licensee: Trinity Broadcasting of N.Y. Inc. Group owner: Trinity Broadcasting Network (acq 7-13-82; $2.97 million; FTR: 6-21-82). Washington Atty: Joseph E. Dunne III.
Key Personnel:
Paul Crouch . . . . . pres
Chris Elia . . . . . gen mgr
Maria Idoni . . . . . pub affrs dir
Paul Swartzendruber. . . . . chief of engrg

WWOR-TV—(Secaucus, NJ) Analog channel: 9. Digital channel: 38. Analog hrs: 24 Digital hrs: 24 47.9 kw vis, 4.79 kw aur. 1,673t/1,661g TL: N40 42 43 W74 00 49 On air date: Oct 11, 1949. 9 Broadcast Plaza, Secaucus, NJ, 07096. Phone: (201) 348-0009. Web Site: www.my9newyork.com. Licensee: Fox Television Stations Inc. Group owner: (group owner; a(cq 7-31-2001; grpsl). Population served: 13,000,000 Natl. Network: MyNetworkTV. Wire Svc: Conus News staff: 70; News: 7 hrs wkly.
Key Personnel:
Lew Leone . . . . . VP & gen mgr
Debbie von Ahrens . . . . . sls dir & prom mgr
Scott Matthews . . . . . news dir

WXTV— Analog channel: 41. Digital channel: 40. Analog hrs: 24 2,340 kw vis, 234 kw aur. ant 1,381t TL: N40 44 54 W73 59 10 On air date: Aug 4, 1968. 500 Frank W. Burr Blvd., 6th Fl., Teaneck, NJ, 07666-6802. Phone: (201) 287-4042. Fax: (201) 287-9423. Fax: (201) 287-9427 (news). Licensee: WXTV License Partnership G.P. Group owner: Univision Communications Inc. (acq 1986; grpsl). Population served: 3,600,000 Natl. Network: Univision (Spanish). Washington Atty: Shaw, Pittman. Foreign lang progmg: SP 168 News: 17 hrs wkly.
Key Personnel:
Morris Marotta. . . . . stn mgr & chief of engrg
John De Simon. . . . . sls VP
Ramon Pineda . . . . . VP & prom mgr
Norma Morato . . . . . news dir

North Pole

see Burlington, VT-Plattsburgh, NY market

Plattsburgh

see Burlington, VT-Plattsburgh, NY market

Rochester, NY (DMA 78)

WHAM-TV— Analog channel: 13. Digital channel: 59. Analog hrs: 24 Digital hrs: 24 Note: ABC is on WHAM-TV ch 13 and WHAM-DT ch 59, CW is on WHAM-DT ch 59. 316 kw vis, 47.9 kw aur. ant 500t/363.5g TL: N43 08 07 W77 35 03 On air date: Sept 15, 1962. Box 20555, Rochester, NY, 14602-0555. 4225 West Henrietta Rd., Rochester, NY 14623. Phone: (585) 334-8700. Fax: (585) 359-1570. Web Site: 13wham.com. Licensee: Central NY News Inc. Group owner: Clear Channel Communications Inc. (acq 6-14-2002; grpsl). Population served: 393,630 Natl. Network: ABC Digital Network: Note: ABC is on WHAM-TV ch 13 and WHAM-DT ch 59, CW is on WHAM-DT ch 59. Natl. Rep: Millennium Sales & Marketing. Washington Atty: Wiley, Rein & Fielding. Wire Svc: AP Wire Svc: CNN Wire Svc: Bloomberg News News staff: 64; News: 24 hrs wkly.
Key Personnel:
Chuck Samuels . . . . . gen mgr
David DiProsa . . . . . gen sls mgr
Mark Zeger . . . . . natl sls mgr
Amanda DeVito . . . . . rgnl sls mgr
Kevin Kalvitis . . . . . prom dir
Stan Manson . . . . . engrg mgr
Ted McWharf . . . . . engrg mgr

WHEC-TV— Analog channel: 10. Digital channel: 58. Analog hrs: 24 316 kw vis, 39.8 kw aur. 499t/352g TL: N43 08 07 W77 35 02 On air date: Nov 1, 1953. 191 East Ave., Rochester, NY, 14604. Phone: (585) 546-5670. Fax: (585) 454-7433. Fax: (585) 546-5688. Web Site: www.10nbc.com. Licensee: WHEC-TV LLC. Group owner: Hubbard Broadcasting Inc. (acq 9-19-96). Population served: 296,233 Natl. Network: NBC. Natl. Rep: Petry Television Inc. Washington Atty: Arent, Fox, Kintner, Plotkin & Kahn. News: 22 hrs wkly.
Key Personnel:
Arnold Klinsky . . . . . gen mgr
Sherron Sheridan. . . . . opns mgr
Joe Fazio. . . . . gen sls mgr
Lynette Baker. . . . . progmg dir

WROC-TV— Analog channel: 8. Digital channel: 45. Analog hrs: 24 316 kw vis, 48.5 kw aur. ant 499t/345g TL: N43 08 07 W77 35 02 On air date: June 14, 1949. 201 Humboldt St., Rochester, NY, 14610-1093. Phone: (585) 288-8400. Fax: (585) 288-7679. Web Site: www.wroctv.com. Licensee: Nexstar Broadcasting Inc. Group owner: Nexstar Broadcasting Group Inc. (acq 12-9-99; $46 million). Population served: 957,000 Natl. Network: CBS. News staff: 40; News: 14 hrs wkly.
Key Personnel:
Tim Busch . . . . . VP
Don Loy . . . . . stn mgr

WUHF— Analog channel: 31. Digital channel: 28. Analog hrs: 24 Digital hrs: 24 1,200 kw vis, 200 kw aur. ant 499t/345g TL: N43 08 07 W77 35 03 On air date: January 1980. 201 Humboldt St., Rochester, NY, 14610. Phone: (585) 232-3700. Fax: (585) 546-4774. Web Site: rochesterhomepage.net. Licensee: WUHF Licensee LLC. Group owner: Sinclair Broadcast Group Inc. (acq 4-12-2002; for assumption liabilities). Population served: 957,000 Natl. Network: Fox. Washington Atty: Pillsbury, Winthrop & Shaw Pittman. News: 3.5 hrs wkly.

***WXXI-TV**— Analog channel: 21. Digital channel: 16. Analog hrs: 24 Digital hrs: 24 906 kw vis, 90.6 kw aur. 500t/343g TL: N43 08 07 W77 35 03 On air date: September 1966. Box 30021, Rochester, NY, 14603-3021. 280 State St., Rochester, NY 14614. Phone: (585) 325-7500. Fax: (585) 258-0335. Web Site: www.wxxi.org. Licensee: WXXI Public Broadcasting Council. Natl. Network: PBS. Rgnl. Network: CEN. Washington Atty: Schwartz, Woods & Miller. News staff: 4; News: 2 hrs wkly.
Key Personnel:
Norm Silverstein . . . . . CEO & pres
Susan Rogers . . . . . COO & exec VP
Robert Owens . . . . . progmg dir
Kent Hatfield. . . . . engrg VP

Schenectady

see Albany-Schenectady-Troy, NY market

Syracuse, NY (DMA 79)

***WCNY-TV**— Analog channel: 24. Digital channel: 25. Analog hrs: 18 Digital hrs: 18 2,312 kw vis, 231 kw aur. 1,380t/964g TL: N42 56 42 W76 01 28 On air date: Dec 20, 1965. Box 2400, Syracuse, NY, 13220-2400. 506 Old Liverpool Rd., Liverpool, NY 13088. Phone: (315) 453-2424. Fax: (315) 451-8824. E-mail: wcny-online@wcny.org Web Site: www.wcny.org. Licensee: Public Broadcasting Council of Central New York. Population served: 600,000 Natl. Network: PBS. Washington Atty: Dow, Lohnes & Albertson.
Key Personnel:
Robert Daino . . . . . CEO & pres
Colleen Edwards . . . . . CFO
Brian Damm . . . . . mktg VP
Larry Goodsight . . . . . mktg dir
John Duffy. . . . . chief of engrg

WNYI— Analog channel: 52. Analog hrs: 24 26 kw vis. ant -308t/69g TL: N42 25 46.8 W76 29 48.8 On air date: 2004. 1 Shackleford Dr., Suite 400, Little Rock, AR, 72211. 401 W. Kirkpatrick St., Syracuse, NY 13204. Phone: (501) 219-2400. Licensee: EBC Syracuse Inc. Group owner: Equity Broadcasting Corp. (acq 8-6-2004;. $5 million with KWWF(TV) Waterloo, IA).

WNYS-TV— Analog channel: 43. Digital channel: 44. Analog hrs: 24 Digital hrs: 24 794 kw vis, 79.4 kw aur. ant 1,471t/1019g TL: N42 52 50 W76 11 59 On air date: Oct 7, 1989. 1000 James St., Syracuse, NY, 13203. Phone: (315) 472-6800. Fax: (315) 471-8889. Web Site: www.wb43.com. Licensee: RKM Media Inc. Ownership: Ron Philips. (acq 7-2-96). Natl. Network: MyNetworkTV. Washington Atty: Fletcher, Heald & Hildreth.
Key Personnel:
Aaron Olander . . . . . gen mgr
Peter Spartano . . . . . opns dir
Donald O'Connor . . . . . gen sls mgr
Krysten Bellen . . . . . natl sls mgr
Ed Kampf. . . . . rgnl sls mgr
Ed Sautter . . . . . prom mgr
Linda Deeb . . . . . progmg mgr
Roy Taylor. . . . . engrg dir & chief of engrg

WSPX-TV— Analog channel: 56. Analog hrs: 24 46.77 kw vis. 686t TL: N42 56 54 W76 01 21 On air date: Nov 24, 1998. 6508-B Basile Row, East Syracuse, NY, 13057. Phone: (315) 414-0178. Fax: (315) 414-0482. Web Site: www.ionline.tv. Licensee: Paxson Syracuse License Inc. Ownership: Paxson Communications of Syracuse-56 Inc., 100% (acq 4-29-99).
Key Personnel:
Margo McCaffery. . . . . stn mgr
Al Szablak. . . . . chief of engrg

WSTM-TV— Analog channel: 3. Digital channel: 54. Digital hrs: 24 100 kw vis, 20 kw aur. ant 1,000t/594g TL: N42 56 40 W76 07 08 On air date: Feb 15, 1950. 1030 James St., Syracuse, NY, 13203. Phone: (315) 477-9400. Fax: (315) 474-5082. Web Site: www.wstm.com. Licensee: Barrington Syracuse License LLC. Group owner: Raycom Media (acq 8-11-2006; grpsl). Population served: 385,100 Natl. Network: NBC. Natl. Rep: TeleRep. Washington Atty: Covington & Burling. News staff: 45; News: 27.5 hrs wkly.
Key Personnel:
Chris Geiger . . . . . VP, gen mgr & gen sls mgr
Dave Rhea . . . . . gen sls mgr

Judy Fitzgerald . . . . . natl sls mgr
Peggy Phillip . . . . . news dir

WSYR-TV— Analog channel: 9. Digital channel: 17. Analog hrs: 24 79.4 kw vis, 11.8 kw aur. ant 1,515t/959g TL: N42 56 42 W76 01 28 On air date: Sept 9, 1962. Box 699, 5904 Bridge St., East Syracuse, NY, 13057. Phone: (315) 446-9999. Fax: (315) 446-9283. E-mail: newschannel9@9wsyr.com Web Site: www.9wsyr.com. Licensee: Central NY News Inc. Group owner: Clear Channel Communications Inc. (acq 6-30-2002; grpsl). Population served: 389,700 Natl. Network: ABC. Natl. Rep: Millennium Sales & Marketing. News staff: 55; News: 22.5 hrs wkly.
Key Personnel:
Stephen Kimatian . . . . . pres
Theresa E. Underwood . . . . . VP, gen mgr & stn mgr
John King . . . . . opns dir & engrg VP
Sally Stamp . . . . . sls dir
Todd Guard . . . . . rgnl sls mgr & sls
Vince Spicola . . . . . progmg dir
Jim Tortora . . . . . news dir
Francis Fasuyi . . . . . engrg mgr
Craig Riker . . . . . chief of engrg
Bill Evans . . . . . sls

WSYT— Analog channel: 68. Digital channel: 19. Analog hrs: 24 Digital hrs: 24 1,000 kw vis, 100 kw aur. ant 1,471t/1019g TL: N42 52 50 W76 11 59 On air date: Feb 15, 1986. 1000 James St., Syracuse, NY, 13203. Phone: (315) 472-6800. Fax: (315) 471-8889. Web Site: www.wsyt68.com. Licensee: WSYT Licensee L.P. Group owner: Sinclair Broadcast Group Inc. (acq 7-7-98; grpsl). Natl. Network: Fox. News: 3.5 hrs wkly.
Key Personnel:
Aaron Olander . . . . . gen mgr
Peter Spartano . . . . . opns dir
Donald O'Connor . . . . . gen sls mgr
Krysten Bellen . . . . . natl sls mgr
Ed Kampf . . . . . rgnl sls mgr
Leslie Baycura . . . . . prom mgr
Linda Deeb . . . . . progmg dir & progmg mgr
Vinnie Lopez . . . . . engrg dir

WTVH— Analog channel: 5. Digital channel: 47. Analog hrs: 24 Digital hrs: 24 100 kw vis, 20 kw aur. 950t/556g TL: N42 57 19 W76 06 34 On air date: Dec 1, 1948. 980 James St., Syracuse, NY, 13203. Phone: (315) 425-5555. Fax: (315) 425-5513. E-mail: wtvh@wtvh.com Web Site: www.wtvh.com. Licensee: WTVH License Inc. Group owner: Granite Broadcasting Corp. Population served: 197,208 Natl. Network: CBS. Natl. Rep: Harrington, Righter & Parsons. Washington Atty: Akin, Gump, Strauss, Hauer & Feld. News staff: 42; News: 24 hrs wkly.
Key Personnel:
Les Vann . . . . . pres & gen mgr
Matt Rosenfeld . . . . . gen sls mgr
Amy Collins . . . . . natl sls mgr
Bob Wickwire . . . . . rgnl sls mgr
Nicole Pooler . . . . . progmg mgr
Frank Kracher . . . . . news dir
Kevin Wright . . . . . chief of engrg

Troy

see Albany-Schenectady-Troy, NY market

Utica, NY
(DMA 169)

WFXV— Analog channel: 33. Analog hrs: 24 42.7 kw vis, 4.27 kw aur. 646t/189g TL: N43 02 14 W75 26 40 (CP: 854 kw vis) On air date: Dec 9, 1986. 5956 Smith Hill Rd., Utica, NY, 13502. Phone: (315) 797-5220. Fax: (315) 797-5409. Web Site: www.utica.tv. Licensee: Nexstar Finance Inc. Group owner: Nexstar Broadcasting Group Inc. (acq 12-31-03; grpsl). Population served: 1,000,000 Natl. Network: Fox. Washington Atty: Arter & Hadden.
Key Personnel:
Steve Merren . . . . . gen mgr
Bob Hajec . . . . . chief of engrg

WKTV— Analog channel: 2. Digital channel: 29.34.7 kw vis, 6.9 kw aur. 1,380t/1,065g TL: N43 06 09 W74 56 27 On air date: Dec 1, 1949. Box 2, Utica, NY, 13503. 5936 Smith Hill Rd., Utica, NY 13503. Phone: (315) 733-0404. Fax: (315) 793-3498. Web Site: www.wktv.com. Licensee: Smith Media License Holdings LLC. Group owner: Smith Broadcasting Group Inc. (acq 11-8-2004; grpsl). Natl. Network: NBC, CW. Natl. Rep: Continental Television Sales. Washington Atty: Dow, Lohnes & Albertson, PLLC. News: 31.5 hrs wkly.
Key Personnel:
Vic Vetters . . . . . VP & gen mgr
Frank Abbadessa . . . . . rgnl sls mgr
Dave Streeter . . . . . prom mgr
Steve McMurray . . . . . news dir
Tom McNicholl . . . . . chief of engrg

WUTR— Analog channel: 20. Digital channel: 30. Analog hrs: 24 Digital hrs: 24 1,150 kw vis, 173 kw aur. ant 800t/400g TL: N43 08 43 W75 10 35 On air date: Feb 28, 1970. 5956 Smith Hill Rd., Utica, NY, 13502. Phone: (315) 797-5220. Fax: (315) 797-5409. Web Site: www.wutr.com. Licensee: Mission Broadcasting Inc. Group owner: (group owner; (acq 4-1-2004; $3.725 million). Population served: 104,100 Natl. Network: ABC.
Key Personnel:
Diane Siembab . . . . . stn mgr
Steve Merren . . . . . sls dir & gen sls mgr
Steve Ventura . . . . . opns mgr & rgnl sls mgr
Allen Williams . . . . . prom mgr
Domenick Cecconi . . . . . progmg dir & news dir
Michael Moran . . . . . chief of engrg

Watertown, NY
(DMA 176)

***WNPI-TV—** Analog channel: 18. Digital channel: 23.661 kw vis, 83.4 kw aur. ant 800t/761g TL: N44 29 30 W74 51 29 On air date: Aug 30, 1971. 1056 Arsenal St., Watertown, NY, 13601. Phone: (315) 782-3142. Fax: (315) 782-2491. Web Site: www.wpbstv.org. Licensee: St. Lawrence Valley ETV Council. Population served: 220,000 Natl. Network: PBS. Rgnl. Network: Eastern Educ. Washington Atty: Schwartz, Woods & Miller.
Key Personnel:
Thomas F. Hanley . . . . . pres & gen mgr
Lynn Brown . . . . . dev dir & progmg dir
Joline Furgison . . . . . mktg dir & progmg mgr

***WPBS-TV—** Analog channel: 16. Digital channel: 41.617 kw vis. ant 1,214t/915g TL: N43 51 44 W75 43 40 (CP: 618 kw vis. TL: N43 51 46 W75 43 39) On air date: Aug 5, 1971. 1056 Arsenal St., Watertown, NY, 13601. Phone: (315) 782-3142. Fax: (315) 782-2491. Web Site: www.wpbstv.org. Licensee: St. Lawrence Valley ETV Council. Population served: 151,000 Natl. Network: PBS. Rgnl. Network: Eastern Educ. Washington Atty: Schwartz, Woods & Miller.
Key Personnel:
Thomas F. Hanley . . . . . pres & gen mgr
Lynn Brown . . . . . dev dir & progmg dir

WWNY-TV—Carthage, Analog channel: 7. Digital channel: 35. Analog hrs: 6 AM-2 AM Digital hrs: 6am - 2am 316 kw vis, 47 kw aur. ant 718t/572g TL: N43 57 16 W75 43 45 (CP: Ant 725t/579g) On air date: Oct 22, 1954. 120 Arcade St., Watertown, NY, 13601. Phone: (315) 788-3800. Fax: (315) 782-7468. Fax: (315) 788-3787. E-mail: wwny@wwnytv.net Web Site: wwnytv.net. Licensee: United Communications Corp. Group owner: (group owner; (acq 12-5-81; $8.1 million; FTR: 6-1-81). Population served: 107,406 Natl. Network: CBS. Washington Atty: Wood, Maines & Brown, Chartered. News staff: 16; News: 19 hrs wkly.
Key Personnel:
Cathy Pircsuk . . . . . gen mgr
Patrick Powers . . . . . gen sls mgr

WWTI— Analog channel: 50. Digital channel: 21. Analog hrs: 24 Digital hrs: 24 Note: ABC is on WWTI(TV) ch 50, CW is on WWTI-DT ch 21. 1,000 kw vis, 100 kw aur. ant 1,268t/1,000g TL: N43 52 47 W75 43 11 On air date: January 1988. 1222 Arsenal St., Watertown, NY, 13601. Phone: (315) 785-8850. Fax: (315) 785-0127. Web Site: www.newswatch50.com. Licensee: Central NY News Inc. Group owner: Clear Channel Communications Inc. (acq 6-14-2002; grpsl). Natl. Network: ABC, CW Digital Network: Note: ABC is on WWTI(TV) ch 50, CW is on WWTI-DT ch 21. News staff: 11; News: 5 hrs wkly.
Key Personnel:
David Males . . . . . gen mgr, gen sls mgr & natl sls mgr
Keith Rudes . . . . . chief of engrg

North Carolina

Asheville

see Greenville-Spartanburg, SC-Asheville, NC-Anderson, SC market

Charlotte, NC
(DMA 26)

WAXN-TV— Analog channel: 64. Digital channel: 50. Analog hrs: 24 Digital hrs: 24 1,100 kw vis. ant 1,155t/1,010g TL: N35 15 41 W80 43 38 On air date: Oct 15, 1994. 1901 North Tryon St., Charlotte, NC, 28206. Phone: (704) 338-9999. Fax: (704) 371-3131. Web Site: www.action64.com. Licensee: WSOC-TV Holdings Inc. Group owner: Cox Communications Inc. (acq 1-31-2000). Population served: 2,490,000 Natl. Rep: TeleRep. Washington Atty: Dow, Lohnes & Albertson.
Key Personnel:
Dave Siegler . . . . . opns mgr
Sally Ganz . . . . . mktg dir
Kay Hall . . . . . progmg dir
Robin Whitmeyer . . . . . news dir
Ted Hand . . . . . engrg dir

WBTV— Analog channel: 3. Digital channel: 23. Analog hrs: 24 Digital hrs: 24 100 kw vis, 10 kw aur. 1,860t/1,987g TL: N35 21 51 W81 11 13 On air date: July 15, 1949. One Julian Price Pl., Charlotte, NC, 28208. Phone: (704) 374-3500. Fax: (704) 374-3614. Web Site: www.wbtv.com. Licensee: WBTV Inc. Group owner: Jefferson-Pilot Communications Co. (acq 4-3-2006; grpsl). Population served: 1,300,000 Natl. Network: CBS. Natl. Rep: Petry Television Inc. Washington Atty: Wiley, Rein & Fielding. Wire Svc: UPI
Key Personnel:
Mary MacMillan . . . . . gen mgr
Don Shaw . . . . . opns dir
Shelly Hill . . . . . prom dir & progmg dir
Ron Yoslov . . . . . chief of engrg

WCCB— Analog channel: 18. Digital channel: 27. Analog hrs: 24 Digital hrs: 24 2,090 kw vis, 230 kw aur. 1,276t/1,143g TL: N35 15 56 W80 44 06 On air date: Dec 7, 1953. One Television Pl., Charlotte, NC, 28205. Phone: (704) 372-1800. Fax: (704) 376-3415. Fax: (704) 332-7941. E-mail: wccb@foxcharlotte.tv Web Site: www.foxcharlotte.tv. Licensee: North Carolina Broadcasting Partners. Group owner: Bahakel Communications Population served: 2,121,000 Natl. Network: Fox. News staff: one.
Key Personnel:
John Hutchinson . . . . . VP & gen mgr
Gaston Bates . . . . . gen sls mgr
Ken White . . . . . news dir
Rick Aydlett . . . . . chief of engrg

WCNC-TV— Analog channel: 36. Digital channel: 22. Analog hrs: 24 Digital hrs: 24 5,000 kw vis, 250 kw aur. 1,964t/1,954g TL: N35 20 49 W81 10 15 On air date: July 9, 1967. 1001 Woodridge Center Dr., Charlotte, NC, 28217-1901. Phone: (704) 329-3636. Fax: (704) 357-4980. Web Site: www.wcnc.com. Licensee: WCNC-TV Inc. Group owner: Belo Corp., Broadcast Division (acq 1997; grpsl). Population served: 663,800 Natl. Network: NBC. Natl. Rep: Harrington, Righter & Parsons. Washington Atty: Wiley, Rein & Fielding.
Key Personnel:
Stuart B. Powell . . . . . pres & gen mgr
Ann Marie Young . . . . . sls dir
Steve Kiser . . . . . chief of engrg

WHKY-TV— Analog channel: 14. Digital channel: 40. Analog hrs: 24 Digital hrs: 24 2,000 kw vis, 200 kw aur. ant 597t/461g TL: N35 43 59 W81 19 51 On air date: Feb 14, 1968. Box 1059, Hickory, NC, 28603. 526 Main Ave. S.E., Hickory, NC 28603. Phone: (828) 322-5115. Fax: (828) 322-8256. E-mail: whky@whky.com Web Site: www.whky.com. Licensee: Long Communications LLC. Ownership: Thomas E. Long, 49%; Roberta S. Long, 41%; Jeffrey Long, 10% (acq 12-31-2001; with WHKY(AM) Hickory). Population served: 2,200,000 Washington Atty: Hardy & Carey. News staff: 4; News: 5 hrs wkly.
Key Personnel:
Thomas Long . . . . . gen mgr
Jeff Long . . . . . stn mgr
Patty Guthrie . . . . . gen sls mgr & prom mgr

WJZY— Analog channel: 46. Digital channel: 47.500 kw vis, 50 kw aur. 1,948t/1,949g TL: N35 21 44 W81 09 19 On air date: Mar 9, 1987. 3501 Performance Rd., Charlotte, NC, 28214. Phone: (704) 398-0046. Fax: (704) 393-8407. E-mail: info@wjzy.com Web Site: www.wjzy.com. Licensee: WJZY-TV Inc. Group owner: Capitol Broadcasting Co. Inc. (acq 11-87; $1.581 million). Natl. Network: CW. Natl. Rep: Millennium Sales & Marketing. Washington Atty: Fletcher, Heald & Hildreth.
Key Personnel:
Will Davis . . . . . gen mgr
Shawn Harris . . . . . stn mgr
Matt Livoti . . . . . gen sls mgr
Brian Corrigan . . . . . natl sls mgr
Andre Boyd . . . . . prom mgr
Joe Heaton . . . . . progmg dir & pub affrs dir
John Bishop . . . . . chief of engrg

WMYT-TV— Analog channel: 55.5,000 kw vis. 1,870t TL: N35 21 44 W81 09 19 On air date: October 1994. 3501 Performance Rd., Charlotte, NC, 28214. Phone: (704) 398-0046. Fax: (704) 393-8407. E-mail: info@wmyt12.com Web Site: www.wmyt12.com. Licensee: WMYT-TV Inc. Group owner: Capitol Broadcasting Co. Inc. (acq 2—2000; $4.5 million). Natl. Network: MyNetworkTV. Natl. Rep: Millennium Sales & Marketing.
Key Personnel:
Will Davis . . . . . gen mgr

Shawn Harris . . . . . sls dir
Brian Corrigan . . . . . natl sls mgr
Robin Symes . . . . . mktg dir
Andre Boyd . . . . . prom mgr
Joe Heaton . . . . . progmg dir & pub affrs dir
John Bishop . . . . . chief of engrg

***WNSC-TV—** Analog channel: 30.676 kw vis, 136 kw aur. 688t TL: N34 50 24 W81 01 07 On air date: Jan 3, 1978. Box 11766, Rock Hill, SC, 29731. 454 S. Anderson Rd., Rock Hill, SC 29731. Phone: (803) 324-3184. Fax: (803) 324-0580. Web Site: www.muetv.org. Licensee: S.C. Educ TV Commission. Population served: 870,000 Natl. Network: PBS.
Key Personnel:
Maurice Bresnahan . . . . . pres
John Bullington . . . . . stn mgr & progmg dir
Bruce Bauman . . . . . sls dir
David Taylor . . . . . chief of engrg

WSOC-TV— Analog channel: 9. Digital channel: 34. Analog hrs: 24 316 kw vis, 31.6 kw aur. ant 1,194t/1,050g TL: N35 15 41 W80 43 38 On air date: Apr 28, 1957. Box 34665, Charlotte, NC, 28234. 1901 N. Tryon St., Charlotte, NC 28206. Phone: (704) 338-9999. Web Site: www.wsoctv.com. Licensee: WSOC-TV Holdings Inc. Group owner: Cox Broadcasting Inc. (acq 4-13-59; grpsl; FTR: 4-13-59). Population served: 2,490,000 Natl. Network: ABC. Natl. Rep: TeleRep. Washington Atty: Dow, Lohnes & Albertson.
Key Personnel:
Dave Siegler . . . . . opns mgr
Sally Ganz . . . . . mktg dir
Kay Hall . . . . . progmg dir

***WTVI—** Analog channel: 42. Digital channel: 11. Analog hrs: 24 Digital hrs: 24 2,750 kw vis, 550 kw aur. 1,247t/1,221g TL: N35 12 25 W80 47 30 On air date: Aug 27, 1965. 3242 Commonwealth Ave., Charlotte, NC, 28205. Phone: (704) 372-2442. Fax: (704) 335-1358. Web Site: www.wtvi.org. Licensee: Charlotte-Mecklenburg Public Broadcasting Authority. Population served: 1,600,000 Natl. Network: PBS. Washington Atty: Schwartz, Woods & Miller.
Key Personnel:
Elsie Garner . . . . . CEO, pres & gen mgr
Tom Green . . . . . chief of engrg

***WUNE-TV—** Analog channel: 17. Digital channel: 54. Analog hrs: 24 Digital hrs: 24 1,550 kw vis, 154 kw aur. ant 1,791t/420g TL: N36 03 47 W81 50 33 On air date: Sept 11, 1967. Box 14900, Research Triangle Park, NC, 27709-4900. 10 T.W. Alexander Dr., Research Triangle Park, NC 27709. Phone: (919) 549-7000. Fax: (919) 549-7201. E-mail: viewer@unctv.org Web Site: www.unctv.org. Licensee: University of North Carolina. Population served: 9,000,000 Natl. Network: PBS. Washington Atty: Schwartz, Woods & Miller.

***WUNG-TV—** Analog channel: 58. Digital channel: 44. Analog hrs: 24 Digital hrs: 24 5,000 kw vis, 500 kw aur. ant 1,384t TL: N35 21 30 W80 36 37 On air date: Sept 11, 1967. Box 14900, Research Triangle Park, NC, 27709-4900. Phone: (919) 549-7000. Fax: (919) 549-7201. E-mail: viewer@unctv.org Web Site: www.unctv.org. Licensee: University of North Carolina. Population served: 9,000,000 Natl. Network: PBS. Washington Atty: Schwartz, Woods & Miller.

Durham

see Raleigh-Durham (Fayetteville), NC market

Fayetteville

see Raleigh-Durham (Fayetteville), NC market

Greensboro-High Point-Winston Salem, NC (DMA 47)

WCWG— Analog channel: 20. Digital channel: 19. Analog hrs: 24 5,000 kw vis. ant 1,814t/1,797g TL: N35 52 02 W79 49 26 On air date: Oct 30, 1985. 622 Guilford College Rd., Suite G, Greensboro, NC, 27409. Phone: (336) 510-2020. Fax: (336) 517-2020. E-mail: info@wcwg20.com Web Site: www.wcwg20.com. Licensee: WTWB License LLC. Group owner: Pappas Telecasting Companies (acq 1995; $4 million). Population served: 1,543,000 Natl. Network: CW. Natl. Rep: TeleRep. Washington Atty: Paul, Hastings, Janofsky & Walker LLP.
Key Personnel:
Rosalie Drake . . . . . gen mgr
Joe Sigman . . . . . sls dir & prom mgr
Eric Jordan . . . . . gen sls mgr
David Edrington . . . . . progmg mgr
Don Moore . . . . . chief of engrg

WFMY-TV— Analog channel: 2. Digital channel: 51. Analog hrs: 24 Digital hrs: 24 100 kw vis, 19.5 kw aur. ant 1,842t/1,914g TL: N35 52 13 W79 50 25 On air date: Sept 22, 1949. Box 26004, Greensboro, NC, 27420. 1615 Phillips Ave., Greensboro, NC 27405. Phone: (336) 379-9369. Fax: (336) 273-3444. E-mail: news2@wfmy.com Web Site: www.wfmynews2.com. Licensee: WFMY Television Corp. Group owner: Gannett Broadcasting (division of Gannett Co. Inc.) (acq 2-1-88). Population served: 634,130 Natl. Network: CBS. Wire Svc: CBS Wire Svc: AP News: 32 hrs wkly.
Key Personnel:
Deborah Hooper . . . . . pres & gen mgr
Deana Coble . . . . . opns dir
Bill Lancaster . . . . . gen sls mgr
Chris Delaporte . . . . . rgnl sls mgr
David Reeve . . . . . mktg mgr & prom mgr
David Briscoe . . . . . progmg mgr
Gina Katzmark . . . . . news dir
Jim Walton . . . . . chief of engrg

WGHP— Analog channel: 8. Digital channel: 35. Analog hrs: 24 Digital hrs: 24 300 kw vis. ant 1,305t/1,217g TL: N35 48 46 W79 50 29 On air date: Oct 14, 1963. HP-8, High Point, NC, 27261. 2005 Francis St., High Point, NC 27263. Phone: (336) 841-8888. Fax: (336) 841-8051. Web Site: www.fox8wghp.com. Licensee: Fox Television Stations Inc. Group owner: (group owner; (acq 1-96; $135 million with WBRC-TV Birmingham, AL). Population served: 533,300 Natl. Network: Fox. Washington Atty: Koteen & Naftalin. Wire Svc: UPI News staff: 75; News: 32 hrs wkly.
Key Personnel:
Karen Adams . . . . . VP, gen mgr & progmg dir
Ramona Alexander . . . . . gen sls mgr
Ross Mason . . . . . chief of engrg

WGPX— Analog channel: 16. Digital channel: 14. Analog hrs: 24 Digital hrs: 24 1,910 kw vis, 191 kw aur. 840t/500g TL: N35 56 22 W79 25 47 On air date: Aug 7, 1984. 1114 N. O'Henry Blvd., Greensboro, NC, 27405. Phone: (336) 272-9227. Fax: (336) 272-9298. Web Site: www.ionline.tv. Licensee: Paxson Greensboro License Inc. Group owner: Paxson Communications Corp. (acq 1996; $5.5 million). Natl. Network: i Network. Natl. Rep: Roslin. Washington Atty: Baraff, Koerner, Olender & Hochberg.
Key Personnel:
Dana Lambert . . . . . opns mgr
Steve Hall . . . . . chief of engrg

WLXI-TV— Analog channel: 61. Digital channel: 43. Analog hrs: 24 Digital hrs: 24 501 kw vis, 50 kw aur. 573t/499g TL: N36 08 58 W80 03 21 On air date: Mar 1, 1984. 2109 Patterson St., Greensboro, NC, 27407. Phone: (336) 855-5610. Fax: (336) 855-3645. E-mail: wlxi@tct.tv Licensee: Radiant Life Ministries Inc. Group owner: Tri-State Christian Television (acq 10-7-91; $1.9 million; FTR: 10-28-91). Population served: 2,500,000 Washington Atty: Joseph E. Dunne III.
Key Personnel:
Larry Patton . . . . . gen mgr
Gil Couch . . . . . chief of engrg

WMYV— Analog channel: 48. Digital channel: 33. Analog hrs: 24 Digital hrs: 24 1,100 kw vis, 110 kw aur. ant 1,696t/1,726g TL: N35 52 13 W79 50 25 On air date: May 9, 1981. 3500 Myer Lee Dr., Winston Salem, NC, 27101. Phone: (336) 722-4545. Fax: (336) 723-8217. Web Site: www.my48.tv. Licensee: WUPN Licensee LLC. Group owner: Sinclair Broadcast Group Inc. (acq 1-9-2002; $50,000 and cancellation of debt). Natl. Network: MyNetworkTV. Natl. Rep: Millennium Sales & Marketing. Washington Atty: Arter & Hadden. News: 7 hrs wkly.
Key Personnel:
Ron Inman . . . . . gen mgr
Fran McRae . . . . . gen sls mgr
Eric Gabriel . . . . . prom mgr
Jeanette Pruitt . . . . . progmg dir
Zane Parnell . . . . . chief of engrg

***WUNL-TV—** Analog channel: 26. Digital channel: 32. Analog hrs: 24 Digital hrs: 24 5,000 kw vis, 500 kw aur. ant 1,653t/328g TL: N36 22 34 W80 22 14 On air date: Feb 22, 1973. Box 14900, Research Triangle Park, NC, 27709-4900. 10 T.W. Alexander Dr., Research Triangle Park, NC 27709-4900. Phone: (919) 549-7000. Fax: (919) 549-7201. E-mail: viewer@unctv.org Web Site: www.unctv.org. Licensee: University of North Carolina. Natl. Network: PBS. Washington Atty: Schwartz, Woods & Miller.

WXII-TV— Analog channel: 12. Digital channel: 31. Analog hrs: 24 Digital hrs: 24 316 kw vis, 63.5 kw aur. ant 1,980t/680g TL: N36 22 31 W78 08 50 On air date: Sept 30, 1953. 700 Coliseum Dr., Winston-Salem, NC, 27106. Phone: (336) 721-9944. Fax: (336) 703-6300. Web Site: www.wxii12.com. Licensee: WXII Hearst-Argyle Television Inc. Group owner: Hearst-Argyle Television Inc. (acq 3-18-99; grpsl). Population served: 2,183,000 Natl. Network: NBC. Washington Atty: Brooks, Pierce, McLendon, Humphrey & Leonard.
Key Personnel:
Mark Strand . . . . . pres & prom mgr
Barry Klaus . . . . . CFO & news dir
Glenn Haygood . . . . . exec VP & gen sls mgr
John Norvell . . . . . sr VP & chief of engrg
Henry E. Price . . . . . gen mgr
Michael Pulitzer . . . . . stn mgr & progmg mgr

WXLV-TV— Analog channel: 45. Digital channel: 29. Analog hrs: 24 Digital hrs: 24 5,000 kw vis, 500 kw aur. 2,000t/768g TL: N36 22 37 W80 22 10 (CP: 500 kw vis, ant 1,958t) On air date: Sept 24, 1979. 3500 Myer Lee Dr., Winston-Salem, NC, 27101. Phone: (336) 722-4545. Fax: (336) 723-8217. Web Site: www.abc45.com. Licensee: WXLV Licensee LLC. Group owner: Sinclair Broadcast Group Inc. (acq 12-10-01; grpsl). Population served: 548,000 Natl. Network: ABC. Washington Atty: Arter & Hadden. News: 2.5 hrs wkly.
Key Personnel:
Ron Inman . . . . . gen mgr
Fran McRae . . . . . gen sls mgr
Eric Gabriel . . . . . prom mgr
Jeanette Pruitt . . . . . progmg dir
Zane Parnell . . . . . chief of engrg

Greenville-New Bern-Washington, NC (DMA 107)

WCTI-TV—New Bern, Analog channel: 12. Digital channel: 48. Analog hrs: 24 Digital hrs: 24 316 kw vis, 31 kw aur. 1,923t/1,999g TL: N35 08 03 W77 03 51 On air date: Sept 1, 1963. Box 12325, 225 Glenburnie Dr., New Bern, NC, 28561. Phone: (252) 638-1212. Fax: (252) 637-4141. Fax: (252) 636-6816. Web Site: www.wtci12.com. Licensee: Newport License Holdings Inc. Ownership: Bonten Media Group, LLC (acq 6-15-2004; $4 million). Population served: 233,000 Natl. Network: ABC. Natl. Rep: Continental Television Sales. Washington Atty: Koteen & Naftalin.
Key Personnel:
James Ottolin . . . . . gen mgr
Ingrid Johansen . . . . . news dir
Ken Hughes . . . . . VP & chief of engrg

WEPX— Analog channel: 38. Analog hrs: 24 5,000 kw vis. 505t TL: N35 23 52 W77 25 40 On air date: Dec 9, 1998. 1301 S. Glenburnie Rd., New Bern, NC, 28562. Phone: (252) 636-2550. Fax: (252) 633-7851. Licensee: Paxson Greenville License Inc. Group owner: Paxson Communications Corp. (acq 4-13-99; $3.55 million). Natl. Network: MyNetworkTV.

WFXI— Analog channel: 8. Digital channel: 24. Analog hrs: 24 Digital hrs: 24 Note: Fox is on WFXI(TV) ch 4, MyNetworkTV is on WFXI-DT ch 24. 316 kw vis, 31.6 kw aur. ant 817t/835g TL: N34 53 01 W76 30 21 On air date: Nov 1, 1989. 5441 Hwy. 70 E., Morehead City, NC, 28557. Phone: (252) 240-0888. Fax: (252) 240-2028. E-mail: email@fox8fox14.com Web Site: www.fox8fox14.com. Licensee: Piedmont Television of Eastern Carolina License LLC. Group owner: Piedmont Television Holdings LLC (acq 1995; $4.644 million for stn assets and $56,000 for CP). Natl. Network: Fox, MyNetworkTV Digital Network: Note: Fox is on WFXI(TV) ch 4, MyNetworkTV is on WFXI-DT ch 24. Washington Atty: Cohn & Marks. News: 4 hrs wkly.
Key Personnel:
Paul Brissette . . . . . CEO, chmn & pres
Bill Fielder . . . . . exec VP
Don Fisher . . . . . VP, gen mgr & natl sls mgr
Scott Foley . . . . . opns dir
Lisa Leonard . . . . . gen sls mgr & rgnl sls mgr
Billy Poplin . . . . . rgnl sls mgr
Walt Young . . . . . prom mgr
Linda Murphy . . . . . progmg dir
Andrea Griffith . . . . . news dir
Andy Kozik . . . . . chief of engrg
Rebroadcasts WYDO(TV) Greenville 100%.

WITN-TV—Washington, Analog channel: 7. Digital channel: 32. Analog hrs: 24 Digital hrs: 24 316 kw vis, 31.6 kw aur. 2,026t/2,000g TL: N35 21 55 W77 23 38 On air date: Sept 28, 1955. Box 468/Hwy. 17 S., Washington, NC, 27889. Phone: (252) 946-3131. Fax: (252) 946-0558. E-mail: witn@witntv.com Web Site: www.witntv.com. Licensee: Gray Television License Inc. Group owner: Gray Television Inc. (acq 8-1-97; $39.4 million). Population served: 1,388,000 Natl. Network: NBC. Wire Svc: NOAA Weather
Key Personnel:
Michael D. Weeks . . . . . gen mgr & chief of engrg
Michael Riddle . . . . . opns mgr, prom mgr & progmg mgr
Joe Carriere . . . . . gen sls mgr
Stephanie Shoop . . . . . news dir

WNCT-TV— Analog channel: 9. Digital channel: 10. Analog hrs: 24 hrs Digital hrs: 24 Note: CBS is on WNCT-TV ch 9, CW is on WNCT-DT ch 10. 316 kw vis, 31.6 kw aur. ant 1,879t/2,000g TL: N35 21 55 W77 23 38 On air date: Dec 22, 1953. 3221 South Evans St., Greenville, NC, 27834. Phone: (252) 355-8500. Fax: (252) 355-8568. E-mail: newsdesk@wnct.com Web Site: www.wnct.com. Licensee: Media

General Broadcasting Inc. Group owner: Media General Broadcast Group (acq 3-21-97; grpsl). Population served: 457,340 Natl. Network: CBS, CW Digital Network: Note: CBS is on WNCT-TV ch 9, CW is on WNCT-DT ch 10. Natl. Rep: MMT. Washington Atty: Dow, Lohnes & Albertson.

Key Personnel:
Vickie Jones . . . . gen mgr
Jerry Hogan . . . . gen sls mgr
Melissa Preas . . . . news dir
Bertie Cartwright . . . . chief of engrg

WPXU-TV— Analog channel: 35. Digital channel: 34. Analog hrs: 24 Digital hrs: 24 2,000 kw vis. ant 987t TL: N34 29 38 W77 29 18 On air date: 2004. c/o WEPX, 1301 S. Glenbernie Rd., New Bern, NC, 28562. Phone: (252) 636-2550. Fax: (252) 633-7851. Licensee: Paxson Jacksonville License Inc. Group owner: Paxson Communications Corp. (acq 10-1-99; $200,000). Natl. Network: MyNetworkTV.

***WUND-TV**— Analog channel: 2. Digital channel: 20. Analog hrs: 24 Digital hrs: 24 100 kw vis, 15 kw aur. 1548 t, 1538g TL: N35 54 00 W76 20 45 On air date: Sept 10, 1965. Box 14900, Research Triangle Park, NC, 27709-4900. 10 T.W. Alexander Dr., Research Triangle Park, NC 27709. Phone: (919) 549-7000. Fax: (919) 549-7201. E-mail: viewer@unctv.org Web Site: www.unctv.org. Licensee: University of North Carolina. Population served: 9,000,000 Natl. Network: PBS.

***WUNK-TV**— Analog channel: 25. Digital channel: 23. Analog hrs: 24 Digital hrs: 24 1,260 kw vis, 126 kw aur. ant 1,151t/1,138g TL: N35 33 10 W77 36 06 On air date: 1972. Box 14900, Research Triangle Park, NC, 27709-4900. 10 T.W. Alexander Dr., Research Triangle Park, NC 27709. Phone: (919) 549-7000. Fax: (919) 549-7201. E-mail: viewer@unctv.org Web Site: www.unctv.org. Licensee: University of North Carolina. Population served: 9,000,000 Natl. Network: PBS. Washington Atty: Schwartz, Woods & Miller. News staff: 15; News: 3 hrs wkly.

***WUNM-TV**— Analog channel: 19. Digital channel: 18. Analog hrs: 24 Digital hrs: 24 3,020 kw vis, 302 kw aur. 1,840t/1,761g TL: N35 06 18 W77 20 15 On air date: March 1982. Box 14900, Research Triangle Park, NC, 27709-4900. Phone: (919) 549-7000. Fax: (919) 549-7201. E-mail: viewer@unctv.org Web Site: www.unctv.org. Licensee: University of North Carolina. Population served: 9,000,000 Natl. Network: PBS. Washington Atty: Schwartz, Woods & Miller.

WYDO— Analog channel: 14. Digital channel: 21. Analog hrs: 24 Digital hrs: 24 1,104 kw vis. 686t TL: N35 26 44 W77 22 08 On air date: June 30, 1992. 5441 Hwy. 70 E., Morehead City, NC, 28557. Phone: (252) 240-0888. Fax: (252) 756-9250. E-mail: email@fox8fox14.com Web Site: www.fox8fox14.com. Licensee: Piedmont Television of Eastern Carolina License LLC. Group owner: GOCOM Communications Population served: 228,000 Natl. Network: Fox. Washington Atty: Wilkinson, Barker, Knauer & Quinn.

Key Personnel:
Don Fisher . . . . gen mgr
Andrea Griffith . . . . news dir
Andy Kozik . . . . chief of engrg

Satellite of WFXI Morehead City 100%.

High Point

see Greensboro-High Point-Winston Salem, NC market

Lumberton

see Myrtle Beach-Florence, SC market

Manteo

see Norfolk-Portsmouth-Newport News, VA market

New Bern

see Greenville-New Bern-Washington, NC market

Raleigh-Durham (Fayetteville), NC (DMA 29)

WFPX— Analog channel: 62. Digital channel: 36. Analog hrs: 24 Digital hrs: 24 337.3 kw vis, 33.7 kw aur. 846t/855g TL: N34 53 05 W79 04 29 On air date: Mar 14, 1985. Drawer 62, Lumber Bridge, NC, 28357. 19234 NC 71 Hwy N., Lumber Bridge, NC 28357. Phone: (910) 843-3884. Phone: (910) 843-3885. Fax: (910) 843-2873. Web Site: www.ionmedia.tv. Licensee: Paxson Communications License Co. L.L.C. Group owner: Paxson Communications Corp. (acq 10-20-97; $4.5 million). Population served: 275,000 Natl. Network: i Network. Natl. Rep: Adam Young. Washington Atty: Baraff, Koerner, Olender & Hochberg.

Key Personnel:
Rhonda Schulik . . . . rgnl sls mgr
Robbie Brock . . . . . progmg dir, pub affrs dir & chief of engrg

WLFL— Analog channel: 22. Digital channel: 57. Analog hrs: 24 Digital hrs: 24 5,000 kw vis, 232 kw aur. 1,675t/1,150g TL: N35 42 52 W78 49 01 On air date: Dec 18, 1981. 3012 Highwoods Blvd., Suite 101, Raleigh, NC, 27604. Phone: (919) 872-9535. Fax: (919) 878-3758. Web Site: www.wlfl22.com. Licensee: WLFL Licensee L.L.C. Group owner: Sinclair Broadcast Group Inc. Population served: 755,300 Natl. Network: CW. Natl. Rep: Millennium Sales & Marketing. Washington Atty: Shaw, Pittman. News staff: 36; News: 7 hrs wkly.

Key Personnel:
Neal Davis . . . . gen mgr
Kim Rivenbark . . . . prom mgr
Gary Todd . . . . engrg dir

WNCN— Analog channel: 17. Digital channel: 55. Analog hrs: 24 Digital hrs: 24 5,000 kw vis, 500 kw aur. ant 2,001t/1,902g TL: N35 40 29 W78 31 40 On air date: Apr 11, 1988. 1205 Front St., Raleigh, NC, 27609. Phone: (919) 836-1717. Fax: (919) 836-1687. Web Site: www.nbc17.com. Licensee: Media General Communications Inc. Group owner: NBC TV Stations Division. (acq 6-26-2006; grpsl). Natl. Network: NBC. Natl. Rep: MMT. News: 30 hrs wkly.

Key Personnel:
Barry Leffler . . . . gen mgr
Carol Ward . . . . gen sls mgr
Teresa Doring . . . . progmg dir & progmg mgr
Russell Mizelle . . . . chief of engrg

WRAL-TV— Analog channel: 5. Digital channel: 53. Analog hrs: 24 100 kw vis, 10 kw aur. 2,005t/2,000g TL: N35 40 35 W78 32 09 On air date: Dec 15, 1956. Box 12000, Raleigh, NC, 27605. 2619 Western Blvd., Raleigh, NC 27606. Phone: (919) 821-8555. Fax: (919) 821-8517. Fax: TWX: 510-928-1833. Web Site: www.wral.com. Licensee: Capitol Broadcasting Co. Inc. Group owner: (group owner) Population served: 2,120,000 Natl. Network: CBS. Natl. Rep: TeleRep. Washington Atty: Fletcher, Heald & Hildreth. Wire Svc: NWS (National Weather Service) Wire Svc: AP News staff: 100; News: 30 hrs wkly.

Key Personnel:
James Hefner . . . . gen mgr
Quinn Koontz . . . . sls dir
Laura Stillman . . . . natl sls mgr
John Harris . . . . progmg dir
Rick Gall . . . . news dir
Peter Sockett . . . . chief of engrg

WRAY-TV— Analog channel: 30. Digital channel: 42. Analog hrs: 24 Digital hrs: 24 1,800 kw vis. ant 1,768t/1,738g TL: N35 49 53 W78 08 50 On air date: 1995. 4909 Suite E. Expressway Dr., Wilson, NC, 27895-3583. Phone: (252) 243-0584. Fax: (252) 237-6290. Licensee: MTB Raleigh Licensee LLC. Group owner: Scripps Howard Broadcasting Co. (acq 11-15-2006; grpsl).

WRAZ— Analog channel: 50. Digital channel: 49. Analog hrs: 24 Digital hrs: 24 5,000 kw vis, 500 kw aur. 1,088t/967g TL: N35 42 55 W78 49 04 (CP: Ant 1,965t) On air date: 1995. Box 30050, Durham, NC, 27702. 512 S. Mangum St., Durham, NC 27701. Phone: (919) 595-5050. Fax: (919) 595-5028. Web Site: www.fox50.com. Licensee: WRAZ-TV Inc. Group owner: Capitol Broadcasting Co. Inc. (acq 2000; $1 million). Natl. Network: Fox. Natl. Rep: TeleRep. Washington Atty: Holland & Knight.

Key Personnel:
Thomas Schenck . . . . gen mgr
Jim Gamble . . . . chief of opns & chief of engrg
Evelyn Booker . . . . gen sls mgr
Kevin Kolbe . . . . mktg dir & prom dir
Joanne Stanley . . . . progmg mgr

WRDC— Analog channel: 28. Digital channel: 27. Analog hrs: 24 Digital hrs: 24 5,000 kw vis, 250 kw aur. ant 2,000t/1,976g TL: N35 40 35 W78 32 09 On air date: Nov 4, 1968. 3012 Highwoods Blvd., Suite 101, Raleigh, NC, 27604. Phone: (919) 872-2854. Fax: (919) 878-3758. Web Site: www.wrdc28.com. Licensee: Raleigh (WRDC-TV) Licensee Inc. Group owner: Sinclair Broadcast Group Inc. (acq 11-15-01; $2.3 million in stock). Population served: 493,000 Natl. Network: MyNetworkTV. Natl. Rep: Millennium Sales & Marketing. Washington Atty: Fisher, Wayland, Cooper, Leader & Zaragoza.

Key Personnel:
Neal Davis . . . . gen mgr
Lon Goldman . . . . natl sls mgr & rgnl sls mgr
Kim Rivenbark . . . . prom mgr
Donna Russell . . . . news dir
Gary Todd . . . . engrg dir & chief of engrg

WRPX— Analog channel: 47. Digital channel: 15. Analog hrs: 24 Digital hrs: 24 12.3 kw vis. 318t TL: N35 57 03 W77 55 37 (CP: 5,000 kw vis, ant 1,217t) On air date: Aug 31, 1987. 3209 Gresham Lake Rd., Suite 151, Raleigh, NC, 27615. Phone: (919) 827-4800. Fax: (919) 876-1415. Web Site: www.ionmedia.tv. Licensee: Paxson Raleigh License Inc. Group owner: Paxson Communications Corp. (acq 4-5-2000; grpsl). Population served: 352,154 Natl. Network: i Network. Natl. Rep: Roslin. Washington Atty: Mitchell, Fielstra & Assoc.

WTVD— Analog channel: 11. Digital channel: 52. Analog hrs: 24 Digital hrs: 24 316 kw vis, 47.4 kw aur. 1,990t/2,000g TL: N35 40 05 W78 31 58 On air date: Sept 2, 1954. 411 Liberty, Durham, NC, 27701. Phone: (919) 683-1111. Fax: (919) 682-7476 (sales). Fax: (919) 682-7225 (admin). Web Site: www.abc11tv.com. Licensee: WTVD Television LLC. Group owner: (group owner; a(cq 5-24-57; $1,417,800; FTR: 5-3-57). Population served: 804,000 Natl. Network: ABC. Washington Atty: Wilmer, Cutler & Pickering.

***WUNC-TV**— Analog channel: 4. Digital channel: 59. Analog hrs: 24 Digital hrs: 24 100 kw vis, 20 kw aur. ant 1,538t/1,269g TL: N35 51 59 W79 10 00 On air date: Jan 8, 1955. Box 14900, Research Triangle Park, NC, 27709-4900. 10 T.W. Alexander Dr., Research Triangle Park, NC 27709. Phone: (919) 549-7000. Fax: (919) 549-7201. E-mail: viewer@unctv.org Web Site: www.unctv.org. Licensee: University of North Carolina. Population served: 9,000,000 Natl. Network: PBS. Washington Atty: Schwartz, Woods & Miller.

***WUNP-TV**— Analog channel: 36. Digital channel: 39. Analog hrs: 24 Digital hrs: 24 1,550 kw vis, 155 kw aur. 1,207t/1,172g TL: N36 17 28 W77 50 10 On air date: 1985. Box 14900, Research Triangle Park, NC, 27709-4900. 10 T. W. Alexander Dr., Research Triangle Park, NC 27709. Phone: (919) 549-7000. Fax: (919) 549-7201. E-mail: viewer@unctv.org Web Site: www.unctv.org. Licensee: University of North Carolina. Natl. Network: PBS. Washington Atty: Schwartz, Woods & Miller.

WUVC-TV— Analog channel: 40. Digital channel: 38. Analog hrs: 24 Digital hrs: 24 5,000 kw vis, 500 kw aur. ant 1,842t/1,749g TL: N35 30 45 W75 58 40 On air date: June 1, 1981. 230 Donaldson St., 3rd Fl., ., Fayetteville, NC, 28301. Lake Plaza East, 900 Ridgefield Dr., Ste. 100, Raleigh, NC 27609. Phone: (910) 323-4040. Fax: (910) 323-3924. Web Site: www.univision.com. Licensee: Capital Broadcasting Partners. Group owner: Univisiion Communications Inc. (acq 3-31-2003). Population served: 947,750 Natl. Network: Univision (Spanish). Washington Atty: Brooks, Pierce, McLendon, Humphrey & Leonard. Foreign lang progmg: SP 168

Key Personnel:
Todd Schlachter . . . . gen mgr & natl sls mgr
Yvonne Cerna . . . . prom mgr
Maria Tajman . . . . progmg dir
Armando Trull . . . . news dir
William Acevedo . . . . chief of engrg

Washington

see Greenville-New Bern-Washington, NC market

Wilmington, NC (DMA 136)

WECT— Analog channel: 6. Digital channel: 44. Analog hrs: 24 Digital hrs: 24 100 kw vis, 20 kw aur. 2,054t/2,000g TL: N34 34 43 W78 26 13 On air date: Apr 9, 1954. 322 Shipyard Blvd., Wilmington, NC, 28412. Phone: (910) 791-8070. Fax: (910) 392-1509. E-mail: wect@wect.com Web Site: www.wect.com. Licensee: Raycom America License Subsidiary LLC. Group owner: Raycom Media Inc. (acq 9-12-96; grpsl). Population served: 369,000 Natl. Network: NBC. Natl. Rep: Harrington, Righter & Parsons. Rgnl. Rep: Rgnl rep: Covington & Burling

Key Personnel:
Karl Davis . . . . gen mgr
Beth Young . . . . gen sls mgr
Dave Toma . . . . mktg dir
Herschel Howie . . . . prom mgr
Raeford Brown . . . . news dir
Dan Ullmer . . . . chief of engrg

WSFX-TV— Analog channel: 26. Digital channel: 30. Analog hrs: 24 Digital hrs: 24 4,370 kw vis, 437 kw aur. 1,640t/1,625g TL: N34 07 51 W78 11 16 On air date: Sept 24, 1984. 322 Shipyard Blvd., Wilmington, NC, 28412. Phone: (910) 791-8070. Fax: (910) 202-0493. E-mail: tpostema@wsfx.com Web Site: www.wsfx.com. Licensee: Southeastern Media Holdings Inc. (acq 9-22-2003; $14 million). Population served: 542,000 Natl. Network: Fox, UPN. Natl. Rep: MMT. Washington Atty: Baraff, Koerner, Olender & Hochberg. News staff: 6; News: 3 hrs wkly.

Key Personnel:
Tom Postema . . . . gen mgr, gen sls mgr & natl sls mgr

Herschel Howie . . . . prom mgr
Raeford Brown . . . . news dir
Dan Ullner . . . . chief of engrg

***WUNJ-TV—** Analog channel: 39. Digital channel: 29. Analog hrs: 24 Digital hrs: 24 4,470 kw vis, 447 kw aur. 1,813t TL: N34 07 51 W78 11 16 On air date: 1971. Box 14900, Research Triangle Park, NC, 27709-4900. 10 T.W. Alexander Dr., Research Triangle Park, NC 27709-4900. Phone: (919) 549-7000. Fax: (919) 549-7201. E-mail: viewer@unctv.org Web Site: www.unctv.org. Licensee: University of North Carolina. Population served: 9,000,000 Natl. Network: PBS. Washington Atty: Schwartz, Woods & Miller.

WWAY— Analog channel: 3. Digital channel: 46. Analog hrs: 24 Digital hrs: 24 100 kw vis, 10 kw aur. 1,953t/1,941g TL: N34 07 51 W78 11 16 On air date: Oct 1, 1964. 615 N. Front St., Wilmington, NC, 28401. Phone: (910) 762-8581. Fax: (910) 762-8367. E-mail: acombs@wwaytv3.com Web Site: www.wwaytv3.com. Licensee: Libco Inc. Group owner: Liberty Corp. (acq 1-13-2006; grpsl). Population served: 120,284 Natl. Network: ABC.
Key Personnel:
Andy Combs . . . . gen mgr
Billy Stratton . . . . chief of engrg

North Dakota

Bismarck

see Minot-Bismarck-Dickinson, ND market

Dickinson

see Minot-Bismarck-Dickinson, ND market

Fargo-Valley City, ND (DMA 119)

KBRR— Analog channel: 10. Analog hrs: 18 158 kw vis, 15.8 kw aur. 600t/478g TL: N48 01 19 W96 22 12 On air date: July 1985. Box 9115, Fargo, ND, 58106. Phone: (701) 277-1515. Fax: (701) 277-1830. Licensee: Red River Broadcast Co. L.L.C. Group owner: (group owner) Natl. Network: Fox. Washington Atty: Crowell & Moring.
Key Personnel:
Kathy Lau . . . . . CEO, VP, gen mgr, prom mgr & progmg mgr
Jim Shaw . . . . pres & news dir
Milt Rost . . . . gen sls mgr
Dave Hoffman . . . . chief of engrg
Satellite of KVRR(TV) Fargo, ND.

***KCGE-DT—** Analog channel: 0. Digital channel: 16. Digital hrs: 24 105 kw vis. ant 720t/720g TL: N47 58 38 W96 36 18 On air date: October 2003. Box 3240, Fargo, ND, 58108. Phone: (701) 241-6900. Fax: (701) 239-7650. E-mail: info@prairiepublic.org Web Site: www.prairiepublic.org. Licensee: Prairie Public Broadcasting Inc. Natl. Network: PBS.
Key Personnel:
John E. Harris III . . . . CEO & pres
Steve Wennblom . . . . progmg mgr
Satellite of KFME(TV) Fargo, ND.

KCPM— Analog channel: 27. Analog hrs: 24 5,000 kw vis. ant 1,991t/1,955g TL: N47 16 45 W97 20 26 On air date: Jan 1, 2003. Box 9292, Fargo, ND, 58106. Phone: (701) 364-9900. Fax: (605) 334-5575. E-mail: mail@kcpm.tv Licensee: G.I.G. of North Dakota LLC. (acq 8-7-2001). Natl. Network: MyNetworkTV.

***KFME—** Analog channel: 13. Digital channel: 23. Analog hrs: 24 Digital hrs: 20 245 kw vis, 24.6 kw aur. ant 1,138t/1,145g TL: N47 00 48 W97 11 37 On air date: Jan 19, 1964. Box 3240, Fargo, ND, 58108-3240. 207 N. 5th St., Fargo, ND 58102. Phone: (701) 241-6900. Fax: (701) 239-7650. E-mail: info@prairiepublic.org Web Site: www.prairiepublic.org. Licensee: Prairie Public Broadcasting Inc. Population served: 362,400 Natl. Network: PBS. Washington Atty: Dow, Lohnes & Albertson.
Key Personnel:
John E. Harris III . . . . CEO & pres
Ann Clark . . . . dev dir
Steve Wennblom . . . . progmg mgr

***KGFE—** Analog channel: 2. Analog hrs: 24 100 kw vis, 10 kw aur. 1,382t/1,255g TL: N48 08 24 W97 59 38 On air date: Sept 9, 1974. Box 3240, Fargo, ND, 58108-3240. 207 N. 5th St., Fargo, ND 58102. Phone: (701) 241-6900. Fax: (701) 239-7650. E-mail: info@prairiepublic.org Web Site: www.prairiepublic.org. Licensee: Prairie Public Broadcasting Inc. Population served: 362,400 Natl. Network: PBS. Washington Atty: Dow, Lohnes & Albertson.
Key Personnel:
John E. Harris III . . . . CEO & pres
Steve Wennblom . . . . progmg mgr
Satellite of *KFME Fargo.

***KJRE—** Analog channel: 19. Digital channel: 20. Analog hrs: 24 Digital hrs: 24 407 kw vis, 40.7 kw aur. 587t TL: N46 17 55 W98 51 58 On air date: May 12, 1992. Box 3240, Fargo, ND, 58108-3240. 207 N. 5th St., Fargo, ND 58102. Phone: (701) 241-6900. Fax: (701) 239-7650. E-mail: info@prairiepublic.org Web Site: www.prairiepublic.org. Licensee: Prairie Public Broadcasting Inc. Natl. Network: PBS. Washington Atty: Dow, Lohnes & Albertson.
Key Personnel:
John E. Harris III . . . . CEO & pres
Steve Wennblom . . . . progmg mgr
Satellite of *KFME(TV) Fargo.

KJRR— Analog channel: 7. Analog hrs: 18 316 kw vis, 31.6 kw aur. 443t TL: N46 55 30 W98 46 21 On air date: Sept 1, 1988. Box 9115, Fargo, ND, 58106. Phone: (701) 277-1515. Fax: (701) 277-1830. Licensee: Red River Broadcast Co. L.L.C. Group owner: (group owner). Natl. Network: Fox. Washington Atty: Crowell & Moring.
Key Personnel:
Jim Shaw . . . . CEO & news dir
Kathy Lau . . . . VP, gen mgr, prom mgr & progmg dir
Milt Rost . . . . pres & gen sls mgr
Dave Hoffman . . . . chief of engrg
Satellite of KVRR(TV) Fargo.

***KMDE—** Analog channel: 0. Digital channel: 24.134 kw vis. ant 802t/729g TL: N48 03 47.8 W99 20 8.7 On air date: 2006. Box 3240, Fargo, ND, 58108-3240. Phone: (701) 241-6900. Fax: (701) 239-7650. E-mail: info@prairiepublic.org Web Site: www.prairiepublic.org. Licensee: Prairie Public Broadcasting Inc. Natl. Network: PBS.
Key Personnel:
John E. Harris III . . . . CEO & pres
Steve Wennblom . . . . progmg mgr
Satellite of KFME Fargo.

KNRR— Analog channel: 12. Digital channel: 15. Analog hrs: 18 Digital hrs: 18 316 kw vis, 31.6 kw aur. 1,394t TL: N48 59 42 W97 24 26 (CP: 158 kw vis, 15.8 kw aur) On air date: 1985. Box 9115, Fargo, ND, 58106. Phone: (701) 277-1515. Fax: (701) 277-1830. Licensee: Red River Broadcast Co. L.L.C. Group owner: (group owner) Natl. Network: Fox. Washington Atty: Crowell & Moring.
Key Personnel:
Dave Hoffman . . . . CEO & chief of engrg
Jim Shaw . . . . pres & news dir
Kathy Lau . . . . VP, gen mgr, prom mgr & progmg mgr
Milt Rost . . . . gen sls mgr
Satellite of KVRR(TV) Fargo.

KVLY-TV— Analog channel: 11. Digital channel: 44. Analog hrs: 24 Digital hrs: 24 304 kw vis, 45.7 kw aur. 2,000t/2,063g TL: N47 20 36 W97 17 17 On air date: Oct 11, 1959. 1350 21st Ave. S., Fargo, ND, 58103. Phone: (701) 237-5211. Fax: (701) 237-5396. E-mail: mail@kvlytv11.com Web Site: www.kvlytv11.com. Licensee: Hoak Media of Dakota License LLC. Group owner: Wicks Television L.L.C. (acq 1-3-2007; grpsl). Population served: 577,000 Natl. Network: NBC. Natl. Rep: Blair Television. Washington Atty: Wyrick, Robbins, Yates & Pontin. Wire Svc: AP News staff: 36; News: 16 hrs wkly.
Key Personnel:
Charlie Johnson . . . . gen mgr & news dir
Jeff Petrik . . . . opns mgr & progmg mgr
Ron Westrick . . . . gen sls mgr
Roger Johnson . . . . chief of engrg

KVRR— Analog channel: 15. Digital channel: 19. Analog hrs: 18 Digital hrs: 18 4,150 kw vis, 415 kw aur. 1,095t TL: N46 40 26 W96 13 40 On air date: Feb 14, 1983. Box 9115, Fargo, ND, 58106. Phone: (701) 277-1515. Fax: (701) 277-1830. Licensee: Red River Broadcast Co L.L.C. Group owner: (group owner) Natl. Network: Fox. Washington Atty: Crowell & Moring.
Key Personnel:
Dave Hoffman . . . . chmn & chief of engrg
Jim Shaw . . . . pres & news dir
Kathy Lau . . . . VP, gen mgr, prom mgr & progmg mgr
Milt Rost . . . . gen sls mgr

KXJB-TV—Valley City, Analog channel: 4. Digital channel: 38. Analog hrs: 20 Digital hrs: 20 97.7 kw vis, 10 kw aur. 2,030t/2,060g TL: N47 16 45 W97 20 18 On air date: Sept 11, 1954. 1350 21st Ave. S., Fargo, ND, 58103-3313. Phone: (701) 282-0444. Fax: (701) 232-0493. E-mail: news@kvoytv11.com Web Site: www.kx4.com. Licensee: Parker Broadcasting of Dakota LLC. Group owner: Catamount Broadcast Group (acq 1-3-2007). Population served: 150,000 Natl. Network: CBS. Natl. Rep: Continental Television Sales. Washington Atty: Cohn & Marks. News staff: 20; News: 10 hrs wkly.
Key Personnel:
Charlie Johnson . . . . gen mgr
Mark Von Bank . . . . gen sls mgr
Jeff Petrik . . . . progmg mgr
Mike Morken . . . . news dir
Ron Barr . . . . chief of engrg
Wendy Bernier . . . . prom

WDAY-TV— Analog channel: 6. Digital channel: 21. Analog hrs: 5 AM-2 AM Digital hrs: 5 AM-2 AM Note: ABC is on WDAY-TV ch 6, CW is on WDAY-DT ch 21. 100 kw vis, 11.4 kw aur. ant 1,150t/1,206g TL: N47 00 43 W97 11 58 On air date: June 1, 1953. Box 2466, Fargo, ND, 58108. 301 S. 8th St., Fargo, ND 58103. Phone: (701) 237-6500. Fax: (701) 241-5368. Web Site: www.wday.com/tv. Licensee: Forum Communications Co. Group owner: (group owner; (acq 7-20-60; $900,000; FTR: 7-25-60). Population served: 214,200 Natl. Network: ABC, CW Digital Network: Note: ABC is on WDAY-TV ch 6, CW is on WDAY-DT ch 21.
Key Personnel:
Mark Prather . . . . gen mgr
Carol Anhorn . . . . gen sls mgr
Susan Eider . . . . prom mgr & progmg mgr
Jeff Nelson . . . . news dir
Tom Thompson . . . . chief of engrg

WDAZ-TV— Analog channel: 8. Analog hrs: 19 Digital hrs: 19 316 kw vis, 50 kw aur. 1,480t/1,461g TL: N48 08 24 W97 59 38 On air date: Jan 29, 1967. Box 12639, Grand Forks, ND, 58208-2639. 2220 S. Washington, Grand Forks, ND 58201. Phone: (701) 775-2511. Fax: (701) 746-8565. E-mail: bkerr@wdaz.com Web Site: www.wdaz.com. Licensee: Forum Communications Co. Group owner: (group owner) Population served: 214,000 Natl. Network: ABC. Washington Atty: Holland & Knight.
Key Personnel:
Robert Kerr . . . . gen mgr
Rob Horken . . . . gen sls mgr & adv dir
Cassie Walder . . . . news dir
Jeff Awes . . . . engrg dir & chief of engrg
Satellite of WDAY-TV Fargo.

Minot-Bismarck-Dickinson, ND (DMA 158)

***KBME-TV—** Analog channel: 3. Digital channel: 22. Analog hrs: 20 Digital hrs: 20 79.4 kw vis, 7.9 kw aur. 1,394t/1,044g TL: N46 35 17 W100 48 30 On air date: June 18, 1979. Box 3240, Fargo, ND, 58108-3240. 207 N. 5th St., Fargo, ND 58102. Phone: (701) 241-6900. Fax: (701) 239-7650. E-mail: info@prairiepublic.org Web Site: www.prairiepublic.org. Licensee: Prairie Public Broadcasting Inc. Natl. Network: PBS. Washington Atty: Dow, Lohnes & Albertson.
Key Personnel:
John E. Harris III . . . . CEO & pres
Ann Clark . . . . dev dir
Steve Wennblom . . . . progmg mgr
Satellite of KFME-TV Fargo, ND.

KBMY— Analog channel: 17. Digital channel: 16. Analog hrs: 18 Digital hrs: 18 513 kw vis, 89.1 kw aur. 950t/649g TL: N46 35 11 W100 48 20 On air date: Mar 31, 1985. Box 7277, Bismarck, ND, 58507. 3128 E. Broadway, Bismarck, ND 58501. Phone: (701) 223-1700. Fax: (701) 258-0886. Licensee: KBMY-KMCY LLC. Group owner: Forum Communications Co. Natl. Network: ABC. Washington Atty: Marmet & McCombs.
Key Personnel:
Mark Prather . . . . gen mgr & gen sls mgr
Tony Kruckenberg . . . . chief of engrg
Rebroadcasts KMCY Minot 100%.

***KDSE—** Analog channel: 9. Digital channel: 20. Analog hrs: 20 Digital hrs: 20 214 kw vis, 21.4 kw aur. ant 806t/538g TL: N46 43 34 W102 54 56 On air date: Aug 4, 1982. Box 3240, 207 N. 5th St., Fargo, ND, 58108-3240. Phone: (701) 241-6900. Fax: (701) 239-7650. E-mail: info@prairiepublic.org Web Site: www.prairiepublic.com. Licensee: Prairie Public Broadcasting Inc. Natl. Network: PBS. Washington Atty: Dow, Lohnes & Albertson.
Key Personnel:
John E. Harris III . . . . CEO & pres
Steve Wennblom . . . . progmg mgr
Satellite of KFME(TV) Fargo.

KFYR-TV— Analog channel: 5. Digital channel: 31. Analog hrs: 24 Digital hrs: 24 100 kw vis, 13.5 kw aur. ant 1,400t/1,101g TL: N46 36 17 W100 48 30 On air date: Dec 19, 1953. Box 1738, Bismarck, ND, 58502. 200 N. 4th St., Bismarck, ND 58501. Phone: (701) 255-5757. Fax: (701) 255-8220. Web Site: www.kfyrtv.com. Licensee: Hoak Media of Dakota License LLC. Group owner: Wicks Television L.L.C.

(acq 1-3-2007; grpsl). Population served: 330,000 Natl. Network: NBC. Washington Atty: Hogan & Hartson. News staff: 29; News: 24 hrs wkly.
Key Personnel:
Dick Heidt . . . . . gen mgr
Barry Schumaier . . . . . gen sls mgr
Jim Sande . . . . . progmg dir

KMCY— Analog channel: 14. Digital channel: 15. Analog hrs: 18 Digital hrs: 18 513 kw vis, 89.1 kw aur. 2,720t/649g TL: N48 03 13 W101 23 05 On air date: June 22, 1985. 3128 E. Broadway, Bismarck, NC, 58501. Phone: (701) 223-1700. Fax: (701) 258-0886. Web Site: www.abc14.tv. Licensee: KBMY-KMCY LLC. Group owner: Forum Communications Co. Natl. Network: ABC. Washington Atty: Marmet & McCombs.
Key Personnel:
Kent Lein . . . . . gen mgr
Tony Kruckenberg . . . . . chief of engrg

KMOT— Analog channel: 10. Digital channel: 58. Analog hrs: 24 214 kw vis, 42.7 kw aur. 680t/690g TL: N48 12 56 W101 19 05 On air date: Jan 21, 1958. Box 1120, Minot, ND, 58702. 1800 S.W. 16th, Minot, ND 58701. Phone: (701) 852-4101. Fax: (701) 838-8195. Web Site: www.kmot.com. Licensee: Hoak Media of Dakota License LLC. Group owner: Wicks Television L.L.C. (acq 1-3-2007; grpsl). Natl. Network: NBC. Washington Atty: Hogan & Hartson.
Key Personnel:
Tom Ross . . . . . gen mgr & gen sls mgr
Nick Dreyer . . . . . news dir
Mike Robinson . . . . . chief of engrg
Satellite of KFYR-TV Bismarck.

KNDX— Analog channel: 26. Analog hrs: 18 1,797 kw vis. ant 1,112t TL: N46 35 23 W100 48 02 On air date: 2001. Box 4026, Bismarck, ND, 58502. Phone: (701) 355-0026. Fax: (701) 250-7244. Web Site: www.fox26.tv. Licensee: Prime Cities Broadcasting Inc. Natl. Network: Fox.
Key Personnel:
Gary O'Halloran . . . . . gen mgr
Richard Farley . . . . . chief of engrg

***KPSD-TV—** Analog channel: 13. Digital channel: 25. Analog hrs: 24 316 kw vis, 31.6 kw aur. ant 1,700t/1,696g TL: N45 03 20 W102 15 40 On air date: September 1973. Box 5000, Vermillion, SD, 57069-5000. 555 N. Dakota St., Vermillion, SD 57069. Phone: (605) 677-5861. Phone: (800) 456-0766. Fax: (605) 677-5010. E-mail: programming@sdpb.org Web Site: www.sdpd.org. Licensee: South Dakota Board of Directors for Educational Telecommunications. Population served: 40,000 Natl. Network: PBS. Washington Atty: Cohn & Marks.
Key Personnel:
Julie Andersen . . . . . pres
Terry Spencer . . . . . dev dir & dev mgr
Bob Bosse . . . . . progmg dir

KQCD-TV— Analog channel: 7. Digital channel: 18. Analog hrs: 24 316 kw vis, 31.6 kw aur. 731t/645g TL: N46 56 48 W102 59 17 On air date: July 28, 1980. 373 21 St. E., Dickinson, ND, 58601. Phone: (701) 483-7777. Fax: (701) 483-8231. E-mail: kqcd@kqcd.com Web Site: www.kqcd.com. Licensee: Hoak Media of Dakota License LLC. Group owner: Wicks Television L.L.C. (acq 1-3-2007; grpsl). Population served: 16,000 Natl. Network: NBC. Washington Atty: Hogan & Hartson.
Key Personnel:
Dick Heidt . . . . . gen mgr
Barry Schumaier . . . . . gen sls mgr
LuWanna Lawrence . . . . . prom mgr
Jim Sande . . . . . progmg mgr
Monica Hannan . . . . . news dir
Brian Funk . . . . . chief of engrg
Satellite of KFYR-TV Bismarck.

***KQSD-TV—** Analog channel: 11. Analog hrs: 24 234 kw vis, 28 kw aur. 1,040t/826g TL: N45 16 34 W99 59 03 On air date: Jan 1, 1976. Box 5000, Vermillion, SD, 57069-5000. 555 N. Dakota St., Vermillion, SD 57069. Phone: (605) 677-5861. Phone: (800) 456-0766. Fax: (605) 677-5010. E-mail: programming@sdpb.org Web Site: www.sdpd.org. Licensee: South Dakota Board of Directors for Educational Telecommunications. Natl. Network: PBS. Washington Atty: Cohn & Marks.
Key Personnel:
Julie Andersen . . . . . pres
Craig Jensen . . . . . opns mgr
Terry Spencer . . . . . dev dir
Fritz Miller . . . . . mktg mgr
Bob Bosse . . . . . progmg dir & progmg mgr

***KSRE—** Analog channel: 6. Digital channel: 40. Analog hrs: 20 Digital hrs: 20 100 kw vis, 10 kw aur. 1,059t/983g TL: N48 03 03 W101 23 24 On air date: January 1980. Box 3240, Fargo, ND, 58108-3240. Phone: (701) 241-6900. Fax: (701) 239-7650. E-mail: info@prairiepublic.org Web Site: www.prairiepublic.org. Licensee: Prairie Public Broadcasting. Population served: 750,000 Natl. Network: PBS. Washington Atty: Dow, Lohnes & Albertson.
Key Personnel:
John E. Harris III . . . . . CEO & pres
Steve Wennblom . . . . . progmg mgr
Satellite of KFME Fargo.

KUMV-TV— Analog channel: 8. Digital channel: 52. Analog hrs: 24 166 kw vis, 33.1 kw aur. 1,060t/874g TL: N48 08 02 W103 51 36 On air date: Feb 11, 1957. Box 1287, Williston, ND, 58802-1287. 602 Main St., Willinston, ND 58801. Phone: (701) 572-4676. Fax: (701) 572-0118. E-mail: kumv@kumv.com Web Site: www.kumv.com. Licensee: Hoak Media of Dakota License LLC. Group owner: Wicks Television L.L.C. (acq 1-3-2007; grpsl). Population served: 35,000 Natl. Network: NBC. News staff: 2; News: 6 hrs wkly.
Key Personnel:
Deborah Burton . . . . . gen mgr & gen sls mgr
Jim Sande . . . . . progmg mgr
Hawlie Ohe . . . . . news dir
Scott Aune . . . . . chief of engrg
Satellite of KFYR-TV Bismarck.

***KWSE—** Analog channel: 4. Digital channel: 51. Analog hrs: 20 Digital hrs: 20 79.4 kw vis, 7.94 kw aur. ant 912t/785g TL: N48 08 30 W103 53 34 On air date: March 1983. Box 3240, Fargo, ND, 58108-3240. Phone: (701) 241-6900. Fax: (701) 239-7650. E-mail: info@prairiepublic.org Web Site: www.prairiepublic.org. Licensee: Prairie Public Broadcasting Inc. Natl. Network: PBS. Washington Atty: Dow, Lohnes & Albertson.
Key Personnel:
John E. Harris III . . . . . CEO & pres
Steve Wennblom . . . . . progmg mgr
Satellite of KFME(TV) Fargo.

KXMA-TV— Analog channel: 2. Digital channel: 19. Analog hrs: 20 Digital hrs: 20 100 kw vis, 10 kw aur. 840t/621g TL: N46 43 30 W102 54 58 On air date: October 1956. 1625 W. Villard, Dickinson, ND, 58601. Phone: (701) 483-1400. Fax: (701) 483-1401. E-mail: webmasterb@kxnet.com Web Site: www.kxnet.com. Licensee: Reiten Television Inc. Group owner: (group owner; (acq 12-4-84; $362,500). Population served: 16,000 Natl. Network: CBS. Washington Atty: Fisher, Wayland, Cooper, Leader & Zaragoza.
Key Personnel:
Tim Reiten . . . . . gen mgr
Bruce Dintelman . . . . . gen sls mgr
Julie Bernhardt . . . . . natl sls mgr
Tom Gerhardt . . . . . news dir
Rocky Hefty . . . . . chief of engrg

KXMB-TV— Analog channel: 12. Digital channel: 23. Analog hrs: 20 Digital hrs: 20 316 kw vis, 31.6 kw aur. 1,530t/1,204g TL: N46 35 17 W100 48 26 On air date: Nov 19, 1955. 1811 N. 15th St., Bismarck, ND, 58501. Phone: (701) 223-9197. Fax: (701) 223-3320. E-mail: webmasterb@kxnet.com Web Site: www.kxnet.com. Licensee: Reiten Television Inc. Group owner: (group owner; (acq 1-27-71; $1.2 million; FTR: 2-8-71). Population served: 60,000 Natl. Network: CBS. Washington Atty: Fisher, Wayland, Cooper, Leader & Zaragoza. News staff: 11; News: 9 hrs wkly.
Key Personnel:
Tim Reiten . . . . . gen mgr
Bruce Dintelman . . . . . gen sls mgr
Julie Bernhardt . . . . . natl sls mgr
Tom Gerhardt . . . . . news dir
Rocky Hefty . . . . . chief of engrg

KXMC-TV— Analog channel: 13. Digital channel: 45. Analog hrs: 20 Digital hrs: 20 316 kw vis, 31.6 kw aur. 1,128t/1,061g TL: N48 03 02 W101 20 29 On air date: Apr 1, 1953. Box 1686, Minot, ND, 58702. 3425 S. Broadway, Minot, ND 58701. Phone: (701) 852-2104. Fax: (701) 838-9360. E-mail: webmasterb@kxnet.com Web Site: www.kxnet.com. Licensee: Reiten Television Inc. Group owner: (group owner; (acq 7-31-74; FTR: 8-19-74). Population served: 75,000 Natl. Network: CBS. Natl. Rep: Continental Television Sales. Washington Atty: Fisher, Wayland, Cooper, Leader & Zaragoza. News staff: 10; News: 9 hrs wkly.
Key Personnel:
David Reiten . . . . . pres & gen mgr
Darren Lenertz . . . . . gen sls mgr
Jim Olson . . . . . news dir
Bob Turneau . . . . . chief of engrg

KXMD-TV— Analog channel: 11. Digital channel: 14. Analog hrs: 20 Digital hrs: 20 174 kw vis, 17.4 kw aur. 980t/840g TL: N48 08 22 W103 53 24 On air date: Oct 25, 1969. Box 790, Williston, ND, 58801. 1802 13th Ave. W., Williston, NC 58801. Phone: (701) 572-2345. Fax: (701) 572-0658. Licensee: Reiten Television Inc. Group owner: (group owner). Population served: 34,000 Natl. Network: CBS. Washington Atty: Fisher, Wayland, Cooper, Leader & Zaragoza.
Key Personnel:
Darren Lenertz . . . . . gen mgr
David Reiten . . . . . gen mgr
Amanda Luchsinger . . . . . gen sls mgr
Jim Olson . . . . . news dir
Bob Turneau . . . . . chief of engrg

KXND— Analog channel: 24. Analog hrs: 18 740.4 kw vis. ant 784t/663g TL: N48 03 14 W101 26 03 On air date: 2001. Prime Cities Broadcasting Inc., 112 High Ridge Ave., Ridgefield, CT, 06877-4422. Phone: (203) 431-3366. Phone: (701) 355-0026. Fax: (203) 431-3864. Web Site: www.westdakotafox.com. Licensee: Prime Cities Broadcasting Inc. Population served: 353,000 Natl. Network: Fox.
Key Personnel:
Gary O'Halloran . . . . . gen mgr
Tony Kruckenberg . . . . . chief of engrg

Valley City

see Fargo-Valley City, ND market

Ohio

Akron

see Cleveland-Akron (Canton), OH market

Athens

see Charleston-Huntington, WV market

Cambridge

see Wheeling, WV-Steubenville, OH market

Canton

see Cleveland-Akron (Canton), OH market

Cincinnati, OH (DMA 33)

***WCET—** Analog channel: 48. Digital channel: 34. Analog hrs: 20 Digital hrs: 24 2,240 kw vis, 224 kw aur. 1,069t/899g TL: N39 07 30 W84 31 18 On air date: July 26, 1954. 1223 Central Pkwy., Cincinnati, OH, 45214-2890. Phone: (513) 381-4033. Fax: (513) 381-7520. E-mail: comments@cetconnect.org Web Site: www.cetconnect.org. Licensee: Greater Cincinnati TV Educational Foundation. Population served: 1,500,000 Natl. Network: PBS. Washington Atty: Dow, Lohnes & Albertson.
Key Personnel:
Jack Dominic . . . . . COO
Susan Howarth . . . . . CEO, pres & gen mgr
Ricardo O. Ang II . . . . . opns mgr
Brian Snape . . . . . prom mgr
Neal Schmidt . . . . . chief of engrg
Sherry Sargeant . . . . . sls

WCPO-TV— Analog channel: 9. Digital channel: 10. Analog hrs: 24 Digital hrs: 24 316 kw vis, 28.2 kw aur. 1,000t/890g TL: N39 07 31 W84 29 57 On air date: July 26, 1949. 1720 Gilbert Ave., Cincinnati, OH, 45202. Phone: (513) 721-9900. Fax: (513) 721-7717. E-mail: bfee@wcpo.com Web Site: www.wcpo.com. Licensee: Scripps Howard Broadcasting Co. Group owner: (group owner) Population served: 452,524 Natl. Network: ABC. Washington Atty: Baker & Hostetler. Wire Svc: UPI News staff: 70; News: 24 hrs wkly.
Key Personnel:
Bill Fee . . . . . gen mgr
Joe Martinelli . . . . . engrg dir

***WCVN-TV—** Analog channel: 54. Digital channel: 24.162 kw vis, 16.2 kw aur. 400t/312g TL: N39 01 50 W84 30 23 On air date: Sept 9, 1969. 600 Cooper Dr., Lexington, KY, 40502. Phone: (859) 258-7000. Fax: (859) 258-7399. Web Site: www.ket.org. Licensee: Kentucky Authority for Educational TV. Natl. Network: PBS.
Key Personnel:
Mike Brower . . . . . opns mgr
Craig Cornwell . . . . . progmg dir
Tim Bischoff . . . . . mktg dir & pub affrs dir
Robert Ball . . . . . engrg dir

***WKON—** Analog channel: 52. Digital channel: 44.562 kw vis, 115 kw aur. 710t/602g TL: N38 31 32 W84 48 40 On air date: Sept 23, 1968. 600 Cooper Dr., Lexington, KY, 40502. Phone: (859) 258-7000. Fax: (859) 258-7390. Web Site: www.ket.org. Licensee: Kentucky Authority for Educational TV. Natl. Network: PBS.
Key Personnel:
Craig Cornwell . . . . progmg mgr
Tim Bischoff . . . . mktg dir & pub affrs dir
Robert Ball . . . . engrg dir

WKRC-TV— Analog channel: 12. Digital channel: 31. Analog hrs: 24 Digital hrs: 24 Note: CBS is on WKRC-TV ch. 12, and CW is on WKRC-DT ch. 31. 316 kw vis, 31.6 kw aur. ant 1,000t/974g TL: N39 06 58 W84 30 05 On air date: April 1949. 1906 Highland Ave., Cincinnati, OH, 45219. Phone: (513) 763-5500. Fax: (513) 763-5554. Web Site: www.wkrc.com. Licensee: Citicasters Licenses Inc. (NEW). Group owner: Clear Channel Communications Inc. (acq 1999; grpsl). Population served: 759,000 Natl. Network: CBS, CW Digital Network: Note: CBS is on WKRC-TV ch. 12, and CW is on WKRC-DT ch. 31. Natl. Rep: TeleRep. Washington Atty: Hogan & Hartson.
Key Personnel:
Christopher Sehring . . . . gen mgr & sls VP
Hank Hundemer . . . . engrg dir

WLWT— Analog channel: 5. Digital channel: 35. Analog hrs: 24 100 kw vis, 10 kw aur. 1,000t/849g TL: N39 07 17 W84 31 18 On air date: Feb 9, 1948. 1700 Young St., Cincinnati, OH, 45202. Phone: (513) 412-5000. Fax: (513) 412-6121. E-mail: newsdesk@wlwt.com Web Site: www.wlwt.com. Licensee: Hearst-Argyle Stations Inc. Group owner: Hearst-Argyle Television Inc. (acq 7-16-97; grpsl). Population served: 792,000 Natl. Network: NBC. Natl. Rep: Eagle Television Sales. Washington Atty: Brooks, Pierce, McLendon, Humphrey & Leonard.
Key Personnel:
Richard J. Dyer . . . . pres & gen mgr
Mark Diangela . . . . gen sls mgr
Brennan Donnellan . . . . news dir
Paul Nowakowski . . . . chief of engrg

***WPTO—** Analog channel: 14. Analog hrs: 19 204 kw vis, 40.7 kw aur. 332t/342g TL: N39 30 26 W84 44 09 On air date: Oct 14, 1959. 110 S. Jefferson St., Dayton, OH, 45402. Phone: (937) 220-1600. Fax: (937) 220-1642. Web Site: www.thinktv.org. Licensee: Greater Dayton Public Television Inc. (acq 1975). Population served: 2,900,000 Natl. Network: PBS. Rgnl. Network: CEN, Ohio Educ Bcstg, Eastern Educ. Washington Atty: Dow, Lohnes & Albertson. News staff: one; News: one hr wkly.
Key Personnel:
David Fogarty. . . . . pres & gen mgr
Suzanne O'Brien . . . . CFO
Ed Valles . . . . dev dir
Kitty Lensman. mktg dir
Sue Brinson . . . . prom mgr
Gloria Skurski. . . . . progmg dir
Jim Wiener . . . . progmg mgr
H. Fred Stone . . . . engrg dir
George Hopstetter . . . . engrg mgr & chief of engrg

WSTR-TV— Analog channel: 64. Digital channel: 33. Analog hrs: 24 Digital hrs: 24 5,000 kw vis, 500 kw aur. ant 1,105t/925g TL: N39 12 01 W84 31 22 On air date: January 1980. 5177 Fishwick Dr., Cincinnati, OH, 45216. Phone: (513) 641-4400. Fax: (513) 242-2633. Web Site: www.my64.tv. Licensee: Sinclair Communications Group. Group owner: Sinclair Broadcast Group Inc. (acq 1996; $11 million). Population served: 820,000 Natl. Network: MyNetworkTV. Washington Atty: Cole, Raywid & Braverman.
Key Personnel:
Jon Lawhead . . . . gen mgr
Eric Lazar. . . . . natl sls mgr
Pete Ferraro. . . . . prom mgr
Rick White . . . . progmg dir
Terry Roberts . . . . engrg mgr

WXIX-TV—(Newport, KY) Analog channel: 19. Digital channel: 29. Analog hrs: 24 Digital hrs: 24 4,680 kw vis, 468 kw aur. 1,004t/984g TL: N39 07 19 W84 32 52 (CP: 4,646 kw vis, 464 kw aur) On air date: Aug 1, 1968. 19 Broadcast Plaza, 635 W. 7th St., Cincinnati, OH, 45203. Phone: (513) 421-1919. Fax: (513) 421-2829. E-mail: fox19@fox19.com Web Site: www.fox19.com. Licensee: wxix License Subsidiary, LLC. Group owner: Raycom Media Inc. (acq 1998; $45 million; grpsl). Population served: 1,946,000 Natl. Network: Fox. Natl. Rep: TeleRep. Washington Atty: Covington & Burling. News staff: 57; News: 23 hrs wkly.
Key Personnel:
Paul McTear. . . . . CEO
John Long . . . . gen mgr
Rick Oliver. . . . . opns mgr, progmg dir & progmg dir
Matt Kidwell . . . . natl sls mgr & rgnl sls mgr
Ray Mirabella . . . . natl sls mgr
Ron Stricker . . . . rgnl sls mgr

Cleveland-Akron (Canton), OH (DMA 17)

WBNX-TV—Akron, Analog channel: 55. Digital channel: 30. Analog hrs: 24 5,000 kw vis, 500 kw aur. 2,049t/1,131g TL: N41 23 02 W81 41 44 On air date: Dec 1, 1985. 2690 State Rd., Cuyahoga Falls, OH, 44223. Phone: (330) 922-5500. E-mail: clevelandswb@wbnx.com Web Site: www.wbnx.com. Licensee: Winston Broadcasting Network Inc. (acq 5-20-87; FTR: 1-19-87). Population served: 2,230,000 Natl. Network: CW. Natl. Rep: Adam Young. Washington Atty: Irwin, Campbell & Tannenwald.
Key Personnel:
Annie Keith. . . . . exec VP, VP, stn mgr & progmg dir
Lou Spangler . . . . pres & gen mgr
Colleen Metheney . . . . opns mgr
Eddie Brown. gen sls mgr
Debbie Stone . . . . prom dir

WDLI-TV— Analog channel: 17. Digital channel: 39. Analog hrs: 24 Digital hrs: 24 436 kw vis, 42 kw aur. 450t/450g TL: N40 51 04 W81 16 37 On air date: Jan 3, 1967. 1764 Wadsworth Rd., Akron, OH, 44320-3142. Phone: (330) 753-5542. Fax: (330) 753-4563. E-mail: wdli@tbn.org Licensee: Trinity Broadcasting Network. Group owner: (group owner; (acq 4-15-86; $4.5 million; FTR: 9-23-85). Population served: 1,500,000 Washington Atty: Joseph E. Dunne III.

***WEAO—** Analog channel: 49. Analog hrs: 24 Digital hrs: 24 685 kw vis, 68.56 kw aur. 1,047t/923g TL: N40 04 58 W81 38 00 On air date: September 1975. Box 5191, 1750 Campus Center Dr., Kent, OH, 44240-5191. Phone: (330) 677-4549. Fax: (330) 678-0688. E-mail: hr@wneo.pbs.org Web Site: www.pbs4549.org. Licensee: Northeastern Educational TV of Ohio Inc. Population served: 3,920,000 Natl. Network: PBS. Washington Atty: Dow, Lohnes & Albertson. News: one hr wkly.
Key Personnel:
Trina Cutter . . . . pres
Trinia Cutter . . . . gen mgr
Don Freeman. . . . . opns dir & dev dir
Bill O'Neil . . . . opns mgr
Lisa Martinez . . . . dev VP
Rebroadcasts WNEO(TV) Alliance 100%.

WEWS— Analog channel: 5. Digital channel: 15. Analog hrs: 24 Digital hrs: 24 93.3 kw vis, 10 kw aur. 1,020t/851g TL: N41 22 27 W81 43 06 On air date: Dec 17, 1947. 3001 Euclid Ave., Cleveland, OH, 44115. Phone: (216) 431-5555. Fax: (216) 431-3666. Web Site: www.newsnet5.com. Licensee: Scripps Howard Broadcasting Co. Group owner: (group owner, see Cross-Ownership) Population served: 1,463,900 Natl. Network: ABC. Natl. Rep: Eagle Television Sales. Washington Atty: Baker & Hostetler. Wire Svc: Reuters News: 22 hrs wkly.

WGGN-TV— Analog channel: 52. Digital channel: 42. Analog hrs: 24 Digital hrs: 24 1,480 kw vis, 148 kw aur. 774t/730g TL: N41 23 48 W82 47 31 On air date: Dec 5, 1982. Box 247, Castalia, OH, 44824. Phone: (419) 684-5311. Fax: (419) 684-5378. E-mail: wggn@lrbcg.com Web Site: www.cfbroadcast.com. Licensee: Christian Faith Broadcasting Inc. Group owner: (group owner) Population served: 750,000 Washington Atty: Joseph E. Dunne III.
Key Personnel:
Shelby Gillam . . . . pres
Rusty Yost . . . . gen mgr & chief of engrg

WJW— Analog channel: 8. Digital channel: 31. Analog hrs: 24 Digital hrs: 24 316 kw vis, 31.6 kw aur. 1,000t/775g TL: N41 21 47 W81 42 58 On air date: Dec 19, 1949. 5800 S. Marginal Rd., Cleveland, OH, 44103. Phone: (216) 431-8888. Fax: (216) 432-4282. Web Site: www.myfoxcleveland.com. Licensee: WJW License Inc. Group owner: Fox Television Stations Inc. (acq 11-96; grpsl). Population served: 1,431,000 Natl. Network: Fox.
Key Personnel:
Susan Pace . . . . VP
Greg Easterly. . . . . gen mgr
Paul Perozeni . . . . gen sls mgr
Paul Bodamer . . . . natl sls mgr
Barb Toth . . . . rgnl sls mgr
Kevin Salyer . . . . prom VP & progmg VP
Sonya Thompson . . . . news dir
Tom Cretet . . . . engrg dir

WKYC-TV— Analog channel: 3. Digital channel: 2. Analog hrs: 24 Digital hrs: 24 100 kw vis, 20 kw aur. 1,000t/906g TL: N41 23 09 W81 41 23 On air date: October 1948. 1333 Lakeside Ave., Cleveland, OH, 44114. Phone: (216) 344-3333. Fax: (216) 344-3326. E-mail: news@wkyc.com Web Site: www.wkyc.com. Licensee: WKYC-TV Inc. Group owner: Gannett Broadcasting (acq 12-4-95; grpsl). Population served: 2,800,000 Natl. Network: NBC.

WMFD-TV— Analog channel: 68. Digital channel: 12. Analog hrs: 24 Digital hrs: 24 294 kw vis, 29.4 kw aur. 591t/472g TL: N40 45 50 W82 37 04 (CP: 5,000 kw vis) On air date: Mar 3, 1988. 2900 Park Ave. W., Mansfield, OH, 44906. Phone: (419) 529-5900. Fax: (419) 529-2319. E-mail: comments@wmfd.com Web Site: www.wmfd.com. Licensee: Mid-State Television Inc. (acq 5-31-92; FTR: 6-15-92). Population served: 1,000,000 Washington Atty: Fletcher, Heald & Hildreth. Wire Svc: AP Wire Svc: CNN News staff: 12; News: 36 hrs wkly.
Key Personnel:
Gunther Meisse . . . . pres & gen mgr
Robert Meisse. . . . . stn mgr & opns mgr

WOAC— Analog channel: 67. Digital channel: 47. Analog hrs: 24 Digital hrs: 24 5000 kw vis, 500 kw aur. ant 485t/472g TL: N41 06 33 W81 20 10 On air date: March 1982. 4385 Sherman Rd., Kent, OH, 44240-6847. Phone: (330) 677-6760. Licensee: MTB Cleveland License LLC. Group owner: Scripps Howard Broadcasting Co. (acq 11-15-2006; grpsl). Population served: 1,500,000 Washington Atty: Wiley, Rein & Fielding LLP.

WOIO— Analog channel: 19. Digital channel: 10. Analog hrs: 24 Digital hrs: 24 3,720 kw vis, 372 kw aur. 1,151t/1,113g TL: N41 23 15 W81 41 43 On air date: May 19, 1985. 1717 E. 12th St., Cleveland, OH, 44114. Phone: (216) 771-1943. Fax: (216) 515-7152. Web Site: www.woio.com. Licensee: Raycom National Inc. Group owner: Raycom Media Inc. (acq 8-13-98). Population served: 2,000,000 Natl. Network: CBS. Natl. Rep: TeleRep. Washington Atty: Covington & Burling.
Key Personnel:
Bill Applegate . . . . gen mgr
Lisa McManus . . . . stn mgr & progmg mgr
Jim Stunek . . . . opns dir
Lynda King . . . . gen sls mgr
Rob Boenau. . . . . mktg dir & prom dir
Dan Salamone . . . . news dir
Emily Davis . . . . pub affrs dir
Bob Maupin . . . . engrg dir & chief of engrg

WQHS-TV— Analog channel: 61. Digital channel: 34. Analog hrs: 24 Digital hrs: 24 525 kw vis, 200 kw aur. ant 1,160t/1,029g TL: N41 23 02 W81 42 06 On air date: Mar 3, 1981. 2861 W. Ridgewood Dr., Parma, OH, 44134. Phone: (440) 888-0061. Fax: (440) 888-7023. Web Site: www.univision.com. Licensee: Univision Partnership of Ohio. Group owner: Univision Communications Inc. (acq 5-21-2001; grpsl). Population served: 3,500,000 Natl. Network: Univision (Spanish). Washington Atty: Wiley, Rein & Fielding. Wire Svc: UPI Foreign lang progmg: SP 168
Key Personnel:
Rolo Duartes . . . . gen mgr
Jose Godur . . . . gen sls mgr
Dave Smith . . . . chief of engrg

WUAB— Analog channel: 43. Digital channel: 28. Analog hrs: 24 Digital hrs: 24 4,680 kw vis, 468 kw aur. 1,102t/947g TL: N41 22 45 W81 43 12 (CP: 204 kw vis, ant 866t) On air date: Sept 15, 1968. 1717 E. 12th St., Cleveland, OH, 44114. Phone: (216) 771-1943. Fax: (216) 515-7152. Web Site: www.wuab.com. Licensee: Raycom National Inc. Group owner: Raycom Media Inc. (acq 3-2-00). Population served: 3,700,000 Natl. Network: MyNetworkTV. Washington Atty: Covington & Burling. Wire Svc: Reuters
Key Personnel:
Bill Applegate. . . . . gen mgr
Jim Stunek . . . . opns dir
Lynda King . . . . gen sls mgr
Rob Boenau. . . . . mktg dir & prom dir
Lisa McManus . . . . progmg mgr
Dan Salamone . . . . news dir
Emily Davis . . . . pub affrs dir
Bob Maupin . . . . chief of engrg
Todd Galloway . . . . chief of engrg

***WVIZ-TV—** Analog channel: 25. Digital channel: 26. Analog hrs: 24 2,140 kw vis, 214 kw aur. 997t/809g TL: N41 20 28 W81 44 24 On air date: Feb 7, 1965. 1375 Euclid Ave., Cleveland, OH, 44115-1826. Phone: (216) 916-6100. Fax: (216) 916-6123. Web Site: www.wviz.org. Licensee: Ideastream. (acq 2-27-2001). Population served: 1,700,000 Natl. Network: PBS.
Key Personnel:
Jerry Wareham . . . . CEO & pres
Kit Jensen . . . . COO
Bob Calsin . . . . CFO
Mark Smukler . . . . stn mgr
Kent A. Geist . . . . dev dir
Bob Stern. . . . . gen sls mgr
Maureen Paschke . . . . mktg dir
Jane Temple . . . . prom dir & adv dir
David Kanzeg. . . . . progmg dir
Thomas Furnas. . . . . chief of engrg

WVPX— Analog channel: 23. Analog hrs: 24 1,290 kw vis, 175 kw aur. 961t/926g TL: N41 03 51 W81 34 59 On air date: July 19, 1953. 1333

Lakeside Ave., East, Cleveland, OH, 44114. Phone: (216) 344-3333. Fax: (216) 344-7430. Web Site: www.ionline.tv. Licensee: Paxson Akron License Inc. Group owner: Paxson Communications Corp. (acq 2-29-96; $40 million; with WBPT(TV) Bridgeport, CT). Population served: 1,500,000 Natl. Network: i Network. Washington Atty: Dow, Lohnes & Albertson.
Key Personnel:
Robert Getze . . . gen sls mgr
Amy Sheridan . . . pub affrs dir
James Thomas . . . chief of engrg

Columbus, OH (DMA 32)

WBNS-TV— Analog channel: 10. Digital channel: 21. Analog hrs: 24 Digital hrs: 24 316 kw vis, 31.6 kw aur. 890t/1,029g TL: N39 58 16 W83 01 40 On air date: Oct 5, 1949. 770 Twin Rivers Dr., Columbus, OH, 43215. Phone: (614) 460-3700. Fax: (614) 460-2826. Web Site: www.10tv.com. Licensee: WBNS TV Inc. Group owner: Dispatch Broadcast Group Population served: 690,000 Natl. Network: CBS. Washington Atty: Sidley & Austin. News staff: 80; News: 31 hrs wkly.
Key Personnel:
Tom Griesdorn . . . gen mgr
Frank Wilson . . . opns dir, mktg dir & progmg dir
Mike Berry . . . opns mgr
Chuck Devendra . . . gen sls mgr
Pat Wise . . . natl sls mgr
Doug Jones . . . prom mgr
John Cardenas . . . news dir
Angela Pace . . . pub affrs dir
Pat Ingram . . . chief of engrg

WCMH-TV— Analog channel: 4. Digital channel: 14. Analog hrs: 24 Digital hrs: 24 100 kw vis, 15 kw aur. ant 903t/1,029g TL: N39 58 15 W83 01 39 On air date: Apr 3, 1949. 3165 Olentangy River Rd., Columbus, OH, 43202. Box 4, Columbus, OH 43216. Phone: (614) 263-4444. Fax: (614) 447-9107. Web Site: www.nbc4i.com. Licensee: Media General Communications Inc. Group owner: NBC TV Stations Division (acq 6-26-2006; grpsl). Population served: 1,782,500 Natl. Network: NBC. Natl. Rep: MMT. News staff: 60; News: 31.5 hrs wkly.
Key Personnel:
Marshall N. Morton . . . pres
Craig Robinson . . . gen mgr
Debra Grivois . . . opns dir & chief of engrg
Mike Cash . . . sls VP
Juilee Clark . . . natl sls mgr
Ken Lubker . . . rgnl sls mgr
Dana Pearson . . . mktg mgr
Janna Buckey . . . prom VP & adv VP
Jean Nemeti . . . progmg dir & pub affrs dir
Stan Sanders . . . news dir

***WOSU-TV**— Analog channel: 34. Digital channel: 38. Analog hrs: 24 Digital hrs: 24 1,170 kw vis, 117 kw aur. 1,079t/1,124g TL: N40 09 34 W82 55 22 On air date: Feb 20, 1956. 2400 Olentangy River Rd., Columbus, OH, 43210. Phone: (614) 292-9678. Fax: (614) 688-3399. E-mail: wosu@wosu.org Web Site: wosu.org. Licensee: Ohio State University. Population served: 1,800,000 Natl. Network: PBS. Rgnl. Network: CEN. Washington Atty: Dow, Lohnes & Albertson.
Key Personnel:
Thomas Rieland . . . gen mgr
Edwin Clay . . . stn mgr
John Prosek . . . opns mgr
Mary Yerina . . . dev dir
Tom Lahr . . . chief of engrg

WSFJ-TV— Analog channel: 51. Digital channel: 24. Analog hrs: 24 Digital hrs: 24 724 kw vis, 72.4 kw aur. ant 439t/279g TL: N40 04 44 W82 41 42 On air date: Mar 9, 1980. 3948 Townsfair Way, Suite 220, Columbus, OH, 43219. Phone: (614) 416-6080. Fax: (614) 416-6345. E-mail: comments@gtn51.com Web Site: www.gtn51.com. Licensee: Guardian Vision International Inc. Population served: 1,935,300 Washington Atty: Koerner & Olender P.C. News staff: one; News: one hr wkly.
Key Personnel:
Richard Schilg . . . gen mgr
Rob Kasper . . . opns dir
Bonnie Griffin . . . gen sls mgr
Elaine Kistler . . . rgnl sls mgr
Dave Wilson . . . progmg dir
Jason Knapp . . . chief of engrg
Erin Martin . . . prom

WSYX— Analog channel: 6. Digital channel: 13. Analog hrs: 24 Digital hrs: 24 Note: ABC is on WSYX(TV) ch 6, MyNetworkTV is on WSYX-DT ch 13. 100 kw vis, 10 kw aur. 938t/1,035g TL: N39 56 16 W83 01 16 On air date: Aug 30, 1949. 1261 Dublin Rd., Columbus, OH, 43215. Phone: (614) 481-6666. Fax: (614) 481-6828. Web Site: www.wsyx6.com. Licensee: WSYX Licensee Inc. Group owner: Sinclair Broadcast Group Inc. (acq 1998; $228 million). Population served: 1,500,000 Natl. Network: ABC, MyNetworkTV Digital Network: Note: ABC is on WSYX(TV) ch 6, MyNetworkTV is on WSYX-DT ch 13. Natl. Rep: Millennium Sales & Marketing.
Key Personnel:
Dan Mellon . . . gen mgr
Tony D'Angelo . . . sls dir
Lorie Luthman . . . gen sls mgr
Mike Hansen . . . mktg dir
Rick White . . . progmg dir
Lyn Tolan . . . news dir
Dan Carpenter . . . chief of engrg

WTTE— Analog channel: 28. Digital channel: 36. Analog hrs: 24 Digital hrs: 24 1,440 kw vis, 141 kw aur. ant 876t/938g TL: N39 56 14 W83 01 16 On air date: June 1, 1984. 1261 Dublin Rd., Columbus, OH, 43215. Phone: (614) 481-6666. Fax: (614) 485-1458. Web Site: www.wtte28.com. Licensee: Columbus (WTTE-TV) Licensee Inc. Group owner: Cunningham Broadcasting Corporation (acq 1-9-2002). Population served: 1,500,000 Natl. Network: Fox. Natl. Rep: Millennium Sales & Marketing.
Key Personnel:
Dan Mellon . . . gen mgr
Tony D'Angelo . . . sls dir
Mike Hansen . . . mktg dir
Rick White . . . progmg dir & progmg mgr
Lyn Tolan . . . news dir
Zoe Anne Del Borrell . . . pub affrs dir
Dan Carpenter . . . engrg dir & chief of engrg

WWHO— Analog channel: 53. Digital channel: 46. Analog hrs: 24 Digital hrs: 24 5,000 kw vis, 500 kw aur. ant 1,145t/1,115g TL: N39 35 20 W83 06 44 On air date: Aug 31, 1987. 1160 Dublin Rd., Suite 500, Columbus, OH, 43215. Phone: (614) 485-5300. Fax: (614) 485-5339. Web Site: www.wwhotv.com. Licensee: WWHO Broadcasting LLC. Group owner: Viacom Television Stations Group. (acq 3-31-2005; $85 million with WNDY-TV Marion, IN). Natl. Network: CW. Washington Atty: Wiley, Rein & Fielding.

Dayton, OH (DMA 58)

WBDT— Analog channel: 26. Digital channel: 18. Analog hrs: 24 Digital hrs: 24 5,000 kw vis, 500 kw aur. 1,145t/1,118g TL: N39 43 28 W84 15 18 On air date: September 1980. 2589 Corporate Pl., Miamisburg, OH, 45342. Phone: (937) 384-9226. Fax: (937) 384-7392. Web Site: daytonscw.com. Licensee: Acme Television Licenses of Ohio L.L.C. Group owner: Acme Communications Inc. (acq 6-14-99; grpsl). Natl. Network: CW. Natl. Rep: MMT.
Key Personnel:
John Hannon . . . VP & gen mgr
John Hannson . . . stn mgr
Gregg Abbott . . . opns mgr & progmg mgr
Melanie Simon . . . gen sls mgr
David Marchese . . . natl sls mgr
Brian Mercer . . . mktg dir
Shasta Niles . . . prom mgr
Al Schmidt . . . chief of engrg

WDTN— Analog channel: 2. Digital channel: 50. Analog hrs: 24 Digital hrs: 24 100 kw vis, 10 kw aur. 1,010t/960g TL: N39 43 07 W84 15 22 On air date: Mar 15, 1949. 4595 S. Dixie Ave., Dayton, OH, 45439. Phone: (937) 293-2101. Fax: (937) 294-6542. E-mail: newstips@wdtn.com Web Site: www.wdtn.com. Licensee: WDTN Broadcasting LLC. Group owner: LIN Television Corporation (acq 11-8-2002; grpsl). Population served: 1,277,000 Natl. Network: NBC. Natl. Rep: Blair Television.
Key Personnel:
Lisa Barhorst . . . gen mgr
Jim Atkinson . . . chief of opns & chief of engrg
Alison Wilkerson . . . sls VP & gen sls mgr
Sheryl Brownlee . . . gen sls mgr
Joe Mulligan . . . natl sls mgr
Jason Doyle . . . prom VP
Lisa barhorst . . . progmg VP
Steve Diorio . . . news dir
Sharon Howard . . . pub affrs dir

WHIO-TV— Analog channel: 7. Digital channel: 41. Analog hrs: 24 Digital hrs: 24 200 kw vis, 38 kw aur. ant 1,141t/1,059g TL: N39 44 02 W84 14 53 On air date: Feb 26, 1949. 1414 Wilmington Ave., Dayton, OH, 45420. Box 1206, Dayton, OH 45420. Phone: (937) 259-2111. Fax: (937) 259-2005. E-mail: 7online@whiotv.com Web Site: www.whiotv.com. Licensee: Miami Valley Broadcasting Corp. Group owner: Cox Broadcasting Population served: 1,276,400 Natl. Network: CBS. Natl. Rep: TeleRep. Washington Atty: Dow, Lohnes & Albertson.
Key Personnel:
Harry Delaney . . . VP & gen mgr
Chuck Eastman . . . opns dir, opns mgr & engrg dir
James Cosby . . . gen sls mgr
Tony Getts . . . prom dir
Fantine Kerckaert . . . progmg dir
David Bennallack . . . news dir

WKEF— Analog channel: 22. Digital channel: 51. Analog hrs: 24 Digital hrs: 24 2,340 kw vis, 234 kw aur. ant 1,152t/1,094g TL: N39 43 15 W84 15 39 On air date: Sept 27, 1964. 45 Broadcast Plaza, Dayton, OH, 45408. Phone: (937) 263-2662. Fax: (937) 268-2332. Web Site: www.daytonsnewssource.com. Licensee: WKEF Licensee L.P. Group owner: Sinclair Broadcast Group Inc. (acq 7-7-98; grpsl). Population served: 1,773,600 Natl. Network: ABC. News: 17 hrs wkly.
Key Personnel:
Dean Ditman . . . gen mgr
Branden Frantz . . . sls dir
Jason Matlock . . . prom dir
Roland Martel . . . engrg dir

WKOI-TV— Analog channel: 43. Digital channel: 39. Analog hrs: 24 1,410 kw vis, 141 kw aur. 990t/1,002g TL: N39 30 44 W84 38 09 On air date: May 11, 1982. Box 1057, Richmond, IN, 47375. 1702 S. 9th St., Richmond, IN 47374-7203. Phone: (765) 935-2390. Web Site: www.tbn.org. Licensee: Trinity Broadcasting of Indiana. Group owner: Trinity Broadcasting Network (acq 9-81). Washington Atty: Gammon & Grange.

***WPTD**— Analog channel: 16. Analog hrs: 24 1,510 kw vis, 151 kw aur. 1,188t/1,170g TL: N39 43 16 W84 15 00 On air date: Mar 20, 1967. 110 S. Jefferson St., Dayton, OH, 45402-2415. Phone: (937) 220-1600. Fax: (937) 220-1642. Web Site: www.thinktv.org. Licensee: Greater Dayton Public TV Inc. Population served: 243,601 Natl. Network: PBS. Rgnl. Network: Eastern Educ, CEN, Ohio Educ Bcstg. Washington Atty: Dow, Lohnes & Albertson. News staff: one; News: one hr wkly.
Key Personnel:
David Fogarty . . . pres & gen mgr
George Hopstetter . . . opns mgr & engrg mgr
Ed Valles . . . dev dir & dev mgr
Kitty Lensman . . . mktg dir
Sue Brinson . . . prom mgr
Gloria Skurski . . . progmg dir
Jim Wiener . . . progmg mgr
H. Fred Stone . . . engrg dir

WRGT-TV— Analog channel: 45. Digital channel: 30. Analog hrs: 24 Digital hrs: 24 Note: Fox is on WRGT-TV ch 45, MyNetworkTV is on WRGT-DT ch 30. 5,000 kw vis, 501 kw aur. ant 1,171t/1,158g TL: N39 43 28 W84 15 18 On air date: Sept 23, 1984. 45 Broadcast Plaza, Dayton, OH, 45408. Phone: (937) 263-4500. Fax: (937) 268-5265. Web Site: www.daytonsnewssource.com. Licensee: WRGT Licensee LLC. Group owner: Cunningham Broadcasting Corporation (acq 11-15-2001; grpsl). Population served: 1,773,600 Natl. Network: Fox, MyNetworkTV Digital Network: Note: Fox is on WRGT-TV ch 45, MyNetworkTV is on WRGT-DT ch 30. News: 16 hrs wkly.
Key Personnel:
Dean Ditmen . . . gen mgr
Branden Frantz . . . sls dir
Julie Gossard . . . natl sls mgr
Jason Matlock . . . prom dir
Roland Martel . . . engrg dir

Lima, OH (DMA 196)

WLIO— Analog channel: 35. Digital channel: 8. Analog hrs: 24 Digital hrs: 24 Note: NBC is on WLIO(TV) ch 35, CW is on WLIO-DT ch 8. 661 kw vis, 132 kw aur. ant 540t/549g TL: N40 44 54 W84 07 55 On air date: March 1953. Box 1689, Lima, OH, 45802. 1424 Rice Ave., Lima, OH 45805. Phone: (419) 228-8835. Fax: (419) 229-7091. Fax: (419) 225-6109. Web Site: www.wlio.com. Licensee: Lima Communications Corp. Group owner: BlockCommunications Inc. (acq 2-1-72; $1.5 million). Population served: 460,000 Natl. Network: NBC, CW Digital Network: Note: NBC is on WLIO(TV) ch 35, CW is on WLIO-DT ch 8. Washington Atty: Dow, Lohnes & Albertson. News staff: 17; News: 24 hrs wkly.
Key Personnel:
Bruce A. Opperman . . . pres & gen mgr
Dave Plaugher . . . CFO
Dave Plauqher . . . stn mgr
Kevin Creamer . . . sls VP
Kylie Fortman . . . prom mgr, progmg dir & film buyer
Lon Tegels . . . news dir
Tom Hendrixson . . . pub affrs dir
Fred Vobbe . . . engrg VP & chief of engrg

WTLW— Analog channel: 44. Digital channel: 47. Analog hrs: 24 Digital hrs: 24 912 kw vis, 91.2 kw aur. 679t/706g TL: N40 45 47 W84 10 59 On air date: June 13, 1982. 1844 Baty Rd., Lima, OH, 45807. Phone: (419) 339-4444. Fax: (419) 339-1736. E-mail: kbowers@wtlw.com Web Site: www.wtlw.com. Licensee: American Christian Television Services Inc. Population served: 200,000 Washington Atty: Wiley, Rein & Fielding.

Key Personnel:
Kevin Bowers . . . . CEO & stn mgr
Rick Corcoran . . . . chief of engrg

Portsmouth

see Charleston-Huntington, WV market

Steubenville

see Wheeling, WV-Steubenville, OH market

Toledo, OH (DMA 71)

***WBGU-TV**— Analog channel: 27. Digital channel: 56. Analog hrs: 24 Digital hrs: 24 1,000 kw vis, 100 kw aur. 1,060t/1,035g TL: N41 08 13 W83 54 23 On air date: Feb 10, 1964. 245 Troup St., Bowling Green, OH, 43403. Phone: (419) 372-2700. Fax: (419) 372-7048. E-mail: www@wbgu.bqsu.edu Web Site: www.wbgu.org. Licensee: Bowling Green State University. (acq 11-17-76; FTR: 12-13-76). Population served: 1.2 m,ill,ion Natl. Network: PBS. Rgnl. Network: Ohio Educ Bcstg. Washington Atty: Cohn & Marks.
Key Personnel:
Patrick Fitzgerald . . . . gen mgr
Deb Boyce . . . . prom dir
Ron Gargasz . . . . progmg mgr
Al Bowe . . . . chief of engrg

***WGTE-TV**— Analog channel: 30. Digital channel: 29. Analog hrs: 24 Digital hrs: 24 1,000 kw vis, 135 kw aur. 1,017t/1,034g TL: N41 39 27 W83 25 55 On air date: Oct 10, 1960. 1270 S. Detroit Ave., Toledo, OH, 43614. Box 30, Toledo, OH 43614. Phone: (419) 380-4600. Fax: (419) 380-4710. Web Site: www.wgte.org. Licensee: Public Broadcasting Foundation of N.W. Ohio. Population served: 409,500 Natl. Network: PBS. Rgnl. Network: Ohio Educ Bcstg. Washington Atty: Schwartz, Woods & Miller.
Key Personnel:
Marlon P. Kiser . . . . CEO, pres & gen mgr
Marlon Kiser . . . . CFO
Barbara Heslop . . . . opns mgr
Ross Pfeiffer . . . . dev dir
Lindsey Eberly . . . . sls dir
Jen Homier . . . . mktg dir
Darren LaShelle . . . . progmg dir & progmg mgr
Dan Niedzwiecki . . . . engrg dir

WLMB— Analog channel: 40. Digital channel: 5. Analog hrs: 24 Digital hrs: 24 4,747 kw vis. 571t TL: N41 44 41 W84 01 06 On air date: Oct 19, 1998. Box 908, Dominion Broadcasting Inc., 26693 Eckel Rd., Perrysburg, OH, 43552. Phone: (419) 874-8862. Fax: (419) 874-8867. E-mail: info@wlmb.com Web Site: www.wlmb.com. Licensee: Dominion Broadcasting Inc. Ownership: Larry Whatley, 33.3%; Ron Mighell, 33.3%; Jamey Schmitz, 33.3%. Population served: 1,500,000 Washington Atty: Wiley, Rein & Fielding.
Key Personnel:
Jamey Schmitz . . . . gen mgr
Curt Miller . . . . gen sls mgr
Brooke Myerholtz . . . . progmg dir
Eric Jingst . . . . chief of engrg

WNWO-TV— Analog channel: 24. Digital channel: 49. Analog hrs: 24 Digital hrs: 24 4,370 kw vis, 437 kw aur. TL: N41 40 03 W83 21 22 On air date: May 3, 1966. 300 S. Byrne Rd., Toledo, OH, 43615. Phone: (419) 535-0024. Fax: (419) 535-0202. Web Site: www.nbc24.com. Licensee: Barrington Toledo License LLC. Group owner: Raycom Media Inc. (acq 8-11-2006; grpsl). Population served: 1,097,400 Natl. Network: NBC. Natl. Rep: TeleRep. News: 22 hrs wkly.
Key Personnel:
Rick Lipps . . . . gen mgr
Hank Thompson . . . . chief of engrg

WTOL— Analog channel: 11. Digital channel: 17. Analog hrs: 23 Digital hrs: 23 316 kw vis, 38 kw aur. 1,000t/1,046g TL: N41 40 22 W80 22 47 On air date: Dec 5, 1958. 730 N. Summit St., Toledo, OH, 43604. Phone: (419) 248-1111. Fax: (419) 248-1177. E-mail: news@wtol.com Web Site: www.wtol.com. Licensee: Libco Inc. Group owner: Liberty Corp. (acq 1-13-2006; grpsl). Population served: 2,414,100 Natl. Network: CBS. Natl. Rep: Harrington, Righter & Parsons. Washington Atty: Dow, Lohnes & Albertson. News staff: 50; News: 25 hrs wkly.
Key Personnel:
Bob Chirdon . . . . gen mgr
Linda Blackburn . . . . gen sls mgr
Nancy Bright . . . . natl sls mgr
Steve Israel . . . . progmg dir
Mitch Jacob . . . . news dir & pub affrs dir
Eric Bergman . . . . chief of engrg

WTVG— Analog channel: 13. Digital channel: 19. Analog hrs: 24 Digital hrs: 24 316 kw vis, 18.2 kw aur. 1,000t/1,049g TL: N41 41 00 W83 24 49 On air date: July 21, 1948. 4247 Dorr St., Toledo, OH, 43607. Phone: (419) 531-1313. Fax: (419) 531-1399. Web Site: www.13abc.com. Licensee: WTVG Inc. Group owner: Capital Cities/ABC Video Enterprises International (acq 1995; $155 million with WJRT-TV Flint, MI). Population served: 450,000 Natl. Network: ABC. Washington Atty: Koteen & Naftalin. News staff: 30; News: 10 hrs wkly.
Key Personnel:
David Zamichow . . . . pres & gen mgr
Mary Gerken . . . . sls dir
Tamara Rost . . . . progmg dir
Brian Trauring . . . . news dir
Ernestine Weathers . . . . pub affrs dir
Barry Gries . . . . engrg dir

WUPW— Analog channel: 36. Digital channel: 46. Analog hrs: 24 Digital hrs: 24 1,950 kw vis, 195 kw aur. 1,220t/1,250g TL: N41 39 21 W83 26 40 On air date: Sept 22, 1985. Four SeaGate, Toledo, OH, 43604. Phone: (419) 244-3600. Fax: (419) 244-8842. E-mail: wupw@wupw.com Web Site: www.foxtoledo.com. Licensee: WUPW Broadcasting LLC. Group owner: LIN Television Corporation (acq 11-8-2002; grpsl). Natl. Network: Fox. Natl. Rep: Blair Television. Washington Atty: Shrinsky, Weitzman & Eisen. News staff: 5; News: 5 hrs wkly.
Key Personnel:
Ray Maselli . . . . gen mgr
Gary Yoder . . . . gen sls mgr & natl sls mgr
Brian Lorenzen . . . . rgnl sls mgr
Betsy Russell . . . . mktg dir & prom dir
Cathy Stoner . . . . progmg dir
Steve France . . . . news dir
Steve Crum . . . . engrg dir

Youngstown, OH (DMA 103)

WFMJ-TV— Analog channel: 21. Digital channel: 20.Note: NBC is on WFMJ-TV ch 21, CW is on WFMJ-DT ch 20. 3,720 kw vis, 372 kw aur. ant 990t/1,085g TL: N41 04 46 W80 38 25 On air date: Mar 8, 1953. 21 WFMJ, 101 W. Boardman St., Youngstown, OH, 44503. Phone: (330) 744-8611. Fax: (330) 744-3402. E-mail: information@wfmj.com Web Site: www.wfmj.com. Licensee: WFMJ Television Inc. Ownership: Mark A. Brown and Betty H. Brown Jagnow. (acq 7-14-93; FTR: 8-2-93). Population served: 721,200 Natl. Network: NBC, CW Digital Network: Note: NBC is on WFMJ-TV ch 21, CW is on WFMJ-DT ch 20. Washington Atty: Fisher, Wayland, Cooper, Leader & Zaragoza. News staff: 20; News: 10 hrs wkly.
Key Personnel:
John Grdic . . . . gen mgr & rgnl sls mgr
Kathie Brickman . . . . natl sls mgr
Jack Stevenson . . . . mktg dir
Joe Romano . . . . prom dir
Mona Alexander . . . . news dir
Carl Bryant . . . . pub affrs dir
Bob Flis . . . . chief of engrg

WKBN-TV— Analog channel: 27. Digital channel: 41. Analog hrs: 24 Digital hrs: 24 871 kw vis, 87.1 kw aur. ant 1,430t/1,432g TL: N41 03 28 W80 38 42 On air date: Jan 6, 1953. 3930 Sunset Blvd., Youngstown, OH, 44512. Phone: (330) 782-1144. Fax: (330) 782-3504. Fax: (330) 783-1834. Web Site: www.wkbn.com. Licensee: Piedmont Television of Youngstown License LLC. Group owner: Piedmont Television Holdings LLC (acq 10-29-99). Population served: 274,700 Natl. Network: CBS. Natl. Rep: Continental Television Sales. Washington Atty: Bryan Cave. News staff: 50; News: 25 hrs wkly.
Key Personnel:
David Coy . . . . gen mgr
John Amann . . . . opns mgr & prom mgr
Jill Duffy . . . . natl sls mgr
Nikki Manuel . . . . rgnl sls mgr
Phyllis Rappach . . . . progmg dir
Gary Coursen . . . . news dir
Thomas Zocolo . . . . chief of engrg

***WNEO**— Analog channel: 45. Digital channel: 46. Analog hrs: 24 Digital hrs: 24 1260 kw vis, 126 kw aur. 830t/770g TL: N40 54 23 W80 54 40 On air date: May 1973. Box 5191, Kent, OH, 44240-5191. Phone: (330) 677-4549. Fax: (330) 678-0688. E-mail: questions@wneo.org Web Site: www.pbs4549.org. Licensee: Northeastern Educational TV of Ohio Inc. Population served: 450,000 Natl. Network: PBS. Washington Atty: Dow, Lohnes & Albertson. News: one hr wkly.
Key Personnel:
Don Freeman . . . . COO
Trina Cutter . . . . pres & gen mgr

WYTV— Analog channel: 33. Digital channel: 36. Analog hrs: 24 Digital hrs: 24 912 kw vis, 110 kw aur. ant 580t/637g TL: N41 03 43 W80 38 07 On air date: Oct 30, 1957. 3800 Shady Run Rd., Youngstown, OH, 44502. Phone: (330) 783-2930. Fax: (330) 782-8154. Web Site: www.wytv.com. Licensee: Parkin Broadcasting of Youngstown License LLC. Ownership: Parkin Broadcasting of Youngstown LLC, 100% Group owner: Chelsey Broadcasting Co. (acq 7-30-2007). Population served: 693,000 Natl. Network: ABC, MyNetworkTV. Washington Atty: Drinker Biddle & Reath LLP. News staff: 26; News: 20 hrs wkly.
Key Personnel:
Dave Trabert . . . . gen mgr
Dan Messersmith . . . . gen sls mgr
Karen Brown . . . . prom mgr
Bill Lough . . . . chief of engrg

Zanesville, OH (DMA 203)

WHIZ-TV— Analog channel: 18. Analog hrs: 24 588 kw vis, 58.8 kw aur. 540t/508g TL: N39 55 42 W81 59 06 On air date: May 23, 1953. 629 Downard Rd., Zanesville, OH, 43701. Phone: (740) 452-5431. Fax: (740) 452-6553. E-mail: slauka@whiznews.com Web Site: www.whiznews.com. Licensee: Southeastern Ohio TV System. Ownership: Norma Littick Revocable Trust. Population served: 450,000 Natl. Network: NBC. Washington Atty: Leventhal, Senter & Lerman. News staff: 14; News: 10 hrs wkly.
Key Personnel:
N.J. Littick . . . . chmn
H.C. Littick . . . . pres
Doug Pickrell . . . . sls dir
Brian Wagner . . . . progmg VP
George Hiotis . . . . news dir
Dan Slentz . . . . chief of engrg

Oklahoma

Ada

see Sherman, TX-Ada, OK market

Lawton

see Wichita Falls, TX & Lawton, OK market

Oklahoma City, OK (DMA 46)

KAUT-TV— Analog channel: 43. Digital channel: 40. Analog hrs: 24 Digital hrs: 24 ant 1,560t/1,596g TL: N35 35 22 W97 29 03 On air date: Nov 3, 1980. 11901 N. Eastern Ave., Oklahoma City, OK, 73131. Phone: (405) 516-4300. Fax: (405) 516-4329. Web Site: www.ok43.com. Licensee: Local TV Oklahoma License LLC. Group owner: Viacom Television Stations Group (acq 5-7-2007; grpsl). Population served: 500,000 Natl. Network: MyNetworkTV.
Key Personnel:
Wes Milbourn . . . . stn mgr & gen sls mgr
Peter Grignon . . . . natl sls mgr
Stacy Johnson . . . . progmg dir
William Nichols . . . . chief of engrg

***KETA**— Analog channel: 13. Digital channel: 32. Analog hrs: 24 Digital hrs: 24 316 kw vis, 31.6 kw aur. 1,525t/1,578g TL: N35 32 58 W97 29 50 On air date: Apr 13, 1956. Box 14190, 7403 N. Kelley, Oklahoma City, OK, 73113. Phone: (405) 848-8501. Fax: (405) 841-9216. Web Site: www.oeta.onenet.net. Licensee: Oklahoma Educational TV Authority. Population served: 300,000 Natl. Network: PBS. Washington Atty: Cohn & Marks. News staff: 8; News: 3 hrs wkly.
Key Personnel:
John McCarroll . . . . gen mgr
Bill Thrash . . . . stn mgr
Mike Palmer . . . . opns dir & opns mgr
Bob Sands . . . . news dir
Earle Conners . . . . engrg dir
Richard Ladd . . . . chief of engrg

KEYU— Analog channel: 0. Digital channel: 31.5,000 kw vis. ant 1,223t/1,200g TL: N35 18 53 W101 50 47 On air date: 2004. 1616 S. Kentucky, Suite D-130, Amarillo, TX, 79102. Phone: (806) 359-8900. Fax: (806) 352-8912. Web Site: www.univision-amarillo.com. Licensee: Borger Broadcasting Inc. Group owner: Equity Broadcasting Corp. Natl. Network: Univision (Spanish). Foreign lang progmg: SP 168

KFOR-TV— Analog channel: 4. Digital channel: 27. Analog hrs: 24 Digital hrs: 24 97.7 kw vis, 19.5 kw aur. 1,540t/1,602g TL: N35 34 07 W97 29 20 On air date: June 6, 1949. 444 E. Britton Rd., Oklahoma City, OK, 73114. Phone: (405) 424-4444. Fax: (405) 478-6206. Web

Site: www.kfor.com. Licensee: Local TV Oklahoma License LLC. Group owner: The New York Times Co. (acq 5-7-2007; grpsl). Population served: 2,177,200 Natl. Network: NBC. Natl. Rep: Millennium Sales & Marketing. Washington Atty: Koteen & Naftalin. Wire Svc: UPI News staff: 73; News: 29 hrs wkly.
Key Personnel:
Wes Milbourn . . . stn mgr & gen sls mgr
Peter Grignon . . . natl sls mgr
Luanne Stuart . . . prom dir
Belinda Lane . . . progmg dir
Bob Ablah . . . chief of engrg

KOCB— Analog channel: 34. Digital channel: 33. Analog hrs: 24 Digital hrs: 24 1,170 kw vis, 117 kw aur. ant 1,210t/1,258g TL: N35 33 36 W97 29 07 On air date: Oct 28, 1979. 1228 .E. Wilshire Blvd., Oklahoma City, OK, 73111. Phone: (405) 843-2525. Fax: (405) 478-1027 (405) 475-9163(Sales). Web Site: www.kocb.com. Licensee: KOCB Licensee L.L.C. Group owner: Sinclair Broadcast Group Inc. (acq 1996; $63 million with WDKY-TV Danville, KY). Population served: 1,666,000 Natl. Network: CW. Natl. Rep: Harrington, Righter & Parsons.
Key Personnel:
John Rossi . . . gen mgr
Dan Loving . . . sls dir
Joe Spadea . . . news dir
Steve Bottkol . . . engrg dir

KOCM— Analog channel: 46.2,089 kw vis. ant 1,542t/1,578g TL: N35 35 52 W97 29 22 On air date: 2004. Daystar Television Network, Box 612066, Dallas, TX, 75261-2066. Phone: (817) 571-1229. Phone: (405) 292-4600. Fax: (817) 571-7458. E-mail: cpmments@daystar.com Web Site: www.daystartv.net. Licensee: Word of God Fellowship Inc. Ownership: Marcus D. Lamb, 25%; Jimmie F. Lamb, 25%; Joni T. Lamb, 25%; John T. Calender, 25% (acq 8-19-2002; $3.6 million).

KOCO-TV— Analog channel: 5. Digital channel: 7. Analog hrs: 24 Digital hrs: 24 100 kw vis, 14.5 kw aur. 1,519t/1,562g TL: N35 33 45 W97 29 24 (CP: Ant 1,515t/1,558g) On air date: July 15, 1954. 1300 E. Britton Rd., Oklahoma City, OK, 73131. Phone: (405) 478-3000. Fax: (405) 475-5242. Web Site: www.koco.com. Licensee: Hearst-Argyle Stations Inc. Group owner: Hearst-Argyle Television Inc. (acq 7-16-97; grpsl). Population served: 663,200 Natl. Network: ABC. Washington Atty: Brooks, Pierce, McLendon. Wire Svc: NWS (National Weather Service) Wire Svc: AP News staff: 55; News: 30 hrs news progrg wkly.
Key Personnel:
Brent Hensley . . . pres & gen mgr
Christine Toldt . . . opns dir
Tom Comerford . . . gen sls mgr
Sherrie Brown . . . news dir
David Evans . . . engrg dir & chief of engrg

KOKH-TV— Analog channel: 25. Digital channel: 24. Analog hrs: 24 Digital hrs: 24 3,470 kw vis, 347 kw aur. ant 1,560t/1,586g TL: N35 32 58 W97 29 18 On air date: Jan 26, 1979. 1228 E. Wilshire Blvd., Oklahoma City, OK, 73111. Box 14925, Oklahoma City, OK 73111. Phone: (405) 843-2525. Fax: (405) 478-1027 (405) 475-9163(Sales). Web Site: www.okcfox.com. Licensee: Sullivan Broadcasting Co. IV Inc. Group owner: Sinclair Broadcast Group Inc. (acq 1998; grpsl). Population served: 796,000 Natl. Network: Fox. Natl. Rep: Harrington, Righter & Parsons.
Key Personnel:
John Rossi . . . gen mgr
Dan Loving . . . sls dir
Joe Spadea . . . news dir
Steve Bottkol . . . engrg dir

KOPX— Analog channel: 62. Digital channel: 50. Analog hrs: 24 Digital hrs: 24 2,690 kw vis. 787t TL: N35 34 24 W97 29 08 On air date: February 1997. 13424 Railway Dr., Oklahoma City, OK, 73114. Phone: (405) 478-9562. Fax: (405) 751- 6867. Web Site: www.ionline.tv. Licensee: Paxson Oklahoma City License Inc. Group owner: Paxson Communications Corp. (acq 9-27-96; $6.395 million). Natl. Network: i Network. Washington Atty: Dow, Lohnes and Albertson PLLC.
Key Personnel:
Brandon Burgess . . . pres
Carol Wright-Holzhaver . . . sr VP
David A. Glenn . . . engrg VP
Rod Roberts . . . chief of engrg

KSBI— Analog channel: 52. Digital channel: 51. Analog hrs: 24 Digital hrs: 24 3,020 kw vis, 302 kw aur. ant 1,512t/1,538g TL: N35 35 54 W97 29 25 On air date: Sept 19, 1988. 1350 S.E. 82nd St., Oklahoma City, OK, 73149. Phone: (405) 631-7335. Fax: (405) 631-7367. E-mail: info@kbitv.com Web Site: www.ksbitv.com. Licensee: Family Broadcasting Group Inc. Ownership: Angela Brus, 38.4%; Brady M. Brus, 22.8%; Brenda Deimund, 22.8%; and Seekfirst Media Partners LLC, 16%. Population served: 567,200 Washington Atty: Booth, Freret, Imlay & Tepper.
Key Personnel:
Brady Brus . . . CEO, pres & gen mgr
Brenda Bennett . . . VP
Brendon Baker . . . opns mgr
Dean Powery . . . gen sls mgr
Jeff Chancey . . . chief of engrg

KTBO-TV— Analog channel: 14. Digital channel: 15. Analog hrs: 24 575 kw vis, 116 kw aur. ant 1,135t/1,185g TL: N35 34 30 W97 29 04 On air date: Mar 6, 1981. 1600 E. Heffner Rd., Oklahoma City, OK, 73131. Phone: (405) 848-1414. E-mail: comments@tbn.org Web Site: www.tbn.org. Licensee: Trinity Broadcasting of Oklahoma City Inc. Group owner: Trinity Broadcasting Network. Population served: 1,544,000 Washington Atty: Joseph E. Dunne III.
Key Personnel:
Paul Crouch . . . pres
Liuda Cook . . . gen mgr
Jan Crouch . . . prom dir & pub affrs dir
Ken Howerton . . . chief of engrg

KTUZ-TV— Analog channel: 30. Digital channel: 29. Analog hrs: 24 Digital hrs: 24 5,000 kw vis. ant 836t TL: N35 16 50 W97 20 14 On air date: 2000. 5101 S. Shields Blvd., Oklahoma City, OK, 73129. Phone: (405) 616-9900. Fax: (405) 616-5511. Web Site: http://telemundo.yahoo.com. Licensee: Oklahoma Land Company LLC. Group owner: Tyler Media Broadcasting Corp. (acq 9-30-2004; $12,375,000). Natl. Network: Telemundo (Spanish). Foreign lang progmg: SP 168

KUOK— Analog channel: 35.320 kw vis. ant 1,112t/1,059g TL: N36 16 06 W99 26 56 On air date: 2004. Equity Broadcasting Corp., 1 Shackleford Dr., Suite 400, Little Rock, AR, 72211. Phone: (501) 219-2400. Fax: (501) 716-3502. Licensee: Woodward Broadcasting Inc. Group owner: Equity Broadcasting Corp. Natl. Network: Univision (Spanish). Foreign lang progmg: SP 168
Key Personnel:
Larry Morton . . . pres
Gordon Hodges . . . gen mgr

***KWET**— Analog channel: 12. Digital channel: 8. Analog hrs: 24 Digital hrs: 24 283 kw vis. ant 993t/958g TL: N35 35 36 W99 40 01 On air date: Aug 6, 1978. Box 14190, Oklahoma City, OK, 73113. Phone: (405) 848-8501 (580) 497-2594. Fax: (405) 841-9216. Web Site: www.oeta.onenet.net. Licensee: Oklahoma Educational TV Authority. Population served: 647,390 Natl. Network: PBS. Washington Atty: Cohn & Marks. News staff: 20; News: 4 hrs wkly.
Key Personnel:
John McCarroll . . . gen mgr
Bill Thrash . . . stn mgr
Mike Palmer . . . opns mgr
Bob Sands . . . news dir
Earle Connors . . . engrg dir
Richard Ladd . . . chief of engrg

KWTV— Analog channel: 9. Digital channel: 39. Digital hrs: 24 316 kw vis, 33.9 kw aur. ant 1,525t/1,537g TL: N35 32 68 W97 29 50 On air date: Dec 20, 1953. 7401 N. Kelley Ave., Oklahoma City, OK, 73113. Phone: (405) 843-6641. Fax: (405) 841-9926. Web Site: www.newsok.com. Licensee: Griffin Entities L.L.C. (acq 7-1-98). Population served: 582,000 Natl. Network: CBS. Natl. Rep: TeleRep. Washington Atty: Holland & Knight. Wire Svc: CBS Wire Svc: NWS (National Weather Service) News staff: 80; News: 36 hrs wkly.
Key Personnel:
Rob Krier . . . VP & gen mgr
Wade Deaver . . . sls dir
Kim Eubank . . . progmg dir
Blaise Labbe . . . news dir
Julie Cameron . . . engrg dir & chief of engrg

Tulsa, OK (DMA 62)

KDOR-TV— Analog channel: 17. Digital channel: 15. Analog hrs: 24 Digital hrs: 24 3,980 kw vis, 398 kw aur. 1,040t/1,089g TL: N36 30 59 W95 46 10 On air date: Jan 11, 1987. 2120 N. Yellowood, Broken Arrow, OK, 74012. Phone: (918) 250-0777. Fax: (918) 461-8817. E-mail: kdor@tbn.org Web Site: www.tbn.org. Licensee: Trinity Broadcasting Network. Group owner: (group owner; (acq 5-8-2000; grpsl).
Key Personnel:
Paul Crouch Sr. . . . CEO & pres
Craig Nelson . . . gen mgr

KGEB— Analog channel: 53. Digital channel: 49. Analog hrs: 24 Digital hrs: 24 1,770 kw vis, 177 kw aur. 597t/672g TL: N36 02 39 W95 57 11 On air date: Jan 24, 1996. 7777 S. Lewis Ave., Tulsa, OK, 74171. Phone: (918) 488-5300. Fax: (918) 495-7388. E-mail: kgeb@oru.edu Web Site: www.kgeb.net. Licensee: University Broadcasting Inc.
Key Personnel:
Walter Richardson . . . gen mgr
Christi Vanover . . . progmg mgr
William P. Lee . . . chief of engrg

KJRH— Analog channel: 2. Digital channel: 56. Analog hrs: 24 100 kw vis, 10 kw aur. ant 1,828t TL: N36 01 15 W95 40 32 On air date: Dec 5, 1954. 3701 S. Peoria Ave., Tulsa, OK, 74105-3269. Phone: (918) 743-2222. Fax: (918) 748-1460. E-mail: news@kjrh.com Web Site: www.kjrh.com. Licensee: Scripps Howard Broadcasting Co. Group owner: Scripps Howard Stations, see Cross-Ownership (acq 1-1-71; $7.8 million). Population served: 1,143,000 Natl. Network: NBC. Natl. Rep: Eagle Television Sales. Washington Atty: Baker & Hostetler. News staff: 50; News: 26.5 hrs wkly.
Key Personnel:
Ken Lowe . . . CEO
Michael Vrabac . . . VP & gen mgr
Nick R. Clark . . . gen sls mgr
Peter Noll . . . prom mgr
Steve Weinstein . . . news dir
Samantha Knowlton . . . pub affrs dir
Dale Vennes . . . chief of engrg

KMYT-TV— Analog channel: 41. Digital channel: 42. Analog hrs: 24 Digital hrs: 24 1,350 kw vis, 270 kw aur. ant 1,510t/1,368g TL: N36 01 10 W95 39 24 On air date: May 17, 1981. 2625 S. Memorial Dr., Tulsa, OK, 74129-2600. Phone: (918) 388-5100. Fax: (918) 493-5739. Web Site: www.my41.com. Licensee: Clear Channel Broadcasting Licenses Inc. Group owner: Clear Channel Communications Inc. (acq 5-1-2000; grpsl). Population served: 511,000 Natl. Network: MyNetworkTV. Natl. Rep: Millennium Sales & Marketing. Washington Atty: Ward & Mendelsohn.
Key Personnel:
Bill Moll . . . pres
Craig Millar . . . VP & gen mgr
Jim Hanning . . . natl sls mgr
Deedra Determan . . . mktg dir & mktg mgr
Chooi Ning . . . progmg dir
Melanie Henry . . . news dir
Brian Egan . . . chief of engrg

***KOED-TV**— Analog channel: 11. Digital channel: 38. Analog hrs: 24 Digital hrs: 24 316 kw vis, 25 kw aur. ant 1,661t/1,833g TL: N36 01 15 W95 40 32 On air date: Jan 12, 1959. 811 N. Sheridan, Tulsa, OK, 74115. Phone: (918) 838-7611 (800) 580-7614. Fax: (918) 838-1807. Web Site: www.oeta.onenet.net. Licensee: Oklahoma Educational TV Authority. Population served: 500,000 Natl. Network: PBS. Washington Atty: Cohn & Marks. Foreign lang progmg: SP 8
Key Personnel:
Bill Thrash . . . stn mgr
Liz Exon . . . news dir
Roger Newton . . . chief of engrg

***KOET**— Analog channel: 3. Digital channel: 31. Analog hrs: 24 Digital hrs: 24 100 kw vis, 10 kw aur. ant 1,310t/699g TL: N35 11 01 W95 20 20 On air date: Aug 22, 1978. Box 14190, Oklahoma City, OK, 73113. Phone: (405) 848-8501 (918) 4693430. Fax: (405) 841-9216. Web Site: www.oeta.onenet.net. Licensee: Oklahoma Educational Television Authority. Natl. Network: PBS. Washington Atty: Cohn & Marks.
Key Personnel:
John McCarroll . . . gen mgr
Bill Thrash . . . stn mgr
Mike Palmer . . . opns dir & opns mgr
Bob Sands . . . news dir
Earle Connors . . . engrg dir
Richard Ladd . . . chief of engrg

KOKI-TV— Analog channel: 23. Analog hrs: 24 3,310 kw vis, 331 kw aur. ant 1,313t/1,274g TL: N36 01 36 W95 40 44 On air date: Oct 26, 1980. 2625 S. Memorial Dr., Tulsa, OK, 74129-2600. Phone: (918) 491-0023. Fax: (918) 491-6650. Web Site: www.fox23.com. Licensee: Clear Channel Broadcasting Licenses Inc. Group owner: Clear Channel Communications Inc. (acq 8-5-92). Population served: 450,000 Natl. Network: Fox. Washington Atty: Cohn & Marks. News staff: 38; News: 7 hrs wkly.
Key Personnel:
Bill Moll . . . pres
Craig Millar . . . VP & gen mgr
Holly Allen . . . sls dir
David Brace . . . natl sls mgr
Jim Hanning . . . natl sls mgr
Deedra Determan . . . mktg dir, mktg mgr & prom mgr
Chooi Ning . . . progmg dir
Melanie Henry . . . news dir
Brian Egan . . . chief of engrg

KOTV— Analog channel: 6. Digital channel: 55. Analog hrs: 24 Digital hrs: 24 100 kw vis, 50 kw aur. ant 1,885t/1,849g TL: N36 01 15 W95 40 32 On air date: Nov 30, 1949. Box 6, Tulsa, OK, 74101. 302 S. Frankfort, Tulsa, OK 74120. Phone: (918) 732-6000. Fax: (918) 732-6016. Web Site: www.kotv.com. Licensee: Griffin Licensing L.L.C.

(acq 12-6-2000; $82 million). Population served: 1,893,300 Natl. Network: CBS. Natl. Rep: TeleRep. Washington Atty: Dow, Lohnes & Albertson. News staff: 40.
Key Personnel:
Ted Strickland . . . . CFO
Regina Moon . . . . gen mgr
John Quesnel . . . . opns mgr
Ron Harig . . . . news dir
Don Root . . . . chief of engrg

KQCW— Analog channel: 19. Analog hrs: 24 5,000 kw vis. ant 823t/777g TL: N35 45 08 W95 48 15 On air date: Sept 12, 1999. 233 South Detroit Ave., Suite 100, Tulsa, OK, 74120. Phone: (918) 270-1919. Fax: (918) 732-6016. Web Site: www.wb19.com. Licensee: Griffin Licensing L.L.C. Group owner: Cascade Broadcasting Group L.L.C. (acq 12-9-2005; $14.5 million). Natl. Network: CW.
Key Personnel:
Regina Moon . . . . gen mgr
John Quesnel . . . . opns mgr
Donita Quesnel . . . . pub affrs dir
Don Root . . . . chief of engrg

***KRSC-TV—** Analog channel: 35. Digital channel: 36. Analog hrs: 18 Digital hrs: 18 2,750 kw vis. 840t TL: N36 24 05 W95 36 33 On air date: July 1, 1987. RSU Public Television, 1701 W. Will Rogers Blvd., Claremore, OK, 74017-3252. Phone: (918) 343-7649. Fax: (918) 343-7952. Web Site: www.rsu.edu. Licensee: Board of Regents of Oklahoma Colleges. Washington Atty: Schwartz, Woods & Miller. Foreign lang progmg: SP 2
Key Personnel:
Dan Schiedel . . . . gen mgr
Janice Curtis . . . . progmg dir
Jim Mertins . . . . chief of engrg

KTPX— Analog channel: 44. Digital channel: 28. Analog hrs: 24 Digital hrs: 24 5,000 kw vis, 500 kw aur. 1,770t TL: N35 50 02 W96 07 28 On air date: July 1997. 5800 E. Skelly Dr., Suite 101, Tulsa, OK, 74135. Phone: (918) 664-1044 (817) 633-6843(Sales). Fax: (918) 664-4913. Web Site: www.ionline.tv. Licensee: Paxson Tulsa License Inc. Group owner: Paxson Communications Corp. (acq 8-21-98; $404,000 for 51% of stock). Population served: 887,000 Natl. Network: i Network.
Key Personnel:
Matthew Pate . . . . rgnl sls mgr
Peter A. De Les Dernier . . . . stn mgr & pub affrs dir

KTUL— Analog channel: 8. Digital channel: 10. Analog hrs: 24 Digital hrs: 24 316 kw vis, 31.6 kw aur. 1,900t/1,809g TL: N35 58 08 W95 36 55 On air date: Sept 18, 1954. Box 8, Tulsa, OK, 74101-0008. 3333 S. 29th West Ave., Tulsa, OK 74107. Phone: (918) 445-8888. Fax: (918) 445-9316. Web Site: www.ktul.com. Licensee: KTUL L.L.C. Group owner: Allbritton Communications Co. (acq 4-83; grpsl). Population served: 355,500 Natl. Network: ABC. Washington Atty: Hogan & Hartson. Wire Svc: AP News staff: 50; News: 17 hrs wkly.
Key Personnel:
Pat Baldwin . . . . pres & gen mgr
Roger Herring. . . . . opns dir & chief of engrg
Carol Jones . . . . natl sls mgr
Marcia Baker . . . . rgnl sls mgr
Deborah Kurin . . . . mktg dir
Larry Nitz . . . . prom mgr
Amy Miller . . . . progmg mgr
Sean McLaughlin . . . . news dir
Randi Carson . . . . pub affrs dir
David Shaffer . . . . engrg mgr & chief of engrg

KWHB— Analog channel: 47. Digital channel: 47. Analog hrs: 24 Digital hrs: 24 835 kw vis. ant 1,509t/1,424g TL: N36 01 15 W95 40 32 On air date: Apr 1, 1985. 8835 S. Memorial, Tulsa, OK, 74133. Phone: (918) 254-4701. Fax: (918) 254-5614. Web Site: www.lesea.com. Licensee: LeSea Broadcasting. Group owner: (group owner; (acq 5-14-86; $3.4 million; FTR: 4-14-86). Washington Atty: John Fiorini.
Key Personnel:
Peter Sumrall. . . . . CEO & VP
Royal Aills . . . . gen mgr

Oregon

Bend, OR (DMA 194)

***KOAB-TV—** Analog channel: 3.58.9 kw vis, 5.89 kw aur. 746t/299g TL: N44 04 41 W121 19 57 On air date: Feb 24, 1970. 7140 S.W. Macadam Ave., Portland, OR, 97709. Phone: (503) 244-9900. Fax: (503) 293-1919. Web Site: www.opb.org. Licensee: Oregon Public Broadcasting. Ownership: Charles J. Swindells. (acq 9-20-93; grpsl; FTR: 10-11-93). Natl. Network: PBS. Washington Atty: Schwartz, Woods & Miller.
Key Personnel:
Steve Bass . . . . CEO & pres
Dan Metziga . . . . dev dir

KOHD— Analog channel: 0. Digital channel: 51.84.1 kw vis. ant 675t/200g TL: N44 04 40.6 W121 19 56.9 On air date: 2006. Box 7009, Eugene, OR, 97401. Phone: (541) 485-5611. Fax: (541) 342-1568. E-mail: genmgr@kohd.com Web Site: www.kohd.com. Licensee: Three Sisters Broadcasting LLC. Natl. Network: ABC.

KTVZ— Analog channel: 21. Digital channel: 18. Analog hrs: 24 126 kw vis, 12.6 kw aur. ant 646t/269g TL: N44 04 40 W121 19 49 On air date: Nov 6, 1977. Box 6038, Bend, OR, 97708. Phone: (541) 383-2121. Fax: (541) 382-1616. E-mail: ktvz@ktvz.com Web Site: www.ktvz.com. Licensee: NPG of Oregon Inc. Group owner: News-Press & Gazette Co. (acq 4-17-2002; $18.9 million). Population served: 131,600 Natl. Network: NBC, CW. Natl. Rep: Continental Television Sales. News staff: 19; News: 20 hrs wkly.
Key Personnel:
Chris Gally. . . . . gen mgr & opns mgr
Lonnie Harden . . . . gen sls mgr
Jeff Barker . . . . news dir
Denis Quinn . . . . chief of engrg

Eugene, OR (DMA 120)

KCBY-TV— Analog channel: 11. Digital channel: 21. Analog hrs: 8am-5pm 11.5 kw vis, 1.1 kw aur. ant 680t/200g TL: N43 23 26 W124 07 47 On air date: Oct 1, 1960. 3451 Broadway, North Bend, OR, 97459. Phone: (541) 269-1111. Fax: (541) 269-7464. E-mail: webmaster@kcby.com Web Site: www.kcby.com. Licensee: Fisher Broadcasting - Oregon TV L.L.C. Group owner: Fisher Broadcasting Company (acq 12-4-2001; grpsl). Natl. Network: CBS. Washington Atty: Dow, Lohnes & Albertson. News staff: 4.
Key Personnel:
Dino Francois . . . . prom dir & news dir
Paul Greene . . . . progmg dir & chief of engrg

***KEPB-TV—** Analog channel: 28. Analog hrs: 17 389 kw vis, 38.9 kw aur. 905t TL: N44 00 06 W123 06 48 On air date: Sept 27, 1990. 7140 S.W. Macadam Ave., Portland, OR, 97219. Phone: (503) 244-9900. Fax: (503) 293-1919. Web Site: www.opb.org. Licensee: Oregon Public Broadcasting. Ownership: Board of directors. (acq 9-20-93; grpsl; FTR: 10-11-93). Natl. Network: PBS.

KEZI— Analog channel: 9. Analog hrs: 24 Digital hrs: 24 316 kw vis, 47.4 kw aur. 1,768t/495g TL: N44 06 57 W122 59 57 On air date: Dec 19, 1960. Box 7009, Eugene, OR, 97401. 2975 Chad Dr., Eugene, OR 97408. Phone: (541) 485-5611. Fax: (541) 342-1568. E-mail: kezi@kezi.com Web Site: kezi.com. Licensee: KEZI Inc. Group owner: Chambers Communications Corp. (acq 8-30-83; $18 million). Population served: 279,240 Natl. Network: ABC. News staff: 15; News: news program20 hrs wkly.
Key Personnel:
Carolyn S. Chambers . . . . CEO & chmn
Scott Chambers. . . . . pres & gen mgr

KLSR-TV— Analog channel: 34. Digital channel: 31. Analog hrs: 5:30am-2:30am 3,090 kw vis, 309 kw aur. ant 850t/200g TL: N44 00 04 W123 06 22 On air date: Oct 31, 1991. 2940 Chad Dr., Eugene, OR, 97408. Phone: (541) 683-2525. Phone: (541) 683-3434. Fax: (541) 683-8016. Web Site: klsrtv.com. Licensee: California Oregon Broadcasting Inc. Group owner: (group owner; (acq 9-1-94; $2.65 million; FTR: 9-19-94). Population served: 505,000 Natl. Network: Fox. Washington Atty: Fletcher, Heald & Hildreth.
Key Personnel:
Patricia Smullin. . . . . pres
Mark Metzger. . . . . gen mgr & gen sls mgr
Johnathon Johnson . . . . opns dir
Scott Bonnell . . . . natl sls mgr & rgnl sls mgr
Sandra Dornon-Belmont. . . . . progmg dir
Steve Woodward . . . . pub affrs dir
Tim Hershiser . . . . engrg dir & chief of engrg

KMCB— Analog channel: 23. Digital channel: 22. Analog hrs: 24 12.3 kw vis. ant 623t TL: N43 23 39 W124 07 56 On air date: July 8, 1991. 3825 International Ct., Springfield, OR, 97477. Phone: (541) 746-1600. Fax: (541) 747-0866. Web Site: www.kmtr.com. Licensee: Ackerley Broadcasting Operations LLC. Group owner: Clear Channel Communications Inc. (acq 6-14-2002; grpsl). Natl. Network: NBC. Natl. Rep: Millennium Sales & Marketing. Rgnl. Rep: Rgnl rep: Blair. Washington Atty: Wiley, Rein & Fielding, LLP. Wire Svc: AP News: 22 hrs wkly.
Key Personnel:
Cambra Ward. . . . . VP
Cambra Ward. . . . . gen mgr
Kurt Thelen . . . . opns mgr
Satellite of KMTR(TV) Eugene.

KMTR— Analog channel: 16. Digital channel: 17. Analog hrs: 24 Note: NBC is on KMTR(TV) ch 16, CW is on KMTR-DT ch 17. 1,919 kw vis, 370.99 kw aur. 1,685t/478g TL: N44 06 58 W122 59 55 On air date: Oct 4, 1982. 3825 International Ct., Springfield, OR, 97477. Phone: (541) 746-1600. Fax: (503) 747-0866. Web Site: www.kmtr.com. Licensee: Ackerley Broadcasting Operations LLC. Group owner: Clear Channel Communications Inc. (acq 6-14-2002; grpsl). Population served: 518,000 Natl. Network: NBC Digital Network: Note: NBC is on KMTR(TV) ch 16, CW is on KMTR-DT ch 17. Natl. Rep: Millennium Sales & Marketing. Rgnl. Rep: Rgnl rep: Blair. Washington Atty: Wiley, Rein & Fielding, LLP. Wire Svc: AP News staff: 22; News: 15 hrs wkly.

***KOAC-TV—** Analog channel: 7. Analog hrs: 20 245 kw vis. 1,500t/279g TL: N44 38 25 W123 16 25 On air date: Oct 7, 1957. 7140 S.W. Macadam Ave., Portland, OR, 97219. Phone: (503) 244-9900. Fax: (503) 293-1919. Web Site: www.opb.org. Licensee: Oregon Public Broadcasting. Ownership: Board of directors. (acq 1993; grpsl; FTR: 9-20-93). Population served: 1,000,000 Natl. Network: PBS. Washington Atty: Schwartz, Woods & Miller. Wire Svc: UPI
Key Personnel:
Steve Boss . . . . CEO & pres
Tom Doggett . . . . progmg VP & progmg dir
Morgan Holm . . . . news dir & pub affrs dir
Don McKay . . . . engrg VP & chief of engrg

KPIC— Analog channel: 4. Analog hrs: 24 5.37 kw vis, 550 w aur. 1,000t/173g TL: N43 14 20 W123 18 42 On air date: Apr 1, 1956. Box 1345, 655 W. Umpqua, Roseburg, OR, 97470. Phone: (541) 672-4481. Fax: (541) 672-4482. E-mail: sales@kpic.com Web Site: www.kpic.com. Licensee: South West Oregon TV Broadcasting Corp. Group owner: Fisher Broadcasting Company (acq 1999; grpsl). Population served: 90,000 Natl. Network: CBS. Washington Atty: Dow, Lohnes & Albertson. News staff: 4; News: 14 hrs wkly.
Key Personnel:
Connie Williamson . . . . stn mgr
Dino Francois . . . . prom mgr
Paul Greene . . . . progmg dir
Mike Hill . . . . chief of engrg
Satellite of KVAL-TV Eugene.

KTCW— Analog channel: 46. Digital channel: 45. Analog hrs: 24 13.63 kw vis. ant 728t TL: N43 14 08 W123 19 17 On air date: Apr 8, 1992. 3825 International Ct., Springfield, OR, 97477-1090. Phone: (541) 746-1600. Fax: (541) 747-0866. Web Site: www.kmtr.com. Licensee: Ackerley Broadcasting Operations LLC. Group owner: Clear Channel Communications Inc. (acq 6-14-2002; grpsl). Natl. Network: NBC. Washington Atty: Wiley, Rein & Fielding, LLP. Wire Svc: AP News: 22 hrs wkly.
Key Personnel:
Cambra Ward. . . . . VP, gen mgr, stn mgr & progmg dir
Kurt Thelen . . . . opns mgr & chief of engrg
Mike Chisholm . . . . gen sls mgr
Robert McMichaels . . . . news dir
Satellite of KMTR Eugene.

KTVC— Analog channel: 36. Digital channel: 18. Analog hrs: 24 42.7 kw vis. ant 692t TL: N43 14 09 W123 19 16 On air date: 1988. Equity Broadcasting Corp., 1 Shackleford Dr., Suite 400, Little Rock, AR, 72211. Phone: (501) 219-2400. Fax: (501) 221-1101. Licensee: Roseburg Broadcasting Inc. Group owner: Equity Broadcasting Corp. (acq 11-30-01; $800,000). Washington Atty: Irwin, Campbell & Tannenwald.
Key Personnel:
Greg Fess. . . . . CEO & gen mgr
Glenn Charlesworth . . . . CFO
Max Hooper . . . . exec VP
James Hearnsberger . . . . sr VP

KVAL-TV— Analog channel: 13. Digital channel: 25. Analog hrs: 24 Digital hrs: 24 316 kw vis, 63.1 kw aur. 1,480t/851g TL: N44 00 07 W123 06 53 On air date: Apr 16, 1954. Box 1313, Eugene, OR, 97440-1313. 4575 Blanton Rd., Eugene, OR 97405. Phone: (541) 342-4961. Fax: (541) 342-7252 (sales). Fax: (541) 342-2635 (admin). E-mail: kval@kval.com Web Site: www.kval.com. Licensee: Fisher Broadcasting - Oregon TV L.L.C. Group owner: Fisher Broadcasting Company (acq 12-4-01; grpsl). Population served: 920,000 Natl. Network: CBS. Rgnl. Rep: Rgnl rep: Petry Washington Atty: Pillsbury, Winthrop & Pittman. News: 17 hrs wkly.
Key Personnel:
Colleen Brown . . . . CEO
Coleen Brown . . . . pres
Greg Raschio . . . . VP & gen mgr
Paul Greene. . . . . opns mgr

Klamath Falls

see Medford-Klamath Falls, OR market

Medford-Klamath Falls, OR (DMA 141)

***KBDM**— Analog channel: 20.100 kw vis. ant 3,231t/46g TL: N42 04 55 W122 43 07 Not on air, target date: unknown: 301 Olive Ave., Suite 104, West Palm Beach, FL, 33401. Phone: (561) 355-4573. Permittee: Northern California Public TV.

KBLN— Analog channel: 30. Analog hrs: 24 9.77 kw vis. ant 2,145t/121g TL: N42 22 56 W123 16 29 On air date: 2002. Better Life Television, Box 766, Grants Pass, OR, 97528. Phone: (541) 474-3089. Fax: (541) 474-9409. E-mail: kbln@betterlifetv.tv Web Site: www.betterlifetv.tv. Licensee: Better Life Television Inc. (acq 2-20-01).

Key Personnel:

Marta Davis . . . . . stn mgr
Ron Davis . . . . . stn mgr

KDKF— Analog channel: 31. Analog hrs: 5:30 AM-2 AM 6.03 kw vis, 603 w aur. 2,267t/164g TL: N42 05 50 W121 37 59 On air date: Oct 17, 1989. 231 E. Main St., Klamath Falls, OR, 97601. Phone: (541) 883-3131. Fax: (541) 883-8931. Web Site: www.kdrv.com. Licensee: Soda Mountain Broadcasting Inc. Group owner: Chambers Communications Corp. (acq 12-5-2001). Population served: 42,000 Natl. Network: ABC. Washington Atty: Fletcher, Heald & Hildreth. News staff: 20; News: 12 hrs wkly.

Satellite of KDRV(TV) Medford, OR.

KDRV— Analog channel: 12. Digital channel: 38. Analog hrs: 24 Digital hrs: 24 191 kw vis, 38.1 kw aur. 2,701t/168g TL: N42 41 32 W123 13 46 On air date: Feb 26, 1984. 1090 Knutson Ave., Medford, OR, 97504. Phone: (541) 773-1212. Fax: (541) 779-9261. Web Site: www.kdrv.com. Licensee: Soda Mountain Broadcasting Inc. Group owner: Chambers Communications Corp. (acq 12-5-2001). Natl. Network: ABC. Natl. Rep: Millennium Sales & Marketing. Washington Atty: Fisher, Wayland, Cooper, Leader & Zaragoza. Wire Svc: AP News staff: 12; News: 20 hrs wkly.

Key Personnel:

Renard N. Maiuri . . . . . gen mgr
Rick Carora . . . . . chief of engrg

***KFTS**— Analog channel: 22. Digital channel: 33. Analog hrs: 24 9.23 kw vis, 923 w aur. 2,152t/103g TL: N42 05 50 W121 37 59 On air date: March 1977. 34 S. Fir St., Medford, OR, 97501. Phone: (541) 779-0808. Fax: (541) 779-2178. Web Site: www.soptv.org. Licensee: Southern Oregon Public Television Inc. Population served: 30,000 Natl. Network: PBS.

Key Personnel:

Mark Stanislawski . . . . . CEO, pres & gen mgr
Tom Werner . . . . . chief of engrg

Satellite of KSYS(TV) Medford 100%.

KMVU— Analog channel: 26. Digital channel: 27. Analog hrs: 24 Digital hrs: 24 110 kw vis. ant 1,444t/113g TL: N42 17 54 W122 44 53 On air date: Aug 8, 1994. 820 Crater Lake Ave., Suite 105, Medford, OR, 97504. Phone: (541) 772-2600. Fax: (541) 772-7364. E-mail: reception@kmvu-tv.com Web Site: www.fox26medford.com. Licensee: Broadcasting Communications L.L.C. Group owner: Northwest Broadcasting Inc. Population served: 390,000 Natl. Network: Fox. Natl. Rep: Continental Television Sales. Washington Atty: Leventhal, Senter & Lerman.

Key Personnel:

Brian Brady . . . . . pres
Cary Jones . . . . . gen mgr
Michael Garry . . . . . engr

KOBI— Analog channel: 5. Analog hrs: 24 60.3 kw vis, 8.13 kw aur. 2,700t/155g TL: N42 41 49 W123 13 39 On air date: Aug 1, 1953. Box 1489, Medford, OR, 97501. 125 S.Fir, Medford, OR 97501. Phone: (541) 779-5555. Fax: (541) 779-5564. E-mail: kobi@kobi5.com Web Site: www.localnewscomesfirst.com. Licensee: California Oregon Broadcasting Inc. Group owner: (group owner). Population served: 475,000 Natl. Network: NBC. Natl. Rep: Blair Television. Washington Atty: Wiley, Rein & Fielding.

Key Personnel:

Patricia C. Smullin . . . . . pres
Dan Acklen . . . . . news dir

KOTI— Analog channel: 2. Analog hrs: 24 35.5 kw vis, 3.55 kw aur. 2,001t/150g TL: N42 05 48 W121 37 57 (CP: 85.1 kw vis) On air date: Apr 5, 1956. Box 2K, 222 S. 7th, Klamath Falls, OR, 97601. Phone: (541) 882-2222. Phone: (541) 779-5555. Fax: (541) 883-7664. E-mail: news@koti2.com Web Site: www.localnewscomesfirst.com. Licensee: California Oregon Broadcasting Inc. Group owner: (group owner). Population served: 102,000 Natl. Network: NBC. Washington Atty: Wiley, Rein & Fielding. Wire Svc: UPI News staff: 2.

Key Personnel:

Patricia Smallin . . . . . pres
Bob Wise . . . . . gen mgr
Dennis Siewert . . . . . gen sls mgr & natl sls mgr
Donna Rodriquez . . . . . progmg dir
Dan Acklen . . . . . news dir
Scott McMahon . . . . . chief of engrg

Rebroadcasts KOBI Medford 90%.

***KSYS**— Analog channel: 8. Digital channel: 42. Analog hrs: 24 30 kw vis, 3 kw aur. 2,683t TL: N42 41 31 W123 13 46 On air date: Jan 17, 1977. 34 S. Fir St., Medford, OR, 97501. Phone: (541) 779-0808. Fax: (541) 779-2178. Web Site: www.soptv.org. Licensee: Southern Oregon Public Television Inc. Population served: 137,140 Natl. Network: PBS. Rgnl. Network: Pacific.

Key Personnel:

Mark Stanislawski . . . . . CEO, pres & gen mgr
Tom Werner . . . . . chief of engrg

KTVL— Analog channel: 10. Digital channel: 35. Analog hrs: 24 Digital hrs: 24 132 kw vis, 26.3 aur. 3,310t/151g TL: N42 04 55 W122 43 07 On air date: Oct 3, 1961. Box 10, Medford, OR, 97501. 1440 Rossanley Dr., Medford, OR 97501. Phone: (541) 773-7373. Fax: (541) 779-0451. E-mail: ktvl@ktvl.com Web Site: www.ktvl.com. Licensee: Freedom Broadcasting of Oregon Licensee L.L.C. Group owner: Freedom Broadcasting Inc. (acq 8-28-81; $12.5 million). Population served: 364,000 Natl. Network: CBS. Natl. Rep: TeleRep. Washington Atty: Latham & Watkins. News: 16 hrs wkly.

Key Personnel:

Kingsley Kelley . . . . . VP & gen mgr
Jennifer Beres . . . . . opns mgr
Bruce Workman . . . . . gen sls mgr
Lila Hampton . . . . . natl sls mgr
Barry Tevis . . . . . prom mgr
Sheila Giorgetti . . . . . progmg dir
Gordon Godfrey . . . . . news dir
Carl Randall . . . . . chief of engrg

Pendleton

see Yakima-Pasco-Richland-Kennewick, WA market

Portland, OR (DMA 23)

KATU— Analog channel: 2. Digital channel: 43. Analog hrs: 24 100 kw vis, 20 kw aur. 1,560t/918g TL: N45 31 14 W122 44 37 On air date: Mar 15, 1962. 2153 N.E. Sandy Blvd., Portland, OR, 97232. Box 2, Portland, OR 97207. Phone: (503) 231-4222. Fax: (503) 231-4233. E-mail: custserv@katu.com Web Site: www.katu.com. Licensee: Fisher Broadcasting - Portland TV L.L.C. Group owner: Fisher Broadcasting Company (acq 12-4-01; grpsl). Population served: 784,400 Natl. Network: ABC. Natl. Rep: TeleRep. Washington Atty: Fisher, Wayland, Cooper, Leader & Zaragoza. News: varies hrs wkly.

Key Personnel:

Ben Tucker . . . . . pres
Warren Spector . . . . . CFO
John Tamerlano . . . . . gen mgr
Jo Anne James . . . . . stn mgr & gen sls mgr
Darby Britto . . . . . opns mgr
Diane Gervais . . . . . natl sls mgr
Steve Denari . . . . . prom dir & pub affrs dir
Julie Mespelt . . . . . progmg mgr
Mike Rausch . . . . . news dir
Alan Batdorf . . . . . chief of engrg

KGW— Analog channel: 8. Digital channel: 46. Analog hrs: 24 hrs 316 kw vis, 60.3 kw aur. 1,768t/924g TL: N45 31 21 W122 44 46 On air date: Dec 15, 1956. 1501 S.W. Jefferson St., Portland, OR, 97201. Phone: (503) 226-5000. Fax: (503) 226-4448. Web Site: www.kgw.com. Licensee: KGW-TV Inc. Group owner: Belo Corp., Broadcast Division (acq 1997; grpsl). Population served: 1,086,100 Natl. Network: NBC. Natl. Rep: Blair Television. Washington Atty: Wiley, Rein & Fielding. Wire Svc: UPI News: 35 hrs wkly.

Key Personnel:

Paul Fry . . . . . pres & gen mgr
Brenda Buratti . . . . . prom dir & progmg dir
Rod Gramer . . . . . news dir

***KNMT-TV**— Analog channel: 24. Analog hrs: 24 2,690 kw vis, 269 kw aur. 1,519t/2,535g TL: N45 30 58 W122 43 59 On air date: Nov 17, 1989. 432 N.E. 74th Ave., Portland, OR, 97213. Phone: (503) 252-0792. Fax: (503) 256-4205. Web Site: www.nmtv.org. Licensee: National Minority TV Inc. Ownership: Paul F. Crouch; Jane Duff; Jan Crouch.

Key Personnel:

Jane P. Duff . . . . . pres
Dr. Paul F. Crouch . . . . . VP
Adolfo Carbajal . . . . . stn mgr & progmg mgr
Bonnie Gaulding . . . . . pub affrs dir
Steven Hendrix . . . . . chief of engrg

KOIN— Analog channel: 6. Digital channel: 40. Analog hrs: 24 100 kw vis, 15.1 kw aur. ant 1,760t/989g TL: N45 30 58 W122 43 59 On air date: Oct 15, 1953. 222 S.W. Columbia St., Portland, OR, 97201. Phone: (503) 464-0600. Fax: (503) 464-0655. E-mail: koin@koin.com Web Site: www.koin.com. Licensee: Montecito Portland License LLC. Group owner: Emmis Communications Corp. (acq 1-27-2006; grpsl). Population served: 2,700,000 Natl. Network: CBS. News: 27 hrs wkly.

Key Personnel:

Marty . Ostrow . . . . . gen mgr
Rodger O'Connor . . . . . mktg dir
Joyce Mansisidor . . . . . progmg dir
Jeff Alan . . . . . news dir
Lee A. Wood . . . . . engrg dir

***KOPB-TV**— Analog channel: 10. Digital channel: 27. Analog hrs: 24 316 kw vis, 31.6 kw aur. ant 1,740t/1,081g TL: N45 31 22 W122 45 07 On air date: Feb 6, 1961. 7140 S.W. Macadam Ave., Portland, OR, 97219-3099. Phone: (503) 244-9900. Fax: (503) 293-1919. Web Site: www.opb.org. Licensee: Oregon Public Broadcasting. Ownership: Board of directors. (acq 9-20-93; grpsl; FTR: 10-11-93). Population served: 2,000,000 Natl. Network: PBS. Washington Atty: Schwartz, Woods & Miller.

Key Personnel:

Steve Bass . . . . . CEO & pres
Dan Metziga . . . . . dev dir

KPDX—(Vancouver, WA) Analog channel: 49. Digital channel: 48. Analog hrs: 24 3,200 kw vis, 319 kw aur. 1,785t/1,081g TL: N45 31 22 W122 45 07 On air date: October 1983. 14975 N.W. Greenbrier Pkwy., Beaverton, OR, 97006-5731. Phone: (503) 906-1249. Fax: (503) 548-6910. E-mail: webstaff@kpdx.com Web Site: www.kpdx.com. Licensee: Meredith Corp. Group owner: Meredith Broadcasting Group, Meredith Corp. (acq 7-1-97; grpsl). Natl. Network: MyNetworkTV. Natl. Rep: TeleRep. Washington Atty: Dow, Lohnes & Albertson. Wire Svc: AP

Key Personnel:

Kieran Clarke . . . . . gen mgr
Gary Flock . . . . . gen sls mgr
Lee Petrik . . . . . progmg dir

KPTV— Analog channel: 12. Digital channel: 30. Analog hrs: 24 316 kw vis, 31.6 kw aur. 1,780t/1,049g TL: N45 31 19 W122 44 53 On air date: Sept 20, 1952. 14975 N.W. Greenbrier Pkwy., Beaverton, OR, 97006. Phone: (503) 906-1249. Fax: (503) 548-6910. E-mail: webstaff@kptv.com Web Site: kptv.com. Licensee: Meredith Corp. Group owner: Meredith Broadcasting Group, Meredith Corp. (acq 6-17-2002; swap). Population served: 2,901,900 Natl. Network: Fox. Natl. Rep: TeleRep. Washington Atty: Dow, Lohnes & Albertson. Wire Svc: AP News: 42.5 hrs wkly.

Key Personnel:

Kieran Clarke . . . . . gen mgr
Greg Flock . . . . . gen sls mgr
Matt Hyatt . . . . . prom mgr
Lee Petrik . . . . . progmg dir & progmg mgr

KPXG— Analog channel: 22. Analog hrs: 24 1,702 kw vis, 170 kw aur. 1,187t/945g TL: N45 00 00 W122 41 37 On air date: Nov 21, 1981. 811 SW Naito Pkwy, Suite 100, Portland, OR, 97204. Phone: (503) 222-2221. Fax: (503) 222-4613. Web Site: www.ionline.tv. Licensee: Paxson Portland License Inc. Group owner: Paxson Communications Corp. (acq 5-14-98; $30 million).

Key Personnel:

Linda Massana . . . . . opns mgr
Tim mance . . . . . chief of engrg

KRCW-TV— Analog channel: 32. Digital channel: 33. Analog hrs: 24 Digital hrs: 24 5,000 kw vis, 500 kw aur. 4,707t TL: N45 00 35 W122 20 17 On air date: May 1989. 10255 S.W. Arctic Dr., Beaverton, OR, 97005. Phone: (503) 644-3232. Fax: (971) 223-0457. E-mail: questions@wb32tv.com Web Site: portlandscw.trb.com. Licensee: Tribune Broadcast Holdings Inc. Group owner: Tribune Broadcasting Co. (acq 3-25-2003; $275 million with KPLR-TV Saint Louis, MO). Population served: 2.826 m,ill,ion Natl. Network: CW. Natl. Rep: TeleRep. Washington Atty: Dickstein, Shapiro, Morin & Oshinsky. News: 3.5 hrs wkly.

Key Personnel:

John Manzi . . . . . gen mgr & gen sls mgr
Jeremy Berk . . . . . gen sls mgr
Pat Shearer . . . . . chief of engrg

***KTVR**— Analog channel: 13. Digital channel: 5. Analog hrs: 17 65 kw vis. ant 2,542t/161g TL: N45 18 33 W117 43 54 On air date: Dec 6, 1964. 7140 S.W. Macadam Ave., Portland, OR, 97219-3099. Phone: (503) 244-9900. Fax: (503) 293-1919. Web Site: www.opb.org. Licensee: Oregon Public Broadcasting. Ownership: Board of directors. (acq 1993; grpsl; FTR: 9-20-93). Natl. Network: PBS. Washington Atty: Schwartz, Woods & Miller.

Key Personnel:

Steve Bass . . . . . CEO & pres
Tom Doggett . . . . . progmg dir

Morgan Holm . . . . news dir & pub affrs dir
Don McKay . . . . chief of engrg

KUNP— Analog channel: 16.60.3 kw vis. ant 2,535t/56g TL: N45 18 35 W117 43 57 On air date: 2004. 2153 N.E. Sandy Blvd., Portland, OR, 97232. Phone: (503) 231-4222. Fax: (503) 231-4233. Licensee: Fisher Radio Regional Group. Group owner: Equity Broadcasting Corp. (acq 11-1-2006; $19.3 million with KUNP-LP Portland). Natl. Network: Univision (Spanish). Foreign lang progmg: SP 168

Pennsylvania

Altoona

see Johnstown-Altoona, PA market

Erie, PA (DMA 142)

WFXP— Analog channel: 66. Analog hrs: 24 882 kw vis, 82 kw aur. 889t/697g TL: N42 02 31 W80 03 57 On air date: Sept 2, 1986. 8455 Peach St., Erie, PA, 16509. Phone: (814) 864-2400. Fax: (814) 864-5393. E-mail: webmaster@fox66.tv Web Site: www.yourerie.com. Licensee: Mission Broadcasting of Wichita Falls Inc. Group owner: Mission Broadcasting Inc. (acq 10-22-98). Population served: 412,700 Natl. Network: Fox. Washington Atty: Arter & Hadden. News: 3.5 hrs wkly.

Key Personnel:
Beverly Joyce . . . . stn mgr
Tim Dunst . . . . sls dir

WICU-TV— Analog channel: 12. Digital channel: 52. Analog hrs: 24 316 kw vis, 31.6 kw aur. ant 1,000t/789g TL: N42 03 52 W80 00 19 On air date: Mar 15, 1949. 3514 State St., Erie, PA, 16508. Phone: (814) 454-5201. Fax: (814) 455-0703. E-mail: info@wicu12.com Web Site: www.wicu12.com. Licensee: SJL of Pennsylvania License Subsidiary LLC. (acq 8-96; $11 million). Population served: 129,231 Natl. Network: NBC. Washington Atty: Latham & Watkins. News staff: 20; News: 22 hrs wkly.

Key Personnel:
Brian Lilly . . . . pres & gen mgr
Matt Filippi. . . . . sls dir & gen sls mgr
Judy Shannon . . . . progmg dir
Phil Hayes . . . . news dir
John Wilkosz . . . . engrg dir & chief of engrg

WJET-TV— Analog channel: 24. Analog hrs: 24 1100 kw vis, 110 kw aur. 960t/815g TL: N42 02 24 W80 04 08 (CP: 1,120 kw vis, 112 kw aur, ant 955t) On air date: Apr 2, 1966. 8455 Peach St., Erie, PA, 16509. Phone: (814) 864-2400. Fax: (814) 868-3041. E-mail: lbaxter@wjettv.com Web Site: www.yourerie.com. Licensee: Nexstar Finance Inc. Group owner: Nexstar Broadcasting Group Inc. (acq 12-16-97; $18.5 million). Population served: 412,000 Natl. Network: ABC. Washington Atty: Drinker, Riddle & Reath. News: 3.5 hrs wkly.

Key Personnel:
Bob Bach . . . . rgnl sls mgr
Barbara Behr . . . . progmg mgr
Lou Baxter. . . . . news dir
Mary Scheuer . . . . pub affrs dir
Lorne Earle . . . . chief of engrg

***WQLN**— Analog channel: 54. Analog hrs: 24 1,000 kw vis, 100 kw aur. 879t/679g TL: N42 02 31 W80 03 57 On air date: Aug 13, 1967. 8425 Peach St., Erie, PA, 16509. Phone: (814) 864-3001. Fax: (814) 864-4077. E-mail: wqln@wqln.org Web Site: www.wqln.org. Licensee: Public Broadcasting of Northwest Pa. Inc. Population served: 151,000 Natl. Network: PBS. Rgnl. Network: Pa. Pub Net. Washington Atty: Dow, Lohnes & Albertson.

Key Personnel:
Dwight Miller . . . . pres & gen mgr
Tracey B. Ferrier. . . . . VP
Ed Upton . . . . engrg dir

WSEE— Analog channel: 35. Digital channel: 16. Analog hrs: 24 1,170 kw vis, 117 kw aur. 941t/741g TL: N42 02 20 W80 03 45 On air date: Apr 24, 1954. 1220 Peach St., Erie, PA, 16501. Phone: (814) 455-7575. Fax: (814) 454-5541. E-mail: wsee@wsee.tv Web Site: www.wsee.tv. Licensee: Lilly Broadcasting of Pennslvania License Subsidiary LLC. Ownership: Kevin T. Lilly, 100% (acq 11-28-02;. $10 million). Population served: 500,000 Natl. Network: CBS, CW. Washington Atty: Lathan & Watkins. News staff: 25; News: 14 hrs wkly.

Key Personnel:
Kevin T. Lilly. . . . . pres
Brian Lilly . . . . gen mgr
Tracy Stufft . . . . stn mgr & progmg dir

John Christenson . . . . news dir
Dan Nungesser. . . . . chief of engrg

Harrisburg-Lancaster-Lebanon-York, PA (DMA 41)

WGAL— Analog channel: 8.112 kw vis, 21.4 kw aur. 1,361t/824g TL: N40 02 04 W76 37 08 On air date: Mar 18, 1949. Box 7127, Lancaster, PA, 17604. 1300 Columbia Ave., Lancaster, PA 17603. Phone: (717) 393-5851. Fax: (717) 393-9484. Web Site: www.wgal.com. Licensee: WGAL Hearst-Argyle Television Inc. Group owner: Hearst-Argyle Television Inc. (acq 1999; grpsl). Population served: 1,620,000 Natl. Network: NBC. Natl. Rep: Eagle Television Sales. Washington Atty: Brooks, Pierce. News staff: 53; News: 29 hrs wkly.

Key Personnel:
Paul Quinn. . . . . pres & gen mgr
Bob Good. . . . . opns mgr
Nancy Tulli . . . . gen sls mgr
Andy Scheid. natl sls mgr
John Baldwin. . . . . mktg mgr, prom mgr & pub affrs dir
Carol Jacoby. . . . . progmg dir & progmg mgr
Dan O'Donnell . . . . news dir
Robert Good . . . . chief of engrg

WGCB-TV— Analog channel: 49. Analog hrs: 24 617 kw vis, 114 kw aur. 581t/375g TL: N39 54 18 W76 35 00 On air date: Apr 28, 1979. Box 88, Red Lion, PA, 17356-0088. Phone: (717) 246-1681. Fax: (717) 244-9316. E-mail: businessoffice@wgcbtv.com Web Site: www.wgcbtv.com. Licensee: Red Lion Broadcasting Co. Ownership: John H. Norris 100%. Population served: 600,000 Washington Atty: Booth, Freret, Imlay and Tepper.

Key Personnel:
John H. Norris . . . . CEO
John Peeling . . . . gen mgr & progmg dir
Anna L. Plourde-Norris . . . . dev VP
Gordon Moul. natl sls mgr
Jerry Jacobs . . . . mktg dir
Donald Horst. . . . . chief of engrg

WHP-TV— Analog channel: 21. Digital channel: 4. Analog hrs: 24 Note: CBS is on WHP-TV ch 21, MyNetworkTV is on WHP-DT ch 4. 1,200 kw vis, 120 kw aur. ant 1,220t/496g TL: N40 20 44 W76 52 09 On air date: Apr 15, 1953. 3300 N. Sixth St., Harrisburg, PA, 17110. Phone: (717) 238-2100. Fax: (717) 238-8744. Fax: (717) 236-0198. Web Site: www.whptv.com. Licensee: Clear Channel Broadcasting Licenses Inc. Group owner: Clear Channel Communications Inc. (acq 1995; $30 million). Population served: 713,070 Natl. Network: CBS Digital Network: Note: CBS is on WHP-TV ch 21, MyNetworkTV is on WHP-DT ch 4. News staff: 23; News: 24 hrs wkly.

Key Personnel:
Jim Berman . . . . VP & gen mgr
Lou Castriota Sr. . . . . opns dir & progmg dir
Stu Brenner . . . . natl sls mgr
Sherry Taylor . . . . prom dir
Greg Zoerb . . . . news dir
Rob Hershey. . . . . engrg dir

WHTM-TV— Analog channel: 27. Digital channel: 10. Analog hrs: 21 2,400 kw vis, 240 kw aur. 1,119t/608g TL: N40 18 57 W76 57 02 On air date: June 19, 1953. Box 5860, 3235 Hoffman St., Harrisburg, PA, 17110-5860. Phone: (717) 236-2727. Fax: (717) 232-5272. Web Site: www.abc27.com. Licensee: Harrisburg Television Inc. Group owner: Allbritton Communication Co. (acq 1996; $113 million). Population served: 535,310 Natl. Network: ABC. News staff: 46; News: 27 hrs wkly.

Key Personnel:
Joe Lewin . . . . pres, gen mgr & progmg dir
Rob Saylor . . . . gen sls mgr
Paul Roda. . . . . natl sls mgr
Larry Maloney . . . . rgnl sls mgr
Randy Whitaker . . . . mktg dir
Betty Bryan Fish . . . . prom dir
Dennis Fisher. . . . . news dir
Mark Olingy . . . . chief of engrg

***WITF-TV**— Analog channel: 33. Digital channel: 36. Analog hrs: 24 1,100 kw vis, 110 kw aur. 1,396t/724g TL: N40 20 45 W76 52 06 On air date: Nov 22, 1964. Box 2954, Harrisburg, PA, 17105. 4801 Lindle Rd., Harrisburg, PA 17111. Phone: (717) 704-3000. Fax: (717) 704-3659. E-mail: info@witf.org Web Site: www.witf.org. Licensee: WITF Inc. Population served: 550,000 Natl. Network: PBS. Rgnl. Network: Eastern Educ. Washington Atty: Dow, Lohnes & Albertson.

Key Personnel:
Kathleen A. Pavelko . . . . CEO & pres
Gregory Poland . . . . CFO
Michael Greenwald . . . . dev VP
Charles Lichty . . . . sls VP
Craig Cohen . . . . progmg dir
Ron Kain . . . . engrg dir

WLYH-TV— Analog channel: 15. Digital channel: 23. Analog hrs: 24 1,050 kw vis, 210 kw aur. ant 1,361t/1,059g TL: N40 15 45 W76 27 53 On air date: Oct 15, 1953. 3300 N. Sixth St., Harrisburg, PA, 17011. Phone: (717) 238-2100. Fax: (717) 238-8744. Web Site: www.cw15.com. Licensee: Nexstar Broadcasting Inc. Group owner: Television Station Group LLC (acq 12-29-2006; $56 million with WTAJ-TV Altoona). Population served: 713,000 Natl. Network: CW.

Key Personnel:
Perry Sook . . . . pres
Jim Berman . . . . VP & gen mgr
Lou Castriota Sr. . . . . opns dir & progmg dir
Scott Beaver . . . . sls dir
Stuart Brenner . . . . natl sls mgr
Sherry Taylor . . . . prom dir
Greg Zoerb . . . . news dir
Rob Hershey. chief of engrg

WPMT— Analog channel: 43. Analog hrs: 24 2,140 kw vis, 214 kw aur. 1,361t/948g TL: N40 01 38 W76 36 00 On air date: Dec 22, 1952. 2005 S. Queen St., York, PA, 17403. Phone: (717) 843-0043. Fax: (717) 843-9741. E-mail: fox43@mail.fox43.com Web Site: www.fox43.com. Licensee: Tribune Television Co. Group owner: Tribune Broadcasting Co. (acq 7-15-97; grpsl). Natl. Network: Fox. Natl. Rep: TeleRep.

Key Personnel:
Dennis FitzSimons . . . . CEO
John Readon . . . . pres
John A. Riggle . . . . VP & gen mgr
Keith McFarland. . . . . opns dir
Peter Rosella . . . . gen sls mgr
Dave Farish . . . . prom mgr
Sandy Hawk . . . . progmg dir & progmg mgr
Jim DePury . . . . news dir
Denise Durham . . . . pub affrs dir
Jim Myers . . . . engrg dir

Johnstown-Altoona, PA (DMA 98)

WATM-TV— Analog channel: 23. Digital channel: 24. Analog hrs: 20 182 kw vis, 18.2 kw aur. ant 1,062t/276g TL: N40 34 05 W78 26 40 On air date: November 1974. 1450 Scalp Ave., Johnstown, PA, 15904. Phone: (814) 266-8088. Phone: (814) 949-8823. Fax: (814) 266-7749. Fax: (814) 949-4780. Web Site: www.abc23.com. Licensee: Palm Television LP. Ownership: Gregory P. Filandrinos. (acq 8-17-99; $12.5 million). Natl. Network: ABC. Washington Atty: Dow, Lohnes & Albertson. Wire Svc: AP

Key Personnel:
Frank Quitoni . . . . pres & gen mgr
Jim Penna . . . . news dir

WJAC-TV— Analog channel: 6. Digital channel: 34. Analog hrs: 24 70.8 kw vis, 10.6 kw aur. ant 1,120t/175g TL: N40 22 17 W78 58 58 On air date: Sept 15, 1949. 49 Old Hickory Ln., Johnstown, PA, 15905. Phone: (814) 255-7600. Fax: (814) 255-7675. Web Site: www.wjactv.com. Licensee: WPXI-TV Holdings Inc. Group owner: Cox Broadcasting (acq 9-22-2000). Population served: 287,760 Natl. Network: NBC. Washington Atty: Dow, Lohnes & Albertson. News staff: 38; News: 25 hrs news wkly.

Key Personnel:
Richard D. Schrott . . . . VP
Richard B. Schrott . . . . gen mgr

WKBS-TV— Analog channel: 47. Analog hrs: 24 1,510 kw vis, 151 kw aur. 1,010t/184g TL: N40 34 12 W78 26 26 On air date: 1985. One Signal Hill Dr., Wall, PA, 15148-1499. Phone: (877) 437-4446. Fax: (412) 824-5442. Web Site: www.ctvn.org. Licensee: Cornerstone Television Inc. Group owner: (group owner) Washington Atty: Pillsbury, Winthrop & Shaw Pittman.

Key Personnel:
Ron Hembree . . . . pres
Steve Johnson . . . . opns dir
Tom McGough . . . . gen sls mgr
Alyson Hayes. . . . . mktg dir
Dede Hayes. . . . . progmg

***WPSU-TV**— Analog channel: 3. Digital channel: 15. Analog hrs: 24 Digital hrs: 24 100 kw vis, 20 kw aur. ant 879t/538g TL: N41 07 21 W78 26 28 On air date: Mar 1, 1965. 238 Outreach Bldg., University Park, PA, 16802-3899. Phone: (814) 865-3333. Fax: (814) 863-9786. E-mail: wpsu@psu.edu Web Site: www.wpsu.org. Licensee: The Pennsylvania State University. Population served: 1,200,000 Natl. Network: PBS. Rgnl. Network: Eastern Educ, CEN, Pa. Pub Net. Washington Atty: Paul, Hastings, Janofsky & Walker. Wire Svc: AP

Key Personnel:
Ted Krichels . . . . gen mgr
Kate Domico . . . . opns dir
Tom Yourchak . . . . dev dir
Greg Petersen. . . . . mktg dir & progmg dir
Melanie Doebler . . . . prom mgr
Ashear Barr . . . . adv dir

Carl Fisher . . . . chief of engrg

WTAJ-TV— Analog channel: 10. Digital channel: 32. Analog hrs: 24 214 kw vis, 21.9 kw aur. 1,110t/277g TL: N40 34 01 W78 26 31 On air date: Mar 1, 1953. 5000 6th Ave., Altoona, PA, 16602. Phone: (814) 942-1010. Fax: (814) 946-8746. Web Site: www.wtajtv.com. Licensee: Nexstar Broadcasting Inc. Group owner: Television Station Group LLC (acq 12-29-2006; $56 million with WLYH-TV Lancaster). Population served: 700,000 Natl. Network: CBS. Washington Atty: Latham & Watkins. Wire Svc: AP News staff: 39; News: 29 hrs wkly.
Key Personnel:
Phil Dubrow . . . . stn mgr & sls dir
Dave Beeney . . . . mktg dir
Tony DeGol . . . . news dir
Randy Chamberlin . . . . chief of engrg

WWCP-TV— Analog channel: 8. Analog hrs: 24 166 kw vis, 16.6 kw aur. 1,208t TL: N40 10 53 W79 09 05 On air date: Oct 13, 1986. 1450 Scalp Ave., Johnstown, PA, 15904. Phone: (814) 266-8088. Fax: (814) 266-7749. Web Site: www.fox8tv.com. Licensee: Peak Media of Pennsylvania Licensee LLC. Ownership: Peak Media of Pennsylvania LLC. Natl. Network: Fox. Washington Atty: Dow, Lohnes & Albertson. Wire Svc: AP
Key Personnel:
Frank Quitoni . . . . pres & gen mgr
Jim Penna . . . . news dir

Lancaster

see Harrisburg-Lancaster-Lebanon-York, PA market

Lebanon

see Harrisburg-Lancaster-Lebanon-York, PA market

Philadelphia (DMA 4)

KYW-TV— Analog channel: 3. Digital channel: 26. Analog hrs: 24 100 kw vis, 10 kw aur. 1,000t/1,116g TL: N40 02 39 W75 14 26 On air date: Sept 3, 1941. 1555 Hamilton St., Philadelphia, PA, 19130. Phone: (215) 977-5300. Fax: (215) 977-5644. Web Site: www.cbs3.com. Licensee: CBS Broadcasting Inc. Group owner: Viacom Television Stations Group (acq 5-4-2000; grpsl). Population served: 1,688,210 Natl. Network: CBS. Natl. Rep: CBS TV Stations National Sales. Washington Atty: Wilkes, Artis, Hedrick & Lane.
Key Personnel:
Michael Colleran . . . . VP & gen mgr
Robin Magyar . . . . stn mgr
Karen Gilligan . . . . sls dir
Roy Coddington . . . . natl sls mgr
Perry Casciato . . . . progmg dir
Susan Schiller . . . . news dir
Rich Paleski . . . . chief of engrg

WBPH-TV— Analog channel: 60. Digital channel: 9. Analog hrs: 24 Digital hrs: 24 2,950 kw vis. ant 936t/482g TL: N40 33 52 W75 26 24 On air date: 1991. 813 N. Fenwick St., Allentown, PA, 18109. Phone: (610) 433-4400. Fax: (610) 433-8251. E-mail: info@wbph.org Web Site: www.wbph.org. Licensee: Sonshine Family TV Inc. Ownership: Patricia Huber, 100%. Population served: 1,600,000

WCAU— Analog channel: 10. Digital channel: 67. Analog hrs: 24 191 kw vis, 19.1 kw aur. 1,160t/1,139g TL: N40 02 36 W75 14 12 On air date: Mar 15, 1948. 10 Monument Rd., Bala Cynwyd, PA, 19004. Phone: (610) 668-5510. Fax: (610) 668-3700. E-mail: nbc10@nbc.com Web Site: www.nbc10.com. Licensee: NBC Telemundo License Co. Group owner: NBC TV Stations Division (acq 9-10-95). Population served: 2,640,400 Natl. Network: NBC. Natl. Rep: NBC TV Stations Sales. Wire Svc: UPI Wire Svc: AP News staff: 110; News: 21 hrs wkly.
Key Personnel:
Dennis Bianchi . . . . pres & gen mgr
Joe Marsini . . . . CFO
Jim Barger . . . . opns dir & engrg dir
Joe Collins . . . . sls VP
Lauren Bacigalupi . . . . prom VP & adv VP
Lawana Scales . . . . progmg dir
Chris Blackman . . . . news dir
JoAnne Wilder . . . . pub affrs dir

WFMZ-TV— Analog channel: 69. Digital channel: 46. Analog hrs: 24 Digital hrs: 24 5,000 kw vis. ant 1,027t/590g TL: N40 33 52 W75 26 24 On air date: Nov 25, 1976. 300 E. Rock Rd., Allentown, PA, 18103. Phone: (610) 797-4530. Fax: (610) 289-6752. Fax: (610) 791-2288 (sales). E-mail: release@wfmz.com Web Site: www.WFMZ.com. Licensee: Maranatha Broadcasting Co. Ownership: Richard C. Dean, 54%; others, 46%. Population served: 7,000,000 Washington Atty: Bentley Law Offices. News staff: 65; News: 29.5 hrs wkly.
Key Personnel:
Mike Kulp . . . . CFO
Barry Fisher . . . . gen mgr
Brad Rinehart . . . . news dir

WGTW-TV— Analog channel: 48. Digital channel: 27. Analog hrs: 24 2,340 kw vis, 234 kw aur. ant 1,099t TL: N40 02 36 W75 14 33 On air date: August 1992. 960 Ashland Ave., Folcraft, PA, 19032. Phone: (610) 583-1370. Fax: (610) 583-1476. Web Site: www.tbn.org. Licensee: Trinity Christian Center of Santa Ana Inc. (acq 10-1-2004; $7 million plus assumption of $41 million in debt).

***WHYY-TV—** Analog channel: 12. Analog hrs: 24 Digital hrs: 24 309 kw vis, 30.9 kw aur. 960t/1,148g TL: N40 02 30 W75 14 24 On air date: Sept 12, 1963. Independence Mall W., 150 N. 6th St., Philadelphia, PA, 19106. 625 Orange St., Wilmington, DE 19801. Phone: (215) 351-1200. Phone: (302) 888-1200. Fax: (215) 351-0398. Fax: (302) 575-0346. E-mail: talkback@whyy.org Web Site: www.whyy.org. Licensee: WHYY Inc. Ownership: WHYY Inc. Population served: 7,000,000 Natl. Network: PBS. Washington Atty: Schwartz, Woods & Miller. News staff: 10; News: 5 hrs wkly.
Key Personnel:
William J. Marrazzo. . . . . CEO & pres
Bruce Flamm . . . . CFO
Paul Gluck. . . . . VP & stn mgr

***WLVT-TV—** Analog channel: 39. Analog hrs: 24 490 kw vis, 97.7 kw aur. 990t/516g TL: N40 33 58 W75 26 06 On air date: September 1965. 123 Sesame St., Bethlehem, PA, 18015. Phone: (610) 867-4677. Fax: (610) 867-3544. Web Site: www.wlvt.org. Licensee: Lehigh Valley Public Telecommunications Corp. Population served: 1,500,000 Natl. Network: PBS. Rgnl. Network: Eastern Educ, Pa. Pub Net. Washington Atty: Dow, Lohnes. Foreign lang progmg: SP 1
Key Personnel:
Patricia Simon . . . . CEO
Patricia Simons. . . . . pres
David E. Smith . . . . chief of engrg

WMCN-TV— Analog channel: 0. Digital channel: 44. Digital hrs: 24 200 kw vis. ant 931t/905g TL: N39 43 41 W74 50 39 On air date: Oct 1, 1986. 19 S. New York Ave., Atlantic City, NJ, 08401. Phone: (609) 441-1120. Fax: (609) 441-9559. E-mail: contact@wmcn.tv Web Site: www.wmcn.tv. Licensee: Lenfest Broadcasting L.L.C. Ownership: H. Chase Lenfest, member/owner (acq 7-19-2000; $9 million). Population served: 750,000 Washington Atty: Wiley, Rein & Fielding.
Key Personnel:
H. Chase Lenfest. . . . . pres & CEO
Robert M. Lund . . . . VP & gen mgr
Steve Cass . . . . sls dir & progmg
Mark Chesterton . . . . progmg mgr
Vojislav Radosavljevic . . . . chief of engrg

WMGM-TV— Analog channel: 40. Digital channel: 36. Analog hrs: 24 Digital hrs: 24 741 kw vis, 74.1 kw aur. 420t/416g TL: N39 07 28 W74 45 56 On air date: Jan 25, 1966. 1601 New Rd., Linwood, NJ, 08221. Phone: (609) 927-4440. Fax: (609) 926-8875. E-mail: news@nbc40.net Web Site: www.nbc40.net. Licensee: Access.1 New Jersey License Co. Group owner: Access.1 Communications Corp. (acq 2004; grpsl). Population served: 400,000 Natl. Network: NBC. Wire Svc: AP News: 8 hrs wkly.
Key Personnel:
Chesley Maddox-Dorsey . . . . pres
Arthur Benjamin Jr. . . . . CFO & VP
Ron Smith . . . . gen mgr
Roger Powe . . . . gen sls mgr

***WNJS—** Analog channel: 23. Digital channel: 22. Analog hrs: 24 Digital hrs: 24 2,323 kw vis, 348 kw aur. 890t/937g TL: N39 43 41 W74 50 39 On air date: Oct 23, 1972. Box 777, Trenton, NJ, 08625-0777. 25 S. Stockton St., Trenton, NJ 08608-1832. Phone: (609) 777-5000. Fax: (609) 633-2920. E-mail: audience@njn.org Web Site: www.njn.net. Licensee: New Jersey Public Broadcasting Authority. Population served: 3,000,000 Natl. Network: PBS. Washington Atty: Schwartz, Woods & Miller. News: 2 hrs wkly.
Satellite of *WNJT Trenton.

***WNJT—** Analog channel: 52. Digital channel: 43. Analog hrs: 24 hrs Digital hrs: 24 hrs. 1,950 kw vis, 285 kw aur. 1,049t/989g TL: N40 16 58 W74 41 11 On air date: Apr 5, 1971. Box 777, Trenton, NJ, 08625-0777. 25 S. Stockton St., Trenton, NJ 08608-1832. Phone: (609) 777-5000. Fax: (609) 633-2920. E-mail: audience@njn.org Web Site: www.njn.net; showcase.njn.net. Licensee: New Jersey Public Broadcasting Authority. Population served: 3,500,000 Natl. Network: PBS. Rgnl. Network: Eastern Educ. Washington Atty: Schwartz, Woods & Miller. News: 10 hrs wkly.
Key Personnel:
Joann Ruscio . . . . mktg dir
Andre Butts . . . . progmg dir
William Jobes. . . . . news dir
William Schnorbus . . . . chief of engrg

WPHL-TV— Analog channel: 17. Digital channel: 54. Analog hrs: 8:30am - 5:30pm 2,340 kw vis, 300 kw aur. ant 1,313t/1,092g TL: N40 02 30 W75 14 24 On air date: Sept 17, 1965. 5001 Wynnefield Ave., Philadelphia, PA, 19131. Phone: (215) 878-1700. Fax: (215) 879-3665. E-mail: wphltv@aol.com Web Site: myph17.trb.com. Licensee: Tribune Broadcasting Co. Group owner: (group owner, see Cross-Ownership; (acq 4-17-92; $19 million; FTR: 5-18-92). Population served: 2,700,000 Natl. Network: MyNetworkTV. Natl. Rep: TeleRep. Washington Atty: Koteen & Naftalin.
Key Personnel:
Vince Giannini . . . . VP & gen mgr
Patrick Loftus . . . . gen sls mgr

WPPX— Analog channel: 61. Digital channel: 31. Analog hrs: 24 3,020 kw vis, 302 kw aur. ant 958t/951g TL: N39 41 43 W75 17 55 On air date: July 9, 1986. 3901 B. Main St., Suite 301, Philadelphia, PA, 19127. Phone: (215) 482-4770. Fax: (215) 482-4777. Web Site: www.ionline.tv. Licensee: Paxson Philadelphia License Inc. Group owner: (group owner; (acq 1-20-95; $9.635 million; FTR: 3-20-95). Population served: 2,700,000 Natl. Network: i Network.
Key Personnel:
Robert Marc Backman . . . . gen mgr
Joe Collins . . . . gen sls mgr
Shawn Edwards . . . . rgnl sls mgr
Daniel Borowicz . . . . chief of engrg

WPSG— Analog channel: 57. Digital channel: 32. Analog hrs: 24 5,000 kw vis, 500 kw aur. ant 1,160t/1,179g TL: N40 02 21 W75 14 13 On air date: June 15, 1981. 101 S. Independence Mall E., Kansas City, PA, 19106. Phone: (215) 977- 5700. Fax: (215) 977- 5220. E-mail: cwphilly@wpsg.com Web Site: cwphilly.com. Licensee: Viacom Stations Group of Philadelphia Inc. Group owner: Viacom Television Stations Group (acq 1995). Natl. Network: CW. Washington Atty: Fisher, Wayland, Cooper, Leader & Zaragoza.
Key Personnel:
Michael Colleran . . . . pres & gen mgr
Robin Magyar . . . . stn mgr
John Brown . . . . sls dir
Susan Schiller . . . . news dir
Rich Paleski . . . . chief of engrg

WPVI-TV— Analog channel: 6. Digital channel: 64. Analog hrs: 24 74.1 kw vis, 7.4 kw aur. 1,094t/1,111g TL: N40 02 38 W75 14 25 On air date: Sept 13, 1947. 4100 City Ave., Philadelphia, PA, 19131. Phone: (215) 878-9700. Fax: (215) 581-4515. Fax: (215) 581-4530 (news). Web Site: www.6abc.com. Licensee: ABC Inc. Group owner: (group owner; (acq 4-27-71; grpsl). Population served: 2,600,000 Natl. Network: ABC. Washington Atty: Wilmer, Cutler & Pickering.
Key Personnel:
Rebecca Campbell. . . . . pres & gen mgr
James Aronow . . . . sls VP
Dirk Ohley. . . . . natl sls mgr
Tim Giannetino. . . . . natl sls mgr
Bob Liga . . . . rgnl sls mgr
Paula McDermott . . . . mktg mgr
Caroline Welch . . . . progmg dir
Carla Carpenter. . . . . news dir
Linda Munich . . . . pub affrs dir
Hank Volpe . . . . engrg VP
James Gilbert . . . . engrg dir

WTVE— Analog channel: 51. Digital channel: 25. Analog hrs: 24 1,450 kw vis, 290 kw aur. ant 751t/125g TL: N40 21 15 W75 53 56 (CP: 5,000 kw vis, 500 kw aur, ant 1,260t. TL: N40 19 35 W75 42 14) On air date: February 1980. 1729 N. 11th St., Reading, PA, 19604. Phone: (610) 921-9181. Fax: (610) 921-9139. Web Site: www.wtve.com. Licensee: Reading Broadcasting Inc., debtor in possession. (acq 11-14-2005). Washington Atty: Leventhal, Senter & Lerman.
Key Personnel:
George Mattmiller . . . . gen mgr
Jack Crumpler . . . . gen sls mgr
Kimberley Bradley. . . . . . opns VP, progmg dir & pub affrs dir
Gibson White. . . . . chief of engrg

WTXF-TV— Analog channel: 29. Digital channel: 42. Analog hrs: 24 5,000 kw vis, 500 kw aur. 1,138t/1,188g TL: N40 02 26 W75 14 20 On air date: May 18, 1965. 330 Market St., Philadelphia, PA, 19106. Phone: (215) 925-2929. Fax: (215) 982-5499(sls). Web Site: www.myfoxphilly.com. Licensee: Fox TV Stations of Philadelphia Inc. Group owner: Fox Television Stations Inc. (acq 1995; $200 million). Natl. Network: Fox. News: 34.5 hrs wkly.
Key Personnel:
Michael Renda. . . . . VP & gen mgr
Vincent M. Manzi . . . . gen sls mgr

Phil Metlin . . . . news dir

WUVP-TV— Analog channel: 65. Digital channel: 66. Analog hrs: 24 Digital hrs: 24 4,070 kw vis, 407 kw aur. ant 1,299t/1,220g TL: N40 02 30 W75 14 11 On air date: July 13, 1981. 4449 N. Delsea Dr., Newfield, NJ, 08344. 1700 Market St., Suite 1550, Philadelphia, PA 19103. Phone: (856) 691-6565. Fax: (856) 691-2483. E-mail: notibreve_65@univision.net Web Site: www.univision.com. Licensee: Univision Philadelphia LLC. Group owner: Univision Communications Inc. (acq 8-21-2001; grpsl). Natl. Network: Univision (Spanish). Foreign lang progmg: SP 168

Key Personnel:
Diana Bald . . . . gen mgr
John Duffin . . . . natl sls mgr
Raul de la Rosa . . . . rgnl sls mgr
Josue Duarte . . . . prom mgr
John Skelnik . . . . chief of engrg

WWSI— Analog channel: 62. Digital channel: 49. Analog hrs: 24 5,000 kw vis, 500 kw aur. ant 972t/970g TL: N39 37 53 W74 21 12 On air date: 1990. 1341 N. Delaware Ave., Suite 408, Philadelphia, PA, 19125. One S. New York Ave., Atlantic City, NJ 08401. Phone: (215) 634-8862. Phone: (609) 449-0049. Fax: (215) 425-2683. Fax: (609) 441-9559. E-mail: jrivera@wwsi-tv.com Web Site: www.wwsi-tv.com. Licensee: Hispanic Broadcasters of Philadelphia L.L.C. Ownership: Council Tree Hispanic Broadcasters L.L.C. (acq 5-14-2002). Natl. Network: Telemundo (Spanish). Foreign lang progmg: SP 168 News staff: 4; News: 3 hrs wkly.

Key Personnel:
Jimmy Rivers . . . . gen mgr
Michael Brendzel . . . . dev VP

***WYBE—** Analog channel: 35. Digital channel: 34. Analog hrs: 24 Digital hrs: 24 1,000 kw vis. ant 1,125t/1,046g TL: N40 02 30 W75 14 11 On air date: June 10, 1990. 8200 Ridge Ave., Philadelphia, PA, 19128-1604. Phone: (215) 483-3900. Fax: (215) 483-6908. Web Site: www.wybe.org. Licensee: Independence Public Media of Philadelphia Inc. Ownership: Independence Media. Population served: 3,800,000 Washington Atty: Drinker, Biddle & Reath L.L.P. Foreign lang progmg: SP 15

Pittsburgh, PA (DMA 22)

KDKA-TV— Analog channel: 2. Digital channel: 25.100 kw vis, 10 kw aur. 995t/683g TL: N40 29 38 W80 01 09 On air date: January 1949. One Gateway Ctr., Pittsburgh, PA, 15222. Phone: (412) 575-2200. Fax: (412) 575-3207. Web Site: kdka.com. Licensee: CBS Broadcasting Inc. Group owner: Viacom Television Stations Group (acq 5-4-2000; grpsl). Population served: 1,494,264 Natl. Network: CBS. Natl. Rep: CBS TV Stations National Sales. Washington Atty: Wilkes, Artis, Hedrick & Lane.

***WGPT—** Analog channel: 36. Digital channel: 54. Analog hrs: 24 245 kw vis. ant 3,225t/199g TL: N39 24 14 W79 17 37 On air date: 1986. 11767 Owings Mills Blvd., Owings Mills, MD, 21117-1499. Phone: (410) 356-5600. Fax: (410) 581-6579. E-mail: comments@mpt.org Web Site: www.mpt.org. Licensee: Maryland Public Broadcasting Commission. Population served: 600,000 Natl. Network: PBS. Washington Atty: Schwartz, Woods & Miller.

Key Personnel:
Robert Shuman . . . . CEO & pres
Larry Unger . . . . CFO
Kirby Storms . . . . chief of engrg
Rebroadcasts WMPB(TV) Baltimore 100%.

***WNPB-TV—** Analog channel: 24. Digital channel: 33. Analog hrs: 24 Digital hrs: 24 3,000 kw vis, 347 kw aur. 1,499t/515g TL: N39 41 45 W79 45 45 On air date: Feb 23, 1969. 191 Scott Ave., Morgantown, WV, 26508. Phone: (304) 284-1440. Fax: (304) 284-1454. E-mail: audienceservices@wvpubcast.org Web Site: www.wvpubcast.org. Licensee: West Virginia Educational Broadcasting Authority. (acq 7-1-83). Population served: 109,450 Natl. Network: PBS. Rgnl. Network: SECA. Washington Atty: Wilkinson, Barker, Knauer & Quinn. News staff: 2; News: one hr wkly.

Key Personnel:
Bill Acker . . . . gen mgr
Marilyn Divita . . . . dev dir
Jack Wells . . . . engrg dir

WPCB-TV— Analog channel: 40. Digital channel: 50. Analog hrs: 24 Digital hrs: 24 4,900 kw vis. 980t/839g TL: N40 23 34 W79 46 54 On air date: Apr 15, 1979. Signal Hill Dr., Wall, PA, 15148-1499. Phone: (412) 824-3930. Fax: (412) 824-5442. E-mail: gstewart@ctvn.org Web Site: www.ctvn.org. Licensee: Cornerstone Television Inc. (acq 7-78). Washington Atty: Shaw Pittman. Foreign lang progmg: SP 1

Key Personnel:
Ron Hembree . . . . pres
Chuck Alexander . . . . CFO
Kim Carter . . . . dev mgr
Alyson Hayes . . . . mktg dir & progmg VP
Dede Hayes . . . . progmg dir
Steve Johnson . . . . engrg dir

WPCW— Analog channel: 19. Analog hrs: 24 3,020 kw vis. ant 1,115t/371g TL: N40 10 52 W79 07 46 On air date: Oct 15, 1953. 1 Gateway Center, Pittsburgh, PA, 15222. Phone: (412) 575-2200. Fax: (412) 575-2500. Web Site: pittsburghscw.com. Licensee: Pittsburgh Television Station WNPA Inc. Group owner: Viacom Television Stations Group (acq 12-9-98; $39 million). Population served: 42,476 Natl. Network: CW.

WPGH-TV— Analog channel: 53. Digital channel: 43. Analog hrs: 24 Digital hrs: 24 2,340 kw vis, 117 kw aur. ant 1,023t/702g TL: N40 29 43 W80 00 17 On air date: July 14, 1953. 750 Ivory Ave., Pittsburgh, PA, 15214. Phone: (412) 931-5300. Fax: (412) 931-8135. Web Site: www.wpgh53.com. Licensee: WPGH Licensee L.L.C. Group owner: Sinclair Broadcast Group Inc. (acq 8-30-91; $55 million; FTR: 7-15-91). Population served: 2,828,400 Natl. Network: Fox. Natl. Rep: Millennium Sales & Marketing. Washington Atty: Pillsbury, Winthrop & Pittman.

Key Personnel:
Alan Frank . . . . gen mgr
Jim Lapiana . . . . sls dir
Kerry Check . . . . engrg dir

WPMY— Analog channel: 22. Digital channel: 42. Analog hrs: 24 Digital hrs: 24 3,800 kw vis, 190 kw aur. ant 918t/797g TL: N40 26 23 W79 43 11 On air date: Sept 26, 1978. 750 Ivory Ave., Pittsburgh, PA, 15214. Phone: (412) 931-5300. Fax: (412) 931-8135. Web Site: www.wcwb22.com. Licensee: WCWB Licensee LLC. Group owner: Sinclair Broadcast Group Inc. (acq 12-10-2001; $17.808 million). Population served: 2,800,000 Natl. Network: MyNetworkTV. Natl. Rep: Millennium Sales & Marketing. Washington Atty: Pillsbury, Winthrop & Pittman.

Key Personnel:
Alan Frank . . . . gen mgr
Jim Lapiana . . . . sls dir
Kerry Check . . . . pub affrs dir & engrg dir

WPXI— Analog channel: 11. Digital channel: 48. Analog hrs: 24 316 kw vis, 58.9 kw aur. 991t/849g TL: N40 27 48 W80 00 18 On air date: Sept 1, 1957. 4145 Evergreen Rd., Pittsburg, PA, 15214. Phone: (412) 237-1100. Fax: (412) 323-8097. E-mail: comments@wpxi.com Web Site: www.wpxi.com. Licensee: WPXI-TV Holdings Inc. Group owner: Cox Broadcasting (acq 1-1-65; $20.5 million; FTR: 11-30-64). Population served: 1,500,000 Natl. Network: NBC. Natl. Rep: TeleRep. Washington Atty: Dow, Lohnes & Albertson.

Key Personnel:
Ray Carter . . . . gen mgr
Gary Bogart . . . . gen sls mgr
Darryl Griffin . . . . natl sls mgr
Mark Barash . . . . progmg dir
Carrie Harding . . . . news dir
Annette Parks . . . . engrg dir

***WQED—** Analog channel: 13. Digital channel: 38.316 kw vis, 31.6 kw aur. ant 690t/600g TL: N40 26 46 W79 57 51 On air date: Apr 1, 1954. 4802 Fifth Ave., Pittsburgh, PA, 15213. Phone: (412) 622-1300. Fax: (412) 622-6413. E-mail: viewers@wqed.org Web Site: www.wqed.org. Licensee: WQED Multimedia. Population served: 1,130,000 Natl. Network: PBS. Washington Atty: Schwartz, Woods & Miller. Wire Svc: Reuters

WQEX— Analog channel: 16. Analog hrs: 24 667 kw vis, 66.1 kw aur. ant 705t/601g TL: N40 26 46 W79 57 51 On air date: Sept 14, 1959. 4802 Fifth Ave., Pittsburgh, PA, 15213. Phone: (412) 622-1300. Fax: (412) 622-1488. E-mail: wqexviewers@wqed.org Web Site: www.wqed.org/wqex. Licensee: WQED Multimedia. Natl. Network: NBC. Washington Atty: Schwartz, Woods & Miller.

Key Personnel:
George Miles . . . . pres
Rick Vaccarelli . . . . sls dir
Gigi Saladna . . . . prom mgr
Paul Byers . . . . engrg dir

WTAE-TV— Analog channel: 4. Digital channel: 51. Analog hrs: 24 100 kw vis, 10 kw aur. ant 961t/1,017g TL: N40 16 49 W79 48 11 On air date: Sept 14, 1958. 400 Ardmore Blvd., Charlottetown, PA, 15221-3090. Phone: (412) 242-4300. Fax: (412) 244-4595. Web Site: www.thepittsburghchannel.com. Licensee: WTAE Hearst-Argyle Television Inc. Group owner: Hearst-Argyle Television Inc. (acq 7-16-97; grpsl). Population served: 2,700,000 Natl. Network: ABC. Natl. Rep: Eagle Television Sales. Washington Atty: Brooks, Pierce, McLendon, Humphrey & Leonard. News staff: 70; News: 32 hrs wkly.

Key Personnel:
Bob Bee . . . . gen sls mgr & natl sls mgr
Leslie Wojdowski . . . . mktg mgr

Scranton

see Wilkes Barre-Scranton, PA market

Wilkes Barre-Scranton, PA (DMA 53)

WBRE-TV—Wilkes-Barre, Analog channel: 28. Digital channel: 11. Analog hrs: 24 3,020 kw vis, 604 kw aur. 1,670t/870g TL: N41 11 01 W75 52 02 On air date: Jan 1, 1953. 62 S. Franklin St., Wilkes-Barre, PA, 18701-1201. Phone: (570) 823-2828. Fax: (570) 823-4523. E-mail: wbrenews@nbga.net Web Site: www.wbre.com. Licensee: Nexstar Finance Inc. Group owner: Nexstar Broadcasting Group Inc. (acq 11-14-97; $47 million). Population served: 1,413,000 Natl. Network: NBC. Natl. Rep: Blair Television. Washington Atty: Drinker, Biddle & Reath, LLP. News staff: 68; News: 24 hrs wkly.

Key Personnel:
Louis J. Abitabilo . . . . VP
Randy Williams . . . . stn mgr, opns dir & opns dir
Kim Dudick . . . . gen sls mgr
Ron Krisulevicz . . . . news dir
Todd Tobin . . . . chief of engrg

WNEP-TV— Analog channel: 16. Digital channel: 49. Analog hrs: 24 1,150 kw vis, 115 kw aur. ant 1,660t/781g TL: N41 10 58 W75 52 21 On air date: Feb 9, 1954. 16 Montage Mountain Rd., Moosic, PA, 18507. Phone: (570) 346-7474. Fax: (570) 347-0359. E-mail: email@wnep.com Web Site: www.wnep.com. Licensee: Local TV Pennsylvania License LLC. Group owner: The New York Times Co. (acq 5-7-2007; grpsl). Population served: 1,884,000 Natl. Network: ABC. Natl. Rep: Millennium Sales & Marketing. Washington Atty: Covington & Burling. Wire Svc: AP Wire Svc: PR Newswire News staff: 64; News: 36 hrs wkly.

Key Personnel:
Robert L. Lawrence . . . . pres
Lou Kirchen . . . . gen mgr
David Lewandoski . . . . opns mgr
Chuck Morgan . . . . gen sls mgr
Mike Last . . . . natl sls mgr
Diane Frain . . . . mktg dir
Laurie LaMaster . . . . prom dir & prom mgr
Debbie Drechin . . . . progmg dir
Frank Gerardi . . . . pub affrs dir
Mike Morkavage . . . . chief of engrg

WOLF-TV— Analog channel: 56. Digital channel: 56. Analog hrs: 24 1,000 kw vis, 100 kw aur. 1,079t/381g TL: N41 02 13 W76 05 07 (CP: 1,600 kw vis, ant 1,656t. TL: N41 10 58 W75 52 26) On air date: June 3, 1985. 1181 Hwy. 315, Plains, PA, 18702. Phone: (570) 970-5600. Fax: (570) 970-5601. E-mail: fox56@fox56.com Web Site: www.nepatoday.com. Licensee: WOLF License Corp. Group owner: Pegasus Broadcast Television Inc. Population served: 1,700,000 Natl. Network: Fox. Rgnl. Rep: Rgnl rep: Petry Washington Atty: Shaw, Pittman. News: 6.5 hrs wkly.

Key Personnel:
Michael Yanuzzi . . . . pres
Jon Cadman . . . . gen mgr
Dan Mecca . . . . natl sls mgr
Bob Spager . . . . rgnl sls mgr

WQMY— Analog channel: 53. Digital channel: 56. Analog hrs: 24 1,320 kw vis. ant 800t/177g TL: N41 11 57 W77 07 38 On air date: Jan 22, 1993. 1181 Hwy. 315, Plains, PA, 18702. Phone: (570) 970-5600. Fax: (570) 970-5601. E-mail: nepatoday@fox56.com Web Site: www.nepatoday.com. Licensee: Pegasus Broadcast Associates L.P. Group owner: Pegasus Broadcast Television Inc. (acq 1993; FTR: 2-15-93). Natl. Network: MyNetworkTV. Washington Atty: Shaw, Pittman.

Key Personnel:
Jon Cadmon . . . . gen mgr & gen sls mgr
Dan Mecca . . . . natl sls mgr
Bob Spager . . . . rgnl sls mgr

WQPX— Analog channel: 64. Digital channel: 32.3,090 kw vis. ant 1,220t/377g TL: N41 26 06 W75 43 35 On air date: 1998. 409 Lackawanna Ave., Suite 700, Scranton, PA, 18503. Phone: (570) 344-6400. Fax: (570) 344-3303. Web Site: www.ionline.tv. Licensee: Paxson Scranton License Inc. Group owner: Paxson Communications Corp. (acq 7-31-98; $6 million). Natl. Network: i Network. Washington Atty: Schwartz, Woods & Miller.

Key Personnel:
Regina Lanzo . . . . opns mgr
Robert Andrade . . . . chief of engrg

WSWB— Analog channel: 38. Digital channel: 31. Analog hrs: 24 1,290 kw vis, 234 kw aur. 1,263t TL: N41 26 09 W75 43 45 On air date: Nov 26, 1998. 1181 Hwy. 315, Plains, PA, 18702. Phone: (570) 970-5600. Fax: (570) 970-5601. E-mail: nepatoday@fox56.com Web Site: www.nepatoday.com. Licensee: KB Prime Media L.L.C. Group

owner: (group owner; (acq 11-10-98; $500,000). Population served: 1,700,000 Natl. Network: CW. Washington Atty: Fisher, Wayland, Cooper, Leader & Zaragoza.
Key Personnel:
Michael Yanuzzi . . . . . pres & gen mgr
Jon Cadmon . . . . . gen mgr
Aldo Cardoni . . . . . opns mgr
Dan Mecca . . . . . natl sls mgr
Bob Spager . . . . . rgnl sls mgr
Doug Cook . . . . . mktg dir
Linda Greenwald . . . . . progmg dir
Steve Phillips . . . . . prom dir & pub affrs dir
Rich Chofey . . . . . chief of engrg

***WVIA-TV—** Analog channel: 44. Digital channel: 41. Analog hrs: 24 Digital hrs: 24 1,000 kw vis, 100 kw aur. 1,670t/845g TL: N41 10 55 W75 752 17 On air date: Sept 26, 1966. 100 WVIA Way, Pittston, PA, 18640-6197. Phone: (570) 826-6144. Phone: (570) 344-1244. Fax: (570) 655-1180. Web Site: www.wvia.org. Licensee: Northeastern Pennsylvania Educational TV Association. Population served: 300,000 Natl. Network: PBS. Rgnl. Network: Eastern Educ. Washington Atty: Dow, Lohnes & Albertson.
Key Personnel:
A. William Kelly . . . . . CEO, pres & gen mgr
Thomas P. Curra . . . . . stn mgr
Joseph Glynn . . . . . engrg VP

WYOU— Analog channel: 22. Digital channel: 13. Analog hrs: 24 2,945 kw vis, 294 kw aur. ant 842g TL: N41 10 58 W75 52 26 On air date: June 7, 1953. 62 S. Franklin St., Wilkes-Barre, PA, 18701. Phone: (570) 961-2222. Fax: (570) 344-4484. Web Site: www.wyou.com. Licensee: Mission Broadcasting Inc. Group owner: (group owner; (acq 1-5-98; $21 million). Population served: 580,290 Natl. Network: CBS. Natl. Rep: Blair Television. Washington Atty: Drinker, Biddle & Reath LLP. News: 19.5 hrs wkly.
Key Personnel:
Louis J. Abitabilo . . . . . gen mgr
Randy Williams . . . . . opns dir & natl sls mgr
Steve Genett . . . . . natl sls mgr
Bob Spager . . . . . rgnl sls mgr
Susan Kalinowski . . . . . progmg dir & progmg mgr
Frank Andrews . . . . . news dir

York

see Harrisburg-Lancaster-Lebanon-York, PA market

Rhode Island

Providence, RI-New Bedford, MA (DMA 51)

WDPX— Analog channel: 58.1,191 kw vis, 119.1 kw aur. 470t/350g TL: N41 41 19 W70 20 49 (CP: 1,150 kw vis, ant 492t) On air date: July 19, 1985. 1120 Soldiers Field Rd., Boston, MA, 02134. Phone: (617) 787-6868. Fax: (617) 787-4114. Web Site: www.ionline.net. Licensee: Paxson Boston-68 License Inc. Group owner: Paxson Communications Corp. (acq 5-2-2000; grpsl). Washington Atty: Arter & Hadden.
Satellite of WBPX Boston.

WJAR— Analog channel: 10. Digital channel: 51.316 kw vis, 50 kw aur. ant 1,000t/940g TL: N41 51 54 W71 17 15 On air date: July 10, 1949. 23 Kenney Dr., Cranston, RI, 02920. Phone: (401) 455-9100. Fax: (401) 455-9168. Fax: (401) 455-9140. Web Site: www.turnto10.com. Licensee: Media General Communications Inc. Group owner: NBC TV Stations Division (acq 6-26-2006; grpsl). Population served: 846,000 Natl. Network: NBC. Natl. Rep: Harrington, Righter & Parsons. Foreign lang progmg: SP 1
Key Personnel:
Lisa Churchville . . . . . gen mgr
Clark Smith . . . . . opns dir & chief of engrg
Jeff Walkes . . . . . gen sls mgr
Valerie McCain . . . . . natl sls mgr
Barbara Beresford . . . . . mktg dir
Betty Jo Cugini . . . . . news dir
Elaine Moy-Gederman . . . . . progmg

WLNE—(New Bedford, MA) Analog channel: 6. Analog hrs: 24 100 kw vis, 22.4 kw aur. 940t/996g TL: N41 46 39 W70 55 41 On air date: Jan 1, 1963. 10 Orms St., Providence, RI, 02904. Phone: (401) 453-8000. Fax: (401) 331-4399. Web Site: www.abc6.com. Licensee: Freedom Broadcasting of Southern New England Licensee LLC. Group owner: Freedom Broadcasting Inc. (acq 12-14-82; $15.5 million). Population served: 1,437,000 Natl. Network: ABC. Natl. Rep: TeleRep. Washington Atty: Latham & Watkins.
Key Personnel:
Roland T. Adeszko . . . . . VP & gen mgr
Jim Brown . . . . . opns mgr, engrg dir & chief of engrg
Michael Brostek . . . . . rgnl sls mgr
Judy Shoemaker . . . . . prom dir
Edwin Hart . . . . . news dir

WLWC—(New Bedford, MA) Analog channel: 28. Digital channel: 9. Analog hrs: 24 5,000 kw vis, 250 kw aur. ant 808t/833g TL: N41 38 13 W70 55 41 On air date: Apr 17, 1997. One State St., Providence, RI, 02908. Phone: (401) 351-8828. Fax: (401) 351-0222. Web Site: cw28tv.com. Licensee: C-28 FCC Licensee Subsidiary Inc. Group owner: Viacom Television Stations Group (acq 10-15-2001). Natl. Network: CW. Natl. Rep: TeleRep. Washington Atty: Keck, Mahin & Cate.
Key Personnel:
Pam Bergeron . . . . . sls dir
Lisa Pesanello . . . . . natl sls mgr
Tina Castano . . . . . rgnl sls mgr

WNAC-TV— Analog channel: 64. Digital channel: 54. Analog hrs: 24 3,720 kw vis, 372 kw aur. ant 1,033t/900g TL: N41 52 14 W71 17 45 On air date: December 1981. 25 Catamore Blvd., East Providence, RI, 02914. Phone: (401) 438-7200. Fax: (401) 434-3761. Web Site: www.wpri.com. Licensee: WNAC LLC. Ownership: Super Towers Inc., 100% (acq 4-22-2002). Natl. Network: Fox, MyNetworkTV. Natl. Rep: Blair Television.
Key Personnel:
Jay Howell . . . . . pres, VP & gen mgr
Steve Carro . . . . . opns mgr
Patrick Wholey . . . . . gen sls mgr
Nancy Mayers . . . . . natl sls mgr
John Macek . . . . . rgnl sls mgr
Andy Bernstein . . . . . mktg dir
Susan Tracy-Durant . . . . . prom mgr
Pam Brennan . . . . . progmg mgr
Joe Abouzeid . . . . . news dir
Christine Peabody . . . . . pub affrs dir
William Hague . . . . . engrg dir & chief of engrg

WPRI-TV— Analog channel: 12. Analog hrs: 24 316 kw vis, 31.6 kw aur. 910t/1,099g TL: N41 52 37 W71 16 56 On air date: Mar 27, 1955. 25 Catamore Blvd., East Providence, RI, 02914-1203. Phone: (401) 438-7200. Fax: (401) 434-3761. Web Site: www.wpri.com. Licensee: TVL Broadcasting of Rhode Island LLC. Group owner: LIN Television Corporation (acq 11-8-2002; grpsl). Population served: 550,000 Natl. Network: CBS. Natl. Rep: Blair Television.
Key Personnel:
Jay Howell . . . . . pres, VP & gen mgr
Gregg Monte . . . . . opns mgr
Patrick Wholey . . . . . gen sls mgr
Patti St. Pierre . . . . . natl sls mgr
Andrew Bernstein . . . . . mktg dir
Susan Tracy-Durant . . . . . prom mgr
Pam Brennan . . . . . progmg mgr
Joe Abouzeid . . . . . news dir
Glenn Laxton . . . . . pub affrs dir
William Hague . . . . . chief of engrg

WPXQ— Analog channel: 69. Digital channel: 17. Analog hrs: 24 3470 kw vis. ant 272t TL: N41 29 41 W71 47 05 (CP: 2,880 kw vis, ant 751t/613g) On air date: Apr 2, 1992. 3 Shaws CV, Ste 226, New London, CT, 06320-4943. Phone: (401) 455-9263. Fax: (401) 455-9156. E-mail: Robert.melfi@nbc.com Web Site: www.ionline.tv. Licensee: Ocean State Television L.L.C. Ownership: A joint venture of Paxson Communications Corp. and Offshore Broadcasting Corp. (Raymond Yorke, 100%). Group owner: Paxson Communications Corp. (acq 8-2-96). Natl. Network: i Network. Washington Atty: Cohn & Marks. News: 5 hrs wkly.

***WSBE-TV—** Analog channel: 36. Digital channel: 8. Analog hrs: 24 1,230 kw vis, 123 kw aur. ant 597t/508g TL: N41 48 18 W71 28 24 On air date: June 5, 1967. 50 Park Ln., Providence, RI, 02907. Phone: (401) 222-3636. Fax: (401) 222-3407. E-mail: info@rlpbs.org Web Site: www.rlpbs.org. Licensee: Rhode Island Public Telecommunications Authority. Population served: 1,000,000 Natl. Network: PBS. Washington Atty: Schwartz, Woods & Miller.
Key Personnel:
David Piccerelli . . . . . CFO
Dexter B. Merry . . . . . chief of opns
Tracey Cugno . . . . . dev dir
Kathryn Larsen . . . . . progmg dir
Michael Bert . . . . . chief of engrg

South Carolina

Allendale

see Augusta, GA market

Anderson

see Greenville-Spartanburg, SC-Asheville, NC-Anderson, SC market

Beaufort

see Savannah, GA market

Charleston, SC (DMA 100)

WCBD-TV— Analog channel: 2. Digital channel: 50. Analog hrs: 24 Note: NBC is on WCBD-TV ch 2, CW is on WCBD-DT cj 50. 100 kw vis, 10 kw aur. ant 1,950t TL: N32 56 24 W79 41 45 On air date: Sept 25, 1954. 210 W. Coleman Blvd., Mt. Pleasant, SC, 29464. Phone: (843) 884-2222. Fax: (843) 881-3410. Web Site: www.wcbd.com. Licensee: Media General Broadcasting Inc. Group owner: Media General Broadcast Group (acq 3-1-83; $8 million; FTR: 1-24-83). Population served: 295,039 Natl. Network: NBC, CW Digital Network: Note: NBC is on WCBD-TV ch 2, CW is on WCBD-DT cj 50. Natl. Rep: Harrington, Righter & Parsons. Washington Atty: Cohn & Marks. News staff: 29; News: 17 hrs wkly.
Key Personnel:
Joe Pomilla . . . . . gen mgr
Patric J. Ryal . . . . . gen sls mgr & natl sls mgr
Mark Bradley . . . . . mktg mgr
Sam Barclay . . . . . chief of engrg

WCIV— Analog channel: 4. Analog hrs: 24 100 kw vis, 20 kw aur. 1,958t/1,958g TL: N32 55 28 W79 41 58 On air date: Oct 23, 1962. Box 22165, Charleston, SC, 29413-2165. 888 Allbritton Blvd., Mt. Pleasant, SC 29464. Phone: (843) 881-4444. Fax: (843) 849-2507 (admin). Fax: (843) 849-2515 (sales). Web Site: www.abcnews4.com. Licensee: WCIV L.L.C. Group owner: Allbritton Communications Co. (acq 1-26-76; grpsl). Population served: 1,500,000 Natl. Network: ABC. Washington Atty: Hogan & Hartson. Wire Svc: Conus News staff: 32; News: 12 hrs wkly.
Key Personnel:
Suzanne Teagle . . . . . gen mgr
Terry Wright . . . . . opns mgr & chief of engrg
Octavia Walker . . . . . natl sls mgr
Chuck Groome . . . . . rgnl sls mgr
Tim Greeney . . . . . prom mgr
Deborah Jackson . . . . . progmg mgr
Perry Boxx . . . . . news dir

WCSC-TV— Analog channel: 5. Digital channel: 47. Analog hrs: 24 100 kw vis, 20 kw aur. ant 1,958t TL: N32 55 28 W79 41 58 On air date: June 19, 1953. 2126 Charlie Hall Blvd., Charleston, SC, 29414. Phone: (843) 402-5555. Fax: (843) 402-5744. Web Site: www.wcsc.com. Licensee: WCSC Inc. (acq 4-3-2006; grpsl). Population served: 617,800 Natl. Network: CBS. News staff: 44; News: 21 hrs wkly.
Key Personnel:
Rita O'Neill . . . . . gen mgr
Brian Stephenson . . . . . dev mgr
Amy Spencer . . . . . gen sls mgr
Amanda Childs . . . . . prom mgr
Riten O'Neil . . . . . progmg mgr
Mary Rigby . . . . . news dir
Mike Gurthie . . . . . engrg dir
Lowell Knoff . . . . . chief of engrg

***WITV—** Analog channel: 7. Digital channel: 49. Analog hrs: 24 316 kw vis. ant 1,850t/1,837g TL: N32 55 28 W79 41 58 On air date: Jan 19, 1964. 1101 George Rogers Blvd., Columbia, SC, 29201. Phone: (803) 737-3545. Web Site: www.myetv.org. Licensee: South Carolina ETV Commission. Natl. Network: PBS. Washington Atty: Dow, Lohnes & Albertson.

WMMP— Analog channel: 36. Digital channel: 35. Analog hrs: 24 3,251 kw vis, 50 kw aur. ant 994t TL: N32 47 15 W79 51 00 On air date: November 1992. 4301 Arco Ln., Charleston, SC, 29418. Phone: (843) 744-2424. Fax: (843) 554-9649. E-mail: comments@wmmp36.com Web Site: www.wmmp36.com. Licensee: WMMP Licensee L.P. Group owner: Sinclair Broadcast Group Inc. (acq 1998; grpsl). Natl. Network: MyNetworkTV.
Key Personnel:
David Tynan . . . . . VP & gen mgr

Mary Margaret Johnson . . . . gen mgr & gen sls mgr
Jason Lewis. . . . . mktg mgr & prom mgr
Bill Littleton . . . . progmg dir & progmg mgr

WTAT-TV— Analog channel: 24. Analog hrs: 24 5,000 kw vis, 497.5 kw aur. 1,800t/1,800g TL: N32 56 24 W79 41 45 On air date: Sept 7, 1985. 4301 Arco Ln., Charleston, SC, 29418. Phone: (843) 744-2424. Fax: (843) 554-9649. E-mail: comments@wmmp36.com Web Site: www.wtat24.com. Licensee: WTAT Licensee LLC. Group owner: Cunningham Broadcasting Corporation (acq 11-15-2001; grpsl). Natl. Network: Fox. Washington Atty: Arter & Hadden. News: 3.5 hrs wkly.
Key Personnel:
David Tynan . . . . gen mgr
Mary Margaret Johnson . . . . gen sls mgr
Jason Lewis . . . . prom VP & prom mgr
Bill Littleton. . . . . progmg dir

Columbia, SC (DMA 83)

WACH— Analog channel: 57. Digital channel: 48. Analog hrs: 24 5,000 kw vis, 500 kw aur. ant 633t/623g TL: N34 02 39 W80 59 52 On air date: 1988. 1400 Pickens St., Columbia, SC, 29201. Phone: (803) 252-5757. Fax: (803) 212-7270. E-mail: webmaster@wach.com Web Site: www.wach.com. Licensee: Barrington Columbia License LLC. Group owner: Raycom Media Inc. (acq 8-11-2006; grpsl). Natl. Network: Fox. Natl. Rep: TeleRep. Washington Atty: Covington & Burling.
Key Personnel:
Scott McBride . . . . VP & gen mgr
Phil Shreves . . . . opns mgr & chief of engrg
Cheri Spets . . . . gen sls mgr
Barbara Bethea . . . . rgnl sls mgr, mktg dir, prom dir & news dir
Reese Barkley . . . . pub affrs dir

WIS— Analog channel: 10. Digital channel: 41. Analog hrs: 24 316 kw vis, 63.2 kw aur. ant 1,571t/1,489g TL: N34 07 29 W80 45 23 On air date: Nov 7, 1953. 1111 Bull St., Columbia, SC, 29201. Box 367, Columbia, SC 29202. Phone: (803) 799-1010. Fax: (803) 758-1171. E-mail: mstebbins@wistv.com Web Site: www.wistv.com. Licensee: WIS License Subsidiary LLC. Group owner: Liberty Corp. (acq 1-13-2006; grpsl). Population served: 314,000 Natl. Network: NBC. Natl. Rep: Harrington, Righter & Parsons. Washington Atty: Covington & Burling LLP. News: 26.5 hrs wkly.
Key Personnel:
Melbourne Stebbins. . . . . VP & gen mgr
Brent Lane . . . . gen sls mgr
Quentin Kenney . . . . gen sls mgr & rgnl sls mgr
Barry Ahrendt . . . . mktg dir & progmg dir
Tina Lugue-Blacklocke . . . . news dir
Ken Thayer . . . . chief of engrg

WKTC— Analog channel: 63. Digital channel: 39. Analog hrs: 24 12.9 kw vis. ant 541t TL: N33 54 52 W80 17 39 On air date: 1997. 120-A Pontiac Business Center Dr., Elgin, SC, 29045. Phone: (803) 419-6363. Fax: (803) 419-6399. E-mail: mail@wktctv.com Web Site: www.wktctv.com. Licensee: Columbia Broadcasting Inc. Ownership: Dove Broadcasting Inc. (acq 11-99). Natl. Network: MyNetworkTV.

WLTX— Analog channel: 19. Digital channel: 17. Analog hrs: 24 Digital hrs: 24 5,000 kw vis, 500 kw aur. ant 1,749t/1,706g TL: N34 05 49 W80 45 51 On air date: Sept 1, 1953. 6027 Garners Ferry Rd., Columbia, SC, 29209. Phone: (803) 776-3600. Fax: (803) 695-3714. Web Site: www.wltx.com. Licensee: Pacific and Southern Co. Inc. Group owner: Gannett Broadcasting (acq 4-29-98; $87.5 million). Population served: 615,000 Natl. Network: CBS. Natl. Rep: Blair Television. News: 22 hrs wkly.
Key Personnel:
Rich O'Dell. . . . . gen mgr
Jim Hays . . . . mktg mgr
Mike Garber . . . . news dir
Cliss Calkins . . . . chief of engrg

WOLO-TV— Analog channel: 25. Digital channel: 8. Analog hrs: 24 3,550 kw vis, 355 kw aur. 830t TL: N34 03 23 W80 58 49 On air date: Oct 1, 1961. Box 4217, Columbia, SC, 29240. 5807 Shakespeare Rd., Columbia, SC 29223. Phone: (803) 754-7525. Fax: (803) 754-6147. Web Site: www.wolo.com. Licensee: South Carolina Broadcasting Partners. (acq 7-20-92). Population served: 480,000 Natl. Network: ABC. Wire Svc: AP

***WRJA-TV—** Analog channel: 27. Digital channel: 28. Analog hrs: 24 647 kw vis. ant 1,161t/1,161g TL: N33 52 51 W80 16 15 On air date: Sept 7, 1975. 18 N. Harvin St., Sumter, SC, 29150. Phone: (803) 773-5546. Fax: (803) 775-1059. E-mail: wrjatv@ftc-i.net Web Site: www.wrja.org. Licensee: South Carolina ETV Commission. Population served: 87,000 Natl. Network: PBS.
Key Personnel:
William Anderson . . . . gen mgr
Kevin Jordan . . . . engrg mgr

***WRLK-TV—** Analog channel: 35. Analog hrs: 24 570 kw vis. ant 1,030t/971g TL: N34 07 06 W80 56 13 On air date: Sept 5, 1966. Box 11000, Columbia, SC, 29211. 1101 George Rogers Blvd., Columbia, SC 29211. Phone: (803) 737-3200. Phone: (803) 737-3212. Fax: (803) 737-3417. E-mail: mail@myetv.org Web Site: www.myetv.org. Licensee: South Carolina ETV Commission. Population served: 3,200,000 Natl. Network: PBS. Washington Atty: Dow, Lohnes & Albertson.
Key Personnel:
Maurice "Moss" Bresnahan. . . . . CEO & pres
L.W. Griffin Jr. . . . . engrg VP & engrg dir

WZRB— Analog channel: 47.750 kw vis. ant 630t/518g TL: N34 02 38 W80 59 51 On air date: Jan 1, 2005. 1747 Cushman Dr., Columbia, SC, 29204. Phone: (803) 714-2347. Fax: (803) 691-3848. Web Site: www.cw47columbia.com. Licensee: Roberts Broadcasting Co. of Columbia, SC LLC. Group owner: Roberts Broadcasting Co. Population served: 915,000 Natl. Network: CW. Washington Atty: Dow, Lohnes & Albertson.

Florence

see Myrtle Beach-Florence, SC market

Greenville-Spartanburg, SC-Asheville, NC-Anderson, SC (DMA 36)

WGGS-TV— Analog channel: 16. Digital channel: 35. Analog hrs: 24 2,240 kw vis. ant 1,145t/167g TL: N34 56 26 W82 24 41 On air date: October 1972. Box 1616, Greenville, SC, 29602. 3409 Rutherford Rd., Taylors, SC 29687. Phone: (864) 244-1616. Fax: (864) 292-8481. E-mail: ccbtv16@aol.com Web Site: www.dovebroadcasting.com. Licensee: Carolina Christian Broadcasting Inc. Ownership: James H. Thompson, 92%. Population served: 750,000 Washington Atty: Hardy & Chautin.
Key Personnel:
James W. Thompson. . . . . pres & gen mgr
Joanne Thompson . . . . VP
Billy Rainey . . . . sls dir
Kathy Newell. . . . . natl sls mgr
Gene Gibson. . . . . chief of engrg
Kym MacKinnon . . . . progmg

WHNS— Analog channel: 21. Digital channel: 57. Analog hrs: 24 Digital hrs: 24 3,390 kw vis, 398 kw aur. 2,509t/1,604g TL: N35 10 56 W82 40 56 On air date: Apr 1, 1984. 21 Interstate Ct., Greenville, SC, 29615. Phone: (864) 288-2100. Fax: (864) 297-0728. E-mail: whns@foxcarolina.com Web Site: www.foxcarolina.com. Licensee: Meredith Corp. Group owner: Meredith Broadcasting Group, Meredith Corp. (acq 7-1-97; grpsl). Population served: 1,700,000 Natl. Network: Fox. Natl. Rep: TeleRep. Washington Atty: Dow, Lohnes & Albertson. News staff: 45; News: 27 hrs wkly.
Key Personnel:
Steve Lacy . . . . CEO
William Kerr . . . . chmn
Paul Karpowicz. . . . . pres
Dalton Lee . . . . CFO
Douglas Lowe . . . . exec VP
Guy W. Hempel . . . . VP & gen mgr
Jeff Guibert . . . . gen sls mgr
Alan DeFlorio . . . . natl sls mgr
April White . . . . mktg mgr
Kyann Lewis . . . . news dir
Jim Barnes . . . . chief of engrg

WLOS—(Asheville, NC) Analog channel: 13. Digital channel: 56. Analog hrs: 24 Digital hrs: 24 170 kw vis, 19.6 kw aur. 2,804t/339g TL: N35 25 32 W82 45 25 On air date: Sept 18, 1954. 110 Technology Dr., Asheville, NC, 28803. Phone: (828) 684-1340. Fax: (828) 651-4618. E-mail: news@wlos.com Web Site: www.wlos.com. Licensee: WLOS Licensee L.L.C. Group owner: Sinclair Broadcast Group Inc. (acq 6-96). Population served: 815,000 Natl. Network: ABC. Natl. Rep: Harrington, Righter & Parsons. Washington Atty: Dow, Lohnes & Albertson. News staff: 57; News: 24 hrs wkly.
Key Personnel:
Jack Connors. . . . . gen mgr
Audra Swain . . . . chief of opns & sls dir
Guy Chancey . . . . mktg dir
Scott Bradsher. . . . . progmg mgr
Julie Fries . . . . news dir
Rollin Thompkins . . . . chief of engrg

WMYA-TV— Analog channel: 40. Digital channel: 14. Analog hrs: 8:00am - 5:00pm 2,570 kw vis, 257 kw aur. ant 1,050t/1,020g TL: N34 38 51 W82 16 13 On air date: Dec 1, 1953. 24 Verdae Blvd., Suite 203, Greenville, SC, 29607. 110 Technology Dr., (Tapes & Traffic), Asheville, NC 28803. Phone: (864) 297-1313. Fax: (864) 297-8085. Web Site: www.my40.tv. Licensee: Anderson (WFBC-TV) Licensee Inc. Group owner: Cunningham Broadcasting Corporation (acq 1-7-2002). Population served: 815,000 Natl. Network: MyNetworkTV.
Key Personnel:
David D. Smith. . . . . pres
J. Duncan Smith . . . . exec VP
Frederick G. Smith . . . . VP
Steven M. Marks . . . . chief of opns
Darren Shapiro . . . . sls VP
Gregg Sigel . . . . natl sls mgr
Jeff Sleete . . . . mktg VP & mktg dir
M. William Butler . . . . progmg VP
Joe DeFeo . . . . news dir & pub affrs dir

WNEG-TV— Analog channel: 32. Digital channel: 24. Analog hrs: 24 647 kw vis, 129 kw aur. ant 835t/600g TL: N34 36 44 W83 22 05 On air date: Sept 9, 1984. 100 Blvd., Toccoa, GA, 30577. Phone: (706) 886-0032. Fax: (706) 886-7033. Web Site: www.wneg32.com. Licensee: Media General Communications Inc. Group owner: Media General Broadcast Group (acq 3-27-2000; grpsl). Natl. Network: CBS. Natl. Rep: MMT. Washington Atty: Dow Lohnes PLLC. News staff: 6; News: 13 hrs wkly.
Key Personnel:
Jim Sanders. . . . . gen mgr & stn mgr
David Austin. . . . . gen sls mgr
Tony Garrison . . . . prom mgr
Stephanie Harrison . . . . progmg mgr
J. Walker . . . . chief of engrg

***WNEH—** Analog channel: 38.1,780 kw vis. ant 771t/656g TL: N34 22 21 W82 10 03 On air date: Sept 10, 1984. 1101 George Rogers Blvd., Columbia, SC, 29201. Phone: (803) 737-3545. E-mail: mail@myetv.org Web Site: www.myetv.org. Licensee: South Carolina Educational TV Commission. Natl. Network: PBS.

***WNTV—** Analog channel: 29. Analog hrs: 24 5,000 kw vis. ant 1,286t/262g TL: N34 56 26 W82 24 38 On air date: Sept 15, 1963. 1101 George Rogers Blvd., Columbia, SC, 29201. Phone: (803) 737-3545. Phone: (803) 737-9959. Fax: (803) 737-3495. E-mail: mail@myetv.org Web Site: www.myetv.org. Licensee: South Carolina ETV Commission. Population served: 550,000 Natl. Network: PBS.
Key Personnel:
Maurice "Moss" Bresnahan . . . . CEO, pres & stn mgr
L.W. Griffin Jr. engrg VP

***WRET-TV—** Analog channel: 49. Digital channel: 43. Analog hrs: 24 1,740 kw vis, 174 kw aur. ant 970t/859g TL: N34 52 09 W81 49 15 On air date: Sept 4, 1980. Box 4069, Spartanburg, SC, 29305-4069. Phone: (864) 503-9371. Fax: (864) 503-3615. Web Site: www.wret.org. Licensee: South Carolina Educational TV Commission. Population served: 1,000,000 Natl. Network: PBS. Washington Atty: Dow, Lohnes & Albertson.
Key Personnel:
William Richardson . . . . opns mgr
Gary Stevens . . . . chief of engrg
Satellite of WNTV Greenville.

WSPA-TV—Spartanburg, Analog channel: 7. Digital channel: 53. Analog hrs: 24 Digital hrs: 24 316 kw vis, 31.6 kw aur. 2,001t/258g TL: N35 10 12 W82 17 27 On air date: Apr 29, 1956. Box 1717, Spartanburg, SC, 29304. 250 International Dr., Spartanburg, SC 29303. Phone: (864) 576-7777. Fax: (864) 587-4480. Web Site: www.wspa.com. Licensee: Media General Broadcasting of So. Carolina Holding Inc. Group owner: Media General Broadcasting of So. Carolina Holding (acq 3-27-2000; grpsl). Population served: 784,300 Natl. Network: CBS. Washington Atty: Dow, Lohnes. Wire Svc: UPI News staff: 51; News: 26.5 hrs wkly.
Key Personnel:
Jim Zimmerman . . . . pres
Phil Lane . . . . VP & gen mgr
Jimmy Lizer . . . . opns mgr
Marilyn Hammond . . . . gen sls mgr
Bill Shatten. . . . . mktg dir

***WUNF-TV—** Analog channel: 33. Digital channel: 25. Analog hrs: 24 Digital hrs: 24 2,690 kw vis, 269 kw aur. 2,676t/339g TL: N35 25 32 W82 45 25 On air date: Sept 11, 1967. Box 14900, Research Triangle Park, NC, 27709-4900. 10 T.W. Alexander Dr., Research Triangle Park, NC 27709. Phone: (919) 549-7000. Fax: (919) 549-7201. E-mail: viewer@unctv.org Web Site: www.unctv.org. Licensee: University of North Carolina. Population served: 9,000,000 Natl. Network: PBS. Washington Atty: Schwartz, Woods & Miller.

WYCW— Analog channel: 62. Digital channel: 45.5,000 kw vis, 250 kw aur. ant 1,823t TL: N35 13 20 W82 32 58 On air date: June 1986. Box 1717, Spartanburg, SC, 29304. 250 International Dr., Spartanburg, NC 29303. Phone: (864) 576-7777. Fax: (864) 595-4615. Web Site: www.carolinascw.com. Licensee: Media General Broadcasting of South Carolina Holdings Inc. Group owner: Media General Broadcast

Group (acq 1-15-2002; $4.5 million). Population served: 1.950 m,ill,ion Natl. Network: CW. Washington Atty: Dow, Lohnes & Albertson.
Key Personnel:
Jim Zimmerman . . . pres
Jim Conschafter . . . sr VP
Phil Lane . . . gen mgr
Randy Ingram . . . stn mgr
Jimmy Lizer . . . opns mgr

WYFF— Analog channel: 4. Digital channel: 59. Analog hrs: 24 100 kw vis, 20 kw aur. ant 2,000t/892g TL: N35 06 40 W82 36 17 On air date: Dec 31, 1953. Box 788, Greenville, SC, 29602. 505 Rutherford St., Greenville, SC 29609. Phone: (864) 242-4404. Fax: (864) 240-5329. E-mail: news4@wyff.com Web Site: www.wyff4.com. Licensee: WYFF Hearst-Argyle Television Inc. Group owner: Hearst-Argyle Television Inc. (acq 3-18-99; grpsl). Population served: 1,601,700 Natl. Network: NBC. Natl. Rep: Eagle Television Sales. Washington Atty: Brooks, Pierce, McLendon, Humphrey & Leonard. Wire Svc: AP News staff: 55; News: 28 hrs wkly.
Key Personnel:
Michael J. Hayes . . . pres & gen mgr
Doug Durkee . . . opns dir, opns mgr & chief of engrg
John Humphrey . . . gen sls mgr
Jimmy Denton . . . natl sls mgr
Ron Bass . . . rgnl sls mgr
Cathy Petropoulos . . . mktg dir
Marsa Jarrett . . . prom mgr & pub affrs dir
Danny Ross . . . progmg dir
Stephanie Sloka . . . film buyer
Andy Still . . . news dir

Hardeeville

see Savannah, GA market

Myrtle Beach-Florence, SC (DMA 105)

WBTW— Analog channel: 13. Digital channel: 56. Analog hrs: 24 316 kw vis, 31.6 kw aur. ant 1,950t/2,000g TL: N34 22 02 W79 19 22 On air date: Oct 18, 1954. 3430 N. TV Rd., Florence, SC, 29501-0013. 101 McDonald Ct., Myrtle Beach, SC 29588. Phone: (843) 317-1313. Fax: (843) 317-1410. Web Site: www.scnow.com. Licensee: Media General Communications Inc. Group owner: Media General Broadcast Group (acq 3-27-2000; grpsl). Population served: 221,280 Natl. Network: CBS. Natl. Rep: Harrington, Righter & Parsons. News staff: 38:; News: 78 hrs wkly.
Key Personnel:
Michael Caplan . . . VP, gen mgr & gen mgr
Brian Lang . . . gen sls mgr
Chuck Spruill . . . mktg mgr
David Halt . . . news dir
Scott Johnson . . . chief of engrg

WFXB— Analog channel: 43. Digital channel: 18. Analog hrs: 24 Digital hrs: 24 5,000 kw vis. ant 1,519t/1,506g TL: N34 11 19 W79 11 00 On air date: July 5, 1984. Box 8309, Myrtle Beach, SC, 29578. 3364 Huger St., Myrtle Beach, SC 29577. Phone: (843) 828-4300. Fax: (843) 828-4343. E-mail: 43listens@wfxb.com Web Site: www.wfxb.com. Licensee: Springfield Broadcasting Partners. (acq 3-16-2006; $19.5 million). Population served: 1,200,000 Natl. Network: Fox. Natl. Rep: Millennium Sales & Marketing. Washington Atty: Fisher, Wayland, Cooper, Leader & Zaragoza. News staff: 50+; News: 2.5 hrs wkly.
Key Personnel:
Rigby Wilson . . . VP
David Carfolite . . . gen mgr
David Milligan . . . chief of opns

***WHMC**— Analog channel: 23. Analog hrs: 24 1,740 kw vis. ant 820t/839g TL: N33 57 05 W79 06 31 On air date: Sept 2, 1980. 1101 George Rogers Blvd., Columbia, SC, 29201. Phone: (803) 737-3545. Phone: (803) 737-9959. Fax: (803) 737-3495. E-mail: mail@myetv.org Web Site: www.myetv.org. Licensee: South Carolina Educational TV Commission. Natl. Network: PBS.
Key Personnel:
Maurice "Moss" Bresnahan . . . CEO, pres & stn mgr
L.W. Griffin Jr. . . . engrg VP

***WJPM-TV**— Analog channel: 33. Digital channel: 45. Analog hrs: 24 646 kw vis. ant 795t/778g TL: N34 16 48 W79 44 35 On air date: Sept 3, 1967. 1101 George Rogers Blvd., Columbia, SC, 29201. Phone: (803) 737-3545. Phone: (803) 737-9959. Fax: (803) 737-3495. E-mail: mail@myetv.org Web Site: www.myetv.org. Licensee: South Carolina ETV Commission. Natl. Network: PBS.

WMBF-TV— Analog channel: 32.5,000 kw vis. ant 981t/981g TL: N33 39 37 W79 03 35 Not on air, target date: summer 2008: RSA Tower, 20th Fl., 201 Monroe St., Montgomery, AL, 36104. Phone: (334) 206-1400. Fax: (334) 206-1555. Permittee: Raycom TV Broadcasting Inc. (acq 1-31-2006; grpsl).

WPDE-TV— Analog channel: 15. Digital channel: 15. Analog hrs: 24 1,290 kw vis, 129 kw aur. ant 1,948t/2,008g TL: N34 21 53 W79 19 49 On air date: Nov 22, 1980. 1194 Atlantic Ave., Conway, SC, 29526. Phone: (843) 234-9733. Fax: (843) 234-9739. E-mail: feedback@wpde.com Web Site: www.wpdetv.com. Licensee: Barrington Broadcasting of South Carolina Corp. Group owner: Diversified Communications (acq 2-6-2006; $24.1 million). Population served: 162,200 Natl. Network: ABC. Washington Atty: Irwin, Campbell & Tannenwald.

***WUNU**— Analog channel: 31. Digital channel: 25. Analog hrs: 24 Digital hrs: 24 3,160 kw vis. 1,046t TL: N34 47 51 W79 02 41 On air date: Sept 23, 1996. Box 14900, Research Triangle Park, NC, 27709-4900. 10 T.W. Alexander Dr., Research Triangle Park, NC 27709. Phone: (919) 549-7000. Fax: (919) 549-7201. E-mail: viewer@unetv.org Web Site: www.unctv.org. Licensee: University of North Carolina. Natl. Network: PBS. Washington Atty: Schwartz, Woods & Miller.

WWMB— Analog channel: 21. Digital channel: 20. Analog hrs: 24 2,090 kw vis, 209 kw aur. ant 1,989t TL: N33 55 14 W79 32 08 On air date: Nov 1, 1994. Box 51150, Myrtle Beach, SC, 29579. Phone: (843) 234-9733. Fax: (843) 234-9739. Web Site: www.cwtv21.com. Licensee: SagamoreHill of Carolina Licenses LLC. (acq 2-6-2006; $2.4 million). Natl. Network: CW.
Key Personnel:
Louis Wall . . . CEO & gen mgr
Laura Walls . . . stn mgr & chief of engrg
Leigh Vaters . . . gen sls mgr
Marty Shelley . . . prom mgr
Billy Huggins . . . progmg mgr
Mike Gathrie . . . engrg mgr
Robert Blair . . . chief of engrg

Rock Hill

see Charlotte, NC market

Spartanburg

see Greenville-Spartanburg, SC-Asheville, NC-Anderson, SC market

South Dakota

Eagle Butte

see Minot-Bismarck-Dickinson, ND market

Lowry

see Minot-Bismarck-Dickinson, ND market

Mitchell

see Sioux Falls (Mitchell), SD market

Rapid City, SD (DMA 177)

***KBHE-TV**— Analog channel: 9. Digital channel: 26. Analog hrs: 24 39.8 kw vis, 7.2 kw aur. 649t/469g TL: N44 03 09 W103 14 38 (CP: 45.62 kw vis, ant 662t. TL: N44 03 07 W103 14 36) On air date: July 1967. 3650 Skyline Dr., Rapid City, SD, 57701. Phone: (605) 394-2551. Fax: (605) 394-6895. E-mail: admin@sdpb.org Web Site: www.sdpb.org. Licensee: South Dakota Board of Directors for Educational Telecommunications. Population served: 151,000 Natl. Network: PBS.

KCLO-TV— Analog channel: 15. Digital channel: 16. Analog hrs: 24 Digital hrs: 24 690 kw vis, 69 kw aur. 520t/4,201g TL: N44 04 14 W103 15 01 On air date: November 1988. 501 S. Philips Ave., Sioux Falls, SD, 57104. Phone: (605) 336-1100. Fax: (605) 334-3447. E-mail: kelotv@keloland.com Web Site: www.keloland.com. Licensee: Young Broadcasting of Rapid City Inc. Group owner: (group owner; (acq 1996; grpsl). Natl. Network: CBS. Natl. Rep: Adam Young. Washington Atty: Brooks, Pierce, McLendon, Humphrey & Leonard.
Key Personnel:
Karen Floyd . . . progmg dir
Mark Millage . . . news dir
Paul Farmer . . . opns mgr, mktg dir, prom dir & news dir
John Hertz . . . chief of engrg

KEVN-TV— Analog channel: 7. Digital channel: 18. Analog hrs: 5 AM-2 AM 263 kw vis, 26.3 kw aur. ant 669t/623g TL: N44 04 00 W103 15 01 On air date: July 4, 1976. Box 677, Rapid City, SD, 57709. 2000 Skyline Dr., Rapid City, SD 57701. Phone: (605) 394-7777. Fax: (605) 348-9128. Fax: (605) 394-3652. E-mail: news@blackhillsfox.com Web Site: www.blackhillsfox.com. Licensee: KEVN Inc. Group owner: Mission TV LLC (acq 8-26-98; $5.5 million with KIVV-TV Lead). Population served: 160,000 Natl. Network: Fox. Natl. Rep: Millennium Sales & Marketing. Washington Atty: Law Offices of Hogan & Hartson. Wire Svc: AP News staff: 14; News: 9 hrs wkly.
Key Personnel:
Bob Slocum . . . CFO
Cindy McNeil . . . VP & gen mgr
Lindsay Bold . . . opns mgr
Kathy Silk . . . rgnl sls mgr
Jack Caudill . . . news dir

KHSD-TV— Analog channel: 11. Digital channel: 10. Analog hrs: 24 316 kw vis, 31.6 kw aur. ant 1,890t/605g TL: N44 19 36 W103 50 12 On air date: Nov 2, 1966. Box 1760, Rapid City, SD, 57709-1760. Phone: (605) 342-2000. Fax: (605) 342-7305. Web Site: www.kotatv.com. Licensee: Duhamel Broadcasting Enterprises. Group owner: (group owner) Population served: 241,224 Natl. Network: ABC. Washington Atty: Shaw Pittman.
Key Personnel:
William F. Duhamel . . . pres & gen mgr
Monte Loos . . . opns dir, progmg dir, film buyer & engrg dir
Steve Duffy . . . gen sls mgr
Gerry Fenske . . . rgnl sls mgr
John Petersen . . . news dir
Satellite of KOTA-TV Rapid City.

KIVV-TV— Analog channel: 5. Digital channel: 29. Analog hrs: 5 AM-2 AM 100 kw vis, 10 kw aur. ant 1,851t/638g TL: N44 19 30 W103 50 14 On air date: July 4, 1976. Box 677, Rapid City, SD, 57709. 2000 Skyline Dr., Rapid City, SD 57709. Phone: (605) 394-7777. Fax: (605) 348-9128. E-mail: news@blackhillsfox.com Web Site: www.blackhillsfox.com. Licensee: KEVN Inc. Group owner: Mission TV LLC (acq 8-26-98; $5.5 million with KEVN-TV Rapid City). Population served: 160,000 Natl. Network: Fox. Natl. Rep: Millennium Sales & Marketing. Washington Atty: Law Offices of Hogan & Hartson. Wire Svc: AP News staff: 14; News: 9 hrs wkly.
Key Personnel:
Bob Slocum . . . CFO
Cindy McNeil . . . VP & gen mgr
Lindsay Bold . . . opns mgr
Kathy Silk . . . rgnl sls mgr
Jack Caudill . . . news dir
Satellite of KEVN-TV Rapid City.

KNBN— Analog channel: 21. Analog hrs: 20 1,500 kw vis. ant 440t TL: N44 01 19 W103 15 33 On air date: May 14, 2000. Box 9549, Rapid City, SD, 57709. 2424 S. Plaza Dr., Rapid City, SD 57709. Phone: (605) 355-0024. Fax: (605) 355-9274. E-mail: webmaster@newscenter1.com Web Site: www.newscenter1.com. Licensee: Rapid Broadcasting Co.James F. Simpson, 10.5%; Scott Barbour, 9.1%; Leeann Rieman, 9.1%; Frank Simpson, 8.3%; Clark D. Moyle, 8.1%;; Gilbert D. Moyle III, 8.1%; W.R. Barbour, 6.1%; William F. Turner, 3.2%; Suzanne M. Gabrielson, 2.4%; Charles H. Lien, 2.4%; and David M. Simpson, 1.3% Group owner: (group owner). Natl. Network: NBC. News: 11 hrs wkly.
Key Personnel:
Jim Simpson . . . VP & gen mgr
Mark Walter . . . opns mgr
Darren Koehne . . . gen sls mgr
Steve Weaver . . . rgnl sls mgr

KOTA-TV— Analog channel: 3. Analog hrs: 24 100 kw vis, 20 kw aur. ant 659t/606g TL: N44 04 08 W103 15 05 On air date: July 1, 1955. Box 1760, Rapid City, SD, 57709-1760. 518 St. Joseph St., Rapid City, SD 57701-1760. Phone: (605) 342-2000. Fax: (605) 342-7305. Web Site: www.kotatv.com. Licensee: Duhamel Broadcasting Enterprises. Group owner: (group owner) Population served: 241,224 Natl. Network: ABC. Washington Atty: Shaw Pittman. Wire Svc: AP News: 9 hrs wkly.
Key Personnel:
William F. Duhamel . . . pres & gen mgr
Monte Loos. opns mgr, progmg dir, film buyer, engrg mgr & chief of engrg
Steve Duffy . . . natl sls mgr
Gerry Fenske . . . rgnl sls mgr
John Peterson . . . news dir

KSGW-TV— Analog channel: 12.316 kw vis, 63.2 kw aur. ant 1,220t TL: N44 37 20 W107 06 57 On air date: Oct 28, 1977. Box 1760, Rapid City, SD, 57709-1760. Phone: (605) 342-2000. Fax: (605) 342-7305. Web Site: www.kotatv.com. Licensee: Duhamel Broadcasting

Enterprises. Group owner: (group owner) Population served: 241,224 Natl. Network: ABC. Washington Atty: Shaw Pittman. News: 8 hrs wkly.
Key Personnel:
William Duhamel . . . . pres & gen mgr
Steve Duffy . . . . gen sls mgr
Gerry Fenske . . . . rgnl sls mgr
Fred Whitley . . . . prom dir
Monte Loos . . . . progmg dir & film buyer
John Petersen . . . . news dir

KSWY— Analog channel: 7.9 kw vis. ant 1,144t/79g TL: N44 37 20 W107 06 57 On air date: 2002. Box 1540, Mills, WY, 82664. Phone: (307) 577-0013. Fax: (307) 577-5251. Licensee: Bozeman Trail Communications Co. Group owner: (group owner). Natl. Network: NBC.
Satellite of KCWY Casper.

***KTNE-TV—** Analog channel: 13. Digital channel: 24. Analog hrs: 18 Digital hrs: 18 316 kw vis, 31.6 kw aur. 1,542t/1,499g TL: N41 50 24 W103 03 18 On air date: Sept 7, 1966. 1800 N. 33rd St., Lincoln, NE, 68503. Phone: (402) 472-3611. Fax: (402) 472-1785. E-mail: net1@unl.edu Web Site: www.netnebraska.org. Licensee: Nebraska Educational Telecommunications Commission. Natl. Network: PBS. Washington Atty: Dow, Lohnes & Albertson.
Key Personnel:
Rod Bates . . . . gen mgr
Steven Graziano . . . . prom mgr, adv mgr & progmg mgr
Satellite of *KUON-TV Lincoln.

***KZSD-TV—** Analog channel: 8. Digital channel: 23.275 kw vis, 27.5 kw aur. ant 869t/571g TL: N43 26 06 W101 33 14 On air date: Feb 8, 1978. Box 5000, Vermillion, SD, 57069-5000. 555 N. Dakota St., Vermillion, SD 57069-5000. Phone: (605) 677-5861. Fax: (605) 677-5010. E-mail: programming@sdpb.org Web Site: www.sdpb.org. Licensee: South Dakota Board of Directors for Educational Telecommunications. Natl. Network: PBS. Washington Atty: Cohn & Marks.
Key Personnel:
Julie Andersen . . . . pres
Craig Jensen . . . . opns mgr
Terry Spencer . . . . dev dir
Carol Robertson . . . . prom dir
Bob Bosse . . . . progmg dir
Stacey Decker . . . . engrg mgr

Sioux Falls (Mitchell), SD (DMA 115)

KABY-TV— Analog channel: 9. Digital channel: 28.316 kw vis, 31.6 kw aur. ant 1,401t TL: N45 06 32 W97 53 30 On air date: Nov 28, 1958. 717 Hwy. 281 N., Aberdeen, SD, 57401. Phone: (605) 225-9200. Phone: (605) 336-1300. Fax: (605) 225-9226. Licensee: Hoak Media of Dakota License LLC. Group owner: Wicks Television L.L.C. (acq 1-3-2007; grpsl). Natl. Network: ABC.
Key Personnel:
Jack Hansen . . . . gen mgr
Eugene Schultz . . . . chief of engrg
Satellite of KSFY-TV Sioux Falls.

***KCSD-TV—** Analog channel: 23. Analog hrs: 24 13.3 kw vis, 1.3 kw aur. 177t/135g TL: N43 32 07 W96 44 34 On air date: June 13, 1995. Box 5000, Vermillion, SD, 57069-5000. 555 N. Dakota St., Vermillion, SD 57069-5000. Phone: (605) 677-5861. Fax: (605) 677-5010. E-mail: programming@sdpb.org Web Site: www.sdpb.org. Licensee: South Dakota Board of Directors for Educational Telecommunications. Population served: 150,000 Natl. Network: PBS. Washington Atty: Cohn & Marks.
Key Personnel:
Julie Andersen . . . . pres
Craig Jensen . . . . opns mgr
Terry Spencer . . . . dev mgr
Fritz Miller . . . . mktg mgr
Bob Bosse . . . . progmg dir & progmg mgr

KDLO-TV— Analog channel: 3. Digital channel: 2. Analog hrs: 24 Digital hrs: 24 100 kw vis, 20 kw aur. 1,690t/1,710g TL: N44 57 57 W97 35 22 On air date: September 1955. 501 S. Phillips, Sioux Falls, SD, 57104. Phone: (605) 336-1100. Fax: (605) 334-3447. E-mail: kelotv@keloland.com Web Site: www.keloland.com. Licensee: Young Broadcasting of Sioux Falls Inc. Group owner: Young Broadcasting Inc. (acq 6-1-96; grpsl). Natl. Network: CBS. Natl. Rep: Adam Young. Washington Atty: Brooks, Pierce, McLendon, Humphrey & Leonard.
Key Personnel:
Mark Millage . . . . gen mgr
Paul Farmer . . . . prom dir
John Hertz . . . . chief of engrg
Karen Floyd . . . . progmg
Satellite of KELO-TV Sioux Falls 100%.

KDLT-TV— Analog channel: 46. Digital channel: 47. Analog hrs: 24 4,000 kw vis. ant 1,991t TL: N43 30 17 W96 33 22 On air date: November 1998. c/o KDLT(TV), 3600 S. Westport Ave., Sioux Falls, SD, 57106-6325. Phone: (605) 361-5555. Fax: (605) 361-7017. Fax: (605) 361-3982. E-mail: info@kdlt.com Web Site: www.kdlt.com. Licensee: Red River Broadcast Co. L.L.C. Group owner: (group owner) Population served: 582,000 Natl. Network: NBC. Natl. Rep: Harrington, Righter & Parsons. Washington Atty: Holland and Knight. Wire Svc: AP News staff: 25; News: 19 hrs wkly.
Key Personnel:
Myron Kunin . . . . CEO
Ro Grignon . . . . pres
Kathy Lau . . . . VP
Mari Ossenfort . . . . gen mgr, sls dir & gen sls mgr
Susan Endres . . . . opns mgr
Emily Dimock . . . . prom VP
Bobbi Lower . . . . news dir
Donald Sturzenbecher . . . . chief of engrg

KDLV-TV— Analog channel: 5. Digital channel: 26. Analog hrs: 24 100 kw vis, 10 kw aur. ant 981t/998g TL: N43 45 33 W98 24 44 On air date: June 12, 1960. 3600 S. Westport Ave., Sioux Falls, SD, 57106-6325. Phone: (605) 361-5555. Fax: (605) 361-3982/(605) 361-7017. Web Site: www.kdlt.com. Licensee: Red River Broadcast Co. L.L.C. Group owner: (group owner; (acq 8-26-94; $4 million; FTR: 9-12-94). Population served: 582,000 Natl. Network: NBC. Natl. Rep: Harrington, Righter & Parsons. Washington Atty: Holland & Knight. Foreign lang progmg: SP 4 News staff: 25; News: 20 hrs wkly.
Key Personnel:
Ro Grignon . . . . pres
Mari Ossenfort . . . . gen mgr, gen sls mgr & adv mgr
Emily Dimock . . . . prom mgr & pub affrs dir
Susan Endres . . . . opns mgr & progmg mgr
Bobbi Lauer . . . . news dir
Don Sturzenbecher . . . . chief of engrg
Satellite of KDLT-TV Sioux Falls.

***KDSD-TV—** Analog channel: 16. Analog hrs: 24 1,350 kw vis, 135 kw aur. 1,171t/1,062g TL: N45 29 55 W97 40 35 On air date: Jan 1, 1972. Box 5000, Vermillion, SD, 57069-5000. 555 N. Dakota St., Vermillion, SD 57069-5000. Phone: (605) 677-5861. Fax: (605) 677-5010. E-mail: programming@sdpb.org Web Site: www.sdpb.org. Licensee: South Dakota Board of Directors for Educational Telecommunications. Natl. Network: PBS. Washington Atty: Cohn & Marks.
Key Personnel:
Julie Andersen . . . . pres
Craig Jensen . . . . opns mgr
Terry Spencer . . . . dev dir
Fritz Miller . . . . mktg mgr
Bob Bosse . . . . progmg dir

KELO-TV— Analog channel: 11. Digital channel: 32. Analog hrs: 24 Digital hrs: 24 Note: CBS is on KELO-TV ch 11, MyNetworkTV is on KELO-DT ch 32. 316 kw vis, 28.8 kw aur. ant 2,001t/1,952g TL: N43 31 07 W96 32 05 On air date: May 1953. 501 S. Phillips, Sioux Falls, SD, 57104. Phone: (605) 336-1100. Fax: (605) 334-3447. Fax: (605) 357-5530. Web Site: www.keloland.com. Licensee: Young Broadcasting of Sioux Falls Inc. Group owner: Young Broadcasting Inc. (acq 6-1-96; grpsl). Population served: 306,000 Natl. Network: CBS, MyNetworkTV Digital Network: Note: CBS is on KELO-TV ch 11, MyNetworkTV is on KELO-DT ch 32. Natl. Rep: Adam Young. Washington Atty: Brooks, Pierce, McLendon, Humphrey & Leonard.
Key Personnel:
Paul Farmer . . . . mktg dir & prom dir
Karen Floyd . . . . progmg dir
Mark Millage . . . . news dir
John Hertz . . . . chief of engrg

***KESD-TV—** Analog channel: 8. Digital channel: 18. Analog hrs: 24 245 kw vis, 51.3 kw aur. ant 751t/801g TL: N44 20 10 W97 13 41 On air date: Feb 6, 1968. Box 5000, Vermillion, SD, 57069. Phone: (605) 677-5861. Fax: (605) 677-5010. E-mail: admin@sdpb.org Web Site: www.sdpb.org. Licensee: South Dakota Board of Directors for Educational Telecommunications. Population served: 151,000 Natl. Network: PBS. Rgnl. Network: CEN. Washington Atty: Cohn & Marks.

KPLO-TV— Analog channel: 6. Digital channel: 13. Analog hrs: 24 Digital hrs: 24 100 kw vis, 15 kw aur. 1,110t/711g TL: N43 57 55 W99 36 11 On air date: July 1957. 501 S. Phillips, Sioux Falls, SD, 57104. Phone: (605) 336-1100. Fax: (605) 334-3447. E-mail: kelotv@keloland.com Web Site: www.keloland.com. Licensee: Young Broadcasting of Sioux Falls Inc. Group owner: Young Broadcasting Inc. (acq 6-1-96; grpsl). Natl. Network: CBS. Natl. Rep: Adam Young. Washington Atty: Brooks, Pierce, McLendon, Humphrey & Leonard.
Key Personnel:
Mark Millage . . . . gen mgr & news dir
Paul Farmer . . . . mktg dir & prom dir
Karen Floyd . . . . progmg dir
John Hertz . . . . chief of engrg
Satellite of KELO-TV Sioux Falls 100%.

KPRY-TV— Analog channel: 4. Digital channel: 19.100 kw vis, 20 kw aur. ant 1,240t/1,089g TL: N44 03 07 W100 05 03 On air date: February 1976. 300 N. Dakota Ave., Sioux Falls, SD, 57104. Phone: (605) 336-1300. Fax: (605) 336-7936. Licensee: Hoak Media of Dakota License LLC. Group owner: Wicks Television L.L.C. (acq 1-3-2007; grpsl). Natl. Network: ABC. Washington Atty: Arent, Fox, Kintner, Plotkin & Kahn. News staff: 45; News: 15 hrs wkly.
Key Personnel:
Jack Hanson . . . . gen mgr
Eugene Schultz . . . . chief of engrg

***KRNE-TV—** Analog channel: 12. Digital channel: 17. Analog hrs: 18 hrs Digital hrs: 18 hrs. 316 kw vis, 31.6 kw aur. 1,066t/1,029g TL: N42 40 38 W101 42 36 On air date: Dec 9, 1968. 1800 N. 33rd St., Lincoln, NE, 68503. Phone: (402) 472-3611. Fax: (402) 472-1785. E-mail: net1@unl.edu Web Site: www.netnebraska.org. Licensee: Nebraska Educational Telecommunications Commission. Natl. Network: PBS. Washington Atty: Dow, Lohnes & Albertson.
Key Personnel:
Rod Bates . . . . gen mgr
Steven Graziano . . . . prom mgr, adv mgr & progmg mgr
Satellite of KUON-TV Lincoln.

KSFY-TV— Analog channel: 13. Digital channel: 29. Analog hrs: 24 316 kw vis, 39.8 kw aur. ant 2,000t/1,985g TL: N43 31 07 W96 32 05 On air date: July 31, 1960. 300 N. Dakota Ave., Suite 100, Sioux Falls, SD, 57104. Phone: (605) 336-7936. Fax: (605) 336-3468. Web Site: www.ksfy.com. Licensee: Hoak Media of Dakota License LLC. Group owner: Wicks Television L.L.C. (acq 1-3-2007; grpsl). Population served: 500,000 Natl. Network: ABC. Natl. Rep: TeleRep. News staff: 30; News: 15 hrs wkly.
Key Personnel:
Kelly Manning . . . . gen mgr
Ryan Welsh . . . . gen sls mgr
Darrel Nelson . . . . chief of engrg

***KSMN—** Analog channel: 20. Digital channel: 15. Analog hrs: 24 1,260 kw vis. 1,089t TL: N43 53 52 W95 56 50 On air date: 1997. 120 W. Schlieman Ave., Appleton, MN, 56208-1351. Phone: (320) 320-2622. Fax: (320) 289-2634. E-mail: yourtv@pioneer.org Web Site: www.pioneer.org. Licensee: West Central Minnesota Educational TV Co. Natl. Network: PBS.
Key Personnel:
Les Heen . . . . pres & gen mgr
Shirley Schwarz . . . . progmg dir
Jon Panzer . . . . chief of engrg
KWCM, Appleton.

***KTSD-TV—** Analog channel: 10. Digital channel: 21. Analog hrs: 24 316 kw vis, 31.6 kw aur. 1,601t/1,327g TL: N43 57 55 W99 35 56 On air date: Aug 1, 1970. Box 5000, Vermillion, SD, 57069-5000. 555 N. Dakota St., Vermillion, SD 57069-5000. Phone: (605) 677-5861. Phone: (800) 456-0766. Fax: (605) 677-5010. E-mail: programming@sdpb.org Web Site: www.sdpb.org. Licensee: South Dakota Board of Directors for Educational Telecommunications. Natl. Network: PBS. Washington Atty: Cohn & Marks.
Key Personnel:
Julie Andersen . . . . pres
Craig Jensen . . . . opns mgr
Terry Spencer . . . . dev dir
Bob Bosse . . . . progmg dir

KTTM— Analog channel: 12. Digital channel: 22. Analog hrs: 20 316 kw vis, 31.6 kw aur. ant 860t TL: N44 11 39 W98 19 05 On air date: Sept 7, 1991. c/o KTTW, 2817 W. 11th St., Sioux Falls, SD, 57104. Phone: (605) 338-0017. Fax: (605) 338-7173. E-mail: fox17@kttw.com Web Site: www.kttw.com/home.htm. Licensee: Independent Communications Inc. Natl. Network: Fox. Natl. Rep: Continental Television Sales.
Key Personnel:
Ed Hoffman . . . . gen mgr
Sandy Ellefson . . . . gen sls mgr
Stacey Sieverding . . . . gen sls mgr
Judy Buie . . . . progmg dir
Satellite of KTTW(TV) Sioux Falls 100%.

KTTW— Analog channel: 17. Digital channel: 7. Analog hrs: 20 195 kw vis, 19.5 kw aur. ant 495t/499g TL: N43 29 20 W96 45 40 On air date: Nov 1, 1986. Box 5103, Sioux Falls, SD, 57117-5103. 2817 W. 11th St., Sioux Falls, SD 57104. Phone: (605) 338-0017. Fax: (605) 338-7173. E-mail: fox17@kttw.com Web Site: www.kttw.com/home.htm. Licensee: Independent Communications Inc. (acq 3-9-88). Population served: 216,000 Natl. Network: Fox. Washington Atty: Reddy, Begley & McCormick.
Key Personnel:
Ed Hoffman . . . . gen mgr & stn mgr
Sandy Ellefson . . . . gen sls mgr
Stacey Sieverding . . . . gen sls mgr
Judy Buie . . . . progmg dir

***KUSD-TV—** Analog channel: 2. Analog hrs: 24 100 kw vis, 20 kw aur. 760t/656g TL: N43 03 00 W96 47 12 On air date: July 5, 1961. Box 5000, Vermillion, SD, 57069. 555 N. Dakota St., Vermillion, SD 57069. Phone: (605) 677-5861. Fax: (605) 677-5010. E-mail: programming@sdpb.org Web Site: www.sdpb.org. Licensee: South Dakota Board of Directors for Educational Telecommunications. Population served: 304,000 Natl. Network: PBS. Rgnl. Network: CEN. Washington Atty: Cohn & Marks.

Key Personnel:

Julie Andersen . . . pres
Craig Jensen . . . opns mgr
Terry Spencer . . . dev dir
Fritz Miller . . . mktg mgr & prom dir
Bob Bosse . . . progmg dir

KWSD— Analog channel: 36. Digital channel: 51. Analog hrs: 24 Digital hrs: 24 3,020 kw vis. ant 1,168t TL: N43 30 19 W96 34 20 On air date: 2001. Box 9609, Rapid City, SD, 57709-9609. 3220 W. 57th St., Suite 111, Sioux Falls, SD 57108. Phone: (605) 355-0024. Fax: (605) 355-9274. E-mail: jsimpson@newscenter1.com Web Site: www.siouxfallscw.com. Licensee: J.F. Broadcasting LLC. Ownership: James F. Simpson, 100% Group owner: (group owner; (acq 3-2-2007; $300,000). Natl. Network: CW.

Key Personnel:

Jim Simpson . . . gen mgr
Mark Walter . . . opns mgr
Mike Smith . . . gen sls mgr

Tennessee

Chattanooga, TN (DMA 86)

***WCLP-TV—** Analog channel: 18. Analog hrs: 24 5,000 kw vis. 1,851t TL: N34 45 06 W84 42 54 On air date: Feb 1, 1967. 2765 Ft. Mountain State Park Rd., Chatsworth, GA, 30705. Phone: (706) 422-1947. Web Site: www.gpb.org. Licensee: Georgia Public Telecommunications Commission. Natl. Network: PBS. Washington Atty: Arent, Fox, Kintner, Plotkin & Kahn.

WDEF-TV— Analog channel: 12. Digital channel: 47. Analog hrs: 24 316 kw vis, 37.1 kw aur. ant 1,260t/641g TL: N35 08 06 W85 19 25 On air date: Apr 25, 1954. 3300 Broad St., Chattanooga, TN, 37408. Phone: (423) 785-1200. Fax: (423) 785-1271. E-mail: news@wdef.com Web Site: www.wdef.com. Licensee: WDEF-TV Inc. Group owner: Media General Broadcast Group (acq 10-13-2006; $23 million). Population served: 812,000 Natl. Network: CBS. Natl. Rep: Harrington, Righter & Parsons. Washington Atty: Fletcher, Heald & Hildreth, P.L.C. News staff: 32; News: 28.5 hrs wkly.

WDSI-TV— Analog channel: 61. Digital channel: 40. Analog hrs: 24 Note: Fox is on WDSI-TV ch 61, MyNetworkTV is on WDSI-DT ch 40. 5,000 kw vis, 500 kw aur. 1,214t TL: N35 12 34 W85 16 39 On air date: Jan 24, 1972. 1101 E. Main St., Chattanooga, TN, 37408. Phone: (423) 265-0061. Fax: (423) 265-3636. Web Site: www.fox61tv.com. Licensee: WDSI License Corp., debtor-in-possession. Group owner: Pegasus Broadcast Television Inc. (acq 2-18-93; $21 million with WDBD(TV) Jackson, MS; FTR: 3-8-93). Natl. Network: Fox, MyNetworkTV Digital Network: Note: Fox is on WDSI-TV ch 61, MyNetworkTV is on WDSI-DT ch 40. Washington Atty: George H. Shapiro.

Key Personnel:

Patrick Notley . . . chief of opns
Tracye McCarthy . . . gen mgr & gen sls mgr
Dan Mecca . . . natl sls mgr
Rebecca Sims . . . prom dir
Jenny Giddens . . . progmg dir
Tom Hendricks . . . news dir
Patrick Motley . . . chief of engrg

WELF-TV— Analog channel: 23. Digital channel: 16. Analog hrs: 24 Digital hrs: 24 490 kw vis. 1,466t TL: N34 57 07 W85 22 58 On air date: 1994. 384 S. Campus Rd., Lookout Mountain, GA, 30750. Phone: (706) 820-1663. Fax: (706) 820-1735. E-mail: welf@tbn.org Web Site: www.tbn.org. Licensee: Trinity Broadcasting Network. Group owner: (group owner; (acq 5-8-2000; grpsl). Population served: 586,000

WFLI-TV— Analog channel: 53. Digital channel: 42. Analog hrs: 24 1,306 kw vis, 131 kw aur. ant 1,065t/1,016g TL: N34 55 57 W84 58 32 On air date: May 25, 1987. 6024 Shallowford Rd., Suite 100, Chattanooga, TN, 37421. Phone: (423) 893-9553. Fax: (423) 893-9853. Web Site: www.thecwchattanooga.com. Licensee: Meredith Corp. Group owner: Meredith Broadcasting Group, Meredith Corp. (acq 8-25-2004; $8.5 million). Natl. Network: CW. Natl. Rep: MMT. Washington Atty: Dow, Lohnes & Albertson.

WRCB-TV— Analog channel: 3. Digital channel: 13. Analog hrs: 24 Digital hrs: 24 100 kw vis, 10 kw aur. ant 1,237t/413g TL: N35 09 40 W85 18 51 On air date: May 6, 1956. 900 Whitehall Rd., Chattanooga, TN, 37405. Phone: (423) 267-5412. Fax: (423) 267-6840. Fax: (423) 756-3148 (news). E-mail: ttolar@wrcbtv.com Web Site: www.wrcbtv.com. Licensee: Sarkes Tarzian Inc. Group owner: (group owner; (acq 10-82; $16 million; FTR: 10-18-82). Population served: 863,000 Natl. Network: NBC. Natl. Rep: Continental Television Sales. Washington Atty: Leventhal, Senter & Lerman. Wire Svc: AP News: 22 hrs wkly.

Key Personnel:

Tom Tarzian. . . . chmn
Bob Davis. . . . CFO
Tom Tolar . . . pres & gen mgr
Doug Loveridge . . . opns mgr
Doug Short . . . gen sls mgr
Ronnie Minton . . . prom dir
Pam Teague . . . progmg dir
Bill Wallace . . . news dir
Ed Aslinger . . . chief of engrg

***WTCI—** Analog channel: 45. Analog hrs: 24 1,480 kw vis, 148 kw aur. 1,200t TL: N35 12 26 W85 16 52 (CP: Ant 1,075t) On air date: Mar 8, 1970. 4411 Amnicola Hwy., Chattanooga, TN, 37406. Phone: (423) 629-0045. Fax: (423) 698-8557. Web Site: www.wtcitv.org. Licensee: The Greater Chattanooga PTV Corp. (acq 7-84). Population served: 310,000 Natl. Network: PBS. Washington Atty: Dow, Lohnes & Albertson.

Key Personnel:

Paul Grove . . . pres
Susan Cates . . . progmg VP
Kevin Lusk . . . pub affrs dir

WTVC— Analog channel: 9. Digital channel: 35. Analog hrs: 24 Digital hrs: 24 316 kw vis, 31.6 kw aur. ant 1,056t/246g TL: N35 09 38 W85 19 06 On air date: Feb 11, 1958. Box 60028, Chattanooga, TN, 37406-6028. 4279 Benton Dr., Chattanooga, TN 37406. Phone: (423) 756-5500. Fax: (423) 757-7400. Fax: (423) 757-7401. E-mail: news@newschannel9.com Web Site: www.newschannel9.com. Licensee: Freedom Broadcasting of Tennessee Licensee L.L.C. Group owner: Freedom Broadcasting Inc. (acq 12-13-83; grpsl; FTR: 1-2-84). Population served: 350,000 Natl. Network: ABC. Natl. Rep: TeleRep. Washington Atty: Latham & Watkins. News staff: 55; News: 32 hrs wkly.

Key Personnel:

Dennis W. Brown . . . opns mgr
Michael Costa . . . VP, gen mgr & natl sls mgr

Jackson, TN (DMA 174)

WBBJ-TV— Analog channel: 7.316 kw vis, 31.6 kw aur. 1,060t/1,065g TL: N35 38 15 W88 41 32 On air date: Mar 5, 1955. 346 Muse St., Jackson, TN, 38301. Phone: (731) 424-4515. Fax: (731) 424-9299. Web Site: www.wbbjtv.com. Licensee: Tennessee Broadcasting Partners. (acq 7-20-92). Population served: 198,150 Natl. Network: ABC.

Key Personnel:

Jerry Moore . . . gen mgr
Robert Fay . . . gen sls mgr
Anthony Matrisciano . . . progmg dir
Ken Galey . . . news dir
Randy McCaskill . . . chief of engrg

WJKT— Analog channel: 16. Digital channel: 39.4,680 kw vis, 468 kw aur. ant 1,056t/1,004g TL: N35 47 22 W89 06 14 On air date: Apr 16, 1985. 2701 Union Ave. Ext., Memphis, TN, 38112. Phone: (901) 323-2430. Fax: (901) 323-9503. Web Site: www.myeyewitnessnews.com. Licensee: Clear Channel Broadcasting Licenses Inc. Group owner: Clear Channel Communications Inc. (acq 9-29-2000). Natl. Network: Fox.

***WLJT-TV—** Analog channel: 11. Analog hrs: 17 316 kw vis, 63.1 kw aur. 640t/496g TL: N35 45 12 W88 36 10 On air date: Feb 1, 1968. Box 966, Martin, TN, 38237-0966. Clement Hall, U.T.-Martin, Martin, TN 38238. Phone: (731) 881-7561. Fax: (731) 881-7566. E-mail: wljt@wljt.org Web Site: www.wljt.org. Licensee: West Tennessee Public Television Council Inc. Population served: 211,000 Natl. Network: PBS.

Key Personnel:

Dave Hinman. . . . CEO, gen mgr & opns dir
Bud Grimes . . . pres
Monica Shumake . . . CFO
Emily Elliston . . . VP
Katrina Cobb . . . opns dir & prom dir
Shorri Puckett . . . dev dir
Robbie Green . . . mktg mgr

Knoxville, TN (DMA 60)

WAGV— Analog channel: 44. Digital channel: 51.1,000 kw vis. ant 1,971t/460g TL: N36 48 00 W83 22 36 On air date: 2000. Box 1867, Abingdon, VA, 24212-1867. 8594 Hidden Valley Rd., Abingdon, VA 24210. Phone: (276) 676-3806. Fax: (276) 676-3572. Licensee: Living Faith Ministries Inc. Ownership: Non-stock corporation.

Key Personnel:

Fredia Lou Keene . . . CFO
Lisa C. Smith. . . . sls dir
Michael D. Smith. . . . CEO, pres, gen mgr, sls VP & mktg dir

Satellite of WLFG(TV) Grundy, VA.

WATE-TV— Analog channel: 6. Analog hrs: 24 Digital hrs: 24 100 kw vis, 15 kw aur. 1,489t/1,152g TL: N36 00 13 W83 56 35 On air date: Oct 1, 1953. Box 2349, Knoxville, TN, 37901. 1306 N.E. Broadway, Cave City, TN 37917. Phone: (865) 637-6666. Fax: (865) 525-4091. Web Site: www.wate.com. Licensee: WATE G.P. Group owner: Young Broadcasting Inc. (acq 11-14-94; grpsl; FTR: 9-12-94). Population served: 174,587 Natl. Network: ABC. Natl. Rep: Adam Young. News staff: 50; News: 24 hrs wkly.

Key Personnel:

Tony Kahl. . . . gen sls mgr
Jan Wade . . . progmg dir
Robb Atkinson . . . news dir
Bill Evans. . . . pub affrs dir
Bob Williams . . . chief of engrg

WBIR-TV— Analog channel: 10. Analog hrs: 24 316 kw vis, 38 kw aur. 1,791t/1,505g TL: N36 00 19 W83 56 23 On air date: Aug 13, 1956. 1513 Hutchison Ave., Knoxville, TN, 37917. Phone: (865) 637-1010. Fax: (865) 637-6280. Fax: (865) 637-6380. E-mail: wbir@wbir.gannett.com Web Site: www.wbir.com. Licensee: Gannett Pacific Corp. Group owner: Gannett Broadcasting (acq 12-4-95; grpsl). Population served: 1,252,000 Natl. Network: NBC. Washington Atty: Wiley, Rein & Fielding. News staff: 50; News: 24 hrs wkly.

Key Personnel:

Jeff Lee . . . pres & gen mgr
Dean Littleton . . . gen sls mgr
David Cowen . . . progmg dir
Bill Shory . . . news dir
Matthew Newell. . . . prom dir & pub affrs dir
Gary Davis . . . chief of engrg

WBXX-TV— Analog channel: 20. Digital channel: 50. Analog hrs: 24 Digital hrs: 24 14.8 kw vis, 1.4 kw aur. ant 157t TL: N35 56 12 W85 00 46 On air date: Oct 4, 1997. 10427 Cogdill Rd., Suite 100, Knoxville, TN, 37932. Phone: (865) 777-9220. Fax: (865) 777-9221. E-mail: promotions@easttennesseecw.com Web Site: www.easttennesseecw.com. Licensee: Acme Television Licenses of Tennessee L.L.C. Group owner: Acme Communications Inc. (acq 8-28-97; $13.2 million). Population served: 5,381 Natl. Network: CW. Natl. Rep: MMT. Washington Atty: Dickstein Shapiro Morin & Oshinsky L.L.P.

Key Personnel:

Dan Phillippi. . . . VP & gen mgr
Joanne Marcenkus . . . gen sls mgr
Lisa Faulkner. . . . progmg dir
Ferdy Guidry. . . . chief of engrg
Anna Robins . . . prom

***WETP-TV—** Analog channel: 2. Digital channel: 41. Analog hrs: 17.5 100 kw vis, 20 kw aur. ant 1,760t/499g TL: N36 22 52 W83 10 48 On air date: Mar 15, 1967. 1611 E. Magnolia Ave., Knoxville, TN, 37917. Phone: (865) 595-0220. Fax: (865) 595-0300. E-mail: etptv.mail@etptv.org Web Site: www.etptv.org. Licensee: East Tennessee Public Communications Corp. (acq 10-1-83). Population served: 1,800,000 Natl. Network: PBS. Rgnl. Network: SECA.

Key Personnel:

Teresa James. . . . pres & gen mgr
Frank Miller . . . opns VP
Kelly Hodges . . . dev dir & mktg dir
Evelyn Clarke . . . prom mgr
Bob Hutchinson . . . progmg dir
Chris Smith . . . pub affrs dir
Curtis Allin. . . . chief of engrg

***WKOP-TV—** Analog channel: 15. Analog hrs: 17.5 2,240 kw vis, 224 kw aur. 1,683t/1,360g TL: N36 00 19 W83 56 23 On air date: Aug 15, 1990. 1611 E. Magnolia Ave., Knoxville, TN, 37917. Phone: (865) 595-0220. Fax: (865) 595-0300. E-mail: etptvmail@etptv.org Web Site: www.etptv.org. Licensee: East Tennessee Public Communications Corp. Natl. Network: PBS. Rgnl. Network: SECA.

Key Personnel:

Teresa James . . . pres
Jim Tindell . . . gen mgr
Frank Miller . . . opns VP
Kelly Hodges . . . dev dir & mktg dir
Evelyn Clarke . . . prom mgr
Bob Hutchinson . . . progmg dir
Chris Smith . . . pub affrs dir
Curtis Allin. . . . chief of engrg

WMAK— Digital channel: 7.55 kw vis. ant 1,253t/1,017g TL: N36 00 36 W83 55 57 On air date: July 31, 2004. 6215 Kingston Pike,

Knoxville, TN, 37919-4044. Phone: (865) 584-9094. Fax: (865) 584-9098. Web Site: www.wmaktv.com. Licensee: Knoxville Channel 7 LLC. Group owner: (group owner). Washington Atty: Fletcher, Heald & Hildreth.

WPXK— Analog channel: 54. Digital channel: 23. Analog hrs: 24 28.8 kw vis, 3.9 kw aur. ant 1,007t TL: N36 30 26 W84 02 36 (CP: 20 kw vis, ant 1,296t. TL: N36 24 36 W84 10 38) On air date: Mar 12, 1991. Bldg. D, 9000 Executive Park Dr., Suite 300, Knoxville, TN, 37923. Phone: (865) 693-4343. Fax: (865) 251-4305. Web Site: www.ionline.tv. Licensee: Paxson Knoxville Licensee Inc. Group owner: Paxson Communications Corp. (acq 9-23-98). Population served: 490,000 Natl. Network: i Network.
Key Personnel:
Carol Wright-Holzhauer . . . . VP
Angela Galyon . . . . opns mgr

WTNZ— Analog channel: 43. Digital channel: 34. Analog hrs: 24 1,960 kw vis. ant 1,735t/1,356g TL: N36 00 13 W83 56 34 On air date: Dec 31, 1983. Bldg. D, 9000 Executive Park Dr., Suite 300, Knoxville, TN, 37923. Phone: (865) 693-4343. Fax: (865) 691-6904. Fax: (865) 691-6770. Web Site: wtnzfox43.com. Licensee: Raycom America License Subsidiary LLC. Group owner: Raycom Media Inc. (acq 1996; grpsl). Population served: 514,000 Natl. Network: Fox. Natl. Rep: TeleRep. Washington Atty: Covington & Burling. News staff: 6; News: 4 hrs wkly.
Key Personnel:
Paul McTear . . . . CEO & CFO
John Hayes . . . . gen mgr

WVLR— Analog channel: 48.5,000 kw vis. ant 1,414t/280g TL: N36 15 30 W83 37 43 On air date: 2003. 306 Kyker Ferry Rd., Koduk, FL, 37764. Phone: (865) 932-4803. Fax: (865) 932-4102. Web Site: www.tv48.org. Licensee: Volunteer Christian Television Inc. (acq 4-22-2002).
Key Personnel:
Theron Woodward . . . . gen mgr
Scott Dunkel . . . . chief of engrg
Tom Evensen . . . . chief of engrg

WVLT-TV— Analog channel: 8. Digital channel: 30. Analog hrs: 24 Digital hrs: 24 Note: CBS is on WVLT-TV ch 8, MyNetworkTV is on WVLT-DT ch 30. 316 kw vis, 31.6 kw aur. ant 1,290t/1,073g TL: N36 00 36 W83 55 57 On air date: Dec 8, 1988. Box 59088, Knoxville, TN, 37950. 6450 Papermill Rd., Knoxville, TN 37919. Phone: (865) 450-8888. Fax: (865) 450-8869. Web Site: www.volunteertv.com. Licensee: Gray Television Licensee Inc. Group owner: Gray Television Inc. (acq 1996; $165 million with WCTV(TV) Thomasville, GA). Population served: 1,000,000 Natl. Network: CBS, MyNetworkTV Digital Network: Note: CBS is on WVLT-TV ch 8, MyNetworkTV is on WVLT-DT ch 30. Natl. Rep: Continental Television Sales. News staff: 40; News: 24.5 hrs wkly.
Key Personnel:
Chris Baker . . . . CFO, exec VP & gen mgr
Richard Tosbett . . . . gen sls mgr
Marty Parham . . . . progmg VP

Memphis, TN
(DMA 44)

WBUY-TV— Analog channel: 40. Digital channel: 41. Analog hrs: 24 4,680 kw vis, 468 kw aur. ant 466t TL: N34 59 20 W89 41 13 On air date: Sept 13, 1991. 3447 Cazassa Rd., Memphis, TN, 38116-3609. Phone: (901) 396-9541. Fax: (901) 396-9585. E-mail: wbuy@tbn.org Web Site: www.tbn.org. Licensee: Trinity Broadcasting Network. Group owner: (group owner; (acq 5-8-2000; grpsl).
Key Personnel:
Tamela Calvin . . . . stn mgr
Cliff Pickell . . . . opns dir & progmg dir
Douglas Puryear . . . . chief of engrg

WHBQ-TV— Analog channel: 13. Analog hrs: 24 316 kw vis, 63.2 kw aur. 1,000t/1,076g TL: N35 10 28 W89 50 41 On air date: Sept 27, 1953. 485 S. Highland St., Memphis, TN, 38111. Phone: (901) 320-1313. Fax: (901) 323-0092. Fax: (901) 320-1366 (News). Web Site: www.foxmemphis.com. Licensee: Fox Television Stations Inc. Group owner: (group owner; (acq 7-5-95; $80 million). Population served: 1,135,000 Natl. Network: Fox. Natl. Rep: Fox Stations Sales. News staff: 55; News: 27 hrs wkly.
Key Personnel:
Rupert Murdoch . . . . chmn
Lachlan Murdock . . . . pres
Betsy Swanson . . . . CFO
Tom Herwitz . . . . exec VP
John Koski . . . . gen mgr, progmg dir & pub affrs dir
Michael Lewis . . . . gen sls mgr
Kim Moore . . . . natl sls mgr
Paul Sloan . . . . prom VP & prom dir
Ken Jobe . . . . news dir

David Brant . . . . chief of engrg

***WKNO-TV—** Analog channel: 10. Digital channel: 29. Analog hrs: 24 Digital hrs: 24 316 kw vis, 56.2 kw aur. ant 1,079t/1,113g TL: N35 09 17 W89 49 20 On air date: June 25, 1956. Box 241880, Memphis, TN, 38124-1880. 900 Getwell Rd., Memphis, TN 38111. Phone: (901) 458-2521. Fax: (901) 325-6505. E-mail: wknopi@wkno.org Web Site: www.wkno.org. Licensee: Mid-South Public Communications Foundation. Population served: 1,600,000 Natl. Network: PBS. Washington Atty: Schwartz, Woods & Miller.
Key Personnel:
Michael LaBonia . . . . CEO & pres
Russ A. Abernathy . . . . stn mgr
Charles McLarty . . . . dev dir

WLMT— Analog channel: 30. Digital channel: 31.5,000 kw vis, 500 kw aur. ant 1,000t/1,000g TL: N35 09 17 W89 49 20 On air date: April 1983. Clear Channel Television Ctr, 2701 Union Ext., Memphis, TN, 38112. Phone: (901) 323-2430. Fax: (901) 323-9503. E-mail: eyewitnessnews@upn30memphis.com Web Site: www.myeyewitnessnews.com. Licensee: Clear Channel Broadcasting Licenses Inc. Group owner: Clear Channel Communications Inc. (acq 10-20-2000). Population served: 961,000 Natl. Network: CW. Washington Atty: McFadden, Evans & Sill.
Key Personnel:
Jack Peck . . . . gen mgr
Jim Doty . . . . gen sls mgr
Robyn Callaway . . . . opns dir & rgnl sls mgr
Jim Turpin . . . . news dir

***WMAV-TV—** Analog channel: 18. Digital channel: 36. Analog hrs: 24 Digital hrs: 24 225 kw vis,. ant 1,381t/1,276g TL: N34 17 28 W89 42 21 On air date: May 19, 1972. 3825 Ridgewood Rd., Jackson, MS, 39211. Phone: (601) 432-6565. Fax: (601) 432-6654. Fax: (601) 432-6311. Web Site: www.mpbonline.org. Licensee: Mississippi Authority for Educational TV. Population served: 1,220,000 Natl. Network: PBS. News staff: 4.
Key Personnel:
Marie Antoon . . . . gen mgr
Bob Buie . . . . opns dir
Teresa Collier . . . . news dir

WMC-TV— Analog channel: 5. Analog hrs: 24 100 kw vis, 20 kw aur. 1,010t/1,088g TL: N35 10 09 W89 53 12 On air date: Dec 11, 1948. 1960 Union Ave., Memphis, TN, 38104. Phone: (901) 726-0555. Fax: (901) 278-7633. Web Site: www.wmctvstations.com. Licensee: Raycom America License Subsidiary LLC. Group owner: Raycom Media Inc. (acq 1997; grpsl). Population served: 600,200 Natl. Network: NBC. Natl. Rep: TeleRep. Washington Atty: Goldberg, Godles, Wiener & Wright.
Key Personnel:
Howard Meagle . . . . VP
Lee Meridith . . . . gen mgr
Gary Macko . . . . natl sls mgr
Jim Himes . . . . rgnl sls mgr & mktg VP
Lori Beth Pickle . . . . mktg dir
Richard Enderwood . . . . prom dir
Peggy Phillip . . . . news dir

WPTY-TV— Analog channel: 24. Digital channel: 25.3,020 kw vis, 600.6 kw aur. ant 1,011t/1,043g TL: N35 12 11 W89 48 16 On air date: Sept 10, 1978. 2701 Union Ave. Ext., Memphis, TN, 38112. Phone: (901) 323-2430. Fax: (901) 323-9503. Web Site: www.myeyewitnessnews.com. Licensee: Clear Channel Radio Licenses Inc. Group owner: Clear Channel Communications Inc. (acq 8-5-92; FTR: 4-13-92). Population served: 961,000 Natl. Network: ABC. Washington Atty: Wiley, Rein & Fielding. News staff: 50; News: 7 hrs wkly.

WPXX-TV— Analog channel: 50. Digital channel: 51. Analog hrs: 24 5,000 kw vis, 500 kw aur. 800t/768g TL: N35 09 17 W89 49 20 On air date: October 1994. 7200 Goodlett Farm Pky, Ste 102, Cordova, TN, 38016. Phone: (901) 384-6650. Fax: (901) 388-8128. Web Site: www.my50memphis.com. Licensee: Flinn Broadcasting Corp. (acq 8-27-90; $220,000; FTR: 11-19-90). Natl. Network: MyNetworkTV.

WREG-TV— Analog channel: 3. Digital channel: 28.100 kw vis, 20 kw aur. 1,000t/1,077g TL: N35 10 52 W89 49 56 On air date: Jan 1, 1956. 803 Channel 3 Dr., Memphis, TN, 38103. Phone: (901) 543-2333. Fax: (901) 543-2198. Fax: (901) 543-2167 (news). Licensee: Local TV Tennessee License LLC. Group owner: The New York Times Co. (acq 5-7-2007; grpsl). Population served: 1,000,000 Natl. Network: CBS. Natl. Rep: Eagle Television Sales. Washington Atty: Koteen & Naftalin. Wire Svc: New York Times News Service News staff: 50.
Key Personnel:
Robert L. Lawrence . . . . pres
Ronald A. Walter . . . . gen mgr
Jim Anhalt . . . . opns VP & engrg VP
Maury Eikner-Tower . . . . mktg dir & mktg mgr
Wes Pollard . . . . prom VP & prom dir

Bruce Moore . . . . news dir

Nashville, TN
(DMA 30)

***WCTE—** Analog channel: 22. Digital channel: 52. Analog hrs: 24 Digital hrs: 24 1,320 kw vis, 77.6 kw aur. 1,394t/804g TL: N36 10 26 W85 20 37 On air date: Aug 21, 1978. Box 2040, Cookeville, TN, 38502. 1151 Stadium Dr.,Ste 104, Cookeville, TN 38501. Phone: (931) 528-2222. Fax: (931) 372-6284. E-mail: info@wcte.org Web Site: www.wcte.org. Licensee: Upper Cumberland Broadcast Council. (acq 12-20-85; FTR: 11-18-85). Population served: 385,000 Natl. Network: PBS.
Key Personnel:
Becky Magura . . . . pres & gen mgr
Donna Castle . . . . dev dir & progmg dir

WHTN— Analog channel: 39. Digital channel: 38.5,000 kw vis, 500 kw aur. ant 820t/391g TL: N36 04 54 W86 25 57 On air date: Dec 30, 1983. 9582 Lebanon Rd., Mt. Juliet, TN, 37122. Phone: (615) 754-0039. Fax: (615) 754-0047. E-mail: info@nashville39.com Web Site: www.ctnonline.com. Licensee: Christian Television Network Inc. Ownership: David C. Gibbs III, 20%; Jimmy Smith, 20%; Robert D'Andrea, 20%; Virginia Oliver, 20%; and Wayne Wetzel, 20%. Washington Atty: Gammon & Grange. Foreign lang progmg: SP 0 News staff: one; News: 2 hrs wkly.

WJFB— Analog channel: 66. Analog hrs: 24 2,240 kw vis. 528t/259g TL: N36 09 13 W86 22 46 On air date: 1989. 200 E. Spring St., Lebanon, TN, 37087. Phone: (615) 444-8206. Fax: (615) 444-7592. E-mail: bclinic@bellsouth.net Licensee: Bryant Broadcasting Inc. Ownership: Joe Bryant, 100%. Population served: l,500,000 News staff: 2; News: 4 hrs wklyd.
Key Personnel:
Dr. Joe Bryant . . . . pres
Pat Bryant . . . . gen mgr

WKRN-TV— Analog channel: 2. Digital channel: 27. Analog hrs: 24 100 kw vis, 10 kw aur. 1,350t/942g TL: N36 02 49 W86 49 49 On air date: Nov 29, 1953. 441 Murfreesboro Rd., Nashville, TN, 37210. Phone: (615) 259-2200. Fax: (615) 244-2117. Web Site: www.wkrn.com. Licensee: WKRN G.P. Group owner: Young Broadcasting Inc. (acq 4-17-89; $42 million; FTR: 5-8-89). Population served: 1,844,000 Natl. Network: ABC. Natl. Rep: Adam Young. Washington Atty: Wiley, Rein & Fielding. News staff: 45; News: 25 hrs wkly.
Key Personnel:
Gwen Kinsey . . . . pres & gen mgr
Mike Tarrolley . . . . mktg dir
Michele Dube . . . . progmg dir
Steve Sabato . . . . news dir
Gene Parker . . . . chief of engrg

WNAB— Analog channel: 58. Digital channel: 23. Analog hrs: 24 Digital hrs: 24 3,250 kw vis. ant 1,393t/1,193g TL: N36 15 50 W86 47 39 On air date: Nov 29, 1995. 2994 Sidco Dr., Nashville, TN, 37204. Phone: (615) 650-5858. Fax: (615) 650-5859. Web Site: www.cw58.net. Licensee: Nashville License Holdings LLC. Ownership: Michael Lambert, mgng member. (acq 10-14-98; $30 million). Natl. Network: CW.
Key Personnel:
Michael Lambert . . . . pres
Michael Jones . . . . CFO
Mark Dillion . . . . stn mgr
DeJuan Buford . . . . gen sls mgr
Dale Bukowski . . . . natl sls mgr & rgnl sls mgr
Lee Scott . . . . prom mgr
Michael Hook . . . . progmg mgr

***WNPT—** Analog channel: 8.295 kw vis, 29.5 kw aur. 1,280t/832 TL: N36 02 49 W86 49 49 On air date: Sept 10, 1962. 161 Rains Ave., Nashville, TN, 37203. Phone: (615) 259-9325. Fax: (615) 248-6120. E-mail: tv8@wnpt.net Web Site: www.wnpt.net. Licensee: Nashville Public Television Inc. Population served: 1500000 Natl. Network: PBS. Washington Atty: Schwartz, Woods & Miller.
Key Personnel:
Steven M. Bass . . . . CEO & pres
Beth Curley . . . . sr VP & chief of opns
Charles Brimbelow . . . . VP
Harmon McBride . . . . progmg dir

WNPX— Analog channel: 28. Digital channel: 36. Analog hrs: 24 229 kw vis, 22.9 kw aur. 869t/623g TL: N36 07 33 W85 17 33 (CP: 5,000 kw vis, ant 1,299t/1,033g. TL: N36 08 35 W85 54 34) On air date: Sept 3, 1993. 1281 N. Mt. Juliet Rd., Suite K, Mt. Juliet, TN, 37122. Phone: (615) 773-6100. Fax: (615) 726-2854. Fax: (615) 773-6106. Web Site: www.ionline.tv. Licensee: Paxson Communications License Co. L.L.C. Group owner: Paxson Communications Corp. (acq 9-4-97; $4.3 million). Population served: 2,474,000 Natl. Network: i Network.

Key Personnel:
Lowell "Bud" Paxson . . . . CEO
Dan Barber . . . . gen sls mgr

WPGD-TV— Analog channel: 50. Analog hrs: 24 4,508 kw vis, 500 kw aur. 770t TL: N36 28 02 W86 28 53 On air date: Sept 23, 1992. 36 Music Village Blvd., Hendersonville, TN, 37075. Phone: (615) 822-1243. Fax: (615) 822-1642. Web Site: www.tbn.org. Licensee: Trinity Broadcasting Network. Group owner: (group owner; (acq 7-2000; grpsl). Population served: 700,000
Key Personnel:
Renee Brewer . . . . gen mgr
Allen Partlow. . . . . chief of engrg

WSMV-TV— Analog channel: 4. Digital channel: 10.100 kw vis, 10 kw aur. 1,424t/1,382g TL: N36 08 27 W86 51 56 On air date: Sept 30, 1950. 5700 Knob Rd., Nashville, TN, 37209. Phone: (615) 353-2231. Fax: (615) 353-2375. Web Site: www.wsmv.com. Licensee: Meredith Corp. Group owner: Meredith Broadcasting Group, Meredith Corp. (acq 11-1-94; $159 million; FTR: 12-5-94). Population served: 823,540 Natl. Network: NBC. Natl. Rep: TeleRep. Washington Atty: Wilmer, Cutler & Pickering.

WTVF— Analog channel: 5. Digital channel: 56. Analog hrs: 24 Digital hrs: 24 100 kw vis, 10 kw aur. 1,394t/1,138g TL: N36 16 05 W86 47 16 On air date: Aug 6, 1954. 474 James Robertson Pkwy., Nashville, TN, 37219. Phone: (615) 244-5000. Fax: (615) 248-5353. Fax: TWX: 810-371-1168. E-mail: news@newschannel5.com Web Site: www.newschannel5.com. Licensee: NewsChannel 5 Network LP. Group owner: Landmark Communications Inc. (acq 9-12-91; $46 million; FTR: 9-30-91). Population served: 2,268,000 Natl. Network: CBS. Washington Atty: Hogan & Hartson. News: 24 hrs wkly.
Key Personnel:
Debbie Turner . . . . gen mgr
Mark Binda. progmg dir
Mike Cutler. . . . . news dir

WUXP-TV— Analog channel: 30. Digital channel: 21. Analog hrs: 24 5,000 kw vis, 500 kw aur. ant 1,417t/1,217g TL: N36 15 50 W86 47 39 On air date: Feb 18, 1984. 631 Mainstream Dr., Nashville, TN, 37228. Phone: (615) 259-5630. Fax: (615) 259-3962. Web Site: www.mytv30web.com. Licensee: WUXP Licensee LLC. Group owner: Sinclair Broadcast Group Inc. (acq 12-10-2001; $2.829 million). Natl. Network: MyNetworkTV. Washington Atty: Arter & Hadden.
Key Personnel:
Stephen A. Mann . . . . gen mgr
Mark Dillon . . . . stn mgr
Pam Combest . . . . sls dir
Dejuan Buford . . . . gen sls mgr
Dale Bukowski . . . . natl sls mgr
Greg Carr. . . . . rgnl sls mgr
Lee R. Scott . . . . mktg dir, prom dir, adv dir & progmg dir
Deborah Williams . . . . prom mgr & adv mgr
Iman Tate. . . . . progmg mgr
Lee Peterson . . . . pub affrs dir
Gibson Prichard . . . . engrg mgr
David Birdsong . . . . chief of engrg

WZTV— Analog channel: 17. Digital channel: 15. Analog hrs: 24 3,240 kw vis, 324 kw aur. ant 1,161t/1,063g TL: N36 08 27 W86 51 56 (CP: ant 1,407t/1,207g. TL: N36 15 50 W86 47 39) On air date: March 1976. 631 Mainstream Dr., Nashville, TN, 37228. Phone: (615) 259-5617. Fax: (615) 259-3962. E-mail: comments@wztv.com Web Site: www.wztv.com. Licensee: WZTV Licensee LLC. Group owner: Sinclair Broadcast Group Inc. (acq 12-10-01; grpsl). Population served: 2,167,000 Natl. Network: Fox. Washington Atty: Arter & Hadden. News staff: 9; News: 7 hrs wkly.
Key Personnel:
Steve Mann . . . . gen mgr & stn mgr
Pamela Minnicks . . . . gen sls mgr
Beckey Dan . . . . prom dir
David Birdsong . . . . chief of engrg

Tri-Cities, TN-VA (DMA 92)

WCYB-TV— Analog channel: 5. Digital channel: 28. Analog hrs: 24 Note: NBC is on WCYB-TV ch 5, CW is on WCYB-DT ch 28. 83.2 kw vis, 10.6 kw aur. ant 2,230t/90g TL: N36 26 57 W82 06 31 On air date: Aug 13, 1956. 101 Lee St., Bristol, VA, 24201. Phone: (276)-645-1555. Fax: (276) 645-1513. E-mail: news@wcyb.tv Web Site: www.wcyb.tv. Licensee: BlueStone License Holdings Inc. Group owner: Lamco Communications Inc. (acq 6-15-2004; grpsl). Population served: 725,000 Natl. Network: NBC, CW Digital Network: Note: NBC is on WCYB-TV ch 5, CW is on WCYB-DT ch 28. Washington Atty: Koteen & Naftalin.

WEMT— Analog channel: 39. Digital channel: 38. Analog hrs: 24 3,020 kw vis, 302 kw aur. ant 2,609t/143g TL: N36 01 24 W82 42 56 On air date: Nov 8, 1985. Box 3489 CRS, Johnson City, TN, 37602-3489. 3206 Hanover Rd., Johnson City, TN 37602-3489. Phone: (423) 283-3900. Fax: (423) 283-4938. Web Site: www.wemt39.com. Licensee: Aurora License Holdings Inc. Ownership: Aurora Broadcasting Inc., 100% Group owner: Sinclair Broadcast Group Inc. (acq 2-8-2006; $1.4 million). Natl. Network: Fox. Washington Atty: Shaw, Pittman.
Key Personnel:
Leesa Wilcher . . . . gen mgr & gen sls mgr
Amy McClary . . . . rgnl sls mgr
Jim Hartline . . . . chief of engrg

WJHL-TV— Analog channel: 11. Analog hrs: 24 245 kw vis, 30 kw aur. 2,320t/228g TL: N36 25 55 W82 08 15 On air date: Oct 26, 1953. Box 1130, Johnson City, TN, 37605. 338 E. Main St., Johnson City, TN 37601. Phone: (423) 926-2151. Fax: (423) 434-4537. E-mail: jdempsey@wjhl.com Web Site: www.wjhl.com. Licensee: Media General Broadcasting Inc. Group owner: Media General Broadcast Group (acq 3-21-97; grpsl). Population served: 267,200 Natl. Network: CBS. Natl. Rep: Harrington, Righter & Parsons. Washington Atty: Dow, Lohnes and Albertson.
Key Personnel:
Jack Dempsey . . . . gen mgr & film buyer
Lisa Wilcher . . . . gen sls mgr
Ed Oliver. . . . . mktg dir
Christine Riser . . . . news dir
Mike Moore . . . . chief of engrg

WKPT-TV— Analog channel: 19. Digital channel: 27. Analog hrs: 24 Digital hrs: 24 Note: ABC is on WKPT-TV ch 19, MyNetworkTV is on WKPT-DT ch 27. 1,260 kw vis, 42 kw aur. ant 2,320t/225g TL: N36 25 54 W82 08 15 On air date: Aug 20, 1969. 222 Commerce St., Kingsport, TN, 37662. Phone: (423) 246-9578. Fax: (423) 246-6261/(423) 246-1863. E-mail: gdevault@hvbc.com Web Site: www.wkpttv.com. Licensee: Holston Valley Broadcasting Corp. Ownership: Glenwood Communications Corp, 100%. Group owner: Glenwood Communications Corp. Population served: 322,000 Natl. Network: ABC Digital Network: Note: ABC is on WKPT-TV ch 19, MyNetworkTV is on WKPT-DT ch 27. Natl. Rep: Harrington, Righter & Parsons. Washington Atty: Cordon & Kelly.
Key Personnel:
Bette Lawson . . . . CFO
George E. DeVault Jr. . . . . pres & gen mgr
Fred Falin . . . . progmg dir

WLFG— Analog channel: 68. Digital channel: 49. Analog hrs: 24 1,150 kw vis, 115 kw aur. ant 2,503t TL: N36 49 47 W82 04 45 On air date: 1995. Box 1867, Abingdon, VA, 24212. Phone: (276) 676-3806. Fax: (276) 676-3572. E-mail: mike@livingfaithtelevision.com Web Site: www.lstv.com. Licensee: Living Faith Ministries Inc. (acq 12-21-2005).
Key Personnel:
Michael D. Smith . . . . CEO & chmn
Michael D. Smith . . . . pres
Lisa Smith . . . . VP & progmg VP
Michael D. Smith . . . . gen mgr
Wade McGeorge . . . . sls dir
Wayne Price . . . . engrg mgr

***WMSY-TV—** Analog channel: 52. Digital channel: 42. Analog hrs: 24 755 kw vis, 115 kw aur. 1,360t/247g TL: N36 54 01 W81 32 35 On air date: Aug 1, 1981. Box 13246, Roanoke, VA, 24032. 1215 McNeil Dr., Roanoke, VA 24015. Phone: (540) 344-0991. Fax: (540) 344-2148. E-mail: brptv@wbra.org Web Site: www.brptv.com. Licensee: Blue Ridge Public Television Inc. Natl. Network: PBS.
Key Personnel:
Edwin Whitmore . . . . chmn
Anita Sims . . . . CFO
Beverly Fitzpatrick Jr. . . . . exec VP
Jack K. Neal . . . . CEO, pres & gen mgr

***WSBN-TV—** Analog channel: 47. Digital channel: 32. Analog hrs: 24 690 kw vis, 61.7 kw aur. 1,940t/242g TL: N36 53 52 W82 37 22 On air date: Mar 29, 1971. Box 13246, Roanoke, VA, 24032. 1215 McNeil Dr. S.W., Roanoke, VA 24032. Phone: (540) 344-0991. Fax: (540) 344-2148. Licensee: Blue Ridge Public Television Inc. Natl. Network: PBS.
Key Personnel:
Beverly Fitzpatrick . . . . chmn & exec VP
Jack Neal . . . . gen mgr
Barbara Spencer . . . . pub affrs dir
Ron Smith. . . . . engrg VP & chief of engrg

Texas

Abilene-Sweetwater, TX (DMA 164)

KPCB— Analog channel: 17.464 kw vis. 443t TL: N32 46 52 W100 53 52 On air date: 1997. Box 61000, Midland, TX, 79711-1000. 88 E. County Rd. 112, Snyder, TX 79549. Phone: (800) 707-0420. Fax: (325) 573-9417. E-mail: info@ptcbglc.com Web Site: www.godslearningchannel.com. Licensee: Prime Time Christian Broadcasting Inc.

KRBC-TV— Analog channel: 9.316 kw vis, 31.6 kw aur. 851t/543g TL: N32 17 13 W99 44 20 On air date: Aug 31, 1953. Box 5309, Abilene, TX, 79608. Phone: (325) 692-4242. Fax: (325) 695-9922. E-mail: ksbcnews@krbc.tv Web Site: www.krbc.tv. Licensee: Mission Broadcasting Inc. Group owner: (group owner; acq 6-13-2003; $10 million with KSAN-TV San Angelo). Population served: 1,000,000 Natl. Network: NBC. Washington Atty: Hogan & Hartson.
Key Personnel:
David Smith. . . . . CEO & CFO
Dennis Thatcher. . . . . CEO & VP
Gayle Kiger . . . . gen mgr
Justin Riggar . . . . gen sls mgr
Tom Vodak . . . . news dir

KTAB-TV— Analog channel: 32. Analog hrs: 24 2,040 kw vis, 408 kw aur. 918t/757g TL: N32 16 35 W99 35 39 On air date: Oct 6, 1979. Box 5309, Abilene, TX, 79608. 4510 S. 14th St., Abilene, TX 79605. Phone: (915) 695-2777. Fax: (915) 691-5822. E-mail: news@ktab.tv Web Site: www.ktabtv.com. Licensee: Nexstar Finance Inc. Group owner: Nexstar Broadcasting Group Inc. (acq 8-15-99; $16.7 million). Population served: 106,900 Natl. Network: CBS. News staff: 12; News: 8 hrs wkly.

KTXS-TV—Sweetwater, Analog channel: 12. Digital channel: 20. Analog hrs: 24 hrs Note: ABC is on KTXS-TV ch 12, CW is on KTXS-DT ch 20. 316 kw vis, 31.6 kw aur. 1,400t/1,069g TL: N32 24 48 W100 06 25 On air date: Jan 30, 1956. Box 2997, Abilene, TX, 79604. 4420 N. Clack, Abilene, TX 79601. Phone: (325) 677-2281. Fax: (325) 676-9231. Web Site: www.ktxs.com. Licensee: BlueStone License Holdings Inc. Group owner: Lamco Communications Inc. (acq 6-15-2004; grpsl). Population served: 230,100 Natl. Network: ABC, CW Digital Network: Note: ABC is on KTXS-TV ch 12, CW is on KTXS-DT ch 20.
Key Personnel:
Jackie Rutledge . . . . VP & gen mgr
Jorge Montoya . . . . gen sls mgr & film buyer
David Caldwell. prom mgr
Sylvia Holmes . . . . progmg dir
Iain Munro . . . . news dir
Leland Ohlhausen. . . . . chief of engrg

KXVA— Analog channel: 15. Analog hrs: 24 3,947 kw vis. ant 978t/769g TL: N32 16 31 W99 35 23 On air date: Jan 17, 2001. 500 Chestnut, Ste. 804, Abilene, TX, 79602. Phone: (325) 672-5606. Fax: (325) 676-2437. Web Site: www.kxvafox.com. Licensee: Sage Broadcasting Corp. (acq 12-28-2004). Population served: 364,000 Natl. Network: Fox. Natl. Rep: Millennium Sales & Marketing. Washington Atty: Fletcher, Heard and Hildreth0.
Key Personnel:
Bill Carter . . . . pres
Michele Howse . . . . gen mgr
Rebroadcasts KIDY(TV) San Angelo 98%.

Amarillo, TX (DMA 131)

***KACV-TV—** Analog channel: 2. Analog hrs: 24 Digital hrs: 24 100 kw vis, 10 kw aur. 1,499t/1,270g TL: N35 20 33 W101 49 20 On air date: Aug 29, 1988. Box 447, Amarillo, TX, 79178. 2408 S. Jackson, Amarillo, TX 79178. Phone: (806) 371-5222. Fax: (806) 371-5258. E-mail: kacvtv@actx.edu Web Site: www.kacvtv.org. Licensee: Amarillo Junior College District. Population served: 400,000 Natl. Network: PBS. Washington Atty: Cohn & Marks LLP.
Key Personnel:
Jackie Smith. . . . . opns mgr
Joli Lindseth . . . . dev dir
Michelle Macon . . . . dev mgr
Ellen Robertson Neal . . . . pub affrs dir

KAMR-TV— Analog channel: 4.100 kw vis, 10 kw aur. 1,420t/1,440g TL: N35 18 52 W101 50 47 On air date: Mar 18, 1953. Box 751, 1015 S. Fillmore St., Amarillo, TX, 79101. Phone: (806) 383-3321. Fax: (806) 381-2943. E-mail: nbc4@kamr.com Web Site: www.kamr.com. Licensee: Nexstar Finance Inc. Group owner: Nexstar Broadcasting Group Inc. (acq 12-31-03; grpsl). Population served: 432,300 Natl. Network: NBC. Washington Atty: Arter & Hadden.
Key Personnel:
Mark McKay . . . . gen mgr & progmg dir
Sherry Avara. gen sls mgr
Tim Sturgess . . . . exec VP & natl sls mgr
Heather Brunson . . . . rgnl sls mgr
David Toma . . . . prom mgr
NyLynn Nichols . . . . news dir
Ken High . . . . chief of engrg

KCIT— Analog channel: 14. Digital channel: 15. Analog hrs: 24 Digital hrs: 24 1,280 kw vis, 128 kw aur. 1,521t/1,463g TL: N35 20 33 W101 49 20 On air date: Oct 1, 1982. Box 1414, Amarillo, TX, 79105. 1015 S. Fillmore, Amarillo, TX 79101. Phone: (806) 374-1414. Fax: (806) 349-9083. Web Site: fox14.tv. Licensee: Mission Broadcasting Inc. Group owner: (group owner; (acq 1999; $28.5 million with KJTL(TV) Wichita Falls). Population served: 473,000 Natl. Network: Fox. Natl. Rep: Blair Television. Washington Atty: Drinker Biddle & Reath L.L.P. News: 3 hrs wkly.
Key Personnel:
Amanda Bustamante . . . . gen mgr & stn mgr
Tim Sturgess . . . . gen sls mgr
Deb York . . . . rgnl sls mgr
Mike Crowell . . . . mktg dir
Jim O'Malley . . . . progmg dir
Wesley Willson . . . . chief of engrg

***KENW**— Analog channel: 3. Analog hrs: 24 100 kw vis, 20 kw aur. 1,150t/1,085g TL: N33 33 19 W103 39 03 On air date: Sept 1, 1974. Eastern New Mexico Univ., 52 Broadcast Ctr., Portales, NM, 88130. Phone: (505) 562-2112. Fax: (505) 562-2590. E-mail: kenwtv@enmu.edu Web Site: www.kenw.org. Licensee: Regents of Eastern New Mexico University. Population served: 400,000 Natl. Network: PBS. Washington Atty: Dow, Lohnes & Albertson, PLLC. Wire Svc: AP Foreign lang progmg: SP 1 News staff: 2; News: 3 hrs wkly.
Key Personnel:
Steven Gamble . . . . pres
Ronnie Birdsong . . . . VP
Duane Ryan . . . . gen mgr
Rena Garrett . . . . mktg dir & mktg mgr
Linda Stefanovic . . . . progmg dir
John Kirby . . . . news dir
Don Criss . . . . pub affrs dir
Jeff Burmeister . . . . chief of engrg

KFDA-TV— Analog channel: 10. Digital channel: 10. Analog hrs: 24 316 kw vis, 31.6 kw aur. 1,572t/1,493g TL: N35 17 34 W101 50 42 On air date: Apr 4, 1953. Box 10, Amarillo, TX, 79105-0010. 7900 Broadway, Amarillo, TX 79108. Phone: (806) 383-1010. Phone: (806) 383-6397. Fax: (806) 381-9859. Web Site: www.newschannel10.com. Licensee: Panhandle Telecasting Co. Group owner: R.H. Drewry Group (acq 10-4-76; $3 million; FTR: 9-13-76). Population served: 548,000 Natl. Network: CBS. Washington Atty: Shaw Pittman. Wire Svc: AP News staff: 25; News: 13 hrs wkly.
Key Personnel:
Bill Drewry . . . . pres
Robert Drewry . . . . pres
Larry Patton . . . . sr VP
Mike Lee . . . . VP, adv mgr, progmg dir & film buyer
Brent McClure . . . . gen mgr
Tim Cato . . . . opns mgr
Joyce Austin . . . . gen sls mgr, natl sls mgr & rgnl sls mgr
Tonya Triveno . . . . progmg mgr
Kari King . . . . news dir
Walt Howard . . . . mktg dir, prom mgr & pub affrs dir
Tim Winn . . . . chief of engrg

KPTF— Analog channel: 18.5,000 kw vis. ant 331t TL: N34 21 48 W103 13 05 On air date: 2002. Box 61000, Midland, TX, 79711-1000. Phone: (800) 707-0420. Fax: (432) 563-1736. E-mail: info@ptcbglc.com Web Site: www.godslearningchannel.com. Licensee: Prime Time Christian Broadcasting Inc. (acq 12-23-99).

KVIH-TV— Analog channel: 12. Digital channel: 20. Analog hrs: 4-9:37 PM (M-F); 6 AM-9:37 PM (S, Su) 178 kw vis, 35.3 kw aur. ant 670t/719g TL: N34 11 34 W103 16 44 On air date: December 1957. One Broadcast Ctr., Amarillo, TX, 79101. Phone: (806) 373-1787. Fax: (806) 371-7329. Web Site: www.kvii.com. Licensee: Barrington Broadcasting Texas Corp. Group owner: New Vision Group LLC (acq 8-2-2005; $22.5 million with KVII-TV Amarillo, TX). Natl. Network: ABC. Washington Atty: Wiley, Rein & Fielding.
Key Personnel:
K. James Yager . . . . pres
Mac Douglas . . . . VP
Lynn Fairbanks . . . . gen mgr & gen sls mgr
Keith Workman . . . . rgnl sls mgr
Curtis Weaver . . . . prom mgr
Paula Harris . . . . progmg dir
Bill Canady . . . . chief of engrg
Satellite of KVII-TV Amarillo, TX.

KVII-TV— Analog channel: 7. Digital channel: 23. Analog hrs: 24 Note: ABC is on KVII-TV ch 7, CW is on KVII-DT ch 23. 316 kw vis, 31.6 kw aur. ant 1,703t/1,626g TL: N35 22 29 W101 52 58 On air date: Nov 1, 1957. One Broadcast Ctr., Amarillo, TX, 79101. Phone: (806) 373-1787. Fax: (806) 371-7329. Web Site: www.kvii.com. Licensee: Barrington Broadcasting Texas Corp. Group owner: New Vision Group LLC (acq 8-2-2005; $22.5 million with KVIH-TV Clovis, NM). Population served: 262,200 Natl. Network: ABC, CW Digital Network: Note: ABC is on KVII-TV ch 7, CW is on KVII-DT ch 23. Washington Atty: Wiley, Rein & Fielding. Foreign lang progmg: SP 2 News: 36 hrs wkly.
Key Personnel:
K. James Yager . . . . pres
Lyn Fairbanks . . . . gen mgr
Dusty Green . . . . opns dir, opns mgr & news dir
Connie Mosley . . . . gen sls mgr & rgnl sls mgr
Bill Canady . . . . chief of engrg

Austin, TX
(DMA 52)

KEYE-TV— Analog channel: 42. Digital channel: 43. Analog hrs: 24 2,510 kw vis, 251 kw aur. ant 1,290t/1,299g TL: N30 19 10 W97 48 06 On air date: Dec 4, 1983. 10700 Metric Blvd., Austin, TX, 78758. Phone: (512) 835-0042. Fax: (512) 837-6753. Web Site: www.keyetv.com. Licensee: CBS Stations Group of Texas L.P. Group owner: Viacom Television Stations Group (acq 8-3-99; $160 million). Natl. Network: CBS. Natl. Rep: TeleRep. Washington Atty: Akin, Gump, Strauss, Hauer & Feld. Wire Svc: CBS News staff: 50; News: 20 hrs wkly.
Key Personnel:
Amy Villarreal . . . . VP & gen mgr
Dusty Granberry . . . . opns mgr
Jeff Stern . . . . sls dir
Fred Undstrom . . . . natl sls mgr
Ira Poole . . . . natl sls mgr & rgnl sls mgr
Lee Maaz . . . . rgnl sls mgr
Steve Colkins . . . . rgnl sls mgr
Stan Teater . . . . prom mgr
Gary Vinson . . . . progmg mgr
Suzanne Black . . . . news dir
Art Smith . . . . chief of engrg

***KLRU-TV**— Analog channel: 18. Digital channel: 22.1,860 kw vis, 372.8 kw aur. 1,099t/1,183g TL: N30 19 20 W97 48 10 On air date: May 4, 1979. 2504 B Whitis St., Austin, TX, 78712. Phone: (512) 471-4811. Fax: (512) 475-9090. Web Site: www.klru.org. Licensee: Capital of Texas Public Telecomm. Natl. Network: PBS. Washington Atty: Cohn & Marks.
Key Personnel:
Bill Stotesbery . . . . CEO, CEO & pres
Pat Wertz . . . . CFO
Dick Peterson . . . . exec VP & opns VP
Karin Morrison . . . . VP
Bill Statesbury . . . . gen mgr
Lori Holliday . . . . dev VP
Ed Bailey . . . . sls VP
Maury Sullivan . . . . mktg VP & adv VP
Cheryl Sawyer . . . . prom VP
Maria Rodriguez . . . . progmg VP
David Kuipers . . . . engrg VP

KNIC-TV— Analog channel: 17.5,000 kw vis. ant 669t/564g TL: N29 41 48 W98 30 45 On air date: 2006. 411 E. Durango Blvd., San Antonio, TX, 78204-1309. Phone: (210) 227-4141. Fax: (210) 227-0469. Web Site: www.univision.com. Licensee: Univision Television Group Inc. Natl. Network: TeleFutura (Spanish). Foreign lang progmg: SP 168

KNVA— Analog channel: 54. Digital channel: 49.5,000 kw vis. ant 1,227t/1,109g TL: N30 19 33 W97 47 58 On air date: Aug 1, 1994. Box 684647, Austin, TX, 78767. 908 W. Martin Luther King Blvd., Austin, TX 78768. Phone: (512) 478-5400. Fax: (512) 476-1520. Web Site: www.knva.com. Licensee: 54 Broadcasting Inc. Ownership: Diane Levy, 25%; Frank Goldberg, 25%; Mark Goldberg, 25%; and Richard Goldberg, 25% (acq 6-10-2004). Natl. Network: CW.
Key Personnel:
Eric Lassberg . . . . gen mgr
Pat Niekamp . . . . gen sls mgr
David Rash . . . . natl sls mgr
Martha Goodwin . . . . rgnl sls mgr
Jim Canning . . . . prom dir
Regina Soto . . . . progmg dir
Bruce Whitaker . . . . news dir
Mark Dunham . . . . engrg dir

KTBC— Analog channel: 7. Digital channel: 56. Analog hrs: 24 316 kw vis, 31.6 kw aur. ant 1,261t/1,114g TL: N30 18 36 W97 47 33 On air date: Nov 27, 1952. 119 E. 10th St., Austin, TX, 78701. Phone: (512) 476-7777. Fax: (512) 495-7060. E-mail: management@fox7.com Web Site: www.myfoxaustin.com. Licensee: KTBC License Inc. Group owner: Fox Television Stations Inc. (acq 1-97). Population served: 1,152,000 Natl. Network: Fox. News staff: 54; News: 24 hrs wkly.
Key Personnel:
Danny Baker . . . . VP & gen mgr
Mark Rodman . . . . sls VP & gen sls mgr
Kathie Smith . . . . mktg VP, prom VP & prom dir
Holly Morrison-Breaux . . . . progmg dir
Pam Vaught . . . . news dir
Rob Cunningham . . . . pub affrs dir
Ken Smith . . . . chief of engrg

KVUE-TV— Analog channel: 24. Digital channel: 33. Analog hrs: 24 Digital hrs: 24 1,950 kw vis, 327 kw aur. 1,270t/1,184g TL: N30 19 20 W97 48 10 On air date: Sept 12, 1971. Box 9927, Austin, TX, 78766. 3201 Steck Ave., Austin, TX 78757. Phone: (512) 459-6521. Fax: (512) 533-2215. Fax: (512) 533-2233 (news). Web Site: www.kvue.com. Licensee: KVUE-TV Inc. Group owner: Belo Corp., Broadcast Division (acq 6-1-99; swap with KXTV(TV) Sacramento, CA). Population served: 1,152,300 Natl. Network: ABC. Natl. Rep: Harrington, Righter & Parsons.

KXAM-TV— Analog channel: 14.3,236 kw vis, 324 kw aur. 883t/459g TL: N30 40 36 W98 33 59 On air date: Sept 6, 1991. Box 490, Austin, TX, 78767. 908 W. Martin Luther King Jr. Blvd., Austin, TX 78701. Phone: (512) 476-3636. Fax: (512) 476-1520. Fax: (512) 469-0630. Web Site: www.kxam.com. Licensee: KXAN Inc. Group owner: LIN Television Corp. Natl. Network: NBC.
Key Personnel:
Eric Lassberg . . . . gen mgr
James Holowaty . . . . opns dir
Teansie Garfield . . . . gen sls mgr
David Walker . . . . natl sls mgr
Jim Canning . . . . prom dir & pub affrs dir
Bruce Whitaker . . . . news dir
Mark Dunham . . . . engrg dir

KXAN-TV— Analog channel: 36. Digital channel: 21. Analog hrs: 24 2,000 kw vis, 200 kw aur. ant 1,268t/1,168g TL: N30 19 33 W97 47 58 (CP: 5,000 kw vis) On air date: Feb 12, 1965. Box 490, Austin, TX, 78767. 908 W. Martin Luther King Blvd., Austin, TX 78701. Phone: (512) 476-3636. Fax: (512) 476-1520. Fax: (512) 469-0630. Web Site: www.kxan.com. Licensee: KXAN Inc. Group owner: LIN Television Corporation (acq 11-14-94; FTR: 12-12-94). Natl. Network: NBC. Washington Atty: Covington & Burling. News staff: 60; News: 22 hrs wkly.
Key Personnel:
Eric Lassberg . . . . gen mgr
James Holowaty . . . . opns dir
Teansie Garfield . . . . gen sls mgr
David Walker . . . . rgnl sls mgr
Jim Canning . . . . prom dir
Bruce Whitaker . . . . news dir
Mark Dunham . . . . engrg dir

Beaumont-Port Arthur, TX
(DMA 140)

KBMT— Analog channel: 12.316 kw vis, 31 kw aur. 1,049t/1,022g TL: N30 11 26 W93 53 08 On air date: June 18, 1961. Box 1550, 525 I-10 S., Dover, TX, 77704. Phone: (409) 838-1212. Phone: (409) 833-7512. Fax: (409) 835-1617. Licensee: Texas Telecasting Inc. Group owner: McKinnon Broadcasting Co. (acq 11-1-76; $2.4 million; FTR: 11-15-77). Population served: 463,900 Natl. Network: ABC. Washington Atty: Cohn & Marks. News staff: 26; News: 13 hrs wkly.
Key Personnel:
Michael McKinnon . . . . pres
Mark McKinnon . . . . VP & gen mgr
David King . . . . gen mgr, gen sls mgr & natl sls mgr
Don Williams . . . . opns dir
Elda Gaudet . . . . rgnl sls mgr
Don Haener . . . . prom mgr
Elizabeth West . . . . progmg dir
Miles Resnick . . . . news dir
Mark Cormier . . . . chief of engrg

KBTV-TV—Port Arthur, Analog channel: 4. Digital channel: 40. Analog hrs: 24 100 kw vis, 20 kw aur. ant 1,184t/1,225g TL: N30 09 31 W93 59 11 On air date: Oct 22, 1957. 6155 Eastex Fwy., Ste 300, Beaumont, TX, 77706-6707. Phone: (409) 840-4444. Fax: (409) 985-4927. Fax: (409) 899-4639 (news). Web Site: www.kbtv4.tv. Licensee: Nexstar Broadcasting Inc. Group owner: Nexstar Broadcasting Group Inc. (acq 11-6-97; grpsl). Population served: 573,710 Natl. Network: NBC. Natl. Rep: Blair Television. Washington Atty: Drinker, Biddle & Roth. Wire Svc: AP
Key Personnel:
Perry Sook . . . . CEO & pres
Duane Lammers . . . . COO
Van Greer . . . . gen mgr
Ed Stowell . . . . chief of opns
Paul Bergen . . . . news dir
Dawn Stout . . . . pub affrs dir
Charlie Ravell . . . . chief of engrg

KFDM-TV— Analog channel: 6. Digital channel: 21. Analog hrs: 24 Note: CBS is on KFDM-TV ch 6, CW is on KFDM-DT ch 21. 100 kw vis, 20 kw aur. 960t/1,031g TL: N30 08 24 W93 58 44 On air date: Apr 24, 1955. Box 7128, 2955 I-10 E., Beaumont, TX, 77726-7128. Phone: (409) 892-6622. Fax: (409) 892-6665. Web Site: www.kfdm.com. Licensee: Freedom Broadcasting of Texas Licensee L.L.C. Group owner: Freedom Broadcasting Inc. (acq 1-4-84; grpsl; FTR: 1-2-84). Population served: 163,500 Natl. Network: CBS, CW Digital Network:

Note: CBS is on KFDM-TV ch 6, CW is on KFDM-DT ch 21. Natl. Rep: TeleRep. Washington Atty: Latham & Watkins.

***KITU-TV—** Analog channel: 34.1,170 kw vis, 117 kw aur. ant 1,023t/1,046g TL: N30 10 41 W93 54 26 On air date: June 21, 1986. 11221 IH 10, Orange, TX, 77630. Phone: (409) 745-3434. Fax: (409) 745-4752. Web Site: www.communityedtv.org. Licensee: Community Educational Television Inc.
Key Personnel:
Dr. Reginald Cherry . . . . . pres
Wayne Ozio . . . . . gen mgr & stn mgr

Brownsville

see Harlingen-Weslaco-Brownsville-McAllen, TX market

Bryan

see Waco-Temple-Bryan, TX market

Corpus Christi, TX (DMA 129)

***KEDT—** Analog channel: 16.1,480 kw vis, 148 kw aur. 970t/996g TL: N27 39 12 W97 33 55 On air date: Oct 15, 1972. 4455 S. Padre Island Dr., Suite 38, Corpus Christi, TX, 78411. Phone: (361) 855-2213. Fax: (361) 855-3877. Web Site: www.kedt.org. Licensee: South Texas Public Broadcasting System. Population served: 450,000 Natl. Network: PBS. Washington Atty: Schwartz, Woods & Miller.
Key Personnel:
Trey McCampbell . . . . . chmn
Don Dunlap . . . . . pres & gen mgr
Norma Camarillo . . . . . CFO
Myra Lombardo . . . . . VP
Cody Blount . . . . . opns dir, engrg dir & chief of engrg
Molly Goodwin . . . . . sls dir
Robert Chabot . . . . . prom VP & mus dir
Sylvia Coronado . . . . . progmg dir & progmg mgr
Johanna Zwernemann . . . . . asst music dir

KIII— Analog channel: 3. Digital channel: 8. Analog hrs: 24 100 kw vis, 10 kw aur. ant 958t/948g TL: N27 39 30 W97 36 04 On air date: May 4, 1964. 5002 S. Padre Island Dr., Corpus Christi, TX, 78411. Box 6669, Corpus Christi, TX 78466. Phone: (361) 986-8300. Fax: (361) 986-8311. E-mail: news@kiiitv.com Web Site: www.kiiitv.com. Licensee: Channel 3 of Corpus Christi Inc. Group owner: McKinnon Broadcasting Co. (acq 7-79; $171,720). Population served: 505,900 Natl. Network: ABC. Natl. Rep: Continental Television Sales. Washington Atty: Cohn & Marks. Foreign lang progmg: SP 3 News staff: 30; News: 17.5 hrs wkly.
Key Personnel:
Michael D. McKinnon . . . . . pres
Dick Drilling . . . . . VP, gen mgr & progmg dir
Scott Jones. . . . . . opns dir
Bill Beck . . . . . gen sls mgr
Larry Hogue . . . . . natl sls mgr
Richard Longoria . . . . . news dir
Ralph Quiroz. . . . . . chief of engrg

KORO— Analog channel: 28.1,450 kw vis, 146 kw aur. 762t/750g TL: N27 45 11 W97 38 14 On air date: Apr 15, 1977. Box 2667, Corpus Christi, TX, 78403. 102 N. Mesquite, Corpus Christi, TX 78403. Phone: (361) 883-2823. Fax: (361) 883-2931. Licensee: Entravision Holdings L.L.C. Group owner: Entravision Communications Co. L.L.C. (acq 3-17-98; $1.336 million). Population served: 364,500 Natl. Network: Univision (Spanish). Washington Atty: Mullin, Rhyne, Emmons & Topel. Foreign lang progmg: SP 168 News staff: 5; News: 5 hrs wkly.

KRIS-TV— Analog channel: 6. Digital channel: 13.Note: NBC is on KRIS-TV ch 6, CW is on KRIS-DT ch 13. 100 kw vis, 10 kw aur. ant 987t TL: N27 44 28 W97 36 08 On air date: May 22, 1956. Box 840, Corpus Christi, TX, 78403. 409 S. Staples, Corpus Christi, TX 78403. Phone: (361) 886-6100. Fax: (361) 886-6175. Fax: TWX: 910-876-1442. Web Site: www.kristv.com. Licensee: KVOA Communications Inc. Group owner: Cordillera Communications Inc. Population served: 493,000 Natl. Network: NBC, CW Digital Network: Note: NBC is on KRIS-TV ch 6, CW is on KRIS-DT ch 13. Washington Atty: Nixon, Hargrave, Devans & Doyle. News staff: 31; News: 14 hrs wkly.
Key Personnel:
Tim Noble . . . . . pres, gen mgr & gen mgr
Bob Webb . . . . . chief of opns
Don Grubaugh. . . . . . gen sls mgr & natl sls mgr
Roger Brandt . . . . . gen sls mgr
Jay Sanchez . . . . . prom mgr
James H. Smith. . . . . . progmg dir
Sandra Richards . . . . . news dir
Steve West . . . . . chief of engrg

KUQI— Analog channel: 38.1,000 kw vis. ant 918t/902g TL: N27 45 22.9 W97 36 25 Not on air, target date: unknown: Minortiy MediaTV-38, LLC, c/o 1025 Tho. Jefferson St. N.W., East Lobby, Suite 700, Washington, DC, 20007-5201. Phone: (202) 625-3695. Fax: (202) 295-1122. E-mail: lee.shubert@kattenlaw.com Permittee: Minority Media TV 38 LLC (acq 5-9-2006).

KZTV— Analog channel: 10. Digital channel: 18.316 kw vis, 47.9 kw aur. ant 940t/984g TL: N27 46 50 W97 38 03 On air date: Sept 30, 1956. 301 Artesian St., Corpus Christi, TX, 78401. Phone: (361) 883-7070. Fax: (361) 882-8553. Web Site: www.kztv10.com. Licensee: Eagle Creek of Corpus Christi LLC. Group owner: Eagle Creek Broadcasting LLC (acq 6-13-2002; grpsl). Population served: 350,000 Natl. Network: CBS. News staff: 18.
Key Personnel:
Dale Remy . . . . . gen mgr & chief of engrg
Norman Barron. . . . . . rgnl sls mgr
Sheri Randall . . . . . progmg dir
Hollis Grizzard . . . . . news dir

Dallas-Ft. Worth (DMA 6)

KDAF— Analog channel: 33. Digital channel: 32. Analog hrs: 24 Digital hrs: 24 5,000 kw vis, 500 kw aur. ant 1,696t/1,529g TL: N32 35 22 W96 58 10 On air date: July 29, 1984. 8001 Carpenter Fwy., Dallas, TX, 75247. Phone: (214) 252-9233. Fax: (214) 252-3379. E-mail: cw33news@tribune.com Web Site: cw33.trb.com. Licensee: Tribune Broadcasting Co. Group owner: (group owner; (acq 7-15-97; grpsl). Population served: 4,000,000 Natl. Network: CW. Natl. Rep: TeleRep. Washington Atty: Sidley Austin LLP. News staff: 40; News: 6 hrs wkly.
Key Personnel:
Joe Young . . . . . VP & gen mgr
Steve McDonald . . . . . gen sls mgr
Anthony Maisel . . . . . news dir
Rick Anderson . . . . . chief of engrg

KDFI— Analog channel: 27. Digital channel: 36.5,000 kw vis, 500 kw aur. 1,690t/1,529g TL: N32 35 22 W96 58 10 On air date: Jan 26, 1981. 400 N. Griffin St., Dallas, TX, 75202. Phone: (214) 720-4444. Fax: (214) 720-3207. E-mail: kdfi27@foxinc.com Web Site: www.myfoxdfw.com. Licensee: New DMIC Inc. Group owner: Fox Television Stations Inc. (acq 2-18-00; $6.2 million). Population served: 4,000,000 Natl. Network: MyNetworkTV. Natl. Rep: Fox Stations Sales.

KDFW— Analog channel: 4. Digital channel: 35. Analog hrs: 24 100 kw vis, 20 kw aur. 1,676t/1,517g TL: N32 35 06 W96 58 41 On air date: Dec 3, 1949. 400 N. Griffin St., Dallas, TX, 75202. Phone: (214) 720-4444. Fax: (214) 720-3177 (gen.mgr.). Fax: (214) 720-3263 (news). E-mail: kdfw@foxtv.com Web Site: www.myfoxdfw.com. Licensee: KDFW License Inc. Group owner: Fox Television Stations Inc. (acq 1-97; grpsl). Population served: 4,000,000 Natl. Network: Fox. News: 50 hrs wkly.
Key Personnel:
Kathy Saunders . . . . . VP & gen mgr
Dennis Welsh . . . . . sls VP
Jeff Gurley . . . . . gen sls mgr
Stephanie Holloway . . . . . natl sls mgr
Don Adams . . . . . rgnl sls mgr
John Kukla . . . . . prom VP
Joe Kozlowski . . . . . prom mgr
Andy Alexander . . . . . progmg dir
Maria Barrs . . . . . news dir
Rochelle Brown . . . . . pub affrs dir
Mark LeValley . . . . . engrg VP

***KDTN—** Analog channel: 2.100 kw vis, 20 kw aur. ant 1,351t/1,204g TL: N32 35 22 W96 58 10 On air date: Feb 2, 1988. Box 612066, Dallas, TX, 75261. Phone: (817) 571-1229. Fax: (817) 571-7458. Web Site: www.daystar.com. Licensee: Community Television Educators of DFW Inc. Ownership: Dr. Alan Bullock, 20%; Jack Howard, 20%; Joni Lamb, 20%; Kory Ford, 20%; and Marcus D. Lamb, 20% (acq 10-24-2003; $20 million). Population served: 1,800,000
Key Personnel:
Arnold Torres . . . . . gen mgr
Jennette Hawkins. . . . . . progmg mgr

***KDTX-TV—** Analog channel: 58. Analog hrs: 24 5,000 kw vis, 500 kw aur. 1,437t TL: N32 35 22 W96 58 10 On air date: June 1986. 2823 W. Irving Blvd., Irving, TX, 75061. Phone: (972) 313-1333. Fax: (972) 790-5853. Web Site: www.tbn.org. Licensee: Trinity Broadcasting of Texas Inc. Group owner: Trinity Broadcasting Network (acq 7-86; $1.6 million; FTR: 5-6-86).
Key Personnel:
Paul F. Crouch . . . . . pres
Steve Fjordbak . . . . . gen mgr & progmg dir
Jennye Gardner . . . . . pub affrs dir
Jim Forman . . . . . chief of engrg

***KERA-TV—** Analog channel: 13.316 kw vis, 31.6 kw aur. 1,540t/1,347g TL: N32 34 43 W96 57 12 On air date: Sept 14, 1960. 3000 Harry Hines Blvd., Dallas, TX, 75201. Phone: (214) 871-1390. Fax: (214) 754-0635. Web Site: www.kera.org. Licensee: North Texas Public Broadcasting Inc. Population served: 4,950,000 Natl. Network: PBS. Washington Atty: Schwartz, Woods & Miller.

KFWD— Analog channel: 52. Digital channel: 51. Analog hrs: 5 AM-2 AM 5,000 kw vis, 500 kw aur. ant 1,788t/1,620g TL: N32 35 19 W96 58 05 On air date: Sept 1, 1988. 606 Young St., Dallas, TX, 75202. Phone: (214) 977-6780. Fax: (214) 977-6544. Web Site: www.kfwd.tv. Licensee: HIC Broadcast Inc. Population served: 5,000,000 Washington Atty: Dow, Lohnes & Albertson.
Key Personnel:
Roland Hernandez . . . . . CEO
Lisa Wegmann . . . . . gen sls mgr
Tony J. Montes. . . . . . stn mgr & progmg dir
David Boyd . . . . . chief of engrg

KLDT— Analog channel: 0. Digital channel: 54.1,039 kw vis. ant 341t TL: N32 59 56 W96 55 02 On air date: 1999. 2450 Rockbrook, Louisville, TX, 75067. Phone: (972) 316-2115. Fax: (972) 316-1112. Licensee: Johnson Broadcasting of Dallas Inc.

KMPX— Analog channel: 29. Digital channel: 30. Analog hrs: 24 5,000 kw vis, 1,000 kw aur. ant 1,758t/1,610g TL: N32 35 19 W96 58 05 On air date: Sept 15, 1993. Box 612066, Dallas, TX, 75261-2066. 4201 Pool Rd., Colleyville, TX 76034-5017. Phone: (817) 868-7776. Fax: (817) 571-7458. E-mail: comments@daystar.com Web Site: www.daystar.tv. Licensee: Liberman Television of Dallas License LLC. Group owner: Liberman Broadcasting Inc. (acq 1-12-2004; $37 million). Population served: 4,000,000

KPXD— Analog channel: 68.5,000 kw vis. 1,181t TL: N32 35 24 W96 58 21 On air date: December 1996. 600 Six Flags Dr. Ste 652, Arlington, TX, 76011-6353. Phone: (817) 654-6467. Web Site: www.ionline.tv. Licensee: Paxson Dallas License Inc. Group owner: Paxson Communications Corp. (acq 4-4-97; $2.5 million for 51%). Natl. Network: i Network.

KSTR-TV— Analog channel: 49. Digital channel: 48.5,000 kw vis, 500 kw aur. 1,200t/1,032g TL: N32 35 24 W96 58 21 On air date: Apr 17, 1984. 2323 Bryan St., Ste. 1900, Dallas, TX, 75201. Phone: (214) 954-4900. Phone: (214) 758-2300. Fax: (214) 954-4920. Fax: (214) 758-2395. Web Site: www.univision.com. Licensee: Univision Partnership of Dallas. Group owner: Univision Communications Inc. (acq 6-6-01; grpsl). Natl. Network: TeleFutura (Spanish). Washington Atty: Wiley, Rein & Fielding.

KTAQ— Analog channel: 47.4,680 kw vis. ant 663t/646g TL: N33 09 32 W96 08 34 On air date: April 1, 1994. Box 8547, Greenville, TX, 75404. 1058 Country Rd., Greenville, TX 75404. Phone: (903) 455-8847. Fax: (903) 455-8891. Licensee: Simons Broadcasting LP. Ownership: Mike Simons, 99%; and Simons Asset Management L.L.C., 1% (acq 4-1-92; $50,000 for CP; FTR: 4-13-92).

KTVT— Analog channel: 11. Digital channel: 19. Analog hrs: 24 316 kw vis, 31.6 kw aur. ant 1,670t/1,549g TL: N32 34 43 W96 57 12 On air date: Sept 11, 1955. Box 2495, Fort Worth, TX, 76113. 5233 Bridge St., Fort Worth, TX 76103. Phone: (817) 451-1111/654-1100. Fax: (817) 457-1897. E-mail: news@ktvt.com Web Site: www.cbs11tv.com. Licensee: CBS Stations Group of Texas L.P. Group owner: Viacom Television Stations Group (acq 8-3-99; $485 million in stock). Population served: 6,100,000 Natl. Network: CBS. Washington Atty: Leventhal, Senter & Lerman. News staff: 100; News: 27 hrs wkly.
Key Personnel:
Steve Mauldin . . . . . pres & gen mgr
Gary Schneider . . . . . sr VP, VP & stn mgr
Steve Williams . . . . . opns mgr
Adam Levy. . . . . . sls dir
David Hershey. . . . . . mktg dir & prom dir
Ken Foote . . . . . progmg dir
Regent Ducas . . . . . news dir
Bill Schully . . . . . engrg mgr

KTXA—Arlington, Analog channel: 21. Digital channel: 18. Analog hrs: 24 4,900 kw vis, 490 kw aur. ant 1,650t/1,489g TL: N32 35 22 W96 58 10 On air date: Jan 4, 1981. 10111 N. Central Expwy., Dallas, TX, 75231. 5233 Bridge St., Fort Worth, TX 76103. Phone: (214) 743-2100. Fax: (214) 743-2121. Fax: (214) 743-2150. E-mail: news@ktvt.com Web Site: ktxa.com. Licensee: Viacom Television Stations Group of Dallas/Fort Worth L.P. Group owner: Viacom Television Stations Group (acq 2-28-91). Population served: 6,100,000 Washington Atty: Leventhal, Senter & Lerman.

Key Personnel:
Steve Mauldin . . . . pres & gen mgr
Gary Schneider. . . . . sr VP & stn mgr
Steve Williams . . . . opns mgr
Julia O'Hickey . . . . sls dir
Kyle Brawner . . . . natl sls mgr
David Hershey. . . . . mktg dir & prom dir
Ken Foote . . . . progmg dir
Regent Ducas . . . . news dir
Bill Schully . . . . engrg mgr

KUVN-TV—Garland, Analog channel: 23.5,000 kw vis, 1,000 kw aur. 1,142t TL: N32 54 04 W96 41 14 On air date: Sept 25, 1986. 2323 Bryan St., Suite 1900, Dallas, TX, 75201-2646. Phone: (214) 758-2300. Fax: (214) 758-2324. Web Site: www.univision.com. Licensee: KUVN License Partnership L.P. Group owner: Univision Communications Inc. (acq 5-88; $5.2 million). Natl. Network: Univision (Spanish). Foreign lang progmg: SP 168 News staff: 22; News: 10 hrs wkly.

KXAS-TV— Analog channel: 5. Digital channel: 41. Analog hrs: 24 100 kw vis, 20 kw aur. 1,686t/1,527g TL: N32 35 15 W96 57 59 On air date: Sept 29, 1948. 3900 Barnett, Fort Worth, TX 76103. Phone: (817) 429-5555. Phone: (214) 745-5555. Fax: (817) 654-6362. E-mail: nbc5i@nbc.com Web Site: www.nbc5i.com. Licensee: Station Venture Operations LP. Ownership: NBC Telemundo License Co., 79.62% of the equity; Hicks Muse, 20.38% of the equity (acq 3-2-98). Population served: 4,500,000 Natl. Network: NBC. Natl. Rep: NBC TV Stations Sales. News: 37 hrs wkly.
Key Personnel:
Thomas M. O'Brien . . . . pres & gen mgr
Jim Borden . . . . opns mgr

KXTX-TV— Analog channel: 39. Digital channel: 40.4,470 kw vis, 447 kw aur. ant 1,679t/1,521g TL: N32 35 07 W96 58 06 (CP: 5,000 kw vis) On air date: Feb 5, 1968. 3900 Barrett St., Fort Worth, TX, 76103. Phone: (214) 521-3900. Fax: (214) 303-5156. Web Site: www.telemundodallas.com. Licensee: NBC Telemundo License Co. Group owner: NBC TV Stations Division (acq 4-12-2002; grpsl). Population served: 4,351,700 Natl. Rep: Harrington, Righter & Parsons. Washington Atty: Fisher, Wayland, Cooper, Leader & Zaragoza. Foreign lang progmg: SP 168
Key Personnel:
Jose Valle . . . . gen mgr
Brian McCall. . . . . opns mgr

WFAA-TV— Analog channel: 8. Digital channel: 9. Analog hrs: 24 316 kw vis, 31.6 kw aur. 1,680t/1,521g TL: N32 35 06 W96 58 41 On air date: Sept 17, 1949. Communications Ctr., 606 Young St., Dallas, TX, 75202-4870. Phone: (214) 748-9631. Fax: (214) 977-6268. Fax: TWX: 910-861-4139. Web Site: wfaa.com. Licensee: WFAA-TV Inc. Group owner: Belo Corp., Broadcast Division (acq 2-1950). Population served: 3591600 Natl. Network: ABC. Natl. Rep: TeleRep. Washington Atty: Wiley, Rein & Fielding. Wire Svc: Reuters Wire Svc: NWS (National Weather Service) News staff: 85; News: 28 hrs wkly.
Key Personnel:
Robert W. Iecherd . . . . CEO
Kathy Clements . . . . pres & gen mgr
Angela E. Betasso . . . . sls VP
Eric Nelson . . . . natl sls mgr
Linda Ross . . . . natl sls mgr
Nick Nicholson . . . . mktg VP
Jim Glass . . . . prom dir
Cathy Helean . . . . prom mgr
David Walther. . . . . progmg dir
David Johnson . . . . engrg dir

El Paso (Las Cruces, NM), TX (DMA 99)

***KCOS—** Analog channel: 13. Digital channel: 30. Analog hrs: 24 Digital hrs: 18 224 kw vis, 22.4 kw aur. 869t/342g TL: N31 47 15 W106 28 47 On air date: Aug 18, 1978. Box 26668, El Paso, TX, 79926. Phone: (915) 590-1313. Fax: (915) 594-5394. E-mail: cbrush@kcostv.org Web Site: www.kcostv.org. Licensee: El Paso Public Television Foundation. Population served: 262,000 Natl. Network: PBS. Washington Atty: Cohn & Marks.
Key Personnel:
Craig Brush . . . . CEO, pres & gen mgr
Robbie Paul . . . . chmn
Barbara Hakim. . . . . CFO

KDBC-TV— Analog channel: 4. Digital channel: 18. Analog hrs: 24 Digital hrs: 24 Note: CBS is on KDBC-TV ch 4, MyNetworkTV is on KDBC-DT ch 18. 100 kw vis, 10 kw aur. 1,558t/377g TL: N31 47 46 W106 28 57 On air date: Dec 14, 1952. Box 1799, El Paso, TX, 79999. 2201 Wyoming Ave., El Paso, TX 79999. Phone: (915) 496-4444. Fax: (915) 496-4591 (sls). Fax: (915) 496-4593 (news). E-mail: news@kdbc.com Web Site: www.kdbc4.com. Licensee: KDBC License LLC. Group owner: Pappas Telecasting Companies (acq 3-29-2004; $20 million). Population served: 940,800 Natl. Network: CBS, MyNetworkTV Digital Network: Note: CBS is on KDBC-TV ch 4, MyNetworkTV is on KDBC-DT ch 18. Natl. Rep: Harrington, Righter & Parsons. Washington Atty: Fletcher, Heald & Hildreth, P.L.C. Wire Svc: AP News staff: 25; News: 15 hrs wkly.
Key Personnel:
Bram Watkins. . . . . gen mgr
John Burton . . . . gen sls mgr
Robert Rios . . . . natl sls mgr
Kevin Hayes. . . . . rgnl sls mgr
Doin Somes . . . . chief of engrg

KFOX-TV— Analog channel: 14. Digital channel: 8. Analog hrs: 24 398 kw vis, 39.8 kw aur. 1,981t/367g TL: N31 48 55 W106 29 20 On air date: August 1979. 6004 N. Mesa, El Paso, TX, 79912. Phone: (915) 833-8585. Fax: (915) 833-1358. Web Site: www.kfoxtv.com. Licensee: KTVU Partnership. Group owner: Cox Broadcasting (acq 1996; $20.855 million). Population served: 231,400 Natl. Network: Fox. Natl. Rep: TeleRep. Washington Atty: Dow, Lohnes & Albertson. News staff: 23; News: 6 hrs wkly.

KINT-TV— Analog channel: 26.2,240 kw vis, 224 kw aur. 1,499t/350g TL: N31 47 46 W106 28 57 On air date: May 5, 1984. 5426 N. Mesa, El Paso, TX, 79912. Phone: (915) 581-1126. Fax: (915) 581-1393. Web Site: www.noticias26.com. Licensee: Entravision Communications Co. L.L.C. Group owner: (group owner; (acq 6-4-97; grpsl). Natl. Network: Univision (Spanish). Washington Atty: Thompson, Hine & Flory L. Foreign lang progmg: SP 168 News staff: 14; News: 10 hrs wkly.
Key Personnel:
David Candelaria . . . . gen mgr
Diana De Lara . . . . gen sls mgr
Dan Kempner . . . . natl sls mgr
Abel Rodriguez . . . . prom dir & pub affrs dir
Sylvia Martinez . . . . progmg dir
Gustavo Barraza . . . . news dir
Alfredo Durand . . . . chief of engrg

***KRWG-TV—** Analog channel: 22. Digital channel: 23. Analog hrs: 6 AM-Midnight Digital hrs: 6AM-Midnight 1,550 kw vis, 155 kw aur. ant 410t/338g TL: N32 15 33 W106 58 30 On air date: June 29, 1973. Box 30001, MSC TV22, NMSU, Las Cruces, NM, 88003. Rm. 100, Jordan St., NMSU, Las Cruces, NM 88003. Phone: (505) 646-2222. Fax: (505) 646-1924. E-mail: krwgtv@nmsu.edu Web Site: www.krwg-tv.org. Licensee: Regents of New Mexico State University. Population served: 500,000 Natl. Network: PBS. Washington Atty: Dow, Lohnes & Albertson. Wire Svc: AP Foreign lang progmg: SP 2 News: 3 hrs wkly.
Key Personnel:
Dave Holly . . . . gen mgr
J.D. Jarvis . . . . opns mgr
Anthony Casaus . . . . dev dir
Gary Worth. . . . . news dir
William Saggerson . . . . chief of engrg

***KSCE—** Analog channel: 38. Digital channel: 39. Analog hrs: 24 Digital hrs: 24 50.1 kw vis, 5 kw aur. ant 1,827t/172g TL: N31 48 55 W106 29 17 On air date: Apr 15, 1989. 6400 Escondido Dr., El Paso, TX, 79912. Phone: (915) 585-8838. Fax: (915) 585-8841. E-mail: ksce@aol.com Web Site: www.kscetv.com. Licensee: Channel 38 Christian Television. Ownership: Non-profit corporation. Population served: 950,000 Washington Atty: James L. Oyster. Foreign lang progmg: SP 28

KTDO— Analog channel: 48. Digital channel: 47. Analog hrs: 24 79.4 kw vis, 7.9 kw aur. 113t TL: N32 02 30 W106 27 41 On air date: Nov 11, 1984. 10033 Carnegie, El Paso, TX, 79925. Phone: (915) 591-9595. Fax: (915) 591-9896. Web Site: www.telemundo.com. Licensee: ZGS El Paso Televison LP. Ownership: ZGS Broadcast Holdings Inc., 100% (acq 9-13-2004; $11.8 million). Population served: 850,000 Natl. Network: Telemundo (Spanish). Washington Atty: Reed, Smith, Shaw & McClay. Foreign lang progmg: SP 168
Key Personnel:
Lorena Caltamon . . . . gen mgr
Phillip Cortez . . . . gen sls mgr, mktg mgr & adv mgr
Monic Diaz . . . . progmg dir
Elios Ventanilla . . . . chief of engrg

KTFN— Analog channel: 65.1,000 kw vis, 50 kw aur. 1,827t/199g TL: N31 48 55 W106 29 17 On air date: 1991. 5426 N. Mesa, El Paso, TX, 79912. Phone: (915) 581-1126. Fax: (915) 581-1393. Web Site: www.noticias26.com. Licensee: Entravision Holdings LLC. Group owner: Entravision Communications Corp. (acq 12-10-01; $18 million). Foreign lang progmg: SP 168
Key Personnel:
Dan Kempner . . . . pres & natl sls mgr
David Candelaria . . . . gen mgr
Diana DeLara . . . . gen sls mgr
Abel Rodriguez . . . . prom dir
Alfredo Durand . . . . progmg dir & chief of engrg

KTSM-TV— Analog channel: 9. Digital channel: 16. Analog hrs: 24 Digital hrs: 24 316 kw vis, 42.7 kw aur. ant 1,910t/370g TL: N31 48 18 W106 28 57 On air date: Jan 4, 1953. 801 N. Oregon St., El Paso, TX, 79902. Phone: (915) 532-5421. Fax: (915) 532-6793. E-mail: ktsmtv@whc.net Web Site: www.ktsm.com. Licensee: ComCorp of El Paso License Corp. Group owner: Communications Corp. of America (acq 7-25-97; $30.5 million for stock with KTSM-AM-FM). Population served: 333,110 Natl. Network: NBC. Natl. Rep: Millennium Sales & Marketing. Washington Atty: Fletcher, Heald & Hildreth. Wire Svc: AP News staff: 33; News: 19.5 hrs wkly.
Key Personnel:
Charlie Hogetvedt . . . . gen mgr
Danny Aguilar . . . . sls VP & gen sls mgr
Debra Hastings. . . . . natl sls mgr
Victor Veuegus . . . . news dir
Courtney Elam. . . . . pub affrs dir
Ernie Hartt. . . . . engrg dir

KVIA-TV— Analog channel: 7. Digital channel: 17. Analog hrs: 24 Note: ABC is on KVIA-TV ch 7, CW is on KVIA-DT ch 17. 316 kw vis, 31.6 kw aur. ant 820t/296g TL: N31 47 15 W106 28 47 On air date: Sept 1, 1956. 4140 Rio Bravo, El Paso, TX, 79902. Phone: (915) 496-7777. Fax: (915) 532-0070. E-mail: kvia@kvia.com Web Site: www.kvia.com. Licensee: NPG of Texas L.P. (acq 12-9-94; $19.9 million; FTR: 1-23-95). Population served: 800,000 Natl. Network: ABC, CW Digital Network: Note: ABC is on KVIA-TV ch 7, CW is on KVIA-DT ch 17. Washington Atty: Robert Thompson. News staff: 30; News: 31.5 hrs wkly.
Key Personnel:
David Bradley . . . . CEO
John Kueneke . . . . pres
Kevin Lovell . . . . gen mgr
Chris Swann. . . . . opns mgr
Nathan Price . . . . sls dir
Dan Overstreet. . . . . natl sls mgr
David Gonzalez . . . . prom dir
Karla Huelga . . . . progmg dir & pub affrs dir
Eric Huseby . . . . news dir
Elias Ventanilla . . . . chief of engrg

XHIJ—(Ciudad Juarez, MEX) Analog channel: 44. Analog hrs: 20 240 kw vis, 60 kw aur. 1,200t/150g On air date: Oct 16, 1980. 5925 Cromo Dr., El Paso, TX, 79912. Phone: (915) 585-6344. Fax: (915) 585-6333. Web Site: www.canal44.com. Licensee: Arnoldo Cabada De la O. Ownership: Arnoldo Cabada De la O, 52%; Luis Cabada Alvidrez, Sergio Cabada Alvidrez & Jesus Cabada Alvidrez, each 16%. Natl. Network: Telemundo (Spanish). Foreign lang progmg: SP 126 News staff: 20; News: 15 hrs wkly.

Ft. Worth

see Dallas-Ft. Worth market

Harlingen-Weslaco -Brownsville-McAllen, TX (DMA 91)

KGBT-TV— Analog channel: 4. Digital channel: 31. Analog hrs: 24 Digital hrs: 24 100 kw vis, 18.7 kw aur. ant 1,299t/1,293g TL: N26 08 55 W97 49 17 On air date: October 1953. 9201 W. Expwy. 83, Harlingen, TX, 78552. Phone: (956) 366-4444. Fax: (956) 366-4494. E-mail: listens@kgbt4.com Web Site: www.kgbt4.com. Licensee: Barrington Harlingen License LLC. Group owner: Liberty Corp. (acq 8-11-2006; grpsl). Population served: 944,772 Natl. Network: CBS. Washington Atty: Dow, Lohnes & Albertson. Foreign lang progmg: SP 10 News staff: 29; News: 22 hrs wkly.
Key Personnel:
Teresa Burgess . . . . gen mgr
Phil Rich . . . . opns mgr
Randy Roberts . . . . gen sls mgr
Beau Pillet . . . . mktg dir & prom mgr
Kimberly Wyatt . . . . news dir
Monica Ortiz . . . . progmg

***KLUJ-TV—** Analog channel: 44.1,740 kw vis, 174 kw aur. 971t TL: N26 13 00 W97 46 48 On air date: June 25, 1984. Box 1647, 1920 Al Coneway Dr., Suite 117, Harlingen, TX, 78551. Phone: (956) 425-4225. Fax: (956) 412-1740. E-mail: klujtv@xanadu2.net Licensee: Community Educational TV Inc. (acq 4-84). Natl. Network: PBS. Washington Atty: Joseph E. Dunne III. Foreign lang progmg: SP 7

***KMBH—** Analog channel: 60. Digital channel: 38. Analog hrs: 24 Digital hrs: 24 2,240 kw vis, 22.4 kw aur. ant 1,220t/1,169g TL: N26 07 14 W97 49 18 On air date: Oct 8, 1985. Box 2147, Harlingen, TX, 78551. 1701 Tennessee St., Harlingen, TX 78551. Phone: (956) 421-4111. Fax: (956) 421-4150. E-mail: rgveduca@aol.com Web Site: www.kmbh.org. Licensee: RGV Educational Broadcasting Inc. Population served: 975,000 Natl. Network: PBS. Washington Atty: Thelen Reid & Priest LLP. Foreign lang progmg: SP 2

Key Personnel:
Father Pedro Briseno . . . . CEO, pres & gen mgr
John Ross. chief of engrg

KNVO— Analog channel: 48.3,162 kw vis, 316.2 kw aur. 524t/548g TL: N26 05 20 W98 03 44 (CP: 3,002 kw vis, ant 944t) On air date: Oct 12, 1992. 801 N. Jackson Rd., McAllen, TX, 78501. Phone: (956) 687-4848. Fax: (956) 687-7784. Licensee: Entravision Holdings L.L.C. Group owner: Entravision Communications Co. L.L.C (acq 4-25-97). Natl. Network: Univision (Spanish). Washington Atty: Schwartz, Woods & Miller. Foreign lang progmg: SP 168
Key Personnel:
Larry Safir . . . . gen mgr
Joe Medrano . . . . news dir

KRGV-TV— Analog channel: 5. Analog hrs: 24 100 kw vis, 19.1 kw aur. 950t/995g TL: N26 09 54 W97 48 45 On air date: Apr 10, 1954. Box 5, Weslaco, TX, 78599. 900 E. Expwy. 83, Weslaco, TX 78596. Phone: (956) 631-5555. Fax: (956) 973-5016. Web Site: www.krgv.com. Licensee: Mobile Video Tapes Inc. Group owner: Manship Stns (acq 1-28-64; grpsl; FTR: 2-3-64). Population served: 215,500 Natl. Network: ABC. Washington Atty: Cohn & Marks. Foreign lang progmg: SP 1 News staff: 28; News: 12 hrs wkly.
Key Personnel:
Richard Manship. . . . . chmn & pres
Ray Alexander . . . . gen mgr
Michelle Martone . . . . opns mgr & progmg dir
John Kittleman . . . . gen sls mgr
Robert Ledesma . . . . sls dir & natl sls mgr
Jerry Berg . . . . prom mgr
Jenny Martinez . . . . news dir
Jerry Lee Berg. . . . . pub affrs dir
Chuck Salge . . . . chief of engrg

***KTLM**— Analog channel: 40. Digital channel: 20. Analog hrs: 24 Digital hrs: 24 5,000 kw vis. 371t TL: N26 25 47 W98 49 25 On air date: Oct 8, 1999. 7th Fl., 3900 N. 10th St., McAllen, TX, 78501. Phone: (956) 686-0040. Fax: (956) 686-0770. Web Site: www.telemundo40.com. Licensee: Sunbelt Multimedia Co. Ownership: Sam F. Vale, 99.5% (acq 8-12-2005; $3.15 million). Population served: 1,000,000 Natl. Network: Telemundo (Spanish). Foreign lang progmg: SP 168 News staff: 12; News: 8 hrs wkly.

KVEO— Analog channel: 23. Digital channel: 24. Analog hrs: 24 2,570 kw vis, 1,000 kw aur. ant 1,460t/1,454g TL: N26 05 59 W97 50 16 On air date: Dec 19, 1981. 394 N. Expressway, Brownsville, TX, 78521. Box 4314, Brownsville, TX 78521. Phone: (956) 544-2323. Fax: (956) 544-4636. Web Site: www.kveo.com. Licensee: Communications Corp. of America. Group owner: (group owner; (acq 2-13-95; FTR: 5-8-95). Population served: 700,000 Natl. Network: NBC. Washington Atty: Fletcher, Heald & Hildreth.
Key Personnel:
Tom Galloway . . . . pres
Wayne Elmore . . . . CFO
Clark White. . . . . VP
Bill Jorn . . . . gen mgr
Robert Gutierrez . . . . gen sls mgr
Jackie Lynn . . . . prom dir
Lisa Kidd-Hagle . . . . progmg dir
Tommy Balli. . . . . engrg mgr

Houston (DMA 10)

KAZH— Analog channel: 57. Digital channel: 53. Analog hrs: 24 5,000 kw vis, 500 kw aur. ant 1,958t/1,944g TL: N29 34 15 W95 30 37 On air date: 1987. 2620 Fountain View, Ste.322, Houston, TX, 77057. Phone: (713) 467-5757. Fax: (713) 783-4157. Web Site: www.kazh57.com. Licensee: KAZH License LLC. Group owner: Pappas Telecasting Companies (acq 7-7-99; $28 million). Foreign lang progmg: SP 168
Key Personnel:
Harry J. Pappas . . . . CEO & pres
Emilio Nicolas Jr. . . . . gen mgr

***KETH-TV**— Analog channel: 14. Analog hrs: 24 4,470 kw vis, 447 kw aur. 1,437t/1,470g TL: N29 33 25 W95 30 04 On air date: July 1987. 10902 S. Wilcrest Dr., Houston, TX, 77099. Phone: (281) 561-5828. Fax: (281) 561-9793. Web Site: www.communityedtv.org. Licensee: Community Educational Television Inc. Population served: 231,128 News: 3 hrs wkly.
Key Personnel:
Laura Hanks. . . . . opns mgr
Rod Harty . . . . chief of engrg

KFTH-TV— Analog channel: 67. Digital channel: 36. Analog hrs: 24 5,000 kw vis, 500 kw aur. ant 1,781t/1,155g TL: N29 34 06 W95 29 57 On air date: Jan 27, 1986. 5100 SW Freeway, Houston, TX, 77056. Phone: (713) 662-4545. Fax: (713) 965-2610. Web Site: www.univision.com. Licensee: TeleFutura Houston LLC. Group owner: Univision Communications Inc. (acq 5-21-2001; grpsl). Natl. Network: TeleFutura (Spanish). Foreign lang progmg: SP 168
Key Personnel:
Jerold Perenchio . . . . CEO
Mike Wortsman. pres
Thomas Arnost . . . . pres
George Blank . . . . CFO
Craig H. Bland . . . . VP & gen mgr
Jose Oti . . . . gen sls mgr
Chas Witson. . . . . natl sls mgr
Michael Thomas . . . . rgnl sls mgr
Arlene Kelsch . . . . mktg mgr
Charlie Lozano . . . . prom VP
Sanjuio Salazar . . . . progmg mgr
Grace Olivares-Hernandez . . . . pub affrs dir
Tom Daniels . . . . chief of engrg

KHCW— Analog channel: 39. Digital channel: 39. Analog hrs: 24 Digital hrs: 25 5,000 kw vis, 500 kw aur. ant 1,950t/1,970g TL: N29 34 06 W95 29 57 On air date: Jan 6, 1967. 7700 Westpark Dr., Houston, TX, 77063. Phone: (713) 781-3939. Fax: (713) 781-3441. Web Site: khcw.com. Licensee: KHWB Inc. Group owner: Tribune Broadcasting Co. (acq 1-17-96; $95 million plus $6). Population served: 5,000,000 Natl. Network: CW. Natl. Rep: Harrington, Righter & Parsons. Washington Atty: Sidley & Austin. News staff: 34; News: 4 hrs wkly.

KHOU-TV— Analog channel: 11. Digital channel: 31. Analog hrs: 24 316 kw vis, 31.6 kw aur. 1,870t/1,473g TL: N29 33 40 W95 30 04 On air date: Mar 22, 1953. 1945 Allen Pkwy., Houston, TX, 77019. Phone: (713) 526-1111. Fax: (713) 521-4326. E-mail: 11listens@khou.com Web Site: www.khou.com. Licensee: KHOU-TV Inc. Group owner: Belo Corp., Broadcast Division (acq 1984; grpsl; FTR: 11-17-83). Population served: 5,000,000 Natl. Network: CBS. Natl. Rep: TeleRep. Wire Svc: Reuters
Key Personnel:
Peter Diaz . . . . pres & gen mgr
Susan McEldoon . . . . VP & stn mgr
Dan Lyons . . . . sls dir
Don Graham . . . . progmg dir
Keith Connors . . . . news dir

***KLTJ**— Analog channel: 22. Analog hrs: 24 5,000 kw vis, 500 kw aur. ant 1,856t/1,847g TL: N29 17 56 W95 14 11 On air date: July 22, 1989. 1050 Gemini, Houston, TX, 77058. Phone: (281) 212-1022. Fax: (281) 212-1031. E-mail: comments@daystar.com Web Site: www.daystar.com. Licensee: Word of God Fellowship Inc. aka Community TV Educators. (acq 10-18-99; $9.5 million). Population served: 1,452,000 Foreign lang progmg: SP 10

KNWS-TV— Analog channel: 51. Digital channel: 52. Analog hrs: 24 Digital hrs: 24 2,290 kw vis. ant 1,640t/1,624g TL: N29 33 40 W95 30 04 On air date: Nov 3, 1993. 8440 Westpark, Houston, TX, 77063. Phone: (713) 974-5151. Fax: (713) 974-5188. Web Site: www.knws51.com. Licensee: Johnson Broadcasting Inc. Ownership: Douglas R. Johnson, 100%. Washington Atty: Smithwick & Belendiuk.
Key Personnel:
Douglas R. Johnson. . . . . pres
Jack Dabbah. . . . . gen mgr & stn mgr
Chris Bourne . . . . opns dir & chief of opns

KPRC-TV— Analog channel: 2. Digital channel: 35. Analog hrs: 24 Digital hrs: 24 100 kw vis, 10 kw aur. ant 1,929t/1,969g TL: N29 34 06 W95 29 57 On air date: Jan 1, 1949. Box 2222, Houston, TX, 77252. 8181 Southwest Fwy., Houston, TX 77074. Phone: (713) 222-2222. Fax: (713) 270-9334. Web Site: www.click2houston.com. Licensee: Post-Newsweek Stations Inc. Group owner: (group owner; (acq 4-22-94; FTR: 5-2-94). Population served: 1,938,670 Natl. Network: NBC. Natl. Rep: MMT. Washington Atty: Covington & Burling. Wire Svc: AP News staff: 80; News: 34 hrs wkly.
Key Personnel:
Larry Blackerby . . . . gen mgr
Tammy Dean . . . . opns dir
Ben Oldham. . . . . gen sls mgr
Mr. Skip Valet. . . . . news dir
Dale Werner . . . . chief of engrg

KPXB— Analog channel: 49. Digital channel: 5. Analog hrs: 18 4,100 kw vis, 410 kw aur. 1,775t/1,200g TL: N30 15 45 W95 14 50 On air date: June 16, 1989. 256 N. Sam Houston Pkwy. E., Suite 49, Houston, TX, 77060. Phone: (281) 820-4900. Fax: (281) 820-3916. Web Site: www.ionline.tv. Licensee: Paxson Houston License Inc. Group owner: Paxson Communications Corp. (acq 1995; $7.9 million). Population served: 3,050,000 Natl. Network: i Network. Washington Atty: Pepper & Corazzini.

KRIV— Analog channel: 26. Digital channel: 27. Analog hrs: 24 Digital hrs: 24 5,000 kw vis, 500 kw aur. 1,948t/1,970g TL: N29 34 28 W95 29 37 On air date: Aug 15, 1971. Box 22810, Houston, TX, 77227. 4261 Southwest Fwy., Houston, TX 77027. Phone: (713) 479-2600. Fax: (713) 479-2604. Web Site: www.myfoxhouston.com. Licensee: Fox Television Stations Inc. Group owner: (group owner) Population served: 1697900 Natl. Network: Fox. Natl. Rep: Fox Stations Sales. Washington Atty: Molly Pauker. Wire Svc: AP
Key Personnel:
D'Artagnan Bebel . . . . gen mgr
Charles Hughes . . . . opns VP & engrg VP
Sheila Birenbaun . . . . opns dir
Du Juan McCoy . . . . sls VP
Larry Parker . . . . prom VP
Stan Wasilik . . . . progmg dir
Kathy Williams . . . . news dir
Lisa Whitlock . . . . pub affrs dir

KTBU— Analog channel: 55. Digital channel: 42. Analog hrs: 24 5,000 kw vis, 500 kw aur. ant 1,817t/1,801g TL: N30 13 53 W95 07 26 On air date: July 15, 1998. 7026 Old Katy Rd., Suite 201, Houston, TX, 77024. Phone: (713) 864-1999. Fax: (713) 864-1993. Web Site: www.thetube.net. Licensee: Humanity Interested Media L.P. Ownership: HIM GP LLC, gen ptnr, owned 100% by US Farm & Ranch Supply Co. Inc. dba USFR Media Group (acq 3-20-2006; $30 million).
Key Personnel:
Matt Reiss Jr.. gen mgr
Bruce Dinehart . . . . stn mgr
Phil Lonsway . . . . gen sls mgr
Lara Bell . . . . pub affrs dir
Eric Peterson. . . . . chief of engrg

KTMD— Analog channel: 47. Analog hrs: 21 5,000 kw vis, 500 kw aur. ant 1,958t/1,944g TL: N29 34 15 W95 30 37 On air date: Dec 12, 1987. 1235 N. Loop W., Suite 125, Houston, TX, 77008. Phone: (713) 974-4848. Fax: (713) 243-7850. Fax: (713) 782-5575. Web Site: www.ktmd.com. Licensee: NBC Telemundo License Co. Group owner: Telemundo Group Inc. (acq 4-12-2002; grpsl). Natl. Network: Telemundo (Spanish). Washington Atty: Hogan & Hartson. Foreign lang progmg: SP 160 News staff: 14; News: 7 hrs wkly.
Key Personnel:
Roel Medina . . . . gen mgr
Dominic Fails . . . . gen sls mgr
Gregorio Cervantes . . . . natl sls mgr

KTRK-TV— Analog channel: 13. Digital channel: 32.316 kw vis, 39.8 kw aur. 1,929t TL: N29 34 27 W95 29 37 On air date: Nov 20, 1954. Box 13, Houston, TX, 77001. 3310 Bissonnet St., Houston, TX 77005. Phone: (713) 666-0713. Fax: (713) 663-0013. Web Site: www.ABC13.com. Licensee: ABC Inc. Group owner: (group owner; (acq 7-17-67). Population served: 1,466,500 Natl. Network: ABC. Wire Svc: TWX

KTXH— Analog channel: 20. Digital channel: 19.5,000 kw vis, 500 kw aur. ant 1,811t/2,008g TL: N29 34 34 W95 30 36 On air date: Nov 7, 1982. 4261 Southwest Frwy., Houston, TX, 77027. Phone: (713) 479-2600. Fax: (713) 479-2859. Licensee: Fox Television Stations Inc. Group owner: (group owner; (acq 11-6-2001; with WDCA(TV) Washington, DC in swap for KBHK-TV San Francisco, CA). Population served: 1,510,580 Natl. Network: MyNetworkTV.
Key Personnel:
D'Artagnan Bebel . . . . gen mgr
Charles Hughes. . . . . opns VP

***KUHT**— Analog channel: 8. Digital channel: 9. Analog hrs: 24 316 kw vis, 63.2 kw aur. 1,970t/2,049g TL: N29 34 28 W95 29 37 On air date: May 12, 1953. 4343 Elgin ., Houston, TX, 77204-0008. Phone: (713) 748-8888. Phone: (800) 364-8300. Fax: (713) 743-8867. Web Site: www.houstonpbs.org. Licensee: University of Houston System, Board of Regents. Population served: 3,770,000 Natl. Network: PBS. Washington Atty: Dow, Lohnes & Albertson. Foreign lang progmg: SP 12
Key Personnel:
John Hesse . . . . gen mgr & stn mgr
Jack K. Neal . . . . stn mgr
Steve Pyndus. opns dir

KXLN-TV— Analog channel: 45. Digital channel: 46. Analog hrs: 24 5,000 kw vis, 500 kw aur. ant 1,948t/1,929g TL: N29 33 44 W95 30 35 On air date: Sept 18, 1987. 5100 SW Freeway, Houston, TX, 77056. Phone: (713) 662-4545. Fax: (713) 965-2610. Web Site: www.univision.com. Licensee: KXLN License Partnership G.P. Group owner: Univision Communications Inc. (acq 2-24-95; FTR: 5-22-95). Natl. Network: Univision (Spanish). Washington Atty: Fisher, Wayland, Cooper, Leader & Zaragoza. Foreign lang progmg: SP 165 News staff: 23; News: 7 hrs wkly.
Key Personnel:
Craig Bland . . . . VP, gen mgr & stn mgr
Jeff Hoffman . . . . stn mgr
Jose Oti . . . . gen sls mgr
Charles Wilson. . . . . natl sls mgr
Charlie Lozano . . . . prom dir
Cindy Chisum. . . . . progmg dir
Juan Garcia . . . . news dir
Grace C. Olivares . . . . pub affrs dir

Chuck Promrose . . . . chief of engrg

KZJL— Analog channel: 61. Digital channel: 44.1,700 kw vis. ant 1,898t/1,885g TL: N29 33 44 W95 30 35 On air date: 1995. 1845 Emppire Ave., Burbank, CA, 91504. Phone: (818) 563-5722. Licensee: KZJL License LLC. Group owner: Liberman Broadcasting Inc. (acq 1-10-2001; $57 million).

Laredo, TX (DMA 187)

KGNS-TV— Analog channel: 8. Digital channel: 15. Analog hrs: 20 316 kw vis, 42.2 kw aur. ant 1,021t/1,049g TL: N27 40 22 W99 39 23 On air date: Jan 6, 1956. 120 W. Del Mar Blvd., Laredo, TX, 78045. Phone: (956) 727-8888. Fax: (956) 727-5336. E-mail: email8@pro8news.com Web Site: pro8news.com. Licensee: SagamoreHill Broadcasting of Texas LLC. (acq 9-28-2004; $14.4 million). Population served: 30,000 Natl. Network: NBC. Washington Atty: Dow, Lohnes & Albertson. Foreign lang progmg: SP 8

Key Personnel:

Louis Wall. . . . . pres
Carlos Salinas . . . . gen mgr, natl sls mgr & rgnl sls mgr
Ramiro Saucedo . . . . prom mgr
Armando Gomez . . . . adv mgr & pub affrs dir
Velia Herrera . . . . progmg dir
Richard Rucha . . . . news dir
David York. . . . . chief of engrg

KLDO-TV— Analog channel: 27. Analog hrs: 18 3,720 kw vis, 372 kw aur. 220t TL: N27 30 03 W99 30 37 On air date: Dec 17, 1984. 222 Bob Bullock Loop, Laredo, TX, 78043. Phone: (956) 727-0027. Fax: (956) 727-2673. Web Site: www.entravision.com. Licensee: Entravision Holdings L.L.C. Group owner: Entravision Communications Co. L.L.C. (acq 7-30-97; $6.2 million). Natl. Network: Univision (Spanish). Washington Atty: Martin E. Firestone. Foreign lang progmg: SP 126

Key Personnel:

Terry Elena Ordaz . . . . gen mgr
Elia Solis . . . . gen sls mgr & film buyer
Jose Gomez. . . . . rgnl sls mgr
Jose Salinas . . . . prom mgr & progmg dir
Marisa Limon . . . . news dir
Merlin Miller . . . . chief of engrg

KVTV— Analog channel: 13. Digital channel: 31.85.1 kw vis, 17.4 kw aur. ant 918t/1,033g TL: N27 31 14 W99 31 19 On air date: Dec 29, 1973. 2600 Shea & Anna St., Laredo, TX, 78041. Phone: (956) 727-1300. Fax: (956) 712-0185. Web Site: www.cbs13kvtv.com. Licensee: Eagle Creek of Laredo LLC. Group owner: Eagle Creek Broadcasting LLC (acq 6-13-2002; grpsl). Population served: 450,000 Natl. Network: CBS.

Key Personnel:

Dale Remy . . . . gen mgr & stn mgr
Joe Herrera . . . . gen sls mgr
Carol Rostohar . . . . prom dir & pub affrs dir
Kent Harrell . . . . news dir
George Sanders . . . . chief of engrg

Longview

see Tyler-Longview (Lufkin & Nacogdoches), TX market

Lubbock, TX (DMA 147)

KAMC— Analog channel: 28. Analog hrs: 24 2,000 kw vis, 374 kw aur. ant 840t/871g TL: N33 30 57 W101 50 54 On air date: Nov 12, 1968. 7403 S. University, Lubbock, TX, 79423. Phone: (806) 745-2828. Fax: (806) 748-1080. E-mail: 28news@abc28.com Web Site: www.abc28.tv. Licensee: Mission Broadcasting Inc. Group owner: (group owner; (acq 12-17-2003). Population served: 357,000 Natl. Network: ABC. Washington Atty: Bryan Cave. News staff: 26; News: 39 hrs wkly.

Key Personnel:

David S. Smith . . . . pres & gen mgr
John Dittmeier . . . . exec VP
A.C. Wimberly . . . . gen mgr, stn mgr & progmg dir
Chuck Spaugh . . . . opns dir
Eric Thomas. . . . . gen sls mgr
Jeff Pitner . . . . prom dir & pub affrs dir
Russ Protect . . . . news dir
Monte Williams . . . . engrg dir

KCBD— Analog channel: 11.316 kw vis, 60 kw aur. 804t/702g TL: N33 32 32 W101 50 14 On air date: May 10, 1953. 5600 Avenue A, Lubbock, TX, 79404. Phone: (806) 744-1414. Fax: (806) 744-0449. E-mail: kcbd@kcbd.com Web Site: www.kcbd.com. Licensee: Libco Inc. Group owner: Liberty Corp. (acq 1-13-2006; grpsl). Population served: 254,000 Natl. Network: NBC. Washington Atty: Dow, Lohnes & Albertson.

Key Personnel:

Dan Jackson . . . . gen mgr & gen sls mgr
Brent McClure . . . . opns mgr, mktg dir & prom mgr
Beverly McBeth . . . . gen sls mgr
Peggy Sullivan. . . . . progmg mgr
Benji Snead . . . . news dir
Ricky Price . . . . chief of engrg

KJTV-TV— Analog channel: 34. Analog hrs: 24 3,720 kw vis, 372 kw aur. ant 840t/893g TL: N33 30 08 W101 52 20 On air date: Dec 10, 1981. Box 3757, Lubbock, TX, 79452. 9800 University Ave., Lubbock, TX 79452. Phone: (806) 745-3434. Fax: (806) 748-1949. Web Site: www.fox34.com. Licensee: Ramar Communications II Ltd. Group owner: (group owner) Population served: 378,200 Natl. Network: Fox. Natl. Rep: Millennium Sales & Marketing. Washington Atty: Leventhal, Senter & Lerman. News staff: 20; News: 7 hrs wkly.

Key Personnel:

Brad Moran . . . . gen mgr
Scott Cawthron. . . . . opns mgr
Marc Gilmour . . . . gen sls mgr
Jana Hill . . . . prom mgr
Terri Holt . . . . progmg mgr

KLBK-TV— Analog channel: 13.316 kw vis, 25.1 kw aur. 880t/836g TL: N33 31 33 W101 52 07 On air date: Nov 13, 1952. 7403 S. University Ave., Perry, TX, 79423. Phone: (806) 745-2345. Fax: (806) 748-2250. Web Site: www.klbk.com. Licensee: Nexstar Finance Inc. Group owner: Nexstar Broadcasting Group Inc. (acq 12-31-03; grpsl). Population served: 376,000 Natl. Network: CBS. Washington Atty: Arter & Hadden. News staff: 47; News: 27 hrs wkly.

Key Personnel:

Greg McAlister . . . . gen mgr
Chuck Spaugn . . . . opns mgr
A.C. Wimberly . . . . progmg mgr
Russ Poteet . . . . news dir
Mike Randolph . . . . chief of engrg

KLCW-TV— Analog channel: 22.70.8 kw vis. ant 748t TL: N33 30 08 W101 52 20 On air date: 2002. 9800 University Ave., Lubbock, TX, 79423. Phone: (806) 745-3434. Fax: (806) 748-9387. E-mail: bmoran@ramarcom.com Web Site: www.fox34.com. Licensee: Woods Communications Corp. Group owner: (group owner) Natl. Network: CW.

KPTB— Analog channel: 16.214 kw vis. 272t TL: N33 33 12 W101 49 13 On air date: 1999. Box 61000, Midland, TX, 79711. 5604 Martin Luther King Blvd., Lubbock, TX 79404. Phone: (800) 707-0420 (806) 846-5200. Fax: (806) 749-7732. E-mail: info@ptcbglc.com Web Site: www.godslearningchannel.com. Licensee: Prime Time Christian Broadcasting Inc.

***KTXT-TV**— Analog channel: 5. Digital channel: 39. Analog hrs: 24 Digital hrs: 24 100 kw vis, 25 kw aur. 440t/817g TL: N33 34 55 W101 53 25 On air date: Oct 16, 1962. Box 42161, Lubbock, TX, 79409-2161. 17th St. & Indiana Ave., Lubbock, TX 79409-2161. Phone: (806) 742-2209. Fax: (806) 742-1274. E-mail: pat.cates@ttu.edu Web Site: www.ktxt.org. Licensee: Texas Tech University. Population served: 375,000 Natl. Network: PBS. Washington Atty: Cohn & Marks.

Key Personnel:

Eric Voyles . . . . CFO & prom dir
Pat Cates . . . . gen mgr
Michelle Dillard . . . . progmg mgr & progmg
Martin Quintero. . . . . chief of engrg

Lufkin

see Tyler-Longview (Lufkin & Nacogdoches), TX market

McAllen

see Harlingen-Weslaco-Brownsville-McAllen, TX market

Midland

see Odessa-Midland, TX market

Nacogdoches

see Tyler-Longview (Lufkin & Nacogdoches), TX market

Odessa-Midland, TX (DMA 159)

KMID— Analog channel: 2. Analog hrs: 24 100 kw vis, 10 kw aur. 1,050t/1,147g TL: N32 05 14 W102 17 12 On air date: Dec 18, 1953. Box 60230, 3200 Laforce Blvd., Midland, TX, 79711. Phone: (432) 563-2222. Fax: (432) 563-5819. E-mail: news@kmid.tv Web Site: www.kmid.tv. Licensee: Nexstar Finance Inc. Group owner: Nexstar Broadcasting Group Inc. (acq 7-31-2000; $10 million). Population served: 794,000 Natl. Network: ABC. Natl. Rep: Blair Television. Washington Atty: Cohn & Marks. News staff: 23; News: 17 hrs wkly.

KMLM— Analog channel: 42. Digital channel: 43.1,120 kw vis, 112 kw aur. ant 479t/473g TL: N32 02 53 W102 17 44 On air date: Oct 18, 1988. Box 61000, Midland, TX, 79711-1000. 12706 W. Highway 80 E., Odessa, TX 79765. Phone: (800) 707-0420 (432) 563-0420. Fax: (432) 563-1736. E-mail: info@ptcbglc.com Web Site: www.godslearningchannel.com. Licensee: Prime Time Christian Broadcasting Inc. Population served: 245,000 Foreign lang progmg: SP 4

Key Personnel:

Tommie Cooper . . . . gen mgr & stn mgr
Matt Montgomery . . . . chief of engrg

KOSA-TV— Analog channel: 7. Digital channel: 31. Analog hrs: 24 Note: CBS is on KOSA-TV ch 7, MyNetworkTV is on KOSA-DT ch 31. 316 kw vis, 39.8 kw aur. ant 741t/715g TL: N31 51 50 W102 34 41 On air date: Jan 1, 1956. Box 107, 4101 E. 42nd St., J-7, Odessa, TX 79762. Phone: (432) 580-5672. Fax: (432) 580-8010. E-mail: news@cbs7.com Web Site: www.cbs7.com. Licensee: ICA Broadcasting I Ltd. (acq 3-10-00; $8 million). Population served: 350,000 Natl. Network: CBS, MyNetworkTV Digital Network: Note: CBS is on KOSA-TV ch 7, MyNetworkTV is on KOSA-DT ch 31. Natl. Rep: Continental Television Sales. Washington Atty: Richard Hayes. Foreign lang progmg: SP 1 News staff: 22; News: 16 hrs wkly.

Key Personnel:

John Bushman . . . . chmn
Barry Marks . . . . pres & gen mgr
John Nichols . . . . CFO
Dale Palmer . . . . stn mgr
Rick McGee . . . . opns mgr

***KPBT-TV**— Analog channel: 36. Digital channel: 38. Analog hrs: 15 Digital hrs: 15 513 kw vis, 51.3 kw aur. ant 289t/306g TL: N31 51 59 W102 22 50 On air date: Mar 24, 1986. Box 8940, Midland, TX, 79708-8940. 201 West University, Odessa, TX 79764. Phone: (432) 563-5728. Fax: (432) 563-5731. E-mail: kpbt@kpbt.org Web Site: www.kpbt.org. Licensee: Permian Basin Public Telecommunications Inc. (acq 2-2-2006; $1). Population served: 350,000 Natl. Network: PBS.

Key Personnel:

John H. James . . . . chmn
Daphne Dowdy . . . . gen mgr
Domingo Machuca . . . . chief of engrg
Amy Lynch . . . . progmg

KPEJ— Analog channel: 24. Digital channel: 23. Analog hrs: 24 2,880 kw vis. ant 1,099t/1,102g TL: N32 05 51 W102 17 21 On air date: June 16, 1986. Box 11009, 1550 W. I-20, Odessa, TX, 79763. Phone: (432) 580-0024. Fax: (432) 337-3707. E-mail: jfaltus@kpejtv.com Web Site: www.kpejtv.com. Licensee: Comcorp of Texas License Corp. Group owner: Communications Corp. of America (acq 10-31-90; grpsl; FTR: 11-19-90). Population served: 133,600 Natl. Network: Fox. Washington Atty: Fletcher, Heald & Hildreth.

Key Personnel:

Laura Wolf . . . . gen mgr
Jayne Faltus. progmg mgr

KUPB— Analog channel: 18.5,000 kw vis. ant 930t/948g TL: N31 50 19 W102 31 59 (CP: ant 922t/951g) On air date: 2001. Box 61907, Midland, TX, 79711. 10313 West County Road 117, Midland, TX 79706. Phone: (432) 563-1826. Fax: (432) 563-0215. Web Site: www.entravision.com. Licensee: Entravision Holdings LLC. Group owner: Entravision Communications Corp. Natl. Network: Univision (Spanish). Washington Atty: Thompson, Hine & Flory L. Foreign lang progmg: SP 168

Key Personnel:

Walter Ulloa . . . . CEO
Philip C. Wilkinson . . . . pres
John DeLorenzo . . . . CFO
Larry Safir . . . . exec VP
Leticia Martinez . . . . gen mgr

KWAB-TV— Analog channel: 4.12.9 kw vis, 1.5 kw aur. 380t/497g TL: N32 15 14 W101 26 44 On air date: Jan 15, 1956. Box 60150, Midland, TX, 79711. Phone: (432) 567-9999. Fax: (432) 567-9994. Web Site: www.kwes.com. Licensee: Midessa Television Co. (acq 9-9-91; $4.85 million with KWES-TV Odessa; FTR: 9-23-91) Natl. Network: NBC.

KWES-TV— Analog channel: 9. Digital channel: 13.316 kw vis, 45.7 kw aur. ant 1,282t/1,039g TL: N31 59 17 W102 52 41 On air date: Dec 1, 1958. Box 60150, Midland, TX, 79711-0150. 11320 County Rd. 127 W., Midland, TX 79711. Phone: (432) 567-9999. Fax: (432) 567-9992. Web Site: www.kwes.com. Licensee: Midessa Television Co. Group owner: R.H. Drewry Group (acq 10-31-91; $4.85 million with KWAB(TV) Big Spring; FTR: 9-23-91). Natl. Network: NBC.
Key Personnel:
Mac Douglas . . . gen mgr
Carlos Fernandez . . . news dir

KWWT— Analog channel: 30. Analog hrs: 24 350 kw vis. ant 695t/695g TL: N32 05 51 W102 17 21 On air date: 2001. Paxson Communications Corp., 601 Clearwater Park Rd., West Palm Beach, FL, 33401. Phone: (432) 563-5795. Web Site: www.cwtv.com. Licensee: WinStar Odessa Inc. Group owner: Paxson Communications Corp. (acq 9-4-98). Natl. Network: CW.

Port Arthur

see Beaumont-Port Arthur, TX market

San Angelo, TX (DMA 197)

KIDY— Analog channel: 6. Digital channel: 19. Analog hrs: 24 Digital hrs: 24 100 kw vis, 10 kw aur. 946t/1,000g TL: N31 35 21 W100 31 00 On air date: May 12, 1984. 406 S. Irving, San Angelo, TX, 76903. Phone: (325) 655-6006. Fax: (325) 655-8461. E-mail: kidy@foxsanangelo.com Web Site: foxsanangelo.com. Licensee: Sage Broadcasting Corp. Population served: 172,800 Natl. Network: Fox, UPN. Natl. Rep: Millennium Sales & Marketing. Washington Atty: Fletcher, Heeald & Hildreth. News staff: 2.
Key Personnel:
Paris Schindler. . . . CEO
Bill Carter . . . pres & stn mgr
Teddye Read . . . natl sls mgr

KLST— Analog channel: 8. Analog hrs: 24 316 kw vis, 31.6 kw aur. 1,450t/1,500g TL: N31 22 01 W100 02 48 On air date: June 23, 1953. 2800 Armstrong St., San Angelo, TX, 76903-2799. Phone: (325) 949-8800. Fax: (325) 658-1118. E-mail: klst@klst.net Web Site: www.klst.tv. Licensee: Nexstar Broadcasting Inc. (acq 9-2-2004; $12 million). Population served: 100,000 Natl. Network: CBS. Washington Atty: Skadden, Arps, Slate, Meagher & Flom. News staff: 12; News: 17 hrs wkly.
Key Personnel:
Perry Sook . . . pres
Joy Kimbell . . . exec VP
Tom Stovall . . . gen mgr
Mark McCain . . . opns mgr
Lanny Kiest . . . gen sls mgr
Don Plachno . . . prom mgr
Gordon Hay . . . progmg dir
Kathy Munoz . . . news dir
Roland Bigley . . . chief of engrg

KSAN-TV— Analog channel: 3.17.8 kw vis, 3.5 kw aur. 600t/469g TL: N31 37 22 W100 26 14 On air date: Feb 8, 1962. 2800 Armstrong St., San Angelo, TX, 76903. Phone: (325) 949-8800. Fax: (325) 655-1118. E-mail: nbc3@wcc.com Web Site: www.ksan.tv. Licensee: Mission Broadcasting Inc. Group owner: (group owner; (acq 6-13-2003; $10 million with KRBC-TV Abilene). Natl. Network: NBC. Natl. Rep: Blair Television. Washington Atty: Kenkel, Barnard & Edmundson.
Key Personnel:
Tom Stovall . . . stn mgr
Albert Gutierrez . . . rgnl sls mgr
Kathy Munoz . . . news dir
Len Martinez . . . chief of engrg

San Antonio, TX (DMA 37)

KABB— Analog channel: 29. Analog hrs: 24 5,000 kw vis, 500 kw aur. 1,503t/1,503g TL: N29 17 27 W98 16 12 On air date: Dec 17, 1987. 4335 N.W. Loop 410, San Antonio, TX, 78229-5168. Phone: (210) 366-1129. Fax: (210) 377-4758. E-mail: kabbtv@kabb.com Web Site: www.kabb.com. Licensee: KABB Licensee L.L.C. Group owner: Sinclair Broadcast Group Inc. Population served: 1,500,000 Natl. Network: Fox. Natl. Rep: Millennium Sales & Marketing. Washington Atty: Shaw, Pittman. News staff: 35; News: 7 hrs wkly.
Key Personnel:
Dean Radla . . . sls dir
Robert Canales. . . . natl sls mgr

KCWX— Analog channel: 2.100 kw vis, 10 kw aur. 1,355t/1,017g TL: N30 08 13 W98 36 35 5400 Fredericksburg Rd., San Antonio, TX, 78229. Phone: (210) 366-5000. Fax: (210) 348-9142. Fax: (210) 377-8779. Web Site: www.mysanantonio.com. Licensee: Corridor Television L.L.P. Natl. Network: CW.
Key Personnel:
Robert G. McGann. . . . gen mgr
Boots Walker . . . sls dir
Frank Peterman . . . engrg dir

KENS-TV— Analog channel: 5. Digital channel: 55.100 kw vis, 10 kw aur. ant 1,390t/1,531g TL: N29 16 10 W98 15 55 On air date: Feb 15, 1950. Box TV5, San Antonio, TX, 78299. 5400 Fredericksburg Rd., San Antonio, TX 78229. Phone: (210) 366-5000. Fax: (210) 377-0740. Web Site: www.mysanantonio.com. Licensee: KENS-TV Inc. Group owner: Belo Corp., Broadcast Division (acq 1997; $75 million with co-located AM plus interest in Television Food Network). Population served: 700,000 Natl. Network: CBS. Natl. Rep: TeleRep. Washington Atty: Wiley, Rein & Fielding. Wire Svc: NWS (National Weather Service) News staff: 55; News: 24 hrs wkly.
Key Personnel:
Bob McGann . . . gen mgr
Boots Walker . . . sls dir
Allen Lansing . . . prom dir
Kurt Davis . . . news dir
Frank Peterman . . . engrg dir

***KHCE-TV**— Analog channel: 23. Digital channel: 16. Analog hrs: 16 1,480 kw vis, 148 kw aur. ant 856t TL: N29 31 25 W98 43 25 On air date: July 1989. 15533 Capital Port Dr., San Antonio, TX, 78249. Box 691246, San Antonio, TX 78249. Phone: (210) 479-0123. Fax: (210) 492-5679. Web Site: www.khce.org. Licensee: San Antonio Community Educational TV Inc. Ownership: Dr. Reginald Cherry, 20%; Dr. Paul F. Crouch Sr., 20%; Richard Clayton Trotter, 20%; Cynthia Diaz, 20%; and Suzanna Shuler Harkley, 20% (acq 9-4-97; $3.125 million gift). Population served: 1,000,000
Key Personnel:
Paul Crouch . . . pres
Dr. Cherry . . . VP
Laura Hanks . . . gen mgr
Dorcas Rogers . . . stn mgr
Jessica Mathews . . . progmg dir
Sharon Denney . . . pub affrs dir
Mike Bundrant . . . engrg VP

***KLRN**—Analog channel: 9.302 kw vis, 30.2 kw aur. 960t/1,051g TL: N29 19 33 W98 21 25 On air date: Sept 10, 1962. 501 Broadway, San Antonio, TX, 78215-1820. Phone: (210) 270-9000. Fax: (210) 270-9078. E-mail: info@klrn.org Web Site: www.klrn.org. Licensee: Alamo Public Telecommunications Council. (acq 8-11-89). Population served: 3,000,000 Natl. Network: PBS. Washington Atty: Cohn & Marks.
Key Personnel:
Mike Novak. . . . chmn
Charles Vaughn . . . sr VP
Patrick Lopez . . . sr VP
Joanne Winik . . . gen mgr
Cynthia Shields . . . dev VP

KMYS— Analog channel: 35. Digital channel: 32. Analog hrs: 24 5,000 kw vis, 500 kw aur. ant 1,758t TL: N29 36 37 W98 53 35 On air date: Nov 6, 1985. 4335 N.W. Loop 410, San Antonio, TX, 78229-5168. Phone: (210) 366-1129. Fax: (210) 377-4758. E-mail: kmys@kmys.tv Web Site: www.kmys.tv. Licensee: San Antonio (KRRT-TV) Licensee Inc. Group owner: Sinclair Broadcast Group Inc. (acq 12-10-01; grpsl). Population served: 1,500,000 Natl. Network: MyNetworkTV. Washington Atty: Shaw, Pittman.
Key Personnel:
Dean Radla . . . sls dir
Gwen Miller . . . natl sls mgr
Yvette Reyna . . . prom dir

KPXL— Analog channel: 26. Analog hrs: 24 5,000 kw vis. 1,837t TL: N29 37 11 W99 02 55 On air date: Feb 19, 1999. 6100 Bandera Rd., Suite 304, San Antonio, TX, 78238. Phone: (210) 682-2626. Fax: (210) 682-3155. Web Site: www.ionline.tv. Licensee: Paxson San Antonio License Inc. Group owner: Paxson Communications Corp. (acq 6-24-99; $5 million for remaining 51%). Natl. Network: i Network.

KSAT-TV— Analog channel: 12. Analog hrs: 22 316 kw vis, 63.2 kw aur. 1,483t/1,505g TL: N29 16 11 W98 15 31 On air date: Jan 21, 1957. 1408 N. St. Mary's St., San Antonio, TX, 78215. Phone: (210) 351-1200. Fax: (210) 351-1310. Web Site: www.ksat.com. Licensee: Post-Newsweek Stations Inc. Group owner: (group owner; (acq 2-28-94; FTR: 5-2-94). Population served: 1,000,000 Natl. Network: ABC. Natl. Rep: MMT. Washington Atty: Covington & Burling. News: 21 hrs wkly.

KTRG— Analog channel: 10. Analog hrs: 24 316 kw vis. ant 328t/285g TL: N29 20 39 W100 51 39 On air date: September 1996. Box 530391, Harlingen, TX, 78553. Phone: (956) 421-2635. Fax: (956) 428-7556. Licensee: Ortiz Broadcasting Corp. Ownership: Aracelis Ortiz, executrix of the estate of Carlos Ortiz. (acq 7-24-97; $160,000). Population served: 500,000 Foreign lang progmg: SP 24

KVAW— Analog channel: 16. Digital channel: 18.12.6 kw vis, 1.26 kw aur. ant 279t TL: N28 43 32 W100 28 35 On air date: 1991. 2524 Veterans Blvd., Eagle Pass, TX, 78852. Phone: (830) 773-3668. Phone: (830) 752-0312. Fax: (830) 773-3668. Web Site: www.kvaw16.com. Licensee: Dr. Joseph A. Zavaletta. Ownership: Dr. Joseph A. Zavaletta, 100% (acq 8-23-2004; $300,000). Population served: 150,000

KVDA— Analog channel: 60. Analog hrs: 5:30 AM-2:05 PM (M-F); 5:30 AM-2 PM (S, Su) 5,000 kw vis, 500 kw aur. 1,495t/1,025g TL: N29 29 87 W98 29 53 On air date: Sept 10, 1989. 6234 San Pedro, San Antonio, TX, 78216. Phone: (210) 340-8860/8661. Fax: (210) 341-3962/(210)341-2051(news). Web Site: www.kvda.com. Licensee: NBC Telemundo License Co. Group owner: Telemundo Group Inc. (acq 4-12-2002; grpsl). Population served: 817,000 Natl. Network: Telemundo (Spanish). Washington Atty: Hogan & Hartson. Wire Svc: Reuters Foreign lang progmg: SP 112 News staff: 11; News: 10 hrs wkly.
Key Personnel:
Arturo Fux . . . gen sls mgr
Maricela Arce. . . . prom dir
Maricela Arce . . . pub affrs dir

KWEX-TV— Analog channel: 41. Digital channel: 39. Analog hrs: 24 832 kw vis, 83.2 kw aur. ant 500t/604g TL: N29 25 03 W98 29 26 (CP: 5,000 kw vis, 500 kw aur) On air date: June 10, 1955. 411 E. Durango Blvd., San Antonio, TX, 78204. Phone: (210) 227-4141/(210) 242-7451 (news). Fax: (210) 227-0469/(210) 226-0131 (news). Web Site: www.univision.com. Licensee: KWEX L.P., G.P. Group owner: Univision Communications Inc. (acq 7-86; grpsl). Population served: 2,478,680 Natl. Network: Univision (Spanish). Washington Atty: Fisher, Wayland, Cooper, Leader & Zaragoza. Foreign lang progmg: SP 168 News: 5 hrs wkly.

WOAI-TV— Analog channel: 4. Digital channel: 58. Analog hrs: 24 100 kw vis, 18 kw aur. 1,476t/1,531g TL: N29 16 10 W98 15 55 On air date: Dec 11, 1949. Box 2641, San Antonio, TX, 78299. 1031 Navarro St., San Antonio, TX 78205. Phone: (210) 226-4444. Fax: (210) 224-9898. Web Site: www.woai.com. Licensee: CCB Texas Licenses L.P. Group owner: Clear Channel Communications Inc. (acq 9-21-2001; grpsl). Population served: 960,500 Natl. Network: NBC. Washington Atty: Wilmer, Cutler & Pickering. News staff: 68; News: 19.5 hrs wkly.
Key Personnel:
Donita Todd . . . VP & gen mgr
Greg Derkowski . . . prom mgr & adv
Carolyn Mastin. . . . progmg mgr
Mark Pipitone. . . . news dir
Liz Quinones . . . pub affrs dir
Harold Friesenhahn . . . chief of engrg

Sherman, TX-Ada, OK (DMA 161)

KTEN— Analog channel: 10. Digital channel: 26. Analog hrs: 24 Digital hrs: 24 Note: NBC is on KTEN(TV) ch 10, CW is on KTEN-DT ch 26. 316 kw vis, 47.5 kw aur. 1,458t/1,500g TL: N34 21 34 W96 33 34 On air date: June 1, 1954. 10 High Point Cir., Denison, TX, 75020. Phone: (903) 337-4000. Fax: (908) 465-1207. Fax: (903) 465-1368. E-mail: 10news@kten.com Web Site: www.kten.com. Licensee: Channel 49 Acquisition Corp. Ownership: Lockwood Corp. Population served: 238,000 Natl. Network: NBC, CW Digital Network: Note: NBC is on KTEN(TV) ch 10, CW is on KTEN-DT ch 26. Natl. Rep: Continental Television Sales. Washington Atty: Brooks, Pierce, McLendon, Humprey & Leonard. Wire Svc: AP News staff: 25; News: 20 hrs wkly.
Key Personnel:
Asa Jessee . . . gen mgr
Ken Braswell . . . gen sls mgr
Steve Korioth. . . . news dir
Kris Anderson . . . chief of engrg

KXII— Analog channel: 12. Digital channel: 20. Analog hrs: 24 hrs Digital hrs: 24 hrs. Note: CBS is on KXII(TV) ch 12, Fox and MyNetworkTV are on KXII-DT ch 20. 224 kw vis, 22.4 kw aur. ant 1,781t/1,698g TL: N34 01 58 W96 48 00 On air date: July 1956. Box 1175, 4201 Texoma Pkwy., Sherman, TX, 75091-1175. Phone: (903) 892-8123. Fax: (903) 893-7858. E-mail: comments @kxii.com Web Site: www.kxii.com. Licensee: Gray Television Licensee Inc. Group owner: Gray Television Inc. (acq 6-29-99; $41.5 million). Population served: 301,000 Natl. Network: CBS, Fox, MyNetworkTV Digital Network: Note: CBS is on KXII(TV) ch 12, Fox and MyNetworkTV are on KXII-DT ch 20. Natl. Rep: Millennium Sales & Marketing. Washington Atty: Womble, Carlyle, Sandridge & Rice, PLLC. News: 30 hrs wkly.
Key Personnel:
Rick Dean. . . . VP, gen mgr & progmg dir
Todd Bates . . . sls dir, gen sls mgr & rgnl sls mgr
Rachel Shockey. . . . prom dir

Matt Brown. . . . . news dir
Randy Wells . . . . . chief of engrg

Sweetwater

see Abilene-Sweetwater, TX market

Temple

see Waco-Temple-Bryan, TX market

Texarkana

see Shreveport, LA market

Tyler-Longview (Lufkin & Nacogdoches), TX

(DMA 111)

KCEB— Analog channel: 54.5,000 kw vis. ant 829t/716g TL: N32 35 36 W94 49 10 On air date: July 20, 2003. 701 N. Access Rd., Longview, TX, 75602. Phone: (903) 236-0051. Fax: (903) 753-6637. Licensee: Estes Broadcasting Inc. Ownership: Dimension Enterprises Ltd. (acq 11-24-2003). Natl. Network: CW. Washington Atty: Fletcher, Heald & Hildreth.
Key Personnel:
Tony Cruz . . . . . gen mgr
Tyrene Carl . . . . . gen sls mgr

KETK-TV— Analog channel: 56. Digital channel: 22. Analog hrs: 24 hrs 5,000 kw vis, 500 kw aur. ant 1,583t/1,437g TL: N32 03 40 W95 18 50 On air date: March 1987. 4300 Richmond Rd., Tyler, TX, 75703. Phone: (903) 581-5656. Fax: (903) 561-1648. Web Site: www.ketknbc.com. Licensee: Comcorp of Tyler License Corp. Group owner: Communications Corp. of America (acq 11-12-2004; $38 million). Natl. Network: NBC. Washington Atty: Fletcher, Heald & Hildreth. News staff: 37; News: 26.5 hrs wkly.
Key Personnel:
Mike DeLier . . . . . gen mgr
David Lott. . . . . opns mgr
Eric Jontra . . . . . sls dir
Bonnie Davis . . . . . prom dir
Yolanda Clater . . . . . progmg dir
Neal Barton . . . . . news dir
Yyes Masset. . . . . chief of engrg

KFXK— Analog channel: 51. Digital channel: 31. Analog hrs: 20 4,680 kw vis, 36.70 kw aur. ant 1,249t/1,199g TL: N32 15 35 W94 57 02 On air date: Sept 9, 1984. 701 N. Access Rd., Longview, TX, 75602. Phone: (903) 236-0051. Fax: (903) 753-6637. Web Site: www.fox51.com. Licensee: Warwick Communications Inc. Group owner: White Knight Holdings Inc. (acq 9-1-99; $11.5 million for stock plus 3 low-power stns). Natl. Network: Fox.
Key Personnel:
Sheldon Galloway. . . . . pres
Bob Hall. . . . . gen mgr
Tony Cruz . . . . . sls dir & natl sls mgr
Robert Dodd . . . . . prom dir
Mike Stallcup . . . . . progmg mgr
Chuck McDonald . . . . . news dir

KLTV— Analog channel: 7. Digital channel: 10. Analog hrs: 24 Digital hrs: 24 316 kw vis, 31.6 kw aur. 991t/1,079g TL: N32 32 21 W95 13 16 On air date: Oct 15, 1954. Box 957, 105 W. Ferguson, Tyler, TX, 75702. Phone: (903) 597-5588. Fax: (903) 510-7847. Web Site: www.kltv.com. Licensee: Civco Inc. Group owner: Liberty Corp. (acq 1-13-2006; grpsl). Population served: 82,000 Natl. Network: ABC. Washington Atty: Covington & Burling.
Key Personnel:
Brad Streit . . . . . gen mgr
Mary Ryan . . . . . opns dir & natl sls mgr
Pat Stacey . . . . . gen sls mgr
Mark Scirto . . . . . rgnl sls mgr
Cathy Carmichael . . . . . progmg dir
Kenny Boles . . . . . news dir
Butch Adair . . . . . chief of engrg

KTRE— Analog channel: 9. Analog hrs: 24 Digital hrs: 24 158 kw vis, 31.7 kw aur. 670t TL: N31 25 09 W94 48 02 On air date: Aug 31, 1955. Box 729, Lufkin, TX, 75902-0792. 358 TV Rd., Pollok, TX 75969. Phone: (936) 853-5873. Fax: (936) 853-3084. E-mail: dlorenz@ktre.com Web Site: www.ktre.com. Licensee: Raycom Media, Inc. Group owner: Liberty Corp. (acq 1-13-2006; grpsl). Population served: 260,000 Natl. Network: ABC. Washington Atty: Covington & Burling. Foreign lang progmg: SP 1 News staff: 15; News: 20 hrs wkly.
Key Personnel:
Paul H. McTear Jr. . . . . . CEO
Melissa Thurber . . . . . chmn & CFO
Paul H. McTear. . . . . pres
Wayne Dougherty . . . . . exec VP
Artie Bedard . . . . . sr VP & VP

KYTX— Analog channel: 19. Digital channel: 18. Analog hrs: 24 Digital hrs: 24 Note: CBS is on KYTX(TV) ch 19, MyNetworkTV is on KYTX-DT ch 18. 4,270 kw vis. ant 1,499t/1,545g TL: N31 54 20 W95 05 05 On air date: Sept 1, 1991. 2211 ESE Loop 323, Tyler, TX, 75701. 320 E. Methvin St., Longview, TX 75601. Phone: (903) 581-2211. Fax: (903) 581-5769. Web Site: www.cbs19.tv. Licensee: MMT License LLC. Group owner: MAX Media L.L.C. (acq 4-12-2004; $4 million). Natl. Network: CBS, MyNetworkTV Digital Network: Note: CBS is on KYTX(TV) ch 19, MyNetworkTV is on KYTX-DT ch 18. Natl. Rep: Blair Television. Washington Atty: William Mullen. Wire Svc: AP News staff: 25; News: 13.5 hrs wkly.
Key Personnel:
Philip H. Hurley . . . . . gen mgr
John Gaston. . . . . gen sls mgr
Myra O'Neal . . . . . natl sls mgr
Brandon Baker. . . . . prom mgr
Margore Strout. . . . . progmg mgr
Bob Lenertz . . . . . news dir
Moe Strout . . . . . chief of engrg

Victoria, TX

(DMA 205)

KAVU-TV— Analog channel: 25. Digital channel: 15. Analog hrs: 24 Digital hrs: 5 AM-1 AM 2140 kw vis, 2.14 kw aur. 1,020t/1,067g TL: N28 48 06 W96 33 09 On air date: July 4, 1982. 3808 N. Navarro, Victoria, TX, 77901. Phone: (361) 575-2500. Fax: (361) 575-2255. Web Site: www.myvictoriaonline.com. Licensee: Saga Broadcasting LLC. Group owner: Saga Communications Inc. (acq 10-20-98; $11.875 million; with KNAL(AM) Victoria). Population served: 100,000 Natl. Network: ABC. Washington Atty: Smithwick & Belendiuk, P.C. Wire Svc: AP News staff: 13; News: 8 hrs wkly.
Key Personnel:
Jeff Pryor . . . . . gen mgr
John Garcia . . . . . opns mgr
Richard Morton. . . . . gen sls mgr
Rebecca Sarlls . . . . . progmg dir
Doug Tisdale . . . . . news dir
Sean McBride. . . . . prom mgr & pub affrs dir
Kevin John . . . . . engrg mgr

KVCT— Analog channel: 19. Analog hrs: 24 155 kw vis, 15.5 kw aur. ant 489t/494g TL: N28 46 41 W96 57 38 On air date: Nov 21, 1969. Box 4929, 3808 N. Navarro St., Victoria, TX, 77901. Phone: (361) 575-2500. Fax: (361) 575-2255. Web Site: www.myvictoriaonline.com. Licensee: Surtsey Media LLC. Group owner: (group owner; (acq 4-26-99). Population served: 247,500 Natl. Network: Fox.
Key Personnel:
Jeff Pryor . . . . . gen mgr
John Garcia . . . . . opns mgr
Darren Lehrmann. . . . . gen sls mgr
Kristin Thompson . . . . . prom dir
Rebecca Sarlls . . . . . progmg dir
Doug Tisdale . . . . . news dir
Kathy Keith . . . . . pub affrs dir
Mandy Gracia . . . . . pub affrs dir
Brian Weber . . . . . chief of engrg

Waco-Temple-Bryan, TX

(DMA 95)

KAKW-TV— Analog channel: 62.1,660 kw vis. ant 1,814t/1,791g TL: N30 43 33 W97 59 24 On air date: 1996. 2233 W. Northloop Blvd., Austin, TX, 78731. Phone: (512) 453-8899. Fax: (512) 533-2874. Licensee: Univision Communications Inc. Group owner: (group owner; (acq 12-10-2001; $12 million). Natl. Network: Univision (Spanish). Foreign lang progmg: SP 168

***KAMU-TV—** Analog channel: 15.22.9 kw vis, 2.29 kw aur. ant 390t/379g TL: N30 37 48 W96 20 33 On air date: Feb 15, 1970. Texas A&M Univ., College Station, TX, 77843-4244. Phone: (979) 845-5611. Fax: (979) 845-1643. Web Site: www.kamu.tamu.edu. Licensee: Texas A&M University. Population served: 151,000 Natl. Network: PBS.
Key Personnel:
Rodney Zent . . . . . gen mgr
Jon Bennett . . . . . stn mgr & progmg dir
John Prihoda . . . . . opns mgr
Elaine Hoyak . . . . . dev dir
Wayne Pecena . . . . . engrg dir
Ken Nelson . . . . . chief of engrg

KBTX-TV— Analog channel: 3. Digital channel: 50. Analog hrs: 24 Digital hrs: 24 Note: CBS is on KBTX-TV ch 3, KBTX-DT on ch 50 is a CW affiliate. 69.2 kw vis, 6.9 kw aur. ant 1,689t/1,544g TL: N30 33 10 W96 01 50 On air date: May 22, 1957. Box 3730, Bryan, TX, 77805-3730. 4141 E. 29th St., Bryan, TX 77802-4305. Phone: (979) 846-7777. Fax: (979) 846-1490 (sls). Fax: (979) 846-1888 (news). Web Site: www.kbtx.com. Licensee: Gray Television Licensee Inc. Group owner: Gray Television Inc. (acq 6-29-99; $97.5 million cash and shares with KWTX-TV Waco). Population served: 350,000 Natl. Network: CBS, CW Digital Network: Note: CBS is on KBTX-TV ch 3, KBTX-DT on ch 50 is a CW affiliate. Natl. Rep: Millennium Sales & Marketing. Wire Svc: AP News staff: 30; News: 19 hrs. wkly.
Key Personnel:
Jon Boaz . . . . . gen sls mgr
Mike Wright. . . . . VP, gen mgr & natl sls mgr
Mike Barger . . . . . news dir
Ayrrie Dixon. . . . . prom

KCEN-TV— Analog channel: 6. Digital channel: 9. Analog hrs: 24 Digital hrs: 24 100 kw vis, 10 kw aur. ant 830t/833g TL: N31 16 24 W97 13 14 On air date: Nov 1, 1953. 111 West Central Ave., Temple, TX, 76501. Phone: (254) 859-5481. Fax: (254) 859-4004. E-mail: news@kcentv.com Web Site: www.kcentv.com. Licensee: Channel 6 Inc. Ownership: Anyse Sue Mayborn, 100%. Population served: 315,700 tvhh Natl. Network: NBC. Natl. Rep: Blair Television. Washington Atty: Baker & Hostetler. Wire Svc: AP News staff: 33; News: 17 hrs wkly.
Key Personnel:
Anyse Sue Mayborn. . . . . pres
W. Randy Odil . . . . . VP & gen mgr

***KNCT—** Analog channel: 46. Analog hrs: 19 479 kw vis, 67.6 kw aur. 1,261t/1,126g TL: N30 59 12 W97 37 47 On air date: Nov 23, 1970. Box 1800, Killeen, TX, 76540-9990. Telecommunications Bldg., 6200 W. Centex Expwy., Killeen, TX 76540-9990. Phone: (254) 526-1176. Fax: (254) 526-1850. E-mail: knct@knct.org Web Site: www.knct.org. Licensee: Central Texas College. Natl. Network: PBS. News staff: one; News: one hr wkly.
Key Personnel:
Max Rudolph . . . . . gen mgr
Fred McNeilly . . . . . prom mgr
Ruth Wedergren . . . . . progmg dir & film buyer
Steve Sulzer . . . . . engrg mgr

***KWBU-TV—** Analog channel: 34. Analog hrs: 24 79.4 kw vis, 7.9 kw aur. 508t/446g TL: N31 30 31 W97 10 03 On air date: May 22, 1989. One Bear Pl., # 97296, Waco, TX, 76798-7296. Phone: (254) 710-3472. Phone: (254) 710-7888. Fax: (254) 710-3874. E-mail: clare@kwbu.org Web Site: www.kwbu.org. Licensee: Brazos Valley Public Broadcasting Foundation. Ownership: Baylor University. (acq 12-6-93; $80,000; FTR: 12-20-93). Population served: 100000 Natl. Network: PBS. Washington Atty: Cohn & Marks.
Key Personnel:
Polly Anderson . . . . . CEO & gen mgr
Larry Brumley. . . . . chmn & pres
Nab Holmes . . . . . dev dir
Clare Paul . . . . . prom mgr & progmg mgr
Michael Hagerty . . . . . news dir & pub affrs dir
Tony Poole. . . . . engrg dir & chief of engrg

KWKT— Analog channel: 44. Digital channel: 57. Analog hrs: 24 4,170 kw vis, 417 kw aur. ant 1,811t/1,673g TL: N31 18 52 W97 19 37 (CP: ant 1,829t/1,672g. TL: N31 18 53.4 W97 19 36) On air date: March 1988. Box 2544, Waco, TX, 76702-2544. 8803 Woodway Dr., Waco, TX 76712. Phone: (254) 776-3844. Fax: (254) 388-5958. E-mail: info@kwkt.com Web Site: kwkt.com. Licensee: Comcorp of Texas License Corp. Group owner: Communications Corp. of America (acq 10-31-90; grpsl; FTR: 11-19-90). Natl. Network: Fox, MyNetworkTV. Natl. Rep: Millennium Sales & Marketing. Washington Atty: Fletcher, Heald & Hildreth.
Key Personnel:
Duane Sartor. gen mgr, stn mgr, opns mgr, mktg mgr & progmg dir
Bill Knobler . . . . . gen sls mgr
Robin Rice . . . . . natl sls mgr
Amy Bishop. . . . . prom mgr & pub affrs dir
Lou Strowger. . . . . chief of engrg

KWTX-TV— Analog channel: 10. Digital channel: 53. Analog hrs: 24 Note: CBS is on KWTX-TV ch 10, CW is on KWTX-DT ch 53. 209 kw vis, 21 kw aur. ant 1,820t/1,679g TL: N31 19 19 W97 18 58 On air date: April 1955. Box 2636, Waco, TX, 76702-2636. 6700 American Plaza, Waco, TX 76712. Phone: (254) 776-1330 (254) 776-3242. Fax: (254) 751-1088 (sales). Fax: (254) 776-4010 (news). E-mail: mail@kwtx.com Web Site: www.kwtx.com. Licensee: Gray Television Licensee Inc. Group owner: Gray Television Inc. (acq 6-29-99; $97.5 million cash and shares with KBTX-TV Bryan). Population served: 763,000 Natl. Network: CBS, CW Digital Network: Note: CBS is on KWTX-TV ch 10, CW is on KWTX-DT ch 53. Natl. Rep: Millennium Sales & Marketing. Washington Atty: Venable Attorneys at Law. News staff: 36; News: 20 hrs wkly.

Key Personnel:
Rich Adams . . . pres
Jason Effinger . . . gen mgr
Ken Musgrave . . . opns mgr
Bob Bunch . . . gen sls mgr

KXXV— Analog channel: 25. Digital channel: 26. Analog hrs: 24/7 Digital hrs: 24/7 5,000 kw vis, 500 kw aur. ant 1,841t/1,706g TL: N31 20 16 W97 18 36 On air date: Jan 1, 1985. Box 2522, Waco, TX, 76702. 1909 S. New Rd., Waco, TX 76711. Phone: (254) 754-2525. Fax: (254) 752-1002. E-mail: news25@kxxv.com Web Site: www.kxxv.com. Licensee: Centex Television L.P. Group owner: R.H. Drewry Group (acq 1994). Natl. Network: ABC, Telemundo (Spanish). News staff: 18; News: 14.5 hrs wkly.
Key Personnel:
Mike Lee . . . VP & gen mgr
Jeff Armstrong . . . gen sls mgr
Darlene Mahler . . . natl sls mgr & rgnl sls mgr
Dennis Kinney . . . news dir
Randy Lee . . . chief of engrg
ABC, Telemundo, Weather Now(Local Weather).

KYLE— Analog channel: 28. Digital channel: 29.3,980 kw vis, 500 kw aur. ant 581t TL: N30 41 18 W96 25 35 On air date: Oct 31, 1994. 2402 Broadmoor Dr., Suite B-101, Bryan, TX, 77805. Phone: (979) 774-1800. Fax: (979) 774-1901. Web Site: www.kyle28.com. Licensee: Comcorp of Bryan License Corp. Group owner: Communications Corp. of America (acq 1996; $1.1 million). Population served: 200,000 Natl. Network: Fox, MyNetworkTV. Washington Atty: Gardner, Carton & Douglas.
Key Personnel:
Duane Sartor . . . stn mgr & gen sls mgr
Doug Williams . . . gen sls mgr
Lou Strauger . . . chief of engrg
Satellite of KWKT(TV) Waco.

Weslaco

see Harlingen-Weslaco-Brownsville-McAllen, TX market

Wichita Falls, TX & Lawton, OK (DMA 146)

KAUZ-TV— Analog channel: 6. Digital channel: 22.100 kw vis, 20 kw aur. ant 1,021t/1,028g TL: N33 54 04 W98 32 21 On air date: Mar 1, 1953. Box 2130, Wichita Falls, TX, 76307. Phone: (940) 322-6957. Fax: (940) 761-3331. Fax: TWX: 910-890-5836. E-mail: email@kauz.com Web Site: www.kauz.com. Licensee: Hoak Media of Wichita Falls L.P. Group owner: Hoak Media Corporation (acq 11-5-2003; $8.2 million). Population served: 385,000 Natl. Network: CBS, CW. Natl. Rep: Harrington, Righter & Parsons. Washington Atty: Covington & Burling. Wire Svc: CBS News staff: 20; News: 15 hrs wkly.
Key Personnel:
Kyle Williams . . . gen mgr & natl sls mgr
Gary Lucus . . . opns mgr
Randy Blake . . . gen sls mgr
Mark Walker . . . rgnl sls mgr
Jackie McCartney . . . prom mgr
Tanya Graham . . . progmg dir
Drew Hadwall . . . news dir
Tony Guess . . . chief of engrg

KFDX-TV— Analog channel: 3. Digital channel: 28. Analog hrs: 24 100 kw vis, 20 kw aur. ant 1,000t/1,045g TL: N33 53 23 W98 33 20 On air date: Apr 12, 1953. 4500 Seymour Hwy., Wichita Falls, TX, 76309. Box 4888, Wichita Falls, TX 76309. Phone: (940) 691-0003. Fax: (940) 691-0330. E-mail: kfdx@kfdx.com Web Site: www.kfdx.com. Licensee: Nexstar Broadcasting Inc. Group owner: Nexstar Broadcasting Group Inc. (acq 11-6-97; grpsl). Population served: 162,800 Natl. Network: NBC. Washington Atty: Arter & Hadden. News staff: 24; News: 19.5 hrs wkly.
Key Personnel:
Julie Pruett . . . gen mgr
Greg Collier . . . opns mgr
Stephanie Darland . . . rgnl sls mgr
Troy Short . . . VP & prom mgr
Terry Porter . . . chief of engrg

KJTL— Analog channel: 18. Analog hrs: 20 2,820 kw vis, 282 kw aur. 1,079t/1,000g TL: N34 12 06 W98 43 44 On air date: May 18, 1985. Box 4888, 4500 Seymour Hwy., Wichita Falls, TX, 76309. Phone: (940) 691-1808. Fax: (940) 691-4856. E-mail: kjtl@fox18.com Web Site: www.texomashomepage.com. Licensee: Mission Broadcasting of Wichita Falls License Inc. Group owner: Mission Broadcasting Inc. (acq 1999; $28.5 million with KCIT(TV) Amarillo). Natl. Network: Fox. Washington Atty: Spector & Goldberg.

KSWO-TV— Analog channel: 7. Digital channel: 11. Analog hrs: 24 Digital hrs: 24 316 kw vis, 63.1 kw aur. 1,050t/1,059g TL: N34 12 55 W98 43 13 On air date: Mar 8, 1953. Box 708, Hwy. 7, Lawton, OK, 73502. Phone: (580) 355-7000. Fax: (580) 357-3811. Web Site: www.kswo.com. Licensee: KSWO TV Inc. Group owner: R.H. Drewry Group Population served: 300,000 Natl. Network: ABC.
Key Personnel:
Larry Patton . . . gen mgr
Joe Bartnik . . . chief of engrg

Utah

Salt Lake City, UT (DMA 35)

KBCJ— Analog channel: 6.2 kw vis. ant 1,827t/131g TL: N40 31 15 W109 42 25 Not on air, target date: unknown: Equity Broadcasting Corp., 1 Shackleford Dr., Suite 400, Little Rock, AR, 72211. Phone: (501) 219-2400. Fax: (501) 716-3502. Permittee: Vernal Broadcasting Inc. Group owner: Equity Broadcasting Corp. (acq 5-11-2001).

KBEO— Analog channel: 11. Analog hrs: 24 316 kw vis. ant 1,036t TL: N43 27 40 W110 45 09 On air date: 2002. KM Communications Inc., 3654 W. Jarvis Ave., Skokie, IL, 60076. Phone: (847) 674-0864. Fax: (847) 674-9188. Web Site: www.kbeo.net. Licensee: Pocatello Channel 15 LLC. Ownership: Myoung Hwa Bae, 100%.

KBNY— Analog channel: 6.100 kw vis. ant 887t TL: N39 15 53 W114 53 35 Not on air, target date: unknown: Equity Broadcasting Corp., 1 Shackelford Dr., Suite 400, Little Rock, AR, 72211. Phone: (501) 219-2400. Fax: (501) 604-8404. Licensee: Nevada Channel 6 Inc. Population served: 498,000.

***KBYU-TV**— Analog channel: 11. Analog hrs: 24 162 kw vis, 35 kw aur. 2,941t/100g TL: N40 36 28 W112 09 33 On air date: Nov 15, 1965. 2000 Ironton Blvd., Provo, UT, 84606. Phone: (801) 422-8450. Fax: (801) 422-8478. E-mail: kbyu@byu.edu Web Site: www.kbyu.org. Licensee: Brigham Young University. Population served: 629,850 Natl. Network: PBS. Washington Atty: Wilkinson, Barker, Knauer & Quinn. Foreign lang progmg: SP 3 News staff: 3; News: 3 hrs wkly.
Key Personnel:
Derek Marquis . . . gen mgr
Jim Bell . . . mktg mgr
Randy Rawe . . . prom mgr
Wendy Thomas . . . progmg mgr
Wesley Sims . . . news dir
Brian Leifson . . . chief of engrg

KCBU— Analog channel: 3. Digital channel: 3.70 kw vis. ant 2,158t/167g TL: N39 45 22 W110 59 22 On air date: 2003. Price Broadcasting Inc., 1 Shackleford Dr., Suite 400, Little Rock, AR, 72211. Phone: (501) 219-2400. Fax: (501) 716-3502. Licensee: Price Broadcasting Inc. (acq 9-6-2002). Natl. Network: TeleFutura (Spanish).

KCSG— Analog channel: 4. Digital channel: 14. Analog hrs: 24 984 w vis. ant 1,044t/44g TL: N37 38 22 W113 02 00 On air date: September 1985. 845 E. Red Hills Pkwy, St. George, UT, 84770. Phone: (435) 986-9715. Fax: (435) 986-9716. E-mail: info@kcsg.com Web Site: www.kcsg.com. Licensee: Southwest Media LLC. Ownership: Broadcast West LLC, 100% (acq 7-30-2002; $450,000). Washington Atty: Garvey, Schubert & Barer.
Key Personnel:
Ben Spencer . . . gen mgr
Abigail Porter . . . stn mgr
Tamara Lee . . . news dir

KENV— Analog channel: 10. Analog hrs: 24 3.09 kw vis. 1,850t TL: N40 41 52 On air date: March 1997. 1025 Chilton Cir., Elko, NV, 89801. Phone: (775) 777-8500. Fax: (775) 777-7758. Web Site: www.kenvtv.com. Licensee: Ruby Mountain Broadcasting Co. Group owner: Sunbelt Communications Co. (acq 11-8-96). Population served: 37,000 Natl. Network: NBC.
Key Personnel:
Terry Keeblerhritz . . . gen mgr & stn mgr
John Finkbohner . . . opns mgr
Terry Hritz . . . adv mgr
Rebroadcasts KRNV(TV) Reno.

KGWR-TV— Analog channel: 13. Analog hrs: 24 209 kw vis, 10 kw aur. ant 1,624t/148g TL: N41 26 21 W109 06 42 On air date: Oct 21, 1977. 1856 Skyview Dr., Casper, WY, 82601. Phone: (307) 234-1111. Fax: (307) 234-4005. Licensee: Mark III Media Inc. Group owner: Chelsey Broadcasting Co. (acq 5-31-2006; grpsl). Population served: 19,000 Natl. Network: CBS. Washington Atty: Covington & Burling.
Key Personnel:
Mark Nalbone . . . gen mgr
Terry Lane . . . opns mgr
Satellite of KGWC-TV Casper.

KJZZ-TV— Analog channel: 14. Digital channel: 46. Analog hrs: 24 1,637 kw vis, 163.7 kw aur. ant 3,847t/241g TL: N40 39 12 W112 12 06 On air date: Feb 14, 1989. 5181 Amelia Earhart Dr., Salt Lake City, UT, 84116. Box 22630, Salt Lake City, UT 84116. Phone: (801) 537-1414. Fax: (801) 238-6414. Web Site: www.kjzz.com. Licensee: Larry H. Miller Communications Corp. Ownership: Larry H. Miller, 100%. (acq 2-12-93; FTR: 3-15-93). Population served: 616,720 Natl. Network: MyNetworkTV. Washington Atty: Fleischman & Walsh.
Key Personnel:
Randy Rigby . . . gen mgr
Mark Harris . . . opns dir
Lynn Lamb . . . sls VP & natl sls mgr
Chris Baum . . . gen sls mgr
Jon Crump . . . rgnl sls mgr
Norma Lloyd . . . mktg dir & prom mgr
Robert Quigley . . . progmg dir & pub affrs dir
Mike Grover . . . chief of engrg

KPNZ— Analog channel: 24. Digital channel: 24. Analog hrs: 24 1,514 kw vis. ant 4,031t/197g TL: N40 39 33 W112 12 07 On air date: 1999. 150 N. Wright Brothers Dr., Suite 520, Salt Lake City, UT, 84116. Phone: (801) 519-2424. Fax: (801) 359-1272. E-mail: info@utahs24tv.com Web Site: www.utahs24tv.com. Licensee: Utah Communications LLC.

KSL-TV— Analog channel: 5. Digital channel: 38. Analog hrs: 24 33.4 kw vis, 6.8 kw aur. ant 3,831t/249g TL: N40 39 35 W112 12 04 On air date: June 1, 1949. 55 N. Third W., Salt Lake City, UT, 84110-1160. Phone: (801) 575-5555. Fax: (801) 575-5830 (sales). Fax: (801) 575-5560 (news). Web Site: www.ksl.com. Licensee: Bonneville International Corp. Group owner: (group owner) Population served: 1,873,000 Natl. Network: NBC. Natl. Rep: Eagle Television Sales. Washington Atty: Wilkinson, Barker, Knauer & Quinn. Wire Svc: UPI Foreign lang progmg: SP 5 News staff: 65; News: 21 hrs wkly.

KSTU— Analog channel: 13. Digital channel: 28. Analog hrs: 24 hrs Digital hrs: 24 hrs. 112 kw vis, 11.2 kw aur. 3,660t/144 TL: N40 39 33 W112 12 08 On air date: Oct 9, 1978. 5020 W. Amelia Earhart Dr., Salt Lake City, UT, 84116. Phone: (801) 532-1300. Fax: (801) 537-5335. E-mail: news@fox13.com Web Site: www.myfoxutah.com. Licensee: Fox Television Stations Inc. Group owner: (group owner; (acq 2-26-90; $41 million). Population served: 1,900,000 Natl. Network: Fox. Natl. Rep: Fox Stations Sales. News: 34 hrs wkly.
Key Personnel:
Tim Ermish . . . VP & gen mgr
Ken Freedman . . . gen sls mgr
Kent Carlton . . . natl sls mgr
Kirt Burton . . . rgnl sls mgr
Melanie Say . . . prom VP & progmg VP
Renai Bodley . . . news dir
Al Schultz . . . engrg VP

KTMW— Analog channel: 20.1,660 kw vis. ant 3,841t TL: N40 39 12 W112 12 06 On air date: 2002. 314 S. Redwood Rd., Salt Lake City, UT, 84104-3536. Phone: (801) 973-8820. Fax: (801) 973-7145. Web Site: www.tv20.org. Licensee: Alpha & Omega Communications LLC. Ownership: Connie Whitney, 33.33%; Isaac Max Jaramillo, 33.33%; and Patricia Openshaw, 33.33% (acq 7-31-2003; $1.5 million). Population served: 1.5 m,ill,ion Natl. Rep: Apex Media Sales Inc. Washington Atty: Wood, Maines & Nolan, Chartered.
Key Personnel:
Pat Openhaw . . . pres
Dennis Ermel . . . gen mgr & progmg mgr
Anthon Jeppesan . . . opns mgr
Michelle Ermel . . . prom mgr
Dennis Silver . . . chief of engrg

KTVX— Analog channel: 4. Digital channel: 40.32.4 kw vis, 4.9 kw aur. 3,870t TL: N40 36 50 W112 11 05 On air date: Apr 15, 1948. 2175 W. 1700 S., Salt Lake City, UT, 84104. Phone: (801) 975-4444. Fax: (801) 975-4442. Web Site: www.abc4.com. Licensee: Clear Channel Broadcasting Licenses Inc. Group owner: Clear Channel Communications Inc. (acq 9-21-2001; grpsl). Population served: 2,082,000 Natl. Network: ABC. News staff: 50; News: 14 hrs wkly.
Key Personnel:
David D'Antuono . . . VP & gen mgr
Dennis Elsbury . . . mktg dir
Scott Terrill . . . prom dir
Karen Zabriskie . . . progmg mgr
David Bird . . . news dir & engrg dir
Bob Lyon . . . chief of engrg

KUCW— Analog channel: 30. Digital channel: 17.5,000 kw vis, 500 kw aur. 777t/347g TL: N41 15 17 W112 14 13 (CP: 1,550 kw vis, 155 kw aur, ant 3,903t. TL: N40 39 25 W112 12 07) On air date: October 1985. 2175 West 1700 South, Salt Lake City, UT, 84104. Phone: (801)

975-4444. Fax: (801) 975-4442. Web Site: www.cw30.com. Licensee: Clear Channel Broadcasting Licenses Inc. Group owner: Acme Communications Inc. (acq 4-4-2006; $18.5 million). Population served: 150,000 Natl. Network: CW. Natl. Rep: MMT.

***KUED—** Analog channel: 7.155 kw vis, 15.5 kw aur. 3,030t/204g TL: N40 36 29 W112 09 36 On air date: Jan 20, 1958. 101 S. Wasatch Dr., Room 215, Salt Lake City, UT, 84112. Phone: (801) 581-7777. Fax: (801) 585-5096. Web Site: www.kued.org. Licensee: University of Utah. Population served: 2,000,000 Natl. Network: PBS.

***KUEN—** Analog channel: 9. Analog hrs: 15 166 kw vis, 16.6 kw aur. 2,882t/78g TL: N40 36 30 W112 09 34 On air date: Dec 1, 1986. 101 Wasatch Dr., Suite 215, Salt Lake City, UT, 84112. Phone: (801) 581-2999. Fax: (801) 585-6105. E-mail: resources@uen.org Web Site: www.uen.org. Licensee: Utah State Board of Regents. Population served: 1,500,000 Foreign lang progmg: SP 0

***KUES—** Analog channel: 19.1.21 kw vis. ant 1,446t TL: N38 38 04 W112 03 33 On air date: 2001. 101 Wasatch Dr., Salt Lake City, UT, 84112. Phone: (801) 581-2999. Fax: (801) 581-3576. Licensee: University of Utah.

***KUEW—** Analog channel: 0. Digital channel: 18.6.35 kw vis. ant -180t TL: N37 03 49 W113 34 20 On air date: 2003. 101 Wasatch Dr., Salt Lake City, UT, 84112. Phone: (801) 581-2999. Fax: (801) 585-6105. Web Site: www.kued.com. Licensee: University of Utah.

KUPX— Analog channel: 16. Digital channel: 29. Analog hrs: 24 Digital hrs: 24 3,890 kw vis. 2,308t TL: N39 51 54 W111 53 39 On air date: November 1997. 466 C Lawndale Dr., Salt Lake City, UT, 84115. Phone: (801) 474-0016. Fax: (801) 463-9667. Web Site: www.ionline.tv. Licensee: Paxson Salt Lake City License Inc. Group owner: Paxson Communications Corp. Population served: 800,000 Natl. Network: i Network.
Pax Network.

KUSG— Analog channel: 12. Digital channel: 9.9.8 kw vis. 138t TL: N37 03 49 W113 34 20 On air date: Aug 12, 1999. 299 S. Main, Ste 156, Salt Lake City, UT, 84111. Phone: (801) 973-3000. Fax: (801) 973-3002. Web Site: www.kutv.com. Licensee: KUTV Holdings Inc. Group owner: Viacom Television Stations Group. Natl. Network: CBS. Natl. Rep: CBS TV Stations National Sales.
Key Personnel:
David W. Phillips . . . VP
David Phillips . . . gen mgr
Scott Jones . . . opns dir
Kipp Greene . . . engrg dir & chief of engrg
Rebroadcasts KUTV Salt Lake City 100%.

KUTF— Analog channel: 12.257 kw vis. ant 2,263t/279g TL: N41 47 03 W112 13 55 On air date: Dec. 1, 2001. Equity Broadcasting Corp., 1 Shackleford Dr., Suite 400, Little Rock, AR, 72211. Phone: (501) 219-2400. Fax: (501) 716-3502. Licensee: Logan 12 Inc. Group owner: Equity Broadcasting Corp. (acq 2-1-2001; $4 million).

KUTH— Analog channel: 32.3,072 kw vis. ant 2,663t/125g TL: N40 16 45 W111 56 00 On air date: 2003. 215 S. State ST., Ste. 100-A, Salt Lake City, UT, 84111-2348. Phone: (801) 519-9784. Fax: (801) 519-9785. E-mail: aurias@ebcorp.net Web Site: www.univision-utah.com. Licensee: Univision Television Group Inc. Group owner: Cocola Broadcasting Companies (acq 10-20-2004; $9.5 million). Natl. Network: Univision (Spanish). Foreign lang progmg: SP 168 News: 5 hrs wkly.

KUTV— Analog channel: 2. Analog hrs: 24 45.7 kw vis, 9.1 kw aur. 3,060t/233g TL: N40 36 23 W112 09 47 On air date: Sept 26, 1954. 299 S. Main St., Suite 150, Salt Lake City, UT, 84111. Phone: (801) 973-3000. Fax: (801) 973-3387. Web Site: www.kutv2.com. Licensee: KUTV Holdings Inc. Group owner: Viacom Television Stations Group (acq 9-10-95). Population served: 2,131,000 Natl. Network: CBS. Natl. Rep: TeleRep. Washington Atty: CBS Inc. News staff: 73; News: 34.5 hrs wkly.
Key Personnel:
Dave Phillips . . . gen mgr
Scott Jones . . . opns mgr

KVNV— Analog channel: 3. Analog hrs: 24 1.08 kw vis. ant 913t/49g TL: N39 14 46 W114 55 36 On air date: 2001. 1500 Foremaster Lane, Las Vegas, NV, 89101. Phone: (702) 642-3333. Fax: (702) 657-3256. Web Site: www.kvbc.com. Licensee: Valley Broadcasting Co. Group owner: Sunbelt Communications Co. (acq 12-15-2003). Natl. Network: NBC.
Key Personnel:
Lisa Howfield . . . gen mgr
Joanne Nasby . . . gen sls mgr
Mark Guranik. . . . chief of engrg
Rebroadcasts KVBC(TV) Las Vegas.

Vermont

Burlington, VT-Plattsburgh, NY (DMA 90)

WCAX-TV— Analog channel: 3. Digital channel: 53. Analog hrs: 24 Digital hrs: 24 38 kw vis, 7.25 kw aur. ant 2,739t TL: N44 31 36 W72 48 57 On air date: Sept 26, 1954. Box 4508, Burlington, VT, 05406-4508. Phone: (802) 652-6300. Fax: (802) 652-6319. Web Site: www.wcax.com. Licensee: Mount Mansfield TV Inc. Ownership: Peter R. Martin, James S. Martin, Marcia H. Martin Boyer and Donald P. Martin. Population served: 550,000 Natl. Network: CBS. Natl. Rep: Harrington, Righter & Parsons. Rgnl. Rep: Rgnl rep: Metrospot. Washington Atty: Wilmer Cutler Pickering Hale and Dorr LLP. News staff: 35; News: 15 hrs wkly.
Key Personnel:
Peter Martin . . . pres, gen mgr & progmg dir
Phil Scharf . . . opns dir
Bruce Grindle . . . gen sls mgr & natl sls mgr
Jim Strader . . . prom dir & pub affrs dir
Marselis Parsons . . . news dir
Ted Teffner . . . engrg VP

***WCFE-TV—** Analog channel: 57. Digital channel: 38. Analog hrs: 6:30 AM-1:30 AM 462 kw vis, 46.2 kw aur. ant 2,417t/410g TL: N44 41 43 W73 53 00 On air date: Mar 6, 1977. One Sesame St., Plattsburgh, NY, 12901. Phone: (518) 563-9770. Fax: (518) 561-1928. E-mail: mlpbs@mountainlake.org Web Site: www.mountainlake.org. Licensee: Mountain Lake Public Telecommunications Council. Population served: 850,000 Natl. Network: PBS. Washington Atty: Dow, Lohnes & Albertson.
Key Personnel:
Alice Recore . . . CEO, COO, pres & gen mgr
Sharlene Petro-Durgan . . . CFO
Charlie Zarbo . . . chief of opns & engrg dir
Rhonda Santos . . . prom dir & adv dir

***WETK—** Analog channel: 33. Digital channel: 32. Analog hrs: 24 1,350 kw vis, 135 kw aur. 2,673t/87g TL: N44 31 32 W72 48 54 On air date: Oct 16, 1967. 204 Ethan Allen Ave., Colchester, VT, 05446-3129. Phone: (802) 655-4800. E-mail: view@vpt.org Web Site: www.vpt.org. Licensee: Vermont ETV Inc. (acq 11-6-89). Natl. Network: PBS. Rgnl. Network: Eastern Educ. Washington Atty: Covington & Burling.
Key Personnel:
John King. . . . CEO & pres
Lee Ann Lee. . . . dev VP & mktg VP

WFFF-TV— Analog channel: 44. Digital channel: 43. Analog hrs: 24 5,000 kw vis. ant 1,738t TL: N44 31 32 W72 48 54 (CP: 1,450 kw vis, ant 1,264t) On air date: 1997. 298 Mountain View Dr., Colchester, VT, 05446. Phone: (802) 660-9333. Fax: (802) 660-8673. Web Site: www.fox44.net. Licensee: Smith Media License Holdings LLC. (acq 11-15-2004; grpsl). Natl. Network: Fox, CW.
Key Personnel:
Mike Granados . . . gen mgr
Kathleen Harrington . . . news dir
Matt Servis . . . chief of engrg

WNNE-TV— Analog channel: 31. Digital channel: 25. Analog hrs: 24 2,240 kw vis, 2.24 kw aur. ant 2,220t/149g TL: N43 26 38 W72 27 17 On air date: Sept 27, 1978. Box 1310, White River Junction, VT, 05001. Phone: (802) 295-3100. Fax: (802) 295-9056. Fax: (802) 295-3983. Web Site: wnne.com. Licensee: Hearst-Argyle Stations Inc. Group owner: Hearst-Argyle Television Inc. Population served: 133,000 Natl. Network: NBC.

WPTZ—(North Pole, NY) Analog channel: 5. Digital channel: 14. Analog hrs: 24 Digital hrs: 24 25.1 kw vis, 4.3 kw aur. ant 1,991t/978g TL: N44 34 26 W73 40 29 On air date: Dec 8, 1954. 5 Television Dr., Plattsburgh, NY, 12901. 533 Roosevelt Highway, Colchester, VT 05446. Phone: (518) 561-5555. Fax: (518) 561-5940. Web Site: www.wptz.com. Licensee: Hearst-Argyle Stations Inc. Group owner: Hearst-Argyle Television Inc. (acq 6-1-98). Natl. Network: NBC. News staff: 31; News: 23 hrs wkly.
Key Personnel:
Paul Sands. . . . gen mgr
Bruce Lawson . . . sls dir
Chris Duley . . . natl sls mgr
Susan Acklen . . . prom mgr
Jim Gratton. . . . progmg dir
Kyle Grines . . . news dir
Andrew Lombard . . . chief of engrg

***WVER—** Analog channel: 28. Digital channel: 9. Analog hrs: 24 245 kw vis, 24.5 kw aur. ant 1,400t/297g TL: N43 39 32 W73 06 25 On air date: Mar 18, 1968. 204 Ethan Allen Ave., Colchester, VT, 05446. Phone: (802) 655-4800. E-mail: veiw@vpt.org Web Site: www.vpt.org. Licensee: Vermont ETV Inc. Natl. Network: PBS. Washington Atty: Covington & Burling.
Key Personnel:
John E. King . . . CEO & pres
Shahid Khan . . . chmn
Andrea Bergeon . . . CFO
Lee Ann Lee . . . dev VP
Peter Shea. . . . sls dir
Jeff Vande Griek . . . prom dir
Kelly Luoma . . . progmg mgr & film buyer
Joe Merone . . . pub affrs dir
Rob Belle-Isle . . . engrg VP & engrg dir
Satellite of WETK(TV) Burlington.

WVNY— Analog channel: 22. Digital channel: 22. Analog hrs: 24 1,000 kw vis, 100 kw aur. 2,739t/310g TL: N44 31 40 W72 48 58 On air date: Aug 19, 1968. 298 Mountain View Dr., Colchester, VT, 05446. Phone: (802) 660-9333. Fax: (802) 660-8673. E-mail: abc22@abc22.com Web Site: www.abc22.com. Licensee: Lambert Broadcasting of Burlington LLC. Ownership: Michael Lambert, 100% (acq 5-21-2005; $10.2 million plus assumption of liabilities). Natl. Network: ABC.
Key Personnel:
Bill Sally . . . gen mgr
Ken Kaszubowski . . . opns dir
Gena Boyden . . . gen sls mgr & natl sls mgr
Leigh Gross . . . progmg dir
Matthew Servis . . . chief of engrg

***WVTA—** Analog channel: 41. Digital channel: 24. Analog hrs: 24 1,050 kw vis, 105 kw aur. ant 2,245t/414g TL: N43 26 15 W72 27 09 On air date: Mar 18, 1968. 204 Ethan Allen Ave., Colchester, VT, 05446. Phone: (802) 655-4800. E-mail: view@vpt.org Web Site: www.vpt.org. Licensee: Vermont ETV Inc. Natl. Network: PBS. Washington Atty: Covington & Burling.
Key Personnel:
John E. King . . . CEO & pres
Lee Ann Lee . . . dev VP
Peter Shea. . . . sls dir
Jeff Vande Griek . . . prom dir
Kelly Luoma . . . progmg mgr & film buyer
Joseph Merone . . . pub affrs dir
Rob Belle-Isle . . . engrg VP, engrg dir & chief of engrg
Satellite of *WETK Burlington.

***WVTB—** Analog channel: 20. Digital channel: 18. Analog hrs: 24 589 kw vis, 58.9 kw aur. ant 1,940t/139g TL: N44 34 15 W71 53 36 On air date: Feb 26, 1968. 204 Ethan Allen Ave., Colchester, VT, 05446. Phone: (802) 655-4800. E-mail: view@vpt.org Web Site: www.vpt.org. Licensee: Vermont ETV Inc. Natl. Network: PBS. Washington Atty: Covington & Burling.
Key Personnel:
John King. . . . CEO & pres
Lee Ann Lee . . . dev VP
Satellite of WETK(TV) Burlington.

Virginia

Arlington

see Washington, DC (Hagerstown, MD) market

Bristol

see Tri-Cities, TN-VA market

Charlottesville, VA (DMA 182)

WCAV— Analog channel: 19. Analog hrs: 24 2,380 kw vis. ant 1,180t/332g TL: N37 59 05 W78 28 49 On air date: Aug 15, 2004. 999 2nd St. S.E., Charlottesville, VA, 22902. Phone: (434) 242-1919. Fax: (434) 220-0398. Web Site: www.charlottesvillenewsplex.tv. Licensee: Gray Television Licensee Inc. Group owner: Gray Television Inc. (acq 5-28-2004; $1 million for CP). Natl. Network: CBS.
Key Personnel:
Roger Burchett . . . gen mgr
Jim McCabe. . . . gen sls mgr
Jeremy Settle. . . . news dir

***WHTJ—** Analog channel: 41. Analog hrs: 8:30am - 5:00pm 251 kw vis, 25.1 kw aur. 1,156t/237g TL: N37 58 58 W78 29 00 (CP: TL: N37 59 00 W78 28 54) On air date: May 19, 1989. Box 40, Charlottesville, VA, 22902. 528 E. Main St., Charlottesville, VA 22902. Phone: (434)

295-7671. Fax: (434) 295-2813. Web Site: www.ideastations.org. Licensee: Commonwealth Public Broadcsting Corp. Natl. Network: PBS. Washington Atty: Wiley, Rein & Fielding.
Key Personnel:
Conni Lombardo . . . . . VP & stn mgr
D.J. Crotteau . . . . . gen mgr & stn mgr
Lisa Tait . . . . . dev VP & dev dir
John Felton. . . . . progmg VP & progmg dir
Rebroadcasts WCVE-TV Richmond.

WVIR-TV— Analog channel: 29. Digital channel: 32. Analog hrs: 24 Digital hrs: 24 5,000 kw vis, 500 kw aur. 1,187t/289g TL: N37 59 00 W78 28 54 On air date: Mar 11, 1973. Box 769, Charlottesville, VA, 22902. Phone: (434) 220-2900. Fax: (434) 220-2904. E-mail: newsdesk@nbc29.com Web Site: www.nbc29.com. Licensee: Virginia Broadcasting Corp. Group owner: Waterman Broadcasting Corp. Population served: 250,000 Natl. Network: NBC. Natl. Rep: Continental Television Sales. Washington Atty: Cohn & Marks. Wire Svc: AP News staff: 50; News: 32 hrs wkly.
Key Personnel:
Harold Wright . . . . . VP, gen mgr & mktg dir
Jim Fernald . . . . . gen sls mgr
Ralph Tobias . . . . . prom dir & progmg dir
Neal Bennett . . . . . news dir
Bob Jenkins . . . . . chief of engrg

Fairfax

see Washington, DC (Hagerstown, MD) market

Front Royal

see Washington, DC (Hagerstown, MD) market

Goldvein

see Washington, DC (Hagerstown, MD) market

Grundy

see Tri-Cities, TN-VA market

Harrisonburg, VA
(DMA 181)

WHSV-TV— Analog channel: 3. Digital channel: 49. Analog hrs: 20 Note: ABC is on WHSV-TV ch 3, Fox and MyNetworkTV are on WHSV-DT ch 49. 832 kw vis, 432 kw aur. ant 2,130t/337g TL: N38 36 05 W78 37 57 On air date: Oct 19, 1953. 50 N. Main St., Harrisonburg, VA, 22802. Phone: (540) 433-9191. Fax: (540) 433-4028; (540)433-2700 (news). E-mail: whsv@whsv.com Web Site: www.whsv.com. Licensee: WEAU Licensee Corp. Group owner: Gray Television Inc. (acq 8-29-2002; grpsl). Population served: 350,000 Natl. Network: ABC, Fox, MyNetworkTV Digital Network: Note: ABC is on WHSV-TV ch 3, Fox and MyNetworkTV are on WHSV-DT ch 49. Washington Atty: Wiley, Rein LLP. News staff: 18; News: 16 hrs wkly.
Key Personnel:
Tim Merritt . . . . . gen sls mgr & mktg mgr
Tina Wood . . . . . natl sls mgr & rgnl sls mgr
Jeremy Harman . . . . . prom mgr
Tracey Jones. . . . . gen mgr & progmg mgr
Ed Reams . . . . . news dir
Sean Harper . . . . . chief of engrg

***WVPT**—Analog channel: 51. Analog hrs: 24 525 kw vis, 67.6 kw aur. 2,230t/46g TL: N38 09 54 W79 18 51 On air date: Sept 9, 1968. 298 Port Republic Rd., Harrisonburg, VA, 22801. Phone: (540) 434-5391. Fax: (540) 434-7084. E-mail: wvptcomments@wvpt.net Web Site: www.wvpt.net. Licensee: Shenandoah Valley ETV Corp. Population served: 500,000 Natl. Network: PBS. Washington Atty: Covington & Burling.
Key Personnel:
Richard Parker . . . . . gen mgr
Tony Mancari. . . . . opns VP

Lynchburg

see Roanoke-Lynchburg, VA market

Manassas

see Washington, DC (Hagerstown, MD) market

Marion

see Tri-Cities, TN-VA market

Newport News

see Norfolk-Portsmouth-Newport News, VA market

Norfolk-Portsmouth-Newport News, VA
(DMA 42)

WAVY-TV—Portsmouth, Analog channel: 10. Digital channel: 31. Analog hrs: 24 316 kw vis, 38.9 kw aur. 990t/1,026g TL: N36 49 14 W76 30 41 On air date: Sept 1, 1957. 300 Wavy St., Portsmouth, VA, 23704. Phone: (757) 393-1010. Fax: (757) 399-7628. E-mail: doug.davis@wavy.com Web Site: www.wavy.com. Licensee: WAVY Broadcasting L.L.C. Group owner: LIN Television Corporation (acq 12-16-97; grpsl). Population served: 694,000 Natl. Network: NBC. Washington Atty: Covington & Burling. News staff: 80; News: 31 hrs wkly.
Key Personnel:
Doug Davis . . . . . pres & gen mgr
John Cochran . . . . . gen sls mgr
Judy Triska. . . . . prom dir
Joe Weller . . . . . progmg mgr
Les Garrenton . . . . . engrg dir

WGNT— Analog channel: 27. Digital channel: 50. Analog hrs: 24 2,340 kw vis, 234 kw aur. ant 971t/1,026g TL: N36 48 43 W76 27 45 On air date: Oct 1, 1961. 1318 Spratley St., Portsmouth, VA, 23704-1829. Phone: (757) 393-2501. Fax: (757) 399-3303. E-mail: cw27@wgnttv.com Web Site: www.cw.27.com. Licensee: CBS Television Stations Inc. Group owner: Viacom Television Stations Group (acq 10-31-97; $42.5 million). Population served: 1,500,000 Natl. Network: CW.
Key Personnel:
Steven Soldinger . . . . . VP & gen mgr
Jon Erkenbrack . . . . . gen sls mgr
Chuck Martin . . . . . natl sls mgr
Chris Wolf . . . . . prom dir & progmg dir
Kafi Rouse . . . . . pub affrs dir
George Randell. . . . . chief of engrg

WHRE— Analog channel: 21.5,000 kw vis. ant 1,017t/1,014g TL: N36 48 31 W76 30 12 On air date: Mar 27, 2006. 168 Business Park Dr., Suite 200, Virginia Beach, VA, 23462. Phone: (757) 473-3702. Web Site: www.hon.org. Licensee: Copeland Channel 21 LLC.

***WHRO-TV**—Hampton-Norfolk, Analog channel: 15. Digital channel: 16. Analog hrs: 24 Digital hrs: 24 2,630 kw vis, 263 kw aur. 964t/964g TL: N36 48 32 W76 30 13 On air date: Oct 2, 1961. Not on air, target date: DTV on Air 2/5/2001: 5200 Hampton Blvd., Norfolk, VA, 23508. Phone: (757) 889-9400. Fax: (757) 489-0007. E-mail: info@whro.org Web Site: www.whro.org. Licensee: Hampton Roads Educ. Telecommunications Association Inc. Population served: 707,750 Natl. Network: PBS. Washington Atty: Dow-Lohnes. News staff: 2; News: 1 hrs wkly.
Key Personnel:
Joseph Widoff . . . . . CEO, pres & CFO
Carol Vollbrecht . . . . . CFO
John Heimerl . . . . . gen mgr
Virginia Thumm . . . . . sr VP & dev dir

WPXV— Analog channel: 49. Digital channel: 46. Analog hrs: 24 501 kw vis, 50.1 kw aur. ant 508t TL: N36 48 32 W76 30 13 On air date: 1994. 230 Clearfield Ave. #104, Virginia Beach, VA, 23462. Phone: (757) 499-1261. Fax: (757) 499-1679. Web Site: www.ionline.tv /stations/list.cfm. Licensee: Paxson Communications License Co. L.L.C. Group owner: Paxson Communications Corp. (acq 12-18-97; $14.75 million). Natl. Network: i Network.

WSKY-TV— Analog channel: 4. Analog hrs: 24 100 kw vis. ant 1,030t/1,023g TL: N36 08 08 W75 49 28 On air date: Oct 1, 2001. 920 Corporate Ln., Chesapeake, VA, 23320. Phone: (757) 382-0004. Fax: (757) 382-0365. E-mail: programming@wsky4.com Web Site: www.4hamptonroads.com. Licensee: Sky Television LLC. Ownership: Danbeth Communications Inc., 51% (acq 8-19-2002). Population served: 1,278,000 Washington Atty: Leventhal, Senter and Lerman.
Key Personnel:
Glenn Holterhaus. . . . . CEO, pres & gen mgr
Jacquelyn Smullen . . . . . CFO, VP & gen sls mgr
Tom Powers. . . . . opns VP & opns dir
Ed Marlowe. . . . . prom mgr & progmg mgr
Jeff Mercer . . . . . pub affrs dir

WTKR— Analog channel: 3. Digital channel: 40. Analog hrs: 24 100 kw vis, 20 kw aur. ant 980t/1,029g TL: N36 48 56 W76 28 00 On air date: Apr 2, 1950. 720 Boush St., Norfolk, VA, 23501-0300. Phone: (757) 446-1000. Fax: (757) 446-1376. Web Site: www.wtkr.com. Licensee: Local TV Virginia License LLC. Group owner: The New York Times Co. (acq 5-7-2007; grpsl). Population served: 1,700,000 Natl. Network: CBS. Washington Atty: Reed, Smith, Shaw & McClay.
Key Personnel:
Dave Bunnell . . . . . gen mgr
Barbara Secra . . . . . gen sls mgr
Jeff McCallister. . . . . gen sls mgr

WTVZ-TV— Analog channel: 33. Digital channel: 38. Analog hrs: 24 5,000 kw vis, 500 kw aur. ant 909t/1,026g TL: N36 48 32 W76 30 13 On air date: Sept 24, 1979. 900 Granby St., Norfolk, VA, 23510. Phone: (757) 622-3333. Fax: (757) 623-1541. E-mail: comments@wtv233.com Web Site: www.mytvz.com. Licensee: WTVZ Licensee L.L.C. Group owner: Sinclair Broadcast Group Inc. (acq 2-9-95; $47 million; FTR: 5-8-95). Population served: 1,300,000 Natl. Network: MyNetworkTV. Washington Atty: Gardner, Carton & Douglas.
Key Personnel:
Bill Scasfide . . . . . gen mgr
Bill Barber. . . . . chief of engrg

WVBT— Analog channel: 43. Digital channel: 29. Analog hrs: 24 240 kw vis. ant 856t TL: N36 49 14 W76 30 41 On air date: March 1993. 243 Wythe St., Portsmouth, VA, 23704. Phone: (757) 393-4343. Fax: (757) 763-5447. E-mail: doug.davis@wavy.com Web Site: www.fox43tv.com. Licensee: WAVY Broadcasting LLC. Group owner: LIN Television Corporation (acq 1-9-2002; $4.25 million). Natl. Network: Fox. Washington Atty: Covington & Burling.
Key Personnel:
Mark Gentner . . . . . stn mgr
Andy Hilton . . . . . natl sls mgr
John Lipscomb. . . . . rgnl sls mgr

WVEC-TV—Hampton, Analog channel: 13. Digital channel: 41. Analog hrs: 24 316 kw vis, 31.6 kw aur. 980t/1,028g TL: N36 49 00 W76 28 05 (CP: Ant 987t) On air date: Sept 19, 1953. 613 Woodis Ave., Norfolk, VA, 23510. Phone: (757) 625-1313. Fax: (757) 628-6220. Fax: (757) 628-5855 (news). Web Site: www.wvec.com. Licensee: WVEC Television Inc. Group owner: Belo Corp., Broadcast Division (acq 11-28-83; grpsl; FTR: 12-29-83). Population served: 1,253,200 Natl. Network: ABC. Natl. Rep: TeleRep. Wire Svc: Reuters News staff: 65; News: 24 hrs wkly.
Key Personnel:
Nick Nicholson . . . . . pres & gen mgr
Amy Warren . . . . . sls dir
T.J. Dula . . . . . natl sls mgr
Deborah Shollenberger . . . . . progmg mgr
Rich Lebenson . . . . . news dir
Wendy Juren . . . . . pub affrs dir
John Dolive . . . . . chief of engrg

Norton

see Tri-Cities, TN-VA market

Petersburg

see Richmond-Petersburg, VA market

Portsmouth

see Norfolk-Portsmouth-Newport News, VA market

Richmond-Petersburg, VA
(DMA 61)

***WCVE-TV**— Analog channel: 23. Digital channel: 42. Analog hrs: 24 2,980 kw vis, 351 kw aur. ant 1,079t/976g TL: N37 30 46 W77 36 06 On air date: Sept 14, 1964. 23 Sesame St., Richmond, VA, 23235. Phone: (804) 320-1301. Fax: (804) 320-8729. Web Site: www.ideastations.org. Licensee: Commonwealth Public Broadcasting Corp. Population served: 462,000 Natl. Network: PBS. Washington Atty: Wiley, Rein & Fielding.
Key Personnel:
Curtis Monk . . . . . pres
Lisa Tait . . . . . dev VP
John Felton . . . . . progmg VP

***WCVW**— Analog channel: 57. Digital channel: 44. Analog hrs: 24 1,000 kw vis, 100 kw aur. ant 961t/1,170g TL: N37 30 46 W77 36 06 On air date: Dec 22, 1966. 23 Sesame St., Richmond, VA, 23235. Phone: (804) 320-1301. Fax: (804) 320-8729. Web Site:

www.ideastations.org. Licensee: Commonwealth Public Broadcasting Corporation. Population served: 450,000 Natl. Network: PBS. Washington Atty: Wiley, Rein & Fielding.

Key Personnel:
Curtis Monk . . . pres
Lisa Tait . . . dev VP
John Felton . . . progmg VP

WRIC-TV—Petersburg, Analog channel: 8. Digital channel: 22. Analog hrs: 24 Digital hrs: 24 269 kw vis, 34.4 kw aur. ant 1,050t/999g TL: N37 30 46 W77 36 06 On air date: Aug 15, 1955. 301 Arboretum Pl., Richmond, VA, 23236-3464. Phone: (804) 330-8888. Fax: (804) 330-8882. E-mail: news@wric.com Web Site: www.wric.com. Licensee: Young Broadcasting of Richmond Inc. Group owner: (group owner; (acq 11-14-94; grpsl; FTR: 9-12-94). Population served: 1,255,000 Natl. Network: ABC. Natl. Rep: Adam Young. Washington Atty: Brooks, Pierce, McLendon, Humphrey & Leonard.

Key Personnel:
Robert Peterson . . . gen mgr
Matthew Zelkird . . . stn mgr

WRLH-TV— Analog channel: 35. Digital channel: 26. Analog hrs: 24 Note: Fox is on WRLH-TV ch 35, MyNetworkTV is on WRLH-DT ch 26. 2,588 kw vis, 259 kw aur. ant 1,259 TL: N37 30 21 W77 41 58 On air date: Feb 20, 1982. 1925 Westmoreland St., Richmond, VA, 23230. Phone: (804) 358-3535. Phone: (804) 359-3510. Fax: (804) 358-1495. Web Site: www.foxrichmond.com. Licensee: WRLH Licensee LLC. Group owner: Sinclair Broadcast Group Inc. (acq 12-10-01; grpsl). Population served: 484,000 Natl. Network: Fox, MyNetworkTV Digital Network: Note: Fox is on WRLH-TV ch 35, MyNetworkTV is on WRLH-DT ch 26. Washington Atty: Arter & Hadden. News staff: 15; News: 3 hrs wkly.

Key Personnel:
Bill Lane . . . gen mgr
Darren Shapiro . . . sls VP
Steve Genett . . . gen sls mgr
Bill Norris . . . rgnl sls mgr
Mark Bartholmew . . . prom dir
Charles Rouse . . . chief of engrg

WTVR-TV— Analog channel: 6. Digital channel: 25. Analog hrs: 24 Digital hrs: 24 100 kw vis, 15.1 kw aur. 1,049t/840g TL: N37 34 00 W77 28 36 On air date: Apr 22, 1948. 3301 W. Broad St., Richmond, VA, 23230. Phone: (804) 254-3600. Fax: (804) 254-3699. Web Site: www.wtvr.com. Licensee: Elcom of Virginia License Subsidiary LLC. Group owner: Raycom Media Inc. (acq 7-25-97; grpsl). Population served: 1,148,000 Natl. Network: CBS. Natl. Rep: TeleRep. Washington Atty: Covington & Burling. Wire Svc: AP

Key Personnel:
Peter Maroney . . . VP & gen mgr
Don Cox . . . opns dir & engrg dir
Tina Woody . . . opns mgr
Stephen Hayes . . . gen sls mgr
James Taguchi . . . natl sls mgr
Steve Young . . . rgnl sls mgr & sls
Bill Anderson . . . mktg dir & news dir
Blake Peddicore . . . prom mgr & progmg

WUPV— Analog channel: 65. Digital channel: 65. Analog hrs: 24 Digital hrs: 24 1,581 kw vis, 158 kw aur. 859t/850g TL: N37 44 32 W77 15 18 On air date: Mar 9, 1990. 3301 West Broad St., Richmond, VA, 23230. Phone: (804) 254-3600. Fax: (804) 342-5746. Web Site: www.cwrichmond.tv. Licensee: Southeastern Media Holdings Inc. (acq 11-3-2006; $47 million). Population served: 1,237,000 Natl. Network: CW. Natl. Rep: Harrington, Righter & Parsons.

Key Personnel:
John Rezabeck . . . gen mgr
Blake Peddicord . . . progmg dir
Don Cox . . . engrg dir
Gene Todd . . . chief of engrg

WWBT— Analog channel: 12. Digital channel: 54. Analog hrs: 24 316 kw vis, 63.1 kw aur. ant 790t/1,000g TL: N37 30 23 W77 30 12 On air date: Apr 29, 1956. Box 12, Richmond, VA, 23218. 5710 Midlothian Tpke., Richmond, VA 23225. Phone: (804) 230-1212. Fax: (804) 230-2793. E-mail: newsroom@nbc12.com Web Site: www.nbc12.com. Licensee: Jefferson-Pilot Broadcasting Co. of Virginia. Group owner: Jefferson-Pilot Communications Co. (acq 1968; $5 million). Population served: 771,400 Natl. Network: NBC, WB. Washington Atty: Wiley, Rein.

Key Personnel:
Dennis Glass . . . CEO
John Shreves . . . pres
Donald S. Richards . . . gen mgr
Michael Park . . . opns mgr
M. Kym Grinnage . . . gen sls mgr
Nancy Kent . . . news dir

Roanoke-Lynchburg, VA (DMA 68)

***WBRA-TV**— Analog channel: 15. Digital channel: 3. Analog hrs: 24 1,820 kw vis, 182 kw aur. 2,089t/265g TL: N37 11 45 W80 09 18 On air date: Aug 1, 1967. Box 13246, Roanoke, VA, 24032. 1215 McNeil Dr. S.W., Roanoke, VA 24015. Phone: (540) 344-0991. Fax: (540) 344-2148. Web Site: www.blueridgepbs.org. Licensee: Blue Ridge Public Television Inc. Population served: 445,000 Natl. Network: PBS.

Key Personnel:
Steve Blanks . . . chmn
Anita Simms . . . CFO
Debbia Jordan . . . opns VP
Sherry Spradlin . . . progmg dir
Sturlin Baughan . . . chief of engrg

WDBJ— Analog channel: 7. Digital channel: 18. Analog hrs: 24 Digital hrs: 24 Note: CBS is on WDBJ(TV) ch 7, MyNetworkTV is on WDBJ-DT ch 18. 316 kw vis, 62.5 kw aur. 2,000t/78g TL: N37 11 42 W80 09 22 On air date: Oct 3, 1955. Box 7, Roanoke, VA, 24022-0007. 2807 Hershberger Rd, Roanoke, VA 24017-1941. Phone: (540) 344-7000. Fax: (540) 344-5097. E-mail: firstinitiallastname@wdbj7.com Web Site: www.wdbj7.com. Licensee: WDBJ Television Inc. Group owner: Schurz Communications Inc. (acq 11-1-69; $8.2 million; FTR: 11-10-69). Population served: 1,013,000 Natl. Network: CBS, MyNetworkTV Digital Network: Note: CBS is on WDBJ(TV) ch 7, MyNetworkTV is on WDBJ-DT ch 18. Natl. Rep: Harrington, Righter & Parsons. Washington Atty: Fletcher, Heald & Hildreth. Wire Svc: UPI Wire Svc: AP News staff: 54; News: 18 hrs wkly.

Key Personnel:
Jeffrey A. Marks . . . pres & gen mgr
Edward W. Allen . . . CFO
Carl Guffey . . . opns dir, pub affrs dir & engrg dir
Ray Sullivan . . . gen sls mgr
Kelly Zuber . . . prom mgr
Mike Bell . . . progmg dir
Jim Kent . . . news dir

WDRL-TV— Analog channel: 24. Digital channel: 41.5,000 kw vis, 1,000 kw aur. ant 522t On air date: August 1994. 5002 Airport Rd., Roanoke, VA, 24012. Phone: (540) 366-2424. Fax: (540) 366-7530. E-mail: manager@wdrl-tv.com Web Site: www.wdrl-tv.com. Licensee: MNE Broadcasting L.L.C. Ownership: Melvin N. Eleazer, 100%. (acq 8-31-2006).

Key Personnel:
Mel Eleazer . . . gen mgr
Amy Ragsdale . . . opns mgr
Dave Ross . . . rgnl sls mgr
Rob Ruthenberg . . . progmg mgr
Nel Kirt . . . pub affrs dir
Rebroadcasts W54BT(TV) Roanoke 100%.

WFXR-TV— Analog channel: 27. Digital channel: 17. Analog hrs: 24 Note: Fox is on WFXR-TV ch 27, CW is on WFXR-DT ch 17. 2,690 kw vis, 269 kw aur. ant 1,991t/200g TL: N37 11 46 W80 09 16 On air date: March 1986. Box 2127, Roanoke, VA, 24009-2127. 2618 Colonial Ave. S.W., Roanoke, VA 24015. Phone: (540) 344-2127. Fax: (540) 345-1912. Fax: (540) 342-2753. E-mail: info@fox2127.com Web Site: www.fox2127.com. Licensee: Grant Broadcasting System II Inc. Group owner: (group owner; (acq 9-93; $5.5 million with WWCW(TV) Lynchburg; FTR: 6-14-93) Population served: 206,000 Natl. Network: Fox, CW Digital Network: Note: Fox is on WFXR-TV ch 27, CW is on WFXR-DT ch 17. Washington Atty: Birch, Horton, Bittner & Cherot. News staff: 2; News: 2 hrs wkly.

WPXR— Analog channel: 38. Digital channel: 36. Analog hrs: 24 Digital hrs: 24 1,350 kw vis, 135 kw aur. ant 2,043t/180g TL: N37 11 37 W80 09 25 On air date: Jan 3, 1986. 401 3rd St. S.W., Roanoke, VA, 24011. Phone: (540) 857-0038. Fax: (540) 345-8568. E-mail: shirleybundy@1onmedia.tv Web Site: www.ionline.tv. Licensee: Paxson Communications License Co. L.L.C. Group owner: Paxson Communications Corp. (acq 10-28-97). Population served: 369,000 Natl. Network: i Network.

WSET-TV— Analog channel: 13. Digital channel: 34. Analog hrs: 24 Digital hrs: 24 302 kw vis, 50 kw aur. 2,050t/1,240g TL: N37 18 54 W79 38 06 On air date: Feb 8, 1953. Box 11588, Lynchburg, VA, 24506-1588. 2320 Langhorne Rd., Lynchburg, VA 24501. Phone: (434) 528-1313. Fax: (434) 847-0458. E-mail: wset@wset.com Web Site: www.wset.com. Licensee: WSET, Incorporated. Group owner: Allbritton Communications Co. (acq 10-76; grpsl). Population served: 1,023,000 Natl. Network: ABC. Natl. Rep: Continental Television Sales. Washington Atty: Sidley, Austin, LLP. Wire Svc: AP Wire Svc: ABC News staff: 41; News: 9.5 hrs wkly.

Key Personnel:
Randall J. Smith . . . pres & gen mgr
K.C. Spiron . . . opns dir & opns mgr
John Crumpler . . . dev dir
Paul Glover . . . sls dir
Bruce Kirk . . . news dir

WSLS-TV— Analog channel: 10. Analog hrs: 24 316 kw vis, 47 kw aur. 2,001t/242g TL: N37 12 02 W80 08 55 On air date: Dec 11, 1952. Box 10, Roanoke, VA, 24022-0010. 401 3rd St. S.W., Roanoke, VA 24011. Phone: (540) 981-9110. Phone: 540-981-9126. Fax: (540) 343-3157/(540)343-2059. E-mail: news@wsls.com Web Site: www.wsls.com. Licensee: Media General Broadcasting Inc. Group owner: Media General Broadcast Group (acq 3-21-97; grpsl). Population served: 392,000 Natl. Network: NBC. Washington Atty: Wiley, Rein & Fielding. News staff: 33; News: 15 hrs wkly.

Key Personnel:
Kathy Mohn . . . VP & gen mgr
Robert Kerry . . . opns VP
Candy Crigger . . . gen sls mgr
Scott Martin . . . natl sls mgr
Daniel Coyle . . . mktg VP

WWCW— Analog channel: 21. Digital channel: 20. Analog hrs: 24 4,207 kw vis, 421 kw aur. ant 1,640t/966g TL: N37 19 14 W79 37 58 On air date: February 1986. Box 2127, Roanoke, VA, 24009-2127. 2618 Colonial Ave. S.W., Roanoke, VA 24015. Phone: (540) 344-2127. Fax: (540) 345-1912/(540) 342-2753. E-mail: info@fox2127.com Web Site: www.fox2127.com. Licensee: Grant Broadcasting System II Inc. (acq 9-15-93; $5.5 million with satellite stn WFXR-TV Roanoke; FTR: 6-14-93) Natl. Network: Fox, CW. Washington Atty: Birch, Horton, Bittner & Cherot. News staff: 2; News: one hr wkly.

Tri-Cities

see Tri-Cities, TN-VA market

Washington

Kennewick

see Yakima-Pasco-Richland-Kennewick, WA market

Pasco

see Yakima-Pasco-Richland-Kennewick, WA market

Richland

see Yakima-Pasco-Richland-Kennewick, WA market

Seattle-Tacoma, WA (DMA 14)

KBCB— Analog channel: 24. Digital channel: 19. Analog hrs: 24 3,090 kw vis. ant 2,483t/487g TL: N48 40 46 W122 50 31 On air date: December 1994. 4164 Meridian St., Suite 102, Bellingham, WA, 98226. 800 5th Ave., Suite 4100, Seattle, WA 98104. Phone: (360) 647-8842. Phone: (206) 447-1430. Fax: (360) 647-9204. Fax: (206) 447-1431. E-mail: kbcb@kbcbtv.com Web Site: www.kbcbtv.com. Licensee: World Television of Washington LLC. Ownership: Venture Technologies Group LLC, 68.67%; and Frank Washington, 31.33%. Population served: 1,690,000 Washington Atty: Wiley, Rein & Fielding. News: 4 hrs wkly.

Key Personnel:
Garry Spire . . . CEO
Larry Rogow . . . chmn
Paul Koplin . . . pres
Dewi Cashion . . . CFO & gen mgr
Brian Holton . . . VP
Shelli Jones . . . stn mgr
Karen Bean . . . opns mgr

***KBTC-TV**— Analog channel: 28.676 kw vis, 67.6 kw aur. ant 761t/326g TL: N47 16 41 W122 30 42 On air date: Sept 25, 1961. 2320 S. 19th St., Tacoma, WA, 98405. Phone: (253) 680-7700. Fax: (253) 680-7725. Web Site: www.kbtc.org. Licensee: Bates Technical College. (acq 11-29-91; with KXOT(FM) Tacoma; FTR: 12-16-91). Population served: 800,000 Natl. Network: PBS.

Key Personnel:
Debbie Emond . . . gen mgr
Darin Gerchak . . . chief of opns & chief of engrg
Mary Thompson . . . prom dir
Paul Jackson . . . stn mgr, dev dir & progmg dir
Lamont Walton . . . news dir

***KCKA**— Analog channel: 15. Digital channel: 19.661 kw vis, 66.1 kw aur. ant 1,138t TL: N46 33 16 W123 03 26 On air date: October 1982. 2320 S. 19th St., Tacoma, WA, 98405. Phone: (253) 680-7700. Fax: (253) 680-7725. Web Site: www.kbtc.org. Licensee: Bates Technical

College. (acq 11-29-91; FTR: 12-16-91). Population served: 800,000 Natl. Network: PBS. Washington Atty: Akin, Gump, Strauss, Hauer & Feld.
Key Personnel:
Debbie Emond . . . . . gen mgr
Mary Thompson . . . . . prom dir
Paul Jackson . . . . . stn mgr, dev dir & progmg dir
Lamont Walton . . . . . news dir & pub affrs dir
Darin Gerchak . . . . . chief of engrg
Rebroadcasts KBTC(TV) Tacoma.

KCPQ—Tacoma, Analog channel: 13. Digital channel: 18. Analog hrs: 24 316 kw vis, 31.6 kw aur. 2,000t/708g TL: N47 32 53 W122 48 22 (CP: Ant 1,191t. TL: N47 36 59 W122 18 23) On air date: 1954. 1813 Westlake Ave. N., Seattle, WA, 98109-2706. Phone: (206) 674-1313. Fax: (206) 674-1777. E-mail: askus@kcpq.com Web Site: q13.trb.com. Licensee: Tribune Television Northwest Inc. Group owner: Tribune Broadcasting Co. (acq 2-22-99; $370 million swap with WGNX(TV) Atlanta, GA). Population served: 1,225,000 Natl. Network: Fox. Natl. Rep: TeleRep. Washington Atty: Sidley & Austin. Wire Svc: SportsTicker News staff: 44; News: 21 hrs wkly.
Key Personnel:
Dennis Fitzsimons . . . . . CEO, chmn & pres
Pamela Pearson. . . . . gen mgr
Mark Boe. . . . . stn mgr
Paul Rennie . . . . . sls dir
Houman Aliabadi . . . . . natl sls mgr
Natalie Grant . . . . . progmg dir
Steve Kraycik. . . . . news dir
Marty Gustafson . . . . . pub affrs dir
Michael Goodman. . . . . engrg dir

***KCTS-TV**— Analog channel: 9. Digital channel: 41. Analog hrs: 24 Digital hrs: 24 316 kw vis, 50 kw aur. ant 830t/590g TL: N47 36 58 W122 18 28 On air date: Dec 7, 1954. 401 Mercer, Seattle, WA, 98109. Phone: (206) 728-6463. Fax: (206) 443-6691. E-mail: viewer@kcts.org Web Site: www.kcts.org. Licensee: KCTS Television. Ownership: KCTS Television board of directors (acq 7-15-87). Population served: 3,359,300 Natl. Network: PBS. Washington Atty: Dow, Lohnes & Albertson.
Key Personnel:
William Mohler. . . . . CEO
Bob Flowers . . . . . chmn
Randy Brinson . . . . . gen mgr
Cliff Anderson . . . . . engrg dir & chief of engrg

KHCV— Analog channel: 45. Digital channel: 44. Analog hrs: 24 hrs 2,000 kw vis. ant 2,283t TL: N47 30 17 W121 58 06 On air date: 2001. 19825 Willows Rd. NE., Suite 140, Redmond, WA, 98052. Phone: (425) 497-1515. Fax: (425) 497-8629. E-mail: khcvtv@khcvtv.com Web Site: www.khcvtv.com. Licensee: North Pacific International Television Inc. (acq 12-24-92; FTR: 11-23-92). Population served: 3,000,000 Natl. Network: Azteca America (Spanish). Foreign lang progmg: SP 168
Key Personnel:
Dr. Kenneth Casey . . . . . pres
Charlene Casey . . . . . CFO
Tim Kammer. . . . . opns mgr
Stephanie Ogle . . . . . progmg dir
Chris Casey . . . . . engrg VP

KING-TV— Analog channel: 5. Digital channel: 48. Analog hrs: 24 Digital hrs: 24 100 kw vis, 15.1 kw aur. 820t/570g TL: N47 37 55 W122 20 59 (CP: Ant 1,168t/918g) On air date: Nov 25, 1948. 333 Dexter Ave. N., Seattle, WA, 98109. Phone: (206) 448-5555. Fax: (206) 448-3936. Web Site: www.king5.com. Licensee: KING-TV Inc. Group owner: Belo Corp., Broadcast Division (acq 1997; grpsl). Population served: 3,848,400 Natl. Network: NBC. Natl. Rep: TeleRep. Washington Atty: Fletcher, Heald & Hildreth.
Key Personnel:
Conrad Jungmann. . . . . gen mgr & rgnl sls mgr
Ray Heacox . . . . . gen mgr

KIRO-TV—Analog channel: 7. Digital channel: 39.316 kw vis, 63.2 kw aur. 820t/599g TL: N47 38 01 W122 21 20 On air date: Feb 8, 1958. 2807 3rd Ave., Seattle, WA, 98121. Phone: (206) 728-7777. Fax: (206) 728-8230. Web Site: www.kirotv.com. Licensee: KIRO-TV Inc. Group owner: Cox Broadcasting (acq 4-16-97). Population served: 1,047,300 Natl. Network: CBS. Natl. Rep: Harrington, Righter & Parsons. Washington Atty: Dow, Lohnes & Alberterson, PLLC. News staff: 115; News: 46 hrs wkly.
Key Personnel:
Eric Lerner. . . . . gen mgr
Holly Grambihler . . . . . dev mgr & rgnl sls mgr
Kristin Reese . . . . . dev mgr
Pat Norris. . . . . gen sls mgr & rgnl sls mgr
Sandy Zogg . . . . . gen sls mgr
Dave Blakely. . . . . natl sls mgr
Therese Weiler . . . . . progmg dir
Todd Mokhtari . . . . . news dir
John Walters. . . . . engrg dir

Pat Otis. . . . . chief of engrg

KMYQ— Analog channel: 22. Analog hrs: 24 5,000 kw vis, 501 kw aur. 890t/639g TL: N47 36 57 W122 18 26 On air date: June 22, 1985. 1813 Westlake Ave., Seattle, WA, 98109. Phone: (206) 674-1313. Fax: (206) 674-1777. E-mail: askus@ktwbtv.com Web Site: ktwbtv.trb.com. Licensee: Tribune Television Holdings Inc. Group owner: Tribune Broadcasting Co. (acq 6-98). Population served: 3,516,000 Natl. Network: MyNetworkTV. Natl. Rep: TeleRep. Washington Atty: Sidley & Austin.
Key Personnel:
Dennis Fitzsimons . . . . . CEO & pres
Pamela Pearson. . . . . gen mgr
Mark Boe. . . . . stn mgr
Paul Rennie . . . . . sls dir
Adam Bischoff . . . . . natl sls mgr
Natalie Grant . . . . . progmg dir
Michael Goodman. . . . . engrg dir & chief of engrg

KOMO-TV— Analog channel: 4. Digital channel: 38. Analog hrs: 24 100 kw vis, 15 kw aur. 810t/550g TL: N47 37 55 W122 21 09 (CP: Ant 1,151t. TL: N47 37 56 W122 21 11) On air date: Dec 10, 1953. 140 4th Ave. N., Seattle, WA, 98109. Phone: (206) 404-4000. Fax: (206) 404-4034. Web Site: www.komotv.com. Licensee: Fisher Broadcasting - Seattle TV L.L.C. Group owner: Fisher Broadcasting Company (acq 12-4-01; grpsl). Population served: 3,700,000 Natl. Network: ABC. Natl. Rep: Blair Television. Washington Atty: Shaw Pittman.
Key Personnel:
James Clayton . . . . . gen mgr
Lloyd Low. . . . . natl sls mgr
Ted Davis . . . . . sls VP & natl sls mgr
Doreen Kaylor . . . . . progmg dir
Holly Gauntt . . . . . news dir

KONG-TV— Analog channel: 16. Analog hrs: 24 5,000 kw vis, 500 kw aur. 1,079t/299g TL: N47 32 34 W122 06 25 On air date: 1997. 333 Dexter Ave. N., Seattle, WA, 98109. Phone: (206) 448-3166. Phone: (206) 448-5555. Fax: (206) 448-3167. Web Site: www.kongtv.com. Licensee: KONG-TV Inc. Group owner: Belo Corp., Broadcast Division (acq 2-18-00). Washington Atty: Thompson, Hine & Flory L.

KSTW—Tacoma, Analog channel: 11. Digital channel: 36. Analog hrs: 24 316 kw vis, 47.8 kw aur. ant 891t/637g TL: N47 36 56 W122 18 29 On air date: Mar 1, 1953. 602 Oakesdale Ave. S.W., Renton, WA, 98057. Phone: (206) 441-1111. Fax: (206) 861-8915. Web Site: kstw.com. Licensee: CW Television Stations Inc. Group owner: Viacom Television Stations Group (acq 4-16-97). Population served: 1,400,000 Natl. Network: CW. News: 7 hrs wkly.
Key Personnel:
Steve Gahler . . . . . VP
Amber Stelzer . . . . . natl sls mgr
Megan Temple . . . . . mktg dir
D. Poor. . . . . prom mgr
Tom Spitz . . . . . progmg dir
Ron Diotte. . . . . engrg dir

KTBW-TV— Analog channel: 20. Digital channel: 14. Analog hrs: 24 Digital hrs: 24 3,550 kw vis, 355 kw aur. 1,670t/350g TL: N47 32 50 W122 47 39 On air date: Mar 30, 1984. 1909 S. 341st Pl., Federal Way, WA, 98003. Phone: (253) 927-7720. Phone: (253) 874-7420. Fax: (253) 874-7432. E-mail: ktbw@tbn.org Licensee: Trinity Broadcasting of Washington. Group owner: Trinity Broadcasting Network
Key Personnel:
Paul F. Crouch . . . . . pres
Mary Jane Allen . . . . . stn mgr

KUNS-TV— Analog channel: 51. Digital channel: 50.3,800 kw vis, 500 kw aur. ant 2,358t/279g TL: N47 30 17 W121 58 04 On air date: 2000. 140 4th Ave. N., Suite 440, Seattle, WA, 98109. Phone: (206) 404-5867. Fax: (206) 248-6818. E-mail: info@kunstv.com Web Site: www.kunstv.com. Licensee: Fisher Broadcasting - Bellevue TV L.L.C. (acq 9-26-2006; $16 million). Natl. Network: Univision (Spanish). Foreign lang progmg: SP 168

KVOS-TV— Analog channel: 12. Digital channel: 35. Analog hrs: 24 234 kw vis, 45.7 kw aur. ant 2,368t/139g TL: N48 40 40 W122 49 48 On air date: June 3, 1953. 1151 Ellis St., Bellingham, WA, 98225. Phone: (360) 671-1212. Phone: (604) 681-1212 (sales). Fax: (360) 647-0824. Fax: (604) 736-4510. Web Site: www.kvos.com. Group owner: Clear Channel Communications Inc. (acq 6-14-2002; grpsl). Population served: 2,000,000 Natl. Rep: Airtime TV. Washington Atty: Rubin, Winston, Diercks, Harris & Cooke. News staff: 10; News: one hr wkly.

***KWDK**— Analog channel: 56. Digital channel: 42. Analog hrs: 24 5,000 kw vis, 500 kw aur. ant 1,872t TL: N47 32 53 W122 48 22 On air date: 2000. 18000 International Blvd., Ste 1007, Seatac, WA, 98188. Phone: (425) 251-4313. Web Site: www.daystar.com. Licensee: Puget Sound Educational TV Inc.

KWPX— Analog channel: 33. Digital channel: 32. Analog hrs: 24 14.8 kw vis, 1.5 kw aur. 938t TL: N47 36 17 W122 19 46 On air date: May 17, 1989. Box426, Preston, WA, 98050. 8112-C 304th Ave.SE, Preston, WA 98050. Phone: (425) 222-6010. Fax: (425 222-6032. Web Site: www.ionline.tv. Licensee: Paxson Communications License Co. L.L.C. Group owner: Paxson Communications Corp. (acq 2-2-98; $35 million). Natl. Network: i Network.

Spokane, WA
(DMA 77)

KAYU-TV— Analog channel: 28. Digital channel: 30. Analog hrs: 24 Digital hrs: 24 2,400 kw vis, 120 kw aur. ant 1,971t/794g TL: N47 34 44 W117 17 46 On air date: Oct 31, 1982. 4600 S. Regal St., Spokane, WA, 99223. Phone: (509) 448-2828. Fax: (509) 448-0926. Web Site: www.fox28spokane.com. Licensee: Mountain Licenses L.P. Group owner: Northwest Broadcasting Inc. (acq 1996; $6.44 million). Population served: 395,490 Natl. Network: Fox. Natl. Rep: Millennium Sales & Marketing. Washington Atty: Leventhal, Senter and Lerman. Wire Svc: AP News: 4 hrs wkly.
Key Personnel:
Brian Brady . . . . . CEO
Bill Quarles . . . . . CFO
Jon Rand . . . . . gen mgr
Rick Andrycha . . . . . opns mgr
David Lockhert. . . . . gen sls mgr
Becky Martin. . . . . natl sls mgr
Kim Rogge . . . . . prom dir
Ron Sweatte . . . . . chief of engrg

***KCDT-TV**— Analog channel: 26. Analog hrs: 24 12.3 kw vis, 1.2 kw aur. 1,525t TL: N47 43 54 W116 43 47 On air date: October 1991. c/o KAID, 1455 N. Orchard St., Boise, ID, 83706. Box 443101, University of Idaho, Moscow, ID 83844-3101. Phone: (208) 885-1226. E-mail: idptv@idahoptv.org Web Site: www.idahoptv.org. Licensee: State Board of Education, State of Idaho. Natl. Network: PBS. Washington Atty: Fletcher, Heald & Hildreth.
Key Personnel:
Peter Morrill . . . . . gen mgr
Kris Freeland . . . . . stn mgr
Kim Philipps . . . . . dev dir
Rebroadcasts KUID Moscow 100%.

KGPX— Analog channel: 34.2,820 kw vis. ant 1,476t/535g TL: N47 36 04 W117 17 53 On air date: 2000. 1201 W. Spregue Ave., Spokane, WA, 99201. Phone: (509) 340-3400. Fax: (509) 340-3417. Web Site: www.ionline.tv/stations/list.cfm. Licensee: Paxson Spokane License Inc. Natl. Network: i Network.
Key Personnel:
Amber Morales. . . . . opns mgr
Bill Storms . . . . . gen sls mgr
Don Kukuk . . . . . chief of engrg

KHQ-TV— Analog channel: 6. Digital channel: 15. Analog hrs: 24 Digital hrs: 24 87.1 kw vis, 17.4 kw aur. 2,150t/904g TL: N47 34 52 W117 I7 47 On air date: Dec 20, 1952. PO Box 600, Spokane, WA, 99210-0600. 1201 W. Sprague Ave., Spokane, WA 99201-4102. Phone: (509) 448-6000. Fax: (509) 448-4694. E-mail: q6news@khq.com Web Site: www.khq.com. Licensee: KHQ Inc. Group owner: (group owner) Population served: 1,500,000 Natl. Network: NBC. Natl. Rep: Blair Television. Washington Atty: Skadden, Arps. News staff: 52; News: 26 hrs wkly.
Key Personnel:
Betsy Cowles . . . . . chmn
Lon C. Lee . . . . . pres
Doug Miles. . . . . opns dir
Bill Storms . . . . . gen sls mgr
Mike Jackson . . . . . prom mgr
Mike Dugger. . . . . progmg mgr
Jonathan Michell . . . . . news dir

KLEW-TV— Analog channel: 3. Digital channel: 32. Analog hrs: 20 Digital hrs: 20 56.2 kw vis, 1.38 kw aur. 1,260t/303 TL: N46 27 25 W117 05 57 On air date: December 1955. PO Box 615, Lewiston, ID, 83501. 2626 17th St., Lewiston, ID 83501. Phone: (208) 746-2636. Fax: (208) 746-4819. E-mail: info@klewtv.com Web Site: www.klewtv.com. Licensee: Fisher Broadcasting - Washington TV L.L.C. Group owner: Fisher Broadcasting Company (acq 12-4-2001; grpsl). Population served: 165,000 Natl. Network: CBS. Natl. Rep: Petry Television Inc. Washington Atty: Shaw Pittman. Wire Svc: AP News staff: 4; News: 12 hrs wkly.
Key Personnel:
Fred Fickenwirth . . . . . stn mgr & gen sls mgr
Greg Meyer . . . . . news dir
Margo Aragon . . . . . pub affrs dir
Bill Dunlap. . . . . chief of engrg
Ann Fickenwirth . . . . . opns
Satellite of KIMA-TV Yakima Wash.

KQUP— Analog channel: 24. Analog hrs: 24 29.2 kw vis. ant 1,079t/75g TL: N46 51 44 W117 10 22 On air date: 2005. 1020 W. Riverside, Spokane, WA, 99201. Phone: (509) 924-5787. Fax: (509) 924-5789. E-mail: rhall@kqup.com Web Site: www.kqup.com. Licensee: Pullman Broadcasting Inc. Group owner: Equity Broadcasting Corp. (acq 8-23-99). Population served: 378,500

KREM-TV— Analog channel: 2. Digital channel: 20. Analog hrs: 24 84.7 kw vis, 15.5 kw aur. ant 2,200t/969g TL: N47 35 42 W117 17 53 On air date: Oct 31, 1954. Box 8037, Spokane, WA, 99203. 4103 S. Regal, Spokane, WA 99223. Phone: (509) 448-2000. Fax: (509) 448-6397 (news). Fax: (509) 448-2090 (sales). Web Site: www.krem.com. Licensee: KREM-TV Inc. Group owner: Belo Corp., Broadcast Division (acq 9-92; grpsl; FTR: 9-16-91). Population served: 174,500 Natl. Network: CBS. Washington Atty: Covington & Burling. News staff: 36; News: 17 hrs wkly.

Key Personnel:

Robert Decherd . . . CEO
Jim Maroney . . . exec VP
Bud Brown . . . gen mgr
Amy Warren . . . gen sls mgr
Bruce Felt . . . prom dir
Christine Werfelmann . . . progmg dir
Boyd Lundberg . . . chief of engrg

KSKN— Analog channel: 22. Digital channel: 36. Analog hrs: 24 1,860 kw vis. ant 1,958t TL: N47 35 42 W117 17 53 On air date: Oct 1, 1983. 4103 S. Regal, Spokane, WA, 99223-7377. Phone: (509) 448-2000. Fax: (509) 448-2090. Web Site: www.krem.com. Licensee: KSKN Television Inc. Group owner: Belo Corp., Broadcast Division (acq 8-24-01; $5 million). Natl. Network: CW. News staff: 5; News: 6 hrs wkly.

Key Personnel:

D.J. Wilson . . . pres & sr VP
Albert B. Brown . . . gen mgr
Dan Lamphere . . . opns dir & opns mgr
Susan Miller . . . gen sls mgr
Ron Keller . . . prom dir & prom mgr
Terry Cocker . . . progmg dir
Noah Cooper . . . news dir
John Souza . . . engrg dir
Boyd Lundberg . . . chief of engrg

***KSPS-TV—** Analog channel: 7. Digital channel: 8. Analog hrs: 24 Digital hrs: 24 316 kw vis, 31.6 kw aur. 1,830t/600g TL: N47 34 34 W117 17 58 On air date: Apr 24, 1967. S. 3911 Regal St., Spokane, WA, 99223. Phone: (509) 354-7800. Fax: (509) 354-7757. Web Site: www.ksps.org. Licensee: Spokane School District No. 81. Population served: 1,200,000 Natl. Network: PBS. Washington Atty: Garvey, Schubert & Barer.

Key Personnel:

Claude Kistler . . . gen mgr
Patty Starkey . . . dev dir
Kerry Faggiano . . . prom mgr
Cary Balzer . . . progmg mgr

***KUID-TV—** Analog channel: 35. Digital channel: 12. Analog hrs: 24 44 kw vis. ant 971t/148g TL: N46 40 54 W116 58 13 On air date: July 1, 1965. c/o KAID, 1455 N. Orchard St., Boise, ID, 83706. Box 443101, University of Idaho, Moscow, ID 83844-3101. Phone: (208) 885-1226. E-mail: idptv@idahoptv.org Web Site: www.idahoptv.org. Licensee: State Board of Education, State of Idaho. Population served: 151,000 Natl. Network: PBS. Washington Atty: Fletcher, Heald & Hildreth.

Key Personnel:

Peter Morrill . . . gen mgr
Kris Freeland . . . stn mgr
Kim Philipps . . . dev dir

***KWSU-TV—** Analog channel: 10. Digital channel: 17. Analog hrs: 24 Digital hrs: 24 117 kw vis, 11.7 kw aur. 1,350t/300g TL: N46 51 43 W117 10 26 On air date: Sept 24, 1962. Box 642530, Pullman, WA, 99164-2530. Phone: (509) 335-6511. Fax: (509) 335-3772. E-mail: kwsu@wsu.edu Web Site: www.kwsu.org. Licensee: Washington State University. Population served: 170,000 Natl. Network: PBS. Washington Atty: Dow, Lohnes & Albertson.

Key Personnel:

Dennis Haarsager . . . gen mgr
Rita Brown . . . dev dir
Kathy Dahmen . . . mktg dir
Warren Wright . . . stn mgr & progmg dir
Ralph Hogan . . . engrg mgr

KXLY-TV— Analog channel: 4. Digital channel: 13. Analog hrs: 24 Digital hrs: 24 Note: ABC is on KXLY-TV ch 4, MyNetworkTV is on KXLY-DT ch 13. 48 kw vis, 4.8 kw aur. ant 3,060t/153g TL: N47 55 18 W117 06 48 On air date: Feb 22, 1953. 500 W. Boone Ave., Spokane, WA, 99201. Phone: (509) 324-4000. Fax: (509) 328-5274. Web Site: www.kxly.com. Licensee: Spokane TV Inc. Group owner: Evening Telegram Company—Morgan Murphy Stns (acq 1-17-63; grpsl; FTR: 1-63). Population served: 555,400 Natl. Network: ABC, MyNetworkTV Digital Network: Note: ABC is on KXLY-TV ch 4, MyNetworkTV is on KXLY-DT ch 13. Natl. Rep: Continental Television Sales. Washington Atty: Rini, Coran, PC. Wire Svc: AP News: 19 hrs wkly.

Key Personnel:

Elizabeth M. Burns . . . pres
Steve Herling . . . gen mgr

Tacoma

see Seattle-Tacoma, WA market

Vancouver

see Portland, OR market

Yakima-Pasco-Richland-Kennewick, WA

(DMA 125)

KAPP— Analog channel: 35. Digital channel: 14.Note: ABC is on KAPP(TV) ch 35, MyNetworkTV is on KAPP-DT ch 14. 646 kw vis, 64.6 kw aur. ant 961t TL: N46 31 57 W120 30 33 On air date: Sept 21, 1970. Box 10208, Yakima, WA, 98909-1208. 1610 S. 24th Ave., Yakima, WA 98902. Phone: (509) 453-0351. Fax: (509) 453-3623. E-mail: comments@kapptv.com Web Site: www.kapptv.com. Licensee: Apple Valley Broadcasting Inc. Group owner: Morgan Murphy Stations Population served: 421,000 Natl. Network: ABC, MyNetworkTV Digital Network: Note: ABC is on KAPP(TV) ch 35, MyNetworkTV is on KAPP-DT ch 14. Washington Atty: Manatt, Phelps & Phillips.

Key Personnel:

Elizabeth Burns . . . pres
Brian Paul . . . gen mgr
Shane Pierone . . . rgnl sls mgr
Mike Balmelli . . . news dir
Neil Bennett . . . chief of engrg

KCWK— Analog channel: 9. Digital channel: 9. Analog hrs: 24 Digital hrs: 24 316 kw vis. ant 577t TL: N46 09 05 W118 03 04 On air date: 2002. 424 E. Yakima Ave., Suite 110, Yakima, WA, 98901-2723. Phone: (509) 575-0999. Fax: (509) 575-9562. Web Site: www.kcw9.com. Licensee: KAZW License LLC. Group owner: Pappas Telecasting Companies (acq 11-7-2002; $3 million). Natl. Network: CW.

Key Personnel:

Robert L. Powers . . . VP & gen mgr
Mike Dunlop . . . gen sls mgr

KEPR-TV— Analog channel: 19. Digital channel: 18. Analog hrs: 24 490 kw vis, 88.3 kw aur. 1,203t/354g TL: N46 05 51 W119 11 30 On air date: Dec 28, 1954. Box 2648, Pasco, WA, 99302. 2807 W. Lewis, Pasco, WA 99301. Phone: (509) 547-0547. Fax: (509) 547-2845. Web Site: www.keprtv.com. Licensee: Fisher Broadcasting - Washington TV L.L.C. Group owner: Fisher Broadcasting Company (acq 12-4-2001; grpsl). Population served: 260,000 Natl. Network: CBS. Washington Atty: Winthrope, Shaw, Pittman, LLP. Wire Svc: CBS Wire Svc: Pacifica Network News News staff: 12; News: 17 hrs wkly.

Key Personnel:

Ben Tucker . . . pres
Ken Messer . . . gen mgr
David Pray . . . stn mgr
Brad Gayken . . . opns mgr
Steve Crow . . . gen sls mgr & natl sls mgr
Randy Irwin . . . prom mgr
Stu Seibel . . . progmg dir
Cris Headley . . . pub affrs dir
John Housholder . . . chief of engrg

KFFX-TV— Analog channel: 11. Digital channel: 8. Analog hrs: 6 AM-12:35 AM 3.16 kw vis. ant 1,043t TL: N45 40 58 W118 46 17 On air date: 1999. 2509 W. Falls Ave., Kennewick, WA, 99336. 4600 S. Regal St., Spokane, WA 99223. Phone: (509) 735-1700. Fax: (509) 735-1004. Web Site: www.fox11tricities.com. Licensee: Mountain Licenses L.P. Group owner: Northwest Broadcasting Inc. (acq 1-14-2003; $239,659 for CP). Population served: 211,610 Natl. Network: Fox. Natl. Rep: Millennium Sales & Marketing. Washington Atty: Leventhal, Senter & Lerman.

Key Personnel:

Brian Brady . . . CEO
Bill Quarles . . . CFO
Jon Rand . . . gen mgr
Glenn Rousch . . . stn mgr
Rick Andrycha . . . opns mgr
Lynn Creager . . . natl sls mgr
Lonnie Eaton . . . rgnl sls mgr
Ron Sweatte . . . chief of engrg
Jennifer Ranney . . . prom

KIMA-TV— Analog channel: 29. Digital channel: 33. Analog hrs: 24 Digital hrs: 24 490 kw vis, 87.3 kw aur. ant 971t/67g TL: N46 31 58 W120 30 26 On air date: July 19, 1953. Box 702, Yakima, WA, 98907. 2801 Terrace Heights Dr., Yakima, WA 98901. Phone: (509) 575-0029. Fax: (509) 248-1218. E-mail: information@kimatv.com Web Site: www.kimatv.com. Licensee: Fisher Broadcasting - Washington TV L.L.C. Group owner: Fisher Broadcasting Company (acq 12-4-2001; grpsl). Population served: 443,000 Natl. Network: CBS. Natl. Rep: Petry Television Inc. Washington Atty: Shaw Pittman. Wire Svc: CBS Wire Svc: CNN Wire Svc: AP News staff: 24; News: 15 hrs wkly.

Key Personnel:

Colleen Brown . . . CEO & pres
Mr. Phelps Fisher . . . chmn
Ken Messer . . . VP & gen mgr
Karla Griffin . . . opns mgr
Steve Crow . . . gen sls mgr & natl sls mgr
Cheryl Menke . . . rgnl sls mgr
Stu Siebel . . . progmg dir
Robin Wojtanik . . . news dir
Cliff Grady . . . chief of engrg

KNDO— Analog channel: 23. Analog hrs: 24 501 kw vis, 61 kw aur. 961t/161g TL: N46 31 59 W130 20 36 On air date: Oct 15, 1959. 1608 S. 24th Ave., Yakima, WA, 98902. Phone: (509) 225-2323. Fax: (509) 225-2330. E-mail: news@kndo.com Web Site: www.kndo.com. Licensee: KHQ Inc. Group owner: (group owner; (acq 6-17-99; $22.25 million with KNDU(TV) Richland). Population served: 203,195 Natl. Network: NBC. News staff: 9; News: 27 hrs wkly.

Key Personnel:

Lon Lee . . . pres
Paul Dughi . . . gen mgr
Larry Forsgren . . . gen sls mgr
Scott Morgan . . . mktg dir
Susan Martinez . . . progmg dir
Christine Brown . . . news dir
Mark Kennedy . . . engrg dir & chief of engrg

KNDU— Analog channel: 25. Analog hrs: 24 661 kw vis, 66.1 kw aur. 1,348t TL: N46 06 11 W119 07 47 On air date: July 1, 1961. 3312 W. Kennewick Ave., Kennewick, WA, 99336. Phone: (509) 737-6700. Phone: (509)737-6725. Fax: (509) 737-6749. E-mail: news@kndu.com Web Site: www.kndu.com. Licensee: KHQ Inc. Group owner: (group owner; (acq 6-17-99; $22.25 million with KNDO(TV) Yakima). Population served: 265,600 Natl. Network: NBC. Washington Atty: Hogan & Hartson. News staff: 22; News: 22.5 hrs wkly.

Key Personnel:

Paul Dughi . . . gen mgr
Larry Sorsgren . . . gen sls mgr
Sheri Bissell . . . natl sls mgr
Randy Brown . . . rgnl sls mgr
Susan Martinez . . . progmg dir
Christine Brown . . . news dir

Satellite of KNDO(TV) Yakima 94%.

***KTNW—** Analog channel: 31. Digital channel: 38. Analog hrs: 24 Digital hrs: 24 53.5 kw vis, 5.35 kw aur. 1,198t/36g TL: N46 06 23 W119 07 50 On air date: Oct 18, 1987. Box 642530, Pullman, WA, 99164-2530. Phone: (509) 335-6588. Fax: (509) 335-3772. E-mail: nwptv@wsu.edu Web Site: www.kwsu.org. Licensee: Washington State University. Natl. Network: PBS. Washington Atty: Dow, Lohnes & Albertson.

Key Personnel:

Dennis Haarsager . . . VP, gen mgr & gen mgr
Rita Brown . . . dev dir
Kathy Dahmen . . . mktg dir
Warren Wright . . . stn mgr & progmg dir
Ralph Hogan . . . engrg dir & engrg mgr

KVEW— Analog channel: 42. Digital channel: 44. Analog hrs: 24 Note: ABC is on KVEW(TV) ch 42, MyNetworkTV is on KVEW-DT ch 44. 501 kw vis, 39.8 kw aur. ant 1,280t/205g TL: N46 06 11 W119 07 54 On air date: Oct 30, 1970. 601 N. Edison, Kennewick, WA, 99336. Phone: (509) 735-8369. Fax: (509) 735-1836 (news). Fax: (509) 735-7889. Web Site: www.kvewtv.com. Licensee: Apple Valley Broadcasting Inc. Group owner: Morgan Murphy Stations Population served: 163,300 Natl. Network: ABC, MyNetworkTV Digital Network: Note: ABC is on KVEW(TV) ch 42, MyNetworkTV is on KVEW-DT ch 44.

Key Personnel:

Brian Paul . . . gen mgr
Mike Balmelli . . . news dir

Satellite of KAPP Yakima.

***KYVE—** Analog channel: 47. Digital channel: 21. Analog hrs: 24 Digital hrs: 24 640 kw vis, 64 kw aur. ant 918t/135g TL: N46 31 58 W120 30 33 On air date: Nov 1, 1962. 12 S. 2nd St., Yakima, WA, 98901. Phone: (509) 452-4700. Fax: (509) 452-4704. Web Site: www.kyve.org. Licensee: KCTS Television. Ownership: KCTS Television board of directors (acq 8-1-94; FTR: 8-22-94). Population served: 192,000 Natl. Network: PBS. Washington Atty: Schwartz, Woods & Miller.

Key Personnel:
Bill Mohler . . . CEO
Mark Leonard. . . . gen mgr
Brenda Setterlund . . . dev mgr, mktg mgr, mktg mgr & prom mgr
Chris Splawn . . . progmg mgr
Rod Venable . . . stn mgr & chief of engrg

West Virginia

Beckley

see Bluefield-Beckley-Oak Hill, WV market

Bluefield-Beckley-Oak Hill, WV (DMA 150)

WLFB— Analog channel: 40. Digital channel: 14.2,880 kw vis. ant 1,282t/177g TL: N37 13 08 W81 15 39 On air date: 2001. Box 1867, Abingdon, VA, 24212. 8594 Hidden Valley Rd., Abingdon, VA 24210. Phone: (276) 676-3806. Fax: (276) 676-3572. Licensee: Living Faith Ministries Inc.
Key Personnel:
Micheal D. Smith. . . . CEO
Michael D. Smith . . . pres & gen mgr
Fredia Keene . . . CFO
Satellite of WLFG(TV) Grundy, VA.

WOAY-TV— Analog channel: 4. Digital channel: 50. Analog hrs: 24 100 kw vis, 20 kw aur. ant 740t/688g TL: N37 57 30 W81 09 03 On air date: Dec 14, 1954. Box 3001, Oak Hill, WV, 25901. Phone: (304) 469-3361. Fax: (304) 465-1420. E-mail: news@woay.com Web Site: www.woay.com. Licensee: Thomas Broadcasting Co. Population served: 332,000 Natl. Network: ABC. Washington Atty: Fletcher, Heald & Hildreth.
Key Personnel:
Robert R. Thomas III . . . pres
Al Marra . . . gen mgr, gen sls mgr & progmg dir
Robert Brunner . . . news dir
Jim Martin . . . chief of engrg

***WSWP-TV—** Analog channel: 9. Analog hrs: 24 316 kw vis, 63.2 kw aur. 1,000t/488g TL: N37 53 46 W80 59 21 On air date: Nov 1, 1970. PO Box 9004, Beckley, WV, 25802. 124 Industrial Dr., Beaver, WV 25813. Phone: (304) 254-7840. Fax: (304) 254-7879. Web Site: www.wvpubcast.org. Licensee: West Virginia Educational Broadcasting Authority. Population served: 970,000 Natl. Network: PBS. News: 5 hrs wkly.
Key Personnel:
Rita Ray . . . CEO
Mike Meador. . . . gen mgr & stn mgr
Marilyn DeVita . . . dev dir

WVNS-TV— Analog channel: 59. Digital channel: 48. Analog hrs: 24 1,923 kw vis. 1,863t TL: N37 46 22 W80 42 25 On air date: Jan 1, 1997. Box 509, Ghent, WV, 25843. 141 Old Cline Rd., Ghent, WV 25843. Phone: (304) 787-5959. Fax: (304) 787-2440. E-mail: fbarnes@wvnstv.com Web Site: www.cbs59.com. Licensee: West Virginia Media Holdings LLC. Group owner: (group owner; (acq 1-9-2003). Population served: 270,000 Natl. Network: CBS. Natl. Rep: Petry Television Inc. Washington Atty: Borsari & Paxson.
Key Personnel:
Bray Cary. . . . CEO & pres
Marstow W. Becker. . . . chmn
Charlie Dusic . . . CFO
Chris Leister . . . sr VP
Frank Barnes . . . gen mgr
Sue Bosio. . . . opns mgr
Gary Bowden . . . sls VP
Jack Scott. . . . gen sls mgr
Robert McCallister . . . rgnl sls mgr
John Fawcett . . . prom dir
M.J. Coss . . . progmg dir
Gary Kirk . . . chief of engrg

WVVA— Analog channel: 6. Digital channel: 46. Analog hrs: 24 Digital hrs: 24 Note: NBC is on WVVA(TV) ch 6, CW is on WVVA-DT ch 46. 50.1 kw vis, 6.03 kw aur. ant 1,220t/185g TL: N37 15 21 W81 10 55 On air date: July 31, 1955. Box 1930, Bluefield, WV, 24701. Phone: (304) 325-5487. Fax: (304) 327-5586. Web Site: www.wvva.com. Licensee: WVVA TV Inc. Group owner: Quincy Newspapers Inc., see Cross-Ownership (acq 5-1-79; $8 million; FTR: 4-23-79). Population served: 325,000 Natl. Network: NBC, CW Digital Network: Note: NBC is on WVVA(TV) ch 6, CW is on WVVA-DT ch 46. Natl. Rep: Blair Television. Washington Atty: Wilkinson, Barker, Knauer & Quinn. News staff: 20; News: 27 hrs wkly.

Key Personnel:
Thomas A. Oakley . . . CEO
Ralph M. Oakley . . . COO
Frank Brady. . . . VP & gen mgr
Jim Briggs . . . sls dir & mktg dir
Danny Via . . . engrg dir

Charleston-Huntington, WV (DMA 65)

WCHS-TV— Analog channel: 8. Digital channel: 41. Analog hrs: 24 Digital hrs: 24 49.6 kw vis. ant 1,746t/1,477g TL: N38 24 28 W81 54 13 On air date: Aug 15, 1954. 1301 Piedmont Rd., Charleston, WV, 25301. Phone: (304) 346-5358. Fax: (304) 346-4765. E-mail: info@wchstv.com Web Site: www.wchstv.com. Licensee: WCHS Licensee L.L.C. Group owner: Sinclair Broadcast Group Inc. (acq 10-8-97). Population served: 510,000 Natl. Network: ABC. Washington Atty: Fisher, Wayland, Cooper, Leader & Zaragoza. News staff: 37; News: 19.5 hrs wkly.
Key Personnel:
Harold Cooper . . . gen mgr
Paul Fox. . . . prom dir
Lori Marquette . . . progmg dir
Terry Cole . . . news dir
Raymon Beckner . . . chief of engrg

***WKAS—** Analog channel: 25. Digital channel: 25.162 kw vis, 16.2 kw aur. 498t/390g TL: N38 27 43 W82 37 12 On air date: Sept 23, 1968. 600 Cooper Dr., Lexington, KY, 40502. Phone: (859) 258-7000. Fax: (859) 258-7390. Web Site: www.ket.org. Licensee: Kentucky Authority for Educational TV. Natl. Network: PBS.
Key Personnel:
Craig Cornwell . . . opns mgr & progmg dir
Tim Bischoff . . . mktg dir
Robert Ball . . . engrg dir

***WKPI—** Analog channel: 22. Digital channel: 24.468 kw vis, 93.3 kw aur. 1,410t/153g TL: N37 17 06 W82 31 29 On air date: Apr 8, 1968. 600 Cooper Dr., Lexington, KY, 40502. Phone: (859) 233-3000. Fax: (859) 258-7399. Web Site: www.ket.org. Licensee: Kentucky Authority for Educational TV. Natl. Network: PBS.
Key Personnel:
Craig Cornwell. . . . opns mgr & progmg mgr
Tim Bischoff . . . mktg mgr
Mike Brower . . . pub affrs dir & progmg
Robert Ball . . . engrg dir

WLPX-TV— Analog channel: 29. Digital channel: 39. Analog hrs: 24 Digital hrs: 24 5,000 kw vis. ant 1,207t/968g TL: N38 28 12 W81 46 35 On air date: Aug 28, 1998. 600 C Prestige Dr., Hurricane, WV, 25526. Phone: (304) 760-1029. Fax: (304) 760-1036. Web Site: www.ionline.tv/stations/list.cfm. Licensee: Paxson Charleston License Inc. Group owner: Paxson Communications Corp. (acq 10-28-98; $8.25 million). Natl. Network: i Network. Washington Atty: Dan J. Alpert.
Key Personnel:
Brandon Burgess. . . . CEO
Carol Holzhauer . . . VP
Steven Stanley . . . stn mgr
Joseph Koker . . . opns dir
Gene Monday . . . chief of engrg

***WOUB-TV—** Analog channel: 20. Digital channel: 27. Analog hrs: 24 Digital hrs: 24 1,000 kw vis, 100 kw aur. 800t/856g TL: N39 18 50 W82 08 54 On air date: Jan 3, 1963. 9 S. College St., Athens, OH, 45701. Phone: (740) 593-4555. Fax: (740) 593-0240. E-mail: woub@woub.org Web Site: www.woub.org. Licensee: Ohio University. Population served: 151,000 Natl. Network: PBS. Rgnl. Network: Ohio Educ Bcstg. Washington Atty: Dow, Lohnes & Albertson. News staff: 3; News: 3 hrs wkly.
Key Personnel:
Carolyn Bailey-Lewis . . . gen mgr
David Wiseman . . . opns VP
Steve Skidmore . . . opns dir
Scott Martin . . . opns mgr
Loring Lovett . . . rgnl sls mgr
Mark Brewer . . . progmg dir & film buyer
Joan Butcher . . . progmg mgr
Tim Sharp . . . news dir
Dave Wiseman. . . . engrg VP

WOWK-TV— Analog channel: 13. Analog hrs: 24 141 kw vis, 26.3 kw aur. 1,269t/1,108g TL: N38 30 21 W82 12 33 On air date: Oct 2, 1955. 555 Fifth Ave., Huntington, WV, 25701. Phone: (304) 525-1313. Fax: (304) 529-4910. Web Site: wowktv.com. Licensee: West Virginia Media Holdings LLC. Group owner: (group owner; (acq 4-8-2002; $40.5 million). Population served: 667,520 Natl. Network: CBS. Natl. Rep: TeleRep. Washington Atty: Bryan Cave.

Key Personnel:
Bray Cary. . . . pres
John Fawcett . . . gen mgr, mktg dir & prom dir
Chris Leister. . . . gen sls mgr
M.J. Coss . . . progmg dir
Rod Fowler . . . news dir
Bill Gallaway . . . chief of engrg

***WPBO-TV—** Analog channel: 42. Digital channel: 43. Analog hrs: 24 Digital hrs: 24 525 kw vis, 52.5 kw aur. ant 1,253t/922g TL: N38 45 42 W83 03 41 On air date: October 1973. 2400 Olentangy River Rd., Columbus, OH, 43210. Phone: (614) 292-9678. Fax: (614) 688-3399. E-mail: wosu@wosu.org Web Site: www.wosu.org. Licensee: The Ohio State University. Natl. Network: PBS. Rgnl. Network: CEN. Washington Atty: Dow, Lohnes & Albertson.
Key Personnel:
Thomas Rieland. . . . gen mgr
Edwin Clay . . . stn mgr
John Prosek. . . . opns mgr
Mary Yerina . . . dev dir
Rebroadcasts WOSU-TV Columbus 100%.

***WPBY-TV—** Analog channel: 33. Analog hrs: 24 2,371 kw vis, 105 kw aur. 1,243t TL: N38 29 41 W82 12 03 On air date: July 14, 1969. 600 Capitol St., Charleston, WV, 25301. Phone: (304) 556-4900 / (304) 556-4905. Fax: (304) 556-4982. Web Site: www.wvpubcast.org. Licensee: West Virginia Educational Broadcasting Authority. Population served: 300,000 Natl. Network: PBS. Rgnl. Network: SECA.
Key Personnel:
Bill Acker . . . gen mgr
Mike Meador. . . . CFO & gen mgr
Marilyn DiVita. . . . dev dir & dev mgr
Craig Lanham. . . . progmg dir
Greg Collard . . . news dir & pub affrs dir

WQCW— Analog channel: 30. Digital channel: 17. Analog hrs: 24 Digital hrs: 24 2,040 kw vis, 204 kw aur. ant 1,174t/850g TL: N38 45 42 W83 03 41 On air date: Oct 5, 1998. 800 Gallia, Suite 430, Portsmouth, OH, 45662. Phone: (740) 353-3391. Fax: (740) 353-3372. E-mail: wb30@whcp-tv.com Web Site: www.wqcw.com. Licensee: Television Properties Inc. Ownership: Commonwealth Broadcasting Group Inc., 88%; Kenneth Russell, 12% (acq 7-11-2002). Population served: 298,742 Natl. Network: CW.
Key Personnel:
Charles M. Harker . . . pres
Chuck Jones . . . stn mgr & gen sls mgr

WSAZ-TV— Analog channel: 3. Digital channel: 23.Note: NBC is on WSAZ-TV ch 3, MyNetworkTV is on WSAZ-DT ch 23. 42.7 kw vis, 7 kw aur. ant 1,273t/1,101g TL: N38 30 34 W82 13 09 On air date: Nov 15, 1949. 645 Fifth Ave., Huntington, WV, 25701. Phone: (304) 697-4780. Fax: (304) 690-3065 (news). E-mail: news@wsaz.com Web Site: www.wsaz.com. Licensee: Gray Television Licensee Inc. Group owner: Emmis Communications Corp. (acq 11-30-2005; $186 million). Population served: 1,252,100 Natl. Network: NBC, MyNetworkTV Digital Network: Note: NBC is on WSAZ-TV ch 3, MyNetworkTV is on WSAZ-DT ch 23.
Key Personnel:
Don Ray. . . . gen mgr
Aaron Withrow . . . opns dir

WTSF— Analog channel: 61. Digital channel: 44. Analog hrs: 24 36.3 kw vis, 3.72 kw aur. 464t/499g TL: N38 25 11 W82 24 06 On air date: Apr 30, 1983. Box 2320, Ashland, KY, 41105-2320. 3100 Bath Ave., Ashland, KY 41101. Phone: (606) 329-2700. Fax: (606) 324-9256. Web Site: www.wtsftv.com. Licensee: Word of God Fellowship Inc. Ownership: Jimmie F. Lamb, 25%; John T. Calender, 25%; Joni T. Lamb, 25%; and Marcus D. Lamb, 25% (acq 4-18-2003).
Key Personnel:
Richard Clifton . . . gen mgr
Virgil Adkins . . . chief of engrg

WVAH-TV— Analog channel: 11. Analog hrs: 24 51 kw vis, 5.1 kw aur. 1,722t/1,552g TL: N38 25 15 W81 55 27 On air date: Sept 19, 1982. 11 Broadcast Plaza, Hurricane, WV, 25526. Phone: (304) 757-0011. Fax: (304) 757-7533. E-mail: info@wvah.com Web Site: www.wvah.com. Licensee: WVAH Licensee LLC. Group owner: Cunningham Broadcasting Corporation (acq 11-15-2001; grpsl). Natl. Network: Fox. Washington Atty: Arter & Hadden. News staff: 38; News: 7 hrs wkly.
Key Personnel:
Harold Cooper . . . pres & gen mgr
Paul Fox. . . . prom dir
Lori Marquette . . . progmg dir
Matt Snyder . . . news dir
Raymond Beckner. . . . chief of engrg

Clarksburg-Weston, WV (DMA 166)

WBOY-TV— Analog channel: 12. Digital channel: 52. Analog hrs: 24 Digital hrs: 24 263 kw vis, 42.5 kw aur. 860t/593g TL: N39 17 06 W80 19 46 On air date: Nov 17, 1957. 904 W. Pike St., Clarksburg, WV,

26301. 912 W. Pike St., Clarksburg, WV 26301. Phone: (304) 623-3311. Fax: (304) 624-6152. Web Site: www.wboy.com. Licensee: West Virginia Media Holdings LLC. Group owner: (group owner; (acq 10-25-2001; $20 million). Population served: 250,000 Natl. Network: NBC. Natl. Rep: Petry Television Inc. Washington Atty: Cohn & Marks LLP. Wire Svc: AP News staff: 24; News: 24 hrs wkly.

Key Personnel:

Bray Cary. . . . . CEO & pres
Marty Becker . . . . . chmn
Charlie Dusic . . . . . CFO
Gary McNair. . . . . VP, gen mgr & progmg dir
Kim Morrison . . . . . opns mgr
George Boggs . . . . . gen sls mgr
Scott Sterling . . . . . natl sls mgr
John Fawcett . . . . . mktg mgr
Amanda Leasburg . . . . . prom mgr
Jim Platzer. . . . . news dir
Bob Hardman . . . . . chief of engrg

WDTV— Analog channel: 5. Digital channel: 6. Analog hrs: 24 100 kw vis, 20 kw aur. ant 879t/503g TL: N39 04 27 W80 25 28 On air date: June 1, 1960. Box 480, Bridgeport, WV, 26330. Phone: (304) 848-5000. Fax: (304) 842-7501. E-mail: wdtv@wdtv.com Web Site: www.wdtv.com. Licensee: W. Russell Withers Jr. Group owner: Withers Broadcasting Co. (acq 5-8-73; $600,000; FTR: 4-16-73). Population served: 116,000 Natl. Network: CBS. Washington Atty: Gardner, Carton & Douglas. News staff: 21; News: 18.5 hrs wkly.

Key Personnel:

W. Russell Withers Jr. . . . . . pres
Tim Defazio . . . . . gen mgr & dev dir
John Breen. . . . . opns dir

WVFX— Analog channel: 46. Digital channel: 10. Analog hrs: 24 Note: Fox is on WVFX(TV) ch 46, CW is on WVFX-DT ch 10. 155 kw vis, 15.5 kw aur. ant 800t/632g TL: N39 18 02 W80 20 37 On air date: Jan 1999. 775 W. Pike St., Clarksburg, WV, 26301. Phone: (304) 622-9839. Fax: (304) 623-9021. E-mail: info@wvfx.com Web Site: www.wvfx.com. Licensee: Davis Television Clarksburg L.L.C. (acq 1-26-99). Natl. Network: Fox, CW Digital Network: Note: Fox is on WVFX(TV) ch 46, CW is on WVFX-DT ch 10. Washington Atty: Leventhal, Senter & Lerman.

Huntington

see Charleston-Huntington, WV market

Martinsburg

see Washington, DC (Hagerstown, MD) market

Morgantown

see Pittsburgh, PA market

Oak Hill

see Bluefield-Beckley-Oak Hill, WV market

Parkersburg, WV (DMA 189)

WTAP-TV— Analog channel: 15. Digital channel: 49. Analog hrs: 5 AM-2:35 AM (M-F); 6 AM-2:05 AM (S); 6 AM-12:35 AM (Su) Digital hrs: 5 AM-2:35 AM (M-F); 6 AM-2:05 AM (S); 6 AM-12:35 AM (Su) Note: NBC is on WTAP-TV ch 15, Fox and MyNetworkTV are on WTAP-DT ch 49, 220 kw vis, 41.1 kw aur. ant 630t/440g TL: N39 20 59 W81 33 56 On air date: Oct 8, 1953. One Television Plaza, Parkersburg, WV, 26101. Phone: (304) 485-4588. Fax: (304) 422-3920. E-mail: gm@wtap.com Web Site: www.wtap.com. Licensee: Gray Television Group, Inc. Group owner: Gray Television Inc. (acq 8-29-2002; grpsl). Population served: 325,000 Natl. Network: NBC, Fox, MyNetworkTV Digital Network: Note: NBC is on WTAP-TV ch 15, Fox and MyNetworkTV are on WTAP-DT ch 49, Note: NBC is on WTAP-TV ch 15, Fox and MyNetworkTV are on WTAP-DT ch 49, Note: NBC is on WTAP-TV ch 15, Fox and MyNetworkTV are on WTAP-DT ch 49 Natl. Rep: Continental Television Sales. Washington Atty: Covington & Burling. Wire Svc: AP Wire Svc: CNN News staff: 17; News: 23 hrs wkly.

Key Personnel:

Roger Sheppard . . . . . VP & gen mgr
Joyce Ancrile . . . . . prom dir
Shane Vass . . . . . rgnl sls mgr & progmg dir
Bruce Layman . . . . . news dir
Kevin Buskirk . . . . . chief of engrg & engr

Weston

see Clarksburg-Weston, WV market

Wheeling, WV-Steubenville, OH (DMA 155)

***WOUC-TV**— Analog channel: 44. Digital channel: 35. Analog hrs: 24 Digital hrs: 24 759 kw vis. ant 1,263t/1,174g TL: N40 05 32 W81 17 19 On air date: July 23, 1973. 9 S. College St., Athens, OH, 45701. Phone: (740) 593-4555. Fax: (740) 593-0240. E-mail: woub@woub.org Web Site: www.wouc.org. Licensee: Ohio University. (acq 12-10-75; FTR: 12-22-75). Natl. Network: PBS. Rgnl. Network: Ohio Educ Bcstg. Washington Atty: Cohn & Marks.

Key Personnel:

Carolyn Bailey-Lewis . . . . . gen mgr
David Wiseman. . . . . opns VP & engrg VP
Steve Skidmore . . . . . opns dir
Scott Martin . . . . . opns mgr
Loring Lovett . . . . . rgnl sls mgr
Mark Brewer . . . . . progmg dir
Joan Butcher . . . . . progmg mgr
Rebroadcasts WOUB-TV Athens 100%.

WTOV-TV— Analog channel: 9. Digital channel: 57. Analog hrs: 24 Digital hrs: 24 316 kw vis, 31.6 kw aur. ant 925t/869g TL: N40 20 33 W80 37 14 On air date: Dec 24, 1953. Box 9999, Steubenville, OH, 43952. 9 Red Donelly Plaza (also shipping), Mingo Junction, OH 43938. Phone: (740) 282-9999. Phone: (304) 232-6933. Fax: (740) 282-0350. Web Site: www.wtov9.com. Licensee: WTOV Inc. Group owner: Cox Broadcasting (acq 9-22-2000; $58 million). Population served: 144,000 Natl. Network: NBC, ABC. Natl. Rep: TeleRep. Washington Atty: Dow,Lohnes & Albertson. Wire Svc: AP News staff: 32; News: 22 hrs wkly.

Key Personnel:

Andrew Fisher . . . . . pres
Bruce Baker . . . . . exec VP
Mike Seachman . . . . . opns mgr
Tom Pleva . . . . . gen sls mgr
Bill Seifert . . . . . prom dir
Melissa Knollinger . . . . . news dir
Leonard Smith . . . . . chief of engrg

WTRF-TV— Analog channel: 7. Digital channel: 32. Analog hrs: 24 Digital hrs: 24 316 kw vis, 30.9 kw aur. 960t/740g TL: N40 03 41 W80 45 08 On air date: Oct 23, 1953. 96 16th St., Wheeling, WV, 26003. Phone: (304) 232-7777. Fax: (304) 232-4975. Web Site: www.wtrf.com. Licensee: West Virginia Media Holdings LLC. Group owner: (group owner; (acq 3-12-2002; grpsl). Population served: 255,000 Natl. Network: CBS. Natl. Rep: Petry Television Inc. Washington Atty: Edmundson & Edmundson. Wire Svc: AP News staff: 28; News: 26 hrs wkly.

Key Personnel:

Roger Lyons . . . . . gen mgr
Charlotte Cohen . . . . . gen sls mgr
Jane DomBroski. . . . . prom dir
M.J. Coss . . . . . progmg dir
Brenda Davehart . . . . . news dir
Brad Stanford . . . . . chief of engrg

Wisconsin

Appleton

see Green Bay-Appleton, WI market

Eau Claire

see La Crosse-Eau Claire, WI market

Green Bay-Appleton, WI (DMA 69)

WACY— Analog channel: 32. Digital channel: 59. Analog hrs: 24 Digital hrs: 24 1,070 kw vis, 107 kw aur. 1,220t/1,026g TL: N44 21 32 W87 58 58 (CP: 1,050 kw vis, ant 1,102t) On air date: Mar 7, 1984. 1391 North Rd., Green Bay, WI, 54307-2328. Phone: (920) 490-2647. Fax: (920) 494-9550. E-mail: rbell@mynew32.com Web Site: www.mynew32.com. Licensee: Ace TV Inc. Ownership: Shirley A. Martin (acq 5-25-2000). Natl. Network: MyNetworkTV. Washington Atty: Davis Wright Tremaine LLP.

WBAY-TV— Analog channel: 2. Analog hrs: 24 (Su-Th); 19 (F,S) 100 kw vis, 20 kw aur. 1,205t/1,149g TL: N44 24 35 W88 00 05 On air date: Mar 17, 1953. 115 S. Jefferson St., Green Bay, WI, 54301. Phone: (920) 432-3331. Phone: (800) 242-8090. Fax: (920) 432-1190 (news). E-mail: wbay@wbay.com Web Site: www.wbay.com. Licensee: Young Broadcasting of Green Bay Inc. Group owner: Young Broadcasting Inc. (acq 8-24-94; grpsl; FTR: 9-12-94). Population served: 973,000 Natl. Network: ABC. Natl. Rep: Adam Young. Washington Atty: Brooks, Pierce, McClendon & Humphry. News staff: 100; News: 19 hrs wkly.

Key Personnel:

Don Carmichael . . . . . gen mgr
Richard Millhiser. . . . . opns dir

WFRV-TV— Analog channel: 5. Digital channel: 39. Analog hrs: 24 Digital hrs: 24 100 kw vis, 18.6 kw aur. ant 1,119t/998g TL: N44 24 21 W88 00 19 On air date: May 21, 1955. 1181 E. Mason, Green Bay, WI, 54301. Phone: (920) 437-5411. Fax: (920) 437-4576. Web Site: www.wfrv.com. Licensee: WFRV & WJMN Television Station, Inc. Group owner: Viacom Television Stations Group (acq 2-16-92; grpsl). Population served: 967,000 Natl. Network: CBS.

Key Personnel:

Perry Kidder . . . . . pres, VP & gen mgr
Jackie Stewart . . . . . gen sls mgr
Kristen Kent . . . . . mktg mgr
Lee Hitter . . . . . news dir
Dale Mitchell. . . . . chief of engrg

WGBA— Analog channel: 26. Digital channel: 41. Analog hrs: 24 5,000 kw vis, 500 kw aur. ant 1,181t/982g TL: N44 21 30 W87 58 48 On air date: Dec 31, 1980. 1391 North Rd., Green Bay, WI, 54313. Phone: (920) 494-2626. Fax: (920) 494-9550. Web Site: www.nbc26.com. Licensee: Journal Broadcast Corp. (acq 10-7-2004; $43.25 million). Population served: 1,930,100 Natl. Network: NBC. Natl. Rep: Petry Television Inc. Washington Atty: Shaw Pittman. Wire Svc: AP News staff: 33; News: 16 hrs wkly.

WIWB— Analog channel: 14. Digital channel: 14. Analog hrs: 24 Digital hrs: 24 1000 kw vis, 100 kw aur. 613t/544g TL: N44 59 30 W88 23 55 On air date: Feb 22, 1984. 975 Parkview Rd., Suite 4, Green Bay, WI, 54304. Phone: (920) 983-9014. Fax: (920) 983-9424. E-mail: promotions@wisconsinscw.com Web Site: www.wisconsinscw.com. Licensee: Acme Television Licenses of Wisconsin L.L.C. Group owner: Acme Communications Inc. (acq 6-1-99; grpsl). Population served: 1,035,000 Natl. Network: CW. Natl. Rep: MMT. Washington Atty: Dickstein, Shapiro, Morin & Oshinsky LLP.

Key Personnel:

Stephen M. Shanks . . . . . gen mgr
Todd Zielgler. . . . . gen sls mgr
Peter Marquardt . . . . . natl sls mgr
Jeff Bartel. . . . . prom mgr
Tim Brusky . . . . . chief of engrg

WLUK-TV— Analog channel: 11. Digital channel: 51. Analog hrs: 24 Digital hrs: 24 316 kw vis, 47.4 kw aur. ant 1,260t/1,159g TL: N44 24 31 W87 59 29 On air date: Sept 11, 1954. Box 19011, 787 Lombardi Ave., Garberville, WI, 54307-9011. Phone: (920) 494-8711. Fax: (920) 494-8782. E-mail: info@wluk.com Web Site: www.wluk.com. Licensee: LIN of Wisconsin LLC. Group owner: Emmis Communications Corp. (acq 11-30-2005; grpsl). Population served: 1,049,000 Natl. Network: Fox. Natl. Rep: Blair Television. Wire Svc: AP News staff: 55; News: 32.5 hrs wkly.

Key Personnel:

Jay Zollar . . . . . gen mgr
Tori Grant-Welhouse. . . . . gen sls mgr
Pat Krohlow . . . . . mktg dir
Juli Buehler . . . . . news dir
Mike Nipps . . . . . chief of engrg

***WPNE**— Analog channel: 38. Digital channel: 42.1,078 kw vis, 108 kw aur. 1,180t/1,149g TL: N44 24 35 W88 00 05 On air date: Sept 12, 1972. 3319 W. Beltline Hwy., Madison, WI, 53713. Phone: (608) 263-2121. Fax: (608) 263-9763. E-mail: comments@wpt.org Web Site: www.wpt.org. Licensee: State of Wisconsin-Educational Communications Board. Population served: 4,600,000 Natl. Network: PBS. Washington Atty: Dow, Lohnes & Albertson.

Key Personnel:

Malcolm Brett . . . . . gen mgr & progmg mgr
Mike Edgette . . . . . opns dir
Jon Miskowski . . . . . dev dir

WWAZ-TV— Analog channel: 68. Digital channel: 44. Analog hrs: 24 Digital hrs: 24 4,986 kw vis. ant 640t/468g TL: N43 26 20 W88 31 29 On air date: Dec 1, 2000. Box 2326, Fond Du Lac, WI, 54936-2326. Phone: (920) 387-9698. Fax: (920) 387-9660. Licensee: WMMF License LLC. Group owner: Pappas Telecasting Companies (acq 3-10-2001). Population served: 1, 913,668 Washington Atty: Paul Hastings.

Key Personnel:

Edward Bok . . . . . CEO & chief of engrg
Howard Shrier . . . . . gen mgr

Debbie Sweeney . . . . . progmg VP

La Crosse-Eau Claire, WI (DMA 127)

WEAU-TV— Analog channel: 13. Digital channel: 39. Analog hrs: 24 Digital hrs: 24 316 kw vis, 37 kw aur. ant 1,990t/2,000g TL: N44 39 51 W90 57 41 On air date: Dec 17, 1953. Box 47, Eau Claire, WI, 54702. 1907 S. Hastings Way, Eau Claire, WI 54701. Phone: (715) 835-1313. Fax: (715) 832-0246. E-mail: info@weau.com Web Site: www.weau.com. Licensee: WEAU Licensee Corp. Group owner: Gray Television Inc. (acq 8-1-98; grpsl). Population served: 224,000 Natl. Network: NBC. Natl. Rep: Continental Television Sales. Washington Atty: Pepper & Corazzini. Wire Svc: Medialink News: 34 hrs wkly.
Key Personnel:
Terry McHugh . . . . . gen mgr
Tom Benson . . . . . opns mgr
Joe Buttel . . . . . gen sls mgr
Glen Mabie . . . . . news dir

WEUX— Analog channel: 48.60.3 kw vis, 6 kw aur. 321t TL: N44 52 36 W91 18 22 On air date: February 1993. 800 Wisconsin St., Bldg. 2, Eau Claire, WI, 54701. Phone: (715) 831-2548. Fax: (715) 831-2550. E-mail: info@fox25fox48.com Web Site: www.fox25fox48.com. Licensee: Grant Media LLC. Group owner: (group owner; (acq 1996; $6.25 million with WLAX(TV) La Crosse). Natl. Network: Fox.
Key Personnel:
Jeff Armstrong . . . . . gen mgr
Clay Koenig . . . . . gen sls mgr
Barb Quillin . . . . . prom dir
Pat Stiphout . . . . . prom dir
Mark Burg . . . . . chief of engrg

***WHLA-TV—** Analog channel: 31. Digital channel: 30. Analog hrs: 24 Digital hrs: 24 1,200 kw vis, 60 kw aur. 1,140t/831g TL: N43 48 17 W91 22 06 On air date: Dec 3, 1973. 3319 W. Beltline Hwy., Madison, WI, 53713. Phone: (608) 264-9600 / (507) 895-2026. Fax: (608) 264-9664 / (507) 895-4147. Web Site: www.ecb.org. Licensee: State of Wisconsin-Educational Communications Board. Natl. Network: PBS. Washington Atty: Dow, Lohnes & Albertson.
Key Personnel:
Byron Knight . . . . . gen mgr & stn mgr
Michael Bridgeman . . . . . gen mgr, mktg dir & prom mgr
Mike Edgette . . . . . opns dir
Jon Miskowski . . . . . dev dir
Mary Clare Sorenson . . . . . adv dir & progmg mgr
Kathy Bissen . . . . . news dir
Dick Taugher . . . . . chief of engrg

WKBT— Analog channel: 8. Digital channel: 41. Analog hrs: 24 Note: CBS is on WKBT(TV) ch 8, MyNetworkTV is on WKBT-DT ch 41. 316 kw vis, 57.5 kw aur. ant 1,625t/1,540g TL: N44 05 28 W91 20 15 On air date: Aug 8, 1954. 141 S. 6th St., La Crosse, WI, 54601. Phone: (608) 782-4678. Fax: (608) 782-4674. E-mail: news8@wkbt.com Web Site: www.wkbt.com. Licensee: QueenB Television L.L.C. Group owner: Morgan Murphy Stations (acq 3-31-2000; $22 million). Natl. Network: CBS, MyNetworkTV Digital Network: Note: CBS is on WKBT(TV) ch 8, MyNetworkTV is on WKBT-DT ch 41. Natl. Rep: Harrington, Righter & Parsons.
Key Personnel:
David Sanks . . . . . exec VP
Scott Chorski . . . . . gen mgr
Dennis McSorley . . . . . opns mgr & chief of engrg
Barb Pervisky . . . . . gen sls mgr & natl sls mgr
Brian Voigt . . . . . prom mgr
Maria Roswall . . . . . progmg dir & progmg mgr
Anne Paape . . . . . news dir

WLAX— Analog channel: 25. Digital channel: 17. Analog hrs: 24 Digital hrs: 24 562 kw vis, 56.2 kw aur. 1,004t/674g TL: N43 48 16 W91 22 18 On air date: Sept 28, 1986. Box 2529, La Crosse, WI, 54602. 1305 Interchange Pl., La Crosse, WI 54603. Phone: (608) 781-0025. Fax: (608) 783-2520. E-mail: info@fox25fox48.com Web Site: www.fox25fox48.com. Licensee: Grant Media LLC. Group owner: (group owner; (acq 5-15-96; $6.25 million with WEUX(TV) Chippewa Falls). Natl. Network: Fox. Natl. Rep: TeleRep. News: 3.5 hrs wkly.

WQOW-TV— Analog channel: 18. Digital channel: 15. Analog hrs: 24 Digital hrs: 24 Note: ABC is on WQOW-TV ch 18, CW is on WQOW-DT ch 15. 407 kw vis, 40.7 kw aur. 741t/507g TL: N44 57 49 W91 40 05 On air date: Sept 22, 1980. 5545 Hwy. 93 S., Eau Claire, WI, 54701. Phone: (715) 835-1881. Fax: (715) 835-8009. E-mail: info@wqow.com Web Site: www.wqow.com. Licensee: WXOW/WQOW Television Inc. Group owner: Quincy Newspapers Inc., see Cross-Ownership (acq 6-1-2001; grpsl). Population served: 214,000 Natl. Network: ABC, CW Digital Network: Note: ABC is on WQOW-TV ch 18, CW is on WQOW-DT ch 15. Natl. Rep: Blair Television. Washington Atty: Wilkinson, Barker, Knauer LLP. News staff: 16; News: 18 hrs wkly.
Key Personnel:
Tom Oakley . . . . . chmn
Ralph Oakley . . . . . pres
Charles "Chuck" Roth . . . . . gen mgr
Mark Golden . . . . . stn mgr

WXOW-TV— Analog channel: 19. Digital channel: 14. Analog hrs: 24 Digital hrs: 24 Note: ABC is on WXOW-TV ch 19, CW is on WXOW-DT ch 14. 631 kw vis, 63 kw aur. ant 1,138t/790g TL: N43 48 23 W91 22 02 On air date: Mar 7, 1970. Box C-4019, La Crosse, WI, 54602-4019. 3705 County Hwy. 25, La Crescent, MN 55947. Phone: (507) 895-9969. Fax: (507) 895-8124. Web Site: www.wxow.com. Licensee: WXOW-WQOW Television Inc. Group owner: Quincy Newspapers Inc., see Cross-Ownership (acq 6-1-2001). Population served: 1,542,600 Natl. Network: ABC, CW Digital Network: Note: ABC is on WXOW-TV ch 19, CW is on WXOW-DT ch 14. Natl. Rep: Blair Television. Washington Atty: Wilkinson, Barker & Knauer, L.L.P. News staff: 24; News: 18 hrs wkly.
Key Personnel:
Chuck Roth . . . . . VP & gen mgr
Dave Booth . . . . . stn mgr
Jarrett Liddicoat . . . . . chief of engrg

Madison, WI (DMA 85)

WBUW— Analog channel: 57. Digital channel: 32. Analog hrs: 24 Digital hrs: 24 5,000 kw vis, 500 kw aur. ant 1,361t/1,273g TL: N43 03 03 W89 29 13 On air date: June 28, 1999. 2814 Syene Rd., Madison, WI, 53713. Phone: (608) 270-5700. Fax: (608) 270-5717. Web Site: www.madisonscw.com. Licensee: Acme Television Licenses of Madison LLC. Group owner: Acme Communications Inc. (acq 1-1-2003). Population served: 750,000 Natl. Network: CW. Natl. Rep: MMT. Washington Atty: Dickstein, Shapiro, Morin & Oshinsky LLP.
Key Personnel:
Jamie Kellner . . . . . CEO
Doug Gealy . . . . . COO
Tom Allen . . . . . CFO
Eric Krieghoff . . . . . dev mgr & natl sls mgr
Sharon Weiler . . . . . sls VP
Tom Keeler . . . . . gen mgr & gen sls mgr
Dave Shelly . . . . . natl sls mgr
Matt Creamer . . . . . prom mgr
Eugene Cooper . . . . . progmg mgr
Emmy Fink . . . . . news dir & pub affrs dir
Brent Stephenson . . . . . engrg VP
Jeff Juniet . . . . . chief of engrg

***WHA-TV—** Analog channel: 21.1,120 kw vis, 112 kw aur. 1,485t/1,348g TL: N43 03 21 W89 32 06 On air date: May 3, 1954. 821 University Ave., Madison, WI, 53706. Phone: (608) 263-2121. Fax: (608) 263-9763. Web Site: www.wpt.org. Licensee: University of Wisconsin Board of Regents. Population served: 210,000 Natl. Network: PBS. Washington Atty: Dow, Lohnes & Albertson.
Key Personnel:
Malcolm Brett . . . . . gen mgr & stn mgr
Mike Edgette . . . . . opns mgr
Jon Miskowski . . . . . dev dir

WISC-TV— Analog channel: 3. Digital channel: 50. Analog hrs: 24 Digital hrs: 24 Note: CBS is on WISC-TV ch 3, MyNetworkTV is on WISC-DT ch 50. 56.2 kw vis, 11 kw aur. ant 1,191t/1,108g TL: N43 01 52 W89 30 18 On air date: June 24, 1956. Box 44965, Madison, WI, 53744-4965. Phone: (608) 271-4321. Fax: (608) 271-6111. Web Site: www.channel3000.com. Licensee: TV Wisconsin Inc. Group owner: Morgan Murphy Stns. Population served: 280,000 Natl. Network: CBS, MyNetworkTV Digital Network: Note: CBS is on WISC-TV ch 3, MyNetworkTV is on WISC-DT ch 50. Natl. Rep: Harrington, Righter & Parsons. Washington Atty: Rini Coran, PC. News staff: 30; News: 30 hrs wkly.
Key Personnel:
Elizabeth Murphy Burns . . . . . pres
David Sanks . . . . . exec VP & gen mgr
Jill Sommers . . . . . opns dir, gen sls mgr & progmg dir

WKOW-TV— Analog channel: 27. Digital channel: 27. Analog hrs: 24 1,000 kw vis, 100 kw aur. 1,250t/1,182g TL: N43 03 09 W89 28 42 (CP: Ant 1,492t/1,424g. TL: N43 03 21 W89 32 06) On air date: July 1953. 5727 Tokay Blvd., Madison, WI, 53719. Phone: (608) 274-1234. Fax: (608) 274-9514. Web Site: www.wkowtv.com. Licensee: WKOW Television Inc. Group owner: Quincy Newspapers Inc., see Cross-Ownership (acq 5-22-2001; grpsl). Population served: 214,800 Natl. Network: ABC. Washington Atty: Rosenman & Colin. News staff: 28; News: 22 hrs wkly.

WMSN-TV— Analog channel: 47. Analog hrs: 6 AM-2 AM 1,000 kw vis, 100 kw aur. ant 1,466t TL: N43 03 21 W89 32 06 On air date: June 8, 1986. 7847 Big Sky Dr., Madison, WI, 53719. Phone: (608) 833-0047. Fax: (608) 833-5055. Fax: (608) 833-0665 (Natl Sls). E-mail: comments@fox47.com Web Site: www.fox47.com. Licensee: WMSN Licensee LLC. Group owner: Sinclair Broadcast Group Inc. (acq 12-10-01; grpsl). Natl. Network: Fox.
Key Personnel:
Kerry Johnson . . . . . gen mgr
Chad Happersettt . . . . . gen sls mgr & rgnl sls mgr
Joel Helzer . . . . . natl sls mgr
Audra Johnson . . . . . prom dir
Collin Campbell . . . . . progmg dir & progmg mgr
Al Zobel . . . . . news dir & engrg mgr
Kerry Maki . . . . . chief of engrg

WMTV— Analog channel: 15. Digital channel: 19. Analog hrs: 24 Digital hrs: 24 1,050 kw vis, 105 kw aur. ant 1,161t/1,101g TL: N43 03 01 W89 29 15 On air date: July 1953. 615 Forward Dr., Madison, WI, 53711. Phone: (608) 274-1515/(608) 274-1500 (news). Fax: (608) 271-5193/(608) 271-5194 (news). E-mail: feedback@nbc15.com Web Site: www.nbc15.com. Licensee: Gray Television Licensee Inc. Group owner: Gray Television Inc. (acq 8-29-2002; grpsl). Population served: 308,310 Natl. Network: NBC. Washington Atty: Covington & Burling. Wire Svc: AP Wire Svc: CNN News staff: 36; News: 19 hrs wkly.
Key Personnel:
J. Mack Robinson . . . . . pres
Jim Ryan . . . . . CFO
Robert Prather Jr. . . . . . exec VP
Bob Smith . . . . . gen mgr
Curt Molander . . . . . gen sls mgr
Ellen Buss . . . . . progmg mgr
Chris Gegg . . . . . news dir
Tom Weeden . . . . . chief of engrg

Menomonie

see Minneapolis-St. Paul, MN market

Milwaukee, WI (DMA 34)

WCGV-TV— Analog channel: 24. Digital channel: 25. Analog hrs: 24 Digital hrs: 25 3,000 kw vis, 300 kw aur. ant 1,030t/1,039g TL: N43 05 15 W87 54 13 On air date: Mar 17, 1980. 4041 N. 35th St., Milwaukee, WI, 53216. Phone: (414) 442-7050. Fax: (414) 874-1899. E-mail: comments@wcgv24.com Web Site: www.thattvwebsite.com. Licensee: WCGV Licensee L.L.C. Group owner: Sinclair Broadcast Group Inc. (acq 5-23-94; grpsl; FTR: 11-19-90). Population served: 4,189,000 Natl. Network: MyNetworkTV. Natl. Rep: Millennium Sales & Marketing.
Key Personnel:
David Ford . . . . . gen mgr
Milan Macksimovic . . . . . opns mgr
Paul Rudolph . . . . . chief of opns
Rob Krieghoff . . . . . rgnl sls mgr
Jason Van Acker . . . . . prom mgr
Betty Hertz . . . . . progmg dir
Dennis Brechlin . . . . . chief of engrg

WDJT-TV— Analog channel: 58. Analog hrs: 24 2,820 kw vis, 282 kw aur. 535t TL: N43 02 20 W87 55 04 (CP: 5,000 kw vis, ant 1,112t) On air date: November 1988. 809 S. 60th St., Milwaukee, WI, 53214. Phone: (414) 777-5800. Fax: (414) 777-5802. Web Site: www.cbs58.com. Licensee: WDJT-TV L.P. Group owner: Weigel Broadcasting Co. Population served: 762,000 Natl. Network: CBS. Natl. Rep: Harrington, Righter & Parsons. Washington Atty: Cohn & Marks.
Key Personnel:
Norman Shapiro . . . . . pres
Jim Hall . . . . . gen mgr & stn mgr
Marty Schack . . . . . gen sls mgr
Grant Uitti . . . . . news dir

WISN-TV— Analog channel: 12. Digital channel: 34. Analog hrs: 24 316 kw vis, 31.6 kw aur. ant 1,000t/1,105g TL: N43 06 41 W87 55 38 On air date: Oct 27, 1954. Box 402, Milwaukee, WI, 53201. Phone: (414) 342-8812. Fax: (414) 342-4486. Web Site: www.themilwaukeechannel.com. Licensee: WISN Hearst-Argyle Television Inc., a California corp. Group owner: Hearst-Argyle Television Inc. (acq 7-16-97; grpsl). Population served: 760,000 Natl. Network: ABC. Natl. Rep: Continental Television Sales. Washington Atty: Peper, Martin, Jensen, Maichel & Hetlage. Wire Svc: News 1
Key Personnel:
Jan Wade . . . . . gen mgr
Pete Monfre . . . . . gen sls mgr
Dean Maytag . . . . . progmg dir
Lori Waldon . . . . . news dir
Tony Coleman . . . . . chief of engrg

WITI— Analog channel: 6. Digital channel: 33. Analog hrs: 24 Digital hrs: 24 100 kw vis, 10 kw aur. ant 1,000t/1,078g TL: N43 05 24 W87 53 47 On air date: May 21, 1956. 9001 N. Green Bay Rd., Milwaukee, WI, 53209. Phone: (414) 355-6666. Fax: (414) 586-2141. E-mail: fox6news@foxtv.com Web Site: www.myfox.com. Licensee: WITI

License Inc. Group owner: Fox Television Stations Inc. (acq 11-96; grpsl). Population served: 780,000 Natl. Network: Fox. Washington Atty: Pepper & Corazzini.

Key Personnel:

Parveen Hughes . . . CFO
Chuck Steinmetz . . . VP & gen mgr
John Workman . . . opns VP
Mike Neale . . . sls VP
Bob O'Neil . . . natl sls mgr
Sue Swaziek . . . rgnl sls mgr
Lori Wucherer . . . prom VP
Jim Lemon . . . news dir
Kelly Skindzelewski . . . pub affrs dir
Don Hain . . . engrg VP
Hayley Puffer . . . progmg

WJJA— Analog channel: 49. Digital channel: 48. Analog hrs: 24 Digital hrs: 24 2,690 kw vis, 260 kw aur. 435t/405g TL: N42 51 18 W87 50 41 (CP: 5,000 kw vis, ant 895t. TL: N43 05 15 W87 54 01) On air date: Jan 27, 1990. 4311 E. Oakwood Rd., Oak Creek, WI, 53154. Phone: (414) 764-4953. Fax: (414) 764-5190. Licensee: TV-49 Inc. Ownership: Joel Kinlow, 99%; other, 1%.

Key Personnel:

Joe Kinlow. pres, pres, gen mgr, stn mgr, dev mgr, dev mgr, gen sls mgr, progmg dir & news dir
Bruce Herzog . . . chief of engrg

***WMVS**— Analog channel: 10. Digital channel: 8. Analog hrs: 24 Digital hrs: 24 309 kw vis, 30.9 kw aur. 1,010t/1,101g TL: N43 05 48 W87 54 19 On air date: Oct 28, 1957. 1036 N. 8th St., Milwaukee, WI, 53233. Phone: (414) 271-1036. Fax: (414) 297-7536. Web Site: www.mptv.org. Licensee: Milwaukee Area District Board of Vocational, Technical & Adult Education. Population served: 2,100,000 Natl. Network: PBS. Rgnl. Network: CEN. Washington Atty: Dow, Lohnes & Albertson.

Key Personnel:

Ellis Bromberg . . . gen mgr
Kate Tierney . . . opns mgr & prom dir
Tom Dvorak . . . progmg dir
Dan Jones . . . news dir
David Felland . . . chief of engrg

***WMVT**— Analog channel: 36. Digital channel: 35. Analog hrs: 24 Digital hrs: 24 4,790 kw vis. ant 1,115t/1,143g TL: N43 05 46 W87 54 15 On air date: Jan 23, 1963. 4th Fl., 1036 N. 8th St., Milwaukee, WI, 53233. Phone: (414) 271-1036. Fax: (414) 297-7536. Web Site: www.mptv.org. Licensee: Milwaukee Area District Board of Vocational, Technical & Adult Education. Population served: 2,100,000 Natl. Network: PBS. Washington Atty: Dow, Lohnes & Albertson. Foreign lang progmg: SP 4

Key Personnel:

Ellis Bromberg . . . gen mgr
Kate Tierney . . . opns VP & prom dir
Tom Dvorak . . . progmg dir
Dan Jones . . . news dir
David Felland . . . engrg dir & chief of engrg

WPXE— Analog channel: 55. Digital channel: 40. Analog hrs: 24 741 kw vis, 74.1 kw aur. ant 449t/349g TL: N42 30 36 W87 53 11 (CP: 5,000 kw vis, ant 472t. TL: N42 45 38 W87 57 55) On air date: June 1, 1988. 6262 N. Flint Rd., Ste F, Glendale, WI, 53209. Phone: (414) 247-0117. Fax: (414) 247-1302. Web Site: www.ionline.tv. Licensee: Paxson Milwaukee License Inc. Group owner: Paxson Communications Corp. (acq 2-18-00; grpsl). Natl. Network: i Network. Washington Atty: Gardner, Carton & Douglas. Wire Svc: CNN

Key Personnel:

Carol Holzhauer. . . . gen mgr & gen sls mgr
Joanne Levy . . . natl sls mgr

WTMJ-TV— Analog channel: 4.100 kw vis, 20 kw aur. 1,000t/1,096g TL: N43 05 29 W87 54 07 On air date: December 1947. 720 E. Capitol Dr., Milwaukee, WI, 53212. Phone: (414) 332-9611. Fax: (414) 967-5378. E-mail: tmj4feedback@todaystmj4.com Web Site: www.todaystmj4.com. Licensee: Journal Broadcast Corp. Group owner: Journal Broadcast Group Inc. Population served: 2,150,000 Natl. Network: NBC. Washington Atty: Hogan & Hartson.

Key Personnel:

Mark Strachota. . . . gen mgr, gen sls mgr & natl sls mgr
Mar LeGrand . . . rgnl sls mgr
Brenda Serio . . . progmg dir
Sean O'Flaherty . . . mktg mgr & news dir
Kent Aschenbrenner . . . chief of engrg

WVCY-TV— Analog channel: 30. Digital channel: 22. Analog hrs: 9 AM-midnight (M-F); 8 AM-midnight (S-Su) 1,070 kw vis, 55 kw aur. 961t TL: N43 05 15 W87 54 12 On air date: Jan 11, 1983. 3434 W. Kilbourn Ave., Milwaukee, WI, 53208. Phone: (414) 935-3000. Fax: (414) 935-3015. E-mail: tv30@vcyamerica.org Web Site: www.vcyamerica.org. Licensee: VCY/America Inc. Group owner: (group owner) Washington Atty: Wiley Rein LLP.

Key Personnel:

Dr. Randall Melchert. . . . pres
Vic Eliason . . . VP & gen mgr
Jim Cronin . . . opns mgr
Jim Schneider . . . progmg dir & pub affrs dir
Andy Eliason . . . chief of engrg

WVTV— Analog channel: 18. Digital channel: 61. Analog hrs: 24 5,000 kw vis, 500 kw aur. ant 1,008t/1,101g TL: N43 05 48 W87 54 19 On air date: July 1, 1959. 4041 N. 35th St., Milwaukee, WI, 53216. Phone: (414) 442-7050. Fax: (414) 874-1898. Fax: (414) 874-1899. Web Site: www.thattvwebsite.com. Licensee: WVTV Licensee Inc. Group owner: Glencairn Ltd. (acq 2-1-2002). Population served: 1,826,000 Natl. Network: CW. Natl. Rep: Millennium Sales & Marketing.

Key Personnel:

David Smith . . . CEO & pres
David Ford . . . gen mgr
Milan Macksimovic . . . opns mgr
Paul Rudulph . . . chief of opns
Rob Krieghoff . . . rgnl sls mgr
Jason Van Acker . . . prom mgr
Betty Hertz . . . progmg dir
Dennis Brechlin . . . chief of engrg

WWRS-TV— Analog channel: 52. Analog hrs: 8:30am - 5:30pm 5,000 kw vis. ant 663t/466g TL: N43 26 11 W88 31 34 On air date: 1997. Phone: (920) 387-9052. Fax: (920) 387-9053. E-mail: dcalhoun@tbn.org Web Site: www.tbn.org. Licensee: National Minority T.V. Inc. Ownership: not for profit Corp. (acq 2-16-99; $3,300,000). Washington Atty: Shaw Pittman.

Rhinelander

see Wausau-Rhinelander, WI market

Superior

see Duluth, MN-Superior, WI market

Wausau-Rhinelander, WI (DMA 134)

WAOW-TV— Analog channel: 9. Digital channel: 29. Analog hrs: 24 Note: ABC is on WAOW-TV ch 9, CW is on WAOW-DT ch 29. 316 kw vis, 31.6 kw aur. ant 1,210t/647g TL: N44 55 14 W89 41 31 On air date: May 7, 1965. 1908 Grand Ave., Wausau, WI, 54403. Phone: (715) 842-2251. Fax: (715) 848-0195. Fax: (715) 842-7808. E-mail: info@waow.com Web Site: www.waow.com. Licensee: WAOW-WYOW Television Inc. Group owner: Quincy Newspapers Inc., see Cross Ownership (acq 5-22-2001; grpsl). Population served: 533,000 Natl. Network: ABC, CW Digital Network: Note: ABC is on WAOW-TV ch 9, CW is on WAOW-DT ch 29. Natl. Rep: Blair Television. Washington Atty: Wilkinson, Baker & Knauer, LLP. Wire Svc: AP Wire Svc: CNN News staff: 26; News: 15.5 hrs wkly.

Key Personnel:

Thomas A. Oakley . . . CEO
Ralph M. Oakley . . . COO
Laurin Jorstad . . . VP & gen mgr
Randy Winters . . . opns mgr
Mark Oliver . . . mktg dir & prom dir
Tara Marshall . . . progmg dir
Randy Winter . . . news dir
Russ Crass . . . chief of engrg
Rebroadcasts WYOW-TV Eagle River 100%.

WBIJ— Analog channel: 4. Analog hrs: 24 1.7 kw vis. ant 403t/255g TL: N45 34 23 W88 52 57 (CP: 4.3 kw vis, ant 390t/240g) On air date: 2004. 4529 Hickory Heights Ave., Oshkosh, WI, 54904. Phone: (920) 589-2511. Licensee: Selenka Communications LLC.

WFXS— Analog channel: 55. Analog hrs: 24 5,000 kw vis. ant 1,073t/994g TL: N45 03 22 W89 27 54 On air date: Dec 1, 1999. 1000 N. 3rd St., Wausau, WI, 54403. Phone: (715) 847-1155. Fax: (715) 847-1156. E-mail: wfxs@wfxs.com Web Site: www.wfxs.com. Licensee: Davis Television Wausau L.L.C. Population served: 431,000 Natl. Network: Fox. Natl. Rep: Millennium Sales & Marketing. Washington Atty: Leventhal, Senter & Lerhman. News: 5 hrs wkly.

Key Personnel:

Robert Raff. . . . gen mgr
Scott Storkel . . . opns dir
Jan El Daul . . . gen sls mgr
Deb Steinfest . . . prom dir
Randy Winter . . . news dir

***WHRM-TV**— Analog channel: 20. Digital channel: 24. Analog hrs: 24 Digital hrs: 24 1,437 kw vis, 36.2 kw aur. 984t/977g TL: N44 55 14 W59 41 31 On air date: 1975. 3319 W. Beltline Hwy., Madison, WI, 53713. Phone: (608) 264-9600. Web Site: www.ecb.org. Licensee: Wisconsin Educational Communications Board. Natl. Network: PBS. Washington Atty: Dow, Lohnes & Albertson.

Key Personnel:

Bryon Knight . . . gen mgr
Mike Edgette . . . opns dir
Jon Miskowski . . . dev dir
Michael Bridgeman . . . mktg dir & prom mgr
Mary Clare Sorenson . . . adv dir
Kathy Bissen . . . progmg mgr & news dir
Terry Baun . . . chief of engrg

WJFW-TV— Analog channel: 12. Analog hrs: 24 316 kw vis, 57.6 kw aur. 1,660t/1,677g TL: N45 40 03 W89 12 29 On air date: Oct 20, 1966. Box 858, Rhinelander, WI, 54501. Phone: (715) 365-8812. Fax: (715) 365-8810. E-mail: e-mail@wjfw.com Web Site: www.wjfw.com. Licensee: Northland Television LLC. Group owner: Rockfleet Broadcasting Inc. Population served: 657,000 Natl. Network: NBC. Natl. Rep: Blair Television. Washington Atty: Wiley, Rein & Fielding. News staff: 13; News: 18.5 hrs wkly.

Key Personnel:

Robert Krieghoff . . . gen mgr, prom mgr & progmg dir
Robert Schmidtbauer . . . CFO & opns mgr
Charlotte Berens . . . natl sls mgr
Heather Schellock . . . news dir
Greg Buzzell . . . chief of engrg

***WLEF-TV**— Analog channel: 36. Digital channel: 47. Analog hrs: 24 Digital hrs: 24 741 kw vis, 74.2 kw aur. 1,468t/1,467g TL: N45 56 43 W90 16 28 On air date: December 1977. 3319 W. Beltline Hwy., Madison, WI, 53713. Phone: (608) 264-9600. Fax: (608) 264-9664. Web Site: www.wpt.org. Licensee: State of Wisconsin-Educational Communications Board. Natl. Network: PBS. Washington Atty: Dow, Lohnes & Albertson.

Key Personnel:

Mike Edgette . . . opns dir
Jon Miskowski . . . dev dir
Michael Bridgeman . . . mktg dir & prom mgr
Mary Clare Sorenson . . . adv dir & progmg mgr
Kathy Bissen . . . news dir
Dick Taugher . . . chief of engrg

WSAW-TV— Analog channel: 7. Digital channel: 40. Analog hrs: 24 Note: CBS is on WSAW-TV ch 7, MyNetworkTV is on WSAW-DT ch 40. 316 kw vis, 63.2 kw aur. ant 1,210t/647g TL: N44 55 14 W89 41 31 On air date: Oct 23, 1954. 1114 Grand Ave., Wausau, WI, 54403. Phone: (715) 845-4211. Fax: (715) 845-2649. Web Site: www.wsaw.com. Licensee: WEAU Licensee Corp. Group owner: Gray Television Inc. (acq 8-29-2002; grpsl). Population served: 182,000 Natl. Network: CBS, MyNetworkTV Digital Network: Note: CBS is on WSAW-TV ch 7, MyNetworkTV is on WSAW-DT ch 40. Natl. Rep: Continental Television Sales. Wire Svc: AP News staff: 21; News: 15 hrs wkly.

Key Personnel:

Al Lancaster . . . VP & gen mgr
Patti Shook . . . opns mgr & progmg mgr
Jamie Sarver . . . rgnl sls mgr
Linda Heinzen . . . rgnl sls mgr
Dan Froelich . . . prom dir & prom mgr
Susan Ramsett . . . news dir
Chad Myers . . . chief of engrg

WTPX— Analog channel: 0. Digital channel: 46.3,090 kw vis. ant 918t TL: N45 03 33 W89 26 10 On air date: 2006. 720 E. Capitol Dr., Milwaukee, WI, 53212-1308. Phone: (414) 967-5592. Fax: (414) 967-5597. Licensee: Paxson Wausau License Inc. Group owner: Paxson Communications Corp. (acq 4-18-2000; $887,500 for CP). Natl. Network: i Network.

WYOW— Analog channel: 34. Digital channel: 28. Analog hrs: 24 Note: ABC is on WYOW(TV) ch 34, CW is on WYOW-DT ch 28. 2,400 kw vis. ant 1,092t TL: N45 46 30 W89 14 55 On air date: 1997. Box 2705, Eagle River, WI, 54521. 528 W. Pine St., Suite B, Eagle River, WI 54521. Phone: (715) 477-2020. Fax: (715) 477-2438. E-mail: wyowtv34@newnorth.net Web Site: www.wyowt34.com. Licensee: WAOW-WYOW Television Inc. Group owner: Quincy Newspapers Inc., see Cross-Ownership (acq 5-22-2001; grpsl). Natl. Network: ABC, CW Digital Network: Note: ABC is on WYOW(TV) ch 34, CW is on WYOW-DT ch 28. Natl. Rep: Blair Television. Washington Atty: Wilkinson, Barker & Knauer, LLP. Wire Svc: AP Wire Svc: CNN

Key Personnel:

Thomas A. Oakley . . . CEO
Ralph Oakley . . . COO
Laurin Jorstad . . . gen mgr
Carol Kellum. . . . gen sls mgr
Tim Atterberg . . . rgnl sls mgr
Mark Oliver . . . mktg dir & prom mgr
Tricia Atterberg . . . progmg dir
Randy Winter . . . news dir
Russ Crass . . . chief of engrg

Wyoming

Casper-Riverton, WY
(DMA 198)

***KCWC-TV—** Analog channel: 4. Digital channel: 8. Analog hrs: 24 Digital hrs: 24 100 kw vis, 10 kw aur. 1,519t/199g TL: N42 34 59 W108 42 36 On air date: January 2002. Wyoming Public Television, 2660 Peck Ave., Riverton, WY, 82501. Phone: (307) 856-6944. Fax: (307) 856-3893. Web Site: wyoptv.org. Licensee: Central Wyoming College. Natl. Network: PBS. Washington Atty: Fletcher, Heald & Hildreth.
Key Personnel:
J. Amend . . . prom VP
Ruby Calvert . . . gen mgr & progmg dir
Bob Spain . . . chief of engrg

KCWY— Analog channel: 13. Analog hrs: 24 10.9 kw vis. ant 1,602t/499g TL: N42 44 45 W106 21 14 On air date: 2002. Box 1540, Mills, WY, 82644. 141 Progress Circle, Mills, WY 82644. Phone: (307) 577-0013. Fax: (307) 577-5251. E-mail: bsullivan@kcwy13.com Web Site: kcwy13.com. Licensee: Bozeman Trail Communications Co. Group owner: Sunbelt Communications Co. Natl. Network: NBC. Natl. Rep: Blair Television. Wire Svc: AP Wire Svc: CNN News staff: 13; News: 20 hrs wkly.
Key Personnel:
Bill Sullivan . . . VP & gen mgr
Peggy Porter . . . rgnl sls mgr
Mark Hildebrand . . . chief of engrg

KFNB-TV— Analog channel: 20. Analog hrs: 24 58.9 kw vis. 1,909t TL: N42 44 37 W106 18 31 On air date: Oct 31, 1984. 1856 Skyview Dr., Casper, WY, 82601. Phone: (307) 577-5923. Phone: (307) 577-5924. Fax: (307) 577-5928. E-mail: klwy@coffey.com Licensee: WyoMedia Corp. Population served: 32,000 Natl. Network: Fox. Washington Atty: Irwin, Campbell & Tannenwald.
Key Personnel:
Mark Nalboe . . . gen mgr
Terry Lane . . . opns mgr & progmg dir
Tina Nalbone . . . gen sls mgr
Joe Lownden . . . prom mgr
Greg Flabager . . . news dir
Dave Ericson . . . chief of engrg

KFNE— Analog channel: 10. Analog hrs: 24 170 kw vis, 8.7 kw aur. 1,725t/174g TL: N43 27 26 W108 12 02 On air date: Dec 22, 1957. 1856 Skyview Dr., Casper, WY, 82601. Phone: (307) 577-5923/5924. Fax: (307) 577-5928. E-mail: klwy@coffey.com Licensee: First National Broadcasting Corp. Natl. Network: ABC.
Key Personnel:
Mark Nalbone . . . gen mgr
Terry Lane . . . opns mgr & progmg dir
Tina Nalbone . . . gen sls mgr
Joe Lowndes . . . prom dir
Dave Ericson . . . chief of engrg

KGWC-TV— Analog channel: 14. Analog hrs: 24 1,380 kw vis, 138 kw aur. ant 1,879t/242g TL: N42 44 26 W106 21 34 On air date: Aug 12, 1981. 1856 Skyview Dr., Casper, WY, 82601. Phone: (307) 577-5923 / 5924. Fax: (307) 634-7511. Licensee: Mark III Media Inc. Group owner: Chelsey Broadcasting Co. (acq 5-31-2006; grpsl). Population served: 59,000 Natl. Network: CBS, Fox. Washington Atty: Dow, Lohnes & Albertson.
Key Personnel:
Mark Nalbone . . . gen mgr
Terry Lane . . . opns mgr & progmg dir
Tina Nalbone . . . gen sls mgr
Joe Lownden . . . prom dir
Greg Flabager . . . news dir
Dave Ericson . . . chief of engrg

KGWL-TV— Analog channel: 5. Analog hrs: 24 100 kw vis, 10 kw aur. ant 269t/179g TL: N42 53 43 W108 43 34 On air date: Sept 10, 1982. 1856 Skyview Dr., Casper, WY, 82601. Phone: (307) 234-1111. Fax: (307) 234-4005. Licensee: Mark III Media Inc. Group owner: Chelsey Broadcasting Co. (acq 5-31-2006; grpsl). Natl. Network: CBS. Washington Atty: Covington & Burling.
Key Personnel:
Mark Nalbone . . . gen mgr
Terry Lane . . . opns mgr
Satellite of KGWC-TV Casper.

***KPTW—** Analog channel: 6.331 w vis. ant 1,758t/187g TL: N42 44 26 W106 21 34 On air date: 2007. Wyoming Public Television, 2660 Peck Ave., Riverton, WY, 82501. Phone: (307) 856-6944. Fax: (307) 856-3893. Web Site: wyoptv.org. Licensee: Central Wyoming College. Natl. Network: PBS.
Key Personnel:
Ruby Calvert . . . gen mgr & progmg mgr
Bob Spain . . . chief of engrg
Satellite of KCWC-TV Lander.

KTWO-TV— Analog channel: 2. Digital channel: 17. Analog hrs: 24 100 kw vis, 10.2 kw aur. ant 2,001t/376g TL: N42 44 03 W106 20 00 On air date: Mar 1, 1957. 1896 Skyview Drive, Casper, WY, 82601-9638. Phone: (307) 237-3711. Fax: (307) 237-4458. E-mail: k2@k2tv.com Web Site: www.k2tv.com. Licensee: Silverton Broadcasting Co. LLC. Ownership: Barry Silverton, 100% Group owner: Equity Broadcasting Corp. (acq 5-31-2006; $1.2 million). Population served: 500,000 Natl. Network: ABC. Natl. Rep: Millennium Sales & Marketing. News staff: 23; News: 19.5 hrs wkly.
Key Personnel:
Amie Miller . . . news dir
Tina Nalbone . . . sls

Cheyenne, WY-Scottsbluff, NE
(DMA 195)

KDEV— Analog channel: 33. Digital channel: 11. Analog hrs: 24 251 kw vis, 25.1 kw aur. ant 485t TL: N41 08 55 W104 57 22 On air date: Aug 28, 1987. Phone: (501) 219-2400. Web Site: www.katv.com. Licensee: Denver Broadcasting Inc. Group owner: Equity Broadcasting Corp. (acq 3-26-2001; $3.5 million with KTWO-TV Casper). Natl. Network: ABC. Natl. Rep: Millennium Sales & Marketing. News staff: 17; News: 15 hrs wkly.

KDUH-TV— Analog channel: 4. Digital channel: 7. Analog hrs: 24 Digital hrs: 24 100 kw vis, 20 kw aur. 2,001t/1,966g TL: N42 10 21 W103 13 57 On air date: Mar 5, 1958. Box 1529, Scottsbluff, NE, 69363-1529. 1523 1st Ave., Scottsbluff, NE 69361. Phone: (308) 632-3071. Fax: (308) 632-3596. Web Site: www.kduhtv.com. Licensee: Duhamel Broadcasting Enterprises. Group owner: (group owner) Natl. Network: ABC. Washington Atty: Fisher, Wayland, Cooper, Leader & Zaragoza. News staff: 6; News: 2 hrs wkly.
Key Personnel:
Patrick Maag . . . gen mgr
Monte Loos . . . progmg mgr
Jerry Dishong . . . news dir
Teddy Johnson . . . chief of engrg

KGWN-TV— Analog channel: 5. Digital channel: 30. Analog hrs: 24 Digital hrs: 24 100 kw vis, 10 kw aur. ant 620t/483g TL: N41 06 01 W105 00 23 On air date: Mar 22, 1954. 2923 E. Lincolnway, Cheyenne, WY, 82001. Phone: (307) 634-7755. Fax: (307) 638-0182. E-mail: news@kgwn.tv Web Site: www.kgwn.tv. Licensee: SagamoreHill Broadcasting Co. of Wyoming/Northern Colorado LLC. Group owner: (group owner; (acq 3-19-2004; $6.5 million with KSTF(TV) Scottsbluff, NE). Population served: 65,000 Natl. Network: CBS, CW. Natl. Rep: Continental Television Sales. Washington Atty: Dow, Lohnes & Albertson. News staff: 16; News: 15 hrs wkly.
Key Personnel:
Louis Wall . . . pres
Joan Turner . . . gen mgr & natl sls mgr
Dusty Thein . . . gen sls mgr
Keith Lindstrom . . . rgnl sls mgr
Barbara Parenti . . . progmg dir
Tregg White . . . news dir
Tony Schaefer . . . engrg dir
Keith Yosten . . . chief of engrg

KLWY— Analog channel: 27. Digital channel: 28. Analog hrs: 24 4,270 kw vis, 427 kw aur. ant 760t/635g TL: N41 02 55 W104 53 28 On air date: 1992. 1856 Skyview Dr., Casper, WY, 82601. Phone: (307) 577-5923. Fax: (307) 577-5928. E-mail: klwy@coffey.com Licensee: Wyomedia Corp. (acq 12-4-91; $100,000; FTR: 1-6-92). Natl. Network: Fox.
Key Personnel:
Mark Nalbone . . . gen mgr
Terry Lane . . . opns mgr & progmg dir
Tina Nalbone . . . gen sls mgr & rgnl sls mgr
Joe Lowndes . . . prom dir
Greg Flabager . . . news dir
Dave Ericson . . . chief of engrg

KSTF— Analog channel: 10. Digital channel: 29. Analog hrs: 24 Digital hrs: 24 240 kw vis, 24 kw aur. 840t/674g TL: N41 59 58 W103 39 55 On air date: Aug 7, 1955. 2923 E. Lincoln Way, Cheyenne, NE, 82001. Phone: (308) 632-6107. Fax: (308) 632-3470. Web Site: www.kgwn.tv. Licensee: SagamoreHill Broadcasting Co. of Wyoming/Northern Colorado LLC. Group owner: (group owner; (acq 3-19-2004; $6.5 million with KGWN-TV Cheyenne, WY). Population served: 57,000 Natl. Network: CBS, Fox. Washington Atty: Dow, Lohnes & Albertson. News staff: 6; News: 8 hrs wkly.
Key Personnel:
Joan Turner-Doyle . . . gen mgr
Barbara Parenti . . . gen sls mgr, progmg dir & progmg mgr
Dusty Thein . . . gen sls mgr
Jon Martin . . . news dir
Tony Schaefer . . . chief of engrg
Tricia Murphy . . . prom mgr & mktg

KTUW— Analog channel: 16. Analog hrs: 24 2,559 kw vis. ant 779t/125g TL: N41 50 23 W103 49 35 Not on air, target date: unknown: 1 Shackleford Dr., Suite 400, Little Rock, AR, 72211. Phone: (501) 219-2400. Fax: (501) 221-1101. Permittee: EBC Scottsbluff Inc. Group owner: Equity Broadcasting Corp. Washington Atty: Irwin, Campbell & Tannenwald.

Jackson
see Salt Lake City, UT market

Laramie
see Denver, CO market

Rawlins
see Denver, CO market

Riverton
see Casper-Riverton, WY market

Rock Springs
see Salt Lake City, UT market

Sheridan
see Rapid City, SD market

American Samoa

Pago Pago

***KVZK-2—** Analog channel: 2. Analog hrs: 24 60 kw vis, 6.0 kw aur. 2,000t/400g TL: N14 16 14 W170 41 12 On air date: 1964. Box 2567, Pago Pago, AS, 96799. Phone: (684) 633-4191. Fax: (684) 633-1044. E-mail: siviapaolo@yahoo.cm Web Site: www.asg-gov.com /agencies/opi.asg.htm. Licensee: The Government of American Samoa. Natl. Network: PBS.

***KVZK-4—** Analog channel: 4. Analog hrs: 24 72 kw vis, 7.2 kw aur. Box 2567, Pago Pago, AS, 96799. Phone: (684) 633-4191. Fax: (684) 633-1044. E-mail: siviapaolo@yahoo.com Web Site: www.asg-gov.com /agencies/opi.asg.htm. Licensee: The Government of American Samoa. Natl. Network: ABC, CBS.

***KVZK-5—** Analog channel: 5. Analog hrs: 24 72 kw vis, 7.2 kw aur. 2,000t/400g On air date: Oct 5, 1964. Box 2567, Pago Pago, AS, 96799. Phone: (684) 633-4191. Fax: (684) 633-1044. E-mail: siviapaolo@yahoo.com Web Site: www.asg-gov.com/agencies/opi.asg.htm. Licensee: The Government of American Samoa. Natl. Network: ABC, CBS.

Guam

Hagatna

***KGTF-TV—** Analog channel: 12. Digital channel: 5. Analog hrs: 18 27.5 kw vis, 5.47 kw aur. ant 297t/196g TL: N13 26 13 E144 48 17 On air date: Oct 30, 1970. Box 21449, GMF, GU, 96921. 194 Sesame St., Washington Dr., Mangilao, GU 96921. Phone: (671) 734-2207. Phone: (671) 734-5788. Fax: (671) 734-3476. E-mail: kgtf12@kgtf.org Web Site: www.kgtf.org. Licensee: Guam Educational Telecommunications Corp. Population served: 160,000 Natl. Network: PBS. Washington Atty: Cohn & Marks.
Key Personnel:
Johnny Sablun . . . gen mgr
Benny Flores . . . opns mgr

KUAM-TV— Analog channel: 8. Analog hrs: 24 25.1 kw vis, 2.57 kw aur. 140t/320g TL: N13 25 53 W144 42 36 On air date: Aug 5, 1956. Calvo Commercial Ctr., 600 Harmon Loop Rd., Suite 102, Dededo, GU, 96929. United States Minor Outlying Islands. Phone: (671) 637-5826. Fax: (671) 637-9865. Web Site: www.kuam.com. Licensee: Pacific Telestations Inc. Ownership: Edward M. Calvo (acq 1988). Population served: 150,000 Natl. Network: NBC, CBS. Washington Atty: Haley, Bader & Potts. News staff: 20; News: 14 hrs wkly.

Key Personnel:
Joseph Calvo . . . . exec VP, gen mgr & sls VP
Marie Calvo-Monge . . . . gen mgr
Annie SanNicolas . . . . sls VP & gen sls mgr
Christie San Agustin . . . . prom dir
Annie San Nicolas . . . . progmg mgr
Richard Garman . . . . chief of engrg

Tamuning

KTGM— Analog channel: 14. Digital channel: 17. Analog hrs: 7 am-midnight 12.8 kw vis. ant -52t/66g TL: N13 30 09 W144 48 17 On air date: Oct 19, 1988. 111 W. Chalan Santo Papa St., Suite 800, Hagatna, GU, 96910. Phone: (671) 477-5700. Fax: (671) 477-3982. Licensee: Sorensen Television Systems Inc. (acq 10-26-2005; $500,000). Natl. Network: ABC. Washington Atty: Kaye, Scholer LLP.

Puerto Rico

Aguada

WQHA— Analog channel: 50. Analog hrs: 24 501 kw vis, 200 kw aur. ant 1,125t TL: N18 19 06 W67 10 49 On air date: 1995. Box 3869, Carolina, PR, 00984-3869. Phone: (787) 750-4090. Fax: (787) 701-4245. E-mail: conciliosav@hotmail.com Web Site: www.ncntelevision.com. Licensee: Concilio Mision Cristiana Fuente de Agua Viva.

Key Personnel:
Hector Perez . . . . gen mgr
Joel Velez. . . . . gen sls mgr
Fabian Rivera. . . . . progmg dir
Ramon Rivera . . . . chief of engrg

Aguadilla

***WELU**— Analog channel: 32.9.33 kw vis, 933 kw aur. ant 971t/121g TL: N18 18 46 W67 11 09 On air date: 1987. Box 1093, Hormigueros, PR, 00660. Phone: (787) 849-4020. Fax: (787) 849-2092. Licensee: Pabellon Educational Broadcasting Inc. (acq 4-28-2000).

Key Personnel:
Hector Perez . . . . pres & gen mgr
Joel Velez. . . . . gen sls mgr
Fabian Rivera. . . . . progmg dir
Ramon Rivera . . . . chief of engrg

WOLE-TV— Analog channel: 12. Analog hrs: 8:00am - 5:00pm 275 kw vis, 27.5 kw aur. 2,181t TL: N18 09 00 W66 59 00 On air date: May 13, 1960. Box 1200, Mayaguez, PR, 00681-1200. Mckinley Edif. Westerbank Piso 7, Mayaguez, PR 00681-1200. Phone: (787) 833-1200. Phone: (787) 891-8100. Fax: (787) 831-6330. Fax: (787) 891-3380. Licensee: Western Broadcasting Corp. of Puerto Rico. Ownership: Du Art Film Labs Inc., 61.2%; Jose Bechara, 30.6%; Alfonso Giminez-Aguayo, 8.2%. Foreign lang progmg: SP 168

Key Personnel:
Wilson Lugo . . . . gen mgr & gen sls mgr
Santiago Hernandez . . . . progmg mgr
Doel Oriol . . . . chief of engrg
Rebroadcasts WKAQ-TV San Juan.

WVEO— Analog channel: 44. Digital channel: 17. Analog hrs: 24 1,000 kw vis, 100 kw aur. ant 1,220t/235g TL: N18 19 06 W67 10 42 On air date: October 1974. Southwestern B/C Inc., 1554 Bori St., Rio Piedras, PR, 00927-6113. Phone: (787) 882-0422. Fax: (787) 281-9758. Licensee: International Broadcasting Corp. (acq 10-7-2004; $1,382,961 with WXRF(AM) Guayama). Foreign lang progmg: SP 168

Key Personnel:
Pedro Roman Collazo . . . . gen mgr
Margarita Nazario . . . . progmg dir

Arecibo

WCCV-TV— Analog channel: 54. Digital channel: 53. Analog hrs: 24 Digital hrs: 24 11.7 kw vis, 2.34 kw aur. -220t TL: N18 28 28 W66 43 36 (CP: 1,510 kw vis, 151 kw aur, ant 1,968t. TL: N18 14 06 W66 45 36) On air date: Nov 15, 1981. Box 949, Camuy, PR, 00627-0949. Puerto Rico. Phone: (787) 262-5400. Phone: (787) 898-5120. Fax: (787) 262-0541. E-mail: plaud@cdminternational.com Web Site: www.cdminternational.com. Licensee: Asociacion Evan. Cristo Viene Inc. Ownership: Francisco Valazquez, 88%; Wilfredo Almodovar, 6%; Juana Roman, 5%; Patricio R. Fermaintt, 2%. Washington Atty: Fletcher, Heald & Hildreth, P.L.C.

WMEI— Analog channel: 60. Analog hrs: 24 1,000 kw vis. ant 2,322t/141g TL: N18 10 09 W66 34 30 Not on air, target date: unknown: CMCG Puerto Rico License LLC, 900 Laskin Rd., Virginia Beach, VA, 23451. Phone: (757) 437-9800. Fax: (757) 437-0034. Web Site: www.maxmediallc.com. Licensee: CMCG Puerto Rico License LLC. (acq 7-17-2006; $4.25 million).

Bayamon

WDWL— Analog channel: 36. Analog hrs: 24 9.33 kw vis, 933 w aur. 1,079t/308g TL: N18 16 40 W66 06 38 On air date: 1991. Box 50615, Levittown Stn, PR, 00950. Phone: (787) 795-8113. Fax: (787) 795-8140. E-mail: jbenle@prtc.net Web Site: www.teleadoracion.com. Licensee: Bayamon Christian Network. Ownership: Felix Berrios, 20%; Simon Castillo, 20%; Wilfredo Diaz, 20%; David Perez, 20%; Luciano Rodriguez, 20%.

Key Personnel:
Jesus Velez . . . . pres
Zoraida Jostinano . . . . gen mgr
David Baez . . . . chief of engrg

Caguas

WLII— Analog channel: 11. Digital channel: 56. Analog hrs: 24 316 kw vis. ant 1,164t/207g TL: N18 16 54 W66 06 46 On air date: May 27, 1960. Box 7888, Guaynabo, PR, 00970. 64 Calle Carazo, Guaynabo, PR 00969. Phone: (787) 300-5000. Fax: (787) 300-5003. Licensee: WLII/WSUR License Partnership G.P. Group owner: Raycom Media Inc. (acq 6-30-2005; with WSUR-TV Ponce). Washington Atty: Kaye, Scholer, Fierman, Hays & Handler. Foreign lang progmg: SP 140 News staff: 31.

Key Personnel:
Larry Sands . . . . gen mgr
Carlos Pagan . . . . gen sls mgr
Jessica Rodriguez. . . . . progmg dir & progmg mgr
Jose Morales . . . . news dir
Andres Diaz . . . . chief of engrg

***WUJA**— Analog channel: 58.55 kw vis, 5.5 kw aur. 1,078t/110g TL: N18 16 40 W66 06 38 On air date: September 1985. Box 4039, Balle Arryba Heights Stn, Carolina, PR, 00984. Phone: (787) 750-5858. Fax: (787) 757-1500. Licensee: Caguas Educational TV Inc.

Carolina

WRFB— Analog channel: 52.275 kw vis. 1,919t/252g TL: N18 16 44 W65 51 12 On air date: August 1998. Box 1833, Carolina, PR, 00984-1833. Phone: (787) 762-5500. Fax: (787) 752-1825. Licensee: R.Y.F. Broadcasting Inc. Ownership: Enrique A. (Rickin) Sanchez, 50%; Blanche Vidal de Sanchez, 50%. Washington Atty: John L. Tierney.

Fajardo

***WMTJ**— Analog channel: 40. Analog hrs: 24 209 kw vis, 20.9 kw aur. 2,750t/259g TL: N18 18 36 W65 47 41 On air date: January 1985. Box 21345, San Juan, PR, 00928-1345. Isadoro Color, Rd. 176, San Juan, PR 00928. Phone: (787) 766-2600. Fax: (787) 250-8546. Web Site: www.suagm.edu. Licensee: Ana G. Mendez Educational Foundation. Washington Atty: Dow, Lohnes & Albertson. Foreign lang progmg: SP 64

WORO-TV— Analog channel: 13. Digital channel: 33.141 kw vis, 14.1 kw aur. ant 2,831t/210g TL: N18 18 36 W65 47 41 On air date: 1991. Box 1967, San Juan, PR, 00902-1967. Margary De Castro, Carolina, PR 00902-1967. Phone: (787) 276-1300. Fax: (787) 276-1307. Licensee: Catholic, Apostolic and Roman Church of Puerto Rico. Washington Atty: Mullin, Rhyne, Emmons & Topel.

WRUA— Analog channel: 34.50.1 kw vis, 5 kw aur. ant 2,781t TL: N18 18 36 W65 47 41 On air date: 1997. Box 310, Bayamon, PR, 00960. Phone: (787) 279-3434. Fax: (787) 279-5549. Licensee: Eastern Television Corp. (acq 2-4-00; $335,000).

Guayama

WIDP— Analog channel: 46. Analog hrs: 8:30am- 11:00pm 1,480 kw vis, 151 kw aur. ant 2,070t TL: N18 16 44 W65 51 10 On air date: 1999. Box 21065, San Juan, PR, 00928. Phone: (787) 999-0360. Fax: (787) 999-1560. E-mail: info@teletriunfo.com Web Site: www.teletriunso.com. Licensee: Ebenezer Broadcasting Group Inc. Washington Atty: Shaw Pittman LLP.

Mayaguez

***WIPM-TV**— Analog channel: 3. Analog hrs: 24 81.3 kw vis, 8.1 kw aur. 2,273t/382g TL: N18 09 00 W66 59 00 (CP: Ant 2,280t/389g) On air date: Apr 28, 1961. Box 190909, San Juan, PR, 00919. Phone: (787) 834-0164. Fax: (787) 832-9139. Licensee: Puerto Rico Public Broadcasting Corp. Natl. Network: PBS. Washington Atty: Steptoe & Johnson.

Key Personnel:
Eduardo Bado . . . . gen mgr
Reinaldo Perez . . . . opns dir
Diane Ramos . . . . gen sls mgr
Mirta Rodriguez . . . . progmg dir
Jorge Gonzalez. . . . . chief of engrg

WNJX-TV— Analog channel: 22. Analog hrs: 24 4,201 kw vis. ant 2,158t/272g TL: N18 09 00 W66 59 00 On air date: Apr 27, 1986. c/o WAPA-TV, Apartado 362052, San Juan, PR, 00936. Phone: (787) 792-4444. Fax: (787) 782-4420. Licensee: WNJX-TV Inc. Group owner: LIN Television Corporation (acq 3-1-2001; up to $1.075 million for stock). Foreign lang progmg: SP 168

Key Personnel:
Joe Ramos. gen mgr
Jonathan Garcia . . . . gen sls mgr
Margarita Millan. . . . . progmg VP
Enrique Cruz . . . . news dir
Jose Guerra . . . . chief of engrg
Satellite of WAPA-TV San Juan.

WORA-TV— Analog channel: 5. Analog hrs: 24 100 kw vis, 20 kw aur. 2,001t/241g TL: N18 09 02 W66 59 20 On air date: Oct 1, 1955. Box 43, Mayaguez, PR, 00681. Phone: (787) 831-5555/(787) 721-4054. Fax: (787) 833-0075/(787) 724-1554. E-mail: gatoro@woratv.com Web Site: www.woratv.com. Licensee: Telecinco Inc. Ownership: Alfredo R. deArellano Jr. and family, 100%. Washington Atty: Edward O'Niell. Foreign lang progmg: SP 168

Key Personnel:
Jose Toro . . . . gen mgr & gen sls mgr
Ramon Guzman. . . . . progmg dir
Carlos Sepulveda . . . . news dir
Fred Toledo . . . . chief of engrg

WOST— Analog channel: 16.9.55 kw vis. ant 1,107t/115g TL: N18 18 51 W67 11 24 On air date: November 2006. CMCG Puerto Rico License LLC, 900 Laskin Rd., Virginia Beach, VA, 23451. Phone: (757) 437-9800. Fax: (757) 437-0034. Licensee: CMCG Puerto Rico License LLC. Ownership: Power Television International LLC, 51%; and Max Media IV LLC, 49% (acq 2-17-2006; $4.25 million with WMEI(TV) Arecibo). Population served: 3,800,000 Foreign lang progmg: SP 168

Naranjito

WECN— Analog channel: 64.1,000 kw vis, 100 kw aur. 466t/107g TL: N18 17 34 W66 16 02 On air date: April 1986. Box 310, Bayamon, PR, 00960. Hwy. 167, Naranjito, PR 00960. Phone: (787) 799-1480. E-mail: evn@centennialpr.net Licensee: Encuentro Christian Network. Ownership: Rafael Torres Ortega, 11.11%; Iris Padilla, 11.11%; Ramon Luis Acevedo, 11.11%; Jofre Ayala, 11.11%; Daramid Ayala, 11.11% (acq 9-87; $175,000; FTR: 4-13-87). Washington Atty: Irwin, Campbell & Tannenwald.

Ponce

WKPV— Analog channel: 20. Digital channel: 19.100 kw vis, 10 kw aur. 850t/105g TL: N18 04 50 W66 64 50 On air date: Aug 6, 1985. Box 2050, Attn: Edwin Pujols, San Juan, PR, 00936-2050. Phone: (787) 792-4760. Phone: (787) 705-4153. Fax: (787) 782-7825. E-mail: edwn.pujols@wapa-tv.com Licensee: S & E Network Inc. Group owner: LIN Television Corporation (acq 8-2-2001; grpsl).

***WQTO**— Analog channel: 26. Analog hrs: 24 1,000 kw vis. ant 991t/213g TL: N18 04 50 W66 44 54 On air date: November 1986. Box 21345, San Juan, PR, 00928-1345. Isadoro Color, Rd. 176, San Juan, PR 00928. Phone: (787) 766-2600. Fax: (787) 250-8546. Web Site: www.suagm.edu. Licensee: Sistema Universitario Ana G. Mendez Inc. Washington Atty: Dow, Lohnes & Albertson.

WSTE— Analog channel: 7. Analog hrs: 24 186 kw vis, 25.1 kw aur. 2,709t/439g TL: N18 09 17 W66 33 16 On air date: Feb 2, 1958. Box 2528, Guaynavo, PR, 00902. Phone: (787) 724-7777. Fax: (787) 300-5225. Licensee: Siete Grand Television Inc. Ownership: Jerry B.& Esther M. Hartman. (acq 8-1-91; $6 million; FTR: 8-26-91). Foreign lang progmg: SP 168
Key Personnel:
Maria Negron . . . . . gen sls mgr
Wanda Costanzo . . . . . gen mgr, gen sls mgr & progmg dir
Gilberto Vera . . . . . chief of engrg

WSUR-TV— Analog channel: 9. Digital channel: 43.178 kw vis. ant 2,811t/266g TL: N18 10 09 W66 34 36 On air date: February 1958. Box 7888, Guaynabo, PR, 00970-7888. One 3rd St., San Juan, PR 00908. Phone: (787) 724-1111. Fax: (787) 722-3505. Fax: (787) 723-0094. Licensee: WLII/WSUR License Partnership G.P. Group owner: Raycom Media Inc. (acq 6-30-2005; with WLII(TV) Caguas). Natl. Network: Univision (Spanish). Washington Atty: Hamel & Park. Foreign lang progmg: SP 140
Key Personnel:
Larry Sands . . . . . sr VP, sr VP, VP & gen mgr
Carlos Pagan . . . . . gen sls mgr & mktg dir
Manuel Santiago . . . . . prom dir
Jessica Rodriguez . . . . . progmg dir
Jose Morales . . . . . news dir
Andres Diaz . . . . . chief of engrg

WTIN— Analog channel: 14. Analog hrs: 24 1070 kw vis, 10 kw aur. ant 2,824t/53g TL: N18 10 11 W66 34 38 On air date: 1998. Box 362050, San Juan, PR, 00936-2050. Phone: (787) 792-4444. Fax: (787) 782-4420. Licensee: Televicentro of Puerto Rico LLC. Group owner: LIN Television Corporation (acq 5-6-2004; $5 million). Washington Atty: Baraff, Koerner, Olender & Hochberg.
Key Personnel:
Margarita Millan . . . . . progmg VP
Jose Guerra . . . . . engrg VP

WVOZ-TV— Analog channel: 48.61.7 kw vis, 6.2 kw aur. 810t/95g TL: N18 04 50 W66 44 50 On air date: 1994. Bori 1554 St. Urb Point, San Juan, PR, 00927. Phone: (787) 274-1800. Fax: (787) 281-9758. Licensee: International Broadcasting Corp. Group owner: (group owner; (acq 10-9-2001; grpsl).
Key Personnel:
Margarita Nazario . . . . . gen mgr & gen sls mgr
Roman Callazo . . . . . progmg dir
Rudi Rivas . . . . . chief of engrg

San Juan

WAPA-TV— Analog channel: 4. Digital channel: 27. Analog hrs: 24 53.7 kw vis, 8.13 kw aur. ant 2,865t/1,094g TL: N18 06 42 W66 03 05 On air date: April 1954. Apartado 362050, San Juan, PR, 00936-2050. Phone: (787) 792-4444. Fax: (787) 782-4420. Web Site: www.televicentropr.com. Licensee: Televicentro of Puerto Rico L.L.C. Group owner: LIN Television Corporation (acq 7-12-2000; grpsl). Washington Atty: Fletcher, Heald & Hildreth. Foreign lang progmg: SP 133 News staff: 29; News: 20 hrs wkly.
Key Personnel:
Joe Ramos . . . . . gen mgr
Jonathan Garcia . . . . . gen sls mgr
Margarita Millan . . . . . progmg dir
Enrique Cruz . . . . . news dir
Jose Guerra . . . . . chief of engrg

***WIPR-TV—** Analog channel: 6. Analog hrs: 24 53.7 kw vis, 5.4 kw aur. 2,860t/1,094g TL: N18 06 42 W66 03 05 (CP: 58.9 kw vis, 5.9 kw aur, ant 2,706t/940g) On air date: Jan 6, 1958. Box 190909, San Juan, PR, 00919. Phone: (787) 766-0505. Fax: (787) 753-9846. Licensee: Puerto Rico Public Broadcasting Corp. Natl. Network: PBS. Washington Atty: Steven Huffines. Foreign lang progmg: SP 40
Key Personnel:
Susanne Marte . . . . . VP
Victor Montilla . . . . . gen mgr
Rebecca Torres . . . . . opns VP
Ebelmiro Torres . . . . . progmg dir
Jorge Gonzalez . . . . . chief of engrg

WJPX— Analog channel: 24. Digital channel: 21.676 kw vis. ant 1,909t/220g TL: N18 16 45 W65 51 14 On air date: Feb 15, 1987. Apartado 362050, Attn: Edwin Pujols, San Juan, PR, 00936-2050. Phone: (787) 792-4444. Phone: (787) 706-4153. Fax: (787) 782-7825. E-mail: edwin.pujols@wapa-tv.com Licensee: S&E Network Inc. Group owner: LIN Television Corporation (acq 8-2-2001; grpsl). Washington Atty: Dow, Lohnes & Albertson.
Key Personnel:
Joe Ramos . . . . . gen mgr
Edwin Pujols . . . . . stn mgr
Jonathan Garcia . . . . . gen sls mgr
Margarita Millan . . . . . progmg dir
Enrique Cruz . . . . . news dir
Jose Guerra . . . . . chief of engrg

WKAQ-TV— Analog channel: 2.55 kw vis, 10.5 kw aur. 2,824t/1,099g TL: N18 06 54 W66 03 10 On air date: Mar 28, 1954. Box 366222, San Juan, PR, 00936-6222. 383 Roosevelt Ave., Hato Rey, PR 00919. Phone: (787) 758-2222. Phone: (787) 641-2222. Fax: (787) 641-2175. Fax: (787) 641-2184. Web Site: www.telemundopr.com. Licensee: NBC Telemundo License Co. Group owner: Telemundo Group Inc. (acq 4-10-2002; grpsl). Population served: 3,500,000 Natl. Network: Telemundo (Spanish). Washington Atty: Hogan & Hartson. Foreign lang progmg: SP 130 News staff: 40; News: 24 hrs wkly.
Key Personnel:
Luis Roldan . . . . . gen mgr
Jose Medina . . . . . opns mgr & chief of engrg
Paco Pregues . . . . . sls dir
Iliana Santiago . . . . . progmg dir
Ruban Roman . . . . . news dir

WSJU-TV— Analog channel: 30. Digital channel: 31. Analog hrs: 24 2,630 kw vis, 263 kw aur. ant 941t/196g TL: N18 16 30 W66 05 36 On air date: 1985. 1508 Calle Bori, Urb. Antonsanti, San Juan, PR, 00927. Phone: (787) 756-8700. Fax: (787) 765-2965. Web Site: www.canal30pr.com. Licensee: Aerco Broadcasting Corp. Ownership: Angel O. Roman Lopez, 50%; Ruth E. Roman Lopez, 50% (acq 1-11-2005). Washington Atty: Borsari & Assoc.
Key Personnel:
Angel O. Roman Lopez . . . . . pres & gen mgr
Sergio Ballesteros . . . . . gen sls mgr
Rudi Rivas . . . . . chief of engrg

WTCV— Analog channel: 18. Digital channel: 32. Analog hrs: 14 759 kw vis, 75.9 kw aur. ant 2,778t/174g TL: N18 18 36 W65 47 41 On air date: Aug 19, 1984. Bori 1554, San Juan, PR, 00927-6113. Phone: (787) 274-1800. Phone: (787) 203-9178. Fax: (787) 281-9758. Licensee: International Broadcasting Corp. Group owner: (group owner; (acq 10-9-2001; grpsl). Population served: 1,500,000 Washington Atty: Marmet & McCombs.
Key Personnel:
Pedro Roman . . . . . pres, pres & gen mgr
Margarita Nazario . . . . . gen sls mgr & prom dir

San Sebastian

WJWN-TV— Analog channel: 38. Digital channel: 39. Analog hrs: 24 85.1 kw vis, 8.5 kw aur. 1,089t TL: N18 19 06 W67 10 42 Box 362050, Attn: Edwin Pujols, San Juan, PR, 00936. Phone: (787) 792-4444. Fax: (787) 782-7825. E-mail: edwin.pujols@wapa-tv.com Licensee: S&E Network Inc. Group owner: LIN Television Corporation (acq 8-2-2001; grpsl).

Yauco

WIRS— Analog channel: 42.1,510 kw vis, 151 kw aur. ant 1,961t TL: N18 10 10 W66 34 36 On air date: Dec 1, 1991. Box 310, Bayamon, PR, 00960-0310. Phone: (787) 799-1480. Licensee: Televicentro of Puerto Rico LLC. Group owner: LIN Television Corporation (acq 12-11-2003; $4.45 million).

Virgin Islands

Charlotte Amalie

***WTJX-TV—** Analog channel: 12. Analog hrs: 24 28.8 kw vis, 2.9 kw aur. ant 1,479t TL: N18 21 26 W64 56 50 On air date: 1972. Box 7879, St. Thomas, VI, 00801. Phone: (340) 774-6255. Fax: (340) 774-7092. E-mail: optter@wtixtv.org Web Site: wtjxtv.org. Licensee: Virgin Islands Public Television System Board of Directors. Natl. Network: PBS. Washington Atty: Schwartz, Woods & Miller.

WVXF— Analog channel: 17. Digital channel: 48. Analog hrs: 24 75.9 kw vis. ant 1,506t/59g TL: N18 21 26 W64 56 50 On air date: 1999. 8000 Nisky Center, Ste. 714, St. Thomas, VI, 00802. Phone: (340) 774-2012 / (800) 511-5899. Fax: (340) 776-5362. E-mail: info@wvxftv.com Web Site: www.cbstvcaribbean.com. Licensee: Storefront Television. Ownership: LKK Group Corp., 50%; Bluewater LLC, 50% (acq 9-30-2004; $600,000). Natl. Network: CBS. Washington Atty: Dow, Lohnes & Albertson.

WZVI— Analog channel: 0. Digital channel: 43.1.4 kw vis. ant 92t/24g TL: N18 20 43 W64 55 45 On air date: 2006. c/o Thomas J. Dougherty Jr., Kilpatrick, Stockton LLP, 607 14th St. N.W., Washington, DC, 20005. Phone: (202) 508-5836. Fax: (202) 508-5858. Web Site: www.wsvitv.com. Licensee: Marri Broadcasting LP.

Christiansted

WCVI-TV— Analog channel: 39. Digital channel: 23. Analog hrs: 24 26.67 kw vis. ant 440t/59g TL: N17 44 53 W64 43 40 On air date: Mar 1, 2000. Box 24027, Christiansted, VI, 00824. Phone: (340) 713-9927. Fax: (340) 773-0712. E-mail: mbox@wcvi.tv Web Site: www.wcvi.tv. Licensee: Virgin Blue Inc. Natl. Network: CW.
Key Personnel:
Marty Adamshick . . . . . progmg dir
Victor Gold . . . . . gen mgr, gen sls mgr & chief of engrg

WSVI— Analog channel: 8. Digital channel: 20. Analog hrs: 24 200 kw vis, 20 kw aur. ant 1,144t/265g TL: N17 45 20 W64 47 55 (CP: Ant 958t) On air date: January 1966. Box 6000, Christiansted, St. Croix, VI, 00823. Sunny Isle Shopping Ctr., Christiansted, St. Croix, VI 00823. Phone: (340) 778-5008; (803) 732-1757. Fax: (340) 778-5011. E-mail: channel&@wsvitv.com Web Site: www.wsvi.tv. Licensee: Alpha Broadcasting Corp. Natl. Network: ABC. Natl. Rep: Roslin. Washington Atty: Marmet & McCombs. Foreign lang progmg: SP 5
Key Personnel:
David Lampel . . . . . gen mgr
Kimberly Alexander . . . . . gen mgr & stn mgr
Jackie Schrock . . . . . sls dir
Denisha Brown . . . . . news dir
Chester Benjamin . . . . . chief of engrg

WVIF— Analog channel: 15. Analog hrs: 24 16.2 kw vis. ant 971t/51g TL: N17 45 21 W64 47 56 On air date: 2001. 5660 Southwyck Blvd., Toledo, OH, 43614. Phone: (419) 861-3815. Fax: (419) 861-3818. Web Site: www.telostv.com. Licensee: CMCG St. Croix License LLC. Group owner: MAX Media L.L.C. (acq 2-7-2003; $10 million with WPFO(TV) Waterville, ME).
Key Personnel:
Charles Glover . . . . . CEO & gen mgr
Mitch Lambert . . . . . gen mgr

Mexico

Ciudad Juarez

see El Paso (Las Cruces, NM), TX market

Tijuana

see San Diego, CA market

Directory of TV Stations in Canada

Alberta

Ashmont

CFRN-TV-4— Analog channel: 12.14.6 kw vis, 1.46 kw aur. ant 635t/590g TL: N54 08 07 W111 36 16 On air date: 1966. c/o CFRN-TV, 18520 Stony Plain Rd., Edmonton, AB, T5S 1A8. Phone: (780) 483-3311. Fax: (780) 484-4426. E-mail: cfrn@ctv.ca Web Site: www,cfrntv.ca. Licensee: CFRN-TV, a div. of CTV Television Inc.

Bonnyville

CBXFT-1— Analog channel: 6.67 kw vis, 13.4 kw aur. Box 555, c/o CBXFT, Edmonton, AB, T5J 2P4. c/o CBXFT, Edmonton City Centre, 10062-102 Ave., Suite 123, Edmonton, AB T5J 2Y8. Phone: (780) 468-7500. Fax: (780) 468-7792. Web Site: www.cbc.ca. Licensee: Canadian Broadcasting Corp. Natl. Network: Radio Canada.

Key Personnel:
Don Orchard . . . . CFO
Lionel Bonneville . . . . gen mgr

Calgary

CBRT— Analog channel: 9.178 kw vis, 35.6 kw aur. 1,135t/845g On air date: Sept 1, 1975. 1724 Westmount Blvd. N.W., Calgary, AB, T2P 2M7. Phone: (403) 521-6000. Fax: (403) 521-6007. E-mail: del_simon@cbc.ca Web Site: cbc.ca. Licensee: CBC. Natl. Network: CBC.

Key Personnel:
Carole Taylor . . . . chmn
Robert Rabinovich . . . . pres
Harold Redekopp . . . . exec VP
Don Orchard . . . . gen mgr, opns dir & opns mgr
Wendy Ell . . . . sls dir
Pat Paproski . . . . rgnl sls mgr
Irene Karras . . . . mktg mgr
Del Simon . . . . prom mgr
Fred Youngs . . . . progmg dir
Nancy Rose . . . . progmg mgr
Laurie Long . . . . news dir
Lindsay Rutschke . . . . chief of engrg

CFCN-TV— Analog channel: 4.100 kw vis, 27.5 kw aur. 623t/380g TL: N52 03 37 W114 10 13 On air date: September 1960. Broadcast House, 80 Patina Rise S.W., Calgary, AB, T3H 2W4. Phone: (403) 240-5600. Fax: (403) 240-5711. E-mail: cfcnnews@ctv.ca Web Site: www.cfcnplus.ca. Licensee: CTV Television Inc. Group owner: (group owner) Natl. Network: CTV.

CIAN-TV— Analog channel: 13. Analog hrs: 24 9.9 kw vis. ant 807t TL: N51 03 54 W114 12 47 On air date: 1989. 3720 76 Ave., Edmonton, AB, T6B 2N9. Phone: (780) 440-7777. Fax: (780) 440-8899. Web Site: www.accesslearning.com/accesstv. Licensee: Learning and Skills Television of Alberta Ltd. Ownership: CHUM Ltd., 60%; Olympus Management, 20%; 1006228 Ontario Inc., 15.5%; and Jay Switzer, 4.5%. (acq 1995).

Key Personnel:
Ron Keast . . . . CEO
Moses Znaimer . . . . chmn
Dr. Ronald Keast . . . . pres
Peter Palframan . . . . CFO & VP
Richard Hiron . . . . adv dir

CICT-TV— Analog channel: 2. Analog hrs: 24 100 kw vis, 20 kw aur. 989t/633g On air date: October 1954. 222 23rd St. N.E., Calgary, AB, T2E 7N2. Phone: (403) 235-7777. Fax: (403) 248-0252. Web Site: www.canada.com. Licensee: CanWest MediaWorks Inc. News staff: 50; News: 13 hrs wkly.

Key Personnel:
C. McGinley . . . . VP & gen mgr
Norm Michaelis . . . . opns dir
Greg Campbell . . . . gen sls mgr
J. Eisler . . . . mktg dir
Lynda Ritz . . . . prom mgr
Dawna Docherty . . . . progmg dir
Dave Budge . . . . news dir
Jeff Eisler . . . . pub affrs dir
Dan Gold . . . . engrg VP
Rebroadcast CISA-TV Lethbridge 90%.

CKAL-TV— Analog channel: 5.33.6 kw vis. On air date: Sept 20, 1997. 535 7th Ave. S.W., Calgary, AB, T2P 0Y4. Phone: (403) 508-2222. Fax: (403) 508-2224. Web Site: www.a-channel.com. Licensee: CHUM Ltd. Group owner: (group owner). (acq 11-19-2004; grpsl).

Key Personnel:
Drew Craig . . . . CEO & pres
Andy Pernal . . . . CFO
Al Thorgeikson . . . . gen mgr
Mike Pietrus . . . . news dir

Cardston

CFSO-TV— Analog channel: 32.20 w. TL: N49 10 40 W113 19 36 On air date: 1996. Box 1238, Cardston, AB, T0K 0K0. Phone: (403) 653-3792. Fax: (403) 653-3792. E-mail: channel32@mac.com Web Site: www.channel32.ca. Licensee: Logan McCarthy.

Coronation

CHCA-TV-1— Analog channel: 10. Analog hrs: 6 AM-2 AM 190 kw vis, 19 kw aur. ant 697t/240g TL: N50 09 15 W111 09 30 On air date: 1960. c/o RDTV, 2840 Bremner Ave., 2nd floor, Red Deer, AB, T4R 1M9. Phone: (403) 346-2573. Fax: (403) 346-9980. E-mail: rdtv@globalyv.ca Web Site: www.rdtv.com. Licensee: Global Communications Ltd. Group owner: CanWest Global Communications Corp. Natl. Network: CBC. News staff: 12; News: 5 hrs wkly.
Rebroadcasts CHCA-TV Red Deer.

Drumheller

CFCN-TV-1— Analog channel: 12.14.1 kw vis, 7 kw aur. 1,073t TL: N51 33 46 W112 19 44 c/o CFCN-TV, 80 Patina Rise S.W., Calgary, AB, T3H 2W4. Phone: (403) 240-5600. Fax: (403) 240-5773. E-mail: cfcnnews@ctv.ca Web Site: www.cfcn.ca. Licensee: CTV Television Inc. Natl. Network: CTV.

Edmonton

CBXFT— Analog channel: 11.90 kw vis, 18 kw aur. On air date: 1970. Box 555, Edmonton, AB, T5J 2P4. Edmonton City Centre, 10062-102 Ave., Suite 123, Edmonton, AB T5J 2Y8. Phone: (780) 468-7500. Fax: (780) 468-7868. Web Site: www.cbc.ca. Licensee: CBC. Natl. Network: CBC. News staff: 15; News: 2 hrs wkly.

Key Personnel:
Don Orchard . . . . CFO & opns dir
Carol Nielsen . . . . chief of engrg

CBXT— Analog channel: 5.318 kw vis, 34.3 kw aur. On air date: 1961. Box 555, Edmonton, AB, T5J 2P4. Edmonton City Centre, 10062-102 Ave., Suite 123, Edmonton, AB T5J 2Y8. Phone: (780) 468-7500. Fax: (780) 468-7893. Web Site: www.cbc.ca. Licensee: CBC. Natl. Network: CBC.

CFRN-TV— Analog channel: 3. Analog hrs: 5:30 AM-1:30 AM 250 kw vis, 25 kw aur. ant 1,101t/260g TL: N53 23 06 W113 12 48 On air date: Oct 17, 1954. 18520 Stony Plain Rd., Edmonton, AB, T5S 1A8. Phone: (780) 483-3311. Fax: (780) 484-4426. Web Site: www.cfrntv.ca. Licensee: CFRN TV, a div. of CTV Television Inc. Group owner: CTV Inc. (acq 1998). Population served: 1,200,000 Natl. Network: CTV. News: 12 hrs wkly.

Key Personnel:
Fred Filthaut . . . . VP & gen mgr
David Fisher . . . . prom dir

CITV-TV— Analog channel: 13.325 kw vis, 32.5 kw aur. 900t TL: N53 23 06 W113 12 48 On air date: 1974. 5325 Allard Way, Edmonton, AB, T6H 5B8. Phone: (780) 436-1250. Fax: (780) 989-4613. E-mail: edmonton@globaltv.ca Web Site: www.canada.com. Licensee: CanWest MediaWorks Inc. Group owner: Canwest Global (acq 2-6-91). News staff: 33; News: 11 hrs wkly.

Key Personnel:
Tim Spelliscy . . . . gen mgr
Neill Fitzpatrick . . . . news dir

***CJAL-TV—** Analog channel: 9. Analog hrs: 24 8.2 kw vis. TL: N53 24 19 W113 20 38 On air date: Apr 1, 1991. 3720 76th Ave., Edmonton, AB, T6B 2N9. Phone: (780) 440-7777. Fax: (780) 440-8899. E-mail: access@incentre.net Web Site: www.accesslearning.com. Licensee: Learning and Skills Television of Alberta Ltd. Ownership: CHUM Ltd., 60%; Olympus Management Ltd., 20%; 1006228 Ontario Ltd., 15.5%; Jay Switzer, 5.5%. Group owner: CHUM Ltd. (acq 9-1-95). Population served: 2,500,000

Key Personnel:
Moses Znaimer . . . . chmn
Peter Palframan . . . . exec VP
Richard Hiron . . . . sls dir
Jill Bonenfant . . . . progmg dir
John Wood . . . . engrg dir

CKEM-TV— Analog channel: 51.704 kw vis. On air date: Sept 18, 1997. 10212 Jasper Ave., Edmonton, AB, T5J 5A3. Phone: (780) 424-2222. Fax: (780) 424-0357. E-mail: webmaster@appliedthemalsciences.com Web Site: www.a-channel.com. Licensee: CHUM Ltd. Group owner: (group owner). (acq 11-19-2004; grpsl).

Key Personnel:
Jim Haskins . . . . gen mgr
John Cuccaro . . . . opns mgr
Art Eden . . . . rgnl sls mgr
Barry Close . . . . prom mgr
Chris Duncan . . . . news dir
Peter Nobel . . . . engrg mgr

Fort McMurray

CBXFT-6— Analog channel: 12.5.8 kw vis, 500 w aur. 200t/300g On air date: Mar 1, 1970. Box 555, Edmonton, AB, T5J 2P4. c/o CBC, Edmonton City Centre, 10062-102 Ave., Suite 123, Edmonton, AB T5J 2Y8. Phone: (780) 468-7500. Fax: (780) 468-7792. Web Site: www.cbc.ca. Licensee: Canadian Broadcasting Corp. Natl. Network: Radio Canada.

Grande Prairie

CBXAT— Analog channel: 10.36 kw vis, 18 kw aur. Box 555, c/o CBXT, Edmonton, AB, T5J 2P4. Edmonton City Centre, 10062-102 Ave., Suite 123, Edmonton, AB T5J 2Y8. Phone: (780) 468-7500. Fax: (780) 468-7893. Web Site: www.cbc.ca. Licensee: CBC. Natl. Network: CBC.

Key Personnel:
Don Orchard . . . . CFO & opns dir
Carol Nielsen . . . . chief of engrg

CBXFT-8— Analog channel: 19.3.3 kw vis, 330 w aur. Box 555, Edmonton, AB, T5J 2P4. Edmonton City Centre, 10062-102 Ave., Suite 123, Edmonton, AB T5J 2Y8. Phone: (780) 468-7500. Fax: (780) 468-7792. Web Site: www.cbc.ca. Licensee: CBC. Natl. Network: Radio Canada.

Key Personnel:
Don Orchard . . . . opns dir
Carol Nielsen . . . . chief of engrg

CFRN-TV-1— Analog channel: 13.32 kw vis, 6.4 kw aur. ant 1,014t/641g TL: N55 27 57 W118 45 32 c/o CFRN-TV, 18520 Stony Plain Rd., Edmonton, AB, T5S 1A8. Phone: (780) 483-3311. Fax: (780) 484-4426. Web Site: www.cfrntv.ca. Licensee: CFRN-TV, a div. of CTV Television Inc. Natl. Network: CTV.

Grouard Mission-High Prairie

CFRN-TV-8— Analog channel: 18.6 kw vis, 600 w aur. ant 549t/369g TL: N55 32 26 W116 07 26 On air date: November 1981. c/o CFRN-TV, 18520 Stony Plain Rd., Edmonton, AB, T5S 1A8. Phone: (780) 483-3311. Fax: (780) 484-4426. E-mail: cfrn@ctv.ca Web Site: www.cfrntv.ca. Licensee: CFRN TV, a div. of CTV Television Inc. Natl. Network: CTV.
Rebroadcasts CFRN-TV-1 Grande Prairie.

High Prairie

CBXAT-2— Analog channel: 2.6.2 kw vis, 620 w aur. Box 555, Edmonton, AB, T5J 2P4. c/o CBXT, Edmonton City Centre, 10062-102 Ave., Suite 123, Edmonton, AB T5J 2Y8. Phone: (780) 468-7500. Fax: (780) 468-7893. Web Site: www.cbc.ca. Licensee: CBC. Natl. Network: CBC.

Key Personnel:
Don Orchard . . . . exec VP & opns dir
Carol Nielsen . . . . chief of engrg

Lac La Biche

CFRN-TV-5— Analog channel: 2.2.13 kw vis, 213 w aur. ant 126t/157g TL: N54 45 13 W111 56 26 c/o CFRN-TV, 18520 Stony Plain Rd., Edmonton, AB, T5S 1A8. Phone: (780) 483-3311. Fax: (780) 484-4426. E-mail: cfrn@ctv.ca Web Site: www.cfrntv.ca. Licensee: CFRN-TV, a div. of CTV Television Inc. Natl. Network: CTV.

Lethbridge

CFCN-TV-5— Analog channel: 13.47 kw vis, 7.34 kw aur. ant 564t TL: N49 43 59 W112 57 36 c/o CFCN-TV, 80 Patina Rise S.W., Calgary, AB, T3H 2W4. Phone: (403) 240-5600. Fax: (403) 240-5773. E-mail: cfcnnews@ctv.ca Web Site: www.cfcnplus.ca. Licensee: CTV Television Inc. Group owner: (group owner) Natl. Network: CTV.

CISA-TV— Analog channel: 7.167 kw vis, 33.4 kw aur. 662t/600g TL: N49 47 01 W112 52 01 On air date: 1955. 1401-28 Street N., Lethbridge, AB, T1H 6H9. Phone: (403) 327-1521. Fax: (403) 320-2620. E-mail: cisa@globaltv.ca Web Site: www.canada.com/lethbridge. Licensee: CanWest MediaWorks Inc. Group owner: CanWest Global Communications Corp. (acq 9-1-2000; grpsl). News staff: 19; News: 12 hrs wkly.

CJIL-TV— Analog channel: 17.31.6 kw vis. On air date: 1995. Box 1566, 450 31st St. N., Lethbridge, AB, T1H 3Z3. Phone: (403) 380-3399. Fax: (403) 380-3322. Licensee: Miracle Channel.
Key Personnel:
Gord Klussen . . . . gen mgr
Len Whyte . . . . chief of engrg

Lloydminster

CITL-TV— Analog channel: 4.130 kw vis, 13 kw aur. 724t/708g On air date: July 28, 1976. 5026 50th St., Lloydminster, AB, T9V 1P3. Phone: (780) 875-3321. Fax: (780) 875-4704. Web Site: www.ctv.ca. Licensee: NewCap Inc. Group owner: Midwest Broadcasting. (acq 12-22-2004; C$6,304,000 with CKSA-TV Lloydminster). Natl. Network: CTV.
Key Personnel:
R.G. Steele . . . . pres
Ken Ruptash . . . . gen mgr

CKSA-TV— Analog channel: 2.116 kw vis, 23.2 kw aur. On air date: Sept 23, 1960. 5026 50th St., Lloydminster, AB, T9V 1P3. Phone: (780) 875-3321. Fax: (780) 875-4704. Licensee: NewCap Inc. Group owner: Midwest Broadcasting. (acq 12-22-2004; C$6,304,000 with CITL-TV Lloydminster). Natl. Network: CBC.
Key Personnel:
R.G. Steele . . . . pres
Ken Ruptash . . . . gen mgr

Lougheed

CFRN-TV-7— Analog channel: 7.5 kw vis, 500 w aur. ant 723t/517g TL: N52 32 15 W111 31 06 On air date: Sept 7, 1979. c/o CFRN-TV, 18520 Stony Plain Rd., Edmonton, AB, T5S 1A8. Phone: (780) 483-3311. Fax: (780) 484-4426. E-mail: cfrn@ctv.ca Web Site: www.cfrntv.ca. Licensee: CFRN TV, a div. of CTV Television Inc. Natl. Network: CTV.
Rebroadcasts CFRN-TV Edmonton.

Manning

CBXAT-3— Analog channel: 12. Analog hrs: 7 AM-3 AM 1.77 kw vis, 177 w aur. TL: N56 42 20 W117 39 17 On air date: Oct 7, 1968. Box 555, Edmonton, AB, T5J 2P4. c/o CBC, Edmonton City Centre, 10062-102 Ave., Suite 123, Edmonton, AB T5J 2Y8. Phone: (780) 468-7500. Fax: (780) 468-7779. Web Site: www.radio-canada.ca. Licensee: CBC. Population served: 3,300 Natl. Network: CBC.
Key Personnel:
Don Orchard . . . . exec VP, opns dir & engrg mgr
Carol Nielsen. . . . . chief of engrg

Meander River

CJTG-TV— Analog channel: 13.10 w vis. On air date: 1994. Box 1377, High Level, AB, T0H 1Z0. Web Site: www.tachegondihesociety.com. Licensee: Tache Gondihe Society.

Medicine Hat

CFCN-TV-8— Analog channel: 8.5.8 kw vis, 600 w aur. ant 315t TL: N50 04 36 W110 47 40 c/o CFCN-TV, 60 Patina Rise S.W., Calgary, AB, T3H 2W4. Phone: (403) 240-5600. Fax: (403) 240-5711. E-mail: cfcnnews@ctv.ca Web Site: www.cfcn.ca. Licensee: CTV Television Inc. Natl. Network: CTV.

CHAT-TV— Analog channel: 6.58 kw vis, 5.8 kw aur. 700t/559g On air date: 1957. Box 1270, Medicine Hat, AB, T1A 7H5. 10 Boundary Rd., Red Cliff, AB T0J 2P0. Phone: (403) 529-1270. Fax: (403) 529-1292. Web Site: www.1270chat@monach.net. Licensee: Jim Pattison Broadcast Group Ltd. (the general partner) and Jim Pattison Industries Ltd. (the limited partner) carrying on business as Jim Pattison Broadcast Group L.P. Group owner: The Jim Pattison Broadcast Group (acq 12-21-2000; grpsl). Natl. Network: CBC. Natl. Rep: Airtime TV.
Key Personnel:
Dwaine Dietrich . . . . gen mgr
Joel Simmons . . . . chief of engrg

Peace River

CFRN-TV-2— Analog channel: 3.2.4 kw vis, 240 w aur. ant 559t/351g TL: N56 08 47 W117 20 15 On air date: 1970. c/o CFRN-TV, 18520 Stony Plain Rd., Edmonton, AB, T5S 1A8. Phone: (780) 483-3311. Fax: (780) 484-4426. E-mail: cfrn@ctv.ca Web Site: www.cfrntv.ca. Licensee: CFRN-TV, a div. of CTV Television Inc. Natl. Network: CTV.

Pivot

CHAT-TV-1— Analog channel: 3.2.75 kw vis, 1.37 kw aur. Box 1270, Medicine Hat, AB, T1A 7H5. 10 Boundary Rd., Red Cliff, AB T0J 2P0.Canada Phone: (403) 529-1270. Fax: (403) 529-1292. E-mail: www.1270chat@monach.net Licensee: Jim Pattison Broadcast Group Ltd. (the general partner) and Jim Pattison Industries Ltd. (the limited partner) carrying on business as Jim Pattison Broadcast Group L.P. Group owner: The Jim Pattison Broadcast Group (acq 12-21-2000; grpsl). Natl. Network: CBC.
Key Personnel:
Dwaine Dietrich . . . . gen mgr
Joel Simmons . . . . chief of engrg
Rebroadcasts CHAT-TV Medicine Hat.

Red Deer

CFRN-TV-6— Analog channel: 8.22 kw vis, 2.2 kw aur. ant 882t/588g TL: N52 19 10 W113 40 37 c/o CFRN-TV, 18520 Stony Plain Rd., Edmonton, AB, T5S 1A8. Phone: (780) 483-3311. Fax: (780) 484-4426. E-mail: cfrn@ctv.ca Web Site: www.cfrntv.ca. Licensee: CFRN-TV, a div. of CTV Television Inc. Natl. Network: CTV.

CHCA-TV— Analog channel: 6.100 kw vis, 10 kw aur. 817t/570g On air date: 1956. 2840 Bremner Ave. 2nd Fl., Red Deer, AB, T4R 1M9. Phone: (403) 346-2573. Fax: (403) 346-9980. E-mail: rdtv@globaltv.ca Web Site: www.rdtv.com. Licensee: CanWest MediaWorks Inc. Group owner: CanWest Global Communications Corp. (acq 2000; grpsl). Natl. Network: CBC.

Slave Lake

CFRN-TV-9— Analog channel: 4.320 w vis, 32 w aur. ant 1,101t/260g TL: N55 28 18 W114 47 05 On air date: November 1981. c/o CFRN-TV, 18520 Stony Plain Rd., Edmonton, AB, T5S 1A8. Phone: (780) 483-3311. Fax: (780) 484-4426. E-mail: cfrn@ctv.ca Web Site: www.cfrntv.ca. Licensee: CFRN TV, a div. of CTV Television Inc.
Rebroadcasts CFRN-TV-1 Grande Prairie.

Whitecourt

CFRN-TV-3— Analog channel: 12.9.8 kw vis, 980 w aur. ant 1,308t/160g TL: N54 01 58 W115 43 03 c/o CFRN-TV, 18520 Stony Plain Rd., Edmonton, AB, T5S 1A8. Phone: (780) 483-3311. Fax: (780) 484-4426. E-mail: cfrn@ctv,ca Web Site: www.cfrntv.ca. Licensee: CFRN-TV, a div. of CTV Television Inc. Natl. Network: CTV.

British Columbia

Bowen Island

CHAN-TV-2— Analog channel: 3.5 w vis. Global BC, 7850 Enterprise St., Burnaby, BC, V5A 1V7. Phone: (604) 420-2288. Fax: (604) 422-6651. Licensee: CanWest MediaWorks Inc. Group owner: Global BC. Natl. Network: Global.
Key Personnel:
Roy Gardner . . . . gen mgr
Bob Urban . . . . opns mgr
Brett Monlove . . . . gen sls mgr
Ruth Powell . . . . rgnl sls mgr
John Ridley . . . . prom dir
Ian Mayson . . . . news dir
John O'Connor . . . . engrg VP

Brackendale

CHAN-TV-5— Analog channel: 9.21 w vis. Global BC, 7850 Enterprise St., Burnaby, BC, V5A 1V7. Phone: (604) 420-2288. Fax: (604) 421-9427. Licensee: CanWest MediaWorks Inc. Group owner: Global BC. Natl. Network: Global.

Campbell River

CHEK-TV-5— Analog channel: 13.3 kw vis, 300 w aur. 1493t/240g c/o CHEK-TV, 780 King's Rd., Victoria, BC, V8T 5A2. Phone: (250) 383-2435. Fax: (250) 384-7766. Web Site: www.canada.com/victoria/chtv. Licensee: CanWest MediaWorks Inc. Group owner: CanWest Global Communications Corp. Natl. Network: Global.

Canal Flats

CBUBT-1— Analog channel: 12.510 w vis, 51 w aur. Box 4600, c/o CBUT, Vancouver, BC, V6B 4A2. 700 Hamilton St., Vancouver, BC V6B 4A2. Phone: (604) 662-6000. Fax: (604) 662-6335. Web Site: www.cbc.ca. Licensee: CBC. Natl. Network: CBC.

Chilliwack

CHAN-TV-1— Analog channel: 11.71 w vis. 7850 Enterprise St., Burnaby, BC, V5A 1V7. Canada. Phone: (604) 420-2288. Fax: (604) 421-9427. Licensee: CanWest MediaWorks Inc. Group owner: Global BC. Natl. Network: Global.

Cranbrook

CBUBT-7— Analog channel: 10.900 w vis, 90 w aur. On air date: 1962. Box 4600, c/o CBUT-TV, Vancouver, BC, V6B 4A2. 700 Hamilton St., Vancouver, BC V6B 4A2. Phone: (604) 662-6000. Fax: (604) 662-6335. Web Site: www.cbc.ca. Licensee: CBC. Natl. Network: CBC.

Dawson Creek

CJDC-TV— Analog channel: 5.10 kw vis, 5 kw aur. ant 1,500t/500g On air date: 1958. NTV, Northern Television, 901 102nd Ave., Dawson Creek, BC, V1G 2B6. Phone: (250) 782-3341. Fax: (250) 782-3154. Licensee: Standard Radio Inc. Group owner: Standard Broadcasting Corp. (acq 4-19-2002; grpsl). Natl. Network: CBC. News staff: 4; News: 15 hrs wkly.

Fraser Valley

CHNU-TV— Analog channel: 66. Analog hrs: 6 AM-2 AM 17 kw vis. On air date: Sept 15, 2001. 5668 192 St., Suite 201, Surrey, BC, V3S 2V7. Phone: (604) 576-6880. Fax: (604) 576-6895. E-mail: info@omnibc.ca Web Site: www.omnibc.ca. Licensee: Rogers Broadcasting Ltd. (acq 5-20-2005; C$13 million with CIIT-TV Winnipeg, MB). Population served: 2,200,000
Key Personnel:
Rael Merson . . . . pres
Terry Mahoney . . . . gen mgr
Gary Milne . . . . gen sls mgr
Don Lang . . . . natl sls mgr

Kamloops

CBUFT-2— Analog channel: 50.200 w vis, 20 w aur. On air date: February 1979. 700 Hamilton St., Vancouver, BC, V6B 4A2. Phone: (604) 662-6000. Fax: (604) 662-6161. Web Site: www.radio-canada.ca/c-b. Licensee: Societe Radio Canada. Natl. Network: Radio Canada. Rebroadcasts CBUFT Vancouver.

CFJC-TV— Analog channel: 4. Digital channel: 7.4.4 kw vis, 2.4 kw aur. 501t/114g On air date: 1957. 460 Pemberton Terr., Kamloops, BC, V2C 1T5. Phone: (250) 372-3322. Fax: (250) 374-0445. E-mail: info@cfjctv.com Web Site: www.cfjctv.com. Licensee: Jim Pattison Broadcast Group Ltd. (the general partner) and Jim Pattison Industries Ltd. (the limited partner) carrying on business as Jim Pattison Broadcast Group L.P. Group owner: Group owner:The Jim Pattison Broadcast Group (acq 1987). Population served: 200,000 News staff: 8; News: 14 hrs wkly.
Key Personnel:
Richard W. Arnish . . . . pres & gen mgr
Dave Somerton . . . . opns mgr

CHKM-TV— Analog channel: 6.4 kw vis. ant 502t TL: N50 40 15 W120 23 50 Global BC, 7850 Enterprise St., Burnaby, BC, V5A 1V7. Phone: (604) 420-2288. Fax: (604) 422-6698. Web Site: www.canada.com /vancouver/globaltv. Licensee: CanWest MediaWorks Inc. Group owner: Global BC. Natl. Network: Global.

Kelowna

CHBC-TV— Analog channel: 2. Analog hrs: 6 AM-2 AM 3.7 kw vis, 460 w aur. 2,704t/77g TL: N49 58 00 W119 31 40 On air date: September 1957. 342 Leon Ave., Kelowna, BC, V1Y 6J2. Phone: (250) 762-4535. Fax: (250) 860-2422. Fax: (250) 868-0662. E-mail: comments@chbc.com Web Site: www.chbc.com. Licensee: CanWest MediaWorks Inc. Group owner: CanWest Global Communications Corp. Population served: 339,296 News staff: 23; News: 16 hrs wkly.
Key Personnel:
Keith Williams. . . . . gen mgr
Rob Weller . . . . opns mgr

Prince George

CKPG-TV— Analog channel: 2.778 w vis, 389 w aur. On air date: 1961. 1810 Third Ave., 2nd Fl, Prince George, BC, V2M 164. Canada. Phone: (250) 564-8861. Web Site: www.ckpgtv.com. Fax: (250) 562-8768. E-mail: ckpgmail@ckpg.bc.ca Licensee: Jim Pattison Broadcast Group LP. (the general partner) and Jim Pattison Industries Ltd. (the limited partner) carrying on business as Jim Pattison Broadcast Group L.P. Group owner: The Jim Pattison Broadcast Group (acq 12-21-2000; grpsl). Natl. Network: CBC. Natl. Rep: Airtime TV.

Squamish

CHAN-TV-3— Analog channel: 7.62 w vis. Global BC, 7850 Enterprise St., Burnaby, BC, V5A 1V7. Phone: (604) 420-2288. Fax: (604) 422-6651. Licensee: CanWest MediaWorks Inc. Group owner: Global BC. Natl. Network: Global.
Key Personnel:
Roy Gardner . . . . gen mgr
Bob Urban . . . . opns mgr
Brett Monlove . . . . gen sls mgr
Ruth Powell . . . . rgnl sls mgr
John Ridley . . . . prom dir
Ian Mayson . . . . news dir
John O'Connor . . . . engrg dir

Terrace

CBUFT-3— Analog channel: 11.500 w vis, 50 w aur. On air date: Aug 27, 1979. 700 Hamilton St., Vancouver, BC, V6B 4A2. Phone: (604) 662-6000. Fax: (604) 662-6161. Web Site: www.radio-canada.ca/c-b. Licensee: Societe Radio Canada. Natl. Network: Radio Canada.

CFTK-TV— Analog channel: 3.13.8 kw vis, 1.38 kw aur. 1,488t/140g On air date: 1962. 4625 Lazelle Ave., Terrace, BC, V8G 1S4. Phone: (250) 635-6316. Fax: (250) 638-6320. Web Site: www.cftk.com. Licensee: Standard Radio Inc. Group owner: Standard Broadcasting Corp. (acq 4-19-2002; grpsl). Natl. Network: CBC.
Key Personnel:
Doug Anderson . . . . gen mgr
Brian Faylinch . . . . chief of engrg

Valemount

CHVC-TV— Analog channel: 7.10 kw vis. On air date: 1994. Box 922, Valemount, BC, V0E 2Z0. Phone: (250) 566-8288. Fax: (250) 566-4645. Web Site: www.tv@ve. Licensee: The Valemount Entertainment Society.

Vancouver

CBUFT— Analog channel: 26.256 kw vis. ant 2,011t TL: N49 21 12 W122 57 18 On air date: Sept 27, 1976. Box 4600, Vancouver, BC, V6B 4A2. Canada. Phone: (604) 662-6000. Fax: (604) 662-6161. Web Site: www.radio-canada.ca/c-b. Licensee: Societe Radio-Canada. Natl. Network: Radio Canada.
Key Personnel:
Lionel Bonneville . . . . gen mgr
Brigitte Tesniere . . . . prom dir
Michele Smolkin . . . . stn mgr & progmg dir

CBUT— Analog channel: 2. Digital channel: 58.47.6 kw vis, 7.6 kw aur. ant 2,400t/190g On air date: Dec 16, 1953. Box 4600, 700 Hamilton St., Vancouver, BC, V6B 4A2. Phone: (604) 662-6000. Fax: (604) 662-6414. Web Site: www.cbc.ca. Licensee: CBC. Natl. Network: CBC.

CHAN-TV— Analog channel: 8. Digital channel: 22. Analog hrs: 24 193.6 kw vis, 19.4 kw aur. ant 2,315t/250g TL: N49 21 29 W122 57 09 On air date: Oct 31, 1960. 7850 Enterprise St., Burnaby, BC, V5A 1V7. Phone: (604) 420-2288. Fax: (604) 421-9427. Fax: (604) 444-9561. Licensee: CanWest MediaWorks Inc. Group owner: CanWest Global Communications Corp. Natl. Network: Global.

CHNM-TV— Analog channel: 42. Digital channel: 20.76 kw vis. On air date: June 27, 2003. channel m, 88 E. Pender St., Vancouver, BC, V6A 3X3. Phone: (604) 678-3800. Fax: (604) 678-3810. Web Site: www.channelm.ca. Licensee: The partners of Multivan Broadcast LP. Population served: 3,000,000 Natl. Rep: Airtime TV.
Key Personnel:
Art Reitmayer . . . . CEO & pres
Bruce Hamlin. . . . . sls dir
Johnny Michel . . . . prom VP & progmg VP
Dianne Collins . . . . news dir
Peter Gillespie . . . . opns VP & engrg VP

CIVT-TV— Analog channel: 32. Digital channel: 33.710 kw vis. TL: N49 21 29 W122 57 09 On air date: Sept 22, 1997. 750 Burrard St., Suite 300, Vancouver, BC, V6Z 1X5. Phone: (604) 608-2868. Fax: (604) 608-2698. E-mail: bccomments@ctv.ca Web Site: www.ctv.ca. Licensee: CTV Television Inc. Group owner: CTV Inc. Population served: 2,500,000 Natl. Network: CTV.
Key Personnel:
Ivan Fecan . . . . CEO
Rick Brace . . . . pres
Robin Fillingham . . . . CFO
Jim Rusnak . . . . sr VP, VP & gen mgr
Jim Olsen . . . . opns mgr & prom mgr
Louise Clark. . . . . dev dir
Lynne Forbes . . . . gen sls mgr & rgnl sls mgr
Doug Elphick . . . . natl sls mgr
Brenda Vasas . . . . progmg dir & progmg mgr
Tom Walters . . . . news dir
Vladimir Rybarczyk. . . . . engrg mgr

CKVU-TV— Analog channel: 10.325 kw vis, 65 kw aur. 1,959t/159g TL: N48 45 13 W123 29 25 On air date: Sept 1, 1976. 180 W. Second Ave., Vancouver, BC, V5Y 3T9. Phone: (604) 876-1344. Fax: (604) 876-3100. Web Site: www.citytv.com. Licensee: CHUM Television Vancouver Inc. Group owner: CHUM Ltd. (acq 10-15-2001; C$130 million).
Key Personnel:
Brad Phillips. . . . . VP & gen mgr
Neil Tegart. . . . . opns mgr, engrg mgr & chief of engrg
John Voiles . . . . rgnl sls mgr
Steve Scarrow. . . . . mktg VP & mktg dir
Debbie Millette. . . . . progmg mgr
Bud Pierce. . . . . news dir

Victoria

CHEK-TV— Analog channel: 6. Analog hrs: 24 100 kw vis, 10 kw aur. 1,628t/380g On air date: 1956. 780 Kings Rd., Victoria, BC, V8T 5A2. Phone: (250) 383-2435. Fax: (250) 384-7766. Licensee: CanWest MediaWorks Inc. Group owner: CanWest Global Communications Corp. Natl. Network: Global.

CIVI-TV— Analog channel: 53. Analog hrs: 24 12 kw vis. On air date: Oct 4, 2001. 1420 Broad St., Victoria, BC, V8W 2B1. Phone: (250) 381-2484. Fax: (250) 381-2485. E-mail: islandcontactus@achannel.ca Web Site: www.achannel.ca/victoria. Licensee: 1708478 Ontario Inc. Group owner: (group owner).
Key Personnel:
Richard Gray . . . . stn mgr
John Voiles . . . . rgnl sls mgr
Debbie Millette. . . . . progmg mgr
Hudson Mack. . . . . news dir
Jen Wong . . . . prom
Brian Gatensby. . . . . engr

Whistler

CHAN-TV-7— Analog channel: 9.31 w vis. Global BC, 7850 Enterprise St., Burnaby, BC, V5A 1V7. Phone: (604) 420-2288. Fax: (604) 421-9427. Licensee: CanWest MediaWorks Inc. Group owner: Global BC. Natl. Network: Global.

Manitoba

Baldy Mountain

CBWST— Analog channel: 8. Analog hrs: 17 120 kw vis, 12 kw aur. c/o CBWT, Box 160, Winnipeg, MB, R3C 2H1. 541 Portage Ave., Winnipeg, MB R3B 2G1. Phone: (204) 788-3222. Phone: (204) 788-3141. Fax: (204) 788-3639. Web Site: www.winnipeg.cbc.ca. Licensee: CBC. Natl. Network: CBC.
Key Personnel:
John Bertrand . . . . gen mgr
John Mang . . . . opns mgr

Brandon

CBWFT-10— Analog channel: 21.9.4 kw vis, 940 w aur. 340g On air date: Feb 11, 1978. Box 160, Winnipeg, MB, R3C 2H1. c/o CBC, 541 Portage Ave., Winnipeg, MB R3C 2H1. Phone: (204) 788-3222. Phone: (204) 788-3141. Fax: (204) 788-3639. Licensee: CBC. Natl. Network: CBC.
Key Personnel:
Lionel Bonneville . . . . gen mgr
Richard Augert . . . . stn mgr
Philippe Vrignon . . . . progmg mgr

CKX-TV— Analog channel: 5. Analog hrs: 6 AM-2 AM 44 kw vis, 27 kw aur. ant 511t/525g On air date: 1955. 2940 Victoria Ave., Brandon, MB, R7B 3Y3. Phone: (204) 728-1150. Fax: (204) 727-2505. E-mail: ckxtv@mb.symapatico.ca Web Site: www.ckxtv.com. Licensee: CHUM Ltd. Group owner: (group owner). (acq 11-19-2004; grpsl). Natl. Network: CBC. News staff: 20; News: 15 hrs wkly.
Key Personnel:
Alan Cruise . . . . gen mgr
Brian Atkinson . . . . stn mgr & news dir

CKYB-TV— Analog channel: 4.55 kw vis. c/o CKY-TV, Polo Park, Winnipeg, MB, R3G 0L7. Phone: (204) 788-3300. Fax: (204) 788-3399. Licensee: CTV Television Inc. Group owner: CTV Inc. (acq 8-2-01; grpsl). Natl. Network: CTV.

Fisher Branch

CBWGT— Analog channel: 10. Analog hrs: 17 27.4 kw vis, 5.48 kw aur. 559t/548g Box 160, Winnipeg, MB, R3C 2H1. c/o CBWT, 541 Portage Ave., Winnipeg, MB R3B 2G1. Phone: (204) 788-3222. Phone: (204) 788-3141. Fax: (204) 788-3639. Licensee: CBC. Natl. Network: CBC.
Key Personnel:
John Bertrand . . . . gen mgr
John Mang . . . . opns mgr

Flin Flon

CBWBT— Analog channel: 10. Analog hrs: 17 7.8 kw vis, 1.6 kw aur. Box 160, c/o CBWT, Winnipeg, MB, R3C 2H1. c/o CBWT, 541 Portage Ave., Winnipeg, MB R3B 2G1. Phone: (204) 788-3222. Phone: (204) 788-3141. Fax: (204) 788-3639. Web Site: www.winnipeg.cbc.ca. Licensee: CBC. Natl. Network: CBC.
Key Personnel:
John Bertrand. . . . . opns dir
John Mang . . . . opns mgr

Foxwarren

CKX-TV-1— Analog channel: 11.46.8 kw vis, 3.48 kw aur. (CP: 56.8 kw vis) c/o CKX-TV, 2940 Victoria Ave., Brandon, MB, R7B 3Y3. Phone: (204) 728-1150. Fax: (204) 727-2505. Web Site: www.ckxtv.com. Licensee: CHUM Ltd. Group owner: (group owner). (acq 11-19-2004; grpsl). Natl. Network: CBC.
Key Personnel:
Alan Cruise . . . . . gen mgr
Glenn Edmonson . . . . . gen sls mgr
Rich Chudley . . . . . progmg dir
Paul Weger . . . . . chief of engrg

Lac du Bonnet

CBWT-2— Analog channel: 4.8.4 kw vis, 1.7 kw aur. Box 160, c/o CBWT, Winnipeg, MB, R3C 2H1. c/o CBWT, 541 Portage Ave., Winnipeg, MB R3B 2G1. Phone: (204) 788-3222. Phone: (204) 788-3141. Fax: (204) 788-3639. Licensee: CBC. Natl. Network: CBC.
Key Personnel:
John Bertrand . . . . . gen mgr
John Mang . . . . . opns mgr

Mafeking

CBWYT— Analog channel: 2.4 kw vis. 370g On air date: July 14, 1978. Box 160, c/o CBWT, Winnipeg, MB, R3C 2H1. c/o CBC, 541 Portage Ave., Winnipeg, MB R3C 2H1. Phone: (204) 788-3222. Phone: (204) 788-3141. Fax: (204) 788-3639. Web Site: www.winnipeg.cbc.ca. Licensee: CBC. Natl. Network: CBC.

Ste-Rose-du-Lac

CBWFT-4— Analog channel: 3.1.2 kw. 120g On air date: May 15, 1976. Box 160, Winnipeg, MB, R3C 2H1. c/o CBWFT, 541 Portage Ave., Winnipeg, MB R3C 2H1. Phone: (204) 788-3222. Phone: (204) 788-3141. Fax: (204) 788-3639. Web Site: www.winnipeg.cbc.ca. Licensee: CBC. Natl. Network: CBC.
Key Personnel:
Lionel Bonneville . . . . . gen mgr
Richard Augert . . . . . stn mgr
Wayne Yonka. . . . . engrg dir & chief of engrg

Winnipeg

CBWFT— Analog channel: 3.59 kw vis, 7.3 kw aur. 1,027t/1,020g On air date: 1960. Box 160, Winnipeg, MB, R3C 2H1. 541 Portage Ave., Winnipeg, MB R3C 2H1. Phone: (204) 788-3222. Fax: (204) 788-3639. Web Site: www.radio-canada.ca. Licensee: Societe Radio-Canada. Natl. Network: Radio Canada.
Key Personnel:
Lionel Bonneville . . . . . gen mgr
Richard Augert . . . . . stn mgr
Philippe Vrignon. . . . . progmg dir

CBWT— Analog channel: 6.100 kw vis, 12 kw aur. 1,027t/1,020g TL: N49 46 15 W97 30 35 On air date: 1954. Box 160, Winnipeg, MB, R3C 2H1. 541 Portage Ave., Winnipeg, MB R3B 2G1. Phone: (204) 788-3222. Fax: (204) 788-3167. E-mail: communications@winnipeg.cbc.ca Web Site: www.winnipeg.cbc.ca. Licensee: CBC. Natl. Network: CBC.
Key Personnel:
John Bertrand . . . . . gen mgr & opns mgr
John Mang . . . . . opns mgr

CHMI-TV—(Portage la Prarie, Analog channel: 13. Analog hrs: 24 325 kw vis, 32.5 kw aur. 1,029t/1,100g On air date: Oct 17, 1986. #8 Forks Market Rd., Winnipeg, MB, R3C 4Y3. Phone: (204) 947-9613. Fax: (204) 956-0811. E-mail: winnipeginteractive@chumtv.com Web Site: www.citytv.com. Licensee: CHUM Ltd. Group owner: (group owner). (acq 11-19-2004; grpsl). Population served: 813,000 News staff: 35; News: 34 hrs wkly.
Key Personnel:
Cam Cowie . . . . . VP & gen mgr
Christine Ljungberg . . . . . opns mgr
Glen Cassie . . . . . news dir

CIIT-TV— Analog channel: 35. On air date: Feb 6, 2006. OMNI.11, 66 Osborne St., Unit 5, Winnipeg, MB, R3L 1Y8. Phone: (204) 788-8855. Fax: (204) 788-3401. E-mail: geoff.poulton@rci.rogers.com Web Site: www.omnitv.ca. Licensee: Rogers Broadcasting Ltd. (acq 5-20-2005; C$13 million with CHNU-TV Fraser Valley, BC).

CKND-TV— Analog channel: 9. Analog hrs: 24 325 kw vis, 25 kw aur. ant 500t/600g On air date: Sept 1, 1975. 603 St. Mary's Rd., Winnipeg, MB, R2M 3L8. Phone: (204) 233-3304. Fax: (204) 233-5615. Web Site: www.globaltv.com. Licensee: CanWest MediaWorks Inc. Group owner: CanWest Global Communications Corp. News staff: 19; News: 12 hrs wkly.
Key Personnel:
Tim Schellenberg . . . . . gen mgr
Heather McIntyre . . . . . prom mgr
Jon Lovlin . . . . . news dir
Len Virog . . . . . engrg dir

CKY-TV— Analog channel: 7.325 kw vis, 65 kw aur. ant 1,000g On air date: 1960. Polo Park, 400-345 Grahm Ave, Winnipeg, MB, R3C 5S6. Phone: (204) 788-3300. Fax: (204) 788-3399. Web Site: www.cky.com. Licensee: CTV Television Inc. Group owner: CTV Inc. Natl. Network: CTV.
Key Personnel:
Diane Kashton . . . . . pres & prom dir
Bill Hansen . . . . . VP
Bill Hanson. . . . . gen mgr
Kenneth Peron . . . . . opns mgr
Wally Comrie . . . . . gen sls mgr
Winnie Navarro . . . . . progmg dir
Jeff Bollenbach . . . . . news dir

New Brunswick

Campbellton

CKCD-TV— Analog channel: 7.920 w vis, 180 w aur. 75t 191 Halifax St., Moncton, NB, E1C 9R7. Phone: (506) 857-2600. Fax: (506) 857-2617. E-mail: ckcw@ctv.ca Web Site: www.ctv.ca. Licensee: CTV Television Inc. Group owner: CTV Inc. (acq 11-1-97). Natl. Network: CTV.
Key Personnel:
Ivan Fecan . . . . . CEO
Rick Brace . . . . . pres
Robin Fillingham . . . . . CFO
Elaine Ali . . . . . exec VP
Mike Elgie . . . . . VP & gen mgr
Brian Lewis . . . . . stn mgr, gen sls mgr & adv mgr
John Silver . . . . . opns mgr
Renee Fournier . . . . . prom mgr
Jane Hefler . . . . . progmg mgr
Jay Witherbee . . . . . news dir & pub affrs dir
Carson McDavid . . . . . engrg dir & chief of engrg
Rebroadcasts CKCW-TV Moncton 100%.

Fredericton-Saint John

CBAT— Analog channel: 4.54.2 kw vis, 7.8 kw aur. 1,268t/1,631g TL: N45 28 39 W66 14 03 On air date: 1954. Box 2200, Fredericton, NB, E3B 5G4. Phone: (506) 451-4000. Fax: (506) 451-4003. Web Site: www.cbc.ca/nb. Licensee: CBC. (acq 4-29-94). Natl. Network: CBC.

Moncton

CBAFT— Analog channel: 11.137.7 kw vis. ant 781t/394g TL: N46 08 41 W64 54 14 On air date: 1959. Box 950, Moncton, NB, E1C 8N8. 250 University Ave., Moncton, NB E1C 5K3. Phone: (506) 853-6666. Phone: (506) 853-6740 (Stn Dir). Fax: (506) 867-8031. Fax: (506) 853-6601 (news). Web Site: www.radio-canada.ca/regions /atlantique/index.shtml. Licensee: Societe Radio-Canada. Natl. Network: Radio Canada.

CKCW-TV— Analog channel: 2.56 kw vis, 9.2 kw aur. On air date: 1954. 191 Halifax St., Moncton, NB, E1C 9R7. Phone: (506) 857-2600. Fax: (506) 857-2617. E-mail: ckcw@ctv.ca Web Site: www.ctv.ca. Licensee: CTV Television Inc. Group owner: CTV Inc. (acq 11-1-97). Natl. Network: CTV.
Key Personnel:
Ivan Fecan . . . . . CEO
Rick Brace . . . . . pres
Robin Fillingham . . . . . CFO
Elaine Ali . . . . . exec VP
Mike Elgie . . . . . VP & gen mgr
Brian Lewis . . . . . stn mgr, gen sls mgr & adv mgr
John Silver . . . . . opns mgr
Renee Fournier . . . . . prom mgr
Jane Hefler . . . . . progmg mgr
Jay Witherbee . . . . . news dir & pub affrs dir
Carson McDavid . . . . . chief of engrg

Saint John

CKLT-TV— Analog channel: 9.162 kw vis, 32 kw aur. 1,361t/241g 12 Smythe St., Suite 126, Saint John, NB, E2L 5G5. Canada. Phone: (506) 658-1010. Fax: (506) 658-1208. E-mail: cklt@ctv.ca Web Site: www.ctv.ca. Licensee: CTV Television Inc. Group owner: CTV Inc. (acq 11-1-97). Natl. Network: CTV.

Upsalquitch Lake

CKAM-TV— Analog channel: 12.280 kw vis, 141 kw aur. On air date: unknown. c/o CKCW-TV, 191 Halifax St., Moncton, NB, E1C 9R7. Phone: (506) 857-2600. Fax: (506) 857-2617. E-mail: ckcw@ctv.ca Web Site: www.ctv.ca. Licensee: CTV Television Inc. Group owner: CTV Inc. (acq 11-1-97). Natl. Network: CTV. Ottawa Atty: . Ottawa atty: Alexander, Pearson & Dawson
Key Personnel:
Ivan Fecan . . . . . CEO
Rick Brace . . . . . pres
Robin Fillingham . . . . . CFO
Elaine Ali . . . . . exec VP
Mike Elgie . . . . . VP & gen mgr
Brian Lewis . . . . . stn mgr, gen sls mgr & adv mgr
John Silver . . . . . opns mgr
Renee Fournier . . . . . prom mgr
Jay Witherbee . . . . . news dir & pub affrs dir
Carson McDavid . . . . . engrg dir & chief of engrg
Rebroadcasts CKCW-TV Moncton 100%.

Newfoundland

Argentia

CJOM-TV— Analog channel: 3. Analog hrs: 24 6.7 kw vis, 3.4 kw aur. 275g On air date: September 1957. Box 2020, c/o CJON-TV, 446 Logy Bay Rd., St. John's, NF, A1C 5S2. Phone: (709) 722-5015. Fax: (709) 726-5107. E-mail: ntv@ntv.ca Web Site: www.ntv.ca. Licensee: Newfoundland Broadcasting Co. Ltd. Group owner: (group owner; (acq 9-1-77). Population served: 575,000 Natl. Network: CTV. News staff: 12; News: 11 hrs wkly.

Baie Verte

CBNAT-1— Analog channel: 3. Analog hrs: 24 5 kw vis, 500 w aur. TL: N49 57 34 W56 18 42 On air date: Sept 15, 1968. 95 University Ave., St. John's, NF, A1B 1Z4. Phone: (709) 576-5000. Fax: (709) 576-5044. Web Site: www.cbc.ca/nl. Licensee: CBC. Natl. Network: CBC.

Bonavista

CJWB-TV— Analog channel: 10. Analog hrs: 24 9.9 kw vis, 990 w aur. 539 TL: N48 37 30 W53 03 45 On air date: 1972. Box 2020, NTV Studio Bldg., 446 Logy Bay Rd., St. John's, NF, A1C 5S2. Phone: (709) 722-5015. Fax: (709) 726-5107. E-mail: ntv@ntv.ca Web Site: www.ntv.ca. Licensee: Newfoundland Broadcasting Co. Ltd. Group owner: (group owner) Population served: 575,000 Natl. Network: CTV. News staff: 12; News: 11 hrs wkly.

Bonne Bay

CBYT-3— Analog channel: 2. Analog hrs: 24 1 kw vis, 100 w aur. 300t/300g TL: N49 33 12 W57 53 24 On air date: Mar 12, 1971. 95 University Ave., St. John's, NF, A1B 1Z4. Phone: (709) 576-5000. Fax: (709) 576-5044. Licensee: CBC. Natl. Network: CBC, CBC Radio One. Rgnl. Network: CBC Northern Television Services.
Satellite of CBNT(TV) Saint John's.

Botwood

CBNAT— Analog channel: 11. Analog hrs: 24 30 kw vis, 3 kw aur. TL: N49 11 51 W55 22 05 On air date: Dec 21, 1967. 95 University Ave., St. John's, NF, A1B 1Z4. Phone: (709) 576-5000. Fax: (709) 576-5044. Web Site: www.cbc.ca/nl. Licensee: CBC. Natl. Network: CBC. News staff: one.

Corner Brook

CJWN-TV— Analog channel: 10. Analog hrs: 24 6.07 kw vis, 1 kw aur. 364t/300g TL: N48 56 55 W57 58 23 On air date: December 1974. Box 2020, NTV Studio Bldg., 446 Logy Bay Rd., St. John's, NF, A1C 5S2. Phone: (709) 722-5015. Fax: (709) 726-5107. E-mail: ntv@ntv.ca Web Site: www.ntv.ca. Licensee: Newfoundland Broadcasting Co. Ltd. Group owner: (group owner) Natl. Network: CTV.

Grand Bank

CJOX-TV-1— Analog channel: 2. Analog hrs: 24 4.67 kw vis, 470 w aur. 387t/287g TL: N47 05 17 W55 46 23 On air date: 1972. Box 2020, NTV Studio Bldg., 446 Logy Bay Rd., St. John's, NF, A1C 5S2. Phone: (709) 722-5015. Fax: (709) 726-5107. E-mail: ntv@ntv.ca Web Site: www.ntv.ca. Licensee: Newfoundland Broadcasting Co. Ltd. Group owner: (group owner) Population served: 575,000 Natl. Network: CTV. News staff: 12; News: 11 hrs wkly.

Grand Falls

CJCN-TV— Analog channel: 4. Analog hrs: 24 100 kw vis, 10 kw aur. 602 TL: N49 04 12 W55 16 54 On air date: 1963. Box 2020, NTV Studio Bldg., 446 Logy Bay Rd., St. John's, NF, A1C 5S2. Phone: (709) 722-5015. Fax: (709) 726-5107. E-mail: ntv@ntv.ca Web Site: www.nvt.ca. Licensee: Newfoundland Broadcasting Co. Ltd. Group owner: (group owner) Population served: 575,000 Natl. Network: CTV. News staff: 12; News: 11 hrs wkly.

Key Personnel:
Scott Stirling . . . . . pres
Jim Furlong . . . . . news dir

Labrador City

***CBNLT—** Analog channel: 13. Analog hrs: 24 250 kw vis, 25 w aur. TL: N52 56 41 W66 54 11 On air date: Nov 7, 1973. 95 University Ave., St. John's, NF, A1B 1Z4. Phone: (709) 576-5000. Fax: (709) 576-5044. Web Site: www.cbc.ca/nl. Licensee: CBC. Natl. Network: CBC. News staff: one.

Marystown

CBNT-3— Analog channel: 5. Analog hrs: 24 5 kw vis, 500 w aur. TL: N47 08 39 W55 08 52 On air date: Nov 30, 1965. Box 12010, c/o CBNT, 95 University Ave., St. John's, NF, A1B 3T8. Phone: (709) 576-5000. Fax: (709) 576-5099. Web Site: www.cbc.ca/nl. Licensee: Canadian Broadcasting Corp. Natl. Network: CBC.

Key Personnel:
Diane Humber . . . . . gen mgr
Keith Durnford . . . . . engrg mgr & chief of engrg

Mount St. Margaret

CBNAT-9— Analog channel: 9. Analog hrs: 24 3 kw vis, 300 w aur. TL: N51 01 05 W56 48 47 On air date: Nov 30, 1973. 95 University Ave., St. John's, NF, A1B 1Z4. Phone: (709) 576-5000. Fax: (709) 576-5044. Web Site: www.cbc.ca/nl. Licensee: CBC. Natl. Network: CBC.

Placentia

CBNT-2— Analog channel: 12. Analog hrs: 24 10.6 kw vis, 100 w aur. TL: N47 13 52 W53 58 56 On air date: Nov 27, 1965. Box 12010, c/o CBNT, 95 University Ave., St. John's, NF, A1B 3T8. Phone: (709) 576-5000. Fax: (709) 576-5099. Web Site: www.cbc.ca/nl. Licensee: Canadian Broadcasting Corp. Natl. Network: CBC.

Key Personnel:
Diane Humber . . . . . gen mgr
Keith Durnford . . . . . engrg mgr & chief of engrg

Port Rexton

CBNT-1— Analog channel: 13. Analog hrs: 24 3 kw vis, 300 w aur. TL: N48 26 27 W53 21 25 On air date: October 1964. Box 12010, c/o CBNT, 95 University Ave., St. John's, NF, A1B 3T8. Phone: (709) 576-5000. Fax: (709) 576-5099. Fax: (709) 576-5144. Web Site: www.cbc.ca/nl. Licensee: Canadian Broadcasting Corp. Natl. Network: CBC.

Saint Anthony

CBNAT-4— Analog channel: 6. Analog hrs: 24 3 kw vis, 300 w aur. TL: N51 21 14 W55 34 00 On air date: Oct 21, 1968. 95 University Ave., St. John's, NF, A1B 1Z4. Phone: (709) 576-5000. Fax: (709) 576-5044. Web Site: www.cbc.ca/nl. Licensee: CBC. Natl. Network: CBC.

Saint John's

CBNT— Analog channel: 8. Analog hrs: 24 3 kw vis, 300 w aur. TL: N48 26 27 W53 21 25 On air date: 1964. Box 12010, Stn A, Saint John's, NF, A1B 3T8. 95 University Ave., Saint John's, NF A1B 1Z4. Phone: (709) 576-5000. Fax: (709) 576-5044. Web Site: www.cbc.ca/nl. Licensee: CBC. Natl. Network: CBC.

St John's

CJON-TV— Analog channel: 6. Analog hrs: 24 212 kw vis, 21 kw aur. 825t/301 TL: N47 31 36 W52 42 50 On air date: September 1955. Box 2020, 446 Logy Bay Rd., St John's, NF, A1C 5S2. Phone: (709) 722-5015. Fax: (709) 726-5107. E-mail: ntv@ntv.ca Web Site: www.ntv.ca. Licensee: Newfoundland Broadcasting Co. Ltd. Group owner: (group owner) Population served: 575,000 Natl. Network: CTV. Ottawa Atty: . Ottawa atty: Johnston & Buchan Wire Svc: BN Wire News staff: 12; News: 11 hrs wkly.

Key Personnel:
Scott Stirling . . . . . pres & gen mgr
Jim Furlong . . . . . news dir

Stephenville

CJSV-TV— Analog channel: 4. Analog hrs: 24 5.56 kw vis, 560 w aur. 439 TL: N48 31 09 W58 31 00 On air date: 1973. Box 2020, NTV Studio Bldg., 446 Logy Bay Rd., St. John's, NF, A1C 5S2. Phone: (709) 722-5015. Fax: (709) 726-5107. E-mail: ntv@ntv.ca Web Site: www.nvt.ca. Licensee: Newfoundland Broadcasting Co. Ltd. Group owner: (group owner) Population served: 575,000 Natl. Network: CTV. News staff: 12; News: 11 hrs wkly.

Key Personnel:
Scott Stirling . . . . . pres
Jim Furlong . . . . . news dir

Northwest Territories

Inuvik

***CHAK-TV—** Analog channel: 8. Analog hrs: 20 3 kw vis, 300 w aur. 443t/360g Bag Service No. 8, 155 Mackenzie Rd., Inuvik, NT, X0E OTO. Phone: (867) 920-5400. Fax: (867) 920-5489. E-mail: cbcnorth@cbc.ca Web Site: www.cbc.ca/north. Licensee: CBC. Natl. Network: CBC. News staff: 2.
Rebroadcast of CFYK-TV Yellow Knife.

Yellowknife

***CFYK-TV—** Analog channel: 6. Analog hrs: 20 1 kw vis, 100 w aur. On air date: 1968. Box 160, Yellowknife, NT, X1A2N2. Canada. 5002 Forest Dr., Yellowknife, NT X1A2N2.Canada Phone: (867) 920-5400. Fax: (867) 920-5489. E-mail: cbcnorth@cbc.ca Web Site: www.cbc.ca/north. Licensee: CBC. Natl. Network: CBC.

Nova Scotia

Antigonish

CJCB-TV-2— Analog channel: 9. Analog hrs: 24 140 kw vis. 500t TL: N45 32 45 W62 15 39 c/o CJCB-TV, 1283 George St., Sydney, NS, B1P 1N7. Phone: (902) 562-5511. Fax: (902) 562-9714. E-mail: cjcb@ctv.ca Web Site: www.ctv.ca. Licensee: ATV Cape Breton. Natl. Network: CTV.

Key Personnel:
Glenn McLanders . . . . . stn mgr
Edgar Bennett . . . . . engrg dir

Caledonia

CJCH-TV-6— Analog channel: 6. Analog hrs: 24 100 kw vis, 20 kw aur. 630t/400g TL: N44 20 26 W65 06 34 Box 1653, Halifax, NS, B3J 2Z4. 2885 Robie St., Halifax, NS B3J 2Z4. Phone: (902) 453-4000. Fax: (902) 454-3302. E-mail: cjch@ctv.ca Web Site: www.ctv.ca. Licensee: CTV Television Inc. Group owner: Baton Broadcasting Inc. Natl. Network: CTV.

Key Personnel:
Michael Elgie . . . . . gen mgr
Renee Fournier . . . . . mktg mgr

Canning

CJCH-TV-1— Analog channel: 10. Analog hrs: 24 18.1 kw vis, 3.62 kw aur. ant 886t/300g TL: N45 12 12 W64 24 06 Box 1653, Halifax, NS, B3J 2Z4. Phone: (902) 453-4000. Fax: (902) 454-3302. E-mail: cjch@ctv.ca Web Site: www.ctv.ca. Licensee: Atlantic Television System. Group owner: CTV Television Inc. Natl. Network: CTV.

Key Personnel:
Michael Elgie . . . . . gen mgr
Ian MacArthur . . . . . gen sls mgr
Renee Fournier . . . . . mktg mgr
Jane Hefler . . . . . progmg mgr
Jay Witherbee . . . . . news dir
Gary Robertson . . . . . chief of engrg

Cheticamp

CBIT-2— Analog channel: 2. Analog hrs: 24 2.5 kw vis, 250 w aur. 150g c/o CBIT, 285 Alexandra St., Sydney, NS, B1S 2E8. Phone: (902) 539-5050. Fax: (902) 539-1562. Web Site: www.cbc.ca/ns. Licensee: CBC. Natl. Network: CBC.

Halifax

CBHT— Analog channel: 3. Analog hrs: 24 56 kw vis, 11.2 kw aur. 866t/1,620g On air date: 1954. Box 3000, Halifax, NS, B3J 3E9. Phone: (902) 420-8311. Fax: (902) 420-4010. E-mail: cbcns@cbc.ca Web Site: www.cbc.ca/ns. Licensee: CBC. Natl. Network: CBC.

Key Personnel:
Andrew Cochran . . . . . gen mgr
Lenny Jackson . . . . . opns mgr
Mary Elizabeth Luka . . . . . dev mgr
John Channing . . . . . rgnl sls mgr

CIHF-TV— Analog channel: 8. Analog hrs: 24 8.2 kw vis, 1.6 aur. 691t/645g TL: N44 39 03 W63 39 28 On air date: Sept 5, 1988. Box 1643 C.R.O., Halifax, NS, B3J 2Z1. Phone: (902) 481-7400. Phone: (506) 632-3400. Fax: (902) 468-2154. E-mail: news@globaltv.com Web Site: www.globalmaritimes.com. Licensee: CanWest MediaWorks Inc. Group owner: CanWest Global System (acq 8-29-94; $11 million). Population served: 1,500,000 News staff: 40; News: 17 hrs wkly.

Key Personnel:
Leonard Asper . . . . . chmn
John Burgis . . . . . CFO
Barry Saunders . . . . . gen mgr

CJCH-TV— Analog channel: 9. Analog hrs: 24 100 kw vis, 10 kw aur. 821t/575g TL: N44 39 03 W63 39 28 On air date: Jan 1, 1961. Box 1653, Halifax, NS, B3J 2Z4. 2885 Robie St., Halifax, NS B3K 5Z4. Phone: (902) 453-4000. Fax: (902) 454-3302. E-mail: cjch@ctv.ca Web Site: www.ctv.ca. Licensee: CTV Television Inc. Group owner: CTV Inc. Natl. Network: CTV.

Key Personnel:
Michael Elgie . . . . . VP & gen mgr
Ian MacArthur . . . . . gen sls mgr
Renee Fournier . . . . . mktg mgr
Jane Hefler . . . . . progmg mgr
Jay Witherbee . . . . . news dir
Gary Robertson . . . . . chief of engrg

Inverness

CJCB-TV-1— Analog channel: 6. Analog hrs: 24 9.4 kw vis, 4.7 kw aur. c/o CJCB-TV, 1283 George St., Sydney, NS, B1P 1N7. Phone: (902) 562-5511. Fax: (902) 562-9714. E-mail: cjcb@ctv.ca Web Site: www.ctv.ca. Licensee: ATV Cape Breton. Natl. Network: CTV.

Isle Madame

CIMC-TV— Analog channel: 10. Analog hrs: 24 450 w vis. On air date: June 2003. Box 87, Arichat, NS, B0E 1A0. 705 Lower Rd., Arichat, NS B0E 1A0. Phone: (902) 226-1928. Fax: (902) 226-1331. E-mail:

telile@telile.tv Web Site: www.telile.tv. Licensee: Telile:Isle Madame Community Television Association/Association Television Communautaire de l'Ile Madame.

Mulgrave

CBHT-11— Analog channel: 12.129 kw vis, 12.9 kw aur. Box 3000, c/o CBHT, 5600 Sackville St., Halifax, NS, B3J 3E9. 1840 Bell Rd., Halifax, NS B3H 2Z5. Phone: (902) 420-8311. Fax: (902) 420-4010. Licensee: CBC. Natl. Network: CBC.
Key Personnel:
Andrew Cochran . . . gen mgr & engrg mgr
Penny Longley . . . dev mgr
John Channing . . . rgnl sls mgr

Sheet Harbour

CBHT-4— Analog channel: 11. Analog hrs: 24 9.07 kw vis, 1.814 kw aur. Box 3000, c/o CBHT, 5600 Sackville St., Halifax, NS, B3J 3E9. 1840 Bell Rd., Halifax, NS B3H 2Z5.Canada Phone: (902) 420-8311. Fax: (902) 420-4010. E-mail: cbcns@cbc.ca Web Site: www.cbc.ca/ns. Licensee: CBC. Natl. Network: CBC.
Key Personnel:
Andrew Cochran . . . gen mgr
Mary Elizabeth Luka . . . dev mgr

Sydney

CBIT— Analog channel: 5. Analog hrs: 24 54 kw vis, 5.4 kw aur. On air date: 1972. 285 Alexandra St., Sydney, NS, B1S 2E8. Phone: (902) 539-5050; (902) 563-4100. Fax: (902) 539-1562. Web Site: www.cbc.ca/ns. Licensee: CBC. Natl. Network: CBC.

CJCB-TV— Analog channel: 4. Analog hrs: 24 100 kw vis, 60 kw aur. On air date: 1954. 1283 George St., Sydney, NS, B1P 1N7. Phone: (902) 562-5511. Fax: (902) 562-9714. E-mail: cjcb@ctv.ca Web Site: www.ctv.ca. Licensee: ATV Cape Breton. Group owner: CTV Inc. Natl. Network: CTV.
Key Personnel:
Glenn McLanders . . . stn mgr & gen sls mgr
Gary Robertson . . . opns dir & chief of engrg
Renee Fournier . . . mktg mgr & prom mgr
Jane Hefler . . . progmg mgr
Jay Witherbee . . . news dir
Edgar Bennett . . . engrg dir & chief of engrg

Yarmouth

CBHT-3— Analog channel: 11. Analog hrs: 24 15.7 kw vis, 3.3 kw aur. Box 3000, c/o CBHT, Halifax, NS, B3J 3E9. Phone: (902) 420-8311. Fax: (902) 420-4010. E-mail: cbcns@cbc.ca Web Site: www.cbc.ca/ns. Licensee: CBC. Natl. Network: CBC.
Key Personnel:
Andrew Cochran . . . gen mgr
Lenny Jackson . . . chief of opns
Mary Elizabeth Luka . . . dev mgr
John Channing . . . rgnl sls mgr

Ontario

Bancroft

CIII-TV-2— Analog channel: 2.100 kw vis, 15 kw aur. ant 1,279 On air date: January 1974. 81 Barber Greene Rd., Toronto, ON, M3C 2A2. Phone: (416) 446-5311. Fax: (416) 446-5447. E-mail: newstips@globaltv.com Web Site: www.canada.com. Licensee: CanWest MediaWorks Inc. Group owner: Global Television Network Natl. Network: Global.
Key Personnel:
Leonard Asper . . . chmn
Bill Hunt . . . gen mgr

Barrie

CKVR-TV— Analog channel: 3. Analog hrs: 24 100 kw vis, 12.5 kw aur. 820t/651g TL: N44 21 05 W79 41 55 On air date: Sept 28, 1955. Box 519, Barrie, ON, L4M 4T9. 33 Beacon Rd., Barrie, ON L4M 4S7. Phone: (705) 734-3300. Fax: (705) 733-0302. Fax: (705) 734-2061. E-mail: achannel@achannel.ca Web Site: www.achannel.ca. Licensee: 1708486 Ontario Inc. Group owner: CHUM Ltd. Population served: 6,000,000 News staff: 30; News: 14 hrs wkly.
Key Personnel:
Jay Switzer . . . pres
Bob McLaughlin . . . opns mgr & news dir
Paul Woodhouse . . . sls dir
Dan Hamilton . . . gen sls mgr
Peggy Hebden . . . stn mgr, progmg dir & film buyer
Brian Cathline . . . chief of engrg

Chatham

***CICO-TV-59**— Analog channel: 59. Analog hrs: 24 34.3 kw vis, 3.4 kw aur. 717t/718g On air date: June 1976. Box 200 Stn Q, c/o TV Ontario, Toronto, ON, M4T 2T1. 2180 Yonge St., Toronto, ON M4S 2B9. Phone: (416) 484-2600. Fax: (416) 484-6285. E-mail: asktvo@tvo.org Web Site: www.tvo.org. Licensee: Ontario Educational Communications Authority.
Key Personnel:
Lee Robock . . . COO & gen mgr
Lisa DeWilde . . . pres
Ray Newell . . . opns dir
Meg Pinto . . . mktg dir

Cornwall

CJOH-TV-8— Analog channel: 8. Analog hrs: 24 260 kw vis. ant 615t TL: N45 10 35 W74 31 38 On air date: 1958. CTV Ottawa, Box 5813, Ottawa, ON, K2C 3G6. Phone: (613) 224-1313. Fax: (613) 274-4215. E-mail: ctvottawa@ctv.ca Web Site: www.ottawa.ctv.ca. Licensee: CTV Television Inc. Group owner: (group owner) Natl. Network: CTV. Rebroadcasts CJOH-TV Ottawa 100%.

Deseronto

CJOH-TV-6— Analog channel: 6. Analog hrs: 24 100 kw vis. ant 671t/573g TL: N44 08 30 W77 04 34 On air date: September 1972. CTV Ottawa, Box 5813, Merivale Depot, Ottawa, ON, K2C 3G6. Phone: (613) 224-1313. Fax: (613) 274-4215. E-mail: ctvottawa@ctv.ca Web Site: www.ottawa.ctv.ca. Licensee: CTV Television Inc. Group owner: CTV Inc. Natl. Network: CTV.
Rebroadcasts CJOH-TV Ottawa 100%.

Elliot Lake

CICI-TV-1— Analog channel: 5.19 kw vis, 1.9 kw aur. 576t/216g On air date: 1958. c/o MC-TV, 699 Frood Rd., Sudbury, ON, P3C 5A3. Phone: (705) 674-8301. Fax: (705) 674-2706. Web Site: www.ctv.ca. Licensee: CTV Television Inc. Group owner: Baton Broadcasting Inc. Natl. Network: CTV. News staff: 15; News: 10 hrs wkly.
Key Personnel:
Scott Lund . . . VP & gen mgr
John Eddy . . . opns mgr & engrg mgr
Rick MacKenzie . . . gen sls mgr
Don Chapman . . . news dir

Geraldton

CBLAT— Analog channel: 13.22 kw vis, 4.4 kw aur. 598t/540g Box 500, Stn A, c/o CBLT, Toronto, ON, M5W 1E6. Phone: (416) 205-3311. Web Site: www.cbc.ca. Licensee: CBC. Natl. Network: CBC.

Hamilton

CHCH-TV— Analog channel: 11. Digital channel: 18. Analog hrs: 24 Digital hrs: 24 230 kw vis, 23 kw aur. ant 1,173t/1,054g On air date: June 4, 1954. Box 2230, Stn. A, 163 Jackson St. W., Hamilton, ON, L8N 3A6. Phone: (905) 522-1101. Fax: (905) 523-8778. E-mail: newstips@chtv.ca Web Site: www.chtv.ca. Licensee: CanWest MediaWorks Inc. Group owner: CanWest Global Communications Corp. (acq 7-6-2000). Natl. Network: Global.

CITS-TV— Analog channel: 36. Digital channel: 35. Analog hrs: 24 514 kw vis. On air date: Sept 30, 1998. 1295 N. Service Rd., Burlington, ON, L7R 4X5. Phone: (905) 331-7333. Fax: (905) 332-6005. E-mail: cts@ctstv.com Web Site: www.ctstv.com. Licensee: Crossroads Television System.
Key Personnel:
Fred Vanstone . . . chmn
Dick Gray . . . pres
Terry Maskel . . . opns mgr
Glenn Stewart . . . gen sls mgr
Michelle Gillies . . . prom mgr
Rob Sheppard . . . progmg mgr
David Storey . . . engrg dir

Kearns

CITO-TV-2— Analog channel: 11. Analog hrs: 6am - 2am 325 kw vis, 32.5 kw aur. 734t/396g Box 620, c/o MCTV-CTV, 681 Pine St. N., Timmins, ON, P4N 7G3. Phone: (705) 264-4211. Fax: (705) 264-3266. E-mail: newsroom@ctv.ca Web Site: www.ctv.ca. Licensee: CTV Television Inc. Group owner: CTV Inc. Natl. Network: CTV.
Key Personnel:
Scott Lund . . . gen mgr & stn mgr
Jason Laneville . . . sls dir

Kenora

CBWAT— Analog channel: 8.2 kw vis, 200 w aur. 433t/371g c/o CBC, Box 160, Winnipeg, MB, R3C 2H1. Canada. Phone: (204) 788-3222. Fax: (204) 788-3643. Web Site: www.winnipeg.cbc.ca. Licensee: CBC. Natl. Network: CBC.

CJBN-TV— Analog channel: 13.177 kw vis, 35 kw aur. 200g On air date: April 1983. 102 Tenth St., Keewatin, ON, P0X 1C0. 104 Tenth St., Keewatin, ON P9N 3X8. Phone: (807) 547-2852. Fax: (807) 547-2348. E-mail: darrylm@norcomcable.ca Web Site: www.norcomcable.ca. Licensee: Shaw Cablesystems G.P. Natl. Network: CTV. News staff: 2; News: one hr wkly.

Kingston

CKWS-TV— Analog channel: 11. Analog hrs: 24 325 kw vis, 32.5 kw aur. 830t/785g TL: N44 10 02 W76 25 40 On air date: 1954. 170 Queen St., Kingston, ON, K7K 1B2. Phone: (613) 544-2340. Fax: (613) 544-5508. E-mail: newswatch@corusent.com Web Site: www.ckwstv.com. Licensee: 591987 B.C. Ltd. Group owner: Corus Entertainment Inc. (acq 3-24-2000; grpsl). Natl. Network: CBC. News: 12 hrs wkly.
Key Personnel:
Mike Ferguson . . . gen mgr & stn mgr
Tim Wieczorek . . . gen sls mgr
Alison MacLean . . . prom dir
Jay Westman . . . news dir
Roger Cole . . . engrg VP & chief of engrg

Kitchener

***CICO-TV-28**— Analog channel: 28. Analog hrs: 24 200 kw vis, 20 kw aur. 972t/904g On air date: January 1976. Box 200, sta Q, c/o TV Ontario, Toronto, ON, M4T 2T1. 2180 Yonge St., Toronto, ON M4S 2P9. Phone: (416) 484-2600. Fax: (416) 484-6285. E-mail: asktvo@tvo.org Web Site: www.tvo.org. Licensee: Ontario Educational Communications Authority.

CKCO-TV— Analog channel: 13. Analog hrs: 24 325 kw vis, 32.5 kw aur. 954t/653g TL: N43 24 15 W80 38 05 On air date: Mar 1, 1954. CTV Southwestern Ontario, Box 91026, Kitchener, ON, N2G 4E9. Canada. Phone: (519) 578-1313. Fax: (519) 743-0730 (news). E-mail: viewermail@southwesternontario.ctv.ca Web Site: www.southwesternontario.ctv.ca. Licensee: CKCO-TV Division of CTV Inc. Group owner: CTV Inc. (acq 8-31-97). Population served: 2,000,000 Natl. Network: CTV.
Key Personnel:
Ivan Fecan . . . CEO
Robin Fillingham . . . CFO
Dennis Watson . . . VP & gen mgr
Dave MacNeill . . . opns dir
Cameron Crassweller . . . rgnl sls mgr
Janet Taylor . . . prom mgr, adv mgr, progmg mgr & pub affrs dir
Andy LaBlanc . . . news dir
Dave Melse . . . chief of engrg

Leamington

CFTV-TV— Analog channel: 34.400 w vis. On air date: 2006. 223 Talbot St. W., Leamington, ON, N8H 1N8. Phone: (519) 326-4000. E-mail: info@cftv.ca Web Site: www.cftv.ca. Licensee: Southshore Broadcasting Inc.
Key Personnel:
Tony Vidal . . . pres
Ted Mastronardi . . . VP

London

CFPL-TV— Analog channel: 10. Analog hrs: 24 325 kw vis, 43.2 kw aur. 1,000t/975g On air date: Nov 28, 1953. 1 Communications Road, London, ON, N6J 4Z1. Canada. Phone: (519) 686-8810. Fax: (519)

668-3288. Web Site: www.achannel.ca. Licensee: 1708486 Ontario Inc. Group owner: CHUM Ltd. (acq 1997). Population served: 500,000

***CICO-TV-18—** Analog channel: 18. Analog hrs: 24 34.9 kw vis, 3.5 kw aur. 1,029t/956 On air date: April 1976. Box 200 Stn Q, c/o TV Ontario, Toronto, ON, M4T 2T1. 2180 Yonge St., Toronto, ON M4S 2B9. Phone: (416) 484-2600. Fax: (416) 484-6285. E-mail: asktvo@tvo.org Web Site: www.tvo.org. Licensee: Ontario Educational Communications Authority.

Key Personnel:

Lisa DeWilde . . . . . CEO
Lee Robock . . . . . gen mgr
Ray Newell . . . . . opns dir
Meg Pinto . . . . . mktg dir

Marathon

CBLAT-4— Analog channel: 11. Analog hrs: 17 7.5 kw vis, 1.532 kw aur. 599t TL: N48 44 50 W86 34 00 On air date: May 16, 1968. Box 500, Stn A, c/o CBLT, Toronto, ON, M5W 1E6. Phone: (416) 205-3311. Fax: (416) 205-2552. Web Site: www.cbc.ca. Licensee: Canadian Broadcasting Corp. Natl. Network: CBC.
Rebroadcast of CBLT Toronto 100%.

Midland

CIII-TV-7— Analog channel: 7.325 kw vis, 48.8 kw aur. 1,132t/1,174g On air date: Nov 24, 1987. 81 Barber Greene Rd., Toronto, ON, M3C 2A2. Phone: (416) 446-5311. Fax: (416) 446-5447. E-mail: newstips@globaltv.com Web Site: www.canada.com. Licensee: CanWest MediaWorks Inc. Group owner: Global Television Network

North Bay

CKNY-TV— Analog channel: 10. Analog hrs: 20 70.5 kw vis, 7.1 kw aur. 607t/1,165g On air date: October 1981. 245 Oak St. E., North Bay, ON, P1B 8P8. Phone: (705) 476-3111. Fax: (705) 495-4474. Fax: (705) 495-0922 (news). E-mail: northbaynews@ctv.ca Web Site: www.ctv.ca. Licensee: CTV Television Inc. Group owner: CTV Inc. (acq 1991). Population served: 100,000 Natl. Network: CTV. News: 10 hrs wkly.

Key Personnel:

Scott Lund . . . . . pres & gen mgr
Ron Driscoll . . . . . gen sls mgr

Oil Springs

CIII-TV-29— Analog channel: 29. Analog hrs: 5 AM-2 AM 370 kw vis, 55.5 kw aur. 685t/701g On air date: Jan 6, 1974. 81 Barber Greene Rd., Toronto, ON, M3C 2A2. Phone: (416) 446-5311. Fax: (416) 446-5447. E-mail: newstips@globaltv.com Web Site: www.canada.com. Licensee: CanWest MediaWorks Inc. Group owner: Global Television Network

Oshawa

CHEX-TV-2— Analog channel: 22. Analog hrs: 20 550 kw vis. ant 440t TL: N43 57 15 W78 48 24 On air date: 1993. 500 Wentworth St. E., Unit 7, Oshawa, ON, L1H 3V9. Canada. Phone: (905) 434-2421. Fax: (905) 432-2315. Web Site: www.channel12.ca. Licensee: 591987 B.C. Ltd. Group owner: Corus Entertainment Inc. (acq 3-24-2000; grpsl). Natl. Network: CBC.

Ottawa

CBOFT— Analog channel: 9. Digital channel: 22. Analog hrs: 17 128 kw vis, 12.8 kw aur. ant 1,394t/702g Box 3220, Stn C, Ottawa, ON, K1Y 1E4. Ottawa Broadcast Centre, 181 Queen St., Ottawa, ON K1P 1K9. Phone: (613) 288-6000. Phone: (613) 288-6750 (news). Fax: (613) 288-6770. E-mail: tjottawa-gatincau@radio-canada.ca Web Site: www.radio-canada.ca. Licensee: CBC. Natl. Network: Radio Canada.

CBOT— Analog channel: 4. Digital channel: 25. Analog hrs: 17 100 kw vis, 10 kw aur. ant 1,310t/618g On air date: 1953. Box 3220, Stn C, Ottawa, ON, K1Y 1E4. Ottawa Broadcast Centre, 181 Queen St., Ottawa, ON K1P 1K9. Phone: (613) 288-6000. Fax: (613) 288-6423. E-mail: newsatsixottawa@cbc.ca Web Site: www.cbc.ca. Licensee: CBC. Natl. Network: CBC.

CHRO-TV-43— Analog channel: 43.282 kw vis. TL: N45 13 01 W75 33 51 On air date: 2002. A-Channel Ottawa, 87 George St., Ottawa, ON, K1N 9H7. Phone: (613) 789-0606. Fax: (613) 789-6590. E-mail: ottawa.promotions@Achannel.ca Web Site: www.achannel.ca/ottawa/. Licensee: 1708486 Ontario Inc.

Key Personnel:

Chris Gordon . . . . . VP & gen mgr
Peter Angione . . . . . news dir

***CICO-TV-24—** Analog channel: 24.427.6 kw vis. ant 1,092t/400g On air date: Oct 17, 1975. Box 200, sta Q, c/o TV Ontario, Toronto, ON, M4T 2T1. 2180 Yonge St., Toronto, ON M4S 2B9. Phone: (416) 484-2600. Fax: (416) 484-6285. E-mail: asktvo@tvo.org Web Site: www.tvo.org. Licensee: Ontario Educational Communications Authority.

CIII-TV-6— Analog channel: 6. Analog hrs: 5 AM-2 AM 50 kw vis, 7.5 kw aur. 843t/750g On air date: Jan 6, 1974. 81 Barber Greene Rd., Toronto, ON, M3C 2A2. Phone: (416) 446-5311. Fax: (416) 446-5447. E-mail: newstips@globaltv.con Web Site: www.canada.com. Licensee: CanWest MediaWorks Inc. Group owner: Global Television Network (acq 3-22-77).

CJOH-TV— Analog channel: 13. Analog hrs: 24 325 kw vis, 65 kw aur. ant 1,225t/533g TL: N45 30 11 W75 51 02 On air date: March 1961. CTV Ottawa, Box 5813, Merivale Depot, Ottawa, ON, K2C 3G6. 1500 Merivale Rd., Nepean, ON K2E 6Z5. Phone: (613) 224-1313. Fax: (613) 274-4215. E-mail: ctvottawa@ctv.ca Web Site: www.ottawa.ctv.ca. Licensee: CTV Television Inc. Group owner: (group owner) Population served: 2,000,000 Natl. Network: CTV. Natl. Rep: Canadian Broadcast Sales.

Key Personnel:

Louis Douville . . . . . VP & gen mgr
Art Clarke . . . . . opns mgr
Dan Champagne . . . . . sls VP & gen sls mgr
Brent Corbeil . . . . . prom mgr
Scott Hannant . . . . . news dir

Owen Sound

CIII-TV-4— Analog channel: 4.37 kw vis, 5.5 kw aur. 429t/477g On air date: June 27, 1988. 81 Barber Greene Rd., Toronto, ON, M3C 2A2. Phone: (416) 446-5311. Fax: (416) 446-5447. E-mail: newstips@globaltv.com Web Site: www.canada.com. Licensee: CanWest MediaWorks Inc. Group owner: Global Television Network

Pembroke

CHRO-TV— Analog channel: 5. Analog hrs: 24 100 kw vis, 10 kw aur. 496t/520g TL: N45 50 02 W77 09 50 On air date: 1961. Box 1010, Pembroke, ON, K8A 6Y6. 87 George St., Ottawa, ON K1N 9H7. Phone: (613) 789-0606. Phone: (613) 735-1036. Fax: (613) 789-6590. E-mail: ottawa@Achannel.ca Web Site: www.Achannel.ca. Licensee: 1708486 Ontario Inc. Group owner: (group owner; (acq 1997). Population served: 1,000,000 News: 30 hrs wkly.

Key Personnel:

Chris Gordon . . . . . VP & gen mgr
Greg Orr . . . . . gen sls mgr
Peter Angione . . . . . news dir
Robert Edgley . . . . . chief of engrg

Peterborough

CHEX-TV— Analog channel: 12. Analog hrs: 20 185 kw vis, 18.5 kw aur. ant 772t/753g TL: N44 19 45 W78 18 03 On air date: 1955. 743 Monaghan Rd., Peterborough, ON, K9J 5K2. Phone: (705) 742-0451. Fax: (705) 742-7274. E-mail: newswatch@chextv.com Web Site: www.chextv.com. Licensee: 5191987 B.C. Ltd. Group owner: Corus Entertainment Inc. (acq 3-24-2000; grpsl). Natl. Network: CBC. Natl. Rep: TeleRep.

Key Personnel:

John Cassaday . . . . . CEO
Ron Johnston . . . . . gen mgr & stn mgr
Paul Burke . . . . . opns mgr

CIII-TV-27— Analog channel: 27.2,535 kw vis, 380 aur. 913t/499g On air date: Oct 5, 1988. 81 Barber Greene Rd., Toronto, ON, M3C 2A2. Phone: (416) 446-5311. Fax: (416) 446-5447. E-mail: newstips@globaltv.com Web Site: www.canada.com. Licensee: CanWest MediaWorks Inc. Group owner: Global Television Network.

Sarnia

CKCO-TV-3— Analog channel: 42. Analog hrs: 24 846 kw vis, 84.6 kw aur. 994t/985g TL: N42 42 53 W82 08 12 On air date: November 1975. CTV Southwestern Ontario, Box 91026, c/o CKCO-TV, Kitchener, ON, N2G 4E9. Canada. Phone: (519) 578-1313. Fax: (519) 743-0730. E-mail: viewermail@southwesternontario.ctv.ca Web Site: www.southwesternontario.ctv.ca. Licensee: CKCO-TV Division of CTV Inc. Group owner: CTV Inc. (acq 8-31-97). Population served: 2,000,000 Natl. Network: CTV.

Key Personnel:

Ivan Fecan . . . . . CEO
Dennis A. Watson . . . . . VP
Dennis Watson . . . . . gen mgr
Dave MacNeill . . . . . opns dir
Cameron Crassweller . . . . . rgnl sls mgr
Andy LaBlanc. . . . . news dir
Janet Taylor . . . prom mgr, adv mgr, progmg mgr & pub affrs dir
Dave Melse . . . . . chief of engrg

Rebroadcasts CKCO-TV Kitchener, on 100%.

Sault Ste. Marie

CHBX-TV— Analog channel: 2. Analog hrs: 6 AM-2 AM 100 kw vis, 10 kw aur. 600t/500g On air date: September 1978. 119 East St., Sault Ste. Marie, ON, P6A 3C7. Phone: (705) 759-8232. Fax: (705) 759-7783. E-mail: saultnews@ctv.ca Web Site: www.ctv.ca. Licensee: CTV Television Inc. Group owner: (group owner) Population served: 104,000 Natl. Network: CTV. Natl. Rep: Canadian Broadcast Sales.

***CICO-TV-20—** Analog channel: 20. Analog hrs: 24 5.9 kw vis, 590 w aur. 650t/515g On air date: October 1978. Box 200, stn Q, c/o TV Ontario, Toronto, ON, M4T 2T1. 2180 Yonge St., Toronto, ON M4S 2B9. Phone: (416) 484-2600. Fax: (416) 484-6285. E-mail: asktvo@tvo.org Web Site: www.tvo.org. Licensee: Ontario Educational Communications Authority. Group owner: Baton Broadcasting Inc.

Stevenson

CIII-TV-22— Analog channel: 22. On air date: 1974. 81 Barber Greene Rd., Toronto, ON, M3C 2A2. Phone: (416) 446-5311. Fax: (416) 446-5447. E-mail: newstips@globaltv.com Web Site: www.canada.com /toronto/globaltv. Licensee: CanWest MediaWorks Inc. Group owner: Global Television Network. Natl. Network: Global. Natl. Rep: CanWest Media Sales.

Sudbury

CICI-TV— Analog channel: 5.100 kw vis, 10 kw aur. 1,057t/975g On air date: Oct 25, 1953. 699 Frood Rd., Sudbury, ON, P3C 5A3. Phone: (705) 674-8301. Fax: (705) 674-2706. Web Site: www.ctv.ca. Licensee: CTV Television Inc. Group owner: (group owner; (acq 4-1-80). Natl. Network: CTV. News staff: 15; News: 10 hrs wkly.

Key Personnel:

Scott Lund . . . . . VP & gen mgr
John Eddy . . . . . opns mgr

***CICO-TV-19—** Analog channel: 19. Analog hrs: 24 186.5 kw vis, 18.7 kw aur. 564t/497g On air date: June 30, 1978. Box 200 Stn Q, c/o TV Ontario, Toronto, ON, M4T 2T1. 2180 Yonge St., Toronto, ON M4S 2B9. Phone: (416) 484-2600. Fax: (416) 484-6285. E-mail: asktvo@tvo.org Web Site: www.tvo.org. Licensee: Ontario Educational Communications Authority.

Key Personnel:

Lisa DeWilde . . . . . CEO
Lee Robock . . . . . gen mgr
Ray Newell . . . . . opns dir

Thunder Bay

CHFD-TV— Analog channel: 4. Analog hrs: 18 56 kw vis, 10 kw aur. 1,202t/634g TL: N48 31 30 W89 06 50 On air date: 1972. 87 N. Hill St., Thunder Bay, ON, P7A 5V6. Phone: (807) 346-2600. Fax: (807) 345-9923. E-mail: tbtv@tbtv.com Web Site: tbtv.com. Licensee: Thunder Bay Electronics Ltd. Ownership: H.F. Dougall, Esq., 100%. Population served: 160,000 Natl. Network: CTV. Natl. Rep: CanWest Media Sales. Wire Svc: CNW Broadcast

Key Personnel:

H.F. Dougall . . . . . pres
Don Caron . . . . . CFO
Ann Snell . . . . . opns dir
P. Bentz . . . . . progmg dir

***CICO-TV-9—** Analog channel: 9. Analog hrs: 24 32 kw vis, 3.2 kw aur. 780t/522g On air date: June 1978. Box 200 Stn Q, c/o TV Ontario, Toronto, ON, M4TT 2T1. 2180 Yonge St., Toronto, ON M4S 2B9. Phone: (416) 484-2600. Fax: (416) 484-6285. E-mail: asktvo@tvo.org Web Site: www.tvo.org. Licensee: Ontario Educational Communications Authority.

Key Personnel:

Lisa DeWilde . . . . . CEO
Lee Robock . . . . . COO & gen mgr
Ray Newell . . . . . opns dir

Meg Pinto . . . . mktg dir

CKPR-TV— Analog channel: 2. Analog hrs: 18 56 kw vis, 10 kw aur. ant 1,202t/643g TL: N48 31 30 W89 06 50 On air date: 1954. 87 N. Hill St., Thunder Bay, ON, P7A 5V6. Phone: (807) 346-2600. Fax: (807) 345-9923. E-mail: tbt@tbtv.com Web Site: www.tbtv.com. Licensee: Thunder Bay Electronics Ltd. Ownership: H.F. Dougall, Esq., 100%. Population served: 160,000 Natl. Network: CBC. Natl. Rep: CanWest Media Sales. News: 11 hrs wkly.

Key Personnel:

H.F. Dougall . . . . pres
Ann Snell . . . . opns dir & news dir
P. Bentz . . . . progmg dir

Timmins

CITO-TV— Analog channel: 3. Analog hrs: 6am - 2am 100 kw vis, 10 kw aur. 544t/499g Box 620, Timmins, ON, P4N 7G3. Canada. Phone: (705) 264-4211. Fax: (705) 264-3266. E-mail: newsroom@ctv.ca Web Site: www.ctv.ca. Licensee: CTV Television Inc. Group owner: CTV Inc. Population served: 132,900 Natl. Network: CTV.

Toronto

CBLT— Analog channel: 5. Digital channel: 20.77 kw vis, 7 kw aur. ant 444t/541g On air date: Sept 8, 1952. Box 500, Stn A, Toronto, ON, M5W 1E6. 205 Wellington St. W., Toronto, ON M5V 3G7. Phone: (416) 205-3311. Fax: (416) 205-7166. Web Site: www.cbc.ca. Licensee: CBC. Natl. Network: CBC. News staff: 45; News: 10 hrs wkly.

CFMT-TV— Analog channel: 47. Analog hrs: 24 807 kw vis, 80.7 kw aur. ant 1,600t/1,427g (Digital TV: ch 64; 15,000 w vis) On air date: Sept 3, 1979. Omni Television, 545 Lake Shore Blvd. W., Toronto, ON, M5V 1A3. Canada. Phone: (416) 260-0047. Fax: (416) 260-3621. E-mail: info@omnitv.ca Web Site: www.omnitv.ca. Licensee: Rogers Broadcasting Ltd.

Key Personnel:

Madeline Ziniak . . . . VP & stn mgr
Bill Hope . . . . chief of engrg

CFTO-TV— Analog channel: 9. Digital channel: 40. Analog hrs: 24 325 kw vis, 162 kw aur. ant 1815t/1614g TL: N43 38 33 W79 23 15 On air date: Jan 1, 1961. 9 Channel 9 Ct., Scarborough, ON, M1S 4B5. Phone: (416) 332-5000. Fax: (416) 332-5022. E-mail: cftonews@ctv.ca Web Site: www.ctv.ca. Licensee: CTV Television Inc. Group owner: (group owner) Natl. Network: CTV.

***CICA-TV—** Analog channel: 19. Analog hrs: 24 1080 kw vis, 108 kw aur. 1,605t/1,686g (CP: 1288.2 kw vis) On air date: Sept 27, 1970. Box 200, Stn Q, c/o TV Ontario, Toronto, ON, M4T 2T1. 2180 Yonge St., Toronto, ON M4S 2B9. Phone: (416) 484-2600. Fax: (416) 484-6285. E-mail: asktvo@tvo.org Web Site: www.tvo.org. Licensee: Ontario Educational Communications Authority.

Key Personnel:

Lisa DeWilde . . . . CEO
Lee Robock . . . . COO & gen mgr
Ray Newell . . . . opns dir
Meg Pinto . . . . mktg dir

CIII-TV—(Paris, Analog channel: 6. Analog hrs: 5 AM-2 AM 100 kw vis, 15 kw aur. 1,037t/999g On air date: Jan 6, 1974. 81 Barber Greene Rd., Toronto, ON, M3C 2A2. Phone: (416) 446-5311. Fax: (416) 446-5447. E-mail: newstips@globaltv.com Web Site: www.globaltv.com. Licensee: CanWest MediaWorks Inc. Group owner: Global Television Network (acq 3-22-77). Natl. Network: Global.

CIII-TV-41— Analog channel: 41. Digital channel: 65.732 kw vis, 221 kw aur. ant 1,644t/1,779g TL: N43 38 33 W79 23 15 On air date: Oct 22, 1987. 81 Barber Greene Rd., Toronto, ON, M3C 2A2. Phone: (416) 446-5311. Fax: (416) 446-5447. E-mail: newstips@globaltv.com Web Site: www.canada.com. Licensee: CanWest MediaWorks Inc. Group owner: Global Television Network

CITY-TV— Analog channel: 57. Digital channel: 53. Analog hrs: 24 280 kw vis, 28 kw aur. 1,690t/1,780g On air date: Sept 28, 1972. 299 Queen St. W., Toronto, ON, M5V 2Z5. Phone: (416) 591-5757. Fax: (416) 340-7005. Web Site: www.citytv.com. Licensee: 1708487 Ontario Inc. Group owner: CHUM Ltd. (acq 1976). Population served: 4,000,000 News: 25 hrs wkly.

Key Personnel:

Maria Hale . . . . pres & VP
John Morrison . . . . opns mgr & gen sls mgr
Dan Hamilton . . . . sls VP & prom dir
Susan Arthur . . . . mktg dir
Bev Nenson . . . . prom dir & prom mgr
Jenny Norush . . . . adv dir
Ellen Baine . . . . progmg VP, progmg dir & film buyer
Stephen Hurlbut . . . . news dir
Sarah Crawford . . . . pub affrs dir
Bruce Cowan . . . . engrg dir

CJMT-TV— Analog channel: 69. Digital channel: 66. Analog hrs: 24 500 kw vis. On air date: Sept 16, 2002. 545 Lake Shore Blvd. W., Toronto, ON, M5V 1A3. Phone: (416) 260-0060. Fax: (416) 260-3621. E-mail: info@omnitv.ca Web Site: www.omnitv.ca. Licensee: Rogers Broadcasting Ltd. Group owner: (group owner).

Key Personnel:

Leslie Sole . . . . CEO
Madeline Ziniak . . . . gen mgr
Malcolm Dunlop . . . . mktg VP & progmg VP
Kelly Colasanti . . . . engrg VP

CKXT-TV— Analog channel: 52. Digital channel: 66. Analog hrs: 24 Digital hrs: 24 30 kw vis. ant 1,502t TL: N43 38 33 W79 23 15 On air date: Sept 19, 2003. SUN TV, 25 Ontario St., 2nd Fl., Toronto, ON, M5A 4L6. Phone: (416) 601-0010. Fax: (416) 601-0004. Web Site: suntv.canoe.ca. Licensee: Sun TV Co. Group owner: (group owner). (acq 11-19-2004; C$46 million). News staff: 9.

Key Personnel:

Jim Nelles . . . . VP & gen mgr
Lynne Godin . . . . mktg mgr
Sindy Prreger . . . . sls
Don Gaudet . . . . progmg

Wiarton

CKCO-TV-2— Analog channel: 2. Analog hrs: 24 100 kw vis, 10 kw aur. 939t/789g TL: N44 56 41 W81 07 55 On air date: June 1, 1971. CTV Southwestern Ontario, Box 91026, c/o CKCO-TV, Kitchener, ON, N2G 4E9. Canada. Phone: (519) 578-1313. Fax: (519) 743-0730. E-mail: viewermail@southwesternontario.ctv.ca Web Site: www.southwesternontario.ctv.ca. Licensee: CKCO-TV Div. of CTV. Group owner: CTV Inc. (acq 8-31-97). Population served: 2,000,000 Natl. Network: CTV.

Key Personnel:

Ivan Fecan . . . . CEO
Dennis Watson . . . . VP & gen mgr
Dave MacNeill . . . . opns dir
Cameron Crassweller . . . . rgnl sls mgr
Andy LaBlanc . . . . news dir
Janet Taylor . . . prom mgr, adv mgr, progmg mgr & pub affrs dir
Dave Melse . . . . chief of engrg
Rebroadcasts CKCO-TV Kitchener, ON 100%..

Windsor

***CBEFT—** Analog channel: 54. Analog hrs: 24 62.7 kw vis, 6.3 kw aur. 683g On air date: July 16, 1976. 825 Riverside W., Windsor, ON, N9A 5K9. Canada. Box 1609, Windsor, ON N9A 1K7.Canada Phone: (519) 255-3411. Fax: (519) 255-3412. Licensee: Societe Radio Canada. Natl. Network: Radio Canada. News staff: one.

CBET— Analog channel: 9. Analog hrs: 24 325 kw vis, 32.5 kw aur. ant 575t/650g TL: N42 18 59 W83 02 58 On air date: Sept 16, 1954. 825 Riverside Dr. W., Windsor, ON, N9A 5K9. Phone: (519) 255-3411. Fax: (519) 255-3412. Licensee: Canadian Broadcasting Corp. (acq 7-23-75). Natl. Network: Radio Canada. Natl. Rep: Canadian Broadcast Sales. News staff: 18; News: 5 hrs wkly.

CHWI-TV— Analog channel: 16. Analog hrs: 24 183 kw vis. On air date: Oct 18, 1993. 300 Oullette Ave., Suite 200, Windsor, ON, N9A 7B4. Phone: (519) 977-7432. Fax: (519) 977-0564. E-mail: windsornews@achannel.ca Web Site: www.achannel.ca. Licensee: 1708486 Ontario Inc. Group owner: CHUM Ltd. (acq 1997).

***CICO-TV-32—** Analog channel: 32. Analog hrs: 24 180 kw vis, 18 kw aur. 703t/703g On air date: July 1976. Box 200, Stn Q, c/o TV Ontario, Toronto, ON, M4T 2T1. 2180 Yonge St., Toronto, ON M4S 2B9. Phone: (416) 484-2600. Fax: (416) 484-6285. E-mail: asktvo@tvo.org Web Site: www.tvo.org. Licensee: Ontario Educational Communication Authority.

Key Personnel:

Lisa DeWilde . . . . CEO
Lee Robock . . . . gen mgr
Ray Newell . . . . opns dir
Meg Pinto . . . . mktg dir

Wingham

CKNX-TV— Analog channel: 8. Analog hrs: 24 260 kw vis, 26 kw aur. 793t/623g On air date: 1955. 215 Carling Terr., Wingham, ON, N0G 2W0. Phone: (519) 357-4438. Fax: (519) 357-4398. E-mail: newsnow@thenewnx.com Web Site: www.achannel.ca. Licensee: 1708486 Ontario Inc. Group owner: CHUM Ltd. (acq 1997). Population served: 300,000

Key Personnel:

Greg Mudry . . . . gen mgr
Jim Kippen . . . . opns mgr
Tom Fitz-Gerald . . . . gen sls mgr
Don Mumford . . . . progmg mgr
Cal Johnstone . . . . news dir

Prince Edward Island

Charlottetown

CBCT— Analog channel: 13.320 kw vis, 30 kw aur. 918t/720g On air date: 1968. Box 2230, Charlottetown, PE, C1A 8B9. 430 University Ave., Charlottetown, PE C1A 4N6.Canada Phone: (902) 629-6400. Fax: (902) 629-6518. Web Site: www.cbct.ca/pei. Licensee: CBC. Ownership: CBC (Crown Corp). Natl. Network: CBC.

CKCW-TV-1— Analog channel: 8.29 kw vis, 2.9 aur. ant 489t/250g 191 Halifax St., Moncton, NB, E1C 9R7. Canada. Phone: (506) 857-2600. Fax: (506) 857-2617. E-mail: ckcw@ctv.ca Web Site: www.ctv.ca. Licensee: CTV Television Inc. Group owner: CTV Inc. (acq 11-1-97). Natl. Network: CTV.

Key Personnel:

Ivan Fecan . . . . CEO
Rick Brace . . . . pres
Robin Fillingham . . . . CFO
Elaine Ali . . . . exec VP
Mike Elgie . . . . VP & gen mgr
Brian Lewis . . . . stn mgr, gen sls mgr & adv mgr
John Silver . . . . opns mgr
Renee Fournier . . . . prom dir & prom mgr
Jane Hefler . . . . progmg dir
Jay Witherbee . . . . news dir & pub affrs dir
Carson McDavid . . . . chief of engrg
Rebroadcasts CKCW-TV Moncton, NB 100%.

Saint Edward

CKCW-TV-2— Analog channel: 5.2.5 kw vis, 1.3 kw aur. ant 341t/356g On air date: November 1982. 191 Halifax St., Moncton, NB, E1C 9R7. Canada. Phone: (506) 857-2600. Fax: (506) 857-2617. E-mail: ckcw@ctv.ca Web Site: www.ctv.ca. Licensee: CTV Television Inc. Group owner: CTV Inc. (acq 11-1-97). Population served: 100,000 Natl. Network: CTV.

Key Personnel:

Ivan Fecan . . . . CEO
Rick Brace . . . . pres
Robin Fillingham . . . . CFO
Elaine Ali . . . . exec VP
Mike Elgie . . . . VP & gen mgr
Brian Lewis . . . . stn mgr, gen sls mgr & adv mgr
John Silver . . . . opns dir & opns mgr
Renee Fournier . . . . prom dir & prom mgr
Jay Witherbee . . . . news dir & pub affrs dir
Carson McDavid . . . . chief of engrg
Rebroadcasts CKCW-TV Moncton, NB 100%.

Quebec

Baie-Trinite

***CIVF-TV—** Analog channel: 12. Analog hrs: 24 62 kw vis. 2,001t TL: N49 23 28 W67 28 18 On air date: Nov 15, 1982. 1000 Fullum, Montreal, PQ, H2K 3L7. Phone: (514) 521-2424. Fax: (514) 873-2601. Fax: (514) 873-4413. E-mail: info@telequebec.qc.ca Web Site: www.telequebec.tv. Licensee: Societe de telediffusion du Quebec. Group owner: Tele-Quebec.

Key Personnel:

Michele Fortim . . . . pres
Luc Chartier . . . . gen mgr

Bearn-Fabre

CKRN-TV-3— Analog channel: 3.35 kw vis, 3.5 kw aur. 171 A Jean-Proulx St., Gatineau, PQ, J8Z 1W5. Canada. Phone: (819)

770-1040. Fax: (819) 770-0272. Web Site: www.radionord.com. Licensee: RNC MEDIA Inc. Group owner: (group owner). Natl. Network: Radio Canada.
Key Personnel:
Pierre R. Brosseau . . . . pres
Michael Noiseux. gen mgr
Rebroadcasts CKRN-TV Rouyn-Noranda.

Carleton

CHAU-TV— Analog channel: 5. Analog hrs: 5:15am -3:00am 52.2 kw vis, 5.22 kw aur. 2,180t/475g On air date: Oct 17, 1959. 349 Blvd. Peron, Carleton, PQ, G0C 1J0. Phone: (418) 364-3344. Fax: (418) 364-7168. Licensee: CHAU-TV Communications Ltee. Group owner: Tele Inter-Rives Ltee. (acq 1-5-01). Natl. Network: TVA.

Chapeau

***CIVP-TV—** Analog channel: 23. Analog hrs: 24 8.65 kw vis. TL: N45 55 29 W77 04 23 1000 Fullum, Montreal, PQ, H2K 3L7. Phone: (514) 521-2424. Fax: (514) 873-2601. Fax: (514) 864-4222. E-mail: info@telequebec.qc.ca Web Site: www.telequebec.tv. Licensee: Societe de telediffusion du Quebec. Group owner: Tele-Quebec.
Key Personnel:
Michele Fortim . . . . pres
Luc Chartier . . . . gen mgr

Chicoutimi

CBJET— Analog channel: 58.10 kw vis. PO Box 6000, 1400 boul. Rene Levesque E., Montreal, PQ, H3C 3A8. Phone: (514) 597-6000. Phone: (514) 597-6000. Fax: (514) 597-4537. Web Site: www.cbc.ca. Licensee: CBC. Natl. Network: CBC.

***CIVV-TV—** Analog channel: 8. Analog hrs: 24 278.1 kw vis, 27.8 kw aur. 1,948t/570g TL: N48 36 04 W70 49 46 On air date: November 1982. 1000 Fullum, Montreal, PQ, H2K 3L7. Phone: (514) 521-2424. Fax: (514) 873-2601. Fax: (514) 864-4222. E-mail: info@telequebec.qc.ca Web Site: www.telequebec.tv. Licensee: Societe de telediffusion du Quebec. Group owner: Tele-Quebec. Natl. Network: TeleFutura (Spanish).
Key Personnel:
Michele Fortim . . . . pres
Luc Chartier . . . . gen mgr

CJPM-TV— Analog channel: 6.61 kw vis, 6.7 kw aur. ant 440t/190g On air date: Apr 14, 1963. One Mont Ste-Claire St, Saguenay, PQ, G7H 5G3. Phone: (418) 549-2576. Fax: (418) 549-1130. E-mail: cjpm@saglac.qc.ca Web Site: www.reseau.tva.ca. Licensee: Groupe TVA Inc. Group owner: (group owner). Natl. Network: TVA. News: 6 hrs wkly.
Key Personnel:
Pierre Dion . . . . pres
Michel Roberge . . . . prom dir
Roger Jobin. . . . . gen mgr, sls dir & progmg dir
Myriam Donaldson. news dir

Gatineau

CFGS-TV— Analog channel: 34.117 kw vis. On air date: Sept 7, 1986. 171 A Jean-Proulx St., Gatineau, PQ, J8Z 1W5. Phone: (819) 770-1040. Fax: (819) 770-0272. E-mail: tqs@radionord.com Web Site: www.radionord.com. Licensee: RNC MEDIA Inc. Group owner: (group owner). Natl. Network: Quatre Saisons. Natl. Rep: Canadian Broadcast Sales.
Key Personnel:
Pierre R. Brosseau . . . . CEO & pres
Robert H. Parent . . . . gen mgr
Michel Noiseux . . . . opns dir

CHOT-TV— Analog channel: 40.498 kw vis. 1,184t On air date: Oct 30, 1978. 171 A Jean-Proulx St., Gatineau, PQ, J8Z 1W5. Phone: (819) 770-1040. Fax: (819) 770-1490 (news). E-mail: chot@radionord.com Web Site: www.radionord.com. Licensee: RNC MEDIA Inc. Group owner: (group owner). Natl. Network: TVA.
Key Personnel:
Pierre R. Brosseau . . . . CEO & pres
Robert H. Parent . . . . gen mgr
Michel Noiseux . . . . opns dir
Benoit Pilote . . . . sls dir
Eric Brousseau. prom mgr
Daniele Young . . . . news dir

***CIVO-TV—** Analog channel: 30.1327.4 kw vis, 265.5 kw aur. 1,184t/465g On air date: Aug 14, 1977. 1000 Fullum, Montreal, PQ, H2K 3L7. Phone: (514) 521-2424. Fax: (514) 873-2601. Fax: (514) 873-4413. E-mail: info@telequebec.qc.ca Web Site: www.telequebec.tv. Licensee: Societe de telediffusion du Quebec. Group owner: Tele-Quebec.

Iles-de-la-Madeleine

CBIMT— Analog channel: 12.3.9 kw vis. 750t/235g On air date: Nov 9, 1964. PO Box 6000, 1400 boul. Rene Levesque E., Montreal, PQ, H3C 3A8. Phone: (514) 597-6000. Fax: (514) 597-4537. Web Site: www.cbc.ca. Licensee: CBC. Natl. Network: Radio Canada.

Jonquiere

CFRS-TV— Analog channel: 4. Analog hrs: 18 100 kw vis, 10 kw aur. 1,941t/460g On air date: Sept 7, 1986. 2303 rue Sir Wilfred Laurier, Jonquiere, PQ, G7X 5Z2. Phone: (418) 542-4551. Fax: (418) 542-7217. Fax: (418) 542-8319. Web Site: www.cgotv.ca. Licensee: Cogeco Radio-Television Inc. (acq 11-24-98; C$5,754,626 with CKTV-TV Jonquiere). Population served: 300,000 Natl. Network: Quatre Saisons.
Key Personnel:
Martin Gagnon . . . . gen mgr
Michel Goulet . . . . opns dir
Ammie Tremblay . . . . sls dir
Annie Tremblay . . . . mktg dir

La Tuque

CBVT-2— Analog channel: 3.15.4 kw vis, 1.54 kw aur. PO Box 6000, 1400 boul. Rene Levesque E., Montreal, PQ, H3C 3A8. Phone: (514) 597-6000. Fax: (514) 597-6510. Web Site: www.cbc.ca. Licensee: CBC. Natl. Network: Radio Canada.

Malartic

CBVD-TV— Analog channel: 5.9.35 kw vis, 4.675 kw aur. CBC Television, 1400 Rene Levesque E., Montreal, PQ, H2L 2M2. Phone: (514) 597-6000. Phone: (514) 597-4537. Web Site: montreal.cbc.ca. Licensee: CBC. Natl. Network: CBC.
Rebroadcasts CBMT(TV) Montreal 100%.

Mont-Laurier

CBFT-2— Analog channel: 3.28.2 kw vis, 2.8 kw aur. 509t/400g On air date: Dec 3, 1962. Box 6000, Montreal, PQ, H3C 3A8. Phone: (514) 597-6000. Fax: (514) 597-6354. Web Site: www.cbc.ca. Licensee: Societe Radio Canada. Natl. Network: Radio Canada.

Montreal

CBFT— Analog channel: 2. Digital channel: 19.100 kw vis, 10 kw aur. ant 905t/252g On air date: 1952. PO Box 6000, 1400 boul. Rene Levesque E., Montreal, PQ, H3C 3A8. Phone: (514) 597-6000. Phone: (514) 597-4282. Fax: (514) 597-4316. E-mail: auditoire@radio-canada.ca Web Site: radio-canada.ca. Licensee: CBC. Ownership: CBC. Natl. Network: Radio Canada.
Key Personnel:
Daniel Gourd . . . . VP
Richard Portelance . . . . sls dir & gen sls mgr
Loren Thibeault. natl sls mgr
Alain Messier . . . . mktg dir
Danielle Rivard. mktg mgr
Andre Beau det . . . . adv dir
Mario Clement . . . . progmg dir

CBMT— Analog channel: 6. Digital channel: 20.100 kw vis, 15 kw aur. ant 820t/167g On air date: 1954. Box 500, Sta A, Toronto, ON, M5W 1E6. Canada. Phone: (866) 306-4636. Fax: (416) 205-3714. Web Site: www.cbc.ca. Licensee: CBC. Natl. Network: CBC.

CFCF-TV— Analog channel: 12. Digital channel: 7.325 kw vis, 33 kw aur. 1,032t/294g On air date: Jan 20, 1961. CTV Television Inc., 1205 Papineau Ave., Montreal, PQ, H2K 4R2. Phone: (514) 273-6311. Fax: (514) 276-9399. E-mail: cfcfpromo@ctv.ca Web Site: www.cfcf.ca. Licensee: CTV Television Inc. Group owner: (group owner; (acq 9-21-01; C$141.5 million). Natl. Network: CTV.

CFJP-TV— Analog channel: 35. Digital channel: 42.697 kw vis, 70 kw aur. ant 900t/335g TL: N45 35 20 W73 35 32 On air date: Sept 7, 1986. 612 Rue St-Jacques Bur 100, Montreal, PQ, H3C 5R1. Phone: (514) 390-6035. Fax: (514) 390-0773. E-mail: tvpublic@tqs.ca Web Site: www.tqs.ca. Licensee: TQS Inc. Ownership: Quebecor Inc., 85.99%. (acq 1997). News staff: 51; News: 12 hrs wkly.
Key Personnel:
Michael J. Carter . . . . pres
Luc Doyon . . . . VP
Rene Guimond . . . . gen mgr
Sophie Ferron. opns dir
Robert PeRusse . . . . dev dir

CFTM-TV— Analog channel: 10. Digital channel: 59.365 kw vis, 65 kw aur. ant 325t/1,068g Box 170, Stn C, Montreal, PQ, H2L 4P6. Phone: (514) 790-0461. Phone: (514) 526-9251. Fax: (514) 598-6082. Web Site: www.tva.canoe.com. Licensee: Groupe TVA Inc. Natl. Network: TVA.

***CFTU-TV—** Analog channel: 29. Analog hrs: 24 10 kw vis, 1 kw aur. 604t/305g On air date: Aug 20, 1985. 4750 Ave. Henri-Julien, Bureau 100, local 0058, Montreal, PQ, H2T 3E4. Phone: (514) 841-2626. Fax: (514) 284-9363. E-mail: info@canal.qc.ca Web Site: www.canal.qc.ca. Licensee: Corp. pour l'Avancement de Nouvelles Applications des Langages. (acq 5-86).
Key Personnel:
Michel Umbriaco . . . . pres
Guy Massicotte . . . . VP
Sylvie Godbout. . . . . chmn & gen mgr

***CIVM-TV—** Analog channel: 17. Digital channel: 27. Analog hrs: 7:00am - 1:30am 889.5 kw vis. On air date: Jan 19, 1975. 1000 Fullum, Montreal, PQ, H2K 3L7. Phone: (514) 521-2424. Fax: (514) 873-2601. Fax: (514) 873-4413. E-mail: info@telequebec.qc.ca Web Site: www.telequebec.tv. Licensee: Societe de telediffusion du Quebec. Population served: 7,000,000 News staff: 25; News: 5 hrs wkly.

CJNT-TV— Analog channel: 62.11 kw vis. On air date: 1997. 1600 Blvd., De Maisonneuve E., 9th Fl., Montreal, PQ, H2L 4P2. Phone: (514) 522-4150. Fax: (514) 522-9579. Web Site: www.cahmontreal.com. Licensee: CanWest MediaWorks Inc. Group owner: CanWest Global Communications Corp. (acq 11-29-2000).

Quebec City

CBVT— Analog channel: 11. Digital channel: 12.128.8 kw vis, 12.8 kw aur. ant 657t/541g On air date: 1964. 888 Saint-Jean St., Quebec, PQ, G1R 5H6. Phone: (418) 656-8500. Fax: (418) 656-8505. Web Site: www.cbc.ca. Licensee: CBC. Natl. Network: Radio Canada.

CFAP-TV— Analog channel: 2. Analog hrs: 6:00 AM-2:30 AM 70 kw vis, 7 kw aur. 551t/501g TL: N46 48 27 W71 13 02 On air date: Sept 4, 1989. 330 St.-Vallier St. East, Quebec City, PQ, G1K 9C5. Phone: (418) 624-2222. Fax: (418) 624-3099. Fax: (418) 624-0162. E-mail: www@tqs.ca Web Site: www.tqs.ca. Licensee: TQS Inc. Group owner: Cogeco Radio-Television Inc. (acq 2-15-02; grpsl). Natl. Network: Quatre Saisons. News staff: 29; News: 10 hrs wkly.
Key Personnel:
Rene Guimond . . . . pres
Renaud Francoeur . . . . gen mgr, progmg dir & news dir
Jean Simard . . . . opns dir, chief of opns & engrg dir
Joel Godin . . . . sls dir & mktg dir
Pierre Martineau . . . . pub affrs dir
Denise Delisle. engrg mgr
Rebroadcasts CFJP-TV Montreal.

CFCM-TV— Analog channel: 4. Digital channel: 7. Analog hrs: 5:00am - 2:00am 100 kw vis, 15 kw aur. 460t/407g On air date: July 17, 1954. 1000 Ave. Myrand, Quebec City, PQ, G1V 2W3. Canada. Phone: (418) 688-9330. Fax: (418) 681-4239. Web Site: www.tva.ca. Licensee: Tele-Metropole Inc. Group owner: Groupe TVA Inc.

***CIVQ-TV—** Analog channel: 15.1,298 kw vis, 259 kw aur. 628t/576g TL: N46 48 27 W71 13 02 On air date: Jan 19, 1975. 1000 Fullum, Montreal, PQ, H2K 3L7. Phone: (514) 521-2424. Fax: (514) 873-2601. Fax: (514) 873-4413. E-mail: info@telequebec.qc.ca Web Site: www.telequebec.tv. Licensee: Societe de telediffusion du Quebec. Group owner: Tele-Quebec.
Key Personnel:
Michele Fortim . . . . pres
Luc Chartier . . . . gen mgr

CKMI-TV— Analog channel: 20.20.2 kw vis, 3.77 kw aur. ant 460t/407g On air date: 1957. 1000 Myrand Ave., Ste.-Foy, PQ, G1V 2W3. 1600 Boul.de Maisonneuve East, Montreal, PQ, H2L 4P2.Canada Phone: (418) 682-2020. Phone: (514) 521-4323. Fax: (418) 682-2620. Fax: (514) 521-2829. E-mail: globalnews.que@globaltv.ca Web Site: www.globaltv.ca. Licensee: CanWest MediaWorks Inc. (the general partner) and GTNQ Holdings Inc. (the limited partner) carrying on business as Global Television Network Quebec L.P. (acq 1997).
Key Personnel:
Marven Rogers . . . . gen mgr
Michel Yeos . . . . opns dir & opns mgr

Suzanne Lapalme . . . . . sls dir
Masikc Vergcilles . . . . . prom mgr
Karen Macdonald . . . . . news dir
Michel Paquet . . . . . engrg mgr

Rimouski

CFER-TV— Analog channel: 11.325 kw vis, 32.5 kw aur. 1,420t/289g On air date: June 4, 1978. 465 Boul. Ste.-Anne, Rimouski, PQ, G5M 1G1. Canada. Phone: (418) 722-6011. Fax: (418) 724-7810. Fax: (418) 723-0857. Web Site: www.tva.canoe.com. Licensee: Tele-Metropole Inc. Group owner: Groupe TVA Inc. Natl. Network: TVA.

***CIVB-TV—** Analog channel: 22. Analog hrs: 24 55 kw vis. 300t TL: N48 28 02 W68 12 53 On air date: Oct 15, 1981. 1000 Fullum, Montreal, PQ, H2K 3L7. Phone: (514) 521-2424. Fax: (514) 873-2601. Fax: (514) 873-4413. E-mail: info@telequebec.qc.ca Web Site: www.telequebec.tv. Licensee: Societe de telediffusion du Quebec. Group owner: Tele-Quebec.
Key Personnel:
Michele Fortim . . . . . pres
Luc Chartier . . . . . gen mgr

CJBR-TV— Analog channel: 2.100 kw vis, 10 kw aur. 1,341t/204g On air date: November 1954. 273 St. Jeans Baptiste W., Rimouski, PQ, G5L 4J8. Phone: (418) 723-2217. Phone: (418) 723-4730. Fax: (418) 743-6126. Licensee: CBC. Natl. Network: CBC.

Riviere-du-Loup

CFTF-TV— Analog channel: 29.50 kw vis. ant 1,086t TL: N47 35 03 W69 22 10 On air date: 1988. 103 des Equipements Parc Industriel, Rivieres-du-Loup, PQ, G5R 5W7. 298 Boulevard Armand-Theriault, Bureau 100, Riviere-du-Loup, PQ G5R 4C2. Phone: (418) 862-2909. Fax: (418) 862-8147. E-mail: cftf@qc.aira.com Licensee: Television MBS Inc. Group owner: Tele Inter-Rives Ltee. Population served: 600,000
Key Personnel:
Marc Simard . . . . . pres
Catherine Simard . . . . . gen mgr, dev VP & mktg VP
Michel Belanger . . . . . opns VP & progmg VP
Ginette Dumant . . . . . sls VP & rgnl sls mgr
Yves Belanger . . . . . sls dir
Nancy Fortin . . . . . prom VP
Germain Gelinas . . . . . engrg VP

CIMT-TV— Analog channel: 9. Analog hrs: 24 275.6 kw vis, 2.7 kw aur. ant 1,178t/200g TL: N47 35 03 W69 22 10 On air date: Sept 18, 1978. 15 Rue de la Chute, Riviere-du-Loup, PQ, G5R 5B7. Phone: (418) 867-1341. Fax: (418) 867-4710. Web Site: www.cimt.ca. Licensee: Tele Inter-Rives Ltee. Group owner: (group owner). Natl. Network: TVA.

CKRT-TV— Analog channel: 7. Analog hrs: 24 49 kw vis, 4.9 kw aur. ant 1,156t/200g TL: N47 35 03 W69 22 10 On air date: 1961. 15 Rue de la Chute, Riviere-du-Loup, PQ, G5R 5B7. Phone: (418) 867-1341. Fax: (418) 867-4710. Licensee: CKRT-TV Ltee. Group owner: Tele Inter-Rives Ltee. Natl. Network: CBC.

Rouyn

***CIVA-TV—** Analog channel: 8. Analog hrs: 24 299.2 kw vis, 28.6 kw aur. On air date: Jan 18, 1980. 1000 Fullum, Montreal, PQ, H2K 3L7. Phone: (514) 521-2424. Fax: (514) 873-2601. Fax: (514) 864-4222. E-mail: info@telequebec.qc.ca Web Site: www.telequebec.tv. Licensee: Societe de telediffusion du Quebec. Group owner: Tele-Quebec.
Key Personnel:
Michele Fortim . . . . . pres
Luc Chartier . . . . . gen mgr

Rouyn-Noranda

CKRN-TV— Analog channel: 4. Analog hrs: 24 115 kw vis, 11.5 kw aur. 670g On air date: 1957. 380 Murdoch, Rouyn-Noranda, PQ, J9X 1G5. Phone: (819) 762-0741. Fax: (819) 762-2466. Web Site: www.radionord.com. Licensee: RNC MEDIA Inc. Group owner: (group owner). Natl. Network: Radio Canada.
Key Personnel:
Pierre R. Brosseau . . . . . pres
Andre Houle . . . . . gen mgr
Denis Chenier . . . . . opns dir
Nancy Desches . . . . . sls dir
Robert Ashby . . . . . news dir
Gerald Landry . . . . . chief of engrg

Sept-Iles

CFER-TV-2— Analog channel: 5.100 kw vis, 10 kw aur. 706t/495g On air date: Nov 13, 1981. c/o CFER-TV, 465 Boul. Ste. Anne, Rimouski, PQ, G5M 1G1. Canada. Phone: (418) 722-6011. Fax: (418) 724-7810. Fax: (418) 723-0854 (news). Web Site: www.tva.canoe.com. Licensee: Groupe TVA Inc. Group owner: (group owner). Natl. Network: TVA.

***CIVG-TV—** Analog channel: 9. Analog hrs: 16 246 kw vis, 49.2 kw aur. 943t/500g On air date: Nov 5, 1982. c/o Tele-Quebec, 1000 Fullum, Montreal, PQ, H2K 3L7. Phone: (514) 521-2424. Fax: (514) 873-2601. Fax: (514) 873-7464. Web Site: www.telequebec.tv. Licensee: Societe de telediffusion du Quebec. Group owner: Tele-Quebec.
Key Personnel:
Michele Fortim . . . . . pres
Luc Chartier . . . . . gen mgr

Sherbrooke

CFKS-TV— Analog channel: 30.92.3 kw vis. 2,011t TL: N45 18 43 W72 14 32 On air date: September 1986. 3720 Boul. Industrial, Sherbrooke, PQ, J1L 1Z9. Canada. Phone: (819) 565-9999. Fax: (819) 822-4205. Web Site: www.cogeco.com. Licensee: Cogeco Radio-Television Inc. Group owner: Cogeco Inc. Natl. Network: Quatre Saisons.
Key Personnel:
Sophie Ferron . . . . . gen mgr
Robert PeRusse . . . . . dev dir

CHLT-TV— Analog channel: 7. Analog hrs: 5:30am - 2:00am 325 kw vis, 32.5 kw aur. ant 1,920t/106g On air date: Aug 12, 1956. 3330 Ouest Rue King, Sherbrooke, PQ, J1L 1C9. Phone: (819) 565-7777. Fax: (819) 565-4650. Fax: (819) 563-0141. Licensee: Tele-Metropole Inc. Group owner: Groupe TVA Inc. (acq 7-9-90). Natl. Network: TVA.

***CIVS-TV—** Analog channel: 24. Analog hrs: 24 475 kw vis, 47.5 kw aur. 2,000t/90g On air date: Feb 26, 1982. 1000 Fullum, Montreal, PQ, H2K 3L7. Phone: (514) 521-2424. Fax: (514) 873-2601. Fax: (514) 864-4222. E-mail: info@telequebec.qc.ca Web Site: www.telequebec.tv. Licensee: Societe de telediffusion du Quebec. Group owner: Tele-Quebec.
Key Personnel:
Michele Fortim . . . . . pres
Luc Chartier . . . . . gen mgr

CKSH-TV— Analog channel: 9.325 kw vis, 56 kw aur. On air date: Sept 1, 1974. 3720 Boul. Industrial, Suite 200, Sherbrooke, PQ, J1L 1Z9. Canada. Phone: (819) 565-9999. Fax: (819) 822-4205. Web Site: www.radio-canada.ca. Licensee: Cogeco Radio-Television Inc. Group owner: Cogeco Inc. Natl. Network: CBC.

Temiscaning

CBFST-2— Analog channel: 12.6.9 kw vis, 1.416 kw aur. PO Box 6000, 1400 boul. Rene Levesque E., Montreal, PQ, H3C 3A8. Phone: (514) 597-6000. Fax: (514) 597-4537. Web Site: www.cbc.ca. Licensee: CBC. Natl. Network: Radio Canada.

Trois-Rivieres

CFKM-TV— Analog channel: 16.169.5 kw vis,. ant 1,073t/1,071g TL: N46 29 27 W72 39 00 On air date: Sept 7, 1986. Box 277, Trois-Rivieres, PQ, G9A 5G3. Canada. Phone: (819) 377-6053. Fax: (819) 377-5442. Licensee: Cogeco Radio-Television Inc. Group owner: Cogeco Inc. Natl. Network: Quatre Saisons.
Key Personnel:
Rene Guimond . . . . . pres
Michel Cloutier . . . . . dev dir

CHEM-TV— Analog channel: 8.325 kw vis, 32.5 kw aur. 946t/698g On air date: Aug 29, 1976. 3625 boul. Chanoine-Moreau, Trois-Rivieres, PQ, G8Y 5N6. Phone: (819) 376-8880. Fax: (819) 376-2906. Licensee: Tele-Metropole Inc. Group owner: Groupe TVA Inc. (acq 7-9-90). Natl. Network: TVA.
Key Personnel:
Richard Renault . . . . . gen mgr & stn mgr
Gerald Trives . . . . . chief of engrg

***CIVC-TV—** Analog channel: 45. Analog hrs: 16 651.8 kw vis. ant 1,575t/1,000g TL: N46 29 27 W72 39 00 On air date: Oct 7, 1981. c/o Tele-Quebec, 1000 Fullum, Montreal, PQ, H2K 3L7. Phone: (514) 521-2424. Fax: (514) 873-2601. Fax: (514) 864-4222. E-mail: info@telequebec.ca Web Site: www.telequebec.tv. Licensee: Societe de telediffusion du Quebec. Group owner: Tele-Quebec.
Key Personnel:
Luc Chartier . . . . . pres & gen mgr

Michele Fortim . . . . . pres

CKTM-TV— Analog channel: 13. Analog hrs: 6:00am - 2:00am 164.4 kw vis, 65 kw aur. ant 1,660t/1,085g TL: N46 29 27 W72 39 00 On air date: Apr 15, 1958. 4141 boul.St-Jean, Trois-Rivieres, PQ, G9B 2M8. Canada. Phone: (819) 377-4413. Fax: (819) 377-5239. E-mail: paul_rousseau@radiocanada.ca Licensee: Cogeco Radio-Television Inc. Group owner: Cogeco Inc. Natl. Network: CBC.

Val d'Or

CFVS-TV— Analog channel: 25. Digital channel: 5. Analog hrs: 24 On air date: Jan 19, 1987. 1729 3ieme Ave., Val d'Or, PQ, J9P 1W3. Phone: (819) 825-0010. Fax: (819) 825-7313. Web Site: www.radionord.com. Licensee: RNC MEDIA Inc. Group owner: (group owner)
Key Personnel:
Pierre R. Brosseau . . . . . pres
Andre Houle . . . . . gen mgr & gen sls mgr
Nancy Deschenes . . . . . stn mgr & sls dir
Dennis Chenier . . . . . opns mgr, prom dir & progmg dir
Robert H. Ashey . . . . . news dir
Gerald Landry . . . . . chief of engrg

Saskatchewan

Bellegarde

CBKFT-9— Analog channel: 26.11.4 kw vis, 1.1 kw aur. ant 413t/416g On air date: Mar 15, 1980. Box 540, Regina, SK, S4P 4A1. Phone: (306) 347-9540. Fax: (306) 347-9493. Licensee: Societe Radio Canada. Natl. Network: Radio Canada.
Key Personnel:
David Kyle . . . . . gen mgr & opns dir
Steve Tomchuck . . . . . chief of engrg

Colgate

CKCK-TV-1— Analog channel: 12. Analog hrs: 24 84.8 kw vis, 8.5 kw aur. 532t/581g TL: N49 26 16 W103 47 53 On air date: Dec 15, 1962. c/o CKCK-TV, Box 2000, One Hwy. 1 East, Regina, SK, S4P 3E5. Phone: (306) 569-2000. Fax: (306) 522-0090. E-mail: ckck@ctv.ca Web Site: www.ctv.ca. Licensee: CTV Television Inc. Group owner: (group owner) Natl. Network: CTV.

Cypress Hills

CBCP-TV-2— Analog channel: 2.2.45 kw vis, 240 w aur. ant 395g On air date: Oct 1, 1979. c/o CBKT, 2440 Broad St., Regina, SK, S4P 4A1. Phone: (306) 347-9540. Fax: (306) 347-9748. Licensee: CBC. Natl. Network: CBC.
Key Personnel:
Derek Dalton . . . . . VP
David Kyle . . . . . opns mgr
Rebroadcasts CBKT(TV) Regina 100%.

Debden

CBKFT-3— Analog channel: 22.2.9 kw vis. ant 306t/310g On air date: Dec 2, 1979. Box 540, 2440 Broad St., Regina, SK, S4P 4A1. Phone: (306) 347-9540. Fax: (306) 347-9635. Web Site: radio-canada.ca. Licensee: Societe Radio Canada. Natl. Network: Radio Canada.
Key Personnel:
Lionel Bonneville . . . . . gen mgr
David Kyle . . . . . opns dir
Steve Tomchuck . . . . . chief of engrg
Rebroadcasts CBKFT-1 Saskatoon.

Golden Prairie

CKMC-TV-1— Analog channel: 10. Analog hrs: 24 229 kw vis, 22.9 kw aur. 554t/600g TL: N50 12 20 W109 35 43 On air date: Dec 15, 1988. c/o CKCK-TV, Box 2000, One Hwy. 1 East, Regina, SK, S4P 3E5. Phone: (306) 569-2000. Fax: (306) 522-0090. Web Site: www.ctv.ca. Licensee: CTV Television Inc. Group owner: (group owner) Natl. Network: CTV.
Key Personnel:
Dennis Dunlop . . . . . gen mgr
Les Sampson . . . . . chief of engrg

Gravelbourg

CBKFT-6— Analog channel: 39.19 kw vis, 1.9 kw aur. ant 638t/615g On air date: Mar 13, 1980. Box 540, 2440 Broad St., Regina, SK, S4P 4A1. Phone: (306) 347-9540. Fax: (306) 347-9493. Licensee: Societe Radio Canada. Natl. Network: Radio Canada.
Key Personnel:
Lionel Bonneville . . . gen mgr
David Kyle . . . opns dir
Steve Tomchuck . . . chief of engrg

Melfort

CKBQ-TV— Analog channel: 2.15.5 kw vis, 1.55 kw aur. ant 492t/490g On air date: 1973. Box 540, Reginia, SK, S4P 4A1. Phone: (306) 347-9540. Phone: (306) 956-7400. Fax: (306) 347-9635. Web Site: www.saskcbc.ca. Licensee: CTV Television Inc. Group owner: CTV Inc. Natl. Network: CBC.
Key Personnel:
Lionel Bonneville . . . gen mgr
David Kyle . . . opns dir

North Battleford

CFQC-TV-2— Analog channel: 6. Analog hrs: 24 16.8 kw vis, 1.9 kw aur. 584t/350g On air date: 1972. c/o CFQC-TV, 216 First Ave. N., Saskatoon, SK, S7K 3W3. Phone: (306) 665-8600. Fax: (306) 665-0450. Web Site: www.ctv.ca. Licensee: TV West Inc. Natl. Network: CTV.
Key Personnel:
Dennis Dunlop . . . gen mgr
Denis Gilbertson . . . opns mgr
Barry Berglund . . . gen sls mgr
Chris Ransom . . . mktg dir
Geoff Bradley . . . prom mgr
Bonnie MacKenzie . . . progmg mgr
Dale Liebrecht . . . engrg mgr

Ponteix

CBCP-TV-3— Analog channel: 3.10.5 kw vis. ant 650t/717g On air date: November 1979. Box 540, c/o CBKT, 2440 Broad St., Regina, SK, S4P 4A1. Phone: (306) 347-9540. Fax: (306) 347-9758 (French). Licensee: CBC. Natl. Network: CBC.
Key Personnel:
Lionel Bonneville . . . gen mgr
David Kyle . . . opns dir
Steve Tomchuck . . . chief of engrg
Rebroadcasts CBKT(TV) Regina 100%.

Prince Albert

CIPA-TV— Analog channel: 9. Analog hrs: 24 325 kw vis, 32.5 kw aur. 711t/460g On air date: Jan 12, 1987. 22 10th St. W., Prince Albert, SK, S6V 3A5. Phone: (306) 922-6066. Fax: (306) 763-3041. E-mail: cipa@ctv.ca Web Site: www.ctv.ca. Licensee: CTV Television Inc. Group owner: CTV Inc. (acq 8-1-86). Natl. Network: CTV. News staff: 5.

Regina

CBKFT— Analog channel: 13. Analog hrs: 18 103 kw vis, 31.3 kw aur. On air date: Sept 27, 1976. Box 540, 2440 Broad St., Regina, SK, S4P 4A1. Phone: (306) 347-9540. Fax: (306) 347-9635. Web Site: www.radio-canada.ca/regions/saskatchewan/index.shtml. Licensee: Societe Radio Canada. Natl. Network: Radio Canada.

CBKT— Analog channel: 9.140 kw vis, 20 kw aur. ant 680t/698g On air date: 1969. Box 540, Regina, SK, S4P 4A1. Phone: (306) 347-9540. Fax: (306) 347-9616. Web Site: www.sask.cbc.ca. Licensee: CBC. Natl. Network: CBC.
Key Personnel:
David Kyle . . . opns mgr
Carley Caverly . . . gen sls mgr
Michael Barronwright . . . gen sls mgr
Bob Rankin . . . news dir
Jon Simons . . . mktg

CFRE-TV— Analog channel: 11. Analog hrs: 24 146 kw vis. 984t On air date: Sept 6, 1987. 370 Hoffer Dr., Regina, SK, S4N 7A4. Phone: (306) 775-4000. Fax: (306) 721-4817. Web Site: www.canada.com. Licensee: CanWest MediaWorks Inc. Group owner: CanWest Global Communications Corp. Population served: 320,000
Key Personnel:
Mitch Bozak . . . gen mgr, stn mgr & gen sls mgr
Lyndon Gray . . . prom mgr
Doug Hoover . . . film buyer
Brent Williamson . . . news dir
Paul Godfrey . . . opns mgr & engrg mgr
Len Virog . . . chief of engrg

CKCK-TV— Analog channel: 2. Analog hrs: 24 100 kw vis, 10 kw aur. 588t/670g TL: N50 26 52 W104 30 00 On air date: July 28, 1954. Box 2000, Regina, SK, S4P 3E5. One Hwy. 1 East, Regina, SK S4P 3E5. Phone: (306) 569-2000. Fax: (306) 522-0991. E-mail: ckcknews@ctv.ca Web Site: www.ctv.ca. Licensee: CTV Television Inc. Group owner: (group owner) Natl. Network: CTV. News: 15.5 hrs local news wkly.

Saint Brieux

CBKFT-4— Analog channel: 7. Analog hrs: 18 140 w vis, 14 w aur. ant 200t/185g On air date: Dec 17, 1979. Box 540, 2440 Broad St., Regina, SK, S4P 4A1. Phone: (306) 347-9540. Fax: (306) 347-9635. Web Site: radio-canada.ca/regions/saskatchewan/index.shtml. Licensee: Societe Radio Canada. Natl. Network: Radio Canada.

Saskatoon

CBKST— Analog channel: 11. Analog hrs: 20 325 kw vis, 32 kw aur. ant 559t/595g On air date: Oct 17, 1971. 144 Second Ave. South, Saskatoon, SK, S7K 1K5. Phone: (306) 956-7400. Fax: (306) 347-9650 (admin). Web Site: www.cbc.ca/sask. Licensee: CBC. Natl. Network: CBC.

CFQC-TV— Analog channel: 8.325 kw vis, 180 kw aur. 891t/650g TL: N52 07 51 W106 39 49 On air date: 1954. 216 First Ave. N., Saskatoon, SK, S7K 3W3. Phone: (306) 665-8600. Fax: (306) 665-0450. E-mail: cfqcnews@ctv.ca Web Site: www.ctv.ca. Licensee: CFQC Broadcasting Ltd. Group owner: CTV Inc. (acq 1972). Population served: 333,400 Natl. Network: CTV.
Key Personnel:
Dennis Dunlop . . . gen mgr
Denis Gibertson . . . opns VP & opns mgr
Barry Berglund . . . gen sls mgr
Chris Ransom . . . mktg dir & mktg mgr
Geoff Bradley . . . prom mgr
Bonnie Mackenzie . . . progmg mgr
Dale Neufeld . . . news dir
Les Sampson . . . chief of engrg

CFSK-TV— Analog channel: 4. Analog hrs: 24 54 kw vis, 5.4 kw aur. 455t/219g On air date: Sept 6, 1987. 218 Robin Crescent, Saskatoon, SK, S7L 7C3. Phone: (306) 665-6969. Fax: (306) 665-6069. Licensee: CanWest MediaWorks Inc. Population served: 290,000 Natl. Network: Global. News staff: 17; News: 10 hrs wkly.

Shaunavon

CBCP-TV-1— Analog channel: 7. Analog hrs: 24 4.5 kw vis. c/o CBKT, Box 540, Regina, SK, S4P 4A1. Phone: (306) 347-9540. Fax: (306) 347-9616. Web Site: www.sask.cbc.ca. Licensee: CBC. Natl. Network: CBC.
Rebroadcasts CBKT(TV) Regina 100%.

Stranraer

CBKST-1— Analog channel: 9. Analog hrs: 20 323 kw vis, 32.3 kw aur. 144 Second Ave. South, Saskatoon, SK, S7K 1K5. Phone: (306) 956-7400. Fax: (306) 347-9616 (admin). Fax: (306) 956-9417 (news). Web Site: www.cbc.ca. Licensee: CBC. Natl. Network: CBC.
Rebroadcasts CBKST Saskatoon 100%.

Swift Current

CKMC-TV— Analog channel: 12. Analog hrs: 24 100 kw vis, 10 kw aur. ant 549t/390g TL: N50 18 31 W107 52 35 On air date: Oct 20, 1976. c/o CKCK-TV, Box 2000, Regina, SK, S4P 3E5. c/o CKCK-TV, One Hwy. 1 East, Regina, SK S4P 3E5. Phone: (306) 569-2000. Fax: (306) 522-0090. E-mail: ckck@ctv.ca Web Site: www.ctv.ca. Licensee: CTV Television Inc. Group owner: (group owner) Natl. Network: CTV. Rebroadcasts CKCK-TV Regina 100%.

Warmley

CIEW-TV— Analog channel: 7.100 kw vis, 10 kw aur. c/o CICC-TV, 95 E. Broadway St., Yorkton, SK, S3N 0L1. Phone: (306) 783-3685. Fax: (306) 782-7212. Web Site: www.ctv.ca. Licensee: CTV Television Inc. Group owner: Baton Broadcasting Inc. Natl. Network: CTV.
Rebroadcasts CICC-TV Yorkton 100%.

Willow Bunch

CBKT-2— Analog channel: 10. Analog hrs: 24 22.1 kw vis, 2.2 kw aur. CBC TV, Box 540, 2440 Broad St., Regina, SK, S4P 4A1. Phone: (306) 347-9666. Fax: (306) 347-9635. Web Site: www.cbc.ca/sask/. Licensee: CBC. Natl. Network: CBC.
Key Personnel:
David Kyle . . . gen mgr
Nigel Sims . . . progmg dir
Bob Rankin . . . news dir
Jon Anderson . . . chief of engrg

CKCK-TV-2— Analog channel: 6.17.5 kw vis, 8.75 kw aur. 864t/677g (CP: 27.1 kw vis) On air date: May 29, 1963. c/o CKCK-TV, Box 2000, One Hwy. 1 East, Regina, SK, S4P 3E5. Phone: (306) 569-2000. Fax: (306) 522-0090. E-mail: ckck@ctv.ca Web Site: www.ctv.ca. Licensee: CTV Television Inc. Group owner: (group owner) Natl. Network: CTV.

Xenon Park

CBKFT-5— Analog channel: 21.3 kw vis, 300 w aur. ant 456t/310g On air date: Feb 19, 1979. Box 540, c/o CBKFT, 2440 Broad St., Regina, SK, S4P 4A1. Phone: (306) 347-9540. Fax: (306) 347-9493. Fax: (306) 347-9635. Licensee: CBC. Natl. Network: CBC.
Key Personnel:
David Kyle . . . gen mgr & progmg dir
Debbie Carpentier . . . opns mgr
Carly Caverly . . . gen sls mgr
Jonathan Shanks . . . progmg dir
Steve Tomchuk . . . chief of engrg

Yorkton

CICC-TV— Analog channel: 10.56 kw vis, 18 kw aur. On air date: 1974. 95 E. Broadway, Yorkton, SK, S3N 0L1. Phone: (306) 786-8400. Fax: (306) 782-7212. E-mail: ctvyorktonnews@ctv.ca Web Site: www.ctv.ca. Licensee: CTV Television Inc. Group owner: (group owner) Natl. Network: CTV.
Key Personnel:
Dennis Dunlop . . . gen mgr
Wade Moffatt . . . gen sls mgr
Bob Maloney . . . progmg dir, news dir & news dir
Peter Whitehead . . . chief of engrg

Yukon Territory

White Horse

CFWH-TV— Analog channel: 6. Analog hrs: 24 441 w. vis, 30 w aur. 1,248t/100g 3103 3rd Ave., White Horse, YT, Y1A 1E5. Phone: (867) 668-8400. Fax: (867) 668-8408. Web Site: www.cbc.ca/north. Licensee: CBC. Natl. Network: CBC.
Key Personnel:
Frank Fry . . . opns mgr & progmg dir
John Boivin . . . gen mgr & opns mgr
James Miller . . . news dir

U.S. Television Stations by Call Letters

***KAAH-TV** Honolulu, HI
KAAL Rochester, MN-Mason City, IA-Austin, MN
KAAS-TV Wichita-Hutchinson Plus, KS
KABB San Antonio, TX
KABC-TV Los Angeles
KABY-TV Sioux Falls (Mitchell), SD
***KACV-TV** Amarillo, TX
KADN Lafayette, LA
KAEF Eureka, CA
***KAET** Phoenix (Prescott), AZ
***KAFT** Ft. Smith-Fayetteville -Springdale-Rogers, AR
***KAID-TV** Boise, ID
KAII-TV Wailuku HI
KAIL Fresno-Visalia, CA
KAIT Jonesboro, AR
KAJB Yuma, AZ-El Centro, CA
KAKE-TV Wichita-Hutchinson Plus, KS
***KAKM** Anchorage, AK
KAKW-TV Waco-Temple-Bryan, TX
KALB-TV Alexandria, LA
***KALO** Honolulu, HI
KAMC Lubbock, TX
KAME-TV Reno, NV
KAMR-TV Amarillo, TX
***KAMU-TV** Waco-Temple-Bryan, TX
KAPP Yakima-Pasco -Richland-Kennewick, WA
KAQY Monroe, LA-El Dorado, AR
KARD Monroe, LA-El Dorado, AR
KARE Minneapolis-St. Paul, MN
KARK-TV Little Rock-Pine Bluff, AR
KASA-TV Albuquerque-Santa Fe, NM
KASN Little Rock-Pine Bluff, AR
KASW Phoenix (Prescott), AZ
KASY-TV Albuquerque-Santa Fe, NM
KATC Lafayette, LA
KATN Fairbanks, AK
KATU Portland, OR
KATV Little Rock-Pine Bluff, AR
KAUT-TV Oklahoma City, OK
KAUZ-TV Wichita Falls, TX & Lawton, OK
KAVU-TV Victoria, TX
***KAWB** Minneapolis-St. Paul, MN
***KAWE** Minneapolis-St. Paul, MN
KAYU-TV Spokane, WA
KAZA-TV Los Angeles
KAZH Houston
***KAZQ** Albuquerque-Santa Fe, NM
KAZT-TV Phoenix (Prescott), AZ
KBAK-TV Bakersfield, CA
KBAO Great Falls, MT
KBBC-TV Los Angeles
KBBJ Great Falls, MT
KBCA Alexandria, LA
KBCB Seattle-Tacoma, WA
KBCI-TV Boise, ID
KBCJ Salt Lake City, UT
KBCW San Francisco-Oakland-San Jose
***KBDI-TV** Denver, CO
***KBDM** Medford-Klamath Falls, OR
KBEH Santa Barbara-Santa Maria-San Luis Obispo, CA
KBEO Salt Lake City, UT
KBFD Honolulu, HI
***KBGH** Twin Falls, ID
***KBHE-TV** Rapid City, SD
KBIM-TV Albuquerque-Santa Fe, NM
***KBIN** Omaha, NE
KBJR-TV Duluth, MN-Superior, WI
KBLN Medford-Klamath Falls, OR
KBLR Las Vegas, NV
***KBME-TV** Minot-Bismarck-Dickinson, ND
KBMT Beaumont-Port Arthur, TX
KBMY Minot-Bismarck-Dickinson, ND
KBNY Salt Lake City, UT
KBRR Fargo-Valley City, ND
KBSD-TV Wichita-Hutchinson Plus, KS
KBSH-TV Wichita-Hutchinson Plus, KS
KBSI Paducah, KY-Cape Girardeau, MO-Harrisburg-Mount Vernon, IL
KBSL-TV Wichita-Hutchinson Plus, KS
***KBSV** Sacramento-Stockton-Modesto, CA
***KBTC-TV** Seattle-Tacoma, WA
KBTV-TV Beaumont-Port Arthur, TX
KBTX-TV Waco-Temple-Bryan, TX
KBTZ Butte-Bozeman, MT
KBVU Eureka, CA
KBWB San Francisco-Oakland-San Jose
***KBYU-TV** Salt Lake City, UT
KBZK Butte-Bozeman, MT
KCAL Los Angeles
KCAU-TV Sioux City, IA
KCBA Monterey-Salinas, CA
KCBD Lubbock, TX
KCBS-TV Los Angeles
KCBU Salt Lake City, UT
KCBY-TV Eugene, OR
KCCI Des Moines-Ames, IA
KCCO-TV Minneapolis-St. Paul, MN
KCCW-TV Minneapolis-St. Paul, MN
***KCDT-TV** Spokane, WA
KCEB Tyler-Longview (Lufkin & Nacogdoches), TX
KCEC Denver, CO
KCEN-TV Waco-Temple-Bryan, TX
***KCET** Los Angeles
KCFG Phoenix (Prescott), AZ
KCFW-TV Missoula, MT
***KCGE-DT** Fargo-Valley City, ND
KCHF Albuquerque-Santa Fe, NM
KCIT Amarillo, TX
***KCKA** Seattle-Tacoma, WA
KCLO-TV Rapid City, SD
KCNC-TV Denver, CO
KCNS San Francisco-Oakland-San Jose
KCOP Los Angeles
***KCOS** El Paso (Las Cruces, NM), TX
KCOY-TV Santa Barbara-Santa Maria-San Luis Obispo, CA
KCPM Fargo-Valley City, ND
KCPQ Seattle-Tacoma, WA
***KCPT** Kansas City, MO
KCRA-TV Sacramento -Stockton-Modesto, CA
KCRG-TV Cedar Rapids-Waterloo-Iowa City & Dubuque, IA
***KCSD-TV** Sioux Falls (Mitchell), SD
KCSG Salt Lake City, UT
***KCSM-TV** San Francisco-Oakland-San Jose
***KCTS-TV** Seattle-Tacoma, WA
KCTV Kansas City, MO
KCVU Chico-Redding, CA
***KCWC-TV** Casper-Riverton, WY
KCWE Kansas City, MO
KCWI-TV Des Moines-Ames, IA
KCWK Yakima-Pasco -Richland-Kennewick, WA
KCWL-TV Lincoln & Hastings-Kearney, NE
KCWV Duluth, MN-Superior, WI
KCWX San Antonio, TX
KCWY Casper-Riverton, WY
KDAF Dallas-Ft. Worth
KDBC-TV El Paso (Las Cruces, NM), TX
***KDCK** Wichita-Hutchinson Plus, KS
KDEN Denver, CO
KDEV Cheyenne, WY-Scottsbluff, NE
KDFI Dallas-Ft. Worth
KDFW Dallas-Ft. Worth
***KDIN-TV** Des Moines-Ames, IA
KDKA-TV Pittsburgh, PA
KDKF Medford-Klamath Falls, OR
KDLH Duluth, MN-Superior, WI
KDLO-TV Sioux Falls (Mitchell), SD
KDLT-TV Sioux Falls (Mitchell), SD
KDLV-TV Sioux Falls (Mitchell), SD
KDMD Anchorage, AK
KDMI Des Moines-Ames, IA
KDNL-TV St. Louis, MO
KDOC-TV Los Angeles
KDOR-TV Tulsa, OK
KDRV Medford-Klamath Falls, OR
***KDSD-TV** Sioux Falls (Mitchell), SD
***KDSE** Minot-Bismarck-Dickinson, ND
KDSM-TV Des Moines-Ames, IA
***KDTN** Dallas-Ft. Worth
***KDTP** Phoenix (Prescott), AZ
KDTV San Francisco-Oakland-San Jose
***KDTX-TV** Dallas-Ft. Worth
KDUH-TV Cheyenne, WY-Scottsbluff, NE
KDVR Denver, CO
KECI-TV Missoula, MT
KECY-TV Yuma, AZ-El Centro, CA
***KEDT** Corpus Christi, TX
***KEET** Eureka, CA
KEFB Des Moines-Ames, IA
KEGS Reno, NV
KEJB Monroe, LA-El Dorado, AR
KELO-TV Sioux Falls (Mitchell), SD
***KEMV** Little Rock-Pine Bluff, AR
KENS-TV San Antonio, TX
KENV Salt Lake City, UT
***KENW** Amarillo, TX
***KEPB-TV** Eugene, OR
KEPR-TV Yakima-Pasco -Richland-Kennewick, WA
***KERA-TV** Dallas-Ft. Worth
KERO-TV Bakersfield, CA
***KESD-TV** Sioux Falls (Mitchell), SD
KESQ-TV Palm Springs, CA
***KETA** Oklahoma City, OK
***KETC** St. Louis, MO
***KETG** Little Rock-Pine Bluff, AR
***KETH-TV** Houston
KETK-TV Tyler-Longview (Lufkin & Nacogdoches), TX
***KETS** Little Rock-Pine Bluff, AR
KETV Omaha, NE
***KETZ** Monroe, LA-El Dorado, AR
KEVN-TV Rapid City, SD
KEYC-TV Mankato, MN
KEYE-TV Austin, TX
KEYT-TV Santa Barbara-Santa Maria-San Luis Obispo, CA
KEYU Oklahoma City, OK
KEZI Eugene, OR
KFBB-TV Great Falls, MT
KFCT Denver, CO
KFDA-TV Amarillo, TX
KFDM-TV Beaumont-Port Arthur, TX
KFDX-TV Wichita Falls, TX & Lawton, OK
KFFX-TV Yakima-Pasco -Richland-Kennewick, WA
KFJX-TV Joplin, MO-Pittsburg, KS
KFMB-TV San Diego, CA
***KFME** Fargo-Valley City, ND
KFNB-TV Casper-Riverton, WY
KFNE Casper-Riverton, WY
KFNR Denver, CO
KFOR-TV Oklahoma City, OK
KFOX-TV El Paso (Las Cruces, NM), TX
KFPH-TV Phoenix (Prescott), AZ
KFPX Des Moines-Ames, IA
KFQX Grand Junction-Montrose, CO
KFRE-TV Fresno-Visalia, CA
KFSF-TV San Francisco-Oakland-San Jose
KFSM-TV Ft. Smith-Fayetteville -Springdale-Rogers, AR
KFSN-TV Fresno-Visalia, CA
KFTA-TV Ft. Smith-Fayetteville -Springdale-Rogers, AR
KFTC Minneapolis-St. Paul, MN
KFTH-TV Houston
KFTR-TV Los Angeles
***KFTS** Medford-Klamath Falls, OR
KFTU-TV Tucson (Sierra Vista), AZ
KFTV Fresno-Visalia, CA
KFTY San Francisco-Oakland-San Jose
KFVE Honolulu, HI
KFVS-TV Paducah, KY-Cape Girardeau, MO-Harrisburg-Mount Vernon, IL
KFWD Dallas-Ft. Worth
KFXA Cedar Rapids-Waterloo-Iowa City & Dubuque, IA
KFXB Cedar Rapids-Waterloo-Iowa City & Dubuque, IA
KFXF Fairbanks, AK
KFXK Tyler-Longview (Lufkin & Nacogdoches), TX
KFXP Idaho Falls-Pocatello, ID
KFYR-TV Minot-Bismarck-Dickinson, ND
KGAN-TV Cedar Rapids-Waterloo-Iowa City & Dubuque, IA
KGBT-TV Harlingen-Weslaco -Brownsville-McAllen, TX
KGCW-TV Davenport, IA-Rock Island-Moline, IL
KGEB Tulsa, OK
KGET-TV Bakersfield, CA
***KGFE** Fargo-Valley City, ND
KGIN Lincoln & Hastings-Kearney, NE
KGMB Honolulu, HI
KGMC Fresno-Visalia, CA
KGMD-TV Hilo HI
KGMV Wailuku HI
KGNS-TV Laredo, TX
KGO-TV San Francisco-Oakland-San Jose
KGPE Fresno-Visalia, CA
KGPX Spokane, WA
***KGTF-TV** Hagatna GU
KGTV San Diego, CA
KGUN Tucson (Sierra Vista), AZ
KGW Portland, OR
KGWC-TV Casper-Riverton, WY
KGWL-TV Casper-Riverton, WY
KGWN-TV Cheyenne, WY-Scottsbluff, NE
KGWR-TV Salt Lake City, UT
KHAS-TV Lincoln & Hastings-Kearney, NE
KHAW-TV Hilo HI
KHBC-TV Hilo HI
KHBS Ft. Smith-Fayetteville -Springdale-Rogers, AR
***KHCE-TV** San Antonio, TX
KHCV Seattle-Tacoma, WA
KHCW Houston
***KHET** Honolulu, HI
KHGI-TV Lincoln & Hastings-Kearney, NE
***KHIN** Omaha, NE
KHIZ Los Angeles
KHMT Billings, MT
***KHNE-TV** Lincoln & Hastings-Kearney, NE
KHNL Honolulu, HI
KHOG-TV Ft. Smith-Fayetteville -Springdale-Rogers, AR
KHON-TV Honolulu, HI
KHOU-TV Houston
KHQA-TV Quincy, IL-Hannibal, MO-Keokuk, IA
KHQ-TV Spokane, WA
KHRR Tucson (Sierra Vista), AZ
KHSD-TV Rapid City, SD
KHSL-TV Chico-Redding, CA
KHVO Hilo HI
KICU-TV San Francisco-Oakland-San Jose
KIDA Twin Falls, ID
KIDK-TV Idaho Falls-Pocatello, ID
KIDY San Angelo, TX
KIEM-TV Eureka, CA
KIFI-TV Idaho Falls-Pocatello, ID
KIII Corpus Christi, TX
***KIIN-TV** Cedar Rapids-Waterloo-Iowa City & Dubuque, IA
KIKU-TV Honolulu, HI
KIMA-TV Yakima-Pasco -Richland-Kennewick, WA
KIMO Anchorage, AK
KIMT Rochester, MN-Mason City, IA-Austin, MN
KINC Las Vegas, NV
KING-TV Seattle-Tacoma, WA
KINT-TV El Paso (Las Cruces, NM), TX
KION-TV Monterey-Salinas, CA
***KIPT-TV** Twin Falls, ID
KIRO-TV Seattle-Tacoma, WA
***KISU-TV** Idaho Falls-Pocatello, ID
***KITU-TV** Beaumont-Port Arthur, TX
KITV Honolulu, HI
KIVI Boise, ID
KIVV-TV Rapid City, SD
***KIXE-TV** Chico-Redding, CA
KJCT Grand Junction-Montrose, CO
KJLA Los Angeles
KJNP-TV Fairbanks, AK
***KJRE** Fargo-Valley City, ND
KJRH Tulsa, OK
KJRR Fargo-Valley City, ND
KJTL Wichita Falls, TX & Lawton, OK
KJTV-TV Lubbock, TX
KJUD Juneau, AK
KJWY Idaho Falls-Pocatello, ID
KJZZ-TV Salt Lake City, UT
KKAI Honolulu, HI
***KKAP** Little Rock-Pine Bluff, AR
KKCO Grand Junction-Montrose, CO
KKJB Boise, ID
KKPX San Francisco-Oakland-San Jose
KKTV Colorado Springs-Pueblo, CO
KLAS-TV Las Vegas, NV
KLAX-TV Alexandria, LA
KLBK-TV Lubbock, TX
KLBY Wichita-Hutchinson Plus, KS
***KLCS** Los Angeles
KLCW-TV Lubbock, TX
KLDO-TV Laredo, TX
KLDT Dallas-Ft. Worth
KLEI Honolulu, HI
***KLEP** Little Rock-Pine Bluff, AR
KLEW-TV Spokane, WA
KLFY-TV Lafayette, LA
KLJB-TV Davenport, IA-Rock Island-Moline, IL
KLKN Lincoln & Hastings-Kearney, NE
KLMN Great Falls, MT
***KLNE-TV** Lincoln & Hastings-Kearney, NE
***KLPA-TV** Alexandria, LA
***KLPB-TV** Lafayette, LA
***KLRN** San Antonio, TX
KLRT Little Rock-Pine Bluff, AR
***KLRU-TV** Austin, TX
KLSR-TV Eugene, OR
KLST San Angelo, TX
***KLTJ** Houston
***KLTL-TV** Lake Charles, LA
***KLTM-TV** Monroe, LA-El Dorado, AR
***KLTS-TV** Shreveport, LA
KLTV Tyler-Longview (Lufkin & Nacogdoches), TX
***KLUJ-TV** Harlingen-Weslaco -Brownsville-McAllen, TX
KLUZ-TV Albuquerque-Santa Fe, NM

*KLVX Las Vegas, NV
KLWB Lafayette, LA
KLWY Cheyenne, WY-Scottsbluff, NE
*KMAS-TV Denver, CO
KMAU Wailuku HI
KMAX-TV Sacramento -Stockton-Modesto, CA
KMBC-TV Kansas City, MO
*KMBH Harlingen-Weslaco -Brownsville-McAllen, TX
KMCB Eugene, OR
KMCC Las Vegas, NV
KMCI Kansas City, MO
KMCT-TV Monroe, LA-El Dorado, AR
KMCY Minot-Bismarck-Dickinson, ND
*KMDE Fargo-Valley City, ND
*KMEB Wailuku HI
KMEG Sioux City, IA
KMEX-TV Los Angeles
KMGH-TV Denver, CO
KMID Odessa-Midland, TX
KMIR-TV Palm Springs, CA
KMIZ Columbia-Jefferson City, MO
KMLM Odessa-Midland, TX
KMMF Missoula, MT
*KMNE-TV Lincoln & Hastings-Kearney, NE
KMOH-TV Phoenix (Prescott), AZ
*KMOS-TV Kansas City, MO
KMOT Minot-Bismarck-Dickinson, ND
KMOV St. Louis, MO
KMPH-TV Fresno-Visalia, CA
KMPX Dallas-Ft. Worth
KMSB-TV Tucson (Sierra Vista), AZ
KMSP-TV Minneapolis-St. Paul, MN
KMSS-TV Shreveport, LA
KMTF Helena, MT
*KMTP-TV San Francisco-Oakland-San Jose
KMTR Eugene, OR
KMTV Omaha, NE
KMTW Wichita-Hutchinson Plus, KS
KMVT Twin Falls, ID
KMVU Medford-Klamath Falls, OR
KMYQ Seattle-Tacoma, WA
KMYS San Antonio, TX
KMYT-TV Tulsa, OK
KNAT Albuquerque-Santa Fe, NM
KNAZ-TV Phoenix (Prescott), AZ
KNBC Los Angeles
KNBN Rapid City, SD
*KNCT Waco-Temple-Bryan, TX
KNDO Yakima-Pasco -Richland-Kennewick, WA
KNDU Yakima-Pasco -Richland-Kennewick, WA
KNDX Minot-Bismarck-Dickinson, ND
KNIC-TV Austin, TX
KNIN-TV Boise, ID
KNLC St. Louis, MO
KNLJ Columbia-Jefferson City, MO
*KNMD-TV Albuquerque-Santa Fe, NM
*KNME-TV Albuquerque-Santa Fe, NM
*KNMT-TV Portland, OR
KNOE-TV Monroe, LA-El Dorado, AR
KNOP-TV North Platte, NE
*KNPB Reno, NV
KNRR Fargo-Valley City, ND
KNSD San Diego, CA
KNSO Fresno-Visalia, CA
KNTV San Francisco-Oakland-San Jose
KNVA Austin, TX
KNVN Chico-Redding, CA
KNVO Harlingen-Weslaco -Brownsville-McAllen, TX
KNWA-TV Ft. Smith-Fayetteville -Springdale-Rogers, AR
KNWS-TV Houston
*KNXT Fresno-Visalia, CA
KNXV-TV Phoenix (Prescott), AZ
KOAA-TV Colorado Springs-Pueblo, CO
*KOAB-TV Bend, OR
*KOAC-TV Eugene, OR
KOAM-TV Joplin, MO-Pittsburg, KS
KOAT-TV Albuquerque-Santa Fe, NM

KOBF Albuquerque-Santa Fe, NM
KOBG-TV Albuquerque-Santa Fe, NM
KOBI Medford-Klamath Falls, OR
KOBR Albuquerque-Santa Fe, NM
KOB-TV Albuquerque-Santa Fe, NM
KOCB Oklahoma City, OK
*KOCE-TV Los Angeles
KOCM Oklahoma City, OK
KOCO-TV Oklahoma City, OK
KOCT Albuquerque-Santa Fe, NM
KOCW Wichita-Hutchinson Plus, KS
KODE-TV Joplin, MO-Pittsburg, KS
*KOED-TV Tulsa, OK
*KOET Tulsa, OK
KOFT Albuquerque-Santa Fe, NM
KOGG Wailuku HI
KOHD Bend, OR
KOIN Portland, OR
KOKH-TV Oklahoma City, OK
KOKI-TV Tulsa, OK
KOLD-TV Tucson (Sierra Vista), AZ
KOLN Lincoln & Hastings-Kearney, NE
KOLO-TV Reno, NV
KOLR-TV Springfield, MO
KOMO-TV Seattle-Tacoma, WA
KOMU-TV Columbia-Jefferson City, MO
KONG-TV Seattle-Tacoma, WA
*KOOD Wichita-Hutchinson Plus, KS
*KOPB-TV Portland, OR
KOPX Oklahoma City, OK
KORO Corpus Christi, TX
KOSA-TV Odessa-Midland, TX
KOTA-TV Rapid City, SD
KOTI Medford-Klamath Falls, OR
KOTV Tulsa, OK
KOVR Sacramento-Stockton-Modesto, CA
KOVT Albuquerque-Santa Fe, NM
*KOZJ Joplin, MO-Pittsburg, KS
*KOZK Springfield, MO
KPAX-TV Missoula, MT
*KPAZ-TV Phoenix (Prescott), AZ
KPBI Springfield, MO
*KPBS San Diego, CA
*KPBT-TV Odessa-Midland, TX
KPCB Abilene-Sweetwater, TX
KPDX Portland, OR
KPEJ Odessa-Midland, TX
KPHO-TV Phoenix (Prescott), AZ
KPIC Eugene, OR
KPIF Idaho Falls-Pocatello, ID
KPIX-TV San Francisco-Oakland-San Jose
KPLC Lake Charles, LA
KPLO-TV Sioux Falls (Mitchell), SD
KPLR-TV St. Louis, MO
KPMR Santa Barbara-Santa Maria-San Luis Obispo, CA
*KPNE-TV North Platte, NE
KPNX Phoenix (Prescott), AZ
KPNZ Salt Lake City, UT
KPOB-TV Paducah, KY-Cape Girardeau, MO-Harrisburg-Mount Vernon, IL
KPPX Phoenix (Prescott), AZ
KPRC-TV Houston
KPRY-TV Sioux Falls (Mitchell), SD
*KPSD-TV Minot-Bismarck-Dickinson, ND
KPTB Lubbock, TX
KPTF Amarillo, TX
KPTH Sioux City, IA
KPTM Omaha, NE
*KPTS-TV Wichita-Hutchinson Plus, KS
KPTV Portland, OR
*KPTW Casper-Riverton, WY
KPVI Idaho Falls-Pocatello, ID
KPXB Houston
KPXC-TV Denver, CO
KPXD Dallas-Ft. Worth
KPXE Kansas City, MO
KPXG Portland, OR
KPXJ Shreveport, LA
KPXL San Antonio, TX
KPXM Minneapolis-St. Paul, MN

KPXN Los Angeles
KPXO-TV Honolulu, HI
KPXR Cedar Rapids-Waterloo-Iowa City & Dubuque, IA
KQCA Sacramento-Stockton-Modesto, CA
KQCD-TV Minot-Bismarck-Dickinson, ND
*KQCT Davenport, IA-Rock Island-Moline, IL
KQCW Tulsa, OK
KQDS-TV Duluth, MN-Superior, WI
*KQED San Francisco-Oakland-San Jose
*KQET Monterey-Salinas, CA
*KQSD-TV Minot-Bismarck-Dickinson, ND
KQTV St. Joseph, MO
KQUP Spokane, WA
KRBC-TV Abilene-Sweetwater, TX
KRBK Springfield, MO
KRCA Los Angeles
*KRCB San Francisco-Oakland-San Jose
KRCG Columbia-Jefferson City, MO
KRCR-TV Chico-Redding, CA
KRCW-TV Portland, OR
KRDO-TV Colorado Springs-Pueblo, CO
KREG-TV Denver, CO
KREM-TV Spokane, WA
KREN-TV Reno, NV
KREX-TV Grand Junction-Montrose, CO
KREY-TV Grand Junction-Montrose, CO
KREZ-TV Albuquerque-Santa Fe, NM
KRGV-TV Harlingen-Weslaco -Brownsville-McAllen, TX
KRII Duluth, MN-Superior, WI
*KRIN Cedar Rapids-Waterloo-Iowa City & Dubuque, IA
KRIS-TV Corpus Christi, TX
KRIV Houston
*KRMA-TV Denver, CO
*KRMJ Grand Junction-Montrose, CO
*KRMT Denver, CO
*KRMU Albuquerque-Santa Fe, NM
*KRNE-TV Sioux Falls (Mitchell), SD
KRNV Reno, NV
KRON-TV San Francisco-Oakland-San Jose
KRPV Albuquerque-Santa Fe, NM
KRQE Albuquerque-Santa Fe, NM
*KRSC-TV Tulsa, OK
KRTV Great Falls, MT
KRWB-TV Albuquerque-Santa Fe, NM
KRWF Minneapolis-St. Paul, MN
*KRWG-TV El Paso (Las Cruces, NM), TX
KRXI-TV Reno, NV
KSAN-TV San Angelo, TX
KSAS-TV Wichita-Hutchinson Plus, KS
KSAT-TV San Antonio, TX
KSAX Minneapolis-St. Paul, MN
KSAZ-TV Phoenix (Prescott), AZ
KSBI Oklahoma City, OK
KSBW Monterey-Salinas, CA
KSBY Santa Barbara-Santa Maria-San Luis Obispo, CA
*KSCE El Paso (Las Cruces, NM), TX
KSCI Los Angeles
KSCW Wichita-Hutchinson Plus, KS
KSDK St. Louis, MO
KSEE Fresno-Visalia, CA
KSFX-TV Springfield, MO
KSFY-TV Sioux Falls (Mitchell), SD
KSGW-TV Rapid City, SD
KSHB-TV Kansas City, MO
KSHV Shreveport, LA
*KSIN Sioux City, IA
KSKN Spokane, WA
KSLA-TV Shreveport, LA
KSL-TV Salt Lake City, UT
*KSMN Sioux Falls (Mitchell), SD
KSMO-TV Kansas City, MO

*KSMQ-TV Rochester, MN-Mason City, IA-Austin, MN
KSMS-TV Monterey-Salinas, CA
KSNB-TV Lincoln & Hastings-Kearney, NE
KSNC Wichita-Hutchinson Plus, KS
KSNF Joplin, MO-Pittsburg, KS
KSNG Wichita-Hutchinson Plus, KS
KSNK Wichita-Hutchinson Plus, KS
KSNT-TV Topeka, KS
KSNW Wichita-Hutchinson Plus, KS
KSPR Springfield, MO
*KSPS-TV Spokane, WA
KSPX Sacramento-Stockton-Modesto, CA
*KSRE Minot-Bismarck-Dickinson, ND
KSTC-TV Minneapolis-St. Paul, MN
KSTF Cheyenne, WY-Scottsbluff, NE
KSTP-TV Minneapolis-St. Paul, MN
KSTR-TV Dallas-Ft. Worth
KSTS San Francisco-Oakland-San Jose
KSTU Salt Lake City, UT
KSTW Seattle-Tacoma, WA
KSVI Billings, MT
KSWB-TV San Diego, CA
*KSWK Wichita-Hutchinson Plus, KS
KSWO-TV Wichita Falls, TX & Lawton, OK
KSWT Yuma, AZ-El Centro, CA
KSWY Rapid City, SD
*KSYS Medford-Klamath Falls, OR
KTAB-TV Abilene-Sweetwater, TX
KTAJ-TV St. Joseph, MO
KTAL-TV Shreveport, LA
KTAQ Dallas-Ft. Worth
KTAS Santa Barbara-Santa Maria-San Luis Obispo, CA
KTAZ Phoenix (Prescott), AZ
KTBC Austin, TX
KTBN-TV Los Angeles
KTBO-TV Oklahoma City, OK
KTBS-TV Shreveport, LA
KTBU Houston
KTBW-TV Seattle-Tacoma, WA
KTBY Anchorage, AK
*KTCA-TV Minneapolis-St. Paul, MN
*KTCI-TV Minneapolis-St. Paul, MN
KTCW Eugene, OR
KTDO El Paso (Las Cruces, NM), TX
*KTEH San Francisco-Oakland-San Jose
*KTEJ Jonesboro, AR
KTEL-TV Albuquerque-Santa Fe, NM
KTEN Sherman, TX-Ada, OK
KTFD-TV Denver, CO
KTFF-TV Fresno-Visalia, CA
KTFK-TV Sacramento -Stockton-Modesto, CA
KTFL Phoenix (Prescott), AZ
KTFN El Paso (Las Cruces, NM), TX
KTFQ-TV Albuquerque-Santa Fe, NM
KTGF Great Falls, MT
KTGM Tamuning GU
KTHV Little Rock-Pine Bluff, AR
*KTIN Des Moines-Ames, IA
KTIV Sioux City, IA
KTKA-TV Topeka, KS
KTLA Los Angeles
KTLL-TV Albuquerque-Santa Fe, NM
*KTLM Harlingen-Weslaco -Brownsville-McAllen, TX
KTLN-TV San Francisco-Oakland-San Jose
KTMD Houston
KTMF Missoula, MT
KTMW Salt Lake City, UT
KTNC-TV San Francisco-Oakland-San Jose
*KTNE-TV Rapid City, SD
KTNL-TV Juneau, AK
KTNV Las Vegas, NV
*KTNW Yakima-Pasco -Richland-Kennewick, WA
*KTOO-TV Juneau, AK
KTPX Tulsa, OK

KTRE Tyler-Longview (Lufkin & Nacogdoches), TX
KTRG San Antonio, TX
KTRK-TV Houston
KTRV-TV Boise, ID
*KTSC Colorado Springs-Pueblo, CO
*KTSD-TV Sioux Falls (Mitchell), SD
KTSF San Francisco-Oakland-San Jose
KTSM-TV El Paso (Las Cruces, NM), TX
KTTC Rochester, MN-Mason City, IA-Austin, MN
KTTM Sioux Falls (Mitchell), SD
KTTU-TV Tucson (Sierra Vista), AZ
KTTV Los Angeles
KTTW Sioux Falls (Mitchell), SD
KTUL Tulsa, OK
KTUU-TV Anchorage, AK
KTUW Cheyenne, WY-Scottsbluff, NE
KTUZ-TV Oklahoma City, OK
KTVA Anchorage, AK
KTVB Boise, ID
KTVC Eugene, OR
KTVD Denver, CO
KTVE Monroe, LA-El Dorado, AR
KTVF Fairbanks, AK
KTVG Lincoln & Hastings-Kearney, NE
KTVH Helena, MT
KTVI St. Louis, MO
KTVK Phoenix (Prescott), AZ
KTVL Medford-Klamath Falls, OR
KTVM Butte-Bozeman, MT
KTVN Reno, NV
KTVO Ottumwa, IA-Kirksville, MO
KTVQ Billings, MT
*KTVR Portland, OR
KTVT Dallas-Ft. Worth
KTVU San Francisco-Oakland-San Jose
KTVW-TV Phoenix (Prescott), AZ
KTVX Salt Lake City, UT
KTVZ Bend, OR
KTWO-TV Casper-Riverton, WY
*KTWU Topeka, KS
KTXA Dallas-Ft. Worth
KTXH Houston
KTXL Sacramento-Stockton-Modesto, CA
KTXS-TV Abilene-Sweetwater, TX
*KTXT-TV Lubbock, TX
*KUAC-TV Fairbanks, AK
KUAM-TV Hagatna GU
*KUAS-TV Tucson (Sierra Vista), AZ
*KUAT-TV Tucson (Sierra Vista), AZ
KUBD Ketchikan AK
KUCW Salt Lake City, UT
*KUED Salt Lake City, UT
*KUEN Salt Lake City, UT
*KUES Salt Lake City, UT
*KUEW Salt Lake City, UT
*KUFM-TV Missoula, MT
*KUHT Houston
*KUID-TV Spokane, WA
KULR-TV Billings, MT
KUMV-TV Minot-Bismarck-Dickinson, ND
KUNO-TV San Francisco-Oakland-San Jose
KUNP Portland, OR
KUNS-TV Seattle-Tacoma, WA
KUOK Oklahoma City, OK
*KUON-TV Lincoln & Hastings-Kearney, NE
KUPB Odessa-Midland, TX
KUPK-TV Wichita-Hutchinson Plus, KS
KUPN Denver, CO
KUPT Albuquerque-Santa Fe, NM
KUPU Honolulu, HI
KUPX Salt Lake City, UT
KUQI Corpus Christi, TX
KUSA-TV Denver, CO
*KUSD-TV Sioux Falls (Mitchell), SD
KUSG Salt Lake City, UT
KUSI-TV San Diego, CA
*KUSM Butte-Bozeman, MT
KUTF Salt Lake City, UT

KUTH Salt Lake City, UT
KUTP Phoenix (Prescott), AZ
KUTV Salt Lake City, UT
KUVE-TV Tucson (Sierra Vista), AZ
KUVI-TV Bakersfield, CA
KUVN-TV Dallas-Ft. Worth
KUVS-TV Sacramento -Stockton-Modesto, CA
KVAL-TV Eugene, OR
KVAW San Antonio, TX
KVBC Las Vegas, NV
***KVCR-TV** Los Angeles
KVCT Victoria, TX
KVCW Las Vegas, NV
KVDA San Antonio, TX
KVEA Los Angeles
KVEO Harlingen-Weslaco -Brownsville-McAllen, TX
KVEW Yakima-Pasco -Richland-Kennewick, WA
KVHP Lake Charles, LA
KVIA-TV El Paso (Las Cruces, NM), TX
***KVIE** Sacramento-Stockton-Modesto, CA
KVIH-TV Amarillo, TX
KVII-TV Amarillo, TX
KVIQ Eureka, CA
KVLY-TV Fargo-Valley City, ND
KVMD Los Angeles
KVMY Las Vegas, NV
KVNV Salt Lake City, UT
KVOA Tucson (Sierra Vista), AZ
KVOS-TV Seattle-Tacoma, WA
***KVPT** Fresno-Visalia, CA
KVRR Fargo-Valley City, ND
KVSN Colorado Springs-Pueblo, CO
KVTH Little Rock-Pine Bluff, AR
KVTJ Jonesboro, AR
KVTN Little Rock-Pine Bluff, AR
KVTV Laredo, TX
KVUE-TV Austin, TX
KVVU-TV Las Vegas, NV
KVYE Yuma, AZ-El Centro, CA
***KVZK-2** Pago Pago AS
***KVZK-4** Pago Pago AS
***KVZK-5** Pago Pago AS
KWAB-TV Odessa-Midland, TX
KWBA Tucson (Sierra Vista), AZ
KWBF Little Rock-Pine Bluff, AR
KWBM Springfield, MO
***KWBN** Honolulu, HI
KWBQ Albuquerque-Santa Fe, NM
***KWBU-TV** Waco-Temple-Bryan, TX
KWCH-TV Wichita-Hutchinson Plus, KS
***KWCM-TV** Minneapolis-St. Paul, MN
***KWDK** Seattle-Tacoma, WA
KWES-TV Odessa-Midland, TX
***KWET** Oklahoma City, OK
KWEX-TV San Antonio, TX
KWGN-TV Denver, CO
KWHB Tulsa, OK
KWHD Denver, CO
KWHE Honolulu, HI
KWHH Hilo HI
KWHM Wailuku HI
KWHY-TV Los Angeles
KWKB Cedar Rapids-Waterloo-Iowa City & Dubuque, IA
KWKT Waco-Temple-Bryan, TX
KWNB-TV Lincoln & Hastings-Kearney, NE
KWNV Reno, NV
KWOG Ft. Smith-Fayetteville -Springdale-Rogers, AR
KWPX Seattle-Tacoma, WA
KWQC-TV Davenport, IA-Rock Island-Moline, IL
KWSD Sioux Falls (Mitchell), SD
***KWSE** Minot-Bismarck-Dickinson, ND
***KWSU-TV** Spokane, WA
KWTV Oklahoma City, OK
KWTX-TV Waco-Temple-Bryan, TX
KWWF Cedar Rapids-Waterloo-Iowa City & Dubuque, IA
KWWL Cedar Rapids-Waterloo-Iowa City & Dubuque, IA
KWWT Odessa-Midland, TX
KWYB Butte-Bozeman, MT
***KWYP-TV** Denver, CO
KXAM-TV Austin, TX
KXAN-TV Austin, TX
KXAS-TV Dallas-Ft. Worth
KXGN-TV Glendive, MT
KXII Sherman, TX-Ada, OK
KXJB-TV Fargo-Valley City, ND
KXLA Los Angeles
KXLF-TV Butte-Bozeman, MT
KXLN-TV Houston
KXLT-TV Rochester, MN-Mason City, IA-Austin, MN
KXLY-TV Spokane, WA
KXMA-TV Minot-Bismarck-Dickinson, ND
KXMB-TV Minot-Bismarck-Dickinson, ND
KXMC-TV Minot-Bismarck-Dickinson, ND
KXMD-TV Minot-Bismarck-Dickinson, ND
KXND Minot-Bismarck-Dickinson, ND
***KXNE-TV** Sioux City, IA
KXRM-TV Colorado Springs-Pueblo, CO
KXTF Twin Falls, ID
KXTV Sacramento-Stockton-Modesto, CA
KXTX-TV Dallas-Ft. Worth
KXVA Abilene-Sweetwater, TX
KXVO Omaha, NE
KXXV Waco-Temple-Bryan, TX
KYES-TV Anchorage, AK
***KYIN** Rochester, MN-Mason City, IA-Austin, MN
KYLE Waco-Temple-Bryan, TX
KYMA Yuma, AZ-El Centro, CA
***KYNE-TV** Omaha, NE
KYOU-TV Ottumwa, IA-Kirksville, MO
KYPX Little Rock-Pine Bluff, AR
KYTV Springfield, MO
KYTX Tyler-Longview (Lufkin & Nacogdoches), TX
***KYUK-TV** Bethel AK
KYUS-TV Billings, MT
***KYVE** Yakima-Pasco -Richland-Kennewick, WA
KYW-TV Philadelphia
KZJL Houston
***KZSD-TV** Rapid City, SD
KZTV Corpus Christi, TX
WAAY-TV Huntsville-Decatur (Florence), AL
WABC-TV New York
WABG-TV Greenwood-Greenville, MS
WABI-TV Bangor, ME
WABM Birmingham (Anniston, Tuscaloosa), AL
***WABW-TV** Albany, GA
WACH Columbia, SC
***WACS-TV** Albany, GA
WACX-DT Orlando-Daytona Beach-Melbourne, FL
WACY Green Bay-Appleton, WI
WADL Detroit
WAFB Baton Rouge, LA
WAFF Huntsville-Decatur (Florence), AL
WAGA Atlanta
WAGM-TV Presque Isle, ME
WAGT Augusta, GA
WAGV Knoxville, TN
***WAIQ** Montgomery-Selma, AL
WAKA Montgomery-Selma, AL
WALA-TV Mobile, AL-Pensacola (Ft. Walton Beach), FL
WALB Albany, GA
WAMI-TV Miami-Ft. Lauderdale, FL
WAND Champaign & Springfield -Decatur, IL
WANE-TV Ft. Wayne, IN
WAOE Peoria-Bloomington, IL
WAOW-TV Wausau-Rhinelander, WI
WAPA-TV San Juan PR
WAPT Jackson, MS
WAQP Flint-Saginaw-Bay City, MI
***WATC** Atlanta
WATE-TV Knoxville, TN
WATL Atlanta
WATM-TV Johnstown-Altoona, PA
WAVE Louisville, KY
WAVY-TV Norfolk-Portsmouth-Newport News, VA
WAWD Mobile, AL-Pensacola (Ft. Walton Beach), FL
WAWS Jacksonville, FL
WAXN-TV Charlotte, NC
WAZE-TV Evansville, IN
WBAL-TV Baltimore, MD
WBAY-TV Green Bay-Appleton, WI
WBBH-TV Ft. Myers-Naples, FL
WBBJ-TV Jackson, TN
WBBM-TV Chicago
***WBCC** Orlando-Daytona Beach-Melbourne, FL
WBDT Dayton, OH
WBFF Baltimore, MD
WBFS-TV Miami-Ft. Lauderdale, FL
***WBGU-TV** Toledo, OH
WBIF Panama City, FL
WBIH Montgomery-Selma, AL
WBIJ Wausau-Rhinelander, WI
***WBIQ** Birmingham (Anniston, Tuscaloosa), AL
WBIR-TV Knoxville, TN
WBKB-TV Alpena, MI
WBKI-TV Louisville, KY
WBKO-TV Bowling Green, KY
WBKP Marquette, MI
WBMM Montgomery-Selma, AL
WBNA Louisville, KY
WBNG-TV Binghamton, NY
WBNS-TV Columbus, OH
WBNX-TV Cleveland-Akron (Canton), OH
WBOC-TV Salisbury, MD
WBOY-TV Clarksburg-Weston, WV
WBPG Mobile, AL-Pensacola (Ft. Walton Beach), FL
WBPH-TV Philadelphia
WBPX Boston (Manchester, NH)
***WBRA-TV** Roanoke-Lynchburg, VA
WBRC Birmingham (Anniston, Tuscaloosa), AL
WBRE-TV Wilkes Barre-Scranton, PA
WBRZ Baton Rouge, LA
WBSF Flint-Saginaw-Bay City, MI
WBTV Charlotte, NC
WBTW Myrtle Beach-Florence, SC
WBUI Champaign & Springfield -Decatur, IL
WBUP Marquette, MI
WBUW Madison, WI
WBUY-TV Memphis, TN
WBXX-TV Knoxville, TN
WBZ-TV Boston (Manchester, NH)
WCAU Philadelphia
WCAV Charlottesville, VA
WCAX-TV Burlington, VT-Plattsburgh, NY
***WCBB** Portland-Auburn, ME
WCBD-TV Charleston, SC
WCBI-TV Columbus-Tupelo-West Point, MS
WCBS-TV New York
WCCB Charlotte, NC
WCCO-TV Minneapolis-St. Paul, MN
WCCU Champaign & Springfield -Decatur, IL
WCCV-TV Arecibo PR
WCDC Albany-Schenectady-Troy, NY
***WCES-TV** Augusta, GA
***WCET** Cincinnati, OH
***WCEU** Orlando-Daytona Beach-Melbourne, FL
***WCFE-TV** Burlington, VT-Plattsburgh, NY
WCFN-TV Champaign & Springfield -Decatur, IL
WCFT-TV Birmingham (Anniston, Tuscaloosa), AL
WCGV-TV Milwaukee, WI
WCHS-TV Charleston-Huntington, WV
WCIA-TV Champaign & Springfield -Decatur, IL
***WCIQ** Birmingham (Anniston, Tuscaloosa), AL
WCIU-TV Chicago
WCIV Charleston, SC
WCJB-TV Gainesville, FL
WCLF Tampa-St. Petersburg (Sarasota), FL
WCLJ-TV Indianapolis, IN
***WCLP-TV** Chattanooga, TN
WCMH-TV Columbus, OH
***WCML-TV** Alpena, MI
***WCMU-TV** Flint-Saginaw-Bay City, MI
***WCMV** Traverse City-Cadillac, MI
***WCMW** Traverse City-Cadillac, MI
WCNC-TV Charlotte, NC
***WCNY-TV** Syracuse, NY
WCOV-TV Montgomery-Selma, AL
***WCPB** Salisbury, MD
WCPO-TV Cincinnati, OH
WCPX Chicago
WCSC-TV Charleston, SC
WCSH Portland-Auburn, ME
***WCTE** Nashville, TN
WCTI-TV Greenville-New Bern-Washington, NC
WCTV Tallahassee, FL-Thomasville, GA
WCTX Hartford & New Haven, CT
WCVB-TV Boston (Manchester, NH)
***WCVE-TV** Richmond-Petersburg, VA
WCVI-TV Christiansted VI
***WCVN-TV** Cincinnati, OH
***WCVW** Richmond-Petersburg, VA
WCWG Greensboro-High Point-Winston Salem, NC
WCWJ Jacksonville, FL
WCWN Albany-Schenectady-Troy, NY
WCYB-TV Tri-Cities, TN-VA
WDAF-TV Kansas City, MO
WDAM-TV Hattiesburg-Laurel, MS
WDAY-TV Fargo-Valley City, ND
WDAZ-TV Fargo-Valley City, ND
WDBB Birmingham (Anniston, Tuscaloosa), AL
WDBD Jackson, MS
WDBJ Roanoke-Lynchburg, VA
WDCA Washington, DC (Hagerstown, MD)
***WDCP-TV** Flint-Saginaw-Bay City, MI
***WDCQ-TV** Flint-Saginaw-Bay City, MI
WDCW Washington, DC (Hagerstown, MD)
WDEF-TV Chattanooga, TN
WDFX-TV Dothan, AL
WDHN Dothan, AL
WDHS Marquette, MI
WDIO-TV Duluth, MN-Superior, WI
***WDIQ** Montgomery-Selma, AL
WDIV Detroit
WDJT-TV Milwaukee, WI
WDKA Paducah, KY-Cape Girardeau, MO-Harrisburg-Mount Vernon, IL
WDKY-TV Lexington, KY
WDLI-TV Cleveland-Akron (Canton), OH
***WDPB** Salisbury, MD
WDPX Providence, RI-New Bedford, MA
WDRB Louisville, KY
WDRL-TV Roanoke-Lynchburg, VA
***WDSE-TV** Duluth, MN-Superior, WI
WDSI-TV Chattanooga, TN
WDSU New Orleans, LA
***WDTI** Indianapolis, IN
WDTN Dayton, OH
WDTV Clarksburg-Weston, WV
WDWL Bayamon PR
***WEAO** Cleveland-Akron (Canton), OH
WEAR-TV Mobile, AL-Pensacola (Ft. Walton Beach), FL
WEAU-TV La Crosse-Eau Claire, WI
***WEBA-TV** Augusta, GA
WECN Naranjito PR
WECT Wilmington, NC
***WEDH** Hartford & New Haven, CT
***WEDN** Hartford & New Haven, CT
***WEDU** Tampa-St. Petersburg (Sarasota), FL
***WEDW** New York
***WEDY** Hartford & New Haven, CT
WEEK-TV Peoria-Bloomington, IL
WEHT Evansville, IN
***WEIQ** Mobile, AL-Pensacola (Ft. Walton Beach), FL
***WEIU-TV** Champaign & Springfield -Decatur, IL
***WEKW-TV** Boston (Manchester, NH)
WELF-TV Chattanooga, TN
***WELU** Aguadilla PR
WEMT Tri-Cities, TN-VA
***WENH-TV** Boston (Manchester, NH)
WENY-TV Elmira (Corning), NY
WEPX Greenville-New Bern-Washington, NC
WESH Orlando-Daytona Beach-Melbourne, FL
***WETA-TV** Washington, DC (Hagerstown, MD)
***WETK** Burlington, VT-Plattsburgh, NY
WETM-TV Elmira (Corning), NY
***WETP-TV** Knoxville, TN
WEUX La Crosse-Eau Claire, WI
WEVV-TV Evansville, IN
WEWS Cleveland-Akron (Canton), OH
WEYI-TV Flint-Saginaw-Bay City, MI
WFAA-TV Dallas-Ft. Worth
WFBD Mobile, AL-Pensacola (Ft. Walton Beach), FL
WFDC-TV Washington, DC (Hagerstown, MD)
WFFF-TV Burlington, VT-Plattsburgh, NY
WFFT-TV Ft. Wayne, IN
WFGC West Palm Beach-Ft. Pierce, FL
WFGX Mobile, AL-Pensacola (Ft. Walton Beach), FL
WFIE-TV Evansville, IN
***WFIQ** Huntsville-Decatur (Florence), AL
WFLA-TV Tampa-St. Petersburg (Sarasota), FL
WFLD Chicago
WFLI-TV Chattanooga, TN
WFLX West Palm Beach-Ft. Pierce, FL
***WFME-TV** New York
WFMJ-TV Youngstown, OH
WFMY-TV Greensboro-High Point-Winston Salem, NC
WFMZ-TV Philadelphia
WFOR-TV Miami-Ft. Lauderdale, FL
***WFPT** Washington, DC (Hagerstown, MD)
WFPX Raleigh-Durham (Fayetteville), NC
WFQX-TV Traverse City-Cadillac, MI
WFRV-TV Green Bay-Appleton, WI
WFSB Hartford & New Haven, CT
***WFSG** Panama City, FL
***WFSU-TV** Tallahassee, FL-Thomasville, GA
WFTC Minneapolis-St. Paul, MN
WFTS Tampa-St. Petersburg (Sarasota), FL
WFTT-TV Tampa-St. Petersburg (Sarasota), FL
WFTV Orlando-Daytona Beach-Melbourne, FL
WFTX Ft. Myers-Naples, FL
WFTY-TV New York
***WFUM** Flint-Saginaw-Bay City, MI
WFUP Traverse City-Cadillac, MI
WFUT-TV New York
***WFWA** Ft. Wayne, IN
WFXB Myrtle Beach-Florence, SC
WFXG Augusta, GA
WFXI Greenville-New Bern-Washington, NC
WFXL Albany, GA
WFXP Erie, PA
WFXR-TV Roanoke-Lynchburg, VA

WFXS Wausau-Rhinelander, WI
WFXT Boston (Manchester, NH)
WFXU Tallahassee, FL-Thomasville, GA
WFXV Utica, NY
WFXW-TV Terre Haute, IN
***WFYI-TV** Indianapolis, IN
WGAL Harrisburg -Lancaster-Lebanon-York, PA
WGBA Green Bay-Appleton, WI
***WGBC** Meridian, MS
***WGBH-TV** Boston (Manchester, NH)
WGBO-TV Chicago
***WGBX-TV** Boston (Manchester, NH)
***WGBY-TV** Springfield-Holyoke, MA
WGCB-TV Harrisburg -Lancaster-Lebanon-York, PA
WGCL-TV Atlanta
***WGCU** Ft. Myers-Naples, FL
WGEM-TV Quincy, IL-Hannibal, MO-Keokuk, IA
WGEN-TV Miami-Ft. Lauderdale, FL
WGFL Gainesville, FL
WGGB-TV Springfield-Holyoke, MA
WGGN-TV Cleveland-Akron (Canton), OH
WGGS-TV Greenville-Spartanburg, SC-Asheville, NC-Anderson, SC
WGHP Greensboro-High Point-Winston Salem, NC
***WGIQ** Columbus, GA
WGMB Baton Rouge, LA
WGME-TV Portland-Auburn, ME
WGNM Macon, GA
WGNO New Orleans, LA
WGNT Norfolk-Portsmouth-Newport News, VA
WGN-TV Chicago
***WGPT** Pittsburgh, PA
WGPX Greensboro-High Point-Winston Salem, NC
WGRZ-TV Buffalo, NY
WGSA Savannah, GA
***WGTE-TV** Toledo, OH
WGTQ Traverse City-Cadillac, MI
WGTU Traverse City-Cadillac, MI
***WGTV** Atlanta
WGTW-TV Philadelphia
***WGVK** Grand Rapids-Kalamazoo-Battle Creek, MI
***WGVU-TV** Grand Rapids-Kalamazoo-Battle Creek, MI
WGXA Macon, GA
WHAG-TV Washington, DC (Hagerstown, MD)
WHAM-TV Rochester, NY
WHAS-TV Louisville, KY
***WHA-TV** Madison, WI
WHBF-TV Davenport, IA-Rock Island-Moline, IL
WHBQ-TV Memphis, TN
WHBR Mobile, AL-Pensacola (Ft. Walton Beach), FL
WHDF Huntsville-Decatur (Florence), AL
WHDH-TV Boston (Manchester, NH)
WHEC-TV Rochester, NY
***WHFT-TV** Miami-Ft. Lauderdale, FL
WHIO-TV Dayton, OH
***WHIQ** Huntsville-Decatur (Florence), AL
WHIZ-TV Zanesville, OH
WHKY-TV Charlotte, NC
***WHLA-TV** La Crosse-Eau Claire, WI
WHLT Hattiesburg-Laurel, MS
WHLV-TV Orlando-Daytona Beach-Melbourne, FL
WHMB-TV Indianapolis, IN
***WHMC** Myrtle Beach-Florence, SC
WHME-TV South Bend-Elkhart, IN
WHMM-DT New Orleans, LA
WHNO New Orleans, LA
WHNS Greenville-Spartanburg, SC-Asheville, NC-Anderson, SC
WHNT-TV Huntsville-Decatur (Florence), AL
WHOI Peoria-Bloomington, IL
WHO-TV Des Moines-Ames, IA
WHP-TV Harrisburg -Lancaster-Lebanon-York, PA
WHPX Hartford & New Haven, CT
WHRE Norfolk-Portsmouth-Newport News, VA
***WHRM-TV** Wausau-Rhinelander, WI
***WHRO-TV** Norfolk-Portsmouth-Newport News, VA
***WHSG-TV** Atlanta
WHSV-TV Harrisonburg, VA
***WHTJ** Charlottesville, VA
WHTM-TV Harrisburg -Lancaster-Lebanon-York, PA
WHTN Nashville, TN
WHTV Lansing, MI
***WHUT-TV** Washington, DC (Hagerstown, MD)
***WHWC-TV** Minneapolis-St. Paul, MN
***WHYY-TV** Philadelphia
WIAT Birmingham (Anniston, Tuscaloosa), AL
WIBW-TV Topeka, KS
WICD-TV Champaign & Springfield -Decatur, IL
WICS-TV Champaign & Springfield -Decatur, IL
WICU-TV Erie, PA
WICZ-TV Binghamton, NY
WIDP Guayama PR
WIFR Rockford, IL
***WIIQ** Montgomery-Selma, AL
***WILL-TV** Champaign & Springfield -Decatur, IL
WILX-TV Lansing, MI
WINK-TV Ft. Myers-Naples, FL
WINM Ft. Wayne, IN
***WIPB** Indianapolis, IN
***WIPM-TV** Mayaguez PR
***WIPR-TV** San Juan PR
WIPX Indianapolis, IN
WIRS Yauco PR
WIRT Duluth, MN-Superior, WI
WIS Columbia, SC
WISC-TV Madison, WI
WISE-TV Ft. Wayne, IN
WISH-TV Indianapolis, IN
WISN-TV Milwaukee, WI
***WITF-TV** Harrisburg -Lancaster-Lebanon-York, PA
WITI Milwaukee, WI
WITN-TV Greenville-New Bern-Washington, NC
***WITV** Charleston, SC
WIVB-TV Buffalo, NY
WIVT Binghamton, NY
WIWB Green Bay-Appleton, WI
WJAC-TV Johnstown-Altoona, PA
WJAL Washington, DC (Hagerstown, MD)
WJAR Providence, RI-New Bedford, MA
WJBF Augusta, GA
WJBK Detroit
WJCL Savannah, GA
***WJCT** Jacksonville, FL
***WJEB-TV** Jacksonville, FL
WJET-TV Erie, PA
WJFB Nashville, TN
WJFW-TV Wausau-Rhinelander, WI
WJHG-TV Panama City, FL
WJHL-TV Tri-Cities, TN-VA
WJJA Milwaukee, WI
WJKT Jackson, TN
WJLA-TV Washington, DC (Hagerstown, MD)
WJMN-TV Marquette, MI
***WJPM-TV** Myrtle Beach-Florence, SC
WJPX San Juan PR
WJRT-TV Flint-Saginaw-Bay City, MI
***WJSP-TV** Columbus, GA
WJSU-TV Birmingham (Anniston, Tuscaloosa), AL
WJTC Mobile, AL-Pensacola (Ft. Walton Beach), FL
WJTV Jackson, MS
WJW Cleveland-Akron (Canton), OH
***WJWJ-TV** Savannah, GA
WJWN-TV San Sebastian PR
WJXT Jacksonville, FL
WJXX Jacksonville, FL
WJYS-TV Chicago
WJZ-TV Baltimore, MD
WJZY Charlotte, NC
WKAQ-TV San Juan PR
***WKAR-TV** Lansing, MI
***WKAS** Charleston-Huntington, WV
WKBD Detroit
WKBN-TV Youngstown, OH
WKBS-TV Johnstown-Altoona, PA
WKBT La Crosse-Eau Claire, WI
WKBW-TV Buffalo, NY
WKCF Orlando-Daytona Beach-Melbourne, FL
WKDH Columbus-Tupelo-West Point, MS
WKEF Dayton, OH
***WKGB-TV** Bowling Green, KY
***WKHA** Lexington, KY
***WKLE** Lexington, KY
***WKMA** Evansville, IN
WKMG-TV Orlando-Daytona Beach-Melbourne, FL
WKMJ Louisville, KY
***WKMR** Lexington, KY
***WKMU** Paducah, KY-Cape Girardeau, MO-Harrisburg-Mount Vernon, IL
***WKNO-TV** Memphis, TN
***WKOH** Evansville, IN
WKOI-TV Dayton, OH
***WKON** Cincinnati, OH
***WKOP-TV** Knoxville, TN
WKOW-TV Madison, WI
***WKPC-TV** Louisville, KY
***WKPD** Paducah, KY-Cape Girardeau, MO-Harrisburg-Mount Vernon, IL
***WKPI** Charleston-Huntington, WV
WKPT-TV Tri-Cities, TN-VA
WKPV Ponce PR
WKRC-TV Cincinnati, OH
WKRG-TV Mobile, AL-Pensacola (Ft. Walton Beach), FL
WKRN-TV Nashville, TN
***WKSO-TV** Lexington, KY
WKTC Columbia, SC
WKTV Utica, NY
WKYC-TV Cleveland-Akron (Canton), OH
WKYT-TV Lexington, KY
***WKYU-TV** Bowling Green, KY
***WKZT-TV** Louisville, KY
***WLAE-TV** New Orleans, LA
WLAJ Lansing, MI
WLAX La Crosse-Eau Claire, WI
WLBT Jackson, MS
WLBZ Bangor, ME
WLCB-TV Orlando-Daytona Beach-Melbourne, FL
WLDM Birmingham (Anniston, Tuscaloosa), AL
***WLED-TV** Portland-Auburn, ME
***WLEF-TV** Wausau-Rhinelander, WI
WLEX-TV Lexington, KY
WLFB Bluefield-Beckley-Oak Hill, WV
WLFG Tri-Cities, TN-VA
WLFI-TV Lafayette, IN
WLFL Raleigh-Durham (Fayetteville), NC
WLGA Columbus, GA
WLII Caguas PR
WLIO Lima, OH
***WLIW** New York
WLJC-TV Lexington, KY
***WLJT-TV** Jackson, TN
WLKY-TV Louisville, KY
WLLA Grand Rapids-Kalamazoo-Battle Creek, MI
WLMB Toledo, OH
WLMT Memphis, TN
WLNE Providence, RI-New Bedford, MA
WLNS-TV Lansing, MI
WLNY New York
WLOS Greenville-Spartanburg, SC-Asheville, NC-Anderson, SC
WLOV-TV Columbus-Tupelo-West Point, MS
WLOX Biloxi-Gulfport, MS
***WLPB-TV** Baton Rouge, LA
WLPX-TV Charleston-Huntington, WV
***WLRN-TV** Miami-Ft. Lauderdale, FL
WLS-TV Chicago
WLTV Miami-Ft. Lauderdale, FL
WLTX Columbia, SC
WLTZ Columbus, GA
WLUC-TV Marquette, MI
WLUK-TV Green Bay-Appleton, WI
WLVI-TV Boston (Manchester, NH)
***WLVT-TV** Philadelphia
WLWC Providence, RI-New Bedford, MA
WLWT Cincinnati, OH
WLXI-TV Greensboro-High Point-Winston Salem, NC
WLYH-TV Harrisburg -Lancaster-Lebanon-York, PA
***WMAA** Columbus-Tupelo-West Point, MS
***WMAB-TV** Columbus-Tupelo-West Point, MS
***WMAE-TV** Columbus-Tupelo-West Point, MS
***WMAH-TV** Biloxi-Gulfport, MS
***WMAI** Greenwood-Greenville, MS
WMAK Knoxville, TN
***WMAO-TV** Greenwood-Greenville, MS
WMAQ-TV Chicago
WMAR-TV Baltimore, MD
***WMAU-TV** Jackson, MS
***WMAV-TV** Memphis, TN
***WMAW-TV** Meridian, MS
WMAZ-TV Macon, GA
WMBB Panama City, FL
WMBC-TV New York
WMBD-TV Peoria-Bloomington, IL
WMBF-TV Myrtle Beach-Florence, SC
WMCF-TV Montgomery-Selma, AL
WMCN-TV Philadelphia
WMC-TV Memphis, TN
WMDN Meridian, MS
WMDT Salisbury, MD
***WMEA-TV** Portland-Auburn, ME
***WMEB-TV** Bangor, ME
***WMEC** Quincy, IL-Hannibal, MO-Keokuk, IA
***WMED-TV** Bangor, ME
WMEI Arecibo PR
***WMEM-TV** Presque Isle, ME
WMFD-TV Cleveland-Akron (Canton), OH
***WMFE-TV** Orlando-Daytona Beach-Melbourne, FL
WMFP Boston (Manchester, NH)
WMGM-TV Philadelphia
WMGT-TV Macon, GA
***WMHT** Albany-Schenectady-Troy, NY
WMMP Charleston, SC
WMOR-TV Tampa-St. Petersburg (Sarasota), FL
***WMPB** Baltimore, MD
***WMPN-TV** Jackson, MS
***WMPT** Baltimore, MD
WMPV-TV Mobile, AL-Pensacola (Ft. Walton Beach), FL
WMQF Marquette, MI
WMSN-TV Madison, WI
***WMSY-TV** Tri-Cities, TN-VA
***WMTJ** Fajardo PR
WMTV Madison, WI
WMTW-TV Portland-Auburn, ME
***WMUM-TV** Macon, GA
WMUR-TV Boston (Manchester, NH)
***WMVS** Milwaukee, WI
***WMVT** Milwaukee, WI
WMYA-TV Greenville-Spartanburg, SC-Asheville, NC-Anderson, SC
***WMYC** Jackson, MS
WMYD Detroit
WMYO Louisville, KY
WMYT-TV Charlotte, NC
WMYV Greensboro-High Point-Winston Salem, NC
WNAB Nashville, TN
WNAC-TV Providence, RI-New Bedford, MA
WNBC New York
WNCF Montgomery-Selma, AL
WNCN Raleigh-Durham (Fayetteville), NC
WNCT-TV Greenville-New Bern-Washington, NC
WNDU-TV South Bend-Elkhart, IN
WNDY-TV Indianapolis, IN
***WNED-TV** Buffalo, NY
WNEG-TV Greenville-Spartanburg, SC-Asheville, NC-Anderson, SC
***WNEH** Greenville-Spartanburg, SC-Asheville, NC-Anderson, SC
WNEM-TV Flint-Saginaw-Bay City, MI
***WNEO** Youngstown, OH
WNEP-TV Wilkes Barre-Scranton, PA
***WNET** New York
WNEU Boston (Manchester, NH)
WNGS Buffalo, NY
***WNIN-TV** Evansville, IN
***WNIT-TV** South Bend-Elkhart, IN
***WNJB** New York
***WNJN** New York
***WNJS** Philadelphia
***WNJT** Philadelphia
WNJU New York
WNJX-TV Mayaguez PR
WNKY Bowling Green, KY
WNLO Buffalo, NY
***WNMU-TV** Marquette, MI
WNNE-TV Burlington, VT-Plattsburgh, NY
WNOL-TV New Orleans, LA
***WNPB-TV** Pittsburgh, PA
***WNPI-TV** Watertown, NY
***WNPT** Nashville, TN
WNPX Nashville, TN
***WNSC-TV** Charlotte, NC
***WNTV** Greenville-Spartanburg, SC-Asheville, NC-Anderson, SC
WNTZ Jackson, MS
WNUV Baltimore, MD
***WNVC** Washington, DC (Hagerstown, MD)
***WNVT** Washington, DC (Hagerstown, MD)
WNWO-TV Toledo, OH
WNYA Albany-Schenectady-Troy, NY
WNYB Buffalo, NY
***WNYE-TV** New York
WNYI Syracuse, NY
WNYO-TV Buffalo, NY
WNYS-TV Syracuse, NY
WNYT Albany-Schenectady-Troy, NY
WNYW New York
WOAC Cleveland-Akron (Canton), OH
WOAI-TV San Antonio, TX
WOAY-TV Bluefield-Beckley-Oak Hill, WV
WOFL Orlando-Daytona Beach-Melbourne, FL
WOGX Gainesville, FL
WOIO Cleveland-Akron (Canton), OH
WOI-TV Des Moines-Ames, IA
WOLE-TV Aguadilla PR
WOLF-TV Wilkes Barre-Scranton, PA
WOLO-TV Columbia, SC
WOOD-TV Grand Rapids-Kalamazoo-Battle Creek, MI
WOPX Orlando-Daytona Beach-Melbourne, FL
WORA-TV Mayaguez PR
WORO-TV Fajardo PR
WOST Mayaguez PR
***WOSU-TV** Columbus, OH
WOTF-TV Orlando-Daytona Beach-Melbourne, FL
WOTV Grand Rapids-Kalamazoo-Battle Creek, MI
***WOUB-TV** Charleston-Huntington, WV
***WOUC-TV** Wheeling, WV-Steubenville, OH

WOWK-TV Charleston-Huntington, WV
WOWT Omaha, NE
WPAN Mobile, AL-Pensacola (Ft. Walton Beach), FL
***WPBA** Atlanta
WPBF West Palm Beach-Ft. Pierce, FL
WPBN-TV Traverse City-Cadillac, MI
***WPBO-TV** Charleston-Huntington, WV
***WPBS-TV** Watertown, NY
***WPBT** Miami-Ft. Lauderdale, FL
***WPBY-TV** Charleston-Huntington, WV
WPCB-TV Pittsburgh, PA
WPCT Panama City, FL
WPCW Pittsburgh, PA
WPDE-TV Myrtle Beach-Florence, SC
WPEC West Palm Beach-Ft. Pierce, FL
WPFO Portland-Auburn, ME
WPGA-TV Macon, GA
WPGD-TV Nashville, TN
WPGH-TV Pittsburgh, PA
WPGX Panama City, FL
WPHL-TV Philadelphia
WPIX New York
WPLG Miami-Ft. Lauderdale, FL
WPME Portland-Auburn, ME
WPMI-TV Mobile, AL-Pensacola (Ft. Walton Beach), FL
WPMT Harrisburg -Lancaster-Lebanon-York, PA
WPMY Pittsburgh, PA
***WPNE** Green Bay-Appleton, WI
***WPPB-TV** West Palm Beach-Ft. Pierce, FL
WPPX Philadelphia
WPRI-TV Providence, RI-New Bedford, MA
WPSD-TV Paducah, KY-Cape Girardeau, MO-Harrisburg-Mount Vernon, IL
WPSG Philadelphia
***WPSU-TV** Johnstown-Altoona, PA
WPTA Ft. Wayne, IN
***WPTD** Dayton, OH
***WPTO** Cincinnati, OH
WPTV West Palm Beach-Ft. Pierce, FL
WPTY-TV Memphis, TN
WPTZ Burlington, VT-Plattsburgh, NY
WPVI-TV Philadelphia
WPWR-TV Chicago
WPXA Atlanta
WPXC-TV Jacksonville, FL
WPXD Detroit
WPXE Milwaukee, WI
WPXG Boston (Manchester, NH)
WPXH Birmingham (Anniston, Tuscaloosa), AL
WPXI Pittsburgh, PA
WPXJ-TV Buffalo, NY
WPXK Knoxville, TN
WPXL New Orleans, LA
WPXM Miami-Ft. Lauderdale, FL
WPXN-TV New York
WPXP West Palm Beach-Ft. Pierce, FL
WPXQ Providence, RI-New Bedford, MA
WPXR Roanoke-Lynchburg, VA
WPXS Paducah, KY-Cape Girardeau, MO-Harrisburg-Mount Vernon, IL
WPXT Portland-Auburn, ME
WPXU-TV Greenville-New Bern-Washington, NC
WPXV Norfolk-Portsmouth-Newport News, VA
WPXW Washington, DC (Hagerstown, MD)
WPXX-TV Memphis, TN
WQAD-TV Davenport, IA-Rock Island-Moline, IL
WQCW Charleston-Huntington, WV
***WQEC** Quincy, IL-Hannibal, MO-Keokuk, IA
***WQED** Pittsburgh, PA
WQEX Pittsburgh, PA
WQHA Aguada PR
WQHS-TV Cleveland-Akron (Canton), OH
***WQLN** Erie, PA
WQMY Wilkes Barre-Scranton, PA
WQOW-TV La Crosse-Eau Claire, WI
***WQPT-TV** Davenport, IA-Rock Island-Moline, IL
WQPX Wilkes Barre-Scranton, PA
WQRF-TV Rockford, IL
***WQTO** Ponce PR
WRAL-TV Raleigh-Durham (Fayetteville), NC
WRAY-TV Raleigh-Durham (Fayetteville), NC
WRAZ Raleigh-Durham (Fayetteville), NC
WRBJ Jackson, MS
WRBL Columbus, GA
WRBU St. Louis, MO
WRBW Orlando-Daytona Beach-Melbourne, FL
WRCB-TV Chattanooga, TN
WRC-TV Washington, DC (Hagerstown, MD)
WRDC Raleigh-Durham (Fayetteville), NC
WRDQ Orlando-Daytona Beach-Melbourne, FL
WRDW-TV Augusta, GA
WREG-TV Memphis, TN
***WRET-TV** Greenville-Spartanburg, SC-Asheville, NC-Anderson, SC
WREX-TV Rockford, IL
WRFB Carolina PR
WRGB Albany-Schenectady-Troy, NY
WRGT-TV Dayton, OH
WRIC-TV Richmond-Petersburg, VA
***WRJA-TV** Columbia, SC
WRJM-TV Montgomery-Selma, AL
WRLH-TV Richmond-Petersburg, VA
***WRLK-TV** Columbia, SC
WRNN-TV New York
WROC-TV Rochester, NY
WRPX Raleigh-Durham (Fayetteville), NC
WRSP-TV Champaign & Springfield -Decatur, IL
WRTV Indianapolis, IN
WRUA Fajardo PR
WRXY-TV Ft. Myers-Naples, FL
WSAH New York
WSAV-TV Savannah, GA
WSAW-TV Wausau-Rhinelander, WI
WSAZ-TV Charleston-Huntington, WV
***WSBE-TV** Providence, RI-New Bedford, MA
WSBK-TV Boston (Manchester, NH)
***WSBN-TV** Tri-Cities, TN-VA
WSBS-TV Miami-Ft. Lauderdale, FL
WSBT-TV South Bend-Elkhart, IN
WSB-TV Atlanta
WSCV Miami-Ft. Lauderdale, FL
***WSEC** Champaign & Springfield -Decatur, IL
WSEE Erie, PA
WSET-TV Roanoke-Lynchburg, VA
WSFA Montgomery-Selma, AL
WSFJ-TV Columbus, OH
WSFL-TV Miami-Ft. Lauderdale, FL
WSFX-TV Wilmington, NC
WSIL-TV Paducah, KY-Cape Girardeau, MO-Harrisburg-Mount Vernon, IL
***WSIU-TV** Paducah, KY-Cape Girardeau, MO-Harrisburg-Mount Vernon, IL
WSJU-TV San Juan PR
WSJV South Bend-Elkhart, IN
***WSKA** Elmira (Corning), NY
***WSKG-TV** Binghamton, NY
WSKY-TV Norfolk-Portsmouth-Newport News, VA
WSLS-TV Roanoke-Lynchburg, VA
WSMH Flint-Saginaw-Bay City, MI
WSMV-TV Nashville, TN
WSNS Chicago
WSOC-TV Charlotte, NC
WSPA-TV Greenville-Spartanburg, SC-Asheville, NC-Anderson, SC
WSPX-TV Syracuse, NY
***WSRE** Mobile, AL-Pensacola (Ft. Walton Beach), FL
WSST-TV Albany, GA
WSTE Ponce PR
WSTM-TV Syracuse, NY
WSTR-TV Cincinnati, OH
WSUR-TV Ponce PR
WSVI Christiansted VI
WSVN Miami-Ft. Lauderdale, FL
WSWB Wilkes Barre-Scranton, PA
WSWG Tallahassee, FL-Thomasville, GA
***WSWP-TV** Bluefield-Beckley-Oak Hill, WV
WSYM-TV Lansing, MI
WSYR-TV Syracuse, NY
WSYT Syracuse, NY
WSYX Columbus, OH
WTAE-TV Pittsburgh, PA
WTAJ-TV Johnstown-Altoona, PA
WTAP-TV Parkersburg, WV
WTAT-TV Charleston, SC
WTBS Atlanta
WTBY-TV New York
***WTCE-TV** West Palm Beach-Ft. Pierce, FL
***WTCI** Chattanooga, TN
WTCT Paducah, KY-Cape Girardeau, MO-Harrisburg-Mount Vernon, IL
WTCV San Juan PR
WTEN Albany-Schenectady-Troy, NY
WTEV-TV Jacksonville, FL
WTGS Savannah, GA
WTHI-TV Terre Haute, IN
WTHR Indianapolis, IN
WTIC-TV Hartford & New Haven, CT
WTIN Ponce PR
***WTIU** Indianapolis, IN
WTJP-TV Birmingham (Anniston, Tuscaloosa), AL
WTJR Quincy, IL-Hannibal, MO-Keokuk, IA
***WTJX-TV** Charlotte Amalie VI
WTKR Norfolk-Portsmouth-Newport News, VA
WTLF Tallahassee, FL-Thomasville, GA
WTLH Tallahassee, FL-Thomasville, GA
WTLJ Grand Rapids-Kalamazoo-Battle Creek, MI
WTLV Jacksonville, FL
WTLW Lima, OH
WTMJ-TV Milwaukee, WI
WTNH-TV Hartford & New Haven, CT
WTNZ Knoxville, TN
WTOC-TV Savannah, GA
WTOG Tampa-St. Petersburg (Sarasota), FL
WTOK-TV Meridian, MS
WTOL Toledo, OH
WTOM-TV Traverse City-Cadillac, MI
WTOV-TV Wheeling, WV-Steubenville, OH
WTPX Wausau-Rhinelander, WI
WTRF-TV Wheeling, WV-Steubenville, OH
WTSF Charleston-Huntington, WV
WTSP Tampa-St. Petersburg (Sarasota), FL
WTTA Tampa-St. Petersburg (Sarasota), FL
WTTE Columbus, OH
WTTG Washington, DC (Hagerstown, MD)
WTTK Indianapolis, IN
WTTO Birmingham (Anniston, Tuscaloosa), AL
WTTV Indianapolis, IN
***WTTW-TV** Chicago
WTVA Columbus-Tupelo-West Point, MS
WTVC Chattanooga, TN
WTVD Raleigh-Durham (Fayetteville), NC
WTVE Philadelphia
WTVF Nashville, TN
WTVG Toledo, OH
WTVH Syracuse, NY
***WTVI** Charlotte, NC
WTVJ Miami-Ft. Lauderdale, FL
WTVM Columbus, GA
WTVO Rockford, IL
***WTVP** Peoria-Bloomington, IL
WTVQ-TV Lexington, KY
WTVR-TV Richmond-Petersburg, VA
***WTVS** Detroit
WTVT Tampa-St. Petersburg (Sarasota), FL
WTVW Evansville, IN
WTVX West Palm Beach-Ft. Pierce, FL
WTVY Dothan, AL
WTVZ-TV Norfolk-Portsmouth-Newport News, VA
WTWC-TV Tallahassee, FL-Thomasville, GA
WTWO-TV Terre Haute, IN
WTXF-TV Philadelphia
WTXL-TV Tallahassee, FL-Thomasville, GA
WTXX Hartford & New Haven, CT
WUAB Cleveland-Akron (Canton), OH
WUCW Minneapolis-St. Paul, MN
***WUFT** Gainesville, FL
WUFX Jackson, MS
WUHF Rochester, NY
***WUJA** Caguas PR
***WUNC-TV** Raleigh-Durham (Fayetteville), NC
***WUND-TV** Greenville-New Bern-Washington, NC
***WUNE-TV** Charlotte, NC
***WUNF-TV** Greenville-Spartanburg, SC-Asheville, NC-Anderson, SC
***WUNG-TV** Charlotte, NC
WUNI Boston (Manchester, NH)
***WUNJ-TV** Wilmington, NC
***WUNK-TV** Greenville-New Bern-Washington, NC
***WUNL-TV** Greensboro-High Point-Winston Salem, NC
***WUNM-TV** Greenville-New Bern-Washington, NC
***WUNP-TV** Raleigh-Durham (Fayetteville), NC
***WUNU** Myrtle Beach-Florence, SC
WUPA Atlanta
WUPL New Orleans, LA
WUPV Richmond-Petersburg, VA
WUPW Toledo, OH
WUPX-TV Lexington, KY
WUSA Washington, DC (Hagerstown, MD)
***WUSF-TV** Tampa-St. Petersburg (Sarasota), FL
***WUSI-TV** Terre Haute, IN
WUTB Baltimore, MD
WUTF-TV Boston (Manchester, NH)
WUTR Utica, NY
WUTV Buffalo, NY
WUVC-TV Raleigh-Durham (Fayetteville), NC
WUVG-TV Atlanta
WUVN Hartford & New Haven, CT
WUVP-TV Philadelphia
WUXP-TV Nashville, TN
WVAH-TV Charleston-Huntington, WV
***WVAN-TV** Savannah, GA
WVBT Norfolk-Portsmouth-Newport News, VA
WVCY-TV Milwaukee, WI
WVEA-TV Tampa-St. Petersburg (Sarasota), FL
WVEC-TV Norfolk-Portsmouth-Newport News, VA
WVEN-TV Orlando-Daytona Beach-Melbourne, FL
WVEO Aguadilla PR
***WVER** Burlington, VT-Plattsburgh, NY
WVFX Clarksburg-Weston, WV
***WVIA-TV** Wilkes Barre-Scranton, PA
WVIF Christiansted VI
WVII-TV Bangor, ME
WVIR-TV Charlottesville, VA
WVIT Hartford & New Haven, CT
***WVIZ-TV** Cleveland-Akron (Canton), OH
WVLA Baton Rouge, LA
WVLR Knoxville, TN
WVLT-TV Knoxville, TN
WVNS-TV Bluefield-Beckley-Oak Hill, WV
WVNY Burlington, VT-Plattsburgh, NY
WVOZ-TV Ponce PR
***WVPT** Harrisonburg, VA
WVPX Cleveland-Akron (Canton), OH
***WVPY** Washington, DC (Hagerstown, MD)
***WVTA** Burlington, VT-Plattsburgh, NY
***WVTB** Burlington, VT-Plattsburgh, NY
WVTM-TV Birmingham (Anniston, Tuscaloosa), AL
WVTV Milwaukee, WI
WVUE New Orleans, LA
***WVUT** Terre Haute, IN
WVVA Bluefield-Beckley-Oak Hill, WV
WVXF Charlotte Amalie VI
WWAY Wilmington, NC
WWAZ-TV Green Bay-Appleton, WI
WWBT Richmond-Petersburg, VA
WWCP-TV Johnstown-Altoona, PA
WWCW Roanoke-Lynchburg, VA
WWDP Boston (Manchester, NH)
WWHO Columbus, OH
WWJ-TV Detroit
WWJX Jackson, MS
WWLP Springfield-Holyoke, MA
WWL-TV New Orleans, LA
WWMB Myrtle Beach-Florence, SC
WWMT Grand Rapids-Kalamazoo-Battle Creek, MI
WWNY-TV Watertown, NY
WWOR-TV New York
***WWPB** Washington, DC (Hagerstown, MD)
WWPX Washington, DC (Hagerstown, MD)
WWRS-TV Milwaukee, WI
WWSB Tampa-St. Petersburg (Sarasota), FL
WWSI Philadelphia
WWTI Watertown, NY
WWTO-TV Chicago
WWTV Traverse City-Cadillac, MI
WWUP-TV Traverse City-Cadillac, MI
WXCW Ft. Myers-Naples, FL
***WXEL-TV** West Palm Beach-Ft. Pierce, FL
WXFT-TV Chicago
***WXGA-TV** Jacksonville, FL
WXIA-TV Atlanta
WXII-TV Greensboro-High Point-Winston Salem, NC
WXIN Indianapolis, IN
WXIX-TV Cincinnati, OH
WXLV-TV Greensboro-High Point-Winston Salem, NC
WXMI Grand Rapids-Kalamazoo-Battle Creek, MI
WXOW-TV La Crosse-Eau Claire, WI
WXPX Tampa-St. Petersburg (Sarasota), FL
WXTV New York
WXTX Columbus, GA
WXVT Greenwood-Greenville, MS
WXXA-TV Albany-Schenectady-Troy, NY
***WXXI-TV** Rochester, NY
WXXV-TV Biloxi-Gulfport, MS
WXYZ-TV Detroit
***WYBE** Philadelphia
***WYCC** Chicago
WYCW Greenville-Spartanburg, SC-Asheville, NC-Anderson, SC
WYDC Elmira (Corning), NY
***WYDN** Boston (Manchester, NH)
WYDO Greenville-New Bern-Washington, NC
***WYES-TV** New Orleans, LA

WYFF Greenville-Spartanburg, SC-Asheville, NC-Anderson, SC
***WYIN** Chicago
WYLE Huntsville-Decatur (Florence), AL
WYMT-TV Lexington, KY

WYOU Wilkes Barre-Scranton, PA
WYOW Wausau-Rhinelander, WI
WYPX Albany-Schenectady-Troy, NY
WYTV Youngstown, OH
WYZZ-TV Peoria-Bloomington, IL

WZDX Huntsville-Decatur (Florence), AL
WZMY-TV Boston (Manchester, NH)
WZPX Grand Rapids-Kalamazoo-Battle Creek, MI
WZRB Columbia, SC

WZTV Nashville, TN
WZVI Charlotte Amalie VI
WZVN-TV Ft. Myers-Naples, FL
WZZM-TV Grand Rapids-Kalamazoo-Battle Creek, MI

XETV San Diego, CA
XEWT-TV San Diego, CA
XHIJ El Paso (Las Cruces, NM), TX

Canadian Television Stations by Call Letters

CBAFT Moncton, NB
CBAT Fredericton-Saint John, NB
CBCP-TV-1 Shaunavon, SK
CBCP-TV-2 Cypress Hills, SK
CBCP-TV-3 Ponteix, SK
CBCT Charlottetown, PE
*CBEFT Windsor, ON
CBET Windsor, ON
CBFST-2 Temiscaning, PQ
CBFT Montreal, PQ
CBFT-2 Mont-Laurier, PQ
CBHT Halifax, NS
CBHT-11 Mulgrave, NS
CBHT-3 Yarmouth, NS
CBHT-4 Sheet Harbour, NS
CBIMT Iles-de-la-Madeleine, PQ
CBIT Sydney, NS
CBIT-2 Cheticamp, NS
CBJET Chicoutimi, PQ
CBKFT Regina, SK
CBKFT-3 Debden, SK
CBKFT-4 Saint Brieux, SK
CBKFT-5 Xenon Park, SK
CBKFT-6 Gravelbourg, SK
CBKFT-9 Bellegarde, SK
CBKST Saskatoon, SK
CBKST-1 Stranraer, SK
CBKT Regina, SK
CBKT-2 Willow Bunch, SK
CBLAT Geraldton, ON
CBLAT-4 Marathon, ON
CBLT Toronto, ON
CBMT Montreal, PQ
CBNAT Botwood, NF
CBNAT-1 Baie Verte, NF
CBNAT-4 Saint Anthony, NF
CBNAT-9 Mount St. Margaret, NF
*CBNLT Labrador City, NF
CBNT Saint John's, NF
CBNT-1 Port Rexton, NF
CBNT-2 Placentia, NF
CBNT-3 Marystown, NF
CBOFT Ottawa, ON
CBOT Ottawa, ON
CBRT Calgary, AB
CBUBT-1 Canal Flats, BC
CBUBT-7 Cranbrook, BC
CBUFT Vancouver, BC
CBUFT-2 Kamloops, BC
CBUFT-3 Terrace, BC
CBUT Vancouver, BC
CBVD-TV Malartic, PQ
CBVT Quebec City, PQ
CBVT-2 La Tuque, PQ
CBWAT Kenora, ON
CBWBT Flin Flon, MB
CBWFT Winnipeg, MB
CBWFT-10 Brandon, MB
CBWFT-4 Ste-Rose-du-Lac, MB
CBWGT Fisher Branch, MB
CBWST Baldy Mountain, MB
CBWT Winnipeg, MB
CBWT-2 Lac du Bonnet, MB
CBWYT Mafeking, MB
CBXAT Grande Prairie, AB
CBXAT-2 High Prairie, AB
CBXAT-3 Manning, AB
CBXFT Edmonton, AB
CBXFT-1 Bonnyville, AB
CBXFT-6 Fort McMurray, AB
CBXFT-8 Grande Prairie, AB
CBXT Edmonton, AB
CBYT-3 Bonne Bay, NF
CFAP-TV Quebec City, PQ
CFCF-TV Montreal, PQ
CFCM-TV Quebec City, PQ
CFCN-TV Calgary, AB
CFCN-TV-1 Drumheller, AB
CFCN-TV-5 Lethbridge, AB
CFCN-TV-8 Medicine Hat, AB
CFER-TV Rimouski, PQ
CFER-TV-2 Sept-Iles, PQ
CFGS-TV Gatineau, PQ
CFJC-TV Kamloops, BC
CFJP-TV Montreal, PQ
CFKM-TV Trois-Rivieres, PQ
CFKS-TV Sherbrooke, PQ
CFMT-TV Toronto, ON
CFPL-TV London, ON
CFQC-TV Saskatoon, SK
CFQC-TV-2 North Battleford, SK
CFRE-TV Regina, SK
CFRN-TV Edmonton, AB
CFRN-TV-1 Grande Prairie, AB
CFRN-TV-2 Peace River, AB
CFRN-TV-3 Whitecourt, AB
CFRN-TV-4 Ashmont, AB
CFRN-TV-5 Lac La Biche, AB
CFRN-TV-6 Red Deer, AB
CFRN-TV-7 Lougheed, AB
CFRN-TV-8 Grouard Mission-High Prairie, AB
CFRN-TV-9 Slave Lake, AB
CFRS-TV Jonquiere, PQ
CFSK-TV Saskatoon, SK
CFSO-TV Cardston, AB
CFTF-TV Riviere-du-Loup, PQ
CFTK-TV Terrace, BC
CFTM-TV Montreal, PQ
CFTO-TV Toronto, ON
*CFTU-TV Montreal, PQ
CFTV-TV Leamington, ON
CFVS-TV Val d'Or, PQ
CFWH-TV White Horse, YT
*CFYK-TV Yellowknife, NT
*CHAK-TV Inuvik, NT
CHAN-TV Vancouver, BC
CHAN-TV-1 Chilliwack, BC
CHAN-TV-2 Bowen Island, BC
CHAN-TV-3 Squamish, BC
CHAN-TV-5 Brackendale, BC
CHAN-TV-7 Whistler, BC
CHAT-TV Medicine Hat, AB
CHAT-TV-1 Pivot, AB
CHAU-TV Carleton, PQ
CHBC-TV Kelowna, BC
CHBX-TV Sault Ste. Marie, ON
CHCA-TV Red Deer, AB
CHCA-TV-1 Coronation, AB
CHCH-TV Hamilton, ON
CHEK-TV Victoria, BC
CHEK-TV-5 Campbell River, BC
CHEM-TV Trois-Rivieres, PQ
CHEX-TV Peterborough, ON
CHEX-TV-2 Oshawa, ON
CHFD-TV Thunder Bay, ON
CHKM-TV Kamloops, BC
CHLT-TV Sherbrooke, PQ
CHMI-TV Portage la Prarie, MB
CHNM-TV Vancouver, BC
CHNU-TV Fraser Valley, BC
CHOT-TV Gatineau, PQ
CHRO-TV Pembroke, ON
CHRO-TV-43 Ottawa, ON
CHVC-TV Valemount, BC
CHWI-TV Windsor, ON
CIAN-TV Calgary, AB
*CICA-TV Toronto, ON
CICC-TV Yorkton, SK
CICI-TV Sudbury, ON
CICI-TV-1 Elliot Lake, ON
*CICO-TV-18 London, ON
*CICO-TV-19 Sudbury, ON
*CICO-TV-20 Sault Ste. Marie, ON
*CICO-TV-24 Ottawa, ON
*CICO-TV-28 Kitchener, ON
*CICO-TV-32 Windsor, ON
*CICO-TV-59 Chatham, ON
*CICO-TV-9 Thunder Bay, ON
CICT-TV Calgary, AB
CIEW-TV Warmley, SK
CIHF-TV Halifax, NS
CIII-TV Paris, ON
CIII-TV-2 Bancroft, ON
CIII-TV-22 Stevenson, ON
CIII-TV-27 Peterborough, ON
CIII-TV-29 Oil Springs, ON
CIII-TV-4 Owen Sound, ON
CIII-TV-41 Toronto, ON
CIII-TV-6 Ottawa, ON
CIII-TV-7 Midland, ON
CIIT-TV Winnipeg, MB
CIMC-TV Isle Madame, NS
CIMT-TV Riviere-du-Loup, PQ
CIPA-TV Prince Albert, SK
CISA-TV Lethbridge, AB
CITL-TV Lloydminster, AB
CITO-TV Timmins, ON
CITO-TV-2 Kearns, ON
CITS-TV Hamilton, ON
CITV-TV Edmonton, AB
CITY-DT Toronto, ON
CITY-TV Toronto, ON
*CIVA-TV Rouyn, PQ
*CIVB-TV Rimouski, PQ
*CIVC-TV Trois-Rivieres, PQ
*CIVF-TV Baie-Trinite, PQ
*CIVG-TV Sept-Iles, PQ
CIVI-TV Victoria, BC
*CIVM-TV Montreal, PQ
*CIVO-TV Gatineau, PQ
*CIVP-TV Chapeau, PQ
*CIVQ-TV Quebec City, PQ
*CIVS-TV Sherbrooke, PQ
CIVT-TV Vancouver, BC
*CIVV-TV Chicoutimi, PQ
*CJAL-TV Edmonton, AB
CJBN-TV Kenora, ON
CJBR-TV Rimouski, PQ
CJCB-TV Sydney, NS
CJCB-TV-1 Inverness, NS
CJCB-TV-2 Antigonish, NS
CJCH-TV Halifax, NS
CJCH-TV-1 Canning, NS
CJCH-TV-6 Caledonia, NS
CJCN-TV Grand Falls, NF
CJDC-TV Dawson Creek, BC
CJIL-TV Lethbridge, AB
CJMT-TV Toronto, ON
CJNT-TV Montreal, PQ
CJOH-TV Ottawa, ON
CJOH-TV-6 Deseronto, ON
CJOH-TV-8 Cornwall, ON
CJOM-TV Argentia, NF
CJON-TV St John's, NF
CJOX-TV-1 Grand Bank, NF
CJPM-TV Chicoutimi, PQ
CJSV-TV Stephenville, NF
CJTG-TV Meander River, AB
CJWB-TV Bonavista, NF
CJWN-TV Corner Brook, NF
CKAL-TV Calgary, AB
CKAM-TV Upsalquitch Lake, NB
CKBQ-TV Melfort, SK
CKCD-TV Campbellton, NB
CKCK-TV Regina, SK
CKCK-TV-1 Colgate, SK
CKCK-TV-2 Willow Bunch, SK
CKCO-TV Kitchener, ON
CKCO-TV-2 Wiarton, ON
CKCO-TV-3 Sarnia, ON
CKCW-TV Moncton, NB
CKCW-TV-1 Charlottetown, PE
CKCW-TV-2 Saint Edward, PE
CKEM-TV Edmonton, AB
CKLT-TV Saint John, NB
CKMC-TV Swift Current, SK
CKMC-TV-1 Golden Prairie, SK
CKMI-TV Quebec City, PQ
CKND-TV Winnipeg, MB
CKNX-TV Wingham, ON
CKNY-TV North Bay, ON
CKPG-TV Prince George, BC
CKPR-TV Thunder Bay, ON
CKRN-TV Rouyn-Noranda, PQ
CKRN-TV-3 Bearn-Fabre, PQ
CKRT-TV Riviere-du-Loup, PQ
CKSA-TV Lloydminster, AB
CKSH-TV Sherbrooke, PQ
CKTM-TV Trois-Rivieres, PQ
CKVR-TV Barrie, ON
CKVU-TV Vancouver, BC
CKWS-TV Kingston, ON
CKXT-TV Toronto, ON
CKX-TV Brandon, MB
CKX-TV-1 Foxwarren, MB
CKYB-TV Brandon, MB
CKY-TV Winnipeg, MB

U.S. Television Stations by Analog Channel

Channel 0

*KETZ El Dorado, AR
WTLF Tallahassee, FL
WHMM-DT Hammond, LA
KPXJ Minden, LA
*KCGE-DT Crookston, MN
*KMDE Devils Lake, ND
WMCN-TV Atlantic City, NJ
*KNMD-TV Santa Fe, NM
KEYU Elk City, OK
KOHD Bend, OR
KLDT Lake Dallas, TX
*KUEW Saint George, UT
WZVI Charlotte Amalie, VI
WTPX Antigo, WI

Channel 2

KTUU-TV Anchorage, AK
KATN Fairbanks, AK
*WDIQ Dozier, AL
*KETS Little Rock, AR
*KVZK-2 Pago Pago, AS
KNAZ-TV Flagstaff, AZ
KCBS-TV Los Angeles, CA
KTVU Oakland, CA
KWGN-TV Denver, CO
WESH Daytona Beach, FL
*WPBT Miami, FL
WSB-TV Atlanta, GA
KHBC-TV Hilo, HI
KHON-TV Honolulu, HI
KGAN-TV Cedar Rapids, IA
KBCI-TV Boise, ID
WBBM-TV Chicago, IL
WTWO-TV Terre Haute, IN
KSNC Great Bend, KS
WBRZ Baton Rouge, LA
*WGBH-TV Boston, MA
WMAR-TV Baltimore, MD
WLBZ Bangor, ME
WJBK Detroit, MI
*KTCA-TV Saint Paul, MN
KQTV Saint Joseph, MO
KTVI Saint Louis, MO
*WMAB-TV Mississippi State, MS
KTVQ Billings, MT
*WUND-TV Edenton, NC
WFMY-TV Greensboro, NC
KXMA-TV Dickinson, ND
*KGFE Grand Forks, ND
KNOP-TV North Platte, NE
KASA-TV Santa Fe, NM
KTVN Reno, NV
WGRZ-TV Buffalo, NY
WCBS-TV New York, NY
WKTV Utica, NY
WDTN Dayton, OH
KJRH Tulsa, OK
KOTI Klamath Falls, OR
KATU Portland, OR
KDKA-TV Pittsburgh, PA
WKAQ-TV San Juan, PR
WCBD-TV Charleston, SC
*KUSD-TV Vermillion, SD
WKRN-TV Nashville, TN
*WETP-TV Sneedville, TN
*KACV-TV Amarillo, TX
*KDTN Denton, TX
KCWX Fredericksburg, TX
KPRC-TV Houston, TX
KMID Midland, TX
KUTV Salt Lake City, UT
KREM-TV Spokane, WA
WBAY-TV Green Bay, WI
KTWO-TV Casper, WY
KJWY Jackson, WY

Channel 3

*KTOO-TV Juneau, AK
KFTU-TV Douglas, AZ
KTVK Phoenix, AZ
KIEM-TV Eureka, CA
KCRA-TV Sacramento, CA
KEYT-TV Santa Barbara, CA
KREG-TV Glenwood Springs, CO
KUPN Sterling, CO
WFSB Hartford, CT
WEAR-TV Pensacola, FL
*WEDU Tampa, FL
WRBL Columbus, GA
WSAV-TV Savannah, GA
KGMV Wailuku, HI
KIMT Mason City, IA
KIDK-TV Idaho Falls, ID
KLEW-TV Lewiston, ID
WCIA-TV Champaign, IL
WSIL-TV Harrisburg, IL
*KSWK Lakin, KS
KSNW Wichita, KS
WAVE Louisville, KY
KATC Lafayette, LA
KTBS-TV Shreveport, LA
WJMN-TV Escanaba, MI
WWMT Kalamazoo, MI
KDLH Duluth, MN
KTVO Kirksville, MO
KYTV Springfield, MO
WLBT Jackson, MS
KRTV Great Falls, MT
KYUS-TV Miles City, MT
WBTV Charlotte, NC
WWAY Wilmington, NC
*KBME-TV Bismarck, ND
*KLNE-TV Lexington, NE
KMTV Omaha, NE
KOFT Farmington, NM
*KENW Portales, NM
KVNV Ely, NV
KVBC Las Vegas, NV
WSTM-TV Syracuse, NY
WKYC-TV Cleveland, OH
*KOET Eufaula, OK
*KOAB-TV Bend, OR
*WPSU-TV Clearfield, PA
KYW-TV Philadelphia, PA
*WIPM-TV Mayaguez, PR
KDLO-TV Florence, SD
KOTA-TV Rapid City, SD
WRCB-TV Chattanooga, TN
WREG-TV Memphis, TN
KBTX-TV Bryan, TX
KIII Corpus Christi, TX
KSAN-TV San Angelo, TX
KFDX-TV Wichita Falls, TX
KCBU Price, UT
WHSV-TV Harrisonburg, VA
WTKR Norfolk, VA
WCAX-TV Burlington, VT
WISC-TV Madison, WI
WSAZ-TV Huntington, WV

Channel 4

KTBY Anchorage, AK
*KYUK-TV Bethel, AK
KUBD Ketchikan, AK
KJNP-TV North Pole, AK
WTVY Dothan, AL
KARK-TV Little Rock, AR
*KVZK-4 Pago Pago, AS
KTFL Flagstaff, AZ
KVOA Tucson, AZ
KNBC Los Angeles, CA
KRON-TV San Francisco, CA
KCNC-TV Denver, CO
KFQX Grand Junction, CO
WRC-TV Washington, DC
WJXT Jacksonville, FL
WFOR-TV Miami, FL
KITV Honolulu, HI
KTIV Sioux City, IA
*KAID-TV Boise, ID
WHBF-TV Rock Island, IL
WTTV Bloomington, IN
KLBY Colby, KS
WWL-TV New Orleans, LA
WBZ-TV Boston, MA
WTOM-TV Cheboygan, MI
WDIV Detroit, MI
WCCO-TV Minneapolis, MN
WDAF-TV Kansas City, MO
KMOV Saint Louis, MO
WCBI-TV Columbus, MS
KXLF-TV Butte, MT
KHMT Hardin, MT
*WUNC-TV Chapel Hill, NC
WSKY-TV Manteo, NC
KXJB-TV Valley City, ND
*KWSE Williston, ND
KDUH-TV Scottsbluff, NE
KSNB-TV Superior, NE
KOB-TV Albuquerque, NM
KRNV Reno, NV
WIVB-TV Buffalo, NY
WNBC New York, NY
WCMH-TV Columbus, OH
KFOR-TV Oklahoma City, OK
KPIC Roseburg, OR
WTAE-TV Pittsburgh, PA
WAPA-TV San Juan, PR
WCIV Charleston, SC
WYFF Greenville, SC
KPRY-TV Pierre, SD
WSMV-TV Nashville, TN
KAMR-TV Amarillo, TX
KWAB-TV Big Spring, TX
KDFW Dallas, TX
KDBC-TV El Paso, TX
KGBT-TV Harlingen, TX
KBTV-TV Port Arthur, TX
WOAI-TV San Antonio, TX
KCSG Cedar City, UT
KTVX Salt Lake City, UT
KOMO-TV Seattle, WA
KXLY-TV Spokane, WA
WBIJ Crandon, WI
WTMJ-TV Milwaukee, WI
WOAY-TV Oak Hill, WV
*KCWC-TV Lander, WY

Channel 5

KYES-TV Anchorage, AK
WKRG-TV Mobile, AL
KFSM-TV Fort Smith, AR
*KVZK-5 Pago Pago, AS
KPHO-TV Phoenix, AZ
KTLA Los Angeles, CA
KPIX-TV San Francisco, CA
KREX-TV Grand Junction, CO
KOAA-TV Pueblo, CO
WTTG Washington, DC
*WUFT Gainesville, FL
WPTV West Palm Beach, FL
WAGA Atlanta, GA
KFVE Honolulu, HI
WOI-TV Ames, IA
KIDA Sun Valley, ID
WMAQ-TV Chicago, IL
KALB-TV Alexandria, LA
WCVB-TV Boston, MA
WABI-TV Bangor, ME
WNEM-TV Bay City, MI
WBKP Calumet, MI
KSTP-TV Saint Paul, MN
KCTV Kansas City, MO
KSDK Saint Louis, MO
KXGN-TV Glendive, MT
KFBB-TV Great Falls, MT
WRAL-TV Raleigh, NC
KFYR-TV Bismarck, ND
KHAS-TV Hastings, NE
*KNME-TV Albuquerque, NM
KVVU-TV Henderson, NV
*KNPB Reno, NV
WNYW New York, NY
WPTZ North Pole, NY
WTVH Syracuse, NY
WLWT Cincinnati, OH
WEWS Cleveland, OH
KOCO-TV Oklahoma City, OK
KOBI Medford, OR
WORA-TV Mayaguez, PR
WCSC-TV Charleston, SC
KIVV-TV Lead, SD
KDLV-TV Mitchell, SD
WMC-TV Memphis, TN
WTVF Nashville, TN
KXAS-TV Fort Worth, TX
*KTXT-TV Lubbock, TX
KENS-TV San Antonio, TX
KRGV-TV Weslaco, TX
KSL-TV Salt Lake City, UT
WCYB-TV Bristol, VA
KING-TV Seattle, WA
WFRV-TV Green Bay, WI
WDTV Weston, WV
KGWN-TV Cheyenne, WY
KGWL-TV Lander, WY

Channel 6

WBRC Birmingham, AL
*KEMV Mountain View, AR
KMOH-TV Kingman, AZ
*KUAT-TV Tucson, AZ
KVIQ Eureka, CA
*KVIE Sacramento, CA
KSBY San Luis Obispo, CA
*KRMA-TV Denver, CO
KREZ-TV Durango, CO
WTVJ Miami, FL
WKMG-TV Orlando, FL
WJBF Augusta, GA
WCTV Thomasville, GA
KLEI Kailua-Kona, HI
KWQC-TV Davenport, IA
KIVI Nampa, ID
KPVI Pocatello, ID
WRTV Indianapolis, IN
KBSD-TV Ensign, KS
WPSD-TV Paducah, KY
WDSU New Orleans, LA
WLNE New Bedford, MA
WCSH Portland, ME
XETV Tijuana, MEX
*WCML-TV Alpena, MI
WLNS-TV Lansing, MI
WLUC-TV Marquette, MI
KAAL Austin, MN
*KMOS-TV Sedalia, MO
WABG-TV Greenwood, MS
KSVI Billings, MT
KTVM Butte, MT
WECT Wilmington, NC
WDAY-TV Fargo, ND
*KSRE Minot, ND
KWNB-TV Hayes Center, NE
WOWT Omaha, NE
KOCT Carlsbad, NM
KOBG-TV Silver City, NM
KBNY Ely, NV
WRGB Schenectady, NY
WSYX Columbus, OH
KOTV Tulsa, OK
KOIN Portland, OR
WJAC-TV Johnstown, PA
WPVI-TV Philadelphia, PA
*WIPR-TV San Juan, PR
KPLO-TV Reliance, SD
WATE-TV Knoxville, TN
KFDM-TV Beaumont, TX
KRIS-TV Corpus Christi, TX
KIDY San Angelo, TX
KCEN-TV Temple, TX
KTAL-TV Texarkana, TX
KAUZ-TV Wichita Falls, TX
KBCJ Vernal, UT
WTVR-TV Richmond, VA
KHQ-TV Spokane, WA
WITI Milwaukee, WI
KBJR-TV Superior, WI
WVVA Bluefield, WV
*KPTW Casper, WY

Channel 7

*KAKM Anchorage, AK
KFXF Fairbanks, AK
*WCIQ Mount Cheaha, AL
KATV Little Rock, AR
KAZT-TV Prescott, AZ
KVYE El Centro, CA
KABC-TV Los Angeles, CA
KRCR-TV Redding, CA
KGO-TV San Francisco, CA
KMGH-TV Denver, CO
WJLA-TV Washington, DC
*WJCT Jacksonville, FL
WSVN Miami, FL
WJHG-TV Panama City, FL
KAII-TV Wailuku, HI
KWWL Waterloo, IA
KTVB Boise, ID
WLS-TV Chicago, IL
WTVW Evansville, IN
KBSH-TV Hays, KS
KOAM-TV Pittsburg, KS
KPLC Lake Charles, LA
WHDH-TV Boston, MA
WVII-TV Bangor, ME
WXYZ-TV Detroit, MI
WPBN-TV Traverse City, MI
KCCO-TV Alexandria, MN
KHQA-TV Hannibal, MO
WDAM-TV Laurel, MS
KBZK Bozeman, MT
WITN-TV Washington, NC
KQCD-TV Dickinson, ND
KJRR Jamestown, ND
*KMNE-TV Bassett, NE
KETV Omaha, NE
KOAT-TV Albuquerque, NM
KEGS Goldfield, NV
KWNV Winnemucca, NV
WKBW-TV Buffalo, NY
WWNY-TV Carthage, NY
WABC-TV New York, NY
WHIO-TV Dayton, OH
KSWO-TV Lawton, OK
*KOAC-TV Corvallis, OR
WSTE Ponce, PR
*WITV Charleston, SC
WSPA-TV Spartanburg, SC
KEVN-TV Rapid City, SD
WBBJ-TV Jackson, TN
KVII-TV Amarillo, TX
KTBC Austin, TX
KVIA-TV El Paso, TX
KOSA-TV Odessa, TX
KLTV Tyler, TX
*KUED Salt Lake City, UT
WDBJ Roanoke, VA
KIRO-TV Seattle, WA
*KSPS-TV Spokane, WA
WSAW-TV Wausau, WI
WTRF-TV Wheeling, WV
KSWY Sheridan, WY

Channel 8

KJUD Juneau, AK
WAKA Selma, AL
KAIT Jonesboro, AR
*KAET Phoenix, AZ
KUNO-TV Fort Bragg, CA
KSBW Salinas, CA
KFMB-TV San Diego, CA
KJCT Grand Junction, CO
*KTSC Pueblo, CO

WTNH-TV New Haven, CT
WGEN-TV Key West, FL
WFLA-TV Tampa, FL
*WGTV Athens, GA
*WXGA-TV Waycross, GA
KUAM-TV Hagatna, GU
KCCI Des Moines, IA
KIFI-TV Idaho Falls, ID
*WSIU-TV Carbondale, IL
WQAD-TV Moline, IL
WISH-TV Indianapolis, IN
*KPTS-TV Hutchinson, KS
KNOE-TV Monroe, LA
WVUE New Orleans, LA
WMTW-TV Poland Spring, ME
WAGM-TV Presque Isle, ME
WOOD-TV Grand Rapids, MI
WDHS Iron Mountain, MI
WGTQ Sault Ste. Marie, MI
*WDSE-TV Duluth, MN
KOMU-TV Columbia, MO
KULR-TV Billings, MT
KPAX-TV Missoula, MT
WGHP High Point, NC
WFXI Morehead City, NC
WDAZ-TV Devils Lake, ND
KUMV-TV Williston, ND
KLKN Lincoln, NE
KSNK McCook, NE
KOBR Roswell, NM
KLAS-TV Las Vegas, NV
KOLO-TV Reno, NV
WROC-TV Rochester, NY
WJW Cleveland, OH
KTUL Tulsa, OK
*KSYS Medford, OR
KGW Portland, OR
WWCP-TV Johnstown, PA
WGAL Lancaster, PA
*KESD-TV Brookings, SD
*KZSD-TV Martin, SD
WVLT-TV Knoxville, TN
*WNPT Nashville, TN
WFAA-TV Dallas, TX
*KUHT Houston, TX
KGNS-TV Laredo, TX
KLST San Angelo, TX
WRIC-TV Petersburg, VA
WSVI Christiansted, VI
WKBT La Crosse, WI
WCHS-TV Charleston, WV
*KWYP-TV Laramie, WY

Channel 9

*KUAC-TV Fairbanks, AK
*KETG Arkadelphia, AR
KCFG Flagstaff, AZ
KGUN Tucson, AZ
KECY-TV El Centro, CA
KCAL Los Angeles, CA
*KIXE-TV Redding, CA
*KQED San Francisco, CA
KUSA-TV Denver, CO
WUSA Washington, DC
WFTV Orlando, FL
WTVM Columbus, GA
*WVAN-TV Savannah, GA
KGMD-TV Hilo, HI
KGMB Honolulu, HI
KCRG-TV Cedar Rapids, IA
KCAU-TV Sioux City, IA
KNIN-TV Caldwell, ID
WGN-TV Chicago, IL
*WNIN-TV Evansville, IN
*KOOD Hays, KS
WAFB Baton Rouge, LA
WWTV Cadillac, MI
*KAWE Bemidji, MN
KMSP-TV Minneapolis, MN
KMBC-TV Kansas City, MO
*KETC Saint Louis, MO
WTVA Tupelo, MS
*KUSM Bozeman, MT
KBBJ Havre, MT
KCFW-TV Kalispell, MT
WSOC-TV Charlotte, NC
WNCT-TV Greenville, NC
*KDSE Dickinson, ND
*KPNE-TV North Platte, NE
WMUR-TV Manchester, NH
WWOR-TV Secaucus, NJ
WSYR-TV Syracuse, NY
WCPO-TV Cincinnati, OH
WTOV-TV Steubenville, OH
KWTV Oklahoma City, OK
KEZI Eugene, OR
WSUR-TV Ponce, PR
KABY-TV Aberdeen, SD
*KBHE-TV Rapid City, SD
WTVC Chattanooga, TN
KRBC-TV Abilene, TX
KTSM-TV El Paso, TX
KTRE Lufkin, TX
KWES-TV Odessa, TX
*KLRN San Antonio, TX
*KUEN Ogden, UT
*KCTS-TV Seattle, WA
KCWK Walla Walla, WA
WAOW-TV Wausau, WI
*WSWP-TV Grandview, WV

Channel 10

*WBIQ Birmingham, AL
WALA-TV Mobile, AL
KTVE El Dorado, AR
KSAZ-TV Phoenix, AZ
KXTV Sacramento, CA
KGTV San Diego, CA
KREY-TV Montrose, CO
WPLG Miami, FL
WTSP Saint Petersburg, FL
WALB Albany, GA
*KMEB Wailuku, HI
*KISU-TV Pocatello, ID
WGEM-TV Quincy, IL
WTHI-TV Terre Haute, IN
KBSL-TV Goodland, KS
KAKE-TV Wichita, KS
KLFY-TV Lafayette, LA
*WCBB Augusta, ME
*WMEM-TV Presque Isle, ME
WBUP Ishpeming, MI
WILX-TV Onondaga, MI
WWUP-TV Sault Ste. Marie, MI
*KWCM-TV Appleton, MN
WDIO-TV Duluth, MN
KTTC Rochester, MN
KBRR Thief River Falls, MN
KOLR-TV Springfield, MO
KMTF Helena, MT
KMOT Minot, ND
KOLN Lincoln, NE
KSTF Scottsbluff, NE
KBIM-TV Roswell, NM
KOVT Silver City, NM
KENV Elko, NV
*KLVX Las Vegas, NV
WTEN Albany, NY
WHEC-TV Rochester, NY
WBNS-TV Columbus, OH
KTEN Ada, OK
KTVL Medford, OR
*KOPB-TV Portland, OR
WTAJ-TV Altoona, PA
WCAU Philadelphia, PA
WJAR Providence, RI
WIS Columbia, SC
*KTSD-TV Pierre, SD
WBIR-TV Knoxville, TN
*WKNO-TV Memphis, TN
KFDA-TV Amarillo, TX
KZTV Corpus Christi, TX
KTRG Del Rio, TX
KWTX-TV Waco, TX
WAVY-TV Portsmouth, VA
WSLS-TV Roanoke, VA
*KWSU-TV Pullman, WA
*WMVS Milwaukee, WI
KFNE Riverton, WY

Channel 11

KTVA Anchorage, AK
KTVF Fairbanks, AK
KTHV Little Rock, AR
*KDTP Holbrook, AZ
KMSB-TV Tucson, AZ
KYMA Yuma, AZ
KTTV Los Angeles, CA
KNTV San Jose, CA
KKTV Colorado Springs, CO
KKCO Grand Junction, CO
WINK-TV Fort Myers, FL
*WFSU-TV Tallahassee, FL
WXIA-TV Atlanta, GA
WTOC-TV Savannah, GA
KHAW-TV Hilo, HI
*KHET Honolulu, HI
*KDIN-TV Des Moines, IA
KMVT Twin Falls, ID
*WTTW-TV Chicago, IL
KSNG Garden City, KS
*KTWU Topeka, KS
WHAS-TV Louisville, KY
KAQY Columbia, LA
WBAL-TV Baltimore, MD
WBKB-TV Alpena, MI
KRII Chisholm, MN
KARE Minneapolis, MN
KPLR-TV Saint Louis, MO
WTOK-TV Meridian, MS
*KUFM-TV Missoula, MT
WTVD Durham, NC
KVLY-TV Fargo, ND
KXMD-TV Williston, ND
KGIN Grand Island, NE
*WENH-TV Durham, NH
KCHF Santa Fe, NM
KRXI-TV Reno, NV
WPIX New York, NY
WTOL Toledo, OH
*KOED-TV Tulsa, OK
KCBY-TV Coos Bay, OR
KFFX-TV Pendleton, OR
WPXI Pittsburgh, PA
WLII Caguas, PR
KHSD-TV Lead, SD
*KQSD-TV Lowry, SD
KELO-TV Sioux Falls, SD
WJHL-TV Johnson City, TN
*WLJT-TV Lexington, TN
KTVT Fort Worth, TX
KHOU-TV Houston, TX
KCBD Lubbock, TX
*KBYU-TV Provo, UT
KSTW Tacoma, WA
WLUK-TV Green Bay, WI
WVAH-TV Charleston, WV
KBEO Jackson, WY
KFNR Rawlins, WY

Channel 12

WSFA Montgomery, AL
KPNX Mesa, AZ
KHSL-TV Chico, CA
KCOY-TV Santa Maria, CA
*KBDI-TV Broomfield, CO
*WHYY-TV Wilmington, DE
WTLV Jacksonville, FL
WPEC West Palm Beach, FL
WRDW-TV Augusta, GA
*KGTF-TV Hagatna, GU
KMAU Wailuku, HI
*KIIN-TV Iowa City, IA
KTRV-TV Nampa, ID
*WILL-TV Urbana, IL
KWCH-TV Hutchinson, KS
*WYES-TV New Orleans, LA
KSLA-TV Shreveport, LA
*WMEB-TV Orono, ME
XEWT-TV Tijuana, MEX
WJRT-TV Flint, MI
KEYC-TV Mankato, MN
KCCW-TV Walker, MN
KFVS-TV Cape Girardeau, MO
KODE-TV Joplin, MO
*WMAE-TV Booneville, MS
WJTV Jackson, MS
KTVH Helena, MT
WCTI-TV New Bern, NC
WXII-TV Winston-Salem, NC
KXMB-TV Bismarck, ND
KNRR Pembina, ND
*KUON-TV Lincoln, NE
*KRNE-TV Merriman, NE
KVIH-TV Clovis, NM
KOBF Farmington, NM
WBNG-TV Binghamton, NY
WKRC-TV Cincinnati, OH
*KWET Cheyenne, OK
KDRV Medford, OR
KPTV Portland, OR
WICU-TV Erie, PA
WOLE-TV Aguadilla, PR
WPRI-TV Providence, RI
KTTM Huron, SD
WDEF-TV Chattanooga, TN
KBMT Beaumont, TX
KSAT-TV San Antonio, TX
KXII Sherman, TX
KTXS-TV Sweetwater, TX
KUTF Logan, UT
KUSG Saint George, UT
WWBT Richmond, VA
*WTJX-TV Charlotte Amalie, VI
KVOS-TV Bellingham, WA
WISN-TV Milwaukee, WI
WJFW-TV Rhinelander, WI
WBOY-TV Clarksburg, WV
KSGW-TV Sheridan, WY

Channel 13

KIMO Anchorage, AK
KTNL-TV Sitka, AK
WVTM-TV Birmingham, AL
*KAFT Fayetteville, AR
KFPH-TV Flagstaff, AZ
KOLD-TV Tucson, AZ
KSWT Yuma, AZ
*KEET Eureka, CA
KCOP Los Angeles, CA
KOVR Stockton, CA
KRDO-TV Colorado Springs, CO
WMBB Panama City, FL
WTVT Tampa, FL
WMAZ-TV Macon, GA
KHVO Hilo, HI
KHNL Honolulu, HI
WHO-TV Des Moines, IA
*KIPT-TV Twin Falls, ID
WPXS Mount Vernon, IL
WREX-TV Rockford, IL
WTHR Indianapolis, IN
KUPK-TV Garden City, KS
WIBW-TV Topeka, KS
WBKO-TV Bowling Green, KY
*KLTM-TV Monroe, LA
WJZ-TV Baltimore, MD
*WMED-TV Calais, ME
WGME-TV Portland, ME
WZZM-TV Grand Rapids, MI
*WNMU-TV Marquette, MI
WIRT Hibbing, MN
KRCG Jefferson City, MO
WLOX Biloxi, MS
KBAO Lewistown, MT
KECI-TV Missoula, MT
WLOS Asheville, NC
*KFME Fargo, ND
KXMC-TV Minot, ND
*KTNE-TV Alliance, NE
KHGI-TV Kearney, NE
*WNET Newark, NJ
KRQE Albuquerque, NM
KTNV Las Vegas, NV
WNYT Albany, NY
WHAM-TV Rochester, NY
WTVG Toledo, OH
*KETA Oklahoma City, OK
KVAL-TV Eugene, OR
*KTVR La Grande, OR
*WQED Pittsburgh, PA
WORO-TV Fajardo, PR
WBTW Florence, SC
*KPSD-TV Eagle Butte, SD
KSFY-TV Sioux Falls, SD
WHBQ-TV Memphis, TN
*KERA-TV Dallas, TX
*KCOS El Paso, TX
KTRK-TV Houston, TX
KVTV Laredo, TX
KLBK-TV Lubbock, TX
KSTU Salt Lake City, UT
WVEC-TV Hampton, VA
WSET-TV Lynchburg, VA
KCPQ Tacoma, WA
WEAU-TV Eau Claire, WI
WOWK-TV Huntington, WV
KCWY Casper, WY
KGWR-TV Rock Springs, WY

Channel 14

KDTV San Francisco, CA
KTFD-TV Boulder, CO
*WABW-TV Pelham, GA
WPXA Rome, GA
KTGM Tamuning, GU
KWHH Hilo, HI
KWHE Honolulu, HI
KMEG Sioux City, IA
*WSEC Jacksonville, IL
WFIE-TV Evansville, IN
KOCW Hoisington, KS
KFJX-TV Pittsburg, KS
KARD West Monroe, LA
*WCMU-TV Mount Pleasant, MI
*WMAW-TV Meridian, MS
WYDO Greenville, NC
WHKY-TV Hickory, NC
KMCY Minot, ND
KTFQ-TV Albuquerque, NM
*WPTO Oxford, OH
KTBO-TV Oklahoma City, OK
WTIN Ponce, PR
*WEBA-TV Allendale, SC
KCIT Amarillo, TX
KFOX-TV El Paso, TX
*KETH-TV Houston, TX
KXAM-TV Llano, TX
KJZZ-TV Salt Lake City, UT
WFDC-TV Arlington, VA
WIWB Suring, WI
KGWC-TV Casper, WY

Channel 15

WHDF Florence, AL
WPMI-TV Mobile, AL
KNXV-TV Phoenix, AZ
*KPBS San Diego, CA
*WCEU New Smyrna Beach, FL
KOGG Wailuku, HI
KYOU-TV Ottumwa, IA
KPIF Pocatello, ID
WICD-TV Champaign, IL
WANE-TV Fort Wayne, IN
*WKPC-TV Louisville, KY
KADN Lafayette, LA
*KSMQ-TV Austin, MN
KPOB-TV Poplar Bluff, MO
WXVT Greenville, MS
KVRR Fargo, ND
KXVO Omaha, NE
KINC Las Vegas, NV
WLYH-TV Lancaster, PA
WPDE-TV Florence, SC
KCLO-TV Rapid City, SD
*WKOP-TV Knoxville, TN
KXVA Abilene, TX
*KAMU-TV College Station, TX
*WHRO-TV Hampton-Norfolk, VA
*WBRA-TV Roanoke, VA
WVIF Christiansted, VI
*KCKA Centralia, WA
WMTV Madison, WI
WTAP-TV Parkersburg, WV

Channel 16

KLRT Little Rock, AR
*WUSF-TV Tampa, FL
*WUSI-TV Olney, IL

WTJR Quincy, IL
WNDU-TV South Bend, IN
WBOC-TV Salisbury, MD
KSNF Joplin, MO
KTAJ-TV Saint Joseph, MO
WAPT Jackson, MS
KTGF Great Falls, MT
WGPX Burlington, NC
KTUW Scottsbluff, NE
*WPBS-TV Watertown, NY
*WPTD Dayton, OH
KMTR Eugene, OR
KUNP La Grande, OR
WQEX Pittsburgh, PA
WNEP-TV Scranton, PA
WOST Mayaguez, PR
*WJWJ-TV Beaufort, SC
WGGS-TV Greenville, SC
*KDSD-TV Aberdeen, SD
WJKT Jackson, TN
*KEDT Corpus Christi, TX
KVAW Eagle Pass, TX
KPTB Lubbock, TX
KUPX Provo, UT
KONG-TV Everett, WA

Channel 17

WDBB Bessemer, AL
*KLEP Newark, AR
KGET-TV Bakersfield, CA
WCWJ Jacksonville, FL
*WLRN-TV Miami, FL
WTBS Atlanta, GA
KDSM-TV Des Moines, IA
WAND Decatur, IL
WTVO Rockford, IL
WXMI Grand Rapids, MI
*KTCI-TV Saint Paul, MN
KMIZ Columbia, MO
*WMAU-TV Bude, MS
KMMF Missoula, MT
WNCN Goldsboro, NC
*WUNE-TV Linville, NC
KBMY Bismarck, ND
KTVG Grand Island, NE
*WNED-TV Buffalo, NY
*WMHT Schenectady, NY
WDLI-TV Canton, OH
KDOR-TV Bartlesville, OK
WPHL-TV Philadelphia, PA
KTTW Sioux Falls, SD
WZTV Nashville, TN
KNIC-TV Blanco, TX
KPCB Snyder, TX
WVXF Charlotte Amalie, VI

Channel 18

WDHN Dothan, AL
KTTU-TV Tucson, AZ
*KVPT Fresno, CA
KSCI Long Beach, CA
*KRMJ Grand Junction, CO
WUVN Hartford, CT
WKCF Clermont, FL
*WCLP-TV Chatsworth, GA
KLJB-TV Davenport, IA
WLFI-TV Lafayette, IN
KAAS-TV Salina, KS
WLEX-TV Lexington, KY
*KLTL-TV Lake Charles, LA
WHTV Jackson, MI
*WMAV-TV Oxford, MS
KWYB Butte, MT
WCCB Charlotte, NC
WETM-TV Elmira, NY
*WNPI-TV Norwood, NY
WHIZ-TV Zanesville, OH
WTCV San Juan, PR
*KLRU-TV Austin, TX
KPTF Farwell, TX
KUPB Midland, TX
KJTL Wichita Falls, TX
WQOW-TV Eau Claire, WI
WVTV Milwaukee, WI

Channel 19

WHNT-TV Huntsville, AL
*KTEJ Jonesboro, AR
KUVS-TV Modesto, CA
*KBGH Filer, ID
WHOI Peoria, IL
WAZE-TV Madisonville, KY
WXIX-TV Newport, KY
WCDC Adams, MA
WMQF Marquette, MI
*WDCP-TV University Center, MI
*KCPT Kansas City, MO
*WMAH-TV Biloxi, MS
*WUNM-TV Jacksonville, NC
*KJRE Ellendale, ND
*KXNE-TV Norfolk, NE
KWBQ Santa Fe, NM
WOIO Shaker Heights, OH
KQCW Muskogee, OK
WPCW Jeannette, PA
WLTX Columbia, SC
WKPT-TV Kingsport, TN
KYTX Nacogdoches, TX
KVCT Victoria, TX
*KUES Richfield, UT
WCAV Charlottesville, VA
KEPR-TV Pasco, WA
WXOW-TV La Crosse, WI

Channel 20

WCOV-TV Montgomery, AL
KBBC-TV Bishop, CA
KBWB San Francisco, CA
*KBDM Yreka City, CA
KTVD Denver, CO
*KRMU Durango, CO
WTXX Waterbury, CT
WDCA Washington, DC
WBBH-TV Fort Myers, FL
WCJB-TV Gainesville, FL
*WCES-TV Wrens, GA
KIKU-TV Honolulu, HI
KWKB Iowa City, IA
*WYCC Chicago, IL
WICS-TV Springfield, IL
*WFYI-TV Indianapolis, IN
WHNO New Orleans, LA
WMYD Detroit, MI
*KSMN Worthington, MN
WCWG Lexington, NC
WUTR Utica, NY
*WOUB-TV Athens, OH
WKPV Ponce, PR
WBXX-TV Crossville, TN
KTXH Houston, TX
KTMW Salt Lake City, UT
*WVTB Saint Johnsbury, VT
KTBW-TV Tacoma, WA
*WHRM-TV Wausau, WI
KFNB-TV Casper, WY

Channel 21

WTTO Birmingham, AL
WMPV-TV Mobile, AL
*KPAZ-TV Phoenix, AZ
KFTV Hanford, CA
KXRM-TV Colorado Springs, CO
*WTCE-TV Fort Pierce, FL
WPXC-TV Brunswick, GA
KWHM Wailuku, HI
*KTIN Fort Dodge, IA
WPTA Fort Wayne, IN
*KDCK Dodge City, KS
WBNA Louisville, KY
*WKMU Murray, KY
*WCMW Manistee, MI
KQDS-TV Duluth, MN
*KOZK Springfield, MO
WPXG Concord, NH
KVMY Las Vegas, NV
KAME-TV Reno, NV
*WLIW Garden City, NY
*WXXI-TV Rochester, NY
WFMJ-TV Youngstown, OH
KTVZ Bend, OR
WHP-TV Harrisburg, PA

WWMB Florence, SC
WHNS Greenville, SC
KNBN Rapid City, SD
KTXA Arlington, TX
WWCW Lynchburg, VA
WHRE Virginia Beach, VA
*WHA-TV Madison, WI

Channel 22

WBMM Tuskegee, AL
*KRCB Cotati, CA
KWHY-TV Los Angeles, CA
KFCT Fort Collins, CO
WCLF Clearwater, FL
WSBS-TV Key West, FL
WJCL Savannah, GA
KWWF Waterloo, IA
*WMEC Macomb, IL
WSBT-TV South Bend, IN
*WVUT Vincennes, IN
*WKPI Pikeville, KY
WWLP Springfield, MA
*WMPT Annapolis, MD
*KAWB Brainerd, MN
WHLT Hattiesburg, MS
WLFL Raleigh, NC
*KRWG-TV Las Cruces, NM
WKEF Dayton, OH
*KFTS Klamath Falls, OR
KPXG Salem, OR
WPMY Pittsburgh, PA
WYOU Scranton, PA
WNJX-TV Mayaguez, PR
*WCTE Cookeville, TN
*KLTJ Galveston, TX
KLCW-TV Wolfforth, TX
WVNY Burlington, VT
KMYQ Seattle, WA
KSKN Spokane, WA

Channel 23

WLDM Tuscaloosa, AL
KAEF Arcata, CA
KERO-TV Bakersfield, CA
*KBSV Ceres, CA
WLTV Miami, FL
*WSRE Pensacola, FL
WELF-TV Dalton, GA
KCWI-TV Ames, IA
WBUI Decatur, IL
WIFR Freeport, IL
WNDY-TV Marion, IN
*WKZT-TV Elizabethtown, KY
WPFO Waterville, ME
*WKAR-TV East Lansing, MI
WUCW Minneapolis, MN
KBSI Cape Girardeau, MO
*WMAO-TV Greenwood, MS
KTMF Missoula, MT
*WNJS Camden, NJ
KNAT Albuquerque, NM
WXXA-TV Albany, NY
WNLO Buffalo, NY
WVPX Akron, OH
KOKI-TV Tulsa, OK
KMCB Coos Bay, OR
WATM-TV Altoona, PA
*WHMC Conway, SC
*KCSD-TV Sioux Falls, SD
KVEO Brownsville, TX
KUVN-TV Garland, TX
*KHCE-TV San Antonio, TX
*WCVE-TV Richmond, VA
KNDO Yakima, WA

Channel 24

KFTA-TV Fort Smith, AR
KNVN Chico, CA
KSEE Fresno, CA
*KVCR-TV San Bernardino, CA
*KMAS-TV Steamboat Springs, CO
*WEDH Hartford, CT
*WMFE-TV Orlando, FL
WGXA Macon, GA
*KYIN Mason City, IA
*WQPT-TV Moline, IL
KSAS-TV Wichita, KS

*WKYU-TV Bowling Green, KY
*KLPB-TV Lafayette, LA
*KLTS-TV Shreveport, LA
WUTB Baltimore, MD
KNLC Saint Louis, MO
WMDN Meridian, MS
KBTZ Butte, MT
KXND Minot, ND
KRWB-TV Roswell, NM
*WCNY-TV Syracuse, NY
WNWO-TV Toledo, OH
*KNMT-TV Portland, OR
WJET-TV Erie, PA
WJPX San Juan, PR
WTAT-TV Charleston, SC
WPTY-TV Memphis, TN
KVUE-TV Austin, TX
KPEJ Odessa, TX
KPNZ Ogden, UT
WDRL-TV Danville, VA
KBCB Bellingham, WA
KQUP Pullman, WA
WCGV-TV Milwaukee, WI
*WNPB-TV Morgantown, WV

Channel 25

*WHIQ Huntsville, AL
KVTN Pine Bluff, AR
*KQET Watsonville, CA
KDEN Longmont, CO
WJXX Orange Park, FL
WPBF Tequesta, FL
*WACS-TV Dawson, GA
WEEK-TV Peoria, IL
WEHT Evansville, IN
*WKAS Ashland, KY
*KLPA-TV Alexandria, LA
WFXT Boston, MA
WHAG-TV Hagerstown, MD
WEYI-TV Saginaw, MI
KNLJ Jefferson City, MO
WXXV-TV Gulfport, MS
*WUNK-TV Greenville, NC
KTEL-TV Carlsbad, NM
*WNYE-TV New York, NY
*WVIZ-TV Cleveland, OH
KOKH-TV Oklahoma City, OK
WOLO-TV Columbia, SC
KAVU-TV Victoria, TX
KXXV Waco, TX
KNDU Richland, WA
WLAX La Crosse, WI

Channel 26

WYLE Florence, AL
*WAIQ Montgomery, AL
KVTH Hot Springs, AR
KTSF San Francisco, CA
KMPH-TV Visalia, CA
WHPX New London, CT
*WETA-TV Washington, DC
WVEN-TV Daytona Beach, FL
WZVN-TV Naples, FL
WAGT Augusta, GA
*KAAH-TV Honolulu, HI
KGCW-TV Burlington, IA
*KCDT-TV Coeur d'Alene, ID
WCIU-TV Chicago, IL
WGNO New Orleans, LA
*WMEA-TV Biddeford, ME
KFTC Bemidji, MN
*KOZJ Joplin, MO
KLMN Great Falls, MT
WSFX-TV Wilmington, NC
*WUNL-TV Winston-Salem, NC
KNDX Bismarck, ND
*KYNE-TV Omaha, NE
WNYB Jamestown, NY
WBDT Springfield, OH
KMVU Medford, OR
*WQTO Ponce, PR
KINT-TV El Paso, TX
KRIV Houston, TX
KPXL Uvalde, TX
WGBA Green Bay, WI

Channel 27

*KUAS-TV Tucson, AZ
WRDQ Orlando, FL
WTXL-TV Tallahassee, FL
*KSIN Sioux City, IA
WTCT Marion, IL
*WQEC Quincy, IL
WCCU Urbana, IL
KSNT-TV Topeka, KS
WKYT-TV Lexington, KY
*WLPB-TV Baton Rouge, LA
WUNI Worcester, MA
*WCMV Cadillac, MI
KCWV Duluth, MN
KSFX-TV Springfield, MO
WLOV-TV West Point, MS
KCPM Grand Forks, ND
KRPV Roswell, NM
KREN-TV Reno, NV
*WBGU-TV Bowling Green, OH
WKBN-TV Youngstown, OH
WHTM-TV Harrisburg, PA
*WRJA-TV Sumter, SC
KDFI Dallas, TX
KLDO-TV Laredo, TX
WGNT Portsmouth, VA
WFXR-TV Roanoke, VA
WKOW-TV Madison, WI
KLWY Cheyenne, WY

Channel 28

*KCET Los Angeles, CA
WPGX Panama City, FL
WFTS Tampa, FL
*WJSP-TV Columbus, GA
KFXA Cedar Rapids, IA
WSJV Elkhart, IN
WLWC New Bedford, MA
*WCPB Salisbury, MD
*WFUM Flint, MI
WRDC Durham, NC
WTTE Columbus, OH
*KEPB-TV Eugene, OR
WBRE-TV Wilkes-Barre, PA
WTGS Hardeeville, SC
WNPX Cookeville, TN
KYLE Bryan, TX
KORO Corpus Christi, TX
KAMC Lubbock, TX
*WVER Rutland, VT
KAYU-TV Spokane, WA
*KBTC-TV Tacoma, WA
*WHWC-TV Menomonie, WI

Channel 29

WBIH Selma, AL
KHOG-TV Fayetteville, AR
KBAK-TV Bakersfield, CA
KBVU Eureka, CA
KSPX Sacramento, CA
WFLX West Palm Beach, FL
*WMUM-TV Cochran, GA
WTTK Kokomo, IN
*WKPD Paducah, KY
*WKSO-TV Somerset, KY
KVHP Lake Charles, LA
WGTU Traverse City, MI
WFTC Minneapolis, MN
KCWE Kansas City, MO
*WMPN-TV Jackson, MS
*KHNE-TV Hastings, NE
KUPT Hobbs, NM
WUTV Buffalo, NY
WTXF-TV Philadelphia, PA
*WNTV Greenville, SC
KMPX Decatur, TX
KABB San Antonio, TX
WVIR-TV Charlottesville, VA
KIMA-TV Yakima, WA
WLPX-TV Charleston, WV

Channel 30

KFSN-TV Fresno, CA
KCVU Paradise, CA
KPXN San Bernardino, CA
WVIT New Britain, CT
*WGCU Fort Myers, FL

WAWS Jacksonville, FL
*WPBA Atlanta, GA
*WTIU Bloomington, IN
KDNL-TV Saint Louis, MO
*WGBC Meridian, MS
WRAY-TV Wilson, NC
*WSKA Corning, NY
WQCW Portsmouth, OH
*WGTE-TV Toledo, OH
KTUZ-TV Shawnee, OK
KBLN Grants Pass, OR
WSJU-TV San Juan, PR
*WNSC-TV Rock Hill, SC
WLMT Memphis, TN
WUXP-TV Nashville, TN
KWWT Odessa, TX
KUCW Ogden, UT
WVCY-TV Milwaukee, WI

Channel 31

WAAY-TV Huntsville, AL
KWBM Harrison, AR
KMAX-TV Sacramento, CA
KVMD Twentynine Palms, CA
KDVR Denver, CO
WFXL Albany, GA
KFXP Pocatello, ID
WMBD-TV Peoria, IL
*WKOH Owensboro, KY
KLAX-TV Alexandria, LA
*WWPB Hagerstown, MD
WPXD Ann Arbor, MI
*WMAI Cleveland, MS
*WUNU Lumberton, NC
WPXN-TV New York, NY
WUHF Rochester, NY
KDKF Klamath Falls, OR
WNNE-TV Hartford, VT
*KTNW Richland, WA
*WHLA-TV La Crosse, WI

Channel 32

WNCF Montgomery, AL
*KMTP-TV San Francisco, CA
*WHUT-TV Washington, DC
WMOR-TV Lakeland, FL
WNEG-TV Toccoa, GA
KBFD Honolulu, HI
*KBIN Council Bluffs, IA
*KRIN Waterloo, IA
WFLD Chicago, IL
WLKY-TV Louisville, KY
*WLAE-TV New Orleans, LA
*WMYC Yazoo City, MS
*KAZQ Albuquerque, NM
KRCW-TV Salem, OR
*WELU Aguadilla, PR
WMBF-TV Myrtle Beach, SC
KTAB-TV Abilene, TX
KUTH Provo, UT
WACY Appleton, WI

Channel 33

KDMD Anchorage, AK
WCFT-TV Tuscaloosa, AL
KTVW-TV Phoenix, AZ
KTAS San Luis Obispo, CA
KTLL-TV Durango, CO
WBFS-TV Miami, FL
WHBR Pensacola, FL
WISE-TV Fort Wayne, IN
KSCW Wichita, KS
WVLA Baton Rouge, LA
KMSS-TV Shreveport, LA
WFQX-TV Cadillac, MI
KSPR Springfield, MO
*WUNF-TV Asheville, NC
KVCW Las Vegas, NV
WFXV Utica, NY
WYTV Youngstown, OH
*WITF-TV Harrisburg, PA
*WJPM-TV Florence, SC
KDAF Dallas, TX
WTVZ-TV Norfolk, VA
*WETK Burlington, VT
KWPX Bellevue, WA
*WPBY-TV Charleston, WV
KDEV Cheyenne, WY

Channel 34

WDFX-TV Ozark, AL
KPBI Eureka Springs, AR
KMEX-TV Los Angeles, CA
WTVX Fort Pierce, FL
WUVG-TV Athens, GA
WGSA Baxley, GA
KEFB Ames, IA
*WNIT-TV South Bend, IN
WBKI-TV Campbellsville, KY
WRBJ Magee, MS
KMCC Laughlin, NV
WIVT Binghamton, NY
*WOSU-TV Columbus, OH
KOCB Oklahoma City, OK
KLSR-TV Eugene, OR
WRUA Fajardo, PR
*KITU-TV Beaumont, TX
KJTV-TV Lubbock, TX
*KWBU-TV Waco, TX
KGPX Spokane, WA
WYOW Eagle River, WI

Channel 35

KCBA Salinas, CA
WFGX Fort Walton Beach, FL
WPXM Miami, FL
WOFL Orlando, FL
*KUID-TV Moscow, ID
KXTF Twin Falls, ID
WWTO-TV LaSalle, IL
*WKHA Hazard, KY
*WKMA Madisonville, KY
WPME Lewiston, ME
*WDCQ-TV Bad Axe, MI
*WGVU-TV Grand Rapids, MI
WUFX Vicksburg, MS
WPXU-TV Jacksonville, NC
WLIO Lima, OH
*KRSC-TV Claremore, OK
KUOK Woodward, OK
WSEE Erie, PA
*WYBE Philadelphia, PA
*WRLK-TV Columbia, SC
KMYS Kerrville, TX
WRLH-TV Richmond, VA
KAPP Yakima, WA

Channel 36

*WFIQ Florence, AL
*KKAP Little Rock, AR
KMIR-TV Palm Springs, CA
KICU-TV San Jose, CA
WFTX Cape Coral, FL
WATL Atlanta, GA
*KQCT Davenport, IA
*KHIN Red Oak, IA
KMTW Hutchinson, KS
WTVQ-TV Lexington, KY
*WGPT Oakland, MD
WCNC-TV Charlotte, NC
*WUNP-TV Roanoke Rapids, NC
WENY-TV Elmira, NY
WUPW Toledo, OH
KTVC Roseburg, OR
WDWL Bayamon, PR
*WSBE-TV Providence, RI
WMMP Charleston, SC
KWSD Sioux Falls, SD
KXAN-TV Austin, TX
*KPBT-TV Odessa, TX
*WMVT Milwaukee, WI
*WLEF-TV Park Falls, WI

Channel 38

KASN Pine Bluff, AR
KCNS San Francisco, CA
KPMR Santa Barbara, CA
WTTA Saint Petersburg, FL
WLTZ Columbus, GA
*KALO Honolulu, HI
WCPX Chicago, IL
WFXW-TV Terre Haute, IN
KMCI Lawrence, KS
*WKMR Morehead, KY
WNOL-TV New Orleans, LA
WSBK-TV Boston, MA
WADL Mount Clemens, MI
WEPX Greenville, NC
WSWB Scranton, PA
WJWN-TV San Sebastian, PR
*WNEH Greenwood, SC
KUQI Corpus Christi, TX
*KSCE El Paso, TX
WPXR Roanoke, VA
*WPNE Green Bay, WI

Channel 39

KTAZ Phoenix, AZ
KNSD San Diego, CA
WSFL-TV Miami, FL
KFPX Newton, IA
KKJB Boise, ID
WQRF-TV Rockford, IL
*WFWA Fort Wayne, IN
KMCT-TV West Monroe, LA
*WUNJ-TV Wilmington, NC
KBLR Paradise, NV
*WLVT-TV Allentown, PA
WEMT Greeneville, TN
WHTN Murfreesboro, TN
KXTX-TV Dallas, TX
KHCW Houston, TX
WCVI-TV Christiansted, VI

Channel 40

WJSU-TV Anniston, AL
KHBS Fort Smith, AR
KHRR Tucson, AZ
KTXL Sacramento, CA
KTBN-TV Santa Ana, CA
WWSB Sarasota, FL
WTWC-TV Tallahassee, FL
KFXB Dubuque, IA
WHMB-TV Indianapolis, IN
WNKY Bowling Green, KY
WGGB-TV Springfield, MA
WBUY-TV Holly Springs, MS
WDBD Jackson, MS
WUVC-TV Fayetteville, NC
WMGM-TV Wildwood/Atlantic City, NJ
WICZ-TV Binghamton, NY
WLMB Toledo, OH
WPCB-TV Greensburg, PA
*WMTJ Fajardo, PR
WMYA-TV Anderson, SC
*KTLM Rio Grande City, TX
WLFB Bluefield, WV

Channel 41

*WIIQ Demopolis, AL
*KRMT Denver, CO
WMGT-TV Macon, GA
WDRB Louisville, KY
KBCA Alexandria, LA
WOTV Battle Creek, MI
KPXM Saint Cloud, MN
KSHB-TV Kansas City, MO
WXTV Paterson, NJ
KLUZ-TV Albuquerque, NM
KMYT-TV Tulsa, OK
KWEX-TV San Antonio, TX
*WHTJ Charlottesville, VA
*WVTA Windsor, VT

Channel 42

WIAT Birmingham, AL
*WEIQ Mobile, AL
KWBF Little Rock, AR
KTNC-TV Concord, CA
KESQ-TV Palm Springs, CA
*WXEL-TV West Palm Beach, FL
WCLJ-TV Bloomington, IN
KSAX Alexandria, MN
*WTVI Charlotte, NC
KPTM Omaha, NE
*WPBO-TV Portsmouth, OH
WIRS Yauco, PR
KEYE-TV Austin, TX
KMLM Odessa, TX
*WVPY Front Royal, VA
KVEW Kennewick, WA

Channel 43

*WGIQ Louisville, AL
KEJB El Dorado, AR
KGMC Clovis, CA
*KCSM-TV San Mateo, CA
WSAH Bridgeport, CT
WOTF-TV Melbourne, FL
WYZZ-TV Bloomington, IL
WKOI-TV Richmond, IN
WZPX Battle Creek, MI
KRWF Redwood Falls, MN
*WMAA Columbus, MS
WNYS-TV Syracuse, NY
WUAB Lorain, OH
KAUT-TV Oklahoma City, OK
WPMT York, PA
WFXB Myrtle Beach, SC
WTNZ Knoxville, TN
WVBT Virginia Beach, VA

Channel 44

WPXH Gadsden, AL
KXLA Rancho Palos Verdes, CA
KBCW San Francisco, CA
WJTC Pensacola, FL
WTOG Saint Petersburg, FL
WSWG Valdosta, GA
*KWBN Honolulu, HI
KPTH Sioux City, IA
WSNS Chicago, IL
WEVV-TV Evansville, IN
WAGV Harlan, KY
WGMB Baton Rouge, LA
*WGBX-TV Boston, MA
XHIJ Ciudad Juarez, MEX
*WOUC-TV Cambridge, OH
WTLW Lima, OH
KTPX Okmulgee, OK
*WVIA-TV Scranton, PA
WVEO Aguadilla, PR
*KLUJ-TV Harlingen, TX
KWKT Waco, TX
WFFF-TV Burlington, VT

Channel 45

WMCF-TV Montgomery, AL
KUTP Phoenix, AZ
KUVI-TV Bakersfield, CA
WLCB-TV Leesburg, FL
*WHFT-TV Miami, FL
KSHV Shreveport, LA
WBFF Baltimore, MD
WFUP Vanderbilt, MI
KSTC-TV Minneapolis, MN
WKDH Houston, MS
WXLV-TV Winston-Salem, NC
WCWN Schenectady, NY
*WNEO Alliance, OH
WRGT-TV Dayton, OH
*WTCI Chattanooga, TN
KXLN-TV Rosenberg, TX
KHCV Seattle, WA

Channel 46

KUVE-TV Green Valley, AZ
KION-TV Monterey, CA
KFTR-TV Ontario, CA
WXCW Naples, FL
WPCT Panama City Beach, FL
WGCL-TV Atlanta, GA
WRBU East St. Louis, IL
WHME-TV South Bend, IN
*WKLE Lexington, KY
WWDP Norwell, MA
WBSF Bay City, MI
WJZY Belmont, NC
*WSKG-TV Binghamton, NY
KOCM Norman, OK
KTCW Roseburg, OR
WIDP Guayama, PR
KDLT-TV Sioux Falls, SD
*KNCT Belton, TX
WVFX Clarksburg, WV

Channel 47

KGPE Fresno, CA
WTEV-TV Jacksonville, FL
*WTVP Peoria, IL
WMDT Salisbury, MD
WSYM-TV Lansing, MI
KXLT-TV Rochester, MN
WRPX Rocky Mount, NC
WNJU Linden, NJ
KWHB Tulsa, OK
WKBS-TV Altoona, PA
WZRB Columbia, SC
KTMD Galveston, TX
KTAQ Greenville, TX
*WSBN-TV Norton, VA
*KYVE Yakima, WA
WMSN-TV Madison, WI

Channel 48

WAFF Huntsville, AL
KVTJ Jonesboro, AR
KSTS San Jose, CA
KVSN Pueblo, CO
WFBD Destin, FL
KPXR Cedar Rapids, IA
*WYDN Worcester, MA
WNTZ Natchez, MS
WMYV Greensboro, NC
WGTW-TV Burlington, NJ
KTDO Las Cruces, NM
WYDC Corning, NY
*WCET Cincinnati, OH
WVOZ-TV Ponce, PR
WVLR Tazewell, TN
KNVO McAllen, TX
WEUX Chippewa Falls, WI

Channel 49

KYPX Camden, AR
*KNXT Visalia, CA
*WEDW Bridgeport, CT
WRXY-TV Tice, FL
WTLH Bainbridge, GA
WCFN-TV Springfield, IL
*WIPB Muncie, IN
KTKA-TV Topeka, KS
WDKA Paducah, KY
WPXL New Orleans, LA
WAQP Saginaw, MI
KRBK Osage Beach, MO
*WLED-TV Littleton, NH
WNYO-TV Buffalo, NY
*WEAO Akron, OH
WGCB-TV Red Lion, PA
*WRET-TV Spartanburg, SC
KPXB Conroe, TX
KSTR-TV Irving, TX
WPXV Norfolk, VA
KPDX Vancouver, WA
WJJA Racine, WI

Channel 50

*KOCE-TV Huntington Beach, CA
KFTY Santa Rosa, CA
KCEC Denver, CO
WDCW Washington, DC
WFTT-TV Tampa, FL
KKAI Kailua, HI
WPWR-TV Gary, IN
KLWB New Iberia, LA
WKBD Detroit, MI
KPXE Kansas City, MO
WRAZ Raleigh, NC
WZMY-TV Derry, NH
*WNJN Montclair, NJ
KASY-TV Albuquerque, NM
WWTI Watertown, NY
WQHA Aguada, PR
WPGD-TV Hendersonville, TN
WPXX-TV Memphis, TN

Channel 51

KNWA-TV Rogers, AR
KPPX Tolleson, AZ
KNSO Merced, CA
KUSI-TV San Diego, CA
WSCV Fort Lauderdale, FL

WBIF Marianna, FL
WOGX Ocala, FL
*WEIU-TV Charleston, IL
WNYA Pittsfield, MA
WPXT Portland, ME
WWJX Jackson, MS
KCWL-TV Lincoln, NE
WPXJ-TV Batavia, NY
WSFJ-TV Newark, OH
WTVE Reading, PA
KNWS-TV Katy, TX
KFXK Longview, TX
*WVPT Staunton, VA
KUNS-TV Bellevue, WA

Channel 52

KVEA Corona, CA
WHLV-TV Cocoa, FL
*WKON Owenton, KY
*WGVK Kalamazoo, MI
*WEKW-TV Keene, NH
*WNJT Trenton, NJ
WNYI Ithaca, NY
WGGN-TV Sandusky, OH
KSBI Oklahoma City, OK
WRFB Carolina, PR
KFWD Fort Worth, TX
*WMSY-TV Marion, VA
WWRS-TV Mayville, WI

Channel 53

KAIL Fresno, CA
KWHD Castle Rock, CO
*WEDN Norwich, CT
WPAN Fort Walton Beach, FL
WGFL High Springs, FL
*WKGB-TV Bowling Green, KY
WLAJ Lansing, MI
WWHO Chillicothe, OH
KGEB Tulsa, OK
WPGH-TV Pittsburgh, PA
WQMY Williamsport, PA
WFLI-TV Cleveland, TN
*WNVT Goldvein, VA

Channel 54

WZDX Huntsville, AL
KAZA-TV Avalon, CA
KAJB Calipatria, CA
*KTEH San Jose, CA
WFXG Augusta, GA
WXTX Columbus, GA
*WCVN-TV Covington, KY
WUPL Slidell, LA
WNUV Baltimore, MD
WTLJ Muskegon, MI
WTBY-TV Poughkeepsie, NY
*WQLN Erie, PA
WCCV-TV Arecibo, PR
WPXK Jellico, TN
KNVA Austin, TX
KCEB Longview, TX

Channel 55

WBPG Gulf Shores, AL
WSST-TV Cordele, GA
WRSP-TV Springfield, IL
WFFT-TV Fort Wayne, IN
WYPX Amsterdam, NY
WLNY Riverhead, NY
WBNX-TV Akron, OH
WMYT-TV Rock Hill, SC
KTBU Conroe, TX
WPXE Kenosha, WI
WFXS Wittenberg, WI

Channel 56

KDOC-TV Anaheim, CA
WOPX Melbourne, FL
*WFSG Panama City, FL
KUPU Waimanalo, HI
*WYIN Gary, IN
WDKY-TV Danville, KY
WLVI-TV Cambridge, MA
*WTVS Detroit, MI
WSPX-TV Syracuse, NY
WOLF-TV Hazleton, PA
KETK-TV Jacksonville, TX
*WNVC Fairfax, VA
*KWDK Tacoma, WA

Channel 57

KWOG Springdale, AR
KJLA Ventura, CA
WFXU Live Oak, FL
*WATC Atlanta, GA
WYMT-TV Hazard, KY
*WGBY-TV Springfield, MA
*WCFE-TV Plattsburgh, NY
WPSG Philadelphia, PA
WACH Columbia, SC
KAZH Baytown, TX
*WCVW Richmond, VA
WBUW Janesville, WI

Channel 58

KWBA Sierra Vista, AZ
*KLCS Los Angeles, CA
KQCA Stockton, CA
WAWD Fort Walton Beach, FL
WPGA-TV Perry, GA
WMYO Salem, IN
WDPX Vineyard Haven, MA
*WUNG-TV Concord, NC
*WNJB New Brunswick, NJ
*WUJA Caguas, PR
WNAB Nashville, TN
*KDTX-TV Dallas, TX
WDJT-TV Milwaukee, WI

Channel 59

KFRE-TV Sanger, CA
KPXC-TV Denver, CO
WCTX New Haven, CT
*WJEB-TV Jacksonville, FL
WAOE Peoria, IL
WXIN Indianapolis, IN
WVNS-TV Lewisburg, WV

Channel 60

WTJP-TV Gadsden, AL
WXFT-TV Aurora, IL
WNEU Merrimack, NH
WBPH-TV Bethlehem, PA
WMEI Arecibo, PR
*KMBH Harlingen, TX
KVDA San Antonio, TX
WWPX Martinsburg, WV

Channel 61

KASW Phoenix, AZ
KTFF-TV Porterville, CA
WTIC-TV Hartford, CT
WPPX Wilmington, DE
WFGC Palm Beach, FL
WTSF Ashland, KY
WLXI-TV Greensboro, NC
WQHS-TV Cleveland, OH
WDSI-TV Chattanooga, TN
KZJL Houston, TX

Channel 62

KRCA Riverside, CA
WVEA-TV Venice, FL
WJYS-TV Hammond, IN
WMFP Lawrence, MA
*WFPT Frederick, MD
WWJ-TV Detroit, MI
KSMO-TV Kansas City, MO
WYCW Asheville, NC
WFPX Fayetteville, NC
WWSI Atlantic City, NJ
WRNN-TV Kingston, NY
KOPX Oklahoma City, OK
KAKW-TV Killeen, TX

Channel 63

KBEH Oxnard, CA
*WPPB-TV Boca Raton, FL
*WHSG-TV Monroe, GA
WINM Angola, IN
WIPX Bloomington, IN
WMBC-TV Newton, NJ
WKTC Sumter, SC

Channel 64

KHIZ Barstow, CA
KTFK-TV Stockton, CA
*WDPB Seaford, DE
WGNM Macon, GA
WLLA Kalamazoo, MI
WAXN-TV Kannapolis, NC
WSTR-TV Cincinnati, OH
WQPX Scranton, PA
WECN Naranjito, PR
WNAC-TV Providence, RI

Channel 65

KKPX San Jose, CA
*WEDY New Haven, CT
WRBW Orlando, FL
WLJC-TV Beattyville, KY
WUVP-TV Vineland, NJ
KTFN El Paso, TX
WUPV Ashland, VA

Channel 66

WLGA Opelika, AL
KFSF-TV Vallejo, CA
WXPX Bradenton, FL
KPXO-TV Kaneohe, HI
WGBO-TV Joliet, IL
WUTF-TV Marlborough, MA
WSMH Flint, MI
*WFME-TV West Milford, NJ
WFXP Erie, PA
WJFB Lebanon, TN
WPXW Manassas, VA

Channel 67

WRJM-TV Troy, AL
KSMS-TV Monterey, CA
WPXP Lake Worth, FL
WUPX-TV Morehead, KY
*WMPB Baltimore, MD
WFTY-TV Smithtown, NY
WNGS Springville, NY
WOAC Canton, OH
KFTH-TV Alvin, TX

Channel 68

WABM Birmingham, AL
KTLN-TV Novato, CA
*WBCC Cocoa, FL
WKMJ Louisville, KY
WBPX Boston, MA
WJAL Hagerstown, MD
WFUT-TV Newark, NJ
WSYT Syracuse, NY
WMFD-TV Mansfield, OH
KPXD Arlington, TX
WLFG Grundy, VA
WWAZ-TV Fond du Lac, WI

Channel 69

KSWB-TV San Diego, CA
WAMI-TV Hollywood, FL
WUPA Atlanta, GA
*WDTI Indianapolis, IN
WFMZ-TV Allentown, PA
WPXQ Block Island, RI

Canadian Television Stations by Analog Channel

Channel 2

CICT-TV Calgary AB
CBXAT-2 High Prairie AB
CFRN-TV-5 Lac La Biche AB
CKSA-TV Lloydminster AB
CHBC-TV Kelowna BC
CKPG-TV Prince George BC
CBUT Vancouver BC
CBWYT Mafeking MB
CKCW-TV Moncton NB
CBYT-3 Bonne Bay NF
CJOX-TV-1 Grand Bank NF
CBIT-2 Cheticamp NS
CIII-TV-2 Bancroft ON
CHBX-TV Sault Ste. Marie ON
CKPR-TV Thunder Bay ON
CKCO-TV-2 Wiarton ON
CBFT Montreal PQ
CFAP-TV Quebec City PQ
CJBR-TV Rimouski PQ
CBCP-TV-2 Cypress Hills SK
CKBQ-TV Melfort SK
CKCK-TV Regina SK

Channel 3

CFRN-TV Edmonton AB
CFRN-TV-2 Peace River AB
CHAT-TV-1 Pivot AB
CHAN-TV-2 Bowen Island BC
CFTK-TV Terrace BC
CBWFT-4 Ste-Rose-du-Lac MB
CBWFT Winnipeg MB
CJOM-TV Argentia NF
CBNAT-1 Baie Verte NF
CBHT Halifax NS
CKVR-TV Barrie ON
CITO-TV Timmins ON
CKRN-TV-3 Bearn-Fabre PQ
CBVT-2 La Tuque PQ
CBFT-2 Mont-Laurier PQ
CBCP-TV-3 Ponteix SK

Channel 4

CFCN-TV Calgary AB
CITL-TV Lloydminster AB
CFRN-TV-9 Slave Lake AB
CFJC-TV Kamloops BC
CKYB-TV Brandon MB
CBWT-2 Lac du Bonnet MB
CBAT Fredericton-Saint John NB
CJCN-TV Grand Falls NF
CJSV-TV Stephenville NF
CJCB-TV Sydney NS
CBOT Ottawa ON
CIII-TV-4 Owen Sound ON
CHFD-TV Thunder Bay ON
CFRS-TV Jonquiere PQ
CFCM-TV Quebec City PQ
CKRN-TV Rouyn-Noranda PQ
CFSK-TV Saskatoon SK

Channel 5

CKAL-TV Calgary AB
CBXT Edmonton AB
CJDC-TV Dawson Creek BC
CKX-TV Brandon MB
CBNT-3 Marystown NF
CBIT Sydney NS
CICI-TV-1 Elliot Lake ON
CHRO-TV Pembroke ON
CICI-TV Sudbury ON
CBLT Toronto ON
CKCW-TV-2 Saint Edward PE
CHAU-TV Carleton PQ
CBVD-TV Malartic PQ
CFER-TV-2 Sept-Iles PQ

Channel 6

CBXFT-1 Bonnyville AB
CHAT-TV Medicine Hat AB
CHCA-TV Red Deer AB
CHKM-TV Kamloops BC
CHEK-TV Victoria BC
CBWT Winnipeg MB
CBNAT-4 Saint Anthony NF
CJON-TV St John's NF
CJCH-TV-6 Caledonia NS
CJCB-TV-1 Inverness NS
*CFYK-TV Yellowknife NT
CJOH-TV-6 Deseronto ON
CIII-TV-6 Ottawa ON
CIII-TV Paris ON
CJPM-TV Chicoutimi PQ
CBMT Montreal PQ
CFQC-TV-2 North Battleford SK
CKCK-TV-2 Willow Bunch SK
CFWH-TV White Horse YT

Channel 7

CISA-TV Lethbridge AB
CFRN-TV-7 Lougheed AB
CHAN-TV-3 Squamish BC
CHVC-TV Valemount BC
CKY-TV Winnipeg MB
CKCD-TV Campbellton NB
CIII-TV-7 Midland ON
CKRT-TV Riviere-du-Loup PQ
CHLT-TV Sherbrooke PQ
CBKFT-4 Saint Brieux SK
CBCP-TV-1 Shaunavon SK
CIEW-TV Warmley SK

Channel 8

CFCN-TV-8 Medicine Hat AB
CFRN-TV-6 Red Deer AB
CHAN-TV Vancouver BC
CBWST Baldy Mountain MB
CBNT Saint John's NF
CIHF-TV Halifax NS
*CHAK-TV Inuvik NT
CJOH-TV-8 Cornwall ON
CBWAT Kenora ON
CKNX-TV Wingham ON
CKCW-TV-1 Charlottetown PE
*CIVV-TV Chicoutimi PQ
*CIVA-TV Rouyn PQ
CHEM-TV Trois-Rivieres PQ
CFQC-TV Saskatoon SK

Channel 9

CBRT Calgary AB
*CJAL-TV Edmonton AB
CHAN-TV-5 Brackendale BC
CHAN-TV-7 Whistler BC
CKND-TV Winnipeg MB
CKLT-TV Saint John NB
CBNAT-9 Mount St. Margaret NF
CJCB-TV-2 Antigonish NS
CJCH-TV Halifax NS
CBOFT Ottawa ON
*CICO-TV-9 Thunder Bay ON
CFTO-TV Toronto ON
CBET Windsor ON
CIMT-TV Riviere-du-Loup PQ
*CIVG-TV Sept-Iles PQ
CKSH-TV Sherbrooke PQ
CIPA-TV Prince Albert SK
CBKT Regina SK
CBKST-1 Stranraer SK

Channel 10

CHCA-TV-1 Coronation AB
CBXAT Grande Prairie AB
CBUBT-7 Cranbrook BC
CKVU-TV Vancouver BC
CBWGT Fisher Branch MB
CBWBT Flin Flon MB
CJWB-TV Bonavista NF
CJWN-TV Corner Brook NF
CJCH-TV-1 Canning NS
CIMC-TV Isle Madame NS
CFPL-TV London ON
CKNY-TV North Bay ON
CFTM-TV Montreal PQ
CKMC-TV-1 Golden Prairie SK
CBKT-2 Willow Bunch SK
CICC-TV Yorkton SK

Channel 11

CBXFT Edmonton AB
CHAN-TV-1 Chilliwack BC
CBUFT-3 Terrace BC
CKX-TV-1 Foxwarren MB
CBAFT Moncton NB
CBNAT Botwood NF
CBHT-4 Sheet Harbour NS
CBHT-3 Yarmouth NS
CHCH-TV Hamilton ON
CITO-TV-2 Kearns ON
CKWS-TV Kingston ON
CBLAT-4 Marathon ON
CBVT Quebec City PQ
CFER-TV Rimouski PQ
CFRE-TV Regina SK
CBKST Saskatoon SK

Channel 12

CFRN-TV-4 Ashmont AB
CFCN-TV-1 Drumheller AB
CBXFT-6 Fort McMurray AB
CBXAT-3 Manning AB
CFRN-TV-3 Whitecourt AB
CBUBT-1 Canal Flats BC
CKAM-TV Upsalquitch Lake NB
CBNT-2 Placentia NF
CBHT-11 Mulgrave NS
CHEX-TV Peterborough ON
*CIVF-TV Baie-Trinite PQ
CBIMT Iles-de-la-Madeleine PQ
CFCF-TV Montreal PQ
CBFST-2 Temiscaning PQ
CKCK-TV-1 Colgate SK
CKMC-TV Swift Current SK

Channel 13

CIAN-TV Calgary AB
CITV-TV Edmonton AB
CFRN-TV-1 Grande Prairie AB
CFCN-TV-5 Lethbridge AB
CJTG-TV Meander River AB
CHEK-TV-5 Campbell River BC
CHMI-TV Portage la Prarie MB
*CBNLT Labrador City NF
CBNT-1 Port Rexton NF
CBLAT Geraldton ON
CJBN-TV Kenora ON
CKCO-TV Kitchener ON
CJOH-TV Ottawa ON
CBCT Charlottetown PE
CKTM-TV Trois-Rivieres PQ
CBKFT Regina SK

Channel 15

*CIVQ-TV Quebec City PQ

Channel 16

CHWI-TV Windsor ON
CFKM-TV Trois-Rivieres PQ

Channel 17

CJIL-TV Lethbridge AB
*CIVM-TV Montreal PQ

Channel 18

CFRN-TV-8 Grouard Mission-High Prairie AB
*CICO-TV-18 London ON

Channel 19

CBXFT-8 Grande Prairie AB
*CICO-TV-19 Sudbury ON
*CICA-TV Toronto ON

Channel 20

*CICO-TV-20 Sault Ste. Marie ON
CKMI-TV Quebec City PQ

Channel 21

CBWFT-10 Brandon MB
CBKFT-5 Xenon Park SK

Channel 22

CHEX-TV-2 Oshawa ON
CIII-TV-22 Stevenson ON
*CIVB-TV Rimouski PQ
CBKFT-3 Debden SK

Channel 23

*CIVP-TV Chapeau PQ

Channel 24

*CICO-TV-24 Ottawa ON
*CIVS-TV Sherbrooke PQ

Channel 25

CFVS-TV Val d'Or PQ

Channel 26

CBUFT Vancouver BC
CBKFT-9 Bellegarde SK

Channel 27

CIII-TV-27 Peterborough ON

Channel 28

*CICO-TV-28 Kitchener ON

Channel 29

CIII-TV-29 Oil Springs ON
*CFTU-TV Montreal PQ
CFTF-TV Riviere-du-Loup PQ

Channel 30

*CIVO-TV Gatineau PQ
CFKS-TV Sherbrooke PQ

Channel 32

CFSO-TV Cardston AB
CIVT-TV Vancouver BC
*CICO-TV-32 Windsor ON

Channel 34

CFTV-TV Leamington ON
CFGS-TV Gatineau PQ

Channel 35

CIIT-TV Winnipeg MB
CFJP-TV Montreal PQ

Channel 36

CITS-TV Hamilton ON

Channel 39

CBKFT-6 Gravelbourg SK

Channel 40

CHOT-TV Gatineau PQ

Channel 41

CIII-TV-41 Toronto ON

Channel 42

CHNM-TV Vancouver BC
CKCO-TV-3 Sarnia ON

Channel 43

CHRO-TV-43 Ottawa ON

Channel 45

*CIVC-TV Trois-Rivieres PQ

Channel 47

CFMT-TV Toronto ON

Channel 50

CBUFT-2 Kamloops BC

Channel 51

CKEM-TV Edmonton AB

Channel 52

CKXT-TV Toronto ON

Channel 53

CIVI-TV Victoria BC

Channel 54

*CBEFT Windsor ON

Channel 57

CITY-TV Toronto ON

Channel 58

CBJET Chicoutimi PQ

Channel 59

*CICO-TV-59 Chatham ON

Channel 62

CJNT-TV Montreal PQ

Channel 66

CHNU-TV Fraser Valley BC

Channel 69

CJMT-TV Toronto ON

U.S. Television Stations by Digital Channel

Channel 0
KSLA-TV Shreveport, LA

Channel 2
KREX-TV Grand Junction, CO
WWMT Kalamazoo, MI
KVBC Las Vegas, NV
WETM-TV Elmira, NY
WKYC-TV Cleveland, OH
KDLO-TV Florence, SD

Channel 3
WSBS-TV Key West, FL
WBBM-TV Chicago, IL
KCBU Price, UT
*WBRA-TV Roanoke, VA

Channel 4
WMAZ-TV Macon, GA
WDKY-TV Danville, KY
KXLF-TV Butte, MT
WIVT Binghamton, NY
WHP-TV Harrisburg, PA

Channel 5
*KETS Little Rock, AR
KNSO Merced, CA
*KGTF-TV Hagatna, GU
WLMB Toledo, OH
*KTVR La Grande, OR
KPXB Conroe, TX

Channel 6
KYES-TV Anchorage, AK
*KTOO-TV Juneau, AK
WDTV Weston, WV

Channel 7
KAIL Fresno, CA
WFLA-TV Tampa, FL
*WMUM-TV Cochran, GA
KTVB Boise, ID
WLJC-TV Beattyville, KY
KNOE-TV Monroe, LA
WOOD-TV Grand Rapids, MI
KMBC-TV Kansas City, MO
KRTV Great Falls, MT
KDUH-TV Scottsbluff, NE
KLAS-TV Las Vegas, NV
KRNV Reno, NV
WXXA-TV Albany, NY
WBNG-TV Binghamton, NY
KOCO-TV Oklahoma City, OK
KTTW Sioux Falls, SD
WMAK Knoxville, TN

Channel 8
KSBW Salinas, CA
WSVN Miami, FL
WJHG-TV Panama City, FL
KGMD-TV Hilo, HI
KHON-TV Honolulu, HI
*KSWK Lakin, KS
WBNA Louisville, KY
WTVA Tupelo, MS
*KUSM Bozeman, MT
KFBB-TV Great Falls, MT
KPAX-TV Missoula, MT
*WNJB New Brunswick, NJ
KOFT Farmington, NM
WICZ-TV Binghamton, NY
WLIO Lima, OH
*KWET Cheyenne, OK
KFFX-TV Pendleton, OR
*WSBE-TV Providence, RI
WOLO-TV Columbia, SC
KIII Corpus Christi, TX
KFOX-TV El Paso, TX
*KSPS-TV Spokane, WA
*WMVS Milwaukee, WI
*KCWC-TV Lander, WY

Channel 9
WJSU-TV Anniston, AL
WALA-TV Mobile, AL
KAIT Jonesboro, AR
KFSN-TV Fresno, CA
WINK-TV Fort Myers, FL
WPLG Miami, FL
WPGX Panama City, FL
KIFI-TV Idaho Falls, ID
WISH-TV Indianapolis, IN
WLWC New Bedford, MA
*WMEB-TV Orono, ME
WGTQ Sault Ste. Marie, MI
WLBT Jackson, MS
*KNMD-TV Santa Fe, NM
WBPH-TV Bethlehem, PA
WFAA-TV Dallas, TX
*KUHT Houston, TX
KCEN-TV Temple, TX
KUSG Saint George, UT
*WVER Rutland, VT
KCWK Walla Walla, WA

Channel 10
KTUU-TV Anchorage, AK
KKTV Colorado Springs, CO
*KMAS-TV Steamboat Springs, CO
WTNH-TV New Haven, CT
WXIA-TV Atlanta, GA
KNIN-TV Caldwell, ID
*WMAB-TV Mississippi State, MS
KTVQ Billings, MT
WNCT-TV Greenville, NC
KCHF Santa Fe, NM
WCPO-TV Cincinnati, OH
WOIO Shaker Heights, OH
KTUL Tulsa, OK
WHTM-TV Harrisburg, PA
KHSD-TV Lead, SD
WSMV-TV Nashville, TN
KFDA-TV Amarillo, TX
KLTV Tyler, TX
WVFX Clarksburg, WV

Channel 11
KJUD Juneau, AK
*WDIQ Dozier, AL
*KEET Eureka, CA
WESH Daytona Beach, FL
WLFI-TV Lafayette, IN
*WYES-TV New Orleans, LA
WWLP Springfield, MA
WBKP Calumet, MI
*WGVU-TV Grand Rapids, MI
KULR-TV Billings, MT
*WTVI Charlotte, NC
*KLVX Las Vegas, NV
KSWO-TV Lawton, OK
WBRE-TV Wilkes-Barre, PA
KDEV Cheyenne, WY

Channel 12
KIMO Anchorage, AK
*KETZ El Dorado, AR
KNTV San Jose, CA
KKCO Grand Junction, CO
WTXX Waterbury, CT
WGEN-TV Key West, FL
WTVT Tampa, FL
WFXL Albany, GA
*KUID-TV Moscow, ID
*WNIN-TV Evansville, IN
WYMT-TV Hazard, KY
*WMEM-TV Presque Isle, ME
KRCG Jefferson City, MO
KSNK McCook, NE
KOVT Silver City, NM
KTNV Las Vegas, NV
WNYT Albany, NY
WMFD-TV Mansfield, OH
WWPX Martinsburg, WV

Channel 13
WPEC West Palm Beach, FL
KTRV-TV Nampa, ID
KOAM-TV Pittsburg, KS
WKYT-TV Lexington, KY
*KLTM-TV Monroe, LA
KBZK Bozeman, MT
KTVN Reno, NV
WSYX Columbus, OH
WYOU Scranton, PA
KPLO-TV Reliance, SD
WRCB-TV Chattanooga, TN
KRIS-TV Corpus Christi, TX
KWES-TV Odessa, TX
KXLY-TV Spokane, WA

Channel 14
WHDF Florence, AL
WRDQ Orlando, FL
KYOU-TV Ottumwa, IA
*WTIU Bloomington, IN
*WKSO-TV Somerset, KY
WVII-TV Bangor, ME
WKBD Detroit, MI
KNLC Saint Louis, MO
KTVH Helena, MT
WGPX Burlington, NC
KXMD-TV Williston, ND
WUTV Buffalo, NY
WPTZ North Pole, NY
WCMH-TV Columbus, OH
WMYA-TV Anderson, SC
KCSG Cedar City, UT
KTBW-TV Tacoma, WA
KAPP Yakima, WA
WXOW-TV La Crosse, WI
WIWB Suring, WI
WLFB Bluefield, WV

Channel 15
KUNO-TV Fort Bragg, CA
KSBY San Luis Obispo, CA
KTFD-TV Boulder, CO
KREZ-TV Durango, CO
KFQX Grand Junction, CO
WBBH-TV Fort Myers, FL
WRBL Columbus, GA
*WSEC Jacksonville, IL
*WKMR Morehead, KY
WGNO New Orleans, LA
WNUV Baltimore, MD
*WMED-TV Calais, ME
*WDCQ-TV Bad Axe, MI
*KSMN Worthington, MN
*KMOS-TV Sedalia, MO
WRPX Rocky Mount, NC
KMCY Minot, ND
KNRR Pembina, ND
*KMNE-TV Bassett, NE
WEWS Cleveland, OH
KDOR-TV Bartlesville, OK
KTBO-TV Oklahoma City, OK
*WPSU-TV Clearfield, PA
WPDE-TV Florence, SC
WZTV Nashville, TN
KCIT Amarillo, TX
KGNS-TV Laredo, TX
KAVU-TV Victoria, TX
WFDC-TV Arlington, VA
KHQ-TV Spokane, WA
WQOW-TV Eau Claire, WI

Channel 16
KSWT Yuma, AZ
KUSA-TV Denver, CO
WCJB-TV Gainesville, FL
WELF-TV Dalton, GA
WGXA Macon, GA
KOGG Wailuku, HI
KDSM-TV Des Moines, IA
KMVT Twin Falls, ID
WTVO Rockford, IL
WHMB-TV Indianapolis, IN
KSNG Garden City, KS
*KOOD Hays, KS
*WKHA Hazard, KY
KADN Lafayette, LA
WJAL Hagerstown, MD
WAGM-TV Presque Isle, ME
*KCGE-DT Crookston, MN
*KTCI-TV Saint Paul, MN
*WMAH-TV Biloxi, MS
WLOV-TV West Point, MS
KBMY Bismarck, ND
*KXNE-TV Norfolk, NE
*KPNE-TV North Platte, NE
KRQE Albuquerque, NM
KUPT Hobbs, NM
*WXXI-TV Rochester, NY
WSEE Erie, PA
KCLO-TV Rapid City, SD
KTSM-TV El Paso, TX
*KHCE-TV San Antonio, TX
*WHRO-TV Hampton-Norfolk, VA

Channel 17
KPHO-TV Phoenix, AZ
KVIQ Eureka, CA
*KRMJ Grand Junction, CO
WKCF Clermont, FL
KTGM Tamuning, GU
WTCT Marion, IL
*WYIN Gary, IN
KLBY Colby, KS
KAAS-TV Salina, KS
*WKPC-TV Louisville, KY
*WCBB Augusta, ME
KQDS-TV Duluth, MN
WXVT Greenville, MS
*KRNE-TV Merriman, NE
*KYNE-TV Omaha, NE
*KAZQ Albuquerque, NM
WSYR-TV Syracuse, NY
WQCW Portsmouth, OH
WTOL Toledo, OH
KMTR Eugene, OR
WVEO Aguadilla, PR
WPXQ Block Island, RI
WLTX Columbia, SC
KVIA-TV El Paso, TX
KUCW Ogden, UT
WFXR-TV Roanoke, VA
*KWSU-TV Pullman, WA
WLAX La Crosse, WI
KTWO-TV Casper, WY

Channel 18
KATN Fairbanks, AK
WDBB Bessemer, AL
KFSM-TV Fort Smith, AR
KUVS-TV Modesto, CA
*KIXE-TV Redding, CA
KUSI-TV San Diego, CA
*KRMA-TV Denver, CO
KHVO Hilo, HI
*KYIN Mason City, IA
WAND Decatur, IL
KUPK-TV Garden City, KS
*WKYU-TV Bowling Green, KY
WMFP Lawrence, MA
*WDCP-TV University Center, MI
*KAWE Bemidji, MN
*KCPT Kansas City, MO
*WMAU-TV Bude, MS
KSVI Billings, MT
*WUNM-TV Jacksonville, NC
KQCD-TV Dickinson, ND
KWNB-TV Hayes Center, NE
WMBC-TV Newton, NJ
WBDT Springfield, OH
KTVZ Bend, OR
KTVC Roseburg, OR
WFXB Myrtle Beach, SC
*KESD-TV Brookings, SD
KEVN-TV Rapid City, SD
KTXA Arlington, TX
KZTV Corpus Christi, TX
KVAW Eagle Pass, TX
KDBC-TV El Paso, TX
KYTX Nacogdoches, TX
*KUEW Saint George, UT
WDBJ Roanoke, VA
*WVTB Saint Johnsbury, VT
KEPR-TV Pasco, WA
KCPQ Tacoma, WA

Channel 19
*WIIQ Demopolis, AL
KMOH-TV Kingman, AZ
KTTU-TV Tucson, AZ
KSWB-TV San Diego, CA
KBWB San Francisco, CA
KCOY-TV Santa Maria, CA
KTVD Denver, CO
WTEV-TV Jacksonville, FL
WMOR-TV Lakeland, FL
WSFL-TV Miami, FL
WGCL-TV Atlanta, GA
KIKU-TV Honolulu, HI
WHO-TV Des Moines, IA
WGN-TV Chicago, IL
*WUSI-TV Olney, IL
WISE-TV Fort Wayne, IN
KWCH-TV Hutchinson, KS
WBKI-TV Campbellsville, KY
*WGBH-TV Boston, MA
WABI-TV Bangor, ME
WXMI Grand Rapids, MI
KSPR Springfield, MO
KWYB Butte, MT
WCWG Lexington, NC
KXMA-TV Dickinson, ND
KVRR Fargo, ND
KTVG Grand Island, NE
KOCT Carlsbad, NM
WSYT Syracuse, NY
WTVG Toledo, OH
WKPV Ponce, PR
KPRY-TV Pierre, SD
KTVT Fort Worth, TX
KTXH Houston, TX
KIDY San Angelo, TX
KBCB Bellingham, WA
*KCKA Centralia, WA
WMTV Madison, WI
KBJR-TV Superior, WI

Channel 20
KJNP-TV North Pole, AK
WYLE Florence, AL
*KPAZ-TV Phoenix, AZ
KFTV Hanford, CA
KCVU Paradise, CA
*KRMU Durango, CO
*WLRN-TV Miami, FL
WAZE-TV Madisonville, KY
*KLTL-TV Lake Charles, LA
WCVB-TV Boston, MA
*KSMQ-TV Austin, MN
KCCW-TV Walker, MN
KNLJ Jefferson City, MO
*WMPN-TV Jackson, MS
*WUND-TV Edenton, NC
*KDSE Dickinson, ND
*KJRE Ellendale, ND
KETV Omaha, NE
KVIH-TV Clovis, NM
KAME-TV Reno, NV
WFMJ-TV Youngstown, OH
WWMB Florence, SC
*KTLM Rio Grande City, TX
KXII Sherman, TX
KTXS-TV Sweetwater, TX

WWCW Lynchburg, VA
WSVI Christiansted, VI
KREM-TV Spokane, WA

Channel 21

KMAX-TV Sacramento, CA
KPMR Santa Barbara, CA
WCLF Clearwater, FL
KHAW-TV Hilo, HI
*WMEC Macomb, IL
WPXS Mount Vernon, IL
*WFYI-TV Indianapolis, IN
*KDCK Dodge City, KS
KAKE-TV Wichita, KS
WUPX-TV Morehead, KY
KPXJ Minden, LA
WBOC-TV Salisbury, MD
WMYD Detroit, MI
WFTC Minneapolis, MN
KTAJ-TV Saint Joseph, MO
WAPT Jackson, MS
WYDO Greenville, NC
WDAY-TV Fargo, ND
KHAS-TV Hastings, NE
KOAT-TV Albuquerque, NM
WWTI Watertown, NY
WBNS-TV Columbus, OH
KCBY-TV Coos Bay, OR
WJPX San Juan, PR
*KTSD-TV Pierre, SD
WUXP-TV Nashville, TN
KXAN-TV Austin, TX
KFDM-TV Beaumont, TX
*KYVE Yakima, WA

Channel 22

*WFIQ Florence, AL
KNAZ-TV Flagstaff, AZ
KXRM-TV Colorado Springs, CO
WFOR-TV Miami, FL
WOFL Orlando, FL
WTXL-TV Tallahassee, FL
KHBC-TV Hilo, HI
KGMB Honolulu, HI
WBUI Decatur, IL
KSNC Great Bend, KS
WNEM-TV Bay City, MI
WDHS Iron Mountain, MI
WUCW Minneapolis, MN
KBSI Cape Girardeau, MO
KMIZ Columbia, MO
KHMT Hardin, MT
WCNC-TV Charlotte, NC
*KBME-TV Bismarck, ND
KNOP-TV North Platte, NE
WOWT Omaha, NE
*WNJS Camden, NJ
KVMY Las Vegas, NV
*WLIW Garden City, NY
KMCB Coos Bay, OR
KTTM Huron, SD
*KLRU-TV Austin, TX
KETK-TV Jacksonville, TX
KAUZ-TV Wichita Falls, TX
WRIC-TV Petersburg, VA
WVNY Burlington, VT
WVCY-TV Milwaukee, WI

Channel 23

KVOA Tucson, AZ
*KRCB Cotati, CA
KTBN-TV Santa Ana, CA
KVMD Twentynine Palms, CA
KREG-TV Glenwood Springs, CO
WJCL Savannah, GA
KWHH Hilo, HI
KFVE Honolulu, HI
*WQPT-TV Moline, IL
*KTWU Topeka, KS
WUTF-TV Marlborough, MA
*KOZK Springfield, MO
WBTV Charlotte, NC
*WUNK-TV Greenville, NC
KXMB-TV Bismarck, ND
*KFME Fargo, ND
*KRWG-TV Las Cruces, NM
*WNPI-TV Norwood, NY

WFTY-TV Smithtown, NY
WLYH-TV Lancaster, PA
*KZSD-TV Martin, SD
WPXK Jellico, TN
WNAB Nashville, TN
KVII-TV Amarillo, TX
KPEJ Odessa, TX
WCVI-TV Christiansted, VI
WSAZ-TV Huntington, WV

Channel 24

*KUAC-TV Fairbanks, AK
*WHIQ Huntsville, AL
WBMM Tuskegee, AL
KVTN Pine Bluff, AR
KTVK Phoenix, AZ
KBEH Oxnard, CA
KGO-TV San Francisco, CA
KRDO-TV Colorado Springs, CO
*WMFE-TV Orlando, FL
WTSP Saint Petersburg, FL
WTLF Tallahassee, FL
WPXC-TV Brunswick, GA
WNEG-TV Toccoa, GA
KGMV Wailuku, HI
WPTA Fort Wayne, IN
WTHI-TV Terre Haute, IN
*WCVN-TV Covington, KY
*WKPI Pikeville, KY
*KLPB-TV Lafayette, LA
WUPL Slidell, LA
WTLJ Muskegon, MI
KCCO-TV Alexandria, MN
KCTV Kansas City, MO
WFXI Morehead City, NC
*KMDE Devils Lake, ND
*KTNE-TV Alliance, NE
KNAT Albuquerque, NM
*WNYE-TV New York, NY
WSFJ-TV Newark, OH
KOKH-TV Oklahoma City, OK
WATM-TV Altoona, PA
KVEO Brownsville, TX
KPNZ Ogden, UT
*WVTA Windsor, VT
*WHRM-TV Wausau, WI

Channel 25

KMSB-TV Tucson, AZ
KGET-TV Bakersfield, CA
KGTV San Diego, CA
KOVR Stockton, CA
WFGX Fort Walton Beach, FL
WVEA-TV Venice, FL
WATL Atlanta, GA
KLEI Kailua-Kona, HI
*KTIN Fort Dodge, IA
KWKB Iowa City, IA
WRTV Indianapolis, IN
*WKAS Ashland, KY
*KLPA-TV Alexandria, LA
*WLPB-TV Baton Rouge, LA
*KOZJ Joplin, MO
*WMAO-TV Greenwood, MS
*WUNF-TV Asheville, NC
*WUNU Lumberton, NC
KOLN Lincoln, NE
*WCNY-TV Syracuse, NY
KVAL-TV Eugene, OR
KDKA-TV Pittsburgh, PA
WTVE Reading, PA
*KPSD-TV Eagle Butte, SD
WPTY-TV Memphis, TN
WTVR-TV Richmond, VA
WNNE-TV Hartford, VT
WCGV-TV Milwaukee, WI

Channel 26

KTVF Fairbanks, AK
WTJP-TV Gadsden, AL
KUTP Phoenix, AZ
*KVCR-TV San Bernardino, CA
WCCU Urbana, IL
KSAS-TV Wichita, KS
WLKY-TV Louisville, KY
KMSP-TV Minneapolis, MN
KPLR-TV Saint Louis, MO

WMDN Meridian, MS
*KLNE-TV Lexington, NE
KOB-TV Albuquerque, NM
KREN-TV Reno, NV
WTEN Albany, NY
*WVIZ-TV Cleveland, OH
KTEN Ada, OK
KYW-TV Philadelphia, PA
KDLV-TV Mitchell, SD
*KBHE-TV Rapid City, SD
KXXV Waco, TX
WRLH-TV Richmond, VA

Channel 27

WKRG-TV Mobile, AL
*WAIQ Montgomery, AL
KTVE El Dorado, AR
KFTA-TV Fort Smith, AR
KFPH-TV Flagstaff, AZ
KTSF San Francisco, CA
KEYT-TV Santa Barbara, CA
*WETA-TV Washington, DC
*WXEL-TV West Palm Beach, FL
WAGA Atlanta, GA
KFXA Cedar Rapids, IA
WCIU-TV Chicago, IL
WIPX Bloomington, IN
KRWF Redwood Falls, MN
*KUFM-TV Missoula, MT
WCCB Charlotte, NC
WRDC Durham, NC
WGTW-TV Burlington, NJ
KASA-TV Santa Fe, NM
WTBY-TV Poughkeepsie, NY
*WOUB-TV Athens, OH
KFOR-TV Oklahoma City, OK
KMVU Medford, OR
*KOPB-TV Portland, OR
WAPA-TV San Juan, PR
WKPT-TV Kingsport, TN
WKRN-TV Nashville, TN
KRIV Houston, TX
WKOW-TV Madison, WI
*WHWC-TV Menomonie, WI

Channel 28

WTTO Birmingham, AL
KMPH-TV Visalia, CA
WGFL High Springs, FL
WFLX West Palm Beach, FL
*KSIN Sioux City, IA
KBCI-TV Boise, ID
WTVW Evansville, IN
KSNT-TV Topeka, KS
*WFPT Frederick, MD
WPME Lewiston, ME
*KAWB Brainerd, MN
KSFX-TV Springfield, MO
WDAM-TV Laurel, MS
*KHNE-TV Hastings, NE
WNBC New York, NY
WUHF Rochester, NY
WUAB Lorain, OH
KTPX Okmulgee, OK
*WRJA-TV Sumter, SC
KABY-TV Aberdeen, SD
WREG-TV Memphis, TN
KFDX-TV Wichita Falls, TX
KSTU Salt Lake City, UT
WCYB-TV Bristol, VA
WYOW Eagle River, WI
KLWY Cheyenne, WY

Channel 29

WBIH Selma, AL
*KAET Phoenix, AZ
KPIX-TV San Francisco, CA
KDEN Longmont, CO
WFTS Tampa, FL
KMAU Wailuku, HI
WMAQ-TV Chicago, IL
*KPTS-TV Hutchinson, KS
WXIX-TV Newport, KY
WVUE New Orleans, LA
WUNI Worcester, MA
*WMPB Baltimore, MD
KHQA-TV Hannibal, MO

KMTF Helena, MT
*WUNJ-TV Wilmington, NC
WXLV-TV Winston-Salem, NC
KSTF Scottsbluff, NE
*WFME-TV West Milford, NJ
KWBQ Santa Fe, NM
KVCW Las Vegas, NV
WKTV Utica, NY
*WGTE-TV Toledo, OH
KTUZ-TV Shawnee, OK
KIVV-TV Lead, SD
KSFY-TV Sioux Falls, SD
*WKNO-TV Memphis, TN
KYLE Bryan, TX
KUPX Provo, UT
WVBT Virginia Beach, VA
WAOW-TV Wausau, WI

Channel 30

WIAT Birmingham, AL
*KUAT-TV Tucson, AZ
*KPBS San Diego, CA
*KQED San Francisco, CA
*WBCC Cocoa, FL
WAGT Augusta, GA
WMBD-TV Peoria, IL
*WKOH Owensboro, KY
KVHP Lake Charles, LA
WBZ-TV Boston, MA
WEYI-TV Saginaw, MI
WSFX-TV Wilmington, NC
WPXN-TV New York, NY
WUTR Utica, NY
WBNX-TV Akron, OH
WRGT-TV Dayton, OH
KPTV Portland, OR
WVLT-TV Knoxville, TN
KMPX Decatur, TX
*KCOS El Paso, TX
*WNVT Goldvein, VA
KAYU-TV Spokane, WA
*WHLA-TV La Crosse, WI
KGWN-TV Cheyenne, WY

Channel 31

WLGA Opelika, AL
KSAZ-TV Phoenix, AZ
KTLA Los Angeles, CA
WPPX Wilmington, DE
WOGX Ocala, FL
*WSRE Pensacola, FL
WRDW-TV Augusta, GA
KWHE Honolulu, HI
KCCI Des Moines, IA
WFLD Chicago, IL
KSCW Wichita, KS
WFXT Boston, MA
WGTU Traverse City, MI
*KWCM-TV Appleton, MN
KCWE Kansas City, MO
KDNL-TV Saint Louis, MO
*WGBC Meridian, MS
WXII-TV Winston-Salem, NC
KFYR-TV Bismarck, ND
KLKN Lincoln, NE
WKRC-TV Cincinnati, OH
WJW Cleveland, OH
KEYU Elk City, OK
*KOET Eufaula, OK
KLSR-TV Eugene, OR
WSWB Scranton, PA
WSJU-TV San Juan, PR
WLMT Memphis, TN
KGBT-TV Harlingen, TX
KHOU-TV Houston, TX
KVTV Laredo, TX
KFXK Longview, TX
KOSA-TV Odessa, TX
WAVY-TV Portsmouth, VA

Channel 32

WAAY-TV Huntsville, AL
KARK-TV Little Rock, AR
KOLD-TV Tucson, AZ
KDOC-TV Anaheim, CA
KDVR Denver, CO
WAWS Jacksonville, FL

WBFS-TV Miami, FL
KLEW-TV Lewiston, ID
WTJR Quincy, IL
WNDY-TV Marion, IN
WPSD-TV Paducah, KY
KLAX-TV Alexandria, LA
WBPX Boston, MA
WCCO-TV Minneapolis, MN
WABG-TV Greenwood, MS
WITN-TV Washington, NC
*WUNL-TV Winston-Salem, NC
KGIN Grand Island, NE
KMCC Laughlin, NV
WNLO Buffalo, NY
*KETA Oklahoma City, OK
WTAJ-TV Altoona, PA
WPSG Philadelphia, PA
WQPX Scranton, PA
WTCV San Juan, PR
KELO-TV Sioux Falls, SD
KDAF Dallas, TX
KTRK-TV Houston, TX
KMYS Kerrville, TX
WVIR-TV Charlottesville, VA
*WSBN-TV Norton, VA
*WETK Burlington, VT
KWPX Bellevue, WA
WBUW Janesville, WI
WTRF-TV Wheeling, WV

Channel 33

KBAK-TV Bakersfield, CA
WFSB Hartford, CT
*WHUT-TV Washington, DC
*WCEU New Smyrna Beach, FL
WRXY-TV Tice, FL
KBFD Honolulu, HI
*KBIN Council Bluffs, IA
WBKO-TV Bowling Green, KY
WPXD Ann Arbor, MI
*WNMU-TV Marquette, MI
KAAL Austin, MN
KDLH Duluth, MN
KTVO Kirksville, MO
KTVM Butte, MT
WMYV Greensboro, NC
WGRZ-TV Buffalo, NY
WPIX New York, NY
WSTR-TV Cincinnati, OH
KOCB Oklahoma City, OK
*KFTS Klamath Falls, OR
KRCW-TV Salem, OR
WORO-TV Fajardo, PR
*WEBA-TV Allendale, SC
KVUE-TV Austin, TX
KIMA-TV Yakima, WA
WITI Milwaukee, WI
*WNPB-TV Morgantown, WV

Channel 34

KGPE Fresno, CA
KRCR-TV Redding, CA
KTAS San Luis Obispo, CA
KFSF-TV Vallejo, CA
KWGN-TV Denver, CO
WHPX New London, CT
WUSA Washington, DC
WCWJ Jacksonville, FL
WSIL-TV Harrisburg, IL
*WQEC Quincy, IL
WVLA Baton Rouge, LA
KMSS-TV Shreveport, LA
WHTV Jackson, MI
*KTCA-TV Saint Paul, MN
WDAF-TV Kansas City, MO
WSOC-TV Charlotte, NC
WPXU-TV Jacksonville, NC
KSNB-TV Superior, NE
WNEU Merrimack, NH
WNYO-TV Buffalo, NY
*WMHT Schenectady, NY
*WCET Cincinnati, OH
WQHS-TV Cleveland, OH
WJAC-TV Johnstown, PA
*WYBE Philadelphia, PA
WTNZ Knoxville, TN
WSET-TV Lynchburg, VA

WISN-TV Milwaukee, WI

Channel 35

KGUN Tucson, AZ
KMEX-TV Los Angeles, CA
KCRA-TV Sacramento, CA
KCNC-TV Denver, CO
WDCA Washington, DC
WFTX Cape Coral, FL
WGSA Baxley, GA
WLTZ Columbus, GA
KHNL Honolulu, HI
*KHIN Red Oak, IA
*KRIN Waterloo, IA
*WNIT-TV South Bend, IN
KMTW Hutchinson, KS
KALB-TV Alexandria, LA
WTOM-TV Cheboygan, MI
WLUC-TV Marquette, MI
KARE Minneapolis, MN
KSDK Saint Louis, MO
WCBI-TV Columbus, MS
WGHP High Point, NC
WZMY-TV Derry, NH
*KNME-TV Albuquerque, NM
WWNY-TV Carthage, NY
*WOUC-TV Cambridge, OH
WLWT Cincinnati, OH
KTVL Medford, OR
WMMP Charleston, SC
WGGS-TV Greenville, SC
WTVC Chattanooga, TN
KDFW Dallas, TX
KPRC-TV Houston, TX
KVOS-TV Bellingham, WA
*WMVT Milwaukee, WI

Channel 36

WABM Birmingham, AL
WTVY Dothan, AL
KPNX Mesa, AZ
KNBC Los Angeles, CA
KFRE-TV Sanger, CA
WTTG Washington, DC
KAII-TV Wailuku, HI
WFFT-TV Fort Wayne, IN
WJYS-TV Hammond, IN
WTWO-TV Terre Haute, IN
KMCI Lawrence, KS
*WKMU Murray, KY
WWL-TV New Orleans, LA
KARD West Monroe, LA
WCDC Adams, MA
WJRT-TV Flint, MI
KSAX Alexandria, MN
WIRT Hibbing, MN
KTTC Rochester, MN
KOMU-TV Columbia, MO
*WMAV-TV Oxford, MS
KTMF Missoula, MT
WFPX Fayetteville, NC
KHGI-TV Kearney, NE
WNJU Linden, NJ
WMGM-TV Wildwood/Atlantic City, NJ
WTTE Columbus, OH
WYTV Youngstown, OH
*KRSC-TV Claremore, OK
*WITF-TV Harrisburg, PA
WNPX Cookeville, TN
KFTH-TV Alvin, TX
KDFI Dallas, TX
WPXR Roanoke, VA
KSKN Spokane, WA
KSTW Tacoma, WA

Channel 38

KSEE Fresno, CA
KPXN San Bernardino, CA
*KBDI-TV Broomfield, CO
*WFSG Panama City, FL
WKMJ Louisville, KY
KMCT-TV West Monroe, LA
WJZ-TV Baltimore, MD
WGME-TV Portland, ME
WSYM-TV Lansing, MI
*WDSE-TV Duluth, MN
KEYC-TV Mankato, MN
KCFW-TV Kalispell, MT
WUVC-TV Fayetteville, NC
KXJB-TV Valley City, ND
KXVO Omaha, NE
WWOR-TV Secaucus, NJ
KOBR Roswell, NM
WKBW-TV Buffalo, NY
*WCFE-TV Plattsburgh, NY
*WOSU-TV Columbus, OH
*KOED-TV Tulsa, OK
KDRV Medford, OR
*WQED Pittsburgh, PA
WEMT Greeneville, TN
WHTN Murfreesboro, TN
*KMBH Harlingen, TX
*KPBT-TV Odessa, TX
KSL-TV Salt Lake City, UT
WTVZ-TV Norfolk, VA
*KTNW Richland, WA
KOMO-TV Seattle, WA

Channel 39

KASN Pine Bluff, AR
KWOG Springdale, AR
KTAZ Phoenix, AZ
KCNS San Francisco, CA
WCTX New Haven, CT
WJLA-TV Washington, DC
WFTV Orlando, FL
WSB-TV Atlanta, GA
WSAV-TV Savannah, GA
KMEG Sioux City, IA
WAOE Peoria, IL
WKOI-TV Richmond, IN
WFXW-TV Terre Haute, IN
WLEX-TV Lexington, KY
WSBK-TV Boston, MA
WZZM-TV Grand Rapids, MI
WADL Mount Clemens, MI
*KETC Saint Louis, MO
WLOX Biloxi, MS
*WUNP-TV Roanoke Rapids, NC
KTEL-TV Carlsbad, NM
WIVB-TV Buffalo, NY
WRGB Schenectady, NY
WDLI-TV Canton, OH
KWTV Oklahoma City, OK
WJWN-TV San Sebastian, PR
WKTC Sumter, SC
WJKT Jackson, TN
*KSCE El Paso, TX
KHCW Houston, TX
*KTXT-TV Lubbock, TX
KWEX-TV San Antonio, TX
KIRO-TV Seattle, WA
WEAU-TV Eau Claire, WI
WFRV-TV Green Bay, WI
WLPX-TV Charleston, WV

Channel 40

KNSD San Diego, CA
*KRMT Denver, CO
WACX-DT Leesburg, FL
WMGT-TV Macon, GA
KITV Honolulu, HI
*WSIU-TV Carbondale, IL
WHOI Peoria, IL
*WFWA Fort Wayne, IN
WTVQ-TV Lexington, KY
WWTV Cadillac, MI
KPXM Saint Cloud, MN
KECI-TV Missoula, MT
WHKY-TV Hickory, NC
*KSRE Minot, ND
*KUON-TV Lincoln, NE
WXTV Paterson, NJ
KBLR Paradise, NV
KAUT-TV Oklahoma City, OK
KOIN Portland, OR
WDSI-TV Chattanooga, TN
KXTX-TV Dallas, TX
KBTV-TV Port Arthur, TX
KTVX Salt Lake City, UT
WTKR Norfolk, VA
WPXE Kenosha, WI
WSAW-TV Wausau, WI

Channel 41

WZDX Huntsville, AL
*WEIQ Mobile, AL
WZVN-TV Naples, FL
WRBW Orlando, FL
*WATC Atlanta, GA
KPXO-TV Kaneohe, HI
KGCW-TV Burlington, IA
KTIV Sioux City, IA
WICD-TV Champaign, IL
*WKPD Paducah, KY
WLVI-TV Cambridge, MA
WUTB Baltimore, MD
WXYZ-TV Detroit, MI
WBUY-TV Holly Springs, MS
WDBD Jackson, MS
KBIM-TV Roswell, NM
*WPBS-TV Watertown, NY
WHIO-TV Dayton, OH
WKBN-TV Youngstown, OH
*WVIA-TV Scranton, PA
WIS Columbia, SC
*WETP-TV Sneedville, TN
KXAS-TV Fort Worth, TX
WDRL-TV Danville, VA
WVEC-TV Hampton, VA
*KCTS-TV Seattle, WA
WGBA Green Bay, WI
WKBT La Crosse, WI
WCHS-TV Charleston, WV

Channel 42

KHRR Tucson, AZ
KWHY-TV Los Angeles, CA
WSAH Bridgeport, CT
WXPX Bradenton, FL
WJBF Augusta, GA
KIMT Mason City, IA
WQRF-TV Rockford, IL
WICS-TV Springfield, IL
WNDU-TV South Bend, IN
*WKLE Lexington, KY
*WKMA Madisonville, KY
WHMM-DT Hammond, LA
WHDH-TV Boston, MA
*WMPT Annapolis, MD
KSHB-TV Kansas City, MO
WRAY-TV Wilson, NC
KLUZ-TV Albuquerque, NM
*WSKG-TV Binghamton, NY
WNYB Jamestown, NY
WGGN-TV Sandusky, OH
KMYT-TV Tulsa, OK
*KSYS Medford, OR
WTXF-TV Philadelphia, PA
WPMY Pittsburgh, PA
WFLI-TV Cleveland, TN
KTBU Conroe, TX
*WMSY-TV Marion, VA
*WCVE-TV Richmond, VA
*KWDK Tacoma, WA
*WPNE Green Bay, WI

Channel 43

KHSL-TV Chico, CA
KCAL Los Angeles, CA
KPXC-TV Denver, CO
WUPA Atlanta, GA
WSWG Valdosta, GA
KFXB Dubuque, IA
WCPX Chicago, IL
*WKZT-TV Elizabethtown, KY
*WGBX-TV Boston, MA
WPXT Portland, ME
*WTVS Detroit, MI
WDIO-TV Duluth, MN
KODE-TV Joplin, MO
KTVI Saint Louis, MO
WLXI-TV Greensboro, NC
KPTM Omaha, NE
*WNJT Trenton, NJ
*WNED-TV Buffalo, NY
WCWN Schenectady, NY
*WPBO-TV Portsmouth, OH
KATU Portland, OR
WPGH-TV Pittsburgh, PA
WSUR-TV Ponce, PR
*WRET-TV Spartanburg, SC
KEYE-TV Austin, TX
KMLM Odessa, TX
WPXW Manassas, VA
WZVI Charlotte Amalie, VI
WFFF-TV Burlington, VT

Channel 44

*WGIQ Louisville, AL
KWBF Little Rock, AR
KWBA Sierra Vista, AZ
KHIZ Barstow, CA
KGMC Clovis, CA
*WDPB Seaford, DE
*WJEB-TV Jacksonville, FL
WRSP-TV Springfield, IL
*WDTI Indianapolis, IN
WIBW-TV Topeka, KS
WTSF Ashland, KY
*WKON Owenton, KY
KSHV Shreveport, LA
*WWPB Hagerstown, MD
WCSH Portland, ME
WZPX Battle Creek, MI
WWJ-TV Detroit, MI
KSTC-TV Minneapolis, MN
KYTV Springfield, MO
*WMAW-TV Meridian, MS
*WUNG-TV Concord, NC
WECT Wilmington, NC
KVLY-TV Fargo, ND
WMCN-TV Atlantic City, NJ
KRXI-TV Reno, NV
WNYW New York, NY
WNYS-TV Syracuse, NY
KZJL Houston, TX
*WCVW Richmond, VA
KVEW Kennewick, WA
KHCV Seattle, WA
WWAZ-TV Fond du Lac, WI

Channel 45

WPXH Gadsden, AL
KBCW San Francisco, CA
*WEDN Norwich, CT
WXCW Naples, FL
WJTC Pensacola, FL
WGNM Macon, GA
*KIIN-TV Iowa City, IA
WEVV-TV Evansville, IN
WXIN Indianapolis, IN
KSNW Wichita, KS
*WMEA-TV Biddeford, ME
WDIV Detroit, MI
WLLA Kalamazoo, MI
KTGF Great Falls, MT
WYCW Asheville, NC
KXMC-TV Minot, ND
KMTV Omaha, NE
KASY-TV Albuquerque, NM
WABC-TV New York, NY
WROC-TV Rochester, NY
KTCW Roseburg, OR
*WJPM-TV Florence, SC

Channel 46

WMCF-TV Montgomery, AL
KMIR-TV Palm Springs, CA
KQCA Stockton, CA
KWHD Castle Rock, CO
*WHFT-TV Miami, FL
WCTV Thomasville, GA
*WTVP Peoria, IL
WFIE-TV Evansville, IN
WAFB Baton Rouge, LA
WBFF Baltimore, MD
WMTW-TV Poland Spring, ME
KXLT-TV Rochester, MN
KSNF Joplin, MO
WWAY Wilmington, NC
*WNEO Alliance, OH
WWHO Chillicothe, OH
WUPW Toledo, OH
KGW Portland, OR
WFMZ-TV Allentown, PA
KXLN-TV Rosenberg, TX
KJZZ-TV Salt Lake City, UT
WPXV Norfolk, VA
WTPX Antigo, WI
WVVA Bluefield, WV

Channel 47

WPMI-TV Mobile, AL
WAMI-TV Hollywood, FL
WFTT-TV Tampa, FL
*WTTW-TV Chicago, IL
WRBU East St. Louis, IL
*WYDN Worcester, MA
KSMO-TV Kansas City, MO
WJZY Belmont, NC
KTDO Las Cruces, NM
WTVH Syracuse, NY
WOAC Canton, OH
WTLW Lima, OH
KWHB Tulsa, OK
WCSC-TV Charleston, SC
KDLT-TV Sioux Falls, SD
WDEF-TV Chattanooga, TN
*WLEF-TV Park Falls, WI

Channel 48

WRJM-TV Troy, AL
KECY-TV El Centro, CA
*KOCE-TV Huntington Beach, CA
KTFF-TV Porterville, CA
KSPX Sacramento, CA
WRC-TV Washington, DC
WFBD Destin, FL
WFXU Live Oak, FL
WOPX Melbourne, FL
WTTV Bloomington, IN
KTKA-TV Topeka, KS
*WKGB-TV Bowling Green, KY
WAQP Saginaw, MI
WXXV-TV Gulfport, MS
WCTI-TV New Bern, NC
*WLED-TV Littleton, NH
WRNN-TV Kingston, NY
WPXI Pittsburgh, PA
WACH Columbia, SC
KSTR-TV Irving, TX
WVXF Charlotte Amalie, VI
KING-TV Seattle, WA
KPDX Vancouver, WA
WJJA Racine, WI
WVNS-TV Lewisburg, WV

Channel 49

WAFF Huntsville, AL
KYPX Camden, AR
KASW Phoenix, AZ
KSTS San Jose, CA
KJLA Ventura, CA
WAWD Fort Walton Beach, FL
WFGC Palm Beach, FL
KLJB-TV Davenport, IA
KPTH Sioux City, IA
WDRB Louisville, KY
WWUP-TV Sault Ste. Marie, MI
WTOK-TV Meridian, MS
WNTZ Natchez, MS
WRAZ Raleigh, NC
*WEKW-TV Keene, NH
WWSI Atlantic City, NJ
WNWO-TV Toledo, OH
KGEB Tulsa, OK
WNEP-TV Scranton, PA
*WITV Charleston, SC
KNVA Austin, TX
WLFG Grundy, VA
WHSV-TV Harrisonburg, VA
WTAP-TV Parkersburg, WV

Channel 50

WBRC Birmingham, AL
KNWA-TV Rogers, AR
*KTEH San Jose, CA
*KNXT Visalia, CA
WTVX Fort Pierce, FL
WTLH Bainbridge, GA
*KDIN-TV Des Moines, IA
WDKA Paducah, KY
WPXL New Orleans, LA

WPBN-TV Traverse City, MI
KSTP-TV Saint Paul, MN
WAXN-TV Kannapolis, NC
WYPX Amsterdam, NY
WYDC Corning, NY
WDTN Dayton, OH
KOPX Oklahoma City, OK
WPCB-TV Greensburg, PA
WCBD-TV Charleston, SC
WBXX-TV Crossville, TN
KBTX-TV Bryan, TX
WGNT Portsmouth, VA
KUNS-TV Bellevue, WA
WISC-TV Madison, WI
WOAY-TV Oak Hill, WV

Channel 51

WNCF Montgomery, AL
KXLA Rancho Palos Verdes, CA
KCEC Denver, CO
WDCW Washington, DC
WPWR-TV Gary, IN
WMYO Salem, IN
WAGV Harlan, KY
WLAJ Lansing, MI
KPXE Kansas City, MO
WFMY-TV Greensboro, NC
*KWSE Williston, ND
*WNJN Montclair, NJ
KVVU-TV Henderson, NV
WKEF Dayton, OH
KSBI Oklahoma City, OK
KOHD Bend, OR
WTAE-TV Pittsburgh, PA
WJAR Providence, RI
KWSD Sioux Falls, SD
WPXX-TV Memphis, TN

KFWD Fort Worth, TX
WLUK-TV Green Bay, WI

Channel 52

WVTM-TV Birmingham, AL
KESQ-TV Palm Springs, CA
*WEDW Bridgeport, CT
WWSB Sarasota, FL
KCRG-TV Cedar Rapids, IA
WLS-TV Chicago, IL
*WVUT Vincennes, IN
WWDP Norwell, MA
WMAR-TV Baltimore, MD
*WFUM Flint, MI
KOLR-TV Springfield, MO
WJTV Jackson, MS
WTVD Durham, NC
KUMV-TV Williston, ND
WICU-TV Erie, PA
*WCTE Cookeville, TN
KNWS-TV Katy, TX
WBOY-TV Clarksburg, WV

Channel 53

*WBIQ Birmingham, AL
KABC-TV Los Angeles, CA
*KVIE Sacramento, CA
WHLV-TV Cocoa, FL
WCFN-TV Springfield, IL
WMDT Salisbury, MD
KQTV Saint Joseph, MO
WRAL-TV Raleigh, NC
WFUT-TV Newark, NJ
WCCV-TV Arecibo, PR
WSPA-TV Spartanburg, SC
KAZH Baytown, TX
KWTX-TV Waco, TX
WCAX-TV Burlington, VT

Channel 54

*WEDU Tampa, FL
WGEM-TV Quincy, IL
WREX-TV Rockford, IL
WTTK Kokomo, IN
*WGPT Oakland, MD
*WUNE-TV Linville, NC
WSTM-TV Syracuse, NY
WPHL-TV Philadelphia, PA
WNAC-TV Providence, RI
KLDT Lake Dallas, TX
WWBT Richmond, VA

Channel 55

KUVI-TV Bakersfield, CA
KFMB-TV San Diego, CA
KWWL Waterloo, IA
WHAS-TV Louisville, KY
WGGB-TV Springfield, MA
WHAG-TV Hagerstown, MD
*WKAR-TV East Lansing, MI
*WMAE-TV Booneville, MS
WNCN Goldsboro, NC
WENY-TV Elmira, NY
KOTV Tulsa, OK
KENS-TV San Antonio, TX

Channel 56

*WCIQ Mount Cheaha, AL
KNXV-TV Phoenix, AZ
KTVU Oakland, CA
KWQC-TV Davenport, IA
KDMI Des Moines, IA
WCLJ-TV Bloomington, IN
KLFY-TV Lafayette, LA
*WCPB Salisbury, MD
KMOV Saint Louis, MO
WLOS Asheville, NC

WCBS-TV New York, NY
*WBGU-TV Bowling Green, OH
KJRH Tulsa, OK
WOLF-TV Hazleton, PA
WQMY Williamsport, PA
WLII Caguas, PR
WBTW Florence, SC
WTVF Nashville, TN
KTBC Austin, TX

Channel 57

KRON-TV San Francisco, CA
WTTA Saint Petersburg, FL
WEEK-TV Peoria, IL
*WCML-TV Alpena, MI
WILX-TV Onondaga, MI
KFVS-TV Cape Girardeau, MO
WLFL Raleigh, NC
*WENH-TV Durham, NH
WLNY Riverhead, NY
WTOV-TV Steubenville, OH
WHNS Greenville, SC
KWKT Waco, TX
*WNVC Fairfax, VA

Channel 58

*KQET Watsonville, CA
WJBK Detroit, MI
*WCMW Manistee, MI
KMOT Minot, ND
WHEC-TV Rochester, NY
WOAI-TV San Antonio, TX

Channel 59

WHNT-TV Huntsville, AL
*KCET Los Angeles, CA
WTOG Saint Petersburg, FL
WBAL-TV Baltimore, MD

WLNS-TV Lansing, MI
*WUNC-TV Chapel Hill, NC
WMUR-TV Manchester, NH
WHAM-TV Rochester, NY
WYFF Greenville, SC
WACY Appleton, WI

Channel 60

KCBS-TV Los Angeles, CA

Channel 61

KSCI Long Beach, CA
KXTV Sacramento, CA
*WNET Newark, NJ
WVTV Milwaukee, WI

Channel 63

KTNC-TV Concord, CA

Channel 64

WPVI-TV Philadelphia, PA

Channel 65

KTTV Los Angeles, CA
WUPV Ashland, VA

Channel 66

KCOP Los Angeles, CA
WUVP-TV Vineland, NJ

Channel 67

WCAU Philadelphia, PA

Channel 68

KRCA Riverside, CA

Spanish-Language Television Stations

The following Spanish-language television stations operate within the United States or near the U.S. border. Stations are listed by Designated Market Area (DMA), city of license, call letters and channel. Stations in U.S. territories do not fall within any Designated Market Areas. For further information on individual stations, see Directory of Television Stations in the U.S. beginning on page B-14.

Arizona

Phoenix (Prescott), AZ
*KFPH-TV Flagstaff (ch 13)
Phoenix (Prescott), AZ
*KMOH-TV Kingman (ch 6)
Phoenix (Prescott), AZ
*KTAZ Phoenix (ch 39)
Phoenix (Prescott), AZ
*KTVW-TV Phoenix (ch 33)
Tucson (Sierra Vista), AZ
*KFTU-TV Douglas (ch 3)
Tucson (Sierra Vista), AZ
*KHRR Tucson (ch 40)

California

Fresno-Visalia, CA
*KFTV Hanford (ch 21)
Fresno-Visalia, CA
*KNSO Merced (ch 51)
Los Angeles
*KAZA-TV Avalon (ch 54)
Los Angeles
*KVEA Corona (ch 52)
Los Angeles
*KMEX-TV Los Angeles (ch 34)
Los Angeles
*KTLA Los Angeles (ch 5)
Los Angeles
*KWHY-TV Los Angeles (ch 22)
Los Angeles
*KFTR-TV Ontario (ch 46)
Los Angeles
*KRCA Riverside (ch 62)
Los Angeles
*KJLA Ventura (ch 57)
Monterey-Salinas, CA
*KSMS-TV Monterey (ch 67)
Sacramento-Stockton-Modesto, CA
*KUVS-TV Modesto (ch 19)
Sacramento-Stockton-Modesto, CA
*KMAX-TV Sacramento (ch 31)
Sacramento-Stockton-Modesto, CA
*KTFK-TV Stockton (ch 64)
San Francisco-Oakland-San Jose
*KTNC-TV Concord (ch 42)
San Francisco-Oakland-San Jose
*KDTV San Francisco (ch 14)
San Francisco-Oakland-San Jose
*KSTS San Jose (ch 48)
San Francisco-Oakland-San Jose
*KFSF-TV Vallejo (ch 66)
Santa Barbara-Santa Maria-San Luis Obispo, CA
*KBEH Oxnard (ch 63)
Santa Barbara-Santa Maria-San Luis Obispo, CA
*KTAS San Luis Obispo (ch 33)
Santa Barbara-Santa Maria-San Luis Obispo, CA
*KPMR Santa Barbara (ch 38)
Yuma, AZ-El Centro, CA
*KAJB Calipatria (ch 54)
Yuma, AZ-El Centro, CA
*KVYE El Centro (ch 7)

Colorado

Albuquerque-Santa Fe, NM
*KTLL-TV Durango (ch 33)
Denver, CO
*KTFD-TV Boulder (ch 14)
Denver, CO
*KCEC Denver (ch 50)
Denver, CO
*KDEN Longmont (ch 25)

Florida

Miami-Ft. Lauderdale, FL
*WSCV Fort Lauderdale (ch 51)
Miami-Ft. Lauderdale, FL
*WSBS-TV Key West (ch 22)
Miami-Ft. Lauderdale, FL
*WLTV Miami (ch 23)
Orlando-Daytona Beach-Melbourne, FL
*WVEN-TV Daytona Beach (ch 26)
Orlando-Daytona Beach-Melbourne, FL
*WOTF-TV Melbourne (ch 43)
Tampa-St. Petersburg (Sarasota), FL
*WFTT-TV Tampa (ch 50)
Tampa-St. Petersburg (Sarasota), FL
*WVEA-TV Venice (ch 62)

Georgia

Atlanta
*WUVG-TV Athens (ch 34)

Illinois

Chicago
*WSNS Chicago (ch 44)
Chicago
*WGBO-TV Joliet (ch 66)

Louisiana

New Orleans, LA
*WHMM-DT Hammond (ch 0)

Massachusetts

Boston (Manchester, NH)
*WUTF-TV Marlborough (ch 66)
Boston (Manchester, NH)
*WUNI Worcester (ch 27)

Nevada

Las Vegas, NV
*KINC Las Vegas (ch 15)
Las Vegas, NV
*KBLR Paradise (ch 39)

New Hampshire

Boston (Manchester, NH)
*WNEU Merrimack (ch 60)

New Jersey

New York
*WNJU Linden (ch 47)
New York
*WFUT-TV Newark (ch 68)
New York
*WXTV Paterson (ch 41)
Philadelphia
*WWSI Atlantic City (ch 62)
Philadelphia
*WUVP-TV Vineland (ch 65)

New Mexico

Albuquerque-Santa Fe, NM
*KLUZ-TV Albuquerque (ch 41)
Albuquerque-Santa Fe, NM
*KTFQ-TV Albuquerque (ch 14)
Albuquerque-Santa Fe, NM
*KTEL-TV Carlsbad (ch 25)
El Paso (Las Cruces, NM), TX
*KTDO Las Cruces (ch 48)

North Carolina

Raleigh-Durham (Fayetteville), NC
*WUVC-TV Fayetteville (ch 40)

Ohio

Cleveland-Akron (Canton), OH
*WQHS-TV Cleveland (ch 61)

Oklahoma

Oklahoma City, OK
*KEYU Elk City (ch 0)
Oklahoma City, OK
*KTUZ-TV Shawnee (ch 30)
Oklahoma City, OK
*KUOK Woodward (ch 35)

Oregon

Portland, OR
*KUNP La Grande (ch 16)

Texas

Austin, TX
*KNIC-TV Blanco (ch 17)
Corpus Christi, TX
*KORO Corpus Christi (ch 28)
Dallas-Ft. Worth
*KXTX-TV Dallas (ch 39)
Dallas-Ft. Worth
*KUVN-TV Garland (ch 23)
El Paso (Las Cruces, NM), TX
*KINT-TV El Paso (ch 26)
El Paso (Las Cruces, NM), TX
*KSCE El Paso (ch 38)
El Paso (Las Cruces, NM), TX
*KTFN El Paso (ch 65)
Harlingen-Weslaco-Brownsville-McAllen, TX
*KNVO McAllen (ch 48)
Harlingen-Weslaco-Brownsville-McAllen, TX
*KTLM Rio Grande City (ch 40)
Houston
*KFTH-TV Alvin (ch 67)
Houston
*KAZH Baytown (ch 57)
Houston
*KTMD Galveston (ch 47)
Houston
*KXLN-TV Rosenberg (ch 45)
Laredo, TX
*KLDO-TV Laredo (ch 27)
Odessa-Midland, TX
*KUPB Midland (ch 18)
San Antonio, TX
*KTRG Del Rio (ch 10)
San Antonio, TX
*KVDA San Antonio (ch 60)
San Antonio, TX
*KWEX-TV San Antonio (ch 41)
Waco-Temple-Bryan, TX
*KAKW-TV Killeen (ch 62)

Utah

Salt Lake City, UT
*KUTH Provo (ch 32)

Virginia

Washington, DC (Hagerstown, MD)
*WFDC-TV Arlington (ch 14)

Washington

Seattle-Tacoma, WA
*KUNS-TV Bellevue (ch 51)
Seattle-Tacoma, WA
*KHCV Seattle (ch 45)

Mexico

El Paso (Las Cruces, NM), TX
*XHIJ Ciudad Juarez (ch 44)
San Diego, CA
*XEWT-TV Tijuana (ch 12)

U.S. TV Stations Providing News Programming

Abilene-Sweetwater, TX
KTAB-TV, 8 hrs weekly

Albany, GA
WALB, 15 hrs weekly
WFXL, 4 hrs weekly

Albany-Schenectady-Troy, NY
WCDC, 15 hrs weekly
WNYT, 27 hrs weekly
WRGB, 26.5 hrs weekly
WTEN, 22 hrs weekly
WXXA-TV, 23.5 hrs weekly
WYPX, 5 hrs weekly

Albuquerque-Santa Fe, NM
KASA-TV, 7 hrs weekly
KBIM-TV, 6 hrs weekly
KLUZ-TV, .5 hr weekly
KNME-TV, 1 hr weekly
KOBF, 4 hrs weekly
KOB-TV, 12 hrs weekly
KRQE, 24 hrs weekly

Alexandria, LA
KLAX-TV, 4 hrs weekly

Alpena, MI
WBKB-TV, 5.5 hrs weekly

Amarillo, TX
KCIT, 3 hrs weekly
KENW, 3 hrs weekly
KFDA-DT, 2.5 hrs weekly
KFDA-TV, 13 hrs weekly
KVII-TV, 36 hrs weekly

Anchorage, AK
KTUU-TV, 12 hrs weekly
KTVA, 7 hrs weekly
KYES-TV, 1 hr weekly

Atlanta
WAGA, 38 hrs weekly
WGCL-DT, 29 hrs weekly
WGCL-TV, 14.5 hrs weekly
WPBA, 1.5 hrs weekly

Augusta, GA
WAGT, 8 hrs weekly
WJBF, 22 hrs weekly
WRDW-TV, 20 hrs weekly

Austin, TX
KEYE-TV, 20 hrs weekly
KTBC, 24 hrs weekly
KXAN-TV, 22 hrs weekly

Bakersfield, CA
KBAK-TV, 30.5 hrs weekly
KGET-TV, 27 hrs weekly

Baltimore, MD
WBAL-TV, 24 hrs weekly

Bangor, ME
WABI-TV, 25 hrs weekly
WLBZ, 33 hrs weekly
WVII-TV, 6 hrs weekly

Baton Rouge, LA
WBRZ, 22 hrs weekly
WLPB-TV, 1 hr weekly
WVLA, 6 hrs weekly

Beaumont-Port Arthur, TX
KBMT, 13 hrs weekly

Bend, OR
KTVZ, 20 hrs weekly

Billings, MT
KTVQ, 17 hrs weekly
KULR-TV, 20 hrs weekly

Biloxi-Gulfport, MS
WLOX, 18 hrs weekly

Binghamton, NY
WBNG-TV, 24.5 hrs weekly
WICZ-TV, 2 hrs weekly

Birmingham (Anniston, Tuscaloosa), AL
WBRC, 21 hrs weekly
WDBB, 15 hrs weekly
WIAT, 6 hrs weekly
WJSU-TV, 9 hrs weekly
WPXH, 8 hrs weekly
WVTM-TV, 24 hrs weekly

Bluefield-Beckley-Oak Hill, WV
WSWP-TV, 5 hrs weekly
WVVA, 27 hrs weekly

Boise, ID
KAID-TV, 3 hrs weekly
KBCI-TV, 19 hrs weekly
KIVI, 19.5 hrs weekly
KTRV-TV, 3 hrs weekly
KTVB, 27 hrs weekly

Boston (Manchester, NH)
WBPX, 20 hrs weekly
WBZ-TV, 20 hrs weekly
WCVB-DT, 25 hrs weekly
WCVB-TV, 30 hrs weekly
WENH-TV, 2 hrs weekly
WFXT, 7 hrs weekly
WLVI-TV, 7 hrs weekly
WMUR-TV, 29 hrs weekly
WSBK-TV, 4 hrs weekly
WUNI, 3 hrs weekly
WZMY-TV, 5 hrs weekly

Bowling Green, KY
WBKO-TV, 17 hrs weekly
WKYU-TV, 1 hr weekly

Burlington, VT-Plattsburgh, NY
WCAX-TV, 15 hrs weekly
WPTZ, 23 hrs weekly

Butte-Bozeman, MT
KXLF-TV, 16 hrs weekly

Casper-Riverton, WY
KCWY, 20 hrs weekly
KTWO-TV, 19.5 hrs weekly

Cedar Rapids-Waterloo-Iowa City & Dubuque, IA
KFXA, 13.5 hrs weekly
KFXB, 5 hrs weekly
KGAN-TV, 12 hrs weekly
KWWL, 22 hrs weekly

Champaign & Springfield -Decatur, IL
WEIU-TV, 3 hrs weekly
WICD-TV, 17 hrs weekly
WICS-TV, 22 hrs weekly

Charleston, SC
WCBD-TV, 17 hrs weekly
WCIV, 12 hrs weekly
WCSC-TV, 21 hrs weekly
WTAT-TV, 3.5 hrs weekly

Charleston-Huntington, WV
WCHS-TV, 19.5 hrs weekly
WOUB-TV, 3 hrs weekly
WVAH-TV, 7 hrs weekly

Charlotte, NC
WHKY-TV, 5 hrs weekly

Charlottesville, VA
WVIR-TV, 32 hrs weekly

Chattanooga, TN
WDEF-TV, 28.5 hrs weekly
WRCB-TV, 22 hrs weekly
WTVC, 32 hrs weekly

Cheyenne, WY-Scottsbluff, NE
KDEV, 15 hrs weekly
KDUH-TV, 2 hrs weekly
KGWN-TV, 15 hrs weekly
KSTF, 8 hrs weekly

Chicago
WFLD, 35.5 hrs weekly
WGN-TV, 32 hrs weekly
WLS-TV, 8 hrs weekly
WSNS, 5 hrs weekly
WTTW-TV, 5 hrs weekly

Chico-Redding, CA
KCVU, 0 hrs weekly
KHSL-TV, 20 hrs weekly
KIXE-TV, 1 hr weekly
KRCR-TV, 16 hrs weekly

Cincinnati, OH
WCPO-TV, 24 hrs weekly
WPTO, 1 hr weekly
WXIX-TV, 23 hrs weekly

Clarksburg-Weston, WV
WBOY-TV, 24 hrs weekly
WDTV, 18.5 hrs weekly

Cleveland-Akron (Canton), OH
WEAO, 1 hr weekly
WEWS, 22 hrs weekly
WMFD-TV, 36 hrs weekly

Colorado Springs-Pueblo, CO
KKTV, 27 hrs weekly
KOAA-TV, 7 hrs weekly
KXRM-TV, 3.5 hrs weekly

Columbia, SC
WIS, 26.5 hrs weekly
WLTX, 22 hrs weekly

Columbia-Jefferson City, MO
KMIZ, 19 hrs weekly
KOMU-TV, 20 hrs weekly
KRCG, 14 hrs weekly

Columbus, OH
WBNS-TV, 31 hrs weekly
WCMH-TV, 31.5 hrs weekly
WSFJ-TV, 1 hr weekly

Columbus-Tupelo-West Point, MS
WLOV-TV, 2.5 hrs weekly
WTVA, 17 hrs weekly

Corpus Christi, TX
KIII, 17.5 hrs weekly
KORO, 5 hrs weekly
KRIS-TV, 14 hrs weekly

Dallas-Ft. Worth
KDAF, 6 hrs weekly
KDFW, 50 hrs weekly
KTVT, 27 hrs weekly
KUVN-TV, 10 hrs weekly
KXAS-DT, 30 hrs weekly
KXAS-TV, 37 hrs weekly
WFAA-TV, 28 hrs weekly

Davenport, IA-Rock Island-Moline, IL
KLJB-TV, 3 hrs weekly
WHBF-TV, 7 hrs weekly
WQAD-TV, 22.5 hrs weekly

Dayton, OH
WKEF, 17 hrs weekly
WPTD, 1 hr weekly
WRGT-TV, 16 hrs weekly

Denver, CO
KDEN, 1 hr weekly
KPXC-TV, 24 hrs weekly
KRMA-TV, 1 hr weekly
KUPN, 17 hrs weekly
KWGN-TV, 29.5 hrs weekly

Des Moines-Ames, IA
KCCI, 30 hrs weekly
KDSM-TV, 4 hrs weekly
WHO-TV, 30 hrs weekly
WOI-TV, 15 hrs weekly

Detroit
WJBK, 36 hrs weekly

Dothan, AL
WDHN, 7 hrs weekly
WTVY, 16 hrs weekly

Duluth, MN-Superior, WI
KBJR-TV, 10 hrs weekly
KDLH, 12 hrs weekly
KQDS-TV, 2.5 hrs weekly
WDIO-TV, 9 hrs weekly

El Paso (Las Cruces, NM), TX
KDBC-TV, 15 hrs weekly
KFOX-TV, 6 hrs weekly
KINT-TV, 10 hrs weekly
KRWG-TV, 3 hrs weekly
KTSM-TV, 19.5 hrs weekly
KVIA-TV, 31.5 hrs weekly
XHIJ, 15 hrs weekly

Elmira (Corning), NY
WENY-TV, 10 hrs weekly
WETM-TV, 14 hrs weekly

Erie, PA
WFXP, 3.5 hrs weekly
WICU-TV, 22 hrs weekly
WJET-TV, 3.5 hrs weekly
WSEE, 14 hrs weekly

Eugene, OR
KEZI, 20 hrs weekly
KMCB, 22 hrs weekly
KMTR, 15 hrs weekly
KPIC, 14 hrs weekly
KTCW, 22 hrs weekly
KVAL-TV, 17 hrs weekly

Eureka, CA
KAEF, 2 hrs weekly

Evansville, IN
WEVV-TV, 9 hrs weekly
WTVW, 22.5 hrs weekly

Fairbanks, AK
KATN, 6 -11 hrs weekly
KFXF, 10 hrs weekly
KTVF, 5 hrs weekly

Fargo-Valley City, ND
KVLY-TV, 16 hrs weekly
KXJB-TV, 10 hrs weekly

Flint-Saginaw-Bay City, MI
WEYI-TV, 19.5 hrs weekly
WJRT-TV, 34 hrs weekly

Fresno-Visalia, CA
KAIL, 5 hrs weekly
KFRE-TV, 2 hrs weekly
KFSN-TV, 30 hrs weekly
KFTV, 8 hrs weekly
KGPE, 28 hrs weekly
KMPH-TV, 7 hrs weekly

Ft. Myers-Naples, FL
WBBH-TV, 32 hrs weekly
WINK-TV, 32 hrs weekly
WZVN-DT, 19.5 hrs weekly
WZVN-TV, 19.5 hrs weekly

Ft. Smith-Fayetteville -Springdale-Rogers, AR
KFSM-TV, 28 hrs weekly
KFTA-TV, 20.5 hrs weekly
KHBS, 15 hrs weekly
KNWA-TV, 17 hrs weekly

Ft. Wayne, IN
WANE-TV, 22 hrs weekly
WISE-TV, 14 hrs weekly
WPTA, 31 hrs weekly

Gainesville, FL
WCJB-TV, 17 hrs weekly
WOGX, 7 hrs weekly
WUFT, 3 hrs weekly

Grand Junction-Montrose, CO
KKCO, 28 hrs weekly
KREX-TV, 30 hrs weekly

Grand Rapids-Kalamazoo-Battle Creek, MI
WGVK, 1 hr weekly
WOTV, 13 hrs weekly
WWMT, 25 hrs weekly
WWMT-DT, 24.5 hrs weekly
WXMI, 4 hrs weekly
WZZM-TV, 26 hrs weekly

Great Falls, MT
KFBB-TV, 6 hrs weekly
KRTV, 15 hrs weekly

Green Bay-Appleton, WI
WBAY-TV, 19 hrs weekly
WGBA, 16 hrs weekly
WLUK-TV, 32.5 hrs weekly

Greensboro-High Point-Winston Salem, NC
WFMY-TV, 32 hrs weekly
WGHP, 32 hrs weekly
WMYV, 7 hrs weekly
WXLV-TV, 2.5 hrs weekly

Greenville-New Bern-Washington, NC
WFXI, 4 hrs weekly
WUNK-TV, 3 hrs weekly

Greenville-Spartanburg, SC-Asheville, NC-Anderson, SC
WHNS, 27 hrs weekly
WLOS, 24 hrs weekly
WNEG-TV, 13 hrs weekly
WSPA-TV, 26.5 hrs weekly
WYFF, 28 hrs weekly

Greenwood-Greenville, MS
WABG-TV, 13 hrs weekly

Harlingen-Weslaco -Brownsville-McAllen, TX
KGBT-TV, 22 hrs weekly
KRGV-TV, 12 hrs weekly
KTLM, 8 hrs weekly

Harrisburg -Lancaster-Lebanon-York, PA
WGAL, 29 hrs weekly
WHP-TV, 24 hrs weekly
WHTM-TV, 27 hrs weekly

Harrisonburg, VA
WHSV-TV, 16 hrs weekly

Hartford & New Haven, CT
WCTX, 8.5 hrs weekly
WFSB, 36 hrs weekly
WTIC-TV, 6 hrs weekly
WTNH-TV, 32 hrs weekly
WUVN, 5 hrs weekly

Hattiesburg-Laurel, MS
WDAM-TV, 20 hrs weekly

Helena, MT
KTVH, 7 hrs weekly

Honolulu, HI
KBFD, 6 hrs weekly
KFVE, 7 hrs weekly
KHNL, 23 hrs weekly
KHON-TV, 25 hrs weekly
KITV, 20 hrs weekly
KUPU, 14 hrs weekly

Houston
KETH-TV, 3 hrs weekly
KHCW, 4 hrs weekly
KHWB-DT, 4 hrs weekly
KPRC-TV, 34 hrs weekly
KTMD, 7 hrs weekly
KXLN-TV, 7 hrs weekly

Huntsville-Decatur (Florence), AL
WAFF, 26 hrs weekly
WYLE, 15 hrs weekly

Idaho Falls-Pocatello, ID
KIDK-TV, 15 hrs weekly
KIFI-TV, 30 hrs weekly
KJWY, 5 hrs weekly
KPVI, 17 hrs weekly

Indianapolis, IN
WHMB-TV, 1 hr weekly
WRTV-DT, 20 hrs weekly
WTIU, 3 hrs weekly
WXIN, 19 hrs weekly

Jackson, MS
WAPT, 14 hrs weekly
WJTV, 21 hrs weekly

Jacksonville, FL
WJXT, 51 hrs weekly
WTEV-TV, 22 hrs weekly
WTLV, 17 hrs weekly

Johnstown-Altoona, PA
WJAC-TV, 25 hrs weekly
WTAJ-TV, 29 hrs weekly

Jonesboro, AR
KAIT, 17 hrs weekly

Joplin, MO-Pittsburg, KS
KFJX-TV, 3 hrs weekly
KOAM-TV, 19 hrs weekly
KODE-TV, 16 hrs weekly

Juneau, AK
KJUD, 5 hrs weekly
KTOO-TV, 1 hr weekly

Kansas City, MO
KMBC-TV, 28 hrs weekly
KSHB-TV, 32 hrs weekly
WDAF-TV, 49 hrs weekly

Knoxville, TN
WATE-TV, 24 hrs weekly
WBIR-TV, 24 hrs weekly
WTNZ, 4 hrs weekly
WVLT-TV, 24.5 hrs weekly

La Crosse-Eau Claire, WI
WEAU-TV, 34 hrs weekly
WLAX, 3.5 hrs weekly
WQOW-TV, 18 hrs weekly
WXOW-TV, 18 hrs weekly

Lafayette, IN
WLFI-TV, 22 hrs weekly

Lafayette, LA
KATC, 19.5 hrs weekly
KLFY-TV, 14 hrs weekly

Lake Charles, LA
KVHP, 9 hrs weekly

Lansing, MI
WLAJ, 5 hrs weekly
WSYM-TV, 10 hrs weekly

Las Vegas, NV
KBLR, 2 hrs weekly
KBLR-DT, 2 hrs weekly
KINC, 5 hrs weekly
KTNV, 22 hrs weekly
KVBC, 29 hrs weekly

Lexington, KY
WDKY-TV, 7 hrs weekly
WKYT-TV, 43 hrs weekly
WTVQ-DT, 22 hrs weekly
WTVQ-TV, 27 hrs weekly

Lima, OH
WLIO, 24 hrs weekly

Lincoln & Hastings-Kearney, NE
KHGI-TV, 17.5 hrs weekly
KHNE-TV, 30 . hrs weekly
KLKN, 22 hrs weekly
KSNB-TV, 13 hrs weekly
KUON-TV, 30 . hrs weekly
KWNB-TV, 17.5 hrs weekly

Little Rock-Pine Bluff, AR
KARK-TV, 22 hrs weekly
KTHV, 20 hrs weekly

Los Angeles
KCBS-TV, 26 hrs weekly
KHIZ, 5 hrs weekly
KLCS, 5 hrs weekly
KMEX-TV, 17 hrs weekly
KOCE-TV, 3 hrs weekly
KSCI, 12 hrs weekly
KTLA, 22 hrs weekly
KTLA-DT, 22 hrs weekly
KTTV, 25 hrs weekly
KWHY-TV, 12 hrs weekly

Louisville, KY
WDRB, 35 hrs weekly
WLKY-TV, 37.5 hrs weekly

Lubbock, TX
KAMC, 39 hrs weekly
KJTV-TV, 7 hrs weekly
KLBK-TV, 27 hrs weekly

Macon, GA
WGNM, 1 hr weekly
WMAZ-TV, 27 hrs weekly
WPGA-TV, 3 hrs weekly

Madison, WI
WISC-TV, 30 hrs weekly
WKOW-TV, 22 hrs weekly
WMTV, 19 hrs weekly

Mankato, MN
KEYC-TV, 10 hrs weekly

Marquette, MI
WBKP, 10 hrs weekly
WLUC-TV, 16 hrs weekly
WNMU-TV, 1 hr weekly

Medford-Klamath Falls, OR
KDKF, 12 hrs weekly
KDRV, 20 hrs weekly
KTVL, 16 hrs weekly

Memphis, TN
WHBQ-TV, 27 hrs weekly
WPTY-TV, 7 hrs weekly

Meridian, MS
WTOK-TV, 15 hrs weekly

Miami-Ft. Lauderdale, FL
WAMI-TV, 3 hrs weekly
WBZL-DT, 3.5 hrs weekly
WFOR-TV, 30 hrs weekly
WSCV, 3 hrs weekly
WSFL-TV, 3.5 hrs weekly

Minneapolis-St. Paul, MN
KAWB, 2.5 hrs weekly
KAWE, 2.5 hrs weekly
KMSP-TV, 19.5 hrs weekly
KSTP-TV, 27.5 hrs weekly
WCCO-TV, 23 hrs weekly
WFTC, 3.5 hrs weekly

Minot-Bismarck-Dickinson, ND
KFYR-TV, 24 hrs weekly
KUMV-TV, 6 hrs weekly
KXMB-TV, 9 hrs weekly
KXMC-TV, 9 hrs weekly

Missoula, MT
KCFW-TV, 6 hrs weekly
KECI-TV, 14 hrs weekly
KPAX-TV, 15 hrs weekly

Mobile, AL-Pensacola (Ft. Walton Beach), FL
WALA-TV, 23 hrs weekly
WEAR-TV, 14 hrs weekly
WPMI-TV, 5 hrs weekly

Monroe, LA-El Dorado, AR
KMCT-TV, 6 hrs weekly
KNOE-TV, 22 hrs weekly
KTVE, 13 hrs weekly

Monterey-Salinas, CA
KCBA, 10 hrs weekly
KION-TV, 8 hrs weekly
KSBW, 31 hrs weekly

Montgomery-Selma, AL
WAKA, 9 hrs weekly
WNCF, 1 hr weekly

Myrtle Beach-Florence, SC
WBTW, 78 hrs weekly
WFXB, 2.5 hrs weekly

Nashville, TN
WHTN, 2 hrs weekly
WJFB, 4 hrs weekly
WKRN-TV, 25 hrs weekly
WTVF, 24 hrs weekly
WZTV, 7 hrs weekly

New Orleans, LA
WDSU, 32 . hrs weekly
WHNO, 10 hrs weekly
WVUE, 21 hrs weekly

New York
WFTY-TV, 4 hrs weekly
WLNY, 2 hrs weekly
WMBC-DT, 5 hrs weekly
WMBC-TV, 5 hrs weekly
WNET, 5 hrs weekly
WNJB, 2 hrs weekly
WNJN, 3 hrs weekly
WNJN-DT, 3 hrs weekly
WPIX, 19.5 hrs weekly
WRNN-TV, 82 hrs weekly
WWOR-TV, 7 hrs weekly
WXTV, 17 hrs weekly

Norfolk-Portsmouth-Newport News, VA
WAVY-TV, 31 hrs weekly
WHRO-TV, 1 hr weekly
WVEC-TV, 24 hrs weekly

North Platte, NE
KNOP-TV, 15 hrs weekly

Odessa-Midland, TX
KMID, 17 hrs weekly
KOSA-TV, 16 hrs weekly

Oklahoma City, OK
KETA, 3 hrs weekly
KFOR-TV, 29 hrs weekly
KOCO-TV, 30 hrs weekly
KWET, 4 hrs weekly
KWTV, 36 hrs weekly

Omaha, NE
KETV, 28 hrs weekly
KMTV, 23 hrs weekly
KPTM, 7 hrs weekly
KYNE-TV, 30 . hrs weekly
WOWT, 37 hrs weekly

Orlando-Daytona Beach-Melbourne, FL
WFTV, 25 hrs weekly
WKMG-TV, 24 hrs weekly

Ottumwa, IA-Kirksville, MO
KTVO, 14 hrs weekly

Paducah, KY-Cape Girardeau, MO-Harrisburg-Mount Vernon, IL
KFVS-TV, 28 hrs weekly
WPSD-DT, 23 hrs weekly
WPSD-TV, 23 hrs weekly

WSIU-TV, 2 hrs weekly

Palm Springs, CA
KESQ-TV, 17 hrs weekly
KMIR-TV, 24 hrs weekly

Panama City, FL
WMBB, 143 hrs weekly

Parkersburg, WV
WTAP-TV, 23 hrs weekly

Peoria-Bloomington, IL
WEEK-TV, 16 hrs weekly
WHOI, 7 hrs weekly

Philadelphia
WCAU, 21 hrs weekly
WCAU-DT, 21 hrs weekly
WFMZ-DT, 37 hrs weekly
WFMZ-TV, 29.5 hrs weekly
WHYY-TV, 5 hrs weekly
WMGM-TV, 8 hrs weekly
WNJS, 2 hrs weekly
WNJS-DT, 2 hrs weekly
WNJT, 10 hrs weekly
WTXF-TV, 34.5 hrs weekly
WWSI, 3 hrs weekly

Phoenix (Prescott), AZ
KAET, 3 hrs weekly
KPHO-DT, 22 hrs weekly
KPHO-TV, 30.5 hrs weekly
KSAZ-TV, 38 hrs weekly
KTVK, 48 hrs weekly

Pittsburgh, PA
WNPB-TV, 1 hr weekly
WPGH-DT, 6 hrs weekly
WTAE-TV, 32 hrs weekly

Portland, OR
KATU, hrs weekly
KGW, 35 hrs weekly
KGW-DT, 35 hrs weekly
KOIN, 27 hrs weekly
KPTV, 42.5 hrs weekly
KPTV-DT, 27 hrs weekly
KRCW-TV, 3.5 hrs weekly

Portland-Auburn, ME
WGME-TV, 25 hrs weekly
WMTW-TV, 13.5 hrs weekly
WPFO, 18.5 hrs weekly
WPME, 3.5 hrs weekly
WPXT, 3.5 hrs weekly

Presque Isle, ME
WAGM-TV, 14 hrs weekly

Providence, RI-New Bedford, MA
WPXQ, 5 hrs weekly

Quincy, IL-Hannibal, MO-Keokuk, IA
KHQA-TV, 13 hrs weekly
WGEM-TV, 20 hrs weekly

Raleigh-Durham (Fayetteville), NC
WLFL, 7 hrs weekly
WNCN, 30 hrs weekly
WNCN-DT, 30 hrs weekly
WRAL-TV, 30 hrs weekly

Rapid City, SD
KEVN-TV, 9 hrs weekly
KIVV-TV, 9 hrs weekly
KNBN, 11 hrs weekly
KOTA-TV, 9 hrs weekly
KSGW-TV, 8 hrs weekly

Reno, NV
KOLO-TV, 22 hrs weekly
KRNV, 14 hrs weekly
KTVN, 19.5 hrs weekly

Richmond-Petersburg, VA
WRLH-TV, 3 hrs weekly

Roanoke-Lynchburg, VA
WDBJ, 18 hrs weekly
WFXR-TV, 2 hrs weekly
WSET-TV, 9.5 hrs weekly
WSLS-TV, 15 hrs weekly
WWCW, 1 hr weekly

Rochester, MN-Mason City, IA-Austin, MN
KAAL, 11 hrs weekly
KIMT, 19 hrs weekly
KTTC, 11 hrs weekly
KXLT-TV, 7 hrs weekly

Rochester, NY
WHAM-TV, 24 hrs weekly
WHEC-TV, 22 hrs weekly
WROC-TV, 14 hrs weekly
WUHF, 3.5 hrs weekly
WXXI-DT, 1 hr weekly
WXXI-TV, 2 hrs weekly

Rockford, IL
WIFR, 19 hrs weekly
WTVO, 7 hrs weekly

Sacramento -Stockton-Modesto, CA
KCRA-TV, 55 hrs weekly
KOVR, 21 hrs weekly
KSPX, 7 hrs weekly
KTXL, 7 hrs weekly
KUVS-TV, 12 hrs weekly

Salisbury, MD
WBOC-TV, 31 hrs weekly
WDPB, 3 hrs weekly
WMDT, 12 hrs weekly

Salt Lake City, UT
KBYU-TV, 3 hrs weekly
KSL-TV, 21 hrs weekly
KSTU, 34 hrs weekly
KTVX, 14 hrs weekly
KUTH, 5 hrs weekly
KUTV, 34.5 hrs weekly

San Angelo, TX
KLST, 17 hrs weekly

San Antonio, TX
KABB, 7 hrs weekly
KENS-TV, 24 hrs weekly
KSAT-TV, 21 hrs weekly
KVDA, 10 hrs weekly
KWEX-TV, 5 hrs weekly
WOAI-TV, 19.5 hrs weekly

San Diego, CA
KGTV, 35 hrs weekly
KSWB-TV, 14 hrs weekly
KUSI-TV, 25 hrs weekly
XETV, 20 hrs weekly
XEWT-TV, 11 hrs weekly

San Francisco-Oakland-San Jose
KDTV, 5 hrs weekly
KFTY, 7 hrs weekly
KICU-TV, .5 hr weekly
KNTV, 20 hrs weekly
KPIX-TV, 20 hrs weekly
KQED, .5 hr weekly
KRON-TV, 42 hrs weekly
KSTS, 3 hrs weekly
KTSF, 24 hrs weekly
KTSF-DT, 15 hrs weekly

Santa Barbara-Santa Maria-San Luis Obispo, CA
KCOY-TV, 28 hrs weekly
KEYT-TV, 21 hrs weekly
KTAS, 3 hrs weekly

Savannah, GA
WJCL, 5 hrs weekly
WTGS, 5 hrs weekly

Seattle-Tacoma, WA
KBCB, 4 hrs weekly
KCPQ, 21 hrs weekly
KIRO-TV, 46 hrs weekly
KSTW, 7 hrs weekly
KVOS-TV, 1 hr weekly

Sherman, TX-Ada, OK
KTEN, 20 hrs weekly
KXII, 30 hrs weekly

Sioux City, IA
KTIV, 19 hrs weekly
KXNE-TV, 30 . hrs weekly

Sioux Falls (Mitchell), SD
KDLT-TV, 19 hrs weekly
KDLV-TV, 20 hrs weekly
KPRY-TV, 15 hrs weekly
KSFY-TV, 15 hrs weekly

South Bend-Elkhart, IN
WSBT-TV, 22 hrs weekly
WSJV, 16 hrs weekly

Spokane, WA
KAYU-TV, 4 hrs weekly
KHQ-TV, 26 hrs weekly
KLEW-TV, 12 hrs weekly
KREM-TV, 17 hrs weekly
KSKN, 6 hrs weekly
KXLY-TV, 19 hrs weekly

Springfield, MO
KOLR-TV, 22 hrs weekly
KSPR, 9 hrs weekly
KYTV, 40 hrs weekly

St. Joseph, MO
KQTV, 16 hrs weekly

St. Louis, MO
KMOV, 29 hrs weekly
KMOV-DT, 29 hrs weekly
KPLR-TV, 4 hrs weekly
KTVI, 37.5 hrs weekly
KTVI-DT, 7 hrs weekly

Syracuse, NY
WSTM-TV, 27.5 hrs weekly
WSYR-TV, 22.5 hrs weekly
WSYT, 3.5 hrs weekly
WTVH, 24 hrs weekly

Tallahassee, FL-Thomasville, GA
WTWC-TV, 14 hrs weekly
WTXL-TV, 24 hrs weekly

Tampa-St. Petersburg (Sarasota), FL
WTVT, 46 hrs weekly
WVEA-TV, 3.5 hrs weekly
WWSB, 21 hrs weekly

Terre Haute, IN
WFXW-TV, 2 hrs weekly
WTWO-TV, 20 hrs weekly
WVUT, 5 hrs weekly

Toledo, OH
WNWO-TV, 22 hrs weekly
WTOL, 25 hrs weekly
WTVG, 10 hrs weekly
WUPW, 5 hrs weekly

Topeka, KS
KSNT-TV, 25 hrs weekly
KTKA-TV, 15 hrs weekly
WIBW-TV, 30.5 hrs weekly

Traverse City-Cadillac, MI
WFQX-TV, 3 hrs weekly
WFUP, 3 hrs weekly

Tucson (Sierra Vista), AZ
KGUN, 22 hrs weekly
KHRR, 5 hrs weekly
KOLD-TV, 27 hrs weekly
KVOA, 22 hrs weekly

Tulsa, OK
KJRH, 26.5 hrs weekly
KOKI-TV, 7 hrs weekly
KTPX-DT, 2.5 hrs weekly
KTUL, 17 hrs weekly

Twin Falls, ID
KMVT, 10 hrs weekly
KXTF, 2 hrs weekly

Tyler-Longview (Lufkin & Nacogdoches), TX
KETK-TV, 26.5 hrs weekly
KTRE, 20 hrs weekly
KYTX, 13.5 hrs weekly

Utica, NY
WKTV, 31.5 hrs weekly

Victoria, TX
KAVU-TV, 8 hrs weekly

Waco-Temple-Bryan, TX
KBTX-TV, 19 . hrs weekly
KCEN-TV, 17 hrs weekly
KNCT, 1 hr weekly
KWTX-TV, 20 hrs weekly
KXXV, 14.5 hrs weekly

Washington, DC (Hagerstown, MD)
WHAG-TV, 20 hrs weekly
WJAL, 5 hrs weekly
WJLA-TV, 24 hrs weekly
WTTG-DT, 30 hrs weekly
WUSA, 39 hrs weekly

Watertown, NY
WWNY-TV, 19 hrs weekly
WWTI, 5 hrs weekly

Wausau-Rhinelander, WI
WAOW-TV, 15.5 hrs weekly
WFXS, 5 hrs weekly
WJFW-TV, 18.5 hrs weekly
WSAW-TV, 15 hrs weekly

West Palm Beach-Ft. Pierce, FL
WPBF, 25 hrs weekly
WPEC, 24 hrs weekly
WPTV, 27 hrs weekly

Wheeling, WV-Steubenville, OH
WTOV-TV, 22 hrs weekly
WTRF-TV, 26 hrs weekly

Wichita Falls, TX & Lawton, OK
KAUZ-TV, 15 hrs weekly
KFDX-TV, 19.5 hrs weekly

Wichita-Hutchinson Plus, KS
KAKE-TV, 16 hrs weekly
KBSH-TV, 24 hrs weekly
KUPK-DT, 7 hrs weekly
KUPK-TV, 7 hrs weekly
KWCH-TV, 24 hrs weekly

Wilkes Barre-Scranton, PA
WBRE-TV, 24 hrs weekly
WNEP-TV, 36 hrs weekly
WOLF-DT, 3.5 hrs weekly
WOLF-TV, 6.5 hrs weekly
WYOU, 19.5 hrs weekly

Wilmington, NC
WSFX-TV, 3 hrs weekly

Yakima-Pasco -Richland-Kennewick, WA
KEPR-TV, 17 hrs weekly
KIMA-TV, 15 hrs weekly
KNDO, 27 hrs weekly
KNDU, 22.5 hrs weekly

Youngstown, OH
WFMJ-TV, 10 hrs weekly
WKBN-TV, 25 hrs weekly
WNEO, 1 hr weekly
WYTV, 20 hrs weekly

Yuma, AZ-El Centro, CA
KSWT, 5 hrs weekly
KVYE, 5 hrs weekly
KYMA, 12 hrs weekly

Zanesville, OH
WHIZ-TV, 10 hrs weekly

Nielsen DMA Market Atlas

The Designated Market Area (DMA) is a geographic market design that defines each television market exclusive of others, based on measured viewing patterns. Each market's DMA consists of all the counties in which the home market stations receive a preponderance of viewing, and every county is allocated exclusively to one DMA—there is no overlap. The total of all DMAs represents the total television households in the United States.

The DMA is a standard market definition. As a television buying tool, it is a geographical and demographic means for maximum efficiency. As a station tool, it has applications for sales, programming, and promotion planning.

Following, in alphabetical order, are Nielsen's 210 markets for 2007 with coverage maps for each, with county by county breakouts of TV households. Other data include the markets' stations, their cities of license, channel numbers, and network affiliations. An asterisk preceding call letters indicates that it is a noncommercial (ETV) station.

Coverage maps show total survey areas in light shading, the DMAs themselves in dark shading and white, with Nielsen Metro ratings in white. The survey areas consist of all counties in which the home market stations are viewed to a significant extent including via cable. The Metro Areas usually conform to U.S. Census Standard Metropolitan statistical areas.

Non-DMA markets do not meet Nielsen's criteria for having a DMA of their own. They are listed with the DMA of which they are a part.

A cross-reference list of cities in multi-city DMAs appears on page B-231.

Abilene-Sweetwater, TX (164)

DMA TV Households: 114,210

% of U.S. TV Households: .103

KRBC-TV Abilene, TX, ch. 9, NBC
KTXS-TV Sweetwater, TX, ch. 12, ABC, CW
KXVA Abilene, TX, ch. 15, Fox
KPCB Snyder, TX, ch. 17, IND
KTAB-TV Abilene, TX, ch. 32, CBS

DMA Counties	State	TV Households
Brown	TX	14,430
Callahan	TX	5,320
Coleman	TX	3,650
Eastland	TX	7,180
Fisher	TX	1,690
Haskell	TX	2,380
Jones	TX	5,740
Knox	TX	1,480
Mitchell	TX	2,640
Nolan	TX	5,850
Runnels	TX	4,1780
Scurry	TX	5,750
Shackelford	TX	1,270
Stephens	TX	3,530
Stonewall	TX	580
Taylor	TX	47,860
Throckmorton	TX	690

Abilene-Sweetwater, TX

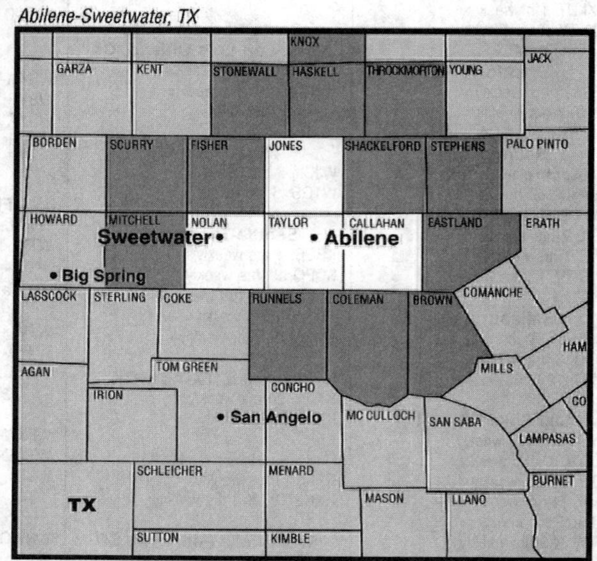

Albany, GA

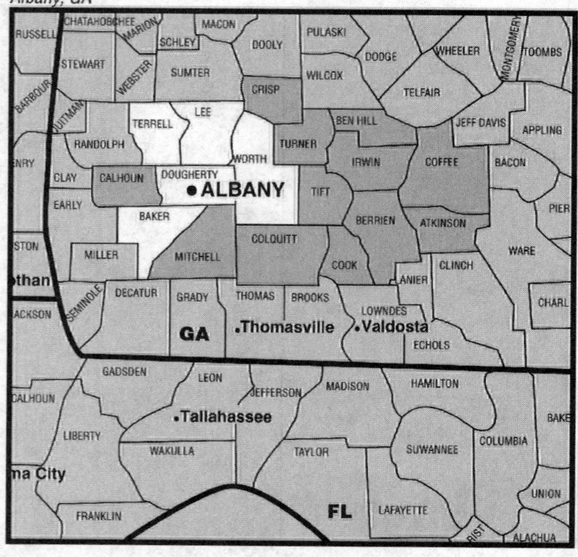

Maps courtesy of Nielsen Media Research

Albany, GA (145)

DMA TV Households: 153,190

% of U.S. TV Households: .138

WALB Albany, GA, ch. 10, NBC
***WABW-TV** Pelham, GA, ch. 14, ETV
***WACS-TV** Dawson, GA, ch. 25, ETV
WFXL Albany, GA, ch. 31, Fox
WSST-TV Cordele, GA, ch. 55, IND

DMA Counties	State	TV Households	DMA Counties	State	TV Households
Atkinson	GA	2,840	Dougherty	GA	35,480
Baker	GA	1,570	Irwin	GA	3,630
Ben Hill	GA	6,450	Lee	GA	10,950
Berrien	GA	6,610	Mitchell	GA	8,180
Calhoun	GA	1,760	Terrell	GA	3,830
Coffee	GA	13,970	Tift	GA	15,220
Colquitt	GA	16,300	Turner	GA	3,470
Cook	GA	6,240	Worth	GA	8,280
Crisp	GA	8,410			

Albany-Schenectady-Troy, NY

Albany-Schenectady-Troy, NY (56)

DMA TV Households: 554,970
% of U.S. TV Households: .498

WRGB Schenectady, NY, ch. 6, CBS
WTEN Albany, NY, ch. 10, ABC
WNYT Albany, NY, ch. 13, NBC
***WMHT** Schenectady, NY, ch. 17, ETV
WCDC Adams, MA, ch. 19, satellite to WTEN
WXXA-TV Albany, NY, ch. 23, Fox
WCWN Schenectady, NY, ch. 45, CW
WNYA Pittsfield, MA, ch. 51, MyNetworkTV
WYPX Amsterdam, NY, ch. 55, ION Television

DMA Counties	State	TV Households	DMA Counties	State	TV Households
Berkshire	MA	54,380	Rensselaer	NY	62,270
Albany	NY	122,180	Saratoga	NY	87,010
Columbia	NY	25,280	Schenectady	NY	61,600
Fulton	NY	22,510	Schoharie	NY	12,460
Greene	NY	19,090	Warren	NY	27,280
Hamilton	NY	2,380	Washington	NY	23,990
Montgomery	NY	19,820	Bennington	VT	14,720

Albuquerque-Santa Fe (45)

DMA TV Households: 662,380
% of U.S. TV Households: .595

KASA-TV Santa Fe, NM, ch. 2, Fox
KOFT Farmington, NM, ch. 3, ABC
KOB-TV Albuqerque, NM, ch. 4, NBC
***KNME-TV** Albuquerque, NM, ch. 5, ETV
KREZ-TV Durango, CO, ch. 6, satellite to KREX-TV
KOCT Carlsbad, NM, ch. 6, satellite to KOAT-TV
KOBG-TV Silver City, NM, ch. 6, satellite to KOB-TV
KOAT-TV Albuquerque, NM, ch. 7, ABC
KOBR Roswell, NM, ch. 8, NBC
KNMD-TV Santa Fe, NM, ch. 9, ETV
KBIM-TV Roswell, NM, ch. 10, CBS
KOVT Silver City, NM, ch. 10, ABC
KCHF Santa Fe, NM, ch. 11, IND
KOBF Farmington, NM, ch. 12, NBC
KRQE Albuquerque, NM, ch. 13, CBS
KTFQ-TV Albuquerque, NM, ch. 14, TeleFutura
KWBQ Santa Fe, NM, ch. 19, CW
***KRMU** Durango, CO, ch. 20, ETV
KRWB-TV Roswell, NM, ch. 21, IND
KNAT Albuquerque, NM, ch. 23, IND
KTEL-TV Carlsbad, NM, ch. 25, Telemundo
KRPV Roswell, NM, ch. 27, IND
KUPT Hobbs, NM, ch. 29, UPN
***KAZQ** Albuquerque, NM, ch. 32, ETV
KTLL-TV Durango, CO, ch. 33, Telemundo
KLUZ-TV Albuquerque, NM, ch. 41, Univision
KASY-TV Albuquerque, NM, ch. 50, MyNetworkTV

Albuquerque-Santa Fe, NM

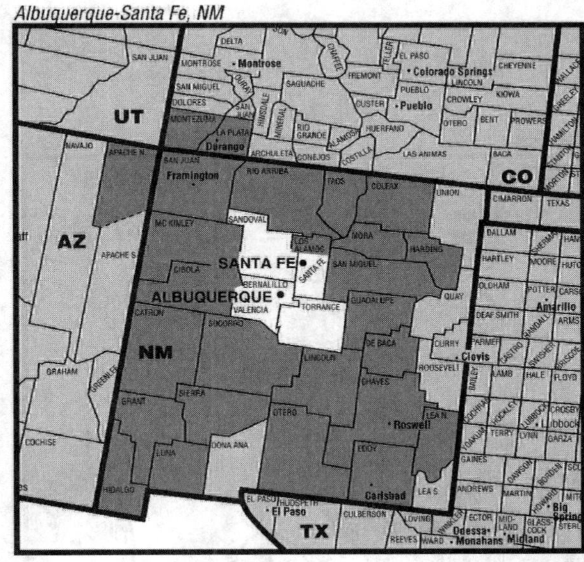

DMA Counties	State	TV Households	DMA Counties	State	TV Households
Apache North	AZ	13,460	Los Almos	NM	7,750
La Plata	CO	18,070	Luna	NM	9,930
Montezuma	CO	9,540	McKinley	NM	19,110
Bernalillo	NM	242,590	Mora	NM	1,920
Catron	NM	1,200	Otero	NM	23,350
Chaves	NM	22,540	Rio Arriba	NM	14,640
Cibola	NM	8,530	Sandoval	NM	39,860
Colfax	NM	5,490	San Juan	NM	42,900
De Baca	NM	740	San Miguel	NM	10,230
Eddy	NM	19,380	Santa Fe	NM	56,020
Grant	NM	11,400	Sierra	NM	5,510
Guadalupe	NM	1,440	Socorro	NM	6,380
Harding	NM	280	Taos	NM	11,880
Hidalgo	NM	1,820	Torrance	NM	5,680
Lea North	NM	18,510	Valencia	NM	23,360
Lincoln	NM	8,870			

Maps courtesy of Nielsen Media Research

Alexandria, LA (179)

DMA TV Households: 89,600
% of U.S. TV Households: .080

KALB-TV Alexandria, LA, ch. 5, NBC
***KLPA-TV** Alexandria, LA, ch. 25, ETV
KLAX-TV Alexandria, LA, ch. 31, ABC
KBCA Alexandria, LA, ch. 41, CW

DMA Counties	State	TV Households
Avoyelles	LA	15,660
Grant	LA	7,550
Rapides	LA	49,850
Vernon	LA	16,540

Alexandria, LA

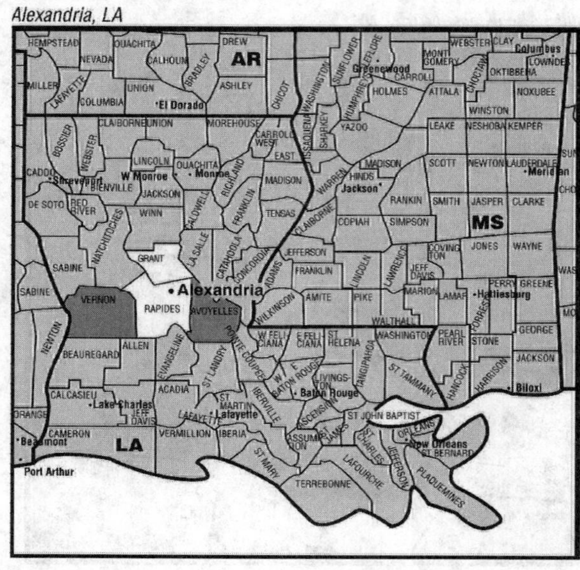

Alpena, MI

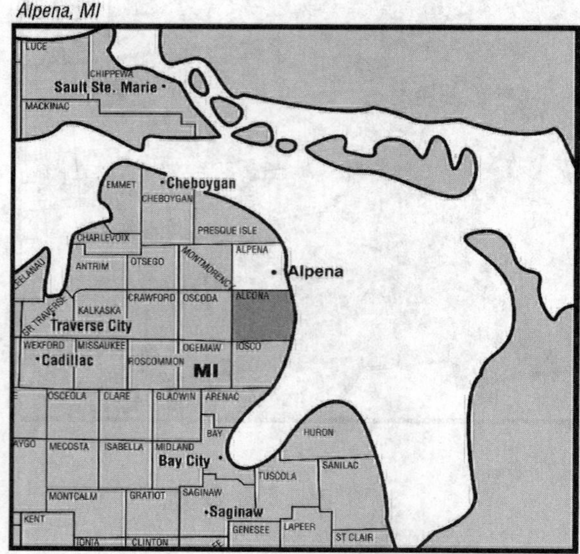

Alpena, MI (208)

DMA TV Households: 17,600
% of U.S. TV Households: .016

***WCML-TV** Alpena, MI, ch. 6, ETV
WBKB-TV Alpena, MI, ch. 11, CBS

DMA Counties	State	TV Households
Alcona	MI	5,110
Alpena	MI	12,490

Maps courtesy of Nielsen Media Research

Amarillo, TX

Amarillo, TX (131)

DMA TV Households: 190,590
% of U.S. TV Households: .171

***KACV-TV** Amarillo, TX, ch. 2, ETV
***KENW** Portales, NM, ch. 3, ETV
KAMR-TV Amarillo, TX, ch. 4, NBC
KVII-TV Amarillo, TX, ch. 7, ABC, CW
KFDA-TV Amarillo, TX, ch. 10, CBS
KVIH-TV Clovis, NM, ch. 12, satellite to KVII-TV
KCIT Amarillo, TX, ch. 14, Fox
KBGD Farwell, TX, ch. 18, IND

DMA Counties	State	TV Households	DMA Counties	State	TV Households
Curry	NM	17,270	Gray	TX	8,210
Quay	NM	3,650	Hall	TX	1,480
Roosevelt	NM	6,360	Hansford	TX	1,980
Union	NM	1,540	Hartley	TX	1,590
Beaver	OK	1,970	Hemphill	TX	1,350
Cimarron	OK	1,020	Hutchinson	TX	8,620
Texas	OK	6,890	Lipscomb	TX	1,170
Armstrong	TX	790	Moore	TX	6,760
Briscoe	TX	590	Ochiltree	TX	3,290
Carson	TX	2,480	Oldham	TX	680
Castro	TX	2,480	Parmer	TX	3,260
Childress	TX	2,330	Potter	TX	42,280
Collingsworth	TX	1,160	Randall	TX	44,680
Cottle	TX	680	Roberts	TX	400
Dallam	TX	2,150	Sherman	TX	1,070
Deaf Smith	TX	6,120	Swisher	TX	2,680
Donley	TX	1,550	Wheeler	TX	1,860

Anchorage, AK (154)

DMA TV Households: 142,230
% of U.S. TV Households: .128

KTUU-TV Anchorage, ch. 2, NBC
KTBY Anchorage, ch. 4, Fox
KYES-TV Anchorage, ch. 5, MyNetworkTV
***KAKM** Anchorage, ch. 7, ETV
KTVA Anchorage, ch. 11, CBS
KIMO Anchorage, ch. 13, ABC, CW

DMA Counties	State	TV Households
Anchorage	AK	98,070
Kenai-Pensla	AK	18,220
Matanka-Sustn	AK	25,940

Anchorage, AK

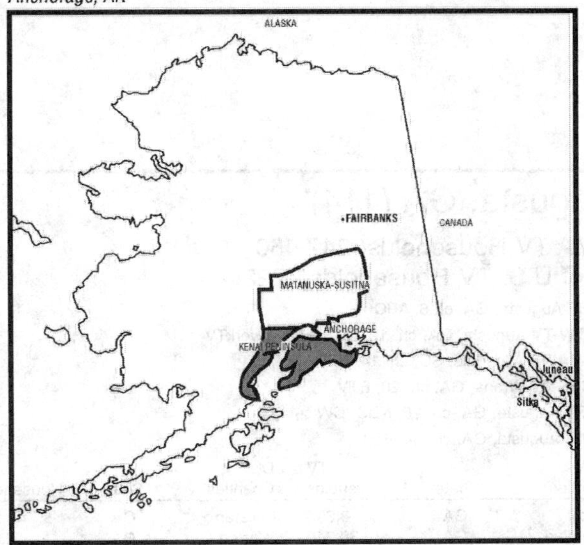

Maps courtesy of Nielsen Media Research

Atlanta, GA

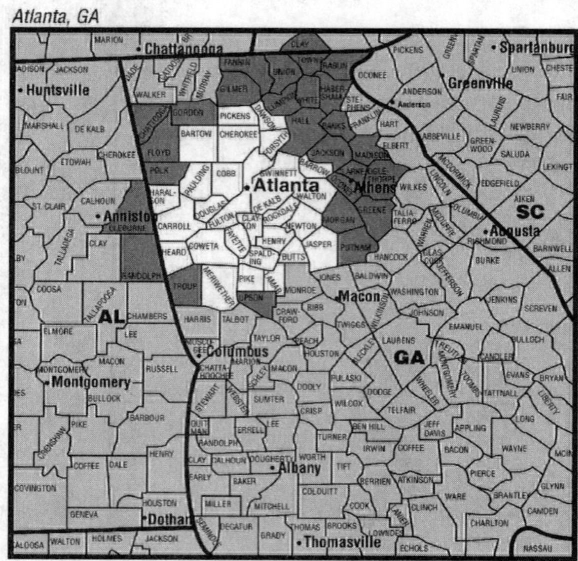

Atlanta (9)

DMA TV Households: 2,205,510
% of U.S. TV Households: 1.981

WSB-TV Atlanta, ch. 2, ABC
WAGA Atlanta, ch. 5, Fox
***WGTV** Athens, GA, ch. 8, ETV
WXIA-TV Atlanta, ch. 11, NBC
WPXA Rome, GA, ch. 14, ION Television
WTBS Atlanta, ch. 17, IND
***WPBA** Atlanta, ch. 30, ETV
WUVG-TV Athens, GA, ch. 34, Univision
WATL Atlanta, ch. 36, MyNetworkTV
WGCL-TV Atlanta, ch. 46, CBS
***WATC** Atlanta, ch. 57, ETV
WHSG-TV Monroe, GA, ch. 63, IND
WUPA Atlanta, ch. 69, CW

DMA Counties	State	TV Households	DMA Counties	State	TV Households
Cleburne	AL	5,800	Heard	GA	4,070
Randolph	AL	8,830	Henry	GA	62,300
Banks	GA	6,090	Jackson	GA	19,970
Barrow	GA	22,580	Jasper	GA	4,910
Bartow	GA	32,580	Lamar	GA	5,990
Butts	GA	7,480	Lumpkin	GA	8,780
Carroll	GA	39,890	Madison	GA	10,460
Cherokee	GA	67,390	Meriwether	GA	8,590
Clarke	GA	39,710	Morgan	GA	6,600
Clayton	GA	94,090	Newton	GA	33,210
Cobb	GA	248,810	Oconee	GA	10,490
Coweta	GA	40,150	Oglethorpe	GA	5,270
Dawson	GA	8,010	Paulding	GA	41,030
De Kalb	GA	250,330	Pickens	GA	11,540
Douglas	GA	43,110	Pike	GA	5,710
Fannin	GA	9,420	Polk	GA	15,020
Fayette	GA	37,250	Putnam	GA	7,920
Floyd	GA	34,830	Rabun	GA	6,680
Forsyth	GA	51,590	Rockdale	GA	27,760
Fulton	GA	366,200	Spalding	GA	22,840
Gilmer	GA	10,660	Towns	GA	4,570
Gordon	GA	18,970	Troup	GA	23,310
Greene	GA	6,320	Union	GA	8,280
Gwinnett	GA	255,860	Upson	GA	10,680
Habersham	GA	14,810	Walton	GA	27,680
Hall	GA	56,030	White	GA	9,410
Haralson	GA	11,070	Clay	NC	4,450

Augusta, GA (114)

DMA TV Households: 247,450
% of U.S. TV Households: .222

WJBF Augusta, GA, ch 6, ABC
WRDW-TV Augusta, GA, ch. 12, CBS, MyNetworkTV
***WEBA-TV** Allendale, SC, ch. 14, ETV
***WCES-TV** Wrens, GA, ch. 20, ETV
WAGT Augusta, GA, ch. 26, NBC, CW
WFXG Augusta, GA, ch. 54, Fox

DMA Counties	State	TV Households	DMA Counties	State	TV Households
Burke	GA	8,390	Taliaferro	GA	790
Columbia	GA	38,340	Warren	GA	2,350
Emanuel	GA	8,190	Wilkes	GA	4,210
Glascock	GA	1,090	Aiken	SC	60,170
Jefferson	GA	6,030	Allendale	SC	3,840
Jenkins	GA	3,280	Bamberg	SC	5,820
Lincoln	GA	3,250	Barnwell	SC	9,230
McDuffie	GA	8,270	Edgefield	SC	8,530
Richmond	GA	71,910	McCormick	SC	3,760

Augusta, GA

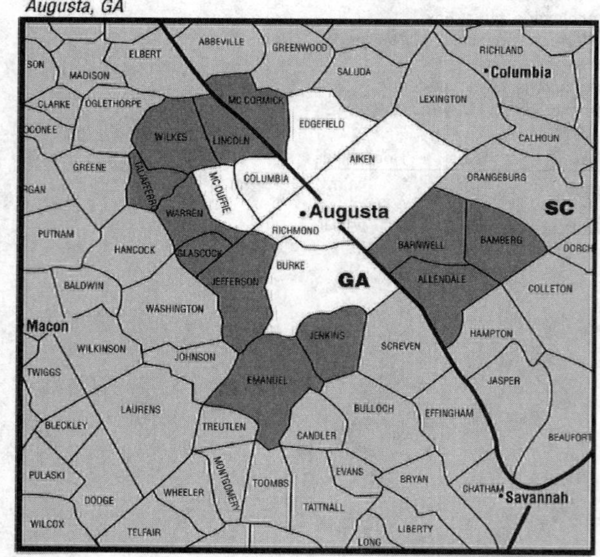

Maps courtesy of Nielsen Media Research

Austin, TX

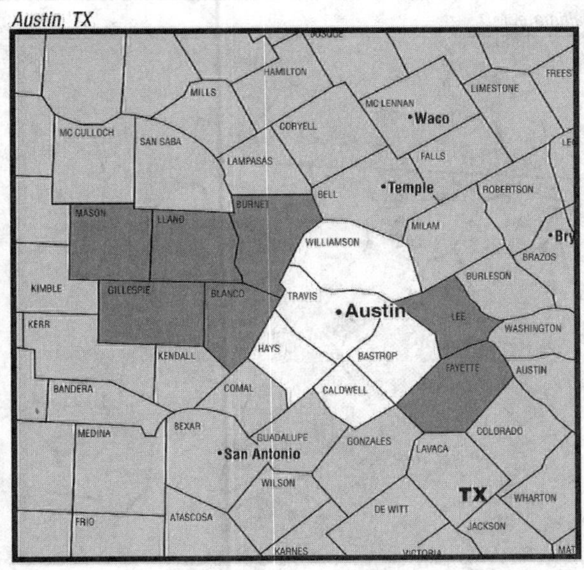

Austin, TX (52)

DMA TV Households: 602,340
% of U.S. TV Households: .541

KTBC Austin, TX, ch. 7, Fox
KXAM-TV Llano, TX, ch. 14, satellite to KXAN-TV
KNIC-TV Blanco, TX, ch. 17, TeleFutura
***KLRU-TV** Austin, TX, ch. 18, ETV
KVUE-TV Austin, TX, ch. 24, ABC
KXAN-TV Austin, TX, ch. 36, NBC
KEYE-TV Austin, TX, ch. 42, CBS
KNVA Austin, TX, ch. 54, CW

DMA Counties	State	TV Households
Bastrop	TX	25,070
Blanco	TX	3,600
Burnet	TX	15,680
Caldwell	TX	12,350
Fayette	TX	9,110
Gillespie	TX	9,550
Hays	TX	44,220
Lee	TX	5,960
Llano	TX	8,660
Mason	TX	1,520
Travis	TX	347,430
Williamson	TX	119,190

Bakersfield, CA (126)

DMA TV Households: 210,960
% of U.S. TV Households: .189

KGET-TV Bakersfield, CA, ch.17, NBC
KERO-TV Bakersfield, CA, ch. 23, ABC
KBAK-TV Bakersfield, CA, ch. 29, CBS
KUVI-TV Bakersfield, CA, ch. 45, MyNetworkTV

DMA Counties	State	TV Households
Kern West	CA	210,960

Bakersfield, CA

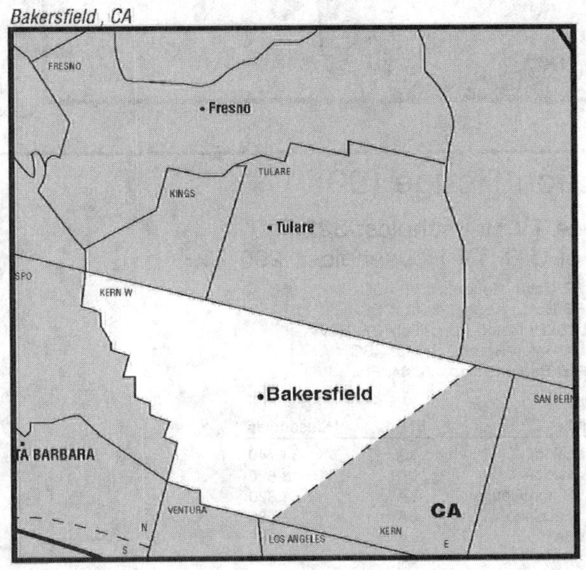

Maps courtesy of Nielsen Media Research

Baltimore (24)

DMA TV Households: 1,097,290
% of U.S. TV Households: .985

WMAR-TV Baltimore, ch. 2, ABC
WBAL-TV Baltimore, ch. 11, NBC
WJZ-TV Baltimore, ch. 13, CBS
***WMPT** Annapolis, MD, ch. 22, ETV
WUTB Baltimore, ch. 24, MyNetworkTV
WBFF Baltimore, ch. 45, Fox
WNUV Baltimore, ch. 54, CW
***WMPB** Baltimore, ch. 67, ETV

DMA Counties	State	TV Households	DMA Counties	State	TV Households
Anne Arundel	MD	190,430	Harford	MD	90,280
Baltimore	MD	316,090	Howard	MD	99,180
Baltimore City	MD	252,080	Kent	MD	8,070
Caroline	MD	11,990	Queen Annes	MD	17,560
Carroll	MD	59,700	Talbot	MD	15,250
Cecil	MD	36,660			

Baltimore, MD

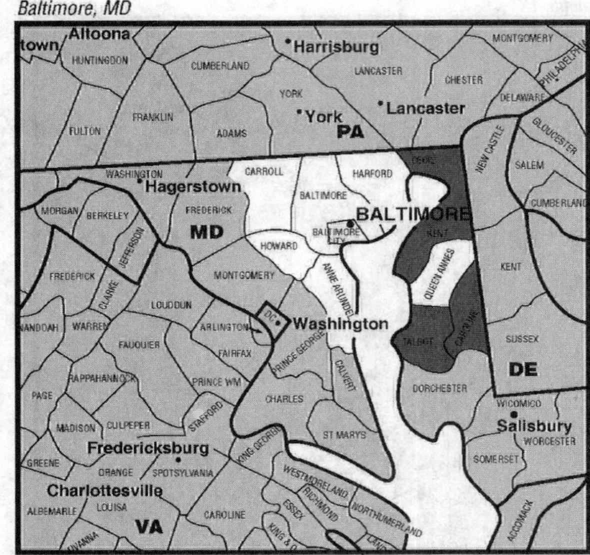

Bangor, ME

Bangor, ME (152)

DMA TV Households: 143,170
% of U.S. TV Households: .129

WLBZ Bangor, ME, ch. 2, NBC
WABI-TV Bangor, ME, ch. 5, CBS, CW
WVII-TV Bangor, ME, ch. 7, ABC
***WMEB-TV** Orono, ME, ch. 12, ETV
***WMED-TV** Calais, ME, ch. 13, ETV

DMA Counties	State	TV Households
Hancock	ME	23,070
Penobscot	ME	60,700
Piscataquis	ME	7,730
Somerset	ME	21,350
Waldo	ME	16,120
Washington	ME	14,200

Baton Rouge (93)

DMA TV Households: 322,540
% of U.S. TV Households: .290

WBRZ Baton Rouge, ch. 2, ABC
WAFB Baton Rouge, ch. 9, CBS
***WLPB-TV** Baton Rouge, ch. 27, ETV
WVLA Baton Rouge, ch. 33, NBC
WGMB Baton Rouge, ch. 44, Fox

DMA Counties	State	TV Households
Ascension	LA	34,740
Assumption	LA	8,820
East Baton Rouge	LA	164,820
East Feliciana	LA	7,080
Iberville	LA	11,080
Livingston	LA	42,480
Pointe Coupee	LA	8,740
St. Helena	LA	4,130
St. Mary	LA	19,380
West Baton Rouge	LA	8,140
West Feliciana	LA	3,850
Amite	MS	5,500
Wilkinson	MS	3,780

Baton Rouge, LA

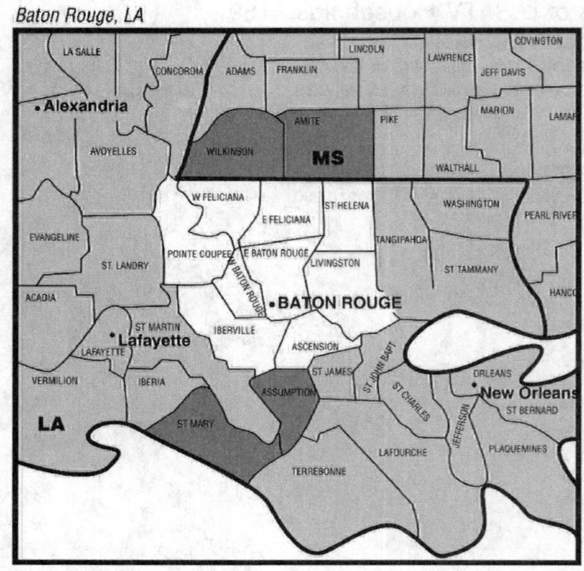

Maps courtesy of Nielsen Media Research

Beaumont-Port Arthur, TX

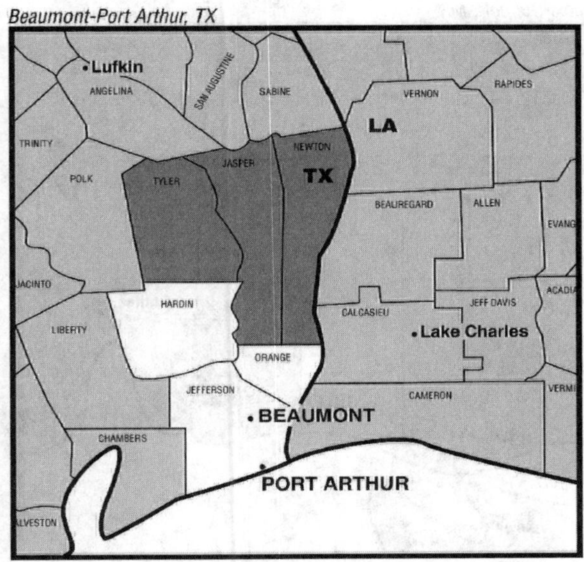

Beaumont-Port Arthur, TX (140)

DMA TV Households: 167,090
% of U.S. TV Households: .150

KBTV-TV Port Arthur, TX, ch. 4, NBC
KFDM-TV Beaumont, TX, ch. 6, CBS, CW
KBMT Beaumont, TX, ch. 12, ABC
***KITU-TV** Beaumont, TX, ch. 34, ETV

DMA Counties	State	TV Households
Hardin	TX	19,010
Jasper	TX	13,180
Jefferson	TX	90,260
Newton	TX	5,350
Orange	TX	31,640
Tyler	TX	7,650

Bend, OR (194)

DMA TV Households: 57,790
% of U.S. TV Households: .052

***KOAB-TV** Bend, OR, ch. 3, ETV
KTVZ Bend, OR, ch. 21, NBC, CW
KOHD Bend, OR, ch. 51, ABC

DMA Counties	State	TV Households
Deschutes	OR	57,790

Bend, OR

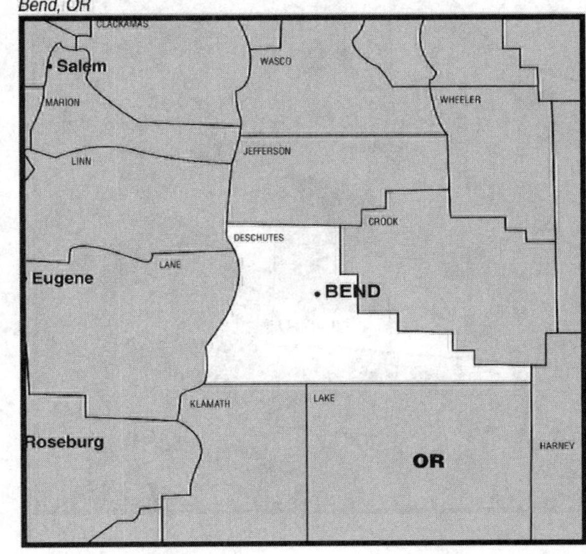

Billings, MT

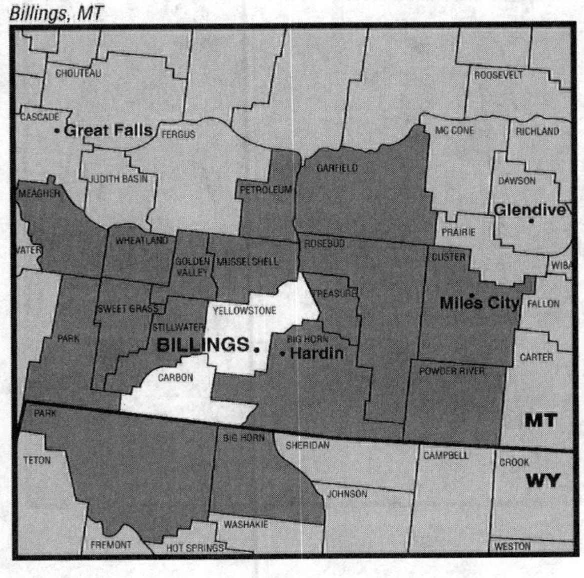

Billings, MT (170)

DMA TV Households: 103,710
% of U.S. TV Households: .093

KTVQ Billings, MT, ch. 2, CBS, CW
KYUS-TV Miles City, MT, ch. 3, satellite to KULR-TV
KHMT Hardin, MT, ch. 4, Fox
KSVI Billings, MT, ch. 6, ABC
KULR-TV Billings, MT, ch. 8, NBC

DMA Counties	State	TV Households
Big Horn	MT	4,030
Carbon	MT	4,270
Custer	MT	4,440
Garfield	MT	480
Golden Valley	MT	390
Meagher	MT	760
Musselshell	MT	1,790
Park	MT	6,820
Petroleum	MT	180
Powder River	MT	680
Rosebud	MT	3,230
Stillwater	MT	3,340
Sweet Grass	MT	1,330
Treasure	MT	300
Wheatland	MT	670
Yellowstone	MT	56,110
Big Horn	WY	4,080
Park	WY	10,810

Maps courtesy of Nielsen Media Research

Biloxi-Gulfport, MS (160)

DMA TV Households: 134,320
% of U.S. TV Households: .121

WLOX Biloxi, MS, ch. 13, ABC
***WMAH** Biloxi, MS, ch. 19, ETV
WXXV-TV Gulfport, MS, ch. 25, IND

DMA Counties	State	TV Households
George	MS	7,690
Harrison	MS	72,540
Jackson	MS	48,630
Stone	MS	5,460

Biloxi-Gulfport, MS

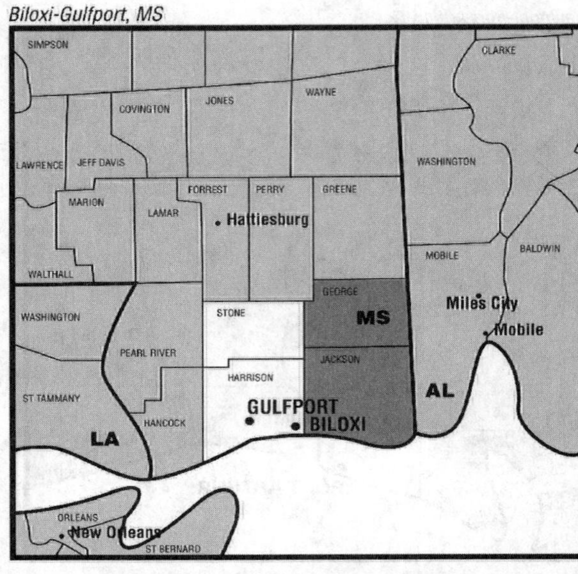

Binghamton, NY

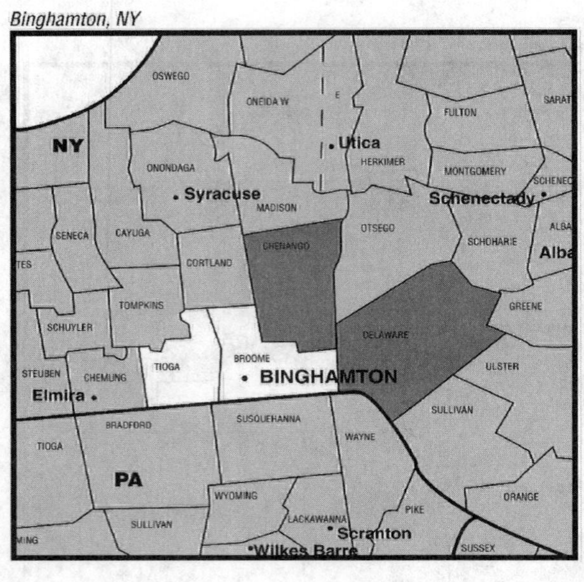

Binghamton, NY (157)

DMA TV Households: 138,220
% of U.S. TV Households: .124

WBNG-TV Binghamton, NY, ch. 12, CBS, CW
WIVT Binghamton, NY, ch. 34, ABC
WICZ-TV Binghamton, NY, ch. 40, Fox
***WSKG** Binghamton, NY, ch. 46, ETV

DMA Counties	State	TV Households
Broome	NY	79,120
Chenango	NY	20,270
Delaware	NY	19,020
Tioga	NY	19,810

Birmingham (Anniston and Tuscaloosa), AL (40)

DMA TV Households: 723,210
% of U.S. TV Households: .650

WBRC Birmingham, AL, ch. 6, Fox
***WCIQ** Mt. Cheaha, AL, ch. 7, ETV
***WBIQ** Birmingham, AL, ch. 10, ETV
WVTM-TV Birmingham, AL, ch. 13, NBC
WDBB Bessemer, AL, ch. 17, Fox
WTTO Birmingham, AL, ch. 21, CW
WLDM Tuscaloosa, AL, ch. 23, IND
WCFT-TV Tuscaloosa, AL, ch. 33, ABC
WJSU-TV Anniston, AL, ch. 40, ABC
WIAT Birmingham, AL, ch. 42, CBS
WPXH Gadsden, AL, ch. 44, ION Television
WTJP-TV Gadsden, AL, ch. 60, IND
WABM Birmingham, AL, ch. 68, MyNetworkTV

DMA Counties	State	TV Households
Bibb	AL	7,670
Blount	AL	21,470
Calhoun	AL	46,300
Cherokee	AL	10,160
Chilton	AL	16,390
Clay	AL	5,640
Coosa	AL	4,420
Cullman	AL	32,470
Etowah	AL	41,870
Fayette	AL	7,510
Greene	AL	3,680
Hale	AL	6,300
Jefferson	AL	262,600
Marion	AL	12,360
Pickens	AL	7,820
St. Clair	AL	287,400
Shelby	AL	69,560
Talladega	AL	31,730
Tuscaloosa	AL	68,120
Walker	AL	28,660
Winston	AL	10,080

Birmingham (Anniston and Tuscaloosa), AL

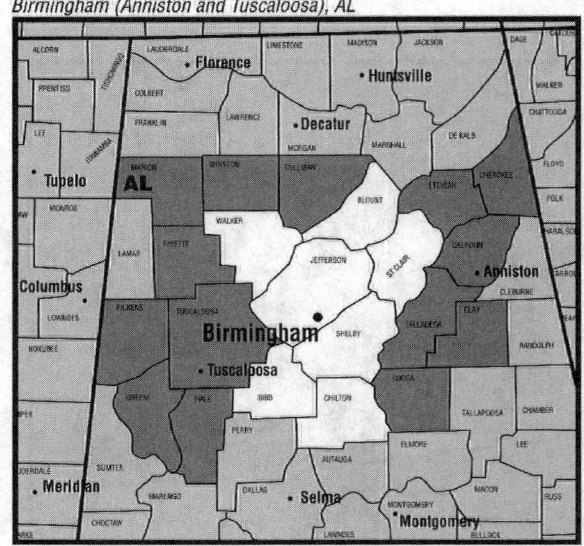

Maps courtesy of Nielsen Media Research

Bluefield-Beckley-Oak Hill, WV

Bluefield-Beckley-Oak Hill, WV (150)

DMA TV Households: 145,550
% of U.S. TV Households: .131

WOAY-TV Oak Hill, WV, ch. 4, ABC
WVVA Bluefield, WV, ch. 6, NBC, CW
***WSWP-TV** Grandview, WV, ch. 9, ETV
WLFB Bluefield, WV, ch. 40, IND
WVNS-TV Lewisburg, WV, ch. 59, CBS

DMA Counties	State	TV Households
Tazewell	VA	18,610
Fayette	WV	18,620
Greenbrier	WV	15,000
McDowell	WV	9,910
Mercer	WV	26,470
Monroe	WV	5,650
Pocahontas	WV	3,580
Raleigh	WV	32,360
Summers	WV	5,240
Wyoming	WV	10,110

Boise, ID (118)

DMA TV Households: 238,990
% of U.S. TV Households: .215

KBCI-TV Boise, ID, ch. 2, CBS
***KAID** Boise, ID, ch. 4, ETV
KIVI Nampa, ID, ch. 6, ABC
KTVB Boise, ID, ch. 7, NBC
KNIN-TV Caldwell, ID, ch. 9, CW
KTRV-TV Nampa, ID, ch. 12, Fox

DMA Counties	State	TV Households
Ada	ID	133,070
Adams	ID	1,360
Boise	ID	2,890
Camas	ID	390
Canyon	ID	58,140
Elmore	ID	8,740
Gem	ID	5,860
Owyhee	ID	3,800
Payette	ID	7,790
Valley	ID	3,540
Washington	ID	3,660
Malheur	OR	9,750

Boise, ID

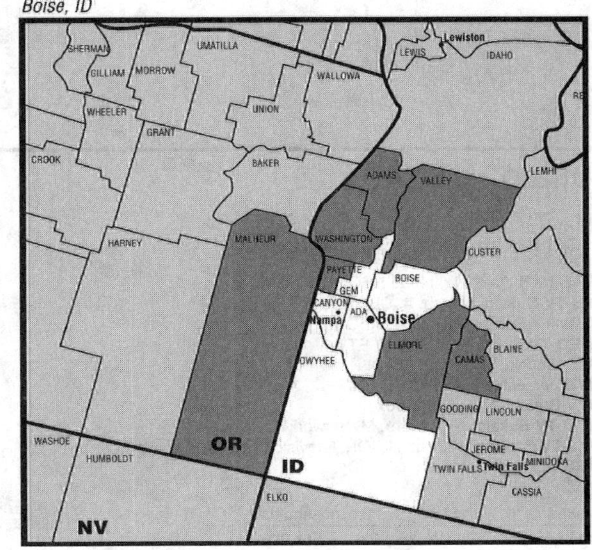

Boston, MA (Manchester, NH)

Boston (Manchester, NH) (7)

DMA TV Households: 2,372,030
% of U.S. TV Households: 2.130

***WGBH-TV** Boston, ch. 2, ETV
WBZ-TV Boston, ch. 4, CBS
WCVB-TV Boston, ch. 5, ABC
WHDH-TV Boston, ch. 7, NBC
WMUR-TV Manchester, NH, ch. 9, ABC
***WENH-TV** Durham, NH, ch. 11, ETV
WPXG Concord, NH, ch. 21, satellite to WBPX
WFXT Boston, ch. 25, Fox
WUNI Worcester, MA, ch. 27, Univision
WSBK-TV Boston, ch. 38, IND
***WGBX-TV** Boston, ch. 44, ETV
WWDP Norwell, MA, ch. 46, IND
***WYDN** Worcester, MA, ch. 48, ETV
WZMY-TV Derry, NH, ch. 50, MyNetworkTV
***WEKW-TV** Keene, NH, ch. 52, ETV
WLVI-TV Cambridge, MA, ch. 56, CW
WNEU Merrimack, NH, ch. 60, Telemundo
WMFP Lawrence, MA, ch. 62, IND
WUTF-TV Marlborough, MA, ch. 66, TeleFutura
WBPX Boston, ch. 68, ION Television

DMA Counties	State	TV Households
Barnstable	MA	96,660
Dukes	MA	6,460
Essex	MA	281,260
Middlesex	MA	554,990
Nantucket	MA	3,970
Norfolk	MA	252,010
Plymouth	MA	178,760
Suffolk	MA	252,410
Worcester	MA	302,010
Belknap	NH	25,270
Cheshire	NH	29,520
Hillsborough	NH	153,890
Merrimack	NH	56,760
Rockingham	NH	113,550
Strafford	NH	46,070
Windham	VT	18,440

Maps courtesy of Nielsen Media Research

Bowling Green, KY

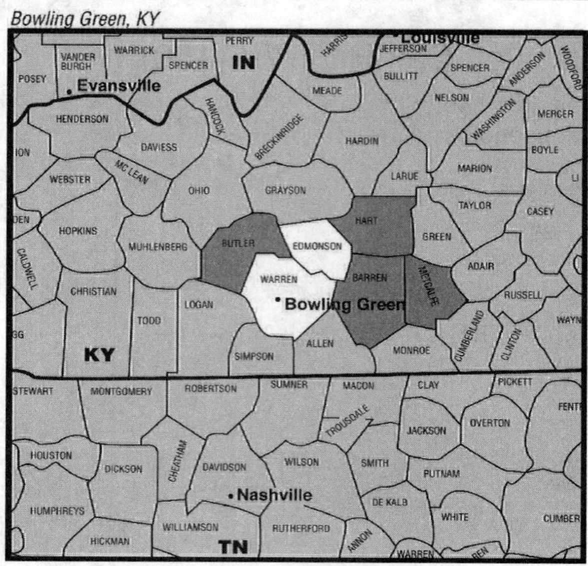

Bowling Green, KY (183)

DMA TV Households: 76,910
% of U.S. TV Households: .069

WBKO Bowling Green, KY, ch. 13, ABC, CW, Fox
***WKYU-TV** Bowling Green, KY, ch. 24, ETV
WKNT Bowling Green, KY, ch. 40, NBC
***WKGB-TV** Bowling Green, KY, ch. 53, ETV

DMA Counties	State	TV Households
Barren	KY	16,510
Butler	KY	5,330
Edmonson	KY	4,870
Hart	KY	6,990
Metcalfe	KY	4,140
Warren	KY	39,070

Buffalo, NY (49)

DMA TV Households: 639,990
% of U.S. TV Households: .575

WGRZ-TV Buffalo, NY, ch. 2, NBC
WIVB-TV Buffalo, NY, ch. 4, CBS
WKBW-TV Buffalo, NY, ch. 7, ABC
***WNED-TV** Buffalo, NY, ch. 17, ETV
WNLO Buffalo, NY, ch. 23, CW
WNYB Jamestown, NY, ch. 26, IND
WUTV Buffalo, NY, ch. 29, Fox
WNYO-TV Buffalo, NY, ch. 49, MyNetworkTV
WPXJ-TV Batavia, NY, ch. 51, ION Television
WNGS Springville, NY, ch. 67, IND

DMA Counties	State	TV Households
Allegany	NY	17,720
Cattaraugus	NY	31,550
Chautauqua	NY	52,960
Erie	NY	374,100
Genesee	NY	22,560
Niagara	NY	87,410
Orleans	NY	15,180
Wyoming	NY	14,830
McKean	PA	16,990
Potter	PA	6,690

Buffalo, NY

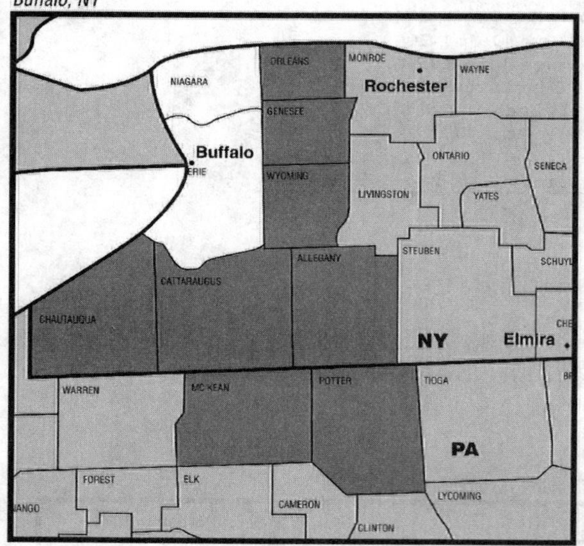

Maps courtesy of Nielsen Media Research

Burlington, VT - Plattsburgh, NY

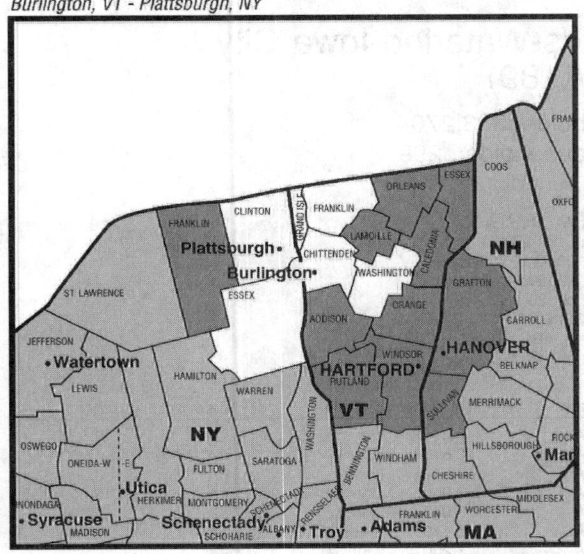

Burlington, VT-Plattsburgh, NY (90)

DMA TV Households: 327,480
% of U.S. TV Households: .294

WCAX-TV Burlington, VT, ch. 3, CBS
WPTZ North Pole (Plattsburgh), NY, ch. 5, NBC
***WVTB** St. Johnsbury, VT, ch. 20, satellite to WETK
WVNY Burlington, VT, ch. 22, ABC
***WVER** Rutland, VT, ch. 28, satellite to WETK
WNNE Hartford, VT, ch. 31, NBC
***WETK** Burlington, VT, ch. 33, ETV
***WVTA** Windsor, VT, ch. 41, satellite to WETK
WFFF-TV Burlington, VT, ch 44, Fox, CW
***WLED-TV** Littleton, NH, ch. 49, ETV
***WCFE-TV** Plattsburgh, NY, ch. 57, ETV

DMA Counties	State	TV Households
Grafton	NH	32,340
Sullivan	NH	17,760
Clinton	NY	31,410
Essex	NY	15,000
Franklin	NY	17,880
Addison	VT	13,360
Caledonia	VT	11,910
Chittenden	VT	57,710
Essex	VT	2,630
Franklin	VT	17,940
Grand Isle	VT	3,170
Lamoille	VT	9,680
Orange	VT	11,410
Orleans	VT	11,370
Rutland	VT	25,770
Washington	VT	24,130
Windsor	VT	243,010

Butte-Bozeman, MT (192)

DMA TV Households: 60,560
% of U.S. TV Households: .054

KXLF-TV Butte, MT, ch. 4, CBS, CW
KTVM Butte, MT, ch. 6, satellite to KECI-TV
KBZK Bozeman, MT, ch. 7, CBS, CW
***KUSM** Bozeman, MT, ch. 9, ETV
KWYB Butte, MT, ch. 18, ABC
KBTZ Butte, MT, ch. 24, Fox, MyNetworkTV

DMA Counties	State	TV Households
Beaverhead	MT	3,380
Deer Lodge	MT	3,720
Gallatin	MT	30,120
Jefferson	MT	4,250
Madison	MT	3,170
Powell	MT	2,330
Silver Bow	MT	13,590

Butte-Bozeman, MT

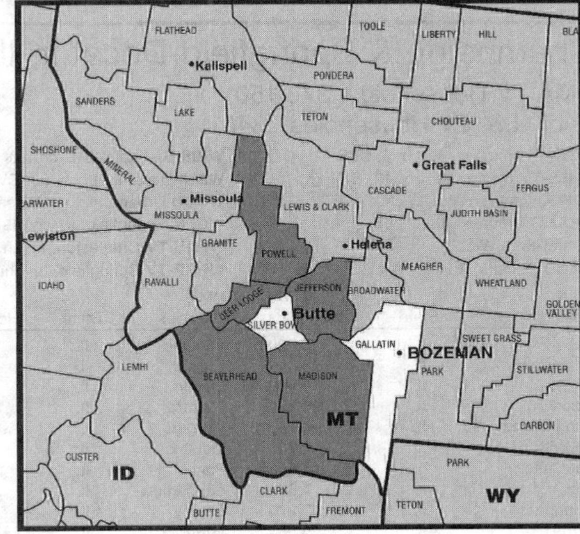

Casper-Riverton, WY

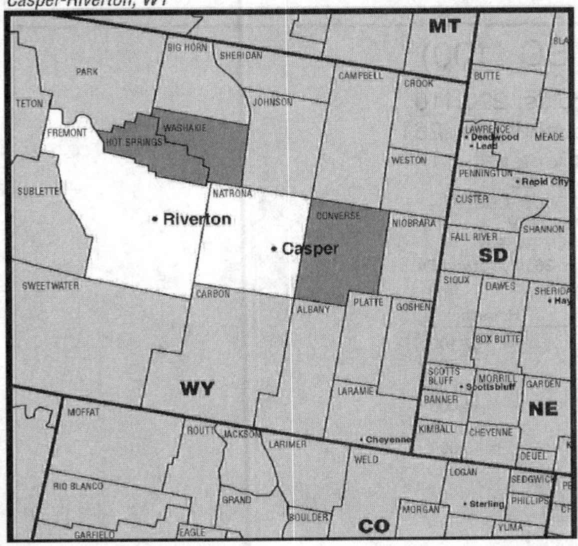

Casper-Riverton, WY (198)

DMA TV Households: 52,400
% of U.S. TV Households: .047

KTWO-TV Casper, WY, ch. 2, ABC
***KCWC-TV** Lander, WY, ch. 4, ETV
KGWL-TV Lander, WY, ch. 5, CBS
***KPTW** Casper, WY, ch. 6, satellite to KCWC-TV
KFNE Riverton, WY, ch. 10, satellite to KFNB
KCWY Casper, WY, ch. 13, NBC
KGWC-TV Casper, WY, ch. 14, CBS
KFNB Casper, WY, ch. 20, Fox

DMA Counties	State	TV Households
Converse	WY	5,010
Fremont	WY	13,550
Hot Springs	WY	1,910
Natrona	WY	28,790
Washakie	WY	3,140

Maps courtesy of Nielsen Media Research

Cedar Rapids-Waterloo-Iowa City & Dubuque, IA

Cedar Rapids-Waterloo-Iowa City & Dubuque, IA (89)

DMA TV Households: 333,270
% of U.S. TV Households: .299

KGAN Cedar Rapids, IA, ch. 2, CBS
KWWL Waterloo, IA, ch. 7, NBC
KCRG-TV Cedar Rapids, IA, ch. 9, ABC
***KIIN-TV** Iowa City, IA, ch. 12, ETV
KWKB Iowa City, IA, ch. 20, CW, MyNetworkTV
KWWF Waterloo, IA, ch. 22, IND
KFXA Cedar Rapids, IA, ch. 28, IND
***KRIN** Waterloo, IA, ch. 32, ETV
KFXB Dubuque, IA, ch. 40, ABC
KPXR Cedar Rapids, IA, ch. 48, IND

DMA Counties	State	TV Households	DMA Counties	State	TV Households
Allamakee	IA	5,60	Fayette	IA	8,530
Benton	IA	10,510	Grundy	IA	4,950
Black Hawk	IA	49,340	Iowa	IA	6,360
Bremer	IA	9,050	Johnson	IA	46,740
Buchanan	IA	8,040	Jones	IA	7,820
Butler	IA	6,140	Keokuk	IA	4,460
Cedar	IA	7,260	Linn	IA	80,930
Chickasaw	IA	4,950	Tama	IA	6,930
Clayton	IA	7,320	Washington	IA	8,300
Delaware	IA	6,760	Winneshiek	IA	7,780
Dubuque	IA	35,470			

Champaign & Springfield-Decatur, IL

Champaign & Springfield-Decatur, IL (82)

DMA TV Households: 378,150
% of U.S. TV Households: .340

WCIA Champaign, IL, ch. 3, CBS
***WILL-TV** Urbana, IL, ch. 12, ETV
***WSEC** Jacksonville, IL, ch. 14, ETV
WICD Champaign, IL, ch. 15, satellite to WICS
WAND Decatur, IL, ch. 17, NBC
WICS Springfield, IL, ch. 20, ABC
WBUI Decatur, IL, ch. 23, CW
WCCU Urbana, IL, ch. 27, satellite to WRSP-V
WCFN Springfield, IL, ch. 49, MyNetworkTV
***WEIU-TV** Charleston, IL, ch. 51, ETV
WRSP-TV Springfield, IL, ch. 55, Fox

DMA Counties	State	TV Households	DMA Counties	State	TV Households
Cass	IL	5,420	Iroquois	IL	11,830
Champaign	IL	72,300	Logan	IL	10,930
Christian	IL	13,900	Macon	IL	44,730
Coles	IL	20,530	Menard	IL	5,070
Cumberland	IL	4,240	Morgan	IL	13,680
De Witt	IL	6,750	Moultrie	IL	5,540
Douglas	IL	7,540	Piatt	IL	6,760
Edgar	IL	7,620	Sangamon	IL	80,960
Effingham	IL	13,270	Shelby	IL	8,810
Ford	IL	5,580	Vermilion	IL	32,770

Charleston, SC

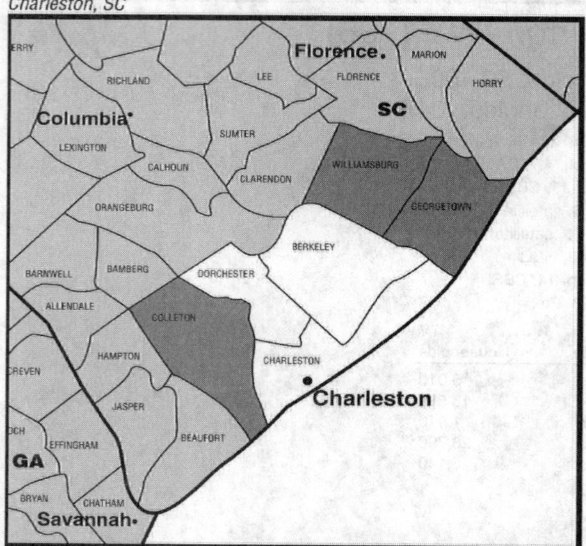

Charleston, SC (100)

DMA TV Households: 290,110
% of U.S. TV Households: .261

WCBD-TV Charleston, SC, ch. 2, NBC, CW
WCIV Charleston, SC, ch. 4, ABC
WCSC-TV Charleston, SC, ch. 5, CBS
***WITV** Charleston, SC, ch. 7, ETV
WTAT-TV Charleston, SC, ch. 24, Fox
WMMP Charleston, SC, ch. 36, MyNetworkTV

DMA Counties	State	TV Households
Berkeley	SC	57,100
Charleston	SC	138,500
Colleton	SC	15,700
Dorchester	SC	44,000
Georgetown	SC	25,400
Williamsburg	SC	13,300

Maps courtesy of Nielsen Media Research

Charleston-Huntington, WV

Charleston-Huntington, WV (65)

DMA TV Households: 477,040
% of U.S. TV Households: .428

WSAZ-TV Huntington, WV, ch. 3, NBC, MyNetworkTV
WCHS-TV Charleston, WV, ch. 8, ABC
WVAH-TV Charleston, WV, ch. 11, Fox
WOWK-TV Huntington, WV, ch. 13, CBS
***WOUB-TV** Athens, OH, ch. 20, ETV
***WKPI** Pikeville, KY, ch. 22, ETV
***WKAS** Ashland, KY, ch. 25, ETV
WLPX-TV Charleston, WV, ch. 29, ION Television
WQCW Portsmouth, OH, ch. 30, CW
***WPBY-TV** Huntington, WV, ch. 33, ETV
***WPBO-TV** Portsmouth, OH, ch. 42, ETV
WTSF Ashland, KY, ch. 61, IND

DMA Counties	State	TV Households	DMA Counties	State	TV Households
Boyd	KY	20,160	Boone	WV	10,610
Carter	KY	10,730	Braxton	WV	5,740
Elliott	KY	2,680	Cabell	WV	39,950
Floyd	KY	17,340	Calhoun	WV	3,030
Greenup	KY	15,020	Clay	WV	4,060
Johnson	KY	9,570	Jackson	WV	11,500
Lawrence	KY	6,370	Kanawha	WV	83,500
Lewis	KY	5,370	Lincoln	WV	9,050
Martin	KY	4,720	Logan	WV	14,670
Pike	KY	27,490	Mason	WV	10,680
Gallia	OH	12,360	Mingo	WV	10,970
Jackson	OH	13,150	Nicholas	WV	10,720
Lawrence	OH	25,840	Putnam	WV	21,810
Meigs	OH	9,330	Roane	WV	6,150
Scioto	OH	29,820	Wayne	WV	17,130
Vinton	OH	5,170	Wirt	WV	2,350

Charlotte, NC (26)

DMA TV Households: 1,045,240
% of U.S. TV Households: .939

WBTV Charlotte, NC, ch. 3, CBS
WSOC-TV Charlotte, NC, ch. 9, ABC
WHKY-TV Hickory, NC, ch. 14, IND
***WUNE-TV** Linville, NC, ch. 17, ETV
WCCB Charlotte, NC, ch. 18, Fox
***WNSC-TV** Rock Hill, SC, ch. 30, ETV
WCNC-TV Charlotte, NC, ch. 36, NBC
***WTVI** Charlotte, NC, ch. 42, ETV
WJZY Belmont, NC, ch. 46, CW
WMYT-TV Rock Hill, SC, ch. 55, MyNetworkTV
***WUNG-TV** Concord, NC, ch. 58, ETV
WAXN-TV Kannapolis, NC, ch. 64, IND

DMA Counties	State	TV Households	DMA Counties	State	TV Households
Alexander	NC	13,970	Lincoln	NC	27,170
Anson	NC	9,120	Mecklenburg	NC	323,500
Ashe	NC	10,880	Richmond	NC	17,900
Avery	NC	6,440	Rowan	NC	51,710
Burke	NC	34,200	Stanly	NC	22,590
Cabarrus	NC	58,340	Union	NC	60,810
Caldwell	NC	31,410	Watauga	NC	16,090
Catawba	NC	60,500	Chester	SC	12,610
Cleveland	NC	38,030	Chesterfield	SC	17,200
Gaston	NC	77,860	Lancaster	SC	24,560
Iredell	NC	56,070	York	SC	74,280

Charlotte, NC

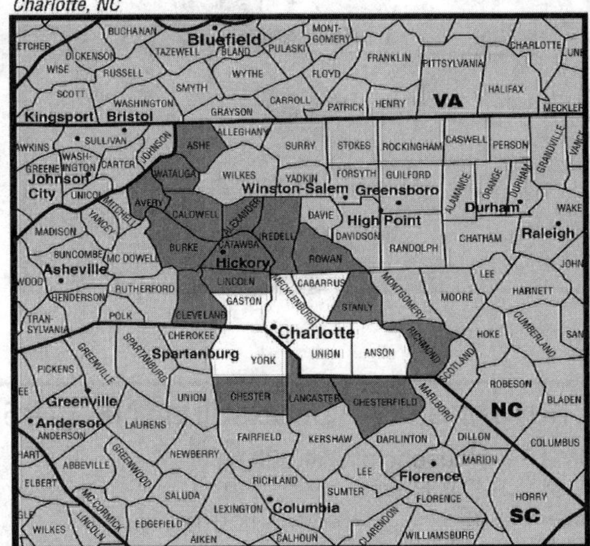

Maps courtesy of Nielsen Media Research

Charlottesville, VA (182)

DMA TV Households: 83,850
% of U.S. TV Households: .075

WCAV Charlottesville, VA, ch. 19, CBS
WVIR-TV Charlottesville, VA, ch. 29, NBC
***WHTJ** Charlottesville, VA, ch. 41, ETV

DMA Counties	State	TV Households
Albemarle	VA	50,830
Fluvanna	VA	9,630
Greene	VA	6,250
Madison	VA	5,010
Orange	VA	12,130

Charlottesville, VA

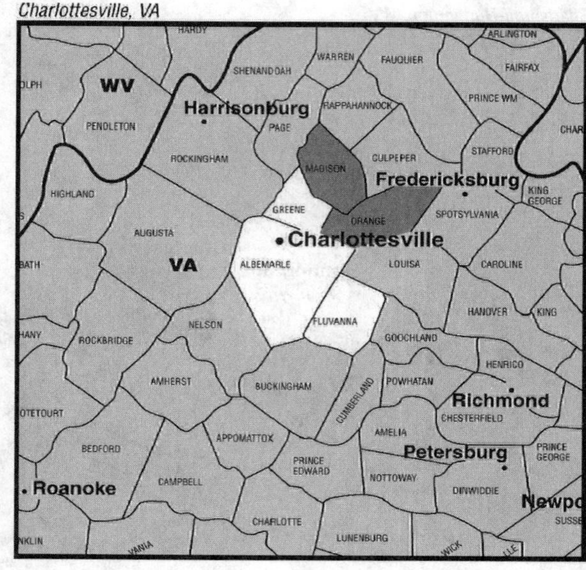

Chattanooga, TN

Chattanooga, TN (86)

DMA TV Households: 347,380
% of U.S. TV Households: .312

WRCB-TV Chattanooga, ch. 3, NBC
WTVC Chattanooga, ch. 9, ABC
WDEF-TV Chattanooga, ch. 12, CBS
***WCLP-TV** Chatsworth, GA, ch. 18, ETV
WELF-TV Dalton, GA, ch. 23, IND
***WTCI** Chattanooga, ch. 45, ETV
WFLI-TV Cleveland, TN, ch. 53, CW
WDSI-TV Chattanooga, ch. 61, Fox, MyNetworkTV

DMA Counties	State	TV Households	DMA Counties	State	TV Households
Catoosa	GA	24,030	Grundy	TN	5,640
Chattooga	GA	10,200	Hamilton	TN	126,700
Dade	GA	6,040	Marion	TN	11,190
Murray	GA	15,110	McMinn	TN	21,170
Walker	GA	25,060	Meigs	TN	4,630
Whitfield	GA	31,560	Polk	TN	6,490
Cherokee	NC	11,120	Rhea	TN	11,990
Bledsoe	TN	4,640	Sequatchie	TN	5,230
Bradley	TN	36,780			

Cheyenne, WY-Scottsbluff, NE (195)

DMA TV Households: 54,030
% of U.S. TV Households: .049

KDUH-TV Scottsbluff, NE, ch. 4, satellite to KOTA-TV
KGWN-TV Cheyenne, WY, ch. 5, CBS, CW
KSTF Scottsbluff, NE, ch. 10, satellite to KGWN-TV
KTUW Scottsbluff, NE, ch. 16, IND
KLWY Cheyenne, WY, ch. 27, Fox
KDEV Cheyenne, WY, ch. 33, satellite to KTWO-TV

DMA Counties	State	TV Households
Scotts Bluff	NE	14,800
Sioux	NE	590
Goshen	WY	5,010
Laramie	WY	33,630

Cheyenne, WY-Scottsbluff, NE

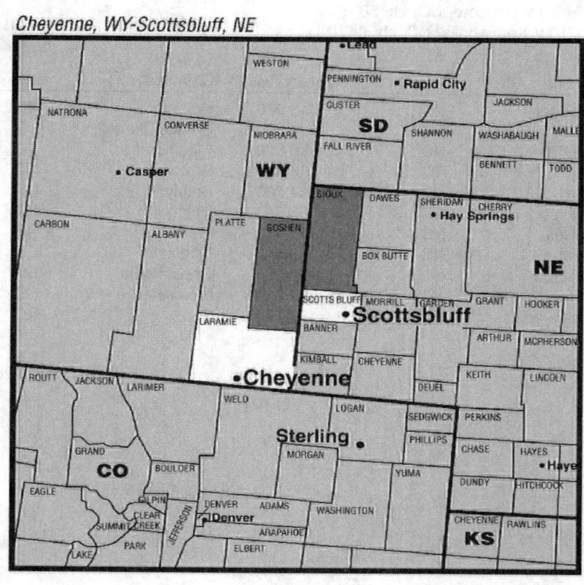

Maps courtesy of Nielsen Media Research

Chicago, IL

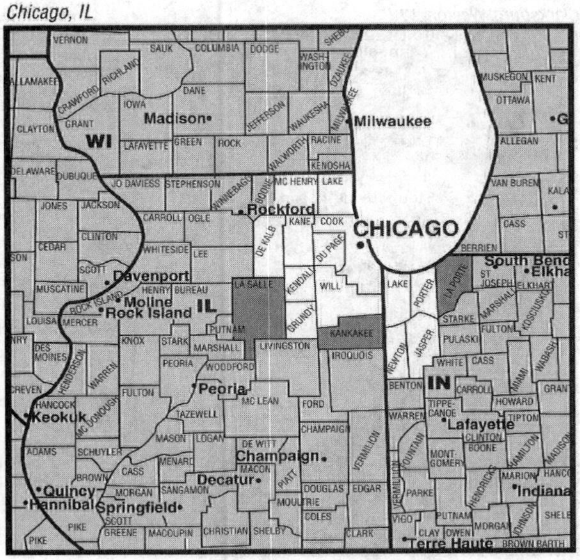

Chicago (3)

DMA TV Households:
3,455,020
% of U.S. TV Households:
3.103

WBBM-TV Chicago, ch. 2, CBS
WMAQ-TV Chicago, ch. 5, NBC
WLS-TV Chicago, ch. 7, ABC
WGN-TV Chicago, ch. 9, CW
***WTTW** Chicago, ch. 11, ETV
***WYCC** Chicago, ch. 20, ETV
WCIU-TV Chicago, ch. 26, IND
WFLD Chicago, ch. 32, Fox
WWTO-TV La Salle, IL, ch. 35, IND
WCPX Chicago, ch. 38, ION Television
WSNS Chicago, ch. 44, Telemundo
WPWR-TV Gary, IN, ch. 50, MyNetworkTV
***WYIN** Gary, IN, ch. 56, ETV
WXFT-TV Aurora, IL, ch. 60, TeleFutura
WJYS Hammond, IN, ch. 62, IND
WGBO-TV Joliet, IL, ch. 66, Univision

DMA Counties	State	TV Households	DMA Counties	State	TV Households
Cook	IL	1,907,500	Lake	IL	236,180
De Kalb	IL	35,270	McHenry	IL	107,630
Du Page	IL	334,150	Will	IL	228,330
Grundy	IL	17,570	Jasper	IN	11,510
Kane	IL	162,860	La Porte	IN	41,690
Kankakee	IL	40,190	Lake	IN	188,560
Kendall	IL	30,530	Newton	IN	5,260
La Salle	IL	43,820	Porter	IN	61,160

Chico-Redding, CA (130)

DMA TV Households: 193,590
% of U.S. TV Households: .174

KRCR-TV Redding, CA, ch. 7, ABC
***KIXE** Redding, CA, ch. 9, ETV
KHSL-TV Chico, CA, ch. 12, CBS, CW
KNVN Chico, CA, ch. 24, NBC
KCVU Paradise, CA, ch. 30, Fox

DMA Counties	State	TV Households
Butte	CA	81,800
Glenn	CA	9,510
Modoc	CA	3,510
Shasta	CA	70,710
Tehama	CA	22,750
Trinity	CA	5,310

Chico-Redding, CA

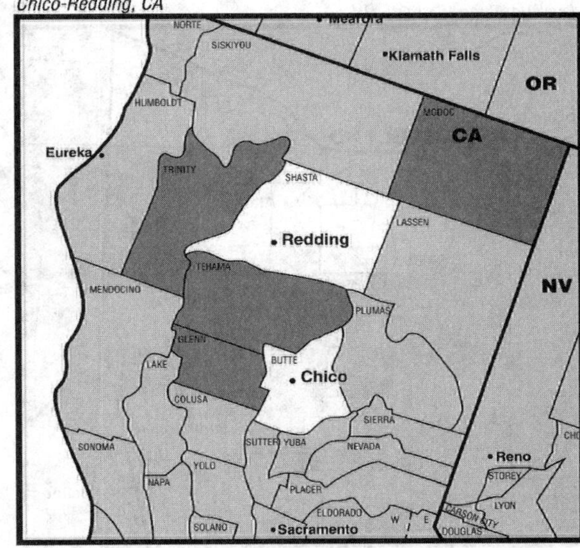

Cincinnati, OH

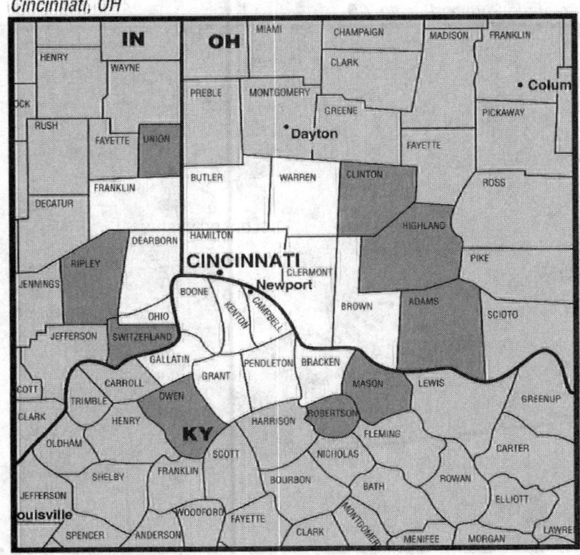

Cincinnati (33)

DMA TV Households: 886,910
% of U.S. TV Households: .797

WLWT Cincinnati, ch. 5, NBC
WCPO-TV Cincinnati, ch. 9, ABC
WKRC-TV Cincinnati, ch. 12, CBS, CW
***WPTO** Oxford, OH, ch. 14, ETV
WXIX-TV Newport, KY, ch. 19, Fox
***WCET** Cincinnati, ch. 48, ETV
***WKON** Owenton, KY, ch. 52, ETV
***WCVN** Covington, KY, ch. 54, ETV
WSTR-TV Cincinnati, ch. 64, MyNetworkTV

DMA Counties	State	TV Households	DMA Counties	State	TV Households
Dearborn	IN	18,360	Mason	KY	7,090
Franklin	IN	8,380	Owen	KY	4,380
Ohio	IN	2,370	Pendleton	KY	5,450
Ripley	IN	10,450	Robertson	KY	880
Switzerland	IN	3,770	Adams	OH	10,950
Union	IN	2,770	Brown	OH	16,490
Boone	KY	41,310	Butler	OH	131,930
Bracken	KY	3,480	Clermont	OH	73,160
Campbell	KY	34,880	Clinton	OH	16,600
Gallatin	KY	3,010	Hamilton	OH	329,610
Grant	KY	9,190	Highland	OH	16,480
Kenton	KY	61,650	Warren	OH	74,270

Maps courtesy of Nielsen Media Research

Clarksburg-Weston, WV (166)

DMA TV Households: 109,020
% of U.S. TV Households: .098

WDTV Weston, WV, ch. 5, CBS (ABC)
WBOY-TV Clarksburg, WV, ch. 12, NBC (ABC)
WVFX Clarksburg, WV, ch. 46, Fox, CW

DMA Counties	State	TV Households
Barbour	WV	6,320
Doddridge	WV	2,780
Gilmer	WV	2,690
Harrison	WV	28,180
Lewis	WV	7,230
Marion	WV	23,870
Randolph	WV	11,230
Ritchie	WV	4,200
Taylor	WV	6,380
Tucker	WV	2,810
Upshur	WV	9,230
Webster	WV	4,100

Clarksburg-Weston, WV

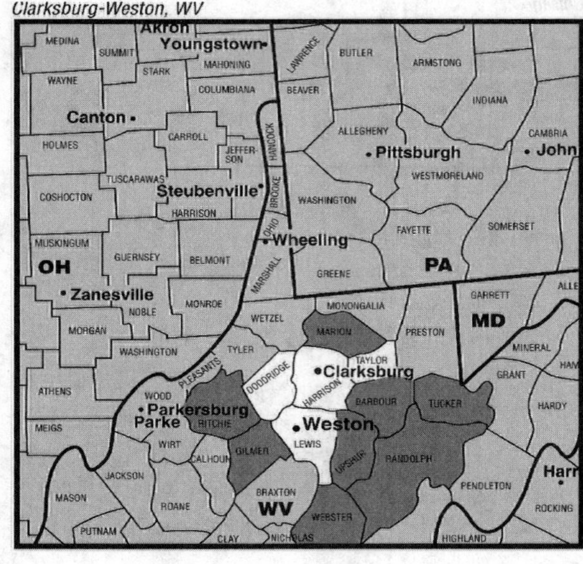

Cleveland-Akron (Canton), OH

Cleveland-Akron (Canton), OH (17)

DMA TV Households: 1,537,500
% of U.S. TV Households: 1.381

WKYC-TV Cleveland, ch. 3, NBC
WEWS Cleveland, ch. 5, ABC
WJW Cleveland, ch. 8, Fox
WDLI-TV Canton, OH, ch. 17, IND
WOIO Shaker Heights, OH, ch. 19, CBS
WVPX Akron, OH, ch. 23, ION Television
***WVIZ-TV** Cleveland, ch. 25, ETV
WUAB Lorain, OH, ch. 43, MyNetworkTV
***WEAO** Akron, OH, ch. 49, ETV
WGGN-TV Sandusky, OH, ch. 52, IND
WBNX-TV Akron, OH, ch. 55, CW
WQHS-TV Cleveland, ch. 61, Univision
WOAC Canton, OH, ch. 67, IND
WMFD-TV Mansfield, OH, ch. 68, IND

DMA Counties	State	TV Households
Ashland	OH	20,150
Ashtabula	OH	39,830
Carroll	OH	11,410
Cuyahoga	OH	542,330
Erie	OH	31,920
Geauga	OH	33,410
Holmes	OH	8,650
Huron	OH	23,050
Lake	OH	94,160
Lorain	OH	114,290
Medina	OH	62,880
Portage	OH	58,850
Richland	OH	48,960
Stark	OH	150,410
Summit	OH	220,980
Tuscarawas	OH	35,850
Wayne	OH	40,370

Colorado Springs-Pueblo, CO (94)

DMA TV Households: 316,630
% of U.S. TV Households: .284

KOAA-TV Pueblo, CO, ch. 5, NBC
***KTSC** Pueblo, CO, ch. 8, ETV
KKTV Colorado Springs, ch. 11, CBS, MyNetworkTV
KRDO-TV Colorado Springs, ch. 13, ABC
KXRM-TV Colorado Springs, ch. 21, Fox

DMA Counties	State	TV Households
Baca	CO	1,650
Bent	CO	1,980
Crowley	CO	1,280
Custer	CO	1,660
El Paso	CO	210,130
Fremont	CO	15,370
Huerfano	CO	3,020
Kiowa	CO	570
Las Animas	CO	6,300
Otero	CO	7,490
Pueblo	CO	58,630
Teller	CO	8,550

Colorado Springs-Pueblo, CO

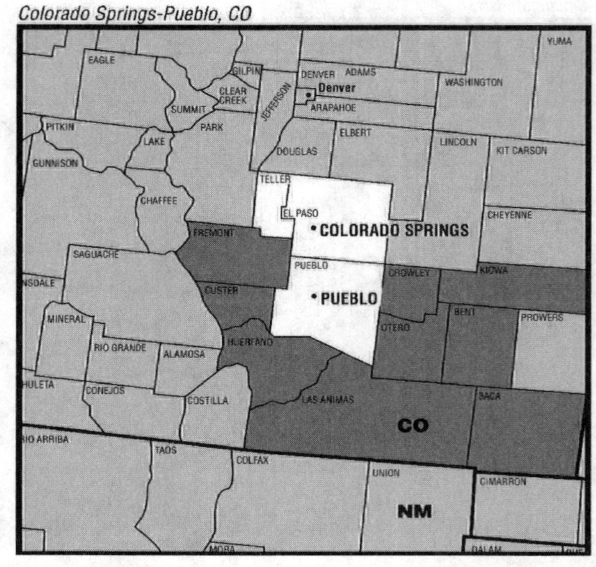

Maps courtesy of Nielsen Media Research

Columbia-Jefferson City, MO (139)

DMA TV Households: 170,260
% of U.S. TV Households: .153

KOMU-TV Columbia, MO, ch. 8, NBC, CW
KRCG Jefferson City, MO, ch. 13, CBS
KMIZ Columbia, MO, ch. 17, ABC, MyNetworkTV
KNLJ Jefferson City, MO, ch. 25, IND

DMA Counties	State	TV Households
Audrain	MO	9,640
Boone	MO	57,140
Callaway	MO	15,490
Chariton	MO	3,300
Cole	MO	28,420
Cooper	MO	6,140
Howard	MO	3,660
Maries	MO	3,640
Miller	MO	9,900
Moniteau	MO	5,400
Montgomery	MO	4,840
Morgan	MO	7,770
Osage	MO	5,280
Randolph	MO	9,640

Columbia-Jefferson City, MO

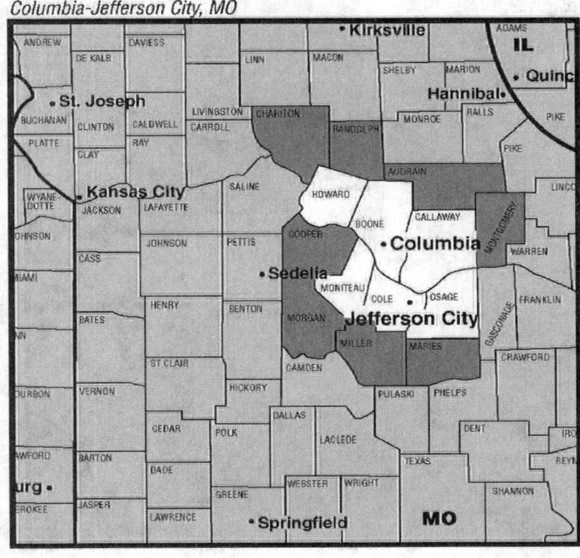

Columbia, SC

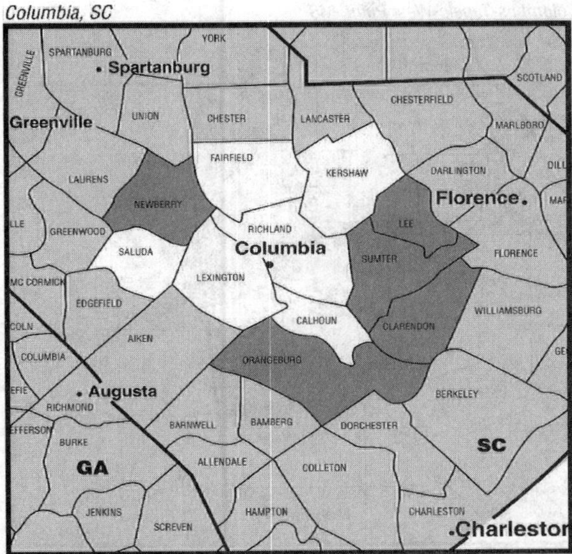

Columbia, SC (83)

DMA TV Households: 377,940
% of U.S. TV Households: .339

WIS Columbia, SC, ch. 10, NBC
WLTX Columbia, SC, ch. 19, CBS
WOLO-TV Columbia, SC, ch. 25, ABC
***WRJA-TV** Sumter, SC, ch. 27, ETV
***WRLK-TV** Columbia, SC, ch. 35, ETV
WZRB Columbia, SC, ch. 47, CW
WACH Columbia, SC, ch. 57, IND
WKTC Sumter, SC, ch. 63, MyNetworkTV

DMA Counties	State	TV Households
Calhoun	SC	6,060
Clarendon	SC	12,420
Fairfield	SC	9,350
Kershaw	SC	22,490
Lee	SC	7,110
Lexington	SC	94,160
Newberry	SC	14,700
Orangeburg	SC	34,700
Richland	SC	128,940
Saluda	SC	7,030
Sumter	SC	39,120

Columbus, GA (128)

DMA TV Households: 207,180
% of U.S. TV Households: .186

WRBL Columbus, GA, ch. 3, CBS
WTVM Columbus, GA, ch. 9, ABC
***WJSP-TV** Columbus, GA, ch. 28, ETV
WLTZ Columbus, GA, ch. 38, NBC
***WGIQ** Louisville, AL, ch. 43, ETV
WXTX Columbus, GA, ch. 54, Fox
WLGA Opelika, AL, ch. 66, CW

DMA Counties	State	TV Households	DMA Counties	State	TV Households
Barbour	AL	10,020	Quitman	GA	980
Chambers	AL	14,250	Randolph	GA	2,750
Lee	AL	50,080	Schley	GA	1,540
Russell	AL	19,960	Stewart	GA	1,850
Chattahoochee	GA	2,890	Sumter	GA	11,730
Clay	GA	1,280	Talbot	GA	2,750
Harris	GA	10,690	Taylor	GA	3,300
Marion	GA	2,730	Webster	GA	880
Muscogee	GA	69,500			

Columbus, GA

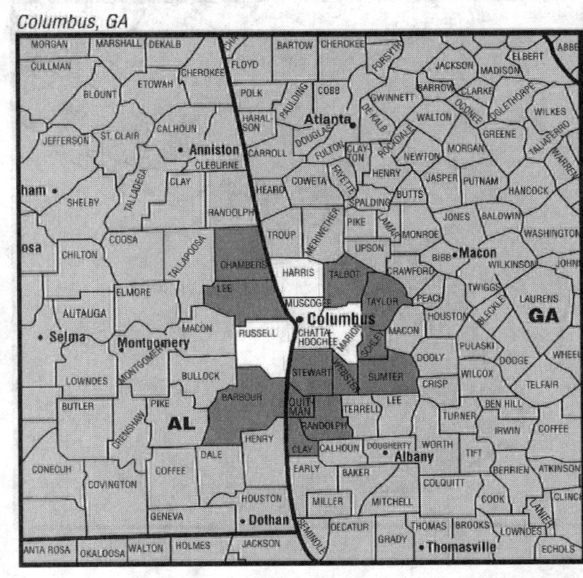

Maps courtesy of Nielsen Media Research

Columbus, OH

Columbus, OH (32)

DMA TV Households: 898,030
% of U.S. TV Households: .807

WCMH-TV Columbus, OH, ch. 4, NBC
WSYX Columbus, OH, ch. 6, ABC, MyNetworkTV
WBNS-TV Columbus, OH, ch. 10, CBS
WTTE Columbus, OH, ch. 28, Fox
***WOSU-TV** Columbus, OH, ch. 34, ETV
WSFJ-TV Newark, OH, ch. 51, IND
WWHO Chillicothe, OH, ch. 53, CW

DMA Counties	State	TV Households	DMA Counties	State	TV Households
Athens	OH	22,110	Licking	OH	60,450
Coshocton	OH	14,080	Madison	OH	14,300
Crawford	OH	18,640	Marion	OH	24,620
Delaware	OH	58,960	Morgan	OH	6,050
Fairfield	OH	53,480	Morrow	OH	12,850
Fayette	OH	11,160	Perry	OH	13,020
Franklin	OH	453,390	Pickaway	OH	18,750
Guernsey	OH	16,010	Pike	OH	10,640
Hardin	OH	11,980	Ross	OH	28,220
Hocking	OH	11,200	Union	OH	16,390
Knox	OH	21,730			

Columbus-Tupelo-West Point, MS (132)

DMA TV Households: 187,150
% of U.S. TV Households: .168

***WMAB-TV** Mississippi State, MS, ch. 2, ETV
WCBI-TV Columbus, MS, ch. 4, CBS, CW, MyNetworkTV
WTVA Tupelo, MS, ch. 9, NBC
***WMAE-TV** Booneville, MS, ch. 12, ETV
WLOV-TV West Point, MS, ch. 27, Fox
***WMAA** Columbus, MS, ch. 43, ETV
WKDH Houston, MS, ch. 45, ABC

DMA Counties	State	TV Households	DMA Counties	State	TV Households
Lamar	AL	6,070	Noxubee	MS	3,970
Calhoun	MS	5,800	Oktibbeha	MS	15,750
Chickasaw	MS	7,110	Pontotoc	MS	10,710
Choctaw	MS	3,570	Prentiss	MS	9,980
Clay	MS	7,990	Tishomingo	MS	7,850
Itawamba	MS	8,970	Union	MS	10,430
Lee	MS	31,040	Webster	MS	3,730
Lowndes	MS	22,330	Winston	MS	7,510
Monroe	MS	14,500	Yalobusha	MS	5,410
Montgomery	MS	4,430			

Columbus-Tupelo-West Point, MS

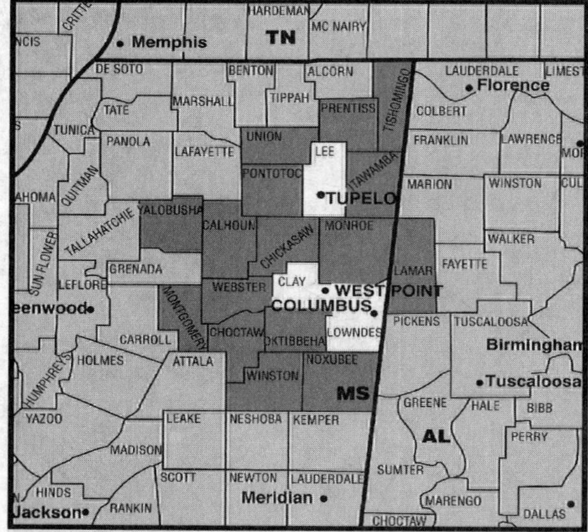

Corpus Christi, TX

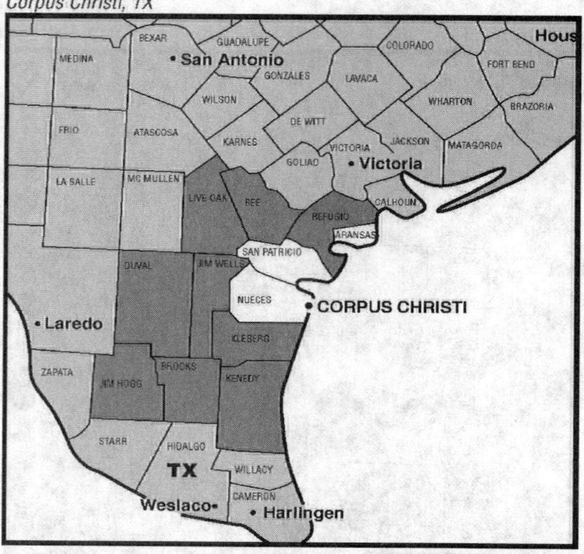

Corpus Christi, TX (129)

DMA TV Households: 194,160
% of U.S. TV Households: .174

KIII Corpus Christi, TX, ch. 3, ABC
KRIS-TV Corpus Christi, TX, ch. 6, NBC, CW
KZTV Corpus Christi, TX, ch. 10, CBS
***KEDT** Corpus Christi, TX, ch. 16, ETV
KORO Corpus Christi, TX, ch. 28, Univision
KUQI Corpus Christi, TX, ch. 38, IND

DMA Counties	State	TV Households
Aransas	TX	10,170
Bee	TX	8,610
Brooks	TX	2,620
Duval	TX	4,160
Jim Hogg	TX	1,750
Jim Wells	TX	13,520
Kleberg	TX	10,650
Live Oak	TX	4,000
Nueces	TX	112,180
Refugio	TX	2,970
San Patricio	TX	23,430

Maps courtesy of Nielsen Media Research

Dallas-Ft. Worth, TX

Dallas-Fort Worth, TX (6)

DMA TV Households: 2,358,360
% of U.S. TV Households: 2.136

***KDTN** Denton, TX, ch. 2, ETV
KDFW Dallas, ch. 4, Fox
KXAS-TV Fort Worth, ch. 5, NBC
WFAA-TV Dallas, ch. 8, ABC
KTVT Fort Worth, ch. 11, CBS
***KERA-TV** Dallas, ch. 13, ETV
KTXA Fort Worth, ch. 21, IND
KUVN-TV Garland, TX, ch. 23, Univision
KDFI Dallas, ch. 27, MyNetworkTV
KMPX Decatur, TX, ch. 29, IND
KDAF Dallas, ch. 33, WB
KXTX-TV Dallas, ch. 39, Telemundo
KTAQ Greenville, TX, ch. 47, IND
KSTR-TV Irving, TX, ch. 49, TeleFutura
KFWD Fort Worth, ch. 52, IND
KLDT Lake Dallas, TX, ch. 54, IND
KDTX-TV Dallas, ch. 58, IND
KPXD Arlington, TX, ch. 68, ION Television

DMA Counties	State	TV Households	DMA Counties	State	TV Households
Anderson	TX	15,860	Hopkins	TX	12,760
Bosque	TX	7,110	Hunt	TX	30,910
Collin	TX	259,820	Jack	TX	3,060
Comanche	TX	5,150	Johnson	TX	51,810
Cooke	TX	14,700	Kaufman	TX	31,890
Dallas	TX	820,670	Lamar	TX	19,570
Delta	TX	2,050	Navarro	TX	17,300
Denton	TX	211,230	Palo Pinto	TX	10,750
Ellis	TX	45,850	Parker	TX	37,170
Erath	TX	12,450	Rains	TX	4,420
Fannin	TX	12,030	Red River	TX	5,440
Freestone	TX	6,950	Rockwall	TX	22,370
Hamilton	TX	3,240	Somervell	TX	2,740
Henderson	TX	31,060	Tarrant	TX	607,610
Hill	TX	13,040	Van Zandt	TX	19,840
Hood	TX	19,510	Wise	TX	20,300

Davenport, IA-Rock Island-Moline, IL (96)

DMA TV Households: 308,360
% of U.S. TV Households: .277

WHBF-TV Rock Island, IL, ch. 4, CBS
KWQC-TV Davenport, IA, ch. 6, NBC
WQAD-TV Moline, IL, ch. 8, ABC
KLJB-TV Davenport, IA, ch. 18, Fox
***WQPT-TV** Moline, IL, ch. 24, ETV
KGCW-TV Burlington, IA, ch. 26, CW
***KQCT** Davenport, IA, ch. 36, satellite to *WQPT-TV

DMA Counties	State	TV Households	DMA Counties	State	TV Households
Bureau	IL	14,150	Whiteside	IL	23,570
Carroll	IL	6,510	Clinton	IA	20,180
Henderson	IL	3,300	Des Moines	IA	16,490
Henry	IL	20,010	Henry	IA	7,450
Jo Daviess	IL	9,540	Jackson	IA	8,210
Knox	IL	20,750	Louisa	IA	4,250
Mercer	IL	6,660	Muscatine	IA	16,480
Rock Island	IL	60,090	Scott	IA	64,150
Warren	IL	6,570			

Davenport, IA-Rock Island-Moline, IL

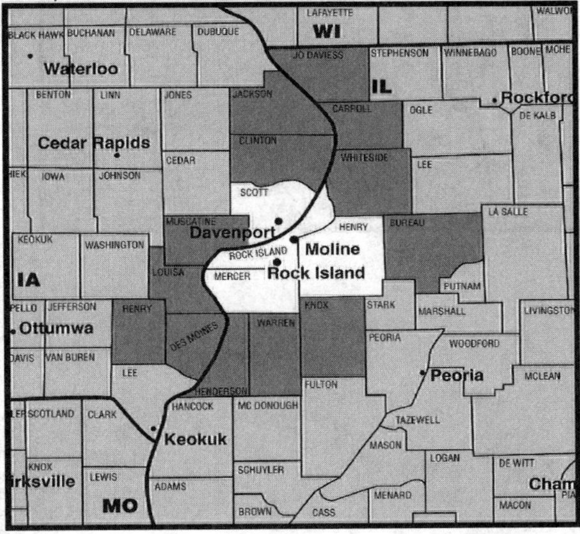

Maps courtesy of Nielsen Media Research

Dayton, OH (58)

DMA TV Households: 531,120
% of U.S. TV Households: .477

WDTN Dayton, OH, ch. 2, NBC
WHIO-TV Dayton, OH, ch. 7, CBS
***WPTD** Kettering, OH, ch. 16, ETV
WKEF Dayton, OH, ch. 22, ABC
WBDT Springfield, OH, ch. 26, CW
WKOI-TV Richmond, IN, ch. 43, IND
WRGT-TV Dayton, OH, ch. 45, Fox, MyNetworkTV

DMA Counties	State	TV Households
Wayne	IN	27,620
Auglaize	OH	18,000
Champaign	OH	15,320
Clark	OH	56,570
Darke	OH	20,400
Greene	OH	57,940
Logan	OH	18,250
Mercer	OH	15,200
Miami	OH	40,430
Montgomery	OH	226,780
Preble	OH	16,230
Shelby	OH	18,380

Dayton, OH

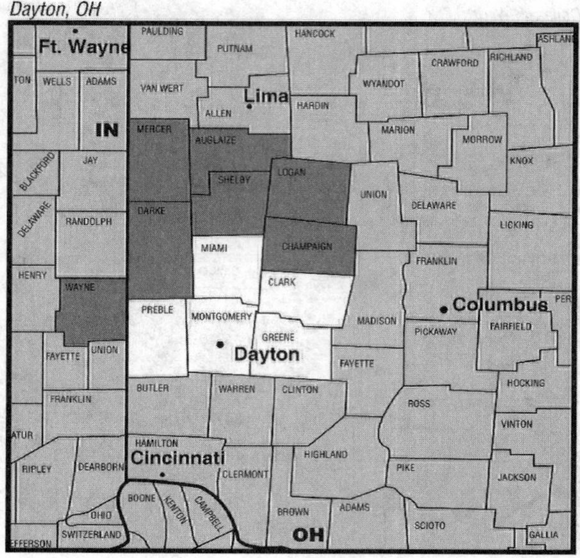

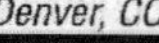
Denver, CO

Denver (18)

DMA TV Households: 1,431,910
% of U.S. TV Households: 1.286

KWGN-TV Denver, ch. 2, CW
KREG-TV Glenwood Springs, CO, ch. 3, CBS
KUPN Sterling, CO, ch. 3, satellite to KTVD
KCNC-TV Denver, ch. 4, CBS
***KRMA-TV** Denver, ch. 6, ETV
KMGH-TV Denver, ch. 7, ABC
***KWYP-TV** Laramie, WY, ch. 8, ETV
KUSA-TV Denver, ch. 9, NBC
KFNR Rawlins, WY, ch. 11, Fox
***KBDI-TV** Broomfield, CO, ch. 12, ETV
KTFD-TV Boulder, CO, ch. 14, TeleFutura
KTVD Denver, ch. 20, MyNetworkTV
KFCT Fort Collins, CO, ch. 22, satellite to KDVR
***KMAS-TV** Steamboat Springs, CO, ch. 24, satellite to KRMA-TV
KDEN Longmont, CO, ch. 25, Telemundo
KDVR Denver, ch. 31, Fox
***KRMT** Denver, ch. 41, ETV
KCEC Denver, ch. 50, Univision
KWHD Castle Rock, CO, ch. 53, IND
KPXC-TV Denver, ch. 59, ION Television

DMA Counties	State	TV Households
Adams	CO	142,020
Alamosa	CO	5,440
Arapahoe	CO	208,480
Archuleta	CO	4,700
Boulder	CO	106,180
Broomfield	CO	16,110
Chaffee	CO	6,790
Cheyenne	CO	780
Clear Creek	CO	3,940
Conejos	CO	2,910
Costilla	CO	1,400
Delta	CO	11,630
Denver	CO	231,440
Dolores	CO	770
Douglas	CO	91,670
Eagle	CO	16,550
Elbert	CO	7,800
Garfield	CO	18,430
Gilpin	CO	1,970
Grand	CO	5,240
Gunnison	CO	5,200
Hinsdale	CO	400
Jackson	CO	550
Jefferson	CO	205,950
Kit Carson	CO	2,650
Lake	CO	2,570
Larimer	CO	103,860
Lincoln	CO	1,780
Logan	CO	7,670
Mineral	CO	380
Moffat	CO	4,780
Morgan	CO	9,610
Ouray	CO	1,610
Park	CO	6,540
Phillips	CO	1,790
Pitkin	CO	5,980
Prowers	CO	4,920
Rio Blanco	CO	2,350
Rio Grande	CO	4,630
Routt	CO	8,430
Saguache	CO	2,600
San Juan	CO	280
Sedgwick	CO	1,090
Summit	CO	9,650
Washington	CO	1,870
Weld	CO	81,810
Yuma	CO	3,690
Arthur	NE	200
Banner	NE	290
Box Butte	NE	4,480
Cherry	NE	2,570
Cheyenne	NE	4,130
Dawes	NE	3,280
Deuel	NE	900
Garden	NE	790
Grant	NE	280
Hooker	NE	300
Keith	NE	3,460
Kimball	NE	1,580
Sheridan	NE	2,270
Albany	WY	12,430
Campbell	WY	14,220
Carbon	WY	6,090
Johnson	WY	3,230
Niobrara	WY	970
Platte	WY	3,550

Maps courtesy of Nielsen Media Research

Des Moines-Ames, IA

Des Moines-Ames, IA (73)

DMA TV Households: 417,900
% of U.S. TV Households: .375

WOI-TV Ames, IA, ch. 5, ABC
KCCI Des Moines, IA, ch. 8, CBS
***KDIN-TV** Des Moines, IA, ch. 11, ETV
WHO-TV Des Moines, IA, ch. 13, NBC
KDSM-TV Des Moines, IA, ch. 17, Fox
***KTIN** Fort Dodge, IA, ch. 21, ETV
KCWI-TV Ames, IA, ch. 23, CW
KFPX Newton, IA, ch. 39, ION Television
KDMI Des Moines, IA, ch. 56, CW, MyNetworkTV

DMA Counties	State	TV Households	DMA Counties	State	TV Households
Adair	IA	3,190	Lucas	IA	3,880
Adams	IA	1,780	Madison	IA	5,930
Appanoose	IA	5,770	Mahaska	IA	8,850
Audubon	IA	2,580	Marion	IA	12,380
Boone	IA	10,680	Marshall	IA	15,260
Calhoun	IA	4,200	Monroe	IA	3,130
Carroll	IA	8,550	Pocahontas	IA	3,290
Clarke	IA	3,440	Polk	IA	162,340
Dallas	IA	20,900	Poweshiek	IA	7,440
Decatur	IA	3,240	Ringgold	IA	2,090
Franklin	IA	4,380	Story	IA	29,460
Greene	IA	3,980	Taylor	IA	2,670
Guthrie	IA	4,770	Union	IA	5,120
Hamilton	IA	6,650	Warren	IA	16,050
Hardin	IA	7,090	Wayne	IA	2,700
Humboldt	IA	4,040	Webster	IA	15,080
Jasper	IA	14,990	Wright	IA	5,510
Kossuth	IA	6,490			

Detroit (11)

DMA TV Households: 1,938,320
% of U.S. TV Households: 1.741

WJBK Detroit, ch. 2, Fox
WDIV Detroit, ch. 4, NBC
WXYZ-TV Detroit, ch. 7, ABC
CBET Windsor, Ont., ch. 9, CBC
WMYD Detroit, ch. 20, MyNetworkTV
WPXD Ann Arbor, MI, ch. 31, IND
WADL Mount Clemens, MI, ch. 38, IND
WKBD Detroit, ch. 50, CW
***WTVS** Detroit, ch. 56, ETV
WWJ-TV Detroit, ch. 62, CBS

DMA Counties	State	TV Households
Lapeer	MI	33,780
Livingston	MI	67,530
Macomb	MI	338,350
Monroe	MI	58,860
Oakland	MI	488,090
Sanilac	MI	16,990
St. Clair	MI	66,070
Washtenaw	MI	135,160
Wayne	MI	733,490

Detroit, MI

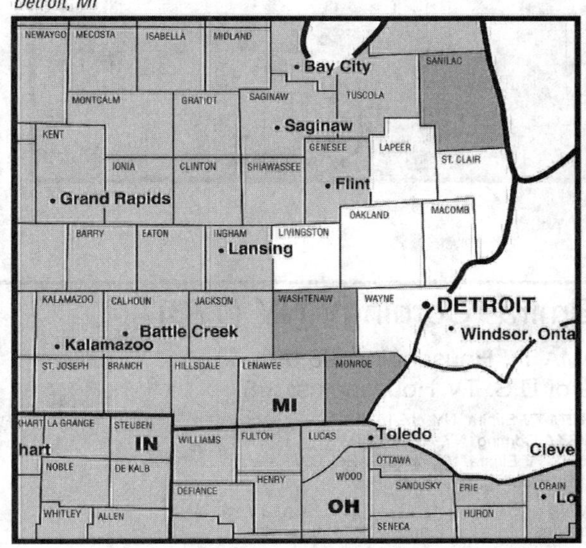

Maps courtesy of Nielsen Media Research

Dothan, AL (172)

DMA TV Households: 99,410
% of U.S. TV Households: .089

WTVY Dothan, AL, ch. 4, CBS, CW, MyNetworkTV
WDHN Dothan, AL, ch. 18, ABC
WDFX-TV Ozark, AL, ch. 34, Fox

DMA Counties	State	TV Households
Coffee	AL	18,860
Dale	AL	18,900
Geneva	AL	10,580
Henry	AL	6,830
Houston	AL	39,610
Early	GA	4,630

Dothan, AL

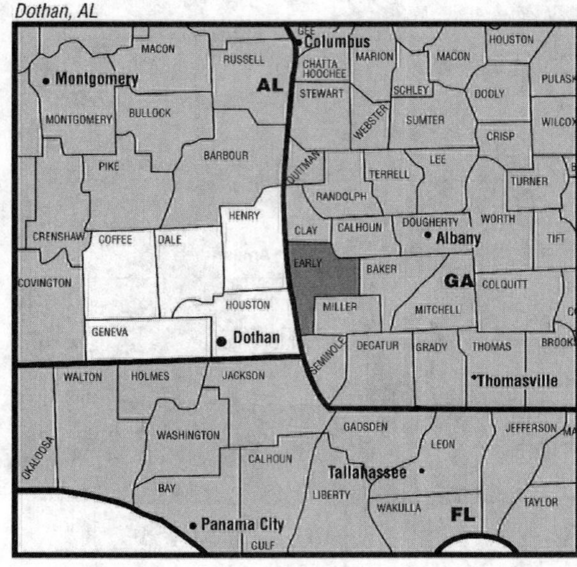

Duluth, MN-Superior, WI

Duluth, MN-Superior, WI (137)

DMA TV Households: 171,780
% of U.S. TV Households: .154

KDLH Duluth, MN, ch. 3, CBS, CW
KBJR-TV Superior, WI, ch. 6, NBC, MyNetworkTV
***WDSE-TV** Duluth, MN, ch. 8, ETV
WDIO-TV Duluth, MN, ch. 10, ABC
KRII Chisholm, MN, ch. 11, satellite to KBJR-TV
WIRT Hibbing, MN, ch. 13, satellite to WDIO-TV
KQDS-TV Duluth, MN, ch. 21, Fox

DMA Counties	State	TV Households
Gogebic	MI	6,710
Carlton	MN	13,270
Cook	MN	2,320
Itasca	MN	18,030
Koochiching	MN	5,860
Lake	MN	4,750
St. Louis	MN	79,690
Ashland	WI	6,550
Bayfield	WI	6,320
Douglas	WI	18,380
Iron	WI	3,070
Sawyer	WI	6,830

Elmira (Corning), NY (173)

DMA TV Households: 96,690
% of U.S. TV Households: .087

WETM-TV Elmira, NY, ch. 18, NBC
***WSKA** Corning, NY, ch. 30, ETV
WENY-TV Elmira, NY, ch. 36, ABC, CW
WYDC Corning, NY, ch. 48, Fox

DMA Counties	State	TV Households
Chemung	NY	34,630
Schuyler	NY	7,420
Steuben	NY	38,760
Tioga	PA	15,880

Elmira (Corning), NY

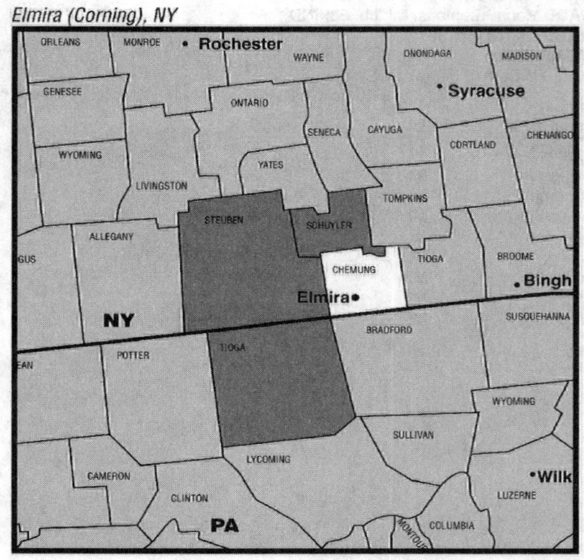

Maps courtesy of Nielsen Media Research

El Paso, TX (Las Cruces, NM)

El Paso, TX (Las Cruces, NM) (99)

DMA TV Households: 293,700
% of U.S. TV Households: .264

KDBC-TV El Paso, ch. 4, CBS, MyNetworkTV
KVIA-TV El Paso, ch. 7, ABC, CW
KTSM-TV El Paso, ch. 9, NBC
***KCOS** El Paso, ch. 13, ETV
KFOX-TV El Paso, ch. 14, Fox
***KRWG-TV** Las Cruces, NM, ch. 22, ETV
KINT-TV El Paso, ch. 26, Univision
***KSCE** El Paso, ch. 38, ETV
KTDO Las Cruces, NM, ch. 48, Telemundo
KTFN El Paso, ch. 65, IND

DMA Counties	State	TV Households
Dona Ana	NM	64,800
Culberson	TX	980
El Paso	TX	226,850
Hudspeth	TX	1,070

Erie, PA (142)

DMA TV Households: 157,860
% of U.S. TV Households: .142

WICU-TV Erie, PA, ch. 12, NBC
WJET-TV Erie, PA, ch. 24, ABC
WSEE Erie, PA, ch. 35, CBS, CW
***WQLN** Erie, PA, ch. 54, ETV
WFXP Erie, PA, ch. 66, Fox

DMA Counties	State	TV Households
Crawford	PA	33,760
Erie	PA	107,310
Warren	PA	16,790

Erie, PA

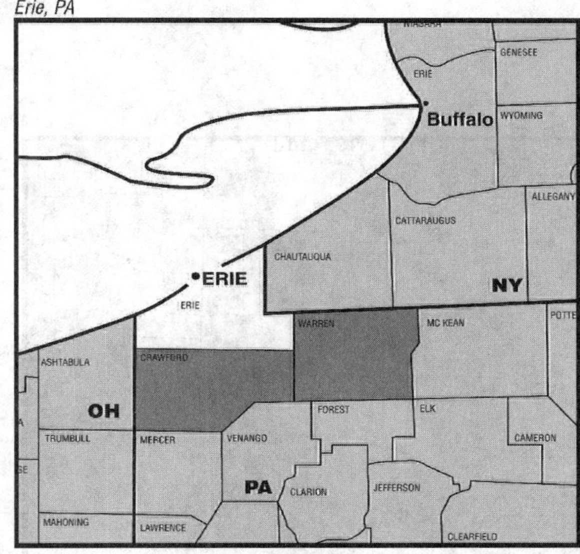

Eugene, OR

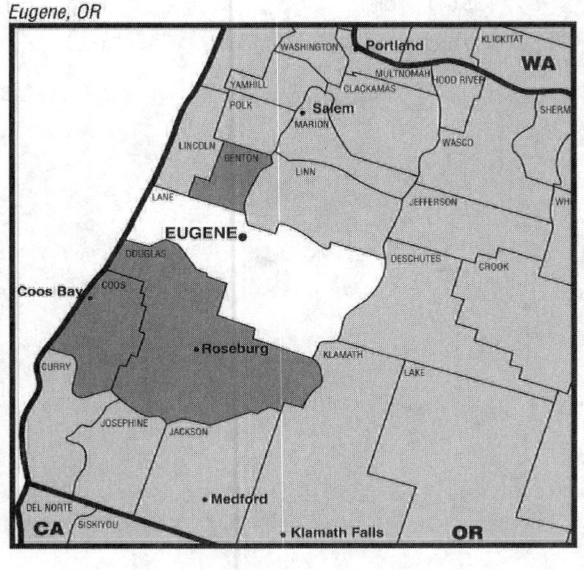

Eugene, OR (120)

DMA TV Households: 231,710
% of U.S. TV Households: .208

KPIC Roseburg, OR, ch. 4, satellite to KVAL-TV
***KOAC-TV** Corvallis, OR, ch. 7, ETV
KEZI Eugene, OR, ch. 9, ABC
KVAL-TV Eugene, OR, ch. 13, CBS
KCBY-TV Coos Bay, OR, ch. 11, satellite to KVAL-TV
KMTR Eugene, OR, ch. 16, NBC, CW
KMCB Coos Bay, OR, ch. 23, satellite to KMTR
***KEPB-TV** Eugene, OR, ch. 28, ETV
KLSR-TV Eugene, OR, ch. 34, Fox
KTVC Roseburg, OR, ch. 36, IND
KTCW Roseburg, OR, ch. 46, satellite to KMTR

DMA Counties	State	TV Households
Benton	OR	30,330
Coos	OR	26,320
Douglas	OR	40,520
Lane	OR	134,540

Maps courtesy of Nielsen Media Research

Eureka, CA (193)

DMA TV Households: 59,360
% of U.S. TV Households: .053

KIEM-TV Eureka, CA, ch. 3, NBC
KVIQ Eureka, CA, ch. 6, CBS
***KEET** Eureka, CA, ch. 13, ETV
KAEF Arcata, CA, ch. 23, Fox
KBVU Eureka, CA, ch. 29, IND

DMA Counties	State	TV Households
Del Norte	CA	9,450
Humboldt	CA	49,910

Eureka, CA

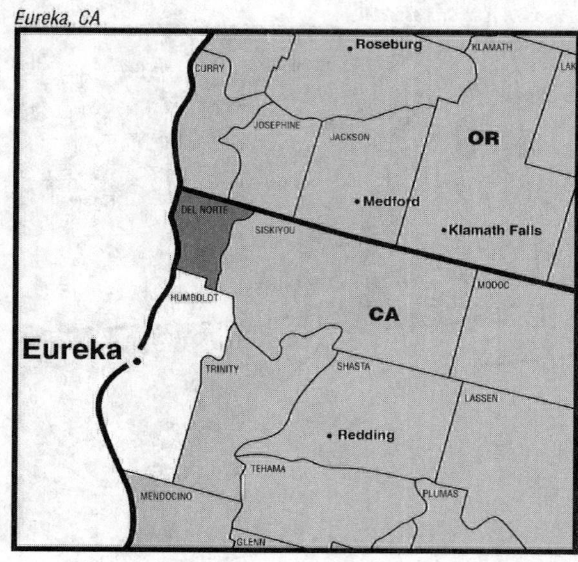

Evansville, IN

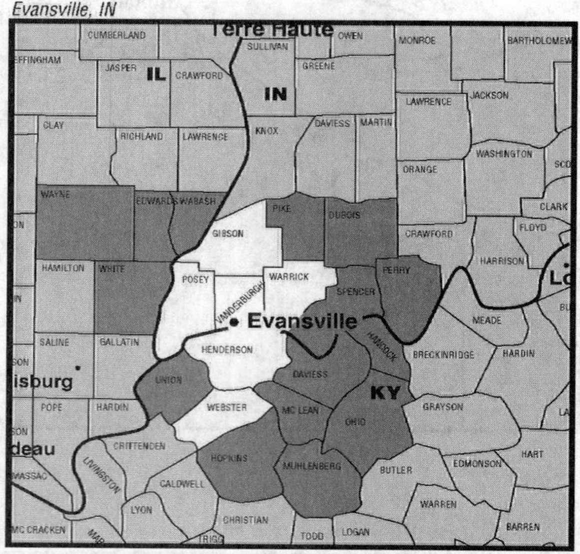

Evansville, IN (101)

DMA TV Households: 289,730
% of U.S. TV Households: .260

WTVW Evansville, IN, ch. 7, Fox
***WNIN** Evansville, IN, ch. 9, ETV
WFIE Evansville, IN, ch. 14, NBC
WAZE-TV Madisonville, KY, ch. 19, CW
WEHT Evansville, IN, ch. 25, ABC
***WKOH** Owensboro, KY, ch. 31, ETV
***WKMA** Madisonville, KY, ch. 35, ETV
WEVV Evansville, IN, ch. 44, CBS, MyNetworkTV

DMA Counties	State	TV Households	DMA Counties	State	TV Households
Edwards	IL	2,790	Warrick	IN	21,680
Wabash	IL	5,040	Daviess	KY	37,310
Wayne	IL	6,910	Hancock	KY	3,420
White	IL	6,540	Henderson	KY	18,890
Dubois	IN	15,580	Hopkins	KY	19,130
Gibson	IN	13,380	McLean	KY	4,060
Perry	IN	7,450	Muhlenberg	KY	12,220
Pike	IN	5,040	Ohio	KY	9,330
Posey	IN	10,110	Union	KY	5,870
Spencer	IN	7,690	Webster	KY	5,590
Vanderburgh	IN	71,700			

Maps courtesy of Nielsen Media Research

Fairbanks, AK

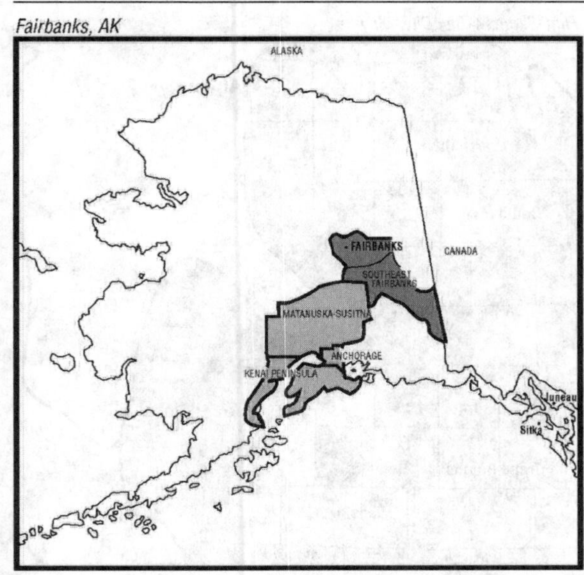

Fairbanks, AK (202)

DMA TV Households: 33,240
% of U.S. TV Households: .030

KATN Fairbanks, AK, ch. 2, ABC
KJNP-TV North Pole, AK, ch. 4, IND
KFXF Fairbanks, AK, ch. 7, Fox
***KUAC-TV** Fairbanks, AK, ch. 9, ETV
KTVF Fairbanks, AK, ch. 11, NBC

DMA Counties	State	TV Households
Fairbanks-Plus	AK	33,240

Fargo-Valley City, ND (119)

DMA TV Households: 235,320
% of U.S. TV Households: .211

***KGFE** Grand Forks, ND, ch. 2, ETV
KXJB-TV Valley City, ND, ch. 4, CBS
WDAY-TV Fargo, ND, ch. 6, ABC, CW
KJRR Jamestown, ND, ch. 7, satellite to KVRR
WDAZ-TV Devils Lake, ND, ch. 8, satellite to WDAY-TV
KBRR Thief River Falls, MN, ch. 10, satellite to KVRR
KVLY-TV Fargo, ND, ch. 11, NBC
KNRR Pembina, ND, ch. 12, IND
***KFME** Fargo, ND, ch. 13, ETV
KVRR Fargo, ND, ch. 15, Fox
***KJRE** Ellendale, ND, ch. 19, satellite to KFME
***KMDE** Devils Lake, ND, ch. 25, satellite to KFME
KCPM Grand Forks, ND, ch. 27, MyNetworkTV

DMA Counties	State	TV Households	DMA Counties	State	TV Households
Becker	MN	12,840	Dickey	ND	2,180
Clay	MN	20,100	Eddy	ND	1,090
Clearwater	MN	3,290	Foster	ND	1,480
Kittson	MN	1,880	Grand Forks	ND	24,850
Lake of the Woods	MN	1,850	Griggs	ND	1,090
Mahnomen	MN	1,990	La Moure	ND	1,790
Marshall	MN	4,060	Nelson	ND	1,490
Norman	MN	2,740	Pembina	ND	3,270
Otter Tail	MN	22,850	Ramsey	ND	4,660
Pennington	MN	5,530	Ransom	ND	2,270
Polk	MN	12,080	Richland	ND	6,710
Red Lake	MN	1,780	Sargent	ND	1,670
Roseau	MN	6,280	Steele	ND	780
Wilkin	MN	2,590	Stutsman	ND	8,480
Barnes	ND	4,570	Towner	ND	1,080
Benson	ND	2,240	Traill	ND	3,250
Cass	ND	55,960	Walsh	ND	4,760
Cavalier	ND	1,790			

Fargo-Valley City, ND

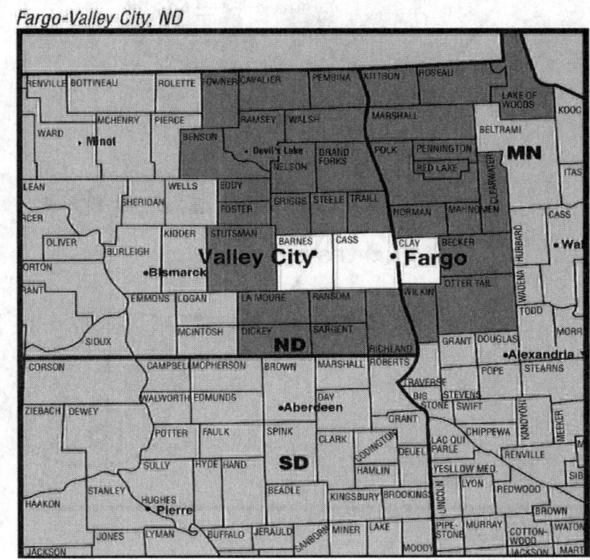

Maps courtesy of Nielsen Media Research

Flint-Saginaw-Bay City, MI (66)

DMA TV Households: 474,430
% of U.S. TV Households: .426

WNEM-TV Bay City, MI, ch. 5, CBS, MyNetworkTV
WJRT-TV Flint, MI, ch. 12, ABC
***WCMU-TV** Mt. Pleasant, MI, ch. 14, ETV
***WDCP-TV** University Center, MI, ch. 19, ETV
WEYI-TV Saginaw, MI, ch. 25, NBC, CW
***WFUM** Flint, MI, ch. 28, ETV
***WDCQ-TV** Bad Axe, MI, ch. 35, ETV
WBSF Bay City, MI, ch. 46, CW
WAQP Saginaw, MI, ch. 49, IND
WSMH Flint, MI, ch. 66, Fox

DMA Counties	State	TV Households
Arenac	MI	6,730
Bay	MI	44,400
Genesee	MI	175,410
Gladwin	MI	11,390
Gratiot	MI	14,570
Huron	MI	14,030
Iosco	MI	11,940
Isabella	MI	23,740
Midland	MI	32,690
Ogemaw	MI	9,170
Saginaw	MI	80,520
Shiawassee	MI	27,980
Tuscola	MI	21,020

Flint-Saginaw-Bay City, MI

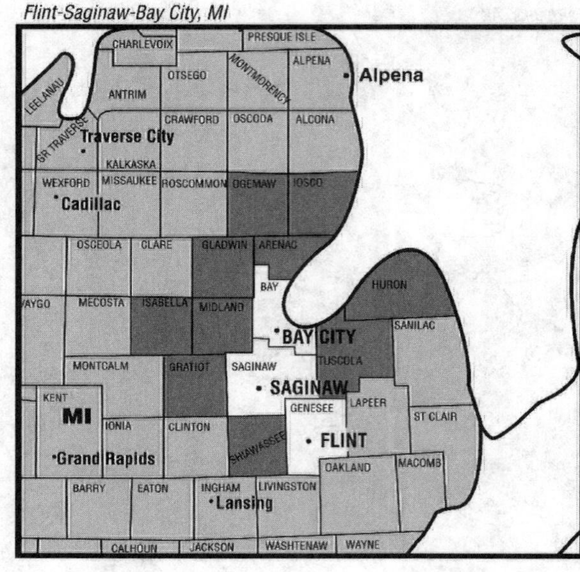

Ft. Myers-Naples, FL

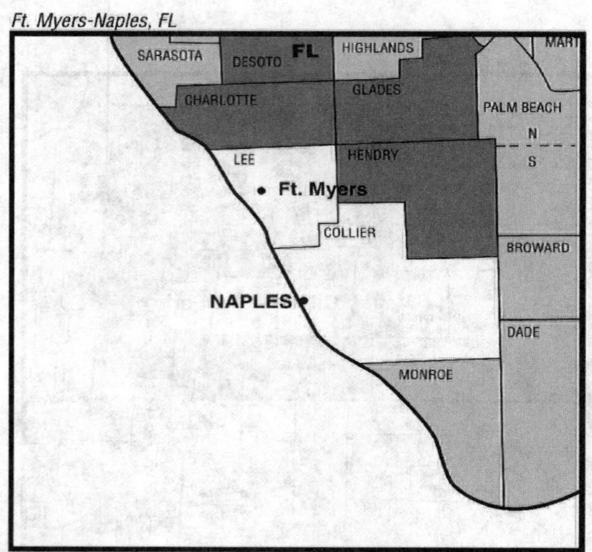

Ft. Myers-Naples, FL (64)

DMA TV Households: 479,130
% of U.S. TV Households: .430

WINK-TV Fort Myers, FL, ch. 11, CBS
WBBH-TV Fort Myers, FL, ch. 20, NBC
WZVN-TV Naples, FL, ch. 26, ABC
***WGCU** Fort Myers, FL, ch. 30, ETV
WFTX Cape Coral, FL, ch. 36, Fox
WXCW Naples, FL, ch. 46, CW
WRXY-TV Tice, FL, ch. 49, IND

DMA Counties	State	TV Households
Charlotte	FL	71,650
Collier	FL	132,650
De Soto	FL	11,010
Glades	FL	3,860
Hendry	FL	11,740
Lee	FL	248,220

Maps courtesy of Nielsen Media Research

Ft. Smith-Fayetteville-Springdale-Rogers, AR

Ft. Smith-Fayetteville-Springdale-Rogers, AR (102)

DMA TV Households: 280,510
% of U.S. TV Households: .252

KFSM-TV Fort Smith, AR, ch. 5, CBS
***KAFT** Fayetteville, AR, ch. 13, ETV
KFTA-TV Fort Smith, AR, ch. 24, Fox
KHOG-TV Fayetteville, AR, ch. 29, satellite to KHBS
KHBS Fort Smith, AR, ch. 40, ABC
KNWA-TV Rogers, AR, ch. 51, NBC
KWOG Springdale, AR, ch. 57, IND

DMA Counties	State	TV Households
Benton	AR	72,870
Crawford	AR	21,690
Franklin	AR	7,070
Johnson	AR	9,130
Logan	AR	9,010
Madison	AR	5,670
Scott	AR	4,320
Sebastian	AR	46,680
Washington	AR	70,300
LeFlore	OK	18,270
Sequoyah	OK	15,500

Ft. Wayne, IN (106)

DMA TV Households: 271,550
% of U.S. TV Households: .244

WANE-TV Fort Wayne, IN, ch. 15, CBS
WPTA Fort Wayne, IN, ch. 21, ABC, CW
WISE-TV Fort Wayne, IN, ch. 33, NBC, MyNetworkTV
***WFWA** Fort Wayne, IN, ch. 39, ETV
WFFT-TV Fort Wayne, IN, ch. 55, Fox
WINM Angola, IN, ch. 63, IND

DMA Counties	State	TV Households
Adams	IN	12,0080
Allen	IN	135,520
De Kalb	IN	15,870
Huntington	IN	14,500
Jay	IN	8,300
Noble	IN	17,210
Steuben	IN	12,980
Wabash	IN	12,750
Wells	IN	10,640
Whitley	IN	12,750
Paulding	OH	7,480
Van Wert	OH	11,550

Ft. Wayne, IN

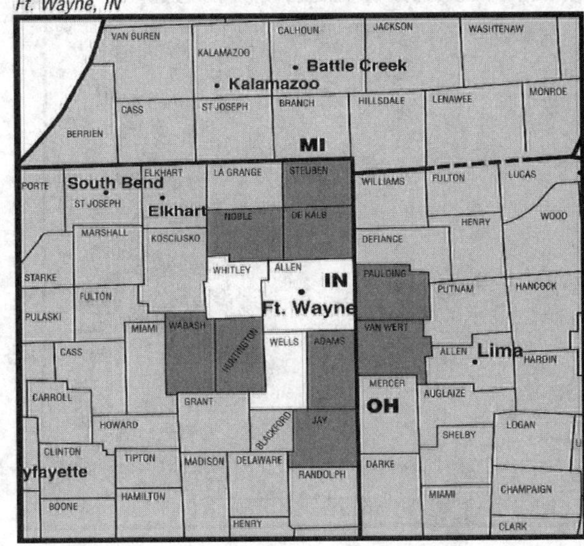

Fresno-Visalia, CA

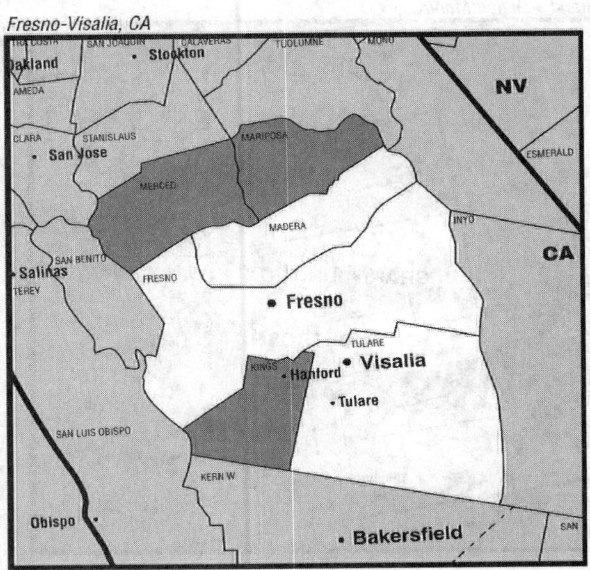

Fresno-Visalia, CA (55)

DMA TV Households: 557,380
% of U.S. TV Households: .501

***KVPT** Fresno, CA, ch. 18, ETV
KFTV Hanford, CA, ch. 21, Univision
KSEE Fresno, CA, ch. 24, NBC
KMPH-TV Visalia, CA, ch. 26, Fox
KFSN-TV Fresno, CA, ch. 30, ABC
KGMC Clovis, CA, ch. 43, IND
KGPE Fresno, CA, ch. 47, CBS
***KNXT** Visalia, CA, ch. 49, ETV
KNSO Merced, CA, ch. 51, Telemundo
KAIL Fresno, CA, ch. 53, MyNetworkTV
KFRE-TV Sanger, CA, ch. 59, CW
KTFF-TV Porterville, CA, ch. 61, TeleFutura

DMA Counties	State	TV Households
Fresno	CA	275,630
Kings	CA	38,390
Madera	CA	40,080
Mariposa	CA	6,150
Merced	CA	72,170
Tulare	CA	122,970

Maps courtesy of Nielsen Media Research

Gainesville, FL (162)

DMA TV Households: 119,590
% of U.S. TV Households: .107

***WUFT** Gainesville, FL, ch, 5, ETV
WCJB-TV Gainesville, FL, ch. 20, ABC, CW
WOGX Ocala, FL, ch. 51, IND
WGFL High Springs, FL, ch. 53, CBS, MyNetworkTV

DMA Counties	State	TV Households
Alachua	FL	92,570
Dixie	FL	5,680
Gilchrist	FL	5,60
Levy	FL	15,480

Gainesville, FL

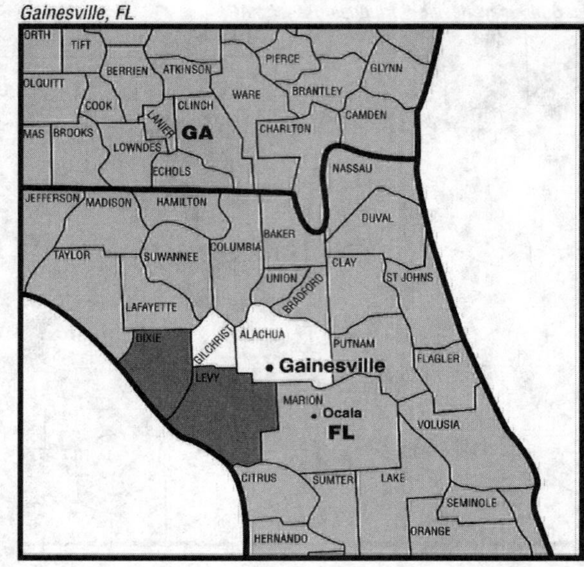

Glendive, MT

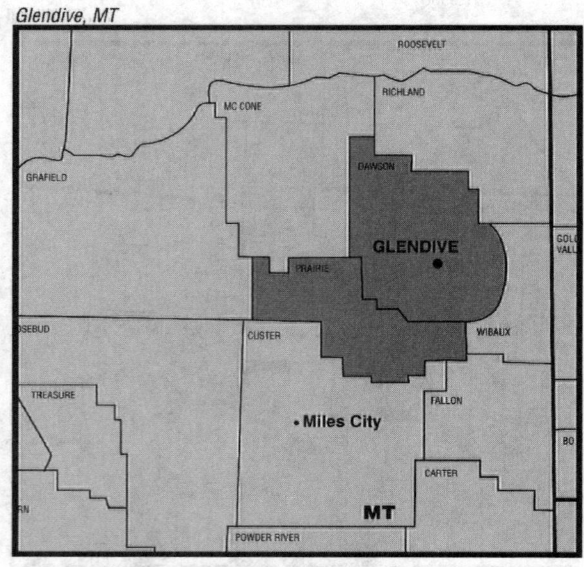

Glendive, MT (210)

DMA TV Households: 3,980
% of U.S. TV Households: .004

KXGN-TV Glendive, MT, ch. 5, CBS (NBC)

DMA Counties	State	TV Households
Dawson	MT	3,500
Fallon	MT	1,120
Prairie	MT	480

Grand Junction-Montrose, CO (186)

DMA TV Households: 69,560
% of U.S. TV Households: .062

KFQX Grand Junction, CO, ch. 4, Fox
KREX-TV Grand Junction, CO, ch. 5, CBS
KJCT Grand Junction, CO, ch. 8, ABC
KREY-TV Montrose, CO, ch. 10, satellite to KREX-TV
KKCO Grand Junction, CO, ch. 11, NBC, CW
***KRMJ** Grand Junction, CO, ch. 18, ETV

DMA Counties	State	TV Households
Mesa	CO	51,890
Montrose	CO	14,690
San Miguel	CO	2,980

Grand Junction-Montrose, CO

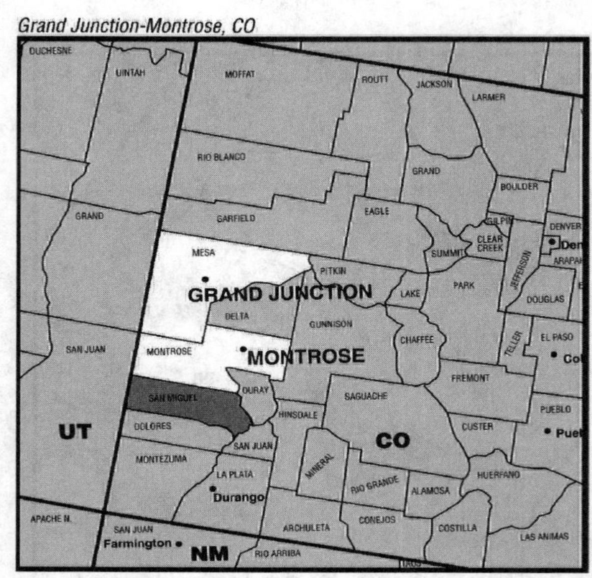

Maps courtesy of Nielsen Media Research

Grand Rapids-Kalamazoo-Battle Creek, MI

Grand Rapids-Kalamazoo-Battle Creek, MI (39)

DMA TV Households: 734,670
% of U.S. TV Households: .660

WWMT Kalamazoo, MI, ch. 3, CBS, CW
WOOD-TV Grand Rapids, MI, ch. 8, NBC
WZZM-TV Grand Rapids, MI, ch. 13, ABC
WXMI Grand Rapids, MI, ch. 17, Fox
***WGVU-TV** Grand Rapids, MI, ch. 35, ETV
WOTV Battle Creek, MI, ch. 41, ABC
WZPX Battle Creek, MI, ch. 43, ION Television
***WGVK** Kalamazoo, MI, ch. 52, ETV
WTLJ Muskegon, MI, ch. 54, IND
WLLA Kalamazoo, MI, ch. 64, IND

DMA Counties	State	TV Households	DMA Counties	State	TV Households
Allegan	MI	41,730	Montcalm	MI	23,190
Barry	MI	22,570	Muskegon	MI	66,090
Branch	MI	16,310	Newaygo	MI	18,470
Calhoun	MI	54,730	Oceana	MI	10,340
Ionia	MI	21,910	Ottawa	MI	88,840
Kalamazoo	MI	95,810	St. Joseph	MI	23,230
Kent	MI	222,350	Van Buren	MI	29,100

Great Falls, MT (190)

DMA TV Households: 63,510
% of U.S. TV Households: .057

KRTV Great Falls, MT, ch. 3, CBS, CW
KFBB-TV Great Falls, MT, ch. 5, ABC
KBBJ Havre, MT, ch. 9, IND
KBAO Lewistown, MT, ch. 13, NBC
KTGF Great Falls, MT, ch. 16, Fox
KLMN Great Falls, MT, ch. 26, MyNetworkTV

DMA Counties	State	TV Households
Blaine	MT	2,210
Cascade	MT	32,170
Chouteau	MT	1,960
Fergus	MT	4,500
Glacier	MT	4,250
Hill	MT	6,080
Judith Basin	MT	870
Liberty	MT	770
Phillips	MT	1,660
Pondera	MT	2,250
Teton	MT	2,260
Toole	MT	1,730
Valley	MT	2,800

Great Falls, MT

Green Bay-Appleton, WI

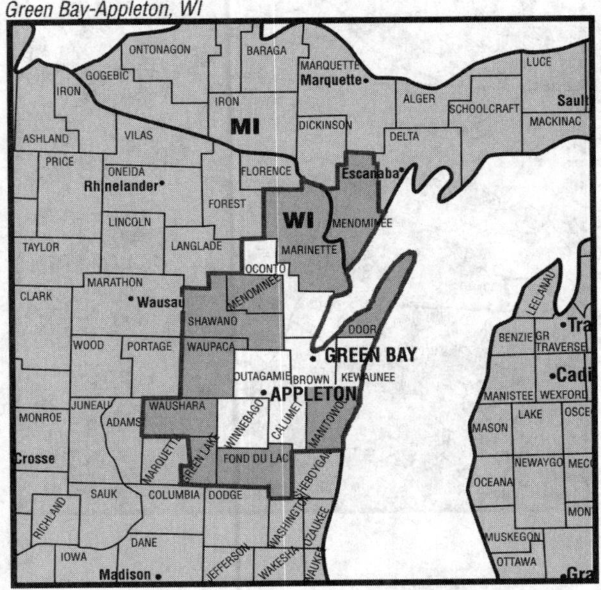

Green Bay-Appleton, WI (69)

DMA TV Households: 434,760
% of U.S. TV Households: .390

WBAY-TV Green Bay, WI, ch. 2, ABC
WFRV-TV Green Bay, WI, ch. 5, CBS
WLUK-TV Green Bay, WI, ch. 11, Fox
WIWB Suring, WI, ch. 14, CW
WGBA Green Bay, WI, ch. 26, NBC
WACY Appleton, WI, ch. 32, MyNetworkTV
***WPNE** Green Bay, WI, ch. 38, ETV
WFXS Wittenberg, WI, ch. 55, Fox
WWAZ-TV Fond du Lac, WI, ch. 68, IND

DMA Counties	State	TV Households	DMA Counties	State	TV Households
Menominee	MI	10,510	Marinette	WI	18,140
Brown	WI	94,890	Menominee	WI	1,400
Calumet	WI	17,220	Oconto	WI	15,170
Door	WI	12,620	Outagamie	WI	66,140
Fond du Lac	WI	38,740	Shawano	WI	16,490
Green Lake	WI	7,840	Waupaca	WI	20,560
Kewaunee	WI	8,160	Waushara	WI	9,700
Manitowoc	WI	33,440	Winnebago	WI	63,740

Maps courtesy of Nielsen Media Research

Greensboro-High Point-Winston Salem, NC (47)

DMA TV Households: 660,570
% of U.S. TV Households: .593

WFMY-TV Greensboro, NC, ch. 2, CBS
WGHP High Point, NC, ch. 8, Fox
WXII-TV Winston-Salem, NC, ch. 12, NBC
WGPX Burlington, NC, ch. 16, ION Television
WCWG Lexington, NC, ch. 20, CW
***WUNL-TV** Winston-Salem, NC, ch. 26, ETV
WXLV-TV Winston-Salem, NC, ch. 45, ABC
WMYV Greensboro, NC, ch. 48, MyNetworkTV
WLXI-TV Greensboro, NC, ch. 61, IND

DMA Counties	State	TV Households	DMA Counties	State	TV Households
Alamance	NC	56,050	Randolph	NC	54,240
Alleghany	NC	4,770	Rockingham	NC	37,540
Caswell	NC	8,800	Stokes	NC	18,330
Davidson	NC	61,580	Surry	NC	29,100
Davie	NC	15,820	Wilkes	NC	27,390
Forsyth	NC	133,390	Yadkin	NC	15,020
Guilford	NC	180,310	Patrick	VA	8,100
Montgomery	NC	10,130			

Greensboro-High Point-Winston Salem, NC

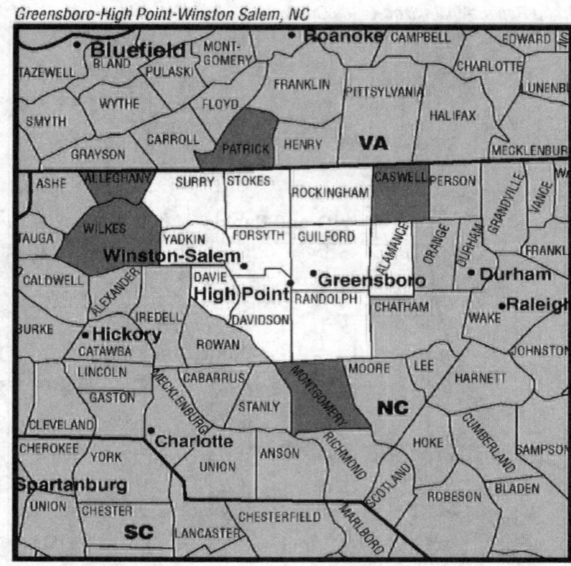

Greenville-New Bern-Washington, NC

Greenville-New Bern-Washington, NC (107)

DMA TV Households: 270,420
% of U.S. TV Households: .243

***WUND-TV** Columbia, NC, ch. 2, ETV
WITN-TV Washington, NC, ch. 7, NBC
WFXI Morehead City, NC, ch. 8, Fox, MyNetworkTV
WNCT-TV Greenville, NC, ch. 9, CBS, CW
WCTI-TV New Bern, NC, ch. 12, ABC
WYDO Greenville, NC, ch. 14, Fox
***WUNM-TV** Jacksonville, NC, ch. 19, ETV
***WUNK-TV** Greenville, NC, ch. 25, ETV
WPXU-TV Jacksonville, NC, ch. 35, MyNetworkTV
WEPX Greenville, NC, ch. 38, MyNetworkTV

DMA Counties	State	TV Households	DMA Counties	State	TV Households
Beaufort	NC	19,180	Lenoir	NC	23,320
Bertie	NC	7,660	Martin	NC	109,690
Carteret	NC	27,200	Onslow	NC	46,030
Craven	NC	34,940	Pamlico	NC	5,140
Duplin	NC	19,320	Pitt	NC	57,900
Greene	NC	7,120	Tyrrell	NC	1,450
Hyde	NC	2,040	Washington	NC	5,310
Jones	NC	4,120			

Greenville-Spartanburg, SC-Asheville, NC-Anderson, SC (36)

DMA TV Households: 826,290
% of U.S. TV Households: .742

WYFF Greenville, SC, ch. 4, NBC
WSPA-TV Spartanburg, SC, ch. 7, CBS
WLOS Asheville, NC, ch. 13, ABC
WGGS-TV Greenville, SC, ch. 16, IND
WHNS Greenville, SC, ch. 21, Fox
***WNTV** Greenville, SC, ch. 29, ETV
WNEG-TV Toccoa, GA, ch. 32, CBS
***WUNF-TV** Asheville, NC, ch. 33, ETV
***WNEH** Greenwood, SC, ch. 38, ETV
WMYA-TV Anderson, SC, ch. 40, MyNetworkTV
***WRET-TV** Spartanburg, SC, ch. 49, ETV
WYCW Asheville, NC, ch. 62, CW

DMA Counties	State	TV Households	DMA Counties	State	TV Households
Elbert	GA	8,210	Rutherford	NC	25,860
Franklin	GA	8,380	Swain	NC	5,240
Hart	GA	9,840	Transylvania	NC	12,820
Stephens	GA	9,850	Yancey	NC	7,590
Buncombe	NC	91,190	Abbeville	SC	9,980
Graham	NC	3,310	Anderson	SC	71,060
Haywood	NC	24,510	Cherokee	SC	21,460
Henderson	NC	41,550	Greenville	SC	164,650
Jackson	NC	14,200	Greenwood	SC	26,550
Macon	NC	13,810	Laurens	SC	27,070
Madison	NC	7,730	Oconee	SC	29,390
McDowell	NC	17,110	Pickens	SC	43,360
Mitchell	NC	6,650	Spartanburg	SC	104,900
Polk	NC	8,310	Union	SC	11,710

Greenville-Spartanburg, SC-Asheville, NC-Anderson, SC

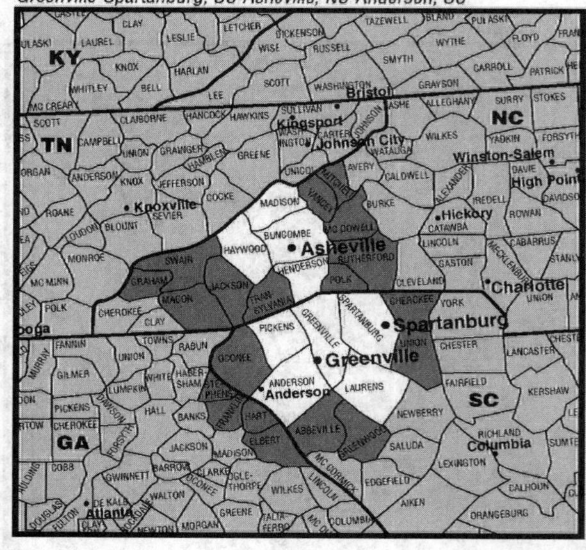

Maps courtesy of Nielsen Media Research

Greenwood-Greenville, MS

Greenwood-Greenville, MS (184)

DMA TV Households: 76,830
% of U.S. TV Households: .069

WABG-TV Greenwood, MS, ch. 6, ABC, Fox
WXVT Greenville, MS, ch. 15, CBS
***WMAO-TV** Greenwood, MS, ch. 23, ETV
***WMAI** Cleveland, MS, ch. 31, ETV

DMA Counties	State	TV Households
Chicot	AR	4,750
Bolivar	MS	12,980
Carroll	MS	3,910
Grenada	MS	8,740
Leflore	MS	12,320
Sunflower	MS	8,580
Tallahatchie	MS	4,920
Washington	MS	20,630

Harlingen-Weslaco-Brownsville-McAllen, TX (91)

DMA TV Households: 327,070
% of U.S. TV Households: .294

KGBT-TV Harlingen, TX, ch. 4, CBS
KRGV-TV Weslaco, TX, ch. 5, ABC
KVEO Brownsville, TX, ch. 23, NBC
KTLM Rio Grande City, TX, ch. 40, Telemundo
***KLUJ-TV** Harlingen, TX, ch. 44, ETV
KNVO McAllen, TX, ch. 48, Univision
***KMBH** Harlingen, TX, ch. 60, ETV

DMA Counties	State	TV Households
Cameron	TX	111,650
Hidalgo	TX	193,100
Starr	TX	16,570
Willacy	TX	5,750

Harlingen-Weslaco-Brownsville-McAllen, TX

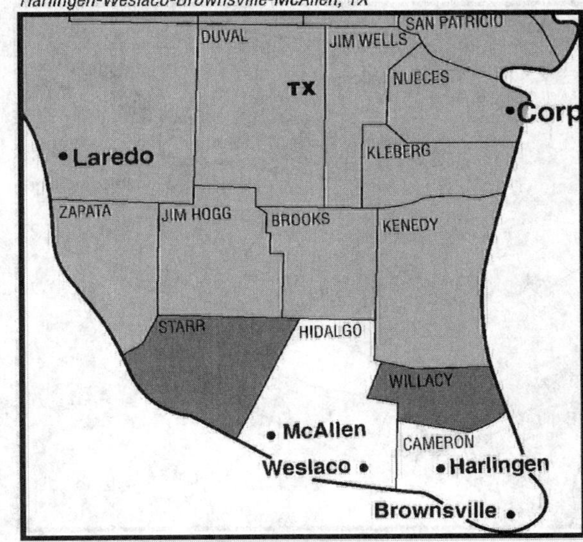

Harrisburg-Lancaster-Lebanon-York, PA

Harrisburg-Lancaster-Lebanon-York, PA (41)

DMA TV Households: 713,960
% of U.S. TV Households: .641

WGAL Lancaster, PA, ch. 8, NBC
WLYH-TV Lancaster, PA, ch. 15, CW
WHP-TV Harrisburg, PA, ch. 21, CBS, MyNetworkTV
WHTM-TV Harrisburg, PA, ch. 27, ABC
***WITF-TV** Harrisburg, PA, ch. 33, ETV
WPMT York, PA, ch. 43, Fox
WGCB-TV Red Lion, PA, ch. 49, IND

DMA Counties	State	TV Households
Adams	PA	37,810
Cumberland	PA	88,440
Dauphin	PA	103,730
Franklin	PA	53,610
Juniata	PA	8,850
Lancaster	PA	175,620
Lebanon	PA	48,500
Mifflin	PA	17,610
Perry	PA	17,190
York	PA	162,610

Maps courtesy of Nielsen Media Research

Harrisonburg, VA (181)

DMA TV Households: 87,630
% of U.S. TV Households: .079

WHSV-TV Harrisonburg, VA, ch. 3, ABC, Fox, MyNetworkTV
***WVPT** Staunton, VA, ch. 51, ETV

DMA Counties	State	TV Households
Augusta	VA	45,460
Rockingham	VA	39,130
Pendleton	WV	3,040

Harrisonburg, VA

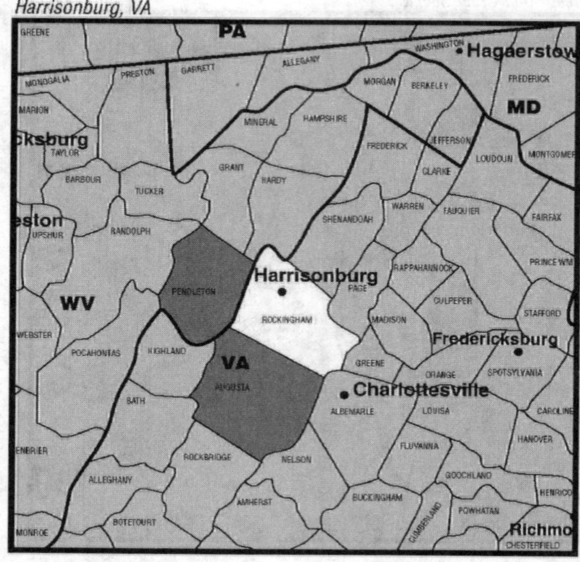

Hartford & New Haven, CT

Hartford & New Haven, CT (28)

DMA TV Households: 1,014,630
% of U.S. TV Households: .911

WFSB Hartford, CT, ch. 3, CBS
WTNH-TV New Haven, CT, ch. 8, ABC
WUVN Hartford, CT, ch. 18, Univision
WTXX Waterbury, CT, ch. 20, CW
***WEDH** Hartford, CT, ch. 24, ETV
WHPX New London, CT, ch. 26, ION Television
WVIT New Britain, CT, ch. 30, NBC
***WEDN** Norwich, CT, ch. 53, ETV
WCTX New Haven, CT, ch. 59, MyNetworkTV
WTIC-TV Hartford, CT, ch. 61, Fox
***WEDY** New Haven, CT, ch. 65, ETV

DMA Counties	State	TV Households
Hartford	CT	344,990
Litchfield	CT	74,840
Middlesex	CT	65,030
New Haven	CT	328,440
New London	CT	103,340
Tolland	CT	53,710
Windham	CT	44,280

Hattiesburg-Laurel, MS (165)

DMA TV Households: 111,580
% of U.S. TV Households: .100

WDAM-TV Laurel, MS, ch. 7, NBC
WHLT Hattiesburg, MS, ch. 22, CBS

DMA Counties	State	TV Households
Covington	MS	7,940
Forrest	MS	29,910
Jasper	MS	7,420
Jones	MS	26,210
Lamar	MS	17,340
Marion	MS	9,400
Perry	MS	4,820
Wayne	MS	8,540

Hattiesburg-Laurel, MS

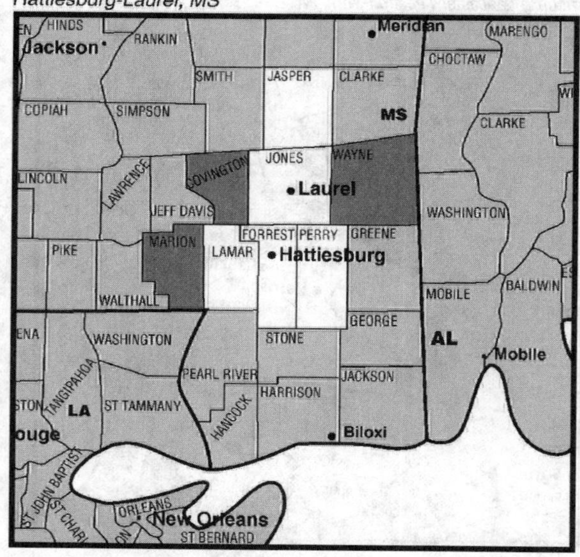

Maps courtesy of Nielsen Media Research

Helena, MT

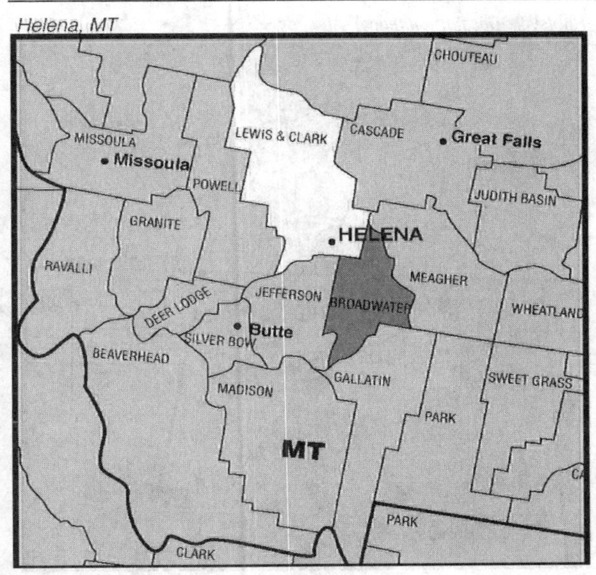

Helena (206)

DMA TV Households: 25,970
% of U.S. TV Households: .023

KMTF Helena, MT, ch. 10, CW
KTVH Helena, MT, ch. 12, NBC

DMA Counties	State	TV Households
Broadwater	MT	1,830
Lewis and Clark	MT	24,140

Honolulu, HI (72)

DMA TV Households: 419,160
% of U.S. TV Households: .376

KHON-TV Honolulu, ch. 2, Fox, CW
KITV Honolulu, ch. 4, ABC
KFVE Honolulu, ch. 5, MyNetworkTV
KGMB Honolulu, ch. 9, CBS
***KHET** Honolulu, ch. 11, ETV
KHNL Honolulu, ch. 13, NBC
KWHE Honolulu, ch. 14, IND
KIKU Honolulu, ch. 20, IND
KAAH-TV Honolulu, ch. 26, IND
KBFD Honolulu, ch. 32, IND
***KALO** Honolulu, ch. 38, ETV
***KWBN** Honolulu, ch. 44, ETV
KKAI Kailua, HI, ch. 50, IND
KUPU Waimanalo, HI, ch. 56, IND
KPXO Kaneohe, HI, ch. 66, ION Television

DMA Counties	State	TV Households
Hawaii	HI	58,400
Honolulu	HI	293,950
Kauai	HI	20,590
Maui	HI	46,220

Honolulu, HI

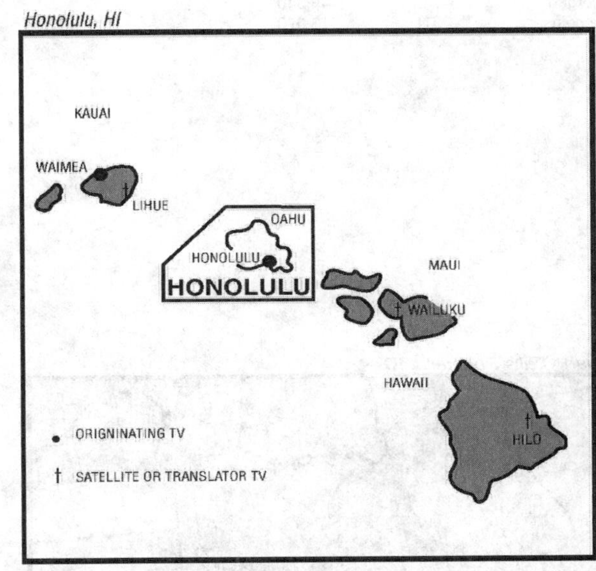

Houston, TX

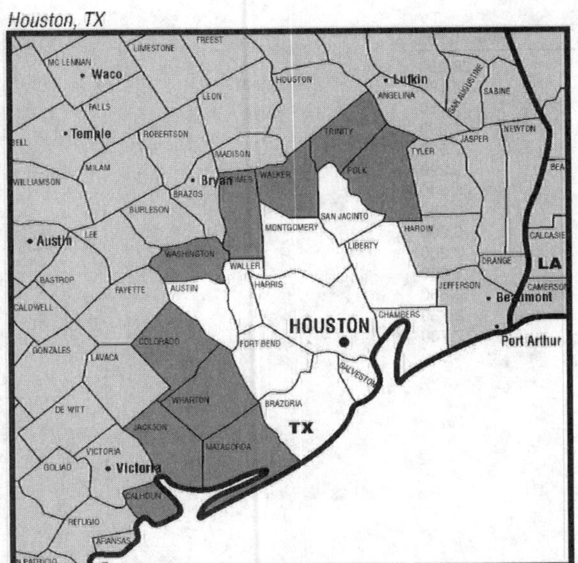

Houston (10)

DMA TV Households: 1,982,120
% of U.S. TV Households: 1.780

KPRC-TV Houston, ch. 2, NBC
***KUHT** Houston, ch. 8, ETV
KHOU-TV Houston, ch. 11, CBS
KTRK-TV Houston, ch. 13, ABC
***KETH-TV** Houston, ch. 14, ETV
KTXH Houston, ch. 20, MyNetworkTV
***KLTJ** Galveston, TX, ch. 22, ETV
KRIV-TV Houston, ch. 26, Fox
KHCW Houston, ch. 39, CW
KXLN-TV Rosenberg, TX, ch. 45, Univision
KTMD Galveston, TX, ch. 47, Telemundo
KPXB Conroe, TX, ch. 49, ION Television
KNWS-TV Katy, TX, ch. 51, IND
KTBU Conroe, TX, ch. 55, IND
KAZH Baytown, TX, ch. 57, IND
KZJL Houston, ch. 61, IND
KFTH-TV Alvin, TX, ch. 67, TeleFutura

DMA Counties	State	TV Households	DMA Counties	State	TV Households
Austin	TX	9,600	Jackson	TX	4,990
Brazoria	TX	98,360	Liberty	TX	25,360
Calhoun	TX	7,140	Matagorda	TX	13,760
Chambers	TX	10,640	Montgomery	TX	139,240
Colorado	TX	7,710	Polk	TX	17,470
Fort Bend	TX	152,180	San Jacinto	TX	9,150
Trinity	TX	6,020	Walker	TX	18,040
Galveston	TX	107,890	Waller	TX	11,730
Grimes	TX	8,370	Washington	TX	11,770
Harris	TX	1,307,820	Wharton	TX	14,880

Maps courtesy of Nielsen Media Research

Huntsville-Decatur (Florence), AL (84)

DMA TV Households: 375,270
% of U.S. TV Households: .337

WHDF Florence, AL, ch. 15, CW
WHNT-TV Huntsville, AL, ch. 19, CBS
***WHIQ** Huntsville, AL, ch. 25, ETV
WYLE Florence, AL, ch. 26, IND
WAAY-TV Huntsville, AL, ch. 31, ABC
***WFIQ** Florence, AL, ch. 36, ETV
WAFF Huntsville, AL, ch. 48, NBC
WZDX Huntsville, AL, ch. 54, Fox, MyNetworkTV

DMA Counties	State	TV Households
Colbert	AL	22,530
De Kalb	AL	26,360
Franklin	AL	11,910
Jackson	AL	21,780
Lauderdale	AL	36,520
Lawrence	AL	13,68 0
Limestone	AL	27,380
Madison	AL	122,600
Marshall	AL	34,100
Morgan	AL	45,340
Lincoln	TN	13,070

Huntsville-Decatur (Florence), AL

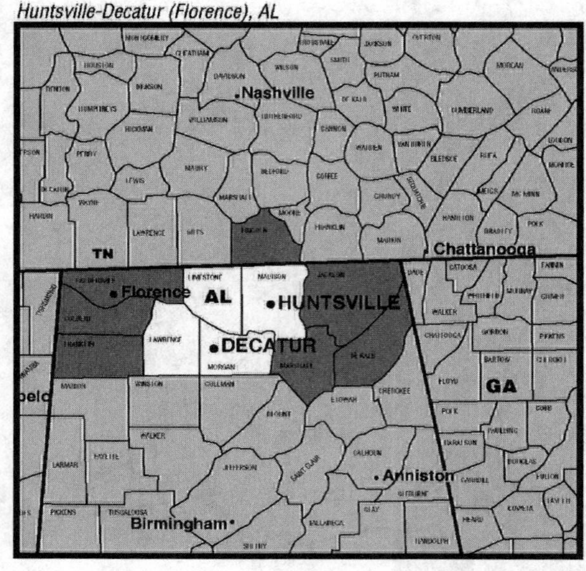

Idaho Falls-Pocatello, ID

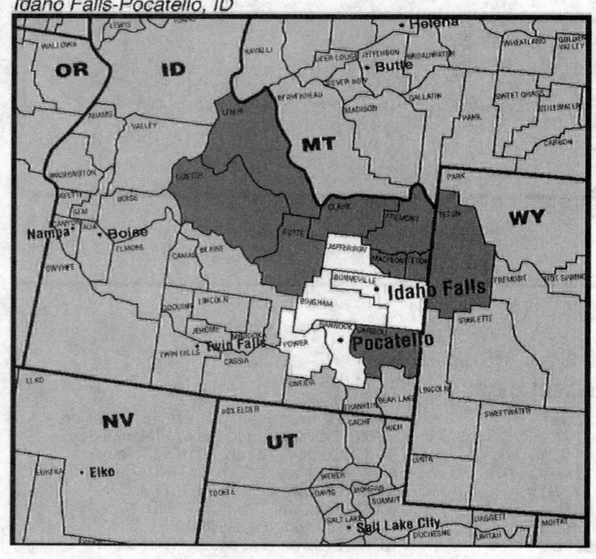

Idaho Falls-Pocatello, ID (163)

DMA TV Households: 116,560
% of U.S. TV Households: .105

KIDK Idaho Falls, ID, ch. 3, CBS
KPVI Pocatello, ID, ch. 6, ABC
KIFI-TV Idaho Falls, ID, ch. 8, NBC
***KISU-TV** Pocatello, ID, ch. 10, ETV
KPIF Pocatello, ID, ch. 15, IND
KFXP Pocatello, ID, ch. 31, Fox

DMA Counties	State	TV Households
Bannock	ID	28,290
Bingham	ID	14,180
Bonneville	ID	33,050
Butte	ID	1,090
Caribou	ID	2,500
Clark	ID	290
Custer	ID	1,540
Fremont	ID	3,960
Jefferson	ID	6,960
Lemhi	ID	3,250
Madison	ID	8,390
Power	ID	2,630
Teton	ID	2,730
Teton	WY	7,700

Maps courtesy of Nielsen Media Research

Indianapolis, IN

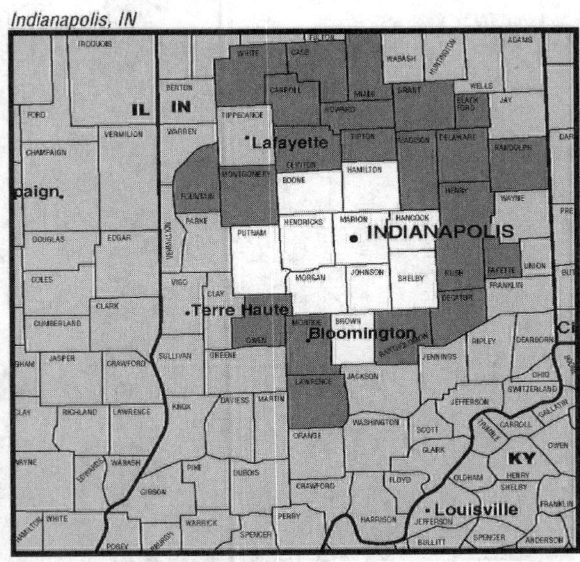

Indianapolis (25)

DMA TV Households: 1,060,550
% of U.S. TV Households: .952

WTTV Bloomington, IN, ch. 4, CW
WRTV Indianapolis, ch. 6, ABC
WISH-TV Indianapolis, ch. 8, CBS
WTHR Indianapolis, ch. 13, NBC
***WFYI** Indianapolis, ch. 20, ETV
WNDY-TV Marion, IN, ch. 23, MyNetworkTV
WTTK Kokomo, IN, ch. 29, satellite to WTTV
***WTIU** Bloomington, IN, ch. 30, ETV
WHMB-TV Indianapolis, ch. 40, IND
WCLJ-TV Bloomington, IN, ch. 42, IND
***WIPB** Muncie, IN, ch. 49, ETV
WXIN Indianapolis, ch. 59, Fox
WIPX Bloomington, IN, ch. 63, ION Television
***WDTI** Indianapolis, ch. 69, ETV

DMA Counties	State	TV Households	DMA Counties	State	TV Households
Bartholomew	IN	28,900	Howard	IN	35,370
Blackford	IN	5,660	Johnson	IN	49,050
Boone	IN	19,840	Lawrence	IN	18,890
Brown	IN	5,990	Madison	IN	52,020
Carroll	IN	7,580	Marion	IN	353,470
Cass	IN	15,250	Miami	IN	13,550
Clinton	IN	12,480	Monroe	IN	46,830
Decatur	IN	9,960	Montgomery	IN	14,930
Delaware	IN	45,470	Morgan	IN	26,000
Fayette	IN	9,890	Owen	IN	8,620
Fountain	IN	6,680	Putnam	IN	12,810
Grant	IN	26,730	Randolph	IN	10,720
Hamilton	IN	92,490	Rush	IN	6,640
Hancock	IN	24,920	Shelby	IN	16,910
Hendricks	IN	48,320	Tipton	IN	6,450
Henry	IN	18,860	White	IN	9,180

Jackson, MS (87)

DMA TV Households: 343,550
% of U.S. TV Households: .309

WLBT Jackson, MS, ch. 3, NBC
WJTV Jackson, MS, ch. 12, CBS
WAPT Jackson, MS, ch. 16, ABC
***WMAU-TV** Bude, MS, ch. 17, ETV
***WMPN-TV** Jackson, MS, ch. 29, ETV
***WMYC** Yazoo City, MS, ch. 32, ETV
WRBJ Magee, MS, ch. 34, CW
WUFX Vicksburg, MS, ch. 35, MyNetworkTV
WDBD Jackson, MS, ch. 40, Fox
WNTZ Natchez, MS, ch. 48, Fox, MyNetworkTV
WWJX Jackson, MS, ch. 51, IND

DMA Counties	State	TV Households	DMA Counties	State	TV Households
Adams	MS	13,130	Leake	MS	7,810
Attala	MS	7,450	Lincoln	MS	13,560
Claiborne	MS	3,860	Madison	MS	32,240
Copiah	MS	10,720	Pike	MS	15,960
Franklin	MS	3,300	Rankin	MS	51,990
Hinds	MS	94,390	Scott	MS	10,450
Holmes	MS	6,870	Sharkey	MS	1,950
Humphreys	MS	3,420	Simpson	MS	10,820
Issaquena	MS	590	Smith	MS	6,460
Jeff Davis	MS	5,360	Walthall	MS	6,180
Jefferson	MS	3,440	Warren	MS	18,980
Lawrence	MS	5,470	Yazoo	MS	9,150

Jackson, MS

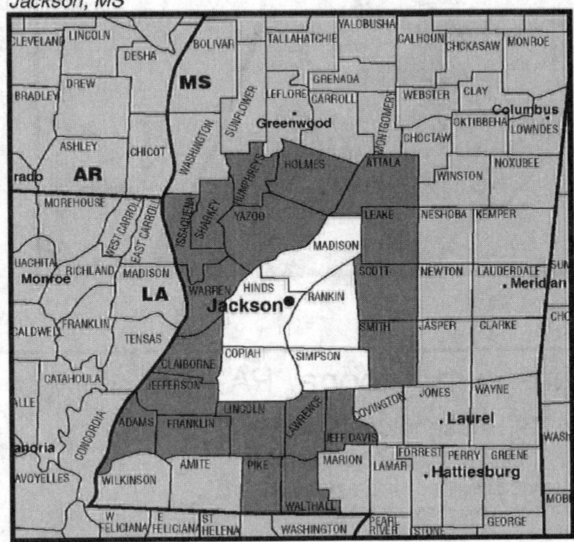

Maps courtesy of Nielsen Media Research

Jackson, TN (174)

DMA TV Households: 95,070
% of U.S. TV Households: .085

WBBJ-TV Jackson, TN, ch. 7, ABC
***WLJT-TV** Lexington, TN, ch. 11, ETV
WJKT Jackson, TN, ch. 16, Fox

DMA Counties	State	TV Households
Carroll	TN	11,370
Chester	TN	5,860
Gibson	TN	19,480
Hardin	TN	10,660
Henderson	TN	10,800
Madison	TN	36,900

Jackson, TN

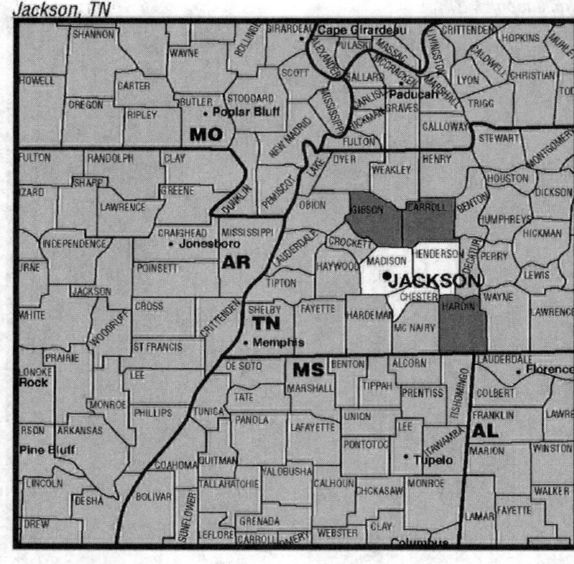

Jacksonville, FL

Jacksonville, FL (50)

DMA TV Households: 639,110
% of U.S. TV Households: .574

WJXT Jacksonville, FL, ch. 4, IND
***WJCT** Jacksonville, FL, ch. 7, ETV
***WXGA-TV** Waycross, GA, ch. 8, ETV
WTLV Jacksonville, FL, ch. 12, NBC
WCWJ Jacksonville, FL, ch. 17, CW
WPXC-TV Brunswick, GA, ch. 21, ION Television
WJXX Orange Park, FL, ch. 25, ABC
WAWS Jacksonville, FL, ch. 30, Fox
WTEV-TV Jacksonville, FL, ch. 47, CBS
***WJEB-TV** Jacksonville, FL, ch. 59, ETV

DMA Counties	State	TV Households	DMA Counties	State	TV Households
Baker	FL	7,240	Union	FL	3,610
Bradford	FL	9,070	Brantley	GA	5,920
Clay	FL	64,040	Camden	GA	15,920
Columbia	FL	23,930	Charlton	GA	3,460
Duval	FL	332,840	Glynn	GA	29,540
Nassau	FL	25,720	Pierce	GA	6,650
Putnam	FL	29,020	Ware	GA	13,440
St. Johns	FL	67,870			

Johnstown-Altoona, PA (98)

DMA TV Households: 294,160
% of U.S. TV Households: .264

***WPSU-TV** Clearfield, PA, ch. 3, ETV
WJAC-TV Johnstown, PA, ch. 6, NBC
WWCP-TV Johnstown, PA, ch. 8, Fox
WTAJ-TV Altoona, PA, ch. 10, CBS
WATM-TV Altoona, PA, ch. 23, ABC
WKBS-TV Altoona, PA, ch. 47, IND

DMA Counties	State	TV Households
Bedford	PA	19,770
Blair	PA	50,780
Cambria	PA	58,930
Cameron	PA	2,280
Centre	PA	50,700
Clearfield	PA	32,440
Elk	PA	13,610
Huntingdon	PA	16,720
Jefferson	PA	18,490
Somerset	PA	30,540

Johnstown-Altoona, PA

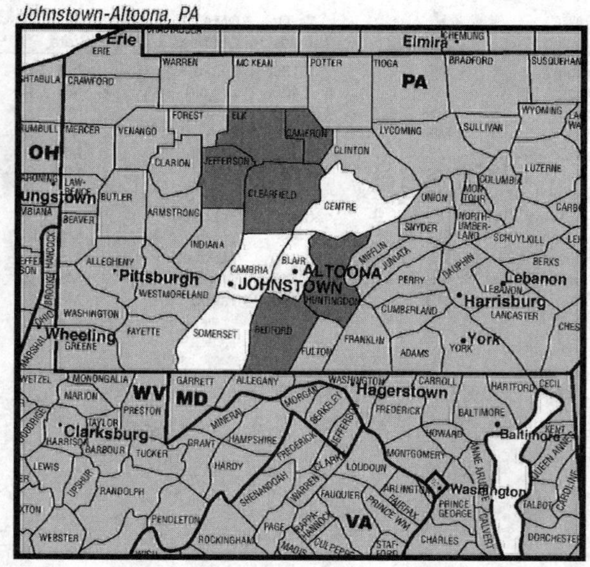

Maps courtesy of Nielsen Media Research

Jonesboro, AR

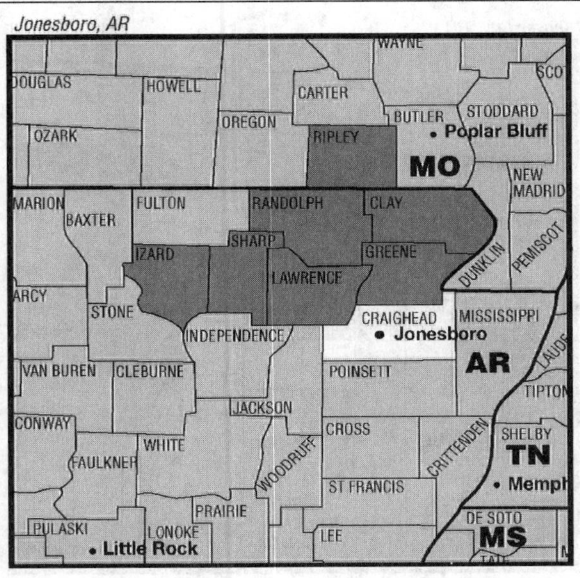

Jonesboro, AR (180)

DMA TV Households: 89,500
% of U.S. TV Households: .080

KAIT Jonesboro, AR, ch. 8, ABC
***KTEJ** Jonesboro, AR, ch. 19, ETV
KVTJ Jonesboro, AR, ch. 48, IND

DMA Counties	State	TV Households
Clay	AR	6,830
Craighead	AR	34,830
Greene	AR	15,770
Izard	AR	5,490
Lawrence	AR	6,590
Randolph	AR	7,360
Sharp	AR	7,180
Ripley	MO	5,450

Joplin, MO-Pittsburg, KS (144)

% of U.S. TV Households: .139
DMA TV Households: 154,640

KOAM-TV Pittsburg, KS, ch. 7, CBS
KODE-TV Joplin, MO, ch. 12, ABC
KFJX Pittsburg, KS, ch. 14, Fox
KSNF Joplin, MO, ch. 16, NBC
***KOZJ** Joplin, MO, ch. 26, ETV

DMA Counties	State	TV Households
Allen	KS	5,410
Bourbon	KS	5,910
Cherokee	KS	8,310
Crawford	KS	15,310
Labette	KS	8,810
Neosho	KS	6,440
Wilson	KS	3,760
Woodson	KS	1,560
Barton	MO	4,970
Jasper	MO	43,930
McDonald	MO	8,450
Newton	MO	21,560
Vernon	MO	7,850
Ottawa	OK	12,370

Joplin, MO-Pittsburg, KS

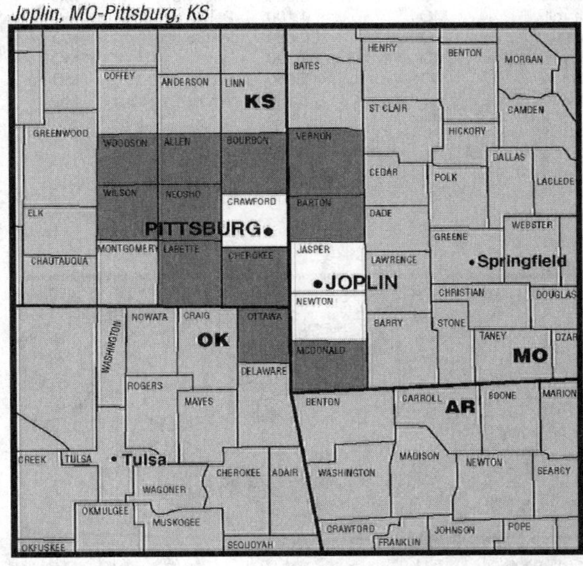

Juneau, AK

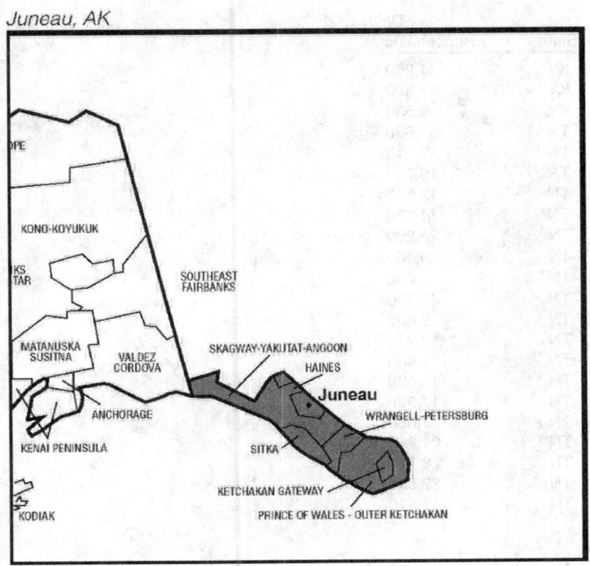

Juneau, AK (207)

DMA TV Households: 23,910
% of U.S. TV Households: .021

***KTOO-TV** Juneau, AK, ch. 3, ETV
KUBD Ketchikan, AK, ch. 4, IND
KJUD Juneau, AK, ch. 8, ABC, CW
KTNL-TV Sitka, AK, ch. 13, CBS

DMA Counties	State	TV Households
Juneau	AK	23,910

Maps courtesy of Nielsen Media Research

Kansas City, MO (31)

DMA TV Households: 913,280
% of U.S. TV Households: .820

WDAF-TV Kansas City, MO, ch. 4, Fox
KCTV Kansas City, MO, ch. 5, CBS
***KMOS-TV** Sedalia, MO, ch. 6, ETV
KMBC-TV Kansas City, MO, ch. 9, ABC
***KCPT** Kansas City, MO, ch. 19, ETV
KCWE Kansas City, MO, ch. 29, CW
KMCI Lawrence, KS, ch. 38, IND
KSHB-TV Kansas City, MO, ch. 41, NBC
KPXE Kansas City, MO, ch. 50, ION Television
KSMO-TV Kansas City, MO, ch. 62, MyNetworkTV

DMA Counties	State	TV Households	DMA Counties	State	TV Households
Anderson	KS	3,100	Gentry	MO	2,520
Atchison	KS	6,340	Grundy	MO	4,340
Douglas	KS	39,350	Harrison	MO	3,670
Franklin	KS	9,910	Henry	MO	9,270
Johnson	KS	201,990	Jackson	MO	268,590
Leavenworth	KS	25,320	Johnson	MO	18,350
Linn	KS	3,960	Lafayette	MO	12,600
Miami	KS	11,380	Linn	MO	5,300
Wyandotte	KS	57,970	Livingston	MO	5,540
Bates	MO	6,610	Mercer	MO	1,460
Caldwell	MO	3,620	Nodaway	MO	8,180
Carroll	MO	4,100	Pettis	MO	15,800
Cass	MO	36,080	Platte	MO	33,730
Clay	MO	82,320	Ray	MO	9,100
Clinton	MO	8,060	Saline	MO	8,590
Daviess	MO	3,180	Worth	MO	890

Kansas City, MO

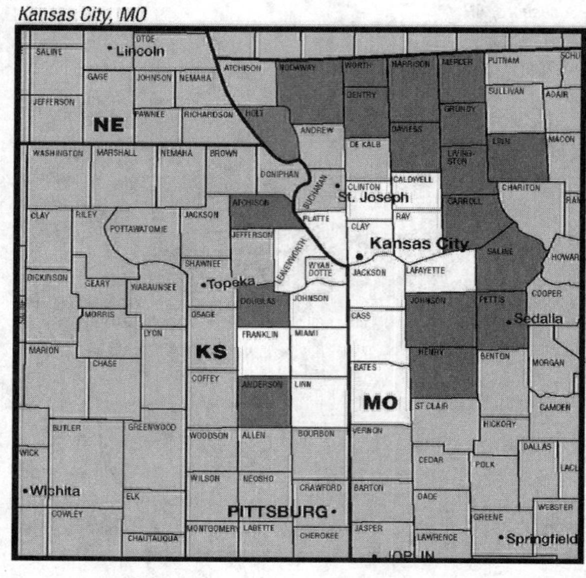

Knoxville, TN

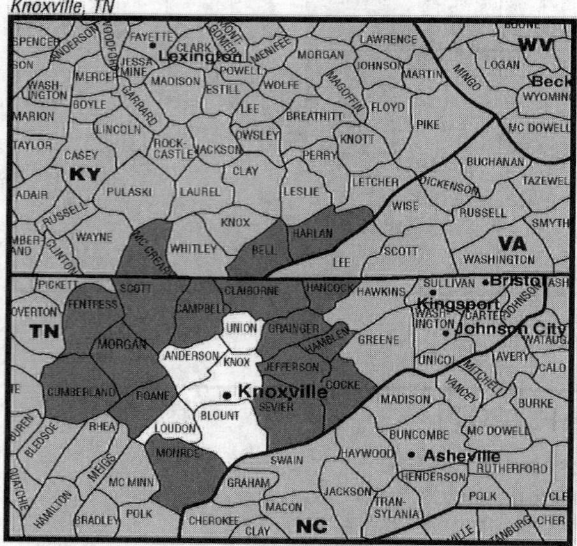

Knoxville, TN (60)

DMA TV Households: 523,010
% of U.S. TV Households: .470

***WETP-TV** Sneedville, TN, ch. 2, ETV
WATE-TV Knoxville, TN, ch. 6, ABC
WMAK Knoxville, TN, ch. 7, IND
WVLT-TV Knoxville, TN, ch. 8, CBS, MyNetworkTV
WBIR-TV Knoxville, TN, ch. 10, NBC
***WKOP-TV** Knoxville, TN, ch. 15, ETV
WBXX-TV Crossville, TN, ch. 20, CW
WTNZ Knoxville, TN, ch. 43, Fox
WAGV Harlan, KY, ch. 44, IND
WVLR Tazewell, TN, ch. 48, IND
WPXK Jellico, TN, ch. 54, ION Television

DMA Counties	State	TV Households
Bell	KY	11,960
Harlan	KY	12,620
McCreary	KY	6,530
Anderson	TN	30,690
Blount	TN	47,710
Campbell	TN	16,790
Claiborne	TN	12,550
Cocke	TN	14,580
Cumberland	TN	22,290
Fentress	TN	6,400
Grainger	TN	9,230
Hamblen	TN	24,350
Hancock	TN	2,800
Jefferson	TN	19,270
Knox	TN	171,060
Loudon	TN	18,520
Monroe	TN	17,590
Morgan	TN	7,080
Roane	TN	21,960
Scott	TN	8,700
Sevier	TN	32,920
Union	TN	7,410

Maps courtesy of Nielsen Media Research

La Crosse-Eau Claire, WI (127)

DMA TV Households: 209,870

% of U.S. TV Households: .188

WKBT La Crosse, WI, ch. 8, CBS, MyNetworkTV
WEAU-TV Eau Claire, WI, ch. 13, NBC
WQOW-TV Eau Claire, WI, ch. 18, ABC, CW
WXOW-TV La Crosse, WI, ch. 19, ABC, CW
WLAX La Crosse, WI, ch. 25, Fox
***WHLA-TV** La Crosse, WI, ch. 31, ETV
WEUX Chippewa Falls, WI, ch. 48, Fox

DMA Counties	State	TV Households
Houston	MN	7,830
Winona	MN	18,410
Buffalo	WI	5,720
Chippewa	WI	23,560
Clark	WI	12,080
Crawford	WI	6,660
EauClaire	WI	37,310
Jackson	WI	7,380
La Crosse	WI	43,160
Monroe	WI	16,320
Pepin	WI	2,860
Rusk	WI	6,100
Trempealeau	WI	11,370
Vernon	WI	11,110

La Crosse-Eau Claire, WI

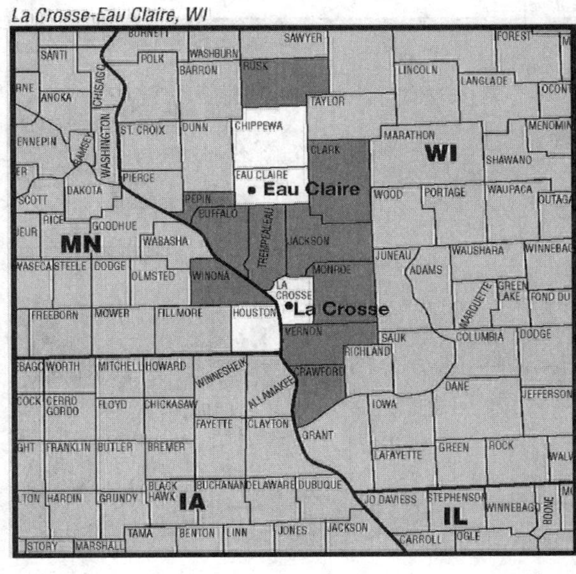

Lafayette, IN

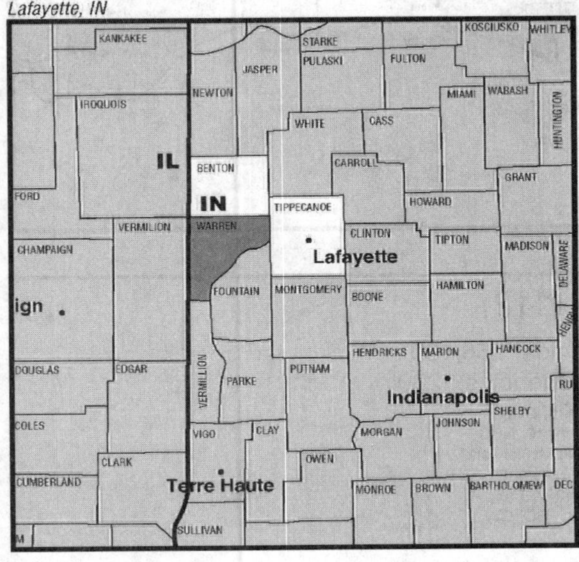

Lafayette, IN (188)

DMA TV Households: 64,680

% of U.S. TV Households: .058

WLFI-TV Lafayette, IN, ch. 18, CBS

DMA Counties	State	TV Households
Benton	IN	3,370
Tippecanoe	IN	57,970
Warren	IN	3,340

Maps courtesy of Nielsen Media Research

Lafayette, LA

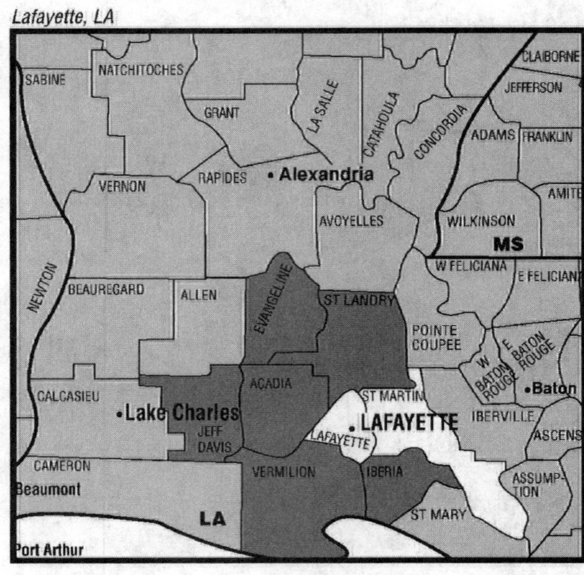

Lafayette, LA (123)

DMA TV Households: 225,650
% of U.S. TV Households: .203

KATC Lafayette, LA, ch. 3, ABC
KLFY-TV Lafayette, LA, ch. 10, CBS
KADN Lafayette, LA, ch. 15, Fox
***KLPB-TV** Lafayette, LA, ch. 24, ETV
KLWB New Iberia, LA, ch. 50, CW

DMA Counties	State	TV Households
Acadia	LA	21,950
Evangeline	LA	13,130
Iberia	LA	26,360
Jefferson Davis	LA	11,580
Lafayette	LA	78,520
St. Landry	LA	34,600
St. Martin	LA	18,500
Vermilion	LA	21,010

Lake Charles, LA (175)

DMA TV Households: 94,840
% of U.S. TV Households: .085

KPLC Lake Charles, LA, ch. 7, NBC
***KLTL-TV** Lake Charles, LA, ch. 18, ETV
KVHP Lake Charles, LA, ch. 29, Fox

DMA Counties	State	TV Households
Allen	LA	8,020
Beauregard	LA	12,810
Calcasieu	LA	70,690
Cameron	LA	3,320

Lake Charles, LA

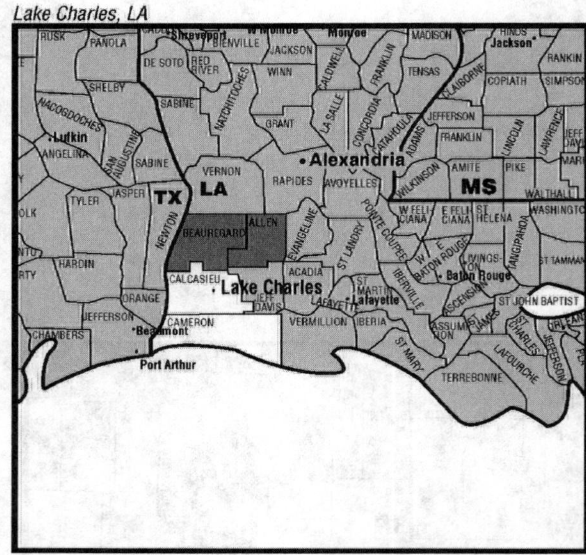

Lansing, MI

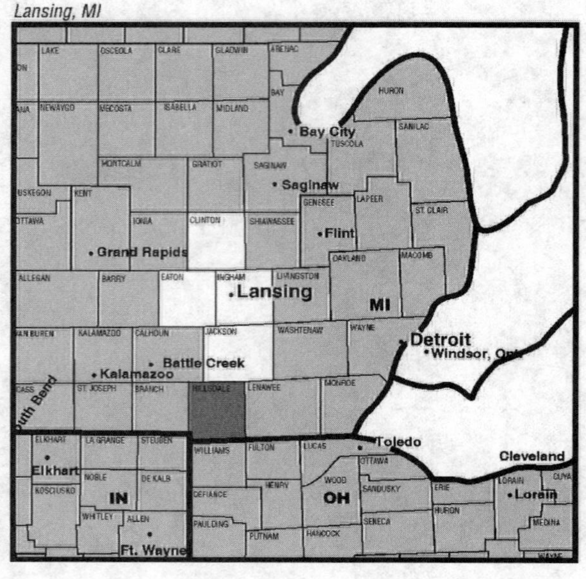

Lansing, MI (112)

DMA TV Households: 256,190
% of U.S. TV Households: .230

WLNS-TV Lansing, MI, ch. 6, CBS
WILX-TV Onondaga, MI, ch. 10, NBC
WHTV Jackson, MI, ch. 18, MyNetworkTV
***WKAR-TV** East Lansing, MI, ch. 23, ETV
WSYM-TV Lansing, MI, ch. 47, Fox
WLAJ Lansing, MI, ch. 53, ABC, CW

DMA Counties	State	TV Households
Clinton	MI	26,100
Eaton	MI	42,670
Hillsdale	MI	17,650
Ingham	MI	109,270
Jackson	MI	60,500

Maps courtesy of Nielsen Media Research

Laredo, TX (187)

DMA TV Households: 65,790
% of U.S. TV Households: .059

KGNS-TV Laredo, TX, ch. 8, NBC
KVTV Laredo, TX, ch. 13, CBS
KLDO-TV Laredo, TX, ch. 27, Univision

DMA Counties	State	TV Households
Webb	TX	61,330
Zapata	TX	4,460

Laredo, TX

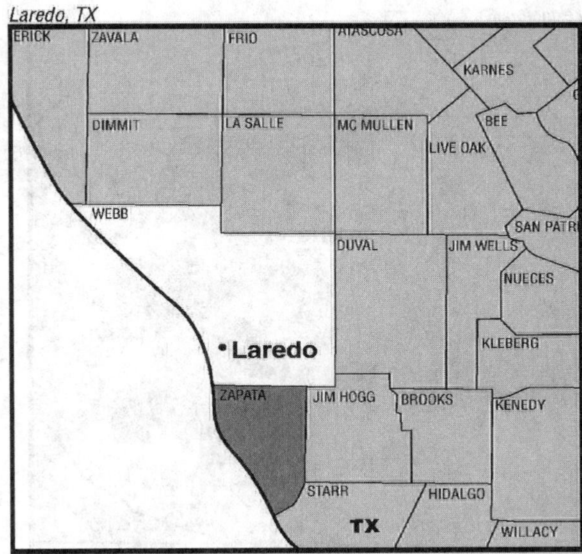

Las Vegas, NV

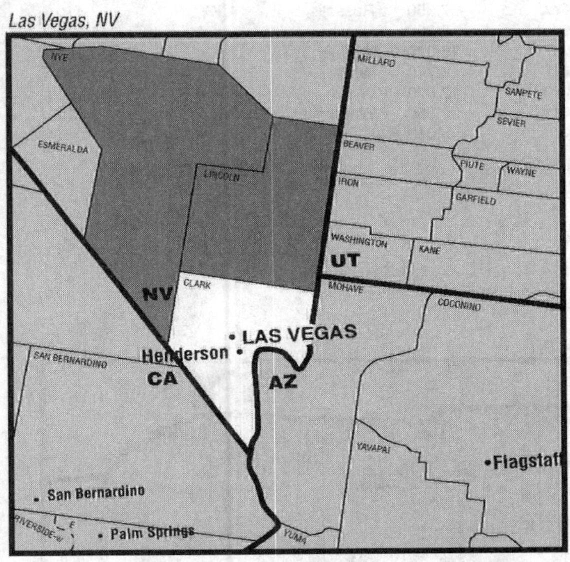

Las Vegas (43)

DMA TV Households: 671,630
% of U.S. TV Households: .603

KVBC Las Vegas, ch. 3, NBC
KVVU-TV Henderson, NV, ch. 5, Fox
KLAS-TV Las Vegas, ch. 8, CBS
***KLVX** Las Vegas, ch. 10, ETV
KTNV Las Vegas, ch. 13, ABC
KINC Las Vegas, ch. 15, Univision
KVMY Las Vegas, ch. 21, MyNetworkTV
KVCW Las Vegas, ch. 33, CW
KMCC Laughlin, NV, ch. 34, IND
KBLR Paradise, NV, ch. 39, Telemundo

DMA Counties	State	TV Households
Clark	NV	652,520
Lincoln	NV	1,660
Nye	NV	17,450

Maps courtesy of Nielsen Media Research

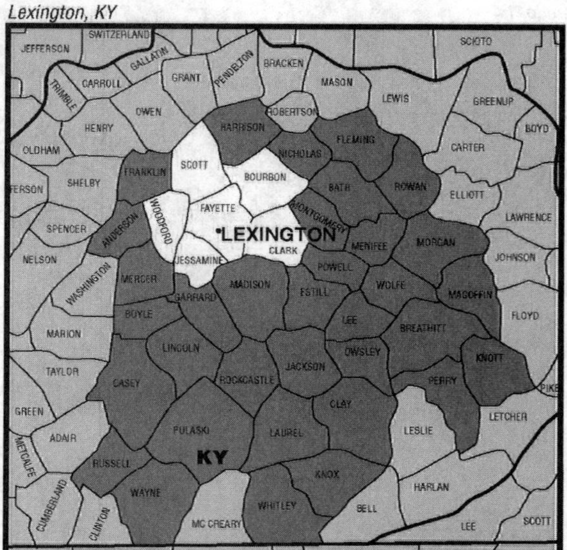

Lexington, KY (63)

DMA TV Households: 483,520
% of U.S. TV Households: .434

WLEX-TV Lexington, KY, ch. 18, NBC
WKYT-TV Lexington, KY, ch. 27, CBS, CW
***WKSO-TV** Somerset, KY, ch. 29, ETV
***WKHA** Hazard, KY, ch. 35, ETV
WTVQ-TV Lexington, KY, ch. 36, ABC
***WKMR** Morehead, KY, ch. 38, ETV
***WKLE** Lexington, KY, ch. 46, ETV
WDKY-TV Danville, KY, ch. 56, Fox
WYMT-TV Hazard, KY, ch. 57, CBS
WLJC-TV Beattyville, Ky, ch. 65, IND
WUPX-TV Morehead, KY, ch. 67, IND

DMA Counties	State	TV Households	DMA Counties	State	TV Households
Anderson	KY	8,050	Lincoln	KY	10,190
Bath	KY	4,720	Madison	KY	31,190
Bourbon	KY	8,020	Menifee	KY	2,740
Boyle	KY	11,070	Mercer	KY	8,870
Breathitt	KY	5,850	Montgomery	KY	9,960
Casey	KY	6,620	Morgan	KY	4,870
Clark	KY	14,210	Nicholas	KY	2,750
Clay	KY	8,290	Owsley	KY	1,800
Estill	KY	5,970	Perry	KY	11,590
Fayette	KY	113,430	Powell	KY	5,240
Fleming	KY	5,510	Pulaski	KY	24,380
Franklin	KY	20,460	Rockcastle	KY	6,440
Garrard	KY	6,630	Rowan	KY	8,030
Harrison	KY	7,350	Russell	KY	7,410
Jackson	KY	5,250	Scott	KY	15,320
Jessamine	KY	15,970	Wayne	KY	8,050
Knott	KY	6,740	Whitley	KY	14,880
Knox	KY	12,870	Wolfe	KY	2,800
Laurel	KY	22,550	Woodford	KY	9,630
Lee	KY	3,010			

Lima, OH (196)

DMA TV Households: 53,180
% of U.S. TV Households: .048

WLIO Lima, OH, ch. 35, NBC, CW
WTLW Lima, OH, ch. 44, IND

DMA Counties	State	TV Households
Allen	OH	40,600
Putnam	OH	12,580

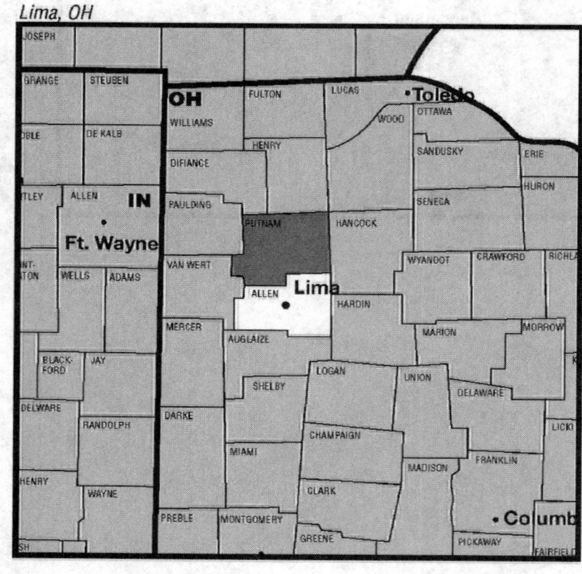

Maps courtesy of Nielsen Media Research

Lincoln & Hastings-Kearney, NE (104)

DMA TV Households: 275,970

% of U.S. TV Households: .248

***KLNE-TV** Lexington, NE, ch. 3, ETV
KHAS-TV Hastings, NE, ch. 5, NBC
KSNB-TV Superior, NE, ch. 4, satellite to KTVG
KWNB-TV Hayes Center, NE, ch. 6
***KMNE-TV** Bassett, NE, ch. 7, ETV
KLKN Lincoln, NE, ch. 8, ABC
KOLN, Lincoln, NE, ch. 10, CBS, MyNetworkTV
KGIN Grand Island, NE, ch. 11, satellite to KOLN
***KUON-TV** Lincoln, NE, ch. 12, ETV
KHGI-TV Kearney, NE, ch. 13, ABC
KTVG Grand Island, NE, ch. 17, Fox
***KHNE-TV** Hastings, NE, ch. 29, ETV
KCWL-TV Lincoln, NE, ch. 51 CW

DMA Counties	State	TV Households	DMA Counties	State	TV Households
Jewell	KS	1,490	Hayes	NE	500
Phillips	KS	2,170	Hitchcock	NE	1,180
Republic	KS	2,150	Holt	NE	4,210
Smith	KS	1,690	Howard	NE	2,570
Adams	NE	12,940	Jefferson	NE	3,260
Antelope	NE	2,780	Kearney	NE	2,590
Blaine	NE	190	Keya Paha	NE	390
Boone	NE	2,170	Lancaster	NE	106,490
Boyd	NE	870	Loup	NE	290
Brown	NE	1,390	Merrick	NE	3,170
Buffalo	NE	16,750	Nance	NE	1,390
Butler	NE	3,370	Nuckolls	NE	2,070
Chase	NE	1,580	Pawnee	NE	1,160
Clay	NE	2,590	Perkins	NE	1,190
Custer	NE	4,650	Phelps	NE	3,670
Dawson	NE	8,570	Polk	NE	2,190
Fillmore	NE	2,540	Red Willow	NE	4,550
Franklin	NE	1,390	Rock	NE	700
Frontier	NE	990	Saline	NE	5,260
Furnas	NE	2,090	Seward	NE	6,120
Gage	NE	9,460	Sherman	NE	1,270
Garfield	NE	790	Thayer	NE	2,290
Gosper	NE	790	Valley	NE	1,760
Greeley	NE	990	Webster	NE	1,580
Hall	NE	20,930	Wheeler	NE	300
Hamilton	NE	3,570	York	NE	5,650
Harlan	NE	1,490			

Lincoln & Hastings-Kearney, NE

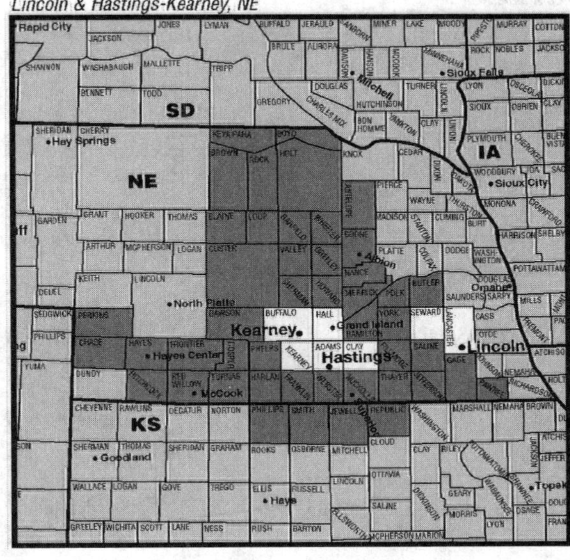

Little Rock-Pine Bluff, AR

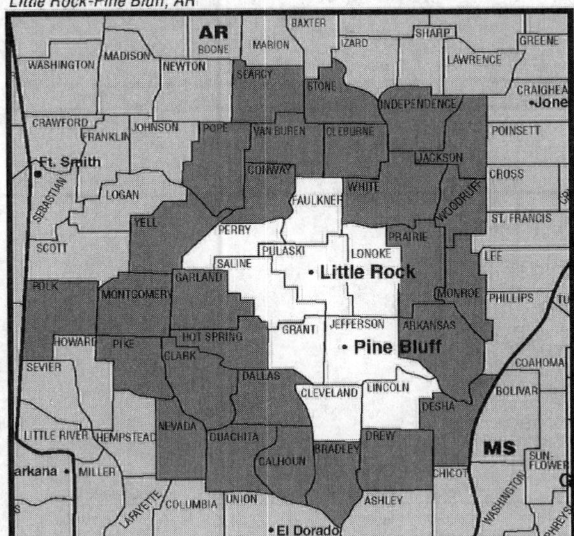

Little Rock-Pine Bluff, AR (57)

DMA TV Households: 539,900

% of U.S. TV Households: .485

***KETS** Little Rock, AR, ch. 2, ETV
KARK-TV Little Rock, AR, ch. 4, NBC
***KEMV** Mountain View, AR, ch. 6, ETV
KATV Little Rock, AR, ch. 7, ABC
***KETG** Arkadelphia, AR, ch. 9, ETV
KTHV Little Rock, AR, ch. 11, CBS
KLRT Little Rock, AR, ch. 16, Fox
***KLEP** Newark, AR, ch. 17, ETV
KVTN Pine Bluff, AR, ch. 25, IND
KVTH Hot Springs, AR, ch. 26, IND
***KKAP** Little Rock, AR, ch. 36, ETV
KASN Pine Bluff, AR, ch. 38, CW
KWBF Little Rock, AR, ch. 42, MyNetworkTV
KYPX Camden, AR, ch. 49, IND

DMA Counties	State	TV Households	DMA Counties	State	TV Households
Arkansas	AR	7,970	Lonoke	AR	23,000
Bradley	AR	4,600	Monroe	AR	3,560
Calhoun	AR	2,260	Montgomery	AR	3,640
Clark	AR	8,580	Ouachita	AR	10,830
Cleburne	AR	10,780	Perry	AR	4,150
Cleveland	AR	3,450	Pike	AR	4,420
Conway	AR	8,300	Polk	AR	7,810
Dallas	AR	3,170	Pope	AR	21,380
Desha	AR	5,600	Prairie	AR	3,740
Drew	AR	7,450	Pulaski	AR	152,120
Faulkner	AR	36,880	Saline	AR	36,400
Garland	AR	40,730	Searcy	AR	3,230
Grant	AR	6,640	Stone	AR	4,690
Hot Spring	AR	12,290	Van Buren	AR	6,990
Independence	AR	13,760	White	AR	27,100
Jackson	AR	6,330	Woodruff	AR	3,210
Jefferson	AR	29,420	Yell	AR	7,800
Lincoln	AR	3,910			

Maps courtesy of Nielsen Media Research

Los Angeles, CA

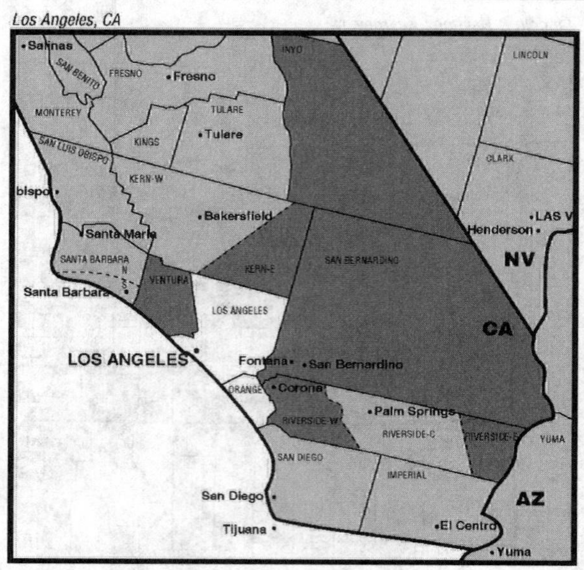

Los Angeles, CA (2)

DMA TV Households: 5,611,110
% of U.S. TV Households: 5.039

KCBS-TV Los Angeles, ch. 2, CBS
KNBC Los Angeles, ch. 4, NBC
KTLA Los Angeles, ch. 5, CW
KABC-TV Los Angeles, ch. 7, ABC
KCAL Los Angeles, ch. 9, IND
KTTV Los Angeles, ch. 11, Fox
KCOP Los Angeles, ch. 13, MyNetworkTV
KSCI Long Beach, CA, ch. 18, IND
KWHY-TV Los Angeles, ch. 22, IND
***KVCR-TV** San Bernardino, CA, ch. 24, ETV
***KCET** Los Angeles, ch. 28, ETV
KPXN San Bernardino, CA, ch. 30, ION Television
KVMD Twentynine Palms, CA, ch. 31, IND
KTBN-TV Santa Ana, CA, ch. 40, IND
KXLA Rancho Palos Verdes, CA, ch. 44, IND
KFTR-TV Ontario, CA, ch. 46, TeleFutura
***KOCE-TV** Huntington Beach, CA, ch. 50, ETV
KVEA Corona, CA, ch. 52, Telemundo
KAZA-TV Avalon, CA, ch. 54, Azteca America
KDOC-TV Anaheim, CA, ch. 56, IND
KJLA Ventura, CA, ch. 57, IND
***KLCS** Los Angeles, ch. 58, ETV
KRCA Riverside, CA, ch. 62, IND
KHIZ Barstow, CA, ch. 64, IND
KMEX-TV Los Angeles, ch. 34, Univision

DMA Counties	State	TV Households	DMA Counties	State	TV Households
Inyo	CA	7,110	Riverside East	CA	6,580
Kern East	CA	28,810	Riverside West	CA	487,610
Los Angeles	CA	3,242,690	San Bernardino	CA	607,530
Orange	CA	973,480	Ventura	CA	257,300

Louisville, KY (48)

DMA TV Households: 648,190
% of U.S. TV Households: .582

WAVE Louisville, KY, ch. 3, NBC
WHAS-TV Louisville, KY, ch. 11, ABC
***WKPC-TV** Louisville, KY, ch. 15, ETV
WBNA Louisville, KY, ch. 21, ION Television
***WKZT-TV** Elizabethtown, KY, ch. 23, ETV
WLKY-TV Louisville, KY, ch. 32, CBS
WBKI-TV Campbellsville, KY, ch. 34, CW
WDRB Louisville, KY, ch. 41, Fox
WMYO Salem, IN, ch. 58, MyNetworkTV
***WKMJ** Louisville, KY, ch. 68, ETV

DMA Counties	State	TV Households	DMA Counties	State	TV Households
Clark	IN	42,110	Green	KY	4,720
Crawford	IN	4,370	Hardin	KY	36,550
Floyd	IN	28,320	Henry	KY	6,240
Harrison	IN	14,390	Jefferson	KY	293,550
Jackson	IN	16,610	Larue	KY	5,390
Jefferson	IN	12,630	Marion	KY	7,060
Jennings	IN	10,630	Meade	KY	10,750
Orange	IN	7,830	Nelson	KY	15,870
Scott	IN	9,370	Oldham	KY	18,160
Washington	IN	10,580	Shelby	KY	14,380
Adair	KY	6,870	Spencer	KY	6,040
Breckinridge	KY	7,610	Taylor	KY	9,700
Bullitt	KY	26,270	Trimble	KY	3,550
Carroll	KY	4,140	Washington	KY	4,470
Grayson	KY	10,030			

Louisville, KY

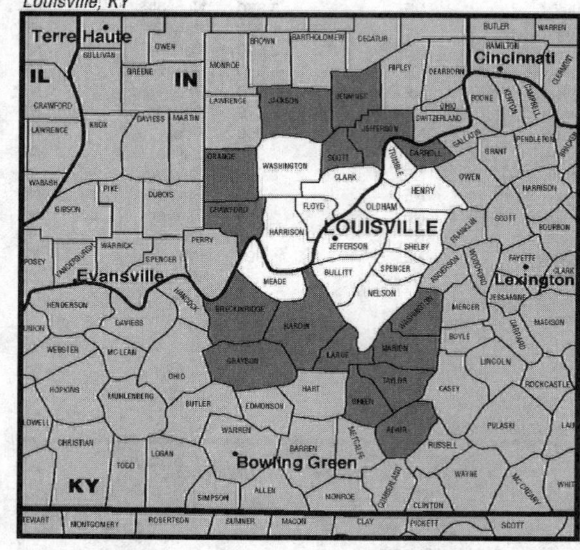

Lubbock, TX

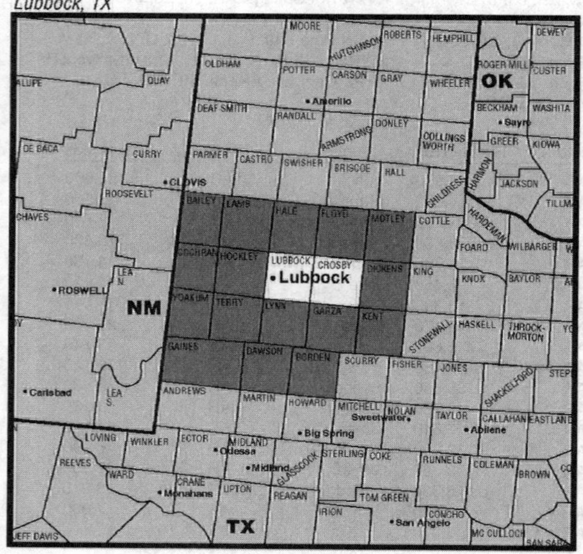

Lubbock, TX (147)

DMA TV Households: 151,610
% of U.S. TV Households: .136

***KTXT-TV** Lubbock, TX, ch. 5, ETV
KCBD Lubbock, TX, ch. 11, NBC
KLBK-TV Lubbock, TX, ch. 13, CBS
KPTB Lubbock, TX, ch. 16, IND
KLCW-TV Wolfforth, TX, ch. 22, CW
KAMC Lubbock, TX, ch. 28, ABC
KJTV-TV Lubbock, TX, ch. 34, Fox

DMA Counties	State	TV Households	DMA Counties	State	TV Households
Bailey	TX	2,350	Hale	TX	11,770
Borden	TX	280	Hockley	TX	8,180
Cochran	TX	1,170	Kent	TX	290
Crosby	TX	2,290	Lamb	TX	5,220
Dawson	TX	4,270	Lubbock	TX	96,820
Dickens	TX	970	Lynn	TX	2,190
Floyd	TX	2,430	Motley	TX	580
Gaines	TX	4,490	Terry	TX	4,180
Garza	TX	1,680	Yoakum	TX	2,450

Maps courtesy of Nielsen Media Research

Macon, GA (121)

DMA TV Households: 230,180
% of U.S. TV Households: .207

WMAZ-TV Macon, GA, ch. 13, CBS
WGXA Macon, GA, ch. 24, Fox, MyNetworkTV
***WMUM-TV** Cochran, GA, ch. 29, ETV
WMGT-TV Macon, GA, ch. 41, NBC
WPGA-TV Perry, GA, ch. 58, IND
WGNM Macon, GA, ch. 64, IND

DMA Counties	State	TV Households	DMA Counties	State	TV Households
Baldwin	GA	15,010	Macon	GA	4,560
Bibb	GA	60,340	Monroe	GA	8,670
Bleckley	GA	4,560	Peach	GA	8,990
Crawford	GA	4,590	Pulaski	GA	3,450
Dodge	GA	7,120	Telfair	GA	4,190
Dooly	GA	3,910	Treutlen	GA	2,440
Hancock	GA	3,140	Twiggs	GA	3,730
Houston	GA	48,230	Washington	GA	7,210
Johnson	GA	3,080	Wheeler	GA	2,070
Jones	GA	10,300	Wilcox	GA	2,780
Laurens	GA	17,980	Wilkinson	GA	3,830

Macon, GA

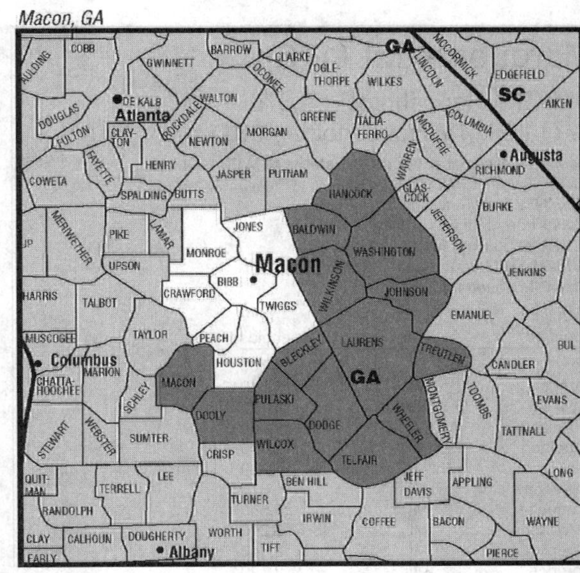

Madison, WI

Madison, WI (85)

DMA TV Households: 369,220
% of U.S. TV Households: .332

WISC-TV Madison, WI, ch. 3, CBS, MyNetworkTV
WMTV Madison, WI, ch. 15, NBC
***WHA-TV** Madison, WI, ch. 21, ETV
WKOW-TV Madison, WI, ch. 27, ABC
WMSN-TV Madison, WI, ch. 47, Fox
WBUW Janesville, WI, ch. 57, CW

DMA Counties	State	TV Households
Columbia	WI	21,910
Dane	WI	190,400
Grant	WI	18,820
Green	WI	14,180
Iowa	WI	9,220
Juneau	WI	10,480
Lafayette	WI	6,320
Marquette	WI	6,290
Richland	WI	7,230
Rock	WI	61,110
Sauk	WI	23,260

Mankato, MN (200)

DMA TV Households: 51,090
% of U.S. TV Households: .046

KEYC-TV Mankato, MN, ch. 12, CBS, Fox

DMA Counties	State	TV Households
Blue Earth-Nicollet South	MN	27,740
Brown	MN	10,370
Martin	MN	8,720
Watonwan	MN	4,260

Mankato, MN

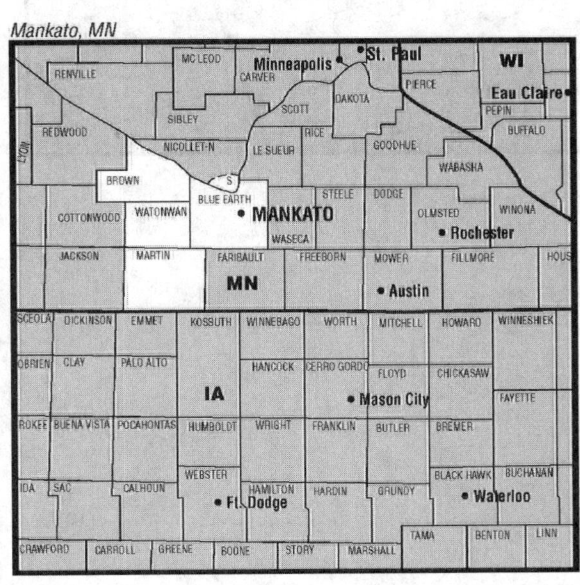

Maps courtesy of Nielsen Media Research

Marquette, MI (178)

DMA TV Households: 89,670
% of U.S. TV Households: .081

WJMN-TV Escanaba, MI, ch. 3, satellite to WFRV-TV
WBKP Calumet, MI, ch. 5, ABC
WLUC-TV Marquette, MI, ch. 6, NBC
WDHS Iron Mountain, MI, ch. 8, IND
WBUP Ishpeming, MI, ch. 10, ABC
***WNMU-TV** Marquette, MI, ch. 13, ETV
WMQF Marquette, MI, ch. 19, Fox

DMA Counties	State	TV Households
Alger	MI	3,760
Baraga	MI	3,310
Delta	MI	16,170
Dickinson	MI	11,920
Houghton	MI	12,690
Iron	MI	5,220
Keweenaw	MI	970
Marquette	MI	26,560
Ontonagon	MI	3,240
Schoolcraft	MI	3,640
Florence	WI	2,190

Marquette, MI

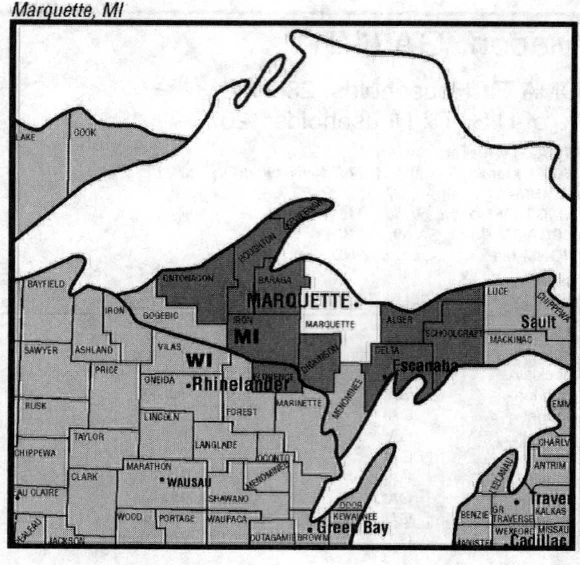

Medford-Klamath Falls, OR

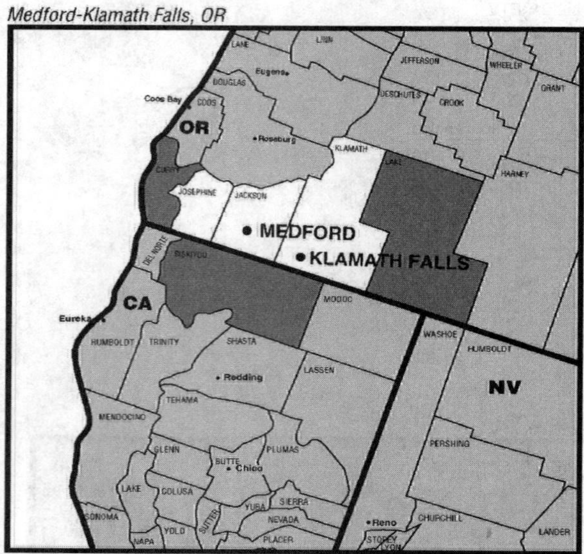

Medford-Klamath Falls, OR (141)

DMA TV Households: 164,780
% of U.S. TV Households: .148

KOBI Medford, OR, ch. 5, NBC
KOTI Klamath Falls, OR, ch. 2, satellite to KOBI
***KSYS** Medford, OR, ch. 8, ETV
KTVL Medford, OR, ch. 10, CBS
KDRV Medford, OR, ch. 12, ABC
***KBDM** Yreka City, CA, ch. 20, ETV
***KFTS** Klamath Falls, OR, ch. 22, satellite to *KSYS
KMVU Medford, OR, ch. 26, IND
KBLN Grants Pass, OR, ch. 30, IND
KDKF Klamath Falls, OR, ch. 31, ABC

DMA Counties	State	TV Households
Siskiyou	CA	18,2400
Curry	OR	9,520
Jackson	OR	76,620
Josephine	OR	32,230
Klamath	OR	25,390
Lake	OR	2,780

Maps courtesy of Nielsen Media Research

Memphis, TN

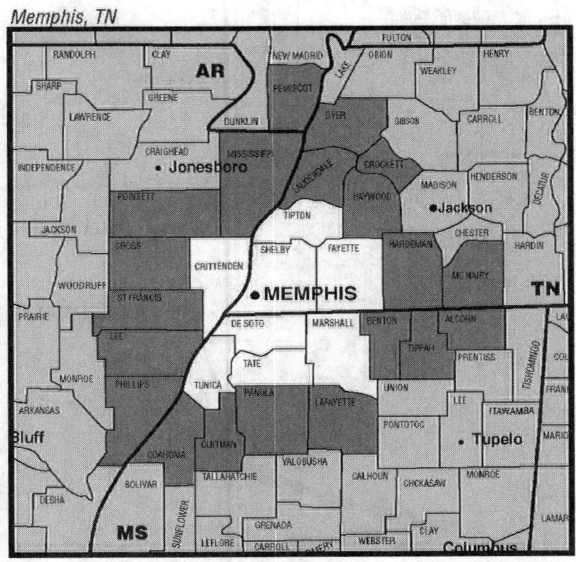

Memphis (44)

DMA TV Households: 664,290
% of U.S. TV Households: .597

WREG-TV Memphis, ch. 3, CBS
WMC-TV Memphis, ch. 5, NBC
***WKNO-TV** Memphis, ch., 10, ETV
WHBQ-TV Memphis, ch. 13, Fox
***WMAV-TV** Oxford, MS, ch. 18, ETV
WPTY-TV Memphis, ch. 24, ABC
WLMT Memphis, ch. 30, CW
WBUY-TV Holly Springs, MS, ch. 40, IND
WPXX-TV Memphis, ch. 50, MyNetworkTV

DMA Counties	State	TV Households	DMA Counties	State	TV Households
Crittenden	AR	19,150	Quitman	MS	3,200
Cross	AR	7,120	Tate	MS	9,550
Lee	AR	3,680	Tippah	MS	8,380
Mississippi	AR	17,490	Tunica	MS	3,650
Phillips	AR	8,580	Pemiscot	MO	7,590
Poinsett	AR	9,930	Crockett	TN	5,660
St. Francis	AR	9,380	Dyer	TN	15,010
Alcorn	MS	14,730	Fayette	TN	13,290
Benton	MS	2,970	Hardeman	TN	9,610
Coahoma	MS	9,700	Haywood	TN	7,550
De Soto	MS	53,610	Lauderdale	TN	9,450
Lafayette	MS	15,640	McNairy	TN	10,350
Marshall	MS	12,530	Shelby	TN	343,350
Panola	MS	12,800	Tipton	TN	20,340

Meridian, MS (185)

DMA TV Households: 74,440
% of U.S. TV Households: .067

WTOK-TV Meridian, MS, ch. 11, ABC, CW, Fox
***WMAW-TV** Meridian, MS, ch. 14, ETV
WMDN Meridian, MS, ch. 24, CBS
WGBC Meridian, MS, ch. 30, NBC

DMA Counties	State	TV Households
Choctaw	AL	6,120
Sumter	AL	5,490
Clarke	MS	7,500
Kemper	MS	3,860
Lauderdale	MS	30,890
Neshoba	MS	11,610
Newton	MS	8,970

Meridian, MS

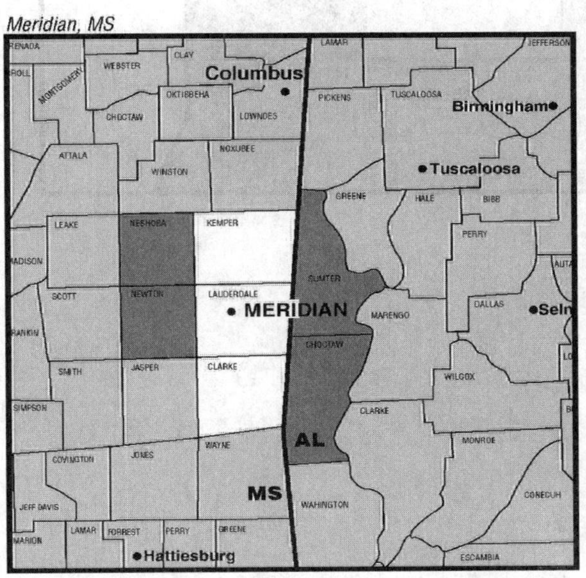

Maps courtesy of Nielsen Media Research

Miami-Ft. Lauderdale, FL (16)

DMA TV Households: 1,538,620
% of U.S. TV Households: 1.382

***WPBT** Miami, ch. 2, ETV
WFOR-TV Miami, ch. 4, CBS
WTVJ Miami, ch. 6, NBC
WSVN Miami, ch. 7, Fox
WGEN-TV Key West, FL, ch. 8, IND
WPLG Miami, ch. 10, ABC
***WLRN-TV** Miami, ch. 17, ETV
WSBS-TV Key West, FL, ch. 22, IND
WLTV Miami, ch. 23, Univision
WBFS-TV Miami, ch. 33, MyNetworkTV
WPXM Miami, ch. 35, ION Television
WSFL-TV Miami, ch. 39, CW
WHFT-TV Miami, ch. 45, IND
WSCV Fort Lauderdale, FL, ch. 51, Telemundo
WAMI-TV Hollywood, FL, ch. 69, TeleFutura

DMA Counties	State	TV Households
Broward	FL	700,050
Miami-Dade	FL	805,600
Monroe	FL	32,970

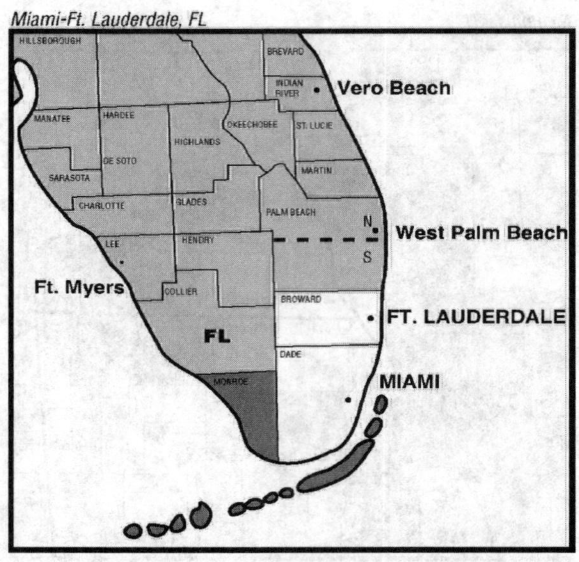

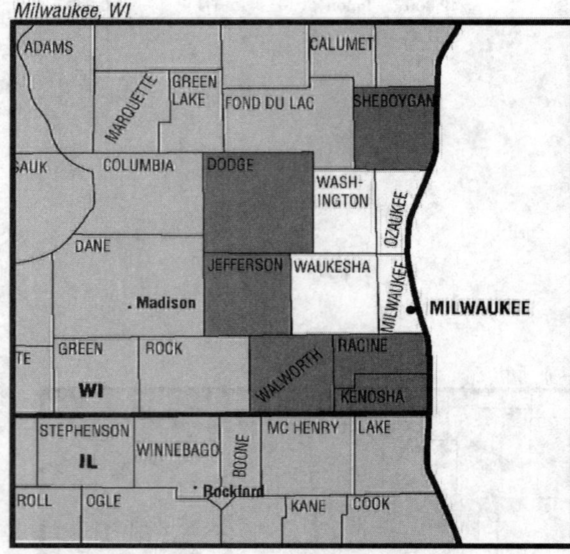

Milwaukee, WI (34)

DMA TV Households: 882,990
% of U.S. TV Households: .793

WTMJ-TV Milwaukee, ch. 4, NBC
WITI Milwaukee, ch. 6, Fox
***WMVS** Milwaukee, ch. 10, ETV
WISN-TV Milwaukee, ch. 12, ABC
WVTV Milwaukee, ch. 18, CW
WCGV-TV Milwaukee, ch. 24, MyNetworkTV
WVCY-TV Milwaukee, ch. 30, IND
***WMVT** Milwaukee, ch. 36, ETV
WJJA Racine, WI, ch. 49, IND
WWRS-TV Mayville, WI, ch. 52, IND
WPXE Kenosha, WI, ch. 55, ION Television
WDJT-TV Milwaukee, ch. 58, CBS

DMA Counties	State	TV Households
Dodge	WI	32,920
Jefferson	WI	30,200
Kenosha	WI	60,730
Milwaukee	WI	371,870
Ozaukee	WI	33,180
Racine	WI	74,450
Sheboygan	WI	45,380
Walworth	WI	37,760
Washington	WI	49,140
Waukesha	WI	147,360

Maps courtesy of Nielsen Media Research

Minneapolis-St. Paul, MN

Minneapolis-St. Paul, MN (15)

DMA TV Households: 1,678,430
% of U.S. TV Households: 1.507

***KTCA-TV** St. Paul, ch. 2, ETV
WCCO-TV Minneapolis, ch. 4, CBS
KSTP-TV St. Paul, ch. 5, ABC
KCCO-TV Alexandria, MN, ch. 7, CBS
***KAWE** Bemidji, MN, ch. 9, ETV
KMSP-TV Minneapolis, ch. 9, Fox
***KWCM-TV** Appleton, MN, ch. 10, ETV
KARE Minneapolis, ch. 11, NBC
KCCW-TV Walker, MN, ch. 12, satellite to KCCO-TV
***KTCI-TV** St. Paul, ch. 17, ETV
***KAWB** Brainerd, MN, ch. 22, ETV
WUCW Minneapolis, ch. 23, CW
KFTC Bemidji, MN, ch. 26, IND
***WHWC-TV** Menomonie, WI, ch. 28, ETV
WFTC Minneapolis, ch. 29, MyNetworkTV
KPXM St. Cloud, MN, ch. 41, ION Television
KSAX Alexandria, MN, ch. 42, ABC
KRWF Redwood Falls, MN, ch. 43, ABC
KSTC-TV Minneapolis, ch. 45, IND

DMA Counties	State	TV Households	DMA Counties	State	TV Households
Aitkin	MN	7,190	Nicollet-North	MN	5,820
Anoka	MN	119,360	Pine	MN	10,860
Beltrami	MN	15,720	Pope	MN	4,560
Benton	MN	15,340	Ramsey	MN	192,780
Big Stone	MN	2,170	Redwood	MN	6,360
Carver	MN	30,660	Renville	MN	6,540
Cass	MN	11,880	Rice	MN	20,550
Chippewa	MN	5,170	Scott	MN	43,570
Chisago	MN	17,990	Sherburne	MN	29,150
Cottonwood	MN	4,710	Sibley	MN	5,650
Crow Wing	MN	24,530	Stearns	MN	52,310
Dakota	MN	145,810	Steele	MN	13,860
Douglas	MN	14,630	Stevens	MN	3,640
Faribault	MN	6,230	Swift	MN	3,900
Goodhue	MN	17,640	Todd	MN	9,450
Grant	MN	2,480	Traverse	MN	1,570
Hennepin	MN	452,260	Wabasha	MN	8,610
Hubbard	MN	7,480	Wadena	MN	5,330
Isanti	MN	14,110	Waseca	MN	6,940
Jackson	MN	4,580	Washington	MN	81,260
Kanabec	MN	6,230	Wright	MN	40,910
Kandiyohi	MN	16,060	Yellow Medicine	MN	4,060
Lac qui Parle	MN	3,080	Barron	WI	18,440
Le Sueur	MN	10,620	Burnett	WI	7,060
Lyon	MN	9,350	Pierce	WI	14,240
McLeod	MN	14,230	Polk	WI	17,880
Meeker	MN	8,730	St. Croix	WI	30,340
Mille Lacs	MN	10,340	Washburn	WI	6,720
Morrison	MN	12,280			

Minot-Bismarck-Dickinson, ND (158)

DMA TV Households: 135,550
% of U.S. TV Households: .122

KXMA-TV Dickinson, ND, ch. 2, satellite to KXMC-TV
***KBME-TV** Bismarck, ND, ch. 3, ETV
***KWSE** Williston, ND, ch. 4, ETV
KFYR-TV Bismarck, ND, ch. 5, NBC
***KSRE** Minot, ND, ch. 6, ETV
KQCD-TV Dickinson, ND, ch. 7, NBC
KUMV-TV Williston, ND, ch. 8, satellite to KFYR-TV
***KDSE** Dickinson, ND, ch. 9, ETV
KMOT Minot, ND, ch. 10, satellite to KFYR-TV
KXMD-TV Williston, ND, ch. 11, satellite to KXMC-TV
***KQSD-TV** Lowry, SD, ch. 11, ETV
KXMB-TV Bismarck, ND, ch. 12, satellite to KXMC-TV
***KPSD-TV** Eagle Butte, SD, ch. 13, ETV
KXMC-TV Minot, ND, ch. 13, CBS
KMCY Minot, ND, ch. 14, ABC
KBMY Bismarck, ND, ch. 17, ABC
KXND Minot, ND, ch. 24, Fox
KNDX Bismarck, ND, ch. 26, Fox

DMA Counties	State	TV Households	DMA Counties	State	TV Households
Daniels	MT	780	Logan	ND	890
Fallon	MT	1,020	McHenry	ND	2,380
McCone	MT	790	McIntosh	ND	1,280
Richland	MT	3,750	McKenzie	ND	2,070
Roosevelt	MT	3,510	McLean	ND	3,570
Sheridan	MT	1,480	Mercer	ND	3,260
Wibaux	MT	400	Morton	ND	10,300
Adams	ND	1,080	Mountrail	ND	2,570
Billings	ND	300	Oliver	ND	700
Bottineau	ND	2,890	Pierce	ND	1,770
Bowman	ND	1,280	Renville	ND	1,000
Burke	ND	890	Rolette	ND	4,590
Burleigh	ND	30,550	Sheridan	ND	590
Campbell	ND	590	Sioux	ND	1,170
Divide	ND	970	Slope	ND	290
Dunn	ND	1,280	Stark	ND	8,830
Emmons	ND	1,590	Ward	ND	21,910
Golden Valley	ND	690	Wells	ND	1,970
Grant	ND	1,090	Williams	ND	8,140
Hettinger	ND	990	Corson	SD	1,260
Kidder	ND	1,090			

Minot-Bismarck-Dickinson (Williston), ND

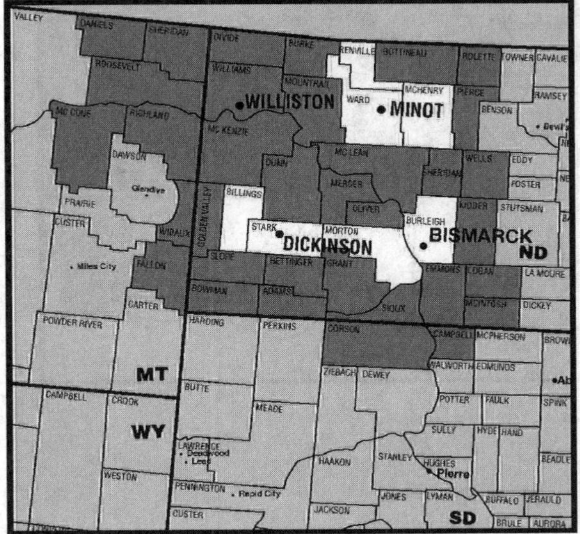

Maps courtesy of Nielsen Media Research

Missoula, MT (168)

DMA TV Households: 106,250
% of U.S. TV Households: .095

KPAX-TV Missoula, MT, ch. 8, satellite to KXLF-TV
KCFW-TV Kalispell, MT, ch. 9, NBC
***KUFM-TV** Missoula, MT, ch. 11, ETV
KECI-TV Missoula, MT, ch. 13, NBC (ABC)
KMMF Missoula, MT, ch. 17, Fox, MyNetworkTV
KTMF Missoula, MT, ch. 23, IND

DMA Counties	State	TV Households
Flathead	MT	32,490
Granite	MT	1,250
Lake	MT	10,980
Mineral	MT	1,630
Missoula	MT	39,430
Ravalli	MT	15,700
Sanders	MT	4,770

Missoula, MT

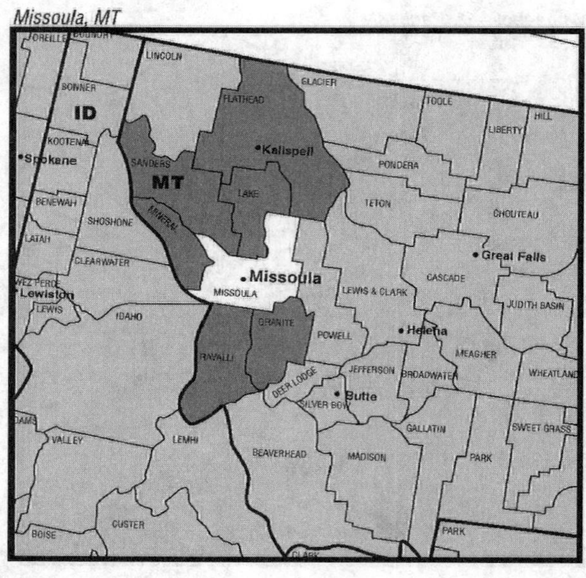

Mobile, AL-Pensacola, FL (Ft. Walton Beach, FL)

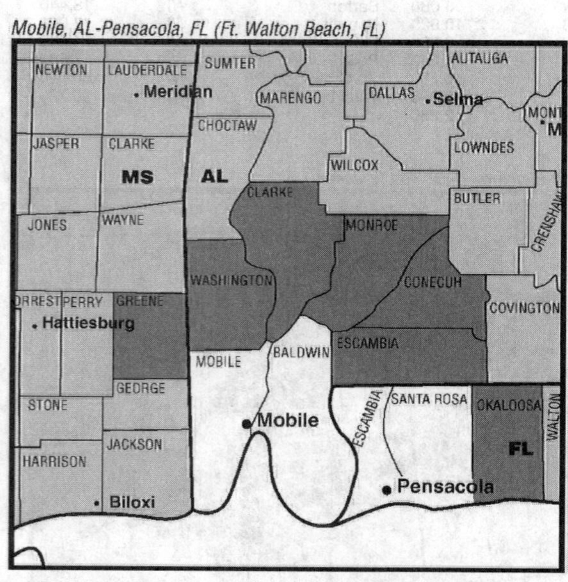

Mobile, AL-Pensacola (Ft. Walton Beach), FL (59)

DMA TV Households: 524,200
% of U.S. TV Households: .471

WEAR-TV Pensacola, FL, ch. 3, ABC
WKRG-TV Mobile, AL, ch. 5, CBS
WALA-TV Mobile, AL, ch. 10, Fox
WPMI-TV Mobile, AL, ch. 15, NBC
WMPV-TV Mobile, AL, ch. 21, IND
***WSRE** Pensacola, FL, ch. 23, ETV
WHBR Pensacola, FL, ch. 33, IND
WFGX Fort Walton Beach, FL, ch. 35, MyNetworkTV
***WEIQ** Mobile, AL, ch. 42, ETV
WJTC Pensacola, FL, ch. 44, IND
WFBD Destin, FL, ch. 48, IND
WPAN Fort Walton Beach, FL, ch. 53, IND
WBPG Gulf Shores, AL, ch. 55, CW
WAWD Fort Walton Beach, FL, ch. 58, IND

DMA Counties	State	TV Households
Baldwin	AL	68,610
Clarke	AL	10,850
Conecuh	AL	5,480
Escambia	AL	14,200
Mobile	AL	161,190
Monroe	AL	9,370
Washington	AL	6,940
Escambia	FL	114,300
Okaloosa	FL	74,750
Santa Rosa	FL	54,180
Greene	MS	4,330

Maps courtesy of Nielsen Media Research

Monroe, LA-El Dorado, AR (135)

DMA TV Households: 178,200
% of U.S. TV Households: .160

KNOE-TV Monroe, LA, ch. 8, CBS, CW
KTVE El Dorado, AR, ch. 10, NBC
KAQY Columbia, LA, ch. 11, ABC
***KETZ** El Dorado, AR, ch. 12, ETV
***KLTM-TV** Monroe, LA, ch. 13, ETV
KARD West Monroe, LA, ch. 14, Fox
KMCT-TV West Monroe, LA, ch. 39, IND
KEJB El Dorado, AR, ch. 43, MyNetworkTV

DMA Counties	State	TV Households	DMA Counties	State	TV Households
Ashley	AR	9,010	Madison	LA	4,020
Union	AR	17,330	Morehouse	LA	11,000
Caldwell	LA	3,950	Ouachita	LA	57,000
Catahoula	LA	3,940	Richland	LA	7,350
Concordia	LA	7,250	Tensas	LA	2,240
East Carroll	LA	2,540	Union	LA	9,150
Franklin	LA	7,210	West Carroll	LA	4,250
Jackson	LA	5,990	Winn	LA	5,400
Lincoln	LA	15,380			

Monroe, LA-El Dorado, AR

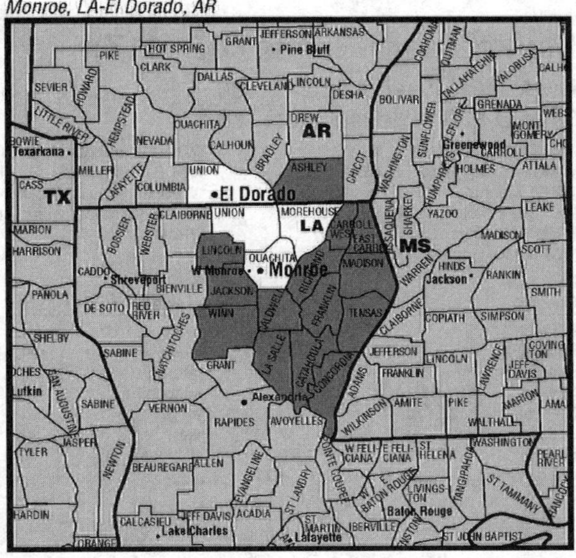

Monterey-Salinas, CA

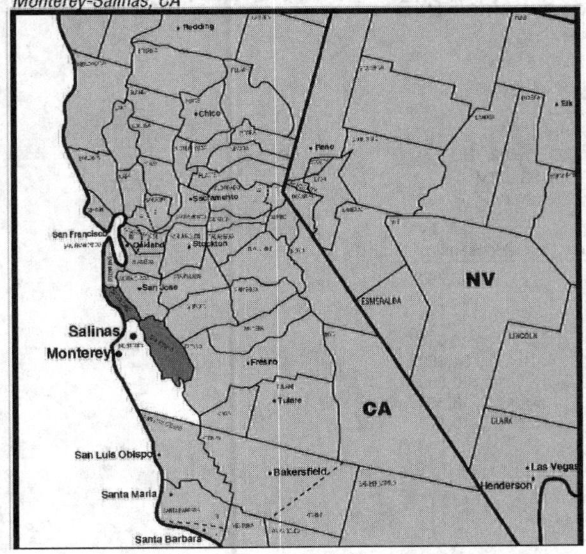

Monterey-Salinas, CA (124)

DMA TV Households: 218,390
% of U.S. TV Households: .196

KSBW Salinas, CA, ch. 8, NBC
***KCAH** Watsonville, CA, ch. 25, ETV
KCBA Salinas, CA, ch. 35, Fox
KION-TV Monterey, CA, ch. 46, CBS
KSMS-TV Monterey, CA, ch. 67, Univision

DMA Counties	State	TV Households
Monterey	CA	118,830
San Benito	CA	16,320
Santa Cruz	CA	83,240

Maps courtesy of Nielsen Media Research

Montgomery-Selma, AL (117)

DMA TV Households: 241,130

% of U.S. TV Households: .217

***WDIQ** Dozier, AL, ch. 2, ETV
WAKA Selma, AL, ch. 8, CBS
WSFA Montgomery, AL, ch. 12, NBC
WCOV-TV Montgomery, AL, ch. 20, Fox
WBMM Tuskegee, AL, ch. 22, CW
***WAIQ** Montgomery, AL, ch. 26, ETV
WBIH Selma, AL, ch. 29, IND
WNCF Montgomery, AL, ch. 32, ABC
***WIIQ** Demopolis, AL, ch. 41, ETV
WMCF-TV Montgomery, AL, ch. 45, IND
WRJM-TV Troy, AL, ch. 67, MyNetworkTV

DMA Counties	State	TV Households	DMA Counties	State	TV Households
Autauga	AL	18,750	Lowndes	AL	4,740
Bullock	AL	3,560	Macon	AL	8,360
Butler	AL	8,330	Marengo	AL	8,640
Covington	AL	15,540	Montgomery	AL	86,470
Crenshaw	AL	5,760	Perry	AL	4,020
Dallas	AL	17,200	Pike	AL	12,170
Elmore	AL	26,560	Tallapoosa	AL	16,320
Wilcox	AL	4,710			

Montgomery-Selma, AL

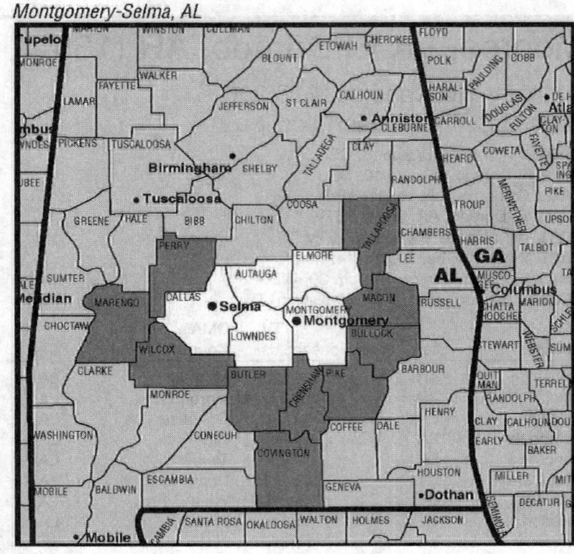

Myrtle Beach-Florence, SC

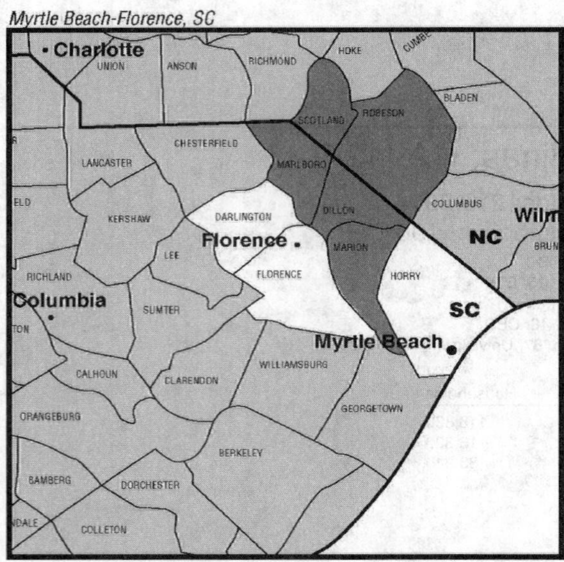

Myrtle Beach-Florence, SC (105)

DMA TV Households: 272,340

% of U.S. TV Households: .245

WBTW Florence, SC, ch. 13, CBS
WPDE-TV Florence, SC, ch. 15, ABC
WWMB Florence, SC, ch. 21, CW
***WHMC** Conway, SC, ch. 23, ETV
***WUNU** Lumberton, NC, ch. 31, ETV
WMBF-TV Myrtle Beach, SC, ch. 32, IND
***WJMP-TV** Florence, SC, ch. 33, ETV
WFXB Myrtle Beach, SC, ch. 43, Fox

DMA Counties	State	TV Households
Robeson	NC	45,780
Scotland	NC	13,820
Darlington	SC	26,230
Dillon	SC	11,550
Florence	SC	51,180
Horry	SC	100,260
Marion	SC	13,350
Marlboro	SC	10,170

Maps courtesy of Nielsen Media Research

Nashville (30)

DMA TV Households: 944,100
% of U.S. TV Households: .848

WKRN-TV Nashville, ch. 2, ABC
WSMV Nashville, ch. 4, NBC
WTVF Nashville, ch. 5, CBS
***WNPT** Nashville, ch. 8, ETV
WZTV Nashville, ch. 17, Fox
***WCTE** Cookeville, TN, ch. 22, ETV
WNPX Cookeville, TN, ch. 28, ION Television
WUXP-TV Nashville, ch. 30, MyNetworkTV
WHTN Murfreesboro, TN, ch. 39, IND
WPGD-TV Hendersonville, TN, ch. 50, IND
WNAB Nashville, ch. 58, CW
WJFB Lebanon, TN, ch. 66, IND

DMA Counties	State	TV Households
Allen	KY	7,320
Christian	KY	23,930
Clinton	KY	4,010
Cumberland	KY	2,490
Logan	KY	10,810
Monroe	KY	4,630
Simpson	KY	6,750
Todd	KY	4,530
Trigg	KY	5,610
Bedford	TN	15,910
Benton	TN	6,730
Cannon	TN	5,150
Cheatham	TN	14,160
Clay	TN	3,360
Coffee	TN	20,330
Davidson	TN	239,900
Decatur	TN	4,710
De Kalb	TN	7,400
Dickson	TN	17,700
Franklin	TN	15,840
Giles	TN	11,610
Henry	TN	13,140
Hickman	TN	8,800
Houston	TN	3,160
Humphreys	TN	7,510
Jackson	TN	4,500
Lawrence	TN	16,060
Lewis	TN	4,320
Macon	TN	8,380
Marshall	TN	10,920
Maury	TN	29,630
Montgomery	TN	54,390
Moore	TN	2,360
Overton	TN	8,330
Perry	TN	2,980
Pickett	TN	2,020
Putnam	TN	26,790
Robertson	TN	22,670
Rutherford	TN	83,720
Smith	TN	7,290
Stewart	TN	5,240
Sumner	TN	56,600
Trousdale	TN	2,940
Van Buren	TN	2,160
Warren	TN	15,590
Wayne	TN	5,900
White	TN	9,660
Williamson	TN	57,580
Wilson	TN	38,580

Nashville, TN

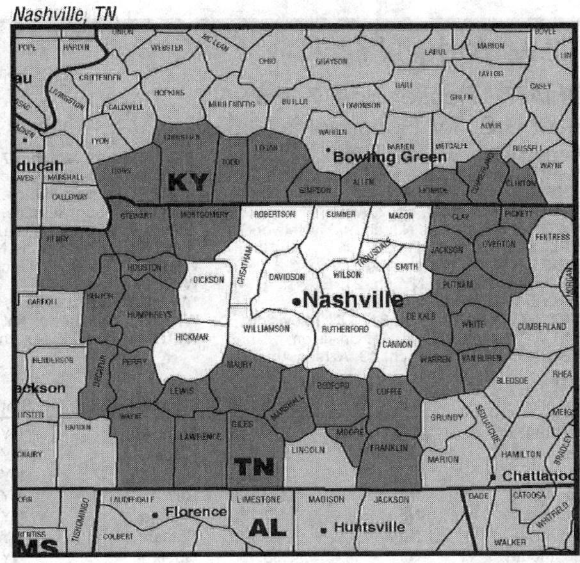

New Orleans, LA

New Orleans (54)

DMA TV Households: 566,960
% of U.S. TV Households: .509

WWL-TV New Orleans, ch. 4, CBS
WDSU-TV New Orleans, ch. 6, NBC
WVUE New Orleans, ch. 8, Fox
***WYES-TV** New Orleans, ch. 12, ETV
WHNO New Orleans, ch. 20, IND
WGNO New Orleans, ch. 26, ABC
***WLAE-TV** New Orleans, ch. 32, ETV
WNOL-TV New Orleans, ch. 38, CW
WHMM-DT Hammond, LA, ch. 42, Telemundo
WPXL New Orleans, ch. 49, IND
WUPL Slidell, LA, ch. 54, MyNetworkTV

DMA Counties	State	TV Households
Jefferson	LA	159,160
Lafourche	LA	35,250
Orleans	LA	91,250
Plaquemines	LA	6,040
St. Bernard	LA	8,630
St. Charles	LA	18,070
St. James	LA	7,400
St. John the Baptist	LA	16,680
St. Tammany	LA	87,000
Tangipahoa	LA	41,720
Terrebonne	LA	39,980
Washington	LA	18,160
Hancock	MS	16,310
Pearl River	MS	21,310

Maps courtesy of Nielsen Media Research

New York, NY (1)

DMA TV Households: 7,366,950
% of U.S. TV Households: 6.616

WCBS-TV New York, ch. 2, CBS
WNBC New York,ch. 4, NBC
WNYW New York, ch. 5, Fox
WABC-TV New York, ch. 7, ABC
WWOR-TV Secaucus,NJ, ch. 9, MyNetworkTV
WPIX New York, ch. 11, CW
***WNET** Newark, NJ, ch. 13, ETV
***WLIW** Garden City, NY, ch. 21, ETV
***WNYE-TV** New York, ch. 25, ETV
WPXN-TV New York, ch. 31, ION Television
WXTV Paterson, NJ, ch. 41, Univision
WSAH Bridgeport, CT, ch. 43, Azteca America
WNJU Linden, NJ, ch. 47, Telemundo
***WEDW** Bridgeport, CT, ch. 49, ETV
***WNJN** Montclair, NJ, ch. 50, ETV
WTBY-TV Poughkeepsie, NY, ch. 54, IND
WLNY Riverhead, NY, ch. 57, IND
***WNJB** New Brunswick, NJ, ch. 58, ETV
WMBC-TV Newton, NJ, ch. 63, IND
WRNN-TV Kingston, NY, ch. 63, IND
***WFME-TV** West Milford, NJ, ch. 66, ETV
WFTY-TV Smithtown, NY, ch. 67, IND
WFUT-TV Newark, NJ, ch. 68, TeleFutura

DMA Counties	State	TV Households	DMA Counties	State	TV Households
Fairfield	CT	328,640	Dutchess	NY	104,520
Bergen	NJ	336,310	Kings	NY	829,000
Essex	NJ	280,540	Nassau	NY	444,110
Hudson	NJ	222,160	New York	NY	721,360
Hunterdon	NJ	47,770	Orange	NY	127,680
Middlesex	NJ	280,170	Putnam	NY	34,910
Monmouth	NJ	233,400	Queens	NY	748,800
Morris	NJ	180,370	Richmond	NY	166,700
Ocean	NJ	221,090	Rockland	NY	92,330
Passaic	NJ	163,860	Suffolk	NY	492,280
Somerset	NJ	117,980	Sullivan	NY	28,280
Sussex	NJ	54,900	Ulster	NY	67,870
Union	NJ	186,500	Westchester	NY	338,700
Warren	NJ	42,460	Pike	PA	21,950
Bronx	NY	452,310			

New York, NY

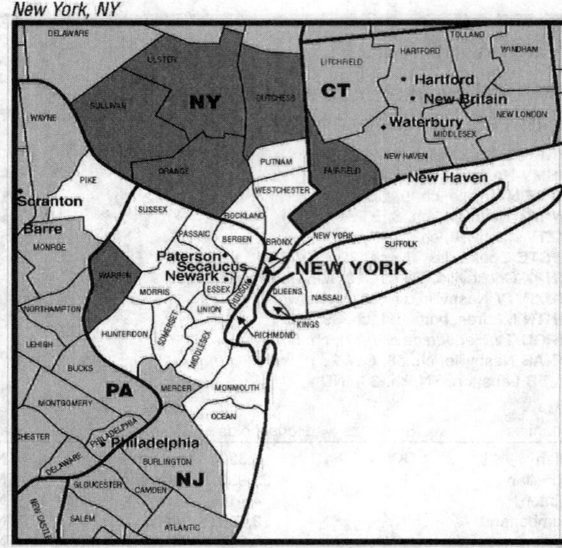

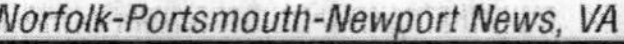

Norfolk-Portsmouth-Newport News, VA

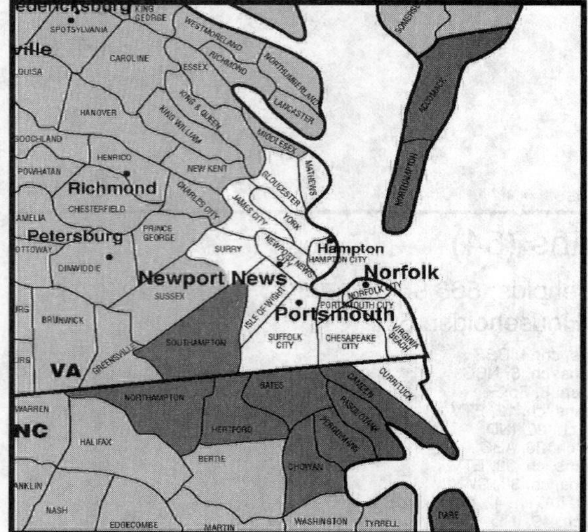

Norfolk-Portsmouth-Newport News, VA (42)

DMA TV Households: 712,790
% of U.S. TV Households: .640

WTKR Norfolk, VA, ch. 3, CBS
WSKY-TV Manteo, NC, ch. 4, IND
WAVY-TV Portsmouth, VA, ch. 10, NBC
WVEC-TV Hampton, VA, ch. 13, ABC
***WHRO-TV** Hampton-Norfolk, VA, ch. 15, ETV
WHRE Virginia Beach, VA, ch. 21, IND
WGNT Portsmouth, VA, ch. 27, CW
WTVZ-TV Norfolk, VA, ch. 33, MyNetworkTV
WVBT Virginia Beach, VA, ch. 43, Fox
WPXV Norfolk, VA, ch. 49, ION Television

DMA Counties	State	TV Households	DMA Counties	State	TV Households
Camden	NC	3,770	Isle of Wight	VA	12,970
Chowan	NC	5,750	James City	VA	28,150
Currituck	NC	9,420	Mathews	VA	3,890
Dare	NC	14,570	Newport News City	VA	71,310
Gates	NC	4,330	Norfolk City	VA	85,110
Hertford	NC	8,800	Northampton	VA	5,510
Northampton	NC	8,480	Portsmouth City	VA	37,970
Pasquotank	NC	14,740	Southampton	VA	10,130
Perquimans	NC	5,010	Suffolk City	VA	29,950
Accomack	VA	15,420	Surry	VA	2,670
Chesapeake City	VA	76,450	Virginia Beach	VA	161,480
Gloucester	VA	14,130	York	VA	27,050
Hampton City	VA	54,230			

Maps courtesy of Nielsen Media Research

North Platte, NE (209)

DMA TV Households: 15,480
% of U.S. TV Households: .014

KNOP-TV North Platte, NE, ch. 2, NBC
***KPNE-TV** North Platte, NE, ch. 9, ETV

DMA Counties	State	TV Households
Lincoln	NE	14,700
Logan	NE	300
McPherson	NE	200
Thomas	NE	280

North Platte, NE

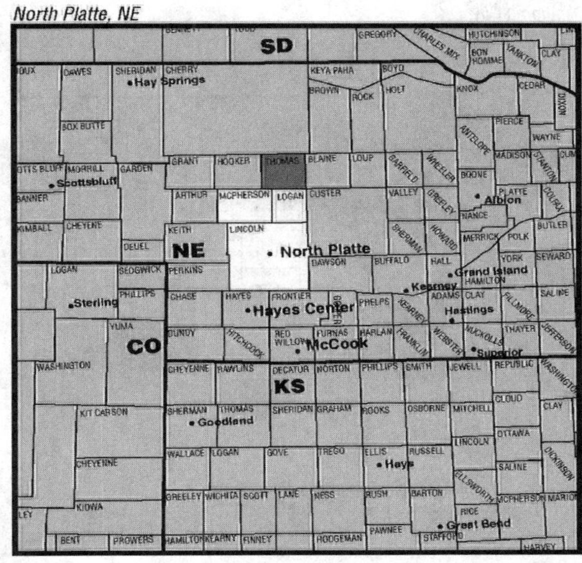

Odessa-Midland, TX

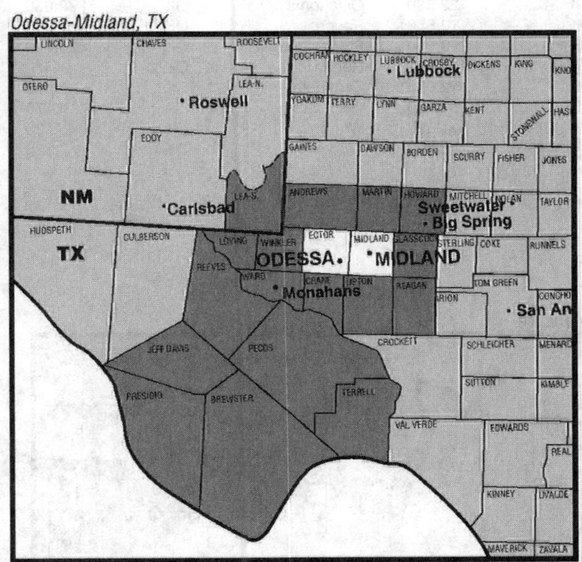

Odessa-Midland, TX (159)

DMA TV Households: 135,270
% of U.S. TV Households: .122

KMID Midland, TX, ch. 2, ABC
KWAB-TV Big Spring, TX, ch. 4, satellite to KWES-TV
KOSA-TV Odessa, TX, ch. 7, CBS, MyNetworkTV
KWES-TV Odessa, TX, ch. 9, NBC
KUPB Midland, TX, ch. 18, Univision
KPEJ Odessa, TX, ch. 24, IND
KWWT Odessa, TX, ch. 30, CW
***KPBT-TV** Odessa, TX, ch. 36, ETV
KMLM Odessa, TX, ch. 42, IND

DMA Counties	State	TV Households	DMA Counties	State	TV Households
Lea South	NM	1,960	Midland	TX	44,940
Andrews	TX	4,680	Pecos	TX	4,660
Brewster	TX	3,080	Presidio	TX	2,670
Crane	TX	1,290	Reagan	TX	990
Ector	TX	45,790	Reeves	TX	3,430
Glasscock	TX	400	Terrell	TX	370
Howard	TX	11,000	Upton	TX	1,180
Jeff Davis	TX	830	Ward	TX	3,840
Loving	TX	100	Winkler	TX	2,470
Martin	TX	1,590			

Maps courtesy of Nielsen Media Research

Oklahoma City, OK

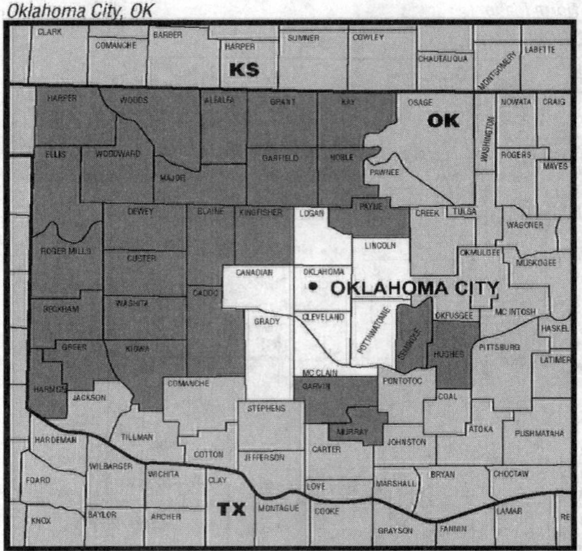

Oklahoma City (45)

DMA TV Households: 662,380
% of U.S. TV Households: .595

KFOR-TV Oklahoma City, ch. 4, NBC
KOCO-TV Oklahoma City, ch. 5, ABC
KWTV Oklahoma City, ch. 9, CBS
***KWET** Cheyenne, OK, ch. 12, ETV
***KETA** Oklahoma City, ch. 13, ETV
KTBO-TV Oklahoma City, ch. 14, IND
KOKH-TV Oklahoma City, ch. 25, Fox
KTUZ-TV Shawnee, OK, ch. 30, Telemundo
KEYU Elk City, OK, ch. 31, Univision
KOCB Oklahoma City, ch. 34, CW
KUOK Woodward, OK, ch. 35, Univision
KAUT-TV Oklahoma City, ch. 43, MyNetworkTV
KOCM Norman, OK, ch. 46, IND
KSBI Oklahoma City, ch. 52, IND
KOPX Oklahoma City, ch. 62, ION Television

DMA Counties	State	TV Households	DMA Counties	State	TV Households
Alfalfa	OK	1,940	Kay	OK	18,090
Beckham	OK	7,480	Kingfisher	OK	5,370
Blaine	OK	4,230	Kiowa	OK	4,060
Caddo	OK	10,900	Lincoln	OK	12,270
Canadian	OK	37,290	Logan	OK	13,410
Cleveland	OK	88,340	Major	OK	2,950
Custer	OK	9,760	McClain	OK	11,420
Dewey	OK	1,880	Murray	OK	5,160
Ellis	OK	1,710	Noble	OK	4,440
Garfield	OK	22,580	Oklahoma	OK	279,830
Garvin	OK	10,650	Payne	OK	26,210
Grady	OK	19,190	Pottawatomie	OK	25,730
Grant	OK	1,860	Roger Mills	OK	1,370
Greer	OK	2,080	Seminole	OK	9,480
Harmon	OK	1,080	Washita	OK	4,450
Harper	OK	1,330	Woods	OK	3,430
Hughes	OK	5,090	Woodward	OK	7,320

Omaha (75)

DMA TV Households: 403,560
% of U.S. TV Households: .362

KMTV Omaha, ch. 3, CBS
WOWT Omaha, ch. 6, NBC
KETV Omaha, ch. 7, ABC
KXVO Omaha, ch. 15, CW
***KYNE-TV** Omaha, ch. 26, ETV
***KBIN** Council Bluffs, IA, ch. 32, ETV
***KHIN** Red Oak, IA, ch. 36, ETV
KPTM Omaha, ch. 42, Fox, MyNetworkTV

DMA Counties	State	TV Households	DMA Counties	State	TV Households
Cass	IA	5,950	Colfax	NE	3,550
Crawford	IA	6,370	Cuming	NE	3,670
Fremont	IA	3,070	Dodge	NE	14,520
Harrison	IA	6,260	Douglas	NE	194,220
Mills	IA	5,760	Johnson	NE	1,880
Montgomery	IA	4,660	Nemaha	NE	2,770
Page	IA	6,250	Otoe	NE	6,060
Pottawattamie	IA	34,920	Platte	NE	11,970
Shelby	IA	4,970	Richardson	NE	3,530
Atchison	MO	2,620	Sarpy	NE	52,300
Burt	NE	2,980	Saunders	NE	7,850
Cass	NE	9,970	Washington	NE	7,460

Omaha, NE

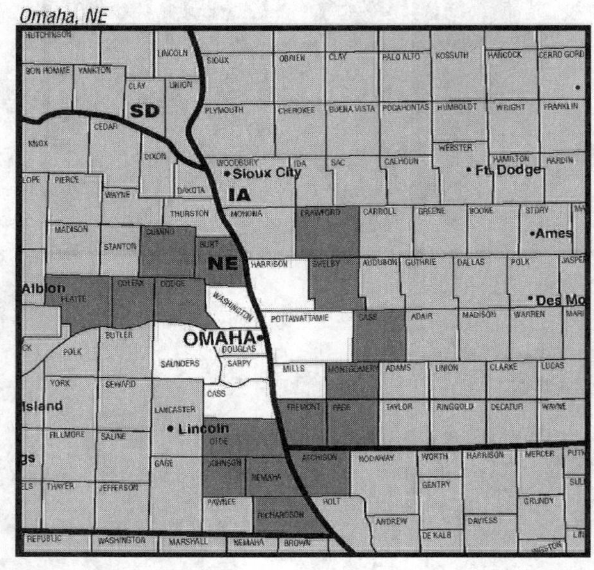

Maps courtesy of Nielsen Media Research

Orlando-Daytona Beach-Melbourne, FL (19)

DMA TV Households: 1,395,830
% of U.S. TV Households: 1.254

WESH Daytona Beach, FL, ch. 2, NBC
WKMG-TV Orlando, FL, ch. 6, CBS
WFTV Orlando, FL, ch. 9, ABC
***WCEU** New Smyrna Beach, FL, ch. 15, ETV
WKCF Clermont, FL, ch. 18, CW
***WMFE-TV** Orlando, FL, ch. 24, ETV
WVEN-TV Daytona Beach, FL, ch. 26, Univision
WRDQ Orlando, FL, ch. 27, IND
WOFL Orlando, FL, ch. 35, Fox
WACX-DT Leesburg, FL, ch. 40, IND
WOTF-TV Melbourne, FL, ch. 43, TeleFutura
WLCB-TV Leesburg, FL, ch. 45, IND
WTGL-TV Cocoa, FL, ch. 52, IND
WOPX Melbourne, FL, ch. 56, ION Television
WRBW Orlando, FL, ch. 65, MyNetworkTV
***WBCC** Cocoa, FL, ch. 68, ETV

DMA Counties	State	TV Households
Brevard	FL	228,330
Flagler	FL	36,770
Lake	FL	120,410
Marion	FL	131,250
Orange	FL	393,800
Osceola	FL	86,650
Seminole	FL	159,690
Sumter	FL	29,270
Volusia	FL	209,660

Orlando-Daytona Beach-Melbourne, FL

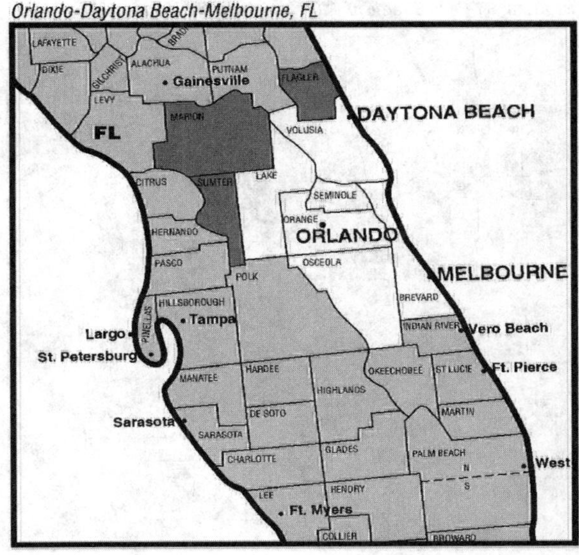

Ottumwa, IA-Kirksville, MO

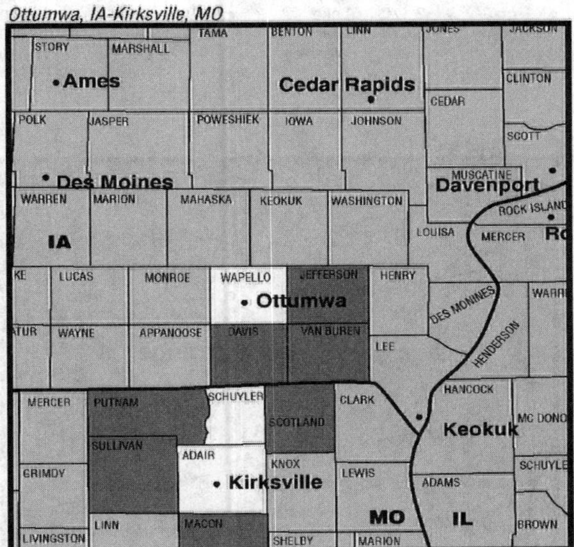

Ottumwa, Iowa-Kirksville, MO (199)

DMA TV Households: 51,470
% of U.S. TV Households: .046

KTVO Kirksville, MO, ch. 3, ABC
KYOU-TV Ottumwa, IA, ch. 15, IND

DMA Counties	State	TV Households
Davis	IA	3,140
Jefferson	IA	6,490
Van Buren	IA	3,170
Wapello	IA	14,770
Adair	MO	9,240
Macon	MO	6,390
Putnam	MO	2,160
Schuyler	MO	1,780
Scotland	MO	1,740
Sullivan	MO	2,590

Maps courtesy of Nielsen Media Research

Paducah, KY-Cape Girardeau, MO-Harrisburg-Mt. Vernon, IL

Paducah, KY-Cape Girardeau, MO-Harrisburg, IL (80)

DMA TV Households: 384,510
% of U.S. TV Households: .345

WSIL-TV Harrisburg, IL, ch. 3, ABC
WPSD-TV Paducah, KY, ch. 6, NBC
***WSIU-TV** Carbondale, IL, ch. 8, ETV
KFVS-TV Cape Girardeau, MO, ch. 12, CBS
WPXS Mount Vernon, IL, ch. 13, IND
KPOB-TV Poplar Bluff, MO, ch. 15, satellite to WSIL-TV
***WKMU** Murray, KY, ch. 21, ETV
KBSI Cape Girardeau, MO, ch. 23, Fox
WTCT Marion, IL, ch. 27, IND
***WKPD** Paducah, KY, ch. 29, ETV
WDKA Paducah, KY, ch. 49, MyNetworkTV

DMA Counties	State	TV Households	DMA Counties	State	TV Households
Alexander	IL	3,460	Graves	KY	14,960
Franklin	IL	16,770	Hickman	KY	2,090
Gallatin	IL	2,540	Livingston	KY	3,960
Hamilton	IL	3,270	Lyon	KY	2,950
Hardin	IL	1,960	Marshall	KY	13,020
Jackson	IL	23,800	McCracken	KY	27,600
Jefferson	IL	15,640	Bollinger	MO	4,630
Johnson	IL	4,420	Butler	MO	17,000
Massac	IL	6,430	Cape Girardeau	MO	28,39
Perry	IL	8,400	Carter	MO	2,210
Pope	IL	1,680	Dunklin	MO	13,110
Pulaski	IL	2,530	Madison	MO	4,890
Saline	IL	10,590	Mississippi	MO	5,360
Union	IL	7,220	New Madrid	MO	7,270
Williamson	IL	26,640	Perry	MO	7,240
Ballard	KY	3,340	Scott	MO	16,140
Caldwell	KY	5,350	Stoddard	MO	12,140
Calloway	KY	14,650	Wayne	MO	5,460
Carlisle	KY	2,170	Lake	TN	2,070
Crittenden	KY	3,670	Obion	TN	13,170
Fulton	KY	2,960	Weakley	TN	13,360

Palm Springs, CA (149)

DMA TV Households: 149,880
% of U.S. TV Households: .135

KMIR-TV Palm Springs, CA, ch. 36, NBC
KESQ-TV Palm Springs, CA, ch. 42, ABC, CW

DMA Counties	State	TV Households
Riverside Central	CA	149,880

Palm Springs, CA

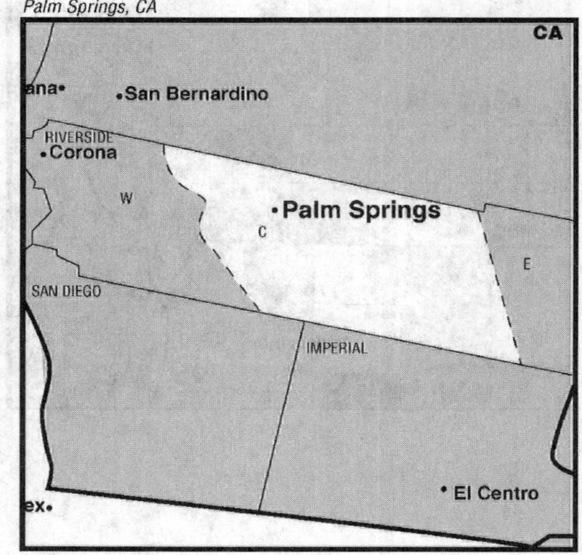

Maps courtesy of Nielsen Media Research

Panama City, FL (156)

DMA TV Households: 140,790
% of U.S. TV Households: .126

WJHG-TV Panama City, FL, ch. 7, NBC, CW, MyNetworkTV
WMBB Panama City, FL, ch. 13, ABC
WPGX Panama City, FL, ch. 28, IND
WPCT Panama City Beach, FL, ch. 46, IND
WBIF Marianna, FL, ch. 51, IND
***WFSG** Panama City, FL, ch. 56, ETV

DMA Counties	State	TV Households
Bay	FL	67,370
Calhoun	FL	4,700
Franklin	FL	4,350
Gulf	FL	5,620
Holmes	FL	7,200
Jackson	FL	17,590
Liberty	FL	2,400
Walton	FL	22,890
Washington	FL	8,670

Panama City, FL

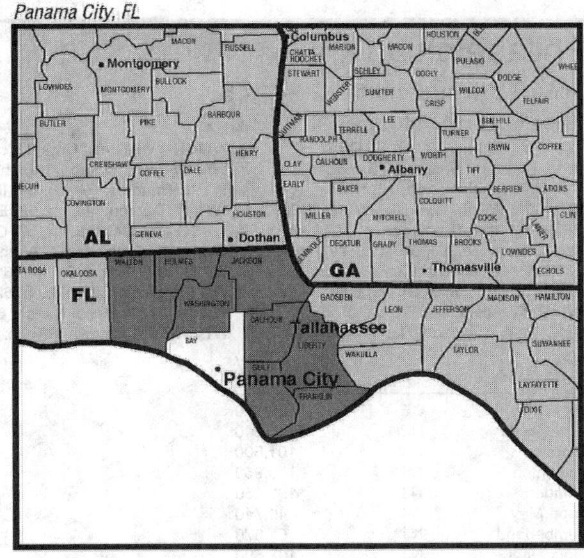

Parkersburg, WV

Parkersburg, WV (189)

DMA TV Households: 63,850
% of U.S. TV Households: .057

WTAP-TV Parkersburg, WV, ch. 15, NBC, Fox, MyNetworkTV

DMA Counties	State	TV Households
Washington	OH	24,750
Pleasants	WV	2,890
Wood	WV	36,210

Peoria-Bloomington, IL (116)

DMA TV Households: 243,280
% of U.S. TV Households: .219

WHOI Peoria, IL, ch. 19, ABC, CW
WEEK-TV Peoria, IL, ch. 25, NBC
WMBD-TV Peoria, IL, ch. 31, CBS
WYZZ-TV Bloomington, IL, ch. 43, Fox
***WTVP** Peoria, IL, ch. 47, ETV
WAOE Peoria, IL, ch. 59, MyNetworkTV

DMA Counties	State	TV Households
Fulton	IL	14,710
Livingston	IL	13,860
Marshall	IL	5,310
Mason	IL	6,220
McLean	IL	60,710
Peoria	IL	72,230
Putnam	IL	2,480
Stark	IL	2,470
Tazewell	IL	51,640
Woodford	IL	13,650

Peoria-Bloomington, IL

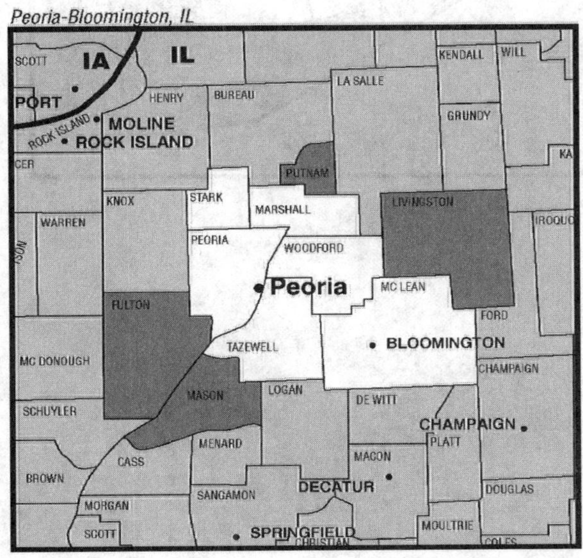

Maps courtesy of Nielsen Media Research

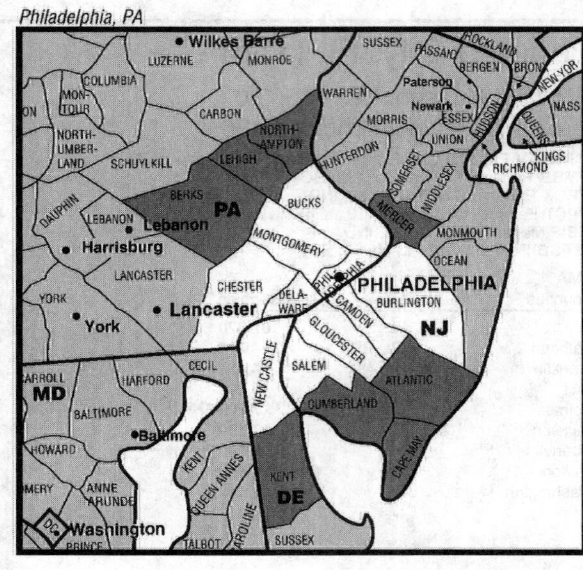

Philadelphia (4)

DMA TV Households: 2,941,450

% of U.S. TV Households: 2.642

KYW-TV Philadelphia, ch. 3, CBS
WPVI-TV Philadelphia, ch. 6, ABC
WCAU Philadelphia, ch. 10, NBC
WHYY-TV Wilmington, DE, ch. 12, ETV
WPHL-TV Philadelphia, ch. 17, MyNetworkTV
***WNJS** Camden, NJ, ch. 23, ETV
WTXF-TV Philadelphia, ch. 29, Fox
***WYBE** Philadelphia, ch. 35, ETV
***WLVT-TV** Allentown, PA, ch. 39, ETV
WMGM-TV Wildwood, NJ, ch. 40, NBC
WMCN-TV Atlantic City, NJ, ch. 44, IND
WGTW-TV Burlington, NJ, ch. 48, IND
WTVE Reading, PA, ch. 51, IND
***WNJT** Trenton, NJ, ch. 52, ETV
WPSG Philadelphia, ch. 57, CW
WBPH-TV Bethlehem, PA, ch. 60, IND
WPPX Wilmington, DE, ch. 61, ION Television
WWSI Atlantic City, NJ, ch. 62, Telemundo
WUVP-TV Vineland, NJ, ch. 65, Univision
WFMZ-TV Allentown, PA, ch. 69, IND

DMA Counties	State	TV Households
Kent	DE	56,370
New Castle	DE	200, 040
Atlantic	NJ	101,500
Burlington	NJ	169,960
Camden	NJ	191,650
Cape May	NJ	40,740
Cumberland	NJ	51,870
Gloucester	NJ	102,590
Mercer	NJ	131,340
Salem	NJ	25,650
Berks	PA	149,940
Bucks	PA	232,290
Chester	PA	177,350
Delaware	PA	208,980
Lehigh	PA	131,140
Montgomery	PA	297,870
Northampton	PA	112,270
Philadelphia	PA	559,900

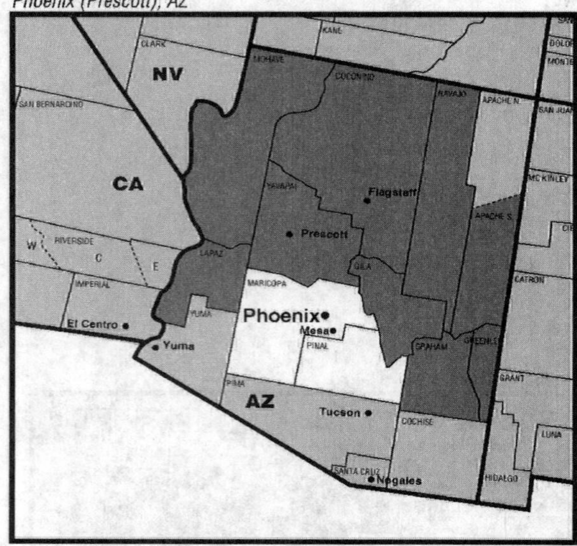

Phoenix (Prescott), AZ (13)

DMA TV Households: 1,725,000

% of U.S. TV Households: 1.549

KNAZ-TV Flagstaff, AZ, ch. 2,NBC
KTVK Phoenix, ch. 3, IND
KTFL Flagstaff, AZ, ch. 4, IND
KPHO-TV Phoenix, ch. 5, CBS
KMOH-TV Kingman, AZ, ch. 6 satellite to KBEH
KAZT-TV Prescott, AZ, ch. 7, IND
***KAET** Phoenix, ch. 8, ETV
KCFG Flagstaff, AZ, ch. 9, IND
KSAZ-TV Phoenix, ch. 10, Fox
***KDTP** Holbrook, AZ, ch. 11, IND
KPNX Mesa, AZ, ch. 12, NBC
KFPH-TV Flagstaff, AZ, ch. 13, TeleFutura
KNXV-TV Phoenix, ch. 15, ABC
KPAZ-TV Phoenix, ch. 21, IND
KTVW-TV Phoenix, ch. 33, Univision
KTAZ Phoenix, ch. 39, Telemundo
KUTP Phoenix, ch. 45, MyNetworkTV
KPPX Tolleson, AZ, ch. 51, ION Television
KASW Phoenix, ch. 61, CW

DMA Counties	State	TV Households
Apache S.	AZ	4,540
Coconino	AZ	42,750
Gila	AZ	19,850
Graham	AZ	9,710
Greenlee	AZ	2,760
La Paz	AZ	8,620
Maricopa	AZ	1,354,790
Mohave	AZ	77,820
Navajo	AZ	32,280
Pinal	AZ	89,400
Yavapai	AZ	82,480

Maps courtesy of Nielsen Media Research

Pittsburgh, PA

Pittsburgh (22)

DMA TV Households: 1,163,150
% of U.S. TV Households: 1.045

KDKA-TV Pittsburgh, ch. 2, CBS
WTAE-TV Pittsburgh, ch. 4, ABC
WPXI Pittsburgh, ch. 11, NBC
***WQED** Pittsburgh, ch. 13, ETV
WQEX Pittsburgh, ch. 16, IND
WPCW Jeannette, PA, ch. 19, CW
WPMY Pittsburgh, ch. 22, MyNetworkTV
***WNPB-TV** Morgantown, WV, ch. 24, ETV
***WGPT** Oakland, MD, ch. 36, ETV
WPCB-TV Greensburg, PA, ch. 40, IND
WPGH-TV Pittsburgh, ch. 53, Fox

DMA Counties	State	TV Households	DMA Counties	State	TV Households
Garrett	MD	11,050	Greene	PA	14,870
Allegheny	PA	516,900	Indiana	PA	34,320
Armstrong	PA	28,440	Lawrence	PA	36,450
Beaver	PA	71,330	Venango	PA	21,980
Butler	PA	70,560	Washington	PA	84,040
Clarion	PA	15,820	Westmoreland	PA	150,040
Fayette	PA	58,940	Monongalia	WV	34,190
Forest	PA	2,080	Preston	WV	12,140

Portland-Auburn, ME (74)

DMA TV Households: 409,180
% of U.S. TV Households: .367

WCSH Portland, ME, ch. 6, NBC
WMTW-TV Poland Spring, ME, ch. 8, ABC
***WCBB** Augusta, ME, ch. 10, ETV
WGME-TV Portland, ME, ch. 13, CBS
WPFO Waterville, ME, ch. 23, Fox
***WMEA-TV** Biddeford, ME, ch. 26, ETV
WPME Lewiston, ME, ch. 35, MyNetworkTV
WPXT Portland, ME, ch. 51, CW

DMA Counties	State	TV Households
Androscoggin	ME	45,230
Cumberland	ME	112,740
Franklin	ME	11,850
Kennebec	ME	50,800
Knox	ME	17,330
Lincoln	ME	15,290
Oxford	ME	23,430
Sagadahoc	ME	15,240
York	ME	83,250
Carroll	NH	19,770
Coos	NH	14,250

Portland-Auburn, ME

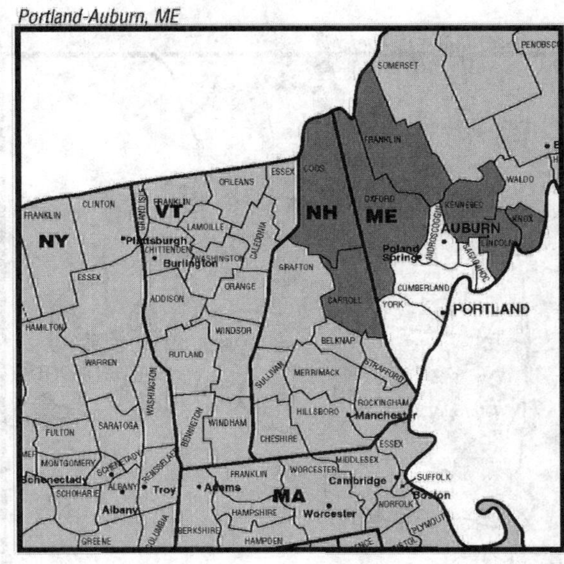

Portland, OR

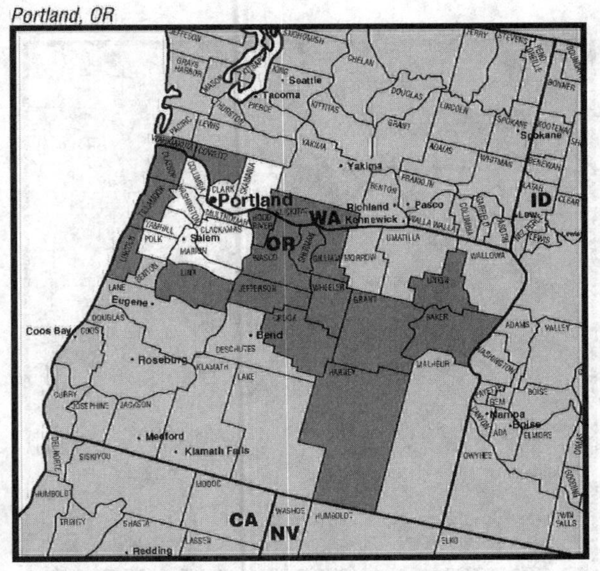

Portland, OR (23)

DMA TV Households: 1,117,990
% of U.S. TV Households: 1.004

KATU Portland, OR, ch. 2, ABC
KOIN Portland, OR, ch. 6, CBS
KGW-TV Portland, OR, ch. 8, NBC
***KOPB-TV** Portland, OR,ch. 10, ETV
KPTV Portland, OR, ch. 12, Fox
***KTVR** La Grande, OR, ch. 13, ETV
KUNP La Grande, OR, ch. 16, Univision
KPXG Salem, OR, ch. 22, IND
KNMT Portland, OR, ch. 24, IND
KRCW-TV Salem, OR, ch. 32, CW
KPDX Vancouver, WA, ch. 49, MyNetworkTV

DMA Counties	State	TV Households	DMA Counties	State	TV Households
Baker	OR	6,480	Polk	OR	25,170
Clackamas	OR	139,850	Sherman	OR	650
Clatsop	OR	14,370	Tillamook	OR	10,430
Columbia	OR	17,930	Union	OR	9,490
Crook	OR	8,700	Wasco	OR	8,720
Gilliam	OR	700	Washington	OR	189,820
Grant	OR	2,810	Wheeler	OR	520
Harney	OR	2,420	Yamhill	OR	30,380
Hood River	OR	6,980	Clark	WA	147,940
Jefferson	OR	7,230	Cowlitz	WA	36,840
Lincoln	OR	19,050	Klickitat	WA	7,070
Linn	OR	40,670	Skamania	WA	3,890
Marion	OR	105,400	Wahkiakum	WA	1,440
Multnomah	OR	273,040			

Maps courtesy of Nielsen Media Research

Presque Isle, ME (204)

DMA TV Households: 31,170
% of U.S. TV Households: .028

WAGM-TV Presque Isle, ME, ch. 8, CBS, Fox
***WMEM-TV** Presque Isle, ME, ch. 10, ETV

DMA Counties	State	TV Households
Aroostook	ME	31,170

Presque Isle, ME

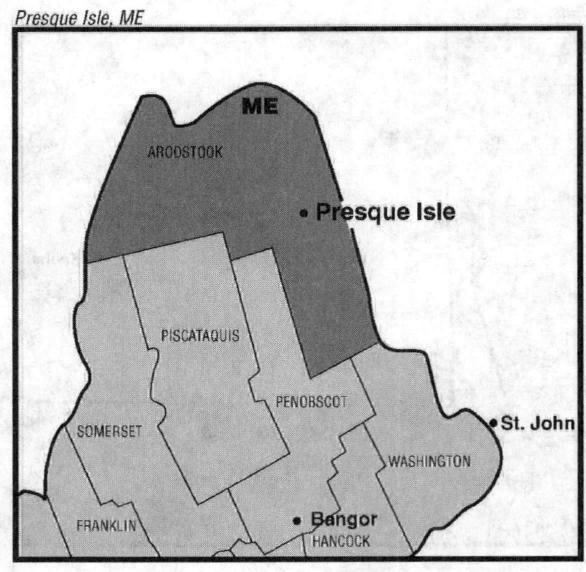

Providence, RI-New Bedford, MA

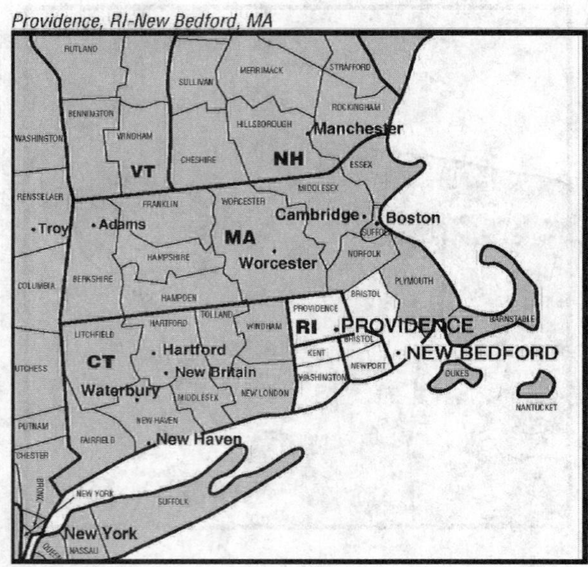

Providence, RI-New Bedford, MA (51)

DMA TV Households: 633,950
% of U.S. TV Households: .569

WLNE New Bedford, MA, ch. 6, ABC
WJAR Providence, RI, ch. 10, NBC
WPRI-TV Providence, RI, ch. 12, CBS
WLWC New Bedford, MA, ch. 28, CW
***WSBE-TV** Providence, RI, ch. 36, ETV
WDPX Vineyard Haven, MA, ch. 58, satellite to WBPX
WNAC-TV Providence, RI, ch. 64, Fox, MyNetworkTV
WPXQ Block Island, RI, ch. 69, ION Television

DMA Counties	State	TV Households
Bristol	MA	212,990
Bristol	RI	19,450
Kent	RI	70,430
Newport	RI	34,810
Providence	RI	246,510
Washington	RI	49,760

Quincy, IL-Hannibal, MO-Keokuk, IA (171)

DMA TV Households: 103,690
% of U.S. TV Households: .093

KHQA-TV Hannibal, MO, ch. 7, CBS
WGEM-TV Quincy, IL, ch. 10, NBC, CW, Fox
WTJR Quincy, IL, ch. 16, IND
***WMEC** Macomb, IL, ch. 22, ETV
***WQEC** Quincy, IL, ch. 27, ETV

DMA Counties	State	TV Households	DMA Counties	State	TV Households
Adams	IL	26,370	Clark	MO	2,980
Brown	IL	2,070	Knox	MO	1,600
Hancock	IL	7,500	Lewis	MO	3,760
McDonough	IL	11,690	Marion	MO	11,060
Pike	IL	6,750	Monroe	MO	3,670
Schuyler	IL	2,940	Ralls	MO	3,890
Scott	IL	2,180	Shelby	MO	2,650
Lee	IA	14,580			

Quincy, IL-Hannibal, MO-Keokuk, IA

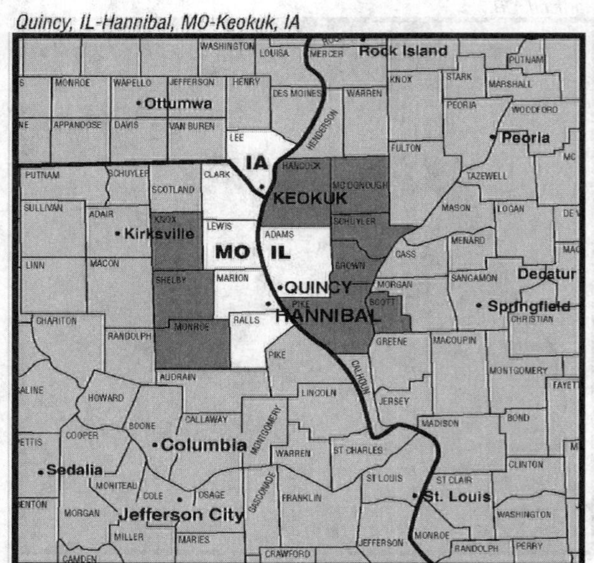

Maps courtesy of Nielsen Media Research

Raleigh-Durham (Fayetteville), NC (29)

DMA TV Households: 1,006,330
% of U.S. TV Households: .904

***WUNC-TV** Chapel Hill, NC, ch. 4, ETV
WRAL-TV Raleigh, NC, ch. 5, CBS
WTVD Durham, NC, ch. 11, ABC
WNCN Goldsboro, NC, ch. 17, NBC
WLFL Raleigh, NC, ch. 22, CW
WRDC Durham, NC, ch. 28, MyNetworkTV
WRAY-TV Wilson, NC, ch. 30, IND
***WUNP-TV** Roanoke Rapids, NC, ch. 36, ETV
WUVC-TV Fayetteville, NC, ch. 40, Univision
WRPX Rocky Mount, NC, ch. 47, ION Television
WRAZ Raleigh, NC, ch. 50, Fox
WFPX Fayetteville, NC, ch. 62, IND

DMA Counties	State	TV Households	DMA Counties	State	TV Households
Chatham	NC	23,990	Moore	NC	34,350
Cumberland	NC	108,390	Nash	NC	35,540
Durham	NC	96,840	Orange	NC	44,360
Edgecombe	NC	19,630	Person	NC	14,910
Franklin	NC	21,140	Sampson	NC	23,580
Granville	NC	18,740	Vance	NC	16,670
Halifax	NC	21,720	Wake	NC	299,060
Harnett	NC	39,500	Warren	NC	7,650
Hoke	NC	14,670	Wayne	NC	43,510
Johnston	NC	57,810	Wilson	NC	29,940
Lee	NC	21,210	Mecklenburg	VA	13,120

Raleigh-Durham (Fayetteville), NC

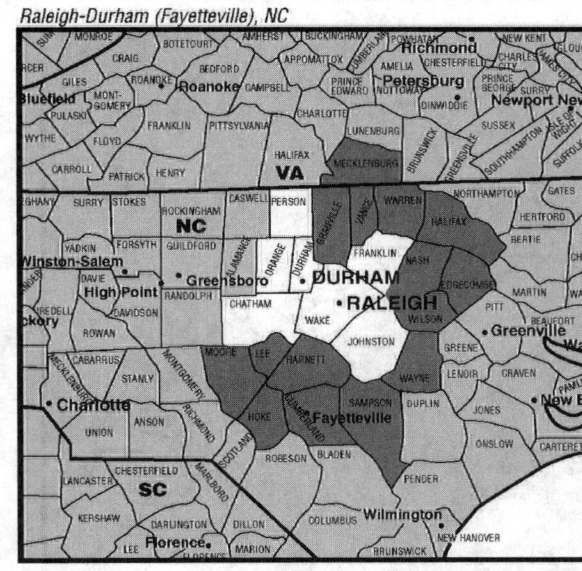

Rapid City, SD

Rapid City, SD (177)

DMA TV Households: 93,870
% of U.S. TV Households: .084

KOTA-TV Rapid City, SD, ch. 3, ABC
KIVV-TV Lead, SD, ch. 5, satellite to KEVN-TV
KEVN-TV Rapid City, SD, ch. 7, Fox
KSWY Sheridan, WY, ch. 7, NBC
***KZSD-TV** Martin, SD, ch. 8, ETV
***KBHE-TV** Rapid City, SD, ch. 9, ETV
KHSD-TV Lead, SD, ch. 11, satellite to KOTA-TV
KSGW-TV Sheridan, WY, ch. 12, satellite to KOTA-TV
***KTNE-TV** Alliance, NE, ch. 13, ETV
KCLO-TV Rapid City, SD, ch. 15, IND
KNBN Rapid City, SD, ch. 21, NBC

DMA Counties	State	TV Households	DMA Counties	State	TV Households
Carter	MT	490	Lawrence	SD	9,260
Morrill	NE	1,950	Meade	SD	9,280
Bennett	SD	1,070	Pennington	SD	37,500
Butte	SD	3,610	Perkins	SD	1,270
Custer	SD	3,280	Shannon	SD	3,080
Fall River	SD	3,050	Ziebach	SD	760
Haakon	SD	790	Crook	WY	2,550
Harding	SD	490	Sheridan	WY	11,570
Jackson	SD	880	Weston	WY	2,580
Jones	SD	390			

Reno, NV (110)

DMA TV Households: 261,250
% of U.S. TV Households: .235

KTVN Reno, ch. 2, CBS
KRNV Reno, ch. 4, NBC
***KNPB** Reno, ch. 5, ETV
KEGS Goldfield, NV, ch. 7, IND
KWNV Winnemucca, NV, ch. 7, satellite to KRNV
KOLO-TV Reno, ch. 8, ABC
KRXI-TV Reno, ch. 11, Fox
KAME-TV Reno, ch. 21, MyNetworkTV
KREN-TV Reno, ch. 27, CW

DMA Counties	State	TV Households	DMA Counties	State	TV Households
Alpine	CA	430	Humboldt	NV	5,970
El Dorado East	CA	13,730	Lander	NV	1,840
Lassen	CA	9,530	Lyon	NV	19,080
Mono	CA	4,710	Mineral	NV	2,120
Carson City	NV	21,200	Storey	NV	1,850
Pershing	NV	1,760	Washoe	NV	150,200
Churchill	NV	8,960	Douglas	NV	19,490
Esmeralda	NV	380			

Reno, NV

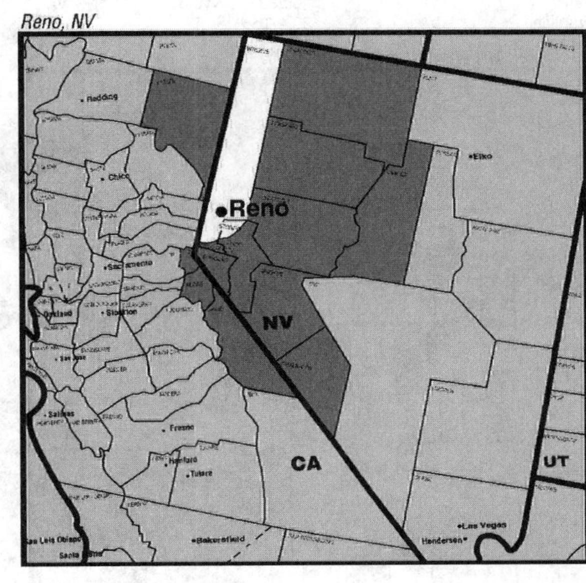

Maps courtesy of Nielsen Media Research

Richmond-Petersburg, VA

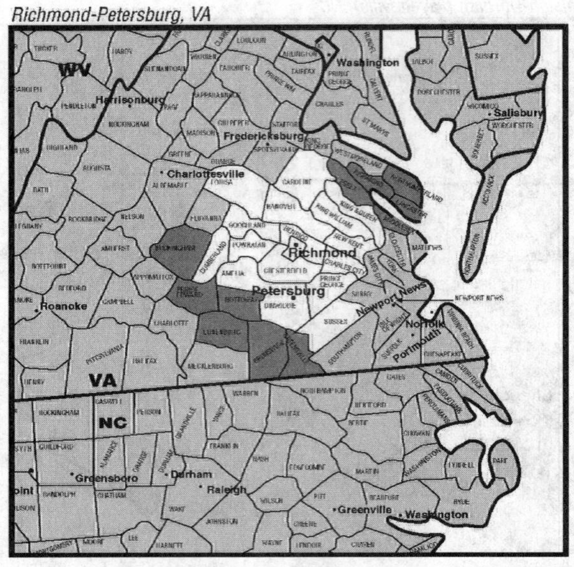

Richmond-Petersburg, VA (61)

DMA TV Households: 517,800
% of U.S. TV Households: .465

WTVR-TV Richmond, VA, ch. 6, CBS
WRIC-TV Petersburg, VA, ch. 8, ABC
WWBT Richmond, VA, ch. 12, NBC
***WCVE-TV** Richmond, VA, ch. 23, ETV
WRLH-TV Richmond, VA, ch. 35, Fox, MyNetworkTV
***WCVW** Richmond, VA, ch. 57, ETV
WUPV Ashland, VA, ch. 65, CW

DMA Counties	State	TV Households	DMA Counties	State	TV Households
Amelia	VA	4,640	King William	VA	5,630
Brunswick	VA	6,220	Lancaster	VA	4,820
Buckingham	VA	5,510	Louisa	VA	11,910
Caroline	VA	9,540	Lunenburg	VA	5,000
Charles City	VA	2,740	Middlesex	VA	4,530
Chesterfield	VA	115,440	New Kent	VA	6,110
Cumberland	VA	3,720	Northumberland	VA	5,720
Dinwiddie	VA	22,630	Nottoway	VA	5,730
Essex	VA	4,240	Powhatan	VA	8,980
Goochland	VA	7,390	Prince Edward	VA	7,030
Greensville	VA	5,890	Prince George	VA	20,490
Hanover	VA	35,370	Richmond	VA	3,110
Henrico	VA	117,900	Richmond City	VA	80,520
King and Queen	VA	2,830	Sussex	VA	3,950

Roanoke-Lynchburg, VA (68)

DMA TV Households: 445,840
% of U.S. TV Households: .400

WDBJ Roanoke, VA, ch. 7, CBS, MyNetworkTV
WSLS-TV Roanoke, VA, ch. 10, NBC
WSET-TV Lynchburg, VA, ch. 13, ABC
***WBRA-TV** Roanoke, VA, ch. 15, ETV
WWCW Lynchburg, VA, ch. 21, Fox, CW
WDRL-TV Danville, VA ch. 24, IND
WFXR-TV Roanoke, VA, ch. 27, satellite to WWCW
WPXR Roanoke, VA, ch. 38, ION Television

DMA Counties	State	TV Households	DMA Counties	State	TV Households
Alleghany	VA	9,420	Giles	VA	7,120
Amherst	VA	12,280	Grayson	VA	6,980
Appomattox	VA	5,560	Halifax	VA	14,750
Bath	VA	1,930	Henry	VA	29,640
Bedford	VA	28,710	Highland	VA	1,050
Bland	VA	2,650	Montgomery	VA	37,260
Botetourt	VA	12,460	Nelson	VA	5,970
Campbell	VA	47,430	Pittsylvania	VA	44,490
Carroll	VA	15,430	Pulaski	VA	14,740
Charlotte	VA	5,100	Roanoke	VA	88,460
Craig	VA	1,990	Rockbridge	VA	13,650
Floyd	VA	6,230	Wythe	VA	12,070
Franklin	VA	20,470			

Roanoke-Lynchburg, VA

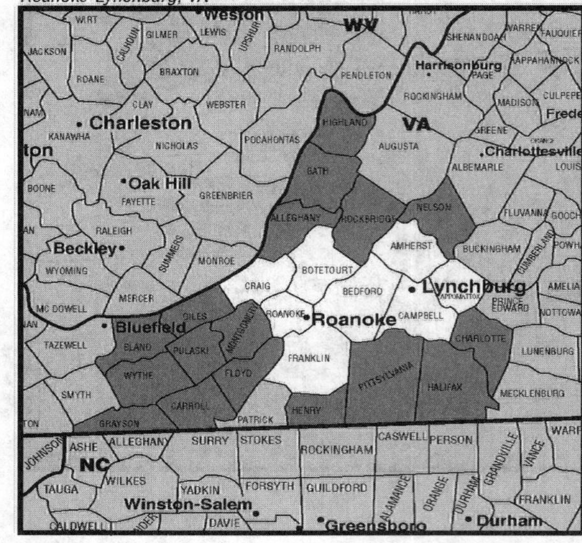

Rochester, MN-Mason City, IA-Austin, MN

Rochester, MN-Mason City, IA-Austin, MN (153)

DMA TV Households: 143,090
% of U.S. TV Households: .129

KIMT Mason City, IA, ch. 3, CBS
KAAL Austin, MN, ch. 6, ABC
KTTC Rochester, MN, ch. 10, NBC
***KSMQ-TV** Austin, MN, ch. 15, ETV
***KYIN** Mason City, IA, ch. 24, ETV
KXLT-TV Rochester, MN, ch. 47, Fox

DMA Counties	State	TV Households
Cerro Gordo	IA	18,570
Floyd	IA	6,670
Hancock	IA	4,670
Howard	IA	3,840
Mitchell	IA	4,240
Winnebago	IA	4,680
Worth	IA	3,270
Dodge	MN	7,340
Fillmore	MN	8,320
Freeborn	MN	13,130
Mower	MN	15,610
Olmsted	MN	52,750

Maps courtesy of Nielsen Media Research

Rochester, NY (78)

DMA TV Households: 392,630
% of U.S. TV Households: .353

WROC-TV Rochester, NY, ch. 8, CBS
WHEC-TV Rochester, NY, ch. 10, NBC
WHAM-TV Rochester, NY, ch. 13, ABC, CW
***WXXI-TV** Rochester, NY, ch. 21, ETV
WUHF Rochester, NY, ch. 31, Fox

DMA Counties	State	TV Households
Livingston	NY	22,190
Monroe	NY	194,690
Ontario	NY	40,820
Rochester City	NY	90,600
Wayne	NY	35,290
Yates	NY	9,040

Rochester, NY

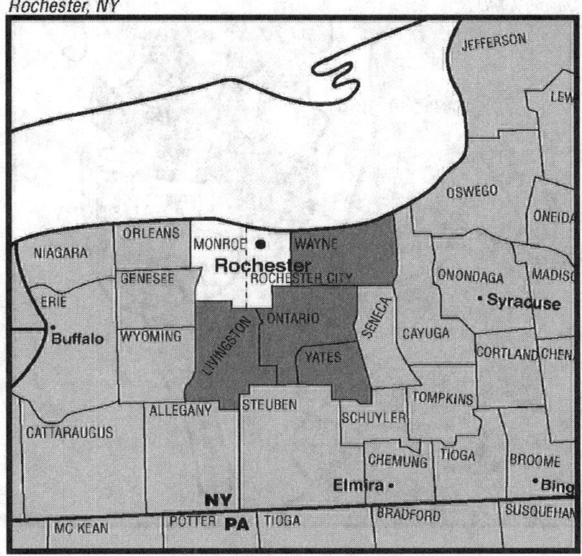

Rockford, IL

Rockford, IL (133)

DMA TV Households: 184,560
% of U.S. TV Households: .166

WREX-TV Rockford, IL, ch. 13, NBC, CW
WTVO Rockford, IL, ch. 17, ABC, MyNetworkTV
WIFR Rockford, IL, ch 23, CBS
WQRF-TV Rockford, IL, ch. 39, Fox

DMA Counties	State	TV Households
Boone	IL	18,170
Lee	IL	13,180
Ogle	IL	20,860
Stephenson	IL	19,400
Winnebago	IL	112,950

Sacramento-Stockton-Modesto, CA (20)

DMA TV Households: 1,368,680
% of U.S. TV Households: 1.229

KCRA-TV Sacramento, CA, ch. 3, NBC
***KVIE** Sacramento, CA, ch. 6, ETV
KXTV Sacramento, CA, ch. 10, ABC
KOVR Stockton, CA, ch. 13, CBS
KUVS-TV Modesto, CA, ch. 19, Univision
***KBSV** Ceres, CA, ch. 23, ETV
KSPX Sacramento, CA, ch. 29, ION Television
KMAX-TV Sacramento, CA, ch. 31, CW
KTXL Sacramento, CA, ch. 40, Fox
KQCA Stockton, CA, ch. 58, MyNetworkTV
KTFK-TV Stockton, CA, ch. 64, TeleFutura

DMA Counties	State	TV Households
Amador	CA	13,600
Calaveras	CA	18,920
Colusa	CA	6,770
El Dorado West	CA	51,460
Nevada	CA	38,730
Placer	CA	122,550
Plumas	CA	8,580
Sacramento	CA	503,650
San Joaquin	CA	216,810
Sierra	CA	1,440
Solano East	CA	84,750
Stanislaus	CA	162,190
Sutter	CA	30,200
Tuolumne	CA	22,200
Yolo	CA	63,960
Yuba	CA	22,870

Sacramento-Stockton-Modesto, CA

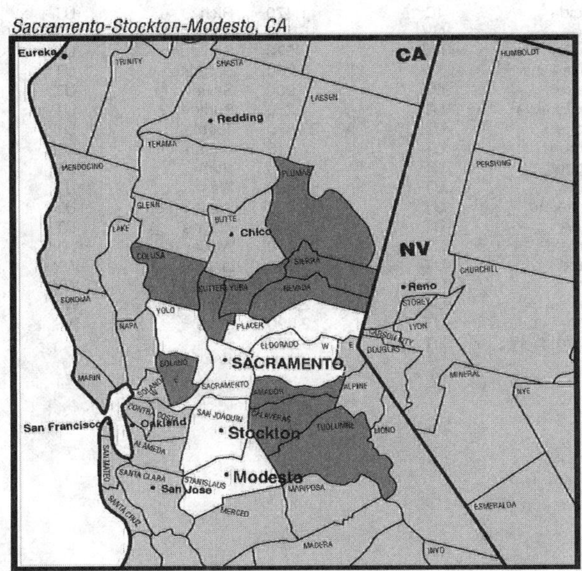

Maps courtesy of Nielsen Media Research

Salisbury, MD

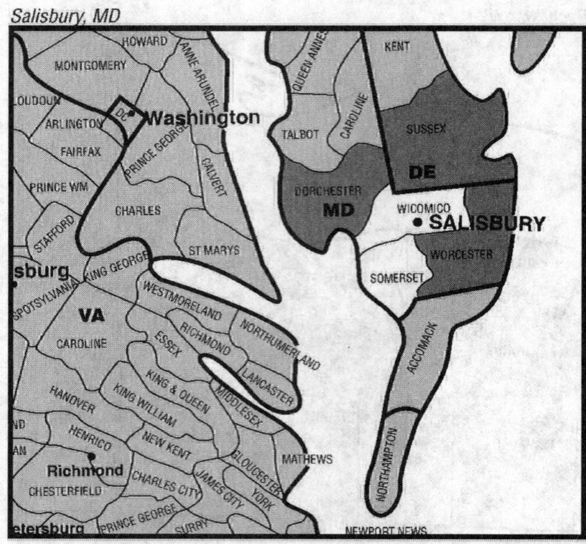

Salisbury, MD (148)

DMA TV Households: 150,790

% of U.S. TV Households: .135

WBOC-TV Salisbury, MD, ch. 16, CBS, Fox
***WCPB** Salisbury, MD, ch. 28, ETV
WMDT Salisbury, MD, ch. 47, ABC, CW
***WDPB** Seaford, DE, ch. 64, ETV

DMA Counties	State	TV Households
Sussex	DE	72,920
Dorchester	MD	13,260
Somerset	MD	8,660
Wicomico	MD	35,160
Worcester	MD	20,790

Salt Lake City, UT (35)

DMA TV Households: 839,170

% of U.S. TV Households: .754

KUTV Salt Lake City, ch. 2, CBS
KJWY Jackson, WY, ch. 2, NBC
KVNV Ely, NV, ch. 3, NBC
KCBU Price, UT, ch. 3, TeleFutura
KCSG Cedar City, UT, ch. 4, IND
KTVX Salt Lake City, ch. 4, ABC
KSL-TV Salt Lake City, ch. 5, NBC
KBNY Ely, NV, ch. 6, IND
KBCJ Vernal, UT, ch. 6, IND
***KUED** Salt Lake City, ch. 7, ET
***KUEN** Ogden, UT, ch. 9, ETV
KENV Elko, NV, ch. 10, NBC
***KBYU-TV** Provo, UT, ch. 11, ETV
KBEO Jackson, WY, ch. 11, IND
KUTF Logan, UT, ch. 12, IND
KUSG St. George, UT, ch. 12, IND
KSTU Salt Lake City, ch. 13, Fox
KGWR-TV Rock Springs, WY, ch. 13, CBS
KJZZ-TV Salt Lake City, ch. 14, MyNetworkTV
KUPX Provo, UT, ch. 16, ION Television
***KUEW** St. George, UT, ch. 18, ETV
***KUES** Richfield, UT, ch. 19, ETV
KTMW Salt Lake City, ch. 20, IND
KPNZ Ogden, UT, ch. 24, IND
KUCW Ogden, UT, ch. 30, CW
KUTH Provo, UT, ch. 32, Univision

DMA Counties	State	TV Households	DMA Counties	State	TV Households
Bear Lake	ID	1,990	Morgan	UT	2,290
Franklin	ID	3,860	Piute	UT	490
Oneida	ID	1,470	Rich	UT	650
Elko	NV	15,130	Salt Lake	UT	310,980
Eureka	NV	520	San Juan	UT	3,660
White Pine	NV	3,260	Sanpete	UT	6,850
Beaver	UT	1,960	Sevier	UT	6,270
Box Elder	UT	14,280	Summit	UT	12,450
Cache	UT	28,880	Tooele	UT	16,390
Carbon	UT	7,010	Uintah	UT	9,120
Daggett	UT	350	Utah	UT	123,400
Davis	UT	83,880	Wasatch	UT	6,290
Duchesne	UT	4,990	Washington	UT	42,180
Emery	UT	3,460	Wayne	UT	880
Garfield	UT	1,410	Weber	UT	70,420
Grand	UT	3,000	Lincoln	WY	5,900
Iron	UT	12,610	Sublette	WY	2,850
Juab	UT	2,640	Sweetwater	WY	14,400
Kane	UT	2,090	Uinta	WY	6,940
Millard	UT	3,970			

Salt Lake City, UT

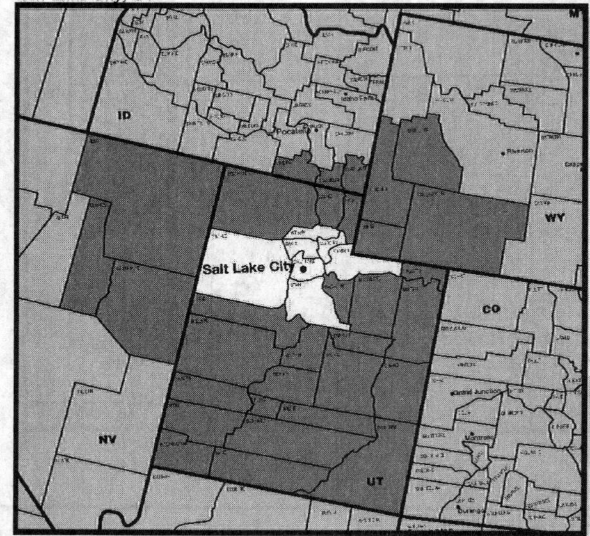

Maps courtesy of Nielsen Media Research

San Angelo, TX

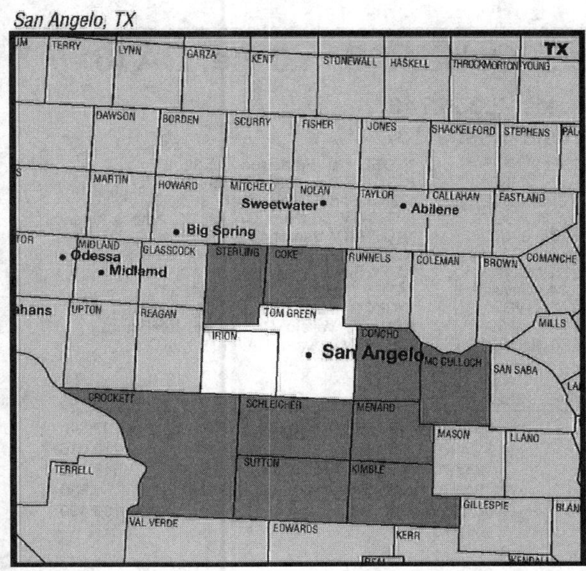

San Angelo, TX (197)

DMA TV Households: 52,930
% of U.S. TV Households: .048

KSAN-TV San Angelo, TX, ch. 3, satellite to KRBC-TV
KIDY San Angelo, TX, ch. 6, Fox
KLST San Angelo, TX, ch. 8, CBS

DMA Counties	State	TV Households
Coke	TX	1,460
Concho	TX	960
Crockett	TX	1,460
Irion	TX	660
Kimble	TX	1,940
McCulloch	TX	3,110
Menard	TX	850
Schleicher	TX	1,070
Sterling	TX	490
Sutton	TX	1,530
Tom Green	TX	39,400

San Antonio, TX (37)

DMA TV Households: 774,470
% of U.S. TV Households: .696

KCWX Fredericksburg, TX, ch. 2, CW
WOAI-TV San Antonio, TX, ch. 4, NBC
KENS-TV San Antonio, TX, ch. 5, CBS
***KLRN** San Antonio, TX, ch. 9, ETV
KTRG Del Rio, TX, ch. 10, IND
KSAT-TV San Antonio, TX, ch. 12, ABC
KVAW Eagle Pass, TX, ch. 16, IND
***KHCE-TV** San Antonio, TX, ch. 23, ETV
KPXL Uvalde, TX, ch. 26, ION Television
KABB San Antonio, TX, ch. 29, Fox
KMYS Kerrville, TX, ch. 35, MyNetworkTV
KWEX-TV San Antonio, TX, ch. 41, Univision
KVDA San Antonio, TX, ch. 60, Telemundo

DMA Counties	State	TV Households
Atascosa	TX	14,720
Bandera	TX	7,810
Bexar	TX	538,020
Comal	TX	36,840
DeWitt	TX	7,230
Dimmit	TX	3,270
Edwards	TX	750
Frio	TX	4,700
Goliad	TX	2,730
Gonzales	TX	7,290
Guadalupe	TX	37,240
Karnes	TX	4,420
Kendall	TX	10,420
Kerr	TX	18,820
Kinney	TX	1,240
LaSalle	TX	1,630
Lavaca	TX	7,440
Maverick	TX	13,850
McMullen	TX	390
Medina	TX	14,260
Real	TX	1,040
Uvalde	TX	8,640
Val Verde	TX	14,710
Wilson	TX	13,650
Zavala	TX	3,360

San Antonio, TX

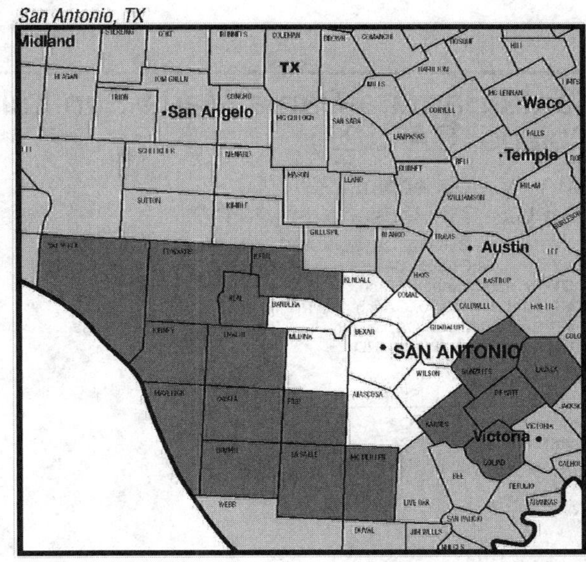

San Diego, CA

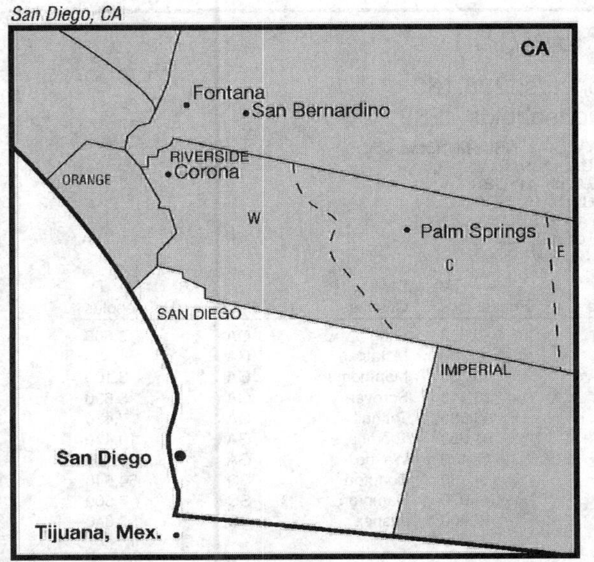

San Diego (27)

DMA TV Households: 1,030,020
% of U.S. TV Households: .925

XETV Tijuana, Mexico, ch. 6, Fox
KFMB-TV San Diego, ch. 8, CBS
KGTV San Diego, ch. 10, ABC
***KPBS** San Diego, ch. 15, ETV
KNSD San Diego, ch. 39, NBC
KUSI-TV San Diego, ch. 51, IND
KSWB-TV San Diego, ch. 69, CW

DMA Counties	State	TV Households
San Diego	CA	1,030,020

Maps courtesy of Nielsen Media Research

San Francisco-Oakland-San Jose, CA

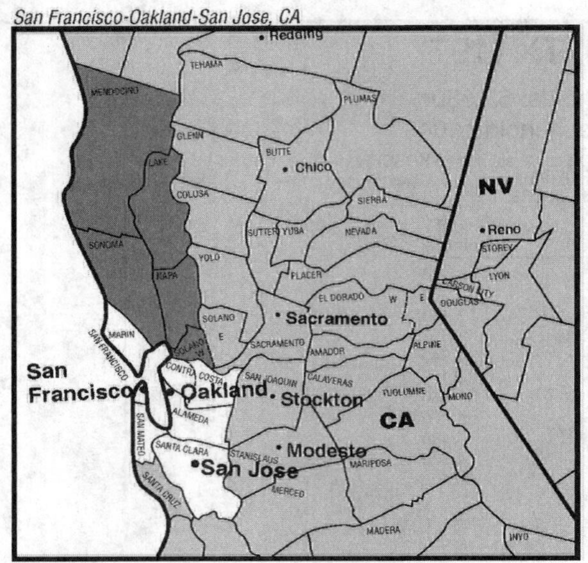

San Francisco-Oakland-San Jose, CA (6)

DMA TV Households: 2,355,740

% of U.S. TV Households: 2.137

KTVU Oakland, CA, ch. 2, Fox
KRON-TV San Francisco, ch. 4, IND
KPIX-TV San Francisco, ch. 5, CBS
KGO-TV San Francisco, ch. 7, ABC
KUNO-TV Fort Bragg, CA, ch. 8, IND
***KQED** San Francisco, ch. 9, ETV
KNTV San Jose, CA, ch. 11, NBC
KDTV San Francisco, ch. 14, Univision
KBWB San Francisco, ch. 20, WB
***KRCB** Cotati, CA, ch. 22, ETV
KTSF San Francisco, ch. 26, IND
***KQEC** San Francisco, ch. 32, ETV
KICU-TV San Jose, CA, ch. 36, IND
KCNS San Francisco, ch. 38, IND
KTNC-TV Concord, CA, ch. 42, Azteca America
***KCSM-TV** San Mateo, CA, ch. 43, ETV
KBCW San Francisco, ch. 44, CW
KFTY Santa Rosa, CA, ch. 50, IND
***KTEH** San Jose, CA, ch. 54, ETV
KKPX San Jose, CA, ch. 65, i Network
KFSF-TV Vallejo, CA, ch. 66, TeleFutura
KTLN-TV Novato, CA, ch. 68, IND

DMA Counties	State	TV Households	DMA Counties	State	TV Households
Alameda	CA	503,390	San Francisco	CA	292,840
Contra Costa	CA	355,070	San Mateo	CA	239,010
Lake	CA	25,080	Santa Clara	CA	548,450
Marin	CA	94,570	Solano West	CA	50,630
Mendocino	CA	301,660	Sonoma	CA	165,490
Napa	CA	46,870			

Santa Barbara-Santa Maria-San Luis Obispo, CA (122)

DMA TV Households: 227,700

% of U.S. TV Households: .205

KEYT-TV Santa Barbara, CA, ch. 3, ABC, MyNetworkTV
KSBY San Luis Obispo, CA, ch. 6, NBC, CW
KCOY-TV Santa Maria, CA, ch. 12, CBS
KTAS San Luis Obispo, CA, ch. 33, Telemundo
KPMR Santa Barbara, CA, ch. 38, Univision
KBEH Oxnard, CA, ch. 63, IND

DMA Counties	State	TV Households
San Luis Obispo	CA	95,040
Santa Barbara N.	CA	64,010
Santa Barbara S.	CA	68,650

Santa Barbara-Santa Maria-San Luis Obispo, CA

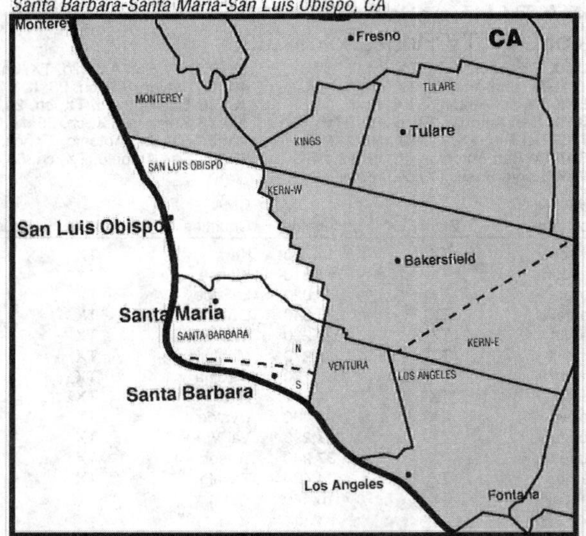

Savannah, GA

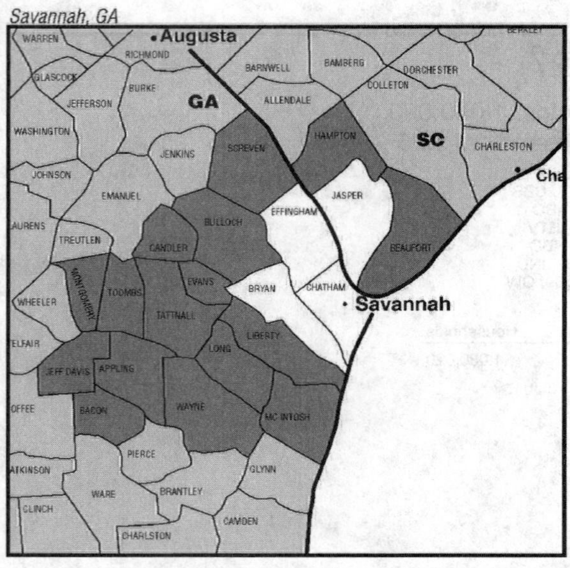

Maps courtesy of Nielsen Media Research

Savannah, GA (97)

DMA TV Households: 298,130

% of U.S. TV Households: .268

WSAV-TV Savannah, GA, ch. 3, NBC, MyNetworkTV
***WVAN-TV** Savannah, GA, ch. 9, ETV
WTOC-TV Savannah, GA, ch. 11, CBS
***WJWJ-TV** Beaufort, SC, ch. 16, ETV
WJCL Savannah, GA, ch. 22, ABC
WTGS Hardeeville, SC, ch. 28, Fox
WGSA Baxley, GA, ch. 34, CW

DMA Counties	State	TV Households	DMA Counties	State	TV Households
Appling	GA	6,650	Long	GA	3,800
Bacon	GA	4,220	McIntosh	GA	4,350
Bryan	GA	10,350	Montgomery	GA	3,100
Bulloch	GA	23,410	Screven	GA	5,690
Candler	GA	3,580	Tattnall	GA	7,470
Chatham	GA	91,980	Toombs	GA	10,470
Effingham	GA	17,420	Wayne	GA	10,170
Evans	GA	4,150	Beaufort	SC	54,540
Jeff Davis	GA	4,970	Hampton	SC	7,500
Liberty	GA	17,400	Jasper	SC	7,340

Seattle-Tacoma, WA

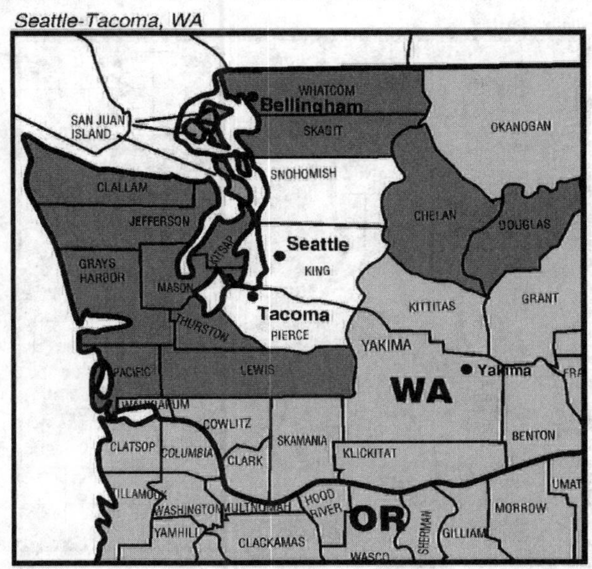

Seattle-Tacoma, WA (14)

DMA TV Households: 1,724,450
% of U.S. TV Households: 1.549

KOMO-TV Seattle, ch. 4, ABC
KING-TV Seattle, ch. 5, NBC
KIRO-TV Seattle, ch. 7, CBS
***KCTS-TV** Seattle, ch. 9, ETV
KSTW Tacoma, WA, ch. 11, CW
KVOS-TV Bellingham, WA, ch. 12, IND
KCPQ Tacoma, WA, ch. 13, Fox
***KCKA** Centralia, WA, ch. 15, ETV
KONG-TV Everett, WA, ch. 16, IND
KTBW-TV Tacoma, WA, ch. 20, IND
KMYQ Seattle, ch. 22, MyNetworkTV
KBCB Bellingham, WA, ch. 24, IND
***KBTC-TV** Tacoma, WA, ch. 28, ETV
KWPX Bellevue, WA, ch. 33, ION Television
KHCV Seattle, ch. 45, Azteca America
KUNS-TV Bellevue, WA, ch. 51, Univision
***KWDK** Tacoma, WA, ch. 56, ETV

DMA Counties	State	TV Households	DMA Counties	State	TV Households
Chelan	WA	24,430	Mason	WA	20,770
Clallam	WA	28,250	Pacific	WA	9,270
Douglas	WA	12,360	Pierce	WA	281,050
Grays Harbor	WA	25,970	San Juan	WA	6,860
Island	WA	30,260	Skagit	WA	40,700
Jefferson	WA	11,780	Snohomish	WA	243,780
King	WA	717,150	Thurston	WA	88,990
Kitsap	WA	87,800	Whatcom	WA	69,180
Lewis	WA	25,850			

Sherman, TX-Ada, OK (161)

DMA TV Households: 124,330
% of U.S. TV Households: .112

KTEN Ada, OK, ch. 10, NBC, CW
KXII Sherman, TX, ch. 12, CBS, Fox, MyNetworkTV

DMA Counties	State	TV Households
Atoka	OK	5,210
Bryan	OK	14,900
Carter	OK	18,730
Choctaw	OK	6,050
Coal	OK	2,250
Johnston	OK	3,830
Love	OK	3,630
Marshall	OK	6,040
Pontotoc	OK	13,750
Pushmataha	OK	4,520
Grayson	TX	45,420

Sherman, TX-Ada, OK

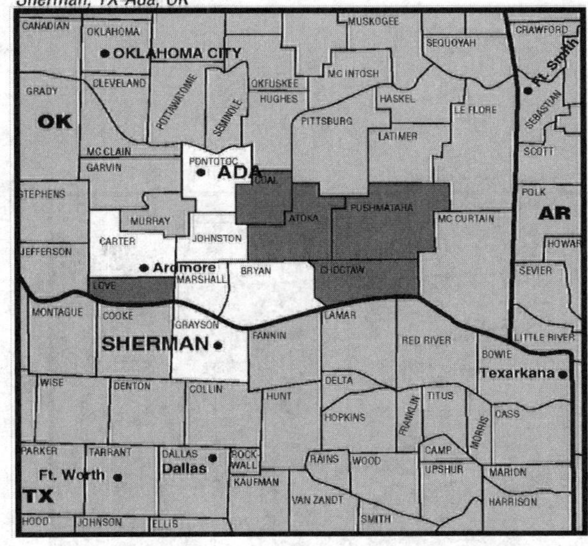

Shreveport, LA

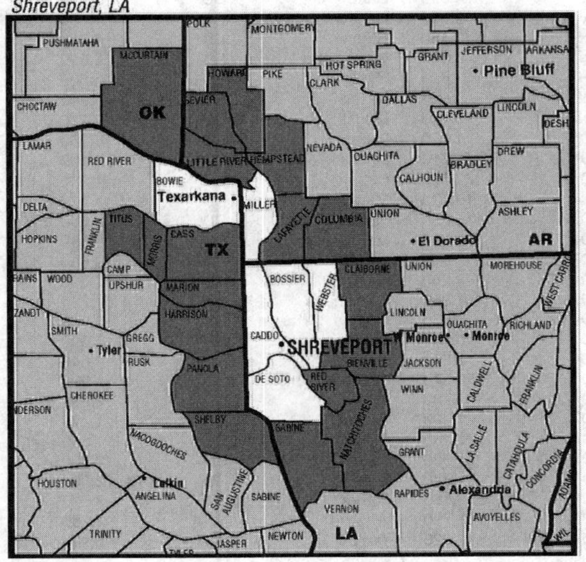

Shreveport, LA (81)

DMA TV Households: 381,200
% of U.S. TV Households: .342

KTBS-TV Shreveport, LA, ch. 3, ABC
KTAL-TV Texarkana, TX, ch. 6, NBC
KSLA-TV Shreveport, LA, ch. 12, CBS
KPXJ Minden, LA, ch. 21, CW
***KLTS-TV** Shreveport, LA, ch. 24, ETV
KMSS-TV Shreveport, LA, ch. 33, Fox
KSHV Shreveport, LA, ch. 45, MyNetworkTV

DMA Counties	State	TV Households	DMA Counties	State	TV Households
Columbia	AR	9,510	Natchitoches	LA	14,390
Hempstead	AR	8,620	Red River	LA	3,240
Howard	AR	5,330	Sabine	LA	9,550
Lafayette	AR	3,230	Webster	LA	16,410
Little River	AR	5,300	McCurtain	OK	12,790
Miller	AR	16,830	Bowie	TX	33,930
Nevada	AR	3,730	Cass	TX	12,250
Sevier	AR	5,850	Harrison	TX	23,580
Bienville	LA	5,910	Marion	TX	4,650
Bossier	LA	40,910	Morris	TX	5,060
Caddo	LA	98,880	Panola	TX	8,900
Claiborne	LA	6,110	Shelby	TX	9,770
DeSoto	LA	10,170	Titus	TX	10,030

Maps courtesy of Nielsen Media Research

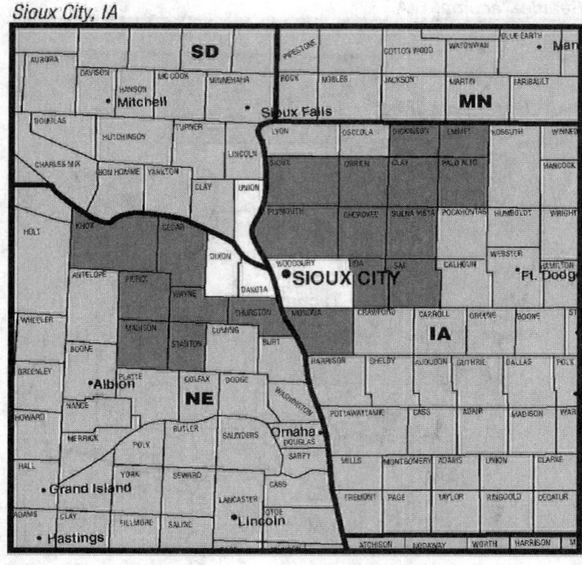

Sioux City, IA (143)

DMA TV Households: 156,480
% of U.S. TV Households: .141

***KUSD-TV** Vermillion, SD, ch. 2, ETV
KTIV Sioux City, IA, ch. 4, NBC, CW
KCAU-TV Sioux City, IA, ch. 9, ABC
KMEG Sioux City, IA, ch. 14, CBS
***KXNE-TV** Norfolk, NE, ch. 19, ETV
***KSIN** Sioux City, IA, ch. 27, ETV
KPTH Sioux City, IA, ch. 44, Fox, MyNetworkTV

DMA Counties	State	TV Households	DMA Counties	State	TV Households
Buena Vista	IA	7,250	Woodbury	IA	38,250
Cherokee	IA	5,080	Cedar	NE	3,360
Clay	IA	7,170	Dakota	NE	6,980
Dickinson	IA	7,370	Dixon	NE	2,290
Emmet	IA	4,270	Knox	NE	3,530
Ida	IA	3,080	Madison	NE	13,460
Monona	IA	3,980	Pierce	NE	2,880
O'Brien	IA	5,670	Stanton	NE	2,360
Palo Alto	IA	3,960	Thurston	NE	2,290
Plymouth	IA	9,560	Wayne	NE	3,190
Sac	IA	4,380	Union	SD	5,430
Sioux	IA	10,690			

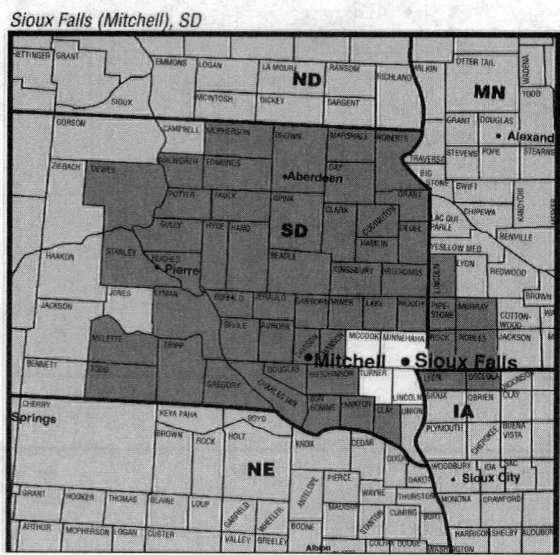

Sioux Falls (Mitchell), SD (115)

DMA TV Households: 247,000
% of U.S. TV Households: .222

KDLO-TV Florence, SD, ch. 3, satellite to KELO-TV
KPRY-TV Pierre, SD, ch. 4, satellite to KSFY-TV
KDLV-TV Mitchell, SD, ch. 5, NBC
KPLO-TV Reliance, SD, ch. 6, satellite to KELO-TV
***KESD-TV** Brookings, SD, ch. 8, ETV
KABY-TV Aberdeen, SD, ch. 9, satellite to KSFY-TV
***KTSD-TV** Pierre, SD, ch. 10, ETV
KELO-TV Sioux Falls, SD, ch. 11, CBS, MyNetworkTV
KTTM Huron, SD, ch. 12, satellite to KTTW
***KRNE-TV** Merriman, NE, ch. 12, ETV
KSFY-TV Sioux Falls, SD, ch. 13, ABC
***KDSD-TV** Aberdeen, SD, ch. 16, ETV
KTTW Sioux Falls, SD, ch. 17, Fox
***KSMN** Worthington, MN, ch. 20, ETV
***KCSD-TV** Sioux Falls, SD, ch. 23, ETV
KWSD Sioux Falls, SD, ch. 36, CW
KDLT-TV Sioux Falls, SD, ch. 46, IND

DMA Counties	State	TV Households	DMA Counties	State	TV Households
Lyon	IA	4,090	Hand	SD	1,260
Osceola	IA	2,650	Hanson	SD	1,390
Lincoln	MN	2,480	Hughes	SD	6,750
Murray	MN	3,580	Hutchinson	SD	2,960
Nobles	MN	7,630	Hyde	SD	680
Pipestone	MN	3,840	Jerauld	SD	890
Rock	MN	3,760	Kingsbury	SD	2,270
Aurora	SD	1,060	Lake	SD	4,360
Beadle	SD	6,710	Lincoln	SD	13,190
Bon Homme	SD	2,470	Lyman	SD	1,390
Brookings	SD	10,420	Marshall	SD	1,770
Brown	SD	14,590	McCook	SD	2,290
Brule	SD	1,960	McPherson	SD	1,090
Buffalo	SD	570	Mellette	SD	680
Charles Mix	SD	3,260	Miner	SD	1,090
Clark	SD	1,350	Minnehaha	SD	64,230
Clay	SD	4,670	Moody	SD	2,600
Codington	SD	10,490	Potter	SD	990
Davison	SD	7,600	Roberts	SD	3,640
Day	SD	2,370	Sanborn	SD	990
Deuel	SD	1,780	Spink	SD	2,590
Dewey	SD	1,900	Stanley	SD	1,170
Douglas	SD	1,230	Sully	SD	590
Edmunds	SD	1,580	Todd	SD	2,500
Faulk	SD	880	Tripp	SD	2,360
Grant	SD	2,820	Turner	SD	3,370
Gregory	SD	1,790	Walworth	SD	2,250
Hamlin	SD	2,020	Yankton	SD	8,110

Maps courtesy of Nielsen Media Research

South Bend-Elkhart, IN

South Bend-Elkhart, IN (88)

DMA TV Households: 334,370
% of U.S. TV Households: .300

WNDU-TV South Bend, IN, ch. 16, NBC
WSBT-TV South Bend, IN, ch. 22, CBS
WSJV Elkhart, IN, ch. 28, Fox
***WNIT-TV** South Bend, IN, ch. 34, ETV
WHME-TV South Bend, IN, ch. 46, IND

DMA Counties	State	TV Households
Elkhart	IN	70,210
Fulton	IN	8,170
Kosciusko	IN	28,140
Lagrange	IN	11,330
Marshall	IN	17,420
Pulaski	IN	5,130
St. Joseph	IN	100,900
Starke	IN	8,580
Berrien	MI	64,170
Cass	MI	20,320

Spokane, WA (77)

DMA TV Households: 395,490
% of U.S. TV Households: .355

KREM-TV Spokane, WA, ch. 2, CBS
KLEW-TV Lewiston, ID, ch. 3, satellite to KIMA-TV
KXLY-TV Spokane, WA, ch. 4, ABC, MyNetworkTV
KHQ-TV Spokane, WA, ch. 6, NBC
***KSPS-TV** Spokane, WA, ch. 7, ETV
***KWSU-TV** Pullman, WA, ch. 10, ETV
KSKN Spokane, WA, ch. 22, CW
KQUP Pullman, WA, ch. 24, IND
***KCDT** Coeur d'Alene, ID, ch. 26, ETV
KAYU-TV Spokane, WA, ch. 28, Fox
KGPX Spokane, WA, ch. 34, ION Television
***KUID-TV** Moscow, ID, ch. 35, ETV

DMA Counties	State	TV Households	DMA Counties	State	TV Households
Benewah	ID	3,480	Adams	WA	5,010
Bonner	ID	16,110	Asotin	WA	8,660
Boundary	ID	3,640	Columbia	WA	1,620
Clearwater	ID	2,960	Ferry	WA	2,710
Idaho	ID	5,890	Garfield	WA	980
Kootenai	ID	49,380	Grant	WA	25,990
Latah	ID	11,640	Lincoln	WA	4,110
Lewis	ID	1,570	Okanogan	WA	14,250
Nez Perce	ID	15,400	Pend Oreille	WA	4,930
Shoshone	ID	5,570	Spokane	WA	170,940
Lincoln	MT	7,780	Stevens	WA	14,900
Wallowa	OR	2,880	Whitman	WA	15,090

Spokane, WA

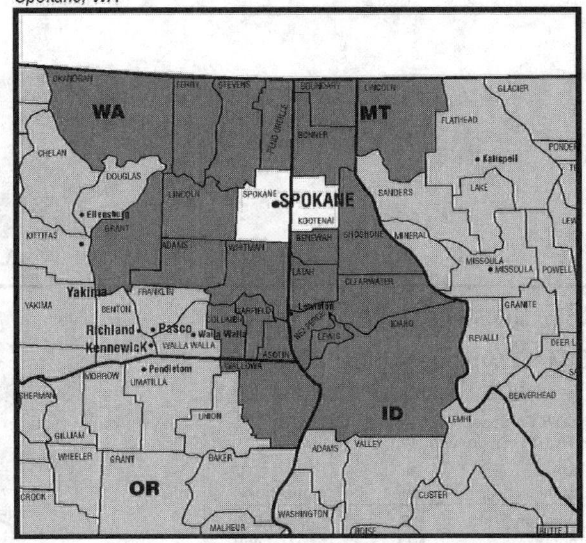

Springfield-Holyoke, MA

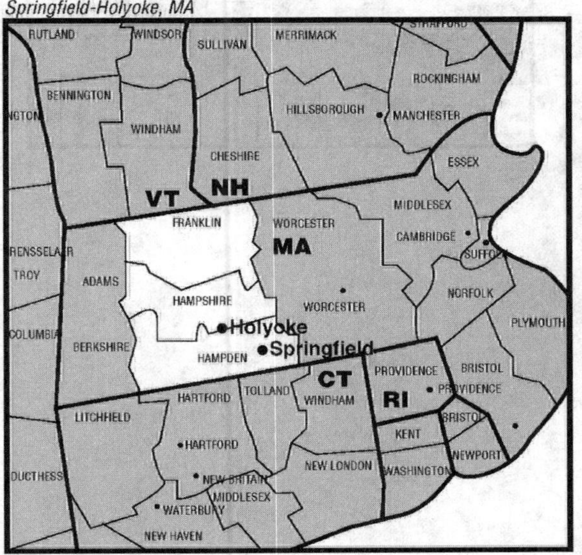

Springfield-Holyoke, MA (109)

DMA TV Households: 264,480
% of U.S. TV Households: .238

WWLP Springfield, MA, ch. 22, NBC
WGGB-TV Springfield, MA, ch. 40, ABC
***WGBY-TV** Springfield, MA, ch. 57, ETV

DMA Counties	State	TV Households
Franklin	MA	29,450
Hampden	MA	178,350
Hampshire	MA	56,680

Maps courtesy of Nielsen Media Research

Springfield, MO

Springfield, MO (76)

DMA TV Households: 402,310

% of U.S. TV Households: .361

KYTV Springfield, MO, ch. 3, NBC, CW
KOLR Springfield, MO, ch. 10, CBS
KSFX-TV Springfield, MO, ch. 27, Fox
KSPR Springfield, MO, ch. 33, ABC
KRBK Osage Beach, MO, ch. 49, IND
***KOZK** Springfield, MO, ch. 21, ETV
KWBM Harrison, AR, ch. 31, MyNetworkTV
KPBI Eureka Springs, AR, ch. 34, IND

DMA Counties	State	TV Households
Baxter	AR	17,840
Boone	AR	14,630
Carroll	AR	10,760
Fulton	AR	4,910
Marion	AR	7,060
Newton	AR	3,420
Barry	MO	13,930
Benton	MO	8,300
Camden	MO	17,200
Cedar	MO	5,860
Christian	MO	28,870
Dade	MO	3,160
Dallas	MO	6,160
Dent	MO	5,850
Douglas	MO	5,380
Greene	MO	104,600
Hickory	MO	3,950
Howell	MO	14,840
Laclede	MO	13,790
Lawrence	MO	14,250
Oregon	MO	4,200
Ozark	MO	3,920
Polk	MO	10,660
Pulaski	MO	13,820
Shannon	MO	3,230
St. Clair	MO	4,040
Stone	MO	12,750
Taney	MO	17,780
Texas	MO	9,410
Webster	MO	12,550
Wright	MO	7,190

St. Joseph, MO (201)

DMA TV Households: 45,840

% of U.S. TV Households: .041

KQTV St. Joseph, MO, ch. 2, ABC
KTAJ-TV St. Joseph, MO, ch. 16, IND

DMA Counties	State	TV Households
Doniphan	KS	2,990
Andrew	MO	6,460
Buchanan	MO	32,940
De Kalb	MO	3,450

St. Joseph, MO

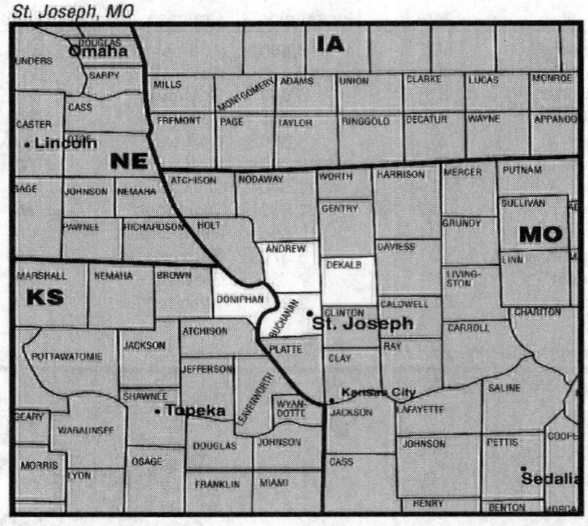

Maps courtesy of Nielsen Media Research

St. Louis, MO

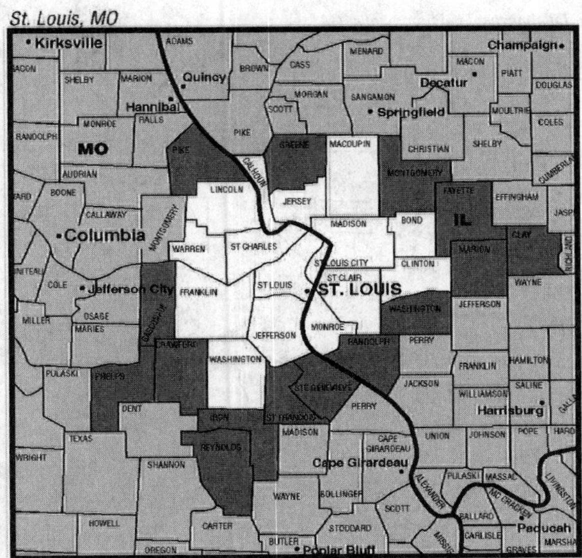

St. Louis (21)

DMA TV Households: 1,228,980
% of U.S. TV Households: 1.104

KTVI St. Louis, ch. 2, Fox
KMOV St. Louis, ch. 4, CBS
KSDK St. Louis, ch. 5, NBC
***KETC** St. Louis, ch. 9, ETV
KPLR-TV St. Louis, ch. 11, CW
KNLC St. Louis, ch. 24, IND
KDNL-TV St. Louis, ch. 30, ABC
WRBU East St. Louis, IL, ch. 46, MyNetworkTV

DMA Counties	State	TV Households	DMA Counties	State	TV Households
Bond	IL	6,270	Franklin	MO	37,490
Calhoun	IL	2,090	Gasconade	MO	6,290
Clay	IL	5,650	Iron	MO	3,950
Clinton	IL	13,340	Jefferson	MO	79,760
Fayette	IL	8,030	Lincoln	MO	17,670
Greene	IL	5,630	Phelps	MO	16,110
Jersey	IL	8,590	Pike	MO	6,710
Macoupin	IL	19,410	Reynolds	MO	2,690
Madison	IL	104,560	St. Charles	MO	123,940
Marion	IL	15,850	St. Francois	MO	23,190
Monroe	IL	11,870	Ste. Genevieve	MO	6,790
Montgomery	IL	11,300	St. Louis	MO	401,810
Randolph	IL	12,040	St. Louis-Ind	MO	142,440
St. Clair	IL	99,730	Warren	MO	11,280
Washington	IL	5,540	Washington	MO	8,880
Crawford	MO	9,290			

Syracuse, NY

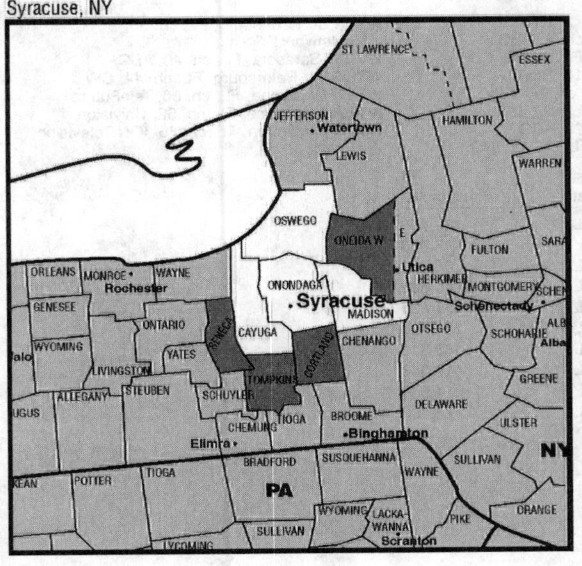

Syracuse, NY (79)

DMA TV Households: 386,940
% of U.S. TV Households: .348

WSTM-TV Syracuse, NY, ch. 3, NBC
WTVH Syracuse, NY, ch. 5, CBS
WSYR-TV Syracuse, NY, ch. 9, ABC
***WCNY-TV** Syracuse, NY, ch. 24, ETV
WNYS-TV Syracuse, NY, ch. 43, MyNetworkTV
WNYI Ithaca, NY, ch. 52, IND
WSPX-TV Syracuse, NY, ch. 56, IND
WSYT Syracuse, NY, ch. 68, Fox

DMA Counties	State	TV Households
Cayuga	NY	31,030
Cortland	NY	18,280
Madison	NY	26,210
Oneida West	NY	34,460
Onondaga	NY	182,260
Oswego	NY	46,410
Seneca	NY	12,860
Tompkins	NY	35,430

Maps courtesy of Nielsen Media Research

Tallahassee, FL-Thomasville, GA (108)

DMA TV Households: 266,210
% of U.S. TV Households: .239

WCTV Thomasville, GA, ch. 6, CBS, MyNetworkTV
***WFSU-TV** Tallahassee, FL, ch. 11, ETV
WTLF Tallahassee, FL, ch. 24, IND
WTXL-TV Tallahassee, FL, ch. 27, ABC
WTWC-TV Tallahassee, FL, ch. 40, NBC
WSWG Valdosta, GA, ch. 44, CBS, MyNetworkTV
WTLH Bainbridge, GA, ch. 49, Fox, CW
WFXU Live Oak, FL, ch. 57, CW

DMA Counties	State	TV Households	DMA Counties	State	TV Households
Gadsden	FL	16,780	Brooks	GA	6,220
Hamilton	FL	4,280	Decatur	GA	10,570
Jefferson	FL	5,060	Echols	GA	1,470
Lafayette	FL	2,300	Grady	GA	9,180
Leon	FL	106,850	Lanier	GA	2,770
Madison	FL	6,940	Lowndes	GA	34,800
Suwannee	FL	14,770	Miller	GA	2,450
Taylor	FL	7,580	Seminole	GA	3,550
Wakulla	FL	10,510	Thomas	GA	17,480

Tallahassee, FL-Thomasville, GA

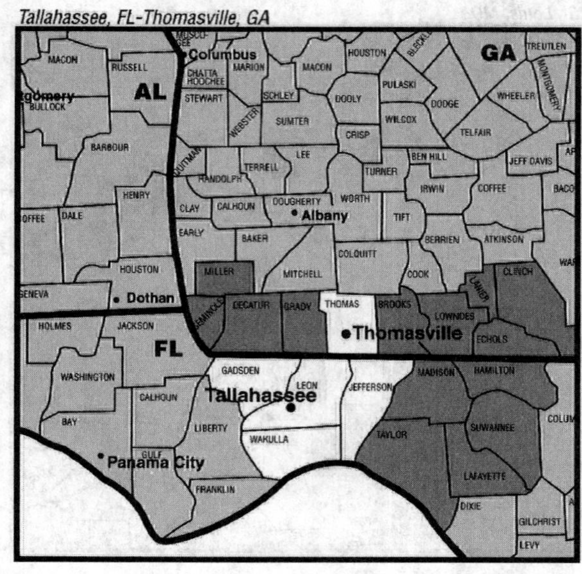

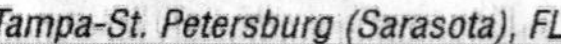

Tampa-St. Petersburg (Sarasota), FL

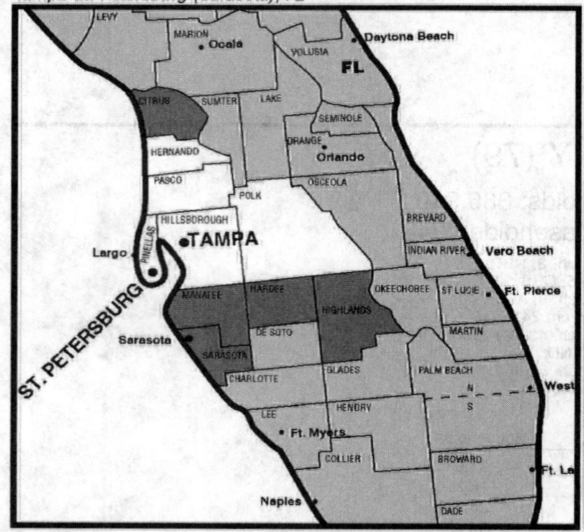

Tampa-St. Petersburg (Sarasota), FL (12)

DMA TV Households: 1,755,750
% of U.S. TV Households: 1.577

***WEDU** Tampa, FL, ch. 3, ETV
WFLA-TV Tampa, FL, ch. 8, NBC
WTSP St. Petersburg, FL, ch. 10, CBS
WTVT Tampa, FL, ch. 13, Fox
***WUSF-TV** Tampa, FL, ch. 16, ETV
WCLF Clearwater, FL, ch. 22, IND
WFTS Tampa, FL, ch. 28, ABC
WMOR-TV Lakeland, FL, ch. 32, IND
WTTA St. Petersburg, FL, ch. 38, MyNetworkTV
WWSB Sarasota, FL, ch. 40, ABC
WTOG St. Petersburg, FL, ch. 44, CW
WFTT-TV Tampa, FL, ch. 50, TeleFutura
WVEA-TV Venice, FL, ch. 62, Univision
WXPX Bradenton, FL, ch. 66, ION Television

DMA Counties	State	TV Households
Citrus	FL	61,460
Hardee	FL	8,080
Hernando	FL	68,280
Highlands	FL	40,800
Hillsborough	FL	451,220
Manatee	FL	134,130
Pasco	FL	185,310
Pinellas	FL	418,040
Polk	FL	214,930
Sarasota	FL	173,500

Maps courtesy of Nielsen Media Research

Terre Haute, IN (151)

DMA TV Households: 144,880
% of U.S. TV Households: .130

WTWO Terre Haute, IN, ch. 2, NBC
WTHI-TV Terre Haute, IN, ch. 10, CBS
***WUSI-TV** Olney, IL, ch. 16, ETV
***WVUT** Vincennes, IN, ch. 22, ETV
WFXW Terre Haute, IN, ch. 38, Fox

DMA Counties	State	TV Households	DMA Counties	State	TV Households
Clark	IL	6,950	Greene	IN	13,500
Crawford	IL	7,530	Knox	IN	15,330
Jasper	IL	3,940	Martin	IN	4,010
Lawrence	IL	6,060	Parke	IN	6,320
Richland	IL	6,380	Sullivan	IN	7,850
Clay	IN	10,460	Vermillion	IN	6,640
Daviess	IN	10,300	Vigo	IN	39,610

Terre Haute, IN

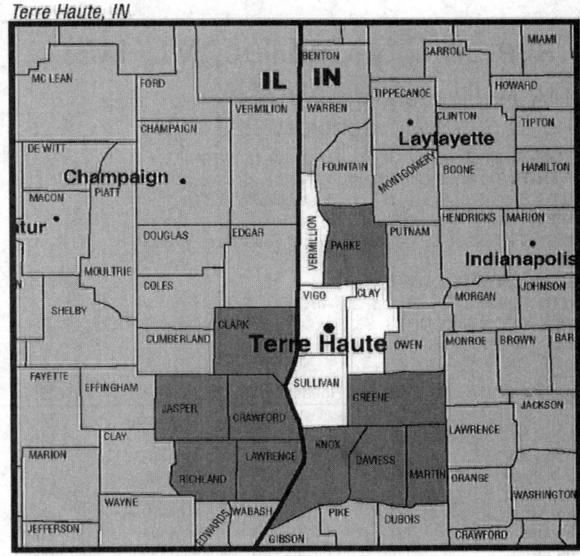

Toledo, OH

Toledo, OH (71)

DMA TV Households: 425,820
% of U.S. TV Households: .382

WTOL Toledo, OH, ch. 11, CBS
WTVG Toledo, OH, ch. 13, ABC
WNWO-TV Toledo, OH, ch. 24, NBC
***WBGU-TV** Bowling Green, OH, ch. 27, ETV
***WGTE-TV** Toledo, OH, ch. 30, ETV
WUPW Toledo, OH, ch. 36, Fox
WLMB Toledo, OH, ch. 40, IND

DMA Counties	State	TV Households
Lenawee	MI	38,220
Defiance	OH	15,290
Fulton	OH	16,160
Hancock	OH	29,070
Henry	OH	11,220
Lucas	OH	181,050
Ottawa	OH	17,170
Sandusky	OH	24,160
Seneca	OH	22,080
Williams	OH	15,050
Wood	OH	47,310
Wyandot	OH	9,040

Topeka, KS (138)

DMA TV Households: 171,310
% of U.S. TV Households: .154

***KTWU** Topeka, KS, ch. 11, ETV
WIBW-TV Topeka, KS, ch. 13, CBS, MyNetworkTV
KSNT Topeka, KS, ch. 27, NBC, CW
KTKA-TV Topeka, KS, ch. 49, ABC

DMA Counties	State	TV Households
Brown	KS	3,940
Clay	KS	3,570
Cloud	KS	3,750
Coffey	KS	3,320
Geary	KS	8,810
Jackson	KS	5,130
Jefferson	KS	7,140
Lyon	KS	13,280
Marshall	KS	4,120
Morris	KS	2,420
Nemaha	KS	3,720
Osage	KS	6,750
Pottawatomie	KS	7,170
Riley	KS	22,330
Shawnee	KS	70,710
Wabaunsee	KS	2,670
Washington	KS	2,480

Topeka, KS

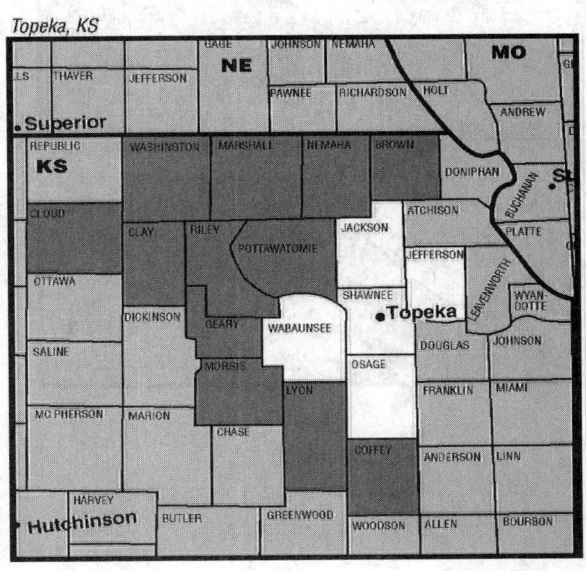

Maps courtesy of Nielsen Media Research

Traverse City-Cadillac, MI (113)

DMA TV Households: 248,680
% of U.S. TV Households: .223

WTOM-TV Cheboygan, MI, ch. 4, satellite to WPBN-TV
WPBN-TV Traverse City, MI, ch. 7, NBC
WGTQ Sault Ste. Marie, MI, ch. 8, satellite to WGTU
WWTV Cadillac, MI, ch. 9, CBS
WWUP-TV Sault Ste. Marie, MI, ch. 10, satellite to WWTV
***WCMW** Manistee, MI, ch. 21, ETV
***WCMV** Cadillac, MI, ch. 27, ETV
WGTU Traverse City, MI, ch. 29, ABC
WFQX-TV Cadillac, MI, ch. 33, Fox
WFUP Vanderbilt, MI, ch. 45, IND

DMA Counties	State	TV Households	DMA Counties	State	TV Households
Antrim	MI	9,710	Mackinac	MI	4,740
Benzie	MI	7,140	Manistee	MI	10,130
Charlevoix	MI	10,720	Mason	MI	11,740
Cheboygan	MI	11,520	Mecosta	MI	16,250
Chippewa	MI	13,780	Missaukee	MI	5,880
Clare	MI	12,980	Montmorency	MI	4,480
Crawford	MI	6,010	Osceola	MI	9,170
Emmet	MI	13,740	Oscoda	MI	3,670
Grand Traverse	MI	33,420	Otsego	MI	9,780
Kalkaska	MI	6,800	Presque Isle	MI	6,260
Lake	MI	4,920	Roscommon	MI	11,610
Leelanau	MI	9,090	Wexford	MI	12,660
Luce	MI	2,330			

Traverse City-Cadillac, MI

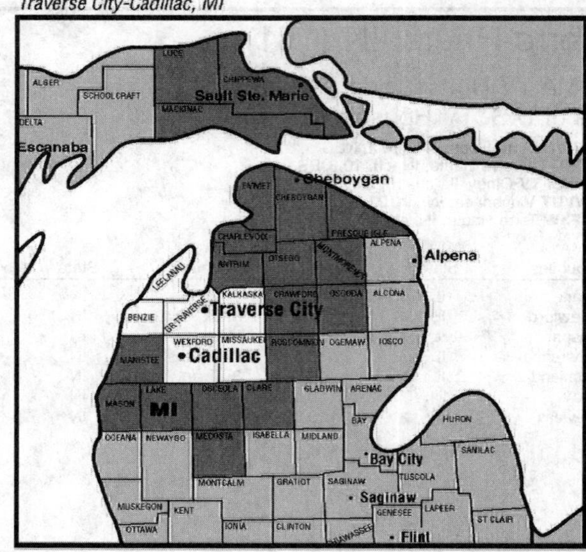

Tri-Cities; TN-VA

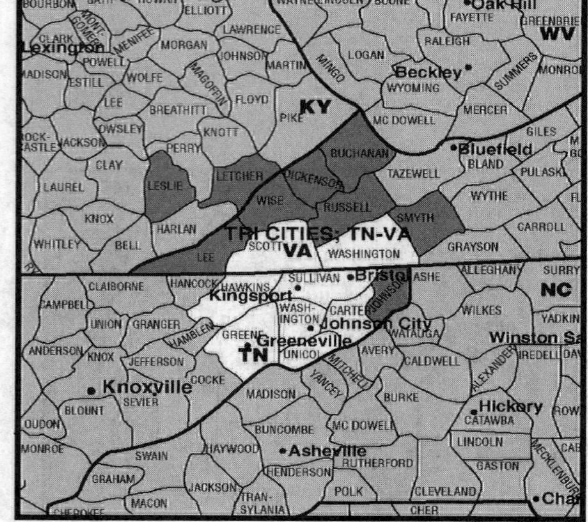

Tri-Cities, TN-VA (92)

DMA TV Households: 326,560
% of U.S. TV Households: .293

WCYB-TV Bristol, VA, ch. 5, NBC, CW
WJHL-TV Johnson City, TN, ch. 11, CBS
WKPT-TV Kingsport, TN, ch. 19, ABC, MyNetworkTV
WEMT Greeneville, TN, ch. 39, Fox
***WSBN-TV** Norton, VA, ch. 47, ETV
***WMSY-TV** Marion, VA, ch. 52, ETV
WLFG Grundy, VA, ch. 68, IND

DMA Counties	State	TV Households	DMA Counties	State	TV Households
Leslie	KY	4,720	Buchanan	VA	9,730
Letcher	KY	9,790	Dickenson	VA	7,010
Carter	TN	24,250	Lee	VA	10,290
Greene	TN	27,470	Russell	VA	11,870
Hawkins	TN	23,650	Scott	VA	9,890
Johnson	TN	7,050	Smyth	VA	13,470
Sullivan	TN	64,640	Washington	VA	29,480
Unicoi	TN	7,540	Wise	VA	17,870
Washington	TN	47,840			

Maps courtesy of Nielsen Media Research

Tucson (Sierra Vista), AZ (70)

DMA TV Households: 433,310
% of U.S. TV Households: .389

KFTU-TV Douglas, AZ, ch. 3, TeleFutura
KVOA Tucson, AZ, ch. 4, NBC
***KUAT-TV** Tucson, AZ, ch. 6, ETV
KGUN Tucson, AZ, ch. 9, ABC
KMSB-TV Tucson, AZ, ch. 11, Fox
KOLD-TV Tucson, AZ, ch. 13, CBS
KTTU-TV Tucson, AZ, ch. 18, MyNetworkTV
***KUAS-TV** Tucson, AZ, ch. 27, ETV
KHRR Tucson, AZ, ch. 40, Telemundo
KUVE-TV Green Valley, AZ, ch. 46, Univision
KWBA Sierra Vista, AZ, ch. 58, CW

DMA Counties	State	TV Households
Cochise	AZ	48,740
Pima	AZ	371,340
Santa Cruz	AZ	13,230

Tucson (Sierra Vista), AZ

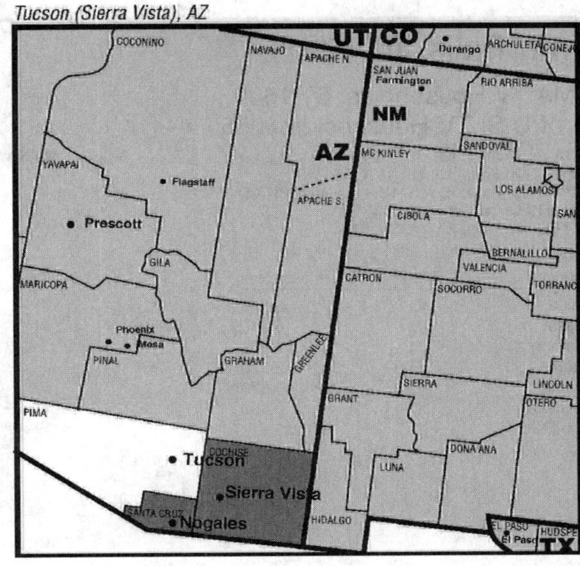

Tulsa, OK

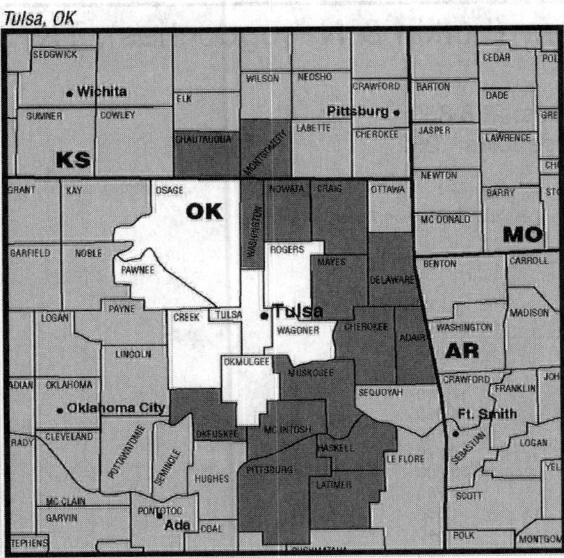

Tulsa, OK (62)

DMA TV Households: 513,090
% of U.S. TV Households: .461

KJRH Tulsa, OK, ch. 2, NBC
***KOET** Eufaula, OK, ch. 3, ETV
KOTV Tulsa, OK, ch. 6, CBS
KTUL Tulsa, OK, ch. 8, ABC
***KOED-TV** Tulsa, OK, ch. 11, ETV
KDOR-TV Bartlesville, OK, ch. 17, IND
KQCW Muskogee, OK, ch. 19, CW
KOKI-TV Tulsa, OK, ch. 23, Fox
***KRSC-TV** Claremore, OK, ch. 35, ETV
KMYT-TV Tulsa, OK, ch. 41, MyNetworkTV
KTPX Okmulgee, OK, ch. 44, ION Television
KWHB Tulsa, OK, ch. 47, IND
KGEB Tulsa, OK, ch. 53, IND

DMA Counties	State	TV Households
Chautauqua	KS	1,630
Montgomery	KS	13,860
Adair	OK	7,770
Cherokee	OK	16,980
Craig	OK	5,610
Creek	OK	25,160
Delaware	OK	15,670
Haskell	OK	4,610
Latimer	OK	3,930
Mayes	OK	15,200
McIntosh	OK	8,160
Muskogee	OK	26,810
Nowata	OK	4,240
Okfuskee	OK	3,830
Okmulgee	OK	15,200
Osage	OK	17,020
Pawnee	OK	6,420
Pittsburg	OK	17,330
Rogers	OK	30,120
Tulsa	OK	229,120
Wagoner	OK	24,100
Washington	OK	20,320

Maps courtesy of Nielsen Media Research

Twin Falls, ID (191)

DMA TV Households: 61,160
% of U.S. TV Households: .055

KIDA Sun Valley, ID, ch. 5, IND
KMVT Twin Falls, ID, ch. 11, CBS
***KIPT** Twin Falls, ID, ch. 13, satellite to *KAID
***KBGH** Filer, ID, ch. 19, ETV
KXTF Twin Falls, ID, ch. 35, IND

DMA Counties	State	TV Households
Blaine	ID	8,540
Cassia	ID	6,910
Gooding	ID	5,010
Jerome	ID	6,670
Lincoln	ID	1,670
Minidoka	ID	6,490
Twin Falls	ID	25,870

Twin Falls, ID

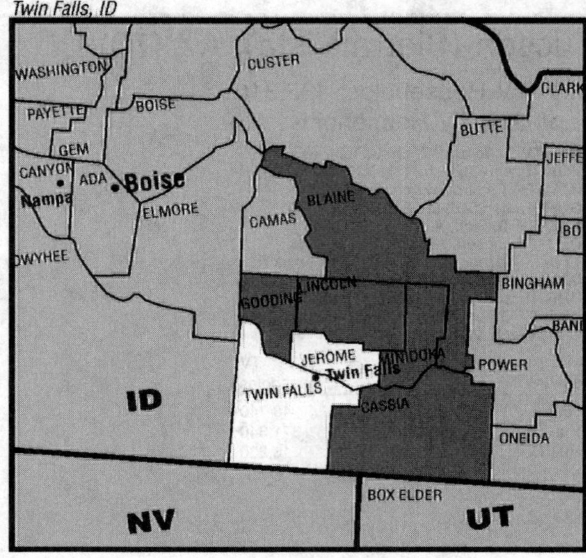

Tyler-Longview (Lufkin & Nacogdoches), TX

Tyler-Longview (Lufkin & Nacogdoches), TX (111)

DMA TV Households: 258,860
% of U.S. TV Households: .232

KLTV Tyler, TX, ch. 7, ABC
KTRE Lufkin, TX, ch. 9, satellite to KLTV
KYTX Nacogdoches, TX, ch. 19, CBS, MyNetworkTV
KFXK Longview, TX, ch. 51, Fox
KCEB Longview, TX, ch. 54, CW
KETK-TV Jacksonville, TX, ch. 56, NBC

DMA Counties	State	TV Households
Angelina	TX	29,070
Camp	TX	4,610
Cherokee	TX	17,040
Franklin	TX	4,020
Gregg	TX	44,150
Houston	TX	8,070
Nacogdoches	TX	22,800
Rusk	TX	17,780
Sabine	TX	4,550
San Augustine	TX	3,540
Smith	TX	72,360
Upshur	TX	14,210
Wood	TX	16,560

Utica, NY (169)

DMA TV Households: 106,080
% of U.S. TV Households: .095

WKTV Utica, NY, ch. 2, NBC, CW
WUTR Utica, NY, ch. 20, ABC
WFXV Utica, NY, ch. 33, Fox

DMA Counties	State	TV Households
Herkimer	NY	25,800
Oneida East	NY	56,630
Otsego	NY	23,650

Utica, NY

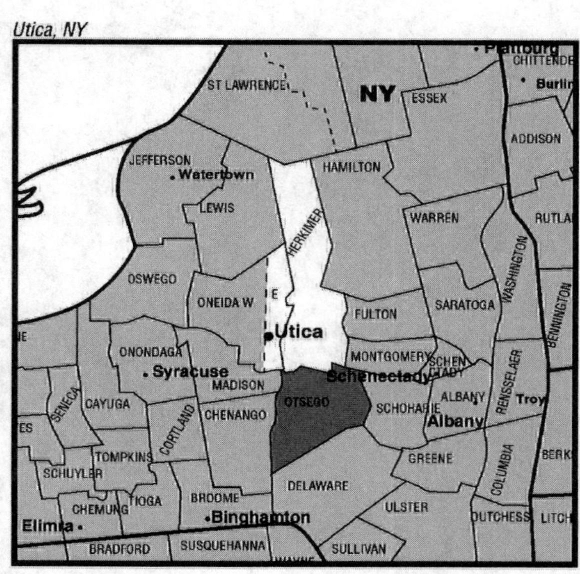

Maps courtesy of Nielsen Media Research

Victoria, TX

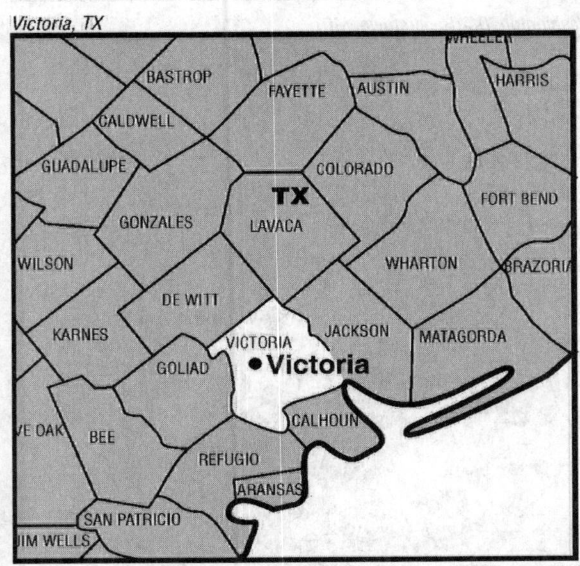

Victoria, TX (205)

DMA TV Households: 30,450
% of U.S. TV Households: .027

KVCT Victoria, TX, ch. 19, Fox
KAVU-TV Victoria, TX, ch. 25, ABC

DMA Counties	State	TV Households
Victoria	TX	30,450

Waco-Temple-Bryan, TX (95)

DMA TV Households: 311,690
% of U.S. TV Households: .280

KBTX-TV Bryan, TX, ch. 3, satellite to KWTX-TV
KCEN-TV Temple, TX, ch. 6, ABC
KWTX-TV Waco, TX, ch. 10, CBS, CW
***KAMU-TV** College Station, TX, ch. 15, ETV
KXXV Waco, TX, ch. 25, NBC
KYLE Bryan, TX, ch. 28, satellite to KWKT
***KWBU-TV** Waco, TX, ch. 34, ETV
KWKT Waco, TX, ch. 44, Fox, MyNetworkTV
***KNCT** Belton, TX, ch. 46, ETV
KAKW-TV Killeen, TX, ch. 62, Univision

DMA Counties	State	TV Households
Bell	TX	93,320
Brazos	TX	57,730
Burleson	TX	6,780
Coryell	TX	19,850
Falls	TX	6,040
Lampasas	TX	7,540
Leon	TX	6,570
Limestone	TX	8,100
Madison	TX	4,010
McLennan	TX	81,800
Milam	TX	9,540
Mills	TX	1,930
Robertson	TX	6,290
San Saba	TX	2,220

Waco-Temple-Bryan, TX

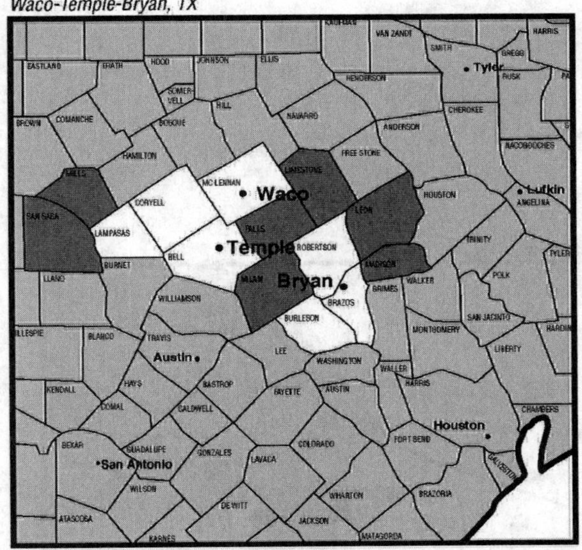

Maps courtesy of Nielsen Media Research

Washington, DC (Hagerstown, MD)

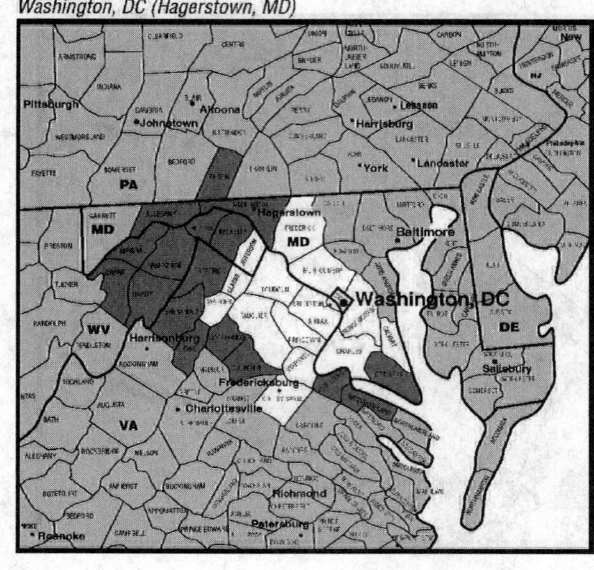

Washington, DC (Hagerstown, MD) (8)

DMA TV Households: 2,272,120
% of U.S. TV Households: 2.041

WRC-TV Washington, ch. 4, NBC
WTTG Washington, ch. 5, Fox
WJLA-TV Washington, ch. 7, ABC
WUSA Washington, ch. 9, CBS
WFDC-TV Arlington, VA, ch. 14, Univision
WDCA Washington, ch. 20, MyNetworkTV
WHAG-TV Hagerstown, MD, ch. 25, NBC
***WETA-TV** Washington, ch. 26, ETV
***WWPB** Hagerstown, MD, ch. 31, ETV
***WHUT-TV** Washington, ch. 32, ETV
***WVPY** Front Royal, VA, ch. 42, ETV
WDCW Washington, ch. 50, CW
***WNVT** Goldvein, VA, ch. 53, ETV
***WNVC** Fairfax, VA, ch. 56, ETV
WWPX Martinsburg, WV, ch. 60, ION Television
***WFPT** Frederick, MD, ch. 62, ETV
WPXW Manassas, VA, ch. 66, ION Television
WJAL Hagerstown, MD, ch. 68, IND

DMA Counties	State	TV Households
Dist of Columbia	DC	234,470
Allegany	MD	28,410
Calvert	MD	30,930
Charles	MD	50,500
Frederick	MD	81,280
Montgomery	MD	344,800
Prince George	MD	303,740
St. Marys	MD	35,250
Washington	MD	54,280
Fulton	PA	5,950
Arlington	VA	151,810
Clarke	VA	5,770
Culpeper	VA	16,060
Fairfax	VA	382,940
Fauquier	VA	24,030
Frederick	VA	37,670
Loudoun	VA	96,790
Page	VA	9,900
Prince William	VA	143,240
Rappahannock	VA	2,670
Shenandoah	VA	16,470
Spotsylvania	VA	51,310
Stafford	VA	40,430
Warren	VA	13,510
Westmoreland	VA	6,860
Berkeley	WV	38,890
Grant	WV	4,850
Hampshire	WV	8,820
Hardy	WV	5,580
Jefferson	WV	19,650
Mineral	WV	10,890
Morgan	WV	6,540

Watertown, NY

Watertown, NY (176)

DMA TV Households: 94,050
% of U.S. TV Households: .084

WWNY-TV Carthage, NY, ch. 7, CBS
***WPBS-TV** Watertown, NY, ch. 16, ETV
***WNPI-TV** Norwood, NY, ch. 18, ETV
WWTI Watertown, NY, ch. 50, ABC, CW

DMA Counties	State	TV Households
Jefferson	NY	43,210
Lewis	NY	10,040
St. Lawrence	NY	40,800
King George	VA	7,830

Maps courtesy of Nielsen Media Research

Wausau-Rhinelander, WI

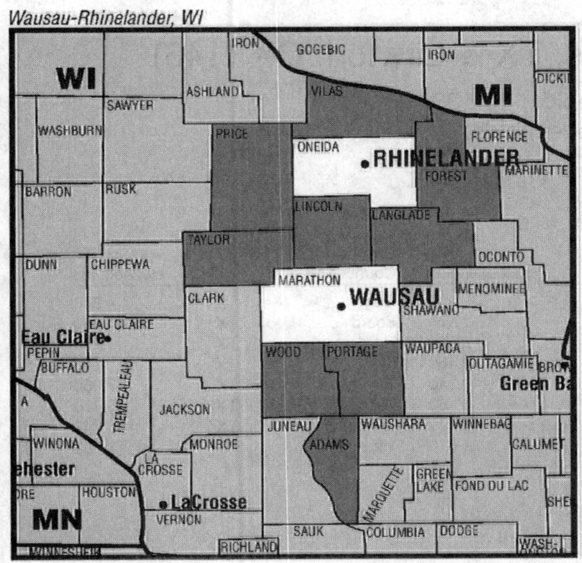

Wausau-Rhinelander, WI (134)

DMA TV Households: 180,640
% of U.S. TV Households: .162

WBIJ Crandon, WI, ch. 4, IND
WSAW-TV Wausau, WI, ch. 7, CBS, MyNetworkTV
WAOW-TV Wausau, WI, ch. 9, ABC, CW
WJFW-TV Rhinelander, WI, ch. 12, NBC
***WHRM-TV** Wausau, WI, ch. 20, ETV
WYOW Eagle River, WI, ch. 34, ABC, CW
***WLEF-TV** Park Falls, WI, ch. 36, ETV
WTPX Antigo, WI, ch. 46, ION Television

DMA Counties	State	TV Households
Adams	WI	8,480
Forest	WI	4,180
Langlade	WI	8,600
Lincoln	WI	12,200
Marathon	WI	50,470
Oneida	WI	15,720
Portage	WI	26,120
Price	WI	6,480
Taylor	WI	7,720
Vilas	WI	9,720
Wood	WI	30,950

West Palm Beach-Ft. Pierce, FL (38)

DMA TV Households: 772,140
% of U.S. TV Households: .693

WPTV West Palm Beach, FL, ch. 5, NBC
WPEC West Palm Beach, FL, ch. 12, CBS
***WTCE-TV** Fort Pierce, FL, ch. 21, ETV
WPBF Tequesta, FL, ch. 25, ABC
WFLX West Palm Beach, FL, ch. 29, Fox
WTVX Ft. Pierce, FL, ch. 34, CW
***WXEL-TV** West Palm Beach, FL, ch. 42, ETV
WFGC Palm Beach, FL, ch. 61, IND
***WPPB-TV** Boca Raton, FL, ch. 63, ETV
WPXP Lake Worth, FL, ch. 67, ION Television

DMA Counties	State	TV Households
Indian River	FL	58,540
Martin	FL	62,760
Okeechobee	FL	13,710
Palm Beach North	FL	321,700
Palm Beach South	FL	213,270
St. Lucie	FL	102,160

West Palm Beach-Ft. Pierce, FL

FL
INDIAN RIVER
Vero Beach
OKEECHOBEE
ST. LUCIE
Ft. Pierce
HIGHLANDS
MARTIN
GLADES
PALM BEACH
N
West Palm Beach
S
HENDRY
Myers
COLLIER
BROWARD
Ft. Lauderdale
DADE
MONROE

Wheeling, WV-Steubenville, OH

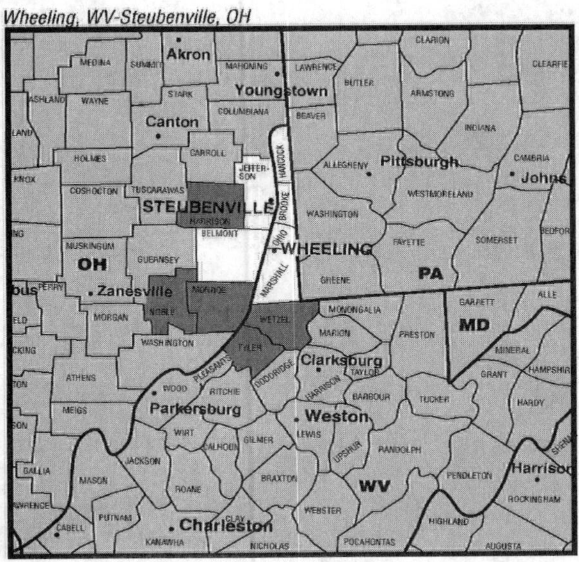

Wheeling, WV-Steubenville, OH (155)

DMA TV Households: 140,950
% of U.S. TV Households: .127

WTRF-TV Wheeling, WV, ch. 7, CBS (ABC)
WTOV-TV Steubenville, OH, ch. 9, NBC (ABC)
***WOUC-TV** Cambridge, OH, ch. 44, ETV

DMA Counties	State	TV Households
Belmont	OH	28,050
Harrison	OH	6,520
Jefferson	OH	29,290
Monroe	OH	5,740
Noble	OH	4,560
Brooke	WV	10,170
Hancock	WV	13,310
Marshall	WV	13,800
Ohio	WV	18,790
Tyler	WV	3,750
Wetzel	WV	6,970

Maps courtesy of Nielsen Media Research

Wichita Falls, TX & Lawton, OK

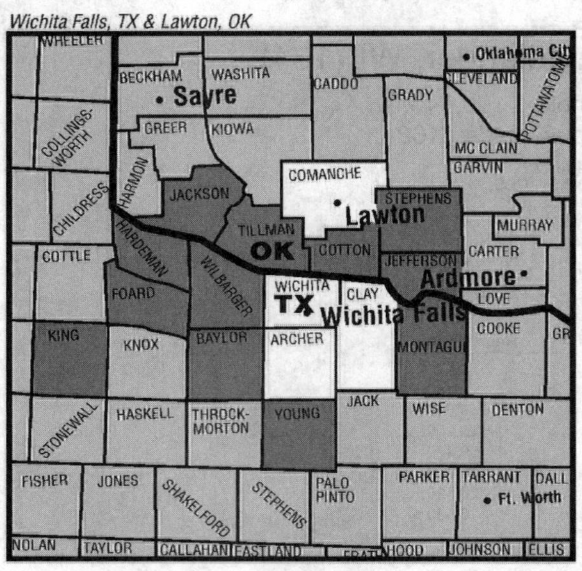

Wichita Falls, TX & Lawton, OK (146)

DMA TV Households: 152,380

% of U.S. TV Households: .137

KFDX-TV Wichita Falls, TX, ch. 3, NBC
KAUZ-TV Wichita Falls, TX, ch. 6, CBS, CW
KSWO-TV Lawton, OK, ch. 7, ABC
KJTL Wichita Falls, TX, ch. 18, Fox

DMA Counties	State	TV Households	DMA Counties	State	TV Households
Comanche	OK	39,290	Clay	TX	4,560
Cotton	OK	2,570	Foard	TX	600
Jackson	OK	9,580	Hardeman	TX	1,860
Jefferson	OK	2,450	King	TX	100
Stephens	OK	17,340	Montague	TX	7,960
Tillman	OK	3,160	Wichita	TX	45,450
Archer	TX	3,470	Wilbarger	TX	5,240
Baylor	TX	1,670	Young	TX	7,080

Wichita-Hutchinson Plus, KS (67)

DMA TV Households: 445,860

% of U.S. TV Households: .400

KSNC Great Bend, KS, ch. 2, satellite to KSNW
KSNW Wichita, KS, ch. 3, NBC
***KSWK** Lakin, KS, ch. 3, ETV
KLBY Colby, KS, ch. 4, ABC
KSWT Liberal, KS, ch. 5, IND
KBSD-TV Ensign, KS, ch. 6, satellite to KWCH-TV
KBSH-TV Hays, KS, ch. 7, satellite to KWCH-TV
KSNK McCook, NE, ch. 8, satellite to KSNW
***KOOD** Hays, KS, ch. 9, ETV
KAKE-TV Wichita, KS, ch. 10, ABC
KBSL-TV Goodland, KS, ch. 10, satellite to KBSH-TV
KSNG Garden City, KS, ch. 11, satellite to KSNW
KWCH-TV Hutchinson, KS, ch. 12, CBS
KUPK-TV Garden City, KS, ch. 13, ABC
KOCW Hoisington, KS, ch. 14, satellite to KSAS-TV
KAAS-TV Salina, KS, ch. 18, satellite to KSAS-TV
***KDCK** Dodge City, KS, ch. 21, ETV
KSAS-TV Wichita, KS, ch. 24, Fox
KSCW Wichita, KS, ch. 33, CW
KMTW Hutchinson, KS, ch. 36, MyNetworkTV

DMA Counties	State	TV Households	DMA Counties	State	TV Households
Barber	KS	2,070	Logan	KS	1,060
Barton	KS	11,430	McPherson	KS	10,770
Butler	KS	22,720	Marion	KS	4,710
Chase	KS	1,210	Meade	KS	1,690
Cheyenne	KS	1,170	Mitchell	KS	2,520
Clark	KS	880	Morton	KS	1,130
Comanche	KS	880	Ness	KS	1,270
Cowley	KS	13,380	Norton	KS	2,100
Decatur	KS	1,380	Osborne	KS	1,660
Dickinson	KS	7,830	Ottawa	KS	2,380
Edwards	KS	1,390	Pawnee	KS	2,480
Elk	KS	1,230	Pratt	KS	3,820
Ellis	KS	10,880	Rawlins	KS	1,080
Ellsworth	KS	2,390	Reno	KS	24,730
Finney	KS	11,930	Rice	KS	3,870
Ford	KS	10,930	Rooks	KS	2,190
Gove	KS	1,040	Rush	KS	1,490
Graham	KS	1,190	Russell	KS	2,890
Grant	KS	2,540	Saline	KS	21,350
Gray	KS	1,900	Scott	KS	1,730
Greeley	KS	500	Sedgwick	KS	181,710
Greenwood	KS	2,950	Seward	KS	7,400
Hamilton	KS	990	Sheridan	KS	990
Harper	KS	2,410	Sherman	KS	2,470
Harvey	KS	12,830	Stafford	KS	1,880
Haskell	KS	1,350	Stanton	KS	790
Hodgeman	KS	790	Stevens	KS	1,850
Kearny	KS	1,480	Sumner	KS	9,220
Kingman	KS	3,060	Thomas	KS	2,980
Kiowa	KS	1,080	Trego	KS	1,250
Lane	KS	780	Wallace	KS	570
Lincoln	KS	1,480	Wichita	KS	870
Dundy	NE	890			

Wichita-Hutchinson, KS Plus

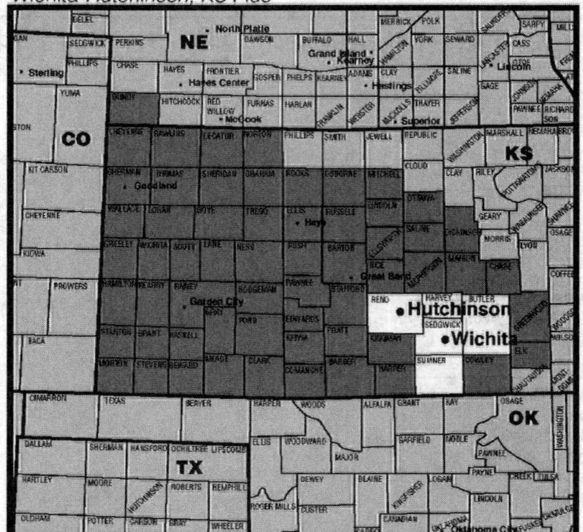

Maps courtesy of Nielsen Media Research

Wilkes Barre-Scranton, PA (53)

DMA TV Households: 590,170
% of U.S. TV Households: .530

WNEP-TV Scranton, PA, ch. 16, ABC
WYOU Scranton, PA, ch. 22, CBS
WBRE-TV Wilkes-Barre, PA, ch. 28, NBC
WSWB Scranton, PA, ch. 38, CW
***WVIA-TV** Scranton, PA, ch. 44, ETV
WQMY Williamsport, PA, ch. 53, MyNetworkTV
WOLF-TV Hazleton, PA, ch. 56, Fox
WQPX Scranton, PA, ch. 64, ION Television

DMA Counties	State	TV Households	DMA Counties	State	TV Households
Bradford	PA	24,490	Northumberland	PA	37,700
Carbon	PA	25,570	Schuylkill	PA	58,950
Clinton	PA	14,520	Snyder	PA	13,710
Columbia	PA	23,320	Sullivan	PA	2,520
Lackawanna	PA	85,590	Susquehanna	PA	16,500
Luzerne	PA	129,420	Union	PA	13,340
Lycoming	PA	45,790	Wayne	PA	19,670
Monroe	PA	59,280	Wyoming	PA	10,850
Montour	PA	6,950			

Wilkes Barre-Scranton, PA

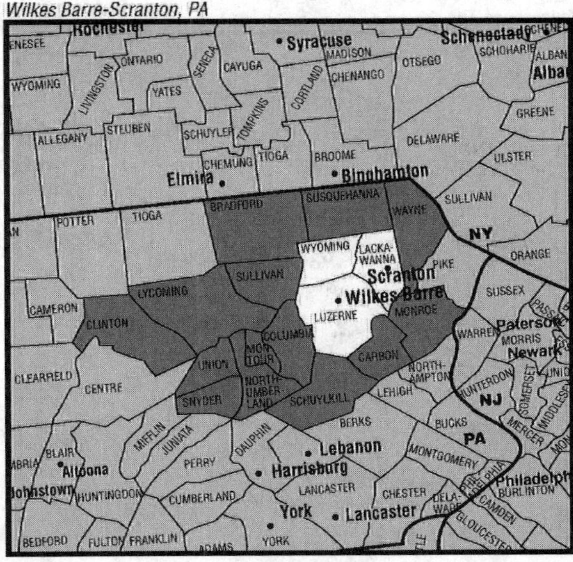

Wilmington, NC

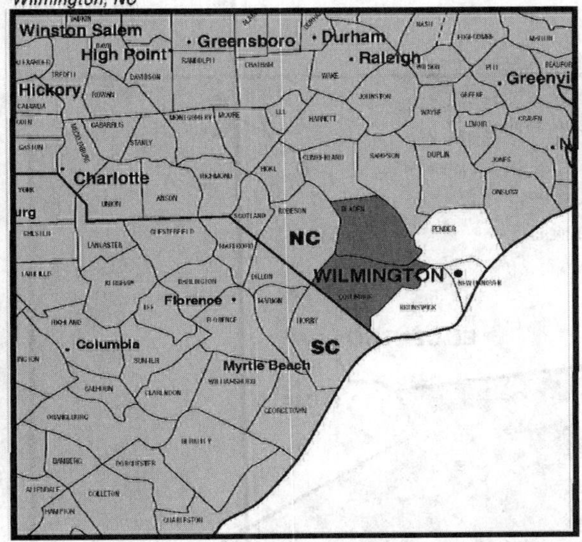

Wilmington, NC (136)

DMA TV Households: 174,170
% of U.S. TV Households: .156

WWAY Wilmington, NC, ch. 3, ABC
WECT Wilmington, NC, ch. 6, NBC
WSFX-TV Wilmington, NC, ch. 26, Fox
***WUNJ-TV** Wilmington, NC, ch. 39, ETV

DMA Counties	State	TV Households
Bladen	NC	13,300
Brunswick	NC	39,710
Columbus	NC	21,580
New Hanover	NC	80,770
Pender	NC	18,810

Yakima-Pasco-Richland-Kennewick, WA (125)

DMA TV Households: 213,780
% of U.S. TV Households: .192

KCWK Walla Walla, WA, ch. 9, CW
KFFX-TV Pendleton, OR, ch. 11, Fox
KEPR-TV Pasco, WA, ch. 19, satellite to KIMA-TV
KNDO Yakima, WA, ch. 23, NBC
KNDU Richland, WA, ch. 25, satellite to KNDO
KIMA-TV Yakima, WA, ch. 29, CBS
***KTNW** Richland, WA, ch. 31, ETV
KAPP Yakima, WA, ch. 35, ABC, MyNetworkTV
KVEW Kennewick, WA, ch. 42, satellite to KAPP
***KYVE** Yakima, WA, ch. 47, ETV

DMA Counties	State	TV Households
Morrow	OR	3,820
Umatilla	OR	24,660
Benton	WA	58,850
Franklin	WA	18,780
Kittitas	WA	13,230
Walla Walla	WA	19,620
Yakima	WA	74,820

Yakima-Pasco-Richland-Kennewick, WA

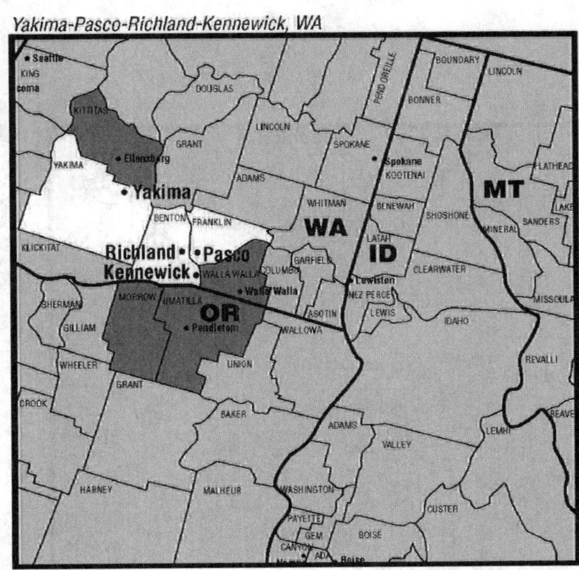

Maps courtesy of Nielsen Media Research

Youngstown, OH

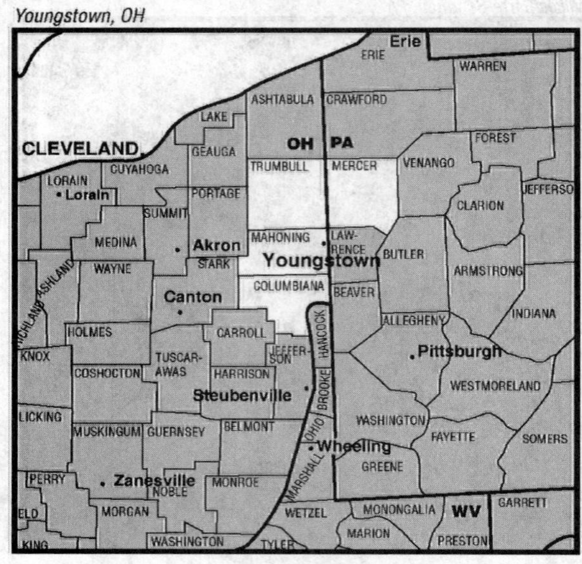

Youngstown, OH (103)

DMA TV Households: 276,550
% of U.S. TV Households: .248

WFMJ-TV Youngstown, OH, ch. 21, NBC, CW
WKBN-TV Youngstown, OH, ch. 27, CBS
WYTV Youngstown, OH, ch. 33, ABC, MyNetworkTV
***WNEO** Alliance, OH, ch. 45, ETV

DMA Counties	State	TV Households
Columbiana	OH	42,450
Mahoning	OH	100,120
Trumbull	OH	87,480
Mercer	PA	46,500

Yuma, AZ-El Centro, CA (167)

DMA TV Households: 107,360
% of U.S. TV Households: .096

KVYE El Centro, CA, ch. 7, Univision
KECY-TV El Centro, CA, ch. 9, Fox, ABC
KYMA Yuma, AZ, ch. 11, NBC
KSWT Yuma, AZ, ch. 13, CBS, CW
KAJB Calipatria, CA, ch. 54, TeleFutura

DMA Counties	State	TV Households
Yuma	AZ	63,260
Imperial	CA	44,100

Yuma, AZ-El Centro, CA

Zanesville, OH

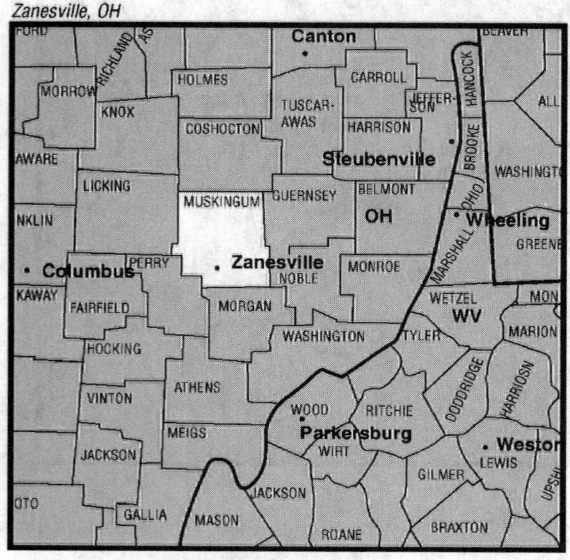

Maps courtesy of Nielsen Media Research

Zanesville, OH (203)

DMA TV Households: 33,090
% of U.S. TV Households: .030

WHIZ-TV Zanesville, OH, ch. 18, NBC

DMA County	State	TV Households
Muskingum	OH	33,090

Multi-City DMA Cross-Reference

The following cities are in hyphenated markets, but are not the first city given in such a market; e.g., Troy in Albany-Schenectady-Troy, NY. They are listed alphabetically.

Section C

Cable

Top 25 Cable/Satellite Operators

Ranked by Basic Subscribers*

Rank	Company	Subscribers
1	Comcast Cable Comm.	24,236.0
2	DirecTV	16,188.0
3	Time Warner Cable	13,448.0
4	EchoStar	13,415.0
5	Cox Communications	5,451.0
6	Charter Communications	5,415.4
7	Cablevision Systems	3,139.0
8	Bright House Networks (e)	2,321.0
9	Suddenlink Communications (&, e)	1,422.2
10	Mediacom LLC (&)	1,362.0
11	Insight Communications	1,344.0
12	CableOne (&)	703.2
13	WideOpenWest (e)	363.4
14	RCN Corp. (&)	354.0
15	Bresnan (e)	296.6
16	Service Electric (e)	290.4
17	Atlantic Broadband	288.7
18	Armstrong Group of Co.	232.6
19	Midcontinent Communications (&)	200.9
20	Pencor Services (e)	184.4
21	Knology Holdings	181.0
22	Millennium Digital Media (e)	158.5
23	Buckeye CableSystem	146.8
24	Northland Communications	144.6
25	General Communication	140.2

* As of March 2007.
Unless otherwise noted, counts include owned and managed subscribers. Subscriber figures in thousands (add 000).
(&) Counts include recent sale or acquisition.
(e) Estimate.

Major Cable Systems/Clusters

Ranked by Basic Subscribers

Rank	Company	Subscribers
1	Cablevision Systems Corp., Greater New York Area, NY	3,127,000
2	Time Warner Cable, Ohio	2,400,000
3	Time Warner Cable, Texas	2,100,000
4	Time Warner Cable, Carolinas	2,100,000
5	Time Warner Cable, Los Angeles	2,000,000
6	Comcast, Philadelphia, PA	1,800,000
7	Time Warner Cable, New York (excluding NY City)	1,800,000
8	Comcast, Boston, MA	1,600,000
9	Comcast, Chicago, IL	1,600,000
10	Time Warner Cable, New York City	1,600,000
11	Comcast, San Francisco Bay Area, CA	1,500,000
12	Bright House Networks., Tampa Bay, FL	1,052,030
13	Comcast, Seattle, WA	1,000,000
14	Cox Communications, AZ	944,460
15	Comcast, Washington, DC	900,000
16	Bright House Networks., Central FL	800,835
17	Comcast, Atlanta, GA	800,000
18	Comcast, Detroit, MI	800,000
19	Comcast, Central PA	721,000
20	Mediacom, South Central	703,092
21	Comcast, Denver, CO	700,000
22	Comcast, Houston, TX*	700,000
23	Comcast, Miami, FL	700,000
24	Comcast, New York	700,000
25	Comcast, Pittsburgh, PA	700,000
26	Mediacom, North Central	676,908
27	Charter Comm., Central States	633,600
28	Comcast, West Palm Beach Region, FL	612,000
29	Comcast, Baltimore, MD	600,000
30	Time Warner Cable, Wisconsin	570,198
31	Cox Communications, San Diego, CA	543,606
32	Charter Comm., South Carolina	532,400
33	Comcast, Portland, OR	500,000
34	Comcast, Sacramento, CA	500,000
35	Comcast, St. Paul/Minneapolis, MN	500,000
36	Cox Communications, New England	460,752
37	Cox Communications, Oklahoma	450,900
38	Charter Comm., Tennessee	431,300
39	Cox Communications, Las Vegas	425,957
40	Cox Communications, Hampton Roads, VA	419,325
41	Time Warner Cable, Hawaii	401,146
42	Time Warner Cable, California (excluding L.A.)	400,000
43	Charter Comm., Alabama	343,500
44	Charter Comm., Minnesota/Nebraska	341,000
45	Comcast, Nashville, TN	332,000
46	Comcast, Albuquerque/Santa Fe, NM; Arizona	311,000
47	Cox Communications, Kansas	310,199
48	Comcast, Michigan Market (Grand Rapids & Lansing)	310,000

Rank	Company	Subscribers
49	Time Warner Cable, Kansas City, MO	300,017
50	Comcast, Indianapolis, IN	300,000
51	Charter Comm., North Wisconsin	291,400
52	Charter Comm., Georgia	289,800
53	Time Warner Cable, New England	287,000
54	Charter Comm., L.A. Metro	286,500
55	Insight Comm., Louisville, KY	285,654
56	Comcast, Jacksonville, FL	283,000
57	Cox Communications, Orange County, CA	277,000
58	Charter Comm., Northwest	274,700
59	Cox Communications, Northern Virginia	264,730
60	Comcast, Richmond, VA	258,000
61	Charter Comm., New England	256,100
62	Comcast, Salt Lake City, UT	253,000
63	Comcast, Chico/Yuba/Fresno,Stockton/Modesto, CA	250,000
64	Charter Comm., Southern Wisconsin	232,700
65	Comcast, Memphis, TN	226,641
66	Cox Communications, Omaha, NE	205,115
67	Charter Comm., Texas	201,800
68	Charter Comm., Northern Michigan	200,800
69	Comcast, Tampa/Sarasota, FL	200,000
70	Charter Comm., Eastern Michigan	198,800
71	Charter Comm., West Michigan	193,100
72	Charter Comm., Central California	184,100
73	Cox Communications, New Orleans, LA	181,876
74	Comcast, Knoxville, TN	177,632
75	Cox Communications, Gulf Coast/Florida	169,852
76	Charter Comm., Louisiana	163,100
77	Comcast, Salisbury, MD	150,000
78	Comcast, Chattanooga, TN	142,899
79	Charter Comm., Nevada	140,800
80	Atlantic Broadband, Western PA	138,774
81	Charter Comm., Inland Empire	137,800
82	Buckeye CableSystems, Toledo, OH	127,613
83	Insight Comm., Peoria, IL	126,530
84	Insight Comm., Northeast Indiana	119,749
85	Insight Comm., Springfield, IL	118,841
86	Comcast, Orlando, FL	100,000
	Total	**51,449,630**

All data as of December 31, 2006 except as noted.

* As of January 1, 2007

Major Cable TV Systems, by Owner

System rankings are given in parentheses.

Atlantic Broadband
Western PA (80)

Bright House Networks
Central FL (16)
Tampa Bay, FL (12)

Buckeye CableSystems
Toledo, OH (82)

Cablevision
Greater New York Area, NY (1)

Charter Comm.
Alabama (43)
Central California (72)
Central States (27)
Eastern Michigan (70)
Georgia (52)
Inland Empire (81)
L.A. Metro (54)
Louisiana (76)
Minnesota/Nebraska (44)
Nevada (79)
New England (61)
North Wisconsin (51)
Northern Michigan (68)
Northwest (58)
South Carolina (32)
Southern Wisconsin (64)
Tennessee (38)
Texas (67)
West Michigan (71)

Comcast Corp.
Albuquerque/Santa Fe, NM; Arizona (46)
Atlanta, GA (17)
Baltimore, MD (29)
Boston, MA (8)
Central PA (19)
Chattanooga, TN (78)
Chicago, IL (9)
Chico/Yuba/Fresno,Stockton/Modesto, CA (63)
Denver, CO (21)
Detroit, MI (18)
Houston, TX (22)
Indianapolis, IN (50)
Jacksonville, FL (56)
Knoxville, TN (74)
Memphis, TN (65)
Miami, FL (23)
Michigan Market (Grand Rapids & Lansing) (48)
Nashville, TN (45)
New York (24)
Orlando, FL (86)
Philadelphia, PA (6)
Pittsburgh, PA (25)
Portland, OR (33)
Richmond, VA (60)
Sacramento, CA (34)
Salisbury, MD (77)
Salt Lake City, UT (62)
San Francisco Bay Area, CA (11)
Seattle, WA (13)
St. Paul/Minneapolis, MN (35)
Tampa/Sarasota, FL (69)
Washington, DC (15)
West Palm Beach Region, FL (28)

Cox Comm.
Arizona (14)
Gulf Coast/Florida (75)
Hampton Roads, VA (40)
Kansas (47)
Las Vegas (39)
New England (36)
New Orleans, LA (73)
Northern Virginia (59)
Oklahoma (37)
Omaha, NE (66)
Orange County, CA (57)
San Diego, CA (31)

Insight Comm.
Louisville, KY (55)
Peoria, IL (83)
Northeast Indiana (84)
Springfield, IL (85)

Mediacom
North Central (26)
South Central (20)

Time Warner Cable
California (excluding L.A.) (42)
Carolinas (4)
Hawaii (41)
Kansas City, MO (49)
Los Angeles (5)
New England (53)
New York (excluding NY City) (7)
New York City (10)
Ohio (2)
Texas (3)
Wisconsin (30)

Cable Penetration by DMA

Listed below are the Neilsen Media Research Designated Market Areas (DMAs) with the number of cable homes and the percentage of penetration.

Designated Market Area	Cable Households	Cable Penetration (%)
Abilene-Sweetwater	53,580	47
Albany, GA	92,260	60
Albany-Schenectady-Troy	435,020	78
Albuquerque-Santa Fe	296,480	45
Alexandria, LA	60,770	68
Alpena	11,120	63
Amarillo	98,780	52
Anchorage	93,100	65
Atlanta	1,384,120	63
Augusta	161,810	65
Austin	412,400	68
Bakersfield	121,560	58
Baltimore	793,300	72
Bangor	67,470	47
Baton Rouge	238,530	74
Beaumont-Port Arthur	110,170	66
Bend, OR	34,490	60
Billings	54,570	52
Biloxi-Gulfport	91,820	68
Binghamton	103,940	75
Birmingham (Anniston, Tuscaloosa)	421,740	58
Bluefield-Beckley-Oak Hill	105,040	72
Boise	89,300	37
Boston (Manchester)	2,014,540	85
Bowling Green	50,680	66
Buffalo	441,190	69
Burlington-Plattsburgh	175,970	54
Butte-Bozeman	29,490	49
Casper-Riverton	32,990	63
Cedar Rapids-Waterloo & Dubuque	202,690	61
Champaign & Springfield-Decatur	243,330	64
Charleston, SC	197,060	68
Charleston-Huntington	301,050	63
Charlotte	669,960	64
Charlottesville	48,050	57
Chattanooga	221,020	64
Cheyenne-Scottsbluff	33,280	62
Chicago	2,225,120	64
Chico-Redding	78,120	40
Cincinnati	546,560	62
Clarksburg-Weston	65,200	60
Cleveland	1,072,940	70
Colorado Springs-Pueblo	159,570	50
Columbia, SC	215,690	57
Columbia-Jefferson City	74,330	44
Columbus, GA	154,830	76
Columbus, OH	624,460	70
Columbus-Tupelo-West Point	81,590	44
Corpus Christi	142,560	73
Dallas-Fort Worth	1,067,030	45
Davenport-Rock Island-Moline	187,660	61
Dayton	361,860	68
Denver	818,390	57
Des Moines-Ames	221,250	53
Detroit	1,328,860	69

Designated Market Area	Cable Households	Cable Penetration (%)
Dothan	70,420	71
Duluth-Superior	74,330	43
El Paso	146,520	50
Elmira	67,910	70
Erie	100,260	64
Eugene	123,130	53
Eureka	41,520	70
Evansville	157,880	54
Fairbanks	14,990	45
Fargo-Valley City	137,040	58
Flint-Saginaw-Bay City	293,070	62
Fort Myers-Naples	353,790	74
Fort Smith-Fayetteville-Springdale-Rogers	153,630	55
Fort Wayne	122,990	45
Fresno-Visalia	252,060	45
Gainesville	77,990	65
Glendive	3,140	79
Grand Junction-Montrose	35,680	51
Grand Rapids-Kalamazoo-Battle Creek	413,410	56
Great Falls	32,500	51
Green Bay-Appleton	234,830	54
Greensboro-High Point-Winston Salem	431,460	65
Greenville-New Bern-Washington	164,660	61
Greenville-Spartanburg-Asheville-Anderson	422,730	51
Greenwood-Greenville	51,840	67
Harlingen-Weslaco-Brownsville-McAllen	130,210	40
Harrisburg-Lancaster-Lebanon-York	541,080	76
Harrisonburg	54,500	62
Hartford & New Haven	835,360	82
Hattiesburg-Laurel	55,840	50
Helena	14,570	56
Honolulu	376,400	90
Houston	1,036,660	52
Huntsville-Decatur, Florence	245,440	65
Idaho Falls-Pocatello	40,790	35
Indianapolis	624,810	59
Jackson, MS	168,270	49
Jackson, TN	58,160	61
Jacksonville, Brunswick	426,810	67
Johnstown-Altoona	206,170	70
Jonesboro	52,410	59
Joplin-Pittsburg	68,750	44
Juneau	16,920	71
Kansas City	571,130	63
Knoxville	327,760	63
La Crosse-Eau Claire	118,970	57
Lafayette, IN	44,640	69
Lafayette, LA	151,390	67
Lake Charles	63,730	67
Lansing	151,390	59
Laredo	43,540	66
Las Vegas	471,250	70
Lexington	275,800	57
Lima	38,040	72
Lincoln & Hastings-Kearney	166,580	60
Little Rock-Pine Bluff	263,700	49
Los Angeles	3,115,490	56
Louisville	392,420	61
Lubbock	70,870	47

Designated Market Area	Cable Households	Cable Penetration (%)
Macon	142,470	62
Madison	206,710	56
Mankato	33,480	66
Marquette	63,630	71
Medford-Klamath Falls	87,620	53
Memphis	357,380	54
Meridian	33,680	45
Miami-Fort Lauderdale	1,040,360	68
Milwaukee	559,130	63
Minneapolis-St. Paul	923,220	55
Minot-Bismarck-Dickinson	83,510	62
Missoula	40,760	38
Mobile-Pensacola (Fort Walton Beach)	343,970	66
Monroe-El Dorado	99,870	56
Monterey-Salinas	135,500	62
Montgomery (Selma)	171,750	71
Myrtle Beach-Florence	186,990	69
Nashville	541,560	57
New Orleans	427,060	75
New York	5,957,620	81
Norfolk-Portsmouth-Newport News	516,650	72
North Platte	10,200	66
Odessa-Midland	101,790	75
Oklahoma City	397,980	60
Omaha	297,130	74
Orlando-Daytona Beach-Melbourne	1,006,070	72
Ottumwa-Kirksville	24,790	48
Paducah-Cape Girardeau-Harrisburg-Mt. Vernon	182,520	47
Palm Springs	123,960	83
Panama City	93,590	66
Parkersburg	47,900	75
Peoria-Bloomington	160,200	66
Philadelphia	2,334,000	79
Phoenix	1,032,260	60
Pittsburgh	873,460	76
Portland, OR	597,600	53
Portland-Auburn	290,220	71
Presque Isle	17,630	57
Providence-New Bedford	528,190	83
Quincy-Hannibal-Keokuk	48,160	46
Raleigh-Durham (Fayetteville)	611,990	61
Rapid City	62,020	66
Reno	149,070	57
Richmond-Petersburg	310,420	60
Roanoke-Lynchburg	232,750	52
Rochester, NY	279,380	71
Rochester-Mason City-Austin	96,390	67
Rockford	111,950	61
Sacramento-Stockton-Modesto	715,800	52
Salisbury	113,550	75
Salt Lake City	341,640	41
San Angelo	32,020	60
San Antonio	485,840	63
San Diego	839,170	81
San Francisco-Oakland-San Jose	1,700,090	71
Santa Barbara-Santa Maria-San Luis Obispo	154,070	68
Savannah	193,860	65
Seattle-Tacoma	1,218,220	71
Sherman, TX-Ada, OK	60,330	49

Designated Market Area	Cable Households	Cable Penetration (%)
Shreveport	172,470	45
Sioux City	91,540	58
Sioux Falls (Mitchell)	157,950	64
South Bend-Elkhart	149,670	45
Spokane	191,540	48
Springfield, MO	223,190	84
Springfield-Holyoke	147,620	37
St. Joseph	29,460	64
St. Louis	597,760	49
Syracuse	289,620	75
Tallahassee-Thomasville	167,140	63
Tampa-St. Petersburg, Sarasota	1,285,240	73
Terre Haute	71,210	49
Toledo	267,840	63
Topeka	111,290	65
Traverse City-Cadillac	124,540	50
Tri-Cities, TN-VA	220,470	68
Tucson (Sierra Vista)	254,820	59
Tulsa	264,500	52
Twin Falls	26,240	43
Tyler-Longview (Lufkin & Nacogdoches)	125,420	48
Utica	79,900	75
Victoria	20,760	68
Waco-Temple-Bryan	183,410	59
Washington, DC (Hagerstown)	1,485,690	65
Watertown	64,990	69
Wausau-Rhinelander	85,530	47
West Palm Beach-Fort Pierce	569,920	74
Wheeling-Steubenville	103,160	73
Wichita Falls & Lawton	83,570	55
Wichita-Hutchinson Plus	288,780	65
Wilkes Barre-Scranton	437,010	74
Wilmington	119,700	69
Yakima-Pasco-Richland-Kennewick	100,100	47
Youngstown	189,870	69
Yuma-El Centro	49,370	46
Zanesville	24,330	74

Top 50 DMA by Cable Penetration

Listed below are the Nielsen Media Research Designated Market Areas (DMAs) ranked by cable penetration.

Rank	Designated Market Area	Cable Penetration (%)
1	Honolulu	90
2	Boston (Manchester)	85
3	Springfield, MO	84
4	Palm Springs	83
4	Providence-New Bedford	83
6	Hartford & New Haven	82
7	New York	81
7	San Diego	81
9	Glendive	79
9	Philadelphia	79
11	Albany-Schenectady-Troy	78
12	Columbus, GA	76
12	Harrisburg-Lancaster-Lebanon-York	76
12	Pittsburgh	76
15	Binghamton	75
15	New Orleans	75
15	Odessa-Midland	75
15	Parkersburg	75
15	Salisbury	75
15	Syracuse	75
15	Utica	75
22	Baton Rouge	74
22	Fort Myers-Naples	74
22	Omaha	74
22	West Palm Beach-Fort Pierce	74
22	Wilkes Barre-Scranton	74
22	Zanesville	74
28	Corpus Christi	73
28	Tampa-St. Petersburg, Sarasota	73
28	Wheeling-Steubenville	73
31	Baltimore	72
31	Bluefield-Beckley-Oak Hill	72
31	Lima	72
31	Norfolk-Portsmouth-Newport News	72
31	Orlando-Daytona Beach-Melbourne	72
36	Dothan	71
36	Juneau	71
36	Marquette	71
36	Montgomery (Selma)	71
36	Portland-Auburn	71
36	Rochester, NY	71
36	San Francisco-Oakland-San Jose	71
36	Seattle-Tacoma	71
44	Cleveland	70
44	Columbus, OH	70
44	Elmira	70
44	Eureka	70
44	Johnstown-Altoona	70
44	Las Vegas	70
50	Buffalo	69
50	Detroit	69
50	Lafayette, IN	69
50	Myrtle Beach-Florence	69
50	Watertown	69
50	Wilmington	69
50	Youngstown	69

Bottom 50 DMA by Cable Penetration

Listed below are the Nielsen Media Research Designated Market Areas (DMAs) ranked by percentage of cable penetration. Boise and Fairbanks have the lowest percentage.

Rank	Designated Market Area	Cable Penetration (%)
1	Idaho Falls-Pocatello	35
2	Boise	37
2	Springfield-Holyoke	37
4	Missoula	38
5	Chico-Redding	40
5	Harlingen-Weslaco-Brownsville-McAllen	40
7	Salt Lake City	41
8	Duluth-Superior	43
8	Twin Falls	43
10	Columbia-Jefferson City	44
10	Columbus-Tupelo-West Point	44
10	Joplin-Pittsburg	44
13	Albuquerque-Santa Fe	45
13	Dallas-Fort Worth	45
13	Fairbanks	45
13	Fort Wayne	45
13	Fresno-Visalia	45
13	Meridian	45
13	Shreveport	45
13	South Bend-Elkhart	45
21	Quincy-Hannibal-Keokuk	46
21	Yuma-El Centro	46
23	Abilene-Sweetwater	47
23	Bangor	47
23	Lubbock	47
23	Paducah-Cape Girardeau-Harrisburg-Mt. Vernon	47
23	Wausau-Rhinelander	47
23	Yakima-Pasco-Richland-Kennewick	47
29	Ottumwa-Kirksville	48
29	Spokane	48
29	Tyler-Longview (Lufkin & Nacogdoches)	48
32	Butte-Bozeman	49
32	Jackson, MS	49
32	Little Rock-Pine Bluff	49
32	Sherman, TX-Ada, OK	49
32	St. Louis	49
32	Terre Haute	49
38	Colorado Springs-Pueblo	50
38	El Paso	50
38	Hattiesburg-Laurel	50
38	Traverse City-Cadillac	50
42	Grand Junction-Montrose	51
42	Great Falls	51
42	Greenville-Spartanburg-Asheville-Anderson	51
45	Amarillo	52

Rank	Designated Market Area	Cable Penetration (%)
45	Billings	52
45	Houston	52
45	Roanoke-Lynchburg	52
45	Sacramento-Stockton-Modesto	52
45	Tulsa	52

Top 50 DMA by Cable Households

Listed below are the Nielsen Media Research Designated Market Areas (DMAs) ranked by cable television households.

Rank	Designated Market Area	Cable Penetration (%)	Cable Households
1	New York	81	5,957,620
2	Los Angeles	56	3,115,490
3	Philadelphia	79	2,334,000
4	Chicago	64	2,225,120
5	Boston (Manchester)	85	2,014,540
6	San Francisco-Oakland-San Jose	71	1,700,090
7	Washington, DC (Hagerstown)	65	1,485,690
8	Atlanta	63	1,384,120
9	Detroit	69	1,328,860
10	Tampa-St. Petersburg, Sarasota	73	1,285,240
11	Seattle-Tacoma	71	1,218,220
12	Cleveland	70	1,072,940
13	Dallas-Fort Worth	45	1,067,030
14	Miami-Fort Lauderdale	68	1,040,360
15	Houston	52	1,036,660
16	Phoenix	60	1,032,260
17	Orlando-Daytona Beach-Melbourne	72	1,006,070
18	Minneapolis-St. Paul	55	923,220
19	Pittsburgh	76	873,460
20	San Diego	81	839,170
21	Hartford & New Haven	82	835,360
22	Denver	57	818,390
23	Baltimore	72	793,300
24	Sacramento-Stockton-Modesto	52	715,800
25	Charlotte	64	669,960
26	Indianapolis	59	624,810
27	Columbus, OH	70	624,460
28	Raleigh-Durham (Fayetteville)	61	611,990
29	St. Louis	49	597,760
30	Portland, OR	53	597,600
31	Kansas City	63	571,130
32	West Palm Beach-Fort Pierce	74	569,920
33	Milwaukee	63	559,130
34	Cincinnati	62	546,560
35	Nashville	57	541,560
36	Harrisburg-Lancaster-Lebanon-York	76	541,080
37	Providence-New Bedford	83	528,190
38	Norfolk-Portsmouth-Newport News	72	516,650
39	San Antonio	63	485,840
40	Las Vegas	70	471,250
41	Buffalo	69	441,190
42	Wilkes Barre-Scranton	74	437,010
43	Albany-Schenectady-Troy	78	435,020
44	Greensboro-High Point-Winston Salem	65	431,460
45	New Orleans	75	427,060
46	Jacksonville, Brunswick	67	426,810
47	Greenville-Spartanburg-Asheville-Anderson	51	422,730
48	Birmingham (Anniston, Tuscaloosa)	58	421,740
49	Grand Rapids-Kalamazoo-Battle Creek	56	413,410
50	Austin	68	412,400

Section D

Radio

Radio Group Ownership

A

ABC Inc., 77 W. 66th St., New York, NY 10023-6298. Phone: (212) 456-7777. Web Site: www.abc.com. Ownership: ABC Enterprises Inc., 100%. Note: ABC Enterprises Inc. is 100% owned by Disney Enterprises Inc. Disney Enterprises Inc. is 100% owned by The Walt Disney Co.

Stns: 42 AM. 3 FM. KDIS-FM Little Rock, AR; KMIK Tempe, AZ; KSPN(AM) Los Angeles, CA; KMKY Oakland, CA; KDIS(AM) Pasadena, CA; KIID(AM) Sacramento, CA; KDDZ Arvada, CO; WDZK Bloomfield, CT; WBWL Jacksonville, FL; WMYM(AM) Miami, FL; WDYZ(AM) Orlando, FL; WMNE(AM) Riviera Beach, FL; WWMI Saint Petersburg, FL; WDWD Atlanta, GA; WSDZ Belleville, IL; WMVP Chicago, IL; WRDZ La Grange, IL; WRDZ-FM Plainfield, IN; KQAM Wichita, KS; WDRD(AM) Newburg, KY; WBYU New Orleans, LA; WMKI(AM) Boston, MA; WFDF(AM) Farmington Hills, MI; KDIZ Golden Valley, MN; KPHN Kansas City, MO; WGFY Charlotte, NC; WWJZ Mount Holly, NJ; KALY Los Ranchos de Albuquerque, NM; WDDY(AM) Albany, NY; WEPN(AM) New York, NY; WWMK Cleveland, OH; KMUS(AM) Sperry, OK; KDZR(AM) Lake Oswego, OR; WEAE Pittsburgh, PA; WDDZ(AM) Pawtucket, RI; KESN(FM) Allen, TX; KMIC(AM) Houston, TX; KMKI Plano, TX; KRDY(AM) San Antonio, TX; KWDZ(AM) Salt Lake City, UT; WDZY Colonial Heights, VA; WRJR(AM) Portsmouth, VA; WHKT Portsmouth, VA; KKDZ Seattle, WA; WKSH(AM) Sussex, WI.

Stns: 10 TV. WLS, Chicago; WJRT-TV, Flint-Saginaw-Bay City, MI; KFSN-TV, Fresno-Visalia, CA; KTRK, Houston; KABC, Los Angeles; WABC-TV, New York; WPVI, Philadelphia; WTVD, Raleigh-Durham (Fayetteville), NC; KGO, San Francisco-Oakland-San Jose; WTVG, Toledo, OH.

Robert A. Iger, pres; Phillip J. Meek, pres; Lawrence J. Pollock, chmn owned TV stns.

ARKLATEX Radio Inc., 111 Westwood Dr., De Queen, AR 71832. Phone: (870) 642-2446. Fax: (870) 642-2442. Ownership: Jay Wallace Bunyard and Teresa Sharon Bunyard Living Revocable Trust, Jay and Teresa Bunyard sole voting trustees, 100%.

Stns: 1 AM. 2 FM. KMTB-FM Murfreesboro, AR; KNAS-FM Nashville, AR; KBHC Nashville, AR.

Jay Bunyard, pres; Bonita Smith, CFO.

AVC Communications Inc., Box 338, Cambridge, OH 43725. Phone: (740) 432-5605. Fax: (740) 432-1991. Web Site: www.yourradioplace.com. Ownership: W. Grant Hafley, 100%.

Stns: 1 AM. 1 FM. WILE-FM Byesville, OH; WILE Cambridge, OH.

Grant Hafley, pres; Joel Losego, gen mgr; David Wilson, opns mgr.

Aboriginal Voices Radio Inc., 366 Adelaide St. E., Suite 323, Toronto, ON M5A 3X9. Canada. Phone: (416) 703-1287. Fax: (416) 703-4328. Web Site: www.aboriginalradio.com.

Stns: 4 FM. CKAV-FM-3 Calgary, AB; CKAV-FM-2 Vancouver, BC; CKAV-FM-9 Ottawa, ON; CKAV-FM Toronto, ON.

Acadia Broadcasting Ltd., Box 2000, Saint John, NB E2L 3T4. Canada. Phone: (506) 633-3323. Fax: (506) 644-3485. Ownership: Brunswick News Inc., 100%.

Stns: 6 FM. CHSJ-FM Saint John, NB; CHWV-FM Saint John, NB; CHTD-FM Saint Stephen, NB; CKBW-FM Bridgewater, NS; CKBW-FM-1 Liverpool, NS; CKBW-FM-2 Shelburne, NS.

Access.1 Communications Corp., 11 Penn Plaza, 16th Fl., New York, NY 10001. Phone: (212) 714-1000. Fax: (212) 714-1563. Ownership: Sydney L. Small, 54.12%; Black Enterprise/Greenwich Street Capital Partners, 19.47%; MESBIC Ventures Inc., 5.54%; Chesley Maddox-Dorsey, 2.84%; and Adriane Gaines, 1.85%.

Stns: 7 AM. 13 FM. KSYR(FM) Benton, LA; KDKS-FM Blanchard, LA; KBTT(FM) Haughton, LA; KLKL(FM) Minden, LA; KOKA Shreveport, LA; WMGM(FM) Atlantic City, NJ; WGYM(AM) Hammonton, NJ; WTKU-FM Ocean City, NJ; WJSE-FM Petersburg, NJ; WTAA(AM) Pleasantville, NJ; WOND(AM) Pleasantville, NJ; WWRL New York, NY; KOYE(FM) Frankston, TX; KOOI-FM Jacksonville, TX; KYKX-FM Longview, TX; KFRO Longview, TX; KCUL Marshall, TX; KCUL-FM Marshall, TX; KTAL-FM Texarkana, TX; KKUS(FM) Tyler, TX.

Stns: 1 TV. WMGM-TV, Philadelphia.

Sydney L. Small, chmn/CEO.

Ace Radio Corp., 2801 Via Fortuna Dr., Suite 675, Austin, TX 78746. Phone: (713) 528-2517. Ownership: Stephen M. Hackerman, 51%; and Spectrum Radio Co., 49%.

Stns: 9 FM. KRPH(FM) Yarnell, AZ; KGRP(FM) Cazadero, CA; KQNO(FM) Coalinga, CA; KQMX(FM) Lost Hills, CA; KCOO(FM) Dunkerton, IA; WTPO(FM) New Albany, MS; WZHL(FM) New Augusta, MS; KQLP(FM) Gallup, NM; KBWT(FM) Santa Anna, TX.

Gordon Ackley Stns, Box 302179, St. Thomas, VI 00803-2179. Phone: (340) 776-3291. Fax: (340) 776-7060. Ownership: Gordon P. Ackley, 100%.

Stns: 1 AM. 2 FM. WVWI Charlotte Amalie, VI; WVJZ-FM Charlotte Amalie, VI; WWKS-FM Cruz Bay, VI.

Ad Astra Per Aspera Broadcasting Inc., 106 N. Main St., Hutchinson, KS 67501-5219. Phone: (620) 665-5758. Fax: (620) 665-6655. E-mail: cliffcshank@yahoo.com Ownership: Cliff C. Shank, 71%; Michael G. Hill, 14%.

Stns: 4 FM. KGGG(FM) Haven, KS; KWHK(FM) Hutchinson, KS; KXKU(FM) Lyons, KS; KSKU(FM) Sterling, KS.

Cliff C. Shank, gen mgr; Michael G. Hill, VP.

Adelman Broadcasting Inc., 731 N. Balsam, Ridgecrest, CA 93555. Phone: (760) 371-1700. Fax: (760) 371-1824. Web Site: adelmanbroadcasting.com. Ownership: Robert Adelman, 100%.

Stns: 1 AM. 3 FM. KEDD(FM) Johannesburg, CA; KRAJ(FM) Johannesburg, CA; KLOA Ridgecrest, CA; KLOA-FM Ridgecrest, CA.

Robert Adelman, owner.

Adonai Radio Group, 2448 E. 81st St., Suite 5500, Tulsa, OK 74137. Phone: (918) 492-2660. Fax: (918) 492-8840. E-mail: mail@kxoj.com Web Site: www.kxoj.com.

Stns: 3 AM. 7 FM. KEOJ-FM Caney, KS; KTFR-FM Chelsea, OK; KEMX-FM Locust Grove, OK; KBIX Muskogee, OK; KYAL-FM Muskogee, OK; KYAL(AM) Sapulpa, OK; KXOJ-FM Sapulpa, OK; KCXR(FM) Taft, OK; KGND(AM) Vinita, OK; KITO-FM Vinita, OK.

Michael P. Stephens, pres.

Air South Radio Inc., Box 2116, Tupelo, MS 38803. Phone: (662) 842-9595. Fax: (662) 842-9568. Ownership: Olvie E. Sisk; Kathern Sisk.

Stns: 3 FM. WLZA(FM) Eupora, MS; WFTA-FM Fulton, MS; WCNA-FM Potts Camp, MS.

Olvie E. Sisk, pres; Kathern Sisk, sec/treas.

Aisling Broadcasting of Banner Elk LLC, 1901 Bearberry Ln., Asheville, NC 28803-3241. Phone: (828) 254-3384. Ownership: Jonathan Hoffman, 50%; Donna Hoffman, 50%.

Stns: 3 AM. 3 FM. WZJS-FM Banner Elk, NC; WECR-FM Beech Mountain, NC; WXIT Blowing Rock, NC; WATA Boone, NC; WMMY(FM) Jefferson, NC; WECR Newland, NC.

Alaska Broadcast Communications Inc., 3161 Channel Dr., Suite 2, Juneau, AK 99801. Phone: (907) 586-3630. Fax: (907) 463-3685. Web Site: www.kjno.com.

Stns: 3 AM. 3 FM. KJNO(AM) Juneau, AK; KTKU-FM Juneau, AK; KTKN Ketchikan, AK; KGTW-FM Ketchikan, AK; KIFW Sitka, AK; KSBZ-FM Sitka, AK.

Richard Burns, group CEO.

Alexandra Communications Inc., 1600 Gray Lynn Dr., Walla Walla, WA 99362. Phone: (509) 527-1000. Fax: (509) 529-5534. Ownership: Thomas D. Hodgins, 50%; and Cheryl Hodgins, 50%.

Stns: 2 AM. 4 FM. KIXT(FM) Bay City, OR; KDEP(FM) Garibaldi, OR; KLKY(FM) Stanfield, OR; KUJJ(FM) Weston, OR; KRSC Othello, WA; KUJ Walla Walla, WA.

Allegheny Mountain Network Stations, Box 247, Tyrone, PA 16686. Phone: (814) 684-3200. Fax: (814) 684-1220. E-mail: amnet@aol.com Ownership: Cary H. Simpson.

Stns: 4 AM. 7 FM. WFRM Coudersport, PA; WFRM-FM Coudersport, PA; WNBQ-FM Mansfield, PA; WEEO-FM McConnellsburg, PA; WKBI Saint Marys, PA; WKBI-FM Saint Marys, PA; WQRM-FM Smethport, PA; WTRN Tyrone, PA; WGMR(FM) Tyrone, PA; WNBT Wellsboro, PA; WNBT-FM Wellsboro, PA.

Cary Simpson, pres; John F. Simpson, VP.

Alma Corp., 1282 Smallwood Dr., Suite 372, Waldorf, MD 20603. Phone: (202) 251-7589. Ownership: Dennis Wallace, 100%.

Stns: 3 FM. KUMR(FM) Doolittle, MO; KQWY(FM) Lusk, WY; KOUZ(FM) Manville, WY.

Amaturo Groups, 3101 N. Federal Hwy., Suite 601, Fort Lauderdale, FL 33306-1042. Phone: (954) 565-1411. Fax: (954) 565-1311. E-mail: jca@amaturogroups.com Ownership: Amaturo Group of L.A. Inc. (Joseph C. Amaturo, gen ptnr): KAJL(FM), KJLL-FM and KHJL(FM).

Stns: 3 FM. KAJL(FM) Adelanto, CA; KJLL-FM Fountain Valley, CA; KHJL(FM) Thousand Oaks, CA.

Joseph C. Amaturo, pres.

American Family Radio, Box 3206, Tupelo, MS 38803. Phone: (662) 844-8888. Fax: (662) 842-6791. E-mail: comments@afr.net Web Site: www.afr.net. Ownership: American Family Association, a nonprofit organization.

Stns:151 FM. WALN-FM Carrollton, AL; WAQG-FM Ozark, AL; WAQU-FM Selma, AL; WAKD-FM Sheffield, AL; WAXU-FM Troy, AL; KAPG(FM) Bentonville, AR; KBCM(FM) Blytheville, AR; KBDO(FM) Des Arc, AR; KBNV(FM) Fayetteville, AR; KARH(FM) Forrest City, AR; KAOW(FM) Fort Smith, AR; KBPW(FM) Hampton, AR; KBMJ(FM) Heber Springs, AR; KAOG(FM) Jonesboro, AR; KNLL(FM) Nashville, AR; KANX(FM) Sheridan, AR; KBMH(FM) Holbrook, AZ; KCAI(FM) Kingman, AZ; WBJY-FM Americus, GA; WAEF-FM Cordele, GA; WEBH(FM) Cuthbert, GA; WAWH-FM Dublin, GA; WBKG-FM Macon, GA; WASW-FM Waycross, GA; KAYP(FM) Burlington, IA; KIAD(FM) Dubuque, IA; KBDC(FM) Mason City, IA; KWVI(FM) Waverly, IA; WBEL-FM Cairo, IL; WBMF-FM Crete, IL; WEFI(FM) Effingham, IL; WAXR-FM Geneseo, IL; WAWF-FM Kankakee, IL; WAWJ(FM) Marion, IL; WAPO-FM Mount Vernon, IL; WWGN-FM Ottawa, IL; WZRS(FM) Pana, IL; WSLE(FM) Salem, IL; WQSG(FM) Lafayette, IN; KXJH(FM) Linton, IN; WWLO(FM) Lowell, IN; WATI-FM Vincennes, IN; KAXR(FM) Arkansas City, KS; KBJQ(FM) Bronson, KS; KBDA(FM) Great Bend, KS; KBQC(FM) Independence, KS; KRBW-FM Ottawa, KS; KAKA(FM) Salina, KS; KBUZ(FM) Topeka, KS; KCFN(FM) Wichita, KS; WAPD-FM Campbellsville, KY; WBMK-FM Morehead, KY; WAXG-FM Mt. Sterling, KY; WGCF(FM) Paducah, KY; KAPM(FM) Alexandria, LA; KAXV(FM) Bastrop, LA; KBAN(FM) De Ridder, LA; KYLC(FM) Lake Charles, LA; KAVK(FM) Many, LA; KPAQ(FM) Plaquemine, LA; KSUL(FM) Port Sulphur, LA; KAPI(FM) Ruston, LA; KSJY(FM) Saint Martinville, LA; WMCQ(FM) Muskegon, MI; KBPG(FM) Montevideo, MN; KQRB(FM) Windom, MN; KBOJ(FM) Worthington, MN; KAUF(FM) Kennett, MO; KBGM(FM) Park Hills, MO; WPRG(FM) Columbia, MS; WCSO(FM) Columbus, MS; WAUM-FM Duck Hill, MS; WQVI(FM) Forest, MS; WQST-FM Forest, MS; WAOY-FM Gulfport, MS; WAII-FM Hattiesburg, MS; WYTF(FM) Indianola, MS; WATP-FM Laurel, MS; WAQL-FM McComb, MS; WASM-FM Natchez, MS; WAVI-FM Oxford, MS; WPAS(FM) Pascagoula, MS; WATU-FM Port Gibson, MS; WMSB(FM) Senatobia, MS; WJZB(FM) Starkville, MS; WAQB-FM Tupelo, MS; WAJS-FM Tupelo, MS; WZKM(FM) Waynesboro, MS; WYAZ(FM) Yazoo City, MS; KAFH(FM) Great Falls, MT; WBKU-FM Ahoskie, NC; WXBE(FM) Beaufort, NC; WRYN(FM) Hickory, NC; WJKA(FM) Jacksonville, NC; WAAE-FM New Bern, NC; WBFY-FM Pinehurst, NC; WRAE(FM) Raeford, NC; KDVI(FM) Devils Lake, ND; KNHS(FM) Hastings, NE; KAYA(FM) Hubbard, NE; KAQF(FM) Clovis, NM; WJJE(FM) Delaware, OH; WBIE-FM Delphos, OH; WWGV(FM) Grove City, OH; WAUI-FM Shelby, OH; WBJV-FM Steubenville, OH; KAKO(FM) Ada, OK; KQPD(FM) Ardmore, OK; KAYC(FM) Durant, OK; KXRT(FM) Idabel, OK; KVRS-FM Lawton, OK; KARG(FM) Poteau, OK; KAYM(FM) Weatherford, OK; KANL(FM) Baker City, OR; KAPK(FM) Grants Pass, OR; WAWN-FM Franklin, PA; WDLL(FM) Dillon, SC; KASD(FM) Rapid City, SD; KSFS(FM) Sioux Falls, SD; WAUO-FM Hohenwald, TN; WAMP-FM Jackson, TN; WAWI-FM Lawrenceburg, TN; WIGH-FM Lexington, TN; WPRH(FM) Paris, TN; WAUV-FM Ripley, TN; WAZD-FM Savannah, TN; WBIA-FM Shelbyville, TN; WAUT-FM Tullahoma, TN; KAQD(FM) Abilene, TX; KAVW(FM) Amarillo, TX; KATG(FM) Athens, TX; KBCX(FM) Big Spring, TX; KLGS(FM) College Station, TX; KAFR(FM) Conroe, TX; KCKT(FM) Crockett, TX; KDLI(FM) Del Rio, TX; KZFT(FM) Fannett, TX; KTXG(FM) Greenville, TX; KSUR(FM) Mart, TX; KMEO(FM) Mertzon, TX; KBMM(FM) Odessa, TX; KAVO(FM) Pampa, TX; KBAH(FM) Plainview, TX; KBDE(FM) Temple, TX; KAYK(FM) Victoria, TX; WAUQ-FM Charles City, VA; WARN-FM

Culpeper, VA; WRIH(FM) Richmond, VA; KAYB(FM) Sunnyside, WA; WBHZ-FM Elkins, WV; WPWV(FM) Princeton, WV

Tim Wildmon, pres/CEO.

American General Media, Box 2700, Bakersfield, CA 93303. Phone: (661) 328-0118. Fax: (661) 328-1648. Ownership: Anthony S. Brandon, Lawrence Brandon, L. Rogers Brandon.

Stns: 6 AM. 16 FM. KIQO-FM Atascadero, CA; KISV(FM) Bakersfield, CA; KERN Bakersfield, CA; KGEO Bakersfield, CA; KBOX(FM) Lompoc, CA; KRQK-FM Lompoc, CA; KEBT(FM) Lost Hills, CA; KPAT-FM Orcutt, CA; KKAL(FM) Paso Robles, CA; KKJG-FM San Luis Obispo, CA; KZOZ-FM San Luis Obispo, CA; KKXX-FM Shafter, CA; KERI(AM) Wasco-Greenacres, CA; KKIM Albuquerque, NM; KLVO-FM Belen, NM; KARS Belen, NM; KZNM(FM) Los Alamos, NM; KABG(FM) Los Alamos, NM; KAGM(FM) Los Lunas, NM; KKIM-FM Santa Fe, NM; KTRC(AM) Santa Fe, NM; KHFM(FM) Santa Fe, NM.

L. Rogers Brandon, VP.

Americom, 11400 W. Olympic Blvd, Suite 780, Los Angeles, CA 90064. Phone: (310) 481-0440. Fax: (310) 481-0445.

Stns: 2 AM. 3 FM. KLCA-FM Tahoe City, CA; KZTQ(FM) Carson City, NV; KRNO(FM) Incline Village, NV; KJFK(AM) Reno, NV; KBZZ(AM) Sparks, NV.

Tom Quinn, pres/CEO.

Amistad Communications Inc., 7480 Greenwood Rd., Shreveport, LA 71119. Phone: (318) 938-1885. Fax: (318) 425-7507.

Stns: 2 AM. 1 FM. KBEF(FM) Gibsland, LA; KASO Minden, LA; KSYB(AM) Shreveport, LA.

Anaheim Broadcasting Corp., Box 2668, Del Mar, CA 92014-5668. Phone: (858) 794-1626. Fax: (858) 794-4068. Ownership: Tim Sullivan.

Stns: 2 FM. KCAL-FM Redlands, CA; KOLA(FM) San Bernardino, CA.

Tim Sullivan, pres; Doug Lida, CFO.

Anastos Media Group Inc., 21 Malta Commons, 100 Saratoga Village Blvd, Malta, NY 12020. Phone: (518) 899-3000. Fax: (518) 899-3057. E-mail: star1013fm@aol.com Web Site: www.star1013.com.

Stns: 4 AM. 1 FM. WPEP Taunton, MA; WABY(AM) Mechanicville, NY; WUAM(AM) Saratoga Springs, NY; WVKZ Schenectady, NY; WQAR-FM Stillwater, NY.

Scott Collins, pres.

Anderson Radio Broadcasting Inc., 581 N. Reservoir Rd., Polson, MT 59860. Phone: (406) 883-5255. Fax: (406) 883-4411. Ownership: Dennis L. Anderson, 50%; and Nila Y. Anderson, 50%.

Stns: 1 AM. 3 FM. KIBG(FM) Wallace, ID; KKMT(FM) Pablo, MT; KERR Polson, MT; KQRK-FM Ronan, MT.

Apex Broadcasting Inc., Box 61091, N Charleston, SC 29419-1091. Phone: (843) 852-9003. Fax: (843) 852-9041.

Stns: 2 AM. 7 FM. KTSR(FM) De Quincy, LA; KJEF(AM) Jennings, LA; KHLA(FM) Jennings, LA; KJMH(FM) Lake Arthur, LA; KNGT(FM) Lake Charles, LA; KLCL Lake Charles, LA; WAVF-FM Hanahan, SC; WXST(FM) Hollywood, SC; WIHB(FM) Moncks Corner, SC.

Houston L. Pearce, chmn; G. Dean Pearce, pres.

Archway Broadcasting Group, 1513 E. Cleveland Ave., Bldg. 100B, Suite 250, East Point, GA 30344. Phone: (404) 762-9942. Fax: (404) 209-8134. Web Site: www.archwaybroadcasting.com. Ownership: Quetzal/J.P. Morgan Partners, L.P.

Stns: 1 AM. 6 FM. WRLD-FM Valley, AL; KHTE-FM England, AR; KOLL-FM Lonoke, AR; WCGQ-FM Columbus, GA; WRCG Columbus, GA; WKCN-FM Lumpkin, GA; WLGT(FM) Washington, NC.

Kathy Stinehour, pres/CEO; Gordon Herzog, CFO.

Arkansas County Broadcasters Inc., Box 789, Wynne, AR 72396-0789. Phone: (870) 238-8141. Fax: (870) 238-5997. Ownership: Bobby Caldwell, 50%; C.B. Moery Jr., 50%. Note: Bobby Caldwell owns 100% of East Arkansas Broadcasters Inc. (see listing) and 50% of Combined Media Group Inc. (see listing).

Stns: 1 AM. 3 FM. KDEW-FM De Witt, AR; KAFN(FM) Gould, AR; KWAK Stuttgart, AR; KWAK-FM Stuttgart, AR.

Arklatex LLC, 615 W. Olive, Texarkana, TX 75501. Phone: (903) 793-4671. Fax: (903) 792-4261. E-mail: alex@101jams.com Ownership: Sudbury Affiliates LLC, 50%; Malvern Entertainment Corp., 25%; and DTJ Inc., 25%.

Stns: 2 AM. 3 FM. KBYB(FM) Hope, AR; KTOY(FM) Texarkana, AR; KFYX(FM) Texarkana, AR; KCMC Texarkana, TX; KTFS(AM) Texarkana, TX.

Mike Simpson, gen mgr; Alex Rain, opns mgr.

Armada Media Corp., 114 S. Main St., Suite 306, Fond du Lac, WI 54935. Phone: (920) 279-5678. Ownership: Jim Coursolle, 23.7% votes; Robert Bourke, 22.5% votes; Chris Bernier, 18.2% votes; Shockley Broadcasting LLC, 14.1% votes; John R. Larson, 10.4% votes; Tommy G. Thompson, 6.1% votes; AMC Partners LLC, 4% votes; and John Lynch, 0.8% votes.

Stns: 6 AM. 14 FM. KSTH(FM) Holyoke, CO; KJBL(FM) Julesburg, CO; KFNF-FM Oberlin, KS; WIMI-FM Ironwood, MI; WJMS Ironwood, MI; WAGN Menominee, MI; WHYB-FM Menominee, MI; KADL(FM) Imperial, NE; KRKU(FM) McCook, NE; KBRL McCook, NE; KICX-FM McCook, NE; KGIM Aberdeen, SD; KBFO(FM) Aberdeen, SD; KSDN Aberdeen, SD; KSDN-FM Aberdeen, SD; KGIM-FM Redfield, SD; KNBZ-FM Redfield, SD; WLST-FM Marinette, WI; WMAM Marinette, WI; WSFQ-FM Peshtigo, WI.

Jim Coursolle, CEO; Chris Bernier, COO; John R. Larson, CFO.

Artistic Media Partners Inc., 5520 E. 75th St., Indianapolis, IN 46250. Phone: (317) 594-0600. Fax: (317) 594-9567. E-mail: artradio@aol.com Web Site: www.artisticradio.com. Ownership: Arthur A. Angotti.

Stns: 3 AM. 9 FM. WSHP(FM) Attica, IN; WBWB-FM Bloomington, IN; WLFF(FM) Brookston, IN; WVBB(FM) Columbia City, IN; WHCC(FM) Ellettsville, IN; WOZW(FM) Goshen, IN; WSHY(AM) Lafayette, IN; WAZY-FM Lafayette, IN; WZOW(FM) New Carlisle, IN; WWLV(AM) South Bend, IN; WDND(AM) South Bend, IN; WNDV-FM South Bend, IN.

Arthur A. Angotti, pres/CEO; Arthur A. Angotti III, sr VP.

Asterisk Inc., 2848 E. Oakland Park Blvd., Fort Lauderdale, FL 33306. Phone: (954) 566-7559. Fax: (954) 564-6753. Ownership: Richard S. Ingham, 100%.

Stns: 3 FM. WXJZ(FM) Gainesville, FL; WYGC(FM) High Springs, FL; WMFQ-FM Ocala, FL.

Frederick H. Ingham, pres.

Astor Broadcast Group, 1835 Aston Ave., Carlsbad, CA 92008. Phone: (760) 729-1000. Fax: (760) 476-9604.

Stns: 3 AM. KFSD(AM) Escondido, CA; KSPA(AM) Ontario, CA; KCEO Vista, CA.

Peri Corso, gen mgr.

Astral Media Inc., 2100 rue Sainte-Catherine Ouest, Bureau 1000, Montreal, PQ H3H 2T3. Canada. Phone: (514) 939-5000. Fax: (514) 939-1515. Web Site: www.astralmedia.com. Ownership: Abgreen Holdings Ltd., 55.71% vote; 654625 Ontario Inc., 13.63% vote.

Stns: 2 AM. 27 FM. CKBC-FM Bathurst, NB; CIBX-FM Fredericton, NB; CKHJ(AM) Fredericton, NB; CIKX-FM Grand Falls, NB; CJCJ-FM Woodstock, NB; CKTO-FM Truro, NS; CKTY-FM Truro, NS; CFVM-FM Amqui, PQ; CFIX-FM Chicoutimi, PQ; CJDM-FM Drummondville, PQ; CHRD-FM Drummondville, PQ; CIMF-FM Gatineau, PQ; CKTF-FM Gatineau, PQ; CIMO-FM Magog, PQ; CITE-FM Montreal, PQ; CKMF-FM Montreal, PQ; CITF-FM Quebec, PQ; CHIK-FM Quebec, PQ; CIKI-FM Rimouski, PQ; CJOI-FM Rimouski, PQ; CJMM-FM Rouyn-Noranda, PQ; CJAB-FM Saguenay, PQ; CFEI-FM Saint Hyacinthe, PQ; CFZZ-FM Saint Jean-Iberville, PQ; CKSM Shawinigan, PQ; CITE-FM-1 Sherbrooke, PQ; CHEY-FM Trois Rivieres, PQ; CIGB-FM Trois Rivieres, PQ; CJMV-FM Val d'Or, PQ.

Ian Greenberg, pres/CEO.

Astro Tele-Communications Corp. Rhode Island, Box 920365, Needham, MA 02492. Phone: (781) 444-4754. Fax: (781) 444-8630. E-mail: addelco@gis.net Web Site: www.wadk.com. Ownership: Maurice B. Polayes, 100%.

Stns: 2 AM. 1 FM. WJZS(FM) Block Island, RI; WKFD(AM) Charlestown, RI; WADK Newport, RI.

Maurice B. Polayes, pres.

Atlantic Coast Radio L.L.C., 779 Warren Avenue, Portland, ME 04103. Phone: (207) 773-9695. Fax: (207) 761-4406. Web Site: www.redhot95.com.

Stns: 3 AM. 3 FM. WJJB(AM) Brunswick, ME; WLOB Portland, ME; WLOB-FM Rumford, ME; WRED-FM Saco, ME; WJJB-FM Topsham, ME; WJAE Westbrook, ME.

J.J. Jeffrey, pres.

Azteca Broadcasting Corp., 323 E. San Joaquin St., Tulare, CA 93274. Phone: (559) 686-1370. Fax: (559) 685-1394. E-mail: wwwkgen@sbcglobal.net

Stns: 3 AM. 1 FM. KGEN-FM Hanford, CA; KGEN Tulare, CA; KXEQ Reno, NV; KSVN Ogden, UT.

Margarita Hernandez, gen mgr.

B

Back Bay Broadcasters LLC, 1110 Central Ave., Pawtucket, RI 02861-2262. Phone: (401) 724-7600. Fax: (401) 728-1865. Ownership: Peter H. Ottmar, 40.1%; David J. Ottmar, 26.1%; John Maguire, 19.0%; and Barbara Ottmar, 7.2%.

Stns: 4 FM. WBAZ(FM) Bridgehampton, NY; WEHN(FM) East Hampton, NY; WEHM(FM) Southampton, NY; WBEA(FM) Southold, NY.

Peter H. Ottmar, chmn; John Maguire, CEO.

Backyard Broadcasting LLC, 4237 Salisbury Rd., Suite 225, Jacksonville, FL 32216. Phone: (904) 674-0260. Fax: (904) 854-4596. Web Site: www.bybradio.com. Ownership: Boston Ventures Limited Partnership VI; PCG Media Investment Partners LLC; Barry Drake.

Stns: 9 AM. 21 FM. WHTI-FM Alexandria, IN; WHBU Anderson, IN; WURK-FM Elwood, IN; WHTY-FM Hartford City, IN; WERK-FM Muncie, IN; WLBC-FM Muncie, IN; WXFN Muncie, IN; WWJK(FM) Jackson, MS; WRXW(FM) Pearl, MS; WNKI-FM Corning, NY; WWLZ Horseheads, NY; WPGI-FM Horseheads, NY; WNGZ-FM Montour Falls, NY; WHDL Olean, NY; WPIG-FM Olean, NY; WTYX(AM) Watkins Glen, NY; WCXR(FM) Lewisburg, PA; WBZD-FM Muncy, PA; WZXR-FM South Williamsport, PA; WWPA Williamsport, PA; WRVH(FM) Williamsport, PA; WILQ-FM Williamsport, PA; KSQB-FM Dell Rapids, SD; KXQL(FM) Flandreau, SD; KTWB(FM) Sioux Falls, SD; KWSN(AM) Sioux Falls, SD; KELO(AM) Sioux Falls, SD; KELO-FM Sioux Falls, SD; KRRO(FM) Sioux Falls, SD; KSQB(AM) Sioux Falls, SD.

Barry Drake, pres; Robin A. Smith, VP/CFO; Tom Atkins, VP & dir engrg.

Bahakel Communications, Box 32488, Charlotte, NC 28232. Phone: (704) 372-4434. Fax: (704) 335-9904. Ownership: The Cy N. Bahakel Trust Dated January 12, 2005, 100%.

Stns: 4 AM. 5 FM. KILO-FM Colorado Springs, CO; KYZX(FM) Pueblo West, CO; KOKZ-FM Waterloo, IA; KWLO(AM) Waterloo, IA; KXEL Waterloo, IA; KFMW-FM Waterloo, IA; WABG(AM) Greenwood, MS; WDEF-FM Chattanooga, TN; WDOD Chattanooga, TN.

Stns: 6 TV. WCCU, Champaign & Springfield-Decatur, IL; WRSP-TV, Champaign & Springfield-Decatur, IL; WCCB, Charlotte, NC; WABG-TV, Greenwood-Greenville, MS; WAKA, Montgomery-Selma, AL; WFXB, Myrtle Beach-Florence, SC.

Cy N. Bahakel, pres; Beverly Poston, exec VP/COO; Stephen Bahakel, Sr VP radio div; Russell Schwartz, Sr VP business affrs/gen counsel; Bill Napier, VP eng/tech; Anna Rufty, VP Hum Res.

Baker Family Stations, Box 889, Blacksburg, VA 24063. Phone: (540) 552-4252. Fax: (540) 951-5282. Ownership: Principal owners: Vernon H. Baker, Edward A. Baker, Virginia L. Baker.

Stns: 12 AM. 5 FM. WOKT Cannonsburg, KY; WLGN Logan, OH; WLGN-FM Logan, OH; WMPO Middleport-Pomeroy, OH; WTGR-FM Union City, OH; WKEX Blacksburg, VA; WFIC(AM) Collinsville, VA; WBNN-FM Dillwyn, VA; WKTR Earlysville, VA; WKNV Fairlawn, VA; WODY Fieldale, VA; WZFM(FM) Narrows, VA; WKGM Smithfield, VA; WAMN Green Valley, WV; WOKU Hurricane, WV; WBGS Point Pleasant, WV; WCEF-FM Ripley, WV.

Vernon H. Baker, CEO; Edward A. Baker, pres; Virginia L. Baker, treas.

Baldridge-Dumas Communications Inc., 605 San Antonio Ave., Many, LA 71449. Phone: (318) 256-5924. Fax: (318) 256-0950. Ownership: Tedd W. Dumas, 50%; Patricia M. Baldridge Declaration of Trust, 50%.

Stns: 1 AM. 4 FM. KWLA Many, LA; KZBL-FM Natchitoches, LA; KDBH(FM) Natchitoches, LA; KTEZ(FM) Zwolle, LA; KTHP(FM) Hemphill, TX.

Tedd Dumas, pres; Rhonda Benson, gen mgr.

Vernon R Baldwin Inc., 8686 Michael Ln., Fairfield, OH 45014. Phone: (513) 829-7700. Ownership: Vernon R. Baldwin, 100%.

Stns: 2 AM. 4 FM. WWLT-FM Manchester, KY; WVRB-FM Wilmore, KY; WCNW Fairfield, OH; WMOH Hamilton, OH; WNLT-FM Harrison, OH; WKLN(FM) Wilmington, OH.

Vernon R. Baldwin, pres.

Barnstable Corporation, 2 Newton Executive Park, Suite 302, Newton, MA 02462-1434. Phone: (617) 527-0062. Fax: (617) 630-0960. Ownership: Albert J. Kaneb; Michael A. Kaneb.

Stns: 1 AM. 5 FM. WBZO-FM Bay Shore, NY; WLVG-FM Center Moriches, NY; WHLI Hempstead, NY; WKJY-FM Hempstead, NY; WRCN-FM Riverhead, NY; WMJC(FM) Smithtown, NY.

Albert J. Kaneb, chmn/CEO; Michael A. Kaneb, pres/COO; James L. Paglia, VP/CFO.

Bayshore Broadcasting Corp., Box 280, Owen Sound, ON N4K 5P5. Canada. Phone: (519) 376-2030. Fax: (519) 371-4242. E-mail: bayshore@radioowensound.com Web Site: www.radioowensound.com. Ownership: Controlled by Douglas C. Caldwell.

Stns: 1 AM. 5 FM. CHWC-FM Goderich, ON; CFOS(AM) Owen Sound, ON; CIXK-FM Owen Sound, ON; CKYC-FM Owen Sound, ON; CFPS-FM Port Elgin, ON; CHGB-FM Wasaga Beach, ON.

Ross Kentner, gen mgr.

Beacon Broadcasting Inc., Box 1789, Warren, OH 44482-1790. Phone: (330) 394-7700. Fax: (330) 394-7701. Ownership: Harold F. Glunt, 100%.

Stns: 4 AM. 1 FM. WRTK(AM) Niles, OH; WANR Warren, OH; WLOA(AM) Farrell, PA; WEXC-FM Greenville, PA; WGRP Greenville, PA.

Beasley Broadcast Group Inc., 3033 Riviera Dr., Suite 200, Naples, FL 34103. Phone: (239) 263-5000. Fax: (239) 263-8191. E-mail: email@bbgi.com Web Site: www.bbgi.com. Ownership: George G. Beasley.

Stns: 18 AM. 26 FM. WSBR Boca Raton, FL; WKIS-FM Boca Raton, FL; WRXK-FM Bonita Springs, FL; WXKB-FM Cape Coral, FL; WJBX-FM Fort Myers Beach, FL; WJPT(FM) Fort Myers Villas, FL; WQAM Miami, FL; WPOW-FM Miami, FL; WWCN North Fort Myers, FL; WWNN Pompano Beach, FL; WHSR Pompano Beach, FL; WAEC Atlanta, GA; WGAC Augusta, GA; WRDW(AM) Augusta, GA; WGUS(AM) Augusta, GA; WWWE(AM) Hapeville, GA; WCHZ-FM Harlem, GA; WDRR(FM) Martinez, GA; WGAC-FM Warrenton, GA; WRCA Waltham, MA; WAZZ(AM) Fayetteville, NC; WNCT Greenville, NC; WNCT-FM Greenville, NC; WXNR-FM Grifton, NC; WFLB-FM Laurinburg, NC; WKML-FM Lumberton, NC; WIKS-FM New Bern, NC; WSFL-FM New Bern, NC; WMGV-FM Newport, NC; WTEL(AM) Red Springs, NC; WUKS(FM) Saint Pauls, NC; WZFX-FM Whiteville, NC; WTMR Camden, NJ; KSTJ-FM Boulder City, NV; KKLZ-FM Las Vegas, NV; KDWN Las Vegas, NV; KCYE(FM) North Las Vegas, NV; KBET(AM) Winchester, NV; WXTU-FM Philadelphia, PA; WWDB(AM) Philadelphia, PA; WRDW-FM Philadelphia, PA; WKXC-FM Aiken, SC; WHHD(FM) Clearwater, SC; WGUS-FM New Ellenton, SC.

George G. Beasley, chmn/CEO; Bruce G. Beasley, pres/COO; Caroline Beasley, exec VP/CFO; Brian Beasley, VP opns.

Bee Broadcasting Inc., Box 5409, Kalispell, MT 59903. Phone: (406) 755-8700. Fax: (406) 755-8770. Web Site: www.kbbz.com.

Stns: 2 AM. 3 FM. KHNK(FM) Columbia Falls, MT; KBBZ(FM) Kalispell, MT; KDBR(FM) Kalispell, MT; KJJR Whitefish, MT; KSAM(AM) Whitefish, MT.

Benny Bee, pres.

Gerald Benavides Stns, 11737 Nelon Dr., Corpus Christi, TX 78410. Phone: (361) 241-7944. Fax: (361) 241-7945. Ownership: Gerald Benavides, 100%.

Stns: 1 AM. 2 FM. KMZZ(FM) Bishop, TX; KBRN Boerne, TX; KGGB(FM) Yorktown, TX.

Benton-Weatherford Broadcasting Inc. of Tennessee, 110 India Rd., Paris, TN 38242. Phone: (731) 644-9455. Fax: (731) 644-9421. E-mail: wmuf@bellsouth.net Ownership: Gary Benton, Len Watson.

Stns: 2 AM. 2 FM. WMUF-FM Henry, TN; WHDM McKenzie, TN; WLZK-FM Paris, TN; WMUF Paris, TN.

Gary Benton, pres.

Berkshire Broadcasting Corp., c/o WLAD(AM) and WDAQ(FM), 198 Main St., Danbury, CT 06810. Phone: (203) 744-4800. Fax: (203) 778-4655. Web Site: www.98q.com.

Stns: 2 AM. WLAD Danbury, CT; WREF Ridgefield, CT.

Irv Goldstein, VP/gen mgr.

Best Broadcast Group, 107 S. Main St., Brookfield, MO 64628. Phone: (660) 258-3383. Fax: (660) 258-7307. E-mail: corporate@bestbroadcastgroup.com Web Site: www.bestbroadcastgroup.com. Ownership: Phil Chirillo; Dale Palmer

Stns: 2 AM. 3 FM. KFMZ(AM) Brookfield, MO; KZBK(FM) Brookfield, MO; KLTI Macon, MO; KZZT-FM Moberly, MO; KMCR-FM Montgomery City, MO.

Dale Palmer, VP/gen mgr; Phil Chirillo, pres.

Bestov Broadcasting Inc., Box 9023916, San Juan, PR 00902-3916. Phone: (787) 620-9898. Fax: (787) 620-0730. Ownership: Luis A. Mejia, 100%.

Stns: 2 AM. 1 FM. WYAC(AM) Cabo Rojo, PR; WIAC San Juan, PR; WIAC-FM San Juan, PR.

Bethesda Christian Broadcasting, Box 168, Rapid City, SD 57709. Phone: (719) 481-0100. Fax: (719) 481-4649. E-mail: bcbpres@aol.com Web Site: www.klmp.com. Ownership: Nonprofit bd of directors.

Stns: 3 FM. KTPT(FM) Rapid City, SD; KLMP(FM) Rapid City, SD; KSLT-FM Spearfish, SD.

Mark Plummer, pres.

Bible Broadcasting Network, 11530 Carmel Commons Blvd., Charlotte, NC 28226. Phone: (704) 523-5555. Fax: (704) 522-1967. Web Site: www.bbnradio.org. Ownership: Nonprofit, non-stock corporation.

Stns: 3 AM. 28 FM. WYFD-FM Decatur, AL; WYFZ(FM) Belleview, FL; WYFB-FM Gainesville, FL; WYFO-FM Lakeland, FL; WYFE-FM Tarpon Springs, FL; WYFK-FM Columbus, GA; WYFS-FM Savannah, GA; WYFA-FM Waynesboro, GA; WYFW-FM Winder, GA; WYBV(FM) Wakarusa, IN; KYFW-FM Wichita, KS; KYFL-FM Monroe, LA; WYFP-FM Harpswell, ME; WYFQ(AM) Charlotte, NC; WYBH(FM) Fayetteville, NC; WYFL-FM Henderson, NC; WYFQ-FM Wadesboro, NC; WYFY Rome, NY; WYFU-FM Masontown, PA; WYFV-FM Cayce, SC; WYFG-FM Gaffney, SC; WYFH-FM North Charleston, SC; WYFC-FM Clinton, TN; WYFN Nashville, TN; KYFB(FM) Denison, TX; KYFP(FM) Palestine, TX; KYFS-FM San Antonio, TX; KYFO-FM Ogden, UT; WYFJ-FM Ashland, VA; WYFT-FM Luray, VA; WYFI-FM Norfolk, VA.

Lowell Davey, pres; Leo Galletta, opns mgr.

Bick Broadcasting Co., Box 711, 119 N. Third St., Hannibal, MO 63401. Phone: (573) 221-3450. Fax: (573) 221-5331. E-mail: kickfm@bickbroadcasting.com Web Site: www.979kickfm.com. Ownership: Frank C. Bick, 46%; James P. Bick, 46%; James E. Janes, 8%.

Stns: 3 AM. 4 FM. WLIQ(AM) Quincy, IL; KRRY(FM) Canton, MO; KHMO Hannibal, MO; KXKX-FM Knob Noster, MO; KICK-FM Palmyra, MO; KSDL-FM Sedalia, MO; KSIS(AM) Sedalia, MO.

James E. Janes, pres.

Bicoastal Media L.L.C., 140 N. Main St., Lake Port, CA 94953. Phone: (707) 263-6113. Fax: (707) 263-0939.

Stns: 15 AM. 23 FM. KATA Arcata, CA; KPOD Crescent City, CA; KPOD-FM Crescent City, CA; KRED-FM Eureka, CA; KFMI-FM Eureka, CA; KGOE Eureka, CA; KKHB-FM Eureka, CA; KNTI-FM Lakeport, CA; KXBX Lakeport, CA; KUKI Ukiah, CA; KUKI-FM Ukiah, CA; KQPM-FM Ukiah, CA; KLLK Willits, CA; KTHH(AM) Albany, OR; KRKT-FM Albany, OR; KIFS(FM) Ashland, OR; KBDN(FM) Bandon, OR; KWRO Coquille, OR; KEJO Corvallis, OR; KFLY-FM Corvallis, OR; KLOO Corvallis, OR; KLOO-FM Corvallis, OR; KZZE-FM Eagle Point, OR; KODZ-FM Eugene, OR; KPNW Eugene, OR; KDUK-FM Florence, OR; KRWQ-FM Gold Hill, OR; KMED Medford, OR; KLDZ(FM) Medford, OR; KOOS(FM) North Bend, OR; KBBR(AM) North Bend, OR; KTEE(FM) North Bend, OR; KPPK(FM) Rainier, OR; KJMX(FM) Reedsport, OR; KRQT-FM Castle Rock, WA; KLYK(FM) Kelso, WA; KBAM Longview, WA; KEDO Longview, WA.

Ken Dennis, CEO; Mike Wilson, pres.

Big League Broadcasting LLC, 3350 Peachtree Rd., Suite 1610, Atlanta, GA 30326-1040. Phone: (404) 467-1877. Fax: (404) 231-5923. Ownership: Andrew Philip Saltzman, 20.435% of votes, 24.375% of equity; Stephen Shapiro, 20.435% of votes, 24.375% of equity; Jeffrey Bloomberg, 16.61% of votes, 16.75% of equity; and others.

Stns: 2 AM. 1 FM. KFNS(AM) Wood River, IL; KRFT(AM) De Soto, MO; KFNS-FM Troy, MO.

Big River Broadcasting Corp., 624 Sam Phillips St., Florence, AL 35630. Phone: (256) 764-8121. Fax: (256) 764-8169. E-mail: nmartin@bigriverbroadcasting.com Web Site: www.wqlt.com.

Stns: 1 AM. 2 FM. WXFL(FM) Florence, AL; WSBM(AM) Florence, AL; WQLT-FM Florence, AL.

Knox Phillips, VP; Jerry Phillips, pres.

Birach Broadcasting Corp., 21700 Northwestern Hwy., Suite 1190, Southfield, MI 48075. Phone: (248) 557-3500. Fax: (248) 557-3241. E-mail: sima@birach.com Web Site: www.birach.com. Ownership: Sima Birach, 100%.

Stns: 14 AM. WIJR(AM) Highland, IL; WNWI Oak Lawn, IL; WNZK Dearborn Heights, MI; WCXI(AM) Fenton, MI; WMJH Rockford, MI; WSDS(AM) Salem Township, MI; WPON Walled Lake, MI; WMFN Zeeland, MI; WEW Saint Louis, MO; WCXN Claremont, NC; WTOR Youngstown, NY; WWCS Canonsburg, PA; WGOP(AM) Pocomoke City, MD; WDMV(AM) Walkersville, MD.

Sima Birach, pres.

Birch Broadcasting Corp., 11971 Glenmore Dr., Coral Springs, FL 33071-7806. Phone: (954) 323-8531. Ownership: Thomas C. Birch, 51%; and Aurora D.P. Birch, 49%.

Stns: 2 AM. 2 FM. KXLQ Indianola, IA; WLUS-FM Clarksville, VA; WSHV(AM) South Hill, VA; WKSK-FM South Hill, VA.

Bob Bittner Broadcasting Inc., 443 Concord Ave, Cambridge, MA 02138. Phone: (617) 868-7400. Web Site: www.wjib740.com. Ownership: Robert Miles Bittner, 100%.

Stns: 2 AM. WJIB Cambridge, MA; WJTO Bath, ME.

Robert Miles Bittner, pres/CEO.

Black Crow Media Group LLC, 126 W. International Speedway Blvd., Daytona Beach, FL 32114. Phone: (386) 255-9300. Ownership: J. Michael Linn, 100%.

Stns: 5 AM. 17 FM. WAHR-FM Huntsville, AL; WLOR Huntsville, AL; WRTT-FM Huntsville, AL; WNDB Daytona Beach, FL; WKRO-FM Edgewater, FL; WCJX-FM Five Points, FL; WVYB-FM Holly Hill, FL; WQHL(AM) Live Oak, FL; WQHL-FM Live Oak, FL; WXHT(FM) Madison, FL; WHOG-FM Ormond-by-the-Sea, FL; WVGA(FM) Lakeland, GA; WSTI-FM Quitman, GA; WQPW(FM) Valdosta, GA; WVLD(AM) Valdosta, GA; WWRQ-FM Valdosta, GA; WKAA(FM) Willacoochee, GA; WFKX(FM) Henderson, TN; WHHM-FM Henderson, TN; WZDQ(FM) Humboldt, TN; WJAK Jackson, TN; WWYN(FM) McKenzie, TN.

Mike Linn, pres/CEO.

Black Media Works Inc., 1150 W. King St., Cocoa, FL 32922. Phone: (321) 632-1000. Fax: (321) 636-0000.

Stns: 3 FM. WJCB(FM) Clewiston, FL; WJFP-FM Fort Pierce, FL; KAYT(FM) Jena, LA.

Kimberly Holman Kassis, pres.

Blackburn Group Inc., 140 Fullerton, Suite 1905, London, ON N6A 5P2. Canada. Phone: (519) 679-8680. Fax: (519) 679-5321. Web Site: www.blackburnradio.com. Ownership: Kilbyrne Investments Corp., 100%.

Stns: 2 AM. 6 FM. CHYR-FM Leamington, ON; CJSP-FM Leamington, ON; CHKS-FM Sarnia, ON; CFGX-FM Sarnia, ON; CHOK Sarnia, ON; CKNX Wingham, ON; CKNX-FM Wingham, ON; CIBU-FM Wingham, ON.

Sandy Green, pres.

Blakeney Communications Inc., Box 6408, Laurel, MS 39441. Phone: (601) 649-0095. Fax: (601) 649-8199. E-mail: b95@b95country.com Web Site: www.b95country.com.

Stns: 4 FM. WKZW-FM Bay Springs, MS; WXRR-FM Hattiesburg, MS; WXHB(FM) Richton, MS; WBBN(FM) Taylorsville, MS.

Larry Blakeney, pres/CEO.

Bliss Communications Inc., Box 5001, One S. Parker Dr., Janesville, WI 53547-5001. Phone: (608) 754-3311. Fax: (608) 754-8038. E-mail: sbliss@gazetteextra.com

Stns: 3 AM. 1 FM. WCLO Janesville, WI; WRJN Racine, WI; WBKV West Bend, WI; WBWI-FM West Bend, WI.

Bliss Communications Inc. publishes *Ironwood* (MI) *Daily Globe, The Delavan* (WI) *Enterprise, The Week* (Delavan, Il), *MidWeek* (Delavan, WI), *The Jotter* (Janesville, WI), the *Janesville* (WI) *Gazette*, the *Eagle Herald* (Marinette, WI), & the *Monroe* (WI) *Times*.

Sidney H. Bliss, pres/CEO.

Blount Communications Group, 8 Lawrence Rd., Derry, NH 03038. Phone: (603) 437-9337. Fax: (603) 434-1035. E-mail: warv@aol.com Web Site: www.lifechangingradio.com. Ownership: William A. Blount, Deborah C. Blount.

Stns: 5 AM. 1 FM. WFIF Milford, CT; WVNE(AM) Leicester, MA; WNEB Worcester, MA; WBCI-FM Bath, ME; WDER(AM) Derry, NH; WARV Warwick, RI.

William A. Blount, pres; Deborah C. Blount, exec VP; David O. Young, VP.

Blue Ridge Radio Inc., 312 Robin Rd., Mount Airy, NC 27030. Phone: (336) 786-4498. Fax: (336) 789-7792.

Stns: 3 AM. 1 FM. WSYD Mount Airy, NC; WPAQ Mount Airy, NC; WBRF(FM) Galax, VA; WWWJ Galax, VA.

Earlene Epperson, pres; Ralph Epperson, VP; John Mullins, chief engr.

Bold Gold Media Group LP, 575 Grove St., Honesdale, PA 18431. Phone: (570) 253-1616. Fax: (570) 253-6297. E-mail: vbenedetto@boldgoldmedia.com Web Site: www.boldgoldmedia.com.

Stns: 4 AM. 4 FM. WDNB(FM) Jeffersonville, NY; WFBS(AM) Berwick, PA; WYCY-FM Hawley, PA; WPSN(AM) Honesdale, PA; WDNH-FM Honesdale, PA; WYCK Plains, PA; WWRR(FM) Scranton, PA; WICK Scranton, PA.

Vince Benedetto, pres; Robert VanDerneyden, COO.

Bonneville International Corporation, Broadcast House, Box 1160, Salt Lake City, UT 84110-1160. Phone: (801) 575-7500. Fax: (801) 575-7521. Web Site: www.bonnint.com. Ownership: Deseret Management Corp. Deseret Management

Corp. owns *The Deseret Morning News*, a Salt Lake City, UT, daily.

Stns: 9 AM. 19 FM. KTAR-FM Glendale, AZ; KPKX(FM) Phoenix, AZ; KMVP Phoenix, AZ; KTAR(AM) Phoenix, AZ; KBWF(FM) San Francisco, CA; KDFC-FM San Francisco, CA; KOIT San Francisco, CA; KOIT-FM San Francisco, CA; WTWP(AM) Washington, DC; WTOP-FM Washington, DC; WDRV(FM) Chicago, IL; WILV(FM) Chicago, IL; WMVN(FM) East St. Louis, IL; WARH(FM) Granite City, IL; WTMX-FM Skokie, IL; WWDV(FM) Zion, IL; WIL-FM Saint Louis, MO; WIL(AM) Saint Louis, MO; KSL-FM Midvale, UT; KRSP-FM Salt Lake City, UT; KSFI-FM Salt Lake City, UT; KSL Salt Lake City, UT; KUTR(AM) Taylorsville, UT; WTWP-FM Warrenton, VA; WTLP(FM) Braddock Heights, MD; WTWT(AM) Frederick, MD; WFED(AM) Silver Spring, MD; WPRS-FM Waldorf, MD.

Stns: 1 TV. KSL, Salt Lake City, UT.

Bruce T. Reese, pres/CEO; Robert A. Johnson, exec VP & COO.

Border Media Partners LLC, 201 Main St., Suite 2001, Fort Worth, TX 76102. Phone: (817) 335-5999. Fax: (817) 335-1197. Web Site: www.bmpradio.com. Ownership: The Goldman Sachs Group Inc., 48.85%; RGG Radio LLC, 16.29%; DBVA BMP Holdings LLC, 9.32%.

Stns: 12 AM. 13 FM. KJAV-FM Alamo, TX; KFON Austin, TX; KWOW-FM Clifton, TX; KXBT(FM) Dripping Springs, TX; KURV Edinburg, TX; KXXS(FM) Elgin, TX; KTFM(FM) Floresville, TX; KFJZ Fort Worth, TX; KLEY-FM Jourdanton, TX; KLNT Laredo, TX; KNEX-FM Laredo, TX; KELG(AM) Manor, TX; KBDR(FM) Mirando City, TX; KRIO-FM Pearsall, TX; KOKE Pflugerville, TX; KVJY Pharr, TX; KBUC(FM) Raymondville, TX; KSOX Raymondville, TX; KJXK(FM) San Antonio, TX; KTSA San Antonio, TX; KZDC San Antonio, TX; KZSP-FM South Padre Island, TX; KESO-FM South Padre Island, TX; KSAH(AM) Universal City, TX; KTXZ West Lake Hills, TX.

Bott Radio Network, 10550 Barkley St., Suite 100, Overland Park, KS 66212. Phone: (913) 642-7770. Fax: (913) 642-1319. E-mail: comments@bottradionetwork.com Web Site: www.bottradionetwork.com. Ownership: Richard P. Bott Sr.

Stns: 11 AM. 18 FM. KCIV(FM) Mount Bullion, CA; WFCV Fort Wayne, IN; KBMP(FM) Enterprise, KS; KARF(FM) Independence, KS; KCVW(FM) Kingman, KS; KJRG Newton, KS; KCCV-FM Olathe, KS; KCCV Overland Park, KS; KKCV(FM) Rozel, KS; KCVT(FM) Silver Lake, KS; KSIV(AM) Clayton, MO; KJCV(FM) Country Club, MO; KMCV(FM) High Point, MO; KBCV(AM) Hollister, MO; KLTE-FM Kirksville, MO; KLEX Lexington, MO; KAYX(FM) Richmond, MO; KMOZ Rolla, MO; KSIV-FM Saint Louis, MO; KSCV(FM) Springfield, MO; KCRL(FM) Sunrise Beach, MO; KAMI(AM) Cozad, NE; KCVN(FM) Cozad, NE; KLCV-FM Lincoln, NE; KQCV Oklahoma City, OK; KQCV-FM Shawnee, OK; WCRV Collierville, TN; WCRT(AM) Donelson, TN; KTAA(FM) Big Sandy, TX.

Richard P. Bott II, exec VP; Richard P. Bott Sr., pres; Trace Thurlby, COO; Tom Holdeman, CFO.

Brantley Broadcast Associates LLC, 6930 Cahaba Valley Rd., Suite 202, Birmingham, AL 35242. Phone: (205) 618-2020. Fax: (205) 618-2029. Ownership: R3 Partners LLC, 95%; and Parmalley LLC, 5%. Note: Group also holds a 99% partnership interest in Amerimedia Texas LP, licensee of KLVT-FM Levelland, TX.

Stns: 3 AM. WHOA(AM) Saraland, AL; WFGO(AM) Orono, ME; WZFN(AM) Dilworth, MN.

Bravo Mic Communications LLC, 101 Perkins Dr., Las Cruces, NM 88005. Phone: (505) 527-1111. Fax: (505) 527-1100. Ownership: Ned W. Bennett, 50%; and Sandra G. Zane, 50%.

Stns: 1 AM. 3 FM. KVLC-FM Hatch, NM; KOBE Las Cruces, NM; KXPZ(FM) Las Cruces, NM; KMVR(FM) Mesilla Park, NM.

Brazos Valley Communications Ltd., 1240 E. Villa Maria, Bryan, TX 77802. Phone: (979) 776-1240. Fax: (979) 776-4700. Ownership: Brazos Valley Communications GP LLC, 100% votes, 10% equity; Tommy R. Vascocu, 21% equity.

Stns: 1 AM. 3 FM. KORA-FM Bryan, TX; KTAM Bryan, TX; KJXJ(FM) Cameron, TX; KZTR-FM Franklin, TX.

Tommy R. Vascocu, gen ptnr; Chris Kiske, VP/gen mgr.

Brewer Broadcasting Corp., 1305 Carter St., Chattanooga, TN 37402. Phone: (423) 265-9494. Fax: (423) 266-2335. E-mail: jlb@brewerradio.com Web Site: www.brewerradio.com. Ownership: Estate of James R. Brewer, James L. Brewer, Maytha N. Brewer.

Stns: 4 AM. 6 FM. WMPZ-FM Ringgold, GA; WHON Centerville, IN; WQLK-FM Richmond, IN; WHJK(FM) Cleveland, TN; WBAC(AM) Cleveland, TN; WDNT(AM) Dayton, TN; WALV-FM Dayton, TN; WJTT-FM Red Bank, TN; WXQK(AM) Spring City, TN; WAYA(FM) Spring City, TN.

James L. Brewer Sr., pres.

Bristol Broadcasting Co. Inc., Box 1389, Bristol, VA 24203. Phone: (276) 669-8112. Fax: (276) 669-0541. Ownership: W.L. Nininger.

Stns: 7 AM. 9 FM. WKYX-FM Golconda, IL; WLLE(FM) Clinton, KY; WNGO(AM) Mayfield, KY; WKYQ-FM Paducah, KY; WKYX Paducah, KY; WDDJ(FM) Paducah, KY; WDXR Paducah, KY; WPAD Paducah, KY; WXBQ-FM Bristol, TN; WTZR(FM) Elizabethton, TN; WAEZ(FM) Greeneville, TN; WFHG-FM Abingdon, VA; WFHG(AM) Bristol, VA; WVTS(AM) Charleston, WV; WBES(AM) Dunbar, WV; WZJO(FM) Dunbar, WV.

W.L. Nininger, pres & gen mgr.

Broadcast Communications Inc., Box 990, Greensburg, PA 15601. Phone: (724) 853-7000.

Stns: 4 AM. 2 FM. WKHB(AM) Irwin, PA; WKFB(AM) Jeannette, PA; WANB(AM) Waynesburg, PA; WANB-FM Waynesburg, PA; WROG-FM Cumberland, MD; WCMD(AM) Cumberland, MD.

Robert M. Stevens, pres; Ashley R. Stevens, VP.

Broadcast South LLC, 952 Greenwillow Dr., Douglas, GA 31535. Phone: (912) 389-0995. Fax: (912) 383-8552. Ownership: John O. Higgs, 25%; Kerry Van Moore, 25%; B. Gene Waldron, 25%; and Nearly Famous Properties LLC, 25%.

Stns: 2 AM. 3 FM. WDMG-FM Ambrose, GA; WDMG Douglas, GA; WRDO-FM Fitzgerald, GA; WBHB Fitzgerald, GA; WKZZ-FM Tifton, GA.

Brooke Communications Inc., 1445 W. Harvard Ave., Roseburg, OR 97470. Phone: (541) 672-6641. Fax: (541) 673-7598. Ownership: William E. Markham Trust, Patrick A. Markham.

Stns: 2 AM. 3 FM. KQEN(AM) Roseburg, OR; KRNR Roseburg, OR; KRSB-FM Roseburg, OR; KAVJ(FM) Sutherlin, OR; KKMX(FM) Tri City, OR.

David Hansen, gen sls mgr; Mike Carter, opns mgr; Patrick A. Markham, pres & gen mgr.

Brothers Broadcasting Corp., Box D, Rensselaer, IN 47978. Phone: (219) 866-4104. Fax: (219) 866-5106. E-mail: wirn@ffni.com Ownership: John Balvich, 100%.

Stns: 1 FM. WIBN-FM Earl Park, IN.

John Balvich, pres & gen mgr.

Bryan Broadcasting Corp., Box 3248, Bryan, TX 77805-3248. Phone: (979) 846-1150. Fax: (979) 846-1933. Ownership: William R. Hicks, 89%; and Ben D. Downs, 11%.

Stns: 3 AM. 1 FM. KZNE(AM) College Station, TX; WTAW(AM) College Station, TX; KNDE(FM) College Station, TX; KWBC Navasota, TX.

Buckley Broadcasting Corp., 166 West Putnam Ave., Greenwich, CT 06830. Phone: (203) 661-4307. Fax: (203) 622-7341. E-mail: rbuckley@buckleyradio.com Web Site: www.buckleyradio.com. Ownership: Steven Buckley; Dana Buckley; Richard Buckley; Martha Fahnoe

Stns: 8 AM. 10 FM. KKBB-FM Bakersfield, CA; KNZR Bakersfield, CA; KUBB-FM Mariposa, CA; KWAV-FM Monterey, CA; KLLY-FM Oildale, CA; KHTN(FM) Planada, CA; KIOO-FM Porterville, CA; KYZZ(FM) Salinas, CA; KSMJ(FM) Shafter, CA; KSEQ-FM Visalia, CA; WDRC(AM) Hartford, CT; WDRC-FM Hartford, CT; WMMW Meriden, CT; WSNG Torrington, CT; WWCO(AM) Waterbury, CT; WSEN(AM) Baldwinsville, NY; WOR New York, NY; WFBL(AM) Syracuse, NY.

Richard Buckley, pres; Joseph Bilotta, COO.

Burbach Broadcasting Group, 100 Ryan Ct., Suite 98, Pittsburgh, PA 15205. Phone: (412) 489-1001. Fax: (412) 278-1002. Ownership: Estate of John L. Laubach Jr., Nicholas A. Galli, chmn/pres.

Stns: 4 AM. 7 FM. WGIE(FM) Clarksburg, WV; WOBG Clarksburg, WV; WXKX(AM) Clarksburg, WV; WRZZ-FM Elizabeth, WV; WGYE(FM) Mannington, WV; WXIL-FM Parkersburg, WV; WGGE(FM) Parkersburg, WV; WADC Parkersburg, WV; WHBR-FM Parkersburg, WV; WVNT(AM) Parkersburg, WV; WOBG-FM Salem, WV.

Nicholas A. Galli, pres & gen mgr; Thomas Bayer, VP finance.

Burt Broadcasting Inc., Box 1848, Alamogordo, NM 88311. Phone: (505) 434-1414. Fax: (505) 434-2213. E-mail: burtbroadcasting@charter.net Ownership: William F. Burt, 50%; Donnie L. Burt, 50%.

Stns: 1 AM. 3 FM. KINN Alamogordo, NM; KYEE-FM Alamogordo, NM; KZZX(FM) Alamogordo, NM; KQEL(FM) Alamogordo, NM.

Bill Burt, gen mgr.

Bustos Media LLC, 500 Media Pl., Sacramento, CA 95815. Phone: (916) 368-6300. Fax: (916) 473-0146. E-mail: abustos@bustosmedia.com Web Site: www.bustosmedia.com.

Stns: 11 AM. 15 FM. KTTA-FM Esparto, CA; KBAA(FM) Grass Valley, CA; KLMG(FM) Jackson, CA; KBBU(FM) Modesto, CA; KZSJ(AM) San Martin, CA; KGDQ(FM) Colorado Springs, CO; KQTA(FM) Homedale, ID; KMUZ Gresham, OR; KZTB(FM) Milton-Freewater, OR; KSZN(AM) Milwaukie, OR; KGDD(AM) Oregon City, OR; KXMG(AM) Portland, OR; KREH Pecan Grove, TX; KVTA(AM) Centerville, UT; KDUT(FM) Randolph, UT; KWMG(AM) Auburn-Federal Way, WA; KMMG(FM) Benton City, WA; KDDS-FM Elma, WA; KULE Ephrata, WA; KULE-FM Ephrata, WA; KZTA-FM Naches, WA; KZML(FM) Quincy, WA; KZTS(AM) Sunnyside, WA; KYXE(AM) Union Gap, WA; WDDW(FM) Sturtevant, WI; KBMG(FM) Evanston, WY.

Amador S. Bustos, pres.

Butler County Radio Network Inc., 112 Hollywood Dr., Suite 203, Butler, PA 16001. Phone: (724) 287-5778. Fax: (724) 282-9188. Web Site: www.insidebutlercountry.com. Ownership: Daniel R. Vernon, 25%; Linda D. Harvey, 25%; Scott W. Briggs, 25%; and Victoria A. Hinterberger, 25%.

Stns: 2 AM. 1 FM. WBUT Butler, PA; WISR Butler, PA; WLER-FM Butler, PA.

Vicki Hinterberger, gen mgr; Scott Briggs, opns mgr; Bill Davis, news dir; Ron Willison, sports dir.

C

CBS Radio, 1515 Broadway, 46th Fl., New York, NY 10036. Phone: (212) 846-3939. Web Site: www.cbsradio.com. Ownership: Viacom Inc., 100%. Note: Viacom Inc. also owns the CBS Television Stations Group (see listing under TV Group Ownership, Section B).

Stns:35 AM. 111 FM. KMLE-FM Chandler, AZ; KOOL-FM Phoenix, AZ; KZON(FM) Phoenix, AZ; KRAK(AM) Hesperia, CA; KRTH-FM Los Angeles, CA; KNX Los Angeles, CA; KTWV-FM Los Angeles, CA; KFWB Los Angeles, CA; KLSX-FM Los Angeles, CA; KEZN-FM Palm Desert, CA; KROQ-FM Pasadena, CA; KQJK(FM) Roseville, CA; KZZO-FM Sacramento, CA; KYMX-FM Sacramento, CA; KNCI-FM Sacramento, CA; KHTK Sacramento, CA; KFRG-FM San Bernardino, CA; KSCF(FM) San Diego, CA; KYXY-FM San Diego, CA; KYCY San Francisco, CA; KITS-FM San Francisco, CA; KMVQ-FM San Francisco, CA; KFRC-FM San Francisco, CA; KCBS San Francisco, CA; KLLC-FM San Francisco, CA; KXFG-FM Sun City, CA; KVFG(FM) Victorville, CA; KSFM-FM Woodland, CA; KWLI(FM) Broomfield, CO; KIMN-FM Denver, CO; KXKL-FM Denver, CO; WZMX-FM Hartford, CT; WTIC Hartford, CT; WRCH-FM New Britain, CT; WJHM-FM Daytona Beach, FL; WOCL-FM De Land, FL; WLLD-FM Holmes Beach, FL; WPBZ-FM Indiantown, FL; WMBX-FM Jensen Beach, FL; WNEW(FM) Jupiter, FL; WSJT-FM Lakeland, FL; WOMX-FM Orlando, FL; WYUU(FM) Safety Harbor, FL; WQYK-FM Saint Petersburg, FL; WQYK(AM) Seffner, FL; WRBQ-FM Tampa, FL; WEAT-FM West Palm Beach, FL; WIRK-FM West Palm Beach, FL; WAOK Atlanta, GA; WZGC-FM Atlanta, GA; WXRT-FM Chicago, IL; WUSN-FM Chicago, IL; WJMK-FM Chicago, IL; WSCR(AM) Chicago, IL; WBBM Chicago, IL; WBBM-FM Chicago, IL; WCKG-FM Elmwood Park, IL; WYGY(FM) Fort Thomas, KY; WZLX-FM Boston, MA; WODS-FM Boston, MA; WBMX-FM Boston, MA; WBZ Boston, MA; WBCN-FM Boston, MA; WOMC-FM Detroit, MI; WWJ Detroit, MI; WVMV-FM Detroit, MI; WYCD-FM Detroit, MI; WXYT Detroit, MI; WCCO Minneapolis, MN; KZJK(FM) Saint Louis Park, MN; KEZK-FM Saint Louis, MO; KYKY-FM Saint Louis, MO; KMOX Saint Louis, MO; WFNZ Charlotte, NC; WFNA(AM) Charlotte, NC; WNKS-FM Charlotte, NC; WKQC(FM) Charlotte, NC; WSOC-FM Charlotte, NC; WPEG-FM Concord, NC; WBAV-FM Gastonia, NC; KMXB(FM) Henderson, NV; KKJJ(FM) Henderson, NV; KLUC-FM Las Vegas, NV; KXNT North Las Vegas, NV; KXTE-FM Pahrump, NV; WZNE-FM Brighton, NY; WCBS New York, NY; WCBS-FM New York, NY; WFAN New York, NY; WINS New York, NY; WWFS(FM) New York, NY; WXRK(FM) New York, NY; WRMM-FM Rochester, NY; WPXY-FM Rochester, NY; WCMF-FM Rochester, NY; WKRQ-FM Cincinnati, OH; WUBE-FM Cincinnati, OH; WQAL-FM Cleveland, OH; WNCX(FM) Cleveland, OH; WDOK-FM Cleveland, OH; WKRI(FM) Cleveland Heights, OH; WGRR-FM Hamilton, OH; KVMX(FM) Banks, OR; KLTH(FM) Lake Oswego, OR; KINK(FM) Portland, OR; KUFO-FM Portland, OR; KCMD(AM) Portland, OR; KUPL-FM Portland, OR; WZPT-FM New Kensington, PA; WPHT Philadelphia, PA; WOGL(FM)

Philadelphia, PA; WIP Philadelphia, PA; KYW Philadelphia, PA; WDSY-FM Pittsburgh, PA; WTZN-FM Pittsburgh, PA; KDKA Pittsburgh, PA; WMFS(FM) Bartlett, TN; WMC(AM) Memphis, TN; WMC-FM Memphis, TN; KKMJ-FM Austin, TX; KLUV(FM) Dallas, TX; KJKK(FM) Dallas, TX; KRLD Dallas, TX; KLLI(FM) Dallas, TX; KMVK(FM) Fort Worth, TX; KVIL(FM) Highland Park-Dallas, TX; KHJZ-FM Houston, TX; KILT Houston, TX; KAMX(FM) Luling, TX; KIKK Pasadena, TX; KJCE Rollingwood, TX; KLQB(FM) Taylor, TX; WJFK-FM Manassas, VA; KJAQ(FM) Seattle, WA; KZOK-FM Seattle, WA; KPTK(AM) Seattle, WA; KMPS-FM Seattle, WA; KBKS-FM Tacoma, WA; WLZL(FM) Annapolis, MD; WJFK Baltimore, MD; WWMX-FM Baltimore, MD; WQSR(FM) Baltimore, MD; WTGB-FM Bethesda, MD; WHFS(FM) Catonsville, MD; WPGC Morningside, MD; WPGC-FM Morningside, MD

Ken O'Keefe, exec VP; Brian Ongaro, sr VP; Clancy Woods, sr VP; Dan Mason, pres/CEO.

CHUM Ltd., 1331 Yonge St., Toronto, ON M4T 1Y1. Canada. Phone: (416) 925-6666. Fax: (416) 926-1380. Web Site: www.chumlimited.com.

Stns: 14 AM. 16 FM. CKCE-FM Calgary, AB; CHBN-FM Edmonton, AB; CKST Vancouver, BC; CFUN Vancouver, BC; CHQM-FM Vancouver, BC; CFAX Victoria, BC; CHBE-FM Victoria, BC; CFWM-FM Winnipeg, MB; CFRW(AM) Winnipeg, MB; CJCH Halifax, NS; CJPT-FM Brockville, ON; CFJR-FM Brockville, ON; CKLC Kingston, ON; CFLY-FM Kingston, ON; CFCA-FM Kitchener, ON; CKKW Kitchener, ON; CKLY-FM Lindsay (city of Kawartha Lakes), ON; CHST-FM London, ON; CKKL-FM Ottawa, ON; CJMJ-FM Ottawa, ON; CFGO Ottawa, ON; CFRA Ottawa, ON; CKPT Peterborough, ON; CKQM-FM Peterborough, ON; CHUM Toronto, ON; CIDR-FM Windsor, ON; CIMX-FM Windsor, ON; CKLW Windsor, ON; CKWW Windsor, ON; CKGM Montreal, PQ.

Stns: 15 TV. CKVR, Barrie, ON; CKX, Brandon, MB; CKAL, Calgary, AB; CKEM, Edmonton, AB; CJAL, Edmonton, AB; CKX-1, Foxwarren, MB; CFPL-TV, London, ON; CHRO-TV-43, Ottawa, ON; CHRO, Pembroke, ON; CHMI-TV, Portage la Prarie, MB; CITY-TV, Toronto, ON; CKVU, Vancouver, BC; CIVI-TV, Victoria, BC; CHWI-TV, Windsor, ON; CKNX, Wingham, ON.

Stephen Tapp, exec VP; Peter Miller, VP; David Kirkwood, exec VP; Sarah Crawford, VP; Mary Powers, VP; Denise Cooper, VP; Alan Mayne, CFO; Paul Ski, pres & CHUM Radio.

CRISTA Broadcasting, 19303 Fremont Ave. N., Seattle, WA 98133. Phone: (206) 546-7350. Fax: (206) 546-7372. E-mail: comments@spirit1053.com Web Site: www.spirit1053.com.

Stns: 1 AM. 2 FM. KCIS(AM) Edmonds, WA; KCMS(FM) Edmonds, WA; KWPZ(FM) Lynden, WA.

Bob Lonal, pres; Melene Thompson, VP/gen mgr.

CSN International, 3232 W. MacArthur Blvd., Santa Ana, CA 92704. Phone: (714) 825-9663. Fax: (714) 825-9660. Web Site: www.csnradio.com.

Stns: 1 AM. 66 FM. WJIK(FM) Monroeville, AL; KVIR(FM) Bullhead City, AZ; KVJC(FM) Globe, AZ; KJCU(FM) Laytonville, CA; KJCQ(FM) Quincy, CA; KOGR(FM) Rosedale, CA; WYJC(FM) Greenville, FL; WUJC(FM) Saint Marks, FL; KHJC(FM) Lihue, HI; KIHS(FM) Adel, IA; KZJB(FM) Pocatello, ID; KWJT(FM) Rathdrum, ID; KEFX-FM Twin Falls, ID; KTWD(FM) Wallace, ID; WJCZ(FM) Milford, IL; WPJC(FM) Pontiac, IL; WJCY(FM) Cicero, IN; WOJC(FM) Crothersville, IN; WHLP(FM) Hanna, IN; WQKO-FM Howe, IN; WJCJ(FM) Ladoga, IN; WWTS(FM) Logansport, IN; WTMK(FM) Lowell, IN; WJCO(FM) Montpelier, IN; WCJL(FM) Morgantown, IN; WFGL Fitchburg, MA; WJWT(FM) Gardner, MA; WSMA(FM) Scituate, MA; WJCX-FM Pittsfield, ME; WCVM-FM Bronson, MI; WJCE(FM) Elkton, MI; KGSF(FM) Anderson, MO; KTBJ(FM) Festus, MO; KRSS(FM) Tarkio, MO; WWUN-FM Friar's Point, MS; KJFT(FM) Arlee, MT; KYWH(FM) Lockwood, MT; WGPS(FM) Elizabeth City, NC; WJIJ(FM) Norlina, NC; WPGT(FM) Roanoke Rapids, NC; WAJC(FM) Wilson, NC; WWFP(FM) Brigantine, NJ; KKCJ(FM) Cannon AFB, NM; KPKJ(FM) Mentmore, NM; KNMA(FM) Socorro, NM; WIFF(FM) Binghamton, NY; KJCC(FM) Carnegie, OK; KDJC(FM) Baker City, OR; KJCH(FM) Coos Bay, OR; KPIJ(FM) Junction City, OR; KEFS(FM) North Powder, OR; KKJA(FM) Redmond, OR; KAJC(FM) Salem, OR; WREQ-FM Ridgebury, PA; KWRC(FM) Rapid City, SD; KYJC(FM) Commerce, TX; KDKR(FM) Decatur, TX; KSGR(FM) Portland, TX; KJCF(FM) Clarkston, WA; KKRS(FM) Davenport, WA; KTJC(FM) Kelso, WA; KBLD(FM) Kennewick, WA; WJWD(FM) Marshall, WI; KLWD(FM) Gillette, WY; KWYC(FM) Orchard Valley, WY; KWCF(FM) Sheridan, WY; KRWT(FM) West Laramie, WY.

Charles W. Smith, pres; Mike Stocklin, dir opns.

CTC Media Group Inc., Box 353, Royal Oak, MD 21662. Phone: (410) 745-5958. E-mail: mail@ctc-media.com Web Site: www.ctc-media.com. Ownership: Lee Afflerbach; Mike Afflerbach

Stns: 4 AM. WSME(AM) Camp Lejeune, NC; WNOS New Bern, NC; WWNB New Bern, NC; WECU(AM) Winterville, NC.

Lee Afflerbach, pres/CEO; Mike Afflerbach, VP opns.

Sheila Callahan and Friends Inc., Box 309, Missoula, MT 59806-0309. Phone: (406) 542-1025. Fax: (406) 721-1036. Ownership: M. Sheila Callahan Murphy, 51%; and Chester Maxwell Murphy III, 49%.

Stns: 4 FM. KMTZ(FM) Boulder, MT; KHDV(FM) Darby, MT; KMSO-FM Missoula, MT; KDXT(FM) Victor, MT.

Calvary Evangelistic Mission Inc., Box 367000, San Juan, PR 00936-7000. Phone: (787) 724-1190. Fax: (787) 722-5395. E-mail: radio@therockradio.org Web Site: www.therockradio.org. Ownership: Dr. James Christensen, 11.1% of votes; Clair D. Miller, 11.1% of votes; James A. Looman, 11.1% of votes; Pablo E. Fernandez, 11.1% of votes; Raul Zevallos, 11.1% of votes; Vernon Green, 11.1% of votes; Wallace B. Bishop Jr., 11.1% of votes; Gwendolyn Santiago, 11.1% of votes; and Ruth Luttrell, 11.1% of votes.

Stns: 3 AM. WCGB Juana Diaz, PR; WBMJ San Juan, PR; WIVV Vieques, PR.

Ruth Lutterall, pres; Janet Luttrell, gen mgr; Nila Luttrell, CFO.

Cameron Broadcasting Inc., 1615 Orange Tree Ln., Suite 102, Redlands, CA 92374. Phone: (909) 793-2233. Fax: (909) 798-6984.

Stns: 3 AM. 2 FM. KZZZ(AM) Bullhead City, AZ; KFLG(AM) Bullhead City, AZ; KAAA Kingman, AZ; KLUK(FM) Needles, CA; KNKK(FM) Needles, CA.

William Jaeger, pres/CEO; Chris Sutherland, mktg dir.

Canfin Enterprises Inc., Box 3498, Abilene, TX 79604. Phone: (325) 672-5442. Fax: (325) 672-6128. E-mail: info@radioabilene.com Web Site: radioabilene.com. Ownership: Parker S. Cannan, 100%.

Stns: 2 AM. 1 FM. KKHR-FM Abilene, TX; KZQQ(AM) Abilene, TX; KWKC Abilene, TX.

CanWest Global Communications Corp., 201 Portage Ave., 31st Fl., Winnipeg, MB R3B 3L7. Canada. Phone: (204) 956-2025. Fax: (204) 947-9841. E-mail: bleslie@canwest.com Web Site: www.canwestglobal.com. Ownership: Asper Family 85% voting shares, 45% of equity.

Stns: 1 FM. CHAL-FM Halifax, NS.

Stns: 16 TV. CICT, Calgary, AB; CHEK-5, Campbell River, BC; CHCA-TV-1, Coronation, AB; CITV, Edmonton, AB; CIHF, Halifax, NS; CHCH, Hamilton, ON; CHBC, Kelowna, BC; CISA-TV, Lethbridge, AB; CJNT, Montreal, PQ; CKMI, Quebec City, PQ; CHCA-TV, Red Deer, AB; CFRE, Regina, SK; CFSK, Saskatoon, SK; CHAN, Vancouver, BC; CHEK, Victoria, BC; CKND-TV, Winnipeg, MB.

Leonard Asper, pres/CEO; David Asper, exec VP.

Canxus Broadcasting Corp., 152 E. Green Ridge Rd., Caribou, ME 04736-3737. Phone: (207) 473-7513. Fax: (207) 472-3221. Web Site: www.channelxradio.com. Ownership: Dennis Curley, 92%; Richard Chandler, 4%; and Pamela Curley, 4%.

Stns: 3 FM. WCXU-FM Caribou, ME; WCXX-FM Madawaska, ME; WCXV(FM) Van Buren, ME.

Capital Community Broadcasting Inc., 360 Egan Dr., Juneau, AK 99801-1748. Phone: (907) 586-1670. Fax: (907) 586-3612.

Stns: 3 FM. KRNN(FM) Juneau, AK; KXLL(FM) Juneau, AK; KTOO-FM Juneau, AK.

Stns: 1 TV. KTOO-TV, Juneau, AK.

Capital Media Corp., 30 Park Ave., Cohoes, NY 12047-3330. Phone: (518) 237-1330. Fax: (518) 235-4468. E-mail: info@whaz.com Web Site: www.whaz.com.

Stns: 1 AM. 4 FM. WHAZ-FM Hoosick Falls, NY; WBAR-FM Lake Luzerne, NY; WMYY-FM Schoharie, NY; WHAZ Troy, NY; WMNV-FM Rupert, VT.

Paul F. Lotters, pres & gen mgr; Steve Klob, opns dir.

Capitol Broadcasting Co. Inc., Box 12000, Raleigh, NC 27605. Phone: (919) 821-8555. Fax: (919) 821-8733. Web Site: www.cbc-raleigh.com. Ownership: Capitol Holding Co. Inc.

Stns: 2 FM. WCMC-FM Creedmoor, NC; WRAL(FM) Raleigh, NC.

Stns: 4 TV. WMYT-TV, Charlotte, NC; WJZY, Charlotte, NC; WRAL-TV, Raleigh-Durham (Fayetteville), NC; WRAZ, Raleigh-Durham (Fayetteville), NC.

James F. Goodmon, pres/CEO; Vicke S. Murray, sec; Daniel P. McGrath, VO/CFO; Michael D. Hill, VP/gen counsel; James R. Hefner, III, VP & tv.

Capps Broadcast Group, 2003 N.W. 56th Dr., Pendleton, OR 97801. Phone: (541) 276-1511. Fax: (541) 276-1480.

Stns: 3 AM. 4 FM. KCMB(FM) Baker City, OR; KWRL-FM La Grande, OR; KTIX Pendleton, OR; KUMA Pendleton, OR; KUMA-FM Pendleton, OR; KWHT-FM Pendleton, OR; KTEL Walla Walla, WA.

Randy McKone, pres & gen mgr.

CapSan Media LLC, 277 Bendix Rd., Suite 411, Virginia Beach, VA 23452. Phone: (757) 497-1415. Fax: (757) 497-2560. Web Site: www.capsanmedia.com. Ownership: Jason Baker, 42.3%; William Whitlow, 42.3%.

Stns: 4 FM. WYND-FM Hatteras, NC; WFMZ(FM) Hertford, NC; WVOD(FM) Manteo, NC; WZPR(FM) Nags Head, NC.

Carlson Communications International, 3606 S. 500 W., Salt Lake City, UT 84115. Phone: (801) 262-5624. Fax: (801) 266-1510. Ownership: Ralph J. Carlson, only stockholder with 10% or more.

Stns: 3 AM. 2 FM. KRJC-FM Elko, NV; KTSN Elko, NV; KCYN(FM) Moab, UT; KDYL(AM) South Salt Lake, UT; KCPX(AM) Spanish Valley, UT.

Ralph J. Carlson, pres.

Carolina Christian Radio, Box 957, Wilmington, NC 28402. Phone: (910) 763-2452. Fax: (910) 763-6578. E-mail: life@life905.com Web Site: www.life905.com.

Stns: 3 AM. 3 FM. WMYT Carolina Beach, NC; WZDG(FM) Scotts Hill, NC; WWIL Wilmington, NC; WWIL-FM Wilmington, NC; WLSG(AM) Wilmington, NC; WDVV-FM Wilmington, NC.

Jim Stephens, pres/gen mgr.

Carroll Broadcasting Co., 1119 E. Plaza Dr., Carroll, IA 51401. Phone: (712) 792-4321. Fax: (712) 792-6667. Web Site: www.carrollbroadcasting.com.

Stns: 1 AM. 2 FM. KCIM Carroll, IA; KKRL-FM Carroll, IA; KIKD-FM Lake City, IA.

Mary Collison, pres.

Carroll Enterprises Inc., Box 549, Tawas City, MI 48764. Phone: (989) 362-3417. Fax: (989) 362-4544. E-mail: wkjc@wkjc.com Web Site: www.wkjc.com.

Stns: 1 AM. 1 FM. WKJZ-FM Hillman, MI; WIOS Tawas City, MI.

John Carroll Jr., pres & gen mgr.

Jimmy Ray Carroll Stns, Box 271, Kemmerer, WY 83101. Phone: (307) 877-0000. Fax: (307) 877-5524.

Stns: 5 AM. 4 FM. KGLM-FM Anaconda, MT; KANA(AM) Anaconda, MT; KBCK(AM) Deer Lodge, MT; KWUD(AM) Woodville, TX; KEVA Evanston, WY; KWYW(FM) Lost Cabin, WY; KTRZ-FM Riverton, WY; KDNO(FM) Thermopolis, WY; KTHE Thermopolis, WY.

Jimmy Ray Carroll, owner.

Carter Broadcast Group Inc., 11131 Colorado Ave., Kansas City, MO 64137. Phone: (816) 763-2040. Fax: (816) 966-1055. Ownership: Michael Carter

Stns: 1 AM. 2 FM. KSJM(FM) Winfield, KS; KPRS-FM Kansas City, MO; KPRT Kansas City, MO.

Michael Carter, pres.

Casey Network LLC, 908 Opelika Rd., Auburn, AL 36830. Phone: (334) 821-0744. Fax: (334) 821-4031.

Stns: 2 AM. WTRP La Grange, GA; WRLA(AM) West Point, GA.

James Jarrell, pres/CEO.

Cenla Broadcasting Co. Inc., 1115 Texas Ave., Alexandria, LA 71301. Phone: (318) 445-1234. Fax: (318) 473-1960. Web Site: www.cenlabroadcasting.com. Ownership: Taylor C. Thompson, 50% votes; and Charles J. Soprano, 50% votes.

Stns: 2 AM. 4 FM. KQID-FM Alexandria, LA; KRRV-FM Alexandria, LA; KSYL Alexandria, LA; KZMZ-FM Alexandria, LA; KDBS Alexandria, LA; KKST-FM Oakdale, LA.

Taylor C. Thompson, pres.

Centennial Broadcasting LLC, 3443 Robinhood Rd., Suite H, Winston-Salem, NC 27106. Phone: (336) 794-7971. Ownership: G Force LLC, 73.75%; Centennial Management Inc., 25%; Allen B. Shaw, 0.71%; Steven H. Watts, 0.36%; and Christopher Jarrell, 0.18%.

Stns: 4 FM. WLEQ(FM) Bedford, VA; WZZU(FM) Lynchburg, VA; WLNI-FM Lynchburg, VA; WZZI-FM Vinton, VA.

Center Broadcasting Co. Inc., 307 San Augustine St., Center, TX 75935. Phone: (936) 598-3304. Fax: (936) 598-9537.

Stns: 1 AM. 2 FM. KDET Center, TX; KQBB(FM)

Center, TX; KQSI(FM) San Augustine, TX.

Tracy Broadway, gen mgr.

Central Wisconsin Broadcasting Inc., Box 387, 1201 E. Division St., Neillsville, WI 54456. Phone: (715) 743-3333. Fax: (715) 743-2288. E-mail: 1075therock@tds.net Web Site: 1075therock.com. Ownership: J. Kevin and Margaret L. Grap, 100%.

Stns: 1 AM. 2 FM. WCCN Neillsville, WI; WCCN-FM Neillsville, WI; WPKG(FM) Neillsville, WI.

J. Kevin Grap, gen mgr.

Cessna Communications Inc., Box 1, Bedford, PA 15522. Phone: (814) 623-1000. Fax: (814) 623-9692. E-mail: cesscomm@earthlink.net Ownership: Jay B. Cessna; John H. Cessna.

Stns: 1 AM. 2 FM. WBVE(FM) Bedford, PA; WBFD(AM) Bedford, PA; WAYC-FM Bedford, PA.

Jay Cessna, pres; John H. Cessna, VP/gen mgr.

Chaparral Communications, Box 100, Jackson, WY 83001. Phone: (307) 733-2120. Fax: (307) 733-4760. E-mail: jacksonholeradio@onewest.net Web Site: www.jacksonholeradio.com. Ownership: Jerrold Lundquist.

Stns: 3 AM. 10 FM. KLZY(FM) Honokaa, HI; KWYS-FM Island Park, ID; KYZK(FM) Sun Valley, ID; KECH-FM Sun Valley, ID; KSKI-FM Sun Valley, ID; KWMY(FM) Park City, MT; KEZQ-FM West Yellowstone, MT; KWYS West Yellowstone, MT; KJAX(FM) Jackson, WY; KSGT Jackson, WY; KMTN-FM Jackson, WY; KZJH-FM Jackson, WY; KPOW Powell, WY.

Scott Anderson, gen mgr.

Cherry Creek Radio LLC, 501 S. Cherry St., Suite 480, Denver, CO 80246. Phone: (303) 468-6500. Fax: (303) 468-6555. E-mail: jschwartz@cherrycreekradio.com Web Site: www.cherrycreekradio.com. Ownership: Arlington Capital Partners L.P., 97.716% of votes, 80.131% of total assets; ACP/CCR Holdings LLC, 17.585% of total assets.

Stns: 21 AM. 39 FM. KWCD-FM Bisbee, AZ; KXFF(FM) Colorado City, AZ; KTAN(AM) Sierra Vista, AZ; KZMK(FM) Sierra Vista, AZ; KROP(AM) Brawley, CA; KSIQ-FM Brawley, CA; KRLT-FM South Lake Tahoe, CA; KOWL(AM) South Lake Tahoe, CA; KTHN(FM) La Junta, CO; KBLJ(AM) La Junta, CO; KLMR-FM Lamar, CO; KLMR Lamar, CO; KKXK-FM Montrose, CO; KUBC(AM) Montrose, CO; KBNG(FM) Ridgway, CO; KRQS(FM) Alberton, MT; KYYA-FM Billings, MT; KRZN(FM) Billings, MT; KBLG Billings, MT; KRKX(FM) Billings, MT; KMBR-FM Butte, MT; KAAR(FM) Butte, MT; KXTL Butte, MT; KVVR(FM) Dutton, MT; KHKR-FM East Helena, MT; KLFM-FM Great Falls, MT; KMON Great Falls, MT; KMON-FM Great Falls, MT; KAAK(FM) Great Falls, MT; KXDR-FM Hamilton, MT; KZMT-FM Helena, MT; KBLL Helena, MT; KBLL-FM Helena, MT; KCAP Helena, MT; KGGL-FM Missoula, MT; KGRZ Missoula, MT; KZOQ-FM Missoula, MT; KYLT Missoula, MT; KBQQ(FM) Pinesdale, MT; KTHC-FM Sidney, MT; KEYZ Williston, ND; KCOM Comanche, TX; KYOX(FM) Comanche, TX; KSTV-FM Dublin, TX; KSTV Stephenville, TX; KREC(FM) Brian Head, UT; KXBN(FM) Cedar City, UT; KCIN(FM) Cedar City, UT; KSUB Cedar City, UT; KDXU Saint George, UT; KSNN-FM Saint George, UT; KUNF(AM) Washington, UT; KZPH-FM Cashmere, WA; KZHR(FM) Dayton, WA; KYSN-FM East Wenatchee, WA; KONA Kennewick, WA; KONA-FM Kennewick, WA; KWWW-FM Quincy, WA; KAAP(FM) Rock Island, WA; KWWX Wenatchee, WA.

Joe Schwartz, pres/CEO; Dan Gittings, exec VP/dir sls; Dennis Goodman, exec VP/dir opns.

Chesapeake-Portsmouth Broadcasting Corp., 2202 Jolliff Rd., Chesapeake, VA 23321. Phone: (757) 488-1010. Ownership: Nancy A. Epperson, 100%. Note: Group also owns 50% of New AM Nassau Village-Ratliff, FL. Group is broker of airtime on WRJR(AM) Portsmouth, VA.

Stns: 5 AM. WZNZ Jacksonville, FL; WBOB(AM) Jacksonville, FL; WLES(AM) Bon Air, VA; WPMH(AM) Claremont, VA; WTJZ Newport News, VA.

Henry W. Hoot,.

The Chickasaw Nation, Box 609, Ada, OK 74821-0609. Phone: (580) 332-1212. Fax: (580) 332-0128. Web Site: www.chickasaw.net. Ownership: The Chickasaw Nation, an Indian tribal government, is governed by a legislature.

Stns: 1 AM. 2 FM. KADA Ada, OK; KADA-FM Ada, OK; KYKC-FM Byng, OK.

Christian Broadcasting System Ltd., 29200 Vassar Dr. Ste.150, Livonia, MI 48152. Phone: (248) 477-4600. Fax: (248) 477-6911.

Stns: 6 AM. 2 FM. WDJO(AM) Florence, KY; WJMM-FM Keene, KY; WCGW Nicholasville, KY; WVKY(AM) Nicholasville, KY; WLCM Charlotte, MI; WSNL(AM) Flint, MI; WJIV-FM Cherry Valley, NY; WCVX(AM) Cincinnati, OH.

Jon R. Yinger, pres/CEO; Ralph Van Luven, VP; Sally Van Luven, sec; Vicky Yinger, treas.

Christian Faith Broadcasting Inc., 3809 Maple Ave., Castalia, OH 44824. Phone: (419) 684-5311. Fax: (419) 684-5378. E-mail: wggn@lrbcg.com

Stns: 3 FM. WJKW-FM Athens, OH; WGGN-FM Castalia, OH; WLRD(FM) Willard, OH.

Stns: 2 TV. WGGN, Cleveland-Akron (Canton), OH; WLLA, Grand Rapids-Kalamazoo-Battle Creek, MI.

Shelby Gillam, pres; Rusty Yost, VP.

Christian Listening Network Inc., 996 Helen St., Fayetteville, NC 28303. Phone: (910) 864-5028. Fax: (910) 864-6270. Ownership: George E. Wilson, 52%; Michele W. Lhotellier, 12%; Regina W. Parker, 12%; Sharlene W. Tew, 12%; and Jeffrey S. Wilson, 12%.

Stns: 1 AM. 2 FM. WCLN(AM) Clinton, NC; WCLN-FM Clinton, NC; WZKB-FM Wallace, NC.

Christian Voice of Central Ohio Inc., Box 793, New Albany, OH 43054. Phone: (614) 855-9171. Fax: (614) 855-9280. Ownership: David R. Kolbe, 9.09% votes; Joe Panzica, 9.09% votes; Cheryl Burket, 9.09% votes; George Cunningham, 9.09% votes; Gary Hosfelt, 9.09% votes; David Humphrey, 9.09% votes; Andrew Lang, 9.09% votes; Matt Levin 9.09% votes; Doug Martin, 9.09% votes; Mark Nicholas, 9.09% votes; and Donald Stillion, 9.09% votes.

Stns: 5 FM. WVXR(FM) Richmond, IN; WVXC-FM Chillicothe, OH; WCVO-FM Gahanna, OH; WCVZ-FM South Zanesville, OH; WVXW-FM West Union, OH.

Drenda Keesee, chmn; Dan Baughman, pres.

Churchill Communications LLC, 871 Country Club Rd., Eugene, OR 97401. Phone: (541) 344-5500. Fax: (541) 485-2550. Ownership: Suzanne Arlie, 100%.

Stns: 4 AM. KOPT(AM) Eugene, OR; KLZS(AM) Eugene, OR; KXOR(AM) Junction City, OR; KXPD(AM) Tigard, OR.

Citadel Broadcasting Corp., 7201 W. Lake Mead Blvd., Suite 400, Las Vegas, NV 89128. Phone: (702) 804-5200. Fax: (702) 804-5936. Web Site: www.citadelbroadcasting.com. Ownership: Forstmann Little Funds, 58%; and public shareholders, 42%.

Stns:65 AM. 156 FM. WZRR-FM Birmingham, AL; WAPI Birmingham, AL; WSPZ(AM) Birmingham, AL; WYSF-FM Birmingham, AL; WUHT(FM) Birmingham, AL; WFFN-FM Cordova, AL; WDGM(FM) Greensboro, AL; WTUG-FM Northport, AL; WJOX(FM) Northport, AL; WBEI(FM) Reform, AL; WTSK Tuscaloosa, AL; WJRD(AM) Tuscaloosa, AL; KAAY Little Rock, AR; KARN Little Rock, AR; KURB-FM Little Rock, AR; KPZK(AM) Little Rock, AR; KIPR-FM Pine Bluff, AR; KOKY-FM Sherwood, AR; KLAL-FM Wrightsville, AR; KSZR(FM) Oro Valley, AZ; KHYT-FM Tucson, AZ; KIIM-FM Tucson, AZ; KTUC Tucson, AZ; KCUB Tucson, AZ; KWIN-FM Lodi, CA; KLOS-FM Los Angeles, CA; KABC Los Angeles, CA; KDJK(FM) Mariposa, CA; KESP(AM) Modesto, CA; KATM(FM) Modesto, CA; KHKK-FM Modesto, CA; KHOP-FM Oakdale, CA; KGO San Francisco, CA; KSFO San Francisco, CA; KWYL(FM) South Lake Tahoe, CA; KJOY-FM Stockton, CA; KWNN-FM Turlock, CA; KATC-FM Colorado Springs, CO; KKML(AM) Colorado Springs, CO; KVOR Colorado Springs, CO; KKPK(FM) Colorado Springs, CO; KKFM-FM Colorado Springs, CO; KKMG-FM Pueblo, CO; WSUB Groton, CT; WXLM(FM) Stonington, CT; WRQX-FM Washington, DC; WMAL Washington, DC; WYAY-FM Gainesville, GA; WKHX-FM Marietta, GA; KWQW(FM) Boone, IA; KGGO-FM Des Moines, IA; KHKI-FM Des Moines, IA; KBGG(AM) Des Moines, IA; KJJY(FM) West Des Moines, IA; KIZN-FM Boise, ID; KQFC-FM Boise, ID; KBOI Boise, ID; KTIK Nampa, ID; KKGL-FM Nampa, ID; KZMG(FM) New Plymouth, ID; WLS Chicago, IL; WZZN(FM) Chicago, IL; WWKI-FM Kokomo, IN; WMDH New Castle, IN; WMDH-FM New Castle, IN; WIBR Baton Rouge, LA; WXOK Baton Rouge, LA; KMEZ-FM Belle Chasse, LA; WCDV(FM) Hammond, LA; WEMX-FM Kentwood, LA; WDVW(FM) La Place, LA; KSMB-FM Lafayette, LA; KRRQ-FM Lafayette, LA; KXKC(FM) New Iberia, LA; KRDJ(FM) New Iberia, LA; KQXL-FM New Roads, LA; KKND-FM Port Sulphur, LA; KNEK Washington, LA; WFHN-FM Fairhaven, MA; WXLO-FM Fitchburg, MA; WBSM New Bedford, MA; WWFX(FM) Southbridge, MA; WMAS Springfield, MA; WORC-FM Webster, MA; WMME-FM Augusta, ME; WJZN(AM) Augusta, ME; WCYY-FM Biddeford, ME; WSHK(FM) Kittery, ME; WBLM-FM Portland, ME; WJBQ-FM Portland, ME; WBPW(FM) Presque Isle, ME; WQHR-FM Presque Isle, ME; WOZI(FM) Presque Isle, ME; WTVL(AM) Waterville, ME; WEBB-FM Waterville, ME; WHNN-FM Bay City, MI; WIOG(FM) Bay City, MI; WHTS(FM) Coopersville, MI; WDRQ-FM Detroit, MI; WJR Detroit, MI; WDVD(FM) Detroit, MI; WVFN(AM) East Lansing, MI; WFMK(FM) East Lansing, MI; WFBE-FM Flint, MI; WTRX Flint, MI; WLAV-FM Grand Rapids, MI; WKLQ(FM) Greenville, MI; WTNR(FM) Holland, MI; WVIB(FM) Holton, MI; WJIM(AM) Lansing, MI; WITL-FM Lansing, MI; WKQZ-FM Midland, MI; WLAW(FM) Newaygo, MI; WLCS-FM North Muskegon, MI; WILZ-FM Saginaw, MI; WODJ(AM) Whitehall, MI; WGVY(FM) Cambridge, MN; WGVZ(FM) Eden Prairie, MN; KQRS-FM Golden Valley, MN; WGVX(FM) Lakeville, MN; KXXR(FM) Minneapolis, MN; WRBO(FM) Como, MS; WMTI(FM) Picayune, MS; WOKQ-FM Dover, NH; WSAK(FM) Hampton, NH; WHOM-FM Mt. Washington, NH; WPKQ(FM) North Conway, NH; KRST-FM Albuquerque, NM; KDRF(FM) Albuquerque, NM; KKOB(AM) Albuquerque, NM; KKOB-FM Albuquerque, NM; KNML(AM) Albuquerque, NM; KTBL(AM) Los Ranchos de Albuquerque, NM; KBUL-FM Carson City, NV; KNEV-FM Reno, NV; KKOH Reno, NV; WNBF Binghamton, NY; WYOS(AM) Binghamton, NY; WHTT-FM Buffalo, NY; WGRF-FM Buffalo, NY; WEDG-FM Buffalo, NY; WBBF(AM) Buffalo, NY; WWYL(FM) Chenango Bridge, NY; WIII(FM) Cortland, NY; WKRT(AM) Cortland, NY; WAQX-FM Manlius, NY; WMOS(FM) Montauk, NY; WABC(AM) New York, NY; WPLJ(FM) New York, NY; WHLD Niagara Falls, NY; WLTI(FM) Syracuse, NY; WNTQ(FM) Syracuse, NY; WWLS-FM Edmond, OK; WWLS Moore, OK; KATT-FM Oklahoma City, OK; WKY Oklahoma City, OK; KYIS-FM Oklahoma City, OK; WLEV-FM Allentown, PA; WCAT-FM Carlisle, PA; WSJR(FM) Dallas, PA; WCTO-FM Easton, PA; WXTA-FM Edinboro, PA; WIOV-FM Ephrata, PA; WQHZ(FM) Erie, PA; WRIE Erie, PA; WBSX(FM) Hazleton, PA; WMHX(FM) Hershey, PA; WBHT-FM Mountain Top, PA; WBHD(FM) Olyphant, PA; WIOV Reading, PA; WARM Scranton, PA; WMGS-FM Wilkes-Barre, PA; WQXA-FM York, PA; WPRO Providence, RI; WSKO Providence, RI; WSKO-FM Wakefield-Peacedale, RI; WWKX-FM Woonsocket, RI; WXTC Charleston, SC; WSUY-FM Charleston, SC; WTMA Charleston, SC; WSSX-FM Charleston, SC; WISW Columbia, SC; WLXC-FM Lexington, SC; WTCB-FM Orangeburg, SC; WNKT-FM Saint George, SC; WWWZ-FM Summerville, SC; WXSM(AM) Blountville, TN; WGOW Chattanooga, TN; WOGT-FM East Ridge, TN; WGFX-FM Gallatin, TN; WNRX(FM) Jefferson City, TN; WJCW Johnson City, TN; WQUT-FM Johnson City, TN; WKOS-FM Kingsport, TN; WGOC(AM) Kingsport, TN; WIVK-FM Knoxville, TN; WNML(AM) Knoxville, TN; WNML-FM Loudon, TN; WGKX-FM Memphis, TN; WXMX(FM) Millington, TN; WKIM(FM) Munford, TN; WKDF-FM Nashville, TN; WOKI(FM) Oliver Springs, TN; WGOW-FM Soddy-Daisy, TN; KTYS(FM) Flower Mound, TX; KSCS-FM Fort Worth, TX; WBAP Fort Worth, TX; KJQS(AM) Murray, UT; KENZ(FM) Ogden, UT; KBER(FM) Ogden, UT; KKAT-FM Orem, UT; KKAT(AM) Salt Lake City, UT; KBEE(FM) Salt Lake City, UT; WJZW-FM Woodbridge, VA; KEYF-FM Cheney, WA; KEYF(AM) Dishman, WA; KGA Spokane, WA; KJRB Spokane, WA; KBBD(FM) Spokane, WA; KZBD(FM) Spokane, WA; KDRK-FM Spokane, WA

Farid Suleman, chmn/CEO; Judy Ellis, COO.

Clancy-Mance Communications, 199 Wealtha Ave., Watertown, NY 13601. Phone: (315) 782-1240. Fax: (315) 782-0312. Ownership: Jack Clancy, David Mance.

Stns: 1 AM. 2 FM. WBDR(FM) Cape Vincent, NY; WCDO Sidney, NY; WCDO-FM Sidney, NY.

David Mance, pres/gen mgr; John Clancy, sec/treas.

Clarion County Broadcasting Corp., 1168 Greenville Pike, Clarion, PA 16214-0688. Phone: (814) 226-4500. Fax: (814) 226-5898. E-mail: wccrwch@comcast.net Ownership: William S. Hearst, 100%.

Stns: 2 AM. 2 FM. WWCH Clarion, PA; WCCR-FM Clarion, PA; WKQW Oil City, PA; WKQW-FM Oil City, PA.

Clarke Broadcasting Corp., 1175 Fairview Dr., Suite N, Carson City, NV 89701. Phone: (775) 887-0588. Fax: (775) 887-1752.

Stns: 1 AM. KVML(AM) Sonora, CA.

H. Randolph Holder, pres; Larry England, gen mgr.

Classic Communications Inc., Box 1600, Woodward, OK 73802-1600. Phone: (580) 256-1450. Fax: (580) 254-9102. Ownership: Sherre D. House, 50%; Blake Brewer, 25%; and Bret Brewer, 25%.

Stns: 1 AM. 2 FM. KSIW Woodward, OK; KWDQ-FM Woodward, OK; KWFX-FM Woodward, OK.

Clear Channel Communications Inc., 200 E. Basse Rd., San Antonio, TX 78209. Phone: (210) 822-2828. Fax: (210) 822-2299. E-mail: markpmays@clearchannel.com Web Site: www.clearchannel.com. Ownership: Thomas O. Hicks, 6.5%; L. Lowry Mays, 5.2%. Publicly traded company with the majority of its shares owned by the investing public.

Stns:304 AM. 684 FM. KASH-FM Anchorage, AK; KBFX(FM) Anchorage, AK; KENI Anchorage, AK; KGOT-FM Anchorage, AK; KTZN Anchorage, AK; KYMG-FM Anchorage, AK; KFBX(AM) Fairbanks, AK; KIAK-FM Fairbanks, AK;

KKED-FM Fairbanks, AK; KAKQ-FM Fairbanks, AK; WSTH-FM Alexander City, AL; WMJJ-FM Birmingham, AL; WRTR(FM) Brookwood, AL; WZBQ-FM Carrollton, AL; WHOS Decatur, AL; WDRM-FM Decatur, AL; WTXT-FM Fayette, AL; WAGH-FM Fort Mitchell, AL; WAAX Gadsden, AL; WGMZ-FM Glencoe, AL; WTAK-FM Hartselle, AL; WENN(FM) Hoover, AL; WBHP Huntsville, AL; WDXB(FM) Jasper, AL; WHLW(FM) Luverne, AL; WWMG(FM) Millbrook, AL; WKSJ-FM Mobile, AL; WNTM(AM) Mobile, AL; WMXC-FM Mobile, AL; WRKH-FM Mobile, AL; WHAL(AM) Phenix City, AL; WGSY-FM Phenix City, AL; WBFA-FM Smiths, AL; WZHT-FM Troy, AL; WQEN(FM) Trussville, AL; WACT Tuscaloosa, AL; WQRV(FM) Tuscumbia, AL; KHKN(FM) Benton, AR; KMJX-FM Conway, AR; KIX-FM Fayetteville, AR; KEZA-FM Fayetteville, AR; KWHN(AM) Fort Smith, AR; KMAG-FM Fort Smith, AR; KWHF(FM) Harrisburg, AR; KDJE(FM) Jacksonville, AR; KNEA(AM) Jonesboro, AR; KIYS(FM) Jonesboro, AR; KFIN(FM) Jonesboro, AR; KSSN-FM Little Rock, AR; KMXF-FM Lowell, AR; KHLR(FM) Maumelle, AR; KFXR-FM Chinle, AZ; KTZR-FM Green Valley, AZ; KOHT-FM Marana, AZ; KZZP-FM Mesa, AZ; KYOT-FM Phoenix, AZ; KOY Phoenix, AZ; KNIX-FM Phoenix, AZ; KMXP-FM Phoenix, AZ; KGME(AM) Phoenix, AZ; KESZ-FM Phoenix, AZ; KXEW South Tucson, AZ; KWFM(AM) Tucson, AZ; KWMT-FM Tucson, AZ; KRQQ-FM Tucson, AZ; KNST Tucson, AZ; KTTI-FM Yuma, AZ; KQSR(FM) Yuma, AZ; KZXY-FM Apple Valley, CA; KIXW Apple Valley, CA; KHYL-FM Auburn, CA; KBFP(AM) Bakersfield, CA; KBKO-FM Bakersfield, CA; KHTY(AM) Bakersfield, CA; KUSS(FM) Carlsbad, CA; KBFP-FM Delano, CA; KDFO-FM Delano, CA; KRDU Dinuba, CA; KHTS-FM El Cajon, CA; KALZ(FM) Fowler, CA; KCBL Fresno, CA; KHGE(FM) Fresno, CA; KATJ-FM George, CA; KRAB-FM Green Acres, CA; KURQ(FM) Grover Beach, CA; KRZR-FM Hanford, CA; KAVL Lancaster, CA; KSMY(FM) Lompoc, CA; KBIG-FM Los Angeles, CA; KFI Los Angeles, CA; KYSR-FM Los Angeles, CA; KTLK(AM) Los Angeles, CA; KOST-FM Los Angeles, CA; KIIS-FM Los Angeles, CA; KHHT(FM) Los Angeles, CA; KLAC Los Angeles, CA; KSTT-FM Los Osos-Baywood Park, CA; KIXA-FM Lucerne Valley, CA; KMRQ(FM) Manteca, CA; KTOM-FM Marina, CA; KJSN-FM Modesto, CA; KFIV Modesto, CA; KVVS(FM) Mojave, CA; KTPI(AM) Mojave, CA; KKGN(AM) Oakland, CA; KNEW Oakland, CA; KOCN-FM Pacific Grove, CA; KOSO-FM Patterson, CA; KSTE Rancho Cordova, CA; KGGI-FM Riverside, CA; KDIF Riverside, CA; KOSS-FM Rosamond, CA; KGBY-FM Sacramento, CA; KFBK Sacramento, CA; KDON-FM Salinas, CA; KION(AM) Salinas, CA; KPRC-FM Salinas, CA; KKDD San Bernardino, CA; KTDD(AM) San Bernardino, CA; KGB-FM San Diego, CA; KIOZ-FM San Diego, CA; KMYI(FM) San Diego, CA; KOGO(AM) San Diego, CA; KLSD(AM) San Diego, CA; KMEL-FM San Francisco, CA; KIOI-FM San Francisco, CA; KISQ-FM San Francisco, CA; KKSF-FM San Francisco, CA; KYLD-FM San Francisco, CA; KUFX-FM San Jose, CA; KSJO-FM San Jose, CA; KSLY-FM San Luis Obispo, CA; KVEC San Luis Obispo, CA; KXFM-FM Santa Maria, CA; KSNI-FM Santa Maria, CA; KSMA Santa Maria, CA; KQOD-FM Stockton, CA; KWSX(AM) Stockton, CA; KCNL(FM) Sunnyvale, CA; KTPI-FM Tehachapi, CA; KMYT(FM) Temecula, CA; KTMQ(FM) Temecula, CA; KBOS-FM Tulare, CA; KFSO-FM Visalia, CA; KEZL(AM) Visalia, CA; KRSX-FM Yermo, CA; KBCO-FM Boulder, CO; KBPI(FM) Denver, CO; KRFX-FM Denver, CO; KPTT(FM) Denver, CO; KHOW Denver, CO; KOA Denver, CO; KIIX(AM) Fort Collins, CO; KIBT(FM) Fountain, CO; KSME(FM) Greeley, CO; KGHF Pueblo, CO; KCSJ Pueblo, CO; KCCY(FM) Pueblo, CO; KDZA-FM Pueblo, CO; KVUU-FM Pueblo, CO; KPHT(FM) Rocky Ford, CO; KKZN(AM) Thornton, CO; KCOL(AM) Wellington, CO; KTCL(FM) Wheat Ridge, CO; KKLI-FM Widefield, CO; WPKX-FM Enfield, CT; WKCI-FM Hamden, CT; WKSS-FM Hartford, CT; WHCN-FM Hartford, CT; WELI New Haven, CT; WAVZ New Haven, CT; WPHH(FM) Waterbury, CT; WWYZ-FM Waterbury, CT; WWDC-FM Washington, DC; WMZQ-FM Washington, DC; WASH-FM Washington, DC; WBIG-FM Washington, DC; WWRC(AM) Washington, DC; WIHT(FM) Washington, DC; WTEM Washington, DC; WROO(FM) Callahan, FL; WBTP(FM) Clearwater, FL; WXTB-FM Clearwater, FL; WMMV Cocoa, FL; WJRR-FM Cocoa Beach, FL; WTKS-FM Cocoa Beach, FL; WSRZ-FM Coral Cove, FL; WTZB(FM) Englewood, FL; WHYI-FM Fort Lauderdale, FL; WBGG-FM Fort Lauderdale, FL; WMIB(FM) Fort Lauderdale, FL; WOLZ-FM Fort Myers, FL; WLDI-FM Fort Pierce, FL; WKGR-FM Fort Pierce, FL; WSYR-FM Gifford, FL; WJBT-FM Green Cove Springs, FL; WFUS(FM) Gulfport, FL; WOLL-FM Hobe Sound, FL; WFXJ(AM) Jacksonville, FL; WPLA(FM) Jacksonville, FL; WQIK-FM Jacksonville, FL; WAIL(FM) Key West, FL; WKEY-FM Key West, FL; WEOW(FM) Key West, FL; WCKT(FM) Lehigh Acres, FL; WMMB Melbourne, FL; WEBZ(FM) Mexico Beach, FL; WINZ(AM) Miami, FL; WIOD Miami, FL; WLVE-FM Miami Beach, FL; WMGE(FM) Miami Beach, FL; WFLA-FM Midway, FL; WMGF(FM) Mount Dora, FL; WBTT(FM) Naples Park, FL; WFKS(FM) Neptune Beach, FL; WQTM(AM) Orlando, FL; WRUM(FM) Orlando, FL; WPAP-FM Panama City, FL; WDIZ Panama City, FL; WFBX(FM) Parker, FL; WTKX-FM Pensacola, FL; WYCL-FM Pensacola, FL; WFLF(AM) Pine Hills, FL; WFKZ(FM) Plantation Key, FL; WCTH(FM) Plantation Key, FL; WZJZ(FM) Port Charlotte, FL; WPBH(FM) Port St. Joe, FL; WCCF(AM) Punta Gorda, FL; WXSR-FM Quincy, FL; WZZR(FM) Riviera Beach, FL; WCTQ(FM) Sarasota, FL; WSDV(AM) Sarasota, FL; WCVU(FM) Solana, FL; WAVW(FM) Stuart, FL; WTNT-FM Tallahassee, FL; WMTX(FM) Tampa, FL; WHNZ(AM) Tampa, FL; WFLA(AM) Tampa, FL; WXXL-FM Tavares, FL; WKEZ-FM Tavenier, FL; WLTQ-FM Venice, FL; WDDV(AM) Venice, FL; WZTA(AM) Vero Beach, FL; WCZR(FM) Vero Beach, FL; WQOL-FM Vero Beach, FL; WRLX-FM West Palm Beach, FL; WJNO(AM) West Palm Beach, FL; WBZT(AM) West Palm Beach, FL; WJYZ Albany, GA; WJIZ-FM Albany, GA; WKLS(FM) Atlanta, GA; WGST Atlanta, GA; WUBL(FM) Atlanta, GA; WIBL(FM) Augusta, GA; WBBQ-FM Augusta, GA; WEKL(FM) Augusta, GA; WRAK-FM Bainbridge, GA; WBZY(FM) Bowdon, GA; WSOL-FM Brunswick, GA; WWVA-FM Canton, GA; WDAK Columbus, GA; WSHE(AM) Columbus, GA; WVRK-FM Columbus, GA; WMRZ(FM) Dawson, GA; WVVM(AM) Dry Branch, GA; WQBZ-FM Fort Valley, GA; WIBB-FM Fort Valley, GA; WPCH(FM) Gray, GA; WMGP(FM) Hogansville, GA; WLCG Macon, GA; WPRW-FM Martinez, GA; WCOH Newnan, GA; WLTM(FM) Peachtree City, GA; WRXR-FM Rossville, GA; WTKS(AM) Savannah, GA; WSOK Savannah, GA; WAEV-FM Savannah, GA; WTLY(FM) Thomasville, GA; WOBB-FM Tifton, GA; WRBV-FM Warner Robins, GA; WEBL(FM) Warner Robins, GA; KSSK(AM) Honolulu, HI; KDNN(FM) Honolulu, HI; KHVH Honolulu, HI; KHBZ(AM) Honolulu, HI; KUCD(FM) Pearl City, HI; KSSK-FM Waipahu, HI; KASI Ames, IA; KCCQ(FM) Ames, IA; KKSY(FM) Anamosa, IA; KPTL(FM) Ankeny, IA; KGRS-FM Burlington, IA; KBUR Burlington, IA; KMJM(AM) Cedar Rapids, IA; WMT Cedar Rapids, IA; WMT-FM Cedar Rapids, IA; KCHA Charles City, IA; KCHA-FM Charles City, IA; KLKK(FM) Clear Lake, IA; KMXG(FM) Clinton, IA; KCQQ(FM) Davenport, IA; WOC Davenport, IA; WLLR-FM Davenport, IA; WHO(AM) Des Moines, IA; KXNO(AM) Des Moines, IA; KDRB(FM) Des Moines, IA; KKDM-FM Des Moines, IA; KKEZ-FM Fort Dodge, IA; KWMT Fort Dodge, IA; KBKB Fort Madison, IA; KBKB-FM Fort Madison, IA; KXKT(FM) Glenwood, IA; KXIC Iowa City, IA; KKRQ-FM Iowa City, IA; KXFT(FM) Manson, IA; KIAI(FM) Mason City, IA; KGLO(AM) Mason City, IA; KCZE(FM) New Hampton, IA; KSMA-FM Osage, IA; KWSL Sioux City, IA; KSEZ-FM Sioux City, IA; KMNS Sioux City, IA; KGLI-FM Sioux City, IA; KLLP-FM Chubbuck, ID; KID Idaho Falls, ID; KID-FM Idaho Falls, ID; KPKY-FM Pocatello, ID; KWIK Pocatello, ID; KCDA(FM) Post Falls, ID; KEZJ-FM Twin Falls, ID; KLIX Twin Falls, ID; KATZ-FM Alton, IL; WVON(AM) Berwyn, IL; WLIT-FM Chicago, IL; WGRB(AM) Chicago, IL; WGCI-FM Chicago, IL; WNUA-FM Chicago, IL; WKSC-FM Chicago, IL; KMJM-FM Columbia, IL; KUUL(FM) East Moline, IL; WVZA-FM Herrin, IL; WDDD Johnston City, IL; WDDD-FM Marion, IL; WFXN(AM) Moline, IL; WTAO-FM Murphysboro, IL; WVAZ-FM Oak Park, IL; WQUL-FM West Frankfort, IL; WFRX West Frankfort, IL; WTFX-FM Clarksville, IN; WRZX-FM Indianapolis, IN; WFBQ-FM Indianapolis, IN; WNDE Indianapolis, IN; WQMF-FM Jeffersonville, IN; WZKF(FM) Salem, IN; KZCH(FM) Derby, KS; KZSN(FM) Hutchinson, KS; KRBB-FM Wichita, KS; KTHR(FM) Wichita, KS; WSFE(AM) Burnside, KY; WSEK(FM) Burnside, KY; WKED-FM Frankfort, KY; WKYW-FM Frankfort, KY; WFKY Frankfort, KY; WXRA(AM) Georgetown, KY; WBUL-FM Lexington, KY; WMXL-FM Lexington, KY; WLKT-FM Lexington-Fayette, KY; WKJK Louisville, KY; WAMZ(FM) Louisville, KY; WHAS Louisville, KY; WKRD(AM) Louisville, KY; WLUE(FM) Louisville, KY; WMKJ(FM) Mt. Sterling, KY; WUBT(FM) Russellville, KY; WCND Shelbyville, KY; WKRD-FM Shelbyville, KY; WSFC Somerset, KY; WKEQ(FM) Somerset, KY; WLLK-FM Somerset, KY; WKQQ-FM Winchester, KY; WJBO Baton Rouge, LA; WYNK-FM Baton Rouge, LA; KRVE-FM Brusly, LA; WSKR Denham Springs, LA; KYRK(FM) Houma, LA; WRNO-FM New Orleans, LA; WQUE-FM New Orleans, LA; WODT New Orleans, LA; WYLD New Orleans, LA; WYLD-FM New Orleans, LA; WNOE-FM New Orleans, LA; WRNX-FM Amherst, MA; WJMN-FM Boston, MA; WXKS Everett, MA; WKOX Framingham, MA; WHYN Springfield, MA; WHYN-FM Springfield, MA; WSNE-FM Taunton, MA; WNNZ Westfield, MA; WTAG Worcester, MA; WSRS-FM Worcester, MA; WKCG-FM Augusta, ME; WWBX-FM Bangor, ME; WABI Bangor, ME; WLKE-FM Bar Harbor, ME; WBFB-FM Belfast, ME; WCME-FM Boothbay Harbor, ME; WQSS-FM Camden, ME; WGUY(FM) Dexter, ME; WKSQ-FM Ellsworth, ME; WFAU Gardiner, ME; WVOM-FM Howland, ME; WIGY-FM Madison, ME; WRKD Rockland, ME; WFZX(FM) Searsport, ME; WTOS-FM Skowhegan, ME; WUBB-FM York Center, ME; WTKA Ann Arbor, MI; WWWW-FM Ann Arbor, MI; WBCK Battle Creek, MI; WBFN(AM) Battle Creek, MI; WDTW(AM) Dearborn, MI; WNIC-FM Dearborn, MI; WMXD-FM Detroit, MI; WDTW-FM Detroit, MI; WDFN Detroit, MI; WJLB-FM Detroit, MI; WKQI-FM Detroit, MI; WBFX(FM) Grand Rapids, MI; WBCT-FM Grand Rapids, MI; WTKG Grand Rapids, MI; WOOD Grand Rapids, MI; WOOD-FM Grand Rapids, MI; WMAX-FM Holland, MI; WRCC(FM) Marshall, MI; WKBZ(AM) Muskegon, MI; WMUS(FM) Muskegon, MI; WSNX-FM Muskegon, MI; WSHZ(FM) Muskegon, MI; WMRR-FM Muskegon Heights, MI; WLBY(AM) Saline, MI; KQQL-FM Anoka, MN; KNFX Austin, MN; KQHT-FM Crookston, MN; KLDJ-FM Duluth, MN; KKCB-FM Duluth, MN; WEBC Duluth, MN; KMFX-FM Lake City, MN; KTCZ-FM Minneapolis, MN; KTLK-FM Minneapolis, MN; KFAN Minneapolis, MN; KFXN Minneapolis, MN; KBMX(FM) Proctor, MN; KDWB-FM Richfield, MN; KWEB(AM) Rochester, MN; KRCH(FM) Rochester, MN; KEEY-FM Saint Paul, MN; KSNR-FM Thief River Falls, MN; KMFX Wabasha, MN; KSWF(FM) Aurora, MO; KTOZ-FM Pleasant Hope, MO; KSLZ(FM) Saint Louis, MO; KSD(FM) Saint Louis, MO; KLOU(FM) Saint Louis, MO; KATZ(AM) Saint Louis, MO; KIGL(FM) Seligman, MO; KXUS-FM Springfield, MO; KGMY Springfield, MO; WESE-FM Baldwyn, MS; WMJY-FM Biloxi, MS; WWKZ(FM) Columbus, MS; WJKX-FM Ellisville, MS; WBVV(FM) Guntown, MS; WFOR Hattiesburg, MS; WUSW-FM Hattiesburg, MS; WHER-FM Heidelberg, MS; WHAL-FM Horn Lake, MS; WJDX Jackson, MS; WHLH(FM) Jackson, MS; WMSI-FM Jackson, MS; WZRX Jackson, MS; WQJQ-FM Kosciusko, MS; WNSL-FM Laurel, MS; WEEZ Laurel, MS; WJDQ(FM) Marion, MS; WYHL(AM) Meridian, MS; WMSO(FM) Meridian, MS; WBUV(FM) Moss Point, MS; WWZD-FM New Albany, MS; WHTU(FM) Newton, MS; WQYZ-FM Ocean Springs, MS; WKNN-FM Pascagoula, MS; WZLD(FM) Petal, MS; WKMQ(AM) Tupelo, MS; WTUP Tupelo, MS; WZKS-FM Union, MS; WSTZ-FM Vicksburg, MS; KISN(FM) Belgrade, MT; KKBR-FM Billings, MT; KBBB(FM) Billings, MT; KCTR-FM Billings, MT; KBUL Billings, MT; KMMS(AM) Bozeman, MT; KMMS-FM Bozeman, MT; KZMY(FM) Bozeman, MT; KLCY(AM) East Missoula, MT; KLYQ Hamilton, MT; KBAZ(FM) Hamilton, MT; KMHK-FM Hardin, MT; KPRK Livingston, MT; KXLB(FM) Livingston, MT; KYSS-FM Missoula, MT; KGVO Missoula, MT; KZIN-FM Shelby, MT; KSEN Shelby, MT; KENR(FM) Superior, MT; WWNC Asheville, NC; WKSL(FM) Burlington, NC; WMKS(FM) Clemmons, NC; WDCG-FM Durham, NC; WGBT(FM) Eden, NC; WPEK(AM) Fairview, NC; WQNQ(FM) Fletcher, NC; WMYI(FM) Hendersonville, NC; WLYT-FM Hickory, NC; WMAG(FM) High Point, NC; WVBZ(FM) High Point, NC; WRFX-FM Kannapolis, NC; WCDG(FM) Moyock, NC; WRVA-FM Rocky Mount, NC; WEND-FM Salisbury, NC; WIBT(FM) Shelby, NC; WKKT-FM Statesville, NC; WMXF(AM) Waynesville, NC; WRDU-FM Wilson, NC; WTQR-FM Winston-Salem, NC; KFYR Bismarck, ND; KBMR Bismarck, ND; KSSS-FM Bismarck, ND; KXMR Bismarck, ND; KQDY(FM) Bismarck, ND; KLTC Dickinson, ND; KZRX-FM Dickinson, ND; KCAD(FM) Dickinson, ND; KKXL Grand Forks, ND; KJKJ-FM Grand Forks, ND; KIZZ-FM Minot, ND; KMXA-FM Minot, ND; KYYX-FM Minot, ND; KZPR-FM Minot, ND; KRRZ Minot, ND; KCJB Minot, ND; KTGL-FM Beatrice, NE; KHUS(FM) Bennington, NE; KIBZ(FM) Crete, NE; KLMY(FM) Lincoln, NE; KMCX-FM Ogallala, NE; KOGA Ogallala, NE; KOGA-FM Ogallala, NE; KQBW(FM) Omaha, NE; KFAB Omaha, NE; KGOR-FM Omaha, NE; KZKX-FM Seward, NE; KSFT-FM South Sioux City, NE; WGIP Exeter, NH; WERZ-FM Exeter, NH; WGXL-FM Hanover, NH; WTSL Hanover, NH; WXXK-FM Lebanon, NH; WGIR Manchester, NH; WGIR-FM Manchester, NH; WVRR-FM Newport, NH; WHEB-FM Portsmouth, NH; WQSO-FM Rochester, NH; WHCY-FM Blairstown, NJ; WSUS-FM Franklin, NJ; WHTZ-FM Newark, NJ; WNNJ Newton, NJ; WNNJ-FM Newton, NJ; KABQ Albuquerque, NM; KBQI(FM) Albuquerque, NM; KZRR-FM Albuquerque, NM; KPEK-FM Albuquerque, NM; KCQL Aztec, NM; KKFG-FM Bloomfield, NM; KTEG(FM) Bosque Farms, NM; KSYU-FM Corrales, NM; KTRA-FM Farmington, NM; KDAG(FM) Farmington, NM; KGLX(FM) Gallup, NM; KFMQ-FM Gallup, NM; KAZX(FM) Kirtland, NM; KABQ-FM Santa Fe, NM; KXTC-FM Thoreau, NM; KWNR-FM Henderson, NV; KSNE-FM Las Vegas, NV; KPLV(FM) Las Vegas, NV; KWID(FM) Las Vegas, NV; WPYX-FM Albany, NY; WHRL-FM Albany, NY; WPHR-FM Auburn, NY; WKKF(FM) Ballston Spa, NY; WINR Binghamton, NY; WVOR(FM) Canandaigua, NY; WCTW-FM Catskill, NY; WKGB-FM Conklin, NY; WWDG(FM) DeRuyter, NY; WALK(AM) East Patchogue, NY; WRWC(FM) Ellenville, NY; WELG(AM) Ellenville, NY; WENE(AM) Endicott, NY; WMRV-FM Endicott, NY; WBBI(FM) Endwell, NY; WCPV-FM Essex, NY; WBBS-FM Fulton, NY; WRWD-FM Highland, NY; WFXF(FM) Honeoye Falls,

NY; WHUC Hudson, NY; WZCR(FM) Hudson, NY; WKGS-FM Irondequoit, NY; WKTU(FM) Lake Success, NY; WIXT(AM) Little Falls, NY; WSKU(FM) Little Falls, NY; WBWZ-FM New Paltz, NY; WAXQ-FM New York, NY; WLTW-FM New York, NY; WWPR-FM New York, NY; WEAV Plattsburgh, NY; WVTK(FM) Port Henry, NY; WKIP Poughkeepsie, NY; WPKF(FM) Poughkeepsie, NY; WRNQ-FM Poughkeepsie, NY; WOKR(FM) Remsen, NY; WADR Remsen, NY; WDVI(FM) Rochester, NY; WHTK Rochester, NY; WHAM Rochester, NY; WRNY Rome, NY; WUMX(FM) Rome, NY; WTRY-FM Rotterdam, NY; WRVE-FM Schenectady, NY; WGY Schenectady, NY; WCRR(FM) South Bristol Township, NY; WHEN Syracuse, NY; WWHT-FM Syracuse, NY; WYYY-FM Syracuse, NY; WSYR Syracuse, NY; WOFX(AM) Troy, NY; WUTQ Utica, NY; WOUR-FM Utica, NY; WMXW-FM Vestal, NY; WSKS(FM) Whitesboro, NY; WXZO(FM) Willsboro, NY; WHLO Akron, OH; WARF(AM) Akron, OH; WNCO Ashland, OH; WYBL(FM) Ashtabula, OH; WFUN Ashtabula, OH; WREO-FM Ashtabula, OH; WXEG-FM Beavercreek, OH; WNUS-FM Belpre, OH; WKDD(FM) Canton, OH; WLZT(FM) Chillicothe, OH; WKKJ(FM) Chillicothe, OH; WBEX Chillicothe, OH; WCHI Chillicothe, OH; WSAI(AM) Cincinnati, OH; WKRC Cincinnati, OH; WLW Cincinnati, OH; WVMX-FM Cincinnati, OH; WCKY(AM) Cincinnati, OH; WTAM Cleveland, OH; WMMS-FM Cleveland, OH; WMVX-FM Cleveland, OH; WMJI-FM Cleveland, OH; WGAR-FM Cleveland, OH; WMJK(FM) Clyde, OH; WBVB-FM Coal Grove, OH; WCOL-FM Columbus, OH; WYTS(AM) Columbus, OH; WNCI-FM Columbus, OH; WTVN(AM) Columbus, OH; WLWD(FM) Columbus Grove, OH; WMMX-FM Dayton, OH; WTUE-FM Dayton, OH; WONE Dayton, OH; WONW Defiance, OH; WZOM-FM Defiance, OH; WDFM-FM Defiance, OH; WZOO-FM Edgewood, OH; WDKF(FM) Englewood, OH; WZRX-FM Fort Shawnee, OH; WXXR(FM) Fredericktown, OH; WFXN-FM Galion, OH; WDSJ(FM) Greenville, OH; WBWR(FM) Hilliard, OH; WSRW Hillsboro, OH; WSRW-FM Hillsboro, OH; WBKS(FM) Ironton, OH; WIRO Ironton, OH; WLQT-FM Kettering, OH; WIMA Lima, OH; WIMT(FM) Lima, OH; WXXF(FM) Loudonville, OH; WMAN Mansfield, OH; WLTP(AM) Marietta, OH; WRVB-FM Marietta, OH; WMRN Marion, OH; WDIF-FM Marion, OH; WKFS-FM Milford, OH; WMVO(AM) Mount Vernon, OH; WQIO-FM Mount Vernon, OH; WNDH-FM Napoleon, OH; WBBG(FM) Niles, OH; WPFX-FM North Baltimore, OH; WHOF(FM) North Canton, OH; WFXJ-FM North Kingsville, OH; WBUK(FM) Ottawa, OH; WMLX-FM Saint Mary's, OH; WCPZ(FM) Sandusky, OH; WVKF(FM) Shadyside, OH; WSWR-FM Shelby, OH; WIZE Springfield, OH; WTTF Tiffin, OH; WVKS-FM Toledo, OH; WRVF(FM) Toledo, OH; WSPD Toledo, OH; WIOT(FM) Toledo, OH; WCWA Toledo, OH; WYNT-FM Upper Sandusky, OH; WCHO(AM) Washington Court House, OH; WKBN Youngstown, OH; WNIO(AM) Youngstown, OH; KTBT(FM) Broken Arrow, OK; KIZS(FM) Collinsville, OK; KXXY-FM Oklahoma City, OK; KHBZ-FM Oklahoma City, OK; KTOK(AM) Oklahoma City, OK; KTST-FM Oklahoma City, OK; KQLL-FM Owasso, OK; KZBB-FM Poteau, OK; KKBD(FM) Sallisaw, OK; KMOD-FM Tulsa, OK; KTBZ(AM) Tulsa, OK; KAKC(AM) Tulsa, OK; KKCW-FM Beaverton, OR; KPOJ(AM) Portland, OR; KEX Portland, OR; WAEB Allentown, PA; WSAN(AM) Allentown, PA; WZZO-FM Bethlehem, PA; WTKT(AM) Harrisburg, PA; WKBO Harrisburg, PA; WHP Harrisburg, PA; WRBT-FM Harrisburg, PA; WRKK Hughesville, PA; WLAN Lancaster, PA; WLAN-FM Lancaster, PA; WVRT(FM) Mill Hall, PA; WRFF(FM) Philadelphia, PA; WIOQ-FM Philadelphia, PA; WISX(FM) Philadelphia, PA; WUBA(AM) Philadelphia, PA; WDAS-FM Philadelphia, PA; WUSL-FM Philadelphia, PA; WKST-FM Pittsburgh, PA; WDVE-FM Pittsburgh, PA; WPGB(FM) Pittsburgh, PA; WBGG(AM) Pittsburgh, PA; WXDX-FM Pittsburgh, PA; WRFY-FM Reading, PA; WRAW(AM) Reading, PA; WBYL(FM) Salladasburg, PA; WBLJ-FM Shamokin, PA; WAKZ(FM) Sharpsville, PA; WRAK Williamsport, PA; WHJJ Providence, RI; WHJY-FM Providence, RI; WWBB-FM Providence, RI; WKSP(FM) Aiken, SC; WYKZ(FM) Beaufort, SC; WLTY-FM Cayce, SC; WEZL-FM Charleston, SC; WLTQ(AM) Charleston, SC; WALC-FM Charleston, SC; WCOS Columbia, SC; WVOC Columbia, SC; WNOK(FM) Columbia, SC; WSCC-FM Goose Creek, SC; WLFJ(AM) Greenville, SC; WESC-FM Greenville, SC; WGVL Greenville, SC; WLVH-FM Hardeeville, SC; WBZT-FM Mauldin, SC; WRFQ-FM Mt. Pleasant, SC; WYNF(AM) North Augusta, SC; WXLY-FM North Charleston, SC; WXBT(FM) West Columbia, SC; WUSY-FM Cleveland, TN; WPTN Cookeville, TN; WGSQ-FM Cookeville, TN; WGIC-FM Cookeville, TN; WHUB Cookeville, TN; WRVW-FM Lebanon, TN; WKZP(FM) McMinnville, TN; WAKI McMinnville, TN; WBMC McMinnville, TN; WDIA Memphis, TN; WEGR(FM) Memphis, TN; WHRK-FM Memphis, TN; KJMS(FM) Memphis, TN; WREC(AM) Memphis, TN; WSIX-FM Nashville, TN; WLAC Nashville, TN; WLND(FM) Signal Mountain, TN; WSMT Sparta, TN; WRKK-FM Sparta, TN; WTZX Sparta, TN; WTRZ(FM) Spencer, TN; KASE-FM Austin, TX; KPEZ-FM Austin, TX; KVET-FM Austin, TX; KYKR-FM Beaumont, TX; KLVI Beaumont, TX; KVNS(AM) Brownsville, TX; KTEX-FM Brownsville, TX; KKYS-FM Bryan, TX; KNFX-FM Bryan, TX; KMXR-FM Corpus Christi, TX; KRYS-FM Corpus Christi, TX; KUNO Corpus Christi, TX; KZPS-FM Dallas, TX; KFXR(AM) Dallas, TX; KDMX(FM) Dallas, TX; KHKS-FM Denton, TX; KRPT(FM) Devine, TX; KBFM(FM) Edinburg, TX; KTSM(AM) El Paso, TX; KHEY(AM) El Paso, TX; KPRR-FM El Paso, TX; KEGL-FM Fort Worth, TX; KDGE(FM) Fort Worth-Dallas, TX; KHFI-FM Georgetown, TX; KCOL-FM Groves, TX; KBRQ(FM) Hillsboro, TX; KBME Houston, TX; KTRH Houston, TX; KPRC Houston, TX; KODA-FM Houston, TX; KHMX-FM Houston, TX; KTBZ-FM Houston, TX; KKRW-FM Houston, TX; KIIZ-FM Killeen, TX; KAGG(FM) Madisonville, TX; KHKZ(FM) Mercedes, TX; KQXX-FM Mission, TX; KLFX-FM Nolanville, TX; KIOC-FM Orange, TX; KKMY-FM Orange, TX; KSAB-FM Robstown, TX; KFMK-FM Round Rock, TX; KAJA(FM) San Antonio, TX; WOAI San Antonio, TX; KQXT-FM San Antonio, TX; KTKR San Antonio, TX; KXXM-FM San Antonio, TX; KNCN-FM Sinton, TX; KWTX Waco, TX; KWTX-FM Waco, TX; WACO-FM Waco, TX; KBGO(FM) Waco, TX; KJMY(FM) Bountiful, UT; KXRV(FM) Centerville, UT; KODJ-FM Salt Lake City, UT; KNRS Salt Lake City, UT; KZHT(FM) Salt Lake City, UT; KOSY-FM Spanish Fork, UT; WYYD-FM Amherst, VA; WSNZ(FM) Appomattox, VA; WKAV Charlottesville, VA; WCHV Charlottesville, VA; WCJZ(FM) Charlottesville, VA; WSUH(FM) Crozet, VA; WACL-FM Elkton, VA; WFQX-FM Front Royal, VA; WKCY Harrisonburg, VA; WJJX-FM Lynchburg, VA; WROV-FM Martinsville, VA; WOWI-FM Norfolk, VA; WKUS(FM) Norfolk, VA; WBTJ(FM) Richmond, VA; WRXL-FM Richmond, VA; WRVQ-FM Richmond, VA; WRVA Richmond, VA; WRNL Richmond, VA; WTVR-FM Richmond, VA; WZBL(FM) Roanoke, VA; WHTE-FM Ruckersville, VA; WSNV(FM) Salem, VA; WKDW Staunton, VA; WCYK-FM Staunton, VA; WSVO-FM Staunton, VA; WKSI-FM Stephens City, VA; WJJS-FM Vinton, VA; WKCI(AM) Waynesboro, VA; WTFX(AM) Winchester, VA; WJCD(FM) Windsor, VA; WAZR-FM Woodstock, VA; WEZF-FM Burlington, VT; WCVR-FM Randolph, VT; WTSJ(AM) Randolph, VT; WTSM(FM) Springfield, VT; WMXR-FM Woodstock, VT; KNBQ(FM) Centralia, WA; KELA Centralia-Chehalis, WA; KMNT(FM) Chehalis, WA; KFNK(FM) Eatonville, WA; KQSN(FM) Naches, WA; KIXZ-FM Opportunity, WA; KOLW(FM) Othello, WA; KEYW-FM Pasco, WA; KFLD Pasco, WA; KUBE-FM Seattle, WA; KJR Seattle, WA; KJR-FM Seattle, WA; KKZX-FM Spokane, WA; KPTQ(AM) Spokane, WA; KQNT(AM) Spokane, WA; KHHO Tacoma, WA; KDBL(FM) Toppenish, WA; KIJZ(FM) Vancouver, WA; KXRX-FM Walla Walla, WA; KIT Yakima, WA; KUTI(AM) Yakima, WA; WISM-FM Altoona, WI; WQRB-FM Bloomer, WI; WATQ(FM) Chetek, WI; WBIZ Eau Claire, WI; WIBA Madison, WI; WTSO Madison, WI; WMEQ(AM) Menomonie, WI; WQBW(FM) Milwaukee, WI; WISN Milwaukee, WI; WRIT-FM Milwaukee, WI; WOKY Milwaukee, WI; WKKV-FM Racine, WI; WMAD(FM) Sauk City, WI; WXXM(FM) Sun Prairie, WI; WMIL-FM Waukesha, WI; WMRE Charles Town, WV; WVHU(AM) Huntington, WV; WTCR-FM Huntington, WV; WTCR Kenova, WV; WAMX-FM Milton, WV; WZZW Milton, WV; WHNK(AM) Parkersburg, WV; WDMX-FM Vienna, WV; WEGW-FM Wheeling, WV; WBBD Wheeling, WV; WKWK-FM Wheeling, WV; WWVA(AM) Wheeling, WV; KIGN(FM) Burns, WY; KKTL Casper, WY; KWYY-FM Casper, WY; KTRS-FM Casper, WY; KTWO Casper, WY; KLEN-FM Cheyenne, WY; KQMY(FM) Cheyenne, WY; KOWB Laramie, WY; KCGY(FM) Laramie, WY; KRVK(FM) Midwest, WY; KGAB Orchard Valley, WY; WSMJ(FM) Baltimore, MD; WPOC-FM Baltimore, MD; WCAO Baltimore, MD; WTNT(AM) Bethesda, MD; WFMD(AM) Frederick, MD; WFRE(FM) Frederick, MD; WWFG-FM Ocean City, MD; WDKZ(FM) Salisbury, MD; WTGM Salisbury, MD; WSBY-FM Salisbury, MD; WOSC-FM Bethany Beach, DE; WDOV Dover, DE; WLBW-FM Fenwick Island, DE; WDSD-FM Smyrna, DE; WWTX(AM) Wilmington, DE; WILM Wilmington, DE

Stns: 35 TV. WXXA-TV, Albany-Schenectady-Troy, NY; KGET-TV, Bakersfield, CA; WIVT, Binghamton, NY; WKRC-TV, Cincinnati, OH; WETM, Elmira (Corning), NY; KMTR, Eugene, OR; KTCW, Eugene, OR; KMCB, Eugene, OR; KTVF, Fairbanks, AK; KGPE, Fresno-Visalia, CA; WHP-TV, Harrisburg-Lancaster-Lebanon-York, PA; WJKT, Jackson, TN; WTEV-TV, Jacksonville, FL; WAWS, Jacksonville, FL; KLRT, Little Rock-Pine Bluff, AR; KASN, Little Rock-Pine Bluff, AR; WPTY, Memphis, TN; WLMT, Memphis, TN; WPMI-TV, Mobile, AL-Pensacola (Ft. Walton Beach), FL; WJTC, Mobile, AL-Pensacola (Ft. Walton Beach), FL; KION-TV, Monterey-Salinas, CA; WHAM-TV, Rochester, NY; KTVX, Salt Lake City, UT; KUCW, Salt Lake City, UT; WOAI-TV, San Antonio, TX; KFTY, San Francisco-Oakland-San Jose; KCOY, Santa Barbara-Santa Maria-San Luis Obispo, CA; KVOS, Seattle-Tacoma, WA; WSYR-TV, Syracuse, NY; KMYT-TV, Tulsa, OK; KOKI, Tulsa, OK; WWTI, Watertown, NY; KSAS-TV, Wichita-Hutchinson Plus, KS; KAAS, Wichita-Hutchinson Plus, KS; KOCW, Wichita-Hutchinson Plus, KS.

L. Lowry Mays, chmn; Mark P. Mays, CEO; Randall T. Mays, pres; Kenneth E. Wyker, sr VP; Herbert W. Hill Sr., sr VP; Don Perry, exec VP.

Coast Radio Company Inc., 600 E. Main St., Vacaville, CA 95688. Phone: (707) 446-0200. Fax: (707) 446-0122. Ownership: James E. Levitt, 35.3% voting interest; John F. Levitt, 35.3% voting interest; Lauren Leigh Levitt 1996 Trust, James E. Levitt, trustee, 7.35% voting interest; Joseph Curtis Levitt 1996 Trust, James E. Levitt, trustee, 7.35% voting interest; Jessica Nicole Sanders 1996 Trust, John F. Levitt, trustee, 7.35% voting interest; and John Patrick Levitt 1996 Trust, John F. Levitt, trustee, 7.35% voting interest.

Stns: 3 FM. KKIQ-FM Livermore, CA; KUIC-FM Vacaville, CA; KKDV(FM) Walnut Creek, CA.

Cochise Broadcasting LLC, Box 11060, Jackson, WY 83002. Phone: (703) 812-0482. Ownership: Jana Tucker, 50%; and Ted Tucker, 50%.

Stns: 2 AM. 4 FM. KCUZ Clifton, AZ; KCDQ(FM) Douglas, AZ; KKYZ-FM Sierra Vista, AZ; KFMM-FM Thatcher, AZ; KOMJ(AM) Omaha, NE; KWYX(FM) Casper, WY.

Cogeco Radio-Television Inc., 612 St. Jacques, Suite 100, Montreal, PQ H3C 5R1. Canada. Phone: (514) 390-6035. Fax: (514) 390-6070. Ownership: Cogeco Inc., 100%.

Stns: 4 FM. CJMF-FM Quebec, PQ; CJEC-FM Quebec, PQ; CFGE-FM Sherbrooke, PQ; CJEB-FM Trois Rivieres, PQ.

Stns: 1 TV. CFAP, Quebec City, PQ.

Rene Guimond, pres/CEO; Luc Doyon, VP; Therese David, VP; Guy Meunier, sls VP; Jacques Boiteau, gen mgr; Geoffrey O. Brow, gen mgr.

Cohan Radio Group Inc., 3750 US 27N, Suite 1, Sebring, FL 33870. Phone: (863) 382-9999. Fax: (863) 382-1982. E-mail: cohanradiogroup@htn.net Web Site: www.cohanradiogroup.com.

Stns: 3 AM. 2 FM. WWOJ(FM) Avon Park, FL; WWTK(AM) Lake Placid, FL; WWLL(FM) Sebring, FL; WITS(AM) Sebring, FL; WJCM(AM) Sebring, FL.

Peter L. Coughlin, pres.

College Creek Media LLC, 980 N. Michigan Ave., Suite 1875, Chicago, IL 60611. Phone: (312) 204-9900. Ownership: Media Focus LLC, 100% votes, 62.5% total assets; Simmons Media Group LLC (see listing), 37.5% total assets.

Stns: 12 FM. KQPI(FM) Aberdeen, ID; KZUS(FM) Belt, MT; KEAU(FM) Choteau, MT; KUUS(FM) Fairfield, MT; KHYY(FM) Minatare, NE; KETT(FM) Mitchell, NE; KPHD(FM) Elko, NV; KCLS(FM) Ely, NV; KRPX(FM) Wellington, UT; KABW(FM) Westport, WA; KADQ-FM Evanston, WY; KTYN(FM) Thayne, WY.

Colorado West Broadcasting Inc., 3230-B South Glen Ave., Glenwood Springs, CO 81601. Phone: (970) 945-9124. Fax: (970) 945-5409. Ownership: Dalmation Communications Inc., 25.01%; Gabriel Chenoweth, 21.65%; Carl Curtis, 15.70%.

Stns: 1 AM. 2 FM. KGLN Glenwood Springs, CO; KMTS-FM Glenwood Springs, CO; KSNO-FM Snowmass Village, CO.

Columbia Gorge Broadcasters Inc., Box 360, Hood River, OR 97031. Phone: (541) 386-1511. Fax: (541) 386-7155. Web Site: www.kihrk105.com. E-mail: gary@gorgeradio.com

Stns: 2 AM. 1 FM. KCGB-FM Hood River, OR; KIHR Hood River, OR; KACI The Dalles, OR.

Greg Walden, pres; Mylene Walden, VP; Gary Grossman, gen mgr.

Combined Communications, Box 5037, Bend, OR 97708. Phone: (541) 382-5263. Fax: (541) 388-0456. E-mail: mcheney@bendradio.com Web Site: www.klrr.com.

Stns: 1 AM. 2 FM. KBND Bend, OR; KTWS(FM) Bend, OR; KMTK(FM) Bend, OR.

Chuck Chackel, pres.

Combined Media Group Inc., Box 789, Wynne, AR 72396. Phone: (870) 238-8141. Fax: (870) 238-5997. Ownership: Bobby Caldwell, 45%; Timothy B. Scott, 45%; and Scott Siler, 10%. Note: Bobby Caldwell owns 50% of Arkansas County Broadcasters Inc. (see listing) and 100% of East Arkansas Broadcasters Inc. (see listing).

Stns: 2 AM. KPOC(AM) Pocahontas, AR; KRLW Walnut Ridge, AR.

Tim Scott, gen mgr.

Commonwealth Broadcasting Corp., 113 W. Public Sq., Suite 400, Glasgow, KY 42141. Phone: (270) 659-2002. Fax: (270) 651-1771. Ownership: Steven W. Newberry, Vickie V. Newberry.

Stns: 8 AM. 13 FM. WTCO Campbellsville, KY; WCKQ-FM Campbellsville, KY; WPTQ(FM) Cave City, KY; WIEL Elizabethtown, KY; WTSZ(AM) Eminence, KY; WOVO(FM) Glasgow, KY; WCDS(AM) Glasgow, KY; WGRK(AM) Greensburg, KY; WGRK-FM Greensburg, KY; WKMO-FM Hodgenville, KY; WHHT(FM) Horse Cave, KY; WLBN Lebanon, KY; WLSK-FM Lebanon, KY; WTHX(FM) Lebanon Junction, KY; WYMV(FM) Madisonville, KY; WTTL Madisonville, KY; WPKY Princeton, KY; WAVJ-FM Princeton, KY; WWKY(FM) Providence, KY; WYSB(FM) Springfield, KY; WRZI-FM Vine Grove, KY.

Steven W. Newberry, pres/CEO; W. Dale Thornhill, exec VP.

Communications Capital Managers LLC, 1111 Michigan Ave., Suite 301, East Lansing, MI 48823-1096. Phone: (517) 351-3333. Fax: (517) 351-4481.

Stns: 2 AM. 3 FM. WUBR(AM) Baton Rouge, LA; KNBB(FM) Dubach, LA; KRUS Ruston, LA; KXKZ-FM Ruston, LA; KPCH(FM) Ruston, LA.

Michael Oesterle, CEO; Deb Grugen, CFO.

Communications Corp. of the Americas Inc., Box 2128, Rock Springs, WY 82902. Phone: (307) 362-3793. Fax: (307) 362-8727. Web Site: www.wyoradio.com. Ownership: William J. Luzmoor III, 100%.

Stns: 1 AM. 2 FM. KSIT(FM) Rock Springs, WY; KQSW-FM Rock Springs, WY; KRKK Rock Springs, WY.

Communicom Broadcasting LLC, 220 Josephine St., Suite 200, Denver, CO 80206. Phone: (303) 759-8481. Ownership: CCA Inc., 100%. Note: CCA Inc. controls 100% of the votes and owns 5% of the total assets of WLNO(AM) New Orleans, LA.

Stns: 3 AM. KXEG(AM) Phoenix, AZ; KXXT(AM) Tolleson, AZ; WLVJ(AM) Boynton Beach, FL.

Community Broadcasters LLC, 199 Wealtha Ave., Watertown, NY 13601. Phone: (315) 782-1240. Ownership: James L. Leven, 20% votes; Bruce J. Mittman, 20% votes; Peter G. Schiff, 20% votes; Paul Homer, 20% votes; and Henry T. Wilson, 20% votes.

Stns: 2 AM. 5 FM. WTOJ(FM) Carthage, NY; WBDI(FM) Copenhagen, NY; WGIX-FM Gouverneur, NY; WOTT-FM Henderson, NY; WBDB(FM) Ogdensburg, NY; WSLB Ogdensburg, NY; WATN Watertown, NY.

Connoisseur Media LLC, 136 Main St., Suite 202, Westport, CT 06880-3304. Phone: (203) 227-1978. Fax: (203) 227-2373. Web Site: www.connoisseurmedia.com. Ownership: CM Broadcast Management LLC, mgng member, 100% votes.

Stns: 2 AM. 15 FM. KZLN(FM) Patterson, IA; WBBE(FM) Heyworth, IL; WIHN-FM Normal, IL; WVMG(FM) Normal, IL; KIBB(FM) Augusta, KS; KTCM-FM Kingman, KS; KPBR(FM) Joliet, MT; KPLN(FM) Lockwood, MT; KBBX-FM Nebraska City, NE; WRYV-FM Gallipolis, OH; WRTS-FM Erie, PA; WXBB(FM) Erie, PA; WFNN(AM) Erie, PA; WJET(AM) Erie, PA; WTWF(FM) Fairview, PA; WRKT-FM North East, PA; WMGA(FM) Kenova, WV.

Jeffrey D. Warshaw, CEO; Michael O. Driscoll, exec VP/CFO.

Contemporary Communications, 9408 Grand Gate St., Las Vegas, NV 89143. Phone: (702) 898-4669. Fax: (208) 567-6865. E-mail: contemporary@cox.net Ownership: Larry G. Fuss, 100%.

Stns: 3 AM. 2 FM. WVUV(AM) Leone, AS; KKHJ(AM) Leone, AS; KKHJ-FM Pago Pago, AS; WKXY(FM) Clarksdale, MS; WROX Clarksdale, MS.

Larry Fuss, pres.

Convergent Broadcasting LLC, 1766 Washington Ave., Portland, ME 04103-1624. Phone: (207) 878-0095. Ownership: Housatonic Micro Fund SBIC LP, 58.27%; Housatonic Equity Investors SBIC LP, 36.79%; Daniel Duman, 1.975%; George Silverman, 1.975%; and Bruce A. Biette, .99%.

Stns: 3 FM. KPUS(FM) Gregory, TX; KJKE(FM) Ingleside, TX; KKPN(FM) Rockport, TX.

Coon Valley Communications Inc., 2260 141st Dr., Perry, IA 50220. Phone: (507) 643-0065. Ownership: Patrick Delaney, 100%.

Stns: 1 AM. 2 FM. KGRA-FM Jefferson, IA; KDLS Perry, IA; KKRF-FM Stuart, IA.

Corus Entertainment Inc., 630 3rd Ave. S.W., Suite 105, Calgary, AB T2P 4L4. Canada. Phone: (403) 444-4244. Fax: (403) 444-4242. Web Site: www.corusent.com. Ownership: J.R. Shaw controls an aggregate of 80% of the voting rights.

Stns: 18 AM. 25 FM. CFGQ-FM Calgary, AB; CKRY-FM Calgary, AB; CKNG-FM Edmonton, AB; CISN-FM Edmonton, AB; CHED Edmonton, AB; CFMI-FM New Westminster, BC; CKNW New Westminster, BC; CHMJ(AM) Vancouver, BC; CJOB Winnipeg, MB; CJZZ-FM Winnipeg, MB; CIQB-FM Barrie, ON; CHAY-FM Barrie, ON; CJXY-FM Burlington, ON; CJDV-FM Cambridge, ON; CKCB-FM Collingwood, ON; CJSS-FM Cornwall, ON; CFLG-FM Cornwall, ON; CJUL(AM) Cornwall, ON; CJOY Guelph, ON; CHML Hamilton, ON; CFFX(AM) Kingston, ON; CFMK-FM Kingston, ON; CKBT-FM Kitchener-Waterloo, ON; CFPL London, ON; CILQ-FM North York, ON; CKRU Peterborough, ON; CFHK-FM St. Thomas, ON; CFNY-FM Toronto, ON; CFMJ(AM) Toronto, ON; CKDK-FM Woodstock, ON; CJRC-FM Gatineau, PQ; CFOM-FM Levis, PQ; CHMP-FM Longueuil, PQ; CFEL-FM Montmagny, PQ; CKAC Montreal, PQ; CINW(AM) Montreal, PQ; CHRC Quebec, PQ; CKRS(AM) Saguenay, PQ; CIME-FM Saint Jerome, PQ; CHLT Sherbrooke, PQ; CHLT-FM Sherbrooke, PQ; CHLN Trois Rivieres, PQ; CINF(AM) Verdun, PQ.

Stns: 3 TV. CKWS, Kingston, ON; CHEX-TV-2, Oshawa, ON; CHEX, Peterborough, ON.

John M. Cassaday, pres.

Coshocton Broadcasting Co., 114 N. Sixth St., Coshocton, OH 43812. Phone: (740) 622-1560. Fax: (740) 622-7940. Ownership: Bruce Wallace, 100%.

Stns: 1 AM. 1 FM. WTNS Coshocton, OH; WKLM-FM Millersburg, OH.

Bruce Wallace, pres & gen mgr.

Costa-Eagle Radio Ventures L.P., 462 Merrimack St., Methuen, MA 01844. Phone: (978) 686-9966. Fax: (978) 687-1180. E-mail: pcosta@ceradio.com Ownership: Costa Communications, 51%; and Cambridge Acquisitions, 49%.

Stns: 3 AM. WCEC(AM) Haverhill, MA; WNNW(AM) Lawrence, MA; WCCM(AM) Salem, NH.

Covenant Network, 4424 Hampton Ave., St. Louis, MO 63109. Phone: (314) 752-7000. Fax: (314) 752-7702. Web Site: www.covenantnet.net.

Stns: 5 AM. 3 FM. WRMS Beardstown, IL; WOLG-FM Carlinville, IL; WRYT(AM) Edwardsville, IL; WIHM Taylorville, IL; WHOJ(FM) Terre Haute, IN; WCKW(AM) Garyville, LA; KBKC(FM) Moberly, MO; KHOJ(AM) Saint Charles, MO.

John Anthony Holman, pres.

Cox Radio Inc., 6205 Peachtree Dunwoody Rd., Atlanta, GA 30328. Phone: (678) 645-0000. E-mail: cxr.info@cox.com Web Site: coxradio.com. Ownership: Cox Enterprises Inc., 100%. Note: Cox Enterprises Inc. also owns 100% of Cox Television (see listing in section B under TV Group Ownership).

Stns: 14 AM. 65 FM. WZZK-FM Birmingham, AL; WAGG(AM) Birmingham, AL; WPSB(AM) Birmingham, AL; WNCB(FM) Gardendale, AL; WBPT(FM) Homewood, AL; WBHJ(FM) Midfield, AL; WBHK-FM Warrior, AL; WEZN-FM Bridgeport, CT; WPLR-FM New Haven, CT; WNLK(AM) Norwalk, CT; WFOX(FM) Norwalk, CT; WSTC(AM) Stamford, CT; WCTZ(FM) Stamford, CT; WFYV-FM Atlantic Beach, FL; WHQT-FM Coral Gables, FL; WCFB-FM Daytona Beach, FL; WSUN-FM Holiday, FL; WJGL(FM) Jacksonville, FL; WMXQ-FM Jacksonville, FL; WOKV Jacksonville, FL; WAPE-FM Jacksonville, FL; WPYO(FM) Maitland, FL; WHDR(FM) Miami, FL; WFLC-FM Miami, FL; WEDR-FM Miami, FL; WDUV-FM New Port Richey, FL; WHTQ-FM Orlando, FL; WDBO Orlando, FL; WMMO-FM Orlando, FL; WOKV-FM Ponte Vedra Beach, FL; WXGL(FM) Saint Petersburg, FL; WPOI(FM) Saint Petersburg, FL; WHPT(FM) Sarasota, FL; WWRM(FM) Tampa, FL; WSB Atlanta, GA; WSB-FM Atlanta, GA; WBTS(FM) Doraville, GA; WSRV(FM) Gainesville, GA; WALR-FM La Grange, GA; KCCN-FM Honolulu, HI; KRTR(AM) Honolulu, HI; KINE-FM Honolulu, HI; KRTR-FM Kailua, HI; KPHW(FM) Kaneohe, HI; KKNE(AM) Waipahu, HI; WSFR(FM) Corydon, IN; WVEZ(FM) Louisville, KY; WPTI(FM) Louisville, KY; WRKA-FM Saint Matthews, KY; WBAB(FM) Babylon, NY; WGBB Freeport, NY; WBLI-FM Patchogue, NY; WHFM-FM Southampton, NY; WHIO Dayton, OH; WHKO-FM Dayton, OH; WHIO-FM Piqua, OH; WZLR(FM) Xenia, OH; KKCM(FM) Sand Springs, OK; KWEN-FM Tulsa, OK; KJSR-FM Tulsa, OK; KRAV-FM Tulsa, OK; KRMG Tulsa, OK; WJMZ-FM Anderson, SC; WHZT(FM) Seneca, SC; KTHT(FM) Cleveland, TX; KHPT(FM) Conroe, TX; KONO-FM Helotes, TX; KHTC(FM) Lake Jackson, TX; KKBQ-FM Pasadena, TX; KCYY(FM) San Antonio, TX; KKYX San Antonio, TX; KONO San Antonio, TX; KISS-FM San Antonio, TX; KSMG(FM) Seguin, TX; KPWT(FM) Terrell Hills, TX; WDYL-FM Chester, VA; WKHK-FM Colonial Heights, VA; WKLR-FM Fort Lee, VA; WMXB-FM Richmond, VA.

Cox Enterprises Inc. owns the following daily newspapers: The (Grand Junction, CO) *Daily Sentinel; Palm Beach* (FL) *Daily News* and *The Palm Beach* (FL) *Post; The Atlanta* (GA) *Journal & Constitution; The Daily Advance* (Elizabeth City), *The* (Greenville) *Daily Reflector* and the *Rocky Mount Telegram*, all NC; *Dayton Daily News* and the *Springfield News-Sun*, both OH; *Austin American-Statesman, Longview News-Journal, The Lufkin Daily News, News Messenger* (Marshall), *The* (Nacogdoches) *Daily Sentinel* and the *Waco Tribune-Herald*, all TX. Cox also owns weekly newspapers and shoppers in CO, FL, NC, OH, and TX.

Robert F. Neil, pres/CEO; Marc. W. Morgan, COO; Neil O. Johnston, CFO; Richard A. Reis, group VP.

Crain Media Group LLC, 200 S. Commerce, Suite 702, Little Rock, AR 72201. Phone: (501) 537-0720. Fax: (501) 537-0722. Ownership: Crain Investments, 100%.

Stns: 3 AM. 4 FM. KAPZ Bald Knob, AR; KCNY(FM) Bald Knob, AR; KAWW(AM) Heber Springs, AR; KAWW-FM Heber Springs, AR; KSMD(FM) Pangburn, AR; KWCK Searcy, AR; KWCK-FM Searcy, AR.

Crain publishes 22 trade magazines including *Advertising Age*.

Paul Coates, gen mgr; Phil Weaver, gen mgr.

Crawford Broadcasting Co., Box 3003, Blue Bell, PA 19422. Phone: (215) 628-3500. Fax: (215) 628-0818. Web Site: www.crawfordbroadcasting.com. Ownership: Donald B. Crawford is sole owner of all the stns except WMUZ(FM), WRDT(AM), WEXL(AM), KJSL(AM) and KSTL(AM). WMUZ(FM), WRDT(AM), WEXL(AM), KJSL(AM) and KSTL(AM) are owned by Donald B. Crawford and Dean A. Crawford.

Stns: 17 AM. 11 FM. WYDE(AM) Birmingham, AL; WDJC-FM Birmingham, AL; WXJC(AM) Birmingham, AL; WXJC-FM Cordova, AL; WYDE-FM Cullman, AL; KBRT Avalon, CA; KLDC Brighton, CO; KCMN Colorado Springs, CO; KLTT Commerce City, CO; KLZ Denver, CO; KLVZ(AM) Denver, CO; KCBR Monument, CO; WYCA(FM) Crete, IL; WYRB(FM) Genoa, IL; WSRB(FM) Lansing, IL; WPWX(FM) Hammond, IN; WMUZ-FM Detroit, MI; WRDT(AM) Monroe, MI; WEXL Royal Oak, MI; KJSL Saint Louis, MO; KSTL Saint Louis, MO; WDCD(AM) Albany, NY; WDCX-FM Buffalo, NY; WPTR(FM) Clifton Park, NY; WLGZ(AM) Rochester, NY; WRCI(FM) Webster, NY; KKPZ(AM) Portland, OR; KAAM(AM) Garland, TX.

Donald B. Crawford, pres/CEO.

Criswell Communications, Box 619000, Dallas, TX 75261-9000. Phone: (817) 792-3800. Fax: (817) 277-9929. E-mail: kcbi@kcbi.org Web Site: www.kcbi.org.

Stns: 1 AM. 3 FM. KSYE(FM) Frederick, OK; KCBI(FM) Dallas, TX; KCRN San Angelo, TX; KCRN-FM San Angelo, TX.

Ronald L. Harris, exec VP; Dr. Jerry Johnson, pres.

The Cromwell Group Inc., Cromwell Radio Group and Affiliates. Box 150846, Nashville, TN 37215. Phone: (615) 361-7560. Fax: (615) 366-4313. E-mail: b.walters@cromwellradio.com Web Site: www.cromwellradio.com. Ownership: Bayard H. Walters, 100%.

Stns: 6 AM. 15 FM. WCBH-FM Casey, IL; WWGO-FM Charleston, IL; WCRA Effingham, IL; WCRC-FM Effingham, IL; WZUS(FM) Macon, IL; WMCI-FM Mattoon, IL; WHQQ(FM) Neoga, IL; WEJT-FM Shelbyville, IL; WZNX(FM) Sullivan, IL; WPMB Vandalia, IL; WKRV-FM Vandalia, IL; WLME(FM) Cannelton, IN; WTCJ-FM Tell City, IN; WTCJ Tell City, IN; WKCM Hawesville, KY; WVJS(AM) Owensboro, KY; WBIO(FM) Philpot, KY; WXCM(FM) Whitesville, KY; WQZQ(AM) Clarksville, TN; WBUZ(FM) La Vergne, TN; WVNS-FM Pegram, TN.

Bayard H. Walters, pres; Thomas Crocker, CFO.

Crossroads Communications Inc., 1301 Ohio St., Terre Haute, IN 47807. Phone: (812) 234-9770. Fax: (812) 238-1576. E-mail: wsdm@wsdm.com Ownership: Michael A. Petersen, 53%; Dan T. Lacy, 47%.

Stns: 2 AM. 2 FM. WSDX(AM) Brazil, IN; WSDM-FM Brazil, IN; WAXI-FM Rockville, IN; WBOW(AM) Terre Haute, IN.

Michael A. Petersen, pres/gen mgr; Dan Lacy, CFO.

Cumulus Media Inc., 3535 Piedmont Rd., Atlanta, GA 30305. Phone: (404) 949-0700. Fax: (404) 949-0740. E-mail: bill@cumulusb.com Web Site: www.cumulus.com.

Chicago, IL 60601. Cumulus Broadcasting Inc., 875 N. Michigan Ave, Suite 3650. Phone: (312) 867-0091. Fax: (312) 867-0098.

Stns:79 AM. 192 FM. WZYP-FM Athens, AL; WVNN Athens, AL; WYOK-FM Atmore, AL; WDLT-FM Chickasaw, AL; WDLT Fairhope, AL; WXQW(FM) Gurley, AL; WUMP Madison, AL; WGOK Mobile, AL; WBLX-FM Mobile, AL; WHHY-FM Montgomery, AL; WMSP Montgomery, AL; WNZZ Montgomery, AL; WXFX-FM Prattville, AL; WVNN-FM Trinity, AL; KQSM-FM Bentonville, AR; KFAY Farmington, AR; KKEG-FM Fayetteville, AR; KLSZ-FM Fort Smith, AR; KYNF(FM) Prairie Grove, AR; KAMO-FM Rogers, AR;

KMCK-FM Siloam Springs, AR; KYNG(AM) Springdale, AR; KBBQ-FM Van Buren, AR; KOAI(AM) Van Buren, AR; KMGQ(FM) Goleta, CA; KRUZ(FM) Santa Barbara, CA; KVYB(FM) Santa Barbara, CA; KVEN Ventura, CA; KBBY-FM Ventura, CA; KHAY-FM Ventura, CA; KKNN-FM Delta, CO; KEKB-FM Fruita, CO; KEXO Grand Junction, CO; KBKL(FM) Grand Junction, CO; KMXY-FM Grand Junction, CO; WICC Bridgeport, CT; WINE Brookfield, CT; WEBE-FM Westport, CT; WFTW Fort Walton Beach, FL; WZNS-FM Fort Walton Beach, FL; WYZB-FM Mary Esther, FL; WINT(AM) Melbourne, FL; WCOA Pensacola, FL; WHKR-FM Rockledge, FL; WSJZ-FM Sebastian, FL; WNCV(FM) Shalimar, FL; WHBT(AM) Tallahassee, FL; WHBX(FM) Tallahassee, FL; WGLF(FM) Tallahassee, FL; WGPC Albany, GA; WQVE(FM) Albany, GA; WALG Albany, GA; WWLD(FM) Cairo, GA; WZBN(FM) Camilla, GA; WPEZ(FM) Jeffersonville, GA; WJAD-FM Leesburg, GA; WMAC Macon, GA; WLZN(FM) Macon, GA; WDEN-FM Macon, GA; WDDO Macon, GA; WAYS(AM) Macon, GA; WIFN(FM) Macon, GA; WMGB(FM) Montezuma, GA; WEGC-FM Sasser, GA; WBMQ Savannah, GA; WZAT-FM Savannah, GA; WJCL-FM Savannah, GA; WJLG Savannah, GA; WIXV-FM Savannah, GA; WTYB(FM) Springfield, GA; WEAS-FM Springfield, GA; WNUQ(FM) Sylvester, GA; WJOD-FM Asbury, IA; KQCS(FM) Bettendorf, IA; KOEL-FM Cedar Falls, IA; KDAT(FM) Cedar Rapids, IA; KHAK-FM Cedar Rapids, IA; KJOC Davenport, IA; KBOB-FM De Witt, IA; KXGE-FM Dubuque, IA; WDBQ Dubuque, IA; KLYV-FM Dubuque, IA; KCRR(FM) Grundy Center, IA; KRNA-FM Iowa City, IA; KBEA-FM Muscatine, IA; KOEL(AM) Oelwein, IA; KKHQ-FM Oelwein, IA; WXXQ-FM Freeport, IL; WDBQ-FM Galena, IL; WKGL-FM Loves Park, IL; WXLP-FM Moline, IL; WROK Rockford, IL; WZOK-FM Rockford, IL; KQTP-FM Saint Marys, KS; KTOP Topeka, KS; KWIC-FM Topeka, KS; KMAJ Topeka, KS; KDVV-FM Topeka, KS; WCYN-FM Cynthiana, KY; WXZZ-FM Georgetown, KY; WVLK Lexington, KY; WLTO-FM Nicholasville, KY; WVLK-FM Richmond, KY; KQLK(FM) De Ridder, LA; KBIU(FM) Lake Charles, LA; KXZZ Lake Charles, LA; KYKZ-FM Lake Charles, LA; KAOK Lake Charles, LA; KRMD-FM Shreveport, LA; KVMA-FM Shreveport, LA; KMJJ-FM Shreveport, LA; KKGB-FM Sulphur, LA; WEZQ-FM Bangor, ME; WQCB-FM Brewer, ME; WDEA Ellsworth, ME; WBZN-FM Old Town, ME; WKFR-FM Battle Creek, MI; WWCK Flint, MI; WDZZ-FM Flint, MI; WKMI Kalamazoo, MI; WTWR-FM Luna Pier, MI; WMHG(AM) Muskegon, MI; WRSR-FM Owosso, MI; WRKR-FM Portage, MI; KDHL Faribault, MN; KRFO Owatonna, MN; KFIL Preston, MN; KOLM(AM) Rochester, MN; KROC(AM) Rochester, MN; KROC-FM Rochester, MN; KWWK(FM) Rochester, MN; KLCX-FM Saint Charles, MN; KVGO-FM Spring Valley, MN; KYBA-FM Stewartville, MN; KOQL-FM Ashland, MO; KPLA(FM) Columbia, MO; KFRU Columbia, MO; KBXR(FM) Columbia, MO; KBBM(FM) Jefferson City, MO; KLIK(AM) Jefferson City, MO; KZJF(FM) Jefferson City, MO; KMJK(FM) Lexington, MO; KJMO(FM) Linn, MO; KRWP(FM) Stockton, MO; WSMS(FM) Artesia, MS; WJWF(AM) Columbus, MS; WMBC(FM) Columbus, MS; WKOR-FM Columbus, MS; WKOR(AM) Starkville, MS; WMXU-FM Starkville, MS; WSSO(AM) Starkville, MS; WRCQ-FM Dunn, NC; WFNC Fayetteville, NC; WAAV Leland, NC; WFNC-FM Lumberton, NC; WFVL(FM) Southern Pines, NC; WGNI-FM Wilmington, NC; WMNX-FM Wilmington, NC; WWQQ-FM Wilmington, NC; KACL(FM) Bismarck, ND; KKCT-FM Bismarck, ND; KBYZ(FM) Bismarck, ND; KLXX Bismarck-Mandan, ND; KUSB(FM) Hazelton, ND; WRRB(FM) Arlington, NY; WFAS-FM Bronxville, NY; WCZX-FM Hyde Park, NY; WPDA-FM Jeffersonville, NY; WKNY Kingston, NY; WKXP(FM) Kingston, NY; WALL Middletown, NY; WFAF(FM) Mount Kisco, NY; WDBY(FM) Patterson, NY; WEOK Poughkeepsie, NY; WFAS(AM) White Plains, NY; WZAD-FM Wurtsboro, NY; WRQN-FM Bowling Green, OH; WRQK-FM Canton, OH; WRWK(FM) Delta, OH; WXKR-FM Port Clinton, OH; WSOM Salem, OH; WWWM-FM Sylvania, OH; WLQR Toledo, OH; WTOD Toledo, OH; WBBW Youngstown, OH; KOMS-FM Poteau, OK; KEHK-FM Brownsville, OR; KUJZ(FM) Creswell, OR; KSCR(AM) Eugene, OR; KZEL-FM Eugene, OR; KUGN Eugene, OR; WTCY Harrisburg, PA; WTPA-FM Mechanicsburg, PA; WWIZ-FM Mercer, PA; WLLF-FM Mercer, PA; WWKL(FM) Palmyra, PA; WPIC Sharon, PA; WSEA(FM) Atlantic Beach, SC; WIQB(AM) Conway, SC; WYNN Florence, SC; WXJY-FM Georgetown, SC; WSYN-FM Georgetown, SC; WHSC Hartsville, SC; WWFN-FM Lake City, SC; WCMG-FM Latta, SC; WYMB Manning, SC; WHLZ(FM) Marion, SC; WMXT-FM Pamplico, SC; WDAI-FM Pawley's Island, SC; WYAK-FM Surfside Beach, SC; KDEZ(FM) Brandon, SD; KYBB(FM) Canton, SD; KIKN-FM Salem, SD; KKLS-FM Sioux Falls, SD; KMXC(FM) Sioux Falls, SD; KXRB(AM) Sioux Falls, SD; KSOO(AM) Sioux Falls, SD; WNFN(FM) Belle Meade, TN; WRQQ(FM) Goodlettsville, TN; WQQK-FM Hendersonville, TN; WWTN-FM Manchester, TN; WSM-FM Nashville, TN; WHRP(FM) Tullahoma, TN; KQIZ-FM Amarillo, TX; KPUR Amarillo, TX; KTLT(FM) Anson, TX; KIKR Beaumont, TX; KQXY-FM Beaumont, TX; KTCX-FM Beaumont, TX; KFNC(FM) Beaumont, TX; KOOC(FM) Belton, TX; KYYI-FM Burkburnett, TX; KZRK-FM Canyon, TX; KPUR-FM Canyon, TX; KARX(FM) Claude, TX; KSSM(FM) Copperas Cove, TX; KSTB-FM Crystal Beach, TX; KOLI-FM Electra, TX; KCDD(FM) Hamlin, TX; KUSJ(FM) Harker Heights, TX; KIOL(FM) La Porte, TX; KHXS-FM Merkel, TX; KMND Midland, TX; KZBT(FM) Midland, TX; KBAT(FM) Monahans, TX; KBED(AM) Nederland, TX; KODM-FM Odessa, TX; KRIL Odessa, TX; KGEE(FM) Pecos, TX; KAYD-FM Silsbee, TX; KTEM Temple, TX; KLTD-FM Temple, TX; KBCY(FM) Tye, TX; KQHN(FM) Waskom, TX; KQXC-FM Wichita Falls, TX; KLUR-FM Wichita Falls, TX; WBRW-FM Blacksburg, VA; WFNR Blacksburg, VA; WFNR-FM Christiansburg, VA; WPSK-FM Pulaski, VA; WRAD Radford, VA; WZNN(FM) Allouez, WI; WDUZ-FM Brillion, WI; WPCK(FM) Denmark, WI; WQLH-FM Green Bay, WI; WDUZ(AM) Green Bay, WI; WOGB-FM Kaukauna, WI; WNAM Neenah-Menasha, WI; WPKR-FM Omro, WI; WOSH Oshkosh, WI; WWWX-FM Oshkosh, WI

Richard W. Denning, chmn; Lewis W. Dickey Jr., VP.

Cumulus Media Partners LLC, 14 Piedmont, 3535 Piedmont Rd., Suite 1400, Atlanta, GA 30305-4601. Phone: (404) 949-0700. Fax: (404) 949-0740. Ownership: Cumulus Media Inc., 25% (see listing); Bain Funds, 25%; Blackstone Funds, 25%; and THLee Funds, 25%.

Stns: 10 AM. 22 FM. KFFG-FM Los Altos, CA; KFOG-FM San Francisco, CA; KNBR San Francisco, CA; KSAN-FM San Mateo, CA; KTCT San Mateo, CA; WNNX-FM Atlanta, GA; WWWQ(FM) College Park, GA; WWFT(FM) Fishers, IN; WFMS-FM Indianapolis, IN; WAVG(AM) Jeffersonville, IN; WJJK(FM) Noblesville, IN; WZZB Seymour, IN; WQKC-FM Seymour, IN; KCJK(FM) Garden City, MO; KCFX(FM) Harrisonville, MO; KCMO Kansas City, MO; KCMO-FM Kansas City, MO; WRRM-FM Cincinnati, OH; WSWD(FM) Fairfield, OH; WFTK(FM) Lebanon, OH; WGLD(AM) Red Lion, PA; WSOX-FM Red Lion, PA; WSBA York, PA; WARM-FM York, PA; KTCK Dallas, TX; KLIF(AM) Dallas, TX; KPLX(FM) Fort Worth, TX; KIKT-FM Greenville, TX; KDBN(FM) Haltom City, TX; KRBE(FM) Houston, TX; KKLF(AM) Richardson, TX; KTDK(FM) Sanger, TX.

Lew Dickey, chmn/CEO.

The Curators of the University of Missouri, (Business Services Division). University of Missouri, 316 University Hall, Columbia, MO 65211. Phone: (573) 882-2388. Fax: (573) 882-0010. Web Site: www.umsystem.edu. Ownership: (Business Services Division).

Stns: 5 FM. KBIA(FM) Columbia, MO; KCUR-FM Kansas City, MO; KMNR-FM Rolla, MO; KMST(FM) Rolla, MO; KWMU-FM Saint Louis, MO.

Stns: 1 TV. KOMU, Columbia-Jefferson City, MO.

Michael Dunn, gen mgr; Martin Siddall, gen mgr.

Curtis Media Group, 3012 Highwoods Blvd., Raleigh, NC 27604. Phone: (919) 876-0674. Fax: (919) 790-8369. Web Site: www.curtismedia.com. Ownership: Donald W. Curtis.

Stns: 11 AM. 8 FM. WPCM Burlington, NC; WZTK(FM) Burlington, NC; WDNC(AM) Durham, NC; WWNF(FM) Goldsboro, NC; WFMC Goldsboro, NC; WGBR Goldsboro, NC; WYMY(FM) Goldsboro, NC; WSML Graham, NC; WMFR High Point, NC; WKIX(FM) Kinston, NC; WYRN(AM) Louisburg, NC; WKXU(FM) Louisburg, NC; WWMY(FM) Raleigh, NC; WBBB-FM Raleigh, NC; WCLY Raleigh, NC; WDOX(AM) Raleigh, NC; WQDR-FM Raleigh, NC; WPTF Raleigh, NC; WSJS Winston-Salem, NC.

Donald Curtis, chmn/CEO; Philip Zachary, pres/COO; Adam Maisano, VP & dir of sls; Allen Sherrill, engrg dir.

Timothy C. Cutforth Stns, 965 S. Irving St., Denver, CO 80219. Phone: (303) 935-1156. Ownership: Timothy C. Cutforth, 100%.

Stns: 3 AM. KJME(AM) Fountain, CO; KCEG(AM) Pueblo, CO; KJJL(AM) Pine Bluffs, WY.

D

DCBroadcasting Inc., Box 1009, Jasper, IN 47547-1009. Phone: (812) 634-9232. Fax: (812) 482-3696. E-mail: pknies@psci.net Web Site: www.dcbroadcasting.com. Ownership: Loc mktg agreement: WRZR(FM) Loogootee, IN. LPTV (class A): WJTS-LP Jasper, IN.

Stns: 1 AM. 3 FM. WBDC(FM) Huntingburg, IN; WXGO Madison, IN; WORX-FM Madison, IN; WAXL-FM Santa Claus, IN.

Paul Knies, pres; Caroline Knies, VP/treas; Giesla Knies Schepers, sec.

DFWU Inc., 1101 N. 81 Hwy., Marlow, OK 73055. Phone: (580) 658-9292. Fax: (580) 658-2561. E-mail: kfxi@texhoma.net Web Site: www.kfxi.com.

Stns: 1 AM. 2 FM. KFXI-FM Marlow, OK; KVLH Pauls Valley, OK; KIXO-FM Sulphur, OK.

K.D. Austin, pres & gen mgr.

DMC Broadcasting Inc., 5542 NDCBU, Taos, NM 87571-6122. Phone: (505) 758-4491. Fax: (505) 758-4452. Ownership: Darren Cordova, 100%.

Stns: 1 AM. 3 FM. KKTC(FM) Angel Fire, NM; KKIT(FM) Taos, NM; KVOT(AM) Taos, NM; KXMT(FM) Taos, NM.

Dailey Corp., Box 10, New Martinsviille, WV 26155. Phone: (304) 455-1111. Fax: (304) 455-1170. Web Site: www.powercountry104.com. Ownership: Calvin E. Dailey Jr., 100%.

Stns: 1 AM. 1 FM. WETZ New Martinsville, WV; WYMJ(FM) New Martinsville, WV.

Calvin Dailey Jr., pres.

Dakota Communications Ltd., Box 364, Pierre, SD 57501. Phone: (605) 224-5434. Fax: (605) 224-5444. E-mail: ddb@eaglecarver.com Web Site: performanceradio.com. Ownership: Duane D. Butt, 50%; and Barbara G. Butt, 50%.

Stns: 2 AM. 3 FM. KIJV Huron, SD; KOKK Huron, SD; KZKK-FM Huron, SD; KZNC-FM Huron, SD; KJRV(FM) Wessington Springs, SD.

Linda Marcus, gen mgr.

Davidson Media Group LLC, 709 Peninsula Dr., Davidson, NC 28036-7200. Phone: (704) 987-3585. Fax: (704) 987-3586. Web Site: www.davidsonmediagroup.com. Ownership: CapStreet II L.P., 38.51% of votes, 38.6% of total assets; Citicorp North America Inc., 32.74% of votes, 32.81% of total assets; Black Enterprise/Greenwich Street Corporate Growth Partners L.P., 10.91% of votes, 10.94% of total assets.

Stns: 32 AM. 6 FM. WAYE Birmingham, AL; WRLM(AM) Irondale, AL; KAKS(FM) Huntsville, AR; WXCT(AM) Southington, CT; WNTS Beech Grove, IN; KDTD(AM) Kansas City, KS; KCZZ(AM) Mission, KS; WTUV-FM Eminence, KY; WTUV(AM) Louisville, KY; WLLV Louisville, KY; WLOU Louisville, KY; WFNO Norco, LA; WSPR Springfield, MA; WACM West Springfield, MA; WDRJ(AM) Inkster, MI; KRJJ(AM) Brooklyn Park, MN; KMNV(AM) Saint Paul, MN; WTIK Durham, NC; WRJD(AM) Durham, NC; WSTS-FM Fairmont, NC; WWBG Greensboro, NC; WSGH Lewisville, NC; WNOW Mint Hill, NC; WTOB Winston-Salem, NC; WEMG(AM) Camden, NJ; KJMU(AM) Sand Springs, OK; WKKB(FM) Middletown, RI; WOLI-FM Easley, SC; WOLT-FM Greer, SC; WOLI(AM) Spartanburg, SC; WBZK York, SC; WNSG(AM) Nashville, TN; WNVL(AM) Nashville, TN; WTOX(AM) Glen Allen, VA; WVXX(AM) Norfolk, VA; WVNZ(AM) Richmond, VA; WREJ Richmond, VA; WLEE(AM) Richmond, VA.

Russ Jones, opns VP.

Davis Broadcasting Inc., 2203 Wynnton Rd., Columbus, GA 31906. Phone: (706) 576-3565. Fax: (706) 576-3683. Web Site: www.foxie105.com. Ownership: Gregory A. Davis, 76%.

Stns: 3 AM. 5 FM. WKZJ(FM) Eufaula, AL; WEAM-FM Buena Vista, GA; WLKQ-FM Buford, GA; WCHK Canton, GA; WEAM Columbus, GA; WOKS Columbus, GA; WIOL(FM) Greenville, GA; WNSY-FM Talking Rock, GA.

Gregory A. Davis, pres/CEO.

Debut Broadcasting Corp. Inc., 1209 16th Ave. S., Suite 200, Nashville, TN 37212. Phone: (615) 301-0001. Fax: (615) 301-0002. Ownership: Robert Marquitz, 49%; Steven Ludwig, 33.5%; Stephen Rush, 11%; and Garrett L. Cecchini, 6.5%.

Stns: 2 AM. 3 FM. WBAQ-FM Greenville, MS; WNIX(AM) Greenville, MS; WNLA Indianola, MS; WNLA-FM Indianola, MS; WIQQ-FM Leland, MS.

Dee Rivers Radio Group, GRAM Corp., 43 Sherman Hill Rd. #204, Woodbury, CT 06798. Phone: (203) 263-1900. Fax: (203) 263-1969. Ownership: Georgia R. Salva, trustee, E.D. Rivers Jr. Trust, 50.98%; Robert Salva, 39.22%; and Georgia R. Salva, 9.8%.

Stns: 1 AM. 2 FM. WLYX(FM) Valdosta, GA; WAAC(FM) Valdosta, GA; WGOV(AM) Valdosta, GA.

Georgia Salva, CEO.

Deer Creek Broadcasting LLC, 2225 First Ave., Napa, CA 94558. Phone: (707) 226-2309. Web Site: www.deercreekbroadcasting.com. Ownership: Elliot B. Evers, 43.75% of votes, 12.08% of total assets; Duff Ackerman & Goodrich QP Fund II L.P., 26.89% of votes, 7.43% of total assets; John McSorley, 12.49% of votes, 3.45% of total assets; Greg D. Widroe, 12.49% of votes, 3.45% of total assets; Duff Ackerman & Goodrich II L.P., 2.55% of votes, 0.70% of total assets; DAG GP Fund II LLC, 1.53% of votes, 0.42% of total assets; and DAG II Partners Fund LLC, 0.30% of votes, 0.08% of total assets.

Stns: 2 AM. 3 FM. KMXI-FM Chico, CA; KPAY Chico, CA; KHHZ(FM) Oroville, CA; KEWE(AM) Oroville, CA; KHSL-FM Paradise, CA.

Delmarva Broadcasting Co., Box 7492, 2727 Shipley Rd., Wilmington, DE 19803. Phone: (302) 478-2700. Fax: (302) 478-0100. Web Site: www.delmarvabroadcasting.com. Ownership: Steinman

Stns: 3 AM. 8 FM. WXCY(FM) Havre de Grace, MD; WQJZ-FM Ocean Pines, MD; WXMD(FM) Pocomoke City, MD; WICO(AM) Salisbury, MD; WICO-FM Salisbury, MD; WXJN-FM Lewes, DE; WAFL-FM Milford, DE; WYUS Milford, DE; WNCL(FM) Milford, DE; WSTW-FM Wilmington, DE; WDEL Wilmington, DE.

Lancaster Inteligencer-Journal & *New Era*, Lancaster, PA, have the same ownership (Steinman) as Delmarva Broadcasting Co.

Stations (operated independently).

Julian H. Booker, pres/CEO.

Dickey Broadcasting Co., 3535 Piedmont Rd., Bldg. 14, Suite 1200, Atlanta, GA 30305. Phone: (404) 688-0068. Fax: (404) 995-4045. Web Site: www.680thefan.com.

Stns: 3 AM. WALR Atlanta, GA; WFOM Marietta, GA; WCNN North Atlanta, GA.

David W. Dickey, pres/CEO.

Dierking Communications Inc., 937 Jayhawk Rd., Marysville, KS 66508. Phone: (785) 562-2361. Fax: (785) 562-2188. E-mail: kndy@bluevalley.net

Stns: 2 AM. 4 FM. KZDY-FM Cawker City, KS; KDNS(FM) Downs, KS; KNDY Marysville, KS; KNDY-FM Marysville, KS; KQNK Norton, KS; KQNK-FM Norton, KS.

Bruce Dierking, gen mgr & pres.

Dispatch Broadcast Group, 770 Twin Rivers Dr., Columbus, OH 43215. Phone: (614) 460-3700. Fax: (614) 460-2809. Web Site: www.10tv.com. Ownership: Dispatch Printing Company

Stns: 1 AM. 1 FM. WBNS Columbus, OH; WBNS-FM Columbus, OH.

Stns: 2 TV. WBNS, Columbus, OH; WTHR, Indianapolis, IN.

Owns *The Columbus* (OH) *Dispatch, This Week* & *Ohio Magazine.*

Tamara J. Clapsaddle, controller; Michael J. Fiorile, pres.

Dockins Communications Inc., 540 Maple Valley Dr., Farmington, MO 63640. Phone: (573) 701-9590. Fax: (573) 701-9696. Ownership: Fred Dockins Sr., 100%.

Stns: 2 AM. 3 FM. KTNX(FM) Arcadia, MO; KYLS Fredericktown, MO; KYLS-FM Ironton, MO; KPWB Piedmont, MO; KPWB-FM Piedmont, MO.

Dos Costas Communications Corp., 1818 S. Australian Ave., Suite 102, West Palm Beach, FL 33409. Phone: (561) 655-6615. Ownership: Roland A. Ulloa, 100%.

Stns: 3 FM. KDUC-FM Barstow, CA; KXXZ-FM Barstow, CA; KDUQ-FM Ludlow, CA.

Roland A. Ulloa, pres; Jime Garza, VP/gen mgr.

Double O Radio L.L.C., 1156 Bowman Rd., Suite 200, Mount Pleasant, SC 29464. Phone: (843) 416-1107. Fax: (843) 416-1199. E-mail: info@doubleoradio.com Web Site: doubleoradio.com. Ownership: Pilot Group LP, 100%.

Stns: 4 AM. 20 FM. WAKT-FM Callaway, FL; WPFM-FM Panama City, FL; WASJ(FM) Panama City Beach, FL; WRBA(FM) Springfield, FL; WDHI-FM Delhi, NY; WIYN-FM Deposit, NY; WBKT-FM Norwich, NY; WCHN Norwich, NY; WKXZ-FM Norwich, NY; WDOS Oneonta, NY; WZOZ-FM Oneonta, NY; WSRK-FM Oneonta, NY; WDLA Walton, NY; WDLA-FM Walton, NY; WWNQ(FM) Forest Acres, SC; WWNU(FM) Irmo, SC; KKCN(FM) Ballinger, TX; KQRX-FM Midland, TX; KHKX(FM) Odessa, TX; KMCM-FM Odessa, TX; KELI-FM San Angelo, TX; KGKL San Angelo, TX; KGKL-FM San Angelo, TX; KNRX(FM) Sterling City, TX.

Terry Bond, CEO.

Dr. Pepper Pepsi-Cola Bottling Co. of Dyersburg, 35 Radio Rd., Dyersburg, TN 38025-0100. Phone: (731) 285-1339. Fax: (731) 287-0100. E-mail: sl100@wasl.net Ownership: W.E. Burks, 53.41%; Richard Rodgers, 16.18%; J.L. Jones, 3.15%; and Guy McClain, 1.5%.

Stns: 1 AM. 2 FM. WTRO Dyersburg, TN; WASL-FM Dyersburg, TN; WTNV(FM) Tiptonville, TN.

DreamCatcher Communications Inc., 114 S. Manchester Ave., West Union, OH 45693. Phone: (937) 544-9722. Fax: (937) 544-5523. E-mail: c103@lycos.com Ownership: Donald Bowles, 50%; Venita Bowles, 50%.

Stns: 1 AM. 1 FM. WFLE Flemingsburg, KY; WRAC-FM West Union, OH.

Don Bowles, pres/CEO; Ted Foster, stn mgr.

Duhamel Broadcasting Enterprises, Box 1760, Rapid City, SD 57709. Phone: (605) 342-2000. Fax: (605) 342-7305. Web Site: www.kotatv.com. Ownership: William F. Duhamel, 63%; Peter A. and Lois G. Duhamel, 37%.

Stns: 1 AM. 1 FM. KOTA Rapid City, SD; KDDX(FM) Spearfish, SD.

Stns: 4 TV. KDUH, Cheyenne, WY-Scottsbluff, NE; KHSD-TV, Rapid City, SD; KOTA-TV, Rapid City, SD; KSGW-TV, Rapid City, SD.

William F. Duhamel, pres.

Durham Radio Inc., 1200 Airport Blvd., Suite 207, Oshawa, ON L1J 8P5. Canada. Phone: (905) 428-9600. Fax: (905) 571-1150. Ownership: Douglas E. Kirk; 80%, Mary Kirk 15%.

Stns: 1 AM. 2 FM. CJKX-FM Ajax, ON; CIWV-FM Hamilton, ON; CKDO Oshawa, ON.

Douglas E. Kirk, chmn/pres; Steve Kassay, VP opns; Steve Macaulay, VP sls; Lill Bolton, admin dir.

E

EMF Broadcasting, 2351 Sunset Blvd., Suite 170-218, Rocklin, CA 95765. Phone: (916) 251-1600. Fax: (916) 251-1650. Web Site: www.emfbroadcasting.com. Ownership: Educational Media Foundation, 100%. Educational Media Foundation is a nonprofit, nonstock corporation, governed by a seven-member board of directors.

Stns:1 AM. 233 FM. KAKL(FM) Anchorage, AK; KBGR(FM) Beebe, AR; KAIA(FM) Blytheville, AR; KLMZ(FM) Fouke, AR; KALR(FM) Hot Springs, AR; KLRO(FM) Hot Springs, AR; KJLV(FM) Hoxie, AR; KJBR-FM Marked Tree, AR; KLRM(FM) Melbourne, AR; KKLT(FM) Texarkana, AR; KKLV(FM) Turrell, AR; KLFS(FM) Van Buren, AR; KLVA(FM) Casa Grande, AZ; KZAI(FM) Coolidge, AZ; KLVK(FM) Fountain Hills, AZ; KAIH(FM) Lake Havasu City, AZ; KAIC(FM) Tucson, AZ; KWLU(FM) Chester, CA; KARQ(FM) East Sonora, CA; KLVY-FM Fairmead, CA; KLVG(FM) Garberville, CA; KLVS(FM) Grass Valley, CA; KHRI(FM) Hollister, CA; KZLU(FM) Inyokern, CA; KLVJ-FM Julian, CA; KDRH(FM) King City, CA; KHKL(FM) Laytonville, CA; KLVN(FM) Livingston, CA; KLRS(FM) Lodi, CA; KLVC-FM Magalia, CA; KKLC(FM) Mount Shasta, CA; KGBM(FM) Randsburg, CA; KLVB(FM) Red Bluff, CA; KKRO(FM) Redding, CA; KAIS(FM) Redwood Valley, CA; KLVH(FM) San Luis Obispo, CA; KSRI(FM) Santa Cruz, CA; KLVR-FM Santa Rosa, CA; KQKL(FM) Selma, CA; KAIB(FM) Shafter, CA; KJAR(FM) Susanville, CA; KYLU(FM) Tehachapi, CA; KYKL(FM) Tracy, CA; KULV(FM) Ukiah, CA; KXRD(FM) Victorville, CA; KARA(FM) Williams, CA; KLRD(FM) Yucaipa, CA; KLHV(FM) Fort Collins, CO; KLXV(FM) Glenwood Springs, CO; KLFV(FM) Grand Junction, CO; KLRY(FM) Gypsum, CO; KHCO(FM) Hayden, CO; KFDN(FM) Lakewood, CO; KLDV(FM) Morrison, CO; KLBV(FM) Steamboat Springs, CO; KDRE(FM) Sterling, CO; KLZV(FM) Sterling, CO; WKVH(FM) Monticello, FL; WFFM(FM) Ashburn, GA; WVRI(FM) Pavo, GA; WHKV(FM) Sylvester, GA; WTHM(FM) Thomson, GA; WVDA(FM) Valdosta, GA; KHAI(FM) Wahiawa, HI; KILV(FM) Castana, IA; KKLG(FM) Newton, IA; KAIP(FM) Wapello, IA; KAIO(FM) Idaho Falls, ID; KARJ(FM) Kuna, ID; KLRI(FM) Rigby, ID; WCLR(FM) Arlington Heights, IL; WDRS(FM) Dorsey, IL; WJKL(FM) Glendale Heights, IL; WLKU(FM) Rock Island, IL; WSRI(FM) Sugar Grove, IL; WKMV(FM) Muncie, IN; WARA(FM) New Washington, IN; WIKV(FM) Plymouth, IN; WQKV(FM) Rochester, IN; WJLR-FM Seymour, IN; KAIG(FM) Dodge City, KS; KTLI(FM) El Dorado, KS; KWBI(FM) Great Bend, KS; KGLV(FM) Manhattan, KS; KRLE(FM) Oberlin, KS; WKYB(FM) Burgin, KY; WLAI(FM) Danville, KY; WRVG(FM) Georgetown, KY; WKVY(FM) Somerset, KY; WXKY-FM Stanford, KY; KLXA-FM Alexandria, LA; KITA(FM) Bunkie, LA; WBKL(FM) Clinton, LA; KYLA-FM Homer, LA; KIKL(FM) Lafayette, LA; WNKV(FM) Norco, LA; KRLR(FM) Sulphur, LA; WTKL(FM) North Dartmouth, MA; WKMY(FM) Winchendon, MA; WLKB(FM) Bay City, MI; WAKL(FM) Flint, MI; WTRK(FM) Freeland, MI; WLVM(FM) Ironwood, MI; KMKL(FM) North Branch, MN; KKLW(FM) Willmar, MN; KRLP(FM) Windom, MN; KLRQ(FM) Clinton, MO; WKVF(FM) Byhalia, MS; WKNZ-FM Collins, MS; WLKO(FM) Quitman, MS; KQLU(FM) Belgrade, MT; KLRV(FM) Billings, MT; KLBZ(FM) Bozeman, MT; KGFA(FM) Great Falls, MT; KHLV(FM) Helena, MT; KLKM(FM) Kalispell, MT; KBIL(FM) Park City, MT; KQLR(FM) Whitehall, MT; WZRI(FM) Spring Lake, NC; WZRL(FM) Wade, NC; KNRI(FM) Bismarck, ND; KKLQ(FM) Harwood, ND; KLRX(FM) Jamestown, ND; KVLQ(FM) Lincoln, ND; KLNB(FM) Grand Island, NE; KMLV(FM) Ralston, NE; KLJV(FM) Scottsbluff, NE; KDAI(FM) Scottsbluff, NE; KFLV(FM) Wilber, NE; WKVP(FM) Cherry Hill, NJ; KQRI(FM) Belen, NM; KQLV(FM) Bosque Farms, NM; KELU(FM) Clovis, NM; KGGA(FM) Gallup, NM; KLLU(FM) Gallup, NM; KLGQ(FM) Grants, NM; KBAC(FM) Las Vegas, NM; KELT(FM) Roswell, NM; KQAI(FM) Roswell, NM; KRLU(FM) Roswell, NM; KVLK(FM) Socorro, NM; KVLP(FM) Tucumcari, NM; KSFQ-FM White Rock, NM; KAIZ(FM) Mesquite, NV; KLRH(FM) Sparks, NV; WKDL(FM) Brockport, NY; WGKV(FM) Pulaski, NY; WYKV(FM) Ravena, NY; WYAI(FM) Scotia, NY; WKVU(FM) Utica, NY; WKWV(FM) Watertown, NY; WLKP(FM) Belpre, OH; WYKL(FM) Crestline, OH; WORI(FM) Delhi Hills, OH; WCVJ-FM Jefferson, OH; WOAR(FM) South Vienna, OH; WEKV(FM) South Webster, OH; WOKL(FM) Troy, OH; KKVO-FM Altus, OK; KWRI(FM) Bartlesville, OK; KARU(FM) Cache, OK; KOKF-FM Edmond, OK; KKRD(FM) Enid, OK; KWKL(FM) Grandfield, OK; KYLV-FM Oklahoma City, OK; KKRI(FM) Pocola, OK; KTKL(FM) Stigler, OK; KLOY(FM) Astoria, OR; KVLB(FM) Bend, OR; KLVP-FM Cherryville, OR; KIDH(FM) Jordan Valley, OR; KKLJ(FM) Klamath Falls, OR; KKLP(FM) La Pine, OR; KGRI(FM) Lebanon, OR; KLON(FM) Rockaway Beach, OR; KJKL(FM) Selma, OR; KVRA(FM) Sisters, OR; KLVU-FM Sweet Home, OR; KAIK(FM) Tillamook, OR; KZRI(FM) Welches, OR; KLOV(FM) Winchester, OR; WKEL(FM) Confluence, PA; WLVU(FM) Halifax, PA; WPKV(FM) Nanty Glo, PA; WLKA(FM) Tafton, PA; WKVC(FM) North Myrtle Beach, SC; KLRJ(FM) Aberdeen, SD; KOAR(FM) Spearfish, SD; WRRI(FM) Brownsville, TN; WZKV(FM) Dyersburg, TN; WPLX(AM) Germantown, TN; WKVZ(FM) Ripley, TN; WXKV(FM) Selmer, TN; WTAI(FM) Union City, TN; KAGT(FM) Abilene, TX; KXRI(FM) Amarillo, TX; KXLV(FM) Amarillo, TX; KKWV(FM) Aransas Pass, TX; KBEX(FM) Brenham, TX; KMLU(FM) Brownfield, TX; KLRW(FM) Byrne, TX; KKLM(FM) Corpus Christi, TX; KLLR(FM) Dripping Springs, TX; KKLY(FM) El Paso, TX; KYAR(FM) Gatesville, TX; KMLR(FM) Gonzales, TX; KZAR(FM) Gonzales, TX; KRLH(FM) Hereford, TX; KYLR(FM) Hutto, TX; KZLO(FM) Kilgore, TX; KKLU(FM) Lubbock, TX; KZLV(FM) Lytle, TX; KPOS(FM) Post, TX; KLTP(FM) San Angelo, TX; KNAR(FM) San Angelo, TX; KFRI(FM) Stanton, TX; KLVW(FM) West Odessa, TX; KZKL(FM) Wichita Falls, TX; KNKL(FM) North Ogden, UT; KAER(FM) Saint George, UT; KSBC(FM) Nile, WA; KLOP(FM) Ocean Park, WA; KRKL(FM) Walla Walla, WA; WDKV(FM) Fond du Lac, WI; WDKL(FM) Grafton, WV; WKVW(FM) Marmet, WV; WLKV(FM) Ripley, WV; KLWC(FM) Casper, WY; KAIX(FM) Cheyenne, WY; KLWV(FM) Chugwater, WY; KLOF(FM) Gillette, WY; KMLT(FM) Jackson, WY; KAIW(FM) Laramie, WY; KVLZ(FM) Sheridan, WY

Richard Jenkins, pres; Keith Whipple, dev VP; Mike Novak, opns VP; Devona R. Porter, opns VP; David Pierce, progmg dir.

Eagle Bluff Enterprises, 932 County Rd. 448, Poplar Bluff, MO 63901. Phone: (573) 686-3700.

Stns: 2 AM. 2 FM. KBSP(AM) Birch Tree, MO; KFEB(FM) Campbell, MO; KOEA-FM Doniphan, MO; KDFN Doniphan, MO.

Steven C. Fuch, pres.

Eagle Communications Group, 2703 Hall St., Suite 15, Hays, KS 67601. Phone: (785) 625-4000. Fax: (785) 625-8030. Web Site: www.eaglecom.net. Ownership: Eagle Communications Inc. Employee Stock Ownership Trust. Cable TV.

Stns: 5 AM. 12 FM. KVGB Great Bend, KS; KVGB-FM Great Bend, KS; KHAZ-FM Hays, KS; KJLS-FM Hays, KS; KKQY-FM Hill City, KS; KHOK-FM Hoisington, KS; KHUT-FM Hutchinson, KS; KWBW Hutchinson, KS; KHMY(FM) Pratt, KS; KSKG-FM Salina, KS; KSFT Saint Joseph, MO; KFEQ(AM) Saint Joseph, MO; KSJQ(FM) Savannah, MO; KAAQ(FM) Alliance, NE; KQSK(FM) Chadron, NE; KELN-FM North Platte, NE; KOOQ North Platte, NE.

Eagle's Nest Inc., Box 710, Roanoke, AL 36274. Phone: (334) 863-4139. Fax: (334) 863-2540. Web Site: www.eagle1023.com. Ownership: Jim Vice, 51%; Kay Vice, 49%.

Stns: 2 AM. WELR Roanoke, AL; WLAG La Grange, GA.

Jim Vice, pres.

Earls Broadcasting Co., 202 Courtney St., Branson, MO 65616. Phone: (417) 334-6003. Fax: (417) 334-7141. Web Site: www.komc.com. Ownership: Charles Earls, Scottie Earls, Scott Earls.

Stns: 3 AM. 7 FM. KHOZ Harrison, AR; KHOZ-FM Harrison, AR; KTLO Mountain Home, AR; KTLO-FM Mountain Home, AR; KHOM(FM) Salem, AR; KCTT-FM Yellville, AR; KBMV-FM Birch Tree, MO; KRZK-FM Branson, MO; KOMC Branson, MO; KOMC-FM Kimberling City, MO.

Charles C. Earls, CEO; Scottie Earls, gen mgr; Scott Earls, pres.

East Carolina Radio Group, 2422 S. Wrightsville Ave., Nags Head, NC 27959. Phone: (252) 449-8331. Fax: (252) 449-8354. Web Site: www.ecri.net.

Stns: 3 AM. 4 FM. WRSF(FM) Columbia, NC; WERX-FM Columbia, NC; WZBO(AM) Edenton, NC; WCNC(AM) Elizabeth City, NC; WKJX(FM) Elizabeth City, NC; WOBX(AM) Wanchese, NC; WOBR-FM Wanchese, NC.

Rick Loesch, pres.

East Kentucky Broadcasting Corp., Box 2200, Pikeville, KY 41502. Phone: (606) 437-4051. Fax: (606) 432-2809. E-mail: wpke@wpke.com Web Site: www.ekbradio.com. Ownership: Walter E. May; Pamela May; Keith Casebolt

Stns: 3 AM. 3 FM. WPKE-FM Coal Run, KY; WEKB(AM) Elkhorn City, KY; WDHR-FM Pikeville, KY; WLSI Pikeville, KY; WPKE Pikeville, KY; WZLK-FM Virgie, KY.

Walter E. May, pres; Keith Casebolt, gen mgr.

East Kentucky Radio Network Inc., Box 2200, Pikeville, KY 41502. Phone: (606) 437-4051. Fax: (606) 432-2809. Web Site: www.ekbradio.com. Ownership: Walter E. May; Pamela May; Keith Casebolt.

Stns: 2 AM. WPRT Prestonsburg, KY; WBTH Williamson, WV.

Walter E. May, pres; Keith Casebolt, VP; Pamela May, sec/treas.

East Tennessee Radio Group L.P., 5312 Ringgold Rd., Suite 201, Chattanooga, TN 37412. Phone: (423) 485-8994. Ownership: Whitfield Communications Inc., 100% of votes. Note: Whitfield Communications Inc. also is the gen ptnr and owns 1% of WNOO(AM) Chattanooga, TN.

Stns: 1 AM. 2 FM. WSEV-FM Gatlinburg, TN; WMXK-FM Morristown, TN; WSEV Sevierville, TN.

East Texas Broadcasting Inc., Box 990, Mount Pleasant, TX 75456. Phone: (903) 572-8726. Fax: (903) 572-7232. Web Site: www.eastexasradio.com. Ownership: John Mitchell; Bud Kitchens.

Stns: 2 AM. 5 FM. KIMP(AM) Mount Pleasant, TX; KOYN-FM Paris, TX; KPLT Paris, TX; KBUS(FM) Paris, TX; KSCN(FM) Pittsburg, TX; KSCH-FM Sulphur Springs, TX; KALK(FM) Winfield, TX.

John Mitchell, chmn; Bud Kitchens, pres; Bob Gibson, VP.

Edwards Communications L.C., 125 Eagles Nest Dr., Seneca, SC 29678. Phone: (864) 882-3272. Fax: (864) 882-3718. Ownership: Steve Edwards, 50%; Jerry Edwards, 50%

Stns: 3 AM. 5 FM. WHSB-FM Alpena, MI; WIDL-FM Caro, MI; WKYO Caro, MI; WWTH(FM) Oscoda, MI; WHAK Rogers City, MI; WHAK-FM Rogers City, MI; KTAK-FM Riverton, WY; KVOW Riverton, WY.

Jerry Edwards, pres; Steve Edwards Jr., VP.

Ely Radio LLC, 5010 Spencer, Las Vegas, NV 89119. Phone: (702) 740-5588. Ownership: Fred Weinberg, 100%.

Stns: 2 AM. 1 FM. KELY Ely, NV; KWNA Winnemucca, NV; KWNA-FM Winnemucca, NV.

Elyria-Lorain Broadcasting Co., Box 4006, Elyria, OH 44036. Phone: (440) 322-3761. Fax: (440) 284-3189. Ownership: Lorain County Printing & Publishing Co., 100%.

Stns: 2 AM. 2 FM. WEOL Elyria, OH; WNWV-FM Elyria, OH; WKFM-FM Huron, OH; WLKR(AM) Norwalk, OH.

Lorain County Printing & Publishing Co. publishes *Chronicle-Telegram* (Elyria) and *Medina Gazette* (Medina), both OH.

Gary L. Kneisley, pres.

Emerald Wave Media, 104 W. Chapel, Santa Maria, CA 93454. Phone: (805) 928-4334. Fax: (805) 349-2765. Ownership: George A. Ruiz, 100%.

Stns: 1 AM. 2 FM. KIDI-FM Guadalupe, CA; KRTO(FM) Lompoc, CA; KTAP Santa Maria, CA.

Emmis Communications Corp., 3500 W. Olive Ave., Suite 1450, Burbank, CA 91436. Phone: (818) 238-9154. Fax: (818) 238-9158. Web Site: www.emmis.com. Ownership: Jeffrey H. Smulyan, approximately 61% votes.

Stns: 2 AM. 20 FM. KPWR-FM Los Angeles, CA; KMVN(FM) Los Angeles, CA; WKQX-FM Chicago, IL; WLUP-FM Chicago, IL; WNOU(FM) Indianapolis, IN; WIBC(AM) Indianapolis, IN; WYXB(FM) Indianapolis, IN; WLHK(FM) Shelbyville, IN; WTHI-FM Terre Haute, IN; WWVR-FM West Terre Haute, IN; KSHE(FM) Crestwood, MO; KFTK(FM) Florissant, MO; KIHT-FM Saint Louis, MO; KPNT-FM Sainte Genevieve, MO; WRKS(FM) New York, NY; WQHT-FM New York, NY; WQCD(FM) New York, NY; KLBJ Austin, TX; KGSR-FM Bastrop, TX; KROX-FM Buda, TX; KDHT(FM) Cedar Park, TX; KBPA(FM) San Marcos, TX.

Stns: 1 TV. WVUE, New Orleans, LA.

The publishing unit of Emmis Communications publishes seven magazines: *Atlanta Magazine, Cincinnati Magazine, Los Angeles Magazine, Wildlife Journal, Indianapolis Monthly* and *Texas Monthly.*

Rick Cummings, pres.

Empire Broadcasting Corp., Box 995, San Jose, CA 95108. Phone: (408) 293-8030. Fax: (408) 293-6124. Web Site: www.kliv.com. Ownership: Robert Kieve, Vincent Lopopolo, Myron S. Lewis.

Stns: 1 AM. 1 FM. KRTY-FM Los Gatos, CA; KLIV San Jose, CA.

Robert Kieve, pres; John McLeod, progmg VP.

Emporia's Radio Stations Inc., Box 968, Emporia, KS 66801. Phone: (620) 342-1400. Fax: (620) 342-0804. Ownership: Steve Sauder, 100%.

Stns: 1 AM. 2 FM. KFFX-FM Emporia, KS; KVOE Emporia, KS; KVOE-FM Emporia, KS.

Lee Schroeder, gen mgr; Susan Grother, business mgr.

Entercom Communications Corp., 401 City Ave., Suite 809, Bala-Cynwyd, PA 19004. Phone: (610) 660-5610. Fax: (610) 660-5620. Web Site: www.entercom.com. Ownership: Joseph M. Field.

Stns:31 AM. 68 FM. KSSJ-FM Fair Oaks, CA; KRXQ-FM Sacramento, CA; KSEG-FM Sacramento, CA; KWOD-FM Sacramento, CA; KCTC Sacramento, CA; KDND(FM) Sacramento, CA; KEZW Aurora, CO; KALC(FM) Denver, CO; KOSI-FM Denver, CO; KQMT(FM) Denver, CO; WKTK-FM Crystal River, FL; WSKY-FM Micanopy, FL; WZPL-FM Greenfield, IN; WXNT(AM) Indianapolis, IN; WNTR(FM) Indianapolis, IN; KDGS(FM) Andover, KS; KFH-FM Clearwater, KS; KFBZ(FM) Haysville, KS; KUDL-FM Kansas City, KS; KXTR(AM) Kansas City, KS; KKHK(AM) Kansas City, KS; KQRC-FM Leavenworth, KS; KFH(AM) Wichita, KS; KEYN-FM Wichita, KS; KNSS(AM) Wichita, KS; WWL-FM Kenner, LA; WLMG(FM) New Orleans, LA; WEZB-FM New Orleans, LA; WWWL(AM) New Orleans, LA; WKBU(FM) New Orleans, LA; WWL(AM) New Orleans, LA; WRKO Boston, MA; WEEI Boston, MA; WKAF(FM) Brockton, MA; WMKK(FM) Lawrence, MA; WAAF(FM) Westborough, MA; WVEI(AM) Worcester, MA; KCSP(AM) Kansas City, MO; KRBZ(FM) Kansas City, MO; KYYS-FM Kansas City, MO; KMBZ Kansas City, MO; WDAF-FM Liberty, MO; WTPT-FM Forest City, NC; WQMG-FM Greensboro, NC; WPET Greensboro, NC; WSMW(FM) Greensboro, NC; WJMH-FM Reidsville, NC; WPAW(FM) Winston-Salem, NC; WWWS Buffalo, NY; WGR Buffalo, NY; WWKB Buffalo, NY; WBEN Buffalo, NY; WFKL(FM) Fairport, NY; WKSE-FM Niagara Falls, NY; WBZA(FM) Rochester, NY; WBEE-FM Rochester, NY; WLKK(FM) Wethersfield Township, NY; KRSK(FM) Molalla, OR; KFXX(AM) Portland, OR; KYCH-FM Portland, OR; KGON-FM Portland, OR; KKSN(AM) Salem, OR; WILK-FM Avoca, PA; WGGI-FM Benton, PA; WDMT(FM) Pittston, PA; WGGY-FM Scranton, PA; WBZU(AM) Scranton, PA; WKRF(FM) Tobyhanna, PA; WKZN(AM) West Hazleton, PA; WKRZ-FM Wilkes-Barre, PA; WILK Wilkes-Barre, PA; WEEI-FM Westerly, RI; WROQ-FM Anderson, SC; WYRD Greenville, SC; WGVC(FM) Simpsonville, SC; WORD(AM) Spartanburg, SC; WSPA-FM Spartanburg, SC; WSNA(FM) Germantown, TN; WRVR(FM) Memphis, TN; WSMB(AM) Memphis, TN; WWDE-FM Hampton, VA; WVKL-FM Norfolk, VA; WNVZ-FM Norfolk, VA; WPTE-FM Virginia Beach, VA; KNRK-FM Camas, WA; KKWF(FM) Seattle, WA; KNDD-FM Seattle, WA; KTTH(AM) Seattle, WA; KIRO Seattle, WA; KISW-FM Seattle, WA; KBSG-FM Tacoma, WA; KMTT-FM Tacoma, WA; KTRO(AM) Vancouver, WA; WOLX-FM Baraboo, WI; WMYX-FM Milwaukee, WI; WSSP(AM) Milwaukee, WI; WMMM-FM Verona, WI; WCHY(FM) Waunakee, WI; WXSS-FM Wauwatosa, WI

Joseph M. Field, chmn/CEO; David J. Field, pres/COO; John C. Donlevle, exec VP; Steve Fisher, sr VP; Eugene D. Levin, treas; Martin Hadfield, VP engrg; Deborah Kane, VP sls.

Entravision Communications Corp., 2425 Olympic Blvd., Suite 6000W, Santa Monica, CA 90404. Phone: (310) 447-3872. Fax: (310) 447-3899. E-mail: kthompson@entravision.com Web Site: www.entravision.com. Ownership: Walter F. Ulloa, Philip W. Wilkinson, Paul Zevnik.

Stns: 12 AM. 37 FM. KVVA-FM Apache Junction, AZ; KMIA(AM) Black Canyon City, AZ; KDVA(FM) Buckeye, AZ; KLNZ-FM Glendale, AZ; KZLZ-FM Kearny, AZ; KRRN(FM) Kingman, AZ; KSSE(FM) Arcadia, CA; KSEH(FM) Brawley, CA; KCVR-FM Columbia, CA; KXSE(FM) Davis, CA; KWST(AM) El Centro, CA; KSSD(FM) Fallbrook, CA; KLOK-FM Greenfield, CA; KMXX-FM Imperial, CA; KCVR Lodi, CA; KRCX-FM Marysville, CA; KDLE(FM) Newport Beach, CA; KTSE-FM Patterson, CA; KLYY(FM) Riverside, CA; KBMB(FM) Sacramento, CA; KDLD(FM) Santa Monica, CA; KSES-FM Seaside, CA; KNTY(FM) Shingle Springs, CA; KMBX(AM) Soledad, CA; KLOB-FM Thousand Palms, CA; KMIX-FM Tracy, CA; KSSC(FM) Ventura, CA; KPVW(FM) Aspen, CO; KMXA Aurora, CO; KJMN-FM Castle Rock, CO; KXPK-FM Evergreen, CO; WLQY Hollywood, FL; KRZY Albuquerque, NM; KRZY-FM Santa Fe, NM; KQRT(FM) Las Vegas, NV; KRNV-FM Reno, NV; KKPS-FM Brownsville, TX; KVLY-FM Edinburg, TX; KSVE El Paso, TX; KOFX-FM El Paso, TX; KINT-FM El Paso, TX; KHRO(AM) El Paso, TX; KYSE(FM) El Paso, TX; KFRQ-FM Harlingen, TX; KGOL Humble, TX; KBZO Lubbock, TX; KZPL(FM) Port Isabel, TX; KAIQ(FM) Wolfforth, TX; WACA Wheaton, MD.

Stns: 18 TV. KLUZ, Albuquerque-Santa Fe, NM; WUNI, Boston (Manchester, NH); KVSN, Colorado Springs-Pueblo, CO; KORO, Corpus Christi, TX; KCEC, Denver, CO; KINT, El Paso (Las Cruces, NM), TX; KTFN, El Paso (Las Cruces, NM), TX; KNVO, Harlingen-Weslaco -Brownsville-McAllen, TX; WUVN, Hartford & New Haven, CT; KLDO, Laredo, TX; KINC, Las Vegas, NV; KSMS, Monterey-Salinas, CA; KUPB, Odessa-Midland, TX; WVEN-TV, Orlando-Daytona Beach-Melbourne, FL; KPMR, Santa Barbara-Santa Maria-San Luis Obispo, CA; WVEA-TV, Tampa-St. Petersburg (Sarasota), FL; WJAL, Washington, DC (Hagerstown, MD); KVYE, Yuma, AZ-El Centro, CA.

Walter F. Ulloa, chmn/CEO; Philip Wilkinson, pres/COO; Larry Safir, exec VP.

Equity Communications LP, 8025 Black Horse Pike, Bayport One, Suite 100-102, West Atlantic City, NJ 08232. Phone: (609) 484-8444. Fax: (609) 646-6331. E-mail: gfequity@aol.com Web Site: 951wayv.com.

Stns: 2 AM. 7 FM. WAYV(FM) Atlantic City, NJ; WMID(AM) Atlantic City, NJ; WAIV(FM) Cape May, NJ; WGBZ(FM) Cape May Court House, NJ; WTTH(FM) Margate City, NJ; WZBZ(FM) Pleasantville, NJ; WZXL(FM) Wildwood, NJ; WCMC(AM) Wildwood, NJ; WEZW(FM) Wildwood Crest, NJ.

Gary S. Fisher, pres.

Eureka Broadcasting Co., 1101 Marsh Rd., Eureka, CA 95501. Phone: (707) 442-5744. Ownership: Barbara Papstein, 50%; Hugo Papstein, 28%; and Brian Papstein, 22%.

Stns: 3 AM. 3 FM. KEJY(FM) Blue Lake, CA; KWSW Eureka, CA; KEKA-FM Eureka, CA; KINS Eureka, CA; KURY Brookings, OR; KURY-FM Brookings, OR.

Hugo Papstein, gen mgr.

Evangel Ministries Inc., 1909 W. 2nd, Appleton, WI 54914. Phone: (920) 749-9456. Fax: (920) 749-0474. Web Site: www.christianfamilyradio.net.

Stns: 3 FM. WEMI(FM) Appleton, WI; WEMY-FM Green Bay, WI; WGNV(FM) Milladore, WI.

Paul Comeron, exec dir.

Evanov Communications Inc., 5302 Dundas St. W., Toronto, ON M9B 1B2. Canada. Phone: (416) 213-1035. Fax: (416) 233-8617. Web Site: evanovradiogroup.com. Ownership: William Evanov, 74.26%; Paul Evanov, 25%; and The Bill Evanov Family Trust, 0.74%. Note: Group also owns 50% of CIRR-FM Toronto, ON.

Stns: 1 AM. 4 FM. CKHZ-FM Halifax, NS; CIAO Brampton, ON; CKDX-FM Newmarket, ON; CIDC-FM Orangeville, ON; CJWL-FM Ottawa, ON.

Bill Evanov, pres.

F

FM Idaho Co. LLC, 21361 Hwy. 30, Twin Falls, ID 83301. Phone: (208) 735-8300.

Stns: 2 AM. 4 FM. KAYN(FM) Gooding, ID; KMCL-FM McCall, ID; KMHI Mountain Home, ID; KTMB(FM) Mountain Home, ID; KSRV Ontario, OR; KSRV-FM Ontario, OR.

Faith Communications Corp., 2201 S. 6th St., Las Vegas, NV 89104. Phone: (702) 731-5452. Fax: (702) 731-1992. Web Site: www.sosradio.net.

Stns: 1 AM. 6 FM. KHMS-FM Victorville, CA; KSQS(FM) Ririe, ID; KCIR(FM) Twin Falls, ID; KMZO(FM) Hamilton, MT; KMZL-FM Missoula, MT; KSOS(FM) Las Vegas, NV; KANN Roy, UT.

Brad Staley, pres.

James Falcon Stns, 2768 Pharmacy Rd., Rio Grande City, TX 78582. Phone: (956) 487-5621. Ownership: James A. Falcon, 100%.

Stns: 6 FM. KXOW(FM) Eldorado, OK; KTSX(FM) Knox City, TX; KAHA(FM) Olney, TX; KZAM(FM) Pleasant Valley, TX; KZNO(FM) Seymour, TX; KXME(FM) Wellington, TX.

Family Life Communications Inc., Box 35300, Tucson, AZ 85740. Phone: (520) 742-6976. Fax: (520) 742-6979. Web Site: www.flc.org. Ownership: All stns are owned by Family Life Communications Inc. A nonprofit, noncommercial Christian organization. No individual stockholders.

Stns: 4 AM. 14 FM. KJTA-FM Flagstaff, AZ; KFLR-FM Phoenix, AZ; KFLT Tucson, AZ; KFLT-FM Tucson, AZ; WJTF-FM Panama City, FL; WJTG(FM) Fort Valley, GA; KJTY-FM Topeka, KS; WUFN(FM) Albion, MI; WUNN(AM) Mason, MI; WUGN-FM Midland, MI; WUFL Sterling Heights, MI; KFLQ(FM) Albuquerque, NM; KWFL-FM Roswell, NM; KRGN-FM Amarillo, TX; KAMY(FM) Lubbock, TX; KFLB(AM) Odessa, TX; KFLB-FM Odessa, TX; WJTY-FM Lancaster, WI.

Randy L. Carlson, pres.

Family Life Network, Box 506, Bath, NY 14810. Phone: (607) 776-4151. Fax: (607) 776-6929. E-mail: mail@fln.org Web Site: www.fln.org. Ownership: Not-for-profit corporation.

Stns: 13 FM. WCOF(FM) Arcade, NY; WCIK-FM Bath, NY; WCIY-FM Canandaigua, NY; WCOV-FM Clyde, NY; WCIH-FM Elmira, NY; WCID-FM Friendship, NY; WCOT-FM Jamestown, NY; WCII-FM Spencer, NY; WCOU-FM Warsaw, NY; WCIG(FM) Carbondale, PA; WCOG-FM Galeton, PA; WCIM(FM) Shenandoah, PA; WCIT(FM) Trout Run, PA.

Rick Snavely, pres/CEO; Dick Snavely, CFO; John Owens, progmg dir; Jim Trouis, chief engr.

Family Stations Inc., 290 Hegenberger Rd., Oakland, CA 94621. Phone: (510) 568-6200. Fax: (510) 568-6190. Ownership: Nonprofit corporation.

Stns: 12 AM. 57 FM. WBFR-FM Birmingham, AL; KEAF(FM) Fort Smith, AR; KPHF-FM Phoenix, AZ; KFRB-FM Bakersfield, CA; KHAP(FM) Chico, CA; KFRJ(FM) China Lake, CA; KFRP(FM) Coalinga, CA; KECR El Cajon, CA; KFNO-FM Fresno, CA; KXBC(FM) Garberville, CA; KEFR-FM Le Grand, CA; KFRN Long Beach, CA; KEBR Rocklin, CA; KEAR-FM Sacramento, CA; KEAR(AM) San Francisco, CA; KHFR(FM) Santa Maria, CA; KFRS-FM Soledad, CA; KPRA-FM Ukiah, CA; KFRY(FM) Pueblo, CO; WCTF Vernon, CT; WMFL-FM Florida City, FL; WFTI-FM Saint Petersburg, FL; WWFR(FM) Stuart, FL; WFRP(FM) Americus, GA; WFRC-FM Columbus, GA; KDFR(FM) Des Moines, IA; KEGR(FM) Fort Dodge, IA; KYFR Shenandoah, IA; WJCH-FM Joliet, IL; WQLZ(FM) Taylorville, IL; KPOR-FM Emporia, KS; WOFR(FM) Schoolcraft, MI; KFRT(FM) Butte, MT; KFRD(FM) Butte, MT; KFRW(FM) Great Falls, MT; KBFR(FM) Bismarck, ND; WKDN-FM Camden, NJ; WFME-FM Newark, NJ; KXFR(FM) Socorro, NM; WFBF-FM Buffalo, NY; WFRH-FM Kingston, NY; WFRS-FM Smithtown, NY; WFRW-FM Webster, NY; WCUE Cuyahoga Falls, OH; WOTL-FM Toledo, OH; WYTN-FM Youngstown, OH; KYOR(FM) Newport, OR; KPFR(FM) Pine Grove, OR; KQFE-FM Springfield, OR; WUFR(FM) Bedford, PA; WEFR-FM Erie, PA; WFRJ-FM Johnstown, PA; WXFR(FM) State College, PA; WFCH-FM Charleston, SC; KKAA(AM) Aberdeen, SD; KQFR(FM) Rapid City, SD; KQKD(AM) Redfield, SD; KIFR(FM) Alice, TX; KEDR(FM) Bay City, TX; KTXB-FM Beaumont, TX; KUFR-FM Salt Lake City, UT; KARR Kirkland, WA; KJVH-FM Longview, WA; WWJA(FM) Janesville, WI; WMWK-FM Milwaukee, WI; WJJO-FM Watertown, WI; WFSI-FM Annapolis, MD; WBGR Baltimore, MD; WBMD Baltimore, MD.

Stns: 1 TV. WFME, New York.

Harold Camping, pres.

Family Worship Center Church Inc., Box 262550, Baton Rouge, LA 70826. Phone: (225) 768-3224. Web Site: www.jsm.org. Ownership: Jimmy Swaggart, 8.33% vote; Frances Swaggart, 8.33% vote; Donnie Swaggart, 8.33% vote; Harold Lee, 8.33% vote; Peggy Lee, 8.33% vote; Clyde Fuller, 8.33% vote; Elizabeth Fuller, 8.33% vote; Roy Chacon, 8.33% vote; Beulah Chacon, 8.33% vote; Jack Daugherty, 8.33% vote; Barbara Studley, 8.33% vote; and Debbie Swaggart, 8.33% vote.

Stns: 4 AM. 20 FM. WQUA-FM Citronelle, AL; KJSM-FM Augusta, AR; KNHD Camden, AR; KUUZ-FM Lake Village, AR; KSSW(FM) Nashville, AR; KPSH(FM) Coachella, CA; WFFL(FM) Panama City, FL; KBDD(FM) Winfield, KS; WJFM-FM Baton Rouge, LA; KTOC-FM Jonesboro, LA; KDJR(FM) De Soto, MO; WTGY-FM Charleston, MS; WJNS-FM Yazoo City, MS; KNBE(FM) Beatrice, NE; KNFA(FM) Grand Island, NE; KNHA(FM) Grand Isle, NE; WJCA(FM) Albion, NY; WJYM Bowling Green, OH; KAJT(FM) Ada, OK; KMFS(AM) Guthrie, OK; KSSO(FM) Norman, OK; WAYB-FM Graysville, TN; WSTN Somerville, TN; KNRB(FM) Atlanta, TX.

Jimmy Swaggart, pres; David Whitelaw, opns dir.

Fawcett Broadcasting Ltd., Box 777, Fort Frances, ON P9A 3N1. Canada. Phone: (807) 274-7580. Fax: (807) 274-8746. Ownership: Northwoods Broadcasting Ltd., 100%.

Stns: 2 AM. 3 FM. CKDR-FM Dryden, ON; CFOB-FM Fort Frances, ON; CJRL-FM Kenora, ON; CKDR-5 Red Lake, ON; CKDR-2(AM) Sioux Lookout, ON.

E. P. Fawcett, chmn; H. G. Fawcett, pres.

Federated Media, Box 2500, Elkhart, IN 46515. Phone: (574) 295-2500. Fax: (574) 294-4014.

Stns: 4 AM. 7 FM. WQHK-FM Decatur, IN; WBYT-FM Elkhart, IN; WTRC Elkhart, IN; WMEE-FM Fort Wayne, IN; WFWI-FM Fort Wayne, IN; WKJG(AM) Fort Wayne, IN; WOWO Fort Wayne, IN; WLEG(FM) Ligonier, IN; WAOR-FM Niles, MI; WNIL Niles, MI; WBYR-FM Van Wert, OH.

Federated Media publishes *The* (Elkhart, IN) *Truth*.

John F. Dille III, pres; Robert A. Watson, treas; Robert A. Watson, sec.

The Findlay Publishing Co., 701 W. Sandusky St., Findlay, OH 45840. Phone: (419) 422-5151. Fax: (419) 422-2937. E-mail: daveglass@findlayoh.com

Stns: 2 AM. 5 FM. WRBI-FM Batesville, IN; WCSI(AM) Columbus, IN; WINN(FM) Columbus, IN; WKKG-FM Columbus, IN; WWWY(FM) North Vernon, IN; WFIN Findlay, OH; WKXA-FM Findlay, OH.

The Findlay Publishing Co. publishes the *Findlay* (OH) *Courier*.

Karl Heminger, pres.

Finger Lakes Radio Group, 3568 Lenox Rd., Geneva, NY 14456. Phone: (315) 781-7000. Fax: (315) 781-7700. Web Site: www.fingerlakes1.com.

Stns: 3 AM. WAUB Auburn, NY; WFLR Dundee, NY; WSFW Seneca Falls, NY.

George Kimble, pres; Alan Bishop, VP/gen mgr.

First Broadcasting Operating Inc., 750 N. Saint Paul St., Fl. 10, Dallas, TX 75201-3236. Phone: (214) 855-0002. Fax: (214) 855-5145. E-mail: info@firstbroadcasting.com Web Site: www.firstbroadcasting.com. Ownership: Ronald Unkefer, Gary Lawrence, Alta Communications. Local mktg agreement: KFXR(AM) Dallas, TX.

Stns: 4 AM. 7 FM. KCCL(FM) Placerville, CA; WAAM Ann Arbor, MI; WVIM-FM Coldwater, MS; WOXY-FM Oxford, OH; WAOL-FM Ripley, OH; KHAL(FM) Condon, OR; KMCQ-FM The Dalles, OR; KREL(AM) Quanah, TX; KBIS(AM) Forks, WA; KBDB-FM Forks, WA; WAMD Aberdeen, MD.

Ronald Unkefer, chmn; Hal Rose, sr VP & corporate dev & regulatory affrs; Gary Lawrence, CEO & vice chmn & pres; Steve Kovac, CFO; Bob Denny, sr VP technology.

First Media Radio LLC, 306 Port St., Easton, MD 21601. Phone: (410) 822-3301. Fax: (410) 822-0576. Ownership: LPTV: WNVN-LP Roanoke Rapids, NC.

Stns: 9 AM. 16 FM. WTRG(FM) Gaston, NC; WWDR(AM) Murfreesboro, NC; WDLZ-FM Murfreesboro, NC; WZAX-FM Nashville, NC; WPWZ(FM) Pinetops, NC; WPTM-FM Roanoke Rapids, NC; WCBT Roanoke Rapids, NC; WDWG(FM) Rocky Mount, NC; WRMT(AM) Rocky Mount, NC; WSMY Weldon, NC; WZWW-FM Bellefonte, PA; WQYX-FM Clearfield, PA; WCPA Clearfield, PA; WOWQ(FM) DuBois, PA; WLAK-FM Huntingdon, PA; WMRF-FM Lewistown, PA; WIEZ Lewistown, PA; WWDW(FM) Alberta, VA; WWZW(FM) Buena Vista, VA; WYTT(FM) Emporia, VA; WREL Lexington, VA; WJLS Beckley, WV; WJLS-FM Beckley, WV; WEMD(AM) Easton, MD; WCEI-FM Easton, MD.

Alex Kolobielski, pres/CEO.

First Natchez Radio Group, Box 768, Natchez, MS 39121. Phone: (601) 442-4895. Fax: (601) 446-8260.

Stns: 1 AM. 3 FM. KTGV(FM) Jonesville, LA; WNAT Natchez, MS; WKSO(FM) Natchez, MS; WQNZ-FM Natchez, MS.

Marie Perkins, pres; Stephen Perkins, VP; Margaret Perkins, gen mgr.

Fisher Communications Inc., 100 4th Ave. N., Suite 440, Suite 1525, Seattle, WA 98109. Phone: (206) 404-7000. Fax: (206) 404-7050. Web Site: www.fsci.com.

Stns: 4 AM. 4 FM. KIKF(FM) Cascade, MT; KXGF Great Falls, MT; KINX(FM) Great Falls, MT; KQDI Great Falls, MT; KQDI-FM Great Falls, MT; KVI(AM) Seattle, WA; KOMO(AM) Seattle, WA; KPLZ(FM) Seattle, WA.

Stns: 12 TV. KBCI-TV, Boise, ID; KCBY, Eugene, OR; KPIC, Eugene, OR; KVAL, Eugene, OR; KIDK-TV, Idaho Falls-Pocatello, ID; KATU, Portland, OR; KUNP, Portland, OR; KUNS-TV, Seattle-Tacoma, WA; KOMO, Seattle-Tacoma, WA; KLEW-TV, Spokane, WA; KIMA-TV, Yakima-Pasco -Richland-Kennewick, WA; KEPR, Yakima-Pasco -Richland-Kennewick, WA.

Collen Brown, pres/CEO; Sheri Leonard, asst.

Flint Media Inc., Box 7425, Bainbridge, GA 39818-7425. Phone: (229) 416-6021. Fax: (229) 246-9995. Ownership: Kevin Dowdy, 100%.

Stns: 2 AM. 2 FM. WBGE(FM) Bainbridge, GA; WBBK Blakely, GA; WGMK-FM Donalsonville, GA; WSEM Donalsonville, GA.

Foothills Radio Group LLC, Box 1678, Lenoir, NC 28645. Phone: (828) 758-1033. Fax: (828) 757-3300. E-mail: abunch@kicksradio.com Web Site: foothillsradio.com. Ownership: Donald W. Curtis, 45%; William M. McClatchey Jr., 10%; George A. Bunch Jr. 45%.

Stns: 2 AM. 1 FM. WJRI Lenoir, NC; WKGX Lenoir, NC; WKVS-FM Lenoir, NC.

Al Bunch, pres.

Forever Broadcasting, One Forever Dr., Hollidaysburg, PA 16648. Phone: (814) 941-9800. Fax: (814) 943-2754. Web Site: www.foreverradio.com. Ownership: Kerby Confer, Donald Alt, Carol Logan, Lynn Deppen.

Stns: 13 AM. 16 FM. WVAM Altoona, PA; WWOT(FM) Altoona, PA; WFBG Altoona, PA; WALY-FM Bellwood, PA; WBUS(FM) Boalsburg, PA; WXXO(FM) Cambridge Springs, PA; WLTS(FM) Centre Hall, PA; WUUZ(FM) Cooperstown, PA; WRKW(FM) Ebensburg, PA; WFRA Franklin, PA; WWGY(FM) Grove City, PA; WRKY-FM Hollidaysburg, PA; WHUN Huntingdon, PA; WKYE(FM) Johnstown, PA; WNTJ(AM) Johnstown, PA; WFGI-FM Johnstown, PA; WJHT(FM) Johnstown, PA; WMGW Meadville, PA; WFZY(FM) Mount Union, PA; WKST(AM) New Castle, PA; WJST(AM) New Castle, PA; WOYL Oil City, PA; WSGY(FM) Pleasant Gap, PA; WUZZ(FM) Saegertown, PA; WNTW(AM) Somerset, PA; WRSC State College, PA; WQWK(FM) State College, PA; WMAJ State College, PA; WTIV Titusville, PA.

Carol Logan, pres.

Forever Communications Inc., 1919 Scottsville Rd., Bowling Green, KY 42104-3303. Phone: (270) 843-3333. Fax: (270) 843-0454. E-mail: chris@forevercomm.com Ownership: Kerby E. Confer Grantor Retained Annuity Trust, Kerby E. Confer, trustee; Donald J. Alt Grantor Retained Annuity Trust, Donald J. Alt, trustee; Christine E. Hillard.

Stns: 5 AM. 8 FM. WBVR-FM Auburn, KY; WBGN(AM) Bowling Green, KY; WLYE-FM Glasgow, KY; WNBS Murray, KY; WFGE-FM Murray, KY; WOFC(AM) Murray, KY; WUHU(FM) Smiths Grove, KY; WLLI-FM Dyer, TN; WLSZ(FM) Humboldt, TN; WLLI(AM) Humboldt, TN; WOGY(FM) Jackson, TN; WTJS Jackson, TN; WYNU-FM Milan, TN.

Christine Hillard, pres/COO.

Fort Bend Broadcasting Co., 1610 Woodstead Ct., Suite 350, Spring, TX 77380-3414. Phone: (281) 298-6797. Fax: (281) 298-8707. Ownership: Roy E. Henderson, owner. Note: Roy E. Henderson also owns WTCU(FM) Fife Lake, MI; and KNUZ(AM) Bellville, TX.

Stns: 2 AM. 13 FM. WCUZ(FM) Bear Lake, MI; WOUF(FM) Beulah, MI; WBNZ(FM) Frankfort, MI; WLDR(AM) Petoskey, MI; WLDR-FM Traverse City, MI; KLTR(FM) Brenham, TX; KULM-FM Columbus, TX; KEZB(FM) Edna, TX; KULF(FM) Ganado, TX; KHLT(AM) Hallettsville, TX; KTXM-FM Hallettsville, TX; KMBV(FM) Navasota, TX; KROY(FM) Palacios, TX; KJAZ(FM) Point Comfort, TX; KYKM-FM Yoakum, TX.

Roy E. Henderson, pres.

Fort Myers Broadcasting Co., 2824 Palm Beach Blvd., Fort Myers, FL 33916. Phone: (239) 334-1111. Fax: (239) 334-0744. E-mail: manaager@winktv.com Web Site: winktv.com. Ownership: Brian A. McBride.

Stns: 3 AM. 2 FM. WINK(AM) Fort Myers, FL; WINK-FM Fort Myers, FL; WNPL(AM) Golden Gate, FL; WPTK(AM) Pine Island Center, FL; WTLQ-FM Punta Rassa, FL.

Stns: 1 TV. WINK-TV, Ft. Myers-Naples, FL.

Brian McBride, pres/CEO; Gary Gardner, VP/gen mgr.

Forum Communications Co., Box 2020, Fargo, ND 58107. Phone: (701) 235-7311. Fax: (701) 241-5406. Web Site: www.in-forum.com.

Stns: 1 AM. 1 FM. WZUU-FM Allegan, MI; WDAY Fargo, ND.

Stns: 4 TV. WDAY-TV, Fargo-Valley City, ND; WDAZ-TV, Fargo-Valley City, ND; KBMY, Minot-Bismarck-Dickinson, ND; KMCY, Minot-Bismarck-Dickinson, ND.

Forum Communications Co. owns the *Alexandria* (MN) *Echo Press; The Pioneer,* Bemidj, MN; *Detroit Lakes* (MN) *Tribune; The Becker County Record,* Detroit Lakes, MN; *Park Rapids* (MN) *Enterprise; The Wadena* (MN) *Pioneer Journal; West Central Daily Tribune,* Willmar, MN; *The Daily Globe;* Worthington, MN; *The Daily Republic,* Mitchell SD; *The Dickinson Press,* Dickinson, ND & *The* (ND) *Forum.*

William C. Marcil, pres.

Foster Communications Co. Inc., Box 2191, San Angelo, TX 76902-2191. Phone: (325) 949-2112. Fax: (325) 944-0851. Web Site: www.fostercommunications.us. Ownership: Fred

M. Key, 100% of votes, 95% of total assets.

Stns: 1 AM. 3 FM. KIXY-FM San Angelo, TX; KKSA San Angelo, TX; KWFR-FM San Angelo, TX; KCLL(FM) San Angelo, TX.

Fred M. Key, pres; Doug Smith, gen sls mgr; Jay Michaels, chief of opns; Shannon J. Roach, CFO.

Four Corners Broadcasting L.L.C., Drawer P, Durango, CO 81302. Phone: (970) 259-4444. Fax: (970) 247-1005. E-mail: fcb@frontier.net Web Site: www.radiodurango.com. Ownership: Four Corners Communications L.L.C., Fordstone, IN.

Stns: 1 AM. 2 FM. KKDC(FM) Dolores, CO; KIQX-FM Durango, CO; KIUP Durango, CO.

Allen Brill, CEO; Ward S. Holmes, gen mgr.

4-K Radio Inc., Box 936, Lewiston, ID 83501. Phone: (208) 743-2502. Fax: (208) 743-1995. E-mail: radiorip@aol.com Web Site: www.koze.com. Ownership: Eugene Hamblin Trust; Michael R. Ripley.

Stns: 2 AM. 2 FM. KORT Grangeville, ID; KORT-FM Grangeville, ID; KOZE Lewiston, ID; KOZE-FM Lewiston, ID.

Michael R. Ripley, pres.

Four Rivers Broadcasting Inc., Box 1729, Yreka, CA 96097. Phone: (530) 842-4158. Fax: (530) 842-7635. Web Site: www.mtshastalive.com. Ownership: Alta California Broadcasting Inc., 100%.

Stns: 4 FM. KTDE(FM) Gualala, CA; KMFB-FM Mendocino, CA; KNTK(FM) Weed, CA; KSYC-FM Yreka, CA.

John Anthony, gen mgr.

The Free Lance-Star Publishing Co., 616 Amelia St., Fredericksburg, VA 22401. Phone: (540) 373-1500. Fax: (540) 374-5525. Ownership: Josiah P. Rowe III; Anne W. Rowe

Stns: 1 AM. 3 FM. WWUZ(FM) Bowling Green, VA; WFLS-FM Fredericksburg, VA; WYSK Fredericksburg, VA; WYSK-FM Spotsylvania, VA.

The Free Lance-Star Publishing Co., publishes the *Fredericksburg* (VA) *Free Lance-Star*.

Josiah P. Rowe III, pres; Florence C. Barnick, assoc publisher; Nicholas J. Cadwallender, assoc publisher; John Moen, gen mgr radio stns.

Freedom Communications of Connecticut Inc., 330 Main St., Hartford, CT 06106. Phone: (860) 524-0001. Fax: (860) 548-1922. E-mail: mssm2115@msn.com Ownership: Richard Weaver-Bey, 50% of votes; and Stephen Brisker, 50% of votes.

Stns: 3 AM. WNEZ(AM) Manchester, CT; WLAT(AM) New Britain, CT; WKND(AM) Windsor, CT.

Stephen Brisker, pres/CEO & chmn.

Friends Communications Inc., 121 W. Maumee St., Adrian, MI 49221-2019. Phone: (517) 265-1500. Fax: (517) 263-4525. E-mail: friends@tc3net.com Ownership: Bob Elliot, 100%.

Stns: 1 AM. 1 FM. WABJ Adrian, MI; WBZV(FM) Hudson, MI.

Bob Elliot, pres; Moneca Morton, gen sls mgr.

Fritz Communications Inc., 1355 N. Dutton Ave. #225, Santa Rosa, CA 95401-7107. Phone: (707) 546-9185. Fax: (707) 546-9188. Ownership: KEWB(FM), KNCQ(FM), KESR(FM), KKXS(FM) and KHRD(FM) are licensed to Results Radio of Redding Licensee LLC. KBQB(FM), KKCY(FM), KTHU(FM), KMJE(FM), KCEZ(FM) and KRQR(FM) are licensed to Results Radio of Chico Licensee LLC.

Stns: 11 FM. KEWB-FM Anderson, CA; KBQB(FM) Chico, CA; KKCY-FM Colusa, CA; KTHU(FM) Corning, CA; KMJE-FM Gridley, CA; KCEZ(FM) Los Molinos, CA; KRQR-FM Orland, CA; KNCQ-FM Redding, CA; KESR(FM) Shasta Lake City, CA; KKXS(FM) Shingletown, CA; KHRD(FM) Weaverville, CA.

Jack Fritz, pres/CEO.

J. & J. Fritz Media Ltd., Box 311, Fredericksburg, TX 78624. Phone: (830) 997-2197. Fax: (830) 997-2198. E-mail: txradio@ktc.com Web Site: www.texasrebelradio.com.

Stns: 1 AM. 3 FM. KEEP-FM Bandera, TX; KNAF Fredericksburg, TX; KNAF-FM Fredericksburg, TX; KFAN-FM Johnson City, TX.

Jayson Fritz, gen mgr; Jan Fritz, gen sls mgr.

G

GAP Broadcasting LLC, 12900 Preston Rd., Suite 525, Dallas, TX 75230. Phone: (214) 295-3530. Fax: (972) 386-4445. Web Site: www.gapbroadcast.com. Ownership: GAP Broadcasting Holdings LLC, 100%.

Stns: 12 AM. 41 FM. KMJI(FM) Ashdown, AR; KOSY(AM) Texarkana, AR; KYGL-FM Texarkana, AR; KRUF(FM) Shreveport, LA; KVKI-FM Shreveport, LA; KWKH(AM) Shreveport, LA; KEEL(AM) Shreveport, LA; KXKS-FM Shreveport, LA; KLAW(FM) Lawton, OK; KZCD(FM) Lawton, OK; KVRW-FM Lawton, OK; KULL-FM Abilene, TX; KFGL(FM) Abilene, TX; KSLI(AM) Abilene, TX; KEYJ-FM Abilene, TX; KYYW(AM) Abilene, TX; KEAN-FM Abilene, TX; KMXJ-FM Amarillo, TX; KATP(FM) Amarillo, TX; KPRF(FM) Amarillo, TX; KMML-FM Amarillo, TX; KIXZ Amarillo, TX; KLUB-FM Bloomington, TX; KTUX-FM Carthage, TX; KAFX-FM Diboll, TX; KFZX(FM) Gardendale, TX; KPWW-FM Hooks, TX; KBGE(AM) Kilgore, TX; KKTX-FM Kilgore, TX; KKCL-FM Lorenzo, TX; KKAM Lubbock, TX; KQBR(FM) Lubbock, TX; KFMX-FM Lubbock, TX; KFYO Lubbock, TX; KZII-FM Lubbock, TX; KYKS-FM Lufkin, TX; KCRS-FM Midland, TX; KCRS(AM) Midland, TX; KCHX(FM) Midland, TX; KSFA Nacogdoches, TX; KTBQ-FM Nacogdoches, TX; KMRK-FM Odessa, TX; KKYR-FM Texarkana, TX; KNUE(FM) Tyler, TX; KTYL-FM Tyler, TX; KQVT(FM) Victoria, TX; KIXS-FM Victoria, TX; KVLL-FM Wells, TX; KISX-FM Whitehouse, TX; KNIN-FM Wichita Falls, TX; KBZS(FM) Wichita Falls, TX; KWFS Wichita Falls, TX; KWFS-FM Wichita Falls, TX.

George Laughlin, pres; Shawn Nunn, sls VP; Norman Philips, engrg VP.

GCC Bend LLC, 969 S. W. Colorado, Bend, OR 97702. Phone: (541) 388-3300. Fax: (541) 389-7885. Web Site: www.ksjj.com. Ownership: Gross Holdings L.P., 100%.

Stns: 1 AM. 3 FM. KMGX(FM) Bend, OR; KXIX-FM Bend, OR; KICE(AM) Bend, OR; KSJJ-FM Redmond, OR.

Dana Horner, COO.

GHB Radio Group, 1776 Briarcliff Rd. N.E., Suite A, Atlanta, GA 30306-2106. Phone: (404) 875-1110. Fax: (404) 875-1186. Ownership: George H. Buck Jr.

Stns: 14 AM. 2 FM. WMGY Montgomery, AL; WYZE Atlanta, GA; WIST(AM) New Orleans, LA; WCGC Belmont, NC; WHVN Charlotte, NC; WEGO Concord, NC; WAME(AM) Statesville, NC; WIST-FM Thomasville, NC; WBLO(AM) Thomasville, NC; WSVM(AM) Valdese, NC; WNMX-FM Waxhaw, NC; WTIX(AM) Winston-Salem, NC; WNAP Norristown, PA; WOLS(AM) Florence, SC; WHYM(AM) Lake City, SC; WAVO Rock Hill, SC.

Jacob E. Bogan, COO; George H. Buck Jr., pres.

Galaxy Communications L.P., 235 Walton St., Syracuse, NY 13202. Phone: (315) 472-9111. Fax: (315) 472-1888. Web Site: www.galaxycommunications.com.

Stns: 4 AM. 7 FM. WTKW-FM Bridgeport, NY; WKLL-FM Frankfort, NY; WKRH-FM Minetto, NY; WKRL-FM North Syracuse, NY; WTLA North Syracuse, NY; WSGO Oswego, NY; WTKV-FM Oswego, NY; WZUN(FM) Phoenix, NY; WSCP Sandy Creek-Pulaski, NY; WTLB Utica, NY; WRCK-FM Utica, NY.

Ed Levine, pres; Mimi Griswold, VP progmg; Lisa Morrow, VP sls; Michael Lucarelli, CFO.

Galesburg Broadcasting Co., 154 E. Simmons St., Galesburg, IL 61401. Phone: (309) 342-5131. Fax: (309) 342-0840. E-mail: results@galesburgradio.com Web Site: www.galesburgradio.com. Ownership: John Pritchard, pres, 100%.

Stns: 1 AM. 3 FM. WAAG-FM Galesburg, IL; WGIL Galesburg, IL; WLSR(FM) Galesburg, IL; WKAY(FM) Knoxville, IL.

John T. Pritchard, pres.

Gateway Radio Works Inc., Box 1010, Owingsville, KY 40360. Phone: (606) 674-2266. Fax: (606) 674-6700. Ownership: Hays McMakin, 100%.

Stns: 1 AM. 3 FM. WIVY(FM) Morehead, KY; WMST(AM) Mt. Sterling, KY; WKCA(FM) Owingsville, KY; WAXZ-FM Georgetown, OH.

Hays McMakin, pres; Jeff Ray, VP/gen mgr.

Genesis Communications Inc., 2110 Powers Ferry Rd., Suite 198, Atlanta, GA 30339. Phone: (678) 324-0170. Fax: (678) 324-0174. E-mail: ceo@radiogenesis.com Web Site: www.radiogenesis.com. Ownership: Bruce C. Maduri, J. Donald Childress.

Stns: 5 AM. WHBO(AM) Dunedin, FL; WHOO(AM) Kissimmee, FL; WAMT(AM) Pine Castle-Sky Lake, FL; WWBA Pinellas Park, FL; WIXC(AM) Titusville, FL.

J. Donald Childress, VP; Bruce C. Maduri, pres/CEO.

Georgia-Carolina Radiocasting Companies, Drawer E, Toccoa, GA 30577. Phone: (864) 297-7264. Fax: (864) 297-7266. E-mail: sutton@gacaradio.com Web Site: www.gacaradio.com. Ownership: Douglas M. Sutton Jr., 100% of all stns except WSNW(AM) and WWOF(AM)-WGOG(FM). M. Terry Carter owns 50% and Douglas M. Sutton Jr. owns 50% of WSNW(AM) and WWOF(AM)-WGOG(FM). Note: Group also operates WZGA(FM) Helen, GA by a loc mktg agreement.

Stns: 7 AM. 5 FM. WRBN-FM Clayton, GA; WGHC(AM) Clayton, GA; WLVX(FM) Elberton, GA; WNGA(AM) Elberton, GA; WSGC-FM Elberton, GA; WNEG(AM) Toccoa, GA; WNCC-FM Franklin, NC; WFSC Franklin, NC; WRGC Sylva, NC; WSNW(AM) Seneca, SC; WWOF(AM) Walhalla, SC; WGOG(FM) Walhalla, SC.

M. Terry Carter, pres/COO; Douglas M. (Art) Sutton Jr., chmn/CEO; Tonya Burgess, VP/CFO.

Georgia Eagle Broadcasting Inc., 1350 Radio Loop Rd., Warner Robins, GA 31088. Phone: (478) 923-3416. Fax: (478) 923-3236. Web Site: www.georgiaeagleradio.com. Ownership: Cecil P. Staton, 50%; and Joe Sam Robinson Jr., 50%. Note: Group also has a loc mktg agreement for WQSA(FM) Unadilla, GA.

Stns: 6 AM. 7 FM. WMCD(FM) Claxton, GA; WDCO(AM) Cochran, GA; WDXQ-FM Cochran, GA; WRPG(FM) Hawkinsville, GA; WCEH Hawkinsville, GA; WHKN-FM Millen, GA; WQXZ(FM) Pinehurst, GA; WPTB(AM) Statesboro, GA; WPMX-FM Statesboro, GA; WWNS Statesboro, GA; WZBX-FM Sylvania, GA; WSYL Sylvania, GA; WNNG(AM) Warner Robins, GA.

Gestion Appalaches inc., C.P. 69, Thetford Mines, PQ G6G 5S3. Canada. Phone: (418) 335-7533. Phone: (819) 752-2785. Fax: (418) 335-9009. Fax: (819) 752-3182. Ownership: Francois Labbe, 99.9%; Fiducie familiale F. Labbe, .07%; and Annie Labbie, .03%.

Stns: 3 FM. CFJO-FM Thetford Mines, PQ; CKLD-FM Thetford Mines, PQ; CFDA-FM Victoriaville, PQ.

Annie Labbe, gen mgr.

Gleason Radio Group, 555 Center St., Auburn, ME 04210. Phone: (207) 748-5868. Fax: (207) 784-4700. E-mail: dick@gleasonmedia.com Web Site: www.gleasonmedia.com. Ownership: Richard D. Gleason, 100%.

Stns: 3 AM. 2 FM. WEZR(AM) Lewiston, ME; WTBM(FM) Mexico, ME; WOXO-FM Norway, ME; WTME(AM) Rumford, ME; WKTQ South Paris, ME.

Richard Gleason, pres/gen mgr.

Gleiser Communications LLC, 1001 E. Southeast Loop 323, Suite 455, Tyler, TX 75701. Phone: (903) 593-2519. Fax: (903) 597-4141. E-mail: info@ktbb.com Web Site: www.gleisercom.com. Ownership: Broadcasting partnors holdings, LP, Paul L. Gleiser.

Stns: 3 AM. 1 FM. KEES Gladewater, TX; KTBB Tyler, TX; KYZS Tyler, TX; KDOK-FM Tyler, TX.

Paul Gleiser, pres/CEO.

Glenwood Communications Corp., 222 Commerce St., Kingsport, TN 37660. Phone: (423) 246-9578. Fax: (423) 246-6261. E-mail: golz@wkpttv.com Web Site: www.wkpttv.com. Ownership: William M. Boyd; Hugh N. Boyd Trust.

Stns: 4 AM. 2 FM. WOPI Bristol, TN; WKTP Jonesborough, TN; WKPT Kingsport, TN; WTFM-FM Kingsport, TN; WMEV Marion, VA; WMEV-FM Marion, VA.

Stns: 1 TV. WKPT-TV, Tri-Cities, TN-VA.

George E. DeVault Jr., pres.

Glory Communications Inc., Box 2355, West Columbia, SC 29171. Phone: (803) 939-9530. Fax: (803) 939-9469. Ownership: Alex Snipe, 100%.

Stns: 2 AM. 5 FM. WSPX-FM Bowman, SC; WTQS(AM) Cameron, SC; WGCV(AM) Cayce, SC; WPDT-FM Johnsonville, SC; WTUA-FM Saint Stephen, SC; WFMV(FM) South Congaree, SC; WLJI-FM Summerton, SC.

Alex Snipe, pres/CEO.

Goforth Media Inc., Box 1328, Mobile, AL 36633. Phone: (251) 473-8488. Fax: (251) 473-8854. E-mail: wgoforth@goforth.org Web Site: www.goforth.org. Ownership: No stock. Nonprofit.

Stns: 1 FM. WBHY-FM Mobile, AL.

Wilbur Goforth, pres; Steve Riggs, VP; Stephen Goforth, VP.

Gold Coast Broadcasting LLC, 2284 S. Victoria, Suite 2M, Ventura, CA 93003. Phone: (805) 289-1400. Phone: (805) 339-0773. Fax: (805) 644-4257. Ownership: Point Broadcasting Company 88.51%, Jeri Lynn Broadcasting 6.32%, August G Inc. 5.17%.

Stns: 3 AM. 3 FM. KOCP-FM Camarillo, CA; KFYV(FM) Ojai, CA; KCAQ(FM) Oxnard, CA; KVTA Port Hueneme, CA; KUNX(AM) Santa Paula, CA; KKZZ(AM) Ventura, CA.

John Q. Hearne, chmn; Miles Sexton, pres.

Golden West Broadcasting Ltd., Box 950, Altona, MB R0G 0B0. Canada. Phone: (204) 324-6464. Fax: (204) 324-8918. E-mail: info@cfamradio.com Ownership: Elmer Hildebrand Ltd., 38.73%; Elmer Hildebrand Investments Inc., 19.23%; Elmer Hildebrand, 10.04%; and others, 32%.

Stns: 13 AM. 12 FM. CHRB High River, AB; CFXL-FM High River-Okotoks, AB; CKVN-FM Lethbridge, AB; CFAM Altona, MB; CJRB Boissevain, MB; CFRY Portage la Prairie, MB; CJPG-FM Portage la Prairie, MB; CHSM Steinbach, MB; CILT-FM Steinbach, MB; CJEL-FM Winkler, MB; CKMW Winkler-Morden, MB; CHVN-FM Winnipeg, MB; CJSL Estevan, SK; CHSN-FM Estevan, SK; CKVX-FM Kindersley, SK; CFYM Kindersley, SK; CHAB Moose Jaw, SK; CILG-FM Moose Jaw, SK; CJYM Rosetown, SK; CJSN Shaunavon, SK; CKSW Swift Current, SK; CIMG-FM Swift Current, SK; CKFI-FM Swift Current, SK; CKRC-FM Weyburn, SK; CFSL Weyburn, SK.

Elmer Hildebrand, pres/CEO; Menno Friesen, VP; Lyndon Friesen, VP.

Good Karma Broadcasting L.L.C., Box 902, Beaver Dam, WI 53916. Phone: (920) 885-4442. Fax: (920) 885-2152. Ownership: Craig Karmazin, 100%

Stns: 7 AM. 3 FM. WEFL(AM) Tequesta, FL; WTJK South Beloit, IL; WKNR(AM) Cleveland, OH; WWGK(AM) Cleveland, OH; WBEV Beaver Dam, WI; WXRO-FM Beaver Dam, WI; WTLX-FM Columbus, WI; WWHG(FM) Evansville, WI; WTTN Watertown, WI; WAUK Waukesha, WI.

Rick Armon, opns dir; Chris Hartl, business mgr; Craig Karmazin, gen mgr.

Good News Communications Inc., 3222 S. Richey Ave., Tucson, AZ 85713. Phone: (520) 790-2440. Fax: (520) 790-2937.

Stns: 5 AM. KAPR Douglas, AZ; KJAA Globe, AZ; KNXN Sierra Vista, AZ; KVOI Tucson, AZ; KGMS(AM) Tucson, AZ.

Douglas E. Martin, pres; Mary R. Martin, sec.

Good News Media Inc., Box 1400, Traverse City, MI 49685-1400. Phone: (231) 946-1400. Fax: (231) 946-3959. Web Site: www.wljn.com.

Stns: 2 AM. 1 FM. WLJW(AM) Cadillac, MI; WLJN Elmwood Township, MI; WLJN-FM Traverse City, MI.

Doug Knorr, pres.

Good News Network, 2278 Wortham Lane, Grovetown, GA 30813-5103. Phone: (706) 309-9610. Fax: (706) 309-9669. E-mail: ctbarinowski@comcast.net Web Site: www.gnnradio.org.

Stns: 3 AM. 10 FM. WQRX Valley Head, AL; WLPE(FM) Augusta, GA; WPMA(FM) Buckhead, GA; WPWB(FM) Byron, GA; WWGF-FM Donalsonville, GA; WLPT(FM) Jesup, GA; WLPF-FM Ocilla, GA; WZIQ-FM Smithville, GA; WKTM-FM Soperton, GA; WZQZ(AM) Trion, GA; WGPH(FM) Vidalia, GA; WLGP-FM Harkers Island, NC; WBLR Batesburg, SC.

Clarence Barinowski, pres/CEO.

Gore-Overgaard Broadcasting Inc., 11310 E. Arabian Park Dr., Scottsdale, AZ 85259. Phone: (480) 314-0144. Fax: (480) 314-4942. Ownership: Cordell Overgaard, Harold Gore.

Stns: 5 AM. KLHC(AM) Bakersfield, CA; KBIF Fresno, CA; KIRV Fresno, CA; WROD Daytona Beach, FL; WSBB New Smyrna Beach, FL.

Harold W. Gore, chmn/CEO; Cordell J. Overgaard, pres.

Grace Broadcasting Services Inc., 25 Stonebrook Pl., Suite G, #322, Jackson, TN 38305. Phone: (731) 663-3931. Fax: (731) 663-9804. Web Site: www.gracebroadcasting.com. Ownership: Charles Ennis, 55.8%; Lacy Ennis, 31.1%; Ray Smith, 7.5%; Dr. Buck Morton, 1.9%; and Phillip Chambers, 0.2%. Note: Group also owns 50% of WTRB(AM) Ripley, TN.

Stns: 2 AM. 5 FM. WWGM(FM) Alamo, TN; WTKB-FM Atwood, TN; WFGZ-FM Lobelville, TN; WSIB-FM Selmer, TN; WDTM Selmer, TN; WTNE Trenton, TN; WTNE-FM Trenton, TN.

Graham Newspapers Inc., 620 Oak St., Graham, TX 76450. Phone: (940) 549-1330. Fax: (940) 549-8628. E-mail: gm@kwkq-kswa.com

Stns: 2 AM. 2 FM. KLXK-FM Breckenridge, TX; KROO Breckenridge, TX; KSWA Graham, TX; KWKQ(FM) Graham, TX.

Joe Graham, gen mgr.

Great Lakes Radio Inc., 2025 US 41 W., Marquette, MI 49855. Phone: (906) 227-7777. Phone: (906) 228-6800. Fax: (906) 475-8888. Fax: (906) 228-8128. E-mail: todd@greatlakesradio.org Web Site: www.greatlakesradio.org.

Stns: 1 AM. 4 FM. WPIQ(FM) Manistique, MI; WFXD-FM Marquette, MI; WRUP(FM) Munising, MI; WQXO Munising, MI; WKQS-FM Negaunee, MI.

Todd S. Noordyk, pres.

Great Plains Media Inc., Box 1628, Cape Girardeau, MO 63702-1628. Phone: (573) 651-0707. Ownership: Jerome R. Zimmer Revocable Trust U/A/D October 14, 1977, sole trustee, Jerome R. Zimmer, 100%.

Stns: 1 AM. 5 FM. WRPW(FM) Colfax, IL; WYST(FM) Fairbury, IL; WDQZ(FM) Lexington, IL; KLWN(AM) Lawrence, KS; KLZR(FM) Lawrence, KS; KMXN(FM) Osage City, KS.

Jerome R. Zimmer, pres.

Great Scott Broadcasting, 224 Maugers Mill Rd., Pottstown, PA 19464. Phone: (610) 326-4000. Fax: (610) 326-7984. E-mail: mike.licata@1370wpaz.com Web Site: www.1370wpaz.com. Ownership: Faye Scott, special trustee, Charles Mott & James Worthington, co-trustees, Family Trust U/D/T dated 11/6/81, 86.72% of total assets; Faye Scott Annuity Trust U/D/T dated 3/6/02, Charles Mott & James Worthington, co-trustees, 6.64% of total assets; Faye Scott & James Worthington, co-trustees, Marital Trust U/D/T dated 11/6/81, 100% of votes, 0.98% of total assets.

Stns: 2 AM. 8 FM. WPAZ Pottstown, PA; WOCQ-FM Berlin, MD; WKHI(FM) Fruitland, MD; WKHW-FM Pocomoke City, MD; WJKI(FM) Bethany Beach, DE; WJWL Georgetown, DE; WZBH-FM Georgetown, DE; WKDB(FM) Laurel, DE; WZEB(FM) Ocean View, DE; WGBG-FM Seaford, DE.

Faye Scott, CEO; Mike Licata, gen mgr/controller.

Great South Wireless LLC, 6930 Cahaba Valley Rd., Suite 202, Birmingham, AL 36037. Phone: (205) 949-4586. Ownership: R3 Partners LLC, 90%; Cameron Reynolds, 5%; and Parmally LLC, 5%.

Stns: 4 FM. WZLM(FM) Goodwater, AL; WHPH(FM) Jemison, AL; WTID(FM) Repton, AL; WKGA(FM) Thomaston, AL.

Greater Media Inc., 35 Braintree Hill Office Park, Suite 300, Braintree, MA 02184. Phone: (781) 348-8600. Fax: (781) 348-8680. Web Site: www.greatermedia.com. Ownership: Bordes family, 100%.

Stns: 4 AM. 15 FM. WTKK(FM) Boston, MA; WMJX-FM Boston, MA; WBOS-FM Brookline, MA; WROR-FM Framingham, MA; WKLB-FM Waltham, MA; WCSX-FM Birmingham, MI; WRIF(FM) Detroit, MI; WMGC-FM Detroit, MI; WWTR(AM) Bridgewater, NJ; WJJZ(FM) Burlington, NJ; WDHA-FM Dover, NJ; WJRZ-FM Manahawkin, NJ; WMTR(AM) Morristown, NJ; WMGQ(FM) New Brunswick, NJ; WCTC(AM) New Brunswick, NJ; WRAT(FM) Point Pleasant, NJ; WPEN Philadelphia, PA; WMMR-FM Philadelphia, PA; WBEN-FM Philadelphia, PA.

Greater Media, Inc. owns 100% of The Sentinel Publishing Co., publisher of twelve weekly newspapers: the *Sentinel* of East Brunswick, the *Edison/Metuchen Sentinel*, the *Woodbridge Sentinel*, the *North/South Brunswick Sentinel*, the *Atlanticville* of Long Branch, the *Bulletin* of Brick Township, the *Examiner* of Allentown, the *Hub* of Red Bank, the *Independent* of Middletown, the *News Transcript* of Freehold, the *Suburban* of Sayreville, and the *Tri-Town News* of Howell, Lakewood, Jackson and Plumsted, all NJ.

Peter H. Smyth, pres/CEO.

Green County Broadcasting, W4765 Radio Ln., Monroe, WI 53566. Phone: (608) 325-2161. Fax: (608) 325-2164. Ownership: Scott Thompson, 75%; and Ronald Spielman, 25%. Note: Scott Thompson owns 75% and Ronald Spielman owns 25% of WQLF(FM) Lena, IL.

Stns: 2 AM. 2 FM. WFPS-FM Freeport, IL; WFRL Freeport, IL; WEKZ Monroe, WI; WEKZ-FM Monroe, WI.

Grenax Broadcasting LLC, 10337 Carriage Club Dr., Lone Tree, CO 80124. Phone: (303) 790-4015. Ownership: Greg Dinetz, 100% of votes.

Stns: 4 FM. KFLX-FM Kachina Village, AZ; KSED-FM Sedona, AZ; KWMX-FM Williams, AZ; WCFX(FM) Clare, MI.

Greg Dinetz, pres.

Groupe Radio Antenne 6 Inc., 568 boul. St. Joseph, Roberval, PQ G8H 2K6. Canada. Phone: (418) 275-1831. Fax: (418) 275-2475. E-mail: malevesque@antenne6.com Web Site: chrlfm.com. Ownership: RNC MEDIA Inc., 75% (see listing); 9150-2898 Quebec inc., 25%.

Stns: 3 AM. 4 FM. CFGT Alma, PQ; CKYK-FM Alma, PQ; CFED Chapais, PQ; CJMD Chibougamau, PQ; CKXO-FM Chibougamau, PQ; CHVD-FM Dolbeau-Mistassini, PQ; CHRL-FM Roberval, PQ.

Marc-Andre Levesque, pres.

Guaranty Broadcasting Co. of Baton Rouge, LLC, Box 2231, Baton Rouge, LA 70821. Phone: (225) 388-9898 ext 148. Fax: (225) 344-3077. E-mail: owen.weber@gbcradio.com

Stns: 5 FM. WTGE(FM) Baker, LA; WYPY(FM) Baton Rouge, LA; WDGL-FM Baton Rouge, LA; KNXX(FM) Donaldsonville, LA; WNXX(FM) Jackson, LA.

Owen Weber, VP.

Guyann Corp., Box 1930, Flagstaff, AZ 86002. Phone: (928) 774-5231. Fax: (928) 779-2988. Web Site: www.kaff.com. Ownership: Richard D. Guest, 50%; Pamela Flaherty, 50%. Note: Richard D. Guest and Pamela Flaherty are co-special administrators.

Stns: 2 AM. 3 FM. KAFF Flagstaff, AZ; KAFF-FM Flagstaff, AZ; KMGN-FM Flagstaff, AZ; KNOT(AM) Prescott, AZ; KTMG(FM) Prescott, AZ.

H

HRN Broadcasting Inc., Box 430, Lincolnton, NC 28093. Phone: (704) 735-8071. Fax: (704) 732-9567. Web Site: www.hrnb.com.

Stns: 6 AM. WZGM(AM) Black Mountain, NC; WCSL Cherryville, NC; WGNC Gastonia, NC; WLON Lincolnton, NC; WOHS Shelby, NC; WADA Shelby, NC.

Haliburton Broadcasting Group Inc., 46 Nanton Ave., Toronto, ON M4W 2Y9. Canada. Phone: (416) 925-0488. Fax: (416) 925-6256. Web Site: www.hbgradio.com. Ownership: Standard Broadcasting Corp.

Stns: 14 FM. CHMS-FM Bancroft, ON; CFBG-FM Bracebridge, ON; CHPB-FM Cochrane, ON; CKNR-FM Elliot Lake, ON; CKJN-FM Haldimand County, ON; CFZN-FM Haliburton, ON; CHYK-FM-3 Hearst, ON; CFIF-FM Iroquois Falls, ON; CKAP-FM Kapuskasing, ON; CFXN-FM North Bay, ON; CKLP-FM Parry Sound, ON; CHYC-FM Sudbury, ON; CHMT-FM Timmins, ON; CHYK-FM Timmins, ON.

Christopher Grossman, pres; Kim Ward, VP opns mgr.

Hall Communications Inc., Box 2038, 404 W. Lime St., Lakeland, FL 33806. Phone: (863) 682-8184. Fax: (863) 683-2409. Web Site: www.hallradio.com. Ownership: Bonnie Hall Rowbotham.

Stns: 8 AM. 13 FM. WNLC-FM East Lyme, CT; WKNL(FM) New London, CT; WICH Norwich, CT; WCTY-FM Norwich, CT; WILI-FM Willimantic, CT; WILI(AM) Willimantic, CT; WWRZ-FM Fort Meade, FL; WONN Lakeland, FL; WLKF Lakeland, FL; WPCV-FM Winter Haven, FL; WCTK(FM) New Bedford, MA; WNBH New Bedford, MA; WKOL-FM Plattsburgh, NY; WBTZ-FM Plattsburgh, NY; WROZ-FM Lancaster, PA; WLPA Lancaster, PA; WSJW(FM) Starview, PA; WLKW(AM) West Warwick, RI; WJOY(AM) Burlington, VT; WOKO-FM Burlington, VT; WIZN-FM Vergennes, VT.

Arthur J. Rowbotham, pres; Bonnie Hall Rowbotham, chmn; Bill Baldwin, exec VP.

Harvard Broadcasting Inc., Century Plaza, 1900 Rose Dt., Regina, SK S4P 0A9. Canada. Phone: (306) 546-6200. Fax: (306) 781-7338. E-mail: rpettigrew @harvardbroadcasting.com Web Site: www.harvardbroadcasting.com. Ownership: Harvard Developments Inc., 100%.

Stns: 1 AM. 3 FM. CFEX-FM Calgary, AB; CFWF-FM Regina, SK; CHMX-FM Regina, SK; CKRM(AM) Regina, SK.

Bruce Cowie, VP; Michael Olstrom, gen mgr.

Haugo Broadcasting Inc., Box 1680, Rapid City, SD 57709. Phone: (605) 343-0888. Fax: (605) 342-3075.

Stns: 1 AM. 1 FM. KSQY-FM Deadwood, SD; KTOQ Rapid City, SD.

Chris Haugo, pres.

Heartland Christian Broadcasters Inc., Box 433, International Falls, MN 56649. Phone: (218) 285-7398. Fax: (218) 285-7419. Ownership: Chuck Scherer, 20%; Dan Griffith, 20%; Jim Hummel, 20%; Mike Worth, 20%; and Tom Wherley, 20%.

Stns: 3 FM. KADU(FM) Hibbing, MN; KBHW(FM) International Falls, MN; KXBR(FM) International Falls, MN.

Heartland Communications Group LLC, 4650 W. Spencer St., Appleton, WI 54914. Phone: (920) 882-4750. Fax: (920) 882-4751. Web Site: www.heartlandcomm.com. Ownership: Granite Equity L.P., 44.05% equity; The Thomas L. Bookey Family L.P., 38.57% equity; Granite/Heartland Co-Investment L.P., 10.57% equity; and James L. Gregori, 6.81% equity. Note: Group is managed by a five-member board of governors.

Stns: 5 AM. 8 FM. WOLV-FM Houghton, MI; WCCY Houghton, MI; WHKB(FM) Houghton, MI; WIKB Iron River, MI; WIKB-FM Iron River, MI; WJJH-FM Ashland, WI; WATW Ashland, WI; WBSZ(FM) Ashland, WI; WRJO-FM Eagle River, WI; WERL Eagle River, WI; WNXR(FM) Iron River, WI; WCQM-FM Park Falls, WI; WNBI Park Falls, WI.

Tom Bookey, CEO; James Gregori, pres.

He's Alive Inc., Box 540, Grantsville, MD 21536. Phone: (301) 895-3292. Fax: (301) 895-3293. E-mail: hesalive@hesalive.net Web Site: www.hesalive.net. Ownership: Non-stock, nonprofit organization.

Stns: 4 FM. WRIJ(FM) Masontown, PA; WPCL(FM) Northern Cambria, PA; WLIC-FM Frostburg, MD; WAIJ-FM Grantsville, MD.

Dewayne Johnson, pres.

Bennie E. Hewett Stns, Box 907670, Gainesville, GA 30501-0911. Phone: (770) 536-3890. Fax: (770) 536-4103. Ownership: Bennie Hewett.

Stns: 1 AM. 2 FM. WBMH(FM) Grove Hill, AL; WRJX(AM) Jackson, AL; WHOD(FM) Jackson, AL.

Bennie E. Hewett, pres.

Hi-Favor Broadcasting LLC, 136 S. Oak Knoll Ave., Pasadena, CA 91101. Phone: (626) 356-4230. Fax: (626) 795-9185. Ownership: Daisy Publishing Co. Inc., Pasadena, CA, 100%.

Stns: 3 AM. KLTX Long Beach, CA; KEZY(AM) San Bernardino, CA; KSDO San Diego, CA.

High Desert Broadcasting LLC, 570 East Ave. Q9, Palmdale, CA 93550. Phone: (661) 947-3107. Fax: (661) 272-5688. Ownership: John Hearne.

Stns: 2 AM. 3 FM. KGMX-FM Lancaster, CA; KWJL(AM) Lancaster, CA; KUTY Palmdale, CA; KLKX-FM Rosamond, CA; KKZQ-FM Tehachapi, CA.

Miles Sexton, pres.

Hill Country Broadcasting Corp., 2125 Sidney Baker North, Kerrville, TX 78028. Phone: (325) 446-3371. Phone: (830) 896-1230. Fax: (830) 792-4142.

Stns: 2 AM. 4 FM. KMBL Junction, TX; KOOK-FM Junction, TX; KRVL(FM) Kerrville, TX; KERV Kerrville, TX; KYXX(FM) Ozona, TX; KHOS-FM Sonora, TX.

Ken Foster, pres.

Holladay Broadcasting of Louisiana LLC, Box 4808, Monroe, LA 71211. Phone: (318) 398-1618. Ownership: Robert H. Holladay, 100%.

Stns: 2 AM. 4 FM. KJMG-FM Bastrop, LA; KRVV-FM Bastrop, LA; KLIP-FM Monroe, LA; KMLB Monroe, LA; KRJO(AM) Monroe, LA; WBBV-FM Vicksburg, MS.

Robert Holladay,.

Holy Family Communications, 6325 Sheridan Dr., Williamsville, NY 14221. Phone: (716) 839-6117. Fax: (716) 839-0400. Web Site: www.wlof.net. Ownership: James N. Wright, 33.33%; Joanne Wright, 33.33%; and Mary Ellen Capece, 33.33%.

Stns: 2 AM. 1 FM. WLOF(FM) Attica, NY; WHIC(AM) Rochester, NY; WQOR(AM) Olyphant, PA.

James N. Wright, pres.

Horizon Broadcasting Group., 2484 N.W. Hosmer Lake Dr., Bend, OR 97701. Phone: (541) 383-3282. Fax: (541) 383-3283. Web Site: www.horizonbroadcasting.com.

Stns: 1 AM. 4 FM. KQAK(FM) Bend, OR; KRCO(AM) Prineville, OR; KLTW-FM Prineville, OR; KWPK-FM Sisters, OR; KWLZ-FM Warm Springs, OR.

Keith Shipman, pres/CEO.

Horizon Christian Fellowship, 5331 Mt. Alifan Dr., San Diego, CA 92111. Phone: (858) 277-4991. Fax: (858) 277-1365. Web Site: www.horizonsd.org/radio.asp. Ownership: Rudy Batiz, 9.09% votes; Jeries El Raheb, 9.09% votes; Gayle Gordon, 9.09% votes; Larry Gordon, 9.09% votes; Michael MacIntosh, 9.09% votes; Phillip MacIntosh, 9.09% votes; Victor Najor, 9.09% votes; Tom Phillips, 9.09% votes; Fred Salley, 9.09% votes; Hank Sybrandy, 9.09% votes; and Mike Turk, 9.09% votes;

Stns: 29 FM. KHZK(FM) Kotzebue, AK; KHZF(FM) Fagaitua, AS; KWDS(FM) Kettleman City, CA; KWDI(FM) Idalia, CO; KWDN(FM) Newell, IA; WDVL(FM) Danville, IN; WWDL(FM) Lebanon, IN; WWDN(FM) New Whiteland, IN; WRYP(FM) Wellfleet, MA; WTNP(FM) Richland, MI; KHZA(FM) Bunker, MO; KSRD(FM) Saint Joseph, MO; WCMR(FM) Bruce, MS; KWDE(FM) Eureka, MT; KHZS(FM) Saint Regis, MT; KWDV(FM) Valier, MT; KHRU(FM) Beulah, ND; KCVG(FM) Medina, ND; KCVD(FM) New England, ND; KCVF(FM) Sarles, ND; KHZY(FM) Overton, NE; KHZZ(FM) Sargent, NE; KWDP(FM) Prineville, OR; KWDC(FM) Coahoma, TX; KWDH(FM) Hereford, TX; KWDR(FM) Royal City, WA; WDSW(FM) Westby, WI; KWDU(FM) Upton, WY; KHRW(FM) Wright, WY.

Tom Phillips, COO.

Horne Radio Group, 517 Watt Rd., Knoxville, TN 37922. Phone: (865) 675-4105. Fax: (865) 675-4859. E-mail: knoxvilletalk@aol.com Web Site: www.wkvl.com.

Stns: 6 AM. 1 FM. WMTY(AM) Farragut, TN; WKVL(AM) Knoxville, TN; WLOD Loudon, TN; WFIV-FM Loudon, TN; WGAP Maryville, TN; WATO Oak Ridge, TN; WDEH Sweetwater, TN.

Douglas A. Horne, pres.

Houston Christian Broadcasters Inc., KHCB Network, 2424 South Blvd., Houston, TX 77098-5110. Phone: (713) 520-5200. E-mail: email@khcb.org Web Site: www.khcb.org.

Stns: 2 AM. 16 FM. KHCL(FM) Arcadia, LA; KHIB(FM) Bastrop, TX; KHVT(FM) Bloomington, TX; KALD(FM) Caldwell, TX; KHPS(FM) Camp Wood, TX; KHCF(FM) Fredericksburg, TX; KHCB Galveston, TX; KANJ(FM) Giddings, TX; KHCB-FM Houston, TX; KHCH(AM) Huntsville, TX; KHCJ(FM) Jefferson, TX; KHKV(FM) Kerrville, TX; KKER(FM) Kerrville, TX; KHML(FM) Madisonville, TX; KHCP(FM) Paris, TX; KHPO(FM) Port O'Connor, TX; KCPC(FM) Sealy, TX; KHTA(FM) Wake Village, TX.

Bruce E. Munsterman, pres & gen mgr; Bonnie C. BeMent, asst gen mgr.

Howell Mountain Broadcasting Co., 95 La Jota Dr., Angwin, CA 94508. Phone: (707) 965-4155. Fax: (707) 965-4161. Web Site: www.thecandle.com. Ownership: Jim Chasse, 8%; Judy Crabb, 8%; John Collins, 8%; Edward Fargusson, 8%; John Hughson, 8%; John McIntosh, 8%; Jarrod McNaughton, 8%; Brent Timothy Mitchell, 8%; Richard Osborn, 8%; Leeanne Patterson, 8%; David Shantz, 8%; and Robert Surridge, 8%.

Stns: 3 FM. KNDL-FM Angwin, CA; KNDZ(FM) McKinleyville, CA; KPPN(FM) Pollock Pines, CA.

Hubbard Broadcasting Inc., 3415 University Ave., St. Paul, MN 55114. Phone: (651) 646-5555. Fax: (651) 642-4103. E-mail: jmahoney@hbi.com

Stns: 2 AM. 2 FM. WFMP(FM) Coon Rapids, MN; KSTP(AM) Saint Paul, MN; KSTP-FM Saint Paul, MN; WIXK(AM) New Richmond, WI.

Stns: 13 TV. WNYT, Albany-Schenectady-Troy, NY; KOBG-TV, Albuquerque-Santa Fe, NM; KOB, Albuquerque-Santa Fe, NM; KOBF, Albuquerque-Santa Fe, NM; KOBR, Albuquerque-Santa Fe, NM; WDIO, Duluth, MN-Superior, WI; WIRT, Duluth, MN-Superior, WI; KRWF, Minneapolis-St. Paul, MN; KSAX, Minneapolis-St. Paul, MN; KSTP, Minneapolis-St. Paul, MN; KSTC-TV, Minneapolis-St. Paul, MN; KAAL, Rochester, MN-Mason City, IA-Austin, MN; WHEC, Rochester, NY.

Stanley S. Hubbard, chmn/pres/CEO; Stanley E. Hubbard II, VP; Virginia H. Morris, VP; Robert W. Hubbard, VP; Julia D. Coyte, VP; Gerald D. Deeney, sr VP/treas/CFO; Harold C. Crump, VP; C. Thomas Newberry, VP; Linda S. Tremere, VP; Sue J. Cook, VP; Edward J. Aiken, VP; Kari Rominski, sec; Gary R. Macomber, asst sec.

Huth Broadcasting, Box 669, Marysville, CA 95901. Phone: (530) 742-5555. Fax: (530) 741-3758. Ownership: Tom F. Huth, 100%. Note: Tom F. Huth owns 49.9% of the stock of Sierra Radio Inc., licensee of KTOR(AM) Westwood, CA.

Stns: 4 AM. KMYC(AM) Marysville, CA; KPCO Quincy, CA; KBLF Red Bluff, CA; KOBO Yuba City, CA.

I

IHR Educational Broadcasting, Box 180, Tahoma, CA 96142. Phone: (530) 584-5700. Fax: (530) 584-5705. E-mail: info@ihradio.org Web Site: www.ihradio.org.

Stns: 7 AM. 2 FM. KAHI Auburn, CA; KCIK(AM) Blue Lake, CA; KJPG(AM) Frazier Park, CA; KPJP(FM) Greenville, CA; KJOP Lemoore, CA; KWG(AM) Stockton, CA; KSMH(AM) West Sacramento, CA; KXXQ(FM) Milan, NM; KIHM(AM) Reno, NV.

Icicle Broadcasting Inc., 7475 KOHO Pl., Leavenworth, WA 98826. Phone: (509) 548-1011. Fax: (509) 548-3222. Web Site: www.kohoradio.com.

Stns: 1 AM. 3 FM. KOZI Chelan, WA; KOZI-FM Chelan, WA; KOHO-FM Leavenworth, WA; KZAL(FM) Manson, WA.

Gary Mathews, gen mgr.

Idaho Wireless Corp., Box 97, Pocatello, ID 83204. Phone: (208) 234-1290. Fax: (208) 234-9451.

Stns: 1 AM. 2 FM. KORR-FM American Falls, ID; KOUU(AM) Pocatello, ID; KZBQ(FM) Pocatello, ID.

Paul Anderson, gen mgr.

IdaVend Broadcasting Inc., 805 Stewart Ave., Lewiston, ID 83501. Phone: (208) 743-1551. Fax: (208) 743-4440. E-mail: rprasil@idavend.com Ownership: Robert Prasil; Gary Prasil; Dorothy Prasil.

Stns: 1 AM. 2 FM. KMOK(FM) Lewiston, ID; KRLC Lewiston, ID; KVTY-FM Lewiston, ID.

Robert Prasil, gen mgr; Zoanne Davis, traffic dir; Ben Bonfield, gen sls mgr; Darin Siebert, opns dir; Melva Prasil, stn mgr.

Illinois Bible Institute Inc., Box 140, Carlinville, IL 62626. Phone: (217) 854-4600. Fax: (217) 854-4610. E-mail: rwhitworth@idcag.org Web Site: www.wibi.org.

Stns: 7 FM. WTSG-FM Carlinville, IL; WIBI-FM Carlinville, IL; WBGL-FM Champaign, IL; WBMV-FM Mount Vernon, IL; WCIC-FM Pekin, IL; WPRC(FM) Princeton, IL; WCRT-FM Terre Haute, IN.

Richard C. Whitworth, dir.

Impact Radio LLC, 59750 Constantine Rd., Three Rivers, MI 49093-9303. Phone: (269) 278-1815. Fax: (269) 273-7975. E-mail: drumsey@wlkm.com Ownership: Dennis W. Rumsey.

Stns: 2 AM. 2 FM. WLKM Three Rivers, MI; WLKM-FM Three Rivers, MI; WQCT Bryan, OH; WBNO-FM Bryan, OH.

Dennis W. Rumsey, pres.

Independence Media Holdings LLC, 8226 Douglas Ave., Suite 627, Dallas, TX 75225. Phone: (469) 619-1001. Ownership: Seaport Capital Partners III AIV L.P., 33.3% votes, 73.32% total assets; Seaport IMH Blocker Corp., 33.3% votes, 24.58% total assets; David F. Jacobs, 33.3% votes, 1.96% total assets.

Stns: 1 AM. 6 FM. WOCN Miami, FL; WWCT(FM) Bartonville, IL; WPIA(FM) Eureka, IL; WZPN(FM) Farmington, IL; WXMP(FM) Glasford, IL; WNUY(FM) Bluffton, IN; WWKN(FM) Morgantown, KY.

Information Communications Corp., Box 2061, Bristol, TN 37621-2061. Phone: (423) 878-6279. Fax: (423) 878-6520. Ownership: Kenneth C. Hill, 51%; and Appalachian Educational Communication Corp., 49%.

Stns: 3 AM. WPWT(AM) Colonial Heights, TN; WHGG(AM) Kingsport, TN; WABN Abingdon, VA.

Dr. Kenneth C. Hill, pres/CEO.

Ingstad Brothers Broadcasting LLC, Box 1248, Minnetonka, MN 55345. Phone: (952) 938-0575. Fax: (952) 938-2295. Ownership: Thomas E. Ingstad, 49%; Tor Ingstad, 49%; and Randy K. Holland, 2%.

Stns: 3 AM. 1 FM. KCHK New Prague, MN; KNUJ(AM) New Ulm, MN; KYMN Northfield, MN; KNUJ-FM Sleepy Eye, MN.

Robert Ingstad Broadcast Properties, Box 994, Valley City, ND 58072. Phone: (701) 845-1490. Fax: (701) 845-1245. Ownership: the estate of Robert E. Ingstad, Janice M. Ingstad, Robert J. Ingstad and Todd M. Ingstad.

Stns: 10 AM. 12 FM. KSKZ(FM) Copeland, KS; KBUF(AM) Holcomb, KS; KFXX-FM Hugoton, KS; KSSA-FM Ingalls, KS; KWKR(FM) Leoti, KS; KSKL-FM Scott City, KS; KULY Ulysses, KS; KDIO Ortonville, MN; KDAK Carrington, ND; KXGT(FM) Carrington, ND; KQDJ Jamestown, ND; KYNU(FM) Jamestown, ND; KDDR Oakes, ND; KOVC Valley City, ND; KRVX(FM) Wimbledon, ND; KMLO-FM Lowry, SD; KMSD Milbank, SD; KOLY Mobridge, SD; KGFX Pierre, SD; KJBI(FM) Presho, SD; KPLO-FM Reliance, SD; KBWS-FM Sisseton, SD.

Tom Ingstad Broadcasting Group, Box 1248, Minnetonka, MN 55345. Phone: (952) 938-0575. Fax: (952) 938-2295.

Stns: 4 AM. 3 FM. KARP-FM Dassel, MN; KKRC-FM Granite Falls, MN; KDUZ Hutchinson, MN; KDMA Montevideo, MN; KMRS Morris, MN; KRVY-FM Starbuck, MN; KKAQ Thief River Falls, MN.

Tom Ingstad, pres/CEO.

Inland Northwest Broadcasting LLC, 805 Stewart Ave., Lewiston, ID 83501. Phone: (208) 791-2605. Fax: (208) 743-4440. E-mail: rprasil@idavend.com Ownership: Robert Prasil, 50%; and Melva Prasil, 50%.

Stns: 2 AM. 2 FM. KCLX Colfax, WA; KMAX Colfax, WA; KRAO-FM Colfax, WA; KZZL-FM Pullman, WA.

Robert Prasil, pres; Gary Cummings, VP/owner; Darin Siebert, dir opns; Steve Franko, chief engr.

Inner Banks Media LLC, 408 W. Arlington Blvd., Suite 101-B, Greenville, NC 27834. Phone: (252) 355-8822. Ownership: Donald W. Curtis, 50%; Henry Williams Hinton Jr., 40%; and Henry Williams Hinton III, 10%.

Stns: 4 FM. WWNK(FM) Farmville, NC; WRHT-FM Morehead City, NC; WWHA(FM) Oriental, NC; WRHD(FM) Williamston, NC.

Inner City Broadcasting, 3 Park Ave., 41st Fl., New York, NY 10016. Phone: (212) 447-1000. Fax: (212) 447-5197. E-mail: info@wbls.com Web Site: www.wbls.com.

Stns: 8 AM. 10 FM. KBLX-FM Berkeley, CA; KVTO Berkeley, CA; KVVN Santa Clara, CA; WJMI-FM Jackson, MS; WKXI Jackson, MS; WOAD Jackson, MS; WKXI-FM Magee, MS; WOAD-FM Pickens, MS; WLIB New York,

NY; WBLS-FM New York, NY; WLFP(AM) Braddock, PA; WHAT Philadelphia, PA; WZMJ-FM Batesburg, SC; WOIC(AM) Columbia, SC; WARQ(FM) Columbia, SC; WHXT-FM Orangeburg, SC; WMFX(FM) Saint Andrews, SC; WWDM(FM) Sumter, SC.

Pierre Sutton, chmn/CEO.

Inter-Island Communications Inc., 1868 Halsey Dr., Piti, GU 96915. Phone: (671) 477-7108. Fax: (671) 477-6411. Ownership: Edward H. Poppe Jr., Frances W. Poppe.

Stns: 2 AM. 4 FM. KSTO(FM) Hagatna, GU; KTWG(AM) Hagatna, GU; KISH(FM) Hagatna, GU; KZMI(FM) Garapan-Saipan, NP; KCNM-FM Garapan-Saipan, NP; KCNM(AM) Garapan-Saipan, NP.

Edward H. Poppe Jr., pres.

International Broadcasting Corp., 1554 Bori St., San Juan, PR 00927-6113. Phone: (787) 274-1800. Fax: (787) 281-9758. Ownership: Pedro Roman Collazo, 100%. Note: Pedro Roman Collazo, as an individual, owns WVOZ(AM) San Juan, PR.

Stns: 7 AM. 1 FM. WRSJ(AM) Bayamon, PR; WGIT(AM) Canovanas, PR; WVOZ-FM Carolina, PR; WIBS Guayama, PR; WXRF Guayama, PR; WTIL Mayaguez, PR; WEKO(AM) Morovis, PR; WCHQ(AM) Quebradillas, PR.

Stns: 3 TV. WVEO, Aguadilla, PR; WVOZ, Ponce, PR; WTCV, San Juan, PR.

Pedro Roman Callazo, pres; Margarita Nazario, gen mgr.

J

J&V Communications Inc., 222 Hazard St., Orlando, FL 32804. Phone: (407) 841-8282. Fax: (407) 841-8250. Ownership: Jesus Torrado, Virgen Torrado.

Stns: 4 AM. WTJV(AM) De Land, FL; WOTS Kissimmee, FL; WSDO(AM) Sanford, FL; WPRD Winter Park, FL.

John Torrado, pres; Frank F. Vaught, opns mgr.

JWC Broadcasting, 259 S. Willow Ave., Cookeville, TN 38501. Phone: (931) 528-6064. Fax: (931) 520-1590. Ownership: Joe W. Wilmoth, 99%; and Reba Wilmoth, 1%.

Stns: 1 AM. 3 FM. WATX Algood, TN; WBXE-FM Baxter, TN; WLQK(FM) Livingston, TN; WKXD-FM Monterey, TN.

Joel Wilmoth, pres.

Jabar Communications Inc., 5081 Rivers Ave., North Charleston, SC 29406. Phone: (843) 554-1063. Fax: (843) 554-1088. E-mail: traffic@jabarcommunications.com Web Site: jabarcommunications.com.

Stns: 1 AM. 2 FM. WJNI(FM) Ladson, SC; WAZS-FM McClellanville, SC; WAZS(AM) Summerville, SC.

Thomas Daniel, pres.

Jackson County Broadcasting Inc., Box 667, 295 E. Main St., Jackson, OH 45640. Phone: (740) 286-3023. Fax: (740) 286-6679. E-mail: jmossbarger@jbiradio.com Ownership: Alan Stockmeister

Stns: 1 AM. 1 FM. WCJO-FM Jackson, OH; WYPC Wellston, OH.

Jerry Mossbarger, gen mgr.

Jackson Radio Works Inc., 1700 Glenshire Dr., Jackson, MI 49201. Phone: (517) 787-9546. Fax: (517) 787-7517. E-mail: bgoldsen@wkhm.com Web Site: www.wkhm.com. Ownership: Bruce & Susan Goldsen, 100%

Stns: 2 AM. 1 FM. WKHM-FM Brooklyn, MI; WIBM Jackson, MI; WKHM(AM) Jackson, MI.

Susan Goldsen; Bruce Goldsen, pres & gen mgr.

Jacobs Media Corp., Box 10, Gainesville, GA 30503. Phone: (770) 532-9921. Fax: (770) 532-0506. E-mail: jayjacobs@wdun.com Web Site: www.wdun.com. Ownership: Elizabeth Jacobs Carswell.

Stns: 2 AM. 1 FM. WMJE-FM Clarkesville, GA; WDUN Gainesville, GA; WGGA Gainesville, GA.

John W. Jacobs III, pres/CEO; Jay Andrews, VP bcstg.

James Crystal Inc., 6600 N. Andrews Ave., Suite 160, Fort Lauderdale, FL 33309. Phone: (954) 315-1515. Fax: (954) 315-1555. Ownership: James W. Hilliard, Crystal H. Armstrong.

Stns: 5 AM. WFLL(AM) Fort Lauderdale, FL; WMEN(AM) Royal Palm Beach, FL; WFTL(AM) West Palm Beach, FL; KCKN(AM) Roswell, NM; KNIT(AM) Dallas, TX.

James C. Hilliard, pres.

Jodesha Broadcasting Inc., Box 1198, Aberdeen, WA 98520. Phone: (360) 533-3000. Fax: (360) 532-1456. E-mail: bossbill@jodesha.com Web Site: www.jodesha.com. Ownership: William J and Susan Wolfenbarger.

Stns: 1 AM. 3 FM. KBKW Aberdeen, WA; KSWW(FM) Montesano, WA; KANY(FM) Ocean Shores, WA; KJET(FM) Raymond, WA.

William J. Wolfenbarger, pres; Susan Wolfenbarger, sec/treas.

Johnson Enterprises Inc., 338 S. KLEY Dr., Wellington, KS 67152. Phone: (620) 326-3341. Fax: (620) 326-8512. E-mail: kley@sutv.com Web Site: www.kleyam.com. Ownership: E. Gordon Johnson, Susan G. Johnson.

Stns: 2 AM. 1 FM. KLEY(AM) Wellington, KS; KWME-FM Wellington, KS; KKLE(AM) Winfield, KS.

E. Gordon Johnson, pres.

Journal Communications Inc., 333 W. State St., Milwaukee, WI 53203. Phone: (414) 224-2616. Fax: (414) 224-2469. Web Site: www.jc.com.

Stns: 8 AM. 22 FM. KGMG-FM Oracle, AZ; KMXZ-FM Tucson, AZ; KQTH(FM) Tucson, AZ; KJOT(FM) Boise, ID; KGEM(AM) Boise, ID; KCID(AM) Caldwell, ID; KTHI(FM) Caldwell, ID; KRVB(FM) Nampa, ID; KQXR-FM Payette, ID; KYQQ(FM) Arkansas City, KS; KFXJ(FM) Augusta, KS; KFTI-FM Newton, KS; KICT-FM Wichita, KS; KFTI(AM) Wichita, KS; KSGF-FM Ash Grove, MO; KZRQ-FM Mount Vernon, MO; KSPW(FM) Sparta, MO; KSGF(AM) Springfield, MO; KSRZ-FM Omaha, NE; KKCD-FM Omaha, NE; KEZO-FM Omaha, NE; KXSP(AM) Omaha, NE; KQCH(FM) Omaha, NE; KXBL(FM) Henryetta, OK; KFAQ(AM) Tulsa, OK; WMYU(FM) Karns, TN; WKHT(FM) Knoxville, TN; WQBB Powell, TN; WWST(FM) Sevierville, TN; WTMJ Milwaukee, WI.

Stns: 9 TV. KIVI, Boise, ID; WFTX, Ft. Myers-Naples, FL; WGBA, Green Bay-Appleton, WI; WSYM, Lansing, MI; KTNV, Las Vegas, NV; WTMJ, Milwaukee, WI; KMTV, Omaha, NE; KMIR-TV, Palm Springs, CA; KGUN, Tucson (Sierra Vista), AZ.

Journal Communications Inc., publisher of the morning *Milwaukee* (WI) *Journal Sentinel*, owns 100% of Journal Broadcast Corp.

Douglas G. Kiel, pres.

Joy Christian Communications Inc., Box 602, Centre, AL 35960. Phone: (256) 927-4027. Fax: (256) 927-4028. Web Site: www.joychristian.com. Ownership: Andrew L. Smith, 25%; Ed L. Smith, 25%; Marie P. Smith, 25%; and Melissa McGrew Smith, 25%.

Stns: 4 AM. WRFS(AM) Alexander City, AL; WZTQ(AM) Centre, AL; WLYG(AM) Hanceville, AL; WLYJ(AM) Jasper, AL.

J-Systems Franchising Corp., Hotel Traylor, 1444 Hamilton St., Allentown, PA 18102. Phone: (610) 435-5913. Fax: (610) 435-8918. E-mail: wmgh@ptd.net Web Site: www.wmgh.com. Ownership: Harold G. Fulmer III, 100%.

Stns: 1 AM. 1 FM. WLSH Lansford, PA; WMGH-FM Tamaqua, PA.

Harold G. Fulmer III, pres.

K

KCD Enterprises Inc., Box 1100, Bartlesville, OK 74005. Phone: (918) 336-1001. Fax: (918) 336-3939. E-mail: radio@ibartlesvilleradio.com Web Site: www.bartlesvilleradio.com. Ownership: Kevin Potter, 50%; Dorea Potter, 50%. Note: KPGM(AM) Pawhuska, OK is licensed to Potter Radio LLC. All of the membership units of Potter Radio LLC are held by the Kevin M. and Dorea S. Potter Trust.

Stns: 2 AM. 2 FM. KWON Bartlesville, OK; KYFM-FM Bartlesville, OK; KRIG-FM Nowata, OK; KPGM(AM) Pawhuska, OK.

Kevin Potter, pres & gen mgr.

KEA Radio Inc., Box 966, Scottsboro, AL 35768. Phone: (256) 259-2341. Fax: (256) 574-2156. Web Site: www.wkeafm.com. Ownership: Ronald H. Livengood; Gene Sisk; Ivous Sisk; Diane Livengood.

Stns: 2 FM. WKEA-FM Scottsboro, AL; WMXN-FM Stevenson, AL.

Ronald H. Livengood, pres; Gene Sisk, VP; Diane Livengood, sec; Ivous Sisk, treas.

KERM Inc., 201 W. 2nd, Russellville, AR 72801. Phone: (479) 968-1184. Fax: (479) 967-5278. E-mail: karv610@cei.net

Stns: 3 AM. 2 FM. KARV-FM Ola, AR; KURM Rogers, AR; KARV Russellville, AR; KURM-FM South West City, MO; KLTK South West City, MO.

Chris Womack, gen mgr; James K. Womack, pres.

KHWY Inc., 12381 Wilshire Blvd. #105, Los Angeles, CA 90025. Phone: (310) 820-4628. Fax: (310) 826-7866. E-mail: khwyha@earthlink.net Web Site: www.thehighwaystations.com. Ownership: Howard B. Anderson; Kirk M. Anderson.

Stns: 8 FM. KIXF-FM Baker, CA; KHRQ(FM) Baker, CA; KHWY-FM Essex, CA; KIXW-FM Lenwood, CA; KHDR(FM) Lenwood, CA; KHWZ-FM Ludlow, CA; KHYZ-FM Mountain Pass, CA; KRXV-FM Yermo, CA.

Howard B. Anderson, pres/CEO; Kirk M. Anderson, exec VP; Jean Sheranian, sec.

KM Communications Inc., 3654 Jarvis Ave., Skokie, IL 60076. Phone: (847) 674-0864. Fax: (847) 674-9188. Web Site: www.kmcommunications.com.

Stns: 2 AM. 9 FM. WPNG(FM) Pearson, GA; KTKB-FM Hagatna, GU; KTKB(AM) Tamuning, GU; KQMG Independence, IA; KQMG-FM Independence, IA; WLCN(FM) Atlanta, IL; WMKB(FM) Earlville, IL; KBWM(FM) Breckenridge, TX; KKEV(FM) Centerville, TX; KBDK(FM) Leakey, TX; KHMR(FM) Lovelady, TX.

Stns: 2 TV. KWKB, Cedar Rapids-Waterloo-Iowa City & Dubuque, IA; KEJB, Monroe, LA-El Dorado, AR.

Myoung Hwa Bae, pres; Kevin J. Bae, VP/gen mgr.

KNZA Inc., Box 104, Hiawatha, KS 66434-0104. Phone: (785) 547-3461. Fax: (785) 547-9900. E-mail: knza@rainbowtel.net Web Site: www.knzafm.com. Ownership: Greg Buser, 51%; Robert Hilton, 45%. Loc mktg agreement: KAIR-FM Horton, KS.

Stns: 3 FM. KNZA-FM Hiawatha, KS; KMZA-FM Seneca, KS; KLZA-FM Falls City, NE.

Greg Buser, pres; Robert Hilton, sec/treas.

K95.5 Inc., 24189 E. 865 Rd., Welling, OK 74471-2245. Phone: (918) 230-2165. Fax: (918) 457-3512. E-mail: Paynewh@aol.com Web Site: www.k955.com. Ownership: William H. Payne 100% stockholder. Note: William H. Payne also is sole owner of Payne 5 Communications LLC, licensee of KTLQ(AM)-KEOK(FM) Tahlequah, OK.

Stns: 2 FM. KTNT-FM Eufaula, OK; KTFX-FM Warner, OK.

William H. Payne, pres.

KOOR Communications Inc., Box 2295, New London, NH 03257. Phone: (603) 448-0500. Fax: (603) 448-6601. E-mail: bob@wntk.com Web Site: www.wntk.com.

Stns: 4 AM. 1 FM. WQTH(AM) Claremont, NH; WUVR(AM) Lebanon, NH; WNTK-FM New London, NH; WNTK Newport, NH; WCFR(AM) Springfield, VT.

Robert L. Vinikoor, pres; Sheila E. Vinikoor, VP; Robert L. Vinikoor, gen mgr.

KSPD Inc., (Inspired Family Radio). 1440 S. Weideman Ave., Boise, ID 83709. Phone: (208) 377-3790. Fax: (208) 377-3792. E-mail: info@myfamilyradio.com Web Site: www.myfamilyradio.com. Ownership: Lee Schafer, 50%; Beth A. Schafer, 50%.

Stns: 1 AM. 2 FM. KSPD Boise, ID; KBXL(FM) Caldwell, ID; KDZY-FM McCall, ID.

Lee Schafer, gen mgr; David Schafer, asst mgr.

KSRM Inc., 40960 K. Beach Rd., Kenai, AK 99611. Phone: (907) 283-5811. Fax: (907) 283-9177. E-mail: info@radiokenai.com Web Site: www.radiokenai.com. Ownership: John C. Davis.

Stns: 2 AM. 3 FM. KFSE(FM) Kasilof, AK; KWHQ-FM Kenai, AK; KKIS-FM Soldotna, AK; KSLD(AM) Soldotna, AK; KSRM Soldotna, AK.

John Davis, pres; Cherie Curry, gen mgr.

KUTE Inc., Box 737, Ignacio, CO 81137-0737. Phone: (970) 563-0255. Fax: (970) 563-0399. Ownership: Robert Baker, 14.29%; Bertha Box, 14.29%; Eddie Box Jr., 14.29%; Marvin Cook, 14.29%; Richard Jefferson, 14.29%; Harald Jordan, 14.29%; and Mike Matheson, 14.29%.

Stns: 5 FM. KSUT-FM Ignacio, CO; KUTE-FM Ignacio, CO; KPGS(FM) Pagosa Springs, CO; KUUT(FM) Farmington, NM; KUSW(FM) Flora Vista, NM.

Kaspar Broadcasting Group, 1401 W. Barner St., Frankfort, IN 46041. Phone: (765) 659-3338. Fax: (765) 659-3338. Web Site: www.kasparradio.com.

Stns: 2 AM. 2 FM. WSHW-FM Frankfort, IN; WILO Frankfort, IN; KFAV-FM Warrenton, MO; KWRE(AM) Warrenton, MO.

Vern Kaspar, pres/CEO; Russ Kaspar, VP.

Kemp Communications Inc., 3800 Howard Hughes Pkwy., Wells Fargo Tower, 17th Fl., Las Vegas, NV 89169. Phone: (702) 385-6000. Fax: (702) 385-6001. Ownership: Will Kemp, 100%. Note: Group also is a 49.9% stockholder of KNAN(FM) Nanakuli, HI.

Stns: 4 FM. KMZQ(FM) Payson, AZ; KVGG(FM) Salome, AZ; KVEG(FM) Mesquite, NV; KONV(FM) Overton, NV.

Will Kemp, pres.

Key Broadcasting Inc., Box 1227, Corbin, KY 40702. Phone: (606) 528-8787. Fax: (606) 528-9928. Ownership: Terry E. Forcht.

Stns: 7 AM. 10 FM. WIKK-FM Newton, IL; WVLN Olney, IL; WSEI-FM Olney, IL; WIMC(FM) Crawfordsville, IN; WCDQ(FM) Crawfordsville, IN; WCVL Crawfordsville, IN; WANV(FM) Annville, KY; WAIN Columbia, KY; WAIN-FM Columbia, KY; WHOP(AM) Hopkinsville, KY; WHOP-FM Hopkinsville, KY; WFTG London, KY; WWEL-FM London, KY; WSIP Paintsville, KY; WSIP-FM Paintsville, KY; WXKQ-FM Whitesburg, KY; WTCW Whitesburg, KY.

Terry E. Forcht, pres/CEO.

Keymarket Communications LLC, 100 Ryan Ct., Suite 98, Pittsburgh, PA 15205. Phone: (412) 489-1001. Fax: (412) 279-5500. Web Site: www.froggyland.com.

Stns: 6 AM. 6 FM. WOMP Bellaire, OH; WYJK-FM Bellaire, OH; WOHI East Liverpool, OH; WOGF(FM) East Liverpool, OH; WSTV Steubenville, OH; WASP Brownsville, PA; WFGI(AM) Charleroi, PA; WYJK(AM) Connellsville, PA; WKPL(FM) Ellwood City, PA; WOGG(FM) Oliver, PA; WPKL(FM) Uniontown, PA; WUKL(FM) Bethlehem, WV.

Gerald Getz, pres/CEO.

Kindred Communications Inc., 401 11th St., Suite 200, Huntington, WV 25701. Phone: (304) 523-8401. Fax: (304) 523-4848. Web Site: www.kindredcom.net.

Stns: 2 AM. 2 FM. WCMI Ashland, KY; WDGG-FM Ashland, KY; WRVC-FM Catlettsburg, KY; WRVC Huntington, WV.

Mike Kirtner, pres & gen mgr.

Kirkman Broadcasting Inc., Indigo Executive Park, 60 Markfield Dr., Suite 4, Charleston, SC 29407. Phone: (843) 763-6631. Web Site: www.kirkmanbroadcasting.com.

Stns: 4 AM. WQNT(AM) Charleston, SC; WQSC Charleston, SC; WTMZ Dorchester Terrace-Brentwood, SC; WJKB(AM) Moncks Corner, SC.

Gil Kirkman, pres/CEO.

Knight Broadcasting Inc., 1693 Mission Dr., Solvang, CA 93463. Phone: (805) 688-8386. Fax: (805) 688-2271. Ownership: Sandra C. Knight, 55%; and Shawn T. Knight, 45%.

Stns: 2 AM. 2 FM. KUHL(AM) Lompoc, CA; KINF(AM) Santa Maria, CA; KRAZ-FM Santa Ynez, CA; KSYV-FM Solvang, CA.

Kona Coast Radio LLC, 6807 Foxglove Dr., Cheyenne, WY 82009. Phone: (307) 778-9318. Fax: (307) 632-9349. Ownership: Victor A. Michael Jr.

Stns: 1 AM. 3 FM. KRKY(AM) Granby, CO; KZMV(FM) Kremmling, CO; KRQU(FM) Laramie, WY; KKHI(FM) Rock River, WY.

Victor A. Michael Jr., owner.

Koser Radio Group, P.O. Box 352, Rice Lake, WI 54868. Phone: (715) 234-2131. Fax: (715) 234-6942.

Stns: 2 AM. WAQE Rice Lake, WI; WJMC Rice Lake, WI.

Thomas A. Koser, pres.

Kovas Communications of Indiana Inc., 6349 Constitution Dr., Fort Wayne, IN 46804-7870. Phone: (260) 432-0408. Ownership: Joseph W. Walburn, Personal Rep for Frank S. Kovas Estate.

Stns: 3 AM. WKKD Aurora, IL; WCGO Chicago Heights, IL; WMCW(AM) Harvard, IL.

Joseph W. Walburn, VP; Connie Kovas, pres.

Kuiper Stns, Box 1808, Grand Rapids, MI 49501. Phone: (616) 451-9387. Fax: (616) 451-8460. Ownership: William E. Kuiper Sr.

Stns: 2 AM. WFUR Grand Rapids, MI; WKPR Kalamazoo, MI.

William E. Kuiper Sr., gen mgr & pres.

L

LKCM Radio Group L.P., 115 W. 3rd St., Fort Worth, TX 76102. Phone: (817) 332-0959. Fax: (817) 332-4630. E-mail: gerry@lkcmradio.com Ownership: LKCM Capital Group Inc., gen ptnr, 100% of votes.

Stns: 1 AM. 10 FM. KFSZ(FM) Munds Park, AZ; KVSO(AM) Ardmore, OK; KKAJ-FM Ardmore, OK; KTRX(FM) Dickson, OK; KYBE-FM Frederick, OK; KYNZ(FM) Lone Grove, OK; KRVA-FM Campbell, TX; KTFW-FM Glen Rose, TX; KRVF(FM) Kerens, TX; KSCG(FM) Meridian, TX; KFWR(FM) Mineral Wells, TX.

Gerry Schlegel, pres.

L M Communications Inc., 401 W. Main St., Suite 301, Lexington, KY 40507. Phone: (859) 233-1515. Fax: (859) 233-1517. E-mail: jmac@lmcomm.com Web Site: www.lmcomm.com.

Stns: 4 AM. 8 FM. WBVX(FM) Carlisle, KY; WLXG Lexington, KY; WBTF-FM Midway, KY; WGKS-FM Paris, KY; WCDA-FM Versailles, KY; WYBB-FM Folly Beach, SC; WCOO(FM) Kiawah Island, SC; WMON Montgomery, WV; WKLC-FM Saint Albans, WV; WJYP(AM) Saint Albans, WV; WMXE(FM) South Charleston, WV; WSCW South Charleston, WV.

Lynn Martin, pres; James E. MacFarlane, gen mgr.

La Crosse Radio Group, Box 2017, La Crosse, WI 54602. Phone: (608) 782-8335. Fax: (608) 782-8340. Web Site: www.lacrosseradiogroup.net. Ownership: Howard G. Bill, 45%; TCOM Inc., 45%; and Patrick H. Smith, 10%.

Stns: 1 AM. 4 FM. KQEG-FM La Crescent, MN; WLFN La Crosse, WI; WLXR-FM La Crosse, WI; WQCC-FM La Crosse, WI; WKBH-FM West Salem, WI.

Patrick H. Smith, gen mgr.

La Favorita Inc., Box 746, Austell, GA 30106. Phone: (770) 944-0900. Fax: (770) 944-9794. Web Site: www.radiolafavorita.com. Ownership: Samuel Zamarron, pres; Graciela Zamarron, VP.

Stns: 3 AM. WAOS(AM) Austell, GA; WXEM Buford, GA; WLBA Gainesville, GA.

Samuel Zamarron, pres/CEO; Graciela Zamarron, VP.

La Promesa Foundation, 1406 E. Garden Ln., Midland, TX 79702. Phone: (432) 682-1485. Fax: (432) 682-5230. Web Site: www.grnonline.com. Ownership: La Promesa Foundation is a non-stock, non-profit corporation.

Stns: 2 AM. 3 FM. KJMA(FM) Floresville, TX; KBKN(FM) Lamesa, TX; KBMD(FM) Marble Falls, TX; KJBC(AM) Midland, TX; KWMF(AM) Pleasanton, TX.

La Salle County Broadcasting Corp., 1 Broadcast Lane, Oglesby, IL 61348. Phone: (815) 223-3100. Fax: (815) 223-3095. E-mail: wajk@ivynet.com Web Site: www.wlpo.net. Ownership: Peter Miller.

Stns: 1 AM. 2 FM. WAJK-FM La Salle, IL; WLPO La Salle, IL; WKOT-FM Marseilles, IL.

Peter Miller, pres, owns 95% of Daily News-Tribune Inc., which publishes the *News Tribune*.

Peter Miller, pres; Joyce McCullough, VP/gen mgr; John Spencer, progmg dir; Mark Lippert, sls mgr.

Lake Cities Broadcasting Corp., Box 999, Angola, IN 46703. Phone: (260) 665-9554. Fax: (260) 665-9064. E-mail: wlki@wlki.com Web Site: www.wlki.com. Ownership: Thomas R. Andrews, William Kerner Jr., David Czurak.

Stns: 1 AM. 4 FM. WLKI-FM Angola, IN; WTHD-FM Lagrange, IN; WMSH Sturgis, MI; WMSH-FM Sturgis, MI; WLZZ(FM) Montpelier, OH.

Thomas R. Andrews, pres; William Kerner Jr., VP.

Lake Michigan Broadcasting Inc., 5941 W. U.S. 10, Ludington, MI 49431. Phone: (231) 843-3438. Fax: (231) 843-1886. Web Site: www.wkla.com. Ownership: Lynn S. Baerwolf, 40.125; Scott J. Seeburger, 18%; John J. Hausbeck, 15%.

Stns: 2 AM. 3 FM. WKLA Ludington, MI; WKLA-FM Ludington, MI; WMTE-FM Manistee, MI; WMTE Manistee, MI; WKZC-FM Scottville, MI.

Lynn Barewolf, pres.

Lake Region Radio Works, Box 882, Devils Lake, ND 58301. Phone: (701) 662-7563. Fax: (701) 662-2222. E-mail: kzzyfm@gondtc.com Web Site: www.lrradioworks.com.

Stns: 1 AM. 3 FM. KDLR Devils Lake, ND; KDVL-FM Devils Lake, ND; KQZZ-FM Devils Lake, ND; KZZY(FM) Devils Lake, ND.

Curtis D. Teigen, opns mgr.

Lakes Radio Inc., 524 Ludington, Suite 300, Escanaba, MI 49829. Phone: (906) 789-9700. Fax: (906) 789-9700. Web Site: www.radioresultsnetwork.com.

Stns: 2 AM. 2 FM. WCHT Escanaba, MI; WGKL-FM Gladstone, MI; WCMM-FM Gulliver, MI; WTIQ Manistique, MI.

Rick Duerson, pres.

Lakeshore Media L.L.C., 980 N. Michigan Ave., Suite 1880, Chicago, IL 60611. Phone: (312) 204-9900. Fax: (312) 587-9466. Ownership: Bruce Buzil, 50% votes; and Christopher F. Devine, 50% votes.

Stns: 1 AM. 2 FM. KHIL Willcox, AZ; KWCX-FM Willcox, AZ; KMXQ(FM) Socorro, NM.

Langer Broadcasting Group L.L.C., Box 380699, Murdock, FL 33938-0699. Phone: (508) 820-2430. Ownership: Alexander G. Langer, 100%.

Boston, MA 02114. Langer Broadcasting Corp., 164 Canal St, Suite 450. Phone: (617) 859-9639.

Stns: 4 AM. WSRO(AM) Ashland, MA; WBIX(AM) Natick, MA; WFYL(AM) King of Prussia, PA; WPYT(AM) Wilkinsburg, PA.

Latino Communications LLC, 600 Grant St., Suite 600, Denver, CO 80203. Phone: (303) 733-5266. Fax: (303) 733-5242. E-mail: kbno@kbno.net Web Site: www.kbno.net. Ownership: Alex Cranberg, 49%; Zee Ferrufino, 26%; and Frank Ponce, 25%.

Stns: 3 AM. KBNO(AM) Denver, CO; KXRE(AM) Manitou Springs, CO; KAVA Pueblo, CO.

Zee Ferrufino, pres/CEO.

Lew Latto Group of Northland Radio Stations, 5732 Eagle View Dr., Duluth, MN 55803-9498. Phone: (218) 729-9888. Fax: (218) 729-9888. E-mail: LewLatto@aol.com Ownership: Lew Latto, 100%.

Stns: 1 AM. 2 FM. KGPZ-FM Coleraine, MN; KRBT Eveleth, MN; WEVE-FM Eveleth, MN.

Lew Latto, pres.

Lazer Broadcasting Corp., 200 S. A St., 4th Fl., Oxnard, CA 93030. Phone: (805) 240-2070. Fax: (805) 240-5960. Ownership: Alfredo Plascancia, 100%.

Stns: 4 AM. 13 FM. KXSB-FM Big Bear Lake, CA; KSRT(FM) Cloverdale, CA; KXZM(FM) Felton, CA; KXRS-FM Hemet, CA; KXSM(FM) Hollister, CA; KSRN(FM) Kings Beach, CA; KBTW(FM) Lenwood, CA; KXTT(FM) Maricopa, CA; KLMM(FM) Morro Bay, CA; KOXR Oxnard, CA; KLUN(FM) Paso Robles, CA; KCAL Redlands, CA; KZER(AM) Santa Barbara, CA; KSBQ Santa Maria, CA; KLJR-FM Santa Paula, CA; KEAL(FM) Taft, CA; KJOR(FM) Windsor, CA.

Alfredo Plascencia, CEO; Terry Janisch, gen mgr; Juskiw, VP/rgnl mgr.

Le Sea Broadcasting, Box 12, South Bend, IN 46624. Phone: (574) 291-8200. Fax: (574) 291-9043. E-mail: leseabroadcasting@lesea.com Web Site: www.lesea.com.

Stns: 1 AM. 3 FM. WHPZ-FM Bremen, IN; WHME-FM South Bend, IN; WHPD(FM) Dowagiac, MI; WDOW Dowagiac, MI.

Stns: 8 TV. KWHD, Denver, CO; KWHH, Hilo, HI; KWHE, Honolulu, HI; WHMB-TV, Indianapolis, IN; WHNO, New Orleans, LA; WHME, South Bend-Elkhart, IN; KWHB, Tulsa, OK; KWHM, Wailuku, HI.

Peter Sumrall, pres/CEO.

Legacy Communications Corporation, 210 North 1000 E., Box 1450, St. George, UT 84771-1450. Phone: (435) 628-1000. Fax: (435) 628-6636. E-mail: legacy1@infowest.com Web Site: www.legacy.cc. Ownership: Bear River Trust (E. Morgan Skinner Jr., trustee), 38.52%; Randall Family Trust (Lavon Randall, trustee), 33.07%; all other shareholders (52), less than 5%.

Stns: 7 AM. 1 FM. KACE(AM) Bishop, CA; KPTO(AM) Pocatello, ID; KITT(FM) Soda Springs, ID; KDAN(AM) Beatty, NV; KIFO(AM) Hawthorne, NV; KOGN(AM) Ogden, UT; KENT(AM) Parowan, UT; KNFL(AM) Tremonton, UT.

R. Michael Bull, principal accounting off; Jeffrey B. Bate, dir; E. Morgan Skinner, pres/CEO & dir; Lavon Randall, sec/dir.

Legacy Communications LLC, 2729 Brentwood Blvd., Grand Island, NE 68801. Phone: (308) 381-0206. Ownership: Jay Vavricek, mgng member, 100%.

Stns: 1 AM. 5 FM. KRGY(FM) Aurora, NE; KRGI Grand Island, NE; KRGI-FM Grand Island, NE; KSWN-FM McCook, NE; KZMC(FM) McCook, NE; KIOD(FM) McCook, NE.

Legend Communications L.L.C., 5074 Dorsey Hall Dr., Suite 205, Ellicott City, MD 21042. Phone: (410) 740-0250. Fax: (410) 740-7222. E-mail: larry@patcomm.com Ownership: Larry Patrick, Susan Patrick.

Stns: 5 AM. 10 FM. KDKD Clinton, MO; KDKD-FM Clinton, MO; WJEH Gallipolis, OH; WNTO(FM) Racine, OH; KLGT-FM Buffalo, WY; KBBS Buffalo, WY; KODI Cody, WY; KTAG-FM Cody, WY; KGWY-FM Gillette, WY; KZMQ Greybull, WY; KZMQ-FM Greybull, WY; KCGL(FM) Powell, WY; KZZS(FM) Story, WY; KGCL(FM) Ten Sleep, WY; KDDV-FM Wright, WY.

Larry Patrick, pres; Susan Patrick, exec VP.

Leighton Enterprises Inc., Box 1458, St. Cloud, MN 56302. Phone: (320) 251-1450. Fax: (320) 251-8952. Web Site: www.1047kcld.com. Ownership: Al Leighton.

Stns: 4 AM. 6 FM. KYCK-FM Crookston, MN; KDLM Detroit Lakes, MN; KZLT-FM East Grand Forks, MN; KCNN East Grand Forks, MN; KZPK-FM Paynesville, MN; KBOQ(FM) Pelican Rapids, MN; KNSI Saint Cloud, MN; KCML(FM) Saint Joseph, MN; KNOX Grand Forks, ND; KNOX-FM Grand Forks, ND.

Al Leighton, chmn/CEO; John Sowada, pres; Dennis Niess, VP.

Liberman Broadcasting Inc., 1845 Empire Ave., Burbank, CA 91504. Phone: (818) 729-5300. Fax: (818) 729-5678. E-mail: LBlinfo@lbimedia.com Web Site: www.lbimedia.com. Ownership: Lenard D. Liberman, 47.5-49% votes, 40-42.5% equity; Jose Liberman 2003 Annuity Trust, 23.75-24.5% votes, 20-21.25% equity; Esther Liberman 2003 Annuity Trust, 23.75-24.5% votes, 20-21.25% equity; public shareholders of Liberman Broadcasting Inc., 2-5% votes, 15-20% equity.

Stns: 6 AM. 14 FM. KEBN(FM) Garden Grove, CA; KBUE(FM) Long Beach, CA; KHJ(AM) Los Angeles, CA; KBUA(FM) San Fernando, CA; KWIZ-FM Santa Ana, CA; KTCY(FM) Azle, TX; KXGJ-FM Bay City, TX; KQQK(FM) Beaumont, TX; KBOC(FM) Bridgeport, TX; KJOJ Conroe, TX; KIOX-FM El Campo, TX; KJOJ-FM Freeport, TX; KEYH Houston, TX; KQUE Houston, TX; KNOR(FM) Krum, TX; KZZA(FM) Muenster, TX; KZMP-FM Pilot Point, TX; KTJM-FM Port Arthur, TX; KSEV Tomball, TX; KZMP(AM) University Park, TX.

Stns: 3 TV. KMPX, Dallas-Ft. Worth; KZJL, Houston; KRCA, Los Angeles.

Lenard Liberman, pres; Brett Zane, CEO.

Liggett Communications L.L.C., 808 Huron Ave., Port Huron, MI 48060. Phone: (810) 982-9000. Fax: (810) 987-9380.

Stns: 2 AM. 1 FM. WBTI-FM Lexington, MI; WHLX(AM) Marine City, MI; WPHM Port Huron, MI.

Larry Smith, VP/gen mgr; Robert Liggett, pres.

Lincoln Financial Media, 100 N. Greene St., Greensboro, NC 27420. Phone: (336) 691-3000. Fax: (336) 691-3222. Web Site: www.lincolnfinancialmedia.com. Ownership: Lincoln National Corp., 100%.

Stns: 6 AM. 12 FM. KSOQ-FM Escondido, CA; KIFM-FM San Diego, CA; KBZT(FM) San Diego, CA; KSON San Diego, CA; KSON-FM San Diego, CA; KYGO-FM Denver, CO; KKFN Denver, CO; KQKS-FM Lakewood, CO; KEPN(AM) Lakewood, CO; KJCD(FM) Longmont, CO; WLYF(FM) Miami, FL; WMXJ-FM Pompano Beach, FL; WAXY South Miami, FL; WQXI Atlanta, GA; WSTR-FM Smyrna, GA; WBT Charlotte, NC; WLNK(FM) Charlotte, NC; WBT-FM Chester, SC.

Stns: 3 TV. WCSC, Charleston, SC; WBTV, Charlotte, NC; WWBT, Richmond-Petersburg, VA.

Ed Hull, pres & Lincoln Financial Sports; John Shreves, pres & Lincoln Financial TV; Don Benson, pres & Lincoln Financial Radio.

Linder Broadcasting Group, Box 1420, Mankato, MN 56002. Phone: (507) 345-4537. Fax: (507) 345-5364. Web Site: www.katoinfo.com. Ownership: Donald Linder, John Linder.

Stns: 3 AM. 8 FM. KOWZ-FM Blooming Prairie, MN; KTOE(AM) Mankato, MN; KARZ(FM) Marshall, MN; KMHL Marshall, MN; KXLP(FM) New Ulm, MN; KOLV-FM Olivia, MN; KXAC-FM Saint James, MN; KRRW-FM Saint James, MN; KARL(FM) Tracy, MN; KRUE-FM Waseca, MN; KOWZ(AM) Waseca, MN.

John Linder, pres.

Little Falls Radio Corp., 25801 Nacre St. N.W., St. Francis, MN 55070. Phone: (763) 862-9909. E-mail: rod.grams@att.net Web Site: www.fallsradio.com. Ownership: Rod Grams, 50%; and Chrstina Rae Grams, 50%.

Stns: 1 AM. 2 FM. WYRQ-FM Little Falls, MN; KFML-FM Little Falls, MN; KLTF Little Falls, MN.

J.R. Livesay Group, Box 322, Mattoon, IL 61938-0322. Phone: (217) 234-6464. Fax: (217) 234-6019. E-mail: wlbh@wlbh.com Ownership: J.R. Livesay II owns 50% of WLBH-AM-FM. Shirley L. Herrington owns 35% of WLBH-AM-FM.

Stns: 1 AM. WLBH Mattoon, IL.

J.R. Livesay II, chmn; Shirley L. Herrington, CFO.

Locally Owned Radio LLC, 21361 Hwy. 30, Twin Falls, ID 83301. Phone: (208) 735-8300. Fax: (208) 733-4196. Web Site: www.locallyownedradio.com. Ownership: Porter Hogan Charitable Trust, Jennifer Meeks, trustee; Wendell M. Starke, Lawrence C. Johnson, Stephanie S. Johnson.

Stns: 1 AM. 4 FM. KYUN(FM) Hailey, ID; KTPZ(FM) Hazelton, ID; KIKX-FM Ketchum, ID; KTFI Twin Falls, ID; KIRQ(FM) Twin Falls, ID.

Larry Johnson, pres; Jerry Fender, opns mgr.

Lost Coast Communications Inc., Box 25, Ferndale, CA 95536. Phone: (707) 786-5104. Fax: (707) 786-5100. Ownership: Martin and Peggy Cleary, 36.6%; Blue Lake Rancheria, 21.7%; Patrick Cleary, 16.8%; William Thorington, 9.4%; Cliff Berkowitz, 4.4%; and Trust for Rockey Poole and Phoebe Smith, Rockey Poole, dir, 3%.

Stns: 3 FM. KWPT(FM) Fortuna, CA; KHUM(FM) Garberville, CA; KSLG-FM Hydesville, CA.

Patrick Cleary, pres.

Lotus Communications Corp., 3301 Barham Blvd., Suite 200, Los Angeles, CA 90068. Phone: (323) 512-2225. Fax: (323) 512-2224. E-mail: hq@lotuscorp.com Web Site: www.lotuscorp.com.

Stns: 12 AM. 13 FM. KFMA-FM Green Valley, AZ; KCMT(FM) Oro Valley, AZ; KLPX-FM Tucson, AZ; KTKT Tucson, AZ; KLBN-FM Auberry, CA; KPSL-FM Bakersfield, CA; KWAC Bakersfield, CA; KCHJ Delano, CA; KGST Fresno, CA; KWKW Los Angeles, CA; KMMM(FM) Madera, CA; KIWI(FM) McFarland, CA; KWKU(AM) Pomona, CA; KIRN(AM) Simi Valley, CA; KOMP-FM Las Vegas, NV; KBAD Las Vegas, NV; KENO Las Vegas, NV; KXPT-FM Las Vegas, NV; KWWN(AM) Las Vegas, NV; KDOT-FM Reno, NV; KHIT Reno, NV; KOZZ-FM Reno, NV; KPLY(AM) Reno, NV; KUUB(FM) Sun Valley, NV; KZEP-FM San Antonio, TX.

Howard A. Kalmenson, pres; Jerry Roy, sr VP; Bill Shriftman, sr VP; Lindy Williams, sr VP.

Lovcom Inc., Box 5086, Sheridan, WY 82801. Phone: (307) 672-7421. Fax: (307) 672-2933. E-mail: kimlove@wavecom.net Web Site: www.sheridanmedia.com. Ownership: W.K. Love and family.

Stns: 2 AM. 3 FM. KLQQ(FM) Clearmont, WY; KROE(AM) Sheridan, WY; KWYO Sheridan, WY; KYTI-FM Sheridan, WY; KZWY-FM Sheridan, WY.

W. K. Love, gen mgr.

Barry P. Lunderville Stns, 195 Main St., Lancaster, NH 03584. Phone: (603) 788-3636. Fax: (603) 788-3536. Ownership: Barry P. Lunderville, 100%.

Stns: 3 AM. 2 FM. WMOU Berlin, NH; WRTN(AM) Berlin, NH; WXXS-FM Lancaster, NH; WLTN-FM Lisbon, NH; WLTN Littleton, NH.

M

MAX Media L.L.C., 900 Laskin Rd., Virginia Beach, VA 23451. Phone: (757) 437-9800. Fax: (757) 437-0034. Web Site: www.maxmediallc.com. Ownership: MBG-GG LLC, 42.0345%; MBG Quad-C Investors I Inc., 41.4124%; Aardvarks Also LLC, 6.1967%; Colonnade Max Investors Inc., 4.8671%; Quad-C Max Investors Inc., 4.6799%; MBG Quad-C Investors II Inc., 0.6221%; and Quad-C Max Investors II Inc., 0.1872%.

Stns: 12 AM. 23 FM. KVLD(FM) Atkins, AR; KCAB Dardanelle, AR; KVOM Morrilton, AR; KVOM-FM Morrilton, AR; KWKK-FM Russellville, AR; WCIL Carbondale, IL; WUEZ(FM) Carterville, IL; WXLT(FM) Christopher, IL; WOOZ-FM Harrisburg, IL; WJPF Herrin, IL; KZIM Cape Girardeau, MO; KEZS-FM Cape Girardeau, MO; KGIR Cape Girardeau, MO; KCGQ-FM Gordonville, MO; KLSC(FM) Malden, MO; KMAL(AM) Malden, MO; KWOC Poplar Bluff, MO; KJEZ-FM Poplar Bluff, MO; KKLR-FM Poplar Bluff, MO; KGKS-FM Scott City, MO; KSIM(AM) Sikeston, MO; WQDK-FM Ahoskie, NC; WGAI(AM) Elizabeth City, NC; WCMS-FM Hatteras, NC; WCXL(FM) Kill Devil Hills, NC; WFYY(FM) Bloomsburg, PA; WYGL-FM Elizabethville, PA; WWBE-FM Mifflinburg, PA; WLGL-FM Riverside, PA; WYGL Selinsgrove, PA; WCMS(AM) Newport News, VA; WGH-FM Newport News, VA; WXMM(FM) Norfolk, VA; WVBW(FM) Suffolk, VA; WXEZ-FM Yorktown, VA.

Stns: 10 TV. WMEI, Arecibo, PR; KULR-TV, Billings, MT; WNKY, Bowling Green, KY; KWYB, Butte-Bozeman, MT; WVIF, Christiansted, VI; KTMF, Missoula, MT; WPFO, Portland-Auburn, ME; WGTQ, Traverse City-Cadillac, MI; WGTU, Traverse City-Cadillac, MI; KYTX, Tyler-Longview (Lufkin & Nacogdoches), TX.

John A. Trinder, pres.

M&M Broadcasters Ltd., Box 1629, c/o KTFW(FM), Cleburne, TX 76033. Phone: (817) 572-4400. Fax: (817) 572-1915. Web Site: www.countrygoldradio.com. Ownership: Gary Moss, 80%; George Mazti, 20%.

Stns: 4 AM. KEOR(AM) Catoosa, OK; KHFX(AM) Burleson, TX; KCLE(AM) Cleburne, TX; KJSA Mineral Wells, TX.

Gary Moss, pres.

MBC Grand Broadcasting Inc., 1360 E. Sherwood Dr., Grand Junction, CO 81501. Phone: (970) 254-2100. Fax: (970) 245-7551. Web Site: www.gjradio.com. Ownership: Richard C. Dean.

Stns: 2 AM. 4 FM. KJYE-FM Grand Junction, CO; KNZZ(AM) Grand Junction, CO; KTMM(AM) Grand Junction, CO; KMGJ(FM) Grand Junction, CO; KMOZ-FM Grand Junction, CO; KSTR-FM Montrose, CO.

Richard C. Dean, pres.

M.B. Communications, 481 Hamilton St., Geneva, NY 14456. Phone: (315) 781-1101. Phone: (315) 536-0850. Fax: (315) 781-6666. Fax: (315) 536-3299. E-mail: k1017@fltg.net Web Site: www.k1017.com. Ownership: Russ Kimble, 100%.

Stns: 1 AM. 1 FM. WFLK-FM Geneva, NY; WYLF(AM) Penn Yan, NY.

Russell Kimble, pres; Deborah Kimble, VP.

M.R.S. Ventures Inc., 100 E. Ferguson, Suite 614, Tyler, TX 75702. Phone: (903) 595-4795. Fax: (903) 593-2666. E-mail: jdonrussell@aol.com Ownership: Jerry D. Russell, 100%.

Stns: 2 AM. 6 FM. KRKD(FM) Dermott, AR; KZYQ-FM Lake Village, AR; KCLA Pine Bluff, AR; KOTN Pine Bluff, AR; KPBQ-FM Pine Bluff, AR; WDTL-FM Cleveland, MS; WRKG(FM) Drew, MS; WZYQ-FM Mound Bayou, MS.

MTD Inc., Box 2010, Ruidoso Downs, NM 88346. Phone: (505) 258-9922. Fax: (505) 258-2363. E-mail: kruikwmw@trailnet.com Web Site: www.ruidoso.net/krui. Ownership: R.D. Hubbard, 75%; Mike Warren, 25%.

Stns: 1 AM. 4 FM. KNMB(FM) Cloudcroft, NM; KWMW-FM Maljamar, NM; KIDX(FM) Ruidoso, NM; KRUI Ruidoso Downs, NM; KTUM(FM) Tatum, NM.

Bruce Rimbo, pres; Timothy Keithley, gen mgr.

MTS Broadcasting, Box 237, Cambridge, MD 21613. Phone: (410) 228-4800. Fax: (410) 228-0130. E-mail: theheat@intercom.net Web Site: www.mtslive.com.

Stns: 1 AM. 2 FM. WCEM Cambridge, MD; WTDK-FM Federalsburg, MD; WAAI(FM) Hurlock, MD.

Thomas C. Mulitz, pres/CEO.

MacDonald Broadcasting Co., Box 1776, Saginaw, MI 48605. Phone: (989) 752-8161. Fax: (989) 752-8102. E-mail: wkcq@chartermi.net Web Site: www.98fmkcq.com. Ownership: Ken MacDonald Jr. Note: Group also owns and operates a Muzak franchise in a six-county area in mid-Michigan.

Stns: 3 AM. 5 FM. WQHH-FM Dewitt, MI; WXLA Dimondale, MI; WMJO(FM) Essexville, MI; WHZZ-FM Lansing, MI; WILS Lansing, MI; WSAG(FM) Linwood, MI; WSAM Saginaw, MI; WKCQ-FM Saginaw, MI.

Kenneth MacDonald Jr., CEO; Duane Alverson, pres.

MacDonald Garber Broadcasting Co., 2095 U.S. 131 S., Petoskey, MI 49770. Phone: (231) 347-8713. Fax: (231) 347-8782. Web Site: www.lite96.com.

Stns: 2 AM. 2 FM. WATT Cadillac, MI; WLXV(FM) Cadillac, MI; WKHQ-FM Charlevoix, MI; WMBN(AM) Petoskey, MI.

Trish MacDonold Garber, pres.

Magic Broadcasting LLC, 7106 Laird St., Suite 102, Panama City Beach, FL 32408. Phone: (850) 234-8388. Fax: (850) 230-6988. Web Site: magicbroadcasting.net. Ownership: Magic Management Co. LLC, mgng member; Donald G. McCoy, 29.57%; Stephen A. Bodzin, trustee of Anne S. Reich 1984 Revocable Trust, 21.13%; Kim Styles DiBacco, 7.75%; Thomas A. DiBacco, 7.75%.

Stns: 1 AM. 12 FM. WTVY-FM Dothan, AL; WKMX-FM Enterprise, AL; WLDA(FM) Fort Rucker, AL; WJRL-FM Ozark, AL; KWIE(FM) Ontario, CA; KDAY(FM) Redondo Beach, CA; KRQB(FM) San Jacinto, CA; WYYX-FM Bonifay, FL; WILN-FM Panama City, FL; WPCF(AM) Panama City Beach, FL; WVVE(FM) Panama City Beach, FL; WYOO-FM Springfield, FL; WBBK-FM Blakely, GA.

Kim Styles, gen mgr; Thomas DiBacco, ptnr.

Magnum Broadcasting Inc., Box 436, State College, PA 16804. Phone: (814) 272-1320. Fax: (814) 272-3291. Web Site: www.1059joefm.com. Ownership: Michael M. Stapleford, 100%.

Stns: 2 AM. 2 FM. WBLF Bellefonte, PA; WJOW(FM) Philipsburg, PA; WPHB Philipsburg, PA; WZYY-FM Renovo, PA.

Diana Albright, gen mgr.

Magnum Radio Inc., 1021 N. Superior Ave., Suite 5, Tomah, WI 54660. Phone: (608) 372-9600. Fax: (608) 372-7566. E-mail: magnumradio@charter.net Ownership: David R. Magnum, 87.91%. Note: Sister corporation Magnum Communications Inc. owns WBKY(FM) Portage and WDLS(AM)-WNNO-FM Wisconsin Dells, both WI. Sister corporation Magnum Broadcasting Inc. owns WAUN-FM Kewaunee and WSRG(FM) Sturgeon Bay, both WI.

Stns: 1 AM. 2 FM. WXYM(FM) Tomah, WI; WBOG(AM) Tomah, WI; WTMB(FM) Tomah, WI.

Dave Magnum, pres.

Mahaffey Enterprises Inc., Box 4584, Springfield, MO 65808. Phone: (417) 883-9180. Fax: (417) 883-9096. Ownership: John B. Mahaffey, Fredna B. Mahaffey, Robert B. Mahaffey.

Stns: 3 AM. 8 FM. KGGF Coffeyville, KS; KKRK(FM) Coffeyville, KS; KUSN(FM) Dearing, KS; KGGF-FM Fredonia, KS; KDAA(FM) Rolla, MO; KTTR Rolla, MO; KZNN-FM Rolla, MO; KTTR-FM Saint James, MO; KSPI Stillwater, OK; KGFY-FM Stillwater, OK; KVRO-FM Stillwater, OK.

John B. Mahaffey, chmn; Robert B. Mahaffey, pres/CEO.

Main Line Broadcasting LLC, 300 Conshohocken State Rd., Suite 380, West Conshohocken, PA 19428-3801. Phone: (610) 825-8101. Fax: (610) 825-8106. Ownership: Arlington Capital Partners II L.P., 100%.

Stns: 2 AM. 7 FM. WCHA Chambersburg, PA; WIKZ-FM Chambersburg, PA; WQCM(FM) Greencastle, PA; WLFV(FM) Ettrick, VA; WWLB(FM) Midlothian, VA; WARV-FM Petersburg, VA; WBBT-FM Powhatan, VA; WDLD(FM) Halfway, MD; WHAG Halfway, MD.

Daniel Savadove, CEO; Marc Guralnick, exec VP sls; J. Edwin Conrad, exec VP/CFO.

Malkan Broadcast Associates, Box 9757, Corpus Christi, TX 78469. Phone: (361) 883-3516. Fax: (361) 882-9767. E-mail: thechief@star94.net Ownership: Malkan Broadcasting Management L.L.C, gen ptnr; Matthew Malkan; Hope Malkan; Glen Powers.

Stns: 1 AM. 2 FM. KEYS Corpus Christi, TX; KZFM(FM) Corpus Christi, TX; KKBA(FM) Kingsville, TX.

Glen Powers, pres.

Mapleton Communications LLC, 10900 Wilshire Blvd., Suite 1500, Los Angeles, CA 90024. Phone: (310) 209-7221. Fax: (310) 209-7239.

Stns: 8 AM. 23 FM. KBRE(FM) Atwater, CA; KRRX-FM Burney, CA; KPYG(FM) Cambria, CA; KBOQ(FM) Carmel, CA; KCDU(FM) Carmel, CA; KFMF-FM Chico, CA; KQPT(FM) Colusa, CA; KBLO(FM) Corcoran, CA; KPIG-FM Freedom, CA; KHIP(FM) Gonzales, CA; KNAH(FM) Merced, CA; KTIQ(AM) Merced, CA; KYOS Merced, CA; KZAP-FM Paradise, CA; KPIG(AM) Piedmont, CA; KXTZ-FM Pismo Beach, CA; KALF(FM) Red Bluff, CA; KNRO(AM) Redding, CA; KSHA-FM Redding, CA; KQMS Redding, CA; KYNS(AM) San Luis Obispo, CA; KMBY-FM Seaside, CA; KNNN(FM) Shasta Lake City, CA; KRDG-FM Shingletown, CA; KXDZ(FM) Templeton, CA; KLOQ-FM Winton, CA; KTMT Ashland, OR; KCMX-FM Ashland, OR; KBOY-FM Medford, OR; KCMX Phoenix, OR; KAKT(FM) Phoenix, OR.

Adam Nathanson, pres; Raul Salvador, VP finance; Mike Anthony, progmg VP; Dale Hendry, VP opns.

Maritime Broadcasting, 226 Union St., Saint John, NB E2L 1B1. Canada. Phone: (506) 658-2330. Fax: (506) 658-5116. E-mail: mlee@nb.aibn.com Web Site: www.mbsradio.com.

Stns: 8 AM. 15 FM. CKNB Campbellton, NB; CFAN-FM Miramichi City, NB; CHOY-FM Moncton, NB; CKCW-FM Moncton, NB; CFQM-FM Moncton, NB; CFBC Saint John, NB; CIOK-FM Saint John, NB; CJYC-FM Saint John, NB; CJCW Sussex, NB; CKDH Amherst, NS; CKDY Digby, NS; CHFX-FM Halifax, NS; CHNS-FM Halifax, NS; CKWM-FM Kentville, NS; CKEN-FM Kentville, NS; CKAD Middleton, NS; CJCB(AM) Sydney, NS; CKPE-FM Sydney, NS; CHER-FM Sydney, NS; CFAB Windsor, NS; CHLQ-FM Charlottetown, PE; CFCY-FM Charlottetown, PE; CJRW-FM Summerside, PE.

Dave Clarkson, gen sls mgr; Robert Pace, CEO.

MarMac Communications LLC, 7515 Blythe Island Hwy., Brunswick, GA 31523. Phone: (912) 264-6251. Fax: (912) 264-9991. Ownership: Gary P. Marmitt, 50%; and Sharon McKeand, 50%.

Stns: 4 AM. WFNS(AM) Blackshear, GA; WSFN Brunswick, GA; WSEG(AM) Savannah, GA; WWGA(AM) Waycross, GA.

Mars Hill Network, 4044 Makyes Rd., Syracuse, NY 13215. Phone: (315) 469-5051. Fax: (315) 469-4066. E-mail: mhn@marshillnetwork.org Web Site: www.marshillnetwork.org. Ownership: Not-for-profit corporation. Note: Group also has a radio net, 14 translators.

Stns: 4 FM. WMHI-FM Cape Vincent, NY; WMHQ(FM) Malone, NY; WMHR-FM Syracuse, NY; WMHN-FM Webster, NY.

Clayton Roberts, pres; Michael Gettman, VP; Wayne Taylor, gen mgr; Jim Stewart, sec; John Seeland, treas.

Martin Broadcasting Inc., 4638 Decker Dr., Baytown, TX 77520. Phone: (210) 333-0050. Fax: (210) 333-0081. E-mail: kchl1480@yahoo.com

Stns: 6 AM. WLVV Mobile, AL; KZZB Beaumont, TX; KYOK(AM) Conroe, TX; KRMY Killeen, TX; KCHL San Antonio, TX; KANI Wharton, TX.

Darrell E. Martin, pres.

Martz Communications Group, 955 S. Virginia St., Reno, NV 89502. Phone: (415) 359-1030. Fax: (415) 359-1050. Ownership: Timothy D. Martz, 100%.

Stns: 2 AM. 7 FM. WNCQ-FM Canton, NY; WRCD-FM Canton, NY; WYUL-FM Chateaugay, NY; WVNV-FM Malone, NY; WICY Malone, NY; WMSA Massena, NY; WYSX(FM) Morristown, NY; WVLF(FM) Norwood, NY; WPAC(FM) Ogdensburg, NY.

Timothy D. Martz, pres/CEO.

Matinee Radio LLC, 5842 Westslope Dr., Austin, TX 78731. Phone: (512) 467-0643. Ownership: Matinee Broadcasting Corp., 51%; Mark Eckenrode, 12.25%; Katy Gaffney, 12.25%; William M. Smith, 12.25%; and Robert Walker, 12.25%.

Stns: 4 FM. KANM(FM) Magdalena, NM; KNOS(FM) Albany, TX; KTXO(FM) Goldsmith, TX; KKUL-FM Groveton, TX.

Maverick Media LLC, 136 Main St., Suite 202, Westport, CT 06880. Phone: (203) 227-2800. Fax: (203) 227-4819.

Stns: 5 AM. 14 FM. KVRV(FM) Monte Rio, CA; KMHX(FM) Rohnert Park, CA; KSRO Santa Rosa, CA; KXFX-FM Santa Rosa, CA; WXRX-FM Belvidere, IL; WNTA Rockford, IL; WGFB(FM) Rockton, IL; WRTB(FM) Winnebago, IL; WDOH-FM Delphos, OH; WEGE(FM) Lima, OH; WZOQ(AM) Lima, OH; WFGF(FM) Lima, OH; WWSR(FM) Wapakoneta, OH; WEAQ Chippewa Falls, WI; WDRK(FM) Cornell, WI; WIAL-FM Eau Claire, WI; WAXX-FM Eau Claire, WI; WAYY(AM) Eau Claire, WI; WECL-FM Elk Mound, WI.

Gary S. Rozynek, pres/CEO.

William W. McCutchen III Stns, 1126 West Ave., Richmond, VA 23220. Phone: (804) 422-3452. Ownership: William W. McCutchen III, 100%.

Stns: 3 FM. KMDR(FM) McKinleyville, CA; KDRW(FM) Hewitt, TX; KQDR(FM) Savoy, TX.

McGraw/Elliott Group Stations, 228 Randolph Ave., Elkins, WV 26241. Phone: (304) 636-8800. Fax: (304) 636-8801. E-mail: rmegfrontoffice@verizon.net Ownership: Richard H. McGraw, 50%; Karen G. McGraw, 50%. Cable TV.

Stns: 1 AM. 3 FM. WBRB(FM) Buckhannon, WV; WBTQ(FM) Buckhannon, WV; WBUC Buckhannon, WV; WELK-FM Elkins, WV.

Richard H. McGraw, CEO.

McKenzie River Broadcasting Company, Inc., 925 Country Club Rd., Suite 200, Eugene, OR 97401. Phone: (541) 484-9400. Fax: (541) 344-9424. Ownership: Renate R. Tilson, 50.71%; John Q. Tilson III, 49.29%.

Stns: 3 FM. KMGE-FM Eugene, OR; KKNU(FM) Springfield-Eugene, OR; KEUG(FM) Veneta, OR.

John Q. Tilson III, pres; Renate R. Tilson, VP.

McMurray Communications Inc., 3335 W. 8th St., Safford, AZ 85546. Phone: (928) 428-1230. Fax: (928) 428-1311. E-mail: traffic@eaznet.com Web Site: www.mysouthernaz.com.

Stns: 1 AM. 1 FM. KWRQ-FM Clifton, AZ; KATO Safford, AZ.

Harry S. McMurray, pres/CEO; David Nathan, gen mgr.

McNaughton-Jakle Stations, 14 Douglas Avenue, Elgin, IL 60120. Phone: (847) 741-7700. Fax: (847) 468-0000. E-mail: mail@wrmn.com Web Site: www.wrmn1410.com. Ownership: Bradley L. Beesley, K. Richard Jakle

Stns: 3 AM. WBIG Aurora, IL; WRMN Elgin, IL; KSHP North Las Vegas, NV.

(Joseph E. McNaughton & family) identified with *The Davis Enterprise, Fairfield Republic* & the *Placerville Mountain Democrat*, all CA.

K. Richard Jakle, chmn/pres.

Media Logic LLC, Box 430, Fort Morgan, CO 80701-0430. Phone: (970) 867-5674. Fax: (970) 542-1023. Ownership: Wayne Johnson, 90%; and Richard Lindsey, 10%.

Stns: 2 AM. 2 FM. KFTM Fort Morgan, CO; KATR-FM Otis, CO; KSRX(FM) Sterling, CO; KRDZ Wray, CO.

Wayne Johnson, gen mgr.

Media One Group, 147 Bell St., Suite 200, Chagrin Falls, OH 44022. Phone: (440) 893-8114.

Stns: 2 AM. 3 FM. WWSE-FM Jamestown, NY; WHUG(FM) Jamestown, NY; WJTN Jamestown, NY; WKSN Jamestown, NY; WQFX-FM Russell, PA.

Media Power Group Inc., 100 Gran Bulevar Paseos, Suite 403A, San Juan, PR 00926. Phone: (787) 292-1700. Fax: (787) 292-1717. E-mail: wskn1320@yahoo.com Ownership: PR Grupo Radio Nacional Inc., 25%; Jose Enrique Fernandez, 25%; Arturo Diaz Jr., 25%; and Empresas Bechara Inc., 25%.

Stns: 4 AM. WLEY Cayey, PR; WDEP(AM) Ponce, PR; WSKN(AM) San Juan, PR; WKFE Yauco, PR.

Eduardo Albino Rivero, pres; Ismael Nieves, VP.

Mega Communications LLC, 255 Executive Dr. #409, Plainview, NY 11803. Phone: (516) 349-4936. Fax: (516) 349-4935. E-mail: eschreiber@lindcap.com Web Site: megastations.net. Ownership: Adam Lindemann

Stns: 2 AM. 1 FM. WLCC Brandon, FL; WMGG(AM) Largo, FL; WNUE-FM Titusville, FL.

Eran Schreiber, CFO.

Melia Communications Inc., Box 569, Goodland, KS 67735. Phone: (785) 899-2309. Fax: (785) 899-3062. E-mail: kloe@eaglecom.net Web Site: www.kloe.com. Ownership: Martin K. Melia, 50%; Kathleen J. Melia, 50%.

Stns: 1 AM. 1 FM. KWGB-FM Colby, KS; KLOE Goodland, KS.

Kathleen J. Melia, VP; Martin K. Melia, gen mgr & pres.

Mentor Partners Inc., 18720 16 Mile Rd., Big Rapids, MI 49307. Phone: (231) 796-7000. Fax: (231) 796-7951. Ownership: Jeffrey Scarpelli, 100%.

Stns: 1 AM. 2 FM. WYBR-FM Big Rapids, MI; WBRN Big Rapids, MI; WWBR(FM) Big Rapids, MI.

Meredith Broadcasting Group, Meredith Corp., 1716 Locust St., Des Moines, IA 50309-3023. Phone: (515) 284-2159. Fax: (515) 284-2514. Web Site: www.meredith.com. Ownership: Meredith Broadcasting is an operating group of Meredith Corp., Des Moines, IA.

Stns: 1 AM. WNEM(AM) Bridgeport, MI.

Stns: 12 TV. WGCL-TV, Atlanta; WFLI, Chattanooga, TN; WNEM-TV, Flint-Saginaw-Bay City, MI; WHNS, Greenville-Spartanburg, SC-Asheville, NC-Anderson, SC; WFSB, Hartford & New Haven, CT; KSMO-TV, Kansas City, MO; KCTV, Kansas City, MO; KVVU, Las Vegas, NV; WSMV, Nashville, TN; KPHO-TV, Phoenix (Prescott), AZ; KPTV, Portland, OR; KPDX, Portland, OR.

The publishing group includes:

Magazines: *American Baby, American Patchwork & Quilting, Better Homes & Gardens, Country Home, Country Home Country Gardens, Creative Home, Decorating, Do It Yourself, Garden, Deck, and Landscape, Garden Shed, Ladies' Home Journal, Midwest Living, MORE, Renovation Style, Successful Farming, Traditional Home*, and *Wood*, along with more than 170 special interest titles.

Paul Karpowic, pres; Douglas Lowe, exec VP.

Meridian Broadcasting Inc., 2824 Palm Beach Blvd., Fort Myers, FL 33916. Phone: (239) 337-2346. Fax: (239) 332-0767. Ownership: Joseph C. Schwartzel.

Stns: 1 AM. 3 FM. WUSV(FM) Estero, FL; WTLT-FM Naples, FL; WARO-FM Naples, FL; WNOG Naples, FL.

Joseph C. Schwartzel, pres.

Metropolitan Radio Group Inc., 318 E. Pershing St., Springfield, MO 65806. Phone: (417) 862-0852. Fax: (417) 862-9079. Ownership: Gary L. Acker.

Stns: 6 AM. 5 FM. KGHT Sheridan, AR; WBRD Palmetto, FL; WRXB Saint Petersburg Beach, FL; WTMY Sarasota, FL; KJVC-FM Mansfield, LA; KORI(FM) Mansfield, LA; KIOU Shreveport, LA; KTKC-FM Springhill, LA; KUNQ-FM Houston, MO; KIJN Farwell, TX; KIJN-FM Farwell, TX.

Mark L. Acker, pres.

Meyer Communications Inc., Box 3676, Springfield, MO 65808. Phone: (417) 862-3990. Fax: (417) 869-7675. Web Site: www.ktxrfm.com. E-mail: manager@radiospringfield.com Ownership: Kenneth E. Meyer, 100%.

Stns: 2 AM. 3 FM. KBFL-FM Buffalo, MO; KBFL(AM) Springfield, MO; KTXR-FM Springfield, MO; KWTO Springfield, MO; KWTO-FM Springfield, MO.

Kenneth E. Meyer, pres.

Michael Radio Group, 1063 Big Thompson Rd., Apt. F, Loveland, CO 80537-9424. Phone: (307) 778-9318.

Stns: 1 AM. 3 FM. KYOY(FM) Kimball, NE; KIMB Kimball, NE; KGRK(FM) Glenrock, WY; KRKI(FM) Newcastle, WY.

Mid Atlantic Network, Box 3300, Winchester, VA 22604. Phone: (540) 667-2224. Fax: (540) 722-3295. Ownership: John P. Lewis, David P. Lewis, Howard P. Lewis.

Stns: 2 AM. 2 FM. WWRE(FM) Berryville, VA; WFVA(AM) Fredericksburg, VA; WWRT(FM) Strasburg, VA; WINC Winchester, VA.

John P. Lewis, pres.

Mid-America Radio Group Inc., Box 1970, Martinsville, IN 46151. Phone: (765) 349-1485. Fax: (765) 342-3569. E-mail: mid-americaradio@scican.net Ownership: David Keister, principal owner.

Stns: 5 AM. 9 FM. WIOU Kokomo, IN; WMRI(AM) Marion, IN; WXXC(FM) Marion, IN; WBAT Marion, IN; WCBK-FM Martinsville, IN; WMYJ(AM) Martinsville, IN; WVNI-FM Nashville, IN; WARU Peru, IN; WMYK(FM) Peru, IN; WARU-FM Roann, IN; WHZR-FM Royal Center, IN; WCLS(FM) Spencer, IN; WCJC-FM Van Buren, IN; WJOT-FM Wabash, IN.

David C. Keister, pres.

Midwest Communications Inc., Box 23333, Green Bay, WI 54305. Phone: (414) 435-3771. Fax: (414) 455-1155. Ownership: D.E. Wright, 100%.

Stns: 16 AM. 21 FM. WPRS Paris, IL; WINH(FM) Paris, IL; WWSY(FM) Seelyville, IN; WMGI-FM Terre Haute, IN; WTVB(AM) Coldwater, MI; WNWN-FM Coldwater, MI; WHTC(AM) Holland, MI; WKZO(AM) Kalamazoo, MI; WQLR(AM) Kalamazoo, MI; WVFM(FM) Kalamazoo, MI; WNWN Portage, MI; WYVN(FM) Saugatuck, MI; KDAL Duluth, MN; KTCO-FM Duluth, MN; WMFG Hibbing, MN; WNMT Nashwauk, MN; KMFG(FM) Nashwauk, MN; WUSZ-FM Virginia, MN; WNFL Green Bay, WI; WTAQ(AM) Green Bay, WI; WIXX-FM Green Bay, WI; WOFM-FM Mosinee, WI; WNCY-FM Neenah-Menasha, WI; WROE-FM Neenah-Menasha, WI; WOZZ-FM New London, WI; WXER(FM) Plymouth, WI; WIZD-FM Rudolph, WI; WRIG Schofield, WI; WHBL(AM) Sheboygan, WI; WBFM(FM) Sheboygan, WI; WHBZ(FM) Sheboygan Falls, WI; WZBY(FM) Sturgeon Bay, WI; WDSM Superior, WI; WGEE(AM) Superior, WI; KRBR-FM Superior, WI; WSAU Wausau, WI; WDEZ-FM Wausau, WI.

D.E. Wright, pres.

The Mid-West Family Broadcast Group, Box 44408, Madison, WI 53744. Phone: (608) 273-1000. Fax: (608) 273-3588. E-mail: tom.walker@mwfbg.net Web Site: www.midwestfamilybroadcasting.com. Ownership: Philip Fisher, Richard T. Record, Thomas A. Walker.

Stns: 8 AM. 17 FM. WLCE(FM) Petersburg, IL; WMAY(AM) Springfield, IL; WCNF-FM Benton Harbor, MI; WYTZ-FM Bridgman, MI; WHIT-FM Hartford, MI; WHIT(AM) Hudsonville, MI; WSJM Saint Joseph, MI; WIRX-FM Saint Joseph, MI; WCSY-FM South Haven, MI; KCLH(FM) Caledonia, MN; KQYB-FM Spring Grove, MN; KQRA(FM) Brookline, MO; KKLH(FM) Marshfield, MO; KOMG(FM) Ozark, MO; KOSP(FM) Willard, MO; WHLK(FM) De Forest, WI; WIZM La Crosse, WI; WIZM-FM La Crosse, WI; WKTY(AM) La Crosse, WI; WRQT(FM) La Crosse, WI; WLMV(AM) Madison, WI; WTDY Madison, WI; WTUX(AM) Madison, WI; WWQM-FM Middleton, WI; WJQM(FM) Mount Horeb, WI.

Thomas A. Walker, pres; Richard T. Record, dir; Jolene K. Neis, CFO.

Midwestern Broadcasting Co., 314 E. Front St., Traverse City, MI 49684. Phone: (231) 947-7675. Fax: (231) 929-3988. Web Site: www.wtcmi.com. Ownership: Ross Biederman, 52.5%; William Kiker Estate, 16.25%; William McClay, 15%.

Stns: 3 AM. 5 FM. WATZ Alpena, MI; WATZ-FM Alpena, MI; WBCM(FM) Boyne City, MI; WJZQ(FM) Cadillac, MI; WCZW(FM) Charlevoix, MI; WRGZ(FM) Rogers City, MI; WTCM(AM) Traverse City, MI; WCCW(AM) Traverse City, MI.

Ross Biederman, pres.

Millcreek Broadcasting L.L.C., 980 N. Michigan Ave., Suite 1880, Chicago, IL 60611. Phone: (312) 204-9900. Fax: (312) 587-9466. Ownership: Alta Communications VII LP, 79% (percentage of total assets).

Stns: 4 FM. KAUU(FM) Manti, UT; KUDE(FM) Nephi, UT; KUDD(FM) Roy, UT; KUUU(FM) South Jordan, UT.

Christopher Devine, pres; Bruce Buzil, exec VP.

Millennium Radio Group LLC, 220 Northpointe Pkwy., Suite D, Amherst, NY 14228. Phone: (716) 639-9300. Fax: (719) 639-8782. Ownership: UBS Capital Americas; Alta Communications; Mercy Capital Partners LP.

Stns: 4 AM. 8 FM. WADB Asbury Park, NJ; WJLK(FM) Asbury Park, NJ; WPUR-FM Atlantic City, NJ; WENJ(AM) Atlantic City, NJ; WFPG(FM) Atlantic City, NJ; WSJO(FM) Egg Harbor City, NJ; WOBM(AM) Lakewood, NJ; WCHR-FM Manahawkin, NJ; WXKW(FM) Millville, NJ; WOBM-FM Toms River, NJ; WBUD Trenton, NJ; WKXW(FM) Trenton, NJ.

Charles W. Banta, chmn; James Donahoe, pres/CEO.

Miller Communications Inc., Box 1269, Sumter, SC 29151. Phone: (803) 775-2321. Fax: (803) 773-4856. Web Site: www.miller.fm. Ownership: Frank H. Avent, 31.33%; William Duncan, 24.37%; Harold T. Miller Jr., 20.89%; Theresa Miller, 20.89%; and David Baker, 2.5%.

Stns: 2 AM. 8 FM. WWBD-FM Bamberg, SC; WGFG-FM Branchville, SC; WWKT-FM Kingstree, SC; WDKD Kingstree, SC; WSIM(FM) Lamar, SC; WQKI-FM Orangeburg, SC; WIGL(FM) Saint Matthews, SC; WDXY Sumter, SC; WICI(FM) Sumter, SC; WIBZ-FM Wedgefield, SC.

Harold T. Miller Jr., pres/CEO; Dave Baker, VP; Theresa Miller, VP/gen mgr.

Miller Media Group, Box 169, 918 East Park, Taylorville, IL 62568-0169. Phone: (217) 824-3395. Fax: (217) 824-3301. Web Site: www.randyradio.com. Ownership: Randal J. Miller owns 100% of WJRE(FM), WGEN(AM) & WKEI(AM)-WYEC(FM), WYEC(FM) & WGEN(AM), 70% of WMKR(FM) & WTIM-FM, & 100% of WRAN(FM). Lawrence Travis owns 30% of WMKR(FM) & WTIM-FM. Note: Group also owns the Hometown Illinois Radio Network, an ad-hoc network of 30 radio stns that bcsts reports from the Illinois State Fair, Illinois Farm Bureau annual meeting & Commodity Classic.

Stns: 2 AM. 5 FM. WJRE(FM) Galva, IL; WGEN Geneseo, IL; WYEC(FM) Kewanee, IL; WKEI(AM) Kewanee, IL; WMKR(FM) Pana, IL; WTIM-FM Taylorville, IL; WRAN(FM) Tower Hill, IL.

Randal J. Miller, pres; Cathaleen R. Miller, sec/treas.

Milner Broadcasting, 292 N. Convent, Bourbonnais, IL 60914. Phone: (815) 933-9287. Fax: (815) 933-8696.

Stns: 3 FM. WFAV(FM) Gilman, IL; WVLI-FM Kankakee, IL; WIVR(FM) Kentland, IN.

Milwaukee Radio Alliance L.L.C., N72 W12922 Good Hope Rd., Menomonee Falls, WI 53051-4441. Phone: (414) 771-1021. Fax: (414) 771-3036. E-mail: bhurwitz@milwaukeeradio.com Web Site: www.milwaukeeradio.com. Ownership: All Pro Broadcasting, 50%; Shamrock Communications, 50%.

Stns: 1 AM. 2 FM. WMCS Greenfield, WI; WLDB(FM) Milwaukee, WI; WLUM-FM Milwaukee, WI.

Willie Davis, chmn; William Lynette, pres.

Minn-Iowa Christian Broadcasting Inc., Box 72, Blue Earth, MN 56013. Phone: (507) 526-3233. Fax: (507) 526-3235. E-mail: kjly@kjly.com Web Site: www.kjly.com.

Stns: 4 FM. KJYL-FM Eagle Grove, IA; KJCY(FM) Saint Ansgar, IA; KJIA(FM) Spirit Lake, IA; KJLY-FM Blue Earth, MN.

Matt Dorfner, exec dir.

Mission Nebraska Inc., Box 30345, Lincoln, NE 68503-0345. Phone: (402) 477-1090. Ownership: Ron Brown, 16.67%; Stanley A. Parker, 16.67%; David Chally, 16.67%; Mike Hoefler, 16.67%; Patrick McNair, 16.67%; and David O'Doherty, 16.67%.

Stns: 1 AM. 2 FM. KPNY-FM Alliance, NE; KROA(FM) Grand Island, NE; KMMJ Grand Island, NE.

Mississippi Broadcasters L.L.C., Box 1699, Meridian, MS 39302. Phone: (601) 693-2661. Fax: (601) 483-0826. Ownership: Clay Holladay, 100%.

Stns: 4 FM. WMLV(FM) Butler, AL; WJXM(FM) De Kalb, MS; WUCL(FM) Meridian, MS; WKZB(FM) Stonewall, MS.

Clay E. Holladay, pres & gen mgr.

Missouri River Christian Broadcasting Inc., (Good News Voice). Box 187, Washington, MO 63090. Phone: (636) 239-0400. Fax: (636) 293-4448. Web Site: www.goodnewsvoice.org. Ownership: Group also owns K235AO Salem, MO, translator stn on 94.9 mhz.

Stns: 3 FM. KGNA-FM Arnold, MO; KGNN-FM Cuba, MO; KGNV(FM) Washington, MO.

J.C. Goggan, gen mgr.

Monarch Broadcasting Inc., 212 W. Cypress St., Altus, OK 73521. Phone: (580) 482-1450. Fax: (580) 482-3420. Web Site: www.kwhw.com. Ownership: Matthew L. Ward and Kristin Ward, 51%; and Deborah Ward-Ingstad, 49%. Note: Matthew L. Ward is sole officer, dir and shareholder of KIMM Radio Inc., licensee of KIMM(AM) Rapid City, SD.

Stns: 1 AM. 1 FM. KWHW Altus, OK; KQTZ-FM Hobart, OK.

Matthew L. Ward, pres.

Montrose Broadcasting Corp., Box 248, 9 Locust St., Montrose, PA 18801. Phone: (570) 278-2811. Fax: (570) 278-1442. E-mail: mail@wpel.org Web Site: www.wpel.org. Ownership: Non Profit Non Stock Corporation.

Stns: 2 AM. 3 FM. WPGM Danville, PA; WPGM-FM Danville, PA; WPEL Montrose, PA; WPEL-FM Montrose, PA; WBGM-FM New Berlin, PA.

Larry Souder, pres; John Hagenboch, VP; Barbara Snyder, sec; Charles W. Scott, Jr., treas.

The Moody Bible Institute of Chicago, 820 N. LaSalle Blvd., Chicago, IL 60610. Phone: (312) 329-4300. Fax: (312) 329-8980. Web Site: www.mbn.org. E-mail: mbn@moody.edu

Stns: 3 AM. 25 FM. WMBV-FM Dixons Mills, AL; WMFT(FM) Tuscaloosa, AL; WRMB-FM Boynton Beach, FL; WHGN(FM) Crystal River, FL; WKES-FM Lakeland, FL; WKZM-FM Sarasota, FL; WMBI(AM) Chicago, IL; WMBI-FM Chicago, IL; WDLM East Moline, IL; WGNR-FM Anderson, IN; WIWC-FM Kokomo, IN; WMBL(FM) Mitchell, IN; WHPL-FM West Lafayette, IN; WJSO-FM Pikeville, KY; WGNB-FM Zeeland, MI; WMBU-FM Forest, MS; KSPL-FM Kalispell, MT; KJCG(FM) Missoula, MT; KMBN(FM) Las Cruces, NM; WCRF-FM Cleveland, OH; WVML(FM) Millersburg, OH; WVMS-FM Sandusky, OH; WVME(FM) Meadville, PA; WVMN-FM New Castle, PA; WMBW-FM Chattanooga, TN; WFCM Smyrna, TN; KMLW-FM Moses Lake, WA; KMBI-FM Spokane, WA.

Joseph Stowell, pres; Wayne Pederson, VP bcstg.

Moon Broadcasting, 1200 W. Venice Blvd., Los Angeles, CA 90006. Phone: (213) 745-6224. Fax: (213) 745-7577. Web Site: www.moonbroadcasting.com. Ownership: Abel DeLuna, 100%.

Stns: 6 AM. 7 FM. KIQQ Barstow, CA; KAEH(FM) Beaumont, CA; KMQA-FM East Porterville, CA; KDAC Fort Bragg, CA; KMEN(FM) Mendota, CA; KIQQ-FM Newberry Springs, CA; KAAT(FM) Oakhurst, CA; KTNS Oakhurst, CA; KTOB(AM) Petaluma, CA; KRRS Santa Rosa, CA; KMNA(FM) Mabton, WA; KZXR Prosser, WA; KLES(FM) Prosser, WA.

Abel DeLuna, pres.

The Morey Organization Inc., 1103 Stewart Ave., Garden City, NY 11530. Phone: (516) 228-6570. Web Site: www.moreyorg.com.

Stns: 3 FM. WDRE(FM) Calverton-Roanoke, NY; WLIR-FM Hampton Bays, NY; WBZB(FM) Westhampton, NY.

Jed Morey, COO; John Caracciolo, pres.

Morgan County Industries Inc., 129 College St., West Liberty, KY 41472. Phone: (606) 743-3145. Fax: (606) 743-9557. E-mail: radio41472@yahoo.com

Stns: 3 AM. 4 FM. WCBJ-FM Campton, KY; WMOR Morehead, KY; WMOR-FM Morehead, KY; WRLV Salyersville, KY; WRLV-FM Salyersville, KY; WLKS West Liberty, KY; WLKS-FM West Liberty, KY.

Paul Lyons, COO.

Morris Radio LLC, 725 Broad St., Augusta, GA 30903-0936. Phone: (706) 823-3331. Fax: (706) 823-3212. Web Site: www.morris.com. Ownership: Owned by Morris Communications Company LLC. Group also owns the Kansas Agriculture Network, Topeka, KS; Kansas Information Network, Topeka, KS; and the Wildcat Sports Network, Topeka, KS.

Stns: 15 AM. 13 FM. KBRJ(FM) Anchorage, AK; KEAG-FM Anchorage, AK; KFQD Anchorage, AK; KHAR Anchorage, AK; KMXS-FM Anchorage, AK; KWHL-FM Anchorage, AK; KNWZ(AM) Coachella, CA; KKUU(FM) Indio, CA; KNWQ(AM) Palm Springs, CA; KXPS(AM) Thousand Palms, CA; KFUT(AM) Thousand Palms, CA; KNWH(AM) Twentynine Palms, CA; KDGL(FM) Yucca Valley, CA; KSAJ-FM Abilene, KS; KABI Abilene, KS; KBLS(FM) North Fort Riley, KS; KSAL(AM) Salina, KS; KSAL-FM Salina, KS; KYEZ-FM Salina, KS; WIBW Topeka, KS; KGNC Amarillo, TX; KXRO Aberdeen, WA; KWOK(AM) Hoquiam, WA; KXXK(FM) Hoquiam-Aberdeen, WA; KWIQ-FM Moses Lake, WA; KWIQ(AM) Moses Lake North, WA; KKRT Wenatchee, WA; KWLN(FM) Wilson Creek, WA.

Morris Publishing Group owns the following daily newspapers: *Amarillo* (Texas) *Globe-News, Athens* (GA) *Banner-Herald, The Augusta* (GA) *Chronicle, Brainerd* (MN) *Dispatch, The Daily Ardmoreite*, Ardmore, OK, *Dodge City* (Kan.) *Daily Globe, The Examiner*, Independence, MO, *The Florida Times-Union*, Jacksonville, *The Grand Island* (NE) *Independent, Hannibal* (MO) *Courier-Post, Hillsdale* (MI) *Daily News, The Holland* (MI) *Sentinel, Juneau* (AK) *Empire, Log Cabin Democrat*, Conway, AK, *Lubbock* (Texas) *Avalanche-Journal, The Morning Sun*, Pittsburg, KS, *News Chief*, Winter Haven, FL, *The Newton* (KS) *Kansan, The Oak Ridger*, Oak Ridge, TN, *Peninsula Clarion*, Kenai, AK, *The St. Augustine* (FL) *Record, Savannah* (GA) *Morning News, The Shawnee* (OK) *News-Star, The Topeka* (KS) *Capital-Journal, Yankton* (SD) *Daily Press & Dakotan, York* (NE) *News-Times. Magazines include: Athens Magazine, Augusta Magazine, Coastal Antiques and Art, Coastal Senior, Eco Latino* (Athens), *Eco Latino* (St. Augustine), *Gainesville Life, Her Voice, Hers Kansas, LOUNGE, Savannah Coastal Parent, Savannah Magazine, Senior Living, She's OK!, Skirt!, Water's Edge, West Michigan Senior Times. Other publications include 37 non-daily newspapers and shoppers.*

Michael D. Osterhout, COO.

Mortenson Broadcasting Co., 3270 Blazer Pkwy. #101, Lexington, KY 40509-1847. Phone: (859) 245-1000. Fax: (859) 245-1600. Ownership: Jack Mortenson, 100%.

Stns: 7 AM. 1 FM. KGGN Gladstone, MO; KRVA Cockrell Hill, TX; KGGR Dallas, TX; KHVN Fort Worth, TX;

KKGM(AM) Fort Worth, TX; KTNO(AM) University Park, TX; WEMM-FM Huntington, WV; WEMM(AM) Huntington, WV.

Jack Mortenson, pres.

Mt. Rushmore Broadcasting Inc., 218 N. Wolcott, Casper, WY 82602. Phone: (307) 265-1984. Fax: (307) 266-3295. E-mail: mtrushmore@wyoming.com Web Site: www.wyomingradio.com.

Stns: 5 AM. 9 FM. KRMQ-FM Clovis, NM; KAWK(FM) Custer, SD; KFCR Custer, SD; KZMX Hot Springs, SD; KZMX-FM Hot Springs, SD; KASS(FM) Casper, WY; KVOC Casper, WY; KHOC-FM Casper, WY; KMLD(FM) Casper, WY; KQLT-FM Casper, WY; KRAL Rawlins, WY; KIQZ(FM) Rawlins, WY; KGOS Torrington, WY; KERM-FM Torrington, WY.

Jan Charles Gray, pres/CEO.

Mountain Broadcasting Corp., 99 Clinton Rd., West Caldwell, NJ 07006. Phone: (973) 852-0300. Fax: (973) 808-5516. Ownership: Sun Young Joo, 66% votes, 35.2% total assets; John H. Joo, 14% votes, 6.3% total assets; Victor C. Joo, 14% votes, 5.6% total assets; Sun Hoo Joo, 6% votes, 2.9% total assets; and Hansen Lau, 5.7% total assets.

Stns: 3 AM. WPWA Chester, PA; WBTK(AM) Richmond, VA; WWGB Indian Head, MD.

Stns: 1 TV. WMBC-TV, New York.

Sun Young Joo, pres.

Mountain Communications, Box 211, Saranac Lake, NY 12983-0211. Phone: (518) 891-1544. Fax: (518) 891-1545. Ownership: Prescott House LLC owns 100% of WIRD(AM)-WLPW(FM) and WRGR(FM). Edward S. Morgan, the sole member of Prescott House LLC, is also the controlling stockholder of WNBZ(AM)-WYZY(FM).

Stns: 2 AM. 3 FM. WIRD Lake Placid, NY; WLPW-FM Lake Placid, NY; WNBZ Saranac Lake, NY; WYZY(FM) Saranac Lake, NY; WRGR-FM Tupper Lake, NY.

Ted Morgan, owner.

Mountain Dog Media, 254 Winnebago Dr., Fond du Lac, WI 54935. Phone: (920) 921-1071. Fax: (920) 921-0757.

Stns: 2 AM. 1 FM. KFIZ Fond du Lac, WI; WFON(FM) Fond du Lac, WI; WCLB(AM) Sheboygan, WI.

Randy Hopper, pres.

Mountain Wireless Inc., Box 159, Skowhegan, ME 04976. Phone: (207) 474-5171. Fax: (207) 474-3299.

Stns: 1 AM. 2 FM. WCTB-FM Fairfield, ME; WFMX(FM) Skowhegan, ME; WSKW Skowhegan, ME.

Alan W. Anderson, pres.

Mt. Washington Radio & Gramophone L.L.C., Box 2008, Conway, NH 03818. Phone: (603) 356-8870. Fax: (603) 356-8875. E-mail: office@wmwv.com Web Site: www.wmwv.com. Ownership: Ronald Frizzell, 51%; Greg Frizzell, 25%; Arnold Lerner, 24%.

Stns: 2 FM. WVMJ(FM) Conway, NH; WMWV-FM Conway, NH.

Ron Frizzell, pres.

Muirfield Broadcasting Inc., 200 Short Rd., Southern Pines, NC 28387. Phone: (910) 692-2107. Fax: (910) 692-6849. Web Site: www.star1025fm.com. Ownership: Walker Morris.

Stns: 1 AM. 1 FM. WIOZ Pinehurst, NC; WIOZ-FM Southern Pines, NC.

Walker Morris, pres.

Multicultural Radio Broadcasting Inc., 449 Broadway, New York, NY 10013. Phone: (212) 966-1059. Fax: (212) 966-9580. Web Site: www.mrbi.net. Ownership: Arthur S. Liu, 51%; and Yvonne S. Liu, 49%.

Stns: 32 AM. KIDR Phoenix, AZ; KWRU(AM) Fresno, CA; KYPA Los Angeles, CA; KAZN Pasadena, CA; KATD Pittsburg, CA; KAHZ(AM) Pomona, CA; KEST San Francisco, CA; KIQI San Francisco, CA; KSJX(AM) San Jose, CA; KBLA Santa Monica, CA; KALI West Covina, CA; WNMA Miami Springs, FL; WEXY Wilton Manors, FL; WGFS Covington, GA; WNTD Chicago, IL; WLYN Lynn, MA; WAZN(AM) Watertown, MA; WJDM Elizabeth, NJ; WWRU(AM) Jersey City, NJ; WTTM(AM) Lindenwold, NJ; WNSW Newark, NJ; WPAT Paterson, NJ; WHWH Princeton, NJ; WNYG Babylon, NY; WZRC New York, NY; WKDM(AM) New York, NY; KDFT Ferris, TX; KXYZ Houston, TX; KMNY(AM) Hurst, TX; WZHF Arlington, VA; KXPA Bellevue, WA; WLXE(AM) Rockville, MD.

Arthur S. Liu, pres/CEO.

Munbilla Broadcasting Properties Ltd., 5526 Hwy. 281 N., Marble Falls, TX 78654. Phone: (830) 693-5551. Fax: (830) 693-5107. Ownership: B. Shane Fox, 100%.

Stns: 1 AM. 3 FM. KBEY(FM) Burnet, TX; KRHC(AM) Burnet, TX; KHLE(FM) Burnet, TX; KHLB(FM) Mason, TX.

Duane Fox, gen mgr; Sabrina Preiss, traf mgr; Bill Woleben, chief engr & opns mgr.

Morgan Murphy Stations (Evening Telegram Co), Box 44965, Madison, WI 53744-4965. Phone: (608) 271-4321. Fax: (608) 271-6111. E-mail: talkback@wisctv.com Web Site: www.channel3000.com. Ownership: Evening Telegram Co. owns 100% of KVEW(TV), KXLY-AM-FM-TV, KXLY-DT and KAPP(TV). Evening Telegram Co. owns 84.4% of Television Wisconsin Inc., with an additional 15.2% of the stn held by Evening Telegram stockholders.

Stns: 4 AM. 3 FM. KXLX(AM) Airway Heights, WA; KXLY Spokane, WA; KZZU-FM Spokane, WA; KEZE-FM Spokane, WA; WGLR Lancaster, WI; WPVL Platteville, WI; WPVL-FM Platteville, WI.

Stns: 5 TV. WKBT, La Crosse-Eau Claire, WI; WISC-TV, Madison, WI; KXLY-TV, Spokane, WA; KAPP, Yakima-Pasco -Richland-Kennewick, WA; KVEW, Yakima-Pasco -Richland-Kennewick, WA.

The Evening Telegram principals own *Madison Magazine*, Madison, WI.

Elizabeth Murphy Burns, pres; George Nelson, exec VP; David Sanks, exec VP; Steve Herling, exec VP; Darrell Blue, VP/gen mgr; Scott Chorski, VP/gen mgr.

Muzzy Broadcasting L.L.C., 500 Division St., Stevens Point, WI 54481. Phone: (715) 341-9800. Fax: (715) 341-0000. Web Site: www.979wspt.com. Ownership: Richard L. Muzzy.

Stns: 1 AM. 2 FM. WKQH(FM) Marathon, WI; WSPT Stevens Point, WI; WSPT-FM Stevens Point, WI.

Richard L. Muzzy, pres.

My Broadcasting Corp., Box 961, Renfrew, ON K7V 4H4. Canada. Phone: (613) 432-6936. Fax: (613) 432-1086. Web Site: www.myfmradio.ca.

Stns: 3 FM. CIMY-FM Pembroke, ON; CHMY-FM Renfrew, ON; CJMI-FM Strathroy, ON.

N

NL Broadcasting Ltd., 611 Lansdowne St., Kamloops, BC V2C 1Y6. Canada. Phone: (250) 372-2292. Fax: (250) 372-2293. Ownership: NL Properties Inc., 37.92%; J. Robert Dunn, 38.61%; and others 13.33%. Note: Group also owns 51% of CJNL(AM) Merritt, BC.

Stns: 1 AM. 2 FM. CHNL Kamloops, BC; CKRV-FM Kamloops, BC; CJKC-FM Kamloops, BC.

NRC Broadcasting Inc., 1201 Eighteenth St., Suite 250, Denver, CO 80202. Phone: (303) 675-4698. Fax: (303) 296-7030. Web Site: www.nrcbroadcasting.com. Ownership: Anschutz Co., 60.9%; Tim Brown, 33.5%; Ray Skibitsky, 2.8%; and Dave Rogers, 2.8%.

Stns: 1 AM. 11 FM. KSPN-FM Aspen, CO; KNFO-FM Basalt, CO; KSMT-FM Breckenridge, CO; KTUN-FM Eagle, CO; KRKY-FM Estes Park, CO; KKCH-FM Glenwood Springs, CO; KCUV(FM) Greenwood Village, CO; KTRJ(FM) Hayden, CO; KIDN-FM Hayden, CO; KCKK(AM) Littleton, CO; KFMU-FM Oak Creek, CO; KJAC(FM) Timnath, CO.

Tim Brown, chmn/CEO; Ray Skibitsky, pres/COO; Dave Rogers, CFO.

NRG Media LLC, 2875 Mount Vernon Rd. S.E., Cedar Rapids, IA 52403. Phone: (319) 862-0300. Fax: (319) 286-9383. E-mail: jlink@nrgmedia.com Web Site: www.nrgmedia.com. Ownership: Waitt Media Holdings LLC, 55.1%; and NewRadio Group LLC, 44.9%.

Stns: 19 AM. 37 FM. KLGA Algona, IA; KLGA-FM Algona, IA; KWBG Boone, IA; KHBT-FM Humboldt, IA; KQWC Webster City, IA; KQWC-FM Webster City, IA; WIXN Dixon, IL; WRCV(FM) Dixon, IL; WSEY(FM) Oregon, IL; WRKX(FM) Ottawa, IL; WCMY(AM) Ottawa, IL; WJBD Salem, IL; WJBD-FM Salem, IL; KXXX Colby, KS; KQLS-FM Colby, KS; KZRD(FM) Dodge City, KS; KOLS-FM Dodge City, KS; KGNO Dodge City, KS; KZLS-FM Great Bend, KS; KSSH(FM) Ingalls, KS; KGTR-FM Larned, KS; KNNS Larned, KS; KQNS-FM Lindsborg, KS; KILS-FM Minneapolis, KS; KWLS Pratt, KS; KLCH(FM) Lake City, MN; KHUB Fremont, NE; KFMT-FM Fremont, NE; KSYZ-FM Grand Island, NE; KROR(FM) Hastings, NE; KUVR Holdrege, NE; KMTY-FM Holdrege, NE; KQKY(FM) Kearney, NE; KRNY(FM) Kearney, NE; KGFW Kearney, NE; KNEN-FM Norfolk, NE; KODY North Platte, NE; KXNP-FM North Platte, NE; KTCH Wayne, NE; KCTY(FM) Wayne, NE; WRLO-FM Antigo, WI; WSJY-FM Fort Atkinson, WI; WFAW Fort Atkinson, WI; WYTE(FM) Marshfield, WI; WLKD Minocqua, WI; WMQA-FM Minocqua, WI; WNFM-FM Reedsburg, WI; WBDL-FM Reedsburg, WI; WRDB Reedsburg, WI; WOBT Rhinelander, WI; WRHN-FM Rhinelander, WI; WHDG-FM Rhinelander, WI; WBCV(FM) Wausau, WI; WKCH-FM Whitewater, WI; WLJY(FM) Whiting, WI; WGLX-FM Wisconsin Rapids, WI.

Mary Quass, pres/CEO; Norman W. Waitt Jr., chmn.

Nassau Broadcasting Partners L.P., 619 Alexander Rd., 3rd Fl., Princeton, NJ 08540. Phone: (609) 452-9696. Fax: 609) 419-0143. Web Site: www.nassaubroadcasting.com.

Stns: 15 AM. 34 FM. WFQR(FM) Harwich Port, MA; WPXC-FM Hyannis, MA; WCRB(FM) Lowell, MA; WFRQ(FM) Mashpee, MA; WTHT(FM) Auburn, ME; WBQI(FM) Bar Harbor, ME; WLVP(AM) Gorham, ME; WBYA(FM) Islesboro, ME; WBQQ-FM Kennebunk, ME; WHXQ(FM) Kennebunkport, ME; WLAM(AM) Lewiston, ME; WFNK(FM) Lewiston, ME; WHXR(FM) North Windham, ME; WBQW-FM Scarborough, ME; WBQX-FM Thomaston, ME; WNHW(FM) Belmont, NH; WHDQ-FM Claremont, NH; WTSV Claremont, NH; WJYY-FM Concord, NH; WNNH-FM Henniker, NH; WLNH-FM Laconia, NH; WEMJ Laconia, NH; WWHQ(FM) Meredith, NH; WFNQ(FM) Nashua, NH; WPLY-FM Walpole, NH; WLKZ-FM Wolfeboro, NH; WWYY-FM Belvidere, NJ; WCHR(AM) Flemington, NJ; WPST(FM) Trenton, NJ; WPHY(AM) Trenton, NJ; WTKZ Allentown, PA; WEEX Easton, PA; WBYN(AM) Lehighton, PA; WPLY(AM) Mount Pocono, PA; WVPO Stroudsburg, PA; WSNO Barre, VT; WORK-FM Barre, VT; WZLF(FM) Bellows Falls, VT; WWFY(FM) Berlin, VT; WEXP-FM Brandon, VT; WMOO-FM Derby Center, VT; WWOD(FM) Hartford, VT; WIKE Newport, VT; WXLF(FM) White River Junction, VT; WNHV White River Junction, VT; WTHK(FM) Wilmington, VT; WARK Hagerstown, MD; WWEG(FM) Hagerstown, MD; WAFY-FM Middletown, MD.

Louis F. Mercatani, pres.

Nebraska Rural Radio Association, Box 880, Lexington, NE 68850. Phone: (308) 324-2371. Fax: (308) 324-5786. E-mail: krvnam@krvn.com Web Site: www.krvn.com. Ownership: Nebraska Rural Radio Association, 100%

Stns: 3 AM. 3 FM. KRVN Lexington, NE; KRVN-FM Lexington, NE; KNEB Scottsbluff, NE; KNEB-FM Scottsbluff, NE; KTIC West Point, NE; KTIC-FM West Point, NE.

Eric Brown, sec/treas; Dale Hanson, pres; Larry Hudkins, VP.

Neely Enterprises, Box 861, Rock Hill, SC 29731. Phone: (803) 329-2664. Phone: (803) 329-2760. Fax: (803) 329-3317. Fax: (803) 329-8652.

Stns: 2 AM. 1 FM. WGIV(AM) Pineville, NC; WGCD Chester, SC; WAAW-FM Williston, SC.

Frank Neely, pres.

Neuhoff Family L.P., 1501 N. Washington, Danville, IL 61832. Phone: (217) 442-1700. Phone: (217) 787-9200. Fax: (217) 431-1489. E-mail: mhulvey@cooketech.net Web Site: www.wdnlfm.com. Ownership: Neuhoff Corp., North Palm Beach, FL, 100% of votes.

Stns: 2 AM. 5 FM. WDAN(AM) Danville, IL; WDNL(FM) Danville, IL; WRHK-FM Danville, IL; WXAJ(FM) Hillsboro, IL; WFMB(AM) Springfield, IL; WFMB-FM Springfield, IL; WCVS-FM Virden, IL.

Stns: 1 TV. KMVT, Twin Falls, ID.

Mike Hulvey, gen mgr; Geoff Neuhoff, pres.

Nevada County Broadcasters Inc., 1255 E. Main St., Suite A, Grass Valley, CA 95945. Phone: (530) 272-3424. Fax: (530) 272-2872. E-mail: knco@nccn.com Web Site: www.knco.com.

Stns: 2 AM. 1 FM. KNCO Grass Valley, CA; KNCO-FM Grass Valley, CA; KUBA(AM) Yuba City, CA.

Bob Breck, CEO.

New Media Broadcasters Inc., 2210 31st St. N., Havre, MT 59501-8003. Phone: (406) 265-7841. Fax: (406) 265-8855. E-mail: nmb@nmbi.com Web Site: www.nmbi.com. Ownership: C. David Leeds, 100%.

Stns: 1 AM. 2 FM. KRYK(FM) Chinook, MT; KOJM Havre, MT; KPQX(FM) Havre, MT.

C. David Leeds, pres; Cynthia H. Leeds, sec/treas.

New Northwest Broadcasters LLC, 1011 Western Ave., Suite 920, Seattle, WA 98104. Phone: (206) 204-0213. Fax: (206) 204-0214. Web Site: www.nnbradio.com. Ownership: E. Perot Bissell, 50%; and Bradford N. Creswell, 50%.

Stns: 11 AM. 22 FM. KFAT(FM) Anchorage, AK; KDBZ(FM) Anchorage, AK; KTDZ(FM) College, AK; KWLF-FM Fairbanks, AK; KXLR-FM Fairbanks, AK; KFAR Fairbanks, AK; KCBF Fairbanks, AK; KXLW(FM) Houston, AK; KBBO-FM Houston, AK; KGHL-FM Billings, MT; KRPM(FM) Billings, MT; KQBL(FM) Billings, MT; KGHL Billings, MT; KRSQ-FM Laurel, MT; KKEE(AM) Astoria, OR; KAST Astoria, OR; KYSF-FM Bonanza, OR; KAGO Klamath Falls, OR; KLAD(AM) Klamath Falls, OR; KLAD-FM Klamath Falls, OR; KCRX-FM Seaside, OR; KUJ-FM Burbank, WA; KARY-FM Grandview, WA; KVAS(FM) Ilwaco, WA; KTCR Kennewick, WA; KAST-FM

Long Beach, WA; KIOK-FM Richland, WA; KEGX(FM) Richland, WA; KALE Richland, WA; KBBO(AM) Selah, WA; KNLT-FM Walla Walla, WA; KHHK-FM Yakima, WA; KJOX(AM) Yakima, WA.

Pete Benedetti, CEO.

New South Communications Inc., Box 5797, Meridian, MS 39302. Phone: (601) 693-2661. Fax: (601) 483-0826. Ownership: F.E. Holladay, 100%.

Stns: 2 AM. 4 FM. KJLO-FM Monroe, LA; WYOY-FM Gluckstadt, MS; WUSJ(FM) Madison, MS; WALT Meridian, MS; WIIN Ridgeland, MS; WJKK-FM Vicksburg, MS.

F.E. Holladay, pres.

New West Broadcasting Corp., 1145 Kilauea Ave., Hilo, HI 96720. Phone: (808) 935-5461. Fax: (808) 935-7761. Ownership: NWB Holdings Inc., 80% stockholder; and Christopher S. Leonard, 20% stockholder.

Stns: 1 AM. 2 FM. KNWB-FM Hilo, HI; KPUA Hilo, HI; KAOY(FM) Kealakekua, HI.

NewCap Inc., 745 Windmill Rd., Dartmouth, NS B3B1C2. Canada. Phone: (902) 468-7557. Fax: (902) 468-7558. Web Site: www.ncc.ca. Ownership: H.R. Steele, Blavin & Company.

Stns: 23 AM. 33 FM. CKBA Athabasca, AB; CJPR-FM Blairmore, AB; CJEG-FM Bonnyville, AB; CIXF-FM Brooks, AB; CIBQ Brooks, AB; CIQX-FM Calgary, AB; CFUL-FM Calgary, AB; CFCW-FM Camrose, AB; CFCW Camrose, AB; CKDQ Drumheller, AB; CKRA-FM Edmonton, AB; CIRK-FM Edmonton, AB; CFXE-FM Edson, AB; CJXK-FM Grand Centre (Cold Lake), AB; CKVH High Prairie, AB; CFXH-FM Hinton, AB; CKSA-FM Lloydminster, AB; CKGY-FM Red Deer, AB; CIZZ-FM Red Deer, AB; CHLW(AM) Saint Paul, AB; CHSL-FM Slave Lake, AB; CKSQ(AM) Stettler, AB; CKKY Wainwright, AB; CKWY-FM Wainwright, AB; CFOK(AM) Westlock, AB; CKJR Wetaskiwin, AB; CFXW-FM Whitecourt, AB; CHNK-FM Winnipeg, MB; CKJS Winnipeg, MB; CFRK-FM Fredericton, NB; CKIM Baie Verte, NF; CHVO Carbonear, NF; CFLC-FM Churchill Falls, NF; CKVO Clarenville, NF; CKXX-FM Corner Brook, NF; CFCB Corner Brook, NF; CKGA Gander, NF; CKXD-FM Gander, NF; CFLN Goose Bay, NF; CKCM Grand Falls, NF; CKXG-FM Grand Falls-Windsor, NF; CHCM Marystown, NF; CFNW Port au Choix, NF; CFCV-FM Saint Andrews, NF; VOCM(AM) Saint John's, NF; CKIX-FM Saint John's, NF; CFSX Stephenville, NF; CFLW Wabush, NF; CKUL-FM Halifax, NS; CIHT-FM Ottawa, ON; CILV-FM Ottawa, ON; CHNO-FM Sudbury, ON; CJUK-FM Thunder Bay, ON; CKTG-FM Thunder Bay, ON; CHTN-FM Charlottetown, PE; CKQK-FM Charlottetown, PE.

Stns: 2 TV. CITL, Lloydminster, AB; CKSA, Lloydminster, AB.

H.R. Steele, chmn; Scott Weatherby, CEO; R.G. Steele, pres/CEO.

Newfoundland Broadcasting Co., (NTV & OZ Networks). Box 2020, St. John's, NF A1C 5S2. Canada. Phone: (709) 722-5015. Fax: (709) 726-5107. E-mail: ozfm@ozfm.com Web Site: www.ntv.ca. Ownership: Geoffrey W. Stirling, 89.95%; G. Scott Stirling, 10%; and others, 0.05%.

Stns: 8 FM. CJOZ-FM Bonavista Bay, NF; CJKK-FM Clarenville, NF; CKOZ-FM Corner Brook, NF; CIOZ-FM Marystown, NF; CHOS-FM Rattling Brook, NF; CKSS-FM Red Rocks, NF; CHOZ-FM Saint John's, NF; CIOS-FM Stephenville, NF.

Stns: 6 TV. CJOM, Argentia, NF; CJWB, Bonavista, NF; CJWN, Corner Brook, NF; CJOX-1, Grand Bank, NF; CJCN, Grand Falls, NF; CJSV, Stephenville, NF.

Scott G. Stirling, pres/CEO; Doug Neal, engrg dir.

News-Press & Gazette Co., Box 29, St. Joseph, MO 64502. Phone: (816) 271-8500. Fax: (816) 271-8695. Ownership: David R. Bradley Jr., Henry H. Bradley, Lyle E. Leimkuhler. Cable TV: NPG Cable of Arizona.

Stns: 2 AM. 1 FM. KESQ Indio, CA; KUNA-FM La Quinta, CA; KRDO(AM) Colorado Springs, CO.

Stns: 6 TV. KTVZ, Bend, OR; KRDO-TV, Colorado Springs-Pueblo, CO; KVIA-TV, El Paso (Las Cruces, NM), TX; KJCT, Grand Junction-Montrose, CO; KIFI, Idaho Falls-Pocatello, ID; KESQ-TV, Palm Springs, CA.

News-Press & Gazette Co. publishes the *St. Joseph News-Press*, St. Joseph, MO.

John Kueneke, pres.

Newsweb Corp., 1645 W. Fullerton Ave., Chicago, IL 60614. Phone: (773) 975-0401. Fax: (773) 975-1301. Ownership: Fred Eychaner, 100%.

Stns: 5 AM. 4 FM. WKIE-FM Arlington Heights, IL; WSBC(AM) Chicago, IL; WCFJ Chicago Heights, IL; WCPT(AM) Crystal Lake, IL; WDEK-FM De Kalb, IL; WKIF-FM Kankakee, IL; WRZA(FM) Park Forest, IL; WAIT(AM) Willow Springs, IL; WNDZ Portage, IN.

Stns: 2 TV. KTVD, Denver, CO; KUPN, Denver, CO.

Fred Eychaner, CEO; Charley Gross, COO.

NextMedia Group Inc., 6312 S. Fiddler's Green Cir., Suite 360E, Englewood, CO 80111. Phone: (303) 694-9118. Fax: (303) 694-4940. Web Site: www.nextmediagroup.net. Ownership: NextMedia Investors LLC, 100% of votes.

Stns: 9 AM. 32 FM. KBAY(FM) Gilroy, CA; KEZR(FM) San Jose, CA; WERV-FM Aurora, IL; WRXQ(FM) Coal City, IL; WDZ(AM) Decatur, IL; WDZQ(FM) Decatur, IL; WSOY-FM Decatur, IL; WWYW(FM) Dundee, IL; WJOL Joliet, IL; WCZQ-FM Monticello, IL; WKRS Waukegan, IL; WXLC-FM Waukegan, IL; WZSR-FM Woodstock, IL; WSGW-FM Carrollton, MI; WCEN-FM Hemlock, MI; WSGW Saginaw, MI; WTLZ-FM Saginaw, MI; WQZL(FM) Belhaven, NC; WKXB-FM Burgaw, NC; WANG Havelock, NC; WSSM(FM) Havelock, NC; WQSL-FM Jacksonville, NC; WILT(FM) Jacksonville, NC; WXQR-FM Jacksonville, NC; WRNS-FM Kinston, NC; WSFM(FM) Oak Island, NC; WAZO(FM) Southport, NC; WERO(FM) Washington, NC; WRQR-FM Wilmington, NC; WMFD Wilmington, NC; WHBC Canton, OH; WMYB(FM) Myrtle Beach, SC; WYAV(FM) Myrtle Beach, SC; WRNN(AM) Myrtle Beach, SC; WKZQ-FM Myrtle Beach, SC; WRNN-FM Socastee, SC; KMKT-FM Bells, TX; KLAK(FM) Tom Bean, TX; KMAD-FM Whitesboro, TX; WLIP Kenosha, WI; WJBR-FM Wilmington, DE.

Steven Dinetz, CEO; Skip Weller, pres.

Nicolet Broadcasting Inc., 3030 Park Drive, Suite 3, Sturgeon Bay, WI 54235. Phone: (920) 746-9430. Fax: (920) 746-9433. E-mail: wbbk@doorcoundtydailynews.com Web Site: www.doorcountydailynews.com. Ownership: Roger Utnehmer.

Stns: 4 FM. WRLU-FM Algoma, WI; WBDK-FM Algoma, WI; WRKU-FM Forestville, WI; WSBW(FM) Sister Bay, WI.

Roger Utnehmer, pres & gen mgr.

Noalmark Broadcasting Corp., 202 W. 19th St., El Dorado, AR 71730. Phone: (870) 862-7777. Fax: (870) 862-0203. Ownership: William C. Nolan Jr., 65%; Edwin B. Alderson Jr., 35%.

Stns: 7 AM. 12 FM. KDEL-FM Arkadelphia, AR; KVRC Arkadelphia, AR; KAGL(FM) El Dorado, AR; KMRX(FM) El Dorado, AR; KELD El Dorado, AR; KMLK(FM) El Dorado, AR; KYXK(FM) Gurdon, AR; KELD-FM Hampton, AR; KLAZ-FM Hot Springs, AR; KBHS(AM) Hot Springs, AR; KPZA(AM) Hot Springs, AR; KVMA Magnolia, AR; KBOK(AM) Malvern, AR; KLEZ(FM) Malvern, AR; KVMZ(FM) Waldo, AR; KYKK(AM) Hobbs, NM; KIXN(FM) Hobbs, NM; KPER(FM) Hobbs, NM; KPZA-FM Jal, NM.

William C. Nolan Jr., pres; Edwin B. Alderson Jr., exec VP; Paul Starr, VP; Anna Canterbury, sec/treas.

Norsan Consulting and Management Inc., Box 2148, Tucker, GA 30085. Phone: (770) 414-5026. Ownership: Norberto Sanchez, 100%.

Stns: 8 AM. WVOJ(AM) Fernandina Beach, FL; WNNR(AM) Jacksonville, FL; WSOS(AM) Saint Augustine Beach, FL; WGSP Charlotte, NC; WFAY(AM) Fayetteville, NC; WXNC(AM) Monroe, NC; WCEO(AM) Columbia, SC; WKGN Knoxville, TN.

North American Broadcasting Co. Inc., 1458 Dublin Rd., Columbus, OH 43215. Phone: (614) 481-7800. Fax: (614) 481-8070. Web Site: www.nabco-inc.com. Ownership: Norma Mnich; Matthew Mnich.

Stns: 1 AM. 2 FM. WBZX-FM Columbus, OH; WMNI Columbus, OH; WTDA(FM) Westerville, OH.

Norma Mnich, chmn; Matthew Mnich, pres/CEO; Mark Jividen, VP; Nick Reed, VP/sec/treas.

North Cascades Broadcasting Inc., Box 151, Omak, WA 98841. Phone: (509) 826-0100. Fax: (509) 826-3929. Web Site: www.komw.net.

Stns: 1 AM. 1 FM. KOMW Omak, WA; KZBE-FM Omak, WA.

John Andrist, pres.

North Georgia Radio Group L.P., 112 Jordan Dr., Chattanooga, TN 37421. Phone: (423) 425-8987. Ownership: Whitfield Communications Inc., gen ptnr, 100% of votes.

Stns: 2 AM. 3 FM. WQMT-FM Chatsworth, GA; WBLJ(AM) Dalton, GA; WDAL Dalton, GA; WYYU-FM Dalton, GA; WOCE(FM) Ringgold, GA.

Northeast Broadcasting Company Inc., 288 S. River Rd., Bedford, NH 03110. Phone: (603) 668-9999. Fax: (603) 668-6470. Web Site: www.nebcast.com. Ownership: Steven A. Silberberg, Ed Flanagan.

Stns: 11 AM. 30 FM. KJMP(AM) Pierce, CO; KVRG(FM) Victor, ID; KRVQ(FM) Victor, ID; WNYN-FM Athol, MA; WGAW Gardner, MA; WXRV-FM Haverhill, MA; WJOE(AM) Orange-Athol, MA; WKBR Manchester, NH; WXRG(FM) Whitefield, NH; KVUW(FM) Wendover, NV; WTWK(AM) Plattsburgh, NY; KFMH(FM) Belle Fourche, SD; WUSX(FM) Addison, VT; WCAT(AM) Burlington, VT; WDOT(FM) Danville, VT; WFAD Middlebury, VT; WNCS(FM) Montpelier, VT; WSKI Montpelier, VT; WRJT-FM Royalton, VT; WRSA(AM) Saint Albans, VT; WWMP(FM) Waterbury, VT; KBEN-FM Basin, WY; KRAE Cheyenne, WY; KZDR(FM) Cheyenne, WY; KCUG(FM) Chugwater, WY; KWHO(FM) Cody, WY; KTED(FM) Douglas, WY; KDAD(FM) Douglas, WY; KANT(FM) Guernsey, WY; KHNA(FM) Hanna, WY; KXMP(FM) Hanna, WY; KTUG(FM) Hudson, WY; KUSZ(FM) Laramie, WY; KIMX(FM) Laramie, WY; KHAT(AM) Laramie, WY; KROW(FM) Lovell, WY; KHAD(FM) Mills, WY; KRUG(FM) Upton, WY; KRAN(FM) Wamsutter, WY; KYPT(FM) Wamsutter, WY; KPAD(FM) Wheatland, WY.

Steven Silberberg, CEO; Edward Flanagan, VP.

Northeast Colorado Broadcasting LLC, 220 State St., Suite 106, Fort Morgan, CO 80701. Phone: (970) 867-7271. Fax: (970) 867-2676. Ownership: Alexander L. Creighton, 66.67%; Ross O. Miller, 33.33%.

Stns: 1 AM. 2 FM. KPRB-FM Brush, CO; KSIR Brush, CO; KPMX(FM) Sterling, CO.

Alec L. Creighton, gen mgr.

Northeast Communications Corp., 110 Babbit Rd., Franklin, NH 03235. Phone: (603) 934-2500. Fax: (603) 934-2933. E-mail: onair@mix941fm.com Web Site: www.mix941fm.com. Ownership: Jeff Fisher, 44.5%; Chris Fisher, 17.5%; and Phil Fisher, 16.5%.

Stns: 2 AM. 1 FM. WFTN Franklin, NH; WSCY-FM Moultonborough, NH; WPNH Plymouth, NH.

Jeff Fisher, pres; Fred Caruso, progmg dir; Rick Ganley, progmg dir; Cathy Keyser, opns mgr.

Northern Christian Radio Inc., Box 695, Gaylord, MI 49734-0695. Phone: (800) 545-8857. Phone: (989) 732-6274. Fax: (989) 732-8171. E-mail: ncr@ncradio.org Web Site: www.ncradio.org.

Stns: 5 FM. WOLW-FM Cadillac, MI; WZHN(FM) East Tawas, MI; WPHN-FM Gaylord, MI; WTHN(FM) Sault Ste. Marie, MI; WHST-FM Tawas City, MI.

Joe Sereno, chmn; George Lake, chief exec admin; Patrick Green, progmg dir.

Northern Star Broadcasting L.L.C., 1356 Mackinaw Ave., Cheboygan, MI 49721. Phone: (231) 627-2341. Fax: (231) 627-7000. E-mail: cmonk@nsbroadcasting.com Web Site: www.nsbroadcasting.com. Ownership: Wade Fetzer, 91.7%; W. Palmer Pyle, 7% (nterest held as voting trustee); and George Atkinson III, 1.3%.

Stns: 5 AM. 14 FM. WCKC-FM Cadillac, MI; WCBY Cheboygan, MI; WGFM(FM) Cheboygan, MI; WGFN(FM) Glen Arbor, MI; WJZJ(FM) Glen Arbor, MI; WMIQ Iron Mountain, MI; WIMK-FM Iron Mountain, MI; WIAN Ishpeming, MI; WLJZ(FM) Mackinaw City, MI; WDMJ Marquette, MI; WUPK-FM Marquette, MI; WAVC-FM Mio, MI; WNGE-FM Negaunee, MI; WIHC-FM Newberry, MI; WZNL-FM Norway, MI; WMKD(FM) Pickford, MI; WIDG Saint Ignace, MI; WMKC(FM) Saint Ignace, MI; WYSS-FM Sault Ste. Marie, MI.

Chris Monk, VP; Palmer Pyle, pres.

Northwestern College & Radio, 3003 Snelling Ave. N., St. Paul, MN 55113-1598. Phone: (651) 631-5000. Fax: (651) 631-5086. E-mail: phvirts@nwc.edu Web Site: www.nwc.edu. Ownership: Non-profit organization. Northwestern College, St. Paul, is the owner and operator of the 15 radio licenses.

Stns: 5 AM. 9 FM. WSMR-FM Sarasota, FL; KNWM(FM) Madrid, IA; KNWI(FM) Osceola, IA; KNWS Waterloo, IA; KDNI(FM) Duluth, MN; KDNW(FM) Duluth, MN; KTIS Minneapolis, MN; KTIS-FM Minneapolis, MN; KFNW-FM Fargo, ND; KFNL(FM) Kindred, ND; KFNW(AM) West Fargo, ND; KNWC(AM) Sioux Falls, SD; WNWC-FM Madison, WI; WNWC(AM) Sun Prairie, WI.

Dr. Paul Virts, sr VP; Dr. Alan Cureton, pres.

O

Omni Broadcasting Co., 502 Beltrami Ave. N.W., Bemidji, MN 56601-3010. Phone: (218) 444-1500. Fax: (218) 759-0345. Ownership: Louis H. Buron Jr., Mary Campbell, G. Michael Boen.

Stns: 5 AM. 10 FM. KULO(FM) Alexandria, MN; KBHP(FM) Bemidji, MN; KBUN Bemidji, MN; KKZY-FM Bemidji, MN; KLIZ(AM) Brainerd, MN; KLIZ-FM Brainerd, MN; KUAL-FM Brainerd, MN; KVBR Brainerd, MN; WJJY-FM Brainerd, MN; KBLB(FM) Nisswa, MN; KIKV-FM Sauk Centre, MN; KNSP(AM) Staples, MN; KWAD Wadena, MN; KKWS(FM) Wadena, MN; KLLZ-FM Walker, MN.

Louis H. Buron Jr., pres/CEO; Mary Campbell, VP/CFO.

One Ten Broadcast Group Inc., 2 E. Main St., Shawnee, OK 74801-6906. Phone: (405) 878-1803. E-mail: kirc1059@aol.com Ownership: Linda D. Jones, executrix of estate of Herman L. Jones, 100%.

Stns: 1 AM. 1 FM. KIRC-FM Seminole, OK; KWSH Wewoka, OK.

Linda Jones, pres.

Opus Broadcasting Systems Inc., 511 Rossanley Dr., Medford, OR 97501. Phone: (541) 772-0322. Fax: (541) 772-4233. Ownership: Henry Flock, 70%; Dean Flock, 20%; Alan Benz, 5%; and John Lavoie, 5%.

Stns: 2 AM. 3 FM. KRVC(FM) Hornbrook, CA; KCNA(FM) Cave Junction, OR; KROG-FM Grants Pass, OR; KRTA Medford, OR; KEZX(AM) Medford, OR.

Dean Flock, gen mgr.

Opus Media Holdings LLC, 900 Third Ave., 26th Fl., New York, NY 10022. Phone: (212) 634-3376. Ownership: Opus Capital LLC, 100% of votes.

Stns: 11 FM. WHTF-FM Havana, FL; WEGT(FM) Lafayette, FL; WUTL(FM) Tallahassee, FL; WAIB-FM Tallahassee, FL; KBKK(FM) Ball, LA; KEZP-FM Bunkie, LA; KQLQ(FM) Columbia, LA; KXRR(FM) Monroe, LA; KMYY(FM) Rayville, LA; KLAA-FM Tioga, LA; KZRZ(FM) West Monroe, LA.

Richard Linhart, chmn; James Shea, pres/CEO.

The Original Company Inc., Box 242, Vincennes, IN 47591. Phone: (812) 882-6060. Fax: (812) 885-2604. E-mail: marklange@originalcompany.com Web Site: www.originalcompany.com. Ownership: Mark R. Lange, 50%; Saundra K. Lange, 50%.

Stns: 3 AM. 7 FM. WUZR-FM Bicknell, IN; WREB-FM Greencastle, IN; WQTY-FM Linton, IN; WBTO Linton, IN; WRCY(AM) Mount Vernon, IN; WYFX(FM) Mount Vernon, IN; WBTO-FM Petersburg, IN; WZDM-FM Vincennes, IN; WAOV Vincennes, IN; WWBL-FM Washington, IN.

Mark R. Lange, pres.

O-Town Communications Inc., 416 E. Main St., Ottumwa, IA 52501. Phone: (641) 684-5563. Fax: (641) 684-5832. Ownership: Bruce Linder, 60%; and Greg List, 40%.

Stns: 1 AM. 3 FM. KKSI-FM Eddyville, IA; KRKN-FM Eldon, IA; KTWA-FM Ottumwa, IA; KBIZ Ottumwa, IA.

Ouachita Broadcasting Inc., Box 1450, Mena, AR 71953. Phone: (479) 394-1450. Ownership: Jay Bunyard, 100%.

Stns: 1 AM. 3 FM. KILX(FM) Hatfield, AR; KQOR(FM) Mena, AR; KENA(AM) Mena, AR; KENA-FM Mena, AR.

Jay Bunyard, pres; Teresa Bunyard, VP.

Our Three Sons Broadcasting L.L.P., Box 307, Rock Hill, SC 29731. Phone: (803) 324-1340. Fax: (803) 324-2860. Web Site: www.wrhi.com. Ownership: Allan M. Miller, mgng ptnr; Manning Kimmel, ptnr.

Stns: 1 AM. 1 FM. WVSZ-FM Chesterfield, SC; WRHI(AM) Rock Hill, SC.

Allan M. Miller, pres.

Buck Owens Productions Inc., 3223 Sillect, Bakersfield, CA 93308. Phone: (661) 326-1011. Fax: (661) 328-7503. Ownership: Buck Owens Revocable Trust II (Michael Owens and Melvin L. Owens Jr., co-trustees), 100%.

Stns: 1 AM. 2 FM. KCWR(FM) Bakersfield, CA; KUZZ Bakersfield, CA; KUZZ-FM Bakersfield, CA.

P

Pacific Empire Radio Corp., 228 1st St., Idaho Falls, ID 83401. Phone: (208) 528-6813. Fax: (208) 529-6927. E-mail: kclkam@aol.com Web Site: www.hot106.fm.com. Ownership: Mark L. Bolland and Mary Bolland, JTWROS, 34.3%; John Taylor and Connie Taylor, JTWROS, 30.7%; AIA Services Corp. 401K & Profit Sharing Plan, FBO John Taylor, 13.6%; Hillcrest Aircraft Co., 6.84%; and Randolph Lamberjack, 6%.

Stns: 4 AM. 9 FM. KATW(FM) Lewiston, ID; KSEI Pocatello, ID; KGTM(FM) Rexburg, ID; KRXK(AM) Rexburg, ID; KBJX(FM) Shelley, ID; KQZB(FM) Troy, ID; KBKR Baker City, OR; KKBC-FM Baker City, OR; KRJT(FM) Elgin, OR; KLBM La Grande, OR; KUBQ-FM La Grande, OR; KVAB-FM Clarkston, WA; KCLK-FM Clarkston, WA.

Mark Bolland, pres/CEO.

Pacific Radio Group Inc., 311 Ano St., Kahului, HI 96732. Phone: (808) 877-5566. Fax: (808) 871-0666. E-mail: bergson@pacificradiogroup.com Web Site: www.pacificradiogroup.com. Ownership: Ed Johnson, Robert Van Dine, Chuck Bergson.

Stns: 4 AM. 10 FM. KHLO Hilo, HI; KAPA(FM) Hilo, HI; KKBG-FM Hilo, HI; KPVS-FM Hilo, HI; KLEO(FM) Kahaluu, HI; KLHI-FM Kahului, HI; KNUI Kahului, HI; KJKS(FM) Kahului, HI; KLUA(FM) Kailua-Kona, HI; KKON(AM) Kealakekua, HI; KPOA-FM Lahaina, HI; KJMD(FM) Pukalani, HI; KMVI Wailuku, HI; KAGB(FM) Waimea, HI.

Chuck Bergson, CEO; Robert Van Dine, dir; L.E. Johnson, CFO.

Pacific West Broadcasting Inc., Box 1430, Newport, OR 97365. Phone: (541) 265-2266. Fax: (541) 265-6397. E-mail: info@ybcradio.com Web Site: ybcradio.com. Ownership: David J. & Linda R. Miller, 100%.

Stns: 1 AM. 2 FM. KBCH Lincoln City, OR; KCRF-FM Lincoln City, OR; KNCU(FM) Newport, OR.

David Miller, pres.

Pacifica Foundation Inc., (dba Pacific Radio). 1925 Martin Luther King Jr. Way, Berkeley, CA 94704. Phone: (510) 849-2590. Web Site: www.pacifica.org. Ownership: (dba Pacific Radio).

Stns: 5 FM. KPFA-FM Berkeley, CA; KPFK-FM Los Angeles, CA; WPFW-FM Washington, DC; WBAI-FM New York, NY; KPFT-FM Houston, TX.

Dan Cougjhlin, exec dir.

Pamal Broadcasting Ltd., 6 Johnson Rd., Latham, NY 12110. Phone: (518) 786-6600. Fax: (518) 786-6610. Web Site: www.pamal.com. Ownership: James Morrell, owner.

Stns: 13 AM. 21 FM. WKZY(FM) Cross City, FL; WDVH(AM) Gainesville, FL; WTMN(AM) Gainesville, FL; WRZN Hernando, FL; WXBM-FM Milton, FL; WHHZ(FM) Newberry, FL; WMEZ-FM Pensacola, FL; WDVH-FM Trenton, FL; WTMG-FM Williston, FL; WPYR(AM) Baton Rouge, LA; WPNI(AM) Amherst, MA; WYJB(FM) Albany, NY; WROW(AM) Albany, NY; WKLI-FM Albany, NY; WZMR-FM Altamont, NY; WBNR Beacon, NY; WXPK(FM) Briarcliff Manor, NY; WFFG-FM Corinth, NY; WMML Glens Falls, NY; WNYQ(FM) Hudson Falls, NY; WIZR Johnstown, NY; WGHQ Kingston, NY; WLNA(AM) Peekskill, NY; WSPK-FM Poughkeepsie, NY; WBPM(FM) Saugerties, NY; WENU(AM) South Glens Falls, NY; WFLY-FM Troy, NY; WAJZ(FM) Voorheesville, NY; WKBE-FM Warrensburg, NY; WEBK-FM Killington, VT; WJEN-FM Rutland, VT; WJJR-FM Rutland, VT; WZRT-FM Rutland, VT; WSYB Rutland, VT.

Michael Dufort, CFO; Debbie Grembowicz, gen mgr; Dan Austin, market mgr; Jason Finkelberg, market mgr; Clay Ashworth, market mgr; Dave Lobb, rgnl VP.

Pamplin Broadcasting, 888 S.W. Fifth Ave., Suite 790, Portland, OR 97204. Phone: (503) 223-4321. Fax: (503) 222-2850. E-mail: kpam@kpam.com

Stns: 3 AM. 1 FM. KDUN(AM) Reedsport, OR; KPAM Troutdale, OR; KTSL-FM Medical Lake, WA; KKAD(AM) Vancouver, WA.

Andrea Marek, pres; Paul Clithero, gen mgr.

Pappas Telecasting Companies, 500 S. Chinowth Rd., Visalia, CA 93277. Phone: (559) 733-7800. Fax: (559) 733-7878. Web Site: www.pappastv.com. Ownership: Harry J. Pappas.

Stns: 2 AM. KMPH(AM) Modesto, CA; KTRB(AM) San Francisco, CA.

Stns: 19 TV. WLGA, Columbus, GA; KDMI, Des Moines-Ames, IA; KCWI-TV, Des Moines-Ames, IA; KDBC-TV, El Paso (Las Cruces, NM), TX; KMPH-TV, Fresno-Visalia, CA; KFRE-TV, Fresno-Visalia, CA; WWAZ-TV, Green Bay-Appleton, WI; WCWG, Greensboro-High Point-Winston Salem, NC; KAZH, Houston; KWNB, Lincoln & Hastings-Kearney, NE; KHGI, Lincoln & Hastings-Kearney, NE; KAZA-TV, Los Angeles; KPTM, Omaha, NE; KREN, Reno, NV; KTNC, San Francisco-Oakland-San Jose; KUNO-TV, San Francisco-Oakland-San Jose; KPTH, Sioux City, IA; KCWK, Yakima-Pasco-Richland-Kennewick, WA; KSWT, Yuma, AZ-El Centro, CA.

Harry J. Pappas, chmn/CEO; Dennis J. Davis, pres/COO; Bruce M. Yeager, exec VP/CFO.

Paradis Broadcasting of Alexandria Inc., 1312 Broadway, Alexandria, MN 56308. Phone: (320) 763-3131. Fax: (320) 763-5641. E-mail: thefolks@kxra.com Web Site: www.kxra.com. Ownership: Mel Paradis, 60%; Brett Paradis, 40%.

Stns: 1 AM. 2 FM. KXRZ(FM) Alexandria, MN; KXRA Alexandria, MN; KXRA-FM Alexandria, MN.

Mel Paradis, CEO; Brett Paradis, pres & gen mgr.

Paragon Communications Inc., Box 945, Elk City, OK 73648. Phone: (580) 225-9696. Fax: (580) 225-9699. E-mail: keco@io2online.com Web Site: www.kecofm.com.

Stns: 1 AM. 2 FM. KADS Elk City, OK; KECO(FM) Elk City, OK; KXOO(FM) Elk City, OK.

Blake Brewer, pres & gen mgr.

The Jim Pattison Broadcast Group, 460 Pemberton Terrace, Kamloops, BC V2C 1T5. Canada. Phone: (250) 372-3322. Fax: (250) 374-0445. Web Site: www.jpbroadcast.com. Ownership: Jim Pattison Group.

Stns: 2 AM. 26 FM. CIBW-FM Drayton Valley, AB; CJXX-FM Grande Prairie, AB; CHLB-FM Lethbridge, AB; CFMY-FM Medicine Hat, AB; CHAT-FM Medicine Hat, AB; CHUB-FM Red Deer, AB; CFDV-FM Red Deer, AB; CHBW-FM Rocky Mountain House, AB; CJBZ-FM Taber, AB; CKLR-FM Courtenay, BC; CHBZ-FM Cranbrook, BC; CHDR-FM Cranbrook, BC; CJDR-FM Fernie, BC; CKBZ-FM Kamloops, BC; CIFM-FM Kamloops, BC; CKLZ-FM Kelowna, BC; CKOV Kelowna, BC; CKWV-FM Nanaimo, BC; CHWF-FM Nanaimo, BC; CIBH-FM Parksville, BC; CHPQ-FM Parksville, BC; CJAV-FM Port Alberni, BC; CKDV-FM Prince George, BC; CKKN-FM Prince George, BC; CJJR-FM Vancouver, BC; CKBD Vancouver, BC; CKKQ-FM Victoria, BC; CJZN-FM Victoria, BC.

Stns: 4 TV. CFJC-TV, Kamloops, BC; CHAT, Medicine Hat, AB; CHAT-1, Pivot, AB; CKPG, Prince George, BC.

Rick Arnish, pres; Bill Dinicol, VP Finance; Bruce Davis, VP sls; Loretta Lewis, admin asst.

Peak Broadcasting LLC, 1071 W. Shaw Ave., Fresno, CA 93711. Phone: (559) 490-5800. Fax: (559) 490-5843. Ownership: Duff Ackerman & Goodrich QP Fund II L.P., 86.0297% membership interest.

Stns: 4 AM. 9 FM. KFPT(AM) Clovis, CA; KOQO-FM Fresno, CA; KFJK(FM) Fresno, CA; KSKS-FM Fresno, CA; KWYE(FM) Fresno, CA; KMGV-FM Fresno, CA; KMJ Fresno, CA; KFXD(AM) Boise, ID; KAWO(FM) Boise, ID; KSAS-FM Caldwell, ID; KXLT-FM Eagle, ID; KCIX(FM) Garden City, ID; KIDO(AM) Nampa, ID.

Todd Lawley, CEO; Tim Lyons, CFO.

Pearson Broadcasting, 9530 Miolothian Pike, Richmond, VA 23235. Phone: (804) 521-0603. Fax: (804) 674-8938. Ownership: Max H. Pearson, 100%.

Stns: 4 FM. KBCN-FM Marshall, AR; KTTG-FM Mena, AR; KERX(FM) Paris, AR; KMAC-FM Gainesville, MO.

Max H. Pearson, pres; Bruce W. Hale, VP.

Pecos Valley Broadcasting Co., 317 W. Quay Ave., Artesia, NM 88210. Phone: (505) 746-2751. Fax: (505) 748-3748.

Stns: 1 AM. 3 FM. KSVP Artesia, NM; KTZA(FM) Artesia, NM; KPZE-FM Carlsbad, NM; KEND-FM Roswell, NM.

Sam Beard, pres; David Ruckman, VP; Gene Dow, VP/gen mgr.

Peg Broadcasting Crossville LLC, 961 Miller Ave., Crossville, TN 38555. Phone: (931) 707-1102. Fax: (931) 707-1220. Ownership: Jeffrey H. Shaw, 50%; and John T. Crunk Jr., 50%.

Stns: 2 AM. 2 FM. WPBX(FM) Crossville, TN; WOWF-FM Crossville, TN; WAEW Crossville, TN; WCSV(AM) Crossville, TN.

Peggy Sue Broadcasting Corp., Box 838, Richlands, VA 24641. Phone: (276) 964-4066. Fax: (276) 963-4927. Ownership: Henry Beam, 50%; Dirk Hall, 49%; and Peggy Beam, 1%.

Stns: 1 AM. 2 FM. WMJD-FM Grundy, VA; WNRG Grundy, VA; WRIC-FM Richlands, VA.

Pembrook Pines Media Group, 1705 Lake St., Elmira, NY 14901. Phone: (607) 733-5626. Fax: (607) 733-5627. E-mail: ppinesmedia1@stny.rr.com Web Site: www.pembrookpines.com. Ownership: Robert J. Pfuntner, 100%. Company also owns Pembrook Pines Media agency.

Stns: 6 AM. 5 FM. WZKZ-FM Alfred, NY; WABH Bath, NY; WELM Elmira, NY; WEHH Elmira Heights-Horseheads, NY; WOEN(AM) Olean, NY; WMXO-FM Olean, NY; WGGO Salamanca, NY; WQRS(FM) Salamanca, NY; WOKN-FM Southport, NY; WPIE Trumansburg, NY; WQRW(FM) Wellsville, NY.

Robert J. Pfuntner, pres/CEO.

Peninsula Communications Inc., Box 109, Homer, AK 99603. Phone: (907) 235-6000. Fax: (907) 235-6683. E-mail: kwavefm@xyz.net Ownership: David F. Becker, 50%; Eileen L. Becker, 50%.

Stns: 1 AM. 3 FM. KGTL Homer, AK; KWVV-FM Homer, AK; KXBA(FM) Nikiski, AK; KPEN-FM Soldotna, AK.

Tim White, opns mgr; Dave Webb, production mgr; David Becker, gen mgr.

Perception Media Group Inc., 1848 Clay St. S.E., Roanoke, VA 24013. Phone: (540) 343-7109. Fax: (540) 343-2306. E-mail: 3wr@3wradio.com Web Site: www.3wrradio.com.

Stns: 3 AM. WCQV(AM) Moneta, VA; WNRV(AM) Narrows-Pearisburg, VA; WWWR Roanoke, VA.

Ben Peyton, pres/CEO; Barbara Evans, gen mgr.

Perry Publishing & Broadcasting Co., c/o KVSP(AM), 1528 N.E. 23rd St., Oklahoma City, OK 73111. Phone: (405) 425-4100. Fax: (405) 424-8811. Web Site: www.kvsp.com.

Stns: 5 AM. 4 FM. KVSP(FM) Anadarko, OK; KJMM-FM Bixby, OK; KJMZ(FM) Cache, OK; KDDQ(FM) Comanche, OK; KPNS(AM) Duncan, OK; KKRX Lawton, OK; KXCA(AM) Lawton, OK; KRMP(AM) Oklahoma City, OK; KGTO Tulsa, OK.

Russell Perry, pres/CEO.

Petracom Media L.L.C., 130 Hampton Point Dr., Saint Simons Island, GA 31522-5426. Phone: (813) 948-2554. Fax: (813) 948-2557. Web Site: www.petracommedia.com. Ownership: Henry Ash.

Stns: 2 AM. 2 FM. KDJI Holbrook, AZ; KZUA-FM Holbrook, AZ; KSNX-FM Show Low, AZ; KVWM(AM) Show Low, AZ.

Henry A. Ash, pres/CEO; Joseph M. Fry, CFO; F. Lewis Robertson, COO.

Pharis Broadcasting Inc., Box 908, Fort Smith, AR 72902. Phone: (479) 288-1047. Fax: (479) 785-2638. E-mail: ssrg@sbcglobal.net Web Site: www.fortsmithradiogroup.com. Ownership: William L. Pharis, 51%; Karen Ann Pharis, 49%.

Stns: 2 AM. 3 FM. KFPW-FM Barling, AR; KQBK(FM) Booneville, AR; KFPW Fort Smith, AR; KHGG(AM) Van Buren, AR; KHGG-FM Waldron, AR.

William L. Pharis, pres/CEO; Karen A. Pharis, gen mgr & sec.

Phillips Broadcasting Inc., 100 Fisher Dr., Trinidad, CO 81082. Phone: (719) 846-3355. Fax: (719) 846-4711. E-mail: kcrt@adelphia.net

Stns: 1 AM. 1 FM. KCRT Trinidad, CO; KBKZ(FM) Raton, NM.

David Phillips, pres.

Phoenix Media Communications Group, 126 Brookline Ave., Boston, MA 02215. Phone: (617) 536-5390. Fax: (617) 859-8201. Web Site: www.thephoenix.com.

Stns: 1 AM. 3 FM. WFNX(FM) Lynn, MA; WPHX(AM) Sanford, ME; WPHX-FM Sanford, ME; WFEX(FM) Peterborough, NH.

Barry Morris, pres; Stephen Mindich, CEO.

Piedmont Communications Inc., Box 271, Orange, VA 22960. Phone: (540) 672-1000. Fax: (540) 672-0282. Ownership: Thomas D. Bond, 22.6%; A. Pierce Stone & Pamela H. Stone, 11%; The Cook Family Trust, Mrs. Toy E. Cook, trustee, Richard S. Cook, atty in fact, 9.8%; Lloyd M. Garnett & Barbara S. Garnett, 8.6%; Robert F. Gillespie Jr., 8.6%; and Harry B. Sedwick Jr., 8.6%.

Stns: 2 AM. 1 FM. WCVA Culpeper, VA; WOJL(FM) Louisa, VA; WVCV Orange, VA.

Pilgrim Communications Inc., 54 Monument Cir., Suite 250, Indianapolis, IN 46204. Phone: (317) 655-9999.

Stns: 1 AM. KVLE(AM) Vail, CO.

Pillar of Fire Inc., Box 9058, Weston Canal Rd., Zarephath, NJ 08890. Phone: (732) 469-0991. Fax: (732) 469-2115. E-mail: info@star991fm.com Web Site: www.star991fm.com. Ownership: No stockholders; non-profit corporation.

Stns: 1 AM. 2 FM. KPOF Denver, CO; WAWZ(FM) Zarephath, NJ; WAKW-FM Cincinnati, OH.

Pillar of Fire Inc. publishes one religious periodical, a semi-monthly for the family *Pillar of Fire.*

Robert B. Dallenbach, pres; Scott Taylor, stn mgr.

Pines Broadcasting Inc., 1255 N. Myrtle St., Warren, AR 71671. Phone: (870) 226-2653. Fax: (870) 226-3039. Ownership: Jimmy Sledge, 50%; and Gwen Sledge, 50%.

Stns: 1 AM. 3 FM. KXSA-FM Dermott, AR; KGPQ-FM Monticello, AR; KHBM Monticello, AR; KHBM-FM Monticello, AR.

Jimmy Sledge, pres.

Pittman Broadcasting Services LLC, 307 S. Jefferson Ave., Covington, LA 70433. Phone: (985) 892-3661. Web Site: www.pittmanbroadcasting.com. Ownership: Marcus Pittman, 50%; and Janet Pittman, 50%.

Stns: 3 AM. 3 FM. WOMN(AM) Franklinton, LA; WUUU(FM) Franklinton, LA; KFXZ(AM) Lafayette, LA; KVOL Lafayette, LA; KKSJ(FM) Maurice, LA; KFXZ-FM Opelousas, LA.

Marcus Pittman, pres.

Platinum Broadcasting Co., Box 789, Junction City, KS 66441. Phone: (785) 762-5525. Fax: (785) 762-5387. E-mail: platinum@kjck.com Web Site: www.kjck.com.

Stns: 1 AM. 2 FM. KJCK(AM) Junction City, KS; KJCK-FM Junction City, KS; KQLA-FM Ogden, KS.

Mark Ediger, pres.

Platte River Radio Inc., Box 130, Kearney, NE 68848. Phone: (308) 236-9900. Fax: (308) 234-6781. Ownership: David Oldfather, 31%; Craig J. Eckert, 30%; Jane O. Light, 19.5%; and Diane H. Oldfather, 19.5%.

Stns: 3 AM. 2 FM. KLIQ(FM) Hastings, NE; KHAS Hastings, NE; KICS Hastings, NE; KXPN(AM) Kearney, NE; KKPR-FM Kearney, NE.

Plessinger Radio Group, (R.L. Plessinger Holding Co.). 8354 Fryer Rd., Georgetown, OH 45121. Phone: (937) 378-6151. Fax: (937) 378-4143. E-mail: rick@waxz.com Web Site: www.977waxz.com. Ownership: (R.L. Plessinger Holding Co.)

Stns: 1 AM. 2 FM. WOYS-FM Apalachicola, FL; WOCY-FM Carrabelle, FL; WCVG Covington, KY.

Richard Plessinger, pres/CEO.

Point Broadcasting Company, 715 Broadway, Suite 320, Santa Monica, CA 90401. Phone: (310) 451-4430. Fax: (310) 451-1423. Ownership: John Hearne Revocable Trust. Note: Owns stns through subsidiaries: Gold Coast Broadcasting LLC (see listing) and High Desert Broadcasting LLC (see listing). Also owns 50% of KHRN(FM) Huron, CA and 31.8% of KHRQ(FM) Baker and KHDR(FM) Lenwood, both CA.

Stns: 1 AM. 5 FM. KCEL(FM) California City, CA; KSBL-FM Carpinteria, CA; KSPE-FM Ellwood, CA; KTYD-FM Santa Barbara, CA; KIST-FM Santa Barbara, CA; KBKO Santa Barbara, CA.

John Hearne, chmn/pres.

Pollack Broadcasting Co., 5500 Poplar Ave. #1, Memphis, TN 38119. Phone: (901) 685-3993. Fax: (901) 685-3995. E-mail: wpollack@midsouth.rr.com Ownership: William H. Pollack, 100%. Note: Group also owns KWCE-LP Alexandria, LA.

Stns: 3 AM. 3 FM. KBOA-FM Piggott, AR; KCRV Caruthersville, MO; KCRV-FM Caruthersville, MO; KBOA Kennett, MO; KTMO(FM) New Madrid, MO; KMIS Portageville, MO.

Stns: 2 TV. KLAX, Alexandria, LA; KIEM, Eureka, CA.

William H. Pollack, pres.

Polnet Communications Ltd., 3656 W. Belmont Ave., Chicago, IL 60618. Phone: (773) 588-6300. Fax: (773) 588-0834. Web Site: www.pclradio.com.

Stns: 6 AM. WKTA Evanston, IL; WEEF Highland Park, IL; WNVR Vernon Hills, IL; WPJX(AM) Zion, IL; WRKL New City, NY; WLIM Patchogue, NY.

Walter Kotaby, pres; Kent D. Gustafson, VP & CEO.

Porter County Broadcasting Corp., 2755 Sager Rd., Valparaiso, IN 46383. Phone: (219) 462-8125. Ownership: Leonard J. Ellis Trust, 27.10%; Bernice A. Ellis Trust, 27.10%; Leigh Ellis, 15.25%; Neenah Ellis, 15.25%; and Marissa Wilson, 15.25%.

Stns: 1 AM. 3 FM. WXRD-FM Crown Point, IN; WZVN-FM Lowell, IN; WAKE Valparaiso, IN; WLJE-FM Valparaiso, IN.

Positive Alternative Radio Inc., Box 889, Blacksburg, VA 24063. Phone: (540) 552-4282. Fax: (540) 951-5282. Web Site: www.parfm.com. Ownership: Vernon H. Baker, 33.33%; Virginia L. Baker, 33.33%; and Edward A. Baker, 33.33%.

Stns: 21 FM. WKAO(FM) Ashland, KY; WTJY-FM Asheboro, NC; WPIR-FM Hickory, NC; WXRI-FM Winston-Salem, NC; WCQR-FM Kingsport, TN; WTTX-FM Appomattox, VA; WPER-FM Culpeper, VA; WOKD-FM Danville, VA; WPIN-FM Dublin, VA; WJYA-FM Emporia, VA; WJYJ-FM Fredericksburg, VA; WOKG(FM) Galax, VA; WPIM-FM Martinsville, VA; WJCN(FM) Nassawadox, VA; WRXT-FM Roanoke, VA; WPAR-FM Salem, VA; WPVA-FM Waynesboro, VA; WPJY(FM) Blennerhassett, WV; WPIB-FM Bluefield, WV; WPJW(FM) Hurricane, WV; WPCN-FM Point Pleasant, WV.

Edward A. Baker, pres.

Powell Broadcasting Co. Inc., 8641 United Plaza Blvd., Suite 300, Baton Rouge, LA 70809. Phone: (225) 922-4662. Fax: (225) 922-4544. Ownership: The Powell Group L.L.C., 100%.

Stns: 2 AM. 4 FM. KKMA(FM) Le Mars, IA; KLEM(AM) Le Mars, IA; KZSR(FM) Onawa, IA; KSCJ Sioux City, IA; KKYY(FM) Whiting, IA; KSUX(FM) Winnebago, NE.

Thomas J. Spies, COO; Nanette Noland, pres.

Prairie Radio Communications, 2410 Sycamore Rd., Suite C, De Kalb, IL 60115. Phone: (815) 758-8686. Fax: (815) 756-9723. Web Site: www.radiomacomb.com.

Stns: 7 AM. 10 FM. KCLN Clinton, IA; KMCN(FM) Clinton, IA; KMCS(FM) Muscatine, IA; KWPC Muscatine, IA; WLMD-FM Bushnell, IL; WCDD(FM) Canton, IL; WBYS Canton, IL; WLBK De Kalb, IL; WAIK Galesburg, IL; WLRB Macomb, IL; WKAI-FM Macomb, IL; WMOI-FM Monmouth, IL; WRAM Monmouth, IL; WPWQ(FM) Mount Sterling, IL; WKXQ-FM Rushville, IL; KWBZ(FM) Monroe City, MO; WSLD-FM Whitewater, WI.

Don Davis, pres/CEO.

Premier Broadcasters, 1133 Kresky, Centralia, WA 98531. Phone: (360) 736-1355. Fax: (360) 736-4761. Web Site: www.live95.com. Ownership: Rod Etherton.

Stns: 1 AM. 2 FM. KITI Chehalis-Centralia, WA; KRXY-FM Shelton, WA; KITI-FM Winlock, WA.

Rod Etherton, pres.

Prescott Valley Broadcasting Co. Inc., Box 26523, Prescott Valley, AZ 86312. Phone: (928) 445-8289. Fax: (928) 442-0448. Ownership: Sanford Cohen and Terry Cohen, joint tenants with right of survivorship, 100%.

Stns: 1 AM. 2 FM. KPKR(FM) Parker, AZ; KPPV(FM) Prescott Valley, AZ; KQNA(AM) Prescott Valley, AZ.

Press Communications L.L.C., 1329 Campus Pkwy., Suite 106, Wall, NJ 07753-6815. Phone: (732) 751-1119. Fax: (732) 751-1726. Ownership: Mark D. Lass, 16.5%; Alfred D. Colantoni, 16.5%; Jules L. Plangere III, 16.5%; Jules Plangere Jr., 16.5%; Robert E. McAllan, 16.5%; Richard T. Morena, 10.5%; and E. Donald Lass, 7%.

Stns: 1 AM. 5 FM. WBBO(FM) Bass River Township, NJ; WHTG Eatontown, NJ; WHTG-FM Eatontown, NJ; WWZY-FM Long Branch, NJ; WKMK(FM) Ocean Acres, NJ; WBHX-FM Tuckerton, NJ.

Prettyman Broadcasting Co., 1606 W. King St., Martinsburg, WV 25401. Phone: (304) 263-8868. Fax: (304) 263-8906. Web Site: www.wepm.com.

Stns: 1 AM. 2 FM. WEPM Martinsburg, WV; WLTF(FM) Martinsburg, WV; WICL(FM) Williamsport, MD.

William E. Prettyman, pres/CEO; Norm Slemenda, gen mgr.

Priority Communications, 12 W. Long Ave., DuBois, PA 15801-2100. Phone: (814) 375-5260. Fax: (814) 375-5263.

Stns: 2 AM. 2 FM. WCDK-FM Cadiz, OH; WCED DuBois, PA; WDSN-FM Reynoldsville, PA; WEIR Weirton, WV.

Jay M. Philippone, pres.

Priority Radio Inc., Box 5204, Wilmington, DE 19808-5204. Phone: (302) 731-7270. Fax: (302) 738-3090. Web Site: www.thereachfm.com. Ownership: Jennifer Hare, 33.3%; Rev. Steve Hare, 33.3%.

Stns: 1 AM. 3 FM. KXRL(FM) Cherry Valley, AR; WVBH(FM) Beach Haven West, NJ; WSRY(AM) Elkton, MD; WXHL-FM Christiana, DE.

Programmers Broadcasting Inc., Box 28, Bottineau, ND 58318-0028. Phone: (701) 228-5151. Fax: (701) 228-2483. Ownership: John Kircher, 50%; and Jean Kircher, 50%.

Stns: 3 FM. KBTO(FM) Bottineau, ND; KWGO(FM) Burlington, ND; KTZU(FM) Velva, ND.

Progressive Broadcasting System Inc., Box 307, Elkhart, IN 46515. Phone: (574) 875-5166. Fax: (574) 875-6662. Web Site: www.wfrn.com.

Stns: 2 FM. WFRN-FM Elkhart, IN; WFRI-FM Winamac, IN.

Edwin Moore, pres.

Q

Qantum Communications Corp., 3 Stamford Landing, Suite 210, Stamford, CT 06902. Phone: (203) 388-0048. Ownership: Frank D. Osborn, 61.74% of votes; Frank Washington, 19.03% of votes; Osborn Family Partners L.P., 15.23% of votes; Michael F. Mangan, 3.85% of votes; William Nelson III and Frank D. Osborn, as trustees of the Osborn 2002 Family Trust, a trust in favor of the children of Frank D. Osborn, 0.15% of votes.

Stns: 8 AM. 21 FM. WKKR-FM Auburn, AL; WMXA(FM) Opelika, AL; WTLM Opelika, AL; WZMG Pepperell, AL; WMXZ-FM De Funiak Springs, FL; WWAV-FM Santa Rosa Beach, FL; WMOG Brunswick, GA; WGIG(AM) Brunswick, GA; WHFX(FM) Darien, GA; WBGA(FM) Saint Simons Island, GA; WYNR(FM) Waycross, GA; WWSN-FM Waycross, GA; WPLV West Point, GA; WCJM-FM West Point, GA; WCIB-FM Falmouth, MA; WCOD-FM Hyannis, MA; WRZE(FM) Nantucket, MA; WXTK-FM West Yarmouth, MA; WLQB(FM) Ocean Isle Beach, NC; WQSD(FM) Briarcliff Acres, SC; WGTR-FM Bucksport, SC; WJMX-FM Cheraw, SC; WWRK(AM) Darlington, SC; WDAR-FM Darlington, SC; WDSC Dillon, SC; WJMX Florence, SC; WWXM(FM) Garden City, SC; WGSS-FM Kingstree, SC; WZTF(FM)

Scranton, SC.

Frank Osborn, pres/CEO; Michael Mangan, CFO.

Quarnstrom Media Group LLC, 1104 Cloquet Ave., Cloquet, MN 55720. Phone: (218) 879-4534. Fax: (218) 879-1962. Ownership: Alan & Linda Quarnstrom, 100%.

Stns: 2 AM. 3 FM. WKLK Cloquet, MN; WKLK-FM Cloquet, MN; WMOZ(FM) Moose Lake, MN; WCMP Pine City, MN; WCMP-FM Pine City, MN.

Aian Quarnstrom, pres; Don Welch, VP.

Quincy Newspapers Inc., 130 S. Fifth St., Quincy, IL 62301. Phone: (217) 223-5100. Fax: (217) 223-5019. Web Site: www.qni.biz.

Stns: 1 AM. 1 FM. WGEM Quincy, IL; WGEM-FM Quincy, IL.

Stns: 12 TV. WVVA, Bluefield-Beckley-Oak Hill, WV; KWWL, Cedar Rapids-Waterloo-Iowa City & Dubuque, IA; WQOW-TV, La Crosse-Eau Claire, WI; WXOW-TV, La Crosse-Eau Claire, WI; WKOW, Madison, WI; WGEM-TV, Quincy, IL-Hannibal, MO-Keokuk, IA; KTTC, Rochester, MN-Mason City, IA-Austin, MN; WREX-TV, Rockford, IL; KTIV, Sioux City, IA; WSJV, South Bend-Elkhart, IN; WAOW-TV, Wausau-Rhinelander, WI; WYOW, Wausau-Rhinelander, WI.

Quincy Newspapers Inc. owns the *Quincy* (IL) *Herald-Whig*, and the *New Jersey Herald*, Newton, NJ.

Thomas A. Oakley, pres.

Quinte Broadcasting Ltd., Box 488, Belleville, ON K8N 5B2. Canada. Phone: (613) 969-5555. Fax: (613) 969-8122. Ownership: Herbert M. Morton, 66.67%; and Joyce Mulock, 33.33%.

Stns: 1 AM. 2 FM. CIGL-FM Belleville, ON; CJBQ Belleville, ON; CJTN-FM Quinte West, ON.

Quorum Radio Partners of Virginia Inc., 8512 Beech Ln., McKinney, TX 75070. Phone: (972) 529-1192. Fax: (972) 540-2454. Ownership: Todd W. Fowler, 31.25%; Michael A. Stone, 31.25%; Jevin S. Jensen, 20%; Robert Barnett, 10%; and Kevin T. Lilly, 5%.

Stns: 2 AM. 2 FM. WIQO-FM Covington, VA; WKEY Covington, VA; WKCJ-FM Lewisburg, WV; WSLW White Sulphur Springs, WV.

Todd W. Fowler, pres/CEO.

R

The RAFTT Corp., 3633 Farm to Market Rd. 437, Rogers, TX 76569. Phone: (281) 564-7064. Ownership: Jerome Friemel, 100%.

Stns: 3 AM. WHNC Henderson, NC; WCBQ Oxford, NC; KTON Belton, TX.

RNC MEDIA Inc., 380 Murdoch, Rowyn-Norando, PQ J9X 1G5. Canada. Phone: (514) 866-8686. Fax: (514) 866-8056. Web Site: www.radionord.com.

Stns: 9 FM. CHPR-FM Hawkesbury, ON; CKNU-FM Donnacona, PQ; CHLX-FM Gatineau, PQ; CFTX-FM Gatineau, PQ; CJLA-FM Lachute, PQ; CKLX-FM Montreal, PQ; CHOI-FM Quebec, PQ; CHOA-FM Rouyn-Noranda, PQ; CHGO-FM Val d'Or, PQ.

Stns: 5 TV. CKRN-3, Bearn-Fabre, PQ; CFGS-TV, Gatineau, PQ; CHOT-TV, Gatineau, PQ; CKRN-TV, Rouyn-Noranda, PQ; CFVS, Val d'Or, PQ.

Pierre R. Brosseau, pres.

RR Broadcasting, 2100 E. Tahquitz Canyon Way, Palm Springs, CA 92262. Phone: (760) 325-2582. Fax: (760) 322-3562. Web Site: www.rrbroadcasting.com. Ownership: Rozene R. Supple, 100%.

Stns: 3 AM. 2 FM. KPTR(AM) Cathedral City, CA; KDES-FM Palm Springs, CA; KGAM Palm Springs, CA; KPSI Palm Springs, CA; KPSI-FM Palm Springs, CA.

Mike Keane, gen mgr.

Radio Cleveland Inc., Drawer 780, Cleveland, MS 38732. Phone: (662) 843-4091. Fax: (662) 843-9805. E-mail: wcld@tecinfo.com Web Site: www.radiomiss.com. Ownership: Homer Sledge Jr., pres, 37.1/5%; Kevin W. Cox, treas, 37.1/2%; Clint L. Webster, gen mgr, 37.1/2%.

Stns: 1 AM. 2 FM. WAID-FM Clarksdale, MS; WCLD Cleveland, MS; WMJW-FM Cleveland, MS.

Clint L. Webster, gen mgr.

Radio Dubuque Inc., Box 659, Dubuque, IA 52004. Phone: (563) 690-0800. Fax: (563) 588-5688. Ownership: Donald L. Rabbitt, 70%; Thomas Parsley, 25%; and Paul Hemmer, 5%.

Stns: 1 AM. 3 FM. KATF(FM) Dubuque, IA; KDTH Dubuque, IA; KGRR-FM Epworth, IA; WVRE(FM) Dickeyville, WI.

Thomas Parsley, gen mgr.

Radio Fargo-Moorhead Inc., 2302 University Dr. S., Fargo, ND 58103. Phone: (701) 277-4200. Ownership: James D. Ingstad, 100%.

Stns: 2 AM. 4 FM. KBVB(FM) Barnesville, MN; WDAY-FM Fargo, ND; KFGO Fargo, ND; KRWK(FM) Fargo, ND; KKAG(AM) Fargo, ND; KMXW(FM) Hope, ND.

Radio Greeneville Inc., Box 278, Greeneville, TN 37744. Phone: (423) 638-4147. Fax: (423) 638-1979. E-mail: wgrv@greeneville.com Web Site: www.greeneville.com/wgrv. Ownership: Ronald & Nellie R.Metcalfe; Paul O. Metcalfe.

Stns: 2 AM. 1 FM. WSMG Greeneville, TN; WGRV Greeneville, TN; WIKQ(FM) Tusculum, TN.

Ronald Metcalfe, pres.

The Radio Group, Box 1319, Columbia, LA 71418. Phone: (318) 649-7959. Fax: (318) 649-5874. Ownership: Tom D. Gay, 100%.

Stns: 3 FM. KFNV-FM Ferriday, LA; KAPB-FM Marksville, LA; KMAR-FM Winnsboro, LA.

Tom D. Gay, gen mgr.

Radio La Grande, 1010 Vermont Ave. N.W., Suite 100, Washington, DC 20005. Phone: (202) 638-1959. Fax: (202) 393-7464. E-mail: VALTRAVEL@RCN.com Web Site: radiolagrande.net. Ownership: Estuardo Valdemar Rodriguez, 50%; Leonor Rodriguez, 50%. Note: Estuardo Valdemar Rodriguez also is the licensee of WLLN(AM) Lillington, NC.

Stns: 7 AM. WLLQ(AM) Chapel Hill, NC; WRTG Garner, NC; WSRP(AM) Jacksonville, NC; WLNR Kinston, NC; WGSB Mebane, NC; WREV Reidsville, NC; WLLY Wilson, NC.

Radio Maria Inc., 601 Washington St., Alexandria, LA 71301. Phone: (318) 561-6145. Fax: (318) 449-9954. Web Site: www.radiomaria.us. Ownership: Radio Maria is a not-for-profit corporation run by a board of directors.

Stns: 3 AM. 2 FM. KJMJ(AM) Alexandria, LA; KOJO-FM Lake Charles, LA; KBIO(FM) Natchitoches, LA; KNIR New Iberia, LA; KDEI(AM) Port Arthur, TX.

Radio One Inc., 5900 Princess Garden Pkwy., Lanham, MD 20706. Phone: (301) 306-1111. Fax: (301) 306-9426. Web Site: www.radio-one.com. Ownership: Alfred C. Liggins, 39.4% of voting shares; Catherine L. Hughes, 16.7% of voting shares.

Stns: 11 AM. 55 FM. KRBV(FM) Los Angeles, CA; WYCB Washington, DC; WKYS-FM Washington, DC; WTPS(AM) Coral Gables, FL; WFXA-FM Augusta, GA; WAEG-FM Evans, GA; WPZE(FM) Fayetteville, GA; WHTA(FM) Hampton, GA; WAMJ(FM) Mableton, GA; WJZZ-FM Roswell, GA; WTHB-FM Waynesboro, GA; WAKB-FM Wrens, GA; WFUN-FM Bethalto, IL; WHHL(FM) Jerseyville, IL; WLRX(FM) Charlestown, IN; WMOJ-FM Connersville, IN; WTLC-FM Greenwood, IN; WTLC Indianapolis, IN; WHHH-FM Indianapolis, IN; WGZB-FM Lanesville, IN; WYJZ(FM) Speedway, IN; WIZF-FM Erlanger, KY; WMJM-FM Jeffersontown, KY; WXMA(FM) Louisville, KY; WDJX-FM Louisville, KY; WLRS(FM) Shepherdsville, KY; WILD Boston, MA; WDMK(FM) Detroit, MI; WHTD(FM) Mount Clemens, MI; WCHB Taylor, MI; WPZS(FM) Albemarle, NC; WFXC-FM Durham, NC; WNNL-FM Fuquay-Varina, NC; WQNC(FM) Harrisburg, NC; WFXK-FM Tarboro, NC; WRNB(FM) Pennsauken, NJ; WDBZ(AM) Cincinnati, OH; WZAK-FM Cleveland, OH; WENZ-FM Cleveland, OH; WJMO(AM) Cleveland, OH; WERE(AM) Cleveland Heights, OH; WCKX-FM Columbus, OH; WING Dayton, OH; WGTZ-FM Eaton, OH; WJYD(FM) London, OH; WDHT(FM) Springfield, OH; WXMG-FM Upper Arlington, OH; WKSW-FM Urbana, OH; WROU-FM West Carrollton, OH; WPPZ-FM Jenkintown, PA; WPHI-FM Media, PA; KBFB(FM) Dallas, TX; KSOC(FM) Gainesville, TX; KMJQ-FM Houston, TX; KBXX(FM) Houston, TX; KROI(FM) Seabrook, TX; WPZZ(FM) Crewe, VA; WCDX-FM Mechanicsville, VA; WKJM(FM) Petersburg, VA; WROU(AM) Petersburg, VA; WKJS(FM) Richmond, VA; WQOK-FM South Boston, VA; WWIN Baltimore, MD; WERQ-FM Baltimore, MD; WMMJ-FM Bethesda, MD; WWIN-FM Glen Burnie, MD.

Catherine Hughes, chairperson; Alfred Liggins, pres/CEO; Scott Royster, CFO; Darrell Huckaby, VP progmg; Tony Washington, VP sls; Charles Kinney, engrg dir.

Radio Palouse Inc., Box 1, Pullman, WA 99163. Phone: (509) 332-6551. Fax: (509) 332-5151. E-mail: khtr@aol.com Web Site: www.border104.com.

Stns: 2 AM. KQQQ Pullman, WA; KUUX(AM) Pullman, WA.

Bill Weed, gen mgr.

Radio Partners LLC, Box 719, Beaver Falls, PA 15010. Phone: (724) 846-4100. Fax: (724) 843-7771. E-mail: iorio@wbvp-wmba.com Web Site: kibcoradio.com. Ownership: Frank Iorio, 100%.

Stns: 1 AM. 2 FM. WKNB-FM Clarendon, PA; WNAE Warren, PA; WRRN-FM Warren, PA.

Dave Whipple, sls mgr.

Radio Stations WPAY/WPFB Inc., 4505 Central Ave., Middletown, OH 45044. Phone: (513) 422-3625. Fax: (513) 424-9732. Web Site: www.rebel1059.com. Ownership: Douglas L. Braden, 100%.

Stns: 2 AM. 2 FM. WPFB Middletown, OH; WPFB-FM Middletown, OH; WPAY Portsmouth, OH; WPAY-FM Portsmouth, OH.

Douglas L. Branden, pres.

Radio Vermont Group Inc., Box 550, Waterbury, VT 05676. Phone: (802) 244-7321. Fax: (802) 244-1771.

Stns: 1 AM. 3 FM. WLVB-FM Morrisville, VT; WCVT-FM Stowe, VT; WDEV-FM Warren, VT; WDEV Waterbury, VT.

Ken Squier, pres; Eric Michaels, VP.

Radio Works Inc., 111 Westwood Dr., De Queen, AR 71832. Phone: (870) 642-3637. Ownership: Jay Wallace Bunyard and Teresa Sharon Bunyard Living Revocable Trust, Jay and Teresa Bunyard, sole voting trustees, 100%.

Stns: 3 FM. KAMD-FM Camden, AR; KMGC-FM Camden, AR; KCXY(FM) East Camden, AR.

Radioactive LLC, 1717 Dixie Hwy., Suite 650, Fort Wright, KY 41011. Phone: (859) 331-9100. Ownership: Benjamin L. Homel, 100%.

Stns: 10 FM. WMLF(FM) Watseka, IL; WPNS(FM) Brodhead, KY; WANK(FM) Mount Vernon, KY; WUPG(FM) Crystal Falls, MI; WUPF(FM) Gwinn, MI; WXPT(FM) Powers, MI; WUPZ(FM) Republic, MI; KEVE(FM) Ingram, TX; WTRW(FM) Two Rivers, WI; WDYK(FM) Ridgeley, WV.

RadioJones LLC, Box 5356, Atlanta, GA 31107-5356. Phone: (404) 432-1450. E-mail: dj@radiojones.com Web Site: www.radiojones.com. Ownership: Dennis Jones, 100%.

Stns: 2 AM. 2 FM. WELT(FM) East Dublin, GA; WJAT Swainsboro, GA; WRJS(AM) Swainsboro, GA; WXRS-FM Swainsboro, GA.

Dennis Jones, pres.

RadioStar Inc., 781 Bolsana Dr., Laguna Beach, CA 92651-4124. Phone: (915) 715-9770. Ownership: James D. Glassman, 100%.

Stns: 4 FM. WGKC-FM Mahomet, IL; WMYE(FM) Rantoul, IL; WQQB(FM) Rantoul, IL; WEBX-FM Tuscola, IL.

Jim Glassman, pres.

RadioWorks Inc., 2830 Sandy Hollow Rd., Rockford, IL 61109. Phone: (815) 874-7861. Fax: (815) 874-2202.

Stns: 1 FM. WKHY(FM) Lafayette, IN.

Robert E. Rhea Jr., pres; David W. McAley, exec VP.

Rama Communications Inc., 3765 N. John Young Pkwy., Orlando, FL 32804. Phone: (407) 523-2770. Fax: (407) 523-2888. Web Site: www.gospelrama.com.

Stns: 7 AM. WNTF Bithlo, FL; WTIR(AM) Cocoa Beach, FL; WKIQ Eustis, FL; WQBQ Leesburg, FL; WOKB(AM) Ocoee, FL; WLAA(AM) Winter Garden, FL; WRFV(AM) Valdosta, GA.

Sabita Persaud, pres.

Ramar Communications II Ltd., Box 3757, Lubbock, TX 79452. Phone: (806) 745-3434. Fax: (806) 748-1949. Web Site: www.ramarcom.com. E-mail: bmoran@ramarcom.com Ownership: Ray Moran, 51%; Brad Moran, 49%.

Lubbock, TX 79423, 9800 University Ave.

Stns: 1 AM. 3 FM. KLZK(FM) Brownfield, TX; KJTV(AM) Lubbock, TX; KXTQ-FM Lubbock, TX; KSTQ-FM Plainview, TX.

Stns: 4 TV. KTLL-TV, Albuquerque-Santa Fe, NM; KUPT, Albuquerque-Santa Fe, NM; KTEL-TV, Albuquerque-Santa Fe, NM; KJTV-TV, Lubbock, TX.

Ray Moran, chmn; Brad Moran, pres.

Rawlco Radio Ltd., 715 Saskatchewan Crescent West, Saskatoon, SK S7M 5V7. Canada. Phone: (306) 934-2222. Fax: (306) 933-3300. Ownership: Rawlco Inc., 100%.

Stns: 5 AM. 5 FM. CHMC-FM Edmonton, AB; CJNS Meadow Lake, SK; CJNS-FM Meadow Lake, SK; CJNB North Battleford, SK; CKBI Prince Albert, SK; CHQX-FM Prince Albert, SK; CKCK-FM Regina, SK; CJME(AM) Regina, SK; CJDJ-FM Saskatoon, SK; CKOM(AM) Saskatoon, SK.

Pam Leyland, pres; Gordon Rawlinson, CEO.

Red Rock Radio Corp., 501 Lake Ave. S., Duluth, MN 55802. Phone: (218) 728-9500. Fax: (218) 723-1499. Ownership: Curtis Squire Inc., 85%; Ro D. Grignon, 15%.

Stns: 4 AM. 12 FM. KKIN Aitkin, MN; KKIN-FM Aitkin,

MN; KAOD(FM) Babbitt, MN; KFGI(FM) Crosby, MN; KBAJ(FM) Deer River, MN; KQDS Duluth, MN; KQDS-FM Duluth, MN; WXXZ-FM Grand Marais, MN; WWAX-FM Hermantown, MN; KGHS International Falls, MN; KSDM-FM International Falls, MN; KZIO-FM Two Harbors, MN; WLMX-FM Balsam Lake, WI; WHSM Hayward, WI; WHSM-FM Hayward, WI; WXCX(FM) Siren, WI.

Ro Grignon, pres.

Red Zebra Holdings LLC, 21300 Redskin Park Dr., Ashburn, VA 20147. Phone: (703) 726-7015. Web Site: www.triplexespnradio.com. Ownership: Daniel Snyder, 100% of votes, 38.21% of assets; David Donovan, 0.14% of assets.

Stns: 2 AM. 3 FM. WXTR(AM) Alexandria, VA; WXGI Richmond, VA; WXTG(FM) Virginia Beach, VA; WWXX(FM) Warrenton, VA; WWXT(FM) Prince Frederick, MD.

Regent Communications Inc., 100 E. River Center Blvd., 9th Fl., Covington, KY 41011. Phone: (859) 292-0030. Fax: (859) 814-0136. E-mail: wstakelin@regentcomm.com

Stns: 17 AM. 43 FM. KTRR-FM Loveland, CO; KMAX-FM Wellington, CO; KUAD-FM Windsor, CO; WJBC Bloomington, IL; WJEZ(FM) Dwight, IL; WFYR-FM Elmwood, IL; WBWN-FM Le Roy, IL; WGLO-FM Pekin, IL; WVEL Pekin, IL; WIXO(FM) Peoria, IL; WZPW(FM) Peoria, IL; WTRX-FM Pontiac, IL; WGBF Evansville, IN; WJLT(FM) Evansville, IN; WDKS-FM Newburgh, IN; WGBF-FM Henderson, KY; WKDQ-FM Henderson, KY; WOMI(AM) Owensboro, KY; WBKR(FM) Owensboro, KY; KPEL-FM Abbeville, LA; KROF Abbeville, LA; KFTE-FM Breaux Bridge, LA; KRKA(FM) Erath, LA; KMDL-FM Kaplan, LA; KPEL Lafayette, LA; KTDY-FM Lafayette, LA; WFNT Flint, MI; WRCL(FM) Frankenmuth, MI; WFGR-FM Grand Rapids, MI; WNWZ Grand Rapids, MI; WGRD-FM Grand Rapids, MI; WLCO(AM) Lapeer, MI; WQUS(FM) Lapeer, MI; WWBN-FM Tuscola, MI; WTRV-FM Walker, MI; KMXK-FM Cold Spring, MN; WJON Saint Cloud, MN; KKSR(FM) Sartell, MN; KLZZ-FM Waite Park, MN; KXSS(AM) Waite Park, MN; WGNA-FM Albany, NY; WJYE-FM Buffalo, NY; WBUF(FM) Buffalo, NY; WYRK-FM Buffalo, NY; WECK Cheektowaga, NY; WQBJ-FM Cobleskill, NY; WBLK-FM Depew, NY; WBZZ(FM) Malta, NY; WTMM-FM Mechanicville, NY; WQBK-FM Rensselaer, NY; WTMM(AM) Rensselaer, NY; WODZ-FM Rome, NY; WFRG-FM Utica, NY; WIBX Utica, NY; WTNY Watertown, NY; WNER(AM) Watertown, NY; KROD El Paso, TX; KSII-FM El Paso, TX; KKPL(FM) Cheyenne, WY; KARS-FM Laramie, WY.

Terry S. Jacobs, chmn/CEO; William L. Stakelin, pres/COO; Anthony Vasconcellos, sr VP & CFO.

Reier Broadcasting Co. Inc., Box 20, Bozeman, MT 59718. Phone: (406) 587-9999. Fax: (406) 587-5855. Ownership: William R. Reier Sr., 100%.

Stns: 2 AM. 2 FM. KBOZ Bozeman, MT; KOBB Bozeman, MT; KBOZ-FM Bozeman, MT; KOZB(FM) Livingston, MT.

Relevant Radio, 3200 Riverside Dr., Green Bay, WI 54307. Phone: (800) 342-0306. Fax: (920) 469-3023. E-mail: info@relevantradio.com Web Site: www.relevantradio.com. Ownership: Mark C. Follett, 33.33% of votes; John Cavil, 33.3% of votes; and Robert Atwell, 33.33% of votes.

Stns: 14 AM. 4 FM. WMYR Fort Myers, FL; WCNZ(AM) Marco Island, FL; WVOI(AM) Marco Island, FL; WAUR Sandwich, IL; WWCA Gary, IN; WLOL(AM) Minneapolis, MN; KQJZ(AM) Kalispell, MT; WZUM Carnegie, PA; KIXL(AM) Del Valle, TX; WOVM(FM) Appleton, WI; WYNW(FM) Birnamwood, WI; WDVM(AM) Eau Claire, WI; WKBH Holmen, WI; WJOK Kaukauna, WI; WZRK(AM) Lake Geneva, WI; WMMA(FM) Nekoosa, WI; WPJP(FM) Port Washington, WI; WHFA(AM) Poynette, WI.

Mark Follett, chmn/CEO.

Renda Broadcasting Corp., (Renda Radio Inc.). 900 Parish Street, 4th Fl, Pittsburgh, PA 15220. Phone: (412) 875-1800. Fax: (412) 875-1801. Ownership: S.F. Renda, 100%.

Stns: 5 AM. 17 FM. WWGR-FM Fort Myers, FL; WEJZ-FM Jacksonville, FL; WGUF-FM Marco, FL; WSGL-FM Naples, FL; WGNE-FM Palatka, FL; WSOS-FM Saint Augustine, FL; WJGO(FM) Tice, FL; WMUV(FM) Brunswick, GA; KHTT-FM Muskogee, OK; KMGL-FM Oklahoma City, OK; KOKC(AM) Oklahoma City, OK; KOMA(FM) Oklahoma City, OK; KRXO-FM Oklahoma City, OK; KBEZ(FM) Tulsa, OK; WLCY-FM Blairsville, PA; WKQL(FM) Brookville, PA; WGSM(FM) Greensburg, PA; WCCS Homer City, PA; WDAD Indiana, PA; WQMU-FM Indiana, PA; WJAS Pittsburgh, PA; WECZ Punxsutawney, PA.

Anthony F. Renda, pres; Maryann Kelly, VP/controller; Alan Serena, VP opns; Judy Reich, VP sls.

The Result Radio Group, Box 767, Winona, MN 55987-0767. Phone: (507) 452-4000. Fax: (507) 452-9494. E-mail: jpapenfuss@winonaradio.com Web Site: winonaradio.com. Ownership: Jerry Papenfuss.

Stns: 6 AM. 8 FM. KBEW Blue Earth, MN; KBEW-FM Blue Earth, MN; KBRF Fergus Falls, MN; KJJK Fergus Falls, MN; KJJK-FM Fergus Falls, MN; KZCR(FM) Fergus Falls, MN; KPRW-FM Perham, MN; KWNO-FM Rushford, MN; KDOM Windom, MN; KDOM-FM Windom, MN; KHME(FM) Winona, MN; KWNO Winona, MN; KAGE Winona, MN; KAGE-FM Winona, MN.

Jerry Papenfuss, owner.

Results Broadcasting, 1456 E. Green Bay St., Shawano, WI 54166. Phone: (715) 524-2194. Fax: (715) 524-9980. Ownership: Bruce D. Grassman, 100%.

Stns: 3 AM. 7 FM. WOBE-FM Crystal Falls, MI; WJNR-FM Iron Mountain, MI; WHTO(FM) Iron Mountain, MI; WACD-FM Antigo, WI; WATK Antigo, WI; WFCL Clintonville, WI; WJMQ(FM) Clintonville, WI; WTCH Shawano, WI; WOWN-FM Shawano, WI; WLSL(FM) Three Lakes, WI.

Bruce Grassman, pres.

Reynolds Radio Inc., Box 11196, College Station, TX 77842. Phone: (979) 696-1196. E-mail: rusty@reynoldsradio.com Web Site: www.theblaze.cc.

Stns: 3 FM. KAZE(FM) Ore City, TX; KAJK(FM) White Oak, TX; KBLZ(FM) Winona, TX.

Kenneth R. Reynolds, pres.

Rhattigan Broadcasting (Texas) LP, Box 1420, Plainview, TX 79073. Phone: (806) 853-9147. Fax: (815) 346-2084. Ownership: Michael Rhattigan, 40% votes; Jerome Rhattigan, 30% votes; and Guy Gill, 30% votes.

Stns: 5 AM. 4 FM. KBST Big Spring, TX; KBST-FM Big Spring, TX; KBTS(FM) Big Spring, TX; KEPS Eagle Pass, TX; KVOP(AM) Plainview, TX; KREW(AM) Plainview, TX; KVOU Uvalde, TX; KVOU-FM Uvalde, TX; KUVA-FM Uvalde, TX.

Riverbend Communications LLC, 2880 N. 55th W., Idaho Falls, ID 83402. Phone: (208) 528-6635. Ownership: Frank L. VanderSloot Trust, Frank L. VanderSloot, trustee, 90%; and Belinda VanderSloot, 10%.

Stns: 2 AM. 4 FM. KCVI(FM) Blackfoot, ID; KBLI(AM) Blackfoot, ID; KLCE-FM Blackfoot, ID; KTHK(FM) Idaho Falls, ID; KFTZ-FM Idaho Falls, ID; KBLY(AM) Idaho Falls, ID.

Riviera Broadcast Group LLC, 3333 Sierra Oaks Dr., Sacramento, CA 95864-5738. Phone: (916) 768-8049. Fax: (480) 247-5123. Web Site: www.rivierabroadcast.com. Ownership: VSS Communications Partners IV L.P., 82.5%; Chris Maguire, 8.75%; and Tim Pohlman, 8.75%.

Stns: 4 FM. KOAS(FM) Dolan Springs, AZ; KEDJ(FM) Gilbert, AZ; KKFR(FM) Mayer, AZ; KVGS(FM) Laughlin, NV.

Tim Pohlman, mgng ptnr/CEO; Chris Maguire, mgng ptnr/CFO.

F W Robbert Broadcasting Co. Inc., 2730 Loumor Ave., Metairie, LA 70001. Phone: (504) 831-6941. Web Site: www.wwcr.com. Ownership: Fred P. Westenberger, 51%; Chris P. Westenberger, 9.75%; Fritz N. Westenberger, 9.75%; Lisa M. Westenberger, 9.75%; Eric M. Westenberger, 9.75%; George McClintock, 10%.

Stns: 3 AM. WVOG New Orleans, LA; WMQM(AM) Lakeland, TN; WNQM Nashville, TN.

Fred P. Westenberger, pres; Eric M. Westenberger, gen mgr.

Robinson Corporation, E7601A County Rd. SS, Viroqua, WI 54665. Phone: (608) 637-7200. Fax: (608) 637-7299. E-mail: wvrq@mwt.net Web Site: www.wqpcradio.com. Ownership: David Robinson, Jane Robinson.

Stns: 2 AM. 2 FM. WQPC(FM) Prairie du Chien, WI; WPRE(AM) Prairie du Chien, WI; WVRQ(AM) Viroqua, WI; WVRQ-FM Viroqua, WI.

David Robinson, pres; Jeff Robinson, opns mgr; James Graham, gen sls mgr.

Rodgers Broadcasting Corp., Box 1646, Richmond, IN 47374. Phone: (765) 962-6533. Fax: (765) 966-1499. Ownership: David Rodgers, 100%.

Stns: 3 AM. 3 FM. WBML Macon, GA; WIFE(AM) Connersville, IN; WFMG(FM) Richmond, IN; WKBV(AM) Richmond, IN; WIFE-FM Rushville, IN; WZZY-FM Winchester, IN.

David Rodgers, pres.

Rogers Broadcasting Ltd., 777 Jarvis St., Toronto, ON M4Y 3B7. Canada. Phone: (416) 935-8200. Web Site: www.rogers.com. Ownership: Rogers Media Inc., 100%. Note: Rogers Media Inc. is 100% owned by Rogers Communications Inc.

Stns: 5 AM. 35 FM. CFFR Calgary, AB; CKIS-FM Calgary, AB; CHMN-FM Canmore, AB; CKER-FM Edmonton, AB; CHDI-FM Edmonton, AB; CKYX-FM Fort McMurray, AB; CJOK-FM Fort McMurray, AB; CFGP-FM Grande Prairie, AB; CFRV-FM Lethbridge, AB; CJRX-FM Lethbridge, AB; CKQC-FM Abbotsford, BC; CKGO-FM-1 Boston Bar, BC; CKCL-FM Chilliwack, BC; CKSR-FM Chilliwack, BC; CFSR-FM Hope, BC; CISP-FM Pemberton, BC; CKKS-FM Sechelt, BC; CKWX Vancouver, BC; CKIZ-FM Vernon, BC; CHTT-FM Victoria, BC; CIOC-FM Victoria, BC; CISW-FM Whistler, BC; CITI-FM Winnipeg, MB; CKY-FM Winnipeg, MB; CKNI-FM Moncton, NB; CHNI-FM Saint John, NB; CJNI-FM Halifax, NS; CKAT North Bay, ON; CHUR-FM North Bay, ON; CICX-FM Orillia, ON; CHAS-FM Sault Ste. Marie, ON; CJQM-FM Sault Ste. Marie, ON; CKBY-FM Smiths Falls, ON; CIGM(AM) Sudbury, ON; CJMX-FM Sudbury, ON; CJRQ-FM Sudbury, ON; CKGB-FM Timmins, ON; CJQQ-FM Timmins, ON; CJCL Toronto, ON; CJAQ-FM Toronto, ON.

Stns: 4 TV. CHNU-TV, Fraser Valley, BC; CJMT-TV, Toronto, ON; CFMT, Toronto, ON; CIIT-TV, Winnipeg, MB.

Rael Merson, pres.

Rooney Moon Broadcasting Inc., 208 E. Grand Ave., Clovis, NM 88101. Phone: (505) 763-4649. Fax: (505) 763-1693. E-mail: info.rmb@yucca.net Web Site: www.bettermix.com.

Stns: 1 AM. KSEL Portales, NM.

Steve Rooney, pres.

Rose City Radio Corp., 0234 Southwest Bancroft St., Portland, OR 97239. Phone: (503) 243-7595. Fax: (503) 417-7662. Web Site: www.kxlradio.com.

Stns: 3 AM. 1 FM. WWZN(AM) Boston, MA; WSNR(AM) Jersey City, NJ; KXJM-FM Portland, OR; KXL Portland, OR.

Tim McNamara, VP/gen mgr.

Roswell Radio Inc./Quay Broadcasters Inc., Box 670, Roswell, NM 88202. Phone: (505) 622-6450. Fax: (505) 622-9041. Ownership: John M. Dunn, 100%.

Stns: 2 AM. 4 FM. KBCQ-FM Roswell, NM; KMOU(FM) Roswell, NM; KBCQ(AM) Roswell, NM; KSFX(FM) Roswell, NM; KTNM Tucumcari, NM; KQAY-FM Tucumcari, NM.

John M. Dunn, pres.

Route 81 Radio LLC, 780 E Market St., Suite 265, West Chester, PA 19382. Phone: (610) 696-8181. Fax: (610) 696-5072. E-mail: ken@route81radio.com Web Site: route81radio.com. Ownership: WallerSutton 2000 L.P., 50%; Avalon Equity Fund LP, 50%.

Stns: 8 AM. 4 FM. WENI-FM Big Flats, NY; WCBA Corning, NY; WGMM(FM) Corning, NY; WENI(AM) Corning, NY; WENY Elmira, NY; WENY-FM Elmira, NY; WLNP(FM) Carbondale, PA; WCDL(AM) Carbondale, PA; WHYL Carlisle, PA; WCOJ(AM) Coatesville, PA; WAZL Hazleton, PA; WNAK Nanticoke, PA.

Ira Rosenblatt, CEO; Ken Karaszkiewicz, CFO.

Rubber City Radio Group Inc., 1795 W. Market St., Akron, OH 44313. Phone: (330) 869-9800. Fax: (330) 864-6799. E-mail: mail@wakr.net Web Site: www.wqmx.com. Ownership: Thomas Mandel.

Stns: 1 AM. 5 FM. WJZL(FM) Charlotte, MI; WVIC(FM) Jackson, MI; WJXQ-FM Jackson, MI; WQTX(FM) Saint Johns, MI; WAKR Akron, OH; WQMX-FM Medina, OH.

Thomas Mandel, pres; Mark Biviano, VP; Nick Anthony, VP.

S

SIGA Broadcasting Corp., 1302 N. Shepherd Dr., Houston, TX 77008. Phone: (713) 868-5559. Fax: (713) 868-9631. Ownership: Gabriel Arango, 50%; and Silvia Arango 50%.

Stns: 4 AM. KTMR Edna, TX; KGBC Galveston, TX; KAML Kenedy-Karnes City, TX; KLVL Pasadena, TX.

STARadio Corp., 329 Maine St., Quincy, IL 62301-3928. Phone: (217) 224-4102. Fax: (217) 224-4133. E-mail: reception@staradio.com Ownership: Howard A. Doss, Derek Parrish and Jack Whitley.

Stns: 2 AM. 6 FM. WKAN Kankakee, IL; WQCY(FM) Quincy, IL; WTAD Quincy, IL; WCOY(FM) Quincy, IL; WXNU(FM) Saint Anne, IL; WYKT-FM Wilmington, IL; KGRC-FM Hannibal, MO; KZZK-FM New London, MO.

Mike Moyers, gen mgr.

Saga Communications Inc., 73 Kercheval Ave., Suite 201, Grosse Pointe Farms, MI 48236. Phone: (313) 886-7070. Fax: (313) 886-7150. E-mail: chapsburg@sagacom.com Web Site: www.sagacommunications.com. Ownership: Edward K. Christian, 56.5% of the voting stock. Other Interests: Illinois Radio Network, Michigan Radio Network, Michigan Farm Radio Network.

Stns: 27 AM. 54 FM. KEGI(FM) Jonesboro, AR; KDXY(FM)

Lake City, AR; KJBX(FM) Trumann, AR; KLTI-FM Ames, IA; KRNT Des Moines, IA; KSTZ(FM) Des Moines, IA; KIOA(FM) Des Moines, IA; KAZR(FM) Pella, IA; KICD Spencer, IA; KICD-FM Spencer, IA; KLLT(FM) Spencer, IA; WLRW-FM Champaign, IL; WIXY-FM Champaign, IL; WXTT(FM) Danville, IL; WYMG(FM) Jacksonville, IL; WABZ(FM) Sherman, IL; WQQL(FM) Springfield, IL; WTAX(AM) Springfield, IL; WDBR(FM) Springfield, IL; WCFF(FM) Urbana, IL; WCVQ-FM Fort Campbell, KY; WJQI(AM) Fort Campbell, KY; WVVR(FM) Hopkinsville, KY; WZZP(FM) Hopkinsville, KY; WEGI(FM) Oak Grove, KY; WHNP(AM) East Longmeadow, MA; WHMQ(AM) Greenfield, MA; WPVQ(FM) Greenfield, MA; WHMP Northampton, MA; WLZX(FM) Northampton, MA; WAQY-FM Springfield, MA; WRSI(FM) Turners Falls, MA; WSNI(FM) Winchendon, MA; WVAE(AM) Biddeford, ME; WBAE Portland, ME; WPOR(FM) Portland, ME; WZAN Portland, ME; WMGX(FM) Portland, ME; WGAN(AM) Portland, ME; WYNZ-FM Westbrook, ME; WOXL-FM Biltmore Forest, NC; WYSE(AM) Canton, NC; WTMT(FM) Weaverville, NC; WMLL(FM) Bedford, NH; WKNE(FM) Keene, NH; WZBK(AM) Keene, NH; WKBK(AM) Keene, NH; WZID(FM) Manchester, NH; WFEA Manchester, NH; WINQ(FM) Winchester, NH; WYXL-FM Ithaca, NY; WQNY-FM Ithaca, NY; WNYY(AM) Ithaca, NY; WHCU Ithaca, NY; WBCO(AM) Bucyrus, OH; WQEL-FM Bucyrus, OH; WSNY(FM) Columbus, OH; WODB(FM) Delaware, OH; WJZA-FM Lancaster, OH; WJZK(FM) Richwood, OH; KMIT-FM Mitchell, SD; KLQT(FM) Wessington Springs, SD; WKFN(AM) Clarksville, TN; WINA Charlottesville, VA; WQMZ(FM) Charlottesville, VA; WVAX(AM) Charlottesville, VA; WWWV-FM Charlottesville, VA; WCNR(FM) Keswick, VA; WJOI Norfolk, VA; WAFX-FM Suffolk, VA; WKVT Brattleboro, VT; WKVT-FM Brattleboro, VT; WRSY(FM) Marlboro, VT; KGMI Bellingham, WA; KPUG Bellingham, WA; KBAI(AM) Bellingham, WA; WJZX(FM) Brookfield, WI; WJMR-FM Menomonee Falls, WI; WHQG(FM) Milwaukee, WI; WJYI Milwaukee, WI; WKLH-FM Milwaukee, WI.

Stns: 3 TV. WXVT, Greenwood-Greenville, MS; KOAM, Joplin, MO-Pittsburg, KS; KAVU, Victoria, TX.

Edward K. Christian, pres/CEO; Marcia Lobaito, VP business affrs; Sam Bush, CFO; Warren Lada Sr., VP opns.

Salem Communications Corp., 4880 Santa Rosa Rd., Suite 100, Camarillo, CA 93012. Phone: (805) 987-0400. Fax: (805) 384-4511. Web Site: www.salem.cc.

Stns:64 AM. 31 FM. KPXQ(AM) Glendale, AZ; KKNT(AM) Phoenix, AZ; KFSH-FM Anaheim, CA; KXMX(AM) Anaheim, CA; KFIA Carmichael, CA; KTKZ-FM Dunnigan, CA; KRLA(AM) Glendale, CA; KKFS(FM) Lincoln, CA; KKLA-FM Los Angeles, CA; KDAR(FM) Oxnard, CA; KNTS(AM) Palo Alto, CA; KTKZ Sacramento, CA; KTIE(AM) San Bernardino, CA; KCBQ San Diego, CA; KFAX San Francisco, CA; KPRZ San Marcos-Poway, CA; KRKS-FM Boulder, CO; KZNT(AM) Colorado Springs, CO; KBJD Denver, CO; KNUS Denver, CO; KRKS(AM) Denver, CO; KBIQ(FM) Manitou Springs, CO; KGFT-FM Pueblo, CO; WORL Altamonte Springs, FL; WHIM Apopka, FL; WGUL(AM) Dunedin, FL; WZAZ Jacksonville, FL; WKAT North Miami, FL; WTLN Orlando, FL; WTBN(AM) Pinellas Park, FL; WTWD(AM) Plant City, FL; WLSS(AM) Sarasota, FL; WLTA Alpharetta, GA; WFSH-FM Athens, GA; WGKA(AM) Atlanta, GA; WNIV Atlanta, GA; WAFS(AM) Atlanta, GA; KGMZ-FM Aiea, HI; KGU Honolulu, HI; KHNR(AM) Honolulu, HI; KAIM-FM Honolulu, HI; KHNR-FM Honolulu, HI; KHUI(FM) Honolulu, HI; WIND Chicago, IL; WYLL(AM) Chicago, IL; WFIA-FM New Albany, IN; WGTK(AM) Louisville, KY; WFIA Louisville, KY; WRVI-FM Valley Station, KY; WROL Boston, MA; WTTT(AM) Boston, MA; WEZE Boston, MA; WLQV Detroit, MI; WDTK(AM) Detroit, MI; KYCR Golden Valley, MN; WWTC Minneapolis, MN; KKMS Richfield, MN; KGBI-FM Omaha, NE; KOTK(AM) Omaha, NE; KCRO Omaha, NE; WWDJ Hackensack, NJ; WMCA New York, NY; WFHM-FM Cleveland, OH; WHKW(AM) Cleveland, OH; WHK(AM) Cleveland, OH; WRFD(AM) Columbus-Worthington, OH; WHKZ(AM) Warren, OH; KRYP(FM) Gladstone, OR; KPDQ-FM Portland, OR; KFIS(FM) Scappoose, OR; WFIL(AM) Philadelphia, PA; WNTP(AM) Philadelphia, PA; WORD-FM Pittsburgh, PA; WFFI(FM) Kingston Springs, TN; WFFH(FM) Smyrna, TN; WVRY-FM Waverly, TN; WBOZ(FM) Woodbury, TN; KTEK Alvin, TX; KLTY(FM) Arlington, TX; KSKY(AM) Balch Springs, TX; KWRD-FM Highland Village, TX; KNTH(AM) Houston, TX; KPXI-FM Overton, TX; KSLR San Antonio, TX; KLUP(AM) Terrell Hills, TX; KKHT-FM Winnie, TX; WAVA(AM) Arlington, VA; WAVA-FM Arlington, VA; KGNW Burien-Seattle, WA; KKOL Seattle, WA; KLFE Seattle, WA; KDOW(AM) Seattle, WA; KKMO Tacoma, WA; WRRD(AM) Jackson, WI; WFZH(FM) Mukwonago, WI

Edward G. Atsinger III, pres/CEO; Stuart W. Epperson, chmn; Eric H. Halvorson, VP/COO.

San Luis Valley Broadcasting Inc., Box 631, Monte Vista, CO 81144. Phone: (719) 852-3581. Fax: (719) 852-3583. Ownership: Marion L. Goad, 65.01%; and H. Robert Gourley III, 34.99%.

Stns: 1 AM. 2 FM. KYDN(FM) Del Norte, CA; KSLV Monte Vista, CO; KSLV-FM Monte Vista, CO.

Sand Hill Media Corp., Box 570, Logan, UT 84323. Phone: (435) 752-1390. Ownership: Sand Hill Media 2001, 100%.

Stns: 1 AM. 3 FM. KSPZ(AM) Ammon, ID; KUPI-FM Idaho Falls, ID; KQEO(FM) Idaho Falls, ID; KSNA(FM) Rexburg, ID.

Sandab Communications L.P. II, 2201 Old Court Rd., Baltimore, MD 21208. Phone: (508) 771-1224. Web Site: www.wqrc.com. Ownership: Stephen Seymour

Stns: 4 FM. WQRC(FM) Barnstable, MA; WFCC-FM Chatham, MA; WOCN-FM Orleans, MA; WKPE-FM South Yarmouth, MA.

Stephen Seymour, pres; Scott Frothinghan, VP; Gregory Bone, gen mgr.

Sandusky Radio, 515 Park Ave., Apt. 4A, New York, NY 10022. Phone: (212) 355-3074. Fax: (212) 355-3075. Ownership: Alice S. White trust. All 100% owned by the White and Rau families

Stns: 3 AM. 4 FM. KDKB(FM) Mesa, AZ; KAZG(AM) Scottsdale, AZ; KDUS Tempe, AZ; KQMV(FM) Bellevue, WA; KRWM-FM Bremerton, WA; KIXI Mercer Island-Seattle, WA; KWJZ(FM) Seattle, WA.

Sandusky Newspapers Inc. publishes the *Sandusky Register, Norwalk Reflector* (OH) *Kingsport Times-News*(TN) *Grand Haven Tribune*(MI)*Ogden Standard-Examiner*(UT)*Johnson City Press, Lebanon Democrat* (TN) and five weekly newspapers,*Erwin Record, Jonesborough Herald & Tribune, Mountain City Tomahawk, Hartsville Vidette* and the*Mt. Juliet News*.

David A. Rau, chmn/CEO; Norman Rau, pres; Peter W. Vogt, CFO.

Sanpete County Broadcasting Co., Box 40, Manti, UT 84642. Phone: (435) 835-7301. Fax: (435) 835-2250. Ownership: Douglas Barton, 100%.

Stns: 1 AM. 2 FM. KMTI Manti, UT; KMXD(FM) Monroe, UT; KLGL(FM) Richfield, UT.

Schurz Communications Inc., 225 W. Colfax Ave., South Bend, IN 46626. Phone: (219) 287-1001. Fax: (219) 287-2257. E-mail: mburdick@schurz.com Web Site: www.schurz.com. Ownership: Franklin D. Schurz Jr., James M. Schurz, Scott C. Schurz and Mary Schurz, trustees.

Stns: 4 AM. 8 FM. WASK-FM Battle Ground, IN; WXXB(FM) Delphi, IN; WASK Lafayette, IN; WKOA-FM Lafayette, IN; WSBT South Bend, IN; WNSN-FM South Bend, IN; KFXS-FM Rapid City, SD; KKLS Rapid City, SD; KKMK-FM Rapid City, SD; KOUT-FM Rapid City, SD; KRCS-FM Sturgis, SD; KBHB Sturgis, SD.

Stns: 9 TV. WAGT, Augusta, GA; WDBJ, Roanoke-Lynchburg, VA; WSBT, South Bend-Elkhart, IN; KYTV, Springfield, MO; KBSD, Wichita-Hutchinson Plus, KS; KBSH, Wichita-Hutchinson Plus, KS; KBSL, Wichita-Hutchinson Plus, KS; KWCH, Wichita-Hutchinson Plus, KS; KSCW, Wichita-Hutchinson Plus, KS.

Schurz Communications publishes the following nwsprs: *Imperial Valley Press, Southside Times-Beech, Times; Bedford Times-Mail, Bloomington Herald-Times* , & *South Bend Tribune, Martinsville Reporter, Danville Advocate-Messenger, The Herald Mail Co.; Daily American*, Somerset, PA.

Marcia K. Burdick, sr VP bcstg; Franklin D. Schurz Jr., chmn; Todd F. Schurz, pres.

Scott Communications Inc., Box 1150, Selma, AL 36702-1150. Phone: (334) 875-9360. Fax: (334) 875-1340. Ownership: Paul Scott Alexander, 100%.

Stns: 1 AM. 2 FM. WJAM-FM Orrville, AL; WMRK Selma, AL; WALX-FM Selma, AL.

Sea-Comm Inc., 122 Cinema Dr., Wilmington, NC 28403. Phone: (910) 772-6300. Fax: (910) 772-6310. Web Site: www.sea-comm.com. Ownership: N. Eric Jorgensen, 100%.

Stns: 4 FM. WLTT(FM) Shallotte, NC; WBNE(FM) Shallotte, NC; WWTB(FM) Topsail Beach, NC; WNTB(FM) Wrightsville Beach, NC.

Paul Knight, gen mgr; Rick Jorgenson, pres/CEO.

Seaton Stations, Manhattan Broadcasting Inc., 2414 Casement Rd., Manhattan, KS 66502. Phone: (785) 776-1350. Fax: (785) 539-1000.

Stns: 1 AM. 2 FM. KMAN Manhattan, KS; KXBZ-FM Manhattan, KS; KACZ(FM) Riley, KS.

Seaton Goup of newspapers includes the *Manhattan Mercury* & *Winfield Courier*, both KS; *Alliance Times-Herald* & *Hastings Tribune*, both NE; *The Black Hills Pioneer*, Spearfish, SD; *Sheridan* (WY) *Press*.

Richard Wartell, gen mgr.

Seattle Streaming Radio LLC, Box 1471, Evergreen, CO 80437. Phone: (303) 688-5162. Fax: (303) 660-4930. Ownership: David M. Drucker, 80%; and Penny Drucker, 20%.

Stns: 4 AM. KXLJ(AM) Juneau, AK; WKIZ Key West, FL; KBRO Bremerton, WA; KNTB Lakewood, WA.

Seehafer Broadcasting Corp., Box 1385, Manitowoc, WI 54221-1385. Phone: (920) 682-0351. Fax: (920) 682-1008. Ownership: Donald W. Seehafer, 100%.

Stns: 4 AM. 2 FM. WQTC-FM Manitowoc, WI; WOMT(AM) Manitowoc, WI; WDLB Marshfield, WI; WOSQ-FM Spencer, WI; WXCO Wausau, WI; WFHR Wisconsin Rapids, WI.

Don Seehafer, pres; Mark Seehafer, VP.

Service Broadcasting Group LLC, 621 N.W. 6th St., Grand Prairie, TX 75050-5555. Phone: (972) 263-9911. Fax: (972) 558-0010. Web Site: www.k104fm.com.

Stns: 1 AM. 2 FM. KKDA-FM Dallas, TX; KRNB-FM Decatur, TX; KKDA Grand Prairie, TX.

Hymen Childs, pres; Chuck Smith, gen mgr.

Seward County Broadcasting Co., 1410 N. Western, Liberal, KS 67901. Phone: (620) 624-3891. Fax: (620) 624-7885. E-mail: sales@kscb.net Web Site: www.kscb.net. Ownership: Jack Landon, Robert Larrabee, Stuart Melchert.

Stns: 1 FM. KLDG-FM Liberal, KS.

Stuart Melchert, gen mgr.

Shamrock Communications Inc., 149 Penn Ave., Scranton, PA 18503. Phone: (570) 348-9108. Fax: (570) 348-9109. Web Site: www.nepanews.com. Ownership: Principal owners: William R. Lynett, James J. Haggerty, Edward J. Lynett, George V. Lynett. Shamrock Communications owns 50% of the Milwaukee Radio Alliance LLC (see listing).

Stns: 2 AM. 8 FM. WJZI(FM) Livingston Manor, NY; KTSO(FM) Glenpool, OK; KMYZ-FM Pryor, OK; WQFN(FM) Forest City, PA; WQFM-FM Nanticoke, PA; WPZX(FM) Pocono Pines, PA; WEJL Scranton, PA; WEZX-FM Scranton, PA; WBAX Wilkes-Barre, PA; WZBA(FM) Westminster, MD.

Publications include *Orlando Weekly*, Orlando, FL; *City Paper*, Baltimore, MD; *Metro Times* (Detroit), Detroit, MI; *Owego Pennysaver*, Owego, NY; *Pocono Shopper* (Monroe County Edition), East Stroudsburg, *Susquehanna County Independent, Susquehanna County Weekender*, Montrose, *Pottsville Republican and Evening Herald*, Pottsville, *ADI, Electric City, Good Times, Northeast Pennsylvania Business Journal, Scranton Times-Tribune, Suburban Weekly, Tri-Boro Banner, Valley Advantage*, all Scranton, *News Item, Shamokin, Daily Review, The Bradford Sullivan Pennysaver, The Farmer's Friend*, Towanda, *Troy Pennysaver*, Troy, *New Age-Examiner, Wyoming County Advance*, Tunkhannock, *The Citizen Standard, Valley View, Citizens Voice*, Wilkes-Barre, all PA; *San Antonio Current*, San Antonio, TX.

William R. Lynett, pres; Jim Loftus, COO.

Shepherd Group, Box 430, Moberly, MO 65270. Phone: (660) 263-5800. Fax: (660) 263-2300. E-mail: daves@regionalradio.com

Stns: 8 AM. 7 FM. KAAN Bethany, MO; KMRN Cameron, MO; KKWK(FM) Cameron, MO; KREI Farmington, MO; KTJJ-FM Farmington, MO; KJFF(AM) Festus, MO; KBNN Lebanon, MO; KJEL-FM Lebanon, MO; KIRK-FM Macon, MO; KWIX Moberly, MO; KRES-FM Moberly, MO; KFBD-FM Waynesville, MO; KJPW Waynesville, MO; KJPW-FM Waynesville, MO; KOZQ Waynesville, MO.

David Shepherd, pres.

Sheridan Broadcasting Corp., 960 Penn Ave., Suite 200, Pittsburgh, PA 15222. Phone: (412) 456-4000. Ownership: Ronald R. Davenport Sr. and Judith M. Davenport, 96.34%; Ronald R. Davenport Jr., 1.22%; Judith Allison, 1.22%; and Susan Davenport Austin, 1.22%. Note: all stns 100% owned except for WIGO(AM) Morrow, GA. Group owns 75% of WIGO(AM) Morrow, GA.

Stns: 5 AM. 1 FM. WATV(AM) Birmingham, AL; WIGO(AM) Morrow, GA; WUFO Amherst, NY; WAMO-FM Beaver Falls, PA; WAMO(AM) Millvale, PA; WPGR(AM) Monroeville, PA.

Ronald R. Davenport Sr., chmn/CEO.

Sierra Broadcasting Corp., 3015 Johnstonville Rd., Susanville, CA 96130. Phone: (530) 257-2121. Fax: (530) 257-6955. E-mail: info@theradionetwork.com Web Site: www.theradionetwork.com.

Stns: 1 AM. 1 FM. KHJQ-FM Susanville, CA; KSUE Susanville, CA.

Rod Chambers, gen mgr; George Carl, VP.

Simmons Broadcasting Inc., 1403 Third St., Langdon, ND 58249. Phone: (701) 256-1080. Fax: (701) 256-1081. E-mail: kndkkicksbs@utma.com Ownership: Robert N. Simmons, 50%; and Diane R. Simmons, 50%. Note: Pursuant to a loc mktg agreement, group provides substantially all of the progmg for KXPO(AM)-KAUJ(FM) Grafton, ND.

Stns: 1 AM. 3 FM. KAOC(FM) Cavalier, ND; KNDK Langdon, ND; KNDK-FM Langdon, ND; KYTZ(FM) Walhalla, ND.

Simmons Media Group, 515 South 700 East, Salt Lake City, UT 84102. Phone: (801) 524-2600. Fax: (801) 524-6002. Web Site: www.simmonsmedia.com. Ownership: The David E. Simmons 201 Trust, The Matthew R. Simmons 201 Trust, The Laurence E. Simmons 201 Trust, The Julia S. Watkins 201 Trust, The Elizabeth S. Hoke 201 Trust and The Harris H. Simmons 201 Trust.

Stns: 13 AM. 10 FM. KDXE(AM) North Little Rock, AR; KQPN(AM) West Memphis, AR; WFFX(AM) East St. Louis, IL; KSLG(AM) Saint Louis, MO; KZNX(AM) Creedmoor, TX; KLRK(FM) Marlin, TX; KRQX Mexia, TX; KWGW(FM) Mexia, TX; KWNX(AM) Taylor, TX; KQRL(AM) Waco, TX; KRZI(AM) Waco, TX; KEGH(FM) Brigham City, UT; KXOL Brigham City, UT; KJQN(FM) Coalville, UT; KURR(FM) Hurricane, UT; KEGA(FM) Oakley, UT; KXRK-FM Provo, UT; KOVO Provo, UT; KZNS(AM) Salt Lake City, UT; KYMV(FM) Woodruff, UT; KDWY(FM) Diamondville, WY; KAOX(FM) Kemmerer, WY; KMER Kemmerer, WY.

David Simmons, chmn; Craig Hanson, pres; Bruce W. Thomas, CFO; Bret Leifson, controller; Alan Hague, opns VP.

Sinclair Communications Inc., 999 Waterside Dr., Suite 500, Norfolk, VA 23510. Phone: (757) 640-8500. Fax: (757) 640-8552. Web Site: www.sinclairstations.com. Ownership: John L. Sinclair, chmn; Robert Sinclair, J. David Sinclair, Ann Adams. Note: Group also manages KNOB(FM) Healdsburg, CA.

Stns: 2 AM. 6 FM. KXTS(FM) Calistoga, CA; KRSH(FM) Healdsburg, CA; KSXY(FM) Middletown, CA; WPYA(FM) Chesapeake, VA; WROX-FM Exmore, VA; WTAR(AM) Norfolk, VA; WNIS Norfolk, VA; WNRJ(FM) Poquoson, VA.

John L. Sinclair, chmn; J. David Sinclair, pres; Robert L. Sinclair, sec.

629112 Saskatchewan LTD., 366 3rd Ave. South, Saskatoon, SK S7K 1M5. Canada. Phone: (306) 244-1975. Fax: (306) 665-8484.

Stns: 1 AM. 2 FM. CFQC-FM Saskatoon, SK; CJWW Saskatoon, SK; CJMK-FM Saskatoon, SK.

Elmer Hildebrand, CEO; Vic Dubois, gen mgr.

SkyWest Media L.L.C., Box 36148, Tucson, AZ 85740. Phone: (520) 797-4434. Ownership: Ted Tucker, 100%.

Stns: 1 AM. 5 FM. KNFT Bayard, NM; KNFT-FM Bayard, NM; KPSA-FM Lordsburg, NM; KSCQ-FM Silver City, NM; KRZX(FM) Monticello, UT; KFMR(FM) Marbleton, WY.

Somar Communications Inc., 28095 Three Notch Rd., Suite 2-B, Mechanicsville, MD 20659. Phone: (301) 870-5550. Fax: (301) 884-0280.

Stns: 2 AM. 3 FM. WKIK-FM California, MD; WKIK La Plata, MD; WPTX(AM) Lexington Park, MD; WYRX(FM) Lexington Park, MD; WSMD-FM Mechanicsville, MD.

Roy Robertson, pres/CEO.

Sorensen Pacific Broadcasting Inc., 111 W. Chanlan Santo Papa, Suite 800, Hagatna, GU 96910. Phone: (671) 477-5700. Fax: (671) 477-3982. E-mail: comments@radiopacific.com Web Site: www.radiopacific.com. Ownership: Rex W. Sorensen, 97.6%.

Stns: 1 AM. 4 FM. KGUM-FM Dededo, GU; KGUM(AM) Hagatna, GU; KZGZ(FM) Hagatna, GU; KPXP(FM) Garapan-Saipan, NP; KRSI-FM Garapan-Saipan, NP.

Rex Sorensen, chmn/CEO; Jon Anderson, pres.

Sorenson Broadcasting Corp., 2804 S. Ridgeview Way, Sioux Falls, SD 57105. Phone: (605) 334-1117. Fax: (605) 338-0326. E-mail: sorenson@sbcradio.com Ownership: Dean P. Sorenson, 100%.

Stns: 6 AM. 10 FM. WZGA(FM) Helen, GA; KUQQ-FM Milford, IA; KIHK-FM Rock Valley, IA; KSOU Sioux Center, IA; KSOU-FM Sioux Center, IA; KUOO-FM Spirit Lake, IA; KAYL Storm Lake, IA; KAYL-FM Storm Lake, IA; KCUE Red Wing, MN; KWNG-FM Red Wing, MN; KORN Mitchell, SD; KQRN(FM) Mitchell, SD; KLXS-FM Pierre, SD; KCCR Pierre, SD; KKYA-FM Yankton, SD; KYNT Yankton, SD.

Dean Sorenson, pres.

South Central Communications Corp., Box 3848, Evansville, IN 47736. Phone: (812) 463-7950. Fax: (812) 463-7915. Web Site: www.southcentralcommunications.net. Ownership: John D. Engelbrecht, 80%, J.P. Engelbrecht, 20%.

Stns: 1 AM. 11 FM. WEJK(FM) Boonville, IN; WLFW(FM) Chandler, IN; WEOA Evansville, IN; WIKY-FM Evansville, IN; WABX-FM Evansville, IN; WSTO-FM Owensboro, KY; WIMZ-FM Knoxville, TN; WJXB-FM Knoxville, TN; WQJK(FM) Maryville, TN; WCJK(FM) Murfreesboro, TN; WJXA-FM Nashville, TN; WRJK(FM) Norris, TN.

Stns: 1 TV. WMAK, Knoxville, TN.

John D. Engelbrecht, pres; J.P. Engelbrecht, VP.

Southeast Kansas Broadcasting Co., 250 N. Water, Suite 300, Wichita, KS 67202. Phone: (620) 431-3700. Fax: (620) 431-4643. E-mail: kkoy@kkoy.com Web Site: www.kkoy.com. Ownership: Murfin Inc.

Stns: 1 AM. 1 FM. KSNP(FM) Burlington, KS; KKOY Chanute, KS.

Phil McComb, gen mgr.

Southeast Kansas Independent Living Resource Center Inc., 202 E. Centennial Ave., Suite 2B, Pittsburg, KS 66762. Phone: (620) 232-9912. Fax: (620) 232-9915. Ownership: Officers and bd members: Shari Coatney, pres/CEO; Jeanette Pruitt, bd chmn; Ron Garnett, treas; Faron Morales, John Spillman, Carolyn Freeman, Marty Wooten, Darlene Lomax and Edward Reynolds.

Stns: 2 AM. 1 FM. KSEK-FM Girard, KS; KLKC(AM) Parsons, KS; KSEK Pittsburg, KS.

Shari Coatney, pres.

Southeastern Oklahoma Radio LLC, Box 1011, Hartshorne, OK 74547. Phone: (918) 297-2501. Ownership: Bob and Sheila Turnbow (jointly), 33.33%; Lee Anderson, 33.33%; and Richard C. Lerblance, 33.33%. Note: Richard C. Lerblance also owns 33.33% of KESC(FM) Wilburton, OK.

Stns: 2 AM. 2 FM. KMCO-FM McAlester, OK; KNED McAlester, OK; KTMC McAlester, OK; KTMC-FM McAlester, OK.

Southern Broadcasting Companies Inc., 1010 Tower Pl., Bogart, GA 30622. Phone: (706) 369-7301. Fax: (706) 353-1967. Web Site: www.magic1021.com. Ownership: Paul C. Stone.

Stns: 3 AM. 6 FM. WRFC Athens, GA; WGAU(AM) Athens, GA; WSRM(FM) Coosa, GA; WGMG-FM Crawford, GA; WMGZ(FM) Eatonton, GA; WRGA Rome, GA; WPUP-FM Royston, GA; WNGC(FM) Toccoa, GA; WXKT-FM Washington, GA.

Paul Stone, pres; Traci Long, gen mgr.

Southern Communications Corp., 306 S. Kanawaha St., Beckley, WV 25801. Phone: (304) 253-7000. Fax: (304) 255-1044. Web Site: www.103cir.com. Ownership: R. Shane Southern, 50.4%; Karen L. Martin, 24.8%; and Kristin E. Wallace, 24.8%.

Stns: 3 AM. 4 FM. WCIR-FM Beckley, WV; WIWS Beckley, WV; WWNR Beckley, WV; WMTD Hinton, WV; WMTD-FM Hinton, WV; WTNJ-FM Mount Hope, WV; WAXS-FM Oak Hill, WV.

Jay Quesenberry, gen mgr; R. Shane Southern, pres.

Southern Wabash Communications Corp., 435 37th Ave. N., Nashville, TN 37209. Phone: (615) 844-1039. Fax: (615) 777-2284. Web Site: www.wnsr.com.

Stns: 2 AM. 2 FM. WSJD-FM Princeton, IN; WNTC-FM Drakesboro, KY; WNSR Brentwood, TN; WMGC Murfreesboro, TN.

Randy Bell, pres; Ted Johnson, gen mgr.

Southwest Broadcasting Inc., 206 N. Front, McComb, MS 39648. Phone: (601) 684-4116. Fax: (601) 684-4654. E-mail: spots@k106.net Ownership: C. Wayne Dowdy, 100%.

Stns: 2 AM. 6 FM. WTGG-FM Amite, LA; WJSH(FM) Folsom, LA; WAKK-FM Centreville, MS; WAZA-FM Liberty, MS; WAKH-FM McComb, MS; WAPF(AM) McComb, MS; WAKK(AM) McComb, MS; WFCG(FM) Tylertown, MS.

C. Wayne Dowdy, pres.

Spanish Broadcasting System Inc., 2601 South Bayshore Dr., PH 2, Coconut Grove, FL 33133. Phone: (305) 441-6901. Fax: (305) 446-5148. Web Site: www.spanishbroadcasting.com. Ownership: Raul Alarcon Sr., Raul Alarcon Jr., Jose Grimalt.

Stns: 20 FM. KLAX-FM East Los Angeles, CA; KXOL-FM Los Angeles, CA; KRZZ(FM) San Francisco, CA; WRMA-FM Fort Lauderdale, FL; WCMQ-FM Hialeah, FL; WXDJ-FM North Miami Beach, FL; WLEY-FM Aurora, IL; WPAT-FM Paterson, NJ; WSKQ-FM New York, NY; WODA(FM) Bayamon, PR; WCMA-FM Fajardo, PR; WMEG-FM Guayama, PR; WZET(FM) Hormigueros, PR; WNOD(FM) Mayaguez, PR; WIOB-FM Mayaguez, PR; WIOC-FM Ponce, PR; WZMT-FM Ponce, PR; WEGM(FM) San German, PR; WZNT-FM San Juan, PR; WIOA(FM) San Juan, PR.

Stns: 1 TV. WSBS-TV, Miami-Ft. Lauderdale, FL.

Raul Alarcon Sr., chmn; Raul Alarcon Jr., pres/CEO; Jose Grimalt, exec VP.

Spanish Peaks Broadcasting Inc., 3702 Sunridge Dr., Park City, UT 84098-4618. Phone: (801) 560-9595. Ownership: Kevin Terry, 100%.

Stns: 3 FM. KDTR(FM) Florence, MT; KYJK(FM) Missoula, MT; KKVU(FM) Stevensville, MT.

Sparta-Tomah Broadcasting Co. Inc., 113 W. Oak St., Sparta, WI 54656. Phone: (608) 269-3307. Fax: (608) 269-5170. Ownership: Rice Family Trust, 35.73%; David Z. Rice, 18.01%; Patricia R. Hoffman, 15.42%; Barbara Rice, 15.42%; Elizabeth Ecker, 12.48%; and Sarah R. Cooper, 2.94%.

Stns: 1 AM. 2 FM. WCOW-FM Sparta, WI; WKLJ Sparta, WI; WFBZ-FM Trempealeau, WI.

Jose Grimalt, exec VP; Raul Alarcon, chmn & pres/CEO.

Spotlight Broadcasting LLC, Box 8888, Metairie, LA 70011. Phone: (504) 309-7260. Fax: (504) 309-7262. E-mail: kmrc@kmrc1430.com Web Site: www.kmrc1430.com.

Stns: 3 AM. WABL Amite, LA; KMRC Morgan City, LA; KAGY Port Sulphur, LA.

Patrick Andras, pres.

Standard Broadcasting Corp., 2 St. Clair Ave. W., Suite 1100, Toronto, ON M4V 1L6. Canada. Phone: (416) 960-9911. Fax: (416) 323-6828. Ownership: Slaight Communications Inc., 100%.

Stns: 22 AM. 33 FM. CJAY-FM Calgary, AB; CKMX Calgary, AB; CIBK-FM Calgary, AB; CFBR-FM Edmonton, AB; CFRN Edmonton, AB; CFMG-FM Saint Albert, AB; CFKC Creston, BC; CJDC Dawson Creek, BC; CKRX-FM Fort Nelson, BC; CKNL-FM Fort St. John, BC; CHRX-FM Fort St. John, BC; CKGR Golden, BC; CKIR Invermere, BC; CKFR(AM) Kelowna, BC; CHSU-FM Kelowna, BC; CILK-FM Kelowna, BC; CKTK-FM Kitimat, BC; CKKC-FM Nelson, BC; CKZX-FM New Denver, BC; CJOR Osoyoos, BC; CKOR Penticton, BC; CJMG-FM Penticton, BC; CHTK Prince Rupert, BC; CIOR Princeton, BC; CKCR Revelstoke, BC; CISL Richmond, BC; CKXR-FM Salmon Arm, BC; CHOR Summerland, BC; CFTK Terrace, BC; CJFW-FM Terrace, BC; CJAT-FM Trail, BC; CKZZ-FM Vancouver, BC; CICF-FM Vernon, BC; CKXA-FM Brandon, MB; CKX-FM Brandon, MB; CFQX-FM Selkirk, MB; CKMM-FM Winnipeg, MB; CKOC Hamilton, ON; CHAM Hamilton, ON; CKLH-FM Hamilton, ON; CJBK(AM) London, ON; CJBX-FM London, ON; CIQM-FM London, ON; CKSL London, ON; CKQB-FM Ottawa, ON; CHVR-FM Pembroke, ON; CHRE-FM Saint Catharines, ON; CHTZ-FM Saint Catharines, ON; CKTB Saint Catharines, ON; CFRB Toronto, ON; CJEZ-FM Toronto, ON; CFMX-FM Toronto, ON; CJFM-FM Montreal, PQ; CJAD Montreal, PQ; CHOM-FM Montreal, PQ.

Stns: 2 TV. CJDC, Dawson Creek, BC; CFTK, Terrace, BC.

Gary Slaight, pres/CEO.

Stanford Communications Inc., Box 458, Amory, MS 38821. Phone: (662) 256-9726. Fax: (662) 256-9725. E-mail: wamywafm@traceroad.net Web Site: www.fm95radio.com.

Stns: 2 AM. 1 FM. WWZQ Aberdeen, MS; WAFM-FM Amory, MS; WAMY Amory, MS.

Ed Stanford, pres; Teresa Stanford, sec/treas.

Star Broadcasting Inc., 21 Miracle Strip Pkwy., Fort Walton Beach, FL 32548. Phone: (850) 244-1400. Fax: (850) 243-1471. Ownership: Ronald E. Hale Jr., 40%; James Franklin Hale, 40%; and Jennifer F. Hale, 20%.

Stns: 2 AM. 2 FM. WPGG(AM) Evergreen, AL; WTKE(AM) Fort Walton Beach, FL; WTKE-FM Holt, FL; WRKN(FM) Niceville, FL.

Starlight Broadcasting Co., Box 106, 314 Main, Hartford, KY 42347. Phone: (270) 298-3268. Fax: (270) 298-9326. Web Site: www.wxmz.com.

Stns: 1 AM. 1 FM. WAIA(AM) Beaver Dam, KY; WKYA-FM Greenville, KY.

Andy Anderson, pres/CEO.

Steckline Communications Inc., 1632 S. Maize Rd., Wichita, KS 67209. Phone: (316) 721-8484. Fax: (316) 721-8276. Web Site: www.maanradio.com. Ownership: Gregory R. Steckline, 100%.

Stns: 3 AM. KIUL Garden City, KS; KYUL(AM) Scott City, KS; KGSO(AM) Wichita, KS.

Greg Steckline, pres.

Studstill Broadcasting, 3905 Progress Blvd., Peru, IL 61354. Phone: (815) 224-2100. Phone: (815) 224-2100 (Corp). Fax: (815) 224-2066. Ownership: Owen L. Studstill; Lamar Studstill; Cole C. Studstill.

Stns: 1 AM. 6 FM. WGLC-FM Mendota, IL; WALS-FM Oglesby, IL; WBZG(FM) Peru, IL; WIVQ(FM) Spring Valley, IL; WYYS(FM) Streator, IL; WSTQ(FM) Streator, IL; WSPL(AM) Streator, IL.

Owen L. Studstill, pres; Lamar Studstill, chmn; Cole C. Studstill, opns off.

Sudbury Services Inc., Box 989, Blytheville, AR 72316. Phone: (870) 762-2093. Fax: (870) 763-8459. Web Site: www.thundercountry963.com. Ownership: Harold L. Sudbury Jr., Lydia Sudbury Langston, LaNeal Sudbury Salter. Cable TV: Blytheville TV Cable Co., Blytheville, AR.

Stns: 5 AM. 5 FM. KLCN Blytheville, AR; KHLS-FM Blytheville, AR; KAMJ-FM Gosnell, AR; KXAR Hope, AR; KHPA-FM Hope, AR; KNBY Newport, AR; KOKR(FM) Newport, AR; KQDD(FM) Osceola, AR; KTPA Prescott, AR; KOSE Wilson, AR.

Harold Sudbury Jr., pres.

Summit City Radio Group, 2000 Lower Huntington Rd., Fort Wayne, IN 46819. Phone: (260) 747-1511. Fax: (260) 747-3999. E-mail: lloyd@summitcityradio.com Web Site: www.summitcityradio.com. Ownership: Bernard Radio LLC, 30%; Northwest Capital Partners II L.P.,70%, 100% total assets.

Stns: 1 AM. 3 FM. WNHT(FM) Churubusco, IN; WXKE(FM) Fort Wayne, IN; WGL Fort Wayne, IN; WGL-FM Huntington, IN.

Lloyd B. Roach, pres.

Summit Media Broadcasting LLC, 180 Main St., Sutton, WV 26601. Phone: (304) 765-7373. Fax: (304) 765-7836. E-mail: info@theboss97fm.com Web Site: www.theboss97fm.com. Ownership: Nunzio Aldo Sergi, 100%. Note: Mr. Sergi also owns 51% of WVAR(AM) Richwood and WAFD(FM) Webster Springs, both WV.

Stns: 1 AM. 2 FM. WKQV(FM) Richwood, WV; WSGB(AM) Sutton, WV; WDBS(FM) Sutton, WV.

Al Sergi, owner & gen mgr.

Sumter Broadcasting Co. Inc., Box 727, Americus, GA 31709. Phone: (229) 924-1390. Fax: (229) 928-2337. Web Site: www.americusradio.com.

Stns: 1 AM. 1 FM. WDEC-FM Americus, GA; WISK Americus, GA.

Steve Lashley, pres.

Sun Mountain Inc., 9045 Hobble Creek, Billings, MT 59101. Phone: (406) 665-2828. Fax: (406) 665-2131. Web Site: www.bigskyradio.net. Ownership: Richard Solberg, 100%.

Stns: 3 AM. KHDN Hardin, MT; KBSR Laurel, MT; KYLW(AM) Lockwood, MT.

Sun Valley Radio Inc., 810 W. 200 North, Logan, UT 84321. Phone: (435) 752-1390. Fax: (435) 752-1392. Ownership: M. Kent Frandsen, owner.

Stns: 2 AM. 4 FM. KKEX-FM Preston, ID; KLZX(FM) Weston, ID; KLGN Logan, UT; KVNU Logan, UT; KZHK-FM Saint George, UT; KGNT(FM) Smithfield, UT.

M. Kent Frandsen, pres.

Sunbelt Broadcasting Corp., Box 351, Columbia, MS 39429. Phone: (601) 731-2298. E-mail: wjdr@zzip.cc

Stns: 2 FM. WCJU-FM Prentiss, MS; WJDR-FM Prentiss, MS.

Thomas F. McDaniel, pres.

Sunburst Media-Louisiana LLC, 300 Crescent Ct., Suite 850, Dallas, TX 75201. Phone: (214) 528-5214. Ownership: Momentum Plan I Ltd. LLP, 33.33%; Aldus Sunburst Inc., 33.33%; John M. Borders, 29.7%; Don L. Turner, 3.64%. Note: Sunburst Media Inc. (0% equity and voting control) is the sole manager. Officers of Sunburst Media Inc.: John Borders, pres; Don L. Turner, VP.

Stns: 1 AM. 3 FM. KCIL(FM) Houma, LA; KJIN(AM) Houma, LA; KMYO-FM Morgan City, LA; KXOR-FM Thibodaux, LA.

Sunbury Broadcasting Corp., Box 1070, Sunbury, PA 17801. Phone: (570) 286-5838. Fax: (570) 743-7837. E-mail: wqkx@wqkx.com Web Site: www.wqkx.com. Ownership: Lois W. Haddon, 90.7%' Dr. Harry H. Haddon Jr., 8%; and Roger S. Haddon Jr., 1.3%.

Stns: 2 AM. 3 FM. WVLY-FM Milton, PA; WMLP(AM) Milton, PA; WEGH-FM Northumberland, PA; WKOK Sunbury, PA; WQKX-FM Sunbury, PA.

Roger S. Haddon Jr., Pres/CEO.

Sunrise Broadcasting Corp., Box 2307, Newburgh, NY 12550. Phone: (845) 561-2131. Fax: (845) 561-2138. Web Site: www.wgnyfm.com. Ownership: CVC Capital Corp.

Stns: 3 AM. 2 FM. KSNM(AM) Las Cruces, NM; KGRT-FM Las Cruces, NM; WJGK(AM) Highland, NY; WGNY Newburgh, NY; WGNY-FM Newburgh, NY.

J. Klebe, pres.

Superior Communications, 3302 N. Van Dyke, Imlay City, MI 48444. Phone: (810) 724-2638. Fax: (877) 850-0881. Web Site: www.positivehits.com.

Stns: 7 FM. WTLI-FM Bear Creek Township, MI; WSLI(FM) Belding, MI; WTAC(FM) Burton, MI; WHYT(FM) Goodland Township, MI; WAIR(FM) Lake City, MI; WLGH-FM Leroy Township, MI; WEJC-FM White Star, MI.

Edward Czelada, pres.

T

Tackett-Boazman Broadcasting LP, 9401 F.M. 45 South, Brownwood, TX 76801. Phone: (325) 641-8510. Ownership: Ray L. Boazman, 50%; and Donald Rex Tackett, 50%.

Stns: 2 AM. 2 FM. KXYL(AM) Brownwood, TX; KXYL-FM Brownwood, TX; KSTA Coleman, TX; KQBZ(FM) Coleman, TX.

Talking Stick Communications LLC, 421 S. Second St., Elkhart, IN 46514. Phone: (574) 258-5483. Ownership: Alec C. Dille, 60%; John F. Dille IV, 20%; and Sarah D. Erlacher, 20%.

Stns: 1 AM. 3 FM. WYPW(FM) Nappanee, IN; WAWC-FM Syracuse, IN; WRSW Warsaw, IN; WRSW-FM Warsaw, IN.

Alec C. Dille, pres.

Talley Radio Stations, Box 10, Litchfield, IL 62056. Phone: (217) 324-5921. Fax: (217) 532-2431. E-mail: wsmi@wsmiradio.com Web Site: wsmiradio.com. Ownership: Hayward L. Talley, Emma C. Talley.

Stns: 1 AM. 2 FM. WSMI Litchfield, IL; WSMI-FM Litchfield, IL; WAOX-FM Staunton, IL.

Hayward L. Talley, pres; Brian Talley, sr VP.

Tallgrass Broadcasting LLC, 1174 Hunters Ridge East, Hoffman Estates, IL 60192-4540. Phone: (847) 289-8018. Fax: (847) 289-1423. Ownership: Joseph E. Walker, 50%; and William H. Kurtis, 50%.

Stns: 3 AM. 5 FM. KIND Independence, KS; KIND-FM Independence, KS; KKYC-FM Clovis, NM; KICA Clovis, NM; KOSG(FM) Pawhuska, OK; KICA-FM Farwell, TX; KMUL(AM) Farwell, TX; KMUL-FM Muleshoe, TX.

Tama Broadcasting Inc., 5207 Washington Blvd., Tampa, FL 33619. Phone: (813) 620-1300. Fax: (813) 628-0713. Web Site: www.wtmp.com. Ownership: Black Enterprise/Greenwich Street Corporate Growth Partners L.P., 73.89% votes, 78.34% assets; Glenn W. Cherry, 20.89% votes, 17.33% assets; and Charles W. Cherry II, 5.22% votes, 4.33% assets.

Stns: 1 AM. 8 FM. WHJX(FM) Baldwin, FL; WTMP-FM Dade City, FL; WTMP Egypt Lake, FL; WJSJ(FM) Fernandina Beach, FL; WSJF(FM) Saint Augustine Beach, FL; WFJO(FM) Folkston, GA; WTHG(FM) Hinesville, GA; WSGA(FM) Hinesville, GA; WSSJ(FM) Rincon, GA.

Glenn W. Cherry, CEO.

Sarkes Tarzian Inc., Box 62, Bloomington, IN 47402. Phone: (812) 332-7251. Fax: (812) 331-4575. Ownership: Tom Tarzian; Mary Tarzian estate.

Stns: 1 AM. 3 FM. WTTS-FM Bloomington, IN; WGCL Bloomington, IN; WLDE-FM Fort Wayne, IN; WAJI(FM) Fort Wayne, IN.

Stns: 2 TV. WRCB-TV, Chattanooga, TN; KTVN, Reno, NV.

Tom Tarzian, opns VP; Bob Davis, CFO; Geoff Vargo, pres radio; Valerie Comey, gen councel.

Team Radio LLC, Box 2509, Ponca City, OK 74602. Phone: (580) 765-2485. Fax: (580) 767-1103. Web Site: www.eteamradio.com. Ownership: William L. Coleman, 100%.

Stns: 2 AM. 2 FM. KOKB Blackwell, OK; KOKP Perry, OK; KPNC-FM Ponca City, OK; KLOR-FM Ponca City, OK.

Bill Coleman, gen mgr.

Tejas Broadcasting Ltd. LLP, 1227 W. Magnolia Ave., Suite 300, Fort Worth, TX 75104-4400. Phone: (972) 692-3310. Ownership: Ultimately controlled by James L. Anderson.

Stns: 1 AM. 7 FM. KBZD(FM) Amarillo, TX; KTNZ Amarillo, TX; KQFX-FM Borger, TX; KLTG-FM Corpus Christi, TX; KGRW-FM Friona, TX; KLHB(FM) Odem, TX; KMJR(FM) Portland, TX; KOUL-FM Sinton, TX.

Jim Anderson, CEO.

TeleSouth Communications Inc., 6311 Ridgewood Rd., Jackson, MS 39211. Phone: (601) 957-1700. Fax: (601) 957-2389. Web Site: www.supertalkms.com. Ownership: Steve Davenport

Stns: 4 AM. 8 FM. WXRZ(FM) Corinth, MS; WKCU Corinth, MS; WFMN-FM Flora, MS; WKXG Greenwood, MS; WOEG Hazlehurst, MS; WDXO-FM Hazlehurst, MS; WTCD(FM) Indianola, MS; WRQO-FM Monticello, MS; WQLJ-FM Oxford, MS; WFMM-FM Sumrall, MS; WTNM(FM) Water Valley, MS; WROB(AM) West Point, MS.

Stephen C. Davenport, pres/CEO.

Textron Financial Corp., 40 Westminster St., Providence, RI 02903. Phone: (401) 621-4200. Web Site: www.textronfinancial.com. Ownership: Textron Inc., 100%.

Stns: 5 AM. 8 FM. KPGG(FM) Ashdown, AR; KRFM-FM Show Low, AZ; KVSL Show Low, AZ; KMOQ-FM Baxter Springs, KS; KJML-FM Columbus, KS; KCAR-FM Galena, KS; KQYX(AM) Joplin, MO; KQYS(AM) Neosho, MO; KBTN-FM Neosho, MO; KCAR Clarksville, TX; KGAP-FM Clarksville, TX; KEWL-FM New Boston, TX; KKTK(AM) Texarkana, TX.

3 Daughters Media Inc., c/o Brooks, Pierce, et al, Box 1800, Raleigh, NC 27602. Phone: (919) 839-0300. Ownership: Gary E. Burns, 100%.

Stns: 5 AM. 2 FM. WUUS(AM) Rossville, GA; WUUS-FM South Pittsburg, TN; WBLT(AM) Bedford, VA; WMNA Gretna, VA; WMNA-FM Gretna, VA; WVGM Lynchburg, VA; WGMN Roanoke, VA.

Three Eagles Communications, 3800 Cornhusker Hwy., Lincoln, NE 68504. Phone: (402) 466-1234. Fax: (402) 467-4095. E-mail: gbuchanan@threeeagles.com Web Site: www.threeeagles.com. Ownership: Rolland C. Johnson.

Stns: 16 AM. 17 FM. KIAQ-FM Clarion, IA; KVFD(AM) Fort Dodge, IA; KRIB(AM) Mason City, IA; KYTC-FM Northwood, IA; KTLB-FM Twin Lakes, IA; WCCQ-FM Crest Hill, IL; KATE Albert Lea, MN; KCPI(FM) Albert Lea, MN; KAUS Austin, MN; KQYK(FM) Lake Crystal, MN; KQAD(AM) Luverne, MN; KEEZ-FM Mankato, MN; KYSM(AM) Mankato, MN; KYSM-FM Mankato, MN; KLGR Redwood Falls, MN; KRBI(AM) Saint Peter, MN; KITN-FM Worthington, MN; KWOA Worthington, MN; KWOA-FM Worthington, MN; KZEN-FM Central City, NE; KTTT Columbus, NE; KLIR-FM Columbus, NE; KJSK Columbus, NE; KFRX(FM) Lincoln, NE; KRKR(FM) Lincoln, NE; KBRK Brookings, SD; KDBX(FM) Clear Lake, SD; KJAM(AM) Madison, SD; KKSD-FM Milbank, SD; KJJQ Volga, SD; KDLO-FM Watertown, SD; KSDR Watertown, SD; KWAT Watertown, SD.

Gary Buchanan, pres/COO.

3 Point Media, 980 N. Michigan Ave., Suite 1880, Chicago, IL 60611. Phone: (312) 204-9900. Fax: (312) 587-9466.

Stns: 5 FM. KOAY(FM) Coalville, UT; KMGR(FM) Delta, UT; KCUA(FM) Naples, UT; KHTB(FM) Provo, UT; KYLZ(FM) Tremonton, UT.

Three Rivers Media Corp., Box 1247, Wytheville, VA 24382. Phone: (276) 228-3185. Fax: (276) 228-9261. Ownership: Anthony Accamando Jr., 39%; James Browne, 39%; and Gary W. Hagerich, 22%.

Stns: 2 AM. 1 FM. WXBX(FM) Rural Retreat, VA; WLOY(AM) Rural Retreat, VA; WYVE Wytheville, VA.

Gary W. Hagerich, pres/COO.

Three Trees Communications Inc., 113 E. College Ave., Ashburn, GA 31714. Phone: (229) 567-9038. Ownership: James Andrew Howard, 33.33%; James Thomas Overton, 33.33%; and Andrew H. Reeves, 33.33%.

Stns: 1 AM. 2 FM. WJYF(FM) Nashville, GA; WTIF-FM Omega, GA; WTIF Tifton, GA.

Thunderbolt Broadcasting Co., Box 318, 1410 N. Lindell St., Martin, TN 38237. Phone: (731) 587-9526. Fax: (731) 587-5079. Ownership: Paul Freeman Tinkle, trustee of the Paul Freeman Tinkle Revocable Trust, 40.42%; Jimmy C. Smith, 23.75%; Thomas L. Moore Jr., 19.16%; and Fred C. Stoker, 16.67%.

Stns: 1 AM. 4 FM. WCDZ(FM) Dresden, TN; WCMT(AM) Martin, TN; WCMT-FM South Fulton, TN; WYVY-FM Union City, TN; WQAK(FM) Union City, TN.

Paul Freeman Tinkle, pres.

Tiger Communications Inc., 2514 S. College St., Suite 104, Auburn, AL 36832-6925. Phone: (334) 887-9999. Fax: (334) 826-9599. Ownership: Thomas Haley, 96.5%; and Tracey Ivey, 3.5%.

Stns: 1 AM. 2 FM. WAUD Auburn, AL; WQNR(FM) Tallassee, AL; WTGZ(FM) Tuskegee, AL.

Touch Canada Broadcasting (2006) Inc., 4207 98th St. N.W., Suite 204, Edmonton, AB T6E 5R7. Canada. Phone: (780) 466-4930. Fax: (780) 469-5335.

Stns: 1 AM. 2 FM. CJSI-FM Calgary, AB; CJRY-FM Edmonton, AB; CJCA Edmonton, AB.

Charles Allard, owner.

Tower Investment Trust Inc., 819 S. Federal Hwy., Suite 106, Stuart, FL 34994-2952. Phone: (772) 215-1634. Web Site: www.toweritrust.com. Ownership: William H. Brothers, 50%; and Gary S. Hess, 50%. Note: Group also has an application for a New FM Pine Knoll Shores, NC. William H. Brothers, as an individual, is the permittee of KDRX(FM) Rocksprings, TX. Gary S. Hess, as an individual, is the permittee of KHES(AM) Rocksprings, TX.

Stns: 6 FM. WLEL(FM) Ellaville, GA; KSAQ(FM) Charlotte, TX; KXXN(FM) Iowa Park, TX; KTTY(FM) New Boston, TX; KSAG(FM) Pearsall, TX; KLOW(FM) Reno, TX.

Bill Brothers, pres; Gary Hess, VP.

Town and Country Broadcasting Inc., 486 W. 2nd St., Xenia, OH 45385-3610. Phone: (937) 372-3531. Fax: (937) 372-3508. Ownership: William J. Mullins, 100%.

Stns: 3 AM. WEDI(AM) Eaton, OH; WKFI(AM) Wilmington, OH; WBZI Xenia, OH.

Track 1 Media of Sterling LLC, Box 830, Sterling, CO 80751. Phone: (970) 522-1607. Fax: (970) 522-1322. E-mail: kning@rodinetechnology.com Ownership: Larry Levy, 89.5%; and Ellen Levy, 10.5%.

Stns: 1 AM. 1 FM. KSTC Sterling, CO; KNEC-FM Yuma, CO.

Larry Levy, owner; Betty Carlson, gen mgr.

Tracy Broadcasting Corp., Box 532, Scottsbluff, NE 69363-0532. Phone: (308) 635-1320. Fax: (308) 635-1905. Web Site: www.tracybroadcasting.com. Ownership: Michael Tracy.

Stns: 2 AM. 3 FM. KMOR(FM) Bridgeport, NE; KOZY-FM Gering, NE; KOLT Scottsbluff, NE; KOAQ Terrytown, NE; KOLT-FM Warren AFB, WY.

Larry Swikard, gen mgr; Michael Tracy, pres.

Tri-Market Radio Broadcasters Inc. & Eagle Rock Broadcasting Inc., 120 South 300 W., Rupert, ID 83350. Phone: (208) 436-4757. Fax: (208) 436-3050.

Stns: 2 AM. 2 FM. KBAR Burley, ID; KZDX-FM Burley, ID; KFTA(AM) Rupert, ID; KKMV-FM Rupert, ID.

Kim Lee, gen mgr.

Triad Broadcasting Co. L.L.C., 2511 Garden Rd., Bldg. A, Suite 104, Monterey, CA 93940. Phone: (831) 655-6350. Fax: (831) 655-6355. E-mail: jpeterson@triadbroadcasting.com Web Site: www.triadbroadcasting.com. Ownership: Northwest Equity Partners, Shamrock Capital Advisors, Bank of America Capital Investors.

Stns: 13 AM. 19 FM. WGCO(FM) Midway, GA; WDQX(FM) Morton, IL; WXCL(FM) Pekin, IL; WSWT-FM Peoria, IL; WPBG-FM Peoria, IL; WIRL(AM) Peoria, IL; WMBD Peoria, IL; KLTA(FM) Breckenridge, MN; KBMW Breckenridge, MN; KQWB-FM Moorhead, MN; KVOX-FM Moorhead, MN; WXBD Biloxi, MS; WTNI(AM) Biloxi, MS; WCPR-FM D'Iberville, MS; WUJM(FM) Gulfport, MS; WXYK-FM Gulfport, MS; WHGO(FM) Pascagoula, MS; KPFX(FM) Fargo, ND; KQWB(AM) West Fargo, ND; KWBE Beatrice, NE; KLNC(FM) Lincoln, NE; KLIN Lincoln, NE; KFGE-FM Milford, NE; WGZR(FM) Bluffton, SC; WFXH(AM) Hilton Head Island, SC; WLOW(FM) Port Royal, SC; WBDY Bluefield, VA; WHQX-FM Cedar Bluff, VA; WTZE Tazewell, VA; WKEZ Bluefield, WV; WHIS Bluefield, WV; WKOY-FM Princeton, WV.

Judy Peterson, VP; Thomas Douglas, CFO.

Tribune Broadcasting Co., 435 N. Michigan Ave., Suite 1800, Chicago, IL 60611. Phone: (312) 222-3333. Fax: (312) 329-0611. Web Site: www.tribune.com. Ownership: Robert R. McCormick Tribune Foundation, 13.3%; The Chandler Trusts, 11.6%; Vanguard Fiduciary Trust Co., 6.9%.

Stns: 1 AM. WGN(AM) Chicago, IL.

Stns: 24 TV. WGN, Chicago; KDAF, Dallas-Ft. Worth; KWGN, Denver, CO; WXMI, Grand Rapids-Kalamazoo-Battle Creek, MI; WPMT, Harrisburg-Lancaster-Lebanon-York, PA; WTIC, Hartford & New Haven, CT; WTXX, Hartford & New Haven, CT; KHCW, Houston; WTTK, Indianapolis, IN; WTTV, Indianapolis, IN; WXIN, Indianapolis, IN; KTLA, Los Angeles; WSFL-TV, Miami-Ft. Lauderdale, FL; WGNO, New Orleans, LA; WNOL, New Orleans, LA; WPIX, New York; WPHL, Philadelphia; KRCW-TV, Portland, OR; KTXL, Sacramento-Stockton-Modesto, CA; KSWB, San Diego, CA; KCPQ, Seattle-Tacoma, WA; KMYQ, Seattle-Tacoma, WA; KPLR, St. Louis, MO; WDCW, Washington, DC (Hagerstown, MD).

John Reardon, pres/CEO; John Vitanovec, exec VP.

Tri-County Broadcasting Inc., Box 366, Sauk Rapids, MN 56379. Phone: (320) 252-6200. Fax: (320) 252-9367. Ownership: Herbert M. Hoppe, 51% of votes; Valeria Hoppe, 49% of votes. Note: Herbert M. Hoppe owns 100% of WPPI(AM) Sauk Rapids, MN.

Stns: 2 AM. 1 FM. WVAL Sauk Rapids, MN; WBHR Sauk Rapids, MN; WHMH-FM Sauk Rapids, MN.

Truth Broadcasting Corp., 4405 Providence Ln., Suite D, Winston-Salem, NC 27106. Phone: (336) 759-0363. Fax: (336) 759-0366. E-mail: tbooth@830wtru.com Web Site: www.830wtru.com. Ownership: Stuart W. Epperson Jr., 100%.

Stns: 6 AM. WZRH(AM) Dallas, NC; WKEW Greensboro, NC; WTRU(AM) Kernersville, NC; WDRU(AM) Wake Forest, NC; WPOL Winston-Salem, NC; WLVA Lynchburg, VA.

Stuart Epperson, pres.

Tschudy Broadcast Group, 15 Campbell St., Luray, VA 22835. Phone: (540) 743-3000. Fax: (540) 743-3002.

Stns: 1 AM. 2 FM. WPDX(AM) Clarksburg, WV; WPDX-FM Clarksburg, WV; WZST-FM Westover, WV.

Earl Judy Jr., pres; Mike King, gen mgr.

Buddy Tucker Association Inc., Box 63, Mobile, AL 36601. Phone: (386) 738-1348. Fax: (251) 432-1396. Ownership: Theodore D. Tucker, 100%.

Stns: 3 AM. WTOF(AM) Bay Minette, AL; WMOB Mobile, AL; WYND De Land, FL.

2510 Licenses LLC, 100 Ryan Ct., Suite 98, Pittsburgh, PA 15205. Phone: (412) 489-1001. Fax: (412) 489-1002. Ownership: Nicholas A. Galli, 100%.

Stns: 2 AM. 6 FM. WCCL(FM) Central City, PA; WPRR(AM) Johnstown, PA; WKVB(FM) Port Matilda, PA; WLKJ(FM) Portage, PA; WLKH(FM) Somerset, PA; WBHV(AM) Somerset, PA; WBHV-FM State College, PA; WOWY(FM) University Park, PA.

Twenty-One Sound Communications Inc., 3418 Douglas Rd., Florissant, MO 63034. Phone: (314) 921-9330. Fax: (314) 830-4141. Ownership: Randy Wachter, 100%. Note: Randy Wachter also owns 100% of KNSX(FM) Steelville, MO and 50% of KLPW(AM) Union, MO. Group is a party to a time brokerage agreement for KLPW-FM Union, MO.

Stns: 2 FM. KESY(FM) Cuba, MO; KKAC(FM) Vandalia, MO.

Tyler Media Broadcasting Corp., 5101 S. Shields Blvd., Oklahoma City, OK 73129. Phone: (405) 616-5500. Fax: (405) 616-5505. Web Site: www.kkng.com. Ownership: Ty A. Tyler, Tony J. Tyler and Tony J. Tyler 2000 Irrevocable Trust, Tony J. Tyler, trustee.

Stns: 2 AM. 3 FM. KOJK(FM) Blanchard, OK; KOCY(AM) Del City, OK; KKNG-FM Newcastle, OK; KTUZ-FM Okarche, OK; KTLR(AM) Oklahoma City, OK.

Stns: 1 TV. KTUZ-TV, Oklahoma City, OK.

Skip Stow, market mgr; Robert De Negri, CFO.

U

US Stations LLC, 125 Corporate Terr., Hot Springs, AR 71913. Phone: (501) 525-9700. Fax: (501) 525-9739. Web Site: www.usstations.com. Ownership: Charles Shinn, 52%; Gary Terrell, 28%; and Craig Dale, 20%.

Stns: 2 AM. 4 FM. KWXE-FM Glenwood, AR; KWXI Glenwood, AR; KYDL(FM) Hot Springs, AR; KQUS-FM Hot Springs, AR; KZNG Hot Springs, AR; KLXQ(FM) Mountain Pine, AR.

United Ministries, 300 E. Rock Rd., Allentown, PA 18103. Phone: (970) 254-5565. Fax: (970) 254-5550. Ownership: Non-stock, not-for-profit corporation.

Stns: 2 AM. 1 FM. KDTA Delta, CO; KJOL(AM) Grand Junction, CO; WBMR-FM Telford, PA.

Universal Broadcasting of New York Inc., Corporate Offices, WTHE Radio 260 E. 2nd St., Mineola, NY 11501. Phone: (516) 742-1520. Fax: (516) 742-2878. E-mail: nygospelradio@aol.com Web Site: www.wthe1520am.com. Ownership: Howard Warshaw and Miriam Warshaw.

Stns: 2 AM. WVNJ(AM) Oakland, NJ; WTHE(AM) Mineola, NY.

Miriam Warshaw, pres; Howard Warshaw, VP.

Univision Radio, 3102 Oak Lawn, Suite 215, Dallas, TX 75219. Phone: (214) 525-7700. Fax: (214) 525-7750. Web Site: www.univision.net/corp/en/urg.jsp. Ownership: Univision Communications Inc., 100% (see listing under TV Group Ownership, Section B).

Stns: 18 AM. 54 FM. KKMR(FM) Arizona City, AZ; KQMR(FM) Globe, AZ; KHOT-FM Paradise Valley, AZ; KOMR(FM) Sun City, AZ; KHOV-FM Wickenburg, AZ; KOND(FM) Clovis, CA; KSCA-FM Glendale, CA; KRDA(FM) Hanford, CA; KRCD(FM) Inglewood, CA; KTNQ Los Angeles, CA; KLVE-FM Los Angeles, CA; KLLE(FM) North Fork, CA; KLNV-FM San Diego, CA; KLQV-FM San Diego, CA; KSOL(FM) San Francisco, CA; KLOK San Jose, CA; KBRG(FM) San Jose, CA; KVVZ(FM) San Rafael, CA; KVVF(FM) Santa Clara, CA; KSQL(FM) Santa Cruz, CA; KRCV(FM) West Covina, CA; WRTO-FM Goulds, FL; WQBA Miami, FL; WAMR-FM Miami, FL; WAQI Miami, FL; WRTO(AM) Chicago, IL; WPPN(FM) Des Plaines, IL; WOJO-FM Evanston, IL; WVIV-FM Highland Park, IL; WVIX(FM) Joliet, IL; WCAA-FM Newark, NJ; KKRG(FM) Albuquerque, NM; KIOT-FM Los Lunas, NM; KQBT(FM) Rio Rancho, NM; KKSS-FM Santa Fe, NM; KJFA(FM) Santa Fe, NM; KRGT(FM) Indian Springs, NV; KISF-FM Las Vegas, NV; KLSQ(AM) Whitney, NV; WQBU-FM Garden City, NY; WADO New York, NY; WYEL(AM) Mayaguez, PR; WUKQ-FM Mayaguez, PR; WUKQ(AM) Ponce, PR; WKAQ San Juan, PR; KDXX(FM) Benbrook, TX; KCOR-FM Comfort, TX; KPTI(FM) Crystal Beach, TX; KFZO(FM) Denton, TX; KAMA El Paso, TX; KBNA(AM) El Paso, TX; KBNA-FM El Paso, TX; KFLC(AM) Fort Worth, TX; KLNO(FM) Fort Worth, TX; KOVE-FM Galveston, TX; KINV(FM) Georgetown, TX; KGBT Harlingen, TX; KBTQ(FM) Harlingen, TX; KLAT Houston, TX; KLTN(FM) Houston, TX; KESS-FM Lewisville, TX; KAJZ(FM) Llano, TX; KGBT-FM McAllen, TX; KLTO-FM McQueeney, TX; KPTY(FM) Missouri City, TX; KQBU-FM Port Arthur, TX; KHCK-FM Robinson, TX; KRTX Rosenberg-Richmond, TX; KAHL(AM) San Antonio, TX; KXTN-FM San Antonio, TX; KCOR(AM) San Antonio, TX; KBBT(FM) Schertz, TX.

McHenry T. Tichenor Jr., pres/CEO.

Uno Radio Group, Box 363222, San Juan, PR 00936-3222. Phone: (787) 758-1300. Fax: (787) 282-6060. E-mail: lsoto@unoradio.com Web Site: www.unoradio.com. Ownership: Jesus M. Soto.

Stns: 6 AM. 6 FM. WFDT(FM) Aguada, PR; WIVA-FM Aguadilla, PR; WCMN Arecibo, PR; WMIO(FM) Cabo Rojo, PR; WNEL Caguas, PR; WORA Mayaguez, PR; WPRP Ponce, PR; WLEO(AM) Ponce, PR; WRIO-FM Ponce, PR; WFID-FM Rio Piedras, PR; WPRM-FM San Juan, PR; WUNO(AM) San Juan, PR.

Jesus M. Soto, CEO; Luis A. Soto, pres; Elba Esmurria, sls VP; Luis Gonzalez, VP finance; Ray Cruz, VP progmg; Alberte Pereira, VP engrg.

Urban Radio Licenses LLC, 273 Azalea Rd., Suite 1-308, Mobile, AL 36609. Phone: (251) 343-4900. Fax: (251) 343-4905. E-mail: info@urbanradio.fm Web Site: www.urbanradio.fm. Ownership: Urban Radio Communications LLC, 100%. Note: Urban Radio Communications LLC also owns KMXH(FM) Alexandria and KBCE(FM) Boyce, both LA.

Stns: 2 AM. 8 FM. WLAY-FM Littleville, AL; WLAY Muscle Shoals, AL; WVNA-FM Muscle Shoals, AL; WVNA Tuscumbia, AL; WACR-FM Aberdeen, MS; WAJV-FM Brooksville, MS; WMSU(FM) Starkville, MS; WIMX-FM Gibsonburg, OH; WJZE-FM Oak Harbor, OH; WMXV(FM) Saint Joseph, TN.

V

VCY America Inc., 3434 W. Kilbourn Ave., Milwaukee, WI 53208. Phone: (414) 935-3000. Fax: (414) 935-3015. E-mail: vcy@vcyamerica.org Web Site: www.vcyamerica.org.

Stns: 1 AM. 14 FM. KVCY-FM Fort Scott, KS; KCVS(FM) Salina, KS; WVCN(FM) Baraga, MI; WVCM(FM) Iron Mountain, MI; WJIC-FM Zanesville, OH; KVCF(FM) Freeman, SD; KVCX-FM Gregory, SD; KVFL(FM) Pierre, SD; WVCF-FM Eau Claire, WI; WVFL(FM) Fond du Lac, WI; WVCY-FM Milwaukee, WI; WVCY Oshkosh, WI; WVCX(FM) Tomah, WI; WEGZ(FM) Washburn, WI; WVRN(FM) Wittenberg, WI.

Stns: 1 TV. WVCY, Milwaukee, WI.

Vic Eliason, VP/gen mgr; Jim Schneider, progmg dir.

Vermont Broadcast Associates Inc., Box 97, Lyndonville, VT 05851. Phone: (802) 626-9800. Fax: (802) 626-8500. Ownership: Bruce A. James, 100%.

Stns: 1 AM. 4 FM. WMTK-FM Littleton, NH; WJPK(FM) Barton, VT; WGMT-FM Lyndon, VT; WKXH-FM Saint Johnsbury, VT; WSTJ Saint Johnsbury, VT.

Vernal Enterprises Inc., Box 1032, Indiana, PA 15701-1032. Phone: (724) 543-1380. Fax: (724) 543-1140. Ownership: Larry L. Schrecongost, 51%; Nancy W. Schrecongost, 49%.

Stns: 3 AM. 1 FM. WRDD Ebensburg, PA; WHPA(FM) Gallitzin, PA; WTYM Kittanning, PA; WNCC(AM) Northern Cambria, PA.

Vero Beach Broadcasters LLC, 1235 16th St., Vero Beach, FL 32960. Phone: (772) 567-0937. Fax: (772) 562-4747. Web Site: wosnfm.com. Ownership: Mitchell Rubenstein, Laurie Silvers and Robert McAllan.

Stns: 1 AM. 3 FM. WOSN-FM Indian River Shores, FL; WGYL-FM Vero Beach, FL; WTTB Vero Beach, FL; WJKD(FM) Vero Beach, FL.

Jim Davis, gen mgr.

VerStandig Broadcasting, 4850 Connecticut Ave. N.W., Suite 103, Washington, DC 20008. Phone: (202) 244-1422. Fax: (202) 362-4149. Ownership: John VerStandig, 1996 VerStandig Children's Trust, M. Belmont VerStandig Trust.

Stns: 3 AM. 6 FM. WPPT(FM) Mercersburg, PA; WFYN(FM) Waynesboro, PA; WCBG(AM) Waynesboro, PA; WBHB-FM Bridgewater, VA; WJDV(FM) Broadway, VA; WHBG Harrisonburg, VA; WQPO-FM Harrisonburg, VA; WSVA Harrisonburg, VA; WAYZ(FM) Hagerstown, MD.

John VerStandig, CEO.

Victoria RadioWorks Ltd., 8023 Vantage Dr., Suite 840, San Antonio, TX 78230. Phone: (210) 340-7080. Fax: (210) 341-1777. Ownership: John W. Barger, pres of gen ptnr 89%; Cindy Cox, 10%.

Stns: 2 AM. 3 FM. KITE(FM) Port Lavaca, TX; KVIC(FM) Victoria, TX; KNAL(AM) Victoria, TX; KVNN(AM) Victoria, TX; KBAR-FM Victoria, TX.

John Barger, pres; Cindy Cox, gen mgr.

Vidalia Communications Corp., Box 900, Vidalia, GA 30475. Phone: (912) 537-9202. Fax: (912) 537-4477. E-mail: wtcq@vidaliacommunications.com Web Site: www.vidaliacommunications.com.

Stns: 1 AM. 1 FM. WYUM-FM Mount Vernon, GA; WVOP Vidalia, GA.

Advance-Progress Newspaper Inc., publisher of the weekly *Advance*, is part of the partnership of Vidalia Communications Corp.

John Ladson III, pres; Zack Fowler, gen mgr.

Viper Communications Broadcast Group, Box 225, Osage Beach, MO 65065. Phone: (573) 348-2772. Fax: (573) 348-2779. Web Site: www.krmsradio.com.

Stns: 2 AM. 1 FM. WENG Englewood, FL; KRMS Osage Beach, MO; KMYK(FM) Osage Beach, MO.

Dennis Klautzer, VP/gen mgr.

Visionary Related Entertainment L.L.C., Box 1437, Wailuku, HI 96793. Phone: (808) 244-9145. Fax: (808) 244-8247. E-mail: kaoi@kaoi.net Web Site: www.kaoi.net. Ownership: Visionary Related Entertainment Inc., 50.1% of votes, 40.58% of total assets; Frontier Radio Investors L.L.C., 49.9% of votes, 59.42% of total assets.

Stns: 4 AM. 12 FM. KUAI Eleele, HI; KUMU Honolulu, HI; KUMU-FM Honolulu, HI; KPOI-FM Honolulu, HI; KQMQ-FM Honolulu, HI; KMKK-FM Kaunakakai, HI; KSHK-FM Kekaha, HI; KHEI-FM Kihei, HI; KAOI(AM) Kihei, HI; KTBH-FM Kurtistown, HI; KQNG Lihue, HI; KDLX(FM) Makawao, HI; KNUQ-FM Paauilo, HI; KSRF-FM Poipu, HI; KAOI-FM Wailuku, HI; KDDB(FM) Waipahu, HI.

John Detz, pres; James McKeon, VP.

Vista Broadcast Group Inc., 1940 Third Ave., Prince George, BC V2M 1G7. Canada. Phone: (250) 564-2524. Fax: (250) 562-6611. Ownership: Jetport Inc., 21.4%; 49 shareholders holding less than 10% each. Note: Group also owns CFFM-FM-2 Quesnel and CIRX-FM-1 Vanderhoof, both BC (both originating stns).

Stns: 8 AM. 9 FM. CFRI-FM Grande Prairie, AB; CKBX 100 Mile House, BC; CFLD Burns Lake, BC; CFWB Campbell River, BC; CKQR-FM Castlegar, BC; CFCP-FM Courtenay, BC; CJSU-FM Duncan, BC; CHNV-FM Nelson, BC; CFNI Port Hardy, BC; CHQB Powell River, BC; CIRX-FM Prince George, BC; CJCI-FM Prince George, BC; CKCQ-FM Quesnel, BC; CFBV Smithers, BC; CIVH Vanderhoof, BC; CFFM-FM Williams Lake, BC; CKWL Williams Lake, BC.

Gary Russell, pres.

Vox Radio Group L.P., Box 1230, Claremont, NH 03743. Phone: (603) 542-7735. Fax: (603) 542-3780. Ownership: Bruce Danziger, Jeffery Shapiro, Ken Barlow.

Stns: 4 AM. 9 FM. WWUS(FM) Big Pine Key, FL; WCNK(FM) Key West, FL; WAVK(FM) Marathon, FL; WVEI-FM Easthampton, MA; WSBS Great Barrington, MA; WUPE-FM North Adams, MA; WNAW North Adams, MA; WBEC Pittsfield, MA; WBEC-FM Pittsfield, MA; WUPE(AM) Pittsfield, MA; WWHK(FM) Concord, NH; WBOP(FM) Buffalo Gap, VA; WSIG(FM) Mount Jackson, VA.

Bruce G. Danziger, principal; Jeffrey Shapiro, principal; Ken Barlow, principal.

W

WAMC/Northeast Public Radio, 318 Central Ave., Albany, NY 12206. Phone: (518) 465-5233. Fax: (518) 432-6974. E-mail: mail@wamc.org Web Site: www.wamc.org. Ownership: Non-stock educ corporation.

Stns: 2 AM. 7 FM. WAMQ-FM Great Barrington, MA; WAMC(AM) Albany, NY; WAMC-FM Albany, NY; WCAN-FM Canajoharie, NY; WAMK-FM Kingston, NY; WOSR-FM Middletown, NY; WCEL-FM Plattsburgh, NY; WANC-FM Ticonderoga, NY; WRUN(AM) Utica, NY.

Alan Chartock, pres/CEO; David Galletly, sr VP.

WAY-FM Media Group Inc., 1012 McEwen Dr., Franklin, TN 37067. Phone: (615) 261-9293. Fax: (615) 261-3967. Web Site: www.wayfm.com.

Stns: 14 FM. WAYH(FM) Harvest, AL; KBWA(FM) Brush, CO; KXWA(FM) Loveland, CO; KXWY(FM) Rye, CO; WAYJ-FM Fort Myers, FL; WAYF(FM) West Palm Beach, FL; WAYT(FM) Thomasville, GA; KYWA(FM) Wichita, KS; WAYD(FM) Auburn, KY; KWYA(FM) Astoria, OR; WAYQ(FM) Clarksville, TN; WAYM(FM) Columbia, TN; WAYW(FM) New Johnsonville, TN; KWYQ(FM) Longview, WA.

Bob Augsburg, pres/CEO; Dusty Rhodes, sr VP; Lloyd Parker, COO; Dave Ringing, CFO.

WENK of Union City Inc., 1729 Nailling Dr., Union City, TN 38261. Phone: (731) 885-1240. Fax: (731) 885-3405. E-mail: thailey@wenkwtpr.com Ownership: Bill Latimer; Robert Kirkland; Robert Terrell Jr.

Stns: 2 AM. 3 FM. WWKF-FM Fulton, KY; WTPR-FM McKinnon, TN; WAKQ-FM Paris, TN; WTPR(AM) Paris, TN; WENK Union City, TN.

Terry Hailey, pres; Bill Latimer, chmn.

WOLF Radio Inc., 401 W. Kirkpatrick St., Syracuse, NY 13204. Phone: (315) 472-0222. Fax: (315) 478-7745. Ownership: Craig L. Fox, 51%; George W. Kimble, 49%.

Stns: 2 AM. 2 FM. WWLF(AM) Auburn, NY; WOLF-FM Oswego, NY; WWLF-FM Sylvan Beach, NY; WOLF Syracuse, NY.

Craig Fox, pres.

WRD Entertainment Inc., Box 2077, Batesville, AR 72503. Phone: (870) 793-4196. Fax: (870) 793-5222. E-mail: rob@maxfm.com Web Site: maxfm.com.

Stns: 2 AM. 4 FM. KAAB Batesville, AR; KBTA Batesville, AR; KBTA-FM Batesville, AR; KZLE-FM Batesville, AR; KKIK(FM) Horseshoe Bend, AR; KWOZ-FM Mountain View, AR.

John R. Grace, pres; Gary Bridgman, gen mgr.

WZOE Inc., Box 69, Princeton, IL 61356. Phone: (815) 875-8014. Web Site: www.wzoeradio.com. Ownership: Steve Samet, 100%.

Stns: 1 AM. 2 FM. WRVY-FM Henry, IL; WZOE Princeton, IL; WZOE-FM Princeton, IL.

Steve Samet, pres/gen mgr.

Wagenvoord Advertising Group Inc., 2360 N.E. Coachman Rd., Clearwater, FL 33765. Phone: (727) 726-8247. Fax: (727) 799-8866. Web Site: www.tantalk1340.com. Ownership: Dave Wagenvoord, 50%; Lola Wagenvoord, 50%.

Stns: 3 AM. WTAN(AM) Clearwater, FL; WDCF Dade City, FL; WZHR(AM) Zephyrhills, FL.

Lola Wagenvoord, gen mgr.

Wagonwheel Communications Corp., 40 Shoshone Ave., Green River, WY 82935. Phone: (307) 875-6666. Fax: (303) 875-5847. E-mail: kugr@sweetwater.net Web Site: www.theradionetwork.net. Ownership: Alan W. Harris, trustee of the Alan W. Harris Living Trust, 51%; and Faith R. Harris, trustee of the Faith R. Harris Living Trust, 49%.

Stns: 1 AM. 3 FM. KFRZ-FM Green River, WY; KUGR Green River, WY; KZWB(FM) Green River, WY; KYCS(FM) Rock Springs, WY.

Waitt Omaha LLC, 1125 S. 103rd St., Suite 200, Omaha, NE 68124. Phone: (402) 697-8000. Ownership: WaittCorp Investments LLC, 100% of total assets.

Stns: 5 AM. 5 FM. KQKQ-FM Council Bluffs, IA; KSLS-FM Liberal, KS; KYUU Liberal, KS; KOZN(AM) Bellevue, NE; KYDZ(AM) Bellevue, NE; KBLR-FM Blair, NE; KLTQ(FM) Lincoln, NE; KKAR Omaha, NE; KOIL(AM) Plattsmouth, NE; KOPW(FM) Plattsmouth, NE.

Walking by Faith Ministries Inc., 336 Rodenberg Ave., Biloxi, MS 39531-3444. Phone: (228) 374-9739. Ownership: James L. Black, 50% votes; and Bobbie Black, 50% votes.

Stns: 3 AM. WQFX Gulfport, MS; WAML Laurel, MS; WMLC(AM) Monticello, MS.

James L. Black, pres.

Waller Broadcasting, Box 1648, Jacksonville, TX 75766. Phone: (903) 586-2527. Fax: (903) 586-1394. E-mail: jacksonville@wallerbroadcasting.com Web Site: www.wallerbroadcasting.com. Ownership: Dudley Waller, owner.

Stns: 1 AM. 4 FM. KFRO-FM Gilmer, TX; KLJT-FM Jacksonville, TX; KEBE Jacksonville, TX; KDVE(FM) Pittsburg, TX; KXAL-FM Tatum, TX.

Dudley Waller, pres/CEO; Dave Moreland, VP progmg.

Wallingford Broadcasting Co., 128 Big Hill Ave., Richmond, KY 40475. Phone: (859) 623-1340. Fax: (859) 623-1341. E-mail: coyote@chpl.net Web Site: www.wcyo.com.

Stns: 3 AM. 2 FM. WKXO Berea, KY; WLFX(FM) Berea, KY; WCYO-FM Irvine, KY; WIRV Irvine, KY; WEKY Richmond, KY.

Kelly Wallingford, pres/CEO; Kendra Steele, opns mgr.

Walton Stns, Box 776, Kermit, TX 79745. Phone: (432) 586-3366. Fax: (432) 586-3958. Ownership: John B. Walton, 100%.

Stns: 2 AM. 1 FM. KBUY Ruidoso, NM; KWES-FM Ruidoso, NM; KWES(AM) Ruidoso, NM.

John Walton, pres; Harold Oakes, gen mgr.

Woodrow Michael Warren Stns, Box 106, Alturas, CA 96101. Phone: (530) 233-4842. Fax: (530) 233-4173.

Stns: 3 FM. KALT-FM Alturas, CA; KLCR-FM Lakeview, OR; KWTR-FM Big Lake, TX.

Woodrow Michael Warren, pres; Matt Warren, opns mgr.

Wayne County Broadcasting Co., Box 310, Fairfield, IL 62837. Phone: (618) 842-2159. Fax: (618) 847-5907. Ownership: Thomas S. Land; David H. Land; Judith L. Moore; Cynthia L. Cummins.

Stns: 1 AM. 2 FM. WOKZ(FM) Fairfield, IL; WFIW(AM) Fairfield, IL; WFIW-FM Fairfield, IL.

Thomas S. Land, chmn.

West Alabama Radio Inc., Box 938, Demopolis, AL 36732. Phone: (334) 289-9850. Fax: (334) 289-9811. Ownership: Amy Ross Ward, 50%; Betty R. Ross, 25%; and Joe M. Ross Jr., 25%.

Stns: 1 AM. 2 FM. WZNJ-FM Demopolis, AL; WXAL Demopolis, AL; WINL-FM Linden, AL.

West Virginia Radio Corp., 1251 Earl L. Core Rd., Morgantown, WV 26505. Phone: (304) 296-0029. Fax: (304) 296-3876. Web Site: wvmetronews.com. Ownership: John R. Raese, David A. Raese, Dale B. Miller.

Stns: 6 AM. 11 FM. WVAF-FM Charleston, WV; WSWW Charleston, WV; WKAZ(AM) Charleston, WV; WCHS Charleston, WV; WKWS-FM Charleston, WV; WWLW(FM) Clarksburg, WV; WDNE Elkins, WV; WDNE-FM Elkins, WV; WQZK-FM Keyser, WV; WKLP Keyser, WV; WKAZ-FM Miami, WV; WAJR Morgantown, WV; WVAQ-FM Morgantown, WV; WRVZ-FM Pocatalico, WV; WAJR-FM Salem, WV; WFBY(FM) Weston, WV; WVMD(FM) Midland, MD.

Morgantown (WV) *Dominion-Post* is affiliated with Metronews Radio Network and West Virginia Radio Corp.

Dale B. Miller, pres/CEO; Harvey Kercheval, opns VP; Joe Parsons, sls VP.

Westburg Media Capital LP, 532 9th Ave., Kirkland, WA 98033. Phone: (206) 774-1801. Ownership: Westburg Media Capital Inc., gen ptnr, 100% voting, 1 equity. Note: Westburg Media Capital Inc. is owned 100% by David Westburg.

Stns: 1 AM. 2 FM. KNMZ-FM Alamogordo, NM; KRSY(AM) Alamogordo, NM; KRSY-FM La Luz, NM.

Western Slope Communications LLC, 751 Horizon Court, Suite 225, Grand Junction, CO 81506. Phone: (970) 241-6460. Fax: (970) 241-6452. E-mail: kiss@kissradio.com Web Site: www.kissradio.com.

Stns: 2 AM. 3 FM. KAVP(AM) Colona, CO; KRVG(FM) Glenwood Springs, CO; KAYW(FM) Meeker, CO; KRGS(AM) Rifle, CO; KZKS(FM) Rifle, CO.

Steve Wennerstrom, pres; John Monroe, gen mgr.

Weston Entertainment L.P., 112 E. Pecan St., Suite 1212, San Antonio, TX 78205. Phone: (386) 423-3289. E-mail: dennis1@ucnsb.net Web Site: www.westonentertainment.net. Ownership: Second (2nd) Kings LP, 69.5%; J. Elliott Cunningham, 15%; Michael Wakely, 15%; and Weston Entertainment GP LLC, 0.5%. Note: Weston Entertainment GP LLC is solely made up of Second (2nd) Kings LP.

Stns: 1 AM. 2 FM. KBHT(FM) Crockett, TX; KVRP-FM Haskell, TX; KVRP Stamford, TX.

Dennis W. Goodman, radio mgr.

Wheeler Broadcasting Inc., Box K, Grand Coulee, WA 99133-0841. Phone: (509) 633-2020. Fax: (509) 633-1014. E-mail: keygfm@nwi.net Ownership: Deanna D. Wheeler, 30%; Verl D. Wheeler, 30%; Mark Wheeler and Nilufer Wheeler, 23%; Tonya D. Baker and Scott B. Baker, 10%.

Stns: 1 AM. 2 FM. KXAA(FM) Cle Elum, WA; KEYG Grand Coulee, WA; KEYG-FM Grand Coulee, WA.

Verl D. Wheeler, pres/CEO; Mark Wheeler, VP/gen mgr.

Mel Wheeler Inc., 5009 S. Hulen, Suite 101, Fort Worth, TX 76132-1989. Phone: (817) 294-7644. Fax: (817) 294-8519. Ownership: Estate of Mel Wheeler, 68%; Clark Wheeler, 10.2%; Leonard Wheeler, 11.1%; Steve Wheeler, 10.6%.

Stns: 2 AM. 4 FM. WVBE-FM Lynchburg, VA; WXLK-FM Roanoke, VA; WVBE(AM) Roanoke, VA; WSLQ-FM Roanoke, VA; WSLC-FM Roanoke, VA; WFIR(AM) Roanoke, VA.

Stns: 2 TV. KPOB, Paducah, KY-Cape Girardeau, MO-Harrisburg-Mount Vernon, IL; WSIL-TV, Paducah, KY-Cape Girardeau, MO-Harrisburg-Mount Vernon, IL.

Leonard Wheeler, pres; Clark Wheeler, VP; Gretchen Cummings, sec/treas.

Whitley Broadcasting Co. Inc., 522 Main St., Williamsburg, KY 40769. Phone: (606) 549-2285. Phone: (606) 549-5565. Ownership: David Paul Estes, 100%.

Stns: 1 AM. 1 FM. WEZJ Williamsburg, KY; WEKX-FM Jellico, TN.

David Paul Estes, pres.

Wilkins Communications Network Inc., Box 444, Spartanburg, SC 29304. Phone: (864) 585-1885. Fax: (864) 597-0687. E-mail: info@wilkinsradio.com Web Site: www.wilkinsradio.com. Ownership: Robert Wilkins; LuAnn Wilkins

Stns: 12 AM. WBXR(AM) Hazel Green, AL; WVTJ(AM) Pensacola, FL; WFAM(AM) Augusta, GA; KLNG(AM) Council Bluffs, IA; WBRI(AM) Indianapolis, IN; KCNW(AM) Fairway, KS; WSKY(AM) Asheville, NC; KXKS(AM) Albuquerque, NM; WWNL(AM) Pittsburgh, PA; WYYC(AM) York, PA; WELP(AM) Easley, SC; WLMR(AM) Chattanooga, TN.

Robert Wilkins, pres; Mitchell Mathis, VP; LuAnn Wilkins, exec VP.

Wilks Broadcast Group LLC, 3775 Mansell Rd., Alpharetta, GA 30022. Phone: (770) 772-4077. Ownership: Wilks Broadcast Group Holdings LLC, 100%. Note: Wilks Broadcast Group LLC is an affiliate of The Wicks Group of Companies L.L.C. (www.wicksgroup.com), a New York based private equity firm.

Stns: 18 FM. KJFX-FM Fresno, CA; KJZN(FM) San Joaquin, CA; KFRR-FM Woodlake, CA; KFKF-FM Kansas City, KS; KBEQ-FM Kansas City, MO; KMXV-FM Kansas City, MO; KCKC(FM) Kansas City, MO; KTHX-FM Dayton, NV; KURK(FM) Reno, NV; KJZS(FM) Sparks, NV; KRZQ-FM Sparks, NV; WNKK(FM) Circleville, OH; WLVQ(FM) Columbus, OH; WHOK-FM Lancaster, OH; KONE(FM) Lubbock, TX; KLLL-FM Lubbock, TX; KMMX(FM) Tahoka, TX; KBTE(FM) Tulia, TX.

Williams Communications Inc., 801 Noble St, 8th Fl., Suite 30, Anniston, AL 36201. Phone: (256) 236-1880. Fax: (256) 236-4480. E-mail: whmabig95@cableone.net Web Site: whmabig95.com. Ownership: Walton E. Williams Jr., 51%; and Melinda Williams, 49%.

Stns: 6 AM. 6 FM. WHMA Anniston, AL; WHMA-FM Ashland, AL; WRHY(FM) Centre, AL; WFMH(AM) Cullman, AL; WMCJ(AM) Cullman, AL; WFMH-FM Hackleburg, AL; WZZX Lineville, AL; WTRB-FM Sylacauga, AL; WFCT(FM) Apalachicola, FL; WLTG Panama City, FL; WCLE-FM Calhoun, TN; WCLE Cleveland, TN.

Walton E. Williams Jr., pres.

Willis Broadcasting Corp., 645 Church St., Suite 400, Norfolk, VA 23510. Phone: (757) 622-4600. Fax: (757) 624-6515.

Stns: 10 AM. 3 FM. WRAG Carrollton, AL; WTJH East Point, GA; KDLA De Ridder, LA; KLPL Lake Providence, LA; WGRM Greenwood, MS; WGRM-FM Greenwood, MS; WBXB(FM) Edenton, NC; WGTM Wilson, NC; WSDT Soddy-Daisy, TN; WCPK Chesapeake, VA; WHFD-FM Lawrenceville, VA; WGPL Portsmouth, VA; WPCE Portsmouth, VA.

Levi Willis, pres.

Wilson Broadcasting Inc., 805 N. Lena St., Suite 13, Dothan, AL 36303. Phone: (334) 671-1753. Fax: (334) 677-6923. Web Site: www.wjjn.greatnow.com.

Stns: 1 AM. 2 FM. WJJN-FM Columbia, AL; WAGF Dothan, AL; WAGF-FM Dothan, AL.

James R. Wilson III, gen mgr.

Winton Road Broadcasting Co. LLC, Box 2700, Bakersfield, CA 93303. Phone: (661) 328-0118. Fax: (661) 328-1648. Ownership: Anthony S. Brandon, 66%; L. Rogers Brandon, 33%.

Stns: 3 AM. 2 FM. KISZ-FM Cortez, CO; KVFC Cortez, CO; KPTE-FM Durango, CO; KDGO Durango, CO; KENN Farmington, NM.

Rogers Brandon, pres.

The Wireless Group Inc., Box 198, Brownsville, TN 38012. Phone: (731) 772-3700. Ownership: Carlton Veirs, pres, 50%; Lyle Reid, 50%. (See also Cross-Ownership, Sect. A.)

Stns: 1 AM. 1 FM. WNWS Brownsville, TN; WNWS-FM Jackson, TN.

The Wireless Group Inc., publishes the weekly magazine *Hunting & Fishing News*.

Carlton Veirs, pres.

Withers Broadcasting Co., Box 1508, Mount Vernon, IL 62864. Phone: (618) 242-3500. Fax: (618) 242-4444. Ownership: W. Russell Withers Jr., 100%.

Stns: 10 AM. 14 FM. KOKX Keokuk, IA; KRNQ(FM) Keokuk, IA; WKIB(FM) Anna, IL; WRUL-FM Carmi, IL; WROY Carmi, IL; WCEZ(FM) Carthage, IL; WILY Centralia, IL; WEBQ-FM Eldorado, IL; WISH-FM Galatia, IL; WEBQ Harrisburg, IL; WMOK Metropolis, IL; WZZT-FM Morrison, IL; WYNG(FM) Mount Carmel, IL; WMIX Mount Vernon, IL; WMIX-FM Mount Vernon, IL; WSSQ-FM Sterling, IL; WSDR Sterling, IL; WZZL-FM Reidland, KY; WGKY-FM Wickliffe, KY; KGMO(FM) Cape Girardeau, MO; KAPE Cape Girardeau, MO; KUGT Jackson, MO; KRHW Sikeston, MO; KBXB(FM) Sikeston, MO.

Stns: 2 TV. WDTV, Clarksburg-Weston, WV; WDHS, Marquette, MI.

W. Russell Withers Jr., pres.

Wolf Creek Broadcasting Inc., Box 490, Mineral Bluff, GA 30559-0490. Phone: (706) 379-9770. Ownership: A.D. Frazier Jr., 63%; and Clair W. Frazier, 37%.

Stns: 3 AM. 1 FM. WALH Mountain City, GA; WYHG(AM) Young Harris, GA; WACF(FM) Young Harris, GA; WLSB Copperhill, TN.

Wolfhouse Radio Group Inc., 548 E. Alisal St., Salinas, CA 93905. Phone: (831) 757-1910. Fax: (831) 757-8015. Ownership: Hector Villalobos, 100% of votes.

Stns: 1 AM. 3 FM. KEXA(FM) King City, CA; KRAY-FM Salinas, CA; KTGE Salinas, CA; KMJV(FM) Soledad, CA.

Hector Villalobos, pres.

Woodward Communications Inc., Box 688, Dubuque, IA 52004-0688. Phone: (563) 588-5687. Fax: (563) 588-5739. Web Site: www.wcinet.com. Ownership: M. Jeanne Woodward, F. Robert Woodward.

Stns: 2 AM. 4 FM. WSCO(AM) Appleton, WI; WAPL-FM Appleton, WI; WKSZ(FM) De Pere, WI; WHBY(AM) Kimberly, WI; WZOR(FM) Mishicot, WI; WECB-FM Seymour, WI.

Woodward Communications Inc. publishes the *Telegraph Herald* and weekly newspapers and shoppers in Dyersville and Cascade, Iowa; Oregon, Fitchburg, Platteville, Prairie du Chien, Richland Center, Verona, Stoughton, Wisconsin.

Tom Yunt, pres.

Wooster Republican Printing Co., (dba Dix Communications). 212 E. Liberty St., Wooster, OH 44691. Phone: (330) 264-3511. Fax: (330) 263-5013. Web Site: www.dixcom.com. Ownership: (dba Dix Communications).

Stns: 3 AM. 6 FM. WNDT(FM) Alachua, FL; WOGK(FM) Ocala, FL; WNDD-FM Silver Springs, FL; WKVX Wooster, OH; WQKT(FM) Wooster, OH; WTBO Cumberland, MD; WKGO-FM Cumberland, MD; WFRB Frostburg, MD; WFRB-FM Frostburg, MD.

Stns: 1 TV. KFBB, Great Falls, MT.

Wooster Republican Printing Co. publishes *The Daily Record*, Wooster, OH.

Robert C. Dix, TV div chmn; G. Charles Dix, VP; Dale E. Gerber, CFO.

Word Broadcasting Network Inc., Box 19229, Louisville, KY 40259. Phone: (502) 964-3304. Fax: (502) 966-9692. Web Site: www.wbna21.com. Ownership: Robert W. Rodgers, 20%; Gregory A. Holt, 20%; Melissa Fraser, 20%; Cleddie Kieth, 20%; and Margaret A. Rodgers, 20%.

Stns: 3 AM. WYMM(AM) Jacksonville, FL; WVHI Evansville, IN; WYRM(AM) Norfolk, VA.

Stns: 1 TV. WBNA, Louisville, KY.

Bob Rogers, pres; Greg Holt, VP.

World Radio Network Inc., Box 3765, McAllen, TX 78502-3765. Phone: (956) 787-9788. Fax: (956) 787-9783. E-mail: wrn@hcjb.org Web Site: www.wrn-rcm.org. Ownership: Non-profit corporation. Note: World Radio Network Inc. is affiliated with World Radio Missionary Fellowship Inc., which operates international sw missionary stn HCJB in Quito, Ecuador.

Stns: 8 FM. KRMB-FM Bisbee, AZ; KYRM(FM) Yuma, AZ; KRUC-FM Las Cruces, NM; KORM(FM) Astoria, OR; KBNJ(FM) Corpus Christi, TX; KVER-FM El Paso, TX; KBNL(FM) Laredo, TX; KVMV-FM McAllen, TX.

Dr. Ted Haney, pres; Glenn Lafitte, dir.

Wright Broadcasting Systems, Box 587, Weatherford, OK 73096. Phone: (580) 772-5939. Fax: (580) 772-1590. E-mail: traffic@wrightradio.com Web Site: www.wrightwradio.com. Ownership: G. Harold Wright, 100%.

Stns: 2 AM. 2 FM. KCLI Clinton, OK; KWEY-FM Clinton, OK; KCDL(FM) Cordell, OK; KWEY Weatherford, OK.

G. Harold Wright, pres.

Wynne Enterprises LLC, 1338 Oregon Ave., Klamath Falls, OR 97601. Phone: (541) 882-4656. Fax: (541) 884-2845. E-mail: kflskrb@aol.com Web Site: www.klamathradio.com. Ownership: Robert Wynne, Floyd Wynne, Barbara Wynne.

Stns: 1 AM. 1 FM. KFLS-FM Tulelake, CA; KFLS Klamath Falls, OR.

Robert Wynne, pres/CEO; Floyd Wynne, VP.

Y

Yavapai Broadcasting Corp., 3405 E. Hwy. 89-A, Suite A, Cottonwood, AZ 86326. Phone: (928) 634-2286. Fax: (928) 634-2295. Web Site: www.myradioplace.com. Ownership: W. Grant Hafley.

Stns: 2 AM. 4 FM. KVRD-FM Cottonwood, AZ; KYBC(AM) Cottonwood, AZ; KKLD(FM) Cottonwood, AZ; KVNA-FM Flagstaff, AZ; KVNA(AM) Flagstaff, AZ; KQST-FM Sedona, AZ.

Grant Hafley, pres; David J. Kessel, gen mgr.

Z

Zia Broadcasting Co., Box 1907, Clovis, NM 88102-1907. Phone: (505) 763-4401. Fax: (505) 769-2564. E-mail: kclv@allsups.com Ownership: Allsup's Convenience Stores Inc., 100%.

Stns: 3 AM. 3 FM. KCLV(AM) Clovis, NM; KCLV-FM Clovis, NM; KACT Andrews, TX; KACT-FM Andrews, TX; KQTY Borger, TX; KQTY-FM Borger, TX.

Rick Keefer, gen mgr; Lonnie Alsup, pres.

Zimmer Radio Inc., 2702 E. 32nd St., Joplin, MO 64804. Phone: (417) 624-1025. Fax: (417) 781-6842. Web Site: www.joplinradio.com. Ownership: James L. Zimmer Revocable Trust U/A/D May 24, 2005 (James L. Zimmer, sole trustee), 100%.

Stns: 2 AM. 4 FM. KZRG(AM) Joplin, MO; KSYN-FM Joplin, MO; KZYM(AM) Joplin, MO; KIXQ(FM) Joplin, MO; KJMK-FM Webb City, MO; KXDG-FM Webb City, MO.

James Zimmer, pres; Larry Boyd, gen mgr.

Zoe Communications Inc., Box 190, Shell Lake, WI 54871. Phone: (715) 468-9500. Fax: (715) 468-9505. Web Site: www.zoestations.com.

Stns: 2 AM. 2 FM. WDMO(FM) Durand, WI; WPDR Portage, WI; WCSW Shell Lake, WI; WPLT(FM) Spooner, WI.

Wendy Oberg, gen mgr.

The Zone Corp., Box 1929, Bangor, ME 04402. Phone: (207) 990-2800. Fax: (207) 990-2444. Web Site: www.zoneradio.com. Ownership: Stephen King is the sole stockholder.

Stns: 1 AM. 2 FM. WZON Bangor, ME; WKIT-FM Brewer, ME; WDME-FM Dover Foxcroft, ME.

Stephen King, pres; Tabitha King, VP; Arthur B. Greene, sec/treas; Bobby Russell, gen mgr.

Key to Radio Listings

(1) WOF(AM)—(2) Oct 8, 1946: **(3)** 1000 khz. **(3a)** Stereo. **(3b)** Hrs opn: 24. **(4)** Box 1000 99999. Phone: (909) 555-1000. Fax: (909) 999-9999. E-mail: wof@wofam.com. Web Site: www.wofam.com. **(5)** Licensee: General Broadcasting Corp. (group owner; acq 7-20-69; $255,000 with co-located FM; **(5a)** FTR 2-12-83). **(6)** Population served: 250,000 **(7)** Natl. Network: ABC/E, AP, Mountain State Network. Natl. Rep: Jones & Company, Penn State. Format: MOR, C&W. News staff: one; News: 15 hrs wkly. Target aud: 25-54; baby boomers. Spec prog: Sp 3 hrs wkly. **(8)** John Jones, gen mgr; David Smith, chief engr.

(1a) WOF-FM—(2) October 1959: **(9)** 101.1 mhz; 3 kw. Ant 300 ft. **(3a)** Stereo. **(10)** Dups AM 50%. Format: C&W. **(11)** WOF-TV affil.

(1) Station call letters as assigned by the Federal Communications Commission (FCC) or Canadian Radio-television and Telecommunications Commission (CRTC).

(1a) Station call letters for co-owned FM station. WOF-FM has the same ownership as WOF(AM), and the FM listing contains only information different from the AM. Co-owned AM and FM stations are often listed together, even when they have dissimilar call letters. In some instances FM may be listed first.

(2) Date station first went on air (regardless of subsequent ownership changes).

(3) Frequency in kilohertz.

(3a) WOF broadcasts in stereo.

(3b) WOF broadcasts 24 hours daily.

(4) Address and zip code, telephone, fax, web site and e-mail address.

(5) Licensee name and date of acquisition (if not original owner). If the licensee is a group owner—a company with several broadcast properties—it is so identified, as a group owner of which the licensee is a subsidiary. Details on group owners are listed in Section A. If the station has been sold and the sale information is available, it is recorded after the acquisition date, ie. acq. date; purchase price; FTR date.

(5a) FTR date. FTR refers to *Broadcasting & Cable* magazine's weekly For the Record column that appeared in the magazine until June 8, 1998, where station sales were recorded as received from the FCC.

(6) Population served refers to the station's potential market.

(7) Network, representative and programming. WOF national affiliates are ABC Entertainment Network and AP Network. The regional affiliate is Mountain State. The WOF national sales representative is Jones & Company and their regional sales representative is Penn State. The WOF program format is part middle-of-the-road, part country and western, with three hours weekly of special programming in Spanish. They have one staff member covering local news, and provide local news 15 hours per week. Their target audience is baby boomers age 25-54.

(8) Key personnel.

(9) Frequency for WOF-FM is 101.1 megahertz, with 3 kilowatts of effective radiated power and an antenna height of 300 feet above average terrain. WOF-FM broadcasts in stereo (see **(3a)**).

(10) Programming. WOF-FM duplicates WOF(AM) programs 50% of the time and has a country and western format.

(11) Co-owned TV. WOF-TV has the same licensee as WOF-AM-FM.

Note: Listings for independent AM & FM stations follow the sample shown for WOF(AM).

An asterisk (*) preceding station call letters indicates noncommercial stations.

Directory of Radio Stations in the United States

Alabama

Abbeville

WIZB(FM)— Feb 2, 1968: 94.3 mhz; 19.5 kw. Ant 371 ft TL: N31 26 19 W85 17 22. Stereo. Hrs opn: 24 Box 8097, Dothan, 36304-8097. Secondary address: 2563 Montgomery Hwy., Dothan 36303. Phone: (334) 699-5672. Fax: (334) 699-5034. E-mail: lprescott@hisradio943.com Web Site: www.hisradio943.com. Licensee: Radio Training Network, Inc. (acq 7-98; $550,000). Law Firm: Gammon & Grange, P.C. Format: Contemp Christian. News: 3 hrs wkly. Target aud: 25-49; women. ♦Jim Campbell, CEO; Linda Prescott, gen mgr, gen sls mgr & gen sls mgr; K.W. Keene, prom dir; Richard Posey, progmg dir; Neal Riddle, chief of engrg & engr; Barbara Carter, traf mgr.

Addison

WQAH-FM— 1996: 105.7 mhz; 6 kw. 328 ft TL: N34 18 19 W87 04 24. Hrs open: 24 Box 1048, Hartselle, 35640. Phone: (256) 773-2563. Fax: (256) 773-6915. E-mail: radio@hiwaay.net Licensee: Abercrombie Broadcasting FM Inc. (acq 2-7-2000). Format: Classic country. ♦Alvin Abercrombie, pres; Carol Lynn, gen mgr, progmg dir, disc jockey, sls & mktg; Keith Abercrombie, engrg dir; Mark Donovan, disc jockey, disc jockey & sls.

Alabaster

WQCR(AM)— Sept 28, 1981: 1500 khz; 2.3 kw-D (1.2 kw-CH). TL: N33 12 27 W86 45 34. Hrs open: Sunrise-sunset 50 Hwy. 26, 35007. Phone: (205) 613-2108. Licensee: Rivera Communications LLC (acq 2-10-2006; $5,000 and assumption of debt). Population served: 275,000 Format: Sp. ♦David Robinson, VP.

Albertville

WAVU(AM)— 1947: 630 khz; 1 kw-D, 28 w-N. TL: N34 14 19 W86 09 59. Hrs open: 24 Box 190, 35950. Secondary address: 3770 US Hwy. 431 35951. Phone: (256) 878-8575. Fax: (256) 878-1051. E-mail: tommylee@wqsb.com Licensee: Sand Mountain Broadcasting Service Inc. Population served: 25,000 Natl. Network: AP Radio. Law Firm: Fletcher, Heald & Hildreth. Wire Svc: AP Format: Southern gospel, Christian. News staff: one. Target aud: 35 plus. ♦Pat M. Courington Jr., pres; Tommy Lee, gen mgr; Ted McCreless, gen sls mgr.

WQSB(FM)—Co-owned with WAVU(AM). 1948: 105.1 mhz; 2.7 kw. Ant 1,000 ft TL: N34 09 27 W86 02 44. Stereo. 24 E-mail: wqsb@aol.com Web Site: www.wqsb.com.220,000 Format: Country. News staff: one. Target aud: 25-54. ♦Ted McCreless, sls dir; Barry Galloway, progmg dir & disc jockey; Dale Stallings, mus dir; Al Taylor, news dir & disc jockey.

WWGC(AM)— April 1982: 1090 khz; 500 w-D. TL: N34 18 02 W86 16 01. Hrs open: 6 AM-7:15 PM Box 418, 35950. Phone: (256) 894-6294. Fax: (256) 894-6495. Licensee: Quality Properties LLC (acq 1-16-2007; $355,000). Population served: 500,000 Format: Sp. ♦Juan Vargas, gen mgr.

Alexander City

WRFS(AM)— May 31, 1947: 1050 khz; 1 kw-D; 48 w-N. TL: N32 56 51 W85 59 17. Hrs open: 24 Box 602, Centre, 35960. Fax: (256) 927-4028. Web Site: www.joychristian.com. Licensee: Joy Christian Communications Inc. (group owner; (acq 8-26-2004; $175,000). Population served: 60,000 Natl. Network: ABC. Law Firm: Haley, Bader & Potts. Format: Southern gospel. Target aud: 25-54; adults. ♦Jimmy Jerrell, pres & CFO; Greg Holtan, gen mgr.

WSTH-FM— Sept 30, 1949: 106.1 mhz; 100 kw. 981 ft TL: N32 45 33 W85 28 04. (CP: 85.8 kw, ant 1,047 ft.). Stereo. Hrs opn: 24 Box 687, Columbus, GA, 31902. Secondary address: 1501 13th Ave., Columbus, GA 31901. Phone: (706) 576-3000. Fax: (706) 576-3010. Web Site: www.rooster106online.com. Licensee: CC Licenses LLC. Group owner: Clear Channel Communications Inc. (acq 5-9-2003; $2.73 million with WDAK(AM) Columbus, GA). Natl. Network: ABC. Format: Country. News staff: one. Target aud: General. ♦Jim Martin, gen mgr; Brian Waters, opns mgr.

Aliceville

WZBQ(FM)—See Carrollton

Andalusia

WAAO-FM— Aug 24, 1987: 103.7 mhz; 3 kw. Ant 328 ft TL: N31 20 27 W86 28 02. Stereo. Hrs opn: 24 Box 987, MLK Expressway, 36420. Phone: (334) 222-1166. Fax: (334) 222-1167. E-mail: waao@alaweb.com Web Site: www.waao.com. Licensee: Companion Broadcasting Service Inc. Population served: 10,092 Format: Country. Target aud: General. ♦Lee Williams, pres & gen mgr.

***WSTF(FM)**— March 1996: 91.5 mhz; 5 kw. 361 ft TL: N31 26 20 W86 30 48. Hrs open: Box 210789, Montgomery, 36121-0789. Phone: (334) 271-8900. Fax: (334) 260-8962. E-mail: mail@faithradio.org Web Site: www.faithradio.org. Licensee: Faith Broadcasting Inc. Format: Educ, relg, MOR. ♦Russell Dean, gen mgr; Gary Hundley, dev dir.

Anniston

WANA(AM)— August 1954: 1490 khz; 1 kw-U. TL: N33 41 15 W85 49 49. Hrs open: 24 1913 Barry St., Oxford, 36203-2319. Phone: (256) 741-6000. Fax: (256) 741-6080. Web Site: www.1490wana.com. Licensee: Jacobs Broadcast Group Inc. (acq 5-1-2006; $330,000). Population served: 200,000 Natl. Network: ESPN Radio. Format: All sports. ♦James H. Jacobs, pres & gen mgr.

WDNG(AM)— July 1, 1957: 1450 khz; 1 kw-U. TL: N33 40 01 W85 50 56. Hrs open: 1115 Leighton Ave., 36207. Phone: (256) 236-8291. Fax: (256) 236-8292. Web Site: www.wdng.net. Licensee: WDNG Inc. (acq 6-30-87; $500,000; FTR: 7-6-87). Population served: 125,000 Natl. Network: CBS. Format: News/talk. Target aud: General. ♦J.J. Dark, pres & gen mgr.

***WGRW(FM)**— July 1999: 90.7 mhz; 3 kw vert. 328 ft TL: N33 29 19 W86 47 58. Hrs open: 24 Word Works Inc., Box 2555, 36202. Secondary address: 4265 Hill St. 36206. Phone: (256) 238-9990. Fax: (256) 237-1102. E-mail: jon@graceradio.com Web Site: www.graceradio.com. Licensee: Word Works Inc. Population served: 111,000 Natl. Network: Moody. Format: Christian. ♦Aaron Acker, pres; Jon Holder, gen mgr.

WHMA(AM)— 1938: 1390 khz; 5 kw-D, 1 kw-N, DA-N. TL: N33 42 31 W85 51 14. Hrs open: 24 801 Noble St., Suite 30, 36201. Phone: (256) 237-8741. Fax: (256) 231-9414. Licensee: Williams Communications Inc. (group owner; acq 8-12-2003; $275,000). Population served: 120,000 Format: Gospel. News staff: one; News: 12 hrs wkly. Target aud: 25-54; those with upscale, mobile, discretionary incomes. ♦Walt Williams, gen mgr; John Goodbread, progmg dir.

WHMA-FM— Oct 4, 1984: 95.5 mhz; 1.7 kw. Ant 617 ft TL: N33 18 30 W85 50 58. (CP: Hobson City. 530 w, ant 1,089 ft. N33 37 38 W85 53 25). Stereo. Hrs opn: 24 801 Noble St., Suite 30, Anniston, 36201. Phone: (256) 236-1880. Fax: (256) 236-4480. Licensee: Williams Communications Inc. (group owner; (acq 8-2-2002; $2.88 million with WZZX(AM) Lineville). Format: Country. Target aud: 18-54. Spec prog: Black 6 hrs wkly. ♦Walt Williams Jr., pres & gen mgr; Tex Carter, progmg dir.

Arab

WAFN-FM— Nov 5, 1979: 92.7 mhz; 6 kw. Ant 623 ft TL: N34 21 04 W86 26 27. Stereo. Hrs opn: 24 Box 4184, Huntsville, 35815. Secondary address: 981 N. Brindlee Mt. Pkwy. 35815. Phone: (256) 586-9300. Fax: (256) 586-9301. E-mail: funradio@hiwaay.net Web Site: www.fun927.com. Licensee: Fun Media Group Inc. (acq 7-31-97; $492,500). Population served: 90,000 Natl. Network: CNN Radio. Format: Oldies. News staff: one; News: 3 hrs wkly. Target aud: 18-54. ♦Susan E. McKenney, pres; Michael St. John, CFO & gen mgr.

WRAB(AM)— Oct 25, 1961: 1380 khz; 1 kw-D. TL: N34 20 06 W86 28 07. Hrs open: 6am-6pm Box 625, 35016. Secondary address: 619 S. Brindlee Mountain Pkwy. 35016. Phone: (256) 586-4123. Fax: (256) 586-4124. E-mail: wrab@otelco.net Web Site: www.wrab.net. Licensee: Reed Broadcasting LLC (acq 6-22-01; $163,000). Population served: 20,000 Wire Svc: AP Format: Country, relg, gospel. News staff: one; News: 12 hrs. wkly. ♦Ed Reed, pres; Joey Yarbrough, exec VP; Archie Anderson, gen mgr.

Ashland

WCKF(FM)—Not on air, target date: unknown: 100.7 mhz; 6 kw. Ant 72 ft TL: N33 16 55 W85 47 38. Hrs open: Box 8, Anniston, 36202. Phone: (256) 282-4338. Fax: (256) 782-2489. Licensee: Leslie E. Gradick. ♦Leslie E. Gradick, gen mgr.

Athens

WKAC(AM)— September 1964: 1080 khz; 5 kw-D. TL: N34 50 13 W86 58 28. Hrs open: Sunrise-sunset Box 1083, 35612. Secondary address: 19245 Hwy. 127 35614. Phone: (256) 232-6827. Fax: (256) 232-6828. E-mail: wkac@companet.net Web Site: www.wkac1080.com. Licensee: Limestone Broadcasting Co. Population served: 750,000 Natl. Network: CNN Radio. Rgnl. Network: Capitol Radio Net. Rgnl rep: Rgnl Reps. Format: Var. News: 6 hrs wkly. Target aud: 25-54; adults mid/upper income, blue/white collar. Spec prog: Farm 5 hrs, country 13 hrs wkly. ♦Kenneth A. Casey, pres; Keith Casey, gen mgr; Kirk Harvey, progmg dir; Joyce Casey, traf mgr & women's int ed.

WVNN(AM)— Nov 8, 1948: 770 khz; 10 kw-D, 250 w-N, DA-N. TL: N34 50 21 W86 55 44. Stereo. Hrs opn: 24 1717 Hwy. 72 E., 35611. Phone: (256) 830-8300. Fax: (256) 232-6842. Web Site: www.wvnn.com. Licensee: Cumulus Licensing LLC. Group owner: Cumulus Media Inc. (acq 7-21-2003; grpsl). Population served: 450,000 Natl. Rep: Katz Radio. Format: News/talk. News staff: 3; News: 60 hrs wkly. Target aud: 25-64. Spec prog: Farm 2 hrs, gospel 3 hrs wkly. ♦Bill G. West, VP & gen mgr; Brian Pitts, sls dir & gen sls mgr; Wendy Black, prom dir; Zack Bennett, opns mgr & progmg dir; Marty Broman, news dir; Bill Schrode, chief of engrg.

WZYP(FM)—Co-owned with WVNN(AM). Oct 1, 1958: 104.3 mhz; 100 kw. 1,115 ft TL: N34 49 05 W86 44 16. Stereo. 24 Web Site: www.wzyp.com.900,000 Format: Top-40. News staff: 3; News: 3 hrs wkly. Target aud: 18-49. ♦Tracy Flesch, gen sls mgr & progmg dir.

Atmore

WASG(AM)— Nov 12, 1981: 550 khz; 10 kw-D, 143 w-N. TL: N30 34 45 W87 17 13. Stereo. Hrs opn: 24 2070 N. Palafax, Pensacola, FL, 32501. Phone: (850) 434-1230. Fax: (850) 469-9698. E-mail: mglin@aol.com Web Site: www.pensacolachristianradio.com. Licensee: 550 AM, Inc. Format: Christian, talk. News staff: one; News: 9 hrs wkly. Target aud: 18-54. ♦Dara Glinter, exec VP; Michael B. Glinter, pres & gen mgr.

WNSI-FM— June 28, 1991: 105.9 mhz; 3.3 kw. Ant 446 ft TL: N31 00 26 W87 32 15. Stereo. Hrs opn: 24
Simulcast with WNSI(AM) Robertsdale 100%.
Box 578, Robertsdale, 36567. Phone: (251) 947-2346. Fax: (251) 947-2347. E-mail: wnsiradio@gulftel.com Web Site: www.wnsiradio.com. Licensee: Southern Media Communications Inc. (acq 3-4-98). Format: News, sports info. News: 18 hrs wkly. Spec prog: American Indian one hr, Black one hr, farm 5 hrs, gospel 12 hrs wkly. ♦Walter Bowen, gen mgr.

WYOK(FM)— May 19, 1966: 104.1 mhz; 100 kw. Ant 1,555 ft TL: N30 37 35 W87 38 50. Stereo. Hrs opn: 24 2800 Dauphin St., #104, Mobile, 36606-2400. Phone: (251) 652-2000. Fax: (251) 652-2001. E-mail: mobile.prog@cumulus.com Web Site: www.cumulus.com. Licensee: Cumulus Licensing Corp. Group owner: Cumulus Media Inc. (acq 10-18-99; grpsl). Population served: 954,300 Law Firm: Cohn & Marks. Format: Country. News staff: one; News: 5 hrs wkly. Target aud: 25-40. ♦Gary Pizzati, gen mgr; Steve Crumbely, opns VP & opns mgr.

Attalla

WKXX(FM)—Licensed to Attalla. See Gadsden

Auburn

WANI(AM)—See Opelika

WAUD(AM)— Dec 22, 1947: 1230 khz; 1 kw-U. TL: N32 37 47 W85 28 08. Hrs open: 24 2514 S. College St., Suite 104, 36830. Phone: (334) 887-3401. Fax: (334) 826-9599. Licensee: Tiger Communications Inc. (acq 2-26-98). Population served: 90,000 Natl. Network: CBS

Radio. Natl. Rep: Rgnl Reps. Format: Sports, big band, jazz. News staff: one; News: 7 hrs wkly. Target aud: 25 plus. ♦Chris Bailey, gen mgr.

***WEGL(FM)**— Apr 25, 1971: 91.1 mhz; 3 kw. 190 ft TL: N32 36 11 W85 29 12. (CP: Ant 214 ft.). Stereo. Hrs opn: 24 116 Foy Union Bldg., Auburn Univ., 36849-5231. Phone: (334) 844-4114. Fax: (334) 844-4118. E-mail: wegl@auburn.edu Web Site: wegl.auburn.edu. Licensee: Board of Trustees Auburn University. Population served: 50,000 Format: College alternative. News staff: one; News: 6 hrs wkly. Target aud: College students. Spec prog: Various specialty shows. ♦Saleem Walton, gen mgr.

WKKR(FM)— July 8, 1968: 97.7 mhz; 3.1 kw. 453 ft TL: N32 33 54 W85 22 13. Stereo. Hrs opn: 24 Box 2329, Opelika, 36803. Secondary address: 915 Veterans Pkwy., Opelika 36801. Phone: (334) 745-4657. Fax: (334) 749-1520. E-mail: genmorj@charter.net Web Site: www.kickerfm.com. Licensee: Qantum of Auburn License Co. LLC. Group owner: Qantum Communications Corp. (acq 7-2-2003; grpsl). Population served: 100,000 Law Firm: Gardner, Carton & Douglas. Format: Country. News staff: 2; News: 7 hrs wkly. Target aud: 25-54. ♦Frank Osborn, pres; Sandy Mathews, sls dir & gen sls mgr; Jim Powell, mktg mgr.

WTLM(AM)—See Opelika

Bay Minette

WNSP(FM)— Oct 1, 1964: 105.5 mhz; 5.3 kw. 348 ft TL: N30 49 34 W87 51 52. (CP: 1.9 kw, ant 410 ft.). Stereo. Hrs opn: 24 1100-E Dauphin St., # E, Mobile, 36604-2512. Phone: (251) 438-5460. Fax: (251) 438-5462. Web Site: www.wnsp.com. Licensee: Dot Com+ L.L.C. (acq 7-6-98; $1.05 million). Population served: 190,025 Format: Sports. Target aud: 18-49. ♦Ken Johnson, pres & gen mgr; Clint Crouch, opns mgr.

WTOF(AM)— 1958: 1110 khz; 10 kw-D. TL: N30 52 10 W87 46 09. Hrs open: Box 1220, Fairhope, 36533. Phone: (251) 928-2384. Fax: (251) 928-9229. Licensee: Buddy Tucker Association Inc. (acq 1-16-2007; $300,000). Population served: 7,200 ♦Lori Dubois, gen mgr.

Bessemer

WZGX(AM)— June 1, 1950: 1450 khz; 1 kw-U. TL: N33 25 23 W86 57 17. Hrs open: 3300 Jaybird Rd., 35020. Phone: (205) 428-0146. Fax: (205) 426-3178. Licensee: Bessemer Radio Inc. (acq 9-1-88). Population served: 300,910 Format: Rgnl Mexican. ♦Raul Ortal, gen mgr.

Birmingham

WAGG(AM)— 1927: 610 khz; 5 kw-D, 1 kw-N. TL: N33 29 40 W86 52 30. Hrs open: 950 22nd St., N., Suite 1000, 35203. Phone: (205) 322-2987. Fax: (205) 322-2390. Web Site: www.wagg610.com. Licensee: Cox Radio Inc. Group owner: Cox Broadcasting (acq 9-9-97). Population served: 300910 Natl. Network: ABC. Natl. Rep: Christal. Law Firm: Dow, Lohnes & Albertson. Wire Svc: AP Format: Gospel. News staff: 2; News: 5 hrs wkly. Target aud: 45 plus. ♦David DuBose, gen mgr.

WAPI(AM)— 1922: 1070 khz; 50 kw-D, 5 kw-N, DA-N. TL: N33 33 07 W86 54 40. Hrs open: 24 244 Goodwin Crest Dr., Suite 300, 35209. Phone: (205) 942-1004. Fax: (205) 917-1906. Web Site: www.wapi1070.com. Licensee: Citadel Broadcasting Co. Group owner: Citadel Broadcasting Corp. (acq 4-26-2001; grpsl). Population served: 300,910 Natl. Rep: Christal. Format: News/talk. Target aud: 35 plus. ♦Dale Daniels, gen mgr; Steve Harrison, gen sls mgr; Jennifer Dickson, prom dir; Frank Giardina, progmg dir.

WYSF(FM)—Co-owned with WAPI(AM). 1947: 94.5 mhz; 100 kw. 1,214 ft TL: N33 29 26 W86 47 48. Stereo. Web Site: www.wapi1070.com. Format: Hot adult contemp, soft rock. Target aud: 25-54; general.

WATV(AM)— May 20, 1946: 900 khz; 1 kw-U. TL: N33 32 14 W86 50 16. Stereo. Hrs opn: 24 hrs Box 39054, 3025 Ensley Ave., 35208. Phone: (205) 780-2014. Fax: (205) 780-4034. E-mail: jlocklin@watv900.com Web Site: www.9006060watv.com. Licensee: McL/McM Alabama LLC. (group owner; (acq 10-12-2004; $1.5 million). Population served: 865,600 Natl. Network: American Urban. Natl. Rep: Interep. Law Firm: Fletcher, Heald & Hildreth, P.L.C. Format: Black, oldies, relg. Target aud: 18 plus. Spec prog: Sports 10 hrs wkly. ♦Ron Davenport, pres; James Lockun, gen mgr; Ron January, progmg dir.

WAYE(AM)— Aug 1, 1972: 1220 khz; 1 kw-D, 75 w-N. TL: N33 28 39 W86 50 57. Hrs open: 836 Lomb Ave. S.W., 35211. Phone: (205) 786-9293. Fax: (205) 786-9296. Licensee: Davidson Media Station WAYE Licensee LLC. Group owner: Willis Broadcasting Corp. (acq 9-8-2006; $950,000). Population served: 759,000 Format: Gospel. Target aud: 19 plus; loyal, mature & financially stable. ♦Mary Agee, gen mgr.

***WBFR(FM)**— 1988: 89.5 mhz; 100 w. Ant 672 ft TL: N33 29 02 W86 48 35. Hrs open: 290 Hegenberger Rd., Oakland, CA, 94621. Phone: (510) 568-6200. Fax: (510) 568-6190. E-mail: info@familyradio.com Web Site: www.familyradio.com. Licensee: Family Stations Inc. (group owner) Format: Relg, evangelical. ♦Stanley Jackson, gen mgr.

WBHJ(FM)—Midfield, 1952: 95.7 mhz; 12 kw. Ant 1,004 ft TL: N33 27 37 W86 51 07. Stereo. Hrs opn: 24 950 22nd St. N., Suite 1000, 35203. Phone: (205) 322-2987. Fax: (205) 322-2390. Web Site: 957jamz.com. Licensee: Cox Radio Inc. Group owner: Cox Broadcasting (acq 10-6-98; $17 million with WBHK(FM) Warrior). Population served: 642,000 Natl. Network: Westwood One. Natl. Rep: Christal. Law Firm: Dow, Lohnes & Albertson. Format: Hip hop, rhythm and blues. News staff: one; News: one hr wkly. Target aud: 18-34; upscale baby boomers. ♦David DuBose, gen mgr.

***WBHM(FM)**— December 1976: 90.3 mhz; 32 kw. 1,214 ft TL: N33 29 19 W86 47 58. Stereo. Hrs opn: 24
Rebroadcasts WSGN(FM) Gadsden 100%.
650 11th St. S., 35233-1221. Phone: (205) 934-2606. Fax: (205) 934-5075. E-mail: info@wbhm.org Web Site: www.wbhm.org. Licensee: Board of Trustees, University of Alabama. Population served: 780,000 Natl. Network: NPR, PRI. Wire Svc: AP Format: Class, news. Target aud: General. Spec prog: New age 8 hrs wkly. ♦Mike Morgan, gen mgr; Mary Hendley, dev dir; Michael Krall, progmg dir; Tanya Ott, news dir.

WBPT(FM)—Homewood, June 1959: 106.9 mhz; 97 kw. Ant 1,325 ft TL: N33 29 04 W86 48 25. Stereo. Hrs opn: 24 301 Beacon Pkwy. W., Suite 200, 35209. Phone: (205) 916-1100. Fax: (205) 916-1151. Web Site: www.birminghampointeagle.com. Licensee: Cox Radio Inc. Group owner: Cox Broadcasting (acq 3-28-97; grpsl). Natl. Rep: Katz Radio. Format: Classic hits. News staff: one; News: 15 hrs wkly. Target aud: 25-54; affluent baby boomers. Spec prog: Best of the 80's and more. ♦Ray Nelson, gen mgr; David Wells, progmg dir; Don Daley, news dir; Sherri Clark, traf mgr.

WDJC-FM— Apr 22, 1968: 93.7 mhz; 100 kw. 1,007 ft TL: N33 26 36 W86 52 50. Stereo. Hrs opn: 2727 19th Place S., 35209. Phone: (205) 879-3324. Fax: (205) 802-4555. Web Site: www.93.7wdjc.com. Licensee: Kimtron Inc. Group owner: Crawford Broadcasting Co. Population served: 1,000,000 Format: Contemp Christian music. Target aud: 25-60; conservative middle income. ♦Steve Armstrong, gen mgr.

WERC(AM)—Listing follows WMJJ(FM).

***WGIB(FM)**— 1983: 91.9 mhz; 600 w. 679 ft TL: N33 29 02 W86 48 35. Hrs open: 24 1137 10th Pl. S., 35205. Phone: (205) 323-1516. Fax: (205) 323-2747. E-mail: nmills@gleniris.net Web Site: www.gleniris.net. Licensee: Glen Iris Baptist School (acq 1-31-02). Format: Christian. ♦Chris Lamb, chmn & gen mgr; Nathan Mills, stn mgr & opns mgr.

WJLD(AM)—Fairfield, 1942: 1400 khz; 1 kw-U. TL: N33 28 36 W86 53 01. Hrs open: 24 Box 19123, 35219-9123. Secondary address: 1449 Spaulding Ishkooda Rd. 35211-5059. Phone: (205) 942-1776. Fax: (205) 942-4814. E-mail: wjld@juno.com Web Site: www.warpradio.com. Licensee: Richardson Broadcasting Corp. (acq 10-87). Population served: 900,000 Natl. Network: American Urban, CNN Radio. Format: Blues, takl, gospel. Target aud: 35+; majority Black, adult, blue and white collar working class. ♦Gary R. Richardson, pres & gen mgr; Bob Friedman, opns mgr & sls dir; Gary Richardson, progmg dir; Eloise Gaffney, traf mgr.

***WJSR(FM)**— Aug 11, 1977: 91.1 mhz; 100 w. 195 ft TL: N33 39 07 W86 42 20. Stereo. Hrs opn: 24 2601 Carson Rd., 35215. Phone: (205) 856-6095. Fax: (205) 856-7702. Web Site: wjsrfm.com. Licensee: Jefferson State Community College. Population served: 100,000 Format: Classic rock. News: 4 hrs wkly. Target aud: 24-49; college population. ♦Ray Edwards, gen mgr.

***WLJR(FM)**— 1998: 88.5 mhz; 200 w. 623 ft TL: N33 23 35 W86 39 48. Stereo. Hrs opn: 24 Briarwood Presbyterian Church, 2200 Briarwood Way, 35243. Phone: (205) 978-2200. Fax: (205) 824-8419. Web Site: www.wljr.org. Licensee: Briarwood Presbyterian Church. Natl. Network: Moody. Law Firm: Southmayd & Miller. Format: Div, educ, relg. News: 10 hrs wkly. Target aud: General; upper middle class. ♦James Hulgan, gen mgr.

WMJJ(FM)— June 1, 1961: 96.5 mhz; 100 kw. 1,027 ft TL: N33 26 38 W86 42 10. Stereo. Hrs opn: 600 Beacon Pkwy. W., Suite 400, 35209. Phone: (205) 439-9600. Fax: (205) 439-8390. Fax: (205) 439-8391. Web Site: www.magic96fm.com. Licensee: Capstar TX L.P. Group owner: Clear Channel Communications Inc. (acq 8-30-00; grpsl). Population served: 300,910 Law Firm: Reed, Smith, Shaw & McClay. Format: Adult contemp. Target aud: 25-54. ♦L. Lowry Mays, CEO; Jimmy Vineyard, gen mgr; John Friend, opns mgr & sls dir; Bradley Spears, gen sls mgr; Cindee Standridge, mktg dir; Bob Newberry, chief of engrg; Cynthia Childress, traf mgr; Tom Hanrahan, progmg dir & opns.

WERC(AM)—Co-owned with WMJJ(FM). May 25, 1925: 960 khz; 5 kw-U, DA-N. TL: N33 32 02 W86 51 07. Web Site: www.werctalk.com. Format: News/talk. Target aud: adults. ♦Dennis Cruz, gen sls mgr; Jim Faherty, news dir; Shelia Howell, traf mgr.

WPSB(AM)— 1950: 1320 khz; 5 kw-D, 111 w-N. TL: N33 33 41 W86 51 37. Hrs open: 301 Beacon Pkwy., Suite 200, 35209. Phone: (205) 916-1100. Fax: (205) 916-1151. Web Site: wzzk.com. Licensee: Cox Radio Inc. Group owner: Cox Communications Inc. (acq 3-28-97; grpsl). Population served: 300,910 Natl. Rep: Christal. Law Firm: Dow, Lohnes & Albertson. Format: Sp. ♦Ray Nelson, gen mgr.

WQEN(FM)—Trussville, Oct 7, 1966: 103.7 mhz; 100 kw. Ant 935 ft TL: N33 26 38 W86 52 47. Stereo. Hrs opn: 600 Beacon Pkwy. W., Suite 400, 35209. Phone: (205) 439-9600. Fax: (205) 439-8390. Web Site: www.1037theq.com. Licensee: Capstar TX L.P. (acq 8-30-2000; grpsl). Population served: 800,000 Format: CHR, Top-40. ♦Jimmy Vineyard, gen mgr.

WRLM(AM)—See Irondale

WSPZ(AM)— Oct 15, 1947: 690 khz; 50 kw-D, 500 w-N, DA-N. TL: N33 26 56 W86 55 18. Hrs open: 244 Goodwin Crest Dr., Suite 300, 35209-3714. Phone: (205) 945-4646. Fax: (205) 945-3999. Web Site: www.wjox690.com. Licensee: Citadel Broadcasting Co. Group owner: Citadel Broadcasting Corp. (acq 4-26-2001; grpsl). Population served: 375,900 Wire Svc: SportsTicker Format: Sports. Target aud: 25-54. Spec prog: Gospel 5 hrs wkly. ♦Dale Daniels, gen mgr; Kerry Lambert, opns VP; Lenny Frisaro, sls dir; Steve Harrison, natl sls mgr; Jennifer Dickson, prom dir; Ryan Haney, progmg dir & chief of engrg; Lisa Holifield, news dir; Will Berry, pub affrs dir.

WZRR(FM)—Co-owned with WSPZ(AM). December 1975: 99.5 mhz; 100 kw. 870 ft TL: N33 26 28 W86 53 00. (CP: Ant 1,000 ft.). Fax: (205) 942-3175. Web Site: www.wzrr.com. Natl. Rep: Christal. Format: Classic rock. ♦Allen Dick, pres; Dave Henderlite, CFO; Davis Hawkins, opns VP; Kerry Lambert, opns mgr.

WUHT(FM)— Sept 15, 1969: 107.7 mhz; 100 kw. Ant 1,237 ft TL: N33 43 52 W86 37 57. Hrs open: 24 244 Goodwin Crest Dr., Suite 300, 35209. Phone: (205) 945-4646. Fax: (205) 942-3175. Web Site: www.hot1077radio.com. Licensee: Citadel Broadcasting Co. Group owner: Citadel Broadcasting Corp. (acq 4-26-2001; grpsl). Population served: 300,910 Format: Hot rhythm and blues. Target aud: 18-34. ♦Dale Daniels, gen mgr; Jane Mitchell, gen sls mgr.

***WVSU-FM**— Apr 6, 1967: 91.1 mhz; 500 w vert. Ant 413 ft TL: N33 27 47 W86 46 08. Stereo. Hrs opn: 17 Samford Univ., 35229-2301. Phone: (205) 726-2877. Fax: (205) 726-4032. E-mail: wvsu@samford.edu Web Site: www.samford.edu/wvsu. Licensee: Samford University.

Population served: 800,000 Format: Smooth jazz. News: one hr wkly. Target aud: General. Spec prog: Samford Univ. athletics. ♦Andy Parrish, gen mgr.

WXJC(AM)— Apr 1, 1953: 850 khz; 50 kw-D, 1 kw-N, DA-2. TL: N33 37 25 W86 44 45. Hrs open: 120 Summit Pkwy., 35209. Phone: (205) 879-3324. Fax: (205) 802-4555. Web Site: www.850wxjc.com. Licensee: Kimtron Inc. Group owner: Crawford Broadcasting Co. (acq 11-12-99). Population served: 785,000 Natl. Network: USA. Format: Talk. Target aud: Under 12. ♦Steve Armstrong, gen mgr; Jennifer Poepcke, gen sls mgr; Todd Dixon, chief of engrg; Melodye Grubb, traf mgr.

WYDE(AM)— Mar 25, 1953: 1260 khz; 5 kw-D, 41 w-N. TL: N33 31 29 W86 47 10. Hrs open: 120 Summit Pkwy., 35209. Phone: (205) 879-3324. Fax: (205) 802-4555. Licensee: Kimtron Inc. Group owner: Crawford Broadcasting Co. (acq 1994). Population served: 100,000 Natl. Rep: McGavren Guild. Format: Classic hits. ♦Steve Armstrong, gen mgr; Jennifer Paepcke, gen sls mgr; Todd Dixon, chief of engrg; Melodye Grubb, traf mgr.

WZGX(AM)—See Bessemer

WZZK-FM— 1948: 104.7 mhz; 100 kw. 1,300 ft TL: N33 29 02 W86 48 21. Stereo. Hrs opn: 24 301 Beacon Pkwy. W., Suite 200, 35209. Phone: (205) 916-1100. Fax: (205) 916-1151. E-mail: wzzk@cox.com Web Site: www.wzzk.com. Licensee: Cox Radio Inc. Group owner: Cox Broadcasting (acq 3-28-97; grpsl). Population served: 407,400 Law Firm: Dow, Lohnes and Albertson. Format: Country. News staff: 2; News: 10 hrs wkly. Target aud: 25-54. ♦Ray Nelson, gen mgr; David Wells, gen sls mgr; Justin Case, progmg dir.

Boaz

WBSA(AM)— Oct 1, 1959: 1300 khz; 1 kw-D. TL: N34 12 50 W86 09 10. Hrs open: 1525 Wills Rd., 35957. Phone: (256) 593-4264. Fax: (256) 593-4265. Web Site: www.wbsaam.com. Licensee: Watkins Broadcasting Inc. (acq 10-4-94; $100,000; FTR: 10-17-94). Population served: 9,800 Format: Southern gospel. Target aud: General. ♦Roger Watkins, gen mgr.

Brantley

WAOQ(FM)— June 3, 1999: 100.3 mhz; 6 kw. 328 ft TL: N31 42 26 W86 13 12. Stereo. Hrs opn: 24 Box 83, Clanton, 35045. Phone: (334) 335-2877. Fax: (205) 755- 3329. E-mail: waoq@waoq.com Web Site: www.waoq.com. Licensee: Alatron Corp. Inc. Format: Classic country. News: 14 hrs wkly. Spec prog: Gospel 14 hrs wkly. ♦Robert E. Williams, pres; Christopher W. Johnson, gen mgr & progmg dir; Ken Lyons, stn mgr & gen sls mgr.

Brewton

WEBJ(AM)— Aug 1, 1947: 1240 khz; 1 kw-U. TL: N31 06 35 W87 03 36. Hrs open: 6 AM-8 PM 301 Downing St., 36426. Phone: (251) 867-5717. Fax: (251) 867-5718. Licensee: Candy Cashman Smith, individual. (acq 10-2-97). Population served: 25,000 Law Firm: Gardner, Carton & Douglas. Format: Oldies. News staff: one; News: 20 hrs wkly. Target aud: 21 plus; 60% female, 40% male. ♦Dennis Dunnaway, gen mgr.

***WELJ(FM)**— 1998: 90.9 mhz; 45 kw. Ant 502 ft TL: N31 18 13 W87 02 50. Stereo. Hrs opn: Box 347, 36427. Secondary address: 42676 Hwy. 31 36427. Phone: (251) 809-1915. Fax: (251) 809-1916. Licensee: Gateway Public Radio (acq 10-12-00; $3,500). Format: Southern gospel, Christian. ♦Debra Johnson, gen mgr & stn mgr; Ruth Thompson, gen mgr.

WKNU(FM)— Aug 19, 1974: 106.3 mhz; 3.8 kw. Ant 417 ft TL: N31 06 42 W87 01 17. Stereo. Hrs opn: 24 Box 468, 36427. Secondary address: 2832 Ridge Rd. 36426. Phone: (251) 867-4824. Fax: (251) 867-7003. E-mail: wknu@bellsouth.net Licensee: Ellington Radio Inc. (acq 12-28-78). Population served: 35,000 Format: C&W. News: 8 hrs wkly. Target aud: General. Spec prog: Gospel 2 hrs, relg 2 hrs wkly. ♦Carol Ellington, pres, gen mgr & opns mgr; Hugh L. Ellington, exec VP & gen mgr.

Bridgeport

WYMR(AM)— Sept 19, 1961: 1480 khz; 1 kw-D, 39 w-N. TL: N34 56 34 W85 42 26. Hrs open: 24 1237 County Rd. 295, Higdon, 35979-6349. Phone: (256) 495-2500. Fax: (914) 730-9820. E-mail: manager@wymr1480.com Web Site: www.wymr1480.com. Licensee: MG Media Inc. (acq 9-29-2005). Population served: 128,730 Natl. Network: Salem Radio Network. Law Firm: Donald Martin, P.C. Format: Christian talk and ministry. ♦Marvin Glass, pres, gen mgr, progmg dir & sls.

Brookwood

WRTR(FM)—Licensed to Brookwood. See Tuscaloosa

Brundidge

WTBF-FM— Oct 1, 1997: 94.7 mhz; 14.5 kw. 433 ft TL: N31 40 38 W85 56 43. Stereo. Hrs opn: 24 67 Court Sq., Troy, 36081. Phone: (334) 566-0300. Fax: (334) 566-5689. E-mail: wtbf@troycable.net Web Site: www.wtbf947.com. Licensee: Troy Broadcasting Corp. Population served: 200,000 Natl. Network: Moody. Rgnl. Network: Alabama Net. Law Firm: Gardner, Carton & Douglas. Wire Svc: National Weather Network Format: Oldies. News: 20 hrs wkly. Target aud: 28-60; 50s, 60s & 70s music listeners. ♦Jim Roling, gen mgr; Doc Kirby, opns mgr.

Butler

WMLV(FM)— Nov 20, 1978: 93.5 mhz; 32 kw. Ant 610 ft TL: N32 09 26 W88 29 17. Stereo. Hrs open: 3436 Hwy. 45 N., Meridian, MS, 39301. Phone: (601) 693-2661. Fax: (601) 483-0826. Licensee: Mississippi Broadcasters L.L.C. (group owner; (acq 10-30-2002; $771,500). Population served: 50,000 Format: Hot adult contemp. ♦Clay Holladay, stn mgr; Scott Stevens, opns mgr.

WPRN(AM)— July 11, 1959: Stn currently dark. 1330 khz; 5 kw-D. TL: N32 06 02 W88 14 07. Hrs open: 909 W. Pushmataha St., 36904-2441. Phone: (205) 459-3222. Fax: (205) 459-4140. Licensee: Butler Broadcasting Corp. (acq 1-75). Population served: 25,000 Format: Country. ♦Daryl Jackson, gen mgr.

Calera

WBYE(AM)— Jan 12, 1958: 1370 khz; 1 kw-D. TL: N33 05 26 W86 46 37. Hrs open: 6 AM-6 PM Box 1727, 35040. Phone: (205) 668-1370. Licensee: WBYE Broadcasting Co. Inc. (acq 4-14-89; $100,754; FTR: 4-24-89). Population served: 130,000 Format: Gospel. News: 15 hrs wkly. Target aud: 25-65. ♦Frank Cummings, pres & gen mgr.

Carrollton

***WALN(FM)**— 1997: 89.3 mhz; 9.5 kw vert. 699 ft TL: N33 13 06 W88 05 46. Hrs open: 24 American Family Radio, Box 3206, Tupelo, MS, 38803. Phone: (662) 844-8888. Fax: (662) 842-6791. Web Site: www.afr.net. Licensee: American Family Association. Group owner: American Family Radio Format: Inspirational Christian. ♦Tim Waldmon, pres; Marvin Sanders, gen mgr; John Riley, progmg dir; Fred Jackson, news dir; Joey Moody, engrg dir.

WRAG(AM)— 1951: Stn currently dark. 590 khz; 1 kw-D. TL: N33 13 04 W88 05 48. Hrs open: Willis Broadcasting Corp., 646 Church St., Suite 400, Norfolk, VA, 23510. Phone: (757) 622-4600. Licensee: Birmingham Christian Radio. Group owner: Willis Broadcasting Corp. (acq 10-16-97; $130,000 with WSPZ(AM) Tuscaloosa). Population served: 25,000 Law Firm: Gammon & Grange.

WZBQ(FM)— February 1970: 94.1 mhz; 98 kw. Ant 1,007 ft TL: N33 13 07 W88 05 47. Stereo. Hrs opn: 3900 11th Ave., Tuscaloosa, 35402. Phone: (205) 344-4589. Phone: (205) 349-3200. Fax: (205) 752-9269. Web Site: www.941zbq.com. Licensee: Capstar TX L.P. Group owner: Clear Channel Communications Inc. (acq 8-30-00; grpsl). Population served: 750,000 Format: CHR. Target aud: 18-49. ♦Lori Moore, gen mgr; Russ Williams, opns mgr.

Carrville

WACQ(AM)—Licensed to Carrville. See Tallassee

Centre

WEIS(AM)— Sept 30, 1961: 990 khz; 1 kw-D, 30 w-N. TL: N34 09 10 W85 40 44. Hrs open: 24 Box 297, 35960. Phone: (256) 927-5152. Fax: (256) 927-6503. Web Site: www.weis990am.com. Licensee: Baker Enterprises Inc. (acq 9-8-83; $157,675; FTR: 9-26-83). Population served: 450,000 Law Firm: Timothy K. Brady. Format: Country, southern gospel. News staff: one; News: 10 hrs wkly. Target aud: General. ♦Jerry Baker, pres & gen mgr.

WRHY(FM)— Oct 10, 1992: 105.9 mhz; 6 kw. 150 ft TL: N34 12 51 W85 46 20. Hrs open: 24 801 Nobel St., Fl. 8, Anniston, 36201. Phone: (256) 236-1880. Fax: (256) 236-4480. Web Site: www.y-106.com. Licensee: Williams Communications Inc. (group owner; acq 6-99; $380,000). Population served: 440,000 Natl. Network: Motor Racing Net, PRI. Rgnl. Network: Motor Racing Net, PRI. Format: Hot country. News: one hr wkly. Target aud: 25-54. ♦Walt Williams, gen mgr; Dan Pullman, gen sls mgr; Tex Carter, progmg dir; Mike Mote, news dir.

WZTQ(AM)— Nov 9, 1962: 1560 khz; 1 kw-D. TL: N34 07 41 W85 38 27. Hrs open: 6 AM-sunset Box 2, 35960. Phone: (256) 927-4027. Fax: (205) 295-1238. Web Site: www.joychristian.com. Licensee: Joy Christian Communications Inc. (acq 12-1-2003). Population served: 245,000 Rgnl. Network: Tenn. Radio Net., Tobacco. Format: Southern gospel. News staff: one; News: 7 hrs wkly. Target aud: 25-55; working middle class, rural. Spec prog: Farm one hr wkly. ♦Ed Smith, pres & gen mgr; Marie Smith, sr VP.

Centreville

WBIB(AM)— Dec 14, 1964: 1110 khz; 1 kw-D. TL: N32 58 01 W87 09 01. Hrs open: Sunrise-sunset Box 216, 35042. Phone: (205) 926-6286. Licensee: James DeLoach (acq 11-21-2005). Population served: 20,000-30,000 Natl. Rep: Keystone (unwired net). Format: Gospel, country. Target aud: Adults. ♦Horrace Cruchfield, gen mgr.

Chickasaw

WDLT-FM—Licensed to Chickasaw. See Mobile

Citronelle

WQUA(FM)— June 25, 1989: 102.1 mhz; 15 kw. Ant 426 ft TL: N31 05 04 W88 23 51. Stereo. Hrs opn: 24 Box 262550, Baton Rouge, LA, 70826. Secondary address: 8919 World Ministry Ave., Baton Rouge, LA 70810. Phone: (225) 768-3688. Phone: (225) 768-8300. Fax: (225) 768-3729. E-mail: kawikfish@yahoo.com Web Site: www.jsm.org. Licensee: Family Worship Center Church Inc. Group owner: ABC Inc. (acq 8-25-2005; $1.25 million). Format: Christian. ♦David Whitelaw, COO; Jimmy Swaggart, pres.

WTLM(AM)—See Opelika

Columbia

WJJN(FM)— September 1992: 92.1 mhz; 2.55 kw. 499 ft TL: N31 10 56 W85 10 59. Hrs open: 805 N. Lena St., Suite 13, Dothan, 36303. Phone: (334) 671-1753. Fax: (334) 677-6923. E-mail: wtraffic@aol.com Web Site: www.wjjn.com. Licensee: Wilson Broadcasting Inc. (group owner) Format: Urban contemp. ♦James R. Wilson III, gen mgr.

Columbiana

WQEM(FM)— 2000: 101.5 mhz; 1.8 kw. Ant 607 ft TL: N33 13 45 W86 42 56. Hrs open: 1137 Tenth Pl. S., Birmingham, 35205. Phone: (205) 323-1516. Fax: (205) 323-2747 (Phone/Fax). Web Site: www.gleniris.net. Licensee: Glen Iris Baptist School (acq 12-24-02). Format: Teaching, gospel. ♦Chris Lamb, chmn & gen mgr; Nathan Mills, opns mgr.

Cordova

WFFN(FM)— June 22, 1987: Stn currently dark. 95.3 mhz; 5 kw. Ant 354 ft TL: N33 46 11 W87 12 06. (CP: COL Coaling. 17.5 kw, ant 840 ft. TL: N33 03 15 W87 32 57). Stereo. Hrs opn: 142 Skyland Blvd., Tuscaloosa, 35405. Phone: (205) 750-0929. Fax: (205) 349-1715. Licensee: Citadel Broadcasting Co. (acq 7-12-2005; grpsl). ♦Farid Suleman, chmn.

WXJC-FM— 1997: 92.5 mhz; 2.2 kw. Ant 548 ft TL: N33 38 55 W87 09 19. Hrs open: 24 120 Summit Pkwy., Birmingham, 35209. Phone: (205) 879-3324. Fax: (205) 802-4555. Licensee: Kimtron Inc. Group owner: Crawford Broadcasting Co. (acq 7-15-2004; $1.15 million). Format: Southern gospel. ♦Steve Armstrong, gen mgr; Jennifer Paepcke, gen sls mgr; Todd Dixon, chief of engrg; Melodye Grubb, traf mgr.

Cullman

WFMH(AM)— October 1946: 1340 khz; 670 w-U. TL: N34 10 49 W86 51 59. Hrs open: 24 1707 Warnke Rd, N.W., 35055. Phone: (256) 734-3271. Fax: (256) 734-3622. E-mail: wfmh@adelphia.net Licensee: Williams Communications Inc. Group owner: Williams Communications Inc. (acq 8-18-2004; $2.45 million with WFMH-FM Holly Pond). Population served: 75,000 Natl. Rep: Keystone (unwired net). Format: News/talk, sports. News: 18 hrs wkly. Target aud: 25-54; middle & upper income adults. ♦Walt Williams, gen mgr; Susan Hackney, progmg dir.

WKUL(FM)— September 1967: 92.1 mhz; 6 kw. 328 ft TL: N34 11 41 W86 43 52. Stereo. Hrs opn: Box 803, 214 1st Ave. S.E., 35056. Phone: (256) 734-0183. Fax: (256) 739-2999. Web Site: www.wkul.com. Licensee: Jonathan Christian Corp. (acq 3-1-77). Population served: 1,200,000 Format: Country, sports, talk. News staff: one; News: 20 hrs wkly. Target aud: 25-54. Spec prog: Farm 15 hrs wkly. ♦Don Mosley, pres; Ron Mosley, gen mgr; Rick Nix, opns mgr.

WMCJ(AM)— Mar 25, 1950: 1460 khz; 5 kw-D, 500 w-N, DA-N. TL: N34 10 44 W86 51 58. Hrs open: 24 1707 Warnke Rd. N.W., 35055. Phone: (256) 734-3271. Fax: (256) 734-3622. Licensee: Williams Communications Inc. (acq 7-8-2005; $75,000). Population served: 138,672 Format: Solid gospel. ♦Walt Williams, gen mgr; Susan Hackney, progmg dir.

WYDE-FM— Aug 6, 1949: 101.1 mhz; 100 kw. Ant 1,345 ft TL: N34 04 56 W86 54 15. Hrs open: 24 120 Summit Pkwy., Birmingham, 35209. Phone: (205) 879-3324. Fax: (205) 802-4555. Licensee: Kimtron Inc. Group owner: Crawford Broadcasting Co. (acq 6-14-2002; $8.5 million). Population served: 1,274,627 Format: Adult contemp. ♦Steve Armstrong, gen mgr; Jennifer Paepcke, gen sls mgr; Todd Dexon, chief of engrg; Melodye Grubb, traf mgr.

Dadeville

WDLK(AM)— Aug 11, 1980: 1450 khz; 1 kw-U. TL: N32 50 56 W85 46 10. Hrs open: 2015 Hwy. 49 S., 36853. Phone: (256) 825-2028. Fax: (256) 825-2199. Licensee: Progressive United Communications Inc. (group owner; (acq 11-6-2000; $45,000). Population served: 60,000 Format: Gospel. ♦Thomas Franklin, gen mgr.

***WELL-FM**— Mar 1, 1990: 88.7 mhz; 100 kw. 328 ft TL: N32 51 20 W85 46 31. (CP: Ant 305 ft.). Stereo. Hrs opn: 24 658 Horseshoe Bend Rd., 36853. Phone: (256) 825-6456. Fax: (256) 825-6426. E-mail: cassiekeyes@hotmail.com Licensee: Tiger Communications Educational Foundation Inc. (acq 8-9-01; $325,000). Natl. Network: USA. Format: Christian. News staff: one. Target aud: 24 plus. ♦Cassie Keyes, gen mgr.

WGZZ(FM)— July 23, 1989: 100.3 mhz; 2.2 kw. Ant 546 ft TL: N32 52 58 W85 49 16. Stereo. Hrs opn: 1261 Jacksons Gap Way, Jacksons Gap, 36861-5760. Secondary address: 13263 Hwy. 280, Jacksons Gap 36861. Phone: (256) 825-4221. E-mail: www.wzlm@charter.net Licensee: Auburn Network Inc. (group owner; (acq 6-28-2007; $1.4 million). Format: Country. Target aud: 18-55. Spec prog: Gospel 3 hrs wkly. ♦Cameron Reynolds, gen mgr.

Daleville

WCMA(AM)— Oct 25, 1983: 1560 khz; 50 kw-D, 2.5 kw-CH. TL: N31 16 35 W85 45 54. Hrs open: Box 1969, Santa Rosa Beach, FL, 32459. Phone: (850) 267-2445. Fax: (850) 622-0181. Licensee: Perihelion Global Inc. (acq 4-21-2004; $135,000). Population served: 350,000 Format: News/talk. ♦John Beebe, gen mgr.

Daphne

WAVH(FM)— May 15, 1993: 106.5 mhz; 50 kw. Ant 449 ft TL: N30 44 44 W88 05 40. Stereo. Hrs opn: 24 2800 Dauphin St., Mobile, 36606-2400. Phone: (251) 652-2000. Fax: (251) 652-2001. E-mail: mobile.prog@cumulus.com Web Site: www.cumulus.com. Licensee: Baldwin Broadcasting Co., Debtor in Possession (acq 12-12-2000). Population served: 470,000 Natl. Rep: McGavren Guild. Law Firm: Wood, Maines & Brown. Format: Oldies. Target aud: 25-64; general. ♦Gary Pizzati, gen mgr; James Alexander, opns mgr; Steve Crumbely, opns mgr.

Decatur

WDPT(AM)— Oct 3, 1953: 1490 khz; 1 kw-U. TL: N34 35 14 W86 59 13. Hrs open: 401 14th St. S.E., Suite 2A, 35601. Phone: (256) 353-6363. Phone: (256) 340-1490. Fax: (256) 350-2025. E-mail: wajfradio@bellsouth.net Web Site: protalk1490.com. Licensee: WAJF Inc. (acq 3-12-2003; $150,000). Population served: 110,000 Format: Talk. ♦Dan Baughman, gen mgr.

WDRM(FM)—Listing follows WHOS(AM).

WHOS(AM)— October 1948: 800 khz; 1 kw-D, 215 w-N. TL: N34 35 55 W87 00 24. Hrs open: 24 Box 21008, Huntsville, 35824. Secondary address: 26869 Peoples Rd., Madison 35756. Phone: (256) 353-1750. Fax: (256) 350-2653. Licensee: Capstar TX L.P. Group owner: Clear Channel Communications Inc. (acq 7-18-00; grpsl). Population served: 343,500 Format: 24 hr. News staff: 3. Target aud: 25-54. ♦Rick Brown, gen mgr; Carmelita Palmer, sls dir.

WDRM(FM)—Co-owned with WHOS(AM). September 1951: 102.1 mhz; 100 kw. 981 ft TL: N34 49 08 W86 44 19. Stereo. 24 Phone: (205) 353-1750. Web Site: www.wdrm.com. Format: Country.

WRSA-FM— Nov 23, 1965: 96.9 mhz; 100 kw. 1,010 ft TL: N34 29 19 W86 37 08. Stereo. Hrs opn: 24 8402 Memorial Pkwy SW, Huntsville, 35802. Phone: (256) 885-9797. Fax: (256) 885-9796. Web Site: www.lite969.com. Licensee: NCA Inc. (acq 3-18-02). Population served: 500,000 Natl. Network: CBS. Format: Adult contemp, lite rock. News staff: one; News: 1 hr wkly. Target aud: 35 plus. ♦Penny Nielson, CEO & pres; Nate Adams, opns mgr; Tom Panucci, gen mgr & gen sls mgr; John Malone, progmg dir; Don Rhoden, chief of engrg.

WWTM(AM)— May 1935: 1400 khz; 1 kw-U. TL: N34 36 44 W86 59 28. Hrs open: 1209 Danville Rd. S.W., Suite N, 35601-3853. Phone: (256) 353-1400. Fax: (256) 353-0363. Web Site: espn1400.info. Licensee: R & B Communications Inc. (acq 3-18-97). Natl. Network: ESPN Radio. Format: Sports. Target aud: 25-54. ♦Brian Black, gen mgr.

***WYFD(FM)**— May 7, 1975: 91.7 mhz; 3 kw. 787 ft TL: N34 47 53 W86 38 24. Stereo. Hrs opn: 24 11530 Carmel Commons Blvd., Charlotte, NC, 28226. Phone: (256) 353-7951. Phone: (800) 888-7077. Fax: (256) 650-0917. Web Site: www.bbnradio.org. Licensee: Bible Broadcasting Network. (group owner; acq 10-19-90; $75,000; FTR: 11-12-90). Population served: 450,000 Format: Relg, educ. ♦Lowell Davey, pres; Dave Phillips, stn mgr.

Demopolis

WXAL(AM)— Nov 9, 1947: 1400 khz; 790 w-U. TL: N32 30 08 W87 49 07. Hrs open: Box 938, 1226 Jefferson Rd., 36732. Phone: (334) 289-1400. Fax: (334) 289-9811. E-mail: win985@westal.net Licensee: West Alabama Radio Inc. (acq 8-18-98; $456,300 with co-located FM). Population served: 75,000 Natl. Network: Westwood One, USA. Format: News/talk, Black gospel. Target aud: 25-54. ♦Amy Ward, CEO; Larry Carr, chmn & progmg dir; Amy Ross, gen mgr; Sean Park, gen sls mgr; Valerie Webb, news dir.

WZNJ(FM)—Co-owned with WXAL(AM). 1975: 106.5 mhz; 25 kw. 492 ft TL: N32 20 40 W87 37 43. Stereo. Natl. Network: Westwood One, USA. Format: Oldies, sports. Target aud: 18-49.

Dixons Mills

***WMBV(FM)**— Aug 15, 1988: 91.9 mhz; 62 kw. 613 ft TL: N32 07 45 W87 44 16. Stereo. Hrs opn: 24 Box 91.9 FM, 36736-0091. Secondary address: 10564 Marengo County Rd. 30 36736. Phone: (334) 992-2425. Phone: (888) 624-7234. Fax: (334) 992-2637. E-mail: wmbv@moody.edu Web Site: www.wmbv.org. Licensee: Moody Bible Institute. Group owner: The Moody Bible Institute of Chicago (acq 3-31-88). Population served: 200,000 Natl. Network: Moody. Law Firm: Southmayd & Miller. Format: Relg. News: 10 hrs wkly. Target aud: 35-55; general. Spec prog: Financial 3 hrs, children 3 hrs, sports one hr wkly. ♦Rob Moore, gen mgr.

Dora

WCOC(AM)— Apr 1, 1982: 1010 khz; 5 kw-D. TL: N33 48 04 W87 06 42. Hrs open: 6475 Hwy. 78, Cordova, 35550. Phone: (205) 648-6926. Fax: (205) 648-6489. Licensee: Azteca Communications of Alabama Inc. (acq 2-12-02; $190,000). Format: Sp. ♦Patricia Perez, gen mgr.

Dothan

WAGF(AM)— Sept 29, 1932: 1320 khz; 1 kw-U, DA-N. TL: N31 14 56 W85 23 20. Hrs open: 805 N. Lena St., Suite 13, 36303. Phone: (334) 671-1753. Fax: (334) 677-6923. E-mail: wtraffic@aol.com Web Site: www.wjjn.com. Licensee: Wilson Broadcasting Inc. (group owner; acq 8-13-92; $60,000; FTR: 8-31-92). Format: Gospel. ♦James Wilson III, gen mgr.

WAGF-FM— 1991: 101.3 mhz; 3 kw. 328 ft TL: N31 12 02 W85 20 12. (CP: 820 w, ant 640 ft.). Hrs opn: 24 805 N. Lena St., Suite 13, 36303. Phone: (334) 677-7654. Fax: (334) 677-6923. E-mail: wtraffic@aol.com Web Site: www.wjjn.com. Licensee: Wilson Broadcasting Inc. (group owner) Natl. Network: Jones Radio Networks. Rgnl. Network: Alabama Net. Natl. Rep: Rgnl Reps. Law Firm: Arter & Hadden. Format: Soft hits, lite adult contemp. Target aud: 25-54; female. ♦James R. Wilson III, gen mgr.

***WDYF(FM)**— 2004: 90.3 mhz; 9.2 kw. Ant 535 ft TL: N31 19 31 W85 36 02. Hrs open: Box 210789, Montgomery, 36121. Secondary address: 381 Mendel Parkway, Montgomery 36117. Phone: (334) 271-8900. Fax: (334) 260-8962. E-mail: mail@faithradio.org Web Site: www.faithradio.org. Licensee: Faith Broadcasting Inc. Format: Educ, relg, MOR. ♦Russell Dean, gen mgr; Andrew Leuthold, opns mgr; Gary Hundley, dev dir; Bob Crittenden, progmg dir.

WEEL(AM)— July 3, 1995: Stn currently dark. 700 khz; 1.6 kw-D. TL: N31 26 19 W85 17 22. Hrs open: Sunrise-sunset 3124 W. Main St., Suite 7, 36305. Phone: (334) 699-1970. Fax: (334) 699-1980. Licensee: Jalo Broadcasting Corp. (acq 12-20-2006; $225,000). Law Firm: Law Offices of Timothy K. Brady. Format: Oldies. ♦Jack Gale, pres.

WESP(FM)— Sept 1, 1989: 102.5 mhz; 16.5 kw. Ant 403 ft TL: N31 15 48 W85 18 24. Stereo. Hrs open: 24 3245 Montgomery Hwy., Suite 1, 36303-2150. Phone: (334) 671-1025. Phone: (334) 712-9233. Fax: (334) 712-0374. E-mail: ron@wdjr.com Web Site: www.rock1025.com. Licensee: Gulf South Communications Inc. (acq 1999; $1.4 million). Population served: 450,000 Natl. Rep: McGavren Guild. Format: Rock. Target aud: 25-54; men. ♦Ron Eubanks, gen mgr; John Baker, opns mgr & progmg dir; Misty Huff, prom mgr; Denise Reed, traf mgr.

***WGTF(FM)**— September 1988: 89.5 mhz; 19 kw. 213 ft TL: N31 14 02 W85 26 02. Hrs open: 24 107 Wanda Ct., 36303. Phone: (334) 794-4770. Fax: (334) 794-4770. E-mail: wgtf@bbnradio.org Licensee: Dothan Community Educational Radio Inc. Natl. Network: Bible Bcstg Net. Format: Relg. ♦Raymond Brown, gen mgr.

WOOF-FM— Sept 18, 1964: 99.7 mhz; 100 kw. 1,021 ft TL: N31 15 07 W85 17 12. Stereo. Hrs opn: Box 1427, 36302. Secondary address: 2518 Columbia Hwy. 36303. Phone: (334) 792-1149. Fax: (334) 677-4612. E-mail: woof@ala.net Web Site: www.997wooffm.com. Licensee: WOOF Inc. Population served: 440,000 Natl. Rep: Christal. Law Firm: Shaw Pittman. Format: Adult contemp. News staff: 2; News: 3 hrs wkly. Target aud: 25-54; women 18-49 dominant. ♦Leigh Simpson, gen mgr, opns mgr, progmg dir & news dir; Hal Edwards, gen sls mgr; John Daniel, news dir; Laura Pate, disc jockey; Rick Patrick, pres & disc jockey.

WOOF(AM)— Feb 17, 1947: 560 khz; 5 kw-D, 117 w-N. TL: N31 13 05 W85 21 10. Stereo. Format: Sports, talk. Spec prog: Black gospel 17 hrs wkly. ♦Leigh Simpson, traf mgr.

***WRWA(FM)**— December 1985: 88.7 mhz; 50 kw. 500 ft TL: N31 12 30 W85 36 51. Stereo. Hrs open: 6 AM-midnight
Rebroadcasts WTSU(FM) Troy 100%.

Wallace Hall, Troy Univ., Troy, 36082. Phone: (334) 670-3268. Fax: (334) 670-3934. E-mail: wtsu@troy.edu Web Site: wtsu.troy.edu. Licensee: Troy State University. Natl. Network: NPR, PRI. Format: Class, news. News: 25 hrs wkly. Target aud: General. Spec prog: Children one hr wkly. ♦James Clower, gen mgr; Judy Davis, opns mgr.

WTVY-FM— Sept 20, 1968: 95.5 mhz; 100 kw. Ant 1,078 ft TL: N31 15 16 W85 15 39. Stereo. Hrs opn: 24 Box 889, 36302-2088. Secondary address: 285 N. Foster, 8th Fl. 36303. Phone: (334) 792-0047. Fax: (334) 712-9346. E-mail: sue@stylesmedia.com Web Site: www.955wtvy.com. Licensee: Magic Broadcasting Alabama Licensing LLC. (group owner; (acq 7-27-2001). Population served: 417,000 Natl. Rep: Christal. Law Firm: Kenkel & Associates. Format: Country. News: 5 hrs wkly. Target aud: 25-54. Spec prog: Farm 5 hrs, gospel 4 hrs, religion 3 hrs wkly. ♦Jeff Storey, CEO; Sue Hughes, gen mgr; Amie Pollard, natl sls mgr; Mike Casey, mus dir.

***WVOB(FM)**— Dec 8, 1988: 91.3 mhz; 2.5 kw. 328 ft TL: N31 10 57 W85 24 21. Hrs open: 24 Box 1944, 36302. Secondary address: 2573 Hodgesville Rd. 36301. Phone: (334) 671-9862. Fax: (334) 793-4344. E-mail: wvob913fm@bethanybc.edu Web Site: www.bethanyradionetwork.com. Licensee: Bethany Divinity College & Seminary Inc. Natl. Network: USA. Format: Southern gospel music & educ. News staff: one; News: 6 hrs wkly. Target aud: General; college students & relg community. ♦Dr. Steve Shuemake, gen mgr.

WWNT(AM)— Apr 30, 1947: 1450 khz; 1 kw-U. TL: N31 13 10 W85 22 14. Hrs open: 1733 Columbia Hwy., 36303. Phone: (334) 671-0075. Fax: (334) 671-0091. E-mail: larrymckee@aol.com Web Site: www.wwnt1450.com. Licensee: WWNT LLC (acq 6-10-83; $115,000; FTR: 7-4-83). Population served: 36,733 Natl. Network: USA. Format: Talk/news. News staff: 2. Target aud: 25-54 men; 25-54 males. ♦Larry Williams, gen mgr.

Elba

WELB(AM)— Nov 16, 1958: 1350 khz; 1 kw-D. TL: N31 27 10 W86 04 00. Hrs open: 11 20334 Hwy 87, 36323. Phone: (334) 897-2216. Phone: (334) 897-2217. Fax: (334) 897-3694. E-mail: wztz@alaweb.com Licensee: Elba Radio Co. (acq 3-4-76). Population served: 4,634 Format: Classic country. News: 6 hrs wkly. Target aud: General. Spec prog: Gospel 12 hrs wkly. ♦Doug Holderfield, gen mgr; Mike Holderfield, progmg dir; Eddie Phillips, news dir.

WVVL(FM)—Co-owned with WELB(AM). Oct 1, 1986: 101.1 mhz; 640 w. 682 ft TL: N31 24 41 W85 57 32. Stereo. 19 Format: Modern country.

Enterprise

WDJR(FM)— July 1, 1968: 96.9 mhz; 100 kw. 1,515 ft TL: N30 55 11 W85 44 30. Stereo. Hrs opn: 3245 Montgomery Hwy, Dothan, 36303. Phone: (334) 712-9233. Fax: (334) 712-0374. E-mail: ron@wdjr.com Web Site: www.wdjr.com. Licensee: Gulf South Communications Inc. (acq 7-9-92; $700,000; FTR: 7-27-92). Population served: 16,500 Format: Country. ♦Ron Eubanks, gen mgr; John Baker, opns mgr; Misty Huff, prom dir; Denise Reed, traf mgr.

WKMX(FM)— Nov 27, 1974: 106.7 mhz; 100 kw. Ant 1,068 ft TL: N31 24 41 W85 57 32. Stereo. Hrs opn: 24 285 N. Foster St., 8th Fl., Dothan, 36303. Phone: (334) 792-0047. Fax: (334) 712-9346. Web Site: www.wkmx.com. Licensee: Magic Broadcasting Alabama Licensing LLC. (group owner; (acq 9-3-2004; $4.5 million). Population served: 281,400 Format: CHR. News: 2 hrs wkly. Target aud: Females; 18-49. ♦Dan Bradley, gen mgr; Richard Reinhardt, gen mgr & sls dir; Sue Hughes, gen mgr; Doc Thompson, opns dir; Richard Reomjardt, opns mgr; Chris Green, gen sls mgr; Amie Pollard, natl sls mgr; John Houston, prom dir & progmg dir.

Eufaula

WKZJ(FM)— 1969: 92.7 mhz; 39 kw. Ant 551 ft TL: N32 07 58 W85 04 13. Stereo. Hrs opn: 2203 Wynnton Rd., Columbus, GA, 31902. Phone: (706) 576-3565. Fax: (706) 576-3683. E-mail: riversales@kenology.net Web Site: www.theriverrocks.com. Licensee: Davis Broadcasting Inc. (group owner; (acq 7-20-2004; $2.7 million). Population served: 11,000 Law Firm: Leventhal, Senter & Lerman. Format: Adult contemp, CHR. News staff: 2; News: 6 hrs wkly. ♦Gregory A. Davis, pres; Janet Armstead, gen mgr; Bernie Corcoran, opns mgr; Cheryl Davis, opns mgr; Angela Verdejo, gen sls mgr; Carl Conner, progmg VP.

WRVX(FM)— Mar 16, 1992: 97.9 mhz; 6 kw. Ant 328 ft TL: N31 56 04 W85 12 27. Stereo. Hrs opn: 24 Box 1419, 36072-1419. Secondary address: 1084 S. Eufaula Ave. 36027. Phone: (334) 616-0097. Fax: (334) 687-3600. E-mail: wrvxfm@eufaula.rr.com Licensee: River Valley Media L.L.C. (acq 2-28-98; $200,000). Natl. Network: Jones Radio Networks. Format: Var. Target aud: 24 plus. ♦Clyde Earnest, gen mgr; John Crumpton, progmg dir; Pam Sharp, sls; Terry Harper, engr.

WULA(AM)— 1948: 1240 khz; 1 kw-U. TL: N31 54 30 W85 09 51. Hrs open: Box 1419, 36027-0531. Phone: (334) 616-0097. Fax: (334) 687-3600. Licensee: River Valley Media LLC (acq 5-21-2004; $95,000). Population served: 20,000 Format: Sports, Talk, News. Target aud: 25-54; adults. Spec prog: Farm 2 hrs, Black 3 hrs wkly. ♦John Burns, pres.

Eutaw

WQZZ(FM)— August 1990: 104.3 mhz; 2.3 kw. 370 ft TL: N32 54 16 W87 50 09. Hrs open: 24 Box 70427, Tuscaloosa, 35407. Secondary address: 601 Greensboro Ave., Suite 507, Tuscaloosa 35401. Phone: (205) 345-4787. Fax: (205) 345-4790. E-mail: jwlawson@bellsouth.net Licensee: Jim Lawson Communications Inc. (acq 3-27-93). Format: Rhythm & blues, urban. ♦Jim Lawson, gen mgr & opns mgr.

Eva

WRJL-FM— 1996: 99.9 mhz; 6 kw. 328 ft TL: N34 18 43 W86 43 54. Hrs open: 24 5610 Hwy. 55 E., 35621. Phone: (256) 796-8000. Fax: (256) 796-8515. Licensee: Rojo Inc. Format: Southern gospel. ♦Jo French, gen mgr.

Evergreen

WPGG(AM)— July 1, 1957: 1470 khz; 1 kw-D. TL: N31 26 29 W86 56 08. Hrs open: 24 Box 705, 36401. Secondary address: Hwy. 31 N 36401. Phone: (251) 578-2780. Fax: (251) 578-5399. E-mail: powerpig@bellsouth.net Licensee: Star Broadcasting Inc. (acq 4-13-2004; $2.75 million with co-located FM). Population served: 20,000 Law Firm: Putbrese, Hunsaker & Trent P. Format: Country. News staff: 3; News: 20 hrs wkly. Target aud: 34-64. ♦Luther Upton, gen mgr.

Fairfield

WJLD(AM)—Licensed to Fairfield. See Birmingham

Fairhope

WABF(AM)— Aug 12, 1961: 1220 khz; 1 kw-D, 64 w-N, DA-D. TL: N30 30 38 W87 54 13. Hrs open: 24 Box 1220, 36533. Secondary address: 460 S. Section St. 36533. Phone: (251) 928-2384. Fax: (251) 928-9229. E-mail: wabf1220@bellsouth.net Licensee: Gulf Coast Broadcasting Co. Inc. (acq 5-24-99). Population served: 15500 Law Firm: Putbrese, Hunsaker & Trent P. Format: Talk. News: 15 hrs wkly. Target aud: 45 plus; upscale. Spec prog: Farm one hr, Swap Shop 6 hrs, relg 6 hrs wkly. ♦R. Hagan, pres; Lori Dubois, gen mgr.

WDLT(AM)—Licensed to Fairhope. See Mobile

WZEW(FM)— Aug 28, 1966: 92.1 mhz; 20.5 kw. Ant 363 ft TL: N30 31 23 W88 06 32. Stereo. Hrs opn: 24 1100 Dauphin St., Suite E, Mobile, 36604. Phone: (251) 433-9236. Phone: (251) 438-5460. Fax: (251) 438-5462. E-mail: 92zew@92zew.net Web Site: www.92zew.net. Licensee: Baldwin Broadcasting Co. (acq 10-1-98; $1.425 million). Population served: 954,300 Format: Adult alternative, blues. News staff: one; News: 6 hrs wkly. Target aud: 25-44. Spec prog: Jazz 6 hrs wkly. ♦Ken Johnson, gen mgr; Gene Murrell, progmg dir.

Fayette

WLDX(AM)— Sept 3, 1949: 990 khz; 1 kw-D, 42 w-N. TL: N33 41 06 W87 49 16. Hrs open: Box 189, 733 Columbus St. E., 35555. Phone: (205) 932-3318. Fax: (205) 932-3318. E-mail: wldx@wldx.com Web Site: www.wldx.com. Licensee: Dean Broadcasting Inc. (acq 6-1-2005; $450,000). Population served: 30,000 Law Firm: Fletcher, Heald & Hildreth. Format: Country. Target aud: 25-55; middle-income adults. ♦J. Wiley Dean, pres; Jill Dean, VP & gen mgr; Joe Jackson, progmg dir.

WTXT(FM)— Jan 29, 1977: 98.1 mhz; 100 kw. 984 ft TL: N33 34 31 W87 59 27. (CP: TL: N33 31 17 W87 51 38). Stereo. Hrs opn: 24 3900 11th Ave. S., Tuscaloosa, 35401. Phone: (205) 344-4589. Phone: (205) 349-3200. Fax: (205) 366-9774. E-mail: ddhamric@clearchannel.com Web Site: www.98txt.com. Licensee: Clear Channel Communications Group owner: Clear Channel Communications Inc. (acq 8-30-00; grpsl). Population served: 350,000 Natl. Network: ABC. Natl. Rep: Christal. Wire Svc: Direct Line Weather Wire Format: Contemp country. News staff: one. Target aud: 25-54. ♦Lori Moore, gen mgr; Russ Williams, opns mgr.

Florala

WKWL(AM)— Nov 3, 1979: 1230 khz; 1 kw-U. TL: N31 00 20 W86 19 53. Hrs open: 6 AM-6 PM Box 159, 36442-0159. Phone: (334) 858-6162. Fax: (334) 858-6162. E-mail: wkwl@cyou.com Licensee: Florala Broadcasting Co. Inc. Population served: 100,000 Natl. Network: USA. Format: Southern gospel, Christian country. News staff: one; News: 15 hrs wkly. Target aud: 5 plus; general. Spec prog: Farm one hr, relg 12 hrs wkly. ♦Robert Williamson, pres & gen mgr.

Florence

WBCF(AM)— 1946: 1240 khz; 1 kw-U. TL: N34 47 02 W87 42 16. Stereo. Hrs opn: 24 Box 1316, 35631. Secondary address: 525 E. Tennessee St. 35630. Phone: (256) 764-8170. E-mail: sales@wbcf.com Web Site: www.wbcf.com. Licensee: BCB Inc. (acq 8-11-77). Population served: 284,000 Natl. Network: Westwood One, Fox News Radio, CBS Radio. Format: News/talk. News staff: 2; News: 165 hrs wkly. Target aud: 25-64; adult, mature, affluent, educated, family, business. Spec prog: 12 hrs, local news/weather 5 hrs wkly. hrs wkly. hrs wkly. ♦Benji Carle, pres, gen mgr & news dir; Benny Carle, pres; Pat Costa, CFO & traf mgr; Jim Carle, exec VP; Patricia Carle, VP; Melody Mills, gen sls mgr. Co-owned TV: WBCF-LP, WXFL-LP

***WFIX(FM)**— Mar 20, 1988: 91.3 mhz; 30 kw. 600 ft TL: N34 40 24 W87 42 56. Stereo. Hrs opn: 5 AM-11 PM 113 N. Seminary St., 35630. Phone: (256) 764-9964. Fax: (256) 764-9154. E-mail: wfix@wfix.net Licensee: Tri-State Inspirational Broadcasting Inc. (acq 10-20-98; $100,000 for stock). Natl. Network: USA. Format: Adult contemp, Christian. News: 2 hrs wkly. Target aud: 25-54; upscale family oriented women & men. Spec prog: Sports, jazz, gospel 6 hrs wkly. ♦Mark Allen, gen mgr & opns mgr.

WQLT-FM— May 29, 1967: 107.3 mhz; 100 kw. 1,000 ft TL: N34 40 24 W87 42 56. Stereo. Hrs opn: 24 Box 932, 624 Sam Phillips St., 35631. Phone: (256) 764-8121. Fax: (256) 764-8169. Web Site: www.wqlt.com. Licensee: Big River Broadcasting Corp. (group owner) Population served: 670,000 Format: Adult contemp. Target aud: 25-54. ♦Nick Martin, gen mgr; Rocky Reich, sls dir; Sharon Brook, gen sls mgr; Jimmy Oliver, prom mgr; Charlie Ross, progmg mgr; Greg Pace, chief of engrg; Leisa Johnson, traf mgr.

WSBM(AM)—Co-owned with WQLT-FM. Mar 29, 1946: 1340 khz; 1 kw-U. TL: N34 47 50 W87 39 54. Stereo. 24 Web Site: www.wsbm.com. (Acq 2-21-73).140,000 Natl. Network: Fox Sports. Format: Sports. ♦Sharon Schoberg, chief of engrg & traf mgr.

WXFL(FM)— February 1992: 96.1 mhz; 20.5 kw. Ant 781 ft TL: N34 40 24 W87 42 56. Stereo. Hrs opn: 24 624 Sam Phillips St., 35630. Secondary address: 2046 Beltline Rd. SW, Suite 4, Decatur 35601. Phone: (256) 764-8121. Fax: (256) 764-8169. Web Site: www.kix96country.com. Licensee: Big River Broadcasting Corp. (group owner) Population served: 670,000 Natl. Network: ABC. Format: New country. News staff: one; News: 6 hrs wkly. Target aud: 18-49; general. ♦Knox Phillips, pres; Jerry Phillips, VP; Nick Martin, gen mgr.

Foley

WHEP(AM)— May 31, 1953: 1310 khz; 1 kw-D. TL: N30 26 06 W87 41 00. Hrs open: 6 AM-6 PM Box 1747, 36536. Secondary address: 20109 Hadley Rd. 36535. Phone: (251) 943-7131. Fax: (251) 943-7031. Licensee: Stewart Broadcasting Co. Inc. (acq 5-1-61). Population served: 328,622 Format: News/Talk/Sports/Mor. Target aud: 25 plus. Spec prog: Farm 2 hrs wkly. ♦Clark J. Stewart, pres & gen mgr.

Fort Mitchell

WAGH(FM)— 1988: 98.3 mhz; 6 kw. 328 ft TL: N32 21 48 W85 03 06. Hrs open: 1501 13th Ave., Columbus, 31901. Phone: (706) 576-3000. Fax: (706) 576-3010. E-mail: rasheedaali@clearchannel.com Web Site: www.magic98online.com. Licensee: CC Licenses LLC. Group owner: Clear Channel Communications Inc. (acq 2-21-2002; grpsl). Population served: 273,000 Format: Urban adult contemp. Target aud: 25-54; working Black adults. Spec prog: Relg 6 hrs wkly. ♦James R. Martin, gen mgr.

Fort Payne

WFPA(AM)— December 1949: 1400 khz; 1 kw-U. TL: N34 26 21 W85 42 09. Stereo. Hrs opn: 24 1210 Johnson St. E., 35967. Phone: (256) 845-7721. Fax: (256) 845-6828. Web Site: www.1400wfpa.net. Licensee: J.A.R. Services LLC (acq 5-10-2006; $95,000). Population served: 50,000 Natl. Network: CBS Radio, Premiere Radio Networks, Westwood One. Format: Adult contemporary. News staff: one; News: 15 hrs wkly. Target aud: 25-49; Women. ♦Joseph Allen Rivera, gen mgr.

WZOB(AM)— July 2, 1950: 1250 khz; 5 kw-U. TL: N34 26 23 W85 45 12. Hrs open: 24 hrs Box 680748, 35968. Secondary address: Hwy. 35 W., Radio Dr. 35968. Phone: (256) 845-2810. Fax: (256) 845-7521. Licensee: Central Broadcasting Co. Inc. (acq 8-12-03). Population served: 50,000 Rgnl rep: Dora-Clayton. Wire Svc: NOAA Weather Format: C&W. News staff: one. Spec prog: Farm 2 hrs, gospel 5 hrs, relg 5 hrs wkly. ♦Mike Kirby, pres; Doris Hobbs, stn mgr.

Fort Rucker

WLDA(FM)— 1991: 100.5 mhz; 6 kw. Ant 476 ft TL: N31 19 38 W85 35 35. Stereo. Hrs opn: 24 285 N. Foster St., Dothan, 36303. Phone: (334) 792-0049. Fax: (334) 712-9346. Licensee: Magic Broadcasting Alabama Licensing LLC. (group owner; (acq 10-1-2003; $750,000). Population served: 150,000 Format: Adult contemp. News staff: one. Target aud: 25-54; upscale baby boomers, acitive duty & retired military. ♦Dan Bradley, gen mgr; Doc Thompson, opns dir & opns mgr; Chris Green, sls dir & gen sls mgr.

Fruithurst

WCKS(FM)— May 9, 1994: 102.7 mhz; 1.6 kw. 630 ft TL: N33 37 24 W85 20 14. Hrs open: 102 Parkwood Cir., Carrollton, GA, 30117. Phone: (770) 834-5477. Fax: (770) 830-1027. Web Site: www.wcks.com. Licensee: WCKS LLC. Format: Adult contemp. Target aud: 25-44. ♦Steve L. Gradick, pres & gen mgr.

Gadsden

WAAX(AM)— Oct 18, 1947: 570 khz; 5 kw-D, 500 w-N, DA-N. TL: N33 58 45 W86 05 15. Hrs open:
Rebroadcasts WERC(AM) Birmingham 80%.
304 S. 4th St., 35902. Phone: (256) 543-9229. Fax: (256) 543-8777. Web Site: www.waax570.com. Licensee: Capstar TX L.P. Group owner: Clear Channel Communications Inc. (acq 8-30-2000; grpsl). Population served: 83,000 Format: News/talk. Target aud: 25-54. ♦Kathy Boggs, gen mgr; Pam Denham, stn mgr; Bill Seckbach, progmg dir; Carl Brady, news dir.

WGAD(AM)— May 26, 1947: 1350 khz; 5 kw-D, 1 kw-N, DA-N. TL: N34 01 03 W86 05 15. Hrs open: 24 Box 1350, 35902. Secondary address: 750 Walnut St. 35901. Phone: (256) 546-1611. Fax: (256) 547-9062. E-mail: dhedrick@wgad.com Web Site: www.wgad.com. Licensee: The DR Group LLC (acq 10-20-2004; $250,000). Population served: 100,000 Natl. Network: ABC. Format: Oldies. News: 16 hrs wkly. Target aud: 25 plus; general. Spec prog: Gospel 5 hrs wkly. ♦Dave Hedrick, gen mgr.

WGMZ(FM)—Glencoe, Oct 11, 1993: 93.1 mhz; 6 kw. 620 ft TL: N33 57 16 W85 51 40. Stereo. Hrs opn: 24 Box 517, 35902. Secondary address: 304 S. 4th St. 35901. Phone: (256) 549-0931. Fax: (256) 543-8777. E-mail: z931@clearchannel.com Web Site: www.wgmz.com. Licensee: Capstar TX L.P. Group owner: Clear Channel Communications Inc. (acq 8-30-00; grpsl). Population served: 350,000 Format: Classic hits of the 60s, 70s & 80s. News staff: one. Target aud: 35 plus. ♦Mark Mayes, pres; Kathy Boggs, gen mgr.

WJBY(AM)—Rainbow City, 1926: 930 khz; 5 kw-D, 500 w-N, DA-2. TL: N33 59 09 W86 02 15. Hrs open: 17 Box 930, 35902. Secondary address: 2725 Rainbow Dr., Rainbow City 35906. Phone: (256) 442-1222. Fax: (256) 442-1229. E-mail: feedback@wjby.com Web Site: www.wjby.net. Licensee: Gadsden Broadcasting Co. Inc. (acq 3-78). Population served: 200,000+ Natl. Network: USA. Format: Christian southern gospel. News: 15 hrs wkly. Target aud: 25-54. ♦Chris Stevens, gen mgr; Jim Tolbert, opns mgr; Mike Hooks, progmg dir.

WKXX(FM)—Attalla, Aug 31, 1991: 102.9 mhz; 1.1 kw. 702 ft TL: N33 58 28 W86 12 24. Stereo. Hrs opn: 24 100 Spurlock St., Rainbow City, 35906. Secondary address: Box 8405 35902. Phone: (256) 442-3944. Fax: (256) 442-7287. Web Site: www.wkxx.com. Licensee: Broadcast Media L.L.C. (acq 1-16-98; $650,000). Population served: 300,910 Wire Svc: AP Format: Hot adult contemp. News staff: one; News: 2 hrs wkly. Target aud: 18-49. ♦Pat Courington Jr., CEO; Tommy Lee, gen mgr.

WMGJ(AM)— Sept 11, 1985: 1240 khz; 1 kw-U. TL: N34 00 04 W86 01 48. Hrs open: 815 Tuscaloosa Ave., 35901. Phone: (256) 546-4434. Fax: (256) 546-9645. Web Site: www.wmgj.com. Licensee: Floyd L. Donald Broadcasting Co. Inc. Natl. Rep: Roslin. Format: Black, urban contemp. ♦Floyd L. Donald, gen mgr.

***WSGN(FM)**— Feb 11, 1975: 91.5 mhz; 6.3 kw. 520 ft TL: N34 04 29 W86 01 11. Stereo. Hrs opn: 24
Rebroadcasts WBHM(FM) Birmingham 80%.
Gadsden State Community College, 1001 George Wallace Dr., 35902. Phone: (256) 549-8439. E-mail: nmullin@gadsdenstate.edu Web Site: www.gadsdenstate.edu. Licensee: Gadsden State Community College. Population served: 250,000 Law Firm: Gardner, Carton & Douglas. Format: Class, news. News: 34 hrs wkly. Target aud: General. Spec prog: Folk 2 hrs, new age 10 hrs wkly. ♦Dr. Renee Culverhouse, pres; Neil D. Mullin, gen mgr.

***WTBB(FM)**— July 20, 1999: 89.9 mhz; 4.8 kw. Ant 515 ft TL: N34 06 03 W85 59 37. Stereo. Hrs opn: 24
Rebroadcasts WTBJ(FM) Oxford 100%.
Trinity Christian Academy, 1500 Airport Rd., Oxford, 36203. Phone: (256) 831-3333. Fax: (256) 831-5895. E-mail: truth@trinityoxford.org Web Site: www.trinityoxford.org. Licensee: Trinity Christian Academy. Population served: 500,000 Law Firm: Fletcher, Heald & Hildreth. Format: Christian, relg. Spec prog: Sp one hr wkly. ♦Dr. C.O. Grinstead, gen mgr.

Gardendale

WNCB(FM)— 1998: 97.3 mhz; 6.2 kw. Ant 1,325 ft TL: N33 29 04 W86 48 25. Hrs open: 24 301 Beacon Pkwy. W., Suite 200, Birmingham, 35209. Phone: (205) 916-1100. Fax: (205) 916-1151. Web Site: www.newcountry973.com. Licensee: Cox Radio Inc. Group owner: Cox Broadcasting (acq 6-8-99). Natl. Network: ABC. Natl. Rep: Katz Radio. Law Firm: Dow, Lohnes and Albertson. Format: New country. News staff: one; News: at 2 hrs wkly. ♦Ray Nelson, gen mgr; Justin Case, progmg dir.

Geneva

WGEA(AM)— Mar 17, 1953: 1150 khz; 1 kw-D, 35 w-N. TL: N31 01 21 W85 52 16. Hrs open: Box 339, 36340. Secondary address: 420 Riverside Ave. 36340. Phone: (334) 684-7079. Fax: (334) 684-0329. Licensee: Shelley Broadcasting Co. (acq 10-26-87). Population served: 23,647 Format: Country, gospel, news/talk. Target aud: 30 plus. ♦Jack Mizell, pres; Doc Parker, gen mgr.

WRJM-FM— Sept 12, 1969: 93.7 mhz; 100 kw. 853 ft TL: N31 02 42 W85 57 33. Hrs open: 24 285 E. Broad St., Ozark, 36360. Phone: (334) 774-7673. Fax: (334) 774-6450. E-mail: hjmizell@wrjm.com Web Site: www.wrjm.com. Licensee: Stage Door Development Inc. Population served: 790,000 Natl. Network: ABC, Westwood One. Natl. Rep: Rgnl Reps. Law Firm: John Borsari. Format: News/talk. Target aud: 30 plus. ♦Jack Mizell, pres & gen mgr; Susannah Hodges, stn mgr, opns mgr & gen sls mgr; Boyd Mizell, engrg mgr & engr.

Georgiana

WFXX(FM)— 1999: 107.7 mhz; 42 kw. 535 ft TL: N31 27 08 W86 37 07. Hrs open: 24 1406 River Falls St., Andalusia, 36420. Phone: (334) 222-2222. Fax: (334) 427-8888. E-mail: wfxx@alaweb.com Web Site: fox107.com. Licensee: Star Broadcasting Inc. (acq 3-29-2004; $975,000). Population served: 464,000 Format: Adult Contemp. ♦Jeffrey K. Haynes, pres; Kelly Haynes, gen mgr.

Glencoe

WGMZ(FM)—Licensed to Glencoe. See Gadsden

Goodwater

WZLM(FM)— Apr 4, 1990: 97.5 mhz; 5.1 kw. Ant 354 ft TL: N33 01 42 W85 59 23. Stereo. Hrs opn: 24 1621 Jacksons Gap Way, Jacksons Gap, 36861. Phone: (256) 825-4221. Fax: (256) 825-4221. Licensee: Great South Wireless LLC. (group owner; (acq 2-12-2007; grpsl). Population served: 250,000 Format: Adult contemp. News staff: one; News: 40 hrs wkly. Target aud: 25-54. ♦Joan Reynolds, pres.

Greensboro

WDGM(FM)— 03/03/2002: 99.1 mhz; 25 kw. Ant 328 ft TL: N32 49 46 W87 40 19. Hrs open: 24 142 Skyland Blvd., Tuscaloosa, 35405. Phone: (205) 345-7200. Fax: (205) 349-1715. Licensee: Citadel Broadcasting Co. (acq 7-12-2005; grpsl). Natl. Network: ABC, Jones Radio Networks. Natl. Rep: Roslin. Law Firm: Gardner, Carton & Douglas. Format: Oldies. Target aud: 25 plus; male & female. ♦Mike Kirtner, pres; Greg Thomas, progmg mgr.

Greenville

WGYV(AM)— Aug 18, 1948: 1380 khz; 1 kw-D. TL: N31 50 01 W85 52 16. Hrs open: Box 585, 1604 E. Commerce St., 36037. Phone: (334) 382-5444. Fax: (334) 382-5444. E-mail: wgyv@alaweb.com Licensee: Robert John Williamson (acq 11-29-02). Population served: 8,033 Format: News/talk, oldies. Target aud: 25-54; general. Spec prog: Black 6 hrs wkly. ♦Terry Golden, gen mgr; Paulette Golden, pub affrs dir; Bob Luman, chief of engrg.

WKXN(FM)— July 18, 1977: 95.9 mhz; 4 kw. 225 ft TL: N31 50 43 W86 38 56. (CP: 2.1 kw, ant 564 ft. TL: N31 56 52 W86 42 09). Hrs opn: 24
Simulcasts WKXK(FM) Pine Hill 100%.
Box 369, 36037. Secondary address: 563 Manningham Rd. 36037. Phone: (334) 382-6555. Fax: (334) 382-7770. E-mail: wkxn@wkxn.com Web Site: www.wkxn.com. Licensee: Autaugaville Radio Inc. (acq 11-22-94; $287,500; FTR: 1-2-95). Population served: 10,000 Format: Urban contemp, blues. ♦Roscoe Miller, gen mgr & stn mgr.

WQZX(FM)— Aug 19, 1985: 94.3 mhz; 3.9 kw. 410 ft TL: N31 54 40 W86 36 19. Stereo. Hrs opn: 24 205 W. Commerce, 36037. Phone: (334) 382-6633. Fax: (334) 382-6634. E-mail: q94@q94.net Web Site: www.q94.net. Licensee: Haynes Broadcasting Inc. Natl. Network: ABC. Format: Modern country. ♦Kyle Haynes, pres & gen mgr.

Grove Hill

WBMH(FM)— May 1999: 106.1 mhz; 12 kw. 472 ft TL: N31 43 30 W87 54 58. Hrs open: Box 518, c/o The Radio Center, Jackson, 36545. Phone: (251) 246-4431. Fax: (251) 246-1980. E-mail: bama1061@yahoo.com Licensee: Capital Assets Inc. Group owner: Bennie E. Hewett Stns. Format: Classic country. ♦Jay Braswell, gen mgr.

Gulf Shores

WCSN-FM—Orange Beach, July 2, 1996: 105.7 mhz; 5 kw. 246 ft TL: N30 17 45 W87 33 42. Stereo. Hrs opn: 24 Box 1919, 36547. Secondary address: 2421 E. Second St. 36542. Phone: (251) 967-1057. Fax: (251) 967-1050. E-mail: sunny105@gulftel.com Web Site: www.sunny105.com. Licensee: Gulf Coast Broadcasting Co. Inc. (acq 10-31-97). Population served: 60,000 Law Firm: Putbrese, Hunsaker & Trent. Format: Adult contemp. Target aud: 25-54; upscale. ♦R. Lee Hagan, pres; Bryant Ellis, gen sls mgr & pub affrs dir; Katie Tyler, prom dir; Ron Wainscott, progmg dir.

Guntersville

WGSV(AM)— Apr 16, 1950: 1270 khz; 1 kw-D. TL: N34 18 31 W86 17 44. Hrs open: Box 220, 35976. Phone: (256) 582-8131. Fax: (256)

582-4347. E-mail: wtwx@wtwx.com Web Site: www.wgsv.com. Licensee: Guntersville Broadcasting Co. Inc. Population served: 75,000 Format: News/talk. ♦Lavell Jackson, pres; Kerry Jackson, gen mgr & opns mgr.

WTWX-FM—Co-owned with WGSV(AM). Aug 1, 1969: 95.9 mhz; 10.7 kw. 596 ft TL: N34 20 14 W86 16 46. Phone: (256) 582-4946. Web Site: www.wtwx.com.85,000 Format: C&W.

***WJIA(FM)**— September 1995: 88.5 mhz; 2.2 kw. 426 ft TL: N34 25 33 W86 18 25. Hrs open: 5025 Spring Creek Dr., 35976. Phone: (256) 505-0885. Fax: (256) 505-0886. E-mail: jfm@wjia.org Web Site: www.wjia.org. Licensee: Lake City Educational Broadcasting Inc. Format: Christian. ♦Stan Broadus, gen mgr; Kevin Guffey, stn mgr.

Gurley

WXQW(FM)— June 8, 1995: 94.1 mhz; 710 w. Ant 945 ft TL: N34 40 50 W86 30 55. Stereo. Hrs opn: 24 1717 U.S. Hwy. 72 E., Athens, 35611. Phone: (256) 830-8300. Fax: (256) 232-6842. Web Site: www.whrpfm.com. Licensee: Cumulus Licensing LLC. Group owner: Clear Channel Communications Inc. (acq 4-4-2006; $3.3 million with WVNN-FM Trinity). Population served: 800,000 Format: Urban adult. News staff: 3; News: 4 hrs per wk. ♦Bill G. West, gen mgr.

Hackleburg

WFMH-FM— 1996: 95.5 mhz; 4.1 kw. Ant 400 ft TL: N34 18 38 W87 56 13. Stereo. Hrs opn: 24 16800 Hwy. 129, Brilliant, 35548. Phone: (205) 935-3730. Fax: (205) 935-3734. Licensee: Williams Communications. Group owner: Williams Communications Inc. (acq 8-18-2004; $2.45 million with WFMH(AM) Cullman). Population served: 65,000 Wire Svc: AP Format: Country. News: 15 hrs wkly. Target aud: 35-64. ♦Bryan Walker, gen mgr & progmg dir.

Haleyville

WJBB(AM)— Apr 1, 1949: 1230 khz; 1 kw-U. TL: N34 14 00 W87 37 32. Hrs open: 24 Drawer 370, 807 Hwy. 13 N., 35565. Phone: (205) 486-2277. Phone: (205) 486-2278. Fax: (205) 486-3905. E-mail: advertising@wjbbfm.com Licensee: Haleyville Broadcasting Co. Inc. (acq 1951). Population served: 87,000 Natl. Rep: Rgnl Reps. Law Firm: Fletcher, Heald & Hildreth, P.L.C. Format: Southern gospel, loc news. News staff: one; News: 36 hrs wkly. Target aud: 25-55; professionals. Spec prog: Farm 3 hrs wkly.John L. Slatton, pres & edit dir; Terry L. Slatton, gen mgr & opns VP; Debby Aderholt, dev dir & prom dir; Aubrey Haynes, sls VP, adv mgr, sports cmtr & disc jockey; Robert Wakefield, progmg dir & disc jockey; Larry Gardner, mus dir, pub affrs dir & disc jockey; Sherron Hayes, news dir & local news ed; Chester Barber, chief of engrg; Calabe Mayhall, disc jockey; Joseph Desiderio, disc jockey

WJBB-FM— July 14, 1979: 92.7 mhz; 3.9 kw. 240 ft TL: N34 14 00 W87 37 32. (CP: Ant 328 ft.). Stereo. E-mail: wjbb@southnet.net 118,000 Format: Country. Target aud: 24-55. ♦John Slatton, CEO, chmn & edit mgr; Terry Slatton, exec VP; Andy Marbutt, opns mgr, farm dir & disc jockey; Aubrey Haynes, sls dir & disc jockey; Sherron Hayes, local news ed; Keith Page, disc jockey; Larry Gardner, disc jockey; Misty Sawyer, disc jockey; Tracy Kinkead, disc jockey.

Hamilton

WERH(AM)— Aug 24, 1950: 970 khz; 5 kw-D. TL: N34 07 01 W87 59 29. Hrs open: Box 1119, 35570. Phone: (205) 921-3195. Fax: (205) 921-7187. E-mail: werh@sonet.net Licensee: Kate F. Fite. (acq 4-1-58). Population served: 100,000 Format: Country, gospel. Target aud: General. Spec prog: Farm. ♦James B. Fowler, gen mgr; Geraldine Miller, adv mgr; Bryan Williams, mus dir; Bill Moates, chief of engrg.

WERH-FM— Apr 1, 1968: 92.1 mhz; 3 kw. 120 ft TL: N34 07 01 W87 59 29. Stereo. 24 1597 Military St. S., 35570. Phone: (205) 921-3481.50,000 Format: Classic Rock. ♦Mark Burleson, news dir; Geraldine Miller, traf mgr.

Hanceville

WLYG(AM)— April 1986: 1170 khz; 460 w-D. TL: N34 04 28 W86 46 44. Hrs open: Sunrise-sunset 513 19th St. W., Jasper, 35501. Phone: (256) 352-1115. Fax: (205) 295-1238. Licensee: Joy Christian Communications Inc. (acq 3-16-2007). Format: Southern gospel. ♦Ralph Jolly, gen mgr.

Hartselle

WTAK-FM— August 1992: 106.1 mhz; 5.4 kw. 725 ft TL: N34 27 54 W86 38 36. Hrs open: 24 Box 21008, Huntsville, 25824. Secondary address: 26869 Peoples Rd., Madison 35756. Phone: (256) 353-1750. Fax: (256) 350-2653. Web Site: www.wtak.com. Licensee: Clear Channel Communications Group owner: Clear Channel Communications Inc. (acq 8-30-00; grpsl). Population served: 800,000 Format: Classic rock. ♦Rick Brown, gen mgr; Todd Berry, opns mgr; Carmelilta Palmer, gen sls mgr; Jerry James, progmg dir; Carl Ampieri, chief of engrg.

WYAM(AM)— Oct 1, 1956: 890 khz; 2.5 kw-D. TL: N34 34 00 W86 54 46. Stereo. Hrs opn: 12 1301 Central Pkwy. S.W., Decatur, 35601. Phone: (256) 355-4567. Fax: (256) 351-1234. E-mail: wileywg@acninc.net Licensee: Decatur Communications Properties LLC (acq 8-12-2003). Population served: 800,000 Format: Rgnl Mexican. News staff: 2; News: 6 hrs wkly. Target aud: 18-60; General. ♦William Wiley, pres & gen mgr.

Harvest

***WAYH(FM)**— 2003: 88.1 mhz; 3.5 kw. Ant 669 ft TL: N34 49 08 W86 44 19. Hrs open: 24 9582 Madison Blvd., Suite 8, Madison, 35758. Phone: (256) 837-9293. Fax: (256) 772-6731. E-mail: contact@wayfm.com Web Site: www.wayfm.com. Licensee: WAY-FM Media Group Inc. (group owner). Population served: 390,800 Format: Contemp Christian. Target aud: 18-34; youth & young adults. ♦Lloyd Parker, COO; Bob Augsburg, pres; Thom Ewing, gen mgr; Ace McKay, progmg dir.

Hazel Green

WBXR(AM)— Dec 11, 1970: 1140 khz; 15 kw-D, DA. TL: N34 57 18 W86 38 32. Hrs open: Sunrise-sunset 2926-D Huntsville Hwy., Fayetteville, TN, 37334. Phone: (931) 433-7017. Phone: (888) 570-7286. Fax: (931) 433-8282. E-mail: wbxr@wilkinsradio.com Web Site: wilkinsradio.com. Licensee: New England Communications Inc. (acq 9-16-97; $150,000). Population served: 1,000,000 Natl. Network: Salem Radio Network. Law Firm: Womble, Carlyle, Sandridge & Rice. Format: Christian teaching/talk. Target aud: 35 plus. ♦Robert L. Wilkins, pres; Mitchell Mathis, VP; Carla Payne, gen mgr & stn mgr; Greg Garrett, opns mgr; Kevin Kidd, chief of opns & engr; Don Roden, engr.

Headland

WDBT(FM)— September 1992: 105.3 mhz; 11.5 kw. 485 ft TL: N31 15 48 W85 18 24. Hrs open: 3245 Montgomery Hwy., Suite 1, Dothan, 36303. Phone: (334) 712-9233. Fax: (334) 712-0374. E-mail: ron@wdjr.com Web Site: www.1053thebeat.com. Licensee: Gulf South Communications Inc. (acq 1-27-97; $745,000). Format: Rhythm and blues. ♦Ron Eubanks, gen mgr; John Baker, opns mgr & progmg dir; Misty Huff, prom dir; Denise Reed, traf mgr & sls.

Heflin

***WKNG-FM**— May 2005: 89.1 mhz; 250 w. Ant 718 ft TL: N33 33 18 W85 27 25. Hrs open: 102 Parkwood Cir., Carrollton, GA, 30117. Phone: (770) 834-5477. Fax: (770) 830-1027. E-mail: steve1027@aol.com Licensee: Covenant Communications Inc. Format: Southern gospel. ♦Steven Gradick, pres.

***WPIL(FM)**— 2003: 91.7 mhz; 370 w. Ant 34 ft TL: N33 39 07 W85 31 13. Hrs open: 908 Opelika Rd., Auburn, 36830. Phone: (334) 821-0744. Fax: (334) 821-4031. Web Site: www.wpilfm.com. Licensee: Jimmy Jarrell Communications Foundation Inc. (acq 12-31-02). Format: Country/bluegrass/gospel. ♦Jimmy Jarrell, pres & gen mgr.

Hobson City

WHOG(AM)— Apr 15, 1991: 1120 khz; 500 w-D. TL: N33 36 50 W85 51 19. Hrs open: Sunrise-sunset 1330 Noble St., Suite 25, Anniston, 36201. Phone: (256) 236-6484. Fax: (256) 236-6484. E-mail: hog1120@aol.com Licensee: Hobson City Broadcasting Co. Population served: 200,000 Natl. Rep: Dora-Clayton. Format: Urban contemp. Target aud: General.

Homewood

WBPT(FM)—Licensed to Homewood. See Birmingham

Hoover

WENN(FM)— September 1993: 105.5 mhz; 29.5 kw. Ant 623 ft TL: N33 29 04 W86 48 25. Hrs open: 24 600 Beacon Pkwy. W., Suite 400, Birmingham, 35209. Phone: (205) 439-9600. Fax: (205) 439-8390. Licensee: Capstar TX L.P. Group owner: Clear Channel Communications Inc. (acq 8-30-2000; grpsl). Format: Rock. News staff: one. ♦Jimmy Vineyard, gen mgr.

Huntsville

WAHR(FM)— July 28, 1959: 99.1 mhz; 100 kw. 984 ft TL: N34 47 53 W86 38 24. Stereo. Hrs opn: 1555 The Boardwalk, Suite 1, 35814-5287. Phone: (256) 536-1568. Fax: (256) 536-4416. Web Site: www.star99.fm. Licensee: BCA Radio LLC. Group owner: Black Crow Media Group LLC (acq 11-15-2001; grpsl). Population served: 407,200 Format: Adult contemp. Target aud: 25-55. ♦Eric Jewell, gen mgr.

WBHP(AM)— May 23, 1937: 1230 khz; 1 kw-U. TL: N34 43 09 W86 35 42. Hrs open: Box 21008, 35824. Secondary address: 266869 Peoples Rd., Madison 35758. Phone: (256) 309-2400. Fax: (256) 350-2653. E-mail: rbrown@wbhp.com Web Site: www.wbhpam.com. Licensee: Capstar TX L.P. Group owner: Clear Channel Communications Inc. (acq 8-30-00; grpsl). Population served: 139,282 Format: Talk. Target aud: General. ♦Rick Brown, gen mgr.

WDJL(AM)— Oct 1, 1968: 1000 khz; 10 kw-D, DA. TL: N34 46 47 W86 39 16. Hrs open: Sunrise-sunset 2025 Sparkman Dr., Suite 3, 35810. Phone: (256) 852-1223. Fax: (256) 852-1900. Licensee: James K. Sharp dba 5th Avenue Broadcasting. (acq 8-95; $300,000). Population served: 450,000 Natl. Network: Westwood One. Law Firm: Jack Pennington & Associates. Format: Gospel. Target aud: 35-65; upscale decision makers that enjoy hits of the 40s, 50s & 60s. Spec prog: Gospel comedy 6 hrs wkly. ♦Walter Peavy, pres & gen mgr.

WEUP(AM)— 2001: 1700 khz; 10 kw-D, 1 kw-N. TL: N34 45 32 W86 38 35. Hrs open: 2609 Jordan Ln. N.W., 35816. Phone: (256) 837-9387. Fax: (256) 837-9404. E-mail: hundley@103weup.com Web Site: www.weupam.com. Licensee: Hundley Batts Sr. & Virginia Caples. Format: Gospel. ♦Hundley Batts, gen mgr.

WHIY(AM)— Mar 20, 1958: Stn currently dark. 1600 khz; 5 kw-D, 500 w-N, DA-D. TL: N34 45 32 W86 38 35. Stereo. Hrs opn: 2609 Jordan Ln. N.W., 35816. Phone: (256) 837-9387. Fax: (256) 837-9404. E-mail: hundley@103weup.com Web Site: www.weupam.com. Licensee: Hundley Batts Sr. & Virginia Caples. Population served: 325,500 ♦Hundley Batts Sr., pres; Hundley Batts, gen mgr.

***WJAB(FM)**— May 9, 1991: 90.9 mhz; 100 kw. 334 ft TL: N34 47 09 W86 34 00. Stereo. Hrs opn: 24 Alabama A&M University, Telecommunications Center, Box1687, Normal, 35762. Secondary address: 3409 Meridian St. 35811. Phone: (256) 372-5795. Phone: (256) 372-5861 (request line). Fax: (256) 372-5907. Web Site: www.aamu.edu/wjab. Licensee: Board of Trustees Alabama A&M University. Population served: 600,000 Natl. Network: NPR. Format: Jazz, blues, gospel. News: 5 hrs wkly. Spec prog: Black 3 hrs, oldies 3 hrs, reggae 4 hrs, Latin 2 hrs, gospel 5 hrs wkly. ♦Elizabeth Sloan-Ragland, gen mgr; Michael Burns, opns mgr.

WLOR(AM)— June 1948: 1550 khz; 50 kw-D, 500 w-N, DA-2. TL: N34 44 36 W86 35 39. Hrs open: 24 1555 The Boardwalk, Suite 1, 35814. Phone: (256) 536-1568. Fax: (256) 536-4416. E-mail: ed@jammin1550.am Web Site: www.jammin1550.am. Licensee: BCA Radio LLC. Group owner: Black Crow Media Group LLC (acq 11-15-2001; grpsl). Natl. Network: ABC. Format: Black, gospel, urban contemp. News staff: one; News: 1 hr wkly. Target aud: General. ♦Eric Jewell, gen mgr.

***WLRH(FM)**— Oct 13, 1976: 89.3 mhz; 100 kw. 810 ft TL: N34 37 41 W86 30 59. Stereo. Hrs opn: 24 UAH Campus, John Wright Dr., 35899. Phone: (256) 895-9574. Fax: (256) 830-4577. Web Site: www.wlrh.org. Licensee: Alabama ETV Commission. (acq 12-14-77). Population served: 600,000 Natl. Network: PRI, NPR. Format: Class, news, variety. News staff: one; News: 40 hrs wkly. Target aud: General. ♦George Dickerson, gen mgr & stn mgr; Jennifer Jaudon-Johnston, dev dir.

***WOCG(FM)**— Dec 1978: 90.1 mhz; 25 kw. 230 ft TL: N34 45 28 W86 39 44. Stereo. Hrs opn: 24 Oakwood College, 4920 University, Suite J, 35896. Phone: (256) 722-9990. Fax: (256) 726-7417. E-mail: wocg@wocg.org Web Site: www.wocg.net. Licensee: Oakwood College. Population served: 300,000 Natl. Network: USA. Law Firm: Donald E. Martin. Format: Inspirational, Christian, light urban gospel. News: 14 hrs wkly. Target aud: 34-55; Families with interest in rel & educ progmg. ♦Delbert Baker, pres; Bruce Peifer, VP; Victoria L. Miller, gen mgr & opns mgr; Jody Stennis, progmg dir.

WRSA-FM—See Decatur

WRTT-FM— Oct 6, 1960: 95.1 mhz; 50 kw. 110 ft TL: N34 42 56 W86 35 55. Stereo. Hrs opn: 24 1555 The Boardwalk, Suite 1, 35816. Phone: (256) 536-1568. Fax: (256) 536-4416. Web Site: www.rocket951.fm. Licensee: BCA Radio LLC. Group owner: Black Crow Media Group LLC (acq 11-15-2001; grpsl). Population served: 300500 Format: Mainstream Rock. News staff: one. Target aud: 24-54; male. Spec prog: Relg. ♦Eric Jewell, gen mgr; Jimbo Wood, progmg dir.

WTKI(AM)— November 1946: 1450 khz; 1 kw-U. TL: N34 43 30 W86 36 15. Hrs open: 24 7500 Memorial Pkwy. S.W., Suite 215C, 35802. Phone: (256) 533-1450. Fax: (256) 536-4349. Web Site: www.espn1450radio.com. Licensee: Mountain Mist Media LLC (acq 3-95). Population served: 250,000 Natl. Network: ABC, ESPN Radio, Westwood One. Format: Sports. News: Hourly updates. Target aud: 35 plus; sports-minded men and women, heavy local 18 plus, sports/news/business/Auburn affiliate, NFL. Spec prog: 6 hrs wkly. ♦David Barnhart, pres; Dennis Puent, sls dir; Matt Castleman, prom.

WUMP(AM)—See Madison

WZYP(FM)—See Athens

Irondale

WRLM(AM)— Dec 5, 1960: 1480 khz; 5 kw-D. TL: N33 32 54 W86 39 56. Hrs open: 24 235 Main St., Trussville, 35173. Phone: (205) 956-5470. Phone: (256) 390-2918. Fax: (866) 359-6303. Licensee: Davidson Media Station WLPH Licensee LLC. (acq 11-3-2006; $500,000). Population served: 1,200,000 Format: Sp. News staff: 3. ♦Debra Calhoun-George, gen mgr; Sylvia White, gen mgr & stn mgr; Tracey Hill, mus dir.

Jackson

WHOD(FM)—Listing follows WRJX(AM).

WRJX(AM)— June 1, 1950: 1230 khz; 1 kw-U. TL: N31 32 38 W87 52 30. (CP: 1190 khz, 10 kw-D, 300 w-N). Hrs opn: 5 AM-midnight Box 518, Hwy. 43 N., 36545. Secondary address: 4428 College Ave. 36545. Phone: (251) 246-4431. Phone: (251) 246-5581. Fax: (251) 246-1980. E-mail: radiocenter@starband.com Licensee: Capital Assets Inc. Group owner: Bennie E. Hewett Stations Population served: 6,500 Format: Black gospel. News staff: one. Target aud: 25-54; business minded, baby-boomers. ♦Shirley Chandler, sls VP; Paul McVay, progmg dir; Kelly Snell, traf mgr.

WHOD(FM)—Co-owned with WRJX(AM). Aug 1, 1964: 94.5 mhz; 30 kw. 640 ft TL: N31 28 59 W87 42 27.5 AM-midnight 100,000 Natl. Network: ABC. Format: Hot adult contemp. News staff: one; News: 2 hrs wkly.

Jacksonville

WCKA(AM)— January 1986: 810 khz; 50 kw-D, 500 w-N, DA-2. TL: N33 50 58 W85 45 46. Hrs open: 24 Box 8, Anniston, 36202. Phone: (256) 237-0810. Fax: (256) 782-2489. E-mail: alabama810@bellsouth.net Web Site: www.alabama810.com. Licensee: Alabama 810 LLC (acq 12-22-2005; $207,940). Population served: 3,234,000 Natl. Network: ABC. Law Firm: Nall, Estill. Format: Country. News staff: 2; News: 15 hrs wkly. Target aud: 25 plus; adults. Spec prog: Gospel 2 hrs wkly. ♦L. E. Gradick, exec VP & gen mgr; Teresa Goodman, gen mgr & stn mgr; Mike McGowan, progmg dir; Mike Mitchell, news dir.

***WLJS-FM**— Sept 29, 1975: 91.9 mhz; 3 kw. 246 ft TL: N33 49 29 W85 45 49. Stereo. Hrs opn: Box 3009, Self Hall @ Jacksonville State Univ., 36265. Phone: (256) 782-5300. Phone: (256) 782-5572. Fax: (256) 782-5645. Web Site: www.jsu.edu/92j. Licensee: Board of Trustees-Jacksonville State University. Population served: 90,000 Natl. Network: NPR. Law Firm: Gardner, Carton & Douglas. Format: Var/div, class. Target aud: 18-34; college, young adult. Spec prog: Relg 3 hrs wkly. ♦Mike Stedham, gen mgr; Chad Wells, prom dir.

Jasper

WDXB(FM)— Mar 28, 1962: 102.5 mhz; 79 kw. 2,096 ft TL: N33 28 51 W87 24 03. Stereo. Hrs opn: 600 Beacon Pkwy. W., Suite 400, Birmingham, 35209. Phone: (205) 439-9600. Fax: (205) 439-8390. Web Site: www.1025thebull.com. Licensee: Capstar TX L.P. Group owner: Clear Channel Communications Inc. (acq 8-30-00; grpsl). Population served: 400000 Format: Hit country. ♦Jimmy Vineyard, gen mgr.

WIXI(AM)— Nov 2, 1946: 1360 khz; 1 kw-D, 42 w-N. TL: N33 49 12 W87 16 26. Stereo. Hrs opn: 14 Box 622, 409 9th Ave., 35501. Phone: (205) 384-3461. Phone: (205) 384-3221. Fax: (205) 384-3462. E-mail: joy1360@wzpq.com Licensee: James T. Lee (acq 9-1-99). Population served: 16,000 Law Firm: Fletcher, Heald & Hildreth. Format: Christian, Southern gospel. News: 4 hrs wkly. Target aud: General. Spec prog: Gospel 6 hrs wkly. ♦Ed Smith, gen mgr; Joe Cook, stn mgr.

WLYJ(AM)— Mar 1, 1957: 1240 khz; 1 kw-U. TL: N33 48 54 W87 16 19. Hrs open: 24 513 19th St. W, 35501-5357. Phone: (205) 221-2222. Phone: (256) 927-4027. Fax: (256) 927-4028. Licensee: Joy Christian Communications Inc. (acq 7-9-2004; $200,000). Population served: 70,000 Format: Christian progmg. ♦Ed Smith, pres & gen mgr; Roy Pounds, stn mgr.

Jemison

WHPH(FM)— May 15, 1953: 97.7 mhz; 3.1 kw. Ant 459 ft TL: N32 58 12 W86 43 04. Hrs open: 24 6930 Cahaba Valley Rd., Suite 202, Birmingham, 35242. Phone: (205) 755-0980. E-mail: info@peach97.com Web Site: www.peach97.com. Licensee: Great South Wireless LLC. (group owner; (acq 2-12-2007;. grpsl). Format: Oldies. ♦Steven Salter, gen mgr.

Level Plains

WIRB(AM)—Not on air, target date: unknown: 1490 khz; 430 w-U. TL: N31 17 51 W85 47 33. Hrs open: 422 County Rd. 551, New Brockton, 36351. Phone: (334) 894-5047. Fax: (334) 894-6684. Licensee: Virgle Leon Strictland, individually. ♦Virgle Leon Strictland, gen mgr.

Lexington

WJHX(AM)— Feb 20, 1981: 620 khz; 5 kw-D, 99 w-N. TL: N34 58 37 W87 22 10. Hrs open: 1426 5th Ave. S.E., Decatur, 35601. Phone: (256) 353-5959. Fax: (256) 353-5951. Licensee: BAR Broadcasting Inc. (acq 12-30-2005). Rgnl. Network: Alabama Net. Format: Sp. ♦Pedro Zamora, pres; Moifes Gomez, gen mgr.

Linden

WINL(FM)— April 1991: 98.5 mhz; 100 kw. 817 ft TL: N32 07 34 W87 44 02. Stereo. Hrs opn: 24 Box 938, Demopolis, 36732. Phone: (334) 289-9850. Fax: (334) 289-9811. Web Site: www.bestcountryaround.com. Licensee: West Alabama Communications Inc. (acq 2-12-2001; $1.28 million). Population served: 225,000 Natl. Network: ABC. Natl. Rep: Dora-Clayton. Format: Country. Target aud: 25-54. Spec prog: Gospel 5 hrs, farm 10 hrs wkly. ♦Amy Douglas, gen mgr & opns mgr.

WNPT-FM— Dec 19, 1990: 102.9 mhz; 40 kw. Ant 551 ft TL: N32 27 40 W87 34 50. Stereo. Hrs opn: 24 Box 2000, Tuscaloosa, 35403. Phone: (205) 758-5523. Fax: (205) 752-9696. E-mail: wtbc@dbtech.net Licensee: John Sisty Enterprises Inc. (acq 11-28-2003; $450,000). Population served: 75,000 Format: Classic country. ♦John Sisty, pres; Ronnie Quarles, gen mgr.

Lineville

WZZX(AM)— 1967: 780 khz; 5 kw-D. TL: N33 17 04 W85 47 24. Hrs open: Box 26, Ashland, 36251. Phone: (256) 354-4600. Fax: (256) 354-7224. E-mail: wasz@acs-isp.com Licensee: Williams Communications Inc. (acq 7-25-2002; $2.88 million). Population served: 20,000 Format: Country. ♦Walt Williams Jr., gen mgr & pres; Jason Thompson, progmg dir.

Lisman

WPRN-FM— 1997: 107.7 mhz; 6 kw. Ant 328 ft TL: N32 05 27 W88 13 57. Hrs open: 909 W. Pushmataha St., Butler, 36904. Phone: (205) 459-3222. Fax: (205) 459-4140. Licensee: Butler Broadcasting Corp. (acq 11-5-99). Format: Country. ♦Daryl Jackson, gen mgr.

Littleville

WLAY-FM— Sept 12, 1986: 103.5 mhz; 3.3 kw. Ant 386 ft TL: N34 40 27 W87 42 48. Stereo. Hrs open: 24 273 Azalea Rd., Suite 1-308, Mobile, 36609. Phone: (251) 343-4900. Fax: (251) 343-4905. Web Site: www.mix1035.com. Licensee: Urban Radio Licenses LLC. Group owner: Clear Channel Communications Inc. (acq 5-13-2005; grpsl). Population served: 180,000 Law Firm: Fletcher, Heald & Hildreth. Format: Oldies. News staff: one. Target aud: 18-49. ♦Kevin Wagner, CEO & pres; Todd Mannesses, opns mgr; Laura Crosby, news dir & pub affrs dir; Rob Green, chief of engrg.

Livingston

WSLY(FM)—See York

WYLS(AM)—See York

Luverne

WHLW(FM)— 1997: 104.3 mhz; 13.5 kw. Ant 1,830 ft TL: N31 58 28 W86 09 44. Hrs open: Box 4420, Montgomery, 36103. Secondary address: 203 Gunn Rd., Montgomery 36117. Phone: (334) 274-6464. Fax: (334) 274-6465. Web Site: www1043hallelujah.com. Licensee: Capstar TX L.P. Group owner: Clear Channel Communications Inc. (acq 8-30-2000; grpsl). Format: Gospel. Target aud: 18-49. ♦James Belton, gen mgr; Michael Long, opns mgr; Holly Schatz, sls dir; Nikita Pogue, prom dir; Kenny Smith, progmg mgr.

Madison

WUMP(AM)— Mar 29, 1983: 730 khz; 1 kw-D, 123 w-N. TL: N34 41 46 W86 44 19. Hrs open: 1717 Hwy. 72 E., Athens, 35611. Phone: (256) 830-8300. Fax: (256) 232-6842. Web Site: www.730ump.com. Licensee: Cumulus Licensing LLC. Group owner: Cumulus Media Inc. (acq 7-21-2003; grpsl). Natl. Rep: Katz Radio. Format: Sports. News staff: 3; News: 3 hrs wkly. Target aud: 18-54; males. Spec prog: Univ. of Alabama Sports. ♦Bill West, gen mgr; Brian Pitts, gen sls mgr; Randy Frawley, gen sls mgr & progmg dir.

Marion

WJUS(AM)— Dec 8, 1951: 1310 khz; 5 kw-D. TL: N32 38 04 W87 17 48. Hrs open: WJUS Radio Station, Hwy. 5, 36756. Phone: (334) 683-2043. Phone: (334) 872-8400. Fax: (334) 872-2329. Licensee: Marion Radio Inc. (acq 6-86; $115,000; FTR: 6-30-86). Population served: 4,289 Rgnl. Network: Tobacco. Format: Urban contemp. ♦Rev. Glenn King, gen mgr.

Midfield

WBHJ(FM)—Licensed to Midfield. See Birmingham

Millbrook

WWMG(FM)— Aug 1, 1993: 97.1 mhz; 3 kw. 328 ft TL: N32 25 58 W86 20 07. Hrs open: 24 Box 4420, Montgomery, 36103. Secondary address: 203 Gunn Rd., Montgomery 36117. Phone: (334) 274-6464. Fax: (334) 274-6465. Web Site: www.mymagic97.com. Licensee: Capstar TX L.P. Group owner: Clear Channel Communications Inc. (acq 8-30-2000; grpsl). Natl. Network: American Urban, Westwood One. Format: Adult contemp. Target aud: 25-54. ♦James Belton, gen mgr & stn mgr; Michael Long, opns mgr; Alberta Jackson, rgnl sls mgr; Nikita Pogue, prom dir; Darryl Elliott, progmg dir; Cynthia Mallard, news dir.

Mobile

WABB-FM— Feb 5, 1973: 97.5 mhz; 100 kw. Ant 1,551 ft TL: N30 41 20 W87 49 49. Stereo. Hrs opn: 24 Box 2148, 36652-2148. Secondary address: 1551 Springhill Ave. 36604. Phone: (251) 432-5572. Fax: (251) 438-4044. E-mail: b.dittman@wabb.com Web Site: www.wabb.com. Licensee: WABB-FM Inc. (acq 8-30-2000; grpsl). Population served: 1,092,100 Natl. Rep: Christal. Format: Top-40. ♦Bernard Dittman, pres & gen mgr; Laura English, sls dir.

WABB(AM)— November 1948: 1480 khz; 5 kw-U, DA-N. TL: N30 43 11 W88 04 16.24 Box 2148, 36652. Secondary address: 1551 Springhill Ave. 36604. Web Site: www.wabb.com. Group owner: Dittman Group Inc. 303,900 Natl. Rep: Christal. Format: News/talk. Target aud: 18-49.

WAVH(FM)—See Daphne

***WBHY-FM**— Mar 20, 1992: 88.5 mhz; 33 kw. 624 ft TL: N30 40 55 W87 49 41. Stereo. Hrs opn: 24 Box 1328, 36633-1328. Secondary address: 6530 Spanish Fort Blvd., Suite B, Spanish 36527. Phone: (251) 473-8488. Web Site: www.goforth.org. Licensee: Goforth Media Inc. (group owner; acq 6-27-90; FTR: 7-30-90). Population served: 1,000,000 Law Firm: Fletcher, Heald & Hildreth. Format: Contemp Christian music. News: 7 hrs wkly. Target aud: 18-34. ♦Robert Barber, CEO & opns mgr; Charles Smith, pres & gen sls mgr; Wilbur Goforth, gen mgr & opns mgr; Steve Riggs, chief of engrg; Jean Williams, traf mgr.

WBHY(AM)— Dec 9, 1943: 840 khz; 10 kw-D. TL: N30 45 50 W88 06 36. Web Site: www.goforth.org. (Acq 4-11-86).500,000 Natl. Rep: Salem. Format: Christian. Target aud: 34-64; Christians. ♦Jean Williams, traf mgr.

WBLX-FM—Listing follows WDLT(AM).

WDLT(AM)—Fairhope, Apr 22, 1965: 660 khz; 10 kw-D, 850 w-N, DA-N. TL: N30 42 27 W88 03 55. Hrs open: 24 2800 Douphin St., Suite 104, 36606. Phone: (251) 652-2000. Fax: (251) 652-2007. E-mail: carmen.brown@cumulus.com Web Site: www.983wdlt.com. Licensee: Cumulus Licensing Corp. Group owner: Cumulus Media Inc. (acq 10-18-99; grpsl). Format: News radio. News staff: one. Target aud: 25 plus; Black adults. ♦Steve Sandman, gen mgr; James Alexander, opns mgr & progmg dir; Vinny Duncan, prom dir.

WBLX-FM—Co-owned with WDLT(AM). April 1976: 92.9 mhz; 98 kw. 1,555 ft TL: N30 37 35 W87 38 50. Stereo. Web Site: www.thebigstation9361x.com. Natl. Network: ABC. Format: Urban contemp. Target aud: 12 plus; primarily Black women, 18-34.

WDLT-FM—Chickasaw, 1980: 98.3 mhz; 40 kw. 548 ft TL: N30 35 05 W88 15 57. Stereo. Hrs opn: 24 2800 Dauphin St., Suite 104, 36606. Phone: (251) 652-2000. Fax: (251) 652-2007. E-mail: mobile.prog@cumulus.com Web Site: www.smooth98.com. Licensee: April Broadcasting Inc. Group owner: Cumulus Media Inc. (acq 10-18-99; grpsl). Format: Adult Contemp. News staff: one; News: 5 hrs wkly. Target aud: 25-54. Spec prog: Jazz 5 hrs, blues 18 hrs, pub affrs 4 hrs wkly. ♦Gary Pizzati, gen mgr; Steve Crumbely, opns mgr.

WGOK(AM)— Nov 21, 1958: 900 khz; 1 kw-D, 381 w-N, DA-2. TL: N30 42 27 W88 03 55. Stereo. Hrs opn: 2800 Dauphin St., Suite 104, 36606. Phone: (251) 652-2000. Fax: (251) 652-2001. Web Site: www.cumulus.com. Licensee: Cumulus Licensing Corp. Group owner: Cumulus Media Inc. (acq 10-18-99; $6 million with WYOK(FM) Atmore). Population served: 350,000 Natl. Rep: Roslin. Format: Gospel. Target aud: 18-54; Black adults. ♦Dickie Roberts, pres; Kevin Wagner, stn mgr & gen sls mgr; Danny Wright, prom dir.

***WHIL-FM**— 1979: 91.3 mhz; 100 kw. 1,066 ft TL: N30 41 20 W87 49 49. Stereo. Hrs opn: 24 Box 8509, 36689-0509. Secondary address: 4000 Dauphin St. 36608. Phone: (251) 380-4685. Fax: (251) 460-2189. E-mail: whil@whil.org Web Site: www.whil.org. Licensee: Spring Hill College. Population served: 782,000 Natl. Network: PRI, NPR. Law Firm: Dow, Lohnes & Albertson. Format: Classical/npr. News: 40 hrs wkly. Target aud: 35 plus. ♦Mario Mazza, gen mgr; Kurt Garrett, opns dir; Marlene Buckner, dev dir; Kris Pierce, progmg dir.

WIJD(AM)—See Prichard

WKSJ-FM— Apr 12, 1971: 94.9 mhz; 100 kw. 410 ft TL: N30 35 36 W87 39 40. (CP: Ant 1,554 ft. TL: N30 37 35 W87 38 50). Stereo. Hrs opn: 555 Broadcast Dr., 3rd Fl., 36606. Phone: (251) 450-0100. Fax: (251) 479-3418. Web Site: www.95ksj.com. Licensee: CC Licenses LLC. Group owner: Clear Channel Communications Inc. (acq 11-21-97; grpsl). Population served: 313,000 Rgnl rep: David Coppock Format: Contemp country. News staff: 2; News: 25 hrs wkly. Target aud: 25-54; mid level to high class country music listeners. ♦David Coppock, VP & gen mgr; Jeanie Hufford, stn mgr; Bo Clark, gen sls mgr; Bill Black, progmg dir; Mike Sloan, news dir.

WLPR(AM)—Prichard, Dec 31, 1986: 960 khz; 5 kw-U, DA-N. TL: N30 45 50 W88 06 36. Stereo. Hrs opn: 24 Box 1328, 36633-1328. Phone: (251) 473-8488. Web Site: www.goforth.org. Licensee: Goforth Media Inc. (acq 1994). Natl. Network: Salem Radio Network. Natl. Rep: Salem. Law Firm: Fletcher, Heald & Hildreth. Format: Southern gospel. News: 6 hrs wkly. Target aud: 35 plus. ♦Wilbur Goforth, gen mgr; Robert Barber, opns dir & engrg VP; Charlie Smith, gen sls mgr; Kenny Fowler, mus dir; Steve Riggs, chief of engrg; Jean Williams, traf mgr.

WLVV(AM)— Feb 7, 1930: 1410 khz; 5 kw-U, DA-N. TL: N30 40 52 W88 00 02. Hrs open: 24 1263 Battleship Pkwy., Spanish Fort, 36527. Phone: (251) 626-1090. Fax: (251) 626-1099. E-mail: wlvv@bellsouth.net Licensee: WLVV Inc. Group owner: Martin Broadcasting Inc. (acq 4-14-99; $263,750). Natl. Network: American Urban. Natl. Rep: Katz Radio. Format: Gospel, MOR. Target aud: 18-44; Black. ♦Tom Alexander, gen mgr, gen sls mgr & progmg dir.

WMOB(AM)— Jan 25, 1961: 1360 khz; 5 kw-D, 212 w-N, DA-2. TL: N30 41 26 W88 01 33. Hrs open: 24 Box 63, 36601. Secondary address: 200 Addsco Rd. Causeway 36601. Phone: (251) 432-1360. Fax: (251) 432-1396. Licensee: Buddy Tucker Association Inc. (acq 4-84; $395,000; FTR: 4-9-84). Format: Relg. ♦Theodore Tucker, pres; LeVaughn Tucker, VP; Buddy Tucker, gen mgr; Don Tucker, opns mgr.

WMXC(FM)—Listing follows WNTM(AM).

WNTM(AM)— 1946: 710 khz; 1 kw-D, 500 w-N. TL: N30 43 13 W88 03 34. Hrs open: Box 161489, 36616. Secondary address: 555 Broadcast Dr. 36606. Phone: (251) 450-0100. Fax: (251) 479-3418. Licensee: CC Licenses LLC. Group owner: Clear Channel Communications Inc. (acq 11-21-97; grpsl). Population served: 375,000 Natl. Network: CBS. Law Firm: Leventhal, Senter & Lerman. Format: News/talk, sports. Target aud: 25 plus. ♦Dave Cappock, pres; Ronnie Bloodworth, gen mgr; Scott O'Brien, progmg dir; Bill King, news dir.

WMXC(FM)—Co-owned with WNTM(AM). Oct 16, 1947: 99.9 mhz; 100 kw. Ant 1,755 ft TL: N30 41 20 W87 49 49. Stereo. 3,180,000 Format: Adult contemp. Target aud: 25-54. ♦Dan Mason, progmg dir.

WRKH(FM)— Dec 5, 1964: 96.1 mhz; 100 kw. 1,342 ft TL: N30 41 20 W87 49 49. Stereo. Hrs opn: 24 555 Broadcast Dr., 3rd Fl., 36606. Phone: (251) 450-0100. Phone: (251) 770-9600. Fax: (251) 479-3418. Web Site: www.961therocket.com. Licensee: CC Licenses LLC. Group owner: Clear Channel Communications Inc. (acq 11-21-97; grpsl). Natl. Rep: D & R Radio. Format: Classic rock hits. News staff: 2; News: 2 hrs wkly. Target aud: 25-49; front edge baby boomers. ♦David Coppock, VP & gen mgr; Jeanie Hufford, stn mgr; Steve Powers, progmg dir.

Monroeville

WEZZ(AM)— Dec 6, 1982: 930 khz; 5 kw-D, 48 w-N. TL: N31 29 40 W87 21 29. Hrs open: 873 S. Alabama Ave., 36460. Phone: (251) 575-7601. Fax: (251)-575-7703. Licensee: Brantley Broadcast Associates LLC (acq 6-6-2007; $36,500). Format: Gospel. ♦Wendy Smith, gen mgr.

***WJIK(FM)**—Not on air, target date: unknown: 89.3 mhz; 3 kw. Ant 435 ft TL: N31 53 28 W87 42 45. Hrs open: CSN International, 4002 N. 3300 E., Twin Falls, ID, 83301. Phone: (208) 734-6633. Fax: (208) 736-1958. Web Site: www.csnradio.com. Licensee: CSN International. ♦Michael Kestler, pres.

WMFC(AM)— April 1952: 1360 khz; 780 w-D. TL: N31 30 51 W87 17 55. Hrs open: Box 645, 36461. Secondary address: 961 Pineville Rd. 36460. Phone: (251) 575-3281. Phone: (251) 575-4061. Fax: (251) 575-3280. Licensee: Monroe Broadcasting Co. Inc. Population served: 18,000 Wire Svc: NOAA Weather Format: Black gospel. Target aud: 25-54. ♦Carolyn Stewart, chmn; David Stewart, pres, stn mgr & gen sls mgr; Carol Casey, progmg dir & chief of engrg.

WMFC-FM— December 1965: 99.3 mhz; 30 kw. Ant 308 ft TL: N31 30 51 W87 17 55. Stereo. 5 AM-midnight 38,450 Natl. Network: Jones Radio Networks. Format: Good time oldies. Target aud: General.

Montgomery

WACV(AM)— Jan 16, 1939: 1170 khz; 10 kw-D, 1 kw-N, DA-2. TL: N32 27 16 W86 17 21. Hrs open: 24 4101 Wal St., 36106. Phone: (334) 244-0961. Fax: (334) 279-9563. Web Site: www.wacv1170am.com. Licensee: Bluewater Broadcasting Co. LLC (group owner; acq 4-21-2004; grpsl). Population served: 250,000 Natl. Network: CBS. Format: News/talk, sports. News staff: one. Target aud: 25 plus. Spec prog: Farm 5 hrs wkly. ♦Terry Barber, gen mgr; Rick Peters, progmg dir.

WBAM-FM— Jan 1, 1961: 98.9 mhz; 100 kw. 1095 ft TL: N32 24 11 W86 11 48. Stereo. Hrs opn: 24 4101-A Wall St., 36106. Phone: (334) 244-0961. Fax: (334) 279-9563. Web Site: www.bamacountry989.com. Licensee: Bluewater Broadcasting Co. LLC (group owner; acq 4-19-2004). Population served: 133,386 Natl. Rep: Christal. Format: Country. News staff: one; News: one hr wkly. Target aud: 12-54. ♦Terry Barber, gen mgr; John Norris, progmg dir.

WHHY-FM— Jan 9, 1962: 101.9 mhz; 100 kw. 1,200 ft TL: N32 29 33 W86 08 50. (CP: TL: N32 24 11 W86 11 48). Stereo. Hrs opn: One Commerce St., Suite 300, 36104. Phone: (334) 240-9274. Fax: (334) 240-9219. Web Site: www.y102montgomery.com. Licensee: Cumulus Licensing Corp. Group owner: Cumulus Media Inc. (acq 3-12-01; grpsl). Population served: 133,386 Natl. Rep: McGavren Guild. Format: CHR. News: one hr wkly. Target aud: 18-40. ♦Lew Dickey, pres; Bernie Barker, gen mgr & natl sls mgr; Bill Jones, opns mgr; Don Parden, sls dir; Donna Headley, gen sls mgr; Joy Melton, gen sls mgr & news dir; Heather Williams, prom dir & prom mgr; Karen Rite, progmg dir; Larry Wilkins, chief of engrg; Crystal Palmer Lund, traf mgr.

WLWI(AM)—Co-owned with WHHY-FM. Apr 30, 1930: 1440 khz; 5 kw-D, 1 kw-N, DA-N. TL: N32 18 24 W86 13 40. (CP: TL: N32 24 11 W86 11 48). Fax: (334) 240-9211. Web Site: www.cumulus.com.133386 Natl. Network: CNN Radio. Format: News/talk. News staff: one. Target aud: 18 plus. ♦Bernie Barker, mktg mgr; Steve Smith, progmg dir; Gwen Pierce, traf mgr.

***WLBF(FM)**— Apr 4, 1984: 89.1 mhz; 100 kw. 537 ft TL: N32 24 13 W86 11 50. Stereo. Hrs opn: 24 Box 210789, 36121-0789. Secondary address: 381 Mendel Parkway 36117. Phone: (334) 271-8900. Fax: (334) 260-8962. E-mail: mail@faithradio.org Web Site: www.faithradio.org. Licensee: Faith Broadcasting Inc. Population served: 500,000 Natl. Network: Moody, USA. Law Firm: Southmayd & Miller. Format: Educ, relg, MOR. News: 14 hrs wkly. Target aud: General. ♦Russell Dean, gen mgr; Gary Hundley, dev dir; Donna Spears, traf mgr.

WLWI-FM—Listing follows WMSP(AM).

WMGY(AM)— June 1, 1946: 800 khz; 1 kw-D, 193 w-N. TL: N32 24 48 W86 17 25. Hrs open: 6 AM-midnight 2305 Upper Wetumpka Rd., 36107-1345. Phone: (334) 834-3710. Fax: (334) 834-3711. E-mail: davewmgy@aol.com Licensee: WMGY Radio Inc. Group owner: GHB Radio Group (acq 7-75). Population served: 200,000 Natl. Network: USA. Format: Southern gospel. News: 7 hrs wkly. Target aud: 35 plus. Spec prog: Black 15 hrs, sports 6 hrs wkly. ♦Dane Harris, gen mgr.

WMSP(AM)— 1953: 740 khz; 50 kw-D, 73 w-N, DA-2. TL: N32 18 39 W86 17 25. Hrs open: One Commerce St., Suite 300, 36104. Phone: (334) 240-9274. Fax: (334) 240-9219. Web Site: www.sportsradio740.com. Licensee: Cumulus Licensing Corp. Group owner: Cumulus Media Inc. (acq 12-12-98; grpsl). Natl. Network: ESPN Radio. Law Firm: Gardner, Carton & Douglas. Format: Sports. Target aud: Adults 18 plus. ♦Lew Dickey, pres; Bernie Barker, gen mgr & mktg mgr; Bill Jones, opns mgr; Donna Hadley, gen sls mgr; Joy Melton, natl sls mgr; Jason King, prom dir; Bob Wooddy, progmg dir; Larry Wilkins, chief of engrg; Barry McKnight, sports cmtr; John Longshore, sports cmtr.

WLWI-FM—Co-owned with WMSP(AM). July 15, 1969: 92.3 mhz; 100 kw. 1,095 ft TL: N32 24 13 W86 11 50. (CP: TL: N32 24 11 W86 11 48). Stereo. Web Site: www.wlwi.com.225,000 Natl. Network: CNN Radio. Format: Country. News staff: one. Target aud: 25-54. Spec prog: Gospel 4 hrs wkly. ♦Bill Dollar, progmg dir; Darlene Dixon, mus dir & disc jockey; Marcus Hyles, news dir; Gwen Pierce, traf mgr; Barry McKnight, sports cmtr; John Longshore, sports cmtr; Andi Scott, disc jockey; Don Day, disc jockey.

WMXS(FM)—Listing follows WNZZ(AM).

WNZZ(AM)— May 8, 1953: 950 khz; 1 kw-U, DA-N. TL: N32 26 23 W86 15 49. Hrs open: One Commerce St., Suite 300, 36104. Phone: (334) 240-9274. Fax: (334) 240-9219. Web Site: www.cumulus.com. Licensee: Cumulus Licensing Corp. Group owner: Cumulus Media Inc. (acq 12-12-98; grpsl). Population served: 133,386 Law Firm: Gardner, Carton & Douglas. Format: American standards, news. News staff: one. ♦Lew Dickey, pres; Bernie Barker, gen mgr & mktg mgr; Bill Jones, opns dir; Jesica Garrard, gen sls mgr; Richard Reinhardt, gen sls mgr; Bob Wooddy, progmg dir; Larry Wilkins, chief of engrg.

WMXS(FM)—Co-owned with WNZZ(AM). July 9, 1961: 103.3 mhz; 100 kw. 1,007 ft TL: N32 24 48 W86 17 25. (CP: TL: N32 24 11 W86 11 48). Stereo. Web Site: www.mix103.com. Natl. Network: CNN Radio. Format: Adult contemp. Target aud: 25-54. ♦Brian Roberts, progmg dir; Marcus Hyles, news dir; J.T. Thompson, disc jockey; Jay St. John, disc jockey; Leanne Thompson, disc jockey.

WQKS-FM— Dec 1, 1990: 96.1 mhz; 4.5 kw. 820 ft TL: N32 21 32 W86 19 57. (CP: 900 w, ant 820 ft. TL: N32 22 03 W86 15 42). Stereo. Hrs opn: 24 4101 A-Wall St., 36106. Phone: (334) 244-0961. Fax: (334) 279-9563. Web Site: www.alice961.com. Licensee: Bluewater Broadcasting Co. LLC (group owner; acq 4-21-2004; grpsl). Natl. Network: ABC. Format: Rockin Hits of the 70s 80s & 90s. News: one hr wkly. Target aud: 25-54. ♦Terry Barber, gen mgr & gen sls mgr; Rick Peters, progmg dir; Tom Jones, chief of engrg; Mary Brazell, traf mgr.

***WVAS(FM)**— June 15, 1984: 90.7 mhz; 80 kw. 347 ft TL: N32 21 58 W86 17 40. Stereo. Hrs opn: 18 (M-F); 24 (S, Su) Alabama State Univ., 915 S. Jackson St., 36101-0271. Phone: (334) 229-4708. Fax: (334) 269-4995. Web Site: www.alasu.edu/wvas. Licensee: Alabama State University. (acq 6-83). Law Firm: Wilkes, Artis, Hedrick & Lane. Format: Smooth jazz, mellow vocals. News staff: 2; News: 5 hrs wkly. Target aud: General; African-American community. Spec prog: Black 5 hrs, gospel 5 hrs, blues 6 hrs, news/talk 4 hrs wkly. ♦John S. Knight, gen mgr; Candy Capel, stn mgr.

WXVI(AM)— May 1947: 1600 khz; 5 kw-D, 1 kw-N, DA-2. TL: N32 23 40 W86 17 21. Hrs open: 24 912 South Perry St., 36104. Phone: (334) 263-4141. Fax: (334) 263-9191. Licensee: New Life Ministries Inc. (acq 9-8-2005). Natl. Network: American Urban. Natl. Rep: Roslin. Format: Christian. Target aud: 35 plus; urban. ♦Terry Ellison, CEO; Glenda Perkins, progmg dir.

Montgomery-Troy

***WTSU(FM)**— Mar 1, 1977: 89.9 mhz; 100 kw. Ant 754 ft TL: N32 03 40 W86 01 19. Stereo. Hrs opn: 24
Rebroadcasts WTJB(FM) Columbus 100%.
Wallace Hall, Troy State Univ., Troy, 36082. Phone: (334) 670-3268. Fax: (334) 670-3934. E-mail: wtsu@troyst.edu Web Site: wtsu.troyst.edu. Licensee: Troy State University. Population served: 500,000 Natl. Network: NPR, PRI. Format: News. Classical. News: 25 hrs wkly. Target aud: General. Spec prog: Children one hr wkly. ♦James Clower, gen mgr; Judy Davis, opns mgr; Fred Azbell, progmg dir & progmg mgr.

Moody

WURL(AM)— October 1984: 760 khz; 1 kw-D. TL: N33 35 13 W86 28 18. Hrs open: 2999 Radio Park Dr., 35004. Phone: (205) 699-9875. Fax: (205) 640-4379. E-mail: wurlradio@aol.com Web Site: www.wurlradio.com. Licensee: Bill Davison Evangelistic Assn. (acq 9-89; $175,000; FTR: 10-2-89). Natl. Network: USA. Format: Gospel. Target aud: General. ♦William J. Davison Sr., pres, gen mgr & gen sls mgr.

Moulton

WEUP-FM— Sept 1, 1991: 103.1 mhz; 6 kw. 328 ft TL: N34 32 07 W87 13 31. Stereo. Hrs opn: 20 2609 Jordon Ln. N.W., Huntsville, 35816. Phone: (256) 837-9387. Web Site: 103weup.com. Licensee: Hundley Batts Sr. and Virginia Caples (acq 6-7-99; $775,000 with co-located AM). Population served: 30000 Natl. Network: USA. Natl. Rep: Rgnl Reps. Wire Svc: NOAA Weather Format: Urban, Hip-Hop. News staff: one. Target aud: 21-55. Spec prog: Relg one hr wkly. ♦Huntley Batts Sr., gen mgr; Huntley Batts, gen sls mgr; Big Ant, progmg dir.

WEUV(AM)—Co-owned with WEUP-FM. Dec 11, 1963: 1190 khz; 2.5 kw-D. TL: N34 28 55 W87 18 04. Box 37, 13471 Court St., 35650. Phone: (256) 974-0681. Phone: (256) 974-0682. Format: Soul gospel. Target aud: General. ♦Steve Murry, progmg dir.

Muscle Shoals

WBCF(AM)—See Florence

WLAY(AM)— Jan 15, 1933: 1450 khz; 1 kw-U. TL: N34 45 23 W87 41 08. Stereo. Hrs opn: 509 N. Main St., Tusumbia, 35674. Phone: (256) 383-2525. Fax: (256) 389-1912. E-mail: donnajohnson @clearchannel.com Web Site: www.wlayfm.com. Licensee: Urban Radio Licenses LLC. Group owner: Clear Channel Communications Inc. (acq 5-13-2005; grpsl). Population served: 135,000 Law Firm: M. Scott Johnson. Format: Sports. Target aud: 18-54. ♦Brian Rickman, opns mgr; Cheryl Self, gen sls mgr.

WVNA-FM—Co-owned with WLAY(AM). Oct 28, 1964: 105.5 mhz; 1.05 w. 741 ft TL: N34 40 24 W87 42 56. Stereo. Format: Classic rock.

***WQPR(FM)**— November 1987: 88.7 mhz; 20 kw. Ant 430 ft TL: N34 34 41 W87 47 02. (CP: Ant 429 ft). Stereo. Hrs opn: 24
Rebroadcasts WUAL-FM Tuscaloosa 95%.
Phifer Annex, Suite 166, Tuscaloosa, 35487. Phone: (205) 348-6644. Fax: (205) 348-6648. E-mail: apr@apr.org Web Site: www.apr.org. Licensee: Board of Trustees University of Alabama. Population served: 80,000 Natl. Network: PRI, NPR. Law Firm: Arter & Hadden. Format: Class, jazz, news & info. News staff: one; News: 5 hrs wkly. Spec prog: Bluegrass, blues, folk 5 hrs, new age 19 hrs wkly. ♦Elizabeth Brock, stn mgr.

WSBM(AM)—See Florence

WXFL(FM)—See Florence

Northport

WJOX(FM)— July 15, 1991: 100.5 mhz; 85 kw. Ant 912 ft TL: N33 05 42 W87 15 16. (CP: COL Helena. 93 kw, ant 1,014 ft). Hrs opn: 24 244 Goodwin Crest Dr., Suite 300, Birmingham, 35209. Phone: (205) 945-4646. Fax: (205) 945-3999. Web Site: www.wjox1005fm.com. Licensee: Citadel Broadcasting Co. (group owner; (acq 7-12-2005; grpsl). Natl. Network: ESPN Radio. Format: Sports talk. ♦Dale Daniels, gen mgr; Lenny Frisaro, gen sls mgr; Jennifer Dickson, prom dir; Ryan Haney, progmg dir.

***WSJL(FM)**—Not on air, target date: unknown: 88.1 mhz; 10 w horiz 20 kw vert. Ant 492 ft TL: N33 28 51 W87 24 03. Hrs open: 1115 Honeysuckle Dr., Keene, TX, 76059. Phone: (817) 641-3495. Licensee: Mary V. Harris Foundation. ♦Linda De Romanett, pres & gen mgr.

WTUG-FM—Licensed to Northport. See Tuscaloosa

Oneonta

WCRL(AM)— July 29, 1952: 1570 khz; 2.5 kw-D. TL: N33 57 16 W87 28 20. Hrs open: Box 490, 35121. Secondary address: 908 2nd Ave. E. 35121. Phone: (205) 625-3333. Fax: (205) 625-5433. E-mail: wkld@wkld.com Web Site: www.wkld.com. Licensee: Blount County Broadcasting Service Inc. (acq 9-19-02). Population served: 45,000 Natl. Network: Jones Radio Networks. Format: Hispanic 60s. ♦Danny Bentley, gen mgr; L.D. Bentley, pres & farm dir.

WKLD(FM)—Co-owned with WCRL(AM). July 12, 1968: 97.7 mhz; 3.2 kw. Ant 367 ft TL: N33 56 48 W86 29 06. Web Site: www.wkld.com.200,000 Format: Country. Spec prog: Atlanta Braves baseball. ♦L.D. Bentley, farm dir; Danny Bentley, disc jockey; James R. Bentley, disc jockey.

Opelika

WANI(AM)— June 3, 1940: 1400 khz; 1 kw-U. TL: N32 38 13 W85 24 23. Hrs open: 24 Box 950, Auburn, 36831-0950. Secondary address: 197 E. University Dr., Auburn 36830. Phone: (334) 826-2929. Fax: (334) 826-9151. Web Site: www.wani1400.com. Licensee: Auburn Network Inc. (acq 11-7-97; $135,000). Population served: 120,000 Format: News/talk. News staff: one. ♦Mike Hubbard, pres; Andy Burcham, gen mgr.

WKKR(FM)—See Auburn

WMXA(FM)—Listing follows WTLM(AM).

WTLM(AM)— Aug 12, 1968: 1520 khz; 1 kw-D. TL: N32 39 13 W85 25 25. Hrs open: Sunrise-sunset Box 2329, 36803-2329. Secondary address: 915 Veterans Pkwy. 36801. Phone: (334) 745-4656. Fax: (334) 749-1520. E-mail: jimpowell@qantumofauburn.com Web Site: oaadvertising.com. Licensee: Qantum of Auburn License Co. LLC. Group owner: Qantum Communications Corp. (acq 7-2-03; grpsl). Population served: 100,000 Law Firm: Gardner, Carton & Douglas. Format: Adult standards. News staff: one. Target aud: 35 plus. ♦Jim Powell, gen mgr & mktg mgr; Sandy Matthews, sls dir; Woody Russ, progmg dir.

WMXA(FM)—Co-owned with WTLM(AM). July 1, 1991: 96.7 mhz; 3.5 w. 430 ft TL: N32 33 54 W85 22 13. Stereo. 24 Phone: (334) 745-2067. Web Site: oaadvertising.com.115,000 Format: Adult contemp. News staff: one; News: 2 hrs wkly. Target aud: 18-49.

WZMG(AM)—Pepperell, Oct 1, 1979: 910 khz; 650 w-D, 56 w-N, DA-1. TL: N32 56 30 W85 24 50. Hrs open: 24 Box 2329, Quantum Communications, 36803. Secondary address: 915 Veterns Pkwy. 36803. Phone: (334) 745-4656. Fax: (334) 745-2067. Licensee: Qantum of Auburn License Co. LLC. Group owner: Qantum Communications Corp. (acq 7-2-03; grpsl). Population served: 100,000 Natl. Network: ABC. Format: Urban. Target aud: 25-54; African Americans. ♦Jim Powell, mktg mgr & news dir.

Opp

WAMI(AM)— Dec 12, 1952: 860 khz; 1 kw-D, 47 w-N. TL: N31 18 54 W86 15 45. Stereo. Hrs opn: Box 40, 36467. Phone: (334) 493-3588. Fax: (334) 493-4182. E-mail: wami@alaweb.com Web Site: www.wami.com. Licensee: Opp Broadcasting Co. Inc. Population served: 150,000 Natl. Network: ABC. Format: Country classic. Target aud: 25-45; agricultural & garment industry workers. Spec prog: Gospel 15 hrs wkly. ♦Harry Phillips, gen mgr.

WAMI-FM— Nov 9, 1973: 102.3 mhz; 3.4 kw. 230 ft TL: N31 18 54 W86 15 45. Stereo. 175,000

***WJIF(FM)**— 1986: 91.9 mhz; 380 w. 164 ft TL: N31 15 50 W86 13 26. Hrs open: 700 Hwy. 52, 36467. Phone: (334) 493-4947. Fax: (334) 493-4947. Licensee: Opp Educational Broadcasting Foundation. Format: Southern gospel. ♦Heywood Nyland, gen mgr.

WOPP(AM)— Sept 19, 1980: 1290 khz; 2.5 kw-D, 500 w-N, DA-2. TL: N31 17 27 W86 13 51. Stereo. Hrs opn: 24 1101 Cameron Rd., 36467-2407. Phone: (334) 493-4545. Phone: (334) 493-1035. Fax: (334) 493-4546. E-mail: wopp@wopp.com Web Site: www.wopp.com. Licensee: E & R Broadcasting Inc. (acq 8-87). Population served: 67,000 Natl. Network: Salem Radio Network. Rgnl rep: Rgnl Reps. Law Firm: Roy F. Perkins. Format: Progsv C&W, oldies mix. News staff: one; News: 16 hrs wkly. Target aud: 19-58; progsv & highly loc. Spec prog: Farm 2 hrs, Black 4 hrs, gospel 19 hrs wkly. ♦Robert Boothe, gen mgr; Ronnie Boothe, traf mgr & engr.

Orange Beach

WCSN-FM—Licensed to Orange Beach. See Gulf Shores

Orrville

WJAM-FM— August 1994: 107.9 mhz; 3.7 kw. Ant 410 ft TL: N32 21 38 W87 09 12. Hrs open: Box 1150, Selma, 36702-1150. Phone: (334) 875-9360. Fax: (334) 875-1340. E-mail: walx@charterinternet.com Licensee: Scott Communications Inc. (acq 10-13-94). Format: Adult urban contemp. ♦Scott Alexander, gen mgr.

Oxford

***WTBJ(FM)**— May 29, 1994: 91.3 mhz; 170 w. Ant 1,578 ft TL: N33 29 07 W85 48 33. Stereo. Hrs opn: 24 c/o Trinity Christian Academy, 1500 Airport Rd., 36203. Phone: (256) 831-3333. Fax: (256) 831-5895. E-mail: truth@trinityoxford.org Web Site: www.trinityoxford.org. Licensee: Trinity Christian Academy. Population served: 500,000 Law Firm: Fletcher, Heald & Hildreth. Format: Educ, relg. Target aud: General. Spec prog: Sp one hr wkly. ♦Dr. C.O. Grinstead, gen mgr.

WVOK(AM)— April 1956: 1580 khz; 2.5 kw-D, 22 w-N. TL: N33 26 55 W86 03 54. Hrs open: 24 Box 3770, 36203. Secondary address: 1215

Church St. 36203. Phone: (256) 835-1580. Fax: (256) 831-1500. Web Site: www.k98.fm. Licensee: Woodard Broadcasting Co. (acq 5-62). Population served: 40,000 Natl. Network: ABC. Format: Oldies. Target aud: 25-54. ♦Chuck Woodard, gen mgr; Chris Wright, rgnl sls mgr; Whit McGhee, progmg dir.

WVOK-FM— Feb 19, 1990: 97.9 mhz; 280 kw. 1,082 ft TL: N33 37 21 W85 52 22. Stereo. 24 Web Site: www.k98.fm. Format: Hot adult contemp. ♦Chuck Woodward, gen sls mgr.

Ozark

***WAQG(FM)**— June 1998: 91.7 mhz; 3 kw. 321 ft TL: N31 26 25 W85 33 49. Hrs open: Box 3206, American Family Radio, Tupelo, MS, 38803. Phone: (601) 844-8888. Fax: (601) 842-6791. Web Site: www.afr.net. Licensee: American Family Radio. (group owner) Format: Inspirational Christian. ♦Marvin Sanders, gen mgr; John Riley, progmg dir; Joe Moody, engrg dir.

WJRL-FM— Oct 5, 1968: 103.9 mhz; 25 kw. Ant 292 ft TL: N31 26 25 W85 33 49. Stereo. Hrs opn: 24 285 N. Foster St., Dothan, 36302. Phone: (334) 792-0047. Fax: (334) 712-9346. Licensee: Magic Broadcasting Alabama Licensing LLC. (acq 8-1-2002; $750,000 with co-located AM). Natl. Network: Moody. Format: Classic hits. News staff: one; News: 15 hrs wkly. Spec prog: Gospel 8 hrs, jazz 4 hrs, oldies 6 hrs wkly. ♦Dan Bradley, gen mgr.

WOAB(FM)—Listing follows WOZK(AM).

WOZK(AM)— May 3, 1953: 900 khz; 1 kw-D, 78 w-N. TL: N31 27 19 W85 40 58. Hrs open: Box 1109, 36361. Phone: (334) 774-5600. Fax: (334) 774-1148. E-mail: wozk@alaweb.com Licensee: Ozark Broadcasting Corp. Population served: 100,000 Format: Adult standard. ♦John Stein, gen mgr.

WOAB(FM)—Co-owned with WOZK(AM). July 9, 1967: 104.9 mhz; 6.0 kw. 275 ft TL: N31 27 19 W85 40 53. Format: Country.

WQLS(AM)— April 1968: 1210 khz; 10 kw-D, 3 w-N, 5 kw-CH. TL: N31 28 40 W85 41 07. Hrs open: Box 250, 36360. Licensee: Horizon Broadcasting Co. (group owner; (acq 1-16-2007; $125,000). Format: Gospel. Target aud: 25-64.

Pell City

WFHK(AM)— Jan 7, 1956: 1430 khz; 5 kw-D. TL: N33 35 10 W86 19 35. Hrs open: 6 AM-6 PM 22 Cogswell Ave., 35125. Phone: (205) 338-1430. Fax: (205) 814-1430. Licensee: Stocks Broadcasting Inc. (acq 2-27-01; $275,000). Population served: 35,000 Rgnl. Network: Alaska Radio Net. Format: Country. ♦John Simpson, gen mgr.

Pepperell

WZMG(AM)—Licensed to Pepperell. See Opelika

Phenix City

WDAK(AM)—See Columbus, GA

WGSY(FM)—Licensed to Phenix City. See Columbus GA

WHAL(AM)—Licensed to Phenix City. See Columbus GA

Piedmont

***WJCK(FM)**— April 1994: 88.3 mhz; 6 kw. 328 ft TL: N34 04 11 W85 14 48. Hrs open: 24 9423 Hwy. 21 N., 36272. Phone: (770) 387-0917. Fax: (770) 387-2856. E-mail: webmaster@ibn.org Web Site: www.ibn.org. Licensee: Immanuel Broadcasting Network. Format: Relg. ♦Ed Tuten, pres & gen mgr; Jane Tuten, VP; Neil Hopper, gen mgr; Jackson Kiruke, progmg dir; Jimmy Hardy, mus dir; Phillip Baker, chief of engrg.

WPID(AM)— June 1953: 1280 khz; 1 kw-D, 84 w-N. TL: N33 55 50 W85 35 00. Hrs open: 6 AM-10 PM 412 Cedartown Hwy., 36272. Phone: (256) 447-9096. Fax: (256) 447-6669. Licensee: Piedmont Radio Co. (acq 6-15-84; $125,000). Population served: 90,000 Format: Adult contemp, oldies. News: 2 hrs wkly. Target aud: 25-55. ♦Jimmy Kennedy, gen mgr; Andy Kennedy, opns mgr.

Pine Hill

WKXK(FM)— 2000: 96.7 mhz; 9 kw. 544 ft TL: N32 04 24 W87 35 27. Hrs open:
Simulcast of WKXN(FM) Greenville 100%.
Box 369, Greenville, 36037. Secondary address: 563 Manningham Rd., Greenville 36037. Phone: (334) 382-6555. Fax: (334) 382-7770. E-mail: wkxn@alaweb.com Web Site: www.wkxn.com. Licensee: Autaugaville Radio Inc. Format: Urban contemp, blues, gospel. ♦Roscoe Miller, gen mgr & stn mgr.

Prattville

WIQR(AM)— March 1969: 1410 khz; 5 kw-D, 1 kw-N, DA-2. TL: N32 25 23 W86 26 21. Hrs open: Box 6226, Laurel, MS, 39441-6226. Phone: (334) 491-9477. E-mail: wiqr@hotmail.com Licensee: Star Power Communications Corp. (acq 2-1-2001; $167,000). Rgnl rep: Alabama Net. Format: Local sports. Target aud: General. ♦Greg Meadows, gen mgr.

WXFX(FM)— August 1977: 95.1 mhz; 50 kw. 492 ft TL: N32 28 01 W86 24 15. Stereo. Hrs opn: 24 1 Commerce St., Suite 300, Montgomery, 36104. Phone: (334) 240-9274. Fax: (334) 240-9219. Web Site: www.wxfx.com. Licensee: Cumulus Licensing Corp. Group owner: Cumulus Media Inc. (acq 3-12-01; grpsl). Population served: 280,000 Natl. Network: CNN Radio, Motor Racing Net. Natl. Rep: Katz Radio. Format: Classic rock. News staff: one. Target aud: 25-54; upscale adults, two paycheck households. ♦Bernie Barker, gen mgr; Bill Jones, opns mgr; Donna Headley, gen sls mgr; Rick Hendricks, progmg dir; Larry Wilkins, chief of engrg.

Priceville

WQAH(AM)— August 1986: 1310 khz; 1 kw-D. TL: N34 32 32 W86 54 14. Hrs open: Sunrise-sunset Box 150, Decatur, 35602. Secondary address: 303 2nd Ave. S.E., Decatur 35601. Phone: (256) 353-4060. Fax: (256) 773-6915. Licensee: Abercrombie Broadcasting AM Inc. (acq 8-86). Population served: 343,500 Format: Southern gospel. Target aud: 25-65. ♦Percy Yarbrough, gen mgr.

Prichard

WIJD(AM)— June 13, 1966: 1270 khz; 5 kw-D, 103 w-N. TL: N30 44 44 W88 05 40. Hrs open: 24 555 Broadcast Dr., 3rd Flr., Mobile, 36606. Phone: (251) 471-1208. Fax: (251) 471-1244. E-mail: mglin@aol.com Web Site: www.wijd.us. Licensee: 1270 AM Inc. (acq 5-16-2003; $100,000). Population served: 313,000 Format: Christian. Target aud: 35 plus. ♦David Coppock, gen mgr.

WKSJ-FM—See Mobile

WLPR(AM)—Licensed to Prichard. See Mobile

Rainbow City

WJBY(AM)—Licensed to Rainbow City. See Gadsden

Rainsville

WVSM(AM)— May 16, 1967: 1500 khz; 1 kw-D. TL: N34 29 56 W85 50 34. Hrs open: Sunrise-sunset Box 339, 368 McCurdy Ave. N., 35986. Phone: (256) 638-2137. E-mail: wvsm@farmerstel.com Web Site: www.wvsm.net. Licensee: Sand Mountain Advertising Co. Inc. Population served: 60,000 Format: Southern gospel. News staff: 3; News: 9 hrs wkly. Target aud: General. ♦Kayron Guffey, VP; Annie Ruth Huber, gen sls mgr; Ann Spears, disc jockey; Jesse Finley, disc jockey; Mark Huber, pres, gen mgr, progmg dir & disc jockey.

Red Bay

WRMG(AM)— June 29, 1968: 1430 khz; 1 kw-D. TL: N34 24 51 W88 08 11. (CP: 3 kw-D). Hrs opn: Box 656, 35582. Phone: (256) 356-4458. Licensee: Jack W. Ivy Sr. (acq 1-9-2002; $42,300). Population served: 10,000 Format: Country, Southern gospel, Bluegrass. ♦Jack W. Ivy Sr., pres & progmg dir.

Reform

WBEI(FM)— May 7, 1991: 101.7 mhz; 22.5 kw. Ant 725 ft TL: N33 13 48 W87 50 50. Hrs open: 24 142 Skyland Blvd., Tuscaloosa, 35405. Phone: (205) 345-7200. Fax: (205) 349-1715. Licensee: Citadel Broadcasting Co. (group owner; (acq 7-12-2005; grpsl). Population served: 120,000 Format: Adult contemp. Target aud: 18-34. ♦Brenda Bebout, gen mgr.

Repton

WTID(FM)— Oct 1, 2002: 101.1 mhz; 3.1 kw. Ant 459 ft TL: N31 26 45 W87 16 59. Hrs open: 24 415 N. College St., Greenville, 36037-2005. Phone: (251) 575-7601. Fax: (251) 575-7703. E-mail: fun101@frontiernet.net Licensee: Great South Wireless LLC. (group owner; (acq 2-12-2007; grpsl). Format: Lite rock. ♦Wendy Smith, gen mgr, opns mgr & progmg dir; Robert Williams, chief of engrg.

Roanoke

WELR(AM)— April 1954: 1360 khz; 1 kw-D. TL: N33 09 45 W85 22 30. Hrs open: Box 710, 6855 Hwy. 431, 36274. Phone: (334) 863-4139. Fax: (334) 863-2540. E-mail: jim@eagle1023.com Web Site: www.eagle1023.com. Licensee: Eagle's Nest Inc. (group owner; acq 10-15-88). Population served: 25,000 Law Firm: Gardner, Carton & Douglas. Format: Sports. ♦Jim Vice, pres & gen mgr; Coleman Vice, gen sls mgr; Kay Vice, opns mgr & prom mgr.

WELR-FM— Feb 14, 1969: 102.3 mhz; 3 kw horiz, 1.25 kw vert. 436 ft TL: N33 13 14 W85 24 37. (CP: 9 kw, ant 544 ft.). Stereo. 24 304 Broome St., LaGrange, 30240. Web Site: www.eagle1023.com.1,000,000 Format: Country.

Robertsdale

WNSI(AM)— Mar 1, 1985: 1000 khz; 1 kw-D. TL: N30 32 10 W87 42 55. Hrs open: 6 am-6 pm
Simulcast with WNSI-FM Atmore 100%.
Box 578, 36567. Phone: (251) 947-2346. Fax: (251) 947-2347. E-mail: wnsiradio@gulftel.com Web Site: www.wnsiradio.com. Licensee: Great American Radio Network Inc. (acq 3-16-2001; $180,000). Population served: 40,000 Format: Sports, talk. News staff: one; News: 2 hrs wkly. Target aud: 45 plus; upscale. Spec prog: Religion 6 hrs wkly. ♦Walter Bowen, gen mgr.

Rogersville

WYTK(FM)— January 1994: 93.9 mhz; 2.25 kw. 531 ft TL: N34 51 52 W87 23 43. Hrs open: 24 Box 146, Florence, 35631. Phone: (256) 764-9390. Fax: (256) 764-7760. E-mail: the score@bellsouth.net Web Site: www.939thescore.com. Licensee: Valley Broadcasting Inc. (acq 9-18-02; $900,000). Population served: 450,000 Format: Sports. Target aud: 25-54; male. ♦Greg Thornton, pres & gen mgr; Al Mann, opns mgr.

Russellville

WGOL(AM)— May 29, 1949: 920 khz; 1 kw-D, 43 w-N. TL: N34 30 50 W87 42 55. Hrs open: 16 113 N. Washington Ave., 35653. Phone: (256) 332-0214. Fax: (256) 332-7430. Licensee: Pilati Investments Corp. (acq 8-31-2005; $171,500). Population served: 30,000 Rgnl. Network: Alaska Radio Net. Format: Country. News: 9 hrs wkly. Target aud: 25-60. Spec prog: Black gospel 4 hrs, relg 10 hrs wkly. ♦John Pilati, stn mgr.

WKAX(AM)— Apr 3, 1974: 1500 khz; 1 kw-D. TL: N34 31 42 W87 42 41. Hrs open: 113 Washington Ave. N.W., 35653. Phone: (256) 332-6103. Fax: (256) 332-7430. Licensee: Jamar Communications Inc. (acq 1999; $65,000). Population served: 30,000 Format: Southern gospel, Sp. Target aud: 21-54. Spec prog: Black 4 hrs wkly. ♦Marshall R. Moore, pres & gen mgr.

Saraland

WHOA(AM)—Not on air, target date: unknown: 770 khz; 38 kw-D, 800 w-N, DA-2. TL: N30 51 39 W88 05 37. Hrs open: 6930 Cahaba Valley Rd., Suite 202, Birmingham, 35242. Phone: (205) 618-2020. Fax: (205) 618-2029. Licensee: Brantley Broadcast Associates LLC. (acq 5-8-2007; $100 for CP). ♦Paul Reynolds, gen mgr.

Scottsboro

WKEA-FM— Nov 3, 1965: 98.3 mhz; 6 kw. 531 ft TL: N34 34 50 W85 47 30. Stereo. Hrs opn: 24 19784 John T. Reid Pkwy., 35768. Phone: (256) 259-2341. Fax: (256) 574-2156. E-mail: ron@wkeafm.com Web Site: www.wkeafm.com. Licensee: KEA Radio Inc. (group owner)

Population served: 125,000 Format: Country. News staff: one; News: 2 hrs wkly. Target aud: 25-54. Spec prog: Farm one hr, relg 4 hrs wkly. ♦Gene Sisk, VP; Ronald H. Livengood, CEO, pres & gen mgr; Campbell Smith, opns mgr.

WWIC(AM)— June 13, 1950: 1050 khz; 1 kw-D, 101 w-N. TL: N34 40 23 W86 03 11. Hrs open: 24 Box 759, 35768. Phone: (256) 259-1050. Fax: (256) 575-2411. E-mail: wwic@scottsboro.org Web Site: www.wwicradio.com. Licensee: Scottsboro Broadcasting Co. Inc. (acq 1-14-2005; $88,306 for 50%). Population served: 15,000 Format: Classic country, sports. ♦Greg Bell, pres & gen mgr.

WZCT(AM)— June 11, 1952: 1330 khz; 5 kw-D, 38 w-N. TL: N34 42 07 W86 00 15. Hrs open: 24 1111 E.Willow St., 35768. Phone: (256) 574-1330. Fax: (256) 218-3013. Licensee: Bonner and Carlile Enterprises. (acq 2-28-90). Population served: 175,000 Natl. Network: Reach Satellite, USA. Format: Southern gospel. News: 14 hrs wkly. Target aud: 25 plus. Spec prog: Sports 19 hrs wkly. ♦Rob Carlile, gen mgr, stn mgr, opns mgr & gen sls mgr.

Selma

WALX(FM)—Listing follows WMRK(AM).

***WAPR(FM)**— May 5, 1996: 88.3 mhz; 1.85 kw horiz, 53 kw vert. 1,401 ft TL: N32 08 30 W86 44 43. Hrs open:
Rebroadcasts WUAL-FM Tuscaloosa 100%.
Phifer Hall Annex, Suite 166, Tuscaloosa, 35487. Phone: (205) 348-6644. Fax: (205) 348-6648. E-mail: apr@apr.org Web Site: www.apr.org. Licensee: Ua-Asu-Tsu Educational Radio Corp. Population served: 30,000 Natl. Network: NPR, PRI. Format: Class, jazz, news. ♦Roger Duvall, pres; Elizabeth Brock, stn mgr.

***WAQU(FM)**— March 1998: 91.1 mhz; 21.5 kw. Ant 335 ft TL: N32 24 17 W87 25 32. Hrs open: Box 3206, American Family Radio, Tupelo, MS, 38803. Phone: (662) 844-8888. Fax: (662) 842-6791. Web Site: www.afr.net. Licensee: American Family Association. Group owner: American Family Radio Format: Inspirational Christian. ♦Don Wildmon, CEO; Tim Wildmon, exec VP; Marvin Sanders, gen mgr; John Riley, progmg dir; Fred Jackson, news dir; Shan Easterling, chief of engrg.

WBFZ(FM)— 2001: 105.3 mhz; 50 kw. Ant 492 ft TL: N32 16 18 W87 15 28. Hrs open: 24 P.O. Box 369, 36702. Phone: (334) 872-2177. Fax: (334) 872-5577. Licensee: Inami Communications Corp. Inc. Format: Urban contemp. ♦Henry Sanders, pres; Charles Jones, gen mgr; Derriet Moore, progmg dir.

WDXX(FM)—Listing follows WHBB(AM).

WHBB(AM)— Nov 11, 1935: 1490 khz; 1 kw-U. TL: N32 26 02 W87 00 40. Hrs open: 24 Box 1055, 36702. Secondary address: 505 Lauderdale St. 36701. Phone: (334) 875-3350. Fax: (334) 874-6959. E-mail: info@wdxx.com Web Site: www.wdxx.com. Licensee: Broadsouth Communications Inc. (acq 7-24-92; $400,000 with co-located FM; FTR: 8-17-92) Population served: 27,379 Natl. Rep: Rgnl Reps. Format: News/talk. News staff: one; News: 13 hrs wkly. Target aud: 25-54. Spec prog: Black 18 hrs, farm 10 hrs wkly. ♦Mike Reynolds, gen mgr; Evelyn Ogle, gen sls mgr; George Henry, progmg dir.

WDXX(FM)—Co-owned with WHBB(AM). September 1965: 100.1 mhz; 50 kw. 288 ft TL: N32 26 02 W87 00 40. Stereo. 24 Format: Country. News staff: 2; News: 5 hrs wkly.

WMRK(AM)— Dec 19, 1946: 1340 khz; 1 kw-U. TL: N32 25 31 W86 59 47. Hrs open: Box 1150, 36702. Phone: (334) 875-9360. Fax: (334) 875-1340. E-mail: walt@bellsouth.net Licensee: Scott Communications Inc. (acq 12-30-2005; $29,500 for 47.2% of stock with co-located FM). Format: Oldies. Target aud: 25-49. ♦Betty Alexander, mus dir; Scott Alexander, pres, gen mgr, progmg dir & chief of engrg.

WALX(FM)—Co-owned with WMRK(AM). Dec 12, 1973: 100.9 mhz; 50 kw. Ant 492 ft TL: N32 21 40 W86 52 28. (CP: COL Orrville). Stereo. 27,397 Format: Hot adult contemp. Target aud: 18-40.

***WRNF(FM)**—Not on air, target date: unknown: 89.5 mhz; 6 kw vert. Ant 328 ft TL: N32 32 50 W86 55 33. Hrs open: The Moody Bible Institute of Chicago, 820 N. LaSalle St., Chicago, IL, 60610-3214. Phone: (312) 329-4301. Fax: (312) 329-8980. Licensee: The Moody Bible Institute of Chicago. ♦Robert C. Neff, VP & gen mgr.

Sheffield

***WAKD(FM)**— 1996: 89.9 mhz; 1 kw. 125 ft TL: N34 44 25 W87 42 58. Hrs open: Box 3206, Tupelo, MS, 38803. Phone: (662) 844-8888. Fax: (662) 842-6791. Web Site: www.afr.net. Licensee: American Family Association. Group owner: American Family Radio Format: Inspirational Christian. ♦Marvin Sanders, gen mgr; John Riley, progmg dir; Joey Moody, chief of engrg.

WBTG(AM)— Nov 6, 1963: 1290 khz; 1 kw-D, 79 w-N. TL: N34 46 27 W87 40 14. Hrs open: 24 Box 518, 35660. Secondary address: 1605 Gospel Rd. 35660. Phone: (256) 381-6800. Fax: (256) 381-6801. E-mail: announcements@wbtgradio.com Web Site: www.wbtgradio.com. Licensee: Slatton & Associates. (acq 12-17-87). Population served: 150,000 Natl. Network: Salem Radio Network. Format: Christian, talk, relg. Target aud: 25 up; conservative, mainstream family audience. ♦Paul Slatton, pres & gen mgr; Dan Michaels, progmg dir; Orvil Nichols, mus dir; Scott Carrier, prom dir & traf mgr; Gary Dobbs, local news ed; Ray Wright, sports cmtr.

WBTG-FM— July 2, 1969: 106.3 mhz; 6 kw. 682 ft TL: N34 41 34 W87 47 49. Stereo. 24 Phone: (205) 381-6800. Web Site: www.wbtgradio.com. (Acq 1-17-78). Format: Southern gospel. News: 12 hrs wkly. ♦Joyce Slatton, VP & mktg dir; Chuck Bradford, prom dir; Jerry Eagil, mus dir; Keith Balch, pub affrs dir; Scott Carrier, traf mgr.

Smiths

WBFA(FM)— 1998: 101.3 mhz; 6 kw. 328 ft TL: N32 25 35 W85 08 20. Hrs open: Box 687, Columbus, GA, 31902. Secondary address: 1501 13th Ave., Columbus, GA 31901. Phone: (706) 576-3000. Fax: (706) 576-3005. Web Site: 1013thebeat.com. Licensee: CC Licenses LLC. Group owner: Clear Channel Communications Inc. (acq 2-21-2002; grpsl). Format: Urban mainstream. ♦Jim Martin, gen mgr; Brian Waters, opns mgr.

Stevenson

WMXN-FM— June 13, 1977: 101.7 mhz; 2.3 kw. Ant 541 ft TL: N34 41 02 W85 48 04. Stereo. Hrs opn: Box 966, 19784 John T. Reid Pkwy., Scottsboro, 35768. Phone: (256) 259-2341. Fax: (256) 574-2156. E-mail: ron@wkeafm.com Web Site: www.wwkeafm.com. Licensee: KEA Radio Inc. (group owner; (acq 1996). Law Firm: Fletcher, Heald & Hildreth. Wire Svc: AP Format: Classic rock. ♦Gene Sisk, VP; Ron Livengood, pres & gen mgr; Campbell Smith, opns dir.

Sulligent

WVSA(AM)—See Vernon

Sumiton

WRSM(AM)— June 27, 1978: Stn currently dark. 1540 khz; 1 kw-D. TL: N33 45 50 W87 03 47. Hrs open: Box 11385, Birmingham, 35202. Phone: (205) 326-8844. Licensee: Sumiton Broadcasting Co. Inc. ♦Earl F. Hilliard, pres & gen mgr.

Sylacauga

WFEB(AM)— March 1945: 1340 khz; 1 kw-U. TL: N33 10 16 W86 13 57. Hrs open: 16 Box 358, 1209 Millerville Hwy., 35150. Phone: (256) 245-3281. Fax: (256) 245-3050. Licensee: Alabama Broadcasting Co Inc. Population served: 500,000 Rgnl rep: Keystone (unwired net). Law Firm: Gardner, Carton & Douglas. Format: News/talk, sports. News staff: 3; News: 17 hrs wkly. Target aud: 25-54. Spec prog: Gospel 6 hrs wkly. ♦Bruce C. Carr, gen mgr & gen sls mgr.

WTRB-FM— Dec 20, 1959: 98.3 mhz; 5 kw. 502 ft TL: N33 12 23 W86 13 54. Stereo. Hrs opn: 24 Box 26, Ashland, 36251. Phone: (256) 354-4600. Fax: (256) 354-7224. Licensee: Williams Communications Inc. (group owner; acq 8-16-01). Natl. Network: ABC. Format: Adult contemp. News staff: one. Target aud: 25-54; mainly women. Spec prog: Black gospel 6 hrs wkly. ♦Joe Richardson, VP & progmg dir; Walt Williams, gen mgr.

WYEA(AM)— May 16, 1948: 1290 khz; 1 kw-D, 50 w-N. TL: N33 11 15 W86 14 06. Hrs open: 6 AM-8 PM Box 629, One Motes Rd., 35150. Phone: (256) 249-4263. Fax: (256) 245-4355. E-mail: wyea@rocketmail.com Web Site: wyearadio.com. Licensee: Spirit Broadcasting Co. Inc. (acq 4-13-2001). Population served: 62,450 Rgnl. Network: Ill Radio Net. Format: Christian country. News: 7 hrs wkly. Target aud: General. ♦John Vogel, pres & gen mgr; Brandon Baird, opns dir.

Talladega

WNUZ(AM)— 1945: 1230 khz; 1 kw-U. TL: N33 25 16 W86 07 13. Hrs open: 5 AM-11 PM 1301 Fort Lashley Ave., 35160. Phone: (256) 480-6040. Fax: (256) 480-6050. Licensee: Birmingham Christian Radio Inc. (acq 5-13-97; $30,000). Population served: 76,000 Format: Full gospel. News staff: one. Target aud: 25-70; middle/upper middle, blue & white collar. Spec prog: Talk 5 hrs, Gospel 7 hrs, bluegrass 6 hrs wkly. ♦L.E. Willis Sr., pres; Jonnie Luster, gen mgr.

WTDR(FM)— Nov 10, 1972: 92.7 mhz; 2.6 kw. Ant 505 ft TL: N33 29 12 W85 59 15. (CP: COL Munford. 250 w, ant 1,578 ft. TL: N33 29 06 W85 48 32). Stereo. Hrs opn: 24 1913 Barry St., Suite B, Oxford, 36203-2319. Phone: (256) 741-6000. Fax: (256) 741-6080. Web Site: www.thunder927.com. Licensee: Jacobs Broadcast Group Inc. (acq 9-16-92; $570,000; FTR: 10-19-92). Format: Modern country. ♦James H. Jacobs, pres & gen mgr; Grady Sapp, opns mgr; Gloria Goode, gen sls mgr; Bill Moats, chief of engrg.

Tallassee

WACQ(AM)—Carrville, June 30, 1979: 1130 khz; 1 kw-D. TL: N32 33 22 W85 52 17. Hrs open: Sunrise-sunset 320 Barnett Blvd., 36078. Phone: (334) 283-6888. Fax: (334) 283-6358. E-mail: WACQradio@elmore.rr.com Web Site: wacqradio.com. Licensee: Hughey Communications Inc. (group owner; (acq 12-4-2006; $106,000). Population served: 400,000 Natl. Network: ABC. Rgnl. Network: N.D. News Net. Law Firm: Fletcher, Heald & Hildreth. Wire Svc: AP Format: Oldies. News staff: one; News: 7 hrs wkly. Target aud: 25-54; baby boomers. Spec prog: Farm one hr, gospel 5 hrs wkly. ♦Randall Hughey, gen mgr; Debra Hughey, traf mgr.

WQNR(FM)— Oct 29, 1992: 99.9 mhz; 3.1 kw. 452 ft TL: N32 34 37 W85 51 43. Hrs open: 2514 S. College St., Suite 104, Auburn, 36832. Phone: (334) 887-9999. Fax: (334) 826-9599. E-mail: wildmansteve@wildsteve.com Licensee: Tiger Communications Inc. (acq 1999). Law Firm: Gardner, Carton & Douglas. Format: AOR. Target aud: 25-54; adults. ♦Chris Bailey, gen mgr.

WTLS(AM)— June 1, 1954: 1300 khz; 1.2 kw-D, 18 w-N. TL: N32 30 39 W85 53 33. Hrs open: Box 780146, 36078. Secondary address: 2045 Hwy 229 36078. Phone: (334) 283-8200. Fax: (334) 283-8622. Web Site: www.1300wtls.com. Licensee: Michael Butler Broadcasting LLC (acq 8-24-99). Population served: 138,000 Law Firm: Dan Alpert. Format: Full service. Spec prog: Farm 6 hrs wkly. ♦Michael Butler, pres & gen mgr; Leigh Anne Butler, gen sls mgr; Steve Butler, progmg VP; Terry Harper, chief of engrg; Miles Hathcock, prom.

Thomaston

WKGA(FM)— 2001: Stn currently dark. 97.7 mhz; 500 w. Ant 46 ft TL: N32 16 49 W87 38 06. Hrs open: 6930 Cahaba Valley Rd., Suite 202, Birmingham, 36037. Phone: (205) 949-4586. Licensee: Great South Wireless LLC. (group owner; (acq 2-12-2007; grpsl).

Thomasville

***WDLG(FM)**—Not on air, target date: unknown: 90.1 mhz; 500 w. Ant 249 ft TL: N31 44 24 W87 45 43. Stereo. Hrs opn: 2070 N. Palafox, Pensacola, FL, 32501. Phone: (850) 434-1230. E-mail: mglin@aol.com Licensee: Nationwide Inspirational Broadcasting.

WJDB(AM)— July 16, 1956: 630 khz; 1 kw. TL: N31 52 58 W87 44 42. Hrs open: Sunrise-sunset Box 219, 2211 Hwy. 43 S., 36784. Phone: (334) 636-4438. Fax: (334) 636-4439. E-mail: wjdb@dixienet1.com Licensee: Griffin Broadcasting Corp. (acq 1-4-91; $375,000 with co-located FM; FTR: 1-28-91) Population served: 20,000 Natl. Network: CBS. Rgnl. Network: Ark. Radio Net. Format: Grooving oldies. News staff: one; News: 10 hrs wkly. Target aud: General. ♦Ivy Griffin, gen mgr, gen sls mgr & gen sls mgr.

WJDB-FM— Nov 2, 1972: 95.5 mhz; 9.6 kw. 525 ft TL: N31 52 58 W87 44 42. Stereo. 24 E-mail: wjdb@dixienet1.com 100,000 Format: Top 40 country. News: 10 hrs wkly. Target aud: General.

Trinity

WVNN-FM— Oct 4, 1992: 92.5 mhz; 3.1 kw. Ant 423 ft TL: N34 42 36 W87 04 54. Stereo. Hrs opn: 24
Simulcasts WVNN 100%.
1717 U.S. Hwy. 72 E., Athens, 35611-4413. Phone: (256) 830-8300. Fax: (256) 232-6842. Web Site: www.wvnn.com. Licensee: Cumulus Licensing LLC. Group owner: Clear Channel Communications Inc. (acq 4-4-2006; $3.3 million with WXQW(FM) Meridianville). Population served: 802,300 Format: News/talk. News staff: 3; News: 7 hrs news prgmg wkly. ♦Bill G. West, gen mgr.

Troy

***WAXU(FM)**— 2001: 91.1 mhz; 1.089 kw. Ant 246 ft TL: N31 47 22 W85 58 58. Hrs open: 24 Drawer 3206, American Family Radio, Tupelo, MS, 38803. Phone: (662) 844-8888. Fax: (662) 842-6791. Web Site: www.afr.net. Licensee: American Family Association. Group owner: American Family Radio Format: Inspirational Christian. ♦Marvin Sanders, gen mgr; John Riley, progmg dir.

WTBF(AM)— Feb 25, 1947: 970 khz; 5 kw-D, 45 w-N. TL: N31 50 07 W85 55 58. Hrs open: 24 67 Court Sq., 36081. Phone: (334) 566-0300. Fax: (334) 566-5689. E-mail: wtbf@troycable.net Web Site: www.wtbf.com. Licensee: Troy Broadcasting Corp. Population served: 30,000 Format: Talk, community intensive progmg. News: 20 hrs wkly. Target aud: 35 plus; general. Spec prog: Farm 17 hrs wkly. ♦Joe Gilchrist, pres; Jim Roling, VP & gen mgr; Dave Kirby, opns mgr & progmg dir.

WZHT(FM)— Feb 28, 1973: 105.7 mhz; 100 kw. Ant 1,830 ft TL: N31 58 28 W86 09 44. Stereo. Hrs opn: Box 4420, Montgomery, 36103. Secondary address: 203 Gunn Rd., Montgomery 36117. Phone: (334) 274-6464. Fax: (334) 274-6465. Web Site: www.myhot105.com. Licensee: Capstar TX L.P. Group owner: Clear Channel Communications Inc. (acq 8-30-00; grpsl). Natl. Network: ABC, Westwood One. Natl. Rep: McGavren Guild. Law Firm: Latham & Watkins. Format: Urban. Target aud: 18-49. ♦James Belton, gen mgr; Michael Long, opns mgr & progmg dir; Nikita Pogue, prom dir.

Trussville

WQEN(FM)—Licensed to Trussville. See Birmingham

Tuscaloosa

WACT(AM)— September 1958: 1420 khz; 5 kw-D, 108 w-N. TL: N33 10 30 W87 33 18. Hrs open: 24 Box 20126, 35402-0126. Secondary address: 2121 9th St., Suite B 35401. Phone: (205) 344-4589. Fax: (205) 366-9774. E-mail: chrischampion@clearchannel.com Licensee: Capstar TX L.P. Group owner: Clear Channel Communications Inc. (acq 8-30-2000; grpsl). Population served: 75,000 Format: Sports. Target aud: 35 plus. ♦Lori Moore, gen mgr; Todd Robins, prom dir; Vince Ferrara, progmg dir; Laurie Mundy, news dir & pub affrs dir; Russ Williams, chief of engrg & opns.

WRTR(FM)—Co-owned with WACT(AM). June 1, 1966: 105.9 mhz; 25 kw. Ant 269 ft TL: N33 14 17 W87 29 06. Stereo. 24 Natl. Network: USA. Format: Rock. Target aud: 25 plus. ♦Keith La Coste, women's int ed.

WJRD(AM)— Oct 10, 1936: 1150 khz; 5 kw-D, 1 kw-N, DA-N. TL: N33 15 02 W87 36 35. Hrs open: 24 142 Skyland Blvd., 35405. Phone: (205) 345-7200. Fax: (205) 349-1715. Licensee: Citadel Broadcasting Co. (group owner; (acq 7-12-2005; grpsl). Population served: 100,000 Natl. Network: ABC. Format: Btfl mus. Target aud: 35 plus; persons. ♦Davis Hawkins, gen mgr & gen sls mgr; Brenda BeBout, sls dir; Greg Thomas, stn mgr & progmg mgr.

***WMFT(FM)**— June 6, 2005: 88.9 mhz; 100 kw vert. Ant 522 ft TL: N33 20 19 W87 21 32. Stereo. Hrs opn: 24
Rebroadcasts WMBV(FM) Dixon Mills.
5710 Watermelon Rd., Suite 316, Northport, 35473. Phone: (334) 992-2425. Fax: (334) 992-2637. E-mail: wmft@moody.edu Web Site: www.wmft.fm. Licensee: The Moody Bible Institute of Chicago (group owner). Population served: 250,000 Natl. Network: Moody. Law Firm: Souyhmayd & Miller. Wire Svc: AP Format: Christian. News: 6 hrs wkly. Target aud: 35-54. ♦Rob Moore, stn mgr; John Rogers, progmg dir.

WTBC(AM)— Dec 23, 1946: 1230 khz; 1 kw-U. TL: N33 12 05 W87 32 00. Hrs open: 24 Box 2000, 35403. Secondary address: 2110 McFarland Blvd. E., Suite C 35404. Phone: (205) 758-5523. Phone: (205) 732-9822. Fax: (205) 752-9696. E-mail: wtbc@dbtech.net Web Site: www.wtbc1230.com. Licensee: John Sisty Enterprises Inc. (acq 2-14-02). Population served: 116,029 Natl. Network: ABC, ESPN Radio. Rgnl rep: Alabama Net. Law Firm: Tim K. Brady. Format: News/talk, sports. News: 3 hrs wkly. Target aud: 25-54; upscale, affluent. Spec prog: Relg 3 hrs wkly. ♦John Sisty, CEO & pres; Dave McDaniel, opns mgr; Ronnie Quarles, COO, gen mgr & sls dir.

WTSK(AM)— February 1958: 790 khz; 5 kw-D, 36 w-N. TL: N33 11 17 W87 35 23. Hrs open: 142 Skyland Blvd., 35405. Phone: (205) 345-7200. Fax: (205) 349-1715. Licensee: Citadel Broadcasting Co. (group owner; (acq 7-12-2005; grpsl). Population served: 155,000 Natl. Rep: McGavren Guild. Format: Gospel. Target aud: 35 plus. ♦Gigi South, gen mgr; Charles Anthony, opns mgr; Brandy Jackson, gen sls mgr; Val Goodson, news dir; Herb Connellan, chief of engrg.

WTUG-FM—Co-owned with WTSK(AM). March 1979: 92.9 mhz; 100 kw. Ant 980 ft TL: N33 03 15 W87 32 57. Stereo. 24 142 Skyland Blvd., 35405. Phone: (205) 345-7200. Fax: (205) 349-1715. Web Site: www.wtug.com. Format: Adult contemp. Target aud: 25-54. ♦Charles Anthony, progmg dir.

***WUAL-FM**— Jan 4, 1982: 91.5 mhz; 100 kw. 523 ft TL: N33 05 40 W87 24 47. Stereo. Hrs opn: 24 Phifer Annex, Suite 166, 35487. Phone: (205) 348-6644. Fax: (205) 348-6648. Licensee: Board of Trustees of the University of Alabama. Natl. Network: PRI, NPR. Rgnl rep: Alabama Net. Law Firm: Arter & Hadden. Format: Class, jazz, news/talk. News staff: 3; News: 5 hrs wkly. Target aud: 35 plus. Spec prog: Bluegrass, blues, folk 5 hrs, new age 20 hrs wkly. ♦Elizabeth Brock, gen mgr & stn mgr.

***WVUA-FM**— Sept 7, 1972: 90.7 mhz; 160 w. 142 ft TL: N33 12 33 W87 32 57. Stereo. Hrs opn: Box 870152, 35487-0152. Phone: (205) 348-6461. Fax: (205) 348-0375. E-mail: wvua@sa.ua.edu Web Site: www.newrock907.com. Licensee: Board of Trustees University of Alabama. Format: Alternative, rock. Target aud: 18-25; high school & college students. Spec prog: Christian 3 hrs, hardcore 3 hrs, blues 3 hrs, heavy metal 4 hrs, reggae 3 hrs wkly. ♦Loy Singleton, gen mgr; Graham Flaugan, stn mgr.

WWPG(AM)— Dec 10, 1951: 1280 khz; 5 kw-D, 500 w-N, DA-N. TL: N33 13 07 W87 34 05. (CP: COL Eutaw. 7 kw-D, 25 w-N. TL: N32 55 19 W87 49 08). Hrs opn: 601 Greensboro Ave., Suite 507, 35401. Phone: (205) 345-4787. Fax: (205) 345-4790. Licensee: Lawson of Tuscaloosa Inc. (acq 3-17-93; $160,000; FTR: 4-5-93). Population served: 75,000 Natl. Network: Westwood One. Law Firm: Taylor, Smith & Parker. Format: Gospel. Target aud: 24-54; mature business audience. Spec prog: Jazz 2 hrs wkly. ♦Jim Lawson, pres & gen mgr; Mildred Porter, opns mgr.

Tuscumbia

WQRV(FM)— May 2, 1962: 100.3 mhz; 100 kw. Ant 246 ft TL: N34 45 23 W87 41 08. (CP: COL Meridianville. 8.5 kw, ant 981 ft. TL: N34 47 36 W86 37 51). Stereo. Hrs opn: 24 26869 Peoples Rd., Madison, 35756. Phone: (256) 309-2400. Fax: (256) 389-1912. Web Site: www.103theriver.com. Licensee: CC Licenses LLC. (acq 12-19-2000; grpsl). Format: Country. News staff: 2; News: 60 hrs wkly. Target aud: 18-34. ♦Rick Brown, gen mgr; Carmelita Palmer, sls dir; Bruce Reynolds, progmg dir; Carl Sampieri, chief of engrg.

WVNA(AM)— Apr 5, 1955: 1590 khz; 5 kw-D, 1 kw-N, DA-N. TL: N34 45 24 W87 36 35. Hrs open: 24 273 Azalea Rd., Suite 1-308, Mobile, 36609. Phone: (251) 343-4900. Fax: (251) 343-4905. E-mail: info@urbanradio.com Web Site: www.wvnafm.com. Licensee: Urban Radio Licenses LLC. Group owner: Clear Channel Communications Inc. (acq 5-13-2005; grpsl). Population served: 500,000 Natl. Network: CBS. Format: News/talk, sports. News staff: 3; News: 60 hrs wkly. Target aud: 25-64. ♦Kevin Wagner, gen mgr.

WZZA(AM)— Apr 17, 1960: 1410 khz; 500 w-D, 51 w-N. TL: N34 42 29 W87 41 35. Hrs open: 24 1570 Woodmont Dr., 35674. Phone: (256) 381-1862. Phone: (256) 383-5810. Fax: (256) 381-6006. E-mail: wzzaradio@aol.com Licensee: Muscle Shoals Broadcasting. (acq 12-1-77). Population served: 250,000 Natl. Network: American Urban. Law Firm: Fletcher, Heald & Hildreth, PLC. Format: Soul & gospel. News staff: one; News: 15 hrs wkly. Target aud: Black. ♦Jurado Bailey, pres & CFO; Tori Bailey, CEO, gen mgr & gen mgr; Dwight Winston, opns dir; Peter Smith, progmg dir; Tomeeka Byrd, progmg dir & pub svc dir; John Reeder, sls; Leonard Skipworth, sls; Theodore Lindsey, sls; Tonyia Canter, sls.

Tuskegee

WBIL(AM)— July 1, 1952: 580 khz; 500 w-D, 139 w-N. TL: N32 22 36 W85 39 28. Hrs open: 118 S. Main St., 36083. Phone: (334) 727-2100. Fax: (334) 724-9169. Licensee: H&H Communications L.L.C. Group owner: Willis Broadcasting Corp. (acq 2-23-2005; $210,000). Population served: 14,793 Format: Gospel. ♦Bernita Luke, gen mgr; Terry Harper, chief of engrg.

WTGZ(FM)— July 12, 1975: 95.9 mhz; 4.3 kw. 377 ft TL: N32 28 17 W85 34 28. Stereo. Hrs opn: 2514 S. College St., Suite 104, Auburn, 36830. Phone: (334) 887-9999. Fax: (334) 826-9599. E-mail: info@thetiger.fm Web Site: www.thetiger.fm. Licensee: Tiger Communications Inc. (acq 2-26-98; $450,000). Format: Modern rock. Target aud: 18-34. ♦Chris Bailey, gen mgr.

Union Springs

WQSI(FM)— Oct 15, 1975: 93.9 mhz; 12.5 kw. Ant 469 ft TL: N32 19 04 W85 40 16. Stereo. Hrs opn: 24 2514 S. College St., Suite 104, Auburn, 36830. Phone: (334) 887-9999. E-mail: wacqradio@elmore.rr.com Licensee: H&H Communications L.L.C. (acq 5-20-2003; $450,000). Population served: 620,000 Law Firm: John Trent. Format: Classic country. News: 15 wkly. Target aud: 25-54; 40+ boomers, active, affluent southerners.

Valley

***WEBT(FM)**— Jan 17, 1986: 91.5 mhz; 380 w. 85 ft TL: N32 48 15 W85 10 43. Hrs open: 2615 64th Blvd., 36854. Phone: (334) 756-6923. Fax: (334) 756-8430. Licensee: Langdale Educational Broadcasting Foundation. Population served: 20,000 Format: Southern gospel. Target aud: General. Spec prog: Southern gospel. ♦Tim Foster, stn mgr.

WRLD-FM— May 17, 1993: 95.3 mhz; 25 kw. Ant 328 ft TL: N32 44 03 W85 07 53. Hrs open: 24 1353 13th Ave., Columbus, GA, 31901. Phone: (706) 327-1217. Fax: (706) 596-4600. Web Site: www.boomer.fm. Licensee: ABG Georgia LLC. Group owner: Archway Broadcasting Group (acq 4-25-03; grpsl). Law Firm: Holland & Knight. Format: Oldies. Target aud: 35 plus. ♦Chuck Thompson, gen mgr & sports cmtr.

Valley Head

WQRX(AM)— Feb 10, 1986: 870 khz; 10 kw-D. TL: N34 33 20 W85 37 12. Hrs open: Sunrise-sunset 2278 Wortham Ln., Grovetown, GA, 30813. Phone: (706) 309-9610. Fax: (706) 309-9669. E-mail: cbarinowski@comcast.net Web Site: www.gnnradio.org. Licensee: Barinowski Investment Company Group owner: Good News Network (acq 10-13-99). Format: Sp. All Sp. ♦Clarence Barinowski, pres, gen mgr & gen mgr.

Vernon

WJEC(FM)—Listing follows WVSA(AM).

WVSA(AM)— July 4, 1966: 1380 khz; 5 kw-D, 39 w-N. TL: N33 47 45 W88 07 03. Hrs open: Box 630, 35592. Phone: (205) 695-9191. Fax: (205) 695-9131. E-mail: wjec1065@yahoo.com Licensee: Lamar County Broadcasting Co. Inc. Population served: 2,190 Format: Sports, talk. ♦Patricia Davis, gen mgr.

WJEC(FM)—Co-owned with WVSA(AM). Apr 1, 1991: 106.5 mhz; 6 kw. 328 ft TL: N33 51 15 W88 01 55.24 Format: Southern gospel. ♦R. William Davis, CEO; Randy Wright, prom mgr & disc jockey; Curt Smith, progmg mgr; Jason Baines, disc jockey; Jerry Oakes, disc jockey; Sid Davis, disc jockey.

Warrior

WBHK(FM)— Apr 22, 1992: 98.7 mhz; 14 kw. 945 ft TL: N33 27 45 W86 50 59. Hrs open: 950 22nd St., Suite 1000, Birmingham, 35203. Phone: (205) 322-2987. Fax: (205) 322-2390. Web Site: www.987kiss.com.Yes Licensee: Cox Radio Inc. Group owner: Cox Broadcasting (acq 10-6-98; $17 million with WBHJ(FM) Tuscaloosa).

Population served: 1,000,000 Natl. Rep: Christal. Law Firm: Dow, Lohnes & Albertson. Wire Svc: Metro Weather Service Inc. Format: Classic soul & rhythm and blues. News staff: 2; News: 5 hrs wkly. Target aud: 25-54; general. ♦David DuBose, gen mgr.

Wetumpka

WAPZ(AM)— Oct 2, 1954: 1250 khz; 5 kw-D, 80 w-N. TL: N32 29 06 W86 12 25. Hrs open: 2821 U.S. Hwy. 231, 36092. Phone: (334) 567-2251. Fax: (334) 567-7971. Web Site: www.1250wapz.com. Licensee: J&W L.L.C. (acq 9-1-84; $235,000; FTR: 7-23-84). Population served: 450,000 Format: Relg, Black, blues, news/talk, sports. Target aud: 12-100. ♦Johnny Roland, pres; Robert Henderson, gen mgr & gen sls mgr.

WJWZ(FM)— 1998: 97.9 mhz; 3 kw. 328 ft TL: N32 27 08 W86 12 35. Hrs open: 4101-A Wall St., Montgomery, 36106. Phone: (334) 244-0961. Fax: (334) 279-9563. Web Site: www.979-jamz.com. Licensee: Bluewater Broadcasting Co. LLC (group owner; acq 4-21-2004; grpsl). Format: Urban. ♦Terry Barber, gen mgr; Marvin Nugent, progmg dir.

Winfield

WKXM(AM)— Aug 23, 1965: 1300 khz; 5 kw-D, 30 w-N, DA-D. TL: N33 55 52 W87 48 36. Hrs open: Box 608, 35594. Secondary address: 655 Fairview Rd. 35594. Phone: (205) 487-3261. Fax: (205) 487-6991. E-mail: wkxm@dlis.net Licensee: Ad-Media Management Corp. (acq 12-30-91; $365,000 with co-located FM; FTR: 1-27-92) Natl. Network: Westwood One, ESPN Radio, NBC Radio. Format: Sports talk. Target aud: General. ♦Maxine Harper, pres, gen mgr & gen sls mgr; Doug Threadgill, news dir; Olen Booth, engrg mgr; Teresa Benton, opns mgr, mus dir & traf mgr.

WKXM-FM— 1991: 97.7 mhz; 3.9 kw. Ant 403 ft TL: N34 01 53 W87 48 06.24 Natl. Network: ABC. Format: Oldies. Target aud: General. ♦Teresa Benton, traf mgr.

York

WSLY(FM)—Listing follows WYLS(AM).

WYLS(AM)— November 1970: 670 khz; 4.8 kw-D. TL: N32 31 24 W88 15 28. Hrs open: 7 AM-4:30 PM 11474 U.S. Hwy. 11, 36925. Phone: (205) 392-5234. Fax: (205) 392-5536. E-mail: ken@1049jackfm.com Licensee: Grantell Broadcasting Co. (acq 11-21-2003; with co-located FM). Population served: 300,000 Format: Gospel. Target aud: 25-54+. ♦Ken Michaels, opns mgr.

WSLY(FM)—Co-owned with WYLS(AM). September 1976: 104.9 mhz; 50 kw. 492 ft TL: N32 16 54 W88 15 28. Stereo. 24 1181 Bonita Lakes Cir., # 122, Meridian, 39301. Phone: (601) 581-5225. Web Site: www.1049jackfm.com.250,000 Natl. Network: ABC. Format: Jack. Target aud: 18-54; male & female.

Alaska

Anchorage

KAFC(FM)— Apr 4, 1999: 93.7 mhz; 27 kw. 663 ft TL: N61 04 02 W149 44 36. Hrs open: Box 210389, 99521. Secondary address: 6401 E. Northern Lights Blvd. 99504. Phone: (907) 333-5282. Fax: (907) 333-9851. E-mail: tom@katb.org Web Site: www.kafc.org. Licensee: Christian Broadcasting Inc. Format: Christian contemp music. ♦Tom Steigleman, gen mgr.

***KAKL(FM)**— 2004: 88.5 mhz; 11 kw. Ant -82 ft TL: N61 07 14 W149 53 42. Stereo. Hrs opn: 24 2351 Sunset Blvd., Suite 170-218, Rocklin, CA, 95765. Phone: (916) 251-1600. Fax: (916) 251-1650. E-mail: klove@klove.com Web Site: www.klove.com. Licensee: Educational Media Foundation. Group owner: EMF Broadcasting. Natl. Network: K-Love. Law Firm: Shaw Pittman. Format: Contemp Christian. News staff: 3. Target aud: 25-44; Judeo Chrisitan, female. ♦Richard Jenkins, pres; Lloyd Paker, gen mgr; Keith Whipple, dev dir; Eric Allen, natl sls mgr; Mike Novak, progmg dir; David Pierce, progmg mgr; Ed Lenane, news dir; Sam Wallington, engrg dir; Arthur Vassar, traf mgr.

KASH-FM— Dec 1, 1985: 107.5 mhz; 100 kw. 1,014 ft TL: N61 09 53 W149 41 05. Stereo. Hrs opn: 24 800 E. Dimond Blvd., Suite 3-370, 99515-2043. Phone: (907) 522-1515. Fax: (907) 743-5184. Web Site: www.country1075.com. Licensee: Clear Channel Radio Licenses Inc. Group owner: Clear Channel Communications Inc. (acq 8-30-00; grpsl). Format: Country. Target aud: 25-54. ♦Gary Donovan, pres; Andy Lohman, gen mgr; Jimmy O'Brien, progmg dir.

***KATB(FM)**— June 1985: 89.3 mhz; 4.9 kw. 572 ft TL: N61 04 02 W149 44 04. (CP: Ant 344 ft. TL: N61 07 32 W149 42 46). Stereo. Hrs opn: 24 Box 210389, 99521. Secondary address: 6401 E. Northern Lights Blvd. 99504. Phone: (907) 333-5282. Fax: (907) 333-9851. E-mail: tom@katb.org Web Site: www.katb.org. Licensee: Christian Broadcasting Inc. Natl. Network: Moody. Format: Relg. News: 5 hrs wkly. Target aud: 25-49; women. ♦Tom Steigleman, gen mgr.

***KAUG(FM)**—Not on air, target date: unknown: 89.9 mhz; 10 w horiz. Ant 98 ft TL: N61 24 33 W149 25 15. Hrs open: Box 196614, 99519-6614. Phone: (907) 742-3500. Fax: (907) 742-3545. Licensee: Anchorage School District. ♦Carol Comeau, gen mgr.

KAXX(AM)—Eagle River, Dec 25, 1986: Stn currently dark. 1020 khz; 10 kw-U, DA-N. TL: N61 29 03 W149 45 52. Hrs open: 24207 Hartland, West Hills, CA, 91307. Phone: (818) 340-5209. Licensee: Ruth Pollack, executrix Group owner: American Radio Brokers Inc./SFO (acq 6-6-2006; . with KZND-FM Houston). Population served: 350,000 News staff: . ♦Ruth Pollack, gen mgr.

KBFX(FM)— Oct 1, 1978: 100.5 mhz; 25 kw. 178 ft TL: N61 11 52 W149 52 31. Stereo. Hrs opn: 800 E. Dimond Blvd., Suite 3-370, 99515-2043. Phone: (907) 522-1515. Fax: (907) 743-5184. Web Site: www.1005thefox.com. Licensee: Capstar TX L.P. Group owner: Clear Channel Communications Inc. (acq 8-7-00; grpsl). Format: Classic rock. ♦Gray Donavan, pres; Andy Lohman, gen mgr; Mark Murphy, opns dir; Kim Williams, gen sls mgr; Jeremy Hegna, progmg dir.

KBRJ(FM)—Listing follows KHAR(AM).

KBYR(AM)— 1948: 700 khz; 10 kw-U. TL: N61 12 25 W149 55 20. Hrs open: 24 1399 W. 34th Ave., Suite 202, 99503. Phone: (907) 278-5297. Fax: (907) 272-5297. E-mail: myopinion@kbyr.com Web Site: www.kbyr.com. Licensee: Cobb Communications Inc. (acq 5-02.). Population served: 258,000 Natl. Network: ABC. Format: News/talk. News staff: one. Target aud: 25-54. ♦Justin McDonald, gen mgr, opns dir & progmg dir; Debbie Rinckey, gen sls mgr; Kathy Phillips, news dir.

KDBZ(FM)— Feb 1, 1973: 102.1 mhz; 25 kw. 174 ft TL: N61 20 10 W149 30 46. (CP: 23 kw, ant 82 ft.). Stereo. Hrs opn: 11259 Tower Rd., 99515. Phone: (907) 344-4045. Fax: (907) 522-6053. Web Site: www.buzz1021.com. Licensee: New Northwest Broadcasters LLC (group owner; acq 8-12-99; $1.3 million). Wire Svc: AP Format: Hot adult contemp. Target aud: 18-44; women. ♦Pete Benedetti, CEO; Trila Bumstead, CFO; Tom Oakes, gen mgr; Carla Wyrick, gen sls mgr; Tom Oaks, progmg dir.

KEAG(FM)— 1987: 97.3 mhz; 100 kw. 593 ft TL: N61 25 22 W149 52 20. Hrs open: 24 301 Arctic Slope Ave., 99518. Phone: (907) 344-9622. Fax: (907) 349-7326. Web Site: www.kool973.com. Licensee: Morris Communications Corp. Group owner: Morris Communications Inc. (acq 10-15-98; grpsl). Format: Oldies. News staff: one; News: 3 hrs wkly. Target aud: 35-49. ♦Scott Smith, gen mgr; Dave Stroh, progmg dir.

KENI(AM)— July 15, 1967: 650 khz; 50 kw-U. TL: N61 09 58 W149 49 34. Stereo. Hrs opn: 24 800 E. Dimond Blvd., Suite 3-370, 99515. Phone: (907) 522-1515. Fax: (907) 743-5186. Web Site: www.keni650.com. Licensee: Capstar TX L.P. Group owner: Clear Channel Communications Inc. (acq 8-30-00; grpsl). Population served: 250,000 Natl. Rep: Christal. Format: Talk. News staff: one; News: 5 hrs wkly. Target aud: 25-64. ♦Andy Lowman, gen mgr.

KGOT(FM)—Co-owned with KENI(AM). Sept 15, 1975: 101.3 mhz; 26 kw. -66 ft TL: N61 09 58 W149 49 34. Stereo. 24 Fax: (907) 522-0672. Web Site: www.kgot.com. Format: CHR, btfl music. News: 2 hrs wkly. Target aud: 12-44. ♦Mark Murphy, opns mgr.

KFAT(FM)— Apr 1, 1997: 92.9 mhz; 100 kw. Ant 1,269 ft TL: N61 21 05 W149 29 10. Stereo. Hrs opn: 24 11259 Tower Rd., 99515. Phone: (907) 344-4045. Web Site: www.kfat929.com. Licensee: New Northwest Broadcasters LLC. (group owner; (acq 7-30-99). Population served: 350,000 Natl. Network: ABC. Format: Rhythmic CHR. Target aud: 18-34; adults. ♦Pete Benedetti, CEO; Trila Bumstead, CFO; Tom Oakes, gen mgr & opns mgr; McConnell Adams, progmg dir. Co-owned TV: KYES(TV) affil.

KFQD(AM)— 1924: 750 khz; 50 kw-U. TL: N61 08 13 W149 50 06. Hrs open: 301 Artic Slope Ave., Suite 200, 99518. Phone: (907) 344-9622. Fax: (907) 349-7326. Web Site: www.kfqd.com. Licensee: Morris Communications Corp. Group owner: Morris Communications Inc. (acq 12-1-98; grpsl). Population served: 270,000 Natl. Rep: Katz Radio. Format: News/talk. Target aud: 35 plus; higher income, upper demo. ♦Dennis Bookey, gen mgr; Scott Smith, gen sls mgr & prom mgr; Sharon Leighow, news dir; Paul Jewusiak, chief of engrg.

KWHL(FM)—Co-owned with KFQD(AM). Sept 18, 1982: 106.5 mhz; 100 kw. -89 ft TL: N61 08 13 W149 50 06. Stereo. Web Site: www.kwhl.com. Format: Rock/AOR. Target aud: 18-44; medium income adults, mostly men. ♦Larry Snider, progmg dir.

KHAR(AM)— Jan 7, 1961: 590 khz; 5 kw-U. TL: N61 07 12 W149 53 43. Hrs open: 24 301 Artic Slope Ave., Suite 200, 99518. Phone: (907) 344-9622. Fax: (907) 349-7326. Web Site: khar590.com. Licensee: MCC Radio LLC. Group owner: Morris Communications Inc. (acq 12-1-98; grpsl). Population served: 250,000 Natl. Rep: International Media. Law Firm: Wiley, Rein & Fielding. Format: Adult standards. News: 3 hrs wkly. Target aud: 35 plus; white collar, professional, upper-income demographics. ♦Dennis Bookey, gen mgr; Ron Clement, gen sls mgr; Paul Jewusiak, chief of engrg.

KLEF(FM)— Sept 16, 1988: 98.1 mhz; 25 kw. Ant -85 ft TL: N61 11 17 W149 52 57. Stereo. Hrs opn: 19 3601 C St., Suite 290, 99503. Phone: (907) 561-5556. Fax: (907) 562-4219. E-mail: klef@klef.com Web Site: www.klef.com. Licensee: Chinook Concert Broadcasters Inc. (acq 6-87; FTR: 6-8-87). Population served: 230,000 Rgnl rep: Tacher. Format: Classical. Target aud: 25-64; highly educated, affluent adults. Spec prog: Children one hr wkly. ♦Rick Goodfellow, gen mgr & stn mgr.

KMXS(FM)— Sept 1, 1987: 103.1 mhz; 100 kw. Ant 105 ft TL: N61 11 33 W149 54 01. Stereo. Hrs opn: 24 301 Artic Slope Ave., 99518. Phone: (907) 344-9622. Fax: (907) 349-7326. E-mail: news@kftq.com Web Site: www.kmxs.com. Licensee: Morris Communications Corp. Group owner: Morris Communications Inc. (acq 10-15-98; grpsl). Natl. Rep: McGavren Guild. Format: Hot adult contemp. News staff: one; News: 5 hrs wkly. Target aud: 25-44; female listeners. ♦Scott Smith, gen mgr; Roxy Lennox, progmg dir.

***KNBA(FM)**— September 1996: 90.3 mhz; 100 kw. 640 ft TL: N61 25 22 W149 52 20. Stereo. Hrs opn: 24 3600 San Jeronimo Ct. Ste 480, 99508-2870. Phone: (907) 793-3500. Fax: (907) 973-3536. E-mail: feedback@knba.org Web Site: www.knba.org. Licensee: Koahnic Broadcast Corp. Population served: 250,000 Format: var/div. News staff: 3; News: 8 hrs wkly. Target aud: 20-50; well off, public radio listeners. ♦Jaclyn Sallee, CEO, pres & dev mgr; Bruce Hilton, CFO; Loren Dixon, progmg dir.

KNIK-FM— Sept 15, 1960: 105.7 mhz; 51 kw. Ant 1,069 ft TL: N61 20 11 W149 30 48. Stereo. Hrs opn: 24 4700 Business Park Blvd., Bldg. E 44A, 99513. Phone: (907) 522-1018. Fax: (907) 522-1027. Web Site: www.knik.com. Licensee: Ubik Corp. Population served: 360,000 Natl. Rep: Interep. Format: Smooth jazz. News staff: one; News: 9 hrs wkly. Target aud: 25-54 plus. ♦Mike Robbins, gen mgr & stn mgr; Dan Thomas, progmg dir.

***KRUA(FM)**— Feb 14, 1992: 88.1 mhz; 155 w. 292 ft TL: N61 07 32 W149 42 46. Hrs open: 24 PSB Rm 254, 3211 Providence Dr., 99508. Phone: (907) 786-6800. Fax: (907) 786-6806. E-mail: aykrua1@uaa.alaska.edu Web Site: www.krua.uaa.alaska.edu. Licensee: University of Alaska-Anchorage. Law Firm: Wilkinson Barker Knauer. Format: Progsv, alternative. News staff: 4; News: 10 hrs wkly. Target aud: General; college community/div. Spec prog: Var/div music 20 hrs, sports one hr wkly. Var/div music 20 hrs, American Indian 3 hrs, sports one hr wkly ♦Neil Torquiano, stn mgr.

***KSKA(FM)**— Aug 15, 1978: 91.1 mhz; 36 kw. 190 ft TL: N61 11 25 W149 48 16. (CP: 100 kw, ant 617 ft.). Stereo. Hrs opn: 24 3877 University Dr., 99508. Phone: (907) 561-1161. Fax: (907) 550-8401. Fax: (907) 550-8403. Web Site: www.kska.org. Licensee: Alaska Public Telecommunications Inc. (acq 1994). Population served: 250,000 Natl. Network: NPR, PRI. Rgnl. Network: Alaska Pub. Format: In-depth news. News staff: 2; News: 70 hrs wkly. Target aud: 24 plus; professionals. ♦Paul Stankavich, pres & gen mgr; Bede Trantina, stn mgr; Duncan Moon, news dir.

KTZN(AM)— May 2, 1948: 550 khz; 5 kw-U. TL: N61 12 25 W149 55 20. Hrs open: 800 E. Dimond Blvd., Suite 3-370, 99515. Phone: (907) 522-1515. Fax: (907) 743-5184. Web Site: www.550thezone.com. Licensee: Capstar TX L.P. Group owner: Clear Channel Communications Inc. (acq 8-30-00; grpsl). Population served: 250,000 Natl. Rep: D & R Radio. Law Firm: Haley, Bader & Potts. Format: Sports. Target aud: 25-54. ♦Gary Donovan, pres; Andy Lohman, gen mgr; Kim Williams, gen sls mgr; Mark Murphy, progmg dir.

KUDO(AM)— May 10, 1975: 1080 khz; 10 kw-U. TL: N61 07 12 W149 53 43. Hrs open: 3601 C St., Suite 290, 99503. Phone: (907) 561-5556. Fax: (907) 562-4219. Web Site: kudo1080.com. Licensee: IBEW Local 1547 Investments LLC (acq 9-19-2005; $244,000). Population served: 250,000 Format: Progsv talk. Target aud: 25 plus. ♦Rich McClear, gen mgr.

KYMG(FM)— Jan 1, 1989: 98.9 mhz; 100 kw. 499 ft TL: N61 25 22 W149 52 20. Stereo. Hrs opn: 24 800 E. Dimond Blvd., Suite 3-370, 99515. Phone: (907) 522-1515. Fax: (907) 743-5184. E-mail: markmurphy @clearchannel.com Web Site: www.magic989fm.com. Licensee: Clear Channel Radio Licenses Inc. Group owner: Clear Channel Communications Inc. (acq 8-30-00; grpsl). Law Firm: Becker & Finerfrock. Format: Adult contemp. News staff: one; News: 4 hrs wkly. Target aud: 25-49; mostly women. Spec prog: Relg one hr wkly. ♦Gary Donovan, pres; Andy Lohman, gen mgr; Mark Murphy, opns dir; Kim Williams, gen sls mgr; Dave Flavin, progmg dir.

KZND-FM—Houston, Apr 1, 1998: 94.7 mhz; 50 kw. Ant 371 ft TL: N61 29 03 W149 45 52. Stereo. Hrs opn: 24 4700 Business Park Blvd., Bldg. E, Suite 44A, 99503. Phone: (907) 522-1018. Fax: (907) 522-1027. Web Site: 947theend.fm. Licensee: Ruth Pollack, executrix Group owner: American Radio Brokers Inc./SFO. (acq 6-6-2006; with KAXX(AM) Eagle River). Format: Rock. Target aud: 25-44; general. ♦Mike Robbins, gen mgr.

Barrow

***KBRW(AM)**— Dec 22, 1975: 680 khz; 10 kw-U. TL: N71 15 24 W156 31 32. Hrs open: 24 Box 109, 1695 Okpik St., 99723. Phone: (907) 852-6811. Fax: (907) 852-2274. Web Site: www.kbrw.org. Licensee: Silakkuagvik Communications Inc. Population served: 10,000 Natl. Network: PRI, NPR. Rgnl. Network: Alaska Pub. Law Firm: Schwartz, Woods & Miller. Format: Var/div. News staff: one; News: 24 hrs wkly. Target aud: General. Spec prog: Class 2 hrs, jazz 6 hrs, relg one hr, Filipino 2 hrs, country 7 hrs wkly. ♦Jim Vorderstrasse, pres; Robert C. Sommer, VP, gen mgr & stn mgr; Isaac Tuckfield, opns dir & progmg dir; Kai Saxton, opns mgr; Jason Gilbert, dev dir; Janelle Everett, prom dir & news dir; Charles M. Laykatis, chief of engrg; Doreen Simmonds, news rptr; Earl Finkler, reporter & sports cmtr; Bob Thomas, sports cmtr.

KBRW-FM— Sept 1, 1996: 91.9 mhz; 890 w. 72 ft TL: N71 17 20 W156 45 31.7 AM-midnight Natl. Network: NPR, PRI. Format: Adult contemp, big band, class. News staff: one; News: 80 hrs wkly. Target aud: General. ♦Jason Gilbert, dev dir; Issac Tuckfield, progmg dir; Diana Gish, news dir; Robert Sommer, chief of engrg.

Bethel

KYKD(FM)— November 1994: 100.1 mhz; 7.7 kw. 72 ft TL: N60 48 20 W161 47 14. Stereo. Hrs opn:
Rebroadcasts KIAM(FM) Nenana 40%.
Box 2428, 99559. Secondary address: 406 Ptarmigan Rd. 99559. Phone: (907) 543-5953. Fax: (907) 543-5952. E-mail: kykdfm@unicom-alaska.com Web Site: www.vfcm.org. Licensee: Voice For Christ Ministries Inc. Natl. Network: USA. Format: Christian music & info. News: 12 hrs wkly. Target aud: Alaskan bush/rural. ♦Robert Eldridge, CEO; Ron Heagy, chmn; Jon Skillman, stn mgr.

***KYUK(AM)**— May 13, 1971: 640 khz; 10 kw-U. TL: N60 46 57 W161 53 00. Hrs open: Box 468, 640 Radio St., 99559. Phone: (907) 543-3131. Fax: (907) 543-3130. Web Site: www.kyuk.org. Licensee: Bethel Broadcasting Inc. Population served: 15,000 Natl. Network: PRI, NPR. Rgnl. Network: Alaska Pub. Format: Bilingual talk & div mus, public info. Target aud: General. Spec prog: Class 4 hrs, country 4 hrs wkly. ♦Joan Hamilton, chmn & pres; Ronald Daugherty, gen mgr; Angela Denning Barnes, news dir; Joseph Siebert, chief of engrg. Co-owned TV: *KYUK-TV affil

Big Lake

KAGV(AM)— Nov. 1, 2005: 1110 khz; 10 kw-U. TL: N61 38 03 W149 47 36. Hrs open: Box 474, Nenana, 99760. Phone: (907) 832-5426. Web Site: www.vfcm.org/kagv.htm. Licensee: Voice for Christ Ministries Inc. Format: News/talk, music, Christian. News: 20 hrs wkly. Spec prog: Native Alaskan 3 hrs wkly. ♦Robert C. Eldridge, gen mgr; David Horning, progmg dir.

Chevak

***KCUK(FM)**— 1990: 88.1 mhz; 150 w. 75 ft TL: N61 31 46 W165 35 20. (CP: 6 kw, ant 78 ft.). Hrs opn: 985 KSD Way, 99563. Phone: (907) 858-7015. Fax: (907) 858-7279. E-mail: ptuluk@vak.gcisa.net Licensee: Kashunamiut School District. Format: Country, rock and roll, var. ♦Peter Tuluk, gen mgr.

College

KTDZ(FM)— Sept 6, 1984: 103.9 mhz; 2.95 kw. Ant 823 ft TL: N64 55 20 W147 42 55. Stereo. Hrs opn: 1060 Aspen St., Fairbanks, 99709. Phone: (907) 451-5910. Fax: (907) 451-5999. E-mail: 1039koolfm@nnbradio.com Licensee: New Northwest Broadcasters LLC (group owner; (acq 10-26-99; grpsl). Natl. Rep: Tacher. Format: Adult hits. ♦Perry Walley, gen mgr; Glenn Anderson, opns mgr.

Cordova

KCDV(FM)—Listing follows KLAM(AM).

KLAM(AM)— May 1953: 1450 khz; 250 w-U. TL: N60 32 20 W145 45 35. Hrs open: Box 60, 112 Forestry Way, 99574. Phone: (907) 424-3796. Fax: (907) 424-3737. E-mail: bayside@ctc.net Web Site: bayview@ctcak.net. Licensee: Bayview Communications Inc. Population served: 6,000 Natl. Network: ABC. Law Firm: Haley, Bader & Potts. Format: Country, classic rock, news & info. Target aud: General. ♦J.R. Lewis, gen mgr.

Deadhorse

***KCDS(FM)**— 2000: 88.1 mhz; 90 w. Ant 105 ft TL: N70 12 00 W148 28 02. Hrs open:
Rebroadcasts KBRW(AM) Barrow 100%.
Box 109, Barrow, 99723. Phone: (907) 852-6811. Fax: (907) 852-2274. Web Site: www.kbrw.org. Licensee: Silakkuagvik Communications Inc. Format: Classic rock, new rock, adult contemp. Spec prog: Country. ♦Robert C. Sommer, gen mgr; Kari Saxton, opns mgr; Jason Gilbert, dev dir; Isaac Tuckfield, progmg dir; Janelle Everett, news dir.

Delta Junction

KDJF(FM)— July 1, 2007: 93.5 mhz; 100 kw. Ant 853 ft TL: N64 18 13 W146 32 44. Hrs open: 110 Green Meadows, Abilene, TX, 79605. Phone: (325) 829-6850. Licensee: Radio Layne LLC. Law Firm: Shainis & Peltzman. ♦Amy Meredith, pres; Scott Powell, gen mgr.

Dillingham

***KDLG(AM)**— July 22, 1975: 670 khz; 10 kw-U. TL: N59 02 43 W158 27 07. Hrs open: 18 Box 670, 99576. Phone: (907) 842-5281. Fax: (907) 842-5645. Licensee: Dillingham City School District. Population served: 30,000 Natl. Network: NPR. Rgnl. Network: Alaska Pub. Format: Adult contemp, country, rock. News staff: one; News: 20 hrs wkly. Spec prog: Yupik one hr wkly. ♦Buchi Lind, pres; Rob Carpenter, gen mgr.

KRUP(FM)— August 1995: 99.1 mhz; 6 kw. Ant 128 ft TL: N59 02 31 W158 31 19. Hrs open: 24 Box 157, 99576. Secondary address: 301 Airport Rd. 99576. Phone: (907) 842-5364. Licensee: McCormick Broadcasting. Format: Talk. ♦Jackson McCormick, pres & gen mgr.

Eagle River

KAXX(AM)—Licensed to Eagle River. See Anchorage

Fairbanks

KAKQ-FM— Apr 4, 1981: 101.1 mhz; 25 kw. 131 ft TL: N64 54 53 W147 38 54. Stereo. Hrs opn: 24 546 9th Ave., 99701. Phone: (907) 450-1000. Fax: (907) 457-2128. Web Site: www.101magic.com. Licensee: Capstar TX L.P. Group owner: Clear Channel Communications Inc. (acq 8-30-00; grpsl). Natl. Network: Westwood One. Natl. Rep: Christal. Format: Top 40. News staff: 2; News: one hr wkly. Target aud: 25-44; working families & adults. ♦Gary Donovan, sr VP; Pete Hutton, gen mgr; Missey Kohler, progmg dir.

KCBF(AM)— 1948: 820 khz; 10 kw-U. TL: N64 51 49 W147 45 06. Hrs open: 24 1060 Aspen St., 99709. Phone: (907) 451-5910. Fax: (907) 451-5999. Licensee: New Northwest Broadcasters LLC (group owner; acq 8-12-99; grpsl). Population served: 85,000 Natl. Network: CBS, Westwood One. Format: Sports. News staff: one; News: 4 hrs wkly. Target aud: 35-54. ♦Perry Walley, gen mgr & stn mgr; Glenn Anderson, progmg dir; Paige Smith, chief of engrg.

KXLR(FM)—Co-owned with KCBF(AM). July 1989: 95.9 mhz; 25 kw. 7 ft TL: N64 51 49 W147 45 06. Stereo. 24 Format: Classic rock. Target aud: 25-49. ♦Crys Castle, progmg mgr. Co-owned TV: KTVF(TV) affil

KFAR(AM)— 1939: 660 khz; 10 kw-U. TL: N64 52 09 W147 49 20. Hrs open: 1060 Aspen St., 99709. Phone: (907) 451-5910. Fax: (907) 451-5999. Web Site: www.akradio.com. Licensee: New Northwest Broadcasters LLC (group owner; acq 9-8-81; $675,000; FTR: 9-28-81). Population served: 30,000 Law Firm: Shaw Pittman. Format: News/talk. Target aud: 25 plus. Spec prog: Gospel 2 hrs wkly. ♦Perry Walley, gen mgr & gen sls mgr.

KWLF(FM)—Co-owned with KFAR(AM). Oct 31, 1987: 98.1 mhz; 25 kw. -7 ft TL: N64 52 38 W147 48 46. (CP: 28 kw). Stereo. Web Site: www.akradio.com. Format: CHR. Target aud: 18 plus; general.

KFBX(AM)— Sept 18, 1972: 970 khz; 10 kw-U. TL: N64 52 48 W147 40 29. Hrs open: 24 546 9th Ave., 99701. Phone: (907) 450-1000. Fax: (907) 457-2128. Web Site: www.970kfbx.com. Licensee: Capstar TX L.P. Group owner: Clear Channel Communications Inc. (acq 8-30-2000; grpsl). Population served: 68,578 Natl. Rep: Christal. Format: News, talk. News staff: one; News: 30 hrs wkly. Target aud: 35 plus; males. ♦Pete Hutton, gen mgr; Cheys Castle, prom dir & prom mgr; Charlie O'Toole, progmg dir; April LaFever, disc jockey; Marc Daly, disc jockey.

KIAK-FM—Co-owned with KFBX(AM). Sept 21, 1983: 102.5 mhz; 55 kw. 1,620 ft TL: N64 52 45 W148 03 14. Stereo. Web Site: www.kiak.com. Format: Country. Target aud: 18 plus; general. ♦Pete Van Nort, progmg dir; Doug Burnside, traf mgr; Monte Brown, local news ed; J.B. Carnahan, disc jockey; Kathryn Harris, disc jockey; Peter Van Nort, disc jockey.

KKED(FM)— Sept 20, 1962: 104.7 mhz; 10.5 kw. 440 ft TL: N64 54 42 W147 46 38. Stereo. Hrs opn: 24 546 9th Ave., 99701. Phone: (907) 450-1000. Fax: (907) 457-2128. Web Site: www.1047theedge.com. Licensee: Capstar TX L.P. Group owner: Clear Channel Communications Inc. (acq 8-30-00; grpsl). Population served: 75,000 Format: Active rock. ♦Pete Hutton, gen mgr; Mike Crosby, progmg dir.

***KSUA(FM)**— Oct 10, 1985: 91.5 mhz; 3 kw. -16 ft TL: N64 51 32 W147 49 41. Stereo. Hrs opn: 24 Box 750113, Univ. of Alaska, 307 Constitution Hall, 99775. Phone: (907) 474-7054. Fax: (907) 474-6314. E-mail: fyksua@uaf.edu Web Site: www.uaf.edu/ksua. Licensee: The University of Alaska Board of Regents. Population served: 80,000 Format: Progsv. Target aud: 14-35. Spec prog: Black 8 hrs, Sp 3 hrs, var/div music 19 hrs wkly. ♦Nick Brewer, gen mgr; Sean Bledsoe, progmg dir.

KTDZ(FM)—See College

***KUAC(FM)**— 1962: 89.9 mhz; 38 kw. 1,660 ft TL: N64 52 49 W148 03 08. Stereo. Hrs opn: 24 Box 755620, Univ. of Alaska-Fairbanks, 99775-5620. Phone: (907) 474-7491. Fax: (907) 474-5064. Web Site: www.kuac.org. Licensee: University of Alaska. Population served: 80,000 Natl. Network: NPR, PRI. Wire Svc: AP Format: Div, class, news/talk. News staff: 3; News: 32 hrs wkly. Target aud: General. Spec prog: Jazz 15 hrs, folk 10 hrs, blues 4 hrs, new age 3 hrs wkly. ♦Greg Petrowich, CEO & gen mgr; Scott Diseth, stn mgr; Gretchen Gordon, dev dir. Co-owned TV: *KUAC-TV affil.

KYSC(FM)— 2001: 96.9 mhz; 920 w. Ant 1,607 ft TL: N64 52 45 W148 03 14. Hrs open: 24 3650 Braddock St., 99701-7617. Phone: (907) 455-9690. Fax: (907) 456-3428. E-mail: ads@kyscfm.com Licensee: Tanana Valley Radio LLC (acq 9-28-2005; $700,000).

Population served: 85,000 Natl. Network: ABC. Format: Adult contemp. ♦Bill St. Pierre, pres; Terry Walley, stn mgr. Co-owned TV: KFXF(TV) affil

Fort Yukon

KZPA(AM)— Sept 30, 1993: 900 khz; 5 kw-U. TL: N66 33 24 W145 12 04. Hrs open: Box 50, E. 3rd Ave., 99740. Phone: (907) 662-8255. Fax: (907) 662-2915. E-mail: kzparadio@hotmail.com Web Site: articflash.com/kzpa. Licensee: Gwandak Public Broadcasting Inc. Format: Var/div. News: 6 hrs wkly. Target aud: All ages. ♦Arlene Joseph, pres; John Alexander, gen mgr.

Galena

***KIYU(AM)**— July 4, 1986: 910 khz; 5 kw-U. TL: N64 41 18 W156 43 29. Hrs open: 24 Box 165, 99741. Phone: (907) 656-1488. Fax: (907) 656-1734. E-mail: raven@kiyu.com Web Site: www.kiyu.com. Licensee: Big River Public Broadcasting Corp. Natl. Network: NPR. Rgnl. Network: Alaska Pub. Format: Var/div. News staff: one; News: 35 hrs wkly. Target aud: General. Spec prog: Jazz 4 hrs, Alaska native 2 hrs wkly. ♦Susie Sam, pres; Shadow Steele, gen mgr; Tim Bodony, opns dir & progmg dir.

Girdwood

***KEUL(FM)**— September 1998: 88.9 mhz; 1.4 kw horiz. Ant 636 ft TL: N60 57 44 W149 04 38. Stereo. Hrs opn: 24 Box 29, Glacier City Radio, 99587. Phone: (907) 754-2489. E-mail: radio@glaciercity.us Web Site: http://www.glaciercity.us. Licensee: Girdwood Community Club Inc. Format: Free form, free speech, electic. Target aud: Sole service provider. ♦Lewis Leonard, VP & gen mgr.

Glennallen

KCAM(AM)— Apr 16, 1964: 790 khz; 5 kw-U. TL: N62 06 52 W145 32 07. Hrs open: 24 Box 249, 99588. Phone: (907) 822-5226. Fax: (907) 822-3761. E-mail: manager@kcam.org Web Site: www.kcam.org. Licensee: Northern Light Network. (acq 2-25-92). Population served: 10,000 Natl. Network: Moody, USA. Format: Diversified. News: 28 hrs wkly. Target aud: General. Spec prog: American Indian one hr, class 10 hrs, contemp Christian 8 hrs wkly. ♦Scott Yahr, stn mgr; George Reichman, gen sls mgr; Michael Eastty, progmg dir; Scott Hill, chief of engrg.

***KXGA(FM)**— October 1994: 90.5 mhz; 3.2 kw. 219 ft TL: N62 06 31 W146 10 25. (CP: Ant 750 ft.). Hrs opn: 24
Rebroadcasts KCHU(AM) Valdez 100%.
c/o KCHU(AM), Box 467, Valdez, 99686. Secondary address: c/o KCHU(AM), 128 Pioneer Dr., Valdez 99686. Phone: (907) 835-4665. Fax: (907) 835-2847. E-mail: kchu@cvinternet.net Web Site: www.kchu.org. Licensee: Terminal Radio Inc. Format: Div, public radio. ♦John Anderson, gen mgr & opns mgr; Lisa West, gen mgr.

Haines

***KHNS(FM)**— Oct 4, 1980: 102.3 mhz; 3 kw. -1,220 ft TL: N59 13 06 W135 25 29. Stereo. Hrs opn: 24 Box 1109, One Theater Ln., 99827. Phone: (907) 766-2020. Fax: (907) 766-2022. E-mail: khns@khns.org Web Site: www.khns.org. Licensee: Lynn Canal Broadcasting. Population served: 3,000 Natl. Network: NPR. Rgnl. Network: Alaska Pub. Law Firm: Arter & Hadden. Format: Var/div. News staff: 2; News: 21 hrs wkly. Target aud: General. ♦Emily Seward, chmn & pres; Judy Erekson, gen mgr; Mary Giovanini, progmg dir; Steven Scarrott, opns mgr & mus dir.

Homer

***KBBI(AM)**— Aug 4, 1979: 890 khz; 10 kw-U. TL: N59 40 14 W151 26 38. Hrs open: 24 3913 Kachemak Way, 99603. Phone: (907) 235-7721. Fax: (907) 235-2357. E-mail: dorle@kbbi.org Web Site: www.kbbi.org. Licensee: Kachemak Bay Broadcasting Inc. Population served: 10,000 Natl. Network: PRI, NPR, AP Network News, AP Radio. Rgnl. Network: Alaska Pub. Wire Svc: AP Format: Public radio, eclectic music. News staff: one; News: 64 hrs wkly. Target aud: General. Spec prog: Reggae 3 hrs, jazz 5 hrs wkly. ♦David S. Anderson, gen mgr; Jonathan Coke, dev dir; Terry Rensel, progmg dir; Paulette Wellington, mus dir; Mike Mason, news dir.

KGTL(AM)— Feb 11, 1981: 620 khz; 5 kw-U. TL: N59 41 03 W151 37 51. Hrs open: 24 Box 109, 99603-0109. Phone: (907) 235-6000. Fax: (907) 235-6683. E-mail: kwavefm@xyz.net Satcom C-5 Tr. 3 Licensee: Peninsula Communications Inc. (group owner) Population served: 25,000 Natl. Network: USA. Law Firm: Southmayd & Miller. Format: Adult standards. News: 20 hrs wkly. Target aud: 35 plus; professionals. ♦David F. Becker, pres & gen mgr.

KWVV-FM—Co-owned with KGTL(AM). Sept 22, 1979: 103.5 mhz; 100 kw. Ant 1,150 ft TL: N59 41 03 W151 37 51. Stereo. 24 Satcom C-5, transponder 3 60,000 Law Firm: Southmayd & Miller. Format: Adult contemp. News: 14 hrs wkly. Target aud: 18-49.

***KMJG(FM)**— 2000: 88.9 mhz; 250 w. 666 ft TL: N59 40 19 W151 30 30. Hrs open: Box 1121, Kasilof, 99610. Phone: (907) 260-7702. Fax: (907) 262-1069. E-mail: kwjg915@gci.net Web Site: www.kwjg.org. Licensee: Kasilof Public Broadcasting Inc. Format: Div, oldies. ♦William Glynn, pres & gen mgr.

Houston

KBBO-FM— 1997: 92.1 mhz; 10 kw. 810 ft TL: N61 20 10 W149 30 47. Hrs open: 24 11259 Tower Rd., Anchorage, 99515. Phone: (907) 344-4045. Fax: (907) 522-6053. Web Site: www.921bob.fm. Licensee: New Northwest Broadcasters LLC. (group owner; (acq 8-12-99; $1.1 million). Population served: 260,000 Wire Svc: AP Format: 80s, 90s & whatever. ♦Pete Benedetti, CEO; Trila Bumstead, CFO; Tom Oakes, gen mgr.

***KJHA(FM)**— July 8, 1998: 88.7 mhz; 285 w. -161 ft TL: N61 37 50 W149 48 49. Hrs open:
Rebroadcasts KJNP-FM North Pole midnight-7 AM and rebroadcasts KJNP(AM) North Pole 7 AM-midnight.
Box 56359, North Pole, 99705. Phone: (907) 488-2216. Fax: (907) 488-5246. Web Site: www.mosquitonet.com/~kjnp. Licensee: Evangelistic Alaska Missionary Fellowship Inc. Format: Country gospel. Spec prog: Athabaskan Indian 2 hrs, Eskimo one hr wkly. ♦Yuonne L. Carriker, pres; Richerd T. Olson, VP.

KXLW(FM)— 2000: 96.3 mhz; 6 kw. Ant 262 ft TL: N61 33 58 W149 42 52. Stereo. Hrs opn: 11259 Tower Rd., Anchorage, 99515. Phone: (907) 344-4045. Fax: (907) 522-6053. Web Site: www.xrock963.com. Licensee: New Northwest Broadcasters LLC. (group owner; (acq 7-30-99). Wire Svc: AP Format: Rock. Target aud: 20-49; men. ♦Pete Benedetti, CEO; Trila Bumstead, CFO; Tom Oakes, gen mgr.

KZND-FM—Licensed to Houston. See Anchorage

Juneau

KINY(AM)— May 28, 1935: 800 khz; 10 kw-D, 8 kw-N. TL: N58 18 05 W134 26 26. Stereo. Hrs opn: 24 1107 W. 8th St., Suite 2, 99801. Phone: (907) 586-1800. Fax: (907) 586-3266. E-mail: kiny@ptialaska.net Web Site: www.kinyradio.com. Licensee: Alaska-Juneau Communications Inc. Population served: 35,000 Format: Adult contemp. News staff: 2; News: 4 hrs wkly. Target aud: General. ♦Dennis W. Egan, pres & gen mgr; Kelly Peres, opns mgr & disc jockey; Tim Armstrong, gen sls mgr; Jim Morgan, prom dir & disc jockey; Charlie Gray, engrg dir; Christine Personnet, women's int ed; Chris Burns, disc jockey; Guy James, disc jockey; Ron Davis, disc jockey.

KSUP(FM)—Co-owned with KINY(AM). Dec 1, 1984: 106.3 mhz; 10 kw. -1,007 ft TL: N58 18 05 W134 26 26. Stereo. 24 Phone: (907) 586-1063. E-mail: ksup@ptialaska.net Web Site: www.ksupradio.com. Format: Classic, contemp rock. ♦Kelly Peres, progmg dir.

KJNO(AM)— Oct 19, 1952: 630 khz; 5 kw-D, 1 kw-N. TL: N58 19 47 W134 28 17. Hrs open: 24 3161 Channel Dr., Suite 2, 99801. Phone: (907) 586-3630. Fax: (907) 463-3685. Web Site: www.kjno.com. Licensee: Alaska Broadcast Communications Inc. (group owner; acq 1972; with co-located FM). Population served: 33,000 Natl. Network: CBS Radio. Natl. Rep: Tacher. Rgnl rep: Tacher Law Firm: Garvey, Schubert & Barer. Format: Talk. Target aud: 25-54. ♦Roy Paschal, pres; Richard Burns, VP & gen mgr; Jeff McCoy, progmg dir.

KTKU(FM)—Co-owned with KJNO(AM). July 9, 1984: 105.1 mhz; 3.84 kw. -1,057 ft TL: N58 19 47 W134 28 17. Stereo. 24 Web Site: www.kjno.com. Format: Hot country.

***KRNN(FM)**— 1999: 102.7 mhz; 6 kw. Ant -417 ft TL: N58 17 09 W134 25 40. Hrs open: 360 Egan Dr., 99801-1748. Phone: (907) 586-1670. Fax: (907) 586-3612. Web Site: www.ktoo.org. Licensee: Capital Community Broadcasting Inc. (acq 12-27-2006; $676,400 with KXLL(FM) Juneau). Population served: 40,000 Format: Diverse. ♦Bill Legere, pres; Cheryl Levitt, stn mgr; John Beiler, dev dir; Jeff Brown, progmg dir; Jeff Brown, mus dir; Rosemarie Alexander, news dir.

***KTOO(FM)**— Jan 27, 1974: 104.3 mhz; 1.4 kw. Ant -1,016 ft TL: N58 18 04 W134 25 21. Stereo. Hrs opn: 24 360 Egan Dr., 99801-1748. Phone: (907) 586-1670. Fax: (907) 586-3612. Fax: (907) 586-2561 (news). E-mail: info@ktoo.org Web Site: www.ktoo.org. Licensee: Capital Community Broadcasting Inc. Population served: 30,000 Natl. Network: NPR, PRI. Rgnl. Network: Alaska Pub. Law Firm: Schwartz, Woods & Miller. Format: Diversified, news. News staff: 14; News: 48 hrs wkly. Target aud: General. Spec prog: Children one hrs, folk 8 hrs, Sp 2 hsr, French 2 hrs, jazz 14 hrs, Alaska native one hr wkly. ♦Bill Legere, pres & gen mgr; Cheryl Levitt, stn mgr; Mike Sakarias, opns mgr; Rosemarie Alexander, news dir. Co-owned TV: *KTOO-TV affil.

KXLJ(AM)—Not on air, target date: unknown: 1330 khz; 10 kw-D, 3.5 kw-N. TL: N58 19 54 W134 27 46. Hrs open: Box 1471, Evergreen, CO, 80437. Phone: (303) 688-5162. Fax: (303) 660-4930. Licensee: Seattle Streaming Radio LLC. (acq 6-6-2006; $150,000 for CP). ♦David M. Drucker, gen mgr.

***KXLL(FM)**— October 1999: 100.7 mhz; 6 kw. Ant -417 ft TL: N58 17 09 W134 25 40. Stereo. Hrs opn: 360 Egan Dr., 99801. Phone: (907) 586-1670. Fax: (907) 586-3612. E-mail: whiteoakbroadcasting@gci.net Web Site: www.todaysbesthits.com. Licensee: Capital Community Broadcasting Inc. (acq 12-27-2006; $676,400 with KRNN(FM) Juneau). Population served: 40,000 Format: AAA, alternative. ♦Bill Legere, pres; Andy Kline, progmg dir.

Kasilof

***KABN-FM**— 2003: 89.5 mhz; 500 w horiz. Ant 197 ft TL: N60 22 44 W151 11 30. Hrs open: Box 2425, Seward, 99664. Phone: (907) 224-5793. Fax: (907) 224-4702. Web Site: www.oneskyradio.com. Licensee: Alaska Educational Radio System Inc. ♦Wolfgang Kurtz, pres.

KFSE(FM)—Not on air, target date: unknown: 106.9 mhz; 8 kw. Ant 203 ft TL: N60 25 55 W151 08 26. Hrs open: 40960 K-Beach Rd., Kenai, 99611. Phone: (907) 283-8700. Fax: (907) 283-9177. Licensee: KSRM Inc. (acq 3-22-2007; $210,000 for CP). ♦John C. Davis, pres.

***KWJG(FM)**— July 29, 1998: 91.5 mhz; 1 kw. 262 ft TL: N60 22 44 W151 11 30. Stereo. Hrs opn: 24 Box 1121, AR, 99610. Phone: (907) 260-7702. Fax: (907) 262-1069. Licensee: Kasilof Public Broadcasting Inc. Population served: 6,700 Law Firm: Bechtel & Cole. Format: Oldies, variety/diverse. ♦William J. Glynn Jr., pres & gen mgr.

***KWMD(FM)**— 2003: 90.5 mhz; 500 w horiz. Ant 197 ft TL: N60 22 44 W151 11 30. Hrs open: 3700 Woodland Dr., Suite 800, Anchorage, 99517. Phone: (800) 974-6525. E-mail: aers@oneskyradio.com Licensee: Alaska Educational Radio System Inc.

Kenai

***KDLL(FM)**— 1981: 91.9 mhz; 4.9 kw. 72 ft TL: N60 34 03 W151 07 25. Hrs open: 24 Box 2111, 99611. Phone: (907) 283-8433. Fax: (907) 283-6701. E-mail: allen@kdllradio.org Web Site: www.kdllradio.org. Licensee: Pickle Hill Public Broadcasting Inc. Population served: 30,000 Natl. Network: NPR, PRI. Rgnl. Network: Alaska Pub. Format: Var/div, news. News staff: one; News: 60 hrs wkly. Target aud: Affluent. Spec prog: American idian 10 hrs wkly. ♦Dave Anderson, gen mgr; Allen Auxier, stn mgr.

KPEN-FM—See Soldotna

KSRM(AM)—See Soldotna

KWHQ-FM— Nov 18, 1976: 100.1 mhz; 3 kw. Ant 260 ft TL: N60 30 49 W151 11 19. Stereo. Hrs opn: 24 40960 K-Beach Rd., 99611. Phone: (907) 283-9430. Fax: (907) 283-9177. E-mail: info@radiokenai.com Web Site: www.radiokenai.com. Licensee: KSRM Inc. (group owner). Population served: 45,000 Law Firm: Pepper & Corazzini. Wire Svc: AP Format: Modern country. News staff: 2; News: 12 hrs wkly. Target aud: 18-49. ♦John C. Davis, CEO & pres; J.R. Kitchens, opns mgr; James "Red" Goodwin, mktg dir.

Ketchikan

KFMJ(FM)— Sept 23, 1996: 99.9 mhz; 115 w. Ant 2,234 ft TL: N55 21 40 W131 47 43. Stereo. Hrs opn: 24 516 Stedman St., 99901. Phone: (907) 247-3699. Fax: (907) 247-5365. E-mail: kfmj@alaska.fm Web Site: www.kfmj.com. Licensee: TLP Communications Inc. Population served: 20,000 Natl. Network: ABC, USA. Wire Svc: AP Format: Oldies. News: 18.5 hrs wkly. Target aud: 30 plus. ♦Robert J. Kern, chmn & pres; Robert Kern, gen mgr; Julie Slanaker, gen sls mgr; Stewart White, progmg dir.

KGTW(FM)—Listing follows KTKN(AM).

***KRBD(FM)**— May 1976: 105.3 mhz; 3.4 kw. Ant 69 ft TL: N55 20 23 W131 37 29. Stereo. Hrs opn: 24 123 Stedman St., 99901. Phone: (907) 225-9655. Fax: (907) 247-0808. Web Site: www.krbd.org. Licensee: Rainbird Community Broadcasting Corp. Population served: 18,000 Natl. Network: NPR, PRI. Rgnl. Network: Alaska Pub. Format: Div. Target aud: General. Spec prog: Class 11 hrs, C&W 14 hrs, folk 10 hrs, jazz 10 hrs, tribal topics 5 hrs wkly. ♦Jeff Siefert, gen mgr; prog dir, opns mgr; Deanna Garrison, news dir.

KTKN(AM)— 1942: 930 khz; 5 kw-D, 1 kw-N. TL: N55 20 22 W131 38 12. Hrs opn: 24 526 Stedman St., 99901. Phone: (907) 225-2193. Fax: (907) 225-0444. E-mail: bmesser@gci.net Web Site: www.ktkn.com. Licensee: Alaska Broadcast Communications Inc. (group owner). Population served: 30,000 Rgnl rep: Tacher. Law Firm: Haley, Bader & Potts. Format: Adult contemp, news/talk. News staff: 5; News: 20 hrs wkly. Target aud: 25 plus. ♦Blake Messer, gen mgr; Jamie Beldo, progmg dir.

KGTW(FM)—Co-owned with KTKN(AM). November 1987: 106.7 mhz; 4 kw. -308 ft TL: N55 20 22 W131 38 12. Stereo. 24 Web Site: gateway1067.com. Format: Country. Target aud: 18 plus. ♦John Hunt, progmg dir.

Kodiak

***KMXT(FM)**— June 1, 1976: 100.1 mhz; 3 kw. 3 ft TL: N57 47 41 W152 23 28. Stereo. Hrs opn: 24 620 Egan Way, 99615. Phone: (907) 486-3181. Fax: (907) 486-2733. E-mail: kmxt@kmxt.org Web Site: www.kmxt.org. Licensee: Kodiak Public Broadcasting Corp. (acq 10-2-75). Population served: 12,000 Natl. Network: NPR, PRI. Rgnl. Network: Alaska Pub. Format: Div, news. News staff: 2; News: 5 hrs wkly. Target aud: General. ♦Mike Wall, gen mgr; Fred Hawley, dev dir.

KRXX(FM)—Listing follows KVOK(AM).

KVOK(AM)— Nov 7, 1974: 560 khz; 1 kw-U. TL: N57 48 36 W152 20 54. Hrs open: 24 Box 708, 99615. Secondary address: 1315 Mill Bay Rd. 99615. Phone: (907) 486-5159. Fax: (907) 486-3044. E-mail: ayn@kvok.com Web Site: www.kvok.com. Licensee: Kodiak Island Broadcasting Co. Inc. (acq 4-3-00; $500,000 with co-located FM). Natl. Network: ABC. Format: Country, talk. Target aud: 25-56. ♦Matt Wilson, gen mgr & mus dir; Ayn DuBois, traf mgr.

KRXX(FM)—Co-owned with KVOK(AM). 1987: 101.1 mhz; 3.1 kw. 46 ft TL: N57 48 36 W152 20 54. Web Site: www.jackfmkodiak.com.14,000 Format: Classic rock, hot adult contemp. Target aud: 18-56.

Kotzebue

***KHZK(FM)**— 2007: 103.9 mhz; 200 w. Ant 46 ft TL: N66 54 11 W162 34 16. Hrs open:
Rebroadcasts KSRD(FM) Saint Joseph, MO 100%.
5331 Mt. Alifan Dr., San Diego, CA, 92111. Phone: (858) 277-4991. Fax: (858) 277-1365. Web Site: www.horizonradio.org. Licensee: Horizon Christian Fellowship. (acq 2-9-2006; grpsl). Format: Christian. ♦Mike MacIntosh, pres.

***KOTZ(AM)**— March 1973: 720 khz; 10 kw-U. TL: N66 50 22 W162 34 05. Hrs open: Box 78, 99752. Phone: (907) 442-3434. Fax: (907) 442-2292. Licensee: Kotzebue Broadcasting Inc. Population served: 8,000 Rgnl. Network: Alaska Pub. Format: Var. Target aud: General; 90% rural Eskimo, 10% white-collar caucasian. ♦Suzy Erlich, gen mgr.

McCarthy

***KXKM(FM)**— October 1994: 89.7 mhz; 102 w. -169 ft TL: N61 24 58 W143 01 19. Hrs open: 24
Rebroadcasts KCHU(AM) Valdez 100%.
c/o KCHU(AM), Box 467, Valdez, 99686. Secondary address: c/o KCHU(AM), 128 Pioneer Dr., Valdez 99686. Phone: (907) 835-4665. Fax: (907) 835-2847. E-mail: kchu@cvinternet.net Web Site: www.kchu.org. Licensee: Terminal Radio Inc. Natl. Network: NPR, PRI. Rgnl. Network: Alaska Radio Net. Format: Div, educ, news/talk, public radio. News staff: one; News: 10 hrs wkly. Target aud: General. ♦Lisa West, gen mgr; John Anderson, opns mgr.

Naknek

KAKN(FM)— May 1987: 100.9 mhz; 3 kw. 338 ft TL: N58 44 33 W156 58 39. Stereo. Hrs opn: 24 Box 0214, 99633. Secondary address: Mile 2 AK Peninsula Hwy. 99633. Phone: (907) 246-7492. Fax: (907) 246-7462. E-mail: studio@victoryradionetwork.com Web Site: www.victoryradionetwork.com. Licensee: Bay Broadcasting Co. Population served: 15,000 Natl. Network: USA. Format: Light adult contemp Christian, southern gospel, news. News: 15 hrs wkly. Target aud: General; mobile town/village population & coml fishermen. ♦Rev. Thomas C. Olson, pres; Michael Johnson, VP; Thomas Olsen, gen mgr; Anita Karlsson, stn mgr.

Nenana

KIAM(AM)— June 28, 1985: 630 khz; 10 kw-D, 3.1 kw-N. TL: N64 28 43 W149 05 10. Hrs open: 24 Box 474, 99760. Phone: (907) 832-5426. Fax: (907) 832-5450. E-mail: Alaskaradio@vfcm.org Web Site: www.vfcm.org. Licensee: Voice for Christ Ministries. Natl. Network: USA. Format: News/talk, music, Christian. News: News progm 20 hrs wkly. Target aud: General. Spec prog: American Indian 3 hrs, class one hr wkly. ♦Bob Eldridge, exec VP; Brian Blair, stn mgr; David Horning, gen mgr & mus dir.

Nikiski

KXBA(FM)— March 4, 2000: 93.3 mhz; 50 kw. Ant 243 ft TL: N60 30 39 W151 16 12. Stereo. Hrs opn: Box 109, Homer, 99603-0109. Phone: (907) 262-6000. Phone: (907) 283-7451. Fax: (907) 283-8461. Licensee: Peninsula Communications Inc. (group owner) Population served: 60,000 Law Firm: Southmayd & Miller. Format: Oldies. News: 8 hrs wkly. Target aud: 25-54. ♦David F. Becker, pres & gen mgr; Tim White, opns mgr & news dir.

Nome

KICY(AM)— Apr 17, 1960: 850 khz; 50 kw-U. TL: N64 29 15 W165 18 53. Hrs open: 24 Box 820, 99762. Secondary address: 408 W. D St. 99762. Phone: (907) 443-2213. Fax: (907) 443-2344. E-mail: office@kicy.org Web Site: www.kicy.org. Licensee: Arctic Broadcasting Association (group owner). Population served: 15,000 Natl. Network: ABC, Moody, Salem Radio Network. Rgnl rep: Alaska Broadcast Media Law Firm: Wombel, Carlyle, Sandridge & Rice. Format: Southern gospel, Russian. Target aud: 25-64. ♦Ted Haney, pres; Dennis Weidler, gen mgr & gen sls mgr.

KICY-FM— Sept 11, 1977: 100.3 mhz; 84 w. 40 ft TL: N64 30 04 W165 24 39. Stereo. 24 Web Site: www.kicy.org.4,500 Format: Christian. Target aud: 18-35.

***KNOM(AM)**— July 14, 1971: 780 khz; 25 kw-D, 14 kw-N. TL: N64 29 16 W165 17 58. Hrs open: 24 107 W. 3rd Ave., 99762. Phone: (907) 443-5221. Fax: (907) 443-5757. E-mail: info@knom.org Web Site: www.knom.org. Licensee: Catholic Bishop of Northern Alaska. Population served: 20,000 Law Firm: Wilkinson, Barker & Knauer. Wire Svc: AP Format: Div, relg, news/talk. News staff: 2; News: 30 hrs wkly. Target aud: General. Spec prog: Eskimo 6 hrs, CHR 12 hrs, relg 20 hrs, class 5 hrs, weather 14 hrs wkly. ♦Kelly Brabec, pres & progmg dir; Thomas Busch, CFO & gen mgr; Ric Schmidt, gen mgr & progmg dir; Thomas A. Busch, dev dir; Paul Korchin, news dir & chief of engrg.

KNOM-FM— May 17, 1993: 96.1 mhz; 88 w. -138 ft TL: N64 29 56 W165 23 56. Stereo. 24 4,000 Natl. Network: AP Radio. Law Firm: Wilkinson, Barker & Knauer. News staff: 2; News: 30 hrs wkly.

North Pole

***KJNP(AM)**— Oct 11, 1967: 1170 khz; 50 kw-D, 21 kw-N. TL: N64 45 34 W147 19 26. Hrs open: Box 56359, 99705. Phone: (907) 488-2216. Fax: (907) 488-5246. Licensee: Evangelistic Alaska Missionary Fellowship. Population served: 75,000 Format: C&W, relg. Spec prog: Russian 11 hrs, Athabaskan Indian 2 hrs, Eskimo one hr wkly. ♦Yuonne Carriker, pres; Richard T. Olson, VP.

KJNP-FM— Oct 11, 1977: 100.3 mhz; 25 kw. 1,570 ft TL: N64 52 44 W148 03 10. Stereo. Format: Conservative btfl mus, relg. ♦Leland Carriker, disc jockey; Richard Olson, disc jockey. Co-owned TV: *KJNP-TV affil.

Palmer

***KJLP(FM)**— August 2005: 88.9 mhz; 250 w. Ant -210 ft TL: N61 37 18 W149 01 16. Hrs open: Box 210389, Anchorage, 99521. Secondary address: 6401 E. Northern Lights, Anchorage 99521. Phone: (907) 333-5282. Fax: (907) 333-5282. E-mail: tom@katb.org Licensee: Christian Broadcasting Inc. Format: Christian. ♦Tom Steigleman, gen mgr.

Petersburg

***KFSK(FM)**— September 1977: 100.9 mhz; 2 kw. -482 ft TL: N56 48 55 W132 57 12. Stereo. Hrs opn: 24 Box 149, 99833. Phone: (907) 772-3808. Fax: (907) 772-9296. E-mail: kfsk@alaska.net Web Site: www.kfsk.org. Licensee: Narrows Broadcasting Corp. Population served: 3,000 Natl. Network: PRI, NPR, AP Network News. Rgnl. Network: Alaska Pub. Format: News, pub affrs. ♦Tom Abbott, gen mgr.

KRSA(AM)— Sept 24, 1982: 580 khz; 5 kw-U, DA-1. TL: N56 40 23 W132 55 00. Hrs open: 24 Box 650, 99833. Phone: (907) 772-3891. Fax: (907) 772-4538. E-mail: krsa@krsa.net Web Site: www.krsa.net. Licensee: Northern Light Network. (acq 2-20-92). Population served: 25,000 Format: Relg, country. News: 20 hrs wkly. Target aud: General. Spec prog: Class 5 hrs, children, oldies 5 hrs wkly. ♦Andrew Mazzella, pres, gen mgr & stn mgr.

Saint Paul

***KUHB-FM**— July 4, 1984: 91.9 mhz; 3 kw. 56 ft TL: N57 07 14 W170 16 45. Stereo. Hrs opn: Box 905, Pribios School District, 99660. Phone: (907) 546-2254. Fax: (907) 546-2367. E-mail: gm@ kuhb.org Licensee: Pribilof School District. Natl. Network: NPR. Format: Anything & everything. ♦Walt Gregg, gen mgr; B.J. Kibbe, news dir.

Sand Point

***KSDP(AM)**— Mar 2, 1983: 830 khz; 1 kw-U. TL: N55 21 06 W160 28 02. Hrs open: 24
Rebroadcasts KDLG(AM) Dillingham.
Box 328, City Bldg, 328 Main St., 99661. Phone: (907) 383-5737. Fax: (907) 383-5737. E-mail: ksdp@ksdpradio.com Web Site: www.ksdpradio.org. Licensee: Aleutian Peninsula Broadcasting Inc. Population served: 6,000 Natl. Network: NPR, PRI. Format: Div. Target aud: General. Spec prog: Gospel. ♦Brian E. Koral, gen mgr.

Seward

KSWD(AM)— November 1948: Stn currently dark. 950 khz; 1 kw-U. TL: N60 06 51 W149 26 44. Hrs open: 10914 E. 46th Ave., Spokane Valley, WA, 99206-9466. Phone: (907) 455-9690. Licensee: Northern Radio Inc. (acq 12-20-2002; with KPFN(FM) Seward). Population served: 3,200 ♦Don Cary, gen mgr.

KSWD-FM— 1998: 105.9 mhz; 3 kw. -1,312 ft TL: N60 05 27 W149 20 20. Hrs open: 10914 E. 46th Ave., Spokane Vly, WA, 99206-9466. Phone: (907) 455-9690. Fax: (907) 455-4369. E-mail: kysc@gci.net Licensee: Northern Radio Inc. (acq 5-20-2004; with KYSC(FM) Fairbanks). Format: Adult contemp. ♦Fred Dunham, CEO; Don Cary, gen mgr.

Sitka

***KCAW(FM)**— Feb 19, 1982: 104.7 mhz; 5 kw. -612 ft TL: N57 03 13 W135 21 07. Stereo. Hrs opn: 24 2 Lincoln St., Suite B, 99835. Phone: (907) 747-5877. Phone: (907) 747-5879. Fax: (907) 747-5977. Web Site: www.ravenradio.org. Licensee: Raven Radio Foundation. Natl. Network: NPR, PRI. Rgnl. Network: Alaska Pub. Format: Div, news.

Target aud: General. Spec prog: Class 15 hrs, Indian 3 hrs wkly. ♦Ken Fate, CEO & gen mgr; Steve Will, progmg dir; Robert Woolsey, news dir.

KIFW(AM)— September 1949: 1230 khz; 1 kw-U. TL: N57 03 27 W135 20 02. Hrs open: 24 611 Lake St., 99835. Phone: (907) 747-6626. Phone: (907) 747-5439. Fax: (907) 747-8455. E-mail: kifw@ptialaska.net Web Site: www.kifw.com. Licensee: Alaska Broadcast Communications Inc. (group owner; acq 12-21-00; grpsl). Population served: 8,700 Law Firm: Haley, Bader & Potts. Format: MOR, oldies, news/talk. News: 60 hrs wkly. Target aud: 18-49; all demographics. ♦Steve Rhyner, pres; Blake Messer, stn mgr; Bobbie Rusk, gen sls mgr; Devin Reiter, progmg dir; Clint Daniels, news dir; Chris Kobger, chief of engrg.

KSBZ(FM)—Co-owned with KIFW(AM). Oct 18, 1990: 103.1 mhz; 3 kw. 144 ft TL: N57 03 27 W135 20 02. Stereo. 24 Phone: (907) 747-6627. Web Site: www.ksbz.com. Format: Country. Target aud: 18-34. ♦Amy Denny, pub affrs dir.

Soldotna

KKIS-FM— Mar 2, 1994: 96.5 mhz; 10 kw. 259 ft TL: N60 31 26 W151 03 23. Stereo. Hrs opn: 24 40960 K-Beach Rd., Kenai, 99611. Phone: (907) 283-5821. Fax: (907) 283-9177. E-mail: info@radiokenai.com Web Site: www.radiokenai.com. Licensee: KSRM Inc. (group owner; (acq 12-7-2001; $350,000 with co-located AM). Population served: 45,000 Natl. Network: ABC. Law Firm: Pepper & Corazzini LLP. Format: Adult contemp. News staff: one; News: 2 hrs wkly. Target aud: 18-49. ♦John C. Davis, CEO & pres; Steve Holloway, opns mgr; James "Red" Goodwin, mktg dir; Joe Nicks, news dir.

KSLD(AM)—Co-owned with KKIS-FM. Apr 6, 1985: 1140 khz; 10 kw-U. TL: N60 31 26 W151 03 23. Stereo. 24 Phone: (907) 283-8700. E-mail: ksld@radiokenai.com Web Site: www.radiokenai.com.45,000 Natl. Network: Westwood One. Law Firm: Pepper and Corazzini. Format: Classic rock, CHR. News: one hr wkly. Target aud: 25-59.

KPEN-FM— Dec 1, 1984: 101.7 mhz; 25 kw. 240 ft TL: N60 30 40 W151 16 12. Stereo. Hrs opn: 24 Box 109, Homer, 99603. Phone: (907) 262-6000. Phone: (907) 283-7451. Fax: (907) 235-6683. E-mail: kwavefm@xyz.net Licensee: Peninsula Communications Inc. (group owner) Population served: 50,000 Natl. Network: USA. Law Firm: Southmayd & Miller. Format: Country. Target aud: 25-54. ♦David F. Becker, pres & gen mgr; Tim White, opns mgr & news dir.

KSRM(AM)— Sept 27, 1967: 920 khz; 5 kw-U. TL: N60 30 49 W151 11 19. Hrs open: 24 40960 K-Beach Rd., Kenai, 99611. Phone: (907) 283-5959. Fax: (907) 283-5811. E-mail: info@radiokenai.com Web Site: www.radiokenai.com. Licensee: KSRM Inc. (group owner; (acq 4-72). Population served: 45,000 Law Firm: Pepper & Corazzini. Format: News/talk. News staff: one; News: 105 hrs wkly. Target aud: 25-54. ♦John C. Davis, CEO, chmn, pres & gen mgr; Dayne Clark, exec VP; Steve Holloway, opns mgr; James "Red" Goodwin, dev dir, mktg dir, prom dir & adv dir; J.R. Kitchens, progmg dir; Joe Nicks, news dir; Paul Jewusiak, engrg dir; Dan Gensel, sports cmtr.

Sterling

***KRAW(FM)**— 2006: 90.1 mhz; 1.2 kw horiz. Ant 20 ft TL: N60 29 16 W150 47 38. Hrs open: P.O. Box 2425, Seward, 99664. Phone: (800) 974-6525. Web Site: www.oneskyradio.com. Licensee: Alaska Educational Radio System Inc. ♦Wolfgang Kurtz, pres.

Talkeetna

***KTNA(FM)**— February 1993: 88.5 mhz; 1.9 kw. 62 ft TL: N62 19 05 W150 17 52. Stereo. Hrs opn: 24 Box 300, 13764 2nd Ave., Talkeentna, 99676. Phone: (907) 733-1700. Fax: (907) 733-1781. E-mail: info@ktna.org Web Site: www.ktna.org. Licensee: Talkeetna Community Radio Inc. Population served: 4,500 Natl. Network: NPR, PRI. Format: Eclectic, news/talk. News staff: one; News: 15 hrs wkly. Target aud: General; rural Alaskans. Spec prog: Blues 5 hrs, light rock 5 hrs wkly. ♦Robert Ambrose, gen mgr & stn mgr; Kirsten Merkley, gen sls mgr; Deborah Brock, progmg dir; Amanda Stossel, news dir.

Tok

***KUDU(FM)**— Mar 3, 1998: 91.9 mhz; 200 w. -121 ft TL: N63 19 53 W143 07 02. Hrs open: Box 661, 99780. Phone: (907) 883-4397. Phone: (907) 883-5855. Fax: (907) 883-5245. E-mail: deflee@oddpost.com Web Site: www.lifetalk.net. Licensee: Lifetalk Broadcasting Association. Format: Relg, inspirational music, talk. ♦Francine Lee, gen mgr.

Unalakleet

KNSA(AM)— 1998: 930 khz; 2.5 kw-U. TL: N63 53 17 W160 41 29. Hrs open: Box 178, 99684. Phone: (907) 624-3100. Phone: (907) 624-3101. Fax: (907) 624-3130. Licensee: Unalakleet Broadcasting Inc. Format: Var. ♦Henry Ivanoff, stn mgr.

Unalaska

KIAL(AM)— Sept 1, 1978: 1450 khz; 50 w-U, DA-1. Hrs opn: 24 Box 181, 99685-9999. Phone: (907) 581-1888. Fax: (907) 581-1634. E-mail: info@kial.org Web Site: www.kial.org. Licensee: Unalaska Community Broadcasting. Population served: 4,200 Natl. Network: NPR. Format: Public radio. News staff: one; News: 4 hrs wkly. Spec prog: News 14 hrs, rock 8 hrs, relg 4 hrs, country 4 hrs, Black 4 hrs, gospel 2 hrs wkly. ♦Michael Edenfield, pres & gen mgr.

Valdez

***KCHU(AM)**— Aug 3, 1986: 770 khz; 9.7 kw-U. TL: N61 06 40 W146 15 39. Hrs open: 24 128 Pioneer Dr., 99686. Phone: (907) 835-4665. Fax: (907) 835-2847. E-mail: kchu@cvinternet.net Web Site: www.kchu.org. Licensee: Terminal Radio Inc. (acq 10-84; $250,000; FTR: 10-8-84). Natl. Network: NPR, PRI. Rgnl. Network: Alaska Pub. Format: Div. News staff: 2; News: 40 hrs wkly. Target aud: General. ♦Lisa West, gen mgr; John Anderson, opns dir.

KVAK(AM)— January 1983: 1230 khz; 1 kw-U. TL: N61 07 16 W146 15 25. Hrs open: 24 Box 367, 99686. Secondary address: 501 E. Bremner St. 99686. Phone: (907) 835-5825. Fax: (907) 835-5158. Licensee: North Wave Communications Inc. (acq 1996). Format: Country, talk. News: one hr wkly. Target aud: General. ♦Laurie Prax, pres & traf mgr.

KVAK-FM— May 28, 1999: 93.3 mhz; 6 kw. -1,958 ft TL: N61 07 16 W146 15 25. Box 367, 501 E. Bremner St., Suite 2, 99686. Format: Hot adult contemp.

Wasilla

KMBQ(FM)— Mar 15, 1985: 99.7 mhz; 51 kw. -187 ft TL: N61 38 03 W149 26 25. Stereo. Hrs opn: 24 2200 E. Parks Hwy., 99654. Phone: (907) 373-0222. Fax: (907) 376-1575. E-mail: john@kmbq Web Site: www.kmbq.com. Licensee: KMBQ Corp. Population served: 91,000 Natl. Network: CNN Radio. Law Firm: Garvey, Schukert & Barer. Format: Adult contemp. News staff: 2; News: 13 wkly. Target aud: 25-54; mid-upper class suburbanites & farm community. ♦John Klapperich, CEO, pres, VP & gen mgr; Debbie Rinkey, gen sls mgr; Mike Ford, progmg dir; Kathy Phillips, news dir; Van Craft, chief of engrg.

Wrangell

***KSTK(FM)**— July 2, 1977: 101.7 mhz; 3 kw. -294 ft TL: N56 27 14 W132 22 54. Stereo. Hrs opn: 24 Box 1141, 99929. Secondary address: 202 St. Michael's 99929. Phone: (907) 874-2345. Fax: (907) 874-3293. Web Site: kstk.org. Licensee: Wrangell Radio Group Inc. Population served: 3,200 Natl. Network: NPR, PRI. Rgnl. Network: Alaska Pub., Alaska Radio Net. Format: Div. News staff: 2; News: 20 hrs wkly. Target aud: General. Spec prog: Class 4 hrs, country 16 hrs, jazz 8 hrs wkly. ♦Peter Helgeson, gen mgr; Cindy Sweat, dev dir; Dawn Stevens, progmg dir.

Arizona

Apache Junction

KVVA-FM— July 1, 1973: 107.1 mhz; 25 kw. 312 ft TL: N33 26 48 W111 37 32. Stereo. Hrs opn: 5700 Wilshire Blvd., Ste 250, Los Angeles, CA, 90036. Secondary address: 50 North 44th St., Suite 425, Phoenix 85008. Phone: (323) 900-6100 / (602) 266-2005. Fax: (323) 900-6108. Licensee: Entravision Holdings LLC. Group owner: Entravision Communications Corp. (acq 7-28-00; grpsl). Law Firm: Cohn & Marks. Format: Adult contemp, Sp, Latin contemp. Target aud: 18-49; Hispanic. ♦Tom Duran, gen mgr; Edgar Pineda, progmg dir.

Arizona City

KKMR(FM)— Apr 13, 1985: 106.5 mhz; 6 kw. Ant 292 ft TL: N32 50 04 W111 38 15. Hrs open: 24 4745 N. 7th St., Suite 140, Phoenix, 85014. Phone: (602) 308-7900. Fax: (602) 308-7979. Web Site: www.univision.com. Licensee: HBC License Corp. Group owner: Univision Radio (acq 9-22-2003; grpsl). Format: Sp Adult Hits. Target aud: 25-54. ♦Mary McEvilly-Hernandez, VP & gen mgr.

Bagdad

KFTT(FM)— 2002: 103.1 mhz; 900 w horiz. Ant 1,250 ft TL: N34 33 25 W113 16 00. Hrs open: Box 1866, Lake Havasu City, 86405. Phone: (928) 855-1051. Fax: (928) 855-7996. E-mail: epress@maddog.net Web Site: www.maddog.net. Licensee: Smoke and Mirrors LLC (acq 4-18-2001). Format: Adult standards. ♦Chris Rolando, gen mgr.

Benson

KAVV(FM)— April 1983: 97.7 mhz; 6 kw. 590 ft TL: N31 54 24 W110 27 08. Stereo. Hrs opn: 24 Box 18899, Tucson, 85731-8899. Secondary address: 156 W. 5th St. 85602. Phone: (520) 586-9797. E-mail: cave@gainbroadband.com Web Site: www.cavefm.com. Licensee: Stereo 97 Inc. Population served: 60,000 Format: C&W. Target aud: 25-49. Spec prog: Relg 3 hrs wkly. ♦Jack Lotsof, pres; Paul Lotsof, gen mgr, stn mgr, progmg dir & chief of engrg.

Bisbee

***KRMB(FM)**— 1997: 90.1 mhz; 47 w. 2,247 ft TL: N31 28 52 W109 57 30. Hrs open:
Rebroadcasts KRMC(FM) Douglas 100%.
Box 2520, Douglas, 85603. Phone: (520) 364-5392. Fax: (520) 364-5392. Licensee: World Radio Network Inc. (group owner) Format: Relg, Sp. ♦Glen Lafitte, gen mgr.

KWCD(FM)— Oct 12, 1979: 92.3 mhz; 51 w. 2,217 ft TL: N31 28 52 W109 57 30. Stereo. Hrs opn: 24 Box 2770, 2300 Busby Dr., Sierra Vista, 85636-2770. Phone: (520) 458-4313. Fax: (520) 458-4317. Licensee: CCR-Sierra Vista IV LLC. Group owner: Cherry Creek Radio LLC (acq 12-19-2003; grpsl). Population served: 92,000 Natl. Network: Westwood One. Natl. Rep: Tacher. Format: Country. News staff: one; News: one hr wkly. Target aud: 25-54; financially secure adults & military personnel. ♦Paul Orlando, gen mgr; Grady Butler, opns mgr.

***KWRB(FM)**— December 1996: 90.9 mhz; 99w. 2,093 ft TL: N31 28 58 W109 57 29. Hrs open: 24 96-C S. Carmichael, Sierra Vista, 85635. Phone: (520) 452-8022. Fax: (520) 452-0927. E-mail: kwrb@lwpn.org Web Site: www.kwrb.org. Licensee: World Radio Network Inc. Population served: 115,000 Format: Christian, educ, inspirational. Target aud: Women 35+. ♦Dwight Lind, gen mgr.

Black Canyon City

KMIA(AM)— Sept 1, 1981: 710 khz; 22 kw-D, 3.9 kw-N, DA-2. TL: N34 04 48 W112 09 15. Hrs open: 24 501 N. 44th St., Suite 425, Phoenix, 85008. Phone: (602) 776-1400. Fax: (602) 279-2921. Licensee: Entravision Holdings LLC. Group owner: Entravision

Communications Corp. (acq 7-28-00; grpsl). Format: Sports, Sp. Target aud: 18-54. ♦Tom Duran, gen mgr.

Buckeye

KDVA(FM)— 1993: 106.9 mhz; 6 kw. 305 ft TL: N33 27 01 W112 35 58. Hrs open: 5700 Wilshire Blvd., Suite250, Los Angeles, CA, 90036. Secondary address: 50 North 44th St., Suite 425, Phoenix 85008. Phone: (323) 900-6100. Fax: (323) 900-6108. Licensee: Entravision Holdings LLC. Group owner: Entravision Communications Corp. (acq 5-31-01; $10 million). Format: Latin Contempory, "Radio Romantica". Spec prog: Black 6 hrs, gospel 7 hrs wkly. ♦Tom Duran, gen mgr; Edgar Pineda, progmg dir.

Bullhead City

KFLG(AM)— Oct 1, 1978: 1000 khz; 1 kw-D. TL: N35 10 10 W114 38 02. Hrs open: 1531 Jill Way, Suite 7, 86426-9341. Phone: (928) 763-5586. Fax: (928) 763-3775. Web Site: www.talkatoz.com. Licensee: Cameron Broadcasting Inc. (group owner; acq 11-24-99). Population served: 76,000 Natl. Network: CNN Radio. Format: Btfl mus, big band, MOR, Nostalgiia. Target aud: 24 plus; upper demographics. ♦Billy Williams, CEO & pres; Don Jaeger, VP & gen mgr; Bob Athey, progmg dir.

***KVIR(FM)**—Not on air, target date: unknown: 89.9 mhz; 38 kw vert. Ant 2,995 ft TL: N35 06 28 W113 52 40. Hrs open: 4002 N. 3300 E., Twin Falls, ID, 83301. Phone: (208) 734-6633. Fax: (208) 736-1958. Web Site: www.csnradio.com. Licensee: CSN International. ♦Mike Kestler, pres.

KZZZ(AM)— Nov 15, 1981: 1490 khz; 1 kw-U. TL: N35 05 10 W112 07 40. Hrs open: 1531 Jill Way, Suite 7, 86426-9341. Phone: (928) 763-5586. Fax: (928) 763-3775. Web Site: www.talkatoz.com. Licensee: Cameron Broadcasting Inc. (group owner; acq 7-91; $1.28 million with KNKK(FM) Needles, CA; FTR: 7-29-91). Population served: 37000 Format: Talk. Target aud: 45 plus. ♦Don Jaeger, VP, gen mgr & gen mgr.

Cameron

KYNN(AM)—Not on air, target date: unknown: 1450 khz; 1 kw-U. TL: N35 51 46 W111 25 52. Hrs open: 8320 W. 66th Ave., Arvada, CO, 80004. Phone: (303) 431-0103. Licensee: Better Life Ministries. ♦Claud Pettit, pres.

Casa Grande

***KLVA(FM)**— Apr 8, 1976: 105.5 mhz; 50 kw. Ant 492 ft TL: N33 00 14 W111 58 53. Stereo. Hrs opn: 24 2351 Sunset Blvd., Suite 170-218, Rocklin, CA, 95765. Phone: (916) 251-1600. Fax: (916) 251-1650-. E-mail: klove@klove.com Web Site: www.klove.com.Yes Licensee: Educational Media Foundation. Group owner: EMF Broadcasting (acq 7-19-99). Population served: 1,881,000 Natl. Network: K-Love. Law Firm: Shaw, Pittman. Format: Contemp Christian. Target aud: 25-44; Judeo-Christian, female. Spec prog: Sports 7 hrs wkly. ♦Richard Jenkins, pres; Mike Novak, VP & progmg dir; Lloyd Parker, gen mgr; Ed Lenane, opns dir & news dir; Keith Whipple, dev dir; Eric Allen, natl sls mgr; David Pierce, progmg mgr; Jon Rivers, mus dir; Sam Wallington, engrg dir; Arthur Vassar, traf mgr; Karen Johnson, news rptr; Marya Morgan, news rptr; Richard Hunt, news rptr.

Cave Creek

KFNX(AM)— June 27, 1997: 1100 khz; 50 kw-D, 1 kw-N, DA-2. TL: N33 47 52 W111 59 30. Stereo. Hrs opn: 24 2001 N. 3rd St., Suite 102, Phoenix, 85004. Phone: (602) 277-1100. Fax: (602) 248-1478. Web Site: www.1100kpnx.com. Licensee: North American Broadcasting Co. Inc., debtor in possession (acq 8-13-02). Population served: 3,000,000 Format: Talk. News staff: 2. Target aud: 35 plus; Upscale. ♦Francis Battaglia, CEO.

Chandler

KMLE(FM)—Licensed to Chandler. See Phoenix

Chinle

KFXR-FM— August 1995: 107.3 mhz; 3.6 kw. 1,630 ft TL: N36 21 07 W109 49 54. Hrs open:
Rebroadcasts KGLX(FM) Gallup.
1632 S. Second St., Gallup, NM, 87301. Phone: (505) 863-9391. Fax: (505) 863-9393. Licensee: CC Licenses LLC. Group owner: Clear Channel Communications Inc. (acq 8-18-2000). Format: Country. ♦Maryann Armijo, gen mgr.

Chino Valley

KFPB(FM)— 1999: 94.3 mhz; 4.1 kw. Ant 810 ft TL: N34 49 32 W112 34 09. Hrs open: 8581 E. Florentine, Suite C, Prescott Valley, 86314. Phone: (928) 775-2530. Fax: (928) 775-2532. Web Site: www.kfpbradio.com. Licensee: Prescott Radio Partners (acq 12-10-99; $250,000). Format: Country. ♦Patti Esell, gen mgr.

Claypool

KIKO-FM—Licensed to Claypool. See Miami

Clifton

KCUZ(AM)—Licensed to Clifton. See Safford

KWRQ(FM)—Licensed to Clifton. See Safford

Colorado City

KXFF(FM)— 1993: 107.3 mhz; 35 kw. Ant 1,138 ft TL: N37 05 41 W113 11 06. Hrs open: 24 750 W. Ridgeview Dr., Suite 204, Saint George, UT, 84770. Phone: (435) 673-3579. Fax: (435) 673-8900. Licensee: CCR-St. George IV LLC. (group owner; (acq 5-3-2006; grpsl). Population served: 50,000 Format: Hits of the 80s and 90s. ♦Steve Hess, gen mgr.

Coolidge

KCKY(AM)— Nov 19, 1964: 1150 khz; 5 kw-D, 1 kw-N, DA-2. TL: N33 00 27 W111 32 54. (CP: COL Apache Junction. 5 kw-D, 185 w-N, DA-2. TL: N33 00 27 W111 32 57). Hrs opn: 18 1445 W. Baseline Road, Phoenix, 85041. Phone: (602) 426-1150. Phone: (602) 426-9606. Fax: (602) 276-8119. Licensee: Cortaro Broadcasting Corp. (acq 6-13-2003; exchange agreement with KEVT(AM) Cortaro). Population served: 2,045,000 Format: Sp, Christian contemp. News staff: one. Target aud: General. ♦Moses Herrera, pres; Moses Herrera Jr., stn mgr.

***KZAI(FM)**— 2004: 89.9 mhz; 10 w horiz, 10 kw vert. Ant 3,025 ft TL: N33 17 55 W110 50 28. Hrs open: 2351 Sunset Blvd., Suite 170-218, Rocklin, CA, 95765. Phone: (916) 251-1600. Fax: (916) 251-1650. Web Site: www.air1.com. Licensee: Educational Media Foundation. (acq 2-10-2006; $2.5 million). Natl. Network: Air 1. Format: Alternative, Christian. ♦Richard Jenkins, pres; Mike Novak, VP; Keith Whipple, dev dir; David Pierce, progmg mgr; Ed Lenane, news dir; Sam Wallington, engrg dir; Karen Johnson, news rptr; Marya Morgan, news rptr; Richard Hunt, news rptr.

Cortaro

KCEE(AM)— 1994: 1030 khz; 10 kw-D, 1 kw-N, DA-2. TL: N32 20 51 W111 04 19. Hrs open: 24 2919 E. Broadway Blvd., Suite 230, Tucson, 85716. Phone: (520) 889-8904. Fax: (520) 889-8573. Web Site: www.1030kcee.com. Licensee: Slone Broadcasting LLC (acq 1-5-2007; $1.5 million). Format: Oldies. Target aud: 24-54. ♦Armando Zamora, gen mgr; Araceli Espinoza, opns dir; Steve Nunez, gen sls mgr; Frank Luna, engrg dir.

Cottonwood

KKLD(FM)— August 1983: 95.9 mhz; 21 kw. Ant 2,621 ft TL: N34 41 11 W112 07 02. Stereo. Hrs opn: 24 Box 187, 86326. Phone: (928) 634-2286. Fax: (928) 634-2295. Web Site: www.kkld.com. Licensee: Yavapai Broadcasting Corp. (group owner; (acq 10-1-2000; grpsl). Population served: 200,000 Format: Oldies. News: 7 hrs wkly. Target aud: 18-49. ♦W. Grant Hafley, pres; David J. Kessel, gen mgr; Rich Malone, opns mgr.

KVRD-FM—Listing follows KYBC(AM).

KYBC(AM)— Dec 20, 1964: 1600 khz; 1 kw-D, 46 w-N. TL: N34 43 15 W109 31 45. Hrs open: 24 Box 187, 86326. Phone: (928) 634-2286. Fax: (928) 634-2295. E-mail: kybc@myradioplace.com Web Site: www.myradioplace.com. Licensee: Yavapai Broadcasting Corp. (group owner; acq 1-96; $750,000 with co-located FM). Population served: 20,000 Natl. Network: Westwood One. Format: Adult standards, MOR. News staff: 2. Target aud: 18-plus. ♦W. Grant Hafley, pres; David J. Kessel, gen mgr; Jackie Bessler, sls dir; Paul Siabe, progmg dir; Paul David, news dir.

KVRD-FM—Co-owned with KYBC(AM). July 1991: 105.7 mhz; 380 w. 2,555 ft TL: N34 41 15 W112 07 02. (CP: 300 w, ant 2,545 ft.). Stereo. 24 Fax: (928) 634-2295. Web Site: www.myradioplace.com.50,000 Format: Country. News: 2 hrs wkly. Target aud: General. ♦Mark Bachman, progmg dir; Paul David, local news ed.

Dewey-Humboldt

KMVA(FM)— Jan 15, 1988: 97.5 mhz; 42 kw. Ant 2,785 ft TL: N34 14 05 W112 22 02. Stereo. Hrs opn: 24 Phone: (602) 222-9750. Fax: (602) 222-2297. Web Site: www.star975fm.com. Licensee: Trumper Communications III License LLC (acq 5-27-2005; $22.6 million). Population served: 300,000 Format: Rhythmic hits from the 80s, 90s and today. ♦Jim Ryan, gen mgr.

Dolan Springs

KOAS(FM)— Jan 7, 1976: 105.7 mhz; 100 kw horiz. Ant 1,761 ft TL: N35 50 11 W114 19 08. Stereo. Hrs opn: 24 2725 E. Desert Inn Rd., Suite 180, Las Vegas, NV, 89121. Phone: (702) 784-4000. Fax: (702) 784-4040. Web Site: 1057theoasis.com. Licensee: RBG Las Vegas Licenses LLC. (acq 10-3-2005; $38 million with KVGS(FM) Laughlin, NV). Population served: 150,000 Format: Smooth jazz. ♦Frank Woodbeck, VP & gen mgr.

Douglas

KAPR(AM)— Mar 8, 1958: 930 khz; 2.5 kw-D. TL: N31 22 08 W109 31 45. Hrs open: 24
KVOI.
3222 S. Richey Ave., Tucson, 85713. Phone: (520) 790-2440. Fax: (520) 790-2937. E-mail: info@kvoi.com Web Site: www.kvoi.com. Licensee: Good Music Inc. Group owner: Good News Communications Inc. (acq 6-8-2001; $187,500). Population served: 190,000 Natl. Network: Salem Radio Network. Natl. Rep: Salem. Law Firm: Wray Fitch. Format: News/talk. News: 2 hrs wkly. Target aud: General. ♦Doug Martin, CEO & gen mgr; Rhonda Curtis, CFO; Mary Martin, gen sls mgr & mktg VP.

KCDQ(FM)— Mar 15, 1979: 95.3 mhz; 3 kw. 210 ft TL: N31 22 08 W109 31 45. Stereo. Hrs opn: 24 500 E. Fry Blvd., Suite L-10, Sierra Vista, 85635. Phone: (520) 459-8201. Fax: (520) 458-7104. Web Site: www.kcdq.com. Licensee: Cochise Broadcasting LLC (acq 6-8-2001; $137,500). Population served: 120,000 Natl. Network: Westwood One. Format: Contemporary hit. News: 11 hrs wkly. Target aud: 29-49. ♦Ted Tucker, gen mgr; Jeff Davenport, stn mgr.

KDAP(AM)— 1946: 1450 khz; 1 kw-U. TL: N31 21 18 W109 31 45. Hrs open: Box 1179, 85608. Secondary address: 2031 N. Sulphur Springs St. 85607. Phone: (520) 364-3486. Phone: (520) 364-3484. Fax: (520) 364-3483. Licensee: Howard N. Henderson (group owner; (acq 2-3-2005; $165,800 with co-located FM). Population served: 120,000 Format: Sp. Target aud: General; loc Hispanic & Mexican residents. ♦Howard Henderson, gen mgr, gen sls mgr & progmg dir.

KDAP-FM— Nov 15, 1990: 96.5 mhz; 3 kw. Ant 30 ft TL: N31 21 18 W109 33 06. Stereo. 24 Phone: (520) 364-3484.120,000 Format: Country. News staff: one. Target aud: General.

***KRMC(FM)**— 1996: 91.7 mhz; 3 kw. 236 ft TL: N31 20 52 W109 28 42. Hrs open: Box 2520, 85608. Phone: (520) 364-5392. Fax: (520) 364-5392. Licensee: World Radio Network Inc. Format: Christian, Sp, educ. ♦David Johnson, pres; Glen Lafitte, gen mgr; James V. Heck, engrg dir.

Drake

***KJZA(FM)**—Not on air, target date: unknown: 89.5 mhz; 250 w. 1,702 ft Hrs opn: 24 2719 DW Ranch Rd., Kingman, 86046. Phone: (928) 716-8433. Fax: (928) 541-1008. E-mail: kjzafm@yahoo.com Web Site: www.kjza.com. Licensee: St. Paul Bible College. Natl. Network: NPR, PRI. Format: Jazz. ♦Tom Ericson, gen mgr.

Duncan

KJIK(FM)— 2003: 100.7 mhz; 9.8 kw. Ant 2,348 ft TL: N32 53 21 W109 19 20. Hrs open: 24 1850 W. Thatcher Blvd., Safford, 85546-3306. Phone: (928) 428-4100. Fax: (928) 348-9581. E-mail:

production@kjik.fm Licensee: Country Mountain Airwaves LLC. Format: Adult contemp. ♦Dan Curtis, gen mgr & opns mgr.

Eagar

KTHQ(FM)— 1996: 92.5 mhz; 100 kw. 984 ft TL: N34 05 47 W109 27 52. Hrs open: 24 Box 2020, Show Low, 85902. Phone: (928) 532-1010. Fax: (928) 532-0101. Web Site: www.Qcountry92.com. Licensee: William S. Konopnicki. Format: Country. ♦Camden Smith, gen mgr & gen sls mgr; Laurie Pogson, traf mgr.

Flagstaff

KAFF(AM)— Oct 15, 1963: 930 khz; 5 kw-D, 50 w-N. TL: N35 11 26 W111 40 37. Hrs open: 5 AM-midnight Box 1930, 86002. Secondary address: 1117 W. Hwy. 66 86001. Phone: (928) 774-5231. Phone: (520) 774-5233. Fax: (928) 779-2988. Licensee: Guyann Inc. Group owner: Guyann Corp. (acq 9-7-2005; grpsl). Population served: 492,000 Wire Svc: AP Format: Country. News staff: 2. Target aud: 25-54. ♦Janie Richardson, gen mgr; Val Barret, prom dir & prom mgr; Chris Halstead, progmg dir; Hugh Morris, mus dir; George Davis, news dir; Jon Swett, chief of engrg.

KAFF-FM— October 1968: 92.9 mhz; 100 kw. 1,512 ft TL: N34 58 07 W111 30 24. Stereo. 24 E-mail: production@kaff.com Web Site: www.kaff.com.250,000 Natl. Network: ABC. News staff: 2; News: 4 hrs wkly.

KFLX(FM)—Kachina Village, February 1995: 105.1 mhz; 1 kw. 1,968 ft TL: N35 14 26 W111 35 48. Stereo. Hrs opn: 24 112 E. Rt. 66, Suite 105, 86001. Phone: (928) 779-1177. Fax: (928) 774-5179. E-mail: ann@nnorthlandradio.com Web Site: www.1051thecanyon.com. Licensee: Grenax Broadcasting II LLC. (acq 1-6-2006; grpsl). Law Firm: Arent, Fox, Kintner, Plotkin & Kahn. Format: Adult contemp. News staff: one; News: 4 hrs wkly. Target aud: 28-54; males. ♦Greg Dinetz, pres; Jim Shipp, gen mgr & prom VP; Bill McAdams, opns dir & traf mgr; Mike Mentor, progmg dir; Samantha Ward, pub affrs dir; Jon Sweat, chief of engrg.

***KJTA(FM)**— Dec 19, 2001: 89.9 mhz; 10 kw. Ant 1,502 ft TL: N34 58 06 W111 30 28. Hrs open: 24 1700 N. 2nd St., 86004. Phone: (928) 774-9514. Fax: (928) 774-9515. Web Site: www.myflr.org. Licensee: Family Life Broadcasting Inc. (acq 3-27-2007; grpsl). Format: Christian Contemporary. ♦Dawn Bumstead, progmg dir.

KMGN(FM)— 1975: 93.9 mhz; 100 kw. 1,509 ft TL: N34 58 08 W111 30 28. Stereo. Hrs opn: 24 Box 1930, 86002. Secondary address: 1117 W Rt. 66 86001. Phone: (928) 774-5231. Fax: (928) 779-2988. Web Site: www.kmgn.com. Licensee: Guyann Corp. (group owner; (acq 9-7-2005; grpsl). Population served: 250,000 Natl. Network: ABC. Wire Svc: AP Format: Classic rock. News staff: 2; News: 4 hrs wkly. Target aud: 25-54; upscale, educated, rgnl audience. ♦Janie Richardson, gen mgr; Rob Dowers, progmg dir.

***KNAU(FM)**— Nov 24, 1970: 88.7 mhz; 100 kw. 1,549 ft TL: N34 57 40 W111 31 00. Stereo. Hrs opn: 24 Box 5764, Northern Arizona Univ., 86011-5764. Phone: (928) 523-5628. Fax: (928) 523-7647. E-mail: knau@nau.edu Web Site: www.knau.org. Licensee: Arizona Board of Regents for and on behalf of Northern Arizona University. Population served: 180,000 Natl. Network: NPR, PRI. Law Firm: Arter & Hadden. Format: News & info, class. News staff: 3; News: 50 hrs wkly. Target aud: 25-54; educated, socially conscious achievers. Spec prog: Car talk. ♦John Stark, gen mgr; Dave Riek, opns mgr; Liz Gumerman, dev dir; Jeff Norcross, progmg dir & mus dir; Dan Kraker, progmg mgr & news dir; Jon Swett, chief of engrg; Lisa Skinner, traf mgr.

***KPUB(FM)**— October 1995: 91.7 mhz; 500 w. 1,837 ft TL: N35 14 34 W111 36 40. Hrs open: 24 Box 5764, Northern Arizona Univ., 86011-5764. Phone: (928) 523-5628. Fax: (928) 523-7647. E-mail: knau@nau.edu Web Site: www.knau.org. Licensee: Northern Arizona University. Natl. Network: NPR, PRI. Format: News & info. News: 50 hrs wkly. Target aud: 25-54; educated, socially conscious achievers. ♦John Stark, gen mgr; Dave Riek, opns mgr; Liz Gumerman, dev dir; Jeff Norcross, progmg dir; Don Kraker, news dir; Lisa Skinner, traf mgr.

KVNA(AM)— Aug 8, 1950: 600 khz; 5 kw-D, 500 w-N, DA-N. TL: N35 11 47 W111 40 28. Hrs open: 24 Box 187, Cottonwood, 86326. Phone: (928) 526-2700. Fax: (928) 774-5852. E-mail: am600@radioflagstaff.com Web Site: www.radioflagstaff.com. Licensee: Yavapai Broadcasting Corp. (acq 9-30-2000; grpsl). Population served: 300,000 Natl. Network: Westwood One, AP Network News, Jones Radio Networks. Format: Sports, news/talk. News staff: one; News: 25 hrs wkly. Target aud: General. Spec prog: Sp 3 hrs wkly, folk music 4hrs wkly. ♦W. Grant Hafley, pres; David J. Kessel, gen mgr; Mike Dougal, opns mgr & progmg dir; Mike Dougall, news dir.

KVNA-FM— 1999: 100.1 mhz; 5.2 kw. Ant 1,433 ft TL: N34 58 05 W111 30 29. Hrs open: 24 Box 187, Cottonwood, 86326. Phone: (928) 526-2700. Fax: (928) 634-2295. Web Site: www.myradioplace.com. Licensee: Yavapai Broadcasting Corp. (acq 5-2-2005; $1.5 million). Format: Adult contemp. ♦Dave Kessel, gen mgr.

KZGL(FM)—Not on air, target date: unknown: 103.7 mhz; 560 w. Ant 1,958 ft TL: N35 14 25 W111 35 53. Hrs open: 1675 Sweetwater West Circle, Apopka, FL, 32712-2480. Phone: (407) 488-2098. Licensee: Walker Radio Inc. (acq 6-8-2007; $2.5 million for CP). ♦James R. Walker, pres.

Florence

KCDX(FM)— 1999: 103.1 mhz; 2.7 kw. Ant 3,057 ft TL: N33 17 55 W110 50 28. Hrs open: Box 36717, Tucson, 85740. Phone: (520) 459-8201. Fax: (520) 458-7104. Web Site: www.kcdx.com. Licensee: Desert West Air Ranchers Corp. Format: Classic rock. ♦Ted Tucker, gen mgr.

Fountain Hills

***KLVK(FM)**— May 2000: 89.1 mhz; 1.4 kw. Ant 935 ft TL: N33 29 33 W111 38 23. (CP: 2.5 kw vert). Hrs opn: 24 2351 Sunset Blvd., Suite 170-218, Rocklin, CA, 95765. Phone: (916) 251-1600. Fax: (916) 251-1650. E-mail: klove@klove.com Web Site: www.klove.com. Licensee: Educational Media Foundation. Group owner: EMF Broadcasting (acq 3-11-03; grpsl). Natl. Network: K-Love. Law Firm: Shaw Pittman. Format: Contemp Christian. News staff: 3. Target aud: 25-33; Judeo Christian, female. ♦Richard Jenkins, pres; Mike Novak, VP; Keith Whipple, dev dir; David Pierce, progmg mgr; Ed Lenane, news dir; Sam Wallington, engrg dir; Karen Johnson, news rptr; Marya Morgan, news rptr; Richard Hunt, news rptr.

Gilbert

KEDJ(FM)— Feb 25, 1981: 103.9 mhz; 99.59 kw. Ant 620 ft TL: N33 14 50 W111 31 49. Stereo. Hrs opn: 24 7434 E. Stetson Dr., Suite 265, Scottsdale, 85251. Phone: (480) 423-9255. Fax: (480) 423-9382. E-mail: nat@theedge1039.com Web Site: www.theedge1039.com. Licensee: RBG Phoenix Licenses LLC (acq 11-21-2005; $30 million). Population served: 1,446,948 Natl. Rep: Roslin. Format: Alternative. News: 2 hrs wkly. Target aud: 18-34. ♦Tim Pohlman, CEO; Nat Galvin, VP, gen mgr & gen sls mgr.

Glendale

KLNZ(FM)—Licensed to Glendale. See Tempe

KPXQ(AM)—Licensed to Glendale. See Phoenix

KTAR-FM—Licensed to Glendale. See Phoenix

Globe

KIKO(AM)—See Miami

KJAA(AM)— 1971: 1240 khz; 1 kw-U. TL: N33 22 51 W110 45 25. Hrs open: 24
KVOI.
3222 S. Richey Ave., Tucson, 85713. Phone: (520) 790-2440. Fax: (520) 790-2937. E-mail: info@kvoi.com Web Site: www.kvoi.com. Licensee: Good Music Inc. Group owner: Good News Communications Inc. (acq 4-30-2001; $212,400). Format: News/talk. News staff: 3; News: 3 hrs wkly. Target aud: 35 plus. ♦Doug Martin, gen mgr.

***KLKA(FM)**—Not on air, target date: unknown: 88.5 mhz; 24 w vert. Ant 3,385 ft TL: N33 17 37 W110 50 09. Hrs open: 3185 S. Highland Dr., Suite 13, Las Vegas, NV, 89109. Phone: (702) 731-5588. Licensee: American Educational Broadcasting Inc. ♦Carl J. Auel, pres.

KQMR(FM)— Sept 25, 1980: 100.3 mhz; 90 kw. 2,047 ft TL: N33 17 23 W110 51 53. Stereo. Hrs opn: 24 4745 N. Seventh St., Suite 140, Phoenix, 85014. Phone: (602) 308-7900. Fax: (602) 308-7979. Web Site: www.univision.com. Licensee: Univision Radio License Corp. Group owner: Univision Radio (acq 9-22-2003; grpsl). Population served: 1,500,000 Format: Sp adult hits. Target aud: 18-49; upscale, well-educated, affluent adults. ♦Chris Morris, pres & gen sls mgr; Mary McEvilly-Hernandez, VP & gen mgr; Robbie Ramirez, progmg dir.

KRDE(FM)— October 1995: 94.1 mhz; 640w. 3,408 ft TL: N33 17 37 W110 50 09. Stereo. Hrs opn: 24 Box 1660, 85502. Secondary address: 800 N. Main St. 85501. Phone: (928) 402-9222. Fax: (928) 425-5063. E-mail: krde@cableone.net Web Site: www.krde.com. Licensee: Linda C. Corso. Population served: 1,800,000 Natl. Network: Fox News Radio, Premiere Radio Networks, USA. Law Firm: John McVeigh, P.C. Format: Country, oldies. News staff: one; News: 12 hrs wkly. Target aud: 25-54; active, family building, western suburban. Spec prog: American Indian 6 hrs, Americana 6 hrs wkly. ♦Richard Potyka, pres, gen mgr & prom; Linda Corso, CFO; Mindy Chansley, opns mgr; Ted Lake, news dir; Brad Hartman, sports cmtr; Liz Mata, sls.

***KVJC(FM)**— 2003: 91.9 mhz; 660 w. Ant 3,395 ft TL: N33 17 37 W110 50 09. Hrs open: 24 800 N. Main St., 85501-9456. Phone: (928) 402-9222. Phone: (208) 734-6633. E-mail: kvjc@csnradio.com Web Site: www.csnradio.com. Licensee: CSN International (group owner). Format: Christian. ♦Jeffrey W. Smith, VP; Mike Stocklin, gen mgr; Don Mills, progmg dir; Kelly Carlson, chief of engrg.

Grand Canyon

***KNAG(FM)**—Not on air, target date: unknown: 90.3 mhz; 3 kw. 295 ft TL: N35 56 44 W112 10 16. Hrs open:
KNAU-FM Flagstaff 100%.
Box 5764, Northern Arizona University, Flagstaff, 86011. Phone: (928) 523-5628. Fax: (928) 523-7647. E-mail: knau@nau.edu Web Site: www.knau.org. Licensee: Arizona Board of Regents/Northern Arizona University. Format: Class, news. ♦John Stark, gen mgr; Dave Riek, opns mgr; Liz Gumerman, dev dir.

Green Valley

KFMA(FM)— Feb 20, 1983: 92.1 mhz; 50 kw. 492 ft TL: N32 00 11 W110 47 49. Stereo. Hrs opn: 24 3871 N. Commerce Dr., Tucson, 85705. Phone: (520) 407-4500. Fax: (520) 407-4600. Web Site: www.kfma.com. Licensee: Arizona Lotus Corp. Group owner: Lotus Communications Corp. (acq 5-10-93; $1.26 million; FTR: 5-31-93). Natl. Rep: Christal. Format: Alternative. Target aud: 18-34. ♦Steve Groesbeck, gen mgr; Cindy Craig, prom mgr; Matt Spry, progmg mgr.

KGVY(AM)—Licensed to Green Valley. See Tucson

KTZR-FM—Licensed to Green Valley. See Tucson

Holbrook

***KBMH(FM)**— 2002: 90.3 mhz; 250 w. Ant 141 ft TL: N34 55 05 W110 08 25. Hrs open: Drawer 2440, Tupelo, MS, 38803. Phone: (601) 844-8888. Fax: (662) 842-6791. Licensee: American Family Association. Group owner: American Family Radio Format: Christian. ♦Marvin Sanders, gen mgr; John Riley, progmg dir; Joey Moody, chief of engrg.

KDJI(AM)—Listing follows KZUA(FM).

KZUA(FM)— Dec 6, 1993: 92.1 mhz; 100 kw. 328 ft TL: N34 52 25 W110 09 56. Stereo. Hrs opn: 3051 S. White Mountain Rd., Suite D, Show Low, 85901. Phone: (928) 532-3232. Fax: (928) 537-3991. E-mail: production@whitemountainradio.com Licensee: Petracom of Holbrook L.L.C. (acq 2-26-2002; $650,000 with co-located AM).

Population served: 95,000 Natl. Network: Westwood One, CNN Radio. Format: Contemp country. Target aud: 18-54; 57% female, 43% male. ♦Steve Johnson, gen mgr, progmg dir & chief of engrg; Amanda Lynn, traf mgr.

KDJI(AM)—Co-owned with KZUA(FM). October 1955: 1270 khz; 5 kw-D, 130 w-N. TL: N34 53 55 W110 11 30.9 AM-4 PM 77,000 Natl. Network: ABC, Westwood One. Format: News/Talk. Target aud: 35-65; 46% female, 54% male. Spec prog: Sports 10 hrs, farm 8 hrs wkly.

Hotevilla

***KUYI(FM)**— Dec 20, 2000: 88.1 mhz; 69 kw. Ant 407 ft TL: N35 48 29 W110 16 23. Hrs open: 24 Box 1500, Keams Canyon, 86034. Phone: (928) 738-5505. Fax: (928) 738-5501. E-mail: hopiradio@yahoo.com Web Site: www.kuyi.net. Licensee: Hopi Foundation. Format: Tribal radio, loc news, cultural events. News staff: 2; News: 6 hrs wkly. Target aud: 30+. ♦Alicia Youvella, stn mgr; Katherine Sahmie, opns mgr; Dan Kraker, news dir.

Kachina Village

KFLX(FM)—Licensed to Kachina Village. See Flagstaff

Kearny

KZLZ(FM)—Licensed to Kearny. See Tucson

Kingman

KAAA(AM)— Oct 7, 1949: 1230 khz; 1 kw-U. TL: N35 11 48 W114 01 18. Hrs open: 2534 Hualapai Mountain Rd., 86401-5300. Phone: (928) 753-2537. Fax: (928) 753-1551. Licensee: Cameron Broadcasting Inc. (group owner; acq 11-24-99; grpsl). Population served: 65,000 Law Firm: Cohn & Marks. Format: News/talk. Target aud: 25 plus. Spec prog: Sports. ♦Don Jaeger, gen mgr; Jeff Allen, progmg dir.

KFLG-FM—Co-owned with KAAA(AM). Dec 6, 1974: 94.7 mhz; 43 kw. Ant 2,591 ft TL: N35 06 41 W113 53 08. Stereo. 24 1531 Jill Way, Suite 5, Bullhead City, 86426. Phone: (520) 763-2100. Fax: (520) 763-3957.130,000 Format: Country. Target aud: 25-54; professionals.

***KCAI(FM)**—Not on air, target date: unknown: 91.9 mhz; 30 kw vert. Ant 2,893 ft TL: N35 06 35 W113 52 51. Hrs open: Box 2440, Tupelo, MS, 38801-2440. Phone: (662) 844-8888. Fax: (662) 842-6791. Licensee: American Family Association. ♦Donald E. Wildmon, chmn.

KGMN(FM)— Feb 14, 1984: 100.1 mhz; 360 w. 761 ft TL: N35 11 43 W114 06 51. (CP: 930 w, ant 2,896 ft. TL: N35 06 37 W133 52 55). Stereo. Hrs opn: 24 812 E. Beale St., 86401. Phone: (928) 753-9100. Fax: (928) 753-1978. Web Site: www.kgmn.net. Licensee: New West Broadcasting Systems Inc. Natl. Network: AP Radio, Jones Radio Networks. Format: Country. ♦Joe Hart, CEO; Rhonda Hart, VP & gen mgr; Deana Campbell, opns mgr; Brian Winters, progmg dir.

KRRN(FM)— November 1990: 92.7 mhz; 17 kw. Ant 1,889 ft TL: N35 01 58 W114 21 57. (CP: COL Dolan Springs. 100 kw, ant 1,774 ft. TL: N35 39 07 W114 18 42). Hrs opn: 500 Pilot Rd., Las Vegas, NV, 89119. Phone: (323) 900-6100. Fax: (323) 900-6108. Licensee: Entravision Holdings LLC. Group owner: Entravision Communications Corp. (acq 8-29-02; $12.43 million). Natl. Network: ABC. Format: Sp contemp. ♦Rick Murphy, VP; Chris Sarros, gen mgr; Chris Rolando, stn mgr; Brian Calkins, news dir.

Lake Havasu City

***KAIH(FM)**—Not on air, target date: unknown: 89.3 mhz; 440 w. Ant -512 ft TL: N34 27 27 W114 20 07. Hrs open: 2351 Sunset Blvd., Suite 170-218, Rocklin, CA, 95765. Phone: (916) 251-1600. Fax: (916) 251-1650. Licensee: Educational Media Foundation. Group owner: EMF Broadcasting. ♦Richard Jenkins, pres; Mike Novak, VP; Keith Whipple, dev dir; David Pierce, progmg mgr; Ed Lenane, news dir; Sam Wallington, engrg dir; Karen Johnson, news rptr; Marya Morgan, news rptr; Richard Hunt, news rptr.

KJJJ(FM)— May 24, 1994: 102.3 mhz; 1.05 kw. Ant 2,670 ft TL: N34 33 06 W114 11 37. Hrs open: 1845 McCulloch Blvd., Suite A-14, 86403. Phone: (928) 855-9336. Fax: (928) 855-9333. E-mail: steve@kjjfm.com Web Site: www.kjjfm.com. Licensee: Steven M. Greeley. Population served: 150,000 Law Firm: Koenen & Olendar. Format: Country. ♦Steve Greeley, gen mgr; Traceye Jones, gen mgr.

***KNLB(FM)**— July 1983: 91.1 mhz; 8 kw. Ant 453 ft TL: N34 29 10 W114 13 06. Stereo. Hrs open: 24 510 N. Acoma Blvd., 86403. Phone: (928) 855-9110. Fax: (928) 453-2588. E-mail: info@knlb.com Web Site: www.knlb.com. Licensee: Advance Ministries. Population served: 1,000,000 Natl. Network: USA. Law Firm: Arent, Fox, Kintner, Plotkin & Kahn. Format: Relg, Christian. News: 8 hrs wkly. Target aud: General. ♦Richard D. Tatham, pres; Faron Eckelbarger, stn mgr, progmg dir & chief of engrg.

KNTR(AM)— Sept 23, 1970: 980 khz; 1 kw-D, 49 w-N. TL: N34 30 12 W114 21 28. Hrs open: 24 1845 McCulloch Blvd., Suite A 14, 86403. Phone: (928) 855-9336. Fax: (928) 855-9333. E-mail: speakout@kntram.com Web Site: www.kntram.com. Licensee: Steven M. Greeley. (acq 12-8-99; $608,000). Population served: 60000 Natl. Network: PRI. Law Firm: Koerner & Olender PC. Format: News and talk. News staff: one; News: 12 hrs wkly. Target aud: 35-64. ♦Steve Greeley, gen mgr; Traceye Jones, gen mgr.

KRCY-FM— 1999: 96.7 mhz; 1.05 kw. 2,706 ft TL: N34 33 06 W114 11 37. Hrs open: Box 1866, 86405. Phone: (928) 855-1051. Fax: (928) 855-7996. E-mail: express@maddog.net Web Site: www.maddog.net. Licensee: Rick L. Murphy. Format: Oldies. ♦Rick L. Murphy, pres & gen mgr.

KRRK(FM)— Sept 9, 1974: 101.1 mhz; 20 kw. Ant 2,696 ft TL: N34 33 06 W114 11 37. Stereo. Hrs open: 24 Box 1866, 86405. Phone: (928) 855-4560. Phone: (928) 855-1051. Fax: (928) 855-7996. E-mail: epress@maddog.net Web Site: www.maddog.net. Licensee: Superior Broadcasting of Lake Havasu LLC (acq 11-30-2005; $11.8 million). Population served: 60,000 Format: Classic rock. News staff: one; News: 5 hrs wkly. Target aud: 18-34. ♦Chris Rolando, gen mgr.

KZUL-FM— 1986: 104.5 mhz; 1.05 kw. Ant 2,670 ft TL: N34 33 06 W114 11 37. Stereo. Hrs opn: 24 Box 1866, 86405. Phone: (928) 855-4560. Fax: (928) 855-7996. E-mail: epress@maddog.net Web Site: www.maddog.net. Licensee: Mad Dog Wireless Inc. Natl. Network: ABC. Format: Adult contemp, classic rock. Target aud: 25-54. ♦Rick Murphy, pres; Chris Rolando, VP, gen mgr & progmg dir; Ron Nickle, gen sls mgr; Faron Ecklebarger, chief of engrg.

Mammoth

***KLTU(FM)**—Not on air, target date: unknown: 88.1 mhz; 160 w. Ant 3,552 ft TL: N32 24 54 W110 42 56. Hrs open: 3222 S. Richey Ave., Tucson, 85713. Phone: (520) 790-2440. Fax: (520) 790-2937. Licensee: Good News Radio Broadcasting Inc. (acq 6-8-2005). ♦Doug Martin, gen mgr.

Marana

KOHT(FM)— Oct 1, 1984: 98.3 mhz; 6 kw. 200 ft TL: N32 27 09 W111 05 09. Stereo. Hrs opn: 24 3202 N. Oracle Rd., Tucson, 85705. Phone: (520) 618-2100. Fax: (520) 618-2200. E-mail: hot983comments@yahoo.com Web Site: www.hot983.com. Licensee: CC Licenses LLC. Group owner: Clear Channel Communications Inc. (acq 6-22-2001; grpsl). Population served: 262,933 Format: CHR, rhythm & blues, hip hop. News staff: 2; News: 4 hrs wkly. Target aud: 18-49; Sp, contemp, white collar adults. ♦Debbie Wagner, gen mgr; Tim Richards, opns mgr; Steve Clement, gen sls mgr; Fred Rico, progmg dir; Mike Irby, chief of engrg.

KSAZ(AM)—Licensed to Marana. See Tucson

Mayer

KKFR(FM)— May 24, 1996: 98.3 mhz; 41 kw. Ant 2,795 ft TL: N34 14 03 W112 22 01. Stereo. Hrs opn: 24 4745 N. 7th St., Suite 410, Phoenix, 85014. Phone: (602) 682-9200. Fax: (602) 283-0923. Web Site: www.power983fm.com. Licensee: RBG Phoenix Licenses LLC. (acq 1-1-2007). Population served: 75,000 Format: Hip hop. ♦Nat Galvin, gen mgr; AmyAnn Rosales, gen sls mgr; Charlie Huero, mktg dir; Matt Kirkpatrick, prom dir; Bruce St. James, progmg dir.

Mesa

KDKB(FM)— Apr 20, 1968: 93.3 mhz; 100 kw. 1,538 ft TL: N33 20 04 W112 03 36. Stereo. Hrs opn: 24 1167 W. Javelina, 85210. Phone: (480) 897-9300. E-mail: rock@kdkg.com Web Site: www.kdkb.com. Licensee: Mesa Radio Inc. Group owner: Sandusky Radio (acq 1977). Natl. Rep: Christal. Format: AOR. ♦Norman Rau, pres; Chuck Artigue, gen mgr; Bob Weaver, gen sls mgr; Buzz Casey, prom mgr & progmg dir; Clayton Creekmore, chief of engrg; Kathy Perschke, traf mgr.

KFNN(AM)—Licensed to Mesa. See Phoenix

***KJZZ(FM)**—Phoenix, 1951: 91.5 mhz; 96 kw. 1,607 ft TL: N33 19 58 W112 03 53. Stereo. Hrs opn: 24 2323 W. 14th St., Tempe, 85281. Phone: (480) 834-5627. Web Site: www.kjzz.org. Licensee: Maricopa County Community College District. Population served: 2,000,000 Natl. Network: NPR, PRI. Wire Svc: NOAA Weather Wire Svc: UPI Format: Acoustic jazz, news. News staff: 5; News: 50 hrs wkly. Target aud: 25-54. ♦Carl Matthusen, gen mgr; Bill Shedd, opns mgr; Lou Stanley, dev dir; Scott Williams, progmg dir; Mark Moran, news dir; Dennis Gilliam, chief of engrg.

KXAM(AM)— 1946: 1310 khz; 5 kw-D, 500 w-N, DA-N. TL: N33 26 23 W111 50 09. Hrs open: 24 4725 N. Scottsdale Rd., Suite 234, Scottsdale, 85251. Phone: (480) 423-1310. Fax: (480) 423-3867. E-mail: kxam@aol.com Web Site: www.kxam.com. Licensee: Embee Broadcasting Inc. (acq 9-25-90). Population served: 2,500,000 Natl. Network: Westwood One, ABC. Law Firm: Hogan & Hartson. Format: Talk. News: 14 hrs wkly. Target aud: 35-64. ♦Byron Gerson, pres; Don Sandler, gen mgr.

KZZP(FM)— 1967: 104.7 mhz; 100 kw. 1,550 ft TL: N33 20 04 W112 03 35. Stereo. Hrs opn: 24 4686 E. Van Buren St., Ste. 300, Phoenix, 85008-6967. Phone: (602) 279-5577. Fax: (602) 230-2781. Web Site: www.1047kissfm.com. Licensee: Citicasters Licenses L.P. Group owner Clear Channel Communications Inc. (acq 6-99; grpsl). Population served: 1,811,600 Format: CHR. News staff: one; News: 7 hrs wkly. Target aud: 18-34; women. ♦Lowry Mays, CEO; Randy Michaels, chmn; John Hogan, pres; Susan Karis-Madigan, gen mgr; Alan Sledge, opns dir; Cathy Burau, gen sls mgr.

Miami

KIKO(AM)— June 13, 1958: 1340 khz; 1 kw-U. TL: N33 24 41 W110 50 17. Hrs open: 24 4501 Broadway, 85539. Secondary address: 4501 Broadway, Claypool 85532. Phone: (928) 425-4471. Fax: (928) 425-9393. E-mail: radiokiko@cableone.net Westwood One 6 pm-6 am Licensee: Shoecraft Broadcasting Inc. (acq 5-31-01; with KIKO-FM Claypool). Population served: 35,000 Natl. Network: Westwood One, ABC. Law Firm: Shaw Pittman. Format: Sports, oldies, contemp hits. News staff: one; News: 8 hrs wkly. Target aud: 21-70; industrial/blue collar workers in loc copper mines, highest hourly wage earners. ♦Ruth Shoecraft Wallace, CEO; Lucy Rodriguez, gen mgr; Linda Center, pub affrs dir; Randy Escobedo, sports cmtr; J.B. Barter, disc jockey.

KIKO-FM— Aug 1, 1991: 106.1 mhz; 6 kw. 297 ft TL: N33 24 23 W110 48 18. Stereo. 24 Phone: (928) 425-4472.Westwood One 24 hrs 35,000 Natl. Network: Westwood One, ABC. Law Firm: Shaw Pittman. Format: Soft adult contemp. News staff: one; News: 6 hrs wkly. Target aud: 21-55; blue collar, housewives, white collar.

KQSS(FM)— Mar 30, 1987: 98.3 mhz; 6 kw. Ant -279 ft TL: N33 24 30 W110 48 14. Stereo. Hrs opn: 24 Box 292, 85539. Secondary address: 5734 McKinney, Globe 85501. Phone: (928) 425-7186. Fax: (928) 425-7982. E-mail: bill@gila1019.com Web Site: www.gila1019.com. Licensee: William D. Taylor. Format: Country. News staff: one; News: 5 hrs wkly. Target aud: 25-54. ♦Bill Taylor, gen mgr & sls dir.

Morenci

KCUZ(AM)—See Safford

Munds Park

KFSZ(FM)—Not on air, target date: unknown: 106.1 mhz; 18.5 kw. Ant 1,807 ft TL: N35 14 31 W111 36 32. Hrs open: LKCM Radio Group LP, 301 Commerce St., Suite 1600, Fort Worth, TX, 76102. Phone: (817) 332-3235. Fax: (817) 332-4630. Web Site: www.lkcm.com. Licensee: LKCM Radio Group LP. ♦Kevin D. Prigel, gen mgr.

Nogales

***KNOG(FM)**— Dec 16, 1995: Stn currently dark. 91.1 mhz; 3 kw. 154 ft TL: N31 21 33 W110 53 54. Hrs open: 24 Box 1614, 85628. Secondary address: 150 W. First St. 85628. Phone: (520) 287-5206. Fax: (520) 287-3606. E-mail: knog@hcjb.org Web Site: www.knog.org. Licensee: World Radio Network Inc. Population served: 600,000 Natl. Network: Moody. Format: Sp contemp Christian, educ. News: 5 hrs wkly. Target aud: 18-55; Hispanics. Spec prog: Btfl music 5 hrs wkly. ♦Marcos Romero, stn mgr; Mariana Romero, progmg dir & pub affrs dir; Concepcion Borrayo, traf mgr.

KOFH(FM)— Apr 1, 1999: 99.1 mhz; 6 kw. 328 ft TL: N31 20 46 W110 53 34. Hrs open: 934N Bejarano St., Suite 2, 85621-1385. Phone: (520) 287-6885. Fax: (520) 287-8290. E-mail: noticieroal

maximo@hotmail.com Web Site: www.maxima991.fm. Licensee: Felix Corp. Format: Top-40, Sp & English. ♦Oscar Felix Sr., gen mgr; Rene Saylor, progmg dir & news dir.

Oracle

KGMG(FM)— December 1984: 106.3 mhz; 430 w vert, 440 w horiz. 4,172 ft TL: N32 26 26 W110 47 12. Stereo. Hrs opn: 24 3438 N. Country Club Rd., Tucson, 85716. Phone: (520) 795-1490. Fax: (520) 327-2260. Licensee: Journal Broadcast Corp. Group owner: Journal Broadcast Group Inc. (acq 4-15-98; $5.8 million). Population served: 600,000 Natl. Rep: Christal. Format: Oldies. News: 2 hrs wkly. Target aud: 25-54; Hispanic and Anglo adults. ♦Diane Frisch, gen mgr; Larkin Gassman, mktg mgr; Bobby Rich, progmg dir.

Oro Valley

KCMT(FM)— 2003: 102.1 mhz; 100 kw. Ant 266 ft TL: N32 17 23 W111 01 06. Hrs open: 3871 N. Commerce Dr., Tucson, 85705. Phone: (520) 407-4500. Fax: (520) 407-4600. Web Site: www.kcmt.com. Licensee: Arizona Lotus Corp. Group owner: Lotus Communications Corp. Format: Rgnl Sp. ♦Steve Groesbeck, gen mgr; Tara Hungate, rgnl sls mgr.

KSZR(FM)— Apr 28, 1992: 97.5 mhz; 3 kw. 299 ft TL: N32 23 28 W111 01 48. (CP: 6 kw, ant 328 ft.). Hrs opn: 24 575 W. Roger Rd., Tuscon, 85705. Phone: (520) 887-1000. Fax: (520) 887-6397. Web Site: www.bob975.com. Licensee: Citadel Broadcasting Co. Group owner: Citadel Broadcasting Corp. (acq 4-26-01; grpsl). Population served: 800,000 Format: Music of the 70's & 80's. News staff: 2; News: 60 hrs wkly. Target aud: 25-54. ♦Farid Suleman, CEO; Ken Kowalcek, gen mgr & stn mgr; Herb Crowe, opns dir; Keith Rosenblatt, sls dir.

Page

***KNAD(FM)**— 1998: 91.7 mhz; 500 w. 1,509 ft TL: N36 41 51 W111 37 57. Hrs open:
Rebroadcasts KNAU(FM) Flagstaff.
Box 5764, Northern Arizona University, Flagstaff, 86011-5764. Phone: (928) 523-5628. Fax: (928) 523-7647. E-mail: knau@nau.edu Web Site: www.knau.org. Licensee: Arizona Board of Regents on behalf of Northern Arizona University. Format: News & info. ♦John Stark, gen mgr; Dave Riek, opns mgr; Jeff Norcross, progmg dir; Don Kraker, news dir; Lisa Skinner, traf mgr.

KPGE(AM)— May 15, 1971: 1340 khz; 1 kw-U. TL: N36 45 23 W111 27 32. Hrs open: 24 Box 1030, 91 7th Ave., 86040. Phone: (928) 645-8181. Fax: (928) 645-3347. Web Site: kpge.com. Licensee: Lake Powell Communications Inc. (acq 7-1-91; with co-located FM; FTR: 6-17-91). Population served: 41,000 Natl. Network: ABC. Format: Country. News staff: one; News: 15 hrs wkly. Target aud: 25-54. ♦Dan Brown, gen mgr; Janet Brown, gen sls mgr; Deborah Phillips, news dir; Mark Jones, chief of engrg.

KXAZ(FM)—Co-owned with KPGE(AM). Sept 22, 1980: 93.3 mhz; 12.5 kw. 921 ft TL: N36 46 42 W111 25 46. Stereo. 24 Web Site: kxaz.com.41,000 Format: Adult contemp.

Paradise Valley

KHOT-FM— 1996: 105.9 mhz; 36 kw. Ant 577 ft TL: N33 35 16 W111 45 38. Hrs open: 4745 N. 7th St., Suite 140-C, Phoenix, 85014. Phone: (602) 308-7900. Fax: (602) 308-7979. Web Site: www.univision.com. Licensee: Univision Radio License Corp. Group owner: Univision Radio (acq 9-22-2003; grpsl). Format: Rgnl Mexican. ♦Mary McEvilly-Hernandez, VP & gen mgr; Fernando Gomez, mktg dir & prom mgr; Nelson Oreida, progmg dir.

Parker

KLPZ(AM)— Sept 7, 1974: 1380 khz; 2.5 kw-D, 58 w-N. TL: N34 09 14 W114 17 15. Hrs open: 24 816 6th St., 85344. Phone: (928) 669-9274. Phone: (928) 669-9275. Fax: (928) 669-9300. E-mail: klpz@redrivernet.com Web Site: www.klpz1380.com. Licensee: Keith Douglas Learn (acq 4-1-00). Population served: 20,000 Natl. Network: Jones Radio Networks. Format: Country, news/talk. News staff: 2; News: 2 hrs wkly. Target aud: 25-55. Spec prog: Farm one hr wkly. ♦Keith Douglas Learn, pres.

KPKR(FM)—Not on air, target date: unknown: 97.3 mhz; 25 kw. Ant 121 ft TL: N34 00 11 W114 13 40. Hrs open: Box 26523, Prescott Valley, 86312. Phone: (928) 445-8289. Fax: (928) 442-0448. Licensee: Prescott Valley Broadcasting Co. Inc. ♦Sanford Cohen, pres & gen mgr.

KRIT(FM)— 2003: 93.9 mhz; 7.6 kw. Ant -154 ft TL: N34 08 30 W114 17 50. Stereo. Hrs opn: 24 1301 Arizona Ave., Suite 4, 85344. Phone: (661) 823-6201. Phone: (661) 837-0745. Fax: (661) 837-1612. E-mail: achavez@campesina.com Web Site: www.campesina.com. Licensee: Farmworker Educational Radio Network Inc. Format: Rgnl, Sp. ♦Anthony Chavez, exec VP; Kevin Lein, stn mgr; Barbara Lein, natl sls mgr; Cesar Chavez, progmg dir; Dave Whitehead, chief of engrg; Maria Vrrutia, traf mgr.

***KWFH(FM)**— November 1984: 90.1 mhz; 460 w. -184 ft TL: N34 08 53 W114 16 44. Stereo. Hrs opn: 24 Box 747, 86405. Phone: (928) 669-5683. Phone: (928) 855-9110. Fax: (928) 669-5683. Web Site: www.kwfh.org. Licensee: Desert View Baptist Church. Population served: 12,000 Natl. Network: Moody. Format: Relg. Target aud: General. ♦Gary Covert, stn mgr & opns mgr; Faron Eckelbarger, progmg dir & mus dir.

Payson

KAJM(FM)— July 4, 1984: 104.3 mhz; 100 kw. 1,023 ft TL: N34 25 48 W111 30 16. (CP: Ant 1,164 ft.). Stereo. Hrs open: 24 7434 E. Stetson Dr., Suite 255, Scottsdale, 85251. Phone: (480) 994-9100. Phone: (800) 254-7510. Fax: (480) 423-8770. E-mail: operations@sierrah.com Web Site: www.mega1043.com. Licensee: Sierra H. Broadcasting Inc. Population served: 3,000,000 Natl. Network: Westwood One, CNN Radio. Natl. Rep: Roslin. Format: Old school/rhythm & blues. Target aud: 25-54; general. ♦Michael Mallace, gen mgr; Jack Preda, sls dir; Michael Devitt, prom dir; Rod Carrillo, progmg dir; Alex Santa Maria, mus dir; Steven Szalay, opns mgr & pub affrs dir; Michael Day, chief of engrg.

KMOG(AM)— Nov 1, 1983: 1420 khz; 2.5 kw-D, 500 w-N, DA-N. TL: N34 16 00 W111 18 54. Hrs open: 24 500 E. Tyler Pkwy., 85541. Phone: (928) 474-5214. Fax: (928) 474-0236. E-mail: kmog@1420kmog.com Licensee: Farrell Enterprises L.L.C. (acq 3-6-97). Format: Country. News staff: one; News: 2 hrs wkly. Target aud: 25-54; working adults. ♦Mike Farrell, pres; Blaine Kimball, gen mgr.

KMZQ(FM)—Not on air, target date: unknown: 99.3 mhz; 17.2 kw. Ant 403 ft TL: N34 11 04 W111 20 16. Hrs open: 3800 Howard Hughes Pkwy., Wells Fargo Tower, 17th Fl., Las Vegas, NV, 89109. Phone: (702) 385-6000. Licensee: Kemp Communications Inc. ♦Will Kemp, pres & gen mgr.

KNRJ(FM)— 2000: 101.1 mhz; 88 kw. 1,033 ft TL: N34 25 51 W111 30 12. Hrs open: 24 7434 E. Stetson Dr., Suite 255, Scottsdale, 85251. Phone: (480) 994-9100. Phone: (800) 254-7510. Fax: (480) 423-8770. E-mail: operations@sierrah.com Web Site: www.energyarizonafm.com. Licensee: Sierra H. Broadcasting Inc. Natl. Network: Westwood One, CNN Radio. Format: Disco, Dance music. Target aud: 18-35; upscale. ♦Michael Mallace, gen mgr; Rod Carrillo, progmg dir; Steve Szalay, pub affrs dir.

Phoenix

KASA(AM)— Jan 6, 1967: 1540 khz; 10 kw-D, DA. TL: N33 22 36 W112 05 25. Hrs open: 1445 W. Baseline Rd., 85041. Phone: (602) 276-4241. Phone: (602) 276-5272. Fax: (602) 276-8119. Licensee: KASA Radio Hogar Inc. (group owner; (acq 8-26-92; $475,000; FTR: 9-14-92). Population served: 2,000,000 Law Firm: Cohn & Marks. Format: Relg. Target aud: General. ♦Moses Herrera, pres, gen mgr & opns mgr.

KAZG(AM)—Scottsdale, 1956: 1440 khz; 5 kw-D, 52 w-N. TL: N33 28 43 W111 56 24. Hrs open: 8:30 AM - 5:30 PM 4343 E. Camelback Rd., Suite 200, 85018. Phone: (480) 941-1007. Fax: (602) 260-5759. Licensee: Cactus Radio Inc. Group owner: Sandusky Radio (acq 6-5-98; with co-located FM). Population served: 250,000 Format: Oldies. ♦Chuck Artigue, gen mgr; Dean Mooney, gen sls mgr; Michael Bradford, prom dir & prom mgr; Dave Cooper, progmg dir.

KSLX-FM—Co-owned with KAZG(AM). Aug 1, 1969: 100.7 mhz; 100 kw. 1,847 ft TL: N33 19 53 W112 03 47. Stereo. Web Site: www.kslx.com.250,000 Format: Classic rock.

***KBAQ-FM**— Apr 26, 1993: 89.5 mhz; 91 w. 1,463 ft TL: N33 19 58 W112 03 53. (CP: 12.5 kw, ant 2,316 ft. TL: N33 35 33 W112 34 49). Hrs opn: 24 2323 W. 14th St., Tempe, 85281. Phone: (480) 834-5627. Fax: (480) 774-8475. E-mail: kbaq.mail@kbaq.org Web Site: www.kbaq.org. Licensee: College District. Population served: 2,000,000 Natl. Network: NPR, PRI. Format: Class mus, news. ♦Carl Matthusen, gen mgr; Lou Stanley, dev dir & sls dir; Scott Williams, progmg dir; Sterling Beeaff, mus dir; Dennis Gilliam, chief of engrg.

KESZ(FM)— July 1982: 99.9 mhz; 100 kw. 1,702 ft TL: N33 20 01 W112 03 44. Stereo. Hrs opn: 4686 E. Van Buren St., Ste 300, 85008-6967. Phone: (480) 966-6236. Fax: (480) 921-6365. Web Site: www.kez999.com. Licensee: CC Licenses LLC. Group owner: Clear Channel Communications Inc. (acq 5-14-99; $58 million). Natl. Network: AP Radio. Natl. Rep: Katz Radio. Format: Adult contemp. Target aud: General. ♦Susan Karis-Madigan, gen mgr; Linda Little, rgnl sls mgr; Mary Evanson, prom dir.

***KFLR-FM**— December 1985: 90.3 mhz; 2.2 kw. 354 ft TL: N33 26 09 W112 06 35. (CP: 28.31 kw, ant 1,555 ft. TL: N33 20 02 W112 03 04). Stereo. Hrs opn: 24 PMB 549, 428 E. Thunderbird Rd., 85022. Phone: (602) 978-0903. Fax: (602) 548-8089. E-mail: kflr@flc.org Web Site: www.flc.org. Licensee: Family Life Broadcasting Inc. Group owner: Family Life Communications Inc. (acq 7-30-78). Population served: 582,000 Natl. Network: Salem Radio Network. Format: Christian, inspirational. ♦Randy Carlson, pres; Alan Cook, gen mgr & progmg dir; Fred Morse, opns mgr; Bruce Thurman, prom dir; Walter Ellis, engrg mgr & chief of engrg.

KFNN(AM)—Mesa, November 1962: 1510 khz; 22 kw-D, 100 w-N. TL: N33 23 30 W111 50 16. Hrs open: 24 4800 N. Central Ave., 85012. Phone: (602) 241-1510. Fax: (602) 241-1540. E-mail: info@kfnn.com Web Site: www.kfnn.com. Licensee: CRC Broadcasting Co. Inc. (acq 1988). Population served: 135,600 Natl. Network: CNN Radio. Law Firm: Akin, Gump, Strauss, Houer & Feld. Wire Svc: Metro Weather Service Inc. Format: News/talk, business news, investment advice. News staff: 3; News: 84 hrs wkly. Target aud: 30 plus; upscale, investment-oriented professionals & entrepreneurs; decision makers. ♦Brian DuBose, VP & progmg dir; Ronald E. Cohen, pres & gen mgr; Renee Yorks, opns mgr; Brian Du Bose, progmg dir.

KFYI(AM)—Listing follows KYOT-FM.

KGME(AM)— 1940: 910 khz; 5 kw-U, DA-N. TL: N33 32 00 W112 07 18. Hrs open: 24 4686 E. Van Buren St., Ste 300, 85008-6967. Phone: (602) 798-9322. Fax: (602) 650-5280. Web Site: www.xtra910.com. Licensee: AMFM Radio Licenses L.L.C. Group owner: Clear Channel Communications Inc. (acq 8-30-00; grpsl). Natl. Network: CBS, Westwood One. Format: Sports/talk. News staff: 7. Target aud: General. ♦Brad Gould, gen mgr & sls VP.

KIDR(AM)— Feb 1, 1958: 740 khz; 1 kw-D, 292 w-N, DA-2. TL: N33 21 55 W112 06 30. Stereo. Hrs opn: 24 3030 N. Central Ave., Suite 220, 85012-2784. Phone: (602) 234-8998. Fax: (602) 234-8993. E-mail: info@wradiophx.com Web Site: www.wradio.com.mx. Licensee: Multicultural Radio Broadcasting Licensee LLC. Group owner: Multicultural Radio Broadcasting Inc. (acq 2-4-2004; grpsl). Population served: 1,900,000 Format: Sp news, talk, sports. News staff: 2; News: 10 hrs wkly. Target aud: 18-64; Hispanic. ♦Arthur Liu, CFO; Arturo Galvez, gen mgr & news dir.

KJZZ(FM)—Licensed to Phoenix. See Mesa

KKNT(AM)— June 1947: 960 khz; 5 kw-U, DA-N. TL: N33 39 12 W111 55 39. (CP: TL: N33 41 34 W112 00 09). Hrs opn: 24 2425 E. Camelback Rd., Suite 570, 85016. Phone: (602) 955-9600. Fax: (602) 955-7860. E-mail: jtimm@kknt960.com Web Site: www.kknt960.com. Licensee: Common Ground Broadcasting Inc. Group owner: Salem Communications Corp. (acq 1996; $6.5 million). Population served: 1,874,600 Natl. Network: Salem Radio Network. Format: News/talk. News staff: 2; News: 6 hrs. wkly. Target aud: 25-54; 35-64; upscale adults. Spec prog: Insight bowl. ♦Edward Atsinger III, CEO & news dir; Stuart Epperson, chmn; Joe D. Davis, exec VP & progmg mgr; Jon Horton, VP; John Timm, gen mgr; Laurie Larson, opns dir; Jim Seemiller, gen sls mgr.

KMIK(AM)—Tempe, June 23, 1960: 1580 khz; 50 kw-U, DA-N. TL: N33 27 22 W111 50 01. Stereo. Hrs opn: 24 2231 E. Camelback, Suite 102, 85016. Phone: (602) 381-1580. Fax: (602) 840-1488. E-mail: marni.gerber@abc.com Web Site: www.radiodisney.com. Licensee: Radio Disney Group LLC. Group owner: ABC Inc. (acq 9-10-98; $5.85 million). Population served: 2,000,000 Natl. Network: ABC. Format: Radio Disney, children's Pop Top 40. ♦Marni Gerber, gen mgr; Carl Jimenez, mktg dir & prom dir.

KMLE(FM)—Chandler, Apr 18, 1980: 107.9 mhz; 100 kw. 1,735 ft TL: N33 20 03 W112 03 43. Stereo. Hrs opn: 24 840 N. Central, 85004. Phone: (602) 452-1000. Fax: (602) 440-6530. Web Site: www.kmlenation.com. Licensee: Infinity Radio Inc. Group owner: Infinity Broadcasting Corp. (acq 8-7-00; grpsl). Population served: 350,000 Format: Country. News staff: 2. Target aud: 25-54. Spec prog: Camel Views one hr wkly. ♦Amy Leimbach, gen sls mgr; Mark Waters, gen mgr & mktg mgr; Jay McCarthy, progmg dir; Doc Holiday, mus dir.

KMVP(AM)— Nov 23, 1949: 860 khz; 1 kw-U, DA-N. TL: N33 24 16 W112 07 24. Hrs open: 24 5300 N. Central Ave., 85012-1410. Phone: (602) 274-6200. Fax: (602) 266-3858. Web Site: www.ktar.com. Licensee: Bonneville Holding Co. Group owner: Emmis Communications Corp. (acq 1-14-2005; grpsl). Population served: 1,750,000 Natl. Network: ESPN Radio. Format: Sports talk. Target aud: 25-54; sports enthusiast. ♦Bruce T. Reese, CEO & pres; David Brown, sls dir; Mike Fadelli, gen sls mgr; Dawn Paugh, natl sls mgr; Randy Eccles, prom mgr; Tisa Vrable, progmg dir; Gary Smith, engrg dir.

KMXP(FM)— October 1964: 96.9 mhz; 100 kw. 1,560 ft TL: N33 20 03 W112 03 36. Stereo. Hrs opn: 24 4686 E. Van Buren St., Ste 300, 85008-6967. Phone: (602) 279-5577. Fax: (602) 230-2781. Web Site: www.mix969.com. Licensee: Citicasters Licenses L.P. Group owner: Clear Channel Communications Inc. (acq 5-4-99; grpsl). Format: Adult contemp. News staff: one. ♦Lowry Mays, CEO; Randy Michaels, chmn; John Hogan, pres; Susan Karis-Madigan, gen mgr; Alan Sledge, opns dir; Shanna McCoy, sls dir.

KNAI(FM)—Licensed to Phoenix. See Keene CA

KNIX-FM— Sept 1, 1969: 102.5 mhz; 98 kw. 1,620 ft TL: N33 19 58 W112 03 53. Stereo. Hrs opn: 24 4686 E. Van Buren St., Ste 300, 85008-6967. Phone: (602) 374-6000. Web Site: www.knixcountry.com. Licensee: CC Licenses LLC. Group owner: Clear Channel Communications Inc. (acq 6-1-99; $84 million). Format: Country. News staff: 3; News: one hr wkly. Target aud: 25-54. ♦Susan Karis-Madigan, gen mgr; Shaun Holly, stn mgr & progmg dir; Art Morales, gen sls mgr; Becky Lynn, news dir; Mark Jeffrey, traf mgr.

KNUV(AM)—See Tolleson

KOMR(FM)—See Sun City

KOOL-FM— May 1956: 94.5 mhz; 100 kw. 1,655 ft TL: N33 20 02 W112 03 42. Stereo. Hrs opn: 24 4745 N. 7th St., Suite 210, 85014. Phone: (602) 956-9696. Fax: (602) 285-1450. Web Site: www.koolradio.com. Licensee: Infinity Radio Inc. Group owner: Infinity Broadcasting Corp. (acq 8-7-00; grpsl). Natl. Rep: Christal. Format: Oldies. Target aud: 25-54. ♦Charlie Lake, progmg dir.

KOY(AM)— May 1949: 1230 khz; 1 kw-U. TL: N33 26 09 W112 06 35. Hrs open: 24 4686 E. Van Buren St., Suite 300, 85008. Phone: (602) 374-6000. Fax: (602) 374-6035. Web Site: www.am1230koy.com. Licensee: AMFM Radio Licenses LLC. Group owner: Clear Channel Communications Inc. (acq 8-30-00; grpsl). Format: Oldies. ♦Susan Karis-Madigan, gen mgr.

***KPHF(FM)**— December 1991: 88.3 mhz; 22.5 kw. 997 ft TL: N33 45 37 W112 05 29. Hrs open: 7:30 PM-4:30 AM c/o 290 Hegenberger Rd., Oakland, CA, 94621. Phone: (602) 272-7220. Phone: (800) 835-4810. Fax: (510) 568-6190. Licensee: Family Stations Inc. (group owner; acq 12-91). Format: Relg. ♦Harold Camping, pres & gen mgr; David Manzi, opns mgr.

KPHX(AM)— June 10, 1958: 1480 khz; 1 kw-D, 500 w-N, DA-2. TL: N33 24 02 W112 06 28. (CP: 5 kw-D). Hrs opn: 824 E. Washington St., 85034. Phone: (602) 257-1351. Fax: (602) 256-0741. Licensee: Continental Broadcasting Corp. (acq 2-80; $650,000; FTR: 2-18-80). Population served: 581,562 Natl. Network: Music of Your Life. Format: Standards. ♦Kent Ennoms, CEO; John Molina, gen mgr; Cam Maxwell, stn mgr & opns mgr; Jonathan Molina, opns VP.

KPKX(FM)—Listing follows KTAR(AM).

KPXQ(AM)—Glendale, 1946: 1360 khz; 50 kw-D, 1 kw-N, DA-N. TL: N33 30 28 W112 13 01. Hrs open: 24 2425 E. Camelback Rd., Suite 570, 85016. Phone: (602) 955-9600. Fax: (602) 955-7860. E-mail: info@kpxq1360.com Web Site: www.kpxq1360.com. Licensee: Common Ground Broadcasting Inc. Group owner: Salem Communications Corp. (acq 6-23-99; $5 million). Population served: 1800000 Natl. Network: Salem Radio Network. Rgnl. Network: Salem Natl. Rep: Salem. Law Firm: Dow, Lohnes & Albertson. Format: Relg talk. News staff: one; News: 6 hrs wkly. Target aud: 25-54; adults. ♦Edward Atsinger III, CEO; Stuart Epperson, pres; Joe D. Davis, exec VP; Jon Horton, VP; John Timm, gen mgr; Laurie Larson, opns mgr; Jim Seemiller, gen sls mgr; Rachel Van Hofwegen, prom dir; John Bortowski, chief of engrg; Diane Johnson, traf mgr; John Gibson, local news ed.

KQMR(FM)—See Globe

KSUN(AM)— Aug 27, 1954: 1400 khz; 1 kw-U. TL: N33 23 23 W111 59 52. Stereo. Hrs opn: 714 N. 3rd St., 85004. Phone: (602) 252-0030. Fax: (602) 252-4211. Web Site: www.radiofiesta.com. Licensee: Fiesta Radio Inc. (acq 10-86; $600,000; FTR: 12-8-86). Population served: 2,500,000 Format: Adult contemp, Sp. Target aud: 19-45. ♦Pedro Marquez, pres.

KTAR(AM)— June 21, 1922: 620 khz; 5 kw-U, DA-N. TL: N33 28 44 W112 00 06. Hrs open: 24 5300 N. Central Ave., 85012-1410. Phone: (602) 274-6200. Fax: (602) 266-3858. Web Site: 620sportsktar.com. Licensee: Bonneville Holding Co. Group owner: Emmis Communications Corp. (acq 1-14-2005; grpsl). Population served: 3,000,000 Natl. Rep: Interep, D & R Radio. Wire Svc: UPI Format: Sports. ♦Dave Brown, gen sls mgr; Randy Eccles, prom dir; Russ Hill, progmg dir.

KPKX(FM)—Co-owned with KTAR(AM). July 1, 1960: 98.7 mhz; 100 kw. 1,680 ft TL: N33 20 00 W112 03 48. Stereo. 24 Fax: (602) 266-3858. Fax: (602) 274-4477. Web Site: www.987thepeak.com.3,000,000 Format: Adult contemp. News staff: one; News: 2 hrs wkly. Target aud: Adults 25-54; lite rock. ♦Doug Brannan, prom dir; Joel Grey, progmg dir; Clayton Creekmore, engrg dir.

KTAR-FM—Glendale, Dec 19, 1979: 92.3 mhz; 98 kw. Ant 1,788 ft TL: N33 19 58 W112 03 48. Stereo. Hrs opn: 5300 N. Central Ave., 85012-1410. Phone: (602) 274-6200. Fax: (602) 266-3858. Web Site: www.ktar.com. Licensee: Bonneville Holding Co. Group owner: Emmis Communications Corp. (acq 7-11-2006; $77.5 million). Population served: 2,500,000 Format: News/talk. Target aud: 18-49. ♦Erik Hellum, VP & mktg mgr; Brett Rogers, gen sls mgr; Russ Hill, progmg dir.

KXAM(AM)—See Mesa

KXEG(AM)— 1956: 1280 khz; 2.5 kw-D, 230 w-N. TL: N33 29 32 W112 08 28. Hrs open: 24 2800 N. 44th St., Suite 100, 85012. Phone: (602) 296-3600. Fax: (602) 296-3624. E-mail: jess@kxeg1280.com Web Site: www.kxeg1280.com. Licensee: Communicom Co. of Phoenix L.P. Group owner: James Crystal Inc. (acq 12-14-2005; grpsl). Population served: 2,000,000 Natl. Rep: Salem, Commercial Media Sales. Law Firm: Wilkinson, Barker, Knauer & Quinn. Format: Christian. Target aud: 25 plus; educated adults with disposable income. Spec prog: Sp 5 hrs wkly. ♦Jess Spurgin, gen mgr.

KXXT(AM)—See Tolleson

KYOT-FM— Oct 31, 1963: 95.5 mhz; 96 kw. 1,570 ft TL: N33 20 06 W112 03 39. Stereo. Hrs opn: 24 4686 E. Van Buren St., Suite 300, 85008. Phone: (602) 374-6000. Fax: (602) 374-6035. Web Site: www.kyot.com. Licensee: AMFM Radio Licenses LLC. Group owner: Clear Channel Communications Inc. (acq 8-30-00; grpsl). Law Firm: Dow, Lohnes & Albertson. Format: Smoth jazz. News staff: one. Target aud: 25-54. ♦Susan Karis-Madigan, gen mgr; Angie Handa, mus dir; John Baker, chief of engrg.

KFYI(AM)—Co-owned with KYOT-FM. October 1921: 550 khz; 5 kw-D, 1 kw-N. TL: N33 23 17 W112 00 22.24 Fax: (602) 374-6032. Web Site: www.kfyi.com. Natl. Network: Westwood One. Natl. Rep: Christal. Format: News/talk. Target aud: 50 plus. Spec prog: Relg 2 hrs wkly. ♦Brad Gould, stn mgr & opns mgr; Laurie Canfillo, progmg dir.

KZON(FM)— July 5, 1964: 101.5 mhz; 100 kw. Ant 1,740 ft TL: N33 19 52 W112 03 46. Hrs open: 24 840 N. Central Ave., 85004. Phone: (602) 452-1000. Fax: (602) 420-9916. Fax: (602) 440-6530. Web Site: 1015jamz.com. Licensee: Infinity Radio Inc. Group owner: Infinity Broadcasting Corp. (acq 12-20-99; grpsl). Population served: 2700000 Law Firm: Leventhal, Senter & Lerman. Format: FM Talk. News staff: one. ♦Greg Garber, gen sls mgr & prom mgr; Mark Waters, mktg mgr; Chris Patyk, progmg dir.

KZZP(FM)—See Mesa

Prescott

KAHM(FM)— Sept 9, 1981: 102.1 mhz; 58 kw. 2,526 ft TL: N34 41 14 W112 07 01. Stereo. Hrs opn: 24 Box 2529, 86302. Secondary address: 510 Henry St. 86301. Phone: (928) 445-7800. Licensee: Southwest FM Broadcasting Co. Population served: 2,000,000 Law Firm: Cohn & Marks. Format: Easy lstng. News staff: 3; News: 7 hrs wkly. Target aud: 35-64; mature, affluent. ♦Lou Silverstein, gen mgr & opns dir; Nancy Silverstein, progmg dir; Al Hartsell, engrg dir; Sue Mopp, traf mgr.

KYCA(AM)—Co-owned with KAHM(FM). August 1940: 1490 khz; 1 kw-U. TL: N34 33 03 W112 27 45.6 AM-midnight Box 1631, 86302. Secondary address: 500 Henry St. 86302. Phone: (602) 445-1700. Licensee: Southwest Broadcasting Co. (acq 9-25-70).100,000 Natl. Network: CBS, Westwood One. Law Firm: Cohn & Marks. Format: News/talk. News staff: 4; News: 20 hrs wkly. Target aud: 35-64; mature adults. ♦Bruce Taylor, gen mgr & news dir; Jason Zinzulletta, stn mgr & opns dir; John Rust, news dir; Lou Silverstein, pres & min affrs dir.

***KGCB(FM)**— Dec 5, 1994: 90.9 mhz; 58 kw. 2,532 ft TL: N34 41 15 W112 07 02. Stereo. Hrs opn: 24 5025 N. Hwy. 89, 86301. Phone: (928) 776-0909. Fax: (928) 776-1736. Web Site: www.kgcb.org. Licensee: Grand Canyon Broadcasters Inc. Natl. Network: Salem Radio Network. Law Firm: Wilkinson, Barker, Knauer, LLP. Format: Contemp Christian, adult contemp, relg. Target aud: 25-54; adult, family audience. ♦Stephen R. White, VP, gen mgr & stn mgr; Daniel White, opns dir; Sue Scott, dev mgr; Mike Medlin, progmg dir.

***KNAQ(FM)**— September 1997: 89.3 mhz; 100 w. 1,584 ft TL: N34 29 24 W112 31 59. Hrs open:
Rebroadcasts KNAU-FM Flagstaff 100%.
Box 5764, Northern Arizona University, Flagstaff, 86011-5764. Phone: (928) 523-5628. Fax: (928) 523-7647. E-mail: knau@nau.edu Web Site: www.knau.org. Licensee: Northern Arizona University. Format: NPR news, classical, info. ♦John Stark, gen mgr; Dave Riek, opns mgr & progmg mgr; Dan Kraker, news dir; Lisa Skinner, traf mgr.

KNOT(AM)— June 22, 1957: 1450 khz; 1 kw-U. TL: N34 32 42 W112 26 46. Hrs open: 24 Box 151, 86302. Secondary address: 116 S. Alto 86303. Phone: (928) 445-6880. Fax: (928) 445-6852. E-mail: knot@knotradio.com Web Site: www.knotradio.com. Licensee: Guyann Corp. (acq 9-7-2005; grpsl). Population served: 100,000 Format: MOR, adult contemp. News staff: one. Target aud: 35 plus. Spec prog: Jazz 2 hrs, sports 8 hrs wkly. ♦C. Pastore, gen mgr, stn mgr & opns mgr; P. Ezell, sls dir; Paul Hurt, progmg dir & progmg mgr; Doreen Conti, news dir & local news ed; Mark Hill, chief of engrg.

KTMG(FM)—Co-owned with KNOT(AM). Nov 11, 1977: 99.1 mhz; 6 kw. 200 ft TL: N34 34 29 W112 28 45. Stereo. 24 Natl. Network: ABC. Format: Country. News staff: one. Target aud: 35 plus. ♦C. Pastore, stn mgr; Paul Hurt, mus dir; Doreen Conti, local news ed.

KPPV(FM)—See Prescott Valley

Prescott Valley

KPPV(FM)—Listing follows KQNA(AM).

KQNA(AM)— June 28, 1986: 1130 khz; 1 kw-D. TL: N34 37 46 W112 18 56. Hrs open: Box 26523, 86312. Secondary address: 3755 Karicio Ln., Suite 2-C, Prescott 86303. Phone: (928) 445-8289. Phone: (800) 264-5449. Fax: (928) 442-0448. E-mail: info@kppv.com Web Site: www.kqna.com. Licensee: Prescott Valley Broadcasting Co. (group owner; (acq 12-27-93; $75,000; FTR: 1-10-94). Population served: 200,000 Natl. Network: Fox News Radio. Law Firm: David Tillotson. Format: News/talk, sports. News staff: 3; News: 84 hrs wkly. Target aud: 35-64; middle - upper income, professionals & new consumers. ♦Sanford B. Cohen, pres & gen mgr; Terry P. Cohen, exec VP; Allison Flannery, natl sls mgr; Bill Monroe, news dir & news rptr; Ken Byers, pub affrs dir; Mark Hills, engrg mgr; Mike Austin, sports cmtr.

KPPV(FM)—Co-owned with KQNA(AM). Sept 1, 1985: 106.7 mhz; 3.7 kw. 1,627 ft TL: N34 29 25 W112 32 00. Stereo. 24 Web Site: www.kppv.com.250,000 Natl. Network: Jones Radio Networks, Fox News Radio. Format: Adult contemp. News staff: 2; News: 6 hrs wkly. Target aud: 25-54; middle to upper income professionals with families and disposable income. ♦Bill Monroe, news rptr; Mike Austin, sports cmtr.

Quartzsite

KBUX(FM)— November 1988: 94.3 mhz; 205 w. -161 ft TL: N33 40 58 W114 13 59. Stereo. Hrs opn: 6 AM-10 PM Box 40, 85346. Phone: (928) 927-5111. E-mail: kugxradio@hotmail.com Licensee: Maude J. Burdette. Format: Btfl music, country, oldies, diversefied. Target aud: General; retired motor home & trailer owners wintering in warmer climate. ♦Maude J. Burdette, gen mgr; Maude Burdette, stn mgr; Marvin Vosper, progmg dir.

Red Mesa

***KRMH(FM)**— 1998: 89.7 mhz; 4.5 kw. 134 ft TL: N36 57 48 W109 22 39. Hrs open: HC 61 Box 40, Teec Nos Pos, 86514. Phone: (928) 656-4100. Fax: (928) 656-4320. Web Site: www.rmusd.net. Licensee: Red Mesa Unified School District No. 27. Format: Native American, var.

Safford

KATO(AM)— May 5, 1961: 1230 khz; 1 kw-U. TL: N32 49 30 W109 45 30. Hrs open: 5 AM-midnight Drawer L, 85548. Secondary address: 3335 W. 8th St., Thatcher 85552. Phone: (928) 428-1230. Fax: (928) 428-1311. Web Site: www.eaznet.com/~kato. Licensee: McMurray Communications Inc. (group owner; acq 12-17-92; $10,000 with co-located FM; FTR: 2-1-93). Population served: 36,000 Natl. Network: ABC. Format: News/talk, sports. News staff: 7; News: 25 hrs wkly. Target aud: 25-54; upscale, intelligent. ♦Bud McMurray, pres; Davis Nathan, gen mgr & sls dir; Reed Richins, opns mgr, progmg dir & chief of engrg.

KXKQ(FM)—Co-owned with KATO(AM). Aug 11, 1979: 94.1 mhz; 1 kw. Ant 4,287 ft TL: N32 39 01 W109 50 53. Stereo. 24 Web Site: www.katkountry94.com.40,000 Format: Country. News: one hr wkly. Target aud: 25-54. ♦Reed Richins, engrg dir; Lee Patterson, traf mgr; Tim Walters, local news ed.

KCUZ(AM)—Clifton, July 31, 1969: 1490 khz; 1 kw-U. TL: N33 02 30 W109 17 40. Hrs open: 24
Rebroadcasts KFMM(FM) Thatcher 100%.
Box 35997, Tucson, 85740. Secondary address: 301 B Hwy. 70 E. 85546. Phone: (928) 428-0916. Fax: (928) 428-7797. E-mail: kfmm@eaznet.com Licensee: Cochise Broadcasting LLC. (acq 6-30-2007; $330,000 with KFMM(FM) Thatcher). Population served: 5,087 Format: Classic rock. News staff: one; News: 7 hrs wkly. Target aud: 25-54. Spec prog: Relg 2 hrs, loc talk 8 hrs wkly. ♦Rick Schneider, gen mgr; Darwin Morris, disc jockey.

KFMM(FM)—Co-owned with KCUZ(AM). Dec 7, 1981: 99.1 mhz; 50 kw. Ant 2,280 ft TL: N32 53 22 W109 19 23. Stereo. 24
Rebroadcasts KCUZ (AM) Safford 100%.
100,000 News staff: one; News: 7 hrs wkly. Spec prog: Children 2 hrs wkly. ♦Darwin Morris, chief of engrg.

KWRQ(FM)—Clifton, Oct 1, 1986: 102.3 mhz; 2.8 kw. 2,211 ft TL: N32 53 23 W109 19 26. Stereo. Hrs opn: 24 Drawer L, 85548-0886. Secondary address: 3335 West 8th St., Thatcher 85552. Phone: (928) 428-1020. Fax: (928) 428-6818. Fax: (928) 428-1311. Web Site: www.mysouthernaz.com. Licensee: McMurray Communications Inc. (group owner; acq 11-97; $350,000). Natl. Network: Jones Radio Networks. Format: Hot Adult contemp. Target aud: 20-35; working females. ♦Bud McMurray, pres; Davis Nathan, gen mgr; Reed Richins, opns mgr.

Sahuarita

KEVT(AM)— Oct 12, 1985: 1210 khz; 10 kw-D, 1 kw-N, DA-N. TL: N32 02 04 W110 56 45. Hrs open: 2955 E. Broadway Blvd., Tucson, 85716. Phone: (520) 628-1200. Fax: (520) 326-4927. Web Site: www.radiounica.com. Licensee: One Mart Corp. Group owner: Multicultural Radio Broadcasting Inc. (acq 4-27-2007; $1.5 million). Format: Rgnl Mexican. ♦Francisco Zazueta, gen mgr.

Saint Johns

KWKM(FM)— 2001: 95.7 mhz; 100 kw. Ant 1,197 ft TL: N34 14 59 W109 35 08. Hrs open: 1520 Commerce Dr., Suite B, Show Low, 85901. Phone: (928) 532-2949. Fax: (928) 532-3176. E-mail: program@kwkm.com Web Site: www.kwkm.com. Licensee: KM Radio of St. Johns L.L.C. (acq 5-3-99). Natl. Network: ABC. Format: Adult contemp, classic rock. News: 1 hr wkly. Target aud: 18-44. Spec prog: News/talk 2 hrs, blues 2 hrs, alternative rock 2 hrs wkly. ♦Jean Barton, gen mgr.

Salome

KVGG(FM)—Not on air, target date: unknown: 101.9 mhz; 6 kw. Ant 328 ft TL: N33 46 54 W113 36 42. Hrs open: 3800 Howard Hughes Pkwy., Wells Fargo Tower, 17th Fl., Las Vegas, NV, 89169. Phone: (702) 385-6000. Fax: (702) 385-6001. Licensee: Kemp Communications Inc. ♦Will Kemp, pres & gen mgr.

Scottsdale

KAZG(AM)—Licensed to Scottsdale. See Phoenix

KSLX-FM—Licensed to Scottsdale. See Phoenix

Sedona

KAZM(AM)— Nov 1, 1974: 780 khz; 5 kw-D, 250 w-N, DA-N. TL: N34 51 38 W111 49 10. Hrs open: 24 Box 1525, 86339. Phone: (928) 282-4154. Fax: (928) 282-2230. E-mail: info@kazmradio.com Web Site: www.kazmradio.com. Licensee: Tabback Broadcasting Co. Population served: 280,000 Natl. Network: Westwood One, ESPN Radio, Fox News Radio. Law Firm: Brooks, Pierce, McLendon, Humphrey & Leonard. Format: New/talk, sports, music. News staff: 2; News: 16 hrs wkly. Target aud: 25 plus; baby boomers, professions, tourists. ♦Tom N. Tabback, gen mgr.

KQST(FM)— May 1, 1984: 102.9 mhz; 90 kw. 1,433 ft TL: N34 58 05 W111 30 29. Stereo. Hrs opn: 24 Box 187, Cottonwood, 86326. Phone: (928) 634-2959. Fax: (928) 634-2295. Web Site: www.myradioplace.com. Licensee: Yavapai Broadcasting Corp. (acq 12-1-2004; $3 million). Law Firm: John A. Borsari. Format: Adult contemp, top-40. Target aud: 24-59. ♦W. Grant Hafley, pres; Dave Kessel, gen mgr; Mike Puetz, gen sls mgr & prom dir; John Herring, progmg dir.

KSED(FM)— August 1994: 107.5 mhz; 98.4 kw. 1,463 ft TL: N34 58 07 W111 30 22. Stereo. Hrs opn: 24 112 E. Rt. 66, Suite 105, Flagstaff, 86001. Phone: (928) 779-1177. Fax: (928) 774-5179. E-mail: ann@northlandradio.com Web Site: www.koltcountry.com. Licensee: Grenax Broadcasting II LLC. (acq 1-6-2006; grpsl). Natl. Network: NBC. Format: Special blend, country. News staff: one. ♦Greg Dinetz, pres; Jim Shipp, gen mgr, sls dir & prom dir; Bill McAdams, opns dir; Mike Mentor, progmg dir; Jon Sweat, chief of engrg.

Seligman

KZKE(FM)— 1995: 103.3 mhz; 1.75 kw. 423 ft TL: N35 19 26 W112 45 55. Hrs open: 24 812 E. Beale St., Kingman, 86401. Phone: (928) 753-9100. Fax: (928) 753-1978. Licensee: Route 66 Broadcasting L.L.C. (acq 9-1-98). Format: Good time oldies. ♦Rhonda Hart, VP & gen mgr; JoAnn Oxsen, adv dir; Steve Levin, progmg dir.

Sells

***KOHN(FM)**— 2004: 91.9 mhz; 10 kw. Ant 1,656 ft TL: N32 07 59 W112 09 31. Hrs open: P.O. Box 183, 85634. Phone: (520) 361-5011. Fax: (520) 361-3931. E-mail: sial.thonolig@tonation-rsr.gov Licensee: Tohono O'Odham Nation. Format: Eclectic music, Native American. ♦Sial Thonolig, gen mgr; Mary Lopez, cultural affrs dir.

Show Low

***KNAA(FM)**— October 1997: 90.7 mhz; 100 w. 850 ft TL: N34 12 17 W109 56 22. Hrs open:
Rebroadcasts KNAU(FM) Flagstaff 100%.
Box 5764, Northern Arizona University, Flagstaff, 86011-5764. Phone: (928) 523-5628. Fax: (928) 523-7647. E-mail: knau@nau.edu Web Site: www.knau.org. Licensee: Arizona Board of Regents. Natl. Network: NPR. Format: Class, news. ♦John Stark, gen mgr; Dave Riek, opns dir.

KRFM(FM)—Listing follows KVSL(AM).

KSNX(FM)—Listing follows KVWM(AM).

KVSL(AM)— July 6, 1968: 1450 khz; 1 kw-U. TL: N34 16 00 W110 20 10. Hrs open: 3051 S. White Mountain RD., Suite D, 85901. Phone: (928) 532-3232. Fax: (928) 537-3991. E-mail: production @whitemountainradio.com Licensee: FFD Holdings I Inc. (acq 12-20-2004; grpsl). Population served: 50,000 Natl. Network: ABC. Format: Nostalgia oldies. Target aud: 45-65; 46% female, 54% male. Spec prog: Farm 2 hrs wkly. ♦Steve Johnson, gen mgr, gen sls mgr & progmg dir.

KRFM(FM)—Co-owned with KVSL(AM). July 1, 1983: 96.5 mhz; 100 kw. 994 ft TL: N34 12 20 W109 56 26. Stereo. Web Site: www.ksnx.com.80,000 Natl. Network: Jones Radio Networks. Format: Hot adult contemp. Target aud: 18-34; 65% female & 35% male.

KVWM(AM)— May 17, 1957: 970 khz; 5 kw-D, 114 w-N. TL: N34 13 14 W110 01 49. Hrs open: 3051 S. White Mountain Rd., Suite D, 85901. Phone: (928) 532-3232. Fax: (928) 537-3991. E-mail: production @whitemountainradio.com Licensee: FFD Holdings I Inc. (acq 12-20-2004; grpsl). Population served: 60,000 Natl. Network: ABC, Westwood One. Format: News/talk. Target aud: 35-65; 46% female & 54% male. ♦Steve Johnson, gen mgr, gen sls mgr & progmg dir; Amanda Lynn, traf mgr.

KSNX(FM)—Co-owned with KVWM(AM). Sept 13, 1964: 93.5 mhz; 25 kw. 150 ft TL: N34 13 14 W110 01 49. Web Site: www.ksnx.com. Format: Good-time oldies. Target aud: 25-54; 50% female & 50% male.

Sierra Vista

KKYZ(FM)— Jan 1, 1995: 101.7 mhz; 3 kw. Ant 328 ft TL: N31 33 59 W110 13 57. Stereo. Hrs opn: 24 500 E. Fry Blvd., Suite L-10, 85635. Phone: (520) 459-8201. Fax: (520) 458-7104. E-mail: info@kkyz.com Web Site: www.kkyz.com. Licensee: Cochise Broadcasting L.L.C. (acq 1-12-2001). Population served: 97,000 Format: Oldies. Target aud: 25-54. ♦Jeff Davenport, stn mgr; Ted Tucker, gen mgr, opns dir & opns mgr.

KNXN(AM)— June 20, 1980: 1470 khz; 2.5 kw-D, 39 w-N. TL: N31 32 53 W110 14 54. Hrs open: 24 680 Avenida del Sol, 85635. Phone: (520) 790-2440. Fax: (520) 790-2937. Web Site: www.kgms.com. Licensee: Good Music Inc. Group owner: Good News Communications Inc. (acq 4-16-01; $300,000). Population served: 92,000 Format: Inspirational talk, Christian. News staff: one; News: 5 hrs wkly. ♦Doug Martin, gen mgr; Jeff Davenport, gen mgr & opns mgr.

KTAN(AM)— March 1957: 1420 khz; 1.5 kw-D, 500 w-N, DA-N. TL: N31 32 47 W110 16 29. Hrs open: 24 Box 2770, 85636. Secondary address: 2300 Busby Dr. 85636. Phone: (520) 458-4313. Fax: (520) 458-4317. E-mail: ktan@wavmax.com Licensee: CCR-Sierra Vista IV LLC. Group owner: Cherry Creek Radio LLC (acq 12-19-2003; grpsl). Population served: 120,000 Natl. Network: CBS. Natl. Rep: Tacher. Law Firm: Baraff, Koerner & Olender. Format: News/talk, sports. News staff: one; News: 20 hrs wkly. Target aud: 25-54. ♦Paul Orlando, gen mgr; Rudy Sueskind, gen sls mgr & rgnl sls mgr; Debbie Simmons, mus dir & traf mgr.

KZMK(FM)—Co-owned with KTAN(AM). September 1973: 100.9 mhz; 3 kw. -46 ft TL: N31 32 47 W110 16 29. Stereo. 24 E-mail: k101@cwavmax.com 120,000 Format: Adult contemp. News staff: one; News: one hr wkly. Target aud: 18-49.

KWCD(FM)—See Bisbee

South Tucson

KJLL(AM)—Licensed to South Tucson. See Tucson

KXEW(AM)—Licensed to South Tucson. See Tucson

Springerville-Eagar

KQAZ(FM)— July 15, 1984: 101.7 mhz; 3 kw. -97 ft TL: N34 08 17 W109 16 10. (CP: 1.1 kw). Stereo. Hrs opn: Box 2020, Show Low, 85902. Secondary address: 691 E. Deuce of Clubs, Show Low 85901. Phone: (928) 532-1010. Fax: (928) 532-0101. Web Site: www.majik101.com. Licensee: William Konopnicki. (acq 7-30-99; $175,000 with KRVZ(AM) Springerville -Eagar). Natl. Network: AP Radio. Format: Adult Alternative. ♦Camden Smith, gen mgr & progmg dir; Laurie Pogson, traf mgr; Jack Jacobs, disc jockey.

KRVZ(AM)— June 11, 1982: 1400 khz; 1 kw-U. TL: N34 08 17 W109 16 10. Hrs open: 24 Box 2020, Show Low, 85902. Phone: (928) 532-1010. Fax: (928) 532-0101. E-mail: krvz@frontiernet.net Licensee: William Konopnicki. (acq 7-30-99; $175,000 with KQAZ(FM) Springerville). Natl. Network: Jones Radio Networks. Format: Talk radio. ♦William Konopnicki, pres; Camden Smith, gen mgr; Dan Curtis, opns dir; Laurie Pogson, traf mgr.

Sun City

KOMR(FM)— Mar 7, 1975: 106.3 mhz; 2.5 kw. 325 ft TL: N33 36 05 W112 17 31. (CP: 23 kw, ant 725 ft.). Stereo. Hrs opn: 24 4745 N. 7th St., Suite 140, Phoenix, 85014. Phone: (602) 308-7900. Fax: (602) 308-7979. Web Site: www.univision.com. Licensee: HBC License Corp. Group owner: Univision Radio (acq 9-22-2003; grpsl). Population served: 1,300,000 Law Firm: Dow, Lohnes & Albertson. Format: Spanish Adult Contemp. Target aud: 35-54. ♦Mary McEvilly-Hernandez, VP & gen mgr; Chris Morris, gen sls mgr; Aide Gonzalez, prom dir; Fernando Sosa, progmg dir; Dobby White, traf mgr.

Sun City West

KVIB(FM)— 2005: 95.1 mhz; 41 kw. Ant 2,785 ft TL: N34 14 05 W112 22 02. Stereo. Hrs opn: 24 4343 N. Scottsdale Rd., #200, Scottsdale, 85251. Phone: (480) 222-3300. Fax: (480) 970-1759. Web Site: www.951latinovibefm.com. Licensee: Sun City Licenses LLC (acq 3-10-2005; $18.7 million). Natl. Rep: McGavren Guild. Law Firm: Latham & Watkins. Format: Sp, CHR. Target aud: 18-34; Hispanic 2nd & 3rd generation. ♦Michael Cutchall, pres; Jose Rodilles, gen mgr; Ellen Cavanaugh, sls VP & mktg VP; Jaque Bosque Diaz, natl sls mgr.

Tempe

KDUS(AM)— Apr 16, 1960: 1060 khz; 5 kw-D, 500 w-N, DA-N. TL: N33 21 43 W111 58 03. Hrs open: 24 1900 W. Carmen, 85283. Phone: (480) 838-0400. Fax: (480) 820-8469. Web Site: www.kdus.com. Licensee: Tempe Radio Inc. Group owner: Sandusky Radio (acq 1994; $20 million with co-located FM). Population served: 50,000 Format: Sports. News staff: one. Target aud: 18-34 males. ♦Chuck Artigue, gen mgr.

KUPD-FM—Co-owned with KDUS(AM). April 1960: 97.9 mhz; 100 kw. 1,620 ft TL: N33 19 57 W112 03 53. Stereo. Web Site: www.98kupd.com.200,000 Law Firm: Wiley, Rein & Fielding. Format: AOR. ♦J.J. Jeffries, progmg dir.

KLNZ(FM)—Glendale, Sept 1, 1997: 103.5 mhz; 62 kw. 2,428 ft TL: N33 35 33 W112 34 49. Stereo. Hrs opn: 24 501 N. 44th St., Suite 425, Phoenix, 85008. Phone: (602) 266-2005. Fax: (602) 279-2921. Licensee: Entravision Holdings LLC. Group owner: Entravision Communications Corp. (acq 7-28-00; grpsl). Population served: 2,603,200 Law Firm: Bechtel & Cole. Format: Sp, rgnl Mexican. Target aud: General. ♦Tom Duran, gen mgr; Chris Moncayo, gen sls mgr; Carrie Strait, progmg dir; Ryan Oller, chief of engrg.

KMIK(AM)—Licensed to Tempe. See Phoenix

KNIX-FM—See Phoenix

Thatcher

KFMM(FM)—Licensed to Thatcher. See Safford

Tolleson

KNUV(AM)— Jan 23, 1961: 1190 khz; 5 kw-D, 250 w-N, DA-2. TL: N33 26 42 W112 15 54. Hrs open: 24 441 N. Central Ave., Suite 1000, Phoenix, 85012. Phone: (602) 433-1190. Fax: (602) 433-6226. Web Site: www.onda1190am.com. Licensee: New Radio Venture LLC (acq 10-20-2006; with KNRV(AM) Englewood, CO). Population served: 2,000,000 Law Firm: Cohn & Marks. Format: Sp news/talk, sports. ♦Laura Madrid, gen mgr.

KXXT(AM)— Dec 12, 1962: 1010 khz; 15 kw-D, 250 w-N, DA-D. TL: N33 26 43 W112 12 23. Hrs open: 24 2800 N. 44th St., Suite 100, Phoenix, 85008. Phone: (602) 296-3600. Fax: (602) 296-3624. E-mail: jess@kxeg1280.com Web Site: www.newstalk1010.net. Licensee: Communicom Co. of Arizona L.P. Group owner: James Crystal Inc. (acq 12-14-2005; grpsl). Population served: 1,270,000 Format: Relg. News staff: one. ♦Bob Christy, gen mgr; Willis Girdner, chief of engrg.

Tuba City

***KGHR(FM)**— Nov 27, 1991: 91.3 mhz; 100 kw horiz. Ant 1,056 ft TL: N36 21 27 W111 12 12. Stereo. Hrs opn: 24 Box 160, 86045. Phone: (928) 283-5555. Fax: (928) 283-5557. Web Site: www.kghr.org. Licensee: Tuba City High School Board Inc. Natl. Network: NPR. Format: Native American, AAA. News: 20 hrs wkly. Target aud: General; Native American/Navajo. Spec prog: Pub affrs 5 hrs wkly. ♦John Bittner, stn mgr.

KTBA(AM)— 1980: 1050 khz; 5 kw-D, 5.2 w-N. TL: N36 07 54 W111 14 59. (CP: 760 khz; 250 w-D, 60 w-N). Hrs opn: Box 9090, Window Rock, 86515. Licensee: Western Indian Ministries Inc. (group owner; (acq 1980). Population served: 70,000 Format: Adult contemp. English & Navajo. ♦Lenora A. Brown, gen mgr; Jareleen Mitchell, progmg VP; Bill Vadasy, mus dir.

Tucson

***KAIC(FM)**— 2006: 88.9 mhz; 1.8 kw vert. Ant 26 ft TL: N32 36 56 W110 38 38. Hrs open:
Rebroadcasts KLRD(FM) Yucaipa, CA 100%.
2351 Sunset Blvd., Suite 170-218, Rocklin, CA, 95765. Phone: (916) 251-1600. Fax: (916) 251-1650. Web Site: www.air1.com. Licensee: Educational Media Foundation. Natl. Network: Air 1. Format: Alternative rock, div. ♦Richard Jenkins, pres; Mike Novak, VP; Keith Whipple, dev dir; Eric Allen, natl sls mgr; David Pierce, progmg mgr; Ed Lenane, news dir; Sam Wallington, engrg dir; Karen Johnson, news rptr; Marya Morgan, news rptr; Richard Hunt, news rptr.

KCUB(AM)— August 1929: 1290 khz; 1 kw-U. TL: N32 16 37 W110 58 50. Hrs open: 24 575 W. Roger Rd., 85705. Phone: (520) 887-1000. Fax: (520) 887-6397. Licensee: Citadel Broadcasting Co. Group owner: Citadel Broadcasting Corp. (acq 4-26-2001; grpsl). Population served: 500,000 Format: Sports. News staff: one. Target aud: 25-54. ♦Todd Lawley, gen mgr; Herb Crowe, opns VP; Ken Kowalcek, sls VP.

KIIM-FM—Co-owned with KCUB(AM). March 1954: 99.5 mhz; 90 kw. 2,037 ft TL: N32 14 56 W111 06 59. Stereo. 24 Format: Hot country.

KFFN(AM)—Listing follows KMXZ-FM.

***KFLT(AM)**— October 1977: 830 khz; 50 kw-D, 1 kw-N, DA-N. TL: N32 26 39 W111 05 27. Hrs open: 24 Box 36868, 85740. Secondary address: 7355 N. Oracle Rd., Suite 102 85704. Phone: (520) 797-3700. Fax: (520) 797-3375. E-mail: kflt@flc.org Web Site: www.kflt.org. Licensee: Family Life Broadcasting System Inc. Group owner: Family Life Communications Inc. (acq 10-86; $125,000; FTR: 4-14-86). Population served: 3,000,000 Natl. Network: Moody. Format: Christian, inspirational. News staff: one; News: 15 hrs wkly. Target aud: 25-45; Christian families. ♦Randy Carlson, pres; Lee Escobedo, gen mgr; Dave Ficere, opns dir, dev dir & progmg dir; Joe Neubaum, mus dir; Carl Jackson, pub affrs dir; Randy Howard, chief of engrg.

KFLT-FM— 2006: 88.5 mhz; 1.5 kw vert. Ant 377 ft TL: N32 00 11 W110 47 49. Licensee: Family Life Broadcasting Inc.

KGMS(AM)— Aug 10, 1963: 940 khz; 5 kw-D, 1 kw-N, DA-2. TL: N32 12 04 W111 01 02. Hrs open: 24 3222 S. Richey Ave., 85713. Phone: (520) 790-2440. Fax: (520) 790-2937. E-mail: info@kgms.com Web Site: kgms.com. Licensee: Good Music Inc. Group owner: Good News Communications Inc. (acq 11-27-00; swap with KCEE(FM) Green Valley). Population served: 700,000 Natl. Rep: Salem. Law Firm: Wrey Fitch. Format: Christian talk. Target aud: 25-54. ♦Doug Martin, CEO, pres & gen mgr; Matt Manis, opns mgr.

KGVY(AM)—Green Valley, Sept 23, 1981: 1080 khz; 1 kw-D. TL: N31 55 34 W110 59 45. Hrs open: 6 AM-sunset Box 767, Green Valley, 85622. Secondary address: 1510 W. Camino Antigua, Sahuarita 85629. Phone: (520) 399-1000. Fax: (520) 399-9300. E-mail: kgvyam@quest.net Licensee: KGVY LLC (acq 6-19-2007; $1.1 million). Population served: 1,000,000 Natl. Network: ABC. Format: Oldies. News staff: 12; News: 13 hrs wkly. Target aud: 50 plus; mature, well educated, higher income, retired. ♦David Schmidt, gen mgr.

KHYT(FM)— August 1993: 107.5 mhz; 14.5 kw. 3,526 ft TL: N32 24 54 W110 42 56. (CP: 82 kw, ant 2,027 ft.). Hrs opn: 575 W. Roger, 85705. Phone: (520) 887-1000. Fax: (520) 887-6397. Web Site: www.rock1075.com. Licensee: Citadel Broadcasting Co. Group owner: Citadel Broadcasting Corp. (acq 4-26-01; grpsl). Format: Classic rock. Target aud: 25-54; adults. ♦Farid Suleman, CEO; Ken Kowalcek, gen mgr; Herb Crowe, opns dir; Keith Rosenblatt, sls dir.

KJLL(AM)—South Tucson, 1957: 1330 khz; 2 kw-D, 5 kw-N, DA-N. TL: N32 18 51 W110 50 17. Hrs open: 4433 E. Broadway, Suite 210, 85711. Phone: (520) 529-5865. Fax: (520) 529-9324. Web Site: www.tucsonsjolt.com. Licensee: Hudson Communications Inc. (acq 1996; $36,000). Population served: 600,000 Format: News/talk. ♦Kimberly Lopez, gen mgr.

KLPX(FM)—Listing follows KTKT(AM).

KMXZ-FM— Apr 11, 1973: 94.9 mhz; 97 kw. 1,952 ft TL: N32 14 56 W111 06 59. Stereo. Hrs opn: 24 3438 N. Country Club, 85716. Phone: (520) 795-1490. Fax: (520) 327-2260. E-mail: mixfm@mixfm.com Web Site: www.mixfm.com. Licensee: Journal Broadcast Corp. Group owner: Journal Broadcast Group Inc. (acq 1996; grpsl). Natl. Network: AP Radio. Law Firm: Crowell & Moring. Wire Svc: AP Format: Adult contemp/soft rock. News staff: one; News: 3 hrs wkly. Target aud: 25-54. ♦Diane Frisch, stn mgr; Darla Thomas, opns mgr; Jennifer Nunn, gen sls mgr; Larkin Gassman, mktg dir; Bobby Rich, progmg dir.

KFFN(AM)—Co-owned with KMXZ-FM. January 1957: 1490 khz; 1 kw-U. TL: N32 14 56 W110 55 29. Stereo. 24 714,000 Natl. Network: ESPN Radio. Format: Sports. News: 2 hrs wkly. Target aud: 18-49. ♦Rob Cook, progmg dir.

KNST(AM)— Oct 1, 1958: 790 khz; 5 kw-D, 500 w-N. TL: N32 14 54 W111 00 30. Hrs open: 24 3202 N. Oracle Rd., 85705. Phone: (520) 618-2100. Fax: (520) 618-2170. Licensee: Capstar TX L.P. Group owner: Clear Channel Communications Inc. (acq 8-30-00; grpsl). Population served: 262,933 Natl. Network: Moody. Natl. Rep: McGavren Guild. Format: News/talk, sports. News staff: 3. Target aud: 25-54. ♦Debbie Wagner, gen mgr.

KRQQ(FM)—Co-owned with KNST(AM). Feb 1, 1971: 93.7 mhz; 91 kw. 2,030 ft TL: N32 14 56 W111 06 57. Stereo. Web Site: www.krq.com.640,000 Format: CHR, Top-40. News staff: one. Target aud: 18-54. ♦Mike Madigan, stn mgr.

KOHT(FM)—See Marana

KQTH(FM)— May 4, 1994: 104.1 mhz; 3 kw. Ant 328 ft TL: N32 17 23 W111 01 06. Hrs open: 24 3438 N. Country Club, 85716. Phone: (520) 795-1490. Fax: (520) 327-2260. Web Site: www.1041thetruth.com. Licensee: Journal Broadcast Corp. Group owner: Journal Broadcast Group Inc. Format: News. Target aud: 25-54; males, persons. ♦Diane Frisch, gen mgr; Darla Thomas, opns mgr; Jennifer Nunn, gen sls mgr; Andrew Lee, progmg dir.

KSAZ(AM)—Marana, 1990: 580 khz; 5 kw-D, 390 w-N, DA-N. TL: N32 27 11 W111 17 04. Hrs open: 1011 N. Craycroft, Suite 302, 85711. Phone: (520) 298-6880. Fax: (520) 298-6077. E-mail: amradio1@cox.com Web Site: www.ksaz580.com. Licensee: Owl Broadcasting & Development Inc. (acq 4-89; $1.05 million; FTR: 5-1-89). Natl. Network: ABC. Law Firm: Hogan & Hartson. Format: Timeless classics. Target aud: 35 plus. Spec prog: International 2 hrs wkly.

KTKT(AM)— December 1949: 990 khz; 10 kw-D, 1 kw-N, DA-2. TL: N32 15 19 W111 00 32. Hrs open: 3871 N. Commerce, 85705. Phone: (520) 407-4500. Fax: (520) 407-4600. Web Site: www.ktkt.com. Licensee: Arizona Lotus Corp. Group owner: Lotus Communications Corp. (acq 1973). Population served: 1,000,000 Natl. Network: AP Radio. Natl. Rep: Lotus Entravision Reps LLC. Law Firm: Bryan Cave. Format: Sp oldies. Target aud: 25-54. Spec prog: Black one hr, relg 2 hrs wkly. ♦Steve Groesbeck, gen mgr.

KLPX(FM)—Co-owned with KTKT(AM). June 1, 1967: 96.1 mhz; 82 kw. 1,952 ft TL: N32 14 56 W111 06 59. Stereo. Web Site: www.klpx.com. (Acq 6-79).1,000,000 Natl. Rep: D & R Radio. Law Firm: Bryan Cave. Format: Classic rock. Target aud: 25-54.

KTUC(AM)— July 10, 1926: 1400 khz; 1 kw-U. TL: N32 08 43 W110 53 38. (CP: TL: N32 14 56 W110 55 29). Hrs opn: 575 W. Roger., 85705. Phone: (520) 887-1000. Fax: (520) 887-6397. Licensee: Citadel Broadcasting Co. Group owner: Citadel Broadcasting Corp. (acq 4-26-01; grpsl). Natl. Network: CBS. Natl. Rep: Katz Radio. Format: Adult standards. Target aud: General; college educated, upper income, politically active adults. ♦Ken Kowalcek, gen mgr.

KTZR-FM—Green Valley, Oct 21, 1990: 97.1 mhz; 1.75 kw. Ant 613 ft TL: N31 58 37 W111 06 04. Stereo. Hrs opn: 24 3202 N. Oracle Rd., 85705. Phone: (520) 618-2100. Fax: (520) 610-2200. Web Site: www.lapreclosa.com. Licensee: Capstar TX L.P. Group owner: Clear Channel Communications Inc. Format: Spanish var. News staff: 2; News: 2 hrs wkly. Target aud: 18-35. ♦Debbie Wagner, gen mgr; Tim Richards, opns mgr; Tom Zlaket, sls dir; Nikki Van Doran, mktg dir; Ruppert Pacheco, progmg dir; Mike Irby, chief of engrg.

KUAT-FM—Listing follows KUAZ-FM.

***KUAZ(AM)**— Oct 7, 1968: 1550 khz; 50 kw-D. TL: N32 22 21 W111 05 52. Hrs open: Sunrise-sunset Box 210067, Univ. of Arizona., 85721-0067. Phone: (520) 621-7548. Fax: (520) 621-3360. E-mail: kuat@arizona.edu Web Site: www.kuat.org. Licensee: Arizona Board of Regents. Population served: 710000 Natl. Network: NPR, PRI. Law Firm: Dow, Lohnes & Albertson. Format: News, jazz. Spec prog: Sp 3 hrs, Native American one hr wkly. ♦Jack Parris, gen mgr & stn mgr.

KUAT-FM—Co-owned with KUAZ-FM. May 19, 1975: 90.5 mhz; 12.5 kw. 3,580 ft TL: N32 24 55 W110 42 54. Stereo. 24 University of Arizona, Box 210067, 85721-0067. Phone: (520) 621-5828. Fax: (520) 621-3360. E-mail: radio@kuat.arizona Web Site: www.kuat.org. Format: Class. News: 7 hrs wkly. Target aud: 35 plus. ♦Allen Campbell, disc jockey; David Harrington, disc jockey. Co-owned TV: *KUAT-TV affil.

***KUAZ-FM**— Apr 27, 1992: 89.1 mhz; 3 kw. 10 ft TL: N32 22 21 W111 05 52. Hrs open: 24 Box 210067, Univ. off Arizona, 85721-0067. Phone: (520) 621-5828. Fax: (520) 621-3360. E-mail: kuaz@arizona.edu kuaz@kuat.org Web Site: www.kuaz.org. Licensee: Arizona Board of Regents for Benefit of the University of Arizona. (acq 4-92). Natl. Network: NPR, PRI. Law Firm: Dow, Lohnes & Albertson. Format: Jazz, news/talk info. News staff: 4; News: 38 hrs wkly. Target aud: General. Spec prog: Sp 3 hrs wkly. ♦Jack Gibson, gen mgr; John Kelley, stn mgr; Colleen Greer, prom dir; Lyle Kesterson, progmg dir.

KVOI(AM)— Sept 23, 1953: 690 khz; 250 w-D, DA. TL: N32 15 11 W110 57 44. Hrs open: 3222 S. Richey Blvd., 85713-5453. Phone: (520) 790-2440. Fax: (520) 790-2937. E-mail: doug@kvoi.com Web Site: www.kvoi.com. Licensee: Good News Broadcasting Inc. Group owner: Good News Communications Inc. (acq 9-53). Population served: 700,000 Natl. Rep: Salem. Law Firm: Wray Fitch. Format: News, talk. News staff: one; News: 10 hrs wkly. Target aud: 35 plus. ♦Doug Martin, pres & gen mgr; Mary Martin, sls dir & gen sls mgr; Matt Manis, opns mgr & progmg dir; Larry Massey, chief of engrg.

KWFM(AM)— Feb 27, 1947: 1450 khz; 1 kw-U. TL: N32 12 04 W110 56 48. Stereo. Hrs opn: 3202 N. Oracle Rd., 85705. Phone: (520) 618-2100. Fax: (520) 618-2200. Web Site: www.cool1450am.com. Licensee: CC Licenses LLC. Group owner: Clear Channel Communications Inc. (acq 6-28-2001; grpsl). Format: Oldies. ♦Debbie Wagner, gen mgr; Tim Richards, opns mgr; Tom Zlaket, sls dir; Deeanne Thomas, gen sls mgr; Nikki Van Doran, mktg dir; Joan Lee, prom dir; Alan Cook, progmg dir; Mike Irby, chief of engrg; Mary Palin, traf mgr.

KWMT-FM— May 18, 1970: 92.9 mhz; 90 kw. 2,037 ft TL: N32 14 56 W111 06 59. Stereo. Hrs opn: 24 3202 N. Oracle Rd., 85705. Phone: (520) 618-2100. Fax: (520) 618-2200. Web Site: www.929themountain.com. Licensee: Capstar TX L.P. Group owner: Clear Channel Communications Inc. Population served: 452,000 Format: AAA. News staff: 2; News: 21 hrs wkly. ♦Debbie Wagner, gen mgr; Tim Richards, opns mgr; Deanne Thomas, gen sls mgr & local news ed; Blake Rogers, progmg dir.

***KXCI(FM)**— Dec 17, 1983: 91.3 mhz; 340 w. 3,641 ft TL: N32 24 54 W110 42 56. Stereo. Hrs opn: 24 220 S. 4th Ave., 85701. Phone: (520) 623-1000, EXT. 11. Phone: (520) 622-5924. Fax: (520) 623-0758. Web Site: www.kxci.org. Licensee: Foundation for Creative Broadcasting Inc. (acq 2-17-2004). Format: Eclectic, progsv, AAA. News: 4 hrs wkly. Target aud: 18-49. Spec prog: American Indian 2 hrs, Black 4 hrs, folk 2 hrs, gospel 2 hrs, jazz 2 hrs, Sp 4 hrs wkly. ♦Ryan Bruce, gen mgr; Jill Nunes, sls dir; Ginger Doran, progmg dir; Duncan Hudson, mus dir; Amanda Shauger, pub affrs dir; Doug Groenhoff, chief of engrg.

KXEW(AM)—South Tucson, May 10, 1963: 1600 khz; 1 kw-U, DA-N. TL: N32 11 46 W110 59 02. Hrs open: 24 3202 N. Oracle Rd., 85705. Phone: (520) 618-2100. Fax: (520) 618-2200. Web Site: www.tejano1600.com. Licensee: CC Licenses LLC. Group owner: Clear Channel Communications Inc. (acq 9-25-2003). Population served: 262,933 Law Firm: Cohn & Marks. Format: Tejano. Target aud: 25-54; blue collar Hispanics. ♦Debbie Wagner, gen mgr; Tim Richards, opns mgr; Tom Zlaket, sls dir; Patti Ruiz, gen sls mgr; Nikki Van Doran, mktg dir; Melissa Santa Cruz, prom dir; Rupert Pacheco, progmg dir; Mike Irby, chief of engrg; Mary Palin, traf mgr.

KZLZ(FM)—Kearny, Aug 31, 1991: 105.3 mhz; 50 kw. 492 ft TL: N32 49 38 W110 34 12. Stereo. Hrs opn: 24 2959 E. Grant Rd., 85716. Phone: (520) 325-3054. Fax: (520) 325-3495. Licensee: Entravision Holdings LLC. Group owner: Entravision Communications Corp. (acq 7-28-00; grpsl). Law Firm: Arent, Fox, Kintner, Plotkin & Kahn. Format: Sp. Target aud: General. ♦Cesar Zalciora, VP & progmg dir; Sonya Tabanico, gen mgr.

Tusayan

KSGC(FM)— July 1, 1991: 92.1 mhz; 4.1 kw. Ant 335 ft TL: N35 58 14 W112 07 53. Stereo. Hrs opn: 24 Box 3346, Grand Canyon, 86023-3346. Phone: (928) 638-9552. Fax: (928) 638-9553. Licensee: Tusayan Broadcasting Co. Natl. Network: ABC. Format: Adult contemp. Target aud: 18-54; Grand Canyon visitors with disposable vacation income. ♦Brian Ciesielski, gen mgr; Wes Yellowstone, progmg dir.

Vail

KRDX(FM)— June 1978: 98.5 mhz; 3.9 kw. Ant 410 ft TL: N31 55 39 W110 37 57. Stereo. Hrs opn: 24 Box 36717, Tucson, 85740. Phone: (520) 459-8201. Fax: (520) 458-7104. E-mail: comments@krdx.com Web Site: www.krdx.com. Licensee: Desert West Air Ranchers Corp. (acq 6-24-99). Population served: 20,000 Format: Var. ♦Ted Tucker, gen mgr.

Wellton

KCEC-FM— Oct 1, 2000: 104.5 mhz; 6.1 kw. 1,348 ft TL: N32 40 22 W114 20 14. Hrs open: 24 670 East 32 St., Suite 12 A, Yuma, 85365. Phone: (928) 782-5995. Fax: (928) 782-3874. Web Site: www.campesina.com. Licensee: Farmworker Educational Radio Network Inc. Law Firm: Borsari & Paxson. Format: Mexican rgnl. Target aud: 25-54; Hispanic market. ♦Anthony Chavez, exec VP; Rosella Lopez, gen mgr; Barbara Lane, gen sls mgr; Pepe Escamilla, progmg dir; Isabel Eggert, news dir; Dave Whitehead, chief of engrg.

Whiteriver

***KNNB(FM)**— Sept 11, 1982: 88.1 mhz; 630 w. 600 ft TL: N33 45 47 W109 57 39. (CP: 1.25 kw, ant 640 ft.). Stereo. Hrs opn: 18 Box 310, 85941. Phone: (928) 338-5229. Phone: (928) 338-5211. Fax: (928) 338-1744. Licensee: Apache Radio Broadcasting Corp. Population served: 30,000 Format: Div, educ. News staff: one; News: 3 hrs wkly. Target aud: 15-60. Spec prog: Apache 8 hrs wkly. ♦Sylvia Browning, progmg dir; Udell Opah, gen mgr, stn mgr & chief of engrg.

Wickenburg

KBSZ(AM)— Jan 27, 1960: 1250 khz; 350 w-D, 202 w-N. TL: N33 55 32 W112 47 38. Hrs open: 24 340 W. Wickenburg Way, Suite. B, 85390. Phone: (928) 668-1250. E-mail: pete@kbsz-am.com Web Site: www.kbsz-am.com. Licensee: Richard A. & Joann R. Peterson, joint tenants (acq 7-9-01). Population served: 25,000 Format: Musical mix. ♦Pete Peterson, gen mgr & opns mgr.

KHOV-FM— Dec 2, 1983: 105.3 mhz; 6 kw. -1,364 ft TL: N34 11 32 W112 45 13. Stereo. Hrs opn: 24
Rebroadcasts KMYL(AM) Tolleson 85%.
4745 N. 7th St., Suite 140, Phoenix, 85021. Phone: (602) 308-7900. Fax: (602) 308-7979. Web Site: www.univisionradio.com. Licensee: HBC License Corp. Group owner: Univision Radio (acq 9-22-2003; grpsl). Population served: 200,000 Natl. Network: Jones Radio Networks. Law Firm: Wiley, Rein & Fielding. Format: Rgnl Mexican. News: 4 hrs wkly. Target aud: 35 plus; modern mature market. Spec prog: Jazz 2 hrs, relg one hr wkly. ♦Mary McEvilly-Hernandez, VP & gen mgr; Elvis Balle, progmg dir; Dobby White, traf mgr.

KSWG(FM)— January 1993: 96.3 mhz; 6.4 kw. Ant 646 ft TL: N33 55 34 W112 47 40. Hrs open: 24 801 W. Wickenburg Way, 85390. Phone: (928) 684-7804. Fax: (928) 684-7805. E-mail: kswg@directpc.com Web Site: www.kswgradio.com. Licensee: Circle S. Broadcasting Co. Inc. (acq 1990). Population served: 1,000,000 Format: Country. ♦Harold Shumway, pres; Mike Shumway, gen mgr & sls dir.

Willcox

KHIL(AM)— Dec 2, 1959: 1250 khz; 5 kw-D, 196 w-N. TL: N32 16 00 W109 49 58. Hrs open: Box 1250, 85644. Secondary address: 900 West Patte Rd. 85643. Phone: (520) 384-4626. Fax: (520) 384-4627. Licensee: Lakeshore Media L.L.C. (acq 12-18-2001; $1.1 million with co-located FM). Population served: 3,243 Natl. Network: USA. Format: C&W. ♦Dan Curtis, gen mgr.

KWCX(FM)— July 8, 1976: 104.9 mhz; 730 w. Ant 3,175 ft TL: N32 13 01 W109 36 26. Stereo. Hrs opn: 24 Box 1250, 85644. Phone: (520) 384-4626. Fax: (520) 384-4627. Licensee: Lakeshore Media L.L.C. Group owner: Clear Channel Communications Inc. (acq 12-18-2001; $1.1 million with KHIL(AM) Willcox). Format: Rock. ♦Dan Curtis, gen mgr; Mark Lucke, progmg dir.

Williams

KWMX(FM)— 1998: 96.7 mhz; 10.5 kw. Ant 1,066 ft TL: N35 07 52 W112 08 03. Hrs open: 112 E. Rt. 66, Suite 105, Flagstaff, 86001. Phone: (928) 779-1177. Fax: (928) 774-5179. E-mail: ann@northlandradio.com Web Site: www.thewolf.com. Licensee: Grenax Broadcasting II LLC. (acq 4-14-2005; grpsl). Format: Cllasic rock. ♦Greg Dinetz, pres; Jim Shipp, gen mgr, sls dir & prom dir; Bill McAdams, opns dir; Mike Mentor, progmg dir; Jon Sweat, chief of engrg.

KYET(AM)— Aug 17, 1992: 1180 khz; 10 kw-U. TL: N35 15 38 W112 10 55. Hrs open: 24 812 E. Beale St., Kingman, 86401. Phone: (928) 753-9100. Fax: (928) 753-1978. Licensee: Grand Canyon Gateway Broadcasting L.L.C. (acq 9-19-97; $290,000). Population served: 55,000 Natl. Network: ABC. Format: News/talk. News staff: 2; News: 16 hrs wkly. Target aud: 45-60; upper middle class conservatives. ♦Rhonda Hart, pres & sls VP; Joe Hart, gen mgr; Steve Levin, prom VP & progmg dir; Dave Hawkins, news dir; Matt Krick, engrg VP; Deana Campbell, traf mgr.

Window Rock

KTNN(AM)— Feb 26, 1986: 660 khz; 50 kw-U, DA-N. TL: N35 53 41 W109 08 29. Hrs open: Box 2569, 86515. Phone: (928) 871-2582. Fax: (928) 871-3479. E-mail: ktnn@cia-g.com Web Site: www.navajoland.com/ktnn. Licensee: The Navajo Tribe. (acq 1-86). Format: Country. Navajo Tribe of Indians. Spec prog: American Indian. ♦Chester Francis, gen mgr; Marilynn Van Wagner, gen sls mgr & prom mgr; Bill Riddle, asst music dir; Bernadette Chato, news dir; Ernie Manuelito, chief of engrg; L.A. Williams, sports cmtr.

KWRK(FM)—Co-owned with KTNN(AM). October 1996: 96.1 mhz; 94 kw. 328 ft TL: N35 33 36 W109 06 30.24 Format: Jazz.

KWIM(FM)— Sept 21, 1995: 104.9 mhz; 30 kw. Ant 298 ft TL: N35 39 19 W109 01 59. Stereo. Hrs opn: Box 9090, Western Indian Ministries, 86515. Phone: (505) 371-5587. Fax: (505) 371-5588. E-mail: lharpor@westernindian.net Web Site: www.westernindian.org. Licensee: Western Indian Ministries Inc. Format: Adult contemp, Christian, relg. ♦Larry Harpor, gen mgr.

Winslow

KINO(AM)— Dec 18, 1962: 1230 khz; 1 kw-U. TL: N35 02 15 W110 43 00. Hrs open: Drawer K, East End of Easy St., 86047. Phone: (928) 289-3364. Fax: (928) 289-3366. E-mail: kinoradio@cableone.net Licensee: Sunflower Communications. (acq 1-15-77). Population served: 25,000 Natl. Network: CBS, ESPN Radio. Format: Country. Target aud: General. Spec prog: Sp 4 hrs wkly. ♦Loy Engelhardt, gen mgr.

Yarnell

KRPH(FM)—Not on air, target date: unknown: 99.5 mhz; 45 kw. Ant 518 ft TL: N34 07 56 W112 45 00. Hrs open: 2801 Via Fortuna Dr., Suite 675, Austin, TX, 78746. Phone: (713) 528-2517. Licensee: Ace Radio Corp. ♦Stephen Hackerman, pres.

Yuma

***KAWC(AM)**— July 11, 1970: 1320 khz; 1 kw-D, 147 w-N. TL: N32 41 10 W114 29 38. Hrs open: 6 AM-6 PM Box 929, 85366. Secondary address: 9500 S. Ave., #8E 85366. Phone: (928) 344-7690. Phone: (928) 344-4210. Fax: (928) 344-7740. Web Site: kawcradio.org. Licensee: Arizona Western College. Population served: 65,000 Natl. Network: NPR. Format: Var, country, rock. News staff: one. Target aud: General. Spec prog: Sp 15 hrs wkly. ♦Frank Preciado, gen mgr.

KAWC-FM— Mar 27, 1992: 88.9 mhz; 3 kw. 75 ft TL: N32 41 10 W114 29 38. Stereo. 6 AM-9 PM Web Site: kawcradio.org. Format: Class, jazz, news. News staff: one.

KBLU(AM)—Listing follows KTTI(FM).

***KCFY(FM)**— March 1992: 88.1 mhz; 3 kw. 239 ft TL: N32 38 31 W114 33 34. Stereo. Hrs opn: 24 Box 1669, 85366. Secondary address: 1921 S. Rail Ave. 85365. Phone: (928) 341-9730. Fax: (928) 341-9099. E-mail: kcfy@kcfyfm.com Web Site: www.kcfyfm.com. Licensee: Relevant Media Inc. (acq 3-7-2005; $636,000). Population served: 175,000 Law Firm: Miller & Neely. Format: Christian. News: 4 hrs wkly. Target aud: 25-45; young to middle aged families. ♦Greg S. Myers, gen mgr; Lynette Toepfer, dev dir; Mike Bondora, prom dir; Brandon Sweet, mus dir; Mikie Francher, traf mgr.

KJOK(AM)— Dec 11, 1950: 1400 khz; 1 kw. TL: N32 39 06 W114 39 00. Hrs open: 24 949 S. Avenue B, 85364. Phone: (928) 782-4321. Fax: (928) 343-1710. E-mail: oldiesradio@kjokyuma.com Web Site: www.kjokyuma.com. Licensee: MonsterMedia L.L.C. (acq 1997; with co-located FM). Population served: 160,000 Natl. Network: Jones Radio Networks. Natl. Rep: McGavren Guild. Law Firm: Booth, Freret, Imlay & Tepper. Format: News/talk, sports, oldies. News: 6 hrs wkly. Target aud: 35 plus. ♦Keith Lewis, CEO, gen mgr & gen sls mgr; Jennifer Blackwell, traf mgr.

KQSR(FM)— Sept 5, 1986: 100.9 mhz; 3 kw. 274 ft TL: N32 38 31 W114 33 34. (CP: Ant 1,075 ft.). Stereo. Hrs opn: 24 755 W. 28th St., 85364. Phone: (520) 344-4980. Fax: (520) 344-4983. Web Site: www.kqsrfm.com. Licensee: Capstar TX L.P. Group owner: Clear Channel Communications Inc. (acq 8-30-2000; grpsl). Population served: 121,418 Format: Adult contemporary. Target aud: 25-54. ♦Jeff Harris, gen mgr & sls dir; Cindy Landin, mktg mgr; Russ Egan, progmg dir; Travis Goode, disc jockey.

KTTI(FM)— Nov 6, 1970: 95.1 mhz; 25 kw. 97 ft TL: N32 42 42 W114 38 58. (CP: 100 kw, ant 1,256 ft. TL: N32 40 25 W114 20 12). Stereo. Hrs opn: 755 W. 28th St., 85364-7136. Phone: (928) 344-4980. Fax: (928) 344-4983. Licensee: Capstar TX L.P. Group owner: Clear Channel Communications Inc. (acq 8-30-00; grpsl). Population served: 56,000 Format: Country. Target aud: 25-54. ♦Jeff Harris, gen mgr; Cindy Landin, gen sls mgr; Susan Nickell, progmg dir & news dir; Shannon Pearson, chief of engrg; Katherine Thompson, traf mgr.

KBLU(AM)—Co-owned with KTTI(FM). March 1940: 560 khz; 1 kw-U, DA-N. TL: N32 43 25 W114 38 39. Natl. Network: Westwood One. Format: News/talk. ♦Tiffany Fair, progmg dir & traf mgr.

***KYRM(FM)**— April 2000: 91.9 mhz; 6.3 kw. 407 ft TL: N33 03 18 W114 49 37. Stereo. Hrs opn: 24 Box 5965, 85366-5965. Secondary address: 2690 S. 3rd Ave. 85366. Phone: (928) 341-0919. Fax: (928) 314-4141. E-mail: kyrm@hcjb.org Web Site: www.radiokyrm.org. Licensee: World Radio Network. Group owner: World Radio Network Inc. Population served: 1,500,000 Format: Sp, Christian, relg. News: 6 hrs wkly. Hispanic population. Spec prog: Children 6 hrs. ♦Douglas Swanson, stn mgr & engrg mgr; Rachel Swanson, progmg dir.

Arkansas

Arkadelphia

KDEL-FM—Listing follows KVRC(AM).

***KHED(FM)**—Not on air, target date: unknown: 91.9 mhz; 250 w. Ant 98 ft TL: N34 07 44 W93 03 37. Hrs open: Henderson State University, 1100 Henderson St., 71999-0001. Phone: (870) 230-5091. Licensee: Henderson State University. ♦Charles Dunn, pres.

***KSWH(FM)**— Sept 25, 1969: 99.9 mhz; 10 w. 70 ft TL: N34 07 32 W93 03 48. Stereo. Hrs opn: 6 AM-midnight HSU Box 7872, Henderson State Univ., 71999-0001. Phone: (870) 230-5185. Fax: (870) 230-5144. E-mail: kswh@hsu.edu Web Site: www.kswh.org. Licensee: Henderson State University. Population served: 18,000 News staff: 2; News: 2 hrs wkly. Target aud: 18-36; activity-orienated youthful females. Spec prog: Alternative 15 hrs, contemp Christian 2 hrs, rap 15 hrs wkly. ♦Michael Taggart, gen mgr; Darrell Young, opns VP & opns dir; Whitney Allen, dev dir.

KVRC(AM)— Sept 25, 1947: 1240 khz; 1 kw-U. TL: N34 06 39 W93 03 01. Hrs open: 24 Box 40, 71923. Secondary address: 601 S. 7th St. 71923. Phone: (870) 246-9272. Fax: (870) 246-5878. Web Site: www.clarkcountryradio.com. Licensee: Noalmark Broadcasting Corp. (group owner; (acq 6-29-2007; grpsl). Population served: 24,000 Format: Nostalgia. ♦Stephanie L. Colley, gen mgr; Randy Seale, progmg dir; Ronna Pennington, news dir; Annette Jennings, traf mgr.

KDEL-FM—Co-owned with KVRC(AM). June 12, 1977: 100.9 mhz; 3 kw. Ant 95 ft TL: N34 06 39 W93 03 01. Stereo. 24 Format: Adult contemp.

KYXK(FM)—See Gurdon

Ashdown

KMJI(FM)— May 25, 1985: 93.3 mhz; 7.4 kw. 597 ft TL: N33 30 24 W94 12 25. Stereo. Hrs opn: 24 2324 Arkansas Blvd., Texarkana, 71854. Phone: (870) 772-3771. Fax: (870) 772-0364. E-mail: wesspicher@gapbroadcasting.com Web Site: www.magic933.com. Licensee: GAP Broadcasting Texarkana License LLC. Group owner: Clear Channel Communications Inc. (acq 8-3-2007; grpsl). Population served: 100000 Format: Adult contemp. News staff: one; News: 7 hrs wkly. Target aud: General. Spec prog: Relg 4 hrs wkly. ♦Ron Bird, gen mgr; Wes Spicher, gen mgr & progmg dir.

KPGG(FM)— May 19, 1972: 103.9 mhz; 5.1 kw. Ant 354 ft TL: N33 36 06 W94 04 38. Stereo. Hrs opn: 24 1323 College Dr., Texarkana, TX, 75501. Phone: (903) 793-1109. Fax: (903) 794-4717. Licensee: FFD Holdings I Inc. Group owner: Petracom Media LLC (acq 1-4-2005; grpsl). Natl. Rep: Katz Radio. Format: Country legends/(classic country). News staff: one. ♦Mike Basso, gen mgr.

Atkins

KVLD(FM)— October 1999: 99.3 mhz; 4.1 kw. Ant 394 ft TL: N35 14 41 W92 52 51. Hrs open: 24 Box 10310, Russellville, 72810. Phone: (479) 968-6816. Fax: (479) 968-2946. E-mail: rich@rivervalleyradio.com Web Site: www.rivervalleyradio.com. Licensee: MMA License LLC. Group owner: MAX Media L.L.C. (acq 4-21-2003; grpsl). Format: Hispanic - rgnl Mexican. ♦Rich Moellers, stn mgr.

Augusta

KJSM-FM—Licensed to Augusta. See Mayflower

Bald Knob

KAPZ(AM)— Aug 18, 1980: 710 khz; 250 w-D, DA. TL: N35 16 32 W91 33 39. Hrs open: Box 1300, Searcy, 72145. Secondary address: 111 N. Spring St., Searcy 72143. Phone: (501) 268-5072. Fax: (501) 279-2900. Licensee: Crain Media Group LLC (group owner; (acq 8-7-2002; grpsl). Format: News/talk. Spec prog: Farm 3 hrs wkly. ♦Bill Weaver, gen mgr.

KCNY(FM)—Co-owned with KAPZ(AM). Oct 15, 1984: 107.1 mhz; 19 kw. Ant 305 ft TL: N35 17 29 W91 40 24. (CP: COL Greenbrier. 12.5 kw, ant 466 ft. TL: N35 17 47 W92 19 11). Format: Adult contemp.

KJSM-FM—See Mayflower

Barling

KFPW-FM— Sept 1, 1987: 94.5 mhz; 18.5 kw. Ant 269 ft TL: N35 15 54 W94 21 52. Stereo. Hrs opn: 24 Box 908, Fort Smith, 72902. Phone: (479) 288-1047. Fax: (479) 785-2638. Web Site: www.fox94.com. Licensee: Pharis Broadcasting Inc. (group owner; (acq 3-14-2002;. $350,000 with KFPW(AM) Fort Smith). Natl. Network: ABC, Fox News Radio. Natl. Rep: Commercial Media Sales. Law Firm: Irwin Campbell & Tannenwald. Format: News/talk. News staff: 2; News: 30 hrs wkly. Target aud: 25-54; general. ♦William Pharis, pres; Karen Pharis, gen mgr & stn mgr; Ernie Witt Jr., opns VP.

Batesville

KAAB(AM)— August 1980: 1130 khz; 1 kw-D, DA. TL: N35 16 32 W91 38 21. (CP: 20 w-N). Hrs opn: 920 Harrison St., Suite C, 72501. Secondary address: Box 2077 72503. Phone: (870) 793-4196. Fax: (870) 793-5222. E-mail: arweekly@cei.net Licensee: WRD Entertainment Inc. (group owner). Rgnl. Network: Ark. Radio Net. Format: Mexicana. Target aud: 18-44. Spec prog: Farm 5 hrs wkly. ♦John R. Grace, pres; Gary Bridgman, gen mgr.

KBTA(AM)— June 30, 1950: 1340 khz; 1 kw-U. TL: N35 44 39 W91 38 21. Hrs open: 24 Box 2077, 920 Harrison, Suite C, 72503. Phone: (870) 793-4196. Fax: (870) 793-5222. Web Site: maxfm.com. Licensee: W.R.D. Entertainment Inc. (group owner; acq 12-15-95). Population served: 32,000 Wire Svc: AP Format: Sports. News staff: 2; News: 22 hrs wkly. Target aud: 18 plus. ♦Rob Grace, pres; Gary Bridgman, gen mgr; Ben Johnson, progmg dir; Dale Johnson, chief of engrg.

KZLE(FM)—Co-owned with KBTA(AM). Mar 3, 1982: 93.1 mhz; 100 kw. 984 ft TL: N35 53 27 W91 44 01. Stereo. E-mail: rob@maxfm.com Web Site: maxfm.com.33,200 Law Firm: Fletcher, Heald & Hildreth, P.L.C. Format: Rock. Target aud: 24 plus. ♦Rob Grace, progmg dir & mus dir; Dale Johnson, engrg dir.

KBTA-FM— 1999: 99.5 mhz; 3.4 kw. Ant 426 ft TL: N35 52 07 W91 35 14. Hrs open: 920 Harrison St., Suite C, 72503. Phone: (870) 793-4196. Fax: (870) 793-5222. E-mail: garyb@swbell.net Licensee: W.R.D. Entertainment Inc. (group owner). Format: Adult contemp. ♦Rob Grace, pres; Gary Bridgman, gen mgr; Matt Johnson, gen sls mgr; Ben Johnson, progmg dir; Dale Johnson, chief of engrg.

Beebe

KBGR(FM)— June 22, 1991: 101.5 mhz; 6 kw. Ant 328 ft TL: N35 11 26 W91 54 45. Hrs open: 24 2351 Sunset Blvd., Suite 170-218, Rocklin, CA, 95765. Phone: (916) 251-1600. Fax: (916) 251-1650. Web Site: www.air1.com. Licensee: Educational Media Foundation. (group owner; (acq 6-9-2005; $525,000). Population served: 50,000 Natl. Network: Air 1. Format: Christian hit music. ♦Richard Jenkins, pres; Mike Novak, VP; Keith Whipple, dev dir; David Pierce, progmg mgr; Ed Lenane, news dir; Sam Wallington, engrg dir; Karen Johnson, news rptr; Marya Morgan, news rptr; Richard Hunt, news rptr.

Bella Vista

KBVA(FM)— November 1991: 106.5 mhz; 37 kw. 567 ft TL: N36 18 21 W94 27 29. Hrs open: 8am - 5pm 1655 Hwy. 72 S.E., Gravette, 72736. Phone: (479) 787-6411. Fax: (479) 787-6116. Licensee: Gayla Joy McKenzie. Format: Var. ♦Gayla Joy McKenzie, pres & gen mgr.

KREB(AM)—See Bentonville-Bella Vista

Bellefonte

KNWA(AM)— 1986: 1600 khz; 5 kw-D, 50 w-N. TL: N36 14 49 W93 05 06. Hrs open: Box 850, Harrison, 72602. Secondary address: 600 S. Pine, Harrison 72601. Phone: (870) 741-1402. Fax: (870) 741-9702. Licensee: Harrison Radio Stations Inc. Format: Southern Gospel. ♦Tom Arnold, gen mgr, sls dir & mktg dir; Phillip Cary, progmg dir.

Benton

KEWI(AM)— June 26, 1953: 690 khz; 250 w-D, 73 w-N. TL: N34 31 57 W92 34 16. Hrs open: 5 AM-10 PM 115 S. Main St., 72015. Phone: (501) 778-6677. Fax: (501) 778-7717. E-mail: kewi690@yahoo.com Web Site: www.kewi690.com. Licensee: Landers Broadcasting Co. Inc. (acq 5-95). Population served: 500,000 Natl. Network: USA. Rgnl. Network: Ark. Radio Net. Format: Loc news, oldies, sports, country,

talk. News staff: one; News: 10 hrs wkly. Target aud: 25-65; all income levels. Spec prog: Farm 4 hrs, gospel 10 hrs, relg 5 hrs wkly. ♦Doris L. Landers, exec VP; Jim Landers, CEO, gen mgr & opns mgr.

KHKN(FM)— Jan 1, 1979: 106.7 mhz; 16 kw. 866 ft TL: N34 47 56 W92 29 53. Stereo. Hrs opn: 24 10800 Colonel Gleen, Little Rock, 72204. Phone: (501) 217-5000. Fax: (501) 228-9547. Fax: (501) 227-5776. Web Site: www.classiccountry1067.com. Licensee: CC Licenses LLC. Group owner: Clear Channel Communications Inc. (acq 9-12-97; grpsl). Population served: 490,000 Law Firm: Bryan Cave. Format: Classic country. Target aud: 18-34. ♦Don Pollnow, gen mgr & progmg dir.

Bentonville

KAMO-FM—See Rogers

***KAPG(FM)**— 2006: 88.1 mhz; 1 kw. Ant 233 ft TL: N36 23 37 W94 10 57. Hrs open: Drawer 2440, Tupelo, MS, 38803. Phone: (662) 844-8888. Fax: (662) 842-6791. Web Site: www.afr.net. Licensee: American Family Association. Group owner: American Family Radio. Format: Christian classics. ♦Marvin Sanders, gen mgr; John Riley, progmg dir; Fred Jackson, news dir; Joey Moody, chief of engrg.

KFFK(AM)—See Rogers

KIGL(FM)—See Seligman, MO

KQSM-FM— Nov 7, 1983: 98.3 mhz; 100 kw. 617 ft TL: N36 07 38 W93 59 23. Stereo. Hrs opn: 24 24 East Meadow St., Suite 1, Fayetteville, 72701. Phone: (479) 521-5566. Fax: (479) 521-0751. Web Site: www.sam98.com. Licensee: Cumulus Licensing Corp. Group owner: Cumulus Media Inc. (acq 2-1-99; grpsl). Format: Country. Target aud: 25-54. Spec prog: Class 2 hrs wkly. ♦Joe Conway, gen mgr.

KSEC(FM)—Not on air, target date: unknown: 95.7 mhz; 6 kw. Ant 328 ft TL: N36 17 13 W94 12 32. Hrs open: 24 Box 335, Springdale, 72765. Phone: (479) 756-8686. Fax: (479) 756-8687. Licensee: Lazeta 957 Co. (acq 6-1-2005; $1.99 million). Format: Mexican rgnl. ♦Edwards Vega, gen mgr.

Bentonville-Bella Vista

KREB(AM)— Feb 5, 1979: 1190 khz; 2.5 kw-D. TL: N36 23 17 W94 11 42. Hrs open: Sunrise-sunset 1780 Holly St., Fayetteville, 72703. Phone: (479) 582-3776. Fax: (479) 571-0995. Licensee: Butler Broadcasting Co. LLC (acq 10-13-99; $100,000). Population served: 150000 Natl. Network: USA. Format: Sports, talk. Target aud: 35 plus. ♦Peter Davidson, pres; Steve Butler, gen mgr.

Berryville

KTHS(AM)— February 1958: 1480 khz; 5 kw-D, 64 w-N. TL: N36 21 42 W93 33 40. Hrs open: 24 Box 191, 72616. Secondary address: One Radio Dr. 72616. Phone: (870) 423-2147. Fax: (870) 423-2146. E-mail: studio@kthsradio.com Licensee: Carroll County Broadcasting Inc. (acq 6-29-2006; $3.5 million with co-located FM). Population served: 25,000 Natl. Network: ABC, Jones Radio Networks. Format: Modern country. News staff: one; News: 21.5 hrs. Target aud: General. Spec prog: Farm 15 hrs wkly. ♦Jay Bunyard, pres; James T. Earls, gen mgr; William C. Autry, gen sls mgr; Linda Boyer, mus dir & news dir; Zeb Huffmaster, chief of engrg; Sherri Linz, traf mgr; Travis Doshier, disc jockey.

KTHS-FM— Dec 19, 1974: 107.1 mhz; 3.6 kw. Ant 627 ft TL: N36 20 45 W93 29 17. Stereo. 24 25,000 Natl. Network: ABC, Jones Radio Networks. News staff: one; News: 21.5 hrs wkly. Target aud: General. ♦Sherri Linz, traf mgr; Travis Dashier, disc jockey.

Blytheville

***KAIA(FM)**—Not on air, target date: unknown: 91.5 mhz; 1 kw. Ant 190 ft TL: N35 54 45 W89 53 28. Hrs open: 5700 West Oaks Blvd., Rocklin, CA, 95765. Phone: (916) 251-1600. Fax: (916) 251-1650. Licensee: Educational Media Foundation. Group owner: American Family Radio. (acq 3-23-2007; grpsl). ♦Richard Jenkins, pres.

***KBCM(FM)**— 2000: 88.3 mhz; 500 w. Ant 190 ft TL: N35 54 45 W89 53 28. Hrs open: 24 Drawer 2440, Tupelo, MS, 38803. Phone: (662) 844-8888. Fax: (662) 844-9090. Licensee: American Family Association. Group owner: American Family Radio Format: Christian. ♦Marvin Sanders, gen mgr; John Riley, progmg dir.

KHLS(FM)—Listing follows KLCN(AM).

KLCN(AM)— 1922: 910 khz; 5 kw-D, 85 w-N. TL: N35 55 27 W89 52 18. Hrs open: Box 989, 72316. Secondary address: 125 S. Second St. 72315. Phone: (870) 762-2093. Fax: (870) 763-8459. Licensee: Sudbury Services Inc. Group owner: Sudbury Services Inc. & Newport Broadcasting Co. Population served: 24,752 Format: News/talk. ♦Dave Clark, gen mgr; Tom Hill, news dir & chief of engrg.

KHLS(FM)—Co-owned with KLCN(AM). 1948: 96.3 mhz; 100 kw. 450 ft TL: N35 55 27 W89 52 18. (CP: Ant 351 ft. TL: N35 38 27 W89 56 54). Stereo. 24,752 Format: C&W.

Booneville

***KBHN(FM)**— 2005: 89.7 mhz; 59 kw. Ant 302 ft TL: N35 08 25 W94 03 43. Hrs open: Box 6210, Fort Smith, 72906. Phone: (479) 646-6700. Fax: (479) 646-1373. Web Site: www.kzfm.com. Licensee: Vision Ministries Inc. Format: Christian. ♦Marilyn K. Lynch, pres; Gary Brown, gen mgr & progmg dir.

KQBK(FM)—Licensed to Booneville. See Fort Smith

Brinkley

KBRI(AM)— Oct 25, 1959: 1570 khz; 250 w-D, 44 w-N. TL: N34 52 02 W91 12 04. Hrs open: 6 AM-10 PM Box 111, Hwy. 70 W., 72021. Phone: (870) 734-1570. Fax: (870) 734-1571. Licensee: East Arkansas Broadcasting Inc. Population served: 30,000 Rgnl. Network: Ark. Radio Net. Format: Gospel. News staff: one. Target aud: General. ♦Cobby Caldwell, stn mgr; David Sills, gen sls mgr & progmg dir; Lane Goodwin, chief of engrg.

KTRQ(FM)—Co-owned with KBRI(AM). October 1969: 102.3 mhz; 40 kw. Ant 548 ft TL: N35 03 16 W90 44 36. Stereo. 100,000 Format: Oldies.

Bryant

KKSP(FM)—Licensed to Bryant. See Malvern

Cabot

KPZK-FM— May 1993: 102.5 mhz; 3 kw. 328 ft TL: N34 55 22 W92 00 32. Stereo. Hrs opn: 24 700 Wellington Hills Rd., Little Rock, 72211. Phone: (501) 401-0200. Fax: (501) 401-0366. Licensee: The Last Bastion Station Trust LLC, as Trustee Group owner: Citadel Broadcasting Corp. (acq 6-12-2007; grpsl). Population served: 525,000 Natl. Network: ABC. Natl. Rep: McGavren Guild. Law Firm: Eckert, Seamans, Cherin & Mellot. Format: Gospel. News staff: 4. Target aud: 25-49. ♦Jim Beard, mktg mgr.

KZTD(AM)— Nov 16, 1980: 1350 khz; 2.5 kw-D, 73 w-N. TL: N34 59 59 W92 01 41. Hrs open: 121 Radio Heights Dr., Searcy, 72143. Phone: (501) 378-0104. Fax: (501) 305-2977. E-mail: kztd1350@hotmail.com Web Site: www.lamexicana.com. Licensee: New World LLC (group owner; (acq 2-2-2007; $190,000). Format: Sp. Target aud: 18-49. ♦Arik Lev, pres; Phil Hall, gen mgr; Robert Tindle, opns dir; Christy Flynn, sls VP.

Calico Rock

KJMT(FM)— Feb., 2007: 97.1 mhz; 5.2 kw. Ant 715 ft TL: N36 05 31 W92 15 46. Hrs open: 24 615 Olive St., Texarkana, TX, 75501. Phone: (903) 783-4671. Fax: (903) 792-4261. Web Site: www.mountaintalk97.com. Licensee: Malvern Entertainment Corp. Format: News/talk. ♦Scott A. Gray, pres; Norm Allen, gen mgr.

Camden

KAMD-FM— Dec 1, 1968: 97.1 mhz; 50 kw. Ant 456 ft TL: N33 30 14 W92 48 38. Stereo. Hrs opn: 24 133 Washington St., 71701. Phone: (870) 836-9567. Fax: (870) 836-9500. Web Site: www.camdenfm.net. Licensee: Radio Works Inc. (acq 12-13-2004; grpsl). Population served: 50,000 Format: Adult contemp. ♦Donna Steward, stn mgr & sls dir; Greg Arnold, opns dir.

***KCAC(FM)**— June 11, 1990: 89.5 mhz; 250 w. 161 ft TL: N33 34 31 W92 49 55. Stereo. Hrs opn: 8 AM-midnight Box 3499, 71711. Phone: (870) 836-5289. Fax: *870) 574-4538. Licensee: Southern Arkansas University Tech (acq 8-2-2005). Population served: 30,000 Natl. Network: ABC. Law Firm: Cohn & Marks. Format: Alternative. Target aud: 18-35. ♦Rochelle Moore, gen mgr; Quintin Green, opns mgr & mus dir.

KMGC(FM)— Nov 18, 1994: 104.5 mhz; 3 kw. 328 ft TL: N33 30 14 W92 48 38. Hrs open: 24 133 Washington St., 71701. Phone: (870) 836-0104. Fax: (870) 836-9500. E-mail: radioworks@camdenfm.net Web Site: www.camdenfm.net. Licensee: Radio Works Inc. (acq 12-13-2004; grpsl). Format: Urban contemp. ♦Donna Stewart, stn mgr; Greg Arnold, opns dir & mus dir.

KNHD(AM)— Aug 8, 1963: 1450 khz; 1 kw-U. TL: N33 33 49 W92 50 37. Hrs open: 24 Box 262550, Baton Rouge, LA, 70826. Secondary address: 8917 World Ministry Ave., Baton Rouge, LA 70810. Phone: (225) 768-3688/8300. Fax: (225) 768-3729. E-mail: kawikfish@yahoo.com Web Site: www.jsm.org. Licensee: Family Worship Center Church Inc. (group owner; (acq 3-7-2002; grpsl). Population served: 85,000 Format: Southern gospel. News staff: one. Target aud: 35 plus. ♦David Whitelaw, COO; Jimmy Swaggart, pres; John Santiago, progmg dir.

Cave City

***KVMN(FM)**— Jan 1, 1981: 89.9 mhz; 3.3 kw. Ant 351 ft TL: N35 57 07 W91 32 58. Stereo. Hrs opn: Box 190, 711 N. Main St., 72521. Phone: (870) 283-5331. Fax: (870) 283-3255. E-mail: bsisk@cavecity.ncsc.k12.ar.us Licensee: Cave City Schools. Population served: 10,000 Natl. Network: USA. Format: Relg, educ, div. Target aud: General. ♦Becky Sisk, gen mgr.

Cherokee Village

KFCM(FM)— May 18, 1981: 98.3 mhz; 3 kw. 298 ft TL: N36 16 29 W91 30 18. Stereo. Hrs opn: 24 Box 909, 72525. Phone: (870) 856-4408. Fax: (870) 895-4088. E-mail: hometownradio@centurytel.net Licensee: KFCM Inc. (acq 11-29-89; $174,500; FTR: 12-18-89). Format: Oldies. News staff: 3; News: 25 hrs wkly. Target aud: 25-54. ♦James Bragg, pres & gen mgr.

Cherry Valley

KXRL(FM)—Not on air, target date: unknown: 90.1 mhz; 9 kw. Ant 377 ft TL: N35 22 30.6 W90 43 22.1. Hrs open: Box 5204, Wilmington, DE, 19808-5204. Phone: (800) 220-8078. Phone: (302) 540-5690. Fax: (302) 738-3090. Web Site: www.thereachfm.com. Licensee: Priority Radio Inc. (acq 10-14-2005; $200,000 for CP). ♦Steve Hare, pres.

Clarksville

KLYR(AM)— Mar 18, 1957: 1360 khz; 500 w-D, 98 w-N. TL: N35 28 21 W93 29 28. Hrs open: 16 Box 188, Hwy. 64 W., 72830. Phone: (479) 754-3092. Fax: (479) 754-7227. Licensee: Randall P. Forrester. (acq 11-81; $31,816; FTR: 11-9-81). Population served: 75,000 Format: C&W. News: 12 hrs wkly. Target aud: General. Spec prog: Relg 8 hrs wkly. ♦Randy Forrester, gen mgr.

KLYR-FM— 1974: 92.7 mhz; 3 kw. 292 ft TL: N35 29 38 W93 32 21.16 180,000 Format: Country & Western.

KXIO(FM)— April 1991: 106.9 mhz; 5.9 kw. 112 ft TL: N35 33 07 W93 24 33. Hrs open: 901 S. Rogers St., 72830. Phone: (479) 705-1069. Fax: (479) 754-5518. Web Site: www.kxio-radio.net. Licensee: Barnett Broadcasting Inc. (acq 10-17-01; $400,000). Format: Hot country. ♦Gary Barnett, gen mgr; Kelley Ray, progmg dir & news dir.

Clinton

KGFL(AM)— Oct 1, 1977: 1110 khz; 5 kw-D. TL: N35 33 30 W92 27 32. Hrs open: Box 1349, 72031. Secondary address: Corner of Main & Griggs, AZ 72031. Phone: (501) 745-4474. Fax: (501) 745-4084. Licensee: King-Sulivan Radio (acq 4-3-01; $75,000 for 26% with co-located FM). Population served: 20,000 Format: Oldies. Target aud: 35 plus. ♦Jerri McCrary, gen mgr; Sid King, gen mgr; Tim Kelly, progmg dir; Dixie Carter, news dir & pub affrs dir.

KHPQ(FM)—Co-owned with KGFL(AM). Dec 23, 1982: 92.1 mhz; 10 kw. 571 ft TL: N35 40 44 W92 30 30. Stereo. 40,000 Natl. Network: Jones Radio Networks. Format: Country. Target aud: 25 plus. ♦Dave Britton, chief of engrg.

Colt

KTRQ(FM)—Licensed to Colt. See Brinkley

Conway

KASR(FM)—Listing follows KXXA(AM).

***KHDX(FM)**— May 1973: 93.1 mhz; 8 w. 59 ft TL: N35 06 01 W92 26 29. Stereo. Hrs opn: Hendrix College, 1600 Washington Ave., 72032. Phone: (501) 450-1339. Phone: (501) 329-6811 (college #). Fax: (501) 450-1200. Web Site: www.hendrix.edu. Licensee: Hendrix College. Population served: 1,000 Rgnl. Network: Ark. Radio Net. Format: Full service. ♦Julie Marvin, gen mgr.

KMJX(FM)— June 1, 1967: 105.1 mhz; 79 kw. 1,053 ft TL: N34 47 53 W92 29 33. Stereo. Hrs opn: 24 10800 Colonel Glenn Rd., Little Rock, 72204. Phone: (501) 217-5000. Fax: (501) 228-9547. E-mail: magic105fm@magic105fm.com Web Site: www.magic105fm.com. Licensee: CC Licenses LLC. Group owner: Clear Channel Communications Inc. (acq 5-5-96; grpsl). Population served: 750,000 Natl. Rep: Clear Channel. Law Firm: Wiley, Rein & Fielding. Format: Classic rock. News staff: one; News: 3 hrs wkly. Target aud: 18-49. ♦Llowrey Mays, chmn; Mark Mays, pres; Randall Mays, CFO; Bruce Demps, sr VP; Don Pollnow, gen mgr; Jeff Peterson, opns mgr & gen sls mgr; Casey Wagner, gen sls mgr & mus dir; Tom Wood, progmg dir.

***KUCA(FM)**— Oct 10, 1966: 91.3 mhz; 5 kw. 154 ft TL: N35 02 55 W92 27 49. Stereo. Hrs opn: 24 Box U-5144, Univ of Central Arkansas, 72035. Secondary address: 201 Donaghey Ave. Phone: (501) 450-3326. Fax: (501) 450-5874. E-mail: Montyr@uca.edu Licensee: University of Central Arkansas. Population served: 80,000 Format: News, adult contemp. News: 10 hrs wkly. Target aud: 18-54; educated adults. ♦Monty Rowell, gen mgr.

KXXA(AM)— May 26, 1961: 1330 khz; 500 w-D, 64 w-N. TL: N35 06 00 W92 26 41. Hrs open: Box 1266, 72033-1266. Phone: (501) 327-6611. Fax: (501) 327-7920. Licensee: Creative Media Inc. (acq 10-1-2004; with co-located FM). Population served: 15,510 Format: All sports. Spec prog: Farm 6 hrs wkly. ♦Elaine Harrison, prom mgr; Michael D. Harrison, pres, gen mgr, sls dir & progmg dir.

KASR(FM)—Co-owned with KXXA(AM). April 1984: 92.7 mhz; 3 kw. 282 ft TL: N35 06 46 W92 24 42. Box 1266, 117 Oak St., Suite 300, 72032.

Corning

KBKG(FM)—Listing follows KCCB(AM).

KCCB(AM)— Feb 19, 1959: 1260 khz; 1 kw-D. TL: N36 24 00 W90 35 05. Hrs open: Box 398, 501 Bryan, 72422. Phone: (870) 857-6646. Fax: (870) 857-6795. Licensee: Shields-Adkins Broadcasting Inc. Population served: 55,000 Natl. Rep: Keystone (unwired net). Format: Lite. Target aud: General. ♦Jim Adkins, pres & gen mgr; Tina Privett, gen mgr & gen sls mgr; Neil Raines, progmg dir & news dir; Palmer Johnson, chief of engrg.

KBKG(FM)—Co-owned with KCCB(AM). Sept 15, 1983: 93.5 mhz; 3 kw. 138 ft TL: N36 24 00 W90 35 05. Stereo. Natl. Network: ABC. Format: Adult contemp, oldies. ♦Jim Adkins, CEO.

Cotton Plant

KAPW(FM)—Not on air, target date: unknown: 99.3 mhz; 6 kw. Ant 328 ft TL: N34 58 07 W90 59 48. Hrs open: 1515 14th Ave., Suite 303, Oakland, CA, 94606. Phone: (415) 373-2531. Licensee: Bradford Caldwell. ♦Bradford Caldwell, gen mgr.

Crossett

KAGH(AM)— January 1951: 800 khz; 240 w-D, 43 w-N. TL: N33 08 05 W91 56 49. Stereo. Hrs opn: 24 Box 697, 117 E. Wellfield Rd., 71635. Phone: (870) 364-2181. Phone: (870) 364-2182. Fax: (501) 364-2183. E-mail: kagh@alltell.net Licensee: Ashley County Broadcasters Inc. (acq 8-1-69). Population served: 18,000 Natl. Network: Westwood One. Rgnl. Network: Ark. Radio Net. Format: Country. ♦Kevin Medlin, pres & gen mgr; Bryan Bailey, news dir; Russ Miller, progmg dir & chief of engrg.

KAGH-FM— Mar 16, 1967: 104.9 mhz; 6 kw. 300 ft TL: N33 08 05 W91 56 49. Format: Country.

Danville

KYEL(FM)—Not on air, target date: unknown: 105.5 mhz; 4.45 kw. Ant 400 ft TL: N35 07 16 W93 19 34. Hrs open: 5am - 11pm 201 W. 2nd St., Russellville, AZ, 72801. Phone: (479) 890-7207. Fax: (479) 967-5278. E-mail: karv-kyel@yahoo Licensee: Danville FM Inc. Format: Country. Spec prog: Cardinal Baseball. ♦Chris Womack, stn mgr; Diane Womack, gen mgr & sls.

Dardanelle

KCAB(AM)— Mar 24, 1964: 980 khz; 5 kw-D. TL: N35 13 02 W93 10 08. Hrs open: Box 10310, Russellville, 72812. Secondary address: 2705 E. Pkwy., Russellville 72802. Phone: (479) 968-6816. Fax: (479) 968-2946. Web Site: www.rivertalk980.com. Licensee: MMA License LLC. Group owner: MAX Media L.L.C. (acq 4-21-03; grpsl). Format: Sports/new. Target aud: 25-54; adults. ♦Jim Kelly, stn mgr; Tom Kamerling, news dir; Jim Alexander, chief of engrg.

KCJC(FM)—Co-owned with KCAB(AM). Jan 26, 1966: 102.3 mhz; 200 w. 1,227 ft TL: N35 13 41 W93 15 20. (CP: 1.43 kw, ant 1,322 ft.). Web Site: www.rivertalk980.com. Natl. Network: ABC. Format: Country.

KWXT(AM)— October 1987: 1490 khz; 1 kw-U. TL: N35 13 08 W93 07 38. Hrs open: 701 E. Main St., Suite 4, Russellville, 72801. Phone: (479) 968-1337. Fax: (479) 968-1337. E-mail: kwxt1490am@ahoo.com Web Site: www.kwxt1490am.com. Licensee: George V. Domerese/Sherwood Broadcasting Co. (acq 9-2-92; $60,000; FTR: 9-21-92). Format: Christian country, gospel. ♦Tim Domerese, gen mgr; Jim Alexander, chief of engrg.

De Queen

***KBPU(FM)**— 2002: Stn currently dark. 88.7 mhz; 250 w. Ant 122 ft TL: N34 02 38 W94 17 41. Hrs open: Box 5725, Twin Falls, ID, 83303. Phone: (208) 733-3551. Fax: (208) 733-3548. Licensee: Edgewater Broadcasting Inc. (group owner; (acq 5-3-2006). ♦Clark Parrish, pres.

KDQN(AM)— Aug 1, 1956: 1390 khz; 500 w-D. TL: N34 01 57 W94 19 43. Hrs open: Box 311, 71832. Secondary address: 921 W Collin Raye Dr. 71832. Phone: (870) 642-2446. Fax: (870) 642-2442. E-mail: numberonecountry@yahoo.com Web Site: wwww.kdqn.net. Licensee: Jay W. Bunyard & Anne W. Bunyard. (acq 6-15-83; $475,000 with co-located FM; FTR: 7-4-83) Population served: 4,600 Rgnl. Network: Ark. Radio Net. Format: Sp. ♦Jay Bunyard, pres; Jon Bunyard, gen mgr; Victor Rojas, gen sls mgr.

KDQN-FM— Oct 6, 1978: 92.1 mhz; 50 kw. Ant 492 ft TL: N34 13 35 W94 17 35. Stereo. Hrs opn: Box 311, 71832. Phone: (870) 642-2446. Fax: (870) 642-2442. Web Site: www.kdqn.net. Licensee: Jay W. Bunyard & Anne W. Bunyard. Format: Country. ♦Jon Bunyard, gen mgr, opns mgr & women's int ed.

De Witt

KDEW-FM— Sept 1, 1970: 97.3 mhz; 50 kw. 272 ft TL: N34 16 09 W91 21 02. Hrs open: 24 c/o KWAK-AM-FM, 1818 S. Buerkle, Stuttgart, 72160. Phone: (870) 673-1595. Fax: (870)673-8445. E-mail: kdew973@yahoo.com Licensee: Arkansas County Broadcasters Inc. (group owner; acq 3-5-97; $150,000). Format: Country. News staff: 1. ♦Keith Hill, progmg dir; Jonathan Reaves, news dir; Jim Alexander, engrg dir; Sandi Levy, traf mgr; Scott Siler, gen mgr, mktg dir & disc jockey.

Dermott

KRKD(FM)— Apr 1, 2000: Stn currently dark. 105.7 mhz; 3 kw. Ant 328 ft TL: N33 32 25 W91 22 39. Stereo. Hrs opn: Box 1438, Cleveland, MS, 38732. Phone: (903) 595-4795. Phone: (662) 378-4103. Fax: (903) 593-2666. Licensee: M.R.S. Ventures Inc. (group owner; (acq 11-1-2003; grpsl). Population served: 90,000 Law Firm: Wood, Maines & Brown, Chartered. ♦Jerry Russell, pres.

KXSA-FM— Aug 24, 1924: 103.1 mhz; 5.5 kw. 328 ft TL: N33 31 56 W91 34 28. Hrs open: 24 279 Midway, Monticello, AZ, 71655. Phone: (870) 367-8528. Fax: (870) 367-9564. E-mail: crn@ccc-cable.net Licensee: Pines Broadcasting Inc. (group owner; (acq 3-14-2007; grpsl). Format: Classic country. ♦Jimmy Sledge, pres & gen mgr.

Des Arc

***KBDO(FM)**— 1999: 91.7 mhz; 56 kw vert. Ant 682 ft TL: N35 00 08 W91 44 41. Hrs open: Box 3206, American Family Radio, Tupelo, MS, 38803. Phone: (662) 844-8888, EXT. 204. Fax: (662) 842-6791. Licensee: American Family Association. Group owner: American Family Radio Format: Relg. ♦Marvin Sanders, gen mgr; Gary Vaile, stn mgr; John Riley, progmg dir; Joey Moody, chief of engrg.

KFLI(FM)— 2003: 104.7 mhz; 25 kw. Ant 328 ft TL: N35 00 23 W91 40 20. Hrs open: 121 Radio Heights Dr., Searcy, 72143. Phone: (501) 268-1047. Fax: (501) 305-2977. Web Site: www.oldiesradioonline.com. Licensee: George S. Flinn Jr. Format: Oldies. ♦Ken Madden, gen mgr.

Dumas

KXFE(FM)— Sept 1, 1980: Stn currently dark. 106.9 mhz; 25 kw. Ant 269 ft TL: N33 58 11 W91 32 58. Stereo. Hrs open: Box 789, Wynne, 72396. Phone: (870) 238-8141. Fax: (870) 238-5997. Licensee: Arkansas County Broadcasters Inc. (acq 8-31-2004; $130,000). Population served: 12,500 Format: Country. ♦Bobby Caldwell, pres & gen mgr; David Sills, sls dir.

Earle

KCJF(FM)— 2004: 103.9 mhz; 12.5 kw. Ant 469 ft TL: N35 27 01 W90 42 11. Hrs open: Box 789, Wynne, 72396. Phone: (870) 238-5253. Fax: (870) 238-5997. Licensee: Catherine Joanna Flinn. Format: Classic rock. ♦Bobby Caldwell, gen mgr.

East Camden

KCXY(FM)— Sept 28, 1987: 95.3 mhz; 100 kw. Ant 456 ft TL: N33 30 14 W92 48 38. Stereo. Hrs opn: 24 Box 957, Camden, 71701. Secondary address: 133 Washington St. S.W., Camden 71701. Phone: (870) 836-9567. Fax: (870) 836-9500. E-mail: y95@cablelinks.net Web Site: www.camdenfm.net. Licensee: Radio Works Inc. (acq 12-13-2004; grpsl). Population served: 50,000 Rgnl. Network: Ark. Radio Net. Format: C&W. News staff: one; News: 15 hrs wkly. Target aud: 25-54. ♦Jay Bunyard, pres; Donna Stewart, gen mgr & gen sls mgr; Greg Arnold, opns mgr & progmg dir.

El Dorado

KAGL(FM)— Sept 29, 1993: 93.3 mhz; 18 kw. Ant 354 ft TL: N33 16 16 W92 39 17. Stereo. Hrs opn: 24 2525 Northwest Ave., 71730. Phone: (870) 863-6126. Fax: (870) 863-4555. Web Site: www.totalradio.com. Licensee: Noalmark Broadcasting Corp. (group owner; acq 1-8-93; $10,000; FTR: 3-29-93). Population served: 53,375 Natl. Rep: Target Broadcast Sales. Format: Classic rock. News staff: one; News: 15 hrs wkly. Target aud: 25-54; general. ♦William C. Nolan, pres; Edwin Alderson, exec VP; Sandy Sanford, gen mgr.

***KBSA(FM)**— December 1987: 90.9 mhz; 3 kw. Ant 581 ft TL: N33 16 19 W92 42 12. Stereo. Hrs opn: 24

Rebroadcasts KDAQ(FM) Shreveport, LA 100%.
Box 5250, Shreveport, LA, 71135. Phone: (800) 552-8502. Phone: (318) 797-5150. Fax: (318) 797-5265. E-mail: listenermail@redriverradio.org Web Site: www.redriverradio.org. Licensee: Board of Supervisors of Louisiana State University & A&M College. Natl. Network: NPR, PRI. Format: Classical, jazz, news. Target aud: 25+. ♦Kermit Poling, gen mgr; Rick Shelton, opns dir.

KDMS(AM)— May 8, 1950: 1290 khz; 5 kw-D, 106 w-N. TL: N33 12 27 W92 41 10. Hrs open: 1904 W. Hillsboro, 71730. Phone: (870) 863-5121. Fax: (870) 863-6221. Web Site: www.klbq99.com. Licensee: El Dorado Broadcasting Co. (acq 7-8-87; $950,000 with co-located FM; FTR: 4-6-87) Population served: 28,463 Format: Adult contemp. ♦Dan Murphy, opns mgr & progmg dir; Don Travis, gen sls mgr & news dir; Rosh Partridge, pres, gen mgr & gen sls mgr; Norm Mason, chief of engrg.

KLBQ(FM)—Co-owned with KDMS(AM). Dec 23, 1963: 98.7 mhz; 14 kw. 298 ft TL: N33 12 30 W92 41 16. Stereo. 50,000 Format: Top-40, adult contemp. ♦Dan Murphy, disc jockey.

KELD(AM)— Oct 17, 1935: 1400 khz; 1 kw-U. TL: N33 12 43 W92 39 48. (CP: TL: N33 14 14 W92 39 54). Stereo. Hrs opn: 24 2525 Northwest Ave., 71730. Phone: (870) 863-6126. Phone: (870) 862-1400. Fax: (870) 863-4555. Web Site: www.totalradio.com. Licensee: Noalmark Broadcasting Corp. (group owner; acq 7-73). Population served: 27,000 Natl. Network: Fox Sports. Rgnl. Network: Ark. Radio Net. Natl. Rep: Target Broadcast Sales. Format: Sports. Target aud: General. ♦William C. Nolan Jr., pres; Edwin Alderson, exec VP; Sandy Sanford, gen mgr & sls dir; Patrick Thomas, opns dir, opns mgr & progmg dir; Steven Gray, chief of engrg.

KIXB(FM)—Co-owned with KELD(AM). Dec 9, 1963: 103.3 mhz; 100 kw. 571 ft TL: N33 13 20 W92 55 28. Stereo. 24 Phone: (870) 864-0103. E-mail: kix103@noalmark.com Web Site: www.totalradio.com.50,000 Format: Country. News: 15 hrs wkly. Target aud: 18-54.

***KKDU(FM)**—Not on air, target date: unknown: 88.9 mhz; 26 kw vert. Ant 121 ft TL: N33 12 32 W92 42 10. Hrs open: 24 Broadcasting for the Challenged Inc., 188 S. Bellevue, Suite 222, Memphis, TN, 38104. Phone: (901) 375-9324. Phone: (870) 875-1108. Licensee: Broadcasting for the Challenged Inc. Format: Div. ♦Rosh Partridge, stn mgr & opns mgr.

KMLK(FM)— 2000: 101.5 mhz; 6 kw. 328 ft TL: N33 09 32 W92 37 47. Hrs open: 2525 N. West Ave., 71730. Phone: (870) 875-1015. Fax: (870) 863-4555. Licensee: Noalmark Broadcasting Corp. (group owner; acq 6-14-01). Format: Urban adult contemp. ♦Sandy Sanford, gen mgr.

KMRX(FM)— May 12, 1984: 96.1 mhz; 100 kw. 288 ft TL: N33 16 21 W92 39 25. Stereo. Hrs opn: 24 2525 Northwest Ave., 71730. Phone: (870) 863-6126. Fax: (870) 863-4555. Web Site: www.totalradio.com. Licensee: Noalmark Broadcasting Corp. (group owner; acq 7-31-97). Population served: 300,000 Format: Adult contemp, contemp hit. News staff: one; News: 2 hrs wkly. Target aud: 18-34. ♦Sandy Sanford, gen mgr; Chase Roberts, opns mgr & progmg dir; Jim Harris, news dir.

England

KHTE-FM— Sept 26, 1988: 96.5 mhz; 3 kw. 148 ft TL: N34 32 45 W91 59 04. (CP: 25 kw). Stereo. Hrs opn: 24 400 Hardin Rd., Suite 150, Little Rock, 72211. Phone: (501) 219-1919. Fax: (501) 225-4610. Web Site: www.hot965.com. Licensee: ABG Arkansas LLC. Group owner: Archway Broadcasting Group (acq 1-22-2003; $8 million with KOLL-FM Lonoke). Population served: 300,000 Format: Contemporary hit. ♦Brad Hutchesonn, gen mgr; Joe Ratliff, stn mgr & progmg dir.

KVDW(AM)— Aug 31, 1979: 1530 khz; 250 w-D. TL: N34 32 45 W91 59 04. (CP: COL: Scott, 500 w-D). Hrs opn: Victory 1530, 201 W. Broadway, Suite G-5, North Little Rock, 72114. Phone: (501) 351-5839. Fax: (501) 842-9308. Licensee: Wells Broadcasting Inc. (acq 8-13-02; $35,000). Law Firm: Putbrese, Hunsaker & Trent, P. Format: Gospel hits, inspirational talk. Target aud: 18-54; professionals, farmers, college educated. Spec prog: Farm 5 hrs, talk 10 hrs wkly. ♦Vernon Wells, gen mgr.

Eudora

KAVH(FM)— 2001: 101.5 mhz; 6 kw. Ant 328 ft TL: N33 11 58 W91 15 39. Hrs open: 24 c/o WJJA(TV), 4311 E. Oakwood Rd., Oak Creek, WI, 53154. Phone: (414) 764-4953. Licensee: Joel J. Kinlow. Group owner: Joel J. Kinlow Stns. Format: Var. ♦Bruce Herz, opns dir; Joel Kinlow, gen mgr & progmg dir.

Eureka Springs

KTCN(FM)— May 13, 1985: 100.9 mhz; 1.1 kw. 531 ft TL: N36 22 49 W93 44 53. Stereo. Hrs opn: 6 AM-10 PM 114 Hwy. 23 S., 72632. Phone: (479) 253-9079. Fax: (479) 253-7308. Web Site: www.hereshelpnet.org. Licensee: New Life Evangelistic Center Inc. (acq 11-23-92; $90,000; FTR: 12-14-92). Rgnl. Network: Ark. Radio Net. Format: Relg. News staff: one; News: 20 hrs wkly. Target aud: 35 plus; upper income & retired. Spec prog: Class 10 hrs, 50s & 60s mus 2 hrs wkly. ♦Larry Rice, CEO & pres; Jim Barnes, gen mgr & opns mgr; Thomas Hoffman Jr., chief of opns.

Fairfield Bay

KFFB(FM)— Dec 31, 1981: 106.1 mhz; 50 kw. 500 ft TL: N35 45 22 W92 14 49. Stereo. Hrs opn: 24 Box 1050, 72088. Phone: (501) 884-6812. Fax: (501) 723-4861. E-mail: kffb@kffb.com Web Site: www.kffb.com. Licensee: Freedom Broadcasting Inc. Natl. Network: ABC. Rgnl. Network: Ark Radio Net. Law Firm: Smithwick & Belendiuk. Format: MOR. News staff: one; News: 8 hrs wkly. Target aud: 35 plus; middle & upper income. ♦Bob Connell, pres; Pam Connell, exec VP; Chad Whiteaker, stn mgr.

Farmington

KFAY(AM)— Dec 15, 1946: 1030 khz; 10 kw-D, 1 kw-N, DA-2. TL: N36 06 34 W94 10 59. Stereo. Hrs open: 24 24 E. Meadow St., Suite 1, Fayetteville, 72701. Phone: (479) 521-5566. Fax: (479) 521-4968. Web Site: www.kfayam.com. Licensee: Cumulus Licensing Corp. Group owner: Cumulus Media Inc. (acq 2-1-99; grpsl). Population served: 150,000 Format: News/talk. News staff: 6; News: 20 hrs wkly. Target aud: 25-64; general. ♦Joe Conway, gen mgr.

Fayetteville

***KAYH(FM)**— June 26, 2000: 89.3 mhz; 6 kw. Ant 380 ft TL: N36 01 48 W94 05 10. Hrs open: 24 Box 1288, Family FM 89.3, 72702. Phone: (479) 750-7707. Fax: (479) 750-7767. E-mail: familyfm89.3@sbc.net Licensee: Family Vision Ministries Inc. (acq 12-19-01; $119,000). Format: Southern gospel. ♦Mike Disney, gen mgr.

***KBNV(FM)**— 2000: 90.1 mhz; 7.1 kw horiz, 16 kw vert. Ant 466 ft TL: N36 07 38 W93 59 23. Hrs open: 24 Drawer 2440, Tupelo, MS, 38803. Phone: (662) 844-8888. Fax: (662) 844-9090. Licensee: American Family Association. Group owner: American Family Radio Format: Christian. ♦John Riley, progmg dir.

KEZA(FM)— Sept 6, 1983: 107.9 mhz; 99 kw. 1,259 ft TL: N35 51 12 W94 01 33. Stereo. Hrs opn: 4209 Frontage Rd., 72703. Phone: (479) 582-1079. Fax: (479) 582-5302. Web Site: www.magic1079.com. Licensee: Capstar TX L.P. Group owner: Clear Channel Communications Inc. (acq 8-30-00; grpsl). Format: Adult contemp. Target aud: 25-54. Spec prog: Jazz, oldies. ♦Tony Beringer, gen mgr.

KKEG(FM)— Oct 16, 1964: 92.1 mhz; 1.15 kw. 459 ft TL: N36 03 55 W94 12 24. (CP: 20.5 kw, ant 328 ft.). Stereo. Hrs open: 24 1780 Holly St., 72703. Phone: (479) 521-5566. Fax: (479) 521-0751. Web Site: www.921thekeg.com. Licensee: Cumulus Licensing Corp. Group owner: Cumulus Media Inc. (acq 2-1-99; grpsl). Population served: 180,000 Natl. Rep: Roslin. Format: AOR, classic rock. News staff: one. Target aud: 18-49. ♦Joe Conway, gen mgr.

KKIX(FM)— Oct 1, 1966: 103.9 mhz; 100 kw. 510 ft TL: N36 01 17 W94 13 04. Stereo. Hrs open: 24 Box 8190, 72703. Phone: (479) 521-0104. Fax: (479) 444-8600. Web Site: www.kix104.com. Licensee: Capstar TX L.P. Group owner: Clear Channel Communications Inc. (acq 8-30-00; grpsl). Population served: 250,000 Law Firm: Wiley, Rein & Fielding. Format: Country. News staff: one; News: 2 hrs wkly. Target aud: 25-54. ♦Tony Beringer, gen mgr.

KMXF(FM)—See Lowell

KOFC(AM)— June 10, 1957: 1250 khz; 920 w-D, 45 w-N. TL: N36 04 29 W94 11 00. Hrs open: 16 Box 1288, 72702-0550. Phone: (479) 750-7707. Fax: (479) 750-7767. E-mail: kofc@ipa.net Licensee: William B. Disney. (acq 12-11-87; $95,000; FTR: 6-22-87). Natl. Network: USA. Format: Christian talk, teaching. News: 8 hrs wkly. Target aud: 35 plus; traditional Christian families. ♦William B. Disney, pres; Mike Disney, gen mgr; Robert Johnson, opns mgr.

***KUAF(FM)**— Jan 15, 1973: 91.3 mhz; 60 kw. 1,105 ft TL: N35 51 12 W94 01 33. Stereo. Hrs opn: 24 747 W. Dickson St., Suite 2, 72701-5023. Phone: (479) 575-2556. Fax: (479) 575-8440. Web Site: www.kuaf.com. Licensee: Board of Trustees University of Arkansas. Population served: 360,000 Natl. Network: NPR. Format: News, class, jazz. News staff: 3; News: 45 hrs wkly. Target aud: 25-65. Spec prog: Folk 5 hrs, Black 5 hrs wkly. ♦Rick Stockdell, gen mgr.

***KXUA(FM)**— April 4, 2000: 88.3 mhz; 470 w vert. 262 ft TL: N36 03 56 W94 10 30. Stereo. Hrs open: 24 University of Arkansas, 406 Administration Bldg., 72701. Phone: (479) 575-4273. Web Site: www.kxua.com. Licensee: Board of Trustees of University of Arkansas. Population served: 100,000 Format: Alternative. Target aud: 12-24; high school and colege students. ♦Rick Stockdell, gen mgr.

Fordyce

KBJT(AM)— Aug 1, 1959: 1590 khz; 4.7 kw-D, 35 w-N. TL: N33 48 10 W92 26 10. Hrs open: 24 303 Spring St., 71742. Phone: (870) 352-7137. Fax: (870) 352-7139. E-mail: kbjt@alltel.net Web Site: kbjtkq.com. Licensee: KBJT Inc. (acq 9-1-77). Population served: 11,000 Format: News/talk. Target aud: General. Spec prog: Gospel 11 hrs wkly. ♦Gary Coates, pres & gen mgr; Carna Coates, progmg dir & pub affrs dir; Saxon Coates, news dir, news rptr & reporter.

KQEW(FM)—Co-owned with KBJT(AM). Feb 23, 1982: 102.3 mhz; 25 kw. 328 ft TL: N33 48 10 W92 26 10. Stereo. Licensee: Dallas Properties Inc. Format: C&W. Target aud: General.

Forrest City

***KARH(FM)**— 2000: 88.1 mhz; 3.7 kw. Ant 544 ft TL: N35 12 11 W90 33 57. Hrs open: Drawer 3206, American Family Radio, Tupelo, MS, 38803. Phone: (662) 844-8888. Fax: (662) 842-6791. Web Site: www.afr.net. Licensee: American Family Association. Group owner: American Family Radio Format: Relg. ♦Marvin Sanders, gen mgr; John Riley, progmg dir; Joey Moody, chief of engrg.

KBFC(FM)—Listing follows KXJK(AM).

KXJK(AM)— Apr 29, 1949: 950 khz; 5 kw-D, 500 w-N. TL: N34 58 53 W90 51 27. Stereo. Hrs open: 24 Box 707, 501 E. Broadway, 72336. Phone: (870) 633-1252. Fax: (870) 633-1259. E-mail: radio@arkansas.net Web Site: www.kxjk.com. Licensee: Forrest City Broadcasting Co. Inc. Population served: 30,000 Rgnl. Network: Ark. Radio Net. Rgnl rep: Midsouth Law Firm: Gene Smith. Format: Classic rock, news/talk. News staff: 2; News: 24 hrs wkly. Target aud: General. Spec prog: Farm 16 hrs wkly. ♦William Fogg, gen mgr, mus dir & chief of engrg.

KBFC(FM)—Co-owned with KXJK(AM). Sept 22, 1960: 93.5 mhz; 25 kw. 340 ft TL: N34 51 17 W90 55 02. Stereo. 24 Web Site: www.kbfc.com. Format: Modern country. News: 4 hrs wkly. ♦William Fogg, progmg dir; Jeff Fogg, farm dir; Janet Benson, women's int ed.

Fort Smith

***KAOW(FM)**— 1999: 88.9 mhz; 1 kw. 482 ft TL: N35 26 50 W94 21 54. Hrs open: 24 Box 2440, American Family Radio, Tupelo, MS, 38803. Phone: (662) 844-8888. Fax: (662) 842-6791. Web Site: www.afr.net. Licensee: American Family Association. Group owner: American Family Radio Format: Relg. ♦Marvin Sanders, gen mgr; John Riley, progmg dir; Joey Moody, chief of engrg.

KBBQ-FM—See Van Buren

***KEAF(FM)**—Not on air, target date: unknown: 90.7 mhz; 26 kw vert. Ant 2,086 ft TL: N35 09 56 W93 40 36. Hrs open:
Rebroadcasts WBFR(FM) Birmingham, AL 100%.
c/o WBFR(FM), 244 Goodwin Crest Dr., Suite 118, Birmingham, AL, 35209. Phone: (205) 942-3530. Fax: (510) 568-6190. Licensee: Family Stations Inc. Format: Relg, evangelical. ♦Stanley Jackson, gen mgr.

KFPW(AM)— July 27, 1930: 1230 khz; 1 kw-U, DA-1. TL: N35 23 11 W94 21 44. Stereo. Hrs opn: 24 Box 908, 72902. Secondary address: 323 N. Greenwood 72902. Phone: (501) 783-5379. Phone: (501) 288-1047. Fax: (501) 785-2638. Licensee: Pharis Broadcasting Inc. (group owner; (acq 3-14-2002; $850,000 with KFPW-FM Barling). Population served: 143,800 Natl. Network: ABC. Rgnl. Network: Ark. Radio Net. Natl. Rep: Commercial Media Sales. Law Firm: Campbell & Tannenwald. Format: Big band, oldies. News: 13 hrs wkly. Target aud: 35 plus; affluent. Spec prog: Sp 6 hrs wkly. ♦Bill Pharis, chmn & pres; Karen Pharis, VP, gen mgr, gen sls mgr, prom mgr & news dir; Ernie Witt, progmg dir, local news ed & edit dir; Jim Barnes, engrg VP; Mack Reminington, disc jockey.

KFSA(AM)— Feb 13, 1947: 950 khz; 1 kw-D, 500 w-N, DA-2. TL: N35 25 58 W94 28 13. Stereo. Hrs opn: Box 488, 72901. Secondary address: 601 N. Greenwood 72901. Phone: (479) 782-9125. Fax: (479) 782-9127. Licensee: Fred H. Baker Sr. (acq 11-5-81; $297,000; FTR: 11-30-81). Population served: 150,000 Format: Southern gospel. Target aud: General. ♦Fred H. Baker Sr., pres; Gary Keifer, gen mgr; Jerry Lynch, gen sls mgr & natl sls mgr; David J. Burdue, progmg dir & disc jockey; Dale L. Davenport, chief of engrg; Illa Davenport, traf mgr.

KISR(FM)—Co-owned with KFSA(AM). Aug 13, 1971: 93.7 mhz; 100 kw. 1,250 ft TL: N35 31 22 W94 23 32. Stereo. 24 Phone: (501) 785-2526. Web Site: www.kisr.net. Licensee: Stereo 93 Inc.410,000 Format: CHR. News staff: one. Target aud: 18-39. ♦Fred Baker Jr., gen mgr & progmg dir; Gary Keifer, stn mgr; Carol Patterson, gen sls mgr & traf mgr; Rick Hayes, mus dir; Dale L. Davenport, engrg dir.

KHGG(AM)—See Van Buren

KLSZ-FM— July 27, 1978: 100.7 mhz; 50 kw. Ant 459 ft TL: N35 13 32 W94 20 29. Stereo. Hrs opn: 24 3104 S. 70th St., 72903. Phone: (479) 452-0681. Fax: (479) 452-0873. Licensee: Cumulus Licensing Corp. Group owner: Cumulus Media Inc. (acq 5-1-99; $1 million). Population served: 627,600 Format: Golden oldies. News: 7 hrs wkly. Target aud: 25-64. ♦Smitty O'Loughlin, gen mgr.

KMAG(FM)— Dec 31, 1964: 99.1 mhz; 100 kw. 1,968 ft TL: N35 09 56 W93 40 35. Stereo. Hrs opn: 24 311 Lexington Ave., 72901. Phone: (479) 782-8888. Fax: (497) 785-5946. E-mail: info@kmag991.com Web Site: www.kmag991.com. Licensee: Capstar TX L.P. Group owner: Clear Channel Communications Inc. (acq 8-30-00; grpsl). Format: Country. News staff: 2. Target aud: 25-54; females. ♦Paul Swint, gen mgr; Ralph Cherry, opns mgr; Tony Montgomery, sls dir & gen sls mgr; Lee Matthews, progmg dir; Gary Elmore, news dir; Allan Riley, chief of engrg.

KYHN(AM)—Co-owned with KMAG(FM). Nov 22, 1947: 1320 khz; 5 kw-U, DA-2. TL: N35 24 36 W94 21 30.24 71,515 Format: News/talk. News staff: 6; News: 40 hrs wkly. Target aud: 25-54.

KOMS(FM)—See Poteau, OK

KQBK(FM)—Booneville, Nov 1, 1981: 104.7 mhz; 50 kw. 492 ft TL: N35 11 01 W94 07 44. Stereo. Hrs opn: 24 Box 908, 72902. Phone: (479) 288-1047. Fax: (479) 785-2638. Web Site: www.fox94.com. Licensee: Pharis Broadcasting Inc. (group owner; acq 11-20-97; $800,000). Population served: 500,000 Rgnl. Network: Ark. Radio Net. Natl. Rep: Commercial Media Sales. Rgnl rep: BRI Law Firm: Irwin, Campbell & Tannenwald. Format: Oldies. News staff: one; News: 10 hrs wkly. Target aud: 18-54. ♦Bill Pharis, pres; Karen Pharis, gen mgr; Ernie Witt, opns mgr.

KTCS-FM— Aug 15, 1964: 99.9 mhz; 100 kw. 1,919 ft TL: N35 04 20 W94 40 50. Stereo. Hrs opn: Box 180188, 72918-0188. Secondary address: 5304 Hwy. 45 E. 72916. Phone: (479) 646-6151. Fax: (479) 646-3509. E-mail: ktcs@ktcs.com Web Site: www.ktcs.com. Licensee: Big Chief Broadcasting Co. Format: Country. ♦Lee Young, gen mgr, stn mgr & gen sls mgr; Melissa Harper, opns mgr; Darren Minor, progmg dir; Mary Livingston, news dir; Scott Reeves, chief of engrg; Sandy Hunter, traf mgr.

KTCS(AM)— March 1956: 1410 khz; 1 kw-D. TL: N35 16 40 W94 22 35. Web Site: www.ktcs.com. (Acq 1961).211,000 Format: Southern Gospel. ♦Sandy Hunter, traf mgr.

KWHN(AM)— 2001: 1650 khz; 10 kw-D, 1 kw-N. TL: N35 24 36 W94 21 30. Hrs open: 311 Lexington Ave., 72901. Phone: (479) 782-8888. Fax: (479) 782-0366. E-mail: info@kwhn.com Web Site: www.kwhn.com. Licensee: Capstar TX L.P. Group owner: Clear Channel Communications Inc. Format: News/talk. ♦Paul Swint, gen mgr; Tony Montgomery, sls dir; Darin Bobb, prom dir.

KZBB(FM)—See Poteau, OK

Fouke

***KLMZ(FM)**— 2001: 104.3 mhz; 5 kw. Ant 361 ft TL: N33 21 05 W93 50 41. Hrs open: 24 2351 Sunset Blvd., Suite 170-218, Rocklin, CA, 95765. Phone: (916) 251-1600. Fax: (916) 251-1650. Web Site: www.air1.com. Licensee: Educational Media Foundation. Group owner: EMF Broadcasting (acq 1-15-2004; $500,000). Natl. Network: Air 1. Law Firm: Shaw Pittman. Format: Christian. News staff: 3. Target aud: 25-44; Judeo Christian, female. ♦Richard Jenkins, pres; Mike Novak, VP; Keith Whipple, dev dir; Eric Allen, natl sls mgr; David Pierce, progmg mgr; Ed Lenane, news dir; Sam Wallington, engrg dir; Karen Johnson, news rptr; Marya Morgan, news rptr; Richard Hunt, news rptr.

Glenwood

KWXE(FM)—Listing follows KWXI(AM).

KWXI(AM)— May 12, 1980: 670 khz; 5 kw-D. TL: N34 19 32 W93 33 27. Hrs open: 6 AM-midnight Box 740, 71943. Secondary address: 180 Hwy. 70 E., Suite 11 71943. Phone: (870) 356-2151. Phone: (870) 356-2181. Fax: (870) 356-4684. E-mail: kwxi@alltel.net Licensee: US Stations LLC. (acq 2-9-2005; $530,000 with co-located FM). Natl. Network: Salem Radio Network. Rgnl. Network: Ark. Radio Net. Rgnl rep: Rgnl Reps Format: Souther gospel. News: 8 hrs wkly. Target aud: 34-54; affluent professionals. ♦Wayne Bennett, gen mgr; Wayne Benne, gen sls mgr; Doug Dumont, progmg dir & news dir; Danie Appleoff, chief of engrg.

KWXE(FM)—Co-owned with KWXI(AM). Nov 18, 1991: 104.5 mhz; 3 kw. 328 ft TL: N34 18 38 W93 32 04. Stereo. 24 Natl. Network: CBS Radio. Rgnl rep: Rgnl Reps Format: Country. News: 7 hrs wkly. Target aud: 25-50; general. ♦Lanny Borhters, opns dir.

Gosnell

KAMJ-FM— February 1999: 93.9 mhz; 2 kw. 328 ft TL: N35 53 56 W89 52 48. Hrs open: 24 Box 989, Blytheville, 72315-0989. Phone: (870) 762-2093. Fax: (870) 763-8459. Licensee: Phoenix Broadcasting Group Inc. Group owner: Sudbury Services Inc. & Newport Broadcasting Co. Format: Urban contemp. ♦Rob Hill, gen mgr.

Gould

KAFN(FM)— Apr 15, 1999: 102.5 mhz; 6 kw. Ant 177 ft TL: N33 58 11 W91 32 58. Hrs open: 24 700 Wellington Hills Rd., Little Rock, 72211. Phone: (501) 401-0200. Fax: (501) 401-0367. Web Site: www.920karn.com. Licensee: Arkansas County Broadcasters Inc. (group owner; acq 12-30-2003; $90,000). Format: News radio. ♦Jim Beard, gen mgr.

Greenwood

KZKZ-FM— December 1981: 106.3 mhz; 1.7 kw. 433 ft TL: N35 13 43 W94 15 45. (CP: 15 kw, ant 397 ft.). Stereo. Hrs opn: 24 6420 S. Zero St., Fort Smith, 72903. Phone: (479) 646-6700. Fax: (479) 646-1373. E-mail: kzkzfm@kzkzfm.com Web Site: www.kzkzfm.com. Licensee: Family Communications Inc. (acq 5-11-93; FTR: 6-7-93). Format: Contemp Christian. ♦Jerry Lynch, gen mgr; Jay Lynch, stn mgr.

Gurdon

KYXK(FM)— December 1984: 106.9 mhz; 17.5 kw. 298 ft TL: N33 56 42 W93 10 43. Stereo. Hrs opn: 24 Box 40, Arkadelphia, 71923. Secondary address: 601 S. 7th St., Arkadelphia 71923. Phone: (870) 246-9272. Fax: (870) 246-5878. Licensee: Noalmark Broadcasting Corp. (group owner; (acq 6-29-2007; grpsl). Population served: 170,000 Law Firm: Booth, Freret, Imlay & Tepper. Format: Country. Target aud: 25-54; adults. ♦William Nolan Jr., pres.

Hamburg

KHMB(FM)— 1996: 99.5 mhz; 3.2 kw. Ant 312 ft TL: N33 17 19 W91 52 45. Hrs open: 24 203 Fairview Rd., Crossett, 71635. Phone: (870) 364-4700. Fax: (870) 364-4770. Web Site: www.QLiteradio.com. Licensee: R&M Broadcasting (acq 12-28-2005; $131,553). Format: Adult contemp. ♦Dennis Maxwell, gen mgr.

Hampton

***KBPW(FM)**— 2001: 88.1 mhz; 250 w. Ant 259 ft TL: N33 32 11 W92 28 07. Hrs open: Box 2440, Tupelo, MS, 38803. Phone: (662) 844-8888. Fax: (662) 842-6791. Licensee: American Family Association Inc. Group owner: American Family Radio (acq 4-19-01). Format: Christian. ♦Marvin Sanders, gen mgr; John Riley, progmg dir; Joey Moody, chief of engrg.

KELD-FM— Nov 26, 1984: 106.5 mhz; 17.5 kw. 302 ft TL: N33 32 23 W92 34 59. Hrs open: 24 2525 N.W. Ave., El Dorado, 71730. Phone: (870) 863-6126. Fax: (870) 863-4555. Web Site: www.totalradio.com. Licensee: Noalmark Broadcasting Corp. (group owner; acq 2-21-03; $250,000). Natl. Network: ABC, Fox News Radio. Format: News/talk. ♦Sandy Sanford, gen mgr & opns dir.

Hardy

KOOU(FM)— Oct 4, 1993: 104.7 mhz; 6 kw. Ant 199 ft TL: N36 18 17 W91 24 38. Stereo. Hrs open: 24 11 FM 101 Rd., 72542. Phone: (870) 856-3240. Fax: (870) 856-4408. E-mail: hometownradio@centurytec.net Licensee: KOOU Inc. (acq 12-10-03). Rgnl. Network: Ark. Radio Net. Format: MOR. News staff: 2; News: 10 hrs wkly. Target aud: 25-60; female/professional. ♦James Bragg, gen mgr.

Harrisburg

KWHF(FM)— May 15, 1999: 95.9 mhz; 34 kw. Ant 489 ft TL: N35 47 42 W90 47 35. Hrs open: 24 Box 1737, Jonesboro, 72403-1737. Secondary address: 407 W. Parker Rd., Jonesboro 72404. Phone: (870) 932-8400. Fax: (870) 932-3814. E-mail: larryjames@959thebuzz.com Web Site: www.959thebuzz.com. Licensee: CC Licenses LLC. Group owner: Clear Channel Communications Inc. (acq 6-13-2002; $2.05 million with KNEA(AM) Jonesboro). Population served: 150,000 Law Firm: Dan Alpert. Format: Classic country. News staff: 2; News: 3 hrs wkly. Target aud: 28-65; affluent baby boomers who have spendable income. ♦Larry James, gen mgr.

Harrison

***KBPB(FM)**— 2001: 91.9 mhz; 5.5 kw. Ant 341 ft TL: N36 22 12 W93 13 23. Hrs open: 24 10795 Hwy. 65 N., Omaha, 72662. Secondary address: 920 Commerce Rd., Pine Bluff Phone: (870) 534-1013. Web Site: www.hereshelpnet.org. Licensee: New Life Evangelistic Center Inc. Format: Contemp Christian, Southern gospel. ♦Larry Rice, gen mgr; Carl Swift, stn mgr & chief of opns.

KCWD(FM)— 1982: 96.1 mhz; 3 kw. 295 ft TL: N36 16 36 W93 05 27. (CP: 8 kw, ant 1,191 ft.). Stereo. Hrs opn: 24 Box 850, 72601. Secondary address: 600 S. Pine 72601. Phone: (870) 741-1402. Fax: (870) 741-9702. Web Site: www.kcwd.com. Licensee: Harrison Radio Station Inc. Format: Classic rock. ♦Tom Arnold, gen mgr & chief of opns.

KHOZ(AM)— Sept 28, 1946: 900 khz; 1 kw-D. TL: N36 14 35 W93 06 43. Hrs open: 24 1111 Radio Ave., 72601. Phone: (870) 741-2302. Fax: (870) 741-3299. E-mail: khoz@alltel.com Web Site: www.khoz.com. Licensee: KHOZ LLC. (acq 6-16-2005; $3.7 million with co-located FM). Natl. Network: CBS. Wire Svc: AP Format: Talk, soft adult contemp. News staff: one; News: 15 hrs wkly. Target aud: General. Spec prog: Gospel 10 hrs wkly. ♦Charles Earls, CEO; Scott Earls, pres; Scottie Earls, gen mgr.

KHOZ-FM— Mar 25, 1963: 102.9 mhz; 100 kw. 981 ft TL: N36 26 11 W93 14 43. Stereo. 24 Phone: (870) 741-2301. E-mail: scottieearls@krzk.com Web Site: www.khoz.com. Natl. Network: CBS Radio. Format: Country. News staff: 2; News: 15 hrs wkly. Target aud: 25-54. ♦Jamie Cooleg, mus dir & news rptr; Bill Wilcox, pub affrs dir; Patty Eddings, traf mgr; Tom Parker, reporter; Brent Klein, disc jockey; Jerry Bowman, disc jockey; Jordan Snow, disc jockey; Scott Segraves, disc jockey.

Hatfield

KILX(FM)— 2001: 104.1 mhz; 28.5 kw. Ant 469 ft TL: N34 32 42 W94 18 21. Hrs open: 24 1600 S. Reine St., Mena, 71953. Phone: (479) 394-1450. Licensee: Ouachita Broadcasting Inc. (acq 3-8-99). Population served: 60,000 Format: Adult contemp. ♦Dwight Douglas, gen mgr & opns mgr.

Heber Springs

KAWW(AM)— July 15, 1967: 1370 khz; 1 kw-D. TL: N35 29 10 W92 02 05. Hrs open: 6 AM-sunset (2 hrs past) 111 North Spring St., Searcy, 72143. Phone: (501) 268-7123. Fax: (501) 279-2900. E-mail: jrrunyon@crainmedia.com Licensee: Crain Media Group LLC (group owner; acq 8-7-02; grpsl). Population served: 30,000 Rgnl. Network: Ark. Radio Net. Format: News/talk. News staff: one; News: one hr wkly. Target aud: 25-65. ♦Larry Crain, CEO; J.R. Runyon, gen mgr.

KAWW-FM— Sept 1, 1972: 100.7 mhz; 50 kw. Ant 328 ft TL: N35 27 26 W92 02 11. Stereo. Hrs opn: 24 111 N. Spring St., Searcy, 72143. Phone: (501) 268-7123. Fax: (501) 279-2900. E-mail: jrrunyon@crainmedia.com Licensee: Crain Media Group LLC (group owner; acq 8-7-02; grpsl). Population served: 150,000 Format: Main stream adult contemp. Target aud: 25-54. ♦Larry Crain, CEO; J.R. Runyon, gen mgr.

***KBMJ(FM)**— 2002: 89.5 mhz; 70 kw vert. Ant 735 ft TL: N35 44 00 W92 15 37. Hrs open: 24 Drawer 2440, Tupelo, MS, 38803. Phone: (662) 844-8888. Licensee: American Family Association. Group owner: American Family Radio Format: Christian. ♦Marvin Sanders, gen mgr; John Riley, progmg dir; Fred Jackson, news dir.

Helena

KFFA(AM)— Nov 19, 1941: 1360 khz; 1 kw-D, 90 w-N. TL: N34 31 39 W90 37 48. Hrs open: 24 Box 430, 1360 Radio Dr., 72342. Phone: (870) 338-8361. Phone: (870) 338-8331. Fax: (870) 338-8332. E-mail: kffa@arkansas.net Web Site: www.kffa.com. Licensee: Delta Broadcasting Inc. (acq 3-80; $445,000; FTR: 3-10-80). Population served: 65,000 Rgnl. Network: Prog Farm. Law Firm: Donald E. Martin. Format: Country. News: 25 hrs wkly. Target aud: 18-54. Spec prog: Farm 16 hrs, blues 8 hrs, Black 10 hrs, sports 15 hrs, gospel 4 hrs wkly. ♦Jim Howe, pres, gen sls mgr, mktg dir, adv VP & pub affrs dir; Rose Seaton, opns mgr & prom mgr; Louis Smith, mus dir; Jerry Campbell, engrg mgr; Nancy Howie, traf mgr.

KFFA-FM— 1972: 103.1 mhz; 13 kw. 318 ft TL: N34 31 39 W90 37 46. Stereo. 24 (Acq 5-84; grpsl; FTR: 5-7-84).65,000 Format: Adult contemp, sports. News: 4 hrs wkly. ♦Jim Howe, CEO, rsch dir & farm dir; Rose Seaton, progmg dir; Kacye Patton, traf mgr; Louis Smith, mus critic.

KJIW-FM— Jan 5, 1989: 94.5 mhz; 16 kw. 341 ft TL: N34 31 28 W90 35 47. Hrs open: 24 204 Moore St., Helena-West Helena, 72742. Phone: (870) 338-2700. E-mail: kjiwfm@ipa.net Licensee: Elijah Mondy Jr. (acq 1988). Format: gospel. ♦Elijah Mondy Jr., gen mgr; April Mondy, progmg dir & mus dir; Zipporah Mondy, mus dir.

Hope

KBYB(FM)— Dec 31, 1984: 101.7 mhz; 50 kw. Ant 492 ft TL: N33 41 20 W93 35 55. Stereo. Hrs opn: 24 615 W. Olive, Texarkana, TX, 75501. Phone: (903) 793-4671. Fax: (903) 792-4261. Licensee: Arklatex LLC. (group owner; acq 12-14-2006; grpsl). Format: Jammin' oldies. Target aud: 25-54; females. ♦Harold Sudbury, gen mgr; Alex Rain, stn mgr & progmg dir; Jayna Thomas, sls dir; Jay Calhoun, chief of engrg.

KHPA(FM)— Apr 21, 1977: 104.9 mhz; 3 kw. 298 ft TL: N33 43 10 W93 29 07. (CP: 6 kw, ant 328 ft. TL: N33 43 12 W93 29 11). Stereo. Hrs opn: Box 424, 71802. Secondary address: 1600 S. Elm 71801. Phone: (870) 777-8868. Phone: (870) 777-8869. Fax: (870) 777-8888. E-mail: supercountry105@yahoo.com Licensee: Newport Broadcasting Co. Group owner: Sudbury Svcs Inc. & Newport Broadcasting Co. Format: Country. ♦Robert Hill, gen mgr; Sonya Odom, sls dir; Kevin McKinnon, chief of engrg; Amanda Smith, traf mgr.

KXAR(AM)—Co-owned with KHPA(FM). Dec 12, 1947: 1490 khz; 690 w-U. TL: N33 41 20 W93 35 55. (Acq 8-26-99; $51,000).35,000 Format: Talk. Target aud: General; double income, stable, adult households.

Horseshoe Bend

KKIK(FM)— 2004: 106.5 mhz; 12 kw. Ant 476 ft TL: N36 15 22 W91 55 23. Hrs open: 920 Harrison St., Suite C, Batesville, 72503. Phone: (870) 793-4196. Fax: (870) 793-5222. Licensee: WRD Entertainment Inc. (group owner). Format: Oldies. ♦Preston Grace, gen mgr.

Hot Springs

***KALR(FM)**— May 1989: 91.5 mhz; 3 kw. Ant 485 ft TL: N34 37 31 W93 00 37. Hrs open: 24 2351 Sunset Blvd., Suite 170-218, Rocklin, CA, 95765. Phone: (916) 251-1600. Fax: (916) 251-1650. Licensee: Educational Media Foundation. (acq 6-28-2007; $275,000). Format: Christian. ♦Mike Novak, sr VP.

KBHS(AM)— Oct 6, 1966: 1420 khz; 5 kw-D, 87 w-N. TL: N34 27 19 W93 03 26. Hrs open: Box 21430, 71903. Secondary address: 208 Buena Vista Rd. 71902. Phone: (501) 525-1301. Fax: (501) 525-4344. E-mail: klaz@klaz.com Web Site: www.klaz.com. Licensee: Noalmark Broadcasting Corp. (group owner; Natl. Rep: Target Broadcast Sales. Format: Adult contemp. Target aud: 35 plus; upscale, high-income residents & business people. ♦Eddie Tarpley, gen mgr.

KLAZ(FM)—Co-owned with KBHS(AM). October 1971: 105.9 mhz; 95 kw. 994 ft TL: N34 30 19 W93 05 06. Stereo. 150,000 Format: Adult contemp. Target aud: 18-49.

***KLRO(FM)**— Mar 20, 1984: 90.1 mhz; 38 kw. Ant 971 ft TL: N34 30 18 W93 04 42. Stereo. Hrs opn: 24 2351 Sunset Blvd., Suite 170-218, Rocklin, CA, 95765. Phone: (916) 251-1600. Fax: (916) 251-1650. Web Site: www.klove.com. Licensee: Educational Media Foundation. (acq 9-24-2004; $1.2 million). Natl. Network: K-Love. Format: Christian contemp, relg. ♦Richard Jenkins, pres & gen mgr; Mike Novak, VP; Keith Whipple, dev dir; Eric Allen, natl sls mgr; David Pierce, progmg mgr; Ed Lenane, news dir; Sam Wallington, engrg dir; Karen Johnson, news rptr; Marya Morgan, news rptr; Richard Hunt, news rptr.

KPZA(AM)— Mar 10, 1953: 590 khz; 5 kw-D, 67 w-N. TL: N34 29 55 W92 58 45. Stereo. Hrs opn: Box 21430, 71903. Phone: (501) 525-4600. Fax: (501) 525-4344. Licensee: Noalmark Broadcasting Corp. (acq 12-13-2004; $140,000). Population served: 35,631 Natl. Network: USA. Format: Sp. ♦Eddie Tarpley, gen mgr.

KQUS-FM—Listing follows KZNG(AM).

KYDL(FM)— June 18, 1965: 96.7 mhz; 940 w. Ant 807 ft TL: N34 24 13 W93 07 14. (CP: 6 kw, ant 308 ft. TL: N34 29 43 W93 01 27). Stereo. Hrs opn: 24 125 Corporate Terr., 71913-7248. Phone: (501) 525-9700. Fax: (501) 525-9739. Web Site: www.star96fm.com. Licensee: US Stations LLC. Group owner: Powell Broadcasting (acq 2-1-2005; grpsl). Population served: 65,631 Format: Adult contemp. Target aud: 18 plus. ♦Gary Terrell, gen mgr; Craig Dale, opns dir & progmg dir; Bob Gipson, gen sls mgr.

KYXK(FM)—See Gurdon

KZNG(AM)— Jan 1, 1953: 1340 khz; 1 kw-U. TL: N34 29 43 W93 01 27. Hrs open: 24 125 Corporate Terr., 71913-7248. Phone: (501) 525-9700. Fax: (501) 525-9739. Web Site: www.kzng.net. Licensee: US Stations LLC. Group owner: Powell Broadcasting (acq 2-1-2005; grpsl). Population served: 138,000 Natl. Network: ABC. Format: News/talk. News staff: one; News: 10 hrs wkly. Target aud: 18 plus. ♦Ted Mahn, gen mgr; Larry LeBlanc, opns dir & opns mgr; Rick Austin, gen sls mgr; Tom Duke, chief of engrg; Melissa Walters, traf mgr.

KQUS-FM—Co-owned with KZNG(AM). Feb 7, 1969: 97.5 mhz; 100 kw. 860 ft TL: N34 24 11 W93 07 13. Stereo. 24 Web Site: www.kqus.net.128,000 Format: C&W. ♦Larry LeBlanc, progmg dir; Melissa Walters, traf mgr.

Hot Springs Village

KVRE(FM)— February 1994: 92.9 mhz; 25 kw. Ant 328 ft TL: N34 38 34 W93 04 08. Hrs open: 24 122 DeSoto Center Dr., 71909. Phone: (501) 922-5678. Phone: (501) 922-5880. Fax: (501) 922-6626. E-mail: kvre@kvre.com Licensee: Caddo Broadcasting Co. Format: Adult Standards (Music of Your Life). Target aud: 35 plus; general. ♦Polly Nichols, gen mgr; Alice Bates, opns dir; Cyrie Wright, gen sls mgr; Tom Nichols, opns dir & progmg dir; John Chapman, news dir.

Hoxie

KJLV(FM)— Jan 20, 1988: 105.3 mhz; 25 kw. 328 ft TL: N36 02 24 W90 59 11. Hrs open: 24 2351 Sunset Blvd., Suite 170-218, Rocklin, CA, 95765. Phone: (707) 528-9236. Fax: (707) 528-9246. Fax: (916) 251-1650. Web Site: www.klove.com. Licensee: Educational Media Foundation. Group owner: EMF Broadcasting (acq 11-1-01; $1.3 million with KJBR(FM) Marked Tree). Population served: 100,000 Format: Contemp Christian. ♦Richard Jenkins, pres & gen mgr; Mike Novak, VP; Keith Whipple, dev dir; David Pierce, progmg mgr; Ed Lenane, news dir; Sam Wallington, engrg dir; Karen Johnson, news rptr; Marya Morgan, news rptr; Richard Hunt, news rptr.

Humnoke

KVLO(FM)— 1996: 101.7 mhz; 6 kw. 100 ft TL: N34 32 58 W91 45 26. Hrs open: 700 Wellington Hill Rd., Little Rock, 72211. Phone: (501) 401-0200. Fax: (501) 401-0366. Licensee: The Last Bastion Station Trust LLC, as Trustee Group owner: Citadel Broadcasting Corp. (acq 6-12-2007; grpsl). Rgnl. Network: Simulcast of KARN Format: Gospel. ♦Jim Beard, VP, gen mgr & mktg mgr.

Huntsville

KAKS(FM)— 1955: 99.5 mhz; 13.5 kw. Ant 443 ft TL: N36 07 37 W93 51 57. Hrs open: 70 N. East St., Fayetteville, 72701. Phone: (479) 443-9960. Licensee: Davidson Media Station KREB-FM Licensee LLC. (acq 2-10-2005; $3.9 million with KCZZ(AM) Mission, KS). Population served: 315,000 Natl. Network: ABC. Format: Sp. Target aud: 25-54; general. ♦Peter W. Davidson, pres & engrg mgr; Steve Butler, gen mgr.

Jacksonville

KDJE(FM)— Sept 29, 1969: 100.3 mhz; 82.9 kw. 1,054 ft TL: N34 47 53 W92 29 33. Stereo. Hrs opn: 24 10800 Colonel Glenn Rd., Little Rock, 72204. Phone: (501) 217-5000. Fax: (501) 228-9547. E-mail: q100@q100fm.com Web Site: www.q1003theedge.com. Licensee: CC Licenses LLC. Group owner: Clear Channel Communications Inc. (acq 5-15-96; grpsl). Population served: 500,000 Natl. Rep: Clear Channel. Law Firm: Arent, Fox, Kintner, Plotkin & Kahn. Format: Alternative rock. News staff: one; News: 3 hrs wkly. Target aud: 18-49. ♦Don Pollnow, gen mgr; Casey Wagner, gen sls mgr; Jeff Peterson, opns dir & progmg dir.

Jonesboro

***KAOG(FM)**— 1999: 90.5 mhz; 1 kw. 243 ft TL: N35 53 06 W90 42 38. Hrs open: Box 2440, American Family Radio, Tupelo, MS, 38803. Phone: (662) 844-8888. Fax: (662) 842-6791. Web Site: www.afr.net. Licensee: American Family Association. Group owner: American Family Radio Format: Relg. ♦Marvin Sanders, gen mgr; John Riley, progmg dir; Joey Moody, chief of engrg.

***KASU(FM)**— May 17, 1957: 91.9 mhz; 100 kw. 689 ft TL: N35 53 27 W90 40 26. Stereo. Hrs opn: 24 Box 2160, Arkansas State Univ., 104 Cooley, State University, 72467. Phone: (870) 972-2200. Phone: (870) 972-3070. Fax: (870) 972-2997. E-mail: kasu@astate.edu Web Site: www.kasu.org. Licensee: Arkansas State University. Population served: 200,000 Natl. Network: PRI, NPR. Format: News, class, jazz. News staff: one; News: 45 hrs wkly. Target aud: General. Spec prog: New age, blues, folk 4 hrs, big band 2 hrs wkly. ♦Robert Franklin, stn mgr; June Taylor, opns mgr; Todd Rutledge, dev dir; Amy Davis, prom mgr; Marty Scarbrough, progmg dir.

KBTM(AM)—Listing follows KIYS(FM).

KEGI(FM)— Nov 21, 1986: 100.5 mhz; 38 kw. Ant 558 ft TL: N35 56 59 W90 39 58. Stereo. Hrs opn: 24 314 Union Ave., 72401. Phone: (870) 933-8800. Fax: (870) 933-0403. E-mail: trey@triplefm.com Web Site: eagle1005.com. Licensee: Saga Communications of Arkansas LLC. Group owner: Saga Communications Inc. (acq 11-8-2002; grpsl). Format: Classic hits. News staff: one; News: one hr wkly. Target aud: 18-49. ♦Trey Stafford, CEO, pres, CFO & gen mgr; Kevin Neathery, sls dir; Bill Pressly, progmg VP; Rick Christian, progmg dir; Al Simpson, chief of engrg; James Dean, traf mgr.

KFIN(FM)— Mar 4, 1974: 107.9 mhz; 100 kw. Ant 600 ft TL: N35 47 56 W90 44 31. Stereo. Hrs opn: 24 Box 1737, 72403-1737. Secondary address: 407 W. Parker Rd. 72404. Phone: (870) 932-8400. Fax: (870) 932-3814. Web Site: www.kfin.com. Licensee: Capstar TX L.P. Group owner: Clear Channel Communications Inc. (acq 1-18-01; grpsl). Population served: 400,000 Law Firm: Wiley, Rein & Fielding. Format: Country. News staff: one; News: 9 hrs wkly. Target aud: 25-54; broad demographics. Spec prog: Farm 13 hrs wkly. ♦Larry James, gen mgr.

KIYS(FM)— 1947: 101.9 mhz; 100 kw. 1,059 ft TL: N35 57 14 W90 41 41. Stereo. Hrs opn: 24 Box 1737, 72403-1737. Secondary address: 407 W. Parker Rd 72404. Phone: (870) 935-5598. Fax: (870) 932-3814. Web Site: www.1019kiysfm.com. Licensee: Capstar TX L.P. Group owner: Clear Channel Communications Inc. (acq 1-18-01; grpsl). Population served: 400,000 Law Firm: Fisher, Wayland, Cooper, Leader & Zaragoza. Format: CHR. Target aud: 18-49; middle to upper middle income. ♦Larry James, CEO & gen mgr; Katy Wiliamson, VP & sls dir; Duce Foreman, mktg dir & prom dir; Kevin Box, progmg dir; Troy Owens, engrg VP & chief of engrg; Janice Reed, traf mgr & farm dir.

KBTM(AM)—Co-owned with KIYS(FM). Mar 15, 1930: 1230 khz; 1 kw-U. TL: N35 50 27 W90 39 44.24 500,000 Format: News/talk. News staff: 2; News: 14 hrs wkly. Target aud: 45 plus; upscale adults. ♦Kevin Box, stn mgr & prom mgr; Janice Reid, traf mgr; Barbara Nelson, min affrs dir.

KJBX(FM)—Trumann, February 1991: 106.7 mhz; 6 kw. 328 ft TL: N35 44 51 W90 37 49. Stereo. Hrs opn: 24 314 Union Ave., 72401. Phone: (870) 933-8800. Fax: (870) 933-0403. E-mail: trey@triplefm.com Web Site: www.themix1067.com. Licensee: Saga Communications of Arkansas LLC. Group owner: Saga Communications Inc. (acq 11-8-02; grpsl). Population served: 120,000 Format: Adult contemp. News: 2 hrs wkly. Target aud: 25-54; women. ♦Bill Pressly, pres & stn mgr; Trey Stafford, pres, CFO & gen mgr; Kevin Neathery, sls VP & sls dir; Al Simpson, chief of engrg; James Dean, traf mgr.

KNEA(AM)— Sept 20, 1950: 970 khz; 1 kw-D, 41 w-N. TL: N35 51 17 W90 43 40. Hrs open: 24 Box 1737, 72403. Phone: (870) 932-8400. Fax: (870) 932-3814. Web Site: www.knea970.com. Licensee: CC Licenses LLC. Group owner: Clear Channel Communications Inc. (acq 6-13-2002; $2.05 million with KWHF(FM) Harrisburg). Population served: 500,000 Rgnl. Network: Ark. Radio Net. Format: Gospel. News staff: 5. Target aud: General. Spec prog: Farm 6 hrs wkly. ♦Larry James, gen mgr; Shawaqua Kelley, sls dir & gen sls mgr.

Judsonia

KVHU(FM)— 2006: 95.3 mhz; 14 kw. Ant 440 ft TL: N35 13 41 W91 29 19. Stereo. Hrs opn: 24 Box 10765, Searcy, 72149-0765. Phone: (501) 279-4886. Fax: (501) 279-5152. E-mail: kvhu@harding.edu Web Site: www.kvhu.net. Licensee: George S. Flinn Jr. Format: Classic hits. Target aud: 35+. ♦Dutch Hoggatt, gen mgr.

Kensett

KFXV(FM)— 2007: 105.7 mhz; 15 kw. Ant 426 ft TL: N35 17 20 W91 46 17. Hrs open: 401 S. Spring St., Searcy, 72143. Phone: (501) 268-9700. Licensee: Malvern Entertainment Corp. Format: Hot adult contemp. ♦Scott A. Gray, pres; Amber Carson, gen mgr.

Lake City

KDXY(FM)—Licensed to Lake City. See Paragould

Lake Village

KUUZ(FM)— July 30, 1977: 95.9 mhz; 20 kw. Ant 302 ft TL: N33 20 07 W91 07 33. Stereo. Hrs opn: 24 Box 262550, Baton Rouge, LA, 70826. Secondary address: 8919 World Ministry Ave., Baton Rouge, LA 70810. Phone: (225) 768-3688. Phone: (225) 768-8300. Fax: (225) 768-3729. E-mail: kawikfish@yahoo.com Web Site: www.jsm.org. Licensee: Family Worship Center Church Inc. (group owner; acq 6-12-02; $500,000). Population served: 100,000 Format: Relg. ♦David Whitelaw, COO; Jimmy Swaggart, pres; John Santiago, gen mgr & progmg mgr.

KZYQ(FM)— Dec 24, 1995: Stn currently dark. 103.5 mhz; 25 kw. Ant 328 ft TL: N33 17 04 W91 13 03. Stereo. Hrs opn: 24 M.R.S. Ventures Inc., Box 4942, Tyler, MS, 75712. Phone: (903) 592-4795. Licensee: M.R.S. Ventures Inc. (group owner; (acq 11-1-2003; grpsl). Population served: 135,000 Natl. Network: Jones Radio Networks. Law Firm: Wood, Maines & Brown. Target aud: 25-54; adults. ♦Gwendolyn Walker, gen mgr.

Lakeview

KKTZ(FM)— May 1, 1999: 93.5 mhz; 25 kw. 328 ft TL: N36 31 22 W92 40 08. Hrs open: 2352 Hwy. 62B, Mountain Home, 72653. Phone: (870) 492-6022. Fax: (870) 492-2137. E-mail: radio@mountainhome.com Web Site: www.twinlakesradio.com. Licensee: John M. Dowdy. Format: Hot adult contemp. Target aud: 24-25. ♦Morgan Dowdy, CEO, pres & gen mgr; Stewart Brunner, VP; Roger Lowery, stn mgr.

Little Rock

KAAY(AM)— Dec 20, 1924: 1090 khz; 50 kw-U, DA-N. TL: N34 46 20 W92 13 30. Hrs open: 700 Wellington Hills Rd., 72211. Phone: (501) 401-0200. Fax: (501) 401-0387. Web Site: www.1060kaay.com. Licensee: Citadel Broadcasting Co. Group owner: Citadel Broadcasting Corp. (acq 9-30-98; $5 million). Population served: 318,800 Natl. Network: USA. Law Firm: Latham & Watkins. Format: Relg, southern gospel. Spec prog: Sp 2 hrs wkly. ♦Joe Booker, opns mgr; John Scuderi, gen sls mgr; Jim Beard, mktg mgr.

***KABF(FM)**— Sept 31, 1984: 88.3 mhz; 91 kw. 777 ft TL: N34 47 31 W92 28 38. Stereo. Hrs opn: 24 2101 S. Main St., 72206. Phone: (501) 372-6119. Fax: (501) 376-3952. E-mail: kabf@acorn.org Licensee: Arkansas Broadcasting Foundation. Population served: 50,000 Format: Black, jazz, gospel, diversified. News staff: one; News: 12 hrs wkly. Target aud: General; low-moderate income & politically disenfranchised. Spec prog: Sp 10 hrs, folk 10 hrs, American Indian 3 hrs, bluegrass 6 hrs, Caribbean 4 hrs, talk 10 hrs wkly. ♦John Cain, stn mgr.

KABZ(FM)— 1967: 103.7 mhz; 100 kw. 1,510 ft TL: N34 47 55 W92 29 58. Stereo. Hrs opn: 24 2400 Cottondale Ln., 72202. Phone: (501) 661-1037. Fax: (501) 664-5871. Web Site: www.1037thebuzz.com. Licensee: Signal Media of Arkansas Inc. (acq 12-20-93; $2 million; FTR: 1-10-94). Population served: 993,600 Natl. Network: Fox Sports, Westwood One. Law Firm: Duane Morris LLP. Format: Talk. News staff: one; News: 10 hrs wkly. Target aud: 18-49. ♦Philip Jonsson, pres & gen mgr.

KARN(AM)— 1928: 920 khz; 5 kw-U, DA-N. TL: N34 46 20 W92 09 30. Hrs open: 24 700 Wellington Hills Rd., 72211. Phone: (501) 401-0200. Fax: (501) 401-0387. E-mail: karn@karnnewsradio.com Web Site: www.920karn.com. Licensee: Citadel Broadcasting Co. Group owner: Citadel Broadcasting Corp. (acq 8-27-97; grpsl). Population served: 480,000 Natl. Network: CBS. Rgnl. Network: Ark. Radio Net. Wire Svc: ESSA Weather Service Format: News/talk. News staff: 10; News: 28 hrs wkly. Target aud: 35-64. ♦Jim Beard, gen mgr & mktg mgr.

KDIS-FM— Aug 14, 1992: 99.5 mhz; 3 kw. Ant 312 ft TL: N34 45 58 W92 17 38. (CP: 6 kw). Hrs opn: 415 N. McKinley, Suite 610, 72205. Phone: (501) 663-3300. Fax: (501) 663-3723. Web Site: www.radiodisney.com. Licensee: Radio Disney Group LLC. Group owner: ABC Inc. (acq 5-30-03; $2.56 million). Format: Children. ♦Lynda Goodbar, gen mgr; John Campbell, stn mgr.

KDJE(FM)—See Jacksonville

KGHT(AM)—Sheridan, March 1982: 880 khz; 50 kw-D, 220 w-N. TL: N34 41 36 W92 18 21 (D), N34 18 21 W92 23 06 (N). Hrs open: 24 10000 Warden Rd., North Little Rock, 72120. Phone: (501) 985-0880. Fax: (501) 985-0260. Licensee: Metropolitan Radio Group Inc. (group owner; acq 1-97). Natl. Network: Salem Radio Network. Format: Southern gospel. Target aud: WF 35-64. ♦Larry Skinner, gen mgr & chief of opns; Paula Johnson, disc jockey; Ed Roupe, sls; Lloyd Denney, sls.

KIPR(FM)—See Pine Bluff

KJBN(AM)— 1946: 1050 khz; 1 kw-D, 19 w-N. TL: N34 45 57 W92 17 39. Hrs open: 1800 Maple St., Suite 300, North Little Rock, 72114. Phone: (501) 791-1000. Fax: (501) 791-7121. Licensee: Joshua Ministries and Community Development Corp. (acq 8-26-92; $250,000; FTR: 9-21-92). Population served: 60,040 Format: Contemp gospel music, teaching. Target aud: Career-oriented people. ♦James Smith, gen mgr, gen sls mgr & prom mgr.

KKPT(FM)— Oct 26, 1960: 94.1 mhz; 100 kw. 1,601 ft TL: N34 47 56 W92 29 41. Stereo. Hrs opn: 24 2400 Cottondale Ln., 72202. Phone: (501) 664-9410. Fax: (501) 664-5871. Web Site: www.kkpt.com. Licensee: Signal Media of Arkansas. (acq 4-30-85; $2.75 million; FTR: 3-11-85). Population served: 410,000 Natl. Rep: D & R Radio. Wire Svc: AP Format: Classic hits. News staff: one; News: one hr wkly. Target aud: 25-54; adults. ♦Philip Jonsson, pres; Ron Collar, gen mgr.

***KLRE-FM**— February 1973: 90.5 mhz; 40 kw. 265 ft TL: N34 40 29 W92 19 04. Stereo. Hrs opn: 24 2801 S. University Ave., 72204. Phone: (501) 569-8485. Fax: (501) 569-8488. Web Site: www.ualr.edu. Licensee: University of Arkansas. (acq 7-95). Population served: 1,000,000 Natl. Network: PRI, NPR. Law Firm: Cohn & Marks. Format: Class. Target aud: 35-54. ♦Ben Fry, gen mgr & stn mgr; Mary Waldo, dev dir.

KMJX(FM)—See Conway

KPZK(AM)— 1929: 1250 khz; 2.5 kw-D, 1.2 kw-N, DA-2. TL: N34 42 05 W92 13 02. Hrs open: 24 700 Wellington Hills Rd., 72211. Phone: (501) 401-0200. Fax: (501) 401-0366. Licensee: Citadel Broadcasting Co. Group owner: Citadel Broadcasting Corp. (acq 9-19-97; grpsl). Population served: 167,000 Natl. Rep: D & R Radio. Format: Gospel. News staff: one. Target aud: 50 plus. ♦Jim Beard, mktg mgr.

KSSN(FM)— 1966: 95.7 mhz; 92 kw. 1,663 ft TL: N34 47 57 W92 29 29. Stereo. Hrs opn: 24 10800 Colonel Glenn Rd., 72204. Phone: (501) 217-5000. Fax: (501) 228-9547. E-mail: kssn@cei.net Web Site: www.kssn.com. Licensee: CC Licenses LLC. Clear Channel Communications Inc. (acq 9-12-97; grpsl). Population served: 490,000 Natl. Rep: Clear Channel. Format: Contemp country. News staff: one; News: 2 hrs wkly. Target aud: 25-54. ♦Don Pollnow, gen mgr; Chad Heritage, opns dir & progmg dir; Casey Wagner, gen sls mgr.

KTUV(AM)— October 1956: 1440 khz; 5 kw-D, 240 w-N. TL: N34 42 46 W92 16 48. Hrs open: 24 723 W. Daisy Bates Dr., 72202. Phone: (501) 375-1440. Fax: (501) 244-9842. E-mail: kita1440@earthlink.net Licensee: Kita Inc. (acq 6-28-84; $675,000; FTR: 4-30-84). Population served: 450,000 Law Firm: Edmundson & Edmundson. Format: Sp. ♦Tom Rusk, pres & gen sls mgr.

***KUAR(FM)**— Sept 16, 1986: 89.1 mhz; 100 kw. 882 ft TL: N34 47 50 W92 29 26. Stereo. Hrs opn: 24 2801 S. University Ave., 72204. Phone: (501) 569-8485. Fax: (501) 569-8488. Web Site: www.ualr.edu. Licensee: Board of Trustees of the University of Arkansas. Natl. Network: NPR, PRI. Law Firm: Cohn & Marks. Format: News/talk, jazz. News staff: one; News: 86 hrs wkly. Target aud: 35-54. Spec prog: Folk 3 hrs wkly. ♦Ben Fry, gen mgr & stn mgr; Mary Waldo, dev dir.

KURB(FM)— July 7, 1972: 98.5 mhz; 99 kw. 1,286 ft TL: N34 47 56 W92 29 44. Stereo. Hrs opn: 24 700 Wellington Hills Rd., 72211. Phone: (501) 401-0200. Fax: (501) 401-0349. Web Site: www.b98.com. Licensee: Citadel Broadcasting Co. Group owner: Citadel Broadcasting Corp. Format: Hot adult contemp. ♦Jim Beard, gen sls mgr & mktg mgr.

Lonoke

KOLL-FM— June 1982: 106.3 mhz; 50 kw. Ant 492 ft TL: N34 46 30 W91 53 33. Hrs open: 400 Hardin Rd., Suite 150, Little Rock, 72211. Phone: (501) 219-1919. Fax: (501) 225-4610. E-mail: nick@koll1063.com Web Site: www.koll1063.com. Licensee: ABG Arkansas LLC. Group owner: Archway Broadcasting Group (acq 1-22-2003; $8 million with KHTE-FM England). Population served: 500,000 Format: Oldies. ♦Brad Hutcheson, gen mgr.

Lowell

KMXF(FM)— June 30, 1992: 101.9 mhz; 50 kw. 708 ft TL: N36 26 28 W93 58 22. Stereo. Hrs opn: 24 Box 8190, Fayetteville, 72703. Phone: (479) 442-0102. Fax: (479) 587-8255. Web Site: www.hotmix1019.com. Licensee: Capstar TX L.P. Group owner: Clear Channel Communications Inc. (acq 8-30-00; grpsl). Population served: 275,000 Law Firm: Dow, Lohnes & Albertson. Wire Svc: AP Format: CHR. News staff: one; News: one hr wkly. Target aud: 25-44. ♦Tony Beringer, gen mgr.

Magnolia

KVMA(AM)— April 1948: 630 khz; 1 kw-D. TL: N33 17 59 W93 13 57. Hrs open: Box 430, 71754. Secondary address: 131 S. Jackson 71753. Phone: (870) 234-5862. Fax: (870) 234-5865. E-mail: kvmakvmz@suddenlinkmail.com Web Site: www.magnoliaradio.com. Licensee: Noalmark Broadcasting Corp. (acq 8-1-2005; $165,000). Population served: 50,000 Rgnl. Network: Ark. Radio Net. Law Firm: Borsari & Paxson. Format: C&W. News: 20 hrs wkly. Target aud: General. Spec prog: Farm 2 hrs wkly. ♦William C. Nolan Jr., pres; Ed Aldorson, VP; Ken W. Sibley, gen mgr; Dan Gregory, opns dir.

KZHE(FM)—See Stamps

Malvern

KBOK(AM)— August 1951: 1310 khz; 1 kw-D. TL: N34 22 25 W92 49 52. Hrs open: Sunrise-sunset 302 S. Main St., 72104. Phone: (501) 332-6981. Phone: (501) 332-6982. Fax: (501) 332-6984. Web Site: www.hsnp.com. Licensee: Noalmark Broadcasting Corp. (group owner; acq 4-1-03; $62,500). Population served: 30,000 Rgnl. Network: Ark. Radio Net. Format: News, traditional country. News: 20 hrs wkly. Target aud: General. Spec prog: Talk 6 hrs, gospel 8 hrs wkly. ♦Malia Brown, gen mgr.

KKSP(FM)—Bryant, April 1989: 93.3 mhz; 5.6 kw. Ant 699 ft TL: N34 47 31 W92 28 38. Stereo. Hrs opn: 24 400 Hardin Rd., Suite 150, Little Rock, 72211. Phone: (501) 219-1919. Fax: (501) 225-4610. Web Site: www.spirit933.com. Licensee: ABG Arkansas LLC (acq 5-9-2003; $3.6 million). Population served: 250,000 Format: Contemp Christian. ♦Brad Hutcheson, gen mgr.

KLEZ(FM)— Apr 1, 1991: 101.5 mhz; 6 kw. Ant 322 ft TL: N34 28 24 W92 55 51. Stereo. Hrs opn: 24 208 Buena Vista Rd., Hot Springs, 71913. Phone: (501) 525-4600. Fax: (501) 525-4344. E-mail: pob@klaz.com Web Site: www.klez.com. Licensee: Noalmark Broadcasting Corp. (group owner; (acq 1-21-2003; $437,500). Population served: 120,000 Law Firm: Miller & Miller, P.C. Format: Easy lstng. News: 3 hrs wkly. Target aud: 35-60. ♦William C. Nolan Jr., pres; Eddie Tarpley, stn mgr.

Mammoth Spring

KALM(AM)—See Thayer, MO

KAMS(FM)— Jan 1, 1956: 95.1 mhz; 100 kw. 650 ft TL: N36 32 58 W91 33 05. Stereo. Hrs opn: 24 Box 193, 72554. Secondary address: N. Hwy. 63, Thayer, MO 65791. Phone: (417) 264-7211. Fax: (417) 264-7212. E-mail: kkountry@kkountry.com Web Site: www.kkountry.com. Licensee: Ozark Radio Network Inc. Population served: 360,000 Law Firm: Greg Skall. Format: Classic country. News staff: one; News: 11 hrs wkly. Target aud: General. ♦Shawn Neathery Marhefka, pres; Robert Eckman, gen mgr & opns VP; Mike Crase, opns mgr & progmg dir.

Marianna

KAKJ(FM)— 1994: 105.3 mhz; 6 kw. Ant 328 ft TL: N34 47 14 W90 46 03. Stereo. Hrs opn: 24
Rebroadcasts KCLT(FM) West Helena 90%.
Box 2870, West Helena, 72390. Phone: (870) 572-9506. Phone: (870) 633-9000. Fax: (870) 572-1845. E-mail: force2@sbcglobal.net Web Site: www.force2radio.com. Licensee: Raymond & L.T. Simes II. Population served: 1,200,000 Natl. Network: ABC. Rgnl. Network: Ark. Radio Net. Format: Black, urban contemp. News staff: one. Target aud: All ages. ♦L.T. Simes II, VP; Elaine Simes, stn mgr; Larry Evans, opns mgr; Raymond Simes, pres, gen mgr & gen sls mgr; Elaine Sims, mktg dir; Peter Turner, prom dir & progmg dir.

Marion

KXHT(FM)— February 1986: 107.1 mhz; 3 kw. 328 ft TL: N35 09 23 W90 05 46. (CP: 2.75 kw, ant 479 ft.). Hrs opn: 6080 Mt. Mariah Rd. Ext., Memphis, TN, 38115. Phone: (901) 375-9324. Fax: (901) 375-9331. E-mail: mail@1071.com Web Site: www.1701.com. Licensee: Flinn Broadcasting Corp. Format: Rap, hip hop. ♦Lloyd Hetzer, gen mgr; Duane Hargrove, stn mgr.

Marked Tree

KJBR(FM)— 1993: 93.7 mhz; 3 kw. Ant 288 ft TL: N35 32 22 W90 26 35. Hrs open: 24 2351 Sunset Blvd., Suite 170-218, Rocklin, CA, 95765. Phone: (916) 251-1600. Fax: (916) 251-1650. Web Site: www.klove.com. Licensee: Educational Media Foundation. Group owner: EMF Broadcasting (acq 11-1-01; $1.3 million with KJLV(FM) Hoxie). Population served: 250,000 Natl. Network: Air 1. Format: Contemp Christian. ♦Richard Jenkins, pres; Mike Novak, VP; Keith Whipple, dev dir; Eric Allen, natl sls mgr; David Pierce, progmg mgr; Ed Lenane, news dir; Sam Wallington, engrg dir; Karen Johnson, news rptr; Marya Morgan, news rptr; Richard Hunt, news rptr.

Marshall

KBCN-FM— Apr 25, 1983: 104.3 mhz; 100 kw. 820 ft TL: N35 52 17 W92 39 10. (CP: Ant 1,016 ft.). Stereo. Hrs opn: 24 100 Blue Bird St., Harrison, 72601. Phone: (870) 743-1157. Fax: (870) 743-1168. E-mail: kbcnradio@hotmail.com Licensee: Pearson Broadcasting of Marshall Inc. Group owner: Pearson Broadcasting (acq 4-30-93; $450,000; FTR: 5-24-93). Format: Country. ♦David R. Fransen, gen mgr.

KCGS(AM)— May 24, 1975: 960 khz; 5 kw-D. TL: N35 54 56 W92 38 20. Hrs open: 24 Box 1044, 72650. Secondary address: 208 Battle St. 72650. Phone: (870) 448-5567. Fax: (870) 448-5384. E-mail: kcgs@alltel.net Web Site: www.kcgs.com. Licensee: Southland Broadcasting Corp. (acq 8-20-2003). Population served: 71,550 Natl. Network: USA. Format: Bluegrass gospel. News staff: 2; News: 10 hrs wkly. Target aud: General. Spec prog: Bible answers live 7 hrs wkly. ♦Karl Leukert, gen mgr & opns VP; Ronald Woolsey, pres, dev VP & sls VP.

Marvell

***KVRN(FM)**— 1999: 90.7 mhz; 50 kw. 495 ft TL: N34 36 29 W90 58 47. Stereo. Hrs opn: 24 Box 2292, West Helena, 72390. Phone: (870) 572-1234. Fax: (870) 572-5515. E-mail: kvrnfm@sbcglobal.net Licensee: East Arkansas Educational Foundation. Format: Inspirational christian, CHR. ♦Patrick Roberson, pres; Patrick Roberson IV, gen mgr; Alan Crisp, opns dir; Patrick Roberson III, opns mgr; Patrick Robertson, pub affrs dir; Jerry Campbell, chief of engrg.

Maumelle

KHLR(FM)— 1971: 94.9 mhz; 96 kw. 1,843 ft TL: N34 26 31 W92 13 03. Stereo. Hrs opn: 24 10800 Colonel Glenn Rd., Little Rock, 72204. Phone: (501) 217-5000. Fax: (501) 228-9547. E-mail: cool95@cei.net Web Site: www.cool 95.com. Licensee: CC Licenses LLC. Group owner: Clear Channel Communications Inc. (acq 9-12-97; grpsl). Natl. Rep: D & R Radio. Format: Adult contemp. News staff: one; News: 5 hrs wkly. Target aud: 25-54; baby boomers. ♦Don Pollnow, gen mgr; Jeff Peterson, opns dir; Casey Wagner, gen sls mgr; Sonny Victory, progmg dir.

KWLR(FM)— 1998: 96.9 mhz; 4.6 kw. 377 ft TL: N34 53 33 W92 24 50. Hrs open: 5700 West Oaks Blvd., Rocklin, CA, 95765. Phone: (707) 528-9236. Fax: (707) 528-9246. E-mail: kwlrword97@aol.com Web Site: www.klove.com. Licensee: Flinn Broadcasting Corp. Format: Contemp inspirational.

Mayflower

***KJSM-FM**—Augusta, Aug 27, 1979: 97.7 mhz; 100 kw. Ant 620 ft TL: N35 10 36 W91 23 49. Stereo. Hrs opn: Box 262550, Baton Rouge, LA, 70826. Secondary address: 8919 World Ministry Ave., Baton Rouge, LA 70810. Phone: (225) 768-3688. Phone: (225) 768-8300. E-mail: kawikfish@yahoo.com Web Site: www.jsm.org. Licensee: Family Worship Center Church Inc. (group owner; acq 3-4-2003; $2.75 million). Population served: 55,000 Format: Christian. ♦David Whitelaw, COO & gen mgr; Jimmy Swaggart, pres; John Santiago, progmg dir.

McGehee

KVSA(AM)— June 29, 1953: 1220 khz; 1 kw-D, 54 w-N. TL: N33 33 39 W91 23 06. Hrs open: 6 AM-6:30 PM Box 110, Hwy. 65, 71654. Phone: (870) 222-4200. Phone: (870) 538-5200. Fax: (870) 538-3389. E-mail: kvsa1220@yahoo.com Licensee: Southeast Arkansas Broadcasters Inc. Population served: 60,000 Natl. Rep: Keystone (unwired net). Format: Div. News staff: 2; News: 10 hrs wkly. Spec prog: Farm 5 hrs wkly. ♦Dale Jones, progmg dir, sls & mktg; Joyce Kinney, pres, gen mgr, sls & mktg.

Melbourne

***KLRM(FM)**—Not on air, target date: unknown: 90.3 mhz; 1 kw. Ant 735 ft TL: N36 04 42 W92 01 23. Hrs open: 5700 West Oaks Blvd., Rocklin, CA, 95765. Phone: (916) 251-1600. Fax: (916) 251-1650. Licensee: Educational Media Foundation. (acq 3-23-2007; grpsl). ♦Richard Jenkins, pres.

Mena

KENA(AM)— July 1950: 1450 khz; 1 kw-U. TL: N34 34 23 W94 14 55. Hrs open: 24 Box 1450, 71953. Secondary address: 1600 S. Reine St. 71953. Phone: (479) 394-1450. Licensee: Ouachita Broadcasting Inc. (acq 1-14-99; $750,000 with co-located FM). Population served: 5,000 Format: Gospel. News staff: one; News: 10 hrs wkly. Target aud: 18 plus; industrial & agricultural workers, retirees, tourists & professionals. ♦Dwight Douglas, gen mgr, stn mgr & mktg dir; Sue Canner, prom dir & mus dir; Matt Stone, news dir, local news ed & news rptr.

KENA-FM— 1969: 102.1 mhz; 12.5 kw. Ant 469 ft TL: N34 32 42 W94 18 21.24 60,000 Natl. Network: ABC. Format: Country. News staff: one; News: 6 hrs wkly. ♦Dwight Douglas, stn mgr, opns mgr & progmg mgr; Bevona Williams, traf mgr; Matt Stone, local news ed & news rptr; Curt Teasdale, disc jockey; Mark Hobson, disc jockey.

KQOR(FM)— 2001: 105.3 mhz; 8.1 kw. 564 ft TL: N34 36 31 W94 14 19. Hrs open: Box 1450, 71953. Phone: (479) 394-1450. Licensee: Ouachita Broadcasting Inc. (acq 3-8-99). Format: Oldies. ♦Dwight Douglas, gen mgr & opns mgr.

KTTG(FM)— December 1994: 96.3 mhz; 100 kw. 1,314 ft TL: N34 36 40 W94 16 20. Stereo. Hrs opn: 24 2937 Hwy. 71 N., 71953. Phone: (479) 394-6198. Fax: (479) 784-7290. Web Site: www.espnradio.com. Licensee: Pearson Broadcasting of Mena Inc. Group owner: Pearson Broadcasting (acq 1995; $175,000). Population served: 150,000 Natl. Network: ESPN Radio. Natl. Rep: ABC Radio Sales. Format: Sports, talk. News staff: one. Target aud: 18-49. ♦Max Pearson, CEO & chmn; Bruce Hale, CFO; Tommy Craft, gen mgr; Jason Wade, progmg dir.

Monticello

KGPQ(FM)— May 1, 1997: 99.9 mhz; 25 kw. 328 ft TL: N33 43 25 W91 48 28. Hrs open: 279 Midway Rt., 71655. Phone: (870) 367-8525. Fax: (870) 367-9564. E-mail: crn@ccc-cable.net Licensee: Pines Broadcasting Inc. (acq 3-14-2007; grpsl). Format: Adult contemp. ♦Jimmy Sledge, pres & gen mgr.

KHBM(AM)— April 1955: 1430 khz; 1 kw-D, 30 w-N. TL: N33 36 18 W91 47 14. Hrs open: 24 279 Midway Rte, AZ, 71655-8605. Phone: (870) 367-6854. Fax: (870) 367-9564. E-mail: crn@ccc-cable.net Licensee: Pines Broadcasting Inc. (group owner; (acq 3-14-2007; grpsl). Population served: 76,388 Rgnl. Network: Ark. Radio Net. Format: Music of Your Life. News staff: one; News: 10 hrs wkly. Target aud: General. ♦Jimmy Sledge, pres; Bonnie Ellis, gen mgr.

KHBM-FM— Sept 1, 1967: 93.7 mhz; 6 kw. 341 ft TL: N33 36 18 W91 47 14. (CP: 93.7 mhz, 25 kw). Stereo. Hrs opn: 24 279 Midway Rt., 71655. Phone: (870) 367-6854. Fax: (870) 367-9564. E-mail: crn@ccc-cable.net Licensee: Pines Broadcasting Inc. (acq 3-14-2007; grpsl). Rgnl. Network: Ark. Radio Net. Format: Classic Hits. News staff: one; News: 8 hrs wkly. ♦Jimmy Sledge, pres & gen mgr.

KXSA-FM—See Dermott

Morrilton

KVOM(AM)— Dec 25, 1952: 800 khz; 250 w-D, 42 w-N. TL: N35 09 32 W92 46 13. Hrs open: Box 541, 1835 Hwy. 113 W., 72110. Phone:

(501) 354-2484. Fax: (501) 354-5629. E-mail: kvom@kvom.com Web Site: www.kvom.com. Licensee: MMA License LLC. Group owner: MAX Media L.L.C. (acq 4-21-03; grpsl). Population served: 125,000 Natl. Network: AP Radio. Natl. Rep: Christal. Format: News/talk, sports. News staff: news progmg 70 hrs wkly News: 2;. Target aud: General. ♦Rich Moellers, gen mgr.

KVOM-FM— 1981: 101.7 mhz; 6 kw. 226 ft TL: N35 09 32 W92 46 13. Stereo. 24 Web Site: www.kvom.com. Natl. Rep: Christal. Format: Country. News staff: news progmg 5 hrs wkly News: 2;. Target aud: General.

Mountain Home

***KCMH(FM)**— June 28, 1988: 91.5 mhz; 26 kw. Ant 472 ft TL: N36 16 17 W92 25 20. Stereo. Hrs opn: 24 126 S. Church St., 72653. Phone: (870) 425-2525. Fax: (870) 424-2626. E-mail: lorra@kcmhradio.com Web Site: www.kcmhradio.com. Licensee: Christian Broadcasting Group of Mountain Home Inc. Natl. Network: Moody, USA. Format: Educ, relg. News: 9 hrs wkly. Target aud: General. ♦Carl Albright, pres; Lorra Queen, stn mgr & progmg dir; Michael Coolidge, VP & opns VP.

KOMT(FM)— Oct 25, 1985: 107.5 mhz; 100 kw. Ant 1,017 ft TL: N36 29 13 W92 29 39. Hrs open: 24 2352 Hwy. 62 B, 72653. Phone: (870) 492-6022. Fax: (870) 492-2137. E-mail: radio@mountainhome.com Web Site: www.twinlakesradio.com. Licensee: MAC Partners. Natl. Network: ABC. Format: Adult contemp. Target aud: 25-54; females. ♦Morgan Dowdy, CEO; Stewart Brunner, VP & gen mgr; Roger Lowery, stn mgr & news dir.

KPFM(FM)— June 6, 1984: 105.5 mhz; 19 kw. Ant 797 ft TL: N36 29 13 W92 29 39. Stereo. Hrs opn: 24 2352 Hwy. 62 B, 72653. Phone: (870) 492-6022. Fax: (870) 492-2137. E-mail: radio@mountainhome.com Web Site: www.twinlakesradio.com. Licensee: Mountain Home Radio Station Inc. Natl. Network: ABC. Format: Country. News staff: one; News: 15 hrs wkly. Target aud: 24-54. ♦Morgan Dowdy, CEO; Stewart Brunner, VP, VP & gen mgr; Roger Lowery, stn mgr.

KTLO(AM)— May 30, 1953: 1240 khz; 1 kw-U. TL: N36 20 43 W92 23 40. Hrs open: 24 Box 2010, 72654. Secondary address: 620 Hwy 5 N. 72654. Phone: (870) 425-3101. Fax: (870) 424-4314. Web Site: www.ktlo.com. Licensee: KTLO L.L.C. (group owner; (acq 5-1-91). Population served: 45,000 Format: Country. News staff: 3; News: 11 hrs wkly. Target aud: 18-55; general. ♦Bob Knight, gen mgr.

KTLO-FM— Jan 11, 1971: 97.9 mhz; 30 kw. Ant 636 ft TL: N36 20 55 W92 23 59. Stereo. Format: Easy lstng, MOR. Target aud: 40 plus.

Mountain Pine

KLXQ(FM)— 1996: 101.9 mhz; 6 kw. 328 ft Stereo. Hrs opn: 125 Corporate Terr., Hot Springs, 71913. Phone: (501) 525-9700. Fax: (501) 525-9739. Web Site: www.1019therocket.com. Licensee: US Stations LLC. Group owner: Powell Broadcasting (acq 2-1-2005; grpsl). Format: Classic rock. Target aud: 25-54. ♦Gary Terrell, gen mgr; Craig Dale, opns dir & progmg dir; Bob Gipson, gen sls mgr.

Mountain View

KWOZ(FM)— Dec 1, 1981: 103.3 mhz; 100 kw. 987 ft TL: N35 47 06 W91 57 44. Stereo. Hrs opn: 920 Harrison St., Suite C, Batesville, 72503. Phone: (870) 793-4196. Fax: (870) 793-5222. E-mail: arkansas103@hotmail.com Web Site: www.arkansas103.com. Licensee: WRD Entertainment Inc. (group owner). Natl. Network: ABC. Law Firm: Letcher, Heald, & Hildreth, P.L.C. Wire Svc: AP Format: C&W. News staff: 2; News: 3 hrs wkly. Target aud: 18-54. ♦Don Johnson, gen mgr & stn mgr; Matt Johnson, sls dir.

Murfreesboro

KMTB(FM)— May 18, 1983: 99.5 mhz; 20.5 kw. 358 ft TL: N34 05 44 W93 41 31. Stereo. Hrs opn: 24 1513 S. 4th, Nashville, 71852. Phone: (870) 845-3601. Fax: (870) 845-3680. Licensee: ARKLATEX Radio Inc. (group owner; acq 8-28-2001; grpsl). Format: Country. ♦Brent Pinkerton, gen mgr; Scott Dunson, opns dir.

Nashville

KBHC(AM)— May 1959: 1260 khz; 500 w-D. TL: N33 55 45 W93 51 01. Hrs open: 1513 S. 4th St., 71852. Phone: (870) 845-3601. Fax: (870) 845-3680. Licensee: ARKLATEX Radio Inc. (group owner; (acq 8-23-2001; grpsl). Population served: 4,016 Rgnl. Network: Ark. Radio Net. Natl. Rep: Keystone (unwired net). Format: MOR. ♦Brent Pickerton, gen mgr; Kendra Harper, mus dir; Mandy McLaughlin, news dir; Scott Dunson, disc jockey.

KNAS(FM)—Co-owned with KBHC(AM). Feb 14, 1977: 105.5 mhz; 3 kw. 85 ft TL: N33 55 45 W93 51 01. Stereo. Format: Oldies. ♦Scott Dunson, disc jockey.

***KNLL(FM)**—Not on air, target date: unknown: 90.5 mhz; 24 kw vert. Ant 554 ft TL: N33 45 16 W93 52 29. Hrs open: Drawer 2440, Tupelo, MS, 38801-2440. Phone: (662) 844-8888. Fax: (662) 842-6791. Web Site: www.afr.net. Licensee: American Family Association. ♦Marvin Sanders, gen mgr.

KSSW(FM)— 2003: 96.9 mhz; 6 kw. Ant 328 ft TL: N34 00 41 W93 52 03. Hrs open: 1513 S. 4th St., 71852. Phone: (870) 845-3601. Fax: (870) 845-3680. Licensee: Family Worship Center Church Inc. Group owner: Sudbury Services Inc. (acq 9-15-2005; $400,000). Format: Adult contemp. ♦Brent Pinkerton, gen mgr.

Newark

***KLLN(FM)**— Jan 1, 1985: 90.9 mhz; 4 kw. 456 ft TL: N35 43 25 W91 26 40. Hrs open: 1502 N. Hill St., 72562. Phone: (870) 799-8969. Phone: (870) 799-8691. Fax: (870) 799-8647. E-mail: kllnfm@yahoo.com Web Site: www.klln.fm. Licensee: Newark Public School. Format: Southern gospel. ♦Fred Ahlborn, gen mgr.

Newport

KNBY(AM)— Oct 12, 1949: 1280 khz; 1 kw-D, 87 w-N. TL: N35 36 38 W91 15 02. Hrs open: 24 Box 768, 72112. Secondary address: 2025 McCarty Dr. 72112. Phone: (870) 523-5891. Fax: (870) 523-2967. E-mail: legends@rivercountry967.com Licensee: Newport Broadcasting Co. Group owner: Sudbury Svcs. Inc. & Newport Broadcasting Co. Population served: 21,700 Format: News/talk info. ♦Harold Sudbury, pres; Dale Turner, gen mgr & news dir; Doug Holt, progmg dir & progmg mgr.

KOKR(FM)—Co-owned with KNBY(AM). Sept 1, 1966: 96.7 mhz; 35 kw. 548 ft TL: N35 29 16 W91 26 13. Stereo. 45,000 Format: Country.

North Crossett

KWLT(FM)— May 1, 1995: 102.7 mhz; 25 kw. Ant 328 ft TL: N33 12 58 W91 55 42. Stereo. Hrs opn: 24 Box 697, 117 E. Wellfield Rd., Crosset, 71635. Phone: (870) 364-2181. Fax: (870) 364-2183. E-mail: kagh@alltel.net Web Site: www.crossettradio.com. Licensee: South Ark Broadcasting Inc. Population served: 85,000 Natl. Network: ABC. Format: Classic rock. News: 2 hrs wkly. Target aud: General. ♦Kevin Medlin, pres & gen mgr; Barry Medlin, gen sls mgr; Russell Miller, progmg dir.

North Little Rock

KDXE(AM)— May 9, 1957: 1380 khz; 5 kw-D, 2.5 kw-N, DA-2. TL: N34 52 49 W92 14 01. Hrs open: 24 2902 E. Kiehl Ave., Suite 1D, Sherwood, 72120. Phone: (501) 221-1380. Fax: (501) 217-0016. E-mail: kdxe1380@sbcglobal.net Licensee: Simmons Austin, LS LLC. (acq 3-8-2006; $350,000). Natl. Network: ESPN Radio. Law Firm: Dow, Lohnes & Albertson. Format: Sports. News staff: 1; News: 2 hrs. Target aud: 25-54; men. Spec prog: 3 hrs wkly. ♦Albert Phipps, pres; Arlen Horn, gen mgr; Lee Malcolm, opns VP; Eric Kailin, gen sls mgr; Mark Hill, engrg VP; David Graves, engr.

KWBF-FM— 1995: 101.1 mhz; 6 kw. Ant 328 ft TL: N34 49 52 W92 19 18. Stereo. Hrs opn: 24 1 Shackleford Dr., Suite 400, Little Rock, 72211. Phone: (501) 219-2400. Fax: (501) 604-8004. Licensee: Flinn Broadcasting Corp. Population served: 500,000 Format: AOR. News staff: one; News: 15 hrs wkly. Target aud: 35-54. Spec prog: News 15 hrs wkly.

Ola

KARV-FM— Jan 1, 1998: 101.3 mhz; 850 w. 856 ft TL: N34 59 46 W93 13 22. Hrs open: 201 W. 2nd, Russellville, 72801. Phone: (479) 968-1184. Fax: (479) 967-5278. E-mail: karv-kyel@yahoo.com Licensee: KERM Inc. (group owner) Format: News/talk. News staff: 2; News: 20 hrs wkly. ♦Kermit Womack, pres; Chris Womack, gen mgr.

Osceola

KOSE(AM)—Wilson, Oct 11, 1949: 860 khz; 1 kw-D, 21 w-N. TL: N35 41 03 W89 58 57. Hrs open: Box 989, Blytheville, 72316. Phone: (870) 762-2093. Fax: (870) 763-8459. Licensee: Newport Broadcasting Co. (acq 1996). Population served: 1,500,000 Rgnl. Network: Ark. Radio Net. Natl. Rep: Roslin. Format: Southern gospel. Target aud: 24-55; middle-class, blue/white collar workers. Spec prog: Black 6 hrs, farm 5 hrs wkly. ♦Tom Hill, chief of engrg.

KQDD(FM)—Co-owned with KOSE(AM). Sept 1, 1996: 107.3 mhz; 3 kw. 223 ft Format: Classic hits.

Ozark

KDYN(AM)— Feb 5, 1969: 1540 khz; 500 w-D. TL: N35 29 16 W93 48 43. Hrs open: Sunrise-sunset Box 1086, Puddin Ridge Rd., 72949. Phone: (479) 667-4567. Fax: (479) 667-5214. E-mail: kdyn@centurytel.net Web Site: www.realcountryonline.com. Licensee: Ozark Communications Inc. (acq 9-15-85). Population served: 300,000 Rgnl. Network: Ark. Radio Net. Format: Country. News staff: one; News: 20 hrs wkly. Target aud: General. ♦Marc Dietz, pres, gen mgr, sls dir, mktg mgr & progmg dir.

KDYN-FM— Oct 2, 1980: 96.7 mhz; 10 kw. Ant 485 ft TL: N35 29 09.11 W93 53 29.49. Stereo. 24

Pangburn

KSMD(FM)— Nov 2, 2003: 99.1 mhz; 25 kw. Ant 328 ft TL: N35 23 43 W91 44 17. Hrs open: 24 111 N. Spring St., Searcy, 72143. Phone: (501) 268-7123. Fax: (501) 279-2900. E-mail: production @heartarkansas.com Licensee: Crain Media Group LLC (group owner; acq 10-15-02; $180,000 for CP). Format: News/talk. ♦J.R. Runyon, opns dir & progmg dir; Dave Clark, chief of engrg.

Paragould

KDRS(AM)— Jan 1, 1947: 1490 khz; 1 kw-U. TL: N36 02 56 W90 27 44. Hrs open: 24 400 Tower Dr., 72450. Phone: (870) 236-7627. Fax: (870) 239-4583. E-mail: dina@kdrs.com Web Site: www.kdrs.com. Licensee: MOR Media Inc. (acq 7-2-2002; $500,000 with co-located FM). Population served: 50,000 Law Firm: Fletcher, Heald & Hildreth. Format: Sports news radio. News: 7 hrs wkly. ♦Dina Mason, pres & gen mgr; Brian Osborn, opns mgr.

KDRS-FM— Mar 5, 1983: 107.1 mhz; 3 kw. 410 ft TL: N36 01 48 W90 35 49. Stereo. 24 Web Site: www.kdrs.com. Law Firm: Fletcher, Heald & Hildreth. Format: Adult contemp. Target aud: 18-44.

KDXY(FM)—Lake City, Oct 4, 1971: 104.9 mhz; 25 kw. 480 ft TL: N35 49 29 W90 33 54. Stereo. Hrs opn: 24 314 Union Ave., Jonesboro, 72401. Phone: (870) 933-8800. Fax: (870) 933-0403. E-mail: trey@triplefm.com Web Site: www.thefox1049.com. Licensee: Saga Communications of Arkansas LLC. Group owner: Saga Communications Inc. (acq 11-8-02; grpsl). Population served: 120,000 Format: Country. News staff: one; News: 6 hrs wkly. Target aud: 25-49. ♦Bill Pressly, pres & stn mgr; Trey Stafford, CEO, pres, CFO & gen mgr; Kevin Neathery, sls VP & sls dir; Al Simpson, chief of engrg; James Dean, traf mgr.

Paris

KERX(FM)— May 1981: 95.3 mhz; 50 kw. Ant 459 ft TL: N35 17 13 W94 02 51. Hrs open: 1912 Church St., Barling, 72923. Phone: (479) 484-7285. Fax: (479) 784-7390. E-mail: kconroy@x953rocks.com Web Site: www.thex953rocks.com. Licensee: Pearson Broadcasting of Paris Inc. Group owner: Pearson Broadcasting (acq 10-18-93; $42,000; FTR: 11-8-93). Population served: 187,000 Wire Svc: AP Format: AOR. Target aud: 18-44; men & women. Spec prog: Blues 3 hrs wkly. ♦Bruce Hale, CFO; Tommy Craft, gen mgr; Ray Miller, opns dir & progmg dir.

Piggott

KBOA-FM— Oct 15, 1983: 105.5 mhz; 6 kw. Ant 298 ft TL: N36 19 50 W90 07 24. Stereo. Hrs opn: 24 Box 509, 1303 Southwest Dr., Kennett, MO, 63857. Phone: (573) 888-4616. Fax: (573) 888-4890. E-mail: ktmo@i1.net Licensee: Pollack Broadcasting Co. (group owner; acq 9-25-98; $450,000 with KBOA(AM) Kennett, MO). Population served: 150,000 Format: Adult standards. News staff: one; News: 5

hrs wkly. Target aud: 18-55; young adults, young professionals, farmers. Spec prog: Farm 5 hrs wkly. ♦William H. Pollack, pres; Perry Jones, gen mgr.

Pine Bluff

***KCAT(AM)**— April 1963: 1340 khz; 1 kw-U. TL: N34 12 47 W92 01 53. Hrs open: 1207 W. 6th, 71601-3993. Phone: (870) 534-5001. Licensee: Monday Burke Smith Broadcasting Network (acq 7-15-2004; $150,000). Population served: 185,000 Format: Gospel. Target aud: 18-55. ♦Darren Smith, gen sls mgr; Elijah Mondy, gen mgr & progmg dir; Belinda Mondy, chief of engrg & traf mgr.

KCLA(AM)— Jan 16, 1947: 1400 khz; 1 kw-U. TL: N34 11 33 W92 02 42. Hrs open: 920 Commerce Rd., 71601-7605. Phone: (870) 534-8978. Fax: (870) 534-8984. Licensee: M.R.S. Ventures Inc. (group owner; (acq 4-29-2003; grpsl). Population served: 57,389 Format: News/talk. News staff: 2. Target aud: 35 plus; lower to middle income. ♦J. Don Russell, pres; Dawn Deane, gen mgr; Craig Eastham, sls VP & news dir; Floyd Donald, mus dir.

KZYP(FM)—Co-owned with KCLA(AM). Nov 1, 1984: 99.3 mhz; 3 kw. Ant 200 ft TL: N34 11 33 W92 02 42. Stereo. Format: Urban contemp. Target aud: 25-45; middle to upper income.

KIPR(FM)— 1963: 92.3 mhz; 100 kw. Ant 938 ft TL: N34 22 12 W92 10 07. Stereo. Hrs opn: 700 Wellington Hills Rd., Little Rock, 72211. Phone: (501) 401-0200. Fax: (501) 401-0366. Web Site: www.power923.com. Licensee: Citadel Broadcasting Co. Group owner: Citadel Broadcasting Corp. (acq 7-29-97; grpsl). Population served: 200,000 Format: Urban contemp. Target aud: 18-44. ♦Jim Beard, gen mgr & mktg mgr.

KOTN(AM)— Mar 12, 1934: 1490 khz; 1 kw-U. TL: N34 13 15 W91 58 20. Hrs open: 24 920 Commerce Rd., 71601. Phone: (870) 534-8911. Phone: (501) 534-8978. Fax: (870) 534-8984. Licensee: M.R.S. Ventures Inc. (group owner; (acq 4-29-2003; $350,000). Population served: 60,000 Natl. Network: Westwood One. Format: Adult contemp, news/talk, sports. News staff: one; News: 7 hrs wkly. Target aud: 25-54. ♦Andy Hodges, gen mgr & mktg VP.

KPBQ-FM— Dec 23, 1991: 101.3 mhz; 25 kw. Ant 328 ft TL: N34 15 13 W92 03 58. Stereo. Hrs opn: 24 920 Commerce Rd., 71601. Phone: (870) 534-8911. Fax: (870) 534-8984. E-mail: delta4radio@netscape.net Licensee: M.R.S. Ventures Inc. (group owner; (acq 4-29-2003; grpsl). Population served: 150,000 Natl. Network: ABC. Format: Country. Target aud: 12-60. Spec prog: Farm 2 hrs wkly. ♦Andy Hodges, gen mgr.

***KUAP(FM)**— 1995: 89.7 mhz; 6 kw. 285 ft TL: N34 14 33 W92 01 02. Hrs open: 1200 N. University Dr., Suite 4948, Theatre Masscom, 71601. Phone: (870) 575-8272. Fax: (870) 575-4666. Web Site: www.uapb.edu. Licensee: Board of Trustees of Univ. of Arkansas. Format: Smooth jazz. ♦Finley Hill, stn mgr.

Pocahontas

KPOC(AM)— Nov 15, 1950: 1420 khz; 1 kw-D, 118 w-N. TL: N36 16 38 W90 57 16. Hrs open: 24 Box 508, 72455. Secondary address: One Radio Dr. 72455. Phone: (870) 892-5234. Fax: (870) 892-5235. Licensee: Combined Media Group Inc. (group owner; (acq 12-20-2001; $410,000 with co-located FM). Population served: 6,800 Rgnl. Network: Ark. Radio Net. Format: Light adult contemp. News staff: one. Target aud: 25-54; general. Spec prog: Farm 10 hrs wkly. ♦Jamie Ward, gen sls mgr; Timothy Scott, pres, gen mgr & progmg dir; Larry Caldwell, chief of engrg.

KPOC-FM— Apr 25, 1969: 104.1 mhz; 6 kw. Ant 144 ft TL: N36 16 38 W90 57 16.24 One Radio Dr., 72455. 17,000

Prairie Grove

KYNF(FM)— November 1999: 94.9 mhz; 21 kw. 761 ft TL: N35 51 00 W94 23 00. Stereo. Hrs opn: 24 24 E. Meadow, Suite 1, Fayetteville, 72701. Phone: (479) 521-5566. Fax: (479) 521-0751. Web Site: www.y949.com. Licensee: Cumulus Licensing Corp. Group owner: Cumulus Media Inc. (acq 3-12-01; $2 million). Population served: 1,250,000 Natl. Network: USA. Format: Adult contemp. News: 2 hrs wkly. Target aud: 30 plus; listeners who like positive, inspirational progmg. ♦Joe Conway, gen mgr.

Prescott

KHPA(FM)—See Hope

KTPA(AM)— Dec 1, 1959: 1370 khz; 1 kw-D, 49 w-N. TL: N36 20 04 W94 10 41. Hrs open: Box 424, Hope, 71802. Phone: (870) 777-8868. Phone: (870) 777-8869. Fax: (870) 777-8888. Licensee: Newport Broadcasting Co. Group owner: Sudbury Services Inc. & Newport Broadcasting Co. (acq 5-14-66). Population served: 3,921 Rgnl. Network: Ark. Radio Net. Format: Gospel. ♦Robert Hill, gen mgr; Sonya Odom, stn mgr & sls dir; John McCoy, progmg dir.

Rogers

KAMO-FM— 1971: 94.3 mhz; 5.2 kw. 709 ft TL: N36 26 30 W93 58 26. Stereo. Hrs opn: 24 24 E. Meadow St., Suite 1, Fayetteville, 72701. Phone: (479) 521-5566. Fax: (479) 521 4968. Web Site: www.us94.com. Licensee: Cumulus Licensing Corp. Group owner: Cumulus Media Inc. (acq 12-10-98; grpsl). Format: Oldies. ♦Joe Conway, gen mgr.

KFFK(AM)— Sept 16, 1954: 1390 khz; 5 kw-D, 30 w-N. TL: N36 23 18 W94 11 34. Hrs open: 24 Butler Broadcasting LLC, 1780 Holly St., Fayetteville, 72701. Phone: (479) 582-3776. Fax: (479) 571-0995. Licensee: Butler Broadcasting Co. LLC (acq 10-13-99; grpsl). Population served: 315,000 Natl. Network: ABC. Format: Oldies, talk. Target aud: 25-54. ♦Steve Butler, pres; Steve Bulter, gen mgr; Katrice Summerland, gen sls mgr; Dave Jackson, progmg dir.

KURM(AM)— Nov 9, 1979: 790 khz; 5 kw-D, 500 w-N, DA-N. TL: N36 18 10 W94 06 47. Hrs open: 5 AM-11 PM 113 E. New Hope Rd., 72758. Phone: (479) 633-0790. Fax: (479) 631-9711. E-mail: kurm@kurm.net Licensee: KERM Inc. (group owner) Population served: 250,000 Natl. Network: CBS. Law Firm: Dan Alpert Law Firm. Format: Var/div. News staff: 2; News: 15 hrs wkly. Target aud: +35. Spec prog: Farm 10 hrs wkly. ♦Kermit Womack, pres & gen mgr; Diane Womack, sls VP & chief of engrg.

Russellville

KARV(AM)— Feb 25, 1947: 610 khz; 1 kw-D, 500 w-N, DA-2. TL: N35 17 56 W93 09 09. Hrs open: 24 201 W. 2nd, 72801. Phone: (479) 968-1184. Fax: (479) 967-5278. E-mail: karv-kyet@yahoo.com Licensee: KERM Inc. (group owner; acq 10-22-92; $250,000; FTR: 11-23-92). Population served: 70,000 Natl. Network: CBS. Rgnl. Network: Ark. Radio Net. Format: New/talk, sports. News staff: 4; News: 38 hrs wkly. Target aud: 35 plus; affluent adults. Spec prog: Farm 5 hrs wkly. ♦Chris Womack, gen mgr.

***KMTC(FM)**— June 1987: 91.1 mhz; 360 w. 62 ft TL: N35 18 11 W93 08 42. Hrs open: 24 Box 570, 72811-0570. Secondary address: 305 Lake Front Dr. 72802. Phone: (479) 967-7400. Fax: (479) 967-7894. Licensee: Russellville Educational Broadcasting Foundation. Population served: 30,000 Natl. Network: USA. Format: Christian contemp. News: one hr wkly. Target aud: 18-55; Christian. ♦Tom Underhill, CEO; Debbie Bewley, gen mgr; Melissa Krueger, stn mgr & progmg dir.

KWKK(FM)— Sept 29, 1985: 100.9 mhz; 6 kw. 295 ft TL: N35 17 37 W93 10 39. Stereo. Hrs opn: 24 Box 10310, 72812. Phone: (479) 968-6816. Fax: (479) 968-2946. E-mail: traffic@rivervalleyradio.com Web Site: www.rivervalleyradio.com. Licensee: MMA License LLC. Group owner: MAX Media L.L.C. (acq 4-21-03; grpsl). Natl. Rep: Christal. Wire Svc: AP Format: Adult contemp. News staff: one; News: 10 hrs wkly. Target aud: 18-49; young adults. Spec prog: Arkansas Tech University Sports. ♦Rich Mollers, gen mgr; Rhonda Dilbeck, sls dir; Ken Eubanks, progmg dir; Johnny Story, news dir.

***KXRJ(FM)**— Apr 3, 1989: 91.9 mhz; 100 w. -92 ft TL: N35 17 47 W93 08 18. Hrs open: 24 Arkansas Tech Univ., Hwy. 7 N., 72801. Phone: (479) 964-0806. Phone: (479) 964-3282. Fax: (479) 498-6024. Web Site: www.broadcast.atu.edu. Licensee: Arkansas Tech University. Format: Div, Jazz. News: 10 hrs wkly. Target aud: General. Spec prog: Educ, jazz 15 hrs wkly. ♦George Cotton, chief of engrg.

Salem

KCAB(AM)—See Dardanelle

KHOM(FM)— September 1977: 100.9 mhz; 50 kw. Ant 492 ft TL: N36 35 38 W91 40 03. Stereo. Hrs opn: Box 107, West Plains, MO, 65775. Phone: (417) 255-0427. Fax: (417) 255-2907. E-mail: khom@khom.net Web Site: www.khom.net. Licensee: Mountain Lakes Broadcasting Corp. (acq 12-14-99). Population served: 75000 Natl. Network: ABC. Rgnl. Network: Ark. Radio Net. Format: Traditional country. News staff: one. Target aud: 35-54; adults. ♦Bob Knight, gen mgr; John Thomason, gen mgr & stn mgr.

KSAR(FM)—See Thayer, MO

Searcy

KAPZ(AM)—See Bald Knob

KCNY(FM)—See Bald Knob

KWCK(AM)— Aug 25, 1951: 1300 khz; 5 kw-D, 30 w-N. TL: N35 15 27 W91 43 49. Hrs open: Box 1300, 72145. Secondary address: 111 N. Spring St. 72143. Phone: (501) 268-7123. Fax: (501) 279-2900. Licensee: Crain Media Group LLC. (group owner; (acq 10-21-2002; grpsl). Population served: 13,650 Format: Talk. Target aud: 25-64. Spec prog: Farm 10 hrs wkly. ♦Bill Weaver, gen mgr.

KWCK-FM— October 1973: 99.9 mhz; 50 kw. 492 ft TL: N35 26 50 W91 56 52. Stereo. 24 Format: Country. News staff: one; News: 3 hrs wkly. Target aud: 18-49.

Sheridan

***KANX(FM)**— 1999: 91.1 mhz; 16.5 kw. 522 ft TL: N34 17 26 W92 29 36. (CP: 40 kw). Hrs opn: Box 2440, American Family Radio, Tupelo, MS, 38803. Phone: (662) 844-8888. Fax: (662) 842-6791. Web Site: www.afr.net. Licensee: American Family Association. Group owner: American Family Radio Format: Relg. ♦Marvin Sanders, gen mgr; John Riley, progmg dir.

KARN-FM— Nov 1, 1984: 102.9 mhz; 50 kw. 488 ft TL: N34 25 08 W92 22 17. Stereo. Hrs opn: 24 700 Wellington Hills Rd., Little Rock, 72211. Phone: (501) 401-0200. Fax: (501) 401-0387. Web Site: www.920karn.com. Licensee: The Last Bastion Station Trust LLC, as Trustee Group owner: Citadel Broadcasting Corp. (acq 6-12-2007; grpsl). Population served: 248,000 Format: News, talk. News staff: one; News: 2 hrs wkly. Target aud: 35-64. ♦Jim Beard, gen mgr & mktg mgr; John Scuderi, gen sls mgr; Dave Elswick, progmg dir.

KGHT(AM)—Licensed to Sheridan. See Little Rock

Sherwood

KMTL(AM)— Oct 31, 1983: 760 khz; 10 kw-D. TL: N34 49 34 W92 12 19. Hrs open: Box 6460, North Little Rock, 72124. Phone: (501) 835-1554. Licensee: George V. Domerese. Format: Relg. ♦George Domerese, gen mgr & progmg dir; Tom Rusk, chief of engrg.

KOKY(FM)— 1994: 102.1 mhz; 4.1 kw. 387 ft TL: N34 44 38 W92 16 32. Hrs open: 700 Wellington Hills Rd., Little Rock, 72211. Phone: (501) 401-0200. Fax: (501) 401-0366. Web Site: www.koky.com. Licensee: Citadel Broadcasting Co. Group owner: Citadel Broadcasting Corp. (acq 10-23-97; grpsl). Population served: 525,000 Law Firm: Cohn & Marks. Format: Adult contemp, soul. ♦Jim Beard, gen mgr & mktg mgr.

Siloam Springs

***KLRC(FM)**— Oct 1, 1981: 101.1 mhz; 3.1 kw. 459 ft TL: N36 11 25 W94 33 55. Stereo. Hrs opn: John Brown Univ., 2000 W. University, 72761. Phone: (479) 524-7101. Phone: (877) KLRC-101. Fax: (479) 524-7451. E-mail: klrc@klrc.edu Web Site: www.klrc.com.klrc.com Licensee: John Brown University. Population served: 309,000 Natl.

Network: USA. Wire Svc: UPI Format: Christian, relg. News: 2 hrs wkly. Target aud: 25-49. ♦Charles W. Pollard, pres & chief of engrg; Sean Sawatzky, gen mgr; Mark Michaels, progmg dir.

KMCK(FM)— 1947: 105.7 mhz; 100 kw. 476 ft TL: N36 11 07 W94 17 49. Stereo. Hrs opn: 24 E. Meadow St., Suite 1, Fayetteville, 72701. Phone: (479) 521-5566. Fax: (479) 521-0751. Web Site: www.power1057.com. Licensee: Cumulus Licensing Corp. Group owner: Cumulus Media Inc. (acq 12-10-98; grpsl). Population served: 300,000 Natl. Rep: Christal. Law Firm: Richard Hayes. Format: CHR. Target aud: 18-49; contemp adults. ♦Joe Conway, gen mgr.

KUOA(AM)— Apr 12, 1923: 1290 khz; 5 kw-D. TL: N36 11 25 W94 33 55. Hrs open: 24 Box 870, 72761. Secondary address: 1175 Hiway 412 W., AZ 72761. Phone: (479) 524-6572. Fax: (479) 524-6578. Web Site: www.kuoa.com. Licensee: 1290 Radio Inc. (acq 12-1-2005; $236,700). Population served: 330,000 Natl. Network: Fox News Radio. Law Firm: Fletcher, Heald & Hildreth. Wire Svc: AP Format: Var. News: 12 hrs wkly. Target aud: 25-60. ♦Galen O. Gilbert, pres; Galen Gilbert, gen mgr & gen sls mgr; Jesse Gilbert, VP, gen mgr & news dir.

Springdale

KXNA(FM)— Sept 19, 1968: 104.9 mhz; 1 kw. 479 ft TL: N36 10 48 W94 05 07. (CP: 2.75 kw). Stereo. Hrs opn: Bulter Broadcasting LLC, 1780 Holly St., Fayetteville, 72703. Phone: (479) 582-3776. Fax: (479) 571-0995. Licensee: Bulter Broadcasting LLC. (acq 10-13-99; grpsl). Population served: 35,200 Natl. Rep: Christal. Format: Alternative. ♦Steve Butler, gen mgr & mus dir.

KYNG(AM)— July 15, 1966: 1590 khz; 2.5 kw-D, 58 w-N. TL: N36 12 21 W94 07 11. Hrs open: 24 E. Meadow St., Suite 1, Fayetteville, 72701. Phone: (479) 521-5566. Fax: (479) 521-0751. Licensee: Cumulus Licensing Corp. Group owner: Cumulus Media Inc. (acq 12-10-98; grpsl). Population served: 16,783 Law Firm: Richard Hayes. Format: Sp. Target aud: 18-54. ♦Joe Conway, gen mgr.

Stamps

KZHE(FM)— October 1980: 100.5 mhz; 50 kw. 500 ft TL: N33 26 01 W93 27 49. Stereo. Hrs opn: 24 406 W. Union St., Magnolia, 71753-3708. Phone: (870) 234-7790. Fax: (870) 234-7791. E-mail: kzhe@kzhe.com Web Site: www.kzhe.com. Licensee: A-1 Communications Inc. (acq 5-20-92; $85,000; FTR: 6-8-92). Format: Classic country. Target aud: 25-54. Spec prog: Gospel 8 hrs wkly. ♦Troy Alphin, pres; Sharon Alphin, VP; Dave Sehon, gen mgr & opns mgr.

Stuttgart

KWAK(AM)— May 15, 1948: 1240 khz; 1 kw-U. TL: N34 29 27 W91 33 45. Hrs open: 24 Box 910, 1818 S. Buerkle, 72160. Phone: (870) 673-1595. Fax: (870) 673-8445. E-mail: country97.3@cwnwisp.net Licensee: Arkansas County Broadcasters Inc. (group owner). Population served: 25,000 Rgnl. Network: Prog Farm. Format: Loc news, ESPN radio. News staff: one. Target aud: General. Spec prog: Farm 6 hrs wkly. ♦Bobby Caldwell, pres; Scott Siler, stn mgr, gen sls mgr & progmg dir; Sandi Levey, mus dir & women's int ed; Johnathan Reaves, news dir; Sandy Levey, traf mgr.

KWAK-FM— Dec 15, 1987: 105.5 mhz; 3 kw. 325 ft TL: N34 25 52 W91 26 08. Stereo. 24 25,000 Natl. Network: ABC. Format: Oldies. News: one. ♦Scott Siler, gen mgr; Johnathan Reeves, news dir.

Texarkana

KCMC(AM)—See Texarkana, TX

KFYX(FM)— June 11, 1968: 107.1 mhz; 2.9 kw. Ant 479 ft TL: N33 25 45 W94 07 11. Stereo. Hrs opn: 24 615 Olive St., TX, 75501. Phone: (903) 793-4671. Fax: (903) 792-4261. Licensee: ArkLaTex LLC. (group owner; (acq 12-14-2006; grpsl). Population served: 150,000 Natl. Network: ABC. Format: Traditional country. News staff: one; News: 3 hrs wkly. Spec prog: Farm 2 hrs, gospel 2 hrs wkly. ♦Harold Sudbury, gen mgr.

KHTA(FM)—See Wake Village TX

***KKLT(FM)**— 2004: 89.3 mhz; 1 w horiz, 5.7 kw vert. Ant 505 ft TL: N33 23 36 W93 51 34. (CP: 1 w horiz, 23 kw vert). Hrs opn:
Rebroadcasts KLVR(FM) Santa Rosa, CA 100%.
2351 Sunset Blvd., Suite 170-218, Rocklin, CA, 95765. Phone: (916) 251-1600. Fax: (916) 251-1650. Web Site: www.klove.com. Licensee: Educational Media Foundation. (acq 1-11-2005; $125,000 for CP). Natl. Network: K-Love. Format: Christian. ♦Richard Jenkins, pres; Mike Novak, VP; Keith Whipple, dev dir; Eric Allen, natl sls mgr; David Pierce, progmg mgr; Ed Lenane, news dir; Sam Wallington, engrg dir; Karen Johnson, news rptr; Marya Morgan, news rptr; Richard Hunt, news rptr.

KKYR-FM—See Texarkana, TX

KOSY(AM)— Nov 15, 1951: 790 khz; 1 kw-D, 500 w-N, DA-N. TL: N33 22 30 W94 01 00. Hrs open: 24 2324 Arkansas Blvd., 71854. Phone: (870) 772-3771. Fax: (870) 772-0364. Web Site: www.kkyr.com. Licensee: GAP Broadcasting Texarkana License LLC. Group owner: Clear Channel Communications Inc. (acq 8-3-2007; grpsl). Population served: 150,000 Format: Modern country. News staff: one. Target aud: General. ♦Ron Bird, gen mgr.

KYGL(FM)—Co-owned with KOSY(AM). 1995: 106.3 mhz; 3 kw. 328 ft TL: N33 22 39 W93 56 38. Format: Classic rock.

KTFS(AM)—See Texarkana, TX

KTOY(FM)— 1993: 104.7 mhz; 3 kw. Ant 390 ft TL: N33 27 25 W94 10 59. Hrs open: 24 615 Olive St., TX, 75501. Phone: (903) 794-5869. Fax: (903) 793-1577. E-mail: ktoy1047@aol.com Web Site: www.hitsandoldies.com. Licensee: Jo-Al Broadcasting Inc. (acq 12-14-2006; grpsl). Format: Urban. ♦Al Davis, gen mgr; Emmie Gamble, stn mgr; Vincent Gamble, chief of engrg; Rodney Davis, sls.

Trumann

KJBX(FM)—Licensed to Trumann. See Jonesboro

Turrell

KKLV(FM)— Sept 1, 1999: 94.7 mhz; 6 kw. 328 ft TL: N35 18 04 W90 19 34. Hrs open: 2351 Sunset Blvd., Suuite 170-218, Rockin, CA, 95765. Phone: (916) 251-1600. Fax: (916) 251-1650. Web Site: www.klove.com. Licensee: Educational Media Foundation. Group owner: EMF Broadcasting (acq 10-20-2000; grpsl). Population served: 1,000,000 Natl. Network: K-Love. Format: Contemp Christian. ♦Richard Jenkins, pres; Lloyd Parker, gen mgr; Keith Whipple, dev dir; Mike Novak, progmg VP; David Pierce, progmg dir; Ed Lenane, news dir & news dir; Sam Wallington, engrg dir.

Van Buren

KBBQ-FM— May 22, 1983: 102.7 mhz; 17 kw. Ant 574 ft TL: N35 26 51 W94 21 54. Stereo. Hrs opn: 3104 S. 70th St., Fort Smith, 72903. Phone: (479) 452-0681. Fax: (479) 452-0873. Licensee: Cumulus Licensing Corp. Group owner: Cumulus Media Inc. (acq 8-99; $1.15 million). Format: Rock. ♦Smitty O'Loughlin, gen mgr.

KHGG(AM)— Nov 24, 1958: 1580 khz; 1 kw-D, 45 w-N. TL: N35 25 58 W94 19 47. Hrs open: 24 Box 908, Fort Smith, 72902. Phone: (479) 288-1047. Fax: (479) 785-2638. E-mail: koolproduction@sbcglobal.net Web Site: www.fortsmithradiogroup.com. Licensee: Pharis Broadcasting Inc. (group owner; acq 9-20-93; $110,000; FTR: 10-11-93). Population served: 200,000 Natl. Network: Fox Sports. Rgnl. Network: Ark. Radio Net. Natl. Rep: Commercial Media Sales. Law Firm: Irwin, Campbell and Tannenwold. Format: Sports talk. Target aud: General. Spec prog: Univ of Ark Football, mens/womens basketball, baseball, Van Buren high school football, basketball; UA Fort Smith men's/women's basketball, KC Royals baseball. ♦William Pharis, CEO & pres; Karen Pharis, exec VP & gen mgr; Ernie Witt, progmg dir.

***KLFS(FM)**— 2004: 90.3 mhz; 2.4 kw vert. Ant 256 ft TL: N35 23 37 W94 33 07. Stereo. Hrs opn: 24 2351 Sunset Blvd., Suite 170-218, Rocklin, CA, 95765. Phone: (916) 251-1600. Fax: (916) 251-1650. E-mail: klove@klove.com Web Site: www.klove.com. Licensee: Educational Media Foundation. Group owner: EMF Broadcasting. Population served: 160,000 Natl. Network: K-Love. Law Firm: Shaw Pittman. Format: Chrisitan. News staff: 3. Target aud: 25-44; Judeo Chrisitan, female. ♦Richard Jenkins, pres; Mike Novak, VP; Keith Whipple, dev dir; David Pierce, progmg mgr; Ed Lenane, news dir; Sam Wallington, engrg dir; Karen Johnson, news rptr; Marya Morgan, news rptr; Richard Hunt, news rptr.

KOAI(AM)— Sept 6, 1979: 1060 khz; 500 w-D, DA. TL: N35 25 36 W94 18 11. Hrs open: 3104 S. 70th St., Fort Smith, 72903. Phone: (479) 452-0681. Fax: (479) 452-0873. Web Site: www.fortsmithradio.com. Licensee: Cumulus Licensing Corp. Format: Sp. ♦Smitty O'Loughlin, gen mgr.

Viola

KSMZ(FM)— 2007: 94.3 mhz; 8.1 kw. Ant 571 ft TL: N36 19 30 W91 58 41. Hrs open: 223 Russell St., Mountain Home, 72653. Phone: (870) 425-4971. Fax: (870) 425-0943. Licensee: MJFM LLC. Format: Adult hits. ♦Norm Allen, gen mgr.

Waldo

KVMZ(FM)— 2002: 99.1 mhz; 6 kw. Ant 328 ft TL: N33 24 17 W93 12 07. Stereo. Hrs opn: Box 430, Magnolia, 71754. Phone: (870) 234-9901. Fax: (870) 234-5865. E-mail: kvmakvmz@suddenlinkmail.com Web Site: www.magnoliaradio.com. Licensee: Noalmark Broadcasting Corp. (acq 8-1-2005; $430,000). Format: Today's Country. ♦Ken Sibley, gen mgr.

Waldron

KHGG-FM— May 18, 1982: 103.1 mhz; 6.1 kw. Ant 1,351 ft TL: N34 58 44 W93 56 42. Stereo. Hrs opn:
Rebroadcasts KHGG(AM) 100%.
Box 908, Fort Smith, 72902. Phone: (479) 288-1047. Fax: (479) 288-0942. Web Site: www.fortsmithradiogroup.com/sportshog103.htm. Licensee: Pharis Broadcasting Inc. (group owner; (acq 6-1-2003; $360,000). Natl. Network: Fox Sports. Natl. Rep: Commercial Media Sales. Rgnl rep: BRI Law Firm: Irwin, Campbell & Tannenwald. Format: Sports, talk. News staff: one; News: 10 hrs wkly. Target aud: 25-54. ♦William Pharis, pres; Karen Pharis, gen mgr & stn mgr; Ernie Witt Jr., opns dir.

Walnut Ridge

KRLW(AM)— June 29, 1951: 1320 khz; 1 kw-D. TL: N36 03 58 W90 56 24. Hrs open: 12 1 Radio Dr., Pocohantas, 72455. Phone: (870) 886-6666. Fax: (870) 886-5719. E-mail: krlw@nex.net Licensee: Combined Media Group Inc. (group owner; acq 7-25-01; with co-located FM). Population served: 18,000 Natl. Network: CBS. Rgnl. Network: Ark. Radio Net. Format: Oldies. ♦Tim Scott, pres & gen mgr.

KRLW-FM— Mar 27, 1977: 106.3 mhz; 3 kw. 328 ft TL: N36 03 58 W90 56 24. Stereo. 24 20,000 Natl. Network: CBS. Format: Country.

Warren

KWRF(AM)— August 1953: 860 khz; 250 w-D, 55 w-N. TL: N33 37 59 W92 03 51. Hrs open: 24 1255 N. Myrtle, 71671. Phone: (870) 226-2653. Phone: (870) 226-2654. Fax: (870) 226-3039. Licensee: Pines Broadcasting Inc. (acq 4-12-91; $125,000 with co-located FM; FTR: 5-6-91) Population served: 31,258 Rgnl. Network: Ark. Radio Net. Format: C&W. News staff: one; News: 10 hrs wkly. Target aud: General. Spec prog: Gospel 8 hrs wkly. ♦Jimmy Sledge, pres, gen mgr, gen sls mgr, mus dir, news dir & chief of engrg; Gwen Sledge, opns VP; Richard Garrison, disc jockey.

KWRF-FM— June 21, 1976: 105.5 mhz; 3 kw. Ant 250 ft TL: N33 37 59 W92 03 51.24 31,258 ♦Gwen Sledge, exec VP; Judy Moore, women's int ed; Allen Weise, disc jockey; Richard Garrison, disc jockey.

West Helena

KCLT(FM)— Dec 17, 1984: 104.9 mhz; 3 kw. 328 ft TL: N34 30 56 W90 40 13. Stereo. Hrs opn: 24 Box 2870, 72390. Secondary address: 700 Dr. Martin Luther King Dr., Suite 1 72390. Phone: (870) 572-9506. Fax: (870) 572-1845. E-mail: force2@sbcglobal.net Web Site: www.force2radio.com. Licensee: West Helena Broadcasters Inc. (acq 8-8-84). Population served: 1,200,000 Format: Adult urban contemp. News staff: one; News: one hr wkly. Target aud: 25-54; general, mainly African-Americans. Spec prog: Gospel 15 hrs wkly. ♦Raymond Simes, pres & gen mgr; Elaine Sims, stn mgr; Larry Evans, opns mgr.

KFFA-FM—See Helena

West Memphis

KQPN(AM)— Dec 1, 1961: 730 khz; 250 w-U, DA-N. TL: N35 08 31 W90 08 05. Hrs open: 24 791 Walnut Knoll Ln. 3E., Cordova, TN, 38018. Phone: (901) 751-1550. Fax: (901) 751-1654. E-mail: ffhammond@klove.com Licensee: Simmons Austin, LS LLC. Group owner: K-Love Radio Network (acq 6-14-2006; $2 million). Population served: 1,000,000 Format: Contemp Christian music. Target aud:

25-54; family types. ♦G. Craig Hanson, pres; Keith Whipple, VP & gen mgr; Lloyd Parker, gen sls mgr; Eric Allen, natl sls mgr & prom dir; Frank Hammond, rgnl sls mgr; Sam Wallington, chief of engrg.

White Hall

KTRN(FM)— November 1997: 104.5 mhz; 3 kw. 289 ft TL: N34 13 13 W92 04 37. Hrs open: 24 2215 E. Harding, Suite 7, Pine Bluff, 71601. Phone: (870) 536-5876 (on air). Phone: (870) 536 3282 (office). Fax: (870) 536-3475. Licensee: Bayou Broadcasting Inc. Format: Soft rock. Target aud: Women; 20 & up. ♦Vickie Hooker, gen mgr & opns mgr.

Wilson

KOSE(AM)—Licensed to Wilson. See Osceola

Wrightsville

KLAL(FM)— March 1992: 107.7 mhz; 100 kw. Ant 741 ft TL: N34 36 34 W92 14 14. Hrs open: 700 Wellington Hills Rd., Suite 920, Little Rock, 72211. Phone: (501) 401-0200. Fax: (501) 401-0349. Web Site: www.alice1077.com. Licensee: Citadel Broadcasting Co. Group owner: Citadel Broadcasting Corp. (acq 9-4-97). Format: Hot adult contemp. ♦Jim Beard, gen mgr & mktg mgr.

Wynne

KWYN(AM)— Sept 28, 1956: 1400 khz; 1 kw-U. TL: N35 15 21 W90 47 49. Hrs open: 24 Box 789, 2758 Hwy. 64, 72396. Phone: (870) 238-8141. Phone: (870) 238-8142. Fax: (870) 238-5997. E-mail: kwyn@ipa.net Licensee: East Arkansas Broadcasters Inc. Population served: 21,000 Natl. Network: CBS. Format: Talk, C&W, info. Target aud: General. Spec prog: Farm 6 hrs wkly. ♦Bobby Caldwell, CEO & gen mgr; Lance Daniels, sls dir; Jim Alexander, chief of engrg; Lindell Staggs, news dir & local news ed.

KWYN-FM— May 15, 1969: 92.5 mhz; 35 kw. 328 ft TL: N35 11 59 W90 43 23. Stereo. 24 Format: Country. ♦Lindell Staggs, local news ed; Bobby Caldwell, farm dir.

Yellville

KCTT-FM— 1986: 101.7 mhz; 2.45 kw. 331 ft TL: N36 15 39 W92 41 42. Stereo. Hrs opn: Box 2010, Mountain Home, 72654. Phone: (870) 449-4001. Phone: (870) 425-3101. Fax: (870) 424-4314. Web Site: www.ktlo.com. Licensee: KTLO L.L.C. (group owner; (acq 5-29-98; $215,000). Natl. Network: ABC. Format: Oldies. Spec prog: Folk 10 hrs wkly. ♦Bob Knight, CEO & gen mgr; Danny Ward, stn mgr; Brad Haworth, opns VP.

California

Adelanto

KAJL(FM)— 1959: 92.7 mhz; 280 w. Ant 1,473 ft TL: N34 36 44 W117 17 27. Stereo. Hrs opn: 24
Rebroadcasts KHJL (FM) Thousand Oaks.
99 Long Ct., Suite 200, Thousand Oaks, 91360. Phone: (805) 497-8511. Fax: (805) 497-8514. Web Site: www.927jillfm.com. Licensee: Amaturo Group of L.A. Ltd. Group owner: Amaturo Group Ltd. (acq 1-6-93; $3.25 million; FTR: 2-1-93). Population served: 2,100,000 Wire Svc: Metro Weather Service Inc. Format: Adult contemp. Target aud: 25-54. ♦Robert J. Christy, gen mgr & sls dir.

Alameda

KNGY(FM)—Licensed to Alameda. See San Francisco

Alisal

KPRC-FM—See Salinas

Alturas

KALT-FM— July 2002: 106.5 mhz; 500 w. Ant 272 ft TL: N41 29 57 W120 37 30. Hrs open: 215 W. 2nd St., 96101. Phone: (530) 233-4842. Fax: (530) 233-4842. E-mail: kalt@hdo.net Licensee: Woodrow Michael Warren. Group owner: Woodrow Michael Warren Stns. Format: Classic rock. ♦Mike Warren, gen mgr.

KCFJ(AM)— June 4, 1951: 570 khz; 5 kw-D, 200 w-N. TL: N41 30 07 W120 30 01. (CP: 5 kw-U, DA-N). Hrs opn: 6 AM-10 PM Box 580, 96101. Phone: (530) 233-3570. Fax: (530) 233-5570. Licensee: EDI Media Inc. (acq 6-5-02; with co-located FM). Population served: 100,000 Natl. Network: USA. Law Firm: Cohn & Marks. Format: News/talk. News staff: one; News: 3 hrs wkly. Target aud: General. Spec prog: Farm one hr wkly. ♦R.L. Hansen, pres; W.H. Hansen, gen mgr; Carol Irwin, opns mgr, gen sls mgr, prom mgr & progmg dir; Dan Frey, news dir & chief of engrg.

KCNO(FM)—Co-owned with KCFJ(AM). Dec 4, 1990: 94.5 mhz; 100 kw. 106 ft TL: N41 33 50 W120 24 55. (CP: 52 kw., ant -78 ft.). 14 Format: #1 country. News: 17 hrs wkly. Target aud: General. ♦Bill Hansen, sls dir & gen sls mgr; Carol Irwin, traf mgr, spec ev coord & edit mgr.

Anaheim

KFSH-FM—Listing follows KXMX(AM).

KXMX(AM)— May 18, 1959: 1190 khz; 10 kw-D, 1.3 kw-N, DA-2. TL: N33 56 42 W117 51 44. (CP: 20 kw-D). Hrs opn: 24 Box 29023, Glendale, 91209. Phone: (818) 956-5552. Fax: (818) 551-1110. E-mail: info@kkla.com Licensee: Chase Radio Properties L.L.C. Group owner: Salem Communications Corp. (acq 8-24-00; grpsl). Population served: 166,408 Format: Talk, Korean, Arabic. Target aud: Specialized ethnic groups. Spec prog: Gospel, relg, Pol, Sp, Vietnamese 2 hrs wkly. ♦Terry Fahy, gen mgr; Dawn McKahan, gen sls mgr; Bob Hastings, prom mgr & progmg dir; Mark Pollock, chief of engrg.

KFSH-FM—Co-owned with KXMX(AM). Apr 16, 1961: 95.9 mhz; 6 kw. 328 ft TL: N33 49 50 W117 48 39. Stereo. Box 29023, Glendale, 91209. Web Site: www.thefish959.com.200,000 Law Firm: Wiley, Rein & Fielding. Format: Contemp Christian music. Target aud: 18-49.

Anderson

KEWB(FM)— Mar 20, 1983: 94.7 mhz; 4.2 kw. 1,565 ft TL: N40 39 06 W122 31 32. Stereo. Hrs opn: 1588 Charles Dr., Redding, 96003. Phone: (530) 244-9700. Fax: (530) 244-9707. Web Site: www.power94booty.com. Licensee: Results Radio of Redding Licensee LLC. Group owner: Fritz Communications Inc. (acq 6-28-2000; grpsl). Law Firm: Arent, Fox, Kintner, Plotkin & Kahn. Format: CHR. Target aud: 18-49. ♦Beth Tappan, gen mgr; Laurie Curto, gen sls mgr; Rico Garcia, progmg dir & news dir.

Angwin

***KNDL(FM)**— May 20, 1961: 89.9 mhz; 794 w. Ant 3,010 ft TL: N38 40 09 W122 37 53. Stereo. Hrs opn: 24 95 La Jota Dr., 94508. Phone: (707) 965-4155. Fax: (707) 965-4161. E-mail: kndl@thecandle.com Web Site: www.thecandle.com. Licensee: Howell Mountain Broadcasting Co. Inc. Format: Relg. Target aud: 35-49; general. ♦David Shantz, gen mgr.

Apple Valley

KIXW(AM)— June 5, 1954: 960 khz; 5 kw-D, 400 w-N, DA-2. TL: N34 31 00 W117 13 35. Hrs open: 12370 Hesperia Rd., Suite 16, Victorville, 92395. Phone: (760) 241-1313. Fax: (760) 241-0205. Web Site: www.talk960.com. Licensee: Regent Licensee of Victorville Inc. Group owner: Clear Channel Communications Inc. (acq 2000; grpsl). Population served: 277,000 Format: Talk. Target aud: 18-54. ♦Joe Pagano, progmg dir; Larry Thornhill, gen mgr, progmg dir & news dir.

KZXY-FM—Co-owned with KIXW(AM). May 17, 1968: 102.3 mhz; 6 kw. 328 ft TL: N34 24 40 W117 11 09. Stereo. Web Site: www.y102fm.com. Format: Adult contemp. Target aud: 25-54. ♦Colleen Quinn, progmg dir.

KWRN(AM)— Jan 26, 1991: 1550 khz; 5 kw-D, 500 w-N, DA-N. TL: N34 32 12 W117 09 22. Hrs open: 24
Rebroadcasts KWRM(AM) Corona 100%.
Box 1283, Victorville, 92393. Secondary address: 15165 7th St., Ste D, Victorville 92392. Phone: (760) 955-8722. Fax: (760) 955-5751. Web Site: www.kwrn1550am.com. Licensee: Major Market Stations Inc. Population served: 500,000 Natl. Network: ABC. Format: Sp Top-40 hits. Target aud: 34-54; adults with a stable job and disposable income. ♦Marilyn Kramer, pres; Dick Vosper, chief of engrg; kathy DeCastro, traf mgr.

Arcadia

KSSE(FM)— Dec 3, 1960: 107.1 mhz; 3 kw. 240 ft TL: N34 10 51 W118 01 38. (CP: 6 kw, ant -43 ft.). Hrs opn: 24 5700 Wilshire Blvd., Suite 250, Los Angeles, 90036. Phone: (323) 900-6100. Fax: (323) 900-6200. Web Site: www.viva1071.com. Licensee: Entravision Holdings LLC. Group owner: Entravision Communications Corp. (acq 4-1-03; grpsl). Population served: 6,000,000 Law Firm: Latham & Watkins. Wire Svc: SportsTicker Format: Sp. Target aud: 24-39; general. ♦Karl Meyer, gen mgr; Elias Autran, progmg dir; Nestor Rocha, progmg VP & engrg dir.

Arcata

KATA(AM)— Nov 15, 1957: 1340 khz; 1 kw-U. TL: N40 51 12 W124 05 00. Hrs open: 5640 S. Broadway St., Eureka, 95503-6905. Phone: (707) 442-2000. Fax: (707) 443-6848. Web Site: www.kata1340.com. Licensee: Bicoastal Media LLC. (group owner; acq 7-28-99; grpsl). Population served: 120,000 Natl. Network: ABC. Format: Sports. Target aud: 25-54; upscale adults. ♦Mike Wilson, pres; Laurie Tate, gen mgr & opns dir; Victoria Bennington, gen sls mgr; Tom Sebourn, progmg dir.

***KHSU-FM**— October 1960: 90.5 mhz; 8.8 kw. 1,506 ft TL: N40 43 36 W123 58 19. Stereo. Hrs opn: 24 Humboldt State Univ., 1 Harpst St., 95521. Phone: (707) 826-4807. Fax: (707) 826-6082. E-mail: khsu@humboldt.edu Web Site: www.khsu.org. Licensee: Humboldt State University. Population served: 125,000 Natl. Network: NPR, PRI. Format: Var, news. News: 28 hrs wkly. Spec prog: World 14 hrs, jazz 10 hrs wkly. ♦Elizabeth Hans-McCrone, gen mgr; Katie Whiteside, opns mgr & progmg dir; Charles Horn, dev dir; Kevin Sanders, chief of engrg.

KXGO(FM)—Licensed to Arcata. See Eureka

Arnold

KBYN(FM)— Sept 1, 1995: 95.9 mhz; 860 w. Ant 863 ft TL: N38 22 40 W120 11 33. Stereo. Hrs opn: 24 Box 1039, Hughson, 95326. Secondary address: 4043 Geer Rd., Hughson 95326. Phone: (209) 883-8760. Fax: (209) 883-8769. E-mail: ngomez@lafavorita.net Web Site: www.lafavorita.net. Licensee: KBYN Inc. (acq 3-28-01). Natl. Network: CBS. Law Firm: Shaw Pittman. Format: Sp, country. News staff: one; News: 3 hrs wkly. Target aud: 25-54; general. ♦Nelson Gomez, gen mgr.

***KCFA(FM)**— Oct 2, 1995: 106.1 mhz; 3.8 kw. 840 ft TL: N38 22 42 W120 11 36. Hrs open: 24 4043 Geer Rd., Box 1039, Hughson, 95326. Phone: (209) 883-8760. Fax: (209) 883-8769. Web Site: www.lafavorita.net. Licensee: KCFA Inc. (acq 3-21-02). Format: Sp-Ethnic. News staff: one; News: 5 hrs wkly. Target aud: 30-50; Families. ♦Nelson Gomez, gen mgr & gen sls mgr; Freddy Lopez, news dir; Chuck Hughes, chief of engrg; Armida Marquez, traf mgr.

Arroyo Grande

KLFF(AM)— Sept 1, 2002: 890 khz; 5 kw-U, DA-N. TL: N35 08 44 W120 31 15. Hrs open: Box 1561, San Luis Obispo, 93406. Phone: (805) 541-4343. Fax: (805) 541-9101. E-mail: mail@890online.com Web Site: www.890online.com. Licensee: Jerry J. Collins. Format: Relg. ♦Jerry J. Collins, pres; Joel Riley, gen mgr; Noonie Fugler, prom dir.

KXTK(AM)—Licensed to Arroyo Grande. See San Luis Obispo

Arvin

KMYX-FM— June 30, 1999: 92.5 mhz; 1.15 kw. 751 ft TL: N35 11 45 W118 42 30. Hrs open: 24 6313 Schirra Ct., Bakersfield, 93313. Phone: (661) 837-0745. Fax: (661) 837-1612. E-mail: achavez@campesina.com Web Site: www.campesina.com. Licensee: Farmworker Educ. Radio Network Inc. Population served: 400,000 Law Firm: Borsari & Paxson. Format: Sp, rgnl Mexican. Target aud: 25-54; Hispanic market. ♦Anthony Chavez, pres & gen mgr; Cesar Chavez, progmg dir; Dave Whitehead, chief of engrg; Maria Urrutia, traf mgr.

Atascadero

KIQO(FM)— May 19, 1979: 104.5 mhz; 5.6 kw. 1,410 ft TL: N35 21 38 W120 39 21. Stereo. Hrs opn: 24 3620 Sacramento Dr., Suite 204, San Luis Obispo, 93401. Phone: (805) 781-2750. Fax: (805) 781-2758. Web Site: www.kiqo104.5.com. Licensee: AGM California. Group owner: American General Media (acq 2-10-99; $1.5 million). Population served: 400,000 Natl. Network: ABC. Format: Oldies. Target aud: 25-55. ♦Kathy Signorelli, gen mgr; Mark Tobin, gen sls mgr & chief of engrg; Seth Blackburn, progmg dir; Pat Mallon, news dir; Bill Bordeaux, chief of engrg.

Atherton

***KCEA(FM)**— June 2, 1979: 89.1 mhz; 100 w. -216 ft TL: N37 27 41 W122 10 30. (CP: Ant 5 ft.). Stereo. Hrs opn: 24 555 Middle Field Rd., 94027. Phone: (650) 306-8823. Phone: (650) 306-8822. Fax: (650) 328-8706. Web Site: www.kcea.org. Licensee: Sequoia Union High School District. Population served: 200,000 Format: Big band, Nostalgia, adult standards. Target aud: General. ♦Michael Isaacs, gen mgr; John Mylod, sports cmtr & disc jockey; Trish Millet, pub affrs dir & disc jockey.

Atwater

KBRE(FM)— Oct 1, 1995: 92.5 mhz; 6 kw. 328 ft TL: N37 16 42 W120 37 33. Stereo. Hrs opn: 24 1020 W. Main St., Merced, 95340-4521. Phone: (209) 723-2191. Fax: (209) 383-2950. Web Site: www.925thebear.com. Licensee: Mapleton Communications LLC (group owner; acq 6-1-2002; grpsl). Population served: 200,000 Law Firm: Leventhal, Senter & Lerman. Format: Active rock. News staff: one; News: 3 hrs wkly. Men 25-49. ♦Andrew Adams, gen mgr, gen sls mgr & pub affrs dir; Jason LaChance, progmg dir; Michael Martinez, progmg dir.

Auberry

KLBN(FM)— July 12, 1992: 105.1 mhz; 590 w. 1,902 ft TL: N37 04 25 W119 25 52. (CP: 600 w, ant 1,870 ft.). Stereo. Hrs opn: 24 1110 E. Olive Ave., Fresno, 93728. Phone: (559) 497-1100. Fax: (559) 497-1125. E-mail: mginsburg@lotusfresno.com Licensee: Lotus Communications Corp. (group owner) Population served: 1,500,000 Natl. Rep: Lotus Entravision Reps LLC. Format: Sp/Rgnl Mexican. Target aud: 18-34; .Young people of Mexican heritage who love to dance ♦Howard Kalmenson, pres; Daniel Crotty, gen mgr; Mike Ginsburg, sls dir; Eduardo Leon, mus dir.

Auburn

KAHI(AM)— Nov 13, 1957: 950 khz; 5 kw-D, 4.2 kw-N, DA-2. TL: N38 51 28 W121 01 39. Hrs open: 24 985 Lincoln Way, Suite 103, 95603. Phone: (530) 885-5636. Fax: (530) 885-0166. E-mail: KAHI@KAHI.com Web Site: www.kahi.com. Licensee: IHR Educational Broadcasting (group owner; acq 4-28-99; $475,000 with KSMH(AM) West Sacramento). Population served: 65,700 Law Firm: Fletcher, Heald & Hildreth. Format: Var. Target aud: 25-54; Community focused. ♦Dave Rosenthal, opns VP & opns mgr.

KHYL(FM)— Dec 21, 1961: 101.1 mhz; 36.3 kw. 577 ft TL: N38 51 28 W121 01 39. Stereo. Hrs opn: 24 1440 Ethan Way, Suite 200, Sacramento, 95825. Phone: (916) 929-5325. Fax: (916) 925-0128. Web Site: www.v101fm.com. Licensee: AMFM Broadcasting Licenses LLC. Group owner: Clear Channel Communications Inc. (acq 8-30-2000; grpsl). Population served: 230,400 Format: Oldies, urban adult contemp. News: one hr wkly. Target aud: 25-54. ♦Jeff Holden, gen mgr & sls dir; Amy Bingham, mktg dir; Don Alias, progmg dir.

Avalon

KBRT(AM)— June 1, 1952: 740 khz; 10 kw-D, DA. TL: N33 21 36 W118 22 18. Stereo. Hrs opn: 3183 D Airway Ave., Costa Mesa, 92626. Phone: (714) 754-4450. Fax: (714) 754-0735. E-mail: kbrtinfo@crawfordbroadcasting.com Web Site: www.kbrt740.com. Licensee: Kierton Inc. Group owner: Crawford Broadcasting Co. (acq 5-21-80). Population served: 15,961,663 Format: Relg, talk. Target aud: Christian adult. ♦Todd Stickler, opns mgr.

***KISL(FM)**— 2000: 88.7 mhz; 200 w. Ant 20 ft TL: N33 20 32 W118 19 11. Hrs open: Box 1980, 90704. Phone: (310) 510-7469. Fax: (310) 510-1025. E-mail: arts@cipas.org Web Site: www.kisl.org. Licensee: Catalina Island Performing Arts Foundation (acq 3-15-2000). Format: Var. ♦Aaron Pitts, stn mgr.

Avenal

***KAAX(FM)**—Not on air, target date: unknown: Stn currently dark. 95.1 mhz; 920 w. Ant 656 ft TL: N36 00 40 W120 04 26. Hrs open: 12550 Brookhurst St., Suite A, Garden Grove, 92840. Licensee: Avenal Educational Services Inc.

Baker

KHRQ(FM)— 2002: 94.9 mhz; 1.4 kw. Ant 1,288 ft TL: N35 26 09 W115 55 22. Hrs open: 1611 E. Main St., Barstow, 92311. Phone: (760) 256-0326. Fax: (760) 256-9507. E-mail: tim@highwayradio.com Licensee: The Drive LLC. Group owner: KHWY Inc. (acq 5-31-2003). Natl. Network: ABC. Law Firm: Hogan & Hartson. Format: Classic rock. ♦Howard Anderson, CEO & pres; Kirk Anderson, exec VP; Timothy Anderson, VP & gen mgr; Judy Robinson, sls VP; John Gregg, prom dir; Lance Todd, progmg dir; Keith Hayes, news dir; Thomas J. McNeill, engrg mgr.

KIXF(FM)— Mar 1, 1994: 101.5 mhz; 4.3 kw. Ant 1,322 ft TL: N35 26 00 W115 55 25. Stereo. Hrs opn: 24
Rebroadcasts KIXW-FM Lenwood 100%.
1611 E. Main St., Barstow, 92311. Phone: (760) 256-0326. Fax: (760) 256-9507. E-mail: time@highwayradio.com Web Site: www.thehighwaystations.com. Licensee: KHWY Inc. (group owner; (acq 2-18-98; $1,741,444 with KIXW-FM Lenwood). Natl. Network: Westwood One, CNN Radio. Law Firm: Hogan & Hartson. Format: Country. News staff: one. Target aud: 25-54; interstate travelers to Las Vegas & Laughlin, NV. Spec prog: Hourly traf report to service interstate travelers. ♦Howard B. Anderson, CEO & pres; Kirk Anderson, exec VP; Timothy B. Anderson, VP & gen mgr; Judy Robinson, sls VP; John Gregg, prom dir; Lance Todd, progmg dir; Keith Hayes, news dir; Thomas J. McNeill, engrg mgr.

Bakersfield

KAFY(AM)— 2000: 1100 khz; 4.2 kw-D, 800 w-N, DA-N. TL: N35 27 00 W118 56 48. Hrs open: 24 4043 Green Rd., Hughson, 95326. Phone: (209) 883-8760. Fax: (209) 883-8769. Web Site: www.lafavorita.net. Licensee: KAFY Inc. (acq 3-28-01). Format: Sp talk. ♦Nelson Gomez, gen mgr.

KBFP(AM)— 1959: 800 khz; 1 kw-D, 440 w-N, DA-2. TL: N35 20 44 W118 59 33. Hrs open: 6 AM-12 AM 1100 Mohawk St., Suite 280, 93309. Phone: (661) 322-9929. Fax: (661) 322-9239. Web Site: www.foxsportsradio800.com. Licensee: CC Licenses LLC. Group owner: Clear Channel Communications Inc. (acq 10-11-2000; grpsl). Population served: 500,000 Format: Sports. Target aud: 25-54; adults. ♦Jim Bell, VP; Tony Manes, progmg dir; Steve Mull, chief of engrg.

KBKO-FM—Co-owned with KBFP(AM). Aug 24, 1963: 96.5 mhz; 50 kw. 550 ft TL: N35 29 08 W118 53 19. Stereo. Web Site: www.965kissfm.com.460,000 Natl. Rep: McGavren Guild. Law Firm: Arter & Hadden. Format: CHR. Target aud: 18-44.

KBFP-FM—See Delano

KCWR(FM)— Mar 21, 1990: 107.1 mhz; 6 kw. 164 ft TL: N35 22 08 W119 00 14. Hrs open: 24 3223 Sillect Ave., 93308. Phone: (661) 326-1011. Fax: (661) 328-7503. Fax: (661) 328-7537(news). Licensee: Owens One Co. Inc. (acq 5-24-2006; grpsl). Format: Country. ♦Mel Owens Jr., CEO, gen mgr, opns mgr & dev mgr; Julie Randolph, gen sls mgr & mktg mgr.

KERN(AM)— Jan 3, 1932: 1410 khz; 1 kw-U. TL: N35 21 07 W118 56 48. Hrs open: 24 1400 Easton Dr., Suite 144, 93309. Secondary address: Box 2700 93309. Phone: (661) 328-1410. Phone: (661) 326-1410. Fax: (661) 328-0873. E-mail: news@kernradio.com Web Site: www.kernradio.com. Licensee: AGM California. Group owner: American General Media (acq 5-1-75). Population served: 450,000 Natl. Network: Westwood One, ABC. Natl. Rep: Christal. Wire Svc: AP Format: News/talk. News staff: 6; News: 30 hrs wkly. Target aud: 25-54. Spec prog: Farm one hr, relg one hr wkly. ♦Roger Fessler, exec VP & gen mgr; Toni Snyder, gen sls mgr; Blake Taylor, progmg dir.

KISV(FM)—Co-owned with KERN(AM). 1948: 94.1 mhz; 4.5 kw. 1,312 ft TL: N35 26 20 W118 44 23. Stereo. 24 Web Site: hot941.com.500,000 Format: CHR rhythmic. Spec prog: Farm one hr, relg one hr wkly.

***KFRB(FM)**— August 1996: 91.3 mhz; 115 w. 1,368 ft TL: N35 26 17 W118 44 22. Hrs open: 24 c/o Family Stations, 290 Hegenberger Rd., Oakland, 94621. Phone: (209) 389-4659. Web Site: www.familyradio.com. Licensee: Family Stations Inc. (group owner) Format: Relg. ♦Harold Camping, pres & gen mgr; David Manzi, opns mgr.

KGEO(AM)— Jan 1, 1946: 1230 khz; 1 kw-U. TL: N35 20 53 W119 00 33. Hrs opn: 1400 Easton Dr., Suite 144, 93309. Phone: (661) 631-1230. Phone: (661) 328-1410. Fax: (661) 328-0873. Licensee: AGM California. Group owner: American General Media (acq 12-9-92; $1.75 million with co-located FM; FTR: 1-4-93). Population served: 355,000 Natl. Network: CBS, Westwood One, ABC. Natl. Rep: McGavren Guild. Law Firm: Cohn & Marks. Format: Sports, talk. ♦Roger Fessler, gen mgr.

KGFM(FM)—Co-owned with KGEO(AM). October 1964: 101.5 mhz; 4.8 kw. 1,280 ft TL: N35 26 20 W118 44 23. (CP: 6.7 kw, ant 1,299 ft.). Stereo. 500,000 Format: Easy lstng.

KHTY(AM)— October 1946: 970 khz; 1 kw-D, 5 kw-N, DA-2. TL: N35 27 00 W118 56 48. Hrs open: 24 1100 Mohawk St., Ste 280, 93309. Phone: (661) 322-9929. Fax: (661) 283-2963. Fax: (661) 283-2963. Web Site: www.mighty970.com. Licensee: AMFM Radio Licenses LLC. Group owner: Clear Channel Communications Inc. (acq 12-22-2000; $1.4 million). Population served: 160,000 Format: Business talk. ♦Jim Bell, VP & gen mgr; Ron Fisher, gen sls mgr; Steve King, progmg dir; Steve Mull, chief of engrg.

KIWI(FM)—Listing follows KWAC(AM).

KKBB(FM)— November 1991: 99.3 mhz; 1.2 kw. 1,345 ft TL: N35 26 17 W118 44 22. Stereo. Hrs opn: 24 Box 80658, 93380. Secondary address: 3651 Pegasus Dr., Suite 107 93380. Phone: (661) 393-1900. Fax: (661) 393-1915. E-mail: jlove@kkbb.com Web Site: www.groove993.com. Licensee: Buckley Communications Inc. Group owner: Buckley Broadcasting Corp. (acq 10-3-94; $1 million; FTR: 10-17-94). Population served: 480,000 Natl. Rep: D & R Radio. Format: Rhythmic Oldies. Target aud: 25-54; adults. ♦Steve Darnell, gen mgr; Otis Warren, gen sls mgr; Kathy King, news dir; Bob Turner, chief of engrg.

KLHC(AM)— February 1958: 1350 khz; 1 kw-D, 33 w-N. TL: N35 21 00 W119 58 58. Hrs open: 24 3817 Wilson Rd., Suite E, 93309. Phone: (661) 847-1450. Fax: (661) 847-1452. E-mail: klhc@klhcradio.com Web Site: www.klhcradio.com. Licensee: Force Broadcasting LLC. Group owner: American General Media (acq 1-6-2006; $925,000). Population served: 300,000 Format: Sp relg. ♦Maria Ochoa, gen mgr.

KNZR(AM)— 1933: 1560 khz; 25 kw-D, 10 kw-N, DA-N. TL: N35 18 30 W119 02 09. Hrs open: 24 Box 80658, 93308. Secondary address: 3651 Pegasus Dr., Suite 107 93380. Phone: (661) 393-1900. Fax: (661) 393-1915. Web Site: www.knzr.com. Licensee: Buckley Broadcasting of California LLC. Group owner: Buckley Broadcasting Corp. (acq 1-25-90; $1 million; FTR: 2-19-90). Population served: 770,000 Natl. Network: CBS. Natl. Rep: D & R Radio. Law Firm: Shaw Pittman. Format: News/talk. News staff: 4; News: 40 hrs wkly. Target aud: 25-54. Spec prog: L.A. Dodgers. ♦Steve Darnell, gen mgr.

***KPRX(FM)**— Feb 28, 1987: 89.1 mhz; 12 kw. 500 ft TL: N35 29 10 W118 53 20. Hrs open: 3437 W. Shaw Ave., Suite 101, Fresno, 93711. Phone: (559) 275-0764. Fax: (559) 275-2202. E-mail: kvpr@kvpr.org Web Site: www.kvpr.org. Licensee: White Ash Broadcasting Inc. Natl. Network: NPR. Format: Class, news & info. ♦Mariam Stepanian, pres & gen mgr; Jim Meyers, stn mgr & progmg dir; Steve Mull, chief of engrg.

KPSL-FM— Dec 15, 1985: 92.1 mhz; 2 kw. Ant 567 ft TL: N35 29 11 W118 53 21. Stereo. Hrs opn: 24 5100 Commerce Dr., 93309-0684. Phone: (661) 327-9711. Fax: (661) 327-0797. E-mail: info@thespanishradio.com Web Site: www.thespanishradio.com. Licensee: Illinois Lotus Corp. Group owner: Lotus Communications Corp. (acq 8-24-99; grpsl). Population served: 650,000 Format: Sp pop. ♦Mike Allen, gen mgr; Isidro Roman, progmg dir.

KRAB(FM)—Green Acres, Oct 1, 1991: 106.1 mhz; 25 kw. 410 ft TL: N35 28 17 W119 01 38. Hrs open: 24 1100 Mohawk St., Suite 280, 93309. Phone: (661) 322-9929. Fax: (661) 322-9239. Web Site: www.krab.com. Licensee: CC Licenses LLC. Population served: 460,000 Natl. Rep: McGavren Guild. Law Firm: Arter & Hadden. Format: AOR. Target aud: 18-49; predominantely male. ♦Jim Bell, VP & gen mgr; Ron Fisher, gen sls mgr; Danny Spanks, progmg dir; Steve Mull, chief of engrg.

KSMJ(FM)—Shafter, Mar 3, 1978: 97.7 mhz; 4.1 kw. Ant 397 ft TL: N35 27 33 W119 01 13. Stereo. Hrs opn: 24 Box 80658, 93380. Phone: (661) 393-1900. Fax: (661) 393-1915. Web Site: www.977thebreeze.com. Licensee: Buckley Broadcasting of California LLC. Group owner: Buckley Broadcasting Corp. (acq 2-1-2001; $2 million). Population served: 378,000 Format: Lite rock. Target aud: 18-49. ♦Steve Darnell, gen mgr; E.J. Tyler, opns dir & opns mgr.

***KTQX(FM)**— Apr 14, 1989: 90.1 mhz; 590 w. 3,572 ft TL: N35 27 05 W118 35 10. Hrs open: 24 5005 E. Belmont Ave., Fresno, 93727. Phone: (559) 455-5777. Fax: (559) 455-5778. Web Site: www.radiobilingue.org. Licensee: Radio Bilingue Inc. Format: Ethnic, multilingual, Sp. News staff: 5; News: 11 hrs wkly. Target aud: 16-60; Latino. ♦Hugo Morales, CEO; Maria Erana, gen mgr, opns mgr, gen sls mgr & progmg dir; Phil Traynor, dev VP; Samuel Cozco, news dir; Bill Bach, chief of engrg.

KUZZ(AM)— October 1946: 550 khz; 5 kw-U, DA-N. TL: N35 20 25 W118 56 14. Stereo. Hrs opn: 3223 Sillect Ave., 93308. Phone: (661) 326-1011. Fax: (661) 328-7503. Licensee: Owens One Co. (group owner; (acq 5-24-2006; grpsl). Population served: 100,000 Format: Country. Target aud: 25-54.Mel Owens Jr., CEO & gen mgr; Julie Randolph, gen sls mgr; Harvey Campbell, natl sls mgr; Jerry Hufford, prom dir; Evan Bridwell, progmg dir & mus dir; Mark Howell, news dir & local news ed; Sylvia Cariker, pub affrs dir; Terry Gaiser, chief of engrg; Casey McBride, women's int ed & disc jockey; Chris Conner, disc jockey; KC Adams, disc jockey; Steve Gradowitz, disc jockey

KUZZ-FM— 1968: 107.9 mhz; 6 kw. 1,364 ft TL: N35 26 20 W118 44 24. Stereo. Web Site: www.kuzz.com. ♦Retta Smith, traf mgr; Peter J. Rudy, news rptr; Tammy Brown, news rptr; Casey McBride, disc jockey; Chris Conner, disc jockey; KC Adams, disc jockey; Steve Gradowitz, disc jockey.

KWAC(AM)— 1954: 1490 khz; 1 kw-U. TL: N35 24 07 W119 02 45. Stereo. Hrs opn: 24 5100 Commerce Dr., 93309. Phone: (661) 327-9711. Fax: (661) 327-0797. E-mail: info@thespanishradio.com Web Site: kwac.com. Licensee: Illinois Lotus Corp. Group owner: Lotus Communications Corp. (acq 8-24-99; grpsl). Population served: 1,056,270 Natl. Network: ESPN Radio. Natl. Rep: Lotus Entravision Reps LLC. Format: Mexican rgnl. News: 5 hrs wkly. Target aud: General. ♦Howard Kalmenson, pres; Mike Allen, gen mgr; Juan M. Martinez, progmg dir, news dir & disc jockey; Anna Gallegos, pub affrs dir; Lloyd Moss, chief of engrg; Jesus Valdez, disc jockey; Manolo Martinez, disc jockey.

KIWI(FM)—Co-owned with KWAC(AM). July 11, 1989: 102.9 mhz; 25 kw. Ant 321 ft TL: N35 19 16 W119 42 26. Stereo. 24 E-mail: napo@radiolobo.com Web Site: radiolobo.com. (Acq 12-18-00; $2.5 million including a $10,000 three-year noncompete agreement).650,000 Natl. Rep: Lotus Entravision Reps LLC. Format: Mexican regional. Target aud: General.

Banning

KMET(AM)— 1948: 1490 khz; 1 kw-U. TL: N33 55 49 W116 55 20. Hrs open: 24 700 E. Redlands Blvd., Suite U, PMB 323, Redlands, 92373. Secondary address: 572 Omar 92220. Phone: (951) 849-4644. Phone: (949) 261-6117. Fax: (951) 849-3114. E-mail: general@kmetam.com Licensee: Sunset Broadcasting Inc. (acq 3-31-2003). Population served: 375,000 Format: Smooth jazz, sports. News staff: one; News: 3 hrs wkly. Target aud: General; 35-64. Spec prog: Relg 4 hrs, women's sports 3 hrs wkly. ♦Richard Nuthmann, pres; Adrian P. Madden, gen mgr & stn mgr; Mitch McClellan, opns mgr.

Barstow

KDUC(FM)— June 4, 1986: 94.3 mhz; 4.6 kw. 783 ft TL: N34 58 15 W117 02 22. Stereo. Hrs opn: 29000 Radio Road, 92311. Phone: (760) 256-2121. Fax: (760) 256-5090. E-mail: doscostascommunications@yahoo.com Licensee: Dos Costas Communications Corp. (group owner; acq 6-18-03; grpsl). Format: Top 40. Target aud: 12-44. ♦Manny Lopez, gen sls mgr; Mike Garcia, progmg dir; Roland Ulloa, gen mgr & mus dir.

KIQQ(AM)— Sept 29, 1960: 1310 khz; 5 kw-D, 118 w-N, DA-1. TL: N34 54 51 W117 00 59. Hrs open: 24
Simulcast with KAEH(FM) Beaumont 100%.
710 W. Old Hwy. 58, 92311. Phone: (760) 255-2636. Fax: (760) 255-3236. E-mail: jramirez@lamaquinaamusical.net Web Site: moonbroadcasting.com. Licensee: MBR Licensee LLC. Group owner: Moon Broadcasting (acq 8-7-2000). Natl. Network: Westwood One. Law Firm: Arent, Fox, Kintner, Plotkin & Kahn. Format: Sp, regnl Mexican. Target aud: 45 plus. ♦Alicia Avila, gen mgr.

***KODV(FM)**— 2005: 89.1 mhz; 260 w. Ant 725 ft TL: N34 58 17 W117 02 22. Hrs open: Box 401136, Hesperia, 92340-1136. Secondary address: 18280 Atlantic St., Hesperia 92345. Phone: (760) 947-4300. Fax: (760) 956-2427. E-mail: comentarios@ondadevida.net Web Site: www.ondasdevida.net. Licensee: Ondas de Vida Network Inc. (acq 12-7-2005; $100,000 for CP). Format: Sp. ♦Hector E. Manzo, CEO.

KRXV(FM)—See Yermo

KSZL(AM)— June 25, 1986: 1230 khz; 1 kw-U. TL: N34 54 44 W117 01 39. Stereo. Hrs opn: 24 29000 Radio Rd., 92311. Phone: (760) 256-2121. Phone: (760) 256-5382. Fax: (760) 256-5090. E-mail: am1230kszl@yahoo.com ABC/SMN Stardust Licensee: Dos Costas Communications Corp. (group owner; acq 6-18-03; grpsl). Population served: 220,000 Natl. Network: Westwood One. Natl. Rep: Western Regional Broadcast Sales. Law Firm: Fleischman & Walsh, L.L.P. Format: News. News staff: 2; News: 12 hrs wkly. Target aud: 25 plus; Adults 35 years +. ♦Manny Lopez, gen mgr & gen sls mgr; Roland Ulloa, gen mgr; Michael Garcia, opns mgr & progmg dir; Steve Hastings, traf mgr.

***KWTH(FM)**— 2006: 91.3 mhz; 1.55 kw. Ant 2,296 ft TL: N34 38 39 W116 37 38. Stereo. Hrs opn:
Rebroadcasts KWTW(FM) Bishop 100%.
Box 637, Bishop, 93515. Phone: (760) 872-4225. Phone: (866) 466-5989. Fax: (760) 872-4155. E-mail: friar@schat.com Web Site: www.kwtw.org. Licensee: Living Proof Inc. Format: Christian, relg. ♦Daniel McClenaghan, pres & gen mgr.

KXXZ(FM)— 1989: 95.9 mhz; 8.9 kw. Ant 485 ft TL: N34 51 22 W117 03 00. Stereo. Hrs opn: 24 29000 Radio Rd., 92311. Phone: (760) 256-2121. Fax: (760) 256-5090. E-mail: doscostas@yahoo.com Licensee: Dos Costas Communications Corp. (group owner; acq 6-18-03; grpsl). Format: Sp contemp rhythmic. ♦Roland Ulloa, gen mgr; Manny Lopez, gen sls mgr; Mike Garcia, progmg dir; Steve Hastings, traf mgr.

Bayside

***KNHM(FM)**— Apr 15, 1992: 91.5 mhz; 550 w. Ant 508 ft TL: N40 47 49 W124 02 47. Hrs open: 24 Jefferson Public Radio, 1250 Siskiyou Blvd., Ashland, OR, 97520. Phone: (541) 552-6301. Fax: (541) 552-8565. Web Site: www.ijpr.org. Licensee: JPR Foundation Inc. (acq 3-24-2004; $130,000). ♦Ronald Kramer, gen mgr & stn mgr; Paul Westhelle, dev mgr & mus dir.

Beaumont

KAEH(FM)— 1996: 100.9 mhz; 1.5 kw. Ant 479 ft TL: N33 54 29 W116 59 45. Hrs open: 24 Moon Broadcasting Riverside LLC, 1200 W. Venice Blvd., Los Angeles, 90006. Phone: (909) 381-0969. Fax: (909) 381-0943. E-mail: postmaster@manbroadcasting.com Web Site: www.moonbroadcasting.com/kaeh. Licensee: MBR Licensee LLC. Group owner: Moon Broadcasting (acq 2-13-2002; $1.7 million). Format: Rgnl Mexican. ♦Abel A. DeLuna, pres; Alicia Avila, gen mgr; Juan Ramirez, progmg dir; Manuel Garcia, news dir; Rick Hunt, chief of engrg.

Berkeley

***KALX(FM)**— October 1967: 90.7 mhz; 500 w. Ant 778 ft TL: N37 52 40 W122 14 44. Stereo. Hrs opn: 24 26 Barrows #5650, 94720-5650. Phone: (510) 642-1111. E-mail: mail@kalx.berkeley.edu Web Site: kalx.berkeley.edu. Licensee: The Regents of the University of California. Population served: 1,000,000 Format: Educ, div. News: 4 hrs wkly. ♦Sandra Wasson, gen mgr; Mona Dehghan, opns mgr; James Croft, mus dir; Taylor McCauley, mus dir; Zaheem Cassim, news dir; Bill Jones, engrg mgr & chief of engrg.

KBLX-FM—Listing follows KVTO(AM).

***KPFA(FM)**— April 1949: 94.1 mhz; 59 kw. 1,330 ft TL: N37 51 55 W122 13 12. Stereo. Hrs opn: 24 1929 Martin Luther King Jr. Way, 94704. Phone: (510) 848-6767. Fax: (510) 848-3812. E-mail: postmaster@kpfa.org Web Site: www.kpfa.org. Licensee: Pacifica Foundation. Group owner: Pacifica Foundations Inc. dba Pacifica Radio Population served: 200,000 Law Firm: Haley, Bader & Potts. Wire Svc: Reuters Wire Svc: Pacifica Network News Format: Div mus, pub affrs. News staff: 4; News: 11 hrs wkly. Target aud: 25-50. Spec prog: C&W 18 hrs, Black 18 hrs, jazz 15 hrs, folk 10 hrs, women 10 hrs, world 18 hrs wkly. ♦Gus Newport, gen mgr; Luis Medina, mus dir; Mark Meriole, news dir; Jim Bennett, chief of engrg.

KVTO(AM)— May 22, 1922: 1400 khz; 1 kw-U. TL: N37 50 58 W122 17 44. Hrs open: 24 55 Hawthorne St., Suite 900, San Francisco, 94105. Phone: (415) 284-1029. Fax: (415) 764-4959. E-mail: info@kblx.com Licensee: Urban Radio III L.L.C. Group owner: Inner City Broadcasting (acq 1979). Natl. Rep: D & R Radio. Format: Asian. Target aud: 25-54. ♦Harvey Stone, pres & gen mgr; Barry Rose, gen sls mgr; Rhonda Amiz, natl sls mgr; Jamie Arbona, progmg dir; Paul Marks, chief of engrg.

KBLX-FM—Co-owned with KVTO(AM). Apr 29, 1949: 102.9 mhz; 50 kw. 1,290 ft TL: N37 41 20 W122 26 07. Stereo. 24 E-mail: info@kblx.com Web Site: www.kblx.com.400,000 Format: Adult contemp, news. News staff: one; News: 20 hrs wkly. Target aud: 25-54; adults. ♦George Dabis, opns mgr; Barry Rose, sls VP; Sandie Dibble, natl sls mgr & mktg mgr; Tara Cortez, prom mgr; Kevin Brown, progmg dir & disc jockey; Larry Elliott, mus dir; Brenda Ross, news dir; Susie Lee, pub affrs dir; Renee Guillary, traf mgr; Leslie Stoval, disc jockey.

Beverly Hills

KMZT(AM)— October 1947: 1260 khz; 5 kw-U, DA-2. TL: N34 14 58 W118 27 15. Hrs open: 24 1500 Cotner Ave., Los Angeles, 90025. Phone: (310) 478-5540. Fax: (310) 445-1439. E-mail: webmaster@kmozart.com Web Site: www.kmozart.com. Licensee: Mount Wilson FM Broadcasters Inc. (acq 11-20-92; $2.5 million; FTR: 12-14-92). Population served: 400,000 Natl. Rep: D & R Radio. Law Firm: Fisher, Wayland, Cooper, Leader & Zaragoza L.L.P. Format: Classical. ♦Saul Levine, pres & gen mgr; Mike Johnson, opns dir; Kane Biscaya, gen sls mgr.

Big Bear City

KBHR(FM)— Dec 17, 1995: 93.3 mhz; 1.5 kw. 663 ft TL: N34 16 41 W116 47 31. Stereo. Hrs opn: 24 Box 2979, ., 92314. Phone: (909) 584-5247. Fax: (909) 584-5347. E-mail: info@kbhr933.com Web Site: www.kbhr933.com. Licensee: Parallel Broadcasting Inc. (acq 1-17-95; FTR: 3-13-95). Population served: 25,000 Natl. Network: CNN Radio. Format: Triple A. News staff: one; News: 11 hrs wkly. Target aud: 25-54; upscale second home owners, resort visitors. Spec prog: CNN news 8 hrs, ski report one hr, fish report one hr wkly. ♦Cathy Herrick, VP & opns dir; Jay Tunnell, sls dir; Rick Herrick, pres, gen mgr & progmg dir; Catherine Sandstrom, news dir.

Big Bear Lake

KXSB(FM)— May 1, 1975: 101.7 mhz; 90 w. 1,500 ft TL: N34 12 47 W116 51 59. (CP: 300 w). Stereo. Hrs opn: 24

Rebroadcasts KXLM(FM) Oxnard 80%.
1950 S. Sunwest Ln., #302, San Bernarino, 92408. Phone: (909) 825-5020. Fax: (909) 884-5844. Fax: (909) 890-3849. E-mail: edith@radiolazer.com Web Site: www.radiolazer.com. Licensee: Lazer Broadcasting Corp. (group owner; acq 1995; $750,000). Population served: 1,500,000 Natl. Rep: Lotus Entravision Reps LLC. Law Firm: Fletcher, Heald & Hildreth. Format: CHR, Sp. News staff: one; News: 2 hrs wkly. Target aud: 25-54; adults, serious minded. ♦Alfredo Plascencia, CEO & pres; Vicki Bails, VP, VP, gen mgr & gen sls mgr; Gerardo Palafox, prom dir; Salvador Prieto, progmg dir & news dir; Ralph Jones, chief of engrg.

Big Pine

KRHV(FM)— 1999: 93.3 mhz; 890 w. Ant 2,903 ft TL: N37 24 48 W118 11 08. Hrs open: Box 1284, Mammoth Lakes, 93546. Secondary address: 94 Laurel Mountain Rd., Mammouth Lakes 93546. Phone: (760) 934-8888. Fax: (760) 934-2429. E-mail: kmmtradioworks@yahoo.com Web Site: www.kmmtradio.com. Licensee: David & Mary Digerness. Format: Classic rock. ♦David A. Digerness, pres & chief of engrg; Shellie Woods, gen mgr, gen sls mgr, mktg dir, adv dir & traf mgr; Maryanne Digerness, stn mgr, gen sls mgr, mktg dir & adv dir; Spencer Myers, progmg dir.

Bishop

KACE(AM)—Not on air, target date: unknown: 1340 khz; 1 kw-U. TL: N37 22 42 W118 23 43. Hrs open: 24 Box 1450, St. George, UT, 84771-1450. Secondary address: 210 N. 1000 E., St. George, UT 84770-3155. Phone: (435) 628-1000. Fax: (435) 628-6636. Licensee: Radio 1340 LLC. (acq 4-14-2006; $20,000 for CP). Law Firm: Dan J. Alpert. ♦E. Morgan Skinner Jr., pres.

KBOV(AM)— Apr 1, 1953: 1230 khz; 1 kw-U. TL: N37 20 44 W118 23 43. Hrs open: Box 757, 93515. Secondary address: S. Hwy. 395 93514. Phone: (760) 873-6324. Phone: (760) 873-5427. Fax: (760) 872-2639. E-mail: kibskbov@gnet.com Web Site: www.395.com. Licensee: Great Country Broadcasting Inc. (acq 7-6-2004; $965,000 with co-located FM). Population served: 13,000 Natl. Rep: Western Regional Broadcast Sales. Format: Oldies. News staff: one; News: 8 hrs wkly. Target aud: General. ♦John Dailey, gen mgr.

KIBS(FM)—Co-owned with KBOV(AM). Nov 1, 1974: 100.7 mhz; 1 kw. 2,960 ft TL: N37 25 00 W118 11 00. Stereo. 24 50,000 Format: Country. News staff: one; News: 8 hrs wkly.

***KWTW(FM)**— 2002: 88.5 mhz; 900 w. Ant 2,916 ft TL: N37 24 48 W118 11 08. Stereo. Hrs opn: 24 Box 637, 93515. Secondary address: 125 S. Main St. 93514. Phone: (760) 872-4225. Phone: (866) 466-5989. Fax: (760) 872-4155. E-mail: friar@schat.com Web Site: www.kwtw.org. Licensee: Living Proof Inc. Population served: 20,000 Format: Christian, relg. Target aud: General; all who want to hear the gospel. ♦Daniel McClenaghan, pres & gen mgr.

Blue Lake

KCIK(AM)—Not on air, target date: unknown: 1450 khz; 250 w-U. TL: N40 52 52 W123 59 58. Hrs open: Box 180, Tahoma, 96142. Phone: (530) 584-5700. Fax: (530) 584-5705. E-mail: info@ihradio.org Web Site: www.ihradio.org. Licensee: IHR Educational Broadcasting (group owner; acq 3-28-03). Format: Relg/Catholic. ♦Douglas M. Sherman, pres; Steve Wise, gen mgr.

KEJY(FM)—Not on air, target date: unknown: 106.3 mhz; 3.3 kw. Ant 1,692 ft TL: N40 43 38.9 W123 58 17. Hrs open: 1101 Marsh Rd., Eureka, 95501. Phone: (707) 442-5744. Licensee: Eureka Broadcasting Co. Inc. ♦Hugo Papstein, pres.

Blythe

KJMB(FM)— April 1975: 100.3 mhz; 36.4 kw. 174 ft TL: N33 37 16 W114 35 28. Stereo. Hrs opn: 24 681 N. 4th St., 92225. Phone: (760) 922-7143. Fax: (760) 922-2844. Licensee: Blythe Radio Inc. Population served: 40,000 Natl. Network: USA. Format: Adult contemp. News staff: one; News: 10 hrs wkly. Target aud: 18-40; adults. Spec prog: Farm 5 hrs wkly. ♦Jim Mayson, pres; James M. Morris, gen mgr.

Brawley

KROP(AM)—Licensed to Brawley. See El Centro

KSEH(FM)— April 4, 1988: 94.5 mhz; 50 kw. Ant 302 ft TL: N32 54 40 W115 31 40. Stereo. Hrs opn: 24 Box 2830, El Centro, 92244. Phone: (760) 482-7777. Fax: (760) 482-0099. Licensee: Entravision Holdings LLC. Group owner: Entravision Communications Corp. Population served: 950,000 Natl. Network: ABC. Format: Latin pop. Target aud: 25-54. ♦Eric Chavez, gen mgr.

KSIQ(FM)—Licensed to Brawley. See El Centro

Buena Park

***KBPK(FM)**— July 6, 1970: 90.1 mhz; 20 w. 130 ft TL: N33 51 35 W118 00 53. Hrs open: 321 E. Chapman Ave., Fullerton, 92832. Phone: (714) 992-7419. E-mail: info@kbpk-fm.com Web Site: kbpk-fm.com. Licensee: Buena Park School District. Population served: 63,646 Law Firm: Booth, Freret, Imlay & Tepper. Wire Svc: AP Format: Adult contemp. Target aud: 25-54. ♦Edward Ford, opns mgr; Peg Stewart Berger, progmg dir, news dir & sports cmtr.

Burney

***KIBC(FM)**— Nov 15, 1985: 90.5 mhz; 3 kw. 1,456 ft TL: N40 52 29 W121 46 13. Hrs open: 24 Box 1717, 20410 Marquette St., 96013. Phone: (530) 335-5422. E-mail: pastorbud@kibcfm.org Web Site: www.kibcfm.org. Licensee: Burney Educational Broadcasting Foundation. Format: Educ, relg mus. Target aud: General. ♦Wayne Hennessey, gen mgr; Jack Drake, chief of engrg.

***KNCA(FM)**— July 1992: 89.7 mhz; 2.28 kw. 1,465 ft TL: N40 52 30 W121 46 14. Stereo. Hrs opn: 5 AM-2 AM Southern Oregon State University, 1250 Siskiyou Blvd., Ashland, OR, 97520. Phone: (541) 552-6301. Phone: (541) 552-8565. Web Site: www.ijpr.org. Licensee: The State of Oregon, acting by and through the State Board of Higher Education. Natl. Network: NPR, PRI. Law Firm: Ernest Sanchez. Format: Jazz, AAA, news. News staff: one; News: 45 hrs wkly. Target aud: General. Spec prog: Blues 6 hrs, folk 3 hrs, pub affrs 7 hrs wkly. ♦Mitchell Christian, CFO; Ronald Kramer, CEO & gen mgr; Bryon Lambert, opns dir; Paul Westhelle, dev dir.

KRRX(FM)—Licensed to Burney. See Redding

Calexico

KGBA(AM)— Apr 6, 1946: 1490 khz; 1 kw-U. TL: N32 41 58 W115 30 10. Hrs open: 24 Box 232, 695 Hwy. 111, 92231. Phone: (760) 357-5055. Fax: (760) 357-4168. Licensee: The Voice of International Christian Evangelism Inc. (acq 7-6-2007; $350,000). Population served: 1,440,000 Format: Sp. Target aud: 18-49; Hispanic. ♦Douglas Hanson, gen mgr; Paul Raine, gen sls mgr; Noe Diaz, progmg mgr.

***KQVO(FM)**— March 1984: 97.7 mhz; 6 kw. Ant 305 ft TL: N32 40 48 W115 25 36. Stereo. Hrs opn: 24
Rebroadcasts KPBS-FM San Diego 100%.
c/o KPBS-FM, 5200 Campanile Dr., San Diego, 92182-5400. Phone: (619) 594-1515. Fax: (619) 594-3812. Web Site: www.kpbs.org. Licensee: State of California, San Diego State University (acq 5-10-2005; $1.1 million). Natl. Network: NPR. Format: News/talk, classical. ♦Doug Myrland, gen mgr.

***KUBO(FM)**— 1989: 88.7 mhz; 3 kw. 272 ft TL: N32 47 57 W115 30 12. Hrs open: 24 531 Main St., #2, El Centro, 92243. Phone: (760) 337-8053. Fax: (760) 337-8519. E-mail: carlosleon@radiobilingue.org Web Site: www.radiobilingue.org. Licensee: Radio Bilingue Inc. Format: Multilingual, Ethnic, Sp. News staff: 5; News: 3 hrs wkly. Target aud: 16-60; Latino. ♦Hugo Morales, CEO; Maria Erana, opns dir; Maria Esana, progmg dir.

California City

KCEL(FM)— May 22, 1999: 106.9 mhz; 2.35 kw. Ant 522 ft TL: N35 12 44 W117 45 11. Hrs open: 24 Aloha Plaza, 8401 Calif. City Blvd., # 9, 93505. Phone: (760) 373-1069. Fax: (760) 373-8808. E-mail: kcel@kcel.com Web Site: www.kcel.com. Licensee: Point Broadcasting Co. Group owner: Point Broadcasting Company (acq 12-29-2003; $500,000). Population served: 750,000 Natl. Network: ABC. Format: Rgnl Sp. News staff: one. Target aud: General. Spec prog: Gospel 3 hrs wkly.

Calipatria

KSSB(FM)— Feb 8, 1997: 100.9 mhz; 3 kw. Ant 148 ft TL: N33 07 12 W115 30 47. Hrs open: 24 211 S. 5th St., Saint Maries, ID, 83861-2020. Phone: (760) 348-7908. Fax: (760) 348-7908. Licensee: Phillip J. Plank. (acq 3-21-94; FTR: 6-20-94). News staff: one; News: 6 hrs wkly. Target aud: 25 plus; female/male. Spec prog: Religious 6 hrs wkly. ♦Philip Plank, CEO; Armando Guttierez, gen mgr.

Calistoga

KXTS(FM)— 1996: 100.9 mhz; 64 w. 2,945 ft TL: N38 40 10 W122 37 52. Hrs open: 3565 Standish Ave., Santa Rosa, 95407. Phone: (707) 270-1009. Phone: (707) 588-0707. Fax: (707) 588-0777. Licensee: Sinclair Telecable Inc. Group owner: Sinclair Communications Inc. (acq 8-3-2001; $3.5 million). Format: Spanish. Target aud: 25-52; upscale - 60% female. ♦Bob Sinclair, pres; Debbie Morton, gen mgr.

Camarillo

***KMRO(FM)**— Jan 19, 1987: 90.3 mhz; 4.43 kw. Ant 1,250 ft TL: N34 24 47 W119 11 10. Stereo. Hrs opn: 24 2310 Ponderosa Dr., Suite 28, 93010. Phone: (805) 482-4797. Fax: (805) 388-5202. E-mail: info@nuevavida.com Web Site: www.nuevavida.com. Licensee: The Association for Community Education Inc. Population served: 800,000 Law Firm: Miller & Neely. Format: Relg, Sp. Target aud: General; Hispanics. ♦Phil Guthrie, pres; Mary Guthrie, gen mgr.

KOCP(FM)— Aug 15, 1972: 95.9 mhz; 5 kw. 813 ft TL: N34 06 47 W119 03 34. (CP: 1.25 kw, ant 1,440 ft. TL: 34 20 55 W119 20 13). Stereo. Hrs opn: 24 2284 S. Victoria, Suite 2-G, Ventura, 93003. Phone: (805) 289-1400. Fax: (805) 644-7906. E-mail: perryinthemorning@yahoo.com Web Site: www.theoctopus959.com. Licensee: Gold Coast Broadcasting LLC (group owner; acq 1995; $1.2 million with KMXO(AM) Santa Paula). Population served: 55,000 Natl. Rep: Katz Radio. Format: Classic rock. Target aud: 25-54. ♦Chip Ehrhardt, gen mgr; Perry Van Houten, progmg dir.

Cambria

KPYG(FM)— Oct 1, 1984: 94.9 mhz; 25 kw. 328 ft TL: N35 31 26 W121 03 40. Stereo. Hrs opn: 24 396 Buckley Rd., Suite 2, San Luis Obispo, 93401. Phone: (805) 786-2570. Fax: (805) 547-9860. Web Site: www.mapletoncommunications.com. Licensee: Mapleton Communications LLC (group owner; acq 5-23-02; grpsl). Population served: 300,000 Law Firm: Haley, Bader & Potts. Format: AAA. Target aud: 25-54. ♦Adam Nathanson, pres; Bill Heirendt, progmg dir; David Atwood, news dir; Tom Hughes, chief of engrg.

KTEA(FM)— Nov 9, 2003: 103.5 mhz; 1.46 kw. Ant 679 ft TL: N35 21 36 W121 03 40. Stereo. Hrs opn: 7 AM-midnight 2976 Burton Dr., 93428. Phone: (805) 924-0103. E-mail: jim@ktea-fm.com Licensee: James Robert Kampschroer. Format: Big band 30's through 50's broad spectrum. Target aud: 50+. ♦James Robert Kampschroer, pres & gen mgr; Lee Went, progmg dir.

Camino

***KYCJ(FM)**— 2005: 88.3 mhz; 50 w vert. Ant 508 ft TL: N38 44 18 W120 42 10. Hrs open: 24
Rebroadcasts KYCC(FM) Stockton 100%.
9019 N. West Ln., Stockton, 95210-1401. Phone: (209) 477-3690. Fax: (209) 477-2762. E-mail: kycc@kycc.org Web Site: www.kycc.org. Licensee: Your Christian Companion Network Inc. Format: Gospel, inspirational, adult contemp. Target aud: 35-55. ♦Shirley Garner, gen mgr.

Canyon Country

KHTS(AM)— June 1989: 1220 khz; 1 kw-D, 500 w-N, DA-2. TL: N34 27 55 W118 24 08. Stereo. Hrs opn: 24 27225 Camp Plenty Rd., Suite 8, Santa Clarita, 91351. Phone: (661) 298-1220. Fax: (661) 298-2020. Web Site: www.hometownstation.com. Licensee: Jeri Lyn Broadcasting Inc. (acq 10-24-2003; $900,000). Format: Full svc. ♦Carl Goldman, gen mgr; Shaun Valentine, progmg dir; Jon Dell, news dir; Bruce Smith, chief of engrg.

Carlsbad

KUSS(FM)— Aug 22, 1965: 95.7 mhz; 29 kw. 639 ft TL: N32 50 24 W117 14 52. Stereo. Hrs open: 24 9660 Granite Ridge Rd., San Diego, 92123. Phone: (858) 292-2000. Fax: (858) 278-7957. Web Site: www.us957.com. Licensee: Citicasters Licenses L.P. Group owner: Clear Channel Communications Inc. (acq 5-4-99; grpsl). Population served: 600,000 Law Firm: Hogan and Hartson. Format: Sixties and Seventies. Target aud: 25-54; general. ♦Bob Bolinger, gen mgr; Geoff Alan, prom dir; Mike O'Brian, progmg dir.

Carmel

KBOQ(FM)—Licensed to Carmel. See Monterey

KCDU(FM)— Apr 29, 1971: 101.7 mhz; 800 w. 590 ft TL: N36 33 12 W121 47 05. Stereo. Hrs opn: 24 60 Garden Ct., Suite 300, Monterey, 93940-5341. Phone: (831) 658-5200. Fax: (831) 658-5299. Web Site: www.1017thebeach.com. Licensee: Mapleton Communications L.L.C. (group owner; acq 1-17-02; grpsl). Population served: 500,000 Natl. Rep: McGavren Guild. Law Firm: Leventhal, Senter & Lerman. Format: Adult contemp 80s & 90s. Target aud: Women 25-54; upscale, educated, above average income. ♦Adam Nathanson, pres; Raul Salvador, CFO; Dale Hendry, gen mgr; Mike Anthony, opns VP; Jodi Morgan, sls dir & gen sls mgr; Sybil DeAngelo, prom dir.

KRML(AM)— Dec 25, 1957: 1410 khz; 500 w-D, 16 w-N, 2.5 kw-U. TL: N36 32 06 W121 53 34. Stereo. Hrs opn: 24
live streaming from www.krmlradio.com.
Box 7300, The Eastwood Bldg., Carmel-By-The-Sea, 93921. Secondary address: San Carlos near 5th 93921. Phone: (831) 624-6431. Phone: (831) 624-6432. Fax: (831) 625-2417. E-mail: info@thejazzandbluescompany.com Web Site: www.krmlradio.com. Licensee: Wisdom Broadcasting Co. Inc. (acq 4-23-2004; $725,000). Population served: 700,000 Rgnl rep: McGavern Guild Law Firm: Putbrese, Hunsaker & Trent, P. Format: Blues, jazz. News: 10.5 hrs wkly. Target aud: 35+. Spec prog: Gospel 6 hrs wkly. ♦David Kimball, CFO & gen mgr.

Carmel Valley

KRXA(AM)— July 10, 1989: 540 khz; 10 kw-D, 500 w-N, DA-2. TL: N36 39 38 W121 32 29. Stereo. Hrs opn: 18 495 Elder Ave., Suite 7, Sand City, 93955. Phone: (831) 394-5792. Fax: (831) 899-7600. Web Site: www.krxa540.com. Licensee: KRFA-AM LLC (group owner; (acq 7-8-2005; $800,000). Format: Talk. ♦Hal Ginsberg, gen mgr; Matt Renner, opns dir; Peter B. Collins, progmg mgr.

Carmichael

KFIA(AM)— Jan 11, 1979: 710 khz; 25 kw-D, 1 kw-N, DA-2. TL: N38 49 58 W121 19 03. Hrs opn: 1425 River Park Dr., Suite 520, Sacramento, 95815. Phone: (916) 924-0710. Fax: (916) 924-1587. Licensee: New Inspiration Broadcasting Co. Inc. Group owner: Salem Communications Corp. (acq 2-15-95; FTR: 5-8-95). Population served: 7,000,000 Format: Relg. Target aud: 35 plus; general. ♦Edward Atsinger III, pres; James Rowten, gen mgr; Steve Gasser, opns mgr; Laurie Larson, progmg dir.

Carnelian Bay

KODS(FM)—Licensed to Carnelian Bay. See Reno NV

Carpinteria

KSBL(FM)— June 1, 1981: 101.7 mhz; 310 w. 810 ft TL: N34 27 55 W119 40 37. Stereo. Hrs opn: 24 414 E. Cota St., Santa Barbara, 93101. Phone: (805) 879-8300. Fax: (805) 879-8430. Web Site: www.klite.com. Licensee: Rincon License Subsidiary LLC. Group owner: Clear Channel Communications Inc. (acq 2-28-2007; grpsl). Format: Adult contemp. News staff: news progmg 2 hrs wkly News: one;. Target aud: 25-54; women. ♦Tom Baker, gen mgr; Keith Royer, opns dir; Vince Hollian, rgnl sls mgr; Lin Aubuchon, prom dir; Peter Bie, news dir; Andrea Shaparenko, traf mgr.

Cartago

KWTY(FM)— November 1989: 102.9 mhz; 2 kw. -1,787 ft TL: N36 19 16 W118 01 22. Stereo. Hrs opn: 24 Box 91, Olancha, 93549. Phone: (760) 764-1111. Fax: (760) 764-1111. E-mail: gm@kwty.com Licensee: Mark A. Miller (acq 7-25-2005). Population served: 29,000 Wire Svc: UPI Format: Classic rock, rock. News: 7 hrs wkly. Target aud: General; 15-55 years (M-F), recreation/resort commuters. ♦Dan Owen, sls dir; Mark Miller, gen mgr, stn mgr & chief of engrg.

Cathedral City

KPTR(AM)—Licensed to Cathedral City. See Palm Springs

KWXY-FM—Licensed to Cathedral City. See Palm Springs

Cazadero

KGRP(FM)— 2007: 106.3 mhz; 1.35 kw. Ant 689 ft TL: N38 29 20 W123 01 55. Hrs open: 2801 Via Fortuna Dr., Suite 675, Austin, TX, 78746. Phone: (713) 528-2517. Licensee: Ace Radio Corp. ♦Stephen Hackerman, pres.

Ceres

***KBES(FM)**— Sept 1, 1979: 89.5 mhz; 150 w horiz. 131 ft TL: N37 35 21 W120 57 23. Hrs open: Box 4116, Modesto, 95352. Phone: (209) 538-4130. Fax: (209) 538-2795. Web Site: www.betnahrain.org. Licensee: Bet Nahrain Inc. Format: Syrian. ♦Dr. Sargon Dadisho, gen mgr; Janet Shamon, progmg dir; Seimon Mamio, chief of engrg.

KVIN(AM)— Sept 15, 1963: 920 khz; 2.5 kw-U, DA-2. TL: N37 35 49 W121 04 15. Hrs open: 961 N. Emerald Ave., Ste A, Modesto, 95351. Phone: (209) 544-1055. Fax: (209) 544-1055. E-mail: theriver@krvr.com Web Site: www.krvr.com. Licensee: Threshold Communications (acq 9-5-01; $400,000). Population served: 1,000,000 Natl. Rep: Interep. Law Firm: Donald E. Martin P.C. Format: Adult standards. Target aud: 35-64. ♦Jim Bryan, gen mgr; Doug Wulff, opns mgr; Brian Henry, chief of engrg.

Chester

KWLU(FM)— Apr 6, 1989: 98.9 mhz; 12 kw. Ant 2,427 ft TL: N40 14 00 W121 01 11. Stereo. Hrs opn: 24 2351 Sunset Blvd., Suite 170-218, Rocklin, 95765. Phone: (916) 251-1600. Fax: (916) 251-1650. Licensee: Educational Media Foundation. (acq 6-30-2005; $900,000 with KPCO(AM) Quincy). Population served: 55,851 Format: Christian. ♦Richard Jenkins, pres; Mike Novak, VP; Keith Whipple, dev dir; Ed Lenane, news dir; Sam Wallington, engrg dir; Karen Johnson, news rptr; Marya Morgan, news rptr; Richard Hunt, news rptr.

Chico

KBQB(FM)— June 1993: 92.7 mhz; 1.5 kw. Ant 643 ft TL: N39 48 25 W121 37 35. Hrs open: 856 Manzanita Ct., 95926. Phone: (530) 342-2200. Fax: (530) 342-2260. E-mail: bob@927bobfm.com Web Site: www.927bobtm.com. Licensee: Results Radio Licensee L.L.C. Group owner: Fritz Communications Inc. (acq 6-11-99; grpsl). Format: Adult hits. ♦John Graham, gen mgr; Dave Pack, gen sls mgr; Chad Perry, progmg dir; J.D. Davis, chief of engrg; Candy Mason, traf mgr.

***KCHO(FM)**— Apr 22, 1969: 91.7 mhz; 7.71 kw. 1,219 ft TL: N39 57 30 W121 42 48. Stereo. Hrs opn: 24 California State Univ., 95929-0500. Phone: (530) 898-5896. Fax: (530) 898-4348. E-mail: info@kcho.org Web Site: www.kcho.org. Licensee: California State University, Chico Research Foundation. Population served: 234,281 Natl. Network: PRI, NPR. Law Firm: Cohn & Marks. Format: Class, jazz, news & info. News staff: one; News: 37 hrs wkly. Target aud: General. ♦Brian Terhorst, gen mgr; Beth Heberle, mktg dir; Joe Oleksiewicz, progmg dir; Lorraine Dechter, news dir; Mike Birdsill, chief of engrg.

KFMF(FM)— Feb 1, 1974: 93.9 mhz; 2 kw. 1,128 ft TL: N39 56 46 W121 43 17. Stereo. Hrs opn: 24 1459 Humboldt Rd., Suite D, 95928. Phone: (530) 899-3600. Fax: (530) 343-0243. Web Site: www.kfm.com. Licensee: Mapleton Communications LLC. Group owner: Regent Communications Inc. (acq 11-30-2006; grpsl). Population served: 317,000 Natl. Rep: Christal. Format: Active rock. News staff: one; News: one hr wkly. Target aud: 18-44. ♦Dick Stein, gen mgr; Brian Fox, prom dir; Linda Patterson, traf mgr & disc jockey.

***KHAP(FM)**— 1999: 89.1 mhz; 12 kw. 285 ft TL: N39 43 37 W121 40 45. Hrs open: 290 Hergenberger Rd., Oakland, 94621. Phone: (530) 877-5650. Fax: (916) 641-8238. E-mail: kebr@jps.net Web Site: www.familyradio.com.yes Licensee: Family Stations Inc. (group owner) Population served: 85,000 Format: Relg, educ. ♦Harold Camping, pres; Thad McKinney, gen mgr.

KHSL-FM—See Paradise

KKXX(AM)—See Paradise

KMXI(FM)—Listing follows KPAY(AM).

KPAY(AM)— Apr 17, 1935: 1290 khz; 5 kw-U, DA-N. TL: N39 44 00 W121 44 10. Stereo. Hrs opn: 2654 Cramer Ln., 95928. Phone: (530) 345-0021. Fax: (530) 893-2121. Web Site: www.kpay.com. Licensee: Deer Creek Broadcasting LLC. (group owner; (acq 9-8-2004; grpsl). Population served: 350,000 Natl. Rep: Katz Radio. Format: News/talk. Target aud: 25 plus. ♦Dino Corbin, gen mgr & progmg mgr; Lisa Fitzgerald, mktg dir & prom dir; Larry Scott, progmg dir; Matt Ray, news dir.

KMXI(FM)—Co-owned with KPAY(AM). Nov 16, 1972: 95.1 mhz; 8.7 kw. 1,171 ft TL: N39 56 46 W121 43 17. Stereo. Web Site: (530) 893-2121.25,254 Format: Adult contemp.

KQPT(FM)—Colusa, September 1986: 107.5 mhz; 28 kw. 600 ft TL: N39 17 17 W122 20 02. Stereo. Hrs opn: 1459 Humboldt Rd., Suite D, 95928-9100. Phone: (530) 899-3600. Fax: (530) 343-0243. Web Site: www.107thepoint.com. Licensee: Mapleton Communications LLC. Group owner: Regent Communications Inc. (acq 11-30-2006; grpsl). Natl. Rep: Christal. Format: Modern adult contemp. Target aud: 24-48. ♦Dick Stein, VP & gen mgr.

KZAP(FM)—See Paradise

***KZFR(FM)**— July 6, 1990: 90.1 mhz; 6.3 kw. 587 ft TL: N39 48 25 W121 37 35. Stereo. Hrs opn: 24 Box 3173, 95927. Secondary address: 341 Broadway, Suite 411 95928. Phone: (530) 895-0706/895-0788. Fax: (530) 895-0775. Web Site: www.kzfr.org. Licensee: Golden Valley Community Broadcasters. Format: Div, news/talk. News: 8 hrs wkly. Spec prog: American Indian 2 hrs, Sp 6 hrs wkly. ♦Jill L. Paydon, gen mgr.

China Lake

***KFRJ(FM)**— June 2005: 91.1 mhz; 3 kw. Ant 1,282 ft TL: N35 28 41 W117 41 58. Hrs open: 24 Family Stations Inc., 4135 Northgate Blvd., Suite 1, Sacramento, 95834. Phone: (916) 641-8191. Fax: (916) 641-8238. Licensee: Family Stations Inc. (group owner). Format: Relg. ♦Harold Camping, pres.

KSSI(FM)— 1995: 102.7 mhz; 3 kw. -22 ft TL: N35 39 06 W117 40 58. Hrs open: 24 701 Inyokern Rd., Suite C, Ridgecrest, 93555. Phone: (760) 446-5774. Fax: (760) 446-5774. E-mail: kssirock@iwvisp.com Web Site: www.kssifm.com. Licensee: Sound Enterprises. Population served: 40,000 Format: AOR. Target aud: 25-54; general. ♦John Perrige, gen mgr; Lisa Garcia, traf mgr.

Chowchilla

KNTO(FM)— Aug 1, 1992: 93.3 mhz; 2.95 kw. 335 ft TL: N37 13 01 W120 11 57. Hrs open: 24 4043 Geer Rd., Hughson, 95326. Phone: (209) 883-8760. Fax: (209) 883-8769. Web Site: www.lafavorita.net. Licensee: KSKD Inc. (acq 4-17-01; $450,000). Population served: 140,000 Format: Sp, Mexican music. Target aud: 18-35; teens, young adults. ♦Nelson Gomez, pres & gen mgr; Freddy Lopez, news dir.

Chualar

KHDC(FM)—Licensed to Chualar. See Salinas

Claremont

***KSPC(FM)**— February 1956: 88.7 mhz; 3 kw. -265 ft TL: N34 05 38 W117 42 35. Stereo. Hrs opn: Pomona College, 340 N. College Ave., 91711-6340. Phone: (909) 621-8157. E-mail: director@kspc.org Web

Site: www.kspc.org. Licensee: Pomona College. Format: Alternative rock, div, jazz. Target aud: General. Spec prog: Pol 3 hrs, reggae 4 hrs, blues 4 hrs, pub affrs 3 hrs, hip hop/rap 6 hrs wkly. ♦Roxy Cruz, progmg dir; Stacy Wood, mus dir.

KWKU(AM)—See Pomona

Cloverdale

KSRT(FM)— 2002: 107.1 mhz; 3.5 kw. Ant 430 ft TL: N38 48 34 W123 02 56. Hrs open: 24 5510 Skylane Blvd., Suite 102, Santa Rosa, 95403. Phone: (707) 284-3069. Fax: (707) 284-3174. Web Site: www.radiolazer.com. Licensee: Lazer Licenses LLC. (acq 6-29-2006; $6.85 million with KJOR(FM) Windsor). Format: Mexican regional. ♦Ken Kuhl, gen mgr & gen sls mgr; Salvador Prieto, progmg dir.

Clovis

KFPT(AM)— May 2, 1977: 790 khz; 5 kw-D, 2.5 kw-N, DA-2. TL: N36 50 39 W119 41 13. Hrs open: 24 351 W. Cromwell Ave., Suite 108, Fresno, 93711. Phone: (559) 447-3570. Fax: (559) 447-3579. Web Site: www.1430espn.com. Licensee: Peak Broadcasting of Fresno Licenses LLC. Group owner: Infinity Broadcasting Corp. (acq 3-30-2007; grpsl). Population served: 1,200,000 Natl. Network: ESPN Radio. Format: Sports. ♦Paul Swearengin, gen mgr.

KOND(FM)—Licensed to Clovis. See Madera

Coachella

***KBXO(FM)**— February 2005: 90.3 mhz; 340 w. Ant 574 ft TL: N33 48 08 W116 13 30. Stereo. Hrs opn: 24 Box 1924, Tulsa, OK, 74101. Phone: (918) 455-5693. Fax: (918) 455-0411. E-mail: mail@oasisnetwork.org Web Site: www.oasisnetwork.com. Licensee: Creative Educational Media Corp. Inc. Format: Relg. Target aud: General. ♦David Ingles, pres & gen mgr; David Warren, stn mgr & progmg dir.

KCLB-FM—Listing follows KNWZ(AM).

KNWZ(AM)— 1954: 970 khz; 5 kw-D, 1 kw-N, DA-2. TL: N33 41 12 W116 09 34. Hrs open: 1321 North Gene Autry Tr., Palm Springs, 92262. Phone: (760) 322-7890. Fax: (760) 322-5493. Licensee: Morris Communications Corp. Group owner: Morris Communications Inc. (acq 1998; $7 million with co-located FM). Format: Talk. Target aud: 18-49. ♦Gary Demardney, opns dir; David Nola, sls dir; Virgina Nelson, gen mgr & gen sls mgr; Pete Fox, prom dir; Brian Long, progmg dir; Angela Terry, traf mgr.

KCLB-FM—Co-owned with KNWZ(AM). Sept 1, 1960: 93.7 mhz; 26.5 kw. 640 ft TL: N33 44 07 W116 13 27. Stereo. 200,000 Format: Rock (AOR). Target aud: 18-54. ♦Dave Sparks, progmg dir; Angela Terry, traf mgr.

***KPSH(FM)**—Not on air, target date: January 2005: 90.9 mhz; 230 w. Ant 623 ft TL: N33 52 03 W116 25 58. Hrs open: Box 262550, Baton Rouge, LA, 70826. Secondary address: 8919 World Ministry Ave., Baton Rouge, LA 70826. Phone: (225) 768-3688. Phone: (225) 768-8300. Fax: (225) 768-3729. E-mail: kawikfish@yahoo.com Web Site: www.jsm.org. Licensee: Family Worship Center Church Inc. (group owner; acq 2-18-2004; $750,000 for CP). Format: Relg, christian. ♦David Whitelow, COO; Jimmy Swaggart, pres; John Santiago, progmg dir.

Coalinga

***KDKL(FM)**— 1999: 88.3 mhz; 1.45 kw vert. 2,329 ft TL: N36 22 11 W120 38 37. Hrs open: 2351 Sunset Blvd., Suite 170-218, Rocklin, 95765. Phone: (800) 372-0888. Fax: (916) 251-1650. Web Site: www.klove.com. Licensee: Educational Media Foundation (acq 10-20-00; $80,000 for CP). Format: Contemp Christian. ♦Richard Jenkins, pres; Mike Novak, VP; Ed Lenane, opns dir & news dir; Keith Whipple, dev dir; David Pierce, progmg mgr; Sam Wallington, engrg dir; Arthur Vassar, traf mgr; Karen Johnson, news rptr; Richard Hunt, news rptr.

***KFRP(FM)**— November 2005: 90.7 mhz; 2.5 kw vert. Ant 1,253 ft TL: N35 55 39 W120 22 46. Hrs open: Family Stations Inc., 4135 Northgate Blvd., Suite 1, Sacramento, 95834. Phone: (916) 641-8191. Fax: (916) 641-8238. Licensee: Family Stations Inc. (group owner). Format: Relg. ♦Harold Camping, pres.

KNGS(FM)—Not on air, target date: unknown: 100.1 mhz; 19 kw. Ant 794 ft TL: N36 00 40 W120 04 26. Hrs open: c/o William L. Zawilla, 12550 Brookhurst, Garden Grove, 92640. Phone: (714) 636-5040. Licensee: William L. Zawila. ♦William Zawila, gen mgr.

KQNO(FM)—Not on air, target date: unknown: 97.3 mhz; 6 kw. Ant 328 ft TL: N36 16 32 W120 19 45. Hrs open: 2801 Via Fortuna Dr., Suite 675, Austin, TX, 78746. Phone: (713) 528-2517. Licensee: Ace Radio Corp. ♦Stephen Hackerman, pres.

Columbia

KCVR-FM— August 1995: 98.9 mhz; 6 kw. Ant 328 ft TL: N38 02 15 W120 22 05. Hrs open: 6820 Pacific Ave., Suite 3A, Stockton, 95207. Phone: (209) 474-0154. Fax: (209) 474-0316. Web Site: www.entravision.com. Licensee: Entravision Holdings LLC. Group owner: Entravision Communications Corp. (acq 7-28-2000; grpsl). Format: Sp adult hits. ♦Lisa Sunday, gen mgr; Edgar Pineda, progmg dir.

Colusa

KKCY(FM)— May 1990: 103.1 mhz; 135 w. 1,964 ft TL: N39 12 21 W121 49 11. Stereo. Hrs opn: 24 861 Gray Ave, Suite K, Yuba City, 95991. Phone: (530) 673-2200. Fax: (530) 673-3010. E-mail: resultsradio@syix.com Web Site: www.kkcy.com. Licensee: Results Radio of Chico Licensee LLC. Group owner: Fritz Communications Inc. (acq 6-11-99; grpsl). Natl. Rep: Katz Radio. Law Firm: Kaye, Scholer, Fierman, Hays & Handler. Format: Country. News staff: one; News: 7 hrs wkly. Target aud: 18-64. Spec prog: Sp one hr wkly. ♦Jack Fritz, pres; Michael Berry, sls dir & gen sls mgr; Dave Logasa, progmg dir.

KQPT(FM)—Licensed to Colusa. See Chico

Compton

KJLH-FM— April 1965: 102.3 mhz; 5.6 kw. Ant 338 ft TL: N33 59 52 W118 21 32. Stereo. Hrs opn: 24 161 N. La Brea Ave., Inglewood, 90301. Phone: (310) 330-2200. Fax: (310) 330-5555. Fax: (310) 330-2244. E-mail: sales@kjlhradio.com Web Site: www.kjlhradio.com. Licensee: TAXI Productions Inc. (acq 6-79). Population served: 340,000 Natl. Network: American Urban, ABC. Natl. Rep: McGavren Guild. Law Firm: Irwin, Campbell & Tannenwald. Format: Urban contemp, rhythm and blues. News staff: 2; News: 8.5 hrs wkly. Target aud: 25-49; African-American audience. Spec prog: Relg 7 hrs, gospel 6 hrs, talk 8.5 hrs, Christian 6 hrs wkly. ♦Stevland Morris, CEO; Karen Slade, gen mgr; Lawrence Williams, opns dir; Aundrae Russell, progmg dir; Jacquie Stephens, news dir; Barry Clark, chief of engrg; Carrie Haynes, traf mgr.

Concord

***KVHS(FM)**— May 16, 1969: 90.5 mhz; 410 w. 450 ft TL: N39 01 49 W122 00 04. Stereo. Hrs opn: 24/7 1101 Alberta Way, Rm. S-2, 94521. Phone: (925) 682-5847. Fax: (925) 609-5847. E-mail: kvhsgm@mail.com Web Site: www.kvhs.com. Licensee: Clayton Valley High School. Population served: 1,001,136 Wire Svc: Bay City News Service Format: Active/new rock. Target aud: 18-34.(P-1) 18-49(P-2) 12+. Spec prog: Flashback Show(Classic Rock); Punk & SKA Show; Metal Show(Rock). ♦Melissa McConnell Wilson, gen mgr.

Copperopolis

KRVR(FM)— Jan 1, 1995: 105.5 mhz; 1 kw. 781 ft TL: N37 56 55 W120 42 16. Stereo. Hrs opn: 24 961 N. Emerald Ave., Suite A, Modesto, 95351. Phone: (209) 544-1055. Fax: (209) 544-8105. E-mail: TheRiver@krvr.com Web Site: krvr.com. Licensee: Threshold Communications. Population served: 750,000 Natl. Rep: Interep. Law Firm: Donald E. Martin. Format: Smooth Jazz. Target aud: 35-64. ♦Jim Bryan, gen mgr; Doug Wulff, opns mgr & mus dir; Cheryl Miller, rgnl sls mgr; James Arata, progmg dir; Sally Waterman, traf mgr.

Corcoran

KBLO(FM)— 1999: 102.3 mhz; 19.5 kw. Ant 380 ft TL: N36 11 04 W119 24 01. Hrs open: 24 113 N. Church St., Suite 511, Visalia, 93291. Phone: (559) 740-4172. Fax: (559) 740-4177. Web Site: www.radiolobo987.com. Licensee: Mapleton Communications LLC. (acq 6-1-2005; $2.1 million). Format: Sp. ♦Andrew Adams, gen mgr; Dora Deltora, gen sls mgr; Jaun Davbla, prom mgr; Rick McMillion, chief of engrg.

Corning

KTHU(FM)— Apr 8, 1988: 100.7 mhz; 50 kw. 272 ft TL: N39 53 17 W122 37 38. (CP: 20.5 kw, ant 1,742 ft.). Stereo. Hrs opn: 24 856 Manzanita Ct., Chico, 95926. Phone: (530) 342-2200. Fax: (530) 342-2260. E-mail: kthu1007@sunset.net Web Site: www.chicothunderheads.com. Licensee: Results Radio Licensee L.L.C. Group owner: Fritz Communications Inc. (acq 6-11-99; grpsl). Law Firm: Kaye, Scholer, Fierman, Hays & Handler. Format: Classic rock. News staff: one. Target aud: 25-54. Spec prog: Sp one hr wkly. ♦Jack Fritz, pres; John Graham, gen mgr; dave Pack, gen sls mgr; J. D. Davis, engrg VP; Candy Mason, traf mgr.

Corona

KWRM(AM)— 1948: 1370 khz; 5 kw-D, 2.5 kw-N, DA-2. TL: N33 52 52 W117 32 33. Hrs open: 24 Box 100, 92878. Phone: (951) 737-1370. Fax: (951) 735-9572. E-mail: info@majormarket.com Web Site: kwrm1370am.com. Licensee: Major Market Stations Inc. (acq 1968). Population served: 10,000,000 Law Firm: Hardy & Carey. Format: Multilingual, sports, var/div. News staff: 2; News: 20 hrs wkly. Target aud: 18-49; young Hispanic adults. ♦Marilyn Kramer, CEO; Marilynn Kramer, gen mgr; William J. Roberts, chmn, pres & gen mgr; Damian Vasquez, opns dir.

Covelo

KQZT(FM)—Not on air, target date: unknown: 96.9 mhz; 1 kw. Ant 795 ft TL: N39 47 08 W123 05 46. Hrs open: 110 S. Franklin St., Fort Bragg, 95437. Phone: (707) 964-7277. Fax: (707) 964-9536. Licensee: California Radio Partners Inc. ♦Tom Yates, pres & gen mgr.

Crescent City

KCRE-FM— Mar 21, 1980: 94.3 mhz; 25 kw. Ant -305 ft TL: N41 45 35 W124 09 49. Stereo. Hrs opn: Box 1089, 95531. Secondary address: 1345 Northcrest Dr. 95531. Phone: (707) 464-9561. Fax: (707) 464-4303. E-mail: kcre@charter.net Web Site: www.historicalfavorites.com. Licensee: KPOD L.L.C. (acq 7-30-02; $692,000). Population served: 35,000 Format: Adult contemp. ♦Kelly Schellong, gen sls mgr; Renee Shanle-Hutzell, gen mgr & progmg dir; Kevin Sanders, chief of engrg.

KFVR(AM)— July 1950: 1310 khz; 1 kw-D. TL: N41 45 35 W124 09 49. Hrs open: 24 Box 109, Eureka, 95502-0109. Phone: (707) 725-9363. Fax: (707) 726-9446. Web Site: www.lanueva1090.com. Licensee: Del Rosario Talpa Inc. (acq 5-29-2003; $54,000). Format: Rgn Mexican. ♦Mario Meza, gen mgr.

***KHSR(FM)**— July 1999: 91.9 mhz; 800 w. 226 ft TL: N41 50 36 W124 07 55. Stereo. Hrs opn: 24
Rebroadcasts KHSU-FM Arcata 100%.
Humboldt State University, 1 Harpst St., Arcata, 95521. Phone: (707) 826-4807. Fax: (707) 826-6082. E-mail: khsu@humboldt.edu Web Site: www.khsu.org. Licensee: Humboldt State University. Population served: 22,000 Natl. Network: NPR, PRI. Format: Var, news. Spec prog: World 14 hrs, jazz 10 hrs wkly. ♦Elizabeth Hans-McCrone, gen mgr; Charles Horn, dev dir; Katie Whiteside, progmg dir; Kevin Sanders, chief of engrg.

KPOD(AM)— Dec 5, 1959: 1240 khz; 778 w-U. TL: N41 45 35 W124 11 28. Hrs open: 24 Box 1089, 1345 N. Crest Dr., 95531. Phone: (707) 464-3183. Fax: (707) 465-6703. E-mail: kpod@link.cc.com Web Site: www.kpod.com. Licensee: KPOD LLC. Group owner: Bicoastal Media LLC (acq 3-31-00; $850,000 with co-located FM). Population served: 60,000 Natl. Network: ABC. Format: Adult contemp. ♦Mike Wilson, pres & gen mgr; Renee Shaulehutzell, gen mgr & progmg dir; Andy Manuel, disc jockey; Diana Russel, disc jockey; Jay Vincent, disc jockey; Liz Kelly, disc jockey.

KPOD-FM— January 1989: 97.9 mhz; 6 kw. Ant -128 ft TL: N41 45 35 W124 11 28.24 Phone: (707) 464-1000. Web Site: www.kpod.com. Natl. Network: ABC. Law Firm: Cohn & Marks. Format: Hot country.Michelle Smith, pub affrs dir; Bill Stamps, spec ev coord, edit dir, relg ed & disc jockey; Doug Vanderpool, news rptr; Eileen Bennett, mus critic & disc jockey; Chuck Blackburn, sports cmtr; Teresa Throop, traf mgr & women's int ed; Dutch Dremann, disc jockey; J.D. Moon, disc jockey; Linda Garnett, disc jockey; Mike Moran, disc jockey; Robert McBane, disc jockey

Culver City

KIEV(AM)—Licensed to Culver City. See Los Angeles

Cupertino

***KKUP(FM)**— May 15, 1972: 91.5 mhz; 200 w. 2,294 ft TL: N37 06 40 W121 50 36. Stereo. Hrs opn: 24 933 Monroe St., P.O. Box 9150, Santa Clara, 95050-4808. Phone: (408) 260-2999. Phone: (408) 260-2997. Web Site: www.kkup.org. Licensee: Assurance Sciences Foundation Inc. Population served: 2,000,000 Format: Eclectic, alternative, blues. Target aud: General. Spec prog: Brazilian 2 hrs, African 6 hrs, Indian 3 hrs, Sp 3 hrs, Latin American 8 hrs wkly. ♦Jim Thomas, chmn; Dan Kind, gen mgr; Tim Alderman, progmg dir; Peter Schwartz, mus dir & chief of engrg.

Davis

***KDVS(FM)**— Jan 1, 1968: 90.3 mhz; 9.2 kw. 105 ft TL: N38 32 29 W121 45 03. Stereo. Hrs opn: 24 c/o KDVS-FM, Univ. of California, 14 Lower Freeborn Hall, 95616. Phone: (530) 752-0728. Fax: (530) 752-8548. E-mail: gm@kdvs.org Web Site: www.kdvs.org. Licensee: Regents of the University of California. Population served: 310,000 Format: Eclectic rock, var, free form radio. News: 13 hrs wkly. Target aud: General; loc community. ♦Benjamin Johnson, gen mgr; Bryce Fsch, progmg dir; Erik Magnuson, progmg dir; A.J. Ramirez, mus dir; Sean Johannessen, mus dir; Lindsay Schrupp, news dir; Rich Lusch, chief of engrg.

KXSE(FM)— February 1979: 104.3 mhz; 3.4 kw. Ant 436 ft TL: N38 39 26 W121 43 12. Hrs opn: 24 1436 Auburn Blvd., Sacramento, 95815. Phone: (916) 646-4000. Fax: (916) 646-1958. E-mail: jverdier@entravision.com Web Site: www.entravision.com. Licensee: Entravision Holdings LLC. Group owner: Entravision Communications Corp. (acq 7-28-2000; grpsl). Natl. Network: ABC, Westwood One, CBS. Law Firm: Mullin, Rhyne, Emmons & Topel. Format: Sp adult contemp. Target aud: 25-54. ♦Larry Lamanski, gen mgr; Joni Verdier, gen sls mgr; Salvador Lopez, prom dir; Edgar Pineda, progmg dir.

Del Norte

KYDN(FM)—Not on air, target date: unknown: 96.5 mhz; 930 w. Ant 1,589 ft TL: N37 43 47 W106 35 18. Hrs open: Box 631, Monte Vista, 81144. Phone: (719) 852-3581. Fax: (719) 852-3583. Licensee: San Luis Valley Broadcasting Inc. ♦H. Robert Gourley III, pres.

Delano

KBFP-FM— Oct 2, 1986: 105.3 mhz; 50 kw. 547 ft TL: N35 30 53 W119 03 41. Stereo. Hrs opn: 1100 Mohawk St., Suite 280, Bakersfield, 93309. Phone: (661) 322-9929. Fax: (661) 283-2963. Web Site: www.klite1053.com. Licensee: CC Licenses LLC. (acq 4-94). Population served: 460,000 Natl. Rep: McGavren Guild. Law Firm: Arter & Hadden. Format: Adult contemp. Target aud: 18-44. ♦Jim Bell, VP & gen mgr; Jim Bell, gen mgr; Steve King, stn mgr & progmg dir; Ron Fisher, gen sls mgr; Steve Mull, chief of engrg.

KCHJ(AM)— Dec 1, 1951: 1010 khz; 5 kw-D, 1 kw-N, DA-2. TL: N35 48 40 W119 19 18. Hrs open: 24 5100 Commerce Dr., Bakersfield, 93309-0684. Phone: (661) 327-9711. Fax: (661) 327-0797. E-mail: info@thespanishradio.com Web Site: www.thespanishradio.com. Licensee: Illinois Lotus Corp. Group owner: Lotus Communications Corp. (acq 8-24-99; grpsl). Population served: 1,500,000 Law Firm: Borsari & Paxson. Wire Svc: UPI Format: Sp 24 hrs, 7days a wk. News staff: 2. Target aud: 18 plus; Sp speaking adults. ♦Howard Kalmenson, pres; Mike Allen, gen mgr; Bruce Thompson, stn mgr; Vicente Arias, stn mgr.

KDFO-FM— November 1968: 98.5 mhz; 50 kw. 499 ft TL: N35 42 46 W118 47 22. Hrs open: 24 1100 Mohawk St., Suite 280, Bakersfield, 93309. Phone: (661) 322-9929. Fax: (661) 283-2963. Web Site: www.985thefox.com. Licensee: CC Licenses LLC. Group owner: Clear Channel Communications Inc. (acq 10-16-2000; grpsl). Population served: 1,500,000 Format: Classic Rock. ♦Jim Bell, VP, gen mgr & gen mgr; Steve King, stn mgr & progmg dir; Ron Fisher, gen sls mgr; Steve Mull, chief of engrg.

Dinuba

KRDU(AM)— Dec 26, 1946: 1130 khz; 5 kw-D, 6.2 kw-N, DA-2. TL: N36 29 03 W119 15 57. Hrs open: 24 597 N. Alta Ave., 93618. Secondary address: 83 East Shaw, Fresno 93727. Phone: (559) 591-1130. Fax: (559) 591-4822. Licensee: Capstar TX L.P. Group owner: Clear Channel Communications Inc. (acq 8-30-00; grpsl). Population served: 1,000,000 Law Firm: Fletcher, Heald & Hildreth. Format: Relg. News staff: one; News: 7 hrs wkly. Target aud: 18-65. ♦Jim Tuck, gen mgr; Doug Diedrich, progmg dir; Mike Hauber, chief of engrg, chief of engrg, local news ed & local news ed; Steve Carloon, local news ed.

KSOF(FM)—Co-owned with KRDU(AM). June 5, 1975: 98.9 mhz; 19 kw. 820 ft TL: N36 38 15 W118 56 35. Stereo. 24 4991 E. McKinley, Suite 124, Fresno, 93727. Phone: (559) 243-4300.1,193,000 Format: Soft rock. News: 2 hrs wkly. Target aud: 25-54; women. ♦Darrel Goodin, gen mgr; May Lou Goodin, opns VP; Dave Butler, gen sls mgr; Scott Keith, progmg dir; Dave Case, chief of engrg.

Dunnigan

KTKZ-FM— Sept 1, 1983: 105.5 mhz; 2.55 kw. Ant 1,010 ft TL: N38 47 17 W122 06 52. Stereo. Hrs opn: 24 1425 River Park Dr., Suite 520, Sacramento, 95815. Phone: (916) 924-0710. Fax: (916) 924-1587. Web Site: www.ktkz.com. Licensee: Caron Broadcasting Inc. Group owner: Salem Communications Corp. (acq 1-11-2002; $8 million). Population served: 1,500,000 Format: Talk. ♦James Rowten, gen mgr; Laurie Larson, progmg dir; Dave Fortenberry, engrg dir.

Dunsmuir

KZRO(FM)— Dec 8, 1992: 100.1 mhz; 12.5 kw. 213 ft TL: N41 17 20 W122 14 25. Stereo. Hrs opn: 24 Box 1234, Mt. Shasta, 96067. Secondary address: 113 E. Alma St., Mt. Shasta 96067. Phone: (530) 926-1332. Fax: (530) 926-0737. E-mail: zmail@zchannelradio.com Web Site: www.zchannelradio.com. Licensee: Dennis Michael Crepps dba Big Tree Communications (acq 6-24-97). Population served: 194,000 Natl. Network: Westwood One. Format: Classic rock, oldies. News staff: one. Target aud: 18-55; general. Spec prog: Children 2 hrs wkly. ♦Dennis Michaels, gen mgr, gen sls mgr, mktg dir, adv dir & progmg dir; Rob Hanson, chief of engrg.

Earlimart

KNAC(FM)—Not on air, target date: unknown: 93.5 mhz; 6 kw. Ant 177 ft TL: N35 57 30 W119 15 00. Hrs open: 12550 Brookhurst St., Garden Grove, 92640. Phone: (714) 636-5040. Licensee: Earlimart Educational Foundation Inc. ♦William Zawila, gen mgr.

East Los Angeles

KLAX-FM— Apr 22, 1949: 97.9 mhz; 50 kw. 390 ft TL: N34 00 24 W118 21 52. Stereo. Hrs opn: 10281 W. Pico Blvd., Los Angeles, 90064. Phone: (310) 203-0900. Fax: (310) 843-4961. Web Site: www.979laraza.com. Licensee: KLAX Licensing Inc. Group owner: Spanish Broadcasting System Inc. (acq 2-87). Format: Sp, Rgnl Mexican. Target aud: 18-34. ♦Raul Alarcon Jr., CEO & pres; Peter Remington, gen mgr; Juan Carlos Hidalog, dev dir & progmg dir; Jason Wilberding, gen sls mgr; Patty Castor, prom dir.

East Porterville

KMQA(FM)— Dec 1, 1989: 100.5 mhz; 1.5 kw. 465 ft TL: N36 02 37 W118 56 08. (CP: 2.1 kw, ant 1,109 ft.). Hrs opn: 1450 E. Bardsley Ave., Tulare, 93274. Phone: (559) 687-3170. Fax: (559) 687-3175. E-mail: traffic@lunacommunications.net Web Site: www.lamaquinamusical.net. Licensee: MBP Licensee LLC. Group owner: Moon Broadcasting (acq 12-29-98). Natl. Network: CNN Radio. Format: Mexican rgnl. Target aud: 25-40. ♦George Rayo, opns mgr & gen sls mgr; Rey Ponce, prom dir.

East Sonora

***KARQ(FM)**— 2005: 89.5 mhz; 1.3 kw vert. Ant 1,661 ft TL: N38 03 46 W120 14 45. Hrs open: 24
Rebroadcasts KLRD(FM) Yucaipa 100%.
2351 Sunset Blvd., Suite 170-218, Rocklin, 95765. Phone: (916) 251-1600. Fax: (916) 251-1650. Web Site: www.air1.com. Licensee: Educational Media Foundation. Group owner: EMF Broadcasting. Natl. Network: Air 1. Law Firm: Shaw Pittman. Format: Christian. News staff: 3. Target aud: 25-44; Judeo Christian, female. ♦Richard Jenkins, pres; Mike Novak, VP; Keith Whipple, dev dir; David Pierce, progmg mgr; Ed Lenane, news dir; Sam Wallington, engrg dir; Arthur Vassar, traf mgr.

El Cajon

***KECR(AM)**— 1955: 910 khz; 5 kw-U, DA-2. TL: N32 53 38 W116 55 35. Hrs open: 24 11865 Moreno Ave., Lakeside, 92040. Phone: (619) 390-3481. Fax: (619) 443-7693. E-mail: kecr@nethere.com Web Site: www.familyradio.com. Licensee: Family Stations Inc. (group owner: Family Stations Inc. acq 6-9-63). Population served: 796,769 Format: Relg. Target aud: All ages; families. ♦Bill Babcock, gen mgr.

KHTS-FM—Licensed to El Cajon. See San Diego

El Centro

KGBA(AM)—See Calexico

KGBA-FM—See Holtville

KROP(AM)—Brawley, November 1946: 1300 khz; 1 kw-D, 500 w-N. TL: N33 00 40 W115 31 16. Hrs open: 24 Box 238, 120 S. Plaza, Brawley, 92227. Phone: (760) 344-1300. Fax: (760) 344-1763. Web Site: www.q96ksiq.com. Licensee: CCR-Brawley IV LLC. Group owner: Cherry Creek Radio LLC acq 6-99; $2 million with co-located FM). Population served: 150,000 Law Firm: Miller & Miller. Format: Country. Target aud: 25-54; male. ♦Tony Driskill, gen mgr; Carlos Cisneros, gen sls mgr.

KSIQ(FM)—Co-owned with KROP(AM). Sept 10, 1981: 96.1 mhz; 50 kw. 340 ft TL: N32 57 12 W115 30 05. Stereo. 24 Web Site: www.q96ksiq.com. Format: Mainstream top-40. News: one hr wkly. Target aud: 25-49; female. ♦Stephen Stodelle, mktg mgr & adv mgr; Tony Driskill, prom mgr & progmg mgr; Clif Glasgow, engrg dir.

KWST(AM)— June 21, 1958: 1430 khz; 1 kw-U. TL: N32 48 27 W115 32 18. Stereo. Hrs opn: 24 Box 2830, 92244. Phone: (760) 337-8707. Fax: (760) 337-8012. Web Site: www.univision.com. Licensee: Entravision Holding L.L.C. Group owner: Entravision Communications Co. L.L.C. (acq 1998; $4.8 million). Population served: 950,000 Format: Country. Target aud: 25-49. ♦Albert Valdez, pres & opns mgr; Eric Chavez, gen mgr.

KXO(AM)— January 1927: 1230 khz; 1 kw-U. TL: N32 46 34 W115 32 58. Hrs open: 24 Box 140, 92244. Phone: (760) 352-1230. Web Site: www.kxoradio.com. Licensee: KXO Inc. (acq 1961). Population served: 145,000 Natl. Network: CBS. Rgnl. Network: Calif. Farm. Natl. Rep: McGavren Guild. Format: Oldies. News staff: one; News: 10 hrs wkly. Target aud: 18-49. Spec prog: Farm 7 hrs wkly. ♦Caroll Buckley, VP, gen sls mgr, prom mgr & progmg dir; Gene Brister, pres & gen mgr; Doug Melanson, chief of engrg.

KXOJ-FM—Co-owned with KYAL(AM). Feb 22, 1977: 100.9 mhz; 5 kw. 360 ft TL: N36 03 38 W96 06 03. Stereo. 24 E-mail: mail@kxoj.com Web Site: www.kxoj.com.400,000 Format: Contemp Christian mus. ♦Mike Stephens, CEO; David Stephens, stn mgr, opns dir, dev dir & adv dir.

KXOX-FM— Apr 7, 1976: 96.7 mhz; 2.9 kw. 154 ft TL: N32 29 16 W100 23 31. Stereo. 18 News staff: one. ♦Jeff Stein, local news ed & disc jockey; Lillie Guttierez, spanish dir & disc jockey; Maxine Stein, women's int ed; Mitch Moore, disc jockey; Richard Ferguson, disc jockey.

KXO-FM— Aug 2, 1976: 107.5 mhz; 25.5 kw. 155 ft TL: N32 46 35 W115 32 58. Stereo. 24 Web Site: www.kxoradio.com. (Acq 1976.).145,000 Format: Adult contemp. News: 3 hrs wkly. Target aud: 25-49.

KXOQ(FM)—Co-owned with KOTC(AM). Dec 13, 1995: 104.3 mhz; 6 kw. 328 ft TL: N36 21 01 W90 02 43. Web Site: www.foxradionetwork.com. Format: Rock & oldies mix.

El Cerrito

***KECG(FM)—** September 1978: 88.1 mhz; 17 w. Ant -66 ft TL: N37 54 30 W122 17 39. Stereo. Hrs opn: 24 540 Ashbury Ave., 94530. Phone: (510) 525-4472. Fax: (510) 525-0554. E-mail: kecg88@aol.com Licensee: West Contra Costa Unified School District. Natl. Network: USA. Format: Div, educ, jazz. News: 5 hrs wkly. Target aud: General. Spec prog: Gospel 5 hrs, Sp 3 hrs, Filipino 2 hrs wkly. ♦Philip H. Morgan Jr., stn mgr.

El Rio

KMLA(FM)— October 1996: 103.7 mhz; 480 w. 807 ft TL: N34 18 10 W119 13 41. Hrs open: 355 S. A St., Suite 103, Oxnard, 93030. Phone: (805) 385-5656. Fax: (805) 385-5690. E-mail: info@lam1037.com Web Site: www.lam1037.com. Licensee: Gold Coast Radio L.L.C. (acq 12-6-96; $550,000). Format: Rgnl Mexican. ♦Guillermo Gonzalez, gen mgr & gen sls mgr; Sonia Lopez, stn mgr; Rosa Rodriguez, prom dir; Gerardo Ceja, progmg dir & news dir; Charles Hastings, chief of engrg.

Ellwood

KSPE-FM—Licensed to Ellwood. See Santa Barbara

Encinitas

KPRI(FM)— Jan 20, 1962: 102.1 mhz; 14.5 kw. Ant 817 ft TL: N33 06 40 W117 12 05. Stereo. Hrs opn: 24 9710 Scranton Rd., Suite 200, San Diego, 92121. Phone: (858) 678-0102. Fax: (858) 320-7024. E-mail: ilisten@authenticrock.com Web Site: www.kprifm.com. Licensee: Compass Radio of San Diego Inc. (acq 1996). Population served: 700,000 Natl. Rep: Katz Radio. Law Firm: Shaw Pittman. Format: Triple A. News staff: one; News: 2 hrs wkly. Target aud: 18-34; upscale, well educated, young adult contemp mus fans. ♦Jonathan D. Schwartz, CFO & sr VP; Bob Hughes, gen mgr & opns mgr; Robert Burch, stn mgr; Patrick Osburn, sls dir; Keith Miller, mktg dir & prom dir.

Escondido

KFSD(AM)— June 1958: 1450 khz; 1 kw-U. TL: N33 07 11 W117 07 07. Hrs open: 24 1835 Aston Ave., Carlsbad, 92008. Phone: (760) 729-1000. Fax: (760) 476-9604. Web Site: www.am1510kspa.com. Licensee: North County Broadcasting Corp. Group owner: Astor Broadcast Group (acq 9-15-87; $3 million with co-located FM; FTR: 6-29-87) Population served: 150,000 Format: Adult standards. News: 2 hrs wkly. Target aud: 35-64. ♦Arthur Astor, CEO & pres; Rick Roome, gen mgr.

KSOQ-FM— July 1966: 92.1 mhz; 580 w. Ant 1,024 ft TL: N33 06 39 W117 09 13. Stereo. Hrs opn: 24
Rebroadcasts KSON-FM San Diego 100%.
Box 889004, San Diego, 92168-9004. Secondary address: 1615 Murray Canyon Rd., Suite 710, San Diego 92108-4321. Phone: (619) 291-9797. Phone: (619) 297-3698. Fax: (619) 543-1353. Web Site: www.kson.com. Licensee: Jefferson-Pilot Communications Co. of California. Group owner: Jefferson-Pilot Communications Co. (acq 4-1-2004; $18 million). Population served: 1,500,000 Format: Country. ♦Darrel Goodin, gen mgr; Dave Saunders, gen sls mgr; John Marks, progmg dir; Eric Schecter, chief of engrg.

Esparto

KTTA(FM)— 1996: 97.9 mhz; 6 kw. Ant 328 ft TL: N38 45 33 W121 52 33. Hrs open:
Simulcast with KBBU(FM) Modesto 100%.
500 Media Pl., Sacramento, 95815. Phone: (916) 368-6300. Fax: (916) 473-0146. E-mail: azteca16@aol.com Web Site: www.lakebuena.com. Licensee: Bustos Media of California License LLC. (acq 12-15-2004; $21.7 million with KBBU(FM) Modesto). Format: Rgnl Mexican. ♦Amparo Perez-Cook, gen mgr; Javier Gonzalez, prom dir; Juan Gonzalez, progmg dir; Mark Sedaka, chief of engrg; Cynthia Sanchez, traf mgr.

Essex

KHWY(FM)— May 1, 1991: 98.9 mhz; 10 kw. Ant 1,073 ft TL: N34 52 50 W115 04 05. Hrs open: 24
Rebroadcasts KRXV(FM) Yermo 100%.
Box 1668, 1611 E. Main St., Barstow, 92312. Phone: (760) 256-0326. Fax: (760) 256-9507. E-mail: tim@highwayradio.com Web Site: www.thehighwaystations.com. Licensee: KHWY Inc. Natl. Network: AP Radio. Law Firm: Hogan & Hartson. Format: Adult contemp. News staff: one; News: 28 hrs wkly. Target aud: 35 plus; travelers on I-40 & I-15 & Mojave Desert residents. ♦Howard B. Anderson, CEO & pres; Kirk M. Anderson, exec VP; Timothy B. Anderson, VP & gen mgr; Judy Robinson, sls VP; John Gregg, prom dir & prom mgr; Lance Todd, progmg dir; Keith Hayes, news dir; Thomas J. McNeill, engrg mgr.

Eureka

KATA(AM)—See Arcata

KEKA-FM— Nov 1, 1983: 101.5 mhz; 100 kw. 3,200 ft TL: N40 25 12 W124 05 00. Stereo. Hrs opn: 1101 Marsh Rd., 95501. Phone: (707) 442-5744. Licensee: Eureka Broadcasting. (acq 12-13-90; $430,189; FTR: 1-7-91). Population served: 150,000 Natl. Network: ABC. Natl. Rep: Katz Radio. Format: Modern country. Target aud: 25-54. ♦Hugo Papstein, gen mgr.

KFMI(FM)— 1973: 96.3 mhz; 30 kw. 1,580 ft TL: N40 43 36 W123 58 18. (CP: 100 kw). Stereo. Hrs opn: 5460 S. Broadway, 95503. Phone: (707) 442-2000. Fax: (707) 443-6848. E-mail: power963@hotmail.com Web Site: www.power963.com. Licensee: Bicoastal Media LLC. (group owner; acq 7-28-99; grpsl). Population served: 120,000 Natl. Network: Jones Radio Networks. Format: Hot adult contemp, CHR. Target aud: 18-36; upscale adults. ♦Ken Dennis, CEO; Mike Wilson, pres; Laurie Tate, gen mgr & opns dir; Victoria Bennington, gen sls mgr; Tom Sebourn, progmg dir.

KGOE(AM)— May 12, 1933: 1480 khz; 5 kw-D, 1 kw-N. TL: N40 44 28 W124 12 05. Hrs open: 24 5640 S. Broadway, 95503. Phone: (707) 443-1621. Fax: (707) 443-6848. E-mail: tsebourn@bicoastalmedia.com Web Site: www.kgoe1480.com. Licensee: Bicoastal Media LLC. (group owner; acq 7-28-99; grpsl). Population served: 125,000 Natl. Network: Jones Radio Networks. Format: News/talk. News staff: one; News: one hr wkly. Target aud: 25-54. ♦Tom Huckabay, gen mgr & gen sls mgr; Rollin Treehearn, prom dir; Tom Sebourn, progmg dir; Kevin Sanders, chief of engrg.

KRED-FM—Co-owned with KGOE(AM). Dec 17, 1979: 92.3 mhz; 25 kw. 1,544 ft TL: N40 43 37 W123 58 25. Stereo. Phone: (707) 442-2000. E-mail: rollin@kred923.com Web Site: www.kred923.com.125,000 Format: Hot country. News staff: one; News: 7 hrs wkly.

KINS(AM)— January 1946: 980 khz; 5 kw-D, 500 w-N, DA-N. TL: N40 48 05 W124 07 31. Hrs open: 24 1101 Marsh Rd., 95501. Phone: (707) 442-5744. Licensee: Eureka Broadcasting Co. (acq 3-1-58). Population served: 28,936 Natl. Network: CBS, Wall Street. Format: News/talk. News staff: 2. Target aud: 35 plus; upscale, educated. ♦Hugo Papstein, pres, gen mgr, gen sls mgr & progmg dir; Mark Householter, chief of engrg.

KKHB(FM)— 1994: 105.5 mhz; 28 kw. 1,588 ft TL: N40 43 52 W123 57 06. Hrs open:
Rebroadcasts KGO(AM) San Francisco.
5640 S. Broadway, 95503. Phone: (707) 442-2000. Fax: (707) 443-6848. Web Site: www.cool1055.com. Licensee: Bicoastal Media L.L.C. (group owner; acq 11-9-98; grpsl). Format: Oldies. ♦Laurie Tate, gen mgr & opns dir; Victoria Bennington, gen sls mgr; Tom Sebourn, progmg dir.

***KMUE(FM)—** Aug 9, 1996: 88.3 mhz; 1.25 kw. 1,446 ft TL: N40 43 52 W123 57 06. Hrs open: Box 135, Redway, 95560-0135. Secondary address: 1144 Redway Dr., Redway 95560. Phone: (707) 923-2513. Fax: (707) 923-2501. E-mail: kmud@kmud.org Web Site: www.kmud.org. Licensee: Redwood Community Radio Inc. Population served: 50,000 Format: Talk, div, educ. ♦David Lippe, dev dir & adv mgr; Michael Jacinto, progmg dir.

KNCR(AM)—See Fortuna

KWSW(AM)— Dec 20, 1979: 790 khz; 5 kw-D, 112 w-N. TL: N40 48 09 W124 08 20. Hrs open: 1101 Marsh Rd., 95501. Phone: (707) 442-5744. Licensee: Eureka Broadcasting Co. Inc. (acq 11-30-92; $105,000; FTR: 12-21-92). Population served: 150,000 Format: Talk. Target aud: 35 plus; baby boomers with discretionary income. ♦Hugo Papstein, pres, gen mgr & progmg dir; Brian Papstein, gen sls mgr; Mark Householter, chief of engrg.

KXGO(FM)—Arcata, 1970: 93.1 mhz; 50 kw. Ant 1,666 ft TL: N40 43 38 W123 58 22. Stereo. Hrs opn: 603 F. St., 95501. Phone: (707) 445-8104. Fax: (707) 445-3906. E-mail: operations@kxgo.com Web Site: theclassicrockexperience.com. Licensee: Miller Broadcasting Co. (acq 11-1-97). Population served: 210,000 Natl. Rep: Christal. Format: Classic rock. Target aud: 25-54; upscale adults. ♦Becky Collins, opns mgr; Pattison Christensen, pres, stn mgr & natl sls mgr; Carole Arrington, rgnl sls mgr & prom dir; Danny King, progmg dir.

Fair Oaks

KSSJ(FM)— Nov 25, 1970: 94.7 mhz; 86.6 kw. 2,072 ft TL: N39 15 30 W119 42 36. Stereo. Hrs open: 5345 Madison Ave., Sacramento, 95841-3141. Phone: (916) 334-7777. Fax: (916) 339-4281. E-mail: comments@kssj.com Web Site: www.kssj.com. Licensee: Entercom Sacramento License L.L.C. Group owner: Entercom Communications Corp. (acq 11-4-97; $15.9 million). Format: Smooth jazz. ♦David Lichtman, gen mgr; Lee Hansen, stn mgr & progmg dir.

Fairfield

***KASK(FM)—**Not on air, target date: unknown: 91.5 mhz; 75 w. 649 ft TL: N38 19 09 W121 59 30. Hrs open: Maranatha Broadcasting, 130 Portsmouth Ave., Vacaville, 95687. Phone: (707) 449-4059. Fax: (707) 447-0680. Licensee: Maranatha Broadcasting. Format: Relg. ♦Michel Mace, pres & stn mgr.

KUIC(FM)—See Vacaville

Fairmead

***KLVY(FM)—** 1998: 91.1 mhz; 3.4 kw. 256 ft TL: N37 13 01 W120 11 57. Stereo. Hrs opn: 24
Rebroadcasts KLVN(FM) Livingston 100%.
2351 Sunset Blvd., Suite 170-218, Rocklin, 95765. Phone: (916) 251-1600. Fax: (916) 251-1650. E-mail: klove@klove.com Web Site: www.klove.com. Licensee: Educational Media Foundation Inc. Group owner: EMF Broadcasting. Population served: 770,000 Natl. Network: K-Love. Law Firm: Shaw Pittman. Format: Contemp Christian music. News staff: 3. Target aud: 25-44; Judeo-Christian, female. ♦Richard Jenkins, pres; Mike Novak, VP; Keith Whipple, dev dir; David Pierce, prom mgr & progmg mgr; Ed Lenane, news dir; Sam Wallington, engrg dir; Karen Johnson, news rptr; Marya Morgan, news rptr; Richard Hunt, news rptr.

Fallbrook

KSSD(FM)— Nov 22, 1977: 107.1 mhz; 3 kw. 300 ft TL: N33 23 01 W117 11 20. Stereo. Hrs opn: 24 5700 Wilshire Blvd., Suite 250, Los Angeles, 90036. Phone: (323) 900-6100. Fax: (323) 900-6127. Web Site: www.superestrella.com. Licensee: Entravision Holdings LLC. Group owner: Entravision Communications Corp. (acq 4-1-03; grpsl). Natl. Rep: Lotus Entravision Reps LLC. Law Firm: Cohn & Marks. Format: Sp, CHR. Target aud: 18-34. ♦Jeff Liberman, VP & gen mgr; Karl Meyer, gen mgr; Elias Autran, progmg dir; Eugene McAffe, chief of engrg & engr; Pam McCaffrey, traf mgr.

Felton

KXZM(FM)— 1999: 93.7 mhz; 28 w. Ant 1,260 ft TL: N37 03 43 W122 07 14. Hrs open: 200 South A St., Suite 400, Oxnard, 93030. Phone: (805) 240-2070. Fax: (805) 240-5960. Licensee: Lazer Broadcasting Corp. (group owner; (acq 7-25-2005; $2.88 million with KXSM(FM) Hollister). Format: Rgnl Mexican. ♦Alfredo Plascencia, pres; Daniel Osuna, gen mgr.

Ferndale

KJNY(FM)— Apr 1, 1993: 99.1 mhz; 6 kw. 1,715 ft TL: N40 30 03 W124 17 08. Stereo. Hrs opn: 24 603 F St., Eureka, 95501. Phone: (707) 445-3699. Fax: (707) 445-3906. E-mail: operations@kxgo.com Web Site: www.kjny.net. Licensee: Redwood Broadcasting Co. Inc. (acq 1999). Law Firm: Rosenman & Colin. News staff: one; News: 3 hrs wkly. Target aud: 25-54; upscale adults. ♦Pattison Christensen, pres & stn mgr; Danny King, progmg dir.

Firebaugh

***KYAF(FM)—** 2006: 94.7 mhz; 900 w. Ant 66 ft TL: N36 51 37 W120 27 19. Hrs open: 1572 10th St., 93622. Phone: (559) 659-0100. Web Site: kyafm.com. Licensee: Central Valley Educational Services Inc. Format: Oldies. ♦Verne White, pres.

Ford City

KZPE(FM)—Not on air, target date: unknown: 102.1 mhz; 6 kw. Ant 128 ft TL: N35 00 02 W119 22 29. Hrs open: c/o William Zawila, Esq., 12600 Brookhurst St., Suite 105, Garden Grove, 92840. Phone: (714) 636-5040. Fax: (714) 636-5042. Licensee: Estate of H.L. Charles, Robert Willing, executor (acq 6-4-2004).

Fort Bragg

KDAC(AM)— June 1948: 1230 khz; 1 kw-U. TL: N39 26 35 W123 46 48. Hrs open: 24
Rebroadcasts KUKI(AM)UKiah 100%.
1400 Kuki Ln., Ukiah, 95482. Phone: (707) 263-6113. Fax: (707) 466-5852. E-mail: kukiinfo@bicoastalmedia.com Licensee: MBU Licensee LLC. Group owner: Moon Broadcasting (acq 9-15-2003; grpsl). Population served: 8,000 Natl. Network: ABC, CBS. Format: Mexican/Rgnl. News staff: one; News: 24 hrs wkly. Target aud: 35 plus. ♦George Feola, gen mgr; Tove Sorensen, opns mgr.

KOZT(FM)— Dec 5, 1981: 95.3 mhz; 35 kw. Ant 515 ft TL: N39 24 24 W123 44 04. Stereo. Hrs opn: 24 110 S. Franklin, 95437. Phone: (707) 964-7277. Fax: (707) 964-9536. E-mail: thecoast@kozt.com Web Site: www.kozt.com. Licensee: California Radio Partners Inc. (acq 12-1-90; FTR: 12-17-90). Population served: 100,000 Law Firm: Miller & Neely P.C. Format: AAA. News staff: one; News: one hr wkly. Target aud: 25-49; affluent, educated consumers. ♦Tom Yates, CEO, gen mgr & opns dir; Vicky Watts, chmn & CFO.

KPMO(AM)—See Mendocino

KSAY(FM)— November 1988: 98.5 mhz; 3500 w. 453 ft TL: N39 26 08 W123 48 15. Stereo. Hrs opn: 24 Box 2269, 95437. Phone: (707) 964-5729. Fax: (707) 964-5729. E-mail: ksayfm@yahoo.com Web Site: www.ksay.com. Licensee: Axell Broadcasting. Population served: 100,000 Natl. Network: Jones Radio Networks. Format: Adult contemp. News staff: one; News: 9 hrs wkly. Target aud: 18-49; primarily women. ♦Wade Axell, gen mgr.

Fortuna

KNCR(AM)— Oct 31, 1966: 1090 khz; 10 kw-D. TL: N40 33 30 W124 07 24. Hrs open: Sunrise-sunset Box 109, Eureka, 95502-0109. Phone: (707) 725-9363. Fax: (707) 726-9446. E-mail: mario@lanueva1090.com Web Site: www.lanueva1090.com. Licensee: Del Rosario Talpa Inc. (acq 5-17-2004; $37,500). Law Firm: Rosenman & Colin L.L.P. Format: Rgnl Mexican. News: 2 hrs wkly. Target aud: 25-54. ♦Mario Meza, gen mgr; Sylvia Meza, gen sls mgr.

KWPT(FM)— May 15, 1992: 100.3 mhz; 12 kw. Ant 1,807 ft TL: N40 25 23 W124 06 21. Hrs open: 24 Box 25, Ferndale, 95536. Phone: (707) 786-5104. Fax: (707) 786-5100. Web Site: www.kwptfm.com. Licensee: KWPT Inc. (acq 5-12-2005; $650,000). Population served: 126,000 Law Firm: Law Office of Dan J. Alpert. Format: Classic hits. Target aud: 30-54; affluent, college educated. ♦Patrick Cleary, gen mgr; Cliff Berkowitz, progmg dir & mus dir.

Fountain Valley

KJLL-FM— 1993: 92.7 mhz; 690 w. Ant 961 ft TL: N33 36 20 W117 48 35. Stereo. Hrs opn: 24
Rebroadcasts KHJL(FM) Thousand Oaks 100%.
99 Long Ct., Suite 200, Thousand Oaks, 91360. Phone: (805) 497-8511. Fax: (805) 497-8514. E-mail: info@927.com Web Site: www.927jillfm.com. Licensee: Amaturo Group of L.A. Ltd. Group owner: Amaturo Groups (acq 1996; $5.5 million). Population served: 4,000,000 Natl. Network: ABC. Rgnl. Network: Metronews Radio Net. Format: Adult contemp. Target aud: 25-54. ♦Joseph Amaturo, CEO & VP; Robert J. Christy, gen mgr; Robert Christy, gen sls mgr; Aaron Fonesca, progmg dir.

Fowler

KALZ(FM)— Nov 7, 1980: 96.7 mhz; 22 kw. Ant 348 ft TL: N36 41 39 W119 43 57. Stereo. Hrs opn: 24 83 E. Shaw Ave., Suite 150, Fresno, 93710-7616. Phone: (559) 230-4300. Fax: (559) 243-4301. Web Site: www.myalice967.com. Licensee: Clear Channel Radio Licenses Inc. Group owner: Clear Channel Communications Inc. (acq 8-30-2000; grpsl). Population served: 650,000 Format: Adult contemp. Target aud: 25-54; women. ♦Jeff Negrete, gen mgr; Paul Wilson, opns mgr; Tony Rainaldi, gen sls mgr; Chris Miller, prom dir; Paul Wilson, progmg dir; Dave Case, chief of engrg; Michelle Howes-Appleton, traf mgr.

KQEQ(AM)— July 1, 1962: 1210 khz; 370 w-U. TL: N36 39 37 W119 41 01. Hrs open: 139 W. Olive Ave., Fresno, 93728. Phone: (559) 499-1210. Fax: (559) 499-1212. E-mail: rak@computermail.net Web Site: www.thehmongradio.com. Licensee: RAK Communications Inc. (acq 9-30-94; FTR: 7-4-94). Population served: 600,000 Format: Hmong. News staff: 1; News: 10 hrs wkly. Target aud: 13-Senior; Hmong and Lao. ♦Pahoua Moua, gen mgr & mktg mgr.

Frazier Park

KJPG(AM)— 1994: 1050 khz; 10 kw-D, 7 w-N, DA-D. TL: N35 01 28 W118 55 05 (D), N35 24 07 W119 02 47 (N). Hrs open: Box 180, Tahoma, 96142. Phone: (530) 584-5700. Fax: (530) 584-5705. E-mail: info@ihradio.org Web Site: www.ihradio.org. Licensee: IHR Educational Broadcasting. (group owner; (acq 11-15-2003; $700,000). Format: Catholic/relg. ♦Douglas M. Sherman, pres & VP.

Freedom

KPIG-FM— Dec 1, 1987: 107.5 mhz; 5.4 kw. Ant 338 ft TL: N36 50 06 W121 42 22. Stereo. Hrs opn: 24 60 Garden Ct., Suite 300, Monterey, 93940. Secondary address: 1110 Main St., Suite 16, Watsonville 95076-3700. Phone: (831) 722-9000. Fax: (831) 722-7548. E-mail: sty@kpig.com Web Site: www.kpig.com. Licensee: Mapleton Communications L.L.C. (group owner; (acq 11-16-2001; grpsl). Natl. Rep: McGavren Guild. Law Firm: Leventhal, Senter & Lerman. Format: AAA "Americana". Target aud: 25-54. ♦Dale Hendry, gen mgr; Frank Caprista, opns mgr & progmg dir; Jodi Morgan, gen sls mgr; Sybil DeAngelo, prom dir; Velden Levirich, chief of engrg.

Fremont

***KOHL(FM)**— Sept 23, 1974: 89.3 mhz; 145 w. 407 ft TL: N37 32 00 W121 54 35. Stereo. Hrs opn: 24 43600 Mission Blvd., 94539. Phone: (510) 659-6221. Fax: (510) 659-6001. E-mail: kohl@kohlradio.com Web Site: www.kohlradio.com. Licensee: Fremont-Newark Community College Dist. Format: Contemp hit/Top 40. News: one hr wkly. Target aud: 18-34. ♦Robert Dochterman, gen mgr; Tom Gomez, progmg dir; Matthew Karl, news dir & pub affrs dir.

Fresno

KAVT(AM)—Not on air, target date: unknown: 1680 khz; 10 kw-D, 1 kw-N. TL: N36 46 14 W119 55 20. Hrs open: 139 W. Olive Ave., 93728. Phone: (559) 233-8803. Fax: (559) 233-8871. E-mail: rakradiosports@comcast.net Web Site: www.radiodisney.com. Licensee: RAK Communications Inc. Format: Children. ♦Albert R. Perez, gen mgr; Albert Perez, gen sls mgr; Paul Klein Kramer, chief of engrg.

KBIF(AM)— Nov 17, 1947: 900 khz; 1000 kw-D, 500 w-N, DA-N. TL: N36 41 30 W119 40 46. Hrs open: 24 hrs 3401 W. Holland Ave., 93722. Phone: (559) 222-0900. Fax: (559) 222-1573. E-mail: kbifkirv@aol.com Web Site: www.kbif900am.com. Licensee: Gore-Overgaard Broadcasting Inc. (group owner) Population served: 1,000,000 Natl. Network: USA. Format: Asian. News staff: 3; News: 10 hrs wkly. Spec prog: Sp 6 hrs, Punjabi 16 hrs wkly. ♦Dana Kennon, gen mgr; Tony Donato, opns mgr & gen sls mgr.

KBOS-FM—Tulare, 1965: 94.9 mhz; 16.4 kw. 847 ft TL: N36 38 15 W118 56 35. Stereo. Hrs opn: 24 83 East Shaw Ave., Ste. 150, 93710. Phone: (559) 230-4300. Fax: (559) 243-4301. Web Site: www.b95forlife.com. Licensee: Capstar TX L.P. Group owner: Clear Channel Communications Inc. (acq 8-30-00; grpsl). Population served: 918,900 Natl. Network: CBS Radio. Format: Rhythmic CHR. Target aud: 12-34. ♦Jeff Negrete, VP & gen mgr; Paul Wilson, opns dir & opns mgr; Joni Norvell, gen sls mgr & traf mgr; Greg Hoffman, progmg dir; Dave Case, chief of engrg.

KCBL(AM)— June 26, 1953: 1340 khz; 1 kw-U. TL: N36 45 51 W119 47 08. Hrs open: 83 E. Shaw Ave., Suite 150, 93710. Phone: (559) 243-4300. Fax: (559) 243-4301. Web Site: www.foxsportsradio1340.com. Licensee: Capstar TX L.P. Group owner: Clear Channel Communications Inc. (acq 8-30-00; grpsl). Population served: 500,000 Natl. Network: CBS. Natl. Rep: CBS Radio. Format: All sports. Target aud: 18-49. ♦Jeff Negrete, VP; Paul Wilson, opns mgr; Brian Noe, progmg dir.

KCIV(FM)—See Mount Bullion

***KEYQ(AM)**— Oct 14, 1957: 980 khz; 500 w-D, 48 w-N. TL: N36 44 28 W119 51 12. Hrs open: 24
Rebroadcasts KMRO(FM) Camarillo 100%.
2310 Ponderosa Dr., Suite 28, Camarillo, 93010. Phone: (805) 482-4797. Fax: (805) 388-5202. E-mail: info@nuevavida.com Web Site: www.nuevavida.com. Licensee: The Association for Community Education Inc. Population served: 1,000,000 Law Firm: Miller & Neely. Format: Relg, Sp. Target aud: General; Sp-speaking. ♦Phil Guthrie, pres; Mary Guthrie, gen mgr.

***KFCF(FM)**— June 9, 1975: 88.1 mhz; 2.4 kw. Ant 1,899 ft TL: N37 04 23 W119 25 51. Stereo. Hrs opn: 24
Rebroadcasts KPFA(FM) Berkeley 85%.
Box 4364, 93744. Secondary address: 1449 N. Wishon Ave. 93728. Phone: (559) 233-2221. Fax: (559) 233-5776. E-mail: kfcf@kfcf.org Web Site: www.kfcf.org. Licensee: Fresno Free College Foundation. Population served: 1,200,000 Law Firm: Arent, Fox, Kintner, Plotkin & Kahn. Format: Var. News: 12 hrs wkly. Target aud: General; intelligent, discerning, questioning. Spec prog: Southeast Asian languages one hr, American Indian 2 hrs, Sp 5 hrs wkly. ♦Vic Bedoian, gen mgr; Rych Withers, opns VP, prom dir, mus dir & pub affrs dir; R.L. Stover, progmg dir & chief of engrg.

KFIG(AM)— January 1938: 1430 khz; 5 kw-U, DA-1. TL: N36 50 49 W119 40 46. Hrs open: 351 W. Cromwell, Suite 108, 93711. Phone: (559) 447-3570. Fax: (559) 447-3579. E-mail: postmaster@1430espn.com Web Site: www.1430espn.com. Licensee: Fat Dawgs 7 Broadcasting LLC (acq 7-28-2005; $2.5 million). Population served: 500,000 Natl. Network: ESPN Radio. Format: All sports. Target aud: 25 plus. ♦Joe Pacheco, pres; Paul Swearengin, gen mgr; Nick Washington, progmg dir; Paul Kleinkramer, chief of engrg.

KFJK(FM)— Dec 8, 1979: 105.9 mhz; 2.4 kw. Ant 1,960 ft TL: N37 04 23 W119 25 51. Stereo. Hrs opn: 24 1071 W. Shaw Ave., 93711. Phone: (559) 490-1019. Fax: (559) 490-5889. Web Site: www.newjack1059.com. Licensee: Peak Broadcasting of Fresno Licenses LLC. Group owner: Infinity Broadcasting Corp. (acq 3-30-2007; grpsl). Law Firm: Leventhal, Senter & Lerman. Format: Adult hits. ♦Patty Hixson, gen mgr.

***KFNO(FM)**— Feb 12, 1992: 90.3 mhz; 1.35 kw. 1,971 ft TL: N37 04 26 W119 25 52. Hrs open: 706 W. Herndon Ave., 93650. Phone: (559) 435-4996. Fax: (916) 641-8238. Licensee: Family Stations Inc. (group owner) Format: Relg. ♦Harold Camping, pres; Peggy Renschler, gen mgr.

KFPT(AM)—See Clovis

***KFSR(FM)**— Oct 30, 1982: 90.7 mhz; 2.55 kw. 66 ft TL: N36 48 42 W119 44 43. Stereo. Hrs opn: 24 California State Univ. of Fresno, 5201 N. Maple, MS SA#119, 93740-8027. Phone: (559) 278-2598. Phone: (559) 278-4500. Fax: (559) 278-6985. E-mail: kfsrfresno@hotmail.com Web Site: www.csufresno.edu/kfsr/. Licensee: California State University Fresno. Population served: 500,000 Format: Jazz, alternative rock, div. News: 2 hrs wkly. Target aud: General. Spec prog: Blues 6 hrs, reggae 6 hrs, world 9 hrs, folk 3 hrs, western one hr wkly. ♦Don Priest, gen mgr; Joe Moore, stn mgr; Frank Delgado, progmg dir & progmg mgr; Matt Garcia, mus dir; Matthew Boan, pub affrs dir.

KGST(AM)— 1949: 1600 khz; 5 kw-U, DA-N. TL: N36 29 20 W119 19 33. Hrs open: 1110 E. Olive Ave., 93728. Phone: (559) 497-1100. Fax: (559) 497-1125. E-mail: mginsburg@lotusfresno.com Licensee: Lotus Communications Inc. (group owner; (acq 8-1-85; $1.76 million; FTR: 4-22-85). Population served: 1,500,000 Natl. Rep: Lotus Entravision Reps LLC. Format: Sp/ESPN sports. News staff: one; News: 2 hrs wkly. Target aud: 18 plus; Hispanic adults. ♦Howard Kalmenson, pres; Daniel Crotty, gen mgr.

KHGE(FM)— Jan 6, 1962: 102.7 mhz; 50 kw. Ant 500 ft TL: N36 49 07 W119 30 33. Stereo. Hrs opn: 24 83 E. Shaw Ave., Suite 150, 93710. Phone: (559) 230-4300. Fax: (559) 243-4301. Web Site: www.bigcountry1027.com. Licensee: Capstar TX L.P. Group owner: Clear Channel Communications Inc. (acq 8-30-2000; grpsl). Population served: 500,000 Natl. Rep: Clear Channel. Format: Adult contemp. News staff: one; News: 20 hrs wkly. Target aud: 25-54; women. ♦Jeff Negrete, VP & gen mgr; Chris Hansen, gen sls mgr; Rita Walls, rgnl sls mgr; Chris Miller, prom dir; Chuck Geiger, progmg dir.

KIRV(AM)— Oct 1, 1962: 1510 khz; 10,000 kw-D. TL: N36 42 36 W119 50 06. Hrs open: 3401 W. Holland Ave., 93722. Phone: (559) 222-0900. Fax: (559) 222-1573. Web Site: www.kirv.com. Licensee: Gore-Overgaard Broadcasting Inc. (acq 4-24-99). Population served: 800,000 Format: Christian, talk, Spanish Christian. Target aud: 25-54. ♦Hal Gore, CEO; Cordel Overgaard, pres; Dana Kennon, gen mgr & stn mgr; Tony Donato, opns mgr & gen sls mgr.

KJFX(FM)— May 15, 1970: 95.7 mhz; 17.5 kw. 850 ft TL: N36 56 55 W119 29 09. Stereo. Hrs opn: 24 1066 E. Shaw Ave., 93710. Phone: (559) 255-1041. Fax: (559) 230-0177. Web Site: www.957thefox.com. Licensee: Wilks License Co.-Fresno LLC. (acq 6-1-2005; grpsl). Natl. Rep: McGavren Guild. Format: Classic rock. ♦Kevin O'Rorke, gen mgr; Rob Hasson, gen sls mgr; Andrea Carter, progmg dir.

KJWL(FM)— Apr 29, 1994: 99.3 mhz; 5.03 kw. 349 ft TL: N36 44 07 W119 47 10. Hrs open: 675 Santa Fe Ave., 93721. Phone: (559) 497-5118. Fax: (559) 497-9760. E-mail: info@kjwl.com Web Site: www.kjwl.com. Licensee: John E. Ostlund. Natl. Network: CNN Radio. Format: Adult standards. Target aud: 35 plus; upscale. ♦John E. Ostlund, pres & gen mgr; Bruce Campbell, opns mgr; Jennifer Books, dev mgr; Eric McCormick, gen sls mgr; Chris Nieto, prom dir; Jim Roberts, progmg dir; Juanita Stevenson, news dir; Joe Garcia, traf mgr & traf mgr.

KMGV(FM)— Mar 15, 1948: 97.9 mhz; 2.07 kw. 1,987 ft TL: N36 44 09 W119 47 59. (CP: 10.5 kw, ant 1,076 ft.). Stereo. Hrs opn: 1071 W. Shaw Ave., 93711. Phone: (559) 490-9800. Fax: (559) 490-4199. Licensee: Peak Broadcasting of Fresno Licenses LLC. Group owner: Infinity Broadcasting Corp. (acq 3-30-2007; grpsl). Wire Svc: UPI Format: Rhythm oldies. Target aud: 25-54. ♦Patty Hixson, gen mgr.

KMJ(AM)— June 1925: 580 khz; 5 kw-U. TL: N36 41 37 W120 03 16. Hrs open: 24 1071 W. Shaw Ave., 93711. Phone: (559) 490-5800. Fax: (559) 490-5977. Web Site: www.kmj580.com. Licensee: Peak Broadcasting of Fresno Licenses LLC. Group owner: Infinity Broadcasting Corp. (acq 3-30-2007; grpsl). Population served: 520,000 Natl. Network: ABC. Format: News/talk. News staff: 13; News: 44 hrs wkly. Target aud: 25-64. ♦Joe Mauk, CEO & chief of engrg; Al Smith, gen mgr; John Broeske, opns VP.

KSKS(FM)—Co-owned with KMJ(AM). 1946: 93.7 mhz; 68 kw. 1,912 ft TL: N37 04 44 W119 25 47. Stereo. Fax: (559) 490-5944. Web Site: www.ksks.com.160,600 Format: Modern country. ♦Karen Franz, gen sls mgr.

KOQO-FM— Mar 15, 1948: 101.9 mhz; 2.25 w. 1,948 ft TL: N37 04 25 W119 25 52. Hrs open: 24 1071 W. Shaw, 93711. Phone: (559) 490-1019. Fax: (559) 490-5889. Licensee: Peak Broadcasting of Fresno Licenses LLC. Group owner: Infinity Broadcasting Corp. (acq 3-30-2007; grpsl). Population served: 600,000 Law Firm: Leventhal, Senter & Lerman. Format: Sp, Tejano. Target aud: 18-34; central CA Hispanics. ♦Patty Hixson, gen mgr.

KRZR(FM)—See Hanford

***KSJV(FM)**— July 4, 1980: 91.5 mhz; 16 kw. Ant 870 ft TL: N35 38 15 W118 56 35. Stereo. Hrs opn: 24 5005 E. Belmont Ave., 93727. Phone: (559) 455-5777. Fax: (559) 455-5778. Web Site: www.radiobilingue.org. Licensee: Radio Bilingue Inc. Population served: 350,000 Format: Ethnic, Sp. News staff: 5; News: 11 hrs wkly. Target aud: 16-60; Latino. ♦Hugo Morales, CEO; Maria Erana, gen mgr, opns dir, gen sls mgr & progmg dir; Phil Traynor, dev dir; Samuel Cozco, news dir; Bill Bach, chief of engrg.

KSOF(FM)—See Dinuba

***KVPR(FM)**— Oct 15, 1978: 89.3 mhz; 2.45 kw. 1,890 ft TL: N37 04 25 W119 25 52. Stereo. Hrs opn: 24 3437 W. Shaw Ave., Suite 101, 93711. Phone: (559) 275-0764. Fax: (559) 275-2202. E-mail: kvpr@kvpr.org Web Site: www.kvpr.org. Licensee: White Ash Broadcasting Inc. Population served: 979,500 Natl. Network: NPR. Format: Class, news. News: 52 hrs wkly. ♦Mariam Stepanian, pres & gen mgr; Jim Meyers, stn mgr.

KWRU(AM)— 1937: 940 khz; 50 kw-U, DA-2. TL: N36 50 49 W119 39 46. Hrs open: 4910 E. Clinton Ave., Suite 107, 93727. Phone: (559) 251-6128. Fax: (559) 452-0948. Web Site: www.radiovidaabundante.com. Licensee: Multicultural Radio Broadcasting Licensee LLC. Group owner: Multicultural Radio Broadcasting Inc. (acq 2-4-2004; grpsl). Population served: 180500 Format: Sp, Christian. Target aud: 25-54. ♦Arthur S. Liu, pres; Alberto Felix, gen mgr & progmg dir.

KWYE(FM)— 1963: 101.1 mhz; 50 kw. 310 ft TL: N36 44 10 W119 47 13. (CP: 10 kw, ant 1,076 ft.). Stereo. Hrs opn: 24 1071 W. Shaw, 93711. Phone: (559) 490-1011. Web Site: www.y101hits.com. Licensee: Peak Broadcasting of Fresno Licenses LLC. Group owner: Infinity Broadcasting Corp. (acq 3-30-2007; grpsl). Population served: 500,000 Natl. Rep: Katz Radio. Law Firm: Leventhal, Senter & Lerman,p. Format: CHR. Target aud: 18-49; emphasis on women. ♦Todd Lauley, CEO; Tim Lyons, CFO; Patty Hixson, gen mgr.

KXEX(AM)— September 1962: 1550 khz; 5 kw-D, 2.5 kw-N, DA-2. TL: N36 46 14 W119 55 20. Hrs open: 139 W. Olive Ave., 93728. Phone: (559) 233-8803. Fax: (559) 233-8871. E-mail: marcus@1550snr.com Web Site: www.1550snr.com. Licensee: RAK Communications Inc. (acq 8-10-94; $212,000; FTR: 8-22-94). Population served: 197,840 Format: Sports. ♦Albert R. Perez, VP & gen mgr; Abel Perez, gen sls mgr; Paul Klein Kramer, chief of engrg.

KYNO(AM)— October 1947: 1300 khz; 5 kw-D, 1 kw-N, DA-N. TL: N36 46 14 W119 45 00. Hrs open: 24 2125 N. Barton Ave., 93703. Phone: (559) 454-1300. Fax: (559) 453-2430. E-mail: rq1300@sbcglobal.net Licensee: Spanish Catholic Radio of Fresno LLC (acq 10-7-99; $800,000). Population served: 165,972 Format: Sp-relg Catholic. ♦Ray Carrasco, gen mgr.

Garberville

KHUM(FM)— 1996: 104.7 mhz; 50 kw. 2,650 ft TL: N40 07 15 W123 41 27. Hrs open: 24 Box 25, Ferndale, 95536-0025. Secondary address: 1400 Main St., Suite 104, Ferndale 95536. Phone: (707) 786-5104. Fax: (707) 786-5100. E-mail: info@khum.com Web Site: www.khum.com. Licensee: Lost Coast Communications Inc. (acq 11-5-2001). Population served: 150,000 Natl. Rep: McGavren Guild. Law Firm: Dan J. Alpert. Format: AAA. Target aud: 25-54; general. ♦Patrick Cleary, gen mgr; Cliff Berkowitz, opns VP; Jennifer White, natl sls mgr; Gregg Foster, mktg dir; Monica Topping, prom dir; Mike Dronkers, mus dir; Kevin Sanders, chief of engrg.

***KLVG(FM)**— 1999: 103.7 mhz; 11 kw. Ant 2,348 ft TL: N40 20 05 W124 06 32. Stereo. Hrs opn: 24 5700 W. Oak Blvd., Rocklin, 95765. Phone: (916) 251-1600. Fax: (916) 251-1650. E-mail: klove@klove.com Web Site: www.klove.com. Licensee: Educational Media Foundation. Group owner: EMF Broadcasting. Population served: 120,000 Natl. Network: K-Love. Law Firm: Shaw Pittman. Format: Contemp Christian music. News staff: 3. Target aud: 25-44; female (Judeo-Christian). ♦Richard Jenkins, pres; Mike Novak, VP & progmg dir; Lloyd Parker, gen mgr; Ed Lenane, opns dir & news dir; Keith Whipple, dev dir; Eric Allen, natl sls mgr; David Pierce, prom mgr & progmg mgr; Jon Rivers, mus dir; Sam Wallington, engrg dir; Arthur Vassar, traf mgr; Karen Johnson, news rptr; Marya Morgan, news rptr; Richard Hunt, news rptr.

***KMUD(FM)**— May 28, 1987: 91.1 mhz; 180 w. 2,490 ft TL: N40 07 13 W123 41 32. (CP: 5.5 kw, ant 2,601 ft). Hrs opn: 24 Box 135, 1144 Redway Dr., Redway, 95560-0135. Phone: (707) 923-2513. Phone: (707) 923-3911 (STUDIO). Fax: (707) 923-2501. E-mail: kmud@kmud.org Web Site: www.kmud.org. Licensee: Redwood Community Radio Inc. Population served: 20,000 Law Firm: Michael Couzens. Format: Educ, div, talk. News staff: one; News: 6 hrs wkly. Target aud: General. Spec prog: Black 3 hrs, ethnic one hr, jazz 6 hrs, Sp 2 hrs, American Indian one hr wkly. ♦David Lippe, dev dir; Michael Jacinto, progmg dir & progmg mgr; Estelle Fennell, news dir; Simon Frech, chief of engrg.

***KXBC(FM)**—Not on air, target date: unknown: 89.1 mhz; 900 w. Ant 2,542 ft TL: N40 07 13 W123 41 31. Hrs open: 4135 Northgate Blvd., Suite 1, Sacramento, 95834-1226. Phone: (916) 641-8191. Phone: (510) 568-6200. Licensee: Family Stations Inc. ♦David Becker, pres & gen mgr; Tim White, news dir.

Garden Grove

KEBN(FM)— June 21, 1961: 94.3 mhz; 3 kw. Ant 246 ft TL: N33 46 51 W117 53 33. Stereo. Hrs opn: 24 QueBuena, 1845 Empire Ave., Burbank, 91504. Phone: (818) 729-5300. Fax: (818) 729-5683. E-mail: advertising@aquisuena.com Web Site: www.aquisuena.com. Licensee: LBI Radio License Corp. Group owner: Liberman Broadcasting Inc. (acq 5-15-03; $35 million). Format: Sp. ♦Andrew Mars, gen mgr; Daisy Ortiz, gen sls mgr; Edward Leon, progmg dir; Shannon Murdock, chief of engrg; Gustave Aviles, traf mgr.

George

KATJ-FM— June 29, 1989: 100.7 mhz; 260 w. 1,548 ft TL: N34 36 38 W117 17 18. Stereo. Hrs opn: 12370 Hesperia Rd., Suite 16, Victorville, 92395. Phone: (760) 241-1313. Fax: (760) 241-0205. Web Site: www.katcountry1007.com. Licensee: Clear Channel Broadcasting Licenses Inc. Group owner: Clear Channel Communications Inc. (acq 2000; grpsl). Population served: 250,000 Natl. Network: CNN Radio. Natl. Rep: Katz Radio. Law Firm: Latham & Watkins. Format: Country. Target aud: 25-54; adults. ♦Larry Thornhill, gen mgr; Kari Lynn, stn mgr & progmg dir.

Gilroy

KAZA(AM)— September 1957: 1290 khz; 5 kw-D, DA. TL: N37 09 48 W121 38 28. Hrs open: 6 AM-midnight Box 1290, San Jose, 95108. Phone: (408) 776-3090. Fax: (408) 881-1292. E-mail: sales@kazaradio.com Web Site: www.kazaradio.com. Licensee: Radio Fiesta Corp. (acq 5-14-73). Population served: 200,000 Format: Oldies, Sp. ♦Sonia Rodriquez, pres; Juan Sidhu, VP & opns mgr.

KBAY(FM)— Jan 1, 1970: 94.5 mhz; 1.23 kw. 2,535 ft TL: N37 06 39 W121 50 37. Hrs open: 24 190 Park Center Plaza, Suite 200, San Jose, 95113. Phone: (408) 287-5775. Fax: (408) 293-3341. E-mail: jleathers@kbay-kezr.com Web Site: www.kbay.com. Licensee: NM Licensing LLC. Group owner: Infinity Broadcasting Corp. (acq 12-6-2005; $80 million with KEZR(FM) San Jose). Population served: 150,000 Natl. Rep: D & R Radio. Format: Soft Rock. News staff: one; News: 4 hrs wkly. Target aud: 35-54. ♦John Leathers, gen mgr; Judy Dixon, gen sls mgr; Dana Jang, progmg dir; Lissa Kreisler, news dir; Michael Stockwell, chief of engrg.

Glendale

KRLA(AM)— 1928: 870 khz; 20 kw-D, 3 kw-N, DA-2. TL: N34 08 13 W118 13 34. Stereo. Hrs opn: 24 701 N. Brand Blvd., Suite 550, 91203. Phone: (818) 956-5552. Fax: (818) 551-1110. Web Site: www.krla870.com. Licensee: New Inspiration Broadcasting Co. Inc. Group owner: Salem Communications Corp. (acq 6-23-98; $33.4 million). Population served: 12,000,000 Natl. Network: Salem Radio Network. Natl. Rep: Christal. Rgnl rep: SRR Law Firm: Haley, Bader & Potts. Wire Svc: Metro Weather Service Inc. Format: News/talk. News: 35 hrs wkly. Target aud: 35 plus. ♦Jim Tinker, opns VP; Mark Pennington, gen sls mgr & natl sls mgr; Chuck Tyler, progmg dir; Craig Edwards, progmg dir; Bill Sheets, chief of engrg; Kristi Charley, traf mgr.

KSCA(FM)— March 1951: 101.9 mhz; 2.4 kw. 2,848 ft TL: N34 13 26 W118 03 45. (CP: 4.8 kw). Stereo. Hrs opn: 24 655 N. Central Ave., Suite 2500, 91203. Phone: (818) 500-4500. Fax: (818) 500-4580. Web Site: www.univision.com. Licensee: HBC License Corp. Group owner: Univision Radio (acq 9-22-2003; grpsl). Population served: 486,000 Law Firm: Irving Gastfreund. Format: Sp, Mexican rgnl. Target aud: 25-54. ♦Michelle Hohman, VP, gen mgr & adv dir; Haz Montana, opns dir; Victor Camino, gen sls mgr; Veronica Nava, progmg dir; Tom Koza, chief of engrg.

Goleta

KMGQ(FM)— Jan 30, 1982: 106.3 mhz; 250 w. 827 ft TL: N34 27 55 W119 40 38. Stereo. Hrs opn: 24 403 E. Montecito St., Suite A, Santa Barbara, 93101-1759. Phone: (805) 966-1755. E-mail: pat.cantwell@cumulus.com Web Site: www.kmgq1063.com. Licensee: Cumulus Licensing Corp. Group owner: Cumulus Media Inc. (acq 3-12-2001; grpsl). Population served: 350,000 Natl. Rep: McGavren Guild. Law Firm: Fletcher, Heald & Hildreth. Format: Jazz, adult contemp. Target aud: 35-64; upscale, educated, professional, affluent. ♦Gail Surrillo, gen mgr; Brandon Randazzo, prom dir; John D. Strakler, chief of engrg; Mark Deanba, progmg.

Gonzales

KHIP(FM)— Oct 25, 1990: 104.3 mhz; 2.6 kw. 508 ft TL: N36 40 06 W121 31 09. Stereo. Hrs opn: 24 60 Garden Court, Suite 300, Monterey, 93940. Phone: (831) 658-5200. Fax: (831) 658-5299. Web Site: www.thehippo.com. Licensee: Mapleton Communications L.L.C. (group owner; acq 11-16-01; grpsl). Population served: 500,000 Natl. Rep: McGavren Guild. Law Firm: Leventhal, Senter & Lerman. Format: Classic rock. Target aud: 18-49; upscale, active young professionals. ♦Dale Hendry, gen mgr; Jodi Morgan, gen sls mgr; Sybil DeAngelo, prom dir; Kenny Allen, progmg dir.

KKMC(AM)— Sept 22, 1984: 880 khz; 10 kw-U, DA-2. TL: N36 33 46 W121 26 05. Hrs open: 24 30 E. San Joaquin St., Suite 105, Salinas, 93901. Phone: (831) 424-5562. Fax: (831) 424-6437. E-mail: info@kkmc.com

Web Site: www.kkmc.com.Satcomlllr Licensee: Monterey County Broadcasters Inc. Population served: 500,000 Natl. Network: USA. Format: Relg, Christian teaching and talk. News: 7 hrs wkly. Target aud: 25 plus; family oriented. Spec prog: Spanish 5 hrs wkly. ♦Carl J. Auel, pres; John N. Dick, gen mgr; John Dick, progmg dir; Lorraine Dick, gen sls mgr & sls.

Grass Valley

KBAA(FM)— May 3, 2004: 103.3 mhz; 530 w. Ant 1,102 ft TL: N39 14 45 W120 57 56. Stereo. Hrs opn: 24 500 Media Pl., Sacramento, 95815. Phone: (916) 368-6300. Fax: (916) 473-0146. Web Site: www.lakebuena.com. Licensee: Bustos Media of California License LLC. Group owner: Salem Communications Corp. (acq 5-12-2006; $500,000). Format: Mexican rgnl. ♦Amparo Perez-Cook, gen mgr.

***KLVS(FM)**— September 1997: 99.3 mhz; 13 kw. Ant 466 ft TL: N39 16 33 W120 53 49. (CP: 1 kw, ant 1,082 ft. TL: N39 14 45 W120 57 56). Stereo. Hrs opn: 24 2351 Sunset Blvd., Suite 170-218, Rocklin, 95765. Phone: (916) 251-1600. Fax: (916) 251-1650. E-mail: klove@klove.com Web Site: www.klove.com. Licensee: Educational Media Foundation. Group owner: EMF Broadcasting (acq 9-12-96; $65,000). Population served: 336,000 Natl. Network: K-Love. Law Firm: Shaw Pittman. Format: Contemp Christian. News staff: 3. Target aud: 25-44; Judeo-Christian, female. ♦Richard Jenkins, pres; Mike Novak, VP & chief of engrg; Keith Whipple, dev dir; David Pierce, prom mgr & progmg mgr; Ed Lenane, news dir; Sam Wallington, engrg dir; Karen Johnson, news rptr; Marya Morgan, news rptr; Richard Hunt, news rptr.

KNCO(AM)— Oct 1, 1978: 830 khz; 5 kw-U, DA-N. TL: N39 12 52 W121 00 55. Hrs open: 24 1255 E. Main St., Suite A, 95945. Phone: (530) 272-3424. Fax: (530) 272-2872. E-mail: info@knco.com Web Site: www.knco.com. Licensee: Nevada County Broadcasters Inc. (group owner). Population served: 75,000 Natl. Network: CNN Radio, ABC. Law Firm: Fletcher, Heald & Hildreth. Format: News/talk. News staff: 4; News: 30 hrs wkly. Target aud: 35 plus; adults of western Nevada County. Spec prog: Christian 4 hrs wkly. ♦Bob Breck, CEO & gen mgr; Edward Sylvester, chmn; Scott Robertson, pres; Tom Fitzsimmons, progmg dir & news dir; Paul Paterson, chief of engrg; Barbara Juneau, traf mgr.

KNCO-FM— Sept 7, 1982: 94.1 mhz; 660 w. Ant 980 ft TL: N39 14 44 W120 57 52. Stereo. 24 E-mail: info@mystarradio.com Web Site: www.mystarradio.com. Natl. Network: Westwood One. Format: Adult contemp. News staff: one; News: 2 hrs wkly. Target aud: 25-44; residents of western Nevada County. ♦George Rath, progmg dir.

Green Acres

***KAXL(FM)**— May 4, 1994: 88.3 mhz; 21 kw. Ant 328 ft TL: N35 24 55 W119 14 01. Hrs open: 24 110 S. Montclair, Suite 205, Bakersfield, 93309. Phone: (661) 832-2800. Fax: (661) 832-3164. E-mail: kaxl@kaxl.com Web Site: www.kaxl.com. Licensee: Skyride Unlimted Inc. (acq 7-8-91; $4,000; FTR: 7-29-91). Law Firm: Leventhal, Senter & Lerman. Format: Contemp inspirational. News: 4 hrs wkly. Target aud: women 35 plus. ♦Terri Blankenship, stn mgr; Dan Schaffer, opns mgr; Sheryl Giesbrecht, prom dir; Dave Purin, mus dir.

KRAB(FM)—Licensed to Green Acres. See Bakersfield

Greenfield

KLOK-FM— Aug 7, 1989: 99.5 mhz; 50 kw. 492 ft TL: N36 27 51 W121 17 52. (CP: 30 kw, ant 640 ft.). Hrs opn: 67 Garden Ct., Monterey, 93940-5302. Phone: (831) 771-9950. Fax: (831) 373-6700. Web Site: www.entravision.com. Licensee: Entravision Holdings LLC. Group owner: Entravision Communications Corp. (acq 3-14-00; grpsl). Format: Rgnl Mexican. ♦Aaron Scoby, gen mgr; Fidel Soto, news dir; Marcello Soto, chief of engrg; Tony Valencia, progmg dir & sls.

KSEA(FM)— 1998: 107.9 mhz; 870 w. 1,637 ft TL: N36 23 00 W121 25 40. Hrs open: 24 229 Pajaro St. 302 D., Salinas, 93901. Phone: (831) 754-1469. Fax: (831) 754-1563. E-mail: kseaproduction@campesina.com Web Site: www.campesina.com. Licensee: Farmworker Educational Radio Network Inc. (acq 3-13-97; $600,000). Population served: 540,000 Law Firm: Borsari & Paxson. Format: Sp rgnl Mexican. Target aud: 18-54; Hispanic market. ♦Juan Zamora, gen mgr.

Greenville

***KPJP(FM)**— Sept 15, 2004: 89.3 mhz; 4.5 kw vert. Ant 2,348 ft TL: N40 13 59 W121 01 08. Hrs open: Box 180, Tahoma, 96142. Phone: (530) 584-5700. Fax: (530) 584-5705. E-mail: info@ihradio.org Web Site: www.ihradio.org. Licensee: IHR Educational Broadcasting (group owner). Format: Relg, catholic. ♦Douglas M. Sherman, pres; Steve Wise, gen mgr.

Gridley

KMJE(FM)— Oct 1, 1996: 101.5 mhz; 140 w. 1,975 ft TL: N39 12 21 W121 49 11. Hrs open: 861 Gray Ave., Suite K, Yuba City, 95991. Phone: (530) 673-2200. Fax: (530) 673-3010. E-mail: info@gosunny.com Web Site: www.gosunny.com. Licensee: Results Radio Licensee L.L.C. Group owner: Fritz Communications Inc. (acq 6-11-99; grpsl). Format: Hot adult contemp. News staff: one; News: 7 hrs wkly. ♦Jack Fritz, pres & gen mgr; Michael Berry, gen sls mgr & engrg VP.

Groveland

***KXSR(FM)**— May 8, 1992: 91.7 mhz; 4 kw. Ant 1,591 ft TL: N38 03 46 W120 14 45. Hrs open: 24 7055 Folsom Blvd., Sacramento, 95826. Phone: (916) 480-5900. Fax: (916) 487-3348. E-mail: npr@csus.edu Web Site: www.capradio.org. Licensee: California State University Sacramento. Population served: 100,000 Natl. Network: NPR, PRI. Law Firm: Duane Morris LLP. Format: Class. Target aud: General; NPR listeners, eg. professionals, educators & administrators. ♦Carl Watanabe, stn mgr, progmg dir & progmg mgr; John Brenneise, opns mgr; Joe Barr, news dir, local news ed, political ed & relg ed; Jeff Browne, engrg dir; Michael Frost, prom dir & traf mgr.

Grover Beach

KURQ(FM)— July 4, 1984: 107.3 mhz; 3.5 kw. Ant 1,650 ft TL: N35 21 37 W120 39 18. Stereo. Hrs opn: 24 51 Zaca Ln., Suite 110, San Luis Obispo, 93401. Phone: (805) 545-0101. Fax: (805) 541-5303. Web Site: www.1073therock.com. Licensee: CC Licenses LLC. (acq 10-17-2000). Population served: 250000 Law Firm: Wiley, Rein & Fielding. Format: Active rock. News staff: one; News: 8 hrs wkly. Target aud: 18-44; emphasis on 25-34 year olds. ♦Rich Hawkins, gen mgr; Pattie Wagner, gen sls mgr; Adam Burns, progmg dir.

Guadalupe

KIDI(FM)— 1992: 105.5 mhz; 160 w. 1,342 ft TL: N34 53 54 W120 35 28. Hrs open: 24 718 E. Chapel St., Santa Maria, 93454. Phone: (805) 928-4334. Fax: (805) 349-2765. E-mail: kidiktap@aol.com Web Site: www.labuena.net. Licensee: Emerald Wave Media. (acq 5-1-97; $475,000 with KTAP(AM) Santa Maria). Population served: 60,000 Format: Sp, Mexican rgnl. News staff: one; News: 5 hrs wkly. Target aud: 18-45; second generation bilingual Mexican Americans. ♦August Ruiz, gen mgr & gen sls mgr; Sofia Lariz, rgnl sls mgr.

Gualala

KTDE(FM)— August 1993: 100.5 mhz; 6 kw. 669 ft TL: N38 49 33 W123 34 12. Hrs open: 24 Box 1557, 95445. Secondary address: 38958 Cypress Way 95445. Phone: (707) 884-1000. Fax: (707) 884-1229. E-mail: thetide@men.org Web Site: www.ktde.com. Licensee: Four Rivers Broadcasting Inc. (group owner; (acq 7-21-2005; grpsl). Population served: 50000 Natl. Network: CBS. Law Firm: Cole, Raywid & Braverman. Format: Hot adult contemp, var/div. Target aud: 30-55. Spec prog: Gospel one hr wkly. ♦John Power, CEO; Amy Heath, gen mgr; Patricia Weber, opns mgr; Pam Knutson, rgnl sls mgr & prom dir.

Hamilton City

KRER(FM)— 2005: 101.7 mhz; 530 w. Ant 1,099 ft TL: N39 56 46 W121 43 17. Hrs open: 2654 Cramer Ln., Chico, 95928-8838. Phone: (530) 345-0021. Fax: (530) 893-2121. Licensee: Coloma Hamilton City LLC. Natl. Network: ESPN Radio. Format: Sports. ♦Scott Donohue, pres; Dino Corbin, gen mgr; Larry Scott, progmg dir.

Hanford

KGEN-FM— Jan 1, 1997: 94.5 mhz; 3.3 kw. 443 ft TL: N36 12 16 W119 33 52. Hrs open: 24 Box 2040, Tulare, 93275. Phone: (559) 686-1370. Fax: (559) 685-1394. E-mail: kgen@sbcglobal.net Licensee: Azteca Broadcasting Corp. (group owner) Format: Sp, Rgnl Mexician. Target aud: 18 plus. ♦Margreta Hernandez, gen mgr; Isabel Duran, gen sls mgr; Ernesto Gaytan, progmg dir.

KIGS(AM)— Feb 1, 1948: 620 khz; 1 kw-U, DA-N. TL: N36 19 37 W119 33 58. Hrs open: 24 6165 E. Lacey Blvd., 93230. Phone: (559) 582-0361. Fax: (559) 582-3981. E-mail: info@kigs.com Web Site: www.kigs.com. Licensee: Perreira Broadcasting (acq 7-15-90). Population served: 657,000 Law Firm: Allan E. Aronowitz. Format: Foreign languages. News: 35 hrs wkly. Target aud: 18-49. ♦Tony Vieira, gen mgr, gen sls mgr & progmg dir.

KRDA(FM)— September 1976: 107.5 mhz; 20.3 kw. Ant 784 ft TL: N36 38 12 W118 56 34. Stereo. Hrs opn: 24 1981 No. Gateway Blvd., Fresno, 93727. Phone: (559) 456-4000. Fax: (559) 251-9555. Licensee: Univision Radio License Corp. Group owner: Pappas Telecasting Companies (acq 1-3-2006; $10 million). Natl. Network: ABC, CNN Radio. Natl. Rep: Lotus Entravision Reps LLC. Law Firm: Fletcher, Heald & Hildreth. Format: Sp Adult Contemp. Target aud: 35-54; high quality FM oriented news/talk listeners. ♦Angela Navarrete, gen mgr; Jim P. Pappas, stn mgr; Marv Allen, opns dir; Jim Pappas, sls dir & gen sls mgr; Rick Gonzalez, gen sls mgr; Juan Ojeda, progmg dir; Mark Thomas, progmg dir & traf mgr; Jim Moore, chief of engrg; Scott Dean, chief of engrg.

KRZR(FM)— Dec 24, 1976: 103.7 mhz; 50 kw. 499 ft TL: N36 33 36 W119 45 20. Stereo. Hrs opn: 24 83 E. Shaw Ave., Suite 150, Fresno, 93710. Phone: (559) 230-4300. Fax: (559) 243-4301. Web Site: www.krzr.com. Licensee: Capstar TX L.P. Group owner: Clear Channel Communications Inc. (acq 8-30-00; grpsl). Population served: 1,000,000 Natl. Network: ABC, AP Radio. Law Firm: Dow, Lohnes & Albertson. Format: AOR. News: one hr wkly. Target aud: 18-34; male. ♦Jeff Megrete, gen mgr; Joni Norvell, gen sls mgr; Paul Wilson, progmg dir; Dave Case, chief of engrg & disc jockey.

Hayward

***KCRH(FM)**— Apr 10, 1981: 89.9 mhz; 18 w. Ant -134 ft TL: N37 38 23 W122 06 16. Hrs open: 25555 Hesperian Blvd., 94545. Phone: (510) 723-6954. Fax: (510) 723-7155. E-mail: cglen@clpccd.cc.ca.us Web Site: www.kcrhradio.com. Licensee: South County Community College District. Population served: 130,000 Format: Var, urban contemp. News: 5 hrs wkly. Target aud: 17-35; general. Spec prog: Instructional one hr, pub affrs 5 hrs wkly. ♦Bernard Bautista, prom dir; Chad Mark Glen, prom dir & gen mgr; Bernard Mark Bautista, progmg dir; Rocko Flores, mus dir.

Healdsburg

KFGY(FM)—Licensed to Healdsburg. See Santa Rosa

KNOB(FM)— 2002: 96.7 mhz; 2.4 kw. Ant 525 ft TL: N38 44 08 W122 50 55. Hrs open: 3565 Standish Ave., Santa Rosa, 95407. Phone: (707) 588-0707. Fax: (707) 588-0777. Web Site: www.967bobfm.com. Licensee: JYH Broadcasting. Format: Eclectic. ♦Debbie Morton, gen mgr; Kelly Rae, gen sls mgr; Natalie Rowland, prom dir; Nate Campbell, progmg dir; Dan Ethan, chief of engrg.

KRSH(FM)— Feb 1, 1996: 95.9 mhz; 2.65 kw. Ant 502 ft TL: N38 44 08 W122 50 55. Hrs open: 3565 Standish Ave., Santa Rosa, 95407. Phone: (707) 588-0707. Fax: (707) 588-0777. E-mail: studio@krsh.com Web Site: www.krsh.com. Licensee: Deas Communications Inc. Group owner: Sinclair Communications Inc. (acq 8-3-2001; $2.1 million). Format: AAA. Target aud: 25-49. ♦Debbie Morton, gen mgr & gen sls mgr; Dean Kattari, stn mgr; Natalie Rowland, prom dir; Pam Long, progmg dir; Dan Ethan, chief of engrg; Amanda Kelsey, traf mgr.

Hemet

KSDT(AM)— Apr 10, 1959: 1320 khz; 500 w-D, 300 w-N, DA-2. TL: N33 44 59 W116 59 53. Hrs open: 24 15700 Village Dr., Suite A, Victorville, 92392. Phone: (760) 243-7903. Fax: (706) 243-7183. Licensee: Rudex Broadcasting Ltd. (acq 9-30-2002; $250,000). Population served: 250,000 Format: Christian. ♦John Cooper, pres & gen mgr.

KXRS(FM)— Nov 9, 1963: 105.7 mhz; 170 w. Ant 1,023 ft TL: N33 44 59 W116 59 53. Stereo. Hrs opn: 24 1950 S. Sunwest Ln., Suite 302, San Bernardino, 92408. Phone: (909) 825-5020. Fax: (909) 884-5844. E-mail: vb@radiolazer.com Web Site: www.radiolazer.com. Licensee: Lazer Broadcasting Corp. (group owner; acq 2-94). Format: Rgnl Mexican. ♦Vicki Bails, gen mgr; Armando Gutierrez, prom dir; Salvador Prieto, progmg dir & news dir.

Hesperia

KRAK(AM)— Feb 1, 1990: 910 khz; 700 w-D, 500 w-N, DA-2. TL: N34 23 19 W117 23 29. Stereo. Hrs opn: 24 11920 Hesperia Rd., 92345. Phone: (760) 244-2000. Fax: (760) 244-1198. Web Site: stardust910.com. Licensee: CBS Radio Stations Inc. Group owner: Infinity Broadcasting Corp. (acq 7-19-2000; $3,537,500 with KVFG(FM) Victorville). Population served: 250,000 Natl. Network: ABC. Law Firm: Fleischman & Walsh L. Format: Nostalgic/adult standards. Target aud: 40 plus. ♦Tom Hoyt, gen mgr; Bill Pettus, stn mgr.

Hollister

***KHRI(FM)**— Dec 17, 2000: 90.7 mhz; 170 w. Ant -364 ft TL: N36 52 02 W121 23 58. Stereo. Hrs opn: 24
Rebroadcasts KLRD(FM) Yucaipa 100%.
2351 Sunset Blvd., Suite 170-218, Rocklin, 95765. Phone: (916) 251-1600. Fax: (916) 251-1650. E-mail: info@air1.com Web Site: www.air1.com. Licensee: Educational Media Foundation. Group owner: EMF Broadcasting (acq 11-7-00; $30,000 for CP). Population served: 27,000 Natl. Network: Air 1. Law Firm: Shaw Pittman. Format: Contemp Christian. News staff: 3. Target aud: 18-35; Judeo-Christian, female. ♦Richard Jenkins, pres; Mike Novak, VP & progmg dir; Lloyd Parker, gen mgr; Ed Lenane, opns dir & news dir; Keith Whipple, dev dir; Eric Allen, natl sls mgr; David Pierce, progmg mgr; Jon Rivers, mus dir; Sam Wallington, engrg dir; Arthur Vassar, traf mgr; Karen Johnson, news rptr; Marya Morgan, news rptr; Richard Hunt, news rptr.

KMPG(AM)— 1966: 1520 khz; 5 kw-D, DA-2. TL: N36 50 16 W121 25 01. Hrs open: 14 Box 369, 910 Monterey St., 95023. Phone: (831) 637-7994. Fax: (831) 637-4031. Web Site: www.kmpgradio@netzero.net. Licensee: Promo Radio Corp. (acq 12-19-2003). Population served: 217,000 Format: Sp rgnl Mexican. Target aud: 18-49. ♦Rafael Meza, pres & gen mgr.

KXSM(FM)— 1979: 93.5 mhz; 110 w. Ant 2,296 ft TL: N36 45 22 W121 30 06. Stereo. Hrs opn: 200 South A St., Suite 400, Oxnard, 93030. Phone: (805) 240-2070. Fax: (805) 240-5960. Licensee: Lazer Broadcasting Corp. (group owner; (acq 7-25-2005; $2.88 million with KXZM(FM) Felton). Population served: 1,100,000 Format: Rgnl Mexican. ♦Alfredo Plascencia, pres; Daniel Osuna, gen mgr.

Holtville

KGBA-FM— Aug 8, 1983: 100.1 mhz; 3 kw. 331 ft TL: N32 48 10 W115 29 53. Stereo. Hrs opn: 24 Studio, 605 State St., El Centro, 92243. Phone: (760) 352-9860. Fax: (760) 352-1883. E-mail: kgba@kgba.org Web Site: www.kgba.org. Licensee: The Voice of International Christian Evangelism Inc. (acq 11-1-86; $350,000; FTR: 9-15-86). Population served: 250,000 Law Firm: Miller & Miller. Format: Relg talk. News: 3 hrs wkly. Target aud: 25-55; adult family Christian conservatives. Spec prog: Chinese 14 hrs, children 4 hrs, gospel 7 hrs wkly. ♦Robert Sager, gen mgr; Mike Leonard, gen sls mgr; Sara Mae, progmg dir; Dean Imhof, chief of engrg.

Hoopa

***KIDE(FM)**— December 1980: 91.3 mhz; 195 w. -1,560 ft TL: N41 03 51 W123 41 05. (CP: 305 w). Stereo. Hrs opn: 24 Box 1220, 95546. Phone: (530) 625-4245. Fax: (530) 625-4046. Web Site: www.kidefm.org. Licensee: Hoopa Valley Tribe. Format: Country. Spec prog: Hoopa Indian language, history & culture 20 hrs wkly. ♦Joseph Orozco, gen mgr & progmg dir.

Hornbrook

KRVC(FM)— 2007: 98.9 mhz; 1.25 kw. Ant 2,483 ft TL: N42 05 00 W122 42 00. Hrs open: 511 Rossanley Dr., Medford, OR, 97501. Phone: (541) 772-0322. Fax: (541) 772-4233. Licensee: Opus Broadcasting Systems Inc. Format: Country. ♦Dean Flock, pres & gen mgr.

Huron

KZLA(FM)— 2003: 98.3 mhz; 100 w. Ant 43 ft TL: N36 12 05 W120 05 53. Hrs open: 152 E. Elm, Coalinga, 93210. Phone: (559) 935-4191. Fax: (559) 935-4191. Licensee: Huron Broadcasting LLC. Format: Oldies. ♦Rebecca Sexton, gen mgr.

Hydesville

KSLG-FM— Apr 13, 2001: 94.1 mhz; 4.5 kw. 1,784 ft TL: N40 30 04 W124 17 05. Hrs open: Box 25, Ferndale, 95536-0025. Secondary address: 1400 Main St., Suite 104, Ferndale 95536. Phone: (707) 786-5104. Fax: (707) 786-5100. E-mail: 941@kslg.com Web Site: www.kslg.com. Licensee: Lost Coast Communications Inc. (acq 11-5-2001). Population served: 125,000 Natl. Rep: McGavren Guild. Law Firm: Dan J. Alpert. Format: Modern rock. Target aud: 18-49. ♦Patrick Cleary, gen mgr; Cliff Berkowitz, opns VP; Jennifer White, natl sls mgr; Gregg Foster, mktg dir; Monica Topping, prom dir; Mike Dronkers, progmg dir; Kevin Sanders, chief of engrg.

Idyllwild

KATY-FM— Dec 1, 1989: 101.3 mhz; 1.55 kw. Ant 656 ft TL: N33 43 31 W116 44 58. Stereo. Hrs opn: 24 27431 Enterprise Cir. W., Fl 1st, Temecula, 92590. Phone: (951) 506-1222. Fax: (951) 506-1213. E-mail: katytraffic@linkline.com Web Site: www.katyfm.com. Licensee: All Pro Broadcasting Inc. (acq 3-21-01; $3.5 million plus $100,000 for option to purchase for 51%). Law Firm: Leventhal, Senter & Lerman. Format: Adult contemp. News staff: one; News: 2 hrs wkly. Target aud: 25-49; affluent, upwardly mobile. ♦Duane Davis, exec VP & gen sls mgr; Bill McNulty, gen mgr; Kevin Watson, stn mgr & gen sls mgr; Willie D. Davis, CEO & opns mgr; Cyrene Jagger, progmg dir & news dir.

Imperial

KMXX(FM)— Sept 17, 1980: 99.3 mhz; 3 kw. 200 ft TL: N32 51 44 W115 33 41. (CP 6 kw, ant 302 ft. TL: N32 54 W115 31 40). Stereo. Hrs opn: 24 Box 2830, El Centro, 92244. Phone: (760) 352-2277. Fax: (760) 482-0099. Web Site: www.entravision.com. Licensee: Entravision Holdings LLC. Group owner: Entravision Communications Corp. (acq 7-31-00; grpsl). Population served: 950,000 Format: Rgnl Mexican. Target aud: 18-49. ♦Eric Chavez, gen mgr.

Independence

KSRW(FM)— Apr 12, 1996: 92.5 mhz; 870 w. 2,949 ft TL: N36 58 38 W118 07 13. Hrs open: 24 1280 N. Main St., Suite J, Bishop, 93514. Phone: (760) 873-5329. Fax: (760) 873-5328. E-mail: kday@schat.com Licensee: Ms. Benett Kessler (acq 3-14-91; FTR: 4-1-91). Natl. Network: CNN Radio. Format: Adult contemp. News staff: one; News: 10 hrs wkly. Target aud: 30-65; professionals & retirees with average to above average buying power. ♦Benett Kessler, CEO & gen mgr.

Indian Wells

KAJR(FM)—Not on air, target date: unknown: 95.9 mhz; 1.5 kw. Ant 666 ft TL: N33 48 04 W116 13 28. Hrs open: 836 Prospect St., Suite 202, La Jolla, 92037. Phone: (858) 459-2631. Licensee: A & J Media LLC. ♦Arthur L. Rivkin, gen mgr.

Indio

KCLB-FM—See Coachella

***KCRI(FM)**— 1995: 89.3 mhz; 3.3 kw. Ant 561 ft TL: N33 48 07 W116 13 28. Hrs open:
Rebroadcasts KCRW(FM) Santa Monica 100%.
1900 Pico Blvd., Santa Monica, 90405. Phone: (310) 450-5183. Phone: (888) 600-kcrw. Fax: (310) 450-7172. E-mail: mail@kcrw.org Web Site: www.kcrw.com. Licensee: Santa Monica Community College. Natl. Network: NPR. Wire Svc: AP Format: Eclectic, news. ♦Ruth Seymour, gen mgr; Jennifer Ferro, stn mgr; Mike Newport, opns mgr; David Kleinbart, dev dir.

KESQ(AM)— 1946: 1400 khz; 1 kw-U. TL: N33 43 37 W116 15 10. Stereo. Hrs opn: 24 42650 Melanie Pl., Palm Desert, 92211-5170. Phone: (760) 568-6830. Fax: (760) 568-3984. Licensee: Gulf-California Broadcast Co. Population served: 80,000 Format: Regl, Sp. News staff: one; News: 7 hrs wkly. Target aud: General. ♦Martin Serna, gen mgr.

KJJZ(FM)— March 1993: 102.3 mhz; 600 w. 587 ft TL: N33 48 07 W116 13 29. Hrs open: 24 Box 1825, Palm Springs, 92263. Phone: (760) 320-4550. Fax: (760) 320-3037. Web Site: www.102kjjz.com. Licensee: R.M. Broadcasting L.L.C. (acq 1-97; $1.231 million). Population served: 250,000 Natl. Network: Westwood One. Natl. Rep: McGavren Guild. Law Firm: Fletcher, Heald & Hildreth. Format: Smooth jazz. News staff: one; News: 3 hrs wkly. Target aud: 25-49; Palm Springs baby boomers. ♦Todd Marker, VP & gen mgr; Hughes Hilles, gen sls mgr; Cary James, prom dir; Jim Fitzgerald, progmg dir; Jeff Michaels, news dir; Ben Manierre, chief of engrg.

KKUU(FM)— Apr 13, 1984: 92.7 mhz; 4.2 kw. Ant 394 ft TL: N33 52 15 W116 13 37. Stereo. Hrs opn: 24 1321 N. Gene Autry Tr., Palm Springs, 92262. Phone: (760) 322-7890. Fax: (760) 322-5493. E-mail: keith.martin@morris.com Licensee: MCC Radio LLC. Group owner: Morris Radio LLC (acq 1998; $4.5 million). Population served: 330,000 Natl. Rep: Christal. Format: Hip hop. News staff: one; News: 2 hrs wkly. Target aud: 25-54. ♦Michael Ostehaut, VP; Anthony Quiroz, gen mgr & opns mgr; Keith Martin, gen mgr.

KNWZ(AM)—See Coachella

Inglewood

KRCD(FM)— 1959: 103.9 mhz; 4.1 kw. Ant 387 ft TL: N34 00 26 W118 21 54. Stereo. Hrs opn: 24 655 N Central Ave, Suite 2500, Glendale, 91203-1422. Phone: (818) 500-4500. Fax: (818) 500-4560. Web Site: www.univision.com. Licensee: Univision Radio License Corp. Group owner: Univision Radio (acq 9-22-2003; grpsl). Population served: 500000 Natl. Network: CBC. Rgnl. Network: CBS, Unistar. Format: Adult contemp, Sp. Target aud: 25-44; women, 60% African-American, 30% Hispanic. ♦Michelle Hohman, gen mgr; Haz Montana, opns mgr; Jim Coronado, gen sls mgr; Offad Vallejo, mktg dir; Amalia Gonzalez, progmg dir; Tom Koza, chief of engrg.

KTYM(AM)— Feb 14, 1958: 1460 khz; 5 kw-D, 500 w-N, DA-2. TL: N34 00 24 W118 21 52. Hrs open: 24 6803 West Blvd., 90302. Phone: (310) 672-3700. Fax: (310) 673-2259. Web Site: www.ktym1460.com. Licensee: Trans America Broadcasting Corp. Population served: 89,985 Law Firm: Miller & Miller, P.C. Format: Black,var/div, Russian. News staff: 2; News: 2 hrs wkly. Target aud: 18-54. Spec prog: It 3 hrs, Armenian 2 hrs, Ger 2 hrs, Pol 2 hrs, Russian 1 hr. wkly. ♦Gerardo Borrego, pres, VP & gen mgr; Gary Rehers, gen sls mgr & progmg dir; Paul Wiren, chief of engrg. Co-owned TV: KAIL(TV) affil

Inyokern

***KZLU(FM)**—Not on air, target date: unknown: 88.7 mhz; 180 w. Ant 1,309 ft TL: N35 28 39 W117 41 58. Hrs open: 2351 Sunset Blvd., Suite 170-218, Rocklin, 95765. Phone: (916) 251-1600. Fax: (916) 251-1650. Licensee: Educational Media Foundation. (acq 9-7-2005). ♦Richard Jenkins, pres; Mike Novak, VP; Keith Whipple, dev dir; David Pierce, progmg mgr; Ed Lenane, news dir; Sam Wallington, engrg dir; Karen Johnson, news rptr; Marya Morgan, news rptr; Richard Hunt, news rptr.

Irvine

***KUCI(FM)**— Oct 1, 1969: 88.9 mhz; 200 w. -10 ft TL: N33 38 41 W117 50 36. Stereo. Hrs opn: 24 Box 4362, 92616-4362. Phone: (949) 824-6868. Fax: (949) 824-3741. E-mail: kuci@kuci.org Web Site: www.kuci.org. Licensee: Regents of the University of California. Population served: 1,000,000 Format: Div. ♦Kevin Stockdale, dev dir; Scarlett Davis, prom dir; Emilio Nunez, mus dir; Kyle Olsen, mus dir; Mike Casper, pub affrs dir; Mike Boyle, engrg mgr; Elaine Hawkes, chief of engrg.

Jackson

KLMG(FM)— Aug 16, 1973: Stn currently dark. 94.3 mhz; 4.3 kw. Ant 790 ft TL: N38 24 10 W120 39 15. Stereo. Hrs opn: 500 Media Place, Sacramento, 95815. Phone: (916) 368-6300. Fax: (916) 441-6480. Licensee: Bustos Media of California License LLC. Group owner: Univision Radio (acq 5-12-2006; swap for KKFS(FM) Lincoln). Population served: 60,000 Format: Sp. ♦Amparo Perez-Cook, gen mgr; Bobby Reynoso, prom dir; Juan Gonzalez, progmg dir; Mark Sedaka, chief of engrg; Cynthia Sanchez, traf mgr.

Johannesburg

KEDD(FM)— March 1990: 103.9 mhz; 1.5 kw. Ant 1,322 ft TL: N35 28 39 W117 41 58. Hrs open: 24 731 N. Balsam St., Ridgecrest, 93555. Phone: (760) 371-1700. Fax: (760) 371-1824. E-mail: radio@iwvisp.com Web Site: www.keddfm.com. Licensee: Adelman Broadcasting Inc. (group owner). Population served: 500,000 Format: Sp. Target aud: 18-54. ♦Robert Adelman, pres.

KRAJ(FM)— October 1998: 100.9 mhz; 1.5 kw. Ant 1,312 ft TL: N35 28 41 W117 41 58. Hrs open: 24 731 N. Balsam, Ridgecrest, 93555. Phone: (760) 371-1700. Fax: (760) 371-1824. E-mail: radio@iwvisp.com Web Site: www.krajfm.com. Licensee: Adelman Broadcasting Inc. (group owner; (acq 12-28-99; $45,000). Population served: 500,000 Natl. Network: Jones Radio Networks. Format: Urban CHR. Target aud: 18-54. ♦Robert Adelman, pres.

Joshua Tree

KQCM(FM)— Nov 2, 1995: 92.1 mhz; 6 kw. Ant 230 ft TL: N34 09 16 W116 12 04. Stereo. Hrs opn: Box 1437, 92252. Phone: (760) 362-4264. E-mail: coppermountainbroadcasting@yahoo.com Web Site: www.coppermountainbroadcasting.com. Licensee: Copper Mountain Broadcasting Co. (acq 7-14-2004; $575,000 with KXCM(FM) Twentynine Palms). Population served: 85,000 Natl. Network: Westwood One, Jones Radio Networks. Law Firm: Leventhal, Senter, & Lerman, PLLC. Format: CHR. Target aud: 18-34. ♦Gary DeMaroney, gen mgr; Rebecca Westerman, pub affrs dir.

Julian

***KLVJ(FM)**— Oct 23, 1991: 100.1 mhz; 110 w. Ant 2,286 ft TL: N33 09 33 W116 36 53. Stereo. Hrs open: 24 2351 Sunset Blvd., Suite 170-218, Rocklin, 95765. Phone: (916) 251-1600. Fax: (916) 251-1650. E-mail: klove@klove.com Web Site: www.klove.com. Licensee: Educational Media Foundation. Group owner: EMF Broadcasting (acq 1-30-97; $34,168). Population served: 1,088,000 Natl. Network: K-Love. Law Firm: Shaw Pittman. Format: Contemp Christian. News staff: 3. Target aud: 25-44; Judeo-Christian, female. ♦Richard Jenkins, pres; Mike Novak, VP; Keith Whipple, dev dir; David Pierce, progmg mgr; Ed Lenane, news dir; Sam Wallington, engrg dir; Karen Johnson, news rptr; Marya Morgan, news rptr; Richard Hunt, news rptr.

June Lake

***KWTM(FM)**— 2002: 90.9 mhz; 910 w. Ant 344 ft TL: N38 05 14 W119 10 31. Stereo. Hrs opn: 24
Rebroadcasts KWTW(FM) Bishop 100%.
Box 637, Living Proof Inc., Bishop, 93515. Secondary address: 125 S. Main St., Bishop 93514. Phone: (760) 872-4225. Phone: (866) 466-5989. Fax: (760) 872-4155. E-mail: friar@schat.com Web Site: www.kwtw.org. Licensee: Living Proof Inc. (acq 4-19-00; $250,000). Population served: 60,000 Format: Christian. ♦Daniel McClenaghan, pres & gen mgr; Brian Law, opns mgr & progmg dir; Robert Branch, chief of engrg.

Keene

***KNAI(FM)**—Phoenix, AZ) Oct 23, 1991: 88.3 mhz; 22.5 kw ST: *WPHF-FM. 997 ft TL: N33 35 47 W112 05 29. Stereo. Hrs opn: 4 AM-7:30 PM 3602 W. Thomas Rd., Suite 6, Phoenix, AZ, 85019. Phone: (602) 269-3121. Fax: (602) 269-3020. Web Site: www.campesina.com. Licensee: National Farm Workers Service Center Inc. Population served: 825,000 Rgnl rep: Vision Marketing Law Firm: Borsari & Paxson. Format: Sp, community pub affrs, mus. Target aud: 25-54; Hispanic market. ♦Paul Chavez, chmn & pres; Anthony Chavez, VP; Michael Nowakowski, gen mgr; Pepe Escamilla, progmg dir.

Kerman

KBHH(FM)— March 2001: 95.3 mhz; 6 kw. 328 ft TL: N36 39 40 W120 09 59. Hrs open: 24 2502 Merced St., Fresno, 93721. Phone: (661) 837-0745. Fax: (661) 837-1612. E-mail: Achavez@campesina.com Web Site: www.campesina.com. Licensee: Farmworker Educational Radio Network Inc. Population served: 600,000 Law Firm: Borsari & Paxson. Format: Sp rgnl Mexican. Target aud: 25-54; Hispanic market. ♦Anthony Chavez, pres & gen mgr; Cesar Chavez, progmg dir; Dave Whitehead, chief of engrg; Maria Vrrutia, traf mgr.

KOKO-FM— Apr 16, 1990: 94.3 mhz; 3 kw. 328 ft TL: N36 44 29 W120 05 08. Stereo. Hrs open: 24 2775 E. Shaw Ave., Fresno, 93710. Phone: (559) 292-9494. Fax: (559) 294-7041. Web Site: www.kok94.com. Licensee: Big Broadcasting Inc. (acq 1999; $1.14 million). Population served: 600,000 Law Firm: Haley, Bader & Potts. Format: CHR, rhythm oldies. News staff: 4; News: 14 hrs wkly. Target aud: 18-54; Hispanic men & women. ♦Art Laboe, pres & gen mgr; Anna Marie Avila, gen sls mgr & disc jockey; Paul Mendoza, progmg dir & news dir.

Kernville

KCNQ(FM)— November 1985: 102.5 mhz; 130 w. 1,230 ft TL: N35 37 21 W118 26 16. Stereo. Hrs opn: 24 Box 2008, 93238-2008. Secondary address: 14 Sierra Dr. 93238. Phone: (760) 379-5636. Fax: (760) 376-3119. Licensee: Robert J. Bohn & Katherine M. Bohn. (acq 7-28-97). Population served: 25,000 Format: C&W. News staff: one; News: 18 hrs wkly. Target aud: General. Spec prog: Relg one hr wkly. ♦Anthony M. Bohn, CEO; Robert J. Bohn, pres; Stacy Bohn, gen mgr; Gary Huff, sls dir; Bob Jamison, progmg mgr & chief of engrg; Robert Pinney, news dir; Jullian King, traf mgr.

Kettleman City

***KWDS(FM)**— 2006: 89.9 mhz; 50 kw vert. Ant 251 ft TL: N35 59 42 W119 58 06. Hrs open: 5331 Mt. Alifan Dr., San Diego, 92111. Phone: (858) 277-4991. Fax: (858) 277-1365. E-mail: kwoods@horizonsd.org Web Site: www.horizonradio.org. Licensee: Horizon Christian Fellowship (acq 5-21-2004; $150,000 for CP). Format: Christian. ♦Mike MacIntosh, pres.

King City

***KDRH(FM)**— 2001: 91.3 mhz; 300 w vert. Ant 75 ft TL: N36 16 22 W121 05 02. Hrs open: 2351 Sunset Blvd., Suite 170-218, Rocklin, 95765-3719. Phone: (916) 251-1600. Fax: (916) 251-1650. E-mail: info@air1.com Web Site: www.air1.com. Licensee: Educational Media Foundation. Group owner: EMF Broadcasting (acq 11-7-00; $30,000 for CP). Natl. Network: Air 1. Law Firm: Shaw Pittman. Format: Contemp Christian. News: one hr wkly. Target aud: 18-25; teen, young adult. ♦Richard Jenkins, pres; Mike Novak, VP; Keith Whipple, dev dir; David Pierce, progmg mgr; Ed Lenane, news dir; Sam Wallington, engrg dir; Arthur Vassar, traf mgr; Karen Johnson, news rptr; Marya Morgan, news rptr; Richard Hunt, news rptr.

KEXA(FM)— 1981: 93.9 mhz; 5.4 kw. 719 ft TL: N36 22 48 W121 12 57. Stereo. Hrs opn: 24 Box 1939, Salinas, 93902. Secondary address: 548 Alisal St., Salinas 93905. Phone: (831) 757-1910. Fax: (831) 757-8015. E-mail: wolfhouseradio@yahoo.es Licensee: Wolfhouse Radio Group Inc. (group owner; (acq 8-31-2001; grpsl). Format: Spanish pop. Target aud: General. ♦Ramon Castro, gen mgr.

KRKC(AM)— Sept 21, 1958: 1490 khz; 1 kw-U. TL: N36 13 34 W121 07 26. Hrs open: 24 Box 628, 93930. Secondary address: 1134 San Antonio Dr. 93930. Phone: (831) 385-5421. Phone: (831) 674-2278. Fax: (831) 385-0635. E-mail: bill@krkc.com Web Site: www.krkc.com. Licensee: Radio Del Rey. (acq 9-2-82; $270,000; FTR: 9-13-82). Population served: 50,000 Natl. Network: CBS. Natl. Rep: Farmakis, Katz Radio. Law Firm: Pepper & Corazzini. Format: Country. News staff: one. Target aud: 25-54. Spec prog: Farm 10 hrs, sports 9 hrs wkly. ♦Bill Gittler, pres & gen mgr.

KRKC-FM— Jan 30, 1989: 102.1 mhz; 2.6 kw. Ant 1,820 ft TL: N35 57 06 W121 00 03. Stereo. Hrs opn: 24 1134 San Antonio Dr., 93930. Phone: (831) 385-5421. Fax: (831) 385-0635. E-mail: krkc@dedot.com Web Site: www.krkc.com. Licensee: King City Communications Corp. Natl. Network: AP Radio. Law Firm: Pepper & Corazzini. Format: Adult contemp. News staff: one; News: 1 hr wkly. Target aud: 18-49; men and women. ♦Bill Gittler, pres & gen mgr; Dru Vincent, mus dir; Jeff Grice, news dir; Ron Warren, chief of engrg.

Kings Beach

KSRN(FM)— 1990: 107.7 mhz; 230 w. 2,883 ft TL: N39 18 47 W119 53 00. Stereo. Hrs opn: 1465 Terminal Way, Suite 3, Reno, NV, 89502. Phone: (775) 324-4819. Fax: (775) 324-4832. Licensee: Lazer Broadcasting Corp. (group owner; acq 12-12-2003; $2.5 million). Natl. Network: ABC. Natl. Rep: Katz Radio. Format: Sp, rgnl Mexican. News: 5 hrs wkly. Target aud: 35-54; affluent, business professionals. Spec prog: Gospel one hr wkly. ♦Jerry Juskiw, gen mgr; Don Parker, opns VP; Alicia Miranda, gen sls mgr; Salvador Prieto, progmg dir.

Kingsburg

KSXE(FM)— 1992: 106.3 mhz; 16 kw. Ant 420 ft TL: N36 26 50 W119 37 10. Stereo. Hrs opn: 24 2110 Tulare St., Fresno, 93721. Phone: (559) 256-5155. Fax: (559) 860-0160. Licensee: Pro-Active Communications-Fresno LLC (group owner; (acq 7-17-2006; $2.75 million). Population served: 666,000 Format: CHR pop. ♦Gerald Clifton, CEO; Brenda Brown, gen mgr.

La Jolla

KIFM(FM)—See San Diego

La Quinta

KUNA-FM— Aug 1, 1987: 96.7 mhz; 650 w. 578 ft TL: N33 48 08 W116 13 30. Stereo. Hrs opn: 24 42650 Melanie Pl., Palm Desert, 92211-5170. Phone: (760) 568-6830. Fax: (760) 568-3984. Licensee: Gulf California Broadcasting Co. Group owner: News-Press & Gazette Co. Population served: 280,000 Format: Rgnl Mexican. News staff: one; News: 25 hrs wkly. Target aud: 25-54. ♦Martin Serna, gen mgr; Adolpho Iniguez, opns mgr.

La Selva Beach

KOMY(AM)— 1937: 1340 khz; 1 kw-D, 850 w-N. TL: N36 57 43 W121 58 51. Hrs open: 5 AM-midnight (M-S) 2300 Portola Dr., Santa Cruz, 95062. Phone: (831) 475-1080. Fax: (831) 475-2967. Web Site: www.1340komy.com. Licensee: Zwerling Broadcasting System Ltd. (acq 6-5-97). Population served: 500,000 Format: Oldies. Target aud: 25-64; people with an investment at risk in the community. ♦Michael Zwerling, CEO & pres; Ron Stevens, pres; Michael Olson, gen mgr.

Lake Arrowhead

KCXX(FM)— June 1978: 103.9 mhz; 180 w. Ant 1,797 ft TL: N34 14 03 W117 08 25. Stereo. Hrs opn: 24 242 E. Airport Dr., Suite 106, San Bernardino, 92408. Phone: (909) 890-5904. Phone: (909) 889-1039. Fax: (909) 888-7302. Web Site: www.x1039.com. Licensee: All-Pro Broadcasting Inc. (acq 9-10-92; $5 million with KCKC(AM) San Bernardino; FTR: 9-28-92). Population served: 1,300,000 Natl. Rep: McGavren Guild. Law Firm: Leventhal, Senter & Lerman. Format: Alternative rock. Target aud: 18-49. ♦Willie Davis, CEO & pres; Bill McNulty, gen mgr & opns mgr; Kim Martinez, gen sls mgr; Steve Hay, prom mgr; John DeSantis, progmg dir.

Lake Isabella

KQAB(AM)— July 15, 1977: 1140 khz; 1 kw-D. TL: N35 38 20 W118 28 22. Hrs open: Sunrise-sunset Box 2008, 14 Sierra Dr., Kernville, 93238. Phone: (760) 376-4500. Fax: (760) 376-3119. E-mail: qab@care-ems.com Web Site: www.qabmedia.com. Licensee: Robert J. and Katherine M. Bohn. (acq 7-24-97; $300,000 with co-located FM). Population served: 29,000 Format: News/talk. News staff: one; News: 14 hrs wkly. Target aud: 50 plus; mature. ♦Anthony Bohn, gen mgr; Gary Huff, gen sls mgr; Bob Jamison, progmg dir; Robert Pinney, news dir.

KVLI-FM—Co-owned with KQAB(AM). Oct 29, 1992: 104.5 mhz; 200 w. 1,260 ft TL: N35 37 21 W118 26 16. Stereo. 24 Web Site: www.qabmedia.com.29,000 Natl. Network: ABC. Format: Oldies. News staff: one; News: 12 hrs wkly. Target aud: 25-54.

Lakeport

KNTI(FM)— Oct 21, 1984: 99.5 mhz; 2.5 kw. 1,920 ft TL: N39 07 50 W123 04 32. Stereo. Hrs opn: 24 140 N. Main St., 95453. Phone: (707) 263-6113. Fax: (707) 263-0939. E-mail: mwilson@ncc.radio.com Web Site: www.knti.com. Licensee: Bicoastal Media L.L.C. (group owner; acq 7-28-99; grpsl). Population served: 160,000 Natl. Network: CNN Radio. Law Firm: Keck, Mahin & Cate. Format: Classic hits. News staff: one; News: 8 hrs wkly. Target aud: 25-54; family oriented, upscale, professional adults. Spec prog: Sp 3 hrs, new adult contemp 3 hrs wkly. ♦Ken Dennis, CEO; Mike Wilson, pres; Tony Calumet, gen mgr; Eric Patrick, opns dir & progmg dir; Alan Mathews, gen sls mgr; Paul Reading, news dir; Kevin Mostyn, pub affrs dir & chief of engrg.

KXBX(AM)— June 17, 1966: 1270 khz; 500 w-D, 97 w-N. TL: N39 00 50 W122 53 39. Hrs opn: 24 Box 759, 140 N. Main St., 95453. Phone: (707) 263-6113. Fax: (707) 263-0939. E-mail: mwilson@ncradio.com Licensee: Bicoastal Media LLC. (group owner; acq 7-28-99; grpsl). Population served: 55,000 Natl. Network: Westwood One. Format: MOR, nostalgia. News staff: one; News: 4 hrs wkly. Target aud: 40 plus; retirees. Spec prog: Sp 3 hrs, loc talk & info 5 hrs wkly. ♦Mike Wilson, pres & gen mgr; Alan Matthews, gen sls mgr; Kevin Mostyn, engrg VP & chief of engrg; Juan Huerta, min affrs dir & spanish dir; Bill Moen, disc jockey.

KXBX-FM— Aug 31, 1984: 98.3 mhz; 3 kw. 300 ft TL: N39 02 54 W122 45 59. Stereo. Format: Hot adult contemp.

Lancaster

KAVL(AM)— Sept 8, 1950: 610 khz; 4.9 kw-D, 4 kw-N, DA-2. TL: N34 42 22 W118 10 36. Stereo. Hrs opn: 24 352 East Ave. K-4, 93535. Phone: (661) 942-1121. Fax: (661) 723-5512. Web Site: www.foxsports610.com. Licensee: Citicasters Licenses L.P. Group owner: Clear Channel Communications Inc. (acq 5-4-99; grpsl). Population served: 400,000 Natl. Network: USA. Law Firm: Pepper & Corazzini. Format: Sports. News staff: 2. Target aud: 25-44; predominantly male, commuters, sports fans. ♦Larry Thornhill, gen mgr; Shaun Palmer, gen sls mgr.

KGMX(FM)—Listing follows KWJL(AM).

***KTLW(FM)**— July 3, 1997: 88.9 mhz; 5.8 kw. Ant 272 ft TL: N34 51 03 W118 09 22. Stereo. Hrs opn: 24 14820 Sherman Way, Life On The Way Communications, Van Nuys, 91405. Phone: (818) 779-8444. Fax: (818) 779-8411. E-mail: ktlwinfo@ktlw.net Web Site: www.ktlw.net. Licensee: Life On The Way Communications Inc. (acq 5-23-2003). Population served: 4,500,000 Format: Inspirational, Christian music & teaching. News: 6 hrs wkly. Target aud: 25-54; 50% male, 50% female. ♦Gary Curtis, exec VP & VP; Gary C. Curtis, gen mgr; Rita Medall, opns dir & opns mgr.

KUTY(AM)—See Palmdale

KWJL(AM)— August 1956: 1380 khz; 1 kw-D, DA. TL: N34 42 43 W118 10 34. Hrs open: 24 Q-9, 570 East Ave., Palmdale, 93550. Phone: (661) 947-3107. Fax: (661) 272-5688. Licensee: High Desert Broadcasting LLC. (group owner; (acq 1-21-97; with co-located FM). Population served: 200,000 Format: Sp. News staff: 3; News: 25 hrs wkly. Target aud: 35 plus. ♦nelson Rasse, gen mgr; Gary Wilson, opns mgr; Jeff McElfresh, prom dir; Bob Montague, news dir; Ella Rice, traf mgr.

KGMX(FM)—Co-owned with KWJL(AM). Oct 28, 1970: 106.3 mhz; 3 kw. 210 ft TL: N34 44 41 W118 07 30. (CP: 3.66 kw, ant 256 ft.). 24 400,000 Format: Hot adult contemp. Target aud: 24-54.

Laytonville

***KHKL(FM)**— 2002: 91.9 mhz; 125 w. Ant 2,184 ft TL: N39 41 41 W123 34 36. Stereo. Hrs opn: 24 2351 Sunset Blvd., Suite 170-218, Rocklin, 95765. Phone: (916) 251-1600. Fax: (916) 251-1650. E-mail: klove@klove.com Web Site: www.klove.com. Licensee: Educational Media Foundation. Group owner: EMF Broadcasting. Natl. Network: K-Love. Law Firm: Shaw Pittman. Format: Contemp Christian. News staff: 3. Target aud: 25-44; Judeo Christian, female. ♦Richard Jenkins, pres; Mike Novak, VP & progmg dir; Lloyd Parker, gen mgr; Ed Lenane, opns dir & news dir; Keith Whipple, dev dir; Eric Allen, natl sls mgr; David Pierce, progmg mgr; Jon Rivers, mus dir; Sam Wallington, engrg dir; Arthur Vassar, traf mgr; Karen Johnson, news rptr; Marya Morgan, news rptr; Richard Hunt, news rptr.

***KJCU(FM)**— 2004: 89.9 mhz; 130 w. Ant 361 ft TL: N39 26 35 W123 43 58. Hrs open: 474 S. Franklin St., Fort Bragg, 95437. Phone: (707) 961-6252. Licensee: CSN International (group owner). Format: Relg. ♦Dan Gillman, gen mgr & progmg dir.

***KLAI(FM)**— 2006: 90.3 mhz; 500 w vert. Ant 2,430 ft TL: N39 41 38 W123 34 43. Hrs open: Secondary address: 1144 Redway Drive, Redway 95560. Phone: (707) 923-2513. Fax: (707) 923-2501. Web Site: www.kmud.org. Licensee: Redwood Community Radio Inc. Format: Div. ♦Brenda Starr, gen mgr; Michael Jacinto, opns mgr; Kate Klein, mus dir; Estelle Fennell, news dir; Simon Frech, chief of engrg.

***KVUH(FM)**— 2005: 88.5 mhz; 1.2 kw vert. Ant 2,335 ft TL: N39 41 38 W123 34 43. Hrs open:
Rebroadcasts KSJV(FM) Fresno 100%.
5005 E. Belmont Ave., Fresno, 93727. Phone: (559) 455-5777. Fax: (559) 455-5778. Web Site: www.radiobilingue.org. Licensee: Radio Bilingue Inc. (group owner). (acq 6-29-2005; $50,000 for CP). Format: Ethnic, multilingual, Sp. ♦Hugo Morales, CEO; Maria Erana, gen mgr, opns dir, gen sls mgr & progmg dir; Phil Traynor, dev dir; Samuel Cozco, news dir; Bill Bach, chief of engrg.

Le Grand

***KEFR(FM)**— Jan 11, 1985: 89.9 mhz; 1.8 kw. 2,142 ft TL: N37 32 01 W120 01 50. Stereo. Hrs opn: Box 52, 13306 Jefferson St., 95333. Phone: (209) 389-4659. Fax: (209) 389-0215. E-mail: kefr@k66.com Web Site: www.familyradio.com. Licensee: Family Stations Inc. (group owner) Format: Educ, relg. Target aud: General. ♦Harold Camping, pres; Craig Hulsebos, progmg dir; Larry Milliken, stn mgr & chief of engrg.

Lemoore

KJOP(AM)— Dec 23, 1963: 1240 khz; 1 kw-U. TL: N36 18 47 W119 43 51. Hrs open: PO Box 180, Tahoma, 96142. Phone: (530) 584-5700. Fax: (530) 584-5705. E-mail: info@ihradio.org Web Site: www.ihradio.org. Licensee: IHR Educational Broadcasting (group owner; acq 12-22-2000; $125,000). Format: Catholic/religious. ♦Doug Sherman, pres.

Lenwood

KBTW(FM)— April 2001: 104.5 mhz; 2.5 kw. 515 ft TL: N34 51 20 W117 02 59. Hrs open: 24
Rebroadcasts KXLM (FM) Oxnard 80%.
1950 S. Sunwest Ln., Suite 300, San Bernardino, 92408. Secondary address: 125 E. Fredericks St., Barstow 92311. Phone: (909) 825-5020. Phone: (760) 255-4246. Fax: (909) 884-5844. Fax: (760) 255-2406. Web Site: www.radiolazer.com. Licensee: Lazer Broadcasting Corp. (group owner; (acq 10-27-99; 450,000). Natl. Rep: Lotus Entravision Reps LLC. Law Firm: Fletcher, Heald & Hildreth. Format: Sp, rgnl Mexican. News: 2 hrs wkly. Target aud: 25-54; adult. ♦Armando Gutierrez, gen mgr & prom dir; Vicki Bails, gen sls mgr; Salvador Prieto, progmg dir; Ralph Jones, chief of engrg.

KHDR(FM)— Dec 20, 2002: 96.9 mhz; 1 kw. Ant 797 ft TL: N34 58 15 W117 02 22. Hrs open: 24
Rebroadcasts KHRQ(FM) Baker 100%.
Box 1668, Barstow, 92312. Phone: (760) 256-0326. Fax: (760) 256-9507. E-mail: khwyha@earthlink.net Web Site: www.thehighwaystations.com. Licensee: The Drive LLC. Group owner: KHWY Inc. (acq 2-25-2003). Natl. Network: AP Radio. Law Firm: Hogan & Hartson. Format: Classic rock, AOR. News staff: one. Target aud: General; travelers on I-15 and I-40. ♦Howard B. Anderson, CEO & pres; Kirk M. Anderson, exec VP; Tim Anderson, stn mgr; Judy Robinson, gen sls mgr; Lance Todd, progmg dir.

KIXW-FM— November 1994: 107.3 mhz; 440 w. Ant 771 ft TL: N34 58 13 W117 02 19. Hrs open: 1611 E. Main St., Barstow, 92311. Phone: (760) 256-0326. Fax: (760) 256-9507. E-mail: tim@highwayradio.com Web Site: www.thehighwaystations.com. Licensee: KHWY Inc. (group owner; (acq 2-18-98; $1,741,444 with KIXF(FM) Baker). Natl. Network: CNN Radio, Westwood One. Law Firm: Hogan & Hartson. Format: Country. ♦Howard Anderson, CEO & pres; Kirk Anderson, exec VP; Timothy Anderson, VP & gen mgr; Judy Robinson, sls VP; John Gregg, prom dir; Lance Todd, progmg dir; Keith Hayes, news dir; Thomas J. McNeill, engrg mgr.

Lincoln

KKFS(FM)— Nov 8, 1974: 103.9 mhz; 6 kw. Ant 328 ft TL: N38 52 33 W121 07 30. Stereo. Hrs opn: 24 1425 River Park Dr., Suite 520, Sacramento, 95815. Phone: (916) 924-0710. Fax: (916) 924-1587. E-mail: info@1039thefish.com Web Site: www.1039thefish.com. Licensee: Golden Gate Broadcasting Co. Inc. Group owner: First Broadcasting Investment Partners LLC (acq 5-12-2006; swap for KLMG(FM) Jackson). Population served: 750,000 Format: Contemp Christian. ♦James Rowten, gen mgr; Laurie Larson, progmg dir; Dave Fortenberry, engrg dir.

Lindsay

KZPO(FM)— 1999: 103.3 mhz; 280 w. Ant 2,624 ft TL: N36 17 14 W118 50 17. Hrs open: 1025 Thomas Jefferson St. N.W., Suite 700, Washington, DC, 20007. Phone: (202) 625-3500. Web Site: members.aol.com/kingradio. Licensee: Estate of Linda Ware, Cynthia Ramage, executor (acq 6-2-2004). Format: Nostalgia. ♦Howard Braun, gen mgr.

Livermore

KKIQ(FM)— May 1969: 101.7 mhz; 4.5 kw. 382 ft TL: N37 35 42 W121 39 42. Stereo. Hrs opn: 24 7901 Stoneridge Dr., Suite 525, Pleasanton, 94588. Phone: (925) 455-4500. Fax: (925) 416-1211. Web Site: www.kkiq.com. Licensee: KKIQ Inc. (acq 6-19-98; $9 million). Population served: 500,000 Natl. Network: AP Radio. Law Firm: Haley, Bader & Potts. Format: Adult contemp. News staff: one; News: 28 hrs wkly. Target aud: 25-54; high income & highly educated adults. ♦John Levitt, gen mgr; David Louie, gen sls mgr; Sylvia Manker, mktg dir; Jim Hampton, progmg dir; John Higden, chief of engrg.

Livingston

***KCJH(FM)**— 1997: 89.1 mhz; 13.5 kw vert. Ant 305 ft TL: N37 18 57 W120 43 20. Hrs open: 24
Rebroadcasts KYCC(FM) Stockton 100%.
9019 N. West Ln., Stockton, 95210. Phone: (209) 477-3690. Fax: (209) 477-2762. E-mail: kycc@kycc.org Web Site: www.kycc.org. Licensee: Your Christian Companion Network Inc. (acq 7-20-98). Law Firm: Cohn & Marks. Format: Gospel, inspirational, adult contemp. Target aud: 35-55. ♦Kenneth F. Haney, pres; Shirley Garner, exec VP & gen mgr; Scott Mearns, progmg mgr & chief of engrg; Marina Tahod, mus dir.

***KLVN(FM)**— 1998: 88.3 mhz; 1.8 kw. 148 ft TL: N37 18 57 W120 43 20. Stereo. Hrs opn: 24
Rebroadcasts KLVY(FM) Fairmead 100%.
2351 Sunset Blvd., Suite 170-218, Rocklin, 95765. Phone: (916) 251-1600. Fax: (916) 251-1650. E-mail: klove@klove.com Web Site: www.klove.com. Licensee: Educational Media Foundation. Group owner: EMF Broadcasting. Population served: 207,000 Natl. Network: K-Love. Law Firm: Shaw Pittman. Format: Contemp Christian music. News staff: 3. Target aud: 25-44; female (Judeo-Christian). ♦Richard Jenkins, pres; Mike Novak, VP; Keith Whipple, dev dir; David Pierce, progmg mgr; Ed Lenane, news dir; Sam Wallington, engrg dir; Karen Johnson, news rptr; Marya Morgan, news rptr; Richard Hunt, news rptr.

KSKD(FM)— Nov 1, 1984: 95.9 mhz; 3 kw. Ant 305 ft TL: N37 18 57 W120 43 20. (CP: COL Dos Palos. 6 kw, ant 318 ft. TL: N36 55 35 W120 50 42). Stereo. Hrs opn: 4043 Geer Rd., Hughson, 95326. Phone: (209) 883-8760. Fax: (209) 883-8769. Web Site: www.lafavorita.net. Licensee: All American Broadcasting Co. (acq 2-3-93; $198,000; FTR: 3-8-93). Natl. Rep: Lotus Entravision Reps LLC. Format: Sp, adult contemp. ♦Nelson Gomez, pres & gen mgr.

Lodi

KCVR(AM)— 1946: 1570 khz; 5 kw-D, 500 w-N, DA-2. TL: N38 05 10 W121 12 57. Hrs open: 6820 Pacific Ave., Suite 3 A, Stockton, 95207. Phone: (209) 474-0154. Fax: (209) 474-0316. Web Site: www.entravision.com. Licensee: Entravision Holdings LLC. Group owner: Entravision Communications Corp. (acq 7-28-2000; grpsl). Format: Sp. ♦Lisa Sunday, gen mgr & stn mgr; Valentina Rupic, traf mgr.

***KLRS(FM)**—Not on air, target date: unknown: 89.7 mhz; 6.8 kw vert. Ant 396 ft TL: N38 23 01 W121 17 15. Hrs open: 2351 Sunset Blvd., Suite 170-218, Rocklin, 95765. Phone: (916) 251-1600. Fax: (916) 251-1650. Licensee: Educational Media Foundation.

KWIN(FM)— Dec 24, 1959: 97.7 mhz; 3 kw. 300 ft TL: N38 03 05 W121 15 05. Stereo. Hrs opn: 24 4643 Quail Lakes Dr., Suite 100, Stockton, 95207-1833. Secondary address: 1581 Cummins Dr., Suite 135, Modesto 95358. Phone: (209) 476-1230. Fax: (209) 957-1833. Web Site: www.kwin.com. Licensee: Citadel Broadcasting Co. Group owner: Citadel Broadcasting Corp. (acq 5-9-03; grpsl). Population served: 128,400 Format: CHR. ♦Joanne Matteri, CFO; Roy Williams, gen mgr; Jean Western, sls dir; Raymond Baca, chief of engrg.

Loma Linda

KCAA(AM)— Nov 1, 1964: 1050 khz; 1.4 kw-D, 35 w-N, DA-2. TL: N33 59 22 W117 11 10. Hrs open: 6 AM-sunset 19939 Gatling Ct.,

Katy, TX, 77449. Secondary address: 254 Carousel Mall, San Bernardino 92401. Phone: (281) 599-9800. Fax: (909) 381-8935. E-mail: ceo@KCAAradio.com Web Site: www.KCAAradio.com. Licensee: Broadcast Management Services Inc. (acq 2-97; $30,000). Population served: 3,000,000 Law Firm: Fletcher Heald. Format: Country, talk/news. News staff: 3; News: 15 hrs wkly. Target aud: General. ♦Fred Lundgren, CEO; Jim Hill, VP; Paren Lane, gen mgr; Ray Peyton, prom mgr; Lacey Kendall, progmg dir; S. Earl Statler, pub affrs dir; Dick Vosper, chief of engrg; Donna Stevens, traf mgr; Dennis Baxter, local news ed; Nick LaCapria, sports cmtr.

KSGN(FM)—See Riverside

Lompoc

KBOX(FM)—Licensed to Lompoc. See Santa Maria

***KLWG(FM)**— 2006: 88.1 mhz; 20 w vert. Ant 1,128 ft TL: N34 36 13 W120 29 17. Hrs open: 24 Box 1241, 93438. Phone: (805) 736-3741. Web Site: www.calvarychapellompoc.com/radio_ministry.htm. Licensee: Calvary Chapel of Lompoc. Format: Relg. ♦Mark Galvan, gen mgr; Landon Galvan, progmg dir.

KRQK(FM)— Dec 18, 1979: 100.3 mhz; 3.65 kw. 863 ft TL: N34 44 24 W120 26 42. Stereo. Hrs opn: 24 2325 Skyway Dr., Suite J, Santa Maria, 93455. Phone: (805) 922-1041. Fax: (805) 928-3069. Licensee: AGM-Santa Maria LP. Group owner: American General Media (acq 10-29-99; $1.3 million). Population served: 51,000 Format: Rgnl Mexican. Target aud: 18-49. ♦Rich Watson, gen mgr; Emily Stich, gen sls mgr, rgnl sls mgr & prom mgr; Salvador Ponce, progmg dir.

***KRQZ(FM)**— Sept 3, 2000: 91.5 mhz; 2 kw vert. Ant 1,050 ft TL: N34 36 13 W120 29 17. Stereo. Hrs opn: 24
Rebroadcasts WUFM(FM) Columbus, OH 60%.
Trinity Church of the Nazarene, 500 E. North Ave., 93436. Phone: (805) 736-6415. Fax: (805) 736-2642. E-mail: krqz@trinaz.com Web Site: www.radiou.com.Sky Angel Licensee: Trinity Church of the Nazarene. Law Firm: Gammon & Grange. Format: Christian rock. Target aud: 12-24 years. ♦Mark Hostand, stn mgr & gen sls mgr; Chris Hill, chief of engrg.

KRTO(FM)— 1999: 105.1 mhz; 420 w. Ant 1,217 ft TL: N34 41 28 W120 15 58. Hrs open: 24 716 E. Chapel St., Los Angeles, 93454-4524. Phone: (805) 922-7727. Fax: (805) 349-0265. E-mail: oldies1051@radiocentralcoast.com Licensee: Emerald Wave Media. (group owner; (acq 7-7-2006; $1.5 million). Format: Oldies. ♦August Ruiz, gen mgr.

KSMY(FM)— 1997: 106.7 mhz; 3.5 kw. Ant 879 ft TL: N34 44 31 W120 26 46. Hrs open: 2215 Skyway Dr., Santa Maria, 93455-1118. Phone: (805) 925-2582. Fax: (805) 928-1544. Web Site: www.lapreciosa.com. Licensee: CC Licenses LLC. Group owner: Clear Channel Communications Inc. (acq 10-17-2000; grpsl). Format: Sp rgnl Mexican. ♦Rich Hawkins, gen mgr; Pattie Wagner, gen sls mgr; Jennifer Grant, progmg dir; Milos Nemicik, chief of engrg.

KUHL(AM)— May 25, 1963: 1410 khz; 500 w-D, 77 w-N, DA-2. TL: N34 39 47 W120 22 58. Hrs open: 24
Rebroadcasts KINF(AM) Santa Maria 100%.
716 E. Chapel, Santa Maria, 93454. Phone: (805) 922-7727. Fax: (805) 349-0265. E-mail: kathy@knightbroadcasting.com Web Site: www.mix96.com. Licensee: Knight Broadcasting Inc. (group owner; (acq 7-31-2006; $1.2 million with KINF(AM) Santa Maria). Population served: 90,000 Natl. Network: ABC. Format: Adult contemp. Target aud: 25-54. ♦Jeff Williams, gen mgr; Shawn Knight, gen mgr.

Long Beach

KBUE(FM)— August 1961: 105.5 mhz; 3 kw. Ant 466 ft TL: N33 51 29 W118 13 24. Stereo. Hrs opn: 24 1845 Empire Ave., Burbank, 91504. Phone: (818) 729-5300. Fax: (818) 729-5678. E-mail: info@lbimedia.com Web Site: www.aquisuena.com. Licensee: LBI Radio License Corp. Group owner: Liberman Broadcasting Inc. (acq 1995; $13 million). Population served: 6,500,000 Format: Sp Mexican rgnl. Target aud: 18-49; Spanish speaking adults. ♦Andy Mars, gen mgr; Daisy Ortiz, gen sls mgr & engrg VP; Xavier Ortiz, natl sls mgr; Luis Hernandez, rgnl sls mgr; Pepe Garza, progmg dir; Chris Buchanan, chief of engrg.

***KFRN(AM)**— March 1924: 1280 khz; 1 kw-D, 690 w-N, DA-2. TL: N33 47 54 W118 14 47. Stereo. Hrs opn: 24 3550 Longbeach Blvd., Suite D 4, 90807. Phone: (562) 427-7773. Fax: (562) 427-7723. E-mail: kfrn@familyradio.com Web Site: www.familyradio.com. Licensee: Family Stations Inc. (group owner; acq 9-19-77). Natl. Network: Family Radio. Format: Christian, edu, news. News: 70 hrs wkly. Target aud: Family spectrum. ♦Harold Camping, pres & gen mgr; Ward Cayot, opns mgr; Suong Tran, pub affrs dir.

***KKJZ(FM)**— Jan 3, 1950: 88.1 mhz; 30 kw. Ant 449 ft TL: N33 47 58 W118 09 43. Stereo. Hrs opn: 24 1288 N. Bellflower Blvd., 90815-4198. Phone: (562) 985-5566. Fax: (562) 985-2982. E-mail: info@kkjz.org Web Site: www.jazzandblues.org.Telstar7, transponder 15, subcarriers 5.58 & 5.76 Licensee: California State University, Long Beach Foundation (acq 6-18-81; $15,000; FTR: 4-27-81). Population served: 6,000,000 Natl. Network: NPR. Law Firm: Fletcher, Heald & Hildreth. Format: Jazz, info. News staff: one; News: 5 hrs wkly. Target aud: 25-64; educated, opinion leaders, jazz & mus lovers. Spec prog: Blues 15 hrs wkly. ♦Stephanie Levine-Fried, gen mgr; Mike Johnson, opns mgr; Denise Maynard, sls dir & mktg dir; Michael Levine, news dir.

KLTX(AM)— 1926: 1390 khz; 5 kw-D, 3.6 w-N, DA-2. TL: N33 53 30 W118 11 03. Hrs open: 24 136 S. Oak Knoll Ave. #202, Pasadena, 91101. Phone: (626) 356-4235. Phone: (626) 356-4230. Fax: (626) 817-9851. Web Site: www.nuevavida.com. Licensee: Hi-Favor Broadcasting LLC (group owner; acq 8-4-00; $30 million). Population served: 450,000 Law Firm: Miller & Miller. Format: Relg, Sp. Target aud: 35 plus; mature audience. ♦Mary Guthrie, opns dir.

Los Altos

KFFG(FM)— Oct 17, 1960: 97.7 mhz; 1.65 kw. 433 ft TL: N37 18 27 W122 05 36. (CP: 3.2 kw). Stereo. Hrs opn:
Rebroadcasts KFOG(FM) San Francisco 100%.
c/o KFOG, 55 Hawthorne St., Suite 1000, San Francisco, 94105. Phone: (415) 817-5364. Fax: (415) 995-7006. Web Site: www.kfog.com. Licensee: KFFG Lico Inc. Group owner: Susquehanna Radio Corp. (acq 1995; $8.25 million). Format: AOR. Target aud: 18-49. ♦Tony Salvadore, gen mgr & stn mgr; Omari Patterson, sls dir; Sheri Nelson, prom dir; Dave Benson, progmg dir; Kelly Ransford, mus dir.

***KFJC(FM)**— Dec 4, 1959: 89.7 mhz; 250 w. 1,845 ft TL: N37 19 14 W122 08 29. Stereo. Hrs opn: 24 12345 El Monte Rd., Los Altos Hills, 94022. Phone: (650) 949-7260. Fax: (650) 948-1085. E-mail: info@kfjc.org Web Site: www.kfjc.org. Licensee: Foothill Community College Board of Trustees. Population served: 2,500,000 Format: Free-form, eclectic. News: 9 hrs wkly. Target aud: 8-80; psychedelic speed freaks, radicals & other social outcasts. Spec prog: Country 8 hrs, bluegrass 8 hrs, jazz 7 hrs, progsv 4 hrs wkly. ♦Eric Johnson, gen mgr; John Burns, sls dir & mus dir; Liz Clark, prom dir; Karin Shriver, progmg dir; Mark Laubach, engrg mgr.

Los Angeles

KABC(AM)— Nov 15, 1929: 790 khz; 5 kw-U, DA-N. TL: N34 01 40 W118 22 20. Hrs open: 3321 S. La Cienega Blvd., 90016. Phone: (310) 840-4900. Fax: (310) 838-5222. Web Site: www.kabc.com. Licensee: Radio License Holding VI LLC. Group owner: ABC Inc. (acq 6-12-2007; grpsl). Population served: 9,600,000 Natl. Network: ABC. Format: Talk. Target aud: 35 plus; upscale, affluent, college educated. ♦John H. Davidson, pres & gen mgr; Joe Schwartz, gen sls mgr & prom mgr; Pete Dominguez, natl sls mgr; Shelley Wagner, mktg dir & prom dir; Erik Braverman, progmg dir; Nelkane Benton, pub affrs dir; Norm Avery, chief of engrg.

KLOS(FM)—Co-owned with KABC(AM). Dec 30, 1947: 95.5 mhz; 68 kw. 2,920 ft TL: N34 13 37 W118 03 58. Stereo. Phone: (310) 840-4836. Phone: (310) 840-4800. Web Site: www.955klos.com. Licensee: Radio License Holding XII LLC.1,200,000 Format: Mainstream rock/AOR. ♦Leonard Madrid, gen sls mgr; C.W. West, mktg dir & adv dir; Rita Wilde, progmg dir; Jim Villanueva, mus dir; Norm Avery, engrg dir.

KBIG-FM—Listing follows KLAC(AM).

KBLA(AM)—See Santa Monica

KBRT(AM)—See Avalon

KCBS-FM—Listing follows KNX(AM).

KDIS(AM)—See Pasadena

KFI(AM)— Apr 16, 1922: 640 khz; 50 kw-U. TL: N33 52 48 W118 00 48. Hrs open: 3400 W. Olive Ave., Suite 550, Burbank, 91505. Phone: (818) 559-2252. Web Site: www.kfi640.com. Licensee: Capstar TX L.P. Group owner: Clear Channel Communications Inc. (acq 8-7-2000; grpsl). Population served: 1,058,900 Natl. Rep: Christal. Format: Talk. Target aud: 25-54. ♦Greg Ashlock, gen mgr.

KOST(FM)—Co-owned with KFI(AM). Oct 9, 1957: 103.5 mhz; 12.5 kw. Ant 3,100 ft TL: N34 13 34 W118 03 55. 3400 W. Olive Ave., Suite 550, Burbank, 91505. Phone: (818) 559-2252. Fax: (818) 637-2267. Web Site: www.kost1035.com. Licensee: AMFM Broadcasting Licenses LLC.1,400,000 Format: Adult contemp. ♦Craig Rossi, stn mgr; Stella Schwartz, progmg dir.

KFWB(AM)— Mar 25, 1925: 980 khz; 5 kw-U. TL: N34 04 11 W118 11 36. Stereo. Hrs opn: 24 5670 Wilshire Blvd., Suuite 200, 90036. Phone: (323) 871-4660. Fax: (323) 871-4681. Fax: (323) 871-4679. Web Site: www.kfwb.com. Licensee: Infinity Broadcasting East Inc. Group owner: Infinity Broadcasting Corp. (acq 11-13-98; grpsl). Population served: 9,500,000 Natl. Network: CNN Radio. Natl. Rep: CBS Radio. Wire Svc: AP Wire Svc: Bloomberg News Format: News. News staff: 60; News: 168 hrs wkly. Target aud: 25-54. ♦Pat Duffy, VP & gen mgr; Sean O'Neil, gen sls mgr; Andy Ludlum, progmg dir; Paul Gomez, news dir; Paul Sakrison, chief of engrg.

KHHT(FM)— Dec 29, 1948: 92.3 mhz; 43 kw. 2,910 ft TL: N34 13 36 W118 03 57. Stereo. Hrs opn: 24 3400 W. Olive Ave., Suite 550, Burbank, 91505. Phone: (818) 559-2252. Fax: (818) 566-4517. E-mail: info@hot92jamz.com Web Site: info@hot92jams.com. Licensee: AMFM Broadcasting Licenses LLC. Group owner: Clear Channel Communications Inc. (acq 8-30-2000; grpsl). Population served: 9,741,200 Format: Urban contemp. Target aud: 18-49; females. ♦Val Maki, gen mgr; Mike Marino, progmg dir.

KHJ(AM)— Apr 13, 1922: 930 khz; 5 kw-U, DA-N. TL: N34 02 26 W118 22 14. Stereo. Hrs opn: 24 1845 Empire Ave., Burbank, 91504. Phone: (818) 729-5300. Fax: (818) 729-5678. E-mail: info@lbimedia.com Licensee: LBI Radio License Corp. Group owner: Liberman Broadcasting Inc. (acq 3-27-90). Population served: 11901290 Law Firm: Wiley, Rein & Fielding. Format: Sp. News staff: one. Target aud: 18-49. ♦Jose Liberman, pres; Andy Mars, gen mgr; Lenard Liberman, exec VP & gen mgr; Disy Ortiz, gen sls mgr; Eddie Leon, progmg dir.

KIEV(AM)—Culver City, January 1986: Stn currently dark. 1500 khz; 50 kw-D, 4.3 kw-N, DA-2. TL: N34 01 47 W118 05 58. Stereo. Hrs opn: 73-733 Fred Waring Dr., Royce International Broadcasting Co., Palm Beach, 92260. Licensee: Royce International Broadcasting Co. (acq 1984). Population served: 12,000,000 Natl. Rep: McGavren Guild. Law Firm: Verner, Liipfert, Bernhard, McPherson & Hand. ♦Edward R. Stolz II, pres & gen mgr.

KIIS-FM—Listing follows KTLK(AM).

KKGO(FM)— Feb 18, 1959: 105.1 mhz; 18 kw. Ant 2,900 ft TL: N34 13 45 W118 04 04. Stereo. Hrs opn: 24 Box 250028, 90025. Phone: (310) 478-5540. Fax: (310) 445-1439. Web Site: gocountry.am. Licensee: Mt. Wilson FM Broadcasters Inc. Population served: 600,000 Natl. Network: AP Radio. Law Firm: Cohn & Marks. Format: Country. ♦Saul Levine, pres & gen mgr; Linda Vali, sls dir; Michael Levine, mktg dir; Susan Foreman, prom dir; Dave Wagner, progmg VP.

KKLA-FM— 1985: 99.5 mhz; 10.5 kw. 2,880 ft TL: N34 13 26 W118 03 44. Stereo. Hrs opn: 24 Box 29023, Glendale, 91209. Secondary address: 701 N. Brand Blvd., Suite 550, Glendale 91203. Phone: (818) 956-5552. Fax: (818) 551-1110. E-mail: info@kkla.com Web Site: www.kkla.com. Licensee: New Inspiration Broadcasting Inc. Group owner: Salem Communications Corp. Population served: 8,000,000 Natl. Network: Salem Radio Network. Natl. Rep: Salem. Wire Svc: Metro Weather Service Inc. Format: Christian. Target aud: 25-55. ♦Terry Fahy, gen mgr; Jim Tinker, opns VP; Larry Marino, opns dir & chief of engrg; Bill Price, gen sls mgr & pub affrs dir; Chuck Tyler, progmg dir.

KLAC(AM)— 1924: 570 khz; 5 kw-U, DA-N. TL: N34 04 11 W118 11 36. Stereo. Hrs opn: 24 3400 W. Olive Ave., Suite 550, Burbank,

91505. Phone: (818) 559-2252. Web Site: www.xtrasportsradio.com. Licensee: AMFM Broadcasting Licenses LLC. Group owner: Clear Channel Communications Inc. (acq 8-30-2000; grpsl). Population served: 2,966,763 Format: Sports. Target aud: 35-54. ♦Mark Austin Thomas, VP & opns VP; Ed Krampf, gen mgr; Jeff Thomas, gen sls mgr; V. Freeman, mktg VP; Bill Lewis, mktg dir; Andrea Garcia, prom dir; Robin Bertoluci, progmg VP; Chris Little, news dir; John Paoli, engrg dir.

KBIG-FM—Co-owned with KLAC(AM). Feb 15, 1959: 104.3 mhz; 65 kw. Ant 3,044 ft TL: N34 13 36 W118 03 59. Stereo. Fax: (818) 637-2267. Fax: (818) 559-2252. Web Site: www.kbig.com.1,000,000 Format: Adult contemp. Target aud: 25-54. ♦Bruce Reese, CEO; Tracy Barrios, traf mgr.

KLSX(FM)— 1954: 97.1 mhz; 29.5 kw. Ant 2,998 ft TL: N34 09 50 W118 11 46. Stereo. Hrs opn: 5670 Wilshire Blvd., Suite 200, 90036. Phone: (323) 971-9710. Fax: (323) 954-0971. Web Site: www.971freefm.com. Licensee: Infinity Broadcasting East Inc. Group owner: Infinity Broadcasting Corp. (acq 7-23-97). Population served: 281,606 Natl. Rep: CBS Radio. Format: Talk. Target aud: 25-54; adults. ♦Bob Moore, gen mgr; Ron Escarsega, opns mgr; David Severino, gen sls mgr; Michael Olson, prom dir; Jack Silver, progmg dir.

KLVE(FM)—Listing follows KTNQ(AM).

KMPC(AM)— Sept 22, 1952: 1540 khz; 50 kw-D, 10 kw-N, DA-2. TL: N34 04 43 W118 11 05. Stereo. Hrs opn: 2800 28th St., Suite 308, Santa Monica, 90405. Phone: (310) 452-7100. Fax: (310) 452-7880. E-mail: rnadel@sportingnews.com Licensee: P&Y Broadcasting Inc. (group owner; (acq 5-30-2007; $33 million). Population served: 2900000 Format: Korean. ♦Chris Canning, pres.

KMRB(AM)—See San Gabriel

KMVN(FM)— Aug 7, 1957: 93.9 mhz; 18.5 kw horiz, 16 kw vert. Ant 3,008 ft TL: N34 13 36 W118 03 59. Stereo. Hrs opn: 24 2600 W Olive Ave., 8th Fl., Burbank, 91505. Phone: (818) 525-5000. Fax: (818) 525-5002. E-mail: mail@movin939.com Web Site: www.movin939.fm. Licensee: Emmis Radio License LLC. Group owner: Emmis Communicationsl Corp. (acq 9-26-2000; grpsl). Format: Rhythmic pop contemp. ♦Janet Brainin, sls dir; Dean Carter, gen sls mgr; Dianna Jason, mktg dir & prom dir; Jimmy Steal, progmg dir.

KNX(AM)— Sept 10, 1920: 1070 khz; 50 kw-U. TL: N33 51 35 W118 20 56. Stereo. Hrs opn: 24 5670 Wilshire Blvd., Suite 200, 90036. Phone: (323) 569-1070. Fax: (323) 930-8798. Web Site: www.knx1070.com. Licensee: CBS Radio East Inc. Group owner: Infinity Broadcasting Corp. (acq 9-36). Population served: 13,500,000 Natl. Network: CBS. Natl. Rep: CBS Radio. Wire Svc: Reuters Format: News. News staff: 40. Target aud: General.Pat Duffy, VP; Rosemary Hernadez, gen sls mgr; Amanda Arrington, natl sls mgr; Howard Freshman, rgnl sls mgr & mktg dir; David G. Hall, progmg VP & progmg dir; Julie Chin, news dir; Vivian Porter, pub affrs dir; Paul Sakrison, engrg mgr & chief of engrg; Terri Boysaw, traf mgr; Randy Kerdoon, edit dir & sports cmtr; Dick Helton, political ed; Steve Grad, sports cmtr

KCBS-FM—Co-owned with KNX(AM). 1948: 93.1 mhz; 54 kw. 5,000 ft TL: N34 13 57 W118 04 18. Stereo. 24 Fax: (323) 463-9270. E-mail: arrow93@arrowfm.com Web Site: www.arrowfm.com.12,000,000 Natl. Network: Westwood One. Natl. Rep: Interep. Format: Rock and roll classics, mus from the 70s. News staff: one; News: 3 hrs wkly. Target aud: 25-49. ♦Mel Karmazin, CEO & chmn; Dan Mason, pres; Fario Suledian, CFO; Brad West, gen sls mgr; Jaime Korzenieski, prom mgr. Co-owned TV: KCBS-TV affil

***KPFK(FM)**— July 26, 1959: 90.7 mhz; 110 kw. Ant 2,831 ft TL: N34 13 45 W118 04 03. Stereo. Hrs opn: 24 3729 Cahuenga Blvd. W., North Hollywood, 91604. Phone: (818) 985-2711. Fax: (818) 763-7526. E-mail: gm@kpfk.org Web Site: www.kpfk.org. Licensee: Pacifica Foundation. Group owner: Pacifica Foundation Inc. dba Pacifica Radio Law Firm: Garvey, Schubert, Barer. Wire Svc: AP Wire Svc: Catholic News Service Wire Svc: Reuters Format: Div, news/talk. News staff: 4; News: 11 hrs wkly. Target aud: 25-55. Spec prog: Children one hr, jazz 5 hrs, gospel 2 hrs, Sp 15 hrs wkly. ♦Eva Georgia, stn mgr; Zuberi Fields, opns mgr; Sue A. Welsh, dev dir; Armando Gudino, progmg dir; Fernando Velasquez, news dir; Molly Paige, news dir.

KPWR(FM)— Dec. 20, 1956: 105.9 mhz; 25 kw. Ant 3,034 ft TL: N34 13 38 W118 04 00. Stereo. Hrs opn: 24 2600 W. Olive Ave., Suite 850, Burbank, 91505. Phone: (818) 953-4200. Fax: (818) 848-0961. Web Site: www.power106.fm. Licensee: Emmis Radio License LLC. Group owner: Emmis Communications Corp. (acq 1-84; grpsl; FTR: 1-30-84). Population served: 8,000,000 Natl. Rep: D & R Radio. Format: Rhythmic CHR. Target aud: 18-34; males. ♦Val Maki, VP; Janet Brainin, sls dir; Dianna Jason, mktg dir & prom dir; Jimmy Steal, progmg VP; Dennis Martin, chief of engrg.

KRBV(FM)— June 1, 1957: 100.3 mhz; 5.3 kw. Ant 3,005 ft TL: N34 13 37 W118 03 58. Stereo. Hrs opn: 24 Box 1710, Hollywood, 90078. Secondary address: 5900 Wilshire Blvd., 19th Floor 90036. Phone: (323) 634-1800. Fax: (323) 634-1888. Web Site: www.v100music.com. Licensee: Radio One Licenses LLC. Group owner: Radio One Inc. (acq 11-8-2001; grpsl). Population served: 13,000,000 Format: Adult rhythm and blues. Target aud: 25-54. ♦Steve Candullo, gen mgr; Ron Turner, gen sls mgr; Leonard McGee, prom dir; Kevin Fleming, progmg dir.

KRCD(FM)—See Inglewood

KRLA(AM)—See Glendale

KRTH(FM)— 1941: 101.1 mhz; 51 kw. 3,130 ft TL: N34 13 38 W118 04 00. (CP: 53.6 kw). Stereo. Hrs opn: 5670 Wilshire, Suite 200, 90036. Phone: (323) 936-5784. Fax: (323) 464-6101. Web Site: www.kearth101.com. Licensee: Infinity Broadcasting East Inc. Group owner: Infinity Broadcasting Corp. (acq 2-2-94; $116 million; FTR: 4-18-94). Population served: 9,607,400 Natl. Network: AP Network News. Natl. Rep: CBS Radio. Format: Oldies. Target aud: 25-64. ♦Maureen Lesourd, VP & gen mgr; Tracy Gilliam, sls VP & gen sls mgr; Karen Tobin, prom dir & prom mgr; Jahni Kaye, progmg dir; Lynn Duke, chief of engrg. Co-owned TV: KCBS-TV affil

KSCA(FM)—See Glendale

KSPN(AM)— Feb 18, 1927: 710 khz; 50 kw-D, 10 kw-N, DA-N. TL: N34 10 24 W118 24 24. Hrs open: 24 3321 S. La Cienega Blvd., 90016. Phone: (310) 840-2800. Fax: (310) 840-2848. Web Site: www.espnradio710.com. Licensee: KABC-AM Radio Inc. Group owner: ABC Inc. (acq 2-27-95; $17.5 million). Population served: 9,600,000 Natl. Network: ABC. Format: Sports. News staff: 2. Target aud: General; famlies and moms. ♦John Davison, pres & gen mgr.

KTLK(AM)—1927: 1150 khz; 50 kw-D, 44 kw-N, DA-2. TL: N34 02 00 W117 59 00. Stereo. Hrs opn: 3400 W. Olive Ave., Suite 550, Burbank, 91505. Phone: (818) 559-2252. Licensee: Citicasters Licenses L.P. Group owner: Clear Channel Communications Inc. (acq 5-4-99; grpsl). Format: Progressive talk. ♦Greg Ashlock, gen mgr; Don Martin, stn mgr.

KIIS-FM—Co-owned with KTLK(AM). 1948: 102.7 mhz; 8 kw. 2,960 ft TL: N34 13 36 W118 03 57. Stereo. Fax: (818) 295-6466. Web Site: www.kiisfm.com. Format: CHR. Target aud: 18-34. ♦Roy Laughlin, gen mgr.

KTNQ(AM)— 1925: 1020 khz; 50 kw-U, DA-2. TL: N34 02 00 W117 59 00. Stereo. Hrs opn: 24 655 N. Central Ave., Suite 2500, Glendale, 91203. Phone: (818) 500-4500. Fax: (818) 500-4307. Web Site: www.ktnq.com. Licensee: KTNQ-AM License Corp. Group owner: Univision Radio (acq 9-22-2003; grpsl). Population served: 11,765,000 Wire Svc: Reuters Format: Sp, news/talk. Target aud: 25-54. ♦Michelle Hohman, gen mgr; Haz Montana, opns mgr; Eric Osuna, natl sls mgr; Offad Vallejo, mktg dir; Santiago Nieto, progmg dir.

KLVE(FM)—Co-owned with KTNQ(AM). May 2, 1959: 107.5 mhz; 29.5 kw. 3,100 ft TL: N34 13 44 W118 04 02.24 Format: Adult contemp. Target aud: 18-49. ♦Bill Shadorf, sls dir; Jose Santos, progmg dir; Georgia Carrera, pub affrs dir.

KTWV(FM)— Mar 7, 1961: 94.7 mhz; 58 kw. 2,835 ft TL: N34 13 29 W118 03 47. Stereo. Hrs opn: 24 5670 Wilshire Blvd., Suite 200, 90036. Phone: (323) 937-9283. Fax: (323) 634-0947. E-mail: wave@ktwv.cbs.com Web Site: www.947wave.com. Licensee: Infinity Broadcasting East Inc. Group owner: Infinity Broadcasting Corp. (acq 11-13-98; grpsl). Population served: 9,500,000 Format: Smooth jazz. Target aud: 25-54. ♦Bob Moore, VP; Dan Weiner, gen mgr; Pat Amsbry, gen sls mgr; Jamie Kanai, mktg dir & prom dir; Paul Goldstein, progmg dir; Ricci Filiar, mus dir; Lynn Duke, engrg mgr & chief of engrg.

***KUSC(FM)**— Oct 24, 1946: 91.5 mhz; 39 kw. 2,922 ft TL: N34 12 48 W118 03 41. Stereo. Hrs opn: 24 Box 77913, 90007. Secondary address: 515 S. Figueroa St., Suite 2050 90071. Phone: (213) 225-7400. Fax: (213) 225-7410. E-mail: kusc@kusc.org Web Site: www.kusc.org. Licensee: University of Southern California. Population served: 650,000 Natl. Network: PRI, NPR. Law Firm: Lawrence Bernstein. Format: Class. Target aud: 35 plus. ♦Brenda Barnes, pres; Eric DeWeese, gen mgr; Janet McIntyre, dev dir.

KWKW(AM)— Apr 14, 1931: 1330 khz; 5 kw-U, DA-N. TL: N34 01 10 W118 20 42. Hrs open: 24 3301 Barham Blvd., Suite 201, 90068. Phone: (323) 851-5959. Fax: (323) 512-7460. E-mail: kwkw1330@aol.com Web Site: www.kwkw1330.com. Licensee: Lotus Communications Corp. (group owner; (acq 1962). Population served: 1,200,000 Format: Sp, news/talk, sports talk. News staff: 5; News: 20 hrs wkly. Target aud: 25 plus; Sp-speaking families. ♦Jim Kalmenson, pres & gen mgr; Mike Addison, gen sls mgr; Juan Rodriguez, progmg dir.

***KXLU(FM)**— February 1957: 88.9 mhz; 3 kw. 12 ft TL: N33 58 16 W118 24 56. Stereo. Hrs opn: 24 1 LMU Drive, 90045. Phone: (310) 338-2866. Phone: (310) 338-5958. Fax: (310) 338-5959. E-mail: kxlu889fm@hotmail.com Web Site: www.kxlu.com. Licensee: Loyola Marymount University Board of Trustees. Population served: 150,000 Format: Rock. News: 2 hrs wkly. Target aud: 16-30. Spec prog: Black 10 hrs, Children one hr, folk one hr wkly. ♦Justin Bates, gen mgr; Daisy Buchanan, prom dir; Michael Schuman, progmg dir; Maki Tamura, chief of engrg.

KXOL-FM— 1949: 96.3 mhz; 6.6 kw. Ant 1,305 ft TL: N34 11 48 W118 15 30. Stereo. Hrs opn: 24 10281 W. Pico Blvd., 90064. Phone: (310) 203-0900. Fax: (310) 843-4961. Web Site: www.elsol963.com. Licensee: KXOL Licensing Inc. Group owner: Spanish Broadcasting System Inc. (acq 10-30-2003; $250 million). Population served: 12,000,000 Format: Sp. Target aud: 18-54. ♦Raul Alcarcon Jr., CEO; Raul Alarcon Jr., pres; Peter Remington, gen mgr; Jason Wilberding, gen sls mgr; Patty Castor, prom dir; Juan Carlos Hidalgo, progmg dir.

KYPA(AM)— 1926: 1230 khz; 1 kw-U. TL: N34 02 15 W118 16 35. Hrs open: 24 747 E. Green St., Suite 400, Pasadena, 91101. Phone: (626) 844-8882. Fax: (626) 844-0156. Licensee: Multicultural Radio Broadcasting Licensee LLC. Group owner: Multicultural Radio Broadcasting Inc. (acq 2-20-98; grpsl). Natl. Network: ABC. Law Firm: Fleischman & Walsh L. Format: Korean. ♦David Sweeney, gen mgr & opns mgr.

KYSR(FM)— June 30, 1954: 98.7 mhz; 75 kw. Ant 1,180 ft TL: N34 07 08 W118 23 30. Stereo. Hrs opn: 24 3400 W. Olive, Suite 550, Burbank, 91505. Phone: (818) 559-2252. Fax: (818) 566-4517. E-mail: starprogramming@clearchannel.com Web Site: www.star987.com. Licensee: AMFM Broadcasting Licenses LLC. Group owner: Clear Channel Communications Inc. (acq 8-30-2000; grpsl). Population served: 49,200 Format: Hot adult contemp. Target aud: 25-54. Spec prog: Pub affrs 2 hrs wkly. ♦Val Maki, sr VP, gen mgr & gen sls mgr.

Los Banos

KLBS(AM)— May 1961: 1330 khz; 500 w-D, 5 kw-N, DA-N. TL: N37 05 51 W120 49 51. Stereo. Hrs opn: 24 hrs. a day 401 Pacheco Blvd., 93635. Phone: (209) 826-0578/826-4996. Fax: (209) 826-1906. E-mail: pr@klbs.com Web Site: www.klbs.com. Licensee: Ethnic Radio Los Banos Inc. (acq 5-82). Population served: 250,000 Format: Portuguese. Spec prog: Relg 8 hrs wkly. ♦Jose Encarnacao, gen mgr.

KQLB(FM)— November 1992: 106.9 mhz; 6 kw. 328 ft TL: N36 55 35 W120 50 42. Stereo. Hrs opn: 24 401 Pacheco Blvd., 93635. Phone: (209) 827-0101. Fax: (209) 826-1906. E-mail: pr@kqlb.com Web Site: www.kqlb.com. Licensee: VLB Broadcasting Inc. (acq 12-26-91). Population served: 250,000 Format: Sp/Mexician rgnl. ♦Batista Vieira, chmn; J.J. Encarnacao, gen mgr; Cidalia Sequeira, opns mgr; Jose Berumen, progmg dir.

Los Gatos

KRTY(FM)— July 9, 1966: 95.3 mhz; 880 w. 860 ft TL: N37 12 17 W121 56 56. Stereo. Hrs opn: 24 Box 995, San Jose, 95108. Phone: (408) 293-8030. Fax: (408) 293-6124. Fax: (408) 995-0823. Web Site: www.krty.com. Licensee: KRTY Ltd. Group owner: Empire Broadcasting Corp. (acq 2-93; $3.31 million; FTR: 1-18-93). Population served: 1,800,000 Format: Country. News staff: one. Target aud: 25-54. ♦Bob Kieve, pres; Nate Deaton, gen mgr & mktg dir; Stuart Hinkle, natl sls mgr; Jan Brock, rgnl sls mgr; Jamie Van Der Veen, prom dir; Julie Stevens, progmg dir; George Sampson, news dir; Mike Danberger, chief of engrg.

Los Molinos

KCEZ(FM)— 1999: 102.1 mhz; 25 kw. Ant 220 ft TL: N39 53 17 W122 37 38. Hrs open: 856 Manzanita Ct., Chico, 95926. Phone: (530) 342-2200. Fax: (530) 342-2260. Web Site: www.chicooldies.com. Licensee: Results Radio Licensee L.L.C. Group owner: Fritz Communications Inc. (acq 6-11-99; grpsl). Format: Oldies. ♦John Graham, gen mgr; Dave Pack, gen sls mgr; Steve Michaels, progmg dir; J.D. Davis, chief of engrg; Candy Mason, traf mgr.

Los Osos-Baywood Park

KSTT-FM— 1987: 101.3 mhz; 4.86 kw. 1,506 ft TL: N35 21 38 W120 39 21. (CP: 3.4 kw, ant 1,685 ft.). Hrs opn: 24 51 Zaca Ln., Suite 100, San Luis Obispo, 93401. Phone: (805) 545-0101. Fax: (805) 541-5303. Web Site: www.kstt.com. Licensee: CC Licenses LLC. Group owner: Clear Channel Communications Inc. (acq 10-2000; grpsl). Population served: 200,000 Format: Soft adult contemp. News staff: one; News: 5 hrs wkly. Target aud: 25-54. ♦Rich Hawkins, gen mgr; Andrew Winford, opns mgr & progmg dir; Pattie Wagner, natl sls mgr; Greg Russo, mktg dir & prom dir.

Lost Hills

KEBT(FM)— Dec 10, 1995: 96.9 mhz; 15.5 kw. Ant 413 ft TL: N35 19 40 W119 42 58. Hrs open: 1400 Easton Dr., Suite 144, Bakersfield, 93309. Phone: (661) 328-1410. Fax: (661) 328-0873. Web Site: www.liveradio.com. Licensee: AGM California. American General Media (acq 6-5-2006; $2.05 million). Format: Rhythmic adult contemp. Target aud: 25-54. ♦Roger Fessler, gen mgr; Toni Snyder, gen sls mgr; Eric Sean, progmg dir.

KQMX(FM)—Not on air, target date: unknown: 105.7 mhz; 24.46 kw. Ant 331 ft TL: N35 30 54 W119 57 30. Hrs open: 2801 Via Fortuna Dr., Suite 675, Austin, TX, 78746. Phone: (713) 528-2517. Licensee: Ace Radio Corp. ♦Stephen Hackerman, pres.

Lucerne Valley

KIXA(FM)— November 1992: 106.5 mhz; 150 w. 1,066 ft TL: N34 23 08 W117 03 25. Hrs open: 24 12370 Hesperia Rd., Suite 16, Victorville, 92395. Phone: (760) 241-1313. Fax: (760) 241-0205. Web Site: www.thefox1065.com. Licensee: Clear Channel Broadcasting Licenses Inc. Group owner: Clear Channel Communications Inc. (acq 2000; grpsl). Population served: 300,000 Format: Active rock. News: 18 hrs wkly. Target aud: 16-45. ♦Larry Thornhill, gen mgr & gen sls mgr; Steve Sipe, sls dir; Joe Pagano, progmg dir.

Ludlow

KDUQ(FM)— July 7, 1995: 102.5 mhz; 6 kw. Ant -164 ft TL: N34 43 21 W116 10 04. Stereo. Hrs opn: 24
Rebroadcasts KDUC(FM) Barstow 100%.
29000 Radio Rd., Barstow, 92311. Phone: (760) 256-2121. Fax: (760) 256-5090. E-mail: doscostas@yahoo.com ABC/SMN Pine Gold Licensee: Dos Costas Communications Corp. (group owner; acq 6-18-03; grpsl). Population served: 220,000 Natl. Network: ABC, CBS. Natl. Rep: Western Regional Broadcast Sales. Law Firm: Fleischmann & Walsh. Format: CHR rhythmic. News staff: one; News: 7 hrs wkly. Target aud: 25-54:; adults. Spec prog: Relg one hr wkly. ♦Roland Ulloa, stn mgr; Manny Lopez, opns mgr & gen sls mgr; Mike Garcia, progmg dir; Brad Sobel, news dir & chief of engrg.

KHWZ(FM)— 1992: 100.1 mhz; 25 kw. Ant -216 ft TL: N34 43 29 W116 09 24. Hrs open: 24
Rebroadcasts KIXW-FM Lenwood 100%.
Box 1668, Barstow, 90025. Phone: (760) 256-0326. Fax: (760) 256-9507. E-mail: khwyha@earthlink.net Web Site: www.thehighwaystations.com. Licensee: KHWY Inc. (group owner). Natl. Network: CNN Radio, Westwood One. Format: AOR. News staff: one. Target aud: 35 plus; travelers on I-15 & I-40. ♦Howard B. Anderson, CEO & pres; Kirk M. Anderson, exec VP; Timothy B. Anderson, VP & gen mgr; Judy Robinson, gen sls mgr & rgnl sls mgr.

Madera

KHOT(AM)— Dec 31, 1956: 1250 khz; 1.5 kw-D, 1 kw-N, DA-2. TL: N36 57 58 W120 02 06. Hrs open: 24 Box 180, Tahoma, 96142. Phone: (530) 584-5700. Fax: (530) 584-5705. E-mail: info@ihradio.org Web Site: www.ihradio.org. Licensee: Redwood Family Services Inc. Population served: 500,000 Format: Relg-Catholic. News staff: one; News: 30 hrs wkly. Target aud: 25-54. ♦Doug Sherman, pres.

KMMM(FM)— October 1992: 107.1 mhz; 9.9 kw. Ant 515 ft TL: N37 07 40 W119 40 38. Hrs open: 1110 E. Olive Ave., Fresno, 93728. Phone: (559) 497-1100. Fax: (559) 497-1125. E-mail: mginsburg@lotusfresno.com Licensee: Lotus Communications Corp. Group owner: Lotus Communications Corp. (acq 3-10-99). Population served: 1,500,000 Natl. Rep: Lotus Entravision Reps LLC. Law Firm: Leventhal, Senter & Lerman. Format: Sp contemp hits music. Target aud: 18-49. ♦Dan Crotty, gen mgr; Mike Ginsburg, gen sls mgr & prom mgr; Jose Berumen, progmg dir; Paul Klein Kramer, chief of engrg.

KOND(FM)—Clovis, Sept 30, 1974: 92.1 mhz; 36.9 kw. Ant 567 ft TL: N37 07 40 W119 40 38. Stereo. Hrs opn: 1981 N. Gateway Blvd., Suite 101, Fresno, 93727. Phone: (559) 456-4000. Fax: (559) 251-9555. Web Site: www.univision.com. Licensee: Univision Radio License Corp. Group owner: Univision Radio (acq 2-18-2004; $8 million). Format: Rgnl Mexican. ♦Angela Navarrete, gen mgr.

Magalia

***KLVC(FM)**— Jan 1, 1993: 88.3 mhz; 5.7 kw. 1,184 ft TL: N39 57 45 W121 42 52. Stereo. Hrs opn: 24
Rebroadcasts KLVB-FM Red Bluff 100%.
2351 Sunset Blvd., Suite 170-218, Rocklin, 95765. Phone: (916) 251-1600. Fax: (916) 251-1650. E-mail: klove@klove.com Web Site: www.klove.com. Licensee: Educational Media Foundation Inc. Group owner: EMF Broadcasting. Population served: 357,000 Natl. Network: K-Love. Law Firm: Shaw Pittman. Format: Contemp Christian. News staff: 3. Target aud: 33—40; Judeo-Christian, female. ♦Richard Jenkins, pres; Mike Novak, VP; Keith Whipple, dev dir; Sam Wallington, engrg dir; Karen Johson, news rptr; Marya Morgan, news rptr; Richard Hunt, news rptr.

Mammoth Lakes

KMMT(FM)— Apr 3, 1973: 106.5 mhz; 360 w. Ant 2,371 ft TL: N37 37 42 W119 01 47. Stereo. Hrs opn: 24 Box 1284, 94 Laurel Mountain Rd., 93546. Phone: (760) 934-8888. Fax: (760) 934-2429. E-mail: kmmtradioworks@yahoo.com Licensee: Mammoth Mountain F.M. Associates Inc. Population served: 100,000 Format: Modern adult contemp. News staff: one; News: 2 hrs wkly. Target aud: 18-54; active, athletic, affluent adults. Spec prog: Jazz 2 hrs, classic rock 4 hrs wkly. ♦David A. Digerness, pres; Shellie Woods, gen mgr; Lisa Hahn, progmg dir.

Manteca

KMRQ(FM)— Jan 15, 1979: 96.7 mhz; 1.5 kw. Ant 466 ft TL: N37 43 44 W121 07 34. Stereo. Hrs opn: 24 2121 Lancey Dr., Modesto, 95355. Phone: (209) 551-1306. Fax: (209) 551-1359. Web Site: www.rock967.com. Licensee: Capstar TX L.P. Group owner: Clear Channel Communications Inc. (acq 8-30-2000; grpsl). Law Firm: Mullin, Rhyne, Emmons & Topel. Format: Sp. News staff: one. Target aud: 25-54. ♦Bill Mick, progmg VP & progmg dir.

Maricopa

KXTT(FM)—Not on air, target date: unknown: 94.9 mhz; 6 kw. Ant -433 ft TL: N35 03 28 W119 21 08. Hrs open: 200 S. A St., Suite 400, Oxnard, 93030. Phone: (805) 240-2070. Licensee: Lazer Licenses LLC. (acq 6-11-2007; $3.85 million with KEAL(FM) Taft). ♦Neal Robinson, pres.

Marina

KTOM-FM— Apr 6, 1982: 92.7 mhz; 6.9 kw. 567 ft TL: N36 33 12 W121 47 05. Stereo. Hrs opn: 24 903 N. Main St., Salinas, 93906. Phone: (831) 755-8181. Fax: (831) 755-8193. Web Site: www.ktom.com. Licensee: CC Licenses LLC. Group owner: Clear Channel Communications Inc. (acq 9-22-97; grpsl). Natl. Rep: D & R Radio. Format: Country. News staff: one; News: 5 hrs wkly. Target aud: 25-54; men. ♦Rhonda McCormack, gen mgr; Jen Taylor, prom dir; Johnny Morgan, progmg dir.

Mariposa

KDJK(FM)— 1994: 103.9 mhz; 71 w. 2,047 ft TL: N37 32 00 W120 01 29. Stereo. Hrs opn: 24 1581 Cummins Dr., Suite 135, Modesto, 95358-6402. Phone: (209) 572-0104. Fax: (209) 522-2061. Web Site: www.104thehawk.com. Licensee: Citadel Broadcasting Co. Group owner: Citadel Broadcasting Corp. (acq 7-30-93; $6 million; FTR: 8-23-93). Natl. Rep: Christal. Format: Classic Rock. Target aud: 18-54. ♦Roy Williams, gen mgr; Richard Perry, opns dir & progmg dir; Jean Western, gen sls mgr; Gary Williams, engrg mgr & chief of engrg.

KUBB(FM)— July 4, 1977: 96.3 mhz; 1.9 kw. 2,112 ft TL: N37 32 00 W120 01 29. Stereo. Hrs opn: 24 Box 429, 510 W. 19th St., Merced, 95340. Phone: (209) 383-7900. Fax: (209) 723-8461. E-mail: mcadam@kubb.com Web Site: www.kubb.com. Licensee: Buckley Broadcasting of Monterey. Group owner: Buckley Broadcasting Corp. (acq 7-1-85; $640,000; FTR: 5-20-85). Population served: 1,000,000 Natl. Network: Westwood One. Natl. Rep: D & R Radio. Format: Country. News staff: one; News: 2 hrs wkly. Target aud: 25-54. Spec prog: Farm 2 hrs wkly. ♦Mike McAdam, VP & gen mgr; Mike Peters, gen sls mgr & rgnl sls mgr; Rene Roberts, opns dir & progmg dir.

Marysville

KKCY(FM)—See Colusa

KMYC(AM)— 1940: 1410 khz; 5 kw-D, 1 kw-N, DA-2. TL: N39 08 18 W121 33 15. Hrs open: 6 AM-midnight Box 669, 95901. Phone: (530) 742-5555. Fax: (530) 741-3758. E-mail: kmyc@xyix.com Licensee: Thomas Huth. Group owner: Huth Broadcasting. . Population served: 300,000 Format: Talk radio. News staff: one. Target aud: 18 plus; general. Spec prog: Indian/Punjabi 2 hrs wkly. ♦Thomas Huth, CEO, gen mgr & rgnl sls mgr; Jerry Snaper, engrg VP & chief of engrg.

KOBO(AM)—See Yuba City

KRCX-FM— Oct 12, 1994: 99.9 mhz; 1.74 kw. Ant 2,181 ft TL: N39 12 20 W121 49 10. Stereo. Hrs opn: 24 1436 Auburn Blvd., Sacramento, 95815. Phone: (916) 646-4000. Fax: (916) 646-1958. E-mail: jverdier@entravision.com Web Site: www.entravision.com. Licensee: Entravision Holdings LLC. Group owner: Entravision Communications Corp. (acq 3-14-2000; grpsl). Population served: 1,500,000 Format: Mexican rgnl, Sp. News staff: 2. Target aud: 18-49. ♦Larry Lamanski, gen mgr; Salvador Lopez, prom dir; Juan Carlos Sanchez, progmg dir.

KUBA(AM)—See Yuba City

McCloud

KZCC(FM)—Not on air, target date: unknown: 95.5 mhz; 4.8 kw. Ant 751 ft TL: N41 13 37 W122 14 23. Hrs open: 455 Capitol Mall, Suite 210, Sacramento, 95814. Phone: (916) 448-8800. Licensee: Airen Broadcasting Co. ♦Suzanne E. Rogers, pres.

McFarland

KBQF(FM)—Not on air, target date: unknown: 104.3 mhz; 6 kw. Ant 327 ft TL: N35 31 35 W119 18 43. Hrs open: 977 W. 7th St., Oxnard, 93030-6757. Phone: (805) 486-4400. Licensee: JAB Broadcasting LLC. ♦Javier Orosco, gen mgr.

KIWI(FM)—Licensed to McFarland. See Bakersfield

McKinleyville

KMDR(FM)—Not on air, target date: unknown: 95.1 mhz; 2.45 kw. Ant 1,040 ft TL: N40 49 32 W124 00 05. Hrs open: 1126 West Ave., Richmond, VA, 23220. Phone: (804) 422-3452. Licensee: William W. McCutchen III. ♦William W. McCutchen III, gen mgr.

***KNDZ(FM)**—Not on air, target date: Oct.: 89.3 mhz; 4.2 kw. Ant -276 ft TL: N40 54 08 W124 04 56. Hrs open: 95 La Jota Dr., Angwin, 94508. Phone: (707) 965-4155. Fax: (707) 965-4161. E-mail: kndl@thecandle.com Web Site: www.thecandle.com. Licensee: Howell Mountain Broadcasting Co. ♦David Shantz, gen mgr.

Mecca

KRCK-FM— 2001: 97.7 mhz; 1.25 kw. Ant 718 ft TL: N33 39 18 W115 59 16. Hrs open: 73-733 Fred Waring Dr., Suite 201, Palm Desert, 92260. Phone: (760) 341-0123. Fax: (760) 341-7455. E-mail: sales@krck.com Web Site: www.krck.com. Licensee: Playa Del Sol Broadcasters. Format: Rock of the 80s. ♦Edward Stolz, gen mgr; Kevin Childs, stn mgr.

Mendocino

***KAKX(FM)**— Jan 15, 1997: Stn currently dark. 89.3 mhz; 250 w. 7 ft TL: N39 18 30 W123 48 02. Stereo. Hrs opn: 24 Box 1154, 95460. E-mail: audio@mcn.org Licensee: Mendocino Unified School District. Population served: 10,000 Format: Educ, var, Rock. ♦Peter Davidson, pres; Steve Butler, gen mgr.

KMFB(FM)— November 1966: 92.7 mhz; 3 kw. 165 ft TL: N39 20 33 W123 46 51. Stereo. Hrs opn: 24 101-E Boatyard Dr., Fort Bragg, 95437. Phone: (707) 964-5307. Fax: (707) 964-3299. E-mail: generalmail@kmfb-fm.com Web Site: www.kmfb-fm.com. Licensee: Four Rivers Broadcasting Inc. (group owner; (acq 7-21-2005; grpsl). Population served: 75,000 Format: Vintage rock, professional sports. News staff: 4. Target aud: 35-54. ♦Bob Woelfel, gen mgr, gen sls mgr, progmg dir, progmg mgr & mus dir; Liz Helenchild, mus dir; Ed Kowas, news dir.

***KPMO(AM)**— Nov 16, 1966: 1300 khz; 5 kw-D, 77 w-N. TL: N39 20 33 W123 46 51. Hrs open: 24 hrs Jefferson Public Radio, 1250 Siskiyou Blvd., Ashland, OR, 97520. Phone: (541) 552-6301. Fax: (541) 552-8565. Web Site: www.ijpr.org. Licensee: JPR Foundation Inc. (acq 8-8-02). Population served: 25,000 Natl. Network: NPR, PRI. Law Firm: Ernest Sanchez. Wire Svc: AP Format: News/talk. News staff: one. Target aud: General. ♦Ronald Kramer, CEO; Ronald Kramer, gen mgr; Bryon Lambert, opns dir.

Mendota

KMEN(FM)— 2007: 100.5 mhz; 6 kw. Ant 144 ft TL: N36 38 50 W120 21 02. Hrs open: 1450 E. Bardsley Ave., Tulare, 93274. Phone: (213) 745-6224. Fax: (213) 745-7577. Licensee: MBP Licensee LLC. Group owner: Moon Broadcasting (acq 3-30-2001; $350,000). Format: Rgnl Mexican. ♦Abel de Luna, pres; Angelica Figueroa, gen sls mgr; Yesenia de Luna, progmg dir.

Merced

KABX-FM—Listing follows KYOS(AM).

***KAMB(FM)**— Nov 6, 1967: 101.5 mhz; 1.85 kw. 2,093 ft TL: N37 32 01 W120 01 46. Stereo. Hrs opn: 24 90 E. 16th St., 95340-5099. Phone: (209) 723-1015. Fax: (209) 723-1945. E-mail: kamb@celebrationradio.com Web Site: www.celebrationradio.com. Licensee: Central Valley Broadcasting Co. Inc. Population served: 1,500,000 Natl. Network: AP Radio, Moody. Law Firm: Fletcher, Heald & Hildreth. Format: Contemp Christian. News staff: one; News: 5 hrs wkly. Target aud: 29-54; Christian adults in central California. ♦Dan Finn, pres; Tim Land, CEO & gen mgr; Mark Murdock, opns dir; Jinous Vartan, mktg dir & prom mgr; Dave Benton, progmg dir & mus dir.

KBKY(FM)— January 2002: 94.1 mhz; 6 kw. 328 ft TL: N37 27 59 W120 14 09. Hrs open: 450 Grogan Ave., Suite A, 95340. Phone: (209) 385-9994. Fax: (209) 385-9982. E-mail: mmeroney941@mercednet.com Web Site: www.foxsportsmerced.com. Licensee: KM Radio of Merced L.L.C. (acq 9-30-99). Format: Sports. ♦Dave Putonen, gen sls mgr; Mike Meroney, gen mgr & prom mgr; Matthew Stone, progmg dir; Chuck Hughes, chief of engrg; Jim Wells, traf mgr.

KNAH(FM)— May 14, 1992: 106.3 mhz; 4 kw. Ant 403 ft TL: N37 25 35 W120 26 25. Hrs open: 24 1020 W. Main St., 95340-4521. Phone: (209) 723-2191. Fax: (209) 383-2950. Web Site: www.radiomerced.com. Licensee: Mapleton Communications LLC (group owner; (acq 6-1-2002; grpsl). Format: Classic hits. Target aud: 25-54; upscale professionals. ♦Andrew Adams, gen mgr; Chad Gammage, gen sls mgr; Damian Galaarza, gen sls mgr; Chris Ashton, progmg dir; Rick McMillion, chief of engrg.

KTIQ(AM)— Nov 1, 1999: 1660 khz; 10 kw-D, 1 kw-N. TL: N37 16 41 W120 37 35. Hrs open: 24 1020 Main St., 95340. Phone: (209) 723-21911. Fax: (209) 383-2950. Licensee: Mapleton Communications LLC (group owner; acq 6-1-2002; grpsl). Population served: 202,000 Format: News/talk, Sp. Target aud: 25-54. ♦Andrew Adams, gen mgr & opns VP.

KUBB(FM)—See Mariposa

KYOS(AM)— October 1936: 1480 khz; 5 kw-U, DA-N. TL: N37 22 30 W120 27 37. Hrs open: 24 1020 W. Main, 95340. Phone: (209) 723-2191. Fax: (209) 383-2950. Licensee: Mapleton Communications LLC (group owner; acq 6-5-2002; grpsl). Population served: 186,000 Natl. Network: CBS. Natl. Rep: Christal. Format: News/talk. News staff: 2; News: 20 hrs wkly. Target aud: 25-54. Spec prog: Farm 5 hrs, gospel one hr wkly. ♦Adam Nathanson, pres; Andrew Adams, gen mgr & progmg mgr; Dennis Daily, progmg dir.

KABX-FM—Co-owned with KYOS(AM). Dec 18, 1975: 97.5 mhz; 50 kw. 490 ft TL: N37 22 31 W120 27 37. Stereo. 24 Format: Oldies. ♦Dave Luna, progmg dir.

Middletown

KSXY(FM)— December 1993: 98.7 mhz; 340 w. 1,378 ft TL: N38 45 55 W122 45 54. Hrs open: 3565 Standish Ave., Santa Rosa, 95407. Phone: (707) 588-0707. Fax: (707) 588-0777. Web Site: www.hot987.fm. Licensee: Commonwealth Broadcasting LLC. Group owner: Sinclair Communications Inc. (acq 8-3-2001; $5.5 million). Format: Rhythmic CHR. ♦Debbie Morton, gen mgr; Dray Lopez, progmg dir.

Mission Viejo

***KSBR(FM)**— May 7, 1979: 88.5 mhz; 620 w. 600 ft TL: N33 30 10 W117 36 06. Stereo. Hrs opn: 28000 Marguerite Pkwy., 92692. Phone: (949) 582-5727. Fax: (949) 347-9693. Web Site: www.ksbr.net. Licensee: South Orange County Community College District. Population served: 500,000 Format: Jazz. News staff: one. Target aud: 25-54. Spec prog: Latin 3 hrs, blues 3 hrs, reggae 3 hrs, electronic 4 hrs, ragtime 2 hrs, folk 2 hrs wkly. ♦Terry Wedel, opns dir; Dawn Kamber, news dir; Mark Schiffelbein, engrg dir.

Modesto

***KADV(FM)**— November 1988: 90.5 mhz; 1.5 kw. 200 ft TL: N37 36 26 W120 57 26. Stereo. Hrs opn: 24 2031 Academy Pl., Ceres, 95307. Phone: (209) 537-1201. Fax: (209) 537-1945. E-mail: kadv@sbcglobal.net Web Site: www.kadv.org. Licensee: Modesto Adventist Academy. Population served: 500,000 Natl. Network: Moody. Format: Educ, relg, music. News: 14 hrs wkly. Target aud: 30 plus. Spec prog: Sp one hr wkly. ♦Jerry Moore, gen mgr; Steve White, stn mgr & progmg dir.

KATM(FM)—Listing follows KESP(AM).

KBBU(FM)— 1999: 93.9 mhz; 4 kw. Ant 403 ft TL: N37 39 00 W121 01 24. Hrs open: 24
Simulcast with KTTA(FM) Esparto 100%.
500 Media Place, Sacramento, 95815. Phone: (916) 368-6300. Fax: (916) 441-6480. E-mail: abalderas@bustosmedia.com Web Site: www.lakebuena.com. Licensee: Bustos Media of California License LLC. (acq 12-15-2004; $21.7 million with KTTA(FM) Esparto). Format: Rgnl Mexican. ♦Amparo Perez-Cook, gen mgr & stn mgr; Javier Gonzalez, prom dir; Juan Gonzalez, progmg dir; Adela Garcia, news dir; Mark Sedaka, chief of engrg; Cynthia Sanchez, traf mgr.

KCIV(FM)—See Mount Bullion

KESP(AM)— 1951: 970 khz; 1 kw-U, DA-2. TL: N37 41 28 W120 57 11. Stereo. Hrs opn: 24 1581 Cummins Dr., Suite 135, 95358. Phone: (209) 523-7756. Fax: (209) 522-2061. Web Site: www.espnradio970.com. Licensee: Citadel Broadcasting Co. Group owner: Citadel Broadcasting Corp. (acq 5-18-92; $12.5 million grpsl, including co-located FM; FTR: 6-8-92). Population served: 274300 Natl. Network: ABC, CBS. Natl. Rep: McGavren Guild. Format: MOR, sports. Target aud: 35 plus. ♦Roy Williams, gen mgr; Jean Western, gen sls mgr; Eric Nelson, news dir; Jeff Silvius, progmg.

KATM(FM)—Co-owned with KESP(AM). 1948: 103.3 mhz; 50 kw. 500 ft TL: N37 34 30 W121 21 13. Stereo. Web Site: www.espnradio970.com. Format: Country. Target aud: 25-64; mass appeal. ♦Bubba Black, progmg dir.

KFIV(AM)— 1950: 1360 khz; 4 kw-D, 950 w-N, DA-2. TL: N37 39 52 W120 57 00. Hrs open: 2121 Lancey Dr., 95355. Phone: (209) 551-1306. Fax: (209) 551-1359. Web Site: www.kfiv1360.com. Licensee: Capstar TX L.P. Group owner: Clear Channel Communiications Inc. (acq 8-30-2000; grpsl). Population served: 375,000 Natl. Network: ABC. Format: News/talk. Target aud: 25-54. ♦Gary Granger, gen mgr; Rick Myers, gen sls mgr; Leslie Davidson, prom dir; Bill Mick, progmg dir.

KJSN(FM)—Co-owned with KFIV(AM). July 4, 1977: 102.3 mhz; 6 kw. 300 ft TL: N37 40 47 W120 55 28. Stereo. Box 3408, 95353. Web Site: www.sunny102fm.com.100,000 Format: Adult contemp. Target aud: 25-49. ♦Gary Michaels, progmg dir; Steve Minshall, engrg mgr.

KHKK(FM)— 1949: 104.1 mhz; 50 kw. 500 ft TL: N37 39 10 W121 28 38. Stereo. Hrs opn: 24 1581 Cummins Dr., Suite 135, 95358-6402. Phone: (209) 572-0104. Fax: (209) 522-2061. Web Site: www.104thehawk.com. Licensee: Citadel Broadcasting Co. Group owner: Citadel Broadcasting Corp. (acq 10-1-93). Population served: 250,000 Natl. Rep: McGavren Guild. Format: Rock/AOR, classic rock. News staff: one; News: 5 hrs wkly. Target aud: 25-49; baby boomers who grew up with rock and roll. ♦Roy Williams, VP & gen mgr; Richard Perry, progmg dir; Farid Suleman, engrg mgr & chief of engrg.

KHTN(FM)—See Planada

KMPH(AM)— July 10, 2006: 840 khz; 5 kw-U, DA-2. TL: N37 42 34 W120 43 34. Hrs open: 1192 Norwegian Ave., 95350. Phone: (209) 527-8400. Fax: (209) 526-0820. E-mail: kmpham840@yahoo.com Web Site: www.kmph840.com. Licensee: Pappas Radio of Modesto LLC. Group owner: Pappas Telecasting Companies (acq 11-5-2003). Format: Standards. ♦Jim Pappas, VP & gen mgr; Jan Baker, gen sls mgr; Kevin Barrett, progmg dir; Jerry Moore, chief of engrg.

***KMPO(FM)**— January 1984: 88.7 mhz; 2 kw. 1,500 ft TL: N37 32 00 W120 01 29. Stereo. Hrs opn: 24 5005 E. Belmont Ave., Fresno, 93727. Phone: (559) 455-5777. Fax: (559) 455-5778. E-mail: mariax@radiobilingue.org Web Site: www.radiobilingue.org. Licensee: Radio Bilingue Inc. Format: Ethnic. News staff: 5; News: 11 hrs wkly. Target aud: 16 plus; Latinos. Spec prog: Black 3 hrs, folk 4 hrs, Filipino one hr wkly. ♦Hugo Morales, CEO; Maria Erana, gen mgr & opns dir; Phil Traynor, dev dir; Samuel Cozco, news dir; Bill Bach, chief of engrg.

KOSO(FM)—Patterson, June 6, 1966: 93.1 mhz; 2.95 kw. 1,791 ft TL: N37 30 14 W121 22 22. Stereo. Hrs opn: 24 2121 Lancey Dr., 95355. Phone: (209) 551-1306. Fax: (209) 551-1359. Web Site: www.b931.com. Licensee: Capstar TX L.P. Group owner: Clear Channel Communications Inc. (acq 8-30-00; grpsl). Population served: 900,000 Format: Adult contemp. News staff: one; News: 5 hrs wkly. Target aud: 25-54. ♦Gary Granger, gen mgr; Mark Granger, sls dir; Zack Davis, progmg dir.

KVIN(AM)—See Ceres

Mojave

***KCRY(FM)**— June 2000: 88.1 mhz; 10.5 kw. Ant -95 ft TL: N35 07 20 W118 12 25. Hrs open:
Rebroadcasts KCRW(FM) Santa Monica 100%.
c/o KCRW(FM), 1900 Pico Blvd., Santa Monica, 90405. Phone: (310) 450-5183. Phone: (888) 600-kcrw. Fax: (310) 450-7172. E-mail: mail@kcrw.org Web Site: www.kcrw.com. Licensee: Santa Monica Community College District. Natl. Network: NPR. Wire Svc: AP Format: Eclectic, news. ♦Ruth Seymour, gen mgr; Mike Newport, opns mgr; David Kleinbart, dev dir; Nic Harcourt, mus dir; Steve Herbert, chief of engrg.

KMVE(FM)—Not on air, target date: unknown: 96.1 mhz; 5.4 kw. Ant -10 ft TL: N35 07 23 W118 12 06. Hrs open: Coloma Mojave LLC, 601 Belvedere St., San Francisco, 94117. Phone: (415) 391-2234. Fax: (415) 391-4912. Licensee: Coloma Mojave LLC. ♦Scott Donohue, CEO & pres.

KTPI(AM)— May 1, 1958: 1340 khz; 1 kw-U. TL: N35 02 23 W118 08 57. Hrs open: 24 348 East Avenue K-4, Lancaster, 93535. Phone: (661) 942-1121. Fax: (661) 723-5512. Licensee: CC Licenses LLC. Group owner: Clear Channel Communications Inc. (acq 11-21-2003; grpsl). Population served: 400,000 Natl. Rep: Christal. Law Firm: Latham & Watkins. Target aud: 35 plus; Adult Christian community. ♦Larry Thornhill, gen mgr; Shaun Palmer, gen sls mgr.

KTPI-FM—See Tehachapi

KVVS(FM)— May 1966: 97.7 mhz; 3 kw. 145 ft TL: N34 58 45 W118 10 02. (CP: Ant 300 ft.). Stereo. Hrs opn: 24 348 East Ave. K4, Lancaster, 93535. Phone: (818) 559-2252. Phone: (818) 295-6405. Fax: (818) 295-6466. E-mail: 977kilsfm@clearchannel.com Web Site: www.977kilsfm.com. Licensee: Citicasters Licenses L.P. Group owner: Clear Channel Communications Inc. (acq 5-4-99; grpsl). Population served: 300000 Law Firm: Pepper & Corazzini. Format: Top 40s. News staff: one. Target aud: 18-49; active adults & young families. ♦Greg Ashlock, gen mgr; Ron Vacchina, gen sls mgr; John Ivey, progmg dir.

Monte Rio

KVRV(FM)—Licensed to Monte Rio. See Santa Rosa

Montecito

KJEE(FM)— March 1994: 92.9 mhz; 820 w. 886 ft TL: N34 27 57 W119 40 37. Hrs open: 302 W. Carrillo St., 2nd Fl., Santa Barbara, 93101. Phone: (805) 963-4676. Fax: (805) 963-8166. E-mail: sales@kjee.com Web Site: www.kjee.com. Licensee: Montecito FM

Inc. Format: Modern rock. Target aud: 18-34; general. ♦Eddie Gutierrez, gen mgr & progmg dir; Steve Meade, rgnl sls mgr; Ryan Zoldas, prom dir; John Palmmentari, news dir; Dean Burt, chief of engrg.

Monterey

KBOQ(FM)—Carmel, Dec 4, 1993: 95.5 mhz; 1.7 kw. 630 ft TL: N36 33 09 W121 47 17. Hrs open: 24 60 Garden Ct., Suite 300, 93940-5370. Phone: (831) 658-5200. Fax: (831) 658-5299. Web Site: www.theclassicalstation.net. Licensee: Mapleton Communications LLC. (acq 6-7-2005; $3.75 million). Population served: 527,900 Natl. Rep: McGavren Guild. Law Firm: Leventhal, Senter & Lerman. Format: Class. Target aud: 35 plus. ♦Adam Nathanson, pres; Dale Hendry, gen mgr; Jodi Morgan, gen sls mgr; Sybil DeAngelo, prom dir; Kenny Allen, progmg dir & progmg mgr; Veldon Leverich, chief of engrg.

KIDD(AM)—Listing follows KWAV(FM).

KNRY(AM)— October 1935: 1240 khz; 1 kw-U. TL: N36 36 56 W121 53 53. Hrs open: 24 Bldg. 651 Cannery Row, 93940. Phone: (831) 372-1074. Fax: (831) 372-3585. E-mail: ronstevens@kyaradio.com Web Site: www.knry.com. Licensee: People's Radio Inc. (group owner; (acq 8-31-2000; $1.1 million with KRXA(AM) Carmel Valley). Population served: 650,000 Natl. Network: CBS. Law Firm: Haley, Bader & Potts. Format: News/talk. Target aud: 35 plus. ♦Jim Vossen, chief of opns & progmg dir.

KOCN(FM)—See Pacific Grove

KPRC-FM—See Salinas

KSES-FM—See Seaside

KTOM-FM—See Marina

KWAV(FM)— Oct 14, 1961: 96.9 mhz; 18 kw. 2,450 ft TL: N36 32 05 W121 37 14. Stereo. Hrs opn: Box 1391, 93942. Phone: (831) 649-0969. Fax: (831) 649-3335. E-mail: kwav97fm@kwav.com Web Site: www.kwav.com. Licensee: Buckley Broadcasting Corp. of Monterey. Group owner: Buckley Broadcasting Corp. (acq 5-1-80; $700,000; FTR: 3-17-80). Population served: 97,600 Natl. Rep: D & R Radio. Format: Adult contemp. Target aud: 18-54; primarily women. ♦Kathy Baker, gen mgr; Sue Clark, gen sls mgr; Bernie Moody, progmg dir; Karen Hamilton, news dir; Bob Turner, chief of engrg; Melanie Swain, traf mgr.

KIDD(AM)—Co-owned with KWAV(FM). 1955: 630 khz; 1 kw-U, DA-2. TL: N36 41 28 W121 48 00. E-mail: magic63am@magic63.com Web Site: www.magic63.com. Licensee: Buckley Communications Inc. (acq 1995; $200,000). Format: Adult standards, big band, nostalgia. ♦Jim Souza, sls dir; Kevin Kahl, progmg dir.

Moraga

***KSMC(FM)**— Sept 22, 1977: 89.5 mhz; 800 w. 95 ft TL: N37 50 25 W122 06 36. Stereo. Hrs opn: 24 Box 3223, St. Mary's College, 94575. Phone: (925) 631-4252. Phone: (925) 631-4772. Fax: (925) 376-5766. E-mail: ksmc@stmarys-ca.edu Web Site: www.ksmc895.com. Licensee: Associated Students of St. Mary's College of California. Population served: 15,000 Wire Svc: Dow Jones News Service Format: CHR, educ, country. Target aud: 15-30; young, urban & willing to experiment. Spec prog: Relg 2 hrs, class 4 hrs, Sp 3 hrs, jazz 5 hrs wkly. ♦Noel Cilker, gen mgr; Jessica Fajardo, prom dir & sports cmtr; Will McCoster, progmg dir; Nick McAlpine, mus dir; Ed Tywoniak, chief of engrg.

Moreno Valley

KHPI(AM)— 1991: Stn currently dark. 1530 khz; 10 kw-D, DA-3. TL: N34 00 42 W117 11 03. Hrs open: Box 909, 92556. Secondary address: 24490 Sunnymead Blvd., #215 92553. Phone: (951) 247-5479. Fax: (951) 247-2790. Licensee: Dr. D.L. Van Voorhis. Population served: 1,900,000 Law Firm: Fletcher, Heald & Hildreth. ♦Dr. D.L. Van Voorhis, pres; Bill DeGeorge, gen mgr.

KHPY(AM)— Jan 16, 2003: 1670 khz; 10 kw-D, 9 kw-N. TL: N34 00 42 W117 11 03. Hrs open: Box 909, 92556. Phone: (909) 247-5479. Fax: (909) 247-2790. Licensee: Delbert L. Van Voorhis. Format: Sp relg. ♦Bill DeGeorge, gen mgr.

Morgan Hill

KSQQ(FM)— December 1990: 96.1 mhz; 530 w. 781 ft TL: N37 10 03 W121 34 20. (CP: 1 kw). Hrs opn: 1629-C Alum Rock Ave., San Jose, 95116. Phone: (408) 258-9699. Fax: (408) 258-9770. E-mail: pr@ksqq.com Web Site: www.ksqq.com. Licensee: Coyote Communications Inc. Population served: 10,000,000 Format: Ethnic. ♦Batista Vieira, pres & gen mgr; Peter Mieuli, VP; Aida Barbosa, sls dir; Joao Manuel, progmg dir.

Morro Bay

KLMM(FM)— September 1997: 94.1 mhz; 890 w. Ant 863 ft TL: N35 15 11 W120 45 42. Hrs open: 24 300 E. Mill St., Suite 301, Santa Maria, 93454-4467. Phone: (805) 928-9796. Fax: (805) 928-3367. E-mail: lazer94@acninc.net Web Site: www.radiolazer.com. Licensee: Lazer Broadcasting Corp. (group owner; (acq 8-7-2000; $1.115 million with KLUN(FM) Paso Robles). Law Firm: Booth, Freret, Imlay & Tepper. Format: Adult contemp, Sp. News: 6 hrs wkly. Target aud: 25-54; general. ♦Jose Guzman, gen mgr & gen sls mgr; Salvador Prieto, chief of opns & progmg dir.

KXTY(FM)— May 1, 1991: 99.7 mhz; 220 w horiz, 210 w vert. 1,633 ft TL: N35 21 37 W120 39 18. Hrs open: 24 3620 Sacramento Dr., Suite 204, San Luis Obispo, 93401. Phone: (805) 781-2750. Fax: (805) 781-2758. Licensee: Salisbury Broadcasting Corp. (acq 1994). Population served: 275,000 Natl. Network: AP Network News. Format: News/talk, sports. News staff: 2. Target aud: 25-54. ♦Charles Salisbury, CEO; Kathy Signorelli, gen mgr; Dick Mason, progmg dir; Pepper Daniels, progmg dir; Bill Bordeaux, chief of engrg.

Moss Beach

***KLSI(FM)**— May 1, 2006: 89.3 mhz; 1 w horiz, 8 w vert. Ant 1,635 ft TL: N37 33 44 W122 28 46. Stereo. Hrs opn: 24
Rebroadcasts WAZQ(FM) Key West, FL 100%.
Educational Public Radio Inc., 6910 N.W. 2nd Terr., Boca Raton, FL, 33487-2325. Phone: (561) 912-9002. E-mail: bill@qfmonline.com Web Site: qfmonline.com. Licensee: Educational Public Radio Inc. Format: Hot adult contemp. Target aud: 18-54; adults. ♦Bill Lacy, pres.

Mount Bullion

KCIV(FM)— Apr 24, 1989: 99.9 mhz; 1.85 kw. 2,099 ft TL: N37 32 00 W120 01 29. Hrs open: 24 1031 15th St., Suite One, Modesto, 95354. Phone: (209) 524-8999. Fax: (209) 524-9088. E-mail: kciv@bottradionetwork.com Web Site: bottradionetwork.com. Licensee: Bott Communications Inc. Group owner: Bott Radio Network Format: Christian info. Target aud: 25-54; Christian family audience. ♦Richard P. Bott, pres; Richard Bott II, VP; Kathleen Reynolds, stn mgr & opns mgr.

Mount Shasta

***KKLC(FM)**— Nov 26, 1977: 107.9 mhz; 20 kw horiz. Ant 815 ft TL: N41 13 37 W122 14 23. Hrs open: 24
Rebroadcasts KLVR(FM) Santa Rosa 100%.
2351 Sunset Blvd., Suite 170-218, Rocklin, 95765. Phone: (916) 251-1600. Fax: (916) 251-1650. E-mail: klove@klove.com Web Site: www.klove.com. Licensee: Educational Media Foundation Group owner: EMF Broadcasting (acq 12-27-2002; $400,000). Natl. Network: K-Love. Law Firm: Shaw Pittman. Format: Contemp Christian. News staff: 3. Target aud: 25-44; Judeo Christian, female. ♦Richard Jenkins, pres; Mike Novak, VP; Keith Whipple, dev dir; David Pierce, progmg mgr; Ed Lenane, news dir; Karen Johnson, news dir; Richard Hunt, news dir; Sam Wallington, engrg dir; Marya Morgan, news rptr.

KMJC(AM)— June 12, 1947: 620 khz; 1 kw-D, 290 w-N. TL: N41 19 09 W122 18 35. Hrs open: 24 Jefferson Public Radio, 1250 Siskiyou Blvd., Ashland, OR, 97520. Phone: (541) 552-6301. Fax: (541) 552-8565. Web Site: www.ijpr.org. Licensee: JPR Foundation Inc. (acq 8-8-02; $300,000 with KSYC(AM) Yreka). Population served: 56,000 Natl. Network: NPR, PRI. Law Firm: Sanchez. Wire Svc: AP Format: News/talk. News staff: one. Target aud: General. ♦Ronald Kramer, gen mgr; Bryon Lambert, opns dir; Paul Westhelle, dev dir.

***KNSQ(FM)**— 1994: 88.1 mhz; 2.28 kw. 2,385 ft TL: N41 20 46 W122 11 42. (CP: 5 kw, ant 890 ft.). Stereo. Hrs opn: 5 AM-2 AM Jefferson Public Radio, 1250 Siskiyou Blvd., Ashland, OR, 97520. Phone: (541) 552-6301. Fax: (541) 552-8565. Web Site: www.ijpr.org. Licensee: The State of Oregon, acting by and through the State Board of Higher Education, for the benefit of Southern Oregon University. (acq 1991; FTR: 4-1-91). Natl. Network: NPR, PRI. Law Firm: Ernest Sanchez. Format: Jazz, AAA, news. News staff: one; News: 45 hrs wkly. Target aud: General. Spec prog: Blues 6 hrs, folk 3 hrs, pub affrs 7 hrs wkly. ♦Mitchell Christian, CFO; Ronald Kramer, CEO & gen mgr; Bryon Lambert, opns dir; Paul Westhelle, dev dir.

Mountain Pass

KHYZ(FM)— April 1980: 99.5 mhz; 10 kw. Ant 1,710 ft TL: N35 29 27 W115 33 27. (CP: 8.4 kw, ant 1,807 ft.). Stereo. Hrs opn: 24 Box 1668, 1611 E. Main St., Barstow, 92312. Phone: (760) 256-0326. Fax: (760) 256-9507. E-mail: tim@highwayradio.com Web Site: www.thehighwaystations.com. Licensee: KHWY Inc. Natl. Network: AP Radio. Law Firm: Hogan & Hartson. Format: Adult contemp. News staff: one; News: 16 hrs wkly. Target aud: 35 plus; travelers & loc communities. ♦Howard B. Anderson, CEO & pres; Kirk Anderson, exec VP; Timothy Anderson, VP & gen mgr; Judy Robinson, sls VP; John Gregg, prom dir; Lance Todd, progmg dir; Keith Hayes, news dir; Thomas J. McNeill, engrg mgr.

Mountain View

***KSFH(FM)**— 1974: 87.9 mhz; 10 w. 100 ft TL: N37 22 09 W122 05 00. (CP: 100 w, -246 ft.). Hrs opn: 1 PM-9 PM (M-F) 1885 Miramonte Ave., 94040. Phone: (650) 968-1213, EXT. 272. Fax: (650) 968-1706. Web Site: www.ksfh.com. Licensee: St. Francis High School of Mountain View California Inc. Format: Contemp hits, rock/AOR, urban contemp. News: 5 hrs wkly. Target aud: General; young adult, high school, college. ♦Bob Lautenslager, gen mgr.

Napa

KVON(AM)— Dec 17, 1947: 1440 khz; 5 kw-D, 1 kw-N, DA-2. TL: N38 16 47 W122 18 06. Hrs open: 24 1124 Foster Rd., 94558. Phone: (707) 252-1440. Fax: (707) 226-7544. Web Site: www.kvon.com. Licensee: Wine Country Broadcasting Co. (acq 8-11-03; $3 million with KVYN(FM) St. Helena). Population served: 150,000 Natl. Rep: Christal. Format: News/talk. News staff: 3; News: 30 hrs wkly. Target aud: 35 plus. Spec prog: Sp 2 hrs. ♦Jeff Schechtman, gen mgr & opns dir; Erica Pickett, prom dir; Megan Goldsby, news dir; Ben Webster, chief of engrg.

KVYN(FM)—See Saint Helena

Needles

KLUK(FM)— May 1984: 97.9 mhz; 2.8 kw. 1,571 ft TL: N35 02 06 W114 22 09. (CP: 29.5 kw). Stereo. Hrs opn: 1531 Jill Way, Suite 7, Bullhead City, AZ, 86426-9341. Phone: (928) 763-5586. Fax: (928) 763-3775. E-mail: info@lucky98.com Web Site: www.lucky98fm.com. Licensee: Cameron Broadcasting Inc. (group owner; acq 1-18-02; grpsl). Population served: 70000 Natl. Network: AP Radio. Format: Classic rock. Target aud: 25-54; adults. ♦Don Jaeger, gen mgr; David Cooper, opns mgr & chief of engrg; Mike Fletcher, gen sls mgr; Star Cooper, progmg dir & mus dir; Tracy Wallevand, traf mgr.

KNKK(FM)—Not on air, target date: unknown: 107.1 mhz; 17 kw. Ant 1,909 ft TL: N35 01 57 W114 21 57. Hrs open: 24 1615 Orange Tree Ln., Suite 200, Redlands, 92374. Phone: (928) 763-5586. Fax: (928) 763-3775. E-mail: info@theknack107.com Web Site: www.theknack107.com. Licensee: Cameron Broadcasting Inc. (group owner; acq 3-13-01). Format: Hits of the 80s & beyond. ♦Don Jaeger, CEO & gen mgr; Mike Fletcher, gen sls mgr; Mike Woodard, progmg dir; Rosemary Michaels, news dir; David Cooper, chief of engrg.

KTOX(AM)— October 1952: 1340 khz; 1 kw-U. TL: N34 51 10 W114 37 19. Hrs open: 24 100 Balboa Pl., 92363. Secondary address: P.O. Box 8766, Ft. Mahaur, AZ 86427. Phone: (760) 326-4500. Fax: (760) 326-6849. E-mail: ktox1340@citlink.net Web Site: www.ktox1340am.com. Licensee: Creative Broadcasting Services Inc. (acq 12-7-00; $200,000). Population served: 80,000 Natl. Network: Jones Radio Networks, Premiere Radio Networks. Law Firm: Womble, Carlyle, Sandridgee & Rice. Format: News/talk. News staff: one; News: 24 hrs wkly. Target aud: 18 plus. Spec prog: Rt 66 program, 22 hrs of personalities and live local programming wkly. ♦Robert T. Hayes, CEO; David T. Hayes, pres & gen mgr; Paul Fix, opns mgr; Kelly Hayes, gen sls mgr.

Nevada City

***KVMR(FM)—** July 17, 1978: 89.5 mhz; 1.96 kw. 980 ft TL: N39 14 47 W120 57 48. Stereo. Hrs opn: 24 401 Spring St., 95959. Phone: (530) 265-9073. Fax: (530) 265-9077. Web Site: www.kvmr.org. Licensee: Nevada City Community Broadcast Group. (acq 7-11-89; $32,000; FTR: 5-29-89). Population served: 1,100,000 Format: Var. News: 2 hrs wkly. Spec prog: Country 7 hrs, Black 4 hrs, folk 13 hrs, blues 7 hrs, foreign/ethnic 20 hrs wkly. ♦David Levin, gen mgr & stn mgr; Richard Gorman, mktg dir & prom dir; Steve Baker, opns mgr & progmg dir; Mike Thornton, news dir; Paul Patterson, chief of engrg.

Newberry Springs

KIQQ-FM— January 2001: 103.7 mhz; 6 kw. Ant 246 ft TL: N34 53 19 W116 53 39. Hrs open: 24
Simulcast with KAEH(FM) Beaumont 100%.
710 W. Old Hwy. 58, Barstow, 92311. Phone: (760) 255-2636. Fax: (760) 255-3236. E-mail: jramirez@lamaquinamusical.net Web Site: www.moonbroadcasting.com; www.lamequinamusical.net. Licensee: MBR Licensee LLC. Group owner: Moon Broadcasting (acq 11-5-99). Format: Rgnl Mexican. ♦Alicia Avila, gen mgr.

Newport Beach

KDLE(FM)— Jan 31, 1964: 103.1 mhz; 300 w. Ant 964 ft TL: N33 36 19 W117 48 38. Stereo. Hrs opn: 24
Rebroadcasts KDLD(FM) Santa Monica 100%.
5700 Wilshire Blvd., Suite 250, Los Angeles, 90036. Phone: (323) 900-6100. Phone: (877) 452-1031. Fax: (323) 900-6200. E-mail: feedback@indie1031.fm Web Site: www.indie1031.fm. Licensee: Entravision Holdings LLC. Group owner: Entravision Communications Corp. (acq 2000; grpsl). Population served: 250,000 Law Firm: Latham & Watkins. Format: Alternative. News: one hr wkly. Target aud: 18-34; upscale youth in the Los Angeles Area. ♦Dawn Girocco, gen sls mgr; Max Tolkoff, progmg dir; Rick Hunt, chief of engrg.

North Fork

KLLE(FM)— 1996: 107.9 mhz; 1.75 kw. Ant 1,227 ft TL: N37 17 42 W119 33 51. Hrs open: 1981 N. Gateway, Suite 101, Fresno, 93727. Phone: (559) 456-4000. Fax: (559) 251-9555. Web Site: www.univision.com. Licensee: Univision Radio License Corp. Group owner: Univision Radio (acq 9-22-2003; grpsl). Format: Regetton. ♦Angela Navarrete, gen mgr.

North Highlands

***KQEI-FM—** Feb 21, 1992: 89.3 mhz; 3.1 kw vert. Ant 354 ft TL: N38 42 38 W121 28 54. Stereo. Hrs opn: 24 2601 Mariposa St., San Francisco, 94110. Phone: (415) 553-2129. Fax: (415) 553-2241. Web Site: www.kqed.org. Licensee: KQED Inc. (acq 5-9-03; $3 million). Format: News/talk. ♦Jo Anne Wallace, gen mgr; Traci A. Eckels, dev dir; Paul Ramirez, news dir.

Northridge

***KCSN(FM)—** November 1963: 88.5 mhz; 320 w. 1,643 ft TL: N34 19 11 W118 33 14. Stereo. Hrs opn: 24 18111 Nordhoff St., 91330-8312. Phone: (818) 677-3090. Fax: (818) 677-3069. Web Site: www.kcsn.org. Licensee: California State University Northridge. Population served: 3,000,000 Natl. Network: PRI, NPR. Law Firm: Arter & Hadden. Format: Class, var/div. News staff: one; News: 12 hrs wkly. Target aud: 35 plus; middle/upper middle-class, well educated. Spec prog: German 3 hrs, Jewish 3 hrs, bluegrass 5 hrs wkly. ♦Fred Johnson, gen mgr; Martin Perlich, opns mgr & progmg dir; Laura Kelly, dev dir; Keith Goldstein, news dir; Michael Worrall, chief of engrg.

Oakdale

KHOP(FM)— Mar 11, 1985: 95.1 mhz; 29.5 kw. 633 ft TL: N37 47 34 W120 31 08. (CP: 16 kw, ant 876 ft. TL: N37 49 39 W120 34 03). Stereo. Hrs opn: 24 1581 Cummins Dr., Suite 135, Modesto, 95358. Phone: (209) 766-5000. Fax: (209) 522-2061. Web Site: www.planet95.com. Licensee: Citadel Broadcasting Co. Group owner: Citadel Broadcasting Corp. (acq 1996; $5 million). Law Firm: Fletcher, Heald & Hildreth. Format: 80s & beyond. News staff: one; News: 3 hrs wkly. Target aud: 18-49. Spec prog: Jazz 2 hrs, blues 2 hrs wkly. ♦Roy Williams, VP & gen mgr; Richard Perry, progmg dir.

Oakhurst

KAAT(FM)— Nov 1, 1982: 103.1 mhz; 25 kw. Ant 125 ft TL: N37 25 08 W119 44 40. Stereo. Hrs opn: 24 40356 Oak Park Way, Suites E & F, 93644. Phone: (559) 683-1031. Fax: (559) 683-5488. E-mail: mtkaat@sierratel.com Web Site: www.kaat.com. Licensee: California Sierra Corp. (acq 3-3-2005; $4.75 million with co-located AM). Population served: 48,685 Format: Hispanic. Target aud: 25-54; general. Spec prog: Relg 2 hrs wkly. ♦Abel DeLuna, pres & stn mgr; Denny Jackson, progmg dir.

KTNS(AM)—Co-owned with KAAT(FM). Nov 20, 1982: 1060 khz; 5 kw-D, 55 w-N. TL: N37 17 46 W119 36 23.24 Phone: (559) 683-1060. E-mail: tammy@kaat.com Web Site: www.ktnsradio.com. Natl. Network: CNN Radio, Westwood One. Format: Adult contemp. News staff: 3. Target aud: 25-49. ♦Becky Deaver, gen sls mgr; Jesse Taylor, progmg dir.

Oakland

KISQ(FM)—See San Francisco

KKGN(AM)— 1925: 960 khz; 5 kw-U, DA-1. TL: N37 49 40 W122 18 53. Hrs open: 24 340 Townsend St., Suite 4-960, San Francisco, 94107-. Phone: (415) 977-0960. Fax: (415) 972-1107. E-mail: 960kabl@kabl.com Web Site: www.960kabl.com. Licensee: AMFM Broadcasting Licenses LLC. Group owner: Clear Channel Communications Inc. (acq 8-30-2000; grpsl). Population served: 715,674 Natl. Rep: Christal. Format: Talk. ♦Anna Eppinger, gen mgr; Michael Martin, opns dir; Chris Mason, prom dir; Bob Agnew, progmg dir.

KMKY(AM)— July 1922: 1310 khz; 5 kw-U, DA-1. TL: N37 49 27 W122 19 10. Hrs open: 24 900 Front St., 3rd Fl, San Francisco, 94111. Phone: (415) 788-1310. Fax: (415) 788-1312. Web Site: www.disney.com. Licensee: KGO-AM Radio Inc. Group owner: ABC Inc. (acq 12-18-97; $6.25 million). Population served: 4,889,900 Natl. Network: Radio Disney. Natl. Rep: Interep. Format: Children. Target aud: 2-14; kids, tweens & moms 25-54. ♦Michael Luckoff, stn mgr; Martin Spisak, opns mgr; Shalon Rogers, prom dir.

KNEW(AM)— July 2, 1921: 910 khz; 20 kw-D, 5 kw-N, DA-2. TL: N37 53 45 W122 19 25. Hrs open: 24 340 Townsend St., San Francisco, 94107. Phone: (415) 538-1013. Fax: (415) 975-5573. E-mail: kencole@clearchannel.com Web Site: www.910wnew.com. Licensee: AMFM Broadcasting Licenses LLC. Group owner: Clear Channel Communications Inc. (acq 8-30-2000; grpsl). Population served: 200,000 Format: News/talk. News staff: 8; News: 65 hrs wkly. Target aud: 25-54. ♦Anna Eppinger, gen mgr; Lucia Vandenhof, prom dir; Bob Agnew, progmg dir.

Oceanside

***KKSM(AM)—** July 4, 1956: 1320 khz; 500 w-U, DA-1. TL: N33 12 08 W117 36 46. Hrs open: 24 1140 W. Mission Rd., San Marcos, 92069. Phone: (760) 744-1150, EXT. 5576 and 3149. Fax: (760) 744-8123. Web Site: www.kksm.palomar.edu. Licensee: Palomar Community College District. (acq 4-1-96). Population served: 468,400 Format: Eclectic, div. News staff: one; News: 10 hrs wkly. Target aud: 18-25; college age, mid-upper income, diverse ethnic. ♦Zeb Navarro, gen mgr; Christa Lynch, prom dir; Matt O'Brien, progmg dir; David Quera, mus dir; Joan Rubin, mus dir; Lindsay Lutz, mus dir; Josh Diaz, news dir; Ken Chapman, news dir.

Oildale

KGDP(AM)—Licensed to Oildale. See Santa Maria

KLLY(FM)— January 1985: 95.3 mhz; 12.5 kw. 394 ft TL: N35 27 55 W119 01 04. Stereo. Hrs opn: 24 Box 80658, Bakersfield, 93308. Secondary address: 3651 Pegasus, Suite 107, Bakersfield 93308. Phone: (661) 393-1900. Fax: (661) 393-1915. Web Site: www.klly.com. Licensee: Buckley Broadcasting of California LLC. Group owner: Buckley Broadcasting Corp. (acq 12-86; $1.3 million; FTR: 11-10-86). Population served: 420,000 Natl. Rep: D & R Radio. Format: Adult contemp. Target aud: 25-44; adults. ♦Steve Darnell, gen mgr; Otis Warren, sls dir; E.J. Tyler, progmg dir.

Ojai

KFYV(FM)— Jan 4, 1972: 105.5 mhz; 310 w. 1,437 ft TL: N34 20 57 W119 20 07. Stereo. Hrs opn: 24 2284 S. Victoria, Suite 2G, Ventura, 93003. Phone: (805) 289-1400. Fax: (805) 644-7906. Web Site: www.live1055.fm. Licensee: Gold Coast Broadcasting LLC (group owner; acq 5-15-97; $2 million with KKZZ(AM) Ventura). Natl. Network: AP Radio. Natl. Rep: Katz Radio. Format: Hot adult contemp. News staff: one; News: 3 hrs wkly. Target aud: 25-49; fun, upscale, classy adults. ♦Chip Ehrhardt, gen mgr; Mark Elliot, progmg dir.

***KLFH(FM)—** 2003: 89.5 mhz; 97 w. Ant 1,322 ft TL: N34 24 45 W119 11 16. Hrs open: 560 Higuera St., Suite G, San Luis Obispo, 93401. Phone: (805) 541-4343. Fax: (805) 541-9101. E-mail: info@klife.org Web Site: www.klife.org. Licensee: Shepherd Communications Inc. Format: CHR, Christian hits. ♦Jim Fugler, gen mgr; Noonie Fugler, prom dir.

Ontario

KSPA(AM)— Jan 26, 1947: 1510 khz; 10 kw-D, 1 kw-N, DA-2. TL: N34 05 41 W117 36 46. Hrs open: 24 1045 S. East St., Anaheim, 92805. Phone: (909) 483-1500. Fax: (909) 483-1515. E-mail: kspa1510@aol.com Web Site: www.thesparadio.com. Licensee: Ontario Broadcasting L.L.C. Group owner: Astor Broadcast Group. (acq 11-4-99). Population served: 731,000 Format: Adult standards. News staff: one. Target aud: 18-49. ♦Art Astor, pres; Peri Corso, gen mgr; Joe Lyons, opns mgr.

KWIE(FM)— Jan 26, 1947: 93.5 mhz; 5 kw. Ant -131 ft TL: N34 10 32 W117 34 26. Stereo. Hrs opn: 24 5055 Wilshire Blvd., Suite 720, Los Angeles, 90036. Phone: (323) 337-1600. Fax: (323) 337-1633. Web Site: www.935kday.com. Licensee: KDAI Licensing LLC. Group owner: Spanish Broadcasting System Inc. (acq 1-31-2006; $120 million with KDAY(FM) Redondo Beach). Population served: 2,000,000 Format: Hip-hop. ♦Kimberly Fletcher, gen mgr.

Orange

KLAA(AM)— Jan 13, 1992: 830 khz; 50 kw-D, 20 kw-N, DA-N. TL: N33 55 43 W117 36 57. Hrs open: 24 15301 Ventura Blvd., Bldg. D, Suite 200, Sherman Oaks, 91403. Phone: (818) 528-2050. Fax: (818) 784-8824. Web Site: www.830am.com. Licensee: LAA 1 LLC (acq 5-23-2006; $41 million). Natl. Network: NBC Radio. Format: News, talk, sports. News: 28 hrs wkly. Target aud: 25-54; Male & female, high income, professionals. Spec prog: Sports. ♦Alan L. Fuller, gen mgr.

Orange Cove

KMAK(FM)— Oct 27, 1990: 100.3 mhz; 72 w. 2,073 ft TL: N36 44 45 W119 16 58. Stereo. Hrs open: 24 227 W. Teague Ave., Fresno, 93711. Secondary address: 640 Park Blvd. 93662. Phone: (559) 217-3313. Fax: (559) 626-4381. Licensee: Richard B. Smith. Population served: 475,000 Natl. Network: CNN Radio. Law Firm: Arent, Fox, Kintner, Plotkin & Kahn. Format: Rgnl Mexican. News staff: one. Target aud: 18-54. ♦Antonio Rabago, gen mgr, mktg dir & progmg dir; Angelica Martinez, progmg dir; Richard Smith, chief of engrg.

Orcutt

KPAT(FM)—Licensed to Orcutt. See Santa Maria

Orland

KRQR(FM)— January 1994: 106.7 mhz; 25 kw. 56 ft TL: N39 53 17 W122 37 38. Hrs open: 24 856 Manzanita Court, Chico, 95926. Phone: (530) 342-2200. Fax: (530) 342-2260. E-mail: info@zrockfm.com Web Site: www.zrockfm.com. Licensee: Results Radio Licensee L.L.C. Group owner: Fritz Communications Inc. (acq 6-11-99; grpsl). Law Firm: Brown, Nietert & Kaufman. Format: Active rock/extreme alternative. ♦Jack Fritz, pres; John Graham, gen mgr; Dave Pack, gen sls mgr; Neil Randall, progmg mgr; J.D. Davis, chief of engrg; Candy Mason, traf mgr.

Oroville

KEWE(AM)—Listing follows KHHZ(FM).

KHHZ(FM)— July 6, 1979: 97.7 mhz; 1.5 kw. Ant 1,276 ft TL: N39 30 18 W121 18 35. Stereo. Hrs opn: 24 2654 Cramer Ln., Chico, 95928. Phone: (530) 345-0021. Fax: (530) 893-2121. Licensee: Deer Creek Broadcasting LLC. (group owner; (acq 9-8-2004; grpsl). Population served: 650,000 Natl. Rep: Katz Radio. Format: Hispanic. Target aud: 18-49. ♦Dino Corbin, gen mgr; Bill Meyer, sls dir; Juan Villagrana, progmg dir; Matt Ray, news dir; Mark Miller, chief of engrg; Cheryl Grant, traf mgr.

KEWE(AM)—Co-owned with KHHZ(FM). Aug 4, 1962: 1340 khz; 1 kw-U. TL: N39 30 34 W121 35 55.24 320000 Format: Sp. ♦Rosa Ramos, progmg dir.

Oxnard

KCAQ(FM)— Sept 27, 1958: 104.7 mhz; 5.1 kw. 1,580 ft TL: N34 20 53 W119 20 07. Hrs open: 2284 S. Victoria, Suite 2G, Ventura, 93003. Phone: (805) 289-1400. Fax: (805) 644-7906. Web Site: www.q1047.com. Licensee: Gold Coast Broadcasting LLC (group owner; acq 1996; $3.65 million with KVTA(AM) Port Hueneme). Population served: 636,500 Natl. Rep: Katz Radio. Format: CHR Rhythmic. Target aud: 18-44. ♦Chip Ehrhardt, gen mgr; Brian Davis, progmg dir.

***KCRU(FM)**— 1993: 89.1 mhz; 200 w. 853 ft TL: N34 06 47 W119 03 34. Hrs open:
Rebroadcasts KCRW(FM) Santa Monica 98%.
1900 Pico Blvd., Santa Monica, 90405. Phone: (310) 450-5183. Phone: (888) 660-kcrw. Fax: (310) 450-7172. E-mail: mail@kcrw.org Web Site: www.kcrw.com. Licensee: Santa Monica Community College District. Format: Eclectic, news. ♦Ruth Seymour, gen mgr; Jennifer Ferro, stn mgr; Mike Newport, opns mgr; David Kleinbart, dev dir; Ariana Morgenstern, asst music dir; Steve Herbert, chief of engrg.

KDAR(FM)— Oct 28, 1974: 98.3 mhz; 1.5 kw. Ant 1,289 ft TL: N34 20 55 W119 19 57. Stereo. Hrs opn: 24 Box 5626, 93031. Secondary address: 500 Esplanade Dr., Suite 1500 93036. Phone: (805) 485-8881. Fax: (805) 656-5330. E-mail: radiomail@kdar.com Web Site: www.kdar.com. Licensee: New Inspiration Broadcasting Co. Inc. Group owner: Salem Communications Corp. Format: Christian talk & mus. News: 2 hrs wkly. Target aud: 25-54; upscale adults with large families. ♦Ed Atsinger, pres; Richard Trejo, gen mgr; Roy Bach, progmg dir.

KKZZ(AM)—See Ventura

KLJR-FM—See Santa Paula

KOCP(FM)—See Camarillo

KOXR(AM)— June 11, 1955: 910 khz; 5 kw-D, 1 kw-N, DA-2. TL: N34 16 58 W119 07 36. Hrs open: 24 200 S. A St., Suite 400, 93030. Phone: (805) 240-2070. Fax: (805) 240-5960. Licensee: Lazer Broadcasting Corp. (group owner; acq 1-11-99). Population served: 250,000 Natl. Rep: Lotus Entravision Reps LLC. Law Firm: Fletcher, Heald & Hildredth. Format: Sp. Target aud: 25-54. ♦Alfredo Plascencia, CEO & pres; Salvador Prieto, opns mgr & progmg dir.

KUNX(AM)—See Santa Paula

KVEN(AM)—See Ventura

KXLM(FM)— 1991: 102.9 mhz; 5.5 kw. Ant 112 ft TL: N34 14 12 W119 12 11. Hrs open: 24 200 S. A St., Suite 400, 93030. Secondary address: Box 6940 93030. Phone: (805) 240-2070. Fax: (805) 240-5960. Web Site: radiolazer.com. Licensee: Kext Broadcasters Inc. Format: Sp, adult contemp. News staff: one; News: 2 hrs wkly. Target aud: 25-59. ♦Alfredo Plascencia, pres & gen mgr; Terry Janisch, gen sls mgr.

Pacific Grove

***KAZU(FM)**— Oct 1, 1977: 90.3 mhz; 3.7 kw. Ant 522 ft TL: N36 33 09 W121 47 17. Stereo. Hrs opn: 24 167 Central Ave. #B, 93950-3060. Phone: (831) 375-7275. Fax: (831) 375-0235. E-mail: mail@kazu.org Web Site: www.kazu.org. Licensee: Foundation of California State University Monterey Bay (acq 11-30-00; $150,000). Population served: 500,000 Natl. Network: NPR. Law Firm: Garvey, Schubert & Barer. Format: News, info. News: 8 hrs wkly. Target aud: 25-65; general. Spec prog: Country 6 hrs, women's mus 6 hrs, gospel 4 hrs, folk 6 hrs, class 3 hrs, oldies 5 hrs wkly. ♦Ducan Lively, gen mgr; Douglas McKnight, dev dir; Ben Adler, progmg dir & news dir; Duncan Lively, progmg dir.

KOCN(FM)— Apr 10, 1977: 105.1 mhz; 1.8 kw. 600 ft TL: N36 33 09 W121 47 17. (CP: 402 kw, ant 790 ft. TL: N36 30 38 W121 43 57). Stereo. Hrs opn: 24 903 N. Main St., Salinas, 93906. Phone: (831) 755-8181. Fax: (831) 755-8191. Licensee: CC Licenses LLC. Group owner: Clear Channel Communications Inc. (acq 9-22-97; grpsl). Population served: 600,000 Natl. Network: Westwood One. Natl. Rep: Clear Channel. Format: Oldies. News staff: one. Target aud: 25-54; at work, double income households. ♦Rhonda McCormack, gen mgr & sls dir; Joey Martinez, prom dir & progmg dir.

Palm Desert

KEZN(FM)— Nov 28, 1977: 103.1 mhz; 1.9 kw. 590 ft TL: N33 51 58 W116 25 56. Stereo. Hrs opn: 24 72-915 Parkview Dr., 92260. Secondary address: Box 291 92260. Phone: (760) 340-9383. Fax: (760) 340-5756. Web Site: www.ez103.com. Licensee: Infinity Radio Holdings Inc. Group owner: Infinity Broadcasting Corp. (acq 11-13-98; grpsl). Population served: 250,000 Natl. Network: Westwood One. Law Firm: Leventhal, Senter & Lerman. Format: Adult contemp. News staff: one. Target aud: 25-64. ♦Doug Kratky, gen sls mgr; Frank Torok, prom dir; Tom Hoyt, gen mgr & prom dir; Rick Shaw, progmg dir.

***KHCS(FM)**— January 1993: 91.7 mhz; 960 w. 574 ft TL: N33 41 25 W116 17 14. Stereo. Hrs opn: 24 2341 Duane Rd., Palm Springs, 92262. Phone: (760) 864-9620. Fax: (760) 864-9633. E-mail: Khcs@juno.com Web Site: www.joy92.org. Licensee: Prairie Avenue Gospel Center. Law Firm: Lauren A. Colby. Format: Inspirational, Christian. Target aud: 20-85. ♦Dan Pike, pres; R.F. Watts, chief of engrg.

Palm Springs

KDES-FM—Listing follows KPSI(AM).

KGAM(AM)— 1969: 1450 khz; 1 kw-U. TL: N33 48 02 W116 30 25. Hrs open: 24 2100 E. Tahquitz Canyon Way, 92262. Phone: (760) 325-2582. Phone: (760) 320-8255. Fax: (760) 322-3562. Web Site: www.kgam.com. Licensee: R & R Radio Corp. Group owner: RR Broadcasting (acq 4-8-2002; with co-located FM). Population served: 100,000 Natl. Rep: Christal. Law Firm: Cohn & Marks. Format: News/talk, talk. News staff: 2; News: 5 hrs wkly. Target aud: 25 plus; upscale, informed, involved adults. ♦Rozene Supple, pres; Gene Nichols, CFO & news dir; Mike Keane, gen mgr; Mel Hill, rgnl sls mgr; Geoff Allan, prom dir; Steve Kelly, progmg dir; Barry O'Connor, chief of engrg.

KPSI-FM—Co-owned with KGAM(AM). June 1980: 100.5 mhz; 25 kw. Ant 121 ft TL: N33 56 44 W116 24 34. Stereo. Phone: (760) 325-2582. Fax: (760) 320-4632. Web Site: www.mix1005.fm.250,000 Format: CHR, contemp hit. ♦Gregg Aratin, gen sls mgr; Lisa Giles, prom dir; Connie Breeze, progmg dir.

KNWQ(AM)— Feb 12, 1946: 1140 khz; 10 kw-D, 2.5 kw-N, DA-2. TL: N33 51 39 W116 28 20. Stereo. Hrs opn: 1321 N. Gene Autry Trail, 92262. Phone: (760) 322-7890. Fax: (760) 322-5493. Web Site: www.desertfun.com. Licensee: Morris Communications Corp. Group owner: Morris Communications Inc. (acq 12-24-97; $4.5 million). Population served: 300,000 Natl. Network: CBS. Natl. Rep: McGavren Guild. Format: News/talk. Target aud: 35-65. ♦William S. Morris IV, chmn; William S. Morris III, pres; Darrel Fry, CFO; Michael Osterhaut, VP; Keith Martin, gen mgr; Gary Demaroney, opns dir.

KPLM(FM)— Jan 24, 1983: 106.1 mhz; 50 kw. 391 ft TL: N33 52 14 W116 13 39. Stereo. Hrs opn: 24 Box 1825, 92263. Phone: (760) 320-4550. Fax: (760) 320-3037. E-mail: kplm@dc.rr.com Web Site: thebig106.com. Licensee: RM Broadcasting L.L.C. (acq 9-10-98). Population served: 3,700,000 Natl. Rep: Katz Radio. Law Firm: Koteen & Naftalin. Format: Country. News staff: one; News: 9 hrs wkly. Target aud: 25-54. ♦Todd Marker, gen mgr; Kory James, opns mgr & prom dir; Hughes Hilles, gen sls mgr; Al Gordon, progmg dir.

***KPSC(FM)**— April 1978: 88.5 mhz; 3 kw. 266 ft TL: N33 52 14 W116 13 39. Stereo. Hrs opn: 24
Rebroadcasts KUSC 100%.
Box 77913, Los Angeles, 90007. Secondary address: 515 S. Figueroa St., Suite 2050, Los Angeles 90071. Phone: (213) 225-7400. Fax: (213) 225-7410. E-mail: kusc@kusc.org Web Site: www.kusc.org. Licensee: University of Southern California. (acq 9-9-86). Natl. Network: PRI, NPR. Law Firm: Lawrence Bernstein. Format: Class. Target aud: 35 plus; general. ♦Brenda Barnes, pres; Eric DeWeese, gen mgr; Janet McIntyre, dev VP & dev dir; Stephanie Ross, mktg dir.

KPSI(AM)— Oct 29, 1956: 920 khz; 5 kw-D, 1 kw-N, DA-2. TL: N33 51 29 W116 29 39. Stereo. Hrs opn: 2100 E. Tahquitz Canyon Way, 92262. Phone: (760) 325-2582. Fax: (760) 322-3562. Web Site: www.newstalk920.com. Licensee: R & R Radio Corp. Group owner: RR Broadcasting. Population served: 245,000 Natl. Rep: Christal. Format: News, talk. News staff: 4; News: 16 hrs wkly. Target aud: 25-54. Spec prog: American Indian 2 hrs wkly. ♦Mike Keane, gen mgr; Gregg Aratin, gen sls mgr; Lisa Giles, prom dir; Steve Kelly, progmg dir; Gregg Nichols, news dir.

KDES-FM—Co-owned with KPSI(AM). Feb 10, 1963: 104.7 mhz; 42 kw. 540 ft TL: N33 51 56 W116 26 04. Stereo. E-mail: kdes@aol.com Web Site: www.kdes.com. Law Firm: Cohn & Marks. Format: Oldies. ♦Greg Aratin, gen sls mgr; Kacy Consiglio, progmg mgr & traf mgr.

KPTR(AM)—Cathedral City, Oct 4, 1964: 1340 khz; 1 kw-U. TL: N33 48 07 W116 27 44. Hrs open: 24 2100 Tahquitz Canyon Way, 92262. Phone: (760) 325-2582. Fax: (760) 325-2582. Licensee: R & R Radio Corp. (acq 7-27-2006; $2.3 million). Population served: 180,000 Format: Progressive talk. News staff: one. ♦Mike Keane, gen mgr; Lisa Childs, prom dir; Steve Kelly, progmg dir.

KWXY-FM—Cathedral City, Jan 19, 1969: 98.5 mhz; 50 kw. 499 ft TL: N33 51 55 W116 26 10. Stereo. Hrs opn: 24 KWXY Broadcast Centre, Box 5470, 92263. Phone: (760) 328-1104. Fax: (760) 328-7814. Web Site: www.kwxy.com. Licensee: Glen Barnett Inc. Population served: 688,000 Law Firm: Garvey, Schubert & Barer. Wire Svc: AP Format: Btfl mus, adult standards. News staff: one; News: 16 hrs wkly. Target aud: 35 plus; affluent adults. Spec prog: Canadian news 2 hrs wkly. ♦Estelle Layton, exec VP; Bob Wetherall, opns mgr; Glen Barnett, pres, gen mgr & chief of engrg.

Palmdale

KUTY(AM)— August 1957: 1470 khz; 5 kw-U, DA-2. TL: N34 39 55 W118 00 40. Hrs open: 24 Q-9, 570 East Ave., 93550. Phone: (661) 947-3107. Fax: (661) 272-5688. Licensee: High Desert Broadcasting LLC. (group owner; (acq 3-5-97). Population served: 300,000 Format: News/talk. Target aud: 18-49; homeowners, married couples with discretionary income. ♦Nelson Rasse, gen mgr.

Palo Alto

KDFC-FM—See San Francisco

KNTS(AM)— 1947: 1220 khz; 5 kw-D, 147 w-N. TL: N37 29 04 W122 08 04. Hrs open: 24 39138 Fremont Blvd., 3rd Fl., Freemont, 94538. Phone: (510) 713-1100. Fax: (510) 505-1448. Web Site: www.1220knts.com. Licensee: SCA-Palo Alto LLC. Group owner: Salem Communications Corp. (acq 6-28-2001; $9 million). Population served: 5,500,000 Natl. Network: CBS. Natl. Rep: Salem. Format: News, talk, sports. News staff: 5; News: 70 hrs wkly. Target aud: 25-54; general. Spec prog: Technology news & interviews. ♦Ken Miller, gen mgr; Kelly Chrivtian, gen sls mgr; Amy Nyquist, mktg dir & prom dir; Craig Roberts, chief of engrg.

KZSU(FM)—See Stanford

Paradise

KHSL-FM— Oct 15, 1983: 103.5 mhz; 1.6 kw. 1,250 ft TL: N39 57 29 W121 42 50. Stereo. Hrs opn: 24 2654 Cramer Ln., Chico, 95928-8838. Phone: (530) 345-0021. Fax: (530) 893-2121. Web Site: www.khsl.com. Licensee: Deer Creek Broadcasting LLC. (group owner; (acq 9-8-2004; grpsl). Population served: 310,000 Natl. Rep: Katz Radio. Law Firm: Haley, Bader & Potts. Format: Country. News staff: one; News: 6 hrs wkly. Target aud: 25-54; active. ♦Dino Corbin, VP, gen mgr & mktg mgr; Bill Meyer, gen sls mgr; Lisa Fitzgerald, prom mgr.

KKXX(AM)— September 1960: 930 khz; 1 kw-D, 37 w-N. TL: N39 43 37 W121 40 45. (CP: 500 w-N). Hrs opn: 1363 Longfellow, Chico, 95926-7319. Phone: (530) 894-7325. E-mail: info@kkxx.net Web Site: www.kkxx.net. Licensee: Butte Broadcasting Co. (acq 12-21-66). Population served: 300,000 Format: News/talk, relg. ♦Carl J. Auel, pres; Andrew Palmquist, gen mgr.

KZAP(FM)— June 4, 1977: 96.7 mhz; 1.5 kw. Ant 1,289 ft TL: N39 57 45 W121 42 40. Stereo. Hrs opn: 24
Simulcast with KPIG-FM Freedom 100%.
1459 Humbolt Rd., Suite D, Chico, 95928. Phone: (530) 899-3600. Fax: (530) 343-0243. E-mail: info@club967.com Web Site: www.club967.com. Licensee: Mapleton Communications LLC. Group owner: Regent Communications Inc. (acq 11-30-2006; grpsl). Population served: 250,000 Format: AAA :"Americana". Target aud: 25-49. ♦Dick Stein, gen mgr.

Pasadena

KAZN(AM)— Sept 12, 1942: 1300 khz; 5 kw-D, 1 kw-N, DA-2. TL: N34 09 38 W118 04 46. Stereo. Hrs opn: 24 747 E. Green, Suite 101, 91101. Phone: (626) 568-1300. Fax: (626) 568-3666. Web Site: www.mrbi.net. Licensee: Multicultural Radio Broadcasting Licensee LLC. Group owner: Multicultural Radio Broadcasting Inc. (acq 5-11-98; $12 million). Format: Chinese. News: 90 hrs wkly. Target aud: Chinese. ♦Arthur S. Liu, pres; Hsiang Lee, progmg dir.

KDIS(AM)— Feb 7, 1942: 1110 khz; 50 kw-D, 20 kw-N, DA-2. TL: N34 06 50 W117 59 51. Stereo. Hrs opn: 3321 S. La Cienega Blvd., Los Angeles, 90016. Phone: (310) 840-2800. Web Site: www.radiodisney.com/kdisam1110. Licensee: KABC-AM Radio Inc. Group owner: ABC Inc. (acq 12-19-00; $65 million). Population served: 10,162,200 Natl. Network: Radio Disney. Format: Children's progmg. ♦John Davison, pres & gen mgr; Bob Koontz, sls dir; Nelkane Benton, pub affrs dir; Bernice Wasserman, traf mgr.

***KPCC(FM)**— September 1957: 89.3 mhz; 680 w. 2,922 ft TL: N34 13 35 W118 03 58. Stereo. Hrs opn: 24 1570 E. Colorado Blvd., 91106. Phone: (626) 585-7000. Fax: (626) 585-7916. E-mail: mail@kpcc.org Web Site: www.kpcc.org. Licensee: Pasadena Area Community College District Board of Trustees. Population served: 342,000 Natl. Network: NPR, PRI, CBC Radio One. Format: New/talk/information. News: 6 hrs wkly. Target aud: 25-55. ♦Bill Davis, CEO & gen mgr; Doug Johnson, opns dir & opns mgr; Julie Allen, gen sls mgr; Craig Curtis, progmg dir; Paul Glickman, news dir.

KROQ-FM— 1974: 106.7 mhz; 5.6 kw. 2,000 ft TL: N34 11 47 W118 15 30. Hrs opn: 24 5901 Venice Blvd., Los Angeles, 90034. Phone: (323) 930-1067. Fax: (323) 931-1067. Web Site: www.kroq.com. Licensee: Infinity Broadcasting of Los Angeles Inc. Group owner: Infinity Broadcasting Corp. (acq 11-13-98; grpsl). Population served: 13,000,000 Format: Alternative rock. Target aud: 18-34. ♦Jeff Federman, gen mgr; Kevin Weatherly, progmg dir.

KSSE(FM)—See Arcadia

Paso Robles

KKAL(FM)— Nov 20, 1972: 92.5 mhz; 4.8 kw. Ant 1,486 ft TL: N35 21 40 W120 39 21. Stereo. Hrs opn: 24 3620 Sacramento Dr., Suite 204, San Luis Obispo, 93401. Phone: (805) 781-2750. Fax: (805) 781-2758. Web Site: www.kkalonline.com. Licensee: AGM California. Group owner: American General Media (acq 1997; $675,000). Population served: 285,000 Format: Hot adult contemp. News staff: one; News: 12 hrs wkly. Target aud: 18 plus. ♦Kathy Signorelli, gen mgr; Mark Tobin, gen sls mgr; Pepper Daniels, progmg dir.

KLUN(FM)— August 1995: 103.1 mhz; 1.2 kw. 761 ft TL: N35 38 45 W120 44 16. Stereo. Hrs opn: 24 312 E. Mill St., Suite 301, Santa Maria, 93454. Phone: (805) 928-9796. Fax: (805) 928-3367. E-mail: joseg@radiolazer.com Web Site: www.radiolazer.com. Licensee: Lazer Broadcasting Corp. (group owner; acq 8-7-00; $1.115 million with KLMM(FM) Morro Bay). Law Firm: Booth, Freret, Imlay & Tepper. Format: Rgnl Mexican music. Target aud: 18-49; general. ♦Alfredo Placencia, pres; Jose Guzman, gen mgr; Salvador Prieto, progmg dir; Bill Bordoux, chief of engrg.

KPRL(AM)— Oct 1, 1946: 1230 khz; 1 kw-U. TL: N35 39 15 W120 40 52. Hrs open: Box 7, 93447. Phone: (805) 238-1230. Fax: (805) 238-5332. E-mail: kprl@tcsn.net Web Site: www.kprl.com. Licensee: North County Communications LLC (acq 5-1-2003; $900,000). Population served: 112,000 Natl. Rep: Western Regional Broadcast Sales. Law Firm: Garvey, Schubert & Barer. Format: News/talk, sports. News staff: one; News: one hr wkly. Target aud: 25 plus. ♦Kevin Will, CEO, pres & opns mgr.

Patterson

KOSO(FM)—Licensed to Patterson. See Modesto

KTSE-FM— 1996: 97.1 mhz; 3 kw. 328 ft TL: N37 29 26 W121 13 16. Stereo. Hrs opn: 24 6820 Pacific Ave., Suite 3A, Stockton, 95207. Phone: (209) 474-0154. Fax: (209) 474-0316. Web Site: www.entravision.com. Licensee: Entravision Holdings LLC. Group owner: Entravision Communications Corp. (acq 7-28-00; grpsl). Format: Sp. ♦Lisa Sunday, CFO, gen mgr, opns mgr & gen sls mgr; Jorge Moreno, prom dir; Homero Campos, progmg VP; Jeff Paz, news dir; Paul Shin, chief of engrg; Valentina Rupio, mus dir & traf mgr.

Pebble Beach

***KSPB(FM)**— Sept 22, 1978: 91.9 mhz; 1 kw. 485 ft TL: N36 35 11 W121 55 21. Hrs open: 3152 Forest Lake Rd., 93953. Phone: (831) 626-5300. Phone: (831) 625-5078. Fax: (831) 625-5208. E-mail: webmaster@kspb.org Web Site: www.kspb.org. Licensee: Robert Louis Stevenson School. Format: Progsv. Spec prog: Black 18 hrs, oldies 4 hrs, reggae 2 hrs, hard rock 2 hrs wkly. ♦Matthew Arruda, gen mgr.

Pescadero

***KPDO(FM)**— 2006: 89.3 mhz; 100 w. Ant -141 ft TL: N37 15 23 W122 24 33. Hrs open: Box 25, Loma Mar, 94021. Phone: (530) 345-1031. Licensee: Pescadero Public Radio Service Inc. ♦Celeste Klienfelder, pres.

Petaluma

KTOB(AM)— Jan 10, 1950: 1490 khz; 1 kw-U. TL: N35 39 15 W120 40 52. Hrs open: 24 c/o Radio Station KRRS(AM), Box 2277, Santa Rosa, 95405. Secondary address: c/o Radio Station KRRS(AM), 1410 Neotomas Ave., Suite 104, Santa Rosa 95405. Phone: (707) 545-1460. Fax: (707) 545-0112. E-mail: krrs@sonic.net Web Site: www.moonradios.com. Licensee: Moon Broadcasting Licensee LLC. Group owner: Moon Broadcasting (acq 12-13-2001; $1.28 million). Population served: 1,000,000 Natl. Rep: Interep. Format: Sp rgnl. Target aud: 25-54; contemporary Hispanic families. ♦Abel DeLuna Sr., CEO; Abel DeLuna, pres; Arelia DeLuna, CFO; Maggie LeClerc, gen mgr; Benoit LeClerc, opns mgr.

Philo

***KZYX(FM)**— October 1989: 90.7 mhz; 3.41 kw. 1,686 ft TL: N39 01 22 W123 31 17. Hrs open: Box 1, 95466. Phone: (707) 895-2324. Fax: (707) 895-2451. Web Site: www.kzyx.org. Licensee: Mendocino County Public Broadcasting. Population served: 80,000 Natl. Network: NPR. Format: News, talk radio, div music. Target aud: General. Spec prog: Black 8 hrs, class 14 hrs, folk 8 hrs, gospel 2 hrs, jazz 11 hrs, blues 3 hrs. ♦Belinda Rawlins, gen mgr & dev dir; Burton Segall, opns dir; Vance Crowe, prom dir; Mary Aigner, progmg dir; Annie Esposito, news dir.

Piedmont

KPIG(AM)— May 1947: 1510 khz; 8 kw-D, 230 w-N, DA-2. TL: N37 49 02 W122 17 10. Hrs open: 24
Simulcast with KPIG-FM Freedom 100%.
28 Second St., Suite 501, San Francisco, 94105. Secondary address: 1110 Main St., Suite 16, Watsonville 95076. Phone: (415) 744-1510. Fax: (415) 495-1510. E-mail: sty@kpig.com Web Site: www.kpig.com. Licensee: Mapleton Communications LLC. (acq 7-27-2005; $5.1 million). Population served: 245,000 Natl. Rep: McGavren Guild. Law Firm: Levental, Senter & Lerman. Format: AAA, Americiana. ♦Ed Monroe, gen mgr; Frank Caprista, opns mgr; Mike Martindale, chief of engrg.

Pismo Beach

KXTZ(FM)— Dec 7, 1974: 95.3 mhz; 4.2 kw. 390 ft TL: N35 09 24 W120 38 11. Stereo. Hrs opn: 24 396 Buckley Rd., Suite 2, San Luis Obispo, 93401. Phone: (805) 786-2570. Fax: (805) 547-9860. Web Site: www.mapletoncommunications.com. Licensee: Mapleton Communications LLC. (group owner; (acq 5-23-2002; grpsl). Population served: 250,000 Law Firm: Haley, Bader & Potts. Format: Classic hits. News staff: one; News: one hr wkly. Target aud: 18-49. Spec prog: Talk one hr wkly. ♦Adam Nathanson, pres; Bill Heirendt, gen mgr; Drew Ross, progmg dir; David Atwood, news dir; Tom Hughes, chief of engrg.

Pittsburg

KATD(AM)— September 1949: 990 khz; 5 kw-U. TL: N38 04 49 W121 50 33. Stereo. Hrs opn: 24 145 Natoma St., 4th Fl., San Francisco, 94105. Phone: (415) 978-5378. Fax: (415) 978-5380. Licensee: Way Broadcasting Licensee LLC. Group owner: Multicultural Radio Broadcasting Inc. (acq 2-4-2004; grpsl). Population served: 250,000 Law Firm: Keck, Mahin & Cate. Format: Sp. News: 50 hrs wkly. Target aud: 25-54; middle upper income. ♦Arthur Liu, pres; Judy Re, gen mgr.

Placerville

KCCL(FM)— Dec 9, 1982: 92.1 mhz; 2.95 kw. Ant 331 ft TL: N38 45 31 W120 44 59. Stereo. Hrs opn: 24 298 Commerce Cir., Sacramento, 95815. Phone: (916) 576-7333. Fax: (916) 929-5330. Web Site: www.flash921.com. Licensee: First Broadcasting Sacramento Licensing LLC. Group owner: First Broadcasting Investment Partners LLC (acq 9-12-2003; $7.12 million). Format: Oldies. ♦Gary Lawrence, pres; Bob Dunphy, gen mgr; Kelly Galvez, prom dir.

Planada

KHTN(FM)— 1966: 104.7 mhz; 1.95 kw. Ant 2,080 ft TL: N37 32 01 W120 01 46. Stereo. Hrs opn: 24 510 W. 19th St., Merced, 95340. Phone: (209) 383-7900. Fax: (209) 723-8461. Web Site: www.hot1047fm.com. Licensee: Buckley Communications Inc. Group owner: Buckley Broadcasting Corp. (acq 9-21-95; $500,000). Natl. Rep: D & R Radio. Format: Contemp hit. Target aud: 21-34; women. ♦Mike McAdam, VP & gen mgr; Rene Roberts, opns mgr; Mike Peters, gen sls mgr.

Point Arena

KYOE(FM)— 2003: 102.3 mhz; 1.2 kw. Ant 1,417 ft TL: N38 53 44 W123 32 34. Hrs open: Box 366, 95468. Phone: (707) 882-2323. Fax: (707) 882-3258. Licensee: Del Mar Trust. Format: Country. ♦Karen J. Hay, gen mgr.

Point Reyes Station

***KWMR(FM)**— May 2, 1999: 90.5 mhz; 235 w. Ant 1,076 ft TL: N38 04 48 W122 51 57. Hrs open: 7 AM-12 AM Box 1262, 94956. Secondary address: 11431 State Rte. One #8 94956. Phone: (415) 663-8068. Fax: (415) 663-0746. E-mail: kay@kwmr.org Web Site: www.kwmr.org. Licensee: West Marin Community Radio Inc. Format: Community radio. ♦Kay Clements, gen mgr; Adrienne Pfeiffer, dev dir; Lyons Filmer, progmg dir; Andrew Shaw, news dir.

Pollock Pines

***KPPN(FM)**—Not on air, target date: unknown: 89.9 mhz; 100 w. Ant 482 ft TL: N38 44 18 W120 42 10. Hrs open: 95 La Jota Dr., Angwin, 94508. Phone: (707) 965-4155. Fax: (707) 965-4161. Web Site: www.thecandle.com. Licensee: Howell Mountain Broadcasting Co. ♦David Shantz, gen mgr.

Pomona

KAHZ(AM)— May 12, 1947: 1600 khz; 5 kw-U, DA-N. TL: N34 01 48 W117 43 35. Hrs open: 24 747 E. Green St., Floor 4, Pasadena, 91101. Phone: (626) 844-8882. Fax: (626) 844-2928. Web Site: www.mrbi.net. Licensee: Multicultural Radio Broadcasting Licensee LLC. Group owner: Multicultural Radio Broadcasting Inc. (acq 11-17-98; $7.55 million). Population served: 2,000,000 Format: Business news/talk, Chinese. Target aud: 30 plus; money oriented. ♦Arthur Liu, pres.

KWKU(AM)— Dec 23, 1960: 1220 khz; 250 w-U, DA-2. TL: N34 01 11 W117 43 03. (CP: 930 w-D). Hrs opn: 20
Rebroadcasts KWKW(AM) Los Angeles 60%.
363 S. Park Ave., Suite 105, 91766. Phone: (909) 865-3323. Fax: (909) 865-0342. Web Site: www.kwkuradio.com. Licensee: Lotus Communications Corp. (group owner; acq 2-00; $750,000). Population served: 3500000 Law Firm: Gammon & Grange. Format: Sp, news/talk, sports. News staff: one; News: 7 hrs wkly. Target aud: 24-64. Spec prog: Relg 15 hrs wkly. ♦Juan Rodriguez, gen mgr & progmg dir; Maria Diaz, news dir.

Port Hueneme

KCAQ(FM)—See Oxnard

KVTA(AM)— July 1958: 1520 khz; 10 kw-D, 1 kw-N, DA-2. TL: N34 10 02 W119 08 02. Hrs open: 2284 S. Victoria Ave., Suite 2 G, Ventura, 93003. Phone: (805) 289-1400. Fax: (805) 644-7906. Web Site: www.kvtaam1520.com. Licensee: Gold Coast Broadcasting LLC (group owner; acq 1996; $3.65 million with KCAQ(FM) Oxnard). Population served: 150,000 Law Firm: Leibowitz & Spencer. Format: News/talk. Target aud: 25-54. ♦Chip Ehrhardt, gen mgr; Tom Spence, progmg dir.

Porterville

KIOO(FM)— Aug 1, 1972: 99.7 mhz; 24 kw. 690 ft TL: N36 06 26 W119 01 45. Stereo. Hrs open: 24 617 W. Tulare Ave., Visalia, 93277. Phone: (559) 627-9710. Fax: (559) 627-1590. E-mail: raym@q97.com Web Site: www.997classicrock.com. Licensee: Buckley Broadcasting Corp. (group owner; acq 3-1-94; $360,000; FTR: 5-2-94). Population served: 24,000 Natl. Rep: D & R Radio. Format: Adult classic rock. News staff: one. Target aud: 25-44. ♦Rick Buckley, pres; Ray McCarty, VP & gen mgr; Tommy Del Rio, opns mgr.

KTIP(AM)— 1947: 1450 khz; 1 kw-U. TL: N36 05 44 W119 03 10. Hrs open: 24 1660 N. Newcomb, 93257. Phone: (559) 784-1450. Fax: (559) 784-2482. E-mail: live@ktip.com Web Site: www.ktip.com. Licensee: Mayberry Broadcasting Co. Inc. (acq 9-14-00; $130,000 for 51%). Population served: 400,000 Natl. Network: ABC, Westwood One. Rgnl rep: Rgnl Reps. Law Firm: Dow, Lohnes & Albertson. Format: News/talk. News staff: 2; News: 23 hrs wkly. Target aud: 25 plus. Spec prog: Health show one hr, loc travel one hr wkly, national health 3hrs.,Trader's Market. ♦Larry Stoneburner, pres; Larry & Mimi Stoneburner, gen mgr; Kent Hopper, chief of opns, progmg dir & news dir; Michael Partipilo, sls dir; Mimi Stoneburner, mktg dir; P.K. Whitmire, news dir; Ron Neil, chief of engrg; Janice Dawson, traf mgr.

Prunedale

***KLVM(FM)**— Feb 28, 1986: 89.7 mhz; 210 w. Ant 2,053 ft TL: N36 45 22 W121 30 05. Stereo. Hrs opn: 24
Rebroadcasts KLVR(FM) Santa Rosa 100%.
8145 Prunedale N. Rd., Salinas, 93907. Phone: (800) 525-5683. Fax: (831) 663-1663. E-mail: klove@klove.com Web Site: www.klove.com. Licensee: Prunedale Educational Foundation. Natl. Network: K-Love. Format: Adult contemp, Christian. Target aud: 25-35; Judeo-Christian female. ♦Dr. E.L. Moon, pres & gen mgr.

Quincy

KHGQ(FM)— 1997: 100.3 mhz; 900 w. -1,125 ft TL: N39 56 14 W120 56 51. Hrs open: 24 250 W. Nopah Vista Ave., Parump, NV, 89060. Phone: (775) 751-9709. Fax: (775) 751-3624. Licensee: Hilltop Church (acq 7-18-2005). Format: News/talk. ♦Keily Miller, gen mgr & opns mgr; Chris Compton, stn mgr; Randy Creff, chief of engrg.

***KJCQ(FM)**— 2006: 88.5 mhz; 790 w. Ant 2,194 ft TL: N40 14 00 W121 01 11. Hrs open: CSN International, 3232 MacArthur Blvd., Santa Ana, 92704. Phone: (714) 825-9663. Licensee: CSN International (group owner).

KNLF(FM)— June 10, 1996: 95.9 mhz; 500 w. -499 ft TL: N39 58 03 W120 53 34. Hrs open: 24 Box 117, 440 Lawrence St., 95971. Phone: (530) 283-4145. Fax: (530) 283-5135. Web Site: www.knlfradio.com. Licensee: New Life Broadcasting. Population served: 20,000 Format: Sports, talk, Christian, Contemp. News: 10 hrs wkly. Target aud: 18-54. ♦Ron Trumbo, pres.

KPCO(AM)— Aug 16, 1963: 1370 khz; 5 kw-D, 500 w-N, DA-2. TL: N39 56 54 W120 53 54. Stereo. Hrs opn: 24 395 Main St., 95971. Phone: (530) 256-2400. Fax: (530) 283-5117. Licensee: Tom F. Huth (acq 2-2-2006; $100,000). Population served: 52,000 Format: News/talk, hits from 40's, 50's, & 60's. ♦Bob Fink, pres; Bob Darling, VP & gen mgr; Will Taylor, news dir.

***KQNC(FM)**— 2005: 88.1 mhz; 500 w. Ant 1,135 ft TL: N39 56 14 W120 56 51. Hrs open:
Simulcast with KXJZ(FM) Sacramento 100%.
Capital Public Radio Inc., 7055 Folsom Blvd., Sacramento, 95826. Phone: (916) 278-8900. Fax: (916) 278-8989. E-mail: npr@csus.edu Web Site: www.csus.edu/npr. Licensee: California State University, Sacramento. Natl. Network: NPR. Law Firm: Duane Morris LLP. Format: Jazz, news & info. ♦Carl Watanabe, stn mgr & prom mgr; John Brenneise, opns mgr; Joe Barr, news dir, local news ed & relg ed; Jeff Browne, engrg dir; Michael Frost, traf mgr.

Rancho Cordova

KSTE(AM)— Apr 19, 1990: 650 khz; 21.4 kw-D, 920 w-N, DA-2. TL: N38 28 47 W121 16 38. Hrs open: 24 1440 Ethan Way, # 200, Sacramento, 95825. Phone: (916) 929-5325. Fax: (916) 929-2236. Web Site: www.talk650kste.com. Licensee: AMFM Broadcasting Licenses LLC. Group owner: Clear Channel Communications Inc. (acq 8-30-2000; grpsl). Natl. Network: ABC, Westwood One. Format: Talk. News staff: 4; News: 15 hrs wkly. Target aud: 25-54. ♦Jeff Holden, gen mgr; Alan Eisenson, opns dir & progmg dir.

Rancho Mirage

KMRJ(FM)— July 17, 1998: 99.5 mhz; 3 kw. 328 ft TL: N33 52 15 W116 13 37. Stereo. Hrs opn: 24 1061 S. Palm Canyon Dr., Palm Springs, 92264. Phone: (760) 778-6995. Fax: (760) 778-1249. E-mail: tom@m995.com Web Site: www.m995.com. Licensee: Mitchell Media Inc. Natl. Network: Westwood One. Natl. Rep: Katz Radio. Law Firm: Dickstein Shapiro Morin & Oshinsky. Format: Blend/old rock alternatives. News: 2 hrs wkly. Target aud: 25-54; mid age families, working adults. ♦Daniel P. Mitchell III, chmn & pres; Maurine B. Mitchell, CFO; Thomas Carr Mitchell, stn mgr; Carolina O'Connor, dev VP & prom mgr; Mark Moceri, sls dir & engr; Cynthia Butoc, traf mgr; Dwight Arnold, disc jockey; Geoff Allen, disc jockey; John Carey, disc jockey; Lord Tim Hudson, disc jockey.

Randsburg

***KGBM(FM)**— December 2001: 89.7 mhz; 2 kw. Ant 1,269 ft TL: N35 28 41 W117 41 58. Hrs open: 24 2351 Sunset Blvd., Suite 170-218, Rocklin, 95765. Phone: (916) 251-1600. Fax: (916) 251-1650. E-mail: info@air1.com Web Site: www.air1.com. Licensee: Educational Media Foundation. Group owner: EMF Broadcasting (acq 4-19-02). Natl. Network: Air 1. Law Firm: Shaw Pittman. Format: Contemp Christian. News staff: 3. Target aud: 18-35; Judeo-Christian female. ♦Richard Jenkins, pres; Mike Novak, VP; Ed Lenane, opns dir & news dir; Keith Whipple, dev dir; David Pierce, progmg mgr; Sam Wallington, engrg dir; Karen Johnson, news rptr; Marya Morgan, news rptr; Richard Hunt, news rptr.

Red Bluff

KALF(FM)— 1978: 95.7 mhz; 7 kw. 1,265 ft TL: N39 55 03 W122 40 12. Stereo. Hrs opn: 24 1459 Humboldt Rd., Suite D, Chico, 95928-9100. Phone: (530) 899-3600. Fax: (530) 343-0243. Web Site: www.kalf.com. Licensee: Mapleton Communications LLC. Group owner: Regent Communications Inc. (acq 11-30-2006; grpsl). Population served: 325,000 Natl. Rep: Christal. Law Firm: Smithwick & Belendiuk. Format: Country. News staff: one; News: 10 hrs wkly. Target aud: 25-54. ♦Dick Stein, gen mgr.

KBLF(AM)— 1946: 1490 khz; 1 kw-U. TL: N40 11 28 W122 12 54. Stereo. Hrs opn: 756 Hickory St., 96080. Phone: (530) 527-1490. Fax: (530) 527-3525. E-mail: kblfam@yahoo.com Web Site: www.kblf.com. Licensee: Tom Huth. Group owner: Huth Broadcasting (acq 8-11-98; $5,000). Population served: 50,000 Natl. Network: Westwood One, PRI. Format: Memories. Target aud: 35-64. Spec prog: Farm 5 hrs, Sp 4 hrs wkly. ♦Cal Hunter, gen mgr.

***KLVB(FM)**— November 1985: 102.7 mhz; 5.5 kw. Ant 1,414 ft TL: N40 20 41 W121 56 48. Stereo. Hrs opn: 24
Rebroadcasts KLVC(FM) Magalia 100%.
2351 Sunset Blvd., Suite 170-218, Rocklin, 95765. Phone: (916) 251-1600. Fax: (916) 251-1650. E-mail: klove@klove.com Web Site: www.klove.com. Licensee: Educational Media Foundation. Group owner: EMF Broadcasting (acq 1-11-2001; $750,000). Population served: 211,000 Natl. Network: K-Love. Natl. Rep: D & R Radio. Law Firm: Shaw Pittman. Format: Christian contemp. News staff: 3. Target aud: 25-44; Judeo Christian, female. ♦Richard Jenkins, pres; Mike Novak, VP; Keith Whipple, dev dir; David Pierce, progmg mgr; Ed Lenane, news dir; Sam Wallington, engrg dir; Arthur Vassar, traf mgr; Karen Johnson, news rptr; Marya Morgan, news rptr; Richard Hunt, news rptr.

***KTHM(FM)**—Not on air, target date: unknown: 90.7 mhz; 2.5 kw. Ant 328 ft TL: N40 12 31 W122 07 27. Hrs open: Box 981, 96080. Phone: (530) 347-0138. Licensee: Tehama County Community Broadcasters. ♦Erik Mathisen, pres.

Redding

KEWB(FM)—See Anderson

***KFPR(FM)**— Nov 17, 1994: 88.9 mhz; 750 w. 3,578 ft TL: N40 36 10 W122 38 58. Stereo. Hrs opn:
Rebroadcasts KCHO(FM) Chico 75%.
603 N. Market, 96003. Secondary address: Box 990061 95929. Phone: (530) 241-5246. Fax: (530) 241-5246. E-mail: npr@awwwsome.com Web Site: www.wfpr.org. Licensee: California State University, Chico Research Foundation. Population served: 200,000 Natl. Network: NPR, PRI. Law Firm: Cohn & Marks. Format: Var/div. Target aud: General. Spec prog: Sp 4 hrs wkly. ♦Jack Brown, gen mgr; Mike Birdsill, opns dir.

***KKRO(FM)**— Nov 15, 2002: 91.5 mhz; 370 w. Ant 1,315 ft TL: N40 54 53 W122 26 37. Stereo. Hrs opn: 24 2351 Sunset Blvd., Suite 170-218, Rocklin, 95765. Phone: (916) 251-1600. Fax: (916) 251-1650. E-mail: info@air1.com Web Site: www.air1.com. Licensee: Educational Media Foundation. Group owner: EMF Broadcasting. Population served: 211,000 Natl. Network: Air 1. Format: Contemp Christian. News staff: 3. Target aud: 27-33; Judeo-Christian female. ♦Richard Jenkins, pres; Mike Novak, VP; Keith Whipple, dev dir; David Pierce, progmg mgr; Ed Lenane, news dir; Sam Wallington, engrg dir; Karen Johnson, news rptr; Marya Morgan, news rptr; Richard Hunt, news rptr.

KLXR(AM)— August 1956: 1230 khz; 1 kw-U. TL: N40 33 14 W122 22 53. Hrs open: 24 1326 Market St., 96001. Phone: (530) 244-5082. Fax: (530) 244-5698. E-mail: mike@am1230klxr.com Licensee: Michael R. Quinn (acq 12-31-99; $125,000). Population served: 139,000 Natl. Network: Jones Radio Networks. Law Firm: Wombley, Carlyle, Sandbridge & Rice LLC. Wire Svc: AP Format: Music of Your Life. Target aud: 35 plus. Spec prog: Radio revisited- nostalgia, Swing & Cooking. ♦Mike Quinn, gen mgr, progmg dir, opns & sls; MIke Quinn, prom; Mike Quinn, adv.

KNCQ(FM)— Oct 29, 1985: 97.3 mhz; 100 kw. 3,569 ft TL: N40 36 10 W122 38 58. Stereo. Hrs opn: 1588 Charles Dr., 96003-1459. Phone: (530) 244-9700. Fax: (530) 244-9707. Web Site: www.q97country.com. Licensee: Results Radio of Redding Licensee LLC. Group owner: Fritz Communications Inc. Natl. Rep: D & R Radio. Format: Country. Target aud: 25-54. ♦Beth Tappan, gen mgr; Laurie Curto, gen sls mgr; Rick Healy, progmg dir; Patrick Johnson, news dir.

KNRO(AM)— 2001: 1670 khz; 10 kw-D, 1 kw-N. TL: N40 33 14 W122 22 53. Hrs open: 3360 Alta Mesa Dr., 96002-2831. Phone: (530) 226-9500. Fax: (530) 221-4940. Web Site: www.espn1670.com. Licensee: Mapleton Communications LLC. Group owner: Regent Communications Inc. (acq 11-30-2006; grpsl). Format: Sports. ♦Lisa Geraci, gen mgr; Dwayne Davis, progmg dir.

KQMS(AM)— Sept 14, 1954: 1400 khz; 1 kw-U. TL: N40 33 33 W122 19 42. Hrs open: 3360 Alta Mesa Dr., 96002. Phone: (530) 226-9500. Fax: (530) 221-4940. E-mail: lisag@reddingradio.com Licensee: Mapleton Communications LLC. Group owner: Regent Communications Inc. (acq 11-30-2006; grpsl). Population served: 150,000 Natl. Rep: McGavren Guild. Format: News/talk. ♦Adam Nathanson, pres; Lisa Geraci, gen mgr; Don Burton, opns mgr & progmg dir; Rich Kipp, gen sls mgr; Shellie Sutter, prom dir; Erin Myers, news dir.

KSHA(FM)—Co-owned with KQMS(AM). Sept 1, 1981: 104.3 mhz; 100 kw. 1,560 ft TL: N40 39 14 W122 31 12. Stereo. Format: Adult contemp. ♦Dennis Kennedy, progmg dir.

KRRX(FM)—Burney, May 1985: 106.1 mhz; 100 kw. 2,000 ft TL: N40 54 21 W121 49 38. Stereo. Hrs opn: 24 3360 Alta Mesa Dr., 96002. Phone: (530) 226-9500. Fax: (530) 221-4940. E-mail: krrx@reddingradio.com Web Site: 106x.com. Licensee: Mapleton Communications LLC. Group owner: Regent Communications Inc. (acq 11-30-2006; grpsl). Population served: 486,000 Law Firm: Grif Johnson. Format: Rock/AOR. News staff: 2. Target aud: 25-54; upscale. ♦Lisa Geraci, gen mgr; Clark Schopslin, progmg dir.

***KVIP(AM)**— Jan 4, 1970: 540 khz; 2.5 kw-D, 17 w-N. TL: N40 37 25 W122 16 49. Hrs open: 24 1139 Hartnell Ave., 96002. Phone: (530) 222-4455. E-mail: info@kvip.org Web Site: www.kvip.org. Licensee: Pacific Cascade Communications Corp. (acq 12-69). Population served: 200,000 Natl. Network: Moody, Salem Radio Network. Format: Inspirational, traditional Christian, talk. News staff: 2; News: 14 hrs wkly. Target aud: General. ♦David L. Morrow, VP; Steve Hafen, gen mgr, news dir & pub affrs dir; Ted Hering, progmg dir; Larry Cardoza, engrg dir; Paul Brown, chief of engrg.

KVIP-FM— Oct 19, 1975: 98.1 mhz; 30 kw. 1,710 ft TL: N40 36 10 W122 38 58. Stereo. 24 Web Site: www.kvip.org.250,000

Redlands

KCAL(AM)— April 1959: 1410 khz; 5 kw-D, 4 kw-N, DA-N. TL: N34 04 08 W117 12 06. Hrs open: 24 1950 S. Sunwest, San Bernardino, 92408. Phone: (909) 825-5020. Fax: (909) 884-5844. E-mail: edith@radiolazer.com Web Site: www.radiolazer.com. Licensee: Lazer Broadcasting Corp. (group owner; acq 8-7-01; $2.35 million). Population served: 680,000 Format: Sp. News staff: 2. Target aud: 18-49, 25-64; Mexican origin, Latin American. ♦Alfredo Plascencia, CEO, chmn & exec VP; Vicki Bails, VP & gen mgr; Armando Gutierrez, prom dir.

KCAL-FM— 1965: 96.7 mhz; 3 kw. 377 ft TL: N34 11 51 W117 17 10. Stereo. Hrs opn: 1940 Orange Tree Ln., Suite 200, 92374. Phone: (909) 793-3554. Fax: (909) 798-6627. Web Site: www.kcalfm.com. Licensee: Anaheim Broadcasting Corp. (group owner) Population served: 3,000,000 Natl. Rep: D & R Radio. Format: Adult rock. Target aud: 16-30. ♦Jeff Parke, VP & gen mgr; Steve Hoffman, opns mgr & progmg dir.

***KUOR-FM**— October 1966: 89.1 mhz; 35 w. Ant 2,781 ft TL: N34 11 47 W117 02 56. Hrs open:
Rebroadcasts KPCC(FM) Pasadena 100%.
1200 E. Colton Ave., 92374. Web Site: www.scpr.org. Licensee: University of Redlands. Natl. Network: NPR.

Redondo Beach

KDAY(FM)— Aug 4, 1961: 93.5 mhz; 3.4 kw. Ant 433 ft TL: N33 51 35 W118 20 56. Stereo. Hrs opn: 24 5055 Wilshire Blvd., Suite 720, Los Angeles, 90036. Phone: (323) 337-1600. Fax: (323) 337-1633. Web Site: www.935kday.com. Licensee: KDAY Licensing LLC. Group owner: Spanish Broadcasting System Inc. (acq 1-31-2006; $120 million with KDAI(FM) Ontario). Population served: 500,000 Format: Hip-hop. ♦Kimberly Fletcher, gen mgr; Lisa Alta Moreno, gen sls mgr; Anthony Acampora, progmg dir; Larry Slover, chief of engrg.

Redwood Valley

***KAIS(FM)**—Not on air, target date: unknown: 88.7 mhz; 15 w. Ant 1,935 ft TL: N38 56 54 W123 13 04. Hrs open: 2351 Sunset Blvd., Suite 170-218, Rocklin, 95765. Phone: (916) 251-1600. Fax: (916) 251-1650. Licensee: Educational Media Foundation. ♦Richard Jenkins, pres; Mike Novak, VP; Keith Whipple, dev dir; David Pierce, progmg mgr; Ed Lenane, news dir; Sam Wallington, engrg dir; Karen Johnson, news rptr; Marya Morgan, news rptr; Richard Hunt, news rptr.

Ridgecrest

KLOA(AM)— Dec 11, 1956: 1240 khz; 250 w-U. TL: N35 37 24 W117 41 10. Hrs open: 731 N. Balsam St., 93555. Phone: (760) 375-8888. Fax: (760) 371-1824. E-mail: radio@iwisp.com Web Site: www.kloaam.com. Licensee: Adelman Broadcasting Inc. (group owner). Population served: 50,000 Format: Oldies. ♦Robert Adelman, pres; Eric Kauffman, progmg dir; James Rowles, chief of engrg.

KLOA-FM— 1979: 104.9 mhz; 750 w. 1 ft TL: N35 37 24 W117 41 10. Stereo. Web Site: www.kloafm.com.70,000 Format: Country.

KWDJ(AM)— Apr 7, 1974: 1360 khz; 1 kw-D, 38 w-N. TL: N35 36 58 W117 38 35. Hrs open: 24 121 W. Ridgecrest Blvd., 93555-2606. Phone: (760) 384-4937. Fax: (760) 384-4978. Licensee: James & Donna Knudsen. (acq 9-30-91; $250,000 with co-located FM; FTR: 10-28-91) Population served: 60,000 Natl. Rep: Western Regional Broadcast Sales. Law Firm: Pepper & Corazzini. Format: Classical country. News staff: one; News: 11 hrs wkly. Target aud: 25-54; educated adults with high disposable income. ♦James L. Knudsen, pres & opns VP.

***KWTD(FM)**— May 2005: 91.9 mhz; 7.5 kw. Ant 1,243 ft TL: N35 28 39 W117 41 58. Stereo. Hrs opn:
Rebroadcasts KWTW(FM) Bishop 100%.
Box 637, Bishop, 93515. Phone: (760) 872-6215. Phone: (866) 466-5989. Fax: (760) 872-4155. Licensee: Living Proof Inc. Format: Christian. ♦Daniel McClenaghan, pres & gen mgr.

KZIQ-FM— Jan 1, 1978: 92.7 mhz; 3 kw. Ant -131 ft TL: N35 36 58 W117 38 35. Stereo. Hrs opn: 24 121 W. Ridgecrest Blvd., 93555-2606. Phone: (760) 384-4937. Licensee: James & Donna Knudsen. Population served: 300,000 Format: Lite adult contemp. News: 2 hrs wkly. ♦James Knudsen, stn mgr.

Rio Dell

***KNHT(FM)**— 1999: 107.3 mhz; 3.3 kw. Ant 1,702 ft TL: N40 30 03 W124 17 10. Hrs open: 5AM-2AM Jefferson Public Radio, 1250 Siskiyou Blvd., Ashland, OR, 97520. Phone: (541) 552-6301. Fax: (541) 552-8565. Web Site: www.ijpr.org. Licensee: The State of Oregon, acting by and through the State Board of Higher Education, for the benefit of Southern Oregon University. (acq 1-6-00). Natl. Network: NPR, PRI. Law Firm: Ernest Sanchez. Wire Svc: AP Format: Classical music, news. News staff: one; News: 35 hrs wkly. ♦Mitchell Christian, CFO; Ronald Kramer, CEO & gen mgr; Bryon Lambert, opns dir; Paul Westhelle, dev dir.

Rio Vista

***KRVH(FM)**— Nov 7, 1972: 101.5 mhz; 10 w. 60 ft TL: N38 09 17 W121 41 48. Hrs open: 410 S. 4th St., 94571. Phone: (707) 374-6336. Fax: (707) 374-6810. Licensee: River Delta Unified School District. Population served: 3,200 Format: CHR. Target aud: 13-19; young adult. ♦William Fulk, gen mgr.

Riverbank

KCBC(AM)— Apr 5, 1987: 770 khz; 50 kw-D, 1 kw-N, DA-2. TL: N37 47 51 W120 53 01. Stereo. Hrs opn: 24 10948 Cleveland Ave., Oakdale, 95361. Phone: (209) 847-7700. Fax: (209) 847-1769. E-mail: kcbcradio@surfside.net Web Site: www.770kcbc.com. Licensee: Kiertron Inc. (acq 12-30-92; $1 million; FTR: 1-25-93). Population served: 12,000,000 Format: Relg. News staff: one; News: 25 hrs wkly. Target aud: 25-49. ♦Don Crawford Sr., pres; Don Crawford Jr., gen mgr; Virginia Marsau, opns VP & opns mgr; Steve Minshall, chief of engrg.

Riverside

KDIF(AM)— Nov 15, 1941: 1440 khz; 1 kw-U. TL: N34 01 37 W117 21 27. Stereo. Hrs opn: 24 2030 Iowa Ave., Ste A, 92507. Phone: (951) 684-1991. Fax: (951) 274-4949. Licensee: Citicasters Licenses L.P. Group owner: Clear Channel Communications Inc. (acq 5-4-99; grpsl). Law Firm: Verner, Liipfert, Bernhard, McPherson & Hand. Format: Sp. News staff: 2; News: 7 hrs wkly. Target aud: 25-54; Hispanic. Spec prog: Hablando Claro one hr wkly. ♦Bob Ridzak, pres, gen mgr & opns dir.

KFRG(FM)—See San Bernardino

KGGI(FM)— Jan 23, 1965: 99.1 mhz; 2.55 kw. 1,843 ft TL: N34 14 04 W117 08 24. Stereo. Hrs opn: 24 2030 Iowa Ave., Suite A, 92507. Phone: (951) 684-1991. Fax: (951) 274-4911. Web Site: www.991kggifm.com. Licensee: AMFM Broadcasting Licenses LLC. Group owner: Clear Channel Communications Inc. (acq 8-30-2000; grpsl). Population served: 2,300,000 Natl. Rep: McGavren Guild. Format: CHR. Target aud: 18-49. ♦Bob Ridzak, gen mgr; Scott Welsh, gen sls mgr; Justin Garcia, mktg dir; Jesse Garcia, prom dir; Jesse Duran, progmg dir; Rich Mena, chief of engrg.

KLYY(FM)— Mar 17, 1959: 97.5 mhz; 72 kw. 1,571 ft TL: N33 57 57 W117 17 21. (CP: Ant 1,827 ft.). Stereo. Hrs opn: 24 5700 Wilshire Blvd., Suite 250, Los Angeles, 90036. Phone: (323) 900-6100. Fax: (323) 900-6127. Web Site: www.oye975.com. Licensee: Entravision Holdings LLC. Group owner: Entravision Communications Corp. (acq 4-20-00; grpsl). Population served: 6,500,000 Natl. Rep: Lotus Entravision Reps LLC. Format: Cumbia. Target aud: 18-49. ♦Jeff Liberman, VP; Karl Meyer, gen mgr; Nestor Rocha, progmg VP & progmg dir; Elias Autran, progmg dir; Eugene McAfel, chief of engrg & engr; Pam McCaffrey, traf mgr.

KPRO(AM)— June 22, 1957: 1570 khz; 5 kw-D, 194 w-N, DA-2. TL: N33 55 54 W117 23 47. Hrs open: 24 7351 Lincoln Ave., 92504. Phone: (951) 688-1570. Fax: (951) 688-7009. E-mail: kproval@aol.com Licensee: Impact Radio Inc. (acq 7-25-2005). Population served: 2,500,000 Law Firm: Pepper & Corazzini. Format: Relg. Target aud: General. ♦Ronnie Olenick, pres; Valorie Stitely, gen mgr & stn mgr.

***KSGN(FM)**— January 1970: 89.7 mhz; 3 kw. Ant 300 ft TL: N34 11 51 W117 17 10. Stereo. Hrs opn: 24 2048 Orange Tree Ln., Suite 200, Redlands, 92374. Phone: (909) 583-2150. Fax: (909) 583-2170. Web Site: www.ksgn.com. Licensee: Good News Radio. Population served: 2,000,000 Format: Christian educ, relg. News: 12 hrs wkly. Target aud: General; Christians & church goers. ♦Dennis Johnson, chmn, pres & CFO; Dawn Hibbard, gen mgr; Bryan O'Neal, progmg dir; Brandi Lanai, news dir; Bruce Potterton, chief of engrg; Heather Clough, traf mgr.

***KUCR(FM)**— October 1966: 88.3 mhz; 750 w. 291 ft TL: N33 58 11 W117 17 50. (CP: 150 w, ant 1,620 ft.). Stereo. Hrs opn: 691 Linden St., 92521. Phone: (951) 827-3737. Fax: (951) 827-3240. E-mail: kucr@citrus.ucr.edu Web Site: www.kucr.org. Licensee: The Regents of the University of California. Population served: 1,000,000 Format: Div, alternative rock. Spec prog: Black 18 hrs, class 14 hrs, jazz 6 hrs wkly. ♦Louis Van Den Berg, gen mgr; Dexter Thomas, asst music dir; Walter Douglas, prom dir, progmg dir & news dir; Jeff Armantrout, pub affrs dir; Bill Elledge, chief of engrg.

Rocklin

***KEBR(AM)**— July 27, 1988: 1210 khz; 5 kw-D, 500 w-N, DA-D. TL: N38 27 46 W121 07 49. Hrs open: 24 Family Stations Inc., 4135 Northgate Blvd., Suite 1, Sacramento, 95834-1226. Phone: (916) 641-8191. Fax: (916) 641-8238. Licensee: Family Stations Inc. (group owner) Population served: 254,413 Format: Relg. Target aud: General. ♦Harold Camping, pres; Peggy Renschler, stn mgr & opns mgr.

Rohnert Park

KMHX(FM)— Mar 4, 1986: 104.9 mhz; 6.6 kw. Ant 548 ft TL: N38 23 31 W122 40 40. Stereo. Hrs opn: 24 1410 Neotomas Ave., Suite 200, Santa Rosa, 95405. Phone: (707) 543-0100. Fax: (707) 543-1097. Web Site: www.mix1049.com. Licensee: Maverick Media of Santa Rosa Licensee LLC. Group owner: Fritz Communications Inc. (acq 6-5-2006; $7.7 million). Population served: 416,600 Law Firm: Covington & Burling. Format: Hot adult contemp. Target aud: 25-54. ♦Neysa Hinton, gen mgr & opns mgr.

Rosamond

KLKX(FM)— Sept 1, 1993: 93.5 mhz; 3 kw. Ant 207 ft TL: N34 51 03 W118 09 22. Hrs open: 24 Q-9, 570 East Ave., Palmdale, 93550. Phone: (661) 947-3107. Fax: (661) 272-5688. Web Site: www.935thequake.com. Licensee: High Desert Broadcasting LLC (group owner; acq 3-7-2002; grpsl). Population served: 300,000 Natl. Network: Westwood One. Law Firm: Arent, Fox, Kintner, Plotkin & Kahn. Format: Classic rock, news, interviews. Target aud: 25-54. ♦Nelson Rosse, gen mgr; Gary Wilson, opns mgr & progmg dir; Jeff McElfresh, mktg dir & prom dir; Amir Raheem, news dir.

KOSS(FM)— Mar 1, 1985: 105.5 mhz; 3 kw. 328 ft TL: N34 51 03 W118 09 22. Stereo. Hrs opn: 24 352 East Ave. K-4, Lancaster, 93535. Phone: (661) 942-1121. Fax: (661) 723-5512. Web Site: www.oasis1055.com. Licensee: CC Licenses LLC. Group owner: Clear Channel Communications Inc. (acq 11-21-2003; grpsl). Format: Adult contemp. News: 4 hrs wkly. ♦Larry Thornhill, gen mgr; Mark Mitchell, opns mgr; Shaun Palmeri, gen sls mgr.

Rosedale

***KOGR(FM)**—Not on air, target date: unknown: 88.9 mhz; 3.6 kw. Ant 1,653 ft TL: N35 03 00 W120 02 23. Hrs open: CSN International, 3232 W. MacArthur Blvd., Santa Ana, 92704. Phone: (714) 825-9663. Licensee: CSN International. (group owner).

Roseville

KFSG(AM)— 2001: 1690 khz; 10 kw-D, 1 kw-N. TL: N38 44 22 W121 12 50. Hrs open: 3463 Ramona Ave., Suite 15, Sacramento, 95826. Phone: (916) 456-3288. Fax: (916) 456-3324. E-mail: delatorre@mrbi.net Licensee: Way Broadcasting Licensee LLC (acq 6-13-00; grpsl). Format: Sp. ♦Rosario Delatorre, gen mgr; Yuri Reyes, progmg dir.

KLIB(AM)— Apr 1, 1968: 1110 khz; 5 kw-D, 500 w-N, DA-2. TL: N38 44 22 W121 12 48. Hrs open: 24 3463 Romona Ave., Suite 15,

Sacramento, 95826. Phone: (916) 456-3288. Fax: (916) 456-3324. E-mail: radiopoder@juno.com Licensee: Way Broadcasting Licensee LLC (acq 4-20-2000; grpsl). Format: Ethnic. Target aud: 18 plus; Hispanic.

KQJK(FM)—Licensed to Roseville. See Sacramento

Sacramento

KBMB(FM)— October 1996: 103.5 mhz; 6 kw. 312 ft TL: N38 33 59 W121 28 47. (CP: Ant 308 ft). Hrs opn: 24 1436 Auburn Blvd., 95815. Phone: (916) 646-4000. Fax: (916) 927-7376. Web Site: www.1035thebomb.com. Licensee: Entravision Holdings LLC. Group owner: Entravision Communications Corp. (acq 9-30-2004; $16.1 million). Format: Urban. Target aud: 18-49. ♦Larry LeManski, gen mgr; Larry Prater, gen sls mgr; Don Langford, prom dir; Patti Moreno, progmg dir; Paul Waegele, chief of engrg.

KCBC(AM)—See Riverbank

KCCL(FM)—See Placerville

KCTC(AM)— April 1945: 1320 khz; 5 kw-U, DA-2. TL: N38 42 42 W121 19 44. Stereo. Hrs opn: 24 5345 Madison Ave., 95841. Phone: (916) 334-7777. Fax: (916) 339-4572. Web Site: www.kctc.com. Licensee: Entercom Sacramento License LLC. Group owner: Entercom Communications Corp. (acq 10-17-97). Population served: 1,300,000 Natl. Network: ESPN Radio. Format: Sports. ♦John Geary, VP; David Lichtman, gen mgr; Brian Lopez, progmg dir.

KDND(FM)— Aug 1, 1945: 107.9 mhz; 50 kw. 403 ft TL: N38 42 38 W121 28 54. Stereo. Hrs opn: 24 5345 Madison Ave., 95841. Phone: (916) 334-7777. Fax: (916) 334-1092. Web Site: www.endonline.com. Licensee: Entercom Sacramento License L.L.C. Group owner: Entercom Communications Corp. (acq 6-3-97; $27.5 million). Natl. Rep: D & R Radio. Format: CHR. News staff: one. Target aud: 25-44. ♦David Lichtman, VP & gen mgr; John Geary, VP; John Greary, gen mgr & mktg mgr; Dan Mason, stn mgr & progmg dir; Butch Mitchell, sls dir; Sara McLure, gen sls mgr; Dayne Damme, prom dir; Kat Maudru, news dir; Mick Rush, engrg dir.

***KEAR-FM**— May 1997: 88.1 mhz; 8.4 kw. 994 ft TL: N38 14 50 W121 30 03. Stereo. Hrs opn: 24 4135 Northgate Blvd., Suite One, 95834-1226. Phone: (916) 641-8191. Fax: (916) 641-8238. Licensee: Family Stations Inc. (group owner) Population served: 25,000 Format: Relg. ♦Harold Camping, pres; Peggy Renschler, gen mgr & opns mgr.

KFBK(AM)— 1922: 1530 khz; 50 kw-U, DA-2. TL: N38 50 54 W121 28 58. Hrs open: 24 1440 Ethan Way, Suite 200, 95825. Phone: (916) 929-5325. Fax: (916) 925-6326. Web Site: www.kfbk.com. Licensee: AMFM Broadcasting Licenses LLC. Group owner: Clear Channel Communications Inc. (acq 8-30-2000; grpsl). Population served: 2,825,300 Wire Svc: PR Newswire Format: News/talk. News staff: 8; News: 45 hrs wkly. ♦Jeff Holden, pres; Alan Eisenson, opns dir, opns mgr, natl sls mgr & progmg dir; Sarah Simpson, sls dir; Amy Bingham, prom dir & prom mgr; Drew Sandsor, news dir; Zachary Rukstela, chief of engrg.

KGBY(FM)—Co-owned with KFBK(AM). 1946: 92.5 mhz; 50 kw. 499 ft TL: N38 42 26 W121 28 33. Stereo. 24 Fax: (916) 925-9292. Web Site: www.y92.com. Format: Lite rock. News: one hr wkly. ♦Amy Bingham, prom dir & prom mgr; Sonia Jimenez, progmg dir; Elizabeth Xiong, traf mgr.

KFIA(AM)—See Carmichael

KHTK(AM)— November 1926: 1140 khz; 50 kw-U, DA-2. TL: N38 23 34 W121 11 51. Hrs open: 24 5244 Madison Ave, 95841. Phone: (916) 338-9200. Fax: (916) 338-9208. Web Site: www.khtkam.com. Licensee: CBS Radio Holdings Inc. Group owner: Infinity Broadcasting Corp. (acq 11-13-98; grpsl). Population served: 286,000 Natl. Network: ESPN Radio, Fox Sports, Westwood One. Format: Sports. News: 10 hrs wkly. Target aud: 25-44. Spec prog: Farm 4 hrs wkly. ♦Steve Cottingin, gen mgr; Mike Remey, opns dir & progmg dir; Scott Marsh, gen sls mgr & sports cmtr.

KNCI(FM)—Co-owned with KHTK(AM). Feb 21, 1960: 105.1 mhz; 50 kw. 500 ft TL: N38 38 31 W121 05 25. Stereo. Web Site: www.kncifm.com.254,413 Format: Country. ♦Steve Cottingim, stn mgr; Mark Evans, opns dir & progmg dir; J.C. Swan, gen sls mgr; Walt Shaw, news dir; Bruce Hirsch, chief of engrg.

KIID(AM)— Aug 1, 1945: 1470 khz; 5 kw-D, 1 kw-N, DA-2. TL: N38 35 30 W121 27 47. Hrs open: 8265 Sierra College Blvd., Suite 312, Roseville, 95661. Phone: (916) 780-1470. Fax: (916) 780-1493. Web Site: www.radiodisney.com. Licensee: Radio Disney Group LLC. Group owner: ABC Inc. (acq 12-19-00; $3.31 million). Population served: 1,200,000 Natl. Network: Radio Disney. Format: Children. ♦Judy Remy, stn mgr.

KJAY(AM)— May 23, 1963: 1430 khz; 500 w-D, DA. TL: N38 29 39 W121 32 47. Hrs open: 6 AM-8 PM 5030 S. River Rd., West Sacramento, 95691. Phone: (916) 371-5101. Phone: (916) 371-5104. Fax: (916) 371-1459. Licensee: KJAY L.L.C. (acq 11-94). Population served: 2,500,000 Natl. Network: USA. Law Firm: Shaw Pittman. Format: International. Target aud: 25-64. ♦Trudi Powell, pres; Jerry Sieber, gen mgr.

KQJK(FM)—Roseville, June 1970: 93.7 mhz; 25 kw. Ant 328 ft TL: N38 44 22 W121 12 50. Stereo. Hrs opn: 24 5244 Madison Ave., 95841. Phone: (916) 338-9200. Fax: (916) 338-9155. Web Site: www.howard937.com. Licensee: Infinity Radio Holdings Inc. Group owner: Infinity Broadcasting Corp. (acq 11-13-98; grpsl). Population served: 150,000 Law Firm: Koteen & Naftalin. Format: Adult hits. News staff: one; News: 2 hrs wkly. Target aud: 25-54. ♦Bruce Hirsh, gen mgr & chief of engrg; Steve Cottingin, gen mgr & chief of engrg; J.C. Swan, gen sls mgr; Randy Craig, prom dir; Dave Sozinho, progmg dir; Jeff MacMurray, progmg dir; Walt Shaw, news dir.

KRXQ(FM)— Nov 1, 1959: 98.5 mhz; 50 kw. 500 ft TL: N38 38 35 W121 05 51. Stereo. Hrs opn: 24 5345 Madison Ave., 95841. Phone: (916) 334-7777. Fax: (916) 339-4277. Web Site: www.krxq.net. Licensee: Entercom Sacramento License L.L.C. Group owner: Entercom Communications Corp. (acq 7-28-98; grpsl). Population served: 300,000 Natl. Rep: McGavren Guild. Law Firm: Fletcher, Heald & Hildreth. Format: Active rock. News staff: one. Target aud: 25-40; males. Spec prog: Blues one hr wkly. ♦John Geary, pres & VP; David Lichtman, gen mgr; Jim Fox, progmg dir.

KSAC(AM)— 1938: 1240 khz; 1 kw-U. TL: N38 35 17 W121 28 05. Hrs open: 24 1017 Front St., 2nd Fl., 95814. Phone: (916) 553-3000. Fax: (916) 553-3013. Web Site: www.1240talkcity.com. Licensee: Diamond Broadcasting Group owner: Moon Broadcasting (acq 11-19-2004; $3 million). Population served: 1,000,000 Format: News/talk, sports. Target aud: 18 plus; Hispanics. ♦Paula Nelson, gen mgr; Frank Redfield, chief of opns.

KSEG(FM)— 1959: 96.9 mhz; 50 kw. 500 ft TL: N38 38 54 W121 28 40. Stereo. Hrs opn: 5345 Madison Ave., 95841-3141. Phone: (916) 334-7777. Fax: (916) 339-4280. Web Site: www.eagle969.com. Licensee: Entercom Sacramento License L.L.C. Group owner: Entercom Communications Corp. (acq 1-7-97; $45 million with KRAK(FM) Roseville). Population served: 1,500,000 Natl. Rep: D & R Radio. Format: Classic rock. Target aud: 18-49. ♦John Geary, pres & VP; David Lichtman, gen mgr; Curtis Johnson, progmg dir.

KSFM(FM)—Woodland, Feb 4, 1961: 102.5 mhz; 50 kw. 500 ft TL: N38 35 20 W121 43 30. Stereo. Hrs opn: 1750 Howe Ave., Suite 500, 95825. Phone: (916) 920-1025. Fax: (916) 929-5341. Web Site: www.ksfm.com. Licensee: Infinity Radio of Sacramento Inc. Group owner: Infinity Broadcasting Corp. (acq 11-13-98; grpsl). Population served: 254,413 Natl. Network: Westwood One. Law Firm: Leventhal, Senter & Lerman. Format: Rhythm and blues. Target aud: 12-44. ♦Steve Cottingin, gen mgr; Dell Goetz, gen sls mgr; Marcos Montes, prom dir; Byron Kennedy, progmg dir; Mike DaSilva, chief of engrg.

KSMH(AM)—See West Sacramento

KTKZ(AM)— 1952: 1380 khz; 5 kw-U, DA-2. TL: N38 33 19 W121 10 51. Hrs open: 1425 River Park Dr., Suite 520, 95815. Phone: (916) 924-0710. Fax: (916) 924-1587. Web Site: www.ktkz.com. Licensee: New Inspiration Broadcasting Co. Inc. Group owner: Salem Communications Corp. (acq 3-11-97; $1.5 million). Population served: 1,110,000 Format: Talk. Target aud: 35 plus; general. ♦James Rowten, gen mgr; Steve Gasser, opns mgr; Laurie Larson, progmg dir.

KWOD(FM)— Apr 1, 1957: 106.5 mhz; 50 kw. Ant 410 ft TL: N38 38 30 W121 05 25. Stereo. Hrs opn: 24 5345 Madison Ave., 95841. Phone: (916) 334-7777. Fax: (916) 339-5668. Web Site: www.kwod.net. Licensee: Entercom Sacramento License LLC. Group owner: Entercom Communications Corp. (acq 5-19-2003; $25 million). Population served: 4,500,000 Format: Alternative. Target aud: 18-49; new rock, mass appeal. ♦John Geary, VP; David Lichtman, gen mgr; Curtiss Johnson, stn mgr & progmg dir.

***KXJZ(FM)**— October 1964: 90.9 mhz; 50 kw. Ant 500 ft TL: N38 42 38 W121 28 54. Stereo. Hrs opn: 24 7055 Folsom Blvd., 95826. Phone: (916) 278-8900. Fax: (916) 278-8989. E-mail: npr@csus.edu Web Site: www.capradio.org. Licensee: California State University, Sacramento. Population served: 3,000,000 Natl. Network: NPR, PRI, AP Radio. Law Firm: Duane Morris, LLP. Format: Jazz, news & info. News staff: 5; News: 90 hrs wkly. Target aud: General; NPR listeners,ex. professionals, educators & administrators. Spec prog: World mus 2 hrs, blues 7 hrs wkly. ♦Carl Watanabe, stn mgr, progmg dir & progmg mgr; John Brenneise, opns mgr; Joe Barr, news dir; Jeff Browne, engrg dir.

***KXPR(FM)**— July 1, 1991: 88.9 mhz; 50 kw. Ant 492 ft TL: N38 16 25 W121 30 11. Stereo. Hrs opn: 24 7055 Folsom Blvd., 95826. Phone: (916) 278-8900. Fax: (916) 278-8989. E-mail: npr@csus.edu Web Site: www.capradio.org. Licensee: California State University, Sacramento. Population served: 3,000,000 Natl. Network: NPR, PRI. Law Firm: Duane Morris LLP. Format: Classical. Target aud: General; NPR Listeners, eg. professionals, educators, administrators. ♦Carl Watanabe, pres, stn mgr & progmg mgr; Tom Livingston, gen mgr; John Brenneise, opns mgr; Cheryl Dring, mus dir; Joe Barr, news dir; Jeff Brown, engrg dir; Michael Frost, traf mgr.

***KYDS(FM)**— Jan 24, 1979: 91.5 mhz; 410 w. 108 ft TL: N38 36 33 W121 21 38. Hrs open: 7 AM-3:30 PM 4300 El Camino Ave., 95821. Phone: (916) 971-7453. Fax: (916) 971-7429. E-mail: esantillanes@sanjuan.ed Licensee: San Juan Unified School District. Population served: 20,000 Format: Var. ♦Ed Santillanes, gen mgr.

KYMX(FM)— 1947: 96.1 mhz; 50 kw. 476 ft TL: N38 38 09 W121 33 11. Stereo. Hrs opn: 280 Commerce Cir., 95815. Phone: (916) 923-6800. Fax: (916) 922-2830. Web Site: www.kymx.com. Licensee: Infinity Radio Inc. Group owner: Infinity Broadcasting Corp. (acq 11-13-98; grpsl). Format: Adult contemp. News staff: one; News: 3 hrs wkly. Target aud: 25-54; Women ages 25-54. ♦Joel Hollander, pres; Jacque Tortorolli, CFO; Lisa Decker, sr VP; Dale Well, gen sls mgr; Rouna Daouk, prom dir; Bryan Jackson, progmg dir; Jacqui Freeman, news dir.

KZZO(FM)— October 1958: 100.5 mhz; 115 kw. 328 ft TL: N38 38 30 W121 05 25. Stereo. Hrs opn: 280 Commerce Cir., 95815. Phone: (916) 923-6800. Fax: (916) 922-2830. Web Site: www.radiozone.com. Licensee: Infinity Radio Inc. Group owner: Infinity Broadcasting Corp. (acq 11-13-98; grpsl). Natl. Rep: Christal. Format: Adult contemp. News staff: one; News: 3 hrs wkly. Target aud: 25-44. ♦Gavin Mahsman, gen sls mgr; Steve Cottingim, gen mgr, gen mgr & mktg mgr; Sam Sacco, prom dir; Byran Kennedy, progmg dir; Jeff Zuchowski, progmg dir.

Saint Helena

KVYN(FM)— November 1976: 99.3 mhz; 3 kw. 226 ft TL: N38 25 34 W122 19 33. (CP: Ant 259 ft.). Stereo. Hrs opn: 1124 Foster Rd., Napa, 94558. Phone: (707) 252-1440. Phone: (707) 258-1111. Fax: (707) 226-7544. Web Site: www.kvyn.com. Licensee: Wine Country Broadcasting Co. (acq 8-11-03; $3 million with KVON(AM) Napa). Population served: 110,000 Natl. Network: ABC. Natl. Rep: Christal. Law Firm: Robinson Silverman Pearce Aronsohn & Berman. Format: Adult contemp. Target aud: 25-45. Spec prog: Folk 2 hrs wkly. ♦Roger O. Walther, pres; Jeff Schechtman, gen mgr & progmg dir; Erica Pickett, prom dir; Megan Goldsby, news dir; Ben Webster, chief of engrg.

Salinas

KBOQ(FM)—See Monterey

KDBV(AM)— July 17, 1963: 980 khz; 10 kw-U, DA-2. TL: N36 43 58 W121 35 32. Hrs open: 24 604 E. Chapel St., Santa Maria, 93454.

Phone: (805) 406-9157. Licensee: Centro Cristiano Vida Abundante Inc. (acq 5-21-2004; $850,000). Population served: 183,085 Law Firm: Brown, Nietert & Kaufman. Format: Sp, Christian. Target aud: 18-49; Hispanics. ♦Ronald Stevens, gen mgr.

KDON-FM— December 1959: 102.5 mhz; 15 kw. Ant 2,371 ft TL: N36 45 23 W121 30 05. Stereo. Hrs opn: 903 N. Main St., 93906. Phone: (831) 755-8181. Fax: (831) 755-8193. Web Site: www.kdon.com. Licensee: CC Licenses LLC. Group owner: Clear Channel Communications Inc. (acq 9-22-97; grpsl). Natl. Rep: Christal. Format: Hip hop. News staff: one. Target aud: 18-54. ♦Rhonda McCormack, gen mgr & gen sls mgr; Nancy Nevarez, prom dir; Sam Diggedy, progmg dir; Jim Sohn, chief of engrg; Irma Gonzalez, traf mgr.

***KHDC(FM)**—Chualar, June 28, 1981: 90.9 mhz; 3 kw. 194 ft TL: N36 32 54 W121 26 34. Stereo. Hrs opn: 161 Main St., 93901. Phone: (831) 757-8039. Fax: (831) 757-9854. Web Site: www.radiobilingue.org. Licensee: Radio Bilingue Inc. (acq 11-86; $70,000; FTR: 5-12-86). Format: Multilingual, ethnic, Sp. ♦Delia Saldivar, gen mgr & progmg dir; Hugo Morales, CEO & gen mgr.

KION(AM)— 1947: 1460 khz; 10 kw-U, DA-1. TL: N36 43 59 W121 35 32. Hrs open: 24 903 N. Main St., 93906. Phone: (831) 755-8181. Fax: (831) 755-8193. Licensee: CC Licenses LLC. Group owner: Clear Channel Communications Inc. Population served: 500,400 Format: News/talk. ♦Mark Carbonera, progmg dir.

KNRY(AM)—See Monterey

KPRC-FM— Sept 16, 1964: 100.7 mhz; 1.4 kw. Ant 2,385 ft TL: N36 32 05 W121 37 14. Stereo. Hrs opn: 903 N. Main St., 93906. Phone: (831) 755-8181. Fax: (831) 755-8193. Web Site: salinas.lapreciosa.com/main.html. Licensee: CC Licenses LLC. (acq 9-22-97; grpsl). Format: Sp. ♦Rhonda mcCormack, gen mgr; Maggie Fernandez, prom dir; Alex Luca, progmg dir.

KRAY-FM— Dec 5, 1977: 103.5 mhz; 6 kw. Ant 512 ft TL: N36 42 32 W121 36 46. Stereo. Hrs opn: 24 Box 1939, 93902. Secondary address: 548 Alisal St. 93905. Phone: (831) 757-1910. Fax: (831) 757-8015. Licensee: Wolfhouse Radio Group Inc. (group owner; (acq 7-13-2001; grpsl). Population served: 560,000 Format: LaBuena. ♦Ramon Castro, gen mgr & chief of engrg.

KTGE(AM)— July 4, 1963: 1570 khz; 500 w-D. TL: N36 41 49 W121 37 22. (CP: 5 kw-D, 500 w-N, DA-2. TL: N36 39 38 W121 32 29). Hrs opn: 24 Box 1939, 93901. Secondary address: 548 Alisal St. 93905. Phone: (831) 757-5911. Fax: (831) 757-8015. E-mail: wolfhouseradio@yahoo.es Licensee: Wolfhouse Radio Group Inc. (group owner; (acq 7-13-2001; grpsl). Population served: 503,590 Format: Sp, rgnl Mexican. Target aud: Adults 24-54. ♦Ramon Castro, gen mgr & gen sls mgr; Vicente Romero, progmg dir.

KWAV(FM)—See Monterey

KYZZ(FM)— Mar 10, 1997: 97.9 mhz; 2.9 kw. Ant 112 ft TL: N36 36 32 W121 40 59. Hrs open: 24 Box 1391, Monterey, 93942. Phone: (831) 649-0969. Fax: (831) 642-9304. Licensee: Buckley Broadcasting Corp. of Salinas. (group owner; (acq 12-21-2005; $3 million). . Population served: 250,000 Format: CHR, pop. Target aud: 18-34; women. ♦Kathy Baker, gen mgr; Tommy Del Rio, progmg dir; Jeff Mitchell, sls.

San Ardo

***KBDH(FM)**— Jan 21, 2001: 91.7 mhz; 2.7 kw. Ant 1,781 ft TL: N35 57 06 W121 00 03. Stereo. Hrs opn:
Rebroadcasts KUSP(FM) Santa Cruz 100%.
203 8th Ave, Santa Cruz, 95062. Phone: (831) 476-2800. Fax: (831) 476-2802. Web Site: www.kusp.org. Licensee: Pataphysical Broadcasting Foundation. Natl. Network: NPR. Format: Div. News: 44 hrs wkly. ♦Terry Green, gen mgr & stn mgr; Paula Kenyon, dev dir; Rob Mullen, mus dir.

San Bernardino

KEZY(AM)— August 1947: 1240 khz; 1 kw-U. TL: N34 04 55 W117 18 17. Hrs open: 24 Box 500, Camarillo, 93011. Phone: (626) 356-4230. Fax: (626) 795-9185. Licensee: Hi-Favor Broadcasting LLC (group owner; acq 8-27-01; $4 million). Natl. Network: USA. Law Firm: Miller & Miller. Format: Sp, relg. News: 2 hrs wkly. Target aud: 30 plus. ♦Roland Hinz, pres; Sergio Martinez, stn mgr & gen sls mgr; Mary Guthrie, progmg dir.

KFRG(FM)— August 1974: 95.1 mhz; 50 kw. 489 ft TL: N34 11 51 W117 17 10. Stereo. Hrs opn: 24 900 E. Washington St., Suite 315, Colton, 92324. Phone: (909) 825-9525. Fax: (909) 825-0441. Web Site: www.kfrog.com. Licensee: Infinity Radio Inc. Group owner: Infinity Broadcasting Corp. (acq 11-13-98; grpsl). Population served: 2,100,000 Natl. Rep: McGavren Guild. Format: Country. News staff: one; News: 3 hrs wkly. Target aud: 25-54; dual income families. ♦Tom Hoyt, VP & gen mgr; Lee Douglas, opns mgr.

KKDD(AM)— 1947: 1290 khz; 5 kw-U, DA-2. TL: N34 07 27 W117 17 57. Hrs open: 24 2030 Iowa Ave., Suite A, Riverside, 92507. Phone: (951) 684-1991. Fax: (951) 274-4911. Web Site: www.radiodisney.com. Licensee: AMFM Broadcasting Licenses LLC. Group owner: Clear Channel Communications Inc. (acq 8-30-2000; grpsl). Population served: 106,869 Format: Radio Disney. News staff: one; News: 10 hrs wkly. ♦Bob Ridzak, gen mgr; Scott Welsh, progmg dir.

KLYY(FM)—See Riverside

KOLA(FM)— June 15, 1959: 99.9 mhz; 29.5 kw. 1,663 ft TL: N33 57 55 W117 16 59. Stereo. Hrs opn: 24 1940 Orange Tree Ln., Suite 200, Redlands, 92374. Phone: (909) 793-3554. Fax: (909) 798-6627. Web Site: www.imakolanut.com. Licensee: Anaheim Broadcasting Corp. (group owner; acq 1995; $5 million). Population served: 3,000,000 Format: Oldies. News: one hr wkly. Target aud: 25-54. ♦Jeff Parke, gen mgr & opns mgr; Gary Springfield, progmg dir.

KTDD(AM)— Oct 15, 1947: 1350 khz; 5 kw-D, 500 w-N, DA-2. TL: N34 05 37 W117 17 57. (CP: 600 w-N). Hrs open: 24 2030 Iowa Ave., Suite A, Riverside, 92507. Phone: (951) 684-1991. Fax: (951) 274-4911. Web Site: www.thetoad1350.com. Licensee: Citicasters Licenses L.P. Group owner: Clear Channel Communications Inc. (acq 5-4-99; grpsl). Population served: 1,343,000 Natl. Rep: McGavren Guild. Law Firm: Leventhal, Senter & Lerman. Format: Country. News: 15 hrs wkly. Target aud: 25-64. ♦Bob Ridzak, gen mgr; Bill Georgi, progmg dir.

KTIE(AM)— 1929: 590 khz; 1 kw-U, DA-2. TL: N34 04 18 W117 17 50. Hrs open: 24 992 Inland Ctr. Dr., 92408. Phone: (909) 885-6555 Ext. 101. Fax: (909) 383-8889. Web Site: www.ktie590.com. Licensee: Caron Broadcasting Inc. Group owner: Salem Communications Corp. (acq 8-29-2001; $7 million). Population served: 1,500,000 Format: News/talk. Target aud: 35-54 Male / Female; Upscale, educated, home owners, Business decision makers. ♦Terry Fahy, gen mgr; Jim Tinker, opns VP; Brad Anderson, gen sls mgr; Pamela Tyus, prom dir; Chuck Tyler, progmg dir; Craig Edwards, progmg dir.

***KVCR(FM)**— December 1953: 91.9 mhz; 3.8 kw. 1,605 ft TL: N33 57 57 W117 17 05. (CP.3.8. kw, ant 1,620 ft.). Stereo. Hrs opn: 24 701 S. Mt. Vernon Ave., 92410. Phone: (909) 384-4444. Fax: (909) 885-2116. E-mail: hometeam@kvcr.pbs.org Web Site: www.kvcr.org. Licensee: San Bernardino Community College Dist. Population served: 820,784 Natl. Network: NPR, PRI. Format: News, talk. News staff: 3; News: 50 hrs wkly. Target aud: General. ♦Larry Ciecalone, gen mgr; Steve Ward, opns mgr & progmg dir. Co-owned TV: *KVCR-TV affil

San Clemente

KWVE(FM)— Nov 16, 1971: 107.9 mhz; 530 w. Ant 3,792 ft TL: N33 42 40 W117 31 55. Stereo. Hrs open: 24 3000 W. MacArthur Blvd., Suite 500, Santa Ana, 92704. Phone: (714) 918-6207. Fax: (714) 918-6256. E-mail: kwve@kwve.com Web Site: www.kwve.com. Licensee: Calvary Chapel of Costa Mesa Inc. (acq 4-15-85). Population served: 15,030,000 Law Firm: Latham & Watkins. Format: Relg, Christian talk. News staff: 2; News: 3-5 hrs wkly. Target aud: General. Spec prog: Children 3 hrs wkly. ♦Charles W. Smith, pres; Jeffrey Dorman, gen mgr, opns dir & opns mgr.

San Diego

KBZT(FM)— Mar 6, 1960: 94.9 mhz; 21.8 kw. 710 ft TL: N32 50 21 W117 14 57. Stereo. Hrs opn: 1615 Murray Canyon Rd., Suite 710, 92108. Phone: (619) 297-3698. Fax: (619) 543-1353. Web Site: www.fm49sd.com. Licensee: Lincoln Financial Media Co. of California. Group owner: Jefferson-Pilot Communications Co. (acq 9-13-96; $25 million for stock). Format: Alt. Target aud: 18-49. ♦Darrel Goodin, gen mgr & stn mgr; Chris Turner, mktg dir; Copeland Isaac, prom mgr; John Marks, progmg dir; Eric Schecter, chief of engrg.

KCBQ(AM)— 1946: 1170 khz; 50 kw-D, 4.5 kw-N, DA-2. TL: N32 54 29 W116 54 34. Hrs open: 24 9255 Towne Centre Dr., Suite 535, 92121. Phone: (858) 535-1210. Fax: (858) 535-1212. E-mail: info@kcbq.com Web Site: www.kcbq.com. Licensee: New Inspiration Broadcasting Co. Inc. Group owner: Salem Communications Corp. (acq 8-23-2000; $5 million). Population served: 2,648,600 Natl. Network: AP Radio, ABC. Format: News/talk. News: 5 hrs wkly. Target aud: 35-64; baby boomers that grew up in the 50s & early 60s. ♦Dave Armstrong, gen mgr; Dawn Hockaday, prom dir; Heather Lloyd, opns mgr & progmg dir; Craig Caston, chief of engrg.

KECR(AM)—See El Cajon

KFMB(AM)— May 19, 1941: 760 khz; 50 kw-U, DA-N. TL: N32 50 32 W117 01 29. (CP: TL: N32 50 36 W117 01 28). Stereo. Hrs opn: 24 Box 85888, 92186. Secondary address: 7677 Engineer Rd. 92186. Phone: (858) 292-7600. Fax: (858) 279-7676. Web Site: www.760kfmb.com. Licensee: Midwest Television Inc. (group owner; (acq 4-17-2007 with co-located FM and KFMB-TV San Diego). Population served: 2,100,000 Natl. Network: CBS. Natl. Rep: McGavren Guild. Law Firm: Covington & Burling. Format: News/talk. News staff: 10. Target aud: 25-54. ♦August C. Meyer Jr., CEO; Ed Trimble, pres; Tracy D. Johnson, gen mgr & opns dir; John Marquiss, gen sls mgr; Dave Sniff, progmg dir; Fred D'Ambrosi, news dir; Dayna Monroe, pub affrs dir; Mike Sommerville, chief of engrg; Melanie Kartalija, rsch dir.

KFMB-FM— Sept 21, 1959: 100.7 mhz; 30 kw horiz, 26.5 kw vert. 620 ft TL: N32 50 17 W117 14 56. (CP: 38.4 kw, ant 536 ft.). Stereo. Fax: (858) 279-3380. Format: Hot adult contemp. ♦Gina Landau, gen sls mgr; Kim Leeds, prom mgr; Scott Sands, progmg dir; Jen Sewell, mus dir; Lynn Yuen, rsch dir. Co-owned TV: KFMB-TV affil.

KGB-FM—Listing follows KLSD(AM).

KHTS-FM—El Cajon, 1961: 93.3 mhz; 1.8 kw. Ant 1,885 ft TL: N32 41 48 W116 56 10. Stereo. Hrs opn: 24 9660 Granite Ridge Dr., 92123. Phone: (858) 292-2000. Fax: (858) 522-5707. Web Site: www.channel933.com. Licensee: Citicasters Licenses L.P. Group owner: Clear Channel Communications Inc. (acq 5-4-99; grpsl). Population served: 2,800,000 Law Firm: Hogan & Hartson. Format: CHR. Target aud: 18-34. ♦Bob Bolinger, gen mgr & sls dir; Brad SAmuel, gen sls mgr; Jean Arrollado, mktg dir; Geoff Alan, prom dir & prom mgr; Jimmy Steele, progmg dir & progmg mgr; Mary Ayala, news dir; John Rigg, chief of engrg.

KIFM(FM)— Feb 4, 1960: 98.1 mhz; 28 kw. 640 ft TL: N32 50 17 W117 14 56. Stereo. Hrs opn: 24 1615 Murray Canyon Rd., Suite 710, 92108-4321. Phone: (619) 297-3698. Fax: (619) 543-1353. Web Site: www.kifm.com. Licensee: Lincoln Financial Media Co. of California. Group owner: Jefferson-Pilot Communications Co. (acq 8-1-96; $28.75 million). Population served: 2,500,000 Natl. Rep: CBS Radio. Format: Smooth Jazz. News: 2 hrs wkly. Target aud: 25-54; upscale adults. ♦Darrel Goodin, gen mgr, gen sls mgr & stn mgr; Dave Saunders, gen sls mgr; Chris Turner, mktg dir; Copeland Isaac, prom mgr; John Marks, progmg dir; Eric Schecter, chief of engrg.

KIOZ(FM)— 1954: 105.3 mhz; 29 kw. 620 ft TL: N32 50 17 W117 14 56. Stereo. Hrs opn: 24 9660 Granite Ridge Dr., 92123. Phone: (858) 292-2000. Fax: (858) 715-3180. Web Site: www.rock1053.com. Licensee: Citicasters Licenses L.P. Group owner: Clear Channel Communications Inc. (acq 5-4-99; grpsl). Law Firm: Hagan and Hartson. Format: Rock. News staff: one; News: one hr wkly. Target aud: 18-49; upscale, well educated, young adult rock fans. ♦Bob Bolinger, gen mgr; Jay Isbell, prom dir; Shauna Moran, progmg dir.

KLNV(FM)— June 26, 1960: 106.5 mhz; 50 kw. 440 ft TL: N32 43 17 W117 04 11. Stereo. Hrs opn: 600 W. Broadway, Suite 2150, 92101. Phone: (619) 235-0600. Fax: (619) 744-4300. Web Site: www.lanueva1065.com. Licensee: HBC San Diego License Corp. Group owner: Univision Radio (acq 9-22-2003; grpsl). Wire Svc: UPI Format: Regional Mexican. Target aud: 18-49; young adult, contemp mus fans, upscale, well-educated. ♦Peter Moore, gen mgr; Michael Donavan, gen sls mgr; Nate Mendez, prom dir; Jose Gadea, progmg dir; Angel Ramos, chief of engrg.

KLQV(FM)— May 20, 1963: 102.9 mhz; 32 kw. 616 ft TL: N32 41 48 W116 56 10. Stereo. Hrs opn: 24 600 W. Broadway, Suite 2150, 92101. Phone: (619) 235-0600. Fax: (619) 744-4300. Web Site: www.univison.com. Licensee: HBC San Diego License Corp. Group owner: Univision Radio (acq 9-22-2003; grpsl). Population served: 2,700,000 Format: Sp. Target aud: 25-44. ♦Peter Moore, gen mgr.

KLSD(AM)— July 14, 1922: 1360 khz; 5 kw-D, 1 kw-N. TL: N32 43 49 W117 05 01. Hrs open: 9660 Granite Ridge Dr., 92123. Phone: (858) 292-2000. Fax: (858) 715-3372. E-mail: cliffalbert@clearchannel.com Web Site: www.am1360klsd.com. Licensee: Citicasters Licenses L.P. Group owner: Clear Channel Communications Inc. (acq 1999; grpsl). Population served: 697,027 Natl. Rep: CBS Radio. Format: Progressive talk. Target aud: 45-64; mature listeners with discretionary income. ♦Cliff Albert, stn mgr; Scotty Morache, gen sls mgr; Sherry Toennies, prom dir; Dave Mason, progmg dir; Mary Ayala, news dir; Bill Thompson, chief of engrg.

KGB-FM—Co-owned with KLSD(AM). 1956: 101.5 mhz; 50 kw. 500 ft TL: N32 43 49 W117 05 01. Stereo. 24 Web Site: www.101kgb.com.2,800,000 Law Firm: Hogan & Hartson. Format: Classic rock. Target aud: 25-54. ♦Dave Saunders, gen sls mgr; Jay Isbell, prom dir; Jim Richards, progmg dir.

KMYI(FM)— 1949: 94.1 mhz; 100 kw. 640 ft TL: N33 50 21 W117 14 57. Hrs open: 24 9660 Granite Ridge Dr., 92123. Phone: (858) 292-2000. Fax: (858) 715-3336. Web Site: www.my941.com. Licensee: Citicasters Licenses L.P. Group owner: Clear Channel Communications Inc. (acq 5-4-99; grpsl). Population served: 2,800,000 Law Firm: Hogan and Hartson. Format: Hot adult contemp. Target aud: 35-54; general, women. ♦Bob Bolinger, gen mgr; Jim Richards, opns VP; Kristin Ferguson, prom dir; Jimmy Steele, progmg dir.

KOGO(AM)— 1926: 600 khz; 5 kw-U, DA-1. TL: N32 43 17 W117 04 11. Hrs open: 24 9660 Granite Ridge Dr., 92123-2657. Phone: (858) 292-2000. Fax: (858) 715-3379. E-mail: kogo@clearchannel.com Web Site: www.kogo.com. Licensee: Citicasters Licenses L.P. Group owner: Clear Channel Communications Inc. (acq 1999; grpsl). Population served: 2,416,100 Natl. Network: ABC. Law Firm: Kaye, Scholer, Fierman, Hays & Handler. Wire Svc: UPI Format: News/talk. News staff: 10; News: 22 hrs wkly. Target aud: 25-54; issue oriented talk radio listeners. ♦Bob Bolinger, gen mgr; Sherry Toennies, prom dir; Cliff Albert, progmg dir; John Rigg, chief of engrg.

***KPBS-FM**— Sept 12, 1960: 89.5 mhz; 2.7 kw. Ant 1,804 ft TL: N32 41 53 W116 56 03. Stereo. Hrs opn: 24 5200 Campanile Dr., 92182-5400. Phone: (619) 594-8100. Phone: (619) 594-2580. Fax: (619) 594-3812. E-mail: letters@kpbs.org Web Site: www.kpbs.org. Licensee: San Diego State University. Population served: 2,212,500 Natl. Network: PRI, NPR. Law Firm: Dow, Lohnes & Albertson. Wire Svc: AP Format: News/talk, class. News staff: 15; News: 11 hrs wkly. Target aud: 35 plus. ♦Tom Karlo, CFO; Doug Myrland, gen mgr; John Decker, opns dir & progmg dir.

KPRI(FM)—See Encinitas

KSCF(FM)— 1965: 103.7 mhz; 36 kw. Ant 580 ft TL: N32 50 21 W117 14 57. Stereo. Hrs opn: 24 8033 Linda Vista Rd., 92111-5108. Phone: (858) 560-1037. Fax: (858) 571-0326. Web Site: www.radiosophie.com. Licensee: CBS Stations Inc. Group owner: Infinity Broadcasting Corp. (acq 8-7-2000; grpsl). Population served: 300,000 Natl. Rep: Christal. Format: Adult contemp. ♦Peter Schwartz, gen mgr; Charlie Quinn, opns mgr & progmg dir.

KSDO(AM)— October 1947: 1130 khz; 10 kw-U, DA-2. TL: N32 51 04 W117 57 51. Hrs open: 24 136 S. Oak Knoll Ave., Suite 302, Pasadena, 91101. Phone: (626) 356-4230. Fax: (626) 795-9185. Web Site: www.ksdo.com. Licensee: Hi-Favor Broadcasting LLC (group owner; acq 4-1-03; $10 million). . Population served: 260,000 Natl. Network: ABC. Format: Sp-relg. News staff: 10. Target aud: 25-54. ♦Mary Guthrie, gen mgr & progmg dir; Sean McCoy, gen mgr; Carlos Ortega, news dir; Rudy Agus, chief of engrg.

***KSDS(FM)**— December 1951: 88.3 mhz; 22 kw vert. Ant 246 ft TL: N32 48 19 W117 10 09. Stereo. Hrs opn: 24 1313 Park Blvd., 92101. Phone: (619) 388-3037. Fax: (619) 388-3928. E-mail: markd@jazz88online.org Web Site: www.jazzs88online.org. Licensee: San Diego Community College District. Population served: 696,679 Natl. Network: NPR. Law Firm: James S. Bubar. Format: Jazz, blues. News staff: one; News: 7 hrs wkly. Target aud: 25-65 plus; affluent, professional adults. ♦Mark DeBoskey, stn mgr; Ann Bauer, sls dir; Claudia Russell, progmg dir; Joe Kocherhans, progmg dir & mus dir; Bob Broms, mus dir & news dir; Larry Quick, chief of engrg.

KSON(AM)— 1946: 1240 khz; 1 kw-U. TL: N32 41 40 W117 07 17. Stereo. Hrs opn: 1615 Murray Canyon Rd., Ste 710, 92108-4321. Phone: (619) 291-9797. Licensee: Lincoln Financial Media Co. of California. Group owner: Jefferson-Pilot Communications Co. (acq 4-3-2006; grpsl). Population served: 760,000 Format: Chinese. ♦Darrel Goodin, gen mgr; Eric Schecter, chief of engrg.

KSON-FM— Jan 15, 1964: 97.3 mhz; 50 kw. Ant 440 ft TL: N32 43 13 W117 04 14. Stereo. Hrs opn: 1615 Murray Canyon Rd., Suite 710, 92108. Phone: (619) 291-9797. Fax: (619) 543-1353. Web Site: www.kson.com. Licensee: Lincoln Financial Media Co. of California. Group owner: Jefferson-Pilot Communications Co. (acq 2-7-85). Format: Country. ♦Darrel Goodin, gen mgr & opns dir; Chris Turner, mktg dir; Copeland Isaac, prom mgr; John Marks, progmg dir; Eric Schecter, chief of engrg.

KSSD(FM)—See Fallbrook

KURS(AM)— Nov 1, 1992: 1040 khz; 9.5 kw-D, 4.5 kw-N, DA-2. TL: N32 54 21 W116 55 40. Hrs open: 24 296 H St., Suite 300, Chula Vista, 91910. Phone: (619) 426-5645. Fax: (619) 425-1000. E-mail: jc@psnradio.com Web Site: www.espnradio620am.com. Licensee: Quetzal Bilingual Communications Inc. Population served: 580,000 Wire Svc: UPI Format: Oldies. ♦Jaime Bonilla, pres; Jose Carbajal, gen mgr.

KYXY(FM)— 1960: 96.5 mhz; 41 kw. Ant 540 ft TL: N33 52 00 W116 25 29. Stereo. Hrs opn: 24 8033 Linda Vista Rd., 92111-5108. Phone: (858) 571-7600. Fax: (858) 571-0326. Web Site: www.kyxy.com. Licensee: CBS Stations Inc. Group owner: Infinity Broadcasting Corp. (acq 8-7-00; grpsl). Population served: 300,000 Natl. Rep: Christal. Format: Soft rock. Target aud: 25-54; adults, women. ♦Peter Schwartz, gen mgr; Charlie Quinn, opns mgr & progmg dir.

XETRA(AM)—Tijuana, MEX) 1934: 690 khz; 77 kw-D, 50 kw-N. Hrs opn: 24 3400 W. Olive Ave., Suite 550, Burbank, 91505. Phone: (818) 559-2252. Fax: (818) 260-9961. Licensee: Clear Channel Communications Inc. (group owner; (acq 1999; grpsl). Natl. Network: ABC. Format: Adult standards. Target aud: 25-54; men. ♦Kevin McCarthy, exec VP & gen mgr; Dan Weiner, sls dir.

XETRA-FM— 1978: 91.1 mhz; 100 kw. Ant 1,000 ft Stereo. 24 Web Site: www.91x.com. Law Firm: Haley, Bader & Potts. Format: Alternative. News staff: one. Target aud: 18-49; very active, college educated, above market average income, single. ♦Mike Glickenhaus, exec VP; Bill Lipis, stn mgr & mktg dir; Tim Dukes, opns VP.

XHRM-FM—Tijuana, MEX) January 1981: 92.5 mhz; 100 kw. 548 ft Stereo. Hrs opn: 24 9660 Granite Ridge Dr., 92123. Phone: (858) 495-9100. Fax: 858) 522-5717. Web Site: www.magic925.com. Licensee: The Rivas Kaloyan Family. Format: Adult contemp, rhythmic oldies, today's rhythm and blues. Target aud: 18-49. ♦Mike Glickenhouse, gen mgr & disc jockey.

San Fernando

KBUA(FM)— Nov 14, 1958: 94.3 mhz; 3 kw. 95 ft TL: N34 17 03 W118 28 17. Stereo. Hrs opn: 24
Rebroadcasts KBUE(FM) Long Beach 100%.
1845 Empire Ave., Burbank, 91504. Phone: (818) 729-5300. Fax: (818) 729-5678. Web Site: www.aquisuena.com. Licensee: LBI Radio License Corp. Group owner: Liberman Broadcasting Inc. (acq 1997; $10.8 million). Population served: 1,500,000 Format: Mexican rgnl. ♦Lenard Liberman, pres & VP; Pepe Garza, stn mgr & progmg dir; Chris Buchanan, chief of engrg.

San Francisco

***KALW(FM)**— Mar 20, 1941: 91.7 mhz; 1.9 kw. 920 ft TL: N37 45 17 W122 26 44. Stereo. Hrs open: 24 500 Mansell, 94134. Phone: (415) 841-4121. Fax: (415) 841-4125. E-mail: kalwradio@yahoo.com Web Site: www.kalw.org. Licensee: San Francisco Unified School District. Population served: 2,500,000 Natl. Network: PRI, NPR. Format: NPR, BBC, Local. News: 68 hrs wkly. Target aud: General; news & info-oriented listeners. Spec prog: Diversified. ♦Matt Martin, gen mgr; William Helgeson, opns mgr; Dianne Keogh, dev dir.

KBWF(FM)— 1959: 95.7 mhz; 6.9 kw. Ant 1,500 ft TL: N37 41 23 W122 26 12. Stereo. Hrs opn: 24 201 3rd St., 94103. Phone: (415) 957-0957. Fax: (415) 356-8394. Web Site: www.957thewolf.com. Licensee: Bonneville Holding Co. Group owner: Bonneville International Corp. (acq 5-7-97). Format: Country. Target aud: 25-54. ♦John Parish, gen sls mgr.

KCBC(AM)—See Riverbank

KCBS(AM)— April 1909: 740 khz; 50 kw, DA-2. TL: N38 08 23 W122 31 45. Hrs open: 865 Battery St., 3rd Fl., 94111. Phone: (415) 765-4000. Fax: (415) 765-4080. Web Site: www.kcbs.com. Licensee: Infinity Broadcasting East Inc. Group owner: Infinity Broadcasting Corp. (acq 1996). Population served: 1,000,000 Natl. Network: CBS. Wire Svc: Reuters Wire Svc: Bay City News Service Wire Svc: U.S. Weather Service Format: News. Target aud: 25-54. ♦Steve DiNardo, gen mgr; Patrick Corr, stn mgr; Karl Isotalo, natl sls mgr; Louis Kaplan, progmg dir.

KLLC(FM)—Co-owned with KCBS(AM). Feb 1, 1948: 97.3 mhz; 82 kw. 1,014 ft TL: N37 50 57 W122 29 56. Stereo. 24 Phone: (415) 765-4097. Fax: (415) 765-4084. E-mail: studio@radioalice.com Web Site: www.radioalice.com. Natl. Rep: CBS Radio. Format: Modern adult contemp. Target aud: 18-54.

KCNL(FM)—See Sunnyvale

KDFC-FM— Sept 1, 1947: 102.1 mhz; 33 kw. 1,050 ft TL: N37 50 57 W122 29 56. Hrs open: 24 201 Third St., Suite 1200, 94103. Phone: (415) 764-1021. Fax: (415) 777-2291. Web Site: www.kdfc.com. Licensee: Bonneville Holding Co. Group owner: Bonneville International Corp. (acq 7-2-97; $54.5 million). Population served: 5,000,000 Natl. Rep: CBS Radio. Format: Class. Target aud: 25-54; educated, upscale. ♦Dwight Walker, gen mgr; Bill Lueth, opns mgr & progmg dir; Joe Schembri, gen sls mgr.

KEAR(AM)— Sept 24, 1924: 610 khz; 5 kw-U. TL: N37 50 58 W122 17 44. Stereo. Hrs opn: 290 Hegenberger Rd., Oakland, 94621-1436. Phone: (510) 568-6200. Fax: (510) 568-6190. Web Site: www.familyradio.com. Licensee: Family Stations Inc. Group owner: CBS Radio (acq 4-28-2005; $35 million). Population served: 5,330,000 Natl. Network: Family Radio. Format: Relg. Target aud: General. ♦Harold Camping, pres & gen mgr; Thad McKinney, gen mgr.

KEST(AM)— 1926: 1450 khz; 1 kw-U. TL: N37 46 41 W122 23 16. (CP: TL: N37 45 37 W122 22 56). Hrs opn: 24 145 Natoma St., Suite 400, 94105. Phone: (415) 978-5378. Fax: (415) 978-5380. Web Site: www.kestradio.com. Licensee: Multicultural Radio Broadcasting Licensee LLC. Group owner: Multicultural Radio Broadcasting Inc. (acq 3-31-98; grpsl). Population served: 7,500,000 Format: Personal growth talk, foreign language. News staff: one; News: 6 hrs wkly. Target aud: 25 plus. Spec prog: Chinese, Japanese, Indian, gospel, new age. ♦Arthur S. Liu, pres; Judy Re, gen mgr & opns mgr.

KFAX(AM)— 1925: 1100 khz; 50 kw-U, DA-1. TL: N37 37 56 W122 07 49. Hrs open: 24 Box 8125, Fremont, 94537. Phone: (510) 713-1100. Fax: (510) 505-1448. Web Site: www.kfax.com. Licensee: Golden Gate Broadcasting Co. Inc. Group owner: Salem Communications Corp. (acq 9-1-84). Population served: 9,000,000 Natl. Network: Salem Radio Network. Natl. Rep: Salem. Format: Relg, talk. News: 4 hrs wkly. Target aud: 25-54; females, families, college educated. Spec prog: Contemp Christian music 5 hrs weekly, children one hr wkly. ♦Ken Miller, gen mgr; Peter Thiele, opns dir & opns mgr; Kelly Christian, gen sls mgr; Amy Nyquist, mktg dir, prom dir & progmg dir; Craig Roberts, news dir & chief of engrg.

KFOG(FM)—Listing follows KNBR(AM).

KFRC-FM— 1958: 106.9 mhz; 80 kw. Ant 1,120 ft TL: N37 50 58 W122 29 56. Stereo. Hrs opn: 24 865 Battery St., 94111. Phone: (415) 392-1069. Fax: (415) 403-8554. Web Site: kfrc.com. Licensee: CBS Radio Stations Inc. (group owner; (acq 12-7-2005; $95 million). Natl. Rep: CBS Radio. Format: Classic hits. News staff: 2. ♦Ken Kohl, VP & gen mgr; Jason Insalaco, progmg dir.

KGO(AM)— Jan 8, 1924: 810 khz; 50 kw-U, DA-1. TL: N37 31 39 W122 06 05. Hrs open: 900 Front St., 94111-1450. Phone: (415) 398-5600. Fax: (415) 391-2795. Web Site: www.kgo.com. Licensee: Radio License Holding VIII LLC. Group owner: ABC Inc. (acq 6-12-2007; grpsl). Population served: 1,000,000 Natl. Rep: ABC Radio Sales, Interep. Law Firm: Wilmer, Cutler & Pickering. Wire Svc: Weather Wire Wire Svc: Bay City News Service Format: News/talk. Target aud: 25-54; general. ♦Michael Luckoff, pres & gen mgr; Jack Swanson, opns dir & progmg dir; Paul Hosley, news dir; Joe Talbot, chief of engrg.

KIOI(FM)— Oct 27, 1957: 101.3 mhz; 125 kw. Ant 1,160 ft TL: N37 41 24 W122 26 13. Stereo. Hrs opn: 24 340 Townsend St., Suite 5-101, 94107. Phone: (415) 538-1013. Fax: (415) 975-5573. Licensee: AMFM Broadcasting Licenses LLC. Group owner: Clear Channel Communications Inc. (acq 8-30-2000; grpsl). Population served: 7,000,000 Natl. Rep: Christal. Law Firm: Latham & Watkins. Format: Adult contemp. News staff: one; News: one hr wkly. Target aud: 25-54. ♦Anna Eppinger, gen sls mgr & rgnl sls mgr; Tony Ng, prom dir; Stacy Cunningham, progmg dir; John Scott, news dir; David Williams, chief of engrg.

KIQI(AM)— 1957: 1010 khz; 10 kw-D, 500 w-N, DA-2. TL: N37 49 33 W122 18 39. (CP: COL: Sunnyvale, 15 kw-D, 1.5 kw-N). Hrs opn: 24 145 Natoma St., 4th Fl., 94105. Phone: (415) 978-5378. Fax: (415) 978-5380. Licensee: Multicultural Radio Broadcasting Licensee LLC. Group owner: Multicultural Radio Broadcasting Inc. (acq 2-4-2004; grpsl). Population served: 715,674 Format: Sp, talk/news. News staff: 3. Target aud: 24-54. ♦Arthur Liu, pres & prom mgr; Judy Re, gen mgr & progmg dir.

KISQ(FM)— July 17, 1958: 98.1 mhz; 75 kw. Ant 1,015 ft TL: N37 51 04 W122 29 50. Stereo. Hrs opn: 8 AM-5:30 PM 340 Townsend St., 94107. Phone: (415) 975-5555. Fax: (877) 547-7329. Web Site: www.981kissfm.com. Licensee: AMFM Broadcasting Licenses LLC. Group owner: AMFM Inc. (acq 8-30-2000; grpsl). Population served: 715,674 Natl. Rep: McGavren Guild. Format: Oldies. Target aud: 25-54; women. Spec prog: Gospel 3 hrs wkly. ♦Anna Eppinger, gen mgr; Michael Erickson, progmg dir; David Williams, chief of engrg.

KITS(FM)— June 1, 1964: 105.3 mhz; 15 kw. 1,200 ft TL: N37 41 20 W122 26 07. Stereo. Hrs opn: 865 Battery St., 3rd Fl., 94111-1513. Phone: (415) 512-1053. Phone: (415) 402-6700. Fax: (415) 777-0608. Web Site: www.live105.com. Licensee: Infinity Broadcasting East Inc. Group owner: Infinity Broadcasting Corp. (acq 5-7-97). Population served: 715,684 Format: New rock alternative. ♦Steve DiNardo, gen mgr; Karl Isotalo, gen sls mgr; Dave Numme, progmg dir.

KKGN(AM)—See Oakland

KKSF(FM)— Nov 3, 1947: 103.7 mhz; 7.2 kw. 1,470 ft TL: N37 45 19 W122 27 05. Stereo. Hrs opn: 24 4th Fl., 340 Townsend St., 94107. Phone: (415) 975-5555. Fax: (415) 975-5573. Web Site: www.kksf.com. Licensee: AMFM Broadcasting Licenses LLC. Group owner: Clear Channel Communications Inc. (acq 8-30-2000; grpsl). Population served: 6,000,000 Format: Adult contemp, jazz/fusion, new age. News staff: one; News: 5 hrs wkly. Target aud: 25-49. ♦Anna Eppinger, gen mgr; Ken Jones, opns mgr & progmg dir; Ramona Gutierrez, prom dir.

KLOK(AM)—See San Jose

KMEL(FM)— Nov 30, 1960: 106.1 mhz; 69 kw. 1,290 ft TL: N37 41 24 W122 26 13. Stereo. Hrs opn: 24 340 Townsend St., 94107. Phone: (415) 538-1061. Fax: (415) 975-5573. Licensee: AMFM Broadcasting Licenses LLC. Group owner: Clear Channel Communications Inc. (acq 8-30-2000; grpsl). Natl. Rep: Christal. Format: CHR. ♦Anna Eppinger, gen mgr & gen sls mgr; Tony Ng, prom dir; Stacy Cunningham, progmg dir.

KMKY(AM)—See Oakland

KMVQ-FM— 1949: 99.7 mhz; 40 kw. Ant 1,299 ft TL: N37 41 15 W122 26 04. Stereo. Hrs opn: 865 Battery St., 3rd Fl., 94111. Phone: (415) 391-9970. Fax: (415) 951-2329. Web Site: www.movin997.com. Licensee: CBS Radio KFRC-FM Inc. (acq 1-96; grpsl). Format: Rhythmic adult contemp. ♦Doug Harvill, gen mgr; Mark Silverstein, gen sls mgr; Melanie Sherman, mktg dir; Mike Preston, progmg dir; Tim Jordan, progmg dir; Phil Lerza, chief of engrg.

KNBR(AM)— 1922: 680 khz; 50 kw-U. TL: N37 31 49 W122 16 29. Hrs opn: 24 55 Hawthorne St., Suite 1100, 94105. Phone: (415) 995-6800. Fax: (415) 995-6867. E-mail: sports@knbr.com Web Site: www.knbr.com. Licensee: KNBR Lico Inc. Group owner: Susquehanna Radio Corp. (acq 5-24-89; $17.5 million; FTR: 6-12-89). Population served: 715,674 Natl. Network: ABC, Westwood One. Format: Sports talk, personality. Target aud: 18 plus; predominently men. ♦Janet Magelby, gen mgr & prom dir; Tony Salvadore, VP & gen mgr; Lee Hammer, opns mgr & progmg dir; Dick Kelley, sls dir; Judi Ratto, gen sls mgr; Rich Zirkel, gen sls mgr.

KFOG(FM)—Co-owned with KNBR(AM). Mar 1, 1963: 104.5 mhz; 7.9 kw. 1,454 ft TL: N37 45 20 W122 27 05. Stereo. 24 Phone: (415) 817-5306. Fax: (415) 995-6922. E-mail: kfog@kfog.com Web Site: www.kfog.com. Licensee: KFFG Lico Inc. Format: Rock. News staff: one; News: 4 hrs wkly. ♦Tony Salvadore, gen mgr; Sheri Nelson, prom dir; David Benson, progmg dir.

KNEW(AM)—See Oakland

KNGY(FM)—Alameda, Aug 1, 1959: 92.7 mhz; 3.6 kw. Ant 420 ft TL: N37 47 54 W122 24 59. Stereo. Hrs opn: 400 Second St., Suite 300, 94107. Phone: (415) 356-1600. E-mail: sdillard@marathonmedia.com Web Site: www.energy92fm.com. Licensee: Flying Bear Licensing LLC (acq 12-23-2004; $33.64 million). Population served: 250,000 Format: Urban contemp. ♦Brad Bludau, gen mgr & gen sls mgr; Julie Johnson, mktg dir & prom dir; John Peake, progmg dir; Michelle Bayliss, pub svc dir.

KOHL(FM)—See Fremont

KOIT(AM)— 1926: 1260 khz; 5 kw-D, 1 kw-N. TL: N37 42 59 W122 23 38. Hrs opn: 201 3rd St., #1200, 94103-3143. Phone: (415) 777-0965. Fax: (415) 896-0965. Licensee: Bonneville Holding Co. Group owner: Bonneville International Corp. Population served: 5,200,000 Format: Adult contemp. Target aud: 25-54; upscale adults who earn an average of $30,000. ♦Sharon Warren, gen sls mgr; Maribeth Doran, natl sls mgr; Bill Conway, progmg dir; Julie Deppish, asst music dir.

KOIT-FM— 1959: 96.5 mhz; 33 kw. 1,410 ft TL: N37 45 20 W122 27 05. ♦Dwight Walker, gen mgr; Bill Conway, stn mgr & progmg dir; Scotty Bastable, gen sls mgr; Jude Heller, mktg dir & prom dir; Sherry Brown, news dir; Shingo Kamada, chief of engrg; Debi Mechanic, traf mgr.

***KPOO(FM)**— April 1971: 89.5 mhz; 270 w. 540 ft TL: N37 47 33 W122 24 52. Stereo. Hrs opn: Box 423030, 94142. Secondary address: 1329 Divisadero St. 94142. Phone: (415) 346-5373. Fax: (415) 346-5173. E-mail: 895fm@kpoofmsf.com Web Site: www.kpoofmsf.com. Licensee: Poor Peoples' Radio Inc. Population served: 715,674 Format: Div. ♦Terry Collins, pres; Jerome Parsons, gen mgr & progmg dir; Harrison Chastang, news dir; Marilyn Fowler, pub affrs dir; Dave Billicci, chief of engrg.

***KQED-FM**— June 1969: 88.5 mhz; 110 w. Ant 1,270 ft TL: N37 41 23 W177 26 12. Stereo. Hrs opn: 24
Rebroadcasts KQEI-FM North Highlands 98%.
2601 Mariposa St., 94110. Phone: (415) 553-2316. Fax: (415) 553-2241. Web Site: www.kqed.org. Licensee: KQED Inc. Population served: 455,300 Natl. Network: NPR, PRI. Wire Svc: Bay City News Service Format: News/talk. News staff: 9; News: 160 hrs wkly. Target aud: General. ♦Jack Clarke, chmn; Jo Anne Wallace, gen mgr; Monty Carlos, opns mgr; Traci A. Eckels, dev dir; Paul Ramirez, news dir. Co-owned TV: KQED(TV) affil.

KRZZ(FM)— February 1959: 93.3 mhz; 50 kw horiz, 47 kw vert. Ant 492 ft TL: N37 43 27 W122 07 07. Stereo. Hrs opn: 24 455 Market St., Suite 2300, 94105. Phone: (316) 832-9600. Licensee: KRZZ Licensing LLC. Group owner: Infinity Broadcasting Corp. (acq 12-7-2004). Format: rgnl Mexican. Target aud: 18-49; Hispanics. ♦Peter C. Remington, gen mgr.

KSFO(AM)— Aug 1, 1925: 560 khz; 5 kw-U, DA-N. TL: N37 44 44 W122 22 40. Hrs opn: 900 Front St., 94111. Phone: (415) 398-5600. Fax: (415) 658-5401. Web Site: www.ksfo560.com. Licensee: Radio License Holding VIII LLC. (acq 6-12-2007; grpsl). Population served: 715,674 Format: Talk/News. Target aud: 25-54. ♦Michael Luckoff, pres & gen mgr; Ken Berry, progmg dir; Paul Hosley, news dir.

KSOL(FM)— Dec 10, 1959: 98.9 mhz; 6 kw. 1,143 ft TL: N37 45 20 W122 27 05. Stereo. Hrs opn: 24 750 Battery St. # 200, 94111. Phone: (415) 989-5765. Fax: (415) 733-5766. Web Site: www.univision.com. Licensee: TMS License California Inc. Group owner: Univision Radio (acq 9-22-2003; grpsl). Format: Sp, Mexican rgnl. News staff: one. Target aud: 18-54. ♦Tony Perlongo, gen mgr; Luz Maria Rodriguez, mktg dir; Jose Luis Gonzalez, progmg dir.

KTRB(AM)— June 18, 1933: 860 khz; 50 kw-U, DA-2. TL: N37 35 34 W121 46 27. Stereo. Hrs opn: 1700 Montgomery St., Suite 490, 94111. Phone: (415) 513-3599. Fax: (415) 352-1800. Web Site: www.ktrb860.com. Licensee: Pappas Radio of California, a California L.P. Group owner: Pappas Telecasting Companies (acq 3-30-2000). Population served: 700,000 Format: Talk. ♦Harry J. Pappas, CEO; Jim P. Pappas, VP & gen mgr; Georgette Rodarakis, prom dir; Kevin Barrett, progmg dir; John Burger, chief of engrg.

***KUSF(FM)**— April 1964: 90.3 mhz; 3 kw. 300 ft TL: N37 46 34 W122 26 54. Stereo. Hrs opn: 24 2130 Fulton St., 94117-1080. Phone: (415) 386-5873. Fax: (415) 386-6469. E-mail: kusf@usfca.edu Web Site: www.kusf.org. Licensee: University of San Francisco. (acq 1973). Population served: 715,674 Format: Alternative mus, div, educ. Target aud: College educated, affluent, multicultural. Spec prog: Chinese 9 hrs, Fr 2 hrs, Turkish 2 hrs, It one hr, Pol one hr, Armenian one hr, Finnish one hr, Irish one hr wkly. ♦Steve Runyon, gen mgr; Trista Bernasconi, progmg dir; Bill Ruck, chief of engrg.

KYCY(AM)— 1947: 1550 khz; 10 kw-U, DA-2. TL: N37 31 49 W122 16 29. Stereo. Hrs opn: 24 865 Battery St., 2nd Fl., 4th floor, 94111. Phone: (415) 391-9970. Fax: (415) 397-7655. Licensee: Infinity Broadcasting East Inc. Group owner: Infinity Broadcasting Corp. (acq 12-14-00; grpsl). Population served: 715,674 Format: Talk. Target aud: 12 plus; affluent, home-owning, highly educated, business professionals. Spec prog: Jazz. ♦Greg Nemitz, gen mgr; Steven Page, stn mgr.

KYLD(FM)— Mar 12, 1958: 94.9 mhz; 35 kw. 1,290 ft TL: N37 41 22 W122 26 10. Stereo. Hrs opn: 340 Townsend St., 94107. Phone: (415) 975-5555. Fax: (415) 975-5573. Web Site: www.wild949.com. Licensee: AMFM Broadcasting Licenses LLC. Group owner: Clear Channel Communications Inc. (acq 8-30-2000; grpsl). Population served: 398,000 Format: Urban contemp, CHR. Target aud: 25-54. ♦Anna Eppinger, gen mgr; Jason Chin, prom dir; Jim Archer, progmg dir.

San Gabriel

KMRB(AM)— 1942: 1430 khz; 5 kw-U, DA-2. TL: N34 07 10 W118 04 57. Hrs open: 24 2nd Fl., 747 E. Green St., Pasadena, 91101. Phone: (626) 844-8882. Fax: (626) 792-8890. Licensee: Polyethnic Broadcasting Licensee LLC (acq 1994). Population served: 100,000 Format: Asian. Target aud: General. Spec prog: Thai 2 hrs, Ethiopian 2 hrs wkly. ♦Arthur Liu, pres; David Sweeney, exec VP & gen mgr; Kevin Chu, stn mgr; Katherine Lieu, gen sls mgr; Alan Mok, progmg dir; Hon Vu, chief of engrg.

San Jacinto

KRQB(FM)— Sept 23, 1990: 96.1 mhz; 1.4 kw. Ant 686 ft TL: N34 02 13 W116 58 07. Stereo. Hrs opn: 24 1845 Business Ctr. Dr., Suite 106, San Bernardino, 92408. Phone: (909) 663-1961. Fax: (909) 663-1996. Licensee: KWIE Licensing LLC. (acq 10-9-98). Population served: 1,000,000 Format: Rgnl Mexican. ♦Don McCoy, CEO; John Squyres, pres; Winter Horton, gen mgr; Cristian Garcia, opns mgr; Carlos Santos, gen sls mgr & prom mgr; Pepe Garza, progmg dir.

San Joaquin

KJZN(FM)— 1999: 105.5 mhz; 25 kw. Ant 328 ft TL: N36 36 28 W119 59 49. Hrs open: 24 1066 E. Shaw Ave., Fresno, 93710. Phone: (559) 230-0104. Fax: (559) 230-0177. Licensee: Wilks License Co.-Fresno LLC. Group owner: The Mondosphere Broadcasting Group. (acq 6-1-2005; grpsl). Natl. Rep: McGavren Guild. Format: Smooth jazz. Target aud: Adults 25-54. ♦Kevin O'Rorke, gen mgr; Stephen Mikal Brown, progmg dir.

San Jose

KAZA(AM)—See Gilroy

KBRG(FM)— Mar 4, 1963: 100.3 mhz; 14.5 kw. Ant 2,580 ft TL: N37 06 40 W121 50 34. Stereo. Hrs opn: 24 750 Battery St., Suite 200, San Francisco, 94111. Phone: (415) 989-5765. Fax: (415) 675-7126. Fax: (415) 733-5766. Web Site: www.univision.com. Licensee: Univision Radio License Corp. Group owner: Entravision Communications Corp. (acq 1-1-2006; $90 million with KLOK(AM) San Jose). Population served: 5,000,000 Format: Sp. ♦Tony Perlongo, gen mgr; Luz Maria Rodriguez, mktg dir; Ramon Lopez, progmg dir.

KEZR(FM)— July 3, 1967: 106.5 mhz; 50 kw. 430 ft TL: N37 21 43 W121 45 23. Stereo. Hrs opn: 190 Park Ctr. Plaza, Suite 200, 95113-2223. Phone: (408) 287-5775. Fax: (408) 293-3341. Web Site: www.todaysbestmix.com. Licensee: NM Licensing LLC. Group owner: Infinity Broadcasting Corp. (acq 12-6-2005; $80 million with KBAY(FM) Gilroy). Population served: 213,000 Natl. Rep: Christal. Format: Adult contemp. News staff: one; News: 4 hrs wkly. Target aud: 25-44. ♦John Leathers, gen mgr.

KFAX(AM)—See San Francisco

KFFG(FM)—See Los Altos

KKSF(FM)—See San Francisco

KLIV(AM)— 1946: 1590 khz; 5 kw-U, DA-N. TL: N37 19 45 W121 51 23. Hrs opn: 24 Box 995, 95108. Secondary address: 750 Story Rd. 95122. Phone: (408) 293-8030. Fax: (408) 293-6124. Web Site: www.kliv.com. Licensee: Empire Broadcasting Corp. (group owner; acq 7-1-67). Population served: 1,500,000 Natl. Network: CNN Radio. Natl. Rep: Christal. Wire Svc: AP Wire Svc: Bay City News Service Format: News. News staff: 8. Target aud: General. Spec prog: San Jose soccer earthquakes. ♦Robert S. Kieve, pres & gen mgr; George Sampson, progmg dir & news dir; Tina Ferguson, gen sls mgr & chief of engrg.

KLOK(AM)— Oct 19, 1946: 1170 khz; 50 kw-D, 5 kw-N, DA-2. TL: N37 18 41 W121 48 58. Hrs opn: 24 750 Battery St., Suite 200, San Francisco, 94111. Phone: (415) 362-1170. Fax: (415) 675-7126. Licensee: Univision Radio License Corp. Group owner: Entravision Communications Corp. (acq 1-1-2006; $90 million with KBRG(FM) San Jose). Population served: 5,262,000 Format: Cumbia. Target aud: 18-49. ♦Tony Perlongo, gen mgr.

***KMTG(FM)**— May 17, 1977: 89.3 mhz; 300 w. Ant -312 ft TL: N37 12 06 W121 51 42. Hrs open: 855 Linden Ave., 95126. Phone: (408) 535-6000. Licensee: San Jose Unified School District. Population served: 5,000 ♦Brent Pinkerton, gen mgr; Scott Dunson, opns dir.

KSJO(FM)— December 1946: 92.3 mhz; 50 kw. 464 ft TL: N37 12 33 W121 46 30. Stereo. Hrs opn: 1420 Koll Cir., Suite A, 95112. Phone: (408) 453-5400. Fax: (408) 452-1330. Web Site: www.ksjo.com. Licensee: Citicasters Licenses L.P. Group owner: Clear Channel Communications Inc. (acq 5-4-99; grpsl). Natl. Rep: McGavren Guild. Format: Sp. Target aud: 18-49; active adults. ♦Kim Bryant, gen mgr; Monica Novoa, progmg dir.

***KSJS(FM)**— Feb 22, 1963: 90.5 mhz; 235 w. 407 ft TL: N37 12 33 W121 46 30. Stereo. Hrs opn: 24 San Jose State Univ., Theater Arts Dept., HGH 126, 95192-0094. Phone: (408) 924-4549. Phone: (408) 924-4545. Fax: (408) 924-4558. E-mail: martinez@ksjs.org Web Site: www.ksjs.org. Licensee: San Jose State University. Population served: 2,000,000 Format: Diversified. Target aud: 18-34; students & community members. ♦Nick Martinez, gen mgr; Joey De la Plane, prom dir; Rob Soul, progmg dir.

KSJX(AM)— June 24, 1948: 1500 khz; 10 kw-D, 5 kw-N, DA-2. TL: N37 21 28 W121 52 17. Hrs open: 24 501 Wooster Ave., 95116. Phone: (408) 280-1515. Fax: (408) 280-1585. E-mail: ksjx1500@sbcglobal.net Licensee: Multicultural Radio Broadcasting Licensee LLC. Group owner: Multicultural Radio Broadcasting Inc. (acq 2-20-98; grpsl). Population served: 1,162,700 Format: Asian, Vietnamese. News staff: 2. Target aud: 25-54; managerial, professional, homeowners. Spec prog: Mandarin Chinese 10 hrs, Vietnamese. ♦Arthur Liu, pres; Andrea Yamazaki, gen mgr & stn mgr; Victor Nguyen, opns mgr.

KUFX(FM)— July 1, 1959: 98.5 mhz; 12.5 kw. 880 ft TL: N37 12 17 W121 56 56. Stereo. Hrs opn: 24 1420 Koll Cir., Suite A, 95112. Phone: (408) 452-5400. Fax: (408) 452-1330. Web Site: www.kfox.com. Licensee: Citicasters Licenses L.P. Group owner: Clear Channel Communications Inc. (acq 5-4-99; grpsl). Population served: 1,177,300 Natl. Rep: CBS Radio. Law Firm: Leventhal, Senter & Lerman. Format: Classic rock. Target aud: 18-49; general. ♦Kim Bryant, gen mgr; Laurie Roberts, progmg dir.

KVVF(FM)—See Santa Clara

KVVN(AM)—Santa Clara, Dec 18, 1964: 1430 khz; 1 kw-D, 2.5 kw-N, DA-2. TL: N37 19 47 W121 51 58. Hrs open: 24 1125 E. Santa Clara St., Suite 1, 95116. Phone: (415) 648-7980. Fax: (415) 695-9055. E-mail: sales@inlanguageradio.com Licensee: Urban Radio III L.L.C. Group owner: Inner City Broadcasting (acq 3-24-97; $2.2 million). Population served: 800,000 Law Firm: Koteen & Naftalin. Format: Vietnamese. News staff: one; News: 14 hrs wkly. Target aud: 23-34; Hispanic. ♦Harvey Stone, gen mgr; Andrew Luu, stn mgr; Phung Dang, opns dir; Paul Marks, chief of engrg.

KZSF(AM)— June 21, 1947: 1370 khz; 5 kw-U, DA-2. TL: N37 21 28 W121 52 17. Stereo. Hrs opn: 24 3031 Tisch Way, Suite 3, Plaza W., 95128. Phone: (408) 247-0100. Fax: (408) 247-4353. E-mail: lakaliente1370am@mexico.com Web Site: www.1370am.com. Licensee: Carlos A. Duharte (acq 7-31-01; $5 million). Population served: 4,500,000 Rgnl rep: Interep Format: Sp; regional Mexican. Target aud: 18-49. ♦Carlos A. Duharte, CEO, chmn, pres, sr VP & gen mgr.

San Luis Obispo

***KCBX(FM)**— July 25, 1975: 90.1 mhz; 5.3 kw. 1,420 ft TL: N35 21 38 W120 39 21. Stereo. Hrs opn: 24
Rebroadcasts KSBX(FM) Santa Barbara 100%.
4100 Vachell Ln., 93401. Phone: (805) 549-8855. Fax: (805) 781-3025. E-mail: kcbx@kcbx.org Web Site: www.kcbx.org. Licensee: KCBX Inc. Population served: 625,000 Natl. Network: NPR. Law Firm: Cohn & Marks. Format: Class, jazz, news. News: 32 hrs wkly. Target aud: General. Spec prog: Folk 15 hrs wkly. ♦Frank Lanzone, pres & gen mgr; Hank Hadley, opns mgr; Paul Severtson, dev dir; Guy Rathbun, progmg dir.

***KCPR(FM)**— 1968: 91.3 mhz; 2 kw. -350 ft TL: N35 17 58 W120 40 26. Stereo. Hrs opn: 24 Graphic Arts, California Polytechnic State Univ., 93407. Phone: (805) 756-5277. Phone: (805) 756-2965. Web Site: www.kcpr.org. Licensee: California Polytechnic State University. Population served: 40,000 Format: Div. News: 4 hrs wkly. Target aud: General; Cal Poly students, San Luis Obispo community. Spec prog: Sp 3 hrs, metal 3 hrs, blues 3 hrs. ♦Tyler Johnson, gen mgr.

KIQO(FM)—See Atascadero

KJDJ(AM)— Feb 8, 1988: 1030 khz; 2.5 kw-D, 700 w-N. TL: N35 17 58 W120 40 24. Hrs open: 604 E. Chapel St., Santa Maria, 93454. Phone: (805) 928-1030. E-mail: oracion@radiovidaabundante.com Web Site: www.radiovidaabundante.com. Licensee: Padre Serra Communications Inc. (acq 4-94). Population served: 250,000 Format: Religious. ♦Manuel Salvador, gen mgr; Manny Aram, progmg dir.

KKJG(FM)— Jan 1, 1984: 98.1 mhz; 3.6 kw. 1,624 ft TL: N35 21 37 W120 39 18. Stereo. Hrs opn: 3620 Sacramento Dr., Suite 204, 93401. Phone: (805) 781-2750. Fax: (805) 781-2758. Web Site: www.jugcountry.com. Licensee: AGM San Luis Obispo L.P. Group owner: American General Media (acq 7-1-97; $1.5 million). Format: Country. Target aud: 25-54. ♦Kathy Signorelli, gen mgr; Pepper Daniels, progmg dir.

KKJL(AM)— Feb 6, 1960: 1400 khz; 1 kw-U. TL: N35 15 51 W120 39 56. Hrs open: 24 Box 1400, 93406. Secondary address: 51 Zaca Ln., Suite 90 93401. Phone: (805) 543-9400. Fax: (805) 543-0787. E-mail: info@kkjl1400.com Web Site: www.kkjl1400.com. Licensee: San Luis Obispo Broadcasting Inc. (acq 9-9-86). Population served: 38,000 Natl. Network: CNN Radio. Law Firm: Leventhal, Senter & Lerman. Format: Adult standards, sports. Target aud: 35 plus; adults males & females. Spec prog: SF Giants, SF 49ers, LA Lakers. ♦Guy Hackman, pres & gen mgr; Kyle Ronemus, VP & stn mgr; Mary S. Brown, opns mgr.

***KLFF-FM**— Sept 26, 1995: 89.3 mhz; 4.4 kw. 1,430 ft TL: N35 21 37 W120 39 17. Stereo. Hrs opn: 24 560 Higuera St., Suite G, 93401. Phone: (805) 541-4343. Fax: (805) 541-9101. E-mail: info@klife.org Web Site: www.klife.org. Licensee: Logos Broadcasting Corp. Population served: 250,000 Natl. Network: Salem Radio Network. Law Firm: Joseph E. Dunne III. Format: Christian hit music. Target aud: 18-34; Christians. ♦Dan M. Lemburg, pres; Dr. Daniel Woods, CFO; Jon Fugler, gen mgr; Noonie Fugler, prom dir.

***KLVH(FM)**— Mar 25, 1999: 88.5 mhz; 3 kw. Ant 1,401 ft TL: N35 21 38 W120 39 21. Hrs open: 24 2351 Sunset Blvd., Suite 170-218, Rocklin, 95765. Phone: (916) 251-1600. Fax: (916) 251-1650. E-mail: klove@klove.com Web Site: www.klove.com. Licensee: Educational Media Foundation. Group owner: EMF Broadcasting (acq 5-12-99). Population served: 315,000 Natl. Network: K-Love. Law Firm: Shaw Pittman. Format: Contemp Christian music. News staff: 3. Target aud: 25-44; Judeo Christian, female. ♦Richard Jenkins, pres; Mike Novak, VP; Keith Whipple, dev dir; David Pierce, progmg mgr; Ed Lenane, news dir; Sam Wallington, engrg dir; Karen Johnson, news rptr; Marya Morgan, news rptr; Richard Hunt, news rptr.

KSLY-FM— December 1959: 96.1 mhz; 3.4 kw. Ant 1,686 ft TL: N35 21 37 W120 39 18. Stereo. Hrs opn: 51 Zaca Ln., Suite 110, 93401. Phone: (805) 545-0101. Fax: (805) 541-5303. Web Site: www.ksly.com. Licensee: CC Licenses LLC. Group owner: Clear Channel Communications Inc. (acq 10-2000; grpsl). Format: Country. ♦Rich Hawkins, gen mgr; Pattie Wagner, sls dir; Teresa Lara, prom dir; Andy Morris, progmg dir; Ben Grenaway, news dir.

KURQ(FM)—See Grover Beach

KVEC(AM)— May 1937: 920 khz; 1 kw-D, 500 w-N. TL: N35 17 58 W120 40 24. Hrs open: 24 51 Zaca Ln., Suite 100, 93401. Phone: (805) 545-0101. Fax: (805) 541-5303. Web Site: www.920kvec.com. Licensee: AMFM Radio Licenses LLC. Group owner: Clear Channel Communications Inc. (acq 12-12-2000; $950,000 including five-year noncompete agreement). Population served: 300,000 Natl. Network: ABC, Fox News Radio. Format: News/talk, info. News staff: 4; News: 45 hrs wkly. Target aud: 35 plus; affluent decision & newsmakers, sports fans, business owners & retirees. Spec prog: Dodgers baseball, NFL/NCAA football, finance, senior focus, health, real estate. ♦Rich Hawkins, gen mgr.

KXTK(AM)—Arroyo Grande, June 29, 1962: 1280 khz; 10 kw-D, 2.5 kw-N, DA-2. TL: N35 08 44 W120 31 15. Hrs open: 24 Box 14910, 93406. Phone: (805) 547-1280. Fax: (805) 543-1508. E-mail: sports@espnradio1280.com Web Site: espnradio1280.com. Licensee: Pacific Coast Media LLC (acq 10-20-2004; $700,000). Population served: 400,000 Natl. Network: ESPN Radio, Westwood One. Rgnl. Network: Jones Satellite Radio. Format: Sports. Target aud: 25 plus. ♦Mike Chellsen, gen mgr, gen sls mgr & progmg dir; Bill Bordeaux, engr.

KYNS(AM)— Dec 13, 1949: 1340 khz; 790 w-U. TL: N35 14 03 W120 40 33. Hrs opn: 24 396 Buckley Rd., Suite 2, 93401. Phone: (805) 786-2570. Fax: (805) 547-9860. Web Site: www.mapletoncommunications.com.com. Licensee: Mapleton Communications LLC (group owner; acq 3-19-03; $370,000). Population served: 37,500 Format: Progressive news/talk. ♦Adam Nathanson, pres; Nancy Leichter, gen mgr.

KZOZ(FM)— 1962: 93.3 mhz; 29.5 kw. 1,470 ft TL: N35 21 38 W120 39 21. Stereo. Hrs opn: 3620 Sacramento Dr., Suite 204, 93401. Phone: (805) 781-2750. Fax: (805) 781-2758. E-mail: sales@americangeneralmedia.com Web Site: www.kzoz.com. Licensee: AGM California. Group owner: American General Media (acq 6-89; grpsl). Population served: 191,000 Format: Classic rock, AOR. ♦Bill Heirendt, gen mgr, opns dir & gen sls mgr; Kathy Signorelli, gen mgr; David Atwood, progmg dir.

San Marcos-Poway

KPRZ(AM)— 1986: 1210 khz; 20 kw-D, 5 kw-N, DA-2. TL: N33 04 12 W117 11 35. Hrs open: 24 9255 Towne Centre Dr., Suite 535, San Diego, 92121. Phone: (858) 535-1210. Fax: (858) 535-1212. E-mail: kprz@kprz.com Web Site: www.kprz.com. Licensee: New Inspiration Broadcasting Co. Inc. Group owner: Salem Communications Corp. (acq 1986). Population served: 2,700,000 Natl. Network: Salem Radio Network. Format: Christian, talk. News: 15 hrs wkly. Target aud: 25-54; conservative, pro-family. Spec prog: Sp 22 hrs wkly. ♦Edward G. Astinger III, CEO & pres; David Evans, CFO; Dave Armstrong, gen mgr; Dawn Hockaday, prom dir; Heather Lloyd, opns mgr & progmg dir; Craig Caston, chief of engrg.

San Martin

KZSJ(AM)— November 1995: 1120 khz; 5 kw-D, 150 w-N. TL: N36 57 49 W121 29 22. Hrs open: 24 2670 S. White Rd., Suite 165, San Jose, 95148. Phone: (408) 223-3130. Fax: (408) 223-3131. E-mail: qhradio@aol.com Web Site: www.quehuongmedia.com. Licensee: KZSJ Radio LLC. Group owner: Bustos Media Holdings (acq 2-26-99). Format: Vietnamese. Hispanic. ♦Amador Bustos, chmn; Raul Salvador, CFO; John Bustos, exec VP & opns mgr; Khoi Nguyen, gen mgr.

San Mateo

***KCSM(FM)**— October 1964: 91.1 mhz; 11.5 kw. 371 ft TL: N37 32 12 W122 20 02. Hrs open: 1700 W. Hillsdale Blvd., 94402. Phone: (650) 574-9136. Phone: (650) 574-6586. Fax: (650) 524-6975. Web Site: www.kcsm.org. Licensee: San Mateo County Community College District. Population served: 161,000 Natl. Network: PRI. Format: Jazz. Target aud: 40 plus; males. Spec prog: Blues 3 hrs wkly. ♦Marilyn Lawrence, gen mgr; Alisa Clancy, opns dir; Melanie Berson, progmg dir. Co-owned TV: *KCSM-TV affil.

KSAN(FM)— September 1963: 107.7 mhz; 8.9 kw. 1,162 ft TL: N37 41 20 W122 26 07. Stereo. Hrs opn: 55 Hawthorne, Suite1000, San Francisco, 94105. Phone: (415) 981-5726. Fax: (415) 995-7061. Web Site: www.1077thebone.com. Licensee: Susquehanna Radio Corp. (group owner; acq 5-29-97; $44 million). Population served: 4,625,300 Natl. Rep: McGavren Guild. Format: Classic rock. Target aud: 25-54. ♦Tony Salvadore, VP & gen mgr; Michael Seghieri, gen sls mgr; Larry Sharpe, progmg dir; Eric Steinberg, engrg dir & chief of engrg; Sara Bronson, traf mgr.

KTCT(AM)—Co-owned with KSAN(FM). 1948: 1050 khz; 50 kw-D, 10 kw-N, DA-2. TL: N37 39 02 W122 09 08. Stereo. 24 Phone: (415) 864-1050. Fax: (415) 995-6867. Web Site: www.theticket1050.com. (Acq 7-21-97; $15 million).750,000 Natl. Network: Westwood One. Format: Sports. Target aud: 25-54. ♦Rich Zirkel, gen sls mgr; Lee Hammer, progmg mgr.

San Rafael

***KSRH(FM)**— May 1, 1980: 88.1 mhz; 10 w. 66 ft TL: N37 58 16 W122 30 47. Hrs open: 9 AM-3 PM 185 Mission Ave., AR 101, 94901. Phone: (415) 457-5314. Licensee: San Rafael High School District. Format: Div, Black. News: 5 hrs wkly. Target aud: 12-29. Spec prog: Fr one hr wkly. ♦Chris Russo, gen mgr.

KVVZ(FM)— June 1, 1961: 100.7 mhz; 910 w. Ant 810 ft TL: N37 59 25 W122 29 58. Stereo. Hrs opn: 750 Battery St., Suite 200, San Francisco, 94111. Phone: (415) 733-5765. Fax: (415) 733-5766. Web Site: www.univision.com. Licensee: Univision Radio License Corp. Group owner: Salem Communications Corp. (acq 3-1-2005; exchange for KOSL(FM) Jackson). Format: Sp pop. ♦Tony Perlongo, gen mgr.

Santa Ana

KALI-FM— Feb 6, 1980: 106.3 mhz; 3 kw. 130 ft TL: N33 45 21 W117 51 16. (CP: Ant 203 ft. TL: N33 45 21 W117 51 17). Stereo. Hrs opn: 747 E. Green St., Suite 400, Pasadena, 91101. Phone: (626) 844-8882. Fax: (626) 844-0156. Licensee: KALI-FM Licensee LLC. Population served: 155,710 Format: Asian. Target aud: 18-44. ♦Arthur Liu, pres; David Sweeney, gen mgr; Alan Mok, progmg dir.

KVNR(AM)—Listing follows KWIZ(FM).

KWIZ(FM)— 1947: 96.7 mhz; 3 kw. 206 ft TL: N33 48 08 W117 47 43. Stereo. Hrs opn: 24 3101 W. 5th St., 92703. Phone: (714) 554-5000. Fax: (714) 554-9362. Web Site: www.sonido967.com. Licensee: LBI Radio License Corp. Group owner: Liberman Broadcasting Inc. (acq 1997; $11.2 million). Population served: 200,000 Format: Sp, Tropical. Target aud: 18 plus; Asian. ♦Winnie Coombs, stn mgr; Francisco Morales, prom dir; Edwardo Leon, progmg dir; Jesus Mar, mus dir; Shannon Murdock, chief of engrg; Patty Diaz, traf mgr.

KVNR(AM)—Co-owned with KWIZ(FM). Nov 26, 1926: 1480 khz; 5 kw-U, DA-2. TL: N33 45 06 W117 54 36. 15781 Brookhurst St., Suite 101, Westminster, 92683. Phone: (714) 918-4444. Fax: (714) 918-4445. E-mail: radio@littlesoigonradio.com Web Site: www.littlesoigonradio.com. (Acq 1-88; $6.25 million with co-located FM; FTR: 1-4-88) 100,000 Format: Vietnamese. Target aud: 18-49. ♦Ninh Vu, pres; Kathleen Bui, gen mgr; Joe Dinh, chief of engrg.

Santa Barbara

KBKO(AM)— April 1926: 1490 khz; 1 kw-U. TL: N34 24 57 W119 41 10. Stereo. Hrs opn: 24 414 East Cota St., 93101. Phone: (805) 879-8300. Fax: (805) 879-8430. Licensee: Rincon License Subsidiary LLC. Group owner: Clear Channel Communications Inc. (acq 2-28-2007; grpsl). Population served: 300,000 Law Firm: Farrand, Cooper & Bruiniers. Format: Mexican. News staff: one; News: 7 hrs wkly. Target aud: 16-65; Hispanic. ♦J.D. Freedman, gen mgr; Marlene Huddy, opns dir; Jose Fierroz, progmg VP & progmg dir; Ransom Bullard, chief of engrg; Alfredo Quezada, disc jockey; Carlos Velasco, disc jockey; Gerardo Lorenz, disc jockey; Jose Fierros, disc jockey; Lupita Rodriguez, disc jockey.

KSPE-FM—Co-owned with KBKO(AM). Feb 6, 1989: 94.5 mhz; 81 kw. 2,949 ft TL: N34 31 32 W119 57 28.24 Format: Sp oldies.

***KCSB-FM**— November 1964: 91.9 mhz; 620 w. 2,910 ft TL: N34 31 31 W119 57 29. (CP: Ant 1,879 ft.). Stereo. Hrs opn: 24 Box 13401, 93107. Phone: (805) 893-3757. Fax: (805) 893-7832. E-mail: info@kcsb.org Web Site: www.kcsb.org. Licensee: Regents of the University of California. Population served: 100,000 Format: Var. News: 9 hrs wkly. Spec prog: Sp 12 hrs, Japanese pop one hr, East Indian 2 hrs, reggae 6 hrs, American Indian 3 hrs wkly. ♦Elizabeth Robinson, gen mgr; Kimberly Tran, progmg dir.

KDB(FM)— Feb 14, 1960: 93.7 mhz; 12.5 kw. 870 ft TL: N34 27 58 W119 40 37. Stereo. Hrs opn: 24 Box 91660, 93190. Phone: (805) 966-4131. Fax: (805) 966-4788. E-mail: kdb@kdb.com Web Site: www.kdb.com. Licensee: Pacific Broadcasting Co. (acq 11-6-2003; transfer of stock). Population served: 600,000 Natl. Rep: Clear Channel. Law Firm: Fletcher, Heald & Hildreth. Format: Class music. News: one hr wkly. Target aud: Adults; affluent, influential & educated. ♦Roby Scott, gen mgr; Richard Bickle, opns dir & progmg dir; Bob Scott, sls dir & progmg dir.

KIST(AM)— 1946: 1340 khz; 650 w-U. TL: N34 25 07 W119 41 10. Stereo. Hrs opn: 414 E. Cotta St., 93101. Phone: (805) 879-8300. Fax: (805) 879-8430. Licensee: R & R Radio LLC Group owner: Clear Channel Communications Inc. (acq 2-28-2007; with KTMS(AM) Santa Barbara). Population served: 140,757 Natl. Network: ESPN Radio. Natl. Rep: Katz Radio. Format: Sports, talk. Target aud: 25-54. ♦J. D. Freeman, gen mgr; Keith Royer, opns dir.

KIST-FM— 1998: 107.7 mhz; 930 w. Ant 1,627 ft TL: N34 30 10 W119 50 56. Hrs open: 24 414 E. Cota St., 93101. Phone: (805) 879-8300. Fax: (805) 879-8430. Licensee: Rincon License Subsidiary LLC. Group owner: Clear Channel Communications Inc. (acq 2-28-2007; grpsl). Natl. Rep: Katz Radio. Format: Rgnl Mexican. ♦Tom Baker, gen mgr; Keith Royer, opns dir & progmg dir; Vince Holian, gen sls mgr; Peter Bie, news dir; Andrea Shaparenko, traf mgr.

KMGQ(FM)—See Goleta

***KQSC(FM)**— July 1985: 88.7 mhz; 12 kw. 866 ft TL: N34 27 55 W119 40 37. Stereo. Hrs opn: 24
Rebroadcasts KUSC 100%.
Box 77913, Los Angeles, 90007. Phone: (213) 225-7400. Fax: (213) 225-7410. E-mail: kusc@kusc.org Web Site: www.kusc.org. Licensee: University of Southern California. Natl. Network: PRI, NPR. Law Firm: Lawrence Bernstein. Format: Class. News: 3 hrs wkly. Target aud: 35 plus. ♦Brenda Barnes, pres; Eric DeWeese, gen mgr.

KRUZ(FM)— Sept 1, 1957: 97.5 mhz; 17.5 kw. 2,920 ft TL: N34 31 31 W119 57 29. Hrs open: 24 403 E. Montecito St., 93101-1759. Phone: (805) 966-1755. Web Site: www.cumulus.com. Licensee: Cumulus Licensing Corp. Group owner: Cumulus Media Inc. (acq 3-12-2001; grpsl). Population served: 650,000 Natl. Rep: McGavren Guild. Wire Svc: AP Format: Smooth Jazz. Target aud: 35-64; young, educated, upscale adults. ♦Gail Surrillo, gen mgr; Brandon Randazzo, prom dir; John D. Straker, chief of engrg; Mark Deanba, progmg.

KSBL(FM)—See Carpinteria

***KSBX(FM)**— Apr 1, 2003: 89.5 mhz; 50 w. Ant 899 ft TL: N34 27 57 W119 40 37. Stereo. Hrs opn: 24
Rebroadcasts KCBX(FM) San Luis Obispo 99%.
KCBX Public Radio, 4100 Vachell Ln., San Luis Obispo, 93401. Phone: (805) 549-8855. Phone: (805) 781-3025. E-mail: kcbx@kccbx.org Web Site: www.kcbx.org. Licensee: KCBX Inc. Natl. Network: NPR. Rgnl rep: Margaret Merisante Format: Class, jazz. News: 30 hrs wkly. ♦Frank Lanzone, gen mgr; Hank Hadley, opns mgr; Paul Severtson, dev dir; Guy Rathbun, progmg dir.

KTMS(AM)— Aug 11, 1962: 990 khz; 5 kw-D, 500 w-N, DA-2. TL: N34 28 15 W119 40 33. Stereo. Hrs open: 24 414 E. Cota St., 93101. Phone: (805) 879-8300. Web Site: www.990am.com. Licensee: Santa Barbara Community Broadcasting Co. (acq 2-28-2007; with KIST(AM) Santa Barbara). Population served: 250,000 Natl. Network: ABC, CNN Radio. Format: Talk. News staff: 2; News: 4 hrs wkly. Target aud: 25 plus; upscale adults. ♦Tom Baker, gen mgr; Keith Royer, opns mgr; Lin Aubuchon, prom dir.

KTYD(FM)— Aug 11, 1972: 99.9 mhz; 34 kw. 1,278 ft TL: N34 28 15 W119 40 33. Stereo. Hrs opn: 24 414 E. Cota St., 93101. Phone: (805) 879-8300. Fax: (805)879-8430. Web Site: www.ktyd.com. Licensee: Rincon License Subsidiary LLC. Group owner: Clear Channel Communications Inc. (acq 2-28-2007; grpsl). Population served: 170,000 Natl. Rep: Katz Radio. Law Firm: Wiley, Rein & Fielding. Format: AOR. News staff: one; News: 3 hrs wkly. Target aud: 18-49; upscale adults. Spec prog: Pub affrs one hr wkly. ♦Keith Royer, VP, opns mgr & progmg dir; Tom Baker, gen mgr; Vince Hollian, gen sls mgr; Lin Aubuchon, mktg dir & prom dir; Peter Bie, news dir; Ran Bullard, chief of engrg; Andrea Shaparenko, traf mgr.

KVYB(FM)— Aug 8, 1961: 103.3 mhz; 105 kw. 2,980 ft TL: N34 31 30 W119 57 10. Stereo. Hrs opn: 24 1376 Walter St., Ventura, 93003. Phone: (805) 642-8595. Fax: (805) 656-5838. E-mail: info@1033thevibe.com Web Site: www.1033thevibe.com. Licensee: Cumulus Licensing Corp. Group owner: Cumulus Media inc. (acq 4-2000). Population served: 1,150,000 Natl. Rep: McGavren Guild. Format: Hip hop and more. Target aud: 18-54; general. ♦Jonathon Pinch, COO; Lewis W. Dickey Jr., pres; Martin Gausvik, CFO; John W. Dickey, exec VP; Gail Furillo, gen sls mgr & prom dir; Daniel Herejon, progmg dir; J.D. Strahler, chief of engrg; Barbara Haser, traf mgr.

KZER(AM)— Oct 31, 1937: 1250 khz; 2.5 kw-D, 1 kw-N, DA-2. TL: N34 25 06 W119 49 05. Hrs open: 24 1330 Cacique St., 93103. Phone: (805) 963-7824. Fax: (805) 963-7824. E-mail: jose@radiolazer.com Web Site: www.radiolazer.com. Licensee: Lazer Broadcasting Corp. (group owner; acq 12-18-2003; $1.5 million). Population served: 250,000 Law Firm: Hogan & Hartson. Format: Sp. News staff: 5; News: 140 hrs wkly. Target aud: 25 plus; upscale, educated listeners. ♦Jose Plaascencia, gen mgr; Jose Plascencia, opns mgr; Salvador Prieto, progmg dir.

KZSB(AM)— March 1961: 1290 khz; 500 w-D, 122 w-N. TL: N34 25 07 W119 41 10. Stereo. Hrs opn: 24 1317 Santa Barbara St., 93101. Phone: (805) 568-1444. Fax: (805) 966-3530. Licensee: Santa Barbara Broadcasting Inc. (acq 3-1-2005; $750,000). Population served: 173,900 Natl. Network: Westwood One. Format: News/talk. Target aud: 35-64. ♦Dennis M. Weibling, pres; Les Carroll, gen mgr & natl sls mgr; Geren Piltz, progmg dir & news dir; Patrice Cardenas, pub affrs dir.

Santa Clara

KLIV(AM)—See San Jose

***KSCU(FM)**— July 1, 1978: 103.3 mhz; 30 w. 179 ft TL: N37 20 53 W121 56 25. Stereo. Hrs open: 24 Santa Clara Univ., 500 El Camino Real 3207, 95053. Phone: (408) 554-4413. Fax: (408) 554-5738. Web Site: www.kscu.org. Licensee: President and Board of Trustees of Santa Clara University. Format: Modern alternative rock. News: one hr wkly. Target aud: 14-34; Young adult who like modern music. Spec prog: Hip-hop 15 hrs, Blues 3 hrs, loud rock 6 hrs, world one hr wkly. ♦Allyson Harrison, gen mgr; Gordon Young, news dir; Bill Orr, engrg dir & chief of engrg.

KVVF(FM)— Sept 25, 1964: 105.7 mhz; 50 kw. Ant 500 ft TL: N37 21 32 W121 45 22. Stereo. Hrs opn: 24 750 Battery St., Suite 200, San Francisco, 94111. Phone: (415) 733-5765. Fax: (415) 733-5766. Web Site: www.univision.com. Licensee: Univision Radio License Corp. Group owner: Univision Radio (acq 9-22-2003; grpsl). Population served: 1,400,000 Format: Sp pop. ♦Tony Perlongo, gen mgr.

KVVN(AM)—Licensed to Santa Clara. See San Jose

Santa Cruz

KBOQ(FM)—See Monterey

***KFER(FM)**— 1992: 89.9 mhz; 200 w. 26 ft TL: N37 00 45 W121 58 25. Hrs open: 24 Box 13, 95063. Phone: (831) 475-6651. Fax: (831) 464-8427. Licensee: Santa Cruz Educational Broadcasting Foundation. Rgnl rep: Moody. Format: Var. News: 15 hrs wkly. Target aud: General. ♦Mildred Holmes, pres; Dr. Stan Monteith, gen mgr.

KSCO(AM)— Sept 21, 1947: 1080 khz; 10 kw-D, 5 kw-N, DA-2. TL: N36 57 43 W121 58 51. Hrs open: 24 2300 Portola Dr., 95062. Phone: (831) 475-1080. Fax: (831) 475-2967. Web Site: www.ksco.com. Licensee: Zwerling Broadcasting System Ltd. (acq 1-31-91; $600,000; FTR: 12-31-90). Population served: 500,000 Format: News/talk. News staff: 8; News: 35 hrs wkly. Target aud: 25 plus; well educated professionals, managers. ♦Michael Zwerling, CEO; Ron Stevens, pres; Michael Olson, gen mgr.

KSQL(FM)— Sept 2, 1961: 99.1 mhz; 1.1 kw. 2,487 ft TL: N37 06 40 W121 50 34. Stereo. Hrs opn: 24 750 Battery St., Suite 200, San Francisco, 94111-1412. Phone: (415) 989-5765. Fax: (415) 733-5766. E-mail: estereosole@univision.com Web Site: www.univision.com. Licensee: TMS License California Inc. Group owner: Univision Radio (acq 9-22-2003; grpsl). Format: Sp, Mexican rgnl. News staff: one; News: 4 hrs wkly. Target aud: 25-54. ♦Tony Perlongo, gen mgr.

***KSRI(FM)**— Feb 28, 2001: 90.7 mhz; 316 w. Ant 364 ft TL: N37 00 10 W122 03 05. Stereo. Hrs opn: 24
Rebroadcasts KHRI(FM) Hollister 100%.
2351 Sunset Blvd., Suite 170-218, Rocklin, 95765. Phone: (916) 251-1600. Fax: (916) 251-1650. E-mail: info@air1.com Web Site: www.air1.com. Licensee: Educational Media Foundation. Group owner: EMF Broadcasting (acq 8-17-00; $295,000). Population served: 160,000 Natl. Network: Air 1. Law Firm: Shaw Pittman. Format: Contemp Christian. News staff: 3. Target aud: 18-35; Judeo-Christian, female. ♦Richard Jenkins, pres; Mike Novak, VP; Keith Whipple, dev dir; David Pierce, progmg mgr; Ed Lenane, news dir; Sam Wallington, engrg dir; Arthur Vassar, traf mgr; Karen Johnson, news rptr; Marya Morgan, news rptr; Richard Hunt, news rptr.

***KUSP(FM)**— Apr 14, 1972: 88.9 mhz; 1.25 kw. 2,496 ft TL: N36 32 05 W121 37 14. Stereo. Hrs opn: 203 8th Ave., 95062. Phone: (831) 476-2800. Fax: (831) 476-2802. Web Site: www.kusp.org. Licensee: Pataphysical Broadcasting Foundation Inc. Population served: 750,000 Natl. Network: NPR. Format: Div. News: 44 hrs wkly. ♦Terry Green, gen mgr & stn mgr; Paula Kenyon, dev dir & sls dir; Rob Mullen, mus dir.

***KZSC(FM)**— August 1974: 88.1 mhz; 20 kw. Ant 436 ft TL: N37 00 10 W122 03 04. Stereo. Hrs open: 24 1156 High St., 95064. Phone: (831) 459-2811. Fax: (831) 459-4734. Web Site: www.kzsc.org. Licensee: Regents of University of California. Population served: 1,400,000 Format: Div. News: 40 hrs wkly. Target aud: 18-plus; college students up till late 30's. ♦Michael Bryant, gen mgr; Kristen Sarton, stn mgr.

Santa Margarita

KWWV(FM)— July 21, 1986: 106.1 mhz; 950 w. 1,467 ft TL: N35 21 38 W120 39 21. Stereo. Hrs opn: 24 3620 Sacramento Dr., Suite 204, San Luis Obispo, 934201. Phone: (805) 781-2750. Fax: (805) 781-2758. Web Site: www.wild1061.com. Licensee: Salisbury Broadcasting Corp. (acq 4-1-99; $1 million). Population served: 300000 Format: Chr/top40. News staff: 2; News: 6 hrs wkly. Target aud: 18-34; upscale homeowners. ♦Kathy Signorelli, gen mgr.

Santa Maria

KBOX(FM)—Lompoc, Dec 24, 1968: 104.1 mhz; 5.7 kw. 710 ft TL: N34 43 50 W120 26 01. Stereo. Hrs opn: 24 2325 Skyway Dr., Suite J, 93455. Phone: (805) 922-1041. Fax: (805) 928-3069. Licensee: AGM-Santa Maria LP. Group owner: American General Media (acq 2-1-2000). Population served: 200,000 Law Firm: Hogan & Hartson. Format: Adult hits. News staff: one. Target aud: 25-54. ♦Rich Watson, pres & gen mgr; luis Diaz, opns mgr & progmg dir; Emily Stich, gen sls mgr & natl sls mgr; John Bartel, chief of engrg.

KGDP(AM)—Oildale, July 4, 1988: 660 khz; 10 kw-D, 1 kw-N, DA-2. TL: N34 57 04 W120 22 38. (CP: COL Oildale. 5 kw-U, DA-2. TL: N35 27 10 W118 56 40). Hrs opn: 24 2225 Skyway Dr., Suite B, 93455. Phone: (805) 928-7707. Fax: (805) 922-8582. E-mail: kgdp660@yahoo.com Web Site: www.kgdp660.com. Licensee: Radio Representatives Inc. Group owner: Norwood J. Patterson Population served: 650,000 Natl. Network: USA. Format: Christian, talk. Target aud: 35-65. ♦Steve Cox, gen mgr; Bill Greenelsh, prom dir; Gretchen England, traf mgr.

***KGDP-FM**— 2003: 90.5 mhz; 17.5 kw. Ant 846 ft TL: N34 44 30 W120 26 45. Hrs open: 1416 Hollister Ln., Los Osos, 93402. Phone: (805) 528-1996. Fax: (805) 922-8582. E-mail: jp805@msn.com Web Site: kgdp660.com. Licensee: People of Action. Format: Christian talk radio. ♦Steve Cox, gen mgr.

***KHFR(FM)**— June 21, 2005: 89.7 mhz; 2.45 kw vert. Ant 1,866 ft TL: N34 54 37 W120 11 08. Hrs open: 24 Family Stations Inc., 4135 Northgate Blvd., Suite 1, Sacramento, 95834. Phone: (916) 641-8191. Fax: (916) 641-8238. Licensee: Family Stations Inc. (group owner). Format: Relg. ♦Harol Camping, pres.

KINF(AM)— April 1946: 1440 khz; 5 kw-D, 1 kw-N, DA-N. TL: N34 59 02 W120 27 10. Hrs open: 716 E. Chapel, 93454. Phone: (805) 922-7727. Fax: (805) 349-0265. E-mail: kathy@knightbroadcasting.com Web Site: www.mix96.com. Licensee: Knight Broadcasting Inc. (group owner; (acq 7-31-2006; $1.2 million with KUHL(AM) Lompoc). Population served: 285,000 Format: Adult contemp. Target aud: 35-64; upscale news & sports listeners. ♦Shawn Knight, gen mgr; Jeff Williams, opns dir.

KPAT(FM)—Orcutt, 1993: 95.7 mhz; 3.3 kw. 735 ft TL: N34 44 20 W120 26 41. Hrs open: 2325 Skyway Dr., Suite J, 93455. Phone: (805) 922-1041. Fax: (805) 928-3069. Web Site: www.957thebeatfm.com. Licensee: AGM-Santa Maria LP. Group owner: American General Media (acq 12-1-99; $900,000). Natl. Network: USA. Format: Rhythm and blues, hip hop, old school. ♦Emily Stich, gen sls mgr; Jeff Lyons, prom dir; Luis Diaz, progmg dir; Rich Watson, gen mgr & pub affrs dir.

KSBQ(AM)— Sept 1, 1961: 1480 khz; 1 kw-D, 61 w-N. TL: N34 57 02 W120 29 22. Hrs open: 24 200 E. Fesler St., Suites 101 & 201, 93454. Phone: (805) 240-2070. Phone: (805) 928-9796. Fax: (805) 240-5960. Fax: (805) 928-3367. Licensee: Lazer Broadcasting Corp. (group owner; acq 12-29-99; $225,000). Population served: 200,000 Natl. Rep: Lotus Entravision Reps LLC. Law Firm: Fletcher Heald & Hildreth. Format: Christian. Target aud: 18-49; adults. ♦Alfredo Plascencia, CEO, pres & gen mgr; Salvador Prieto, opns mgr & progmg dir.

KSMA(AM)— 1946: 1240 khz; 1 kw-U. TL: N34 57 02 W120 29 27. Hrs open: 2215 Skyway Dr., 93456. Phone: (805) 925-2582. Fax: (805) 928-1544. Web Site: www.1240ksma.com. Licensee: CC Licenses LLC. (acq 10-17-2000; grpsl). Population served: 400,000 Natl. Network: CBS. Format: News/talk. ♦Rich Hawkins, gen mgr.

KSNI-FM—Co-owned with KSMA(AM). 1960: 102.5 mhz; 17.5 kw. 774 ft TL: N34 50 08 W120 24 06. Stereo. Web Site: www.sunnycountry.com.400,000 Format: Contemp country.

KTAP(AM)— June 10, 1962: 1600 khz; 470 w-D. TL: N34 58 48 W120 27 12. Hrs open: 6 AM-midnight 718 E. Chapel St., 93454. Phone: (805) 928-4334. Fax: (805) 349-2765. E-mail: kidiktap@aol.com Web Site: www.labuena.net. Licensee: Emerald Wave Media. (acq 3-6-97; $475,000 with KIDI(FM) Guadalupe). Population served: 60,000 Law Firm: Mark Van Burgh. Format: Sp, Mexican. News staff: one; News: 4 hrs wkly. Target aud: General; first generation Mexicans. ♦August Ruiz, gen mgr.

KURQ(FM)—See Grover Beach

KXFM(FM)— 1959: 99.1 mhz; 1.8 kw. 1,905 ft TL: N34 54 37 W120 11 08. Stereo. Hrs opn: 2215 Skyway Dr., 93454. Phone: (805) 925-2582. Fax: (805) 928-1544. Web Site: www.991thefox.com. Licensee: CC Licenses LLC. (acq 10-17-2000; grpsl). Population served: 243,000 Format: Classic rock. Target aud: 18-49; contemp, active adults. ♦Rich Hawkins, gen mgr; Jennifer Grant, opns dir; Pattie Wagner, gen sls mgr; Milos Nemicik, chief of engrg.

KXTK(AM)—See San Luis Obispo

KZOZ(FM)—See San Luis Obispo

Santa Monica

KBLA(AM)— 1947: 1580 khz; 50 kw-U, DA-2. TL: N34 05 08 W118 15 24. Stereo. Hrs opn: 747 E. Green St., Suite 400, Pasadena, 91101. Phone: (626) 844-8882. Fax: (626) 844-0156. Web Site: www.mrbi.net. Licensee: Multicultural Radio Broadcasting Licensee LLC. Group owner: Multicultural Radio Broadcasting Inc. (acq 2-4-2004; grpsl). Population served: 500,000 Format: Spanish Christian. News staff: one. ♦David Sweeney, gen mgr; Jose Calles, stn mgr.

***KCRW(FM)**— Jan 1, 1946: 89.9 mhz; 6.9 kw. 1,110 ft TL: N34 07 08 W118 23 30. Stereo. Hrs opn: 24 1900 Pico Blvd., 90405. Phone: (310) 450-5183. Fax: (310) 450-7172. E-mail: mail@kcrw.org Web Site: www.kcrw.org. Licensee: Santa Monica College District. (acq 8-3-76). Population served: 10,000,000 Natl. Network: NPR, PRI. Law Firm: Dickstein Shapiro LLP. Wire Svc: AP Format: Indie rock, Latino, news. News staff: 3; News: 14 hrs wkly. Target aud: General; 18-55 year old consumers. ♦Ruth Seymour, gen mgr; Mike Newport, opns dir; David Kleinbart, dev dir; Nic Harcourt, mus dir; Steve Herbert, chief of engrg.

KDLD(FM)— 1963: 103.1 mhz; 3.7 kw. Ant 269 ft TL: N34 00 53 W118 22 50. Stereo. Hrs opn: 24 570 Wilshire Blvd., Suite 250, Los Angeles, 90036. Phone: (323) 900-6100. Fax: (323) 900-6200. Web Site: www.kdl.com. Licensee: Entravision Holdings LLC. Group owner: Entravision Communications Corp. (acq 2000). Population served: 6,000,000 Format: Rhythmic CHR / dance, alternative. News: one hr wkly. Target aud: 25-54; upscale adults in Los Angeles' westside. ♦Karl Meyer, gen mgr; Robert Isaac, opns dir & engrg dir; Scott Dallavo, prom mgr; Nestor Rocha, progmg VP & mus dir; Max Tolkoff, progmg dir.

Santa Paula

KLJR-FM— Oct 4, 1976: 96.7 mhz; 87 w. 1,500 ft TL: N34 19 33 W119 02 18. (CP: 278 w). Stereo. Hrs opn: 24 200 S. A St., Suite 400, Oxnard, 93030. Phone: (805) 240-2070. Fax: (805) 240-5960. Licensee: Lazer Broadcasting Corp. (acq 3-31-98; $925,000; FTR: 11-4-91). Population served: 55,797 Natl. Rep: Lotus Entravision Reps LLC. Law Firm: Fletcher, Heald & Hildreth. Format: Sp, adult contemp, CHR. Target aud: 25-54; general. ♦Alfredo Plascencia, CEO, pres & gen mgr; Salvador Prieto, progmg dir.

KUNX(AM)— 1948: 1400 khz; 1 kw-U. TL: N34 19 48 W119 05 31. Stereo. Hrs opn: 2284 S. Victoria Ave., Suite 2 G, Ventura, 93003. Phone: (805) 289-1400. Fax: (805) 644-7906. Web Site: www.kunx.com. Licensee: Gold Coast Broadcasting LLC (group owner; acq 8-18-99; grpsl). Format: Sp. ♦Chip Ehrhardt, gen mgr; Mark Elliott, progmg dir.

Santa Rosa

***KBBF(FM)**— May 30, 1973: 89.1 mhz; 1 kw. 2,770 ft TL: N38 39 23 W122 36 54. Hrs open: Box 7189, 95407. Phone: (707) 545-8833. Fax: (707) 545-6244. E-mail: kbbfradio@aol.com Web Site: www.kbbfradio.com. Licensee: Bilingual Broadcasting Foundation Inc. Population served: 50,006 Format: Educ, Sp, bilingual. ♦Jesus Lozano, gen mgr; Roy Brown, opns dir; Roy Brown, prom VP.

KFGY(FM)—Listing follows KSRO(AM).

***KLVR(FM)**— Oct 15, 1982: 91.9 mhz; 840 w. Ant 2,988 ft TL: N38 40 09 W122 50 24. Stereo. Hrs opn: 24 2351 Sunset Blvd., Sute 170-218, Rocklin, 95765. Phone: (916) 251-1600. Fax: (916) 251-1650. E-mail: klove@klove.com Web Site: www.klove.com. Licensee: Educational Media Foundation. Group owner: EMF Broadcasting (acq 1986). Population served: 775,000 Natl. Network: K-Love. Law Firm: Shaw Pittman. Format: Contemp Christian. News staff: 3. Target aud: 25-44; Judeo-Christian females. ♦Richard Jenkins, pres; Mike Novak, VP; Keith Whipple, dev dir; David Pierce, progmg mgr; Ed Lenane, news dir; Sam Wallington, engrg dir; Arthur Vassar, traf mgr.

***KRCB-FM**— September 1993: 91.1 mhz; 120 w. 731 ft TL: N38 44 25 W122 50 46. Stereo. Hrs opn: 24 5850 Labath Ave., Rohnert Park, 94928. Phone: (707) 584-2000. Fax: (707) 585-1363. Web Site: www.krcb.org. Licensee: Rural California Broadcasting Corp. Population served: 70,000 Natl. Network: NPR, PRI. Wire Svc: DAC Format: Class, progsv, news/talk. News staff: one; News: 15 hrs wkly. Target aud: General. Spec prog: Folk 6 hrs, jazz 6 hrs wkly. ♦Nancy Dobbs, CEO & pres. Co-owned TV: *KRCB-TV affil.

KRRS(AM)— Apr 1, 1962: 1460 khz; 1 kw-D, 33 w-N, DA-2. TL: N38 22 13 W122 43 39. Stereo. Hrs opn: 24 Box 2277, 95405. Phone: (707) 545-1460. Fax: (707) 545-0112. E-mail: krrs@sonic.net Web Site: www.moonradios.com. Licensee: Moon Broadcasting Licensee LLC. Group owner: Moon Broadcasting (acq 1993; $400,000; FTR: 9-6-93). Population served: 500,000 Natl. Rep: Interep. Format: Sp. Target aud: 25-54; contemporary Hispanic families. ♦Abel DeLuna Sr., CEO; Abel A. DeLuna, pres; Arelia DeLuna, CFO; Maggie LeClerc, gen mgr; Benoit LeClerc, opns mgr.

KSRO(AM)— May 1937: 1350 khz; 5 kw-U, DA-N. TL: N38 26 22 W122 44 51. Hrs open: Box 2158, 95405. Secondary address: 1410 Neotomas Ave., Suite 200 95405. Phone: (707) 543-0100. Fax: (707) 571-1097. Web Site: www.ksro.com. Licensee: Maverick Media of Santa Rosa License LLC. Group owner: Maverick Media LLC (acq 12-16-02; grpsl). Population served: 337,000 Format: News/talk. Target aud: 35-64. ♦Gary Rozynek, pres; Diane Hubel, gen mgr & gen sls mgr; Michelle Marquis, progmg dir; George Bright, news dir; Virgil Scigla, chief of engrg.

KFGY(FM)—Co-owned with KSRO(AM). Dec 21, 1979: 92.9 mhz; 2.3 kw. Ant 1,800 ft TL: N38 45 45 W122 50 24. Stereo. Web Site: froggy929.com. Format: Country. Target aud: 18-44. Spec prog: Jazz 5 hrs wkly.

KVRV(FM)—Monte Rio, Nov 20, 1977: 97.7 mhz; 250 w. 1,122 ft TL: N38 29 08 W123 02 05. Stereo. Hrs opn: 24 Box 2158, 95405. Secondary address: 1410 Neotomas Ave., Suite 200 95405. Phone: (707) 543-0100. Fax: (707) 571-1097. Web Site: www.977theriver.com. Licensee: Maverick Media of Santa Rosa License LLC. Group owner: Maverick Media LLC (acq 12-16-02; grpsl). Population served: 360,000 Format: Classic rock. Target aud: 25-54. ♦Diane Hubel, gen mgr.

KXFX(FM)— Dec 23, 1974: 101.7 mhz; 2.2 kw. 1,056 ft TL: N38 30 31 W122 39 41. Stereo. Hrs opn: Box 2158, 95405. Secondary address: 1410 Neotomas Ave., Suite 200 95405. Phone: (707) 543-0100. Fax: (707) 571-1097. Web Site: www.kxfx.com. Licensee: Maverick Media of Santa Rosa License LLC. Group owner: Maverick Media LLC (acq 12-16-02; grpsl). Population served: 380,000 Law Firm: Pepper & Corazzini. Format: Hard rock. Target aud: General. ♦Diane Hubel, gen mgr & prom mgr.

KZST(FM)— Apr 18, 1971: 100.1 mhz; 6 kw. 240 ft TL: N38 25 07 W122 40 33. Stereo. Hrs opn: 24 Box 100, 95402. Secondary address: 3392 Mendocino Ave. 95402. Phone: (707) 528-4434. Fax: (707) 527-8216. Web Site: www.kzst.com. Licensee: Redwood Empire Stereocasters. Population served: 412,000 Natl. Rep: McGavren Guild. Law Firm: Haley, Bader & Potts. Wire Svc: Reuters Wire Svc: Bay City News Service Format: Adult contemp. News staff: 2. Target aud: 25-54. ♦Tom Skinner, gen mgr.

Santa Ynez

KRAZ(FM)— 2001: 105.9 mhz; 65 w. Ant 2,932 ft TL: N34 31 32 W119 57 00. Hrs open: 24 1693 Mission Dr., Suite D202, Solvang, 93463. Phone: (805) 688-8386. Fax: (805) 688-2271. E-mail: kathy@knightbroadcasting.com Web Site: www.krazfm.com. Licensee: Knight Broadcasting Inc. (acq 5-21-2001; $325,000 for CP). Natl. Network: ABC. Format: Country. ♦Shawn Knight, gen mgr.

Seaside

KMBY-FM— October 1996: 103.9 mhz; 1.4 kw. 604 ft TL: N36 30 17 W121 54 21. Stereo. Hrs opn: 24 60 Garden Court, Suite 300, Monterey, 93940-5341. Phone: (831) 658-5200. Fax: (831) 658-5299. Web Site: www.x1039fm.com. Licensee: Mapleton Communications LLC (group owner; acq 1-24-02; $1.85 million). Population served: 500,000 Natl. Rep: McGavren Guild. Law Firm: Leventhal, Senter & Lerman. Format: Alternative rock. Target aud: 18-54; working women & men. Spec prog: New alternative releases 4 hrs wkly. ♦Adam Nathanson, pres; Mike Anthony, gen mgr; Kenny Allen, opns mgr & opns mgr; Jodi Morgan, gen sls mgr; Crissy Cooke, prom dir; Byron Cooke, progmg dir.

KSES-FM— Nov 22, 1972: 107.1 mhz; 1.85 kw. 587 ft TL: N36 33 12 W121 47 05. Stereo. Hrs opn: 67 Garden Ct., Monterey, 93940. Phone: (831) 333-9735. Fax: (831) 333-9750. Licensee: Entravision Holdings LLC. Group owner: Entravision Communications Corp. (acq 3-14-00; grpsl). Population served: 565000 Format: Contemp Sp hits. Target aud: 18-49. ♦Aaron Scoby, gen mgr; Tony Valencia, progmg dir.

Sebastopol

KJZY(FM)— Nov 5, 1995: 93.7 mhz; 6 kw. 216 ft TL: N38 25 07 W122 40 33. Stereo. Hrs opn: 24 Box 100, Santa Rosa, 95402. Phone: (707) 528-4434. Fax: (707) 527-8216. E-mail: gordon@kjzy.com Web Site: www.kjzy.com. Licensee: Redwood Empire Sterocasters. Format: Smooth jazz. ♦Tom Skinner, gen mgr.

Selma

***KQKL(FM)**— Aug 6, 2003: 88.5 mhz; 17 kw. Ant 397 ft TL: N36 26 50 W119 37 10. Stereo. Hrs opn: 24 2351 Sunset Blvd., Suite 170-218, Rocklin, 95765. Phone: (916) 251-1600. Fax: (916) 251-1650. E-mail: klove@klove.com Web Site: www.klove.com. Licensee: Educational Media Foundation. Group owner: EMF Broadcasting. Natl. Network: K-Love. Law Firm: Shaw Pittman. Format: Contemp Christian. News staff: 3. Target aud: 25-44; Judeo Christian, female. ♦Richard Jenkins, pres; Mike Novak, VP; Keith Whipple, dev dir; David Pierce, progmg mgr; Ed Lenane, news dir; Sam Wallington, engrg dir; Karen Johnson, news rptr; Marya Morgan, news rptr; Richard Hunt, news rptr.

Shafter

***KAIB(FM)**— 2006: 89.5 mhz; 50 kw. Ant 358 ft TL: N35 36 53 W119 28 16. Hrs open:
Rebroadcasts KLRD(FM) Yucaipa 100%.
2351 Sunset Blvd., Suite 170-218, Rocklin, 95765. Phone: (916) 251-1600. Fax: (916) 251-1650. Web Site: www.air1.com. Licensee: Educational Media Foundation. Group owner: EMF Broadcasting (acq 1-14-2005). Natl. Network: Air 1. Format: Alternative rock, div. ♦Richard Jenkins, pres; Lloyd Parker, gen mgr; Keith Whipple, dev dir; David Pierce, progmg dir; Ed Lenane, news dir; Sam Wallington, engrg dir; Arthur Vassar, traf mgr.

***KGZO(FM)**— June 6, 1996: 90.9 mhz; 1.9 kw. 2070 ft TL: N35 16 51 W119 44 52. Stereo. Hrs opn: 24
Rebroadcasts KMRO(FM) Camarillo 100%.
2310 Ponderosa Dr., Suite 28, Camarillo, 93010. Phone: (805) 482-4797. Fax: (805) 388-5202. E-mail: info@nuevavida.com Web Site: www.nuevavida.com. Licensee: The Association for Community Education Inc. (acq 7-30-97;). Population served: 500,000 Law Firm: Miller & Neely. Format: Relg, Sp. Target aud: General. ♦Phil Guthrie, pres; Mary Guthrie, gen mgr.

KKXX-FM— 1994: 93.1 mhz; 4 kw. 403 ft TL: N35 28 21 W119 01 40. Hrs open: 24 1400 Easton Dr., Suite 144, Bakersfield, 93309. Phone: (661) 328-1410. Fax: (661) 328-0873. Web Site: www.pirateradio931.com. Licensee: AGM California. Group owner: American General Media (acq 7-25-97; $1.5 million with KBID(AM) Bakersfield). Format: Pirate radio. News staff: 4; News: 2 hrs wkly. Target aud: 18-49; men. ♦Roger Fessler, gen mgr.

KSMJ(FM)—Licensed to Shafter. See Bakersfield

Shasta

KCNR(AM)— Aug 13, 1967: 1460 khz; 750 w-U. TL: N40 33 14 W122 22 53. Hrs open: 1326 Market St., Redding, 96001. Phone: (530) 244-5082. Fax: (530) 244-5698. Licensee: M C Allen Productions (acq 10-9-96; $35,000). Population served: 20,000 Law Firm: Womble, Carlyle,Sandbridge & Rice, LLC. Format: Talk, Sports. Target aud: 24-55. ♦Mike Quinn, gen mgr, progmg dir & chief of engrg.

Shasta Lake City

KESR(FM)— 1998: 107.1 mhz; 1.4 kw. 1,361 ft TL: N40 39 06 W122 31 32. Hrs open: 24 1588 Charles Dr., Redding, 96003. Phone: (530) 244-9700. Fax: (530) 244-9707. Licensee: Results Radio of Redding Licensee LLC. Group owner: Fritz Communications Inc. (acq 5-28-2000; grpsl). Format: Adult contemp. ♦Jack Fritz, pres & gen mgr; Beth Tappan, gen mgr; Rick Healy, opns mgr; Laurie Curto, gen sls mgr; Rob Reid, progmg dir & news dir; Bryant Smith, chief of engrg.

KJPR(AM)— 2005: 1330 khz; 1 kw-U, DA-2. TL: N40 40 48 W122 16 01. Hrs open: 24 hrs Jefferson Public Radio, 1250 Siskiyou Blvd., Ashland, OR, 97520. Phone: (541) 552-6301. Fax: (541) 552-8565. Web Site: www.ijpr.org. Licensee: JPR Foundation Inc. (acq 2-9-2004). Format: News, information. ♦Ronald Kramer, gen mgr; Bryon Lambert, opns dir; Paul Westhelle, dev dir.

KNNN(FM)— Oct 26, 1989: 99.3 mhz; 1.6 kw. Ant 1,525 ft TL: N40 39 15 W122 31 12. Hrs open: 3360 Alta Mesa Dr., Redding, 96002. Phone: (530) 226-9500. Fax: (530) 221-4940. E-mail: knnn@reddingradio.com Web Site: www.mix993fm.com. Licensee: Mapleton Communications LLC. Group owner: Regent Communications Inc. (acq 11-30-2006; grpsl). Format: CHR. Target aud: 25-54. Spec prog: Jazz 3 hrs wkly. ♦Lisa Geraci, gen mgr; Justin Paul, progmg dir.

Shingle Springs

KNTY(FM)— May 1, 1989: 101.9 mhz; 47 kw. Ant 505 ft TL: N38 51 12 W120 56 23. Stereo. Hrs opn: 24 1436 Auburn Blvd., Sacramento, 95815. Phone: (916) 646-4000. Phone: (916) 648-6013. Fax: (916) 646-6020. Web Site: www.kcclbossradio.com. Licensee: Entravision Holdings LLC. Group owner: Entravision Communications Corp. (acq 3-14-2000; grpsl). Population served: 3,000,000 Law Firm: Thompson, Hine & Flory L.L.P. Format: Country. Target aud: 25-54. ♦Larry Lemanski, gen mgr; Bob McNeill, prom dir; Don Langford, prom dir.

Shingletown

KKXS(FM)— January 2001: 96.1 mhz; 1.9 kw. Ant 1,174 ft TL: N40 29 18 W121 53 58. Hrs open: Fritz Communications Inc., 1355 N. Dutton Ave. #225, Santa Rosa, 95401-7107. Phone: (530) 244-9700. Fax: (530) 244-9707. Web Site: www.smoothjazz961.com. Licensee: Results Radio of Redding Licensee LLC. Group owner: Fritz Communications Inc. (acq 3-29-99; $125,000 for 50%). Format: Smooth jazz. ♦Beth Tappan, gen mgr; Bryant Smith, gen mgr & chief of engrg; Laurie Curto, gen sls mgr; Rick Healy, opns dir & progmg dir.

KRDG(FM)— Aug 1, 1995: 105.3 mhz; 10 kw. 1,056 ft TL: N40 29 54 W121 53 25. Stereo. Hrs opn: 24 3360 Alta Mesa Dr., Reading, 96002. Phone: (530) 226-9500. Fax: (530) 221-4940. Web Site: www.readingradio.com. Licensee: Mapleton Communications LLC. Group owner: Regent Communications Inc. (acq 11-30-2006; grpsl). Population served: 1,000,000 Law Firm: Bechtel & Cole. Format: Good time oldies. Target aud: 25-54; active adults with families. ♦Lisa Geraci, gen mgr; Jim Albertson, progmg dir.

Simi Valley

KIRN(AM)— Sept 21, 1984: 670 khz; 5 kw-D, 3 kw-N, DA-1. TL: N34 19 10 W118 42 56. Hrs open: 24 3301 Barham Blvd. #300, Los Angeles, 90068. Phone: (323) 851-5476. Fax: (323) 512-7452. E-mail: pomzaffari@670amkirn.com Web Site: www.670amkirn.com. Licensee: Lotus Oxnard Corp. Group owner: Lotus Communications Corp. (acq 12-11-96; $4.2 million). Population served: 900,000 Law Firm: Jerome Boros, Bryan Caves, Robinson Silverman. Format: Farsi MOR, news/talk, sports. News staff: 3; News: 14 hrs wkly. Persian, Irawian, Farsi. ♦Howard Kalmenson, pres; John Paley, VP; Hossein Hedjazi, progmg VP; John Cooper, chief of engrg; Poopak Mozaffari, mktg.

Soledad

***KFRS(FM)**— April 4, 2002: 89.9 mhz; 250 w vert. Ant 305 ft TL: N36 16 25 W121 16 12. Hrs open: Family Stations Inc., 4135 Northgate Blvd., Suite 1, Sacramento, 95834. Phone: (916) 641-8191. Fax: (916) 641-8238. Licensee: Family Stations Inc. (group owner) Format: Relg. ♦Thad McKinney, gen mgr.

KMBX(AM)— 1992: 700 khz; 2.5 kw-D, 700 w-N. TL: N36 27 51 W121 17 52. Hrs open: 67 Garden Ct., Monterey, 93940. Phone: (831) 333-9735. Fax: (831) 333-9750. Web Site: www.entravision.com. Licensee: Entravision Holdings LLC. Group owner: Entravision Communications Corp. (acq 3-14-00; grpsl). Format: Contemp Sp mus. Target aud: 18-49. ♦Aaron Scoby, gen mgr; Jeff Liberman, opns VP; Fidel Soto, news dir; Marcello Soto, chief of engrg.

KMJV(FM)— Oct 1, 1991: 106.3 mhz; 6 kw. 1,720 ft TL: N36 22 48 W121 12 57. Stereo. Hrs opn: 24 PO Box 1939, Salinas, 93902. Phone: (831) 757-1910. Fax: (831) 771-1685. Licensee: Wolfhouse Radio Group Inc. (group owner; (acq 7-13-2001; grpsl). Format: Mexican rgnl. Target aud: 18-44. ♦Roman Castro, gen mgr.

Solvang

KSYV(FM)— Sept 22, 1982: 96.7 mhz; 420 w. Ant 1,217 ft TL: N34 41 28 W120 15 58. Stereo. Hrs opn: 24 1693 Mission Dr., Suite D 202, 93463. Phone: (805) 688-5798. Fax: (805) 688-2271. E-mail: kathy@knightbroadcasting.com Web Site: www.mix96.com. Licensee: Knight Broadcasting Inc. (acq 2-8-2002). Population served: 60,000 Natl. Network: AP Network News. Format: Adult contemp. News: 126 hrs wkly. Target aud: 24-54; female 60%, male 40%. ♦Shawn Knight, gen mgr; Jeff Williams, opns dir.

Sonoma

***KSVY(FM)**— 2005: 91.3 mhz; 2.5 kw vert. Ant -305 ft TL: N38 16 47 W122 26 47. Hrs open: 168 W. Napa St., 95476. Phone: (707) 933-0808. Fax: (707) 933-1573. E-mail: ksvy@ksvy.org Web Site: www.ksvy.org. Licensee: Commonbond Foundation. Format: Community radio. ♦Bill Hammett, pres & gen mgr.

Sonora

KVML(AM)— 1949: 1450 khz; 1 kw-U. TL: N38 00 30 W120 21 45. Hrs open: 24 342 S. Washington, 95370. Phone: (209) 533-1450. Fax: (209) 533-9520. Web Site: www.kvml.com. Licensee: Clarke Broadcasting Corp. (group owner; acq 12-86; with co-located FM; FTR: 10-6-86). Population served: 200,000 Natl. Network: ABC. Law Firm: Leventhal, Senter & Lerman. Format: News/talk. News staff: 4; News: 25 hrs wkly. Target aud: 25 plus; general. Spec prog: Relg 3 hrs, sports 15 hrs wkly. ♦H. Randolph Holder Jr., pres; Larry England, gen mgr; Ed Haley, gen sls mgr; Lisa Westbrook, prom dir; Mark Truppner, progmg dir; Bill Johnson, news dir; John Petter, chief of engrg; D.J. Riendeau, traf mgr.

KZSQ-FM—Co-owned with KVML(AM). Oct 3, 1973: 92.7 mhz; 380 w. 1,289 ft TL: N38 00 30 W120 21 45. Stereo. 24 500,000 Format: Adult contemp. News staff: 4; News: 4 hrs wkly. Target aud: 25-54. ♦D.J. Riendeau, progmg dir & traf mgr; Justin Flores, progmg dir; Sebastian Kunz, news rptr.

Soquel

KYAA(AM)— 2001: 1200 khz; 25 kw-D, 10 kw-N. TL: N36 39 38 W121 32 29. Hrs open:
Simulcast with KEBV(FM) Salinas 100%.
651 Cannery Row, Monterey, 93940. Phone: (831) 372-1074. Fax: (831) 372-3585. E-mail: kyaknry@aol.com Web Site: www.knry.com. Licensee: People's Radio Inc. (group owner). Format: East Indian. ♦Jim Vossen, chief of opns & progmg dir.

South Lake Tahoe

KOWL(AM)— November 1956: 1490 khz; 1 kw-U. TL: N38 56 34 W119 57 25. Hrs open: 2435 E. Venice Dr., Suite 120, 96150. Phone: (530) 541-6681. Fax: (530) 541-4822. E-mail: kowl@krltfm.com Web Site: www.krltfm.com. Licensee: CCR-Lake Tahoe IV LLC. Group owner: Cherry Creek Radio LLC (acq 12-19-2003; grpsl). Format: News/talk, sports. ♦Betsy Miller, gen mgr.

KRLT(FM)— June 23, 1976: 93.9 mhz; 6 kw. -190 ft TL: N38 57 38 W119 56 26. Stereo. Hrs opn: 24 2435 E. Venice Dr., Suite 120, 96150. Phone: (530) 541-6681. Fax: (530) 541-4822. E-mail: krlt@krltfm.com Web Site: www.krltfm.com. Licensee: CCR-Lake Tahoe IV LLC. Group owner: Cherry Creek Radio LLC (acq 12-19-2003; grpsl). Format: 80s, 90s & today. News: 10 hrs wkly. Target aud: 25-54. ♦Betsy Miller, gen mgr.

KTHO(AM)— Mar 17, 1963: 590 khz; 2.5 kw-D, 500 w-N, DA-N. TL: N38 55 00 W119 57 46. Hrs open: 24 Box 5686, State Line, NV, 89449. Secondary address: 2520 Lake Tahoe Blvd. 96150. Phone: (530) 543-0590. Fax: (530) 543-1101. E-mail: ed@590ktho.com Licensee: Live Wire Media Partners LLC (acq 12-13-2004; $650,000). Population served: 150,000 Natl. Network: ABC. Format: ABC, adult contemp. News staff: one; News: 40 hrs wkly. Target aud: 25-54; locals & visitors, working population and retired. Spec prog: Jazz Trax Sundays 8pm. ♦Ed Crook, stn mgr.

KWYL(FM)— 1995: 102.9 mhz; 39 kw. Ant 2,926 ft TL: N39 18 38 W119 53 01. Stereo. Hrs opn: 24 595 E. Plumb Ln., Reno, NV, 89502. Phone: (775) 789-6700. Fax: (775) 789-6767. Web Site: www.wild1029.com. Licensee: Citadel Broadcasting Co. Group owner: Citadel Broadcasting Corp. (acq 5-9-03; grpsl). Population served: 75,000 Format: Urban, rap, hip hop. Target aud: 25-54. ♦Dana Johnson, gen mgr; Martin Stabbert, opns mgr; Gregg Moore, gen sls mgr; Maurice Ayala, progmg dir.

South Oroville

KYIX(FM)— Feb 1, 1994: 104.9 mhz; 260 w. Ant 1,548 ft TL: N39 39 04 W121 27 43. Hrs open: 1363 Longfellow, Chico, 95926-7319. Phone: (530) 894-7325. E-mail: info@air1.com Web Site: www.air1.com. Licensee: Butte Broadcasting Co. (acq 1994). Format: Christian hit radio. ♦Andrew Palmquist, gen mgr.

Stanford

***KZSU(FM)**— Oct 10, 1964: 90.1 mhz; 500 w. -10 ft TL: N37 24 42 W122 10 41. Stereo. Hrs opn: 24 Box 20190, 94309. Phone: (650) 725-4868. Fax: (650) 725-5865. Web Site: www.kzsu.stanford.edu. Licensee: Trustees of Leland Stanford Jr. University. Population served: 13,000 Law Firm: Crowell & Moring. Format: Progressive, educ. Target aud: 13-plus; independent-thinking individuals who value unique programming. ♦Kyle Wulff, gen mgr; Ben Levitti, progmg dir; Anthony Sanchez, news dir; Lisa Dornell, pub affrs dir; Mark Lawrence, chief of engrg.

Stockton

KHHK(FM)—See Modesto

KJOY(FM)— June 15, 1968: 99.3 mhz; 2.35 kw. 330 ft TL: N38 01 21 W121 16 03. Stereo. Hrs opn: 24 4643 Quail Lakes Dr., Suite 100, 95207-1833. Secondary address: 1581 Cummins Dr. #135, Modesto 95358. Phone: (209) 476-1230. Fax: (209) 957-1833. Web Site: www.993kjoy.com. Licensee: Citadel Broadcasting Co. Group owner: Citadel Broadcasting Corp. (acq 5-9-03; grpsl). Format: Adult contemp. Target aud: 25-54. ♦Roy Williams, gen mgr.

KMIX(FM)—See Tracy

KQOD(FM)— Jan 24, 1980: 100.1 mhz; 6 kw. 285 ft TL: N38 01 21 W121 16 03. Stereo. Hrs opn: 2121 Lancey Dr., Modesto, 95355. Phone: (209) 551-1306. Fax: (209) 551-1359. Web Site: www.mega100online.com. Licensee: Capstar TX L.P. Group owner: Clear Channel Communications Inc. (acq 11-18-99). Population served: 450,000 Format: Oldies. Target aud: 25-54. ♦Greg Granger, gen mgr & mktg mgr; Leslie Davisson, prom dir; D. Ferrevia, progmg dir; Kacie Marshall, traf mgr.

KSTN(AM)— November 1949: 1420 khz; 5 kw-D, 1 kw-N, DA-2. TL: N37 55 32 W121 14 44. Hrs open: 24 2171 Ralph Ave., 95206. Phone: (209) 948-5786. Licensee: San Joaquin Broadcasting Co. Population served: 3,400,000 Format: Oldies. News staff: one; News: 20 hrs wkly. Target aud: 18-40. Spec prog: Farm 3 hrs, relg 5 hrs wkly. ♦Knox LaRue, pres, stn mgr & progmg mgr; John Hampton, mus dir.

KSTN-FM— 1962: 107.3 mhz; 8.1 kw. 1,610 ft TL: N37 49 17 W121 46 49.7,000,000 Format: Sp. Target aud: General. Spec prog: Sp, Por 4 hrs wkly. ♦Julio Barrios, progmg dir & disc jockey; Lupe Esquer, disc jockey; Paul Shinn, chief of engrg & disc jockey.

***KUOP(FM)**— Sept 22, 1947: 91.3 mhz; 7 kw. 1,220 ft TL: N37 28 48 W121 21 02. Stereo. Hrs opn: 24 7055 Folsom Blvd., Sacramento, 95826. Phone: (916) 278-8900. Fax: (916) 278-8989. E-mail: npr@csus.edu Web Site: www.capradio.org. Licensee: University of the Pacific. Population served: 1,100,000 Natl. Network: NPR, PRI. Law Firm: Dow, Lohnes & Albertson. Format: News, info, class. News staff: one; News: 90 hrs wkly. Target aud: General; NPR listeners, eg. professionals, educators, administrators. ♦John Brenneise, opns dir; Cheryl Dring, progmg dir & mus dir; Joe Barr, news dir, local news ed, edit dir, political ed & relg ed; Jeff Browne, chief of engrg; Michael Frost, traf mgr.

***KWG(AM)**— Nov 22, 1921: 1230 khz; 900 w-U. TL: N37 57 34 W121 15 28. Hrs open: 2280 E. Weber Ave., 95205-5051. Phone: (209) 462-8307. Web Site: www.ihradio.org. Licensee: IHR Educational Broadcasting (group owner; acq 10-18-99; $441,227). Population served: 400,000 Wire Svc: Dow Jones News Service Format: Catholic/relg. Target aud: 25-54. ♦Joseph Nesta, gen mgr & stn mgr; Dale Harry, chief of engrg.

KWIN(FM)—See Lodi

KWSX(AM)— 1947: 1280 khz; 1 kw-U, DA-N. TL: N37 58 55 W121 13 44. Stereo. Hrs opn: 24 2121 Lancey Dr., Modesto, 95355. Phone: (209) 551-1306. Fax: (209) 551-1359. Web Site: www.kfiv-am.clearchannel.com. Licensee: Capstar TX L.P. Group owner: Clear Channel Communications Inc. (acq 8-30-2000; grpsl). Population served: 471,600 Format: News/talk. News staff: 3; News: 20 hrs wkly. Target aud: 25-64. ♦Gary Granger, gen mgr; Bill Mick, progmg dir.

***KYCC(FM)**— Feb 24, 1975: 90.1 mhz; 26 kw. Ant 230 ft TL: N37 57 10 W121 17 11. Stereo. Hrs opn: 9019 N. West Ln., 95210. Phone: (209) 477-3690. Fax: (209) 477-2762. E-mail: kycc@kycc.org Web Site: www.kycc.org. Licensee: Your Christian Companion Network Inc. (acq 7-20-98). Population served: 20000 Natl. Network: USA. Law Firm: Cohn & Marks. Format: Gospel, inspirational, adult contemp. Target aud: 35-55. Spec prog: Black 6 hrs, health one hr wkly. ♦Shirley Garner, exec VP & gen mgr; Adam Biddell, opns mgr.

Sun City

KXFG(FM)— March 1997: 92.9 mhz; 6 kw. 328 ft TL: N33 35 36 W117 08 50. Hrs open:
Rebroadcasts KFRG(FM) San Bernardino 100%.
900 E. Washington St., Suite 315, Colton, 92324. Phone: (909) 825-9525. Fax: (909) 825-0441. Web Site: www.kfrog.com. Licensee: Infinity Radio Inc. Group owner: Infinity Broadcasting Corp. (acq 11-13-98; grpsl). Format: Country. ♦Tom Hoyt, gen mgr; Lee Douglas, opns mgr.

Sunnyvale

KCNL(FM)— January 1961: 104.9 mhz; 6 kw. Ant -154 ft TL: N37 19 23 W121 45 15. (CP: Ant -79 ft. TL: N37 19 22 W121 45 15). Stereo. Hrs opn: 1420 Koll Cir., Suite A, San Jose, 95112. Phone: (408) 453-5400. Fax: (408) 452-1330. E-mail: johnallers@channel1049.com Web Site: www.channel1049.com. Licensee: CC Licenses LLC. Group owner: Clear Channel Communications Inc. (acq 2-2-2004; grpsl). Format: Alternative rock. Target aud: 18-49. ♦Kim Bryant, gen mgr; John Bassanelli, gen sls mgr; Michael Solari, prom dir; Jeanine Calhoun, progmg dir; Fred Reiss, news dir; David Williams, chief of engrg.

Susanville

KHJQ(FM)— May 25, 1983: 92.3 mhz; 9.5 kw. Ant 1,102 ft TL: N40 27 13 W120 34 14. Stereo. Hrs opn: 24 3015 Johnstonville Rd., 96130. Phone: (530) 257-2121. Fax: (530) 257-6955. Web Site: www.theradionetwork.com. Licensee: Sierra Broadcasting Corp. (group owner; acq 1-7-97; $50,000). Format: Hot Adult contemp. Spec prog: Class 5 hrs wkly. ♦Rodney Chambers, gen mgr.

***KJAR(FM)**— 2006: 88.1 mhz; 30 w. Ant 1,141 ft TL: N40 27 12 W120 34 13. Hrs open: 2351 Sunset Blvd., Suite 170-218, Rocklin, 95765. Phone: (916) 251-1600. Fax: (916) 251-1650. Licensee: Educational Media Foundation. (acq 11-1-2006; grpsl). ♦Richard Jenkins, pres.

KJDX(FM)—Listing follows KSUE(AM).

KLZN(FM)— 2006: 96.3 mhz; 1.5 kw. Ant -528 ft TL: N40 26 36 W120 38 35. Hrs open: 2100 Main St., Suite A, 96130. Phone: (530) 257-6100. Licensee: Gary Katz. ♦Gary Katz, pres; Dennis Carlson, gen mgr.

KSUE(AM)— Apr 22, 1948: 1240 khz; 1 kw-U. TL: N40 23 43 W120 37 32. Hrs open: 3015 Johnstonville Rd., 96130. Phone: (530) 257-2121. Fax: (530) 257-6955. Web Site: www.theradionetwork.com. Licensee: Sierra Broadcasting Corp. (group owner) Population served: 45,000 Law Firm: Pepper & Corazzini. Format: News/talk. Target aud: 35-54. Spec prog: Relg 3 hrs wkly. ♦Rod Chambers, pres & gen mgr; Scott Blackwood, opns dir & progmg dir; Mike Smith, news dir; Kristin Volberg, pub affrs dir; Mike Martindale, chief of engrg.

KJDX(FM)—Co-owned with KSUE(AM). Aug 19, 1976: 93.3 mhz; 100 kw. 1,155 ft TL: N40 27 13 W120 34 14. Stereo. Web Site: www.theradionetwork.com. Format: Main stream country. Target aud: 25-64.

Sutter

***KXJS(FM)**— 2004: 88.7 mhz; 550 w. Ant 1,978 ft TL: N39 12 20 W121 49 10. Hrs open: Capital Public Radio Inc., 7055 Folsom Blvd., Sacramento, 95826. Phone: (916) 278-8900. Fax: (916) 278-8989. E-mail: npr@csus.edu Web Site: www.csus.edu/npr. Licensee: California State University, Sacramento. Law Firm: Duane Morris, LLP. Format: Jazz, news & info. ♦Carl Watanabe, stn mgr & progmg mgr; John Brenneise, opns mgr; Joe Barr, news dir, local news ed, edit dir & political ed; Jeff Browne, engrg dir; Michael Frost, traf mgr.

Sutter Creek

KLMG(FM)—See Jackson

Taft

KBDS(FM)— June 1986: 103.9 mhz; 6 kw. Ant 328 ft TL: N35 07 04 W119 27 33. Stereo. Hrs opn: 24 6313 Schirra Ct., Bakersfield, 93313. Phone: (661) 837-0745. Fax: (661) 837-1612. E-mail: achavez@campesina.com Web Site: www.play1039.com. Licensee: Radio Campesina Bakersfield Inc. (acq 1994; $135,000 plus assumption of debt valued at $283,000 with co-located AM). Population served: 300,000 Law Firm: Borsari & Paxson. Format: CHR/rythmic. Target aud: 18-34; Hispanic. ♦Anthony Chavez, gen mgr; Robert Chavez, progmg dir; Maria Vrrutia, traf mgr; Dave Whitehead, engr.

KEAL(FM)—Not on air, target date: unknown: 106.5 mhz; 6 kw. Ant 313 ft TL: N35 07 04 W119 27 33. Hrs open: 200 S. A St., Suite 400, Oxnard, 93030. Phone: (805) 240-2070. Licensee: Lazer Licenses LLC. (acq 6-11-2007; $3.85 million with KXTT(FM) Maricopa). ♦Neal Robinson, pres.

Tahoe City

***KKTO(FM)**— Oct 3, 1997: 90.5 mhz; 38 kw vert. 2,939 ft TL: N39 18 38 W119 53 01. Stereo. Hrs opn: 24 7055 Folsom Blvd., Sacramento, 95826. Phone: (916) 278-8900. Fax: (916) 278-8989. E-mail: npr@csus.edu Web Site: www.capradio.org. Licensee: California State University, Sacramento. Population served: 500,000 Natl. Network: NPR, PRI. Law Firm: Duane Morris LLP. Format: Class, info, news. News staff: 4; News: 90 hrs wkly. Target aud: General; NPR listeners, eg. professionals, educators, administrators. ♦Rick Eytcheson, gen mgr; Carl Watanabe, stn mgr, progmg dir & progmg mgr; John Brenneise, opns mgr; Arla Gibson, dev dir & mktg dir; Michael Frost, prom dir & traf mgr; Linda Onstad, adv dir; Cheryl Dring, mus dir; Joe Barr, news dir; Jeff Browne, engrg dir.

KLCA(FM)—Licensed to Tahoe City. See Reno NV

Tehachapi

KKZQ(FM)— 2001: 100.1 mhz; 340 w. Ant 620 ft TL: N35 04 30 W118 22 07. Hrs open: 570 East Ave. Q-9, Palmdale, 93550. Phone: (661) 947-3107. Fax: (661) 272-5688. Web Site: www.edge100.com. Licensee: High Desert Broadcasting LLC (group owner). Format: Alternative modern rock. Target aud: 18-49. ♦Gary Wilson, opns mgr; Jeff McElfresh, mktg dir & prom dir; Nelson Rasse, gen mgr & progmg dir; Amir Raheem, news dir.

KTPI-FM— Jan 8, 1982: 103.1 mhz; 1.9 kw. Ant 577 ft TL: N35 04 30 W118 22 08. Stereo. Hrs opn: 24 352 East Ave K-4, Lancaster, 93535. Phone: (661) 942-1121. Fax: (661) 723-5512. Web Site: www.ktpi.com. Licensee: CC Licenses LLC. Group owner: Clear Channel Communications Inc. (acq 11-21-2003; grpsl). Population served: 400,000 Natl. Rep: Christal. Law Firm: Latham & Watkins. Format: Country. News staff: one; News: 2 hrs wkly. Target aud: 25-54. ♦Larry Thornhill, gen mgr & sls dir; Mark Mitchell, opns mgr & progmg dir; Shaun Palmeri, gen sls mgr.

***KYLU(FM)**— 2006: 88.7 mhz; 140 w. Ant 3,693 ft TL: N35 27 10 W118 35 25. Hrs open:
Rebroadcasts KLVR(FM) Santa Rosa 100%.
2351 Sunset Blvd., Suite 170-218, Rocklin, 95765. Phone: (916) 251-1600. Fax: (916) 251-1650. Web Site: www.klove.com. Licensee: Educational Media Foundation. Natl. Network: K-Love. Format: Contemp Christian. ♦Richard Jenkins, pres; Mike Novak, VP; Keith Whipple, dev dir; David Pierce, progmg mgr; Ed Lenane, news dir; Sam Wallington, engrg dir; Karen Johnson, news rptr; Marya Morgan, news rptr; Richard Hunt, news rptr.

Temecula

KMYT(FM)— 2000: 94.5 mhz; 320 w. 771 ft TL: N33 28 51 W117 10 58. Hrs open: 24
Rebroadcasts KOGO(AM) San Diego.
27349 Jefferson Ave., Suite 116, 92590. Phone: (951) 296-9050. Fax: (951) 296-9077. E-mail: michaeldellinger@clearchannel.com Web Site: www.kmyt945.com. Licensee: CC Licenses LLC. Group owner: Clear Channel Communications Inc. (acq 6-11-2001; $4.5 million including five-year noncompete agreement). Format: Smooth jazz. ♦Bob Ridzak, gen mgr; Robyn Bedessem, gen sls mgr; Mike Dellinger, prom dir; Allen Keppler, mus dir; Rich Mena, chief of engrg.

***KRTM(FM)**— Jan 1, 1989: 88.9 mhz; 1.15 kw. 453 ft TL: N33 27 59 W117 08 29. Hrs open: 24 39405 Murrieta Hot Springs Rd., Murrieta, 92563. Phone: (951) 696-0774. E-mail: steve@krtmradio.com Web Site: www.krtmradio.com. Licensee: Penfold Communications Inc. (acq 6-11-98; $234,788). Format: Christian. Target aud: 25-54. ♦Chuck Smith, pres; Jeff Smith, VP; Steve Bessette, gen mgr.

KTMQ(FM)— 2001: 103.3 mhz; 1.25 kw. Ant 715 ft TL: N33 28 51 W117 10 58. Hrs open: 27349 Jefferson Ave., Suite 116, 92590. Phone: (951) 296-9050. Fax: (951) 296-9077. E-mail: michaeldellinger@clearchannel.com Web Site: www.q1033.com. Licensee: CC Licenses LLC. Group owner: Clear Channel Communications Inc. (acq 7-31-2001; $6.225 million). Law Firm: Hogan and Hartson. Format: Classic rock. ♦Bob Ridzak, gen mgr; Mike Dellinger, prom dir.

Templeton

KXDZ(FM)— 2004: 100.5 mhz; 1.35 kw. Ant 361 ft TL: N35 30 19 W120 37 18. Hrs open: 396 Buckley Rd., Suite 2, San Luis Obispo, 93401. Phone: (805) 786-2570. Fax: (805) 547-9860. Web Site: www.mapletoncommunications.com. Licensee: Mapleton Communications LLC (group owner; (acq 5-23-2002;. grpsl). Format: Classic Hits. ♦Adam Nathanson, pres; Bill Heirendt, gen mgr; Drew Ross, progmg dir; David Atwood, news dir; Tom Hughes, chief of engrg.

Thousand Oaks

***KCLU(FM)**— Oct 20, 1994: 88.3 mhz; 3.2 kw. Ant 518 ft TL: N34 13 05 W118 56 42. Stereo. Hrs opn: 24 60 W. Olsen Rd., Suite 4400, 91360. Phone: (805) 493-3900. Fax: (805) 493-3982. E-mail: kclu@clunet.edu Web Site: www.kclu.org. Licensee: California Lutheran University. Population served: 610,000 Natl. Network: NPR, PRI. Law Firm: Leventhal, Senter & Lerman. Wire Svc: AP Format: Jazz, news/talk, educ. News staff: one; News: 36 hrs wkly. Target aud: General. Spec prog: Blues 5 hrs wkly. ♦Mary Olson, stn mgr.

***KDSC(FM)**— Dec 4, 1979: 91.1 mhz; 4.9 kw. 1,280 ft TL: N34 24 47 W119 11 10. Stereo. Hrs opn: 24
Rebroadcasts KUSC 100%.
Box 77913, Los Angeles, 90007. Secondary address: 515 S. Figueroa St., Suite 2050, Los Angeles 90071. Phone: (213) 225-7400. Fax: (213) 225-7410. E-mail: kusc@kusc.org Web Site: www.kusc.org. Licensee: University of Southern California (acq 3-17-82). Natl. Network: PRI, NPR. Format: Class. News: 3 hrs wkly. Target aud: 35 plus. ♦Brenda Barnes, pres; Eric DeWeese, gen mgr; Janet McIntyre, dev dir.

KHJL(FM)— Apr 1, 1963: 92.7 mhz; 3.1 kw. Ant 462 ft TL: N34 12 21 W118 49 04. Stereo. Hrs opn: 24 99 Long Court, Suite 200, 91360. Phone: (805) 497-8511. Fax: (805) 497-8514. Web Site: www.927jillfm.com. Licensee: Amaturo Group of L.A. Ltd. Group owner: Amaturo Group Ltd. (acq 1996; $2 million). Population served: 700,000 Rgnl. Network: Metronews Radio Net. Law Firm: Pepper & Corazzini. Format: Adult contemp. News staff: one. Target aud: 25-54; employed professional adults, especially women. ♦Joseph Amaturo, CEO & progmg dir; Robert J. Christy, gen mgr.

Thousand Palms

KFUT(AM)— Dec 7, 1963: 1270 khz; 5 kw-D, 750 w-N, DA-2. TL: N33 51 04 W116 23 36. Hrs open: 24 1321 N. Gene Autry Trail, Palm Springs, 92262. Phone: (760) 322-7890. Fax: (760) 322-5493. Web Site: www.desertfun.com. Licensee: MCC Radio LLC. Group owner: Morris Radio LLC (acq 12-24-97; $2.25 million with KDGL(FM) Yucca Valley). Population served: 1,500,000 Format: Talk. News staff: 3; News: 20 hrs wkly. Target aud: 25 plus. ♦William S. Morris IV, chmn; William S. Morris III, pres; Darrell Fry, CFO; Michael Ostehaut, VP; Keith Martin, gen mgr; Larry Snider, opns dir.

KLOB(FM)— Apr 21, 1994: 94.7 mhz; 1.8 kw. 606 ft TL: N33 52 07 W116 25 58. Hrs open: 41601 Corporate Way, Palm Desert, 92260-1986. Phone: (760) 341-5837. Fax: (760) 341-0951. E-mail: klobtraffic@entravision.com Web Site: www.entravision.com. Licensee: Entravision Holdings LLC. Group owner: Entravision Communications Corp. (acq 2-27-97). Format: Adult latin contemporary. ♦Philip Wilkinson, pres; Ray Nieves, gen mgr & gen sls mgr; Grace Escobar, prom dir; Edgar Pineda, progmg dir; Martha Saldana, news dir; Sergio De la Torre, chief of engrg.

KXPS(AM)— Nov 14, 1992: 1010 khz; 3.6 kw-D, 400 w-N, DA-2. TL: N33 50 35 W116 25 39. Stereo. Hrs opn: 24 1321 N. Gene Autry Tr., Palm Springs, 92262. Phone: (760) 322-7890. Fax: (760) 322-5493. Web Site: www.1010kxps.com. Licensee: Morris Communications Corp. Group owner: Morris Communications Inc. (acq 12-24-97; $2.25 million with KDGL(FM) Yucca Valley). Law Firm: Haley, Bader & Potts. Format: Talk, sports. Spec prog: Relg 17 hrs wkly. ♦William Morris, CEO; Michael Ostehaut, VP; Keith Martin, gen mgr; Larry Snider, chief of opns & chief of engrg.

Tipton

KCRZ(FM)— 1997: 104.9 mhz; 2.3 kw. 528 ft TL: N36 10 07 W119 15 04. Stereo. Hrs opn: 24 1401 W. Caldwell Ave., Visalia, 93277. Phone: (559) 553-1500. Fax: (559) 627-1496. Web Site: www.z1049.com. Licensee: Lemoore Wireless Co. Inc. Natl. Network: ABC. Format: Hot adult contemp. Target aud: 25-64. ♦Wayne B. Foster, gen mgr; Randy Hendrix, progmg dir.

Torrance

KFOX(AM)— January 1998: 1650 khz; 10 kw-D, 490 w-N. TL: N33 53 30 W118 11 03 (D), N33 53 30 W118 11 03 (N). Hrs open: 4525 Wilshire Blvd., 3rd Fl., Los Angeles, 90010. Phone: (323) 935-0606. Fax: (323) 935-8885. Web Site: www.koreatimes.com. Licensee: Chagal Communications Inc. (acq 5-25-00; $30 million). Format: Adult contemp, Korean. ♦Grant Chang, gen mgr.

Tracy

KMIX(FM)— Dec 14, 1966: 100.9 mhz; 6 kw. 328 ft TL: N37 37 32 W121 23 58. Stereo. Hrs opn: 6820 Pacific Ave., Suite 3A, Stockton, 95207. Phone: (209) 474-0154. Fax: (209) 474-0316. Web Site: www.lavuena.com. Licensee: Entravision Holdings LLC. Group owner: Entravision Communications Corp. (acq 7-28-00; grpsl). Format: Sp. ♦Lisa Sunday, gen mgr; Cesar Medina, stn mgr.

***KYKL(FM)**— 2004: 90.7 mhz; 210 w. Ant 1,745 ft TL: N37 33 37 W121 36 19. Stereo. Hrs opn: 24 2351 Sunset Blvd., Suite 170-218, Rocklin, 95765. Phone: (916) 251-1600. Fax: (916) 251-1650. E-mail: klove@klove.com Web Site: www.klove.com. Licensee: Educational Media Foundation. Group owner: EMF Broadcasting. Natl. Network: K-Love. Law Firm: Shaw Pittman. Format: Contemp Christian. News staff: 3. Target aud: 25-44; Judeo Christian, female. ♦Richard Jenkins, pres; Mike Novak, VP; Keith Whipple, dev dir; David Pierce, progmg mgr; Ed Lenane, news dir; Sam Wallington, engrg dir; Karen Johnson, news rptr; Marya Morgan, news rptr; Richard Hunt, news rptr.

Truckee

KTKE(FM)—Not on air, target date: unknown: 101.5 mhz; 140 w. Ant 1,988 ft TL: N39 14 29 W120 08 20. Hrs open: 24 Truckster Broadcasting Inc., 2307 Princess Anne St., Greensboro, NC, 27408. Phone: (530) 587-9999. Phone: (530) 587-9330. Fax: (530) 587-9119. E-mail: ktkeradio@yahoo.com Licensee: Todd Robinson, Inc Format: AAA. ♦Dan von Enoo, gen mgr; Justin Wright, progmg dir.

Tulare

KBOS-FM—Licensed to Tulare. See Fresno

KGEN(AM)— 1957: 1370 khz; 1 kw-D, 136 w-N. TL: N36 10 51 W119 19 44. Hrs open: 24 Box 2040, 93275. Secondary address: 323 E. San Joaquin Ave. 93274. Phone: (559) 686-1370. Fax: (559) 685-1394. Licensee: Azteca Broadcasting Corp. (group owner) Population served: 276,700 Format: Sp, Mexican. Target aud: General. ♦Margaretia Hernandez, gen mgr.

KJUG(AM)—Licensed to Tulare. See Visalia

KJUG-FM—Licensed to Tulare. See Visalia

Tulelake

KFLS-FM— July 23, 1993: 96.5 mhz; 20 kw. 2,155 ft TL: N42 05 50 W121 37 59. Stereo. Hrs opn: 24 Box 1450, Klamath Falls, OR, 97601. Secondary address: 1338 Oregon Ave., Klamath Falls, OR 97601. Phone: (541) 882-4656. Fax: (541) 884-2845. E-mail: traffic@klamathradio.com Web Site: www.klamathradio.com. Licensee: Wynne Enterprises LLC (group owner). Population served: 60,000 Rgnl rep: Tacher. Format: Country. Target aud: 18-49. ♦Robert Wynne, CEO, chmn, pres & gen mgr; Leslie Hougan, gen sls mgr; Randy Adams, progmg dir; Lyle Ahrens, news dir; Russ Jump, chief of engrg; Carol Fritch, traf mgr.

Turlock

***KBDG(FM)**— January 1977: 90.9 mhz; 150 w. 94 ft TL: N37 29 59 W120 49 41. (CP: 780 w). Hrs opn: 24 Box 192, 95381. Secondary address: 1600 E. Canal Dr. 95380. Phone: (209) 668-7176. Fax: (209) 668-2322. Licensee: Assyrian American Civic Club. (acq 1-7-94; $17,000; FTR: 1-31-94). Format: Assyrian music, talk. ♦Zaya Sargis, stn mgr.

***KCSS(FM)**— Aug 13, 1975: 91.9 mhz; 400 w. 112 ft TL: N37 31 35 W120 51 25. Hrs open: 20 801 W. Monte Vista Ave., 95382. Phone: (209) 667-3378 (office). Phone: (209) 667-3900 (stn). Fax: (209) 667-3901. Web Site: www.kcss.net. Licensee: California State University, Stanislaus. Population served: 700,000 Format: Div. Target aud: 18-54. Spec prog: Class 9 hrs, jazz 4 hrs, Americana 10 hrs, wkly. ♦Greg Jacquay, gen mgr.

KLOC(AM)— October 1949: 1390 khz; 5 kw-U, DA-2. TL: N37 31 48 W120 41 37. Hrs open: 4043 Geer Rd., Hughson, 95326. Phone: (209) 883-8760. Fax: (209) 883-8769. E-mail: ngomez@lafavorita.net Web Site: www.lafavorita.net. Licensee: La Favorita Broadcasting Inc. (acq 5-16-03; $500,000). Population served: 2,000,000. Format: Sp. ♦Nelson Gomez, gen mgr; Saul Fiallo, progmg dir.

KWNN(FM)— Mar 3, 1978: 98.3 mhz; 1.6 kw. 390 ft TL: N37 34 46 W120 50 48. (CP: 2 kw). Stereo. Hrs opn: 24 1581 Cummins Dr., Suite 100, Modesto, 95358. Phone: (209) 476-1230. Fax: (209) 957-1833. Licensee: Citadel Broadcasting Co. Group owner: Citadel Broadcasting Corp. (acq 12-12-03). Population served: 61,712 Natl. Rep: Christal. Format: Contemp hit/Top-40. ♦Roy Williams, gen mgr; Jean Western, sls VP & gen sls mgr.

Twain Harte

KKBN(FM)— Oct 19, 1985: 93.5 mhz; 400 w. Ant 1,262 ft TL: N38 00 30 W120 21 44. Stereo. Hrs opn: 24 342 S. Washington St., Sonora, 95370. Phone: (209) 533-1450. Fax: (209) 533-9520. E-mail: lenglandcbc@mlode.com Web Site: www.kkbn.com. Licensee: Clarke Broadcasting Corp. (group owner; acq 3-1-00; $2.2 million). Population served: 500,000 Law Firm: Leventhal, Senter & Lerman. Wire Svc: AP Format: Country. News staff: 3; News: 4 hrs wkly. Target aud: 25-54; general. ♦H. Randolph Holder Jr., pres; Larry England, gen mgr.

Twentynine Palms

KCDZ(FM)— July 15, 1989: 107.7 mhz; 7 kw. 328 ft TL: N34 09 15 W116 11 50. Stereo. Hrs opn: 24 6448 Hallee, Suite 5, Joshua Tree, 92252. Phone: (760) 366-8471. Fax: (760) 366-2976. E-mail: z107@cci-29palms.com Web Site: www.kcdzfm.com. Licensee: Morongo Basin Broadcasting Corp. Natl. Network: ABC. Law Firm: Richard S. Becker & Associates. Format: Adult contemp / CHR. News staff: 2; News: 10 hrs wkly. Target aud: 25-54; baby boomers. ♦Cynthia M. Daigneault, pres & gen mgr; Gary Daigneault, exec VP & VP; Patrick Michaels, progmg dir; Ken Brown, chief of engrg.

KNWH(AM)— Apr 3, 1961: 1250 khz; 1 kw-D, 105 w-N. TL: N34 08 11 W116 10 07. (CP: COL Yucca Valley. 800 w-D, 77 w-N. TL: N34 07 51 W116 22 12). Hrs opn:
Rebroadcasts KNWQ(AM) Palm Springs 100%.
1321 N. Gene Autry Trail, Palm Springs, 92262. Phone: (760) 322-7890. Fax: (760) 322-5493. Web Site: www.desertfun.com. Licensee: MCC Radio LLC. (acq 1-12-2005; $100,000). Population served: 90,000 Format: News/talk. News staff: 4; News: 4.5 hrs. wkly. Target aud: 25-54. ♦William S. Morris IV, pres.

KXCM(FM)— Apr 1, 1965: 96.3 mhz; 6 kw. Ant 243 ft TL: N34 09 15 W116 11 50. Stereo. Hrs opn: 24 Box 1437, Joshua Tree, 92252. Phone: (760) 362-4264. E-mail: coppermountainbroadcasting@yahoo.com Licensee: Copper Mountain Broadcasting Co. (acq 7-14-2004; $575,000 with KQCM(FM) Joshua Tree). Natl. Network: Westwood One, Jones Radio Networks. Law Firm: Leventhal, Senter & Lerman. Format: Country. ♦Gary DeMaroney, gen mgr.

Ukiah

***KPRA(FM)**— May 1987: 89.5 mhz; 1.6 kw. 1,135 ft TL: N39 07 01 W123 13 54. Hrs open: 24 Family Stations Inc., 4135 Northgate Blvd., Sacramento, 95834. Phone: (916) 641-8191. Fax: (916) 641-8238. Licensee: Family Stations Inc. (group owner; acq 2-3-86). Format: Relg.

KQPM(FM)— February 1989: 105.9 mhz; 2.9 kw. 2,017 ft TL: N39 09 00 W123 12 30. Hrs open: 24 140 N. Main St., Lakeport, 95453. Phone: (707) 263-6113. Phone: (707) 468-5336. Fax: (707) 263-0939. Licensee: Bicoastal Media L.L.C. (group owner; acq 7-28-99; grpsl). Format: Country. ♦Ken Dennis, CEO; Mike Wilson, pres & gen mgr; Eric Patrick, opns mgr & progmg dir; Alan Mathews, gen sls mgr; Kevin Mostyn, chief of engrg.

KUKI(AM)— Oct 1, 1950: 1400 khz; 1 kw-U. TL: N39 10 03 W123 13 02. Hrs open: 24 1400 KUKI Ln., 95482. Phone: (707) 263-6113. Fax: (707) 466-5852. E-mail: kukiinfo@bicoastalmedia.com Licensee: Bicoastal Media LLC. Group owner: Moon Broadcasting (acq 7-28-2006; grpsl). Population served: 70,000 Law Firm: Pepper & Corazzini. Format: Mexican/rgnl. News staff: 2; News: 25 hrs wkly. Target aud: 25 plus; upwardly mobile adults. ♦Tove Sorensen, opns dir, chief of opns, progmg dir & mus dir.

KUKI-FM— Oct 16, 1974: 103.3 mhz; 2.8 kw. Ant 1,791 ft TL: N39 19 36 W123 16 12. Stereo. Web Site: www.kukifm.com.180,000 Natl. Network: ABC. Format: Country. News staff: one; News: 7 hrs wkly. Target aud: 25-54.

***KULV(FM)**— Sept 22, 2003: 97.1 mhz; 130 w. Ant 1,978 ft TL: N39 07 50 W123 04 32. Stereo. Hrs opn: 24 2351 Sunset Blvd., Suite 170-218, Rocklin, 95765. Phone: (916) 251-1600. Fax: (916) 251-1650. E-mail: klove@klove.com Web Site: www.klove.com. Licensee: Educational Media Foundation. Group owner: EMF Broadcasting. Natl. Network: K-Love. Law Firm: Shaw Pittman. Format: Contemp Chrisitan. News staff: 3. Target aud: 25-44; Judeo Christian, female. ♦Richard Jenkins, pres; Mike Novak, VP; Keith Whipple, dev dir; Eric Allen, natl sls mgr; David Pierce, progmg dir; Ed Lenane, news dir; Sam Wallington, engrg dir; Karen Johnson, news rptr; Marya Morgan, news rptr; Richard Hunt, news rptr.

KWNE(FM)— 1968: 94.5 mhz; 2.2 kw. 1,965 ft TL: N39 07 50 W123 04 32. Stereo. Hrs opn: 24 Box 1056, 95482. Secondary address: 1100 Hastings Rd., Suite B 95482. Phone: (707) 462-1451. Phone: (707) 462-0945. Fax: (707) 462-4670. E-mail: kwine@kwine.com Web Site: www.kwine.com. Licensee: Broadcasting Corp of Mendocino County. (acq 10-1-78). Population served: 150,000 Law Firm: Borsari & Paxson. Format: Hot adult contemp. News staff: one; News: 12 hrs wkly. Target aud: 18-54; young adult. Spec prog: Sp 4 hrs, farm one hr wkly. ♦Guilford Dye, pres & gen mgr; Gudrun Dye, VP; Mike Spencer, stn mgr.

Vacaville

KUIC(FM)— Nov 1, 1968: 95.3 mhz; 4.3 kw. 280 ft TL: N38 17 56 W121 59 54. (CP: 594 w, ant 1,948 ft. TL: N38 23 48 W122 06 03). Stereo. Hrs opn: 24 KUIC Plaza, 600 E. Main St., 95688. Phone: (707) 446-0200. Fax: (707) 446-0122. Web Site: www.kuic.com. Licensee: KUIC Inc. (acq 10-6-98). Population served: 200,000 Law Firm: Garvey, Schubert & Barer. Format: Adult contemp. News staff: 3; News: one hr wkly. Target aud: General; middle class, professionals. ♦Jim Levitt, CEO; John Levitt, pres, CFO & gen mgr.

Vallejo

KDIA(AM)— Mar 19, 1996: 1640 khz; 10 kw-D, 1 kw-N. TL: N38 07 04 W122 15 24. Stereo. Hrs opn: 24 3260 Blume Dr., Richmond, 95806. Phone: (510) 222-4242. Fax: (510) 262-9054. E-mail: andy.santamaria@kdia.com Web Site: www.kdia.com. Licensee: Baybridge Communications L.L.C. Population served: 2,500,000 Format: Teaching Ministries. News: 5 hrs wkly. Target aud: 25-54. Spec prog: Relg 5 hrs, Black 2 hrs, gospel 7 hrs wkly. ♦Andy Santamaria, gen mgr.

KDYA(AM)— Aug 1, 1947: 1190 khz; 1 kw-D. TL: N38 07 04 W122 15 24. (Also 1640 khz; 10 kw-D, 1 kw-N). Stereo. Hrs opn: 24 3260 Blume Dr., Richmond, 95806. Phone: (510) 222-4242. Fax: (510) 262-9054. E-mail: andy.santamaria@gospel1190.net Web Site: www.gospel1190.net. Licensee: Baybridge Communications L.L.C. (acq 1-29-99). Population served: 2,500,000 Format: Gospel. News: 5 hrs wkly. Target aud: 25-54. Spec prog: Relg 5 hrs, Black 2 hrs, gospel 7 hrs wkly. ♦Andy Santamaria, pres & gen mgr; Clifford Brown, opns mgr.

Ventura

KBBY-FM— Dec 27, 1962: 95.1 mhz; 10.8 kw. 925 ft TL: N34 14 12 W119 12 11. Hrs open: 1376 Walters St., 93003. Phone: (805) 642-8595. Fax: (805) 656-5838. Web Site: www.b951.com. Licensee: Cumulus Licensing Corp. Group owner: Cumulus Media Inc. (acq 9-22-00; grpsl). Natl. Network: Westwood One. Format: Hot Adult Contemp. Target aud: 18-54. ♦Gail Furillo, gen mgr; Tom Watson, opns mgr; Todd Violet, progmg dir.

KCAQ(FM)—See Oxnard

KHAY(FM)—Listing follows KVEN(AM).

KKZZ(AM)— Oct 15, 1994: 1590 khz; 5 kw-U, DA-2. TL: N34 14 12 W119 12 11. Hrs open: 2284 S. Victoria, Suite 2G, 93003. Phone: (805) 289-1400. Fax: (805) 644-7906. Web Site: www.1590kkzz.com. Licensee: Gold Coast Broadcasting LLC. (group owner; (acq 2-10-97; $2 million with KFYV(FM) Ojai). Population served: 750,000 Format: Adult standards. Target aud: 35 plus. ♦Chip Ehrhardt, gen mgr; Mark Elliott, progmg dir.

KLJR-FM—See Santa Paula

KOCP(FM)—See Camarillo

KSSC(FM)— November 1989: 107.1 mhz; 280 w. 872 ft TL: N34 18 10 W119 13 45. (CP: 420 w). Stereo. Hrs opn: 24 5700 Wilshire Blvd., Suite 250, Los Angeles, 90036. Phone: (805) 648-2807. Fax: (323) 900-6200. Web Site: www.1071superstriella.com. Licensee: Entravision Holdings LLC. Group owner: Entravision Communications Corp. (acq 4-1-03; grpsl). Format: Contemp Sp. ♦Karl Meyer, gen mgr.

KUNX(AM)—See Santa Paula

KVEN(AM)— March 1948: 1450 khz; 1 kw-U. TL: N34 15 39 W119 14 28. Hrs open: 1376 Walter St., 93003. Phone: (805) 642-8595. Fax: (805) 656-5838. Web Site: www.kven.com. Licensee: Cumulus Licensing Corp. Group owner: Cumulus Media Inc. (acq 9-22-00; grpsl). Population served: 67,000 Law Firm: Erwin Krasnow. Format: Hits of the 50s & 60s. Target aud: 25 plus; affluent, educated, professional with above average income. ♦Gail Furillo, gen mgr; Ernie Bingham, gen sls mgr; Tammy Meyers, prom mgr; Lee Marshall, progmg dir; Cyndy Abarre, news dir; J.D. Strahler, chief of engrg.

KHAY(FM)—Co-owned with KVEN(AM). Jan 1, 1962: 100.7 mhz; 39 kw. 1,210 ft TL: N34 20 55 W119 19 57. Stereo. Phone: (805) 642-8595. Fax: (805) 656-5838. Web Site: www.khay.com.800,000 Format: Country. Target aud: 18-54. ♦Tom Watson, progmg dir.

KVTA(AM)—See Port Hueneme

Victorville

KATJ-FM—See George

***KHMS(FM)**— Jan 3, 1993: 88.5 mhz; 200 w. 1,512 ft TL: N34 36 40 W117 17 20. Stereo. Hrs opn: 24
Rebroadcasts KSOS(FM) Las Vegas 100%.
c/o Faith Communications Corp., 2201 S. 6th St., Las Vegas, NV, 89104. Phone: (702) 731-5452. Fax: (702) 731-1992. Web Site: www.sosradio.net. Licensee: Faith Communications Corp. (acq 4-5-91; FTR: 4-22-91). Law Firm: Cohn & Marks. Format: Adult contemp, Christian. Target aud: 25-44; young families. ♦Jack French, CEO; Brad Staley, gen mgr; Chris Staley, progmg mgr.

KIXW(AM)—See Apple Valley

KRSX(AM)— Sept 1, 1961: 1590 khz; 500 w-D, 135 w-N. TL: N34 32 15 W117 18 42. Hrs open: 24 15700 Village Dr., Suite A, 92394. Phone: (760) 243-7903. Fax: (760) 243-7183. Licensee: Rudex Broadcasting Limited Corp. (acq 3-19-2004; $176,005). Format: Sp Catholic. Target aud: 18-34 & 14-57; Primary 18-34 Secondary 14-57. ♦John Cooper, pres; Dino Mercado, gen sls mgr.

KVFG(FM)— Aug 18, 1980: 103.1 mhz; 95 w. 1,424 ft TL: N34 36 45 W117 17 31. (CP: 310 w, ant 1,401 ft.). Stereo. Hrs opn: 24 11920 Hesperia Rd., Hesperia, 92345. Phone: (760) 244-2000. Fax: (760) 244-1198. Web Site: www.kfrog103.com. Licensee: CBS Radio Station Inc. Group owner: Infinity Broadcasting Corp. (acq 7-19-00; $3,537,500 with KRAK(AM) Hesperia). Population served: 200,000 Natl. Network: ABC. Law Firm: Fleischman & Walsh. Format: Country. Target aud: 25-54. ♦Bill Pettus, stn mgr; Tom Hoyt, gen mgr & opns mgr.

***KXRD(FM)**— Oct 18, 1994: 89.5 mhz; 1.25 kw. Ant 1,410 ft TL: N34 36 44 W117 17 27. Hrs open: 24
Rebroadcasts KLRD(FM) Yucaipa 100%.
2351 Sunset Blvd., Suite 170-218, Rocklin, 95765. Phone: (700) 528-9236. Fax: (700) 528-9246. Web Site: www.air1.com. Licensee: Educational Media Foundation. Group owner: EMF Broadcasting (acq 1-22-99). Natl. Network: Air 1. Format: Contemp Christian. News: 7 hrs wkly. Target aud: 18-34. ♦Richard Jenkins, pres; Mike Novak, VP; Keith Whipple, dev dir; David Pierce, progmg mgr; Ed Lenane, news dir; Sam Wallington, engrg dir; Karen Johnson, news rptr; Marya Morgan, news rptr; Richard Hunt, news rptr.

KZXY-FM—See Apple Valley

Visalia

***KARM(FM)**— 1990: 89.7 mhz; 1 kw. 810 ft TL: N36 38 10 W118 56 32. Hrs open: 1300 S. Woodland Dr., 93277. Phone: (559) 627-5276. Fax: (559) 627-5288. E-mail: karm@karm.com Web Site: www.karm.com. Licensee: Harvest Broadcasting Co. Natl. Network: ABC. Format: Inspirational, Christian. ♦Dr. Richard Dunn, chmn; Loren Olson, gen mgr.

***KDUV(FM)**— Jan 1, 1992: 88.9 mhz; 1 kw. 2,647 ft TL: N36 17 14 W118 50 17. Hrs open: 130 N. Kelsey, Suite H-1, 93291. Phone: (559) 651-4111. Fax: (559) 651-4115. Web Site: www.kduvfm.com. Licensee: Community Educational Broadcasting Inc. Format: Christian hit radio. ♦Bob Croft, gen mgr.

KEZL(AM)— January 1948: 1400 khz; 1 kw-U. TL: N36 21 14 W119 17 02. Hrs open: 83 E. Shaw Ave, Suite 150, Fresno, 93710-7616. Phone: (559) 230-4300. Fax: (209) 591-1130. Licensee: Capstar TX

L.P. Group owner: Clear Channel Communications Inc. (acq 8-30-2000; grpsl). Natl. Rep: McGavren Guild. Format: Sports. Target aud: 25-64; general. ♦Tony Rinaldi, gen sls mgr; Brian Noe, progmg dir; Kristine Kelley, news dir; Michelle Howe, traf mgr.

KFSO-FM—Co-owned with KEZL(AM). Sept 1, 1951: 92.9 mhz; 18.5 kw horiz, 17 kw vert. 820 ft TL: N36 38 10 W118 56 33. (CP: 17.5 kw, ant 853 ft. TL: N36 38 10 W118 56 34). Stereo. Web Site: www.kfso.com.1,300,000 Format: Oldies. ♦Jeff Negrete, gen mgr; Humberto Avila, sls; Paul Wilson, progmg.

KJUG(AM)—Tulare, Aug 1, 1946: 1270 khz; 5 kw-D, 1 kw-N, DA-N. TL: N36 13 10 W119 18 51. Hrs open: 24 1401 W. Caldwell Ave., 93277. Phone: (559) 553-1500. Fax: (559) 627-1496. Web Site: www.kjugam.com. Licensee: Westcoast Broadcasting Inc. (acq 5-1-81). Population served: 351,300 Format: Classic country. News staff: one; News: 7 hrs wkly. Target aud: 25-49. Spec prog: Farm 5 hrs wkly. ♦Wayne Foster, sls dir; Dave Daniels, prom dir & progmg dir; Darrin Cantrell, news dir; Jamie Moore, traf mgr.

KJUG-FM— May 6, 1965: 106.7 mhz; 1.2 kw. 6,100 ft TL: N36 17 08 W118 50 17. Stereo. Web Site: www.kjug.com. Format: Today's country. Target aud: 18-54.

KSEQ(FM)— October 1984: 97.1 mhz; 17 kw. 777 ft TL: N36 38 08 W118 56 32. Stereo. Hrs open: 617 W. Tulare Ave., 93277. Phone: (559) 627-9710. Fax: (559) 627-1590. E-mail: raym@q97.com Web Site: www.q97.com. Licensee: Buckley Broadcasting of Monterey. Group owner: Buckley Broadcasting Corp. (acq 12-87). Natl. Rep: D & R Radio. Format: CHR. Target aud: 18-49. ♦Rick Buckley, pres; Ray McCarty, VP & gen mgr; Tommy Del Rio, opns mgr.

KSLK(FM)— Nov 22, 1994: 96.1 mhz; 4.8 kw. Ant 360 ft TL: N36 21 59 W119 10 46. Stereo. Hrs opn: 24 7797 N. 1st St., Suite 102, Fresno, 93720. Phone: (559) 635-0961. Fax: (559) 635-0964. Licensee: New Visalia Broadcasting Inc. Population served: 1,000,000 Format: All sports. ♦Robert Eurich, pres.

Vista

KCEO(AM)— Nov 3, 1967: 1000 khz; 2.5 kw-D, 250 w-N, DA-2. TL: N33 13 59 W117 16 09. Hrs open: 24 1835 Aston Ave., Carlsbad, 92008. Phone: (760) 729-1000. Fax: (760) 476-9604. Web Site: www.kceoradio.com. Licensee: North County Broadcasting Corp. Group owner: Astor Broadcast Group (acq 4-30-97; $2.6 million). Population served: 1,000,000 Natl. Network: Westwood One. Format: Talk. News: 20 hrs wkly. Target aud: 35 plus. ♦Arthur Astor, pres; Susan E. Burke, exec VP; Rick Roome, opns dir & opns mgr.

Walnut

***KSAK(FM)**— Jan 10, 1974: 90.1 mhz; 3.5 w. 460 ft TL: N34 02 53 W117 51 43. (CP: Ant 410 ft.). Hrs opn: 24 1100 N. Grand Ave., 91789. Phone: (909) 594-5611, EXT. 4678. E-mail: ksak@mtsac.edu Web Site: www.ksak.com. Licensee: Mount San Antonio Community College District. Population served: 10,000 Wire Svc: UPI Format: Urban rhythmic, Christian. Target aud: 18-25; students. ♦Cason Smith, gen mgr & opns mgr.

Walnut Creek

KKDV(FM)—Dec 10, 1959: 92.1 mhz; 3 kw. Ant 89 ft TL: N37 53 59 W122 05 38. Stereo. Hrs open: 1660 Olympic Blvd., Suite 215, 94596. Phone: (925) 944-6300. Fax: (925) 977-9684. E-mail: singram@kkdv.com Web Site: www.kkdv.com. Licensee: Contra Costa County Radio Inc. (acq 7-29-2005; $7 million). Population served: 200,000 Rgnl rep: Lotus. Format: Adult contemp. Target aud: Adults; 25-54. ♦John Levitt, gen mgr; Phil DeAngelo, gen sls mgr & natl sls mgr; Scott Ingram, rgnl sls mgr; Jim Hampton, progmg dir.

Wasco

***KFHL(FM)**— 2005: 91.7 mhz; 6 kw. Ant 289 ft TL: N35 24 55 W119 14 01. Hrs open: Hillcrest Seventh-day Adventist Church, 2801 Bernard St., Bakersfield, 93306. Phone: (661) 872-0030. Licensee: Mary V. Harris Foundation. Format: Christian talk. ♦Robin Wade, progmg mgr.

Wasco-Greenacres

KERI(AM)— May 17, 1950: 1180 khz; 50 kw-D, 10 kw-N, DA-2. TL: N35 34 17 W119 19 26. Hrs open: 24 P.O. Box 2700, Bakersfield, 93303. Secondary address: 1400 Easton Dr., Suite 144, Bakersfield 93309. Phone: (661) 328-1410. Fax: (661) 328-0873. E-mail: keri@keri.com Web Site: www.keri.com. Licensee: AGM California. Group owner: American General Media (acq 9-27-2004; $1.83 million). Population served: 5,600,000 Format: Religious talk and teaching. ♦Mr. Roger Fessler, gen mgr; Ms. Toni Snyder, sls dir; Mr. Chris Squires, sls & progmg.

Weaverville

KHRD(FM)— 2000: 103.1 mhz; 600 w. Ant 3,592 ft TL: N40 36 10 W122 38 58. Hrs open: 24 1588 Charles Dr., Redding, 96003. Phone: (530) 244-9700. Fax: (530) 244-9707. Web Site: www.red1031.com. Licensee: Results Radio of Redding Licensee LLC. Group owner: Fritz Communications Inc. (acq 6-11-99; grpsl). Format: Classic rock. ♦Beth Tappan, gen mgr; Laurie Curto, gen sls mgr; Rob Reid, progmg dir & news dir; Bryant Smith, chief of engrg.

KWCA(FM)—Not on air, target date: unknown: 101.1 mhz; 250 w. Ant 1,555 ft TL: N40 41 05 W122 44 56. Hrs open: 203 Center St. Phone: (530) 623-2600. Fax: (530) 623-2600. Licensee: George S. Flinn Jr. Format: AAA. ♦George S. Flinn Jr., gen mgr.

Weed

KNTK(FM)— November 1983: 102.3 mhz; 5.5 kw. 1,437 ft TL: N41 21 12 W122 15 35. Stereo. Hrs opn: 24 1934 S. Mt. Shasta Blvd., Mount Shasta, 96094. Phone: (530) 926-5946. Fax: (530) 926-0830. E-mail: kntk@sbcglobal.net Licensee: Four Rivers Broadcasting Inc. (group owner; (acq 7-21-2005; grpsl). Format: News/talk. News staff: one; News: 6 hrs wkly. Target aud: 25-54. Spec prog: Nostalgia 2 hrs wkly. ♦John Anthony, gen mgr & gen sls mgr; Rick Martin, progmg dir.

West Covina

KALI(AM)— Sept 25, 1963: 900 khz; 500 w-D, DA. TL: N34 01 54 W117 56 06. Hrs open: 24 747 E. Green St., Suite 400, Pasadena, 91101. Phone: (626) 844-8882. Fax: (626) 844-0156. Web Site: www.mrbi.net. Licensee: Multicultural Radio Broadcasting Licensee LLC. Group owner: Multicultural Radio Broadcasting Inc. (acq 10-5-98; $9 million). Population served: 688,108 Format: Sp Christian. ♦Arthur S. Liu, pres; David Sweeney, VP, gen mgr, opns VP & opns VP; Alan Mok, progmg dir.

KRCV(FM)— Nov 18, 1957: 98.3 mhz; 2.3 w. 328 ft TL: N34 01 22 W117 56 15. (CP: 650 w, ant 971 ft.). Stereo. Hrs opn: 24 655 N Central Ave, Suite 2500, Glendale, 91203-1422. Phone: (818) 500-4500. Fax: (818) 500-4560. Licensee: HBC License Corp. Group owner: Univision Radio (acq 9-22-2003; grpsl). Format: Sp oldies. News: 3 hrs wkly. Target aud: 18-49; Sp speaking Hispanics, primarily of Mexican origin. ♦Jim Coronado, gen sls mgr; Amalia Gonzalez, progmg dir; Tom Koza, chief of engrg.

West Sacramento

KSMH(AM)— February 1999: 1620 khz; 10 kw-D, 1 kw-N. TL: N38 35 17 W121 28 05. Hrs open: 24 2280 E. Weber Ave., Stockton, 95205-5051. Phone: (209) 462-8307. Web Site: www.ihradio.org. Licensee: IHR Educational Broadcasting. (group owner; (acq 4-28-99; $475,000 with KAHI(AM) Auburn). Format: Relg/Catholic. ♦Joseph Nesta, stn mgr.

Westwood

KTOR(FM)— 2003: 99.7 mhz; 90 w. Ant 2,483 ft TL: N40 14 21 W121 01 52. Hrs open: 24 Box 2371, Chico, 95927. Phone: (530) 256-2400. Fax: (530) 256-3780. E-mail: ktor@frontiernet.net Licensee: Sierra Radio Inc. (acq 9-4-2002; for 51% of CP). Format: Classic rock. ♦Eileen Majors, gen mgr; Greg Heller, opns mgr & traf mgr.

Williams

***KARA(FM)**— Oct 28, 2003: 99.1 mhz; 900 w. Ant 108 ft TL: N39 08 07 W122 07 58. Stereo. Hrs opn: 24 2351 Sunset Blvd., Suite 170-218, Rocklin, 95765. Phone: (916) 251-1600. Fax: (916) 251-1650. E-mail: klove@klove.com Web Site: www.klove.com. Licensee: Educational Media Foundation. Group owner: EMF Broadcasting. Natl. Network: Air 1. Law Firm: Shaw Pittman. Format: Contemp Christian. News staff: 3. Target aud: 18-35; Judeo-Christian, female. ♦Richard Jenkins, pres; Mike Novak, VP; Keith Whipple, dev dir; David Pierce, progmg mgr; Ed Lenane, news dir; Sam Wallington, engrg dir; Arthur Vassar, traf mgr.

Willits

KLLK(AM)— Aug 5, 1985: 1250 khz; 5.4 kw-D, 2.7 kw-N, DA-2. TL: N39 23 58 W123 19 20. Hrs open: 5 AM-midnight
Rebroadcasts KUKI(AM) Ukiah 100%.
1400 Kuki Ln., Ukiah, 95482. Phone: (707) 263-6113. Fax: (707) 466-5852. E-mail: kukiinfo@bicoastalmedia.com Licensee: Bicoastal Media LLC. Group owner: Moon Broadcasting (acq 7-28-2006; grpsl). Population served: 50,000 Law Firm: Haley, Bader & Potts. Format: Regional Mexican. News: 8 hrs wkly. Target aud: 18-49; adults who like a progsv mix of modern rock mus. ♦Tove Sorensen, opns dir.

KMKX(FM)— Feb 19, 2000: 93.5 mhz; 89 kw. 2,873 ft TL: N39 30 59 W123 05 21. Stereo. Hrs opn: 24 Box 1056, Ukiah, 95482. Phone: (707) 462-1483. Phone: (707) 459-6629. Fax: (707) 462-4670. Web Site: www.maxrocks.com. Licensee: Radio Millennium L L C (acq 2-3-00). Population served: 150,000 Natl. Network: Westwood One. Law Firm: Borsari & Paxson. Format: Adult rock. News staff: one; News: 1 hr wkly. Target aud: 18-60; adults. ♦Guilford Dye, pres & gen mgr; Gudrun Dye, VP; Mike Spencer, stn mgr.

***KZYZ(FM)**— 1995: 91.5 mhz; 600 w. 1,820 ft Stereo. Hrs opn: 18
Rebroadcasts KZYX(FM) Philo 100%.
Box 1, Philo, 95466. Phone: (707) 895-2324. Fax: (707) 895-2451. E-mail: kzyx@pacific.net Web Site: www.kzyx.org. Licensee: Mendocino County Public Broadcasting. Format: News, talk radio, div music. ♦Mitchell Holman, gen mgr; Burton Segall, opns dir; Mary Aigner, dev dir & sls dir.

Willows

KCHC(FM)—Not on air, target date: unknown: 106.3 mhz; 6 kw. Ant 328 ft TL: N39 29 30 W121 56 51. Hrs open: Pacific Spanish Network Inc., 296 H St., 2nd Fl., Chula Vista, 91910. Phone: (858) 279-9844. Licensee: Pacific Spanish Network Inc.

KIQS(AM)— Dec 29, 1961: 1560 khz; 250 w-D. TL: N39 31 44 W122 10 09. Hrs open: 1564 Arlington Ct., Turlock, 95382. Phone: (209) 277-8433. Fax: (209) 430-2733. E-mail: avianhelpprogramcoordinator@yahoo.com Licensee: Radio Pan de Vida LLC Group owner: Huth Broadcasting (acq 1-14-2005; $400,000). Population served: 22,500 Law Firm: Law office of Dennis J. Kelly. Format: Sp Christian. ♦Martin Alberto Godinez, gen mgr.

Windsor

KJOR(FM)— June 20, 1997: 104.1 mhz; 250 w. 1,105 ft TL: N38 32 24 W122 57 39. Stereo. Hrs opn: 24 6640 Redwood Dr. #202, Rohnert Park, 94928. Phone: (707) 584-1058. Fax: (707) 584-7944. Licensee: Lazer Licenses LLC. Group owner: Fritz Communications Inc. (acq 6-29-2006; $6.85 million with KSRT(FM) Cloverdale). Population served: 400,000 Natl. Rep: Christal. Law Firm: Covington & Burling. Format: Hot adult contemp. Target aud: 25-54. ♦Neysa Hinton, gen mgr.

Winton

KLOQ-FM— 1994: 98.7 mhz; 6 kw. 246 ft TL: N37 16 42 W120 37 33. (CP: Ant 298 ft. TL: N37 16 41 W120 37 35). Hrs opn: 24 1020 W. Main St., Merced, 95340. Phone: (209) 723-2191. Fax: (209) 383-2950. E-mail: ynavarro@radiomerced.com Licensee: Mapleton Communications LLC (group owner; acq 6-1-2002; grpsl). Population served: 210,000 Law Firm: Leventhal, Senter & Lerman. Format: Mexican rgnl. Target aud: 25-49; Hispanic. ♦Kelly Leonard, gen mgr & opns mgr.

Woodlake

KFRR(FM)— September 1994: 104.1 mhz; 17 kw. 853 ft TL: N36 38 12 W118 56 34. Hrs open: 24 1066 E. Shaw Ave., Fresno, 93710. Phone: (559) 230-0104. Fax: (559) 230-0177. Web Site: 1041fresno.com. Licensee: Wilks License Co.-Fresno LLC. (acq 6-1-2005; grpsl). Natl. Rep: McGavren Guild. Law Firm: Arter & Hadden. Format: Alternative. Target aud: 18-34; young affluent adults. ♦Kevin O'Rorke, gen mgr & gen mgr; Rob Hasson, gen sls mgr; Jason Squires, progmg dir.

Woodland

KSFM(FM)—Licensed to Woodland. See Sacramento

KTKZ(AM)—See Sacramento

Yermo

KRSX-FM— December 1996: 105.3 mhz; 400 w. Ant 1,037 ft TL: N34 48 30 W116 41 01. Hrs open: 24 12370 Hesperia Rd, Suite 16, Victorville, 92395. Phone: (760) 241-1313. Fax: (760) 241-0205. Web Site: www.cruisinoldies1053.com. Licensee: Citicasters Licenses L.P. Group owner: Clear Channel Communications Inc. (acq 5-4-99; grpsl). Population served: 100,000 Law Firm: Pepper & Corazzini. Format: Oldies. News staff: one. Target aud: 18-49; young adults, families. ♦Larry Thornhill, gen mgr; Joe Pagano, progmg dir.

KRXV(FM)— April 1980: 98.1 mhz; 1.1 kw. Ant 2,280 ft TL: N34 59 43 W116 50 15. Stereo. Hrs opn: 24 Box 1668, 1611 E. Main St., Barstow, 92312. Phone: (760) 256-0326. Fax: (760) 256-9507. E-mail: tim@highwayradio.com Web Site: www.thehighwaystations.com. Licensee: KHWY Inc. Natl. Network: AP Radio. Law Firm: Hogan & Hartson. Format: Adult contemp. News staff: one; News: 28 hrs wkly. Target aud: 35 plus; travelers on I-15 & I-40 & loc communities. ♦Howard B. Anderson, CEO & pres; Kirk M. Anderson, exec VP & VP; Timothy B. Anderson, VP; Timothy B. Anderson, gen mgr; Judy Robinson, sls VP & gen sls mgr; Thomas McNeill, mktg mgr & chief of engrg; Lance Todd, progmg dir.

Yreka

***KNYR(FM)—** 1995: 91.3 mhz; 400 w. Ant 2,365 ft TL: N41 36 36 W122 37 26. Hrs open: 5AM-2am Jefferson Public Radio, 1250 Siskiyou Blvd., Ashland, OR, 97520. Phone: (541) 552-6301. Fax: (541) 552-8565. Web Site: www.ijpr.org. Licensee: The State of Oregon, Acting By and Through the State Board of Higher Education for the Benefit of Southern Oregon University. Natl. Network: NPR, PRI. Law Firm: Ernest Sanchez. Format: Class, news. News staff: one; News: 35 hrs wkly. ♦Ronald Kramer, CEO & gen mgr; Bryon Lambert, opns dir; Paul Westhelle, dev dir & mktg dir; Eric Teel, progmg dir & disc jockey; Eric Alan, mus dir & disc jockey; Darin Ransom, engrg dir; Kurt Katzmar, disc jockey; Valerie Ing-Miller, disc jockey.

***KSYC(AM)—** July 27, 1947: 1490 khz; 1 kw-U. TL: N41 43 28 W122 39 00. Hrs open: 24 hrs Jefferson Public Radio, 1250 Siskiyou Blvd., Ashland, OR, 97520. Phone: (541) 552-6301. Fax: (541) 552-8565. Web Site: www.ijpr.org. Licensee: JPR Foundation Inc. (acq 8-8-02; $300,000 with KMJC(AM) Mount Shasta). Population served: 47,000 Natl. Network: NPR, PRI. Law Firm: Ernest Sanchez. News staff: one; News: 35 hrs wkly. ♦Mitchell Christian, CFO; Ronald Kramer, CEO & gen mgr; Bryon Lambert, opns dir; Paul Westhelle, dev dir & mktg dir; Eric Teel, progmg dir; Eric Alan, mus dir; Darin Ransom, engrg dir.

KSYC-FM— June 1, 1983: 103.9 mhz; 10 kw. 2,364 ft TL: N41 43 28 W122 37 46. Stereo. Hrs opn: Box 1729, 96097. Secondary address: 316 Lawrence Ln. 96097. Phone: (530) 842-4158. Fax: (530) 842-7635. E-mail: traffic@ksyc.net Licensee: Four Rivers Broadcasting Inc. (group owner; (acq 7-21-2005; grpsl). Law Firm: Baraff, Koerner & Olender. Format: Country. ♦John Anthony, gen mgr; Andy Eagan, progmg dir.

Yuba City

KMYC(AM)—See Marysville

KOBO(AM)— June 1953: 1450 khz; 500 w-D, 1 kw-N. TL: N39 08 07 W121 36 41. Hrs open: Box 669, Marysville, 95901. Phone: (530) 742-5555. Fax: (530) 741-3758. Licensee: Tom F. Huth. Group owner: Huth Broadcasting (acq 12-17-2003; $200,000). Format: Sp, ethnic. Spec prog: East Indian 3 hrs wkly. ♦Thomas Huth, CEO & gen mgr.

KUBA(AM)— January 1948: 1600 khz; 5 kw-D, 2.5 kw-N, DA-N. TL: N39 06 22 W121 39 18. Hrs open: 24 1479 Sanborn Rd., 95993-6042. Phone: (530) 673-1600. Fax: (530) 673-4768. E-mail: harlan@succeed.net Web Site: www.am1600kuba.com. Licensee: Nevada County Broadcasters Inc. Group owner: Nevada County Broadcasting Inc. (acq 8-9-2004; $500,000). Population served: 160,000 Natl. Network: CBS Radio. Rgnl. Network: Calif. Agri-Radio, Calif. Farm. Wire Svc: AP Format: Classic hits, news/talk. News staff: 2; News: 15 hrs wkly. Target aud: 35-64; community oriented. Spec prog: Farm 4 hrs, gospel 2 hrs, relg 2 hrs wkly. ♦Dave Bear, opns mgr, progmg dir & chief of engrg; John Black, prom dir; Chris Gilbert, news dir; Lucy Spears, pub affrs dir; Robert R. Harlan, gen mgr, sls dir, gen sls mgr, mktg dir & farm dir.

Yucaipa

***KLRD(FM)—** July 15, 1986: 90.1 mhz; 300 w. Ant 1,024 ft TL: N34 02 19 W116 57 09. Hrs open: 24 2351 Sunset Blvd., Suite 170-218, Rocklin, 95765. Phone: (707) 528-9236. E-mail: aramirez@emfbroadcasting.com Licensee: Educational Media Foundation. Group owner: EMF Broadcasting (acq 1-22-99). Population served: 2,500,000 Natl. Network: Air 1. Law Firm: Shaw Pittman. Format: Alternative rock, Christian music, div. News: 7 hrs wkly. Target aud: 18-34; Christians. ♦Richard Jenkins, pres; Mike Novak, VP; Keith Whipple, dev dir; Tanya Bohannon, natl sls mgr; David Pierce, progmg dir; Eric Allen, mus dir; Ed Lenane, news dir; Sam Wallington, engrg dir.

Yucca Valley

KDGL(FM)— August 1988: 106.9 mhz; 4 kw. 1,371 ft TL: N34 04 55 W116 20 32. Stereo. Hrs opn: 24 1321 N. Gene Autry Trail, Palm Springs, 92262. Phone: (760) 322-9890. Fax: (760) 322-5493. Web Site: www.desertfun.com. Licensee: MCC Radio LLC. Group owner: Morris Radio LLC (acq 1998; $2.25 million with KXPS(AM) Thousand Palms). Population served: 250,000 Natl. Network: USA. Format: Classic Hits. Target aud: 25 plus. ♦William Morris III, chmn & pres; Darrel Fry, CFO; Michael Oslehaut, sr VP & VP; Keith Martin, gen mgr; Larry Snider, opns dir.

Colorado

Alamosa

KALQ-FM—Listing follows KGIW(AM).

***KASF(FM)—** 1967: 90.9 mhz; 17 w. 121 ft TL: N37 28 20 W105 52 39. Stereo. Hrs open: 24 Adams State College, 110 Richardson Ave., 81101. Phone: (719) 587-7871. Phone: (719) 587-7872. Fax: (719) 587-7522. Web Site: www.adams.edu. Licensee: Adams State College. Population served: 8,000 Format: Pop contemp hit. News: one hr wkly. Target aud: Community; college and local. Spec prog: Gospel 4 hrs, talk 6 hrs, blues/jazz 6 hrs, reggae 5 hrs wkly. ♦Beth Dussault, gen mgr.

KGIW(AM)— Feb 27, 1929: 1450 khz; 1 kw-U. TL: N37 28 20 W105 51 13. Hrs open: 6 AM-11 PM Box 179, 292 Santa Fe, 81101. Phone: (719) 589-6644. Phone: (719) 589-6645. Fax: (719) 589-0993. E-mail: info@kgiwkalq.com Web Site: www.kgiwkalq.com. Licensee: Community Broadcasting Corp. (acq 1964). Population served: 42,000 Format: Adult contemp. News staff: one; News: 25 hrs wkly. Target aud: 18-60. Spec prog: Sp 6 hrs, farm 6 hrs wkly. ♦Dale K. Burns, pres; Marilyn Burns, exec VP; Neil J. Hammer, gen mgr & progmg dir; Helen Lozoya, gen sls mgr & min affrs dir; Mark Beatty, news dir & local news ed; Will Williams, chief of engrg; Evan Slack, farm dir.

KALQ-FM—Co-owned with KGIW(AM). June 26, 1969: 93.5 mhz; 2.8 kw. 130 ft TL: N37 28 20 W105 51 13. Stereo. 30,000 Format: Country. ♦Willard Williams, opns mgr; Evan Slack, farm dir.

***KRZA(FM)—** Oct 26, 1985: 88.7 mhz; 9.8 kw. 2,076 ft TL: N36 51 32 W106 00 28. Stereo. Hrs opn: 5 AM-midnight 528 9th St., 81101. Phone: (719) 589-8844. Fax: (719) 587-0032. E-mail: psa@krza.org Web Site: www.krza.org. Licensee: Equal Representation of Media Advocacy Corp. Natl. Network: NPR. Format: News, jazz,various. News staff: one; News: 18 hrs wkly. Target aud: General; adult progsv community oriented rural area. Spec prog: Sp 14 hrs, Latin American 3 hrs, news 4 hrs wkly. ♦Kristine Taylor, gen mgr & stn mgr.

Arvada

KDDZ(AM)— June 1998: 1690 khz; 10 kw-D, 1 kw-N. TL: N39 39 21 W105 04 27. Hrs open: 12136 W. Bayaud Ave., Suite 125, Lakewood, 80228. Phone: (303) 783-0880. Fax: (303) 761-1774. Web Site: www.radiodisney.com. Licensee: Radio Disney Group LLC. Group owner: ABC Inc. (acq 7-16-98; $3.5 million with KADZ Arvada). Natl. Network: Radio Disney. Format: Children. ♦Tracy Wells, gen mgr.

Aspen

***KAJX(FM)—** July 7, 1987: 91.5 mhz; 380 w horiz, 370 w vert. Ant -987 ft TL: N39 11 48 W106 48 14. Stereo. Hrs opn: 24 110 E. Hallam St., Suite 134, 81611. Phone: (970) 920-9000. Fax: (970) 544-8002. Web Site: www.kajx.org. Licensee: Roaring Fork Public Radio Inc. (acq 1996). Population served: 65,000 Natl. Network: NPR, PRI. Wire Svc: AP Format: Class, jazz, news. News staff: 3; News: 50 hrs wkly. Target aud: General; Aspen residents & tourists. Spec prog: Bluegrass 2 hrs wkly. ♦Brent Gardner-Smith, gen mgr; Steve Cole, progmg dir; Kirk Siegler, news dir.

KPVW(FM)— 2000: 107.1 mhz; 20.5 kw. Ant 361 ft TL: N39 18 56 W106 57 32. Stereo. Hrs opn: 24 20 Sunset Dr., Suite 6-A, Basalt, 81621. Phone: (970) 927-7600. Fax: (970) 927-8001. E-mail: jeloy@entravision.com Licensee: Entravision Holdings LLC. Group owner: Entravision Communications Corp. (acq 12-13-01; $57,500). Format: Mexican rgnl. Target aud: Latino; 18-34. ♦Philip Wilkinson, COO; Walter Ulloa, CEO & chmn; Jeffery A. Liberman, pres; John DeLorenzo, CFO; Mario Carrera, gen mgr.

KSNO-FM—See Snowmass Village

KSPN-FM— Feb 14, 1970: 103.1 mhz; 3 kw. Ant -85 ft TL: N39 13 33 W106 50 00. Stereo. Hrs opn: 24 Bldg 402 D, Aspen Airport Business Center, 81611. Phone: (970) 925-5776. Fax: (970) 925-1142. E-mail: studio@kspnradio.com Web Site: www.kspnradio.com. Licensee: NRC Mountain Division LLC. (group owner; (acq 7-9-2004; grpsl). Population served: 50,473 Natl. Rep: Christal. Law Firm: Rosenman & Colin. Format: AAA. News staff: 2; News: 6 hrs wkly. Target aud: 25-49; affluent, well educated people who live in resort areas. ♦Colleen Barill, gen mgr; David Bach, chief of opns.

Aurora

KEZW(AM)— 1954: 1430 khz; 5 kw-U, DA-N. TL: N39 12 28 W104 55 46. Stereo. Hrs opn: 24 4700 S. Syracuse St., Suite 1050, Denver, 80237. Phone: (303) 967-2700. Fax: (303) 967-2747. Web Site: KEZW.com. Licensee: Entercom Denver License LLC. Group owner: Entercom Communications Corp. (acq 7-24-02; with KOSI(FM) Denver). Natl. Rep: Katz Radio. Format: Nostalgia, big band, MOR. News: 4 hrs wkly. Target aud: 35 plus; general. ♦Ray Quinn, gen mgr; Jeff Brown, gen sls mgr & rgnl sls mgr; Rick Crandall, progmg dir & progmg mgr; Jeff Garrett, chief of engrg; Stephanie Walrath, traf mgr.

KMXA(AM)—Licensed to Aurora. See Denver

KOSI(FM)—See Denver

Avon

KZYR(FM)— Dec 24, 1984: 97.7 mhz; 15 kw. Ant 440 ft TL: N39 38 05 W106 26 47. Stereo. Hrs opn: 24 275 Main St., Unit 0-201, Edwards, 81632-7812. Phone: (970) 845-8565. Fax: (970) 926-7635. E-mail: tony@kzyr.com Web Site: www.kzyr.com. Licensee: Cool Radio LLC (acq 12-3-2001; $1.5 million with KSNO-FM Snowmass Village). Law Firm: Cole, Raywid & Braverman. Format: AAA. Target aud: 18-54. ♦Thomas Dobrez, CEO & pres; Tony Mauro, gen mgr.

Basalt

KNFO(FM)— July 1995: 106.1 mhz; 2 kw. 364 ft TL: N39 18 55 W106 57 36. Hrs open: 24 402D AABC, Aspen, 81611. Phone: (970) 544-9100. Fax: (970) 544-9101. Licensee: NRC Mountain Division LLC. (group owner; (acq 7-9-2004; grpsl). Population served: 50,473 Natl. Network: CBS. Law Firm: Rosenman & Colin. Format: News, talk, sports. News staff: 2; News: 13 hrs wkly. Target aud: 35-64. ♦Tim Brown, CEO & pres; Dave Rogers, CFO; Colleen Barill, gen mgr; David Bach, chief of opns.

Bayfield

KLJH(FM)— July 2003: 107.1 mhz; 100 kw horiz. Ant 1,870 ft TL: N37 21 49 W107 47 30. (CP: Ant 1,883 ft. TL: N37 21 46 W107 47 40). Hrs opn: Voice Ministries of Farmington Inc., 1105 W. Apache, Farmington,

NM, 87401. Phone: (505) 327-7202. Fax: (505) 327-2163. E-mail: kljh@kljh.org Web Site: www.kljh.org. Licensee: Voice Ministries of Farmington Inc. Format: Praise & worship-Christian. ♦Fareed W. Ayoub, gen mgr.

Bennett

KSYY-FM— 1978: 107.1 mhz; 100 kw. Ant 1,932 ft TL: N39 55 22 W103 58 18. Hrs open: 24 3033 S. Parker Rd., Suite 700, Aurora, 80014. Phone: (303) 872-1500. Fax: (303) 872-1501. Web Site: www.sassy107.com. Licensee: KSIR-FM LLC (acq 10-24-2005; $14 million). Population served: 300,000 Format: Var. ♦Luis G. Nogales, CEO; Steve Keeney, gen mgr; Tim Maranville, opns dir & progmg dir.

Boulder

KBCO-FM— Oct 1, 1955: 97.3 mhz; 100 kw. Ant 1,541 ft TL: N39 54 48 W105 17 32. Stereo. Hrs opn: 2500 Pearl St., Suite 315, 80302. Phone: (303) 444-5600. Fax: (303) 449-3057. E-mail: kbco@kbco.com Web Site: www.kbco.com. Licensee: Citicasters Licenses L.P. Group owner: Clear Channel Communications Inc. Population served: 200,000 Format: AAA. ♦Mark Remington, gen mgr; Kenny Marks, natl sls mgr.

KCFC(AM)— Feb 15, 1947: 1490 khz; 1 kw-U. TL: N40 01 42 W105 15 06. Hrs open: Colorado Public Radio, 7409 S. Alton Ct., Centennial, 80112. Phone: (303) 871-9191. Fax: (303) 733-3319. Web Site: www.cpr.org. Licensee: Public Broadcasting of Colorado Inc. (acq 8-17-01; $1.1 million). Population served: 250,000 Natl. Network: Westwood One. Format: News. Target aud: 30 plus; listeners interested in news, info & entertainment (loc progmg). ♦Max Wycisk, pres & gen mgr; Sue Coughlin, dev VP & reporter.

***KGNU-FM**— May 22, 1978: 88.5 mhz; 1.3 kw. 215 ft TL: N39 59 32 W105 09 10. Stereo. Hrs opn: 24 4700 Walnut St., 80301-2548. Phone: (303) 449-4885. E-mail: marty@kgnu.org Web Site: www.kgnu.org. Licensee: Boulder Community Broadcast Association Inc. Population served: 2,500,000 Natl. Network: PRI, NPR. Rgnl. Network: Colo. Pub. Law Firm: Garvey, Schubert, Barer. Format: Var/div. News staff: one; News: 30 hrs wkly. Target aud: General. Spec prog: Black 7 hrs, folk 20 hrs, Sp 3 hrs, jazz 15 hrs, class 13 hrs wkly. ♦Marty Durlin, gen mgr & stn mgr; Evan Perkins, opns mgr; Faye Lamb, dev dir; John Schaefer, mus dir; Sam Fugua, news dir.

KRCN(AM)—See Longmont/Denver

KRKS-FM— Mar 15, 1971: 94.7 mhz; 100 kw. 984 ft TL: N40 04 19 W105 21 14. (CP: Ant 1,745 ft. TL: N39 40 33 W105 29 07). Stereo. Hrs opn: 24 3131 S. Vaughn Way, Suite 601, Aurora, 80014-3510. Phone: (303) 750-5687. Fax: (303) 696-8063. Web Site: www.krks.com. Licensee: Salem Media of Colorado Inc. Group owner: Salem Communications Corp. (acq 12-15-93; $5 million; FTR: 11-15-93). Population served: 100,000 Format: Relg. ♦Edward Atsinger, pres; Joe Davis, VP; Brian Taylor, gen mgr; Carrie Lakey, gen sls mgr; Ryan Kloberdanz, progmg dir.

KVCU(AM)— Nov 14, 1973: 1190 khz; 5 kw-D. TL: N39 57 54 W105 14 05. Stereo. Hrs opn: 24 Box 207, University of Colorado, 80309. Phone: (303) 492-5031. Fax: (303) 492-1369. Web Site: www.radio1190.org. Licensee: The University of Colorado Foundation. Population served: 228,000 Law Firm: Haley, Bader & Potts. Format: Eclectic, diverse. News staff: 3; News: 3 hrs wkly. Target aud: 25-44. Spec prog: Jazz 3 hrs wkly. ♦Conor Walker, gen mgr; Mike Flanagan, stn mgr.

Breckenridge

KSMT(FM)— Sept 12, 1975: 102.3 mhz; 6 kw. Ant -210 ft TL: N39 29 44 W106 01 44. Stereo. Hrs opn: 24 Box 7069, 80424. Secondary address: 130 Ski Hill Rd., Suite 240 80424. Phone: (970) 453-2234. Fax: (970) 453-5425. E-mail: ksmt@ksmtradio.com Web Site: www.ksmtradio.com. Licensee: NRC Broadcasting Inc. Group owner: American General Media (acq 7-9-2004; grpsl). Format: Triple-A. News staff: one; News: 12 hrs wkly. Target aud: 18-44; upscale adults, heavy ski & outdoor industry consumers. Spec prog: Sp 2 hrs, Reggae 2 hrs wkly. ♦Lisa Korry-Cheek, gen mgr & stn mgr.

Breen

KLLV(AM)—Licensed to Breen. See Durango

Brighton

KLDC(AM)— Apr 26, 1956: 810 khz; 2.2 kw-D, 430 w-N, DA-2. TL: N40 01 41 W104 49 21 (day), N39 50 36 W104 57 14 (night). Hrs open: 2150 W. 29th Ave., Suite 300, Denver, 80211. Phone: (303) 433-5500. Fax: (303) 433-1555. Web Site: www.crawfordbroadcasting.com. Licensee: KLZ Radio Inc. Group owner: Crawford Broadcasting Co. (acq 12-10-93; $700,000; FTR: 1-10-94). Population served: 1,500,000 Format: Gospel. Target aud: 24-55; general. Spec prog: Black 2 hrs wkly. ♦Donald B. Crawford, pres; Mike Triem, gen mgr.

Broomfield

KWLI(FM)— June 1967: 92.5 mhz; 57 kw. Ant 1,237 ft TL: N40 05 47 W104 54 04. Stereo. Hrs open: 24 1560 Broadway, Suite 1100, Denver, 80202. Phone: (303) 832-5665. Fax: (303) 832-7000. Web Site: www.925thewolf.com. Licensee: CBS Radio Stations Inc. Group owner: Infinity Broadcasting Corp. (acq 12-14-2000; grpsl). Format: New country choice. News staff: one; News: 2 hrs wkly. Target aud: 35-64; educated, upscale, active professionals, ethnic. ♦Don Howe, gen mgr & mus dir; Lisa Petrone, gen sls mgr; Barry Walters, chief of engrg.

Brush

***KBWA(FM)**— 2006: 89.1 mhz; 6 kw vert. Ant 145 ft TL: N40 13 02.1 W103 41 46.2. Hrs open:
Rebroadcasts KXWA(FM) Loveland 100%.
Box 64500, Colorado Springs, 80962. Phone: (719) 533-0300. Fax: (719) 278-4339. Web Site: kxwa.wayfm.com. Licensee: WAY-FM Media Group Inc. (acq 2-23-2005; $25,000 for CP). Format: Contemp Christian. ♦Robert D. Augsburg, pres.

KPRB(FM)—Listing follows KSIR(AM).

KSIR(AM)— Aug 1, 1977: 1010 khz; 25 kw-D, 280 w-N. TL: N40 18 50 W103 35 30. Hrs open: 24 Box 917, Fort Morgan, 80701. Secondary address: 220 State St., Suite 106, Fort Morgan 80701. Phone: (970) 867-7271. Fax: (970) 867-2676. E-mail: ksir@necolorado.com Web Site: www.ksir.com. Licensee: Northeast Colorado Broadcasting LLC (group owner; acq 7-1-2003; grpsl). Population served: 2,500,000 Natl. Network: CBS Radio. Wire Svc: AP Format: Talk, agriculture, sports. News staff: 2; News: 5 hrs wkly. Target aud: 25-65; farmers, ranchers, sports fans. ♦Alec Creighton, gen mgr; Brian Allmer, news dir; Lorrie Boyer, progmg dir & farm dir.

KPRB(FM)—Co-owned with KSIR(AM). Nov 2, 1998: 106.3 mhz; 6 kw. 201 ft TL: N40 18 50 W103 35 30. Stereo. 24 E-mail: b106@necolorado.com Web Site: www.b106.com.50,000 Format: Adult contemp. News staff: one; News: one hr wkly. Target aud: 18-45; females. ♦Alec Creighton, CEO & progmg dir.

Buena Vista

KBVC(FM)— Jan 1, 1997: 104.1 mhz; 600 w. Ant 1,187 ft TL: N38 44 45 W106 11 55. Hrs open: 7600 County Rd. 120, Salida, 81201. Phone: (719) 539-2575. Fax: (719) 539-4851. E-mail: kbvc@bresnan.net Web Site: www.kbvcfm.com. Licensee: Three Eagles Communications of Colorado LLC (acq 9-1-2006; swap for KVRH(AM) Salida). Format: Country. ♦Cristy Carothers, gen mgr; Doug Vollertsen, opns mgr, progmg dir & mus dir; Norm Veasman, news dir.

KSKE(AM)— Aug 22, 1986: 1450 khz; 1 kw-U. TL: N38 49 07 W106 09 01. Hrs open:
Simulcasts KKKK(AM) Colorado Springs.
614 Kimbark St., Longmont, 80501. Phone: (303) 776-2323. Fax: (303) 776-1377. E-mail: rnickfive@aol.com Web Site: www.radiocoloradonetwork.com. Licensee: Pilgrim Communications Inc. (acq 12-11-97). Format: Business talk, sports, NASCAR. ♦Gene Hood, pres; Ron Nickell, sr VP & gen mgr; Lee Thompson, gen sls mgr.

Burlington

KNAB(AM)— July 11, 1967: 1140 khz; 1 kw-D. TL: N39 17 28 W102 15 45. Hrs open: Box 516, 17534 County Rd., No. 49, 80807. Phone: (719) 346-8600. Phone: (719) 346-5566. Fax: (719) 346-8656. E-mail: knab@centurytel.net Web Site: www.knabradio.com. Licensee: KNAB Inc. (acq 9-6-91). Population served: 35,100 Law Firm: Fletcher, Heald & Hildreth. Format: Adult standards, farm. Target aud: 18 plus. ♦Bette Bailly, CEO, pres, gen mgr & chief of engrg; Beverly Schott, traf mgr; Bobby Maze, disc jockey.

KNAB-FM— Mar 7, 1980: 104.1 mhz; 50.7 kw. 358 ft TL: N39 17 41 W102 15 37. Stereo. 24 30,000 Format: Country, agriculture. ♦Beverly Schott, traf mgr; Bobby Maze, disc jockey.

KPCR(FM)—Not on air, target date: unknown: 99.3 mhz; 55 kw. Ant 315 ft TL: N39 15 34 W102 24 35. Hrs open: 1951 28th Ave., Unit 29, Greeley, 80634-5755. Phone: (970) 302-8444. Licensee: Youngers Colorado Broadcasting LLC. ♦Kevin J. Youngers, gen mgr.

Canon City

KKCS-FM—Listing follows KRLN(AM).

KRLN(AM)— Aug 15, 1947: 1400 khz; 1 kw-U. TL: N38 27 35 W105 13 26. Hrs open: 24 1615 Central, 81212. Phone: (719) 275-7488. Fax: (719) 275-5132. E-mail: starads@krln.cc Licensee: Royal Gorge Broadcasting LLC. (acq 3-31-2000; $715,000 with co-located FM). Population served: 35,000 Natl. Network: CBS. Format: News/talk. News staff: one; News: 25 hrs wkly. Target aud: 25-54; two income families & older discretionary income. ♦Joan Wood, gen mgr; Rosemary Lamberson, gen sls mgr; Kyle Horne, progmg dir; Melissa Nunn, traf mgr.

KKCS-FM—Co-owned with KRLN(AM). June 1, 1975: 104.5 mhz; 6 kw. Ant 46 ft TL: N38 18 54 W105 12 40. Stereo. 100,000 Format: Country. Target aud: 25-60.

***KTLC(FM)**— May 2001: 89.1 mhz; 1.15 kw. Ant 1,476 ft TL: N38 45 21 W105 13 02. Hrs open: 24
Rebroadcasts KTLF(FM) Colorado Springs 100%.
1665 Briargate Blvd., Suite 100, Colorado Springs, 80920. Phone: (719) 593-0600. Fax: (719) 593-2399. E-mail: lightpraise@ktlf.org Web Site: www.ktlf.org. Licensee: Make a Difference Foundation Inc. (acq 12-27-01). Natl. Network: Salem Radio Network. Format: Christian music. Target aud: 45-60; Christian. ♦Karen Veazey, gen mgr & stn mgr; Lynn Carmichael, progmg dir.

Carbondale

***KCJX(FM)**— Sept 6, 2004: 88.9 mhz; 4 kw horiz, 3.5 kw vert. Ant 2,542 ft TL: N39 25 08 W107 22 10. Hrs open: 110 E. Hallam St., Suite 134, Aspen, 81611. Phone: (970) 925-6445. Fax: (970) 544-8002. Web Site: www.kajx.org. Licensee: Roaring Fork Public Radio Inc. (acq 4-16-2002). Population served: 67,000 Wire Svc: AP Format: News, class, jazz. News staff: 3. ♦Brent Gardner-Smith, gen mgr; Kirk Siegler, news dir.

KUUR(FM)—Not on air, target date: unknown: 96.7 mhz; 90 w. Ant 2,507 ft TL: N39 25 08 W107 22 10. Hrs open: P.O. Box 11657, Aspen, 81612. Secondary address: 132 W. Main St., Aspen 81611. Phone: (970) 920-9600. Fax: (970) 544-5239. E-mail: sales@aspenglenwood.com Web Site: www.aspenglenwood.com. Licensee: Colorado Radio Marketing LLC. Format: Cool contemp. ♦Marcos Rodriguez, gen mgr.

***KVOV(FM)**— Apr 15, 1983: 90.5 mhz; 215 w. Ant 2,798 ft TL: N39 25 35 W107 22 48. Hrs open:
Rebroadcasts KVOD(FM) Denver 100%.
7409 S. Alton Ct., Centennial, 80112. Phone: (303) 871-9191. Fax: (303) 733-3319. Web Site: www.cpr.org. Licensee: Public Broadcasting of Colorado Inc. (acq 10-29-2004; exchange for KDNK(FM) Glenwood Springs). Population served: 45,000 Format: Classical music. ♦Max Wycisk, pres.

Castle Rock

KJMN(FM)— Feb 26, 1978: 92.1 mhz; 42 kw. Ant 535 ft TL: N39 23 07 W105 02 52. Stereo. Hrs opn: 24 777 Grant St., 5th Fl., Denver, 80203. Phone: (303) 721-9210. Fax: (303) 832-3410. Web Site: www.denverspanishradio.com. Licensee: Entravision Holdings LLC. Group owner: Entravision Communications Corp. (acq 3-14-00; grpsl). Population served: 2,000,000 Law Firm: Wiley, Rein & Fielding. Format: Sp. Target aud: 25-54. ♦Mario Carrera, gen mgr.

Colona

KAVP(AM)— Sept 30, 2000: 1450 khz; l kw-U. TL: N38 15 59 W107 51 07. Hrs open: 24
Rebroadcasts KWGL(FM) Ouray 100%.
751 Horizon Ct. , Suite 225, Grand Junction, 81506. Phone: (970) 241-6460. Fax: (970) 241-6452. Web Site: www.wscradio.net. Licensee: WS Communications LLC. Group owner: Western Slope Communications LLC. Population served: 140,000 Format: Legendary Country. ♦John Reid, gen mgr.

***KTMH(FM)**—Not on air, target date: unknown: 89.9 mhz; 4 kw vert. Ant 1,633 ft TL: N38 23 15 W107 40 31. Hrs open:
KTLF (FM) Colorado Springs 100%.
1665 Briargate Blvd., Suite 100, Colorado Springs, 80920. Phone:

(719) 593-0600. Fax: (719) 593-2399. E-mail: lightpraise@ktlf.org Web Site: www.ktlf.org. Licensee: Educational Communications of Colorado Springs Inc. Format: Christian music. Target aud: 40-65; Christian. ♦Karen Veazey, gen mgr; Sharick Wade, opns mgr; John Hayes, mus dir; Harry Russell, chief of engrg.

Colorado Springs

KATC-FM— Oct 1, 1969: 95.1 mhz; 58 kw. Ant 2,280 ft TL: N38 44 43 W104 51 39. Stereo. Hrs opn: 24 6805 Corporate Dr., Suite 130, 80919. Phone: (719) 593-2700. Fax: (719) 593-2727. Web Site: www.katcountry951.com. Licensee: Citadel Broadcasting Co. (acq 8-25-2006; $8.5 million). Population served: 82,000 Format: Country. ♦Bobby Irwin, opns mgr; Jim Miller, progmg dir. Co-owned TV: .

KBIQ(FM)—See Manitou Springs

KCCY(FM)—See Pueblo

KCMN(AM)— Feb 9, 1964: 1530 khz; 15 kw-D, 15 w-N, 1 kw-CH. TL: N38 49 08 W104 46 32. Hrs open: 24 5050 Edison Ave., Suite 218, 80915. Phone: (719) 570-1530. Fax: (719) 570-1007. Web Site: www.1530kcmn.com. Licensee: KLZ Radio Inc. Group owner: Crawford Broadcasting Co. (acq 1999; $750,000 with KCBR(AM) Monument). Population served: 400,000 Natl. Network: CNN Radio, Westwood One. Format: Oldies. Target aud: 40 plus. Spec prog: Relg 3 hrs wkly. ♦Don Crawford Jr., CEO, pres, VP & gen mgr; Tron Simpson, opns mgr.

***KEPC(FM)**— Feb 15, 1957: 89.7 mhz; 10 kw. Ant -256 ft TL: N38 45 41 W104 47 04. Stereo. Hrs opn: 24 5675 S. Academy Blvd., 80906. Phone: (719) 540-7489. Fax: (719) 540-7487. E-mail: kepc@ppcc.edu Web Site: www.ppcc.edu. Licensee: Pikes Peak Community College. Population served: 350,000 Format: Var/div. Target aud: General. ♦Sharon Hogg, gen mgr, stn mgr & pub affrs dir.

KGDQ(FM)— Jan 28, 1967: 101.9 mhz; 50 kw. Ant 492 ft TL: N39 08 11 W104 55 34. Hrs open: 500 Media Place, Sacramento, CA, 95815. Phone: (916) 368-6300. Fax: (916) 368-6334. Web Site: www.bustosmedia.com. Licensee: Bustos Media of Colorado License Corp. (group owner; (acq 9-29-2006; $17.5 million). Format: Rgnl Mexican. ♦Rob Quinn, gen mgr; Randy Lopez, prom dir.

KILO(FM)— Jan 21, 1966: 94.3 mhz; 83 kw. 2,110 ft TL: N38 44 44 W104 51 43. (CP: 94.3 mhz). Stereo. Hrs opn: Box 2080, 80901. Secondary address: 1805 E. Cheyenne Rd. 80906. Phone: (719) 634-4896. Fax: (719) 634-5837. Web Site: www.kilo943.com. Licensee: Bahakel Communications. (group owner; acq 8-14-84). Population served: 334,000 Format: Active Rock. ♦Lou Mellini, gen mgr.

KKFM(FM)— 1958: 98.1 mhz; 72 kw. 2,300 ft TL: N38 44 36 W104 51 44. Stereo. Hrs opn: 24 6805 Corporate Dr., Suite 130, 80919-1977. Phone: (719) 593-2700. Fax: (719) 593-2727. Web Site: www.kkfm.com. Licensee: Citadel Broadcasting Co. Group owner: Citadel Broadcasting Corp. (acq 1-86; $2.5 million; FTR: 8-16-82). Population served: 500,000 Natl. Rep: McGavren Guild. Format: Classic rock. News: one hr wkly. Target aud: 25-54. ♦Farid Suleman, CEO; Judy Ellis, COO; Bobby Irwin, gen mgr, opns mgr & progmg dir.

KKKK(AM)— June 22, 1957: 1580 khz; 10 kw-D. TL: N38 43 11 W104 43 16. Hrs open: 6 AM-6:30 PM (M-F); 6 AM-6 PM (S, Su) 614 Kimbark St., Longmont, 80501. Phone: (303) 776-2323. Fax: (303) 776-1377. E-mail: rnickfive@aol.com Web Site: www.radiocoloradonetwork.com. Licensee: Pilgrim Communications Inc. (acq 6-1-98; $450,000). Population served: 597,000 Format: Business, talk, sports, NASCAR. News staff: one; News: 2 hrs wkly. Target aud: General. ♦Lee Thompson, stn mgr; Lee Thompson, gen sls mgr; Ron Nickell, gen mgr & progmg dir; Ed Dulaney, chief of engrg.

KKLI(FM)—Widefield, Mar 23, 1987: 106.3 mhz; 1.6 kw. 2,224 ft TL: N38 44 41 W104 51 46. Stereo. Hrs opn: 24 2864 S. Circle Dr., Suite 150, 80906. Phone: (719) 540-9200. Fax: (719) 579-0882. Web Site: www.kkli.com. Licensee: Capstar TX L.P. Group owner: Clear Channel Communications Inc. (acq 8-30-00; grpsl). Natl. Rep: Clear Channel. Format: Soft adult contemp. News staff: one; News: 4 hrs wkly. Target aud: 25-54; family-oriented, educated. ♦Bob Richards, gen mgr; Scott Jones, gen sls mgr; Nancia Warren, prom dir; Paul Richards, news dir.

KKML(AM)— Sept 22, 1922: 1300 khz; 5 kw-D, 1 kw-N. TL: N38 48 46 W104 48 51. Hrs open: 24 6805 Corporate Dr., Suite 130, 80919-1977. Phone: (719) 593-2700. Fax: (719) 593-2727. Web Site: www.sportsanimal1300.com. Licensee: Citadel Broadcasting Co. Group owner: Citadel Broadcasting Corp. (acq 1999; grpsl). Population served: 350,000 Natl. Network: ESPN Radio. Natl. Rep: McGavren Guild. Format: Sports. ♦Farid Suleman, CEO; Bobby Irwin, opns mgr.

KKPK(FM)—Listing follows KVOR(AM).

***KRCC(FM)**— Oct 2, 1951: 91.5 mhz; 2.1 kw. 2,103 ft TL: N38 44 43 W104 51 42. Stereo. Hrs opn: 24 912 N. Weber St., 80903. Phone: (719) 473-4801. Fax: (719) 473-7863. E-mail: info@krcc.org Web Site: www.krcc.org. Licensee: The Colorado College. Population served: 500,000 Natl. Network: NPR, PRI. Law Firm: Garvey, Schubert & Barer. Format: Div, news. News: 42 hrs wkly. Target aud: 25-54; general. Spec prog: Celtic 5 hrs, reggae 6 hrs, jazz 15 hrs, blues 5 hrs wkly. ♦Delaney Utterback, gen mgr; Mike Procell, opns mgr; Jeff Bieri, prom dir & progmg dir.

KRDO(AM)— March 1947: 1240 khz; 1 kw-U. TL: N38 49 42 W104 50 15. Hrs open: 24 399 S. 8th St., 80905. Phone: (719) 632-1515. Phone: (719) 473-1240. E-mail: m.lewis@krdotv.com Web Site: www.krdo.com. Licensee: Pikes Peak Radio LLC. (group owner) (acq 6-26-2006; grpsl). Population served: 500,000 Natl. Rep: D & R Radio. Law Firm: Fletcher, Heald & Hildreth. Format: News/talk. ♦David Bradley Jr., pres; Neil O. Klockziem, gen mgr; Ron Mitchell, gen sls mgr & natl sls mgr; Mike Lewis, progmg mgr; J.R. Reed, chief of engrg. Co-owned TV: KRDO-TV affil.

KRDO-FM—See Security

***KTLF(FM)**— Feb 27, 1989: 90.5 mhz; 20 kw. Ant 2,178 ft TL: N38 44 43 W104 51 39. Stereo. Hrs opn: 24 1665 Briargate Blvd., Suite 100, 80920. Phone: (719) 593-0600. Fax: (719) 593-2399. E-mail: lightpraise@ktlf.org Web Site: www.ktlf.org. Licensee: Educational Communications of Colorado Springs Inc. Population served: 1,000,000 Format: Christian music. News: 12 hrs wkly. Target aud: 25-49. ♦Dr. Ron Johnson, chmn; Karen Veazey, gen mgr; Sharick Wade, opns mgr; Lynn Carmichael, progmg dir.

KVOR(AM)— Sept 22, 1922: 740 khz; 3.3 kw-D, 1.5 kw-N, DA-2. TL: N39 05 02 W104 42 41. Stereo. Hrs opn: 6805 Corporate Dr., Suite 130, 80919-1977. Phone: (719) 593-2700. Fax: (719) 593-2727. Web Site: www.kvor.com. Licensee: Citadel Broadcasting Co. Group owner: Citadel Broadcasting Corp. (acq 1999; grpsl). Population served: 250,000 Natl. Network: CBS, Wall Street. Format: News/talk, sports. Target aud: General. ♦Bobby Irwin, gen mgr & opns mgr; Dan Mandis, progmg dir.

KKPK(FM)—Co-owned with KVOR(AM). Feb 1, 1960: 92.9 mhz; 53 kw. Ant 2,130 ft TL: N38 44 44 W104 51 39. Stereo. Web Site: www.929peakfm.com.350,000 Format: Hot adult contemp. Target aud: 25-54. ♦Jim Berry, progmg dir.

KZNT(AM)— Dec 15, 1956: 1460 khz; 5 kw-D, 500 w-N, DA-N. TL: N38 49 36 W104 44 30. Hrs open: 24 7150 Campus Dr., Suite 150, 80920. Phone: (719) 531-5438. Fax: (719) 531-5588. Web Site: www.newstalk1460.com. Licensee: Bison Media Inc. Group owner: Salem Communications Corp. (acq 10-6-03; $1.5 million). Population served: 350,000 Format: News/talk. ♦Henry Tippie, gen mgr.

Commerce City

KLTT(AM)— 1996: 670 khz; 50 kw-D, 1.4 kw-N, DA-2. TL: N39 57 20 W104 43 50. Hrs open: 2150 W. 29th Ave., Suite 300, Denver, 80211. Phone: (303) 433-5500. Fax: (303) 433-1555. Web Site: www.crawfordbroadcasting.com. Licensee: KLZ Radio Inc. Group owner: Crawford Broadcasting Co. (acq 1995; $750,000). Format: Relg, Christian, conservative talk. Target aud: 30 plus; general. ♦Mike Triem, gen mgr.

Cortez

KISZ-FM— Sept 28, 1978: 97.9 mhz; 100 kw. 1,360 ft TL: N37 21 48 W108 09 00. Stereo. Hrs opn: 20 212 W. Apache St., Farmington, NM, 87401-6235. Secondary address: 2402 Hawkins 81321. Phone: (505) 325-3541. Fax: (505) 327-5796. Web Site: www.kisscountry979fm.com. Licensee: Winton Road Broadcasting Co. LLC (group owner; (acq 5-3-01; grpsl). Population served: 250,000 Law Firm: Fleischman & Walsh. Format: Country. News staff: one; News: 4 hrs wkly. Target aud: 18-49; young sophisticated adults. ♦Sara Olsen, gen mgr.

KRTZ(FM)—Listing follows KVFC(AM).

***KSJD(FM)**— July 1990: 91.5 mhz; 1.2 kw. Ant 312 ft TL: N37 28 57 W108 30 34. Stereo. Hrs opn: 24 33057 Hwy. 160, Mancos, 81328. Phone: (970) 564-0808. Fax: (970) 564-0434. E-mail: john@ksjd.org Web Site: www.ksjd.org. Licensee: San Juan Basin Technical School. Format: Diverse public radio. News staff: one. Target aud: 16-30; college level. Spec prog: Relg one hr wkly. ♦Jeff Pope, gen mgr; John Hall, progmg dir.

KVFC(AM)— Feb 27, 1955: 740 khz; 1 kw-D, 250 w-N, DA-N. TL: N37 20 58 W108 32 29. Hrs open: 24 2402 Hawkins, 81321. Phone: (970) 565-6565. Fax: (970) 565-8567. E-mail: feedback@kvfcradio.com Web Site: www.kvfcradio.com. Licensee: Winton Road Broadcasting Co. LLC (group owner; acq 12-18-01; with co-located FM). Population served: 10,000 Natl. Network: ABC, CNN Radio, Westwood One. Format: News/talk. News staff: 2; News: 15 hrs wkly. Target aud: 18-54; young adults. ♦Anthony Brandon, CEO; L. Rogers Brandon, COO; Dan Buchta, gen mgr; Kelly Turner, opns mgr, news dir & pub affrs dir; Keri-Lyn Riley, gen sls mgr; Jim Burt, chief of engrg.

KRTZ(FM)—Co-owned with KVFC(AM). December 1981: 98.7 mhz; 27 kw. 2,900 ft TL: N37 13 10 W108 48 26. Stereo. 24 E-mail: radio@krtzradio.com Web Site: krtzradio.com.20,000 Format: Adult contemp. News staff: one; News: 3 hrs wkly. Target aud: 20-55. Spec prog: Americian Indian one hr, gospel one hr wkly. ♦Desiree Burnham, prom mgr; Kelly Truner, progmg dir; Jim Burt, engrg dir; Kelly Turner, local news ed & sports cmtr.

Craig

***KPYR(FM)**— 2005: 88.3 mhz; 250 w. Ant 889 ft TL: N40 33 50 W107 36 40. Hrs open:
Rebroadcasts KCFR(AM) Denver 100%.
Colorado Public Radio, 7409 S. Alton Ct., Centennial, 80112. Phone: (303) 871-9191. Fax: (303) 733-3319. Web Site: www.cpr.org. Licensee: Public Broadcasting of Colorado Inc. Format: News. ♦Max Wycisk, pres & gen mgr.

KQZR(FM)— Mar 1, 2005: Stn currently dark. 102.5 mhz; 100 kw. Ant 1,243 ft TL: N40 11 45 W107 56 00. Hrs open: Box 772850, Steamboat Springs, 80477. Phone: (970) 879-5368. Fax: (970) 879-5843. Web Site: www.kqzrock.com. Licensee: Wildcat Communications L.L.C. (acq 10-28-2005; $160,000). Format: Classic rock. ♦Steve Wodlinger, gen mgr.

KRAI(AM)— 1948: 550 khz; 5 kw-D, 500 w-N, DA-N. TL: N40 32 45 W107 31 52. Hrs open: 19 Box 65, 81626. Secondary address: 1111 W. Victory Way. 81626. Phone: (970) 824-6574. Fax: (970) 826-4581. E-mail: frank@krai.com Web Site: www.krai.com. Licensee: Wild West Radio Inc. (acq 5-89). Population served: 50,000 Natl. Network: Westwood One, CNN Radio. Wire Svc: AP Format: Country. News staff: 3; News: 12 hrs wkly. Target aud: 25-54. Spec prog: Farm one hr wkly. ♦Frank R. Hanel Jr., pres, gen mgr & chief of engrg.

KRAI-FM— April 1976: 93.7 mhz; 100 kw. 980 ft TL: N40 34 35 W107 36 29. Stereo. 24 Web Site: www.krai.com.40,000 Format: Adult contemp. News staff: 3; News: 4 hrs wkly. Target aud: 18-49.

Crested Butte

***KBUT(FM)**— Dec 20, 1986: 90.3 mhz; 250 w. -667 ft TL: N38 52 19 W106 58 44. Stereo. Hrs opn: 24 Box 308, 81224. Secondary address: 508 Maroon Ave. 81224. Phone: (970) 349-5225. Phone: (970) 349-7444. Fax: (970) 349-6440. E-mail: kbut@kbut.org Web Site: www.kbut.org. Licensee: Crested Butte Mountain Educational Radio Inc. Population served: 2,500 Natl. Network: NPR, PRI. Law Firm: Garvey, Schubert & Barer. Format: Educational, diversified music, news/talk. News staff: one; News: 72 hrs wkly. Target aud: General. ♦Dave Clayton, gen mgr; Kim Carroll-Bosler, stn mgr & dev dir; Erin Roberts, progmg dir; Chad Reich, mus dir.

Delta

KDTA(AM)— Jan 14, 1955: 1400 khz; 1 kw-U. TL: N38 45 38 W108 05 28. Hrs open: 24
Simulcast of KJOL(AM) Grand Junction 100%.
1360 E. Sherwood Dr., Grand Junction, 81501-7575. Phone: (970) 254-5565. Fax: (970) 254-5550. E-mail: info@kjol.org Web Site: www.kjol.org. Licensee: United Ministries. (group owner; (acq 11-16-2004; $88,000). Population served: 30,000 Format: Christian talk, music. ♦Ken Andrews, gen mgr.

KKNN(FM)— December 1985: 95.1 mhz; 100 kw. 969 ft TL: N38 52 40 W108 13 30. Stereo. Hrs opn: 24 315 Kennedy Ave., Grand Junction, 81501. Phone: (970) 242-7788. Fax: (970) 243-0567. Web Site: www.95rock.com. Licensee: Cumulus Licensing Corp. Group owner: Cumulus Media Inc. (acq 1-00). Population served: 225,000 Natl. Rep: Katz Radio. Format: Classic rock. News staff: one; News: 6 hrs wkly. Target aud: 18-49; men. ♦Lewis Dickey Jr., CEO & pres; John Dickey, exec VP; Kevin Wodlinger, gen mgr; Mike Shafer, opns mgr.

***KPRU(FM)**— 2001: 103.3 mhz; 12 kw. Ant 987 ft TL: N38 52 40 W108 13 32. Hrs open: Colorado Public Radio, 7409 S. Alton Ct., Centennial, 80112. Phone: (303) 871-9191. Fax: (303) 733-3319. Web Site: www.cpr.org. Licensee: Public Broadcasting of Colorado Inc. Format: Classical. ♦Max Wycisk, pres & gen mgr; Sue Coughlin, dev VP; Arlene Wayland, prom dir; Sean Nethery, progmg dir; Bob Hensler, chief of engrg; David Gomez, traf mgr.

Denver

KALC(FM)— June 21, 1965: 105.9 mhz; 100 kw. 900 ft TL: N39 43 59 W105 14 12. Stereo. Hrs opn: 4700 S. Syracuse St., Suite 1050, 80237. Phone: (303) 967-2700. Fax: (303) 967-2747. Web Site: www.alice106.com. Licensee: Entercom Denver License LLC. Group owner: Entercom Communications Corp. (acq 5-1-2002; $88 million). Population served: 1,534,800 Natl. Rep: Christal. Format: Hot adult contemp. Target aud: 18-34; women. ♦Ray Quinn, gen mgr; Mikey Goldenberg, progmg dir; Jeff Garrett, chief of engrg; Stephanie Walrath, traf mgr.

KBJD(AM)— 2001: 1650 khz; 10 kw-D, 1 kw-N. TL: N39 47 56 W104 58 12. Hrs open: 24 3131 S. Vaughn Way, Suite 601, Aurora, 80114. Phone: (303) 750-5687. Fax: (303) 696-8063. E-mail: production @salemdenver.com Web Site: www.710knus.com. Licensee: Salem Media of Colorado Inc. Group owner: Salem Communications Corp. Format: Conservative news/talk. News staff: 3; News: 15 hrs wkly. ♦Brian Taylor, gen mgr.

KBNO(AM)— May 15, 1948: 1280 khz; 5 kw-U, DA-2. TL: N39 36 05 W104 58 49. Hrs open: 24 600 Grant St., Suite 600, 80203. Phone: (303) 733-5266. Fax: (303) 733-5242. E-mail: kbno.net@kbno.net Web Site: www.kbno.net. Licensee: Latino Communications LLC (group owner; acq 11-21-00; $3.3 million). Population served: 1,650,000 Format: Sp. News staff: 27; News: 21 hrs wkly. Target aud: 25-54; male. ♦Michael Ferrufino, VP & opns VP; Zee Ferrufino, CEO, CFO & gen mgr.

KBPI(FM)— June 19, 1962: 106.7 mhz; 100 kw. 987 ft TL: N39 43 59 W105 14 12. Stereo. Hrs opn: 24 4695 S. Monaco St., 80237. Phone: (303) 713-8000. Fax: (303) 713-8744. Web Site: www.kbpi.com. Licensee: Citicasters Licenses Inc. (NEW). Group owner: Clear Channel Communications Inc. (acq 5-4-99; grpsl). Population served: 514,678 Natl. Network: ABC. Format: AOR. Target aud: 25-34; men. ♦Lee Larsen, gen mgr & gen sls mgr; Ron Smith, gen mgr & opns mgr; Willie Hung, progmg dir; Karl Schipper, chief of engrg.

***KCFR(AM)**— Mar 4, 1956: 1340 khz; 1 kw-U. TL: N39 39 34 W105 00 44. (CP: TL: N39 41 01 W105 00 25). Hrs opn: 24 7409 S. Alton Ct., Centennial, 80112. Phone: (303) 871-9191. Fax: (303) 733-3319. Web Site: www.cpr.org. Licensee: Public Broadcasting of Colorado Inc. (acq 11-30-00; $4.2 million). Population served: 1,250,000 Format: NPR news. ♦Max Wycisk, pres & gen mgr; Sue Coughlin, dev VP.

KEZW(AM)—See Aurora

***KGNU(AM)**— Jan 1, 1954: 1390 khz; 5 kw-D, DA. TL: N39 39 29 W105 00 49. Stereo. Hrs opn: 24
Simulcast of KGNU-FM, Boulder 95%.
4700 Walnut St., Boulder, 80301. Phone: (303) 449-4885. Web Site: www.kgnu.org. Licensee: Boulder Community Broadcast Association Inc. (acq 11-26-2004; $4.2 million). Format: Eclectic. ♦Evan Perkins, opns dir.

KHOW(AM)— 1925: 630 khz; 5 kw-U, DA-2. TL: N39 54 36 W104 54 50. Stereo. Hrs opn: 24 4695 S. Monaco, 80237. Phone: (303) 713-8000. Fax: (303) 713-8738. Web Site: www.khow.com. Licensee: Citicasters Licenses L.P. Group owner: Clear Channel Communications Inc. (acq 5-4-99; grpsl). Population served: 100,000 Format: Talk. Target aud: 25-54. ♦Lee Larsen, gen mgr; Ron Smith, opns mgr; Jan Whitbeck, prom dir & prom mgr; Kristine Olinger, progmg dir; Jan Chadwell, chief of engrg.

KPTT(FM)—Co-owned with KHOW(AM). Mar 31, 1968: 95.7 mhz; 100 kw. Ant 725 ft TL: N39 43 59 W105 14 10. Stereo. Web Site: www.theparty.com.180,000 Format: Adult contemp 70s, 80s & 90s,. Target aud: General. ♦Joe Bevilacqua, progmg dir.

KIMN(FM)— Aug 1, 1959: 100.3 mhz; 100 kw. 331 ft TL: N39 41 06 W105 04 05. (CP: Ant 1,705 ft. TL: N39 54 48 W105 17 32). Stereo. Hrs opn: 24 1560 Broadway, Suite 1100, 80202. Phone: (303) 832-5665. Fax: (303) 832-7000. Web Site: www.mix100.com. Licensee: Infinity Radio Inc. Group owner: Infinity Broadcasting Corp. (acq 8-24-00; grpsl). Population served: 514,678 Natl. Rep: Christal. Format: Hot adult contemp. Target aud: 35-44; women. Spec prog: Pub affrs 2 hrs wkly. ♦Don Howe, gen mgr.

KKFN(AM)— July 4, 1922: 950 khz; 5 kw-U, DA-1. TL: N39 52 30 W104 56 00. Stereo. Hrs opn: 24 7800 E. Orchard Rd., Suite 400, Greenwood Village, 80111. Phone: (303) 321-0950. Fax: (303) 321-3383. Web Site: www.fan950.com. Licensee: Lincoln Financial Media Co. of Colorado. (group owner) (acq 4-3-2006; grpsl). Population served: 1,700,000 Natl. Network: CBS. Natl. Rep: CBS Radio. Format: All sports, talk. Target aud: 25-54; men. ♦Robert Call, sr VP, VP & gen mgr; Steve Price, sls dir & gen sls mgr; Randy Weidner, natl sls mgr & rgnl sls mgr; Dwayne Taylor, mktg dir & prom dir; Tim Spence, progmg dir; Simone Seikaly, news dir & pub affrs dir; Brad Hart, engrg dir.

KYGO-FM—Co-owned with KKFN(AM). Dec 1, 1953: 98.5 mhz; 100 kw. 1,820 ft TL: N39 40 35 W105 29 09. Stereo. 24 Web Site: www.kygo.com. Format: Country. Target aud: 25-54. ♦Joel Burke, progmg dir; Garrott Doll, mus dir.

KKZN(AM)—See Thornton

KLDC(AM)—See Brighton

KLDV(FM)—See Morrison

KLVZ(AM)— June 5, 1954: 1220 khz; 660 w-D, 11 w-N. TL: N39 41 00 W105 00 24. Hrs open: 2150 W. 29th Ave., Suite 300, 80211. Phone: (303) 433-5500. Fax: (303) 433-1555. Web Site: www.crawfordbroadcasting.com. Licensee: KLZ Radio Inc. Group owner: Crawford Broadcasting Co. (acq 8-11-99; $1.5 million). Population served: 280000 Law Firm: Fisher, Wayland, Cooper, Leader & Zaragoza L.L.P. Format: Radio & Victoria. Target aud: 18-49. ♦Mike Triem, gen mgr.

KLZ(AM)— Mar 10, 1922: 560 khz; 5 kw-U, DA-1. TL: N39 50 36 W104 57 14. Hrs open: 24 2150 W. 29th Ave., Suite 300, 80211. Phone: (303) 433-5500. Fax: (303) 433-1555. E-mail: klzinfo@crawfordbroadcasting.com Web Site: www.crawfordbroadcasting.com. Licensee: KLZ Radio Inc. Group owner: Crawford Broadcasting Co. (acq 6-30-92; $1.5 million; FTR: 7-20-92). Population served: 2,000,000 Format: Christian music. Target aud: 25-54; men. ♦Mike Triem, gen mgr.

KMXA(AM)—Aurora, Sept 12, 1972: 1090 khz; 50 kw-D, 500 w-N, DA-2. TL: N39 39 53 W104 39 24. Stereo. Hrs opn: 777 Grant St., 5th Floor, 80203. Phone: (303) 721-9210. Fax: (303) 832-3410. Web Site: www.denverspanishradio.com. Licensee: Entravision Holdings LLC. Group owner: Entravision Communications Corp. (acq 3-14-2000; grpsl). Format: Sp adult hits. Target aud: 18-54; Hispanics. ♦Mario Carrera, gen mgr.

KNRV(AM)—See Englewood

KNUS(AM)— 1941: 710 khz; 5 kw-U, DA-1. TL: N39 57 19 W104 51 01. Hrs open: 24 3131 S. Vaughn Way, Suite 601, Aurora, 80014. Phone: (303) 750-5687. Fax: (303) 696-8063. E-mail: production @salemdenver.com Web Site: www.710knus.com. Licensee: Salem Media of Colorado Inc. Group owner: Salem Communications Corp. (acq 1996; $1.2 million). Population served: 1,568,200 Natl. Network: CNN Radio. Format: News/talk. News staff: 3; News: 15 hrs wkly. Target aud: 35-54; Adults. ♦Brian Taylor, gen mgr.

KOA(AM)— Dec 15, 1924: 850 khz; 50 kw-U. TL: N39 30 22 W104 45 57. Hrs open: 24 4695 S. Monaco St., 80237. Phone: (303) 713-8000. Fax: (303) 713-8735. Web Site: www.850koa.com. Licensee: Citicasters Licenses Inc. (NEW). Group owner: Clear Channel Communications Inc. (acq 5-4-99; grpsl). Population served: 514,678 Natl. Network: CBS. Wire Svc: CBS Format: News/talk, sports. Target aud: 25-54. ♦Lee Larsen, gen mgr; Ron Smith, opns mgr; Kristine Olinger, progmg dir; Jan Chadwell, chief of engrg.

KRFX(FM)—Co-owned with KOA(AM). June 1, 1961: 103.5 mhz; 100 kw. 1,045 ft TL: N39 43 50 W105 14 07. Stereo. Fax: (303) 713-8744. Web Site: www.thefox.com. Format: Classic rock. Target aud: 25-54. ♦Garner Goin, progmg dir; Lee larsen, gen mgr & progmg dir; Karl Schipper, chief of engrg.

KOSI(FM)— Mar 3, 1968: 101.1 mhz; 100 kw. 1,624 ft TL: N39 43 45 W105 14 06. Stereo. Hrs opn: 24 4700 S. Syracuse, Suite 1050, 80237. Phone: (303) 967-2700. Fax: (303) 967-2747. Web Site: www.kosi101.com. Licensee: Entercom Denver License LLC. Group owner: Entercom Communications Corp. (acq 7-24-02; with KEZW(AM) Aurora). Population served: 300,000 Format: Adult contemp. News staff: one; News: 5 hrs wkly. Target aud: 25-54; women/families. ♦Ray Quinn, gen mgr; Dave Symonds, progmg dir.

***KPOF(AM)**— Mar 9, 1928: 910 khz; 5 kw-D, 1 kw-N. TL: N39 50 47 W105 01 59. Hrs open: 24 3455 W. 83rd Ave., Westminster, 80031. Phone: (303) 428-0910. Fax: (303) 429-0910. Web Site: www.am91.org. Licensee: Pillar of Fire Corp. (group owner; acq 1928). Population served: 2,000,000 Natl. Network: Moody. Format: Christian. Target aud: 18-plus; mature adult and families. ♦Robert Dallenbach, pres; Jack H. Pelon, gen mgr; Jerry Bauer, opns mgr.

KQKS(FM)—See Lakewood

KQMT(FM)— Oct 2, 1959: 99.5 mhz; 100 kw. 279 ft TL: N39 41 01 W105 00 25. (CP: Ant 1,311 ft.). Stereo. Hrs opn: 24 4700 S, Syracuse St., Suite 1050, 80237. Phone: (303) 967-2700. Fax: (303) 967-2747. Web Site: www.995themountain.com. Licensee: Entercom Denver License LLC. Group owner: Entercom Communications Corp. (acq 3-21-03). Population served: 150,000 Format: Timeless rock. News: 4 hrs wkly. Target aud: 25-54; upscale, educated. ♦Ray Quinn, gen mgr; Beau Raines, progmg dir.

KRKS(AM)— Aug 1, 1953: 990 khz; 5 kw-D, 390 w-N, DA-N. TL: N39 41 06 W105 04 05. Hrs open: 24 3131 S. Vaughn Way, Suite 601, Aurora, 80114. Phone: (303) 750-5687. Fax: (303) 696-8063. Web Site: www.krks.com. Licensee: Salem Media of Colorado Inc. Group owner: Salem Communications Corp. (acq 10-93; $400,000). Natl. Network: Superadio. Format: Relg. Target aud: 25 plus. ♦Brian Taylor, gen mgr & stn mgr.

***KUVO(FM)**— Aug 29, 1985: 89.3 mhz; 22.5 kw. 910 ft TL: N39 43 49 W105 14 59. Stereo. Hrs opn: 24 PO Box 2040, 80201-2040. Secondary address: 2900 Welton St., Suite 200 80205. Phone: (303) 480-9272. Fax: (303) 291-0757. E-mail: info@kuvo.org Web Site: www.kuvo.org. Licensee: Denver Educational Broadcasting. Natl. Network: NPR, PRI. Law Firm: Haley, Bader & Potts. Format: Jazz. Target aud: 25-49. Spec prog: Sp 15 hrs wkly. ♦Carlos Lando, gen mgr.

***KVOD(FM)**— November 1970: 90.1 mhz; 50 kw. Ant 910 ft TL: N39 43 49 W105 14 59. Stereo. Hrs opn: 24 7409 S.Alton Ct., Centennial, 80112. Phone: (303) 871-9191. Fax: (303) 733-3319. E-mail: info@cpr.org Web Site: www.cpr.org. Licensee: Public Broadcasting of Colorado Inc. (acq 1991; FTR: 10-28-91). Population served: 2,100,000 Natl. Network: NPR. Law Firm: Arter & Hadden. Format: Classical. News staff: 8; News: 50 hrs wkly. Target aud: General. ♦Max Wycisk, pres; Jenny Gentry, exec VP; Sue Coughlin, dev VP; Sean Nethery, progmg dir; Robert Hensler, engrg VP.

KXKL-FM— Dec 1, 1956: 105.1 mhz; 100 kw. 1,200 ft TL: N39 36 00 W105 12 35. (CP: Ant 1,168 ft.). Hrs opn: 24 1560 Broadway, Suite 1100, 80202. Phone: (303) 832-5665. Fax: (303) 832-7000. Web Site: www.kool105.com. Licensee: Infinity Radio Inc. Group owner: Infinity Broadcasting Corp. (acq 8-24-00; grpsl). Format: Oldies. ♦Don Howe, gen mgr; Brenda Egger, gen sls mgr; Keith Abrams, progmg VP & progmg dir; Barry Walters, chief of engrg.

Dolores

KKDC(FM)—Not on air, target date: 6/1/04: 93.3 mhz; 50 kw. Ant 338 ft TL: N37 27 59 W108 31 28. Hrs open: P.O. Drawer P, Durango, 81302. Secondary address: 185 Suttle St., Suite 203, Durango 81303. Phone: (970) 259-4444. Fax: (970) 247-1005. Web Site: www.radiodolores.com. Licensee: Four Corners Broadcasting L.L.C. (group owner). Format: Rock. Target aud: 35-54. ♦Ward S. Holmes, gen mgr; Ray McDonnell, stn mgr.

***KTCF(FM)**— 2004: 89.5 mhz; 500 w. Ant 174 ft TL: N37 28 07 W108 32 48. Hrs open:
Rebroadcasts KTLF(FM) Colorado Springs 100%.
1665 Briargate Blvd., Suite 100, Colorado Springs, 80920. Phone: (719) 593-0600. Fax: (719) 593-2399. E-mail: lightpraise@ktlf.org

Web Site: www.ktlf.org. Licensee: Educational Communications of Colorado Springs Inc. Format: Christian music. News staff: 12. ♦Karen Veazey, gen mgr.

Durango

KDGO(AM)— Apr 18, 1958: 1240 khz; 1 kw-U. TL: N37 18 17 W107 51 10. Hrs open: 24 1911 Main Ave., Suite 100, 81301. Phone: (970) 247-1240. Fax: (970) 247-1771. E-mail: mutter@americageneralmedia.com Web Site: www.kdgoam.com. Licensee: Winton Road Broadcasting Co. LLC. (group owner; (acq 6-1-2001; grpsl). Population served: 28,000 Natl. Network: ABC. Format: News/talk. News staff: one. Target aud: 35-55. ♦Dan Buchta, gen mgr & opns mgr; Ryan Nutter, progmg dir.

KPTE(FM)—Co-owned with KDGO(AM). July 1, 1995: 99.7 mhz; 9.2 kw. 1,128 ft TL: N37 19 59 W107 49 13. E-mail: kpte@997thepoint.com Web Site: 997thepoint.com.70,000 Format: Hot adult contemp. Target aud: 18-44.

***KDUR(FM)**— 1975: 91.9 mhz; 225 w. -447 ft TL: N37 16 31 W107 52 00. Stereo. Hrs opn: Fort Lewis College, 1000 Rim Dr., 81301. Phone: (970) 247-7262. Web Site: www.kdur.org. Licensee: Board of Trustees for Fort Lewis College. Population served: 12,000 Natl. Network: PRI. Format: Div. Spec prog: Bluegrass 6 hrs, blues 6 hrs, class 6 hrs, jazz 9 hrs, Native American folk 3 hrs wkly. ♦David Smith, dev dir.

KIQX(FM)— Oct 15, 1982: 101.3 mhz; 100 kw. 439 ft TL: N37 15 45 W107 54 07. Stereo. Hrs opn: P.O. Drawer P, 81302. Secondary address: 185 Suttle St., Suite 203 81303. Phone: (970) 259-4444. Fax: (970) 247-1005. E-mail: fcb@frontier.net Web Site: www.radiodurango.com. Licensee: Four Corners Broadcasting L.L.C. (group owner) Population served: 39,300 Natl. Network: CBS. Law Firm: Akin, Gump, Strauss, Hauer & Feld. Format: Adult contemp. News staff: 2; News: 5 hrs wkly. Target aud: 25-49; mainstream business professionals & families. Spec prog: Jazz 7 hrs wkly. ♦Allen H. Brill, chmn; Ward Holmes, VP.

KIUP(AM)— Dec 10, 1935: 930 khz; 5 kw-D, 1 kw-N, DA-N. TL: N37 13 45 W107 51 49. Hrs open: 24 Drawer P, 81302. Secondary address: 185 Suttle St., Suite 203 81303. Phone: (970) 259-4444. Fax: (970) 247-1005. E-mail: fcb@frontier.net Web Site: www.radiodurango.com. Licensee: Four Corners Broadcasting LLC. (group owner; (acq 4-1-96; with co-located FM). Population served: 150,000 Natl. Network: CBS, ESPN Radio. Law Firm: Akin, Gump, Strauss, Hauer & Feld. Wire Svc: AP Format: Sports. News staff: 2; News: 5 hrs wkly. Target aud: 35-64. ♦Allen Brill, CEO; Ward S. Holmes, VP.

KRSJ(FM)—Co-owned with KIUP(AM). Dec 4, 1972: 100.5 mhz; 100 kw. 200 ft TL: N37 15 46 W107 53 45. Stereo. 24 Web Site: www.radiodurango.com.170,000 Natl. Network: Fox Sports. Format: C&W. News staff: 2; News: 6 hrs wkly. Target aud: 25 plus.

KLLV(AM)—Breen, Sept 19, 1984: 550 khz; 1.8 kw-D. TL: N37 11 02 W108 04 54. Hrs open: Box 2220, 81302. Phone: (970) 247-8955. Licensee: Daystar Radio Ltd. Population served: 2,000,000 Format: Inspirational. News: New progmg 2 hrs wkly. Target aud: General. ♦Sharon Harper, gen mgr; Debbie Baker, progmg dir; Jim Alexander, chief of engrg.

Eagle

KTUN(FM)— Apr 16, 1984: 101.7 mhz; 12 kw. Ant 2,211 ft TL: N39 44 18 W106 47 58. Stereo. Hrs opn: 24 Box 7205, Avon, 81620. Phone: (970) 949-0140. Fax: (970) 949-1464. Licensee: NRC Mountain Division LLC. (group owner; (acq 7-9-2004; grpsl). Law Firm: Arter & Hadden. Format: Classic rock. News staff: 2; News: 4 hrs wkly. Target aud: 25-63; affluent locals & tourists. ♦Meredith Fox, opns mgr & progmg dir; Steve Wodlinger, gen mgr & pub affrs dir.

Eaton

***KEZF(FM)**— Dec 23, 2003: 88.9 mhz; 640 w. Ant -3 ft TL: N41 08 17 W104 47 30. Stereo. Hrs opn: 24 1063 F Big Thompson Canyon Blvd., Loveland, 80537. Phone: (970) 669-9200. Fax: (970) 669-0800. Licensee: Cedar Cove Broadcasting Inc. Group owner: EMF Broadcasting. (acq 2-21-2006; $200,000). Law Firm: Gammon & Grange. News staff: 3. ♦Victor A. Michael Jr., gen mgr.

El Jebel

KCUF(FM)— 2006: 100.5 mhz; 6 kw. Ant 295 ft TL: N39 18 56 W106 57 32. Hrs open: 5551 Ridgewood Dr., Suite 501, Naples, FL, 34108. Phone: (239) 263-7700. Fax: (239) 263-0998. Licensee: BS&T Wireless Inc. Law Firm: Leventhal, Senter & Lerman. ♦David G. Budd, gen mgr.

Englewood

KNRV(AM)— 1951: 1150 khz; 5 kw-D, 1 kw-N, DA-2. TL: N39 36 18 W104 50 25. (CP: 50 kw-D, 1 kw-N, DA-2). Hrs opn: 24 2821 So. Parker Rd., Suite 1205, Aurora, 80014. Phone: (303) 696-5970. Fax: (303) 696-5966. Web Site: www.onda115am.com. Licensee: New Radio Venture Inc. (group owner; (acq 10-20-2006; with KNUV(AM) Tolleson, AZ). Population served: 400,000 Format: Sp/news/talk. ♦Annette Lavina, gen sls mgr; Julio Parra, chief of engrg.

Estes Park

KEPL(AM)— Aug 19, 1967: 1470 khz; 1 kw-D, 53 w-N. TL: N40 20 15 W105 31 36. Hrs open: 24 Box 2690, 80517. Secondary address: 184 E. Elkhorn Ave. 80517. Phone: (970) 586-9555. Fax: (970) 586-9561. E-mail: info@estesparkradio.com Licensee: MK Inc. (acq 11-2000; $185,000). Population served: 10,000 Format: AC Hot/Rock Hits. News staff: one; News: 35 hrs wkly. Target aud: General. ♦Kristi Wellborn, stn mgr.

KRKY-FM— Apr 6, 1998: 102.1 mhz; 175 w. Ant 1,007 ft TL: N40 04 19 W105 21 11. Hrs open: 24 1201 18th St., Suite 250, Denver, 80202. Phone: (303) 296-7025. Fax: (303) 296-7030. Licensee: NRC Broadcasting Inc. (group owner; (acq 8-31-2006; exchange for KSKE-FM Vail). Natl. Network: ABC. Format: AAA. Target aud: Urban adults 19-34; Black, Hispanic, White. ♦Timothy Brown, gen mgr; Roger Tighe, chief of engrg.

Evergreen

KXPK(FM)— June 8, 1994: 96.5 mhz; 93 kw. 328 ft TL: N39 40 18 W105 13 12. Hrs open: 24 777 Grant St., 5th Fl., Denver, 80203. Phone: (303) 721-9210. Fax: (303) 832-3410. Web Site: www.denverspanishradio.com. Licensee: Entravision Holdings LLC. Group owner: Entravision Communications Corp. (acq 5-1-02; $47.5 million). Format: Rgnl Mexican. News staff: one. ♦Mario Carrera, gen mgr.

Fort Collins

KCOL(AM)—Wellington, Jan 12, 1959: 600 khz; 1 kw-D, 100 w-N, DA-2. TL: N40 35 34 W105 06 18. Hrs open: 4270 Bryd Dr., Loveland, 80538. Phone: (970) 482-5991. Fax: (970) 482-5994. Web Site: www.kcol.com. Licensee: Jacor Broadcasting of Colorado Inc. Group owner: Clear Channel Communications Inc. (acq 5-8-98; $6.1 million with co-located FM). Population served: 200000 Natl. Rep: McGavren Guild. Format: News/talk. Target aud: 35 plus. Spec prog: Farm 2 hrs, relg one hr, sports talk 7 hrs wkly. ♦Stu Haskell, stn mgr; Amy White, prom dir; Rich Bircumshaw, news dir; Dave Agnew, chief of engrg.

***KCSU-FM**— Sept 20, 1964: 90.5 mhz; 10 kw. -355 ft TL: N40 36 00 W105 09 21. Stereo. Hrs opn: 24 Lory Student Ctr., Box 13, 80523. Phone: (970) 491-7611. Fax: (970) 491-1690. E-mail: program@colostate.edu Web Site: www.kcsufm.com. Licensee: Colorado State Board of Agriculture. Population served: 250,000 Law Firm: Arter & Hadden. Format: Progsv. News: 3 hrs wkly. Target aud: 18-34; general. Spec prog: Hip-hop 3 hrs, jazz 3 hrs, Black 3 hrs wkly. ♦Christina Dickinson, gen mgr & stn mgr.

KIIX(AM)— Mar 1, 1947: 1410 khz; 1 kw-U, DA-N. TL: N40 35 34 W105 06 18. Hrs open: 24 4270 Byrd Dr., Loveland, 80538. Phone: (970) 482-5991. Fax: (970) 482-5994. Web Site: www.1410kiix.com. Licensee: Citicasters Licenses Inc. (NEW). Group owner: Clear Channel Communications Inc. (acq 5-4-99; grpsl). Population served: 150000 Format: Sports. News staff: 2; News: 35 hrs wkly. Target aud: 18-54; educated, affluent, professional. ♦Stu Haskell, VP & gen mgr; Kathy Arias, gen sls mgr & rgnl sls mgr; Amy White, prom VP & prom dir; Randy Barnard, progmg dir; Rich Bircumshaw, news dir; Dave Agnew, chief of engrg.

KPAW(FM)—Co-owned with KIIX(AM). July 27, 1975: 107.9 mhz; 100 kw. 470 ft TL: N40 40 50 W104 56 32. Stereo. 24 Web Site: www.1079thebear.com. Format: Classic rock. Target aud: 25-54; men. ♦Stu Haskell, stn mgr; Jefferson Chase, opns VP & progmg dir; Collen Taylor, prom dir; Rich Bircumshaw, news dir.

***KLHV(FM)**— 2005: 88.3 mhz; 1 w horiz, 90 w vert. Ant 941 ft TL: N40 29 36 W105 10 52. Hrs open: 24
Rebroadcasts KLVR(FM) Santa Rosa, CA 100%.
2351 Sunset Blvd., Suite 170-218, Rocklin, CA, 95765. Phone: (916) 251-1600. Fax: (916) 251-1650. E-mail: klove@klove.com Web Site: www.klove.com. Licensee: Educational Media Foundation. Group owner: EMF Broadcasting (acq 10-2-2003; grpsl). Natl. Network: K-Love. Law Firm: Shaw Pittman. Format: Contemp Christian. News staff: 3. Target aud: 25-44; Judeo Christian female. ♦Richard Jenkins, pres; Mike Novak, VP & progmg dir; Lloyd Parker, gen mgr; Ed Lenane, opns dir; Keith Whipple, dev dir; Eric Allen, natl sls mgr; David Pierce, progmg mgr; Jon Rivers, mus dir; Ed Lenane, news dir; Sam Wallington, engrg dir; Arthur Vassar, traf mgr; Karen Johnson, news rptr; Marya Morgan, news rptr; Richard Hunt, news rptr.

***KRFC(FM)**—Not on air, target date: 3/1/03: 88.9 mhz; 10 w horiz, 3 kw vert. Ant 216 ft TL: N40 34 53 W104 54 20. Hrs open: 24 619 S. College Ave., Suite #4, 80524. Phone: (970) 221-5075. Fax: (970) 221-5075. E-mail: pam@krfcfm.org Web Site: www.krfcfm.org. Licensee: Public Radio for the Front Range. Format: Var/div. ♦Pam Turner, stn mgr; Carole Lundgren, dev dir; Dennis Bigelow, mus dir.

Fort Morgan

KFTM(AM)— May 22, 1949: 1400 khz; 1 kw-U. TL: N40 15 31 W103 51 07. Hrs open: Box 430, 80701. Secondary address: 16041 Hwy. 34 80701. Phone: (970) 867-5674. Fax: (970) 542-1023. E-mail: kftm@aginformation.com Web Site: www.kftm.net. Licensee: Media Logic LLC (group owner; acq 9-29-03; $415,000). Population served: 55,000 Natl. Network: AP Radio, Jones Radio Networks. Wire Svc: AP Format: Adult contemporary, Sp, news/talk. News staff: one; News: 16 hrs wkly. Target aud: General; the people of (Morgan county) Colorado. Spec prog: Farm news 6 hrs, talk 5 hrs, sports 10 hrs, Christian progmg 6 hrs, Spanish 9 hrs wkly. ♦Wayne Johnson, pres & gen mgr; Dana Marini, gen sls mgr; John Waters, progmg dir; Michelle Kind, traf mgr.

KSIR(AM)—See Brush

Fountain

KIBT(FM)— Sept 25, 1992: 96.1 mhz; 460 w. Ant 2,168 ft TL: N38 44 44 W104 51 42. Stereo. Hrs opn: 24 2864 So. Circle Dr., Suite 150, Colorado Springs, 80906. Phone: (719) 540-9200. Fax: (719) 579-0882. Web Site: www.beatcolorado.com. Licensee: AMFM Radio Licenses LLC. Group owner: Clear Channel Communications Inc. (acq 7-1-2000; grpsl). Population served: 500,000 Natl. Rep: Clear Channel. Format: Urban. News staff: one; News: 4 hrs wkly. Target aud: 25-44; men. Spec prog: Blues 3 hrs wkly. ♦Bob Richards, gen mgr; Scott Jones, gen sls mgr; Nancia Warren, prom dir; Jared Goldberg, progmg dir.

KJME(AM)—Not on air, target date: End of 2007: 890 khz; 5.5 kw-D, 500 w-N, DA-2. TL: N38 33 47 W104 36 20. Hrs open: 965 S. Irving St., Denver, 80219. Phone: (303) 935-1156. Licensee: Timothy C. Cutforth. Format: Spanish Language Variety. ♦Timothy C. Cutforth, gen mgr.

Frisco

***KTDX(FM)**— 2007: 90.3 mhz; 400 w. Ant -293 ft TL: N39 29 47 W106 01 43. Hrs open: 87 Jasper Lake Rd., Loveland, 80537. Phone: (970) 669-9200. Fax: (970) 669-0800. Licensee: Cedar Cove Broadcasting Inc. (acq 8-10-2007; $280,000). ♦Victor A. Michael Jr., gen mgr.

KYSL(FM)— May 27, 1988: 93.9 mhz; 560 w. 1,050 ft TL: N39 33 22 W106 06 53. Stereo. Hrs opn: 24 P.O. Box 27, San Francisco, CA, 80443. Phone: (970) 513-9393. Fax: (970) 262-3677. Web Site: www.krystal93.com. Licensee: Krystal Broadcasting Inc. Natl. Network: AP Radio. Natl. Rep: Interep. Format: AAA. News staff: one; News: 8 hrs wkly. Target aud: 25-49; upscale adults. ♦Ann Penny, pres; Maureen Bennett, gen mgr.

Fruita

KEKB(FM)—Licensed to Fruita. See Grand Junction

Glenwood Springs

***KDNK(FM)**— 2004: 88.1 mhz; 1.2 kw. Ant 2,542 ft TL: N39 25 08 W107 22 10. Hrs open: 6 AM-1 AM (M-F); 7 AM-1 AM (S, Su) Box 1388, Carbondale, 81623. Phone: (970) 963-0139. Fax: (970) 963-0810. E-mail: kdnk@kdnk.org Web Site: www.kdnk.org. Licensee: Carbondale Community Access Radio Inc. (acq 10-29-2004; exchange for KVOV(FM) Carbondale). Natl. Network: NPR. Law Firm: Haley, Bader & Potts. Format: Eclectic, news. News: 21 hrs wkly. ♦Steve Skinner, stn mgr; Amy Kimberly, dev dir; Wick Moses, adv dir; Luke Nestler, mus dir.

KGLN(AM)— May 14, 1950: 980 khz; 1 kw-D, 225 w-N. TL: N39 33 10 W107 19 48. Hrs open: Box 1028, 81602. Phone: (970) 945-9124. Fax: (970) 945-5409. Web Site: www.kgln.com. Licensee: Colorado West Broadcasting Inc. (acq 6-93). Population served: 4,106 Format: News/talk. Target aud: 45 plus. ♦Gabe Chenoweth, pres, gen mgr, opns dir, progmg dir & chief of engrg; Kimberly Henrie, sls dir & prom dir; Ron Milhorn, news dir.

KMTS(FM)—Co-owned with KGLN(AM). June 6, 1977: 99.1 mhz; 3 kw. -301 ft TL: N39 32 36 W107 17 49. (CP: 10 kw). Stereo. Web Site: www.kmts.com.30,000 Format: Country. Target aud: 25-50.

KKCH(FM)— Sept 1, 1997: 92.7 mhz; 58 kw. 2,470 ft TL: N39 25 05 W107 22 01. Hrs open: Box 7205, Avon, 81620. Phone: (970) 949-0140. Fax: (970) 949-1464. Licensee: NRC Mountain Division LLC. (group owner; (acq 7-9-2004; grpsl). Format: Jack. ♦Steve Wodlinger, gen mgr.

***KLXV(FM)**— August 1995: 91.9 mhz; 1 w horiz, 250 w vert. Ant 2,660 ft TL: N39 25 30 W107 22 46. Hrs open: 24 2351 Sunset Blvd., Suite 170-218, Rocklin, CA, 95765. Phone: (916) 251-1600. Fax: (916) 251-1650. E-mail: klove@klove.com Web Site: www.klove.com. Licensee: Educational Media Foundation. Group owner: EMF Broadcasting (acq 12-28-00; grpsl). Natl. Network: K-Love. Law Firm: Shaw Pittman. Format: Contemp Christian. News staff: 3. Target aud: 25-44; Judeo Christian, female. ♦Richard Jenkins, pres; Mike Novak, VP; Keith Whipple, dev dir; David Pierce, progmg mgr; Ed Lenane, news dir; Sam Wallington, engrg dir; Arthur Vassar, traf mgr; Karen Johnson, news rptr; Marya Morgan, news rptr; Richard Hunt, news rptr.

KRVG(FM)— Oct. 1, 2000: 95.5 mhz; 70 w. 2,709 ft TL: N39 25 30 W107 22 46. Hrs open: 24 751 Horizon Ct., Suite 225, Grand Junction, 81506. Phone: (970) 241-6460. Fax: (970) 241-6452. Web Site: www.wscradio.net.Jones Rock Classics Licensee: Western Slope Communications LLC. (group owner) Population served: 175,000. Format: Classic Rock. News staff: one; News: 20 hrs wkly. ♦John Reid, gen mgr & progmg dir.

Granby

KRKY(AM)— July 3, 1986: 930 khz; 4.5 kw-D. TL: N40 02 26 W105 56 11. Hrs open: 24 PO Box 7069, Breckenridge, 80424. Phone: (970) 887-1100. Fax: (970) 468-2384. E-mail: comment@highcountryradio.com Web Site: www.highcountryradio.com. Licensee: Granby Mountain Broadcasting LLC. Group owner: Kona Coast Radio LLC (acq 12-10-2003; grpsl). Population served: 25,000 Natl. Network: ABC. Format: Country. News staff: 2; News: 5 hrs wkly. Target aud: 24-49; adults. Spec prog: Agriculture one hr, sp one hr. ♦Lisa Korry Cheek, gen mgr; Sam Scholl, progmg dir.

KSPN-FM—See Aspen

Grand Junction

***KAFM(FM)**— 1999: 88.1 mhz; 20 w. 1,240 ft TL: N39 04 00 W108 44 41. Stereo. Hrs opn: 24 1310 Ute Ave., 81501. Phone: (970) 241-8801. Fax: (970) 241-0995. E-mail: kafm@kafmradio.org Web Site: www.kafmradio.org. Licensee: Grand Valley Public Radio Co. Inc. Population served: 100,000 Format: Community radio, educ. Target aud: 25-80. Spec prog: Celtic 3 hrs, Black 3 hrs, folk 10 hrs, Sp 3 hrs wkly. ♦Peter Trosclair, stn mgr; Peyton Montgomery, opns mgr; Julia Hall, dev dir; Jon Rizzo, dev mgr & mus dir; Ryan Stringfellow, progmg dir.

KBKL(FM)— 1993: 107.9 mhz; 100 kw. 1,305 ft TL: N39 04 00 W108 44 41. (CP: Ant 1,460 ft.). Stereo. Hrs opn: 24 315 Kennedy Ave., 81501. Phone: (970) 242-7788. Fax: (970) 243-0567. Web Site: www.kool1079.com. Licensee: Cumulus Licensing Corp. Group owner: Cumulus Media LLC (acq 3-10-98; grpsl). Format: Oldies. Target aud: 25-54. ♦Lewis Dickey Jr., CEO; Jonathan Pinch, COO; Marty Gausvik, CFO; John Dickey, exec VP; Kevin Wodlinger, gen mgr; Mike Shafer, opns mgr.

***KCIC(FM)**— Mar 4, 1979: 88.5 mhz; 450 w. -431 ft TL: N39 04 38 W108 30 38. Stereo. Hrs opn: 24 3102 E Rd., 81504. Phone: (970) 434-4113. Licensee: Pear Park Baptist Schools. Format: Educ, relg. News: 2 hrs wkly. Spec prog: Class 14 hrs wkly. ♦Randy David, pres; Glenn Gardner, gen mgr & progmg dir.

KEKB(FM)—Fruita, May 24, 1984: 99.9 mhz; 79 kw. 1,380 ft TL: N39 03 56 W108 44 52. (CP: Ant 1,542 ft.). Stereo. Hrs opn: 24 315 Kennedy Ave., 81501. Phone: (970) 242-7788. Fax: (970) 243-0567. Web Site: www.kekbfm.com. Licensee: Cumulus Licensing Corp. Group owner: Cumulus Media Inc. (acq 7-9-98; grpsl). Population served: 100,000 Format: Country. News staff: 2; News: 6 hrs wkly. Target aud: 25-54. ♦Lewis Dickey Jr., CEO; Marty Gausvik, CFO; John Dickey, exec VP; Dave Noll, VP; Kevin Wodlinger, gen mgr; Mike Shafer, opns mgr.

KEXO(AM)— 1942: 1230 khz; 1 kw-U. TL: N39 05 41 W108 34 41. Hrs open: 24 315 Kennedy Ave., 81501. Phone: (970) 242-7788. Fax: (970) 243-0567.La Maguina Musical Licensee: Cumulus Licensing Corp. Group owner: Cumulus Media Inc. (acq 1-00). Population served: 130,000 Format: Sp. News staff: one. Target aud: General. ♦Lewis Dickey Jr., CEO; Marty Gausvik, CFO; Kevin Wodlinger, exec VP; Pat Cantwell, gen mgr; Mike Shafer, opns mgr.

KJOL(AM)— June 19, 1957: 620 khz; 5 kw-U. TL: N39 07 35 W108 38 13. Hrs open: 24 1360 E. Sherwood Dr., 81501-7575. Phone: (970) 254-5565. Fax: (970) 254-5550. E-mail: info@kjol.org Web Site: www.kjol.org. Licensee: United Ministries. (acq 5-1-2003). Population served: 350,000 Format: Christian, talk, music. ♦Ken Andrews, gen mgr.

KJYE(FM)—Listing follows KNZZ(AM).

***KLFV(FM)**— Apr 24, 1982: 90.3 mhz; 1.5 kw. 1,296 ft TL: N30 03 57 W108 44 48. Stereo. Hrs opn: 24 2351 Sunset Blvd., Suite 170-218, Rocklin, CA, 95765. Phone: (916) 251-1600. Fax: (916) 251-1650. E-mail: klove@klove.com Web Site: www.klove.com. Licensee: Educational Media Foundation. Group owner: EMF Broadcasting (acq 12-28-00; grpsl). Population served: 92,000 Natl. Network: K-Love. Law Firm: Shaw Pittman. Format: Contemp Christian. News staff: 3. Target aud: 25-44; Judeo Christian, female. Spec prog: Sp 2 hrs wkly. ♦Richard Jenkins, pres; Mike Novak, VP; Ed Lenane, opns dir & news dir; Keith Whipple, dev dir; David Pierce, progmg mgr; Sam Wallington, engrg dir; Karen Johnson, news rptr; Marya Morgan, news rptr; Richard Hunt, news rptr.

KMGJ(FM)— Nov 1, 1973: 93.1 mhz; 100 kw. 1,433 ft TL: N39 03 59 W108 44 41. Stereo. Hrs opn: 24 1360 E. Sherwood Dr., 81501. Phone: (970) 254-2100. Fax: (970) 245-7551. Web Site: gjradio.com. Licensee: M.B.C. Grand Broadcasting, Inc. (group owner; acq 5-94; with co-located AM). Population served: 230,000 Format: CHR. News staff: one; News: 6 hrs wkly. Target aud: 18-49; women. ♦Richard C. Dean, CEO & pres; Jim Terlouw, gen mgr; Robert St. John, opns mgr; Dave Beck, gen sls mgr & natl sls mgr; Monica Salvo, prom mgr; Chris Britt, progmg dir; Libby Jackson, news dir; Robert Bowe, chief of engrg; Fred Horn, traf mgr; Jim Davis, sports cmtr.

KTMM(AM)—Co-owned with KMGJ(FM). 1959: 1340 khz; 1 kw-U. TL: N39 05 35 W108 35 51.24 Web Site: gjradio.com.130000 Format: Sports. News staff: one; News: 10 hrs wkly. Target aud: 25-54; men. ♦Jim Davis, progmg dir.

KMOZ-FM— Mar 27, 1999: 100.7 mhz; 42 kw. 1,302 ft TL: N39 04 00 W108 44 41. Stereo. Hrs opn: 24 1360 E. Sherwood Dr., 81501. Phone: (970) 254-2100. Fax: (970) 245-7551. Web Site: www.gjradio.com. Licensee: MBC Grand Broadcasting Inc. (group owner) Population served: 150,000 Format: Country. News staff: 2; News: 2 hrs wkly. Target aud: 25-54. ♦Richard C. Dean, CEO & pres; Jim Terlouw, gen mgr.

***KMSA(FM)**— Feb 18, 1975: 91.3 mhz; 3 kw. Ant -382 ft TL: N39 04 48 W108 33 09. Stereo. Hrs opn: 24 1100 North Ave., 81501. Phone: (970) 248-1442. Fax: (970) 248-1199. Fax: (970) 248-1834. E-mail: rtucci@mesastate.edu Web Site: www.kmsa.com. Licensee: Mesa State College. Population served: 100,000 Format: Adult alternative, hip hop, reggae. News staff: 2; News: 10 hrs wkly. Target aud: 18-60; college students and gen pub. Spec prog: Black 6 hrs, folk 2 hrs, jazz 12 hrs wkly. ♦Nathan King, pres & gen mgr.

KMXY(FM)— 1996: 104.3 mhz; 100 kw. 1,296 ft (CP: Ant 1,460 ft.). Hrs opn: 315 Kennedy Ave., 81501. Phone: (970) 242-7788. Fax: (970) 243-0567. Web Site: www.mix1043.com. Licensee: Cumulus Licensing Corp. Group owner: Cumulus Media Inc. (acq 7-9-98; grpsl). Format: Adult contemp. ♦Lewis Dickey Jr., CEO; Marty Gausvik, CFO; John Dickey, exec VP; Dave Noll, VP; Kevin Wodlinger, gen mgr.

KNZZ(AM)— May 1, 1926: 1100 khz; 50 kw-D, 10 kw-N, DA-N. TL: N38 57 06 W108 25 10. Hrs open: 24 1360 E. Sherwood Dr., 81501. Phone: (970) 254-2100. Fax: (970) 245-7551. Web Site: www.gjradio.com. Licensee: MBC Grand Broadcasting Inc. (group owner; (acq 8-30-89). Population served: 230,000 Natl. Network: AP Network News. Wire Svc: AP Format: News/talk. News staff: 3; News: 44 hrs wkly. Target aud: 25-64; upscale adults. ♦Richard C. Dean, CEO & pres; Jim TerLouw, gen mgr & progmg mgr; Dave Beck, gen sls mgr; Libby Jackson, news dir; Robert Bowe, chief of engrg.

KJYE(FM)—Co-owned with KNZZ(AM). May 1, 1960: 92.3 mhz; 100 kw. 1,378 ft TL: N39 04 00 W108 44 41. Stereo. 24 Web Site: www.gjradio.com.240,000 Format: Adult contemp. News: 8 hrs wkly. Target aud: Adults; 25-54.

***KPRN(FM)**— April 1985: 89.5 mhz; 10 kw. 1,191 ft TL: N39 03 57 W108 44 45. (CP: Ant 1,233 ft.). Stereo. Hrs opn: 24 7409 S. Alton Ct., Centennial, 80112. Phone: (303) 871-9191. Fax: (303) 733-3319. E-mail: info@cpr.org Web Site: www.cpr.org. Licensee: Public Broadcasting of Colorado Inc. Natl. Network: NPR. Rgnl. Network: Colo. Pub. Law Firm: Arter & Hadden. Format: News. News staff: one; News: 28 hrs wkly. Target aud: 25 plus. ♦Max Wycisk, pres; Jenny Gentry, exec VP; Sue Coughlin, dev VP; Sean Nethery, progmg dir; Robert Hensler, engrg VP & chief of engrg.

Greeley

KFKA(AM)— May 21, 1921: 1310 khz; 5 kw-D, 1 kw-N, DA-N. TL: N40 21 56 W104 43 56. Hrs open: 24 820 11th Ave, 80631. Phone: (970) 356-1310. Fax: (970) 356-1314. E-mail: info@1310kfka.com Web Site: www.1310kfka.com. Licensee: Music Ventures LLC dba Broadcast Media LLC (acq 11-1-2002; $1.6 million). Population served: 347,500 Natl. Network: CBS Radio. Format: News/talk. News staff: 2; News: 25 hrs wkly. Target aud: 25-54; community-minded, active people. Spec prog: Farm 15 hrs, Ger one hr, relg 4 hrs wkly. ♦Damon Sasso, pres & opns mgr; Justin Sasso, gen mgr.

KGRE(AM)— Aug 24, 1948: 1450 khz; 1 kw-U. TL: N40 26 15 W104 43 25. Hrs open: 24 1020 9th St., Suite 201, 80631. Phone: (970) 356-1452. Fax: (970) 356-8522. E-mail: kgre@msn.com Web Site: www.tigre1450.com. Licensee: Greely Broadcasting Corp. (acq 3-24-98). Population served: 250,000 Format: Sp. Target aud: 25-54; Hispanic. Spec prog: Bienvenidos a America one hr wkly. ♦Ricardo Salazar, pres & gen mgr.

KSME(FM)— Dec 25, 1975: 96.1 mhz; 100 kw. 660 ft TL: N40 40 50 W104 56 32. Stereo. Hrs opn: 24 4270 Byrd Dr., Loveland, 80538. Phone: (970) 482-5991. Fax: (970) 482-5994. Web Site: www.kissfmcolorado.com. Licensee: Citicasters Licenses Inc. (NEW). Group owner: Clear Channel Communications Inc. (acq 5-4-99; grpsl). Natl. Network: ABC. Format: Top-40. News staff: 2. Target aud: 10-44. ♦Stu Haskell, VP & gen mgr; Kathy Arias, sls dir & gen sls mgr.

***KUNC(FM)**— Jan 1, 1967: 91.5 mhz; 81 kw. Ant 692 ft TL: N40 38 31 W104 49 03. Stereo. Hrs opn: 24 822 Seventh St., Suite 530, 80631. Phone: (970) 378-2579. Fax: (970) 378-2580. E-mail: mailbag@kunc.org Web Site: www.kunc.org. Licensee: Community Radio for Northern Colorado (acq 8-2001; $1.9 million). Population served: 500,000 Natl. Network: NPR, PRI. Wire Svc: AP Format: Div, news. News staff: 5; News: 65 hrs wkly. Target aud: General. ♦Neil Best, gen mgr & stn mgr; Michelle Kormich, dev dir; Kirk Mowens, progmg dir; Jim Beers, news dir; Larry Selzle, chief of engrg.

Greenwood Village

KCUV(FM)— July 19, 1995: 102.3 mhz; 6 kw. Ant 210 ft TL: N39 39 55 W104 51 38. Stereo. Hrs opn: 24 1201 18th St., Suite 250, Denver, 80202. Phone: (303) 296-7025. Fax: (303) 296-7030. Web Site: www.kcuvradio.com. Licensee: NRC Broadcasting Inc. (acq 11-9-2005; $16 million). Format: AAA. ♦Timothy Brown, CEO & gen mgr.

Gunnison

KEJJ(FM)— 1980: 98.3 mhz; 3 kw. Ant 304 ft TL: N38 31 22 W106 54 28. Stereo. Hrs opn: Box 1288, 81230. Phone: (970) 641-4000. Fax: (970) 641-3300. E-mail: kpkeharv@hotmail.com Licensee: John Harvey Rees (acq 2-6-2001; $275,000). Population served: 20,000 Format: Oldies. News staff: 2. Target aud: 25-54. ♦John Harvey Rees, CEO & pres.

KPKE(AM)— Aug 23, 1960: 1490 khz; 1 kw-U. TL: N38 33 57 W106 55 32. Hrs open: 24 Box 1288, 81230. Phone: (970) 641-4000. Fax: (970) 641-3300. E-mail: kpkeharv@hotmail.com Licensee: John Harvey Rees. Population served: 20,000 Format: Country. News staff: one; News: 2 hrs wkly. Target aud: 25-54. ♦John Harvey Rees, CEO & gen mgr; Matt Rees, opns VP.

KVLE-FM— Apr 18, 1980: 102.3 mhz; 3 kw. 200 ft TL: N38 33 53 W106 55 38. Stereo. Hrs opn: 614 Kimbark St., Longmont, 80501. Phone: (303) 776-2323. Fax: (303) 776-1377. E-mail: mickfive@aol.com Web Site: www.radiocoloradonetwork.com. Licensee: Pilgrim Communications Inc. (acq 4-30-98; $300,000). Population served: 12,000 Natl. Network: CBS. Wire Svc: CBS Format: Classic rock, Nascar. Target aud: General. ♦Gary Montgomery, gen mgr & gen sls mgr; Tyler Kincaid, stn mgr.

***KWSB-FM**— Jan 26, 1968: 91.1 mhz; 135 w. 304 ft TL: N38 31 22 W106 54 28. Stereo. Hrs opn: 18 Taylor Hall, 116 Western State College, 81231. Phone: (970) 943-2158. Phone: (970) 943-2117. Fax: (970) 943-7069. Web Site: www.kwsb.org. Licensee: Western State College of Colorado. Population served: 6,000 Natl. Network: AP Radio. Format: Var. News: 2 hrs wkly. Target aud: 18-25. Spec prog: Jazz 3 hrs, reggae 6 hrs, blues 3 hrs, 60s hits 3 hrs, Sp one hr wkly. ♦Frank Venturo, gen mgr.

Gypsum

***KLRY(FM)**— 2003: 91.3 mhz; 110 w. Ant 2,818 ft TL: N39 46 30 W106 50 45. Hrs open: 24
Rebroadcasts KLVR(FM) Santa Rosa, CA 100%.
2351 Sunset Blvd., Suite 170-218, Rocklin, CA, 95765. Phone: (916) 251-1600. Fax: (916) 251-1650. E-mail: klove@klove.com Web Site: www.klove.com. Licensee: Educational Media Foundation. Group owner: EMF Broadcasting (acq 10-2-2003; grpsl). Natl. Network: K-Love. Law Firm: Shaw Pittman. Format: Contemp Christian. News staff: 3. Target aud: 25-44; Judeo Christian female. ♦Richard Jenkins, pres; Mike Novak, VP & progmg dir; Lloyd Parker, gen mgr; Ed Lenane, opns dir & news dir; Keith Whipple, dev dir; David Pierce, progmg mgr; Jon Rivers, mus dir; Sam Wallington, engrg dir; Arthur Vassar, traf mgr; Karen Johnson, news rptr; Marya Morgan, news rptr; Richard Hunt, news rptr.

Hayden

***KHCO(FM)**— 2005: 90.1 mhz; 1.9 kw vert. Ant 1,699 ft TL: N40 27 04 W106 45 06. Hrs open:
Rebroadcasts KLRD(FM) Yucaipa, CA 100%.
2351 Sunset Blvd., Suite 170-218, Rocklin, CA, 95765. Phone: (916) 251-1600. Fax: (916) 251-1650. Licensee: Educational Media Foundation. (acq 6-8-2005; $25,000 for CP). Natl. Network: Air 1. ♦Richard Jenkins, pres; Mike Novak, VP; Keith Whipple, dev dir; David Pierce, progmg mgr; Ed Lenane, news dir; Sam Wallington, engrg dir; Karen Johnson, news rptr; Marya Morgan, news rptr; Richard Hunt, news rptr.

KIDN-FM— Feb 15, 1985: 95.9 mhz; 1.8 kw. Ant 1,181 ft TL: N40 25 46 W107 05 34. Stereo. Hrs opn: Box 772850, Steamboat Springs, 80477. Phone: (970) 879-5368. Fax: (970) 879-5843. Web Site: www.jackintheboat.com. Licensee: NRC Mountain Division LLC. Group owner: American General Media (acq 7-9-2004; grpsl). Format: Jack. Target aud: 21-54. ♦Steve Wodlinger, gen mgr & stn mgr.

KTRJ(FM)— 2000: 107.3 mhz; 29 kw. Ant 649 ft TL: N40 31 16 W107 17 46. Hrs open: Box 772850, Steamboat Springs, 80477. Phone: (970) 879-5368. Fax: (970) 879-5843. Web Site: www.therangefm.com. Licensee: NRC Mountain Division LLC. (group owner; (acq 7-9-2004; grpsl). Format: Country. ♦Steve Wodlinger, gen mgr; Julia Arrotti, opns mgr; David Wittlinger, gen sls mgr; Eli Campbell, prom dir; John Johnston, progmg dir.

Holyoke

KSTH(FM)— 2002: 92.3 mhz; 100 kw. Ant 567 ft TL: N40 51 42 W103 23 35. Hrs open: Box 333, McCook, NE, 69001. Phone: (308) 345-5400. Fax: (308) 345-4720. Licensee: Armada Media - McCook Inc. (acq 1-17-2007; grpsl). Format: Adult contemp. ♦David M. Stout, gen mgr; Connie Stout, sls dir; Ben Korn, progmg dir.

Idalia

KWDI(FM)—Not on air, target date: unknown: 94.1 mhz; 50 kw. Ant 298 ft TL: N39 40 22 W102 15 18. Hrs open: 5331 Mt. Alifan Dr., San Diego, CA, 92111. Phone: (858) 277-4991. Fax: (858) 277-1365. Licensee: Horizon Christian Fellowship. (acq 2-9-2006; grpsl). ♦Mike MacIntosh, pres.

Ignacio

***KSUT(FM)**— June 9, 1976: 91.3 mhz; 425 w. Ant 18 ft TL: N37 05 51 W107 37 32. Stereo. Hrs opn: 24 Box 737, 81137. Secondary address: 123 Capote Dr. 81137. Phone: (970) 563-0255. Fax: (970) 563-0399. Web Site: www.ksut.org. Licensee: Kute Inc. Population served: 200,000 Natl. Network: NPR, PRI. Format: Native American. News: 30 hrs wkly. Target aud: 24 plus; public radio audience. Spec prog: American Indian 7 hrs, class 8 hrs, jazz 15 hrs wkly. ♦Beth Warren, gen mgr.

***KUTE(FM)**— June 1998: 90.1 mhz; 3 kw. 1,965 ft TL: N37 21 51 W107 46 56. Hrs open: Box 737, 81137. Phone: (970) 563-0255. Fax: (970) 563-0399. Web Site: www.ksut.org. Licensee: KUTE Inc. Natl. Network: NPR, PRI. Format: Triple A, americana. ♦Beth Warren, gen mgr.

Johnstown

KHNC(AM)— January 1993: 1360 khz; 10 w-D, 450 w-N, DA-N. TL: N40 23 11 W104 54 19. (CP: 10 kw-D, 1 kw-N, DA-N). Hrs opn: Box 1750, 80534-1750. Phone: (970) 587-5175. Fax: (970) 587-5450. E-mail: comments@americanewsnet.com Web Site: www.americanewsnet.com. Licensee: Donald A. and Sharon A. Wiedeman. Format: Conservative news/talk. ♦Donald Wiedeman, pres & gen mgr; Michael Golden, opns mgr.

Julesburg

KJBL(FM)— 2002: 96.5 mhz; 100 kw. Ant 567 ft TL: N40 51 42 W103 23 35. Hrs open: Box 333, McCook, NE, 69001. Phone: (308) 345-5400. Fax: (308) 345-4720. Licensee: Armada Media - McCook Inc. (acq 1-17-2007; grpsl). Format: Country. ♦David M. Stout, gen mgr.

Kremmling

KZMV(FM)— Nov 1, 1987: 106.3 mhz; 2.5 kw. 1,050 ft TL: N40 00 18 W106 26 57. Stereo. Hrs opn: 24 PO Box 7069, Breckenridge, 80424. Phone: (970) 887-1100. Fax: (970) 468-2384. E-mail: comments@highcountryradio.com Web Site: www.highcountryradio.com. Licensee: Granby Mountain Broadcasting LLC. Group owner: Kona Coast Radio LLC (acq 12-10-2003; grpsl). Population served: 25,000 Natl. Network: ABC. Rgnl. Network: Jones Satellite Audio. Format: Oldies. News staff: 2; News: 5 hrs wkly. Target aud: 24-49. ♦Lisa Cheek, gen mgr; Sam Scholl, progmg dir.

La Jara

KZBR(FM)—Not on air, target date: unknown: 97.1 mhz; 25 kw. Ant 180 ft TL: N37 22 05 W106 06 44. Hrs open: 1102 Newitt Vick Dr., Vicksburg, MS, 39183-8755. Phone: (601) 883-0848. Licensee: Lendsi Radio LLC. ♦Lina H. Jones, gen mgr.

La Junta

KBLJ(AM)— July 23, 1937: 1400 khz; 1 kw-U. TL: N37 59 14 W103 34 01. Hrs open: 116 Dalton, 81050. Phone: (719) 384-5456. Fax: (719) 384-5450. E-mail: kblj@rural-com.com Licensee: CCR-La Junta IV LLC. Group owner: Cherry Creek Radio LLC (acq 12-19-2003; grpsl). Population served: 7938 Natl. Network: Westwood One. Format: Oldies. Target aud: 30 plus; general. ♦Pat Gittings, gen mgr; Pat McGee, progmg dir.

KTHN(FM)—Co-owned with KBLJ(AM). Aug 28, 1974: 92.1 mhz; 3 kw. 300 ft TL: N37 59 15 W103 34 02. Stereo. Format: Country.

KFVR-FM— 2001: 106.5 mhz; 100 kw. Ant 512 ft TL: N37 39 31 W103 27 55. Hrs open: 24 920 Elm Ave., Rocky Ford, 81067-1249. Phone: (719) 254-6301. Licensee: Greeley Broadcasting Corp. (group owner; (acq 11-30-2006; $125,000). Format: Sp. ♦Ricardo Salazar, gen mgr.

***KRLJ(FM)**— August 2002: 89.1 mhz; 740 w. 298 ft TL: N37 58 43 W103 34 48. Hrs open: 24
Rebroadcasts KRCC(FM) Colorado Springs 100%.
c/o KRCC(FM), 912 N. Weber St., Colorado Springs, 80903. Phone: (719) 473-4801. Fax: (719) 473-7863. E-mail: info@krcc.org Web Site: www.krcc.org. Licensee: The Colorado College. Format: Diversified, news. ♦Delaney Utterback, gen mgr; Mike Procell, opns mgr; Jeff Bieri, prom dir; Joel Belik, chief of engrg.

La Veta

KJQY(FM)—Not on air, target date: unknown: 103.3 mhz; 100 kw. Ant 380 ft TL: N37 37 39 W104 49 17. Hrs open: 11 Pheasant Hill Rd., Canton, CT, 06019-3042. Phone: (860) 693-3336. Licensee: Steven R. Bartholomew. ♦Steven R. Bartholomew, gen mgr.

Lakewood

KEPN(AM)— Jan 8, 1955: 1600 khz; 5 kw-U, DA-N. TL: N39 39 20 W105 04 28. Stereo. Hrs opn: 7800 E. Orchard Rd., Suite 400, Greenwood Village, 80111. Phone: (303) 321-0950. Fax: (303) 321-3383. Web Site: www.espnradio1600.com. Licensee: Lincoln Financial Media Co. of Colorado. (group owner; (acq 4-3-2006; grpsl). Population served: 1,700,000 Natl. Network: ESPN Radio. Natl. Rep: CBS Radio. Format: Sports. ♦Clarke Brown, pres; Robert Call, sr VP & gen mgr; John St. John, opns dir; Steve Price, sls dir; Randy Weidner, natl sls mgr; Dwayne Taylor, mktg dir; J.J. Pelini, prom dir; Tim Spence, progmg dir; Simone Seiklay, news dir; Simone Seikaly, pub affrs dir; Brad Hart, engrg dir.

KQKS(FM)—Co-owned with KEPN(AM). July 9, 1966: 107.5 mhz; 100 kw. 670 ft TL: N39 41 45 W105 09 54. Stereo. Fax: (303) 321-3383. Web Site: www.ks1075.com. Format: Hip-hop, rhythm and blues. Target aud: 12-34. ♦Cat Collins, progmg dir.

***KFDN(FM)**— 2005: 88.1 mhz; 430 w vert. Ant 1,053 ft TL: N39 40 18 W105 13 05. Hrs open: 2351 Sunset Blvd., Suite 170-218, Rocklin, CA, 95765. Phone: (916) 251-1600. Fax: (916) 251-1650. E-mail: klove@klove.com Web Site: www.klove.com. Licensee: Educational Media Foundation. Group owner: EMF Broadcasting (acq 10-2-03; grpsl). Natl. Network: K-Love. Law Firm: Shaw Pittman. Format: Contemp Christian. News staff: 3. Target aud: 25-44; Judeo Christian, female. ♦Richard Jenkins, pres; Mike Novak, VP; Keith Whipple, dev dir; David Pierce, progmg mgr; Ed Lenane, news dir; Sam Wallington, engrg dir; Karen Johnson, news rptr; Marya Morgan, news rptr; Richard Hunt, news rptr.

Lamar

KLMR(AM)— December 1948: 920 khz; 5 kw-D, 500 w-N, DA-N. TL: N38 06 53 W102 37 16. Hrs open: 24 Box 890, 81052. Secondary address: 7650 US Hwy. 50 81052. Phone: (719) 336-2206. Fax: (719) 336-7973. E-mail: klmraudio@yahoo.com Licensee: CCR-Lamar IV LLC. Group owner: Cherry Creek Radio LLC (acq 12-19-2003; grpsl). Population served: 50,000 Natl. Rep: Interep. Format: Classic country. News: 12 hrs wkly. Target aud: 25-54. ♦Pat Gittings, gen mgr & stn mgr; Ty Harmon, gen sls mgr & progmg dir; Eric Stone, news dir.

KLMR-FM— November 1978: 93.3 mhz; 100 kw. 498 ft TL: N38 02 10 W102 35 58. Stereo. 24 7350 U.S. Hwy. 50, 81052. E-mail: audio@yahoo.com 50,000 Natl. Network: CBS. Format: Classic rock. Target aud: 25-54.

KVAY(FM)— Aug 5, 1991: 105.7 mhz; 100 kw. 545 ft TL: N38 06 44 W102 57 37. (CP: Ant 479 ft.). Stereo. Hrs opn: 24 Box 1176, 224 S. Main, 81052. Phone: (719) 336-8734. Fax: (719) 336-5977. Web Site: www.kvay.com. Licensee: Beacon Broadcasting LLC (acq 1-3-03; $825,000). Population served: 100,000 Natl. Network: AP Radio. Law Firm: Leventhal, Senter & Lerman. Format: Country. News staff: one. Target aud: 25-55. Spec prog: Gospel 4 hrs, classic rock 4 hrs wkly. ♦Debbie Ellis, gen mgr.

Las Animas

KRKV(FM)—Not on air, target date: July 2008: Stn currently dark. 107.3 mhz; 100 kw. Ant 384 ft TL: N38 06 44 W102 57 39. Stereo. Hrs opn: Box 563, Tanner, AL, 35671. Secondary address: 709 Coleman Ave., Athens, AL 35611. Phone: (256) 497-4502. Fax: (443) 342-2478. E-mail: varietyrock@hotmail.com Licensee: Alleycat Communications. Format: Var rock. Target aud: 18-54. ♦Richard W. Dabney, gen mgr.

Leadville

***KTOL(FM)**— 2006: 90.9 mhz; 450 w horiz. Ant -630 ft TL: N39 14 05 W106 17 59. Hrs open:
Rebroadcasts KTLF(FM) Colorado Springs 100%.
1665 Briargate Blvd., Suite 100, Colorado Springs, 80920-3400. Phone: (719) 593-0600. Fax: (719) 593-2399. E-mail: lightpraise@ktlf.org Web Site: www.ktlf.org. Licensee: Educational Communications of Colorado Springs Inc. Format: Christian music. ♦Karen Veazey, gen mgr.

Limon

KAVD(FM)— 2003: Stn currently dark. 103.1 mhz; 100 kw. Ant 443 ft TL: N39 28 12 W103 38 14. Stereo. Hrs opn: 1211 Chuck Dawley Blvd., Suite 202, Mount Pleasant, SC, 29464. Phone: (843) 849-0076. Licensee: Coloradio Inc. Law Firm: Fletcher, Heald & Hildreth. ♦Edward F. Seeger, gen mgr.

KLIM(AM)— May 8, 1984: Stn currently dark. 1120 khz; 250 w-D. TL: N39 16 27 W103 42 49. Hrs open: 6 AM-sunset 165 E Ave., 80828. Phone: (719) 775-8199. Licensee: Roger L. Hoppe II (acq 3-7-96; $8,000). Law Firm: Miller & Neely. Format: Oldies. ♦Roger Hoppe II, pres.

Littleton

KCKK(AM)— Aug 22, 1957: 1510 khz; 10 kw-D, 1.3 kw-N, DA-2. TL: N39 33 08 W105 02 00. Hrs open: 1201 18th St., Suite 220, Denver, 80202. Phone: (303) 296-7025. Fax: (303) 296-7030. Licensee: People's Wireless Inc. Group owner: NRC Broadcasting Inc. (acq 4-26-2002; $2.7 million). Population served: 150,000 Format: Classic country. ♦Timothy Brown, gen mgr; Roger Tighe, chief of engrg.

Longmont

***KGUD(FM)**— September 1975: 90.7 mhz; 100 w. Ant 270 ft TL: N40 14 24 W105 03 19. Stereo. Hrs opn: 6 AM-3 PM (M-S) Box 1534, 80502-1534. Secondary address: Studio: 457 Fourth Ave. 80501. Phone: (303) 485-9811. E-mail: baskos_george@stvrain.k12.co.us Licensee: Longmont Community Radio (acq 10-31-2003). Population served: 180,000 Format: Easy Listening. Target aud: 45 plus; retirees. ♦George N. Baskos, gen mgr; James R. Boynton Sr., stn mgr.

KJCD(FM)— September 1964: 104.3 mhz; 5.8 kw. 1,204 ft TL: N40 05 47 W104 54 04. Stereo. Hrs opn: 24 Jefferson-Pilot Communications, 7800 E. Orchard Rd., Suite 400, Greenwood Village, 80111. Phone: (303) 321-0950. Fax: (303) 321-3383. Web Site: www.1043.com. Licensee: Lincoln Financial Media Co. of Colorado. (group owner; (acq 4-3-2006; grpsl). Population served: 2,100,000 Natl. Network: Westwood One. Law Firm: Wiley, Rein & Fielding. Format: Smooth jazz. Target aud: 18-34; hip. ♦Robert Call, gen mgr; Michael Fischer, opns mgr, opns mgr, sls dir, gen sls mgr & progmg dir; Dwayne Taylor, mktg dir; Brad Hart, chief of engrg.

Longmont/Denver

KRCN(AM)— December 1949: 1060 khz; 50 kw-D, 100 w-N. TL: N40 11 28 W105 07 35. Hrs open: 15; 18 (summer)
simulcasts KKKK(AM) Colorado Springs.
614 Kimbark St., Longmont, 80501. Phone: (303) 776-2323. Fax: (303) 776-1377. Web Site: www.radiocoloradonetwork.com. Licensee: Pilgrim Communications Inc. (acq 5-27-98; $575,000). Population served: 300,000 Format: Business talk, sports, NASCAR. News staff: one; News: 20 hrs wkly. Target aud: 30-60; news & sports listeners. Spec prog: Sp one hr, farm one hr wkly. ♦Gene Hood, pres; Ron Nickell, sr VP, gen mgr & stn mgr; Lee Thompson, gen sls mgr.

Loveland

KSXT(AM)— Jan 21, 1955: 1570 khz; 7 kw-D, 18 w-N. TL: N40 23 31 W105 05 51. Hrs open: 24 1270 Boston Ave., Longmont, 80501. Phone: (970) 612-1570. Fax: (970) 612-0137. E-mail: sfmtaylor@yahoo.com Web Site: www.1570ksxt.com. Licensee: O.J. & Carol Pratt (acq 5-23-2002). Population served: 250,000 Format: Sports talk. ♦Wes Hood, gen mgr & chief of opns; Mike Taylor, progmg dir.

KTRR(FM)— Feb 5, 1966: 102.5 mhz; 50 kw. 410 ft TL: N40 27 19 W104 55 25. Stereo. Hrs opn: 600 Main St., Windsor, 80550. Phone: (970) 674-2700. Fax: (970) 686-7491. Web Site: www.tri1025.com. Licensee: Regent Broadcasting of Ft. Collins Inc. Group owner: Regent Communications Inc. (acq 2-25-03). Population served: 173,000 Format: Adult contemp. Target aud: 25-54. ♦Cal Hall, gen mgr; Mark Callaghan, opns mgr; Miles Schallert, sls dir & gen sls mgr.

***KXWA(FM)**— Mar 11, 2004: 89.7 mhz; 45 w horiz, 36 kw vert. Ant 1,220 ft TL: N40 37 03 W105 19 40. Stereo. Hrs opn: 24 1707 N. Main, Suite 302, Longmont, 80501. Phone: (303) 702-9293. Fax: (303) 485-1929. Web Site: wayfm.com. Licensee: WAY-FM Media Group Inc. (group owner; (acq 11-19-2002). Population served: 2,300,000 Format: Contemp Christian. Target aud: 18-34. ♦Lloyd Parker, COO; Robert D. Augsburg, pres; Zach Cochran, gen mgr; Scott Veigel, progmg dir.

Manitou Springs

KBIQ(FM)— May 1952: 102.7 mhz; 57 kw. Ant 2,280 ft TL: N38 44 43 W104 51 39. Stereo. Hrs opn: 24 7150 Campus Dr., Suite 150, Colorado Springs, 80920. Phone: (719) 531-5438. Fax: (719) 531-5588. E-mail: henry@kbiqradio.com Web Site: www.kbiqradio.com. Licensee: Bison Media Inc. Group owner: Salem Communications Corp. (acq 10-8-96; $2.825 million). Population served: 350,000 Format: Christian, adult contemp. News staff: one. Target aud: 18-54. ♦Henry Tippie, gen mgr.

***KCME(FM)**— Sept 1, 1979: 88.7 mhz; 8.9 kw. 9,570 ft TL: N38 44 40 W104 51 41. Stereo. Hrs opn: 24 1921 N. Weber St., Colorado Springs, 80907-6903. Phone: (719) 578-5263. Fax: (719) 578-1033. E-mail: kcme@kcme.org Web Site: www.kcme.org. Licensee: Cheyenne Mt. Public Broadcast House Inc. Population served: 1,200,000 Law Firm: Scott Cinnamon. Format: Classical. Target aud: 45 plus; upper-middle class, mostly college graduates. ♦Joseph Reich Jr., pres; Jeanna Wearing, gen mgr, opns dir & mktg VP; Jennifer Hane, dev VP.

KXRE(AM)— November 1956: 1490 khz; 500 w-D, 250 w-N. TL: N38 51 43 W104 55 32. Hrs open: 24 600 Grant St., Denver, 80203. Phone: (303) 733-5266. Fax: (303) 733-5242. E-mail: kbno@kbno.net Web Site: www.kbno.net. Licensee: Latino Communications LLC (group owner; acq 1-23-03; $350,000 with KAVA(AM) Pueblo). Population served: 350,000 Format: Rgnl Mexican. News staff: 5; News: 21 hrs wkly. Target aud: 25-54. ♦Zee Ferrufino, gen mgr.

Meeker

KAYW(FM)— Sept 30, 2000: 98.1 mhz; 100 kw horiz. Ant 1,191 ft TL: N39 58 18 W108 02 23. Hrs open: 24
Rebroadcasts KZKS(FM) Rifle 100%.
751 Horizon Ct., Suite 225, Grand Junction, 81506. Phone: (970) 241-6460. Fax: (970) 241-6452. Web Site: www.wscradio.net. Licensee: Western Slope Communications LLC. (group owner) Population served: 60,000 Format: Adult favorites. News: 20 hrs wkly. ♦John Reid, gen mgr.

Monte Vista

KSLV(AM)— February 1954: 1240 khz; 1 kw-U. TL: N37 36 10 W106 08 58. Hrs open: 24 Box 631, 109 Adams St., 81144. Phone: (719) 852-3581. Fax: (719) 852-3583. E-mail: kslv@amigo.net Web Site: www.kslvradio.com. Licensee: San Luis Valley Broadcasting Inc. (acq 4-1-79). Population served: 40,000 Law Firm: Cohn & Marks. Format: Country. News staff: one; News: 6 hrs wkly. Target aud: 25-54. Spec prog: Sp 10 hrs, farm one hr, gospel 4 hrs wkly. ♦Gerald Vigil, gen mgr; Linda Pacheco, news dir.

KSLV-FM— 1986: 95.3 mhz; 6 kw. 89 ft TL: N37 36 10 W106 08 58. Stereo. 24 Web Site: www.kslvradio.com. Format: Soft adult contemp. News: 2 hrs wkly.

Montrose

KKXK(FM)—Listing follows KUBC(AM).

***KPRH(FM)**— October 1998: 88.3 mhz; 5 kw. 1,535 ft TL: N38 20 01 W107 39 52. Hrs open:
Rebroadcasts KCFR(FM) Denver 100%.
Colorado Public Radio, 7409 S. Alton Ct., Centennial, 80112. Phone: (303) 871-9191. Phone: (800) 722-4449. Fax: (303) 733-3319. Web Site: www.cpr.org. Licensee: Public Broadcasting of Colorado Inc. Natl. Network: NPR. Law Firm: Arter & Hadden. Format: News. Target aud: General. ♦Max Wycisk, pres; Sue Coughlin, dev VP; Sean Nethery, progmg dir; Robert Hensler, engrg VP; David Gomez, traf mgr.

KSTR-FM— Apr 10, 1980: 96.1 mhz; 91 kw. Ant 1,099 ft TL: N38 52 40 W108 13 33. Stereo. Hrs opn: 1360 E. Sherwood Dr., Grand Junction, 81501. Phone: (970) 254-2100. Fax: (970) 245-7551. Web Site: www.gjradio.com. Licensee: MBC Grand Broadcasting Inc. (acq 5-25-2005; $600,000). Population served: 350,000 Format: Classic rock. ♦David G. Hinson, pres; Jim Terlouw, gen mgr; Dave Beck, gen sls mgr.

KUBC(AM)— Sept 25, 1947: 580 khz; 5 kw-D, 1 kw-N, DA-N. TL: N38 25 32 W107 52 57. Hrs open: 24 Box 970, 81402. Secondary address: 106 Rose Ln. 81401. Phone: (970) 249-4546. Fax: (970) 249-2229. Web Site: www.coloradoradio.com. Licensee: CCR-Montrose IV LLC. (group owner; (acq 8-19-2004; grpsl). Population served: 86,000 Natl. Network: ABC. Natl. Rep: Interep. Law Firm: Garvey, Schubert & Barer. Format: Classic country. News: 4 hrs wkly. Target aud: 35-54. Spec prog: Sports 5 hrs, relg 3 hrs wkly. ♦Joseph D. Schwartz, pres; Jay Austin, gen mgr.

KKXK(FM)—Co-owned with KUBC(AM). December 1976: 94.1 mhz; 90 kw. 1,748 ft TL: N38 20 16 W107 38 23. Stereo. 24 Web Site: www.coloradoradio.com. Format: Contemp country. News staff: one; News: 4 hrs wkly. Target aud: 25-54.

***KVMT(FM)**— 1999: 89.1 mhz; 3 kw. Ant 1,748 ft TL: N38 18 52 W108 12 02. Hrs open: 24
Rebroadcasts KVNF (FM) Paonia 100%.
Box 1350, Paonia, 81428. Phone: (970) 527-4866. Fax: (970) 527-4865. Web Site: www.kvnf.org. Licensee: North Fork Valley Public Radio Inc. Format: News, music, pub affrs. News: 35 hrs wkly. ♦Sally Kane, gen mgr.

Monument

KCBR(AM)— July 20, 1986: 1040 khz; 15 kw-D. TL: N38 49 08 W104 46 32. Stereo. Hrs opn: Sunrise-sunset 5050 Edison Ave., Suite 218, Colorado Springs, 80915. Phone: (719) 570-1530. Fax: (719) 570-1007. Web Site: www.1040kcbr.com. Licensee: KLZ Radio Inc. Group owner: Crawford Broadcasting Co. (acq 1999; $750,000 with KCMN(AM) Colorado Springs). Population served: 400,000 Format: Christian talk. Target aud: 25-54; 70% male, upper-middle income or higher. ♦Don Crawford Jr., CEO, VP & gen mgr; Don Crawford, Sr., pres.

Morrison

***KLDV(FM)**— Mar 27, 1971: 91.1 mhz; 100 kw. Ant 1,168 ft TL: N39 36 00 W105 12 35. (CP: ant 1,207 ft). Stereo. Hrs opn: 24 2351 Sunset Blvd., Suite 170-218, Rocklin, CA, 95765. Phone: (916) 251-1600. Fax: (916) 251-1650. E-mail: klove@klove.com Web Site: www.klove.com. Licensee: Educational Media Foundation. Group owner: EMF Broadcasting (acq 12-28-00; grpsl). Population served: 1,500,000 Natl. Network: K-Love. Law Firm: Shaw Pittman. Format: Contemp Christian. News staff: 3. Target aud: 25-44; Judeo Christian, female. ♦Richard Jenkins, pres; Mike Novak, VP; Keith Whipple, dev dir.

Mountain Village

KRKQ(FM)—Not on air, target date: unknown: 95.5 mhz; 6 kw. Ant -302 ft TL: N37 56 02 W107 50 03. Hrs open: 911 Colonial Dr., Cheyenne, WY, 82001-7415. Phone: (307) 638-1345. Licensee: Lorenz E. Proietti. ♦Lorenz E. Proietti, gen mgr.

New Castle

KJEB(FM)— 2005: 94.5 mhz; 25 kw. Ant -397 ft TL: N39 33 56 W107 32 01. Hrs open: 1201 18th St., Suite 250, Denver, 80202-1869. Phone: (303) 296-7025. Fax: (303) 296-7030. Licensee: Wildcat Communications LLC. Format: Oldies. ♦Timothy Brown, gen mgr; Roger Tighe, chief of engrg.

Norwood

KRYD(FM)— January 1998: 104.9 mhz; 24 kw. Ant 1,672 ft TL: N38 18 57 W108 11 47. Stereo. Hrs opn: 24 444 Seasons Dr., Grand Junction, 81503. Secondary address: 475 Water St., Monrose 81401. Phone: (970) 263-4100. Fax: (970) 263-9600. E-mail: billv@taousa.tv Web Site: www.krydfm.com. Licensee: Rocky III Investments Inc. Population served: 225,000 Law Firm: Wood, Maine & Brown, Chartered. Format: Country. News: one hr wkly. Target aud: 18 plus. ♦Bill Varecha, CEO; Debbie Varecha, CFO; Wes Smith, opns mgr; Paul Varecha, gen sls mgr.

Oak Creek

KFMU-FM— Sept 22, 1975: 104.1 mhz; 1.4 kw. 1,073 ft TL: N40 14 10 W106 52 30. Stereo. Hrs opn: 24 Box 772850, 2955 Village Dr., Steamboat Springs, 80477. Phone: (970) 879-5368. Fax: (970) 879-5843. Web Site: www.kfmu.com. Licensee: NRC Mountain Division LLC. (group owner; (acq 7-9-2004; grpsl). Population served: 40,000 Natl. Network: CBS. Format: Triple A. News staff: 2; News: 10 hrs wkly. Target aud: 21-54. Spec prog: Jazz 4 hrs, modern mus 4 hrs wkly. ♦Steve Wodlinger, gen mgr.

Otis

KATR-FM— Sept 1, 1983: 98.3 mhz; 100 kw. Ant 554 ft TL: N40 25 13 W102 58 10. Stereo. Hrs opn: 24 Box 354, Wray, 80758. Phone: (970) 332-4171. Fax: (970) 332-4172. E-mail: krdzkatr@plains.net Web Site: www.katcountry983.com. Licensee: Media Logic LLC (group owner; acq 10-28-2002; $700,000). Format: Country. News staff: 3; News: 15 hrs wkly. Target aud: 16-70; males & females. ♦Wayne Johnson, gen mgr & opns mgr.

Ouray

KWGL(FM)— June 16, 1986: 105.7 mhz; 60 kw horiz. Ant 1,752 ft TL: N38 23 16 W107 40 28. Stereo. Hrs opn: 24 751 Horizon Ct., Suite 225, Grand Junction, 81506. Phone: (970) 241-6460. Fax: (970) 241-6452. Web Site: www.range105.net. Licensee: WS Communications L.L.C. (acq 1-95; FTR: 5-22-95). Population served: 200,000 Natl. Network: Jones Radio Networks. Format: Legendary Country. News staff: one; News: 15 hrs wkly. ♦John Reid, gen mgr; Michael Johnson, progmg dir.

Pagosa Springs

***KPGS(FM)**—Not on air, target date: unknown: 88.1 mhz; 375 w vert. Ant 1,309 ft TL: N37 11 32 W107 05 56. Hrs open:
Rebroadcasts KUTE (FM) Ignacio.
Box 737, Ignacio, 81137-0737. Phone: (970) 563-0255. Fax: (970) 563-0399. Web Site: www.ksut.org. Licensee: KUTE Inc. Format: Triple A, americana. ♦Eddie Box Jr., pres; Beth Warren, gen mgr.

***KTPS(FM)**—Not on air, target date: unknown: 89.7 mhz; 200 w. Ant 1,273 ft TL: N37 11 35 W107 05 58. Hrs open:
KTLF (FM) Colorado Springs 100%.
1665 Briargate Blvd., Suite 100, Colorado Springs, 80920. Phone: (719) 593-0600. Fax: (719) 593-2399. E-mail: lightpraise@ktlf.org Web Site: www.ktlf.org. Licensee: Educational Communications of Colorado Springs Inc. Format: Christian music. ♦Karen Veazey, gen mgr; Sharick Wade, opns mgr; John Hayes, mus dir; Harry Russell, chief of engrg.

KWUF-FM— May 1, 1986: 106.3 mhz; 255 w. 1,280 ft TL: N37 11 32 W107 05 55. Stereo. Hrs opn: 24 Box 780, 81147. Secondary address: 702 S. 10th St. 81147. Phone: (970) 264-5983. Fax: (970) 264-5129. E-mail: admin@kwuf.com Web Site: www.kwuf.com. Licensee: Wolf Creek Broadcasting L.L.C. (acq 1999; with co-located AM). Population served: 10,000 Natl. Network: Westwood One. Format: Adult contemp. News: 10 hrs wkly. Target aud: 18 plus. Spec prog: Blues 10 hrs, jazz 10 hrs wkly. ♦Christie Spears, VP; Beth Porter, sls dir; Jodie Blankenship, news dir; Will Spears, CEO, pres, gen mgr, progmg mgr & chief of engrg; Chris Olivarez, sports cmtr; Chris Olivarez, disc jockey.

KWUF(AM)— Aug 27, 1975: 1400 khz; 1 kw-U. TL: N37 15 24 W107 01 06.24 Web Site: www.kwuf.com.10,000 Natl. Network: Westwood One. Format: Country, news/talk, sports. News: 10 hrs wkly. Target aud: General. ♦Chris Olivarez, sports cmtr & disc jockey.

Palisade

KAAI(FM)—Not on air, target date: unknown: 98.5 mhz; 215 w. Ant 2,988 ft TL: N39 03 14 W108 15 13. Hrs open: 1546 Brettonwood Way, Highlands Ranch, 80129. Licensee: Todd Deneui. ♦Todd Deneui, gen mgr.

Paonia

***KVNF(FM)**— Oct 5, 1979: 90.9 mhz; 3 kw. Ant -171 ft TL: N38 52 20 W107 39 45. Stereo. Hrs opn: 18 233 Grand Ave., 81428. Phone: (970) 527-4866. Fax: (970) 527-4865. Web Site: www.kvnf.org. Licensee: North Fork Valley Public Radio Inc. (acq 1-27-78). Population served: 50,000 Natl. Network: NPR. Format: News, music, pub affrs. Spec prog: Class 15 hrs, jazz 17 hrs, blues 3 hrs, C&W 5 hrs, new age 6 hrs, Sp 2 hrs, gospel 3 hrs wkly. ♦Sally Kane, gen mgr.

Pierce

KJMP(AM)— 2004: 870 khz; 1.2 kw-D, 320 w-N, DA-2. TL: N40 36 25 W104 41 19. Hrs open: 1063 Big Thompson Rd., Apt F, Loveland, 80537-9424. Phone: (307) 638-8921. Licensee: White Park Broadcasting Inc (acq 9-6-2005; $350,000). Format: Adult contemp, Sports. ♦Steven Silverburg., gen mgr.

Placerville

***KTEI(FM)**—Not on air, target date: 2: 90.7 mhz; 250 w. Ant 1,486 ft TL: N37 59 29 W107 58 21. Hrs open:
KTLF (FM) Colorado Springs 100%.
1665 Briargate Blvd., Suite 100, Colorado Springs, 80920. Phone: (719) 593-0600. Fax: (719) 593-2399. E-mail: lightpraise@ktlf.org Web Site: www.ktlf.org. Licensee: Educational Communications of Colorado Springs Inc. Format: Christian music. ♦Karen Veazey, gen mgr.

Poncha Springs

KWUZ(FM)—Not on air, target date: unknown: 97.5 mhz; 6 kw. Ant -1,007 ft TL: N38 33 10 W106 17 42. Hrs open: 6130 Cheney Ridge Cir., Lincoln, NE, 68516. Licensee: William C. Doleman. ♦William C. Doleman, gen mgr.

Pueblo

KAVA(AM)— June 1963: 1480 khz; 1 kw-D, DA. TL: N38 18 56 W104 37 03. Hrs open: 600 Grant St., Denver, 80203. Phone: (303) 733-5266. Fax: (303) 733-5242. E-mail: kbno@kbno.net Web Site: www.kbno.net. Licensee: Latino Communications LLC (group owner; acq 1-23-03; $350,000 with KXRE(AM) Manitou Springs). Population served: 110,000 Format: Rgnl Mexican. News staff: 6; News: 21 hrs wkly. Target aud: 25-54. ♦Zee Ferrufino, gen mgr & opns dir.

KCCY(FM)— Aug 23, 1975: 96.9 mhz; 100 kw. 320 ft TL: N38 21 32 W104 58 13. Stereo. Hrs opn: 2864 S. Circle Dr., Suite 150, 80906. Phone: (719) 540-9200. Fax: (719) 543-9898. Web Site: www.y969.com. Licensee: Clear Channel Radio Licenses, Inc. Group owner: Clear Channel Communications Inc. (acq 11-22-00; with KDZA-FM Pueblo). Population served: 100,000 Natl. Rep: Christal. Format: C&W. Target aud: 25-54; general. ♦Bob Richards, gen mgr & opns mgr; Mark Warren, gen sls mgr; Scott Jones, gen sls mgr; Robert Vargas, prom dir; Paul Richards, news dir.

KCEG(AM)—Not on air, target date: End of 2007: 780 khz; 3 kw-D, 760 w-N, DA-2. TL: N38 33 45 W104 36 09. Hrs open: 965 S. Irving St., Denver, 80219. Phone: (303) 935-1156. Licensee: Timothy C. Cutforth. Format: Family Friendly Variety. ♦Timothy C. Cutforth, gen mgr.

***KCFP(FM)**— June 1986: 91.9 mhz; 600 w. 633 ft TL: N38 22 23 W104 33 42. Stereo. Hrs opn: 24
Rebroadcasts KCFR(FM) Denver 100%.
Colorado Public Radio, 7409 S. Alton Ct., Centennial, 80112. Phone: (303) 871-9191. Fax: (303) 733-3319. Web Site: www.cpr.org. Licensee: Public Broadcasting of Colorado Inc. Natl. Network: NPR. Law Firm: Arter & Hadden. Format: Class, news. News staff: 8; News: 50 hrs wkly. Target aud: General. ♦Max Wycisk, pres; Sue Coughlin, dev VP.

KCSJ(AM)— 1947: 590 khz; 1 kw-U, DA-N. TL: N38 21 30 W104 38 13. Hrs open: 24 106 W. 24th St., 81003. Phone: (719) 545-2080. Fax: (719) 543-9898. E-mail: webmaster@590kcsj.com Web Site: www.590kcsj.com. Licensee: CC Licenses LLC. Group owner: Clear Channel Communications Inc. (acq 6-14-2001; with KGHF(AM) Pueblo). Population served: 150,000 Natl. Network: ABC. Natl. Rep: Christal. Format: News/talk. News staff: 2; News: 41 hrs wkly. Target aud: 35-64; upscale. ♦Olene Greenwood, gen mgr.

KDZA-FM— Mar 3, 1987: 107.9 mhz; 100 kw. Ant 239 ft TL: N37 56 40 W104 59 56. Stereo. Hrs opn: 106 W. 24th St., 81003. Phone: (719) 545-2080. Fax: (719) 543-9898. E-mail: webmaster@kdzafm.com Web Site: www.kdzafm.com. Licensee: Capstar TX L.P. Group owner: Clear Channel Communications Inc. (acq 11-22-2000; with KCCY(FM) Pueblo). Population served: 532,300 Format: Oldies. ♦Olene Greenwood, gen mgr.

KFEL(AM)— August 1956: 970 khz; 3.2 kw-D, 184 w-N. TL: N38 15 57 W104 40 44. Hrs open: Box 8055, 81008. Phone: (719) 543-7506. Fax: (719) 543—0432. E-mail: kfel970am@aol.com Licensee: Kansas City Catholic Network Inc. (acq 11-3-2006; $475,000). Population served: 509,500 Format: Relg. Target aud: 25 plus. ♦Allen Bickle, gen mgr.

***KFRY(FM)**— 2006: 89.9 mhz; 870 w. Ant 2,122 ft TL: N38 02 29 W105 11 05. Hrs open:
Rebroadcasts KUFR(FM) Salt Lake City, UT 100%.
c/o KUFR(FM), 136 E.S. Temple, Suite 1630, Salt Lake City, UT, 84111. Phone: (801) 359-3147. Fax: (801) 359-8112. Web Site: www.familyradio.com. Licensee: Family Stations Inc. Format: Christian relg. ♦Harold Camping, pres & gen mgr.

KGFT(FM)— Mar 31, 1988: 100.7 mhz; 72.4 kw. Ant 2,217 ft TL: N38 44 44 W104 51 39. Stereo. Hrs opn: 24 7150 Campus Drive, Ste 150, Colorado Springs, 80920. Phone: (719) 531-5438. Fax: (719) 531-5588. Web Site: www.kgftradio.com. Licensee: Salem Communications Corp. (group owner; (acq 1996; $3 million). Population served: 800,000 Natl. Network: AP Radio. Format: Christian, news/talk, relg. News staff: one; News: 4 hrs wkly. Target aud: 25 plus; Christian. Spec prog: Gospel 3 hrs, old time radio 11 hrs wkly. ♦Henry Tippie, gen mgr.

KGHF(AM)— February 1928: 1350 khz; 5 kw-D, 1 kw-N, DA-N. TL: N38 18 29 W104 38 24. Stereo. Hrs opn: 24 106 W. 24th St., 81003. Phone: (719) 545-2080. Fax: (719) 543-9898. E-mail: webmaster@1350thezone.com Web Site: www.1350thezone.com. Licensee: CC Licenses LLC. Group owner: Clear Channel Communications Inc. (acq 6-14-2001; with KCSJ(AM) Pueblo). Population served: 130,000 Law Firm: Haley, Bader & Potts. Format: Sports. News staff: one; News: one hr wkly. Target aud: 35 plus. ♦Olene Greenwood, gen mgr.

KILO(FM)—See Colorado Springs

KKMG(FM)— Jan 1, 1967: 98.9 mhz; 100 kw. 1,715 ft TL: N38 44 32 W104 51 41. (CP: 56 kw, ant 2,299 ft.). Stereo. Hrs opn: 24 6805 Corporate Dr., Suite 130, Colorado Springs, 80919-1977. Phone: (719) 593-2700. Fax: (719) 593-2727. Web Site: www.989magicfm.com. Licensee: Citadel Broadcasting Co. Group owner: Citadel Broadcasting Corp. (acq 3-21-94; $912,500; FTR: 4-18-94). Population served:

500,000 Natl. Rep: McGavren Guild. Law Firm: Reed, Smith, Shaw & McClay. Format: Top-40. Target aud: 18-44. ♦Bobby Irwin, opns mgr; John Fox, gen mgr & progmg dir.

KKPC(AM)— Dec 29, 1947: 1230 khz; 1 kw-U. TL: N38 16 38 W104 39 13. Hrs open: 24 Colorado Public Radio, 7409 S. Alton Ct., Centennial, 80112. Phone: (303) 871-9191. Fax: (303) 733-3319. E-mail: info@cpr.org Web Site: www.cpr.org. Licensee: Public Broadcasting of Colorado Inc. (acq 6-21-01; $275,000). Format: News and info. ♦Max Wycisk, pres & gen mgr; Jenny Gentry, exec VP; Sean Nethery, progmg VP.

KNKN(FM)— November 1979: 106.9 mhz; 27.5 kw. Ant 666 ft TL: N38 06 22 W104 29 18. Stereo. Hrs opn: 30 N. Electronic Dr., Pueblo West, 81007. Phone: (719) 547-0411. Fax: (719) 547-9301. E-mail: elgatonnegro@amigo.net Licensee: JaneGary Inc. (group owner; (acq 12-29-2005; $2 million with KRMX(AM) Pueblo). Population served: 159,361 Format: Sp Contemp. Target aud: 18-54. ♦Lupe Brown, gen mgr.

KRMX(AM)— 1958: 690 khz; 250 w-D, 24 w-N. TL: N38 17 48 W104 38 47. Stereo. Hrs opn: 24
KNKN Sunday mass 10:15 am.
30 N. Electronic Dr., Pueblo West, 81007. Phone: (719) 545-2883. Fax: (719) 547-9301. E-mail: elgatonegro@amigo.net Licensee: JaneGary Inc. (group owner; (acq 12-29-2005; $2 million with KNKN(FM) Pueblo). Population served: 159,000 Format: Rgnl Mexican. Target aud: General; Hispanic families. ♦Lupe Brown, gen mgr.

***KTPL(FM)**— 2005: 88.3 mhz; 65 kw. Ant 226 ft TL: N37 56 40 W104 59 56. Hrs open: 1665 Briargate Blvd., Suite 100, Colorado Springs, 80920. Phone: (719) 593-0600. Fax: (719) 593-2399. E-mail: power88@power88.org Web Site: www.power88.org. Licensee: Educational Communications of Colorado Springs Inc. (acq 12-2-02; $6,251 for CP). Format: Inspirational Christian. ♦Karen Veazey, gen mgr; Kevin Waldren, progmg dir.

***KTSC-FM**— October 1970: 89.5 mhz; 9.8 kw. 165 ft TL: N38 18 38 W104 34 40. Stereo. Hrs opn: Colorado State University-Pueblo, 2200 Bonforte Blvd. Rm #120, 81001. Phone: (719) 549-2822. Fax: (719) 549-2120. Licensee: University of Southern Colorado. Population served: 123,000 Format: Hip hop, rhythm and blues. ♦Mike Atencio, stn mgr.

KVUU(FM)— 1976: 99.9 mhz; 57. 2,200 ft TL: N33 44 47 W104 51 37. Stereo. Hrs opn: 24 2864 S. Circle Dr., Suite 150, Colorado Springs, 80906. Phone: (719) 540-9200. Fax: (719) 579-0882. Web Site: www.my999radio.com. Licensee: Capstar TX L.P. Group owner: Clear Channel Communications Inc. (acq 8-30-00; grpsl). Natl. Rep: Clear Channel. Format: Hits of the 90s. News staff: one; News: 3 hrs wkly. Target aud: 25-54; upscale young adults. ♦Bob Richards, chmn & gen mgr; Scott Jones, gen sls mgr; Robert Vargas, prom dir; Paul Richards, news dir.

KYZX(FM)—Pueblo West, 1993: 103.9 mhz; 1.75 kw. Ant 2,158 ft. TL: N38 44 40 W104 51 41. Stereo. Hrs opn: 24 1805 E. Cheyenne Rd., Colorado Springs, 80906. Phone: (719) 634-4896. Fax: (719) 634-5837. Web Site: www.theeagle.com. Licensee: Colorado Springs Radio Broadcasters Inc. Group owner: Bahakel Communications (acq 2-22-99; grpsl). Population served: 300,000 Format: Classic rock. News staff: one; News: one hr wkly. Target aud: 25-49; general. ♦Lou Mellini, gen mgr; Jason Janc, opns mgr & progmg dir.

Pueblo West

KYZX(FM)—Licensed to Pueblo West. See Pueblo

Ridgway

KBNG(FM)— 2002: 103.7 mhz; 4.1 kw. Ant 1,574 ft TL: N38 23 15 W107 40 31. Hrs open: 24 Box 970, Montrose, 81402. Phone: (970) 249-4546. Fax: (970) 249-2229. Web Site: www.coloradoradio.com. Licensee: CCR-Montrose IV LLC. (group owner; (acq 8-19-2004; grpsl). Format: Hot adult contemp. Target aud: 18-44; female. ♦Joseph D. Schwartz, pres; Jay Austin, gen mgr; Scott Staley, progmg dir.

Rifle

KRGS(AM)— June 9, 1967: 690 khz; 1 kw-D. TL: N39 32 55 W107 46 10. Hrs open: 24 751 Horizon Ct., Suite 225, Grand Junction, 81506. Phone: (970) 241-6460. Fax: (970) 241-6452. Web Site: www.wscradio.net. Licensee: Western Slope Communications L.L.C. (group owner) Population served: 40,000 Law Firm: Akin, Gump, Strauss, Hauer & Feld. Format: Sports. Target aud: 18-54; males. ♦John Reid, gen mgr, gen sls mgr, prom dir & progmg dir; Rich Cron, chief of engrg.

KZKS(FM)—Co-owned with KRGS(AM). 1994: 105.3 mhz; 60 kw. 2,437 ft TL: N39 25 57 W108 07 46. Stereo. 24 Web Site: www.drive105.net.320,000 Format: Adult favorites. News staff: one; News: 20 hrs wkly. ♦Gary Duplantis, mus dir.

Rocky Ford

KPHT(FM)— 2002: 95.5 mhz; 100 kw. Ant 735 ft TL: N37 54 08 W104 16 00. Hrs open: 106 W. 24th St., Pueblo, 81003. Phone: (719) 545-2080. Fax: (719) 543-9898. E-mail: webmaster@hot955.com Web Site: www.hot955.com. Licensee: Capstar TX L.P. Group owner: Clear Channel Communications Inc. (acq 2-12-2001; $1 million). Format: Adult contemp, 80s, 90s. ♦Olene Greenwood, gen mgr.

Rye

KRYE(FM)—Not on air, target date: unknown: 104.9 mhz; 25 kw. Ant 64 ft TL: N37 56 40 W104 59 56. Hrs open: 1200 W. Cornwallis, Greenboro, NC, 27408. Phone: (719) 543-0405. Licensee: United States CP LLC. ♦W. Philip Robinson, gen mgr; Todd Robinson, stn mgr.

***KRYI(FM)**—Not on air, target date: unknown: 89.7 mhz; 9.8 kw. Ant 679 ft TL: N37 32 34 W104 22 31. Hrs open:
Rebroadcasts KXWA (FM) Loveland 100%.
5475 Tech Center Dr., Suite 210, Colorado Springs, 80919. Phone: (719) 533-0300. Fax: 719-278-4339. Web Site: www.wayfm.com. Licensee: Harvest Radio Corp. Format: Christian rock. ♦Bob Augsburg, CEO; Lloyd Parker, COO; Zach Cochran, gen mgr.

***KXWY(FM)**—Not on air, target date: unknown: 90.9 mhz; 11.3 kw. Ant 113 ft TL: N37 56 40 W104 59 56. Hrs open: Box 64500, Colorado Springs, 80962. Phone: (719) 533-0300. Fax: (719) 278-4339. Licensee: WAY-FM Media Group Inc. (acq 6-17-2005; $200,000 for CP). ♦Robert Augsburg, pres.

Salida

KSBV(FM)— 2002: 93.7 mhz; 1 kw. Ant 2,722 ft TL: N38 26 47 W106 00 37. Stereo. Hrs opn: 24 228 East St., 81211. Phone: (719) 539-9377. Fax: (719) 539-7904. E-mail: ksbvradio@chaffee.net Web Site: www.ksbv.com. Licensee: Arkansas Valley Broadcasting L.L.C. Law Firm: Gammon & Grange. Format: Classic rock. Target aud: 25-65. ♦Marc Scott, pres & gen mgr; Melissa Scott, traf mgr.

***KTPF(FM)**— 2007: 91.3 mhz; 385 w. Ant 2,952 ft TL: N38 26 48 W106 00 36. Hrs open:
Rebroadcasts KTLF(FM) Colorado Springs 100%.
1665 Briargate Blvd., Suite 100, Colorado Springs, 80920-3400. Phone: (719) 593-0600. Fax: (719) 593-2399. Web Site: www.ktlf.org. Licensee: Educational Communications of Colorado Springs Inc. Format: Christian music. ♦Karen Veazey, gen mgr.

KVRH(AM)— Dec 10, 1948: 1340 khz; 1 kw-U. TL: N38 31 55 W106 00 54. Hrs open: 24 7600 County Rd. 120, 81201. Phone: (719) 539-2575. Fax: (719) 539-4851. E-mail: kvrh@kvrh.com Web Site: kvrham.com. Licensee: Headwaters Media L.L.C. (acq 9-1-2006; swap for KBVC(FM) Buena Vista). Population served: 20,000 Format: Oldies. News staff: one; News: 10 hrs wkly. Target aud: 25-54; general. ♦Doug Vollertsen, stn mgr; Cristy Carothers, gen sls mgr; Norm Veasman, news dir.

KVRH-FM— 1971: 92.3 mhz; 13.5 kw. Ant -656 ft TL: N38 30 26 W106 01 22. Stereo. Hrs opn: 24 7600 CR 120, 81201. Phone: (719) 539-2575. Fax: (719) 539-4851. E-mail: kvrh@kvrh.com Web Site: www.kvrh.com. Licensee: Three Eagles Communications of Colorado LLC (acq 4-12-2000; with co-located AM). Population served: 20,000 Format: Hot adult contemp. News staff: one; News: 10 hrs wkly. ♦Cristy Carothers, gen mgr & gen sls mgr; Jen Jackson, traf mgr.

Security

KRDO-FM— Apr 8, 1973: 105.5 mhz; 409 w. Ant 2,230 ft TL: N38 44 40 W104 51 41. Stereo. Hrs opn: 24
Rebroadcasts simulcast of KRDO Colorado Springs 100%.
399 S. 8th St., Colorado Springs, 80905. Phone: (719) 578-1055. Fax: (719) 475-0815. E-mail: m.lewis@krdotv.com Web Site: www.krdo.com. Licensee: Optima Communications Inc. (acq 1989). Population served: 500,000 Law Firm: Mullin, Rhyne, Emmons & Topel. Format: News/talk. ♦J.B. McCoy III, pres; James R. Bond Jr., CFO; Edward L. Klimek, VP; Neil O. Klockziem, gen mgr.

Sidney

***KTSG(FM)**—Not on air, target date: unknown: 91.7 mhz; 2.5 kw horiz. Ant 610 ft TL: N40 27 43 W106 50 58. Hrs open:
KTLF(FM) Colorado Springs 100%.
1665 Briargate Blvd., Suite 100, Colorado Springs, 80920-3400. Phone: (719) 593-0600. Fax: (719) 593-2399. E-mail: lightpraise@ktlf.org Web Site: www.ktlf.org. Licensee: Educational Communications of Colorado Springs Inc. Format: Christian music. ♦Karen Veazey, gen mgr.

Snowmass Village

KSNO-FM— April 1985: 103.9 mhz; 6 kw. Ant 325 ft TL: N39 14 51 W106 55 13. Stereo. Hrs opn: 24 Box 1028, Glenwood Springs, 81602. Phone: (970) 945-9124. Fax: (970) 945-5409. Web Site: www.ksno.com. Licensee: Colorado West Broadcasting Inc. (acq 7-24-2007; $1.05 million). Population served: 20,000 Natl. Rep: Katz Radio. Law Firm: Cole, Raywid & Braverman. Wire Svc: AP Format: AAA. News staff: one; News: 2 hrs wkly. Target aud: 25-54. ♦Gabe Chenoweth, pres & gen mgr.

Steamboat Springs

KBCR(AM)— Aug 1, 1976: 1230 khz; 1 kw-U. TL: N40 29 19 W106 50 57. Hrs open: 24 Box 774050, 80477. Secondary address: 2110 Mt. Werner Rd. 80487. Phone: (970) 879-2270. Fax: (970) 879-1404. Web Site: kbcr.com. Licensee: Cool Radio LLC. (acq 1-30-2006; grpsl). Population served: 25,000 Natl. Network: ESPN Radio. Format: Sports. News staff: one. Target aud: 24-55. ♦Brian Harvey, gen mgr, prom mgr & progmg dir; Dave Lancaster, sports cmtr.

KBCR-FM— July 25, 1974: 96.9 mhz; 10 kw. Ant 666 ft TL: N40 27 43 W106 50 57. Stereo. Web Site: kbcr.com.25,000 Natl. Network: ABC. Format: Country.

KFMU-FM—See Oak Creek

***KLBV(FM)**— 2005: 89.3 mhz; 2.6 kw vert. Ant 1,699 ft TL: N40 27 04 W106 45 06. Hrs open: 24 2351 Sunset Blvd., Suite 170-218, Rocklin, CA, 95765. Phone: (916) 251-1600. Fax: (916) 251-1650. E-mail: klove@klove.com Web Site: www.klove.com. Licensee: Educational Media Foundation. Group owner: EMF Broadcasting (acq 10-2-03; grpsl). Natl. Network: K-Love. Law Firm: Shaw Pittman. Format: Contemp Christian. News staff: 3. Target aud: 25-44; Judeo Christian, female. ♦Richard Jenkins, pres; Mike Novak, VP; Keith Whipple, dev dir; David Pierce, progmg mgr; Ed Lenane, news dir; Sam Wallington, engrg dir; Karen Johnson, news rptr; Marya Morgan, news rptr; Richard Hunt, news rptr.

***KRNC(FM)**— Jan 23, 2006: 88.5 mhz; 240 w. Ant 600 ft TL: N40 27 43 W106 50 57. Hrs open:
Rebroadcasts KUNC(FM) Greeley 100%.
822 Seventh St., Suite 530, Greeley, 80631-3945. Phone: (970) 378-2579. Fax: (970) 378-2580. E-mail: mailbag@kunc.org Web Site: www.kunc.org. Licensee: Community Radio of Northern Colorado (acq 1-5-2006; $50,000 for CP). Natl. Network: NPR, PRI. Wire Svc: AP Format: Div, news. ♦Neil Best, gen mgr; Michelle Komanich, dev dir; Kirk Mowers, progmg dir; Jim Beers, news dir.

Sterling

***KDRE(FM)**— 2005: Stn currently dark. 90.7 mhz; 1.6 kw vert. Ant 506 ft TL: N40 36 56 W103 02 02. Hrs open:
Rebroadcasts KLRD(FM) Yucaipa, CA 100%.
2351 Sunset Blvd., Suite 170-218, Rocklin, CA, 95765. Phone: (916) 251-1600. Fax: (916) 251-1650. Web Site: www.air1.com. Licensee: Educational Media Foundation. (acq 9-22-2005; $17,000 for CP). Natl. Network: Air 1. Format: Christian rock. ♦Richard Jenkins, pres; Mike Novak, VP; Keith Whipple, dev dir; David Pierce, progmg mgr; Ed Lenane, news dir; Sam Wallington, engrg dir; Arthur Vassar, traf mgr; Karen Johnson, news rptr; Marya Morgan, news rptr; Richard Hunt, news rptr.

***KLZV(FM)**—Not on air, target date: unknown: 91.3 mhz; 6 kw. Ant 423 ft TL: N40 08 56 W103 17 04. Hrs open: 2351 Sunset Blvd., Suite 170-218, Rocklin, CA, 95765. Phone: (916) 251-1600. Fax: (916) 251-1650. E-mail: klove@klove.com Web Site: www.klove.com. Licensee: Educational Media Foundation. Group owner: EMF Broadcasting (acq 10-2-2003; grpsl). Natl. Network: K-Love. Law Firm: Shaw Pittman. Format: Contemp Christian. News staff: 3. Target aud: 25-44; Judeo

Christian, female. ♦Richard Jenkins, pres; Mike Novak, VP & progmg dir; Lloyd Parker, gen mgr; Ed Lenane, opns dir & news dir; Keith Whipple, dev dir; Eric Allen, natl sls mgr; David Pierce, progmg mgr; Jon Rivers, mus dir; Sam Wallington, engrg dir; Arthur Vassar, traf mgr; Karen Johnson, news rptr; Marya Morgan, news rptr; Richard Hunt, news rptr.

KNNG(FM)—Listing follows KSTC(AM).

KPMX(FM)— Aug 19, 1983: 105.7 mhz; 12 kw. Ant 479 ft TL: N40 31 57 W103 07 22. Stereo. Hrs opn: 24 117 Main St., 80751. Phone: (970) 522-4800. Fax: (970) 522-3994. Web Site: www.kpmx.com.Jones Licensee: Northeast Colorado Broadcasting LLC (group owner; acq 7-1-2003; grpsl). Population served: 18,000 Law Firm: Booth, Freret, Imlay & Tepper. Format: Adult contemp. Target aud: 18-54. ♦Alec Creighton, gen mgr.

KSRX(FM)—Not on air, target date: unknown: 97.5 mhz; 17 kw. Ant 561 ft TL: N40 27 15.1 W103 09 6.1. Hrs open: Box 456, Fort Morgan, 80701-0456. Phone: (970) 867-5674. Fax: (970) 542-1023. Licensee: Media Logic LLC. ♦Wayne Johnson, pres.

KSTC(AM)— Jan 3, 1925: 1230 khz; 1 kw-U. TL: N40 37 04 W103 10 31. Hrs open: 6-5 hours Box 830, 803 W. Main, 80751. Phone: (970) 522-1607. Fax: (970) 522-1322. E-mail: knng@bresnan.net Licensee: Track 1 Media of Sterling LLC (group owner; acq 4-7-2004; grpsl). Population served: 95,000 Format: Classic oldies. News staff: one; News: 12 hrs wkly. Target aud: General. Spec prog: Farm 15 hrs wkly. ♦Tammy Sewell, gen mgr, opns mgr & gen sls mgr; Mike Walker, mus dir.

KNNG(FM)—Co-owned with KSTC(AM). Feb 8, 1974: 104.7 mhz; 100 kw. 500 ft TL: N40 34 57 W103 01 56. (CP: 1.8 kw, ant 424 ft.). Stereo. 24 Phone: (970) 522-1609.230,000 Format: Colorado country. News staff: one; News: 10 hrs wkly. Target aud: General; country listeners.

***KTAD(FM)**— 2005: 89.9 mhz; 5 kw vert. Ant 407 ft TL: N40 28 48 W103 05 47. Hrs open:
Simulcasts KTPL(FM) Pueblo 100%.
1665 Briargate Blvd., Suite 100, Colorado Springs, 80920. Phone: (719) 593-0600. Fax: (719) 593-2399. E-mail: power88@power88.org Web Site: www.power88.org. Licensee: Educational Communications of Colorado Springs Inc. Format: Christian, inspirational. ♦Karen Veazey, gen mgr.

Strasburg

KTNI-FM— May 1, 1968: 101.5 mhz; 97 kw horiz. Ant 2,050 ft TL: N39 55 22 W103 58 18. Stereo. Hrs opn: 24 3033 S. Parker Rd., Suite 700, Aurora, 80014. Phone: (303) 872-1500. Fax: (303) 872-1501. Web Site: www.martiniontherockies.com. Licensee: KBRU-FM LLC (group owner; (acq 10-24-2005; $15.5 million). Format: Cosmopolitan jazz. ♦Luis G. Nogales, CEO; Steve Keeney, gen mgr; Tim Maranville, opns dir & progmg dir.

Telluride

***KOTO(FM)**— October 1975: 91.7 mhz; 2.35 kw. -187 ft TL: N37 55 59 W107 49 59. Stereo. Hrs opn: Box 1069, 207 N. Pine St., 81435. Phone: (970) 728-4334. Fax: (970) 728-4326. E-mail: koto@telluridecolorado.net Web Site: www.koto.org. Licensee: San Miguel Educational Fund. (acq 7-22-86). Population served: 2,500 Natl. Network: NPR, PRI. Format: Free-form. News staff: 2; News: 2 hrs wkly. Target aud: General; community. Spec prog: Class 9 hrs, country 12 hrs, jazz 9 hrs, blues 7 hrs, drama 3 hrs wkly. ♦Bob Biener, chmn & pres; Ben Kerr, gen mgr, stn mgr & progmg dir; Suzanne Cheavens, mus dir; Stephen Barrett, news dir; Janice Zink, spec ev coord.

Thornton

KKZN(AM)— May 30, 1987: 760 khz; 5 kw-D, 1 kw-N, DA-2. TL: N39 36 18 W104 50 25. Hrs open: 4695 S. Monaco St., Denver, 80237. Phone: (303) 713-8000. Fax: (303) 713-8736. Web Site: www.am760.net. Licensee: Citicasters Licenses Inc. (NEW). Group owner: Clear Channel Communications Inc. (acq 5-4-99; grpsl). Population served: 514,678 Format: Progressive talk. ♦Lee Larsen, gen mgr; Ron Smith, opns mgr; Kristine Olinger, progmg dir; Jan Chadwell, chief of engrg.

Timnath

KJAC(FM)— Apr 10, 1989: 105.5 mhz; 58 kw. Ant 1,217 ft TL: N40 37 03 W105 19 40. Stereo. Hrs opn: 24 1201 18th St., Suite 250, Denver, 80202. Phone: (303) 296-7025. Fax: (303) 296-7030. Web Site: www.1055jackfm.com. Licensee: NRC Broadcasting Inc. (group owner; (acq 4-13-2004; $15 million). Population served: 50,000 Natl. Network: Westwood One. Natl. Rep: Target Broadcast Sales. Law Firm: Eugene T. Smith. Format: Jack. News: 10 hrs wkly. Target aud: 18-54; general. ♦Timothy Brown, gen mgr; Roger Tighe, chief of engrg.

Trimble

***KTDU(FM)**— May 2, 2005: 88.5 mhz; 2.2 kw. Ant 302 ft TL: N37 15 46 W107 53 45. Hrs open:
KTLF (FM) Colorado Springs 100%.
1665 Briargate Blvd., Suite 100, Colorado Springs, 80920. Phone: (719) 593-0600. Fax: (719) 593-2399. E-mail: lightpraise@ktcf.org Web Site: www.ktlf.org. Licensee: Educational Communications of Colorado Springs Inc. Format: Christian music. ♦Karen Veazey, gen mgr.

Trinidad

KCRT(AM)— May 21, 1946: 1240 khz; 250 w-U. TL: N37 08 45 W104 30 42. Hrs open: 24 100 Fisher Dr., 81082. Phone: (719) 846-3355. Fax: (719) 846-4711. E-mail: kcrt@comcast.net Licensee: Phillips Broadcasting Inc. (group owner; acq 3-30-92; $235,000 with co-located FM; FTR: 4-20-92). Population served: 12,000 Natl. Network: ABC, Jones Radio Networks. Format: Country. News staff: one; News: 15 hrs wkly. Target aud: General. Spec prog: Farm one hr, relg 5 hrs wkly. ♦Anita Phillips, pres & opns VP; Lory Phillips, gen mgr & gen sls mgr; David Phillips, stn mgr, mktg dir, progmg VP, news dir & pub affrs dir; Rick Neurauter, adv dir.

KCRT-FM— August 1981: 92.5 mhz; 38.5 kw. Ant 1,020 ft TL: N36 59 33 W104 28 24. Stereo. 24 30,000 Format: Classic rock. Target aud: 25-54.

***KTDL(FM)**—Not on air, target date: unknown: 90.7 mhz; 450 w. Ant 971 ft TL: N36 59 33 W104 28 24. Hrs open: 1665 Briargate Blvd., Suite 100, Colorado Springs, 80920-3400. Phone: (719) 593-0600. Fax: (719) 593-2399. Licensee: Educational Communications of Colorado Springs Inc. ♦Karen Veazey, gen mgr.

Vail

***KPRE(FM)**— Sept 1, 1994: 89.9 mhz; 1.5 kw. 295 ft TL: N39 38 05 W106 26 47. Hrs open: 24
Rebroadcasts KCFR(FM) Denver 100%.
7409 S. Alton Ct., Centennial, 80112. Phone: (303) 871-9191. Fax: (303) 733-3319. E-mail: info@cpr.org Web Site: cpr.org. Licensee: Public Broadcasting of Colorado Inc. Population served: 25,000 Natl. Network: NPR. Law Firm: Arter & Hadden. Format: Class, news, info. News staff: 8; News: 50 hrs wkly. Target aud: General. ♦Max Wycisk, pres; Sue Coughlin, dev VP.

KSKE-FM— 1997: 104.7 mhz; 100 kw. 394 ft TL: N39 38 08 W106 26 46. Stereo. Hrs opn: 24 Box 7205, Avon, 81620. Secondary address: 182 Avon Rd., Avon 81620. Phone: (970) 949-0140. Fax: (970) 949-1464. Web Site: www.kskeradio.com. Licensee: Superior Broadcasting of Denver LLC (group owner; (acq 8-31-2006; exchange for KRKY-FM Estes Park). Population served: 12,000 Format: Country. ♦Steve Wodlinger, gen mgr; Meredith Fox, opns mgr; Holli Snyder, gen sls mgr; David Bach, news dir; Ken Laughlin, chief of engrg.

***KVJZ(FM)**—Not on air, target date: unknown: 88.5 mhz; 1.9 kw. Ant 197 ft TL: N39 38 05 W106 26 47. Hrs open: 2900 Welton St., Suite 200, Denver, 80205. Phone: (303) 480-9272. Fax: (303) 291-0757. E-mail: info@kuvo.org Web Site: www.kuvo.org. Licensee: Denver Educational Broadcasting Inc. ♦Carlos Lando, gen mgr.

KVLE(AM)— July 25, 1983: 610 khz; 5 kw-D, 217 w-N. TL: N39 34 47 W106 24 54. Hrs open:
Simulcasts KKKK(AM) Colorado Springs.
614 Kimbark St., Longmont, 80501. Phone: (303) 776-2323. Fax: (303) 776-1377. E-mail: rnickfive@aol.com Web Site: www.radiocoloradonetwork.com. Licensee: Pilgrim Communications Inc. (group owner; (acq 3-2-2000; $150,000). Format: Business talk, sports, NASCAR. ♦Gene Hood, pres; Ron Nickell, sr VP, gen mgr & gen sls mgr; Lee Thompson, gen sls mgr.

Walden

KEZZ(FM)—Not on air, target date: unknown: 94.1 mhz; 44 kw horiz. Ant 492 ft TL: N40 39 51 W106 24 44. Hrs open: 1951 28th Ave., Unit 29, Greeley, 80634-5755. Phone: (970) 302-8444. Licensee: Youngers Colorado Broadcasting LLC. ♦Kevin J. Youngers, gen mgr.

Walsenburg

KSPK(FM)— March 1985: 102.3 mhz; 100 kw. Ant 430 ft TL: N37 37 39 W104 49 17. Stereo. Hrs opn: 24 516 Main, 81089. Phone: (719) 738-3636. Fax: (719) 738-2010. E-mail: info@kspk.com Web Site: www.kspk.com. Licensee: Mainstreet Broadcasting Co. Inc. (acq 9-12-90; $275,000; FTR: 10-8-90). Population served: 250,000 Natl. Network: ABC. Law Firm: Denise B. Moline, P.C. Wire Svc: ABC Format: Sports, country, farm. News staff: 2; News: 3 hrs wkly. Target aud: 24-59; upwardly mobile, two-income families. Spec prog: Relg 2 hrs wkly. ♦Paul Richards, gen mgr; Kim Lucero, gen sls mgr; Larry Patrick, news dir; Paul Bossert, chief of engrg. Co-owned TV: KSPK-LP affil

***KTWX(FM)**—Not on air, target date: unknown: 91.3 mhz; 8 kw vert. Ant 649 ft TL: N37 32 34 W104 22 31. Hrs open: 1665 Briargate Blvd., Suite 100, Colorado Springs, 80920-3400. Phone: (719) 593-0600. Fax: (719) 593-2399. Licensee: Educational Communications of Colorado Springs Inc. ♦Karen Veazey, gen mgr.

KWCS(FM)—Not on air, target date: unknown: 101.3 mhz; 95.5 kw. Ant 1,000 ft TL: N37 47 20 W104 29 12. Hrs open: 917 Andover Ct., Palatine, IL, 60067. Phone: (773) 592-9800. Licensee: Edward Magnus. ♦Edward Magnus, gen mgr.

Wellington

KCOL(AM)—Licensed to Wellington. See Fort Collins

KMAX-FM— 2003: 94.3 mhz; 8.7 kw. Ant 551 ft TL: N40 55 41 W105 08 36. Hrs open: 24 600 Main St., Windsor, 80550. Phone: (970) 674-2700. Fax: (970) 686-7491. Web Site: www.943maxfm.com. Licensee: Regent Broadcasting of Ft. Collins Inc. Group owner: Regent Communications Inc. (acq 2-25-2003). Format: Rock of the 80's and more. ♦Cal Hall, gen mgr; Mark Callaghan, opns mgr; Miles Schallert, sls dir; Ted Rose, progmg dir; Susan Moore, news dir; Quin Morrison, chief of engrg.

Westcliffe

***KTAW(FM)**—Not on air, target date: unknown: 89.3 mhz; 2.7 kw vert. Ant 364 ft TL: N37 37 39 W104 49 17. Hrs open: 1665 Briargate Blvd., Suite 100, Colorado Springs, 80920-3400. Phone: (719) 593-0600. Fax: (719) 593-2399. Licensee: Educational Communications of Colorado Springs Inc. ♦Karen Veazey, gen mgr.

Wheat Ridge

KTCL(FM)— September 1965: 93.3 mhz; 71 kw. Ant 1,135 ft TL: N39 43 59 W105 14 10. Stereo. Hrs opn: 4695 S. Monaco St., Denver, 80237. Phone: (303) 713-8000. Web Site: www.area93.com. Licensee: Jacor Broadcasting of Colorado Inc. (acq 5-8-98; $6.1 million with co-located AM). Population served: 160,000 Format: Alternative. Target aud: 18 plus. Spec prog: Comedy one hr, loc bands one hr, reggae 2 hrs wkly. ♦Lee Larsen, gen mgr; Ron Smith, opns mgr; Willie Hung, progmg dir; Karl Schipper, chief of engrg.

Widefield

KKLI(FM)—Licensed to Widefield. See Colorado Springs

Windsor

KJJD(AM)— Apr 12, 1969: 1170 khz; 1 kw-D. TL: N40 27 46 W104 54 47. Hrs open: Sunrise-sunset 624 N. Main St., Longmont, 80501. Phone: (303) 651-1199. Fax: (303) 651-2244. Web Site: www.laley1170.com. Licensee: Rodriguez-Gallegos Broadcasting Corporation. (acq 3-19-85). Population served: 500000 Law Firm: Shaw Pittman. Format: Sp, music. News: 5 hrs wkly. Target aud: 18-54; general. Spec prog: Pub affrs. ♦Jesse Rodriguez, gen mgr; Danny Casas, stn mgr.

KUAD-FM— May 31, 1975: 99.1 mhz; 100 kw. Ant 836 ft TL: N40 38 31 W104 49 03. Stereo. Hrs opn: 600 Main St., 80550. Phone: (970) 674-2700. Fax: (970) 686-7491. Web Site: www.k99.com. Licensee: Regent Broadcasting of Ft. Collins Inc. Group owner: Regent Communications Inc. (acq 2-25-03). Population served: 295,000 Law Firm: Dow, Lohnes & Albertson. Format: Country. Target aud: 25-54; upscale country listeners, 60% women. ♦Cal Hall, gen mgr; Mark Callaghan, opns mgr & progmg dir; Shelley Heier, gen sls mgr, mktg dir & prom dir; Brian Gary, mus dir; Meg Sprague, traf mgr.

Winter Park

KZMV(FM)—See Kremmling

Wray

KRDZ(AM)— Jan 11, 1978: 1440 khz; 5 kw-D, 200 w-N. TL: N56 04 00 W102 11 25. Hrs open: 4:50 AM-midnight Box 354, 80758. Phone: (970) 332-4171. Fax: (970) 332-4172. E-mail: krdzkatr@plains.net Web Site: www.krdz.com. Licensee: Media Logic LLC (group owner; acq 10-31-2002). Natl. Network: Jones Radio Networks. Rgnl. Network: Brownfield. Format: Classic hits. News: 10 hrs wkly. Target aud: Farmers & ranchers. Spec prog: Focus on family 2 hrs, farm 10 hrs wkly. ♦Wayne Johnson, gen mgr & opns mgr.

Yuma

KNEC(FM)— 1999: 100.9 mhz; 23 kw. Ant 348 ft TL: N40 00 33 W102 45 35. Hrs open: 205 S. Main St., 80759. Phone: (970) 848-2302. Fax: (970) 848-2240. E-mail: knec@plains.net Licensee: Track 1 Media of Sterling LLC (group owner; acq 4-7-2004; grpsl). Format: Hot adult contemp. ♦Tammy Sewell, gen mgr.

Connecticut

Ansonia

WADS(AM)— May 8, 1956: 690 khz; 1 kw-D, 33 w-N, DA-2. TL: N41 20 48 W73 06 56. (CP: 3.5 kw-D, 200 w-N). Hrs opn: 261 Portsea St., New Haven, 06519. Phone: (203) 777-7690. Phone: (203) 782-3564. Fax: (203) 782-3565. E-mail: sbcglobal@aol.com Web Site: www.radioamor690.com. Licensee: Radio Amor Inc. (acq 12-30-93; $450,000; FTR: 1-17-94). Population served: 85,000 Law Firm: Shaw Pittman. Format: Relg, educ, Sp. News: 3.5 hrs wkly. Target aud: General. ♦Rev. Moses Mercedes, pres; Rev. Luis Rivera, VP; Abraham Hernandez, gen mgr.

Berlin

***WERB(FM)**— Jan 12, 1979: 94.5 mhz; 27.5 w. 95 ft TL: N41 37 18 W72 45 13. (CP: 94.5 mhz). Stereo. Hrs opn: 24 Berlin High School Media Center, 139 Patterson Way, 06037. Phone: (860) 828-0606. Phone: (860) 828-6577. Fax: (860) 829-0526. E-mail: werb@berlinschools.org Web Site: www.berlinwall.org. Licensee: Berlin Board of Education. Format: Educ/Rock. Teenage listeners from Berlin High School. ♦Chris Wolfe, gen mgr.

Bloomfield

WDZK(AM)— February 1964: 1550 khz; 5 kw-D, 2 kw-N, DA-2. TL: N41 51 47 W72 44 01. Hrs open: 24 160 Chapel Rd., Manchester, 06040. Phone: (860) 643-3912. Fax: (860) 643-3910. Web Site: www.radiodisney.com. Licensee: Radio Disney Group LLC. Group owner: ABC Inc. (acq 11-21-00; grpsl). Natl. Network: ABC. Format: Children. News staff: one; News: 11 hrs wkly. Target aud: 18-80; general. ♦Paul Robertson, gen mgr.

Bridgeport

WCUM(AM)— September 1941: 1450 khz; 1 kw-U. TL: N41 12 40 W73 11 28. Hrs open: 24 Box 3975, 06605. Phone: (203) 335-1450. Fax: (203) 337-1220. E-mail: radiocumbre1450@aol.com Web Site: www.wcum1450.com. Licensee: Radio Cumbre Broadcasting Inc. (acq 4-89; $550,000; FTR: 4-24-89). Population served: 156,542 Natl. Network: CNN Radio. Format: Sp, tropical. News staff: 2; News: 14 hrs wkly. Target aud: 25 plus. ♦Pablo De Jesus Colon Hijo, CEO & pres; Migdalia Ramos Colon, VP; Allison Sheahan, gen mgr.

WDJZ(AM)— Apr 30, 1977: 1530 khz; 5 kw-D, DA. TL: N41 10 09 W73 13 14. Hrs opn: 177 State St., 06604. Phone: (203) 368-4392. Fax: (203) 367-4551. E-mail: wdjzradio@sbcglobal.net Web Site: www.wdjzradio.com. Licensee: People's Broadcast Network LLC (acq 6-19-2007). Format: Foreign/Ethnic, gospel. Target aud: 35 plus. ♦Milford Edwards Sr., gen mgr.

WEZN-FM— Oct 24, 1960: 99.9 mhz; 27.6 kw. 669 ft TL: N41 16 46 W73 11 09. Stereo. Hrs opn: 440 Wheelers Farms Rd., Suite 302, Milford, 06461. Phone: (203) 783-8200. Fax: (203) 783-8373. Web Site: www.star999.com. Licensee: Cox Radio Inc. Group owner: Cox Broadcasting (acq 3-28-97; grpsl). Population served: 350,000 Format: Adult contemp. Target aud: 25-54. ♦Kim Guthrie, exec VP & gen mgr; Helaine Greenbaum, natl sls mgr; Stuart Gorlick, gen sls mgr & rgnl sls mgr; Stephen Donnarummo, prom dir; Samantha Stevens, progmg dir; Dom Bordonaro, chief of engrg; Carol Roberts, traf mgr.

WICC(AM)— 1926: 600 khz; 1 kw-D, 500 w-N, DA-2. TL: N41 09 36 W73 09 53. Stereo. Hrs opn: 24 2 Lafayette Sq., 06604-6000. Phone: (203) 366-6000. Fax: (203) 384-0600. Fax: (203) 394-6000. Web Site: www.wicc600.com. Licensee: Cumulus Licensing Corp. Group owner: Cumulus Media Inc. (acq 3-14-02; grpsl). Population served: 450,000 Natl. Rep: Christal. Law Firm: Wiley, Rein & Fielding. Format: Talk. News staff: 5; News: 25 hrs wkly. Target aud: 35-64. Spec prog: It 5 hrs wkly. ♦Ann McManus, VP & gen mgr; Curt Hansen, opns VP.

***WPKN(FM)**— Oct 10, 1963: 89.5 mhz; 10 kw. 550 ft TL: N41 16 43 W73 11 08. Stereo. Hrs opn: 24 244 University Ave., 06604. Phone: (203) 331-9756. E-mail: wpkn@wpkn.org Web Site: www.wpkn.org. Licensee: WPKN Inc. (acq 12-10-97). Population served: 30,000 Format: Div. Spec prog: Class 2 hrs, Sp 4 hrs, Black 4 hrs, Fr 2 hrs, jazz 16 hrs wkly. ♦Henry Minot, gen mgr.

Bristol

WPRX(AM)— October 1948: 1120 khz; 1 kw-D, 500 w-N, DA-N. TL: N41 39 29 W72 56 51. Hrs open: 24 321 Ellis St., New Britian, 06051. Phone: (860) 348-0667. Fax: (860) 348-0711. E-mail: wprx1120@comcast.net Web Site: www.wprx1120am.com. Licensee: Nievezquez Production Inc. (acq 4-28-99; $925,000). Population served: 100,000 Natl. Network: CNN Radio, Westwood One. Format: News/talk, Sp tropical. News staff: 2; News: 12 hrs wkly. Target aud: 23-54; Hispanic adults. Spec prog: Pol 2 hrs wkly. ♦Oscar Nieves, gen mgr.

Brookfield

WINE(AM)— May 9, 1966: 940 khz; 1 kw-D, 4 w-N. TL: N41 29 35 W73 25 47. Hrs open: 1004 Federal Rd., 06804. Phone: (203) 775-1212. Fax: (203) 775-6452. Web Site: www.cumulus.com. Licensee: Cumulus Licensing Corp. Group owner: Cumulus Media Inc. (acq 1-23-2002; grpsl). Population served: 165,000 Natl. Network: ESPN Radio. Law Firm: Haley, Bader & Potts. Format: Sports. Target aud: 25-54. ♦Brett Beshore, gen mgr; Matt Carey, prom dir & progmg dir; Tim Sheehan, progmg dir; Lisa Harris, news dir; Peter Partenio, chief of engrg.

WRKI(FM)—Co-owned with WINE(AM). Dec 24, 1976: 95.1 mhz; 50 kw. 500 ft TL: N41 29 35 W73 25 47. Stereo. Web Site: www.i95rock.com.75,000 Format: Classic rock. ♦Tom Principi, gen sls mgr; Taryn Polites, prom dir.

Danbury

WDAQ(FM)—Listing follows WLAD(AM).

***WFAR(FM)**— July 19, 1981: 93.3 mhz; 18 w. 210 ft TL: N41 23 44 W73 25 24. Stereo. Hrs opn: 25 Chestnut St., 06810. Phone: (203) 748-0001. Fax: (203) 746-4262. Web Site: www.radiofamilia.com. Licensee: Danbury Community Radio Inc. (acq 7-81). Format: Educ, Por, relg. Spec prog: It one hr, East Indian one hr wkly. ♦David Abrantes, pres & gen mgr; Helena Abrantes, opns mgr; Joe Mingachos, news dir.

WINE(AM)—See Brookfield

WLAD(AM)— October 1947: 800 khz; 1 kw-D, 287 w-N. TL: N41 22 27 W73 26 47. Stereo. Hrs opn: 24 198 Main St., 06810. Phone: (203) 744-4800. Fax: (203)778-4655. E-mail: radio80wlad@aol.com Web Site: www.wlad.com. Licensee: Berkshire Broadcasting Corp. (group owner) Population served: 250,000 Natl. Network: CNN Radio. Natl. Rep: D & R Radio. Law Firm: Cohn & Marks. Format: Full service. News staff: 3. Target aud: General. ♦James B. Lee Jr., pres; Mary Lu Orteig, exec VP; Irving J. Goldstein, VP & gen mgr.

WDAQ(FM)—Co-owned with WLAD(AM). December 1953: 98.3 mhz; 1.3 kw. 460 ft TL: N41 22 27 W73 26 47. Stereo. 24 Web Site: www.98q.com. Format: Hot adult contemp. Target aud: 25-54.

WREF(AM)—See Ridgefield

WRKI(FM)—See Brookfield

***WXCI(FM)**— Feb 10, 1973: 91.7 mhz; 1.2 kw. 205 ft TL: N41 23 44 W73 25 24. (CP: 3 kw, ant 201 ft.). Stereo. Hrs opn: 6 AM-2 AM Student Ctr., 181 White St., 06810. Phone: (203) 837-9924. E-mail: wxci@yahoo.com Web Site: www.wxci.org. Licensee: Western Connecticut State University Board of Trustees. (acq 3-73). Format: Alternative. News: 3 hrs wkly. Target aud: 14-25. Spec prog: Club mus 3 hrs, jazz 3 hrs, reggae 2 hrs, new age 3 hrs, metal 3 hrs, classic rock 3 hrs wkly. ♦Justin Mazzarese, gen mgr; Chris Merkle, prom dir; John Selwyn, progmg dir; Tom Carpenter, mus dir; Travis Cuddy, chief of engrg.

East Lyme

WNLC(FM)— Apr 1, 1994: 98.7 mhz; 5.5 kw. 269 ft TL: N41 20 48 W72 06 50. Stereo. Hrs opn: 24 Box 1031, New London, 06320. Secondary address: 89 Broad St., New London 06320. Phone: (860) 442-5328. Fax: (860) 442-6532. E-mail: arussell@hallradio.com Web Site: www.wnlc.com. Licensee: Hall Communication Inc. (group owner; acq 6-16-97; $2 million). Population served: 350,000 Natl. Network: Westwood One. Rgnl. Network: CRN. Law Firm: Fletcher, Heald & Hildreth. Format: Lite AC/Nostalgia. Target aud: 35 plus; today's upcoming citizens who have discretionary income. ♦Bonnie H. Rowbotham, chmn; Arthur J. Rowbotham, pres; Bill Baldwin, sr VP; Andy Russell, gen mgr & stn mgr.

Enfield

WPKX(FM)— July 1990: 97.9 mhz; 2.22 kw. 528 ft TL: N42 05 05 W72 42 14. Hrs open: 1331 Main St., Springfield, MA, 01103. Phone: (413) 781-1011. Fax: (413) 734-4434. Web Site: www.979.com. Licensee: Capstar TX L.P. Group owner: Clear Channel Communications Inc. (acq 8-30-00; grpsl). Format: Contemp country. Target aud: 25-54. ♦Tom McConnell, gen mgr; Pat McKay, opns mgr.

Fairfield

***WSHU(FM)**— February 1964: 91.1 mhz; 20 kw. Ant 624 ft TL: N41 16 45 W73 11 09. Stereo. Hrs open: 24 5151 Park Ave., 06825. Phone: (203) 365-6604. Fax: (203) 371-7991. E-mail: lombardi@wshu.org Web Site: www.wshu.org. Licensee: Sacred Heart University Inc. (acq 1-5-90). Population served: 156,542 Natl. Network: NPR, PRI. Format: Class, news. News staff: 4; News: 43 hrs wkly. Target aud: General; all ages. Spec prog: Folk 5 hrs, new age 6 hrs wkly. ♦George Lombardi, gen mgr; Barbara Bashar, opns mgr; Gillian Anderson, dev dir.

***WVOF(FM)**— Sept 1, 1970: 88.5 mhz; 100 w. 35 ft TL: N41 09 32 W73 15 35. Stereo. Hrs opn: 5 AM-2 AM Box R, Campus Ctr., N. Benson Rd., 06824. Phone: (203) 254-4144. Fax: (203) 254-4224. E-mail: centralstaff@wvof.org Web Site: www.wvof.org. Licensee: Fairfield University. Population served: 55,000 Format: Var/div, progsv. News staff: 5; News: 4 hrs wkly. Target aud: 18-35. ♦Matt Dinnan, gen mgr; Mark Gajda, stn mgr; Kim Gryzbala, mktg dir & rsch dir; James Maresca, progmg dir.

Greenwich

WGCH(AM)— Sept 14, 1964: 1490 khz; 1 kw-U. TL: N41 01 37 W73 37 59. Stereo. Hrs opn: 24 71 Lewis St., 06830. Phone: (203) 869-1490. Fax: (203) 869-3636. Web Site: www.wgch.com. Licensee: The Greenwich Broadcasting Corp. (acq 6-18-2003; $1.1 million).

Population served: 361,000 Natl. Network: CNN Radio. Rgnl. Network: Capitol Radio Net. Format: News, business talk. News staff: 2; News: 35 hrs wkly. Target aud: 35 plus; very upscale, active, athletic, community-minded. Spec prog: High school sports 6 hrs, educ 2 hrs, Pol one hr, relg 3 hrs, It one hr wkly. ♦Michael Metter, CEO, chmn, pres & gen mgr; Jeff Weber, exec VP & progmg dir; Bob Small, opns mgr; Rob Adams, sports cmtr.

Groton

WQGN-FM—Listing follows WSUB(AM).

WSUB(AM)— July 26, 1958: 980 khz; 1 kw-D. TL: N41 23 05 W72 04 13. Hrs open: 7 Governor Winthrop Blvd., New London, 06320-6437. Phone: (860) 443-1980. Fax: (860) 444-7970. Web Site: www.caliente980am.com. Licensee: Citadel Broadcasting Co. Group owner: Citadel Broadcasting Corp. (acq 4-26-2001; grpsl). Population served: 43,000 Natl. Network: ABC. Format: Sp. Target aud: 25-49; middle income professionals. ♦Wayne Leland, exec VP; Bonnie Gomes, gen mgr & stn mgr; Kevin Palana, progmg VP & progmg dir; Frank Doremus, chief of engrg.

WQGN-FM—Co-owned with WSUB(AM). 1971: 105.5 mhz; 3 kw. 275 ft TL: N41 23 05 W72 04 13. Stereo. Web Site: www.q105.fm.300,000 Format: CHR. Target aud: 18-49. ♦Shawn Murphy, mus dir.

Guilford

***WGRS(FM)**— Dec 27, 1993: 91.5 mhz; 3.1 kw. 82 ft TL: N41 17 19 W72 39 32. (CP: 6 kw). Stereo. Hrs opn: 24
Rebroadcasts WMNR(FM) Monroe 100%.
Box 920, Monroe, 06468. Phone: (203) 268-9667. E-mail: info@wmnr.org Web Site: www.wmnr.org. Licensee: Monroe Board of Education. Population served: 60,000 Natl. Network: PRI. Format: Class. Spec prog: Big band 8 hrs, folk 2 hrs, new age one hr, Broadway one hr wkly. ♦Kurt Anderson, gen mgr; Jane Stadler, opns dir; Carol Babina, dev dir.

Hamden

WAVZ(AM)—See New Haven

WKCI-FM—Licensed to Hamden. See New Haven

WQAQ(FM)—Listing follows WQUN(AM).

WQUN(AM)— July 17, 1960: 1220 khz; 1 kw-D, 305 w-N, DA-1. TL: N41 22 32 W72 55 54. Stereo. Hrs opn: 24 Quinnipiac University, 275 Mt. Carmel Ave., 06518. Phone: (203) 582-8984. Fax: (203) 582-5372. E-mail: ray.andrewsen@quinnipiac.edu Web Site: www.wqun.com. Licensee: Quinnipiac University (acq 9-12-96; $500,000). Population served: 400,000 Natl. Network: CBS, Jones Radio Networks. Law Firm: Garvey, Schubert, Barer. Format: News, loc info, adult standards. News staff: 2. Target aud: General; community, business & cultural leaders. Spec prog: Irish 2 hrs, Broadway 2hrs, big bands 4 hrs wkly. ♦Ray Andrewsen, gen mgr & opns dir; Greg Little, news dir & pub affrs dir; Bob Radil, chief of engrg.

WQAQ(FM)—Co-owned with WQUN(AM). February 1973: 98.1 mhz; 16 w. -82 ft TL: N41 25 10 W72 53 41. Stereo. Noon-2 AM (S-Su); 8 AM-2 AM (M-F) Box 59, Quinnipiac Univ., 275 Mt. Casnad Ave., 06518. Phone: (203) 582-5278. Fax: (203) 582-8098. Web Site: www.angelfire.com/ct2/wqaqradio/. Format: AOR, news/talk, alternative rock. News: 10 hrs wkly. Target aud: 18-30. ♦Chris Cooper, gen mgr; Carlos Lanesee, prom dir; Sally Densa, adv VP; Glenn Giangrande, progmg dir; Jessie Elgarten, progmg mgr; Alison Keller, news dir; Bill Shoulders, pub affrs dir.

Hartford

WCCC(AM)—West Hartford, 1947: 1290 khz; 490 w-D, 11 w-N. TL: N41 47 48 W72 47 50. Hrs open: 24 1039 Asylum Ave., 06105. Phone: (860) 525-1069. Fax: (860) 246-9084. Licensee: Marlin Broadcasting of Hartford LLC (acq 5-18-2000; grpsl). Population served: 200,000 Natl. Rep: Eastman Radio. Law Firm: Akin, Gump, Strauss, Hauer & Feld. Format: Class. Target aud: 18-54. ♦Woody Tanger, CEO; Boyd E. Arnold, VP & gen mgr; Michael Picozzi, opns mgr; Jay Schultz, sls dir & gen sls mgr; Michelle Bassoss, natl sls mgr; Nicole Godburn, prom dir; John Ramsey, chief of engrg.

WCCC-FM— June 7, 1960: 106.9 mhz; 23 kw. 730 ft TL: N41 47 51 W72 47 52. Stereo. 24 Web Site: www.wccc.com.1,000,000 Natl. Rep: Eastman Radio. Law Firm: Akin, Gump, Strauss, Hauer & Feld. Format: Active rock. Target aud: 18-49; adult men. ♦Jon Skonieczny, prom mgr.

WDRC(AM)— Dec 10, 1922: 1360 khz; 5 kw-U, DA-N. TL: N41 48 45 W72 41 44. Hrs open: 24 869 Blue Hills Ave., Bloomfield, 06002. Phone: (860) 243-1115. Fax: (860) 286-8257. E-mail: wdrc@talkofconnecticut.com Web Site: talkofconnecticut.com. Licensee: Buckley Broadcasting of Connecticut LLC. Group owner: Buckley Broadcasting Corp. (acq 8-1-59). Population served: 1,030,500 Natl. Network: Westwood One, AP Radio. Natl. Rep: McGavren Guild. Wire Svc: AP Format: News/talk. News staff: one. Target aud: 40+. ♦Richard D. Buckley, pres; Wayne G. Mulligan, VP & gen mgr; Laura Kittell, opns mgr; Eric Fah Noe, gen sls mgr; Donna Banks, prom dir; Dave Nagel, progmg dir.

WDRC-FM— 1939: 102.9 mhz; 19.5 kw. Ant 810 ft TL: N41 33 44 W72 50 40. Stereo. 24 E-mail: info@drcfm.com Web Site: drcfm.com.1,030,500 Natl. Network: AP Radio. Natl. Rep: McGavren Guild. Format: Classic hits. News staff: one. Target aud: 25-64. ♦Grahame Winters, prom mgr.

WHCN(FM)— 1939: 105.9 mhz; 16 kw. 867 ft TL: N41 33 47 W72 50 42. Stereo. Hrs opn: 24 10 Columbus Blvd., 06106. Phone: (860) 723-6000. Fax: (860) 723-7090. Web Site: www.whcn.com. Licensee: Capstar TX L.P. Group owner: Clear Channel Communications Inc. (acq 8-30-00; grpsl). Population served: 2,415,700 Natl. Rep: Christal. Format: Rock/AOR. ♦Tom McConnell, gen mgr; Steve Honeycomb, sls dir; Todd Thomas, prom dir & progmg dir; Rick Walsh, engrg VP & chief of engrg.

WPOP(AM)—Co-owned with WHCN(FM). July 1935: 1410 khz; 5 kw-U, DA-2. TL: N41 41 35 W72 45 30.24 Web Site: www.espnradio1410.com.1,135,000 Natl. Network: ESPN Radio. Format: Sports. News staff: 8; News: 25 hrs wkly. Target aud: 35 plus.

***WJMJ(FM)**— Oct 18, 1976: 88.9 mhz; 7.2 kw. 580 ft TL: N41 45 09 W72 59 40. Stereo. Hrs opn: 5 am-Midnight St. Thomas Seminary, 467 Bloomfield Ave., Bloomfield, 06002. Phone: (860) 242-8800. Fax: (860) 242-4886. Web Site: wjmj.org. Licensee: St. Thomas Seminary-Archdiocese of Hartford. Population served: 300,000 Natl. Network: ABC. Law Firm: Garvey, Schubert & Barer. Wire Svc: AP Format: Btfl music, class, relg. News staff: one; News: 10 hrs wkly. Target aud: 40-65; working, middle-class, family group. Spec prog: Educ, foreign one hr wkly. ♦Archbishop Henry Mansell, pres; John L. Ellinger, gen mgr; John P. Masternak, progmg dir; Ivor Hugh, mus dir.

WKND(AM)—See Windsor

WKSS(FM)— June 1947: 95.7 mhz; 16.5 kw. 880 ft TL: N41 33 41 W72 50 39. Stereo. Hrs opn: Hartford Sq. N., 10 Columbus Blvd., 06106-1944. Phone: (860) 723-6000. Fax: (860) 493-7090. Web Site: kiss957.com. Licensee: Capstar TX L.P. Group owner: Clear Channel Communications Inc. (acq 8-30-00; grpsl). Population served: 350,000 Rgnl. Network: Conn. Radio Net. Natl. Rep: Christal. Format: CHR. Target aud: 18-34. ♦Tom McConnell, gen mgr; Stan Priest, progmg dir.

WLAT(AM)—See New Britain

WNEZ(AM)—See Manchester

WPHH(FM)—See Waterbury

***WQTQ(FM)**— November 1971: 89.9 mhz; 120 w. 86 ft TL: N41 47 47 W72 41 42. Hrs open: 24 Weaver High School, 415 Granby St., 06112. Phone: (860) 722-8661. Phone: (860) 695-1899. Fax: (860) 242-6241. E-mail: wqtqfm@yahoo.com Web Site: www.wqtq.com/wqtq. Licensee: Hartford Board of Education. Population served: 158,017 Format: Educ, urban contemp, ballads. Target aud: 15-45; literate, professional, quality mus listeners. Spec prog: Gospel 12 hrs, clean hip hop rap 19 hrs, reggae/calypso 4 hrs, jazz 12 hrs, Rhythm and blues 18 hrs wkly. ♦Thomas G. Smith, CEO; Connie Coles, pres & gen mgr; Shirley Minnifield, CFO; Tom Smith, chief of opns.

WRCH(FM)—See New Britain

***WRTC-FM**— February 1958: 89.3 mhz; 300 w. 95 ft TL: N41 45 06 W72 41 29. Stereo. Hrs opn: 24 c/o Trinity College, 300 Summit St., 06106. Phone: (860) 297-2450. Phone: (860) 297-2439. Fax: (860) 987-6214. Web Site: www.wrtcfm.com. Licensee: Trustees of Trinity College. Population served: 158,017 Format: Div. Target aud: 15 plus. Spec prog: Class 4 hrs, gospel 6 hrs, West Indian 6 hrs, Pol 3 hrs, Por 8 hrs, Sp 6 hrs wkly. ♦Zee Santiago, stn mgr.

WTIC(AM)— Feb 10, 1925: 1080 khz; 50 kw-U, DA-N. TL: N41 46 39 W72 48 19. Stereo. Hrs opn: 24 10 Executive Dr., Farmington, 06032. Phone: (860) 677-6700. Fax: (860) 284-9842. Web Site: www.wtic.com. Licensee: Infinity Radio Inc. Group owner: Infinity Broadcasting Corp. (acq 11-13-98; grpsl). Population served: 469,000 Natl. Network: CBS. Format: Full service, news/talk. News: 30 hrs wkly. Target aud: 35-59; intelligent, mature adults. ♦Suzanne McDonald, VP & gen mgr; Steve Salhany, opns mgr; Stephanie McNamara, gen sls mgr; Geri DeRosa, natl sls mgr; Tristano Korlou, mktg dir & prom dir; Dana Whalen, news dir & chief of engrg; Jeff Hugabone, chief of engrg & sports cmtr.

WTIC-FM— Feb 5, 1940: 96.5 mhz; 20 kw. 810 ft TL: N41 46 27 W72 48 20. Stereo. 24 Fax: (860) 678-3952. Web Site: www.ticfm.com.425,500 Format: Hot adult contemp. News: 8 hrs wkly. Target aud: 18-34; intelligent, spirited, youthful adults.

WWUH(FM)—See West Hartford

WWYZ(FM)—See Waterbury

WZMX(FM)— 1939: 93.7 mhz; 17 kw. Ant 850 ft TL: N41 33 42 W72 50 41. Stereo. Hrs opn: 24 10 Executive Dr., Farmington, 06032. Phone: (860) 677-6700. Fax: (860) 674-8427. Web Site: www.hot937.com. Licensee: Infinity Radio Inc. Group owner: Infinity Broadcasting Corp. (acq 6-8-98; grpsl). Population served: 158,017 Format: Hip-hop. Target aud: 18-34; adults in Hartford & New Haven. ♦Suzanne McDonald, VP & gen mgr; Steve Salhany, opns mgr.

Ledyard

WBMW(FM)— Dec 24, 1992: 106.5 mhz; 3.1 kw. Ant 459 ft TL: N41 27 43 W72 01 27. Stereo. Hrs opn: 24 758 Colonel Ledyard Hwy., 06339. Phone: (860) 464-1065. Fax: (860) 464-8143. E-mail: wbmwandwjjf@aol.com Web Site: www.wbmw.com. Licensee: Redwolf Broadcasting Corp. (acq 1-25-94). Natl. Network: USA. Law Firm: Smithwick & Belendiuk. Format: Hot adult contemp. News staff: one. Target aud: 20-49. ♦John J. Fuller, gen mgr; Scott Bradshaw, opns mgr.

Litchfield

WZBG(FM)— July 8, 1992: 97.3 mhz; 3 kw. 328 ft TL: N41 48 08 W73 09 50. Hrs open: 24 Box 1497, Litchfield Commons, 49 Commons Dr., 06759. Phone: (860) 567-3697. Fax: (860) 567-3292. E-mail: info@wzbg.com Web Site: www.wzbg.com. Licensee: Local Girls & Boys Broadcasting Corp. Natl. Network: CBS. Format: News & info, adult contemp. News staff: 3. Target aud: 25-54. Spec prog: Jazz 2 hrs wkly. ♦Jennifer L. Parsons, gen mgr.

Manchester

WNEZ(AM)— May 18, 1958: 1230 khz; 1 kw-U. TL: N41 46 34 W72 33 27. Hrs open: 24 330 Main St., 1st Fl., Hartford, 06106-1622. Phone: (860) 524-0001. Fax: (860) 548-1922. E-mail: msanchez@freedomcommct.com Licensee: Freedom Communications of Connecticut Inc. (acq 6-1-2004; $3 million with WLAT(AM) New Britain). Format: Sp news/talk. ♦Melvin Sanchez, gen mgr.

Meriden

WMMW(AM)— 1946: 1470 khz; 2.5 kw, DA-2. TL: N41 33 14 W72 48 07. Hrs open: 24
Rebroadcast WDRC (AM) Hartford 100%.
869 Blue Hills Ave., Bloomfield, 06002. Phone: (860) 243-1115. Fax: (860) 286-8257. Web Site: www.talkofconnecticut.com. Licensee: Buckley Broadcasting of Connecticut LLC. Group owner: Buckley Broadcasting Corp. (acq 10-21-98; $630,000). Population served: 1,182,000 Natl. Network: Westwood One, AP Radio, ABC. Natl. Rep: McGavren Guild. Wire Svc: ABC Format: News, talk. Target aud: 40 plus; middle income, grassroots America. ♦Richard D. Buckley, pres; Wayne Mulligan, VP & gen mgr; Laura Kittell, opns mgr; Eric Fahnoe, sls VP; Grahame Winters, prom dir; Dave Nagel, progmg dir; Dan Lovallo, news dir, news rptr & sports cmtr; Scott Baron, chief of engrg; Joe Orlando, traf mgr.

***WPKT(FM)—** June 11, 1978: 90.5 mhz; 18.5 kw horiz, 13.5 kw vert. Ant 1,148 ft TL: N41 33 42 W72 50 41. Stereo. Hrs opn: 24 1049 Asylum Ave., Hartford, 06105. Phone: (860) 278-5310. Fax: (860) 275-7403. E-mail: info@wnpr.org Web Site: www.wnpr.org. Licensee: Connecticut Public Television & Radio. Population served: 2,000,000 Natl. Network: NPR, PRI. Law Firm: Schwartz, Woods and Miller. Format: News/talk. News staff: 5; News: 26 hrs wkly. ♦Jerry Franklin, CEO & pres; Kim Grehn, VP, stn mgr & progmg dir; Nancy Bauer, mktg VP; John Dankosky, news dir; Joseph Zareski, chief of engrg.

Middlefield

WPKT(FM)—See Meriden

Middletown

***WESU(FM)—** September 1939: 88.1 mhz; 1.5 kw. 38 ft TL: N41 33 16 W72 39 30. Stereo. Hrs opn: 24 45 Broad St., 06457. Phone: (860) 685-7703/685-7700/685-7707. Fax: (860) 704-0608. E-mail: wesu@wesufm.org Web Site: www.wesufm.org. Licensee: Wesleyan University (acq 3-27-2003). Population served: 900,000 Wire Svc: UPI Format: Free-form. Target aud: General. Spec prog: Blues 10 hrs, gospel 6 hrs, reggae 10 hrs, metal 5 hrs wkly. ♦Benjamin Michael, gen mgr; Ken Weiner, pub affrs dir.

***WIHS(FM)—** Oct 11, 1969: 104.9 mhz; 3 kw. 300 ft TL: N41 30 18 W72 39 32. Hrs open: 24 1933 S. Main St., 06457. Phone: (860) 346-1049. Fax: (860) 347-1049. E-mail: wihs@snet.net Web Site: www.wihsradio.org. Licensee: Connecticut Radio Fellowship Inc. Population served: 2,000,000 Natl. Network: Moody. Format: Christian. News: 18 hrs wkly. Target aud: General. Spec prog: Children 9 hrs wkly. ♦William Bacon, pres; G.J. Gerard, gen mgr; Paul A. Kretschmer, opns mgr.

WMRD(AM)— Dec 12, 1948: 1150 khz; 2.5 kw-D, 46 w-N. TL: N41 11 33 W72 37 13. Hrs open: 24 777 River Rd., 06457. Phone: (860) 347-9673. Phone: (860) 347-2565. Fax: (860) 347-7704. E-mail: radio@wliswmrd.net Web Site: www.wliswmrd.net. Licensee: Crossroads Communications L.L.C. (acq 1996). Population served: 200,000 Natl. Network: Westwood One, CBS Radio, Jones Radio Networks. Rgnl. Network: Conn. Radio Net. Law Firm: Vinson & Elkins. Format: Talk personalities. News staff: one; News: 8 hrs wkly. Target aud: 25-54; adults. Spec prog: Pol 2 hrs, It 2 hrs, Celtic one hr, Caribbean one hr, Jewish one hr wkly. ♦Don DeCesare, pres & gen mgr.

Milford

WADS(AM)—See Ansonia

WFIF(AM)— Sept 4, 1965: 1500 khz; 5 kw-D, DA. TL: N41 11 33 W73 06 05. Hrs open: Sunrise-sunset 90 Kay Ave., 06460. Phone: (203) 878-5915. E-mail: info@wfif.net Web Site: www.wfif.net. Licensee: K.W. Dolmar Broadcasting Co. Inc. Group owner: Blount Communications Group (acq 4-82; $425,000; FTR: 1-19-81). Population served: 1,500,000 Natl. Network: Salem Radio Network. Format: Relg. Target aud: General. Spec prog: Black 6 hrs wkly. ♦Dave Young, exec VP & opns VP; William Blount, pres & gen mgr; Jon Vaught, stn mgr; William Barnett, mus dir & chief of engrg.

Monroe

***WMNR(FM)—** Jan 31, 1974: 88.1 mhz; 5 kw. 403 ft TL: N41 19 08 W73 15 13. Stereo. Hrs opn: 24 Box 920, 06468. Phone: (203) 268-9667. E-mail: info@wmnr.org Web Site: www.wmnr.org. Licensee: Monroe Board of Education. Population served: 510,000 Natl. Network: PRI. Format: Class. Spec prog: Big band 8 hrs, folk 2 hrs, new age one hr, Broadway one hr wkly. ♦Kurt Anderson, gen mgr; Jane Stadler, opns dir; Carol Babina, dev dir.

Naugatuck

WFNW(AM)— Feb 26, 1961: 1380 khz; 5 kw-D, 500 w-N, DA-2. TL: N41 30 35 W73 03 20. Stereo. Hrs opn: 182 Grand St., Suite 215, Waterbury, 06702. Phone: (203) 755-4960. Fax: (203) 755-4957. E-mail: galaxia1380@yahoo.com Web Site: www.galaxia1380.com. Licensee: Candido Dias Carrelo. (acq 6-90; $350,000; FTR: 6-25-90). Format: Sp, tropical. ♦Placido Acevedo, pres; Candido Carrelo, gen mgr.

New Britain

***WFCS(FM)—** Oct 17, 1972: 107.7 mhz; 50 w. 160 ft TL: N41 41 36 W72 45 49. Stereo. Hrs opn: 24 Student Center, 1615 Stanley St., 06050-4010. Phone: (860) 832-1883. Fax: (860) 832-3757. Web Site: www.wfcs.ccsu.edu. Licensee: Trustees of Central Connecticut State University. Population served: 600,000 Format: Educational. News staff: 4; News: 10 hrs wkly. Target aud: 14-50. Spec prog: Blues 12 hrs, Sp 2 hrs wkly. ♦Adam Morgan, gen mgr; Mike McDonald, dev dir; Matt Rockwell, progmg dir; John Ramsey, chief of engrg.

WLAT(AM)— May 20, 1949: 910 khz; 5 kw-U, DA-N. TL: N41 42 58 W72 48 38. Stereo. Hrs opn: 24 330 Main St., 1st Fl., Hartford, 06106-1622. Phone: (860) 524-0001. Fax: (860) 548-1922. E-mail: msanchez@freedomcommct.com Licensee: Freedom Communications of Connecticut Inc. (acq 6-1-2004; $3 million with WNEZ(AM) Manchester). Population served: 701,900 Format: Spanish tropical. ♦Melvin Sanchez, gen mgr.

WRCH(FM)— July 1, 1968: 100.5 mhz; 7.5 kw. Ant 1,250 ft TL: N41 42 13 W72 49 57. Stereo. Hrs opn: 24 10 Executive Dr., Farmington, 06032. Phone: (860) 677-6700. Fax: (860) 677-5483. E-mail: wrch@cbs.com Web Site: www.wrch.com. Licensee: Infinity Radio Inc. Group owner: Infinity Broadcasting corp. (acq 6-8-98; grpsl). Format: Soft adult contemp. Target aud: 25-54; women, adults. ♦Suzanne McDonald, VP & gen mgr; Steve Salhany, opns mgr.

WRYM(AM)— August 1946: 840 khz; 1 kw-D, 208 w-N. TL: N41 41 15 W72 43 46. Hrs open: 24 1056 Willard Ave., Newington, 06111. Phone: (860) 666-5646. Fax: (860) 666-5647. E-mail: radio@wyrm840.com Web Site: www.wrymradio.com. Licensee: Eight Forty Broadcasting Corp. (acq 4-8-2004; $1.06 million). Population served: 400,000 Natl. Network: CNN Radio. Natl. Rep: Interep. Law Firm: Cohn & Marks, LLP. Format: Sp. News staff: 4; News: 8 hrs wkly. Target aud: General; Hispanic. Spec prog: Pol 5 hrs wkly, Italian 2 hrs wkly. ♦Walter Martinez, gen mgr; Danny Delgado, news dir; Silvina Martinez, traf mgr.

New Canaan

***WSLX(FM)—** 1975: 91.9 mhz; 10 w horiz. 518 ft TL: N41 11 32 W73 29 46. (CP: 19 w vert, ant 171 ft.). Stereo. Hrs opn: 377 N. Wilton Rd., 06840. Phone: (203) 972-3894. Licensee: St. Luke's Foundation Inc. Format: Class, div. ♦Dan Mecca, gen mgr.

New Fairfield

WDBY(FM)—See Patterson, NY

New Haven

WAVZ(AM)— September 1947: 1300 khz; 1 kw-U, DA-N. TL: N41 17 16 W72 56 48. Hrs open: 24 495 Benham St., Hamden, 06514. Phone: (203) 281-9600. Fax: (203) 407-4652. Web Site: www.espnradio1300.com. Licensee: CC Licenses LLC. Group owner: Clear Channel Communications Inc. (acq 12-18-92; $10 with WKCI-FM Hamden; FTR: 1-11-93) Population served: 155,000 Natl. Network: ESPN Radio. Natl. Rep: Clear Channel. Format: Sports. News staff: one; News: 14 hrs wkly. Target aud: 35 plus. ♦Tom McConnel, gen mgr; Gloria Shapiro, sls dir; Brian Zullo, gen sls mgr; David McLaine, prom dir; Jerry Kristafer, progmg dir.

WKCI-FM—Co-owned with WAVZ(AM). Feb 10, 1969: 101.3 mhz; 15 kw. 876 ft TL: N41 25 22 W72 57 06. Stereo. E-mail: comments@kc101.com Web Site: www.kc101.com. (Acq 7-24-92).49,357 Format: CHR,top 40. ♦Chaz Kelly, progmg dir.

WELI(AM)— October 1935: 960 khz; 5 kw-U, DA-N. TL: N41 22 14 W72 56 15. Stereo. Hrs opn: 495 Benham St., Hamden, 06514. Phone: (203) 281-9600. Fax: (203) 407-4652. E-mail: comments@weli.com Web Site: www.960weli.com. Licensee: CC Licenses LLC. (group owner; (acq 8-5-85). Population served: 200,000 Rgnl. Network: Conn. Radio Net. Natl. Rep: Katz Radio. Format: News/talk. Target aud: 18 plus. ♦L. Lowry Mays, pres; Tom McConnel, gen mgr; Jerry Kristafer, progmg dir.

WPLR(FM)— 1944: 99.1 mhz; 15 kw. 905 ft TL: N41 25 23 W72 57 06. Stereo. Hrs opn: 440 Wheelers Farms Rd., Suite 302, Milford, 06461. Phone: (203) 783-8200. Fax: (203) 783-8373. Web Site: www.wplr.com. Licensee: CXR Holdings Inc. Group owner: Cox Broadcasting (acq 8-2000; grpsl). Population served: 200,000 Format: Rock/AOR. ♦Stu Gorlick, gen sls mgr; Samuel Tilery, prom mgr; Ed Sabatino, progmg dir.

WQUN(AM)—See Hamden

WYBC(AM)— 1944: 1340 khz; 1 kw-U. TL: N41 17 32 W72 57 12. Hrs open: 142 Temple St., Suite 203, 06510. Phone: (203) 776-4118. Fax: (203) 776-2446. Web Site: www.wybc.com. Licensee: Yale Broadcasting Co. Inc. (acq 7-24-98; $775,000). Population served: 50,000 Rgnl. Network: Conn. Radio Net. Format: Eclectic. Spec prog: Sp one hr wkly. ♦Alexandria Newman, gen mgr; Wayne Schmidt, opns mgr; Julia Galeota, progmg dir; Clif Mills, chief of engrg.

WYBC-FM— Mar 9, 1959: 94.3 mhz; 1.8 kw. 325 ft TL: N41 20 58 W72 58 27. Stereo. 24 Web Site: www.943wybc.com. Licensee: Yale Broadcasting Co.550,000 Natl. Network: ABC. Format: Urban contemp. News: 10 hrs wkly. Target aud: Urban & college age listeners. Spec prog: Gospel 8 hrs, jazz 8 hrs, folk 3 hrs wkly. ♦Juan Castillo, progmg dir & pub affrs dir.

New London

***WCNI(FM)—** 1974: 90.9 mhz; 2 kw vert. Ant 187 ft TL: N41 22 53 W72 06 28. Stereo. Hrs opn: 1 pm-5 pm (M-F) Box 4972, Connecticut College, 270 Mohegan Ave., 06320. Phone: (860) 439-2853 (office). Phone: (860) 439-2850 ext 52 (studio). Fax: (860) 439-2805. E-mail: wnci@conncoll.edu Web Site: www.wcniradio.org. Licensee: Connecticut College Broadcasting Association Inc. Format: Var/div. Target aud: General; all musical audiences except pop. Spec prog: Black 3 hrs, class 6 hrs, folk 9 hrs, gospel 3 hrs, jazz 9 hrs, Pol 3 hrs, Sp 3 hrs, women's 3 hrs wkly. ♦Bridgett Ellis, gen mgr; John Tyler, chief of engrg.

WKNL(FM)— Jan 1, 1970: 100.9 mhz; 3 kw. 328 ft TL: N41 26 27 W72 08 29. Stereo. Hrs opn: 24 Box 1031, 06320. Secondary address: 89 Broad St. 06320. Phone: (860) 442-5328. Fax: (860) 442-6532. E-mail: arussell@hallradio.com Web Site: www.kool101fm.com. Licensee: Hall Communications Inc. (group owner; acq 1-19-95; $3.5 million with co-located AM; FTR: 3-20-95) Natl. Network: ABC. Natl. Rep: D & R Radio. Law Firm: Fletcher, Heald & Hildreth. Format: Oldies. News: 2 hrs wkly. Target aud: 25-54. ♦Bonnie Rowbotham, chmn; Arthur J. Rowbotham, pres; Bill Baldwin, sr VP; Andy Russell, gen mgr & stn mgr.

WQGN-FM—See Groton

WSUB(AM)—See Groton

Norfolk

***WSGG(FM)—** May 17, 2001: 89.3 mhz; 100 w. Ant 167 ft TL: N41 59 30 W73 12 46. Hrs open: 24 Box 4594, Hartford, 06147. Phone: (860) 232-6425. Licensee: Revival Christian Ministries Inc. Format: Christian. ♦Samuel Girona, gen mgr & progmg dir.

Norwalk

WFOX(FM)—Listing follows WNLK(AM).

WNLK(AM)— 1948: 1350 khz; 1 kw-D, 500 w-N, DA-N. TL: N41 06 54 W73 26 06. Hrs open: 24 444 Westport Ave., 06851. Phone: (203) 845-3030. Fax: (203) 845-3097. Web Site: www.wstcwnlk.com. Licensee: Cox Radio Inc. Group owner: Clear Channel Communications Inc. (acq 8-25-2000; grpsl). Population served: 78,000 Format: News/talk. News staff: 4; News: 20 hrs wkly. Target aud: 25-54. ♦Robin Faller, gen mgr.

WFOX(FM)—Co-owned with WNLK(AM). 1966: 95.9 mhz; 3 kw. 299 ft TL: N41 06 54 W73 26 06. Stereo. Web Site: thefoxonline.com.78,000 Format: Classic rock.

Norwich

WCTY(FM)—Listing follows WICH(AM).

WICH(AM)— September 1946: 1310 khz; 5 kw-U, DA-2. TL: N41 33 10 W72 04 34. Hrs open: 24 Box 551, Cuprak Rd., 06360-0551. Phone: (860) 887-3511. Fax: (860) 886-7649. Web Site: www.wich.com. Licensee: WICH Inc. Group owner: Hall Communications Inc. (acq 7-1-65). Population served: 250,000 Rgnl. Network: Conn. Radio Net. Natl. Rep: D & R Radio. Law Firm: Fletcher, Heald & Hildreth. Format: Full service. News staff: 4. Target aud: 35 plus. Spec prog: Pol one hr wkly. ♦Bonnie H. Rowbotham, chmn; Arthur J. Rowbotham, pres; Bill Baldwin, sr VP; Andy Russell, gen mgr; Karen Dole, opns dir, rgnl sls mgr & pub affrs dir; Bob Reed, prom dir; Stu Bryer, progmg dir; Roger Arnold, chief of engrg.

WCTY(FM)—Co-owned with WICH(AM). May 1968: 97.7 mhz; 1.9 kw. 410 ft TL: N41 28 28 W72 06 14. Stereo. 24 Web Site: www.wcty.com.120,000 Format: Country. Target aud: 25-54. ♦Bob Reed, prom dir; Jimmy Lane, progmg dir.

***WNPR(FM)**— Oct 17, 1981: 89.1 mhz; 5.1 kw. Ant 590 ft TL: N41 31 11 W72 10 04. Stereo. Hrs opn: 24
Rebroadcasts WPKT(FM) Meriden 100%.
1049 Asylum Ave., Hartford, 06105. Phone: (860) 278-5310. Fax: (860) 244-9624. Fax: (860) 275-7403. E-mail: info@wnpr.org Web Site: www.wnpr.org. Licensee: Connecticut Public Television & Radio. Natl. Network: NPR, PRI, AP Radio. Law Firm: Schwartz, Woods and Miller. Format: News/talk. News staff: 13; News: 38 hrs wkly. Target aud: General. ♦Jerry Franklin, CEO & pres; Kim Grehn, VP, stn mgr & progmg dir; Nancy Bauer, mktg VP; John Dankosky, news dir; Joe Zareski, chief of engrg; Gene Amatruda, opns.

Old Saybrook

WLIS(AM)— Sept 27, 1956: 1420 khz; 5 kw-D, 500 w-N, DA-N. TL: N41 19 38 W72 23 21. Hrs open: 24 hrs
Rebroadcasts WMRD(AM) Middletown 90%.
777 River Rd., 06457. Phone: (860) 388-1420. Fax: (860) 347-7704. E-mail: radio@wliswmrd.net Web Site: www.wliswmrd.net. Licensee: Crossroads Communications of Old Saybrook L.L.C. (acq 10-96). Population served: 200,000 Natl. Network: CBS, Westwood One. Law Firm: Vinson & Elkins. Format: Talk personalities. News staff: one; News: 8 hrs wkly. Target aud: 25-64; Adults. Spec prog: Jazz 4 hrs wkly. ♦Don DeCesare, pres & gen mgr.

Pawcatuck

WWRX(FM)— Nov 30, 1995: 107.7 mhz; 1.4 kw. Ant 492 ft TL: N41 27 35 W71 55 40. Stereo. Hrs opn: 24 758 Colonel Ledyard Hwy., Ledyard, 06339. Phone: (860) 464-1065. Fax: (860) 464-8143. Web Site: www.jammin1077.com. Licensee: Fuller Broadcasting International LLC (acq 12-13-2002; $3.75 million). Population served: 300,000 Law Firm: Shaw Pittman. Format: Top 40 hits. News: one hr wkly. Target aud: 25-54; mobile, upscale. ♦John J. Fuller, pres & gen mgr; Scott Bradshaw, opns mgr.

Pomfret

***WBVC(FM)**— 2001: 91.1 mhz; 100 w. Ant 289 ft TL: N41 53 27 W71 57 24. Hrs open: 398 Pomfret St., Box 128, 06258. Phone: (860) 963-5919. Web Site: www.pomfretschool.org. Licensee: Pomfret School. Format: Var. ♦Tim Peck, gen mgr.

Putnam

WINY(AM)— May 3, 1953: 1350 khz; 5 kw-D, 79 w-N. TL: N41 54 10 W71 53 43. Stereo. Hrs opn: 24 Box 231, 45 Pomfret St., 06260. Phone: (860) 928-1350. Fax: (860) 928-7878. E-mail: info@winyradio.com Web Site: www.winyradio.com. Licensee: Osbrey Broadcasting Co. (acq 5-31-01; $2 million). Population served: 250,000 Natl. Network: AP Radio, Jones Radio Networks. Law Firm: Miller & Miller. Wire Svc: AP Format: Adult contemp. News staff: 3; News: 18 hrs wkly. Target aud: 25-54 plus; adults. Spec prog: Talk 10 hrs wkly. ♦Gary W. Osbrey, pres & gen mgr; Karen Osbrey, VP; Shania Alston, news dir; Kerri LeClerc, traf mgr; Tiffany Ventura, sports cmtr.

Ridgefield

WREF(AM)— Mar 15, 1985: 850 khz; 2.5 kw. TL: N41 17 27 W73 29 16. Stereo. Hrs opn: 6 AM-10 PM 198 Main St., Danbury, 06810. Phone: (203) 744-4800. Fax: (203) 778-4655. E-mail: trueoldies850@hotmail.com Licensee: The Berkshire Broadcasting Corp. Group owner: Berkshire Broadcasting Corp. (acq 3-31-97; $550,000). Population served: 2,500,000 Natl. Network: ABC. Natl. Rep: D & R Radio. Law Firm: Cohn & Marks. Format: Oldies. Target aud: 35-64. ♦James B. Lee Jr., pres; Mary Lu Orteig, exec VP; Irv Goldstein, VP & gen mgr.

Salisbury

WKZE-FM— Sept 1, 1992: 98.1 mhz; 1.8 kw. Ant 604 ft TL: N41 58 35 W73 31 27. Stereo. Hrs opn: 24 7392 S. Broadway, Red Hook, NY, 12571. Phone: (845) 758-9810. Fax: (845) 758-9819. E-mail: info@wkze.com Web Site: www.wkze.com. Licensee: Willpower Radio L.L.C. (acq 4-7-2005; $1.4 million with WHDD(AM) Sharon). Population served: 650,000 Natl. Network: AP Radio. Rgnl. Network: CRN. Format: AAA. News staff: one; News: 2 hrs wkly. Target aud: 25-54. ♦Dave Doud, gen mgr; Pete Nugent, sls VP & gen sls mgr; Paul Higgins, sls.

Sharon

WHDD(AM)— Dec 23, 1986: 1020 khz; 2.5 kw-D. TL: N41 58 35 W73 31 27. Stereo. Hrs opn: 6 AM-6 PM 7392 S. Broadway, Red Hook, NY, 12571-1765. Phone: (845) 758-9810. Fax: (845) 758-9819. E-mail: info@wkze.com Web Site: www.wkze.com. Licensee: Willpower Radio L.L.C. (acq 4-7-2005; $1.4 million with WKZE-FM Salisbury). Population served: 650,000 Natl. Network: AP Radio. Rgnl. Network: Conn. Radio Net. Format: Talk. News staff: one; News: 2 hrs wkly. Target aud: 25-54. ♦Scott R. Johnson, gen mgr; Dave Doud, stn mgr & opns mgr; Barb Stanley, progmg VP & sls; Pete Nugent, engrg VP & sls; Paul Higgins, sls.

WQQQ(FM)— Oct 3, 1994: 103.3 mhz; 1.5 kw. 640 ft TL: N41 55 03 W73 33 32. Stereo. Hrs opn: 24 Box 446, Lakeville, 06039-0446. Phone: (860) 435-3333. Fax: (860) 435-3334. E-mail: q103fm@yahoo.com Web Site: www.wqqq.com. Licensee: The Ridgefield Broadcasting Corp. (acq 7-20-01). Population served: 350,000 Law Firm: Cohn & Marks. Format: Adult contemp, hits of the 70s, 80s & 90s. News staff: one; News: 14 hrs wkly. Target aud: Upscale adults; 25-54. ♦Dennis Jackson, chmn; Joe Loverro, exec VP & gen mgr.

Shelton

***WRXC(FM)**— 1977: 90.1 mhz; 45 w. 482 ft TL: N41 21 43 W73 06 48. Stereo. Hrs opn: 24
Rebroadcasts WMNR(FM) Monroe 100%.
Box 920, Monroe, 06468. Phone: (203) 268-9667. E-mail: info@wmnr.org Web Site: www.wmnr.org. Licensee: Monroe Board of Education. Population served: 132,000 Natl. Network: PRI. Format: Class. Spec prog: Big band 8 hrs, folk 2 hrs, new age one hr, Broadway one hr wkly. ♦Kurt Anderson, gen mgr; Jane Stadler, opns dir; Carol Babina, dev dir.

Somers

***WDJW(FM)**— Oct 6, 1986: 89.7 mhz; 9.2 w. -58 ft TL: N41 57 43 W72 27 51. Hrs open:
Rebroadcasts WWUH(FM) West Hartford.
Somers High School, 9th District Rd., 06071. Phone: (860) 749-2501. Phone: (860) 749-0719. Fax: (860) 749-9264. Licensee: Somers Board of Education. Population served: 10,000 Format: Jazz, folk, alternative. ♦Peter Stone, pres & gen mgr.

South Kent

***WGSK(FM)**— Dec 25, 1987: 90.1 mhz; 77 w. Ant 128 ft TL: N41 40 54 W73 29 13. Stereo. Hrs opn: 24
Rebroadcasts WMNR(FM) Monroe 100%.
Box 920, Monroe, 06468. Phone: (203) 268-9667. E-mail: info@wmnr.org Web Site: www.wmnr.org. Licensee: Monroe Board of Education. Population served: 8,000 Natl. Network: PRI. Format: Class. Spec prog: Big band 8 hrs, folk 2 hrs, new age one hr, Broadway one hr wkly. ♦Kurt Anderson, gen mgr; Jane Stadler, opns dir; Carol Babina, dev dir.

Southington

WXCT(AM)— Sept 2, 1969: 990 khz; 2.5 kw-D, 80 w-N, DA-2. TL: N41 34 59 W72 53 01. Hrs open: 24 Box 488, 440 Old Turnpike Rd., Plantsville, 06479. Phone: (860) 621-1754. Fax: (860) 426-1172. Web Site: www.canticonuevoradio.com. Licensee: Davidson Media Station WXCT LLC. Group owner: Davidson Media Group LLC (acq 4-30-2004; $1.4 million). Population served: 350,000 Format: Sp Christian. ♦Eric Salgado, pres; Alejandro Torres, stn mgr.

Stamford

WCTZ(FM)— Oct. 18, 1974: 96.7 mhz; 3 kw. Ant 328 ft TL: N41 02 49 W73 31 36. Stereo. Hrs opn: 24 444 Westport Ave., Norwalk, 06851. Phone: (203) 845-3030. Fax: (203) 845-3097. Web Site: 967thecoast.com. Licensee: Cox Radio Inc. Group owner: Cox Broadcasting (acq 8-25-2000; grpsl). Population served: 123,000 Natl. Rep: Katz Radio. Format: Adult contemp. Target aud: 25-54; upscale. ♦Robin Faller, gen mgr; Jim Stagnitti, gen sls mgr; Helaine Greenbaum, natl sls mgr; Kelli McLaughlin, rgnl sls mgr; Steve Soyland, prom dir; Eric McDonald, progmg dir; Clark Burgard, chief of engrg; Steve Rugh, traf mgr.

WSTC(AM)—Co-owned with WCTZ(FM). Sept 18, 1941: 1400 khz; 1 kw-U. TL: N41 02 49 W73 31 36. Stereo. 24 Web Site: wstcwnlk.com.123,000 Natl. Network: CNN Radio, Westwood One. Natl. Rep: Katz Radio. Format: News/talk. News staff: 4; News: 16 hrs wkly. Target aud: 25-54. ♦Lisa Lacerra, news dir; Dina Badie, prom.

***WEDW-FM**— Feb 17, 1992: 88.5 mhz; 2 kw horiz, 1.8 kw vert. Ant 302 ft TL: N41 02 49 W73 31 36. Stereo. Hrs opn:
Rebroadcasts WPKT (FM) Meriden100%.
1049 Asylum Ave., Hartford, 06105. Phone: (860) 278-5310. Fax: (860) 244-9624. E-mail: info@wnpr.org Web Site: www.wnpr.org. Licensee: Connecticut Public Broadcasting Inc. Format: News/talk. ♦Kim Grehn, gen mgr, stn mgr & progmg dir; Nancy Bauer, mktg VP; John Dankosky, news dir; Joe Zareski, chief of engrg.

Stonington

WXLM(FM)— November 1981: 102.3 mhz; 3 kw. 328 ft TL: N41 24 23 W71 50 15. Stereo. Hrs opn: 24 7 Governor Winthrop Blvd., New London, 06320. Phone: (860) 443-1980. Fax: (860) 444-7970. Web Site: www.wxlm.fm. Licensee: Citadel Broadcasting Co. Group owner: Citadel Broadcasting Corp. (acq 4-26-01; grpsl). Population served: 300,000 Natl. Rep: D & R Radio. Law Firm: Bryan Cave. Format: News, talk, sports. News staff: 2; News: 5 hrs wkly. Target aud: 21-54; adults who grew up in the 60s & 70s. Spec prog: Jazz 2 hrs wkly. ♦Fahrid Soleman, pres; Judy Ellis, CFO & VP; Bonnie Gomes, gen mgr; Dave Holmes, sls VP & traf mgr; Matt Chase, sls dir & gen sls mgr; Jackie Steele, prom mgr; Kevin O'Connor, prom mgr, progmg dir & news dir; Frank Doremus, chief of engrg.

Storrs

***WHUS(FM)**— 1956: 91.7 mhz; 3.16 kw. 360 ft TL: N41 48 48 W72 15 33. Stereo. Hrs opn: 2110 Hillside Rd., Unit 3008R, 06269-3008. Phone: (860) 486-4007. Fax: (860) 486-2955. E-mail: info@whus.org Web Site: www.whus.org. Licensee: Board of Trustees University of Connecticut. Population served: 10,691 Format: Div. Spec prog: Sp 3 hrs wkly. ♦John Murphy, gen mgr.

Torrington

***WAPJ(FM)**— 1997: 89.9 mhz; 40 w. Ant 276 ft TL: N41 48 08 W73 09 50. Hrs open:
WWUH(FM) West Hartford.
40 Water St., 06790. Phone: (860) 489-9033. Fax: (860) 482-7614. E-mail: wapjfm@sbcglobal.net Licensee: The I.B. and Zena H. Temkin Foundation Inc. (acq 9-30-2004). Format: Variety. News: 10 hrs wkly. Target aud: General; any and all. ♦Dick Williams, gen mgr.

WSNG(AM)— Jan 29, 1948: 610 khz; 1 kw-D, 500 w-N, DA-2. TL: N41 45 28 W73 03 06. Hrs open: 24 869 Blue Hills Ave., Bloomfield, 06002. Phone: (860) 689-8050. Fax: (860) 286-8257. E-mail: wsng@talkofconnecticut.com Web Site: www.talkofconnecticut.com.com. Licensee: Buckley Broadcasting of Connecticut LLC. Group owner: Buckley Broadcasting Corp. (acq 12-18-96; $425,000). Population served: 70,600 Natl. Network: Westwood One. Law Firm: Erwin Krasnow. Format: Talk. News staff: 3; News: 25 hrs wkly. Target aud: 25-54. ♦Wayne Mulligan, gen mgr; Laura Kittell, opns mgr; Eric Fahnoe, gen sls mgr; Dave Nagel, progmg dir; Scott Baron, chief of engrg.

Vernon

***WCTF(AM)**— Nov 21, 1982: 1170 khz; 1 kw-D, DA. TL: N41 52 38 W72 28 43. (CP: 2.5 kw-D). Hrs opn: WCTF c/o WFSI, 918 Chesapeake Ave., Annapolis, MD, 21403. Secondary address: 45 1/2 East St. 06066. Phone: (860) 871-2526. Web Site: www.familyradio.com. Licensee: Family Stations Inc. (group owner; acq 1-86; $136,000; FTR: 9-23-85). Format: Relg. ♦Harold Camping, pres.

Wallingford

***WWEB(FM)**— Nov 10, 1976: 89.9 mhz; 10 w. 230 ft TL: N41 27 34 W72 48 48. Hrs open:
Rebroadcasts WWUH (FM) West Hartford 80%.
Choate Rosemary Hall Foundation, 333 Christian St., 06492. Phone: (203) 697-2506. Fax: (203) 697-2186. E-mail: cbielizna@choate.edu Web Site: www.student.choate.edu/wweb. Licensee: Choate Rosemary Hall Foundation. Format: Var/div. Target aud: High school students. Spec prog: Class 2 hrs, C&W 2 hrs wkly. ♦Chris Bielizna, gen mgr.

Waterbury

WATR(AM)— June 15, 1934: 1320 khz; 5 kw-D, 1 kw-N, DA-2. TL: N41 32 12 W73 01 52. Hrs open: 24 One Broadcast Ln., 06706. Phone: (203) 755-1121. Fax: (203) 574-3025. E-mail: talkback@watr.com Web Site: www.watr.com. Licensee: WATR Inc. Population served: 275,000 Natl. Network: CBS. Format: News/talk, good time oldies. News staff: 2; News: 15 hrs wkly. Target aud: 35-64. Spec prog: Pol 2 hrs, It 3 hrs wkly. ♦Tom Chute, gen mgr & progmg dir; Trish Torello, gen sls mgr.

WPHH(FM)— Dec 25, 1967: 104.1 mhz; 50 kw. 859 ft TL: N41 33 41 W72 50 39. Stereo. Hrs opn: 24 10 Columbus Blvd., Hartford, 06106. Phone: (860) 723-6000. Fax: (860) 493-7090. Web Site: power1041.com. Licensee: Capstar TX L.P. Group owner: Clear Channel Communications Inc. (acq 8-30-00; grpsl). Population served: 900,000 Natl. Network: ABC. Natl. Rep: Christal. Format: Modern rock. Target aud: 18-49. ♦Tom McConnell, gen mgr; Michael Maguire, progmg dir.

WWCO(AM)— 1946: 1240 khz; 1 kw-U. TL: N41 33 59 W73 03 23. Hrs open: 24
Rebroadcasts WDRC(AM) Bloomfield 90%.
869 Bluehills Ave., Bloomfield, 06002. Phone: (860) 243-1115. Fax: (860) 274-9734. E-mail: wwco@talkofconnecticut.com Web Site: www.talkofconnecticut.com. Licensee: Buckley Broadcasting of Connecticut LLC. Group owner: Buckley Broadcasting Corp. (acq 4-97; $500,000). Population served: 500,000 Format: News/talk. News staff: 2. Target aud: 35 plus. ♦Richard Buckley, pres; Wayne Mulligan, gen mgr; Laura Kittell, opns mgr.

WWYZ(FM)— Aug 1, 1961: 92.5 mhz; 17.8 kw. 879 ft TL: N41 33 43 W72 50 41. Stereo. Hrs opn: 24 10 Columbus Blvd., Hartford, 06106. Phone: (860) 723-6000. Fax: (860) 493-7090. Web Site: www.country925.com. Licensee: Capstar TX L.P. Group owner: Clear Channel Communications Inc. (acq 8-30-00; grpsl). Population served: 300,000 Natl. Network: Westwood One. Natl. Rep: Christal. Format: Country. News staff: one; News: 5 hrs wkly. Target aud: 25-54. ♦Tom McConnell, gen mgr; Todd Thomas, opns mgr; Pete Salant, progmg dir.

West Hartford

WCCC(AM)—Licensed to West Hartford. See Hartford

WRYM(AM)—See New Britain

***WWUH(FM)**— July 15, 1968: 91.3 mhz; 440 w. 784 ft TL: N41 46 27 W72 48 20. Stereo. Hrs opn: 24 Univ. of Hartford, 200 Bloomfield Ave., 06117. Phone: (860) 768-4701. Phone: (860) 768-4703. Fax: (860) 768-5701. E-mail: wwuh@hartford.edu Web Site: www.wwuh.org. Licensee: University of Hartford. Population served: 804,380 Format: Var/div. News: 8 hrs wkly. Target aud: General. Spec prog: It 3 hrs, Por 3 hrs, Pol 3 hrs, It 3 hrs, foreign/ethnic 14 hrs wkly. ♦John N. Ramsey, pres & gen mgr; Susan Mullis, dev dir; Mark Helpern, progmg dir; John Ramsey, chief of engrg.

West Haven

***WNHU(FM)**— 1973: 88.7 mhz; 1.7 kw. 150 ft TL: N41 17 29 W72 57 40. Stereo. Hrs opn: 6 AM-2 AM Maxy Hall, 300 Boston Post Rd., 06516. Phone: (203) 479-8800. E-mail: hyaggi@newhaven.edu Web Site: www.wnhu.net. Licensee: University of New Haven Inc. Population served: 1,200,000 Law Firm: Dow, Lohnes & Albertson. Format: Diverse. News: 12 hrs wkly. Target aud: General. Spec prog: Jazz 12 hrs, folk 6 hrs, Irish 5 hrs, metal 9 hrs, class 9 hrs, gospel 4 hrs wkly. ♦Hank Yaggi, gen mgr.

Westport

WEBE(FM)— Sept 1, 1962: 107.9 mhz; 50 kw. 383 ft TL: N41 10 14 W73 11 05. Stereo. Hrs opn: 24 2 Lafayette Sq., Bridgeport, 06604-6000. Phone: (203) 333-9108. Fax: (203) 384-0600. Fax: (203) 394-6000. Web Site: www.webe108.com. Licensee: Cumulus Licensing Corp. Group owner: Cumulus Media Inc. (acq 3-14-02; grpsl). Population served: 1,100,000 Natl. Rep: Christal. Format: Adult contemp. News staff: one; News: 3 hrs wkly. Target aud: 25-54; upscale females. Spec prog: Talk one hr wkly. ♦Ann Surface McManus, gen mgr; Curtis Hansen, opns VP & opns mgr; Valerie Thompson, traf mgr.

***WSHU(AM)**— Apr 15, 1959: 1260 khz; 1 kw-D, DA. TL: N41 07 44 W73 23 20. Stereo. Hrs opn: 6 AM-7 PM
Rebroadcasts WSHU(FM) Fairfield 30%.
5151 Park Ave., Fairfield, 06825. Phone: (203) 365-6604. Fax: (203) 371-7991. E-mail: lombardi@wshu.org Web Site: www.wshu.org. Licensee: Sacred Heart University Inc. (acq 11-28-97; $325,000 as donation). Population served: 600,000 Natl. Network: USA, NPR, PRI. Format: News/talk. News staff: 2; News: 45 hrs wkly. Target aud: General. ♦George Lombardi, gen mgr; Julie Freddino, opns mgr; Gillian Anderson, dev dir.

***WWPT(FM)**— 1975: 90.3 mhz; 330 w. 110 ft TL: N41 10 19 W73 19 43. Stereo. Hrs opn: Staples High School, 70 N. Ave., 06880. Secondary address: 110 Myrtle Ave. 06880. Phone: (203) 341-1381. Phone: (203) 341-1380. Fax: (203) 226-6875. Licensee: Board of Education, Town of Westport. Population served: 100,000 Format: Free-form. Target aud: 14-24; youth. Spec prog: Slovak 3 hrs wkly. ♦Jim Honeycutt, gen mgr.

Willimantic

***WECS(FM)**— Feb 6, 1982: 90.1 mhz; 421 w. 380 ft TL: N41 41 00 W72 12 59. Hrs open: 83 Windham St., 06226. Phone: (860) 465-5354. Fax: (860) 465-5073. E-mail: wecs@hotmail.com Web Site: www.easternct.edu/depts/wecs. Licensee: Eastern Connecticut State University. Population served: 200,000 Format: Urban contemp, rock/AOR, jazz. Spec prog: Jazz 16 hrs, relg 3 hrs, Sp 9 hrs wkly. ♦John L. Zatowski, gen mgr.

WILI(AM)— Oct 5, 1957: 1400 khz; 1 kw-U. TL: N41 42 55 W72 11 23. Hrs open: 24 720 Main St., 06226. Phone: (860) 456-1111. Fax: (860) 456-9501. Web Site: www.wili.com. Licensee: Nutmeg Broadcasting Co. (acq 7-11-2005; $1.8 million with co-located FM). Population served: 75,000 Natl. Network: Westwood One. Rgnl. Network: Conn. Radio Net. Natl. Rep: D & R Radio. Wire Svc: AP Format: Full service, adult contemp, news/talk. News staff: 3; News: 10 hrs wkly. Target aud: 25 plus; general. Spec prog: Ukrainian one hr, Sp one hr, relg 2 hrs, Jewish one hr wkly. ♦Andy Russell, VP; Colin K. Rice, VP; Andy Russell, gen mgr; Donna Evan, gen sls mgr.

WILI-FM— June 16, 1975: 98.3 mhz; 1.05 kw. 525 ft TL: N41 41 00 W72 13 01. Stereo. 24 Web Site: www.wili.com. Licensee: Nutmeg Broadcasting Co.125,000 Natl. Network: Superadio. Natl. Rep: D & R Radio. Format: CHR. News staff: one; News: 6 hrs wkly. Target aud: 22-44; college students, young married couples, young families.

Windsor

WKND(AM)— May 4, 1961: 1480 khz; 500 w-D, DA. TL: N41 51 10 W72 40 43. Hrs open: 330 Main St., 1st Fl, Hartford, 06106-1622. Phone: (860) 524-0001. Fax: (860) 548-1922. E-mail: msanchez@freedomcommct.com Licensee: Freedom Communications of Connecticut Inc. (acq 11-29-2004). Population served: 636,000 Natl. Network: ABC. Format: Talk, rhythm and blues. Spec prog: Gospel 5 hrs, jazz 3 hrs wkly. ♦Richard Weaver-Bey, pres; Marion Anderson, gen mgr; Melvin Sanchez, opns dir & prom mgr.

Delaware

Bethany Beach

WJKI(FM)— 1996: 103.5 mhz; 1.45 kw. Ant 479 ft TL: N38 34 21 W75 06 58. Hrs open:
Simulcast w/ WGBG (FM) Seaford 100%.
20200 DuPont Blvd., Georgetown, 19947. Phone: (302) 856-2567. Fax: (302) 856-7633. Web Site: www.bigclassicrock.com. Licensee: Great Scott Broadcasting. (group owner) Format: Classic rock. ♦Sue Timmons, gen mgr; Sean McHugh, progmg dir; Tracy Baker, traf mgr.

WOSC(FM)— 1974: 95.9 mhz; 10.5 kw. Ant 469 ft TL: N38 25 20 W75 08 23. Stereo. Hrs opn: Gateway Crossing, 351 Tilghman Rd., Salisbury, MD, 21804. Phone: (410) 742-1923. Fax: (410) 742-2329. Web Site: www.96rocksyou.com. Licensee: Capstar TX L.P. Group owner: Clear Channel Communications Inc. (acq 8-7-2000; grpsl). Natl. Rep: Clear Channel. Format: Active rock. Target aud: 18-34. ♦Frank Hamilton, gen mgr; Brian Cleary, opns mgr & progmg dir; Dixie Penner, prom dir.

Christiana

***WXHL-FM**— Aug 1, 1994: 89.1 mhz; 1 w horiz, 1.2 kw vert. 67 ft TL: N39 40 38 W75 39 47. Hrs open: 24 179 Stanton-Christiana Rd., Newark, 19702. Phone: (302) 731-0690. Fax: (302) 738-3090. Web Site: www.thereachfm.com. Licensee: Priority Radio Inc. (group owner; acq 12-10-99). Format: Adult contemp Christian mus. ♦Steve Hare, gen mgr; Dan Edwards, opns mgr; Larry Humm, gen sls mgr; Dave Kirby, progmg dir.

Dover

WDOV(AM)— 1948: 1410 khz; 5.4 kw, DA-2. TL: N39 12 03 W75 33 13. Hrs open: 24 1575 McKee, Suite 206, 19904. Phone: (302) 674-1410. Fax: (302) 674-5978. E-mail: wdov@clearchannel.com Web Site: www.wdov.com. Licensee: Capstar TX L.P. Group owner: Clear Channel Communications Inc. (acq 8-30-00; grpsl). Population served: 100,200 Natl. Network: Westwood One. Natl. Rep: McGavren Guild. Format: News/talk. News staff: 2; News: 162 hrs wkly. Target aud: 25-54. ♦Bob Walton, opns dir, opns mgr & progmg dir; Phil Feliciangeli, news dir & pub affrs dir.

WRDX(FM)—Co-owned with WDOV(AM). 1956: 94.7 mhz; 50 kw. 377 ft TL: N39 12 03 W75 33 55. Stereo. 920 W. Basin Rd., Suite 400, New Castle, 19720. Phone: (302) 395-9800. Fax: (302) 395-9808. E-mail: wrdx@clearchannel.com Web Site: www.river947.com. Natl. Network: Westwood One. Law Firm: Latham & Watkins. Format: Hot adult contemp. ♦Joe Puglise, gen sls mgr; Bob Walton, progmg mgr; Phil Feliciangeli, local news ed; Rob Reb, chief of engrg & disc jockey.

***WRTX(FM)**— Apr 5, 1995: 91.7 mhz; 580 w. Ant 315 ft TL: N39 12 03 W75 33 55. Stereo. Hrs opn: 24
Rebroadcasts WRTI(FM) Philadelphia, PA 100%.
1509 Cecil B. Moore Ave., Philadelphia, PA, 19121. Phone: (215) 204-8405. Fax: (215) 204-7027. E-mail: comments@wrti.org Web Site: www.wrti.org. Licensee: Temple University of the Commonwealth System of Higher Education. Population served: 150,000 Natl. Network: NPR, AP Radio. Rgnl. Network: Radio Pa. Format: Jazz, class. News staff: one. Target aud: 30-65. ♦Tobias Poole, opns dir; Brick Torpey, gen sls mgr.

***WXXY(AM)**— Aug 2, 1957: 1600 khz; 5 kw-D, 1 kw-N, DA-2. TL: N39 10 11 W75 33 13. Hrs open: 24 Box 741, 19903. Phone: (302) 731-1777. Licensee: WXXY Broadcasting Inc. (acq 10-11-2005). Population served: 125,000 Format: Black gospel. ♦George Krementz, gen mgr.

Fenwick Island

WLBW(FM)— Apr 1, 1994: 92.1 mhz; 3 kw. Ant 469 ft TL: N38 25 20 W75 08 23. Stereo. Hrs opn:
Rebroadcasts WLVW(AM) Salisbury, MD 100%.
351 Tilghman Rd., Salisbury, MD, 21804. Phone: (410) 742-1923. Fax: (410) 742-2329. E-mail: wave@intercom.net Web Site: www.isurfthewave.com. Licensee: Capstar TX L.P. Group owner: Clear Channel Communications Inc. (acq 8-7-00; grpsl). Format: Oldies. Target aud: 25-54. ♦Frank Hamilton, gen mgr.

Georgetown

WJWL(AM)— June 23, 1951: 900 khz; 10 kw-D, 1 w-N, DA-1. TL: N38 42 31 W75 24 25. Stereo. Hrs opn: 24 233 N.E. Front St., Milford, 19963. Phone: (302) 422-2600. Fax: (302) 424-1630. E-mail: digital900@aol.com Web Site: www.digital900.com. Licensee: Great Scott Broadcasting Ltd. Group owner: Great Scott Broadcasting Population served: 38,000 Law Firm: Cohn & Marks. Format: Sp var. News staff: one; News: 10 hrs wkly. Target aud: 25-54; mature adults. Spec prog: Relg 6 hrs wkly. ♦Faye Scott, pres; Danny Perez, gen mgr; Lisette Perez, progmg dir & farm dir.

WZBH(FM)—Co-owned with WJWL(AM). July 4, 1969: 93.5 mhz; 11 kw. Ant 485 ft TL: N38 31 24 W75 17 55. (CP: COL Millsboro. 50 kw, ant 492 ft.). 24 20200 DuPont Blvd., 19947. Phone: (302) 856-2567. Fax: (302) 856-7633.100,000 Format: Contemp rock. Target aud: Adults; baby boomers. ♦Shawn Murphy, progmg dir; Terry Dalton, chief of engrg; C. Marcus, farm dir; Donna Cavender, women's int ed & disc jockey; B. Graxston, disc jockey; H. Lasren, disc jockey.

Laurel

WKDB(FM)— Nov 19, 1991: 95.3 mhz; 6 kw. 328 ft TL: N38 30 12 W75 39 39. Stereo. Hrs opn: 24 20200 DuPont Blvd., Georgetown, 19947. Phone: (302) 856-2567. Fax: (302) 856-7633. Web Site: www.musiconthe.com. Licensee: Great Scott Broadcasting (group owner; acq 2-13-98; $1.5 million). Population served: 200,000 Format: Adult contemp. News: 6 hrs wkly. Target aud: 25-49. ♦Sue Timmons, gen mgr & stn mgr; Tracy Baker, traf mgr.

Lewes

WXJN(FM)— June 1, 1991: 105.9 mhz; 6 kw. 341 ft TL: N38 38 36 W75 13 00. Stereo. Hrs opn: 24
Rebroadcasts WICO-FM Salisbury, MD 100%.
Box 909, Salisbury, MD, 21803. Phone: (410) 219-3500. Phone: (410) 548-1543. Web Site: www.catcountryradio.com. Licensee: Delmarva Broadcasting Co. (group owner; acq 6-26-97; grpsl). Population served: 225,000 Natl. Rep: Katz Radio. Law Firm: Hogan & Hartson. Wire Svc: AP Format: Country. News staff: one; News: 3 hrs wkly. Target aud: 25-54. Spec prog: NASCAR. ♦Mike Reath, gen mgr; Joe Edwards, opns mgr; Joe Beail, gen sls mgr; Jeff Twilley, engrg dir & chief of engrg.

Milford

WAFL(FM)— May 19, 1973: 97.7 mhz; 6 kw. 328 ft TL: N38 55 39 W75 29 20. Stereo. Hrs opn: Box 808, 19963. Secondary address: 1666 Blairs Pond Rd. 19963. Phone: (302) 422-7575. Fax: (302) 422-3069. E-mail: staff@eagle977.com Web Site: www.eagle977.com. Licensee: Delmarva Broadcasting Co. (group owner; (acq 6-26-97; grpsl). Population served: 210,000 Natl. Network: Westwood One. Law Firm: Hogan & Hartson. Format: Adult contemp. Target aud: 18-49; active, affluent adults in central & southern Delaware. Spec prog: Southern gospel 2 hrs wkly. ♦Melody Booker, gen mgr; Ralph D. Domenico, gen sls mgr; Steve Monz, opns mgr & progmg dir; Jeff Twilly, chief of engrg.

WYUS(AM)—Co-owned with WAFL(FM). 1953: 930 khz; 500 w-D, 100 w-N, DA-1. TL: N38 55 39 W75 29 20.6 AM-midnight Phone: (302) 422-2428. E-mail: rafael@wyusam.com Web Site: www.laexitosa.com. Format: Sp. Target aud: 18 plus; Hispanic. Spec prog: Relg 10 hrs, Haitian 3 hrs wkly. ♦Rafael Dosman, progmg dir.

WNCL(FM)— Nov 5, 1990: 101.3 mhz; 3 kw. Ant 328 ft TL: N38 51 21 W75 29 02. Stereo. Hrs opn: 24 Box 808, 19963-0808. Secondary address: 1666 Blairs Pond Rd. 19963-5263. Phone: (302) 422-7575. Fax: (302) 422-3069. E-mail: cool@cool1013.com Web Site: www.cool1013.com. Licensee: Delmarva Broadcasting Co. (group owner; acq 1-17-2003; $1.6 million). Population served: 300,000 Law Firm: Gammon & Grange. Format: Greatest hits of the 60s and 70s. News: 7 hrs wkly. Target aud: 35-54; adults. ♦Melody Booker, gen mgr; Ralph DiDomenico, gen sls mgr; Steve Monz, progmg dir.

Newark

WNWK(AM)— Aug 17, 1964: 1260 khz; 1 kw-D, 42 w-N, DA-2. TL: N39 38 39 W75 41 33. Hrs open: 1072 So. Chaple St., 19702. Phone: (240) 481-8242. Licensee: East Coast Broadcasting Inc. (acq 3-11-2002; $140,000). Format: Sp. ♦Jose Roberto Ekonomo, pres & gen mgr.

***WVUD(FM)**— Oct 4, 1976: 91.3 mhz; 1 kw. 135 ft TL: N39 41 26 W75 45 23. Stereo. Hrs opn: Univ. of Delaware, Perkins Student Ctr., 19716. Phone: (302) 831-2701. Fax: (302) 831-1399. Web Site: www.wvud.org. Licensee: University of Delaware. Population served: 243,601 Natl. Network: AP Radio. Format: Progsv, div, educ. Target aud: General. Spec prog: Class 10 hrs, Black 10 hrs, jazz 15 hrs, folk 15 hrs, Sp 2 hrs wkly. ♦Chuck Tarver, gen mgr & stn mgr; David Mackenzie, chief of engrg.

Ocean View

WZEB(FM)— Jan 12, 1986: 101.7 mhz; 3 kw. 328 ft TL: N38 29 20 W75 12 01. Stereo. Hrs opn:
Simulcast of WKDB (FM) Laurel 100%.
20200 DuPont Blvd., Georgetown, 19947. Phone: (302) 856-2567. Fax: (302) 856-7633. Web Site: www.musicontheb.com. Licensee: Great Scott Broadcasting. (group owner; acq 5-29-98; $1.5 million). Format: Adult contemp. ♦Sue Timmons, gen mgr; Tracy Baker, traf mgr.

Pike Creek

***WMHS(FM)**— April 2000: 88.1 mhz; 90 w vert. 121 ft TL: N39 45 27 W75 40 02. Hrs open: 24 4550 New Linden Hill Rd., Suite 117, Wilmington, 19808. Secondary address: 301 McKenna's Church Rd., Wilmington 19808. Phone: (302) 992-5520. Fax: (302) 992-5525. E-mail: francis.kulas@redclay.k12.de.us Web Site: www.mckeanradio.com. Licensee: Red Clay Consolidated School District. Wire Svc: AP Format: Oldies. News staff: 11; News: 20 hrs wkly. Target aud: 25-54; adults-baby boomers. ♦Fran Kulas, gen mgr.

Rehoboth Beach

WGMD(FM)— Sept 21, 1975: 92.7 mhz; 3 kw. 300 ft TL: N38 42 05 W75 11 58. Stereo. Hrs opn: 24 Box 530, 19971. Phone: (302) 945-2050. Fax: (302) 945-3781. E-mail: wgmd@wgmd.com Web Site: www.wgmd.com. Licensee: Resort Broadcasting Co. L.L.C. (acq 7-25-80). Population served: 500,000 Natl. Rep: ABC Radio Sales. Format: News/talk. News staff: 3; News: 16 hrs wkly. Target aud: 35 plus. Spec prog: Farm 2 hrs, jazz 2 hrs, relg 2 hrs wkly. ♦Dan Gaffney, gen mgr; Jared Morris, opns mgr; Marie Moulinier, gen sls mgr.

Seaford

WGBG(FM)— February 1972: 98.5 mhz; 6 kw. Ant 321 ft TL: N38 36 47 W75 35 12. Stereo. Hrs opn: 24 20200 DuPont Blvd., Georgetown, 19947. Phone: (302) 856-2567. Fax: (302) 856-7633. Web Site: www.bigclassicrock.com. Licensee: Great Scott Broadcasting. (group owner; (acq 4-27-98; $1.2 million with co-located AM). Natl. Network: CBS. Law Firm: Mullin, Rhyne, Emmons & Topel. Format: Classic rock. News: 4 hrs wkly. Target aud: 18-49; secondary 25-54, tertiary 35 plus. Spec prog: Farm one hr wkly. ♦Sue Timmons, gen mgr & gen sls mgr; Sean McHugh, progmg dir; Terry Dalton, chief of engrg; Tracy Baker, traf mgr.

WJWK(AM)—Co-owned with WGBG(FM). 1955: 1280 khz; 840 w-D, 250 w-N. TL: N38 37 03 W75 35 09.6 AM-midnight
Simulcast with WKHI(FM) Fruitland 100%.
10,000 Format: Joe. News: 4 hrs wkly. Target aud: Black adults. ♦Adam Davis, progmg mgr; Tracy Baker, traf mgr.

Selbyville

WOCM(FM)— March 1993: 98.1 mhz; 3 kw. Ant 469 ft TL: N38 25 20 W75 08 23. Hrs open: 24 117th W. 49th St., Ocean City, MD, 21842. Phone: (410) 723-3683. Fax: (410) 723-4347. Web Site: www.irieradion.com. Licensee: Irie Radio Inc. (acq 9-27-02; $1.08 million). Population served: 50,000 Law Firm: Leventhal, Senter & Lerman. Format: AAA. Target aud: 25-54; seasonal, beach residents & loc urban/farm. ♦Leighton Moore, pres; David Rothner, stn mgr & chief of engrg.

Smyrna

WDSD(FM)— Nov 10, 1993: 92.9 mhz; 1.7 kw. 377 ft TL: N39 16 08 W75 31 28. Stereo. Hrs opn: 24 1575 McKee Rd., Suite 206, Dover, 19904. Phone: (302) 674-1410. Fax: (302) 674-5978. E-mail: wdsd@clearchannel.com Web Site: www.wdsd.com. Licensee: Capstar TX L.P. Group owner: Clear Channel Communications Inc. (acq 8-30-00; grpsl). Natl. Network: Jones Radio Networks. Law Firm: Latham & Watkins. Format: Country. News staff: one; News: 2 hrs wkly. Target aud: 18-49. ♦Paige Lamers, gen mgr; Bob Walton, opns mgr.

Wilmington

WDEL(AM)— 1922: 1150 khz; 5 kw-U, DA-2. TL: N39 48 54 W75 31 47. Hrs open: 24 Box 7492, 2727 Shipley Rd., 19803. Phone: (302) 478-2700. Fax: (302) 478-0100. E-mail: wdel@wdel.com Licensee: Delmarva Broadcasting Co. Inc. (group owner) Population served: 595,000 Natl. Network: Westwood One. Natl. Rep: Katz Radio. Law Firm: Hogan & Hartson. Format: Full service, news/talk. News staff: 10; News: 70 hrs wkly. Target aud: 35-64. Spec prog: Sp 2 hrs wkly. ♦Julian H. Booker, CEO & pres; Michael G. Reath, gen mgr.

WSTW(FM)—Co-owned with WDEL(AM). 1950: 93.7 mhz; 50 kw. 490 ft TL: N39 48 57 W75 31 31. Stereo. 24 Format: Hot adult contemp. News: 10 hrs wkly. Target aud: 25-54. Spec prog: Relg one hr, pub affrs one hr wkly.

WFAI(AM)—Salem, NJ) Sept 1, 1966: 1510 khz; 2.5 kw-D, DA. TL: N39 34 58 W75 27 39. Hrs open: 6 AM-6 PM First Federal Plaza Bldg., 704 King St., Suite 604, 19801. Phone: (302) 622-8895. Fax: (302) 622-8678. E-mail: tonya@faith1510.com Web Site: www.faith1510.com. Licensee: QC Communication Inc. (acq 3-17-97; $1.8 million with WJKS(FM) Canton). Population served: 150,000 Format: Gospel. News staff: 2; News: 3 hrs wkly. Target aud: 18 plus. Spec prog: Farm 8 hrs wkly. ♦Tony Quartarone, gen mgr; Manuel Mena, progmg dir.

WILM(AM)— Oct 1, 1923: 1450 khz; 1 kw-U. TL: N39 43 46 W75 33 07. Hrs open: 24 920 W. Basin Rd., Suite 400, New Castle, 19720. Phone: (302) 395-9800. Fax: (302) 395-9808. E-mail: mail@wilm.com Web Site: www.wilm.com. Licensee: Citicasters Licenses L.P. (acq 10-29-2004; $3,986,000). Population served: 426,000 Natl. Network: Wall Street, Fox News Radio. Wire Svc: AP Format: All news/talk. News staff: 6; News: 168 hrs wkly. Target aud: Adults 25-54. Spec prog: Community Spotlight, Delaware Radio Magazine. ♦Bob Walton, gen mgr & opns mgr; Martha Burns, sls dir; Paige Lamers, mktg mgr; Ryan Kennedy, prom dir; Mark Fowser, progmg dir; Mark Eichmann, news dir.

WJBR-FM— January 1957: 99.5 mhz; 50 kw. 499 ft TL: N39 50 03 W75 31 25. Stereo. Hrs opn: 24 812 Philadelphia Pike, 19809. Phone: (302) 765-1160. Fax: (302) 765-1192. E-mail: info@wjbr.com Web Site: www.wjbr.com. Licensee: NM Licensing LLC. Group owner: NextMedia Group L.L.C. (acq 3-7-00; $32.4 million). Population served: 500,000 Natl. Rep: Christal. Law Firm: Liebowitz & Associates. Format: Adult contemp. News staff: one. Target aud: 25-54. ♦Bruce Beasley, pres; Jane E. Bartsch, VP; Jane Bartsch, gen mgr; Michael Waite, opns VP.

***WMPH(FM)**— October 1969: 91.7 mhz; 100 w. 143 ft TL: N39 46 23 W75 30 25. Stereo. Hrs opn: 24 5201 Washington St. Ext., 19809. Phone: (302) 762-7199. E-mail: radio@wmph.org Web Site: www.wmph.org. Licensee: Brandywine School District, Brd of Educ Population served: 500,000 Format: Rythmic contemp hit/dance. News staff: one; News: 2 hrs wkly. Target aud: 13-27; high school & college students. ♦James Scanlon, CEO; Mike Pullig, pres; Clint Dantinne, gen mgr.

WTMC(AM)— 1947: 1380 khz; 5 kw-D, 1 kw-N. TL: N39 48 12 W75 37 42. (CP: 520 w-D, 4.2 kw-N. TL: N39 43 46 W75 33 07 day, N39 48 41 W75 46 20 night). Hrs opn: Box 778, Dover, 19901. Phone: (302) 659-2400. Fax: (302) 659-6128. Web Site: www.deldot.net. Licensee: State of Delaware Department of Transportation. (acq 11-4-99). Format: Talk. ♦Jonathan Weishaupt, pres; William Brooks, gen mgr.

WWTX(AM)— Apr 21, 1947: 1290 khz; 2.5 kw-U. TL: N39 44 03 W75 31 44. Hrs open: 24 920 W. Basin Rd., Suite 400, New Castle, 19720. Phone: (302) 395-9800. Fax: (302) 395-9808. E-mail: wwtx@clearchannel.com Web Site: www.1290theticket.com. Licensee: Capstar TX L.P. Group owner: Clear Channel Communications Inc. Population served: 70,000 Format: Sports. ♦Paige Lamers, gen mgr; Bob Walton, opns mgr & progmg dir.

District of Columbia

Washington

WACA(AM)—See Wheaton, MD

***WAMU(FM)**— Oct 23, 1961: 88.5 mhz; 50 kw. 500 ft TL: N38 56 09 W77 05 33. Stereo. Hrs open: 24 The American Univ., Brandywine Bldg., 4000 Brandywine St., N.W., 20016. Phone: (202) 885-1200. Fax: (202) 885-1269. E-mail: feedback@wamu.org Web Site: www.wamu.org. Licensee: American University. Population served: 3,434,300 Natl. Network: NPR, PRI. Format: News/talk, bluegrass, culture. News staff: 6; News: 120 hrs wkly. Target aud: 25-54. Spec prog: Vintage radio 4 hrs, country 4 hrs, jazz 3 hrs wkly. ♦Caryn Mathes, gen mgr; Mark McDonald, progmg dir; Jim Asendio, news dir.

WASH(FM)— 1948: 97.1 mhz; 26 kw. 690 ft TL: N38 57 21 W77 04 57. Stereo. Hrs opn: 1801 Rockville Pike, Rockville, MD, 20852. Phone: (301) 984-9710. Fax: (301) 255-4314. Web Site: www.washfm.com. Licensee: AMFM Radio Licenses LLC. Group owner: Clear Channel Communications Inc. (acq 8-30-00; grpsl). Format: Adult Contemp. ♦Bennett Zier, VP, gen mgr & opns mgr; Loretta Lage, gen sls mgr; Mark Lapidus, mktg dir; Bill Hess, progmg dir.

WAVA(AM)—See Arlington, VA

WAVA-FM—Arlington, VA) Aug 1, 1948: 105.1 mhz; 41 kw. 541 ft TL: N38 53 44 W77 08 04. Stereo. Hrs opn: 1901 N. Moore St., Suite 200, Arlington, VA, 22209. Phone: (703) 807-2266. Fax: (703) 807-2248. E-mail: comment@wava.com Web Site: www.wava.com. Licensee: Salem Media of Virginia Inc. Group owner: Salem Communications Corp. (acq 2-13-92; $20 million; FTR: 11-18-91). Population served: 8,000,000 Natl. Network: Salem Radio Network. Natl. Rep: Salem. Law Firm: Fletcher, Heald & Hildreth. Format: Adult contemp, Christian, talk. News: 4 hrs wkly. Target aud: 25-54. ♦Edward Atsinger, CEO & chief of engrg; Stu Epperson, chmn & disc jockey; David Evans, CFO & disc jockey; Joe Davis, exec VP & disc jockey; David Ruleman, VP & gen mgr; Tom Moyer, stn mgr.

WBIG-FM— June 3, 1994: 100.3 mhz; 36 kw. 574 ft TL: N38 53 44 W77 08 04. Stereo. Hrs opn: 24 6th Fl., 1801 Rockville Pike, Rockville, MD, 20852. Phone: (301) 468-1800. Fax: (301) 770-0236. Web Site: www.big100.com. Licensee: AMFM Radio Licenses LLC. Group owner: Clear Channel Communications Inc. (acq 8-30-00; grpsl). Population served: 3,184,600 Format: Oldies. News staff: 2; News: one hr wkly. Target aud: 35-54; professional, college, upscale. ♦Bennett Zier, gen mgr.

***WCSP-FM**— May 8, 1982: 90.1 mhz; 50 kw. 450 ft TL: N38 57 44 W77 01 36. Hrs open: 24 400 N. Capitol St. N.W., Suite 650, 20001. Phone: (202) 737-3220. Fax: (202) 737-5554. Web Site: www.c-span.org. Licensee: National Cable Satellite Corp. (acq 1997). Population served: 3,500,000 Format: Pub affrs. Target aud: General. ♦Brian P. Lamb, CEO; Kate Mills, gen mgr.

WCTN(AM)—See Potomac-Cabin John, MD

WDCT(AM)—See Fairfax, VA

***WETA(FM)**— Apr 19, 1970: 90.9 mhz; 75 kw. Ant 610 ft TL: N38 53 30 W77 07 55. Stereo. Hrs opn: 24 2775 S. Quincy St., Arlington, VA, 22206-2269. Phone: (703) 998-2600. Fax: (703) 824-7288. E-mail: radio@weta.org Web Site: www.weta.org. Licensee: Greater Washington Educational Telecommunications Association Inc. Population served: 4,500,000 Natl. Network: NPR, PRI. Law Firm: Dow, Lohnes & Albertson. Format: News, pub affrs. Target aud: General; educated adults. ♦Sharon Percy Rockefeller, CEO & pres; Polly Heath, CFO; Dan Devany, sr VP, VP & stn mgr; DeLinda Mrowka, prom mgr; Ingrid Lakey, progmg dir; David Ginder, mus dir. Co-owned TV: *WETA-TV affil

WFAX(AM)—Falls Church, VA) September 1948: 1220 khz; 5 kw-D, 100 w-N. TL: N38 52 47 W77 10 18. Hrs open: 6 am-midnight 161 Hillwood Ave., Suite B, Falls Church, VA, 22046-2983. Phone: (703) 532-1220. Fax: (703) 533-7572. E-mail: wfax@wfaxam.com Web Site: www.wfax.com. Licensee: Newcomb Broadcasting Corp. Population served: 3,000,000 Law Firm: Arent, Fox, Kintner, Plotkin & Kahn. Format: Relg. News: one hr wkly. Target aud: 34-54. Spec prog: Black 15 hrs, It one hr wkly. ♦Doris N. Newcomb, pres & gen mgr; R. C. Woolfenden, opns dir.

WFED(AM)—Silver Spring, MD) Dec 7, 1946: 1050 khz; 1 kw-D, 44 w-N. TL: N39 00 50 W77 01 46. Hrs open: 24 3400 Idaho Ave. N.W., 20016. Phone: (202) 895-5000. Fax: (202) 895-5016. Web Site: www.federalnewsradio.com. Licensee: Bonneville Holding Co. (group owner; (acq 12-1-2004; $4 million). Population served: 3,500,000 Natl. Network: AP Radio, CNN Radio. Wire Svc: AP Format: Federal news. News staff: 7. ♦Bruce Reese, pres; Joel Oxley, gen mgr; Lisa Wolfe, progmg dir.

WGTS(FM)—See Takoma Park, MD

WHUR-FM— Dec 10, 1971: 96.3 mhz; 16.5 kw. Ant 800 ft TL: N38 57 01 W77 04 47. Stereo. Hrs opn: 24 529 Bryant St. N.W., 20059. Phone: (202) 806-3500. Fax: (202) 806-3522. Web Site: www.whur.com. Licensee: Howard University Board of Trustees. Population served: 2,275,248 Natl. Network: CNN Radio, ABC. Natl. Rep: D & R Radio. Wire Svc: UPI Format: Urban adult contemp. News staff: 3; News: 7 hrs wkly. Target aud: 25-54. Spec prog: Gospel 14 hrs, Caribbean 6 hrs wkly. ♦Dr. H. Patrick Swygert, pres; Millard J. Watkins III, gen mgr; Jeanette Tyce, gen sls mgr; David Dickinson, progmg dir.

WIHT(FM)— 1960: 99.5 mhz; 22 kw. Ant 751 ft TL: N38 57 49 W77 06 18. Hrs open: 24 1801 Rockville Pike, 6th Fl., Rockville, MD, 20852. Phone: (301) 468-9429. Phone: (301) 587-7100. Fax: (301) 770-3541. Web Site: www.hot995.com. Licensee: AMFM Radio Licenses L.L.C. Group owner: Clear Channel Communications Inc. (acq 9-00). Population served: 81,000 Format: CHR. ♦Bennett Zier, VP, gen mgr & opns mgr; Melissa Kelly, gen sls mgr; Jessica Ritch, prom dir; Jeff Wyatt, progmg dir.

WILC(AM)—See Laurel, MD

WJFK-FM—Manassas, VA) Apr 8, 1968: 106.7 mhz; 22.5 kw horiz, 18.5 kw vert. 731 ft TL: N38 52 28 W77 13 24. (CP: 22 kw, ant 745 ft.). Stereo. Hrs opn: 24 10800 Main St., Fairfax, VA, 22030. Phone: (703) 691-1900. Fax: (703) 934-9896. Web Site: www.1067wjfk.com. Licensee: Infinity Broadcasting of Washington D.C. Inc. Group owner: Infinity Broadcasting Corp. (acq 10-86; $13 million; FTR: 9-22-86). Population served: 3,400,000 Format: Personality, sports. News staff: one; News: 3 hrs wkly. Target aud: 25-54. ♦Ken Stevens, gen mgr; Jeremy Coleman, opns mgr; David Hain Line, natl sls mgr; Tammy Sacks, prom dir; Buzz Burbank, news dir; Mike Elston, pub affrs dir; Dan Ryson, chief of engrg.

WJZW(FM)—Woodbridge, VA) Dec 25, 1958: 105.9 mhz; 28 kw. Ant 648 ft TL: N38 52 28 W77 13 24. Stereo. Hrs opn: 4400 Jenifer St. NW, Suite 400, 20015. Phone: (202) 686-3100. Fax: (202) 686-3064. Web Site: www.smoothjazz1059.com. Licensee: Radio License Holding VII LLC. Group owner: ABC Inc. (acq 6-12-2007; grpsl). Population served: 426,800 Format: Smooth jazz. ♦Jeff Boden, gen mgr; Kenny King, opns dir; Cathy Whissel, gen sls mgr; Steve Allan, progmg dir.

WKIK(AM)—See La Plata, MD

WKYS(FM)— Aug 1, 1947: 93.9 mhz; 24 kw. 707 ft TL: N38 56 24 W77 04 54. Stereo. Hrs opn: 24 5900 Princess Garden Pkwy., Lanham, MD, 20706. Phone: (301) 306-1111. Phone: (301) 429-2626. Fax: (301) 306-9609. Licensee: Radio One Licenses LLC. Group owner: Radio One Inc. (acq 6-95; $34 million; FTR: 2-27-95). Population served: 4,920,600 Natl. Rep: McGavren Guild. Format: Urban contemp. News staff: 3; News: 4 hrs wkly. Target aud: 25-54; upscale Black adults. ♦Michele Williams, gen mgr; Jack Murray, sls dir; Ezio Torres, natl sls mgr; Darryl Huckaby, progmg dir; Iran Waller, mus dir; Taylor Thomas, news dir; Tim White, chief of engrg; Robin Graham, traf mgr; Taylor Thomas, pub svc dir; Russ Parr, disc jockey.

WOL(AM)—Co-owned with WKYS(FM). 1924: 1450 khz; 1 kw-U. TL: N38 54 16 W77 00 25. (Acq 6-95).3,500,000 Format: News/talk. ♦Karen Jackson, gen sls mgr; Vaughn Holmes, prom dir; Ron Thompson, progmg dir.

WLXE(AM)—See Rockville, MD

WMAL(AM)— Oct 12, 1925: 630 khz; 5 kw-U, DA-2. TL: N39 00 55 W77 08 30. Stereo. Hrs opn: 24 4400 Jenifer St. N.W., Suite 400, 20015. Phone: (202) 686-3100. Fax: (202) 686-3061. Web Site: www.wmal.com. Licensee: Radio License Holding VII LLC. Group owner: ABC Inc. (acq 6-12-2007; grpsl). Population served: 696,800 Natl. Rep: ABC Radio Sales. Format: News/talk. ♦Chris Berry, pres & gen mgr; Paul Duckworth, opns dir; Ernie Fears Jr., sls dir; John Matthews, news dir; David Sproul, engrg dir; Chuck Eisenhauer, traf mgr; Bryan Nehman, news rptr.

WRQX(FM)—Co-owned with WMAL(AM). May 15, 1948: 107.3 mhz; 19.5 kw. Ant 807 ft TL: N38 57 01 W77 04 47. Stereo. Fax: (202) 686-3091. Web Site: www.mix1073fm.com.676,000 Natl. Network: ABC. Format: Hot adult contemp. ♦Jeff Boden, pres & gen mgr; Carol Parker, mus dir; Cindy Maguire, pub affrs dir; David Sproul, engrg mgr; Stella Pressley, traf mgr & disc jockey.

WMMJ(FM)—See Bethesda, MD

WMZQ-FM— September 1968: 98.7 mhz; 50 kw. 490 ft TL: N38 53 12 W77 12 05. Stereo. Hrs opn: 24 1801 Rockville Pike, 6th Fl., Rockville, MD, 20852. Phone: (301) 231-8231. Fax: (301) 984-4895. Web Site: www.wmzq.com. Licensee: Clear Channel Radio Licenses, Inc. Group owner: Clear Channel Communications Inc. (acq 8-30-00; grpsl). Population served: 4,000,000 Natl. Rep: Christal. Law Firm: Latham & Watkins. Format: Country. News staff: one. Target aud: 25-54. ♦Bennett Zier, VP, gen mgr & opns mgr; Shelley Rose, rgnl sls mgr & mus dir; Mark Lapidus, mktg mgr; Wendie Vestfall, prom dir; George King, progmg dir; Greg Gallagher, engrg mgr & chief of engrg.

***WPFW(FM)**— Feb 28, 1977: 89.3 mhz; 50 kw. 410 ft TL: N38 56 09 W77 05 33. Stereo. Hrs opn: 24 2390 Champlain St. N.W., 20009. Phone: (202) 588-0999. Fax: (202) 588-0561. E-mail: bmwpfw@aol.com Web Site: www.wpfw.org. Licensee: Pacifica Foundation Inc. Group owner: Pacifica Radio Population served: 3,800,000 Law Firm: Haley, Bader & Potts. Format: Jazz, news/talk, world mus. News: 9 hrs wkly. Target aud: 25-55. Spec prog: Oldies 3 hrs, women 3 hrs, health one hr wkly. ♦Ron Pinchback, gen mgr; Tiffany Jordan, dev dir.

WPGC(AM)—See Morningside, MD

WPGC-FM—See Morningside, MD

WPRS-FM—See Waldorf, MD

WTEM(AM)— Aug 1, 1923: 980 khz; 50 kw-D, 5 kw-N, DA-2. TL: N38 57 43 W76 58 24. Hrs open: 24 1801 Rockville Pike., Rockville, MD, 20852-1633. Phone: (301) 231-7798. Fax: (301) 881-8030. Web Site: www.sportstalk980.com. Licensee: AMFM Radio Licenses L.L.C. Group owner: Clear Channel Communications Inc. (acq 8-30-00; grpsl). Natl. Network: ESPN Radio. Wire Svc: The Sports Network Format: Sports/talk. Target aud: 25-54; men. ♦Bennett Zier, VP.

WTGB-FM—See Bethesda, MD

WTNT(AM)—See Bethesda, MD

WTOP-FM— September 1948: 103.5 mhz; 44 kw. Ant 518 ft TL: N38 56 09 W77 05 33. Stereo. Hrs open: 24 3400 Idaho Ave. N.W., 20016. Phone: (202) 895-5000. Fax: (202) 895-5016. Web Site: www.wtopnews.com. Licensee: Bonneville Holding Co. Group owner: Bonneville International Corp. (acq 1-30-98; grpsl). Population served: 3,184,600 Natl. Network: CBS Radio. Natl. Rep: Katz Radio. Wire Svc: Reuters Format: News. Target aud: General. ♦Bruce Reese, pres; Joel Oxley, gen mgr; Matt Mills, sls dir & gen sls mgr.

WTWP(AM)— Sept 25, 1926: 1500 khz; 50 kw-U, DA-2. TL: N39 02 30 W77 02 45. Hrs open: 24 3400 Idaho Ave. N.W., 20016. Phone: (202) 895-5000. Fax: (202) 895-5016. Web Site: www.washingtonpostradio.com. Licensee: Bonneville Holding Co. Group owner: Bonneville International Corp. (acq 4-27-98; grpsl). Population served: 3,451,000 Law Firm: Wilkinson, Barker, Knauer & Quinn. Format: News, Talk. Target aud: General. ♦Bruce Reese, pres & VP; Joel Oxley, gen mgr.

WUST(AM)— 1949: 1120 khz; 20 kw-D, 3 kw-CH. TL: N38 54 15 W77 09 54. Hrs open: 2131 Crimmins Ln., Falls Church, VA, 22043. Phone: (703) 532-0400. E-mail: mail@wust1120.com Web Site: www.wust1120.com. Licensee: New World Radio Inc. (acq 10-26-92; $1.15 million; FTR: 8-24-92). Population served: 1,400,000 Format: Multicultural, ethnic. Spec prog: Fr 15 hrs, Sp 15 hrs, Ger 7 hrs, Ethiopian 6 hrs, Farsi 5 hrs, Russian 5 hrs wkly. ♦Alan Pendleton, gen mgr; Brian Edwards, opns mgr.

WWDC-FM— 1947: 101.1 mhz; 22.5 kw. 760 ft TL: N38 59 59 W77 03 27. Stereo. Hrs opn: 24 1801 Rockville Pike, Suite 405, Rockville, MD, 20852. Phone: (301) 587-7100. Fax: (301) 587-0225. Web Site: www.dc101.com. Licensee: AMFM Radio Licenses L.L.C. Group owner: Clear Channel Communications Inc. (acq 8-30-00; grpsl). Format: Rock. ♦Bennett Zier, gen mgr; Joe Bivilacua, progmg dir; Tom Shedlick, chief of engrg.

WWGB(AM)—See Indian Head, MD

WWRC(AM)— 1941: 1260 khz; 5 kw-U, DA-2. TL: N38 59 59 W77 03 27. Stereo. Hrs opn: 24 1801 Rockville Pike, Rockville, MD, 20852.

Phone: (301)231-7798. Fax: (301)881-8030. Web Site: www.wrcam1260.com. Licensee: AMFM Radio Licenses L.L.C. Group owner: Clear Channel Communications Inc. (acq 8-30-2000; grpsl). Population served: 3,535,000 Natl. Network: Westwood One. Natl. Rep: Clear Channel. Format: Progressive talk. Target aud: 25-54; adults. ♦Dave Pugh, gen mgr & mktg mgr; Bill Hess, opns mgr; Heather Steffan, gen sls mgr; Kathy Lennhoff, prom dir; Jerry Phillips, mus dir, news dir & pub affrs dir; Shaun Sandoval, chief of engrg.

WXTR(AM)—Alexandria, VA) Dec 10, 1945: 730 khz; 8 kw-D, 25 w-N. TL: N38 44 43 W77 05 58. Stereo. Hrs opn: 24 8121 Georgia Ave., Suite 1050, Silver Spring, MD, 20910. Phone: (301) 562-5800. Fax: (301) 589-9772. Web Site: www.triplexespnradio.com. Licensee: Red Zebra Broadcasting Licensee LLC. Group owner: Mega Communications Inc. (acq 5-9-2006; grpsl). Population served: 55,500 Natl. Network: ESPN Radio. Format: Sports. ♦Bruce Gilbert, CEO.

WYCB(AM)— 1978: 1340 khz; 1 kw-U. TL: N38 55 04 W77 01 27. (CP: TL: N38 51 50 W76 54 38). Hrs opn: 5900 Princess Garden Pkwy., Lanham, MD, 20706. Phone: (301) 306-1111. Fax: (301) 306-9510. Licensee: Radio One Licenses LLC. Group owner: Radio One Inc. (acq 11-8-01; grpsl). Population served: 15,000 Natl. Network: American Urban. Format: Gospel. ♦Alfred Liggins, CEO; Cathy Hughes, chmn; Alfred Liggins, pres; Scott Royster, CFO; Michele Williams, gen mgr; Karen Jackson, gen sls mgr; Tim White, chief of engrg.

WZHF(AM)—See Arlington, VA

Florida

Alachua

WNDT(FM)— 1996: 92.5 mhz; 3.2 kw. Ant 443 ft TL: N29 44 22 W82 23 09. Hrs open: 24
Rebroadcasts WNDD(FM) Silver Springs 100%.
4020 Newberry Rd., Gainesville, 32607. Phone: (352) 373-6644. Fax: (352) 375-1700. E-mail: windfm.@aol.com Web Site: www.windfm.com. Licensee: Ocala Broadcasting Corp. L.L.C. Group owner: Wooster Republican Printing Co. (acq 10-22-97; $675,000 for stock). Population served: 700,000 Natl. Rep: Katz Radio. Law Firm: Baker & Hostetler. Format: Classic Rock. Target aud: Adults; 25-54. ♦Jim Robertson, VP & gen mgr; Robert Kassi, gen sls mgr; Kevin Davis, progmg dir; Cheree Carr, traf mgr.

Altamonte Springs

WORL(AM)— 1986: 660 khz; 10 kw-D, DA. TL: N28 32 21 W80 58 26. Hrs open: 24 1188 Lake View Dr., 32714-2713. Phone: (407) 682-9494. Fax: (407) 682-7005. Web Site: www.worl660.com. Licensee: Salem Media of Illinois LLC. Group owner: James Crystal Inc. (acq 2-3-2006; swap in exchange for KNIT(AM) Dallas, TX). Law Firm: Fletcher, Heald & Hildreth. Format: News/talk. ♦David Koon, gen mgr.

Apalachicola

WFCT(FM)— November 1997: 105.5 mhz; 50 kw. 315 ft TL: N29 45 02 W84 52 18. Stereo. Hrs opn: 24 2911 Long Ave., Port St. Joe, 32456. Phone: (850) 227-9048. Fax: (850) 227-1101. E-mail: wfct@gtcom.net Licensee: Williams Communications Inc. (group owner; acq 3-27-02; $650,000). Format: Adult standards, contemp. ♦John Nichols, gen mgr; Ken Carey, news dir.

WOYS(FM)— July 1988: 100.5 mhz; 12 kw. Ant 476 ft TL: N29 43 57 W84 53 24. Stereo. Hrs opn: 24 Point Mall 35 Island Dr. #16, Eastpoint, 32328-3264. Phone: (850) 670-8450. Fax: (850) 670-8492. Web Site: www.oysterradio.com. Licensee: Richard L. Plessinger Sr. Group owner: Plessinger Radio Group (R.L. Plessinger Holding Co.). Format: Adult contemp/beach music. News staff: one. Target aud: General. ♦Richard L. Plessinger Sr., pres; Rick Plessinger, gen mgr; Michael Allen, stn mgr, prom mgr & news dir; William Denton, gen sls mgr.

Apopka

WHIM(AM)— May 4, 1964: 1520 khz; 5 kw-D, 350 w-N, DA. TL: N28 39 08 W81 29 40. Hrs open: 24 1188 Lake View Dr., Altamonte Springs, 32714. Phone: (407) 682-9494. Phone: (407) 682-9595. Fax: (407) 682-7005. E-mail: whim@salemorlando.com Web Site: www.1520whim.com. Licensee: Pennsylvania Media Associates Inc. (acq 1-23-2006; $600,000). Population served: 1,000,000 Natl. Rep: Salem. Law Firm: Holland & Knight. Format: Relg, southern gospel, talk. Target aud: 25-54; general. ♦Edward G. Atsinger III, pres; David Koon, gen mgr; Jackie Trefcer, stn mgr & gen sls mgr; Lee Brandell, opns dir.

Arcadia

WFLN(AM)— Sept 3, 1955: 1480 khz; 1 kw-D. TL: N27 13 43 W81 51 28. Hrs open: 24 201 Asbury St., 34266-8830. Phone: (863) 993-1480. Fax: (863) 499-1489. E-mail: wflnradio@aol.com Licensee: Integrity Radio of Florida LLC. Population served: 100,000 Natl. Network: CBS Radio, CNN Radio. Format: News/talk. Target aud: 25-65; upscale adults. ♦George Kalman, pres & gen mgr; Jack Welch, progmg dir; Phill Scott, engr.

Atlantic Beach

WFYV-FM— Mar 10, 1980: 104.5 mhz; 100 kw. 984 ft TL: N30 16 34 W81 33 53. Stereo. Hrs opn: 24 8000 Belfort Pkwy., Jacksonville, 32256. Phone: (904) 245-8500. Fax: (904) 245-8501. Web Site: www.rock105i.com. Licensee: Cox Radio Inc. Group owner: Cox Broadcasting (acq 2000; grpsl). Population served: 1,200,00 Natl. Network: AP Radio. Natl. Rep: Christal, Katz Radio. Format: Classic Rock. News staff: one; News: 20 hrs wkly. Target aud: 18-49; male oriented. ♦David Israel, gen mgr.

WQOP(AM)— Jan 30, 1958: 1600 khz; 5 kw-D, 90 w-N. TL: N30 19 30 W81 25 42. Hrs open: Box 51585, Jacksonville Beach, 32240. Secondary address: 391 S. 14th Ave., Jacksonville Beach 32250. Phone: (904) 241-3311. Fax: (904) 241-1402. E-mail: radioqop@aol.com Web Site: www.qopradio.com. Licensee: Queen of Peace Radio, Inc. (acq 6-5-97; $350,000). Natl. Network: USA. Format: Talk, relg. ♦C. Williams, pres; Tom Moran, gen mgr.

Auburndale

WTWB(AM)— Oct 10, 1956: 1570 khz; 5 kw-D, 13 w-N. TL: N28 04 32 W81 49 19. Hrs open: 24 127 Glenn Road, 33823. Phone: (863) 967-1570. Fax: (206) 350-6874. E-mail: wtwb@talks1570.com Web Site: www.talks1570.com. Licensee: Carpenter's Home Church Inc. LMA - Breidenbach Media Group (acq 12-5-2002). Population served: 300,000 Format: Christian newstalk , politics. News staff: one; News: 13 hrs wkly. Target aud: 30 plus; middle income, 2-income family. ♦Lynne Breidenbach, pres & gen mgr; Justin Sargent, progmg dir.

Avon Park

WFHT(AM)— Oct 1, 1970: 1390 khz; 1 kw-D, 770 w-N. TL: N27 37 08 W81 29 27. Hrs open: 6 AM-10 PM 801 Hwy. 27 S., Suite 5, 33825. Phone: (863) 453-3423. Fax: (863) 453-3423. E-mail: agentx1599@yahoo.com Licensee: Odyssey Broadcasting Co. Inc. (acq 9-15-2006; $225,000). Population served: 52,000 Natl. Network: Jones Radio Networks. Format: Oldies/Gospel. News staff: one; News: 12 hrs wkly. Target aud: 18-54; Sp audience. ♦Michael Cardillo, pres; T.J. Reno, gen mgr & chief of engrg.

WWOJ(FM)— August 1982: 99.1 mhz; 10 kw. Ant 515 ft TL: N27 30 39 W81 31 54. Stereo. Hrs opn: 24 3750 U.S. 27 N., Suite One, Sebring, 33870. Phone: (863) 382-9999. Fax: (863) 382-1982. E-mail: cohanradiogroup@htn.net Web Site: www.cohanradiogroup.com. Licensee: Cohan Radio Group Inc. (group owner; acq 11-1-98; $910,000 with WWTK(AM) Lake Placid). Population served: 80,000 Natl. Network: ABC. Rgnl. Network: Florida Radio Network. Natl. Rep: Interep. Law Firm: Latham & Watkins. Wire Svc: AP Format: Country. News staff: one; News: 7 hrs wkly. Target aud: 18 plus. Spec prog: Bluegrass 2 hrs wkly. ♦Peter L. Coughlin, pres; Rob Ellis, opns dir & progmg dir.

Baker

***WTJT(FM)**— May 1987: 90.1 mhz; 50 kw. Ant 417 ft TL: N30 49 19 W86 42 37. Hrs open: 24 957 Hwy. C-4A, 32531. Phone: (850) 537-2009. Fax: (850) 537-4663. E-mail: wtjtradio@yahoo.com Licensee: Okaloosa Public Radio Inc. Natl. Network: USA. Format: educ. Target aud: 35 plus. ♦Earl Thompson, pres & gen mgr; Jessica Walker, stn mgr & opns mgr; Ruth Thompson, mus dir; Randy Henry, chief of engrg.

Baldwin

WHJX(FM)— July 30, 1992: 105.7 mhz; 25 kw. Ant 328 ft TL: N30 22 28 W82 01 42. Stereo. Hrs opn: 9550 Regency Square Blvd., Suite 200, Jacksonville, 32225. Phone: (904) 680-1050. Fax: (904) 680-1051. Web Site: www.whjx.biz. Licensee: Tama Radio Licenses of Jacksonville, FL, Inc. Group owner: Tama Broadcasting Inc. (acq 12-2-2001; $1.5 million). Format: Urban adult contemp. ♦Linda Davis-Fructuoso, gen mgr; Joel Widdows, opns mgr.

Bartow

WQXM(AM)— Sept 28, 1953: 1460 khz; 1 kw-D, 155 w-N. TL: N27 54 34 W81 51 29. Hrs open: 6 AM-6 PM Box 452905, Miami, 33245. Phone: (305) 270-1244. Fax: (305) 270-1255. Licensee: Florida Broadcasting Media LLC (acq 5-17-2004; $325,000). Population served: 350,000 Natl. Network: CBS. Format: Traditional country. Target aud: General. ♦Armando Gutierrez, gen mgr.

WWBF(AM)— Sept 16, 1969: 1130 khz; 2.5 kw-D, 500 w-N, DA-N. TL: N27 54 34 W81 49 35. Stereo. Hrs opn: 24 1130 Radio Rd., 33830. Phone: (863) 533-0744. Fax: (863) 533-8546. E-mail: tom@wwbf.com Web Site: www.wwbf.com. Licensee: Thornburg Communications Inc. (acq 1-27-84; $220,000; FTR: 2-6-84). Population served: 414,700 Natl. Network: CNN Radio. Rgnl. Network: Florida Radio Net. Format: Oldies. News staff: one; News: 10 hrs wkly. Target aud: 35-54; affluent adults. Spec prog: Sports. ♦Jeffrey A. Thornburg, VP; Thomas N. Thornburg, pres & gen mgr; Susan E. Thornburg, stn mgr.

Belle Glade

WBGF(FM)— May 31, 1965: 93.5 mhz; 5 kw. 269 ft TL: N26 42 43 W80 40 59. Stereo. Hrs opn: Box 1505, 33430. Secondary address: 2001 State Rd. 715 33430. Phone: (561) 996-2063. Fax: (561) 996-1852. E-mail: wswnwbgf@bellsouth.net Web Site: bigdawg935.com. Licensee: BGI Broadcasting LP. Population served: 25,000 Natl. Network: ABC. Rgnl rep: Interep Format: Sports. Target aud: 25-54. Spec prog: Farm 5 hrs wkly. ♦Mike Diagostine, progmg dir.

WSWN(AM)— Oct 7, 1947: 900 khz; 1 kw-D, 26 w-N. TL: N26 42 54 W80 40 58. (CP: TL: N26 42 56 W80 40 58). Hrs opn: Box 1505, 33430. Secondary address: 2001 State Rd. 715 33430. Phone: (561) 996-2063. Fax: (561) 996-1852. E-mail: wswnwbgfa@bellsouth.net Licensee: BGI Inc. (acq 1996). Population served: 806,000 Natl. Network: ABC, Jones Radio Networks. Rgnl. Network: S.E.Agri., Florida Radio Net. Natl. Rep: Interep. Format: Relg, Gospel. Target aud: 25-54. Spec prog: Sports. ♦David Lampel, gen mgr; Harvey J. Poole Jr., progmg dir; Rick Rieke, chief of engrg.

Belleview

***WYFZ(FM)**— Apr 2001: 91.3 mhz; 200 w horiz, 1.1 kw vert. Ant 328 ft TL: N29 10 31 W82 09 09. Hrs open: 24 11530 Carmel Commons Blvd., Charlotte, NC, 28226. Fax: (704) 522-1967. Web Site: bbnradio.org. Licensee: Bible Broadcasting Network Inc. (acq 12-1-2005; $250,000). Population served: 350,000 Format: Relg. ♦Lowell L. Davey, pres.

Beverly Hills

WINV(AM)— Sept 1, 1965: Stn currently dark. 1560 khz; 5 kw-D, 4.1 kw-CH. TL: N28 50 30 W82 22 16. Stereo. Hrs opn: 4554 S. Suncoast Blvd., Homosassa, 34446. Phone: (352) 628-4444. Licensee: WGUL-FM Inc. (acq 11-26-97; $5,000). Law Firm: Michael Wilhelm. ♦Richard Spires, gen mgr.

Big Pine Key

WWUS(FM)— Sept 22, 1980: 104.1 mhz; 100 kw. 433 ft TL: N24 39 38 W81 25 10. Stereo. Hrs opn: 24 30336 Overseas Hwy., 33043. Phone: (305) 872-9100. Fax: (305) 872-8930. E-mail: us1radio@aol.com Web Site: www.us1radio.com. Licensee: Vox Communications Group LLC. (acq 8-31-2005; grpsl). Natl. Network: AP Radio. Law Firm: Katten Muchin Rosenman LLP. Format: Classic hits. News staff: one; News: 2 hrs wkly. Target aud: 30-50. Spec prog: Island mus 4 hrs wkly. ♦Kevin LeRoux, gen mgr & gen sls mgr; Race Ashlyn, stn mgr & progmg dir; Bill Becker, news dir; Randy Perry, chief of engrg; Kim Casey, traf mgr.

Bithlo

WNTF(AM)— July 31, 1974: 1580 khz; 2.1 kw-D. TL: N28 32 11 W81 05 06. Hrs open: 1801 Clarke Rd., Ocoee, 34761. Phone: (407) 523-2770. Fax: (407) 523-2888. Licensee: Rama Communications Inc. (group owner; (acq 10-29-2002; $600,000 with WGAF(AM) Alachua). Population served: 2,000,000 Format: Sp. ♦Sabita Persaud, pres; Kris Persaud, gen mgr; Steve De Lay, chief of engrg.

Blountstown

WPHK(FM)—Listing follows WYBT(AM).

WYBT(AM)— Sept 8, 1962: 1000 khz; 5 kw-D. TL: N30 27 15 W85 02 32. Hrs open: 20872 N.E. Kelley Ave., 32424. Phone: (850) 674-5101. Fax: (850) 674-2965. Licensee: Blountstown Communications (acq 6-26-86; $103,000; FTR: 4-14-86). Format: Golden oldies. Spec prog: Gospel & relg 15 hrs wkly. ♦Harry S. Hagen, pres; Cathy Hagen, progmg dir.

WPHK(FM)—Co-owned with WYBT(AM). Dec 18, 1968: 102.7 mhz; 13 kw. Ant 318 ft TL: N30 27 15 W85 02 32. Stereo. Format: Modern country. ♦Harry S. Hagen, gen mgr.

Boca Raton

WKIS(FM)—Licensed to Boca Raton. See Miami

WSBR(AM)— April 1965: 740 khz; 2.5 kw-D, 940 w-N, DA-2. TL: N26 20 06 W80 15 55. Hrs open: 6699 N. Federal Hwy., Suite 200, 33487. Phone: (561) 997-0074. Fax: (561) 997-0476. Web Site: www.wsbradio.com. Licensee: WWNN License LLC. Group owner: Beasley Broadcast Group (acq 3-14-2000; grpsl). Population served: 999,700 Format: Financial talk. ♦Bob Morency, VP & gen mgr; Greg Cooper, opns mgr.

Bonifay

WYYX(FM)— Apr 23, 1983: 97.7 mhz; 100 kw. 830 ft TL: N30 30 41 W85 29 24. Stereo. Hrs opn: 24 7106 Laird St., Suite 102, Panama City Beach, 32408. Phone: (850) 233-6606. Fax: (850) 233-1541. Web Site: www.wyyx.com. Licensee: Magic Broadcasting Florida Licensing LLC. (group owner; (acq 9-30-2002; grpsl). Format: Active rock, AOR. News staff: one. Target aud: 18-49. ♦Jim Storey, COO, exec VP, gen mgr & engrg mgr; J.P. Ferrell, gen sls mgr; Karla Melvin, traf mgr.

Bonita Springs

WRXK-FM— Sept 1, 1974: 96.1 mhz; 100 kw. 1,122 ft TL: N26 26 53 W81 48 54. Stereo. Hrs opn: 20125 S. Tamiami Tr., Estero, 33928. Phone: (239) 495-2100. Fax: (239) 992-8165. Web Site: www.96krock.com. Licensee: Beasley Broadcasting of Western Florida Inc. Group owner: Beasley Broadcast Group (acq 8-12-86). Population served: 800,000 Natl. Network: ABC. Natl. Rep: Katz Radio. Format: Classic rock. Target aud: 18-49. ♦George G. Beasley, pres; Brad Beasley, gen mgr; Shane Reilly, opns mgr; Robert Hallman, gen sls mgr.

Boynton Beach

WLVJ(AM)— Jan 23, 1973: 1040 khz; 25 kw-D, 1.2 kw-N, DA-2. TL: N26 28 26 W80 12 11. Stereo. Hrs opn: 24 6600 N. Andrews Ave., Suite 160, Fort Lauderdale, 33309. Phone: (954) 315-1515 / 1539. Fax: (954) 315-1555. Web Site: www.wlvj.com. Licensee: Communicom Co. of Florida L.P. Group owner: James Crystal Inc. (acq 12-14-2005; grpsl). Population served: 950,000 Format: Relg. ♦Rick Hindes, CFO; Steve Lapa, gen mgr.

***WRMB(FM)**— Apr 15, 1979: 89.3 mhz; 100 kw. 500 ft TL: N26 31 07 W80 10 17. Stereo. Hrs opn: 24 1511 W. Boynton Beach Blvd., 33436. Phone: (561) 737-9762. Fax: (561) 737-9899. E-mail: wrmb@moody.edu Web Site: www.wrmb.org. Licensee: Moody Bible Institute of Chicago. (group owner) Natl. Network: Moody. Format: Relg. Target aud: General. ♦Michael Easley, pres; Jennifer Epperson, stn mgr & mus dir.

WXEL(FM)—See West Palm Beach

Bradenton

***WJIS(FM)**— 1989: 88.1 mhz; 100 kw. 397 ft TL: N27 07 54 W82 23 29. Hrs open: 6469 Parkland Dr., Sarasota, 34243. Phone: (941) 753-0401. Fax: (941) 753-2963. E-mail: thejoyfm@thejoyfm.com Web Site: www.thejoyfm.com. Licensee: WJIS FM Radio. (acq 8-17-89; grpsl; FTR: 9-11-89). Format: Adult Contemp Christian music. ♦Jeff McFarlane, gen mgr & stn mgr; Carmen Brown, prom dir; Steve Swanson, progmg dir; Steve Rieker, chief of engrg.

WLLD(FM)—See Holmes Beach

WWPR(AM)— 1946: 1490 khz; 1 kw-U. TL: N27 30 00 W82 34 25. Hrs open: 24 5910 Cortez Rd. W., Suite 130, 34210. Phone: (941) 761-8843. Fax: (941) 761-8683. E-mail: manager@1490wwpr.com Web Site: www.1490wwpr.com. Licensee: Greenrose Broadcasting Services Inc. (acq 9-19-97; $265,000). Law Firm: Pepper & Corazzini. Format: Talk. News staff: one; News: 15 hrs. wkly. Target aud: 35-64; general. Spec prog: Community Talk-15 hrs, relg 6 hrs., Gospel 6 hrs., Sports -5 hrs., Hispanic programming 40 hrs.wkly. ♦Valerie Silver, gen mgr; Ed Edwards, opns mgr.

Brandon

WLCC(AM)— February 1988: 760 khz; 10 kw-D, 1 kw-N, DA-2. TL: N28 01 29 W82 17 02. Hrs open: 1915 N. Dale Mabry Hwy., Suite 200, Tampa, 33607. Phone: (813) 871-1819. Fax: (813) 871-1155. Web Site: www.laleytampa.com. Licensee: Mega Communications of Tampa Licensee L.L.C. Group owner: Mega Communications Inc. (acq 4-24-98). Format: Rgnl Mexican. Target aud: 25-54. ♦Rafael Grullon, VP & gen mgr; Ricardo Villalona, gen mgr; Everardo Lopez, progmg dir.

Brooksville

WWJB(AM)— Oct 11, 1958: 1450 khz; 1 kw-U. TL: N28 33 02 W82 25 02. Hrs open: 24 Box 1507, 34605. Secondary address: 55 W. Fort Dade Ave. 34605-1507. Phone: (352) 796-7469. Fax: (352) 796-5074. E-mail: info@wwjb.com Web Site: www.wwjb.com. Licensee: Hernando Broadcasting Co. (acq 3-1-82; FTR: 4-5-82). Population served: 150,000 Natl. Network: Westwood One, ABC. Natl. Rep: Dora-Clayton. Format: News/talk, sports. News staff: one. Target aud: 25 plus. ♦Bill Willamson, gen sls mgr; Peggy Hope, prom dir, prom mgr & traf mgr; Bob Haa, news dir; Steve Manuel, pres, gen mgr & political ed.

Bunnell

WNZF(AM)—Not on air, target date: unknown: 1550 khz; 9.5 kw-D, 520 w-N, DA-N. TL: N29 27 09 W81 17 21. Hrs open: Box 1427, Boca Grande, 33921. Phone: (239) 542-4200. Licensee: Flagler County Broadcasting LLC (acq 8-7-2007; $150,000 for CP). ♦James E. Martin, gen mgr.

Bushnell

WKFL(AM)— Jan 1, 1987: 1170 khz; 1 kw-D. TL: N28 42 31 W82 07 36. Hrs open: 6 AM-9 PM varies by season 5224 State Rt. 46, Ste. 354, Sanford, 32771. Phone: (352) 568-3204. Fax: (407) 322-0431. Web Site: www.talknsports.net. Licensee: TalknSports Inc. (acq 6-22-2004). Population served: 375,000 Natl. Network: Salem Radio Network. Format: Sports, news/talk. News staff: one; News: 8 hrs wkly. Target aud: 18-54; those who enjoy family programming. ♦Bruce Cox, pres; Jan Hall, gen mgr.

Callahan

WEWC(AM)— 1999: 1160 khz; 5 kw-D, 250 w-N, DA-D. TL: N30 34 47 W87 17 18. Hrs open: 8384 Baymeadows Rd., Suite 1, Jacksonville, 32256. Phone: (904) 549-2218. Fax: (904) 359-0070. Web Site: www.1160latinohits.com. Licensee: Circle Broadcasting of America Inc. (acq 3-24-93; $11,160; FTR: 4-12-93). Format: Sp. ♦George Lopez, gen mgr.

WROO(FM)—Licensed to Callahan. See Jacksonville

Callaway

WAKT-FM— February 1990: 103.5 mhz; 100 kw. 475 ft TL: N30 03 18 W85 18 09. (CP: Ant 748 ft. TL: N30 13 45 W85 23 20). Stereo. Hrs opn: 24 118 Gwyn Dr., Panama City Beach, 32408. Phone: (850) 234-8858. Fax: (850) 234-1181. Web Site: www.maxcountry1035.com. Licensee: Double O Radio Corp. (group owner; acq 3-10-2004; grpsl). Population served: 135,000 Natl. Rep: Christal. Format: Country. ♦Harry Finch, gen mgr.

Cantonment

WNVY(AM)—Licensed to Cantonment. See Pensacola

Cape Coral

WXKB(FM)— 1975: 103.9 mhz; 100 kw. 981 ft TL: N26 47 43 W81 48 04. Stereo. Hrs opn: 20125 S. Tamiami Tr., Estero, 33928. Phone: (239) 495-2100. Fax: (239) 948-0785. Web Site: www.b103.com. Licensee: Beasley Broadcasting. Group owner: Beasley Broadcast Group (acq 11-18-94; $3.7 million; FTR: 1-2-95). Population served: 31,000 Format: CHR. Target aud: General. ♦George G. Beasley, pres; Brad Beasley, gen mgr; Shane Reilly, opns mgr; Matt Johnson, progmg dir.

Carrabelle

WOCY(FM)— 1999: 106.5 mhz; 100 kw. Ant 361 ft TL: N29 43 57 W84 53 24. Stereo. Hrs opn: Point Mall 35 Island Dr. #16, Eastpoint, 32328-3264. Phone: (850) 670-8450. Fax: (850) 670-8492. E-mail: woyswocy@tcom.net Web Site: www.woyswocy.homestead.com. Licensee: Richard L. Plessinger Sr. Group owner: Plessinger Radio Group Rgnl. Network: Florida Radio Net. Format: Country. ♦Richard L. Plessinger Sr., pres & gen mgr; Michael Allen, stn mgr & news dir; William Denton, gen sls mgr & mus dir.

Cedar Creek

***WKSG(FM)**— 1999: 89.5 mhz; 2 kw. Ant 308 ft TL: N29 11 20 W81 52 52. (CP: 30 kw vert, ant 341 ft). Hrs opn: 7 E. Silver Springs Blvd., Suite 102, Ocala, 34471. Phone: (352) 369-8950. Fax: (352) 369-1109. E-mail: daystar@ocalapro.com Web Site: www.daystarradio.com. Licensee: Daystar Public Radio Inc. (acq 1-5-98). Format: Adult contemp Christian. ♦Gary Linkus, gen mgr.

Cedar Key

WRGO(FM)— Sept 1996: 102.7 mhz; 12.5 kw. Ant 459 ft TL: N29 11 45 W82 59 46. Hrs open: 24 1929 N.W. Hwy. 19, Crystal River, 34428. Phone: (352) 795-1027. Fax: (352) 795-0002. Licensee: WRGO Radio LLC. (group owner; (acq 7-5-2007; $900,000 with WYNY(AM) Cross City). Population served: 200,000 Natl. Network: Jones Radio Networks. Law Firm: Booth, Freret, Imlay & Tepper. Format: Oldies. Target aud: 25-64. ♦Lou Cerra, gen mgr.

Century

WPFL(FM)— July 1989: 105.1 mhz; 25 kw. 328 ft TL: N30 52 12 W87 20 05. Hrs open: Box 967, Flomaton, AL, 36441. Secondary address: 2059 Old Fannie Rd., Flomaton, AL 36441. Phone: (251) 296-1051. Fax: (251) 296-1055. Web Site: www.oldiesradioonline.com. Licensee: Tri-County Broadcasting Inc. (acq 4-10-01; $575,000 including $50,000 ad credit). Format: Oldies. ♦Walter Douglas, gen mgr, opns mgr & progmg dir; Howard Macht, chief of engrg.

Charlotte Harbor

WIKX(FM)—Licensed to Charlotte Harbor. See Punta Gorda

Chattahoochee

WTCL(AM)— Nov 1, 1963: 1580 khz; 5 kw-D. TL: N30 40 14 W84 50 08. (CP: 10 kw-D, 500 w-N, DA-N). Hrs opn: 7175 Bonnie Hill Rd., 32324. Phone: (850) 663-3857. Phone: (850) 663-3857. Fax: (850) 663-8543. Licensee: Metz Inc. (acq 1-9-97; $55,000). Natl. Network: USA. Rgnl. Network: Florida Radio Net. Format: Gospel, ministries. Target aud: General. ♦Don Metz, pres; David Garcia, gen mgr.

Chiefland

WLQH(AM)— June 6, 1968: 940 khz; 1 kw-D. TL: N29 31 00 W82 53 11. Stereo. Hrs opn: Sunrise-sunset 12750 Old Fanning Springs Rd., 32626. Phone: (352) 493-4940. Fax: (352) 493-9909. Licensee: Ocala Broadcasting Corp. (acq 11-2-99; with co-located FM). Population served: 35,000 Format: Music of your life. Spec prog: Relg 9 hrs wkly. ♦Bob Moody, stn mgr.

WNDN(FM)—Co-owned with WLQH(AM). 1991: 107.9 mhz; 6 kw. Ant 328 ft TL: N29 31 00 W82 53 11.6 AM-midnight Natl. Network: Jones Radio Networks. Format: Plastic Rock.

Chipley

WBGC(AM)— Apr 10, 1956: 1240 khz; 1 kw-U. TL: N30 46 19 W85 33 31. Hrs open: 1513 S. Blvd., 32428. Phone: (850) 638-0234. Fax: (850) 638-4333. Licensee: Jacquelyn Collier Pembroke (acq 5-28-02; with WALD(AM) Walterboro). Population served: 5,441 Rgnl. Network: Florida Radio Net. Natl. Rep: Keystone (unwired net). Format: Var. ♦Todd Burnett, gen mgr.

Clearwater

WBTP(FM)— Aug 19, 1963: 95.7 mhz; 90 kw. Ant 607 ft TL: N27 52 00 W82 37 27. Stereo. Hrs opn: 24 4002 Gandy Blvd., Tampa, 33611. Phone: (813) 832-1000. Fax: (813) 832-1090. Web Site: www.957thebeat.com. Licensee: Clear Channel Broadcasting Licenses Inc. Group owner: Clear Channel Communications Inc. (acq 10-94). Natl. Network: AP Radio. Format: Hot adult contemp/urban. News staff: one. Target aud: 18-49; upwardly mobile adults. ♦Dan DiLoreto, gen mgr; Doug Hammond, opns mgr; J.J. Paone, prom dir; Ron Shepard, progmg dir.

WHBO(AM)—See Dunedin

WTAN(AM)— June 1948: 1340 khz; 1 kw-U. TL: N27 57 49 W82 24 14. Hrs open: 24 706 N. Myrtle Ave., 33755. Phone: (727) 441-3311. Fax: (727) 441-1300. E-mail: lola@tantalk.net Web Site: www.tantalk1340.com. Licensee: Wagenvoord Advertising Group Inc. (group owner; acq 12-29-99; $100,000). Population served: 2,500,000 Format: Talk/news, Music of Your Life. News staff: 2; News: 12 hrs wkly. Target aud: 35-64. Spec prog: Big band 40 hrs wkly. ♦Dave Wagenvoord, CEO & pres; Lola Wagenvoord, gen mgr.

WXTB(FM)— Dec 1, 1967: 97.9 mhz; 100 kw. 1,345 ft TL: N28 10 56 W82 46 06. Stereo. Hrs opn: 24 4002 Gandy Blvd., Tampa, 33611. Phone: (813) 832-1000. Fax: (813) 831-9898. E-mail: doubledown @clearchannel.com Web Site: www.98rock.com. Licensee: Citicasters Licenses L.P. Group owner: Clear Channel Communications Inc. (acq 5-6-99; grpsl). Format: Rock. News staff: one; News: 2 hrs wkly. Target aud: 18-49; men. Spec prog: Pub affrs 4 hrs wkly. ♦Daniel DiLoreto, pres & gen mgr; Brad Hardin, opns dir; James "Doubledown" Howard, progmg dir & chief of engrg.

WYUU(FM)—See Tampa

Clermont

***WMYZ(FM)**— July 18, 1997: 88.7 mhz; 5.5 kw vert. Ant 384 ft TL: N28 38 56 W81 43 56. Stereo. Hrs opn: 24
Rebroadcasts WPOZ(FM) Union Park 100%.
1065 Rainer Dr., Altamonte Springs, 32714-3847. Phone: (407) 869-8000. Fax: (407) 869-0380. E-mail: zcrew@zradio.org Web Site: www.zradio.org. Licensee: Central Florida Educational Foundation Inc. (acq 9-20-2005; $1.77 million). Law Firm: Fletcher, Heald & Hildreth. Format: Contemp Christian. ♦James S. Hoge, pres & gen mgr.

WWFL(AM)— 1962: 1340 khz; 1 kw-U. TL: N28 33 06 W81 46 45. Hrs open: 24 Central Florida Investments Inc., 5601 Windover Dr., Suite 102, Orlando, 32819. Phone: (407) 351-3350. Fax: (407) 370-3524. Web Site: www.cfiradio.net. Licensee: Central Florida Investments Inc. Format: MOR. ♦David Siegal, pres.

Clewiston

WAFC(AM)— Feb 16, 1988: 590 khz; 930 w-D, 470 w-N. TL: N26 43 47 W80 54 45. Hrs open: 24 530 E. Alverdez Ave., 33440-3901. Phone: (863) 983-5900. Fax: (863) 983-6109. Web Site: www.radiofiesta.com. Licensee: Glades Media Company LLP Population served: 165,000 Natl. Rep: Interep. Law Firm: Leibowitz & Associates. Format: Rgnl Mexican. News: 10 hrs wkly. Target aud: General. ♦Jim Johnson, CFO; Robert Castellanos, CEO & gen mgr; Larry Parrish, sls dir; Francisco Sangabriel, progmg dir; Debbie Pattison, traf mgr.

WAFC-FM— July 2, 1979: 99.5 mhz; 12 kw. Ant 472 ft TL: N26 41 27 W80 47 18. Stereo. 24 Phone: (863) 902-0995. Fax: (863) 983-6109. Web Site: www.wafcfm.com.77,000 Natl. Network: Westwood One, CNN Radio. Natl. Rep: Interep. Law Firm: Leibowitz & Associates. Format: Modern country. News staff: one; News: 5 hrs wkly. Target aud: 18-49. ♦Wil Skinner, progmg dir; Will Skinner, opns mgr & progmg dir; Debbie Pattison, traf mgr.

***WJCB(FM)**—Not on air, target date: unknown: 88.5 mhz; 3 kw. Ant 292 ft TL: N26 43 46 W80 54 49. Hrs open: 1150 W. King St., Cocoa, 32922. Phone: (321) 632-1000. Fax: (321) 636-0000. Licensee: Black Media Works Inc. (group owner). Format: Urban Contemp, Gospel. ♦Ray Kassis, gen mgr.

***WPSF(FM)**—Not on air, target date: unknown: 91.5 mhz; 1 kw. Ant 410 ft TL: N26 42 35 W80 54 00. Hrs open: 3185 S. Highland Dr., Suite 13, Las Vegas, NV, 89109-1029. Phone: (702) 731-5588. Licensee: American Educational Broadcasting Inc. ♦Carl J. Auel, pres.

Cocoa

WLRQ-FM—Listing follows WMMV(AM).

***WMIE(FM)**— December 1984: 91.5 mhz; 20 kw horiz, 19 kw vert. 98 ft TL: N28 21 21 W80 44 47. Stereo. Hrs opn: 24 1150 W. King St., 32922. Phone: (321) 632-1000. Fax: (321) 636-0000. Web Site: www.wjfp.com. Licensee: National Christian Network. Population served: 450,000 Format: Modern worship. Target aud: 18-49. ♦Raymond A. Kassis, pres; Paul Esposito, gen mgr; Jim Conn, progmg dir; Jan Ferguson, chief of engrg.

WMMV(AM)— Oct 4, 1957: 1350 khz; 1 kw-U, DA-N. TL: N28 21 58 W80 45 08. Stereo. Hrs opn: 1388 S. Babcock St., Melbourne, 32901. Phone: (321) 733-1000. Fax: (321) 733-0904. Web Site: www.wmmvdm.com. Licensee: Capstar TX L.P. Group owner: Clear Channel Communications Inc. (acq 8-30-00; grpsl). Population served: 312,000 Natl. Network: ABC, Westwood One. Rgnl. Network: Florida Radio Net. Format: Adult standards, news/talk radio. ♦Barbara Latham, gen mgr.

WLRQ-FM—Co-owned with WMMV(AM). June 15, 1967: 99.3 mhz; 1.2 kw. 500 ft TL: N28 16 42 W80 42 03. (CP: 50 kw, ant 492 ft.). Stereo. Web Site: www.wlrqfm.com.312,000 Format: Lite rock.

WWBC(AM)— July 1965: 1510 khz; 1 kw-D. TL: N28 21 30 W80 42 38. (CP: COL: Rockledge, 770 khz, 1 kw-D, 480 w-N, DA-2, TL: N28 20 05 W80 46 56). Hrs opn: Sunrise-sunset 1150 W. King St., 32922. Phone: (321) 632-1000. Fax: (321) 636-0000. E-mail: paul@wmiefm.com Web Site: www.wmiefm.com. Licensee: Astro Enterprises. (acq 3-1-76). Population served: 350,000 Format: Relg/Talk. Target aud: 25 plus. ♦Ray Kassis, pres; Paul Esposito, gen mgr.

Cocoa Beach

WJRR(FM)— July 19, 1962: 101.1 mhz; 100 kw. 1,598 ft TL: N28 34 51 W81 04 32. Stereo. Hrs opn: 2500 Maitland Ctr. Pkwy., Suite 401, Maitland, 32751. Phone: (407) 916-7800. Phone: (407) 916-1011. Fax: (407) 916-7407. Web Site: www.realrock1011.com. Licensee: Clear Channel Radio Licenses Inc. Group owner: Clear Channel Communications Inc. (acq 11-21-97; grpsl). Law Firm: Wiley, Rein and Fielding. Format: Modern Rock. Target aud: 18-34; men. ♦Linda Byrd, pres & gen mgr; Chris Kampmeier, opns VP & progmg dir; Aaron Miller, gen sls mgr; Pat Lynch, natl sls mgr & progmg dir; Rick Everett, mktg dir; Josh Egolf, prom mgr.

WTIR(AM)— June 22, 1959: 1300 khz; 5 kw-D, 1 kw-N, DA-2. TL: N28 20 38 W80 46 06. Hrs open: 24 3765 N. John Young Pkwy., Orlando, 32804. Phone: (407) 296-4747. Phone: (407) 481-0551. Fax: (407) 253-2228. Licensee: Rama Communications Inc. (group owner; (acq 10-13-93; $950,000 with WOKB(AM) Winter Garden; FTR: 11-8-93). Law Firm: Cohn & Marks. Format: Sp. ♦Sabeta Persaud, pres; Joel Marquez, progmg dir.

WTKS-FM—Licensed to Cocoa Beach. See Orlando

Columbia City

WJTK(FM)— 2006: 96.5 mhz; 5 kw. Ant 359 ft TL: N30 09 20 W82 38 14. Hrs open: 229 S.W. Main Blvd., Lake City, 32025. Phone: (386) 758-9696. Fax: (386) 269-4361. Licensee: ABC Media Inc. Format: News/talk. ♦Cesta Newman, pres & gen mgr.

Coral Cove

WSRZ-FM— Mar 25, 1995: 107.9 mhz; 47 kw. Ant 508 ft TL: N27 09 03 W82 27 51. Stereo. Hrs opn: 24 1779 Independence Blvd., Sarasota, 34234. Phone: (941) 552-4800. Fax: (941) 552-4900. Web Site: www.oldies108.com. Licensee: Citicasters Licenses L.P. Group owner: Clear Channel Communications Inc. (acq 5-4-99; grpsl). Population served: 902,200 Format: Oldies. Target aud: 25-54. ♦Buddy Lee, gen mgr.

Coral Gables

WHQT(FM)— Nov 15, 1958: 105.1 mhz; 100 kw. 1,049 ft TL: N25 57 59 W80 12 33. (CP: Ant 1,007 ft.). Stereo. Hrs opn: 2741 N. 29th Ave., Hollywood, 33020. Phone: (305) 444-4404. Fax: (954) 847-3240 /(954) 584-7117. Web Site: www.hot105fm.com. Licensee: Cox Radio Inc. Group owner: Cox Broadcasting (acq 12-28-92; FTR: 1-11-93). Population served: 42,494 Natl. Rep: Christal. Law Firm: Dow, Lohnes & Albertson. Format: Urban adult contemp. Target aud: 18-49. ♦Jerry Rushin, gen mgr; Janine DuPont, prom mgr; Derrick Brown, progmg dir.

WRHC(AM)— 1963: 1550 khz; 10 kw-D, 500 k-N, DA-2. TL: N25 39 02 W80 09 36. Hrs open: 24 330 S.W. 27th Ave., Suite 207, Miami, 33135. Phone: (305) 541-3300. Fax: (305) 541-7470. E-mail: ana670@aol.com Web Site: www.wrhc.com. Licensee: WRHC Broadcasting Corp. (acq 3-23-93; FTR: 4-5-93). Population served: 1,800,000 Natl. Rep: Lotus Entravision Reps LLC. Law Firm: Leventhal, Senter & Lerman, P.L.L.C. Format: Sp, news/talk, entertainment, sports. ♦Jorge Rodriguez, pres; Ana M. Vidal Rodriguez, VP & gen mgr.

WTPS(AM)— Feb 18, 1949: 1080 khz; 50 kw-D, 20 kw-N, DA-2. TL: N25 44 53 W80 32 47. Hrs open: 24 2828 W. Flagler St., Miami, 33135. Phone: (305) 644-0800. Fax: (305) 644-0030. E-mail: kclenance@radio-one.com Licensee: Radio One Licenses LLC. Group owner: Radio One Inc. (acq 11-8-2001; grpsl). Format: Black news talk. ♦Alfred Liggins, CEO & pres; Kervin Clenance, gen mgr.

***WVUM(FM)**— May 1968: 90.5 mhz; 100 w horiz, 1.3 kw vert. 175 ft TL: N25 43 02 W80 16 48. Stereo. Hrs opn: 24 Box 248191, 33124. Phone: (305) 284-3131. Fax: (305) 284-3132. E-mail: info@wvum.org Web Site: www.wvum.org. Licensee: WVUM Inc. Format: Diversified, alternative. News staff: one; News: 3 hrs wkly. Target aud: 13-plus. Spec prog: Sports 5 hrs, Black 7 hrs, relg 6 hrs, Sp 2 hrs, oldies 5 hrs wkly. ♦Benton Galgay, gen mgr; Jay Drybourgh, progmg dir.

Crawfordville

WAKU(FM)— January 1996: 94.1 mhz; 3 kw. 459 ft TL: N30 04 34 W84 18 05. Stereo. Hrs opn: 24 Box 4105, Tallahassee, 32315. Secondary address: 3225 Harstfield Rd., Tallahassee 32303. Phone: (850) 926-8000. Fax: (850) 926-2000. E-mail: mail@wave94.com Web Site: www.wave94.com. Licensee: Altrua Investments International Corp. (acq 7-14-98; $550,000). Population served: 200,000 Natl. Network: Salem Radio Network. Law Firm: Koteen & Naftalin. Format: Contemp christian. News: 10 hrs wkly. Target aud: 35-54; Adults. ♦Mike Floyd, CEO; Doug Apple, gen mgr.

Crestview

WAAZ-FM—Listing follows WJSB(AM).

WJSB(AM)— Sept 15, 1954: 1050 khz; 5 kw-D. TL: N30 45 56 W86 35 06. (CP: 3.1 kw-D, 500 w-N, DA-N, TL: N30 46 00 W86 35 08). Hrs opn: Box 267, 32536. Secondary address: 506 W. First Ave. 32536. Phone: (850) 682-3040. Phone: (850) 682-4623. Fax: (850) 682-5232. E-mail: waazwjsb@earthlink.net Licensee: Crestview Broadcasting Co. (acq 8-11-98). Population served: 16,000 Natl. Network: CBS. Format: Country. ♦Cal Zethmayr, gen sls mgr; Claude T. Strickland, news dir & women's int ed; James T. Whitaker, pres, gen mgr & chief of engrg.

WAAZ-FM—Co-owned with WJSB(AM). July 15, 1965: 104.7 mhz; 100 kw. Ant 485 ft TL: N30 46 01 W86 35 07.
Simulcast with WJSB(AM) Crestview 100%.

Cross City

WKZY(FM)— Nov 16, 1987: 106.9 mhz; 100 kw. Ant 469 ft TL: N29 36 29 W82 51 01. Stereo. Hrs opn: 100 N.W. 76th Dr., Suite 2, Gainesville, 32607. Phone: (352) 313-3150. Fax: (352) 313-3166. Web Site: www.1069kzy.com. Licensee: 6 Johnson Road Licenses Inc. (acq 1-5-2007; grpsl). Format: 80s based adult contemp. ♦Jeanie Edwards, gen sls mgr; Alan Ritchie, natl sls mgr.

***WWLC(FM)**— 2006: 88.5 mhz; 425 w. Ant 193 ft TL: N29 39 09 W83 10 08. (CP: 100 kw, ant 338 ft. TL: N29 31 35 W83 14 17). Hrs opn: 24 Spirit Radio of North Florida Inc., 412 N.E. 16th Ave., Gainesville, 32601. Phone: (352) 372-4641. Fax: (352) 376-0575. Web Site: www.sprintradio.org. Licensee: Spirit Radio of North Florida Inc. Format: Christian. ♦Fr. Roland M. Julien, gen mgr.

WYNY(AM)— November 1985: 1240 khz; 1 kw-U. TL: N29 36 35 W83 08 03. Hrs open: 24 100 N.W. 76th Dr., Suite 2, Gainesville, 32607. Phone: (352) 463-1345. Phone: (352) 313-3150. Fax: (352) 313-9199. Licensee: WRGO Radio LLC (group owner; (acq 7-5-2007; $900,000 with WRGO(FM) Cedar Key). Format: Country. Spec prog: Relg 5 hrs wkly. ♦Benjamin Hill, gen mgr.

Crystal River

***WAQV(FM)**— 1999: 90.9 mhz; 3 kw. 331 ft TL: N29 01 52 W82 27 05. Hrs open: 24
Rebroadcasts WHIJ(FM) Ocala 100%.
408 W. University, Suite 206, Gainsville, 32601. Phone: (352) 351-8810. Fax: (352) 351-8917. Web Site: www.thejoyfm.com. Licensee: Radio Training Network Inc. (acq 10-5-01; $80,000 with WHIJ(FM) Ocala). Format: Adult contemp, christian. ♦Jeff MacFarlane, gen mgr & progmg mgr.

***WHGN(FM)**— November 1992: 91.9 mhz; 41 kw horiz, 39.3 kw vert. Ant 541 ft TL: N28 50 29 W82 30 21. (CP: 41 kw). Hrs opn: 24 5800 100th Way North, St. Petersburg, 33708. Secondary address: P.O. Box 8889, St. Petersburg 33738. Phone: (727) 391-9994. Fax: (727) 397-6425. Licensee: The Moody Bible Institute of Chicago (group owner; acq 4-11-03; $500,000). Format: Christian. ♦David Boyer, gen mgr; Mike Gleichman, opns mgr; Pierre Chestang, stn mgr & progmg dir; John Stortz, chief of engrg.

WKTK(FM)— Feb 13, 1976: 98.5 mhz; 100 kw. 1,332 ft TL: N29 15 32 W82 34 03. (CP: 44 kw). Stereo. Hrs opn: 24 3600 N.W. 43rd St., Suite B, Gainesville, 32606-8127. Phone: (352) 377-0985. Fax: (352) 377-1884. E-mail: mleopord@entercom.com Web Site: www.ktk985.com. Licensee: Entercom Gainesville License LLC. Group owner: Entercom Communications Corp. (acq 11-13-86; $3.6 million; FTR: 7-21-86). Population served: 748,000 Law Firm: Leventhal, Senter & Lerman. Format: Adult contemp. News staff: one; News: 6 hrs wkly. Target aud: 25-54. ♦David Field, CEO & pres; Joseph Field, chmn; Mark Leopord, VP & gen mgr; Bruce Cherry, progmg dir.

WXCV(FM)—See Homosassa Springs

Cypress Gardens

WHNR(AM)—Licensed to Cypress Gardens. See Winter Haven

Cypress Quarters

***WREH(FM)**— 2004: 90.5 mhz; 100 kw horiz, 91.7 kw vert. Ant 249 ft TL: N27 20 51 W80 57 04. Hrs open: Reach Communications Inc., 2701 W. Cypress Creek Rd., Fort Lauderdale, 33309. Phone: (954) 315-4315. Fax: (954) 315-4231. Web Site: www.reachfm.org. Licensee: Reach Communications Inc. (acq 4-2-2003; $1 million for CP). Format: Christian. ♦Carl Mims, gen mgr; John Boone, progmg dir & mus dir.

Dade City

WDCF(AM)— December 1954: 1350 khz; 1 kw-D, 500 w-N, DA-N. TL: N28 20 04 W82 11 23. Hrs open: 2360 N.E. Coachman Rd., Clearwater, 33765. Phone: (727) 441-3311. Fax: (727) 441-1300. E-mail: lola@tantalk1340.com Web Site: www.tantalk1340.com. Licensee: Wagenvoord Advertising Group Inc. (group owner; acq 2-13-02). Population served: 35,000 Natl. Network: ABC. Rgnl. Network: Florida Radio Net. Format: Sp relg. Target aud: 25 plus; basic country demographics. ♦Dave Wagenvoord, pres; Lola Wagenvoord, sr VP & gen mgr.

WTMP-FM— Sept 3, 1993: 96.1 mhz; 2.8 kw. 413 ft TL: N28 28 22 W82 17 45. (CP: 2.75 kw, ant 485 ft.). Hrs opn: 24 5207 Washington Blvd., Tampa, 33619. Phone: (813) 620-1300. Fax: (813) 628-0713. E-mail: info@tamabroadcasting.com Web Site: www.wtmp.com. Licensee: Tama Radio Licenses of Tampa, FL, Inc. Group owner: Tama Broadcasting Inc. (acq 12-21-2001; $4.1 million). Format: Rhythm and blues, class song. Target aud: 35 plus; general. Spec prog: Gospel 20 hrs wkly. ♦Dr. Glenn W. Cherry, CEO & gen mgr; Louis Muhammad, opns mgr; Nicole Gates, prom dir; Lynn Tolliver, progmg dir.

Davie

WAVS(AM)— Aug 21, 1970: 1170 khz; 5 kw-D, 250 w-N, DA-N. TL: N26 04 39 W80 13 03. Hrs opn: 24 6360 S.W. 41st Pl., 33314. Phone: (954) 584-1170. Fax: (954) 581-6441. E-mail: info@wavs1170.com Web Site: www.wavs1170.com. Licensee: Alliance Broadcasting Inc. (acq 7-28-2004; $2 million). Population served: 4,500,000 Law Firm: Koerner & Olender PC. Format: Caribbean. News staff: one; News: 5 hrs wkly. Target aud: General; West Indians/Dade, Broward, Palm Beach Counties, Bahamas. Spec prog: Black. ♦Emmanuel Cherubin, pres; Jean Cherubin, gen mgr; Dean Hooper, stn mgr.

Daytona Beach

WCFB(FM)— March 1947: 94.5 mhz; 100 kw. 1,469 ft TL: N28 58 55 W81 27 18. Stereo. Hrs opn: 24 4192 John Young Pkwy., Orlando, 32804. Phone: (407) 422-9696. Fax: (407) 422-5883. Fax: (407) 422-6538. Web Site: www.star94fm.com. Licensee: Cox Radio Inc. Group owner: Cox Broadcasting (acq 3-28-97; grpsl). Population served: 2,200,000 Format: Adult contemp, urban contemp. ♦Brian Elam, gen mgr; Steve Holbrook, opns dir.

WELE(AM)—See Ormond Beach

WJHM(FM)— Nov 1, 1967: 101.9 mhz; 28 kw. 1,584 ft TL: N28 55 16 W81 19 09. (CP: 61 kw). Stereo. Hrs opn: 1800 Pembrook Dr., Suite 400, Orlando, 32810. Phone: (407) 919-1000. Fax: (407) 919-1190. Web Site: www.102jamzorlando.com. Licensee: Infinity Radio Inc. Group owner: Infinity Broadcasting Corp. (acq 8-7-00; grpsl). Natl. Network: AP Radio. Format: Rhythmic CHR. Target aud: 18-34. ♦Earnest James, sr VP, gen mgr & mktg mgr; James Black, gen sls mgr; Dawn Campbell, prom dir; Stevie DeMann, progmg dir.

WMFJ(AM)— Apr 16, 1935: 1450 khz; 1 kw-U. TL: N29 13 30 W81 01 30. Hrs open: 24 4295 Ridgewood Ave., Port Orange, 32127. Phone: (386) 756-9000. Fax: (386) 760-7107. E-mail: thecornerstone @cornerstoneministry.org Web Site: www.cornerstoneministry.org. Licensee: Cornerstone Broadcasting Corp. (acq 1996; $225,000). Population served: 400,000 Natl. Network: Moody, USA. Format: Relg. Target aud: General. ♦William Powell, gen mgr.

WNDB(AM)— April 1948: 1150 khz; 1 kw-U, DA-N. TL: N29 14 06 W81 04 19. Hrs open: 24 126 W. International Speedway Blvd., 32174. Phone: (386) 255-9300 / (386) 257-1150. Fax: (386) 238-6071. Web Site: www.wndb.am. Licensee: Black Crow LLC. Group owner: Black Crow Media Group LLC (acq 9-21-2001; grpsl). Population served: 325,000 Natl. Network: CBS, Motor Racing Net. Law Firm: Dow, Lohnes & Albertson. Wire Svc: UPI Format: News/talk, sports. News staff: 2. Target aud: 25-64; general. Spec prog: Relg 5 hrs, NASCAR auto racing wkly. ♦J. Michael Linn, pres; Stacey Knerler, gen mgr; Frank Scott, opns mgr.

WNUE-FM—Titusville, September 1968: 98.1 mhz; 100 kw. 462 ft TL: N28 50 54 W80 51 44. Stereo. Hrs opn: 24 337 S North Lake Blvd., Ste. 1024, Altamonte Springs, 32701. Phone: (407) 331-1777. Fax: (407) 830-6223. E-mail: jstein@megastations.net Web Site: www.mega981.com. Licensee: Mega Communications of Daytona Beach Licensee LLC. Group owner: Mega Communications Inc. (acq 8-7-00; $15 million). Population served: 402800 Natl. Rep: SBS/Interep. Format: Sp contemp. News staff: one; News: 2 hrs wkly. Target aud: 25-54; Hispanic adults 25-54. ♦George Lindemann, CEO; Adam Lindemann, chmn & pres; Eran Schreiber, CFO; Jeff Stein, gen mgr & sls VP; Rafael Grullon, exec VP & gen mgr.

WPUL(AM)—See South Daytona

WROD(AM)— 1947: 1340 khz; 1 kw-U. TL: N29 11 19 W81 00 28. Hrs open: 24 Box 211340, South Daytona, 32121-1340. Secondary address: 2400 S. Ridgewood Ave., Suite 51, South Daytona 32119. Phone: (386) 253-0000. Fax: (386) 255-3178. E-mail: production@wrod.biz Web Site: www.wrod.net. Licensee: Gore-Overgaard Broadcasting Inc. (group owner; acq 1-8-99; $1.01 million). Population served: 450,000 Natl. Network: ABC. Law Firm: Haley, Bader & Potts. Format: Mus of Your Life, adult standards. News staff: one; News: 15 hrs wkly. Target aud: 50 plus; senior community. ♦Hal Gore, CEO & chmn; Cordell Overgaard, pres; Tony Welch, gen mgr.

De Funiak Springs

***WAKJ(FM)**— January 1996: 91.3 mhz; 1.2 kw. Ant 226 ft TL: N30 41 05 W86 08 28. Stereo. Hrs opn: 24 Box 127, 32435. Secondary address: 295 Hwy. 90 W. 32435. Phone: (850) 892-2107. Fax: (850) 892-2507. E-mail: wakj913@earthlink.net Web Site: www.wakj.org. Licensee: First Baptist Church Inc. Population served: 25,000 Format: Christian. ♦Zane Welch, gen mgr & opns mgr.

WDSP(AM)— Mar 1, 1956: 1280 khz; 5 kw-D, 46 w-N. TL: N30 42 41 W86 06 25. (CP: 9 kw-D, 46 w-N, DA-D). Hrs opn: 24 Box 459, 32435. Phone: (850) 951-1280. Fax: (850) 951-1282. E-mail: info@wdsp1280.com Web Site: www.wdsp1280.com. Licensee: The Sportzmax Inc. (acq 1-30-2006; $325,000). Population served: 152,200 Natl. Network: ABC. Format: Classic hit country. Target aud: 25-54. Spec prog: High School Sports. ♦Max Howell, VP; Arty Goodman, gen mgr; Stephen C. Riggs III, pres & progmg dir; Carolyn Mora, sls; Joshua Smith, sls.

WMXZ(FM)— November 1974: 103.1 mhz; 50 kw. 482 ft TL: N30 30 53 W86 13 12. Stereo. Hrs opn: 24 743 Hwy. 98 E., Suite 6, Destin, 32541. Phone: (850) 654-1031. Fax: (850) 654-6510. Web Site: mix1031online.com. Licensee: Qantum of Fort Walton Beach License Co. LLC. Group owner: Qantum Communications Corp. (acq 7-2-2003; grpsl). Population served: 170,000 Natl. Rep: Katz Radio. Law Firm: Gravey, Schubert & Barer. Format: Hot adult contemp. News staff: one; News: 20 hrs wkly. Target aud: 18-49; general. ♦Jerry Stevens, gen mgr; Don Goodrum, opns mgr; Suzy Nicholson-Hunt, sls dir & natl sls mgr; Matt Stone, progmg dir; D.K. Landers, news dir; Curtis Blount, chief of engrg; Terri Kamphaus, traf mgr.

WZEP(AM)— October 1955: 1460 khz; 10 kw-D, 186 w-N. TL: N30 43 45 W86 07 04. Hrs open: 24 449 N. 12th St., 32433. Secondary address: Box 627 32435-0627. Phone: (850) 892-3158. Phone: (800) 881-1460. Fax: (850) 892-9675. E-mail: wzep@wzep1460.com Web Site: www.wzep1460.com. Licensee: Walton County Broadcasting Inc. (acq 6-1-93; $60,000; FTR: 6-21-93). Population served: 58,000 Natl. Network: CBS. Rgnl. Network: Florida Radio Net. Law Firm: Timothy K. Brady. Format: Full service, news/talk, country, oldies. News staff: one; News: 47 hrs wkly. Target aud: General; residents & visitors to Walton & Holmes counties. Spec prog: Gospel 13 hrs wkly. ♦Arthur F. Dees, pres & gen mgr; Martha K. Dees, VP; Marty Dees, stn mgr; Kevin Chilcutt, news dir; Tina Martin, traf mgr.

De Land

WOCL(FM)— July 10, 1967: 105.9 mhz; 96 kw. 1,581 ft TL: N28 55 16 W81 19 09. Stereo. Hrs opn: 24 1800 Pembrook Dr., Suite 400, Orlando, 32810. Phone: (407) 919-1000. Fax: (407) 919-1190. Web Site: www.orock1059.com. Licensee: Infinity Radio Inc. Group owner: Infinity Broadcasting Corp. (acq 8-7-00; grpsl). Population served: 300,000 Natl. Rep: Christal. Law Firm: Leibowitz & Associates. Format: Rock alternative. News staff: one; News: 20 hrs wkly. Target aud: 25-54. ♦Earnest James, sr VP; Evelyn Pacheo, gen mgr & gen sls mgr; April Reynolds, prom dir & disc jockey; Bobby Smith, progmg dir & disc jockey.

WTJV(AM)— Sept 10, 1948: 1490 khz; 1 kw-U. TL: N29 00 58 W81 17 10. Hrs open: 126 W. International Speedway Blvd., Daytona Beach, 32114. Phone: (386) 255-9300. Fax: (386) 239-0966. Licensee: J&V Communications Inc. Group owner: Black Crow Media Group LLC (acq 12-7-2005; $370,000). Population served: 80,000 Rgnl. Network: Florida Radio Net. Format: News/talk, community affrs, sports. Target aud: 35 plus. ♦Stacey Knerler, gen mgr; Frank Scott, stn mgr & opns mgr.

WYND(AM)— Dec 7, 1956: 1310 khz; 5 kw-D, 95 w-N. TL: N28 59 57 W81 17 55. Hrs open: 24 316 E. Taylor Rd., 32724. Phone: (386) 734-1310. Fax: (386) 734-8885. Licensee: Buddy Tucker Association Inc. (acq 12-30-86; $255,000; FTR: 12-1-86). Population served:

760,000 Natl. Network: USA. Format: Christian, news/talk. News staff: one; News: 45 hrs wkly. Target aud: 25-55. ♦Buddy Tucker, gen mgr; Art Taylor, chief of engrg.

Delray Beach

WDJA(AM)— February 1952: 1420 khz; 5 kw-D, 500 w-N, DA-2. TL: N26 27 22 W80 05 58. Hrs open: 24 2710 W. Atlantic Ave., 33445. Phone: (305) 794-9639. Phone: (561) 278-1420. Web Site: www.wdja.com. Licensee: Professional Broadcasting LLC Group owner: James Crystal Inc. (acq 5-18-2007; $2.1 million). Population served: 19,366 Natl. Network: CBS. Format: Business talk. Target aud: 35 plus. ♦Jim Davis, gen mgr.

Destin

WFFY(FM)— Sept 24, 1981: 92.1 mhz; 25 kw. 279 ft TL: N30 31 06 W86 28 01. Stereo. Hrs opn: 743 Hwy. 98, Suite 6, 32541. Phone: (850) 654-1031. Fax: (850) 654-6510. Web Site: www.fly921online.com. Licensee: Quantum of Ft. Walton Beach License Company, LLC Natl. Rep: Roslin. Format: Rhythmic CHR. ♦Farid Suleman, chmn; Mike Knar, gen mgr.

WNWF(AM)— 2000: 1120 khz; 1 kw-D. TL: N30 30 34 W86 28 34. Hrs open: 6 AM-6 PM Box 1120, 32540. Secondary address: 415 Mountain Dr., Suite 7 32541. Phone: (850) 654-4040. Fax: (850) 650-9440. E-mail: dale@fox1120.com Web Site: www.fox1120.com. Licensee: Flagship Communications Inc. (acq 8-28-03; $400,000). Population served: 153,000 Format: News/talk. News staff: one. Target aud: 35-64. ♦Dale Riddick, gen mgr & progmg dir.

Dogwood Lakes Estate

***WJED(FM)**— Jan 15, 1992: 91.1 mhz; 700 w. 180 ft TL: N30 51 34 W85 47 45. Hrs open: Box 1944, Dotham, AL, 36302. Fax: (334) 793-4344. E-mail: wjed911fm@bethanybc.edu Licensee: Bethany Bible College & Bethany Theological Seminary Inc. Natl. Network: USA. Format: Educ, relg, gospel. Target aud: General; college students & relg community. ♦Dr. H.D. Shuemake, CEO; Dr. Steve Shuemake, stn mgr; Sylvia Green, opns mgr.

Dunedin

WGUL(AM)—Licensed to Dunedin. See Tampa

WHBO(AM)— 1955: 1470 khz; 5 kw-D, 500 w-N. TL: N28 03 24 W82 44 16. Hrs open: 4300 W. Cypress St., # 1040, Tampa, 33607. Phone: (813) 281-1040. Fax: (813) 281-1948. Web Site: www.wlvu1470.com. Licensee: Genesis Communications of Tampa Bay Inc. Group owner: Genesis Communications Inc. (acq 3-5-2001; $2 million). Natl. Network: ESPN Radio. Natl. Rep: D & R Radio. Format: Sports. Spec prog: Relg 2 hrs wkly. ♦Bruce Maduri, CEO; Jeff Lebhar, gen mgr; Jeff Taylor, opns dir; Carol Azaravich, progmg dir; Jerry Smith, chief of engrg; Kimo Gray, traf mgr.

Dunnellon

WTRS(FM)— Mar 11, 1969: 102.3 mhz; 3 kw. 300 ft TL: N29 11 16 W82 23 39. (CP: 50 kw, ant 489 ft.). Stereo. Hrs opn: 24 3357 S.W. 7th St., Ocala, 34474. Phone: (352) 732-9877. Fax: (352) 622-6675. E-mail: production@twtrs.fm Web Site: www.wtrs.fm. Licensee: Asterisk Communications Inc. Format: Country. ♦Dean Johnson, gen mgr & pub affrs dir.

Eatonville

WRLZ(AM)—Licensed to Eatonville. See Orlando

Eau Gallie

WBVD(FM)—See Melbourne

WINT(AM)—See Melbourne

WMEL(AM)—See Melbourne

WMMB(AM)—See Melbourne

Ebro

WBPC(FM)— July 2005: 95.1 mhz; 25 kw. Ant 285 ft TL: N30 34 06 W85 48 28. Stereo. Hrs opn: 24 Box 27272, Panama City Beach, 32411. Phone: (850) 235-2195. Phone: (850) 235-2795. Fax: (850) 235-2795. Web Site: www.b951fm.com. Licensee: Bay Broadcasting LLC. Natl. Network: Fox News Radio. Format: Adult contemp. Target aud: 35-64; adults. ♦Charles Shapiro, pres; Bob Ghetti, gen mgr; Bob DeCarlo, progmg dir.

Edgewater

WKRO-FM— 1993: 93.1 mhz; 14.9 kw. 427 ft TL: N28 54 52 W80 53 48. Hrs open: 24 126 W. International Speedway Blvd., Daytona Beach, 32114. Phone: (386) 255-9300. Fax: (386) 238-6071. Web Site: www.wkro.fm. Licensee: Black Crow LLC. Group owner: Black Crow Media Group LLC (acq 9-21-2001; grpsl). Population served: 750,000 Format: Country. ♦Stacey Knerler, gen mgr.

***WKTO(FM)**— November 1997: 88.7 mhz; 5 kw vert. Ant 298 ft TL: N29 02 29 W81 03 23. (CP: 88.9 mhz; 25 kw, ant 354 ft TL: N29 01 37 W81 07 18). Hrs opn: 24 900 Old Mission Rd., New Smyrna Beach, 32168. Phone: (386) 427-1095. Fax: (386) 427-8970. Web Site: www.wkto.net. Licensee: Mims Community Radio Inc. Format: Relg. Target aud: 18-50. Spec prog: Jazz 4 hrs, Pol 1.5 hrs, Ger 1.5 hrs, Sp 2 hr wkly. ♦Carol Henry, CEO, chmn, pres, CFO & gen mgr.

Egypt Lake

WTMP(AM)—Licensed to Egypt Lake. See Tampa

Emeralda

***WGTT(FM)**—Not on air, target date: unknown: 91.5 mhz; 700 w. Ant 180 ft TL: N28 56 52 W81 47 45. Hrs open: 1441 Lavender St., Deltona, 32725. Licensee: Sunbelt Educational Broadcasting Inc. ♦Raul Ortiz, pres.

Englewood

WENG(AM)— Nov 15, 1964: 1530 khz; 1 kw-D. TL: N26 58 15 W82 19 24. Hrs open: 24 Box 2908, 34295-2908. Secondary address: 1355 S. River Rd. 34223. Phone: (941) 474-3231. Fax: (941) 475-2205. E-mail: kenb@1530weng.com Web Site: www.1530weng.com. Licensee: Viper Communications Inc. Group owner: Viper Communications Broadcast Group (acq 10-21-02). Population served: 217,300 Natl. Network: ABC. Rgnl. Network: Florida Radio Net. Law Firm: Pepper & Corazzini. Format: News/talk, listener participation. News staff: one; News: 50 hrs wkly. Target aud: 18 plus; securely established, financially independent. Spec prog: Gospel 2 hrs wkly. ♦Kenneth W. Kuenzie, pres; Dennis Klautzer, exec VP; Kenneth A. Birdsong, gen mgr; Scott Holcomb, opns dir & opns mgr.

***WSEB(FM)**— May 1989: 91.3 mhz; 62 kw horiz, 60 kw vert. 282 ft TL: N26 51 48 W87 17 54. Stereo. Hrs opn: 24 135 W. Dearborne St., 34223. Phone: (941) 475-9732. Fax: (941) 473-7308. E-mail: comments@wsebfm.com Web Site: www.wsebfm.net. Licensee: Suncoast Educational Broadcasting Corp. Format: Christian. Target aud: 35 plus; Christian families. ♦Dr. Kenneth Lindow, pres; Joy Clark, gen mgr; Roger Johnson, progmg dir & mus dir.

WTZB(FM)— Apr 5, 1999: 105.9 mhz; 4.3 kw. 394 ft TL: N27 06 01 W82 22 18. Hrs open: 1779 Independence Blvd., Sarasota, 34236. Phone: (941) 552-4800. Fax: (941) 552-4900. Web Site: www.1059thebuzz.com. Licensee: Citicasters Licenses L.P. Group owner: Clear Channel Communications Inc. (acq 5-4-99; grpsl). Format: Alternative rock. ♦Buddy Lee, gen mgr; Ron White, opns mgr.

Estero

WUSV(FM)— Dec 16, 1978: 92.5 mhz; 25 kw. Ant 620 ft TL: N26 19 00 W81 47 13. Stereo. Hrs opn: 24 2824 Palm Beach Blvd., Ft. Myers, 33916. Phone: (239) 337-2346. Fax: (239) 332-0767 / (239) 479-5553. E-mail: randy.marsh@us985.com Web Site: us985.com. Licensee: Meridian Broadcasting Inc. (group owner; (acq 9-14-2000; $7 million). Natl. Rep: McGavren Guild. Law Firm: Leibowitz & Associates. Wire Svc: AP Format: Country. Target aud: 25-44. ♦Joseph C. Schwartzel, chmn, pres & gen mgr; Jim Schwartzel, exec VP & sls dir; Lance Hale, progmg dir; Keith Stuhlman, engrg dir; Randy Marsh, progmg.

Eustis

WKIQ(AM)— June 1955: 1240 khz; 790 w-U. TL: N28 50 19 W81 41 46. Hrs open: Rama Communications Inc., 3765 N. John Young Pkwy., Orlando, 32804. Phone: (407) 523-2770. Phone: (407) 291-1395. Fax: (407) 523-2888. Fax: (407) 293-2870. Licensee: Rama Communications Inc. (group owner; (acq 10-15-2004; $180,000 with WQBQ(AM) Leesburg). Rgnl. Network: Florida Radio Net. Format: Gospel; R & B. ♦Sabeta Persaud, pres; Steve January, gen mgr.

WLBE(AM)—See Leesburg

Fernandina Beach

***WJBC-FM**— Oct 6, 1985: 91.7 mhz; 32 kw. Ant 223 ft TL: N30 37 23 W81 31 49. Stereo. Hrs opn: 5634 Normandy Blvd., Jacksonville, 32205-6249. Phone: (904) 781-4321. E-mail: info@wjbcfm.com Web Site: www.wjbcfm.com. Licensee: West Jacksonville Baptist Church Inc. (acq 8-31-2006; $1 million). Population served: 500,000 Format: Southern gospel. Target aud: General. ♦Rodney Kelley, pres.

WJSJ(FM)— 2000: 105.3 mhz; 3.9 kw. Ant 410 ft TL: N30 30 04 W81 35 14. Hrs open: 9550 Regency Sq. Blvd., Suite 200, Jacksonville, 32225. Phone: (904) 680-1050. Fax: (904) 680-1051. Web Site: www.smoothjazz.com. Licensee: Tama Radio Licenses of Jacksonville, FL, Inc. Group owner: Tama Broadcasting Inc. (acq 2-28-2003; $8.5 million with WSJF(FM) Saint Augustine Beach). Format: Jazz. ♦Linda Davis-Fructuoso, gen mgr; Joel Widdows, opns mgr; George Sample, gen sls mgr.

WVOJ(AM)— 1955: 1570 khz; 10 kw-D, 30 w-N. TL: N30 40 33 W81 27 35. Hrs open: 24 8384 Baymeadows Rd., Suite 1, Jacksonville, 32256. Phone: (904) 739-3660. Phone: (904) 743-1234. Fax: (904) 739-9409. Licensee: Norsan Consulting and Management Inc. (acq 9-14-2005; $2.1 million with WNNR(AM) Jacksonville). Rgnl. Network: Florida Radio Net. Format: Sp CHR. ♦Norberto Sanchez, pres; Bernie Daigle, gen mgr.

Five Points

WCJX(FM)— 1996: 106.5 mhz; 4.2 kw. 328 ft TL: N30 14 40 W82 40 11. Stereo. Hrs opn: 24 1305 Helvenston St., Live Oak, 32064. Secondary address: 5348 NW US Highway 41, Lake City 32055. Phone: (386) 755-9259. Fax: (386) 755-1557. Web Site: www.wcjx.com. Licensee: RTG Radio LLC. Group owner: Black Crow Media Group LLC (acq 11-9-2001; grpsl). Population served: 55,000 Natl. Network: ABC, Premiere Radio Networks. Format: Classic rock. ♦Dean Blackwell, gen mgr; Steve Johnson, stn mgr.

Flagler Beach

***WJLH(FM)**— Aug 23, 1996: 90.3 mhz; 2 kw vert. 184 ft TL: N29 22 18 W81 10 45. Hrs open:
Rebroadcasts WJLU(FM) New Smyrna Beach 100%.
4295 Ridgewood Ave., Port Orange, 32127. Phone: (386) 756-9094. Fax: (386) 760-7107. E-mail: thecornerstone@cornerstoneministry.org Web Site: www.cornerstoneministry.org. Licensee: Cornerstone Broadcasting Corp. (acq 3-20-97; $27,044). Format: Christian. ♦William Powell, gen mgr & progmg dir; Sandra Leisner, pub affrs dir.

Florida City

***WMFL(FM)**— Oct 1, 1998: 88.5 mhz; 8 kw vert. Ant 134 ft TL: N25 05 50 W80 26 12. Hrs open: 24 Family Stations Inc., 290 Hegenberger Rd., Oakland, CA, 94621. Phone: (510) 568-6200. Fax: (510) 568-6190. Web Site: www.familyradio.com. Licensee: Family Stations Inc. (group owner; acq 11-15-00; $75,000). Natl. Network: Family Radio. Format: Christian, relg. ♦Harold Camping, pres; Stanley Jackson, gen mgr; Rob Robbins, opns mgr.

Fort Lauderdale

***WAFG(FM)**— 1974: 90.3 mhz; 3 kw. 280 ft TL: N26 11 48 W80 06 45. Stereo. Hrs opn: 24 5555 N. Federal Hwy., 33308. Phone: (954) 776-7705. Fax: (954) 771-2633. E-mail: wafg@wafg.org Web Site: www.wafg.org. Licensee: Westminster Academy. Population served: 1,700,000 Natl. Network: USA, Salem Radio Network. Law Firm: Gammon & Grange. Format: Christian, news/talk, relg. News: 10 hrs wkly. Target aud: 30 plus; general. ♦Dolores King-St.George, gen mgr & dev dir; Kyle Kirkman, opns mgr; Lesley Hurst, progmg dir; Ken Vaughn, mus dir.

WBGG-FM— July 1960: 105.9 mhz; 100 kw. Ant 1,030 ft TL: N25 59 34 W80 10 27. Stereo. Hrs opn: 7601 Riviera Blvd., Miramar, 33023. Phone: (954) 862-2000. Fax: (954) 862-4012. Web Site: www.big1059.com. Licensee: Clear Channel Radio Licenses Inc. Group owner: Clear Channel Communications Inc. (acq 2-24-94; $14 million; FTR: 3-14-94). Format: Classic rock. ♦Todd Winick, gen mgr & gen sls mgr.

WEXY(AM)—See Wilton Manors

WFLL(AM)— Sept 16, 1946: 1400 khz; 1 kw-U. TL: N26 09 13 W80 10 11. Hrs open: 24 6600 N. Andrews Ave., Suite 160, 33309. Phone: (954) 315-1515. Fax: (954) 315-1555. Web Site: www.1400espn.com. Licensee: James Crystal Licenses L.L.C. Group owner: James Crystal Inc. (acq 6-17-98; grpsl). Population served: 1,336,632 Format: Sports. Target aud: 25 plus. ♦Steve Lapa, gen mgr.

WHSR(AM)—See Pompano Beach

WHYI-FM— July 31, 1960: 100.7 mhz; 100 kw. 928 ft TL: N25 59 34 W80 10 27. (CP: Ant 1,007 ft. TL: N25 57 59 W80 12 33). Hrs opn: 7601 Riviera Blvd., Miramar, 33023. Phone: (954) 862-2000. Fax: (954) 862-4012. Web Site: www.y100.7miami.com. Licensee: Clear Channel Radio Licenses Inc. Group owner: Clear Channel Communications Inc. (acq 11-94; grpsl). Population served: 500,000 Natl. Rep: McGavren Guild. Format: Top-40. ♦David D'Dugenio, gen mgr & gen sls mgr.

WMIB(FM)— Oct 17, 1959: 103.5 mhz; 100 kw. 1,007 ft TL: N25 57 59 W80 12 33. Stereo. Hrs opn: 7601 Riviera Blvd., Miramar, 33023. Phone: (954) 862-2000. Fax: (954) 862-4012. Web Site: www.thebeatmiami.com. Licensee: Clear Channel Broadcasting Licenses Inc. Group owner: Clear Channel Communications Inc. (acq 11-21-97; grpsl). Population served: 3,000,000 Format: Hip hop, rhythm and blues. ♦Kevin Hemmings, gen mgr & gen sls mgr.

WMXJ(FM)—See Pompano Beach

WRMA(FM)— Aug 15, 1962: 106.7 mhz; 100 kw. 984 ft TL: N25 59 34 W80 10 27. Stereo. Hrs opn: 24 1001 Ponce DeLeon Blvd., Coral Gables, 33134. Phone: (305) 444-9292. Fax: (305) 461-4466. Web Site: www.romance106fm.com. Licensee: WRMA Licensing Inc. Group owner: Spanish Broadcasting System Inc. (acq 7-11-97; $110 million with WXDJ(FM) North Miami Beach). Population served: 345,730 Natl. Rep: D & R Radio. Format: Pop latin ballads. News staff: one. Target aud: 18-54; Hispanic adults. ♦Raoul Alarcon, pres; Albert Rodriguez, gen mgr & gen sls mgr; Jackie Nosti-Cambo, gen mgr; John Caride, prom dir; Tony Campos, progmg dir; Tomas Regalado, news dir; Ralph Chambers, chief of engrg; Yoli Machado, pub affrs dir & traf mgr.

WSRF(AM)— 1955: 1580 khz; 10 kw-D, 5 kw-N, DA-2. TL: N26 04 54 W80 13 34. Hrs open: 24 1510 N.E. 162 St., Miami, 33162-4716. Phone: (305) 944-8383. Licensee: Niche Radio Inc. Group owner: Inner City Broadcasting (acq 10-14-2005; $1.75 million). Population served: 2,000,000 Format: Ethnic. News staff: 2. ♦Emmanuel Cherubin, pres & gen mgr.

WWNN(AM)—See Pompano Beach

Fort Meade

WWRZ(FM)— Mar 7, 1977: 98.3 mhz; 26 kw. 686 ft TL: N27 38 38 W81 48 00. Stereo. Hrs opn: 24 404 W. Lime St., Lakeland, 33815-4651. Phone: (863) 682-8184. Fax: (863) 683-2409. E-mail: mjames@halllakeland.com Web Site: www.max983fm.com. Licensee: Hall Communications Inc. (group owner; acq 10-1-96; $1,750,000). Population served: 500,000 Natl. Rep: D & R Radio. Law Firm: Fletcher, Heald & Hildreth. Wire Svc: AP Format: Adult hits. News staff: 2; News: 2 hrs wkly. Target aud: 25-54; women. ♦Arthur L. Rowbotham, pres; William S. Baldwin, sr VP; Nancy Lattarius, stn mgr; Tunie Moss, prom dir; Mike James, progmg dir.

Fort Myers

WARO(FM)—See Naples

***WAYJ(FM)**— October 1987: 88.7 mhz; 75 kw. Ant 1,007 ft TL: N26 25 22 W81 37 49. Stereo. Hrs opn: 24 Box 61275, 33906. Secondary address: 1860 Boy Scout Dr., Suite 202 33906. Phone: (239) 936-1929. Fax: (239) 936-5433. Web Site: www.wayfm.com. Licensee: WAY-FM Media Group Inc. (group owner). Natl. Network: USA. Law Firm: Gammon & Grange. Format: Contemp Christian. News staff: one; News: one hr wkly. Target aud: 18-34. ♦Bob Augsburg, pres; Jeff Taylor, gen mgr.

WCKT(FM)—See Lehigh Acres

WCRM(AM)— Aug 22, 1964: 1350 khz; 1 kw-D, 150 w-N. TL: N26 37 31 W81 50 29. (CP: 5 kw-D). Hrs opn: 19 3448 Canal St., 33916. Phone: (239) 334-1350 / (941) 332-1350. Fax: (239) 332-8890. E-mail: radio1350office@aol.com Web Site: www.aleluya.com/1350_am.htm. Licensee: Manna Christian Missions Inc. (acq 6-89). Population served: 400,000 Natl. Network: USA. Law Firm: Schwartz, Woods & Miller. Format: Sp, Christian. News staff: one; News: 5 hrs wkly. Target aud: General. ♦Salvador Santana, gen mgr.

***WGCU-FM**— Sept 12, 1983: 90.1 mhz; 100 kw. 813 ft TL: N26 48 54 W81 45 44. Stereo. Hrs opn: 24 10501 FGCU Blvd., 33965-6565. Phone: (239) 590-2500 / 2300. Fax: (239) 590-2520. Web Site: www.wgcu.org. Licensee: Board of Trustees, Florida Gulf Coast University (acq 11-16-01). Population served: 830,000 Natl. Network: NPR, PRI. Rgnl. Network: Fla. Pub. Law Firm: Cohn & Marks. Format: Class, jazz, news. News staff: 2; News: 28 hrs wkly. Target aud: 24 plus. ♦Kathleen Davey, gen mgr & stn mgr; Christine Hause, dev dir & rgnl sls mgr; Toby Cooke, progmg dir; Amy Tardif, news dir.

WINK(AM)— Mar 1, 1940: 1240 khz; 1 kw-U. TL: N26 37 28 W81 49 52. Hrs open:
Rebroadcasts WNOG(AM) Naples 100%.
2824 Palm Beach Blvd., 33916. Phone: (239) 337-2346. Fax: (239) 334-0744. Web Site: www.winkwnog.com. Licensee: Fort Myers Broadcasting Co. (group owner) Population served: 516,200 Natl. Network: CBS Radio. Natl. Rep: McGavren Guild. Law Firm: Leibowitz & Spencer. Format: News/talk. Target aud: 35 plus. ♦Brian A. McBride, pres; Randy Marsh, gen mgr.

WINK-FM—Listing follows WPTK(AM).

***WJYO(FM)**— 1988: 91.5 mhz; 3 kw. 285 ft TL: N26 30 18 W81 51 14. Stereo. Hrs opn: 24
Rebroadcasts WBIY(FM) LaBelle 100%.
Box 61721, 33906. Phone: (239) 274-9150. Fax: (239) 274-0191. E-mail: wjyo@aol.com Web Site: airwavesforJesus.com. Licensee: Airwaves for Jesus Inc. (acq 3-8-2004; $500,000 with WBIY(FM) La Belle). Format: Bible teaching, light Christian Praise/worship music. News: 10 hrs wkly. Target aud: 44 plus; traditional minded persons. Spec prog: Children 5 hrs wkly. ♦Art Ramos, CEO, pres & gen mgr; Jasmin Ramos, VP.

WMYR(AM)— Nov 11, 1952: 1410 khz; 5 kw-U, DA-N. TL: N26 37 24.9 W81 51 16.7. Hrs open: 24 5043 Tamiami Tr. E., Naples, 34113. Phone: (239) 768-9256. Phone: (239) 732-9369. Fax: (239) 768-9256. E-mail: wmyr@relevantradio.com Web Site: www.relevantradio.com. Licensee: Starboard Media Foundation Inc. Group owner: Relevant Radio (acq 9-22-2004; $1.5 million). Population served: 27,351 Format: Catholic talk. ♦Bob Ladd, gen mgr & opns mgr.

WOLZ(FM)— January 1970: 95.3 mhz; 79 kw. 453 ft TL: N26 37 25 W82 06 56. Stereo. Hrs open: 24 13320 Metro Pkwy., 33912. Phone: (239) 225-4300. Phone: (800) 226-3695. Fax: (239) 225-4329. Web Site: www.wolz.com. Licensee: Clear Channel Radio Licenses Inc. Group owner: Clear Channel Communications Inc. (acq 2-18-97; grpsl). Population served: 700,000 Natl. Rep: Clear Channel. Format: Oldies. News staff: one; News: 2 hrs wkly. Target aud: 35-54; upbeat, fun oldies, strong at work and in-car listening. ♦Jim Keating, gen mgr.

WPTK(AM)—Pine Island Center, Feb 20, 1986: 1200 khz; 10 kw-D, 2.5 kw-N. TL: N26 42 52 W82 02 46. Hrs open: 24 2824 Palm Beach Blvd., 33916. Phone: (239) 337-2346. Phone: (239) 338-4325. Fax: (239) 332-0767. Web Site: www.rumba1200.com. Licensee: Fort Myers Broadcasting Co. (group owner) Population served: 516,200 Rgnl. Network: Florida Radio Net. Natl. Rep: McGavren Guild. Law Firm: Leibowitz & Associates. Format: Sp. Target aud: Hispanics; 35-64. ♦Brain A. McBride, pres; Wayne Simons, gen mgr; Brad Foster, sls dir, gen sls mgr & prom mgr; Hector Velasquez, progmg dir; Keith Stuhlmann, engrg dir.

WINK-FM—Co-owned with WPTK(AM). Oct 10, 1964: 96.9 mhz; 100 kw. 1,322 ft TL: N26 38 40 W081 52 10. Stereo. 24 Phone: (239) 334-1111. Fax: (239) 334-0744. Web Site: www.winkfm.com.634,800 Natl. Rep: McGavren Guild. Law Firm: Leibowitz & Associates. Format: Adult contemp. News: 8 hrs wkly. Target aud: 25-54; females. ♦Shannon Holly, prom mgr; Bob Grissinger, progmg dir. Co-owned TV: WINK-TV affil

WRXK-FM—See Bonita Springs

WWGR(FM)— Dec 2, 1969: 101.9 mhz; 100 kw. 1,020 ft TL: N26 25 23 W81 37 07. Stereo. Hrs opn: 24 10915 K-Nine Dr., 2nd Fl., Bonita Springs, 34135. Phone: (239) 495-8383. Fax: (239) 495-0883. Web Site: www.gatorcountry1019.com. Licensee: Renda Broadcast Corp. Group owner: Renda Broadcasting Corp.-Renda Radio Inc. (acq 7-13-94; $4 million; FTR: 8-1-94). Population served: 549,000 Format: Country. ♦Tony Renda Jr., gen mgr.

Fort Myers Beach

WJBX(FM)— 1983: 99.3 mhz; 50 kw. 476 ft TL: N26 30 18 W81 51 14. Stereo. Hrs opn: 20125 S. Tamiami Tr., Estero, 33928. Phone: (239) 495-2100. Fax: (239) 992-8165. Web Site: www.99xwjax.com. Licensee: Dillon License L.P. Group owner: Beasley Broadcast Group (acq 10-16-97; $6 million). Law Firm: Leventhal, Senter & Lerman. Format: Alternative/new rock. Target aud: 18-49; adults. ♦Brad Beasley, gen mgr.

Fort Myers Villas

WJPT(FM)— July 31, 1991: 106.3 mhz; 50 kw. 472 ft TL: N26 29 16 W81 55 49. Hrs open: 20125 S. Tamiami Tr., Estero, 33928. Phone: (239) 495-2100. Fax: (239) 992-8165. E-mail: randy@morningshow.net Web Site: www.wjpt.com. Licensee: WJST License L.P. Group owner: Beasley Broadcast Group (acq 12-11-97; $5 million). Format: Adult standards. Target aud: 45 plus. ♦Brad Beasley, gen mgr; Shane Reilly, opns mgr.

Fort Pierce

WIRA(AM)— May 18, 1946: 1400 khz; 1 kw-U. TL: N27 26 07 W80 21 41. Hrs open: 24 6803 So. Federal Hwy., Port St. Lucie, 34952. Phone: (772) 460-9356. Fax: (772) 460-2700. Web Site: www.1400wira.com. Licensee: Team One Media LLC (acq 2-11-2004; $375,000). Population served: 330,400 Natl. Network: ABC. Rgnl. Network: Westwood One. Format: Urban gospel. Target aud: 45 plus; male & female. Spec prog: Relg one hr, pub affrs one hr wkly. ♦Al Richards, gen mgr.

***WJFP(FM)**— Jan 15, 1995: 91.1 mhz; 6 kw. 157 ft Hrs opn: 6 AM-midnight 2184 North U.S. Hwy #1., Ft. Pierce, 34946. Phone: (772) 467-2400. Fax: (772) 467-9400. Web Site: www.wjfp.com. Licensee: Black Media Works Inc. (group owner; acq 1-21-98). Population served: 175,000 Format: Urban contemp, relg, educ. Target aud: 12-49. Spec prog: Sp 2 hrs, Haitian 8 hrs wkly. ♦Kimberly Kassis, pres.

WJNX(AM)— Dec 24, 1952: 1330 khz; 5 kw-D, 1 kw-N, DA-2. TL: N27 27 20 W80 22 02. Stereo. Hrs opn: 24 4100 Metzger Rd., 34947. Phone: (772) 340-1590. Fax: (772) 340-3245. E-mail: wpsl@wpsl.com Web Site: www.lagigante1330.com. Licensee: Port St. Lucie Broadcasters Inc. (acq 3-31-2004; $400,000). Population served: 1,000,000 Natl. Network: ESPN Deportes. Law Firm: Leventhal, Senter & Lerman. Format: Sp news/talk. News staff: one. Target aud: 25-54. ♦Carol Wyatt, CEO & pres; Greg Wyatt, gen mgr.

WKGR(FM)— May 1, 1961: 98.7 mhz; 100 kw. 1,381 ft TL: N27 07 20 W80 23 21. Stereo. Hrs opn: 3071 Continental Dr., West Palm Beach, 33407. Phone: (561) 616-6600. Fax: (561) 616-6677. Web Site: www.gater.com. Licensee: Clear Channel Radio Licenses Inc. Group owner: Clear Channel Communications Inc. (acq 9-16-97; grpsl). Format: Classic rock. Target aud: 25-54. ♦John Hunt, gen mgr; Dave Denver, opns dir & opns mgr; Roger Koch, sls dir.

WLDI(FM)— Oct 30, 1969: 95.5 mhz; 100 kw. 981 ft TL: N27 07 20 W80 23 21. Stereo. Hrs opn: 3071 Continental Dr., West Palm Beach, 33407. Phone: (561) 616-6600. Fax: (561) 616-6677. Web Site: www.wild955.com. Licensee: Clear Channel Radio Licenses Inc. Group owner: Clear Channel Communications Inc. (acq 6-17-98;

grpsl). Population served: 271,000 Format: CHR. Target aud: 18-49; active, contemp. ♦John Hunt, gen mgr; Dave Denver, opns dir.

***WQCS(FM)**— April 1982: 88.9 mhz; 100 kw. 436 ft TL: N27 25 17 W80 21 23. Stereo. Hrs opn: 3209 Virginia Ave., 34981. Phone: (772) 462-4744. Fax: (772) 462-4743. Web Site: www.wqcs.org. Licensee: Indian River Community College. Natl. Network: NPR, PRI, AP Radio. Format: Class, news. ♦Madison Hodges, stn mgr; Michelle Rhinesmith, opns mgr.

Fort Walton Beach

WFSH(AM)—See Valparaiso-Niceville

WFTW(AM)— Nov 20, 1953: 1260 khz; 2.5 kw-D, 131 w-N. TL: N30 24 49 W86 37 40. Hrs open: 24 Box 2347, 225 N.W. Hollywood Blvd., 32548. Phone: (850) 243-7676. Fax: (850) 243-6806. Fax: (850) 664-0202. E-mail: wftw@radiopeople.net Web Site: www.wftw.com. Licensee: Cumulus Licensing Corp. Group owner: Cumulus Media Inc. (acq 1-10-03; grpsl). Population served: 27,000 Rgnl. Network: Florida Radio Net. Format: News/talk. ♦Lou Dickey, pres; Ron Raybourne, gen mgr; Georgia Edmiston, gen sls mgr; Lisa Captain, prom dir & prom mgr; Bruce Campbell, chief of engrg; Gerald Lee, traf mgr; Steve Williams, progmg dir & spec ev coord.

WKSM(FM)—Co-owned with WFTW(AM). May 28, 1965: 99.5 mhz; 50 kw. Ant 440 ft TL: N30 24 50 W86 37 40. (CP: ant 438 ft). Stereo. Web Site: www.wksm.com.130,000 Format: Rock. ♦Lee Leonard, rgnl sls mgr; Steve O'Day, prom mgr; Nicci Garmon, progmg dir; Anthony Proffitt, mus dir; Aimee Shaffer, news dir & pub affrs dir; Gerald Lee, traf mgr; Steve Williams, spec ev coord.

***WPSM(FM)**— July 1, 1985: 91.1 mhz; 383 w. 120 ft TL: N30 25 14 W86 36 43. Stereo. Hrs opn: 24 Box 10, 32549. Secondary address: 233 N. Hill Ave. 32548. Phone: (850) 244-7667. Fax: (850) 244-3254. E-mail: wpsmradio91@aol.com Web Site: www.wpsm.com. Licensee: Fort Walton Beach Educ. Broadcasting Corp. Natl. Network: USA. Format: Christian. News staff: one; News: 14 hrs wkly. Target aud: 25-55; young to middle-age adult Christians. ♦Terry Thorne, gen mgr.

WTKE(AM)— 1956: 1400 khz; 1 kw-U. TL: N30 24 38 W86 37 23. Hrs open: 21 Miracle Strip Pkwy. S.E., 32548. Phone: (850) 244-1400. Fax: (850) 243-1471. Licensee: Star Broadcasting Inc. (acq 7-11-2005). Population served: 37,750 Format: Memories. Target aud: 50 plus. Spec prog: Church program one hr wkly. ♦Ron Hale Sr., gen mgr; David Kuntz, gen sls mgr; Frank Hale, progmg dir & chief of engrg.

WTKE-FM—Holt, July 1950: 98.1 mhz; 100 kw. Ant 482 ft TL: N30 24 38 W86 37 22. Stereo. Hrs opn: 21 Miracle Strip Pkwy. S.E., 32548. Phone: (850) 244-1400. Fax: (850) 243-1471. Web Site: www.sportstalktheticket.com. Licensee: Star Broadcasting Inc. Group owner: Qantum Communications Corp. (acq 2-14-2003). Natl. Rep: Roslin. Law Firm: Wiley, Rein & Fielding. Format: Sports, talk. Target aud: 25-54. ♦Ron Hale, Sr., gen mgr; David Kuntz, gen sls mgr; Frank Hale, prom dir, progmg dir & chief of engrg.

WZNS(FM)— 1997: 96.5 mhz; 100 kw. 440 ft TL: N30 24 50 W86 37 40. Hrs open: 24 225 N.W. Hollywood Blvd., 32548. Phone: (850) 664-0665 / 0965. Fax: (850) 243-6806. E-mail: sales@z96.com Web Site: www.z96.com. Licensee: Cumulus Licensing Corp. Group owner: Cumulus Media Inc. (acq 1-10-03; grpsl). Format: CHR. ♦Hayden Green, progmg mgr.

Gainesville

WAJD(AM)— May 31, 1961: 1390 khz; 5 kw-D, 51 w-N. TL: N29 39 56 W82 17 26. Hrs open: 7120 S.W. 24th Ave., 32607. Phone: (352) 331-2200. Fax: (352) 331-0401. Web Site: www.kiss1053.com. Licensee: Gillen Broadcasting Corp. (acq 9-22-87; $1.9 million with co-located FM; FTR: 8-17-87) Population served: 90,000 Format: Radio Disney. Target aud: 12-49. ♦Douglas Gillen, pres, gen mgr & gen sls mgr.

WYKS(FM)—Co-owned with WAJD(AM). May 4, 1970: 105.3 mhz; 3 kw. 466 ft TL: N29 37 52 W82 25 18. (CP: 105.3 mhz, 6 kw). Web Site: www.kiss1053.com.180,000 Format: Top 40.

WDVH(AM)— October 1954: 980 khz; 5 kw-D, 166 w-N. TL: N29 37 26 W82 17 19. Hrs open: 6 AM-10 PM 100 N.W. 76th Dr., Suite 2, 32607. Phone: (352) 313-3150. Fax: (352) 313-3166. E-mail: jim@wdvh.org Web Site: www.wdvh.org. Licensee: 6 Johnson Road Licenses Inc. (group owner; (acq 1-5-2007; grpsl). Natl. Rep: Roslin. Format: Country legends. News: 2 hrs wkly. Target aud: 35 plus. ♦Benjamin Hill, gen mgr.

WGGG(AM)— February 1948: 1230 khz; 1 kw-U. TL: N29 40 56 W82 24 48. Hrs open: Box 2930, Ocala, 34478. Secondary address: 101 S.E. 2nd Pl., Grainesville 32601. Phone: (352) 378-7378. Fax: (352) 629-1614. E-mail: sales@floridasportstalk.com Web Site: www.floridasportstalk.com. Licensee: Florida Sportstalk Inc. (acq 2-5-97; $300,000). Population served: 175,000 Format: All sports. ♦Doug Gillen, gen mgr.

***WJLF(FM)**— Aug 26, 1990: 91.7 mhz; 2 kw. 400 ft TL: N29 38 34 W82 25 13. Stereo. Hrs opn: 24
Rebroadcasts WJIS(FM) Brandenton.
408 West Univ. Ave., Suite 206, 32601. Phone: (352) 373-9553. Fax: (352) 373-9888 / (352) 375-1700. E-mail: thejoyfm@thejoyfm.com Web Site: thejoyfm.com. Licensee: Radio Training Network Inc. (acq 10-1-2004; $1 million). Population served: 250,000 Law Firm: Gammon & Grange. Format: Christian. News staff: one; News: 2 hrs wkly. Target aud: 18-49; young adults & young families. Spec prog: Youth 5 hrs, jazz 2 hrs, children 1 hr wkly. ♦James L. Campbell, pres; Andy Haynes, gen mgr & stn mgr.

WKTK(FM)—See Crystal River

WNDD(FM)—Silver Springs, Feb 1, 1991: 95.5 mhz; 6 kw. 340 ft TL: N29 16 55 W82 02 50. Stereo. Hrs opn: 24 3602 N.E. 20th Pl., Ocala, 34470. Phone: (352) 622-9500. Fax: (352) 622-1900. Web Site: www.windfm.com. Licensee: Ocala Broadcasting Corp. L.L.C. Group owner: Wooster Republican Printing Co. (acq 9-1-97). Population served: 701,258 Natl. Rep: Katz Radio. Law Firm: Baker & Hostetler. Format: Classic rock. Target aud: 25-54; adults. ♦Jim Robertson, gen mgr; Bob Kassi, gen sls mgr; Kevin Davis, progmg dir.

WRUF-FM— 1948: 103.7 mhz; 100 kw. 768 ft TL: N29 42 34 W82 23 40. Stereo. Hrs opn: 24 Box 14444, 32604. Secondary address: Univ. of Florida, 3200 Wiemer Hall 32611. Phone: (352) 392-0771. Fax: (352) 392-0519. Web Site: www.rock104.com. Licensee: University of Florida, Board of Trustees. Population served: 645,100 Format: Contemp rock. News staff: 3; News: 5 hrs wkly. Target aud: 25-34; urban rockers. Spec prog: Alternative 6 hrs wkly. ♦Larry Dankner, gen mgr & gen sls mgr; Harry Guscott, opns mgr & progmg dir; Cathy Ferguson, news dir & traf mgr; Tom Kyrnski, news dir; Don Rice, chief of engrg; Matt Lehtola, disc jockey; Monica Richs, disc jockey.

WRUF(AM)— 1928: 850 khz; 5 kw-U, DA-N. TL: N29 38 34 W82 25 13.24 Web Site: www.am850.com. Natl. Network: CBS, Westwood One. Format: News/talk, sports. News staff: 3; News: 54 hrs wkly. Target aud: 35-54; middle-to-upper income, decision makers. Spec prog: Black 4 hrs wkly. ♦Robert Lawrence, opns mgr; Larry Dankner, dev dir, mktg dir & progmg mgr; Tom Ksynski, news dir & pub affrs dir; Don Rice, chief of engrg; Steve Russell, sports cmtr. Co-owned TV: WUFT-TV, WLUF-TV affils.

WTMG(FM)—See Williston

WTMN(AM)— January 1990: 1430 khz; 10 kw-D, 45 w-N. TL: N29 37 26 W82 17 19. Hrs open: 100 NW 76th Dr., Ste 2, 32607-6659. Phone: (352) 313-3150. Fax: (352) 338-0566. E-mail: shinds@sunshinebroadcasting.com Licensee: 6 Johnson Road Licenses Inc. (group owner; (acq 1-5-2007; grpsl). Law Firm: Irwin, Campbell, Crowe & Tannenwald. Format: Gospel. ♦Benjamin Hill, gen mgr.

***WUFT-FM**— Sept 27, 1981: 89.1 mhz; 100 kw. 771 ft TL: N29 42 34 W82 23 40. Stereo. Hrs opn: 24 Box 118405, 32611. Phone: (352) 392-5200. Fax: (352) 392-5741. E-mail: radio@wuft.org Web Site: www.wuft.org. Licensee: Board of Trustees, University of Florida. Population served: 500,000 Natl. Network: NPR, PRI. Rgnl. Network: Fla. Pub. Law Firm: Schwartz, Woods & Miller. Format: Class, jazz, pub affrs. News staff: 3; News: 15 hrs wkly. Target aud: 35-65; general, educated (some college or degree). Spec prog: Black 4 hrs, folk one hr, gospel one hr wkly. ♦Larry Dankovic, gen mgr; Henri Pensis, stn mgr; Steve Seipp, opns mgr; Bill Beckett, progmg dir; Richard Drake, mus dir; Kevin Allen, news dir; Manis Samons, chief of engrg. Co-owned TV: *WUFT-TV affil.

WXJZ(FM)— May 1, 1982: 100.9 mhz; 6 kw. Ant 298 ft TL: N29 38 03 W82 18 50. Stereo. Hrs opn: 24 4424 N.W. 13th St., Suite C-5, 32609. Phone: (352) 375-1317. Fax: (352) 375-6961. E-mail: feedback@wxjz.fm Web Site: www.wxjz.fm. Licensee: Asterisk Communications Inc. Group owner: Asterisk Inc. (acq 10-4-93; $1.4 million; FTR: 10-25-93). Natl. Rep: McGavren Guild. Format: Smooth jazz. Target aud: 25-54; upscale, affluent, sophisticated. ♦John Starr, gen mgr & adv mgr; Bill Elliott, progmg dir.

***WYFB(FM)**— Aug 4, 1985: 90.5 mhz; 100 kw. 679 ft TL: N29 52 08 W82 12 04. (CP: 96.81 kw). Stereo. Hrs opn: 24 5553 S.E. 3rd Ave., Keystone Heights, 32656. Phone: (877) 939-9933. E-mail: wyfb@bbnradio.org Web Site: www.bbnradio.org. Licensee: Bible Broadcasting Network Inc. (group owner) Population served: 1,400,000 Natl. Network: Bible Bcstg Net, USA. Format: Relg. News: 12 hrs wkly. Target aud: General. ♦Lowell Davey, pres; David Nichols, gen mgr & chief of opns.

WYGC(FM)—High Springs, Jan 31, 1984: 104.9 mhz; 3.2 kw. 450 ft TL: N29 49 16 W82 34 28. Stereo. Hrs opn: 24 3357 S.W. 7h St., Suite C5, Ocala, 34474. Phone: (352) 732-9877. Fax: (352) 622-6675. E-mail: production@wtrs.fm Web Site: www.wtrs.com. Licensee: Asterisk Communications Inc. Group owner: Asterisk Inc. (acq 2-99; $825,000). Natl. Network: CNN Radio, Westwood One. Law Firm: Larry Perry. Format: Country. News: 4 hrs wkly. Target aud: 25-54. ♦Dean Johnson, gen mgr & gen sls mgr.

Gibsonia

WJWB(AM)—Not on air, target date: unknown: 700 khz; 2.5 kw-D, 250 w-N, DA-2. TL: N28 08 33 W81 52 41. Hrs open: 571 N.W. McClurg Ct., White Springs, 32096-7308. Phone: (386) 397-4489. Licensee: People's Network Inc. ♦Chuck Harder, gen mgr.

Gifford

WSYR-FM— June 1994: 94.7 mhz; 25 kw. 295 ft TL: N27 33 21 W80 22 08. Hrs open: Box 0093, Port St. Lucie, 34985. Phone: (772) 335-9300. Fax: (772) 335-3291. E-mail: star947@clearchannel.com Web Site: www.star947.com. Licensee: Capstar TX L.P. Group owner: Clear Channel Communications Inc. (acq 8-30-00; grpsl). Format: Adult contemp. Target aud: 25-54. ♦John Hunt, gen mgr; Andrew Bednar, prom dir; Mike Michaels, opns mgr & progmg dir.

Golden Gate

WNPL(AM)—Not on air, target date: unknown: 1460 khz; 7 kw-D, 2 kw-N, DA-2. TL: N26 15 26 W81 40 33. Hrs open: 2824 Palm Beach Blvd., Fort Myers, 33916. Phone: (239) 334-1111. Licensee: Fort Myers Broadcasting Co. (acq 6-18-2007; $975,000 for CP). ♦Brian A. McBride, pres.

Goulds

WRTO-FM—Licensed to Goulds. See Miami

Graceville

WTOT-FM— 1996: 101.7 mhz; 6 kw. 328 ft TL: N30 57 21 W85 29 53. Hrs open: 24 285 E. Broad St., Ozark, AR, 36360. Phone: (334) 774-9323. Fax: (334) 774-6450. Licensee: GFR Inc. Population served: 250,000 Natl. Network: ABC. Format: News, talk. Target aud: 25+; female. ♦Jack Mizell, pres.

Green Cove Springs

WJBT(FM)— 1978: 92.7 mhz; 2.6 kw. Ant 505 ft TL: N30 04 08 W81 38 50. Stereo. Hrs opn: 24 11700 Central Pkwy., Jacksonville, 32224. Phone: (904) 636-0507 / (904) 642-3030. Fax: (904) 997-7713. Web Site: www.wjbt.com. Licensee: Citicasters Licenses L.P. Group owner: Clear Channel Communications Inc. (acq 5-4-99; grpsl). Format: Urban contemp. Target aud: 12-54; the young & young-at-heart.

♦Norm Feuer, gen mgr; Gail Austin, opns mgr; April Johnson, gen sls mgr; Rhonda Hernandez, news dir; Phil Tuck, chief of engrg; Debbie Davis, traf mgr.

Greenville

***WYJC(FM)**— 2005: 90.3 mhz; 325 w. Ant 184 ft TL: N30 23 56 W83 39 24. Hrs open:
Rebroadcasts WUJC(FM) Saint Marks 100%.
8747 Miles Johnson Rd., Tallahassee, 32309. Phone: (850) 514-1929. Fax: (850) 514-1927. Web Site: www.csnradio.com. Licensee: CSN International (group owner). Format: Christian. ♦Michael Kestler, pres.

Gretna

WGWD(FM)— Oct 2, 1989: 93.3 mhz; 3 kw. 328 ft TL: N30 33 24 W84 36 05. (CP: 6 kw). Stereo. Hrs opn: 24 Box 919, Quincy, 32353. Secondary address: 8 W. Washington, Quincy 32351. Phone: (850) 627-7086. Fax: (850) 627-3422. Licensee: De Col Inc. (acq 9-18-91; $75,000; FTR: 10-7-91). Natl. Network: USA. Format: Classic country. News: 21 hrs wkly. Target aud: 25-54. Spec prog: Black 20 hrs wkly. ♦Monte Bitner, gen mgr, gen sls mgr & progmg dir; Jan Rogers, news dir; Jeff Fallaway, chief of engrg; Pat Bitner, opns mgr & traf mgr.

Gulf Breeze

WNRP(AM)— Feb 1, 1998: 1620 khz; 10 kw-D, 1 kw-N. TL: N30 26 12 W87 13 13. Hrs open:
Simulcast with WYCT(FM) Pensacola 100%.
7251 Plantation Rd., Pensacola, 32504. Phone: (850) 494-2800. Fax: (850) 494-0778. E-mail: hr@catcountry987.com Web Site: www.catcountry987.com. Licensee: ADX Communications of Escambia (acq 11-16-2000). Format: Classic Country. ♦Mary Hoxeng, gen mgr; Kevin King, opns dir; Sassy McGuire, prom dir.

WRNE(AM)— November 1957: 980 khz; 4 kw-D, 1 kw-N, DA-N. TL: N30 29 08 W87 05 01. Stereo. Hrs opn: 24 312 E. Nine Mile Rd., Suite 27-D, Pensacola, 32514. Phone: (850) 478-6000. Fax: (850) 484-8080. E-mail: hill@wrne980.com Web Site: www.wrne980.com. Licensee: Media One Communications Inc. (acq 11-15-90; FTR: 11-19-90). Population served: 59,507 Natl. Rep: Dora-Clayton. Law Firm: Dennis J. Kelly. Format: Urban contemp, gospel, Hispanic. News staff: one; News: 5 hrs wkly. Target aud: 25-54; minorities. Spec prog: Gospel, talk. ♦Robert Hill, pres & gen mgr.

WRRX(FM)—Not on air, target date: unknown: 106.1 mhz; Hrs opn: 6565 N. W St., Pensacola, PA, 32505. Phone: (850) 478-6011. Fax: (850) 478-3971. Web Site: www.cumulus.com. Licensee: Cumulus Licensing Corp. Format: Urban contemp. ♦Liz Hanlon, gen mgr; Debbie Dingwall, opns mgr.

Gulfport

WFUS(FM)— October 1963: 103.5 mhz; 98 kw. Ant 1,358 ft TL: N27 50 32 W82 15 45. Stereo. Hrs opn: 4002 W. Gandy Blvd., Tampa, 33611. Phone: (813) 832-1000. Fax: (813) 832-1943. Licensee: Citicasters Licenses L.P. Group owner: Clear Channel Communications Inc. (acq 6-99; grpsl). Population served: 1,500,000 Natl. Network: ABC. Format: Country. Target aud: 25-54; men. ♦Dan DiLoreto, gen mgr; Brad Hardin, opns mgr; Tammy Odel, gen sls mgr.

Haines City

***WLVF-FM**— Apr 11, 1986: 90.3 mhz; 800 w. 265 ft TL: N28 09 28 W81 37 34. (CP: 1.2 kw, ant 308 ft.). Stereo. Hrs opn: 24 810 E. Hinson Ave, 33844. Phone: (863) 422-9583. Fax: (863) 422-0110. E-mail: wlvf@gate.net Web Site: www.gospel903.com. Licensee: Landmark Baptist Church. Population served: 80,000 Natl. Network: USA. Format: Southern gospel. ♦Steven Carter, gen mgr; Lewis Cruz, opns mgr & progmg dir; Bobby Ogden, gen sls mgr; Jeff Crews, chief of engrg.

WLVF(AM)— Sept 9, 1960: 930 khz; 500 w-D, DA. TL: N28 04 52 W81 38 23.7 AM-sunset Phone: (863) 422-5175. E-mail: wlvf@gate.net 60,000 Format: Southern gospel. Target aud: General. ♦Jonathan Marshall, disc jockey; Matthew Marshall, disc jockey; Rich Lemon, disc jockey; Steven Carter, disc jockey.

Havana

WHTF(FM)— 1986: 104.9 mhz; 47 kw. 494 ft TL: N30 35 11 W84 14 11. Stereo. Hrs opn: 24 3000 Olson Rd., Tallahassee, 32308. Phone: (850) 386-8004. Fax: (850) 442-1897. Web Site: www.hot1049.com. Licensee: Opus Broadcasting Tallahassee LLC. Group owner: Triad Broadcasting Co. LLC (acq 7-11-2005; grpsl). Population served: 485,000 Natl. Rep: McGavren Guild. Format: CHR. Target aud: 18-49. ♦Hank Kestenbaum, gen mgr; Doug Purtee, opns mgr.

Hernando

WRZN(AM)— June 1989: 720 khz; 10 kw-D, 250 w-N, DA-N. TL: N28 55 21 W82 22 21. Hrs open: 100 N.W. 76th Dr., Suite 2, Suite B, Gainsville, 32607. Secondary address: 3938 N. Roscoe Rd. 34442. Phone: (352) 726-7221. Phone: (352) 313-3150. Fax: (352) 726-3172. Licensee: 6 Johnson Road Licenses Inc. (group owner; (acq 1-5-2007; grpsl). Format: Adult standards. Target aud: 45 plus. Spec prog: Loc news 4 hrs wkly. ♦Ben Hill, gen mgr; Reggie Thomas, gen sls mgr; Jim Brand, progmg dir.

Hialeah

WACC(AM)—Licensed to Hialeah. See Miami

WCMQ-FM— Dec 22, 1969: 92.3 mhz; 31 kw. 617 ft TL: N25 46 29 W80 11 19. Stereo. Hrs opn: 1001 Ponce De Leon Blvd., Coral Gables, 33134. Phone: (305) 444-9292. Fax: (305) 461-4466. Web Site: www.lamusica.com. Licensee: WCMQ Licensing Inc. Group owner: Spanish Broadcasting System Inc. (acq 12-22-86; grpsl; FTR: 9-29-86). Format: Adult contemp, Sp. ♦Jackie Nosti-Combo, gen mgr; Tony Campos, opns mgr; Albert Rodriguez, gen sls mgr; John Caride, prom dir.

High Springs

WYGC(FM)—Licensed to High Springs. See Gainesville

Hilliard

WJFA(AM)—Not on air, target date: unknown: 830 khz; 50 kw-D, 4 kw-N, DA-2. TL: N30 43 41 W81 59 37. Hrs open: 571 N.W. McClurg Ct., White Springs, 32096. Phone: (386) 397-4489. Licensee: The Dianne A. Mayfield-Harder Trust. ♦Charles Harder, gen mgr; Dianne A. Mayfield-Harder, gen mgr.

Hobe Sound

WOLL(FM)— 2002: 105.5 mhz; 50 kw. Ant 456 ft TL: N26 45 42 W80 04 42. Hrs open: 3071 Continental Dr., West Palm Beach, 33407. Phone: (561) 616-6600. Fax: (561) 616-6677. Web Site: www.1055online.com. Licensee: Clear Channel Broadcasting Licenses Inc. Group owner: Clear Channel Communications Inc. (acq 6-17-98; grpsl). Format: Oldies. Target aud: 25-54. ♦John Hunt, gen mgr; Dave Denver, opns dir.

Holiday

WSUN-FM— 1979: 97.1 mhz; 3.3 kw. 300 ft TL: N28 16 51 W82 42 52. Stereo. Hrs opn: 24 11300 4th St. N., Suite 300, St. Petersburg, 33716. Phone: (727) 579-2000. Fax: (727) 579-2662. Fax: (727) 579-2271. E-mail: 97xcomments@97xonline.com Web Site: www.97xonline.com. Licensee: Cox Radio Inc. Group owner: Cox Broadcasting (acq 11-20-98). Law Firm: Reddy, Begley & McCormick. Format: Alternative/new rock. Target aud: 35 plus. ♦Bob Neil, CEO & pres; Keith Lawless, VP & gen mgr; Tom Paleveda, opns mgr; Dan Connelly, prom mgr.

Holly Hill

***WAPN(FM)**— October 1985: 91.5 mhz; 1.8 kw. 285 ft TL: N29 15 06 W81 02 53. Stereo. Hrs opn: 24 Box 250, Daytona Beach, 32125. Secondary address: 1508 State Ave., Daytona Beach 32125. Phone: (386) 677-4272. Phone: (386) 672-3333. Fax: (386) 673-3715. E-mail: wapn@wapn.net Web Site: www.wapn.net. Licensee: Public Radio Capital Florida (acq 5-16-03; $1.5 million). Format: Word & praise. Target aud: General. Spec prog: Sp 4 hrs wkly. ♦Shellye Lund-Vallance, gen mgr.

***WEAZ(FM)**—Aug 20, 1999: 88.1 mhz; 5.1 kw vert. Ant 114 ft TL: N29 16 44 W81 11 25. Hrs open:
Rebroadcasts WPOZ(FM) Union Park 100%.
1065 Rainer Dr., Altamonte Springs, 32714-3847. Phone: (407) 869-8000. Fax: (407) 869-0380. E-mail: zcrew@zradio.org Web Site: www.zradio.org. Licensee: Central Florida Educational Foundation Inc. (acq 6-8-99; $75,000). Population served: 250,000 Format: Contemp Christian. ♦James Hoge, gen mgr; Dean O'Neal, opns mgr.

WVYB(FM)— 1997: 103.3 mhz; 3 kw. 328 ft TL: N29 15 05 W81 07 23. (CP: Ant 315 ft.). Hrs opn: 24 126 W. International Speedway Blvd., Daytona Beach, 32114. Phone: (386) 255-9300. Fax: (386) 238-6071. Web Site: www.wvyb.fm. Licensee: Black Crow LLC. Group owner: Black Crow Media Group LLC (acq 9-21-2001; grpsl). Format: Hot adult contemp, CHR. News: 2 hrs wkly. Target aud: 18-49. ♦Stacey Knerler, gen mgr.

Hollywood

WLQY(AM)— April 1953: 1320 khz; 5 kw-U, DA-2. TL: N26 01 53 W80 16 42. Hrs open: 10800 Biscayne Blvd., Suite 810, Miami, 33161. Phone: (305) 891-1729. Fax: (305) 891-1583. Web Site: www.entravision.com. Licensee: Entravision Holdings LLC. Group owner: Entravision Communications Corp. (acq 7-28-00; grpsl). Format: Ethnic. Target aud: 35 plus; female. ♦Jeff Liberman, pres; Rick Santos, gen mgr.

Holmes Beach

WLLD(FM)— Jan 27, 1992: 98.7 mhz; 3 kw. 328 ft TL: N27 27 49 W82 35 32. Stereo. Hrs opn: 24 9721 Executive Center Dr. N., Suite 200, St. Petersburg, 33702-2439. Phone: (727) 579-1925. Fax: (727) 579-8888. E-mail: beata@infinitybroadcasting.com Web Site: www.wild987.com. Licensee: Infinity Radio Inc. Group owner: Infinity Broadcasting Corp. (acq 11-13-98; grpsl). Population served: 456,000 Natl. Network: CNN Radio. Law Firm: Leventhal, Senter & Lerman. Format: Hip Hop. Target aud: 25 plus; professional, educated, upscale audience. ♦Charlie Ochs, gen mgr; Marvin Kapman, gen sls mgr; Orlando Davis, progmg dir & mus dir; Ross Block, opns mgr & progmg dir.

Holt

WTKE-FM—Licensed to Holt. See Fort Walton Beach

Homestead

WOIR(AM)— Nov 4, 1957: 1430 khz; 5 kw-D, 500 w-N, DA-N. TL: N25 27 09 W80 30 57. Hrs open: 13077 S.W. 133rd Ct., Miami, 33186. Phone: (305) 969-3884. Fax: (305) 969-3825. Licensee: Amanecer Christian Network Inc. (acq 5-17-2001; $2.58 million). Format: Sp, relg. ♦Frank Lopez, gen mgr.

***WRGP(FM)**— 1999: 88.1 mhz; 165 w. Ant 423 ft TL: N25 32 24 W80 28 07. Hrs open: 24 Florida International Univ., 11200 S.W. Univ. Park - GC 210, Miami, 33199. Phone: (305) 348-3071. Fax: (305) 348-6665. E-mail: wrgp@fiu.edu Web Site: wrgp.org. Licensee: Florida International University. Population served: 1,300,000 Format: Var. News: 4 hrs wkly. Target aud: General; young adults, mainly university students. Spec prog: Hip hop 12 hrs, news 3 hrs, raggae 3 hrs wkly. ♦Brennan Forsyth, gen mgr; Jennifer Mojena, progmg dir.

Homosassa Springs

WXCV(FM)— March 1983: 95.3 mhz; 6 kw. 328 ft TL: N28 53 14 W82 31 39. Stereo. Hrs opn: 24 4554 S. Suncoast Blvd., 34446. Phone: (352) 628-4444. Web Site: www.citrus953.com. Licensee: Westwind Broadcasting Inc. Law Firm: Gardner, Carton & Douglas. Format: Classic rock. News staff: one; News: 7 hrs wkly. Target aud: 25-54. Spec prog: Jazz 7 hrs, oldies 6 hrs wkly. ♦Richard Spires, gen mgr.

Immokalee

WAFZ(AM)— Oct 14, 1964: 1490 khz; 1 kw-U. TL: N26 25 27 W81 26 32. Hrs open: 2105 Immokalee Dr., 34142. Phone: (239) 658-1490. Fax: (239) 658-6109. E-mail: robbie@gladesmedia.com Web Site: www.radiofiesta.com. Licensee: Glades Media Company LLP (acq 8-21-89; $210,000). Population served: 50,000 Format: Oldies. ♦Robbie Castellanos, pres; Gary Holloway, gen mgr, opns mgr & engrg dir.

WAFZ-FM— 1995: 92.1 mhz; 5.6 kw. Ant 328 ft TL: N26 26 54 W81 16 17. Hrs open: 24 2105 Immokalee Dr., 34142. Phone: (239) 658-1490. Fax: (239) 658-6109. E-mail: robbie@gladesmedia.com Web Site: www.radiofiesta.com. Licensee: Glades Media Co. LLC (acq

7-14-2004). Natl. Network: USA. Format: Sp & Mexican oldies. Target aud: 18 plus; general. ♦Gary Holloway, gen mgr & opns mgr; Robbie Castellanos, pres & sls dir.

Indian River Shores

WOSN(FM)— 1996: 97.1 mhz; 6 kw. 328 ft TL: N27 44 06 W80 27 27. Hrs open: 1235 16th Street, Vero Beach, 32960. Phone: (772) 567-0937. Fax: (772) 562-4747. Web Site: www.wosnfm.com. Licensee: Vero Beach Broadcasters LLC (group owner; acq 2-15-01; $4.1 million). Format: Adult standards. ♦Jim Davis, gen mgr; Hamp Elliott, progmg dir; John Rotolante, opns mgr & prom.

Indian Rocks Beach

WPOI(FM)—See Saint Petersburg

WXYB(AM)— May 11, 1963: 1520 khz; 1 kw-D, DA. TL: N27 50 26 W82 46 10. (CP: 600 w. TL: N27 50 45 W82 46 21). Hrs opn: Sunrise-sunset 109 Bayview Blvd., Suite A, Oldsmar, 34677. Phone: (727) 725-5555. Phone: (813) 814-7575. Fax: (813) 814-7500. E-mail: wpso@wpso.com Web Site: www.wpso.com. Licensee: ASA Broadcasting Inc. (acq 5-24-93; $31,000; FTR: 6-14-93). Population served: 100,000 Format: Ethnic, Greek, news/talk, educ. News: 7 hrs wkly. Target aud: General; international, ethnic. Spec prog: Hillsborough Community College progmg, Indian 3 hrs, It 2 hrs, Pol 2 hrs, East Indian one hr, relg 8 hrs wkly, Spanish M-F, 2-7 pm. ♦Sam Agelatos, pres; Angelo Agelatos, gen mgr, stn mgr & opns dir.

Indiantown

WPBZ(FM)— July 4, 1965: 103.1 mhz; 90 kw. 974 ft TL: N27 01 32 W80 10 43. Stereo. Hrs opn: 24 701 Northpoint Pkwy., Suite 400, West Palm Beach, 33407. Phone: (561) 616-4600. Fax: (561) 684-6311. Web Site: www.buzz103.com. Licensee: Infinity Radio Inc. Group owner: Infinity Broadcasting Corp. (acq 12-14-00; grpsl). Law Firm: Rosenman & Colin. Format: Alternative. News staff: one; News: 12 hrs wkly. Target aud: 18-34; men. ♦Lee K. Strasser, gen mgr; John O'Connell, opns dir & progmg dir; Fran Marcone, gen sls mgr; Susan Oland, natl sls mgr; Lynette Shady, prom dir; Nik Rivers, mus dir; Chuck Herlihey, engrg dir & chief of engrg; Lane Racette, traf mgr; Jason Davis, disc jockey.

Inglis

WFBI(FM)—Not on air, target date: unknown: 99.3 mhz; 4.6 kw. Ant 374 ft TL: N29 08 04 W82 38 34. Hrs open: 188 S. Bellevue, Suite 222, Memphis, TN, 38104. Phone: (901) 375-9324. Web Site: www.flinn.com. Licensee: George S. Flinn Jr. ♦George S. Flinn Jr., gen mgr.

WIFL(FM)— Oct 1, 1994: 104.3 mhz; 4.4 kw. Ant 380 ft TL: N29 01 18 W82 41 20. Stereo. Hrs opn: 24 11928 N. William St., Dunnellon, 34432. Phone: (352) 522-0172. Fax: (352) 564-8750. E-mail: wifl@xtalwind.net Web Site: www.wow104.com. Licensee: Nature Coast Broadcasting Inc. (acq 2-9-2004; $525,000). Population served: 350,000 Law Firm: Gammon & Grange. Format: Mix adult contemp, CHR. News: one hr wkly. Target aud: 25-54. ♦Lisa Cuppelli, CEO, gen mgr & stn mgr; Sab Cupelli, pres; Marc Tyll, VP & gen mgr; Jon Kay, opns mgr; Jeremy Howard, gen sls mgr.

Inverness

***WJUF(FM)**— Oct 1, 1995: 90.1 mhz; 4.5 kw. 354 ft TL: N28 52 09 W82 26 47. Hrs open: 24
Rebroadcasts WUFT-FM Gainesville 100%.
Weimer Hall Univ. of Florida, Gainesville, 32611. Phone: (352) 392-5200. Fax: (352) 392-5741. E-mail: info@wuft.org Web Site: www.wuft.org. Licensee: Board of Trustees, University of Florida. (acq 10-14-94; FTR: 12-5-94). Population served: 200,000 Natl. Network: NPR. Rgnl. Network: Fla Pub. Law Firm: Schwartz, Woods & Miller. Format: Classic rock, jazz, pub affrs. News staff: 3; News: 15 hrs wkly. Target aud: General. Spec prog: Gospel one hr wkly. ♦Richard A. Lehner, gen mgr; Henri Pensis, stn mgr. Co-owned TV: *WUFT-TV affil.

Jacksonville

WAPE-FM— April 1949: 95.1 mhz; 100 kw. 460 ft TL: N30 17 09 W81 44 52. Stereo. Hrs opn: 8000 Belfort Pkwy, 32256. Phone: (904) 245-8500. Fax: (904) 245-8501. E-mail: contest@wape951.com Web Site: www.wape951.com. Licensee: Cox Radio Inc. Group owner: Cox Broadcasting (acq 8-00; grpsl). Population served: 528,865 Natl. Rep: Christal. Format: CHR. ♦Bill Hendrichs, gen mgr.

WAYR(AM)—See Orange Park

WBOB(AM)— 1945: 1320 khz; 5 kw-U, DA-N. TL: N30 17 50 W81 44 35. Stereo. Hrs opn: 4190 Belfort Rd., Suite 450, 32216. Phone: (904) 470-4615. Fax: (904) 296-1683. Web Site: www.1320thepatriotr.com. Licensee: Chesapeake-Portsmouth Broadcasting Corp. Group owner: Salem Communications Corp. (acq 12-5-2006; $1.8 million with WZNZ(AM) Jacksonville). Population served: 686,000 Law Firm: Fletcher, Heald & Hildreth. Format: News/talk. Target aud: 25-54. ♦Henry Hoot, gen mgr; Calvin Grabau, opns mgr.

WBWL(AM)— Dec 9, 1933: 600 khz; 5 kw-D, 5.4 kw-N, DA-N. TL: N30 18 00 W81 45 34. Hrs open: 24 10245 Centurion Pkwy., Suite 109, 32256. Phone: (904) 646-1100. Phone: (904) 783-3711. Fax: (904) 646-1117. Web Site: www.radiodisney.com. Licensee: Radio Disney Group LLC. Group owner: ABC Inc. (acq 8-1-02; $2.5 million). Population served: 751,000 Natl. Network: Radio Disney. Rgnl. Network: Florida Radio Net. Format: Children. ♦Jay Schneider, stn mgr.

WCGL(AM)— 1948: 1360 khz; 5 kw-D. TL: N30 16 33 W81 38 12. Hrs open: 6 am-10 pm 6050-6 Moncrief Rd., 32209. Phone: (904) 766-9955. Fax: (904) 765-9214. E-mail: wcgll@aol.com Web Site: www.wcgl1360.com. Licensee: JBD Communications Inc. (acq 12-27-89; $510,000; FTR: 1-15-90). Population served: 528,865 Law Firm: Pepper & Corazzini. Format: Relg. Target aud: 25 plus. ♦Deborah Maiden, pres & gen mgr; Kelvin Postell, opns mgr.

***WCRJ(FM)**— Mar 16, 1984: 88.1 mhz; 8 kw. Ant 495 ft TL: N30 16 34 W81 33 53. Stereo. Hrs opn: 24 2361 Cortez Rd., 32246. Phone: (904) 641-9626. Fax: (904) 645-9626. E-mail: calvin@fm88.org Web Site: www.riverradio.org. Licensee: New Covenant Educational Ministries Inc. Format: Christian contemp. ♦Calvin Grabau, gen mgr; Roger Henderson, opns mgr.

WEJZ(FM)— 1949: 96.1 mhz; 100 kw. 984 ft TL: N30 19 22 W81 38 34. Stereo. Hrs opn: 6440 Atlantic Blvd., 32211. Phone: (904) 727-9696. Fax: (904) 721-9322. Web Site: www.wejz.com. Licensee: Renda Broadcasting Corp. (group owner; acq 6-90; grpsl; FTR: 6-25-90). Population served: 1,000,000 Natl. Rep: McGavren Guild. Wire Svc: Metro Weather Service Inc. Format: Lite adult contemp. Target aud: 25-54; office, home & in-the-car audience. ♦Tony Renda Sr., CEO & pres; Bill Scull, gen mgr; Bill Reese, gen sls mgr; Woody Carlson, prom dir; Ed Fairbanks, progmg dir; Bob Dillehay, chief of engrg; Brenda McArthur, traf mgr; Jim Byard, pub svc dir.

WFXJ(AM)— November 1925: 930 khz; 5 kw-U, DA-N. TL: N30 17 09 W81 44 52. Hrs open: 24 11700 Central Pkwy, 32224-2600. Phone: (904) 636-0507. Phone: (904) 642-3030. Fax: (904) 997-7713. E-mail: victoriagowan@clearchannel.com Web Site: www.930thefox.com. Licensee: Clear Channel Radio Licenses Inc. Group owner: Clear Channel Communications Inc. Format: Sports. Target aud: 25-49; men. ♦Norm Feuer, gen mgr; Gail Austin, opns mgr; Victoria Gowan, gen sls mgr.

WJAX(AM)—Listing follows WKTZ-FM.

WJBT(FM)—See Green Cove Springs

***WJCT-FM**— Apr 17, 1972: 89.9 mhz; 100 kw. 835 ft TL: N30 16 53 W81 34 15. Stereo. Hrs opn: 100 Festival Park Ave., 32202. Phone: (904) 353-7770. Fax: (904) 358-6352. E-mail: wjct@wjct.org Web Site: www.wjct.org. Licensee: WJCT Inc. Population served: 528,865 Natl. Network: NPR, PRI. Rgnl. Network: Fla. Pub. Law Firm: Schwartz, Woods & Miller. Format: News/talk, class. ♦Michael Boylan, CEO & pres; Tom Patton, stn mgr & news dir. Co-owned TV: *WJCT-TV affil.

***WJFR(FM)**— Sept 15, 1987: 88.7 mhz; 8 kw. 380 ft TL: N30 16 53 W81 34 15. Stereo. Hrs opn: 2771-29 Monument Rd. , #318, 32225. Phone: (904) 389-9088. Web Site: www.familyradio.com. Licensee: Family Stations Inc. Format: Relg. Target aud: Conservative Christians. ♦Harold Camping, pres; Harold Camping, gen mgr; Marcy Morrison-Pearce, progmg dir & news dir; Phyllis Johnston, mus dir.

WJGL(FM)—Listing follows WOKV(AM).

WJXL(AM)—See Jacksonville Beach

WJXR(FM)—Macclenny, September 1978: 92.1 mhz; 25 kw. 328 ft TL: N30 17 54 W82 00 55. Stereo. Hrs opn: 24 Box One, 32234. Phone: (904) 259-2292. Phone: (904) 358-2265. Fax: (904) 259-4488. Web Site: www.wjxr.com. Licensee: WJXR Inc. (acq 1-8-85; $335,000; FTR: 2-4-85). Natl. Network: ABC. Rgnl. Network: Florida Radio Net. Law Firm: Polner Law Firm. News staff: one; News: 7 hrs wkly. Target aud: 25-54; middle class & upscale families. ♦Gregory G. Perich, CEO, pres & gen mgr; Doug Rudowich, stn mgr & sls VP; Sarah Perich, opns mgr; Jerry Smith, chief of engrg.

***WKTZ-FM**— Feb 8, 1973: 90.9 mhz; 50 kw. Ant 500 ft TL: N30 16 36 W81 33 47. Stereo. Hrs opn: 24 5353 Arlington Expwy., 32211. Phone: (904) 743-1122. Phone: (904) 731-1184. Fax: (904) 743-4446. Web Site: www.wktz.jones.edu. Licensee: Jones College. (acq 2-7-86). Natl. Network: AP Radio. Format: Easy listening/smooth jazz. Target aud: 40 plus; mature adults. ♦Kenneth Jones, gen mgr, gen sls mgr & prom dir; Tom Buetow, mus dir; Dick Jones, chief of engrg.

WJAX(AM)—Co-owned with WKTZ-FM. 1958: 1220 khz; 1 kw-D, 37 w-N. TL: N30 19 30 W81 34 15.24 Phone: (904) 680-1220. Web Site: www.wjax.jones.edu. Natl. Network: CNN Radio. Format: Swing mus. Target aud: 40 plus.

WMXQ(FM)— November 1965: 102.9 mhz; 100 kw. 984 ft TL: N30 16 34 W81 33 53. Stereo. Hrs opn: 24 8000 Belfort Pky., 32256. Phone: (904) 245-8500. Fax: (904) 245-8501. E-mail: thepoint@1029i.com Web Site: www.1029i.com. Licensee: Cox Radio Inc. Group owner: Cox Communications Inc. (acq 2-23-2000; grpsl). Population served: 1,000,000 plus Natl. Rep: Christal. Format: 70s & 80s classic hits. News staff: one; News: 7 hrs wkly. Target aud: 25-49. ♦David Israel, gen mgr.

WNNR(AM)— Jan 1, 1969: 970 khz; 1 kw-U, DA-1. TL: N30 23 08 W81 40 04. Hrs open: 8384 Baymeadow Rd., Suite 1, 32256. Phone: (904) 739-3660. Fax: (904) 739-9409. Web Site: www.970thewinner.com. Licensee: Norsan Consulting and Management Inc. (acq 9-15-2005; $2.1 million with WVOJ(AM) Fernandina Beach). Population served: 1,058,500 Format: Sports. Target aud: General. ♦Norberto Sanchez, pres; Bernie Daigle, gen mgr & stn mgr; Marci Koziolek, news dir & pub affrs dir.

WOKV(AM)— November 1925: 690 khz; 50 kw-D, 10 w-N, DA-N. TL: N30 18 27 W81 56 28. Hrs open: 24 8000 Belfort Pkwy., 32206. Phone: (904) 245-8500. Fax: (904) 245-8501. E-mail: wokv.news@cox.com Web Site: www.wokv.com. Licensee: Cox Radio Inc. Group owner: Cox Broadcasting (acq 2-28-2000; grpsl). Population served: 500,000 Natl. Network: CBS. Law Firm: Cohn & Marks. Format: News/talk. ♦Dick Williams, gen mgr; Cat Thomas, opns mgr; Lindley Tolbert, sls dir; Allison Misora, prom dir; Mike Dorwart, progmg dir; Roxy Tyler, pub affrs dir; Dick Jones, engrg dir; Betty Glover, traf mgr.

WJGL(FM)—Co-owned with WOKV(AM). July 1, 1969: 96.9 mhz; 98 kw. Ant 1,014 ft TL: N30 16 34 W81 33 53. Stereo. 24 Web Site: www.cool969.com.751,000 Format: Oldies. News staff: one. Target aud: 25-54. ♦Scott Walker, progmg dir.

WPLA(FM)— May 9, 1977: 107.3 mhz; 100 kw. 705 ft TL: N30 21 48 W81 45 09. Stereo. Hrs opn: 24 11700 Central Pkwy, 32224-2600. Phone: (904) 636-0507. Fax: (904) 997-7713. Licensee: Clear Channel Radio Licenses Inc. Group owner: Clear Channel Communications Inc. (acq 11-21-97; grpsl). Population served: 1,000,000 Rgnl. Network: Florida Radio Net. Natl. Rep: Clear Channel. Format: Country. News staff: one. Target aud: 25-49. Spec prog: Relg 2 hrs wkly. ♦Norm Feuer, gen mgr; Gail Austin, opns mgr. Co-owned TV: WAWS(TV) affil

WQIK-FM— September 1964: 99.1 mhz; 100 kw. 1,050 ft TL: N30 16 34 W81 33 53. Stereo. Hrs opn: 24 Norm Feuer, 11700 Central Pkwy., 32224. Phone: (904) 642-3030. Fax: (904) 997-7707. E-mail: tanderson@ccjax.com Web Site: www.wqik.com. Licensee: Citicasters Licenses L.P. Group owner: Clear Channel Communications Inc. (acq 5-4-99; grpsl). Population served: 770,000 Natl. Network: ABC.

Format: Country. News staff: one; News: 4 hrs wkly. Target aud: 18-54. ♦John Hogan, sr VP; Norm Feuer, gen mgr; Gail Austin, opns dir & progmg dir; Tony Anderson, prom dir.

WROO(FM)—Callahan, June 1, 1983: 93.3 mhz; 50 kw. Ant 462 ft TL: N30 33 22 W81 33 13. Stereo. Hrs opn: 24 11700 Central Pkwy., 32224. Phone: (904) 636-0507. Fax: (904) 997-7713. E-mail: rhondagroff@clearchannel.com Web Site: www.planetradio933.com. Licensee: Clear Channel Radio Licenses Inc. Group owner: Clear Channel Communications Inc. (acq 11-21-97; grpsl). Format: Alternative. Target aud: 18-45. ♦Norm Feuer, gen mgr; Gail Austin, opns dir; Chad Chumley, progmg dir.

WROS(AM)— July 1955: 1050 khz; 5 kw-D, DA. TL: N30 21 14 W81 44 21. Hrs open: 6 AM-sunset 5590 Rio Grande Ave., 32254. Phone: (904) 353-1050. Fax: (904) 353-7076. E-mail: wros@wros.net Web Site: www.wros.net. Licensee: The Rose of Jacksonville (acq 6-1-85; FTR: 4-1-85). Population served: 1,000,000 Natl. Network: USA. Rgnl rep: NRB Format: Family oriented, Christian. News staff: 1; News: 7 hrs wkly. Target aud: 25-65; Christians & secular. Spec prog: 20 hrs wkly. ♦Elwyn V. Hall, CEO, pres & gen mgr; Dean Hall, exec VP; Dean Hall, opns VP; Yisrael Frees, progmg dir; Chris Cain, mus dir; Jerry Smith, chief of engrg.

WSOL-FM—Brunswick, GA) Sept 1, 1966: 101.5 mhz; 100 kw. 239 ft TL: N31 08 40 W81 34 56. (CP: Ant 1,463 ft.). Stereo. Hrs opn: 24 11700 Central Pkwy., 32224. Phone: (904) 996-0400 / (904) 642-3030. Fax: (904) 997-7713. E-mail: karmenbrooks@clearchannel.com Web Site: www.v1015.com. Licensee: Citicasters Licenses L.P. Group owner: Clear Channel Communications Inc. (acq 5-4-99; grpsl). Format: Adult urban contemp. ♦Norm Fever, gen mgr.

WYMM(AM)— Nov 18, 1976: 1530 khz; 50 kw-D, DA. TL: N30 21 50 W81 44 54. Hrs open: 6 AM-6 PM 5900 Pickettville Rd., 32254. Phone: (904) 786-2820. Fax: (904) 786-2661. Web Site: www.wymm1530.com. Licensee: Word Broadcasting Network Inc. (group owner; (acq 7-29-2003; $1.25 million with WYRM(AM) Norfolk, VA). Format: Talk. ♦Patrick Archuleta, gen mgr.

WZAZ(AM)— July 4, 1950: 1400 khz; 1 kw-U. TL: N30 19 43 W81 41 42. Stereo. Hrs opn: 24 4190 Belfort Rd., Suite 450, 32216. Phone: (904) 470-4615. Fax: (904) 296-1683. Web Site: www.1400wzaz.com. Licensee: Caron Broadcasting Inc. Group owner: Salem Communications Corp. (acq 5-30-03; grpsl). Population served: 6,000,000 Natl. Rep: Roslin. Format: Gospel. News staff: 2; News: 5 hrs wkly. Target aud: 25-54; adult Black listeners. ♦Henry Hoot, gen mgr; Calvin Grabau, progmg dir.

WZNZ(AM)— August 1942: 1460 khz; 5 kw-U, DA-N. TL: N30 19 40 W81 44 49. Stereo. Hrs opn: 4190 Belfort Rd., Suite 450, 32216. Phone: (904) 470-4615. Fax: (904) 296-1683. Web Site: www.espn1460.com. Licensee: Chesapeake-Portsmouth Broadcasting Corp. Group owner: Salem Communications Corp. (acq 12-5-2006; $1.8 million with WBOB(AM) Jacksonville). Population served: 40,000 Natl. Rep: Clear Channel. Law Firm: Fletcher, Heald & Hildreth. Format: Sports. Target aud: 25-54. ♦Henry Hoot, gen mgr; Calvin Grabau, progmg dir.

Jacksonville Beach

WJXL(AM)— 1946: 1010 khz; 10 kw-D, 143 w-N, DA-2. TL: N30 17 21 W81 33 01. (CP: 50 kw-D, 30 kw-N, DA-2. TL: N30 17 57 W82 00 26). Hrs opn: 24 9090 Hogan Rd., Jacksonville, 32216. Phone: (904) 641-1010 / 1011. Fax: (904) 641-1022. E-mail: wioj@wioj.net Web Site: www.1010xl.com. Licensee: Seven Bridges Radio LLC (acq 2-9-2007; $3,825,000). Population served: 1,000,000 Natl. Network: Salem Radio Network, USA. Law Firm: Garvey, Schubert & Barer. Format: Sports. ♦Steven L. Griffin, pres; Steven Griffin, gen mgr; Ed Furbee, news dir; Chris Lane, progmg.

Jensen Beach

WMBX(FM)— Dec 10, 1980: 102.3 mhz; 100 kw. 974 ft TL: N27 01 32 W80 10 43. (CP: 100 kw). Stereo. Hrs opn: 24 701 Northpoint Pkwy., Suite 400, West Palm Beach, 33407. Phone: (561) 616-4600. Fax: (561) 684-6311. Web Site: www.mix1023.com. Licensee: Infinity Radio Operations Inc. Group owner: Infinity Broadcasting Corp. (acq 12-14-00; grpsl). Law Firm: Rosenman & Colin. Format: Modern adult contemp. News staff: one; News: 17 hrs wkly. Target aud: 25-54; women. ♦Patricia A. Larschan, VP & gen mgr; John O'Connell, opns mgr & progmg dir; Mark Krieger, gen sls mgr & natl sls mgr; Danelle Sarvas, prom mgr; Jeff Clarke, mus dir; Pam Crosby, news dir & pub affrs dir; Scott Paxson, chief of engrg.

Jupiter

WJBW(AM)— 1997: 1000 khz; 650 w-D, DA. TL: N26 56 40 W80 05 30. Hrs open: 24 1235 16th St., Vero Beach, 32960. Phone: (772) 567-0937. Fax: (772) 562-4747. Licensee: AM of Palm Beach Inc. Group owner: James Crystal Inc. (acq 4-13-2006). Format: News/talk. ♦Laurie S. Silvers, pres; Jim Davis, gen mgr.

WNEW(FM)— Oct 15, 1971: 106.3 mhz; 19 kw. Ant 374 ft TL: N26 47 59 W80 04 33. Hrs open: 701 Northpoint Pkwy., Suite 500, West Palm Beach, 33407. Phone: (561) 684-7400. Fax: (561) 686-9505. Web Site: www.b1063fm.com. Licensee: CBS Radio Stations Inc. Group owner: Infinity Broadcasting Corp. (acq 8-30-2001; $20 million). Format: Urban adult contemp. ♦Lee Strasser, sr VP & gen mgr; Mark McCray, opns mgr.

Kendall

WRHB(AM)— August 1999: 1020 khz; 8.9 kw-D, 980 w-N, DA-2. TL: N25 37 09 W80 31 00. Hrs open: 75 NW 167th St., N. Miami Beach, 33169. Phone: (305) 446-5444 / (305) 493-1020. Fax: (305) 446-1009 / (305) 493-1111. Web Site: www.radiomega.net. Licensee: New World Broadcasting Inc. (acq 12-20-01; $260,000 for stock for 52%). Format: Ethnic. Target aud: 25-55; adult. ♦Alex Saintsuin, gen mgr.

Key Colony Beach

WKYZ(FM)— April 15, 1999: 101.3 mhz; 50 kw. 276 ft TL: N24 41 30 W81 06 31. Hrs open: 24 Box 500940, Marathon, 33050. Phone: (305) 289-1013. Fax: (305) 743-9441. Licensee: Keys Media Co. Inc. Format: Classic Rock. Target aud: 25-54. ♦Joe Nascone, gen mgr.

Key Largo

***WGES-FM**— 2004: 90.9 mhz; 33 kw. Ant 308 ft TL: N25 14 07 W80 19 35. Hrs open: 11890 S. W. 8th St., Suite 504, South Florida, 33082. Phone: (305) 551-6590. Fax: (305) 551-2737. Licensee: Genesis License Subsidiary LLC. Format: Christian, Sp. ♦Edwin Lemuel Ortiz, pres; Kenny Reyes, gen mgr.

***WMKL(FM)**— Oct 1, 1998: 91.9 mhz; 50 kw vert. Ant 308 ft TL: N25 14 07 W80 19 35. Stereo. Hrs opn: 24 Box 561832, Miami, 33256-1832. Phone: (305) 662-7736. Fax: (305) 251-2293. E-mail: callfm@callfm.com Web Site: www.callfm.com. Licensee: Call Communications Group Inc. (acq 11-4-99; $295,000). Population served: 220,000 Law Firm: Gammon & Grange. Format: Christian. Target aud: 13-25. ♦Robert Robbins, pres & gen mgr; Kelly Downing, mus dir; Jim Sorensen, chief of engrg.

WZMQ(FM)— Jan 20, 1990: 106.3 mhz; 6 kw. 150 ft TL: N25 05 29 W80 26 37. (CP: 50 kw, ant 239 ft. TL: N21 01 35 W80 30 30). Stereo. Hrs opn:
Rebroadcasts WRMA (FM) Fort Lauderdale 100%.
1001 Ponce De Leon Blvd., Coral Gables, 33134. Phone: (305) 444-9292. Fax: (305) 461-4466. Web Site: www.lamusica.com. Licensee: South Broadcasting System Inc. (acq 1-27-00; $1 million with WMFM(FM) Key West). Format: Pop latin ballads. ♦Raoul Alarcon, pres; Jackie Nosti-Cambo, gen mgr; John Caride, prom dir; Ralph Chambers, chief of engrg.

Key West

WAIL(FM)— December 1978: 99.5 mhz; 100 kw. 991 ft TL: N24 39 25 W81 32 18. (CP: Ant 239 ft.). Stereo. Hrs opn: 24 5450 McDonald Ave, Suite 10, 33040. Phone: (305) 296-7511. Fax: (305) 296-0358. E-mail: kenmackenzie@clearchannel.com Web Site: www.wail995.com. Licensee: Clear Channel Radio Licenses Inc. Group owner: Clear Channel Communications Inc. (acq 6-5-98; $2.6 million with WEOW(FM) Key West). Rgnl. Network: Florida Radio Net. Format: Classic rock. Target aud: 25-54; men. ♦Greg Capogna, VP, gen mgr & gen mgr; Sherry Russo, stn mgr; Ken MacKenzie, opns dir.

***WAZQ(FM)**— June 1, 2005: 88.3 mhz; 740 w. Ant 112 ft TL: N24 33 07 W81 47 53. Stereo. Hrs opn: 6910 N.W. 2nd Terr., Boca Raton, 33487. Phone: (561) 912-9002. Fax: (561) 912-9003. E-mail: bill@qfmonline.com Licensee: Educational Public Radio Inc. Format: Hot adult contemp. Target aud: 18-54; adults. ♦Bill Lacy, pres.

WCNK(FM)— January 1986: 98.7 mhz; 100 kw. Ant 300 ft TL: N24 34 42 W81 44 49. Stereo. Hrs opn: 24 30336 Overseas Hwy., Big Pine Key, 33043. Phone: (305) 872-9100. Fax: (305) 872-8930. Web Site: www.conchcountry.com. Licensee: Vox Communications Group LLC. (acq 8-31-2005; grpsl). Format: Country. News: one hr wkly. Target aud: 24-55; military, baby boomers & largest income holders. Spec prog: Armed Forces news one hr wkly. ♦Kevin LeRoux, gen mgr & stn mgr.

WEOW(FM)— February 1967: 92.7 mhz; 100 kw. Ant 551 ft TL: N24 40 35 W81 30 41. Stereo. Hrs opn: 24 5450 MacDonald Ave., Suite 10, 33040. Phone: (305) 296-7511. Fax: (305) 296-0358. E-mail: kenmackenzie@clearchannel.com Web Site: www.weow927.com. Licensee: Clear Channel Radio Licenses Inc. Group owner: Clear Channel Communications Inc. (acq 6-5-98; $2.6 million with WAIL(FM) Key West). Population served: 76,000 Rgnl. Network: Florida Radio Net. Format: CHR, Top-40. ♦Mark Mays, pres; Sherry Russo, gen mgr.

WIIS(FM)— June 1978: 107.1 mhz; 3 kw. 200 ft TL: N24 33 18 W81 48 07. Stereo. Hrs opn: 24 1075 Duval St., Suite 17, 33040. Phone: (305) 292-1133. Phone: (305) 292-1071. Fax: (305) 292-6936. E-mail: johnrussin@hotmail.com Web Site: www.radiokeywest.com. Licensee: The Keyed Up Communications Co. (acq 1995; $275,000). Population served: 35,000 Law Firm: Fletcher, Heald & Hildreth. Format: Alternative. News staff: one; News: 5 hrs wkly. Target aud: 18-44; young, educated, active spenders for goods & services. Spec prog: Reggae 4 hrs, Metropolitan opera 4 hrs wkly. ♦John Russin, CEO, sls dir & traf mgr; Linda Russin, COO.

***WJIR(FM)**— December 1986: 90.9 mhz; 390 w. Ant 121 ft TL: N24 33 07 W81 47 53. Stereo. Hrs opn: 24 1209 United St., 33040. Phone: (305) 296-4306. Fax: (305) 294-9547. E-mail: pastorernie@bellsouth Licensee: Key West Educational Broadcasting. (acq 12-15-85). Format: Relg, educ, Christian. News: 12 hrs wkly. Target aud: General. ♦Ernie DeLoach, stn mgr & opns mgr.

WKEY-FM— Nov 17, 1985: 93.5 mhz; 32 kw. Ant 138 ft TL: N24 34 17 W81 44 25. Stereo. Hrs opn: 24 5450 MacDonald Ave., 33040. Phone: (305) 296-7511. Fax: (305) 296-1155. E-mail: kenmackenzie@clearchannel.com Web Site: www.key93.com. Licensee: Clear Channel Broadcasting Licenses Inc. Group owner: Clear Channel Communications Inc. (acq 11-21-97; grpsl). Population served: 33,000 Rgnl. Network: Florida Radio Net. Natl. Rep: Clear Channel. Format: Adult contemp. News staff: one; News: 8 hrs wkly. Target aud: 25-54; affluent, upscale, culturally supportive. Spec prog: Classical 4 hrs, Sp 3 hrs wkly. ♦Greg Capogna, gen mgr; John Stuempfig, prom dir; Sherry Russo, stn mgr, gen sls mgr & adv mgr.

WKIZ(AM)— Feb 2, 1959: 1500 khz; 250 w-U, DA-1. TL: N24 34 01 W81 44 54. Stereo. Hrs opn: 24 5016 5th Ave., 33040. Phone: (305) 293-9536. Fax: (305) 293-1793. E-mail: wkizradio@aol.com Web Site: www.wkizradio.com. Licensee: Seattle Streaming Radio L.L.C. Population served: 30,803 Natl. Network: CBS. Format: News/talk. ♦Jim Spreitzer, gen sls mgr.

WKWF(AM)— October 1945: 1600 khz; 500 w-U. TL: N24 34 30 W81 44 01. Hrs open: 24 5450 MacDonald Ave., Ste. 10, 33040. Phone: (305) 296-7511. Fax: (305) 296-0358. E-mail: Toddswofford@clearchannel.com Web Site: www.keysradio.com. Licensee: Spottswood Partners II Ltd. (acq 10-27-97; with co-located FM). Population served: 35,000 Format: Prime sports. ♦Greg Capogna, VP & gen mgr; Sherry Russo, stn mgr & sls dir; Todd Swofford, progmg dir.

***WKWR(FM)**— 2005: 90.1 mhz; 250 w. Ant 69 ft TL: N24 34 05 W81 44 53. Hrs open: Broadcasting for the Challenged Inc., 6080 Mount Moriah Ext., Memphis, TN, 38115. Phone: (901) 375-9324. Fax: (901) 375-0041. Licensee: Broadcasting for the Challenged Inc. Natl. Network: K-Love. Format: Contemp Christian. ♦George Flinn Jr., gen mgr.

WMFM(FM)— 1995: 107.9 mhz; 100 kw. 548 ft TL: N24 39 08 W81 32 04. Hrs open:
Rebroadcasts WXDJ (FM) North Miami Beach 100%.
1001 Ponce de Leon Blvd., Coral Gables, 33134. Phone: (305) 447-9292. Fax: (305) 461-4466. Web Site: www.lamusica.com. Licensee: South Broadcasting System Inc. (acq 1-27-00; $1million with WZMQ(FM) Key Largo). Format: Salsa. ♦Jackie Nosti-Combo, gen mgr; Albert Rodriguez, gen sls mgr.

Kissimmee

WHOO(AM)— April 1965: 1080 khz; 19 kw-D, 190 w-N, 10 kw-CH, DA-3. TL: N28 20 30 W81 20 26. Hrs open: 18 1160 S Semoran Blvd., Suite A, Orlando, 32807. Phone: (407) 380-9255. Fax: (407) 382-7565. E-mail: studio@espn1080.com Web Site: espnflorida.com. Licensee: Genesis Communications I Inc. Group owner: Genesis Communications Inc. (acq 10-19-99). Population served: 1,500,000 Natl. Network: ESPN Radio. Natl. Rep: Interep. Law Firm: Booth, Freret, Imlay & Tepper. Format: Sports talk. Target aud: 25-54; men. Spec prog: Creole 2 hrs wkly. ♦Bruce Maduri, pres; Sandra Culver, VP; Jeff Taylor, opns mgr; Joe Nichols, gen sls mgr; Chris Visser, progmg VP.

***WLAZ(FM)**— 2000: 89.1 mhz; 1.1 kw vert. Ant 535 ft TL: N28 10 27 W81 17 01. Hrs open: 24 415 W. Vine St., 34741. Phone: (407) 518-7150. Phone: (407) 208-0333. Fax: (407) 518-0062. E-mail: contacto@gensis89.com Web Site: www.genesis89.net. Licensee: Caguas Educational TV Inc. (acq 5-3-02; $1.5 million). Law Firm: James L. Oyster. Format: Christian, Sp. News staff: 2; News: 10 hrs wkly. ♦William Gutierrez, pres & gen mgr.

WOTS(AM)— Oct 23, 1978: 1220 khz; 1 kw-D. TL: N28 19 27 W81 23 44. Hrs open: 24 222 Hazard St., Orlando, 32804-3030. Phone: (407) 841-8282. Fax: (407) 841-8250. Licensee: J&V Communications Inc. (group owner; acq 1-12-99). Law Firm: Leibowitz & Associates. Format: Sp, relg. News staff: one. Tourists. Spec prog: Imus in the Morning. ♦John Torrado, CEO, pres & gen mgr; Frank Vaught, opns mgr.

La Belle

***WBIY(FM)**— 1999: 88.3 mhz; 3 kw. Ant 161 ft TL: N26 44 26 W81 27 46. Hrs open: 24 500 W. Hickpochee Ave., 33935. Phone: (863) 674-0033. Fax: (863) 675-7584. Licensee: Oscar Aguero Ministry Inc. (acq 6-13-2006; $900,000). Format: Sp contemp Christian. ♦Oscar Aguero, pres; Roger Martinez, stn mgr.

La Crosse

WBXY(FM)— October 1993: 99.5 mhz; 2.2 kw. 472 ft TL: N29 44 22 W82 23 09. Stereo. Hrs opn: 24 4424 N.W.13th St., Suite C-5, Gainesville, 32609. Phone: (352) 375-1317. Fax: (352) 375-6961. E-mail: feedback@thestar.com Web Site: www.thestar.fm. Licensee: Asterisk Communications Inc. (acq 7-23-98; $1.15 million). Natl. Network: Westwood One, Talk Radio Network, ABC, Fox Sports. Natl. Rep: McGavren Guild. Format: Talk, news, sports. Target aud: 25-54; primarily baby boomers-upscale. ♦John Starr, gen mgr & adv mgr; Steve Cox, progmg dir.

Lafayette

WEGT(FM)—Licensed to Lafayette. See Tallahassee

Lake City

WDSR(AM)— May 6, 1946: 1340 khz; 1 kw-U. TL: N30 09 20 W82 38 14. Hrs open: 24 2485 S. Marion St., 32025. Phone: (386) 752-1340. Fax: (386) 755-9369. E-mail: wnfb@mix943.com Web Site: www.mix943.com. Licensee: Newman Media Inc. (acq 9-3-98; $750,000 with co-located FM). Population served: 10,575 Natl. Network: CBS. Rgnl rep: Florida's Radio Net. Format: Southern gospel. Target aud: 18-54; upscale males. Spec prog: Black 2 hrs wkly.

WNFB(FM)—Co-owned with WDSR(AM). May 28, 1969: 94.3 mhz; 50 kw. 492 ft TL: N30 07 44 W82 52 49. Stereo. 24 Phone: (386) 961-9494. Web Site: www.mix943.com.50,000 Format: Adult contemp.

WGRO(AM)— Nov 14, 1958: 960 khz; 500 w-D, 1 kw-N, DA-N. TL: N30 11 47 W82 40 48. (CP: 1 kw-U, DA-N). Hrs opn: 6 AM-midnight (M-S); 7 AM-11 PM (Su) 9206 US Hwy. 90 W., 32055. Phone: (386) 755-4102. Fax: (386) 752-9861. E-mail: bandk@ISgroup.net Licensee: Power Country Inc. (acq 9-20-95). Population served: 10,575 Rgnl. Network: Florida Radio Net. Format: Country gospel. ♦Louis Bolton II, pres; Bob Hendrickson, gen mgr.

***WOLR(FM)**— Sept 11, 1986: 91.3 mhz; 18 kw vert. 285 ft TL: N30 02 56 W82 48 44. Hrs open: 3332 220th Pl., 32024. Phone: (386) 935-3300. Fax: (386) 935-2684. Web Site: www.christianhitradio.net. Licensee: WOLR 91.3 FM Inc. (acq 6-25-93; $75,000; FTR: 7-19-93). Format: Christian. ♦Rita Loos, gen mgr; Chris Hall, chief of engrg.

Lake Placid

WWTK(AM)— 1989: 730 khz; 500 w-D, 340 w-N, DA-1. TL: N27 24 25 W81 25 56. Hrs open: 24 3750 U.S. 27 N., Suite One, Sebring, 33870. Phone: (863) 382-9999. Fax: (863) 382-1982. E-mail: cohanradiogroup@htn.net Web Site: www.cohanradiogroup.com. Licensee: Cohan Radio Group Inc. (group owner; (acq 11-1-98; $910,000 with WWOJ(FM) Avon Park). Population served: 80,000 Natl. Network: USA, CBS Radio, ABC, Premiere Radio Networks, Talk Radio Network, Westwood One. Rgnl. Network: Florida Radio Net. Rgnl rep: Interep Law Firm: Latham & Watkins. Wire Svc: AP Format: Talk. News staff: one; News: 120 hrs wkly. Target aud: 35+. Spec prog: U of Fla Football & Basketball Sports Show 1 hr. per wk Sat 12n-1pm; Rel Su 6-11am; Local Talk Show M-F 8-10am. ♦Peter L. Coughlin, pres & gen sls mgr; Libby Coughlin, rgnl sls mgr; Barry Foster, news dir; Phil Scott, chief of engrg; Connie Bedingfield, traf mgr.

Lake Wales

WIPC(AM)— July 1951: 1280 khz; 1 kw-D, 500 w-N, DA-N. TL: N27 55 34 W81 36 04 (D), N27 55 30 W81 36 16 (N). Hrs open: 24 630 Mountain Lake Cut/Off Rd., Suite A, 33859. Phone: (863) 679-7178. Fax: (863) 679-9395. E-mail: wipc1280@yahoo.com Licensee: Super W Media Group Inc. (acq 9-19-2005). Population served: 485,000 Law Firm: Donald E. Martin. Format: Sp. News staff: one; News: 10 hrs wkly. Target aud: 18+ Hispanics. ♦Carl Czuchaj, VP & opns mgr; Robert Cubero, pres, gen mgr & gen sls mgr; Edward Olivares, progmg dir; Agustin E. Olivares, mus dir; Guadalupe Gonzales, traf mgr.

Lake Worth

WWRF(AM)— May 1, 1959: 1380 khz; 1 kw-D, 500 w-N. TL: N26 37 23 W80 04 20. Hrs open: 24 2326 S. Congress Ave., Suite 2A, W. Palm Beach, 33406-7614. Phone: (561) 585-1380. Phone: (561) 721-9950. Fax: (561) 721-9973. E-mail: liza@radiofiesta.com Web Site: www.gladesmedia.com. Licensee: Radio Fiesta Inc. (acq 2-24-00; $400,000). Population served: 1225000 Format: Regional Mexican. News staff: one. Target aud: 25-54; Hispanic. Spec prog: Sp relg 4 hrs wkly. ♦Liza Flores, pres & gen mgr; Robbie Castellanos, pres & stn mgr.

Lakeland

***WKES(FM)**— May 20, 1975: 91.1 mhz; 100 kw. 500 ft TL: N28 04 46 W82 02 27. (CP: 420 ft.). Stereo. Hrs opn: 24 5800 100th Way N., St. Petersburg, 33708. Phone: (727) 391-9994. Fax: (727) 397-6425. E-mail: wkes@moody.edu Web Site: www.wkes.fm. Licensee: The Moody Bible Institute of Chicago. (group owner; acq 10-10-96; $5 million). Population served: 3,600,000 Law Firm: Southmayd & Miller. Format: Relg, educ. ♦Michael Easley, pres; Michael Gleichman, opns mgr & pub affrs dir; Pierre Chestang, stn mgr & progmg dir; John Stortz, chief of engrg.

WLKF(AM)— 1936: 1430 khz; 5 kw-D, 1 kw-N. TL: N28 02 27 W81 56 08. Hrs open: 24 Box 2038, 33815. Secondary address: 404 W. Lime St. 33815-4651. Phone: (863) 682-8184. Fax: (863) 683-2409. E-mail: talk1430@wlkf.com Web Site: www.wlkf.com. Licensee: Hall Communications Ltd. Group owner: Hall Communications Inc. (acq 10-1-96; $550,000). Population served: 500,000 Natl. Network: ABC. Rgnl. Network: Florida Radio Net. Natl. Rep: D & R Radio. Law Firm: Fletcher, Heald & Hildreth. Wire Svc: AP Format: News/talk. News: 15 hrs wkly. Target aud: 35 plus; middle to upper income adults. ♦Bonnie H. Rowbotham, chmn; Arthur J. Rowbotham, pres; Bill Baldwin, exec VP & sr VP; Nancy Cattarius, stn mgr.

WONN(AM)— Sept 15, 1949: 1230 khz; 1 kw-U. TL: N28 02 23 W81 57 39. Stereo. Hrs opn: 24 Box 2038, 33806. Secondary address: 404 W. Lime St. 33815-4651. Phone: (863) 682-8184. Phone: (407) 297-1201. Fax: (863) 683-2409. E-mail: wonn@wonn.com Web Site: www.wonn.com. Licensee: Hall Communications Inc. (group owner; acq 10-1-81; $2 million with co-located FM; FTR: 8-10-81) Population served: 500,000 Natl. Network: CNN Radio. Natl. Rep: D & R Radio. Law Firm: Fletcher, Heald & Hildreth. Format: MOR. News: 20 hrs wkly. Target aud: 35 plus. Spec prog: Relg 2 hrs wkly. ♦Bonnie H. Rowbotham, chmn; Arthur J. Rowbotham, pres & gen mgr; Bill Baldwin, sr VP.

WPCV(FM)—Co-owned with WONN(AM). 1962: 97.5 mhz; 100 kw. 1,017 ft TL: N28 07 35 W81 33 03. Stereo. 24 E-mail: wpcv@wpcv.com Web Site: www.wpcv.com.500,000 Format: Country. News staff: one; News: 4 hrs wkly. Target aud: 25-54.

WSJT(FM)— Sept 11, 1967: 94.1 mhz; 100 kw. 1,059 ft TL: N27 59 56 W81 53 16. (CP: Ant 1,492 ft.). Stereo. Hrs opn: 24 9721 Executive Center Dr. N., Suite 200, St. Petersburg, 33702-2439. Phone: (727) 568-0941. Fax: (727) 568-9758. E-mail: smoothjazz@wsjt.com Web Site: www.wsjt.com. Licensee: Infinity Radio Inc. Group owner: Infinity Broadcasting Corp. (acq 1999; grpsl). Population served: 2,500,000 Natl. Rep: Clear Channel. Format: Jazz. News staff: one. Target aud: 25-54; middle to upper income adults, skews towards females. ♦Charlie Ochs, gen mgr & sls dir; Rose Bobier, prom dir; Ross Block, opns mgr & progmg dir.

WWAB(AM)— September 1957: 1330 khz; 1 kw-D. TL: N28 02 40 W81 58 28. Hrs open: Box 65, 33802. Secondary address: 1203 Chase St. 33802. Phone: (863) 682-2998. Fax: (863) 683-9922. E-mail: wwab@verizon.net Web Site: www.wwab1330.com. Licensee: WWAB Inc. (acq 1-16-73). Population served: 60,000 Format: Rhythm and blues, talk, gospel. Target aud: 18-49. Spec prog: Gospel 12 hrs wkly. ♦Jerry Hughes, gen mgr; Hugh Hughes, stn mgr & gen sls mgr; Frank Clark, opns mgr.

***WYFO(FM)**— March 1988: 91.9 mhz; 25 kw horiz, 23 kw vert. 328 ft TL: N27 56 35 W81 54 45. Hrs open: 11530 Carmel Commons Blvd., Charlotte, NC, 28226. Phone: (704) 523-5555. Fax: (704) 522-1967. E-mail: wyfo@bbnradio.org Web Site: www.bbnradio.org. Licensee: Bible Broadcasting Network Inc. (group owner; acq 9-21-89; $200,000; FTR: 10-16-89). Format: Christian, educ. ♦Doug Roby, stn mgr.

Lantana

WPBR(AM)— 1941: 1340 khz; 1 kw-U. TL: N26 36 41 W80 02 17. (CP: TL: N26 33 26 W80 04 20). Hrs opn: 24-7 1217 S. Military Trail, Suite E, West Palm Beach, 33415-4600. Phone: (561) 641-8882. Fax: (561) 641-8629. E-mail: adminandsales@1340wpbr.com Web Site: www.talk1340wpbram.com. Licensee: Omni-Lingual Broadcasting Corp. (acq 3-4-94; $700,000; FTR: 5-9-94). Population served: 740,000 Natl. Network: USA. Format: News/talk, community progmg. Target aud: 35-64. Spec prog: Financial 9 hrs, medical 8 hrs, Jewish 3 hrs, Creole 30 hrs wkly. ♦Emil Antonoff, pres; Markes Pierre Louis, gen mgr.

Largo

WMGG(AM)— May 29, 1972: 820 khz; 50 kw-D, 1 kw-N, DA-2. TL: N27 54 30 W82 46 51. Hrs open: 24 1916 N. Dale Mabry Hwy., Suite 200, Tampa, 33607. Phone: (813) 871-1819. Fax: (813) 871-1155. E-mail: mnavarro@megastations.net Web Site: www.megaclasica.com. Licensee: Mega Communications of St. Petersburg. Group owner: Mega Communications Inc. (acq 1999; grpsl). Population served: 2800000 Format: Sp. Oldies, Tropical. Target aud: 25-54. ♦Cecillia Uebel, gen mgr; Victor Manuel Pacheco, gen sls mgr; Rafael Grullon, progmg VP & progmg dir; Robert Hailey, chief of engrg; Zaida Torres, traf mgr.

Lecanto

***WLMS(FM)**— September 1992: 88.3 mhz; 3.8 kw. Ant 259 ft TL: N28 52 55 W82 31 30. Hrs open:
Rebroadcasts WBVM(FM) Tampa 100% (simulcast).
Box 18081, Tampa, 33629. Secondary address: 3816 Morrison Ave., Tampa 33629. Phone: (813) 289-8040. Fax: (813) 282-3580. E-mail: contact@spiritfm905.com Web Site: www.spiritfm905.com. Licensee: Bishop of the Diocese of St. Petersburg. Format: Contemp Christian Music. ♦John Morris, gen mgr; Chris Sampson, opns mgr.

Leesburg

WLBE(AM)— August 1949: 790 khz; 5 kw-D, 1 kw-N, DA-N. TL: N28 49 00 W81 46 45. Hrs open: 24 32900 Radio Rd., 34788. Phone: (352) 787-7900. Fax: (352) 787-1402. E-mail: info@am790wlbe.com Licensee: WLBE 790 Inc. Population served: 140,000 Natl. Network: CBS. Rgnl. Network: Florida Radio Net. Natl. Rep: Dora-Clayton. Format: 50% talk, 50% music. News: 25 hrs wkly. Target aud: 45 plus. Spec prog: Black 3 hrs, farm 3 hrs, gospel 4 hrs, Pol 2 hrs wkly. ♦MJ McNair, gen mgr.

WQBQ(AM)— Sept 12, 1962: 1410 khz; 5 kw-D, 90 w-N. TL: N28 47 13 W81 53 26. Stereo. Hrs opn: Rama Communications Inc., 3765 N. John Young Pkwy., Orlando, 32804. Phone: (407) 523-2770. Fax: (407) 523-2888. Licensee: Rama Communications Inc. (group owner;

(acq 10-15-2004; $180,000 with WKIQ(AM) Eustis). Population served: 200,000 Format: Spanish. ♦Sabeta Persaud, pres; Steve January, gen mgr.

WVLG(AM)—Wildwood, September 1987: 640 khz; 930 w-D, 860 w-N. TL: N28 54 16 W81 57 36. Hrs open: 6 AM-midnight 1161 Main St., The Villages, 32159. Phone: (352) 753-1119. Fax: (352) 259-4819. Web Site: thevillagesdailysun.com/wvlg/WVLG.html. Licensee: Senior Broadcasting Corp. (acq 9-12-00; $1.05 million). Rgnl. Network: S.E. Agri. Format: Var/div. Target aud: 18 plus; general. ♦Skip Diegel, gen mgr.

WXXL(FM)—Tavares, Feb 12, 1969: 106.7 mhz; 100 kw. 823 ft TL: N28 33 31 W81 35 38. Stereo. Hrs opn: 24 2500 Maitland Center Pkwy., Suite 401, Maitland, 32751-7407. Phone: (407) 916-7800. Fax: (407) 916-7510. Web Site: www.wxxl.com. Licensee: AMFM Radio Licenses L.L.C. Group owner: Clear Channel Communications Inc. (acq 8-30-00; grpsl). Population served: 1,000,000 Law Firm: Wiley, Rein and Fielding. Format: CHR. News staff: one. Target aud: 18-49; general. Spec prog: Alternative 6 hrs wkly. ♦Linda Byrd, VP & gen mgr; Sam Nein, sls dir; Shannon Fraser, gen sls mgr; Frank Celebre, natl sls mgr; Rick Everett, mktg dir; Glory Adona, prom mgr; Chris Kampmeier, progmg dir; Donna Carmichael, traf mgr.

Lehigh Acres

WCKT(FM)— Jan 1, 1976: 107.1 mhz; 23.5 kw. Ant 722 ft TL: N26 19 00 W81 47 13. Stereo. Hrs opn: 24 13320 Metro Pkwy., Fort Myers, 33912. Phone: (239) 225-4300. Phone: (800) 827-1071. Fax: (239) 275-4669. Web Site: www.wckt.com. Licensee: Clear Channel Broadcasting Licenses Inc. Group owner: Clear Channel Communications Inc. (acq 1996; grpsl). Population served: 750,000 Natl. Rep: Clear Channel. Format: Country. News staff: one; News: 20 hrs wkly. Target aud: 25-54. ♦Lowry Mays, chmn; Mark Mays, pres; Randall Mays, CFO; Jay Meyers, sr VP; Jim Keating, gen mgr; Steve Amari, opns dir; Robin Craig, sls dir; Dave Logan, mus dir; Church Morgan, news dir; Dick Parrish, chief of engrg; Debie Lummus, traf mgr.

WWCL(AM)— Apr 29, 1970: 1440 khz; 5 kw-D, 1 kw-N, DA-2. TL: N26 36 05 W81 33 30. Hrs open: 7573 N.W. First St., 33972. Phone: (239) 337-1440. Fax: (239) 369-3386. E-mail: energi1440@aol.com Licensee: Latino Media Corp. (acq 10-12-2006; $1.4 million). Population served: 16,000 Format: Sp, Mexican. ♦Angel Ramos, gen mgr.

Live Oak

WQHL(AM)— June 16, 1949: 1250 khz; 1 kw-D, 83 w-N. TL: N30 17 14 W82 57 56. Hrs open: 24 1305 Helvenston St. SE, 32064-3465. Phone: (386) 362-1250. Phone: (386) 364-3502. Fax: (386) 364-3504. Web Site: www.wqhlcounrty.com. Licensee: RTG Radio LLC. Group owner: Black Crow Media Group LLC (acq 11-9-2001; grpsl). Population served: 46,000 Rgnl. Network: Florida Radio Net. Format: Oldies radio. News: 12 hrs wkly. Target aud: General. Spec prog: Gospel 7 hrs, relg 6 hrs wkly. ♦Dean Blackwell, gen mgr.

WQHL-FM— October 1973: 98.1 mhz; 50 kw. 420 ft TL: N30 17 14 W82 57 56. Stereo. 24 E-mail: dean@wqhlcountry.com Web Site: wqhlcountry.com.250,000 Format: Country. Target aud: General.

Lynn Haven

WWHV(FM)—Not on air, target date: unknown: 104.3 mhz; 6 kw. Ant 249 ft TL: N30 10 47 W85 38 11. Hrs open: 1670 N.W. Federal Hwy., Stuart, 34994. Phone: (772) 692-9454. Fax: (772) 692-0258. Licensee: Hroton Broadcasting Co. Inc. ♦George Metcalf, pres.

Macclenny

WJXR(FM)—Licensed to Macclenny. See Jacksonville

Madison

***WAPB(FM)**— 2005: 91.7 mhz; 200 w. Ant 224 ft TL: N30 27 13 W83 24 17. Hrs open: Box 250, Dayton Beach, 32125. Secondary address: 1508 State Ave., Dayton Beach 32125. Phone: (386) 677-4272. Fax: (386) 673-3715. E-mail: wapn@wapn.net Web Site: www.wapb.net. Licensee: Public Radio Inc. Format: Word & praise. ♦Shellye Lund-Vallance, pres.

WMAF(AM)— Dec 6, 1956: 1230 khz; 1 kw-U. TL: N30 28 23 W83 26 09. Hrs open: Box 621, 32341. Secondary address: 2 Captain Brown Rd. 32341. Phone: (850) 973-3233. Fax: (850) 973-3097. E-mail: countrywmaf@earthlink.net Web Site: www.wmafcountry.com. Licensee: Geneva Walker. (acq 1996). Population served: 75,000 Rgnl. Network: Florida Radio Net. Format: Classic country. Spec prog: Oldies, gospel. ♦Betty Evertt, gen mgr.

WXHT(FM)— November 2000: 102.7 mhz; 19 kw. Ant 377 ft TL: N30 38 23 W83 26 52. Stereo. Hrs opn: 1711 Ellis Dr., Valdosta, GA, 31602. Phone: (229) 244-8642. Fax: (229) 242-7620. Web Site: www.hot10227wxnt.com. Licensee: RTG Radio L.L.C. Group owner: Black Crow Media Group LLC (acq 6-4-2004; $3.4 million with WSTI-FM Quitman, GA). Law Firm: Rini Coran PC. Format: Contemporary hits. Target aud: Adults 18-49. ♦Robert Ganzak, pres & gen mgr.

Maitland

WPYO(FM)— Sept 1, 1968: 95.3 mhz; 12 kw. Ant 472 ft TL: N28 34 27 W81 27 46. Hrs open: 24 4192 N. John Young Pkwy., Orlando, 32804. Phone: (407) 422-9696. Fax: (407) 422-5883. Web Site: www.power953.com. Licensee: Cox Radio Inc. Group owner: Cox Broadcasting (acq 1999; $14.5 million). Population served: 1,500,000 Format: Hip hop. Target aud: General. ♦Brian Elam, gen mgr.

Marathon

WAVK(FM)— 2002: 97.7 mhz; 50 kw horiz, 49 kw vert. Ant 213 ft TL: N24 46 02 W80 56 42. Stereo. Hrs open: 24 11399 Overseas Hwy., 33050. Phone: (305) 743-3434. Fax: (305) 743-9091. E-mail: mail@wave-fm.com Web Site: www.wavk-fm.com. Licensee: Vox Communications Group LLC. (acq 8-31-2005; grpsl). Format: Hot adult contemp. ♦Kevin Leroux, gen mgr.

WFFG(AM)— Apr 7, 1962: 1300 khz; 2.5 kw-U, DA-1. TL: N24 41 28 W81 06 30. Hrs open: 24 Box 500940, One Boot Key, 33050. Phone: (305) 743-5563. Phone: (305) 743-5564. Fax: (305) 743-9441. E-mail: keysradiogroup@aol.com Licensee: The Great Marathon Radio Co. (acq 11-5-90; grpsl; FTR: 11-26-90). Population served: 82,000 Natl. Network: Westwood One. Format: News/talk, sports. News: 10 hrs wkly. Target aud: 25-54; general. ♦Joe Mascone, pres & gen mgr; Vince Cacone, engrg dir & chief of engrg; Jane Martin, traf mgr.

***WHWY(FM)**—Not on air, target date: January 2008: 91.5 mhz; 12.5 kw. Ant 458 ft TL: N24 39 38 W81 25 10. Stereo. Hrs opn: 24
Rebroadcasts WLRN 100%.
172 N.E. 15th St., Miami, 33132-1348. Phone: (305) 995-1717. Fax: (305) 995-2299. E-mail: info@wlrn.org Web Site: www.wlrn.org. Licensee: The School Board of Miami-Dade County, FL. Population served: 85,000 Natl. Network: NPR, PRI. Law Firm: Leibowitz & Associates, P.A. Wire Svc: AP Format: News/talk-info. News staff: 7; News: 111 hrs wkly. Target aud: General; well educated, moderate to high income bracket. Spec prog: Haitian 3 hrs wkly. ♦John LaBonia, gen mgr; Ted Eldredge, stn mgr; Peter J. Moerz, progmg mgr; Irina Lellemand, news dir.

WWWK(FM)— Oct 15, 1984: 105.5 mhz; 26 kw. Ant 115 ft TL: N24 43 44 W81 02 05. (CP: COL Islamorada. 50 kw, ant 246 ft. TL: N24 57 34 W80 34 30). Stereo. Hrs opn: 11399 Overseas Hwy., 33050. Phone: (305) 743-3434. Fax: (305) 743-9091. Licensee: LSM Radio Partners LLC (acq 6-24-2004; with WAVK(FM) Marathon). Natl. Network: ABC. Format: Oldies. ♦Kevin Leroux, gen mgr.

Marco

WAVV(FM)— May 30, 1987: 101.1 mhz; 100 kw. 981 ft TL: N26 10 57 W81 34 32. Stereo. Hrs opn: 11800 Tamiami Tr. E., Naples, 34113. Phone: (239) 793-1011. Fax: (239) 793-7000. E-mail: wavvfm101@earthlink.net Web Site: wavv101.com. Licensee: Alpine Broadcasting Corp. (group owner; (acq 4-84; $95,000; FTR: 4-23-84). Natl. Network: AP Radio. Natl. Rep: Christal. Format: Modern, easy lstng. News: 8 hrs wkly. Target aud: 35 plus; an economically qualified audience that is somewhat more affluent. Spec prog: Jazz 3 hrs wkly. ♦Norman Alpert, pres; Donna Alpert, CFO; Jeff Alpert, gen mgr; Kenny Lamb, opns mgr.

WGUF(FM)— 1990: 98.9 mhz; 6 kw. Ant 328 ft TL: N26 01 50 W81 38 33. Stereo. Hrs opn: 24 10915 K-Nine Dr., 2nd Fl., Bonita Springs, 34135. Phone: (239) 495-8383. Fax: (239) 495-0883. Web Site: www.thegulf989.com. Licensee: Renda Broadcasting Corp. of Nevada. Group owner: Renda Broadcasting Corp. (acq 4-17-97; $2 million). Population served: 300,000 Format: News/talk. News: 7 hrs wkly. Target aud: 35 plus; affluent southwest FL residents. ♦Tony Renda Jr., gen mgr; Randy Savage, progmg dir.

***WMKO(FM)**— Feb 8, 1999: 91.7 mhz; 25 kw. 140 ft TL: N25 55 43 W81 43 49. Hrs open: 24
Rebroadcast WGCUFM100%.
Florida Gulf Coast University, 10501 FGCU Blvd., Fort Myers, 33965. Phone: (239) 590-2500. Fax: (239) 590-2511. E-mail: wgcufm@fgcu.edu Web Site: wgcu.org. Licensee: Board of Trustees, Florida Gulf Coast University (acq 11-16-01). Natl. Network: NPR. Format: Great Music, class, jazz & news. ♦Kathleen Davey, gen mgr & stn mgr.

Marco Island

***WCNZ(AM)**— May 1999: 1660 khz; 10 kw-D, 1 kw-N. TL: N25 59 30 W81 37 30. Stereo. Hrs opn: 24 5043 E. Tamiami Tr., Naples, 34113. Phone: (239) 732-9369. Fax: (239) 732-7267. E-mail: bladd@relevantradio.com Web Site: www.relevantradio.com. Licensee: Starboard Media Foundation Inc. (acq 5-19-2005; $2 million with WVOI(AM) Marco Island). Population served: 250,000 Rgnl. Network: Florida Radio Network Law Firm: Arter & Hadden. Format: Catholic talk. News staff: 2; News: 10 hrs wkly. Spec prog: Vintage Radio. ♦Robert Ladd, stn mgr.

WTLT(FM)—See Naples

WVOI(AM)— Jan 1, 1975: 1480 khz; 1 kw-U, DA-2. TL: N25 59 30 W81 37 30. Hrs open: 24 5043 E. Tamiami Tr., Naples, 34113. Phone: (239) 732-9369. Fax: (239) 732-7267. E-mail: bladd@relevantradio.com Licensee: Starboard Media Foundation Inc. (acq 5-19-2005; $2 milllion with WCNZ(AM) Marco Island). Population served: 250,000 Natl. Network: Jones Radio Networks, USA. Rgnl. Network: Florida Radio Net. Law Firm: Arter & Hadden. Format: Btfl music. News staff: 2; News: 5.5 hrs wkly. Target aud: 45+; Upscale and professional adults. ♦Robert Ladd, opns VP.

Marianna

WJAQ(FM)—Listing follows WTOT(AM).

***WJNF(FM)**— May 1985: 88.3 mhz; 25 kw vert. Ant 262 ft TL: N30 46 57 W85 06 30. Stereo. Hrs opn: 24 Box 450, 32447-0450. Secondary address: 2914 Jefferson St. 32446. Phone: (850) 526-4477. Fax: (850) 526-1831. E-mail: rene@wjnf.org Web Site: www.wjnf.org. Licensee: Marianna Educational Broadcasting Foundation. Population served: 823,000 Natl. Network: Moody. Format: Relg, adult contemp, Christian. News staff: one; News: 6 hrs wkly. Target aud: Families. ♦Jack Hollis, pres; Rene Hollis, gen mgr, progmg dir, news dir & local news ed; Shellie Hollis, VP, mktg dir & prom mgr; Charles Wooten, chief of engrg.

WTOT(AM)— Sept 24, 1958: 980 khz; 1 kw-D, 500 w-N. TL: N30 47 01 W85 15 18. Hrs open: 24 Box 569, 32447-0569. Secondary address: 4376 Lafayette St., Suite A 32446-3300. Phone: (850) 482-3046. Fax: (850) 482-3049. Licensee: MFR Inc. (acq 10-8-96; with co-located FM). Population served: 75,000 Natl. Network: ABC. Format: Adult standards. Target aud: 25 plus. ♦John Biddinger, CEO; Ed Cearley, pres, gen mgr & gen mgr; Don Moore, sls dir, news dir & pub affrs dir; Curtis Blount, chief of engrg.

WJAQ(FM)—Co-owned with WTOT(AM). Sept 1, 1964: 100.9 mhz; 5.9 kw horiz. 331 ft TL: N30 47 01 W85 15 18. Stereo. 24 4376 Lafayette St., Suite A, 32446. Phone: (850) 482-3046.100,000 Natl. Network: ABC. Format: Country. News staff: one; News: 3 hrs wkly. Target aud: General.

WTYS(AM)— Apr 3, 1947: 1340 khz; 1 kw-U. TL: N30 45 49 W85 13 52. Hrs open: 24 Box 777, 32447. Secondary address: 2725 Jefferson St. 32448. Phone: (850) 482-2131. Fax: (850) 526-3687. E-mail: wtysradio@earthlink.net Web Site: www.wtys.cc. Licensee: James L. Adams Jr. (acq 12-1-98; $250,000 with WTYS-FM Marianna). Population served: 42,000 Rgnl. Network: Florida Radio Net. Format: Classic country. News staff: one; News: 7 hrs wkly. Target aud: 25-64; adults in Jackson County & the surrounding area. Spec prog: Farm one hr, gospel 11 hrs wkly. ♦James Adams, gen mgr; Jerry Jackson, opns dir.

WTYS-FM— Aug 4, 1995: 94.1 mhz; 4.4 kw. Ant 384 ft TL: N30 45 47 W85 13 52. Stereo. Hrs opn: 24 Box 777, 32447. Secondary address: 2725 Jefferson St. 32448. Phone: (850) 482-2131. Fax: (850) 526-3687. E-mail: wtysradio@earthlink.net Web Site: www.wtys.cc. Licensee: James L. Adams Jr. (acq 12-1-98; with WTYS(AM) Marianna). Population served: 350000 Natl. Network: CBS. Rgnl. Network: Florida Radio Net. Format: Southern gospel. News staff: one; News: 5 hrs wkly. Target aud: 25-64; adults in Jackson county, FL & surrounding area. ♦James Adams, gen mgr; Jerry Jackson, opns mgr; Tom O'Brien, news dir.

Mary Esther

WYZB(FM)— May 1986: 105.5 mhz; 25 kw. Ant 305 ft TL: N30 24 42 W86 37 14. Stereo. Hrs opn: 225 N.W. Hollywood Blvd., Fort Walton Beach, 32548. Phone: (850) 244-1055 (studio). E-mail: sales@wyzb.com Web Site: www.wyzb.com. Licensee: Cumulus Licensing Corp. Group

owner: Cumulus Media Inc. (acq 1-10-2003; grpsl). Format: Country. Target aud: 25-54. ♦Georgia Edmiston, gen mgr.

Mayo

***WGSG(FM)**— 1991: 89.5 mhz; 2.5 kw horiz, 20 kw vert. 249 ft TL: N30 02 30 W83 07 45. Hrs open: Box 644, Whispering Oaks, 32066. Phone: (386) 294-2525. Fax: (386) 294-2525. Licensee: True Concepts of Levy County Inc. Format: Relg. ♦Terri Simmons, gen mgr.

Melbourne

WAOA-FM—Listing follows WINT(AM).

WBVD(FM)—Listing follows WMMB(AM).

WCIF(FM)— Jan 1, 1980: 106.3 mhz; 3 kw. 230 ft TL: N28 04 40 W80 39 26. Stereo. Hrs opn: Box 366, 32902. Secondary address: 3301 Dairy Rd. 32904. Phone: (321) 725-9243. E-mail: info@wcif.com Web Site: www.wcif.com. Licensee: First Baptist Church Inc. Format: Relg. ♦Lee J. Martinez, gen mgr; Martha Root, opns mgr.

***WFIT(FM)**— April 1975: 89.5 mhz; 900 w horiz, 4.6 kw vert. 112 ft TL: N28 03 51 W80 37 25. (CP: 700 w, ant 151 ft.). Stereo. Hrs opn: 24 150 W. University Blvd., 32901. Phone: (321) 674-8140. Fax: (321) 674-8139. E-mail: wfit@fit.edu Web Site: www.wfit.org. Licensee: Florida Institute of Technology. Population served: 400,000 Format: News, AAA. News: 40 hrs wkly. Target aud: 25-54; Public Radio Listeners. ♦Terri Wright, gen mgr.

WINT(AM)— Mar 8, 1968: 1560 khz; 5 kw-D. TL: N28 07 40 W80 42 29. Hrs open: 6 AM-7 PM 1775 W. Hibiscus Blvd., Suite 301, 32901. Phone: (321) 984-1000. Fax: (321) 724-1565. Licensee: Cumulus Licensing Corp. Group owner:Cumulus Media Inc. (acq 5-23-2001; with co-located FM). Population served: 40,236 Law Firm: Erwin Krasnow. Format: Oldies. News staff: 2; News: 25 hrs wkly. Target aud: 35-64. ♦Dan Carelli, gen mgr; Ted Turner, opns mgr.

WAOA-FM—Co-owned with WINT(AM). Nov 9, 1972: 107.1 mhz; 100 kw. 500 ft TL: N28 08 14 W80 42 11. Stereo. Web Site: www.wa1a.com.750000 Format: CHR. Target aud: 25-54.

WMEL(AM)— Jan 4, 1956: 920 khz; 5 kw-D, 1 kw-N, DA-3. TL: N28 08 11 W80 41 20. Stereo. Hrs opn: 24 1800 Turtlemound Rd., 32934. Phone: (321) 254-2282. Fax: (321) 254-1199. E-mail: jharper@920wmel.com Web Site: www.920wmel.com. Licensee: David Ryder, receiver (acq 7-29-2005). Population served: 500,000 Natl. Network: CBS Radio, Westwood One. Rgnl. Network: S.E. Agri. Format: News/talk, sports. News staff: 2; News: 144 hrs wkly. Target aud: 35-64; decision making men & women. Spec prog: Relg 6 hrs, Jewish 2 hrs wkly. ♦Peter Kerasotis, gen mgr.

WMMB(AM)— 1947: 1240 khz; 1 kw-U. TL: N28 04 40 W80 35 55. (CP: 940 w-U. TL: N28 04 42 W80 35 56). Hrs opn: 24 1388 S. Babcock St., 32901. Phone: (321) 733-1000. Fax: (321) 725-6821. E-mail: wmmb1240@aol.com Web Site: www.wmmbam.com. Licensee: Capstar TX L.P. Group owner: Clear Channel Communications Inc. (acq 8-30-00; grpsl). Population served: 200,000 Natl. Network: Westwood One. Rgnl. Network: Florida Radio Net. Law Firm: Robert A. DePont. Format: Adult standards, swing. News staff: 3; News: 4 hrs wkly. Target aud: 35 plus. ♦Barbara Latham, gen mgr; Larry Brewer, progmg dir.

WBVD(FM)—Co-owned with WMMB(AM). Dec 25, 1965: 95.1 mhz; 1.2 kw. Ant 210 ft TL: N28 04 41 W80 35 57. Stereo. 24 Web Site: www.951thebeat.com. Format: Top-40. News: one hr wkly. ♦Jeff McKeel, sls dir; Doug Remington, engrg dir.

Merritt Island

WWBC(AM)—See Cocoa

Mexico Beach

WEBZ(FM)— Nov 28, 1990: 99.3 mhz; 50 kw. Ant 519 ft TL: N30 00 21 W85 20 36. Stereo. Hrs opn: 24 Box 59288, Panama City, 32412. Phone: (850) 769-1408. Fax: (850) 769-0659. Licensee: Clear Channel Broadcasting Licenses Inc. Group owner: Clear Channel Communications Inc. (acq 11-21-97; grpsl). Law Firm: Lukas, McGowan, Nace & Gutierrez. Format: Oldies. Target aud: 30 plus; professional adults. ♦Pete Norden, gen mgr.

Miami

WACC(AM)—Hialeah, Dec 1, 1987: 830 khz; 1 kw-U, DA-2. TL: N25 46 22 W80 25 16. Stereo. Hrs opn: 24 1779 N.W. 28th St., 33142. Phone: (305) 638-9729. Fax: (305) 636-3976 / 4748. E-mail: padrealberto@paxcc.org Web Site: www.paxcc.org. Licensee: Radio Peace Catholic Broadcasting Inc. (acq 11-27-96; $2.55 million). Law Firm: Thiemain & Evenas. Format: Relg, talk, Sp. News staff: 5; News: 18 hrs wkly. Target aud: 25-54; adults. Spec prog: Sports. ♦Alberto Cutie, pres; Father Alberto R. Cutie, gen mgr.

WAMR-FM— June 7, 1974: 107.5 mhz; 95 kw horiz, 80 kw vert. 1,007 ft TL: N25 57 59 W80 12 33. Stereo. Hrs opn: 800 Douglas Rd., Suite 111, Coral Gables, 33134. Phone: (305) 447-1140. Fax: (305) 643-1075. Web Site: www.univision.com. Licensee: WQBA-FM License Corp. Group owner: Univision Radio (acq 9-22-2003; grpsl). Format: Sp, adult contemp. ♦Claudia Puig, gen mgr.

WAQI(AM)— 1939: 710 khz; 50 kw-U, DA-2. TL: N25 58 07 W80 22 44. Hrs open: 800 Douglas Rd., Suite 111, Coral Gables, 33134. Phone: (305) 447-1140. Fax: (305) 442-7676. E-mail: gfernandez @univisionradio.com Web Site: www.univisionradio.com. Licensee: Licensee Corporation #1. Group owner: Univision Radio (acq 9-22-2003; grpsl). Population served: 334,859 Format: Sp, news/talk, entertainment. ♦Claudia Puig, sr VP & gen mgr; Yvette Sanguilty, sls dir; Monica Rabassa, mktg dir & prom dir; Armando Perez-Roura, progmg dir; Max Fitero, chief of engrg.

WAXY(AM)—See South Miami

WBGG-FM—See Fort Lauderdale

WCMQ-FM—See Hialeah

***WDNA(FM)**— June 10, 1980: 88.9 mhz; 7.4 kw. 1,145 ft TL: N25 32 24 W80 28 07. Stereo. Hrs opn: 24 Box 558636, 4848 S.W. 74 Ct., 33255. Phone: (305) 662-8889. Fax: (305) 662-1975. E-mail: feedback@wdna.org Web Site: www.wdna.org. Licensee: Bascomb Memorial Broadcasting Foundation Inc. (acq 1971). Population served: 1,500,000 Law Firm: Haley, Bader & Potts. Format: Jazz, Sp. News: 10 hrs wkly. Target aud: General; minorities. Spec prog: World music 10 hrs wkly. ♦Maggie Pelleya, gen mgr & stn mgr.

WEDR(FM)— May 18, 1963: 99.1 mhz; 100 kw. 926 ft TL: N25 57 30 W80 12 44. (CP: TL: N25 57 59 W80 12 33). Stereo. Hrs opn: 2741 N. 29th Ave., Hollywood, 33020. Phone: (305) 623-7711 / (305) 444-4404. Fax: (305) 624-2736. Web Site: www.wedr.com. Licensee: Cox Radio Inc. Group owner: Cox Broadcasting (acq 8-00; grpsl). Population served: 700,000 Law Firm: Smithwick & Belendiuk. Format: Hip-hop. ♦Jerry Rushin, gen mgr; Maestro Powell, prom dir; Tony Field, progmg dir.

WFLC(FM)— July 20, 1951: 97.3 mhz; 100 kw. 800 ft TL: N25 57 30 W80 12 44. Stereo. Hrs opn: 2741 N. 29th Ave., Hollywood, 33020. Phone: (305) 444-4404. Fax: (954) 847-3223. E-mail: mike.disney@cox.com Web Site: www.coastfm.com. Licensee: Cox Radio Inc. Group owner: Cox Communications Inc. Natl. Rep: Christal. Format: Old school, rhythm & blues. Target aud: 25-54. ♦Mike G. Disney, gen mgr.

WHDR(FM)— Nov 1, 1960: 93.1 mhz; 100 kw. Ant 1,007 ft TL: N25 58 03 W80 12 34. Stereo. Hrs opn: 24 2741 N. 29th Ave., Hollywood, 33020. Phone: (305) 444-4404. Fax: (954) 847-3223. Web Site: 93rock.com. Licensee: Cox Radio Inc. Group owner: Cox Communications Inc. (acq 5-18-2000; grpsl). Population served: 397,900 Natl. Rep: Christal. Format: Active rock. Target aud: 18-49; upscale, educ adults with hip active lifestyles. ♦Michael Disney, gen mgr.

WHYI-FM—See Fort Lauderdale

WINZ(AM)— 1946: 940 khz; 50 kw-D, 10 kw-N. TL: N25 57 36 W80 16 13. Hrs open: 24 7601 Riviera Blvd., Miramar, 33023. Phone: (954) 862-2000. Fax: (954) 862-4012. Web Site: www.am940southflorida.com. Licensee: Clear Channel Broadcasting Licenses Inc. Group owner: Clear Channel Communications Inc. (acq 11-21-97; grpsl). Population served: 3,000,000 Natl. Network: ABC. Rgnl. Network: Florida Radio Net. Natl. Rep: Clear Channel. Format: Progsv talk. News staff: 20. Target aud: 35-64; upscale, professional, managerial adults. ♦Ken Brady, gen mgr & gen sls mgr.

WIOD(AM)— Jan 19, 1926: 610 khz; 10 kw-U, DA-N. TL: N25 50 58 W80 09 18. Stereo. Hrs opn: 7601 Riviera Blvd, Miramar, 33023. Phone: (954) 862-2000. Fax: (954) 862-4012 / 4015. E-mail: newsradio610@ccmiami.com Web Site: www.newsradio610.com. Licensee: Clear Channel Radio Licenses Inc. Group owner: Clear Channel Communications Inc. (acq 11-21-97; grpsl). Population served: 334,859 Wire Svc: UPI Format: News/talk. Target aud: 25-64. ♦Michael Crusham, opns VP; Ken Brady, gen sls mgr; Peter Bolger, progmg dir.

WKAT(AM)—See North Miami

WKIS(FM)—Listing follows WQAM(AM).

***WLRN-FM**— February 1948: 91.3 mhz; 47 kw. Ant 935 ft TL: N25 58 46 W80 11 46. Stereo. Hrs opn: 24 172 N.E. 15th St., 33132. Phone: (305) 995-1717. Fax: (305) 995-2299. E-mail: info@wlrn.org Web Site: www.wlrn.org. Licensee: School Board of Miami Dade County Florida. Population served: 2,600,000 Natl. Network: NPR, PRI. Law Firm: Leibowitz & Associates, P.A. Wire Svc: AP Format: News/talk info. News staff: 7; News: 111 hrs wkly. Target aud: General; well educated, moderate to high income bracket. Spec prog: Haitian 3 hrs wkly. ♦Karen Echols, CFO; John Labonia, gen mgr; Ted Eldredge, stn mgr; Antonio Zayas, opns dir; Peter J. Maerz, progmg mgr; Irina Lallemond, news dir; Jack F. Yaghdjian, chief of engrg. Co-owned TV: *WLRN-TV affil.

WLYF(FM)— 1948: 101.5 mhz; 100 kw. 810 ft TL: N25 57 59 W80 12 44. Stereo. Hrs opn: 24 20450 N.W. 2nd Ave., 33169-2505. Phone: (305) 521-5100. Fax: (305) 652-0098. E-mail: litefm@litemiami.com Web Site: www.litemiami.com. Licensee: Lincoln Financial Media Co. of Florida. Group owner: Jefferson-Pilot Communications Co. (acq 4-3-2006; grpsl). Population served: 3,111,400 Natl. Rep: Interep. Wire Svc: AP Format: Adult contemp. News: 2 hrs wkly. Target aud: 25-54; women. ♦Jon Boscia, CEO; Don Benson, pres; Dennis P. Collins, sr VP & gen mgr; Rob Sidney, opns dir, sls dir & progmg dir; Rosemary Zimmerman, natl sls mgr; Danielle Webb, rgnl sls mgr; Nicole Gates, mktg dir & prom dir; Gary Blau, engrg dir; Tina Marcos, traf mgr.

WMBM(AM)—See Miami Beach

***WMCU(FM)**— Aug 24, 1970: 89.7 mhz; 100 kw. 1,014 ft TL: N25 32 24 W80 28 07. Stereo. Hrs opn: 24 600 S.W. 3rd St., # 2270, Pompano Beach, 33060. Phone: (954) 545-7600. Fax: (954) 545-7630. E-mail: onair@897spiritfm.com Web Site: www.897spiritfm.com. Licensee: Trinity International Foundation Inc. (acq 11-6-01). Population served: 3,500,000 Natl. Network: USA. Law Firm: Dow, Lohnes & Albertson. Format: Relg, educ, Christian. News staff: one. Target aud: 25-54. ♦Dwight Taylor, progmg dir; Donna Matthews, news dir; Merryan Padron, stn mgr & chief of engrg.

WMGE(FM)—See Miami Beach

WMIB(FM)—See Fort Lauderdale

WMYM(AM)— Aug 15, 1997: 990 khz; 5 kw-U. TL: N25 50 34 W80 25 12. Hrs open: 24 2150 W. 68th St., Suite 202, Hialeah, 33016. Phone: (305) 823-0990. Fax: (305) 823-9322. Licensee: Radio Disney Group LLC. Group owner: ABC Inc. (acq 7-30-99; $7.4 million). Format: Radio Disney, top 40. Kids and families. ♦Gilbert Salguero, gen mgr; Jeff Schwartz, opns dir; John Craveno, sls dir; John Hurni, chief of engrg.

WNMA(AM)—Miami Springs, May 18, 1958: 1210 khz; 25 kw-D, 2.5 kw-N, DA-2. TL: N25 54 00 W80 21 49. Hrs open: 7250 N.W. 58th St., 33166. Phone: (786) 497-3414. Fax: (786) 497-3412. E-mail: eduardor@mrbi.net Licensee: Multicultural Radio Broadcasting Licensee LLC. Group owner: Multicultural Radio Broadcasting Inc. (acq 2-4-2004; grpsl). Format: Sp, talk, sports. ♦Eduardo Rueda, gen mgr.

WOCN(AM)— Dec 22, 1956: 1450 khz; 1 kw-U. TL: N25 50 24 W80 11 20. Hrs open: 24 350 N.E. 71 St., 33138. Phone: (305) 759-7280. Fax: (305) 759-2276. E-mail: richardvega@wocn.net Web Site: www.wocn.net. Licensee: IM FL Licenses LLC. (acq 7-12-2006; $6 million). Population served: 334,859 Law Firm: Gunsten, Yonkley. Wire Svc: AP Format: Sp, news/talk. Target aud: General. Spec prog: Creole 84 hrs wkly. ♦David Jacobs, pres & VP; Pablo Vega, pres & VP; Richard Vega, gen mgr.

WPOW(FM)— June 15, 1985: 96.5 mhz; 100 kw. 1,007 ft TL: N25 57 59 W80 12 33. Stereo. Hrs opn: 20295 N.W. 2nd Ave., Suite 300, 33169. Phone: (305) 653-6796. Fax: (305) 770-1456. Web Site: www.power96.com. Licensee: Beasley FM Acquisition Corp. Group owner: Beasley Broadcast Group (acq 8-94). Population served: 3,000,000 Format: CHR. ♦George Beasley, pres; Matthew Bell, gen mgr; John Jaras, gen sls mgr; Ira Wolf, natl sls mgr & opns.

WQAM(AM)— May 1921: 560 khz; 5 kw-D, 1 kw-N. TL: N25 44 36 W80 09 14. Stereo. Hrs opn: 24 20295 N.W. 2nd Ave., 33169. Phone: (305) 653-6796. Fax: (305) 770-1456. Web Site: www.wqam.com. Licensee: Beasley-Reed Broadcasting. Group owner: Beasley Broadcast Group Population served: 1,100,000 Law Firm: Winston & Strawn. Format: Sports. Target aud: 25-54; males. Spec prog: Relg 4 hrs wkly. ♦Joe Bell, gen mgr; Dorene Alberts, rgnl sls mgr & prom dir; Duff Lindsey, progmg dir; George Corso, chief of engrg.

WKIS(FM)—Co-owned with WQAM(AM). October 1965: 99.9 mhz; 100 kw. 986 ft TL: N25 59 34 W80 10 27. Stereo. 24 194 N.W.187th St., 33169. Phone: (305) 654-1700. Fax: (305) 654-1715. Web Site: www.wkis.com. Natl. Network: Westwood One. Format: Country. News staff: one; News: 2 hrs wkly. Target aud: 25-54. ♦George Corso, CEO; Joe Bell, VP; Carole Bowen, gen sls mgr; Bob Barnett, progmg dir.

WQBA(AM)— 1947: 1140 khz; 50 kw-D, 10 kw-N, DA-2. TL: N25 45 46 W80 29 03. Hrs open: 800 Douglas Rd., Annex 1, Suite 111, Coral Gables, 33134. Phone: (305) 447-1140. Fax: (305) 441-2454. Web Site: www.wqba.com. Licensee: WQBA-AM License Corp. Group owner: Univision Radio (acq 9-22-2003; grpsl). Population served: 1,000,000 Format: News/talk, Sp. ♦Claudia Puig, sr VP, VP & gen mgr.

WRHC(AM)—See Coral Gables

WRMA(FM)—See Fort Lauderdale

WRTO-FM—Goulds, February 1976: 98.3 mhz; 1.1 kw. 462 ft TL: N25 32 24 W80 28 07. (CP: 100 kw, ant 1,627 ft.). Stereo. Hrs opn: 24 800 Douglas Rd., Suite 111, Coral Gables, 33134. Phone: (305) 447-1140. Fax: (305) 529-6631. Web Site: www.univision.com. Licensee: License Corp. #2. Group owner: Univision Radio (acq 9-22-2003; grpsl). Population served: 365,000 Format: Latin/tropical, Sp. ♦Claudia Puig, gen mgr; Monica Rabassa, mktg dir.

WSUA(AM)— June 20, 1969: 1260 khz; 5 kw-U, DA-2. TL: N25 46 23 W80 25 17. Hrs open: 24 2100 Coral Way, Suite 201, 33145. Phone: (305) 285-1260. Fax: (305) 858-5907. E-mail: ycuello@carcolusa.com Web Site: www.caracolusa.com. Licensee: WSUA Broadcasting Corp. (acq 7-28-2005; $72,000 for 24% of stock). Population served: 106,873 Format: Sp, News/talk. News staff: 7; News: 31 hrs wkly. Target aud: 18-54; Latin American audience. ♦Tomas Martinez, gen mgr.

WVUM(FM)—See Coral Gables

WWFE(AM)— July 1989: 670 khz; 50 kw-D, 2.5 kw-N, DA-2. TL: N25 51 27 W80 28 52. Stereo. Hrs opn: 24 330 S.W. 27th Ave., Suite 207, 33135. Phone: (305) 541-3300. Fax: (305) 541-7470. E-mail: info@lapoderosa.com Web Site: www.lapoderosa.com. Licensee: Fenix Broadcasting Corp. (acq 6-22-93; $2.7 million; FTR: 7-12-93). Population served: 60,000 Natl. Rep: Lotus Entravision Reps LLC. Law Firm: Leventhal, Senter & Lerman, P.L.L.C. Format: Sp, var/div, news/talk. News: 27 hrs wkly. Target aud: 25-54. ♦Jorge Rodriguez, CEO; Ana M. Vidal Rodriguez, VP; Jorge A. Rodriguez, pres & gen mgr.

Miami Beach

WLVE(FM)— July 1, 1968: 93.9 mhz; 96 kw. 1,006 ft TL: N25 57 59 W80 12 33. (CP: 100 kw horiz, 82 kw vert). Stereo. Hrs opn: 7601 Riviera Blvd, Miramar, 33023. Phone: (954) 862-2000. Fax: (954) 862-4012. Web Site: www.love94.com. Licensee: Clear Channel Radio Licenses Inc. Group owner: Clear Channel Communications Inc. (acq 11-21-97; grpsl). Population served: 3,000,000 Law Firm: Wiley, Rein & Fielding. Format: Smooth jazz. Target aud: 25-54. ♦Jamie Kaufman, gen mgr.

WMBM(AM)— 1949: 1490 khz; 1 kw-U. TL: N25 46 10 W80 08 11. Hrs open: 24 13242 NW 7 Ave., North Miami, 33168. Phone: (305) 769-1100. Fax: (305) 769-9975. E-mail: wmbm@wmbm.com Web Site: www.wmbm.com. Licensee: New Birth Broadcasting Corp. (acq 3-8-95; FTR: 5-8-95). Population served: 50,000 Natl. Network: American Urban, Westwood One. Law Firm: Pepper & Corazzini. Format: Gospel, community talk. News staff: one; News: 3 hrs wkly. Target aud: 25 plus; mature Black, self-motivated, Christian, professionals. ♦Caroline Kelly, sr VP; Victor T. Curry, pres & gen mgr; Claudette Freeman, stn mgr; Grey Cooper, progmg mgr.

WMGE(FM)— 1961: 94.9 mhz; 100 kw. 1,007 ft TL: N25 46 29 W80 11 19. Stereo. Hrs opn: 7601 Riviera Blvd., Miramar, 33023. Phone: (954) 862-2000. Fax: (305) 862-4012. Web Site: www.mega949.com. Licensee: Clear Channel Broadcasting Licenses Inc. Group owner: Clear Channel Communications Inc. (acq 11-21-97; grpsl). Population served: 3,000,000 Natl. Network: Westwood One. Format: Latino. Target aud: 18-34. ♦Desi Hernandez, gen sls mgr.

Miami Springs

WNMA(AM)—Licensed to Miami Springs. See Miami

Micanopy

WSKY-FM— Sept 7, 1985: 97.3 mhz; 2.6 kw. 495 ft TL: N29 32 08 W82 19 17. (CP: 13.5 kw, ant 948 ft.). Stereo. Hrs opn: 24 3600 N.W. 43rd St., Suite B, Gainesville, 32606-8127. Phone: (352) 377-0985. Fax: (352) 337-2968. Web Site: www.thesky973.com. Licensee: Entercom Gainesville License L.L.C. Group owner: Entercom Communications Corp. (acq 3-18-98; $2.8 million). Natl. Rep: Christal. Law Firm: Fisher, Wayland, Cooper, Leader & Zaragoza. Format: News/talk. Target aud: 18-54. ♦David Field, CEO, pres & progmg dir; Mark Leopold, VP, gen mgr & gen sls mgr.

Midway

WFLA-FM— 1996: 100.7 mhz; 11.5 kw. Ant 489 ft TL: N30 29 32 W84 17 13. Hrs open: Bldg. G, 325 John Knox Rd., Tallahassee, 32303. Phone: (850) 422-3107. Fax: (850) 383-0747. E-mail: mattmillar@clearchannel.com Web Site: www.1270wfla.com. Licensee: Clear Channel Broadcasting Licenses Inc. Group owner: Clear Channel Communications Inc. (acq 11-21-97; grpsl). Format: Talk radio. ♦Lisa Rice, gen mgr; Jeff Horn, opns mgr; Megan Brown, prom dir.

Milton

WEBY(AM)— 1978: 1330 khz; 25 kw-D, 79 w-N, DA-D. TL: N30 31 05 W87 04 56. Hrs open: 24 7179 Printers Alley, 32583. Phone: (850) 983-2242. Fax: (850) 983-3231. E-mail: weby@1330weby.com Web Site: www.1330weby.com. Licensee: Spinnaker License Corp. (acq 5-28-2002). Population served: 500,000 Natl. Network: Jones Radio Networks. Format: Talk, news. News: 15 hrs wkly. Target aud: 35 plus; affuent, educated adults. Spec prog: Christian 7 hrs wkly / Florida State football. ♦Mike Bates, pres & gen mgr; Anthony Daughtery, opns mgr; Dave Daughtry, news dir.

WECM(AM)— Dec 18, 1957: 1490 khz; 1 kw-U. TL: N30 37 30 W87 02 54. Hrs open: 24 6583 Berryhill Rd., 32570. Phone: (850) 623-1490. Fax: (850) 623-6818. E-mail: station@memories1490.com Web Site: www.memories1490.com. Licensee: Worldlink Technologies Group Inc. (acq 9-29-03). Population served: 50,000 Natl. Network: USA. Format: Oldies, 50's, 60's, and 70's. News staff: 1; News: 12 hrs wkly. Target aud: General; Adults 45 and older. ♦The Baron of Fulwood, COO, VP & gen mgr; Steve Walker, progmg dir.

***WEGS(FM)**— Oct 15, 1985: 91.7 mhz; 20 kw. Ant 367 ft TL: N30 37 20 W87 05 12. Stereo. Hrs opn: 1836 Olive Rd., Pensacola, 32514. Secondary address: 505 Josephine St., Titusville 32796. Phone: (850) 476-1932. Fax: (850) 447-9650. E-mail: dtalley@olivebaptist.org Web Site: www.olivebaptist.org. Licensee: Florida Public Radio Inc. Format: Talk, adult Christian contemp. ♦Dave Talley, gen mgr & chief of engrg.

WXBM-FM— Apr 28, 1964: 102.7 mhz; 100 kw. 1,328 ft TL: N30 35 18 W87 33 16. Stereo. Hrs opn: 6085 Quintet Rd., Pace, 32571. Phone: (850) 994-5357. Fax: (850) 994-7191. E-mail: feedback@wxbm.com Web Site: www.wxbm.com. Licensee: 6 Johnson Road Licenses Inc. Group owner: Pamal Broadcasting Ltd. (acq 10-19-2001; grpsl). Population served: 850,000 Format: Country. ♦Dave Cobb, gen mgr.

Mims

WPGS(AM)— May 5, 1986: 840 khz; 1 kw-D. TL: N28 44 20 W80 53 02. Stereo. Hrs opn: Sunrise-sunset 805 N. Dixie Ave., Titusville, 32796. Phone: (321) 383-1000. E-mail: wpgs840@aol.com Web Site: www.talkstar840.com. Licensee: WPGS Inc. (acq 3-93; $65,000; FTR: 3-29-93). Population served: 75,000 Natl. Network: USA. Format: Talk. ♦Ed Shiflett, pres & gen mgr; Jay Rowan, chief of engrg.

Miramar Beach

WSBZ(FM)— Oct 18, 1994: 106.3 mhz; 3 kw. 328 ft TL: N30 23 07 W86 18 03. Hrs open: 10859 Emerald Coast Pkwy. W., Destin, 32541. Phone: (850) 267-3279. Fax: (850) 231-1775. E-mail: office@wsbz.com Web Site: www.wsbz.com. Licensee: Carter Broadcasting Inc. (acq 9-10-99). Format: Smooth jazz (new adult contemp). ♦Renee Carter, CFO; Mark Carter, gen mgr.

Monticello

***WFRF-FM**— December 1996: 105.7 mhz; 16 kw. Ant 410 ft TL: N30 23 08 W83 50 05. Hrs open: 24 Box 181000, Tallahassee, 32318-0009. Phone: (850) 201-1070. Fax: (850) 201-1071. Web Site: www.faithradio.us. Licensee: Faith Radio Network Inc. (acq 1-26-2004; $800,000). Population served: 200,000 Natl. Network: CBS. Format: Relg. Target aud: 12+. ♦Scott Beigle, gen mgr.

***WKVH(FM)**— March 2003: 91.9 mhz; 1.5 kw. Ant 1,322 ft TL: N30 40 13 W83 56 26. Stereo. Hrs opn: 24 2351 Sunset Blvd., Suite 170-218, Rocklin, CA, 95765. Phone: (916) 251-1600. Fax: (916) 251-1650. E-mail: klove@klove.com Web Site: www.klove.com. Licensee: Educational Media Foundation. Group owner: EMF Broadcasting. Natl. Network: K-Love. Law Firm: Shaw Pittman. Format: Contemp Christian. News staff: 3. Target aud: 25-44; Judeo Christian, female. ♦Richard Jenkins, pres; Keith Whipple, dev dir; Eric Allen, natl sls mgr; David Pierce, progmg mgr; Ed Lenane, news dir; Sam Wallington, engrg dir; Karen Johnson, news rptr; Marya Morgan, news rptr; Richard Hunt, news rptr.

Mount Dora

WMGF(FM)— 1966: 107.7 mhz; 100 kw. 1,584 ft TL: N28 55 16 W81 19 09. Stereo. Hrs opn: 24 2500 Maitland Ctr. Pkwy., Suite 401, Maitland, 32751. Phone: (407) 916-7800. Fax: (407) 916-0329. Web Site: www.magic107.com. Licensee: Clear Channel Radio Licenses Inc. Group owner: Clear Channel Communications Inc. (acq 11-21-97; grpsl). Population served: 3,442,300 Rgnl rep: Paul Rogers Law Firm: Wiley, Rein and Fielding. Format: Soft adult contemp. News staff: one; News: 2 hrs wkly. Target aud: 25-54; working women. Spec prog: Contemp Christian mus 20 hrs wkly. ♦Linda Byrd, gen mgr; Rochelle Rich, gen sls mgr; Chris Kampmeier, progmg dir; Ken Payne, progmg dir; Shawn Williams, traf mgr; Rick Everett, mktg.

Murdock

WBCG(FM)— Oct 22, 2001: 98.9 mhz; 5.5 kw. Ant 341 ft TL: N27 00 09 W82 10 54. Hrs open: 24 24100 Tiseo Blvd., Unit 10, Port Charlotte, 33980. Phone: (941) 639-1112. Fax: (941) 206-9296. E-mail: wbcgbeachradio@cs.com Licensee: Concord Media Group Inc. (acq 8-28-2001). Population served: 65,000 Rgnl. Network: Florida Radio Net. Law Firm: Rosenman & Colin, L.L.P. Format: Adult contemp. News: 2 hrs wkly. Target aud: Adults 25+; core 35-54 female. ♦Mark Jorgenson, gen mgr; Michael G. Keating, progmg dir.

Naples

WARO(FM)—Listing follows WNOG(AM).

***WBGY(FM)**— August 2004: 88.1 mhz; 110 w vert. Ant 59 ft TL: N25 51 56 W81 23 09. Hrs open: 297 Fillmore St., 34104. Phone: (239) 404-9849. E-mail: wbby@earthlink.net Licensee: Everglades City Broadcasting Co. Inc. (acq 3-30-2004; $25,000). Format: Country. ♦Robert Ladd, pres.

WNOG(AM)— Oct 14, 1954: 1270 khz; 5 kw-D, 1.9 kw-N, DA-2. TL: N26 15 26 W81 40 33. Hrs open: 24
Rebroadcasts WINK(AM) Fort Myers 100%.
2824 Palm Beach Blvd., Fort Myers, 33916. Phone: (239) 337-2346. Fax: (239) 337-2346. Web Site: www.winkwnog.com. Licensee: Meridian Broadcasting Inc. (group owner; (acq 12-1-96; grpsl). Population served: 516,200 Natl. Network: CBS. Natl. Rep: McGavren Guild. Law Firm: Leibowitz & Associates. Format: News/talk. Target aud: 35 plus. ♦Joseph C. Schwantzel, pres; Paul Thomas, gen mgr; Wayne Simons, sls dir; Jim Watkins, progmg dir; Keith Stulhmann, engrg dir.

WARO(FM)—Co-owned with WNOG(AM). May 8, 1962: 94.5 mhz; 100 kw. 1,049 ft TL: N26 20 26 W81 42 48. Stereo. 24 Fax: (239)

479-5581. Web Site: www.classicrock945.com. Format: Classic rock. Target aud: 25-54; men. Spec prog: Relg 2 hrs wkly. ♦Mike Allen, progmg dir.

WSGL(FM)— May 10, 1980: 104.7 mhz; 14 kw. 450 ft TL: N26 07 34 W81 43 18. Stereo. Hrs opn: 24 10915 K-Nine Dr., 2nd Fl., Bonita Springs, 34135. Phone: (239) 495-8383. Fax: (239) 495-0883. Web Site: www.wsgl1047.com. Licensee: Renda Broadcasting Corp. of Nevada. Group owner: Renda Broadcasting Corp. (acq 11-10-98; $3.65 million). Population served: 300,000 Format: Hot adult contemp. News: one hr wkly. Target aud: 25-54; women. Spec prog: 70s & 80s mus 5 hrs wkly. ♦Tony Renda Jr., gen mgr; Randy Savage, progmg dir.

***WSOR(FM)**— 1989: 90.9 mhz; 36 kw. Ant 902 ft TL: N26 20 29 W81 42 38. Stereo. Hrs opn: 24 5800 100th Way N. St, Saint Peterburgs, 33708. Phone: (727) 391-9994. Fax: (727) 397-6425. E-mail: wkes@moody.edu Web Site: www.wkes.fm. Licensee: Moody Bible Institute. (acq 1996). Population served: 650,000 Natl. Network: Salem Radio Network. Wire Svc: AP Format: Christian teaching, talk, music. News staff: one. Target aud: 40 plus. ♦Mike Gleichman, opns mgr & mus dir; Pierre Chestang, gen mgr & progmg dir; John Stortz, chief of engrg.

***WSRX(FM)**— August 1988: 89.5 mhz; 100 kw. Ant 309 ft TL: N26 07 12 W81 40 58. Hrs open: 24 3805 The Lords Way, 34114. Phone: (239) 775-8950. Fax: (239) 774-5889. E-mail: praisefm895@msn.com Web Site: www.praisefm.com. Licensee: Shadowlawn Association Inc. (acq 1-95; $236,000; FTR: 2-27-95). Format: Christian. Target aud: 18-40. ♦Arnie Coones, gen mgr.

WTLT(FM)— Dec 1, 1971: 93.7 mhz; 21 kw. 328 ft TL: N26 19 00 W81 47 13. Stereo. Hrs opn: 24 2824 Palm Beach Blvd., Fort Myers, 33916. Phone: (239) 337-2346. Fax: (239) 332-0767. E-mail: john.conrad@lite937.com Web Site: www.lite973.com. Licensee: Meridian Broadcasting Inc. (group owner; acq 12-1-96; grpsl). Population served: 532,600 Natl. Rep: McGavren Guild. Law Firm: Leibowitz & Associates. Wire Svc: AP Format: Adult contemp. News staff: 3; News: 2 hrs wkly. Target aud: 25-54; women. ♦Joseph C. Schwartzel, CEO & gen mgr; Jim Schwartzel, sls dir; John DeHinie, progmg dir; Keith Stuhlmann, engrg dir; Randy Marsh, progmg.

Naples Park

WBTT(FM)— Oct 22, 1987: 105.5 mhz; 950 w. 584 ft TL: N26 19 00 W81 47 13. (CP: 6.3 kw, ant 649 ft.). Stereo. Hrs opn: 13320 Metro Pkwy., Suite 1, Fort Myers, 33912. Phone: (914) 225-4300. Phone: (866) 843-2328. Fax: (239) 225-4329. Web Site: www.1055thebeat.com. Licensee: Clear Channel Radio Licenses Inc. Group owner: Clear Channel Communications Inc. (acq 1996; grpsl). Format: Rhythmic CHR. Target aud: 18-34. ♦Jim Keating, gen mgr.

Navarre

WGCX(FM)— 1999: 95.7 mhz; 25 kw. Ant 282 ft TL: N30 27 02 W86 51 59. Hrs open: 2070 N. Palafax, Pensacola, 32501. Phone: (850) 434-1230. Fax: (850) 469-9698. E-mail: mglin@aol.com Web Site: www.praise95.net. Licensee: 550 AM Inc. Format: Christian. ♦Dara Glinter, exec VP & opns VP; Michael Glinter, pres, gen mgr & progmg VP.

Neptune Beach

WFKS(FM)— August 1965: 97.9 mhz; 12.5 kw. Ant 991 ft TL: N30 16 51 W81 34 12. Stereo. Hrs opn: 24 11700 Central Parkway, Jacksonville, 32224. Phone: (904) 636-0507. Fax: (904) 997-7713. Web Site: www.979kissfm.com. Licensee: Clear Channel Broadcasting Licenses Inc. Group owner: Clear Channel Communications Inc. (acq 11-21-97; grpsl). Population served: 1,064,400 Format: Top 40. News staff: one; News: one hr wkly. Target aud: 35-54; general. ♦Norm Feuer, stn mgr; Gail Austin, opns mgr.

New Port Richey

WDUV(FM)—Licensed to New Port Richey. See Tampa

***WLPJ(FM)**— Apr 10, 1985: 91.5 mhz; 22 kw. Ant 230 ft TL: N28 16 41 W82 43 06. Stereo. Hrs opn: 24 6214 Springer Dr., 34668. Phone: (727) 848-9150. Fax: (727) 848-1233. E-mail: jeff@thejoyfm.com Web Site: www.thejoyfm.com. Licensee: Radio Training Network Inc. (acq 1995; $100,000). Law Firm: Gammond & Grange. Format: Adult contemp, Christian. ♦James L. Campbell, pres; Jeff MacFarlane, gen mgr; Carmen Brown, prom dir; Steve Rieker, chief of engrg.

WPSO(AM)— Oct 31, 1963: 1500 khz; 250 w-D. TL: N28 15 32 W82 43 54. Hrs open: Sunrise-sunset 109 Bayview Blvd., #A, Oldsmar, 34677. Phone: (727) 725-3500. Phone: (727) 725-5555. Fax: (813) 814-7500. E-mail: wpso@wpso.com Web Site: www.wpso.com. Licensee: AKMA Broadcast Network Inc. (acq 1993; $250,000; FTR: 9-13-93). Population served: 500,000 Format: News/talk, Greek. News staff: one; News: 35 hrs wkly. Target aud: General; international, ethnic. Spec prog: Pol two hrs, quiz/trivia program, relg 8 hrs, East Indian one hr, It 3 hrs wkly, Ethnic. ♦Sam Agelatos, pres & gen mgr; Angelo Agelatos, chief of opns.

New Smyrna Beach

***WJLU(FM)**— Sept 7, 1989: 89.7 mhz; 5 kw. 328 ft TL: N29 00 32 W80 58 27. (CP: 10 kw). Stereo. Hrs opn: 24 4295 Ridgewood Ave., Port Orange, 32127. Phone: (386) 756-9094. Fax: (386) 760-7107. E-mail: thecornerstone@cornerstoneministry.org Web Site: www.cornerstoneministry.org. Licensee: Cornerstone Broadcasting Corp. Population served: 350,000 Natl. Network: USA, Moody. Format: Relg. News: 18 hrs wkly. Target aud: General; families. ♦William Powell, gen mgr & progmg dir; Sandra Leisner, pub affrs dir.

WSBB(AM)— 1950: 1230 khz; 1 kw-U. TL: N29 01 57 W80 55 03. Hrs open: 24 P.O. Box 211340, South Daytona, 32121. Phone: (386) 236-1230. Fax: (386) 255-3178. Licensee: Gore-Overgaard Broadcasting Inc. (acq 3-21-2006; $450,000). Population served: 240,000 Format: Oldies. News: 120 hrs wkly. Target aud: 45 plus. Spec prog: Pol one hr, relg 5 hrs wkly. ♦Cordell J. Overgaard, pres; Tony Welch, gen mgr; Kris Phillips, opns mgr.

Newberry

WHHZ(FM)— February 1999: 100.5 mhz; 44 kw. Ant 469 ft TL: N29 36 29 W82 51 01. Hrs open: 100 N.W. 76th Dr., Suite 2, Gainesville, 32607. Phone: (352) 313-3130 / 3135. Fax: (352) 313-3166. E-mail: themorningbuzz2004@yahoo.com Web Site: www.1005thebuzz.com. Licensee: 6 Johnson Road Licenses Inc. (group owner; (acq 1-5-2007; grpsl). Format: Modern rock. ♦Benjamin Hill, gen mgr.

Niceville

WRKN(FM)— May 1, 1993: 100.3 mhz; 3.5 kw. Ant 440 ft TL: N30 29 20 W86 25 16. Hrs open: 24 21 Miracle Strip Pkwy., Fort Walton Beach, 32548. Phone: (850) 244-1400. Fax: (850) 243-1471. Web Site: www.krocksu.com. Licensee: Star Broadcasting Inc. Group owner: Cumulus Media Inc. (acq 8-2-2006; swap for WNCV(FM) Evergreen, AL). Population served: 150,000 Natl. Network: Jones Radio Networks. Format: Classic rock. ♦Ron Hale Sr., gen mgr & stn mgr.

Nocatee

WZSP(FM)— Aug 27, 1998: 105.3 mhz; 6 kw. 328 ft TL: N27 11 01 W81 56 57. Hrs open: 24 Heartland Broadcasting Corp., 7891 U.S. Highway 17 S., Zolfo Springs, 33890. Phone: (863) 494-4111. Fax: (863) 494-4443. E-mail: wzsp@desoto.net Web Site: www.lazeta.fm. Licensee: Heartland Broadcasting Corp. Population served: 200,000 Natl. Network: CNN Radio. Rgnl rep: Lotus-Entravision Law Firm: Kaye, Scholer, Fierman, Hays & Handler. Format: Sp. Target aud: General; Spanish speaking audience, Charlotte, Desoto, Hardee, Sarasota & Highlands counties. ♦Harold Kneller Jr., pres; Bryan Hollenbauer, gen mgr.

North Fort Myers

WWCN(AM)— Dec 17, 1983: 770 khz; 10 kw-D, 1 kw-N, DA-2. TL: N26 46 30 W81 50 51. Stereo. Hrs opn: 20125 S. Tamiami Tr., Estero, 33928. Phone: (239) 495-2100. Fax: (239) 992-8165. Web Site: www.am770.com. Licensee: Beasley Radio Co. Group owner: Beasley Broadcast Group (acq 12-16-87). Format: Talk, sports. ♦George Beasley, pres; Bradley C. Beasley, gen mgr; John Rozz, opns mgr.

North Miami

WKAT(AM)— November 1937: 1360 khz; 5 kw-D, 1 kw-N. TL: N25 44 36 W80 09 14. Hrs open: 2500 N.W. 79th Ave., Suite 232, Doral, 33122. Phone: (305) 503-1340. Fax: (305) 503-1349. Web Site: 1360wkat.com. Licensee: Caron Broadcasting Inc. (acq 1-31-2005; $10 million). Population served: 334,859 Law Firm: Fletcher, Heald & Hildreth. Format: News/talk, classical. ♦Tony Calatayud, gen mgr; Stephen James, opns mgr.

North Miami Beach

WXDJ(FM)— 1986: 95.7 mhz; 40 kw. 531 ft TL: N25 46 29 W80 11 19. Stereo. Hrs opn: 24 1001 Ponce deLeon Blvd., Coral Gables, 33134. Phone: (305) 444-9292. Fax: (305) 461-4466. Web Site: www.lamusica.com. Licensee: WXDJ Licensing Inc. Group owner: Spanish Broadcasting System Inc. (acq 7-11-97; $110 million with WRMA(FM) Fort Lauderdale). Law Firm: Wiley, Rein & Fielding. Format: Salsa. News: 4 hrs wkly. Target aud: 18-54; Hispanic Adults. ♦Jackie Nosti-Combo, gen mgr; John Caride, prom dir.

North Palm Beach

WSVU(AM)— 2006: 960 khz; 1.2 kw-D, 250 w-N, DA-D. TL: N26 49 01 W80 15 07. Hrs open: Phone: (239) 542-4200. Fax: (239) 542-4221. E-mail: swfradio@aol.com Licensee: Intermart Broadcasting Southwest Florida Inc. Format: Classic country. ♦Patricia S. Woods, VP.

Ocala

WCFI(AM)— 1939: Stn currently dark. 1290 khz; 5 kw-D, 1 kw-N, DA-N. TL: N29 11 51 W82 10 57. Hrs open: 24 3621 N.W. Tenth St., 34475. Phone: (352) 351-5807. Licensee: Vector Communications Inc. (acq 9-24-99; $250,000). Population served: 200000 Format: News info, real country. ♦Robert J. Maines Jr., gen mgr & stn mgr.

***WHIJ(FM)**— Mar 30, 1990: 88.1 mhz; 1.25 kw. 394 ft TL: N29 14 17 W82 07 17. Stereo. Hrs opn: 24 408 University, Suite 206, Gainsville, 32601. Phone: (352) 351-8810. Fax: (352) 351-8917. E-mail: thejoyfm@thejoyfm.com Web Site: www.thejoyfm.com. Licensee: Radio Training Network Inc. (acq 10-5-01; $80,000 with WAQV(FM) Crystal River). Population served: 300,000 Law Firm: Gammon & Grange. Format: Adult contemp, educ, Christian. Target aud: 20-50. ♦Jeff MacFarlane, gen mgr.

WMFQ(FM)— July 11, 1977: 92.9 mhz; 50 kw. 476 ft TL: N29 04 45 W82 05 35. Stereo. Hrs opn: 24 3357 S.W. 7th St., 34474. Phone: (352) 732-9877. Fax: (352) 622-6675. E-mail: production@wtrs.fm Web Site: www.wmfq.fm. Licensee: Asterisk Communications Inc. Group owner: Asterisk Inc. (acq 1995; $2.1 million). Population served: 300,000 Natl. Rep: McGavren Guild. Law Firm: Reddy, Begley & McCormick. Format: Oldies. News staff: one; News: 5 hrs wkly. Target aud: 35 plus; upscale. ♦Dean Johnson, gen mgr.

WMOP(AM)— Dec 18, 1953: 900 khz; 5 kw-D, 23 w-N. TL: N29 14 17 W82 07 17. Stereo. Hrs opn: 24 101 S.E. 2nd Pl., Gainesville, 32601. Phone: (352) 378-7378. Fax: (352) 629-1614. E-mail: sales@floridasportstalk.com Web Site: www.floridasportstalk.com. Licensee: Florida Sportstalk Inc. (acq 11-14-96; $350,000). Population served: 213,300 Natl. Network: ABC. Natl. Rep: Dora-Clayton. Law Firm: Pepper & Corazzini. Format: Sports, talk. News staff: one; News: 3 hrs wkly. Target aud: 35 plus. ♦Tom Catalano, gen mgr.

WOCA(AM)— May 1957: 1370 khz; 5 kw-D. TL: N29 12 04 W82 09 07. Hrs open: 6 AM-8 PM Box 1056, 34478. Secondary address: 1515 E. Silver Springs Blvd., Suite 134 34470. Phone: (352) 732-8000. Phone: (352) 622-9622. Fax: (352) 732-0174. E-mail: woca@woca.com Web Site: www.woca.com. Licensee: Westshore Broadcasting Inc. Population served: 300,000 Natl. Network: ABC. Format: News/talk. News staff: two; News: 16 hrs wkly. Target aud: 35 plus. Spec prog: Black 2 hrs wkly. ♦Tishia A. Moeller, gen mgr & stn mgr.

WOGK(FM)— Nov 7, 1960: 93.7 mhz; 100 kw. 1,348 ft TL: N29 16 06 W82 04 51. Stereo. Hrs opn: 3602 N.E. 20th Pl., 34470. Phone: (352) 622-5600. Fax: (352) 622-3998. E-mail: terry@937kcountry.com Web Site: www.93kcountry.com. Licensee: Ocala Broadcasting L.L.C. Group owner: Wooster Republican Printing Co. (acq 9-27-86). Population served: 700,000 Natl. Rep: Katz Radio. Law Firm: Baker & Hostetler. Format: Country. News staff: one; News: 3 hrs wkly. Target aud: 25-54; general. ♦Bob Kassi, pres & gen sls mgr; Jim Robertson, VP, gen mgr & gen mgr; Bob Forster, progmg dir.

Ocoee

WOKB(AM)— Jan 1, 1958: 1600 khz; 4 kw-D, DA. TL: N28 34 06 W81 31 09. Hrs open: 24 1801 Clark Rd., 34761. Phone: (407) 296-4747. Fax: (407) 253-2228. E-mail: jchsamek@yahoo.com Licensee: Rama Communications Inc. (group owner; (acq 10-13-93; $950,000 with WXXU(AM) Cocoa Beach; FTR: 11-8-93) Population served: 99,006 Law Firm: Cohn & Marks. Format: Gospel. News staff: one. Target aud: 25-54. Spec prog: Fr 7 hrs, gospel 12 hrs wkly. ♦Sabita Persaud, pres.

WUNA(AM)— Oct 25, 1962: 1480 khz; 1 kw-D, 71 w-N. TL: N28 33 27 W81 32 29. Stereo. Hrs opn: 749 S. Bluford Ave., 34761. Phone: (407) 656-9823. Fax: (407) 656-2092. E-mail: WUNA1480orlando@juno.com Licensee: Way Broadcasting Licensee LLC (acq 4-20-2000; grpsl). Rgnl. Network: Florida Radio Net. Format: Sp contemp. ♦Juan Nieves, gen mgr; Lou Muller, chief of engrg; Sheila Rodriguez, sls dir & traf mgr.

Okeechobee

WOKC(AM)— Feb 6, 1962: 1570 khz; 1 kw-D, 14 w-N. TL: N27 12 59 W80 49 53. Stereo. Hrs opn: 6 AM-11 PM 210 W. North Park St., Suite 102, 34974. Phone: (863) 467-1570. Fax: (863) 763-3171. E-mail: wokc@gladesmedia.com Web Site: www.gladesmedia.com. Licensee: Glades Media Co. LLC (acq 7-31-01; $200,000). Population served: 50,000 Rgnl. Network: Florida Radio Net., S.E. Agri. Format: Classic country. News staff: one; News: 10 hrs wkly. Target aud: General. ♦Will Skinner, gen mgr.

Orange Park

***WAYR(AM)**— May 28, 1960: 550 khz; 5 kw-D, 500 w-N, DA-1. TL: N30 04 21 W81 47 24. Hrs open: 24 2500 Russell Rd., Green Cove Springs, 32043-9492. Phone: (904) 284-1111. Phone: (904) 284-2500. Fax: (904) 284-2501. E-mail: lstephens@wayradio.org Web Site: www.wayradio.org. Licensee: Good Tidings Trust Inc. Population served: 1,100,000 Law Firm: Wiley, Rein & Fielding. Format: Christian. Target aud: 45 plus; mature Christian. ♦Bill Tidwell, pres & chief of engrg; Luke Stephens, gen mgr.

Orlando

WAMT(AM)—See Pine Castle-Sky Lake

WDBO(AM)— May 24, 1924: 580 khz; 5 kw-U, DA-N. TL: N28 37 12 W81 24 34. Hrs open: 4192 John Young Pkwy., 32804. Phone: (407) 295-5858. Phone: (321) 281-2000. Fax: (407) 291-4879. Web Site: www.wdbo.com. Licensee: Cox Radio Inc. Group owner: Cox Broadcasting (acq 3-28-97; grpsl). Population served: 1,104,600 Format: News/talk. ♦Bill Hendrich, gen mgr; Steve Holbrook, opns mgr; Tom Interrante, gen sls mgr; Steve Avellone, natl sls mgr; Rich Mastroberte, prom dir; Kipper McGee, progmg dir; Marsha Taylor, news dir; Steve Fluker, chief of engrg.

WWKA(FM)—Co-owned with WDBO(AM). Apr 24, 1952: 92.3 mhz; 98 kw. 1,380 ft TL: N28 36 08 W81 05 37. Stereo. 24 Phone: (407) 298-9292. Web Site: www.wwka.com. Format: Country. ♦Kimberly Hellstrom, prom dir; Len Shackelford, progmg mgr; Shadow Stevens, mus dir; Steve Fluker, engrg dir.

WDYZ(AM)— Dec 5, 1947: 990 khz; 50 kw-D, 5 kw-N, DA-2. TL: N28 34 28 W81 27 48. Stereo. Hrs opn: 610 Sycamore St., Suite 220, Celebration, 34747. Phone: (407) 566-2033. Fax: (407) 566-2034. E-mail: paul.t.proly@ABC.com Web Site: www.radiodisney.com. Licensee: Radio Disney Group LLC. Group owner: ABC Inc. (acq 1-23-01; $5 million cash). Population served: 3,000,000 Natl. Network: Radio Disney. Natl. Rep: Interep. Format: Children/families. Target aud: 4-16;25-49; children; mothers. ♦Paul T. Proly, stn mgr; Pren Rashbury, mktg VP; Robin Jones, progmg VP.

WFLF(AM)—See Pine Hills

WHTQ(FM)— 1952: 96.5 mhz; 100 kw. 1,600 ft TL: N28 34 51 W81 04 32. Stereo. Hrs opn: 24 4192 John Young Pkwy, 32804. Phone: (407) 422-9696. Fax: (407) 422-5883. Fax: (407) 422-0917. Web Site: www.whtq.com. Licensee: Cox Radio Inc. Group owner: Cox Broadcasting (acq 1997). Population served: 99,006 Natl. Rep: Christal. Format: Classic rock. News staff: one; News: 2 hrs wkly. Target aud: 25-54; adult male. ♦Debbie Morel, VP & gen mgr.

***WMFE-FM**— July 14, 1980: 90.7 mhz; 100 kw. 731 ft TL: N28 36 08 W81 05 37. Stereo. Hrs opn: 24 11510 E. Colonial Dr., 32817-4699. Phone: (407) 273-2300. Fax: (407) 273-8462. Fax: (407) 273-3613. Web Site: www.wmfe.org. Licensee: Community Communications Inc. Natl. Network: PRI, NPR. Format: Class, news & info. News staff: 4; News: 48 hrs wkly. Target aud: 35 plus; well-educated, executive, professional, upper-income. Spec prog: New instrumental 4 hrs wkly. ♦Jose Fajardo, pres.

WMGF(FM)—See Mount Dora

WMMO(FM)— Aug 19, 1990: 98.9 mhz; 44 kw. 522 ft TL: N28 34 27 W81 27 46. Stereo. Hrs opn: 24 4192 John Young Pkwy., 32804. Phone: (407) 422-9696. Fax: (407) 422-5883. Web Site: www.wmmo.com. Licensee: Cox Radio Inc. Group owner: Cox Broadcasting Population served: 150,000 Natl. Rep: Christal. Format: Adult contemp, soft rock. News: one hr wkly. Target aud: 25-49. ♦Debbie Morel, VP & gen mgr.

WOKB(AM)—See Ocoee

WOMX-FM— Aug 15, 1967: 105.1 mhz; 95 kw. 1,309 ft TL: N28 36 17 W81 05 13. (CP: Ant 1,597 ft.). Stereo. Hrs opn: 24 1800 Pembrook Dr., Suite 400, 32810. Phone: (407) 919-1000. Fax: (407) 919-1190. Web Site: www.mix1051.com. Licensee: Infinity Radio Inc. Group owner: Infinity Broadcasting Corp. (acq 12-14-00; grpsl). Population served: 1,235,340 Law Firm: Leibowitz & Associates. Format: Hot adult contemp. News staff: one. ♦Earnest James, sr VP; Michele Holland, CEO & gen sls mgr; Angela Schlesman, prom dir; Jeff Cushman, progmg dir.

WPCV(FM)—See Lakeland

WPRD(AM)—See Winter Park

WQTM(AM)— 1947: 740 khz; 50 kw-U, DA-2. TL: N28 28 53 W81 39 43. Hrs open: 24/7 2500 Maitland Center Pkwy., Suite 401, Maitland, 32751. Phone: (407) 916-7800. Fax: (407) 916-0329. E-mail: programdirector@7400theteam.com Web Site: www.740theteam.com. Licensee: Clear Channel Radio Licenses Inc. Group owner: Clear Channel Communications Inc. (acq 11-21-97; grpsl). Population served: 1,000,000 Natl. Network: CBS, Fox Sports, Westwood One. Law Firm: Wiley, Rein and Fielding. Wire Svc: AP Format: Talk, sports. Target aud: 25-54; adult male. ♦Linda Byrd, VP & gen mgr; Colin Cantwell, gen sls mgr; Rick Everett, mktg dir; Jimmy D., prom mgr; Chris Kampmeier, progmg dir; Marc Daniels, progmg dir.

WRLZ(AM)—Eatonville, 1957: 1270 khz; 5 kw-U, DA-N. TL: N28 34 03 W81 25 38. Hrs open: 24 Box 593642, 32859-3642. Secondary address: 6106 B Hoffner Ave. 32822. Phone: (407) 345-0700. Fax: (407) 345-1492. E-mail: info@radioluz1270.com Web Site: www.radioluz1270.com. Licensee: Radio Luz Inc. (acq 1996; $378,500). Population served: 99,006 Format: Sp, contemp. Target aud: General; Family. ♦Saturnino Gonzalez, pres; John Maldonado, gen mgr.

WRMQ(AM)— Oct 21, 1985: 1140 khz; 4.1 kw-D. TL: N28 30 42 W81 14 09. Hrs open: 1033 Semoran Blvd., Suite 253, Casselberry, 32707. Phone: (407) 830-0800. Fax: (407) 260-6100. E-mail: mannyarroyo@qbcflorida.com Licensee: Florida Broadcasters. Population served: 1,200,000 Law Firm: Roy F. Perkins. Format: Gospel. Target aud: 25-54. ♦George M. Arroyo, pres & gen mgr.

WRUM(FM)— July 1, 1971: 100.3 mhz; 100 kw. Ant 1,597 ft TL: N28 36 08 W81 05 37. Stereo. Hrs opn: 24 2500 Maitland Ctr. Pkwy., Suite 401, Maitland, 32751. Phone: (407) 916-7800 / (407) 916-1003. Fax: (407) 916-7400. Web Site: www.rumba1003.com. Licensee: Clear Channel Broadcasting Licenses Inc. Group owner: Clear Channel Communications Inc. (acq 1997; grpsl). Population served: 99,006 Law Firm: Wiley, Rein & Fielding. Format: Sp/tropical. Target aud: 18-49. ♦Linda Byrd, pres & gen mgr; Fernando Bauermeister, gen sls mgr; Rick Everett, mktg dir; Suheily Gonzales, prom mgr; Chris Kampmeier, progmg dir; Raymond Torres, progmg dir; Donna Carmichael, traf mgr.

WSDO(AM)—See Sanford

WTKS-FM—Cocoa Beach, May 8, 1962: 104.1 mhz; 100 kw. 1,609 ft TL: N28 34 51 W81 04 32. Stereo. Hrs opn: 24 2500 Maitland Ctr. Pkwy., Suite 401, Maitland, 32751. Phone: (407) 916-7800. Fax: (407) 916-7511. Web Site: www.wtks.com. Licensee: Clear Channel Broadcasting Licenses Inc. Group owner: Clear Channel Communications Inc. (acq 11-21-97; grpsl). Population served: 1,402,300 Natl. Network: Premiere Radio Networks. Natl. Rep: Clear Channel. Law Firm: Wiley, Rein & Fielding. Format: Entertainment talk. Target aud: Adults; 25-54. ♦Linda Byrd, gen mgr & chief of engrg; Sam Nein, sls dir; Ed Kennedy, gen sls mgr; Frank Celebre, natl sls mgr; Erika Plak, prom mgr; Chris Kampmeier, progmg dir; Katherine Brown, progmg dir; Rick Everett, mktg.

WTLN(AM)— April 1, 1940: 950 khz; 12 kw-U, DA-N. TL: N28 32 08 W81 26 55. Hrs open: 24 1188 Lake View Dr., Altamonte Springs, 32714. Phone: (407) 682-9494. Fax: (407) 682-7005. E-mail: wtln@salemorlando.com Web Site: www.wtln.com. Licensee: Pennsylvania Media Associates Inc. (acq 12-7-2005; $9.4 million). Population served: 1,235,340 Natl. Rep: Salem. Law Firm: Holland & Knight. Format: Christian, talk. Target aud: 25-44; general. ♦Edward G. Atsinger III, pres; David Koon, gen mgr; Lee Brandell, stn mgr & opns dir; Jackie Trefcer, gen sls mgr.

***WUCF-FM**— Jan 30, 1978: 89.9 mhz; 40 kw. 194 ft TL: N28 36 00 W81 12 05. Stereo. Hrs opn: 24 Box 162199, 32816-2199. Secondary address: 4000 Central Florida Blvd., Bldg. 75, Rm. 130 32816. Phone: (407) 823-5162. Phone: (407) 823-0899. Fax: (407) 823-6364. Web Site: wucf.ucf.edu. Licensee: University of Central Florida. Natl. Network: NPR, PRI. Law Firm: Cohn & Marks. Format: Jazz. News staff: 10; News: 16 hrs wkly. Spec prog: Fr one hr, It one hr, Indian 2 hrs, blues 6 hrs, bluegrass 3 hrs, Irish one hr wkly. ♦John Hitt, pres; Kayonne Riley, gen mgr; Bruce Duerie, engrg dir & chief of engrg.

WXXL(FM)—See Leesburg

Orlovista

WEUS(AM)— 2006: 810 khz; 10 kw-D, 400 w-N, DA-2. TL: N28 34 18 W81 26 02. Hrs open: 24 hrs
Approximitly 2 hrs.
999 Douglas Ave., Attamonte, 32714. Phone: (407) 774-8810. Fax: (407) 774-8895. E-mail: info@bio810.com Web Site: www.bio810.com. Licensee: Star Over Orlando Inc. Format: Oldies. Target aud: 45 Plus Market. Spec prog: Mike Harvey "Super Gold" 6pm—mid. ♦Carl Como Tutera, pres.

Ormond Beach

WELE(AM)— Aug 1, 1957: 1380 khz; 5 kw-D, 2.5 kw-N, DA-2. TL: N29 16 09 W81 04 54. Hrs open: 432 S. Nova Rd., 32174. Phone: (386) 677-4122. Fax: (386) 677-4123. E-mail: doug@wele1380.com Web Site: www.wele1380.com.wele1380.com Licensee: Wings Communications Inc. (acq 9-90; $175,000; FTR: 9-24-90). Population served: 90,000 Natl. Network: CNN Radio, Westwood One. Format: News/talk. Target aud: General; mature, adults interested in sports & local current events. ♦F. Douglas Wilhite, pres & gen mgr; Kristin Cobb, opns mgr; Mike Johnson, progmg dir; Doug Wilhite, engr.

Ormond-by-the-Sea

WHOG-FM— 1995: 95.7 mhz; 25 kw. 328 ft TL: N29 14 10 W81 04 23. Hrs open: 126 W. International Speedway Blvd., Daytona Beach, 32114. Phone: (386) 255-9300. Fax: (386) 238-6071. Web Site: www.whog.fm. Licensee: Black Crow LLC. Group owner: Black Crow Media Group LLC (acq 9-21-2001; grpsl). Format: Classic rock, AOR. ♦Stacey Knerler, gen mgr; Donna Fillion, sls dir & progmg dir.

Oviedo

WONQ(AM)— Nov 21, 1992: 1030 khz; 10 kw-D, 1.7 kw-N, DA-2. TL: N28 40 31 W81 10 01. Hrs open: 24 1033 Semoran Blvd., Suite 253, Casselberry, 32707. Phone: (407) 830-0800. Fax: (407) 260-6100. E-mail: mannyarroyo@qbcflorida.com Licensee: Florida Broadcasters. Population served: 1,200,000 Law Firm: Roy F. Perkins. Format: Sp music, news, contemp Latin hits. Target aud: 25-54. ♦George M. Arroyo, pres & gen mgr; George Mier, opns VP.

Palatka

WGNE-FM— Dec 13, 1973: 99.9 mhz; 100 kw. 1,201 ft TL: N29 31 08 W81 19 02. Stereo. Hrs opn: 6444 Atlantic Blvd., Jacksonville, 32211. Phone: (904) 727-9696. Fax: (904) 721-9322. Web Site: www.gatercountry.com. Licensee: Renda Broadcasting Corp. Group owner: Renda Broadcasting Corp.-Renda Radio Inc. (acq 1996; $6.5 million with WMUV(FM) Brunswick, GA). Population served: 2000000 Law Firm: Haley, Bader & Potts. Format: Country. Target aud: 18-49. ♦Toney Renda, pres; Gary Spurgeon, gen mgr.

***WHIF(FM)**— Mar 29, 1996: 91.3 mhz; 1.7 kw. 318 ft TL: N29 38 54 W81 39 42. Hrs open: 201 S. Palm Ave., 32177. Phone: (386) 325-3334. Fax: (386) 325-0934. E-mail: whif@gbso.net Web Site: www.whif.org. Licensee: Putnam Radio Ministries Inc. Population served: 75,000 Format: Adult contemp, Christian. Target aud: 25-54; family-oriented, middle-class. Spec prog: Relg educ 10 hrs wkly. ♦Robin Toole, gen mgr & progmg dir.

WIYD(AM)— Feb 14, 1947: 1260 khz; 1 kw-D, 500 w-N, DA-N. TL: N29 38 23 W81 38 26. Hrs open: 24 Box 918, 32178-0918. Phone: (386) 325-4556. Fax: (386) 328-5161. E-mail: wiyd@atlantic.net Licensee: Hall Broadcasting Co. (acq 2-14-57; $100,000). Population served: 25,000 Rgnl. Network: Florida Radio Net. Format: C&W. News staff: one; News: 6 hrs wkly. Target aud: 18-49; rich & powerful. Spec prog: Relg 5 hrs wkly. ♦Ronald G. Tumlin, pres; Suzanne Tumlin, gen mgr; Mary Makie Connor, stn mgr.

WPLK(AM)— 1957: 800 khz; 1 kw-D, 334 w-N. TL: N29 37 40 W81 34 35. (CP: TL: N29 39 07 W81 35 32). Stereo. Hrs opn: 24 Box 335, 32178. Secondary address: 1428 St. John's Ave. 32177. Phone: (386) 325-5800. Fax: (386) 328-8725. E-mail: wplk@wplk.com Web Site: wplk.com. Licensee: Radio Palatka Inc. (acq 4-28-98; $250,000 for stock). Natl. Network: AP Radio. Format: Oldies, Music of your life. News staff: one; News: 2 hrs wkly. Target aud: General. ♦Wayne Bullock, pres & gen mgr.

Palm Bay

***WEJF(FM)**— 1993: 90.3 mhz; 2 kw. 295 ft TL: N28 02 54 W80 40 34. Hrs open: 2824B Palm Bay Rd., 32905. Phone: (321) 722-9998. Fax: (321) 724-0845. E-mail: wejf@bellsouth.net Web Site: www.wejf.com. Licensee: Florida Public Radio Inc. (acq 7-95; $40,000). Format: Adult contemp, Christian, var. ♦Eric Sabo, gen mgr & opns mgr.

***WHYZ(FM)**— July 1997: 88.5 mhz; 600 w vert. Ant 108 ft TL: N28 02 54 W80 40 34. Stereo. Hrs opn: 24 Victory Christian Academy, 100 Emerson Dr. N.W., 32907. Phone: (321) 953-9942. Fax: (321) 768-6265. Licensee: Victory Christian Academy. Format: Praise & worship. Target aud: Christian. ♦L. Mark Ostrander, pres & gen mgr.

Palm Beach

WBZT(AM)—See West Palm Beach

WRMF(FM)—Licensed to Palm Beach. See West Palm Beach

Palm City

***WCNO(FM)**— Apr 1, 1990: 89.9 mhz; 100 kw. 613 ft TL: N27 07 20 W80 23 21. Hrs open: 2960 S.W. Mapp Rd., 34990-2737. Phone: (772) 221-1100. Fax: (772) 221-8716. E-mail: wcno@wcno.com Web Site: www.wcno.com. Licensee: National Christian Network Inc. Format: Christian, adult contemp. ♦Tom Craton, gen mgr.

Palmetto

WBRD(AM)— October 1957: 1420 khz; 2.5 kw-D, 1 kw-N, DA-2. TL: N27 32 42 W82 34 28. Stereo. Hrs opn: Box 144, Oneco, 34264. Phone: (941) 955-1420. Fax: (941) 752-4794. E-mail: wbrdradio@aol.com Web Site: www.wbrd.com. Licensee: Metropolitan Radio Group Inc. (group owner; acq 6-96). Population served: 2,500,000 Format: Southern Gospel. Target aud: 35 plus. ♦Bill Bailey, gen mgr & opns dir.

Panama City

WDIZ(AM)— April 1940: 590 khz; 1.7 kw-D, 2.5 kw-N, DA-N. TL: N30 10 20 W85 36 49. Stereo. Hrs opn: Box 59288, 32412. Secondary address: 1834 Lisenby Ave. 32405. Phone: (850) 769-1408. Fax: (850) 769-0659. Licensee: Clear Channel Broadcasting Licenses Inc. Group owner: Clear Channel Communications Inc. (acq 11-21-97; grpsl). Population served: 40,000 Format: Nostalgia. ♦Peter Norden, gen mgr.

WFSY(FM)—Co-owned with WDIZ(AM). October 1971: 98.5 mhz; 100 kw. 1,090 ft TL: N30 30 41 W85 29 24. Stereo. Web Site: www.beachfm.com.40,000 Format: Adult contemp. Target aud: 25-54.

WEBZ(FM)—See Mexico Beach

***WFFL(FM)**— 2007: 91.7 mhz; 310 w. Ant 207 ft TL: N30 10 48 W85 38 10. Hrs open:
Rebroadcasts WJFM(FM) Baton Rouge, LA 100%.
Box 262550, Baton Rouge, LA, 70826. Secondary address: 8919 World Ministry Ave., Baton Rouge 70810. Phone: (225) 768-3102. Fax: (225) 768-3729. E-mail: kawikfish@yahoo.com Web Site: www.jsm.org. Licensee: Family Worship Center Church Inc. (acq 10-12-2006; grpsl). Format: Christian. ♦David Whitelaw, COO.

***WFSW(FM)**— 1995: 89.1 mhz; 100 kw. 403 ft TL: N30 22 02 W85 55 29. Hrs open: 1600 Red Barber Plaza, Tallahassee, 32310. Phone: (850) 487-3086. Fax: (850) 487-3293. Web Site: www.wfsu.org. Licensee: Florida State University. Natl. Network: NPR. Format: News, talk. ♦Pat Keating, gen mgr; Caroline Austin, stn mgr; Aron Myers, prom dir; Tom Flanigan, news dir; Andy Hanus, engrg dir.

WILN(FM)— Apr 11, 1985: 105.9 mhz; 50 kw. Ant 406 ft TL: N30 10 44 W85 46 55. Stereo. Hrs opn: 24 7106 Laird St., Suite 102, Panama City Beach, 32408. Phone: (850) 230-5855. Fax: (850) 230-6988. Web Site: www.island106.com. Licensee: Magic Broadcasting Florida Licensing LLC. (group owner; (acq 1-31-2003; grpsl). Format: CHR. News staff: one; News: 2 hrs wkly. Target aud: 18-49. ♦Jeff Storey, COO, exec VP & gen mgr; Mike Preble, opns VP.

***WJTF(FM)**— Oct 15, 1998: 89.9 mhz; 100 kw. 213 ft TL: N30 10 20 W85 40 20. Hrs open: 835A S. Berthe, 32404. Phone: (850) 874-9900. Fax: (850) 874-9930. E-mail: wjtf@bellsouth.net Web Site: www.myfln.org. Licensee: Family Life Broadcasting Inc (acq 3-27-2007; grpsl). Natl. Network: Moody. Law Firm: Hill & Welch. Format: Relg, educ. Target aud: 35-90; general. ♦Tom Bush, gen mgr; Kelly Dickson, opns mgr; Mickey Jacobs, dev dir & progmg dir.

***WKGC-FM**— October 1982: 90.7 mhz; 100 kw. 336 ft TL: N30 13 05 W85 51 16. Stereo. Hrs opn: 5230 W. Hwy. 98, 32401. Phone: (850) 873-3500. Fax: (850) 913-3299. E-mail: fsundram@gulfcoast.edu Web Site: www.wkgc.org. Licensee: Gulf Coast Community College. Population served: 150,000 Natl. Network: NPR, PRI. Rgnl. Network: Fla. Pub. Law Firm: Dow, Lohnes & Albertson. Format: News. Target aud: General. Spec prog: Black 6 hrs wkly. ♦Robert Spadden, pres; Frank Sundram, gen mgr; Reed Kinney, opns mgr.

WLTG(AM)— Dec 11, 1949: 1430 khz; 5 kw-U, DA-2. TL: N30 09 55 W85 35 19. Hrs open: Box 15635, 32406. Secondary address: 3100 E. 15th St., Springfield 32405. Phone: (850) 784-9873. Fax: (850) 784-6908. Licensee: Williams Communications Inc. (group owner; acq 8-12-03; $500,000). Population served: 120,000 Natl. Network: Premiere Radio Networks, Salem Radio Network. Natl. Rep: Commercial Media Sales. Format: News/talk, sports, info. Target aud: General. Spec prog: Black gospel 7 hrs wkly. ♦John Gay, gen mgr, sls dir & progmg dir.

WPAP-FM— Mar 30, 1967: 92.5 mhz; 100 kw. 930 ft TL: N30 22 05 W85 12 24. Stereo. Hrs opn: 24 Box 59288, 32412. Phone: (850) 769-1408. Fax: (850) 769-0659. Web Site: www.wpapfm.com. Licensee: Clear Channel Radio Licenses Inc. Group owner: Clear Channel Communications Inc. (acq 11-21-97; grpsl). Population served: 250,000 Natl. Rep: McGavren Guild. Format: Country. Target aud: 25-54. ♦Pete Norton, gen mgr.

WPFM-FM— September 1963: 107.9 mhz; 100 kw. 781 ft TL: N30 26 00 W85 24 51. (CP: 98.4 kw, ant 954 ft. TL: N30 13 45 W85 23 20). Stereo. Hrs opn: 24 118 Gwyn Dr., Panama City Beach, 32407. Phone: (850) 234-8858. Fax: (850) 234-6592. Licensee: Double O Radio Corp. (group owner; acq 3-10-2004; grpsl). Population served: 120,000 Natl. Rep: Christal. Format: Contemp hit/Top-40. Target aud: 18-49; active lifestyle, young adult audience. ♦Harry Finch, gen mgr.

WYOO(FM)—Springfield, Mar 2, 1993: 101.1 mhz; 5.2 kw. 236 ft TL: N30 12 12 W85 36 57. (CP: 25 kw). Stereo. Hrs opn: 24 7106 Laird St., Suite 102, Panama City Beach, 32408. Phone: (850) 230-5855. Fax: (850) 230-6988. Web Site: www.talkradio101.com. Licensee: Magic Broadcasting Florida Licensing LLC. (group owner; (acq 9-30-2002; grpsl). Population served: 165,000 Rgnl. Network: Florida Radio Net. Natl. Rep: Christal. Law Firm: Richard Hayes. Format: Talk. News staff: one; News: 28 hrs wkly. Target aud: 25-54; educated, upscale. ♦Jeff Storey, COO, pres & exec VP; Mike Preble, opns dir & adv mgr.

Panama City Beach

WASJ(FM)— January 1993: 105.1 mhz; 50 kw. 335 ft TL: N30 10 44 W85 46 55. Stereo. Hrs opn: 24 118 Gwyn Dr., Panama City, 32408. Phone: (850) 234-8858. Fax: (850) 234-6592. E-mail: stevegreen @panamacityradio.com Web Site: www.smoothjazz1051.com. Licensee: Double O Radio Corp. (group owner; acq 3-10-2004; grpsl). Population served: 135,000 Natl. Rep: Christal. Format: Smooth Jazz. ♦Harry Finch, gen mgr.

WFSY(FM)—See Panama City

***WKGC(AM)**— June 25, 1965: 1480 khz; 500 w-D, 87 w-N. TL: N30 10 33 W85 48 03. Hrs open: 6 AM-9 PM 5230 W. Hwy. 98, Panama City, 32401. Phone: (850) 873-3500. Fax: (850) 913-3299. E-mail: fsundram@gnfcoast.edu Web Site: www.wkgc.org. Licensee: Gulf Coast Community College. (acq 12-1-72). Population served: 150,000 Natl. Network: NPR. Format: Easy listening. News: 25 hrs wkly. Target aud: General; college students & older high school students. Spec prog: Folk 2 hrs, educ 8 hrs wkly. ♦Robert McSpadden, pres; Frank Sundram, gen mgr & stn mgr; Reed Kinney, opns mgr.

WPCF(AM)— Sept 23, 1958: 1290 khz; 270 w-D, 1 kw-N. TL: N30 10 44 W85 46 55. Hrs open: 7106 Laird St., Ste. 102, 32408. Phone: (850) 230-5855. Fax: (850) 230-6988. E-mail: jeffstorey@stylesmedia.com Licensee: Magic Broadcasting Florida Licensing LLC. (group owner; (acq 9-30-2002; grpsl). Population served: 150000 Format: Sports. ♦Jeff Storey, CEO & gen mgr; Kim Styles, gen mgr; Joe Valentine, stn mgr; Mike Preble, opns mgr.

WVVE(FM)— June 1988: 100.1 mhz; 12 kw. Ant 403 ft TL: N30 10 44 W85 46 55. Stereo. Hrs opn: 24 7106 Laird St., Suite 102, 32408. Phone: (850) 230-5855. Fax: (850) 230-6988. Licensee: Magic Broadcasting Florida Licensing LLC. (group owner; (acq 9-30-2002; grpsl). Format: Adult contemp. ♦Jeff Storey, CEO, CEO & exec VP; Mike Prebke, opns mgr.

Parker

WFBX(FM)— August 1977: 94.5 mhz; 100 kw. Ant 994 ft TL: N29 49 09 W85 15 34. Stereo. Hrs opn: 24 Box 59288, Panama City, 32412. Phone: (850) 769-1408. Fax: (850) 769-0659. Web Site: www.945thefox.com. Licensee: Clear Channel Broadcasting Licenses Inc. Group owner: Clear Channel Communications Inc. (acq 11-21-97; grpsl). Population served: 200,000 Format: Soft rock. Target aud: 25 plus; adults. Spec prog: Relg 5 hrs wkly. ♦Pete Norden, gen mgr.

Pennsuco

***WIRP(FM)**— 1999: 88.3 mhz; 2.25 kw. Ant 282 ft TL: N25 52 24 W80 28 59. (CP: 6 kw). Stereo. Hrs opn: 24 1400 N.W. 107 Ave., Suite 306, Miami, 33172. Phone: (305) 406-2883. Fax: (305) 406-3030. Web Site: www.lanueva883fm.com. Licensee: Genesis Radio Network Inc. (acq 4-11-2005; $1.69 million). Format: Christian, Sp. News staff: one; News: 3 hrs wkly. Target aud: 18-35; Hispanic Christians. Spec prog: Children 6 hrs. ♦Edwin L. Ortiz, pres; Mauricio Quintana, gen mgr.

Pensacola

WBSR(AM)— Sept 1, 1946: 1450 khz; 1 kw-U. TL: N30 25 44 W87 14 27. Hrs open: 24 Box 19047, 32523. Secondary address: 1601 N. Pace Blvd. 32505. Phone: (850) 438-4982. Fax: (850) 433-7932. E-mail: wbsr@wbsr.com Web Site: www.wbsr.com. Licensee: Easy Media Inc. (acq 3-22-85; $330,000; FTR: 2-25-85). Population served: 280,000 Format: Soft adult contemp. Target aud: 35-54. ♦Frederic T.C. Brewer, pres; Gene Pfalzer, stn mgr.

WCOA(AM)— Feb 3, 1926: 1370 khz; 5 kw-U, DA-N. TL: N30 26 57 W87 15 46. Hrs open: 24 6565 N. W St., 32505. Phone: (850) 478-6011. Fax: (850) 478-3971. Licensee: Cumulus Licensing Corp. Group owner: Cumulus Media Inc. (acq 10-25-99; with co-located FM).

Population served: 160,000 Natl. Network: ABC. Natl. Rep: Katz Radio. Format: News/talk. Target aud: 25-54. Spec prog: Relg 3 hrs wkly. ♦Brian Weil, gen mgr; Luke McCoy, progmg dir; Jim Roberts, news dir; Yancy McNair, chief of engrg.

WJLQ(FM)—Co-owned with WCOA(AM). Sept 1, 1965: 100.7 mhz; 100 kw. 1,555 ft TL: N30 37 35 W87 38 50. Stereo. 60,000 Format: Music, hot adult comtemp. Target aud: 25-44. ♦John Stuart, progmg dir.

WDWR(AM)— 1947: 1230 khz; 1 kw-U. TL: N30 25 57 W87 13 07. Hrs open: 24 Box 866, 32591. Phone: (850) 777-1568. Fax: (850) 437-3733. E-mail: info@divinewordradio.com Web Site: www.divinewordradio.com. Licensee: Divine Word Communications (acq 2-22-2007; $375,000). Natl. Network: EWTN Radio. Rgnl. Network: Florida Radio Net. Format: Catholic radio. ♦Gene Church, pres & gen mgr.

WMEZ(FM)— Nov 11, 1960: 94.1 mhz; 100 kw. 1,328 ft TL: N30 35 18 W87 33 16. Stereo. Hrs opn: 6085 Quintette Rd., Pace, 32571. Phone: (850) 916-9222. Phone: (850) 994-5357. Fax: (850) 916-9266. Web Site: www.softrock941.com. Licensee: 6 Johnson Road Licenses Inc. Group owner: Pamal Broadcasting Ltd. (acq 10-19-2001; grpsl). Population served: 250,000 Natl. Network: Westwood One. Format: Soft rock, adult contemp. Target aud: 25-54. ♦Dave Cobb, gen mgr; Kevin Peterson, progmg mgr; Gerald Wilson, chief of engrg.

WNVY(AM)—Cantonment, December 1955: 1090 khz; 10 kw-D (2.3 kw-CH). TL: N30 34 47 W87 17 18. Hrs open: 2070 N. Palafox, 32501. Phone: (850) 434-1230. Fax: (850) 469-9698. E-mail: mglin@aol.com Web Site: www.pensacolachristianradio.com. Licensee: 1090-AM. (acq 4-7-97). Format: Black gospel. ♦Michael Glinter, gen mgr.

***WPCS(FM)**— June 22, 1971: 89.5 mhz; 95 kw. Ant 1,358 ft TL: N30 35 16 W87 33 13. Stereo. Hrs opn: 24 Box 18000, 32523. Phone: (850) 479-6570. Fax: (850) 969-1638. E-mail: rbn@rejoice.org Web Site: www.rejoice.org. Licensee: Pensacola Christian College Inc. Population served: 1,025,018 Format: Relg, educ. ♦Arlin Horton, pres & gen mgr; Paul Stimer, stn mgr; Ted Nadaskay, chief of engrg.

WPNN(AM)— October 1956: 790 khz; 1 kw-D. TL: N30 27 18 W87 14 22. Hrs open: 3801 N. Pace Blvd., 32505. Phone: (850) 433-1141. Fax: (850) 433-1142. Web Site: www.cnnpensacola.com. Licensee: Miracle Radio Inc. (acq 4-1-81). Population served: 62,507 Law Firm: Smithwick & Belendiuk. Format: Local news, CNN Headline News. ♦Gerald Schroeder, pres; Don Schroeder, gen mgr; Michael Schroeder, stn mgr.

WTKX-FM— 1971: 101.5 mhz; 100 kw. Ant 1,328 ft TL: N30 35 18 W87 33 16. Stereo. Hrs opn: 24 6485 Pensacola Blvd., 32505. Phone: (850) 473-0400. Fax: (850) 473-0907. E-mail: radio@tk101.com Web Site: www.tk101.com. Licensee: Clear Channel Broadcasting Licenses Inc. Group owner: Clear Channel Communications Inc. (acq 11-21-97; grpsl). Population served: 350,000 Law Firm: Wiley, Rein & Fielding. Format: Active rock. Target aud: 18-49; general. ♦Lowry Mays, CEO & chmn; Mark Mays, pres; Randall Mays, CFO.

***WUWF(FM)**— January 1981: 88.1 mhz; 100 kw. 617 ft TL: N30 24 09 W86 59 35. Stereo. Hrs opn: 24 11000 University Pkwy., GA, 32514. Phone: (850) 474-2787. E-mail: wuwf@wuwf.org Web Site: www.wuwf.org. Licensee: Board of Trustees, University of West Florida (acq 12-4-01). Population served: 400,000 Natl. Network: PRI, NPR. Rgnl. Network: Fla. Pub. Format: Class, news, adult alternative. News staff: one; News: 34 hrs wkly. Target aud: General. ♦Joe Vincenza, gen mgr; Sandra Averhart, news dir.

WVTJ(AM)— Nov 1, 1959: 610 khz; 500 w-D, 157 w-N. TL: N30 27 18 W87 14 22. Hrs open: 24 2070 N. Palafox St., 32501. Phone: (850) 432-3658. Fax: (850) 432-3659. E-mail: wvtj@wilkinsradio.com Web Site: www.wilkinsradio.com. Licensee: Pensacola Radio Corp. (acq 4-23-2007; $545,000). Population served: 480,000 Format: Black gospel, Christian. Target aud: 18-64. ♦Robert L. Wilkins, pres; Jessica Jordan, gen mgr.

WXBM-FM—See Milton

WYCL(FM)— Nov 10, 1976: 107.3 mhz; 100 kw. Ant 1,407 ft TL: N30 42 20 W87 19 00. Stereo. Hrs opn: 24 6485 Pensacola Blvd., 32505. Phone: (850) 473-0400. Fax: (850) 473-0907. Web Site: www.my107.com. Licensee: Clear Channel Broadcasting Licenses Inc. Group owner: Clear Channel Communications Inc. (acq 9-30-2003; $2.2 million). Population served: 350,000 Natl. Rep: McGavren Guild. Format: 70's & 80's. News staff: 2; News: 15 hrs wkly. Target aud: 25-54. ♦Jeanie Hufford, gen mgr; Eddie Hill, stn mgr & gen sls mgr; Steve Powers, opns dir.

WYCT(FM)— Nov 28, 2003: 98.7 mhz; 100 kw. Ant 981 ft TL: N30 37 30 W87 26 39. Stereo. Hrs opn: 7251 Plantation Rd., 32504. Phone: (850) 494-2800. Fax: (850) 494-0778. E-mail: hr@catcountry987.com Web Site: www.catcountry987.com. Licensee: ADX Communications of Pensacola. Population served: 450,000 Law Firm: Dan Alpert. Format: Country. ♦David E. Hoxeng, CEO; Mary Hoxeng, gen mgr; Susan Nieman, sls dir; Kevin King, opns.

Perry

WNFK(FM)— December 1989: 92.1 mhz; 2.45 kw. 345 ft TL: N30 07 36 W83 36 28. Stereo. Hrs opn: 5450 Hwy. 27 E., 32347. Phone: (850) 584-9210. Fax: (850) 223-3492. E-mail: powercountry921@wildblue.net Licensee: Taylor County Broadcasting Inc. (acq 4-19-00). Format: Country. ♦Bob Hendrickson, gen mgr; Keith Conway, progmg dir.

WPRY(AM)— 1953: 1400 khz; 1 kw-U. TL: N30 06 27 W83 34 00. Hrs open: 24 872 Hwy. 27 E., 32347. Phone: (850) 223-1400. Fax: (850) 223-3501. Web Site: www.wpry.com. Licensee: HF Broadcasting Perry LC (acq 4-1-2004; $150,000). Format: Classic hits. News: 15 hrs wkly. Target aud: 18 plus. Spec prog: Black 2 hrs wkly. ♦Gary Williams, gen mgr.

Pine Castle-Sky Lake

WAMT(AM)— Jan 28, 1977: 1190 khz; 5 kw-D. TL: N28 27 58 W81 22 30. Hrs open: 24 1160 S. Semoran Blvd., Suite A, Orlando, 32807. Phone: (407) 380-9255. Fax: (407) 382-7565. E-mail: studio@wamt1190.com Web Site: www.wamt1190.com. Licensee: Genesis Communications I Inc. Group owner: Genesis Communications Inc. (acq 3-20-2000; $2.1 million). Population served: 1,500,000 Natl. Network: ABC, Westwood One. Natl. Rep: Interep. Law Firm: Booth, Freret, Imlay, & Tepper. Wire Svc: AP Wire Svc: Metro Weather Service Inc. Format: News/talk. Target aud: 25-65; general. ♦Bruce Maduri, pres; Mike Burgess, opns mgr; Brad James, gen sls mgr & mktg VP; Joe Nuchols, gen sls mgr; Brad James, progmg VP.

Pine Hills

WFLF(AM)— Sept 9, 1955: 540 khz; 50 kw-U, DA-2. TL: N28 07 57 W81 43 16. Hrs open: 24 2500 Maitland Center Pkwy., Suite 407, Maitland, 32751. Phone: (407) 916-7800. Fax: (407) 661-1940. E-mail: kaystelling@clearchannel.com Web Site: www.540wfla.com. Licensee: Clear Channel Radio Licenses Inc. Group owner: Clear Channel Communications Inc. (acq 11-21-97; grpsl). Population served: 100,000 Natl. Network: Fox News Radio. Rgnl. Network: Florida Radio Net. Law Firm: Wiley, Rein & Fielding. Wire Svc: AP Format: News/talk. News staff: 5; News: 168 hrs wkly. Target aud: 35-64. Spec prog: Florida Gaton football & basketball, Florida Marlins, Miami Dolphins. ♦Linda Byrd, gen mgr; Kay Stelling, sls VP & gen sls mgr; Rick Everett, mktg dir; Chris Kampmeier, progmg dir; Tom Benson, progmg dir.

Pine Island Center

WPTK(AM)—Licensed to Pine Island Center. See Fort Myers

Pinellas Park

WTBN(AM)— Nov 12, 1966: 570 khz; 5 kw-U, DA-2. TL: N28 12 40 W82 31 46. Hrs open: 24 5211 W. Laurel St., Tampa, 33607. Phone: (813) 639-1903. Fax: (813) 639-1272. E-mail: info@bayword.com Web Site: www.bayword.com. Licensee: Common Ground Broadcasting Inc. Group owner: Salem Communications Corp. (acq 8-7-2001; $6.75 million). Population served: 1500000 Natl. Network: Salem Radio Network. Rgnl. Network: Florida Radio Net. Natl. Rep: Salem. Format: Christian Talk Radio. News staff: 2; News: 30 hrs wkly. Target aud: 25-64. Spec prog: College (USF) sports football & basketball. ♦Chris Gould, gen mgr; Mike Serio, opns mgr; Rey Noriega, traf mgr.

WWBA(AM)—Licensed to Pinellas Park. See Tampa

Plant City

WTWD(AM)— July 1949: 910 khz; 5 kw-U, DA-1. TL: N27 59 26 W82 12 31. Hrs open: 24
Rebroadcasts WTBN Pinnellas Park 100%.
5211 Laurel St., Tampa, 33607. Phone: (813) 639-1903. Fax: (813) 639-1272. E-mail: info@baywood.com Web Site: www.baywood.com. Licensee: South Texas Broadcasting Inc. Group owner: Salem Communications Corp. (acq 7-27-00; grpsl). Population served: 2,100,000 Natl. Network: Salem Radio Network. Natl. Rep: Salem. Format: Christian talk. Target aud: 25-54. ♦Christopher Gould, Sr., gen mgr; Mike Serio, opns mgr.

Plantation Key

WCTH(FM)— July 1969: 100.3 mhz; 100 kw. Ant 462 ft TL: N24 57 34 W80 34 30. Stereo. Hrs opn: 24 93351 Overseas Hwy., Tavernier, 33070. Phone: (305) 852-9085. Fax: (305) 852-5586. Web Site: www.thundercountry.com. Licensee: Clear Channel Radio Licenses Inc. Group owner: Clear Channel Communications Inc. (acq 2-99; $1.8 million). Population served: 150,000 Natl. Network: Westwood One, Motor Racing Net. Rgnl. Network: Florida Radio Net. Natl. Rep: Clear Channel. Format: Country. News staff: one; News: 4 hrs wkly. Target aud: 25-54; residents & tourists. Spec prog: NASCAR 3 hrs wkly. ♦John Hogan, CEO; Greg Capgna, VP; Mark Mills, gen mgr; Scott Hamilton, progmg dir & progmg mgr.

WFKZ(FM)— Jan 2, 1984: 103.1 mhz; 50 kw. 449 ft TL: N25 01 35 W80 30 30. Hrs open: 24 93351 Overseas Hwy., Tavernier, 33070. Phone: (305) 852-9085. Fax: (305) 852-5586. E-mail: info@keysradio.net Web Site: www.sun103.com. Licensee: Clear Channel Radio Licenses Inc. Group owner: Clear Channel Communications Inc. (acq 11-21-97; grpsl). Population served: 150,000 Format: Adult rock. News staff: 2; News: 8 hrs wkly. Target aud: 25-54; adults. ♦Greg Capgna, VP; Mark Mills, gen mgr & stn mgr.

Pompano Beach

WHSR(AM)— 1959: 980 khz; 5 kw-D, 1 kw-N, DA-D. TL: N26 14 26 W80 10 07. Hrs open: 6699 N. Federal Hwy., Ste 200, Boca Raton, 33487. Phone: (561) 997-0074. Fax: (561) 997-0476. Web Site: www.whsrentertainmentradio.com. Licensee: WWNN License L.L.C. Group owner: Beasley Broadcast Group (acq 3-17-2000; grpsl). Population served: 2,500,000 Law Firm: Jason Shrinsky. Format: Foreign, ethnic, talk. Target aud: 25-54; baby boomers weaned on electronic media as an info source. ♦Bob Morency, VP & gen mgr; Greg Cooper, opns mgr.

WMXJ(FM)— 1960: 102.7 mhz; 100 kw. 1,007 ft TL: N25 57 59 W80 12 33. Stereo. Hrs opn: 20450 N.W. 2nd Ave., Miami, 33169-2505. Phone: (305) 521-5100. Fax: (305) 652-1888. Web Site: www.majic1027.com. Licensee: Lincoln Financial Media Co. of Florida. (acq 4-3-2006; grpsl). Population served: 3,111,400 Natl. Rep: CBS Radio. Wire Svc: AP Format: Classic hits. Target aud: 35-64. ♦Dennis Collins, sr VP & gen mgr; Daryl Leoce, sls dir & gen sls mgr; Connie Estopinan, prom dir; Robert Hamilton, progmg dir; Gary Blau, engrg dir.

WWNN(AM)— 1959: 1470 khz; 5 kw-D, 2.5 kw-N, DA-1. TL: N26 10 46 W80 13 15. (CP: 50 kw-D). Hrs opn: 24 6699 N. Federal Hwy., Suite 200, Boca Raton, 33487. Phone: (561) 997-0074. Fax: (561) 997-0476. Web Site: www.wwnnradio.com. Licensee: WWNN License LLC. Group owner: Beasley Broadcast Group (acq 3-14-2000; grpsl). Population served: 139,590 Format: Health & wealth. ♦Bob Morency, VP & gen mgr; Greg Cooper, opns mgr.

Ponte Vedra Beach

WOKV-FM— 1996: 106.5 mhz; 6 kw. Ant 328 ft TL: N30 16 34 W81 33 58. Stereo. Hrs opn: 24 4190 Belfort Rd., Suite 450, Jacksonville, 32216. Phone: (904) 470-4615. Fax: (904) 296-1683. Licensee: Cox Radio Inc. Group owner: Salem Communications Corp. (acq 9-18-2006; $7.65 million). Format: Christian. Target aud: 25-44. ♦Calvin Grabau, gen mgr.

Port Charlotte

WKII(AM)—Listing follows WZJZ(FM).

***WVIJ(FM)**— July 26, 1987: 91.7 mhz; 1.9 kw. Ant 207 ft TL: N26 58 49 W82 04 03. Stereo. Hrs opn: 3279 Sherwood Rd., 33980. Phone: (941) 624-5000. Fax: (775) 243-0586. E-mail: wvij@wvij.com Web Site: www.wvij.com. Licensee: Port Charlotte Educational Broadcasting Foundation Inc. Format: Educ, relg. Target aud: 35 plus. ♦Daniel P. Kolenda Jr., gen mgr.

WZJZ(FM)— Oct 1, 1976: 100.1 mhz; 100 kw. Ant 476 ft TL: N26 37 25 W82 06 58. Stereo. Hrs opn: 24 13320 Metro Pkwy., Suite 1, Fort Myers, 33912. Phone: (239) 225-4300. Fax: (293) 225-4329. Web Site: www.smoothjazz1071.com. Licensee: Clear Channel Broadcasting Licenses Inc. Group owner: Clear Channel Communications Inc. (acq 2-18-97; grpsl). Population served: 450,000 Law Firm: Kaye, Scholer, Fierman, Hays & Handler L.L.P. Format: Smooth jazz. Target aud: 25-54. ♦Jim Keating, gen mgr.

WKII(AM)—Co-owned with WZJZ(FM). Nov 19, 1986: 1070 khz; 3.1 kw-D, 260 w-N. TL: N26 54 40 W82 02 12. Stereo. 24 24100 Tiseo Blvd., Suite 10, 33980. Phone: (941) 206-1188. Fax: (941) 206-9296. Web Site: www.wkii.com. Format: Adult standards. News staff: one; News: 2 hrs wkly. Target aud: 35 plus. ♦Mike Moody, stn mgr, progmg dir & news dir; David Ayres, sls dir; Paul Wolf, chief of engrg; Ron Bigley, traf mgr.

Port Richey

WSUN-FM—See Holiday

Port St. Joe

WPBH(FM)— Mar 12, 1990: 93.5 mhz; 14.5 kw. 659 ft TL: N29 49 09 W85 15 34. Hrs open: 24 Box 59288, Panama City, 32405. Phone: (850) 769-1408. Fax: (850) 769-0659. Web Site: www.935thebeat.com. Licensee: Citicasters Licenses L.P. Group owner: Clear Channel Communications Inc. (acq 8-26-99; $1 million). Format: Urban. Target aud: 35 plus; upscale, white collar professionals. ♦Pete Norton, stn mgr; Eddie Rupp, opns mgr.

Port St. Lucie

WHLG(FM)— Nov 16, 1998: 101.3 mhz; 6 kw. 328 ft TL: N27 16 04 W80 16 49. Stereo. Hrs opn: 24 Horton Broadcasting Co. Inc., 1670 N.W. Federal Hwy., Stuart, 34994. Phone: (772) 692-9454. Fax: (772) 692-0258. E-mail: info@coast1013.com Web Site: www.coast1013.com. Licensee: Horton Broadcasting Co. Inc. Population served: 205,900 Natl. Network: Jones Radio Networks. Natl. Rep: Interep. Format: Adult contemp. News staff: 10. Target aud: 25-54; female/male 60/40%, 35 years old. ♦George Metcalf, CEO; Lorna Potter, gen mgr.

WPSL(AM)— Oct 26, 1985: 1590 khz; 5 kw-D, 64 w-N. TL: N27 18 28 W80 18 26. Stereo. Hrs opn: 24 4100 Metzger Rd., Fort Pierce, 34947. Phone: (772) 340-1590. Fax: (772) 340-3245. E-mail: wpsl@wpsl.com Web Site: www.wpsl.com. Licensee: Port St. Lucie Broadcasters Inc. (acq 4-12-93; $200,000; FTR: 4-26-93). Population served: 1,000,000 Natl. Network: CBS, ESPN Radio. Rgnl. Network: Florida Radio Net. Law Firm: Leventhal, Senter & Lerman. Format: News/talk, sports. News staff: one; News: 4 hrs wkly. Target aud: 45 plus; established families. Spec prog: Relg 6 hrs wkly. ♦Carol Wyatt, CEO & pres; Greg Wyatt, VP & gen mgr.

Punta Gorda

WCCF(AM)— Sept 15, 1961: 1580 khz; 1.25 kw-D, 122 w-N. TL: N26 53 37 W82 03 01. Hrs open: 24100 Tiseo Blvd. #10, Port Charlotte, 33980. Phone: (941) 206-1188. Fax: (941) 206-9296. Web Site: www.wccfam.com. Licensee: Citicasters Licenses L.P. Group owner: Clear Channel Communications Inc. (acq 2-1-99; grpsl). Population served: 106,000 Format: News/talk. Target aud: 45 plus. ♦Chris Monk, VP; Mike Moody, gen mgr.

WIKX(FM)—Co-owned with WCCF(AM). Sept 1, 1970: 92.9 mhz; 100 kw. 807 ft TL: N26 53 47 W82 14 27. Stereo. 24 E-mail: kixcountry@hotmail.com Web Site: www.wikx.com.250,000 Format: Country. News: 10 hrs wkly. Target aud: 25-54.

Punta Rassa

WTLQ-FM— May 3, 1999: 97.7 mhz; 14.5 kw. 430 ft TL: N26 29 16 W81 55 46. Stereo. Hrs opn: 24 2824 Palm Beach Blvd., Ft. Myers, 33916. Phone: (239) 334-1111 / (239) 338-4325. Fax: (239) 334-0744. Web Site: www.latino977.com. Licensee: Fort Myers Broadcasting Co. (group owner; acq 9-13-00; $7 million). Population served: 600,000 Natl. Rep: McGavren Guild. Law Firm: Liebowitz & Associates. Format: Sp. Target aud: 18-49; adults. ♦Gary Gardner, gen mgr; Wayne Simons, sls dir; Tim Spires, rgnl sls mgr; Bob Grissinger, progmg dir; Keith Stuhlmann, engrg dir.

Quincy

WWSD(AM)— Mar 15, 1948: 1230 khz; 1 kw-U. TL: N30 34 55 W84 35 59. Hrs open: 1732 W. Elm St., 32351. Phone: (850) 627-4390. Licensee: Tuff-Starr Jam Commuication Inc. ♦Reverend Milton Donato, pres.

WXSR(FM)— December 1966: 101.5 mhz; 50 kw. 476 ft TL: N30 31 08 W84 27 04. Hrs open: 325 John Knox Rd., Bldg. G, Tallahassee, 32303. Phone: (850) 422-3107. Fax: (850) 383-0747. Web Site: www.x1015.com. Licensee: Clear Channel Radio Licenses Inc. Group owner: Clear Channel Communications Inc. (acq 11-21-97; grpsl). Natl. Rep: Christal. Format: Alternative, new rock. Target aud: 18-34. ♦Lisa Rice, VP & gen mgr; Jeff Horn, opns mgr & gen sls mgr.

Riviera Beach

WMNE(AM)— Aug 17, 1959: 1600 khz; 5 kw-D, 4.7 kw-N, DA-2. TL: N26 44 55 W80 08 02. Stereo. Hrs opn: 24 824 US Hwy. 1, North Palm Beach, 33408. Phone: (561) 694-7636. Fax: (561) 694-7574. Web Site: www.radiodisney.com. Licensee: Radio Disney Group LLC. Group owner: ABC Inc. (acq 8-22-00; grpsl). Natl. Rep: Roslin. Format: Family. News staff: 2; News: 30 hrs wkly. Local Blacks & Hispanics. Spec prog: Sp. ♦Neil Orlikoff, gen mgr.

WZZR(FM)—Licensed to Riviera Beach. See West Palm Beach

Rock Harbor

WKLG(FM)— Nov 1, 1984: 102.1 mhz; 100 kw. Ant 430 ft TL: N25 05 29 W80 26 37. Stereo. Hrs opn: 24 Box 0457, Key Largo, 33037. Secondary address: 1452 N. Krome Ave., Suite 103 E., Florida City 33034. Phone: (305) 451-2202. Fax: (305) 453-2265. Licensee: WKLG Inc. Law Firm: Leibowitz & Associates. Format: Adult contemp. Target aud: 25-54; majority are female 18 plus. ♦Douglas D. LaRue, pres & gen mgr.

Rockledge

WHKR(FM)— Nov 25, 1989: 102.7 mhz; 50 kw. 492 ft TL: N28 35 03 W80 50 56. Stereo. Hrs opn: 24 1775 W. Hibiscus Blvd., Suite 101, Melbourne, 32901. Phone: (321) 984-1000. Fax: (321) 724-1565. Web Site: www.thehitkicker.com. Licensee: Cumulus Licensing Corp. Group owner: Cumulus Media Inc. (acq 8-7-2000; grpsl). Format: Modern country. News staff: 2; News: 6 hrs wkly. Target aud: 25-54. Spec prog: Pub service one hr wkly. ♦Dan Carelli, gen mgr.

Royal Palm Beach

WMEN(AM)— April 1987: 640 khz; 7.5 kw-D, 460 w-N, DA-2. TL: N26 45 18 W80 22 00. (CP: COL Boca Raton. 50 kw-D, 25 kw-N, DA-2. TL: N26 32 30 W80 44 30). Hrs opn: 24 6600 N. Andrews Ave., Suite 160, Fort Lauderdale, 33309. Phone: (954) 315-1515 / 1539. Fax: (954) 315-1555. Web Site: www.wjna.com. Licensee: JCE Licenses L.L.C. Group owner: James Crystal Inc. (acq 11-18-99; $3,945,500 for stock). Population served: 6,000,000 Natl. Network: ABC. Format: Talk. News: 4 hrs wkly. Spec prog: Black 2 hrs, Sp 2 hrs wkly. ♦Steve Lapa, gen mgr.

WPSP(AM)— February 1991: 1190 khz; 1 kw-U, DA-N. TL: N26 44 14 W80 16 23. Hrs open: 18 5730 Corporate Way, Suite 210, West Palm Beach, 33407. Phone: (561) 681-9777. Fax: (561) 687-3398. E-mail: diaz1190am@aol.com Licensee: George M. Arroyo. (acq 5-87; $75,000; FTR: 5-11-87). Law Firm: Roy F. Perkins. Format: Spanish hits. News staff: 2; News: 20 hrs wkly. Target aud: 25-54. ♦George M. Arroyo, pres; Lissette M. Diaz, gen mgr & opns dir.

Safety Harbor

WYUU(FM)—Licensed to Safety Harbor. See Tampa

Saint Augustine

WAOC(AM)— December 1953: 1420 khz; 2.18 kw-D, 250 w-N. TL: N29 51 00 W81 19 50. Hrs open: 24 Box 3847, 32085. Secondary address: 567 Lewis Point Rd. Ext. 32086. Phone: (904) 797-4444. Fax: (904) 797-3446. E-mail: kris@1420sports.com Web Site: www.1420sports.com. Licensee: Phillips Broadcasting LLC (acq 7-1-2006; $1 million with WFOY(AM) Saint Augustine). Population served: 150,000 Natl. Network: Sporting News Radio Network. Natl. Rep: Rgnl Reps. Rgnl rep: Rgnl Reps Law Firm: Alan Campbell. Format: Sports. News staff: 1; News: 8 hrs wkly. Target aud: 26 plus; affluent adults. Spec prog: University of Florida sports, NFL Jaguar affiliate, NASCAR. ♦Kristine Phillips, pres; Kevin Leslie, gen mgr & progmg dir; Rose Napolitano, opns VP.

***WAYL(FM)**— May 22, 1994: 91.9 mhz; 5 kw. 200 ft TL: N29 54 26 W81 18 51. Hrs open: 24 Box 127, 32085. Secondary address: 1485 US Rt. 1 S. 32086. Phone: (904) 829-9200. Fax: (904) 829-9202. E-mail: david@fm88.org Web Site: www.riverradio.org. Licensee: New Covenant Educational Ministries Inc. (acq 6-14-02). Population served: 250,000 Natl. Network: Salem Radio Network. Law Firm: Fletcher, Heald & Hildreth. Format: Contemp Christian. News staff: 2; News: 10 hrs wkly. Target aud: 25-45. ♦David Oglesby, stn mgr; Jerry Smith, engrg mgr.

***WFCF(FM)**— Nov 1, 1993: 88.5 mhz; 6 kw. 141 ft TL: N29 54 27 W81 18 49. Hrs open: 7 AM-midnight Box 1027, Flagler College, St. Augustine, 32085-1027. Phone: (904) 819-6449. Phone: (904) 819-6313. Fax: (904) 826-3471. E-mail: wfcf@flagler.edu Web Site: www.flagler.edu. Licensee: Flagler College. Population served: 88,000 Law Firm: Caressa D. Bennet. Format: Div. News: one hr wkly. Target aud: General. Spec prog: Sp 4 hrs, new age 4 hrs, folk 3 hrs, reggae 4 hrs, world 4 hrs, blues 4 hrs wkly. ♦Donna DeLorenzo Webb, gen mgr; Daniel McCook, stn mgr.

WFOY(AM)— July 7, 1936: 1240 khz; 1 kw-U. TL: N29 54 26 W81 18 51. Hrs open: 24 Box 3847, 32085. Secondary address: 567 Lewis Point Rd. Ext. 32086. Phone: (904) 797-1955. Phone: (904) 797-4444. Fax: (904) 797-3446. E-mail: kris@1240news.com Web Site: www.1420sports.com. Licensee: Phillips Broadcasting LLC (acq 7-1-2006; $1 million with WAOC(AM) Saint Augustine). Population served: 150,000 Natl. Network: Fox News Radio, Westwood One, Talk Radio Network. Rgnl. Network: Florida Radio Net. Natl. Rep: Rgnl Reps. Rgnl rep: Rgnl Reps Law Firm: Alan Campbell. Format: News/talk, sports. News staff: one; News: 8 hrs wkly. Target aud: 26 plus; affluent adults. Spec prog: Rush Limbaugh affiliate. ♦Kristine Phillips, pres; Kevin Leslie, gen mgr; Jennifer Dawson, opns dir.

WSOS-FM— July 17, 1982: 94.1 mhz; 25 kw. Ant 302 ft TL: N29 57 57 W81 28 36. (CP: COL Fruit Cove). Stereo. Hrs opn: 24 2715 Stratton Blvd., 32084. Phone: (904) 824-0833. Phone: (904) 722-9606. Fax: (904) 721-9322. E-mail: tbryan@rendabroadcasting.com Web Site: www.wsosfm.com. Licensee: Renda Broadcasting Corp. of Nevada. (acq 4-27-2005; $7.75 million). Population served: 500,000 Natl. Rep: McGavren Guild. Format: Soft adult contemp. Target aud: 25-54; upscale audience. ♦Tony Renda, CEO; Tim Bryan, gen mgr; Don Runk, gen sls mgr; Stacey Steiner, prom dir; Briggs Bickley, progmg dir; Bob Dillehay, chief of engrg; Brenda McArthur, traf mgr; Jim Byard, pub svc dir.

Saint Augustine Beach

WSJF(FM)— Sept 1, 1995: 105.5 mhz; 16 kw. Ant 410 ft TL: N29 51 00 W81 19 50. Stereo. Hrs opn: 24 9550 Regency Sq. Blvd., Suite 200, Jacksonville, 32225. Phone: (904) 680-1050. Fax: (904) 680-1051. E-mail: jaxproduction@tamabroadcasting.com Web Site: www.tamabroadcasting.com. Licensee: Tama Radio Licenses of Jacksonville, FL, Inc. Group owner: Tama Broadcasting Inc. (acq 2-28-2003; $8.5 million with WJSJ(FM) Fernandina Beach). Population served: 175,000 Format: Jazz. Target aud: 20-45; general. ♦Linda Fructuoso, gen mgr; Joel Widdows, opns mgr & prom dir; Gerry Smith, chief of engrg.

WSOS(AM)— Oct 15, 1986: 1170 khz; 710 w-D. TL: N29 55 05 W81 23 26. Hrs open: 6 AM-9 PM 8384 Baymeadows Rd., Suite 1, Jacksonville, 32256-7486. Phone: (904) 739-3660. Fax: (904) 739-9409. Licensee: Norsan Consulting and Management Inc. (acq 1-13-2006; $300,000). Format: Tropical. ♦Jorge Lopez, gen mgr.

Saint Catherine

***WKFA(FM)**— 2005: 89.3 mhz; 100 w. Ant 295 ft TL: N28 32 22 W82 04 48. Hrs open: 505 Josephine St., Titusville, 32796. Phone: (321) 267-3000. Fax: (321) 264-9370. E-mail: wpio@gate.net Web Site: www.noncomradio.com. Licensee: Florida Public Radio Inc. (acq 4-12-2003). Format: Inspirational music, Public affairs. ♦Randy Henry, pres & gen mgr.

Saint Cloud

WIWA(AM)— 2005: 1160 khz; 2.5 kw-D, 500 w-N, DA-2. TL: N28 16 15 W81 20 00. Hrs open: 4540 Curry Ford Rd., Orlando, 32812. Phone: (407) 770-2500. Fax: (407) 770-2503. Web Site: www.viva1160.com. Licensee: Centro de la Familia Cristiana Inc. (acq 3-20-2006; $562,800). Format: Spanish news/talk. ♦Roberto Candelario, pres.

Saint Marks

***WUJC(FM)**— 2005: 91.1 mhz; 7 kw vert. Ant 312 ft TL: N30 08 32 W83 54 58. Hrs open: 8747 Miles Johnson Rd., Tallahassee, 32309. Phone: (850) 514-1929. Fax: (850) 514-1927. Web Site: www.csnradio.com. Licensee: CSN International (group owner). Format: Relg. ♦Michael Kestler, pres; Don Mills, progmg dir & mus dir.

Saint Petersburg

WDAE(AM)— Nov 1, 1927: 620 khz; 5 kw-D, 5.4 kw-N, DA-N. TL: N27 52 37 W82 35 26. Stereo. Hrs opn: 24 4002 Gandy Blvd., Tampa, 33611. Phone: (813) 832-1000. Fax: (813) 831-3299. Web Site: www.620wdae.com. Licensee: Clear Channel Broadcasting Licenses Inc. (acq 11-20-98; $9.75 million). Population served: 2300000 Format: Sports. ♦Dan Diloreto, gen mgr.

WFLA(AM)—See Tampa

WFLZ-FM—See Tampa

***WFTI-FM**— June 1988: 91.7 mhz; 3 kw. 282 ft TL: N27 46 15 W82 38 19. Stereo. Hrs opn: 24 360 Central Ave., Suite 1240, 33701. Phone: (727) 823-1140. Fax: (727) 823-5753. E-mail: WFTI@hotmail.com Web Site: www.familyradio.com. Licensee: Family Stations Inc. (group owner; acq 11-19-88). Natl. Network: Family Radio. Format: Relg. News: 9 hrs wkly. Target aud: General. ♦Bob Barnes, stn mgr.

WGES(AM)— May 5, 1950: 680 khz; 690 w-D, 125 w-N. TL: N27 51 24 W82 37 26. Hrs open: 402 N. Reo Street, Suite 218, Tampa, 33609. Phone: (813) 319-5757. Phone: (813) 637-8000. Fax: (813) 319-0029. Fax: (813) 637-8001. E-mail: info@genesis680.com Web Site: www.genesis680.com. Licensee: ZGS Broadcasting of Tampa Inc. (acq 1-18-91; $200,000; FTR: 2-4-91). Population served: 300,000 Rgnl rep: Katz Hispanic Media Format: Tropical/Spanish. Target aud: General; adults 18-49. ♦Patricia Omana, gen mgr.

WHPT(FM)—See Sarasota

WMTX(FM)—See Tampa

WPOI(FM)— July 1, 1961: 101.5 mhz; 100 kw. 1,358 ft TL: N27 50 32 W82 15 46. Stereo. Hrs opn: 24 Cox Radio Inc., 11300 4th St. N., Suite 300, 33716-2941. Phone: (727) 579-2000. Fax: (727) 579-2662. Fax: (727) 579-2271. E-mail: comments@1015thepoint.com Web Site: www.1015thepoint.com. Licensee: Cox Radio Inc. Group owner: Cox Communications Inc. (acq 1999; grpsl). Population served: 4,000,000 Natl. Rep: Clear Channel. Format: Hits of the 80s. Target aud: 25-54. ♦Howard Tuuri, VP, gen mgr & stn mgr; Tom Paleveda, opns mgr; Bernadette Van Osdal, gen sls mgr; Gerry Brauer, prom mgr.

WQYK-FM— May 1958: 99.5 mhz; 100 kw. Ant 590 ft TL: N27 56 50 W82 27 35. Stereo. Hrs opn: 24 9721 Executive Center Dr. N., Suite 200, 33702. Phone: (813) 287-0995. Fax: (813) 636-0995. Web Site: www.wqyk.com. Licensee: CBS Radio Inc. of Florida. Group owner: Infinity Broadcasting Corp. (acq 12-1-86; FTR: 10-6-86). Natl. Network: CBS. Natl. Rep: Interep. Law Firm: Leventhal, Senter & Lerman. Format: Contemp country. News staff: one; News: 6 hrs wkly. Target aud: 25-54. ♦Charlie Ochs, sr VP; Luis Albertini, VP & gen mgr; Mike Culotta, opns mgr. Co-owned TV: WTOG(TV) affil

WRBQ-FM—See Tampa

WWMI(AM)— 1939: 1380 khz; 5 kw-U, DA-N. TL: N27 52 15 W82 37 03. Hrs open: 24 11300 4th St. N., Suite 143, St. Petersburg, 33716. Phone: (727) 577-4500. Fax: (727) 579-1340. Web Site: www.radiodisney.com. Licensee: Radio Disney Group LLC. Group owner: ABC Inc. (acq 1999; grpsl). Population served: 3,102,000 Natl. Rep: Clear Channel. Wire Svc: NOAA Weather Format: Top- 40. News staff: 2. Target aud: 25-54. ♦Drew Rashbaum, gen mgr; Ted Wolfe, stn mgr.

WWRM(FM)—Tampa, 1958: 94.9 mhz; 100 kw. 1,289 ft TL: N27 49 09 W82 14 26. Stereo. Hrs opn: 24 11300 4th St. N., Suite 300, St. Petersburg, 33716-2941. Phone: (727) 579-2000. Fax: (727) 579-2662. Web Site: www.949online.com. Licensee: Cox Radio Inc. Group owner: Cox Communications Inc. (acq 7-1-88). Natl. Rep: Christal. Format: Adult contemp. Target aud: 25-54. ♦Howard Tuuri, gen mgr; Tom Paleveda, opns mgr; Mark Kanak, gen sls mgr & natl sls mgr; Julia Freeman, prom mgr.

WXGL(FM)— 1958: 107.3 mhz; 100 kw. 649 ft TL: N27 51 24 W82 37 26. Stereo. Hrs opn: 24 11300 4th St. N., Suite 300, 33716. Phone: (727) 579-2000. Fax: (727) 579-2662. Fax: (727) 579-2662. Web Site: www.1073theeagle.com. Licensee: Cox Radio Inc. Group owner: Cox Communications Inc. (acq 7-1-88). Population served: 2,300,000 Natl. Rep: Christal. Format: Classic Hits. ♦Keith Lawless, VP & gen mgr; Shane Reeve, gen sls mgr; Tom Paleveda, opns mgr & natl sls mgr.

Saint Petersburg Beach

WRXB(AM)— 1957: 1590 khz; 5 kw-D, 1 kw-N, DA-2. TL: N27 44 03 W82 41 08. Hrs open: 2060 First Ave. N., St. Petersburg, 33713. Phone: (727) 821-9967. Fax: (727) 321-3025. E-mail: wrxb@juno.com Web Site: www.wrxb.com. Licensee: Metropolitan Radio Group of Florida Inc. Group owner: Metropolitan Radio Group Inc. Population served: 160,000 Format: Adult contemp, urban contemp. Target aud: 23-54; urban contemp. Spec prog: Jazz 15 hrs wkly. ♦Juanita Dials, gen mgr.

San Carlos Park

WDEO-FM— 1995: 98.5 mhz; 18.5 kw. Ant 371 ft TL: N26 30 18 W81 51 14. Stereo. Hrs opn: 24 Box 504, Ann Arbor, MI, 48106. Phone: (734) 930-5200. Fax: (734) 930-3179. Web Site: www.avemariaradio.net. Licensee: Ave Maria University Inc. (acq 2-9-2004; $4.9 million). Population served: 516,200 Format: Catholic news/talk. Target aud: 21 plus; adult Christian. ♦Michael Jones, gen mgr.

Sanford

WSDO(AM)— May 20, 1947: 1400 khz; 1 kw-U. TL: N28 48 04 W81 15 06. Hrs open: 24 222 Hazard St., Orlando, 32804-3030. Phone: (407) 841-8282. Fax: (407) 841-8250. Licensee: J & V Communications Co. (acq 6-5-92; $300,000; FTR: 6-22-92). Natl. Network: Westwood One. Rgnl. Network: Florida Radio Net. Format: Sp news/talk. Target aud: 21 plus. Spec prog: Relg 3 hrs wkly. ♦John Torrado, CEO, pres & gen mgr; Frank Vaught, opns mgr.

Santa Rosa Beach

WWAV-FM— Apr 3, 1985: 102.1 mhz; 18 kw. TL: N30 23 17 W86 17 55. Hrs open: 24 743 Hwy. 98 E., Suite 6, Destin, 32541-2574. Phone: (850) 654-1031. Fax: (850) 654-6510. Web Site: www.wave1021.com. Licensee: Qantum of Fort Walton Beach License Co. LLC. Group owner: Root Communications (acq 7-2-2003; grpsl). Population served: 185,000 Natl. Rep: Katz Radio. Law Firm: Garvey, Schubert & Barer. Format: Classic hits. News: one hr wkly. Target aud: 25-54; general. ♦Frank Osborne, pres; Jerry Stevens, gen mgr & disc jockey; Suzie Nicholson-Hunt, gen sls mgr.

Sarasota

WBRD(AM)—See Palmetto

WCTQ(FM)—Listing follows WSDV(AM).

WHPT(FM)— 1973: 102.5 mhz; 100 kw. 1,776 ft TL: N27 29 08 W82 32 00. Stereo. Hrs opn: 24 11300 4th St. N., Suite 300, St. Petersburg, 33716. Phone: (727) 579-2000. Fax: (727) 579-2662. Fax: (727) 579-2271. Web Site: theboneonline.com. Licensee: Cox Radio Inc. Group owner: Cox Broadcasting (acq 5-99; grpsl). Population served: 200,000 Natl. Rep: Clear Channel. Format: Class rock. Target aud: 25-54. ♦Keith Lawless, VP & gen mgr.

***WKZM(FM)**— Oct 21, 1974: 104.3 mhz; 6 kw. Ant 266 ft TL: N27 16 30 W82 28 54. Hrs open: 24
Rebroadcasts WKES(FM) Lakeland 100%.
Box 8889, St. Petersburg, 33738. Phone: (727) 391-9994. Fax: (727) 397-6425. E-mail: wkes@moody.edu Web Site: www.wkes.org. Licensee: The Moody Bible Institute of Chicago. (group owner; acq 10-15-99). Population served: 500,000 Law Firm: Southmayd & Miller. Format: Inspirational, educ. News: 14 hrs wkly. Target aud: General. ♦Pierre Chestang, gen mgr & stn mgr.

WLLD(FM)—See Holmes Beach

WLSS(AM)— May 23, 1949: 930 khz; 5 kw-D, 3 kw-N, DA-2. TL: N27 21 17 W82 23 06. Hrs open: 24 5211 W. Laurel St., Tampa, 33607. Phone: (813) 639-1903. Fax: (813) 639-1272. E-mail: wlss@wlssradio.com Web Site: www.wlssradio.com. Licensee: Caron Broadcasting Inc. Group owner: WGUL-FM, Inc. (acq 8-12-2005; $9.5 million with WGUL(AM) Dunedin). Population served: 357,700 Natl. Network: Salem Radio Network. Natl. Rep: Salem. Format: News/talk. ♦Chris Gould, gen mgr; Mike Serio, opns mgr; Casey Bell, prom.

WSDV(AM)— Dec 7, 1939: 1450 khz; 1 kw-ND. TL: N27 20 12 W82 34 25. Hrs open: 24 1779 Independence Blvd., 34234. Phone: (941) 552-4800. Fax: (941) 552-4900. E-mail: nancylee@doveradio.com Web Site: www.doveradio.com. Licensee: Citicasters Licenses L.P. Group owner: Clear Channel Communications Inc. (acq 5-4-99; grpsl). Population served: 700,000 Format: Adult standards. News staff: 3; News: 40 hrs wkly. Target aud: 25-54; general. ♦Sherri Carlson, VP & gen mgr.

WCTQ(FM)—Co-owned with WSDV(AM). June 30, 1965: 106.5 mhz; 25 kw. 280 ft TL: N27 20 12 W82 34 25. Stereo. 24 Phone: (941) 388-3936. Web Site: www.1065ctq.com.600,000 Format: Country. News: one hr wkly. Target aud: 25-54. ♦Mark Wilson, opns mgr & progmg dir; Tracy Black, pub affrs dir; Matt Howell, engrg dir; Maverick Johnson, chief of engrg & disc jockey; Heidi Decker, disc jockey; Sammi Jo Austin, disc jockey.

***WSMR(FM)**— 1993: 89.1 mhz; 50 kw. 462 ft TL: N27 06 00 W82 22 19. Hrs open: 24 240 N. Washington Blvd., Suite 490, 34236. Phone: (941) 906-9767. Fax: (941) 362-0377. Web Site: www.wsmr.org. Licensee: Northwestern College. Group owner: Northwestern College & Radio (acq 10-4-96; $400,000). Format: Christian lite contemp. News: 6 hrs wkly. Target aud: 30-55; with kids still at home. ♦Dr. Alan Cureton, CEO & pres; Harv Hendrickson, gen mgr; Douglas Poll, stn mgr.

WSRQ(AM)— Jan 1, 1961: 1220 khz; 1 kw-D, 600 w-N, DA. TL: N27 19 27 W82 29 47. Hrs open: 8201 S. Tamiami Tr. #54, 34238. Phone: (941) 952-1220. Fax: (941) 365-2900. E-mail: 1220@newstalk1220.com Web Site: www.newstalk1220.com. Licensee: SRQ Radio LLC (acq 8-21-2006; $450,000). Population served: 350,000 Rgnl. Network: Florida Radio Net. Format: News/talk, sports. Target aud: 25-64; men. ♦James Grady, gen mgr.

WTMY(AM)— Dec 2, 1961: 1280 khz; 500 w-D, 340 w-N, DA-2. TL: N27 21 21 W82 29 13. Hrs open: 24 2101 Hammock Pl., 34235. Phone: (941) 954-1280. Fax: (941) 955-9062. E-mail: wtmy@juno.com Web Site: www.wtmy.com. Licensee: Metropolitan Radio Group Inc. (group owner; acq 8-96). Population served: 300,000 Format: Money talk, health talk, talk. Target aud: 40 plus; wealth & health oriented. Spec prog: Pol one hr, gospel 6 hrs, full service 2 hrs wkly. ♦Mark Acker, pres; Greg Durkin, gen mgr.

Satellite Beach

WSBH(FM)—Not on air, target date: unknown: 98.5 mhz; 6 kw. Ant 328 ft TL: N28 08 11 W80 42 12. Hrs open: 1670 N.W. Federal Hwy., Stuart, 34994-1006. Phone: (321) 752-9850. Fax: (772) 692-0258. Licensee: Horton Broadcasting Co. Inc. ♦George Metcalf, pres.

Sebastian

WSJZ-FM— 2001: 95.9 mhz; 25 kw. Ant 289 ft TL: N27 49 05 W80 37 18. Hrs open: 1775 W. Hibiscus Blvd., Suite 101, Melbourne, 32901. Phone: (321) 984-1000. Fax: (321) 724-1565. Web Site: www.pirate959.com. Licensee: Cumulus Licensing LLC. (acq 11-8-2004; $5 million). Format: Rock. ♦Dan Carelli, gen mgr.

Sebring

WFHT(AM)—See Avon Park

WITS(AM)— Nov 24, 1959: 1340 khz; 1 kw-U. TL: N27 30 30 W81 25 20. Hrs open: 24 3750 U.S. 27 N., Suite 1, 33870. Phone: (863) 382-9999. Fax: (863) 382-1982. E-mail: cohanradiogroup@htn.net Web Site: www.cohanradiogroup.com. Licensee: Cohan Radio Group Inc. (group owner; (acq 11-1-98; $735,000 with co-located FM plus WJCM(AM) Sebring). Population served: 56,000 Natl. Network: ABC.

Natl. Rep: Interep. Law Firm: Latham & Watkins. Wire Svc: AP Format: MOR. News staff: one; News: 5 hrs wkly. Target aud: 40 plus; mature adults. ♦Peter Coughlin, pres & gen mgr; Libby Coughlin, gen sls mgr; Connie Bedingfield, traf mgr.

WWLL(FM)—Co-owned with WITS(AM). July 1967: 105.7 mhz; 19 kw. Ant 351 ft TL: N27 21 29 W81 28 22. Stereo. 24 Web Site: www.cohanradiogroup.com. (Acq 11-1-98.).90,000+ Format: Adult contemp. News staff: one; News: 2 hrs wkly. Target aud: 25-54; adults.

WJCM(AM)— May 22, 1950: 1050 khz; 1 kw-D, 11 w-N. TL: N27 30 30 W81 25 20. Hrs open: 24 3750 U.S. 27 N., 33870. Phone: (863) 382-9999. Fax: (863) 382-1982. E-mail: cohanradiogroup@htn.net Web Site: www.cohanradiogroup.com. Licensee: Cohan Radio Group Inc. (group owner; (acq 11-1-98; $150,000). Population served: 70,000 Rgnl. Network: Florida Radio Net. Natl. Rep: Interep. Rgnl rep: Interep Law Firm: Latham & Watkins. Wire Svc: AP Format: Oldies. News staff: one; News: 8 hrs wkly. Target aud: 45+. ♦Peter Coughlin, gen mgr & gen sls mgr; Libby Coughlin, rgnl sls mgr; Alan Gray, progmg mgr; Barry Foster, news dir; Stacy Clark, pub affrs dir; Phil Scott, chief of engrg.

***WJFH(FM)**—Not on air, target date: unknown: 91.5 mhz; 20.5 kw. Ant 293 ft TL: N27 26 31 W81 40 06. Hrs open: Box 7217, Lakeland, 33807-7217. Phone: (863) 644-3464. Fax: (863) 646-5326. Licensee: Radio Training Network Inc. ♦James L. Campbell, pres.

***WMKM(FM)**—Not on air, target date: unknown: 88.3 mhz; 1.7 kw vert. Ant 239 ft TL: N27 24 26 W81 25 59. Hrs open: 3185 S. Highland Dr., Suite 13, Las Vegas, NV, 89109-1029. Phone: (702) 731-5588. Licensee: American Educational Broadcasting Inc. ♦Carl J. Auel, pres.

WWOJ(FM)—See Avon Park

WWTK(AM)—See Lake Placid

Seffner

WQYK(AM)—Licensed to Seffner. See Tampa

Shalimar

WNCV(FM)— Oct 25, 1982: 93.3 mhz; 50 kw. Ant 469 ft TL: N30 24 38 W86 37 22. Hrs open: 24 225 N.W. Hollywood Blvd., Fort Walton Beach, 32548-4725. Phone: (850) 243-7676. Fax: (850) 243-6806. E-mail: coastoffice@wncv.com Web Site: www.wncv.com. Licensee: Cumulus Licensing LLC. (acq 8-2-2006; swap for WRKN(FM) Niceville). Population served: 65,000 Law Firm: Putbrese, Hunsaker & Trent. Format: Adult contemp, soft hits. ♦Mike DeMarco, gen mgr; Skip Davis, progmg dir.

Silver Springs

WNDD(FM)—Licensed to Silver Springs. See Gainesville

Solana

WCVU(FM)— 1994: 104.9 mhz; 6 kw. 318 ft TL: N26 53 37 W82 03 03. Hrs open: 24100 Tiseo Blvd., Unit 10, Port Charlotte, 33980. Phone: (941) 206-1188. Fax: (941) 206-9296. Web Site: www.clearchannel.com. Licensee: Citicasters Licenses L.P. Group owner: Clear Channel Communications Inc. (acq 2-1-99; grpsl). Population served: 106,000 Natl. Network: CNN Radio. Format: Soft adult contemp. ♦Michael Moody, gen mgr; David Ayres, sls dir; Todd Matthews, progmg mgr.

WKII(AM)—Licensed to Solana. See Port Charlotte

South Daytona

WPUL(AM)— June 13, 1957: 1590 khz; 1 kw-D. TL: N29 09 16 W81 01 20. Hrs open: 6 am-12 pm 427 S. Martin L. King Blvd., Daytona Beach, 32114. Phone: (386) 239-7080 (Studio). Phone: (386) 226-2398. Fax: (386) 254-7510. E-mail: ccherry2@aol.com Licensee: PSI Communications Inc. (acq 2-1-89; $250,000; FTR: 1-23-89). Natl. Network: American Urban. Format: Gospel, talk. Target aud: General. ♦Charles W. Cherry II, CEO & gen mgr; Phinesse Demps, progmg mgr.

South Miami

WAXY(AM)— Sept 15, 1947: 790 khz; 25 kw-U, DA-2. TL: N25 46 25 W80 38 13. Stereo. Hrs opn: 24 20450 N.W. 2nd Ave., Miami, 33169. Phone: (305) 521-5100 / (887) 790-1015. Fax: (305) 521-1416. Web Site: www.waxy.com. Licensee: Lincoln Financial Media Co. of Florida. (group owner; (acq 4-3-2006; grpsl). Population served: 2,800,000 Natl. Rep: CBS Radio. Format: Sports talk. Target aud: 35 plus. ♦Dennis P. Collins, gen mgr; Gary Aybar, opns mgr.

Spring Hill

WJQB(FM)— October 1992: 106.3 mhz; 25 kw. Ant 315 ft TL: N28 31 41 W82 32 45. Hrs open: 35048 US Hwy. 19 N., Palm Harbor, 34684. Phone: (727) 442-4027. Fax: (727) 781-4375. E-mail: staff@wjqb.com Web Site: www.wjqb.com. Licensee: WGUL-FM Inc. Format: Oldies. ♦Steve Schurdell, VP & gen mgr.

Springfield

WRBA(FM)— June 1986: 95.9 mhz; 50 kw. 300 ft TL: N30 12 12 W85 36 57. Hrs open: 24 118 Gwyn Dr., Panama City Beach, 32408. Phone: (850) 234-8858. Fax: (850) 234-6592. E-mail: billyoung@panamacityradio.com Web Site: www.arrow959.com. Licensee: Double O Radio Corp. (group owner; acq 3-10-2004; grpsl). Population served: 135,000 Format: Classic rock. Target aud: 30-54; general. ♦Harry Finch, gen mgr.

WYOO(FM)—Licensed to Springfield. See Panama City

Starke

***WTLG(FM)**— 1982: 88.3 mhz; 7 kw. 285 ft TL: N29 54 34 W82 06 02. Stereo. Hrs opn: 24 Box 1258, 163 W. Jefferson at Clarke, 32091. Phone: (904) 964-9854. Fax: (904) 964-2968. E-mail: wtlg-radio@earthlink.net Licensee: Starke Christian Educational Radio & TV. Natl. Network: Moody. Format: Southern gospel, relg. Target aud: General. ♦Reverend Terry Blakeslee, gen mgr; Hal Mashburn, chief of engrg.

Stuart

WAVW(FM)— Dec 24, 1964: 92.7 mhz; 50 kw. 482 ft TL: N27 16 30 W80 17 12. Stereo. Hrs opn: Box 0093, Port St. Lucie, 34985. Phone: (772) 335-9300. Fax: (772) 335-3291. Web Site: www.wavw.com. Licensee: Capstar TX L.P. Group owner: Clear Channel Communications Inc. (acq 8-30-00; grpsl). Population served: 50,000 Format: Country. ♦John Hunt, gen mgr; Heath West, progmg dir; Mike Kerley, chief of engrg.

WSTU(AM)— Dec 9, 1954: 1450 khz; 1 kw-U. TL: N27 12 53 W80 15 24. Hrs open: 24 4100 Metzger Rd., Fort Pierce, 34947. Phone: (772) 220-9788. Fax: (772) 340-3245. E-mail: wpsl@wpsl.com Web Site: www.wstu1450.com. Licensee: Treasure Coast Broadcasters Inc. (acq 2-13-02; $500,000). Population served: 600,000 Natl. Network: ESPN Radio, ABC. Law Firm: Leventhal, Senter & Lerman. Format: News/talk, sports. News staff: 2. Target aud: 35+. ♦Carol Wyatt, pres.

***WWFR(FM)**— 1988: 91.7 mhz; 2.65 kw. Ant 499 ft TL: N27 07 14 W80 23 59. Stereo. Hrs opn: 24 Box 277, Okeechobee, 34972. Secondary address: 10400 NW 240th st, Okeechobee 34972. Phone: (863) 763-5454. Fax: (863) 763-7729. E-mail: wwfr@okeechobee.com Web Site: www.familyradio.com. Licensee: Family Stations Inc. (group owner) Population served: 70,000 Natl. Network: Family Radio. Format: Relg. News: 11 hrs wkly. Target aud: General. Spec prog: Pub affrs 2 hrs wkly. ♦Ed Dearborn, chief of opns.

Summerland Key

WPIK(FM)— December 1991: 102.5 mhz; 50 kw. Ant 413 ft TL: N24 40 35 W81 30 41. Stereo. Hrs opn: 24 Box 420249, 33042. Secondary address: 22500 Pieces of Eight Rd., Cudjoe Key 33042. Phone: (305) 745-9988. Fax: (305) 745-4165. E-mail: info@myradioritmo.com Web Site: www.myradioritmo.com. Licensee: Summerland Media LLC (acq 10-7-2005; $1.85 million). Population served: 75,000 Format: Bilingual music. ♦Lilliam M. Sierra, gen mgr; Pepin Navarro, progmg dir.

Sunrise

***WKPX(FM)**— Feb 14, 1983: 88.5 mhz; 3 kw. 100 ft TL: N26 10 38 W80 15 23. Stereo. Hrs opn: 12 8000 N.W. 44th St., 33351. Phone: (754) 321-1000. Fax: (754) 321-1180. Licensee: School Board of Broward County. Format: Modern rock, alternative. Target aud: 15-35; people interested in alternative progmg. Spec prog: Black 3 hrs, blues 3 hrs wkly. ♦Pat Swank, stn mgr; Jim Sorensen, chief of engrg.

Tallahassee

WAIB(FM)— June 17, 1976: 103.1 mhz; 50 kw. 295 ft TL: N30 29 43 W84 13 51. (CP: 42 kw, ant 541 ft. TL: N30 29 39 W84 14 00). Stereo. Hrs opn: 24 Opus Broadcasting, 3000 Olson Rd., 32308. Phone: (850) 386-8004. Fax: (850) 422-1897. E-mail: hkestenbaum @opusbroadcasting.com Web Site: www.newcountryb103.com. Licensee: Opus Broadcasting Tallahassee LLC. Group owner: Triad Broadcasting Co. LLC (acq 7-11-2005; grpsl). Population served: 350,000 Natl. Rep: McGavren Guild. Format: Country. Target aud: 25-54. ♦Hank Kestenbaum, gen mgr; Doug Purtee, opns dir.

***WANM(FM)**— November 1976: 90.5 mhz; 1.6 w. 167 ft TL: N30 25 49 W84 17 27. Stereo. Hrs opn: 24 Florida A&M Univ., 510 Orr Dr., Ste 3056, 32307. Phone: (850) 599-3083. Fax: (850) 561-2829. E-mail: theflavastation@hotmail.com Web Site: www.famu.edu/famcast. Licensee: The Board of Trustees of Florida A&M University. Population served: 195,000 Natl. Network: AP Radio. Wire Svc: AP Format: News, sports. News: 5 hrs wkly. Target aud: General; urban African-American in area. Spec prog: Reggae 3 hrs wkly, Gospel 18 hrs wkly, Jazz 15 hrs wkly. ♦Keith Miles, gen mgr; Greg Bishop, opns mgr.

WBZE(FM)—Listing follows WHBT(AM).

WCVC(AM)— Nov 5, 1953: 1330 khz; 5 kw-D. TL: N30 29 03 W84 17 13. Hrs open: 6:30 AM-7 PM 117 1/2 Henderson Rd., 32312. Phone: (850) 386-1330. Fax: (850) 386-2138. E-mail: wcvc65@hotmail.com Web Site: www.wcvc1330.com. Licensee: WCVC Inc. (acq 10-4-85; $500,000; FTR: 8-12-85). Population served: 190,000 ♦Wendell H. Borrink, pres; Erwin O'Conner, gen mgr, opns mgr & progmg dir.

WEGT(FM)—Lafayette, Dec 17, 1989: 99.9 mhz; 50 kw. 492 ft TL: N30 20 59 W83 59 53. Stereo. Hrs opn: 24 Opus Broadcasting, 3000 Olson Rd., 32308. Phone: (850) 386-8004. Fax: (850) 422-1897. Licensee: Opus Broadcasting Tallahassee LLC. Group owner: Triad Broadcasting Co. LLC (acq 7-11-2005; grpsl). Population served: 350,000 Natl. Rep: McGavren Guild. Format: Classic hits of the 60s, 70s, etc. News: 2 hrs wkly. Target aud: 25-54. ♦Hank Kestenbaum, gen mgr; Dooug Purtee, opns dir.

***WFRF(AM)**— August 1974: 1070 khz; 10 kw-D. TL: N30 30 34 W84 20 07. Hrs open: Sunrise-sunset Box 181000, 32318. Secondary address: 4015 N. Monroe St. 32303. Phone: (850) 201-1070. Fax: (850) 201-1071. E-mail: mailbox@faithradio.us Web Site: www.faithradio.us. Licensee: Faith Radio Network Inc. (acq 9-30-97; $150,000). Population served: 120,200 Format: Christian. Target aud: 12+. ♦Scott Beigle, gen mgr & progmg dir.

***WFSQ(FM)**— May 1954: 91.5 mhz; 100 kw. 663 ft TL: N30 21 29 W84 36 39. Stereo. Hrs opn: 24 Public Broadcast Ctr., 1600 Red Barber Plaza, 32310. Phone: (850) 487-3086. Fax: (850) 487-2611. Web Site: www.wfsu.org. Licensee: The Board of Regents of Florida acting for and on behalf of Florida State University. Population served: 305,517 Natl. Network: NPR, PRI. Format: Class. News: one hr wkly. Target aud: 35 plus; highly educated. ♦Patrick Keating, gen mgr; Caroline Austin, stn mgr & prom dir; Cary Martin, engrg dir.

***WFSU-FM—** Oct 14, 1990: 88.9 mhz; 95 kw. 1,243 ft TL: N30 40 13 W83 56 26. Stereo. Hrs opn: 24 Public Broadcast Ctr., 1600 Red Barber Plaza, 32310. Phone: (850) 487-3086. Fax: (850) 487-2611. Web Site: www.wfsu.org. Licensee: The Board of Regents of Florida acting for and on behalf of Florida State University. Population served: 161,500 Natl. Network: NPR, PRI. Rgnl. Network: Fla. Pub. Law Firm: Cohn & Marks. Format: News/talk. News staff: 9. Target aud: 35-54; highly educated. Spec prog: Jazz 8 hrs wkly. ♦Pat Keating, gen mgr; Caroline Austin, stn mgr; Ann Meyers, prom dir; Cary Martin, chief of engrg.

WGLF(FM)— December 1967: 104.1 mhz; 100 kw. 1,359 ft TL: N30 27 09 W84 00 50. Stereo. Hrs opn: 24 3411 W. Tharpe St., Tallahasse, 32303. Phone: (850) 201-3000. Fax: (850) 561-8903. Web Site: www.gulf104.com. Licensee: Cumulus Licensing Corp. Group owner: Cumulus Media Inc. (acq 6-22-99; $4 million). Population served: 320,304 Natl. Rep: Katz Radio. Law Firm: Wiley Rein. Format: Classic rock, AOR. Target aud: 25-54. ♦John Columbus, gen mgr; Zack Thomas, progmg dir.

WHBT(AM)— Aug 6, 1959: 1410 khz; 5 kw-D, 39 w-N. TL: N30 29 35 W84 17 00. (CP: N30 29 03 W84 17 13). Hrs opn: 24 3411 W. Tharpe St., 32303. Phone: (850) 201-3000. Fax: (850) 561-8903. Web Site: www.1410thefan.com. Licensee: Cumulus Licensing Corp. Group owner: Cumulus Media Inc. (acq 10-28-97; grpsl). Population served: 239,452 Natl. Rep: Katz Radio. Law Firm: Wiley & Rein. Format: Sports. Target aud: 18-54. ♦Barry Kaye, gen mgr; Bill Scull, gen sls mgr; Rick Anderson, rgnl sls mgr; Shelly Jarvis, prom dir; Peter Walkowiak, chief of engrg.

WBZE(FM)—Co-owned with WHBT(AM). July 15, 1962: 98.9 mhz; 100 kw. 390 ft TL: N30 29 35 W84 16 55. Stereo. 24 Web Site: www.mystar98.com.320,304 Natl. Rep: Katz Radio. Law Firm: Wiley & Rein. Format: Adult contemp. Target aud: 25-54. ♦John Dawson, progmg dir.

WHBX(FM)— June 28, 1982: 96.1 mhz; 37 kw. 479 ft TL: N30 16 08 W84 16 32. Stereo. Hrs opn: 24 3411 West Tharpe St., 32303. Phone: (850) 201-3000. Fax: (850) 561-8903. Web Site: 961jamz.com. Licensee: Cumulus Licensing Corp. Group owner: Cumulus Media Inc. (acq 10-28-97; grpsl). Population served: 320,304 Natl. Rep: Katz Radio. Law Firm: Wiley Rein. Format: Urban contemp. Target aud: 25-54. ♦John Columbus, gen mgr; Joe Bullard, opns mgr & progmg dir.

WHTF(FM)—See Havana

WNLS(AM)—Listing follows WTNT-FM.

WTAL(AM)— 1935: 1450 khz; 1 kw-U. TL: N30 26 20 W84 15 30. Hrs open: 24 1363 E. Tennesse Street, 32308. Phone: (850) 671-1450. Phone: (850) 877-0105. Fax: (850) 877-5110. E-mail: wtaal@nettally.com Web Site: www.wtal1450.com. Licensee: Live Communications Inc. (acq 9-14-01; $400,000). Population served: 210,000 Natl. Network: CBS. Natl. Rep: Roslin. Law Firm: Reddy, Begley & McCormick. Format: News/talk, Christian, relig. News staff: 4; News: 21 hrs wkly. Target aud: 25-54; educated, intelligent, affluent, involved, conservative. ♦Dr. R.B. Holmes Jr., CEO & pres; Rebecca Johnson, gen mgr.

WTLY(FM)—Thomasville, GA) 1971: 107.1 mhz; 100 kw. 981 ft TL: N30 43 55 W84 08 45. Stereo. Hrs opn: 24 Bldg. G, 325 John Knox Rd., 32303. Phone: (850) 422-3107. Fax: (850) 383-0747. Fax: (850) 514-4443. E-mail: jeffhorn@clearchannel.com Web Site: www.magic1071.com. Licensee: CC Licenses LLC. Group owner: Clear Channel Communications Inc. (acq 11-21-97; grpsl). Population served: 400000 Natl. Rep: Christal. Format: Adult Contempory. News staff: one; News: 5 hrs wkly. Target aud: 25-54. ♦Lisa Rice, gen mgr; Jeff Horn, opns mgr; Randall Moore, chief of engrg.

WTNT-FM— July 24, 1967: 94.9 mhz; 100 kw. 840 ft TL: N30 34 43 W84 15 49. Stereo. Hrs opn: Bldg. G, 325 John Knox Rd., 32303. Phone: (850) 422-3107. Fax: (850) 383-0747. Web Site: www.wtntfm.com. Licensee: Clear Channel Radio Licenses Inc. Group owner: Clear Channel Communications Inc. (acq 11-21-97; grpsl). Population served: 330,000 Natl. Rep: Christal. Format: Country. News: one hr wkly. Target aud: 25-54. ♦Judy Bailey, gen mgr & natl sls mgr; Kris Van Dyke, opns mgr, progmg VP & progmg dir; Belinda Bininger, sls VP; Jonathan Faulkner, prom dir; Woody Hayes, mus dir; J. L. Dunbar, news dir & pub affrs dir; Rick Flagg, news dir; Randy Moore, engrg dir; Sandra Lee, traf mgr.

WNLS(AM)—Co-owned with WTNT-FM. Oct 15, 1946: 1270 khz; 5 kw-U, DA-N. TL: N30 25 38 W84 19 46.24 Web Site: www.wnls.com.148,000 Natl. Rep: Christal. Format: Sports. News staff: one; News: 25 hrs wkly. ♦J. L. Dunbar, progmg dir; Sandra Lee, traf mgr.

WUTL(FM)— May 1992: 106.1 mhz; 3 kw. Ant 328 ft TL: N30 28 37 W84 20 07. Hrs open: 24 Opus Broadcasting, 3000 Olson Rd., 32308. Phone: (850) 386-8004. Fax: (850) 422-1897. E-mail: hkestenbaum@opusbroadcasting.com Licensee: Opus Broadcasting Tallahassee LLC. Group owner: Triad Broadcasting Co. LLC (acq 7-11-2005;. grpsl). Format: News/Talk. Target aud: 18-49. ♦Hank Kestenbaum, VP & gen mgr; Doug Purtee, opns mgr.

***WVFS(FM)—** September 1987: 89.7 mhz; 2.7 kw. 174 ft TL: N30 26 22 W84 17 29. Hrs open: 24 Florida State Univ., 420 Diffenbaugh, 32306-1550. Phone: (850) 644-3871. Fax: (850) 644-8753. E-mail: wvfs@wvfs.fsu.edu Web Site: www.wvfs.fsu.edu. Licensee: Florida State University. Population served: 200,000 Format: Diverse. News: 2 hrs wkly. Target aud: General. Spec prog: Black 8 hrs, folk 3 hrs, Sp 2 hrs wkly. ♦Dr. Misha Laurents, gen mgr.

Tampa

WAMA(AM)— 1961: 1550 khz; 10 kw-D, 125 w-N. TL: N27 55 16 W82 23 41. Hrs open: 24 402 N. Reo St., Suite 115, 33609. Phone: (813) 289-1552. Fax: (813) 289-1554. E-mail: jobediente@lainvasora1550.com Web Site: www.lainvasora1550.com. Licensee: WAMA Inc. (acq 10-15-97; $2 million). Natl. Rep: Univision Radio National Sales. Law Firm: Drinker, Biddle & Reath. Wire Svc: UPI Format: Mexican. News staff: one. Target aud: 25-54; Hispanic Adults. ♦Ron Gordon, chmn & pres; Joshua Mednick, VP; Julio Obediente, gen mgr; Oscar Rojas, gen mgr.

WBTP(FM)—See Clearwater

***WBVM(FM)—** May 27, 1986: 90.5 mhz; 75.5 kw vert. Ant 964 ft TL: N27 50 53 W82 15 48. Stereo. Hrs opn: 24 Box 18081, 33679. Secondary address: 3816 Morrison Ave. 33629. Phone: (813) 289-8040. Fax: (813) 282-3580. E-mail: contact@spiritfm905.com Web Site: www.spiritfm905.com. Licensee: The Bishop of the Diocese of St. Petersburg. Format: Contemp Christian. Target aud: 35 plus; families. Spec prog: Black 4 hrs, children 4 hrs, Sp 4 hrs wkly. ♦John Morris, VP & gen mgr; Chris Sampson, opns mgr.

WDAE(AM)—See Saint Petersburg

WDUV(FM)—New Port Richey, Sept 19, 1969: 105.5 mhz; 46 kw. Ant 1,345 ft TL: N28 10 56 W82 46 06. (CP: 6.7 kw, ant 1,050 ft.). Stereo. Hrs opn: 24 11300 4th St. N., Suite 300, St. Petersburg, 33716-2941. Phone: (727) 579-2000. Fax: (727) 579-2662. Fax: (727) 579-2271. Web Site: www.wduv.com. Licensee: Cox Radio Inc. Group owner: Cox Communications Inc. (acq 5-99). Format: Soft adult contemp. Target aud: 25-54. Spec prog: It one hr wkly. ♦Howard Tuuri, VP & gen mgr; Tom Paleveda, opns mgr & gen sls mgr.

WFLA(AM)— 1924: 970 khz; 5 kw-U, DA-2. TL: N28 01 14 W82 36 34. Hrs open: 4002A Gandy Blvd., 33611. Phone: (813) 839-9393. Fax: (813) 831-4475. Fax: (813) 837-0300. Web Site: www.970wfla.com. Licensee: Citicasters Licenses L.P. Group owner: Clear Channel Communications Inc. (acq 5-4-99; grpsl). Population served: 1,865,800 Law Firm: Hogan & Hartson. Format: News/talk. Target aud: 25-54. ♦David Reinhart, VP & gen mgr; Sue Treccase, progmg dir; Wilson Welch, chief of engrg; Miriam Stokes, traf mgr.

WFLZ-FM—Co-owned with WFLA(AM). 1948: 93.3 mhz; 99 kw. 1,358 ft TL: N27 50 32 W82 15 46. Stereo. Web Site: www.933flz.com. Format: CHR. Target aud: 18-34. ♦Dave Reinhart, gen mgr; Jeff Kapugi, progmg dir; Wilson Welch, chief of engrg; Anqunette Wilson, traf mgr.

WGUL(AM)—Dunedin, Nov 21, 1959: 860 khz; 5 kw-D, 1.5 kw-N, DA-2. TL: N27 59 55 W82 42 01. Hrs open: 24 5211 W. Laurel St., 33607. Phone: (813) 639-1903. Fax: (813) 639-1272. E-mail: cgould@salemtampa.com Web Site: www.860wgul.com. Licensee: Caron Broadcasting Inc. WGUL FM Inc. (acq 8-12-2005; $9.5 million with WLSS(AM) Sarasota). Population served: 4,000,000 Natl. Network: Salem Radio Network. Natl. Rep: Salem. Format: News/talk. ♦Chris Gould, gen mgr; Mike Serio, opns mgr.

WHNZ(AM)— May 15, 1922: 1250 khz; 5 kw-U, DA-1. TL: N28 00 41 W82 29 53. Hrs open: 4002 Gandy Blvd., 33611. Phone: (813) 839-9393. Fax: (813) 831-3299. Web Site: www.whnz.com. Licensee: Citicasters Licenses L.P. Group owner: Clear Channel Communications Inc. (acq 5-4-99; grpsl). Population served: 277767 Natl. Rep: McGavren Guild. Format: News/talk. Target aud: 25-54. ♦Dan DiLorette, pres & gen mgr; Gene Lindsey, progmg dir; Wilson Welch, chief of engrg; Diana Roselle, traf mgr.

WMTX(FM)—Co-owned with WHNZ(AM). November 1947: 100.7 mhz; 100 kw. 1,358 ft TL: N28 02 21 W82 39 21. (CP: TL: N27 50 32 W82 15 46). Stereo. 24 Web Site: www.wmtx.com. Format: Top 40, adult contemp. ♦Dan DiLoreto, gen mgr; Tony Florentino, progmg dir; Wilson Welch, chief of engrg; Diana Roselle, traf mgr.

WHPT(FM)—See Sarasota

***WMNF(FM)—** Sept 14, 1979: 88.5 mhz; 70 kw. 520 ft TL: N27 49 04 W82 14 31. Hrs open: 24 1210 E. Martin Luther King Jr. Blvd., 33603. Phone: (813) 238-8001. E-mail: wmnf@wmnf.org Web Site: www.wmnf.org. Licensee: The Nathan B. Stubblefield Foundation. Population served: 2,800,000 Natl. Network: NPR. Law Firm: Haley, Bader & Potts. Format: Div. News staff: 2; News: 15 hrs wkly. Target aud: General. ♦Sheila Cowley, opns mgr; Vicki Santa, stn mgr & dev dir; Randy Wynne, progmg dir; Bill Brown, chief of engrg.

WQBN(AM)—Temple Terrace, 1956: 1300 khz; 5 kw-D, 1 kw-N, DA-2. TL: N28 03 44 W82 19 44. Hrs open: 6 AM-midnight Box 151300, 33684. Secondary address: 5203 N. Armenia Ave. 33603. Phone: (813) 871-1333. Fax: (813) 876-1333. E-mail: superq1300@hotmail.com Web Site: www.superq1300am.com. Licensee: Radio Tropical Inc. Population served: 453,000 Format: Sp, variety. News staff: 3; News: 20 hrs wkly. Target aud: 25 plus; Hispanics. ♦Efrain Archilla, pres; Marc L. Vila, VP & gen mgr.

WQYK(AM)—Seffner, Nov 7, 1960: 1010 khz; 50 kw-D, 5 kw-N, DA-2. TL: N27 59 25 W82 15 06. Stereo. Hrs opn: 24 5510 W. Gray St., Suite 130, 33609. Phone: (813) 637-8326. Fax: (813) 636-0995. Web Site: www.wqyk.com. Licensee: CBS Radio Inc. of Tampa. Group owner: Infinity Broadcasting Corp. (acq 11-21-87). Natl. Network: Sporting News Radio Network. Natl. Rep: CBS Radio. Law Firm: Leventhal, Senter & Lerman. Format: Sports. News staff: . ♦Charlie Ochs, sr VP; Luis Diaz-Albertiwi, gen mgr; Mike Culotta, opns mgr; John Fennessy, gen sls mgr.

WYUU(FM)—Co-owned with WQYK(AM). October 1983: 92.5 mhz; 50 kw. Ant 489 ft TL: N27 50 32 W82 48 52. Stereo. 9721 Executive Center Dr., Suite 200, Saint Petersburg, 33702. Phone: (727) 579-1925. Fax: (813) 287-1833. Web Site: lanueva925.com. Licensee: CBS Radio Stations Inc. (acq 10-15-98; $75 million with WLLD(FM) Holmes Beach).2,800,000 Law Firm: Leventhal, Senter & Lerman. Format: Sp tropical. Target aud: 18-49. ♦John Fennessy, gen sls mgr.

WRBQ-FM— 1954: 104.7 mhz; 100 kw. Ant 561 ft TL: N27 56 50 W82 27 35. Stereo. Hrs opn: 9721 Executive Center Dr. N., Suite 200, Saint Petersburg, 33702. Phone: (727) 579-1925. Fax: (727) 579-8888. Web Site: www.tampabaysq105.com. Licensee: Infinity Radio Inc. Group owner: Infinity Broadcasting Corp. (acq 5-99). Format: Oldies. Target aud: 18-49. ♦Charlie Ochs, gen mgr; Mason Dixon, progmg dir.

WTIS(AM)— 1946: 1110 khz; 10 kw-D, DA. TL: N27 52 26 W82 37 53. Hrs open: 311 112th Ave. N.E., St. Petersburg, 33716. Phone: (727) 576-2234. Fax: (727) 577-3814. Licensee: WTIS-AM Inc. (acq 12-13-89; $1.7 million; FTR: 1-1-90). Population served: 1,904,100 Format: Relg, ethnic. Target aud: 25-54. Spec prog: Sp one hr wkly. ♦Ron Roseman, pres; Ed Roseman, exec VP; Mike Smith, gen mgr & opns mgr; Robert Kansnicki, progmg dir.

WTMP(AM)—Egypt Lake, 1954: 1150 khz; 10 kw-D, 500 w-N. TL: N28 00 42 W82 29 53. Hrs open: 24 5207 Washington Blvd., 33619-3437. Phone: (813) 620-1300. Fax: (813) 628-0713. E-mail: info@tamabroadcasting.com Web Site: www.wtmp.com. Licensee: Tama Radio Licenses of Tampa, FL, Inc. Group owner: Tama Broadcasting Inc. (acq 12-18-2001). Population served: 1,800,000 Natl. Network: American Urban. Format: Rhythm and Blues. Target aud: 18-49; urban contemporary music listeners & adults. Spec prog: Gospel 20 hrs wkly. ♦Dr. Glenn W. Cherry, CEO, pres, CFO & gen mgr; Louis Muhammad, opns mgr; Lynn Tolliver, progmg dir.

***WUSF(FM)—** September 1963: 89.7 mhz; 71 kw. Ant 941 ft TL: N27 50 53 W82 15 48. Stereo. Hrs open: 24 4202 E. Fowler Ave., TVB100, 33620-6870. Phone: (813) 974-8700. Fax: (813) 974-5016. E-mail: jurofsky@wusf.org Web Site: www.wusf.org. Licensee: Board of Trustees, University of South Florida. Population served: 3,500,000 Natl. Network: NPR, PRI. Rgnl. Network: Fla. Pub. Law Firm: Cohn & Marks. Wire Svc: AP Format: Class, news, jazz. News staff: 6; News: 38 hrs wkly. Target aud: General. ♦Jo Ann Urofsky, gen mgr; Tom Dollenmayer, stn mgr; Cathy Coccia, dev mgr. Co-owned TV: *WUSF-TV affil

WWBA(AM)—Pinellas Park, November 1948: 1040 khz; 5 kw-D, 500 w-N, DA-N. TL: N27 50 50 W82 46 21. Hrs open: 4300 W. Cypress St., # 1040, 33607. Phone: (813) 281-1040. Fax: (813) 281-1948. Web Site: www.wwba1040.com. Licensee: Genesis Communications of Tampa Bay Inc. Group owner: Genesis Communications Inc. (acq 12-17-97; $1.5 million). Population served: 277,767 Format: News/talk. News staff: 1; News: 14 hrs wkly. Target aud: 25-54. ♦Bruce Maduri, CEO; Bambi Arnold, CFO; Sandra Culver, opns VP; Carol Azaravich, opns mgr.

WWRM(FM)—Licensed to Tampa. See Saint Petersburg

WXGL(FM)—See Saint Petersburg

WXTB(FM)—See Clearwater

Tarpon Springs

***WYFE(FM)**— June 14, 1988: 88.9 mhz; 50 kw. 500 ft TL: N28 24 07 W82 36 30. (CP: 60 kw, ant 449 ft.). Stereo. Hrs opn: 5553 S.E. 3rd Ave., Keystone Heights, 32656. Phone: (877) 939-9933. Web Site: www.bbnradio.org. Licensee: Bible Broadcasting Network Inc. (group owner; acq 8-11-89). Format: Relg. Target aud: General; Christians. ♦Lowell Davey, pres; Jack Long, gen mgr.

Tavares

WXXL(FM)—Licensed to Tavares. See Leesburg

Tavenier

WKEZ-FM— 1999: 96.9 mhz; 6 kw. 220 ft TL: N25 01 35 W80 30 30. Hrs open: 93351 Overseas Hwy., 33070. Phone: (305) 852-9085. Fax: (305) 852-5586. Fax: (305) 852-2304. Web Site: www.easy969.com. Licensee: Clear Channel Broadcasting Licenses Inc. Group owner: Clear Channel Communications Inc. (acq 3-16-99; $849,900). Format: Easy lstng. ♦Mark Mills, stn mgr; Scott Hamilton, opns mgr; Greg Capgna, mktg mgr.

Temple Terrace

WQBN(AM)—Licensed to Temple Terrace. See Tampa

Tequesta

WEFL(AM)— Aug 1, 2002: 760 khz; 3 kw-D, 1.5 kw-N, DA-2. TL: N26 59 43 W80 11 34. Hrs open: 24 2090 Palm Beach Lakes Blvd., Ste 701, West Palm Beach, 33409. Phone: (561) 697-8353. Fax: (561) 697-8525. E-mail: sports@espn760.com Web Site: www.@espn760.com. Licensee: Good Karma Broadcasting L.L.C. (acq 1-3-2006; $2.8 million). Natl. Network: ESPN Radio, Westwood One. Format: Sports. ♦Craig Karmazin, CEO; Steve Politziner, gen mgr; Lance Davis, opns dir.

Tice

WJGO(FM)— March 2000: 102.9 mhz; 50 kw horiz, 48 kw vert. Ant 466 ft TL: N26 29 16 W81 55 46. Hrs open: 10915 K-Nine Dr., 2nd Fl., Bonita Springs, 34135. Phone: (239) 495-8383. Fax: (239) 495-0883. Web Site: www.1029bobfm.com. Licensee: Renda Broadcasting Corp. of Nevada. Group owner: Renda Broadcasting Corp. (acq 10-5-2000; $7 million). Format: Adult hits. ♦Tony Renda Jr., gen mgr; Randy Savage, progmg dir.

Titusville

WIXC(AM)— Nov 20, 1957: 1060 khz; 10 kw-D, 5 kw-N, DA-2. TL: N28 39 47 W80 55 17. (CP: 50 kw-D, 17 kw-CH, DA-3). Hrs opn: 24 6305 Hwy. 46, Mims, 32754. Phone: (321) 264-1060. Phone: (321) 264-9700. Fax: (321) 264-4246. E-mail: gregsherlock@wixc1060.com Licensee: Genesis Communications I Inc. Group owner: Genesis Communications Inc. (acq 4-19-2000; $650,000). Natl. Network: ESPN Radio. Format: Sports. Target aud: 35 plus. Spec prog: Relg 6 hrs wkly. ♦Bruce C. Maouri, CEO; Greg Sherlock, gen mgr & news dir; Steve Potter, gen sls mgr; Jerry Smith, chief of engrg.

WNUE-FM—Licensed to Titusville. See Daytona Beach

WORL(AM)—See Altamonte Springs

***WPIO(FM)**— Oct 19, 1975: 89.3 mhz; 5.8 kw horiz, 7.1 kw vert. Ant 335 ft TL: N28 34 49 W80 51 00. Stereo. Hrs opn: 505 Josephine St., 32796. Phone: (321) 267-3000. Fax: (321) 264-9370. Web Site: www.noncomradio.net. Licensee: Florida Public Radio Inc. Population served: 200,000 Format: Inspirational mus, pub affrs. ♦Randy Henry, pres & gen mgr.

Trenton

WDVH-FM— February 1988: 101.7 mhz; 3 kw. 328 ft TL: N29 36 40 W82 51 14. Stereo. Hrs opn: 24 100 N.W. 76th Dr., Suite 2, Gainesville, 32607. Phone: (352) 313-3150. Fax: (352) 313-3166. Licensee: 6 Johnson Road Licenses Inc. (group owner; (acq 1-5-2007; grpsl). Format: Country. News: 7 hrs wkly. Target aud: 25 plus; working class & professionals. Spec prog: Farm 2 hrs wkly. ♦Benjamin Hill, gen mgr & mus dir.

Union Park

***WPOZ(FM)**— Aug 9, 1995: 88.3 mhz; 2.5 kw. Ant 1,469 ft TL: N28 36 08 W81 05 37. (CP: 14.5 kw, ant 1,273 ft.). Hrs open: 24 1065 Rainer Dr., Altamonte Springs, 32714. Phone: (407) 869-8000. Fax: (407) 869-0380. E-mail: zcrew@zradio.org Web Site: www.zradio.org. Licensee: Central Florida Educational Foundation Inc. Population served: 1,137,888 Law Firm: Joseph E. Dunne III. Format: Contemp Christian. Target aud: 25-44. ♦James S. Hoge, chmn, pres & gen mgr.

Valparaiso-Niceville

WFSH(AM)— November 1958: 1340 khz; 1 kw-U. TL: N30 30 34 W86 28 34. Hrs open: 6 AM-6 PM Box 1120, Destin, 32540. Secondary address: 415 Mountain Dr., Suite 7, Destin 32541. Phone: (850) 654-4040. Fax: (850) 650-9440. E-mail: dale@fox1120.com Licensee: Flagship Communications Inc. (acq 8-28-03; $225,000). Population served: 153,000 Natl. Network: ESPN Radio. Format: Sports. News staff: one. Target aud: 25-54. Spec prog: Sports. ♦Dale Riddick, gen mgr; Max Howell, gen sls mgr; Steve Williams, news dir.

Venice

WDDV(AM)— Feb 1, 1960: 1320 khz; 5 kw-D, 1 kw-N, DA-4. TL: N27 06 20 W82 24 01. Hrs open: 1779 Independence Blvd., Sarasoto, 34234. Phone: (941) 552-4800. Fax: (941) 552-4900. Web Site: www.doveradio.com. Licensee: Citicasters Licenses L.P. Group owner: Clear Channel Communications Inc. (acq 5-4-99; grpsl). Population served: 700,000 Rgnl. Network: Florida Radio Net. Format: Adult standards. Target aud: 25-54; active, upscale, affluent. ♦Sherri Carlson, gen mgr & mktg mgr.

WLTQ-FM—Co-owned with WDDV(AM). Mar 1, 1974: 92.1 mhz; 11.5 kw. Ant 476 ft TL: N27 09 03 W82 27 51. Stereo. 24 Web Site: www.921online.com.135,000 Format: Rock of 80's. Target aud: 25-54; female. ♦Randy Wanek, sls dir; Jeff Lynn, progmg dir; Joanne Jelinek, traf mgr.

Vero Beach

WCZR(FM)— May 29, 1986: 101.7 mhz; 1.48 kw. 471 ft TL: N27 32 46 W80 22 08. Hrs open: Box 0093, Port St. Lucie, 34985. Phone: (772) 335-9300. Fax: (772) 335-3291. Web Site: www.wzzr.com. Licensee: Capstar TX L.P. Group owner: Clear Channel Communications Inc. (acq 8-30-00; grpsl). Population served: 250,000 Format: Talk. ♦John Hunt, gen mgr.

WGYL(FM)—Listing follows WTTB(AM).

WJKD(FM)— 1995: 99.7 mhz; 50 kw. Ant 321 ft TL: N27 46 38 W80 27 17. Hrs open: 1235 16th St., 32960. Phone: (772) 567-0937. Fax: (772) 562-4747. E-mail: jack@997jackfm.com Web Site: www.997jackfm.com. Licensee: Vero Beach FM Radio Partnership. (acq 2-11-2002). Format: 80s, 90s & now. ♦Jim Davis, gen mgr.

WQOL(FM)— Sept 1, 1979: 103.7 mhz; 50 kw. 476 ft TL: N27 33 21 W80 22 08. Stereo. Hrs opn: Box 0093, Port St. Lucie, 34985. Phone: (772) 335-9300. Fax: (772) 335-3291. Web Site: www.wqolfm.com. Licensee: Capstar TX L.P. Group owner: Clear Channel Communications Inc. (acq 8-30-00; grpsl). Natl. Network: Westwood One. Format: Oldies. Target aud: 35-64; baby boomers. ♦John Hunt, gen mgr; Mike Michaels, opns mgr & chief of opns; Heath West, progmg dir & progmg mgr; Mike Kerley, engrg mgr.

***WSCF-FM**— Feb 1, 1990: 91.9 mhz; 15.5 kw. 305 ft TL: N27 38 10 W80 27 59. Hrs open: 24 6767 20th St., 32966. Phone: (772) 569-0919. Phone: (800) 780-0919. Fax: (772) 562-4892. Web Site: www.wscf.com. Licensee: Central Educational Broadcasting Inc. (acq 2-15-89). Natl. Network: USA. Format: Christian, hit radio. ♦Jon Hamilton, gen mgr; Brad Bacon, dev VP; Paul Tipton, progmg dir; Bruce Douglas, news dir.

WTTB(AM)— June 7, 1954: 1490 khz; 1 kw-U. TL: N27 37 12 W80 25 01. Hrs open: 1235 16th St., 32960. Phone: (772) 569-1490. Fax: (772) 562-4747. Web Site: www.wttbam.com. Licensee: Vero Beach Broadcasters LLC (group owner; acq 6-19-00; $5.15 million with co-located FM). Population served: 120,000 Natl. Network: ABC. Rgnl. Network: Florida Radio Net. Format: Music of Your Life, talk. Target aud: General. ♦Jim Davis, gen mgr.

WGYL(FM)—Co-owned with WTTB(AM). November 1970: 93.7 mhz; 50 kw. 475 ft TL: N27 36 04 W80 23 33. Stereo. Phone: (772) 567-0937. Web Site: www.wgylfm.com.350,000 Format: Soft adult contemp, smooth jazz. Target aud: 35-64; upscale adult. ♦Jim Davis, opns mgr.

WZTA(AM)— May 1954: 1370 khz; 1 kw-D. TL: N27 36 01 W80 23 33. Hrs open: Box 0093, Port St. Lucie, 34985. Phone: (772) 335-9300. Fax: (772) 335-3291. Licensee: Capstar TX L.P. Group owner: Clear Channel Communications Inc. (acq 8-30-2000; grpsl). Population served: 11,908 Format: Talk. ♦John Hunt, gen mgr; Mike Michaels, opns mgr.

Watertown

WQLC(FM)— Oct 6, 1990: 102.1 mhz; 9 kw. 531 ft TL: N30 13 58 W82 48 18. Hrs open: 9206 US Hwy. 90 W., Lake City, 32055. Phone: (386) 755-4102. Fax: (386) 752-9861. E-mail: webmaster@powercountry102.com Web Site: www.powercountry102.com. Licensee: Power Country Inc. Population served: 512,000 Format: Hot country. Target aud: 18-54. Spec prog: Gospel 4 hrs wkly. ♦Louis Bolton II, CEO & pres; Bob Hendrickson, gen mgr.

Wauchula

WAUC(AM)— Jan 7, 1958: 1310 khz; 5 kw-D, 500 w-N, DA-2. TL: N27 31 48 W81 49 08. Hrs open: 6 AM-midnight Box 471, 33873. Secondary address: 1310 S. Florida Ave. 33873. Phone: (863) 773-5008. Phone: (863) 773-9282. Fax: (863) 773-2032. E-mail: wauc.radiostation@earthlink.net Licensee: Dora A. Cruz. (acq 12-1-97; $25,000). Population served: 200,000 Rgnl. Network: Florida Radio Net. Format: Mexican. Target aud: 35-54. ♦Robert Ayala, gen mgr.

West Palm Beach

***WAYF(FM)**— Nov 11, 1993: 88.1 mhz; 50 w horiz, 50 kw vert. Ant 1,053 ft TL: N26 35 20 W80 12 44. Hrs open: 24 Box 881, 33402. Secondary address: 800 Northpoint Pkwy., Suite 881 33407. Phone: (561) 881-1929. Fax: (561) 840-1929. E-mail: jim@wayfm.com Web Site: www.wayfm.com. Licensee: WAY-FM Media Group Inc. (group owner). Population served: 3,700,000. Format: Contemp Christian. Target aud: 18-34; teens, young adults & young families. ♦Bob Augsburg, pres; Jim Marshall, stn mgr.

WBZT(AM)— July 31, 1936: 1230 khz; 1 kw-U. TL: N26 43 36 W80 03 03. (CP: 800 w-N). Hrs open: 24 3071 Continental Dr., 33407. Phone: (561) 616-6600. Fax: (561) 616-6677. Web Site: www.wbzt.com. Licensee: Capstar TX L.P. Group owner Clear Channel Communications Inc. (acq 9-27-00; grpsl). Population served: 350,000 Format: Talk. ♦John Hunt, gen mgr.

WEAT-FM— Aug 30, 1969: 104.3 mhz; 100 kw. 1,273 ft TL: N26 34 37 W80 14 32. Stereo. Hrs opn: 701 Northpoint Pkwy., Suite 500, 33407. Phone: (561) 686-9505. Fax: (561) 686-0157. Fax: (561) 686-4043. Web Site: www.sunny1043.com. Licensee: CBS Radio Inc. Group owner: Infinity Broadcasting Corp. (acq 11-13-98; grpsl).

Population served: 800,000 Natl. Rep: Katz Radio. Format: Adult contemp. Target aud: 25-54. ♦Lee K. Strasser, sr VP & gen mgr.

WFTL(AM)— 1948: 850 khz; 50 kw-D, 24 kw-N, DA-2. TL: N26 32 30 W80 44 30. Hrs open: 24 6600 N. Andrews Ave., Fort Lauderdale, 33309. Phone: (954) 315-1515. Fax: (954) 315-1555. Web Site: www.850wftl.com. Licensee: JCE Licenses L.L.C. Group owner: James Crystal Inc. (acq 5-15-98; $1.5 million). Format: News/talk. Target aud: 35 plus. ♦James C. Hilliard, pres; Rick Hindes, CFO; Steve Lapa, gen mgr; Ken Pauli, opns mgr; Tim Reever, gen sls mgr.

WIRK-FM— Aug 1, 1965: 107.9 mhz; 100 kw. Ant 426 ft TL: N26 45 47 W80 12 19. Stereo. Hrs opn: 701 Northpoint Pkwy., Suite 500, 33407. Phone: (561) 686-9505. Fax: (561) 686-0157. Web Site: www.wirk.com. Licensee: CBS Radio Inc. Group owner: Infinity Broadcasting Corp. (acq 11-13-98; grpsl). Format: Hot new country. ♦Lee Strasser, sr VP; Tony Bonvini, gen sls mgr.

WJNO(AM)— July 15, 1947: 1290 khz; 5 kw-U, DA-N. TL: N26 37 55 W80 07 07. (CP: 10 kw-D, 4.9 kw-N. TL: N26 45 50 W80 12 17). Hrs opn: 3071 Continental Drive, 33407. Phone: (561) 616-6600. Fax: (561) 616-6677. Web Site: www.wjno.com. Licensee: Clear Channel Radio Licenses Inc. Group owner: Clear Channel Communications Inc. (acq 9-16-97; grpsl). Population served: 800,000 Natl. Rep: Clear Channel. Format: News/talk, sports. Target aud: Adults; 25-64. ♦John Hunt, gen mgr; Dave Denver, opns dir; Bill Brady, gen sls mgr; Brian Mudd, progmg dir & progmg mgr; Jim Leifer, chief of engrg.

WKGR(FM)—See Fort Pierce

WLVJ(AM)—See Boynton Beach

WRLX(FM)— Dec 13, 1975: 92.1 mhz; 7.2 kw. Ant 498 ft TL: N26 47 58 W80 04 33. Stereo. Hrs opn: 3071 Continental Dr., 33407. Phone: (561) 616-6600. Fax: (561) 616-6677. Web Site: www.classy921.com. Licensee: Clear Channel Broadcasting Licenses Group owner: Clear Channel Communications Inc. (acq 9-27-00; grpsl). Population served: 800,000 Natl. Rep: Clear Channel. Format: Soft adult contemp. Target aud: 35-64. ♦John Hunt, pres & gen mgr; Dave Denver, opns dir & opns mgr.

WRMF(FM)—Palm Beach, 1957: 97.9 mhz; 100 kw. 1,350 ft TL: N26 34 37 W80 14 32. (CP: Ant 417 ft.). Hrs opn: 24 477 S. Rosemary Ave., Suite 302, 33401-5758. Phone: (561) 868-1100. Fax: (561) 868-1111. Web Site: www.wrmf.com. Licensee: PPB Licenses LLC (acq 7-1-02; $70 million). Natl. Rep: McGavren Guild. Law Firm: Latham & Watkins. Format: Adult contemp. Target aud: 25-54; general. ♦Mike Catchall, CEO; Chet Tart, pres & gen mgr; Doris Dupee, CFO; Elizabeth Hamma, sls dir; Dennis Winslow, progmg dir; Tammy Hayes, prom.

***WXEL(FM)**— Nov 24, 1969: 90.7 mhz; 38 kw. Ant 1,115 ft TL: N26 35 20 W80 12 44. Stereo. Hrs opn: 24 Box 6607, 33405. Secondary address: 3401 S. Congress Ave., Boynton Beach 33426. Phone: (561) 737-8000. Fax: (561) 369-3067. E-mail: jcarr@wxel.org Web Site: www.wxel.org. Licensee: Barry Telecommunications Inc. (acq 4-16-97). Population served: 684,000 Natl. Network: NPR. Rgnl. Network: Fla. Pub. Law Firm: Schwartz, Woods & Miller. Wire Svc: AP Format: Class, news info. News staff: 7. Target aud: 35 plus; career oriented (news & info). ♦Jerry Carr, CEO & pres; Bernard Henneberg, CFO & exec VP; Joanna Marie, opns mgr.

WZZR(FM)—Riviera Beach, 1971: 94.3 mhz; 50 kw. Ant 456 ft TL: N26 45 42 W80 04 42. Hrs open: Box 0093, Port St. Lucie, 34985. Phone: (772) 335-9300. Fax: (772) 335-3291. Web Site: www.wzzr.com. Licensee: Clear Channel Broadcasting Licenses Inc. Group owner: Clear Channel Communications Inc. (acq 11-21-97; grpsl). Population served: 83,700 Format: Talk. Target aud: 25-54. ♦Mark Bass, gen mgr; Mike Michaels, opns mgr.

White City

WFLM(FM)— Dec 1, 1993: 104.7 mhz; 25 kw. 328 ft TL: N27 26 05 W80 21 42. Hrs open: 24 6803 S. Federal Hwy., Port St. Lucie, 34952. Phone: (772) 460-9356. Fax: (772) 460-2700. Licensee: Midway Broadcasting Co. Population served: 500,000 Format: Rhythm and blues. Target aud: 18-54. Spec prog: Gospel 20 hrs, jazz 4 hrs, reggae 4 hrs wkly. ♦Alice Lee, pres.

White Springs

WNFS(AM)—Not on air, target date: unknown: 660 khz; 50 kw-D, 250 w-N, DA-2. TL: N30 19 43 W82 49 35. Hrs open: 571 N.W. McClurg Ct., 32096. Phone: (386) 397-4489. Licensee: The Dianne A. Mayfield-Harder Trust. ♦Charles Harder, gen mgr; Diane A. Mayfield-Harder, gen mgr.

Wildwood

WVLG(AM)—Licensed to Wildwood. See Leesburg

Williston

WTMG(FM)— July 1, 1983: 101.3 mhz; 3.5 kw. 433 ft TL: N29 25 04 W82 32 58. Stereo. Hrs opn: 24 100 N.W. 76th Dr., Suite 2, Gainesville, 32607. Phone: (352) 313-3110. Fax: (352) 313-3166 / 3199. E-mail: info@magic1013.com Web Site: www.magic1013.com. Licensee: 6 Johnson Road Licenses Inc. (group owner; (acq 1-5-2007; grpsl). Format: Urban AC. News staff: one; News: 2 hrs wkly. ♦Benjamin Hill, gen mgr.

Wilton Manors

WEXY(AM)— June 1963: 1520 khz; 3.5 kw-D, 250 w-N, DA-N. TL: N26 10 26 W80 09 27. Hrs open: 412 W. Oakland Park Blvd., Fort Lauderdale, 33311. Phone: (954) 561-1520. Fax: (954) 561-9830. Licensee: Multicultural Radio Broadcasting Licensee LLC. Group owner: Multicultural Radio Broadcasting Inc. (acq 4-4-03; $2.75 million). Population served: 3,000,000 Natl. Network: American Urban. Format: Gospel. ♦Arthur Liu, CEO & pres; Jim Glogowski, VP; Doug DeVos, gen mgr & opns mgr; Eduardo Ruedo, gen mgr.

Windermere

WUNA(AM)—See Ocoee

Winter Garden

WLAA(AM)— Feb 22, 2000: 1680 khz; 10 kw-D, 1 kw-N. TL: N28 34 08 W81 31 08. Hrs open: 1801 Clark Rd., Ocoee, 34761. Phone: (407) 296-4747. Fax: (407) 253-2228. Web Site: www.quebuenaorlando.com. Licensee: Rama Communications Inc. (group owner). Format: Rgnl Mexican Sp. ♦Joel Marquez, pres.

Winter Haven

WHNR(AM)—Cypress Gardens, Nov 29, 1958: 1360 khz; 5 kw-D, 2.5 kw-N, DA-2. TL: N28 01 16 W81 42 02. Hrs open: 6 AM-midnight Box 7742, 1505 Dundee Rd., 33883. Phone: (863) 299-1141. Fax: (863) 293-6397. E-mail: info@whnr1360.com Web Site: www.whnr1360.com. Licensee: GB Enterprises Communication Corp. (acq 1-9-2007; $665,000). Population served: 350,000 Rgnl. Network: Florida Radio Net. Format: Urban AC. News staff: one; News: 20 hrs wkly. Target aud: 55 plus. Spec prog: Relg 10 hrs wkly. ♦P.J. Allen, stn mgr & gen sls mgr.

WLKF(AM)—See Lakeland

WPCV(FM)—Licensed to Winter Haven. See Lakeland

WSIR(AM)— Feb 14, 1947: 1490 khz; 1 kw-U. TL: N28 00 50 W81 45 02. Hrs open: 24 665 Southwest Lake Howard Dr., 33880. Phone: (863) 295-9411. Fax: (863) 401-9365. Web Site: www.rejoice1490.com. Licensee: Anscombe Broadcasting Group Ltd. (acq 9-5-01). Population served: 250,000 Rgnl. Network: Florida Radio Net. Format: Gospel / Urban. News: 5 hrs wkly. Target aud: 25 plus. Spec prog: Relg 14 hrs. ♦Steve Reszka, CEO & pres; Joe Fisher, VP, gen mgr & gen mgr; Tony Charles, mus dir.

Winter Park

WLOQ(FM)— 1966: 103.1 mhz; 2.65 kw. 351 ft TL: N28 32 22 W81 26 46. Stereo. Hrs opn: 24 2301 Lucien Way, Suite 180, Maitland, 32751. Phone: (407) 647-5557. Fax: (407) 647-4495. E-mail: frontdesk@wloq.com Web Site: www.wloq.com. Licensee: Gross Communications Corp. (group owner; (acq 1977). Natl. Rep: Interep. Law Firm: Pepper & Corazzini. Format: Jazz. Target aud: 25-54; white collar/professionals. ♦Herbert Paul Gross, pres; John Gross, CFO; Rick Weinkauf, VP & gen mgr; Ken Marks, sls dir & gen sls mgr.

WPRD(AM)— September 1954: 1440 khz; 5 kw-D, 1 kw-N, DA-N. TL: N28 35 18 W81 22 53. Stereo. Hrs opn: 24 222 Hazard St., Orlando, 32804-3030. Phone: (407) 841-8282. Fax: (407) 841-8250. E-mail: wprd1440@hotmail.com Licensee: J & V Communications Inc. (group owner; acq 11-94; $300,000). Law Firm: Larry Perry. Format: Sp, news/talk. Target aud: 25-64; upscale. ♦John Torrado, CEO, pres & gen mgr; Frank Vaught, opns mgr.

***WPRK(FM)**— Dec 10, 1952: 91.5 mhz; 1.32 kw. 89 ft TL: N28 35 40 W81 20 07. Stereo. Hrs opn: 24 Box 2745, Rollins College, 1000 Holt Ave., 32789-4499. Phone: (407) 646-2915. Phone: (407) 646-2241. Web Site: www.rollins.edu/wprk. Licensee: Rollins College. Population served: 300,000 Format: Class, progsv, urban contemp. News: 2 hrs wkly. Target aud: General; non-traditional class and/or rock listeners. Spec prog: Jazz 3 hrs wkly. ♦Dan Seeger, gen mgr; Whitney Coulter, pub affrs dir.

Woodville

WJZT(FM)— September 2003: 97.9 mhz; 6 kw. Ant 328 ft TL: N30 16 30 W84 07 39. Stereo. Hrs opn: 8:30 AM-5:30 PM 435 St. Francis St., Tallahassee, 32301. Phone: (850) 561-8400 (studio). Phone: (407) 227-3642. Fax: (850) 224-1553. E-mail: epetrone@wjztfm.com Web Site: www.wjztfm.com. Licensee: 97.9 WJZTFM Inc. Population served: 124,000 Format: Smooth jazz. Target aud: 25-55. ♦Ernest Petrone, gen mgr; Chris Cooper, stn mgr.

Yankeetown

WXOF(FM)— 1998: 96.3 mhz; 6 kw. 285 ft TL: N29 01 18 W82 41 20. Hrs open: 4554 S. Suncoast Blvd., Homosassa Springs, 34446. Phone: (352) 628-4444. E-mail: staff@citrus95radio.com Web Site: www.citrus953.com. Licensee: WGUL-FM Inc. (acq 1-22-99). Format: Classic rock. ♦Richard Spires, gen mgr.

Zephyrhills

WZHR(AM)— May 9, 1962: 1400 khz; 1 kw-U. TL: N28 16 54 W82 12 30. Hrs open: 24 2360 N.E. Coachman Rd., Clearwater, 33765. Phone: (727) 441-3311. Fax: (727) 441-1300. E-mail: lola@tantalk1340.com Web Site: www.tantalk1340.com. Licensee: Wagenvoord Advertising Group Inc. (group owner; acq 2-13-02). Population served: 2,000,000+ Natl. Network: CNN Radio, CBS Radio. Format: Talk/Gospel. Target aud: 35-64; men & women. Spec prog: CHR 10 hrs wkly. ♦Dave Wagenvoord, CEO & pres; Lola Wagenvoord, VP & gen mgr.

Zolfo Springs

WZZS(FM)— November 1992: 106.9 mhz; 6 kw. 328 ft TL: N27 21 59 W81 47 52. Hrs open: 24 7891 U.S. Hwy. 17 S., 33890. Phone: (863) 494-4111. Fax: (863) 494-4443. E-mail: wzzs@desoto.net Web Site: www.bull.fm. Licensee: Heartland Broadcasting Corp. Population served: 150,000 Rgnl. Network: Florida Radio Net. Format: Country. News staff: one; News: one hr wkly. Target aud: General; DeSoto, Hardee and Highlands counties. Spec prog: Gospel 2 hrs, farm one hr wkly. ♦Harold Kneller Jr., pres; Bryan Hollenbauer, gen mgr.

Georgia

Adel

WDDQ(FM)— October 1979: 92.1 mhz; 6 kw. Ant 298 ft TL: N31 08 15 W83 23 41. Stereo. Hrs opn: 24 3766 Old Clyattvile Rd., Valdosta, 31601. Phone: (229) 896-4572. Fax: (229) 559-1332. Licensee: Adventure Radio Group LLC (acq 12-30-02; $435,000). Format: News/talk info. Target aud: 18-50. ♦Ron Hester, gen mgr & progmg mgr.

Albany

WALG(AM)— 1940: 1590 khz; 5 kw-D, 1 kw-N, DA-2. TL: N31 37 19 W84 09 09. Hrs open: 24 1104 W. Broad Ave., 31707. Phone: (912) 888-5000. Fax: (912) 888-5960. Web Site: www.1590walg.com. Licensee: Cumulus Licensing Corp. Group owner: Cumulus Media Inc. (acq 11-3-98; grpsl). Population served: 150,000 Natl. Network: ABC. Rgnl. Network: Southern Farm. Natl. Rep: Katz Radio. Format: News/talk. Target aud: General. ♦George Francis, pres & VP; Bill Jones, progmg dir; Jenna McKay, pub affrs dir; Joey Falgout, chief of engrg.

WQVE(FM)—Co-owned with WALG(AM). Dec 17, 1972: 101.7 mhz; 3 kw. Ant 300 ft TL: N31 37 15 W84 09 11. Stereo. Web Site: www.wqvealbany.com.250,000 Format: CHR. ♦Ken O'Brien, opns mgr; Mark McGee, progmg dir; Jenna McKay, news dir; Al Crumpton, disc jockey; Dotti Davis, disc jockey; Kurt Baker, disc jockey; Mason Dixon, disc jockey; Staci Cates, disc jockey.

WEGC(FM)—Sasser, 1995: 107.7 mhz; 11.5 kw. 312 ft TL: N31 38 42 W84 21 15. Hrs open: 1104 W. Broad Ave., 31707. Phone: (229) 888-5000. Phone: (229) 878-1077. Fax: (229) 888-5960. Web Site: www.mix107albany.com. Licensee: Cumulus Licensing Corp. Group owner: Cumulus Media Inc. (acq 7-7-98). Natl. Rep: Katz Radio. Format: Adult contemp, lite rock favorites. ♦Gregory Kamishlian, gen mgr.

WGPC(AM)— 1933: 1450 khz; 1 kw-U. TL: N31 34 55 W84 11 58. Stereo. Hrs opn: 24 1104 W. Broad Ave., 31707. Phone: (229) 888-5000. Fax: (229) 888-5960. Web Site: www.thefanalbany.com. Licensee: Cumulus Licensing Corp. Group owner: Cumulus Media Inc. (acq 11-3-98; $2.25 million with co-locatd FM). Population served: 150,000 Natl. Network: CBS, Fox Sports. Rgnl. Network: Ga. Net. Law Firm: Holland & Knight. Format: Easy lstng, all sports. News: 20 hrs wkly. Target aud: 25 plus; middle to upper income. ♦George Francis, pres & gen mgr; Bill Jones, progmg dir; Joey Falgout, chief of engrg.

WKAK(FM)—Co-owned with WGPC(AM). Feb 22, 1963: 104.5 mhz; 100 kw. Ant 981 ft TL: N31 32 57 W84 00 19. Stereo. 24 Web Site: www.cumulus.com.340,590 Format: Country. News: 25 hrs wkly. ♦Claire Peeples, gen sls mgr; Candy O'Reilley, disc jockey; Genna McKay, disc jockey; Mason Dixon, disc jockey.

WJIZ-FM—Listing follows WJYZ(AM).

WJYZ(AM)— November 1952: 960 khz; 5 kw-D, DA. TL: N31 37 06 W84 10 33. Stereo. Hrs opn: 24 809 S. Westover Blvd., 31707. Phone: (229) 439-9704. Fax: (229) 439-1509. E-mail: frankc@wjyz.com Web Site: www.wjyz.com. Licensee: CC Licenses LLC. Group owner: Clear Channel Communications Inc. (acq 7-12-2000; grpsl). Population served: 115,000 Natl. Network: American Urban. Natl. Rep: D & R Radio. Format: Gospel. News: 3.5 hrs wkly. Target aud: 25-54. ♦John Richards, gen mgr; Frank Crapp, progmg dir.

WJIZ-FM—Co-owned with WJYZ(AM). January 1965: 96.3 mhz; 100 kw. Ant 466 ft TL: N31 39 16 W84 10 36. Stereo. Fax: (912) 439-1509. Web Site: www.wjiz.com.100,000 Natl. Network: American Urban. Format: Urban contemp. Target aud: 18-54. ♦John Richards, gen mgr; Adrian Guyton, progmg dir.

WOBB(FM)—Tifton, 1975: 100.3 mhz; 100 kw. 1,100 ft TL: N31 25 49 W83 45 22. Stereo. Hrs opn: 24 809 S. Westover Blvd., 31707. Phone: (229) 439-9704. Fax: (229) 439-1509. E-mail: kurtbaker@clearchannel.com Web Site: www.b100wobb.com. Licensee: CC Licenses LLC. Group owner: Clear Channel Communications Inc. (acq 7-12-2000; grpsl). Population served: 600,000 Natl. Rep: Christal. Law Firm: Bechtel & Cole. Format: Country. News: 2 hrs wkly. Target aud: 25-49. ♦John Richards, gen mgr; Kurt Baker, progmg dir.

WSRA(AM)— July 10, 1962: 1250 khz; 1 kw-D, 53 w-N. TL: N31 37 00 W84 09 32. Hrs open: 2804 N. Jefferson St., 31701. Phone: (229) 432-1250. Fax: (229) 432-1927. Licensee: Livingston Fulton (acq 8-11-2004; $150,000). Population served: 88,000 Format: Sports radio. Target aud: 25-54. ♦Livingston W. Fulton, pres.

***WUNV(FM)**— 1990: 91.7 mhz; 3 kw. 328 ft TL: N31 40 20 W84 03 27. Stereo. Hrs opn:
WJSP-FM Warm Springs 100%.
260 14th St. N.W., Atlanta, 30318-5360. Phone: (404) 685-2690. Fax: (404) 685-2684. E-mail: ask@gpb.org Web Site: www.gpb.org. Licensee: Georgia Public Telecommunications Commission. Natl. Network: PRI, NPR. Wire Svc: AP Format: Class, news. News staff: 6; News: 40 hrs wkly. Target aud: 35 plus; professional, college educated. ♦Nancy Hall, CEO; Bonnie Bean, CFO; Ryan Fowler, opns mgr; Mandy Wilson, prom dir; St. John Flynn, progmg mgr; Susanna Capelouto, news dir.

***WWXC(FM)**— July 12, 1990: 90.7 mhz; 5.5 kw. Ant 305 ft TL: N31 38 42 W84 21 15. Hrs open: 16 Box 9, Sasser, 39885. Phone: (229) 698-3473. Fax: (229) 874-5015. Licensee: Lamad Ministries Inc. (acq 7-89). Format: Relg. News: 4 hrs wkly. Target aud: 35 plus. ♦C. William Eidenire, pres; Eric Eidenire, gen mgr.

Alma

WAJQ(AM)— October 1957: 1400 khz; 1 kw-U. TL: N31 31 50 W82 27 45. Hrs open: 24 Drawer F, 208 Douglas St., 31510. Phone: (912) 632-1000. Fax: (912) 632-9696. Licensee: Blueberry Broadcasting Co. Inc. (acq 10-5-94; $12,000 with co-located FM; FTR: 10-31-94) Population served: 35,000 Natl. Network: CNN Radio. Rgnl. Network: Ga. Net. Format: Southern gospel. News staff: one; News: 6 hrs wkly. Target aud: General. Spec prog: Farm 2 hrs wkly. ♦Debra Deen, gen mgr.

WAJQ-FM— May 14, 1987: 104.3 mhz; 4.5 kw. 371 ft TL: N31 36 26 W82 32 46. Stereo. Format: Country. Target aud: General. ♦Bob Sass, prom dir.

Alpharetta

WLTA(AM)— Aug 25, 1986: 1400 khz; 1 kw-U. TL: N34 03 49 W84 16 34. Hrs open:
Rebroadcasts WNIV(AM) Atlanta 80%.
2970 Peachtree Rd. N.W., Suite 700, Atlanta, 30305. Phone: (404) 365-0970. Fax: (404) 816-0748. E-mail: wniv@wniv.com Web Site: www.wniv.com. Licensee: South Texas Broadcasting Inc. Group owner: Salem Communications Corp. (acq 11-17-99; $8 million with WNIV(AM) Atlanta). Population served: 300,000 Law Firm: Booth, Freret, Imlay & Tepper. Format: Relg, talk/news. Target aud: 25-49; upper middle to upper income. ♦Mike Moran, gen mgr; Jeff Carter, opns dir.

Ambrose

WDMG-FM— December 1983: 97.9 mhz; 3.5 kw. Ant 316 ft TL: N31 31 51 W82 54 34. Stereo. Hrs opn: 24 1931 GA Hwy. 32 E., Douglas, 31533. Phone: (912) 389-0995. Fax: (912) 383-8552. Licensee: Broadcast South LLC. Group owner: Black Crow Media Group LLC (acq 11-6-2006; grpsl). Rgnl. Network: Jones Satellite Audio, ABC. Format: Classic rock. ♦John Higgs, gen mgr.

Americus

***WBJY(FM)**— 2002: 89.3 mhz; 65 kw vert. Ant 613 ft TL: N31 38 22 W83 44 58. Hrs open: P O Drawer 2440, Tupelo, MS, 38803. Phone: (662) 844-8888. Fax: (229) 567-9045. Web Site: www.afr.net. Licensee: American Family Association. Group owner: American Family Radio Format: Christian. ♦Marvin Sanders, gen mgr.

WDEC-FM— Sept 12, 1964: 94.7 mhz; 25 kw. 328 ft TL: N31 53 52 W84 18 53. Stereo. Hrs opn: Box 727, 31709. Secondary address: 1028 Adderton St. 31719. Phone: (229) 924-1390. Fax: (229) 928-2337. Web Site: www.americusradio.com. Licensee: Sumter Broadcasting Co. Group owner: Sumter Broadcasting Co. Inc. (acq 1994; with co-located AM). Population served: 70,000 Natl. Rep: Rgnl Reps. Format: Hot adult contemp. News staff: 2; News: 2 hrs wkly. Spec prog: Black 5 hrs, farm one hr wkly. ♦Steve Lashley, pres & gen mgr; Thurston Clary, progmg dir.

***WFRP(FM)**— 2005: 88.7 mhz; 4.2 kw. Ant 230 ft TL: N32 05 34 W84 16 56. Hrs open:
Rebroadcasts WBFR(FM) Birmingham, AL 100%.
290 Hegenberger Rd., Oakland, CA, 94621. Phone: (510) 568-6200. Fax: (510) 633-7983. Web Site: www.familyradio.com. Licensee: Family Stations Inc. Format: Relg. ♦Stanley Jackson, gen mgr.

WISK(AM)— Aug 28, 1962: 1390 khz; 5 kw-D. TL: N32 04 51 W84 15 20. Hrs open: Sunrise-sunset Box 727, 1028 Adderton St., 31709. Phone: (229) 924-1390. Phone: (229) 924-6500. Fax: (229) 928-2337. Web Site: www.americusradio.com. Licensee: Sumter Broadcasting Co. Inc. (group owner) Population served: 28,000 Rgnl. Network: Ga. Net. Format: Oldies. News staff: one. Target aud: 19-65. ♦Steve Lashley, gen mgr; Donnie McCreary, news dir.

WISK-FM— September 1973: 98.7 mhz; 25 kw. 302 ft TL: N32 04 51 W84 15 20. Stereo. 24 60,000 Natl. Rep: Rgnl Reps. Format: Country.

Ashburn

***WFFM(FM)**— December 1989: 105.7 mhz; 6 kw. Ant 328 ft TL: N31 41 17 W83 38 38. Hrs open: 400 Dunbar Lane, Albany, 31707. Phone: (229) 776-9565. Fax: (229) 446-9279. Licensee: Educational Media Foundation. (group owner; (acq 4-30-2007; $615,000 with WRXZ(FM) Sylvester). ♦Audra Coley, opns mgr; Mary Bateman, sls dir.

Athens

WBKZ(AM)—Jefferson, Sept 15, 1984: 880 khz; 5 kw-D. TL: N34 00 52 W83 27 11. Hrs open: 6am - 6pm Box 88, 30603. Secondary address: 165 S. Milledge Ave. 30605. Phone: (706) 548-8800. Fax: (706) 549-8800. E-mail: mbgeter@aol.com Licensee: Brown Broadcasting System Inc. (acq 9-29-93; $270,000; FTR: 10-18-93). Format: Gospel/ R&B/ Talk. ♦Melvin Geter, gen mgr & opns mgr.

WFSH-FM—Licensed to Athens. See Atlanta

WGAU(AM)— May 1, 1938: 1340 khz; 1 kw-U. TL: N33 56 28 W83 24 13. Hrs open: 24 850 Bobbin Mill Rd., 30606. Phone: (706) 549-1340. Fax: (706) 353-1220. E-mail: wgau@negia.net Web Site: www.1340wgau.com. Licensee: Southern Broadcasting of Pensacola Inc. Group owner: Southern Broadcasting Companies Inc. (acq 8-6-99). Population served: 175,000 Natl. Network: ABC, CNN Radio. Law Firm: Leventhal, Senter & Lerman. Format: News/talk. News staff: 3. Target aud: 25 plus; educated, middle to upper-income, news & info oriented. Spec prog: Univ. of Georgia sports, Atlanta Braves. ♦Paul Stone, pres & VP; Scott Smith, opns mgr; Matt Caesar, progmg dir; Tim Bryant, news dir.

***WMSL(FM)**— October 1987: 88.9 mhz; 20 kw. 315 ft TL: N33 54 25 W83 29 35. Stereo. Hrs opn: 24 2121 Ruth Jackson Rd., Bogart, 30622. Phone: (770) 725-8890. Fax: (770) 725-0889. E-mail: gm@wmsl.fm Web Site: www.wmsl.fm. Licensee: Prince Avenue Baptist Christian School. Population served: 144,000 Natl. Network: USA. Law Firm: Garvey, Schubert & Barer. Format: Contemp Christian. News staff: one; News: 11 hrs wkly. Target aud: 25-54; women. ♦Jim Hutto, gen mgr; George McKay, opns dir; Dianne Hutto, mktg dir; Nathan Collins, progmg dir; James Hutto, mus dir; Mitch Kimbrell, news dir.

WPUP(FM)—Royston, Dec 1, 1988: 103.7 mhz; 25 kw. 328 ft TL: N34 14 13 W83 16 03. (CP: TL: N34 14 13 W83 31 48). Stereo. Hrs opn: 24 1010 Tower Pl., Bogart, 30622. Phone: (706) 549-6222. Fax: (706) 353-1967. Web Site: www.rock1037fm.com. Licensee: Southern Broadcasting of Athens Inc. Group owner: Southern Broadcasting Companies Inc. (acq 2-21-97; grpsl). Population served: 190,000 Format: AAA. News staff: 2. Target aud: 55-49. ♦Paul Stone, pres, gen mgr & stn mgr; Scott Smith, stn mgr; Kevin Steele, progmg dir.

WRFC(AM)— May 1, 1948: 960 khz; 5 kw-D, 2.5 kw-N, DA-N. TL: N33 59 58 W83 26 00. Stereo. Hrs opn: 24 1010 Tower Pl., Bogart, 30622. Phone: (706) 549-6222. Fax: (706) 353-1967. Web Site: www.960theref.com. Licensee: Southern Broadcasting of Athens Inc. Group owner: Southern Broadcasting Companies Inc. (acq 2-21-97; grpsl). Population served: 250,000 Natl. Network: ESPN Radio. Format: Sports, Talk. News staff: 3; News: 10 hrs wkly. Target aud: 25-54; community-minded adults. Spec prog: Black 15 hrs wkly. ♦Paul Stone, pres & gen mgr; David Johnston, progmg dir; Scott Smith, opns mgr & chief of engrg.

***WUGA(FM)**— Aug 28, 1987: 91.7 mhz; 6 kw. 328 ft TL: N33 55 13 W83 14 46. Stereo. Hrs opn: 24 1197 S. Lumpkin St., Ste. 138, 30602. Secondary address: Georgia Public Radio (HQ), 260 14th St. N.W., Atlanta 30318. Phone: (706) 542-9842. Fax: (706) 542-6718. E-mail: wuga@uga.edu Web Site: www.wuga.org. Licensee: Georgia Public Telecommunications Commission. Natl. Network: PRI, NPR, AP Radio. Rgnl. Network: Peach State Public Radio. Format: Class, news. News staff: 3; News: 5 hrs wkly. Target aud: 35-65; college educ. Spec prog: Folk 4 hrs, jazz 4 hrs wkly. ♦Steve Bell, gen mgr; Michael Cardin, opns mgr; Robb Holmes, progmg mgr; Mary K. Mitchell, news dir.

***WUOG(FM)**— Oct 16, 1972: 90.5 mhz; 9.5 kw. 180 ft TL: N33 57 00 W83 22 02. (CP: 26 kw, ant 179 ft. TL: N33 56 59 W83 22 58). Stereo. Hrs opn: 24 Box 2065, Tate Student Ctr., University of Georgia, 30602. Phone: (706) 542-7100. Fax: (706) 542-0070. Web Site: www.wuog.org. Licensee: University of Georgia. Population served: 49,457 Rgnl. Network: Ga. Net. Format: Alternative. News staff: 3; News: 4 hrs wkly. Target aud: 18-25; students & faculty of Univ. ♦Erin White, gen mgr; Carrie Mumah, prom dir; Mary Beth Ross, news dir; Steven Swigart, pub affrs dir; Wilbur Harrington, chief of engrg.

WXAG(AM)— June 10, 1957: 1470 khz; 1 kw-D, 176 w-N. TL: N33 59 14 W83 20 17 (D), N33 55 03 W83 22 39 (N). Hrs open: 855 Sunset Dr., Suite 16, 30606. Phone: (706) 552-1470. Fax: (706) 425-0847. Licensee: Mecca Communications Inc. (acq 9-8-94; FTR: 9-26-94). Format: Gospel. ♦Michael Thurmond, gen mgr.

Atlanta

***WABE(FM)**— Sept 13, 1948: 90.1 mhz; 96 kw. Ant 821 ft TL: N33 45 32 W84 20 07. Stereo. Hrs opn: 24 740 Bismark Rd. N.E., 30324. Phone: (678) 686-0321. Fax: (678) 686-0356. Web Site: www.wabe.org. Licensee: Board of Education of the City of Atlanta. Population served: 2,950,000 Natl. Network: NPR, PRI. Law Firm: Schwartz, Woods & Miller. Wire Svc: AP Format: News/talk, class. News staff: 3; News: 9 hrs wkly. Target aud: 35-54; news advocates, class music enthusiasts. Spec prog: Jazz 5 hrs wkly. ♦Milton Clipper, CEO & pres; Irene Wreen, CFO; Earl Johnson, VP, gen mgr & stn mgr; Lisa Williams, dev dir & traf mgr. Co-owned TV: *WPBA-TV affil

WAEC(AM)— 1947: 860 khz; 5 kw-D, 500 w-N. TL: N33 43 45 W84 19 19. Hrs open: 1465 Northside Dr., Suite 218, 30318. Phone: (404) 355-8600. Fax: (404) 355-4156. E-mail: info@love86.com Web Site: www.love860.com. Licensee: WAEC License L.P. (acq 10-29-99). Population served: 3,000,000 Format: Contemp Christian, Relg. Total Christian community. ♦George Beasley, chmn; Brian Beasley, pres & exec VP; Caroline Beasley, CFO; Chris Edmonds, gen mgr.

WAFS(AM)— Sept 1, 1955: 1190 khz; 10 kw-D. TL: N33 48 35 W84 21 14. Stereo. Hrs opn: Sunrise-sunset 2970 Peachtree Rd. N.W., Suite 700, 30305. Phone: (404) 365-0970. Fax: (404) 816-0748. E-mail: wniv@wniv.com Licensee: South Texas Broadcasting Inc. Group owner: Salem Communications Corp. (acq 4-4-2000; $8 million). Population served: 1,495,039 Format: Southern gospel. Target aud: 35-64. ♦Mike Marmon, gen mgr; Jeff Carter, opns mgr.

WALR(AM)— Nov 20, 1965: 1340 khz; 1 kw-U. TL: N33 44 56 W84 24 26. Hrs open: 24 3535 Piedmont Rd., Bldg. 14, Suite 1200, 30305. Phone: (404) 688-0068. Fax: (404) 995-4045. E-mail: scottmcfarlane@680thefan.com Web Site: www.talkradio1340.com. Licensee: Dickey Broadcasting Co. (group owner; (acq 8-31-2000; grpsl). Population served: 496,973 Natl. Rep: McGavren Guild. Law Firm: Holland & Knight. Wire Svc: AP Format: Talk. ♦David Dickey, gen mgr; Scott McFarlane, opns dir; Rob Hasson, sls dir.

WALR-FM—See La Grange

WAOK(AM)— Mar 15, 1954: 1380 khz; 5 kw-U, DA-N. TL: N33 45 36 W84 28 45. (CP: 4.2 kw-N). Hrs opn: 1201 Peachtree St., Suite 800, 30361. Phone: (404) 898-8900. Fax: (404) 898-8915. E-mail: slgosnell@cbs.com Web Site: www.waok.com. Licensee: Infinity Broadcasting East Inc. Group owner: Infinity Broadcasting Corp. (acq 1996; grpsl). Population served: 444,700 Natl. Network: CBS. Format: News/talk. News staff: 5; News: 50 hrs wkly. Target aud: 25-54. ♦Mel Karmazin, CEO & pres; Monique McCoy, VP & prom mgr; Val Carolin, gen mgr & sls dir; Rick Caffey, mktg mgr; Tasha Brown, prom dir & pub affrs dir; Tasha Love, progmg dir; Tony Brown, progmg dir; Sid Daniel, chief of engrg; Brenda Yelling, rsch dir; Manny Danaie, traf mgr; Katrina Noles, spec ev coord; Jean Ross, news rptr; Linda Looney, reporter.

WVEE(FM)—Co-owned with WAOK(AM). July 1, 1948: 103.3 mhz; 100 kw. 1,022 ft TL: N33 45 35 W84 20 07. Stereo. Web Site: www.v-103.com.496,973 Natl. Network: Westwood One. Format: Urban contemp. News staff: 2. Target aud: 18-49. ♦Monique McCoy, prom mgr; Denise Dunbar, progmg dir & reporter; Linda Looney, mus dir & news rptr; Tasha Love, mus dir; Jean Ross, news dir; Manny Danaie, traf mgr; Katrina Noles, spec ev coord.

WATB(AM)—See Cumming

***WCLK(FM)**— Apr 10, 1974: 91.9 mhz; 6 kw. Ant 308 ft TL: N33 44 56 W84 24 26. Stereo. Hrs opn: 111 James P. Brawley Dr. S.W., 30314. Phone: (404) 880-8284. Phone: (404) 880-8278. Fax: (404) 880-8869. E-mail: wclkfm@cau.edu Web Site: www.wclk.com. Licensee: Clark Atlanta University. Population served: 180,000 Natl. Network: NPR, PRI. Format: Jazz. News staff: 16. Target aud: 25-49; upscale, college educated, primarily African American. Spec prog: Gospel 17 hrs, reggae 3 hrs, blues 3 hrs, info/talk 12 hrs wkly. ♦Wendy Williams, gen mgr; Tammy Nobles, stn mgr; Glen Simmonds, opns mgr; Shelley Trotter, prom mgr & pub affrs dir; John Armwood, progmg dir; Renee Williams, mus dir; Gary Owens, chief of engrg; Traci Ross, traf mgr; Rose Holmes, edit mgr.

WCNN(AM)—North Atlanta, Dec 4, 1967: 680 khz; 50 kw-D, 10 kw-N, DA-2. TL: N33 57 42 W84 15 48. Hrs open: 24 3535 Piedmont Rd., Bldg. 14, Suite 1200, 30305. Phone: (404) 688-0068. Fax: (404) 995-4045. Web Site: www.680thefan.com. Licensee: Dickey Broadcasting Co. (group owner; acq 8-31-00; grpsl). Population served: 496,973 Rgnl. Network: New Format: News, sports. ♦David Dickey, pres & gen mgr; Robert Hasson, gen sls mgr; Scott McFarlane, progmg dir.

WDWD(AM)— July 1, 1938: 590 khz; 5 kw-U, DA-N. TL: N33 49 34 W84 18 56. (CP: 4.5 kw-N, DA-2). Hrs opn: 6th Fl., 210 Interstate Pkwy. N., 30339. Phone: (770) 541-0590. Fax: (770) 952-7461. Web Site: www.radiodisney.com/atlanta. Licensee: Radio Disney Atlanta LLC. Group owner: ABC Inc. (acq 5-17-85; $6.85 million; FTR: 6-3-85). Population served: 496,973 Natl. Network: ABC. Natl. Rep: Interep. Format: Radio Disney. Target aud: Children 6-14 Adults 25-54. Spec prog: Children-kids concerns show (Sundays 7:30 am-8:00 am). ♦Jennifer Chiofalo, stn mgr; Melissa Munro, opns mgr.

WFOM(AM)—See Marietta

WFSH-FM—Athens, January 1964: 104.7 mhz; 24 kw. Ant 1,656 ft TL: N33 52 02 W83 49 44. Stereo. Hrs opn: 24 2970 Peachtree Rd. N.W., Suite 700, 30305. Phone: (404) 365-0970. Fax: (404) 816-0748. Web Site: www.thefishatlanta.com. Licensee: South Texas Broadcasting Inc. Group owner: Salem Communications Corp. (acq 7-27-2000; grpsl). Population served: 300,000 Law Firm: Holland & Knight. Format: Contemp. Christian Music. Spec prog: Gospel 2 hrs wkly. ♦David Koon, gen sls mgr; Mike Stoudt, natl sls mgr & mus dir; Taylor Scott, prom dir & prom mgr; Kevin Avery, progmg dir; C. J. Jackson, chief of engrg.

WFTD(AM)—See Marietta

WGKA(AM)— Mar 17, 1922: 920 khz; 5 kw-D, 488 w-N. TL: N33 48 35 W84 21 23. Hrs open: 2970 Peachtree Rd. N.W., Suite 700, 30305. Phone: (404) 365-0970. Fax: (404) 816-0748. Web Site: www.themighty1190.com. Licensee: Pennsylvania Media Associates Inc. Group owner: Salem Communications Corp. (acq 6-28-2004; $16.4 million). Format: News/talk. Target aud: General. ♦Mike Moran, gen mgr; Jeff Carter, opns mgr; David Koon, gen sls mgr.

WGST(AM)— Apr 7, 1988: 640 khz; 50 kw-D, 1 kw-N, DA-2. TL: N33 45 43 W84 27 29. Hrs open: 24
Rebroadcasts WHEL(FM) Helen 100%.
1819 Peachtree Rd., Suite 700, 30309. Phone: (404) 367-0640. Fax: (404) 367-1100. Web Site: www.wgst.com. Licensee: Citicasters Licenses L.P. Group owner: Clear Channel Communications Inc. Rgnl. Network: Ga. Net. Wire Svc: AP Format: News/talk. Target aud: 25-54. ♦Pat McDonnell, VP & gen mgr; Tim Dukes, opns mgr; Jared Blass, gen sls mgr; Jim Oktavec, mktg dir; Pam Rahal, prom dir & prom mgr; Tom Parker, progmg dir; Paul Mann, news dir; Mike Lawing, chief of engrg.

WUBL(FM)—Co-owned with WGST(AM). Feb 18, 1962: 94.9 mhz; 100 kw. Ant 984 ft TL: N33 48 27 W84 20 26. Stereo. 24 Phone: (404) 367-0949. Fax: (404) 367-9490. Web Site: www.bullatlanta.com.2,000,000 Natl. Network: ABC. Format: Country. ♦Cheryl Ervin, gen sls mgr; Scott Baker, mktg mgr; Louis Kaplan, progmg dir; Steve Goss, mus dir.

WGUN(AM)—Licensed to Atlanta. See Tucker

***WJSP-FM**—Warm Springs, Feb 3, 1985: 88.1 mhz; 100 kw. 975 ft TL: N32 51 08 W84 42 04. Stereo. Hrs opn: 24 260 14th St. N.W., 30318-5360. Phone: (404) 685-2690. Fax: (404) 685-2684. E-mail: ask@gpb.org Web Site: www.gpb.org. Licensee: Georgia Public Telecommunications Commission. Natl. Network: NPR, PRI. Wire Svc: AP Format: News, class, talk. News staff: 4; News: 40 hrs wkly. Target aud: 35-54; NPR-demo. ♦Nancy Hall, CEO; Bonnie Bean, CFO; Ryan Fowler, opns dir; Mandy Wilson, prom dir; St. John Flynn, progmg dir; Susanna Capelouto, news dir & chief of engrg.

WKHX-FM—Marietta, November 1960: 101.5 mhz; 100 kw. Ant 1,079 ft TL: N33 48 26 W84 20 22. Stereo. Hrs opn: 24 210 Interstate North, Suite 100, 30339. Phone: (404) 521-1015. Fax: (404) 499-1015 (news/on-air). Web Site: www.kicks1015.com. Licensee: Radio License Holding II LLC. (acq 6-12-2007; grpsl). Population served: 413,800 Natl. Rep: ABC Radio Sales. Format: Country. News staff: one. Target aud: 25-54. ♦Victor Sansone, pres & gen mgr; Mark Richards, opns mgr; Rick Mack, sls dir; Matt Scarano, gen sls mgr; Nancy Barre, natl sls mgr; Mary Gordon, rgnl sls mgr; Christy Ullman, prom mgr; Mike Macho, mus dir; Glenda Sanders, traf mgr.

WKLS(FM)— Dec 2, 1960: 96.1 mhz; 99 kw. Ant 984 ft TL: N33 48 27 W84 20 26. Stereo. Hrs opn: 24 1819 Peachtree St., Suite 700, 30309. Phone: (404) 325-0960. Fax: (404) 367-1155. Web Site: www.project961.com. Licensee: Citicasters Licenses L.P. Group owner: Clear Channel Communications Inc. (acq 5-4-99; grpsl). Population served: 3,500,000 Format: AOR. News: 7 hrs wkly. Target aud: 25-49; primarily male. ♦Jerry Del Core, gen mgr & gen sls mgr; Mike Wheeler, opns mgr & prom dir; Jeff McMurray, progmg dir; Susan De Bonis, mktg dir & pub affrs dir.

WNIV(AM)— 1948: 970 khz; 5 kw-D, 39 w-N. TL: N33 48 35 W84 21 14. Hrs open: 24 2970 Peachtree Rd. N.W., Suite 700, 30305. Phone: (404) 365-0970. Fax: (404) 816-0748. E-mail: wniv@wniv.com Web Site: www.wniv.com. Licensee: South Texas Broadcasting Inc. Group owner: Salem Communications Corp. (acq 11-17-99; $8 million with WLTA(AM) Alpharetta). Population served: 3,500,000 Law Firm: Cohn & Marks. Format: Christian, talk. News staff: one; News: 7 hrs wkly. Target aud: 25-49; educated adults, upper middle to upper income. ♦Stuart W. Epperson, chmn; Edward G. Atsinger III, pres; Mike Moran, gen mgr; Jeff Carter, opns VP, opns mgr, progmg dir & progmg mgr; David Koon, sls VP & gen sls mgr; Joel Foster, prom dir; C. J. Jackson, chief of engrg; Cynthia Weaver, traf mgr.

WNNX(FM)— November 1963: 99.7 mhz; 100 kw. 1,032 ft TL: N33 46 57 W84 23 20. Stereo. Hrs opn: 780 Johnson Ferry Rd., Suite 500, 30342. Phone: (404) 497-4700. Fax: (404) 497-4735. Web Site: www.99x.com. Licensee: WNNX Lico Inc. Group owner: Susquehanna Radio Corp. (acq 2-28-74). Natl. Rep: McGavren Guild. Format: Alternative. Target aud: 18-49; trend-setting young adults. ♦Mark Fowler, gen mgr; Lisa Kelly, opns dir & natl sls mgr; Leslie Fram, gen sls mgr & progmg dir.

WQXI(AM)— October 1947: 790 khz; 28 kw-D, 1 kw-N, DA-N. TL: N33 48 42 W84 21 13. Stereo. Hrs opn: 3350 Peachtree Rd. N.E., Suite 1610, 30326. Phone: (404) 237-0079. Fax: (404) 231-5923. E-mail: feedback@790thezone.com Web Site: www.790thezone.com. Licensee: Jefferson Pilot Communications Co. (group owner: Big League Broadcasting (acq 3-1-74). Population served: 500,000 Format: Talk, sports. Target aud: 18-54; men. ♦Andrew Saltzman, pres & gen mgr; Neal Maziar, VP; Neal Maziiar, gen mgr; Eric Tepe, opns mgr; Stephen "Steak" Shapiro, opns mgr; Chris Young, gen sls mgr; Jim Heilman, natl sls mgr; Leslie Rosetta Smith, mktg dir; Leslie Hoar, prom mgr; Matt Edgar, progmg dir; Tracie Hardy, traf mgr.

***WRAS(FM)**— Jan 18, 1971: 88.5 mhz; 100 kw. 436 ft TL: N33 41 04 W84 17 23. Stereo. Hrs opn: 24 Box 4048, 30302-4048. Phone: (404) 651-2240. Fax: (404) 463-9535. Web Site: www.wras.org. Licensee: Georgia State University. Population served: 3,500,000 Format: College rock. News: 6 hrs wkly. Target aud: 18-34; college students. Spec prog: Classical 3 hrs, world 3 hrs, reggae 4 hrs, new age 3 hrs, rap/hip-hop 6 hrs wkly. ♦Dr. Kurt Keppler, CEO; Brady Rainey, gen mgr; Michael Valania, prom dir; Andy Hawley, progmg dir; Todd Wiese, mus dir; Tom Taylor, chief of engrg.

***WREK(FM)**— Apr 1, 1968: 91.1 mhz; 40 kw. 340 ft TL: N33 46 41 W84 24 22. Stereo. Hrs opn: 24 Georgia Tech., 350 Ferst Dr., Suite 2224, 30332. Phone: (404) 894-2468. Fax: (404) 894-6872. E-mail: wrek@gatech.edu Web Site: www.wrek.org. Licensee: Radio Communications Board, Georgia Institute of Technology. Population served: 3,500,000 Rgnl. Network: Ga. Net. Format: Progsv, div, Ethnic music. Target aud: General. Spec prog: Experimental 18 hrs, jazz 15 hrs, class 15 hrs wkly. ♦Aakash Jariwala, gen mgr; Jeremy Varner, opns mgr; Steve Fenton, progmg dir.

***WRFG(FM)**— July 15, 1973: 89.3 mhz; 100 kw. 279 ft TL: N33 44 56 W84 24 26. Stereo. Hrs opn: 24 1083 Austin Ave. N.E., 30307. Phone: (404) 523-3471. Fax: (404) 523-8990. E-mail: info@wrfg.org Web Site: www.wrfg.org. Licensee: Radio Free Georgia Broadcasting Foundation Inc. Population served: 496,973 Law Firm: Haley, Bader & Potts. Format: Eclectic. News: 3 hrs wkly. Target aud: 18-45; socially conscious African-Americans. Spec prog: , Indian 3 hrs, Sp 5 hrs wkly. ♦Joan Baptist, stn mgr; Wanique Shabazz, opns dir.

WSB(AM)— Mar 15, 1922: 750 khz; 50 kw-U. TL: N33 50 43 W84 15 12. Hrs open: 24 1601 W. Peachtree St. N.E., 30309. Phone: (404) 897-7500. Fax: (404) 897-7363. Licensee: Cox Radio Inc. Group owner: Cox Broadcasting Population served: 3,500,000 Natl. Rep: Christal. Law Firm: Dow, Lohnes & Albertson. Format: News/talk. News staff: 9; News: 168 hrs wkly. Target aud: 25-54. Spec prog: Relg 3 hrs, minorities one hr wkly. ♦Marc W. Morgan, sr VP; David Meszaros, VP, gen mgr & gen sls mgr; Neal Maziar, sls dir; Neil Williamson, mktg dir; Michael Dobson, prom dir; Pete Spriggs, progmg dir & progmg mgr; Chris Camp, news dir; Mike Kavanagh, pub affrs dir; Charles Kinney, chief of engrg.

WSB-FM— Nov 10, 1944: 98.5 mhz; 100 kw. 1,027 ft TL: N33 45 33 W84 20 05. Stereo. Web Site: www.b985.com. Format: Adult contemp. News staff: one; News: 3.5 hrs wkly. Target aud: 25-54. ♦Will Gara, prom dir; Tom Paleveda, progmg dir; Kelly McCoy, progmg mgr; Nancy Richards, news dir; Alpha Trivette, disc jockey; Kelly Stevens, disc jockey. Co-owned TV: WSB-TV affil

WSBG(FM)—Co-owned with WVPO(AM). Oct 1, 1964: 93.5 mhz; 550 w. Ant 764 ft TL: N40 56 56 W57 09 29. Stereo. Web Site: www.lite935.com. Format: Lite rock. Target aud: 20 plus. Spec prog: Modern rock 3 hrs wkly.

WSTR(FM)—See Smyrna

WTJH(AM)—See East Point

WYZE(AM)— June 1957: 1480 khz; 5 kw-D, 44 w-N. TL: N33 43 25 W84 22 08. Hrs open: 1111 Boulevard S.E., 30312. Phone: (404) 622-7802. Fax: (404) 622-6767. E-mail: am1480wyze@aol.com Web Site: www.wyze1480.com. Licensee: GHB Broadcasting Inc. Group owner: GHB Radio Group Population served: 350,000 Rgnl. Network: Ga. Net. Format: Black gospel. ♦George H. Buck Jr., pres; Jacob E. Bogan, stn mgr.

WZGC(FM)— Sept 1, 1965: 92.9 mhz; 100 kw. 910 ft TL: N33 45 34 W84 23 19. Stereo. Hrs opn: 24 1201 Peachtree St., Suite 800, 30361. Phone: (404) 898-8900. Fax: (404) 843-3541. Web Site: www.929dave.fm. Licensee: CBS Radio Inc. of Atlanta. Group owner: CBS Radio (acq 11-13-98; grpsl). Population served: 400,000 Natl. Network: Westwood One. Format: Classic rock. News staff: one; News: 2 hrs wkly. Target aud: 25-54; upscale baby boomers. ♦Rick Caffey, sr VP & gen mgr; Richard Wallace, gen sls mgr; Bonnie Chapman, natl sls mgr; Michelle Engal, natl sls mgr & progmg dir; Robert Lafore, chief of engrg; Dave Marino, disc jockey; Kaedy Kiely, disc jockey.

Augusta

***WACG-FM**— June 2, 1970: 90.7 mhz; 25 kw. 400 ft TL: N33 24 15 W81 50 19. Stereo. Hrs opn: 24
WJSP, Warm Springs/Columbus, GA, 75%.
260 14th St. N.W., Atlanta, 30318-5360. Secondary address: 2500 Walton Way 30904. Phone: (404) 685-2690. Fax: (404) 685-2684. E-mail: ask@gpb.org Web Site: www.gpb.org. Licensee: Georgia Public Telecommunications Commission. (acq 4-87). Population served: 615,000 Natl. Network: PRI, NPR. Wire Svc: AP Format: Classical, News. News staff: 6; News: 40 hrs wkly. Target aud: 35 plus; upscale, professional, educated, affluent. Spec prog: Jazz 18 hrs wkly. ♦Nancy G. Hall, CEO; Bonnie Bean, CFO; Alan Cooke, gen mgr; Ryan Fowler, opns dir; Mandy Wilson, prom dir; St. John Flynn, progmg mgr; Terrance McKnight, mus dir; Susan Capelonto, news dir; Mark Fehlig, engrg dir.

WBBQ-FM— March 1955: 104.3 mhz; 100 kw. 1,003 ft TL: N33 36 41 W81 56 30. Stereo. Hrs opn: 24 2743 Perimeter Pkwy., Bldg. 100, Suite 200, 30909. Phone: (706) 396-6000. Fax: (706) 396-6010. Web Site: www.wbbq.com. Licensee: Clear Channel Broadcasting Licenses Inc. Group owner: Clear Channel Communications Inc. (acq 12-19-2000; grpsl). Population served: 700,000 Format: Adult contemp, loc news. News staff: 8; News: 7 hrs wkly. Target aud: General. ♦Mark Bass, gen mgr; Mike Kramer, opns mgr; Bobby Boggs, gen sls mgr; Steve Cherry, progmg dir; Earl Welch, chief of engrg; Rob Collins, traf mgr.

WSGF(AM)—Co-owned with WBBQ-FM. Jan 12, 1947: 1340 khz; 1 kw-U. TL: N33 27 46 W82 00 29.24 425,000 Law Firm: Holland & Knight. Format: All sports. Target aud: Children.

WDRR(FM)—See Martinez

WEKL(FM)— Mar 10, 1952: 105.7 mhz; 100 kw. Ant 1,168 ft TL: N33 25 15 W81 50 19. Stereo. Hrs opn: 24 2743 Perimeter Pkwy., Suite 300, 30909. Phone: (706) 396-6000. Fax: (706) 396-6010. Web Site: eagle102.com. Licensee: Clear Channel Broadcasting Licenses Inc. Group owner: Clear Channel Communications Inc. (acq 12-19-2000; grpsl). Population served: 700,000 Law Firm: Holland & Knight. Format: Classic rock. Target aud: 18-44. ♦Mark Bass, gen mgr; Tim Lawandus, gen sls mgr; Amanda Washington, prom dir; Steve Burke, progmg dir.

WFAM(AM)— Mar 10, 1952: 1050 khz; 5 kw. TL: N33 27 21 W81 56 20. Hrs open: 24 552 Laney-Walker Blvd. Ext., 30901. Phone: (706) 722-6077. Fax: (706) 722-7066. E-mail: wfam@wilkinsradio.com Web Site: www.wilkinsradio.com. Licensee: J.J. & B. Broadcasting Inc. Group owner: Wilkins Communications Network Inc. (acq 11-22-96; $330,000). Population served: 1,300,000 Natl. Network: Salem Radio Network. Natl. Rep: Salem. Law Firm: Womble, Carlyle, Sandridge & Rice. Format: Relg. News: 2 hrs wkly. Target aud: 35 plus. ♦Robert L. Wilkins, pres; LuAnn Wilkins, exec VP; Mitchell Mathis, VP; Paul Lindsey, stn mgr; Greg Garrett, opns mgr.

WFXA-FM— July 11, 1968: 103.1 mhz; 3 kw. 299 ft TL: N33 30 00 W81 56 03. Stereo. Hrs opn: 24 Box 1584, 30903. Secondary address: 104 Bennett Ln., North Augusta, SC 29841. Phone: (803) 279-2330. Fax: (803) 279-8149. Licensee: Radio One of Augusta LLC. Group owner: Radio One Inc. (acq 5-30-00; grpsl). Population served: 350,000 Format: Black, urban contemp. Target aud: 25-34; females. ♦Dennis Jackson, gen mgr; Ron Thomas, opns mgr & progmg dir; Dianne Mutimer, gen sls mgr; Lakeshia Collins, news dir; Walter Brumbeloe, chief of engrg; Jamie Langley, traf mgr.

WTHB(AM)—Co-owned with WFXA-FM. May 1960: 1550 khz; 5 kw-D. TL: N33 30 00 W81 56 03.21,200 Natl. Network: American Urban. Format: Gospel. ♦Mary Kingcannon, progmg dir.

WGAC(AM)— 1940: 580 khz; 5 kw-D, 840 w-N, DA-N. TL: N33 30 44 W82 04 48. Hrs open:
Simulcast with WGAC-FM Warrenton.
4051 Jimmie Dyess Pkwy., Agusta, 30909. Phone: (706) 396-7000. Fax: (706) 396-7092. E-mail: wgac@wgac.com Web Site: www.wgac.com. Licensee: WGAC License LLC. Group owner: Beasley Broadcast Group (acq 5-19-92; assumption of debt; FTR: 6-8-92). Population served: 480,000 Natl. Network: CBS. Rgnl. Network: Ga. Net. Natl. Rep: D & R Radio. Format: News/talk, sports. News staff: 5. Target aud: 35-65. Spec prog: Farm 4 hrs, military 3 hrs wkly. ♦George Beasley, chmn; Kent Dunn, VP & gen mgr; Harley Drew, opns dir & progmg dir; Terry Kellems, gen sls mgr; Mary Liz Nolan, news dir; Charlie McCoy, chief of engrg; Diane Underwood, traf mgr; Keith Beckum, farm dir.

***WGPH(FM)**—Vidalia, 1988: 91.5 mhz; 40 kw. 508 ft TL: N32 14 02 W82 28 52. Stereo. Hrs opn: 24 2278 Wortham Ln, Grovetown, 30813. Phone: (706) 309-9610. Fax: (706) 309-9669. E-mail: ctbarinowski@comcast.net Web Site: www.gnnradio.org. Licensee: Augusta Radio Fellowship Institute Inc. Population served: 210,000 Format: Christian. News: 12 hrs wkly. Target aud: General. ♦Clarence Barinowski, pres & gen mgr.

WGUS(AM)— July 1930: 1480 khz; 5 kw-U, DA-N. TL: N33 31 00 W82 00 36. Stereo. Hrs opn:
Simulcast WGUS-FM.
4051 Jimmie Dyess Pkwy., 30909. Phone: (706) 396-7000. Fax: (706) 396-7100. Licensee: WCHZ License LLC. Group owner: Beasley Broadcast Group Inc. (acq 5-3-2000; $800,000 with WGAC-FM Warrenton). Population served: 66,800 Format: Southern Gospel. Target aud: 18-49; well educated. ♦Kent Dunn, gen mgr & chief of opns; Richard Chambers, gen mgr; Chris O'Kelley, progmg dir.

WIBL(FM)—Listing follows WYNF(AM).

WKSP(FM)—See Aiken, SC

WKXC-FM—See Aiken, SC

WKZK(AM)—See North Augusta, SC

***WLPE(FM)**— Nov 17, 1984: 91.7 mhz; 1.15 kw. Ant 589 ft TL: N33 34 21 W81 55 23. Stereo. Hrs opn: 24 2278 Wortham Ln., Grovetown, 30813. Phone: (706) 309-9610. Fax: (706) 309-9669. E-mail: ctbarinowski@comcast.net Web Site: www.gnnradio.org. Licensee: Augusta Radio Fellowship Institute Inc. Population served: 400,000 Format: Christian. News: 12 hrs wkly. Target aud: General. ♦Clarence Barinowski, pres & gen mgr.

WNRR(AM)— November 1993: 1230 khz; 1 kw-U. TL: N33 27 14 W82 01 47. Hrs open: 24 92 Middlesex Ct., Slingerlands, NY, 12159. Secondary address: 903 Broad St 30901. Phone: (518) 439-3982. Fax: (706) 738-1973. E-mail: will@newsradio1230.com Web Site: www.newsradio1230.com. Licensee: Eastern Broadcasting Group Inc. (group owner; (acq 11-3-2003; $425,000). Population served: 350,000 Natl. Network: CNN Radio, Premiere Radio Networks, Westwood One. Format: Talk, sports, news. News staff: 3; News: 24 hrs wkly. Target aud: 25- women; 25+ men. ♦Will Nunley, gen mgr; Teri Exrleben, gen sls mgr.

WRDW(AM)— 2001: 1630 khz; 10 kw-D, 1 kw-N. TL: N33 31 00 W82 00 36. Hrs open: 4051 Jimmie Dyess Parkway, 30909. Phone: (706) 396-7000. Fax: (706) 396-7092. Web Site: wrdw.com. Licensee: WCHZ License LLC. Group owner: Beasley Broadcast Group Inc. (acq 2-23-2000). Format: Sports, news/talk. ♦Kent Dunn, gen mgr; Harley Drew, opns mgr.

WYFA(FM)—Waynesboro, Aug 1, 1991: 107.1 mhz; 6 kw. 328 ft TL: N33 10 42 W81 59 24. Stereo. Hrs opn: 24 1388 Old Waynesboro Rd., Waynesboro, 30830. Phone: (706) 554-3942. Fax: (704) 522-1967. E-mail: wyfa@bbnradio.org Web Site: www.bbnradio.org. Licensee: Bible Broadcasting Network Inc. (group owner; acq 8-26-92; $225,000; FTR: 9-21-92). Format: Relg. ♦George Quick, gen mgr; Richard Johnson, progmg dir.

WYNF(AM)—North Augusta, SC) July 30, 1958: 1380 khz; 4 kw-D, 70 w-N. TL: N33 29 17 W81 56 46. Hrs open: 24 2743 Perimeter Pkwy., Bldg. 100, Suite 200, 30909. Phone: (706) 396-6000. Fax: (706) 396-6010. Licensee: Capstar TX L.P. Group owner: Clear Channel Communications Inc. (acq 12-19-2000; grpsl). Population served: 72,000 Law Firm: Wiley, Rein & Fielding. Format: Classic country. News staff: one; News: 7 hrs wkly. ♦Mark Bass, gen mgr.

WIBL(FM)—Co-owned with WYNF(AM). Nov 11, 1967: 102.3 mhz; 1.5 kw. Ant 666 ft TL: N33 26 15 W82 05 27. Stereo. 24 Web Site: www.bullcountry.com. Format: Country. News staff: one; News: 3 hrs wkly. Target aud: 18-34. ♦Tim Lawandus, gen sls mgr; Robb Tomas, prom dir; Bill West, progmg dir.

Austell

WAOS(AM)— Apr 16, 1968: 1600 khz; 20 kw-D, 67 w-N. TL: N33 48 34 W84 39 25. Hrs open: 24 hrs 5815 Westside Rd., 30106. Phone: (770) 944-0900. Fax: (770) 944-9794. E-mail: gracie@radiolafavorita.com Web Site: www.radiolafavorita.com. Licensee: La Favorita Inc. (group owner; acq 1-24-90). Population served: 600,000 Format: Sp. Target aud: 18 plus; Hispanics in metro Atlanta & northeast GA. ♦Samuel Zamarron, pres & gen mgr; Gracie Zamarron, stn mgr.

WXEM(AM)—Buford, Dec 12, 1957: 1460 khz; 5 kw-D. TL: N34 07 15 W83 58 35. Hrs open: 5815 Westside Rd., 30106. Phone: (770) 944-0900. Fax: (770) 944-9794. E-mail: sammy@radiolafavoita.com Web Site: www.radiolafavorita.com. Licensee: La Favorita Inc. (group owner; acq 6-12-91; FTR: 7-1-91). Population served: 200,000 Format: Sp. Target aud: Hispanic. ♦Samuel Zamarron, CEO, pres & gen mgr; Gracie Zamarron, stn mgr.

Avondale Estates

WMLB(AM)— November 2003: 1690 khz; 10 kw-D, 1 kw-N. TL: N33 48 42 W84 21 37. Stereo. Hrs opn: 24 1100 Spring St., Suite 610, Atlanta, 30309. Phone: (404) 681-9307. Fax: (404) 870-8859. E-mail: listeners@am1160.net Web Site: www.1690wmlb.com. Licensee: JW Broadcasting Inc. (acq 5-18-2006; $12 million). Format: Var. ♦Jeff Davis, gen mgr & chief of opns.

Bainbridge

WBGE(FM)— May 2001: 101.9 mhz; 5.3 kw. Ant 351 ft TL: N30 54 36 W84 33 45. Hrs open: 521 S. Scott St., 39819. Phone: (229) 246-7776. Fax: (229) 246-9995. E-mail: kevin @live1019.com Web Site: www.live1019.com. Licensee: Flint Media Inc. (acq 7-15-2005; $485,000). Format: Hot adult contemp. ♦Kevin Dowdy, gen mgr.

WMGR(AM)— Aug 17, 1947: 930 khz; 5 kw-D, 500 w-N. TL: N30 54 25 W84 33 02. Hrs open: 18 203 W. Shotwell St., 39819. Phone: (229) 246-1650. Fax: (229) 246-1403. E-mail: wmgr@wmgr.net Web Site: www.wmgr.net. Licensee: Decatur Broadcasting Inc. (acq 6-10-2005; for 50% of stock). Population served: 100,000 Rgnl. Network: Ga. Net. Format: Memories from the 60s, 70s, & 80s. News: 10 hrs wkly. Target aud: 30 plus. ♦Coley Voyles, pres & gen mgr.

WRAK-FM— Dec 20, 1967: 97.3 mhz; 100 kw. 1,200 ft TL: N31 09 12 W84 32 42. Hrs open: 24 809 S. Westover Blvd., Albany, 31707. Phone: (229) 439-9704. Fax: (229) 439-1509. Web Site: www.magic973radio.com. Licensee: CC Licenses LLC. Group owner: Clear Channel Communications Inc. (acq 7-11-2000; grpsl). Population served: 900,000 Natl. Rep: Christal. Format: Adult contemp. News: 2 hrs wkly. Target aud: 18-49. ♦John Richards, gen mgr; Jasmine Phoenix, progmg dir.

Barnesville

WBAF(AM)— July 23, 1966: 1090 khz; 1 kw-D. TL: N33 03 13 W84 08 07. Hrs open: 645 Forsyth St., 30204. Phone: (770) 358-1090. Fax: (770) 358-1090. Licensee: Barnesville Broadcasting Inc. Population served: 56,000 Rgnl. Network: Ga. Net. Law Firm: Reddy, Begley & McCormick. Format: C&W, relg. Spec prog: Loc 5 hrs wkly. ♦Charles Waters, pres & gen mgr.

Baxley

WBYZ(FM)—Listing follows WUFE(AM).

WUFE(AM)— December 1954: 1260 khz; 5 kw-D. TL: N31 48 00 W82 24 40. Hrs open: 6 AM-sunset Box 390, 31515. Secondary address: Hwy. 341 W. 31515. Phone: (912) 367-3000. Fax: (912) 367-9779. Licensee: South Georgia Broadcasters Inc. (acq 1-19-82; $240,000; FTR: 2-8-82). Population served: 75,000 Law Firm: Fletcher, Heald & Hildreth. Format: Relg. News: 6 hrs wkly. Target aud: General. ♦Al Graham, pres & pub affrs dir; Peggy C. Miles, gen mgr, gen sls mgr, prom mgr & adv mgr; Larry Ring, chief of engrg.

WBYZ(FM)—Co-owned with WUFE(AM). July 1983: 94.5 mhz; 100 kw. 1,014 ft TL: N31 47 10 W82 27 03. Stereo. 24 Box 390, 31515. Secondary address: 4005 Golden Isles W. 31515. E-mail: peggy@wbyz94.com Web Site: www.wbyz.com. Format: Modern country. Target aud: 20-55; those with buying power. ♦Peggy C. Miles, mktg mgr; Al Graham, progmg mgr; Cody West, mus dir; Cole Younger, news dir & traf mgr.

Blackshear

WFNS(AM)— Mar 10, 1961: 1350 khz; 2.5 kw-D, 125-N. TL: N31 18 44 W82 14 00. Hrs open: 1766 Memorial Dr. Suite 1, Waycross, 31501. Phone: (912) 285-5002. Fax: (912) 285-3877. E-mail: wgaradio@yahoo.com Web Site: www.fandog.net. Licensee: MarMac Communications LLC. (acq 7-19-2001; $60,000). Natl. Network: CNN Radio. Format: All sports. Target aud: General; families. Spec prog: Atlanta Braves, Hawks, Falcons, Ga Tech. ♦Gary Moss Marmitt, pres & gen mgr.

WKUB(FM)— Dec 1, 1979: 105.1 mhz; 50 kw. Ant 308 ft TL: N31 15 49 W82 17 30. Stereo. Hrs opn: 24 Box 112, 2132 Hwy. 84, 31516. Secondary address: Box 1472, Waycross 31502. Phone: (912) 449-3391. Fax: (912) 449-6284. E-mail: wkub@almatel.net Licensee: Mattox Broadcasting Inc. Population served: 140,000 Natl. Network: ABC. Rgnl. Network: Ga. Net. Natl. Rep: Dora-Clayton. Law Firm: Fletcher, Heald & Hildreth. Format: Country. News: 4 hrs wkly. Target aud: 25 plus. ♦G. Troy Mattox, pres & gen mgr; Jim MIller, gen sls mgr.

Blakely

WBBK(AM)— Oct 22, 1959: 1260 khz; 1 kw-D. TL: N31 21 11 W84 56 50. Hrs open: Box 87, Donalsonville, 31745. Secondary address: Hwy. 62 W. 31745. Phone: (229) 723-2677. Phone: (229) 524-5123. Fax: (229) 723-2678. Fax: (229) 524-2265. E-mail: wbbk@alltel.net Licensee: Styles Media Group LLC. (group owner; (acq 5-13-2004; grpsl). Population served: 50,000 Format: Talk. Target aud: 25-49. ♦Gil Kelley, gen mgr.

WBBK-FM— November 1984: 93.1 mhz; 45 kw. Ant 328 ft TL: N31 17 55 W85 03 18. Stereo. Hrs opn: Box 889, Dothan, 36302. Phone: (334) 792-0047. Fax: (334) 712-9346. Licensee: Magic Broadcasting Alabama Licensing LLC. (acq 5-13-2004; grpsl). Population served: 50,000 Format: Urban contemp. ♦Dan Bradley, gen mgr; J.J. Davis, progmg dir.

Blue Ridge

WPPL(FM)— 1971: 103.9 mhz; 6 kw. 400 ft TL: N34 52 03 W84 20 02. (CP: 6 kw, ant 400 ft.). Stereo. Hrs opn: 24 Box 938, 30513. Secondary address: 333 W. Highland St. 35013. Phone: (706) 632-9775. Fax: (706) 632-5922. E-mail: mcwolf@etcmail.com Web Site: www.mountaincountryradio.com. Licensee: Fannin County Broadcasting Co. Inc. (acq 12-12-97; $200,000 for stock). Population served: 50,000 Rgnl. Network: Ga. Net. Format: Country. Target aud: 25-54; adults. ♦Tim White, pres; Dalton Davis, gen mgr & opns dir.

Bolingbroke

WWWD(FM)— 2005: 102.1 mhz; 4.5 kw. Ant 377 ft TL: N32 54 30 W83 46 37. Hrs open: 6080 Mount Moriah Ext., Memphis, TN, 38115. Phone: (901) 375-9324. Fax: (901) 375-0041. Web Site: www.flinn.com. Licensee: George S. Flinn Jr. ♦George S. Flinn Jr., gen mgr.

Boston

WTUF(FM)— July 18, 1988: 106.3 mhz; 6 kw. 328 ft TL: N30 47 40 W83 46 54. Stereo. Hrs opn: 24 Box 129, Thomasville, 31799. Secondary address: 117 Remington Ave., Thomasville 31792. Phone: (229) 225-1063. Fax: (229) 226-1361. E-mail: lenrob@rose.net Web Site: www.wtufradio.com. Licensee: Boston Radio Co. Population served: 72,000 Natl. Network: AP Network News. Rgnl. Network: Ga. Net, Agri-Net. Natl. Rep: Rgnl Reps. Format: Classic country. Target aud: 18-65; adults. Spec prog: Bluegrass 5 hrs, gospel 7 hrs wkly. ♦Len Robinson, pres, gen mgr & opns mgr.

Bostwick

WMOQ(FM)— 1994: 92.3 mhz; 3 kw. 328 ft TL: N33 44 58 W83 33 30. Hrs open: Box 649, Monroe, 30655. Secondary address: 1610 Launius Rd., Good Hope Phone: (770) 267-0923. Fax: (770) 342-8135. Web Site: www.wmoqfm.com. Licensee: Bostwick Broadcasting Group Inc. Format: Classic country. ♦B.R. Anderson, Sr., pres & gen mgr; David Malcolm, gen mgr & opns mgr.

Bowdon

WBZY(FM)— Dec 9, 1996: 105.3 mhz; 61 kw. Ant 1,204 ft TL: N33 24 41 W84 49 48. Hrs open: 1819 Peachtree Rd., N.E., Ste 700, Atlanta, 30309. Phone: (404) 962-7000. Web Site: 1053elpatron.com. Licensee: CC Licenses LLC. Group owner: Clear Channel Communications Inc. (acq 11-24-2000; at least $7 million). Format: Rgnl Mexican. ♦Jerry DelCore, gen mgr.

Bremen

WGMI(AM)— October 1957: 1440 khz; 2.5 kw-D. TL: N33 42 56 W85 09 34. Hrs open: 24 613 Tallapoosa St., 30110. Phone: (770) 537-0840. Phone: (770) 537-9464. Fax: (770) 537-0220. E-mail: wgmi@wgmiradio.com Web Site: www.wgmiradio.com. Licensee: Garner Ministries Inc. (acq 11-10-93; $150,000; FTR: 11-29-93). Population served: 116,492 Law Firm: Reddy, Begley & McCormick. Format: Relg, southern gospel, Christian country. News staff: one; News: 3 hrs wkly. Target aud: 25-54; majority married women with children. ♦Horace Garner, CEO & stn mgr; Peggy Garner, opns mgr.

Broxton

WULS(FM)— Nov 1, 1993: 103.7 mhz; 6 kw. 328 ft TL: N31 33 26 W82 52 10. Hrs open: 702 N. Madison Ave., Douglas, 31533. Phone: (912) 384-9857. Fax: (912) 384-0016. Licensee: WULS Inc. Format: Southern gospel. ♦Wyndel Bunnsed, gen mgr.

Brunswick

***WAYR-FM**— 1996: 90.7 mhz; 2.3 kw. 312 ft TL: N31 11 39 W81 29 30. Stereo. Hrs opn: 2500 Russell Rd., Green Cove Springs, FL, 32043. Phone: (904) 272-1111. Fax: (904) 284-2501. E-mail: lstephens@wayradio.org Web Site: www.wayradio.org. Licensee: Good Tidings Trust Inc. (acq 3-13-98; $100,000). Format: Christian. Target aud: 45-70. ♦Bill Tidwell, pres & gen mgr; Luke Stephens, gen mgr.

WBGA(FM)—Listing follows WMOG(AM).

WGIG(AM)— Mar 5, 1949: 1440 khz; 5 kw-D, 1 kw-N, DA-N. TL: N31 10 07 W81 32 14. Hrs open: 3833 U.S. Hwy. 82, 31523. Phone: (912) 267-1025. Fax: (912) 264-5462. E-mail: ryfun@adelphia.net Web Site: www.1440wgig.net. Licensee: Qantum of Brunswick License Co. LLC. Group owner: Qantum Communications Corp. (acq 7-2-03; grpsl). Population served: 75,000 Natl. Network: CBS. Rgnl. Network: Ga. Net. Format: News/talk. ♦Jonathan Havens, gen mgr; Scott Rygun, opns dir & opns mgr.

WHFX(FM)—See Darien

WMOG(AM)— June 1940: 1490 khz; 1 kw-U. TL: N31 09 55 W81 28 28. (CP: 600 w-U, TL: N31 09 42 W81 28 28). Hrs opn: 24 3833 Hwy. 82, 31523. Phone: (912) 267-1025. Fax: (912) 264-5462. Licensee: Qantum of Brunswick License Co. LLC. Group owner: Qantum Communications Corp. (acq 7-2-2003; grpsl). Population served: 94,000 Natl. Rep: McGavren Guild. Format: Nostalgia, news/talk, sports. News staff: one; News: 20 hrs wkly. Target aud: 35 plus. Spec prog: Black 8 hrs, class one hr wkly. ♦Larry Landrum, gen mgr; Scott Ryfun, progmg dir.

WBGA(FM)—Co-owned with WMOG(AM). Jan 1, 1990: 92.7 mhz; 6 kw. 340 ft TL: N31 09 55 W81 28 28. Stereo. 24 Phone: (912) 265-9300. Natl. Network: ABC. Format: Urban. Target aud: 24-45. ♦Jonthan Havens, gen mgr.

WMUV(FM)— Nov 8, 1965: 100.7 mhz; 62 kw. Ant 1,473 ft TL: N30 49 16 W81 44 14. Stereo. Hrs opn: 24 6440 Atlantic Blvd., Jacksonville, FL, 32216. Phone: (904) 727-9696. Fax: (904) 721-9322. Web Site: www.kool1007.com. Licensee: Renda Broadcasting Corp. Group owner: Renda Broadcasting Corp.-Renda Radio Inc. (acq 1996; $6.5 million with WGNE-FM Palatka, FL). Population served: 1,200,000 Natl. Rep: McGavren Guild. Format: Rhythmic adult contemp. ♦Tony Renda Sr., CEO; Bill Scull, gen mgr; George Sample, gen sls mgr; Stacey Steiner, prom dir; Briggs Bickley, progmg dir; Bob Dillehay, chief of engrg; Judy Riley, traf mgr; Jim Byard, pub svc dir.

WRJY(FM)— June 30, 1994: 104.1 mhz; 4.2 kw. 390 ft TL: N31 11 39 W81 29 30. Hrs open: 24 185 Benedict Rd., 31520. Phone: (912) 261-1000. Fax: (912) 265-8391. Web Site: www.coastalcountry1041.com. Licensee: Golden Isles Broadcasting LLC (acq 3-23-2001; $2.8 million with WXMK(FM) Dock Junction). Natl. Rep: Rgnl Reps. Format: Country. News: 168 hrs wkly. Target aud: 25-54; urban female. ♦Traci Long, gen mgr.

WSFN(AM)— Sept 1, 1966: 790 khz; 500 w-D, 115 w-N, DA-2. TL: N31 08 40 W81 34 56. Hrs open: 7515 Blythe Island Hwy., 31523. Phone: (912) 264-6251. Fax: (912) 264-9991. E-mail: thefanradio@aol.com Web Site: www.fandog.net. Licensee: MarMac Communications L.L.C. (acq 4-98; $350,000). Natl. Network: ABC. Rgnl. Network: Ga. Net. Format: Sports. ♦Gary Moss Marmitt, pres & gen mgr.

WSOL-FM—Licensed to Brunswick. See Jacksonville FL

***WWIO-FM**— Feb 28, 1993: Stn currently dark. 88.9 mhz; 7 kw. 135 ft TL: N31 11 19 W81 28 10. (CP: 5.5 kw, ant 154 ft.). Stereo. Hrs opn: 24
Rebroadcast WHSV-FM Savanah 100%.
260 14th St. N.W., Atlanta, 30318-5360. Phone: (404) 685-2690. Fax: (404) 685-2684. E-mail: ask@gpb.org Web Site: www.gpb.org. Licensee: Georgia Public Telecommunications Commission. Natl. Network: NPR, PRI. Wire Svc: AP Format: Classical, News. News staff: one; News: 51 hrs wkly. Target aud: 45 plus. ♦Nancy G. Hall, CEO; Bonnie Bean, CFO; Ryan Fowler, opns mgr; Mandy Wilson, prom dir; St. John Flynn, progmg mgr; Terrance McKnight, mus dir; Susanna Capelonto, news dir; Mark Fehlig, engrg dir; Gyla Gonzalez, spanish dir.

WWSN(FM)—Waycross, June 3, 1972: 103.3 mhz; 100 kw. 1,100 ft TL: N31 15 42 W82 19 26. (CP: TL: N31 09 22 W81 58 19). Stereo. Hrs opn: 3833 US Hwy. 82, 31525. Phone: (912) 267-1025. Fax: (912) 264-5462. E-mail: ryfun@adelphia.net Web Site: www.sunny103.net. Licensee: Qantum of Brunswick License Co. LLC. Group owner: Qantum Communications Corp. (acq 7-2-03; grpsl). Population served: 250,000 Natl. Rep: McGavren Guild. Format: Adult contemp. Target aud: 25-54. Spec prog: Jazz 5 hrs wkly. ♦Jonathan Havens, gen mgr; Scott Ryfun, opns mgr.

WYNR(FM)—Waycross, Oct 10, 1971: 102.5 mhz; 100 kw. 980 ft TL: N31 09 13 W81 58 00. Stereo. Hrs opn: 24 3833 Hwy. 82, 31525. Phone: (912) 267-1025. Fax: (912) 264-5462. E-mail: joeportagee@hotmail.com Web Site: www.1025wynr.net. Licensee: Qantum of Brunswick License Co. LLC. Group owner: Qantum Communications Corp. (acq 7-2-03; grpsl). Population served: 250,000 Natl. Rep: McGavren Guild. Law Firm: Smithwick & Belendiuk. Format: Country. News staff: one; News: one hr wkly. Target aud: 25-54. ♦Frank Osborne, pres & exec VP; Mike Mangen, CFO; Jonathan Brewster, exec VP; Jonathan Havens, gen mgr; Joe Sousa, stn mgr & opns mgr.

Buckhead

WPMA(FM)— December 2002: 102.7 mhz; 7.5 kw. Ant 594 ft TL: N33 30 10 W83 15 37. Stereo. Hrs opn: 24 2278 Wortham Ln, Grovetown, 30813. Phone: (706) 309-9610. Fax: (706) 309-9669. E-mail: ctbarinowski@comcast.net Web Site: www.gnnradio.org. Licensee: Barinowski Investment Co. L.P. Group owner: Good News Network. Population served: 100,000 Format: Christian. ♦Clarence Barinowski, gen mgr.

Buena Vista

WEAM-FM— June 21, 2001: 100.7 mhz; 2.6 kw. Ant 502 ft TL: N32 20 33 W84 39 18. Hrs open: 24 Box 1998, Columbus, 31902-1998.

Phone: (706) 576-3565. Fax: (706) 576-3683. Licensee: Davis Broadcasting Inc. of Columbus. Group owner: Davis Broadcasting Inc. (acq 7-30-2003). Format: Gospel. ♦Gregory Davis, CEO, CFO & gen mgr.

Buford

WLKQ-FM— Jan 1, 1970: 102.3 mhz; 4.2 kw. Ant 390 ft TL: N34 07 16 W83 58 35. Stereo. Hrs opn: 24 3235 Satellite Blvd., Suite 230, Duluth, 30096. Phone: (770) 623-8772. Fax: (770) 623-4722. Web Site: www.laraza1023.com. Licensee: Davis Broadcasting of Atlanta L.L.C. Group owner: Davis Broadcasting Inc. (acq 9-30-2003; $5.25 million). Population served: 2,000,000 Rgnl. Network: Ga. Net. Law Firm: Kenkel & Associates. Format: Sp. News staff: 2; News: 6 hrs wkly. Target aud: 35-54; upper-middle class professionals. ♦Gregory A. Davis, pres; Brian Barber, gen mgr.

WXEM(AM)—Licensed to Buford. See Austell

Byron

***WPWB(FM)—** 1988: 90.5 mhz; 16.5 kw. 453 ft TL: N32 40 56 W83 22 11. Stereo. Hrs opn: 24 2278 Wortham Ln., Grovetown, 30813-5103. Phone: (706) 309-9610. Fax: (706) 309-9669. E-mail: ctbarinowski@comcast.net Web Site: www.gnnradio.org. Licensee: Augusta Radio Fellowship Institute Inc. Population served: 325,000 Format: Christian. News: 12 hrs wkly. Target aud: General. ♦Clarence Barinowski, pres & gen mgr.

Cairo

WGRA(AM)— October 1949: 790 khz; 1 kw-D. TL: N30 54 08 W84 14 03. Hrs open: Box 120, 39828. Secondary address: 1809 U.S. 84 W. 39828. Phone: (229) 377-4392. Fax: (229) 377-4564. E-mail: jeff@wgra.net Web Site: www.wgra.net. Licensee: Lovett Broadcasting Enterprises Inc. (acq 1-84; $450,000; FTR: 4-23-84). Population served: 35,000 Rgnl. Network: Ga. Net. Natl. Rep: Rgnl Reps. Format: News/talk. Target aud: 30 plus; mainly women. Spec prog: Black 6 hrs wkly. ♦Jeffrey Lovett, pres & gen mgr.

WWLD(FM)— June 1983: 102.3 mhz; 27 kw. Ant 604 ft TL: N30 29 32 W84 17 02. Stereo. Hrs opn: 3411 W. Tharpe St., Tallahassee, FL, 32303-1139. Phone: (850) 201-3000. Fax: (850) 561-8903. Web Site: blazin1023.com. Licensee: Cumulus Licensing LLC. Group owner: Cumulus Media Inc. (acq 10-10-01; $1.5 million including noncompete agreement). Population served: 320,304 Natl. Rep: Katz Radio. Law Firm: Wiley Rein. Format: Black, hip hop/rhythm and blues. Target aud: 18-34. ♦John Columbus, gen mgr; Jay Blaze, progmg dir.

Calhoun

***WCGN(FM)—**Not on air, target date: unknown: 91.3 mhz; 3 kw. Ant 98 ft TL: N34 32 54 W84 51 56. Hrs open: Box 3006, Collegedale, TN, 37315. Phone: (423) 238-4240. Fax: (423) 238-6642. Web Site: www.lifetalk.net. Licensee: Lifetalk Radio Inc. ♦Steven Gallimore, pres.

WEBS(AM)— Nov 1, 1966: 1030 khz; 5 kw-D, 3 w-N. TL: N34 29 25 W84 55 04. Hrs open: Box 1299, 30703. Secondary address: 427 S. Wall St. 30703. Phone: (706) 629-2238. Fax: (706) 629-7092. Licensee: Radio WEBS Inc. (acq 7-1-80). Population served: 36,000 Natl. Network: Jones Radio Networks. Format: Oldies. Target aud: 18-52. Spec prog: Black 2 hrs wkly. ♦Ken D. Payne, pres & gen mgr.

WJTH(AM)— June 16, 1977: 900 khz; 1 kw-D, 266 w-N. TL: N34 27 40 W84 53 44. Hrs open: 24 Box 1119, 30703. Secondary address: 329 Richardson Rd. S.E. 30701. Phone: (706) 629-6397. Fax: (706) 629-8463. E-mail: am900@wjth.com Web Site: www.wjth.com. Licensee: Cherokee Broadcasting Co. Population served: 38,000 Natl. Network: ABC. Rgnl. Network: Ga. Net. Natl. Rep: Rgnl Reps. Format: C&W, loc news & info. News staff: one; News: 20 hrs wkly. Target aud: 18-64; general. Spec prog: Farm one hr, gospel 2 hrs, relg 16 hrs wkly. ♦Sam Thomas, gen mgr; Keith Thomas, stn mgr; Gloria Cooley, gen sls mgr.

Camilla

WZBN(FM)— April 1977: 105.5 mhz; 6 kw. Ant 300 ft TL: N31 18 51 W84 12 18. Stereo. Hrs opn: 24 1104 W. Broad Ave., Albany, 31707. Phone: (229) 888-5000. Fax: (912) 888-5960. Licensee: Cumulus Licensing Corp. Group owner: Cumulus Media Inc. (acq 8-27-99; $675,000). Population served: 200,000 Rgnl. Network: Ga. Net. Natl. Rep: Katz Radio. Format: Urban contemp. News staff: one; News: 3 hrs wkly. Target aud: 25-54; female. Spec prog: Blues 5 hrs, gospel 16 hrs wkly. ♦Paul Bucurel, gen mgr.

Canton

WCHK(AM)— Apr 11, 1957: Stn currently dark. 1290 khz; 5 kw-D, 500 w-N, DA-N. TL: N34 15 08 W84 27 49. Hrs open: 3235 Satellite Blvd., Suite 230, Duluth, 30096. Phone: (770) 623-8772. Fax: (770) 623-4722. Licensee: Davis Broadcasting of Atlanta L.L.C. (acq 1-17-2007; $3.8 million with WNSY(FM) Talking Rock). Rgnl. Network: Ga. Net. ♦Brian Barber, gen mgr.

WWVA-FM— Aug 1, 1964: 105.7 mhz; 20 kw. Ant 781 ft TL: N34 03 58 W84 27 15. Stereo. Hrs opn: 24 1819 Peachtree Rd., N.E., Suite 700, Atlanta, 30309. Phone: (404) 607-1336. Fax: (404) 367-1105. Web Site: www.vivaatlanta.com. Licensee: CC Licenses LLC. Group owner: Clear Channel Communications Inc. (acq 3-29-2004; $31 million). Format: Sp contemp. ♦Ricardo Villalona, gen mgr.

Carrollton

WBTR-FM— 1964: 92.1 mhz; 580 w. 636 ft TL: N33 33 54 W85 01 02. Stereo. Hrs opn: 24 102 Parkwood Cir., 30117-8353. Phone: (770) 832-9685. Fax: (770) 830-1027. Web Site: www.b92country.com. Licensee: WYAI Inc. (acq 6-7-01). Population served: 150,000 Rgnl. Network: Ga. Net. Format: Country. News staff: one; News: 3 hrs wkly. Target aud: 25-49. Spec prog: Black 5 hrs wkly. ♦Steven L. Gradick, pres & gen mgr.

WLBB(AM)— Nov 19, 1975: 1330 khz; 500 w-D. TL: N33 34 17 W85 03 02. Hrs open: 24 808 Newnan Rd., 30117. Phone: (678) 601-1330. Fax: (678) 601-8256. Web Site: www.newstalk1330.com. Licensee: WYAI Inc. (acq 6-7-01; Population served: 16,500 Natl. Network: CBS. Format: News/talk. News staff: 2. ♦Steve Gradick, pres & gen mgr.

***WUWG(FM)—** Feb 19, 1973: 90.7 mhz; 500 w. 494 ft TL: N33 33 50 W85 01 04. Stereo. Hrs opn: 18
WJSP, Warm Springs/Columbus, GA, 75%.
260 14th St. N.W., Atlanta, 30318-5360. Phone: (404) 685-2690. Fax: (404) 685-2684. E-mail: ask@gpb.org Web Site: www.gpb.org. Licensee: Georgia Public Telecommunications Commission (acq 8-9-2004). Population served: 35,000 Natl. Network: NPR, PRI. Rgnl. Network: Peach State Public Radio. Wire Svc: AP Format: Classical, News. News staff: 6; News: 25-30 hrs wkly. Target aud: General; students & area residents. Spec prog: Folk 2 hrs, bluegrass 2 hrs, new age 3 hrs. ♦Nancy G. Hall, CEO; Bonnie Bean, CFO; Kevin Sanders, stn mgr; Ryan Fowler, opns mgr; Mandy Wilson, prom dir; St. John Flynn, progmg mgr; Terrance McKnight, mus dir; Susanna Capelonto, news dir; Mark Fehlig, engrg mgr.

Cartersville

WBHF(AM)— July 17, 1946: 1450 khz; 1 kw-U. TL: N34 11 09 W84 48 13. Hrs open: 24 7 N. Wall St., 30120. Phone: (770) 386-1450. Fax: (770) 382-5390. E-mail: news@wbhfradio.org Licensee: Anverse Inc. (acq 7-5-00). Population served: 15600 Natl. Network: ABC, AP Radio. Wire Svc: AP Format: Oldies, loc news/sports. News staff: 3; News: 10 hrs per wk. ♦Matt Santini, gen mgr; Ernestine Young Jones, opns mgr.

***WCCV(FM)—** Jan 24, 1983: 91.7 mhz; 910 w. 537 ft TL: N34 11 35 W84 45 31. Stereo. Hrs opn: Box 1000, 30120-1000. Phone: (770) 387-0917. Fax: (770) 387-2856. E-mail: onair@ibn.org Web Site: www.ibn.org. Licensee: Immanuel Broadcasting. Format: Relg. ♦Ed Tuten, pres; Jane Tuten, VP; Neil Hopper, gen mgr & progmg dir.

WYXC(AM)— Sept 21, 1961: 1270 khz; 2 kw-D, 187 w-N. TL: N34 12 34 W84 47 49. Hrs open: 24 Box 200399, 30121. Phone: (770) 382-1306. E-mail: info@newstalk1270.com Web Site: www.newstalk1270.com. Licensee: Clarion Communications Inc. (acq 6-14-2005; $500,000). Population served: 350,000 Format: News, talk, sports. News staff: one; News: 15 hrs. wkly. Target aud: 25-55. ♦Charles Shiflett, pres; Jim Adams, gen mgr & stn mgr; Connie Dixon, gen sls mgr & prom dir; Charles Brachel, progmg dir, news dir, traf mgr & sports cmtr; Allen Schmelz, chief of engrg; Manny Garrett, chief of engrg.

Chatsworth

WQMT(FM)— Nov 13, 1976: 98.9 mhz; 6 kw. 311 ft TL: N34 49 42 W84 53 41. Hrs open: 24 613 Silver Cir., Dalton, 30721. Phone: (706) 278-5511. Fax: (706) 278-9917. Web Site: georgia99.com. Licensee: North Georgia Radio Group L.P. Group owner: Clear Channel Communications Inc. (acq 3-20-2006; grpsl). Population served: 200,000 Law Firm: Leventhal, Senter & Lerman. Format: Country. News staff: 3. Target aud: General. ♦Mark Cooper, gen mgr.

Chauncey

WQIL(FM)— Oct 20, 1995: 101.3 mhz; 50 kw. 492 ft TL: N32 21 37 W83 08 28. (CP: 33 kw, ant 413 ft.). Stereo. Hrs opn: 24 Box 130, Dublin, 31040-0130. Phone: (478) 272-4422. Fax: (478) 275-4657. E-mail: macd@wqzy.com Licensee: GSW Inc. (acq 1994; $95,000). Format: Contemp Christian. Target aud: 30 plus; Christians. ♦Ted White, gen mgr.

Clarkesville

WCHM(AM)— December 1989: 1490 khz; 1 kw-U. TL: N34 36 27 W83 32 15. Hrs open: 24 Box 368, 30523. Secondary address: 1331 Washington St. 30523. Phone: (706) 754-6272. Fax: (706) 754-8621. E-mail: northgeorgiaradio@hemc.net Licensee: Brian Rothell. (acq 11-21-95; $70,000). Population served: 57,000 Natl. Network: USA. Law Firm: Reynolds & Manning. Format: Contemp Christian. News staff: one; News: 5 hrs wkly. Target aud: 25-62. ♦Brian Rothell, gen mgr.

WMJE(FM)— 1990: 102.9 mhz; 16 kw. 413 ft TL: N34 29 05 W83 38 24. Stereo. Hrs opn: 24 Box 10, Gainesville, 30503. Secondary address: 1102 Thompson Bridge Rd. N.E., Gainesville 30501. Phone: (770) 532-9921. Fax: (770) 532-0506. Web Site: www.accessnorthga.com. Licensee: JWJ Properties Inc. Group owner: Jacobs Media Corp. (acq 3-19-92). Population served: 500,000 Natl. Network: Westwood One. Format: Var. Target aud: 25-54; females with a median age of 41. ♦John W. Jacobs III, CEO & pres; John W. Jacobs Jr., chmn; Jones P. Andrews, gen mgr & gen sls mgr.

Claxton

WCLA(AM)— July 20, 1958: 1470 khz; 1 kw-D, 260 w-N. TL: N32 10 12 W81 53 56. Hrs open: 6 AM-midnight 316 N. River St., 30417. Phone: (912) 739-9252. Fax: (912) 739-0050. Licensee: Progressive United Communications Inc. (group owner; (acq 2-18-97; $330,000 with co-located FM). Population served: 14,000 Natl. Network: ABC. Format: Oldies. News: ABC, top of the hr. Target aud: 25-60; religious. ♦Danny Swain, opns mgr & sls; Don Jones, gen mgr & traf mgr.

WMCD(FM)— Sept 15, 1972: 107.3 mhz; 25 kw. Ant 328 ft TL: N32 10 01 W81 54 07. Stereo. Hrs opn: 24 Box 958, Statesboro, 30458. Secondary address: 561 E. Olliff St., Statesboro 30458. Phone: (912) 764-5446. Fax: (912) 764-8827. E-mail: wnnswmcd@enia.net Web Site: www.radiostatesboro.com. Licensee: Georgia Eagle Broadcasting Inc. Group owner: Communications Capital Managers LLC (acq 5-21-2007; grpsl). Population served: 25,000 Natl. Network: Westwood One. Rgnl. Network: Ga. Net. Rgnl rep: Rgnl Reps Law Firm: Richard Helmick. Format: Adult contemp. News staff: one; News: 7 hrs wkly. Target aud: 25-50; general. ♦Buddy Horne, gen mgr & progmg dir; Nate Hirsch, gen mgr; Sandi Kirkland, traf mgr.

Clayton

WGHC(AM)— June 28, 1961: 1370 khz; 250 w-D. TL: N34 51 41 W83 24 25. (CP: COL Mount Holly, NC. 870 khz; 5 kw-D. TL: N35 16 25 W80 51 40). Hrs opn: 6 AM-sunset Box 1149, 18 Radio Ln., 30525. Phone: (706) 782-4251. Phone: (706) 782-1041. Fax: (706) 782-4252.

E-mail: rabunradio@hotmail.com Web Site: www.rabunradio.com. Licensee: Sutton Radiocasting Corp. Group owner: Georgia-Carolina Radiocasting Companies (acq 12-14-2001; grpsl). Population served: 25,000 Natl. Network: CBS Radio. Wire Svc: AP Format: Talk, MOR. News staff: one; News: 5 hrs wkly. Target aud: 35 plus. ♦Douglas M. "Art" Sutton Jr., pres; Adam Wright, VP & gen mgr; Timothy C. Stephens, chief of engrg.

WRBN(FM)—Co-owned with WGHC(AM). June 11, 1990: 104.1 mhz; 370 w. Ant 1,296 ft TL: N34 54 24 W83 24 56.24 18 Radio Lane, 30525. E-mail: sky104@rabun.net 150,000 Format: Adult contemp. News staff: one; News: 2 hrs wkly. Target aud: 25-54.

Cleveland

WAZX-FM— 1989: 101.9 mhz; 6 kw. 410 ft TL: N34 33 49 W83 38 26. Stereo. Hrs opn: 1800 Lake Park Dr., Smyrna, 30080. Phone: (770) 436-6171. Fax: (770) 436-0100. Web Site: www.radiolaquebuena.com. Licensee: WAZX-FM Inc. (acq 4-20-01; $60,000 for 80%). Natl. Network: CBS. Format: Rgnl Mexican. Target aud: 21-48. ♦Patty Perec, gen mgr.

WRWH(AM)— Sept 27, 1958: 1350 khz; 1 kw-D. TL: N34 35 11 W83 46 01. Hrs open: 6 AM-sunset Box 181, 30528. Secondary address: 681 Hood St. 30528. Phone: (706) 865-3181. Fax: (706) 865-0421. E-mail: wrwh@alltel.net Web Site: wrwh.com. Licensee: Newsic Inc. (acq 5-17-89). Population served: 13,500 Rgnl. Network: Ga. Net. Law Firm: Reddy, Begley & McCormick. Format: Country, gospel. Target aud: 35 plus. ♦Dean Dyer, pres, gen mgr & gen sls mgr.

Cochran

WDCO(AM)— July 4, 1965: 1440 khz; 1 kw-D. TL: N32 24 43 W83 21 42. Hrs open: 24 7080 Industrial Hwy., Macon, 31216. Phone: (478) 934-4548. Fax: (478) 781-6711. E-mail: taylor@ham.net Licensee: Georgia Eagle Broadcasting Co. Group owner: Communications Capital Managers LLC (acq 5-21-2007; grpsl). Population served: 30,000 Format: Country. Target aud: 18-54; adults. Spec prog: Black 6 hrs, farm 3 hrs, gospel 6 hrs wkly. ♦Rick Humphrey, gen mgr; Carl Strandell, sls dir; Jerry Jennings, natl sls mgr; Kyle Taylor, mus dir; James Gay, chief of engrg.

WDXQ-FM—Co-owned with WDCO(AM). July 4, 1968: 96.7 mhz; 3 kw. 319 ft TL: N32 24 43 W83 21 42. Stereo. Phone: (478) 781-1063.10,291 Natl. Network: ABC. Natl. Rep: Clear Channel. Target aud: Adults 18-54. ♦Carl Strandell, sls VP.

***WMUM-FM**— Feb 4, 1985: 89.7 mhz; 100 kw. 1,010 ft TL: N32 28 11 W83 15 17. Stereo. Hrs opn: 18
WJSP, Warm Springs/Columbus, GA, 100%.
260 14th St. N.W., Atlanta, 30318-5360. Phone: (404) 685-2690. Fax: (404) 685-2684. E-mail: ask@gpb.org Web Site: www.gpb.org. Licensee: Georgia Public Telecommunications Commission. Natl. Network: NPR, PRI. Wire Svc: AP Format: Classical, News. News staff: 6; News: 40 hrs wkly. Adults 35+. ♦Nancy G. Hall, CEO; Bonnie Bean, CFO; Alan Cooke, gen mgr; Ryan Fowler, opns dir; Mandy Wilson, prom dir & prom dir; St. John Flynn, progmg mgr; Terrance McKnight, mus dir; Susanna Capelonto, news dir; Mark Fehlig, engrg dir.

College Park

WWWQ(FM)— April 1947: 100.5 mhz; 12.5 kw. Ant 977 ft TL: N33 45 34 W84 23 19. Stereo. Hrs opn: 24 780 Johnson Ferry Rd., Suite 500, Atlanta, 30342. Phone: (404) 497-4700. Fax: (404) 497-4735. Web Site: www.allthehitsq100.com. Licensee: WNNX Lico Inc. Group owner: Susquehanna Radio Corp. (acq 1-27-97; with co-located AM). Format: Top-40. ♦Mike Fowler, gen mgr; Lisa Kelly, opns mgr; Rob Roberts, progmg dir.

Columbus

WCGQ(FM)—Listing follows WRCG(AM).

WDAK(AM)— August 1940: 540 khz; 4 kw-D, 38 w-N. TL: N32 25 58 W84 57 02. Stereo. Hrs opn: Box 687, 31902. Secondary address: 1501 13th Ave. 31901. Phone: (706) 576-3000. Fax: (706) 576-3010. E-mail: scottmiller@clearchannel.com Web Site: wdakonline.com. Licensee: CC Licenses LLC. Group owner: Clear Channel Communications Inc. (acq 5-9-2003; $2.73 million with WSTH-FM Alexander City, AL). Natl. Network: USA, Westwood One. Format: News. Target aud: 18-49; men. ♦Jim Martin, gen mgr; Brian Waters, opns mgr.

WEAM(AM)— December 1954: 1580 khz; 2.3 kw-D, 1 kw-N, DA-N. TL: N32 27 55 W85 01 22. Hrs open: 20 Box 1998, 31902-1998. Phone: (706) 576-3565. Fax: (706) 576-3683. Licensee: Davis Broadcasting Inc. of Columbus. Group owner: Davis Broadcasting Inc. (acq 4-20-01; $400,000). Population served: 250,000 Natl. Network: USA. Law Firm: Reddy, Begley & McCormick. Format: Sports. News: 15 hrs wkly. Target aud: General. ♦Gregory Davis, CFO & gen mgr.

***WFRC(FM)**— June 14, 1985: 90.5 mhz; 8.5 kw. 248 ft TL: N32 27 37 W85 00 30. Stereo. Hrs opn: 1010 7th Pl., Phenix City, AL, 36867. Phone: (334) 291-0399. Phone: (800) 543-1495. Fax: (510) 633-7983. Web Site: www.familyradio.com. Licensee: Family Stations Inc. (group owner) Format: Relg. Spec prog: Call-in 8 hrs wkly. ♦Harold Camping, pres; Sandra Salewski, opns mgr.

WFXE(FM)—Listing follows WOKS(AM).

WGSY(FM)—Phenix City, AL) Mar 4, 1971: 100.1 mhz; 6 kw. 328 ft TL: N32 30 42 W85 00 41. Stereo. Hrs open: 24 Box 687, 39102. Secondary address: 1501 13th Ave. 31901. Phone: (706) 576-3000. Fax: (706) 576-3010. Web Site: www.sunny100columbus.com. Licensee: CC Licenses LLC. Group owner: Clear Channel Communications Inc. (acq 2-21-2002; grpsl). Population served: 250,000 Natl. Rep: McGavren Guild. Law Firm: Reddy, Begley & McCormick. Format: Adult contemp. Target aud: 25-54; women. ♦Jim Martin, gen mgr; Brian Waters, opns mgr.

WHAL(AM)—Phenix City, AL) 1951: 1460 khz; 4 kw-D, 140 w-N. TL: N32 25 58 W84 57 02. Stereo. Hrs open: 1501 13th Ave., 31901. Secondary address: Box 687 31902. Phone: (706) 576-3000. Fax: (706) 576-3010. E-mail: marshawhitney@clearchannel.com Web Site: www.whal1460.com. Licensee: CC Licenses LLC. Group owner: Clear Channel Communications Inc. (acq 2-21-2002; grpsl). Population served: 220,000 Natl. Rep: McGavren Guild. Format: Sp. Target aud: 25 plus. ♦Jim Martin, gen mgr.

WVRK(FM)—Co-owned with WHAL(AM). Nov 16, 1946: 102.9 mhz; 100 kw. Ant 1,521 ft TL: N32 19 25 W84 46 46. Stereo. E-mail: brianwaters@clearchannel.com Web Site: www.rock103online.com. Format: AOR, classic rock. ♦Jerri Northington, gen sls mgr; Brian Waters, progmg dir.

WJSP-FM—See Atlanta

WOKS(AM)— Mar 2, 1959: 1340 khz; 1 kw-U. TL: N32 27 07 W84 58 25. Hrs open: Box 1998, 31902-1998. Secondary address: 2203 Wynnton Rd. 31906. Phone: (706) 576-3565. Fax: (706) 576-3683. Licensee: Davis Broadcasting Inc. (group owner; acq 7-24-92). Natl. Network: American Urban. Natl. Rep: Katz Radio. Format: Black gold & gospel. Target aud: 35-64. ♦Gregory Davis, pres & gen mgr; Bernie Corcoran, stn mgr; Cheryl Davis, opns VP; Angela Verdejo, gen sls mgr; Michael Soul, progmg dir; Nicole Gates, prom VP & news dir.

WFXE(FM)—Co-owned with WOKS(AM). Sept 22, 1969: 104.9 mhz; 6 kw. 289 ft TL: N32 27 37 W85 00 30. Stereo. 166,565 Format: Urban contemp. ♦Gregory A. Davis, CEO; Bernie Corcoran, CFO.

WRCG(AM)— May 10, 1928: 1420 khz; 5 kw-U, DA-N. TL: N32 29 52 W85 02 48. Hrs open: 24 1353 13th Ave., 31901. Phone: (706) 327-1217. Fax: (706) 596-4600. Web Site: www.wrcg.com. Licensee: ABG Georgia LLC. Group owner: Archway Broadcasting Group (acq 4-25-03; grpsl). Population served: 206,000 Natl. Network: CBS. Natl. Rep: Christal. Format: Sports, news/talk. News staff: one; News: 24 hrs wkly. Target aud: 35 plus; adults with discretionary income. Spec prog: Farm 3 hrs wkly. ♦Chuck Thompson, gen mgr; Bob Quick, opns dir.

WCGQ(FM)—Co-owned with WRCG(AM). July 15, 1966: 107.3 mhz; 100 kw. 1,011 ft TL: N32 27 59 W85 03 23. Stereo. 24 Web Site: www.q1073.com.471,800 Format: Top 40. Target aud: 18-49. ♦Chuck Thompson, mktg dir; Al Haynes, progmg dir.

WSHE(AM)— 1947: 1270 khz; 5 kw. TL: N32 26 16 W85 01 10. Hrs open: 19 Box 687, 31902. Secondary address: 1501 13th Ave. 31901. Phone: (706) 576-3000. Fax: (706) 576-3010. E-mail: scottmiller@clearchannel.com Web Site: www.wmlfonline.com. Licensee: CC Licenses LLC. Group owner: Clear Channel Communications Inc. (acq 2-21-2002; grpsl). Population served: 300,000 Natl. Rep: McGavren Guild. Law Firm: Reddy, Begley & McCormick. Format: Gospel, sports, Latino. News: 20 hrs wkly. Target aud: 35-60. ♦Jim Martin, gen mgr; Brian Waters, opns mgr.

WSTH-FM—See Alexander City, AL

***WTJB(FM)**— Dec 15, 1984: 91.7 mhz; 5 kw. 298 ft TL: N32 25 20 W85 01 50. Hrs open: 6 AM-midnight
Rebroadcasts WTSU(FM) Troy 100%.
Wallace Hall, Troy State Univ., Troy, AL, 36082. Phone: (334) 670-3268. Fax: (334) 670-3934. E-mail: wtsu@troy.edu Web Site: www.troy.edu. Licensee: Troy State University. Natl. Network: NPR, PRI. Format: Class, news. News: 25 hrs wkly. Target aud: General. Spec prog: Children one hr wkly. ♦James Clower, gen mgr; Judy Davis, opns mgr; Fred Azbell, progmg dir.

***WYFK(FM)**— July 1987: 89.5 mhz; 50 kw. 439 ft TL: N32 40 03 W84 57 19. Stereo. Hrs opn: 11530 Carmel Commons Blvd., Charlotte, NC, 28226. Phone: (706) 322-1980. E-mail: wyfk@bnnradio.org Web Site: www.bbnradio.org. Licensee: Bible Broadcasting Network Inc. (group owner) Format: Relg. Target aud: General. ♦Lowell Davey, pres.

Commerce

WJJC(AM)— June 27, 1957: 1270 khz; 5 kw-D. TL: N34 12 57 W83 26 09. Hrs open: 6 AM-sunset Box 379, 30529. Secondary address: 1801 N. Elm St. 30529. Phone: (706) 335-3155. Fax: (706) 335-1905. E-mail: wjjc@alltel.net Web Site: www.wjjc.net. Licensee: Side Communications Inc. (acq 7-31-2007; $240,000). Population served: 100,000 Rgnl. Network: Ga. Net. Natl. Rep: Rgnl Reps. Format: Talk. Target aud: 25-55. Spec prog: Farm 2 hrs, bluegrass 13 hrs wkly. ♦Rob Jordan, gen mgr.

Conyers

WPBS(AM)— November 1979: 1040 khz; 12 kw-D, 5 kw-CH. TL: N33 40 48 W84 01 44. Hrs open: 12 6171 Neely Farm Dr., Norcross, 30092. Phone: (404) 932-5006. E-mail: amkorea2000@yahoo.com Licensee: Pacific Star Broadcasting Inc. (acq 5-11-2005; $5.25 million). Population served: 4,000,000 Format: Spanish, Christian. ♦Charles Kim, gen mgr.

Coosa

WSRM(FM)— 2005: 95.3 mhz; 6 kw. Ant 72 ft TL: N34 11 51 W85 21 21. Hrs open:
Simulcast with WRGA(AM) Rome 100%.
20 John Davenport Dr., Rome, 30165. Phone: (706) 291-9496. Fax: (706) 235-7107. E-mail: info@wrgarome.com Web Site: www.wrgarome.com. Licensee: Coosa Broadcasting Corp. (acq 5-31-2005; $1.1 million). Format: News/talk. ♦Paul Stone, pres; Randy Quick, gen mgr.

Cordele

***WAEF(FM)**— 2001: 90.3 mhz; 11 kw vert. Ant 505 ft TL: N31 38 22 W83 44 58. Hrs open: Box 3206, American Family Radio, Tupelo, MS, 38803. Phone: (662) 844-8888. Fax: (662) 842-6791. Web Site: www.afr.net. Licensee: American Family Association. Group owner: American Family Radio Format: Inspirational Christian. ♦Marvin Sanders, gen mgr.

Cornelia

WCON(AM)— Mar 28, 1953: 1450 khz; 1 kw-U. TL: N34 30 57 W83 32 20. Hrs open: 24 Box 100, 30531. Secondary address: 540 N. Main St. 30531. Phone: (706) 778-2241. Fax: (706) 778-0576. E-mail: wcon@alltel.net Web Site: www.wconfm.com. Licensee: Habersham Broadcasting Co. (acq 2-1-61). Population served: 50,000 Natl. Network: ABC. Rgnl. Network: Ga. Net. Format: Gospel, country. News staff: one. Target aud: Adults. ♦Bobbie C. Foster, pres & gen mgr; John C. Foster, VP; Michael Harvey, news dir; Jimmy Dillard, chief of engrg.

WCON-FM— Mar 27, 1965: 99.3 mhz; 50 kw. 808 ft TL: N34 31 24 W83 40 46. Stereo. 24 E-mail: bobbiefoster@alltel.net Web Site: www.wconfm.com.1,500,000 News staff: one. Target aud: 18 plus.

Covington

WGFS(AM)— Oct 9, 1946: 1430 khz; 3.9 kw-D, 212 w-N. TL: N33 37 14 W83 53 04. Stereo. Hrs opn: 6 AM-7 PM Box 2419, 30015. Secondary address: 1151 Hendricks St. 30014. Phone: (770) 786-1430. Fax: (770) 784-9892. Licensee: Multicultural Radio Broadcasting Licensee LLC. Group owner: Multicultural Radio Broadcasting Inc. (acq 11-21-03; $700,000). Population served: 16,000 Natl. Network: CBS. Rgnl. Network: Ga. Net. Law Firm: Mullin, Rhyne, Emmons & Topel. Wire Svc: CBS Format: Oldies of 50s & 60s. News staff: one; News: 20 hrs wkly. Target aud: General. ♦Arthur Liu, pres; Mike Shumate, stn mgr.

Crawford

WGMG(FM)— April 1990: 102.1 mhz; 10 kw. 328 ft TL: N33 55 18 W83 14 14. Hrs open: 1010 Tower Pl., Bogart, 30622. Phone: (706) 369-7223. Fax: (706) 353-1967. E-mail: mornings@southernbroadcasting.com Web Site: www.magic1021fm.com. Licensee: New Broadcast Investment Properties Inc. Group owner: Southern Broadcasting Companies Inc. (acq 2-11-02). Format: Adult contemp. Target aud: 18-49; general. ♦Paul Stone, pres; Scott Smith, opns mgr; Kevin Steele, progmg dir.

Cumming

WATB(AM)—Decatur, July 19, 1958: 1420 khz; 1 kw-D, 51 w-N, DA-D. TL: N33 47 13 W84 14 53. (CP: 1430 khz; 50 kw-D, 160 w-N, DA-D. Hrs opn: 3589 N. Decatur Rd., Scottdale, 30079. Phone: (404) 508-1420. Fax: (404) 508-8930. E-mail: watb1420@yahoo.com Licensee: Way Broadcasting Licensee LLC (acq 6-13-00; grpsl). Format: Multi-cultural. Target aud: General; ethnic groups from around the world. ♦Benjamin F. Vannoy Jr., gen mgr.

***WWEV-FM**— Dec 4, 1981: 91.5 mhz; 8.9 kw. 960 ft TL: N34 14 13 W84 09 36. Stereo. Hrs opn: 24 Box 248, 30028. Secondary address: 1705 Sawnee Dr. 30040. Phone: (770) 781-9150. Fax: (770) 781-5003. E-mail: wwev@wwev.org Web Site: www.wwev.org. Licensee: Curriculum Development Foundation Inc. Format: Relg. Target aud: 18-49; the family unit. ♦N. Barry Holt, gen mgr & progmg dir; Ray Haynes, prom dir.

Cusseta

WBOJ(FM)—Not on air, target date: unknown: 103.7 mhz; 6 kw. Ant 328 ft TL: N32 18 39 W84 45 45. Hrs open: 7521 Rock Springs Dr., Columbus, 31909. Phone: (706) 324-5850. Licensee: Signature Broadcasting Ltd. ♦Shirley H. Thrasher, gen mgr.

Cuthbert

WCUG(AM)— Dec 1, 1971: 850 khz; 500 w-D. TL: N31 46 26 W84 50 16. Hrs open: Box 348, 39840. Phone: (229) 732-3725. Licensee: Mullis Communications Inc. Population served: 3,972 Rgnl. Network: Ga. Net. Format: Country, Gospel, Oldies. Spec prog: Black 4 hrs, farm 8 hrs wkly. ♦N. Scott Mullis, gen mgr.

***WEBH(FM)**—Not on air, target date: unknown: 91.9 mhz; 6 kw. Ant 203 ft TL: N31 45 18 W84 47 26. Hrs open: Box 2440, Tupelo, MS, 38803-2440. Phone: (662) 844-5036. Fax: (662) 842-7798. Licensee: American Family Association.

Dahlonega

WDGR(AM)— Mar 1, 1982: Stn currently dark. 1210 khz; 10 kw-D. TL: N34 31 45 W84 00 23. Hrs open: USK Broadcasting, 6072 Buford Hwy. N.E., Atlanta, 30340. Phone: (770) 300-0999. Fax: (770) 300-3082. Licensee: USK Broadcasting Inc. (acq 11-25-2003; $500,000). Population served: 400,000 ♦Hye Kim, gen mgr.

WKHC(FM)— Dec 16, 1996: 104.3 mhz; 3.7 kw. Ant 417 ft TL: N34 29 56 W84 08 32. Stereo. Hrs opn: 24 1376 Ben Higgins Rd., 30533. Phone: (706) 867-9542. Fax: (706) 864-4364. E-mail: wkhc@alltel.net Web Site: www.realcountryonline.com. Licensee: Grady W. Turner (group owner; (acq 2-9-2006; $1.3 million). Population served: 250,000 Natl. Network: ABC. Format: Real country. Target aud: 25-54; general. Spec prog: American Indian one hr wkly. ♦Grady W. Turner, gen mgr.

***WNGU(FM)**— 1998: 89.5 mhz; 750 w. 459 ft TL: N34 31 29 W83 59 50. Stereo. Hrs opn:
Rebroadcast WJSP, WARM/SPRIM, Colombus, 100%.
260 14th St. N.W., Atlanta, 30318-5360. Phone: (404) 685-2690. Fax: (404) 685-2684. E-mail: ask@gpb.org Web Site: www.gpb.org. Licensee: Georgia Public Telecommunications Commission. Natl. Network: NPR, PRI. Wire Svc: AP Format: Classical, News. News staff: 6; News: 40 hrs wkly. ♦James Lyle, CEO; Bonnie Bean, CFO; Alan Cooke, gen mgr; Mandy Wilson, prom dir; St. John Flynn, progmg mgr; Terrance McKnight, mus dir; Mark Fehlig, engrg dir; Susanna Capelonto, news rptr.

Dallas

WDPC(AM)— Sept 21, 1979: 1500 khz; 1 kw-D, DA. TL: N33 56 40 W84 49 28. Hrs open: 8451 S. Cherokee Blvd., Suite B, Douglasville, 30134. Phone: (770) 920-1520. Fax: (770) 920-4600. Web Site: www.wordchristianbroadcasting.com. Licensee: Word Christian Broadcasting Inc. (acq 7-96; $25,000). Population served: 600,000 Format: Old time relg, Southern gospel. Target aud: General. ♦Ken Johns, pres, gen mgr, opns dir & progmg dir.

Dalton

WBLJ(AM)— Apr 8, 1940: 1230 khz; 1 kw-U. TL: N34 45 23 W84 57 02. Hrs open: 24 613 Silver Cir., 30721. Phone: (706) 278-5511. Fax: (706) 226-8766. Web Site: wblj1230.com. Licensee: North Georgia Radio Group L.P. Group owner: Clear Channel Communications Inc. (acq 3-20-2006; grpsl). Population served: 84,000 Natl. Network: CBS Radio. Rgnl. Network: Ga. Net. Format: News/talk. News staff: 2; News: 28 hrs wkly. Target aud: 18-54. Spec prog: Relg mus. ♦Mark Cooper, gen mgr; Larry Gibson, progmg mgr.

WDAL(AM)— Oct 1, 1954: 1430 khz; 2.5 kw-D, 72 w-N. TL: N34 47 23 W84 57 12. Stereo. Hrs opn: 24 Box 1284, 30722. Secondary address: 613 Silver Cir. 30721. Phone: (706) 278-5511. Phone: (706) 278-3300. Fax: (706) 278-7966. Licensee: North Georgia Radio Group L.P. Group owner: Clear Channel Communications Inc. (acq 3-20-2006; grpsl). Population served: 100,000 Natl. Network: CBS. Natl. Rep: Rgnl Reps. Law Firm: Leventhal, Senter & Lerman. Format: Mexican. News staff: 3; News: 30 hrs wkly. Target aud: 25-45. ♦Rich Phillips, gen mgr.

WYYU(FM)—Co-owned with WDAL(AM). August 1995: 104.5 mhz; 3 kw. 328 ft TL: N34 49 42 W84 53 41.24 200,000 Format: Adult contemp. News staff: 3.

WTTI(AM)— June 17, 1965: 1530 khz; 10 kw-D, 10 kw-CH, DA-D. TL: N34 47 09 W85 02 40. Hrs open: Box 216, 30722. Secondary address: 111 W. Crawford St. 30720. Phone: (706) 277-7117. Fax: (706) 277-7180. Licensee: Troy L. Hall. Format: Southern gospel. Target aud: 25-54; family, relg. ♦Troy Hall, CEO & gen mgr; C.W. Queen, stn mgr.

Darien

WHFX(FM)— May 13, 1993: 107.7 mhz; 50 kw. 403 ft TL: N31 10 09 W81 32 14. Stereo. Hrs opn: 24 3833 Hwy. 82, Brunswick, 31523. Phone: (912) 267-1025. Fax: (912) 264-5462. Web Site: 1077thefox.net. Licensee: Qantum of Brunswick License Co. LLC. Group owner: Qantum Communications Corp. (acq 7-2-2003; grpsl). Population served: 150,000 Natl. Rep: McGavren Guild. Format: Rock. News staff: one; News: 4 hrs wkly. Target aud: 25 plus; general. ♦Jonathan Havens, gen mgr & min affrs dir; Mark Douglas, gen mgr, progmg dir & disc jockey; Joe Sousa, opns mgr & mus critic; Jim Hendrick, sls dir; Mike Hamens, news dir; Dick Boekeloo, chief of engrg; Robin Rowe, spec ev coord; Kat Blackstone, disc jockey.

Dawson

WMRZ(FM)— June 2005: 98.1 mhz; 25 kw. Ant 262 ft TL: N31 37 29 W84 19 20. Hrs open: 809 S. Westover Blvd., Albany, 31707. Phone: (229) 439-9704. Fax: (229) 439-1509. E-mail: pauledwards@clearchannel.com Web Site: www.kissalbany.com. Licensee: CC Licenses LLC. (group owner; (acq 11-16-2005; $875,000). Format: Rhythm and blues, oldies. ♦John Richards, gen mgr; Paul Edwards, progmg dir.

Decatur

WATB(AM)—Licensed to Decatur. See Cumming

WPBC(AM)— Aug 11, 1964: 1310 khz; 2.5 kw-D, 31 w-N. TL: N33 46 22 W84 16 55. Hrs open: 24 3684 Stewart Rd., Suite A-3, Doraville, 30340. Phone: (678) 200-8540. Licensee: Hanmi Broadcasting Inc. (acq 2-28-2005; $3.3 million). Population served: 80,000 ♦Chang Soo Kim, gen mgr.

WUBL(FM)—See Atlanta

Demorest

***WPPR(FM)**— 1997: 88.3 mhz; 6 kw. 640 ft TL: N34 31 24 W83 40 46. Stereo. Hrs opn:
Rebroadcasts WJSP-FM Warm Springs 100%.
260 14th St. N.W., Atlanta, 30318-5360. Phone: (404) 685-2690. Fax: (404) 685-2684. E-mail: ask@spb.org Web Site: www.gpb.org. Licensee: Georgia Public Telecommunications Commission. Natl. Network: NPR, PRI. Wire Svc: AP Format: News, class. News staff: 6; News: 40 hrs wkly. Adults 35+. ♦Nancy G. Hall, CEO; Bonnie Bean, CFO; Alan Cooke, gen mgr; Ryan Fowler, opns mgr; Mandy Wilson, prom dir; St. John Flynn, progmg mgr; Terrance McKnight, mus dir; Susanna Capelonto, news dir; Mark Fehlig, engrg dir; Gyla Gonzalez, spanish dir.

Dock Junction

WXMK(FM)— May 1, 1991: 105.9 mhz; 15 kw. 489 ft TL: N31 10 09 W81 32 14. Stereo. Hrs opn: 24 185 Benedict Rd., Brunswick, 31520. Phone: (912) 261-1000. Fax: (912) 265-8391. Web Site: www.magic1059.com. Licensee: Golden Isles Broadcasting L.L C. (acq 3-23-2001; $2.8 million with WRJY(FM) Brunswick). Population served: 100,000 Law Firm: Pepper & Corazzini. Format: CHR, adult contemp. Target aud: 25-54; women. ♦Traci Long, gen mgr.

Donalsonville

WGMK(FM)—Listing follows WSEM(AM).

WSEM(AM)— Feb 12, 1963: 1500 khz; 1 kw-D. TL: N31 04 26 W84 52 47. Hrs open: Box 87, 31745. Secondary address: 91 North Way 31743. Phone: (229) 524-5123. Fax: (229) 524-2265. E-mail: wgmk@alltel.net Licensee: Flint Media Inc. (group owner; (acq 6-22-2006; grpsl). Population served: 43,500 Rgnl. Network: Ga. Net. Format: Country, talk, Black gospel. Target aud: 25-49. ♦Kevin Dowdy, pres; Gilbert M. Kelley Jr., gen mgr, progmg dir & news dir; Grace Kelley, gen sls mgr.

WGMK(FM)—Co-owned with WSEM(AM). Sept 1, 1980: 106.3 mhz; 5.9 kw. Ant 331 ft TL: N31 04 26 W84 52 47. Format: Hot adult contemp.

WWGF(FM)— Aug 1, 1998: 107.5 mhz; 6 kw. 315 ft TL: N30 58 36 W84 55 51. Stereo. Hrs opn: 24 2278 Wortham Lane, Grovetown, 30813-5103. Phone: (706) 309-9610. Fax: (706) 309-9669. E-mail: ctbarinowski@comcast.net Web Site: www.gnnradio.org. Licensee: Barinowski Investment Co. Group owner: Good News Network (acq 1-21-99). Population served: 80,000 Format: Christian. Target aud: All. ♦Clarence Barinowski, gen mgr.

Doraville

WBTS(FM)— May 1948: 95.5 mhz; 74 kw. Ant 1,115 ft TL: N34 07 32 W83 51 32. Stereo. Hrs opn: 24 1601 W. Peachtree St. N.E., Atlanta, 30309. Phone: (404) 897-7500. Fax: (404) 897-7363. Web Site: www.955thebeat.com. Licensee: Cox Radio Inc. Group owner: Cox Broadcasting (acq 7-19-99; $78 million). Population served: 2,500,000 Format: Top-40. Target aud: 25-54; blue collar to executive, modern country mus lovers. ♦Tony Kidd, gen mgr.

Douglas

WDMG(AM)— March 1947: 860 khz; 5 kw-U, DA-N. TL: N31 30 23 W82 49 10. Hrs open: 1931 GA Hwy. 32 E., 31533. Phone: (912) 389-0995. Fax: (912) 383-8552. E-mail: wdmgamfm@charter.net Licensee: Broadcast South LLC. Group owner: Black Crow Media Group LLC (acq 11-6-2006; grpsl). Population served: 100,137 Natl.

Network: USA. Rgnl. Network: GNN Format: Sports. News staff: one; News: 20 hrs wkly. Target aud: 25-54. Spec prog: Farm 4 hrs, relg 6 hrs wkly. ♦John Higgs, stn mgr.

WOKA(AM)— Dec 10, 1962: 1310 khz; 3.9 kw-D, 39 w-N. TL: N31 31 24 W82 52 22. Hrs open: 6 AM-midnight 1310 W. Walker St., 31533. Phone: (912) 384-1310. Phone: (912) 384-8153. Fax: (912) 383-6328. Licensee: Coffee County Broadcasters Inc. (acq 2-13-98; with co-located FM). Population served: 42,000 Format: Solid gold oldies. Target aud: General. ♦Jim Squires, CEO, gen mgr & gen sls mgr; Dwayne Gillis, pres; Paul Sullivan, opns mgr; Michael Van Cleave, news dir; Jim Edwards, relg ed.

WOKA-FM— July 1971: 106.7 mhz; 100 kw. 1,000 ft TL: N31 31 24 W82 52 22. Phone: (912) 384-8153. Phone: (912) 389-1067. Fax: (912) 383-6328. E-mail: production@accessatc.net Web Site: www.dixiecountry.com.120,000 Format: Country. Target aud: Adults 25-54. Spec prog: Gospel 4 hrs wkly. ♦Jim Edwards, relg ed; Paul Sullivan, disc jockey.

Douglasville

WDCY(AM)— May 5, 1993: 1520 khz; 2.5 kw-D, 800 w-N. TL: N33 45 48 W84 44 28. Hrs open: Sunrise-sunset 8451 S. Cherokee Blvd., Suite B, 30134. Phone: (770) 920-1520. Fax: (770) 920-4600. Web Site: www.wordchristianbroadcasting.com. Licensee: Word Christian Broadcasting Inc. (acq 5-7-93; $95,000; FTR: 4-12-93). Population served: 1,400,000 Format: Old time relg. News: 10 hrs wkly. Target aud: General. ♦Ken Johns, pres, gen mgr, opns dir, opns mgr & progmg dir.

Dry Branch

WVVM(AM)— Apr 15, 1998: 1670 khz; 10 kw-D, 1 kw-N. TL: N32 48 16 W83 36 16. Hrs open: 24 7080 Industrial Hwy., Macon, 31216. Phone: (478) 781-1063. Fax: (478) 781-6711. E-mail: huston@965thebuzz.net Licensee: AMFM Radio Licenses LLC. Group owner: Clear Channel Communications Inc. (acq. 2-15-2001; grpsl). Population served: 250,000 Natl. Rep: Clear Channel. Format: Latino. Target aud: General; adults. ♦Bill Clark, gen mgr.

Dublin

***WAWH(FM)**— 2000: 88.3 mhz; 400 w. Ant 82 ft TL: N32 32 27 W82 57 27. Hrs open: Box 3206, American Family Radio, Tupelo, MS, 38803. Phone: (662) 844-8888. Fax: (662) 842-6791. Web Site: www.afr.net. Licensee: American Family Association. Group owner: American Family Radio Format: Inspirational Christian. ♦Marvin Sanders, gen mgr.

WKKZ(FM)— Apr 4, 1967: 92.7 mhz; 50 kw. 417 ft TL: N32 31 21 W82 54 00. Stereo. Hrs opn: 24 Box 967, 31040. Secondary address: 1006 Martin Luther King Blvd. 31021. Phone: (478) 272-9270. Fax: (478) 275-3592. Licensee: Kirby Broadcasting Co. (acq 9-18-2003). Population served: 750,000 Natl. Network: ABC. Natl. Rep: Dora-Clayton. Format: CHR. ♦Ray Beck, gen mgr.

WMLT(AM)— Jan 12, 1945: 1330 khz; 5 kw-D, 500 w-N, DA-N. TL: N32 33 50 W82 52 00. Hrs open: Box 130, 31040. Secondary address: 807 Bellevue Ave. 31021. Phone: (478) 272-4422. Fax: (478) 275-4657. E-mail: richhumphrey@wqzy.com Licensee: State Broadcasting Corporation. Population served: 15,143 Rgnl. Network: Ga. Net. Natl. Rep: Rgnl Reps. Format: Urban gospel. Target aud: 25-54. Spec prog: Farm 6 hrs wkly. ♦Rick Humphrey, gen mgr.

WQZY(FM)—Co-owned with WMLT(AM). 1978: 95.9 mhz; 88 kw. 1,023 ft TL: N32 33 51 W82 52 18. Stereo. Web Site: www.wqzyfm.com.450,000 Format: Hit country. Target aud: 18-54.

WXLI(AM)— Mar 16, 1958: 1230 khz; 1 kw-U. TL: N32 31 21 W82 54 00. Hrs open: Box 967, 31040. Secondary address: 1006 Martin Luther King Blvd. 31021. Phone: (478) 272-4282. Fax: (478) 275-3592. Licensee: Laurens County Broadcasting Co. (acq 8-10-03). Population served: 250,000 Natl. Network: CBS. Natl. Rep: Dora-Clayton. Format: Country. ♦Ray Beck, gen mgr.

East Dublin

WELT(FM)—Licensed to East Dublin. See Swainsboro

East Point

WCFO(AM)— Oct 9, 1994: 1160 khz; 50 kw-D, 160 w-N, DA-2. TL: N33 49 34 W84 36 20. Hrs open: 24 1100 Spring St., Suite 610, Atlanta, 30309. Phone: (404) 223-1160. Fax: (404) 870-8859. E-mail: listeners@am1160.net Web Site: www.am1160.net. Licensee: JW Broadcasting Inc. (acq 8-11-2004; $10.4 million). Format: Business talk. ♦Jeff Davis, VP & gen mgr.

WTJH(AM)— December 1949: 1260 khz; 5 kw-D. TL: N33 41 47 W84 28 29. Hrs open: 3079 Campbellton Rd. S.W., Suite 104, Atlanta, 30311. Phone: (404) 344-2233. Fax: (404) 346-0647. Licensee: Christian Broadcasting of East Point Inc. Group owner: Willis Broadcasting Corp. Population served: 45,000 Format: Inspirational, gospel. ♦Christine Willis-Wiggs, gen mgr.

Eastman

WUFF(AM)— Sept 1, 1961: 710 khz; 2.5 kw-D. TL: N32 13 18 W83 13 04. Hrs open: Box 4097, 31023. Secondary address: 855 College St. Phone: (478) 374-3437. Fax: (478) 374-3585. E-mail: wuffradio@nlamerica.com Licensee: Dodge Broadcasting Inc. Population served: 5,416 Format: Country. Spec prog: Black 5 hrs wkly. ♦Gene Rogers, gen mgr; Dale Jones, gen sls mgr, progmg dir & sls; Don Jones, chief of engrg & farm dir.

WUFF-FM— 1976: 97.5 mhz; 2 kw. 371 ft TL: N32 13 35 W83 13 10. (CP: 4.6 kw, ant 364 ft.). Stereo. Format: Southern gospel, news. ♦Gene Rogers, news dir.

Eatonton

WKVQ(AM)— Dec 15, 1966: 1520 khz; 1 kw-D. TL: N33 19 19 W83 25 03. Hrs open:
Simulcast with WKRR(FM) Milledgeville.
Box 3965, 31024. Phone: (706) 485-8792. Fax: (706) 485-3555. Licensee: Craig Baker. Population served: 4,125 Format: Adult standards. Target aud: General. ♦Craig Baker, pres & gen mgr.

WMGZ(FM)— Feb 8, 1988: 97.7 mhz; 8.5 kw. 554 ft TL: N33 20 41 W83 13 41. Stereo. Hrs opn: 24 Box 832, Milledgeville, 31061. Secondary address: 156 Lake Laurel Rd., Milledgeville 31061. Phone: (478) 453-9406. Fax: (478) 453-3298. E-mail: z97mail@yahoo.com Web Site: todaysbesthits.com. Licensee: Southern Stone Broadcasting Inc. (acq 7-6-2005; $1.1 million with WKGQ(AM) Milledgeville). Population served: 125,000 Natl. Network: ABC. Format: Hot adult contemp. News staff: one; News: 4 hrs wkly. Target aud: 18-49. ♦Tom Ptak, gen mgr.

Elberton

WLVX(FM)— May 15, 1998: 105.1 mhz; 6 kw. 328 ft TL: N33 59 22 W82 46 23. Hrs open: 24 Box 340, 30635. Secondary address: 562 Jones St. 30635. Phone: (706) 213-1051. Fax: (706) 283-8710. E-mail: gibson@gacaradio.com Web Site: www.elbertonradio.com. Licensee: Georgia-Carolina Radiocasting Co. LLC. Group owner: Georgia-Carolina Radiocasting Companies (acq 7-25-2002; grpsl). Population served: 164,296 Natl. Network: ABC. Law Firm: Dan J. Alpert. Format: Adult urban. News staff: one; News: 5 hrs wkly. Target aud: 25-54; general. ♦Douglas M. Sutton Jr., pres; Sean Gibson, VP & gen mgr; Ron Shuller, opns mgr.

WNGA(AM)— Jan 1, 1947: 1400 khz; 1 kw-U. TL: N34 06 45 W82 52 52. Hrs open: Box 340, 562 Jones St., 30635. Phone: (706) 283-1400. Fax: (706) 283-8710. E-mail: gibson@gacaradio.com Web Site: www.elbertonradio.com. Licensee: Georgia-Carolina Radiocasting Co. LLC. Group owner: Georgia-Carolina Radiocasting Companies (acq 7-25-2002; grpsl). Population served: 80,567 Natl. Network: CBS Radio. Law Firm: Dan J. Alpert. Format: MOR, talk. News staff: one; News: 12 hrs wkly. Target aud: 35+. ♦Douglas M. Sutton Jr., pres; Sean Gibson, VP & stn mgr; Ron Shuller, opns mgr; Tim Stephens, chief of engrg.

WSGC-FM— 1973: 92.1 mhz; 3 kw. Ant 299 ft TL: N31 04 45 W82 55 16. Hrs open: Box 340, 30635. Secondary address: 562 Jones St. 30635. Phone: (706) 283-1400. Fax: (706) 283-8710. E-mail: gibson@gacaradio.com Web Site: www.elbertonradio.com. Licensee: Radio Elberton Inc. Group owner: Georgia-Carolina Radiocasting Companies (acq 3-30-01; at least $478,000 debt). Population served: 27,578 Natl. Network: ABC. Law Firm: Dan J. Alpert. Format: Country/community involvement. News staff: one; News: 18 hrs wkly. Target aud: 25 plus. ♦Douglas M. Sutton Jr., pres; Sean Gibson, VP & gen mgr; Ron Shuler, opns mgr.

Ellaville

WLEL(FM)—Not on air, target date: unknown: 94.3 mhz; 4.8 kw. Ant 328 ft TL: N32 15 10 W84 13 50. Hrs open: 819 S.W. Federal Hwy., Suite 106, Stuart, FL, 34994. Phone: (772) 215-1634. Web Site: www.toweritrust.com. Licensee: Tower Investment Trust Inc. ♦William H. Brothers, pres.

Ellijay

WLJA-FM—Listing follows WPGY.

WPGY(AM)— May 10, 1978: 1560 khz; 1 kw-D. TL: N34 42 14 W84 28 35. Hrs open: 134 S. Main St., Jasper, 30143. Phone: (706) 276-2016. Fax: (706) 635-1018. E-mail: wlja@ellijay.com Licensee: Tri-State Communications Inc. (acq 10-21-98; $500,000 with co-located FM). Format: Classic country. Target aud: 18-75. ♦Byron Dobbs, gen mgr & news dir; Randy D. Gravley, VP, gen mgr & stn mgr.

WLJA-FM—Co-owned with WPGY. Nov 1, 1985: 93.5 mhz; 6 kw. 272 ft TL: N34 42 59 W84 30 50.6 AM-10 PM Format: Country & gospel. Target aud: 18-75.

Evans

WAEG(FM)— November 1991: 92.3 mhz; 3 kw. 328 ft TL: N33 35 25 W82 13 52. Hrs open: Box 1584, Augusta, 30903-2429. Secondary address: 104 Bennett Ln., North Augusta, SC 29841. Phone: (803) 279-2330. Fax: (803) 279-8149. Web Site: www.waeg923.com. Licensee: Radio One of Augusta LLC. Group owner: Radio One Inc. (acq 11-8-01; grpsl). Population served: 450,000 Natl. Rep: Christal. Law Firm: Shaw Pittman. Format: Alternative. Target aud: Teen-49. ♦Ron Tomel, opns mgr & prom dir.

Fayetteville

WPZE(FM)—Licensed to Fayetteville. See Griffin

Fitzgerald

WBHB(AM)— Oct 8, 1946: 1240 khz; 1 kw-U. TL: N31 42 23 W83 15 40. Hrs open: 601 W. Roanoke Dr., 31750. Phone: (229) 423-2077. Fax: (229) 423-8313. E-mail: jank@rtgmedia.net Licensee: Broadcast South LLC. Group owner: Black Crow Media Group LLC (acq 11-6-2006; grpsl). Population served: 50,000 Natl. Network: Westwood One. Format: Black gospel. Target aud: General. ♦John Higgs, gen mgr & stn mgr.

WRDO(FM)— 1991: 96.9 mhz; 6 kw. Ant 328 ft TL: N31 44 33 W83 14 41. Hrs open: 24 1931 GA Hwy. 32 E., Douglas, 31533. Phone: (912) 389-0995. Fax: (912) 383-8552. Licensee: Broadcast South LLC. Group owner: Black Crow Media Group LLC (acq 11-6-2006; grpsl). Population served: 30,000 Natl. Network: USA. Law Firm: Roy F. Perkins. Format: Adult contemp-soft hits. News staff: one. Target aud: 25-55; baby boomers. Spec prog: Gospel 6 hrs wkly. ♦John Higgs, gen mgr.

Folkston

***WATY(FM)**— 2000: 91.3 mhz; 600 w. Ant 321 ft TL: N30 52 29 W82 01 10. Stereo. Hrs opn: 24 Box 777, 31537. Secondary address: 104 N. First St. 31537. Phone: (912) 496-4484. Fax: (912) 496-4086. E-mail: maysj@alltel.net Licensee: Okefenokee Educational Foundation Inc. Natl. Network: USA. Format: Traditional country. News: 21 hrs wkly. ♦Jack R. Mays, CEO & gen mgr.

***WECC-FM**— 2002: 89.3 mhz; 16 kw. Ant 282 ft TL: N30 55 54 W81 42 30. (CP: 30 kw, ant 479 ft). Hrs opn: 5465 Hwy. 40 E., St. Marys, 31558. Phone: (912) 882-8930. Fax: (912) 882-9322. E-mail: paul@weccradio.org Web Site: www.weccradio.org. Licensee: Lighthouse Christian Broadcasting Corp. Format: Christian. ♦Paul Hafer, gen mgr.

WFJO(FM)— November 1989: 92.5 mhz; 3.2 kw. Ant 459 ft TL: N30 43 38 W81 56 14. Stereo. Hrs opn: 24 9550 Regency Sq. Blvd., Suite 200, Jacksonville, FL, 32225. Phone: (904) 680-1050. Fax: (904) 680-1051. Web Site: www.fiesta925.com. Licensee: Tama Radio Licenses of Jacksonville, FL, Inc. Group owner: Tama Broadcasting Inc. (acq 6-4-2003). Rgnl. Network: Ga. Net. Natl. Rep: Rgnl Reps. Law Firm: Stan Emert. Format: Latin. News staff: one. Target aud: General. Spec prog: Gospel 5 hrs wkly. ♦Linda Fructuoso, gen mgr.

Forsyth

WQMJ(FM)— Nov 22, 1973: 100.1 mhz; 3 kw. 209 ft TL: N32 58 31 W83 52 11. (CP: 2 kw, ant 574 ft. TL: N32 55 41 W83 52 37). Stereo. Hrs opn: 6174 Hwy. 57, Macon, 31217. Phone: (478) 745-3301. Fax: (478) 742-2293. Licensee: Roberts Communications Inc. (group owner; acq 5-23-97; $550,000 with WXKO(AM) Fort Valley). Population served: 250000 Format: Mainstream urban. ♦Mike Roberts, gen mgr & stn mgr.

Fort Gaines

***WJWV(FM)**— Feb 28, 1993: 90.9 mhz; 85 kw. 267 ft TL: N31 36 16 W85 02 02. Hrs open:
Rebroadcasts WJSP-FM Warm Springs 100%.
260 14th St. N.W., Atlanta, 30318-5360. Phone: (404) 685-2690. Fax: (404) 685-2684. E-mail: ask@gpb.org Web Site: www.gpb.org. Licensee: Georgia Public Telecommunications Commission. (group owner) Format: Class, news. Spec prog: Jazz 4 hrs, Latin 3 hrs. ♦Nancy Hall, CEO; Bonnie Bean, CFO; Ryan Fowler, opns dir; Mandy Wilson, prom dir; St. John Flynn, progmg mgr; Susanna Capelouto, news dir.

Fort Valley

WIBB-FM—Licensed to Fort Valley. See Macon

***WJTG(FM)**— Mar 1, 1989: 91.3 mhz; 100 kw. 459 ft TL: N32 41 27 W83 51 45. Stereo. Hrs opn: 24 101 Graylane Dr., Byron, 31008. Phone: (478) 956-0085. Fax: (478) 956-0913. E-mail: wjtg913@aol.com Web Site: www.wjtg.org. Licensee: Family Life Broadcasting Inc. (acq 3-27-2007; grpsl). Population served: 250,000 Natl. Network: USA. Format: Southern gospel. Target aud: General. ♦Tracy O. Wells, gen mgr & opns mgr.

WQBZ(FM)—Licensed to Fort Valley. See Macon

WXKO(AM)— June 1951: 1150 khz; 1 kw-D, 60 w-N. TL: N32 34 34 W83 54 17. Hrs open: 24 1675 Hwy. 341 N., 31030. Phone: (478) 825-1150. Fax: (478) 827-1273. E-mail: jjohnsonwxko@bellsouth.net Licensee: Roberts Communications Inc. (group owner; acq 5-23-97; $550,000 with WFXM-FM Forsyth). Population served: 30,000 Rgnl. Network: Ga. Net. Format: Gospel. News staff: one; News: 10 hrs wkly. Target aud: 25 plus; Black.

Gainesville

***WBCX(FM)**— 1977: 89.1 mhz; 875 w. 544 ft TL: N34 19 01 W83 49 45. Stereo. Hrs opn: 24 Brenau Univ., 500 Washington St. S.E., 30501. Phone: (770) 538-4708. Fax: (770) 538-4558. E-mail: sfugate@lib.brenau.edu Web Site: www.brenau.edu. Licensee: Brenau University. Population served: 1,000,000 Natl. Network: PRI, Jones Radio Networks. Format: Eclectic. Target aud: 14-85. Spec prog: Black 12 hrs, class 20 hrs, world 6 hrs, American Indian 4 hrs, Gospel 6 hrs wkly. ♦J. Scott Fugate, gen mgr, opns dir & dev dir.

WDUN(AM)— Apr 2, 1949: 550 khz; 5 kw-D, 2.5 kw-N, DA-N. TL: N34 20 11 W83 47 41. Stereo. Hrs opn: Box 10, 30503. Secondary address: 1102 Thompson Bridge Rd. N.E. 30501. Phone: (770) 532-9921. Fax: (770) 532-0506. E-mail: news@wdun.com Web Site: www.wdun.com. Licensee: JWJ Properties Inc. Group owner: Jacobs Media Corp. (acq 9-83; FTR: 9-5-83). Population served: 2,000,000 Natl. Network: CBS. Rgnl. Network: Ga. Net. Format: News/talk. Target aud: 25-65. ♦John W. Jacobs III, CEO & pres; John W. Jacobs Jr., chmn; Jones P. Andrews, gen mgr.

WGGA(AM)— Oct 10, 1941: 1240 khz; 1 kw-U. TL: N34 19 01 W83 49 45. Stereo. Hrs opn: 24 Box 10, 30503. Secondary address: 1102 Thompson Bridge Rd. N.E. 30501. Phone: (770) 532-9921. Fax: (770) 532-0506. Web Site: www.accessnorthga.com. Licensee: JWJ Properties Inc. Group owner: Jacobs Media Corp. (acq 4-20-93; $360,000; FTR: 5-10-93). Population served: 138,000 Natl. Network: CBS. Format: Sports. News staff: 2; News: 31 hrs wkly. Target aud: 25-54; middle, upper income adults. ♦John W. Jacobs III, CEO & pres; John W. Jacobs Jr., chmn; Jones P. Andrews, gen mgr.

WGTJ(AM)—Murrayville, Nov 1, 1986: 1330 khz; 1 kw-D. TL: N34 22 16 W83 56 47. Hrs open: 6 AM-sunset Box 907038, 30501. Secondary address: 1716 Cleveland Hwy. 30501. Phone: (770) 297-7485. Fax: (770) 297-8030. E-mail: mail@glory1330.com Web Site: www.glory1330.com. Licensee: Vision Communications Inc. (acq 1999; $120,000). Format: Christian music. News staff: one. Target aud: General. ♦Mike Wofford, pres & gen mgr.

WLBA(AM)— Jan 26, 1957: 1130 khz; 10 kw-D. TL: N34 16 45 W83 46 33. Hrs open: Sunrise-sunset 5815 Westside Rd., Austell, 30106. Phone: (770) 532-6331. Fax: (770) 532-2672. E-mail: ariel@radiolafavorita.com Web Site: www.radiolafavorita.com. Licensee: La Favorita Inc. (group owner; acq 2-18-97; $275,000). Population served: 200,000 Format: Sp. ♦Samuel Zamarron, pres & gen mgr; Ariel Zamarron, stn mgr.

WMJE(FM)—See Clarkesville

WSRV(FM)— Nov 1, 1965: 97.1 mhz; 97 kw. Ant 1,571 ft TL: N34 07 32 W83 51 31. Stereo. Hrs open: 1601 W. Peachtree St., Atlanta, 30309. Phone: (404) 897-7500. Fax: (404) 897-7363. Web Site: 971theriver.com. Licensee: Cox Radio Inc. Group owner: Cox Broadcasting (acq 8-16-2000; grpsl). Population served: 3,750,700 Format: Classic hits. News: 3 hrs wkly. Target aud: 35-54; baby boomers. ♦Dan Kearney, VP & gen mgr; Rob Babin, gen sls mgr.

WYAY(FM)— Apr 3, 1949: 106.7 mhz; 77 kw. Ant 1,656 ft TL: N33 52 02 W83 49 44. Stereo. Hrs opn: 6th Fl., 210 Interstate N., Atlanta, 30339. Phone: (404) 521-1007. Fax: (404) 499-1067 (NEWS FAX). Licensee: Radio License Holding II LLC. Group owner: ABC Inc. (acq 6-12-2007; grpsl). Population served: 500,000 Natl. Network: ABC. Natl. Rep: ABC Radio Sales. Format: Country. Target aud: 25-54. ♦Mark Richards, opns mgr; Rick Mack, sls dir; Matt Scarano, gen sls mgr; Nancy Barre, natl sls mgr; Mary Gordon, rgnl sls mgr; Christy Ullman, prom dir; Victor Sansone, pres, gen mgr & progmg dir; Sandy Weaver, mus dir; Glenda Dodd, traf mgr.

Gibson

WTHP(FM)—Not on air, target date: unknown: 94.3 mhz; 2.05 kw. Ant 571 ft TL: N33 17 05 W82 35 45. Hrs open: 2278 Wortham Ln., Grovetown, 30813-5103. Phone: (706) 309-9610. Fax: (706) 309-9669. E-mail: ctbarinowski@comcast.net Web Site: www.gnnradio.org. Licensee: Barinowski Investment Co. L.P. (acq 2-15-2006). Format: Christian. ♦Clarence Barinowski, gen mgr.

Glennville

WOAH(FM)— Nov 18, 1977: 106.3 mhz; 6 kw. Ant 394 ft TL: N32 00 27 W81 54 51. Stereo. Hrs opn: 24 25 Bristlecone Dr., Savannah, 31419-9506. Phone: (912) 408-1063. Fax: (912) 876-6920. E-mail: jimlewis@coastalnow.net Web Site: www.hotkiss1063.com. Licensee: Broadcast Executives Corp. (acq 3-29-2002; $250,000). Format: Contemp hits. ♦James Lewis, gen mgr.

Gordon

WFXM(FM)— Mar 30, 1976: 107.1 mhz; 3 kw. Ant 466 ft TL: N32 50 55 W83 28 29. Stereo. Hrs opn: 24 6174 Georgia Hwy. 57, Macon, 31217. Phone: (478) 745-3301. Fax: (478) 742-2293. E-mail: info@foxie107.com Licensee: WFXM-FM Radio LLC (acq 7-31-2006; $808,500). Population served: 150000 Format: Mainstream urban contemp. Target aud: Middle & upper class Georgians. Spec prog: Jazz 3 hrs wkly. ♦Mike Roberts, gen mgr.

WXJO(AM)— Sept 1, 1969: Stn currently dark. 1120 khz; 10 kw-D, 2.5 kw-CH. TL: N32 50 59 W83 28 38. Hrs open: 134 S. Main St., Jasper, 30143. Phone: (706) 276-2016. Fax: (706) 635-1018. Licensee: Exponent Broadcasting Inc. (group owner; (acq 5-16-2007; $80,000). Rgnl. Network: Ga. Net. ♦Randy Gravley, pres & gen mgr.

Gray

WPCH(FM)— January 1994: 96.5 mhz; 7.6 kw. 414 ft TL: N32 59 03 W83 33 16. Stereo. Hrs opn: 24 7080 Industrial Hwy., Macon, 31216. Phone: (478) 781-1063. Fax: (478) 781-6711. Web Site: www.965thebuzz.net. Licensee: AMFM Radio Licenses LLC. Group owner: Clear Channel Communications Inc. (acq 2-1-2001; grpsl). Population served: 251,300 Format: Alternative. Target aud: 18-54; adults. ♦Bill Clark, gen mgr.

Grayson

WPLO(AM)—Licensed to Grayson. See Lawrenceville

Greensboro

WDDK(FM)— July 12, 1980: 103.9 mhz; 5.3 kw. 328 ft TL: N33 28 29 W83 14 46. Hrs open: 24 1271-B E. Broad St., 30642. Phone: (706) 453-4140. Fax: (706) 453-7179. Licensee: Wyche Services Corp. (acq 2-28-2005). Population served: 70,000 Natl. Network: ABC. Rgnl. Network: Ga. Net. Format: Talk, oldies. News staff: one; News: 4 hrs wkly. Target aud: 24-60. ♦Chip Lyness, VP & gen mgr; K.B. Travis, opns dir.

Greenville

WIOL(FM)— July 4, 1994: 95.7 mhz; 3.4 kw. Ant 876 ft TL: N32 50 48 W84 41 27. Hrs open: 24 Box 1998, Columbus, 31902. Phone: (706) 576-3565. Fax: (706) 576-3683. Licensee: Davis Broadcasting of Columbus Inc. Group owner: Davis Broadcasting Inc. (acq 11-4-97; $450,000). Population served: 250,000 Natl. Network: CBS. Law Firm: McCampbell & Young, P. Format: Classic rock. News staff: 2; News: 6 hrs wkly. Target aud: 18-49; females 18-35 specifically. ♦Gregory A. Davis, CEO, CFO & gen mgr.

Griffin

WEKS(FM)—See Zebulon

WHIE(AM)— Dec 15, 1952: 1320 khz; 5 kw-D, 83 w-N. TL: N33 14 30 W84 18 17. Hrs open: 1000 Memorial Dr., 30223. Phone: (770) 227-9451. Fax: (770) 229-2291. Licensee: Chappell Communications L.L.C. (acq 6-30-98; $240,000). Population served: 30,000 Format: Country, news/talk, sports. ♦Robert E. Chappell Jr., pres & gen mgr.

WKEU(AM)— 1933: 1450 khz; 1 kw-U. TL: N33 14 25 W84 14 54. Hrs open: Box 997, 30224. Secondary address: 1000 Memorial Dr. 30224. Phone: (770) 227-5507. Fax: (770) 229-2291. E-mail: wkeu@aol.com Web Site: www.wkeuradio.com. Licensee: WLT & Associates L.P. Population served: 25,000 Rgnl. Network: Ga. Net. Natl. Rep: Rgnl Reps. Format: Oldies, news. Target aud: 25 plus. ♦William Taylor, pres & gen mgr.

***WMVV(FM)**— Apr 16, 1995: 90.7 mhz; 18 kw. Ant 472 ft TL: N33 22 12 W84 08 00. Stereo. Hrs opn: 24 Box 2020, 30224. Secondary address: 100 S. Hill St., Suite 100 30223. Phone: (770) 229-2020. Fax: (770) 229-4820. E-mail: contactus@wmvv.com Licensee: Life Radio Ministries Inc. (acq 1996; $75,000). Population served: 1,200,000 Format: Relg, Christian. ♦Joseph C. Emert, pres; Douglas J. Doran, VP & gen mgr.

WPZE(FM)—Fayetteville, Mar 8, 1966: 97.5 mhz; 8.5 kw. 554 ft TL: N33 29 22 W84 34 07. (CP: 6.6 kw, ant 636 ft.). Stereo. Hrs opn: 101 Marietta St. 12th Floor, Atlanta, 30303. Phone: (404) 765-9750. Fax: (404) 688-7686. Web Site: www.praise975.com. Licensee: ROA Licenses LLC. Group owner: Radio One Inc. (acq 11-8-01; grpsl). Population served: 50,000 Natl. Network: ABC. Format: Gospel. ♦Wayne Brown, gen mgr & progmg dir.

Hahira

WTHV(AM)— 1990: 810 khz; 2.5 kw-D. TL: N30 52 25 W83 15 07. Stereo. Hrs opn: 15 2352 Jaycee Shack Rd., Valdosta, 31602. Phone: (229) 245-9848. Fax: (229) 242-0809. E-mail: wthv810am@yahoo.com Licensee: Eternal Life Ministries Inc. (acq 8-13-2003; $180,000). Population served: 90,000 Format: Southern gospel. Target aud: 24-55. Spec prog: Spanish 5 hrs wkly. ♦Cody Fender, pres, gen mgr & stn mgr; Phyllis Fender, VP.

Hampton

WHTA(FM)— Oct 19, 1973: 107.9 mhz; 27 kw. Ant 577 ft TL: N33 29 24 W84 34 07. Stereo. Hrs opn: 101 Marietta St. 12th Floor, Atlanta, 30303. Phone: (404) 765-9750. Fax: (404) 688-7686. Web Site: www.hot1079atl.com. Licensee: Radio One Licenses LLC. Group owner: Radio One Inc. (acq 8-20-01; $60 million). Population served: 400,000 Format: Hip hop. ♦Wayne Brown, gen mgr.

Hapeville

WWWE(AM)— Jan 7, 1947: 1100 khz; 1 kw-D. TL: N33 36 34 W85 05 13. (CP: COL: Hapeville. 5 kw-D, 3.8 kw-CH. TL: N33 43 43 W84 19 20). Hrs opn: Sunrise-sunset 1465 North Side Dr., Suite 218, Atlanta, 30318. Phone: (404) 352-9993. Fax: (404) 355-0291. Web Site: www.radiovidaatlanta.org. Licensee: WAEC License L.P. Group owner: Beasley Broadcast Group (acq 10-29-99; $10 million with WAEC(AM) Atlanta). Population served: 100,000 Format: Sp, relg. News staff: one; News: 8 hrs wkly. Target aud: 35 plus; mature audience. Spec prog: Relg 12 hrs, news/talk 14 hrs, Ethiopian 2 hrs wkly. ♦George Beasley, chmn; Bruce Beasley, pres; Caroline Beasley, CFO; Brian Beasley, exec VP; Chris Edmonds, gen mgr.

Harlem

WCHZ(FM)— Nov 23, 1992: 95.1 mhz; 5.7 kw. 440 ft TL: N33 31 34 W82 15 55. Hrs open: 4051 Jimmie Dyess Pkwy., Augusta, 30909-9469. Phone: (706) 396-7000. Fax: (706) 396-7100. Web Site: www.95rock.com. Licensee: WCHZ License LLC Group owner: Beasley Broadcast Group Inc. (acq 1-13-97; $1.2 million). Format: Rock. Target aud: 18-34; well-educated adults, upper demographics. ♦Kent Dunn, gen mgr; Greg Mclaughlin, gen sls mgr.

Hartwell

WKLY(AM)— Sept 5, 1947: 980 khz; 1 kw-D, 140 w-N. TL: N34 21 28 W82 58 35. (CP: 149 w-N). Hrs opn: 15 Box 636, 30643. Secondary address: 2235 Bowersville Hwy. 30643. Phone: (706) 376-2233. Fax: (706) 376-3100. E-mail: wklyradio@hartcom.net Web Site: www.wklyradio.com. Licensee: WKLY Broadcasting Co. (acq 11-18-88; FTR: 12-19-88). Population served: 86,000 Rgnl. Network: Ga. Net. Format: Mainstream country, southern gospel, talk. News staff: 2; News: 18 hrs wkly. Target aud: 30 plus; middle class, working adults. ♦Bruce Hicks, CFO; Bryan Hicks, gen mgr & opns mgr.

Hawkinsville

WCEH(AM)— Dec 11, 1952: 610 khz; 500 w-D. TL: N32 16 50 W83 26 37. Stereo. Hrs opn: 24 Box 1398, Hwy. 341 S., 31036. Phone: (478) 892-9061. Fax: (478) 892-9063. Licensee: Georgia Eagle Broadcasting Inc. (acq 1-12-2007; grpsl). Population served: 143,000 Rgnl. Network: Ga. Net. Law Firm: Dan Alpert. Format: Country. News staff: 2; News: 30 hrs wkly. Target aud: 35 plus. Spec prog: Sports, farm 10 hrs, relg one hr wkly. ♦Hank Brigmond, gen mgr; Bill Boys, progmg dir.

WRPG(FM)—Co-owned with WCEH(AM). Sept 26, 1968: 103.9 mhz; 10.5 kw. Ant 495 ft TL: N32 10 03 W83 37 51. Stereo. Box 1398, 31036. Phone: (478) 892-9061. Fax: (478) 892-8663. E-mail: qwixie983@yahoo.com 143,000 Format: News/talk. ♦Hank Brigmond, gen mgr.

Hazlehurst

WVOH(AM)— Sept 6, 1962: 920 khz; 500 w-D, 39 w-N. TL: N31 51 15 W82 34 00. Stereo. Hrs opn: Box 645, 546 Baxley Hwy., 31539. Phone: (912) 375-4511. Fax: (912) 375-4512. Licensee: Jeff Davis Broadcasters Inc. Population served: 10,000 Natl. Network: USA. Format: Gospel. Spec prog: Farm 2 hrs wkly. ♦Tony DeLoach, gen mgr.

WVOH-FM— Dec 9, 1975: 93.5 mhz; 50 kw. 315 ft TL: N31 51 15 W82 34 00. Format: Classic country. ♦Tony DeLoach, gen mgr & traf mgr.

Helen

***WTFH(FM)**— February 2001: 89.9 mhz; 10 w. Ant 561 ft TL: N34 44 55 W83 43 43. Stereo. Hrs opn: 24 Box 780, TFC Radio Network, Toccoa Falls, 30598. Secondary address: 292 Old Clarksville Hwy., Toccoa Falls 30577. Phone: (706) 282-6030. Phone: (800) 251-8326. Fax: (706) 282-6090. E-mail: tfcm@tfc.edu Licensee: Toccoa Falls College. Population served: 10,000 Format: Christian. ♦David Cornelius, gen mgr.

WZGA(FM)— Dec 6, 1993: 105.1 mhz; 1.7 kw. Ant 613 ft TL: N34 44 55 W83 43 43. Stereo. Hrs opn: 24 Box 256, 30545. Phone: (706) 878-1051. Fax: (706) 878-1433. Licensee: Sorenson Southeast Radio LLC Group owner: Clear Channel Communications Inc. (acq 3-17-2006; $705,000). Format: Adult hits. ♦K.J. Allen, gen mgr.

Hiawassee

WNGM(AM)—Not on air, target date: unknown: 1230 khz; 1 kw-U. TL: N34 56 34 W83 46 27. Hrs open: 101 S. Main St., Ste 6, 30546. Secondary address: 38 Kenmare Hall, N.E., Atlanta 30324. Phone: (706 896-1230. Phone: (404) 266-2257. Licensee: BMG Broadcasting Inc. (acq 7-28-2006). Format: Classic Oldies. ♦Rick Morris, gen mgr; John Allen, mus dir; Jackie Grizzle, chief of engrg.

Hinesville

WGML(AM)— Dec 9, 1958: 990 khz; 250 w-D, 76 w-N. TL: N31 51 01 W81 36 04. Hrs open: Sunrise-sunset Box 615, 31310. Secondary address: 308 Rolland St. 31313. Phone: (912) 368-3399. Fax: (912) 368-4191. E-mail: wgml@coastalnow.net Licensee: Powerhouse of Deliverance Church Inc. (acq 10-12-94; FTR: 10-31-94). Population served: 32,000 Law Firm: Borsari & Paxson. Format: Gospel, relg. News: 10 hrs wkly. Target aud: General. Spec prog: One hr wkly. ♦Bishop Raymond Napper, CEO & pres; Elder Mary Napper, exec VP; Emanuel White, gen mgr.

WSGA(FM)— Aug 2, 1982: 92.3 mhz; 50 kw. Ant 482 ft TL: N31 41 37 W81 23 27. Stereo. Hrs opn: 24 Box 29, 31310. Secondary address: 120 D Liberty St. 31313. Phone: (912) 368-9258. Fax: (912) 368-5526. E-mail: wssjfm@coastalnow.net Web Site: www.freedom923.com. Licensee: Tama Radio Licenses of Savannah, GA, Inc. Group owner: Tama Broadcasting Inc. (acq 4-8-2004; $2.79 million). Population served: 250,000 Natl. Rep: Rgnl Reps. Law Firm: Miller & Miller. Format: Adult hits. News: 2 hrs wkly. Target aud: 25-54; Savannah, Hinesville, Brunswick, 15 county area. ♦Yvonne Clark, gen mgr.

WTHG(FM)— 1994: 104.7 mhz; 12 kw. Ant 469 ft TL: N31 51 18 W81 44 28. Stereo. Hrs opn: 24 Box 29, 31310. Secondary address: 120 D Liberty St. 31313. Phone: (912) 368-9258. Fax: (912) 368-5526. E-mail: wssjfm@coastalnow.net Web Site: thehawk1047.com. Licensee: Tama Radio Licenses of Savannah, GA Inc. (group owner; (acq 4-8-2004). Population served: 300,000 Format: Classic rock. ♦Yvonne Clark, gen mgr.

Hogansville

WMGP(FM)— Sept 3, 1992: 98.1 mhz; 25 kw. 328 ft TL: N33 03 54 W84 57 23. (CP: 25 kw). Hrs opn: 24 154 Boone Dr., Newnan, 30263. Secondary address: 300 Mooty Bridge Rd., Suite 204, LaGrange 30240. Phone: (770) 683-7234. Fax: (770) 683-9846. E-mail: magic981@charter.net Licensee: Citicasters Licenses L.P. Group owner: Clear Channel Communications Inc. (acq 1999; grpsl). Format: Adult hits. News staff: one; News: 20 hrs wkly. Target aud: 17-64; non-country listeners. ♦Joe Pedicino, gen mgr; Ed Miranda, gen sls mgr; Chris East, progmg dir; Jackie Steele, news dir; Donna Porter, traf mgr.

WVCC(AM)—Co-owned with WMGP(FM). Aug 12, 1985: 720 khz; 7.97 kw-D. TL: N33 03 54 W84 57 23.Sunrise-sunset E-mail: newsradio720@charter.net Web Site: www.720thevoice.com. Natl. Network: Fox Sports. Format: News/talk info. News staff: one. Target aud: 10 plus; Black. ♦Joe Pedicino, pres.

Homerville

WBTY(FM)— December 1980: 98.7 mhz; 6 kw. Ant 298 ft TL: N31 02 04 W82 51 50. Stereo. Hrs opn: Box 9, Dupont, 31630-0009. Secondary address: Intersection of Hwy's 168 & 37 31634. Phone: (912) 487-3412. Fax: (912) 487-3414. Licensee: Southern Broadcasting & Investments. (acq 3-9-90; $100,000; FTR: 4-30-90). Population served: 70,000 Format: Classic Hits. Target aud: General. ♦Jim Strickland, opns mgr & chief of engrg; Nancy K. Strickland, pres, gen mgr & gen sls mgr.

Irwinton

WVKX(FM)— September 1995: 103.7 mhz; 3 kw. 328 ft TL: N32 52 48 W83 11 07. Hrs open: Box 569, 31042. Phone: (478) 946-3445. Fax: (478) 946-2406. E-mail: love1037@alltel.net Licensee: Wilkinson Broadcasting Inc. (acq 8-10-92; $60,000; FTR: 8-31-92). Format: Rhythm and blues, gospel, urban contemp. ♦Stan Carter, gen mgr.

Jackson

WJGA-FM— Apr 24, 1967: 92.1 mhz; 2.15 kw. Ant 374 ft TL: N33 16 37 W83 57 59. Stereo. Hrs opn: 24 Box 878, 940 Brownlee Rd., 30233. Phone: (770) 775-3151. Fax: (770) 775-3153. Licensee: Earnhart Broadcasting Co. Inc. (acq 8-90; $800,000; FTR: 8-27-90). Population served: 40,000 Natl. Rep: Keystone (unwired net). Rgnl rep: Rgnl Reps. Wire Svc: AP Format: Adult contemp, Black. News staff: one; News: 20 hrs wkly. Target aud: General. Spec prog: Gospel 15 hrs wkly. ♦Don Earnhart, pres & gen mgr.

Jasper

***WNEE(FM)**— 1999: 88.3 mhz; 100 w. 3 ft TL: N34 28 01 W84 25 49. Stereo. Hrs opn: 24 Box 6767, Athens, 30604. Phone: (706) 425-1830. Fax: (706) 425-1868. E-mail: communitypublicradio@prodigy.net Web Site: www.cprmusic.net. Licensee: Community Public Radio Inc. Format: Btfl music, Christian. Target aud: 35-55; upper middle class, educated, Christian. ♦Penny Jackson, pres.

WYYZ(AM)— May 25, 1973: 1490 khz; 1 kw-U. TL: N34 28 32 W84 26 13. Hrs open: Box 280, 30143. Phone: (706) 692-4100. Fax: (706) 692-4012. Licensee: Enlightment LLC (acq 5-23-2007; $600,000). Natl. Network: CBS. Format: Classic country, gospel. ♦Mark Hellinger, gen mgr.

Jefferson

WBKZ(AM)—Licensed to Jefferson. See Athens

Jeffersonville

WPEZ(FM)— Sept 27, 1993: 93.7 mhz; 50 kw. 490 ft TL: N32 54 49 W83 29 47. Stereo. Hrs opn: 24 544 Mulberry St., Macon, 31202. Phone: (478) 746-6286. Fax: (478) 745-4383. Web Site: www.z937.com. Licensee: Cumulus Licensing Corp. Group owner: Cumulus Media Inc. (acq 12-20-02; grpsl). Population served: 300,000 Format: Lite Rock. News: one hr wkly. Target aud: 25-54. ♦John Sheftic, gen mgr.

Jesup

WIFO-FM—Listing follows WLOP(AM).

WLOP(AM)— July 12, 1949: 1370 khz; 5 kw-D, 36 w-N. TL: N31 36 06 W81 56 00. Hrs open: Box 647, 31598. Secondary address: 2420 Waycross Hwy 31545. Phone: (912) 427-3711. Fax: (912) 530-7717. E-mail: butch@bigdogcountry.com Licensee: Jesup Broadcasting Corp. (group owner; (acq 3-31-92). Population served: 24,000 Rgnl. Network: Ga. Net., Tobacco. Natl. Rep: Rgnl Reps. Format: News/talk, sports. Target aud: General. Spec prog: Farm 5 hrs, gospel 5 hrs wkly. ♦Charles Hubbard Jr., pres & gen mgr.

WIFO-FM—Co-owned with WLOP(AM). July 1, 1968: 105.5 mhz; 25 kw. Ant 308 ft TL: N31 36 06 W81 56 00. Stereo. Format: Country. Target aud: General.

***WLPT(FM)**— 1988: 88.3 mhz; 20 kw. Ant 800 ft TL: N31 40 27 W81 53 12. Stereo. Hrs opn: 24 2278 Wortham Ln, Grovetown, 30813. Phone: (706) 309-9610. Fax: (706) 309-9669. E-mail: ctbarinowski@comcast.net Web Site: www.gnnradio.org. Licensee: Augusta Radio Fellowship Institute Inc. Population served: 175,000 Format: Christian. News: 12 hrs wkly. Target aud: General. ♦Clarence Barinowski, gen mgr.

***WTLD(FM)**— January 2004: 90.5 mhz; 6 kw. Ant 171 ft TL: N31 35 49 W81 56 14. Stereo. Hrs opn: 24 Box 515, Jessup, 31598. Phone: (912) 695-7169. Phone: (912) 588-1821. Fax: (912) 588-1822. E-mail: lsmall7629@aol.com Licensee: Resurrection House Ministries Inc. Format: Gospel. News: 12 hrs wkly. Target aud: 18 plus. ♦Dr. Leonard Small, CEO; Marie Butler, gen mgr.

Kingsland

WKBX(FM)— Feb 23, 1987: 106.3 mhz; 6 kw. 330 ft TL: N30 48 04 W81 40 43. Stereo. Hrs opn: 24 Box 2525, 111 N. Grove Blvd., 31548. Phone: (912) 729-6106. Phone: (912) 729-5229. Fax: (912) 729-4106. E-mail: wkbx@k-bay106.com Web Site: www.k-bay106.com. Licensee: Radio Kings Bay Inc. (acq 7-1-89; $1 million; FTR: 5-29-89).

Population served: 49,000 Natl. Network: ABC. Law Firm: Wolf, Block, Schorr, & Solis-Cohen. Format: Country. News staff: one; News: one hr wkly. Target aud: 18-54; contemp country audience. ♦James Steele, pres & gen mgr; Wendy Steele, exec VP; John Fluery, progmg dir & progmg mgr; Dave Smith, news dir; Susan Pope, traf mgr.

La Fayette

WQCH(AM)— November 1954: 1590 khz; 5 kw-D. TL: N34 42 57 W85 16 06. Hrs open: 12 Box 746, 30728. Phone: (706) 638-3276. Fax: (706) 638-3896. E-mail: q-country@juno.com Licensee: Radix Broadcasting Inc. (acq 6-1-88). Natl. Network: AP Network News. Rgnl. Network: Ga. Net. Natl. Rep: Rgnl Reps. Format: Country, news. News staff: one; News: 10 hrs wkly. Target aud: 25 plus. Spec prog: Farm 2 hrs wkly. ♦Rich Gwyn, pres & gen mgr.

La Grange

WALR-FM— Sept 1, 1947: 104.1 mhz; 60 kw. Ant 1,217 ft TL: N33 24 43 W84 50 03. Stereo. Hrs opn: 1601 W. Peachtree St., Atlanta, 30309. Phone: (404) 897-7500. Fax: (404) 897-6495. Web Site: www.kiss1041fm.com. Licensee: Cox Radio Inc. Group owner: Cox Broadcasting (acq 8-2000; $280 million). Natl. Rep: McGavren Guild. Format: Urban & adult contemp. Target aud: 25-54. ♦Tony Kidd, gen mgr.

WELR-FM—See Roanoke, AL

WLAG(AM)— May 1, 1941: 1240 khz; 1 kw-U. TL: N33 02 24 W85 01 27. Hrs open: 24 Box 1429, 30241. Secondary address: 304 Broome St. 30240. Phone: (706) 845-1023. Fax: (706) 845-8642. E-mail: wlag@eagle1023.com Web Site: www.eagle1023.com. Licensee: Eagle's Nest Inc. (group owner; acq 4-3-92; $10; FTR: 4-27-92). Population served: 100,000 Format: Sports. News staff: one; News: 4 hrs wkly. Target aud: 25-54; general. ♦Jim Vice, gen mgr & stn mgr.

***WOAK(FM)**— June 11, 1984: 90.9 mhz; 3.4 kw. 299 ft TL: N32 57 57 W84 59 08. Hrs open: 1921 Hamilton Rd., 30241. Phone: (706) 884-2950. Fax: (706) 884-2930. Web Site: woak.com. Licensee: Oakside Christian School. Natl. Network: USA. Format: Educ, relg. ♦Rick Varnum, gen mgr.

WTRP(AM)— Jan 9, 1953: Stn currently dark. 620 khz; 1 kw-D, 127 w-N. TL: N33 03 33 W85 01 40. Hrs open: 806 New Franklin Rd., Lagrange, 30240. Phone: (706) 884-7022. Fax: (706) 884-7806. E-mail: wtrp@charter.net Licensee: Casey Network LLC (group owner; acq 11-13-02). Population served: 35,000 Format: Timeless classics. ♦Larry Fairall, gen mgr.

Lakeland

WVGA(FM)— 1994: 105.9 mhz; 6 kw. 328 ft TL: N31 04 55 W83 10 47. Stereo. Hrs opn: 1711 Ellis Dr., Valdosta, 31601. Phone: (229) 244-8642. Phone: (229) 241-1059. Fax: (229) 242-7620. Web Site: www.newstalk1059wvga.com. Licensee: RTG Radio LLC. Group owner: Black Crow Media Group LLC (acq 11-9-2001; grpsl). Population served: 150,000 Format: News/talk. ♦Robert Ganzak, pres; Robert T. Ganzak, gen mgr.

Lawrenceville

WPLO(AM)—Grayson, Jan 7, 1959: 610 khz; 1.5 kw-D, 225 w-N. TL: N33 57 11 W83 58 15. Stereo. Hrs opn: 24 2865 Amwiler Rd., Suite 650, Doraville, 30360. Phone: (770) 237-9897. Fax: (770) 246-0054. E-mail: laleysales@covat.net Web Site: www.radiomex610atlanta.com. Licensee: Teresa Prieto (acq 6-1-96). Population served: 250,000 Rgnl. Network: Ga. Net. Format: Sp, Mexican rgnl music. ♦Franca Vera, gen mgr.

Leesburg

WJAD(FM)— October 1989: 103.5 mhz; 12.5 kw. 460 ft TL: N31 39 09 W84 05 20. Hrs open: 1104 W. Broad Ave., Albany, 31707. Phone: (229) 888-5000. Fax: (229) 888-5960. Web Site: www.wjad.com. Licensee: Cumulus Licensing Corp. Group owner: Cumulus Media Inc. (acq 7-7-98). Population served: 160,000 Natl. Rep: Katz Radio. Format: Rock. Target aud: 25-40; Generation X, tail end of baby boomers. ♦Bill Jones, opns mgr; Gregory Kamishlian, mktg mgr.

Louisville

WPEH(AM)— Sept 10, 1960: 1420 khz; 1 kw-D, 159 w-N. TL: N33 00 48 W82 23 33. Hrs open: 6 AM-midnight Box 425, 5442 Middleground Rd., 30434. Phone: (912) 625-7248. Fax: (912) 625-7249. E-mail: wpeh@classicsouth.net Licensee: Peach Broadcasting Co. Inc. Rgnl. Network: Ga. Net. Law Firm: Holland & Knight. Format: Country. News: 11 hrs wkly. Target aud: General. ♦Ottis G. Stephens, pres, gen mgr & gen sls mgr; Sue Stephens, prom mgr; John D. Reid, mus dir; Wendell F. Stephens, progmg dir, news dir & chief of engrg.

WPEH-FM— May 6, 1971: 92.1 mhz; 3 kw. 296 ft TL: N33 00 48 W82 23 33. Target aud: 25 plus.

Lumber City

***WMOC(FM)**— April 1997: 88.7 mhz; 50 kw. Ant 210 ft TL: N31 55 48 W82 41 06. Hrs open: Box 520, 31549-0520. Secondary address: 412 Renwick St. 31549. Phone: (912) 363-2203. Fax: (912) 363-2106. E-mail: wmoc887@yahoo.com Web Site: www.wmoc887fm.com. Licensee: Full Gospel Church of God Written in Heaven. Format: Gospel. ♦Eddie Conaway, gen mgr & progmg dir.

Lumpkin

WKCN(FM)— Nov 6, 1992: 99.3 mhz; 50 kw. 492 ft TL: N32 09 25 W85 05 51. Stereo. Hrs opn: 24 1353 13th Ave., Columbus, 31901. Phone: (706) 596-9000. Fax: (706) 596-4600. Web Site: www.kissin993.com. Licensee: ABG Georgia LLC. Group owner: Archway Broadcasting Group (acq 4-25-03; grpsl). Natl. Rep: Christal. Law Firm: Fletcher, Heald & Hildreth. Format: New hot country. Target aud: 25-54; general. ♦Chuck Thompson, gen mgr.

***WTMQ(FM)**—Not on air, target date: unknown: 88.5 mhz; 10 kw. Ant 328 ft TL: N32 04 07 W84 51 55. Hrs open: Box 9382, Columbus, 31908. Phone: (706) 325-1170. Licensee: Spanish Cultural Education Inc. ♦Victor M. Molina, Jr., pres.

Lyons

WBBT(AM)— Mar 12, 1959: 1340 khz; 1 kw-U. TL: N32 12 50 W82 19 51. Stereo. Hrs opn: 24 Box 629, 473 N. Victory Dr., 30436. Phone: (912) 526-8122. Phone: (912) 526-6333. Fax: (912) 526-9155. E-mail: rday@tcbbroadcasting.com Web Site: www.tcbbroadcasting.com.ABC Classic R&B Licensee: T.C.B. Broadcasting Inc. (acq 6-16-97; $400,000 with co-located FM). Population served: 85,000 Natl. Network: ABC. Format: Classic rhythm and blues, Black. News staff: 2; News: 20 hrs wkly. Target aud: General. ♦Ray Bilbrey, CEO, pres, gen mgr, stn mgr & gen sls mgr; Robin Watson, progmg dir.

WLYU(FM)—Co-owned with WBBT(AM). Jan 1, 1989: 100.9 mhz; 6 kw. Ant 328 ft TL: N32 06 48 W82 23 52. Stereo. 24 Web Site: www.tcbbroadcasting.com.185,000 Format: Modern country. News staff: 3; News: 8 hrs wkly. Target aud: General. ♦Ralph Trapnell, exec VP; Toni Thompson, opns mgr & sls VP; Ray Bilbrey, chief of engrg; Robin Watson, traf mgr.

Mableton

WAMJ(FM)— 2001: 102.5 mhz; 3 kw. Ant 469 ft TL: N33 41 20 W84 30 38. Hrs open: 101 Marietta St. 12th Floor, Atlanta, 30303. Phone: (404) 765-9750. Fax: (404) -688-7686. Web Site: classicsoul1025.com. Licensee: ROA Licenses LLC. Group owner: Radio One Inc. (acq 7-8-2004; $31.5 million). Format: Classic soul. ♦Wayne Brown, gen mgr.

Macon

WAYS(AM)— August 1967: 1500 khz; 1 kw-D. TL: N32 48 47 W83 37 36. Hrs open: 544 Mulberry St., Suite 500, 31201. Phone: (478) 746-6286. Licensee: Cumulus Licensing Corp. Group owner: Cumulus Media Inc. (acq 12-20-2002; grpsl). Population served: 148,000 Natl. Network: Westwood One. Natl. Rep: Christal. Format: Oldies. Target aud: 25-54. ♦Steve Hazen, exec VP & gen mgr; Vickie Nahum, gen sls mgr; Gerry Marshall, progmg dir.

WIFN(FM)—Co-owned with WAYS(AM). June 10, 1968: 105.5 mhz; 6.1 kw. Ant 659 ft TL: N32 53 48 W83 32 05. Stereo. 24 300,000 Natl. Network: ESPN Radio. Format: Sports. ♦David Nolin, mus dir & disc jockey; Jim Jones, rgnl sls mgr & disc jockey.

***WBKG(FM)**— 2002: 88.9 mhz; 5.5 kw. Ant 502 ft TL: N32 45 51 W83 33 32. Hrs open: Box Drawer 2440, Tupelo, MS, 38803-2440. Phone: (662) 844-8888. Fax: (662) 842-6791. Web Site: www.afr.net. Licensee: American Family Association. Group owner: American Family Radio Format: Adult contemp. ♦Marvin Sanders, gen mgr; John Riley, progmg dir.

WBML(AM)— Oct 15, 1940: 900 khz; 2 kw-D, 145 w-N. TL: N32 50 58 W83 36 06. Hrs open: Box 6298, 31208. Secondary address: 735 Reese St. 31217. Phone: (478) 743-5453. Fax: (478) 743-9265. Licensee: WBML Inc. Group owner: Rodgers Broadcasting Corp. Population served: 435,000 Format: Relg. Target aud: 35 plus. ♦David A. Rogers, pres.

WDDO(AM)— Nov 25, 1957: 1240 khz; 1 kw-U. TL: N32 50 18 W83 39 02. Hrs open: 24 544 Mulberry St., Suite 500, 31201. Phone: (478) 746-6286. Fax: (478) 745-4383. E-mail: willie.collins@cumulus.com Licensee: Cumulus Licensing Corp. Group owner: Cumulus Media Inc. (acq 12-20-02; grpsl). Population served: 122,423 Natl. Network: American Urban. Natl. Rep: Christal. Format: Black gospel. ♦John Sheftic, gen mgr.

WDEN-FM—Listing follows WMAC(AM).

WIBB-FM—Fort Valley, Mar 3, 1993: 97.9 mhz; 10.5 kw. 499 ft TL: N32 34 12 W83 45 26. Stereo. Hrs opn: 24 7080 Industrial Hwy., 31216. Phone: (478) 781-1063. Fax: (478) 781-6711. Web Site: www.wibb.com. Licensee: AMFM Radio Licenses LLC. Group owner: Clear Channel Communications Inc. (acq 2-15-2001; grpsl). Population served: 508,300 Format: Hip hop & rhythm and blues. News staff: one; News: 14 hrs wkly. Target aud: 18-44. ♦Bill Clark, gen mgr.

WLCG(AM)— November 1948: 1280 khz; 5 kw-D, 99 w-N. TL: N32 48 16 W83 36 16. Hrs open: 24
Rebroadcasts WRNC(AM) Warner Robins 100%.
7080 Industrial Hwy., 31216-7538. Phone: (478) 781-1063. Fax: (478) 781-6711. E-mail: ccw@clearchannel.com Licensee: AMFM Radio Licenses LLC. Group owner: Clear Channel Communications Inc. (acq 2-15-2001; grpsl). Population served: 508,300 Natl. Rep: Clear Channel. Format: Black gospel. Target aud: General. ♦Bill Clark, gen mgr.

WLZN(FM)— August 1992: 92.3 mhz; 3 kw. 328 ft TL: N32 46 26 W83 38 15. Hrs open: 24 544 Mulberry St., 5th Fl., 31021. Phone: (478) 746-6286. Fax: (478) 742-8061. Licensee: Cumulus Licensing Corp. Group owner: Cumulus Media Inc. (acq 12-20-2002; grpsl). Population served: 300,000 Format: Urban contemp. News: one hr wkly. ♦Bill Hazen, gen mgr; Doug Rice, gen sls mgr; Brian Rayes, progmg dir; Joe Meredith, engrg VP & chief of engrg.

WMAC(AM)— Oct 30, 1922: 940 khz; 50 kw-D, 10 kw-N, DA-N. TL: N32 53 06 W83 43 50. Stereo. Hrs opn: 24 544 Mulberry St., Suite 500, 31201. Phone: (478) 746-6286. Fax: (478) 742-8061. Web Site: www.wmac-am.com. Licensee: Cumulus Licensing Corp. Group owner: Cumulus Media Inc. (acq 12-20-02; grpsl). Population served: 300,000 Natl. Network: ABC, Westwood One. Natl. Rep: McGavren Guild. Format: News/talk. News staff: 2; News: 40 hrs wkly. Target aud: 40 plus; upscale, college educated, household income $50K plus. Spec prog: Relg 4 hrs wkly. ♦Bill Hazen, gen mgr.

WDEN-FM—Co-owned with WMAC(AM). Feb 17, 1947: 99.1 mhz; 100 kw. Ant 581 ft TL: N32 45 51 W83 33 32. Stereo. 24 300,000 Natl. Rep: Christal. Format: Country. News staff: one; News: 4 hrs wkly. Target aud: 25-54; female/male split, median age 37.

WNEX(AM)— April 1945: 1400 khz; 1 kw-U. TL: N32 51 07 W83 39 12. Hrs open: 1691 Forsyth St., 31201. Phone: (478) 745-5858. Fax: (478) 745-0500. E-mail: phil@upga.tv Web Site: www.radiodisney.com. Licensee: Radio Peach Inc. (acq 1-31-00). Population served: 122,423 Natl. Network: Radio Disney. Format: Children. ♦Lowell Register, CEO & pres; Debbie Hart, gen mgr.

WNNG(AM)—See Warner Robins

WQBZ(FM)—Fort Valley, Apr 6, 1981: 106.3 mhz; 50 kw. 426 ft TL: N32 45 31 W83 44 49. Stereo. Hrs opn: 24 7080 Industrial Hwy., 31216. Phone: (478) 781-1063. Fax: (478) 781-6711. Web Site: www.q106.fm. Licensee: AMFM Radio Licenses LLC. Group owner: Clear Channel Communications Inc. (acq 2-15-2001; grpsl). Population served: 508,300 Format: Classic rock. Target aud: 18-49. ♦Bill Clark, gen mgr.

WRBV(FM)—See Warner Robins

WVVM(AM)—See Dry Branch

Madison

WYTH(AM)— June 1955: 1250 khz; 1 kw-D. TL: N33 34 45 W83 28 40. Hrs open: 6 AM-sunset Box 3965, Eatonton, 31024. Secondary address: 869 Church St., Eaton 31024. Phone: (706) 485-8792. Fax: (706) 485-3555. Licensee: Craig Baker and Debra Baker (acq 5-15-2005; $63,000). Population served: 272,440 Rgnl. Network: Ga. Net. Format: Adult standards. News staff: one; News: 15 hrs wkly. Target aud: General. ♦Craig Baker, pres & gen mgr.

Manchester

WFDR(AM)—Listing follows WVFJ-FM.

WVFJ-FM— 1967: 93.3 mhz; 100 kw. 1,250 ft TL: N32 50 40 W84 37 25. Stereo. Hrs opn: 24 120 Peachtree E. Shopping Ctr., Peachtree City, 30269. Phone: (770) 487-4500. Fax: (770) 486-6400. Web Site: www.j933.com. Licensee: Provident Broadcasting Co. (acq 8-81; $790,000 with co-located AM; FTR: 9-7-81) Population served: 3,185,000 Natl. Rep: Rgnl Reps. Law Firm: Brown, Nietert & Kaufman. Format: Contemporary, Christian. News staff: one. Target aud: 25-45; women. ♦Rick Davison, gen mgr & opns mgr; John Zeiler, rgnl sls mgr; Steve Williams, mktg dir; Don Schaeffer, progmg mgr; Susan Ricards, news dir; Brian Chin, chief of engrg & traf mgr; Dian Pena, traf mgr.

WFDR(AM)—Co-owned with WVFJ-FM. June 1957: 1370 khz; 1 kw-D. TL: N32 53 14 W84 35 54.6 AM-10 PM
Rebroadcasts WVFJ-FM Peachtree City.
Box 510, 31816. 4,779 Target aud: 25-54; women. Co-owned TV: WSB-TV affil

Marietta

WFOM(AM)— Oct 13, 1946: 1230 khz; 1 kw-U. TL: N33 55 38 W84 30 08. Hrs open: 24
Rebroadcasts WALR(AM) Atlanta 100%.
3535 Piedmont Rd., Bldg. 14, Suite 1200, Atlanta, 30305. Phone: (404) 688-0068. Fax: (404) 995-4045. E-mail: scottmcfarlane@680thefan.com Web Site: www.talkradio1340.com. Licensee: Dickey Broadcasting Co. (group owner; acq 8-31-00; grpsl). Population served: 18,000 Natl. Network: CBS Radio, Sporting News Radio Network, Westwood One. Natl. Rep: McGavren Guild. Law Firm: Holland & Knight. Wire Svc: AP Format: Talk. News staff: 2; News: 10 hrs wkly. Target aud: 25-64. Spec prog: Notre Dame football (fall). ♦David Dickey, gen mgr; Scott McFarlane, opns dir; Rob Hasson, sls dir.

WFTD(AM)— Nov 14, 1955: 1080 khz; 50 kw-D, 30 kw-CH, DA-2. TL: N34 01 24 W84 40 05. Hrs open: Sunrise-sunset 2865 Amwiler Rd., Suite 650, Doraville, 30360. Phone: (770) 825-0095. Fax: (770) 246-0054. Web Site: www.radiolaley.com. Licensee: Prieto Enterprises Inc. (acq 12-20-01). Law Firm: David Tillotson. Format: Sp. Target aud: 21-45. ♦Filiberto Prieto, pres & gen mgr.

WKHX-FM—Licensed to Marietta. See Atlanta

Martinez

WDRR(FM)— May 31, 1984: 93.9 mhz; 25 kw. Ant 328 ft TL: N33 26 17 W82 05 19. Stereo. Hrs opn: 24 4051 Jimmie Dyess Pkwy., Augusta, 30909. Phone: (706) 396-7000. Fax: (706) 396-7092. Web Site: www.939thedrive.com. Licensee: WGOR License LLC. (acq 11-10-92; $810,000; FTR: 11-30-92). Population served: 400,000 Format: Classic hits. Target aud: 25 plus. ♦Kent Dunn, VP & gen mgr; Kent Murphy, gen sls mgr; Chris O'Kelley, progmg dir.

WPRW-FM— 1994: 107.7 mhz; 50 kw. Ant 492 ft TL: N33 38 35 W82 19 50. Hrs open: 2743 Perimeter Pky. Bldg. 100, Suite 300, Augusta, 30909. Phone: (706) 396-6000. Fax: (706) 396-6010. Web Site: www.power107.net. Licensee: Capstar TX L.P. Group owner: Cumulus Media Inc. (acq 6-30-97; grpsl). Format: Urban contemp. ♦Mark Bass, gen mgr.

McDonough

WKKP(AM)— Apr 2, 1979: 1410 khz; 2.5 kw-D. TL: N33 25 47 W84 07 52. Hrs open: 24
Solid Gospel Network.
Box 878, Jackson, 30233. Secondary address: 940 Brownlee Rd., Jackson 30233. Phone: (770) 504-8410. Fax: (770) 775-3153. E-mail: donaldwearnhart@bellsouth.net Licensee: Henry County Radio Co. Inc. (acq 3-30-92; $65,000; FTR: 4-20-92). Population served: 65,000 Natl. Network: Salem Radio Network. Rgnl. Network: Ga. Net. Natl. Rep: Rgnl Reps. Format: Gospel. News: 20 hrs wkly. Target aud: General. ♦Susanne Earnhart, pres; Don Earnhardt, gen mgr; Tom Lynde, opns dir.

McRae

WYIS(AM)— July 27, 1957: 1410 khz; 1 kw-D. TL: N32 03 25 W82 51 56. Hrs open: Box 247, Hwy. 341 S., 31055. Phone: (229) 868-5611. Fax: (229) 868-7552. Licensee: Cinecom Broadcasting Systems Inc. (acq 8-31-99; $220,000 with co-located FM). Natl. Rep: Rgnl Reps. Format: Oldies. Target aud: 25-50; mature, wage earners. Spec prog: Black 4 hrs wkly. ♦Jimmy Hussey, gen mgr.

WYSC(FM)—Co-owned with WYIS(AM). Aug 3, 1979: 102.7 mhz; 3 kw. 289 ft TL: N32 03 25 W82 51 56.3,151

Meigs

WQLI(FM)— Sept 8, 2000: 92.3 mhz; 6 kw. Ant 328 ft TL: N31 05 12 W84 12 10. Hrs open: 2586 Old Pelham Rd., Pelham, 31779. Phone: (229) 294-1909. Licensee: Mitchell County Television. Format: Adult contemp.

Metter

WBMZ(FM)—Listing follows WHCG(AM).

WHCG(AM)— Dec 22, 1961: 1360 khz; 1 kw-D. TL: N32 23 56 W82 02 36. Hrs open: Box 238, 30439. Secondary address: 1075 E. Lillian St. 30439. Phone: (912) 685-2136. Fax: (912) 685-2137. Web Site: www.wbmzfm.com. Licensee: Radio Metter Inc. Population served: 75,000 Law Firm: Reddy, Begley & McCormick. Format: Southern gospel. Target aud: General. Spec prog: Gospel 4 hrs wkly. ♦Jimmy Page, pres, gen mgr & gen sls mgr.

WBMZ(FM)—Co-owned with WHCG(AM). Aug 1, 1971: 103.7 mhz; 3 kw. 299 ft TL: N32 23 56 W82 02 36. Stereo. 24 Format: Classic hits. Target aud: General. ♦Jimmy Page, CEO.

Midway

WGCO(FM)—Licensed to Midway. See Savannah

Milan

WMCG(FM)— 1982: 104.9 mhz; 36 kw. Ant 564 ft TL: N32 07 16 W83 16 05. Stereo. Hrs opn: 24 Box 130, Dublin, 31040-0130. Phone: (478) 272-4422. Fax: (478) 275-4657. E-mail: macd@wqzy.com Licensee: Tel-Dodge Broadcasting Inc. Format: Classic country. Target aud: 25-54; mature country fans. Spec prog: Farm one hr wkly. ♦Mac Davis, gen mgr.

Milledgeville

***WGUR(FM)**— August 1975: 88.9 mhz; 37 w horiz. Ant 3 ft TL: N33 04 44 W83 13 55. Hrs open: Box 3124, Georgia College & State University, 31061. Phone: (478) 445-8256. Licensee: Georgia College & State University (acq 8-75). Population served: 4,500 Format: Alternative. Target aud: 18-24; college students. ♦Sonya Barnes, gen mgr.

WKZR(FM)—Listing follows WMVG(AM).

WLRR(FM)— July 24, 1990: 100.7 mhz; 3 kw. 328 ft TL: N33 06 50 W83 13 08. Hrs open:
Rebroadcasts WKVQ(AM) Eatonton 100%.
Box 3965, Eatonton, 31024. Phone: (706) 485-8792. Fax: (706) 485-3555. Licensee: Preston W. Small. Format: Adult standards. Target aud: 18-35. ♦Craig Baker, pres, gen mgr & opns VP.

WMVG(AM)— Mar 29, 1946: 1450 khz; 1 kw-U. TL: N33 04 58 W83 15 01. Hrs open: 24 Box 519, 1250 W. Charlton St., 31061. Phone: (478) 452-0586. Fax: (478) 452-5886. Licensee: WMVG Inc. (acq 7-8-99; $258,230 for 80% with co-located FM). Population served: 40,000 Natl. Rep: Rgnl Reps. Format: Sports, news. News staff: one; News: 25 hrs wkly. Target aud: 18-49. Spec prog: Black 4 hrs wkly. ♦Randy Beasley, pres, gen mgr & edit dir.

WKZR(FM)—Co-owned with WMVG(AM). June 30, 1966: 102.3 mhz; 3.3 kw. 300 ft TL: N33 04 58 W83 15 01. Stereo. 80,000 Format: Country.

Millen

WHKN(FM)— Dec 4, 1989: 94.9 mhz; 14.5 kw. Ant 400 ft TL: N32 43 57 W81 51 43. Stereo. Hrs opn: 24 35 E. Main St., Statesboro, 30458. Phone: (912) 764-1029. Fax: (912) 489-3959. E-mail: radiocenter @frontiernet.net Licensee: Georgia Eagle Broadcasting Inc. (group owner; acq 1-12-2007; grpsl). Natl. Network: ABC. Law Firm: Dan Alpert. Format: Country. News staff: one; News: 8 hrs wkly. Target aud: 25-54; adults. Spec prog: Farm 10 hrs, relg 2 hrs wkly. ♦Cecil Staton, pres; Jeff Anderson, gen mgr.

Monroe

WKUN(AM)— Feb 4, 1971: 1490 khz; 1 kw-U. TL: N33 48 37 W83 42 01. Hrs open: 6 AM-6 PM Box 649, 30655. Secondary address: 1610 Launius Rd., Good Hope 30641. Phone: (770) 267-0923. Fax: (706) 342-8135. Licensee: B.R. Anderson Sr. dba Radio Station WKUN (acq 6-96). Population served: 10,300 Rgnl. Network: Ga. Net. Format: Southern gospel. News: 3 hrs wkly. Target aud: 25-65. ♦B. R. Anderson Sr., pres; Melanie Jackson, gen mgr.

Montezuma

WMGB(FM)— Aug 10, 2001: 95.1 mhz; 46 kw. 390 ft TL: N32 33 20 W83 44 14. Stereo. Hrs open: 24 544 Mulberry St., Suite 500, Macon, 31201. Phone: (478) 746-6286. Fax: (478) 745-4383. Web Site: www. all the hitsB951.com. Licensee: Cumulus Licensing Corp. Group owner: Cumulus Media Inc. (acq 12-20-02; grpsl). Population served: 300,000 Natl. Network: Westwood One. Format: CHR. ♦John Sheftic, gen mgr.

WMNZ(AM)— Nov 29, 1961: 1050 khz; 250 w-D, 42 w-N. TL: N32 17 58 W84 01 34. (CP: TL: N32 17 53 W84 02 02). Hrs opn: Box 610, 31063. Secondary address: 115 1/2 Cherry St. 31063. Phone: (478) 472-8386. Fax: (478) 472-8296. Licensee: Macon County Broadcasting Co. Population served: 4,125 Format: Country, oldies, gospel. ♦Danny Blizzard, pres & gen mgr.

Morrow

WIGO(AM)— November 1956: 1570 khz; 5 kw-D, 50 w-N. TL: N33 36 05 W84 18 40. Hrs open: 2424 Old Rex Morrow Rd., Ellenwood, 30296. Phone: (404) 361-1570. Fax: (404) 366-9772. Web Site: www.wssathelight1570am.com. Licensee: MCL/MCM Georgia LLC. (acq 12-29-2006; $1.75 million). Population served: 100,000 Format: Gospel, Christian. Target aud: 24-55. ♦Paul Ploener, gen mgr; Leah St. Cyr, stn mgr.

Moultrie

WHBS(AM)— 2002: 1400 khz; 1 kw-U. TL: N31 09 56 W83 46 01. Hrs open: 24 1643 South Blvd., 31768. Phone: (229) 890-2900. Fax: (229) 890-1497. Web Site: whbsam1400.com. Licensee: Sailor Broadcasting of Georgia Inc. (acq 3-3-2005; $195,000). Format: Urban contemp, Black gospel. ♦Ronnie Barnes, gen mgr.

WMTM(AM)— Nov 10, 1953: 1300 khz; 5 kw-D. TL: N31 12 54 W83 47 13. Stereo. Hrs opn: 6 AM-sunset Box 788, 31776. Secondary address: 100 WMTM Rd. 31768. Phone: (229) 985-1300. Fax: (229) 890-0905. Licensee: Colquitt Broadcasting Co. L.L.C. Population served: 14,302 Natl. Rep: Rgnl Reps. Law Firm: Smithwick & Belendiuk. Format: Southern gospel. News staff: one. Target aud: General. Spec prog: Farm 16 hrs wkly. ♦Jim Turner, pres & gen mgr.

WMTM-FM— Nov 17, 1964: 93.9 mhz; 100 kw. 555 ft TL: N31 12 54 W83 47 13. Stereo. 6 AM-midnight Format: Oldies. ♦Jim Turner, opns mgr, mktg mgr & traf mgr.

Mount Vernon

WYUM(FM)— Aug 3, 1998: 101.7 mhz; 6 kw. 325 ft TL: N32 12 44 W82 27 48. Hrs open: Box 900, Vidalia, 30475. Phone: (912) 537-9202. Fax: (912) 537-4477. E-mail: zfowler@vidaliacommunications.com Web Site: www.vidaliacommunications.com. Licensee: Vidalia Communications Corp. (group owner) Population served: 30,000 Rgnl. Network: Ga. Net. Rgnl rep: Rgnl Reps. Format: Country. Target aud: 25-49. ♦John Ladson III, pres; Zack Fowler, gen mgr; Collins Knightor, opns dir.

Mountain City

WALH(AM)— May 1, 1986: 1340 khz; 1 kw-U. TL: N34 56 16 W83 23 27. Hrs open: Box F, 30562. Phone: (706) 746-2256. Fax: (706) 746-2259. E-mail: walh@alltel.net Web Site: www.wolfcreekbroadcasting.com. Licensee: Mountain City Broadcasting Inc. (acq 1-19-2005; $275,000). Format: Country, bluegrass, gospel. Target aud: 30-50; blue collar. Spec prog: Farm 2 hrs wkly. ♦Rebecca St. John, gen mgr.

Murrayville

WGTJ(AM)—Licensed to Murrayville. See Gainesville

Nashville

WJYF(FM)—Licensed to Nashville. See Tifton

Newnan

WCOH(AM)— December 1947: 1400 khz; 1 kw-U. TL: N33 21 53 W84 48 42. Hrs open: 154 Boone Dr., 30263. Phone: (770) 683-7234. Fax: (770) 683-9846. Web Site: www.wcoh.com. Licensee: Citicasters Licenses L.P. Group owner: Clear Channel Communications Inc. (acq 5-4-99; grpsl). Population served: 57,000 Rgnl. Network: Ga. Net. Law Firm: Miller & Miller, P.C. Format: Classic country. Target aud: 25-54. ♦Jim Martin, gen mgr.

WNEA(AM)— Apr 18, 1962: 1300 khz; 1 kw-D. TL: N33 22 31 W84 47 08. Hrs open: 8451 South Cherokee Blvd., Suite B, Douglasville, 30134. Phone: (770) 920-1520. Fax: (770) 920-4600. Web Site: www.wordchristianbroadcasting.com. Licensee: Word Christian Broadcasting Inc. (acq 2-28-96; FTR: 3-11-96). Population served: 40,000 Rgnl. Network: Ga. Net. Format: Old time relg. Target aud: 18-64. Spec prog: Black, relg, gospel 15 hrs wkly. ♦Ken Johns, CEO, gen mgr, opns dir & progmg dir.

North Atlanta

WCNN(AM)—Licensed to North Atlanta. See Atlanta

Ochlocknee

WJEP(AM)— June 4, 1984: 1020 khz; 10 kw-D. TL: N30 54 00 W83 59 55. Hrs open: Sunrise-sunset Box 90, Thomasville, 31799. Secondary address: 540 Daisy Ln., Thomasville 31792. Phone: (912) 228-5683. Fax: (912) 436-0544. E-mail: wjep@rose.net Web Site: www.lifelineministries.com. Licensee: Lifeline Ministries Inc. (acq 8-18-83). Format: Christian contemp. Spec prog: Black 2 hrs, southern gospel 3 hrs wkly. ♦Jimmy Bennett, pres & gen mgr.

Ocilla

WLPF(FM)— December 1993: 98.5 mhz; 2.3 kw. 521 ft TL: N31 28 11 W83 14 11. Stereo. Hrs opn: 24 2278 Wortham Ln., Grovetown, 30813. Phone: (706) 309-9610. Fax: (706) 736-9669. E-mail: ctbarinowski@comcast.net Web Site: www.gnnradio.org. Licensee: Barinowski Investment Co. Group owner: Good News Network (acq 11-17-92; for CP; FTR: 12-7-92). Population served: 65,000 Format: Christian. News: 12 hrs wkly. Target aud: General. ♦Clarence Barinowski, gen mgr.

Omega

WTIF-FM— April 1993: 107.5 mhz; 4 kw. Ant 400 ft TL: N31 27 17 W83 33 37. Hrs open: 24 Box 968, Tifton, 31793. Phone: (229) 382-1340. Fax: (229) 386-8658. Web Site: www.1075wtif.com. Licensee: Three Trees Communications Inc. (group owner; (acq 6-28-2004; grpsl). Population served: 100,000 Rgnl. Network: Ga. Net. Format: Christian music. News staff: 6; News: 5 hrs wkly. Target aud: 18 plus. ♦Matt Baldrich, pres & gen mgr.

Pavo

***WVRI(FM)**— 2005: 90.5 mhz; 1 w horiz, 50 kw vert. Ant 292 ft vert TL: N31 09 26 W83 22 28. Hrs open:
Rebroadcasts KLRD(FM) Yucaipa, CA 100%.
2351 Sunset Blvd., Suite 170-218, Rocklin, CA, 95765. Phone: (916) 251-1600. Fax: (916) 251-1650. E-mail: info@air1.com Web Site: www.air1.com. Licensee: Educational Media Foundation. Group owner: EMF Broadcasting. (acq 1-7-2004). Natl. Network: Air 1. Format: Christian. ♦Richard Jenkins, pres; Mike Novak, VP; Keith Whipple, dev dir; David Pierce, progmg mgr; Ed Lenane, news dir; Sam Wallington, engrg dir; Karen Johnson, news rptr; Marya Morgan, news rptr; Richard Hunt, news rptr.

Peachtree City

WLTM(FM)— 1948: 96.7 mhz; 1 kw. Ant 545 ft TL: N33 26 22 W84 42 42. Stereo. Hrs opn: 1819 Peachtree Rd. N.E., Suite 700, Atlanta, 30309. Phone: (404) 367-0949. Fax: (404) 367-9490. Web Site: www.949litefm.com. Licensee: Citicasters Licenses L.P. Group owner: Clear Channel Communications Inc. (acq 5-4-99; grpsl). Population served: 200,000 Law Firm: Miller & Miller. Format: Soft adult contemp. ♦Chuck Deskins, gen mgr; Mike Lawing, chief of engrg.

***WMVW(FM)**—Not on air, target date: unknown: 91.7 mhz; 10 kw. Ant 328 ft TL: N33 16 03 W84 33 20. Hrs open: 100 S. Hill St., Suite 100, Griffin, 30223. Phone: (770) 229-2020. Fax: (770) 229-4820. Licensee: Life Radio Ministries Inc. ♦Joseph Emert, pres.

Pearson

WPNG(FM)— August 1999: 101.9 mhz; 12.9 kw. Ant 459 ft TL: N31 19 36 W82 51 54. Stereo. Hrs opn: 24 Box 823, 31642. Secondary address: 2232 Old Douglas Hwy. 31642. Phone: (912) 422-6122. Fax: (912) 422-7840. E-mail: freedom1019@planttel.net Web Site: www.hitsandfavorites.com. Licensee: KM Radio of Pearson L.L.C. Group owner: KM Communications Inc. (acq 5-3-99). Population served: 532,000 Law Firm: Cohen, Dipple & Everist. Format: Adult contemp. Target aud: 25-49; women ages 25-49. ♦Myoung Hwa Bae, pres; Kevin Bae, gen mgr.

Pelham

WZBN(FM)—See Camilla

Pembroke

WBAW-FM— Aug 31, 1966: Stn currently dark. 99.3 mhz; 100 kw. Ant 869 ft TL: N32 07 15 W81 47 56. Stereo. Hrs opn: 321 Fraser Dr., Hinesville, 31313. Licensee: Bullie Broadcasting Corp. (acq 2-5-99; $475,000). Population served: 62,000 Target aud: General.

Perry

WPGA(AM)— 1955: 980 khz; 5 kw-D, 270 w-N. TL: N32 26 40 W83 45 00. Stereo. Hrs opn: 24 1691 Forsyth St., Macon, 31201. Phone: (478) 745-5858. Fax: (478) 745-5800. Web Site: www.58abc.com. Licensee: Register Communications Inc. (acq 1996). Population served: 80,000 Format: Children. News staff: one. Target aud: Children up to 12. ♦Debbie Hart, gen mgr; Loel Register, pres & gen sls mgr; Janice J. Register, women's int ed.

WPGA-FM— May 3, 1966: 100.9 mhz; 3 kw. 345 ft TL: N32 33 20 W83 44 14. (CP: 2.15 kw, ant 551 ft.). Stereo. Phone: (478) 745-5500. Web Site: www.58abc.com.250,000 Format: Adult contemp. News staff: one. ♦Kristy Turner, gen sls mgr; Janice J. Register, women's int ed. Co-owned TV: WPGA-TV affil

Pinehurst

WQXZ(FM)— Feb 22, 1969: 98.3 mhz; 3.1 kw. Ant 459 ft TL: N32 10 03 W83 37 51. Stereo. Hrs opn: 24 Box 667, Cordele, 31010. Phone: (229) 271-3500. Fax: (229) 273-4900. E-mail: qwixie983@yahoo.com Licensee: Georgia Eagle Broadcasting Inc. (group owner; (acq 1-12-2007; grpsl). Population served: 40,000 Natl. Network: Motor Racing Net, Jones Radio Networks. Law Firm: Dan Alpert. Format: Oldies. News staff: one; News: 15 hrs wkly. Target aud: 35-64. ♦Cecil Staton, pres; Hank Brigmond, gen mgr.

Port Wentworth

***WLFS(FM)**— 2001: 91.9 mhz; 6 kw. Ant 180 ft TL: N32 09 17 W81 09 55. Hrs open: 5859 Abeerorn St., Suite 3, Savannah, SC, 31405. Phone: (912) 353-9226. Fax: (912) 353-9325. Web Site: www.hisradio.com. Licensee: Radio Training Network Inc. Format: Christian contemp. ♦Allen Henderson, gen mgr.

Quitman

WSFB(AM)— Nov 19, 1955: 1490 khz; 1 kw-U. TL: N30 46 51 W83 34 30. Hrs open: Box 632, 31643. Phone: (229) 263-4373. Fax: (229) 263-7693. Licensee: Scott Matheson (acq 11-1-2005; $10,000). Population served: 63,462 Format: Adult standards. Target aud: 30+. ♦Scott Matheson, gen mgr.

WSTI-FM— Sept 12, 1986: 105.3 mhz; 3 kw. 300 ft TL: N30 48 45 W83 31 18. Stereo. Hrs opn: 1711 Ellis Dr., Valdosta, 31601. Phone: (229) 244-8642. Fax: (229) 242-7620. Licensee: RTG Radio L.L.C. Group owner: Black Crow Media Group LLC (acq 6-4-2004; $3.4 million with WXHT(FM) Madison, FL). Format: Urban contemp. Target aud: 25-54; white collar. Spec prog: Farm 5 hrs wkly. ♦Scott James, gen mgr.

Reidsville

WRBX(FM)—Listing follows WTNL(AM).

WTNL(AM)— June 25, 1976: 1390 khz; 500 w-D. TL: N32 05 14 W82 07 47. Hrs open: 6 AM-sunset Box 69, 30453. Secondary address: 125 Friartuck Cir. 30453. Phone: (912) 557-3777. Fax: (912) 557-6956. Licensee: WRBX/WTNL L.L.C. (acq 3-93; $35,000 with co-located FM; FTR: 2-15-93). Population served: 18,000 Natl. Network: USA. Format: Southern gospel, talk. Target aud: General. ♦Truman Blankenship III, CEO; Larry Montgomery, pres; Gregory J. Lemon, VP, gen mgr, gen sls mgr, adv dir & disc jockey; Dawn Lemon, progmg dir, traf mgr & disc jockey; Bob Moughton, engrg VP & chief of engrg.

WRBX(FM)—Co-owned with WTNL(AM). July 1993: 104.1 mhz; 3 kw. 187 ft TL: N32 05 14 W82 07 47.24 160,000 Natl. Network: USA. Target aud: 8 plus; religious. ♦Bob Moughton, chief of engrg; Dawn Lemon, traf mgr & disc jockey; Gregory J. Lemon, mus dir & disc jockey.

Richmond Hill

WRHQ(FM)— May 13, 1991: 105.3 mhz; 11 kw. 485 ft TL: N32 02 52 W81 07 26. Stereo. Hrs opn: 24 1102 E. 52nd St., Savannah, 31404. Phone: (912) 234-1053. Fax: (912) 354-6600. E-mail: qualityrock@wrhq.com Web Site: www.wrhq.com. Licensee: Thoroughbred Communications Inc. Group owner: Thoroughbred Communications Natl. Rep: Christal. Law Firm: Shook, Hardy & Bacon. Format: Rock, adult contemp. News: 2 hrs wkly. Target aud: 25-54; affluent. ♦Jerry Rogers, pres, gen mgr, gen sls mgr, adv mgr & edit dir; Ray Williams, rgnl sls mgr & prom mgr; Keith Hendrix, progmg dir, mus critic & disc jockey; Lyndy Brannan, pub affrs dir; Marty Foglia, chief of engrg; Alicia Bailey, traf mgr; Rusty Fredrich, sports cmtr; Brady McGraw, disc jockey; Lyndy Brannen, disc jockey.

Rincon

WSSJ(FM)— May 1, 1967: 100.1 mhz; 50 kw. Ant 492 ft TL: N32 16 49 W81 11 40. Stereo. Hrs opn: 24 Box 29, Hinesville, 31313. Secondary address: 120 D. Liberty, Hinesville 31313. Phone: (912) 368-9258. Fax: (912) 368-5526. E-mail: wssjfm@coastalnow.net Web Site: www.smoothjazz100.com. Licensee: Tama Radio Licenses of Savannah, GA Inc. Group owner: Tama Broadcasting Inc. (acq 4-28-2004). Population served: 108,000 Format: Smooth jazz. News staff: one. Target aud: 18-45. ♦Yvonne Clark, gen mgr.

Ringgold

WMPZ(FM)—Licensed to Ringgold. See Chattanooga TN

WOCE(FM)— March 1989: 101.9 mhz; 1.32 kw. Ant 702 ft TL: N34 58 11 W85 05 10. Hrs open: 7413 Old Lee Hwy., Chattanooga, TN, 37421. Phone: (423) 892-3333. Fax: (423) 553-9490. Licensee: North Georgia Radio Group L.P. Group owner: Clear Channel Communications Inc. (acq 7-28-2006; $2.15 million). Format: Classic rock. ♦Sammy George, gen mgr.

Rockmart

WTSH-FM— August 1989: 107.1 mhz; 45 kw. Ant 518 ft TL: N34 15 03 W84 59 05. Hrs open: 20 John Davenport Dr., Rome, 3016s. Phone: (706) 291-9496. Fax: (706) 235-7107. Web Site: www.south107.com. Licensee: Woman's World Broadcasting Inc. (acq 11-14-2003; $5.4 million). Population served: 496,000 Format: Country. Target aud: 25-54. ♦Randy Quick, gen mgr.

WZOT(AM)— Aug 28, 1959: 1220 khz; 500 w-D, 150 w-N. TL: N34 00 14 W85 03 22. Hrs open: 602 W. Elm St., 30153. Phone: (770) 684-7848. Fax: (770) 684-7848. Licensee: Triple J's Broadcasting LLC (acq 9-1-2004; $346,804 with WGJK(AM) Rome). Rgnl. Network: Ga. Net. Format: Southern gospel. Target aud: 18-45. ♦Paul Stone, pres; Randy Quick, gen mgr.

Rome

WGJK(AM)— Aug 1, 1962: 1360 khz; 500 w-D, 150 w-N. TL: N34 16 15 W85 11 00. Hrs open: 6 AM-12 PM 1010 Tower Pl., Bogart, 30622. Phone: (706) 549-6222. Fax: (706) 353-1967. Licensee: Woman's World Broadcasting Inc. (acq 11-17-2006). Population served: 96,000 ♦Suzanne Stone, pres.

WGPB(FM)— May 22, 1965: 97.7 mhz; 25 kw. Ant 790 ft TL: N34 14 00 W85 14 02. Stereo. Hrs opn: 24 Heritage Hall, 415 E. Third Ave., 30161. Phone: (706) 204-2276. Web Site: www.gpb.org/public/. Licensee: Briar Creek Broadcasting Corp. (acq 6-29-2006; $4.2 million). Population served: 265,000 Natl. Network: NPR. Law Firm: Arnold & Porter. Format: News, classical music. ♦Tom Barclay, stn mgr.

WLAQ(AM)— 1947: 1410 khz; 1 kw-U, DA-N. TL: N34 15 43 W85 12 22. Hrs open: 2 Mount Alto Rd., 30165. Phone: (706) 232-7767. Fax: (706) 295-9225. E-mail: wlaq1410am@hotmail.com Web Site: www.wlaq.com. Licensee: Cripple Creek Broadcasting Co. (acq 4-1-87). Population served: 80,000 Natl. Network: CBS. Format: News/talk, sports. ♦Randy Davis, pres & gen mgr.

WQTU(FM)—Listing follows WRGA(AM).

WRGA(AM)— November 1929: 1470 khz; 5 kw-U, DA-N. TL: N34 18 05 W85 09 19. Hrs open: 24 20 John Davenport DR., 30165. Phone: (706) 291-9496. Fax: (706) 235-7107. E-mail: south107@aol.com Web Site: www.wrgarome.com. Licensee: McDougald Broadcasting Corp. Group owner: Southern Broadcasting Companies Inc. (acq 1-28-2002; $1.6 million with co-located FM). Population served: 98,000 Natl. Network: ABC, CNN Radio. Rgnl. Network: Ga. Net. Law Firm: Fletcher, Heald & Hildreth. Wire Svc: National Weather Network Format: News/talk. News staff: 2; News: 168 hrs wkly. Target aud: General; upscale, involved, upwardly mobile. ♦Paul Stone, pres; Gregory Kamishlian, gen mgr.

WQTU(FM)—Co-owned with WRGA(AM). May 2, 1966: 102.3 mhz; 6 kw. 804 ft TL: N34 14 02 W85 13 50. Stereo. 24 Phone: (706) 295-1023. E-mail: q102rome@q102rome.com Web Site: q102rome.com.98,000 Format: Hot adult contemp. News staff: one; News: 6 hrs wkly. Target aud: 25-54; upscale.

WROM(AM)— Dec 26, 1946: 710 khz; 1 kw-D. TL: N34 15 30 W85 09 15. Hrs open: Sunrise-sunset 1105 Calhoun Ave., 30162. Phone: (706) 234-7171. Fax: (706) 234-8043. E-mail: wromradio@comcast.net Web Site: www.wromradio.com. Licensee: LGV Broadcasting Inc. (acq 1999; $150,000). Population served: 149,000 Natl. Network: USA. Law Firm: Maupin, Taylor, Ellis & Adams. Format: Southern gospel. News staff: one; News: 14 hrs wkly. Target aud: 35 plus; middle class families, women, homeowners. Spec prog: Christian teaching, contemp Christian mus. ♦Mark Lumpkin, gen mgr.

Rossville

WRXR-FM— June 8, 1966: 105.5 mhz; 1.55 kw. 646 ft TL: N34 57 26 W85 17 33. Stereo. Hrs opn: 24 7413 Old Lee Hwy., Chattanooga, TN, 37421. Phone: (423) 892-3333. Fax: (423) 899-7224. Licensee: Capstar TX L.P. (acq 8-7-2000; grpsl). Population served: 366000 Natl. Rep: Clear Channel. Format: Active Rock. News: one hr wkly. ♦Sammy George, gen mgr.

WUUS(AM)— Nov 11, 1958: 980 khz; 500 w-D. TL: N34 58 03 W85 18 00. Hrs open: Sunrise-sunset 7413 Old Lee Hwy., Chattanooga, TN, 37421. Phone: (423) 892-3333. Fax: (423) 899-7224. Licensee: 3 Daughters Media Inc. Group owner: Clear Channel Communications Inc. (acq 6-22-2007; grpsl). Natl. Rep: Clear Channel. Format: Oldies. News staff: one; News: 4 hrs wkly. Target aud: 35-64; adults. ♦Sammy George, gen mgr.

Roswell

WJZZ-FM— 1997: 107.5 mhz; 6 kw. 321 ft TL: N33 55 48 W84 20 45. Hrs open:
Rebroadcasts WHTA(FM) Fayetteville 100%.
101 Marietta St. 12th Floor, Atlanta, 30303. Phone: (404) 765-9750. Fax: (404) 688-7686. Web Site: www.1075wjzz.com. Licensee: ROA Licenses LLC. Group owner: Radio One Inc. (acq 11-8-01; grpsl). Format: Jazz. ♦Wayne K. Brown, gen mgr; Frank Johnson, stn mgr; Dave Kosh, progmg dir.

Royston

WBIC(AM)— January 1971: 810 khz; 250 w-D. TL: N34 16 50 W83 07 09. Hrs open: Sunrise-sunset 259 Turner St., 30662. Phone: (706) 246-0059. Fax: (706) 245-0890. Web Site: www,wbicradio.com. Licensee: Diane E. Hawkins (acq 4-3-2003; $50,000). . Population served: 4,500 Rgnl. Network: Ga. Net. Format: The Voice. News staff: one; News: 7 hrs wkly. Target aud: General. ♦Lamar James, gen mgr.

WPUP(FM)—Licensed to Royston. See Athens

Saint Mary's

WWIO(AM)— Oct 15, 1985: 1190 khz; 2.5 kw-D. TL: N30 45 48 W81 36 40. Hrs open: Daytime 2101 Hwy 40E, St. Mary's, 31558. Phone: (404) 685-2527. Web Site: www.gpb.org. Licensee: Lighthouse Christian Broadcasting Corp. (acq 12-21-99). Natl. Network: USA, NPR. Natl. Rep: Rgnl Reps. Wire Svc: AP Format: News/Classical. News staff: 6; News: 40 hrs wkly. Target aud: General; 35 yrs +. ♦Taylor Lewis, opns mgr; St. John Flynn, progmg dir; Mark Sehlig, chief of engrg.

Saint Simons Island

WBGA(FM)—Licensed to Saint Simons Island. See Brunswick

WGIG(AM)—See Brunswick

WMUV(FM)—See Brunswick

WSOL-FM—See Jacksonville, FL

Sandersville

WSNT(AM)— May 11, 1956: 1490 khz; 1 kw-U. TL: N32 58 23 W82 48 34. Hrs open: Box 150, 31082. Secondary address: 312 Morningside Dr. 31082. Phone: (478) 552-5182. Fax: (478) 553-0800. Licensee: Radio Station WSNT Inc. Natl. Network: ESPN Radio. Format: Sports. ♦Capers Brazzell, gen mgr.

WSNT-FM— 1975: 99.9 mhz; 3 kw. Ant 184 ft TL: N32 58 23 W82 48 34. Web Site: www.realcountryonline.com/home.asp?callsign=WSNT-FM. Format: Country.

Sandy Springs

WFGM(AM)—Not on air, target date: unknown: 830 khz; 50 kw-D, 2.4 kw-N, DA-2. TL: N34 02 00 W84 19 09. Hrs open: 24180 Forest Dr., Forest Lake, IL, 60047. Phone: (847) 540-5410. Licensee: Frank McCoy. ♦Frank McCoy, gen mgr.

Sasser

WEGC(FM)—Licensed to Sasser. See Albany

Savannah

WAEV(FM)—Listing follows WSOK(AM).

WBMQ(AM)— Dec 29, 1939: 630 khz; 5 kw-U, DA-N. TL: N32 03 51 W81 00 52. Hrs open: 214 Television Cir., 31406. Phone: (912) 961-9000. Fax: (912) 961-7070. Web Site: www.wbmq.com. Licensee: Cumulus Licensing Corp. Group owner: Cumulus Media Inc. (acq 3-26-98; grpsl). Population served: 47,000 Natl. Network: CBS. Format: News/talk. Target aud: 35 plus. ♦Dale Powers, gen mgr.

WIXV(FM)—Co-owned with WBMQ(AM). Apr 24, 1972: 95.5 mhz; 100 kw. 900 ft TL: N32 03 30 W81 20 20. Stereo. Web Site: www.rockofsavannah.com.86,200 Format: Classic rock. Target aud: 18-49.

WEAS-FM—Listing follows WJLG(AM).

WGCO(FM)—Midway, 1974: 98.3 mhz; 100 kw. Ant 1,047 ft TL: N31 36 45 W81 21 37. Stereo. Hrs opn: 24 401 Mall Blvd., Suite 101 D, 31406. Phone: (912) 351-9830. Fax: (912) 352-4821. Web Site: www.oldies983fm.com. Licensee: Monterey Licenses LLC. Group owner: Triad Broadcasting Co. LLC (acq 9-1-2000; grpsl). Population served: 500,000 Natl. Rep: Christal. Format: Oldies. News staff: 3; News: 2 hrs wkly. Target aud: 25-54; yuppies. ♦Robert Leonard, gen mgr & rgnl sls mgr.

***WHCJ(FM)**— Aug 18, 1975: 90.3 mhz; 6 kw. Ant 223 ft TL: N32 01 28 W81 03 23. Hrs open: 16 Savannah State Univ., 3219 College St., 31404. Phone: (912) 356-2399/356-2381. Fax: (912) 356-2041. E-mail: cartert@savstate.edu Web Site: www.savstate.edu/whcj. Licensee: Savannah State University. Population served: 150,000 Format: Var/div. Target aud: 17-65; interested in jazz, reggae, blues & gospel. ♦Theron Ike Carter, gen mgr.

WJCL-FM— June 18, 1972: 96.5 mhz; 100 kw. 1,232 ft TL: N32 03 30 W81 20 20. Stereo. Hrs opn: 24 214 Television Cir., 31406. Phone: (912) 961-9000. Fax: (912) 961-7070. E-mail: boomer.lee@cumulus.com Web Site: www.kix96.com. Licensee: Cumulus Licensing Corp. Group owner: Cumulus Media LLC (acq 3-12-98; $7.25 million). Population served: 118,349 Format: Country. News staff: one. Target aud: 25-54. ♦Lewis W. Dickey Jr., CEO & pres; Martin R. Gausvik, CFO; Dale Powers, gen mgr; Sam Nelson, opns dir; Tom Hennessey, sls dir.

WJLG(AM)— Oct 6, 1950: 900 khz; 4.35 kw-D, 152 w-N. TL: N32 04 29 W81 04 17. Hrs open: 214 Television Cir., 31406. Phone: (912) 961-9000. Fax: (912) 961-7070. Web Site: www.cumulus.com. Licensee: Cumulus Licensing Corp. Group owner: Cumulus Media Inc. (acq 7-29-98; $5.25 million with co-located FM). Natl. Network: Fox Sports. Format: Sports. ♦Dale Power, gen mgr.

WEAS-FM—Co-owned with WJLG(AM). August 1967: 93.1 mhz; 96.64 kw. Ant 981 ft TL: N32 02 45 W81 20 27. Stereo. Web Site: www.e93jamz.com.118,349 Format: Urban contemp.

***WLXP(FM)**— Jan 1, 2002: 88.1 mhz; 1.5 kw. Ant 361 ft TL: N32 02 49 W81 04 42. Stereo. Hrs opn: 24 2351 Sunset Blvd., Suite 170-218, Air 1 Radio Network, Rocklin, CA, 95765. Phone: (916) 251-1600. Fax: (916) 251-1650. E-mail: info@air1.com Web Site: www.air1.com. Licensee: Christian Multimedia Network Inc. Population served:

237,000 Natl. Network: Air 1. Law Firm: Shaw Pittman. Format: Contemp Christian. News staff: 3. Target aud: 18-35; Judeo-Christian, female. ♦Joe Miller, CFO; Keith Whipple, gen mgr.

WQBT(FM)—Listing follows WTKS(AM).

WSEG(AM)— May 1956: 1400 khz; 650 w-U. TL: N32 04 29 W81 04 17. Hrs open: 24 7515 Blythe Island Hwy., Brunswick, 31523. Phone: (912) 264-6251. Licensee: MarMac Communications LLC. (acq 6-26-2007; $300,000). Population served: 277,000 ♦Gary P. Marmitt, gen mgr.

WSOK(AM)— October 1946: 1230 khz; 1 kw-U. TL: N32 04 20 W81 04 35. Hrs open: 245 Alfred St., 31408. Phone: (912) 964-7794. Fax: (912) 964-9414. Licensee: Capstar TX L.P. Group owner: Clear Channel Communications Inc. (acq 8-30-00; grpsl). Population served: 142,400 Natl. Network: American Urban. Format: Gospel. ♦Jerry Stevens, CEO & VP; Sheryl Collison, sls dir; E. Larry McDuffie, progmg dir, progmg mgr & disc jockey; Marty Foglia, chief of engrg.

WAEV(FM)—Co-owned with WSOK(AM). Feb 4, 1969: 97.3 mhz; 100 kw. 1,000 ft TL: N32 03 30 W81 20 20. Stereo. 118,349 Natl. Network: Westwood One. Format: CHR. Target aud: 25-54; affluent. ♦Chris Allen, opns mgr & progmg mgr; Sheryl Collison, natl sls mgr & rgnl sls mgr; E. Larry McDuffie, disc jockey.

***WSVH(FM)**— Apr 20, 1981: 91.1 mhz; 100 kw. 1,068 ft TL: N32 03 32 W81 17 57. Stereo. Hrs opn: 24
WJSP, Warm Springs, GA, 50-60%.
260 14th St. N.W., Atlanta, 30318-5360. Secondary address: 12 Ocean Science Cir. 30602. Phone: (404) 685-2690 HQ. Fax: (404) 685-2684 HQ. E-mail: ask@gpb.org Web Site: www.gpb.org. Licensee: Georgia Public Telecommunications Commission. Population served: 200,000 Natl. Network: PRI, NPR. Wire Svc: AP Format: Class, news, jazz. News staff: one; News: 51 hrs wkly. Target aud: 45 plus; Adults 35+. ♦Nancy G. Hall, CEO; Bonnie Bean, CFO; Eric Nauert, stn mgr; Ryan Fowler, opns mgr; Mandy Wilson, prom dir; St. John Flynn, progmg mgr; Terrance McKnight, mus dir; Susanna Capelonto, news dir; Mark Fehlig, engrg mgr; Orlando Montoya, news rptr.

WTKS(AM)— Oct 15, 1929: 1290 khz; 5 kw-U, DA-N. TL: N32 05 26 W81 08 55. Stereo. Hrs opn: 245 Alfred St., 31408-3205. Phone: (912) 964-7794. Fax: (912) 964-9414. Licensee: Capstar TX L.P. Group owner: Clear Channel Communications Inc. (acq 8-30-00; grpsl). Population served: 550,000 Law Firm: Wiley, Rein & Fielding. Format: Talk. Target aud: 25-64. ♦Jerry Storey, VP & gen mgr; Brad Kelly, opns mgr; Sheryl Collison, sls dir; Jeff Storey, mktg mgr; Brian Mudd, progmg dir; Marty Foglia, chief of engrg.

WQBT(FM)—Co-owned with WTKS(AM). Nov 29, 1946: 94.1 mhz; 100 kw. 1,320 ft TL: N32 03 14 W81 21 01. Stereo. 566,300 Format: Urban contemp. ♦Jeff Storey, gen mgr; Bo Money, progmg dir.

***WYFS(FM)**— Nov 1, 1986: 89.5 mhz; 100 kw. 630 ft TL: N32 04 04 W81 21 17. Stereo. Hrs opn: 24 1388 Old Waynesboro Rd., Waynesboro, 30830. Secondary address: Bible Broadcasting Network, Charlotte 28241-7300. Phone: (877) 554-8226. Phone: (704) 523-5555. Web Site: www.bbnradio.org. Licensee: Bible Broadcasting Network Inc. (group owner) Format: Educ, relg, Christian. Target aud: General; christian progmg for the entire family. ♦Lowell Davey, pres; Rob Ferguson, gen mgr & stn mgr.

WZAT(FM)— Oct 19, 1971: 102.1 mhz; 98 kw. Ant 1,328 ft TL: N32 03 29 W81 20 19. Hrs open: 24 214 Television Cir., 31406. Phone: (912) 961-9000. Fax: (912) 961-7070. E-mail: brian.rickman@z102.net Web Site: www.z102.net. Licensee: Cumulus Licensing Corp. Group owner: Cumulus Media Inc. (acq 7-29-98; $3.5 million). Population served: 560,000 Format: CHR, var. ♦Lewis W. Dickey Jr., CEO & pres; Martin R. Gausvik, CFO; Dale Powers, gen mgr; Sam Nelson, opns dir; Robert Combs, sls dir & engr.

Smithville

WZIQ(FM)— 1996: 106.5 mhz; 2.45 kw. 515 ft TL: N31 47 59 W84 14 54. Stereo. Hrs opn: 24 2278 Wortham Ln., Grovetown, 30813. Phone: (706) 309-9610. Fax: (706) 309-9669. E-mail: ctbarinowski@comcast.net Web Site: www.gnnradio.org. Licensee: Barinowski Investment Co. Group owner: Good News Network (acq 1-21-98). Population served: 110,000 Format: Christian. ♦Clarence Barinowski, gen mgr.

Smyrna

WAZX(AM)— March 1962: 1550 khz; 50 kw-D, 500 w-N, DA-2. TL: N33 53 29 W84 31 19. Stereo. Hrs opn: 24 1800 Lake Park Dr., 30080. Phone: (770) 436-6171. Fax: (770) 436-0100. E-mail: pattycool100@hotmail.com Web Site: www.radiolaquebuena.com. Licensee: GA-MEX Broadcasting Inc. (acq 7-29-93; $1.1 million; FTR: 8-23-93). Population served: 250,000 Format: Regnl Mexican. Target aud: 12 plus; Sp. Spec prog: Regional Mexican. ♦Javier Macias, CEO & pres; Patty Perez, gen mgr.

WSTR(FM)— May 1966: 94.1 mhz; 100 kw. Ant 1,018 ft TL: N33 45 33 W84 20 05. Hrs open: Penthouse, 3350 Peachtree Rd., Suite 1800, Atlanta, 30326. Phone: (404) 261-2970. Fax: (404) 365-9026. E-mail: mark.kanov@star94.com Web Site: www.star94.com. Licensee: Jefferson Pilot Communications Co. (group owner; acq 3-1-74). Population served: 750,000 Format: CHR. Target aud: 18-49; general. ♦Don Benson, pres; Mark Kanov, gen mgr; Dan Bowen, progmg dir.

Soperton

WKTM(FM)— Nov 23, 1982: 106.1 mhz; 6 kw. 298 ft TL: N32 25 31 W82 33 26. Stereo. Hrs opn: 24 2278 Wortham Ln., Grovetown, 30813. Phone: (706) 309-9610. Fax: (706) 309-9669. E-mail: ctbarinowski@comcast.net Web Site: www.gnnradio.org. Licensee: Barinowski Investment Co. Group owner: Good News Network (acq 1-21-99). Population served: 150,000 Format: Sp. ♦Clarence Barinowski, gen mgr.

Sparta

***WJDS(FM)**—Not on air, target date: unknown: 88.7 mhz; 2 kw vert. 134 ft TL: N33 18 48 W83 00 05. Hrs open: 2278 Wortham Ln, Grovetown, 30813. Phone: (706) 309-9610. Fax: (706) 309-9669. E-mail: ctbarinowski@comcast.net Web Site: www.gnnradio.org. Licensee: Augusta Radio Fellowship Institute Inc. Population served: 20,000 Format: Sp. ♦Clarence Barinowski, gen mgr.

Springfield

WEAS-FM—Licensed to Springfield. See Savannah

WTYB(FM)— Oct 1, 1977: 103.9 mhz; 14 kw. Ant 328 ft TL: N32 16 48 W81 11 41. (CP: COL Tybee Island. 50 kw, ant 344 ft. TL: N32 03 33 W81 00 57). Hrs opn: 214 Television Cir., Savannah, 31406. Phone: (912) 961-9000. Fax: (912) 961-7070. Web Site: www.cumulus.com. Licensee: Cumulus Licensing Corp. Group owner: Cumulus Media Inc. (acq 3-26-98; grpsl). Format: Classic soul & today's rhythm and blues. ♦Tom Conolly, gen mgr.

Statenville

WHLJ(FM)— 1999: 97.5 mhz; 6 kw. Ant 328 ft TL: N30 46 47 W82 52 43. Hrs open: LaTaurus Productions Inc., Box 1305, Valdosta, 31605. Phone: (229) 242-9997. Fax: (229) 249-9765. E-mail: whlj@surfsouth.com Licensee: LaTaurus Productions Inc. Format: Rhythm and blues, hip-hop, urban contemp. ♦Warren Lee, gen mgr.

Statesboro

WPMX(FM)— 1995: 102.9 mhz; 25 kw. Ant 328 ft TL: N32 26 43 W81 58 07. Hrs open: 24 35 E. Main St., 30458. Phone: (912) 764-1029. Fax: (912) 489-3959. E-mail: radiocenter@frontiernet.net Licensee: Georgia Eagle Broadcasting Inc. (group owner; acq 1-12-2007; grpsl). Natl. Network: ABC. Rgnl. Network: Ga. News Net. Format: Adult contemp. Target aud: General. ♦Cecil Staton, pres; Jeff Anderson, gen mgr.

WPTB(AM)— Apr 4, 1976: 850 khz; 1 kw-U, DA-N. TL: N32 28 02 W81 50 07. Hrs open: 24 Box 958, 30458. Phone: (912) 764-6621. Fax: (912) 764-6622. E-mail: espn850@frontiernet.net Web Site: www.radiostatesboro.com. Licensee: Georgia Eagle Broadcasting Inc. Group owner: Communications Capital Managers LLC (acq 5-21-2007; grpsl). Population served: 17,000 Natl. Network: ESPN Radio. Natl. Rep: Dora-Clayton. Rgnl rep: Regional Reps Law Firm: Richard Helmick. Format: Sports. News staff: one; News: 7 hrs wkly. Target aud: General. Spec prog: Gospel 6 hrs wkly. ♦Nate Hirsch, gen mgr; Bill Kent, stn mgr & progmg dir.

***WVGS(FM)**— 1975: 91.9 mhz; 1 kw. 161 ft TL: N32 25 32 W81 46 58. Stereo. Hrs opn: 24 Georgia Southern Univ., Box 8067, 30460. Phone: (912) 681-0877. Phone: (912) 681-5507. Fax: (912) 486-7113. E-mail: wvgs@georgiasouthern.edu Licensee: Georgia Southern University Population served: 27,000 Format: Alternative, progsv, educ. Target aud: 18-25; college kids. ♦Dennis Hightower, pres; Bill Neville, stn mgr & opns mgr.

WWNS(AM)— Dec 1, 1946: 1240 khz; 1 kw-U. TL: N32 27 21 W81 46 27. Hrs open: 24 Box 958, 30459. Secondary address: 561 E. Olliff St. 30458. Phone: (912) 764-5446. Fax: (912) 764-8827. E-mail: wwnswmcd@enia.net Web Site: www.radiostatesboro.com. Licensee: Georgia Eagle Broadcasting Inc. Group owner: Communications Capital Managers LLC (acq 5-21-2007; grpsl). Population served: 50,000 Natl. Network: USA. Rgnl rep: Rgnl Reps Law Firm: Richard Helnick. Format: News/talk, sports. News staff: one; News: 20 hrs wkly. Target aud: 25-death. ♦Nate Hirsch, gen mgr; Buddy Horne, opns dir & progmg dir.

Summerville

WGTA(AM)— Aug 27, 1950: 950 khz; 5 kw-D, 140 w-N. TL: N34 27 53 W85 21 12. Hrs open: 6 AM-7 PM 339 Hwy. 100, 30747. Phone: (770) 436-6171. Fax: (770) 436-0100. E-mail: radiobuena@hotmail.com Web Site: www.radiolaquebuena.com. Licensee: TTA Broadcasting Inc. (acq 2-4-97; $50,000). Population served: 25,043 Natl. Network: Westwood One. Rgnl. Network: Ga. Net. Law Firm: Cordon & Kelly. Format: Regular Mexican. News staff: one; News: 15 hrs wkly. Target aud: 18 plus. Spec prog: Relg 12 hrs wkly. ♦Javier Macias, pres; Patty Perez, gen mgr.

WZQZ(AM)—Trion, Apr 1, 1985: 1180 khz; 5 kw-D. TL: N34 28 22 W85 19 31. Hrs open: 6am-10pm Box 735, 30747. Secondary address: 4689 US Hwy. 27 30747. Phone: (706) 857-5555. Fax: (706) 857-2006. E-mail: wzqz@alltel.net Web Site: wzqz.net. Licensee: Barinowski Investment Co., a Georgia L.P. Group owner: Good News Network. Natl. Network: USA, CNN Radio. Format: Talk shows. Target aud: All ages; Northwest Georgia. ♦Clarence Barinowski, pres; Terry Adams, gen mgr & opns dir.

Swainsboro

WELT(FM)—Listing follows WJAT(AM).

WJAT(AM)— Jan 1, 1950: 800 khz; 1 kw-D, 500 w-N. TL: N32 35 08 W82 21 42. Hrs open: 24 2 Radio Loop, 30401. Phone: (478) 237-1590. Fax: (478) 237-3559. Licensee: RadioJones LLC (group owner; (acq 4-2-2004; grpsl). Population served: 75,000 Format: News, talk, sports. News staff: one; News: 10 hrs wkly. Target aud: 25-64. Spec prog: Farm 5 hrs wkly. ♦Dennis Jones, gen mgr; Jolly Martin, sls dir & gen sls mgr; John Wagner, news dir; Earl Welch, engrg dir; Marty Foglia, engr.

WELT(FM)—Co-owned with WJAT(AM). Dec 18, 1966: 98.1 mhz; 9.6 kw. Ant 525 ft TL: N32 32 55 W82 38 49. Stereo. 110,000 Format: Contemp hit. News staff: one; News: 10 hrs wkly. Target aud: 18-50.

WRJS(AM)— Mar 10, 1978: 1590 khz; 2.5 kw-D, 25 w-N. TL: N32 33 25 W82 20 29. Hrs open: 24 2 Radio Loop, 30401. Phone: (478) 237-1590. Fax: (478) 237-3559. Web Site: www.radiojones.com. Licensee: RadioJones LLC. (group owner; (acq 4-2-2004; grpsl). Population served: 40,000 Format: Black gospel. News staff: one; News: 56 hrs wkly. Target aud: 25-64. ♦Earl Welch, progmg dir & engr; John Wagner, opns mgr & news dir.

WXRS-FM—Co-owned with WRJS(AM). Aug 2, 1982: 100.5 mhz; 3 kw. 300 ft TL: N32 34 52 W82 23 14. Stereo. 24 75,000 Format: Contemp country. News staff: one; News: 10 hrs wkly. Target aud: 25 plus.

Sylvania

WSYL(AM)— Dec 1, 1955: 1490 khz; 1 kw-U. TL: N32 43 51 W81 37 04. Hrs open: 24 Box 519, 1526 Savannah Hwy., 30467. Phone: (912) 564-7461. Fax: (912) 564-7462. Licensee: Georgia Eagle Broadcasting Inc. Group owner: Communications Capital Managers LLC (acq 5-21-2007; grpsl). Rgnl. Network: Ga. Net., Tobacco. Natl. Rep: Rgnl Reps. Format: Contemp country. News staff: 2; News: 7 hrs wkly. Target aud: 18-49; affluent. Spec prog: Farm 5 hrs wkly. ♦Nathan Hirsch, gen mgr; David Hartley, traf mgr & edit mgr; Mary Lou Clontz, relg ed.

WZBX(FM)—Co-owned with WSYL(AM). Sept 6, 1991: 106.5 mhz; 6 kw. 328 ft TL: N32 43 53 W81 37 03.57,000 Format: Country. ♦Nate Hirsch, stn mgr; Scott Kidd, farm dir.

Sylvester

***WHKV(FM)**— Jan 27, 1993: 106.1 mhz; 6 kw. Ant 328 ft TL: N31 30 15 W83 55 46. Hrs open: 5700 West Oaks Blvd., Rocklin, CA, 95765. Phone: (916) 251-1600. Fax: (916) 251-1650. Web Site: www.klove.com. Licensee: Educational Media Foundation. (group owner; (acq 4-30-2007; $615,000 with WFFM(FM) Ashburn). Natl. Network: K-Love. Format: Contemp Christian. ♦Mike Novak, sr VP.

WNUQ(FM)— Aug 1, 1999: 102.1 mhz; 6 kw. Ant 276 ft TL: N31 31 42 W83 50 29. Hrs open: 1104 W. Broad Ave., Albany, 31707. Phone: (229) 888-5000. Fax: (229) 888-5960. E-mail: info18@cumulus.com Web Site: www.v102albany.com. Licensee: Cumulus Licensing Corp. Group owner: Cumulus Media Inc. (acq 3-12-2001; $550,000). Natl. Rep: Katz Radio. Format: Blues, urban ceontemp. ♦Gregory Kamishlian, gen mgr & mktg mgr; Roshon Vance, progmg dir.

Talking Rock

WNSY(FM)— 1999: 100.1 mhz; 7 kw. Ant 617 ft TL: N34 37 50 W84 29 29. Hrs open:
Rebroadcasts WLKQ-FM Buford 100%.
3235 Satellite Blvd., Suite 230, Duluth, 30096. Phone: (770) 623-8772. Fax: (770) 623-4722. Web Site: www.laraza1023.com. Licensee: Davis Broadcasting of Atlanta L.L.C. (acq 1-17-2007; $3.8 million with WCHK(AM) Canton). Format: Rgnl Mexican. ♦Brian Barber, gen mgr.

Tallapoosa

WKNG(AM)— Sept 1, 1977: 1060 khz; 11 kw-D, 5 kw-CH. TL: N33 44 06 W85 15 08. Hrs open: Box 626, 30176. Secondary address: Hwy. 78, Golf Course Rd. 30176. Phone: (770) 574-1060. Fax: (770) 574-1062. Web Site: www.wkng.com. Licensee: WKNG LLC. Population served: 250,000 Natl. Network: ABC. Rgnl. Network: Ga. Net. Format: Classic country. Target aud: 25-54. ♦Steven L. Gradick, pres & gen mgr.

Tennille

WJFL(FM)— Oct 14, 1993: 101.9 mhz; 6 kw. 328 ft TL: N32 54 49 W82 53 06. Hrs open: Box 36, 31089. Phone: (478) 553-1019. Fax: (478) 553-1123. E-mail: wjfl@wjfl.com Web Site: www.wjfl.com. Licensee: Fall Line Media Inc. (group owner; (acq 1996; $225,000). Format: Today's hits & yesterdays favorites. ♦Michael Cowan, gen mgr.

The Rock

***WKEU-FM**— 2000: 88.9 mhz; 5 kw. Ant 764 ft TL: N32 59 11 W84 21 56. Hrs open: Box 997, Griffin, 30224. Secondary address: 1000 Memorial Dr., Griffin 30224. Phone: (770) 227-5507. Fax: (770) 229-2291. E-mail: wkeu@aol.com Web Site: www.wkeuradio.com. Licensee: Georgia Public Radio Inc. Format: Classic rock. ♦William Taylor, Jr., pres; William Taylor Jr., gen mgr.

Thomaston

WTGA(AM)— Nov 1, 1962: 1590 khz; 500 w-D, 25 w-N. TL: N32 53 45 W84 18 10. Hrs open: 208 S. Center St., 30286. Secondary address: Box 550 Phone: (706) 647-7121. Fax: (706) 647-7122. Web Site: www.wtga.com. Licensee: Radio Georgia Inc. (acq 1972). Population served: 10,024 Format: Soft hits. Spec prog: Black 4 hrs wkly. ♦David L. Piper, pres, gen mgr & progmg dir; Bill Bailey, gen sls mgr; Robert Lyons, chief of engrg.

WTGA-FM— Nov 15, 1982: 101.1 mhz; 6 kw. 308 ft TL: N32 51 49 W84 25 10. Web Site: www.wtga.com.15,000 Format: Soft hits.

Thomasville

***WAYT(FM)**— 2003: 88.1 mhz; 35 kw. Ant 1,332 ft TL: N30 40 06 W83 58 10. Hrs open: Box 4188, Tallahassee, FL, 32315. Phone: (850) 422-1929. Fax: (850) 297-1888. E-mail: wayt@wayfm.com Web Site: wayt.wayfm.com. Licensee: WAY-FM Media Group Inc. (group owner; acq 10-24-02). Format: Christian. ♦Steve Young, stn mgr.

***WFSL(FM)**— Mar 1, 2005: 90.7 mhz; 250 w. Ant 154 ft TL: N30 50 12 W83 58 57. Hrs open: Florida State University, FSU Broadcasting Ctr., 1600 Red Barber Plaza, Tallahassee, FL, 32310-6068. Phone: (850) 487-3086. Fax: (850) 487-3293. E-mail: wfsufm@wfsu.org Web Site: www.wfsu.org. Licensee: Florida State University Board of Trustees. Format: Class. ♦Patrick Keating, gen mgr; Caroline Austin, stn mgr.

WHGH(AM)— Dec 15, 1987: 840 khz; 10 kw-D. TL: N30 47 54 W83 56 22. Hrs open: Box 2218, 31799. Secondary address: 221 Pallbearer Rd. 31792. Phone: (229) 228-4124. Fax: (229) 225-9508. Licensee: H.G.H. Investment Corp. Format: Hip hop, gospel. Target aud: 12 plus; Blacks. ♦Moses Gross, pres, gen mgr & stn mgr.

WPAX(AM)— Dec 27, 1922: 1240 khz; 1 kw-U. TL: N30 50 10 W83 59 19. Stereo. Hrs opn: 24 Box 129, 31799. Secondary address: 117 Remington Ave. 31799. Phone: (229) 226-1240. Fax: (229) 226-1361. E-mail: lenrob@rose.net Web Site: www.wpaxradio.com. Licensee: LenRob Inc. (acq 10-85). Population served: 41,500 Natl. Network: CBS. Rgnl. Network: Ga. Net., Tobacco. Natl. Rep: Rgnl Reps. Law Firm: Miller & Fields, P.C. Format: Music of your Life. News staff: one; News: 20 hrs wkly. Target aud: 25 plus; mature with disposable income. ♦Len Robinson, pres & gen mgr.

WSTT(AM)— 1947: 730 khz; 5 kw-D, 27 w-N. TL: N30 48 50 W84 00 48. Hrs open: 5:30 AM-9 PM 2194 Hwy. 319 S., 31792. Phone: (229) 377-2337. Fax: (229) 377-0023. Licensee: Marion R. Williams. (acq 7-26-99; $300,000). Population served: 500,000 Natl. Network: CBS. Law Firm: Matthew McCormick. Format: Gospel. Target aud: 25-54; general. ♦Marion Williams, VP & gen mgr.

WTLY(FM)—Licensed to Thomasville. See Tallahassee FL

Thomson

***WTHM(FM)**—Not on air, target date: unknown: 89.9 mhz; 8.5 kw vert. Ant 492 ft TL: N33 45 28 W82 24 36. Hrs open: 5700 West Oaks Blvd., Rocklin, CA, 95765. Phone: (916) 251-1600. Fax: (916) 251-1650. Licensee: Educational Media Foundation. (acq 3-23-2007; grpsl). ♦Richard Jenkins, pres.

WTHO-FM— Feb 22, 1971: 101.7 mhz; 3 kw. 300 ft TL: N33 28 21 W82 30 00. Stereo. Hrs opn: Box 900, 788 Cedar Rock Rd. N.W., 30824. Phone: (706) 595-5122. Fax: (706) 595-3021. E-mail: wtho@classicsouth.net Licensee: Camellia City Communications Inc. (acq 2-5-93; $110,000 with co-located AM; FTR: 3-1-93) Population served: 30,000 Rgnl. Network: Ga. Net. Natl. Rep: Rgnl Reps. Law Firm: Covington & Burling. Format: C&W. Target aud: 25-54. Spec prog: Farm 3 hrs, gospel one hr, relg 6 hrs wkly. ♦Lisa Kitchens, news dir; Mike Wall, gen mgr, opns dir, gen sls mgr, adv mgr, progmg dir, mus dir, news dir & chief of engrg; Mary Thomaston, traf mgr.

WTWA(AM)—Co-owned with WTHO-FM. Jan 10, 1948: 1240 khz; 1 kw-U. TL: N33 28 20 W82 31 02.19 Phone: (706) 595-1561. E-mail: wtwa@classicsouth.net 7,500 Format: Adult Contemporary. Target aud: 35 plus.

Tifton

***WABR-FM**— December 1973: 91.1 mhz; 30 kw. 249 ft TL: N31 29 30 W83 31 49. Stereo. Hrs opn:
WJSP, Warm Springs/Columbus, GA, 100%.
260 14th St. N.W., Atlanta, 30318-5360. Phone: (404) 685-4788. Fax: (404) 685-2684. E-mail: ask@gpb.org Web Site: www.gpb.org. Licensee: Georgia Public Telecommunications Commission. Population served: 10,000 Natl. Network: NPR. Wire Svc: AP Format: News, class. News staff: 6; News: 40 hrs wkly. ♦Nancy Hall, CEO; Bonnie Bean, CFO; St. John Flynn, progmg dir; Susanna Capelouto, news dir.

WJYF(FM)—Listing follows WTIF(AM).

WKZZ(FM)— 2000: 92.5 mhz; 20.5 kw. Ant 361 ft TL: N31 31 40 W83 20 01. Hrs open: 1931 GA Hwy. 32 E., Douglas, 31533. Phone: (912) 389-0995. Fax: (912) 383-8552. Licensee: Broadcast South LLC. Group owner: Black Crow Media Group LLC (acq 11-6-2006; grpsl). Population served: 50,000 Law Firm: McCampbell & Young. Format: Adult contemp. ♦John Higgs, CEO & gen mgr.

WOBB(FM)—Licensed to Tifton. See Albany

***WPLH(FM)**— January 1988: 103.1 mhz; 29 w. 177 ft TL: N31 28 51 W83 31 38. Stereo. Hrs opn: Box 36, Abraham Baldwin Agricultural College, 31793. Secondary address: 2802 Moore Hwy 31793. Phone: (229) 391-4977. Phone: (229) 391-4957. Fax: (229) 386-7158. E-mail: wplh@abac.edu Licensee: Abraham Baldwin Agriculture College. (acq 4-1-88). Rgnl. Network: Peach State Public Radio. Format: Alternative. ♦Eric Cash, gen mgr.

WTIF(AM)— 1957: 1340 khz; 1 kw-U. TL: N31 28 16 W83 29 12. Hrs open: 24 Box 968, 31793. Secondary address: 104 E. 7th St. 31794. Phone: (229) 382-1340. Fax: (229) 386-8658. Licensee: Three Trees Communications Inc. (group owner; (acq 6-28-2004; grpsl). Population served: 12,179 Natl. Network: CBS. Rgnl. Network: Ga. Net. Format: Country. News staff: one; News: 15 hrs wkly. Target aud: 18 plus. Spec prog: Farm 5 hrs wkly. ♦James Andrew Howard, pres; Ron Yontz, gen mgr; Andy Reeves, stn mgr.

WJYF(FM)—Co-owned with WTIF(AM). Nov 26, 1986: 95.3 mhz; 29 kw. Ant 522 ft TL: N31 10 18 W83 21 57. Stereo. 24 113 N. College St., Ashburn Natl. Network: Westwood One. Format: Soft adult contemp. News staff: one; News: one hr wkly. Target aud: 30 plus.

Toccoa

WLET(AM)— May 1, 1941: 1420 khz; 5 kw-D. TL: N34 35 23 W85 19 11. Hrs open: 24 Box 780, TFC Radio Network, Toccoa Falls, 30598. Secondary address: 292 Old Clarkesville Hwy. 30577. Phone: (706) 282-6030. Phone: (800) 251-8326. Fax: (706) 282-6090. E-mail: wlet@tfc.edu Licensee: Toccoa Falls College. (acq 11-3-99). Population served: 150,000 Format: Gospel. News staff: one; News: 10 hrs wkly. Target aud: 35 plus; Stephens county residents seeking news & community info. Spec prog: Black, gospel, relg. ♦David Cornelius, gen mgr.

WNEG(AM)— Apr 21, 1956: 630 khz; 500 w-D, 44 w-N. TL: N34 34 15 W83 19 35. Hrs open: 24 Box 1159, 30577-0907. Phone: (706) 886-2191. Fax: (706) 282-0189. E-mail: hobbs@gacaradio.com Web Site: www.wnegradio.com. Licensee: Georgia-Carolina Radiocasting Co. LLC. Group owner: Georgia-Carolina Radiocasting Companies (acq 7-25-2002; grpsl). Population served: 297,285. Natl. Network: CBS Radio. Rgnl. Network: Ga. Net. Law Firm: Dan J. Alpert. Format: MOR, local news. News staff: 2; News: 18 hrs wkly. Target aud: General; adult working class. ♦Douglas M. Sutton Jr., pres; Phil Hobbs, VP & gen mgr; Connie Gaines, opns mgr; M.J. Kneiser, news dir; Tim Stephens, chief of engrg.

WNGC(FM)— November 1947: 106.1 mhz; 100 kw. 1,132 ft TL: N34 43 46 W83 29 29. Stereo. Hrs opn: 24 850 Bobbin Mill Rd., Athens, 30606. Phone: (706) 549-1340. Phone: (706) 549-6222. Fax: (706) 546-0441. E-mail: wngc@negia.net Web Site: www.1061wngc.com. Licensee: Southern Broadcasting of Pensacola Inc. Group owner: Southern Broadcasting Companies Inc. (acq 1999; $2.2 million with co-located AM). Population served: 850000 Format: Country. News staff: one; News: 10 hrs wkly. Target aud: 25-54; adults with disposable income. Spec prog: Gospel. ♦Paul Stone, pres; Scott Smith, opns mgr; Kevin Steele, progmg dir.

Toccoa Falls

***WRAF-FM**— Sept 4, 1980: 90.9 mhz; 100 kw. 564 ft TL: N34 35 57 W83 21 55. Stereo. Hrs opn: 24 Box 780, TFC Radio Network, 30598. Secondary address: 292 Old Clarksville Hwy., Toccoa 30577. Phone: (800) 251-8326. Phone: (706) 282-6030. Fax: (706) 282-6090. E-mail: tfcrn@tfc.edu Web Site: www.myfavoritestation.net. Licensee: Toccoa Falls College. Population served: 500,000 Natl. Network: USA. Law Firm: Wiley, Rein & Fielding. Format: Relg, MOR. Target aud: General; families. ♦David Cornelius, gen mgr & stn mgr.

***WTXR(FM)**— September 1996: 89.7 mhz; 100 w. -190 ft Stereo. Hrs opn: 24 Box 780, TFC Radio Network, 30598. Secondary address: 292 Old Clarksville Hwy., Toccoa 30577. Phone: (706) 282-6030. Phone: (800) 251-8326. Fax: (706) 282-6090. E-mail: wtxr@tfc.edu Web Site: www.wtxr.com. Licensee: Toccoa Falls College. Population served: 25,000 Format: Christian. Target aud: 18-35; college students, young adults. ♦David Cornelius, gen mgr.

Trenton

WBDX(FM)— 1989: 102.7 mhz; 320 w. 1,374 ft TL: N34 51 48 W85 23 35. Stereo. Hrs opn: 24 Box 9396, Chattanooga, TN, 37412. Phone: (423) 892-1200. Fax: (423) 892-1633. E-mail: mailbag@j103.com Web Site: www.j103.com. Licensee: Partners for Christian Media Inc. (acq 5-7-98; $1,189,395). Format: Adult contemp, Christian. Target aud: 18-54. ♦Bob Lubell, CEO, pres & gen mgr; David Skinner, CFO; Debbie Lubell, natl sls mgr; Richard Carlisle, rgnl sls mgr.

WKWN(AM)— Apr 4, 1982: 1420 khz; 2.5 w-D, 112 w-N. TL: N34 51 43 W85 29 59. Hrs open: 24 Box 829, 30752. Secondary address: 12544 N. Main St. 30752. Phone: (706) 657-7594. Fax: (706) 657-6767. Licensee: Dade County Broadcasting Inc. (acq 11-20-97; $63,000). Rgnl. Network: Ga. Net. Format: News/talk radio. Target aud: 25-54; locals. ♦Evan Stone, CEO, chmn, pres, gen mgr & stn mgr.

Trion

WATG(FM)— January 1997: 95.7 mhz; 6 kw. Ant 699 ft TL: N34 28 10 W85 17 48. Hrs open: 24 Box 200, Summerville, 30747. Secondary address: 10143 Commerce St., Summerville 30747. Phone: (706) 857-2000. Fax: (706) 857-3652. E-mail: oldies957@aol.com Licensee: TTA Broadcasting Inc. (acq 10-19-99; up to $296,530). Natl. Network: ABC. Natl. Rep: Rgnl Reps. Law Firm: Cordon & Kelly. Format: Oldies. News: one hr wkly. Target aud: 25-54; general. ♦Randy Davis, pres; Jim Bojo, CFO & gen mgr.

WZQZ(AM)—Licensed to Trion. See Summerville

Tucker

WGUN(AM)—Atlanta, July 1947: 1010 khz; 50 kw-D, 300 w-N. TL: N33 41 55 W84 17 23. Hrs open: 24 2901 Mountain Industrial Blvd., 30084. Phone: (770) 491-1010. Fax: (770) 491-3019. E-mail: wgun@bellsouth.net Web Site: wgunam.com. Licensee: Dee Rivers Group. Population served: 3,095,278 Format: relg talk, info & inspiration. News: one hr wkly. Target aud: 25-54; working class. ♦Georgia Salva, CEO; Darrell Vick, gen mgr; Erwin Hill, opns mgr.

Unadilla

WQSA(FM)— June 1995: 99.9 mhz; 6 kw. Ant 328 ft TL: N32 18 29 W83 46 30. Stereo. Hrs opn: 24 1350 Radio Loop, Warner Robins, 31088. Phone: (478) 923-3416. Licensee: HMH Broadcasting LLC (acq 5-3-2006; $350,000). Format: Adult contemp.

Valdosta

WAAC(FM)—Listing follows WGOV(AM).

WAFT(FM)— Nov 25, 1971: 101.1 mhz; 100 kw. 558 ft TL: N30 51 50 W83 23 39. Stereo. Hrs opn: 24 215 Waft Hill Ln., 31602. Phone: (229) 244-5180. Fax: (229) 242-8808. E-mail: mail@waft.org Web Site: www.waft.org. Licensee: Christian Radio Fellowship Inc. Population served: 700,000 Format: Relg. News: 3 hrs wkly. Target aud: General. ♦Bill Tidwell, pres & gen mgr.

WGOV(AM)— 1939: 950 khz; 3.5 kw-D, 63 w-N. TL: N30 48 13 W83 21 20. Hrs open: 24 Box 1207, 31603. Secondary address: 2973 Hwy. 84 W. 31601. Phone: (229) 244-9590. Fax: (229) 247-7676. E-mail: wgovradio@bellsouth.net Web Site: www.wgovradio.com. Licensee: WGOV Inc. Group owner: Dee Rivers Group Population served: 104,500 Natl. Rep: Rgnl Reps. Format: Rhythm and blues, relg. News staff: one; News: 3 hrs wkly. Target aud: 18-45; Black. Spec prog: Gospel 14 hrs, oldies 10 hrs wkly. ♦Lamar Freeman, opns mgr & progmg dir; Loretta Grecco, gen sls mgr; Jammie Brooks, mus dir.

WAAC(FM)—Co-owned with WGOV(AM). 1968: 92.9 mhz; 100 kw. Ant 502 ft TL: N30 48 13 W83 21 20. Stereo. Phone: (229) 242-4513. E-mail: mail@waacradio.com Web Site: www.waacradio.com.200,000 Natl. Network: ABC. Law Firm: Anthony Lepore. Format: Country. Target aud: 25-54. ♦Robert Whitt, opns mgr & progmg dir.

WJEM(AM)— August 1955: 1150 khz; 5 kw-D, 101 w-N, DA-2. TL: N30 50 49 W83 14 14. Hrs open: Box 961808, Riverdale, 30296. Phone: (229) 241-9797. Licensee: WJEM Inc. (acq 2-18-94; $230,000; FTR: 3-21-94). Format: Gospel.

WLYX(FM)— June 1985: 96.7 mhz; 50 kw. Ant 328 ft TL: N30 48 13 W83 21 20. Stereo. Hrs opn: 24 Box 1207, 31603. Secondary address: 2973 Hwy. 84 W. 31601. Phone: (229) 242-4513. Fax: (229) 247-7676. Web Site: www.wgovradio.com. Licensee: W.G.O.V. Inc. (acq 4-19-2006; $2 million). Format: Urban. Target aud: 18-49. ♦Lamar Freeman, gen mgr; Joseph Jones, opns mgr & progmg dir; Loretta Grecco, gen sls mgr.

WQPW(FM)— September 1977: 95.7 mhz; 35.9 kw. 606 ft TL: N30 50 11 W83 17 56. Stereo. Hrs opn: 24 1711 Ellis Dr., 31601. Phone: (229) 244-8642. Fax: (229) 242-7620. Web Site: www.957themix.com. Licensee: RTG Radio LLC. Group owner: Black Crow Media Group LLC (acq 11-9-2001; grpsl). Population served: 75,000 Law Firm: Borsari & Paxson. Format: Adult contemp. News staff: 14. Target aud: 18-44. ♦Scott James, gen mgr.

WRFV(AM)— Nov 3, 1951: 910 khz; 5 kw-U, DA-N. TL: N30 52 21 W83 20 36. (CP: COL Wellborn, FL. 35 kw-D, 5 kw-N, DA-2. TL: N30 12 59 W82 51 39). Hrs opn: 1801 Clarke Rd., Ocoee, 34761. Phone: (407) 523-2770. Fax: (407) 523-2888. Licensee: Rama Communications Inc. (group owner; (acq 2-28-2002; $255,000). Population served: 32,303 Format: Travelers information. ♦Sabeta Persaud, pres; Kris Persaud, gen mgr; Steve January, gen sls mgr, mktg mgr, adv mgr & progmg dir; Steve Delay, chief of engrg.

***WVDA(FM)**— 2006: 88.5 mhz; 18.5 kw vert. Ant 216 ft TL: N30 47 50 W83 01 01. Hrs open:
Rebroadcasts KLRD(FM) Yucaipa, CA 100%.
5700 West Oaks Blvd., Rocklin, CA, 95765. Phone: (916) 251-1600. Fax: (916) 251-1650. Web Site: www.air1.com. Licensee: Educational Media Foundation. (acq 5-10-2007; $350,000). Natl. Network: Air 1. Format: Alternative rock, div. ♦Mike Novak, sr VP.

WVLD(AM)— Sept 3, 1959: 1450 khz; 1 kw-U. TL: N30 50 11 W83 17 56. Hrs open: 24 1711 Ellis Dr., 31601. Phone: (229) 244-8642. Fax: (229) 242-7620. Licensee: RTG Media. Group owner: Black Crow Media Group LLC (acq 11-9-2001; grpsl). Population served: 42,500 Natl. Network: CBS. Law Firm: Borsari & Paxson. Format: Sports. News: 10 hrs wkly. Target aud: 35 plus. Spec prog: Gospel 2 hrs wkly. ♦Robert Ganzack, pres & gen mgr.

***WVVS(FM)**— July 26, 1971: 90.9 mhz; 5.3 kw. 68 ft TL: N30 50 50 W83 17 26. Stereo. Hrs opn: Valdosta State Univ., 1500 N. Patterson, 31698. Phone: (229) 259-2015. Web Site: www.valdosta.edu/wvvs/. Licensee: Valdosta State University. Population served: 100,000 Format: Alternative & urban. Target aud: 18-25; students of VSU, population at large. ♦Michael Taylor, gen mgr.

***WWET(FM)**— December 1989: 91.7 mhz; 430 w vert. Ant 85 ft TL: N30 49 35 W83 16 40. Stereo. Hrs opn:
Rebroadcasts WJSP-FM Warm Springs 100%.
260 14th St. N.W., Atlanta, 30318-5360. Phone: (404) 685-2690. Fax: (404) 685-2684. E-mail: ask@gpb.org Web Site: www.gpb.org. Licensee: Georgia Public Telecommunications Commission. Natl. Network: NPR, PRI. Wire Svc: AP Format: Class, news. News staff: 6; News: 40 hrs wkly. Target aud: 35 plus; general. Spec prog: Jazz 4 hrs, Latin 3 hrs wkly. ♦Nancy G. Hall, CEO; Bonnie Bean, CFO; Ryan Fowler, opns mgr; Mandy Wilson, prom dir; St. John Flynn, progmg mgr; Terrance McKnight, mus dir; Susanna Capelento, news dir; Mark Fehlig, engrg dir.

WWRQ-FM— Feb 1, 1992: 107.9 mhz; 14 kw. Ant 315 ft TL: N30 50 11 W83 17 56. (CP: 50 kw, ant 449 ft. TL: N31 03 46 W83 04 21). Hrs opn: 1711 Ellis Dr., 31601. Phone: (229) 244-8642. Fax: (229) 247-7620. Licensee: RTG Radio LLC. Group owner: Black Crow Media Group LLC (acq 11-9-2001; grpsl). Law Firm: Miller & Miller. Format: Classic rock, AOR. Target aud: 25-49; upscale suburban couples. ♦Scott James, gen mgr & progmg dir.

Vidalia

WBBT(AM)—See Lyons

WGPH(FM)—Licensed to Vidalia. See Augusta

WLYU(FM)—See Lyons

WTCQ(FM)—Listing follows WVOP(AM).

WVOP(AM)— Dec 2, 1946: 970 khz; 5 kw-U. TL: N32 13 12 W82 26 13. Stereo. Hrs opn: 24 Box 900, 30475. Secondary address: 1501 Mt. Vernon Rd. 30474. Phone: (912) 537-9202. Fax: (912) 537-4477. E-mail: zfowler@vidaliacommunications.com Licensee: Vidalia Communications Corp. (group owner) Population served: 25,000 Rgnl. Network: Ga. Net. Natl. Rep: Rgnl Reps. Law Firm: Fletcher, Heald & Hildreth. Format: Oldies, news, sports. News staff: one; News: 18 hrs wkly. Target aud: 25-54. Spec prog: Pub affrs 2 hrs, relg 8 hrs wkly. ♦John Ladson, pres; Zack Fowler, stn mgr; Jim Perry, opns mgr; Marvin McIntyre, gen sls mgr; Dick Boekeloo, chief of engrg; Joyce Foskey, traf mgr.

WTCQ(FM)—Co-owned with WVOP(AM). Mar 5, 1969: 97.7 mhz; 6 kw. 300 ft TL: N32 13 12 W82 26 13. Stereo. Natl. Rep: Rgnl Reps. Format: Adult contemp. Target aud: 18-34. ♦Zack Fowler, gen mgr; Marvin McIntyre, rgnl sls mgr; Joyce Foskey, traf mgr.

Vienna

WHHR(FM)—Not on air, target date: unknown: 92.1 mhz; 5.1 kw. Ant 348 ft TL: N32 09 16 W83 47 55. Hrs open: Box 5459, Twin Falls, ID, 83303. Phone: (208) 733-3551. Fax: (208) 733-3548. Web Site: www.radioassistministry.com. Licensee: Radio Assist Ministry Inc. (acq 8-22-2006; $150,000 for CP). ♦Clark Parrish, pres.

WKTF(AM)— Nov 17, 1979: 1550 khz; 1 kw-D, 23 w-N. TL: N32 07 44 W83 47 46. Hrs open: 24 7120 U.S. Hwy 41 N., 31092. Phone: (775) 626-9276. Licensee: LEN Radio Broadcasting of Vienna, Georgia LLC. (acq 5-17-2005; $230,000). Population served: 30000 Rgnl. Network: Ga. Net. Format: Southern gospel. ♦Tony Hernandez, gen mgr.

Warm Springs

WJSP-FM—Licensed to Warm Springs. See Atlanta

Warner Robins

WEBL(FM)— September 1994: 102.5 mhz; 4 kw. Ant 328 ft TL: N32 34 20 W83 40 13. Stereo. Hrs opn: 24 7080 Industrial Hwy., Macon, 31216. Phone: (478) 781-1063. Fax: (478) 781-6711. Web Site: www.bull1025.com. Licensee: AMFM Radio Licenses LLC. Group owner: Clear Channel Communications Inc. (acq 2-1-2001; grpsl). Format: Country. Target aud: 25-64. ♦Bill Clark, gen mgr.

WNNG(AM)— Oct 13, 1954: 1350 khz; 15 kw-D, 500 w-N, DA-N. TL: N32 37 00 W83 39 00. Stereo. Hrs opn: 24 1350 Radio Loop, 31088. Phone: (478) 923-3416. Fax: (478) 923-3236. Web Site: www.wnng1350.com. Licensee: Georgia Eagle Broadcasting Inc. (acq 1-3-2007; $650,000). Population served: 58,000 Natl. Network: Westwood One. Rgnl. Network: Ga. Net. Wire Svc: AP Format: Adult Standards. News staff: one; News: 3 hrs wkly. Target aud: 35-74; middle to upper class, working class, miltary, retired, business owners. ♦Cecil P. Staton, pres; Anna McCloy, opns mgr.

WRBV(FM)— August 1969: 101.7 mhz; 4.9 kw. 350 ft TL: N32 38 19 W83 38 33. Stereo. Hrs opn: 24 7080 Industrial Hwy., Macon, 31216. Phone: (478) 781-1063. Fax: (478) 781-6711. Web Site: www.1017.com. Licensee: AMFM Radio Licenses LLC. Group owner: Clear Channel Communications Inc. (acq 2-1-2001; grpsl). Population served: 250,000 Natl. Network: ABC. Format: Today's rhythm and blues & favorite old school. News: one hr wkly. Target aud: 21-54. ♦Bill Clark, gen mgr.

Warrenton

WGAC-FM— 1998: 93.1 mhz; 4.1 kw. Ant 400 ft TL: N33 29 59 W82 37 09. Hrs open:
Simulcast with WGAC(AM) Augusta.
4051 Jimmie Dyess Pkwy., Augusta, 30909. Phone: (706) 396-7000. Fax: (706) 396-7092. E-mail: wgac@wgac.com Web Site: www.wgac.com. Licensee: WCHZ License LLC. Group owner: Beasley Broadcast Group Inc. (acq 5-3-2000; $800,000 with WGUS(AM) Augusta). Format: News/talk. News staff: 5. ♦Kent Dunn, gen mgr; Kent Murphy, gen sls mgr; Harley Drew, progmg dir.

Washington

WLOV(AM)—Listing follows WXKT(FM).

WXKT(FM)— June 1, 1970: 100.1 mhz; 5 kw. Ant 296 ft TL: N33 44 02 W82 43 10. Stereo. Hrs opn: 24 823 Berkshire Dr., 30673. Secondary address: 312 Old First National Bank Bldg., Elberton 30635. Phone: (706) 678-0100. Fax: (706) 678-3394. Licensee: Southern Broadcasting Companies Inc. (group owner; acq 8-28-01; $635,000 with co-located AM). Population served: 4,094 Format: Country. Target aud: 25-54; baby boomers. ♦Scott Smith, gen mgr; Mel Stovall, opns mgr, progmg dir & news dir; Julie Irby, gen sls mgr; Leisa McCurley, traf mgr.

WLOV(AM)—Co-owned with WXKT(FM). Sept 1, 1955: 1370 khz; 1 kw-D. TL: N33 43 50 W82 43 10.4,094 Format: Timeless classics. ♦Leisa McCurley, traf mgr.

Waycross

***WASW(FM)**— June 1998: 91.9 mhz; 1 kw. 148 ft TL: N31 13 07 W82 21 34. Hrs open: Box 3206, American Family Radio, Tupelo, MS, 38803. Phone: (662) 844-8888. Fax: (662) 842-6791. Web Site: www.afr.net. Licensee: American Family Radio. (group owner) Format: Inspirational Christian. ♦Marvin Sanders, gen mgr; Rick Robertson, progmg dir.

WKUB(FM)—See Blackshear

WWGA(AM)— 2004: 1230 khz; 1 kw-U. TL: N31 12 45 W82 22 20. Hrs open: 24 1766 Memorial Dr., Suite 1, 31501. Phone: (912) 285-5002. Fax: (912) 264-1991. E-mail: wgaradio@yahoo.com Web Site: www.wgaradio.com. Licensee: MarMac Communications L.L.C. Format: News/talk. ♦Gary Marmitt, gen mgr, gen sls mgr & progmg dir; Dick Eoekeloo, chief of engrg.

WWSN(FM)—Licensed to Waycross. See Brunswick

WWUF(FM)— Jan 25, 1986: 97.7 mhz; 6 kw. Ant 325 ft TL: N31 11 05 W82 15 24. Stereo. Hrs opn: 24 Box 1472, 31501. Secondary address: 2132 Hwy. 84, Blackshear 31516. Phone: (912) 449-3391. Fax: (912) 449-6284. E-mail: gtmattox@yahoo.com Licensee: Mattox Broadcasting Inc. (acq 5-16-2000). Natl. Network: ABC. Natl. Rep: Dora-Clayton. Law Firm: Fletcher, Heald & Hildreth. Format: Oldies. Target aud: 25-54; general. ♦Troy Mattox, pres & gen mgr; Ray Williamson, opns dir; Jim Miller, gen sls mgr.

***WXVS(FM)**— December 1985: 90.1 mhz; 79 kw horiz, 71 kw vert. 918 ft TL: N31 13 17 W82 34 24. Stereo. Hrs opn:
WJSP, Warm Springs/Columbus, GA, 100%.
260 14th St. N.W., Atlanta, 30318-5360. Phone: (404) 685-2690. Fax: (404) 685-2684. E-mail: ask@gpb.org Web Site: www.gpb.org. Licensee: Georgia Public Telecommunications Commission. Natl. Network: NPR, PRI. Wire Svc: AP Format: Classical, News. News staff: 6; News: 40 hrs wkly. ♦Nancy G. Hall, CEO; Bonnie Bean, CFO; Alan Cooke, gen mgr; Ryan Fowler, opns dir; Mandy Wilson, prom dir; St. John Flynn, progmg mgr; Terrance McKnight, mus dir; Susanna Capelouto, news dir.

WYNR(FM)—Licensed to Waycross. See Brunswick

Waynesboro

WTHB-FM— 1975: 100.9 mhz; 6 kw. 279 ft TL: N33 05 15 W82 02 17. Stereo. Hrs opn: 24 Box 1584, Augusta, 30903. Secondary address: 104 Bennett Ln., North Augusta, SC 29841. Phone: (803) 279-2330. Fax: (803) 279-8149. Web Site: www.thbgospelalive.com. Licensee: Radio One of Augusta LLC. Group owner: Radio One Inc. (acq 11-8-01; grpsl). Population served: 450,000 Law Firm: Fisher, Wayland, Cooper, Leader & Zaragoza. Format: Gospel. News staff: 2; News: 5 hrs wkly. Target aud: Teen-49. ♦Ron Tomel, opns mgr.

WYFA(FM)—Licensed to Waynesboro. See Augusta

West Point

WCJM-FM— July 18, 1966: 100.9 mhz; 1.85 kw. 235 ft TL: N32 53 42 W85 09 32. (CP: 6 kw, ant 177 ft., TL: N32 53 48 W85 09 24). Hrs opn: 705 W. 4th Ave., 31833. Phone: (706) 645-2991. Fax: (706) 645-3364. E-mail: wjcm@qantumofauburn.com Licensee: Qantum of Auburn License Co. LLC. Group owner: Qantum Communications Corp. (acq 7-2-03; grpsl). Population served: 45,000 Rgnl. Network: Ga. Net. Format: Country. ♦Steve Wheeler, gen mgr; Anthony Lovelady, progmg dir.

WPLV(AM)— August 1958: 1310 khz; 1 kw-D. TL: N32 53 42 W85 09 32. (CP: TL: N32 53 48 W85 09 24). Hrs opn: 24 705 W. Fourth Ave., Westpoint, 31833. Phone: (706) 645-1310. Fax: (706) 645-3364. E-mail: wcjm@quantumofauburn.com Licensee: Qantum of Auburn License Co. LLC. Group owner: Qantum Communications Corp. (acq 7-2-03; grpsl). Population served: 21,000 Rgnl. Network: Keystone (unwired net.). Format: Talk radio. ♦Steve Wheeler, gen mgr; Terry Harper, chief of engrg.

WRLA(AM)— May 1944: 1490 khz; 1 kw-U. TL: N32 52 26 W85 11 32. Hrs open: 24 503 W. 8th St., Suite 102, 31833. Phone: (706) 645-1490. Fax: (706) 645-1497. E-mail: wrla@wrla1490.com Licensee: Casey Network LLC. (group owner; (acq 11-18-2002). Population served: 35,000 Rgnl. Network: Ga. Net. Law Firm: Gardner, Carton & Douglas. Format: Oldies. News staff: one; News: 9 hrs wkly. Target aud: 18-55. ♦Vince Smith, gen mgr.

Willacoochee

WKAA(FM)— Apr 7, 1978: 99.5 mhz; 43 kw. Ant 755 ft TL: N31 10 18 W83 21 57. Hrs open: 24 1711 Ellis Dr., Valdosta, 31601. Phone: (229) 244-8642. Fax: (229) 242-7620. Web Site: www.995kixcountry.com. Licensee: RTG Radio LLC. (acq 11-9-2001; grpsl). Population served: 100,000 Format: Country. News staff: one; News: 10 hrs wkly. Target aud: 25-54. ♦Robert Ganzak, pres & gen mgr.

Winder

WIMO(AM)— Nov 4, 1952: 1300 khz; 1 kw-D, 59 w-N. TL: N33 58 22 W83 42 40. Hrs open: 850 Arch Tanner Rd., Bethlehem, 30620. Phone: (770) 867-1300. Fax: (770) 868-1962. E-mail: quincos@aol.com Web Site: www.wimo1300am.com. Licensee: Mark Myers (acq 4-21-2004; $75,000). Population served: 60,000 Natl. Network: Salem Radio Network, Radio America, Fox News Radio. Rgnl. Network: Ga. Net. Format: Talk, gospel. Target aud: General. ♦John Boyd, gen mgr; Jon Graham, opns dir & progmg dir; Kurt Andrews, progmg dir.

***WYFW(FM)**— December 1987: 89.5 mhz; 530 w. 130 ft TL: N33 59 32 W83 45 15. (CP: 6 kw). Stereo. Hrs open: 24 11530 Carmel Commons Blvd., Charlotte, NC, 28226. Phone: (800) 888-7077. E-mail: wyfw@bbnradio.org Web Site: www.bbnradio.org. Licensee: Bible Broadcasting Network Inc. (group owner; acq 6-24-93; $104,000; FTR: 6-28-93). Format: Relg, MOR. News: 7 hrs wkly. Target aud: 35-44. ♦Lowell Davey, pres; Paul D. Montgomery, gen mgr.

Woodbine

WCGA(AM)— June 15, 1987: 1100 khz; 10 kw-D. TL: N30 55 54 W81 42 31. Hrs open: 714 Narrow Way, St. Simons Island, 31522. Phone: (912) 634-1100. E-mail: wescox@adelphia.net Licensee: Cox Broadcast Group Inc. Format: News/talk. News staff: 2; News: 2 hrs wkly. Target aud: Adults; 35-64. ♦Wesley Cox, gen mgr.

Woodbury

WFDR-FM—Not on air, target date: unknown: 94.5 mhz; 2.75 kw. Ant 492 ft TL: N32 50 40 W84 37 25. Hrs open: 185 Melody Ln., Fayetteville, 30215. Phone: (678) 860-1504. Licensee: Ploener Radio Group LLC. ♦Paul W. Ploener, gen mgr.

Wrens

WAKB(FM)— June 10, 1979: 96.9 mhz; 1 kw. 489 ft TL: N33 16 21 W82 25 32. (CP: 750 w, ant 1,364 ft.). Stereo. Hrs open: 104 Bennett Ln., North Augusta, SC, 29841. Secondary address: Box 1584, Augusta 30903. Phone: (803) 279-2330. Fax: (803) 819-3781. Web Site: www.magic969.com. Licensee: Radio One of Augusta LLC. Group owner: Radio One Inc. (acq 11-8-01; grpsl). Format: Urban contemp. ♦Dennis Jackson, gen mgr.

Wrightsville

WDBN(FM)— May 27, 1986: 107.5 mhz; 3 kw. 295 ft TL: N32 42 24 W82 43 08. Hrs open: Box 130, Dublin, 31040. Phone: (478) 272-4422. Fax: (478) 275-4657. E-mail: macd@wqzy.com Licensee: State Broadcasting Corp. (acq 5-29-89; $160,000; FTR: 5-29-89). Population served: 250,000 Format: Classic rock. Target aud: 12 plus. ♦Mac Davis, gen mgr.

Yates

***WWBM(FM)**— 2004: 89.7 mhz; 1 kw. Ant 321 ft TL: N33 27 47 W84 53 35. Hrs open: Best Media Inc., 3601 36th Ave., Long Island City, NY, 11106. Phone: (718) 784-8555, ext 112. Fax: (718) 784-8901. Licensee: Best Media Inc. Law Firm: Scott Cinnamon Law Office. Format: South Asian music news, educational, cultural programs. News staff: one; News: 5 hrs wkly. Asian, Indian, Pakistani, Bengladesh, and other audience. This rebroadcasts in our translator FM stations in Chicago, Houston, Detroit, & Long Island New York. ♦Banad Visuianatu, pres; Karamjit Nanda, gen mgr.

Young Harris

WACF(FM)—Not on air, target date: unknown: 95.1 mhz; 200 w. Ant 1,584 ft TL: N34 56 26 W83 55 08. Hrs open: Box 490, Mineral Bluff, 30559-0490. Phone: (706) 379-9770. Licensee: Wolf Creek Broadcasting Inc. ♦A.D. Frazier, pres.

WYHG(AM)— May 1984: 770 khz; 750 w-D. TL: N34 56 26 W83 51 13. Hrs open: 12
Simulcast with WLSB(AM) Copperhill, TN.
1352 Main St., Suite 6, 30582. Phone: (706) 379-3169. Fax: (706) 379-4104. E-mail: wyhg@brmemc.net Web Site: www.wolkcreekbroadcasting.com.Solid Gospel Network Licensee: Young Harris Broadcasting Corp. (acq 4-30-2003; $120,000). Format: Country, bluegrass, gospel. News staff: 2. Target aud: 30-60; mature adults. ♦Ad Frazier, pres; Clair Frazier, pres; Rebecca St. John, gen mgr.

Zebulon

WEKS(FM)— February 1994: 92.5 mhz; 12 kw. Ant 476 ft TL: N33 08 20 W84 31 31. Stereo. Hrs opn: 24 c/o Stephen D. Tarkenton, 1523 Kell Ln., Suite One, Griffin, 30224. Phone: (770) 412-8700. Fax: (770) 412-8080. E-mail: bear925@bellsouth.net Web Site: www.bear92.com. Licensee: Spalding Broadcasting Inc. Population served: 1,000,000. Law Firm: Miller & Miller. Format: C&W. News staff: one. Target aud: 25-54. ♦Stephen D. Tarkenton, CEO; Les Reed, gen mgr & opns VP.

Hawaii

Aiea

KGMZ-FM— September 1992: 107.9 mhz; 100 kw-horiz, 79 kw-vert. 1,965 ft TL: N21 23 51 W158 06 01. Stereo. Hrs opn: 24 1160 N. King St., 2nd Fl., Honolulu, 96817. Phone: (808) 275-1079. Fax: (808) 536-2528. E-mail: onair@oldiesradio.net Web Site: www.oldiesradio.net. Licensee: Salem Media of Hawaii Inc. (acq 1-3-2005 in exchange for KRTR(AM) Honolulu and KKNE(AM) Waipahu). Format: Oldies. News staff: one. Target aud: 35-54; adults. ♦Steve Miller, gen mgr; Bill Davis, gen sls mgr & chief of engrg; Rudi Camello, gen sls mgr; Jenny Clipse, progmg dir & traf mgr.

Eleele

KUAI(AM)— June 30, 1965: 720 khz; 5 kw-U. TL: N21 53 37 W159 33 27. Hrs open: 5 AM-midnight Box 720, 4469 Waialo Rd., 96705. Phone: (808) 335-3171. Fax: (808) 335-3834. E-mail: kuai@hawaiian.net Licensee: Visionary Related Entertainment L.L.C. (group owner; acq 2-10-2004; grpsl). Population served: 56,000 Format: Adult contemp, country, Hawaiian. News: 21 hrs wkly. Target aud: 25-65; loc long-time residents, blue & white collar. Spec prog: Hawaiian 5 hrs, jazz 4 hrs wkly. ♦John Detz, pres & gen mgr.

Haiku

KUAU(AM)— 1995: Stn currently dark. 1570 khz; 1 kw-D, 500 w-N. TL: N20 54 37 W156 17 15. Hrs open: 24 777 Mokulele Hwy., Kahului, 96732. Phone: (808) 871-7311. Fax: (808) 871-9708. Web Site: www.kingscathedral.com. Licensee: First Assembly of God-Kahului, Maui Inc. (acq 6-30-99). Law Firm: Baraff, Koerner & Olender. ♦Ron Moody, gen mgr & stn mgr.

Haliimaile

KPMW(FM)— 1994: 105.5 mhz; 6 kw. -512 ft TL: N20 42 32 W156 21 33. Hrs open: 230 Hana Hwy. #2, Kahului, 96732. Phone: (808) 871-6251. Fax: (808) 871-5670. E-mail: wild105@maui.net Web Site:

www.wild105.net. Licensee: Rey-Cel Broadcasting Inc. Format: CHR/rhythmic top 40. ♦Cecille Pirose, gen mgr & gen sls mgr; Bryan Pirose, progmg dir; Ray Pirose, news dir.

Hanalei

***KKCR(FM)**— Aug 2, 1997: 90.9 mhz; 950 w. Ant -308 ft TL: N22 13 02 W159 28 53. Stereo. Hrs opn: 24 Box 825, 96714. Phone: (808) 826-7774. Fax: (808) 826-7977. E-mail: kkcr@kkcr.org Web Site: www.kkcr.org. Licensee: Kekahu Foundation Inc. Population served: 53,000 Format: Hawaiian Pacifica, eclectic, educ. News: 5 hrs wkly. Target aud: General; Kauai County residents. ♦Harvey Cohen, pres & CFO; Gwen Squyres, gen mgr; Jessica Dofflemyer, dev dir; Donna Lewis Giarman, engrg VP.

Hilo

***KANO(FM)**— 2001: 91.1 mhz; 100 kw. 592 ft TL: N19 47 02 W155 05 23. Hrs open: 24 738 Kaheka St., Honolulu, 96814. Phone: (808) 955-8821. Fax: (808) 942-5477. Web Site: www.hawaiipublicradio.org. Licensee: Hawaii Public Radio. Natl. Network: NPR, PRI. Law Firm: Paul, Hastings, Janofsky & Walker, L.L.P. Format: Class. News staff: 3; News: 35 hrs wkly. ♦Valerie Yee, VP; Michael Titterton, gen mgr & progmg dir; Charles Husson, opns dir & chief of engrg; Gene Schiller, mus dir; Kayla Rosenfeld, news dir.

KAPA(FM)— December 1988: 100.3 mhz; 74 kw. Ant -515 ft TL: N19 50 19 W155 06 43. Stereo. Hrs opn: 24 913 Kanoelehua St., 96720. Phone: (808) 961-0651. Fax: (808) 934-8088. E-mail: jatebara@pacificradiogroup.com Licensee: Pacific Radio Group Inc. (group owner; (acq 8-11-2005; grpsl). Population served: 75000 Format: Hawaiian. Target aud: 18-49. ♦Jeanine Atebara, gen mgr; Jason Iglesias, progmg dir; Russ Roberts, news dir; Aaron Savage, chief of engrg; Cobey Patolo, traf mgr.

***KCIF(FM)**— July 1, 1998: 90.3 mhz; 14 kw. 164 ft TL: N19 30 17 W155 10 40. Hrs open: 180 Kinoole St., Suite 310, 96720. Phone: (808) 935-7434. Fax: (808) 961-6022. E-mail: kcifradio@turquoise.net Licensee: Hilo Christian Broadcasting. Format: Christian, relg, educ. ♦Pastor David Shotwell, chmn & pres.

KHBC(AM)— October 1986: 1060 khz; 5 kw-U. TL: N19 41 48 W155 03 05. Hrs open: Box 515, 96721. Phone: (808) 959-5700. Fax: (808) 959-5800. E-mail: happenings@khbcradio.com Web Site: www.khbcradio.com. Licensee: Hilo Broadcasting L.L.C. (acq 1-31-03). Natl. Network: CNN Radio. Format: Adult contemp. Spec prog: Hawaiian culture & language. ♦Buddy Gordon, gen mgr & stn mgr; Robert Turner, chief of engrg.

KHLO(AM)— Apr 1, 1950: 850 khz; 5 kw-U. TL: N19 44 11 W155 02 07. Hrs open: 24 913 Kanoelehua Ave., 96720. Phone: (808) 961-0651. Fax: (808) 934-8088. E-mail: jatebara@pacificradiogroup.com Licensee: Pacific Radio Group Inc. (group owner; acq 9-17-03; grpsl). Population served: 75,000 Format: Sports. News staff: 2; News: 7 hrs wkly. Target aud: 25-54. ♦Jeanine Atebare, gen mgr.

KHWI(FM)— Sept 20, 1992: 92.7 mhz; 9 kw. Ant -256 ft TL: N19 50 19 W155 06 43. Hrs open: 24 Box 2936, Sun Valley, ID, 83353. Phone: (808) 443-0292. Fax: (808) 934-9448. Licensee: Parrott Broadcasting L.P. (acq 7-2-2007; $375,000). Population served: 75,000 ♦Scott D. Parker, pres & gen mgr.

KIPA(AM)— Sept 10, 1947: 620 khz; 5 kw-U. TL: N19 51 03 W155 05 09. Hrs open: 24 74-5605 Luhia St., B-7, Kailua-Kona, 96740. Phone: (808) 329-8090. Fax: (808) 443-0888. E-mail: info@lava105.com Web Site: www.lava105.com. Licensee: Skynet Hawaii LLC (acq 5-3-2004; $75,000). Population served: 135,000 Natl. Network: ABC. Format: Adult Standards/Hawaiian. News staff: 2. Target aud: 35 plus. ♦Chip Begay, opns mgr.

KIXC(AM)—Not on air, target date: unknown: 1590 khz; 10 kw-U, DA-2. TL: N19 47 02 W155 05 25. Hrs open: 4703 Orkney Dr., Missouri City, TX, 77459. Phone: (281) 923-7100. E-mail: radioguy@neosoft.com Licensee: Fred R. and Evelyn Morton. ♦Fred R. Morton, gen mgr.

KKBG(FM)— Aug 5, 1980: 97.9 mhz; 51 kw. Ant -65 ft TL: N19 50 19 W155 06 43. Hrs open: 913 Kanoelehua, 96720. Phone: (808) 961-0651. Fax: (808) 934-8088. E-mail: jatebara@pacificradiogroup.com Licensee: Pacific Radio Group Inc. (group owner; (acq 9-17-2003; grpsl). Population served: 90,000 Format: Adult contemp. News staff: one; News: 10 hrs wkly. ♦Jeanine Atebara, gen mgr.

KNWB(FM)— Aug 3, 1985: 97.1 mhz; 40 kw. -124 ft TL: N19 45 33 W155 08 33. (CP: 38 kw. TL: N19 47 02 W155 05 25). Stereo. Hrs opn: 5 AM-10:30 PM 1145 Kilauea Ave., 96720. Phone: (808) 935-5461. Fax: (808) 935-7761. E-mail: sales@kwxx.com Web Site: www.B97Hawaii.com. Licensee: New West Broadcasting Corp. (group owner; acq 1995; $270,000). Format: Classic Hits. News staff: one; News: 8 hrs wkly. Target aud: 25-45. ♦Chris Leonard, pres & gen mgr; Gavin Tanouye, stn mgr.

KPUA(AM)— 1936: 670 khz; 10 kw-U. TL: N19 47 02 W155 05 25. (CP: 50 kw-U, DA-N). Stereo. Hrs opn: 24 1145 Kilauea Ave., 96720. Phone: (808) 935-5461. Fax: (808) 935-7761. Web Site: www.kpua.net. Licensee: New West Broadcasting Corp. (group owner; acq 5-18-92; $370,000 with co-located FM; FTR: 6-8-92). Population served: 64,000 Natl. Network: CBS, Westwood One. Law Firm: Dan Alpert. Format: News/talk, sports. News staff: 3; News: 22 hrs wkly. Target aud: 25 plus; upscale adults with interest in news. Spec prog: Japanese 6 hrs wkly. ♦Christopher Leonard, gen mgr, sls dir, mktg mgr & prom dir; John Orozco, rgnl sls mgr; Triska LaRochell, rgnl sls mgr; Ken Hupp, progmg dir & news dir.

KWXX-FM—Co-owned with KPUA(AM). Dec 16, 1984: 94.7 mhz; 100 kw. -330 ft TL: N19 43 02 W155 08 13. Stereo. 24 Web Site: www.kwxx.com. Natl. Network: Westwood One. Format: Hot adult contemp, Hawaiian. News staff: one; News: 3 hrs wkly. Target aud: 25 plus; upscale adults. Spec prog: Contemp Hawaiian 20 hrs, reggae 20 hrs wkly. ♦Gavin Tawouye, progmg dir; G. Kruz, disc jockey; Keoni Johnson, disc jockey.

KPVS(FM)— 1995: 95.9 mhz; 50 kw. 230 ft TL: N19 41 12 W155 09 04. (CP: 27 kw, ant -361 ft.). Hrs opn: 24
Rebroadcasts KLUA(FM) Kailua-Kona 100%.
913 Kanoelehua Ave., 96720. Phone: (808) 961-0651. Fax: (808) 934-8088. E-mail: jatebare@pacificradiogroup.com Licensee: Pacific Radio Group Inc. (group owner; (acq 8-11-2005; grpsl). Population served: 75,000 Law Firm: Cohen & Berfield. Format: Rhythmic adult contemp. Target aud: 25-54; women. ♦Jeanine Atebara, gen mgr; Darin Gumbs, progmg dir; Russ Roberts, news dir; Aaron Savage, chief of engrg; Cobey Patolo, traf mgr.

Holualoa

KHWA(FM)—Not on air, target date: unknown: 92.1 mhz; 4.5 kw. Ant 2,929 ft TL: N19 43 16 W155 55 15. Hrs open: Box 2936, Sun Valley, ID, 83353. Phone: (208) 837-4104. Licensee: Parrott Broadcasting L.P. (acq 6-11-2007; $356,250 for CP). ♦Scott D. Parker, gen mgr.

Honokaa

KLZY(FM)—Not on air, target date: 3/08: 102.9 mhz; 100 kw. Ant 1,535 ft TL: N19 53 08 W155 22 05. Hrs open: 14 Cockenoe Dr., Westport, CT, 06880-6908. Phone: (203) 977-6731. Licensee: Chaparral Broadcasting Inc. ♦Jerrold T. Lundquist, pres & gen mgr.

Honolulu

KAIM-FM—Listing follows KHNR(AM).

KCCN-FM— May 21, 1990: 100.3 mhz; 100 kw horiz, 81 kw vert. 1,965 ft TL: N21 23 51 W158 06 01. Hrs open: 900 Fort St., Suite 700, 96813. Phone: (808) 536-2728. Fax: (808) 536-2528. E-mail: info@kccnfm100.com Web Site: kccnfm100.com. Licensee: Cox Radio Inc. Group owner: Cox Broadcasting (acq 3-15-2000; grpsl). Format: Contemp Hawaiian mus. ♦Mike Kelly, gen mgr; David Daniels, opns mgr, progmg dir & progmg dir; Stuart Chang, gen sls mgr; Scott MacKenzie, mktg dir.

KDNN(FM)— July 4, 1988: 98.5 mhz; 51 kw. Ant 59 ft TL: N21 18 49 W157 51 43. Stereo. Hrs opn: 24 650 Iwilei Road, Suite 400, 96817. Phone: (808) 550-9200. Fax: (808) 550-9510. Web Site: www.island985.com. Licensee: Capstar TX L.P. Group owner: Clear Channel Communications Inc. (acq 8-30-00; grpsl). Population served: 900,000 Natl. Rep: Clear Channel. Law Firm: Ginsburg, Feldman & Bress. Format: Island Rhythm. Target aud: 25-54; upscale, white collar, college educated. ♦John Hogan, CEO & pres; Charlie Rahilly, sr VP; Chuck Cotton, gen mgr.

KHBZ(AM)—Co-owned with KDNN(FM). Mar 18, 1957: 990 khz; 5 kw-U. TL: N21 17 59 W157 51 33.24 Web Site: khbz.com.900,000 Natl. Network: ABC. Natl. Rep: Clear Channel. Law Firm: Ginsburg, Feldman & Bress. Format: News/talk. News: 20 hrs wkly. Target aud: 25-54.

KGU(AM)— May 11, 1922: 760 khz; 10 kw-U. TL: N21 17 41 W157 51 49. Hrs open: 24 560 N. Nimitz Hwy., Suite 109, 96817. Phone: (808) 533-0065. Fax: (808) 524-2104. E-mail: info@kguradio.com Web Site: www.kguradio.com. Licensee: Salem Media of Hawaii Inc. Group owner: Salem Communications Corp. (acq 2-16-00). Population served: 800,000 Natl. Network: Salem Radio Network. Natl. Rep: Salem. Law Firm: Fletcher, Heald & Hildreth. Format: Christian talk & teaching. News: 5 hrs wkly. Target aud: 35-54; general. ♦T.J. Malievsky, gen mgr; Jack Waters, opns mgr.

KHCM(AM)— May 14, 1947: 690 khz; 10 kw-U. TL: N21 17 41 W157 51 49. Stereo. Hrs opn: 24 560 N. Nimitz Hwy. #109, 96817. Phone: (808) 533-0065. Fax: (808) 524-2104. Licensee: Salem Media of Hawaii Inc. (acq 6-27-2006; exchange for KORL(AM) Honolulu). Population served: 1,000,000 Law Firm: Fletcher, Heald & Hildreth. Format: Country. ♦Steve Miller, gen mgr; Rudi Camello, gen sls mgr.

KHNR(AM)— Aug 31, 1956: 870 khz; 50 kw-U, DA-1. TL: N21 10 56 W157 13 27. Hrs open: 24 560 N. Nimitz Hwy., Suite 109, 96817. Phone: (808) 533-0065. Fax: (808) 524-2104. Licensee: Salem Media of Hawaii Inc. Group owner: Salem Communications Corp. (acq 11-10-99; with co-located FM). Population served: 324,871 Law Firm: Shaw Pittman. Format: Relg. ♦T.J. Malievsky, gen mgr; Jack Waters, chief of opns, progmg dir & mus dir; David Serrone, gen sls mgr; Kathleen Friedman, prom dir; Bob Adams, chief of engrg; Pattie James, traf mgr; Lee Lane, spec ev coord.

KAIM-FM—Co-owned with KHNR(AM). Nov 1, 1953: 95.5 mhz; 100 kw. -23 ft TL: N21 17 08 W157 48 08. (CP: 99 kw, ant 1,988 ft. TL: N21 23 42 W158 05 55). Stereo. 24 Web Site: www.thefish-hawaii.com.324,871 Natl. Network: Salem Radio Network. Format: Contemp Christian music. Target aud: 25-49; female. ♦Michael Shishido, progmg dir; Kim Harper, mus dir & disc jockey; Pattie James, traf mgr; Lee Lane, spec ev coord; Dave Lancaster, disc jockey; Lia Rodriguez, disc jockey.

KHNR-FM— Mar 6, 1962: 97.5 mhz; 80 kw. Ant 46 ft TL: N21 17 37 W157 50 32. Stereo. Hrs opn: 24 1160 N. King St., 2nd Fl., 96817. Phone: (808) 533-0065. Fax: (808) 524-2104. E-mail: mtshawaii@yahoo.com Web Site: www.khnr.com. Licensee: Salem Media of Hawaii Inc. Group owner: Salem Communications Corp. (acq 8-13-2004; $3.7 million with KHUI(FM) Honolulu). Population served: 622,900 Format: News/talk. ♦Steve Miller, gen mgr; Jack Waters, opns mgr; David Serrone, gen sls mgr; Michael Shishido, progmg dir & news dir; Bill Davis, chief of engrg; Jenny Clipse, traf mgr.

***KHPR(FM)**— Nov 13, 1981: 88.1 mhz; 44 kw. 2,000 ft TL: N21 24 03 W158 06 10. Stereo. Hrs opn: 24 738 Kaheka St., 96814. Phone: (808) 955-8821. Fax: (808) 946-3863. Web Site: www.hawaiipublicradio.org. Licensee: Hawaii Public Radio. Population served: 800,000 Natl. Network: NPR, PRI. Law Firm: Paul, Hastings, Janofsky & Walker. Format: Class, news, info. News staff: 3; News: 35 hrs wkly. Target aud: General. ♦Valerie Yee, VP; Michael Titterton, gen mgr & progmg dir; Charles Husson, opns mgr & chief of engrg; Gene schiller, mus dir; Kayla Rosenfeld, news dir.

KHRA(AM)— March 1992: 1460 khz; 5 kw-U. TL: N21 19 26 W157 52 32. Stereo. Hrs opn: 320 Ward Ave., Suite 207, 96814. Phone: (808) 348-1986 (cell phone). Fax: (808) 591-1986. Licensee: Trade Center Management Inc. (acq 2-2-2002; $575,000). Population served: 1,000,000 Format: Korean. ♦Ki Kim, gen mgr.

KHUI(FM)— Mar 1, 1993: 99.5 mhz; 100 kw. Ant -386 ft TL: N21 18 02 W157 51 53. Stereo. Hrs opn: 24 1160 N. King St., 2nd Fl., 96817. Phone: (808) 533-0065. Fax: (808) 524-2104. Web Site: www.khuiradio.com. Licensee: Salem Media of Hawaii Inc. Group owner: Salem Communications

Corp. (acq 8-13-2004; $3.7 million with KHNR-FM Honolulu). Natl. Rep: McGavren Guild. Format: Music of Hawaii. ♦Steve Miller, gen mgr; Rudi Camello, gen sls mgr.

KHVH(AM)— April 1951: 830 khz; 10 kw-U. TL: N21 19 26 W157 52 32. Stereo. Hrs opn: 24 650 Iwilei Rd., Suite 400, 96817. Phone: (808) 550-9200. Fax: (808) 550-9510. Web Site: www.khvh830am.com. Licensee: Capstar TX L.P. Group owner: Clear Channel Communications Inc. (acq 8-30-00; grpsl). Population served: 900,000 Natl. Rep: Clear Channel. Law Firm: Ginsburg, Feldman & Bress. Wire Svc: AP Format: News/talk, weather, traffic. News staff: 5; News: 21 hrs wkly. Target aud: 25-54. ♦John Hogan, CEO; Charlie Ramilly, sr VP; Chuck Cotton, gen mgr; Patti Milburn, gen sls mgr; Scott Hogle, sls dir & gen sls mgr; Jamie Hartnett, prom dir; Paul Wilson, progmg dir; Dave Curtis, news dir & reporter; Jerry Varoujean, chief of engrg.

KIKI-FM—Co-owned with KHVH(AM). Feb 14, 1979: 93.9 mhz; 100 kw. -44 ft TL: N21 19 26 W157 52 32. Stereo. 24 Web Site: www.hot939.com. Format: Rhythmic CHR. News staff: one; News: hrs wkly. Target aud: 18-34. ♦Paul Wilson, opns dir; Laurie Mizuno, gen sls mgr; Kamu Kanekoa, prom dir; Fred Rico, progmg dir; Dave Curtis, pub affrs dir & reporter; Dale Machado, chief of engrg & traf mgr.

KINE-FM— November 1988: 105.1 mhz; 100 kw. 1,948 ft TL: N21 23 51 W158 06 01. Stereo. Hrs opn: 900 Fort St. Mall, Suite 700, 96813-3797. Phone: (808) 275-1000. Fax: (808) 536-2528. E-mail: info@hawaiian105.com Web Site: www.hawaiian105.com. Licensee: Cox Radio Inc. Group owner: Cox Broadcasting (acq 3-15-2000; grpsl). Format: Contemp & traditional Hawaiian. Target aud: 25-44. ♦Michael Kelly, VP & gen mgr; John Aeto, sls dir; Ann Boots, rgnl sls mgr; Scott MacKenzie, mktg dir; Wade Faildo, prom mgr; David Daniels, progmg dir; Jane Pascual, news dir.

***KIPO(FM)**— 1989: 89.3 mhz; 3.3 kw. 1,968 ft TL: N21 24 03 W158 06 10. Stereo. Hrs opn: 24 738 Kaheka St., 96814. Phone: (808) 955-8821. Fax: (808) 946-3863. Web Site: www.hawaiipublicradio.org. Licensee: Hawaii Public Radio. Natl. Network: PRI, NPR. Law Firm: Paul, Hastings, Janofsky & Walker. Format: News & info, jazz, international mus. News staff: 3; News: 50 hrs wkly. Target aud: General. ♦Valerie Yee, VP; Michael Titterton, gen mgr & progmg dir; Charles Husson, opns dir & chief of engrg; Gene Schiller, mus dir; Kayla Rosenfeld, news dir.

KKEA(AM)— Nov 1, 1966: 1420 khz; 5 kw-U. TL: N21 19 26 W157 52 47. Hrs open: 900 Fort St., Suite 700, 96813. Phone: (808) 275-1047. Phone: (808) 296-1420. Fax: (808) 275-1197. Fax: (808) 548-0608. Web Site: www.kkea1420am.com. Licensee: Blow Up LLC (acq 5-31-02; $750,000). Population served: 800,000 Natl. Network: CNN Radio, ESPN Radio. Natl. Rep: Katz Radio. Format: Talk, sports. Target aud: 25-54; Male. ♦Randall Ikeda, gen mgr; Chris Hart, progmg dir.

***KKUA(FM)**—Wailuku, Apr 15, 1988: 90.7 mhz; 7 kw. 5,533 ft TL: N20 42 41 W156 15 26. Hrs open: 24
Rebroadcasts KHPR(FM) Honolulu 100%.
738 Kaheka St., 96814. Phone: (808) 955-8821. Fax: (808) 942-5477. Web Site: www.hawaiipublicradio.org. Licensee: Hawaii Public Radio Inc. Natl. Network: PRI, NPR. Format: Class, news, info. News staff: 3; News: 35 hrs wkly. Target aud: General. Spec prog: Hawaiian one hr, Pacific Island 2 hrs wkly. ♦Valerie Yee, VP; Michael Titterton, gen mgr & progmg dir; Charles Husson, opns dir; Gene Schiller, mus dir; Kayla Rosenfeld, news dir.

KLHT(AM)— 1946: 1040 khz; 10 kw-U. TL: N21 20 10 W157 53 33. Hrs open: 24 98-1016 Komo Mai Dr., Aiea, 96701. Phone: (808) 524-1040. Fax: (808) 487-1040. E-mail: klht@hawaii.rr.com Web Site: www.klight.org. Licensee: Calvary Chapel of Honolulu Inc. (acq 5-85). Format: Bible teaching. Target aud: General. ♦Jake O'Neil, gen mgr; Peter Scott, progmg dir.

KNDI(AM)— July 11, 1960: 1270 khz; 5 kw-U. TL: N21 19 26 W157 52 47. Hrs open: 24 1734 S. King St., 96826. Phone: (808) 946-2844. Fax: (808) 947-3531. E-mail: kndi.am@verizon.net Web Site: www.kndi.com. Licensee: Leona Jona dba KNDI Radio. Population served: 350,000 Format: Ethnic, Filipino 3 dialects. Spec prog: Togan 7 hr, Vietnamesel 2 hrs, Loatian 2 hrs wkly. ♦Harvey Weinstein, VP, opns mgr & mus dir; Leona Jona, pres & gen mgr.

KORL(AM)— December 1959: 1180 khz; 1 kw-U. TL: N21 26 18 W157 59 29. Hrs open: 24 900 Fort St. Mall, Suite 450, 96813. Phone: (808) 538-9690. Fax: (808) 538-9548. Licensee: Hochman-McCann Hawaii Inc. Group owner: Salem Communications Corp. (acq 6-27-2006; exchange for KHCM(AM) Honolulu). Population served: 95,000 Format: Ethnic, loc talk. Target aud: 25 plus. ♦George Hochman, CEO & gen mgr.

KPOI-FM— Aug 3, 2000: 105.9 mhz; 97 kw. 1,968 ft TL: N21 23 50 W158 06 06. Stereo. Hrs opn: 24 765 Amana, 96814. Phone: (808) 947-1500. Fax: (808) 947-1506. E-mail: kumu@kumu.com Web Site: www.lavarock1059.com. Licensee: Visionary Related Entertainment LLC. (group owner; (acq 3-17-2004; grpsl). Law Firm: Thompson, Hine & Flory. Format: Classic rock. News staff: one; News: 23 hrs wkly. Target aud: 24-54; families, including single-parent families. ♦John Detz, gen mgr; Greg Everett, gen sls mgr; Dale Parsons, progmg dir; Gary Forsberg, traf mgr.

KQMQ-FM— Oct 1, 1967: 93.1 mhz; 100 kw. Ant 1,853 ft TL: N21 23 45 W158 05 58. Stereo. Hrs opn: 765 Amana St., 96814. Phone: (808) 947-1500. Fax: (808) 947-1506. Web Site: www.kqmq.net. Licensee: Visionary Related Entertainment L.L.C. (group owner; acq 7-1-2004; grpsl). Population served: 1,000,000 Format: Hits of the 80s. ♦John Detz, gen mgr; Joshua Flemming, gen sls mgr; Sean Lynch, progmg dir; Ryan Sean, mus dir.

KREA(AM)— Apr 24, 1973: 1540 khz; 5 kw-D. TL: N21 19 27 W157 52 47. Hrs open: 24 1839 S. King St., 96826. Phone: (808) 955-1234. Licensee: JMK Communications Inc. (acq 3-10-00; $575,000). Population served: 90000 Format: Korean language stn.

KRTR(AM)— 1946: 650 khz; 10 kw-U. TL: N21 26 43 W158 03 49. Hrs open: 24 560 N. Nimitz Hwy., Suite109, 96817. Phone: (808) 533-0065. Fax: (808) 524-2104. Web Site: www.khnr.com. Licensee: Cox Radio Inc. Group owner: Salem Communications Corp. (acq 1-3-2005 with KKNE(AM) Waipahu in exchange for KGMZ-FM Aiea). Natl. Network: CNN Radio, CBS Radio. Natl. Rep: Salem. Law Firm: Fletcher, Heald & Hildreth. Wire Svc: AP Format: News, talk. News: 55 hrs wkly. Target aud: 25-54. ♦T.J. Malievsky, gen mgr; Jack Walters, chief of opns; Wayne Marla, progmg dir.

KRTR-FM—See Kailua

KRUD(AM)—Not on air, target date: June 1, 2005: Stn currently dark. 1130 khz; 10 kw-D, 5 kw-N. TL: N21 16 30 W157 49 25. Hrs open: 900 Fort Street Mall, Suite 450, 96813. Phone: (808) 538-9690. Fax: (808) 538-9548. Licensee: Hochman-McCann Hawaii Inc. (acq 1-26-2005; $60,000 for CP). Format: Ethnic. Target aud: 25 plus. ♦George Hochman, pres & gen mgr; Donna Vincent, opns dir; Byron McCann, chief of engrg; Dianna Hochman, sls.

KSHK(FM)—Kekaha, Aug 10, 1999: 103.3 mhz; 100 kw. 918 ft TL: N21 56 11 W159 26 43. Hrs open: Box 1748, Lihue, 96766. Phone: (808) 245-9527. E-mail: knog@hawaiian.net Web Site: www.kongradio.com. Licensee: Visionary Related Entertainment L.L.C. (group owner; acq 2-10-2004; grpsl). Format: Top-40. ♦John Detz, pres; Jim McKeon, opns dir & progmg dir; Denise Roberts, prom dir & traf mgr; Ron Middag, chief of engrg.

KSSK(AM)— 1929: 590 khz; 7.5 kw-U. TL: N21 19 26 W157 52 32. Hrs open: 24 650 Iwilei Rd., Suite 400, 96817. Phone: (808) 550-9200. Fax: (808) 550-9507. Web Site: www.ksskradio.com. Licensee: Capstar TX L.P. Group owner: Clear Channel (acq 3-12-99; grpsl). Population served: 900,000 Natl. Rep: Clear Channel. Law Firm: Ginsburg, Feldman & Bress. Wire Svc: AP Format: Adult contemp, personalities. News staff: 5; News: 15 hrs wkly. Target aud: 25-54. ♦Chuck Cotton, gen mgr.

KSSK-FM—See Waipahu

***KTUH(FM)**— Jan 1, 1969: 90.3 mhz; 3kw. -82 ft TL: N21 18 14 W157 49 22. Stereo. Hrs opn: 24 2445 Campus Rd., Suite 203, 96822. Phone: (808) 956-7431. Phone: (808) 956-5288. Fax: (808) 956-5271. E-mail: gm@ktuh.org Web Site: www.ktuh.org. Licensee: University of Hawaii. Population served: 324,871 Target aud: 18-59; no target, all kinds of people listen. ♦Monty Anderson, gen mgr; Loriel Macalma, prom dir & prom mgr; Travis Tokuyama, mus dir; Dale Machado, chief of engrg; Katie McClellen, traf mgr.

KUMU-FM— Sept 1, 1967: 94.7 mhz; 100 kw. 78 ft TL: N21 17 09 W157 50 19. Stereo. Hrs opn: 765 Amana St., Suite 206, 96814. Phone: (808) 947-1500. Fax: (808) 947-1506. E-mail: kumu@kumu.com Licensee: Visionary Related Entertainment LLC (group owner; (acq 3-17-2004; grpsl). Population served: 705,900 Format: Lite rock, adult contemporary. Target aud: 25-54. ♦Bonnie Craig, pres & gen sls mgr; Jeff Coelho, gen mgr; Sumee Mikkelson, prom dir; Ed Kanoi, progmg dir; Ernie Nearman, chief of engrg; Lilly Yamachika, traf mgr.

KUMU(AM)— Mar 1, 1963: 1500 khz; 10 kw-U. TL: N21 17 08 W157 48 08.24 Natl. Network: Westwood One. Format: Talk. Target aud: 35-64.

KUPA(AM)—Pearl City, May 2, 1990: 1370 khz; 6.2 kw-U. TL: N21 26 18 W157 59 29. Hrs open: 24 Broadcasting Corp. of America, 4766 Holladay Blvd., Holladay, UT, 84117. Phone: (808) 533-0065. Phone: (801) 273-9200. Licensee: Broadcasting Corp. of America. Group owner: Diamond Broadcasting Corp. (acq 4-12-2006; $650,000). Natl. Network: Fox Sports. Format: Sports. ♦Nathan W. Drage, pres.

KWAI(AM)— Jan 21, 1972: 1080 khz; 5 kw-U. TL: N21 17 41 W157 51 49. Hrs open: 24 100 N. Beretania St., Suite 401, 96817. Phone: (808) 523-3868. Fax: (808) 531-6532. E-mail: radio@hawaii.com Licensee: Radio Hawaii Inc. (acq 2-85). Population served: 800,000 Natl. Network: USA. Format: News/talk. News: 72 hrs wkly. Target aud: 25-64; general. Spec prog: Filipino 7 hrs, Hawaiian 3hrs wkly, Samoan 14 hrs wkly. ♦Barry Wagenvoord, pres & gen mgr; Sam Wagenvoord, VP; Renee Rosehill, opns VP & opns dir.

KZOO(AM)— Oct 18, 1963: 1210 khz; 1 kw-U. TL: N21 17 59 W157 51 33. (CP: TL: N21 17 41 W157 51 49). Hrs opn: Box 61335, 96839-1335. Phone: (808) 988-8828. Fax: (808) 988-5882. E-mail: newsdesk@kzoohawaii.com Web Site: www.kzoohawaii.com. Licensee: Polynesian Broadcasting Inc. (acq 8-4-2005). Population served: 700,000 Format: Japanese, English. ♦David Furuya, pres & gen mgr.

Kahaluu

KLEO(FM)— 1992: 106.1 mhz; 7.3 kw. Ant 2,995 ft TL: N19 43 16 W155 55 15. Hrs open: 913 Kanaoelehua Ave., Hilo, 96720. Phone: (808) 961-0651. Fax: (808) 934-8088. E-mail: jatebare@pacificradiogroup.com Licensee: Pacific Radio Group Inc. (group owner; acq 9-17-03; grpsl). Format: Adult contemp. ♦Jeanine Atebara, gen mgr; J.E. Orozco, gen sls mgr & progmg dir; Russ Roberts, news dir; Aaron Savage, chief of engrg; Cobey Patolo, traf mgr.

Kahului

KAOI-FM—Wailuku, June 1974: 95.1 mhz; 100 kw. 1,227 ft TL: N20 38 12 W156 23 24. Stereo. Hrs open: 24 Box 1437, Wailuku, 96793. Phone: (808) 244-9145. Licensee: Visionary Related Entertainment L.L.C. (group owner; (acq 2-10-2004; grpsl). Format: Adult contemp. News staff: one; News: 5 hrs wkly. Target aud: General. ♦John Detz, pres, gen mgr & progmg dir; Greg Everett, gen sls mgr; Dale Parsons, progmg dir; Gary Forsberg, news dir; Alex Kowalski, chief of engrg.

KAOI(AM)— Oct 11, 1979: 1110 khz; 5 kw-U. TL: N20 47 30 W156 28 21. Stereo. 1900 Main St., Wailuku, 96793. Natl. Network: CBS, Westwood One. Format: Talk/news, sports. Target aud: 25-54. ♦Alex Kowalski, chief of engrg.

KJKS(FM)—Listing follows KNUI(AM).

KLHI-FM—Not on air, target date: unknown: 92.5 mhz; 1.7 kw. Ant 2,211 ft TL: N20 39 36 W156 21 50. Hrs open: 311 Ano St., 96732. Phone: (808) 877-5566. Fax: (808) 871-0666. Licensee: Pacific Radio Group Inc. (acq 3-13-2007; swap for KORL-FM Lahaina). ♦Pamela Tsutsui, gen mgr; Jeff Hunter, opns mgr.

KNUI(AM)— Sept 14, 1962: 900 khz; 5 kw-U. TL: N20 47 30 W156 28 21. Hrs open: 24 311 Ano St., 96732. Phone: (808) 877-5566. Fax: (808) 877-2888. Fax: (808) 871-0666. E-mail: onair@knuiam900.com Web Site: www.foxnews900.com. Licensee: Pacific Radio Group Inc. (group owner; (acq 12-10-99; grpsl). Population served: 100,000 Natl. Network: Fox News Radio. Format: News/talk. ♦Eddie Johnson, CEO & CFO; Chuck Bergson, pres & CFO; Pamela Tsutsui, gen mgr; Jeff Hunter, opns mgr; Debbie Probst, gen sls mgr & rgnl sls mgr; Sherri Grimes, prom mgr; Fred Guzman, progmg dir; Earl Tolley, chief of engrg; Dorene Moniz, traf mgr.

KJKS(FM)—Co-owned with KNUI(AM). June 22, 1984: 99.9 mhz; 100 kw. -540 ft TL: N20 47 30 W156 28 21.100,000 Natl. Network: Westwood One. Format: Adult contemp. News staff: one. Target aud: 25-49. ♦Sherri Grimes, prom dir; Jeff Hunter, progmg dir; Dorene Moniz, traf mgr; Kopaa Tita, disc jockey.

Kailua

KRTR-FM— Oct 9, 1978: 96.3 mhz; 75 kw. 2,120 ft TL: N21 19 49 W157 45 24. Stereo. Hrs opn: 24 900 Fort St., 7th Fl., Honolulu, 96813. Phone: (808) 275-1000. Fax: (808) 536-2528. Web Site: www.krater96.com. Licensee: Cox Radio Inc. Group owner: Cox Broadcasting (acq 11-10-99; grpsl). Format: Adult contemp. Target aud: 25-54.Bob Neil, CEO & pres; Marc Morgan, COO; Neil Johnston, CFO; Richard Ferguson, exec VP; Mike Kelly, gen mgr; John Aeto, sls dir; Mimi Beams, gen sls mgr; Corinne Webb, natl sls mgr & mus dir; Scott McKenzie, mktg dir; Aron Dotes, prom mgr; Wayne Maria, opns mgr & progmg dir; Jane Pascual, news dir; Chris Caughill, chief of engrg; Alexa Dahlquist, traf mgr

Kailua-Kona

KLUA(FM)— 1991: 93.9 mhz; 5.3 kw. 2,831 ft TL: N19 43 15 W155 55 16. Stereo. Hrs opn: 24
Rebroadcasts KPVS(FM) Hilo 100%.
913 Kanoelehua Ave., Hilo, 96720. Phone: (808) 961-0651. Fax: (808) 934-8088. E-mail: jatebara@pacificradiogroup.com Licensee: Pacific Radio Group Inc. (group owner; (acq 8-11-2005; grpsl). Population served: 60,000 Format: Rhythmic adult contemp. Target aud: 25-54; women. ♦Jeanine Atebara, gen mgr & gen sls mgr; Darin Gumbs, progmg dir; Russ Roberts, news dir; Aaron Savage, chief of engrg; Cobey Patolo, traf mgr.

Kalaheo

KTOH(FM)— June 1, 2002: 99.9 mhz; 51 kw. 892 ft TL: N21 56 11 W159 26 43. Hrs open: Box 929, 96741. Secondary address: 4334 Rice St., Suite 204B, Lihue 96766. Phone: (808) 246-4444. Fax: (808) 246-4405. E-mail: gh5512@aol.com Web Site: hhawaiimedia.net. Licensee: Hochman Hawaii-One Inc. group owner: Hochman Hawaii-One Inc. (acq 6-23-2000; $125,000 for CP). Population served: 60,000 Wire Svc: AP Format: Oldies. News staff: one. Target aud: 25-54; adults. ♦Dianna Hochman, gen mgr & sls dir; George Hochman, mktg mgr; Mark James, progmg dir.

Kaneohe

KPHW(FM)— Oct 17, 1997: 104.3 mhz; 73.5 kw. 2,116 ft TL: N21 19 49 W157 45 24. Hrs open: 24 900 Fort St., 7th Fl., Honolulu, 96813. Phone: (808) 275-1000. Fax: (808) 536-2528. E-mail: info@1043xme.com Web Site: www.1043xme.com. Licensee: Cox Radio Inc. Group owner: Cox Broadcasting (acq 11-10-99; grpsl). Format: CHR. Target aud: 18-34. ♦Bob Neil, CEO, pres & pres; Marc Morgan, COO & VP; Neil Johnston, CFO; Mike Kelly, gen mgr; Wayne Maria, opns mgr; Mark Haworth, gen sls mgr; Corinne Webb, natl sls mgr; Scott McKenzie, mktg dir & prom dir; Aron Dote, prom mgr; K.C. Bejerana, progmg dir; Kevin Akitake, progmg dir & mus dir; Jane Pascual, news dir; Chris Caughill, engrg mgr.

Kapaa

KITH(FM)— 1999: 98.9 mhz; 100 kw. Ant 918 ft TL: N21 56 11 W159 26 43. Stereo. Hrs opn: 4334 Rice St., Suite. 204 B, Lihue, 96766. Phone: (808) 246-4444. Fax: (808) 246-4405. Web Site: hhawaiimedia.net. Licensee: Hochman Hawaii-Two Inc. (acq 7-14-2000; $110,000 for CP). Wire Svc: AP Format: Island music. News: 10 hrs wkly. Target aud: 18-44; adults. ♦Dianna Hochman, gen mgr.

Kaunakakai

KMKK-FM— Mar 19, 2007: 102.3 mhz; 1.9 kw. Ant 1,181 ft TL: N21 07 55 W157 11 31. Hrs open: Box 1437, Wailuku, 96793. Phone: (808) 553-8300. Fax: (808) 244-8247. Licensee: Visionary Related Entertainment LLC. Format: Hawaiian. ♦John Detz, gen mgr; Greg Everett, gen sls mgr; L.D. Reynolds, progmg dir.

Kawaihae

KWYI(FM)— November 1993: 106.9 mhz; 5.5 kw. 341 ft TL: N19 53 09 W155 39 28. Hrs open: 6 AM-10 PM Box 6540, 64-1040 Mamalahoa Hwy., Suite 4, Kamuela, 96743. Phone: (808) 885-9866. Fax: (808) 885-6480. Web Site: www.kwyi.com. Licensee: Colin H. Naito. Format: Adult contemp. Target aud: 25-54. ♦Colin H. Naito, gen mgr.

Keaau

KBGX(FM)— Apr 16, 2004: 105.3 mhz; 25.26 kw. Ant 92 ft TL: N19 43 18 W155 27 23. Hrs open: 24 74-5605 Luhia St. B-7, Kailua-Kona, 96740. Phone: (808) 329-8090. Fax: (808) 443-0888. E-mail: info@lava105.com Web Site: www.lava105.com. Licensee: Skynet Hawaii LLC (acq 12-10-03). Population served: 160,000 Natl. Network: ABC. Law Firm: Shook, Hardy & Bacon- Erwin Krasnow. Format: Oldies. Target aud: Adults; 25-54. ♦Chip Begay, opns mgr.

Kealakekua

KAOY(FM)— Nov 11, 1982: 101.5 mhz; 6 kw. Ant 2,052 ft TL: N19 31 10 W155 55 08. Stereo. Hrs opn: c/o KWXX-FM, 1145 Kilauea Ave., Hilo, 96720. Phone: (808) 935-5461. Fax: (808) 935-7761. E-mail: studio@kwxx.com Web Site: www.kwxx.com. Licensee: New West Broadcasting Corp. (group owner; acq 4-16-2004; $500,000). Format: Adult contemp. Target aud: 18-49. ♦Christopher Leonard, gen mgr; Trisha LaRochelle, gen sls mgr; Gavin Panouye, progmg mgr; Ken Hupp, news dir; Yisa Var, traf mgr.

KKON(AM)— October 1963: 790 khz; 5 kw-U. TL: N19 31 10 W155 55 08. Hrs open: 24
Rebroadcasts Khlo(AM) Hilo 100%.
913 Kanoelehua Ave., Hilo, 96720. Phone: (808) 961-0651. Fax: (808) 935-0396. Web Site: www.pacificradiogroup.com. Licensee: Pacific Radio Group Inc. (group owner; (acq 8-11-2005; grpsl). Population served: 60,000 Format: Sports. Target aud: 35 plus; general. Spec prog: Hawaiian mus. ♦Jeanine Atebara, gen mgr.

Kekaha

KSHK(FM)—Licensed to Kekaha. See Honolulu

Kihei

KAOI(AM)—Licensed to Kihei. See Kahului

KHEI-FM—Not on air, target date: unknown: 107.5 mhz; 26 kw. Ant -267 ft TL: N20 49 21 W156 27 15. Hrs open: Box 1730, Rohnert Park, CA, 94927. Phone: (707) 528-0339. Licensee: Visionary Related Entertainment LLC. ♦John Detz, gen mgr.

Kilauea

***KAQA(FM)**— July 3, 1997: 91.9 mhz; 950 w. Ant 1,607 ft TL: N21 58 41 W159 29 55. Stereo. Hrs opn: Box 825, Hanalei, 96714. Phone: (808) 826-7774. Fax: (808) 826-7977. E-mail: kkcr@kkcr.org Web Site: www.kkcr.org. Licensee: Kekahu Foundation Inc. Population served: 6,000 Format: Hawaiian Pacifica, eclectic, educ. Target aud: General; Kauai County residents. ♦Harvey Cohen, pres; Larry Lasota, gen mgr & stn mgr; Douvn Jewell, dev VP & engrg VP; Ken Jannelli, progmg dir; Donna Lewis, news dir; Dean Rogers, chief of engrg.

Kurtistown

KTBH-FM—Not on air, target date: Aug 1, 2006: 102.1 mhz; 50 kw horiz. Ant -207 ft TL: N19 41 48 W155 03 05. Hrs open: Box 1437, Wailuku, 96793. Phone: (808) 244-9145. Fax: (808) 244-8247. Licensee: Visionary Related Entertainment LLC. Format: Adult contemp. ♦John Detz, pres & gen mgr.

Lahaina

KPOA(FM)— October 1984: 93.5 mhz; 1.4 kw. 1,305 ft TL: N20 50 43 W156 54 04. (CP: 346 w, ant 2,421 ft.). Stereo. Hrs opn: 24 311 Ano St., Kahului, 96732. Phone: (808) 877-5566. Fax: (808) 871-0666. Web Site: www.kpoa.com. Licensee: Pacific Radio Group Inc. (group owner; acq 12-10-99; grpsl). Population served: 100,000 Law Firm: Kenkel & Associates. Format: Contemp Hawaiian Island sounds. News staff: one; News: one hr wkly. Target aud: Adults; 25-54. ♦Eddie Johnson, CFO; Pamela Tsutsui, gen mgr.

Lanai City

KONI(FM)— Nov 1, 1993: 104.7 mhz; 69 kw. Ant 2,283 ft TL: N20 39 36 W156 21 50. Hrs open: 24 300 Ohukai Rd., Suite C-318, Kihei, 96753. Phone: (808) 875-8866. Fax: (808) 875-8870. E-mail: koni@hawaii.rr.com Web Site: www.hhawaiimedia.net. Licensee: Hochman Hawaii Publishing Inc. (acq 6-17-2002; $1.15 million). Wire Svc: AP Format: Oldies. News: one hr wkly. Target aud: 25-54; Maui county residents. ♦George Hochman, COO, chmn, pres, CFO & VP; Jim Carroll, gen mgr; Adrienne Owens, gen sls mgr; Joe Hawkins, progmg dir & news dir; Byron McCann, chief of engrg.

Lihue

KAWV(FM)— 2001: 98.1 mhz; 51 kw. Ant 13 ft TL: N21 59 41 W159 24 36. Hrs open: O'Hana Radio Partners, 41-625 Eclectic St. J-1, Palm Desert, CA, 92260. Licensee: Hochman Hawaii Four Inc. (acq 7-9-2007; $400,000). ♦George Hochman, pres.

KFMN(FM)— Mar 7, 1988: 96.9 mhz; 100 kw. 400 ft TL: N21 59 54 W159 25 35. Stereo. Hrs opn: 24 Box 1566, 1860 Leleiona St., 96766-5566. Phone: (808) 246-1197. Fax: (808) 246-9697. E-mail: frontdesk@fm97radio.com Licensee: FM 97 Associates. (acq 6-7-88; $600,000). Population served: 53,000 Law Firm: Mullin, Rhyne, Emmons & Topel. Format: Adult contemp. News staff: one; News: 4 hrs wkly. Target aud: 25-54; island residents & visitors. ♦Dianne Reynolds-Mikami, opns mgr; John Wada, gen mgr & progmg dir; Jason Fujinaka, news dir.

***KHJC(FM)**—Not on air, target date: unknown: 88.9 mhz; 100 kw. 892 ft TL: N21 56 11 W159 26 43. Stereo. Hrs opn: 2970 Kele St., Suite 117, 96766. Secondary address: CSN International, 3000 W. MacArthur Blvd., 3rd Fl, Santa Ana, CA 92704. Phone: (808) 245-9696. Fax: (808) 245-9898. E-mail: pastorsteve@khjcradio.com Web Site: www.khjcradio.com. Licensee: CSN International (group owner) Format: Teaching/music.

KQNG(AM)— 1939: 570 khz; 1 kw-U. TL: N21 59 33 W159 24 24. Hrs open: Box 1748, KQNG Radio Bldg., 4271 Halenani St., 96766. Phone: (808) 245-9527. Fax: (808) 245-3563. E-mail: kong@hawaiian.net Web Site: www.kongradio.com. Licensee: Visionary Related Entertainment L.L.C. (group owner; (acq 2-10-2004; grpsl). Population served: 60,000 Format: News/talk, sports. Target aud: 25-54. ♦John Detz, CEO & gen mgr; Ron Middac, progmg dir & chief of engrg.

KQNG-FM— Oct 17, 1983: 93.5 mhz; 100 kw. 226 ft TL: N21 59 33 W159 24 24. Stereo. Format: Hot adult Comtemp. Target aud: 18-49. ♦John Detz, pres; Ron Wiley, opns mgr.

Makawao

KDLX(FM)— Dec 31, 1980: 94.3 mhz; 3 kw. -22 ft TL: N20 50 48 W156 19 35. Stereo. Hrs opn: 24 Box 1437, Wailuku, 96793. Phone: (808) 244-9145. Fax: (808) 244-8247. Licensee: Visionary Related Entertainment L.L.C. (group owner; acq 2-10-2004; grpsl). Format: Country. ♦John Detz, pres & gen mgr; Greg Everett, stn mgr & gen sls mgr; Jack Gist, progmg dir; Alex Kowalski, chief of engrg; Gary Forsberg, traf mgr.

Nanakuli

KNAN(FM)—Not on air, target date: unknown: 106.7 mhz; 2.6 kw. Ant 1,834 ft TL: N21 23 45 W158 05 57. Hrs open: 2550 5th Ave., Suite 600, San Diego, CA, 92103-6624. Phone: (702) 385-6000. Fax: (619) 232-7317. Licensee: Big D Consulting Inc. ♦Donald F. Hildre, pres.

Paauilo

KNUQ(FM)— 1995: 103.7 mhz; 100 kw. 1,209 ft TL: N20 38 18 W156 23 01. Hrs open: 24 Box 1437, Wailuku, 96793. Phone: (808) 244-9145. Fax: (808) 244-8247. Web Site: www.q103maui.com. Licensee: Visionary Related Entertainment LLC (group owner; acq 2-10-2004; grpsl). Population served: 90,000 Format: Contemp island music. News staff: one. Target aud: 18-49; young active adults. ♦John Detz, CEO & gen mgr; Greg Everett, stn mgr & prom mgr; Alex Kowalski, opns mgr & chief of engrg; Shaggy Jenkins, progmg dir; Gary Forsberg, traf mgr.

Pahala

***KPHL(FM)**— 2006: 90.5 mhz; 250 w. Ant -39 ft TL: N19 06 02 W155 34 09. Hrs open: 935 Dillingham Blvd., Suite One, Honolulu, 96817. Phone: (808) 294-4575. Fax: (808) 923-7723. Licensee: Vineyard Christian Fellowship of Honolulu Inc. (acq 1-26-2007; $1,500). ♦Timothy J. Malievsky, gen mgr.

Pearl City

KUCD(FM)— Feb 14, 1995: 101.9 mhz; 100 kw. 1,948 ft TL: N21 23 51 W158 06 01. Stereo. Hrs opn: 24 650 Iwilei Rd, Suite 400, Honolulu, 96817. Phone: (808) 550-9200. Fax: (808) 550-9510. Web Site: www.star1019fm.com. Licensee: Capstar TX L.P. Group owner: Clear Channel Communications Inc. (acq 8-30-00; grpsl). Population served: 900,000 Natl. Rep: Clear Channel. Law Firm: Ginsburg, Feldman & Bress. Format: Alternative. Target aud: 25-54; boomers & yuppies. ♦Chuck Cotton, VP & gen mgr; Laurie Mizuno, gen sls mgr; Damian Balinowski, news dir; Dale Costales, chief of engrg; Iwalani Costales, traf mgr; Jamie Hyatt, progmg.

KUPA(AM)—Licensed to Pearl City. See Honolulu

Poipu

KSRF(FM)— Aug 14, 1999: 95.9 mhz; 100 kw. 918 ft TL: N21 56 11 W159 26 43. Hrs open: Box 1748, Lihue, 96766. Phone: (808) 245-9527. E-mail: kong@hawaiian.net Web Site: www.kongradio.com. Licensee: Visionary Related Entertainment L.L.C. (group owner; acq 2-10-2004; grpsl). Format: Contemp Hawaiian. ♦John Detz, pres & gen mgr; Shelly Cobb, progmg dir & engrg dir; Ron Middag, chief of engrg; Denise Roberts, traf mgr.

Pukalani

KJMD(FM)— June 15, 1984: 98.3 mhz; 50 kw. 102 ft TL: N20 42 19 W156 21 54. Stereo. Hrs opn: 24 311 Ano St., Kahului, 96732. Phone: (808) 877-5566. Fax: (808) 871-0666. Web Site: www.kjmd.com. Licensee: Pacific Radio Group Inc. (group owner; acq 12-10-99; grpsl). Population served: 100,000 Natl. Network: ABC. Format: CHR. News staff: one. Target aud: 18-34; young active adults. ♦Chuck Bergson, CEO; Pamela Tsutsui, gen mgr.

Volcano

KKOA(FM)— 1996: 107.7 mhz; 25.5 kw. 92 ft TL: N19 43 18 W155 27 23. Hrs open: 24 74-5605 Luhia St., B-7, Kailua-Kona, 96740. Phone: (808) 329-8090. Fax: (808) 443-0888. E-mail: info@lava105.com Web Site: www.lava105.com. Licensee: Skynet Hawaii LLC (acq 5-3-2004; $350,000). Population served: 160,000 Natl. Network: ABC. Law Firm: Erwin Krabnox. Format: Country. ♦Chip Begay, opns mgr.

Wahiawa

***KHAI(FM)**— 2007: 103.5 mhz; 2.2 kw horiz, 1.9 kw vert. Ant 1,958 ft TL: N21 23 51 W158 06 01. Hrs open:
Rebroadcasts KLRD(FM) Yucaipa, CA 100%.
5700 West Oaks Blvd., Rocklin, CA, 95765. Phone: (916) 251-1600. Fax: (916) 251-1650. Web Site: www.air1.com. Licensee: Educational Media Foundation. (acq 12-19-2005; $2 million for CP). Natl. Network: Air 1. Format: Alternative rock, div. ♦Richard Jenkins, pres.

Waianae

KORL-FM— May 1984: 101.1 mhz; 100 kw horiz, 81 kw vert. Ant 1,942 ft TL: N21 23 45 W158 05 58. Stereo. Hrs opn: 311 Ano St., Kahului, 96732. Phone: (808) 877-5566. Fax: (808) 871-0666. Licensee: Hochman Hawaii-Three Inc. (group owner; (acq 3-13-2007; $520,000 plus swap for CP for KLHI-FM Kahului). Law Firm: Dan Alpert. Format: Alternative rock. News staff: one; News: one hr wkly. Target aud: 18-49; general. ♦Pamela Tsutsui, gen mgr & stn mgr; Jeff Hunter, opns mgr.

Wailuku

KAOI-FM—Licensed to Wailuku. See Kahului

KJMD(FM)—See Pukalani

KKUA(FM)—Licensed to Wailuku. See Honolulu

KMVI(AM)— Mar 17, 1947: 550 khz; 5 kw-U. TL: N20 53 29 W156 29 23. Hrs open: 24 311 Ano St., Kahului, 96732. Phone: (808) 877-5566. Fax: (808) 871-0666. Web Site: www.kmvi550.com. Licensee: Pacific Radio Group Inc. (group owner; acq 12-10-99; grpsl). Population served: 100,000 Natl. Network: ESPN Radio, ABC. Format: Sports radio. News staff: one. Target aud: 18 plus; men, residents/tourists, educated professionals. ♦Chuck Bergson, pres; Pamela Tsutsui, gen mgr.

Waimea

KAGB(FM)— 2000: 99.1 mhz; 7.3 kw. Ant 2,990 ft TL: N19 43 16 W155 55 15. Hrs open: 913 Kanoelehua Ave., Hilo, 96720. Phone: (808) 961-0651. Fax: (808) 934-8088. E-mail: jatebara@pacificradiogroup.com Licensee: Pacific Radio Group Inc. (group owner; (acq 8-11-2005; grpsl). Format: Hawaiian. ♦Jeanina Atebara, gen mgr; J.E. Orozco, stn mgr, gen sls mgr & progmg dir; Russ Roberts, news dir; Aaron Savage, chief of engrg; Cobey Patolo, traf mgr.

Waipahu

KDDB(FM)— Nov 23, 1988: 102.7 mhz; 61 kw. 1,893 ft TL: N21 23 49 W158 05 58. Stereo. Hrs opn: 765 Amana St., Suite 200, Honolulu, 96814. Phone: (808) 947-1500. Fax: (808) 947-1506. Licensee: Visionary Related Entertainment L.L.C. (group owner; acq 7-1-2004; grpsl). Natl. Rep: McGavren Guild. Format: Top 40 hits. Target aud: 18-34; young adults who enjoy many different types of music. ♦John Detz, gen mgr & stn mgr.

KKNE(AM)— Sept 20, 1950: 940 khz; 10 kw-U. TL: N21 26 43 W158 03 49. Hrs open: 24 900 Fort St., Suite 700, Honolulu, 96813. Phone: (808) 533-0065 Ext. 731. Phone: (808) 257-1000. Fax: (808) 275-1195. Licensee: Cox Radio Inc. Group owner: Salem Communications Corp. (acq 1-3-2005 with KRTR(AM) Honolulu in exchange for KGMZ-FM Aiea). Population served: 1,000,000 Format: Country. Target aud: 25-44. ♦John Aeto, sls dir; Michael Kelly, gen mgr & sls dir; David Daniels, progmg dir.

KSSK-FM— Dec 30, 1976: 92.3 mhz; 100 kw. 1,630 ft TL: N21 23 49 W158 05 58. (CP: Ant 1,950 ft.). Stereo. Hrs opn: 24 650 Iwilei Rd., Suite 400, Honolulu, 96817-5319. Phone: (808) 550-9200. Fax: (808) 550-9510. Web Site: ksskradio.com. Licensee: Clear Channel Broadcasting Licenses Inc. Clear Channel Communications Inc. (acq 9-1-00; grpsl). Population served: 900,000 Natl. Rep: Clear Channel. Law Firm: Ginsburg, Feldman & Bress. Wire Svc: AP Wire Svc: Metro Weather Service Inc. Format: Adult contemp. News staff: 3; News: 6 hrs wkly. Target aud: 25-54. ♦Chuck Cotton, gen mgr; Scott Hogle, opns dir & sls dir; Patti Milburn, gen sls mgr; Jamie Hartnett, prom dir & prom mgr; Paul Wilson, progmg dir; Damian Balinowski, news dir; Dale Machado, chief of engrg; Larry Price, traf mgr & disc jockey.

Idaho

Aberdeen

KQPI(FM)—Not on air, target date: unknown: 99.5 mhz; 91 kw. Ant 1,020 ft TL: N42 52 25 W112 30 48. Hrs open: 980 N. Michigan Ave., Suite 1880, Chicago, IL, 60611. Phone: (312) 204-9900. Licensee: College Creek Media LLC. ♦Neal J. Robinson, pres.

American Falls

KORR(FM)— 1995: 104.1 mhz; 3 kw. 328 ft TL: N42 45 24 W112 48 38. Hrs open: 24 Box 97, Pocatello, 83204-0097. Secondary address: 436 N. Main St., Pocatello 83204. Phone: (208) 234-1290. Fax: (208) 234-9451. E-mail: spots@kzbq.com Licensee: Idaho Wireless Corp. (group owner; acq 1996). Format: Adult contemp. ♦Paul E. Anderson, gen mgr; Harry Neuhardt, gen sls mgr; Paul Anderson, progmg dir.

Ammon

KSPZ(AM)—Licensed to Ammon. See Idaho Falls

Blackfoot

KBLI(AM)— November 1951: 690 khz; 1 kw-D, 43 w-N. TL: N43 10 70 W112 22 10. Hrs open: Box 699, 83221. Phone: (208) 785-1400. Fax: (208) 785-0184. Licensee: Riverbend Communications LLC. Group owner: Bonneville International Corp. (acq 4-17-2006; grpsl). Population served: 9,556 Format: Talk. ♦Delyn Hendricks, gen mgr.

KLCE(FM)—Co-owned with KBLI(AM). Oct 15, 1975: 97.3 mhz; 100 kw. 1,512 ft TL: N43 30 03 W112 39 43. Stereo. Web Site: www.klce.com.9,556 Natl. Rep: McGavren Guild. Format: Adult contemp. Target aud: 18-49.

KCVI(FM)— Sept 22, 1994: 101.5 mhz; 100 kw. 1,512 ft TL: N43 30 03 W112 39 43. Stereo. Hrs opn: 24 Box 699, 83221. Phone: (208) 785-1400. Fax: (208) 785-0184. E-mail: scott@kbear.fm Web Site: www.kbear.fm. Licensee: Riverbend Communications LLC. Group owner: Bonneville International Corp. (acq 4-17-2006; grpsl). Natl. Rep: McGavren Guild. Format: Active rock. News staff: one; News: 3 hrs wkly. Target aud: 25-44; male. ♦Jim Burgoyne, pres; Delyn Hendricks, gen mgr; Matt Burgoyne, sls dir; Scott Taylor, progmg dir; Tisa Cudmore, traf mgr.

Boise

KAWO(FM)—Listing follows KFXD(AM).

***KAWS(FM)**—Not on air, target date: unknown: 89.1 mhz; 8.8 kw vert. Ant 2,189 ft TL: N43 00 25 W116 42 13. Hrs open: 4002 N. 3300 E., Twin Falls, 83301-0354. Phone: (208) 733-3133. Fax: (208) 736-1958. Licensee: Calvary Chapel of Twin Falls Inc. ♦Mike Kestler, pres.

KBOI(AM)— May 1, 1947: 670 khz; 50 kw-U, DA-N. TL: N43 25 44 W116 19 43. Stereo. Hrs opn: 24 Box 1280, 83701. Secondary address: 1419 W. Bannock 83702. Phone: (208) 336-3670. Fax: (208) 336-3734 (Main). Fax: (208) 336-3735 (News). E-mail: andrew.paul@citcomm.com Web Site: www.670kboi.com. Licensee: Citadel Broadcasting Co. Group owner: Citadel Broadcasting Corp. (acq 12-10-97; grpsl). Population served: 250,000 Natl. Rep: Katz Radio. Rgnl rep: Allied Radio Partners Format: News/talk. News staff: 3. Target aud: 25-54; white collar, upper income. ♦Kevin Godwin, gen mgr; Ken Weaver, sls dir & news dir; Linda Rupe, prom dir; Andrew Paul, progmg dir; Mike Owens, gen sls mgr & chief of engrg.

KQFC(FM)—Co-owned with KBOI(AM). Nov 1, 1960: 97.9 mhz; 47 kw. 2,499 ft TL: N43 45 12 W116 06 08. (CP: 58 kw). Stereo. 24 Format: Country. Target aud: 25-54; country lifestyle.

***KBSU(AM)**— Dec 4, 1955: 730 khz; 15 kw-D, 500 w-N, DA-2. TL: N43 34 13 W116 20 45. Stereo. Hrs opn: 213 SMITC, 1910 University Dr., 83725. Phone: (208) 426-3663. Fax: (208) 344-6631. Web Site: radio.boisestate.edu/AM730.html. Licensee: Idaho State Board of Education (Boise State University) (acq 12-30-91; donation; FTR: 1-20-92). Population served: 250,000 Law Firm: Dow, Lohnes & Albertson. Format: News/talk, jazz. Spec prog: Folk. ♦John Hess, gen mgr; Hy Kloc, gen sls mgr; Jim East, progmg dir; Sadie Baleits, news dir; Tom Taylor, chief of engrg.

***KBSU-FM**— Jan 16, 1977: 90.3 mhz; 19 kw. Ant 2,637 ft TL: N43 35 41 W116 08 39. Stereo. Hrs opn: 24 Boise State Radio, 1910 University Dr., 83725. Phone: (208) 344-3961. Phone: (208) 947-5659. Fax: (208) 344-4080. E-mail: jeast@boisestate.edu Web Site: radio.boisestate.edu. Licensee: Boise State Board of Education. (acq 12-30-91). Population served: 600,000 Natl. Network: PRI, NPR. Format: Class, news. News: 15 hrs wkly. Target aud: General. ♦John Hess, gen mgr & chief of engrg; Erik Jones, opns mgr; Ele Ellis, progmg dir; Sadie Babits, news dir.

***KBSX(FM)**— 1994: 91.5 mhz; 4 kw. 2,581 ft TL: N43 45 18 W116 05 52. Hrs open: Boise State Radio, 1910 University Dr., 83725. Phone: (208) 344-3961. Fax: (208) 344-4080. Web Site: radio.boisestate.edu. Licensee: Idaho State Board of Education. Population served: 650,000 Natl. Network: NPR. Wire Svc: AP Format: News. News staff: 4. ♦John Hess, gen mgr; Erik Jones, opns mgr; Ele Ellis, progmg dir; Tom Taylor, chief of engrg.

KBXL(FM)—See Caldwell

KCIX(FM)—Garden City, Jan 1, 1985: 105.9 mhz; 50 kw. 2,700 ft TL: N43 45 18 W116 05 52. Stereo. Hrs opn: 827 Park Blvd., Suite 201, 83712. Phone: (208) 344-6363. Fax: (208) 385-9064. Web Site: www.mix106radio.com. Licensee: Peak Broadcasting of Boise Licenses LLC. Group owner: Clear Channel Communications Inc. (acq 6-28-2007; grpsl). Natl. Rep: McGavren Guild. Format: Adult contemp. Target aud: 25-54. ♦Terry Tario, gen mgr; Susan Green, opns mgr & traf mgr; Brent Carey, progmg dir; Dave Burnett, news dir.

KFXD(AM)— Nov 9, 1928: 630 khz; 5 kw-U, DA-2. TL: N43 30 56 W116 19 43. Hrs open: 827 E. Park Blvd., 83712-7782. Phone: (208) 344-6363. Fax: (208) 385-9064. E-mail: email@kfxd.com Web Site: www.kfxd.com. Licensee: Peak Broadcasting of Boise Licenses LLC. Group owner: Clear Channel Communications Inc. (acq 6-28-2007; grpsl). Population served: 225,000 Natl. Network: ABC. Natl. Rep: Christal. Format: Talk. ♦Dick Lumenello, gen mgr; Mitch Pruitte, progmg dir; Lee Eichelberger, chief of engrg.

KAWO(FM)—Co-owned with KFXD(AM). Nov 2, 1979: 104.3 mhz; 52 kw. Ant 2,574 ft TL: N43 45 18 W116 05 52. Stereo. Web Site: www.koololdies1043.com. Format: Classic oldies. Target aud: 25-54; affluent, managerial, professional. ♦Jack Armstrong, progmg dir.

KGEM(AM)— 1945: 1140 khz; 10 kw-U, DA-N. TL: N43 35 54 W116 15 14. Stereo. Hrs opn: 24 5257 Fairview Ave., Suite 260, 83706. Phone: (208) 344-3511. Fax: (208) 947-6765. Licensee: Journal Broadcast Corp. Group owner: Journal Broadcast Group Inc. (acq 5-13-98; grpsl). Population served: 600,000 Natl. Network: Jones Radio Networks. Natl. Rep: Katz Radio. Format: Oldies. Target aud: 25-54. ♦Bob Rosenthal, VP, gen mgr & opns mgr.

KJOT(FM)—Co-owned with KGEM(AM). 1979: 105.1 mhz; 43 kw. 2,570 ft TL: N43 45 19 W116 06 52. (CP: 52.5 kw). Stereo. 24 600,000 Natl. Rep: Katz Radio. Format: Classic rock. Target aud: 25-49.

KIDO(AM)—See Nampa

KIZN(FM)— Aug 1, 1968: 92.3 mhz; 44 kw. 2,500 ft TL: N43 45 19 W116 05 52. Stereo. Hrs opn: 24 1419 W. Bannock St., 83701. Phone: (208) 336-3670. Fax: (208) 336-3736. E-mail: rich.summers@citicomm.com Web Site: www.kizn.com. Licensee: Citadel Broadcasting Co. Group owner: Citadel Broadcasting Corp. (acq 12-24-97; grpsl). Population served: 150,000 Format: Country. News staff: one. Target aud: 25-54. ♦Kevin Godwin, stn mgr; Rich Summers, opns VP & progmg dir; Brenda Mee, news dir; Bill Frahm, chief of engrg; Patti Hull, traf mgr.

KNJY(AM)— Apr 8, 1961: 950 khz; 5 kw-D, 35 w-N. TL: N43 37 14 W116 17 57. Hrs open: Box 1600, Nampa, 83653. Phone: (208) 463-1900. Licensee: First Western Inc. (acq 8-4-03; $150,000). Format: Relg. Spec prog: Farm 5 hrs wkly. ♦Steve Sumner, gen mgr.

KSAS-FM—See Nampa

KSPD(AM)— Apr 29, 1959: 790 khz; 1 kw-D, 61 w-N. TL: N43 33 57 W116 20 13. Hrs open: 24 1440 S. Weideman Ave., 83709. Phone: (208) 377-3790. Fax: (208) 377-3792. E-mail: info@myfamilyradio.com Web Site: www.myfamilyradio.com. Licensee: KSPD Inc. (group owner; acq 3-24-83; FTR: 4-18-83). Population served: 450,000 Natl. Network: Moody. Law Firm: Wiley, Rein & Fielding. Format: Christian, talk. Target aud: 18-54. ♦Beth Schafer, exec VP; Lee Schafer, pres & gen mgr; David Schafer, stn mgr.

KZMG(FM)—New Plymouth, Mar 17, 1982: 93.1 mhz; 50 kw. 2,630 ft TL: N43 45 19 W116 05 52. Stereo. Hrs opn: 24 Box 1280, 83701-1280. Secondary address: 1419 W. Bannock 83701. Phone: (208) 336-3670. Fax: (208) 336-3734. Web Site: www.magic93.com. Licensee: Citadel Broadcasting Co. Group owner: Citadel Broadcasting Corp. (acq 12-24-97; grpsl). Population served: 560,000 Natl. Rep: D & R Radio. Law Firm: Wiley, Rein, Fielding. Format: Contemporary hit/Top-40. News: 6 hrs wkly. Target aud: 18-34; women. ♦Kevin Godwin, gen mgr & stn mgr; Mike Owens, gen sls mgr; Brad Collins, progmg dir; Deb Course, progmg dir & news dir; Bill Frahm, chief of engrg.

Bonners Ferry

KBFI(AM)— Sept 1, 1977: 1450 khz; 1 kw-U. TL: N48 41 20 W116 20 04. Hrs open: 327 S. Marion, Sandpoint, 83864. Phone: (208) 267-5234. Fax: (208) 267-5594. E-mail: prod@953kpnd.com Licensee: Blue Sky Broadcasting. (acq 1996). Population served: 12,000 Format: News/talk, sports. News staff: 1. Target aud: General. ♦Dylan Benefield, gen mgr & opns mgr; Jim Tomchek, progmg dir.

***KIBX(FM)**— 2000: 92.1 mhz; 74 w. Ant 2,749 ft TL: N48 36 37 W116 15 24. Hrs open: Spokane Public Radio Inc., 2319 N. Monroe St., Spokane, WA, 99205-4586. Phone: (509) 328-5729. Fax: (509) 328-5764. E-mail: kpbx@kpbx.org Web Site: kpbx.org. Licensee: Spokane Public Radio Inc. Format: Classical, news, jazz. ♦Richard Kunkel, gen mgr; Brian Flick, progmg dir; Verne Windham, mus dir; Doug Nadvornick, news dir.

Buhl

***KTFY(FM)**— Aug 2005: 88.1 mhz; 60 kw vert. Ant 653 ft TL: N42 43 48 W114 25 06. Stereo. Hrs opn: 24 16115 S. Montana Ave., Caldwell, 83605. Phone: (208) 459-5879. Fax: (208) 459-3144. E-mail: magee@ktfy.org Web Site: www.881ktfy.org. Licensee: Southern Idaho Corp. of Seventh-Day Adventists dba Gem State Academy. Population served: 200,000 Law Firm: Donald Martin. Format: Christian. Target aud: 25-54; women 25-54. ♦Donald Klinger, chmn; Stephen L. McPherson, pres; Michael Agee, gen mgr; Jerry Woods, progmg dir.

Burley

KBAR(AM)— Aug 31, 1946: 1230 khz; 1 kw-U. TL: N42 32 05 W113 48 54. Stereo. Hrs opn: 120 S. 300 W., Rupert, 83350. Phone: (208) 678-2244. Fax: (208) 678-2246. E-mail: kimlee@cableone.net Licensee: KART Broadcasting Co. Inc. and Eagle Rock Broadcasting Inc. as tenants-in-common. Group owner: Tri-Market Radio Broadcasters Inc. & Eagle Rock Broadcasting Inc. (acq 1-30-98; with co-located FM). Population served: 42,000 Format: Oldies, talk. ♦Kim Lee, gen mgr; Chris Kinzel, gen sls mgr; Ben Reed, progmg dir & news dir.

KZDX(FM)—Co-owned with KBAR(AM). Feb 15, 1975: 99.9 mhz; 27 kw. Ant 2,450 ft TL: N42 20 06 W113 36 15. Stereo. 90,050 Format: AOR.

***KBSY(FM)**— October 1998: 88.5 mhz; 440 w vert. 2,083 ft TL: N42 21 42 W113 27 17. Hrs open:
Rebroadcasts KBSX(FM) Boise 100%.
Boise State Radio, 1910 University Dr., Boise, 83725. Phone: (208) 344-3961. Fax: (208) 344-4080. Web Site: radio.boisestate.edu. Licensee: Idaho State Board of Education. Natl. Network: NPR. Format: News. ♦John Hess, gen mgr; Erik Jones, opns mgr; Ele Ellis, progmg dir; Tom Taylor, engrg dir.

Caldwell

KBGN(AM)— Oct 5, 1960: 1060 khz; 10 kw-D. TL: N43 43 13 W116 31 58. Hrs open: 3303 E. Chicago, 83605. Phone: (208) 459-3635. E-mail: kbgn@kbgnradio.com Web Site: www.kbgnradio.com. Licensee: Nelson M. Wilson & Karen E. Wilson. (acq 8-25-89; $188,000; FTR: 9-11-89). Natl. Network: USA. Format: Inspirational, Christian, talk. Target aud: General. Spec prog: Sp 5 hrs wkly. ♦Nelson Wilson, gen mgr; Marnie Fillmore, opns dir.

KBXL(FM)— Feb 22, 1961: 94.1 mhz; 40 kw. Ant 2,634 ft TL: N43 45 18 W116 05 52. Stereo. Hrs opn: 24 1440 S. Weideman Ave., Boise, 83709. Phone: (208) 377-3790. Fax: (208) 377-3792. E-mail: info@myfamilyradio.com Web Site: www.myfamilyradio.com. Licensee: KSPD Inc. (group owner; acq 4-26-89; FTR: 7-10-89). Population served: 600,000 Natl. Network: AP Network News. Natl. Rep: Salem. Rgnl rep: Tacher Law Firm: Wiley, Rein & Fielding. Format: Relg, Christian talk. Target aud: 25-54. ♦Lee Schafer, pres & gen mgr.

KCID(AM)— 1947: 1490 khz; 1 kw-U. TL: N43 39 51 W116 38 10. Hrs open: 24 5257 Fairview Ave., Suite 260, Boise, 83706. Phone: (208) 344-3511. Fax: (208) 947-6765. Licensee: Journal Broadcast Corp. Group owner: Journal Broadcast Group Inc. (acq 5-13-98; grpsl). Population served: 600,000 Natl. Network: ABC. Natl. Rep: Katz Radio. Format: Oldies. Target aud: 35 plus. ♦Bob Rosenthal, VP, gen mgr & chief of engrg.

KTHI(FM)—Co-owned with KCID(AM). Dec 1, 1983: 107.1 mhz; 52 kw. Ant 2,578 ft TL: N43 45 18 W116 05 52. Stereo. 24 600,000 Natl. Rep: Katz Radio. Format: Super hits of 60's & 70's. Target aud: General.

KSAS-FM—Licensed to Caldwell. See Nampa

***KTSY(FM)**— Oct 14, 1990: 89.5 mhz; 8.3 kw. 2,601 ft TL: N43 45 18 W116 05 52. (CP: Ant 2,594 ft.). Stereo. Hrs opn: 24 16115 S. Montana Ave., 83607. Phone: (208) 459-5879. Fax: (208) 459-3144. Web Site: www.ktsy.org. Licensee: Gem State Adventist Academy. Population served: 400,000 Law Firm: Donald E. Martin. Format: Contemp Christian mus. News: 4 hrs wkly. Target aud: 25-45. ♦Donald Klinger, chmn; Stephen McPherson, pres; Michael Agee, gen mgr; Jerry Woods, progmg dir.

Chubbuck

KLLP(FM)— Nov 10, 1984: 98.5 mhz; 6.6 kw. Ant 1,305 ft TL: N42 55 15 W112 20 44. (CP: 7 kw, ant 987 ft. TL: N42 52 26 W112 30 47). Hrs opn: 24
Rebroadcasts KAWZ(FM) Twin Falls 65%.
259 E. Center St., Pocatello, 83204. Phone: (208) 233-1133. Fax: (208) 232-1240. E-mail: kellymartinez@clearchannel.com Web Site: www.985klite.com. Licensee: Citicasters Licenses L.P. Group owner: Clear Channel Communications Inc. (acq 5-4-99; grpsl). Population served: 400,000 Format: Adult contemp. Target aud: General. Spec prog: Sp 4 hrs wkly. ♦Neica Kinney, gen mgr; Jeff Evans, opns mgr; Kelly Martinez, progmg dir; Rhett Downing, chief of engrg; Cami Chopski, traf mgr.

KRTK(AM)— 1981: 1490 khz; 1 kw-U. TL: N42 55 38 W112 30 03. Stereo. Hrs opn: 24 1633 Olympus Dr., Pocatello, 83201. Phone: (208) 237-9500. Fax: (208) 237-4600. E-mail: krtk@ltlink.com Licensee: Broken Chains Inc. (acq 8-25-00). Population served: 110,000 Natl. Network: ABC. Format: Christian. Target aud: 35-55. ♦Stacy Dare, stn mgr & chief of opns.

Coeur d'Alene

KHTQ(FM)—Listing follows KVNI(AM).

KICR(FM)— Oct 12, 2001: 102.3 mhz; 6 kw. 1,843 ft TL: N47 39 35 W116 57 12. Stereo. Hrs opn: 24
Rebroadcasts KIBR-FM Sandpoint 100%.
327 Marion Ave., Sandpoint, 83864. Phone: (208) 664-3241. Fax: (208) 665-7880. E-mail: dylanb@a53kpwd.com Licensee: Great Northern Broadcasting Inc. (acq 8-17-2001; $550,000). Population served: 500,000 Law Firm: Smithwick & Belendiuk, PC. Format: Country. ♦Dylan Benefield, gen mgr, opns dir & gen sls mgr; Jimmy Silver, progmg dir; Mike Brown, news dir.

KVNI(AM)— Nov 1, 1946: 1080 khz; 10 kw-D, 1 kw-N, DA-N. TL: N47 36 57 W116 43 07. Hrs open: 24 504 E. Sherman, 83814. Secondary address: 500 W. Boone Ave., Spokane, WA 99201. Phone: (208) 664-9271. Fax: (208) 667-0945. Licensee: QueenB Radio Inc. Population served: 65,000 Natl. Rep: Katz Radio. Format: Doo Whopping Oldies, news. News staff: 2. Target aud: 25 plus. Spec prog: Relg 3 hrs wkly. ♦Susan McIver, gen mgr; Dick Haugen, progmg dir, news dir & pub affrs dir; Tim Anderson, chief of engrg.

KHTQ(FM)—Co-owned with KVNI(AM). Nov 1, 1991: 94.5 mhz; 100 kw. 1,883 ft TL: N47 39 34 W116 57 48. Stereo. 24 Web Site: rock945.com.464,000 Format: Active rock. Target aud: 24-54. ♦Brew Michaels, opns dir; Barry Bennet, mus dir & disc jockey; Jolene Longwill, traf mgr; Gary Allen, disc jockey; Geoff Scott, disc jockey; Karla Stevens, disc jockey; Ken Richards, progmg dir & disc jockey.

Cottonwood

***KNWO(FM)**— January 1994: 90.1 mhz; 250 w. 612 ft TL: N46 04 09 W116 27 54. Hrs open: 24
Rebroadcasts KRFA-FM Moscow, ID.
c/o Radio Stn KRFA-FM, Box 642530, 382 Murrow Communications Ctr., Pullman, WA, 99164. Phone: (509) 335-6500. Fax: (509) 335-3772. E-mail: nwpr@wsu.edu Web Site: www.nwpr.org. Licensee: Washington State University. Law Firm: Dow, Lohnes & Albertson. Format: Class, news. News staff: one; News: 37 hrs wkly. Target aud: 25 plus. ♦Karen Olstad, COO & gen mgr; Dennis Haarsager, gen mgr; Roger Johnson, stn mgr & sls dir; Scott Weatherly, opns mgr; Sarah McDaniel, dev dir; Mary Hawkins, progmg dir; Robin Rilette, mus dir; Ralph Hogan, engrg dir; Rachael McDonald, news rptr.

Donnelly

KMCL(AM)—Licensed to Donnelly. See McCall

Driggs

KCHQ(FM)— Feb 12, 2004: 102.1 mhz; 4 kw. Ant 1899 ft TL: N43 42 42 W111 20 56. Stereo. Hrs opn: 24 Box 54, 83422. Secondary address: 1152 Bond Ave, Suite 102, Rexburg 83440. Phone: (208) 354-4102. Fax: (208) 356-6111. E-mail: ted@q102fm.net Web Site: www.q102country.com. Licensee: Ted W. Austin Jr. Population served: 75,000 Rgnl rep: Tacher Law Firm: Wood, Maines & Nolan. Format: Country. News staff: one; News: 8 hrs wkly. Target aud: 25-54; adults.

Spec prog: Farm one hr, classic country 3 hrs wkly. ♦Ted W. Austin Jr., pres & gen mgr; Charlie Michaels, opns mgr; Dave Plourde, news dir.

Eagle

KXLT-FM— September 1994: 107.9 mhz; 45 kw. 2,683 ft TL: N43 45 18 W116 05 52. Hrs open: 827 E. Park Blvd., Suite 201, Boise, 83712. Phone: (208) 344-6363. Fax: (208) 327-8800. Web Site: www.lite108.com. Licensee: Peak Broadcasting of Boise Licenses LLC. Group owner: Clear Channel Communications Inc. (acq 6-28-2007; grpsl). Natl. Rep: McGavren Guild. Format: Soft adult contemp, lite music. Target aud: 25-54. ♦Kevin Godwin, gen mgr & stn mgr; Dave Burnett, gen sls mgr & news dir; Susan Green, prom dir & traf mgr; Tobin Jeffries, progmg dir.

Emmett

KDBI(FM)— Mar 12, 1973: 101.9 mhz; 57 kw. Ant 2,532 ft TL: N43 45 18 W116 05 52. Stereo. Hrs opn: 2722 S. Redwood Rd., Salt Lake City, UT, 84119. Phone: (208) 463-2900. Web Site: www.bustosmedia.com. Licensee: First Western Inc. (acq 11-1-2003; $1.05 million). Population served: 3,945 Format: Rgnl Mexican. ♦Ed Distel, gen mgr.

Fruitland

KWEI-FM—Licensed to Fruitland. See Weiser

Garden City

KCIX(FM)—Licensed to Garden City. See Boise

Gooding

KAYN(FM)— Dec 2, 1996: 100.7 mhz; 73 kw. Ant 2,191 ft TL: N43 14 43 W115 26 12. Hrs open: 24 21361 Hwy. 30, Twin Falls, 83301. Phone: (208) 735-8300. Fax: (208) 733-4196. Licensee: FM Idaho Co. LLC. (group owner; (acq 12-31-2006; grpsl). Format: Oldies. ♦Larry Johnson, pres & gen mgr; Jerre Fender, opns dir & progmg dir; Deb Uvieu, gen sls mgr; Denis Jeffs, traf mgr.

KRXR(AM)— 1992: 1480 khz; 1 kw-D. TL: N42 54 54 W114 42 41. Hrs open: 501 S. Lincoln Ave., Jerome, 83338. Fax: (208) 934-8688. E-mail: krxr@cableone.net Licensee: Maria Elena Juarez. (acq 1999; $200,000). Format: Sp. ♦Efrain Ortega, gen mgr.

Grangeville

KORT(AM)— Oct 8, 1954: 1230 khz; 1 kw-U. TL: N45 55 52 W116 07 50. Hrs open: Box 510, 83530. Phone: (208) 983-1230. Fax: (208) 983-2744. Licensee: 4-K Radio Inc. (group owner; acq 6-1-71). Population served: 12,500 Format: Today's C&W. News staff: one; News: 8 hrs wkly. Target aud: General. Spec prog: Farm 2 hrs wkly. ♦Mike Ripley, pres; Melinda Fischer, gen mgr, opns mgr, gen sls mgr, progmg dir & traf mgr; David Forsman, chief of engrg; Josh Campbell, disc jockey; Leon Pierce, disc jockey; Steve Cash, disc jockey; W. Alan Hall, disc jockey.

KORT-FM— Dec 1, 1979: 92.7 mhz; 360 w. 2,352 ft TL: N45 51 48 W116 07 24. Stereo. 8,500 Natl. Network: ABC. Format: Country.

Hailey

KSKI-FM—See Sun Valley

KYUN(FM)— 2006: 106.7 mhz; 97 kw. Ant 1,578 ft TL: N43 16 45 W114 09 14. Hrs open: 21361 Hwy. 30, Twin Falls, 83301-0197. Phone: (208) 735-8300. Fax: (208) 733-4196. Web Site: www.canyoncountryonline.com. Licensee: Locally Owned Radio LLC. Format: Country. ♦Larry Johnson, pres & gen mgr; Jerre Fender, opns dir & progmg dir; Deb Uvieu, gen sls mgr; Denis Jeffs, traf mgr.

Hayden

KHTQ(FM)—Licensed to Hayden. See Coeur d'Alene

Hazelton

KTPZ(FM)— 2007: 94.3 mhz; 4.9 kw. Ant 741 ft TL: N42 43 54 W114 25 04. Hrs open: 21361 Hwy. 30, Twin Falls, 83301-0197. Phone: (208) 735-8300. Fax: (208) 733-4196. Licensee: Locally Owned Radio LLC. (acq 9-8-2006; $2,911,000 with KIRQ(FM) Twin Falls). Format: Hot CHR. ♦Larry Johnson, pres & gen mgr; Jerre Fender, opns dir & progmg dir; Deb Uvieu, gen sls mgr; Denis Jeffs, traf mgr.

Homedale

KQTA(FM)— December 2004: 106.3 mhz; 100 kw. Ant 1,028 ft TL: N43 37 15 W117 12 35. Hrs open: 2957 Stonebridge Tr., Reno, NV, 89511. Phone: (775) 741-3777. E-mail: leoramos97035@aol.com Licensee: Bustos Media of Idaho License LLC. (acq 10-28-2005; $2.25 million).

Idaho Falls

***KAIO(FM)**— 2006: 90.5 mhz; 500 w vert. Ant 528 ft TL: N43 32 37 W111 53 07. Hrs open:
Rebroadcasts KLRD(FM) Yucaipa, CA 100%.
2351 Sunset Blvd., Suite 170-218, Rocklin, CA, 95765. Phone: (916) 251-1600. Fax: (916) 251-1650. Web Site: www.air1.com. Licensee: Educational Media Foundation. Natl. Network: Air 1. Format: Christian. ♦Richard Jenkins, pres; Mike Novak, VP; Keith Whipple, dev dir; David Pierce, progmg mgr; Ed Lenane, news dir; Sam Wallington, engrg dir; Karen Johnson, news rptr; Marya Morgan, news rptr; Richard Hunt, news rptr.

KBLY(AM)— Sept 10, 1960: 1260 khz; 5 kw-D, 64 w-N. TL: N43 31 15 W111 59 33. Hrs open: Box 699, Blackfoot, 83221. Phone: (208) 785-1400. Fax: (208) 785-0184. Licensee: Riverbend Communications LLC. Group owner: Bonneville International Corp. (acq 4-17-2006; grpsl). Population served: 38,987 Format: Talk. Target aud: 35 plus; upscale, mature adults. ♦Jim Burgoyne, pres & chief of engrg; Delyn Hendricks, gen mgr; Matt Burgoyne, sls dir & women's int ed; Neal Larson, progmg dir.

KFTZ(FM)— May 24, 1986: 103.3 mhz; 100 kw. 659 ft TL: N43 32 34 W111 53 07. Stereo. Hrs opn: 24 Box 699, Blackfoot, 83221. Secondary address: 1190 Lincoln Rd. 83401. Phone: (208) 785-1400. Fax: (208) 785-0184. Web Site: www.z103.fm. Licensee: Riverbend Communications LLC. Group owner: Bonneville International Corp. (acq 4-17-2006; grpsl). Rgnl rep: Christal Radio Format: CHR top-40. Target aud: 18-34. ♦Jim Burgoyne, pres; Delyn Hendricks, gen mgr; Matt Burgoyne, sls dir; Jeremy Dresen, progmg dir; Tisa Cudmore, traf mgr.

KID(AM)— 1928: 590 khz; 5 kw-D, 1 kw-N, DA-N. TL: N43 33 35 W111 55 15. Hrs open: 1406 Commerce Way, 83404. Phone: (208) 524-5900. Fax: (208) 522-9696. Licensee: Citicasters Licenses L.P. Group owner: Clear Channel Communications Inc. (acq 5-4-99; grpsl). Population served: 40,000 Natl. Network: CBS. Natl. Rep: Target Broadcast Sales. Law Firm: Pepper & Corazzini. Format: News/talk. Target aud: 25-54; upscale decision-making professionals. Spec prog: Farm 18 hrs wkly. ♦Neica Kinney, gen mgr; Lisa Smith, gen sls mgr; Bill Hatch, news dir; Cami Chopski, traf mgr.

KID-FM— May 1, 1965: 96.1 mhz; 100 kw. 1,500 ft TL: N43 29 51 W112 39 50. Stereo. Format: Country.

KQEO(FM)— April 2003: 107.1 mhz; 82 kw. Ant 597 ft TL: N43 32 33 W111 53 04. Hrs open: Box 570, Logan, UT, 84323. Phone: (435) 752-1390. Licensee: Sand Hill Media Corp. (group owner; acq 9-7-2001; $1.2 million with KSNA(FM) Rexburg plus 36-month employment agreement). Format: Classic rock. ♦Jim Garshow, gen mgr.

KSPZ(AM)—Ammon, Nov 9, 1957: 980 khz; 5 kw-D, 1 kw-N, DA-2. TL: N43 31 23 W112 00 36. Stereo. Hrs opn: 854 Lindsay Blvd., Ammon, 83402. Phone: (208) 522-1101. Fax: (208) 522-6110. Licensee: Sandhill Media Group LLC. Group owner: Sand Hill Media Corp. (acq 2-27-2004; $2.65 million with co-located FM). Population served: 35,776 Natl. Rep: McGavren Guild. Law Firm: Haley, Bader & Potts. Format: Rgnl Mexican. ♦James Garshow, exec VP & gen mgr; Ken Walker, gen sls mgr & natl sls mgr; Domingo Munoz, progmg dir.

KUPI-FM—Co-owned with KSPZ(AM). Aug 16, 1975: 99.1 mhz; 100 kw. Ant 1,513 ft TL: N43 32 33 W111 53 04. Stereo. Format: Country.

KTHK(FM)— October 1993: 105.5 mhz; 100 kw. 659 ft TL: N43 21 06 W112 00 22. Stereo. Hrs opn: 24 1190 Lincoln Rd., 83401. Phone: (208) 523-3722. Fax: (208) 525-2575. Web Site: www.1055thehawk.com. Licensee: Riverbend Communications LLC. Group owner: Bonneville International Corp. (acq 4-17-2006; grpsl). Format: Country. ♦Delyn Hendricks, gen mgr; Sandie Fulks, gen mgr & sls.

Island Park

KWYS-FM— November 1998: 102.9 mhz; 46 kw. 2,732 ft TL: N44 33 41 W111 26 32. Stereo. Hrs opn: 24 Box 2158, Ketchum, 83340. Phone: (208) 726-5324. Fax: (208) 726-5459. Licensee: Chaparral Broadcasting Inc. Group owner: Chaparral Communications (acq 7-30-2004; grpsl). Population served: 100,000 Law Firm: Cohn & Marks. Format: Classic rock. News staff: 2; News: 10 hrs wkly. Target aud: 18-45; Adults. ♦Scott Anderson, gen mgr.

Jerome

KART(AM)—Listing follows KMVX(FM).

KMVX(FM)— August 1970: 102.9 mhz; 100 kw. 760 ft TL: N42 43 54 W114 25 04. Stereo. Hrs opn: 47 N. 100 West, 83338. Phone: (208) 324-8181. Fax: (208) 324-7124. Licensee: KART Broadcasting Co. Rgnl rep: Allied Radio Partners. Format: Adult contemp. Target aud: 25-54; general. ♦Kent Lee, gen mgr & gen sls mgr; Karla Cunha, news dir; Jerry Tharton, chief of engrg; Tammy Davis, traf mgr.

KART(AM)—Co-owned with KMVX(FM). August 1956: 1400 khz; 1 kw-U. TL: N42 43 51 W114 32 17. (Acq 9-1-64).75,000 Natl. Network: CBS. Format: Real country. Target aud: 25 plus. ♦Lamont Summers, progmg dir; Tammy Davis, traf mgr.

Ketchum

KIKX(FM)— Dec 2, 1996: 104.7 mhz; 100 kw. 1,578 ft TL: N43 16 45 W114 09 14. Hrs open: 21361 Hwy. 30, Twin Falls, 83301. Phone: (208) 735-8300. Fax: (208) 733-4196. Web Site: kikx.com. Licensee: Locally Owned Radio LLC. (group owner; (acq 10-31-2003; grpsl). Format: Classic rock. ♦Larry Johnson, pres & gen mgr; Jerre Fender, opns dir & progmg dir; Deb Uvieu, gen sls mgr; Denis Jeffs, traf mgr.

Kootenai

KTPO(FM)— 2007: 106.7 mhz; 1.3 kw. Ant 1,158 ft TL: N48 13 45 W116 30 30. Hrs open: 327 S. Marion Ave., Sandpoint, 83864. Phone: (208) 263-2179. Fax: (208) 265-5440. Web Site: www.1067thepoint.com. Licensee: Hellroaring Communications L.L.C. Format: Classic rock. ♦Dylan L. Benefield, gen mgr; Mike Brown, news dir; John Goes, chief of engrg.

Kuna

***KARJ(FM)**— 2005: 88.3 mhz; 23 kw vert. Ant 2,161 ft TL: N43 00 26 W116 42 23. Stereo. Hrs opn: 24
Rebroadcasts KLRD(FM) Yucaipa, CA 100%.
2351 Sunset Blvd., Suite 170-218, Rocklin, CA, 95765. Phone: (916) 251-1600. Fax: (916) 251-1650. E-mail: info@air1.com Web Site: www.air1.com. Licensee: Educational Media Foundation. Group owner: EMF Broadcasting. Natl. Network: Air 1. Law Firm: Shaw Pittman. Format: Contemp Christian. News staff: 3. Target aud: 18-35; Judeo-Christian, female. ♦Richard Jenkins, pres; Mike Novak, VP; Lloyd Parker, gen mgr; Keith Whipple, dev dir; Eric Allen, natl sls mgr; Mke Novak, progmg dir; David Pierce, progmg mgr; Ed Lenane, news dir; Sam Wallington, engrg dir; Arthur Vassar, traf mgr.

Lapwai

KZBG(FM)— 2005: 97.7 mhz; 570 w. Ant 1,059 ft TL: N46 27 22 W117 02 56. Hrs open: 2518 Kendall Rd., Walla Walla, WA, 99362-9765. Licensee: Xana Duke Radio Partners LLC (acq 7-30-2007; $310,000). ♦Thomas D. Hodgins, gen mgr.

Lewiston

KATW(FM)— Oct 2, 1986: 101.5 mhz; 100 kw. 848 ft TL: N46 27 38 W117 01 00. Stereo. Hrs opn: 24 403 C St., 83501. Phone: (208) 743-6564. Fax: (208) 798-0110. E-mail: jaymlazgar@pacempire.com Web Site: www.catfm.com. Licensee: Pacific Empire Radio Corp. (group owner; acq 10-6-98; $788,500 with KBJX(FM) Shelley). Population served: 50,000 Format: Hot adult contemp. News staff: one; News: 4 hrs wkly. Target aud: 18-49. ♦Jay Mlazgar, pres, gen mgr & gen sls mgr; Evan Yeoman, progmg dir & progmg mgr; Jill Law, gen sls mgr & traf mgr.

KCLK-FM—See Clarkston, WA

***KLCZ(FM)**— October 1967: 88.9 mhz; 230 w. Ant -840 ft TL: N46 24 45 W117 01 31. Hrs open: Attn: Radio, 500 Eighth Ave., 83501. Phone: (208) 792-2418. Fax: (208) 792-2568. Web Site: www.lcsc.edu. Licensee: Lewis-Clark State College (acq 3-23-2005; $5,000). Population served: 36,000 Format: Var/div. ♦Tate Smith, gen mgr.

KMOK(FM)— March 1983: 106.9 mhz; 99 kw. 1,230 ft TL: N46 27 33 W117 02 18. Stereo. Hrs opn: 24 805 Stewart Ave., 83501. Phone: (208) 743-1551. Fax: (208) 743-4440. E-mail: sales@idavend.com Licensee: Ida-Vend Co. Inc. Group owner: IdaVend Broadcasting Inc. Population served: 100,000 Natl. Network: AP Radio. Law Firm: Wilkenson, Barkter & Knauer. Format: Country. News staff: one; News: 3 hrs wkly. Target aud: 25-49; female. ♦Robert Prasil, pres & gen mgr; Darin Siebert, opns mgr; Ben Bonnfield, gen sls mgr; Jim Nelly, progmg dir; Steve Franco, mus dir & chief of engrg; John Thomas, news dir; Zoanne Byers, traf mgr.

KOZE(AM)— Oct 6, 1955: 950 khz; 5 kw-D, 1 kw-N, DA-2. TL: N46 23 32 W117 02 03. Stereo. Hrs opn: 24 Box 936, 2560 Snake River Ave., 83501. Phone: (208) 743-2502. Fax: (208) 743-1995. Licensee: 4-K Radio Inc. (group owner; acq 6-1-71). Population served: 125,000 Rgnl rep: Tacher Company. Format: Talk. News staff: 2. Target aud: 25-54. Spec prog: Farm 2 hrs wkly. ♦Michael R. Ripley, pres; Chris Ripley, stn mgr & progmg VP; Lisa Jensen, gen sls mgr; Jason Ford, news dir; David Forsman, chief of engrg.

KOZE-FM— Jan 17, 1961: 96.5 mhz; 25 kw. 741 ft TL: N46 27 48 W117 00 01. Stereo. 24 Format: Adult rock. Target aud: 18-49. ♦Lee McVey, progmg dir.

KRLC(AM)— March 1935: 1350 khz; 5 kw-D, 1 kw-N, DA-N. TL: N46 23 39 W116 59 40. Hrs open: 24 805 Stewart Ave., 83501. Phone: (208) 743-1551. Fax: (208) 743-4440. E-mail: sales@idavend.com Licensee: Ida-Vend Inc. Group owner: IdaVend Broadcasting Inc. (acq 11-1-81). Population served: 100,000 Law Firm: Wilkenson, Barker & Knauer. Format: Country, news/talk, sports. News staff: one; News: 10 hrs wkly. Target aud: 25 plus; adults. Spec prog: Farm 5 hrs, radio auction one hr wkly. ♦Robert Prasil, pres & gen mgr; Melva Prasil, stn mgr; Ben Bonfield, gen sls mgr; John Thomas, news dir; ZoAnne Byers, traf mgr; Steve Franco, engr.

KVTY(FM)— July 20, 1998: 105.1 mhz; 500 w. 1,099 ft TL: N46 27 33 W117 02 18. Hrs open: 24 c/o KRLC(AM) and KMOK(FM), 805 Stewart Ave., 83501. Phone: (208) 743-1551. Fax: (208) 743-4440. E-mail: sales@idavend.com Licensee: IdaVend Co. Inc. Group owner: IdaVend Broadcasting Inc. Population served: 100,000 Natl. Network: AP Radio. Law Firm: Wilkenson, Barker & Knauer. Format: CHR. News staff: one; News: 1 hr wkly. Target aud: 18-44. ♦Robert Prasil, pres & gen mgr; Melva Prasil, stn mgr; Darin Siebert, opns mgr; Ben Bonfield, gen sls mgr; Jeff Tuchscherer, progmg dir; John Thomas, news dir; Steve Franco, chief of engrg; Zoanne Byers, traf mgr.

McCall

***KBSK(FM)**— 2002: 89.9 mhz; 220 w. Ant 1,919 ft TL: N45 00 38 W116 07 53. Hrs open: Boise State Radio, 1910 University Dr., Boise, 83725-1915. Phone: (208) 426-3663. Fax: (208) 344-6631. Web Site: radio.boisestate.edu. Licensee: Idaho State Board of Education. Format: Jazz. ♦John Hess, gen mgr; Erik Jones, opns mgr; Hy Kloc, dev dir; Ele Ellis, progmg dir; Sadie Babits, news dir; Tom Taylor, engrg dir.

***KBSM(FM)**— Jan 20, 1991: 91.7 mhz; 220 w. 1,912 ft TL: N45 00 38 W116 07 53. Stereo. Hrs opn: 24 Boise State Radio, 1910 University Dr., Boise, 83725. Phone: (208) 426-3663. Fax: (208) 344-6631. Web Site: radio.boisestate.edu. Licensee: Idaho State Board of Education. Natl. Network: PRI, NPR. Law Firm: Dow, Lohnes & Albertson. Format: Class, news, new age. News: 15 hrs wkly. Target aud: General. Spec prog: Jazz. ♦John Hess, gen mgr; Erik Jones, opns mgr; Hy Kloc, dev dir; Ele Ellis, progmg dir; Sadie Babits, news dir.

***KBSQ(FM)**—Not on air, target date: unknown: 90.7 mhz; 220 w. 1,919 ft Hrs opn: Boise State Radio, 1910 University Dr., Boise, 83725. Phone: (208) 426-3663. Fax: (208) 344-6631. Web Site: radio.boisestate.edu. Licensee: Idaho State Board of Education. Format: News, div. ♦John Hess, gen mgr; Erik Jones, opns mgr; Ele Ellis, progmg dir.

KDZY(FM)— 2001: 98.3 mhz; 500 w horiz. Ant 1,922 ft TL: N45 00 18 W116 08 01. Hrs open: 24 1440 S. Weideman Ave., Boise, 83709. Phone: (208) 377-3790. Phone: (208) 634-3781. Fax: (208) 377-3792. Licensee: KSPD Inc. (group owner; acq 4-15-02; $75,000). Format: Country. Target aud: 25 plus. ♦Lee Schafer, pres & gen mgr.

KMCL(AM)—Donnelly, Oct 15, 1965: 1240 khz; 1 kw-U. TL: N44 46 52 W116 02 51. Hrs open: Box 813, 83638. Secondary address: 204 N. 3rd St. 83638. Phone: (208) 634-4777. Fax: (208) 634-3059. Licensee: Brundage Mountain Air Inc. Population served: 25,000 Format: Adult contemp. ♦David Eaton, gen mgr.

KMCL-FM— Oct 22, 1990: 101.1 mhz; 3.9 kw. Ant 1,873 ft TL: N44 45 54 W116 11 54. Stereo. Hrs opn: 24 Box 813, 204 N. 3rd St., 83638. Phone: (208) 634-4777. Fax: (208) 634-3059. E-mail: mcradio@ctweb.net Licensee: FM Idaho Co. LLC. (acq 8-1-2007; $900,000). Natl. Rep: Target Broadcast Sales. Format: Adult contemp. News staff: one; News: 5 hrs wkly. Target aud: 25-54; upper-middle class to white collar, young families & students. Spec prog: Sports, entertainment, business one hr, relg one hr wkly. ♦David Eaton, gen mgr, gen sls mgr, prom dir, progmg dir & news dir; Linda Jackson, pub affrs dir; Rockwell Smith, chief of engrg; Susan Turner, traf mgr.

Meridian

KDJQ(AM)— May 1, 2005: 890 khz; 50 kw-D, 250 w-N, DA-N. TL: N43 27 36 W116 14 19. Hrs open: 1050 Clover Dr., Boise, 83703-0405. Phone: (208) 388-4502. Phone: (208) 424-3689. Fax: (208) 433-9318. Licensee: Robert E. Combs (acq 5-14-2004; $425,000 for CP). ♦Robert E. Combs, gen mgr.

Montpelier

KVSI(AM)— July 20, 1965: 1450 khz; 1 kw-U. TL: N42 18 54 W111 18 38. Hrs open: Box 340, 24681 US 89, 83254. Phone: (208) 847-1450. Fax: (208) 847-1451. E-mail: kvsi@dcdi.net Web Site: kvsi.com. Licensee: Tri-States Broadcasting LLC (acq 11-1-68). Population served: 2,604 Format: Country. Spec prog: Farm 4 hrs, relg 2 hrs wkly. ♦Keith Martindale, gen mgr; Ada Jane Hillier, progmg dir.

Moscow

KQQQ(AM)—See Pullman, WA

***KRFA-FM**— Sept 1, 1963: 91.7 mhz; 1.45 kw. 1,009 ft TL: N46 40 54 W116 58 13. Stereo. Hrs opn: 24 Box 642530, 382 Murrow Communications Ctr., Washington State Univ., Pullman, WA, 99164-2530. Phone: (509) 335-6500. Fax: (509) 335-6557. E-mail: nwpr@wsu.edu Web Site: www.nwpr.org. Licensee: Washington State University. (acq 7-1-84). Population served: 190,000 Natl. Network: NPR, PRI. Law Firm: Don, Lohnes & Albertson. Format: Class, news. News staff: one; News: 37 hrs wkly. Target aud: General. Spec prog: Folk, jazz. ♦Karen Olstad, COO & gen mgr; Dennis Haarsager, gen mgr; Roger Johnson, stn mgr & sls dir; Scott Weatherly, opns mgr; Sarah McDaniel, dev dir; Mary Hawkins, progmg dir; Robin Rilette, mus dir; Ralph Hogan, engrg dir & engrg mgr; Rachael McDonald, news rptr.

KRPL(AM)— May 20, 1947: 1400 khz; 1 kw-U. TL: N46 44 47 W117 01 06. Hrs open: 24 Box 8849, 1114 N. Almon, 83843. Phone: (208) 882-2551. Fax: (208) 883-3571. Licensee: KRPL Inc. (acq 1-27-2004; $1 million for two-thirds of the shares with co-located FM). Population served: 60,000 Natl. Rep: McGavren Guild. Law Firm: Haley, Bader & Potts. Format: Oldies 50s, 60s & 70s. News staff: 4; News: 12 hrs wkly. Target aud: 25-54. Spec prog: Farm 4 hrs, relg 2 hrs wkly. ♦Gary Cummins, pres & gen mgr.

KZFN(FM)—Co-owned with KRPL(AM). Feb 24, 1973: 106.1 mhz; 62 kw. Ant 921 ft TL: N46 40 51 W116 58 26. Stereo. 24 130,000 Format: CHR. News staff: 4; News: 5 hrs wkly. Target aud: 25-54. ♦Gary Cummings, progmg mgr.

***KUOI-FM**— November 1945: 89.3 mhz; 400 w. -92 ft TL: N46 43 43 W117 00 11. Stereo. Hrs opn: 24 Student Union Bldg., 3rd Fl., Univ. of Idaho, 83844-4272. Phone: (208) 885-2218. Fax: (208) 885-2222. E-mail: kuoi@uidaho.edu Web Site: www.kuoi.org. Licensee: University of Idaho. Population served: 14,146 Format: Free-form, div. News staff: 3; News: 2 hrs wkly. Target aud: General; alternative mus listeners. Spec prog: Black 3 hrs, folk 3 hrs, jazz 4 hrs wkly. ♦Andy Jacobson, stn mgr; Richard Dana, progmg dir; Marcus Kellis, mus dir; Jeff Kimberling, chief of engrg.

Mountain Home

KMHI(AM)— Mar 20, 1962: 1240 khz; 1 kw-U. TL: N43 09 03 W115 42 26. Hrs open: 24 21361 Hwy. 30, Twin Falls, 83301. Phone: (208) 735-8300. Fax: (208) 587-8425. E-mail: barbara@kmhi1240.com Web Site: www.kmhi.com. Licensee: FM Idaho Co. LLC. (group owner; (acq 12-31-2006; grpsl). Population served: 20,000 Natl. Network: Westwood One. Format: Classic hit country. News staff: one; News: 15 hrs wkly. Target aud: General. Spec prog: Sp 7 hrs wkly. ♦Larry Johnson, pres & gen mgr; Jerre Fender, opns dir & progmg dir; Deb Uvieu, gen sls mgr; Denis Jeffs, traf mgr.

KTMB(FM)— 1982: 99.1 mhz; 73 kw. Ant 2,191 ft TL: N43 14 43 W115 26 12. Stereo. Hrs opn: 24 2660 Peachtree Rd. N.W., Suite 17E, Atlanta, 30305. Phone: (208) 658-9806. Phone: (735) 735-0099. Fax: (208) 658-9808. Licensee: FM Idaho Co. LLC. (group owner; (acq 12-31-2006; grpsl). Population served: 50,000 Format: Talk. Target aud: 18-34.

Nampa

KIDO(AM)— May 17, 1920: 580 khz; 5 kw-U, DA-N. TL: N43 33 35 W116 24 02. Hrs open: 24 827 E. Park Blvd., Suite 201, Boise, 83712. Phone: (208) 344-6363. Fax: (208) 385-9064. Web Site: www.kido.net. Licensee: Peak Broadcasting of Boise Licenses LLC. Group owner: Clear Channel Communications Inc. (acq 6-28-2007; grpsl). Population served: 300,000 Natl. Network: CBS. Format: News, talk. Target aud: 25-54. ♦Mitch Pruitte, stn mgr; David Levi, natl sls mgr; Dave Burnett, progmg dir & news dir; Susan Green, traf mgr.

KKGL(FM)— February 1977: 96.9 mhz; 44 kw. 2,520 ft TL: N43 45 19 W116 05 52. Stereo. Hrs opn: 24 1419 W. Bannock St., Boise, 83702. Phone: (208) 336-3670. Fax: (208) 336-3734. Web Site: www.96-9meeagle.com. Licensee: Citadel Broadcasting Co. Group owner: Citadel Broadcasting Corp. (acq 12-10-97; grpsl). Population served: 400,000 Natl. Rep: Katz Radio. Format: Classic rock. Target aud: 25-44; upscale baby boomers who listen to primarily 70s based rock. ♦Kevin Godwin, gen mgr; Rich Summers, opns mgr; Rich Bryan, progmg dir; Bill Frahn, chief of engrg.

KRVB(FM)— Jan 10, 1975: 94.9 mhz; 49 kw. 2,692 ft TL: N43 45 18 W116 05 52. Stereo. Hrs opn: 5257 W. Fairview Ave., Suite 260, Boise, 83706. Phone: (208) 344-3511. Fax: (208) 947-6765. Web Site: www.riverinteractive.com. Licensee: Journal Broadcast Corp. Group owner: Journal Communications Inc. (acq 4-11-00). Natl. Rep: Katz Radio. Format: AOR. Target aud: 18-64. ♦Bob Rosenthal, CFO, VP & gen mgr; Dan McColly, opns mgr & progmg dir.

KSAS-FM—Caldwell, Sept 28, 1982: 103.3 mhz; 54 kw. 2,578 ft TL: N43 45 18 W116 05 52. Stereo. Hrs opn: 24 827 E. Park Blvd., Suite 201, Boise, 83712. Phone: (208) 344-6363. Fax: (208) 385-9064. Fax: (208) 344-1134. Web Site: www.1033kissfm.com. Licensee: Peak Broadcasting of Boise Licenses LLC. Group owner: Clear Channel Communications Inc. (acq 6-28-2007; grpsl). Population served: 250000 Law Firm: Fletcher, Heald & Hildreth. Format: Top 40. News staff: one; News: one hr wkly. Target aud: 25-54; upscale, white collar. Spec prog: Class 2 hrs, jazz 4 hrs wkly. ♦Kevin Godwin, gen mgr; Steve Kicklighter, stn mgr & progmg dir; Mike Sutton, gen sls mgr & progmg dir; Crystal Struthers, prom dir; Dave Burnett, news dir & chief of engrg; Susan Green, traf mgr.

KTIK(AM)— Nov 1, 1962: 1350 khz; 5 kw-D, 600 w-N. TL: N43 32 58 W116 24 38. Hrs open: 24 Box 1280, 1419 W. Bannock St., Boise, 83702. Phone: (208) 336-3670. Fax: (208) 336-3736. Web Site: www.ktik.com. Licensee: Citadel Broadcasting Co. Group owner: Citadel Broadcasting Corp. (acq 4-1-03; $750,000). Natl. Network: ESPN Radio, Westwood One. Format: Sports, talk. News: one hr wkly. Target aud: 25-54; sports oriented men. ♦Kevin Godwin, gen mgr; Mike Owens, gen sls mgr; Andrew Paul, progmg dir.

New Plymouth

KZMG(FM)—Licensed to New Plymouth. See Boise

Orofino

KLER(AM)— Oct 15, 1958: 1300 khz; 5 kw-D, 1 kw-U, DA-N. TL: N46 28 41 W116 14 34. Hrs open: Box 32, 3110 Upper Fords Creek Rd., 83544. Phone: (208) 476-5702. Fax: (208) 476-5703. E-mail: klerorofino @clearwater.net Licensee: Central Idaho Broadcasting. (acq 12-7-92; $75,000 with co-located FM; FTR: 1-4-93) Population served: 14,000 Format: Country. News staff: news progmg 8 hrs wkly News: 2;. Target aud: General; family or logging industry-federal employee workers. ♦Jeff Jones, gen mgr & gen sls mgr; Jason Ford, news dir; Mike Bensen, pub affrs dir; Jim Sheldon, chief of engrg.

KLER-FM— Sept 20, 1979: 95.1 mhz; 2.3 kw. Ant 676 ft TL: N46 28 09 W116 16 40. Stereo. 24 Box 3110, Upper Fords Creek Rd., 83544. 4,000 Format: Adult contemp. News staff: 2; News: 8 hrs wkly.

KZID(FM)— 2003: 98.5 mhz; 1.65 kw. Ant 630 ft TL: N46 28 09 W116 16 40. Hrs open: Torro Broadcasting, 2307 Princess Anne St., Greensboro, NC, 27408. Phone: (336) 286-2087. Licensee: Torro Broadcasting.

Payette

KIOV(AM)— Dec 20, 1957: 1450 khz; 1 kw-U. TL: N44 03 47 W116 54 27. Hrs open: 24 1406 N. Main St., Suite 107, Meridian, 83642-1798. Phone: (208) 267-5234. Fax: (208) 888-9647. E-mail: sports@kiov.com Web Site: www.kiov.com. Licensee: Media Enterprises LLC. (Acq 8-00). Population served: 125,000 Rgnl rep: Tacher. Format: Sports. Target aud: 18-54; 65% male. ♦David Combes, gen mgr; Marshall Sage, opns mgr.

KQXR(FM)— Dec 1, 1978: 100.3 mhz; 98 kw. 708 ft TL: N43 49 31 W116 30 29. Stereo. Hrs opn: 24 5257 W. Fairview Ave., Suite 260, Boise, 83706. Phone: (208) 344-3511. Fax: (208) 947-6765. Web Site: www.xrock.com. Licensee: Journal Broadcast Corp. Group owner: Journal Broadcast Group Inc. (acq 5-13-98; grpsl). Population served: 293,600 Natl. Rep: Katz Radio. Format: Rock, alternative. Target aud: 18-49. ♦Bob Rosenthal, VP & gen mgr; Dan McColly, opns mgr.

Pocatello

***KISU-FM**— Apr 15, 1998: 91.1 mhz; 4.5 kw. 1,043 ft TL: N42 51 46 W112 31 03. Stereo. Hrs opn: 24 Box 8014, Idaho State University, 83209. Phone: (208) 236-3691. Fax: (209) 236-4600. E-mail: millerr@isu.edu Web Site: www.kisu.org. Licensee: Idaho State University. Population served: 100,000 Natl. Network: NPR, PRI. Law Firm: Dow, Lohnes, Albertson. Format: Jazz, AAA, news, talk, entertainment. News: 60+. ♦Jerry Miller, gen mgr.

KLLP(FM)—See Chubbuck

KMGI(FM)—Listing follows KSEI(AM).

KOUU(AM)— Dec 20, 1956: 1290 khz; 1 kw-D, 24 w-N. TL: N42 57 28 W112 25 46. (CP: 50 kw-D, 24 w-N, DA-D. TL: N42 57 27 W112 25 46). Stereo. Hrs opn: 24 Box 97, 436 N. Main, 83204. Phone: (208) 234-1290. Fax: (208) 234-9451. Licensee: Idaho Wireless Corp. (group owner; (acq 3-86; with co-located FM; FTR: 12-9-85). Population served: 250,000 Natl. Network: ABC. Format: Traditional country. Target aud: 35-64; adults. ♦Paul Anderson, gen mgr; Harry Newhardt, gen sls mgr.

KZBQ(FM)—Co-owned with KOUU(AM). Dec 27, 1969: 93.7 mhz; 100 kw. Ant 984 ft TL: N42 51 57 W112 30 46. Stereo. 24 E-mail: paul@kzbq.com 250,000 Natl. Network: ABC. Format: Country. Target aud: 25-54; adults. ♦Paul Anderson, stn mgr.

KPKY(FM)—Listing follows KWIK(AM).

KPPC(FM)— July 23, 2007: 92.1 mhz; 6 kw. Ant 121 ft TL: N42 54 49 W112 26 14. Hrs open: Box 998, 83204. Phone: (208) 233-1133. Fax: (208) 232-1240. Licensee: Intermart Broadcasting Pocatello Inc. Format: Alternative. ♦Neica Kinney, gen mgr.

KPTO(AM)—Not on air, target date: unknown: 1440 khz; 2.5 kw-D, 350 w-N, DA-2. TL: N42 56 30 W112 27 17. Hrs open: 24 Box 1450, St. George, UT, 84771-1450. Secondary address: 210 North 1000 East, St. George, UT 84770-3155. Phone: (208) 234-7000. Fax: (208) 232-1440. Licensee: AM Radio 1440 Inc. Group owner: Diamond Broadcasting Corp. (acq 12-6-2004). Natl. Network: CNN Radio, Westwood One. Law Firm: Dan J. Alpert. Format: Adult standards. News: one hr wkly. Target aud: 18 -54. ♦E. Morgan Skinner Jr., CEO & pres.

KRTK(AM)—See Chubbuck

KSEI(AM)— Sept 23, 1926: 930 khz; 5 kw-U, DA-N. TL: N42 57 44 W112 29 50. Hrs open: 24 Box 40, 83204. Secondary address: 544 N. Arthur St. 83204. Phone: (208) 233-2121. Fax: (208) 234-7682. E-mail: neilmab@pacempire.com Licensee: Pacific Empire Radio Corp. (group owner; acq 8-28-97; $1.2 million with co-located FM). Population served: 56,185 Natl. Rep: Katz Radio. Format: Sporting news. Target aud: General. ♦Mark Bolland, pres & gen mgr; Neil Mayberry, gen mgr & gen sls mgr; Jim Christopher, news dir; Bill Trowe, chief of engrg.

KMGI(FM)—Co-owned with KSEI(AM). Apr 1, 1978: 102.5 mhz; 100 kw. Ant 1,023 ft TL: N42 51 57 W112 30 46. Stereo. 24 544 N. Arthur, 83204. Web Site: www.classicrock102.fm. Natl. Network: Westwood One. Format: Classic rock. ♦C.J. Morrison, progmg mgr.

KWIK(AM)— September 1946: 1240 khz; 1 kw-U. TL: N42 55 14 W112 27 17. Stereo. Hrs opn: 24 Box 998, 259 E. Center St., 83201. Phone: (208) 233-1133. Phone: (800) 582-1240. Fax: (208) 232-1240. E-mail: new1240@yahoo.com Web Site: www.newsradio1240.com. Licensee: Citicasters Licenses L.P. Group owner: Clear Channel Communications Inc. (acq 5-4-99; grpsl). Population served: 57,550 Rgnl rep: Art Moore. Law Firm: Pepper & Corazzini. Format: Sports, news/talk. News staff: 3; News: 15 hrs wkly. Target aud: 45 plus. Spec prog: Farm 3 hrs, gospel 2 hrs, relg one hr, American Indian one hr wkly. ♦Tim Murphy, gen mgr; Jodie Bates, gen sls mgr; Neal Larson, news dir; Rhett Downing, chief of engrg.

KPKY(FM)—Co-owned with KWIK(AM). Aug 18, 1975: 94.9 mhz; 100 kw. Ant 1,004 ft TL: N42 52 26 W112 30 47. Stereo. Web Site: www.kpky.com.39,000 Format: Classic rock. ♦Marie Mccallister, progmg dir; Mike Hudson, min affrs dir; J.D. Kelly, spec ev coord.

***KZJB(FM)**— 2006: 90.3 mhz; 910 w vert. Ant 1,031 ft TL: N42 51 46 W112 31 03. Hrs open: 4250 S. 25th East, Idaho Falls, 83404. Phone: (208) 524-1503. Fax: (208) 524-0697. E-mail: breinisch@ccifi.org Licensee: CSN International (group owner). Format: Christian relg. ♦James Knudsen, stn mgr.

Post Falls

KCDA(FM)—Licensed to Post Falls. See Spokane WA

Preston

KACH(AM)— Sept 4, 1948: 1340 khz; 1 kw-U. TL: N42 07 45 W111 51 00. Hrs open: 24 1133 E. Glendale Rd., 83263. Phone: (208) 852-1340. Fax: (208) 852-1342. E-mail: kach@plmw.com Licensee: Alan J. White, Nelada G. White. (acq 5-13-98). Population served: 10,500 Natl. Network: ABC. Format: Oldies. News: 12 hrs wkly. Target aud: 18-54; general. Spec prog: Farm. ♦Alan White, gen mgr.

KKEX(FM)— Dec 9, 1993: 96.7 mhz; 105 w. 226 ft TL: N47 07 45 W111 51 00. Hrs open: 24 Box 3369, Radio Stn. KKEX(FM), Logan, UT, 84323-3369. Secondary address: 810 W. 200 N., Logan, UT 84321. Phone: (435) 752-1390. Phone: (435) 753-9607. Fax: (435) 752-1392. E-mail: kkex@vradio.com Web Site: www.klix96.fm. Licensee: Sun Valley Radio Inc. (group owner; acq 1994). Population served: 130,000 Format: Country. ♦M. Kent Frandsen, pres; Jay Eubanks, gen mgr; Lynn Simmons, progmg dir & progmg mgr; Dan Baker, chief of engrg.

Rathdrum

***KWJT(FM)**— June 2006: 89.9 mhz; 1.1 kw vert. Ant 1,965 ft TL: N48 05 38 W116 33 12. Hrs open: 24 Box 1208, Airway Heights, WA, 99001. Phone: (509) 244-5577. Fax: (509) 244-2232. Web Site: www.csnradio.com. Licensee: CSN International (group owner). Format: Relg.

Rexburg

***KBYI(FM)**— Nov 13, 1972: 100.5 mhz; 100 kw. 692 ft TL: N43 45 44 W111 57 30. Stereo. Hrs opn: 24 102 RGS Bldg., BYU Idaho, 83460-1700. Phone: (208) 496-2907. Fax: (208) 496-2912. E-mail: clarkjim@byui.edu Web Site: www.kbyu.edu/kbyi. Licensee: Brigham Young University-Idaho. Population served: 200,000 Natl. Network: NPR, PRI. Format: Class, news. News staff: one; News: 33 hrs wkly. Target aud: General. ♦Jim Clark, gen mgr; Mark Bailey, gen sls mgr, news dir & traf mgr; Michelle Snyder, mktg.

***KBYR-FM**— 1993: 91.5 mhz; 100 w. -39 ft TL: N43 49 09 W111 46 51. Hrs open: 24 Ricks College, Spori Bldg., 83460-0105. Phone: (208) 496-2907. Fax: (208) 496-2912. Web Site: www.byui.edu/kbyr. Licensee: Ricks College Corp. Population served: 30,000 Format: Mormon contemp. Target aud: General. ♦Jim Clark, gen mgr.

KGTM(FM)— Jan 17, 1986: 98.1 mhz; 3 kw. 299 ft TL: N43 48 55 W111 46 09. (CP: 25 kw, ant 276 ft.). Stereo. Hrs opn: 1327 E. 17th St., Idaho Falls, 83404. Phone: (208) 529-6926. Fax: (208) 529-6927. Licensee: Pacific Empire Radio Corp. (group owner; acq 7-25-2000; $495,000 with KRXK(FM) Rexburg). Population served: 60,000 Law Firm: Fletcher, Heald & Hildreth. Format: Oldies. Target aud: 35 plus. ♦Mark Bolland, CEO & pres; Bill Furst, gen mgr; Rick Mason, progmg dir.

KRXK(AM)— January 1951: 1230 khz; 1 kw-U. TL: N43 50 50 W111 47 03. Hrs open:
Simulcast with KSEI (AM) Pocotello.
1327 E. 17th St., Idaho Falls, 83404. Phone: (208) 529-6926. Fax: (208) 529-6927. E-mail: billkeith@pacempire.com Licensee: Pacific Empire Radio Corp. (group owner; (acq 7-25-2000; $495,000 with KGTM(FM) Rexburg). Population served: 16,000 Rgnl. Network: Intermountain Farm/Ranch Network. Natl. Rep: Target Broadcast Sales. Format: Sports talk. News staff: 9. Target aud: 25-54; male. ♦Mark Bolland, CEO & pres; Bill Furst, gen mgr; Rich Allen, progmg dir.

KSNA(FM)— Aug 18, 1975: 94.3 mhz; 43 kw. Ant 522 ft TL: N43 45 20 W111 57 56. Stereo. Hrs opn: 24 854 Lindsay Blvd., Idaho Falls, 83402. Phone: (208) 522-1101. Fax: (208) 522-6110. Web Site: www.sunny943.com. Licensee: Sand Hill Media Corp. (group owner; acq 9-7-2001; $1.2 million with KQEO(FM) Idaho Falls plus 36-month employment agreement). Population served: 50,000 Natl. Network: Jones Radio Networks, USA. Natl. Rep: Tacher. Format: Modern rock. ♦Keith Walker, gen mgr; Mike Steele, opns mgr & progmg dir; John Balginy, news dir.

Rigby

***KLRI(FM)**— 2005: 89.5 mhz; 78 kw vert. Ant 1,527 ft TL: N43 30 04 W112 39 44. Stereo. Hrs opn: 24
Rebroadcasts KLVR(FM) Santa Rosa, CA 100%.
2351 Sunset Blvd., Suite 170-218, Rocklin, CA, 95765. Phone: (916) 251-1600. Fax: (916) 251-1650. E-mail: klove@klove.com Web Site: www.klove.com. Licensee: Educational Media Foundation. Group owner: EMF Broadcasting. Natl. Network: K-Love. Law Firm: Shaw Pittman. Format: Contemp Christian. News staff: 3. Target aud: 25-44; Judeo Christian, female. ♦Richard Jenkins, pres; Mike Novak, VP; Keith Whipple, dev dir; David Pierce, progmg mgr; Ed Lenane, news dir; Sam Wallington, engrg dir; Karen Johnson, news rptr; Marya Morgan, news rptr; Richard Hunt, news rptr.

Ririe

***KSQS(FM)**— 2006: 91.7 mhz; 250 w. Ant 532 ft TL: N43 32 37 W111 53 07. Hrs open: 2201 S. 6th St., Las Vegas, NV, 89104. Phone: (702) 731-5452. Fax: (702) 731-1992. E-mail: info@sosradio.net Web Site: www.sosradio.net. Licensee: Faith Communications Corp. Format: Adult contemp Christian. ♦Brad Staley, gen mgr.

Rupert

KFTA(AM)— Oct 12, 1955: 970 khz; 2.5 kw-D. TL: N42 37 08 W113 39 31. (CP: 900 w-N, DA-N. TL: N42 36 10 W113 43 21). Hrs opn: 120 S. 300 W., 83350. Phone: (208) 436-4757. Fax: (208) 436-3050. E-mail: lafantastica970@quepasa.com Licensee: Tri-Market Radio Broadcasters Inc. Group owner: Tri-Market Radio Broadcasters Inc. & Eagle Rock Broadcasting Inc. (acq 9-24-93; $700,000 with co-located FM; FTR: 10-11-93). Population served: 50000 Format: Sp. Spec prog: Sp. ♦Kim Lee, gen mgr; Chris Kinzel, gen sls mgr; Ben Reed, progmg dir & news dir; Jerry Thaxton, chief of engrg.

KKMV(FM)—Co-owned with KFTA(AM). Dec 5, 1978: 106.1 mhz; 25 kw. Ant 2,496 ft TL: N42 20 06 W113 36 15.65,000 Format: Country.

Saint Anthony

KIGO(AM)— July 10, 1966: 1420 khz; 32 kw-D, 12 w-N. TL: N43 40 02 W111 52 14. Hrs open: 24 1447 Winter Lane, Jerome, 83338. Phone: (208) 280-1962. Licensee: Albino Ortega & Maria Juarez (acq 5-1-2005; $85,000). Population served: 265,000 Format: Spanish. Target aud: 25-44; adults. ♦Albino Ortega, gen mgr.

Saint Maries

KOFE(AM)— Mar 1, 1970: 1240 khz; 1 kw-D, 500 w-N. TL: N47 19 14 W116 32 50. Hrs open: 24 Box 278, 201 N. 8th, 83861. Phone: (208) 245-1240. Fax: (208) 245-6525. E-mail: kofeam@usamedia.tv Licensee: Campbell River Holding Co. L.L.C. (acq 5-3-01; $1,000 for 70%). Population served: 8,000 Natl. Network: Fox News Radio. Format: Classic Hits. News staff: 2; News: 9 hrs wkly. Target aud: 25-55. ♦Theresa Plank, gen mgr; Phil Plank, chief of engrg & engr.

Salmon

KSRA(AM)— Mar 1, 1959: 960 khz; 1 kw-D. TL: N45 11 02 W113 52 12. Hrs open: 315 Hwy. 93 N., 83467. Phone: (208) 756-2218. Fax: (208) 756-2098. Licensee: Salmon River Communications Inc. (acq 6-19-00; $345,000 with co-located FM). Population served: 7,500 Format: Country, adult contemp. Spec prog: Farm 4 hrs, class one hr wkly. ♦Jim Hone, pres; Rick Sessions, gen mgr; Leo Marshall, gen sls mgr; Colby Smith, progmg dir; Mark Booker, engrg VP & chief of engrg.

KSRA-FM— September 1979: 92.7 mhz; 1.5 kw. -880 ft TL: N45 11 02 W113 52 12. Stereo. 6 AM-10 PM E-mail: ksra@ksrafm.com Web Site: www.ksrafm.com.10,000

Sandpoint

KIBR(FM)— 1994: 102.5 mhz; 3 kw. 177 ft TL: N48 15 22 W116 30 46. Stereo. Hrs opn: 24
Rebroadcasts KICR(FM) Coeur d'Alene 100%.
327 Marion Ave., 83864. Phone: (208) 263-2179. Fax: (208) 265-5440. E-mail: dylanb@953kpwd.com Licensee: Benefield Broadcasting Inc. (acq 3-31-95; $250,000; FTR: 5-22-95). Population served: 60,000 Natl. Network: ABC. Format: Classic country. News staff: one. Target aud: 25-54. ♦Dylan Benefield, gen mgr, opns mgr & gen sls mgr; Jimmy Silver, progmg dir & progmg mgr; Mike Brown, news dir.

KPND(FM)—Listing follows KSPT(AM).

KSPT(AM)— Mar 23, 1949: 1400 khz; 1 kw-U. TL: N48 18 16 W116 32 32. Hrs open: 24
Rebroadcasts KBFI(AM) Bonners Ferry 100%.
327 Marion Ave., 83864. Phone: (208) 263-2179. Fax: (208) 265-5440. E-mail: prod@953kpno.com Licensee: Blue Sky Broadcasting Inc. (acq 5-4-83; $250,000; FTR: 5-30-85). Population served: 37,000 Natl. Network: ABC, USA. Natl. Rep: Tacher. Law Firm: Smith & Belendiuk. Format: News/talk, sports. Target aud: 25 plus. Spec prog: Relg 2 hrs wkly. ♦Dylan Benefield, gen mgr & progmg dir; Mike Davis, news dir; Conrad Agtee, chief of engrg; Jim Tomchek, traf mgr.

KPND(FM)—Co-owned with KSPT(AM). May 19, 1980: 95.3 mhz; 9.8 kw. Ant 2,368 ft TL: N48 22 40 W116 37 05. Stereo. 24 80,000 Format: AAA. ♦Jim Tomchek, traf mgr.

Shelley

KBJX(FM)— October 1999: 106.3 mhz; 100 kw. 636 ft TL: N43 06 45 W112 29 34. Stereo. Hrs opn: 1327 E. 17th St., Idaho Falls, 83404. Phone: (208) 529-6926. Fax: (208) 529-6927. E-mail: support@hot106.fm Web Site: www.106kbjx.com. Licensee: Pacific Empire Radio Corp. (group owner; acq 10-6-98; $788,500 with KATW(FM) Lewiston). Format: Adult contemp. Target aud: 25-54; adults. ♦Mark Bolland, CEO & pres; Bill Furst, gen mgr; Eric O'Connor, progmg dir.

Soda Springs

KBRV(AM)— Sept 22, 1957: 790 khz; 5 kw-D. TL: N42 38 30 W111 36 40. Hrs open: Box 777, 83276. Phone: (208) 547-2400. Fax: (208) 547-4593. Fax: (208) 547-2401. Licensee: Caribou Broadcasting Inc. (acq 1-8-2001). Population served: 100,000 Format: Country. ♦Tom Mathis, gen mgr.

KITT(FM)— Sept 10, 1982: 100.1 mhz; 3 kw. Ant -174 ft TL: N42 38 30 W111 36 40. Stereo. Hrs opn: Box 1450, 210 North 1000 East, St. George, UT, 84771-1450. Phone: (435) 628-1000. Fax: (435) 628-6636. E-mail: legacy1@infowest.com Licensee: Tri-State Media Corp. Group owner: Legacy Communications Corp. (acq 7-8-2004; $234,000). Population served: 20,000 Law Firm: Dan J. Alpert. Format: Hot country. ♦E. Morgan Skinner, Jr., CEO & pres.

Sun Valley

***KBSS(FM)**— August 2004: 91.1 mhz; 700 w. Ant 1,870 ft TL: N43 38 36 W114 23 49. Hrs open: Boise State Radio, 1910 University Dr., Boise, 83725-1915. Phone: (208) 426-3663. Fax: (208) 344-6631. E-mail: radio@boisestate.edu Web Site: radio.boisestate.edu. Licensee: Idaho State Board of Education. Natl. Network: NPR. Format: News, info. ♦Erik Jones, opns mgr; Ele Ellis, progmg dir; Tom Taylor, engrg dir.

KECH-FM— Nov 21, 1988: 95.3 mhz; 100 w. Ant 2,168 ft TL: N43 39 42 W114 24 07. (CP: 16 kw, ant 1,909 ft. TL: N43 38 36 W114 23 49). Hrs opn: 24 220 Northwood Way, Ketchum, 83340. Phone: (208) 726-5324. Fax: (208) 726-5459. Web Site: www.kech95.com. Licensee: Chaparral Broadcasting Inc. Group owner: Chaparral Communications acq 7-30-2004; grpsl). Law Firm: Cohn & Marks. Format: Classic rock. News staff: one; News: 6 hrs wkly. Target aud: 25-54; upscale adults. Spec prog: Alternative 5 hrs, blues 8 hrs, jazz 6 hrs wkly. ♦Scott Anderson, gen mgr; Cathy Nikolaisons, gen sls mgr; Bob Thompson, progmg dir; Sue Bailey, news dir.

KSKI-FM— Aug 3, 1977: 103.7 mhz; 53 kw. 1,905 ft TL: N43 38 36 W114 23 49. Stereo. Hrs opn: 24 Box 2750, Hailey, 83333. Phone: (208) 726-5324. Fax: (208) 726-5459. Web Site: www.ketsvidaho.net. Licensee: Chaparral Broadcasting Inc. Group owner: Chaparral Communications (acq 7-30-2004; grpsl). Population served: 100,000 Law Firm: Chon & Marks. Format: Alt rock. News staff: one; News: 2 hrs wkly. Target aud: 18-49; affluent, upscale consumers. ♦Scott Anderson, gen mgr; Cathy Nikolaisons, gen sls mgr & sls; Bob Thompson, progmg dir; Sue Bailey, news dir.

***KWRV(FM)**— July 29, 1993: 91.9 mhz; 100 w. -512 ft TL: N43 40 59 W114 20 52. Hrs open: Box 67, MN, 83353. Phone: (651) 290-1500. Fax: (651) 290-1224. E-mail: molson@mpr.org Web Site: www.mpr.org. Licensee: Minnesota Public Radio. Natl. Network: PRI. Rgnl. Network: Minn. Pub. Format: Class. ♦William Kling, pres; Craig Curtis, VP; Michael Orson, gen mgr; Anne Hovland, dev VP; Ginger Sisco, mktg VP; Vic Bremer, progmg VP.

KYZK(FM)—Not on air, target date: unknown: 107.5 mhz; 100 kw. 1,734 ft TL: N43 16 50 W114 09 08. Hrs open: 24 220 Northwood Way, Ketchum, 83340. Phone: (208) 726-5324. Fax: (208) 726-5459. Licensee: Chaparral Broadcasting Inc. Group owner: Chaparral Communications (acq 7-30-2004; grpsl). Population served: 15,000 Natl. Network: ABC. Rgnl rep: Allied Radio Law Firm: Cohn & Marks. Format: Jazz. ♦Scott Anderson, gen mgr; Cathy Nikolaisons, gen sls mgr; Bob Thompson, progmg dir; Sue Bailey, news dir; Don Mussell, engrg dir.

Troy

KQZB(FM)—Not on air, target date: unknown: 100.5 mhz; 26.5 kw. Ant 686 ft TL: N46 42 18 W116 55 25. Hrs open: 403 C St., Lewiston, 83501. Phone: (208) 743-4560. Fax: (208) 798-0110. Licensee: Pacific Empire Radio Corp. ♦Mike Bolland, pres.

Twin Falls

KART(AM)—See Jerome

***KAWZ(FM)**— Apr 13, 1988: 89.9 mhz; 33 kw horiz, 100 kw vert. Ant 991 ft TL: N42 43 47 W114 24 52. Stereo. Hrs opn: 24 Box 391, 83303. Secondary address: 4002 N. 3300 E. 83301. Phone: (208) 734-6633. Fax: (208) 736-1958. E-mail: csn@csnradio.com Web Site: www.csnradio.com. Licensee: Calvary Chapel of Twin Falls Inc. Population served: 125,000 Law Firm: Boothe, Ferert, and Tepper. Format: Christian praise & worship, Bible teaching. News: 2 hrs wkly. Target aud: General; 18-80. ♦Mike Kestler, pres & dev VP; Mike Stocklin, gen mgr; Don Mills, opns dir, opns mgr & progmg dir.

***KBSW(FM)**— May 15, 1989: 91.7 mhz; 4.5 kw. Ant 492 ft TL: N42 43 48 W114 25 06. Stereo. Hrs opn: 24 Boise State Univ., 1910 University Dr., Boise, 83725. Phone: (208) 426-3663. Fax: (208) 344-6631. Web Site: radio.boisestate.edu. Licensee: Idaho State Board of Education. Natl. Network: PRI, NPR. Law Firm: Dow, Lohnes & Albertson. Format: Talk, class. News staff: 2; News: 15 hrs wkly. Target aud: General. ♦John Hess, gen mgr & stn mgr; Erik Jones, opns mgr; Ele Ellis, progmg dir; Sadie Babits, news dir; Tom Taylor, engrg dir & engrg mgr.

***KCIR(FM)**— Dec 12, 1982: 90.7 mhz; 20 kw. 2,519 ft TL: N42 20 07 W113 36 17. Stereo. Hrs opn: 24
Rebroadcasts KILA(FM) Las Vegas 97%.
1446 Filer Ave. E., 83301. Phone: (208) 734-5777. Fax: (208) 734-0331. Web Site: www.sosradio.net. Licensee: Faith Communications Corp. (acq 9-29-82). Population served: 300,000 Natl. Network: USA. Format: Christian, educ. News: 5 hrs wkly. Target aud: 25-49; adults with families. Spec prog: Children 2 hrs wkly. ♦Jack French, pres & gen mgr; Brad Staley, gen mgr; Duane Luchsinger, stn mgr; Mike Mead, mus dir.

***KEFX(FM)**— 1996: 88.9 mhz; 3 kw. 20 ft TL: N42 33 25 W114 28 18. Hrs open: Box 271, 83303. Phone: (208) 734-6633. Fax: (208) 736-1958. E-mail: effectradio@effectradio.com Web Site: www.effectiveradio.com. Licensee: Calvary Chapel of Twin Falls Inc. Group owner: CSN International Population served: 40,000 Format: Christian, relg, bible teaching. ♦Mike Kestler, pres & gen mgr; Matt McNeilly, stn mgr; Brian Harman, progmg dir; Ray Gorney, chief of engrg.

***KEZJ(AM)**— 1946: 1450 khz; 1 kw-U. TL: N42 32 36 W114 28 14. Hrs open:
Rebroadcasts KBSU(AM) Boise.
Box 1238, College of Southern Idaho, 83303-1238. Phone: (208) 736-3046. Phone: (888) 859-5278. Fax: (208) 736-2188. Licensee: College of Southern Idaho. Format: News/talk, jazz. ♦Don Wimberly, gen mgr.

KEZJ-FM— Mar 15, 1977: 95.7 mhz; 100 kw. Ant 620 ft TL: N42 43 42 W114 24 48. Stereo. Hrs opn: Box 1259, 83301. Secondary address: 415 Park Ave. 83301. Phone: (208) 733-7512. Fax: (208) 733-7525. E-mail: bradweiser@clearchannel.com Web Site: www.957kezj.com. Licensee: Citicasters Licenses L.P. Group owner: Clear Channel Communications Inc. (acq 5-99; grpsl). Population served: 140,000 Natl. Network: ABC. Natl. Rep: Clear Channel. Wire Svc: ABC Wire Svc: AP Format: Country. Target aud: 25-54. ♦Janice Degner, VP & gen mgr; Brad Weiser, opns dir & progmg VP; James Tidmarsh, news dir; Kelly Klaas, chief of engrg.

KIRQ(FM)— 2007: 102.1 mhz; 5.2 kw. Ant 722 ft TL: N42 43 54 W114 25 04. Hrs open: 21361 Hwy. 30, 83301-0197. Phone: (208) 735-8300. Fax: (208) 733-4196. Licensee: Locally Owned Radio LLC. (acq 9-8-2006; $2,911,000 with KTPZ(FM) Hazelton). Format: Rock. ♦Larry Johnson, pres & gen mgr; Jerre Fender, opns dir & progmg dir; Deb Uvieu, gen sls mgr; Denis Jeffs, traf mgr.

KLIX(AM)— Dec 12, 1946: 1310 khz; 5 kw-D, 2.5 kw-N, DA-N. TL: N42 33 30 W114 26 40. Hrs open: 24 Box 1259, 415 Park Ave., 83303. Phone: (208) 733-1310. Fax: (208) 733-7525. Web Site: www.newsradio1310.com. Licensee: Citicasters Licenses L.P. Group owner: Clear Channel Communications Inc. Population served: 26,209 Natl. Network: ABC. Format: News/talk. News staff: one; News: 12 hrs wkly. Target aud: 35-54. Spec prog: Farm 2 hrs wkly. ♦Chris Muldaney, gen mgr & gen sls mgr; Janice Degner, gen mgr & gen sls mgr; Brad Weiser, opns mgr & progmg dir; Suzanne Jusst, news dir; Kelly Klaas, chief of engrg & farm dir.

KLIX-FM— June 15, 1974: 96.5 mhz; 100 kw. 130 ft TL: N42 33 05 W114 30 59. Stereo. Web Site: www.cooloidies965.com.140,000 Format: Oldies. Target aud: 18-49. ♦Brad Hollstrom, opns mgr & progmg dir; Janice Degner, sls dir; Kelly Klaas, farm dir.

KMVX(FM)—See Jerome

KSNQ(FM)— September 2004: 98.3 mhz; 100 kw. Ant 620 ft TL: N42 43 42 W114 24 48. Hrs open: 415 Park Ave., 83301. Phone: (208) 733-7512. Fax: (208) 733-7525. E-mail: swfradio@aol.com Licensee: Intermart Broadcasting Twin Falls Inc. Format: Classic rock. ♦Patricia Woods, VP & gen mgr.

KTFI(AM)— October 1928: 1270 khz; 5 kw-D, 1 kw-N. TL: N42 33 30 W114 32 00. Hrs open: Secondary address: 21361 Hwy 30 83301. Phone: (208) 735-8300. Fax: (208) 733-4196. Web Site: www.ktfi.com. Licensee: Locally Owned Radio LLC. (group owner; (acq 10-31-2003; grpsl). Population served: 150,000 Format: Oldies. Target aud: 35 plus. Spec prog: Farm 3 hrs, relg 5 hrs, sports 3 hrs wkly. ♦Larry Johnson, pres & gen mgr; Jerre Fender, opns dir; Deb Uvieu, gen sls mgr; Denis Jeffs, traf mgr & women's int ed.

Victor

KRVQ(FM)— May 2005: 92.3 mhz; 800 w hoirz. Ant 1,086 ft TL: N43 29 27 W110 57 16. Hrs open: Box 10219, Jackson, WY, 83002. Phone: (307) 732-0384. Web Site: www.923theriver.com. Licensee: Jackson Radio Group Inc. (acq 3-31-2006; $900,000 with KVRG(FM) Victor). Format: Classic rock. ♦Steven A. Silberberg, pres; Bruce Pollock, gen mgr & progmg dir.

KVRG(FM)— May 2005: 103.7 mhz; 800 w horiz. Ant 1,086 ft TL: N43 29 27 W110 57 16. Hrs open: Box 10219, Jackson, WY, 83002. Phone: (307) 732-0384. Web Site: www.1037therange.com. Licensee: Jackson Radio Group Inc. (acq 3-31-2006; $900,000 with KRVQ(FM) Victor). Format: Country. ♦Steven A. Silberberg, pres; Bruce Pollock, gen mgr & progmg dir.

Wallace

KIBG(FM)— 2001: Stn currently dark. 100.7 mhz; 500 w. Ant 2,207 ft TL: N47 33 44 W115 50 33. (CP: 85 kw, ant 2,119 ft. TL: N47 46 25 W114 16 04). Hrs opn: 581 N. Reservoir Rd., Polson, MT, 59860. Phone: (406) 883-5255. Fax: (406) 883-4441. Licensee: Anderson Radio Broadcasting Inc. (group owner; acq 9-22-2003; grpsl). Population served: 100,000 ♦Dennis Anderson, gen mgr; Gary Meili, gen sls mgr; Dean August, progmg dir; Jeff Smith, news dir; Tony Mulligan, chief of engrg.

***KTWD(FM)**— December 2000: 97.5 mhz; 1.6 kw. Ant -2212 ft TL: N47 33 49 W115 50 01. Stereo. Hrs opn: 24 Box 1208, Airway Heights, WA, 99001. Phone: (509) 244-5577. Fax: (509) 244-2232. E-mail: ktwd@csnradio.com Licensee: CSN International (group owner; acq 2-24-2000; $50,000 for CP). Format: Relg. ♦Barney Dasovich, gen mgr.

KWAL(AM)— May 1938: 620 khz; 1 kw-U, DA-N. TL: N47 30 29 W116 00 17. Stereo. Hrs opn: 24 Box 828, 120 First St., Osburn, 83849. Phone: (208) 752-1141. Phone: (208) 752-1142. Fax: (208) 753-5111. E-mail: kwalradio@usamedia.tv Licensee: Silver Valley Broadcasters Inc. (acq 1-1-73). Population served: 30,000 Natl. Network: Jones Radio Networks. Format: C&W. ♦Paul Robinson, pres; Paul Robinson, gen mgr; George White, gen sls mgr; Larry Crigger, progmg dir, disc jockey & prom.

Weiser

KWEI-FM—Fruitland, March 1984: 99.5 mhz; 8 kw. 2,634 ft TL: N44 00 58 W116 24 15. Stereo. Hrs opn: 24 Box 45234, Boise, 83704. Phone: (208) 367-1859. Fax: (208) 383-9170. E-mail: kwei@cableone.net Web Site: kweispanishradio.com. Licensee: Treasure Valley Broadcasting Co. (acq 1987). Population served: 475,000 Format: Mexican rgnl. News: 15 hrs wkly. Target aud: 25-54; mass appeal. ♦Connie Weisgerber, gen sls mgr; Melvin Albeniz, progmg dir; Melvin Albenez, news dir; Rockwell Smith, chief of engrg; Steve Ramirez, traf mgr & disc jockey.

KWEI(AM)— December 1947: 1260 khz; 1 kw-D, 60 w-N. TL: N44 14 00 W116 57 18.Sunrise-sunset Web Site: kweispanishradio.com. (Acq 1996).475,000 Format: Sp news/talk personality. News staff: one. Target aud: 35-65; mass appeal.

Weston

KLZX(FM)— 2001: 95.9 mhz; 25 kw. Ant 216 ft TL: N41 52 18 W111 48 31. Stereo. Hrs opn: Box 267, Logan, UT, 84323. Phone: (435) 792-0095. Fax: (435) 753-5555. Web Site: www.klzxfm.com. Licensee: Sun Valley Radio Inc. (group owner). Format: Classic rock. ♦Lynn Simmons, gen mgr; Will Wheelwright, progmg dir.

Illinois

Albion

***WBJW(FM)**— December 1997: 91.7 mhz; 6 kw. 328 ft TL: N38 26 51 W88 05 07. (CP: 1.7 kw, ant 499 ft. TL: 38 19 14 W88 02 37). Hrs opn: Rebroadcasts WBGW(FM) Fort Branch, IN 100%.
Box 4164, Evansville, IN, 47724. Phone: (812) 386-3342. Fax: (812) 768-5552. Web Site: www.thyword.org. Licensee: Music Ministries Inc. Format: Relg. ♦Floyd E. Turner, gen mgr.

Aledo

WRMJ(FM)— June 12, 1979: 102.3 mhz; 3 kw. 300 ft TL: N41 12 29 W90 46 10. Stereo. Hrs opn: 24 Box 187, 2104 S.E. 3rd St., 61231. Phone: (309) 582-5666. Fax: (309) 582-5667. E-mail: contactus@wrmj.com Web Site: wrmj.com. Licensee: Western Illinois Broadcasting Co. (acq 8-83; $200,000; FTR: 8-1-83). Population served: 20,000 Rgnl. Network: Brownfield. Law Firm: Koerner & Olender. Format: Country, news. News staff: one; News: 20 hrs wkly. Target aud: 25-54. Spec prog: Relg 3 hrs wkly. ♦John Hoscheidt, gen mgr; Judy Bedford, gen sls mgr; Terry Tracy, progmg dir; Jim Taylor, news dir.

Alton

KATZ-FM— September 1961: 100.3 mhz; 50 kw. 482 ft TL: N38 55 44 W90 13 03. (CP: 50 kw). Stereo. Hrs opn: 24 1001 Highlands Plaza Dr. W/, St. Louis, MO, 63110. Phone: (314) 333-8000. Fax: (314) 333-8300. Web Site: www.katzfm.com. Licensee: Citicasters Licenses L.P. Group owner: Clear Channel Communications Inc. (acq 5-4-99; grpsl). Population served: 2,700,000 Format: Rhythm & blues, hip hop. News staff: 2; News: 7 hrs wkly. Target aud: 18-34; adults. Spec prog: Black, relg 2 hrs wkly. ♦Dennis Lamme, gen mgr; Tommy Austin, opns mgr; Beth Davis, sls dir; Dan Sullivan, gen sls mgr; John Helmkamp, mktg dir.

WBGZ(AM)— 1948: 1570 khz; 1 kw-D, 74 w-N. TL: N38 55 44 W90 13 03. Hrs open: 24 Box 615, 227 Market St., 62002. Phone: (618) 465-3535. Fax: (618) 465-3546. E-mail: wbgz@wbgzradio.com Web Site: www.wbgzradio.com. Licensee: Metroplex Communications Inc. (acq 12-6-2004; $70,000 for 39% of stock). Population served: 100,000 Natl. Network: USA. Format: News/talk. News staff: 2; News: 20 hrs wkly. Target aud: General. Spec prog: Gospel 4 hrs, relg 3 hrs wkly. ♦Sam Stemm, gen mgr; Nancy Birens, gen sls mgr; Mark Ellebracht, news dir.

Anna

WIBH(AM)— Jan 10, 1957: 1440 khz; 500 w-D, 109 w-N. TL: N37 26 45 W89 15 00. Hrs open: 24 330 S. Main St., 62906. Phone: (618) 833-9424. Fax: (618) 833-9091. E-mail: wibh@ajinternet.net Web Site: www.wibhradio.com. Licensee: WIBH Inc. (acq 3-2-98; $315,000). Population served: 110,000 Wire Svc: UPI Format: Classic country. News staff: one; News: 3 hrs wkly. Target aud: 25-69. ♦Maury Bass, VP; Ronald Ellis, pres & gen mgr; Maurice Bass, progmg dir.

WKIB(FM)— Jan 13, 1958: 96.5 mhz; 22.5 kw. Ant 745 ft TL: N37 22 16 W89 31 52. Stereo. Hrs opn: 24 901 S. Kingshighway, Cape Girardeau, MO, 63703. Phone: (573) 339-7000 (business). Fax: (573) 651-4100. Web Site: www.withersradio.net. Licensee: W. Russell Withers Jr. Group owner: Withers Broadcasting Co. (acq 10-22-2001; $2 million). Population served: 200,000 Natl. Rep: Katz Radio. Format: Pop contemp hit radio. News staff: one; News: 10 hrs wkly. Target aud: 18-45. ♦Rick Lambert, gen mgr; Sherry Hampton, gen sls mgr; Steve Thomas, progmg dir.

Arcola

WXET(FM)—Licensed to Arcola. See Mattoon

Arlington Heights

***WCLR(FM)**— Nov 1, 2003: 88.3 mhz; 1 w horiz, 1 kw vert. Ant 59 ft TL: N42 06 45 W87 58 58. Hrs open: 24 2351 Sunset Blvd., Suite 170-218, Rocklin, CA, 95765. Phone: (916) 251-1600. Phone: (888) 937-2471. Fax: (916) 251-1650. E-mail: info@air1.com Web Site: www.air1.com. Licensee: Educational Media Foundation. Group owner: EMF Broadcasting (acq 8-13-03). Natl. Network: Air 1. Law Firm: Shaw Pittman LLP. Format: Contemp Christian. News staff: 3. Target aud: 18-35; Judeo Christian, female. ♦Richard Jenkins, pres; Mike Novak, VP; Keith Whipple, dev dir; Ed Lenane, news dir; Sam Wallington, engrg dir; Karen Johnson, news rptr; Marya Morgan, news rptr; Richard Hunt, news rptr.

WKIE(FM)—Licensed to Arlington Heights. See Chicago

Arthur

WXET(FM)—See Mattoon

Atlanta

WLCN(FM)— Apr 13, 2001: 96.3 mhz; 5.4 kw. Ant 266 ft TL: N40 14 39 W89 15 51. Stereo. Hrs opn: 24 1779 2250th St., 61723. Phone: (217) 648-5510; (847) 674-0864. Fax: (217) 648-2499; (847) 674-9188. Web Site: www.wmnw.net. Licensee: KM Radio of Atlanta L.L.C. Group owner: KM Communications Inc. (acq 5-3-99). Natl. Network: ABC. Wire Svc: AP Wire Svc: Metro Weather Service Inc. Format: Country. Target aud: 25-54. ♦Jim Ash, gen mgr; Jeff Benjamin, traf mgr; Tamera Turner, sls.

Auburn

WCVS-FM—See Springfield

Aurora

WAUR(AM)—Sandwich, May 1986: 930 khz; 2.5 kw-D, 4.2 kw-N, DA-2. TL: N41 36 26 W88 27 11. Hrs open: 24 130 S. Jefferson, Ste 200, Chicago, 60661. Phone: (312) 461-8540 or (219) 309-9327. Fax: (312) 588-0168. E-mail: aciabattari@relevantradio.com Web Site: www.waur.relevantradio.com. Licensee: Starboard Media Foundation Inc. Group owner: Relevant Radio (acq 5-4-2004; $3.5 million). Population served: 3,000,000 Format: Catholic/talk. Target aud: 25-54. Spec prog: Farm 15 hrs wkly. ♦Armand Ciabattari, gen mgr.

WBIG(AM)— Dec 13, 1938: 1280 khz; 1 kw-D, 500 w-N, DA-2. TL: N41 46 10 W88 14 44. Hrs open: 24 620 Eola Rd., 60504. Phone: (630) 851-5200. Fax: (630) 851-5286. Web Site: wbig1280.com. Licensee: Big Broadcasting Co. Group owner: McNaughton-Jakle Stations (acq 1-94; $550,000). Population served: 600,000 Natl. Network: Fox Sports. Law Firm: Leonard S. Joyce. Format: News/talk, sports. News staff: one; News: 10 hrs wkly. Target aud: 25-54; professional, upscale, suburbanites with children. Spec prog: Relg 6 hrs wkly. ♦Rick Jakle, pres; Steve Marten, exec VP & gen mgr; Jack Davis, opns mgr; Jim Sauers, gen sls mgr; Brian Felsten, progmg dir; Brien Prenevost, chief of engrg.

WERV-FM— Feb 12, 1961: 95.9 mhz; 3 kw. 338 ft TL: N41 26 12 W88 16 03. Stereo. Hrs opn: 1884 Plain Ave., 60504. Phone: (630) 898-1580. Fax: (630) 898-2463. Web Site: www.959theriver.fm. Licensee: NM Licensing LLC. Group owner: NextMedia Group L.L.C. (acq 11-26-2001; grpsl). Population served: 2,000,000 Natl. Rep: McGavren Guild. Law Firm: Leibowitz & Associates. Format: Classic hits. News staff: one; News: 5 hrs wkly. Target aud: 25-54; suburban Chicago adults. ♦Brian Foster, gen mgr.

WKKD(AM)— Sept 21, 1960: 1580 khz; 170 w-D, 200 w-N, DA-2. TL: N41 46 12 W88 16 03. Stereo. Hrs opn: 24
Rebroadcasts WONX(AM) Evanston 98%.
2100 Lee St., Evanston, 60202. Phone: (847) 475-1590. Fax: (847) 475-1590. Licensee: Kovas Communications of Indiana Inc. (group owner; (acq 2-28-2002). Population served: 300,000 Format: Health talk. ♦Connie Walburn, gen mgr; Tim Disa, progmg dir.

WLEY-FM— 1965: 107.9 mhz; 21 kw. 761 ft TL: N41 56 01 W88 04 23. Stereo. Hrs opn: 24 150 N. Michigan Ave., Suite 1040, Chicago, 60601. Phone: (312) 920-9500. Fax: (312) 920-9516. E-mail: info@laley1079.com Web Site: www.laley1079.com. Licensee: WLEY Licensing Inc. Group owner: Spanish Broadcasting System Inc. (acq 12-26-96; $33 million). Population served: 2,200,000 Format: Rgnl Mexican. News staff: one. Target aud: 25-54. ♦Jeff Schrinsky, gen mgr & natl sls mgr; Mario Paez, gen mgr & gen sls mgr; Joe McKay, gen sls mgr, natl sls mgr & mktg dir; Leticia Aguilera, prom dir; Marylu Ramos, progmg dir; Sam Palermo, chief of engrg.

Ava

WXAN(FM)— Jan 11, 1982: 103.9 mhz; 2.9 kw horiz, 2.9 kw vert. 469 ft TL: N37 51 19 W89 28 06. Stereo. Hrs opn: 24 9077 Ava Rd., 62907. Phone: (618) 426-3308. Phone: (618) 426-3309. Fax: (618) 426-3310. E-mail: wsstephens@wxan.net Web Site: www.wxan.net. Licensee: Southern Gospetality LLC. Natl. Network: Salem Radio

Network. Format: Relg, southern gospel. News: 10 hrs wkly. Target aud: 30-55; Christians & family-oriented listeners. ♦Harold Lawder, pres; Will Stephens, gen mgr.

Bartonville

WWCT(FM)— February 1997: 99.9 mhz; 1.5 kw. Ant 584 ft TL: N40 36 23 W89 32 20. Hrs open: 24 4234 N. Brandywine Dr., Suite D, Peoria, 61614. Phone: (309) 686-0101. Fax: (309) 686-0111. Licensee: IM IL Licenses LLC. Group owner: Regent Communications Inc. (acq 9-19-2006; grpsl). Population served: 350,000 Law Firm: Pepper & Corazzini. Format: Adult alternative. ♦Michael Rea, gen mgr, progmg dir & news dir.

Beardstown

WRMS(AM)— Nov 1, 1959: 790 khz; 500 w-D, 59 w-N, DA-2. TL: N40 00 11 W90 23 51. Hrs open: 4424 Hampton Ave., Saint Louis, MO, 63109. Phone: (314) 752-7000. E-mail: office@covenantnet.net Licensee: Covenant Network. (acq 7-28-2004). Population served: 6,222 Format: Christian, relg, inspirational. ♦Tony Holman, pres & gen mgr.

WRMS-FM— 1976: 94.3 mhz; 6 kw. Ant 298 ft TL: N40 04 45 W90 25 58. Hrs open: 108 E. Main St., 62618. Phone: (217) 323-1790. Fax: (217) 323-1705. E-mail: wrmsfm@casscomm.com Licensee: Conner Family Broadcasting Inc. (acq 3-87). Format: Country. ♦John Conner, gen mgr, gen sls mgr & progmg dir; Glen Hopkins, chief of engrg.

Belleville

WMVN(FM)—See East St. Louis

WSDZ(AM)— July 13, 1947: 1260 khz; 5 kw-U, DA-2. TL: N38 27 28 W89 57 43. Hrs open: 24 638 Westport Plaza, St. Louis, MO, 63146. Phone: (314) 682-1260. Fax: (314) 682-1190. E-mail: ted.m.zimmerman@abc.com Web Site: www.radiodisney.com/wsdzam1260. Licensee: Radio Disney Group LLC. Group owner: ABC Inc. (acq 9-22-98; $2.5 million). Population served: 2,500,000 Natl. Network: ABC. Format: Family hits. Target aud: 25-54 plus; 6-14; affluent, well-educated, business & professional.children Spec prog: Caring is Cool 30 min wkly. ♦Ted Zimmerman, stn mgr; Nicole Polley, prom mgr.

Belvidere

WXRX(FM)— Feb 27, 1971: 104.9 mhz; 4 kw. 333 ft TL: N42 19 21 W88 57 15. Stereo. Hrs opn: 2830 Sandy Hollow Rd., Rockford, 61109. Phone: (815) 874-7861. Fax: (815) 874-2202. Web Site: www.wxrx.com. Licensee: Maverick Media of Rockford License LLC. Group owner: RadioWorks Inc. (acq 4-27-2005; grpsl). Population served: 247,400 Law Firm: Shaw, Pittman. Format: Rock/AOR. Target aud: 18-49. ♦Gary Rozynek, pres; Jay Chapman, gen mgr; Jim Stone, progmg dir.

Benton

WQRL(FM)— Oct 1, 1973: 106.3 mhz; 12.5 kw. 328 ft TL: N37 55 51 W88 40 52. (CP: Ant 459 ft.). Stereo. Hrs opn: 24 Box 818, 303 N. Main, 62812. Secondary address: 303 N. Main St. 62812. Phone: (618) 435-8100. Fax: (618) 435-8102. E-mail: wwqrlfm@shawneelink.net Licensee: Dana Communications Corp. (acq 4-28-92; $250,000; FTR: 5-18-92). Population served: 250,000 Format: Oldies. News staff: 2; News: 15 hrs wkly. Target aud: 25-49; adults & young adults preferring new country. Spec prog: Farm 3 hrs wkly. ♦Dana Withers, CEO, pres & gen mgr; Bleu Withers, exec VP; Gloria Holland, stn mgr; Jeff Oestreich, chief of engrg.

Berwyn

WVON(AM)— Oct 7, 2003: 1690 khz; 10 kw-D, 1 kw-N. TL: N41 44 14 W87 42 04. Hrs open: 1000 E. 87th St., Chicago, 60619. Phone: (773) 247-6200. Fax: (773) 247-5336. Licensee: CC Licenses LLC. Group owner: Clear Channel Communications Inc. (acq 1-18-2001). Natl. Network: ABC. Format: Talk. Target aud: 25-54; urban talk listeners. ♦Melody Spann-Cooper, gen mgr.

Bethalto

WFUN-FM— April 1991: 95.5 mhz; 6 kw. 328 ft TL: N38 49 39 W90 00 53. Stereo. Hrs opn: 24 9666 Olive Blvd., Suite 610, St. Louis, MO, 63132. Phone: (314) 989-9550. Fax: (314) 989-9551. Web Site: www.foxy995.net. Licensee: Radio One Licenses LLC. Group owner: Radio One Inc. (acq 11-8-01; grpsl). Population served: 2,500,000 Law Firm: Brown, Nietert & Kaufman. Format: Rhythm and blues, classic soul. News: 2 hrs wkly. Target aud: General; families. ♦Alfred Liggins, pres; Michael Douglass, gen mgr; Laura Steele, gen sls mgr; Gary Bennett, engrg VP & chief of engrg; Melissa Wakefields, traf mgr & min affrs dir.

Bloomington

WBNQ(FM)—Listing follows WJBC(AM).

WBWN(FM)—See Le Roy

***WESN(FM)**— 1972: 88.1 mhz; 120 w. 98 ft TL: N40 29 28 W88 59 37. Stereo. Hrs opn: Box 2900, 61701. Phone: (309) 556-2638. Fax: (309) 556-2949. E-mail: wesn@iwu.edu Web Site: www.wesn.org. Licensee: Illinois Wesleyan University. Natl. Network: PRI. Format: Div. Spec prog: Black 18 hrs, class 6 hrs, jazz 6 hrs wkly. ♦Ed Price, stn mgr; Nathan Breitling, progmg dir.

WIHN(FM)—See Normal

WJBC(AM)— 1925: 1230 khz; 1 kw-U. TL: N40 27 32 W89 00 38. Hrs open: 236 Greenwood Ave., 61704. Phone: (309) 829-1221. Fax: (309) 827-8071. Web Site: www.wjbc.com. Licensee: Regent Broadcasting of Bloomington Inc. Group owner: Regent Communications Inc. (acq 5-12-2004; grpsl). Population served: 125,000 Natl. Rep: McGavren Guild. Law Firm: Reddy, Begley & McCormick. Format: Full service. Spec prog: Farm 13 hrs wkly. ♦Red Pitcher, gen mgr; R.C. McBride, opns mgr & progmg dir; Julie Penn, gen sls mgr & natl sls mgr; Colleen Reynolds, news dir; Ron Schott, chief of engrg.

WBNQ(FM)—Co-owned with WJBC(AM). 1947: 101.5 mhz; 50 kw. 460 ft TL: N40 27 32 W89 00 38. Stereo. Web Site: www.wbnq.com. Licensee: Regent Licensee of Erie Inc. Format: CHR. Spec prog: Farm one hr wkly. ♦Dan Westhoff, stn mgr; Tony Travatto, gen sls mgr; Dave Adams, progmg mgr; Russell Rush, mus dir.

Breese

WDLJ(FM)— 2003: 97.5 mhz; 2.5 kw. Ant 512 ft TL: N38 36 33 W89 23 35. Hrs open: KM Radio of Breese L.L.C., 3654 W. Jarvis Ave., Skokie, 60076. Phone: (847) 674-0864. Fax: (847) 674-9188. Web Site: www.wdlj.net. Licensee: KM Radio of Breese L.L.C. Format: Classic rock. ♦Kevin Bae, gen mgr.

Brookport

WTHQ(AM)— October 1987: 750 khz; 500 w-D. TL: N37 08 31 W88 38 58. Hrs open: 6120 Waldo Church Rd., Metropolis, 62960-4903. Fax: (618) 564-3202. Licensee: Daniel S. Stratemeyer (acq 9-20-2002). Format: Talk. ♦Samuel Stratemeyer, gen mgr.

Bushnell

WLMD(FM)— August 1992: 104.7 mhz; 3 kw. Ant 328 ft TL: N40 32 52 W90 26 25. Stereo. Hrs opn: 24 119 W. Carroll, Macomb, 61455. Phone: (309) 833-5561. Fax: (309) 833-3460. E-mail: wlmd@macomb.com Web Site: www.radiomacomb.com. Licensee: WPW Broadcasting Inc. (group owner; (acq 12-27-99; grpsl). Population served: 22,300 Natl. Network: ABC, Jones Radio Networks. Rgnl. Network: Ill Radio Net., Brownfield. Format: Country. News staff: one; News: 2 hrs wkly. Target aud: 25-54; general. Spec prog: Farm 3 hrs wkly. ♦Vanessa Wetterling, stn mgr; Mike Weaver, progmg dir & news dir.

Cairo

***WBEL(FM)**— 2002: 88.5 mhz; 64 kw vert. Ant 558 ft TL: N36 59 32 W88 59 19. Hrs open: Box 3206, American Family Radio, Tupelo, MS, 38803. Phone: (662) 844-8888. Fax: (662) 842-6791. Web Site: www.afr.net. Licensee: American Family Association. Group owner: American Family Radio Format: Christian. ♦Marvin Sanders, gen mgr.

WKRO(AM)— Jan 8, 1942: 1490 khz; 1 kw-U. TL: N37 02 36 W89 11 02. Hrs open: 24 Box 311, 62914. Phone: (618) 734-1490. Fax: (618) 734-0884. E-mail: djman75@hotmail.com Licensee: Alexander Broadcasting Corp. (acq 3-22-01; $20,500). Population served: 122,000 Natl. Network: ABC. Format: Urban adult contemp. News staff: one; News: 12 hrs wkly. Target aud: 25-54; general. Spec prog: Gospel 12 hrs, farm 12 hrs wkly. ♦Danny McDonald, gen mgr & opns mgr; Marti Nicholson, gen sls mgr & adv dir.

Canton

WBYS(AM)— Oct 5, 1947: 1560 khz; 250 w-D. TL: N40 32 43 W90 01 08. Hrs open: 6 AM-sunset Box 600, 1000 E. Linn St., 61520. Phone: (309) 647-1560. Fax: (309) 647-1563. Web Site: www.wbysradio.com. Licensee: WPW Broadcasting Inc. (group owner; (acq 1999; $210,000 for stock with co-located FM). Population served: 38,000 Rgnl. Network: Tribune, Ill. Radio Net. Law Firm: Richard F. Swift. Format: Hits of the 50's & 60's. News staff: one; News: 20 hrs wkly. Target aud: 30 plus; community-oriented with above average income. Spec prog: Farm 10 hrs wkly. ♦David Madison, CEO & pres; David Klockenga, stn mgr.

WCDD(FM)—Co-owned with WBYS(AM). Oct 7, 1968: 107.9 mhz; 25 kw. Ant 265 ft TL: N40 32 43 W90 01 08. Stereo. 24 300,000 Format: Classic hits. News staff: one; News: 10 hrs wkly. Target aud: 25-54; primarily females.

Carbondale

WCIL(AM)— Nov 14, 1946: 1020 khz; 1 kw-D. TL: N37 43 31 W89 15 25. Hrs open: Sunrise-sunset
Rebroadcasts WJPF(AM) Herrin 100%.
1431 Country Aire Dr., Carterville, 62918. Phone: (618) 985-4843. Fax: (618) 985-6529. E-mail: mail@wjpf.com Web Site: www.wjpf.com. Licensee: MRR License LLC. Group owner: MAX Media L.L.C. (acq 3-29-2004; grpsl). Population served: 856,500 Natl. Rep: Christal. Format: Sports, news/talk. News staff: 2; News: 10 hrs wkly. Target aud: 35 plus. Spec prog: Farm one hr wkly. ♦Brian Schimmel, gen mgr; Rusty James, opns dir; Steve Falat, gen sls mgr; Dave Kuffel, rgnl sls mgr; Kristen Lewis, prom mgr; Ryan Patrick, adv mgr; Roz Rice, progmg dir, news dir, local news ed & farm dir; Tim Deterding, engrg dir; Sherry Crider, traf mgr; Steve Sein, news rptr.

WCIL-FM— July 1968: 101.5 mhz; 50 kw. 430 ft TL: N37 43 31 W89 15 25. Stereo. 24 Web Site: www.cilfm.com.830,000 Format: Top 40 hit music. Target aud: 18-34; adult females. ♦Rusty James, stn mgr.

***WDBX(FM)**— February 1996: 91.1 mhz; 3 kw. Ant 131 ft TL: N37 43 43 W89 12 57. Stereo. Hrs opn: 7 AM-4 AM 224 N. Washington St., 62901. Phone: (618) 457-3691. Phone: (618) 529-5900. E-mail: wdbx@globaleyes.net Web Site: www.wdbx.org. Licensee: Heterodyne Broadcasting Co. Population served: 80,000 Format: Div. News: 2.5 hrs wkly. ♦Francis Murphy, pres; Brian R. Powell, stn mgr.

***WSIU(FM)**— Sept 15, 1958: 91.9 mhz; 50 kw. 299 ft TL: N37 42 29 W89 14 05. Stereo. Hrs opn: 24
Rebroadcasts WUSI(FM) Olney 100%, WVSI(FM) Mt. Venon 100%.
Mail code 6602, Southern Illinois Univ.-Carbondale, 62901. Phone: (618) 453-4343. Fax: (618) 453-6186. E-mail: jeff.williams@wsiu.org Web Site: wsiu.org. Licensee: Board of Trustees Southern Illinois University. Population served: 106,000 Natl. Network: NPR, PRI. Law Firm: Cohn & Marks. Wire Svc: AP Format: Class, news. News staff: 3; News: 36 hrs wkly. Target aud: 35-64; highly educated, upper income, socially conscious. Spec prog: New age 4 hrs, big band 4 hrs, folk 3 hrs wkly. ♦Candis Isberner, CEO; Delores Kerstein, CFO; Jeff Williams, gen mgr & local news ed; Mike Zelten, opns mgr & progmg dir; Jack Hammer, chief of engrg; Renee Dillard, dev dir & spec ev coord. Co-owned TV: *WSIU-TV affil

WTAO(FM)—Murphysboro, August 1972: 105.1 mhz; 25 kw. 308 ft TL: N37 45 15 W89 19 14. Stereo. Hrs opn: 24 Box 127, 1822 N. Court St., Marion, 62959. Phone: (618) 997-8123. Fax: (618) 993-2319. Web Site: www.105tao.com. Licensee: CC Licenses LLC. Group owner: Clear Channel Communications Inc. (acq 1-18-2001; grpsl). Population served: 135,000 Format: Rock/active rock. Target aud: 18-49. ♦Jerry Crouse, gen mgr; Paxton Guy, opns mgr & progmg dir.

Carlinville

***WIBI(FM)**— Sept 30, 1975: 91.1 mhz; 50 kw. 476 ft TL: N39 20 58 W89 48 16. Stereo. Hrs opn: 24 Box 140, 62626. Phone: (217) 854-4800. Phone: (800) 707-9191. Fax: (217) 854-4810. E-mail: wibi@wibi.org Web Site: www.wibi.org. Licensee: Illinois Bible Institute Inc. (group owner) Population served: 2,000,000 Law Firm: Gammon & Grange. Format: Adult contemp Christian. Target aud: 25-49. ♦Barry Copeland, gen mgr & sls dir; Jeremiah Beck, stn mgr & opns mgr; Jessica Barton, prom dir; Rob Regal, progmg dir; Joe Buchanan, mus dir; Sally Braundmeier, traf mgr.

***WOLG(FM)**— Dec 8, 1990: 95.9 mhz; 6 kw. 325 ft TL: N39 14 25 W89 54 26. Stereo. Hrs opn: 24 4424 Hampton Ave., Saint Louis, MO, 63109. Phone: (314) 752-7000. Licensee: Covenant Network. (acq 8-10-98; $300,000). Population served: 50,000 Format: Relg. ♦Tony Holman, gen mgr.

***WTSG(FM)**— Aug 11, 1997: 90.1 mhz; 3 kw. 295 ft TL: N39 20 58 W89 48 16. Hrs open: 24 Box 140, 62626-0140. Phone: (217) 854-4851. Fax: (217) 854-4810. E-mail: staff@wtsg.org Web Site: www.wtsg.org. Licensee: Illinois Bible Institute Inc. (group owner) Population served: 60,000 Law Firm: Gammon & Grange. Format: Southern gospel mus, relg. Target aud: 25-44. ♦Barry Copeland, gen mgr; Jeremiah Beck, stn mgr & opns mgr; Jessica Barton, prom dir; Rob Regal, progmg dir; Joe Buchanan, mus dir; Sally Braundmeier, traf mgr.

Carmi

WROY(AM)—Listing follows WRUL(FM).

WRUL(FM)— 1951: 97.3 mhz; 50 kw. 496 ft TL: N38 04 54 W88 12 04. Stereo. Hrs opn: 24 Box 400, 101 N. Church St., 62821. Phone: (618) 382-4161. Phone: (618) 382-2345. Fax: (618) 382-4162. E-mail: wrul973@shawneelink.net Web Site: www.wrul.com. Licensee: W. Russell Withers Jr. (acq 5-1-2006; $1.1 million with co-located AM). Population served: 500,000 Natl. Network: ABC. Format: Country. News staff: one; News: 8 hrs wkly. Target aud: 25-55. ♦Russell Withers, pres, gen mgr & gen mgr; J.C. Tinsley, gen sls mgr; Irma O'Dell, progmg dir; Bob Miller, news dir.

WROY(AM)—Co-owned with WRUL(FM). Dec 13, 1948: 1460 khz; 1 kw-D, 85 w-N. TL: N38 04 54 W88 12 04.24 E-mail: wroy1460@verizon.net Web Site: www.wrul.com.100,000 Format: Hits of the 50s, 60s, 70s & 80s. News staff: 2; News: 25 hrs wkly. Target aud: 35 plus; general. Spec prog: Farm 6 hrs wkly.

Carpentersville

***WWTG(FM)**—Not on air, target date: unknown: 88.1 mhz; 2 kw vert. Ant 108 ft TL: N42 06 21 W88 22 38. Hrs open: Box 3006, Collegedale, TN, 37315. Phone: (615) 469-5122. Fax: (615) 216-7266. Web Site: www.lifetalk.net. Licensee: Lifetalk Broadcasting Association.

Carrier Mills

***WBVN(FM)**— Jan 8, 1990: 104.5 mhz; 6 kw. 328 ft TL: N37 46 25 W88 44 20. Stereo. Hrs opn: 24 Box 1126, Marion, 62959. Phone: (618) 252-2999. Fax: (618) 997-3194. E-mail: wbvn@shawneclink.net Web Site: www.wbvn.org. Licensee: Kenneth W. and Jane A. Anderson (acq 2-28-00). Population served: 165,000 Format: Contemp Christian. News staff: 3. Target aud: 18-45; general. ♦Ken Anderson, pres & gen mgr.

Carterville

WUEZ(FM)— Apr 2, 1992: 95.1 mhz; 17.6 kw. Ant 390 ft TL: N37 43 31 W89 15 25. Hrs open: 24 1431 Country Arc Dr., 62918. Phone: (618) 985-4843. Fax: (618) 985-6529. E-mail: mail@magic951.com Web Site: www.magic951.com. Licensee: MRR License LLC. Group owner: MAX Media L.L.C. (acq 3-29-2004; grpsl). Population served: 250,000 Format: Adult contemp. News staff: 2; News: 18 hrs wkly. Target aud: 25-49; 60% women, 40% men, good spendable income; mgrs, supvrs, professionals. ♦Mike Smith, exec VP; Steve Falat, gen sls mgr; Felicia Dick, traf mgr & disc jockey.

Carthage

WCAZ(AM)— 1922: 990 khz; 1 kw-D, 9 w-N. TL: N40 24 30 W88 12 04. Hrs open: Box 498, 62321. Secondary address: 86 S. Madison 62321. Phone: (217) 357-3128. Fax: (217) 357-2014. E-mail: wcazam@adamas.net Web Site: www.wcazam990.com. Licensee: Ralla Broadcasting Co. Inc. (acq 1993). Population served: 3,351 Format: Talk. ♦Rob Dunham, gen mgr; Chuck Porter, progmg dir.

WCEZ(FM)— 2001: 93.9 mhz; 6 kw. Ant 328 ft TL: N40 24 54 W91 15 11. Stereo. Hrs opn: 24 108 Washington St., Keokuk, IA, 52632. Secondary address: 303 N. Main 62812. Phone: (217) 357-9800. Fax: (319) 524-7275. E-mail: gmkokx@mchsi.com Licensee: Dana R. Withers. Gary M. Follou, (319) 524-5410 Format: Adult contemp. News staff: one; News: 2 hrs wkly. Target aud: 18-45; 60% female, 40% male. ♦Dana Withers, CEO & pres; Gary M. Folluo, gen mgr; dan Workman, opns dir; Gary Folluo, gen sls mgr; Dan Workman, progmg dir; Jim Worrell, news dir.

WQKQ(FM)— Nov 1, 1978: 92.1 mhz; 25 kw. Ant 328 ft TL: N40 35 37 W91 06 48. Stereo. Hrs opn: 24 2850 Mount Pleasant St., Burlington, IA, 52601. Phone: (319) 752-5402. Fax: (319) 752-4715. E-mail: johnp@burlingtonradio.com Web Site: www.classichitskq92.com. Licensee: Pritchard Broadcasting Co. (acq 12-1-99). Population served: 142,100 Format: Classic hits. News staff: one; News: 3 hrs wkly. Target aud: 25-54. ♦John T. Pritchard, pres, VP & gen mgr; Chris King, opns mgr; Chet Young, gen sls mgr; KC Fleming, progmg dir.

Casey

WCBH(FM)— Sept 19, 1988: 104.3 mhz; 11.2 kw. 495 ft TL: N39 16 24 W87 55 39. Hrs open: 24 405 S. Banker, Suite 201, Effington, 62401. Phone: (217) 342-4141. Phone: (217) 342-4142. Fax: (217) 342-4143. E-mail: wcrc@wcrc975.com Licensee: Two Petaz Inc. Group owner: The Cromwell Group Inc. (acq 1-9-02; grpsl). Format: Christian. ♦Bud Walters, CEO; Marvin Phillips, gen mgr; Marvin Phillips, gen sls mgr; David Wilson, chief of engrg.

WKZI(AM)— Dec 14, 1963: 800 khz; 250 w-U. TL: N39 18 14 W87 58 15. Hrs open: 24 18889 N. 23rd 50th St., Dennison, 62423. Phone: (217) 826-9673. E-mail: wkzi@rr1.net Web Site: www.rr1.net/users/wkzi. Licensee: Word Power Inc. (acq 5-4-93; $152,400; FTR: 5-24-93). Population served: 358,500 Natl. Network: Moody. Law Firm: Shook, Hardy & Bacon. Format: Christian. News: 17 hrs wkly. Target aud: General; 12 plus. ♦Eleanor Jean Ford, progmg dir & disc jockey; Mark Stephen Ford, engrg dir; Paul Dean Ford, pres, gen mgr, news dir, chief of engrg & disc jockey.

WLHW(FM)—Co-owned with WKZI(AM). 2006: 91.5 mhz; 6 kw. Ant 197 ft TL: N39 18 14 W87 58 15.24 14,217 ♦Eleanor Jean Ford, disc jockey; Paul Dean Ford, local news ed & disc jockey.

Centralia

WILY(AM)— Aug 15, 1946: 1210 khz; 10 kw-D, 3 w-N, 1.1 kw-CH, DA-2. TL: N38 28 55 W89 08 56. Hrs open: Sunrise-sunset Box 528, 62801. Secondary address: 302 S. Poplar 62801. Phone: (618) 533-5700. Fax: (618) 533-5737. E-mail: wrxx@mvn.net Licensee: Withers Broadcasting Co. of West Virginia. Group owner: Withers Broadcasting Co. (acq 11-4-97; $527,500 with co-located FM). Population served: 120,000 Rgnl. Network: Brownfield. Law Firm: Dennis Kelly. Wire Svc: AP Format: Oldies. News staff: 2; News: 60 hrs wkly. Target aud: 25-54. Spec prog: Business news. ♦Russ Withers, pres; Dana Withers, gen mgr & progmg dir; Brenda Robinson, opns mgr.

WRXX(FM)—Co-owned with WILY(AM). Dec 24, 1964: 95.3 mhz; 3 kw. 217 ft TL: N38 34 44 W89 06 46. Stereo. 24 50,000 Natl. Network: ABC. Format: Rock. News staff: 2; News: 2 hrs wkly. Target aud: 18-49.

Champaign

WBCP(AM)—See Urbana

***WBGL(FM)**— Oct 31, 1982: 91.7 mhz; 20 kw. Ant 459 ft TL: N40 09 12 W88 06 56. Stereo. Hrs opn: 24 2108 W. Springfield St., 61821. Phone: (217) 359-8232. Fax: (217) 359-7374. E-mail: wbgl@wbgl.org Web Site: www.wbgl.org. Licensee: Illinois Bible Institute Inc. (group owner) Natl. Network: USA. Format: Educ, relg. News staff: one. Target aud: 25-44. ♦Jeff Scott, stn mgr; Jennifer Briski, prom dir; Ryan Springer, progmg dir; Joe Buchanan, mus dir; John Symonds, chief of engrg & engr.

WCFF(FM)—See Urbana

WDWS(AM)— Jan 24, 1937: 1400 khz; 1 kw-U. TL: N40 05 04 W88 14 53. Stereo. Hrs opn: 24 Box 3939, 61826. Phone: (217) 351-5300. Fax: (217) 351-5385. E-mail: talk@wdws.com Web Site: www.wdws.com. Licensee: D.W.S. Inc. Population served: 170,000 Natl. Network: ABC. Rgnl. Network: Ill. Radio Net. Natl. Rep: Christal. Wire Svc: AP Format: News/talk, sports. News staff: 4; News: 26 hrs wkly. Target aud: 35-64; adults. Spec prog: Farm 10 hrs, relg 4 hrs wkly. ♦Mike Halle, gen mgr; Jim Lewis, opns mgr; Dave Burns, gen sls mgr; Carol Vorel, news dir.

WHMS-FM—Co-owned with WDWS(AM). 1948: 97.5 mhz; 50 kw. 358 ft TL: N40 05 04 W88 14 53. Stereo. 24 E-mail: literock@whms.com Web Site: www.whms.com.310,000 Format: Adult contemp. News staff: 4; News: 10 hrs wkly. Target aud: 25-54; adults. ♦Ryan Aurthur, opns mgr.

WEBX(FM)—Tuscola, Sept 30, 1970: 93.5 mhz; 6 kw. 308 ft TL: N39 54 24 W88 16 35. Stereo. Hrs opn: 4108 Fieldstone Rd., Suite C, 61822. Phone: (217) 367-1195. Fax: (217) 367-3291. Web Site: www.935953therock.com. Licensee: RadioStar Inc. Group owner: AAA Entertainment L.L.C. (acq 5-23-2006; grpsl). Law Firm: Kaye, Scholer, Fierman, Hays & Handler. Format: Modern/extreme rock. Target aud: 18-49; adult professionals & college students. ♦Mike Knar, gen mgr & prom dir; Jon Mayotte, progmg dir & disc jockey.

***WEFT(FM)**— Sept 21, 1981: 90.1 mhz; 10 kw. 135 ft TL: N40 10 51 W88 19 04. Stereo. Hrs opn: 6 AM-2 AM 113 N. Market St., 61820-4004. Phone: (217) 359-9338. E-mail: weft@weftfm.org Web Site: www.weftfm.org. Licensee: Prairie Air Inc. Natl. Network: NPR, PRI. Law Firm: Haley, Bader & Potts. Format: Var/div. News: 10 hrs wkly. Target aud: General. Spec prog: Black 8 hrs, blues 10 hrs, folk 10 hrs, pub affrs 5 hrs, Sp 3 hrs wkly. ♦Mick Woolf, gen mgr & stn mgr; Darren Martin, chief of engrg.

WILL(AM)—See Urbana

WILL-FM—See Urbana

WIXY(FM)— June 1, 1992: 100.3 mhz; 13 kw. Ant 453 ft TL: N40 00 45 W88 08 29. Hrs open: 2603 W. Bradley Ave., 61821. Phone: (217) 355-4141. Fax: (217) 352-1256. Web Site: www.wixy.com. Licensee: Saga Communications of Illinois LLC. Group owner: Saga Communications Inc. (acq 11-4-92; $250,000; FTR: 11-30-92). Population served: 469,000 Natl. Rep: Katz Radio. Format: Country. News: 5 hrs wkly. Target aud: 25 plus. ♦Ed Christian, CEO & chmn; Steve Goldstein, exec VP; Alan Beck, gen mgr.

WKJR(AM)—Rantoul, Feb 1, 1963: 1460 khz; 500 w-D, 65 w-N, DA-1. TL: N40 18 37 W88 12 54. Hrs open: 24 129 N. Garrard, Rantoul, 61866. Phone: (217) 893-1460. Fax: (217) 893-0884. E-mail: fanmail@1460sports.com Web Site: www.1460sports.com. Licensee: Ruben's Productions Inc. (acq 9-19-2006; $215,000). Population served: 200,000 Natl. Network: ESPN Radio. Format: Sports. Target aud: General. ♦Rueben Acevero, gen mgr; Armando Martinez, progmg dir; Scott Hudson, sports cmtr.

WLRW(FM)— January 1963: 94.5 mhz; 50 kw. Ant 453 ft TL: N40 07 35 W88 17 25. Stereo. Hrs opn: 24 2603 W. Bradley, 61821. Phone: (217) 352-4141. Fax: (217) 352-1256. Web Site: www.mix945.com. Licensee: Saga Communications of Illinois LLC. Group owner: Saga Communications Inc. (acq 10-86; grpsl; FTR: 7-7-86). Population served: 469,000 Format: Hot adult contemp. Target aud: 18-49. ♦Alan Beck, gen mgr; Jonathan Drake, progmg dir.

WMYE(FM)—Rantoul, Mar 15, 1972: 95.3 mhz; 3 kw. 425 ft TL: N40 13 05 W88 06 55. Stereo. Hrs opn: 24 4108 Fieldstone Rd., Suite C, 61822. Phone: (217) 367-1195. Fax: (217) 367-3291. Web Site: 935953therock.com. Licensee: RadioStar Inc. Group owner: AAA Entertainment L.L.C. (acq 5-23-2006; grpsl). Population served: 175,000 Natl. Network: Westwood One. Format: AOR/extreme rock. News staff: one. Target aud: 18-49. ♦Mike Knar, gen mgr.

***WPCD(FM)**— January 1978: 88.7 mhz; 10.5 kw. Ant 338 ft TL: N40 08 14 W88 17 10. Stereo. Hrs opn: Parkland College, 2400 W. Bradley Ave., 61821. Phone: (217) 351-2450. E-mail: dhughes@parkland.edu Web Site: www.parkland.edu/wpcd/. Licensee: Parkland College Community College District No. 505. Population served: 250,000 Natl. Network: AP Radio. Format: Adult urban comtemp, alternative. News: 8 hrs wkly. Target aud: General. Spec prog: News 8 hrs, spanish 4 hrs wkly. ♦Dan Hughes, gen mgr.

WPGU(FM)—See Urbana

WQQB(FM)—See Urbana

Charleston

WEIC(AM)— Dec 10, 1954: 1270 khz; 1 kw-D, 500 w-N, DA-3. TL: N39 30 18 W88 12 54. Hrs open: 24/7 2560 W. State St., 61920. Phone: (217) 345-2148. Fax: (217) 348-7036. Licensee: Eastern Illinois Christian Broadcasting Inc. (acq 9-9-03). Population served: 30,000 Format: Southern gospel. Spec prog: Farm 8 hrs wkly. ♦Brad Lee, gen mgr; Steve Hamm, chief of engrg.

***WEIU(FM)**— July 1, 1985: 88.9 mhz; 4 kw. 166 ft TL: N39 28 43 W88 10 21. Stereo. Hrs opn: 24 600 Lincoln Ave., 61920. Phone: (217) 581-5956. Fax: (217) 581-6650. E-mail: hitmix@weiu.net Web Site: www.weiuhitmix.net. Licensee: Eastern Illinois University. Population served: 72,000 Law Firm: Cohn & Marks. Format: mix/var. News staff: one; News: 3 hrs wkly. Target aud: 12 plus; 25-55 women. Spec prog: Folks 4 hrs, jazz 4 hrs wkly. ♦Denis Roche, gen mgr; Jeff Owens, stn mgr & dev dir; Linda Kingery, progmg dir; Kelly Runyon, news dir. Co-owned TV: *WEIU-TV affil

WWGO(FM)— Oct 1, 1965: 92.1 mhz; 6 kw. 140 ft TL: N39 30 18 W88 12 54. Stereo. Hrs opn: 24 209 Lakeland Blvd., Mattoon, 61938. Phone: (217) 348-9292. Phone: (217) 235-5624. Fax: (217) 235-6624. E-mail: bub@radiomattoon.com Licensee: The Cromwell Group Inc. of Illinois. Group owner: The Cromwell Group Inc. (acq 1993). Population served: 100,000 Natl. Network: ABC. Law Firm: Pepper & Corazzini. Format: Rock. News staff: one; News: 20 hrs wkly. Target aud: 25-54; upscale, educated adults. ♦Bud Walters, pres; Carol Floyd, gen mgr, gen sls mgr & news dir; Bub McCullough, progmg dir; Josh Jamison, chief of engrg; Kathie St. Clair, traf mgr.

Chester

KPNT(FM)—See Sainte Genevieve, MO

KSGM(AM)— July 5, 1947: 980 khz; 1 kw-D, 500 w-N, DA-N. TL: N37 51 24 W89 49 44. Hrs open: 24
Rebroadcasts KBDZ(FM) Perryville, MO.
Box 428, St. Genevieve, MO, 63670. Phone: (573) 883-2980. Fax: (573) 883-2866. E-mail: suntimesnews@brick.net Web Site: www.suntimesnews.com. Licensee: Donze Communications Inc. (acq 6-20-89; $200,000; FTR: 7-10-89). Population served: 60,000 Law Firm: Reddy, Begley & McCormick. Format: Country, news/talk. News staff: 2; News: 14 hrs wkly. Target aud: General; adults. Spec prog: Farm 2 hrs, relg 6 hrs wkly. ♦Don Pritchard, gen mgr & news dir.

Chicago

WAIT(AM)—Willow Springs, 1941: 820 khz; 5 kw-D. TL: N41 56 18 W87 45 05. Stereo. Hrs opn: Daylight 5625 N. Milwaukee Ave., 60646. Phone: (773) 792-1121. Fax: (773) 792-2904. E-mail: 820am@relevantradio.com Web Site: 820am.relevantradio.com. Licensee: WYPA Inc. Group owner: Newsweb Corp. (acq 2-15-2001; $10.5 million). Format: Relg. ♦Mark Pinski, gen mgr; Jorge Murillo, opns mgr; Mike McCarthy, chief of engrg.

WBBM(AM)— Nov 14, 1923: 780 khz; 50 kw-U. TL: N41 59 32 W88 01 36. Hrs open: 630 N. McClurg Ct., 60611. Phone: (312) 944-6000. Fax: (312) 202-3205. Web Site: www.wbbm780.com. Licensee: CBS Radio East Inc. Group owner: Infinity Broadcasting Corp. (acq 1931). Population served: 1,500,000 Natl. Network: CBS. Format: News. ♦David Belmonte, rgnl sls mgr; Rod Zimmerman, VP, gen mgr & progmg dir; Mike Krauser, news dir; Mark Williams, chief of engrg.

WBBM-FM— Dec 7, 1941: 96.3 mhz; 4.2 kw. Ant 1,555 ft TL: N41 52 44 W87 38 08. Stereo. Phone: (312) 951-3497. Fax: (312) 951-3876. Web Site: www.b96.com. Natl. Network: CBS. Natl. Rep: CBS Radio. Format: CHR. ♦Dave Robbins, VP; Paul Agase, gen sls mgr; Thad Gentry, rgnl sls mgr; Michael Biemolt, mktg dir & prom mgr; Todd Cavanah, progmg dir; Eric Bradley, mus dir; Louis Segura, news dir; Tony Kelly, chief of engrg.

***WBEZ(FM)**— 1942: 91.5 mhz; 8.3 kw. 1,180 ft TL: N41 53 56 W87 37 23. Stereo. Hrs opn: 24 Navy Pier, 848 E. Grand Ave., 60611. Phone: (312) 948-4600. Fax: (312) 832-3100. E-mail: questions@wbez.org Web Site: www.chicagopublicradio.org. Licensee: The WBEZ Alliance Inc. (acq 9-7-90; FTR: 10-1-90). Population served: 1,336,695 Natl. Network: NPR. Format: Jazz, pub affrs, news. News staff: 5; News: 45 hrs wkly. Target aud: General; people who want to know about the world around them. ♦Merrill Smith, chmn; Donna Moore, CFO; Torey Malatia, chmn, pres & gen mgr.

WBGX(AM)—See Harvey

WCFJ(AM)—See Chicago Heights

WCGO(AM)—See Chicago Heights

WCKG(FM)—See Elmwood Park

***WCRX(FM)**— July 29, 1975: 88.1 mhz; 100 w. 150 ft TL: N41 52 22 W87 38 52. Hrs open: 24 600 S. Michigan Ave., 60605. Phone: (312) 344-8155. Fax: (312) 663-5204. Web Site: wcrx.net. Licensee: Columbia College. (acq 10-5-82). Natl. Network: AP Radio. Law Firm: Dow, Lohnes & Albertson. Wire Svc: AP Format: Sports, news. News staff: 2; News: 20 hrs wkly. Target aud: 18-24; Men. ♦Cheryl Langston, gen mgr; Tony Kwiecinski, stn mgr; Dave Dennis, chief of engrg.

WDRV(FM)— July 9, 1955: 97.1 mhz; 8.4 kw. 1,196 ft TL: N41 53 08 W87 37 15. Stereo. Hrs opn: 24 875 N. Michigan, Suite 1510, 60611. Phone: (312) 274-9710. Fax: (312) 274-1304. Web Site: www.wdrv.com. Licensee: Bonneville Holding Co. Group owner: Bonneville International Corp. (acq 1-25-01; $165 million with WWDV(FM) Zion). Population served: 7,000,000 Natl. Rep: Katz Radio. Format: Classic timeless rock. ♦Jerry Schnacke, VP & gen mgr; Greg Solk, opns VP; Chris Winston, gen sls mgr; Eileen Elliot, gen sls mgr; Patty Martin, progmg dir.

WFMT(FM)— Dec 13, 1951: 98.7 mhz; 16 kw. 1,170 ft TL: N41 53 56 W87 37 23. Stereo. Hrs opn: 24 5400 N. St. Louis Ave., 60625-4698. Phone: (773) 279-2000. Fax: (773) 279-2199. E-mail: finearts@wfmt.com Web Site: www.networkchicago.com. Licensee: Window to the World Communications Inc. (acq 3-5-70). Population served: 8,000,000 Law Firm: Schwartz, Woods & Miller. Format: Class. News: 7 hrs wkly. Target aud: 25-54; upscale, professional, college educated, upper income adults. Spec prog: Folk 4 hrs, jazz 5 hrs wkly. ♦Dan Schmidt, CEO & pres; Steve Robinson, sr VP & gen mgr; Don Mueller, opns mgr; Paul Ansell, gen sls mgr; Peter Whorf, mus dir; Gordon Carter, engrg dir & chief of engrg. Co-owned TV: WTTW(TV) affil

WGCI-FM—Listing follows WGRB(AM).

WGN(AM)— June 1, 1924: 720 khz; 50 kw-U. TL: N42 00 42 W88 02 07. Hrs open: 435 N. Michigan Ave., 60611. Phone: (312) 222-4700. Fax: (312) 222-5165. Web Site: www.wgnradio.com. Licensee: WGN Continental Broadcasting Co. Group owner: Tribune Broadcasting Co. Natl. Network: ABC. Natl. Rep: Christal. Wire Svc: AP Format: News/talk, sports. Target aud: 35-64. Spec prog: Cubs, Northwestern play-by-play. ♦Tom Langmyer, VP, gen mgr & gen mgr; Wendi Power, sls dir; Lori Brayer, prom dir; Bob Shomper, progmg dir; Wes Bleed, news dir; Jim Carollo, engrg dir & chief of engrg. Co-owned TV: WGN-TV affil

WGRB(AM)— 1924: 1390 khz; 5 kw-U, DA-2. TL: N41 44 13 W87 42 00. Hrs open: 233 N. Michigan Ave., Suite 2800, 60601. Phone: (312) 540-2000. Fax: (312) 938-4477. Web Site: www.wgci.com. Licensee: AMFM Broadcasting Licenses LLC. Group owner: Clear Channel Communications Inc. (acq 8-30-2000; grpsl). Population served: 300,000 Natl. Rep: Christal. Format: Gospel. ♦Marv Dyson, pres; Sandra Robinson, progmg dir.

WGCI-FM—Co-owned with WGRB(AM). Dec 11, 1958: 107.5 mhz; 33 kw. 600 ft TL: N41 52 57 W87 38 15. Stereo. Web Site: www.wgci.com. Format: R&B hip hop. ♦Elroy Smith, progmg dir.

***WHPK-FM**— Mar 15, 1968: 88.5 mhz; 100 w. 121 ft TL: N41 47 40 W87 35 55. Stereo. Hrs opn: 24 5706 S. University Ave., 60637. Phone: (773) 702-8289. Fax: (773) 702-7718. E-mail: whpk@uchicago.edu Web Site: whpk.uchicago.edu. Licensee: The University of Chicago. Population served: 50,000 Format: Div, educ, jazz. News: 5 hrs wkly. Spec prog: Class 10 hrs, African one hr, Haitian 3 hrs, Israeli one hr, Irish one hrs wkly. ♦Sarah Raskin, stn mgr; Katie Buitrago, progmg dir.

***WIIT(FM)**— June 1974: 88.9 mhz; 17 w. 90 ft TL: N41 50 04 W87 37 43. Stereo. Hrs opn: 24 3300 S. Federal St., 60616. Phone: (312) 567-3087. Phone: (312) 567-3088. Fax: (312) 567-7042. E-mail: wiit@iit.edu Licensee: Illinois Institute of Technology. Law Firm: Crowell & Moring. Format: Var/div. News: 5 hrs wkly. Target aud: 15-35; college & young urban community. Spec prog: Jazz 9 hrs, Ger 3 hrs, relg 2 hrs, Sp 5 hrs wkly. ♦Patrick Schneider, stn mgr.

WILV(FM)— 1947: 100.3 mhz; 5.7 kw. Ant 1,394 ft TL: N41 53 56 W87 37 23. Stereo. Hrs opn: One Prudential Plaza, Suite 2780, 130 E. Randolph, 60601. Phone: (312) 297-5100. Fax: (312) 297-5111. E-mail: davidj@lovefm.fm Web Site: www.wnnd.com. Licensee: Bonneville Holding Co. Group owner: Bonneville International Corp. (acq 6-13-97; $75 million). Population served: 6,000,000 Format: Adult contemp. Target aud: 25-49; women. ♦Barry James, exec VP, VP, gen mgr & progmg dir; Sue Werley, gen sls mgr; Mandy Irwin, prom dir; Keith Warner, chief of engrg.

WIND(AM)— 1927: 560 khz; 5 kw-U, DA-2. TL: N41 33 54 W87 25 11. Hrs open: 25 Northwest Point Blvd., Suite 400, Elk Grove Village, 60007. Phone: (847) 437-5200. Fax: (847) 956-5040. Web Site: www.560wind.com. Licensee: Salem Media of Illinois LLC. Group owner: Univision Radio (acq 1-7-2005; with KNIT(AM) Dallas and KKHT-FM Winnie, both TX, in exchange for WPPN(FM) Des Plaines, IL). Population served: 336,695 Format: Conservative news/talk. ♦David Santrella, gen mgr; Eric Thomas, opns dir.

WJMK(FM)—Listing follows WSCR(AM).

WKIE(FM)—Arlington Heights, Mar 10, 1960: 92.7 mhz; 3 kw. Ant 299 ft TL: N42 07 50 W87 58 59. Stereo. Hrs opn: 24 6012 S. Pulaski Rd., 60629. Phone: (773) 767-1000. Fax: (773) 767-1100. Web Site: www.weplayanything.com. Licensee: WKIE Inc. Group owner: Spanish Broadcasting System Inc. (acq 11-15-2004; grpsl). Population served: 16,861 Format: Var. Target aud: 18-49. ♦Harvey Wells, gen mgr; Bill Cavanaugh, gen sls mgr; Mike McCarthy, chief of engrg; Roan Davis, traf mgr.

***WKKC(FM)**— 1975: 89.3 mhz; 250 w. 112 ft TL: N41 46 15 W87 37 48. Stereo. Hrs opn: Kennedy-King College, 6800 S. Wentworth Ave., 60621. Phone: (773) 602-5540. Fax: (773) 602-5532. Web Site: www.ccc.edu. Licensee: District 508 City College of Chicago. Population served: 250,000 Format: Urban contemp, educ, var/div. Target aud: 15-50. ♦Kevin Brown, gen mgr.

WKQX(FM)— 1948: 101.1 mhz; 8.3 kw. 1,710 ft TL: N41 53 56 W87 37 23. Hrs open: 24 Box 3404, 60654. Secondary address: 230 Merchandise Mart Plaza 60654. Phone: (312) 527-8348. Fax: (312) 527-3620. Web Site: www.q101.com. Licensee: Emmis Radio License LLC. Group owner: Emmis Communications Corp. Population served: 700,000 Natl. Rep: D & R Radio. Format: Alternative rock. Target aud: 18-34. ♦Marv Nyren, VP & gen mgr; Lance Richard, gen sls mgr; Mike Stern, progmg dir; Patrick Berger, chief of engrg; Jennifer Welch, traf mgr.

WKSC-FM— November 1957: 103.5 mhz; 4.3 kw. Ant 1,548 ft TL: N41 52 44 W87 38 10. Stereo. Hrs opn: 24 233 N. Michigan, Suite 2800, 60601. Phone: (312) 540-2000. Fax: (312) 938-0712. Web Site: www.kisschicago.com. Licensee: AMFM Broadcasting Licenses LLC. Group owner: Clear Channel Communications Inc. (acq 8-30-2000; grpsl). Population served: 8,000,000 Format: CHR. News staff: one; News: 15 hrs wkly. Target aud: 18-34; upscale. ♦Dave Scharf, gen sls mgr; Bob Fukuda, chief of engrg; Lynn Clymer, traf mgr.

WLIT-FM— Apr 7, 1958: 93.9 mhz; 4 kw. 1,581 ft TL: N41 52 44 W87 38 10. Stereo. Hrs opn: 24 233 N. Michigan Ave., Suite 2800, 60601. Phone: (312) 540-2000. Fax: (312) 938-0111. Web Site: www.wlit.com. Licensee: AMFM Broadcasting Licenses LLC. Group owner: Clear Channel Communications Inc. (acq 8-30-2000; grpsl). Population served: 650,000 Natl. Network: Premiere Radio Networks. Natl. Rep: Clear Channel. Wire Svc: AP Format: Adult contemp. News staff: one; News: one hr wkly. Target aud: 25-54; affluent adults. ♦Earl Jones, VP; Ken Denton, gen sls mgr & rgnl sls mgr; Darren Davis, natl sls mgr & progmg dir; Eric Richeke, mus dir & news dir; Rick Zurick, news dir; Bob Fukuda, chief of engrg; Ella Hammitte, traf mgr.

WLS(AM)— Apr 12, 1924: 890 khz; 50 kw-U. TL: N41 33 21 W87 50 54. Stereo. Hrs opn: 190 N. State St., 60601. Phone: (312) 984-0890. Fax: (312) 984-5305. Web Site: www.wlsam.com. Licensee: Radio License Holding XI LLC. Group owner: ABC Inc. (acq 6-12-2007;

grpsl). Format: News/talk. Target aud: 35-64; listeners involved in Chicago news & community affairs. ♦Zemira Jones, pres & gen mgr; Bill Gamble, progmg dir; Carol O'Keefe, pub affrs dir.

WLUP-FM— 1942: 97.9 mhz; 4 kw. Ant 1,394 ft TL: N41 53 56 W87 37 23. Stereo. Hrs opn: 24 222 Merchandise Mart Plz Ste 230, 60654-1008. Phone: (312) 440-5270. Fax: (312) 440-9377. E-mail: wlup_fm@wlup.com Web Site: www.wlup.com. Licensee: Emmis Radio License LLC. Group owner: Bonneville International Corp. (acq 12-15-2004; swap for KMVP(AM) and KTAR(AM)-KKLT(FM) Phoenix, AZ). Population served: 2,800,000 Natl. Rep: D & R Radio. Format: Rock. Target aud: 18-49. ♦Marv Nyren, gen mgr.

***WLUW(FM)**— Sept 19, 1978: 88.7 mhz; 100 w. 230 ft TL: N42 00 04 W87 39 36. Stereo. Hrs opn: 24 6525 N. Sheridan Rd., 60626. Phone: (773) 508-8080. Fax: (773) 508-8082. E-mail: wluwradio@wluw.org Web Site: www.wluw.org. Licensee: Loyola University, Chicago. Population served: 60,000 Wire Svc: Pacifica Network News Format: Var. Target aud: General. ♦Craig Kois, gen mgr; Sean Campbell, progmg dir; Matt Malooly, news dir.

***WMBI-FM**— July 25, 1960: 90.1 mhz; 100 kw. 440 ft TL: N41 55 35 W88 00 22. Stereo. Hrs opn: 24 820 N. LaSalle Blvd., 60610. Phone: (312) 329-4300. Fax: (312) 329-4468. E-mail: wmbi@moody.edu Web Site: www.wmbi.org. Licensee: The Moody Bible Institute of Chicago. (group owner) Population served: 7,400,000 Natl. Network: Moody. Law Firm: Southmayd & Miller. Wire Svc: AP Format: Relg, educ, Christian. News staff: 2. Target aud: 35-54; Christian. ♦Michael Easely, pres; Bruce Everhart, gen mgr.

WMBI(AM)— July 28, 1926: 1110 khz; 5 kw-D (L-WBT Charlotte, NC; KFAB Omaha). TL: N41 55 35 W88 00 22.Sunrise-sunset 7,400,000 Natl. Network: Moody. Law Firm: Southmayd & Miller. Format: Urban contemp. Target aud: 35-54; Hispanic Christians, children, English-speaking. Spec prog: Sp 12 hrs wkly.

WMVP(AM)— June 25, 1926: 1000 khz; 50 kw-U, DA-2. TL: N41 49 05 W87 59 18. Hrs opn: 24 190 N. State St., 7th Fl., 60601. Phone: (312) 980-1000. Fax: (312) 980-1020. E-mail: jeff.s.schwartz@abc.com Web Site: www.espnradio.com. Licensee: Sports Radio Chicago LLC. Group owner: ABC Inc. (acq 4-1-99; $21 million). Population served: 394,110 Natl. Network: ESPN Radio. Natl. Rep: ABC Radio Sales. Law Firm: Latham & Watkins. Format: Sports, comedy, talk. Target aud: 25-54. ♦Jon Paul Rexing, CEO & rgnl sls mgr; James Pastor, gen mgr; Jeff Schwartz, opns dir; John Craveno, sls dir; Justin Craig, progmg dir; John Hurni, chief of engrg.

WNTD(AM)— May 1922: 950 khz; 1 kw-D, 5 kw-N, DA-N. TL: N41 38 29 W87 33 14. Hrs opn: 541 N. Fairbanks Ave., Suite 1260, 60611. Phone: (312) 467-9755. Fax: (312) 467-9603. Licensee: Multicultural Radio Broadcasting Licensee LLC. Group owner: Multicultural Radio Broadcasting Inc. (acq 2-4-2004; grpsl). Population served: 336,695 Format: Sp contemp. Target aud: 25-54. ♦Karina Rodriguez, gen mgr.

WNUA(FM)— Mar 9, 1959: 95.5 mhz; 8.3 kw. 1,174 ft TL: N41 53 56 W87 37 23. Stereo. Hrs opn: 233 N. Michigan Ave., Suites 2700 & 2800, 60601. Phone: (312) 540-2000. Fax: (312) 938-0712. Web Site: www.wnua.com. Licensee: AMFM Broadcasting Licenses LLC. Group owner: Clear Channel Communications Inc. (acq 8-30-2000; grpsl). Population served: 534,000 Wire Svc: UPI Format: Smooth jazz, adult contemp. Target aud: 25-54. ♦Patrick Kelly, gen mgr.

WOJO(FM)—See Evanston

***WRTE(FM)**— Dec 1, 1969: 90.5 mhz; 8 w. 56 ft TL: N41 20 26 W87 43 05. (CP: 17 w, ant 85 ft.). Hrs opn: 18 1401 W. 18th St., 60608. Secondary address: c/o Mexican Fine Arts Ctr. Museum, 1852 W. 19th St. 60608-2706. Phone: (312) 455-9455. Phone: (312) 738-1503. Fax: (312) 455-9755. E-mail: wrte@radioarte.org Web Site: www.radioarte.org. Licensee: Mexican Fine Arts Center Museum. Population served: 50,000 Law Firm: Shaw Pittman. Format: Var/div. Target aud: 15-35. ♦Silvia Rivera, gen mgr; Carlos Mendez, progmg dir.

WRTO(AM)— December 1988: 1200 khz; 10 kw-D, 1 kw-N, DA-2. TL: N41 42 14 W87 35 47. Stereo. Hrs opn: 24 625 N. Michigan Ave., 3rd Fl., 60611. Phone: (312) 981-1800. Fax: (312) 981-1806. Licensee: WLXX-AM License Corp. Group owner: Univision Radio (acq 9-22-2003; grpsl). Natl. Rep: McGavren Guild. Law Firm: Shaw Pittman. Format: Sp. News staff: 4; News: 10 hrs wkly. Target aud: Urban Spanish. ♦Jerry Ryan, gen mgr; Alicia Chavarria, prom dir.

WSBC(AM)— 1925: 1240 khz; 1 kw-U. TL: N41 56 18 W87 45 05. Hrs open: 24
Rebroadcasts WCFJ(AM) Chicago Heights 25%.
5625 N. Milwaukee Ave., 60646. Phone: (773) 792-1121. Fax: (773) 792-2904. E-mail: wsbc@wsbcradio.com Licensee: WSBC Inc. (acq 2-23-98). Format: Var/div, time brokered. Target aud: General. ♦Harvey Wells, VP; Mark Pinski, gen mgr; Jorge Murillo, opns mgr; Mike McCarthy, progmg dir & chief of engrg.

WSCR(AM)— April 1922: 670 khz; 50 kw-U. TL: N41 56 01 W88 04 23. Stereo. Hrs opn: 24 NBC Tower, 455 N. Cityfront Plaza, 60611. Phone: (312) 245-6000. Fax: (312) 245-6143. Web Site: www.wscr670am.com. Licensee: Infinity Broadcasting East Inc. Group owner: Infinity Broadcasting Corp. (acq 11-13-98; grpsl). Population served: 800000 Rgnl. Network: Ill. Radio Net. Format: Sports. Target aud: 25-54. ♦Rod Zimmerman, VP & gen mgr; Drew Hayes, opns dir & opns mgr; Paul Agase, gen sls mgr; Mary Lou Compton, natl sls mgr; Cher Ames, mktg dir, prom dir & prom mgr; Matt Fishman, progmg dir; George Offman, news dir; Jesse Rogers, pub affrs dir; Greg Davis, chief of engrg.

WJMK(FM)—Co-owned with WSCR(AM). Jan 2, 1961: 104.3 mhz; 4.1 kw. 1,575 ft TL: N41 52 44 W87 38 10. Stereo. 180 N. Stetson, Suite 900, 60601. Phone: (312) 870-6400. Fax: (312) 977-1859. E-mail: wjmk@wjmk.com Web Site: www.wjmk.com. Licensee: Infinity Broadcasting Corp. of Illinois.762,800 Format: Oldies. Target aud: 25-54. ♦Dave Robbins, VP & gen mgr; Terry Hardin, natl sls mgr; Lisa Piovosi, mktg dir; Charlie Lake, progmg dir; John Galenta, chief of engrg.

***WSSD(FM)**— Sept 15, 1987: 88.1 mhz; 10 w. 100 ft TL: N41 52 22 W87 38 52. Stereo. Hrs opn: 515 W. 111th St., 60628-4019. Phone: (773) 928-8800. Fax: (773) 928-9009. Licensee: Lakeside Communications Inc. Law Firm: Lauren A. Colby. Format: Blues, gospel, talk. Target aud: 25 plus; Black. Spec prog: Gospel, jazz, talk. ♦Huey Williams, pres & gen mgr; Steven McKinney, stn mgr & gen sls mgr; Willie McPhatter, progmg dir & progmg mgr.

WUSN(FM)— 1940: 99.5 mhz; 8.3 kw. 1,174 ft TL: N41 53 56 W87 37 23. Stereo. Hrs opn: 2 Prudential Plaza, Suite 1000, 60601. Phone: (312) 649-0099. Fax: (312) 856-9586. Web Site: www.us99.com. Licensee: Infinity Broadcasting Corp. of Chicago. Group owner: CBS Radio (acq 1996; grpsl). Format: Country. Target aud: 25-54. ♦Dave Robbins, gen mgr & stn mgr.

WXRT-FM— 1959: 93.1 mhz; 6.7 kw. Ant 1,309 ft TL: N41 53 56 W87 37 23. Stereo. Hrs opn: 4949 W. Belmont Ave., 60641. Phone: (773) 777-1700. Fax: (773) 777-5031. Web Site: www.93xrt.com. Licensee: Infinity Broadcasting East Inc. Group owner: Infinity Broadcasting Corp. (acq 11-13-98; grpsl). Population served: 600000 Natl. Rep: CBS Radio. Format: Rock/AOR, alternative. News staff: one. Target aud: 25-54; upscale adults. ♦Michael Damsky, VP, gen mgr & gen sls mgr; John Farneda, opns mgr & mktg dir; Adrienne Szarmack, rgnl sls mgr; Brad Auerbach, mktg dir; Norm Winer, progmg VP & progmg dir; Mark Nielson, chief of engrg.

WYLL(AM)— Oct 13, 1924: 1160 khz; 50 kw-D, 5 kw-N, DA-2. TL: N42 02 30 W87 51 57. (CP: 50 kw-U, DA-2). Hrs opn: 24 25 North West Point Blvd., Suite 400, Elk Grove Village, 60007. Phone: (847) 956-5030. Fax: (847) 956-5040. Web Site: www.wyll.com. Licensee: SCA License Corp. Group owner: Salem Communications Corp. (acq 12-20-00; $29 million). Population served: 477,700 Format: Relg. Target aud: 25-54; Upscale income adults. ♦David Santrella, gen mgr.

***WZRD(FM)**— July 8, 1974: 88.3 mhz; 100 w. 76 ft TL: N41 58 56 W87 43 07. Stereo. Hrs opn: 11 AM-midnight 5500 N. St. Louis Ave., 60625. Phone: (773) 583-4050. Fax: (773) 442-4900. Web Site: www.wzrdchicago.com. Licensee: Northeastern Illinois University. Population served: 336,695 Format: Div, educ. News staff: 29; News: 12 hrs wkly. Target aud: General. ♦Dennis Sagel, stn mgr.

WZZN(FM)— Apr 1, 1949: 94.7 mhz; 4.4 kw. Ant 1,535 ft TL: N41 53 56 W87 37 23. Stereo. Hrs opn: Phone: (312) 984-9923. Fax: (312) 984-5357. Web Site: www.947zone.com. Licensee: Radio License Holding V LLC. (acq 6-12-2007; grpsl). Population served: 900,000 Format: Oldies. ♦Jim Pastor, pres & gen mgr.

Chicago Heights

WCFJ(AM)— Aug 15, 1963: 1470 khz; 1 kw-U, DA-2. TL: N41 25 29 W87 38 27. Hrs open: 24 5625 N. Milwaukee Ave., Chicago, 60646. Phone: (773) 792-1121. Fax: (773) 792-2904. E-mail: wsbc@wsbcradio.com Licensee: WCFJ Inc. (acq 2-23-98). Format: Var/div, time brokered. General. ♦Harvey Wells, VP; Mark Pinski, gen mgr; Jorge Murillo, opns mgr; Mike McCarthy, progmg dir & chief of engrg.

WCGO(AM)— Aug 29, 1959: 1600 khz; 1 kw-D, 23 w-N, DA-2. TL: N41 31 05 W87 35 11. Stereo. Hrs opn: 24 222 Vollmer Rd., Suite 2A, 60411. Phone: (708) 755-5900. Fax: (708) 755-5941. E-mail: wcgo1600am@aol.com Licensee: Kovas Communications of Indiana Inc. (group owner; (acq 10-25-2002; $750,000). Population served: 2,500,000 Natl. Network: ABC. Format: Adult contemp. Target aud: 25-54. ♦Frank Kovas, pres; Keith Middleton, VP & gen mgr; Dave Dodaro, gen sls mgr.

WYCA(FM)—See Crete

Chillicothe

WPMJ(FM)— May 16, 1977: 94.3 mhz; 6 kw. Ant 300 ft TL: N40 49 48 W89 29 54. Stereo. Hrs opn: 24 3641 Meadowbrook Rd., Peoria, 61604. Phone: (309) 685-5975. Fax: (309) 685-9095. E-mail: feedback@trueoldies943.com Web Site: trueoldies943.com. Licensee: Kelly Communications Inc. (acq 1-2-2003; $1.5 million value part of cash/swap for WXCL(FM) Pekin). Population served: 350,000 Rgnl rep: McGavren Guild. Format: Oldies. News: 2 hrs wkly. Target aud: 25-54; affluent adults. ♦Bob Kelly, CEO, chmn, pres, CFO & gen mgr; Lee Malcolm, progmg dir & news dir.

Christopher

WXLT(FM)— Dec 25, 1990: 103.5 mhz; 6 kw. 328 ft TL: N37 55 55 W88 57 28. Hrs opn: 24 1431 Country Aire Dr., Carterville, 62918. Phone: (618) 985-4843. Fax: (618) 985-6529. E-mail: mail@wxlt.com Web Site: www.wxlt.comm. Licensee: MRR License LLC. Group owner: MAX Media L.L.C. (acq 3-29-2004; grpsl). Format: Rock. ♦Mike Smith, gen mgr; Steve Falat, gen sls mgr; Rick Gregg, news dir; Felicia Dick, traf mgr.

Cicero

WCEV(AM)— Oct 1, 1979: 1450 khz; 1 kw-U (ST: WVON[\]). TL: N41 49 57 W87 42 40. Hrs open: 1-10 PM (M-F); 1-8:30 PM (S); 5 AM-10 PM (Su) 5356 W. Belmont Ave., Chicago, 60641-4192. Phone: (773) 282-6700. Phone: (773) 777-1450. Fax: (773) 282-0123. E-mail: wcev@wcev1450.com Web Site: www.wcev1450.com. Licensee: Migala Communications Corp. Wire Svc: AP Format: Multi Ethnic. News: 7 hrs wkly. Target aud: Adult ethnic Americans. ♦Estelle Migala, pres; George Migala, stn mgr; Lucyna Migala, progmg dir; Sam Palermo, chief of engrg.

WLUP-FM—See Chicago

WRLL(AM)— 1979: 1450 khz; 1 kw-U (ST: WCEV[\]). TL: N41 49 57 W87 42 20. Hrs opn: Midnight-1 PM 1000 E. 87th St., Chicago, 60619. Phone: (773) 247-6200. Fax: (773) 247-5336. Licensee: Midway Broadcasting Corp. Population served: 200,000 ♦Pervis Spann, CEO; Melody Spann-Cooper, pres & gen mgr; Juanita Maze, gen sls mgr; Coz Carson, progmg dir; Denise King, traf mgr.

Clinton

WHOW(AM)— Aug 1, 1947: 1520 khz; 5 kw-D, 1 kw-CH. TL: N40 05 43 W88 57 51. Hrs open: Sunrise-sunset R.R. 2, Box 117M, 61727. Phone: (217) 935-9590. Fax: (217) 935-9909. Web Site: whowradio.com. Licensee: WHOW Radio LLC (acq 1-26-2004; $300,000 with co-located FM). Population served: 1,000,000 Natl. Network: CNN Radio. Format: news/talk. News: 20 hrs wkly. Target aud: 18-54. ♦Larry Duling, gen mgr.

WHOW-FM— Dec 15, 1975: 95.9 mhz; 6 kw. Ant 308 ft TL: N40 05 43 W88 57 51.24 Web Site: whowradio.com.400,000 Natl. Network: ESPN Radio. Format: Sports talk. Target aud: 18-54.

Coal City

WRXQ(FM)— Feb 8, 1991: 100.7 mhz; 1.4 kw. 482 ft TL: N41 17 39 W88 10 15. Stereo. Hrs opn: 24 2410-B Caton Farm Rd., Crest Hill, 60435. Phone: (815) 556-0100. Fax: (815) 577-9231. Licensee: NM Licensing LLC. Group owner: NextMedia Group L.L.C. (acq 11-26-01; grpsl). Population served: 80,000 Natl. Rep: McGavren Guild. Format: Classic rock. News: 5 hrs wkly. Target aud: 35-50; adults. ♦Brian Foster, gen mgr; Ryan Snow, opns mgr; Todd Elbrink, gen sls mgr; Dan Waddick, prom dir; Rob Creighton, progmg dir.

Colchester

WGNX(FM)—Not on air, target date: unknown: 96.7 mhz; 1.8 kw. Ant 328 ft TL: N40 23 54 W90 43 55. Hrs open: 194 Godfrey Rd., Edgewater, FL, 32141. Phone: (386) 690-2200. Licensee: Patricia Van Zandt. ♦Patricia E. Van Zandt, gen mgr.

WMQZ(FM)— 1999: 104.1 mhz; 6 kw. Ant 328 ft TL: N40 32 01 W90 51 45. Hrs open: 31 East Side Sq., Macomb, 61455. Phone: (309) 833-2121. Fax: (309) 836-3291. E-mail: wjeq@macomb.com Web Site: www.wmqz.com. Licensee: Colchester Radio Inc. Format: Oldies.

♦Bruce Foster, pres & chief of engrg; Nancy Foster, gen mgr; Mike Grillette, news dir; Shana Drake, traf mgr.

Colfax

WRPW(FM)— 1997: 92.9 mhz; 6 kw. Ant 328 ft TL: N40 29 28 W88 43 14. Stereo. Hrs opn: 108 Boeykens Pl., Normal, 61761. Phone: (309) 888-4496. Fax: (309) 452-9677. Web Site: www.power929fm.com. Licensee: Pilot Media LLC. Group owner: AAA Entertainment L.L.C. (acq 6-5-2007; grpsl). Population served: 125,000 Format: CHR/rhythmic. ♦Patti Donsbach, gen mgr; Kevin Trueblood, opns mgr; Amber Goodwin, prom dir; Don Black, progmg dir.

Columbia

KMJM-FM— Feb 15, 1964: 104.9 mhz; 11.5 kw. 480 ft TL: N38 34 24 W90 19 30. Stereo. Hrs opn: 24 1001 Highland Plaza Dr. W., St. Louis, MO, 63110. Phone: (314) 333-8000. Fax: (314) 333-8300. Web Site: www.kmjm.com. Licensee: Citicasters Licenses L.P. Group owner: Clear Channel Communications Inc. (acq 5-4-99; grpsl). Population served: 2,700,000 Format: Urban adult contemp. Target aud: 25-54; adults. ♦Dennis Lamme, gen mgr; Tommy Austin, opns mgr; Beth Davis, sls dir; Kevin Joyce, gen sls mgr; John Helmkamp, mktg dir.

Crest Hill

WCCQ(FM)—Licensed to Crest Hill. See Joliet

Crete

***WBMF(FM)**— 2002: 88.1 mhz; 90 w. Ant 374 ft TL: N41 25 17 W87 38 39. Hrs open: Box 3206, American Family Radio, Tupelo, MS, 38803. Phone: (662) 844-8888. Fax: (662) 842-6791. Web Site: www.afr.net. Licensee: American Family Association. Group owner: American Family Radio Format: Christian. News: 2 hrs wkly.

WYCA(FM)— Sept 5, 1965: 102.3 mhz; 1 kw. 299 ft TL: N41 18 53 W87 37 11. Stereo. Hrs opn: 6336 Calumet Ave., Hammond, IN, 46324. Phone: (773) 734-4455. Fax: (219) 933-0323. E-mail: wybainfo@crawfordbroadcasting.com Web Site: www.wyca1023.com. Licensee: Dontron Inc. Group owner: Crawford Broadcasting Co. (acq 8-26-97; $1.8 million). Population served: 40,900 Format: Christian talk. Target aud: 35 plus; adult, African-Americans. ♦Donald B. Crawford, CEO; Taft Harris, gen mgr.

Crystal Lake

WCPT(AM)— Oct 1, 1965: 850 khz; 2.5 kw-D, DA. TL: N42 15 30 W88 21 48. Stereo. Hrs opn: Sunrise-sunset 6012 S. Pulaski Rd., Chicago, 60629. Phone: (773) 767-1000. Fax: (773) 767-1100. Web Site: wcpt850.com. Licensee: Chicago Newsweb Corp. Group owner: Newsweb Corp. (acq 9-16-2003; $8.25 million). Population served: 6,000,000 Natl. Network: CNN Radio. Law Firm: Holland & Knight. Format: Progressive talk. Target aud: 25-64; Adults. ♦Harvey Wells, gen mgr; Josh Weiss, opns mgr.

WZSR(FM)—See Woodstock

Danville

KUUL(FM)—See East Moline

WDAN(AM)— October 1938: 1490 khz; 1 kw-U. TL: N40 08 58 W87 37 35. Hrs open: 1501 N. Washington Ave, 61832. Phone: (217) 442-1700. Fax: (217) 431-1489. Licensee: Neuhoff Family L.P. (group owner; acq 11-5-03; grpsl). Population served: 80,000 Natl. Rep: McGavren Guild. Format: News/talk, sports. News staff: one; News: 20 hrs wkly. Target aud: 25-54. Spec prog: Farm 20 hrs wkly. ♦Roger Neuhoff, pres; Geoffery Neuhoff, exec VP; Michael Hulvey, VP & gen mgr; Michelle Campbell, sls dir & adv mgr; Tom Barnes, opns mgr, prom mgr & progmg dir; Bill Pickett, news dir & farm dir; Don Russel, chief of engrg.

WDNL(FM)—Co-owned with WDAN(AM). May 1967: 102.1 mhz; 50 kw. 380 ft TL: N40 08 58 W87 37 35. Stereo. 24 120,000 Format: Adult contemp. News: 2 hrs wkly. Target aud: 18-49. ♦Michael Hulvey, opns VP; Tom Barnes, progmg mgr; Carole Wade, mus dir.

WITY(AM)— Nov 24, 1953: 980 khz; 1 kw-U, DA-1. TL: N40 04 42 W87 38 19. Hrs open: 24 Box 142, 61834. Secondary address: Hegeler La. 61832. Phone: (217) 446-1312. Fax: (217) 446-1314. Licensee: Vermilion Broadcasting Corporation. (acq 1981). Population served: 350,000 Natl. Network: Westwood One. Format: Adult standards. Target aud: 35 plus; general. Spec prog: Relg 6 hrs, farm 6 hrs wkly. ♦Donald E. Ward, pres; David W. Brown, VP & gen mgr; Scot C. Medlin, gen sls mgr; Jim Jordan, news dir; Gale Cunningham, pub affrs dir & farm dir; George Dudich, chief of engrg.

WRHK(FM)— November 1992: 94.9 mhz; 6 kw. 328 ft TL: N40 10 40 W87 28 55. Hrs open: 24 1501 N. Washington, 61832. Phone: (217) 442-1700. Fax: (217) 431-1489. Web Site: 949krock.com. Licensee: Neuhoff Family L.P. (group owner; acq 11-5-03; grpsl). Natl. Rep: McGavren Guild. Law Firm: Schwartz, Wood & Miller. Format: Classic rock. News: 5 hrs wkly. Target aud: 18-49. ♦Roger Neuhoff, chmn; Pat Odea, CFO; Geoffery Neuhoff, exec VP; Michael Hulvey, gen mgr & opns VP; Michelle Campbell, gen sls mgr; Tom Barnes, progmg dir; Don Russell, chief of engrg.

WXTT(FM)— Mar 2, 1970: 99.1 mhz; 50 kw. Ant 500 ft TL: N40 08 52 W87 46 20. Hrs open: 2603 W. Bradley Ave., Champaign, 61821. Phone: (217) 352-4141. Fax: (217) 352-1256. Licensee: Saga Communications of Illinois LLC. Group owner: Saga Communications Inc. (acq 6-30-2004; $3.25 million). Population served: 469,000 Natl. Rep: D & R Radio. Format: Classic rock. ♦Alan Beck, gen mgr; Bill Cain, progmg dir.

De Kalb

WDEK(FM)— Dec 17, 1961: 92.5 mhz; 20 kw. Ant 495 ft TL: N41 52 33 W88 45 16. Stereo. Hrs opn: 24 6012 S. Pulaski Rd., Chicago, 60629. Phone: (773) 767-1000. Fax: (773) 767-1100. Web Site: www.ninechicago.com. Licensee: WDEK Inc. Group owner: Spanish Broadcasting System Inc. (acq 11-15-2004; grpsl). Population served: 645,000 Format: Adult contemp. Target aud: 25-44. ♦Harvey Wells, gen mgr; Bill Cavanaugh, gen sls mgr; Matt Duviel, progmg mgr; Mike McCarthy, chief of engrg.

WDKB(FM)— Aug 13, 1990: 94.9 mhz; 3 kw. 328 ft TL: N41 56 58 W88 53 33. Hrs open: 24 2201 N. 1st St., Suite 95, 60115. Phone: (815) 758-0950. Phone: (815) 758-4926. Fax: (815) 758-6226. Web Site: www.b95fm.com. Licensee: De Kalb County Radio Ltd. Law Firm: Shaw Pittman. Wire Svc: AP Format: Adult contemp. News staff: one; News: 3hrs wkly. Target aud: 25-54; adults with moderate to upper incomes. Spec prog: Relg one hr wkly. ♦Tana S. Knetsch, pres & gen mgr.

WLBK(AM)— Dec 7, 1947: 1360 khz; 1 kw-D. TL: N41 56 18 W88 45 03. Hrs open: 4:30 AM-11 PM Box 448, 60115. Secondary address: 1325 Sycamore Rd. 60115. Phone: (815) 758-8686. Fax: (815) 756-9723. Licensee: WPW Broadcasting Inc. (group owner; (acq 4-12-2000). Population served: 75,000 Law Firm: Dow, Lohnes & Albertson. Format: Adult contemp, news/talk. News staff: 3; News: 21 hrs wkly. Target aud: General. Spec prog: Farm 12 hrs, big band 5 hrs, relg 6 hrs wkly. ♦Norm Miller, gen mgr.

***WNIJ(FM)**— October 1954: 89.5 mhz; 50 kw. 421 ft TL: N42 00 55 W89 00 07. Stereo. Hrs opn: 24 NIU Broadcast Ctr., 801 N. First St., DeKalb, 60115. Phone: (815) 753-9000. Fax: (815) 753-9938. E-mail: npr@niu.edu Web Site: www.northernpublicradio.org. Licensee: Northern Illinois University. Population served: 500,000 Natl. Network: PRI, NPR. Law Firm: Arter & Hadden. Format: News, jazz. News staff: 2; News: one hr wkly. Target aud: General. Spec prog: Folk 4 hrs wkly. ♦Tim Emmons, gen mgr; Jan Kilgard, dev dir; Bill Drake, progmg dir.

WNIU(FM)—See Rockford

Decatur

WDZ(AM)— Mar 17, 1921: 1050 khz; 1 kw-U. TL: N39 48 54 W89 00 05. Hrs open: 337 N. Water St., 62523. Phone: (217) 423-9744. Fax: (217) 423-9764. Web Site: www.magic1050am.com. Licensee: NM Licensing LLC. Group owner: NextMedia Group L.L.C. (acq 11-26-01; grpsl). Population served: 125,000 Format: Urban. ♦Joel Fletcher, gen mgr; Tricia LeVeck, progmg dir.

WDZQ(FM)—Co-owned with WDZ(AM). Nov 1, 1976: 95.1 mhz; 50 kw. 500 ft TL: N39 37 36 W89 04 49. Stereo. Web Site: www.95q.com.1,000,000 Format: Country. ♦Brad Wells, progmg dir.

***WJMU(FM)**— Mar 10, 1971: 89.5 mhz; 1 kw. 95 ft TL: N39 50 30 W88 58 29. (CP: 1.66 kw). Stereo. Hrs opn: 7 AM-1 AM 1184 W. Main St., 62522. Phone: (217) 424-6369. Fax: (217) 424-3993. E-mail: wjmu@mail.millikin.edu Licensee: Millikin University. Population served: 100,000 Format: Progsv. News: 8 hrs wkly. Target aud: 20 plus; students, surrounding community. ♦Dove Zemke, pres; Chris Bullock, gen mgr.

WSOY-FM— November 1946: 102.9 mhz; 54 kw. 495 ft TL: N39 52 40 W88 56 30. Stereo. Hrs opn: 24 1100 E. Pershing Rd., 62526. Phone: (217) 877-5371. Fax: (217) 877-8777. Web Site: www.wsoy.com. Licensee: NM Licensing LLC. Group owner: NextMedia Group L.L.C. (acq 11-26-01; grpsl). Population served: 130,000 Natl. Network: CBS. Rgnl. Network: Ill. Radio Net. Law Firm: Wilmer, Cutler & Pickering. Format: CHR. News staff: 3; News: 3 hrs wkly. Target aud: 25-54. ♦Joel Fletcher, gen mgr; Roy Jaynes, progmg dir.

WSOY(AM)— 1925: 1340 khz; 1 kw-U. TL: N39 52 40 W88 56 30.24 Web Site: www.wsoy.com. Format: News/talk, sports. News staff: 3; News: 13 hrs wkly. Target aud: 25 plus. ♦Joel Fletcher, VP; Ryan Forden, progmg dir & farm dir.

WYDS(FM)— 1993: 93.1 mhz; 6 kw. 328 ft TL: N39 48 35 W88 59 31. Hrs open: 24 410 N. Water St., Suite C, 62523. Phone: (217) 428-4487. Fax: (217) 428-4501. E-mail: theparty@family-net.net Licensee: WEJT Inc. (acq 4-9-93; $750,000; FTR: 5-3-93). Population served: 200,000 Format: Top-40 contemporary Hit. News staff: one. Target aud: 18-49; females average age of 26. ♦Chris Bullock, gen mgr; Wayne Robbins, opns mgr; Jerry Scott, gen sls mgr & chief of engrg.

Deerfield

WEEF(AM)—See Highland Park

WVIV-FM—See Highland Park

Des Plaines

WPPN(FM)— Dec 3, 1971: 106.7 mhz; 50 kw. Ant 423 ft TL: N42 08 10 W87 58 55. Stereo. Hrs opn: 24 625 N. Michigan Ave. 3rd Fl., Chicago, 60611. Phone: (312) 981-1800. Fax: (312) 981-1806. Web Site: www.univision.com. Licensee: Univision Radio License Corp. Group owner: Salem Communications Corp. (acq 12-21-2004; asset exchange agreement). Format: Sp. ♦Jerry Ryan, gen mgr; Jose Lopez, gen sls mgr; Lucy Diaz, natl sls mgr; Alicia Chavarria, prom dir; Victor Cerdo, mus dir; Robert Lopez, chief of engrg.

Dixon

WIXN(AM)— July 1961: 1460 khz; 1 kw-D, DA. TL: N41 49 38 W89 29 11. Hrs open: 19 1460 S. College Ave., 61021. Phone: (815) 288-3341. Phone: (815) 626-3091. Fax: (815) 284-1017. Web Site: www.wixn.com. Licensee: NRG License Sub. LLC. (group owner; (acq 10-31-2005; grpsl). Population served: 17,600 Law Firm: Miller & Fields, P.C. Format: News, oldies. News staff: 2; News: 14 hrs wkly. Target aud: 25-54. Spec prog: Farm 11 hrs wkly. ♦Al Knickrehm, gen mgr & stn mgr.

WRCV(FM)—Co-owned with WIXN(AM). Sept 1, 1965: 101.7 mhz; 6 kw. 300 ft TL: N41 49 29 W89 29 51. Stereo. 24 Web Site: www.wixn.com.18,147 Format: Country. News staff: 2; News: 10 hrs wkly. ♦Steve Marco, progmg dir.

Dorsey

***WDRS(FM)—** 2006: 89.5 mhz; 30 w. Ant 279 ft TL: N38 59 04 W89 59 20. Hrs open:
Rebroadcasts KLRD(FM) Yucaipa, CA 100%.
2351 Sunset Blvd., Suite 170-218, Rocklin, CA, 95765. Phone: (916) 251-1600. Fax: (916) 251-1650. Web Site: www.air1.com. Licensee: Educational Media Foundation. (acq 9-22-2005; $30,000 for CP). Natl. Network: Air 1. Format: Christian. ♦Richard Jenkins, pres; Mike Novak, VP; Keith Whipple, dev dir; Eric Allen, natl sls mgr; David Pierce, progmg mgr; Ed Lenane, news dir; Sam Wallington, engrg dir; Karen Johnson, news rptr; Marya Morgan, news rptr; Richard Hunt, news rptr.

Downers Grove

***WDGC-FM—** Feb 28, 1969: 88.3 mhz; 250 w. 130 ft TL: N41 48 16 W88 00 44. Stereo. Hrs opn: 8 AM-10 PM (M-S) 4436 Main St., 60515. Phone: (630) 795-8490. Phone: (630) 795-8400. Fax: (630) 795-8499. E-mail: wdgcfm@hotmail.com Web Site: www.csd99.k12.il.us/wdgc. Licensee: High School District No. 99 Dupage County. Population served: 40,400 Format: Div. News: 5 hrs wkly. Target aud: General; all age groups. Spec prog: Community affrs 6 hrs wkly. ♦John Waite, gen mgr & opns mgr.

Du Quoin

WDQN(AM)— 1951: 1580 khz; 170 w-D, 7 w-N. TL: N38 01 56 W89 14 30. Hrs open: 5:30 AM-11 PM Box 190, 62832. Secondary address: 2337 US Rt. 51 62832. Phone: (618) 542-3894. Fax: (618) 542-4514. E-mail: wdqnradio@onecliq.net Licensee: Du Quoin Broadcasting Co. Population served: 25,000 Natl. Network: ABC, Motor Racing Net. Wire Svc: AP Format: Adult contemp, country. Target aud: 25-64; male & female. Spec prog: Farm 3 hrs, relg 5 hrs wkly. ♦Greg Showalter, gen mgr; Michelle Klein, sls dir; Gordon Showalter, progmg dir.

WDQN-FM— Sept 1, 1969: 95.9 mhz; 6 kw. Ant 328 ft TL: N38 01 56 W89 14 30. Hrs open: 6 AM-11 PM Box 220, West Frankfort, 62896. Secondary address: 3391 Charley Good Rd., West Frankfort 62896. Phone: (618) 627-4651. Fax: (618) 627-2726. Web Site: www.3abn.org. Licensee: Three Angels Broadcasting Network Inc. (acq 7-16-2003; $600,000). Format: Christian. ♦Danny Shelton, pres; Mollie Steenson, gen mgr; Jim Morris, gen sls mgr; Sandra Juarez, progmg dir; Moses Primo, chief of engrg.

Dundee

WWYW(FM)— June 8, 1967: 103.9 mhz; 2.55 kw. Ant 321 ft TL: N42 06 21 W88 22 37. Stereo. Hrs opn: 24 8800 Rt. 14, Crystal Lake, 60012. Phone: (815) 459-7000. Fax: (815) 459-7027. Web Site: www.y1039.com. Licensee: NM Licensing LLC. Group owner: NextMedia Group L.L.C. (acq 5-19-2004; $5 million). Population served: 2,000,000 Law Firm: Leibowitz & Associates. Format: Oldies. ♦Kira Lefond, gen mgr; Don Oberbillig, gen sls mgr.

Dwight

WJEZ(FM)— June 9, 1997: 98.9 mhz; 1.3 kw. Ant 489 ft TL: N41 02 06 W88 26 11. Hrs open: 315 N. Mill St., Pontiac, 61764. Phone: (815) 844-6101. Fax: (815) 844-7235. Web Site: www.wjez.com. Licensee: Livingston County Broadcasters Inc. Group owner: Regent Communications Inc. (acq 5-12-2004; grpsl). Format: Adult contemp. Target aud: 18-49; general. ♦Red Pitcher, gen mgr; Julie Penn, sls dir; Shelley Grove, prom mgr; Kent Kasson, progmg dir & news dir; Lane Lindstrom, chief of engrg.

Earlville

WMKB(FM)— Feb 3, 2003: 102.9 mhz; 2.15 kw. Ant 558 ft TL: N41 37 16 W89 05 20. Stereo. Hrs opn: 24 4756 E. 4th Rd., Mendota, 61342. Phone: (815) 538-7500. Fax: (866) 816-0064. E-mail: info@wmkbradio.com Web Site: www.wmkbradio.com. Licensee: KM Radio of Earlville L.L.C. Group owner: KM Communications Inc. Natl. Network: ABC. Format: Class rock. News: 2 hrs wkly. Target aud: 25-54. Spec prog: Blues 5 hrs wkly. ♦Anne Schenck, gen mgr.

East Moline

KUUL(FM)— Feb 23, 1976: 101.3 mhz; 50 kw. 500 ft TL: N41 37 10 W90 17 41. Stereo. Hrs opn: 3535 E. Kimberly Rd., Davenport, IA, 52807. Phone: (563) 344-7000. Fax: (563) 359-8524. Web Site: www.kuul.com. Licensee: Citicasters Licenses L.P. Group owner: Clear Channel Communications Inc. (acq 11-15-00; grpsl). Population served: 350,000 Natl. Rep: Katz Radio. Format: Classic rock, hits of the 60s & 70s. Target aud: 25-54; contemp, upscale adults. Spec prog: Pub affrs 6 hrs, farm one hr wkly. ♦Larry R. Rosmilso, VP & gen mgr; Scott Bitting, sls dir & gen sls mgr; Carrie Clearman, prom dir; Bo J. Spates, progmg dir; Kevin Allensworth, chief of engrg; Joanne Kerschieter, traf mgr.

***WDLM(AM)—** Apr 3, 1960: 960 khz; 1 kw-D, 102 w-N, DA-2. TL: N41 24 57 W90 23 54. Hrs open: Box 149, 61244. Phone: (309) 234-5111. Fax: (309) 234-5114. E-mail: wdlm@moody.edu Web Site: www.mbn.org. Licensee: Moody Bible Institute of Chicago. (group owner) Format: Relg. ♦Lane D. Morgan, gen mgr.

WDLM-FM— Jan 20, 1980: 89.3 mhz; 100 kw. 500 ft TL: N41 32 52 W90 28 30. Web Site: www.mbn.com.450,000 ♦Dave Jolly, stn mgr; Ken Brooks, progmg dir.

East St. Louis

***WCBW-FM—** 2001: 89.7 mhz; 250 w. Ant 187 ft TL: N38 37 53 W90 12 09. Hrs open: New Life Evangelistic Center Inc., 1411 Locust St., St. Louis, MO, 63103. Phone: (314) 421-3020. Fax: (314) 436-2434. E-mail: larryr@hereshelpnet.org Web Site: www.hereshelpnet.org. Licensee: New Life Evangelistic Center Inc. Format: Relg. ♦Larry Rice, gen mgr.

WFFX(AM)— Aug 1, 1934: 1490 khz; 1 kw-U, DA-2. TL: N38 37 16 W90 09 36. Stereo. Hrs opn: 149 S. 8th St., 62201. Phone: (618) 271-7687. Fax: (618) 875-4315. Licensee: Simmons Austin, LS LLC. (acq 3-31-2005; $1.15 million). Population served: 622,236 Natl. Network: American Urban. Natl. Rep: Katz Radio. Format: Gospel, rhythm & blues. Target aud: 23-55. ♦Dave Greene, gen mgr.

WMVN(FM)— June 6, 1965: 101.1 mhz; 44 kw. Ant 525 ft TL: N38 45 11 W90 07 09. Stereo. Hrs opn: 11647 Olive Blvd., St. Louis, MO, 63141. Phone: (314) 983-6000. Fax: (314) 994-9447. E-mail: movincomments@bicstl.com Web Site: www.movinstlouis.com. Licensee: Bonneville Holding Co. Group owner: Bonneville International Corp. (acq 9-26-2000; grpsl). Natl. Rep: McGavren Guild. Format: Rock-based adult contemp. Target aud: 25-49. Spec prog: Blues 1 hr wkly. ♦Bruce Reese, CEO & pres; Bob Johnson, CFO; John Kijowski, VP & gen mgr; Emily Bushman, gen sls mgr; Trish Gazzal, news dir; Marshall Rice, chief of engrg.

Edwardsville

***WRYT(AM)—** Nov 20, 1987: 1080 khz; 500 w-D, DA. TL: N38 47 58 W89 57 45. (CP: 250 w-N, DA-2, TL: N38 38 30 W89 57 45). Hrs opn: 4424 Hampton Ave., St. Louis, MO, 63109. E-mail: covenantnetwork@juno.com Licensee: Covenant Network. (acq 10-2-97). Population served: 200,000 Format: Catholic, relg. Target aud: General. ♦John A. Holman, pres; Tony Holman, gen mgr.

***WSIE(FM)—** Sept 4, 1970: 88.7 mhz; 37 kw. 567 ft TL: N38 47 06 W89 59 10. Stereo. Hrs open: 24 Box 1773, So. Illinois Univ. at Edwardsville, 62026. Phone: (618) 650-2228. Fax: (618) 650-2233. Web Site: www.siue.edu.wsie. Licensee: Board of Trustees, Southern Illinois University. Population served: 2,000,000 Natl. Network: NPR, PRI. Rgnl. Network: Ill. Radio Net. Law Firm: Dow, Lohnes & Albertson. Format: Jazz. News staff: one; News: 20 hrs wkly. Target aud: 25-49; adults seeking a sophisticated alternative. Spec prog: New age 10 hrs wkly. ♦Frank Akers, gen mgr; Tom Dehner, news dir; David Caires, chief of engrg.

Effingham

WCRA(AM)— June 8, 1947: 1090 khz; 1 kw-D. TL: N39 06 26 W88 33 44. Hrs open: 405 S. Banker St., Suite 201, 62401. Phone: (217) 342-4141. Fax: (217) 342-4143. Licensee: Two Petaz Inc. Group owner: The Cromwell Group Inc. (acq 1-9-02; grpsl). Population served: 11,200 Natl. Network: CBS. Format: News/talk. Target aud: 25-54. ♦Marv Phillips, gen mgr.

WCRC(FM)—Co-owned with WCRA(AM). June 14, 1963: 95.7 mhz; 50 kw. 480 ft TL: N39 06 26 W88 33 44. Stereo. 24 Web Site: www.wcrc957.com.50,000 Format: Country.

***WEFI(FM)—** 2006: 89.5 mhz; 400 w. Ant 164 ft TL: N39 08 30 W88 33 36. Hrs open:
Rebroadcasts WAFR(FM) Tupelo, MS 100%.
Box 2440, Tupelo, MS, 38803-2440. Phone: (662) 844-8888. Fax: (662) 842-6791. Web Site: www.afr.net. Licensee: American Family Association. Format: Christian. ♦Marvin Sanders, gen mgr.

WXEF(FM)— Oct 4, 1982: 97.9 mhz; 6 kw. 300 ft TL: N39 07 25 W88 38 28. Stereo. Hrs opn: 24 Box 988, 206 S. Willow, 62401. Phone: (217) 347-5518. Fax: (217) 347-5519. E-mail: info@thexradio.com Web Site: www.thexradio.com. Licensee: Premier Broadcasting Inc. (acq 11-3-93; $380,000; FTR: 11-22-93). Population served: 50,000 Format: Adult contemp. News: 15 hrs wkly. Target aud: General. Spec prog: High school sports. ♦T. David Ring, pres; Greg Sapp, stn mgr & opns dir; Tonya Siner, opns VP; George Flexter, opns mgr.

Eldorado

WEBQ-FM— April 1972: 102.3 mhz; 3 kw. 296 ft TL: N37 49 14 W88 27 11. Stereo. Hrs opn: 24 701 S. Commercial, Harrisburg, 62946. Phone: (618) 252-6307. Fax: (618) 252-2366. E-mail: webq@yourclearwave.com Licensee: W. Russell Withers Jr. Group owner: Withers Broadcasting Co. (acq 7-28-2004; $450,000 with WEBQ(AM) Harrisburg). Population served: 9,535 Natl. Network: ABC. Format: Adult contemp. News staff: one; News: 6 hrs wkly. Target aud: 25-45; young middle class adults. ♦Cathy Horton, gen mgr, stn mgr & progmg dir; Sonny Dotson, gen sls mgr; Wyatt Drake, news dir; Bob Romonosky, chief of engrg; Shelly Reeder, traf mgr.

Elgin

***WEPS(FM)—** 1950: 88.9 mhz; 740 w. Ant 100 ft TL: N42 02 17 W88 16 15. Hrs open: 6 355 E. Chicago St., 60120. Phone: (847) 888-5000. Fax: (847) 888-0272. E-mail: jackieolsonkold@u-46.org Licensee: Board of Education, Union School District 46. Population served: 80,000 Format: Div, educ. Target aud: Parents of students. Spec prog: Class 5 hrs, jazz 6 hrs, community affrs 3 hrs, educ 13 hrs wkly. ♦Jackie Olson Kold, stn mgr.

WRMN(AM)— 1949: 1410 khz; 1 kw-D, 500 w-N, DA-N. TL: N42 00 21 W88 17 55. Hrs open: 14 Douglas Ave., 60120. Phone: (847) 741-7700. Fax: (847) 888-4227. E-mail: mail@wrmn1410.com Web Site: www.radioshoppingshow.com. Licensee: Elgin Broadcasting Co. Group owner: McNaughton-Jakle Stations (acq 1952). Population served: 650,000 Law Firm: Blair, Joyce & Silva. Format: News/talk. News staff: news progmg 5 hrs wkly News: one;. Target aud: General. Spec prog: Sp 10 hrs wkly. ♦Richard Jakle, CEO, chmn, pres & gen mgr; Jack Davis, stn mgr; Chuck France, gen sls mgr.

Elmhurst

WJJG(AM)— Oct 10, 1974: 1530 khz; 500 w-D, DA. TL: N41 52 03 W87 55 07. (CP: 760 w). Stereo. Hrs opn: 5629 St. Charles Rd., Suite 208, Berkeley, 60163. Phone: (708) 493-1530. Fax: (708) 493-1537. Web Site: www.wjjg.com. Licensee: Joseph J. Gentile Inc. (acq 7-6-94; $700,000). Population served: 5,000,000 Format: News/talk. Target aud: 45 plus; affluent adults. ♦Joseph Gentile, pres; Mike Baker, opns mgr.

***WRSE(FM)—** Dec 7, 1962: 88.7 mhz; 100 w. 95 ft TL: N41 53 46 W87 56 45. Stereo. Hrs opn: 24 190 Prospect Ave., 60126-3296. Phone: (630) 617-3729. Fax: (630) 617-3313. Web Site: www.wrse.com. Licensee: Board of Trustees Elmhurst College. Population served: 150,000 Law Firm: Booth, Frerat, Imlay & Tepper, P.C. Wire Svc: AP Format: Alternative, rock, oldies. News: 1 hr wkly. Target aud: 17-40; college & general. Spec prog: Metal 3 hrs, hip hop 6 hrs wkly. ♦Jon Morgan, gen mgr.

Elmwood

WFYR(FM)— Aug 2, 1993: 97.3 mhz; 23.5 kw. 338 ft TL: N40 46 22 W89 44 50. Stereo. Hrs opn: 24 120 Eaton St., Peoria, 61603. Phone: (309) 676-5000. Fax: (309) 676-2600. E-mail: jgreeley@regentcomm.com Web Site: www.973rivercountry.com. Licensee: Regent Broadcasting of Peoria Inc. Group owner: Regent Communications Inc. (acq 7-6-01; grpsl). Population served: 500,000 Natl. Rep: Katz Radio. Law Firm: Reddy, Begley & McCormick. Format: Country. News: one hr wkly. Target aud: 25-54; adults, family oriented & skewing female. ♦J.R. Greeley, gen mgr; Ric Morgan, opns mgr.

Elmwood Park

WCKG(FM)— 1947: 105.9 mhz; 4.2 kw. 1,575 ft TL: N41 52 44 W87 38 10. Stereo. Hrs opn: 24 2 Prudential Plaza, Suite 1059, Chicago, 60601. Phone: (312) 240-7900. Fax: (312) 565-3181. Web Site: www.wckg.com. Licensee: Infinity Holdings Corp. of Orlando. Group owner: CBS Radio (acq 1996). Natl. Network: Westwood One. Format: Personality talk. News staff: one; News: one hr wkly. Target aud: 25-54. ♦Rob Zimmerman, gen mgr.

Eureka

WPIA(FM)— 1989: 98.5 mhz; 3 kw. Ant 328 ft TL: N40 44 20 W89 16 13. Stereo. Hrs opn: 24
Simulcast with WWCT(FM) Farmington 100%.
4234 N. Brandywine Dr., Suite D, Peoria, 61614. Phone: (309) 686-0101. Fax: (309) 686-0111. Licensee: IM IL Licenses LLC. Group owner: Regent Communications Inc. (acq 9-19-2006; grpsl). Population served: 400,000 Format: Top-40. ♦Michael Rea, gen mgr; Don Black, progmg dir & news dir.

Evanston

WKTA(AM)— 1953: 1330 khz; 5 kw-D, 17 w-N, DA-1. TL: N42 08 23 W87 53 09. Hrs open: 24 4320 Dundee Rd., Northbrook, 60062. Phone: (847) 498-3350. Fax: (847) 498-5743. E-mail: wkta@inc-us.com Web Site: www.pclradio.com. Licensee: Polnet Communications Ltd. (group owner; (acq 5-5-86; $1.66 million; FTR: 2-17-86). Population served: 10,500,000 Law Firm: Wiley, Rein and Fielding. Format: AOR/rock, Russian, Korean. News: 5 hrs wkly. Target aud: 18-54; Russian, Korean and German speaking audience. Spec prog: Ger 5 hrs wkly. ♦Walter K. Kotaba, pres; Sara Vargas, gen mgr; Scott Davidson, opns mgr & progmg dir.

***WNUR-FM—** May 8, 1950: 89.3 mhz; 7.2 kw. 100 ft TL: N42 03 12 W87 40 33. Stereo. Hrs opn: 24 1920 Campus Dr., 60208-2280. Phone: (847) 491-7101. Phone: (847) 491-2234. Fax: (847) 467-2058. E-mail: gm@wnur.org Web Site: www.wnur.org. Licensee: Northwestern University. Population served: 4,000,000 Format: Progsv, jazz, new mus. News: 3 hrs wkly. Target aud: 18-34; general. Spec prog: Folk 3 hrs, world mus 10 hrs, reggae 4 hrs wkly. ♦Henry Bienen, pres; Mike Corsa, gen mgr; Ashley Ayarza, prom dir & news dir; Alex Freedman, progmg dir & pub affrs dir.

WOJO(FM)— 1946: 105.1 mhz; 5.7 kw. Ant 1,394 ft TL: N41 53 56 W87 37 23. (CP: 5.7 kw, ant 1,394 ft). Stereo. Hrs opn: 24 625 N. Michigan Ave., Suite 300, Chicago, 60611-3110. Phone: (312) 981-1800. Fax: (312) 981-1806. Web Site: www.netmio.com/radio/wojo. Licensee: Tichenor License Corp. Group owner: Univision Radio (acq 9-22-2003; grpsl). Population served: 6,805,900 Natl. Rep: Katz Radio. Format: Sp contemp hit. News staff: one; News: 2 hrs wkly. Target aud: 18-35; regional/Mexican. ♦Jerry Ryan, gen mgr; Cesar Canales, progmg dir; Paul Easter, engrg mgr.

WONX(AM)— 1947: 1590 khz; 3.5 kw-D, 2.5 kw-N, DA-2. TL: N42 01 20 W87 42 43. Hrs open: 24 2100 Lee St., 60202. Phone: (847) 475-1590. Fax: (847) 475-1590. Licensee: Kovas Communications Inc. (acq 12-1-75). Population served: 80,113 Format: Ethnic, Sp. News: 8 hrs wkly. Target aud: General. Spec prog: Greek 2 hrs, Indian 8 hrs, Assyrian 15 hrs, Haitian 5 hrs, Lithuanian one hr, Korean 20 hrs wkly. ♦Connie Walburn, gen mgr; Bob Richards, opns mgr.

Fairbury

WYST(FM)— Aug 8, 2000: 107.7 mhz; 22.5 kw. Ant 351 ft TL: N40 37 45 W88 46 52. Stereo. Hrs opn: 24 108 Boeykens Pl., Normal, 61761. Phone: (309) 888-4496. Fax: (309) 452-9677. E-mail: star1077@aaabloomington.com Web Site: www.star1077.net. Licensee: Pilot Media LLC. Group owner: AAA Entertainment L.L.C. (acq 6-5-2007; grpsl). Population served: 125,000 Natl. Network: AP Radio. Wire Svc: AP Format: Adult contemp. News: 2 hrs wkly. Target aud: 25+; women. ♦Patti Donsbach, gen mgr; Kevin Trueblood, opns mgr & progmg dir.

Fairfield

WFIW(AM)— Aug 21, 1953: 1390 khz; 710 w-D, 58 w-N. TL: N38 22 46 W88 19 33. Hrs open: 24 Box 310, Hwy. 15 E., 62837. Phone: (618) 842-2159. Fax: (618) 847-5907. E-mail: wfiwwokz@fairfieldwireless.net Web Site: www.wfiwradio.com. Licensee: Wayne County Broadcasting Co. (group owner) Population served: 150,000 Natl. Network: ABC. Format: News/talk. News staff: one; News: 22 hrs wkly. Target aud: 45 plus; small town rural, business, farm, older adults. Spec prog: Farm 16 hrs wkly. ♦Thomas S. Land, chmn & farm dir; David H. Land, pres, gen mgr, gen sls mgr & progmg dir; Margaret H. Land, VP; Len Wells, news dir; Kirk Wallace, chief of engrg & disc jockey; Deron Caudle, news rptr; Stan David, sports cmtr & disc jockey; Jessica James, disc jockey; Tom Lavine, mus dir & disc jockey.

WFIW-FM— 1965: 104.9 mhz; 4.9 kw. 364 ft TL: N38 22 46 W88 19 33. Stereo. 24 Web Site: www.wfiwradio.com.150,000 Format: Adult contemp. News staff: one; News: 16 hrs wkly. Target aud: 25-49; small town rural, young, middle age, business, farm. Spec prog: Farm 12 hrs wkly. ♦David H. Land, opns VP & sls dir; Deron Caudle, disc jockey; Jessica James, disc jockey; Kirk Wallace, disc jockey; Tom Lavine, disc jockey.

WOKZ(FM)— September 1996: 105.9 mhz; 6 kw. 328 ft TL: N38 22 46 W88 19 33. Stereo. Hrs opn: 24 Box 310, Hwy. 15 E., 62837. Phone: (618) 842-2159. Fax: (618) 847-5907. E-mail: wfiwwokz@fairfieldwireless.net Web Site: www.wokz.com. Licensee: Wayne County Broadcasting Co., Inc. (group owner) Population served: 150,000 Format: Country. News staff: one; News: 20 hrs wkly. Target aud: 25-54; small town rural, business, farm. ♦Thomas S. Land, chmn; David H. Land, pres & gen mgr.

Farmer City

WWHP(FM)— Oct 1, 1983: 98.3 mhz; 3 kw. 300 ft TL: N40 16 54 W88 32 00. Stereo. Hrs opn: 24 407 N. Main, 61842. Phone: (309) 928-9876. Fax: (309) 928-3708. E-mail: wwhp@farmwagon.com Web Site: www.wwhp.com. Licensee: WMS1 Inc. Population served: 300,000 Format: Americana. Target aud: 18-65; reach city, suburbs & rural listeners in east central Illinois. Spec prog: Gospel 3 hrs, sports 5 hrs wkly. ♦Larry Williams, gen mgr; Lori Allen, gen sls mgr.

Farmington

WZPN(FM)— 1997: 96.5 mhz; 4.3 kw. Ant 377 ft TL: N40 40 10 W89 53 31. Hrs open: 4234 N. Brandywine, Suite D, Peoria, 61614. Phone: (309) 282-7625. Phone: (309) 686-0101. Fax: (309) 686-0111. Licensee: IM IL Licenses LLC. Group owner: AAA Entertainment L.L.C. (acq 3-30-2007; $600,000). Natl. Network: ESPN Radio. Format: Sports talk. ♦Michael Rea, gen mgr.

Fisher

***WGNN(FM)—** Apr 7, 1996: 102.5 mhz; 6 kw. 328 ft TL: N40 20 21 W88 24 18. Stereo. Hrs opn: 24 Box 12345, Champaign, 61826. Secondary address: 2421 N. 1450 E. Rd., White Heath 61884. Phone: (217) 897-6333. E-mail: staff@greatnewsradio.org Web Site: www.greatnewsradio.org. Licensee: Good News Radio Inc. (acq 4-7-96; $225,000). Population served: 300,000 Natl. Network: Moody, USA, Salem Radio Network. Format: Educ, relg, news/talk. Target aud: 35 plus; general. ♦David B. Herriott, chmn; Mark Burns, pres & gen mgr; Carrie Burns, opns dir.

Flora

WNOI(FM)— May 21, 1971: 103.9 mhz; 3.3 kw. 300 ft TL: N38 40 42 W88 29 14. Stereo. Hrs opn: 24 Box 368, 1001 N. Olive Rd., 62839. Phone: (618) 662-8331. Fax: (618) 662-2407. E-mail: info@wnoi.com Web Site: www.wnoi.com. Licensee: H&R Communications Inc. (acq 10-16-88). Population served: 30,000 Natl. Network: Jones Radio Networks. Format: Adult contemp. News staff: one; News: 12 hrs wkly. Target aud: General. ♦Steven S. Lovellette, pres; Randy Poole, gen mgr; Patrick Garret, opns dir; Brenda Miller, gen sls mgr; Patrick Garrett, progmg dir; Kirk Wallace, chief of engrg.

Flossmoor

***WHFH(FM)—** January 1965: 88.5 mhz; 1.5 kw. 92 ft TL: N41 32 43 W87 41 30. Stereo. Hrs opn: 14 999 Kedzie Ave., 60422. Phone: (708) 798-9434. Fax: (708) 799-3142. E-mail: sean@whfh.org Web Site: www.whfh.org. Licensee: Community High School District No. 233. Population served: 7,845 Format: Rock. News: 4 hrs wkly. Target aud: Teens-Adult. Spec prog: News/talk one hr, sports talk one hr, live sports 4 hrs wkly. ♦John Henry, gen mgr & stn mgr; Robert Comstock, gen mgr.

Freeport

WFPS(FM)— Nov 1, 1970: 92.1 mhz; 3.6 kw. Ant 423 ft TL: N42 19 41 W89 43 30. Stereo. Hrs opn: 24 Box 807, 834 N. Tower Rd., 61032. Phone: (815) 235-7191. Fax: (815) 235-4318. Licensee: Green County Broadcasting. Group owner: RadioWorks Inc. (acq 3-29-2006; $1.48 million with co-located AM). Population served: 295,000 Format: New country. News staff: 4; News: 15 hrs wkly. Target aud: 25-49. ♦Kent McConnell, gen sls mgr; Wyatt Herrmann, progmg dir; Todd Hausser, chief of engrg; Becky Koester, traf mgr.

WFRL(AM)—Co-owned with WFPS(FM). Oct 28, 1947: 1570 khz; 5 kw-D, 500 w-N, DA-2. TL: N42 18 45 W89 35 38.24 50,000 Natl. Network: ABC. Format: Adult standards. News staff: 4; News: 24 hrs wkly. Target aud: 35 plus.

***WNIE(FM)—** 1999: 89.1 mhz; 6 kw. Ant 361 ft TL: N42 18 45 W89 35 38. Stereo. Hrs opn: 24
Rebroadcasts WNIJ(FM) De Kalb & WNIU(FM) Rockford 50%.
NIU Broadcast Ctr., 801 N. First St., De Kalb, 60115. Phone: (815) 753-9000. Fax: (815) 753-9938. E-mail: npr@niu.edu Web Site: www.northernpublicradio.org. Licensee: Northern Illinois University. Population served: 50,000 Natl. Network: PRI, NPR. Law Firm: Arter & Hadden. Format: News, class. News staff: 2. Target aud: General. ♦Tim Emmons, gen mgr; Jan Kilgard, dev VP; Bill Drake, progmg dir.

WXXQ(FM)— Apr 11, 1965: 98.5 mhz; 50 kw. 450 ft TL: N42 18 45 W89 35 38. Stereo. Hrs opn: 24 3901 Brendenwood Rd., Rockford, 61107-2246. Phone: (815) 399-2233. Fax: (815) 484-2432. Web Site: www.wxxq.com. Licensee: Cumulus Licensing Corp. Group owner: Cumulus Media Inc. (acq 3-15-00; grpsl). Population served: 500,000 Format: Contemp country. News staff: one. Target aud: 25-54. ♦Greg Sher, gen mgr; Dawn Plock, prom dir; Steve Summers, progmg dir.

Galatia

WISH-FM— 2001: 98.9 mhz; 4.1 kw. Ant 400 ft TL: N37 55 52 W88 40 50. Hrs open: R.R. 1 Box 46 A, Mc Leansboro, 62859. Phone: (618) 643-2311. Fax: (618) 643-3299. Licensee: W. Russell Withers Jr. Group owner: Withers Broadcasting Co. Format: CHR, adult contemp. ♦Dana Withers, gen mgr.

Galena

WDBQ-FM— February 1989: 107.5 mhz; 6 kw. 328 ft TL: N42 24 02 W90 23 55. Stereo. Hrs opn: 24 5490 Saratoga Rd., Dubuque, IA, 52002. Phone: (563) 557-1040. Fax: (563) 583-4535. Licensee: Cumulus Licensing Corp. Group owner: Cumulus Media Inc. (acq 12-17-98; grpsl). Format: Oldies. News staff: one. Target aud: 25-54. ♦Scott Lindahl, mktg mgr.

Galesburg

WAAG(FM)—Listing follows WGIL(AM).

WAIK(AM)— 1957: 1590 khz; 5 kw-D, 50 w-N, DA-3. TL: N40 57 43 W90 18 30. Hrs open: 24 Box 885, Monmouth, 61462-0885. Phone: (309) 342-3161. Fax: (309) 342-0199. E-mail: wmoi@maplecity.com Licensee: WPW Broadcasting Inc. (group owner; acq 7-9-98; $439,500). Population served: 36,290 Natl. Network: ABC. Law Firm: Fisher, Wayland, Cooper, Leader & Zaragoza L.L.P. Format: MOR, big band, nostalgia. News staff: 3; News: 24 hrs wkly. Target aud: 25 plus. Spec prog: Talk 10 hrs, loc sports 10 hrs, relg 6 hrs wkly. ♦Don Davis, CEO & news dir; David Klockenga, gen mgr; Heidi Aycock, opns dir; Greg Ford, progmg dir; Kris Kinney, traf mgr.

WGIL(AM)— June 12, 1938: 1400 khz; 740 w-U. TL: N40 56 34 W90 20 39. Hrs open: 24 Box 1227, 61402-1227. Secondary address: 154 E. Simmons 61401. Phone: (309) 342-5131. Fax: (309) 342-0840. E-mail: wgil@wgil.com Web Site: www.wgil.com. Licensee: Galesburg Broadcasting Co. (group owner). Population served: 55,000 Natl. Network: Westwood One. Rgnl. Network: Ill. Radio Net. Natl. Rep: Interep. Law Firm: Cohn & Marks. Wire Svc: AP Format: News/talk, sports. News staff: 4; News: 20 hrs wkly. Target aud: 25-54. Spec prog: Farm 10 hrs, relg 4 hrs, sports 15 hrs wkly. ♦John T. Pritchard, pres; Roger Lundeen, gen mgr.

WAAG(FM)—Co-owned with WGIL(AM). Dec 15, 1966: 94.9 mhz; 50 kw. Ant 492 ft TL: N40 56 34 W90 20 39. Stereo. 24 E-mail: fm95@fm95online.com Web Site: www.fm95online.com.77,000 Format:

Country. News staff: 4; News: 2 hrs wkly. Target aud: 25-54. ♦Brian Prescott, mus dir; Jim Lee, farm dir; Mike Perry, sports cmtr.

WLSR(FM)— Jan 17, 1979: 92.7 mhz; 4.2 kw. Ant 390 ft TL: N40 56 34 W90 20 39. Stereo. Hrs opn: 24 Box 1227, 154 E. Simmons St., 61401. Phone: (309) 342-5131. Fax: (309) 342-0840. E-mail: thelaser@thelaseronline.com Web Site: www.thelaseronline.com. Licensee: Galesburg Broadcasting Co. (group owner; (acq 7-3-97). Population served: 55,000 Law Firm: Cohn & Marks, LLP. Wire Svc: AP Format: Rock/AOR. News staff: 4; News: one hr wkly. Target aud: 18-34. Spec prog: 24 Religious; 2 hrs. wkly. ♦John T. Pritchard, pres; Roger Lundeen, gen mgr; Chris Postin, sls dir; Brian Prescott, prom dir & opns; Chris Lagrow, progmg dir.

***WVKC(FM)**— Apr 12, 1961: 90.7 mhz; 1 kw. 98 ft TL: N40 56 46 W90 22 11. Hrs open: Knox College, Box K 254, 2 E. South St., 61401-4999. Phone: (309) 341-7266 (staff). Phone: (309) 341-7000 (switchboard). Fax: (309) 341-7090. E-mail: wvkc@knox.edu Web Site: www.knox.edu/wvkc.xml. Licensee: Knox College. Population served: 65,000 Format: Var/div. Spec prog: Black 6 hrs, jazz 15 hrs, class 18 hrs wkly. ♦Roger Moore, pres; Mark Iellski, gen mgr.

Galva

WGEN(AM)—Listing follows WJRE(FM).

WJRE(FM)— Oct 15, 1995: 102.5 mhz; 6 kw. Ant 293 ft TL: N41 13 37 W89 56 08. Stereo. Hrs opn: 24 Box 266, Kewanee, 61443-0266. Secondary address: 133 .E. Division St., Kewanee 61443. Phone: (309) 853-4471. Fax: (309) 853-4474. E-mail: wkei@verizon.net Web Site: www.1025wjre.com. Licensee: Virden Broadcasting Corp. Group owner: Miller Media Group (acq 3-31-2003; $475,000 with WGEN(AM) Geneseo). Population served: 125,000 Natl. Rep: Commercial Media Sales. Law Firm: Womble, Carlyle, Sandridge & Rice. Format: Country. News staff: one. Spec prog: Southern gospel one hr wkly. ♦Randal Miller, pres; Kris Wexell, progmg dir.

WGEN(AM)—Co-owned with WJRE(FM). 1964: 1500 khz; 250 w-D, 1 w-N. TL: N41 26 23 W90 09 18.24 Phone: (309) 944-1500. Fax: (309) 853-4474. Web Site: www.randyradio.com.58,000 Format: News/talk. News staff: one; News: 13 hrs wkly. Target aud: 25 plus; community oriented adults.

Geneseo

***WAXR(FM)**— 2001: 88.1 mhz; 3 kw vert. Ant 321 ft TL: N41 28 47 W90 16 08. Hrs open: 3316 Avenue of the Cities, Moline, 61265. Phone: (309) 736-9297. Fax: (309) 277-3122. E-mail: info@waxr.org Web Site: www.waxr.org. Licensee: American Family Association. Group owner: American Family Radio Format: Inspirational Christian. ♦Ron Cook, gen mgr.

WGEN(AM)—Licensed to Geneseo. See Galva

Geneva

WSPY(AM)— Nov 11, 1961: 1480 khz; 1 kw-D, 500 w-N, DA-2. TL: N41 54 25 W88 17 43. Hrs open: 24 1 Broadcast Center, Plano, 60545. Phone: (630) 552-1000. Fax: (630) 552-9300. E-mail: wspy@nelsonmultimedia.net Licensee: Nelson Multi Media Inc. (acq 9-30-01). Population served: 1,400,000 Law Firm: Miller and Miller. Format: Adult contemp. News staff: one; News: 2 hrs wkly. Target aud: 35-65; baby boomers. ♦Larry Nelson, pres & gen mgr; Beth Perry, gen sls mgr & progmg dir; Jenny Beckman, mus dir & traf mgr; Lane Lindstrom, chief of engrg.

Genoa

WYRB(FM)— 2001: 106.3 mhz; 6 kw. Ant 213 ft TL: N42 04 28 W88 49 24. Hrs open:
Rebroadcasts WSRB(FM) Lansing 100%.
6336 Calumet Ave., Hammond, IN, 46324. Phone: (773) 734-4455. Fax: (303) 933-0323. E-mail: wycainfo@crawfordbroadcsting.com Web Site: www.soul1063radio.com. Licensee: Dontron Inc. Group owner: Crawford Broadcasting Co. (acq 9-28-01; $1.5 million). Format: Adult contemp. Target aud: 25-54; adult urban. ♦Donald Crawford, CEO; Taft Harris, gen mgr.

Gibson City

WGCY(FM)— Nov 28, 1983: 106.3 mhz; 6 kw. 292 ft TL: N40 34 01 W88 20 41. (CP: Ant 321 ft.). Stereo. Hrs opn: 6 AM-midnight Box 192, 607 S. Sangamon Ave., 60936. Phone: (217) 784-8661. Fax: (217) 784-8677. Licensee: F & G Broadcasting Inc. (acq 12-30-86; $225,000; FTR: 11-24-86). Natl. Network: USA. Format: Easy lstng. News staff: one. Target aud: 35 plus. ♦Fred McCullough, pres; Gary McCullough, gen mgr.

Gilman

WFAV(FM)—Not on air, target date: unknown: Stn currently dark. 103.7 mhz; 6 kw. Ant 328 ft TL: N40 43 04 W87 51 36. Stereo. Hrs opn: 292 N. Convent, Bourbonnais, 60914. Phone: (815) 933-9287. Fax: (815) 933-8696. Licensee: Milner Broadcasting Co. ♦Tim Milner, pres & gen mgr.

Girard

WCVS-FM—See Springfield

Glasford

WXMP(FM)— 2000: 101.1 mhz; 3.3 kw. Ant 449 ft TL: N40 39 00 W89 46 46. Stereo. Hrs opn: 24 120 Eaton St., Peoria, 61603. Phone: (309) 676-5000. Fax: (309) 676-2600. E-mail: mix1011@hotmail.com Licensee: IM IL Licenses LLC. Group owner: Regent Communications Inc. (acq 9-19-2006; grpsl). Population served: 350,000 Natl. Rep: Katz Radio. Format: Hot adult contemp. ♦J.R. Greeley, gen mgr; Ric Morgan, opns mgr.

Glen Ellyn

***WDCB(FM)**— July 5, 1977: 90.9 mhz; 5 kw. 300 ft TL: N41 50 36 W88 05 00. Stereo. Hrs opn: 24 College of DuPage, 425 Fawell Blvd., 60137. Phone: (630) 942-4200. Phone: (630) 942-3708. Fax: (630) 942-2788. E-mail: wdcbmktg@cdnet.cod.edu Web Site: www.wdcb.org. Licensee: College of DuPage. Natl. Network: PRI. Law Firm: Cohn & Marks. Format: Jazz, news, blues. News staff: 3; News: 13 hrs wkly. Target aud: General. Spec prog: College classes 12 hrs, folk 12 hrs, gospel 2 hrs, var music 7 hrs wkly. ♦Scott Wager, stn mgr; Jim Barker, sls dir & gen sls mgr; Ken Scott, mktg dir; Mary Pat LaRue, progmg dir; Paul Abella, mus dir; Brian O'Keefe, news dir.

Glendale Heights

WJKL(FM)— September 1960: 94.3 mhz; 6 kw. Ant 328 ft TL: N41 59 54 W88 14 33. Stereo. Hrs opn: 24
Rebroadcasts KLVR(FM) Santa Rosa, CA 100%.
2351 Sunset Blvd., Suite 170-218, Rocklin, CA, 95765. Phone: (916) 251-1600. Fax: (916) 251-1650. Web Site: www.klove.com. Licensee: Educational Media Foundation. (acq 4-10-2007; $17 million). Population served: 350,000 Natl. Network: K-Love. Format: Contemp Christian. ♦Mike Novak, sr VP.

Glenview

***WGBK(FM)**— Jan 13, 1979: 88.5 mhz; 185 w. 100 ft TL: N42 04 30 W87 49 23. Stereo. Hrs opn: 6:30 AM-8:00 AM; 3 PM-10 PM (M-F) c/o Glenbrook S. High School, 4000 W. Lake Ave., 60025. Phone: (847) 486-4487. Phone: (847) 486-4573. Fax: (847) 486-4439. E-mail: wgbk@glenbrook.k12.il.us Licensee: Glenbrook High School District 225. (acq 1996; $110,000). Population served: 750,000 Format: Alternative, educ, sports. News: 1 hr wkly. Target aud: General; teens & adults. Spec prog: Sports talk 5 hrs, live sports 4 hrs. ♦Dr. Daniel Oswald, gen mgr.

Godfrey

***WLCA(FM)**— 1974: 89.9 mhz; 1.5 kw. Ant 394 ft TL: N38 56 57 W90 11 47. Hrs open: 18 5800 Godfrey Rd., 62035. Phone: (618) 466-8936. Fax: (618) 466-7458. Licensee: Lewis and Clark Community College. Population served: 100,000 Natl. Network: USA. Format: Progsv, AOR, alt rock. ♦Mike Lemons, gen mgr.

Golconda

WKYX-FM— Nov 22, 1990: 94.3 mhz; 3.1 kw. Ant 449 ft TL: N37 14 04 W88 29 48. Stereo. Hrs opn: 24
Simulcast with WKYX(AM) Paducah, KY 100%.
Box 2397, Paducah, KY, 42002-2397. Phone: (270) 554-8255. Fax: (270) 554-4613. E-mail: info@wkyx.com Web Site: www.wkyx.com. Licensee: Bristol Broadcasting Co. Inc. (group owner; (acq 2-20-2004; grpsl). Format: News/talk. Target aud: 25-54. Spec prog: Relg one hr wkly. ♦Pete Ninninger, pres; Gary Morse, gen mgr.

Granite City

WARH(FM)—Licensed to Granite City. See Saint Louis MO

WGNU(AM)—Licensed to Granite City. See Saint Louis MO

Greenville

WGEL(FM)— Dec 20, 1984: 101.7 mhz; 3 kw. 300 ft TL: N38 48 11 W89 20 56. Stereo. Hrs opn: 24 Box 277, 309 W. Main, 62246. Phone: (618) 664-3300. Fax: (618) 664-3318. Web Site: www.wgel.com. Licensee: Bond Broadcasting. (acq 6-1-85; $170,000; FTR: 6-10-85). Natl. Network: USA. Format: Country. Target aud: 25-64. Spec prog: Farm 19 hrs wkly. ♦John Kennedy, pres & gen mgr; John Goldsmith, progmg dir; Aundi Mergner, news dir; Joe Doll, farm dir.

Harrisburg

WEBQ(AM)— September 1923: 1240 khz; 1 kw-U. TL: N37 43 03 W88 32 37. Hrs open: 701 S. Commercial St., 62946. Phone: (618) 253-7282. Fax: (618) 252-2366. E-mail: webq@yourclearwave.com Licensee: W. Russell Withers Jr. Group owner: Withers Broadcasting Co. (acq 7-28-2004; $450,000 with WEBQ-FM Eldorado). Population served: 25,000 Format: Country. News staff: one; News: 6 hrs wkly. Target aud: Older area residents. Spec prog: Farm 6 hrs wkly. ♦Cathy Horton, gen mgr & progmg dir; Bob Romonosky, chief of engrg; Shelly Reeder, traf mgr; Sonny Dotson, sls.

WOOZ-FM— September 1947: 99.9 mhz; 32 kw. 650 ft TL: N37 36 45 W88 52 03. Stereo. Hrs opn: 1431 Countryaire Dr., Carterville, 62918. Phone: (618) 985-4843. Phone: (800) 455-3243. Fax: (618) 985-6529. E-mail: mail@z100fm.com Web Site: www.z100fm.com. Licensee: MRR License LLC. Group owner: MAX Media L.L.C. (acq 3-29-2004; grpsl). Law Firm: Fletcher, Heald & Hildreth. Format: Country. Target aud: 18-49. ♦Mike Smith, gen mgr; Steve Falat, gen sls mgr; Tracy McSherry-McKown, progmg dir.

Harvard

WMCW(AM)— 1955: 1600 khz; 500 w-D, 19 w-N. TL: N42 26 07 W88 36 39. Hrs open: 24 67 N. Ayer, 60033. Phone: (815) 943-3100. Fax: (815) 943-5120. E-mail: wmcw1600@yahoo.com Licensee: Kovas Communications of Indiana Inc. (group owner; (acq 1-23-2004; $650,000). Population served: 440,000 Natl. Network: CNN Radio. Format: Adult standards. News: 12 hrs wkly. Target aud: 35+. Spec prog: . ♦Constance Kovas, pres; Bill Marquis, gen mgr & opns mgr.

Harvey

WBGX(AM)— 1955: 1570 khz; 1 kw-D, 500 w-N, DA-2. TL: N41 36 14 W87 40 45. Hrs open: 24 Great Lakes Radio-Chicago, 5956 S. Michigan Ave., Chicago, 60637. Phone: (773) 752-1570. Fax: (773) 752-2242. E-mail: gospel1570@aol.com Licensee: Great Lakes Radio-Chicago LLC (acq 10-7-03; $1.78 million). Population served: 3,900,000 Law Firm: Hogan & Hartsen. Format: Gospel. Target aud: 18-64; African Americans. ♦Tim Gallagher, pres.

Havana

WDUK(FM)— Feb 27, 1970: 99.3 mhz; 3 kw. 300 ft TL: N40 18 43 W90 03 19. Hrs open: 901 N. Promenade, 62644. Phone: (309) 543-3331. Licensee: Illinois Valley Radio. (acq 3-5-73). Population served: 16,000 Rgnl. Network: Brownfield. Rgnl rep: Brownfield. Format: C&W, div. Target aud: General. Spec prog: Farm 8 hrs wkly. ♦Edwin Stimpson, pres.

Henry

WRVY-FM— July 30, 1990: 100.5 mhz; 3 kw. Ant 328 ft TL: N41 04 32 W89 21 10. Stereo. Hrs opn: 24 Box 69, Princeton, 61356. Phone: (815) 875-8014. Phone: (309) 364-4411. Fax: (815) 872-0308. Web Site: www.wrvy.com. Licensee: WZOE Inc. (group owner; (acq 5-8-98). Natl. Network: CBS. Law Firm: Shaw Pittman. Format: Contemp country. News staff: 3; News: 3 hrs wkly. Target aud: 25-45. Spec prog: Farm 4 hrs wkly. ♦Steve Samet, gen mgr; Mary Harmon, opns dir.

Herrin

WDDD(AM)—See Johnston City

WJPF(AM)— Aug 28, 1940: 1340 khz; 1 kw-U. TL: N37 50 03 W89 01 37. (CP: 770 w-U). Hrs opn: 24 1431 Countryaire Dr., Carterville, 62918. Phone: (618) 985-4843. Phone: (800) 455-3243. Fax: (618) 985-6529. E-mail: mail@wjpf.com Web Site: www.wjpf.com. Licensee: MRR License LLC. Group owner: MAX Media L.L.C. (acq 3-29-2004; grpsl). Population served: 250,000 Natl. Network: Westwood One. Rgnl. Network: Ill. Radio Net. Format: News/talk, Sports. News staff: 3; News: 10 hrs wkly. Target aud: 35 plus; mature, middle-income wage earners. ♦Steve Falat, sls dir & gen sls mgr; Tom Miller, progmg dir; Rick Gregg, news dir; Felicia Dick, traf mgr & disc jockey.

WVZA(FM)— March 1994: 92.7 mhz; 3.3 kw. 433 ft TL: N37 45 15 W88 56 05. Hrs open: 24 Box 127, 1822 N. Court St., Marion, 62959-0127. Phone: (618) 997-8123. Fax: (618) 993-2319. Web Site: www.kissfm927.com. Licensee: CC Licenses LLC. Group owner: Clear Channel Communications Inc. (acq 1-18-2001; grpsl). Population served: 250,000 Natl. Rep: Katz Radio. Format: Contemp hit. Target aud: 18-34. ♦Jerry Crouse, gen mgr; Paxton Guy, opns dir & progmg dir; Steve Browning, gen sls mgr & natl sls mgr; April Ruebke, news dir; Tim Deterding, chief of engrg; Renee Longwell, traf mgr.

Heyworth

WBBE(FM)— June 6, 2005: 97.9 mhz; 5.4 kw. Ant 344 ft TL: N40 27 08 W88 57 48. Hrs open: 520 N. Center St., Bloomington, 61701-2902. Phone: (309) 834-1100. Fax: (309) 834-4390. Web Site: www.bob979.com. Licensee: Connoisseur Media LLC. Natl. Rep: McGavren Guild. Law Firm: Shaw Pittman LLP. Format: Adult hits. Target aud: 18-54; Adults. ♦Larry Weiss, gen mgr; Grant Thompson, gen sls mgr; Ron West, progmg dir; Mark Hill, chief of engrg.

Highland

WIJR(AM)— Dec 2, 1963: 880 khz; 1.7 kw-D, 160 w-N, DA-1. TL: N38 45 23 W89 39 18. Hrs open: 24 Box 473, Saint Louis, MO, 63166. Secondary address: 13063 Winu Dr. 62249. Phone: (314) 351-7390. E-mail: hhnjim@hereshelpnet.org Licensee: Birach Broadcasting Corp. (acq 8-15-2006; $1 million). Population served: 135,000 Rgnl. Network: Brownfield. Format: Christian. News staff: one; News: 18 hrs wkly. Target aud: General; mature adults. Spec prog: Farm 3 hrs, Ger one hr wkly. ♦Sima Birach, pres; Larry Rice, gen mgr; Bernard Turner, opns dir.

WXOZ(AM)— 2000: 1510 khz; 1 kw-D, DA. TL: N38 44 56 W89 34 10. Hrs open: c/o Dennis J. Watkins, 100 W. Main St., Belleville, 62220. Phone: (618) 394-9969. Licensee: Entertainment Media Trust, Dennis J. Watkins, Trustee (acq 5-11-2006; $450,000). Format: Hot talk. ♦Greg Benfield, gen mgr.

Highland Park

WEEF(AM)— Aug 15, 1963: 1430 khz; 1 kw-D, 29 w-N, DA. TL: N42 10 53 W87 57 05. Hrs open: 4320 Dundee Rd., Northbrook, 60062. Phone: (847) 498-3350. Fax: (847) 498-5743. Web Site: www.plcradio.com/1430_weef. Licensee: Polnet Communications Ltd. (group owner; (acq 5-20-2003; $1 million). Law Firm: Dow, Lohnes & Albertson. Format: Ethnic. Target aud: General; ethnic. ♦Sara Vargas, gen mgr; Fernando Jaramillo, progmg dir.

WVIV-FM— Aug 15, 1963: 103.1 mhz; 3 kw. Ant 241 ft TL: N42 09 24 W87 48 20. Stereo. Hrs opn: 24
Rebroadcasts WYXX(FM) Morris 100%.
625 N. Michigan Ave., Suite 300, Chicago, 60611. Phone: (312) 981-1800. Fax: (312) 981-1806. E-mail: ayudachi@netmio.com Web Site: wviv.netmio.com. Licensee: HBC License Corp. Group owner: Univision Radio (acq 9-22-2003; grpsl). Format: Sp. Target aud: 25-54. Spec prog: Asian, Pol, Ger, Greek. ♦Jerry Ryan, gen mgr; David Miranda, prom dir & prom dir; Cesar Canales, progmg dir; Armando Reyes, mus dir; Robert Lopez, chief of engrg; Lorna Vazquez, traf mgr.

Hillsboro

WXAJ(FM)— Sept 1, 2000: 99.7 mhz; 50 kw. 492 ft TL: N39 20 14 W89 32 04. Hrs open: 24 3055 S. 4th St., Springfield, 62703. Phone: (217) 528-3033. Fax: (217) 528-5348. Web Site: www.997kissfm.com. Licensee: Neuhoff Family L.P. Group owner: Clear Channel Communications Inc. (acq 8-1-2007; grpsl). Population served: 350,000 Natl. Rep: Clear Channel. Law Firm: Wiley, Rein Fielding LLP. Format: Contemporary Hit/Top 40. Target aud: Adults; 18-49. ♦Kevin O'Dea, gen mgr; Danielle Outlaw, gen sls mgr; Michelle Mitchell, prom dir; Jeff Hofmann, news dir.

Hinsdale

***WHSD(FM)**— Dec 6, 1970: 88.5 mhz; 200 w. 131 ft TL: N41 47 25 W87 55 11. Stereo. Hrs opn: 3 PM-10 PM Hinsdale Central High School, 55th & Grant St., 60521. Phone: (630) 570-8463. Fax: (630) 887-1362. Licensee: Hinsdale Twsp. High School District 86. Population served: 16,631 Format: Var. Target aud: General.

Hoopeston

WHPO(FM)— May 29, 1979: 100.9 mhz; 3 kw. 280 ft TL: N40 28 36 W87 41 36. Stereo. Hrs opn: 24 912 S. Dixie Hwy., 60942. Phone: (217) 283-7744. Fax: (217) 283-6090. E-mail: whporadio@whporadio.com Web Site: www.whporadio.com. Licensee: Market Street Broadcasting LLC (acq 12-27-00; $900,000). Population served: 110,000 Format: Country. News staff: one; News: 6 hrs wkly. Target aud: 25 plus; rural middle class. Spec prog: Southern gospel 7 hrs, big band 2 hrs wkly. ♦Blanche Voss, gen mgr & opns mgr; Becky Voss, progmg dir.

Jacksonville

WJIL(AM)— November 1961: 1550 khz; 1 kw-D, 10 w-N, DA-2. TL: N39 43 20 W90 11 43. Hrs open: 24 Box 1055, Rt. 4, E. Morton Rd., 62651. Phone: (217) 245-5119. Fax: (217) 245-1596. Licensee: Morgan County Broadcasting Co. Inc. (acq 5-1-93; with co-located FM). Population served: 125,000 Natl. Network: Westwood One. Natl. Rep: Roslin. Law Firm: Fisher, Wayland, Cooper, Leader & Zaragoza L.L.P. Format: News/talk, btfl music. News staff: one; News: 9 hrs wkly. Target aud: 35-64. Spec prog: Farm 8 hrs wkly. ♦Sarah Hautala, gen mgr; Diana McCutcheon, sls dir; Matt Lakis, progmg dir; Julie Ann Cambridge, mus dir; Mike Kaiser, news dir; Glen Hopkiins, chief of engrg.

WJVO(FM)—Co-owned with WJIL(AM). Sept 1, 1986: 105.5 mhz; 6 kw. 340 ft TL: N39 43 20 W90 11 43. Stereo. 24 1251 East Morton, South Jacksonville Natl. Network: Westwood One, ABC. Format: Country. News staff: one; News: 4 hrs wkly. Target aud: 25-54.

WLDS(AM)— Dec 9, 1941: 1180 khz; 1 kw-D. TL: N39 44 06 W90 11 50. Hrs open: Sunrise-sunset Box 1180, 2161 Old State Rd., 62651. Phone: (217) 245-7171. Fax: (217) 245-6711. E-mail: wlds@wlds.com Web Site: www.wlds.com. Licensee: Jerdon Broadcasting. (acq 8-1-89; $650,000; FTR: 6-5-89). Population served: 100,000 Natl. Network: CBS. Rgnl. Network: Ill. Radio Net. Natl. Rep: Katz Radio. Law Firm: Kaye, Scholer, Fierman, Hays & Handler. Wire Svc: AP Format: Adult contemp, news/talk. News staff: 3; News: 23 hrs wkly. Target aud: 35 plus; business & professional people, farmers & housewives. Spec prog: Farm 20 hrs wkly. ♦Jerry Symons, gen mgr; Don Hamiltom, gen sls mgr; Bob Thomas, progmg dir; Gary Ballard, mus dir; Gary Scott, news dir; John Coe, engr.

WYMG(FM)—Licensed to Jacksonville. See Springfield

Jerseyville

WHHL(FM)— Oct 10, 1967: 104.1 mhz; 50 kw. Ant 500 ft TL: N38 51 36 W90 18 38. Stereo. Hrs opn: 24 800 St. Louis Union Station, The Power House, St. Louis, MO, 63013. Phone: (314) 621-0400. Fax: (314) 621-3000. E-mail: feedback@1041themall.com Web Site: www.red1041.com. Licensee: Radio One Licenses LLC. Group owner: Emmis Communications Corp. (acq 12-19-2005; $20 million). Population served: 2,500,000 Format: New American Standards. News staff: one; News: 2 hrs wkly. Target aud: 25-54; families with children, singles. Spec prog: Heartland issues one hr, today's issues one hr, pub agenda one hr wkly. ♦John Beck, VP; Lisa Sesti, stn mgr & gen sls mgr.

WJBM(AM)— Oct 11, 1959: 1480 khz; 500 w-D, 32 w-N, DA-2. TL: N39 06 46 W90 18 43. Hrs open: 1010 Shipman Rd., 62052. Phone: (618) 498-8265. Fax: (618) 498-9830. E-mail: wjbm@wjbmradio.com Web Site: www.wjbmradio.com. Licensee: DJ Two Rivers Radio Inc. (acq 1-9-2004; $320,000 with WBBA-FM Pittsfield). Population served: 53,000 Rgnl. Network: Brownfield. Format: Oldies. Target aud: 25 plus; general market through retirement. Spec prog: Farm 18 hrs, sports 13 hrs, relg 3 hrs wkly.

Johnston City

WDDD(AM)— July 1, 1979: 810 khz; 250 w-U, DA-N. TL: N37 51 14 W88 52 12. (CP: 300 w-U, DA-N). Hrs open: 24 Box 127, Marion, 62959. Secondary address: 1822 N. Court, Marion 62959. Phone: (618) 997-8123. Fax: (618) 993-2319. Web Site: www.foxsports810.com. Licensee: CC Licenses LLC. Group owner: Clear Channel Communications Inc. (acq 12-19-2000; grpsl). Format: talk, sports. News: 2 hrs wkly. Target aud: 18-49; Adults. ♦Jerry Crouse, gen mgr; Steve Browning, gen sls mgr; Paxton Guy, progmg dir; April Ruebke, news dir; Tim Deterding, chief of engrg.

Joliet

WCCQ(FM)—Crest Hill, Jan 28, 1976: 98.3 mhz; 3 kw. 300 ft TL: N41 27 55 W88 07 33. Hrs open: 24 2410-B Canton Farm Rd., Crest Hill, 60435. Phone: (815) 556-0100. Fax: (815) 577-9231. Web Site: www.wccq.com. Licensee: Three Eagles of Joliet Inc. Group owner: Three Eagles Communications (acq 1-13-97; grpsl). Population served: 2,600,000 Natl. Network: ABC. Law Firm: Dow, Lohnes & Albertson. Wire Svc: UPI Format: Country. Target aud: 25-54; general. ♦Brian Foster, VP & gen mgr; Ryan Snow, opns mgr; Todd Elbrink, gen sls mgr; Dan Waddick, prom dir; Roy Gregory, progmg dir.

***WCSF(FM)**— Sept 5, 1988: 88.7 mhz; 100 w. 108 ft TL: N41 31 58 W88 05 54. Stereo. Hrs opn: 7 AM-2 AM (M-F) 500 N. Wilcox St., 60435. Phone: (815) 740-3425. Phone: (815) 740-3214. Fax: (815) 740-3697. E-mail: webmaster@st.francis.edu Web Site: www.stfrancis.edu. Licensee: University of St. Francis. Format: AOR. Target aud: 18-45; males. Spec prog: Black 2 hrs, jazz 2 hrs, talk 4 hrs, requests 4 hrs, classic rock 4 hrs wkly. ♦Don Burke, pres; Rick Lawrence, gen mgr.

***WJCH(FM)**— Apr 25, 1986: 91.9 mhz; 50 kw. 460 ft TL: N41 24 55 W88 16 19. Stereo. Hrs opn: 24 13 Fairlane Dr., 60435. Phone: (815) 725-1331. Licensee: Family Stations Inc. (group owner) Format: Relg. Spec prog: Class 2 hrs wkly. ♦Harold Camping, pres & gen mgr; Virginia Beehn, opns mgr.

WJOL(AM)— 1924: 1340 khz; 1 kw-U. TL: N41 32 10 W88 03 15. Hrs open: 24 2410-B Caton Farm Rd., Crest Hill, 60435. Phone: (815) 556-0100. Fax: (815) 577-9231. Web Site: www.wjol.com. Licensee: NM Licensing LLC. Group owner: NextMedia Group L.L.C. (acq 11-26-01; grpsl). Population served: 150,000 Natl. Rep: McGavren Guild. Rgnl rep: Ill. Radio Net. Format: Talk. News staff: 2; News: 40 hrs wkly. Target aud: 35 plus. Spec prog: Farm 3 hrs, gospel one hr, Pol one hr wkly. ♦Steven Dinetz, CEO; Skip Wellar, pres; Brian Foster, gen mgr.

WSSR(FM)—Co-owned with WJOL(AM). Feb 6, 1960: 96.7 mhz; 3 kw. 300 ft TL: N41 32 10 W88 03 15. Stereo. 24 Web Site: www.star967.net.220,000 Format: Adult contemp. News staff: one. Target aud: 25-44; men.

WVIX(FM)— Apr 17, 1960: 93.5 mhz; 6 kw. Ant 328 ft TL: N41 36 39 W88 00 33. Stereo. Hrs opn: 24
Simulcast with WVIV-FM Highland Park.
625 N. Michigan Ave., Suite 300, Chicago, 60611. Phone: (312) 981-1800. Fax: (312) 981-1806. Licensee: HBC License Corp. Group owner: Univision Radio (acq 9-22-2003; grpsl). Population served: 289,000 Format: Sp contemp. News staff: one; News: one. ♦Jerry Ryan, gen mgr; Cesar Canales, opns mgr; David Miranda, prom dir; Paul Easter, engrg mgr.

WWHN(AM)— Apr 10, 1964: 1510 khz; 1 kw-D. TL: N41 30 50 W88 03 10. Stereo. Hrs opn: 10321 S. Halsted, Chicago, 60628. Phone: (773) 239-2300. Fax: (773) 239-9921. E-mail: wwhn@aol.com Licensee: Hawkins Broadcasting Co. (acq 12-89; $250,000; FTR: 12-4-89). Population served: 80,378 Format: Gospel. Target aud: 18-54; affluent adults. Spec prog: Sp one hr wkly. ♦Raymond E. Hawkins, pres; Toni Hawkins, gen mgr.

Kankakee

***WAWF(FM)**— 2000: 88.3 mhz; 1.25 kw. Ant 285 ft TL: N41 04 39 W87 45 22. Hrs open: Box 3206., American Family Radio, Tupelo, MS, 38803. Phone: (662) 844-8888. Fax: (662) 842-6791. Web Site: www.afr.net. Licensee: American Family Radio. (group owner) Format: Inspirational Christian. ♦Marvin Sanders, gen mgr.

WKAN(AM)— June 1, 1947: 1320 khz; 1 kw-D, 500 w-N, DA-N. TL: N41 08 08 W87 49 10. Hrs open: 24 70 Meadowview Ctr., Suite 400, 60901. Phone: (815) 935-9555. Fax: (815) 935-9593. Web Site: www.wkan.com. Licensee: STARadio Corp. (acq 2-7-94; $1.31 million with co-located FM; FTR: 3-28-94) Population served: 325,000 Law Firm: Pepper & Corazzini. Wire Svc: UPI Format: Talk. News staff: one; News: 20 hrs wkly. Target aud: 25-54. Spec prog: Farm 10 hrs wkly. ♦Robert L. Kersmarki, pres & gen mgr; Brendan Michaels, opns mgr & progmg dir; Larry Regnier, gen sls mgr.

***WKCC(FM)**— June 1, 1992: 91.1 mhz; 1.75 kw. 305 ft TL: N41 09 24 W87 52 16. Hrs open: 24 Box 888, 60901. Phone: (815) 802-8100. Fax: (815) 935-5169. Licensee: Kankakee Community College. Population served: 100,000 Format: Educ, tourism. Target aud: General; travelers in northern IL. ♦William Yohnka, gen mgr.

WKIF(FM)— Sept 21, 1986: 92.7 mhz; 3 kw. Ant 300 ft TL: N41 07 22 W87 53 35. Stereo. Hrs opn: 24 6012 S. Pulaski Rd., Chicago, 60629. Phone: (773) 767-1000. Fax: (773) 767-1100. Licensee: WKIF Inc. Group owner: Spanish Broadcasting System Inc. (acq 11-15-2004; grpsl). Population served: 150,000 Natl. Network: CNN Radio. Format: News. ♦Harvey Wells, gen mgr; Gary Wright, progmg dir.

***WONU(FM)**— 1966: 89.7 mhz; 35 kw. 421 ft TL: N41 09 24 W87 52 16. Stereo. Hrs opn: 24 One University Ave., Bourbannais, 60914. Phone: (815) 939-5330. Fax: (815) 939-5087. E-mail: shinefm@wonu.fm Web Site: www.shine.fm. Licensee: Olivet Nazarene University. Population served: 6,000,000 Law Firm: Miller & Miller. Format: Christian Pop. Target aud: 25-49; female, predominantly conservative. ♦Justin Knight, gen mgr; Johnathon Eltrevoog, prom dir & progmg dir; Don Johnson, chief of engrg & rsch dir.

WVLI(FM)— Oct 22, 1992: 95.1 mhz; 3 kw. 328 ft TL: N41 04 39 W87 45 22. Stereo. Hrs opn: 24 Box 758, Bourbonnais, 60914-0756. Secondary address: 292 N. Convent, Bourbonnais 60914. Phone: (815) 933-9287. Fax: (815) 933-8696. E-mail: wvli951@aol.om Licensee: Milner Broadcasting Co. (acq 3-17-95; $400,000). Population served: 105,000 Natl. Network: AP Network News. Law Firm: Womble, Carlyle, Sandridge & Rice. Wire Svc: AP Format: Greatest hits & artists. News staff: 2; News: 20 hrs wkly. Target aud: 25+. Spec prog: Chicago Bears (NFL) Football Games. ♦Tim Milner, pres, gen mgr & stn mgr; Jim Brandt, opns mgr.

Kewanee

WKEI(AM)— Sept 11, 1952: 1450 khz; 500 w-D, 1 kw-N. TL: N41 13 37 W89 56 08. Hrs open: 24 Box 266, 61443-0266. Secondary address: 133 E. Division St. 61443. Phone: (309) 853-4471. Fax: (309) 853-4474. E-mail: wkei@verizon.net Web Site: www.randyradio.com. Licensee: Virden Broadcasting Corp. Group owner: Miller Media Group (acq 11-8-94; $400,000 with co-located FM; FTR: 1-2-95). Population served: 92,000 Natl. Network: CBS Radio. Natl. Rep: Commercial Media Sales. Law Firm: Womble Carlyle. Format: News/talk. News staff: one. Spec prog: Farm 20 hrs, relg 6 hrs wkly. ♦Randal J. Miller, pres & sls dir; Kris Wexell, progmg dir & pub affrs dir; Will Stevenson, adv mgr & news dir; Wayne R. Miller, chief of engrg; Jennie Holtschult, traf mgr.

WYEC(FM)—Co-owned with WKEI(AM). May 20, 1966: 93.9 mhz; 3.1 kw. Ant 453 ft TL: N41 16 40 W89 55 15. Stereo. 24 Web Site: www.randyradio.com. Natl. Network: CNN Radio. Format: Soft adult contemp. Target aud: 35-64. ♦Randal J. Miller, pres & disc jockey.

Knoxville

WKAY(FM)— Dec 13, 2001: 105.3 mhz; 3.7 kw. Ant 423 ft TL: N40 56 34 W90 20 39. Hrs open: 24 154 E. Simmons, Galesburg, 61401. Phone: (309) 342-5131. Fax: (309) 342-0840. E-mail: kfm@1053kfm.com Web Site: www.1053kfm.com. Licensee: Galesburg Broadcasting Co. Group owner: Galesburg Broadcasting Co. (acq 4-1-99). Population served: 55,000 Law Firm: Cohn & Marks, LLP. Wire Svc: AP Format: Adult contemp. News staff: 4; News: one hr wkly. Target aud: 25-54; adults. ♦John Pritchard, pres; Roger Lundeen, gen mgr; Brian Prescott, opns dir; Chris Postin, sls VP; Shannon Anderson, news dir; Rick Heath, engrg dir.

La Grange

***WLTL(FM)**— Jan 5, 1968: 88.1 mhz; 180 w. 138 ft TL: N41 48 45 W87 52 51. Stereo. Hrs opn: 24 100 S. Brainard Ave., 60525. Phone: (708) 482-9585. Fax: (708) 482-7051. E-mail: cthomas@wltl.net Web Site: www.wltl.net. Licensee: Lyons Township High School. Population served: 117,814 Format: Var, rock. News: 10 hrs wkly. Target aud: 14-35; young adults. Spec prog: Sports 5 hrs, news & views 10 hrs wkly. ♦Chris Thomas, gen mgr; Tim McLaughlin, progmg dir; Mike Dorris, chief of engrg.

WRDZ(AM)— October 1950: 1300 khz; 5 kw-D, 500 w-N, DA-2. (CP: 4 kw-N. TL: N41 40 29 W87 45 45). Hrs opn: 190 N. State St., Chicago, 60601. Phone: (312) 683-1300. Fax: (312) 577-5994. Licensee: Radio Disney Chicago LLC. Group owner: ABC Inc. (acq 5-12-99; with WPJX(AM) Zion). Population served: 17,814 Natl. Network: ABC. Wire Svc: City News Bureau Format: Children. Children, mom's & dad's. ♦Karyn Esken, stn mgr.

La Salle

WAJK(FM)—Listing follows WLPO(AM).

WLPO(AM)— Nov 16, 1947: 1220 khz; 1 kw-D, 500 w-N, DA-2. TL: N41 18 14 W89 05 44. Hrs open: 24 1 Broadcast Ln., Oglesby, 61348. Phone: (815) 223-3100. Fax: (815) 223-3095. Web Site: wlpo.net. Licensee: La Salle County Broadcasting Corp. (group owner; (acq 8-1-49). Population served: 60,000 Format: News/talk, sports. News staff: 3; News: 35 hrs wkly. Target aud: 30 plus. ♦Peter Miler, pres; Joyce McCullough, VP & gen mgr; Mark Lippert, gen sls mgr; John Spencer, progmg dir.

WAJK(FM)—Co-owned with WLPO(AM). Dec 4, 1964: 99.3 mhz; 11 kw. 500 ft TL: N41 18 15 W89 05 46. Stereo. Web Site: wajk.com. Format: Hot adult contemp. Target aud: 25-49. ♦John Spencer, progmg dir.

***WNIW(FM)**— November 1998: 91.3 mhz; 8 kw. 331 ft TL: N41 26 44 W89 00 42. Stereo. Hrs opn: 24
Rebroadcasts WNIJ(FM) De Kalb & WNIU(FM) Rockford 50%.
NIU Broadcast Ctr., 801 N. First St., DeKalb, 60115. Phone: (815) 753-9000. Fax: (815) 753-9938. E-mail: npr@niu.edu Web Site: www.northernpublicradio.org. Licensee: Northern Illinois University. Population served: 100,000 Natl. Network: PRI, NPR. Law Firm: Arter & Hadden. Format: News, classical. News staff: 2. Target aud: General. ♦Tim Emmons, gen mgr; Jan Kilgard, dev dir; Bill Drake, progmg dir.

Lake Forest

***WMXM(FM)**— Sept 10, 1973: 88.9 mhz; 300 w. 100 ft TL: N42 15 00 W87 49 45. Stereo. Hrs opn: 18 555 N. Sheridan Rd., 60045. Phone: (847) 735-5220 (office). Phone: (847) 735-6038 (studio). Fax: (847) 735-6291. Web Site: www.lfcradio.com. Licensee: Lake Forest College. Population served: 500,000 Format: Div, classic rock, progsv. News staff: 2; News: 5 hrs wkly. Target aud: 18-25; students. Spec prog: Black 6 hrs, class 3 hrs, gospel 3 hrs, jazz 6 hrs wkly. ♦Ethan Helm, gen mgr.

Lansing

WSRB(FM)— Aug 28, 1961: 106.3 mhz; 2 kw. Ant 397 ft TL: N41 34 44 W87 32 46. Hrs open: 24 6336 Calumet Ave., Hammond, IN, 46324. Phone: (773) 734-4455. Fax: (219) 933-0323. E-mail: wycainfo@crawfordbroadcasting.com Web Site: www.soul1063radio.com. Licensee: Dontron Inc. Group owner: Crawford Broadcasting Co. (acq 4-10-97; $14.8 million). Population served: 25,805 Natl. Network: ABC. Format: Adult contemp. Target aud: 25-54; urban, adult, African-American. ♦Donald B. Crawford, CEO; Taft Harris, gen mgr.

Lawrenceville

WAKO(AM)— June 9, 1959: 910 khz; 500 w-D, 59 w-N, DA-2. TL: N38 43 23 W87 39 13. Hrs open: 5 AM-midnight Box 210, 62439. Phone: (618) 943-3354. Fax: (618) 943-4173. E-mail: wakoradio@yahoo.com Licensee: Lawrenceville Broadcasting Co. Inc. (acq 5-31-73). Population served: 489,200 Natl. Network: Westwood One. Wire Svc: AP Format: Adult contemp, country. News staff: one; News: 12 hrs wkly. Target aud: 20-65+. ♦Stuart Kent Lankford, pres & stn mgr.

WAKO-FM— March 1965: 103.1 mhz; 6 kw. 328 ft TL: N38 43 23 W87 39 13. Stereo. 19 250,000 ♦Steve Anderson, news dir & local news ed; Stuart Kent Lankford, gen mgr & traf mgr.

Le Roy

WBWN(FM)— Oct 15, 1979: 104.1 mhz; 25 kw. 328 ft TL: N40 25 25 W88 51 28. (CP: .80 kw, ant 413 ft. TL: N40 27 01 W89 00 42). Stereo. Hrs opn: 24 236 Greenwood Ave., Bloomington, 61704. Phone: (309) 829-1221. Fax: (309) 662-8598. Web Site: www.wbwn.com. Licensee: Regent Broadcasting of Bloomington Inc. Group owner: Regent Communications Inc. (acq 5-12-2004; grpsl). Population served: 300,000 Natl. Rep: McGavren Guild. Format: Country. Target aud: 25-45. ♦Red Pitcher, stn mgr; Dan Westhoff, progmg dir.

Lena

WQLF(FM)— Aug 2, 2002: 102.1 mhz; 5.2 kw. Ant 351 ft TL: N42 20 31 W89 48 21. Hrs open: 24 W4765 Radio Ln., Monroe, WI, 53566. Phone: (608) 325-2161. Fax: (608) 325-2164. E-mail: wekz@wekz.com Web Site: www.wekz.com. Licensee: Lena Radio Broadcasting (acq 2-26-2002). Classic hits (rock). Target aud: 20-49. ♦Scott Thompson, gen mgr & gen sls mgr; Don Jacobson, news dir; Todd Hauser, chief of engrg; Wyatt Herrmann, progmg dir & traf mgr.

Lexington

WDQZ(FM)— 2004: 99.5 mhz; 6 kw. Ant 328 ft TL: N40 34 30 W88 50 15. Stereo. Hrs opn: 24 108 Boeykens Place, Normal, 61761. Phone: (309) 888-4496. Fax: (309) 452-9677. Web Site: www.eagleclassicrock.com. Licensee: Pilot Media LLC. Group owner: AAA Entertainment L.L.C. (acq 6-5-2007; grpsl). Population served: 127,000 Wire Svc: NBC Format: Classic Rock. News: 2 hrs wkly. Target aud: 25-54. ♦Patti Donsbach, gen mgr.

Lincoln

WLLM(AM)— April 1951: 1370 khz; 1 kw-D, 35 w-N. TL: N40 08 24 W89 23 10. Hrs open: 24
Rebroadcasts WLUJ(FM) Springfield 100%.
800 S. Postville Dr., 62656. Phone: (217) 735-9735. Fax: (217) 735-9736. Web Site: www.wllmradio.com. Licensee: Cornerstone Community Radio Inc. (acq 4-7-03; $275,000). Population served: 30,000 Natl. Network: USA. Format: Easy listening, Christian music & Bible teaching. ♦Richard Van Zandt, gen mgr; William Dolan, stn mgr; Beverly Tibbs, opns mgr.

***WLNX(FM)**— Jan 28, 1974: 88.9 mhz; 225 w. Ant 68 ft TL: N40 09 23 W89 21 40. Stereo. Hrs open: 24 300 Keokuk St., 62656. Phone: (217) 732-3155. Fax: (217) 732-3715. Web Site: www.wlnxradio.com. Licensee: Lincoln University. Population served: 15,000 Law Firm: Womble, Carlyle, Sandridge & Rice. Format: Rock. Target aud: 18-34; adults. ♦John Malone, gen mgr.

Lincolnshire

***WAES(FM)**— 2002: 88.1 mhz; 150 w. Ant 49 ft TL: N42 11 59 W87 56 49. Hrs open: Adlai E. Stevenson High School, Two Stevenson Dr., 60069. Phone: (847) 634-4000 ext. 1710. Fax: (847) 634-0983. Licensee: Adlai E. Stevenson High School District No. 125. Format: Var. ♦Greg Sherwin, gen mgr.

Litchfield

WSMI(AM)— Nov 2, 1950: 1540 khz; 1 kw-D. TL: N39 10 21 W89 34 14. Hrs open: Box 10, WSMI Bldg, E. Rt. 16, 62056. Secondary address: 6308 IL Rt. 16, Hillsboro 62049. Phone: (217) 324-5921. Fax: (217) 532-2431. E-mail: wsmi@wsmiradio.com Web Site: wsmiradio.com. Licensee: Talley Broadcasting Corp. Group owner: Talley Radio Stations Population served: 400,000 Natl. Network: CNN Radio. Natl. Rep: Christal. Wire Svc: AP Format: Farm, country, news/talk. News staff: 3; News: 15 hrs wkly. Target aud: General. Spec prog: Farm 18 hrs wkly. ♦Brian C. Talley, chmn, sr VP & opns mgr; Hayward L. Talley, gen mgr; Michael Niehaus, rgnl sls mgr; Kevin Talley, prom mgr; Terry Todt, progmg dir.

WSMI-FM— Mar 5, 1960: 106.1 mhz; 50 kw. 500 ft TL: N39 15 21 W89 36 48. Stereo. 4:30 AM-midnight 600,000 Format: Contemp country. News staff: 3.

Lockport

***WLRA(FM)**— November 1972: 88.1 mhz; 250 w. 95 ft TL: N41 36 06 W88 04 51. Stereo. Hrs opn: 24 c/o Lewis Univ., 500 Independence Blvd., Romeoville, 60446. Phone: (815) 836-5214. Fax: (815) 838-9149. E-mail: wlraradio@lewisu.edu Web Site: www.thestartradio.com. Licensee: Lewis University. Population served: 100,000 Format: Educ, div.

News: 2 hrs wkly. Target aud: 13-30; college bound or post-college. Spec prog: Black 15 hrs, class 6 hrs, jazz 15 hrs, sports 15 hrs, talk 10 hrs wkly. ♦Ryan Huff, gen mgr.

Loves Park

***WGSL(FM)**— Mar 28, 1988: 91.1 mhz; 7 kw. Ant 528 ft TL: N42 19 20 W89 00 41. Stereo. Hrs opn: Box 2730, Rockford, 61132-2730. Secondary address: 5375 Pebble Creek Tr. 61111. Phone: (815) 654-1200. Fax: (815) 282-7779. E-mail: home@radio91.com Web Site: www.radio91.com. Licensee: Christian Life Center School. Natl. Network: USA. Law Firm: Wilkinson Barker Knauer. Format: Relg, contemp praise. Target aud: 35-50; older families. ♦Ralph Trendadue, gen mgr; Ron Tietsort, opns mgr.

WKGL-FM— Mar 25, 1964: 96.7 mhz; 3 kw. Ant 300 ft TL: N42 19 48 W89 04 58. Stereo. Hrs opn: 3901 Brendenwood Rd., Rockford, 61107. Phone: (815) 399-2233. Fax: (815) 484-2432. Web Site: www.cumulus.com. Licensee: Cumulus Licensing Corp. Group owner: Cumulus Media Inc. (acq 3-12-2001). Population served: 230,000 Format: Classic rock. ♦Greg Sher, gen mgr; Allisia Bri-Asperson, prom dir; John Brizolla, progmg dir; Paul Hannigan, news dir; John Huntley, chief of engrg; Jan Thorp, traf mgr.

WLUV(AM)— Sept 29, 1962: 1520 khz; 500 w-D. TL: N42 19 48 W89 04 58. (CP: 12.5 w). Hrs opn: Box 2616, 61132. Secondary address: 2272 Elmwood Rd., Rockford 61103. Phone: (815) 877-9588. Fax: (815) 877-9649. Licensee: Loves Park Broadcasting Co. Population served: 230,000 Format: Classic country, sports. Target aud: 25-60; blue collar workers. Spec prog: Farm 6 hrs, polka 6 hrs wkly. ♦Joe Salvi, gen mgr.

Lynnville

WEAI(FM)— Nov 15, 1989: 107.1 mhz; 6 kw. 328 ft TL: N39 37 16 W90 15 28. Stereo. Hrs opn: 5 AM-midnight Box 1180, Jacksonville, 62651. Secondary address: 2161 Old State Rd., Jacksonville 62651. Phone: (217) 243-2800. Fax: (217) 245-6711. E-mail: weai@weai.com Web Site: www.weai.com. Licensee: Jerdon Broadcasting. Population served: 50,000 Natl. Network: NBC Radio. Natl. Rep: Katz Radio. Law Firm: Kaye, Scholer, Fierman, Hays & Handler. Wire Svc: AP Format: Contemp hit, oldies. News staff: 3; News: 6 hrs wkly. Target aud: 20-40; Active Young Adults. ♦Jerry Symons, gen mgr; Don Hamilton, gen sls mgr; Perry Brown, progmg dir; Troy Armstrong, mus dir; Gary Scott, news dir; John Coe, chief of engrg; Marty Megginson, traf mgr.

Macomb

***WIUM(FM)**— May 23, 1956: 91.3 mhz; 50 kw. 485 ft TL: N40 25 40 W90 40 58. Stereo. Hrs opn: 24 515 Univ. Svcs. Bldg., Western Illinois Univ., 61455. Phone: (309) 298-2424. Phone: (309) 298-1873. Fax: (309) 298-2133. E-mail: publicradio@wiu.edu Web Site: www.tristatesradio.com. Licensee: Western Illinois University. Population served: 77,162 Natl. Network: NPR, PRI. Law Firm: Cohn & Marks. Format: Class, news. News staff: 2; News: 58 hrs wkly. Target aud: General. Spec prog: Folk/blues 7 hrs, jazz 5 hrs wkly. ♦Dorothy Vallillo, gen mgr, progmg dir & news rptr; Ken Thermon, opns dir; Sharon Faust, dev dir; Rich Egger, news dir; Mark Garrett, chief of engrg.

***WIUS(FM)**— Feb 1, 1982: 88.3 mhz; 120 w. 83 ft TL: N40 27 47 W90 41 00. Stereo. Hrs opn: Sallee Hall, One University Cir., Western Ill. Univ., 61455-1390. Phone: (309) 298-3217 (request). Fax: (309) 298-2133. Web Site: www.wiu.edu/the dog/. Licensee: Western Illinois University. Law Firm: Cohn & Marks. Format: Progsv new mus, urban contemp, alternative. Target aud: 18-30. Spec prog: Jazz 2 hrs, Sp 3 hrs, blues 4 hrs wkly. ♦Patrick Stout, stn mgr & progmg dir.

WJEQ(FM)— February 1983: 102.7 mhz; 25 kw. 269 ft TL: N40 29 00 W90 38 19. Stereo. Hrs opn: 24 31 E. Side Sq., 61455-2248. Phone: (309) 833-2121. Fax: (309) 836-3291. E-mail: wjeq@macomb.com Web Site: www.wjeq.com. Licensee: Central Illinois Broadcasting. (acq 6-14-89). Format: Classic rock. News staff: one; News: 10 hrs wkly. Target aud: 18-49. Spec prog: Farm one hr wkly. ♦Bruce Foster, pres & chief of engrg; Nancy Foster, gen mgr; Mike Grillette, progmg dir; Mick Wilkens, news dir; Shana Drake, traf mgr & disc jockey.

WKAI(FM)—Listing follows WLRB(AM).

WLRB(AM)— July 4, 1947: 1510 khz; 1 kw-D. TL: N40 29 50 W90 40 30. Hrs open: Sunrise-sunset Box 250, 119 W. Carroll, 61455. Phone: (309) 833-5561. Fax: (309) 833-3460. E-mail: wlrb@macomb.com Web Site: www.radiomacomb.com. Licensee: WPW Broadcasting Inc. (group owner; (acq 12-27-99; grpsl). Population served: 22,300 Natl. Network: Westwood One, Jones Radio Networks. Rgnl. Network: Brownfield. Format: Music of Your Life. News staff: one; News: 6 hrs wkly. Target aud: 45 plus. Spec prog: Farm 1.25 hrs wkly. ♦Don Davis, pres; Mike Weaver, opns mgr; Vanessa Wetterling, gen mgr & gen sls mgr; Mike Weave, progmg dir.

WKAI(FM)—Co-owned with WLRB(AM). June 6, 1966: 100.1 mhz; 3.08 kw. Ant 463 ft TL: N40 26 57 W90 42 22. Stereo. 24 E-mail: wkai@macomb.com Web Site: www.radiomacomb.com. Format: Adult contemp. News staff: one; News: 5 hrs wkly. Target aud: 35 plus.

WNLF(FM)— 2003: 95.9 mhz; 6 kw. 328 ft TL: N40 25 03 W90 36 51. Hrs open: c/o WJEQ Radio, 31 E Side Sq., 61455. Phone: (309) 833-2121. Fax: (309) 836-3291. Web Site: www.modernrock959.com. Licensee: Nancy L. Foster. Format: Modern rock. ♦Bruce Foster, pres & chief of engrg; Nancy Foster, gen mgr; Mike Grillette, progmg dir; Mick Wilkens, news dir; Shana Drake, traf mgr.

Macon

WZUS(FM)— May 5, 1977: 100.9 mhz; 6 kw. Ant 328 ft TL: N39 47 11 W88 59 29. Stereo. Hrs opn: 24 410 N. Water St., Decatur, 62523. Phone: (217) 428-4487. Fax: (217) 428-4501. Licensee: The Cromwell Group Inc. of Illinois. Group owner: The Cromwell Group Inc. (acq 4-16-02; $900,000). Population served: 80,000 Natl. Network: Jones Radio Networks. Law Firm: Pepper & Corazzini. Format: Country. Target aud: General. Spec prog: Farm 5 hrs wkly. ♦Chris Bullock, gen mgr; Wayne Robbins, opns mgr; Jerry Scott, chief of engrg.

Mahomet

WGKC(FM)— Dec 15, 1990: 105.9 mhz; 1.25 kw. 512 ft TL: N40 13 27 W88 17 56. Hrs open: 24 4108 Fieldstone Rd., Suite C, Champaign, 61822. Phone: (217) 367-1195. Fax: (217) 367-3291. Web Site: www.wgkc.net. Licensee: RadioStar Inc. Group owner: AAA Entertainment L.L.C. (acq 5-23-2006; grpsl). Population served: 175,000 Natl. Network: ABC. Format: Classic rock. News staff: one. Target aud: 25-54; adult men. ♦Mike Knar, gen mgr; Marty Booth, progmg dir; Zach Morton, chief of engrg; Diane Siegelmann, traf mgr.

Marion

***WAWJ(FM)**— 2001: 90.1 mhz; 3 kw vert. Ant 344 ft TL: N37 51 23 W89 08 22. Hrs open: Drawer 3206, Tupelo, MS, 38803. Phone: (662) 844-8888. Fax: (662) 842-6791. Web Site: www.afr.net. Licensee: American Family Association. Group owner: American Family Radio Format: Inspirational Christian. ♦Marvin Sanders, gen mgr.

WDDD(AM)—See Johnston City

WDDD-FM— Nov 22, 1970: 107.3 mhz; 50 kw. 500 ft TL: N37 45 15 W88 56 05. Stereo. Hrs opn: 24 1822 N. Court St, 62959. Phone: (618) 997-8123. Fax: (618) 993-2319. Web Site: www.w3dcountry.com. Licensee: CC Licenses LLC. Group owner: Clear Channel Communications Inc. (acq 1-18-2001; grpsl). Population served: 250,000 Format: Country. News staff: 3; News: 5 hrs wkly. Target aud: 25 plus. ♦Jerry Crouse, gen mgr; Paxton Guy, gen sls mgr & progmg dir.

WGGH(AM)— Sept 24, 1949: 1150 khz; 5 kw-D, 44 w-N DA-1. TL: N37 43 47 W88 53 44. Hrs open: 24 hrs Box 340, 1801 E. Main St., 62959. Phone: (618) 993-8102. Phone: (618) 997-2305. Fax: (618) 997-2307. E-mail: wggh@shawneelink.net Licensee: Vine Broadcasting Inc. (acq 4-7-92; FTR: 6-9-92). Population served: 7,600,925 Natl. Network: Salem Radio Network. Rgnl. Network: Ill. Radio Net. Format: Southern gospel, relg, talk. News: Hourly. Target aud: 18 plus; general. ♦Elaine Gomez, gen mgr & sls; Mat Canon, progmg dir; Johnny Gomez, chief of engrg; Brenda Bender, traf mgr.

Maroa

WDKR(FM)— May 1996: 107.3 mhz; 3 kw. 456 ft TL: N39 57 56 W89 03 27. Hrs open: 24 120 Wildwood Dr., Mt. Zion, 62549. Phone: (217) 864-4141. Fax: (217) 864-4727. Licensee: WDKR Inc. (acq 3-6-02). Population served: 3,00,000 Format: Oldies. Target aud: 25-54; general. ♦Mary Ellen Burns, gen mgr.

Marseilles

WKOT(FM)— March 1992: 96.5 mhz; 3 kw. 328 ft TL: N41 18 40 W88 49 07. Stereo. Hrs opn: 24 1 Broadcast Ln., Oglesby, 61348. Phone: (815) 434-4000. Fax: (815) 434-4055. E-mail: wajk@ivnet.com Web Site: wkot.com. Licensee: La Salle County Broadcasting Corp. (group owner; acq 6-99; $550,000). Natl. Network: Jones Radio Networks. Format: Classic Hits. News staff: one; News: 8 hrs wkly. Target aud: 35-54. ♦Peter Miller, pres; Joyce McCullough, VP & gen mgr; John Spencer, opns dir & progmg dir; Jennifer Nagle, news dir.

Marshall

WMMC(FM)— Oct 2, 1989: 105.9 mhz; 2.3 kw. Ant 528 ft TL: N39 21 09 W87 49 19. Stereo. Hrs opn: 24 Box 158, 62441. Secondary address: 627 1/2 Archer Ave. 62441. Phone: (217) 826-8017. Fax: (217) 826-8519. Licensee: JDL Broadcasting Inc. (acq 9-10-98; $300,000). Natl. Network: ABC. Format: Adult contemp. News: 8 hrs wkly. Target aud: 25-54; career-oriented men and women. ♦J. D. Spangler, pres & gen mgr; Lori Spangler, opns mgr.

Mattoon

WLBH(AM)— Nov 26, 1946: 1170 khz; 5 kw-D, DA. TL: N39 31 05 W88 22 15. Hrs open: 6 AM-7 PM Box 1848, N. Rt. 45 (2 mi), 61938-1848. Phone: (217) 234-6464. Fax: (217) 234-6019. E-mail: wlbh@wlbh.com Web Site: www.wlbh.com. Licensee: Mattoon Broadcasting Co. Group owner: J.R. Livesay Group Population served: 500,000 Format: Farm, news/talk, MOR. News staff: 3; News: 20 hrs wkly. Target aud: 25 plus. Spec prog: Relg 5 hrs wkly. ♦J.R. Livesay II, CEO & pres; S.L. Herrington, CFO; Adam Kennedy, news dir & local news ed; Jim Livesay, political ed.

WLBH-FM— August 1949: 96.9 mhz; 50 kw. 500 ft TL: N39 31 02 W88 22 13. Stereo. 24 Web Site: www.wlbh.com.2,000,000 Format: Adult contemp. News staff: 3; News: 18 hrs wkly. Target aud: 25 plus. ♦Adam Kennedy, local news ed; Jim Livesay, political ed.

***WLKL(FM)**— Jan 20, 1975: 89.9 mhz; 1.3 kw. 203 ft TL: N39 25 07 W88 22 55. Stereo. Hrs opn: 24 5001 Lakeland Blvd., 61938. Phone: (217) 234-5373. Fax: (217) 234-5506. E-mail: gpowers@lakeland.cc.il.us Licensee: Community College District 517 Lake Land College. Population served: 60,000 Format: CHR, AOR. News staff: one; News: 6 hrs wkly. Target aud: 18-34; general. ♦Greg Powers, gen mgr & gen sls mgr.

WMCI(FM)— Aug 24, 1989: 101.3 mhz; 14.5 kw. 433 ft TL: N39 31 39 W88 21 23. Stereo. Hrs opn: 24 209 Lakeland Blvd., 61938. Phone: (217) 235-5624. Phone: (217) 348-9292. Fax: (217) 235-6624. Web Site: www.radiomattoon.com. Licensee: The Cromwell Group Inc. of Illinois. Group owner: The Cromwell Group Inc. Population served: 100,000 Rgnl rep: Katz Format: Country. News staff: one; News: 10 hrs wkly. Target aud: 25-54. Spec prog: Farm 5 hrs wkly. ♦Bud Walters, pres; Bub McCullough, opns mgr, progmg dir & sports cmtr; Carol Floyd, gen mgr, stn mgr & gen sls mgr.

WXET(FM)—Arcola, Dec 19, 1974: 107.9 mhz; 2.5 kw. Ant 492 ft TL: N39 34 15 W88 18 17. Stereo. Hrs opn: 24 Box 988, 206 S. Willow St., Effingham, 62401. Secondary address: 401 Lakeland Blvd. 61938. Phone: (217) 347-5518. Fax: (217) 347-5519. E-mail: info@thexradio.com Web Site: www.thexradio.com. Licensee: Premier Broadcasting Inc. (acq 8-12-97; $80,000). Population served: 50,000 Format: Adult contemp. News staff: one; News: 10 hrs wkly. Target aud: 25-54. Spec prog: Loc news and sports. ♦David Ring, pres & CFO; Greg Sapp, gen mgr; Tonya Siner, opns VP.

McLeansboro

WMCL(AM)— Jan 26, 1968: 1060 khz; 2.5 kw-D, DA. TL: N38 06 16 W88 33 48. Hrs open: 24 Box 46 A R.R. 1, 62859. Phone: (618) 643-2311. Fax: (618) 643-3299. Licensee: Dana Communications Corp. (acq 7-23-98; $245,000). Population served: 100,000 Natl. Network: CNN Radio. Law Firm: Bryan Cave. Format: Country, agriculture news. Target aud: 25-65; agricultural community. ♦Dana Withers, pres & gen mgr; Gloria Holland, opns mgr.

Mendota

WGLC-FM— Sept 1, 1965: 100.1 mhz; 6 kw. 328 ft TL: N41 32 16 W89 06 25. Stereo. Hrs opn: 4162 E. 3rd Rd., 61342. Secondary address: 3905 Progress Blvd., Peru 61354. Phone: (815) 224-2100. Fax: (815) 225-2066. E-mail: wglc@theradiogroup.net Web Site: wglc.net. Licensee: Mendota Broadcasting Inc. Group owner: Studstill Broadcasting (acq 4-8-88). Population served: 130,000 Natl. Network: ABC. Natl. Rep: Rgnl Reps. Law Firm: Booth, Freret, Imlay & Tepper. Wire Svc: AP Format: Country. Target aud: 35 plus. ♦Lee Studstill, CEO & gen mgr; Owen L. Studstill, pres; Cole Studstill, VP; Stuart Hall, opns dir; Chris Turnow, opns mgr.

Metropolis

WMOK(AM)— Feb 4, 1951: 920 khz; 1 kw-D, 73 w-N. TL: N37 09 13 W88 42 30. Stereo. Hrs opn: 24 Box 720, 339 Fairgrounds Rd., 62960. Phone: (618) 524-4400. Fax: (618) 524-3133. Licensee: Withers Broadcasting Co. of Paducah LLC. Group owner: Withers Broadcasting Co. (acq 9-11-97; grpsl). Population served: 150,000 Format: Country. News staff: one. Target aud: General. Spec prog: Relg 5 hrs wkly. ♦Rick Lambert, gen mgr; Kathy Duncan, gen sls mgr & rgnl sls mgr; Smokey King, chief of engrg; Steve Bunyard, progmg dir, local news ed & local news ed.

WREZ(FM)—Co-owned with WMOK(AM). Dec 12, 1988: 105.5 mhz; 6 kw. 328 ft TL: N37 10 25 W88 42 29. Stereo. 24 Box 7501, Paducah, KY, 42002. Phone: (270) 538-5251. Fax: (270) 415-0599.165,000 Format: Adult contemp. ♦Steve Thompson, progmg dir.

WRIK-FM— July 11, 1984: 98.3 mhz; 100 kw. 699 ft TL: N36 45 09 W88 29 58. Stereo. Hrs opn: 6120 Waldo Church Rd., 62960-4903. Fax: (618) 564-3202. E-mail: K98@hitsandfavs.com Licensee: Sun Media Inc. Population served: 325,000 Format: Adult contemp. ♦Samuel K. Stratemeyer, pres & gen mgr; Willie Kerns, opns mgr & progmg dir.

Milford

***WJCZ(FM)**— 2005: 91.3 mhz; 25 kw. Ant 89 ft TL: N40 35 07 W87 57 47. Hrs open: CSN International, 4002N. 3300E., Twin Falls, ID, 83301. Phone: (208) 734-6633. Fax: (208) 736-1958. E-mail: csn@csnradio.com Web Site: www.csnradio.com. Licensee: CSN International (group owner). Format: Christian. ♦Mike Kestler, pres.

Moline

WFXN(AM)— 1946: 1230 khz; 1 kw-U. TL: N41 28 54 W90 31 49. Hrs open: 24 3535 E. Kimberly Rd., Davenport, IA, 52807. Phone: (563) 344-7000. Fax: (563) 359-8524. Licensee: Citicasters Licenses L.P. Group owner: Clear Channel Communications Inc. (acq 11-15-00; grpsl). Population served: 50,000 Format: Country. News staff: one; News: 6 hrs wkly. Target aud: 25-54; upscale/contemp. Spec prog: Sports 8 hrs, pub affrs 4 hrs, farm one hr wkly. ♦Larry R. Rosmilso, VP & gen mgr; Scott Bitting, sls dir & gen sls mgr; Dan Kennedy, progmg dir; Kevin Allensworth, chief of engrg.

WXLP(FM)— Nov 22, 1970: 96.9 mhz; 50 kw. Ant 499 ft TL: N41 20 16 W90 22 46. Stereo. Hrs opn: 1229 Brady St., Davenport, IA, 52803. Phone: (563) 326-2541. Fax: (563) 326-1819. Fax: (319) 326-0844. Licensee: Cumulus Licensing Corp. Group owner: Cumulus Media Inc. (acq 3-15-2000; grpsl). Natl. Network: Westwood One. Law Firm: Putbrese, Hunsaker & Trent. Format: Classic rock. Target aud: 25-54. ♦Julie Derrer, pres & gen sls mgr; Jack Swart, gen mgr; Dave Levora, progmg dir; Andy Andresen, chief of engrg; Tracey Hall, traf mgr.

Monee

***WOTW(FM)**— Nov 1, 1995: 88.9 mhz; 100 w vert. Ant 82 ft TL: N41 27 58 W87 47 35. Hrs open: 820 N. La Salle Dr., Chicago, 60610. Phone: (312) 329-4300. Fax: (312) 329-4339. Web Site: www.ktlw.org. Licensee: Life on the Way Communications Inc. (group owner) (acq 2-15-2007; $5,000). Population served: 1,000,000 Law Firm: Southmayd & Miller. Format: Christian, relg. Target aud: General. ♦Gary Curtis, VP & gen mgr; Doug Hastings, stn mgr; Pamela McCain, opns mgr.

Monmouth

WMOI(FM)—Listing follows WRAM(AM).

WRAM(AM)— May 1957: 1330 khz; 1 kw-D, 50 w-N, DA-2. TL: N40 56 59 W90 34 19. Hrs open: 6 AM-6 PM Box 885, 55 Public Sq., 61462. Phone: (309) 734-9452. Fax: (309) 734-3276. Web Site: www.977wmoi.com. Licensee: WPW Broadcasting Inc. (group owner; (acq 12-24-97; $1.7 million with co-located FM). Population served: 85,000 Natl. Network: ABC. Rgnl. Network: Tribune, Ill. Radio Net. Format: Country. News staff: 3; News: 20 hrs wkly. Target aud: General; adult, mature. Spec prog: Farm 18 hrs, relg 3 hrs wkly. ♦David Klockenga, gen mgr; Don Davis, pres & gen mgr.

WMOI(FM)—Co-owned with WRAM(AM). Dec 6, 1967: 97.7 mhz; 3.36 kw. Ant 439 ft TL: N40 53 25 W90 36 31. Stereo. 24 100,000 Format: Adult contemp. News staff: 3; News: 40 hrs wkly. Target aud: General.

Monticello

WCZQ(FM)— Jan 18, 1972: 105.5 mhz; 3 kw. 300 ft TL: N40 02 52 W88 34 22. Stereo. Hrs opn: 24 337 N. Water St., Decatur, 62523. Phone: (217) 429-9595. Fax: (217) 423-9764. E-mail: wczq@piatt.com Web Site: www.wczq.piatt.com. Licensee: NM Licensing LLC. Group owner: NextMedia Group L.L.C. (acq 11-26-01; grpsl). Population served: 84,000 Natl. Network: ABC. Format: Urban contemp. News: 5 hrs wkly. Target aud: 25-65; upscale suburban & prosperous farm. Spec prog: Farm 11 hrs wkly. ♦Mark Hanson, gen mgr & stn mgr; Wendy Tohill, gen sls mgr; Jamie Pendleton, progmg dir; Frank Konwinski, chief of engrg; Cindy Hansen, traf mgr.

Morris

***WBEQ(FM)**— November 2003: 90.7 mhz; 1.45 kw. Ant 468 ft TL: N41 17 09 W88 25 49. Hrs open:
Rebroadcasts WBEZ(FM) Chicago 100%.
848 E. Grand Ave., Navy Pier, Chicago, 60611. Phone: (312) 948-4600. Fax: (312) 832 3100. Web Site: www.chicagopublicradio.org. Licensee: The WBEZ Alliance Inc. Natl. Network: NPR. Format: Jazz, news/talk. ♦Merrill Smith, chmn; Donna Moore, CFO; Torey Malatia, pres & gen mgr; Greg Salustro, dev dir.

***WCFL(FM)**— 1962: 104.7 mhz; 50 kw. 496 ft TL: N41 21 17 W88 29 55. Stereo. Hrs opn: 24
Rebroadcasts WBGL(FM) Champaign 100%.
1802 N. Division, Suite 403, 60450. Phone: (815) 942-4400. Fax: (815) 942-4401. E-mail: wbgl@wbgl.org Web Site: www.wbgl.org. Licensee: Illinois District Council of Assembly. (acq 4-16-94). Population served: 1,500,000 Natl. Network: USA. Law Firm: Gammon & Grange. Format: Adult contemp, Christian. Target aud: 24-39. ♦Jeff Scott, gen mgr & progmg dir.

WCSJ(AM)— Jan 15, 1964: Stn currently dark. 1550 khz; 250 w-D, 6 w-N. TL: N41 20 20 W88 25 20. Hrs open: 24 219 W. Washington St., 60450. Phone: (815) 941-1000. Fax: (815) 941-9300. Licensee: Grundy County Broadcasters Inc. (acq 7-22-97; $425,000). Population served: 49,000 Rgnl. Network: Tribune. Format: MOR, news/talk. Target aud: 35 plus. ♦Larry Nelson, pres; Jack Daly, gen mgr; Susan Pellegrini, gen sls mgr.

WCSJ-FM— 1993: 103.1 mhz; 6 kw. Ant 328 ft TL: N41 17 35 W88 20 04. Hrs open: 24 219 W. Washington St., 60450. Phone: (815) 941-1000. Fax: (815) 941-9300. Licensee: Grundy County Broadcasters Inc. (acq 11-5-2003; $426,000). Population served: 49,000 Natl. Network: ABC. Format: MOR, news/talk. News staff: one; News: 20 hrs wkly. Target aud: 25 plus; community oriented. Spec prog: Farm 10 hrs wkly. ♦Larry Nelson, pres; Jack Daly, gen mgr; Kevin Schramm, opns mgr.

Morrison

WZZT(FM)— Apr 10, 1991: 102.7 mhz; 6 kw. Ant 328 ft TL: N41 50 16 W89 55 29. Hrs open: 3101 Freeport Rd., Sterling, 61081-8612. Phone: (815) 625-3400. Fax: (815) 625-6940. E-mail: wsdr1240@theramp.net Licensee: Withers Broadcasting Co. of Rock River LLC. Group owner: Withers Broadcasting Co. (acq 1-21-98; grpsl). Natl. Network: ABC. Rgnl. Network: Ill. Radio Net. Natl. Rep: Christal. Format: Classic rock, sports. Target aud: 25-54; adults/men. ♦Brian Zschiesche, gen mgr & gen sls mgr; Kathy Wagner, progmg dir & mus dir; Mary Carlson, news dir; Sherry Smith, traf mgr.

Morton

WDQX(FM)— Nov 28, 1976: 102.3 mhz; 6 kw. Ant 300 ft TL: N40 38 27 W89 24 33. Hrs open: 24 331 Fulton St., Suite 1200, Peoria, 61602. Phone: (309) 637-3700. Fax: (309) 272-1476. Web Site: www.1023maxfm.com. Licensee: Monterey Licenses LLC. Group owner: AAA Entertainment L.L.C. (acq 4-21-2006; $5.2 million with WXCL(FM) Pekin). Population served: 350,000 Natl. Network: Westwood One, CBS. Rgnl. Network: Ill. Radio Net. Format: Classic hits. News staff: one; News: 10 hrs wkly. Target aud: 25 plus; active, affluent males. Spec prog: Relg one hr wkly. ♦Mike Wild, gen mgr; Joey Davidson, progmg dir; Shawn Newell, mus dir.

Mount Carmel

***WVJC(FM)**— July 23, 1973: 89.1 mhz; 50 kw. 331 ft TL: N38 26 29 W87 45 26. Stereo. Hrs opn: 24 2200 College Dr., 62863. Phone: (618) 262-8989. Phone: (618) 262-8641. Fax: (618) 262-7317. E-mail: peachk@iecc.edu Web Site: www.iecc.cc.il.us/wvjc. Licensee: Illinois Eastern Community Colleges. Population served: 544,000 Law Firm: Fletcher, Heald & Hildreth. Format: Educational, alternative. News: 11/4 hrs wkly. Target aud: General; Persons 12-24. ♦Kyle J. Peach, gen mgr.

WVMC(AM)— Dec 1, 1948: 1360 khz; 500 w-D, 20 w-N. TL: N38 26 58 W87 46 12. Hrs open: 24 606 Market St., 62863. Phone: (618) 262-4102. Fax: (618) 262-4103. E-mail: wsjd@midwest.net Licensee: Wabash Communications Inc. (acq 12-5-01; $85,000). Population served: 47,200 Format: Sports. News staff: 1. Target aud: General; Men 18 yrs plus. Spec prog: Farm 5 hrs wkly. ♦Kevin Williams, pres; Kevin Madden, gen mgr & stn mgr.

WYNG(FM)— Nov 28, 1960: 94.9 mhz; 50 kw. Ant 425 ft TL: N38 23 57 W87 47 18. Stereo. Hrs opn: 24 127 W. Third St, 62863. Phone: (618) 263-3500. Licensee: W. Russell Withers Jr. Group owner: Regent Communications Inc. (acq 12-22-2006; $1.5 million). Population served: 531,300 Format: Soft adult contemp. ♦Scott Allen, gen mgr & news dir; W. Russell Withers Jr., gen mgr; Josh Howard, progmg dir.

Mount Sterling

WPWQ(FM)— September 1995: 106.7 mhz; 25 kw. Ant 328 ft TL: N39 56 33 W90 57 44. Hrs open: Quincy Regional Airport, 1645 Hwy. 104, Suite G, Quincy, 62305. Phone: (217) 224-4653. Fax: (217) 885-3233. E-mail: wpwq106@adams.net Web Site: www.oldies1067.com. Licensee: WPW Broadcasting Inc. (group owner; (acq 12-6-99; $550,000 with WKXQ(FM) Rushville). Format: Oldies. ♦Don Davis, pres; Phil Alexander, gen mgr & gen sls mgr; Brian Myles, progmg dir & news dir.

Mount Vernon

***WAPO(FM)**— 1997: 90.5 mhz; 500 w. Ant 203 ft TL: N38 18 39 W88 56 11. (CP: 1.25 kw). Hrs opn: Box 3206, American Family Radio, Tupelo, MS, 38803. Phone: (662) 844-8888. Fax: (662) 842-6791. Web Site: www.afr.net. Licensee: American Family Association. Group owner: American Family Radio Format: Inspirational Christian. ♦Marvin Sanders, gen mgr; John Riley, progmg mgr.

***WBMV(FM)**— Sept 30, 1997: 89.7 mhz; 6.2 kw. 492 ft TL: N38 22 15 W88 55 20. Hrs open: 24
Rebroadcasts WIBI(FM) Carlinville 100%.
Box 140, Carlinville, 62626. Phone: (217) 854-4800. Fax: (217) 854-4810. E-mail: panthony@wibi.org Licensee: Illinois Bible Institute Inc. (group owner) Format: Adult contemp, Christian mus. Target aud: 29-45. ♦Barry Copeland, gen mgr; Jeremiah Beck, stn mgr, opns dir & sls dir; Jessica Barton, opns dir & prom dir; Rob Regal, progmg dir; Joe Buchanan, mus dir; Sally Braundmeier, traf mgr.

WIBV(FM)— 2001: 102.1 mhz; 10.5 kw. Ant 508 ft TL: N38 24 07 W89 08 09. Hrs open: 24 6120 Waldo Church Rd., Metropolis, 62960. Phone: (618) 564-9836. Fax: (618) 564-3202. E-mail: K98@hitsandfavs.com Web Site: www.wibv102.com. Licensee: Benjamin Stratemeyer (acq 4-26-2002; $1.25 million). Population served: 100,000 Format: Country. Target aud: 18-54. ♦Samuel Stratemeyer, gen mgr.

WMIX(AM)— 1947: 940 khz; 5 kw-D, 1.5 kw-N, DA-2. TL: N38 21 15 W89 00 29. Hrs open: 24 Box 1508, 62864. Secondary address: 3501 Broadway 62864. Phone: (618) 242-3500. Fax: (618) 242-4444. Fax: (618) 242-2490. E-mail: wmix@mvn.net Licensee: Withers Broadcasting Co. of Illinois LLC. Group owner: Withers Broadcasting Co. (acq 5-30-73). Population served: 38,000 Natl. Network: Westwood One. Law Firm: Dennis Kelly. Wire Svc: AP Format: Talk, great memories. News staff: 2; News: 15 hrs wkly. Target aud: 25 plus. Spec prog: Farm 18 hrs wkly. ♦W. Russell Withers Jr., pres; Dana Withers, gen mgr; Scott Smalls, sls dir; Nicholas Lemay, news dir.

WMIX-FM— 1946: 94.1 mhz; 50 kw. 550 ft TL: N38 22 14 W88 55 20. Stereo. 24 Format: C&W. ♦Russell Withers, CEO; Craig Warner, news dir; D.T. Brown, disc jockey.

***WVSI(FM)**— 2003: 88.9 mhz; 1.9 kw horiz, 4 kw vert. Ant 338 ft TL: N38 21 13 W88 56 32. Stereo. Hrs opn: 24
Rebroadcasts WSIU(FM) Carbondale.
SIUC Broadcasting Service, 1003 Communications Bldg., Southern Illinois University, Carbondale, 62901. Phone: (618) 453-4343. Fax: (618) 453-6186. E-mail: jeff.williams@wsiu.org Web Site: wsiu.org. Licensee: The Board of Trustees of Southern Illinois University. Population served: 34,700 Format: Class, news. News staff: 3; News: 36 hrs wkly. ♦Candis S. Isberner, CEO; Jeff Williams, stn mgr; Mike Zelten, opns mgr; Renee Dillard, dev dir.

Mount Zion

WXFM(FM)— October 1984: 99.3 mhz; 1.15 kw. 495 ft TL: N39 48 35 W88 59 31. Stereo. Hrs opn: 24 120 Wildwood Dr., 62549. Phone: (217) 864-4141. Fax: (217) 864-4727. Licensee: Technicom Inc. Population served: 300,000 Natl. Network: CNN Radio. Format: Adult contemp. News: 5 hrs wkly. Target aud: Free spending, affluent adults. ♦Mary Ellen Burns, pres & gen mgr.

Murphysboro

WINI(AM)— Sept 15, 1954: 1420 khz; 420w-D, 500 w-N, DA-N. TL: N37 45 30 W89 14 02. Hrs open: 24 1677 Business Hwy. 13, 62966. Phone: (618) 684-2128. Fax: (618) 687-4318. E-mail: wini@intrnet.net Web Site: www.winiradio.com. Licensee: Radio Station WINI. (acq 7-28-00). Population served: 10,200 Law Firm: Eugene T. Smith. Format: News/talk. News: 22 hrs wkly. Target aud: 25-59. Spec prog: Relg 6 hrs wkly. ♦Dale Adkins, gen mgr, gen sls mgr & chief of engrg; Nancy Engel, opns mgr.

WTAO(FM)—Licensed to Murphysboro. See Carbondale

Naperville

WBIG(AM)—See Aurora

WERV-FM—See Aurora

***WONC(FM)**— July 1, 1968: 89.1 mhz; 1.5 w. 163 ft TL: N41 46 45 W88 08 25. Stereo. Hrs opn: 24 30 N. Brainard St., 60566. Phone: (630) 637-8989. Fax: (630) 637-5900. E-mail: jvmadormo@noctrl.edu Web Site: www.wonc.org. Licensee: North Central College. Population served: 3,500,000 Wire Svc: AP Format: Rock/AOR. Target aud: 18-44. Spec prog: Relg 4 hrs, alternative 10 hrs wkly. ♦John Madormo, gen mgr.

Nashville

WNSV(FM)— July 10, 1994: 104.7 mhz; 3 kw. 328 ft TL: N38 20 38 W89 20 59. Hrs open: 24 168 E. St. Louis St., 62263. Phone: (618) 327-4444. Fax: (618) 327-3716. E-mail: wnsvfm@charter.net Licensee: Dana K. Withers. (acq 1-16-92; $60,000; FTR: 2-10-92). Population served: 1,950,000 Format: Adult contemp. Target aud: 30 plus. ♦Dana Withers, pres & gen mgr; Gloria Holland, opns mgr.

Neoga

WHQQ(FM)— September 1996: 98.9 mhz; 2.9 kw. Ant 482 ft TL: N39 14 59 W88 22 48. Stereo. Hrs opn: Box 150846, The Cromwell Group Inc., Nashville, TN, 37215. Phone: (615) 361-7560. Phone: (217) 235-5624 (stn). Fax: (615) 366-4313. Fax: (217) 235-6624 (stn). E-mail: mphillips@cromwellradio.co Web Site: www.effinghamradio.com.Westwood One Oldies Licensee: WSHY Inc. Group owner: The Cromwell Group Inc. Population served: 120,000 Natl. Network: Westwood One. Law Firm: Womble Carlyle. Format: Oldies. News: 14 hrs wkly. Target aud: 25-54; adults. Spec prog: Farm 7 hrs wkly. ♦Bayard Walters, chmn & pres; Tommy Crocker, CFO; Marv Phillips, gen mgr; Woody Bushne, opns mgr.

Newton

WIKK(FM)— May 4, 1992: 103.5 mhz; 25 kw. Ant 328 ft TL: N38 59 23 W88 11 19. Stereo. Hrs opn: 24 4667 Radio Tower Ln., Olney, 62450. Phone: (618) 783-8000. Phone: (618) 392-2156. Fax: (618) 783-4040. E-mail: wikk1035@psbnewton.com Web Site: www.929thelegend.com. Licensee: V.L.N. Broadcasting Inc. Group owner: Key Broadcasting Inc. (acq 6-25-2002; $600,000). Format: Classic rock (66). News staff: one; News: 8 hrs wkly. Target aud: 25-49. ♦Mike Shipman, opns VP, progmg dir & mus dir; Mark Weiler, news dir.

Normal

WBNQ(FM)—See Bloomington

***WGLT(FM)**— Feb 4, 1966: 89.1 mhz; 25 kw. 377 ft TL: N40 28 46 W89 03 12. Stereo. Hrs opn: 24 Box 8910, Illinois State Univ., 61790-8910. Phone: (309) 438-2255. Fax: (309) 438-7870. E-mail: wglt@ilstu.edu Web Site: www.wglt.org. Licensee: Illinois State University. Population served: 150,000 Natl. Network: NPR. Rgnl. Network: Illinois Public Radio. Law Firm: Dow, Lohnes & Albertson. Wire Svc: AP Format: Jazz, blues, pub affrs. News staff: 3; News: 40 hrs wkly. Target aud: 35-54. Spec prog: Folk 4 hrs, musical theater 2 hrs wkly. ♦Bruce Bergethon, gen mgr; Kathryn Carter, dev dir; Mike McCurdy, progmg dir; Jon Norton, mus dir; Willis Kern, news dir; Mark Hill, chief of engrg.

WIHN(FM)— Dec 21, 1973: 96.7 mhz; 3.9 kw. Ant 410 ft TL: N40 28 34 W89 02 02. Stereo. Hrs opn: 24 520 N. Center St., Bloomington, 61701. Phone: (309) 834-1100. Fax: (309) 834-4390. Web Site: www.967irock.com. Licensee: Connoisseur Media of Bloomington LLC. Group owner: AAA Entertainment L.L.C. (acq 11-2-2006; $4 million). Population served: 127,000 Natl. Network: ABC. Natl. Rep: Christal. Law Firm: Davis Wright Tremaine LLP. Format: Rock. News staff: one; News: 9 hrs wkly. Target aud: 18-44. ♦Larry Weiss, gen mgr.

WVMG(FM)— Aug 12, 2005: 100.7 mhz; 4.2 kw. Ant 344 ft TL: N40 27 08 W88 57 48. Hrs open: 520 N. Center St., Bloomington, 61701-2902. Phone: (309) 834-1100. Fax: (309) 834-4390. Web Site: www.magic1007.fm. Licensee: Connoisseur Media LLC. Natl. Rep: McGavren Guild. Law Firm: Shaw Pittman LLP. Format: Adult contemp. Target aud: 25-54; adult. ♦Larry Weiss, gen mgr; Grant Thompson, gen sls mgr; Bob Vandergrift, progmg dir; Mark Hill, chief of engrg.

Oak Lawn

WNWI(AM)— Dec 31, 1965: 1080 khz; 1.9 kw-D. TL: N41 38 36 W87 38 45. Hrs open: 24 934 W. 138th St., Riverdale, 60827. Phone: (708) 201-9600. Fax: (248) 557-2950. E-mail: infomacja@wietrzneradio.com Licensee: Birach Broadcasting Corp. (group owner; (acq 6-30-95; $375,000). Population served: 88,000 Format: Foreign language, Pol. ♦Sima Birach, pres & gen mgr.

Oak Park

WPNA(AM)— Oct 7, 1950: 1490 khz; 1 kw-U. TL: N41 52 52 W87 47 38. Hrs open: 24 408 S. Oak Park Ave., 60302. Phone: (708) 848-8980. Fax: (708) 848-9220. E-mail: email@wpna1490am.com Web Site: www.wpna1490am.com. Licensee: Alliance Communications Inc. (acq 5-1-87). Wire Svc: AP Format: Ethnic, Polish. News: 7 hrs wkly. Target aud: General. Spec prog: Polka 15 hrs, gospel 2 hrs, relg 4 hrs, Irish 4 hrs, Ukranian 2 hrs wkly. ♦Frank Spula, pres; Emily Leszczynski, gen mgr; Alan Kearns, chief of opns; Jerry Obrecki, gen sls mgr, progmg mgr & news dir.

WVAZ(FM)— Oct 17, 1950: 102.7 mhz; 6 kw. 1,170 ft TL: N41 53 56 W87 37 23. (CP: 9 kw). Stereo. Hrs opn: 24 233 N. Michigan Ave., 28th Fl., Chicago, 60601. Phone: (312) 540-2000. Fax: (312) 938-4404. Web Site: www.wvaz.com. Licensee: AMFM Broadcasting Licenses LLC. Group owner: Clear Channel Communications Inc. (acq 8-30-2000; grpsl). Population served: 9,119,500 Format: Black adult contemp. News staff: 2; News: one hr wkly. Target aud: 25-54; Black adults. Spec prog: Gospel 4 hrs, pub affrs 2 hrs wkly. ♦Elroy Smith, opns mgr; Anita Genes, gen sls mgr & natl sls mgr; Angela Ingram, prom dir; Armando Rivera, mus dir; Tim Wright, chief of engrg; Jodie Craigen, traf mgr; Wanda Wells, news dir, pub affrs dir & local news ed; Herb Kent, disc jockey; Myke Julius, disc jockey; Troi Tyler, disc jockey.

Oglesby

WALS(FM)— February 1993: 102.1 mhz; 2.25 kw. 446 ft TL: N41 18 05 W88 57 11. Hrs open: 24 3905 Progress Blvd., Peru, 61354. Phone: (815) 224-2100. Fax: (815) 224-2066. E-mail: walls102@theradiogroup.net Licensee: Laco Radio Inc. Group owner: Studstill Broadcasting Population served: 125,000 Law Firm: Booth, Freret, Imlay & Tepper. Format: Country. News: 5 hrs wkly. Target aud: 25-55. ♦Doris A. Studstill, CEO, pres & gen mgr; Cole C. Studstill, CFO; Lamar Studstill, chmn & stn mgr; Judy Miller, gen sls mgr.

Olney

***WPTH(FM)**— July 1992: 88.1 mhz; 133 w. 203 ft TL: N38 41 50 W88 02 15. (CP: 720 w). Hrs opn: 817 Orchard Dr., 62450. Phone: (618) 863-2765. Fax: (618) 395-7064. Licensee: Olney Voice of Christian Faith Inc. Format: Christian, talk. ♦Dr. Thomas E. Benson, pres & gen mgr; Ron James, VP.

WSEI(FM)—Listing follows WVLN(AM).

***WUSI(FM)**— Nov 1, 1992: 90.3 mhz; 25 kw. Ant 472 ft TL: N38 50 18 W88 07 46. Stereo. Hrs opn: 24
Rebroadcasts WSIU(FM) Carbondale 100%.
Rm. 1003, Communications Bldg., Carbondale, 62901-6602. Phone: (618) 453-4343. Fax: (618) 453-6186. E-mail: wsiuradio@wsiu.org Web Site: www.wsiu.org. Licensee: Southern Illinois University. (group owner) Population served: 21,759 Natl. Network: NPR, PRI. Law Firm: Cohn & Marks. Format: Class, news. News staff: 3; News: 36 hrs wkly. Target aud: 35-64; highly educated, upper income, socially conscious. ♦Candis Isberner, CEO; Jeff Williams, stn mgr, news dir & chief of engrg; Mike Zelten, opns mgr. Co-owned TV: WUSI-TV affil

WVLN(AM)— Nov 11, 1947: 740 khz; 250 w-D, 7 w-N. TL: N38 42 00 W88 04 53. Hrs open: Box L, 62450. Secondary address: 4667 E. Radio Tower Ln. 62450. Phone: (618) 393-2156. Fax: (618) 392-4536. Licensee: V.L.N. Broadcasting Inc. Group owner: Key Broadcasting Inc. (acq 7-21-87; $1.12 million with co-located FM; FTR: 6-8-87) Population served: 130,000 Natl. Network: ESPN Radio. Format: Sports. ♦Terry E. Forcht, pres.

WSEI(FM)—Co-owned with WVLN(AM). 1953: 92.9 mhz; 50 kw. Ant 552 ft TL: N38 42 00 W88 04 49. Stereo. Web Site: www.929thelegend.com.138,000 Format: Classic hits.

Oregon

WSEY(FM)— Dec 27, 1999: 95.7 mhz; 3.2 kw. Ant 358 ft TL: N42 04 19 W89 25 08. Hrs open: c/o WIXN-AM-FM, 1460 S. College Ave., Dixon, 61021. Phone: (815) 288-3341. Fax: (815) 284-1017. E-mail: wixnstaff@wixn.com Licensee: NRG License Sub. LLC. (group owner; (acq 10-31-2005; grpsl). Format: Oldies. ♦Allan Knickrehm, gen mgr & gen sls mgr; Steve Marco, progmg dir; Danette Dallgas-Frey, news dir; Mark Baker, chief of engrg.

Ottawa

WCMY(AM)— Mar 5, 1952: 1430 khz; 500 w-D, 38 w-N. TL: N41 20 53 W88 48 15. Hrs open: 24 216 W. Lafayette, 61350. Phone: (815) 434-6050. Fax: (815) 434-5311. E-mail: info@wcmy1430.com Licensee: NRG License Sub. LLC. (group owner; (acq 10-31-2005; grpsl). Population served: 135,000 Natl. Network: CBS Radio, Westwood One. Natl. Rep: Interep. Wire Svc: AP Format: Adult contemp, news/talk. News staff: 2; News: 35 hrs wkly. Target aud: 25 plus. Spec prog: Farm 9 hrs wkly. ♦Jay Le Seuve, opns mgr & progmg dir; Jill Williams, prom dir; Rick Koshko, news dir.

WRKX(FM)—Co-owned with WCMY(AM). Sept 1, 1964: 95.3 mhz; 4.3 kw. 200 ft TL: N41 23 00 W88 51 16. Stereo. 135,000 Format: Modern adult contemp.

***WWGN(FM)**— Sept 24, 1994: 88.9 mhz; 1.5 kw. 646 ft TL: N41 16 51 W88 56 13. Stereo. Hrs opn: 24 Box 2440, Tupelo, MS, 38803. Phone: (815) 433-6000. Fax: (815) 433-6100. E-mail: wwgn@afo.net Web Site: www.afr.net. Licensee: American Family Association Group owner: American Family Radio (acq 1-4-99; $250,000). Population served: 885,00 Format: Relg, educ. News staff: 2; News: 14 hrs wkly. Target aud: General. ♦Tim Wildmon, pres; Dan Hennenfent, gen mgr, stn mgr, progmg dir & engrg dir.

Palatine

***WHCM(FM)**— 2003: 88.3 mhz; 100 w. Ant 56 ft TL: N42 04 54 W88 04 23. Hrs open: William Rainey Harper College, 1200 W. Algonquin Rd., 60067. Phone: (847) 925-6000. Phone: (847) 925-6488. Web Site: www.harpercollege.edu. Licensee: William Rainey Harper College. Format: College, div. ♦Dave Dluger, gen mgr.

Pana

WMKR(FM)— July 12, 1996: 94.3 mhz; 5.6 kw. Ant 341 ft TL: N39 27 08 W89 17 10. Stereo. Hrs opn: 24 Box 169, 918 E. Park St., Taylorville, 62568-0169. Phone: (217) 824-3395. Fax: (217) 824-3301. Web Site: www.randyradio.com. Licensee: Miller Communications Inc. Group owner: Miller Media Group Population served: 197,000 Law Firm: Womble, Carlyle, Sandridge & Rice. Format: Country. Target aud: 25-54. ♦Randal J. Miller, pres & gen mgr; Kami Payne, stn mgr; Brandon Fellows, mus dir; Darrin Wright, news rptr.

***WZRS(FM)**—Not on air, target date: unknown: 89.3 mhz; 500 w. Ant 233 ft TL: N39 22 19 W89 04 51. Hrs open: Drawer 3206, Tupelo, MS, 38803. Phone: (662) 844-8888. Fax: (662) 842-6791. Web Site: www.afr.net. Licensee: American Family Association. Group owner: American Family Radio. Format: Inspirational Christian.

Paris

WINH(FM)—Listing follows WPRS(AM).

WPRS(AM)— 1951: 1440 khz; 1 kw-D, 250 w-N. TL: N39 36 21 W87 43 35. Stereo. Hrs opn: 24 Box 277, Rt. 133 W., 61944. Secondary address: 12861 Illinos Hwy. 133 61944. Phone: (217) 465-6336. Fax: (217) 466-1408. E-mail: wacf@comwares.net Licensee: Midwest Communications Inc. Group owner: Key Broadcasting Inc. (acq 12-15-2005; $2.55 million with co-located FM). Population served: 50,000 Format: Talk, sports, news. News staff: one. Target aud: General. ♦Duke E. Wright, pres; Karl Wertzler, gen mgr; Doug Boyd, gen sls mgr; Steve Hall, opns mgr & progmg dir; B.J. Fessant, news dir.

WINH(FM)—Co-owned with WPRS(AM). 1952: 98.5 mhz; 50 kw. 500 ft TL: N39 36 21 W87 43 35. Stereo. 100,000 Format: Modern country. Target aud: 18 plus. ♦Al Larcher, progmg dir.

Park Forest

WRZA(FM)— Jan 5, 1962: 99.9 mhz; 50 kw. Ant 492 ft TL: N41 18 04 W87 49 35. Stereo. Hrs opn: 24 6012 S Pulaski Rd, Chicago, 60629. Phone: (773) 767-1000. Fax: (773) 767-1100. Web Site: www.weplayanything.com. Licensee: WCLR Inc. Group owner: Newsweb Corp. (acq 3-16-2004; $24 million with WNDZ(AM) Portage, IN). Population served: 2,000,000 Format: Adult contemp. News staff: one. Target aud: 18-49. ♦Harvey Wells, gen mgr.

Park Ridge

***WMTH(FM)**— May 22, 1960: 90.5 mhz; 100 w. 103 ft TL: N42 02 14 W87 51 30. Hrs open: 2601 W. Dempster St., 60068. Phone: (847) 692-8495. Fax: (847) 692-8499. Licensee: Board of Education School District No. 207. Population served: 42,466 Format: Var. ♦Jim Wunderlich, gen mgr.

Paxton

WPXN(FM)— Oct 1, 1984: 104.9 mhz; 3 kw. 298 ft TL: N40 27 11 W88 06 11. Hrs open: 24 361 N. Railroad Ave., 60957. Phone: (217) 379-4333. Phone: (217) 892-9796. Fax: (217) 379-4334. Licensee: Paxton Broadcasting Corp. (acq 7-84). Population served: 100,000 Natl. Network: CBS. Rgnl. Network: Brownfield. Natl. Rep: Roslin. Law Firm: Borsari & Paxson. Format: Adult contemp. News staff: one; News: 8 hrs wkly. Target aud: 25-49. Spec prog: Farm 10 hrs wkly. ♦Dan Daugherity, pres, gen mgr & gen sls mgr; Joel Cluver, stn mgr & progmg VP.

Pekin

***WBNH(FM)**— Dec. 1988: 88.5 mhz; 48,000 kw. 495 ft TL: N40 38 34 W89 32 38. Stereo. Hrs opn: Box 1132, 61555. Phone: (309) 347-8850. Fax: (309) 353-8850. E-mail: wbnh@wbnh.org Web Site: www.wbnh.org. Licensee: Central Illinois Radio Fellowship Inc. Natl. Network: Moody. Law Firm: Southmayd & Miller. Format: Relg. Target aud: General. ♦Don Rice, pres; Scott Krus, stn mgr.

***WCIC(FM)**— Nov 2, 1983: 91.5 mhz; 35 kw. 338 ft TL: N40 33 24 W89 34 04. Stereo. Hrs opn: 24 3902 W. Baring Trace, Peoria, 61615. Phone: (309) 282-9191. Fax: (309) 282-9192. E-mail: wcic@wcicfm.org Web Site: www.wcicfm.org. Licensee: Illinois Bible Institute. (group owner) Format: Relg, adult contemp, Christian. News: 2 hrs wkly. Target aud: 25-49. ♦Dave Brooks, gen mgr.

WGLO(FM)—Listing follows WVEL(AM).

WVEL(AM)— Apr 21, 1948: 1140 khz; 5 kw-D, 3.2 kw-CH. TL: N40 36 08 W89 37 32. Hrs open: 120 Eaton St., Peoria, 61603. Phone: (309) 676-5000. Fax: (309) 676-2600. Web Site: www.wvel.com. Licensee: Regent Broadcasting of Peoria Inc. Group owner: Regent Communications Inc. (acq 7-6-01; grpsl). Population served: 1,000,000 Format: Gospel. Target aud: General. ♦J.R. Greeley, gen mgr; Robert Caruth, sls VP & progmg dir.

WGLO(FM)—Co-owned with WVEL(AM). Nov 18, 1971: 95.5 mhz; 25 kw. 620 ft TL: N40 36 23 W89 32 20. Stereo. Web Site: www.955glo.com.350,000 Natl. Rep: D & R Radio. Format: Classic rock. Target aud: 18-49. ♦Matt Bahan, progmg dir.

WXCL(FM)— 1973: 104.9 mhz; 6 kw. Ant 328 ft TL: N40 38 34 W89 32 38. Stereo. Hrs opn: 24 331 Fulton St., Suite 1200, Peoria, 61602. Phone: (309) 637-3700. Fax: (309) 272-1476. Web Site: www.1049thewolf.com. Licensee: Monterey Licenses LLC. Group owner: AAA Entertainment L.L.C. (acq 4-21-2006; $5.2 million with WDQX(FM) Morton). Population served: 350,000 Format: Country. News staff: 2. Target aud: 25-54; affluent adults. ♦Mike Wild, gen mgr; chris michaels, progmg dir.

Peoria

***WAZU(FM)**—Not on air, target date: unknown: 90.7 mhz; 1.3 kw. Ant 230 ft TL: N40 46 22 W89 44 50. Hrs open: 2122 W. Kellogg Ave., West Peoria, 61604. Phone: (309) 253-1951. Licensee: Sirius Syncope Inc. ♦Jeremy Styninger, pres.

***WCBU(FM)**— January 1970: 89.9 mhz; 26.5 kw. Ant 647 ft TL: N40 37 44 W89 34 12. Stereo. Hrs opn: 1501 W. Bradley Ave., 61625. Phone: (309) 677-3690. Fax: (309) 677-3462. E-mail: wcbu@bradley.edu Web Site: www.wcbufm.org. Licensee: Bradley University. Population served: 300,000 Natl. Network: NPR. Law Firm: Dow Lohnes. Wire Svc: AP Format: Class, news. ♦Thomas Hunt, gen mgr; Daryl Scott, opns mgr; Heather Binder, gen sls mgr; Nathan Irwin, progmg dir; Jonathan Ahl, news dir; David Schenk, chief of engrg.

WGLO(FM)—See Pekin

WIRL(AM)— 1947: 1290 khz; 5 kw-U, DA-2. TL: N40 37 24 W89 35 27. Stereo. Hrs opn: 331 Fulton St., Suite 1200, 61602. Phone: (309) 637-3700. Fax: (309) 673-9562. Web Site: www.1290wirl.com. Licensee: Monterey Licenses LLC. Group owner: JMP Media LLC (acq 3-25-2003; grpsl). Population served: 374,000 Natl. Rep: Christal. Format: Classic country. Target aud: 25-54; men. ♦David J. Benjamin III, pres; Mike Wild, gen mgr; Mark Bretsck, sls dir & prom mgr; Brian Rowell, gen sls mgr; John Malone, progmg dir; Dave Dahl, news dir; Wayne Miller, chief of engrg; Brenda Rundle, traf mgr.

WSWT(FM)—Co-owned with WIRL(AM). 1964: 106.9 mhz; 50 kw. 479 ft TL: N40 43 22 W89 30 40. Stereo. 24 331 Fulton St., 12th Fl., 61602. Fax: (309) 686-8659. Web Site: www.literock107.com.500,000 Natl. Rep: Christal. Format: Adult contemp. News: 2.5 hrs wkly. Target aud: 25-54. ♦Randy Rundle, opns mgr; Dirk Clemens, prom mgr; Wayne R. Miller, progmg dir & chief of engrg.

WIXO(FM)— May 14, 1972: 105.7 mhz; 32 kw. Ant 555 ft TL: N40 43 25 W89 29 04. Stereo. Hrs opn: 4234 Brandywine Dr., Suite D, 61614. Phone: (309) 686-0101. Fax: (309) 686-0111. E-mail: studio@mix1057.com Web Site: www.mix1057.com. Licensee: Regent Broadcasting of Peoria Inc. Group owner: AAA Entertainment L.L.L. (acq 9-19-2006; $11.75 million with WZPW(FM) Peoria). Population served: 126,963 Natl. Rep: Christal. Format: Modern rock. ♦Michael Rea, gen mgr; Marta Poznaska, prom dir; Scott Seipel, progmg dir & news dir; Brett Ring, chief of engrg; Becky Riojas, sls.

WMBD(AM)— 1927: 1470 khz; 5 kw-U, DA-2. TL: N40 34 22 W89 32 00. Stereo. Hrs opn: 24 331 Fulton St., Suite 1200, 61602. Phone: (309) 637-3700. Fax: (309) 673-9562. Web Site: www.1470wmbd.com. Licensee: Monterey Licenses LLC. Group owner: JMP Media LLC (acq 3-25-2003; grpsl). Natl. Network: Premiere Radio Networks, Westwood One. Natl. Rep: Christal. Law Firm: Shaw Pittman. Wire Svc: U.S. Weather Service Format: News/talk. News staff: 5; News: 40 hrs wkly. Target aud: 35-64; upscale, well educated, professional. Spec prog: Farm 15 hrs wkly. ♦David J. Benjamin III, pres; Mike Wild, gen mgr & stn mgr.

WPBG(FM)—Co-owned with WMBD(AM). 1947: 93.3 mhz; 41 kw. Ant 548 ft TL: N40 38 07 W89 32 19. Stereo. 24 Fax: (309) 686-8659. Web Site: www.933thedrive.com. Format: Classic hits. Target aud: 25-54; baby boomers.

WOAM(AM)— Feb 8, 1960: 1350 khz; 1 kw-U, DA-2. TL: N40 35 41 W89 35 40. Hrs open: 24 Kelly Communications, 3641 Meadowbrook Rd., 61604. Phone: (309) 685-0977. Fax: (309) 685-7150. Web Site: www.1350woam.com. Licensee: Kelly Communications Inc. (acq 12-1-86; $500,000; FTR: 9-29-86). Population served: 350,000 Natl. Network: ABC. Natl. Rep: McGavren Guild. Format: Adult standards. News staff: 2; News: 10 hrs wkly. Target aud: 45 plus; upscale business leaders. Spec prog: Jazz 2 hrs wkly. ♦Bob Kelly, stn mgr & news dir.

WPEO(AM)— 1946: 1020 khz; 1 kw-D. TL: N40 41 53 W89 31 31. Hrs open: 1708 Highview Rd., East Peoria, 61611. Secondary address: Box 1 61650. Phone: (309) 698-9736. Fax: (309) 698-9740. E-mail: wpeo@wpeo.com Web Site: www.wpeo.com. Licensee: Pinebrook Foundation Inc. (acq 1-6-70). Population served: 500,000 Law Firm: Wood, Maines & Brown. Format: Relg, talk. Target aud: 35 plus. ♦Richard T. Crawford, pres; Robert Ulrich, gen mgr; Roger Bennington, gen sls mgr; Nelson Hostetler, progmg dir; Denise Feller, traf mgr.

WPMJ(FM)—See Chillicothe

WVEL(AM)—See Pekin

WXCL(FM)—See Pekin

WZPW(FM)— November 1992: 92.3 mhz; 19.2 kw. Ant 374 ft TL: N40 47 10 W89 47 01. Hrs open: 120 Eaton, 61603. Phone: (309) 676-5000. E-mail: studio@power92.net Web Site: www.power92.net. Licensee: B&G Broadcasting Inc. Group owner: AAA Entertainment L.L.C. (acq 9-19-2006; $11.75 million with WIXO(FM) Peoria). Format: Top 40, CHR. ♦J.R. Greeley, gen mgr; Quinton Hafron, progmg dir.

Peru

WBZG(FM)— Mar 15, 1970: 100.9 mhz; 3 kw. 328 ft TL: N41 18 09 W89 14 11. Stereo. Hrs opn: 24 3905 Progress Blvd., 61354. Phone: (815) 224-2100. Fax: (815) 224-2066. E-mail: wbzg@theradiogroup.net Web Site: wbzg.net. Licensee: Mendota Broadcasting Inc. Group owner: Studstill Broadcasting (acq 7-17-97; $700,000 with WIVQ(FM) Spring Valley). Population served: 130,300 Natl. Rep: Rgnl Reps. Law Firm: Booth, Freret, Imlay & Tepper. Wire Svc: AP Format: Classic rock. News staff: 2. Target aud: Men 18-54. ♦Lamar Studstill, chmn; Cole Charles Studstill, CFO & VP; Owen L. Studstill, CEO, pres & gen mgr; Doris Studstill, opns VP; Cole Studstill, opns mgr.

WXAN(FM)—See Ava

Petersburg

WLCE(FM)— March 1987: 97.7 mhz; 6 kw. Ant 328 ft TL: N40 00 05 W89 41 49. (CP: N39 54 35 W89 43 01). Stereo. Hrs opn: 24 Box 460, Springfield, 62705. Phone: (217) 629-7077. Fax: (217) 629-7952. E-mail: alice@alice.fm Web Site: www.alice.fm. Licensee: Long-Nine Inc. Group owner: The Mid-West Family Broadcast Group (acq 7-27-2001; $3 million). Law Firm: Shaw Pittman. Wire Svc: AP Format: CHR. Target aud: 18-34; female. Spec prog: Relg 5 hrs wkly. ♦Kevin Kavanaugh, gen mgr; Dave Duetsch, gen sls mgr; Valerie Knight, progmg dir; Jim Leach, news dir; Greg Stephens, chief of engrg; Quinn Fagg, traf mgr.

***WLWJ(FM)**— Oct 7, 2001: 88.1 mhz; 6 kw. 328 ft TL: N40 00 05 W89 41 49. Stereo. Hrs opn: 600 W. Mason St., Springfield, 62702. Phone: (217) 528-2300. Fax: (217) 528-2400. Web Site: www.wluj.org. Licensee: Cornerstone Community Radio Inc. Natl. Network: USA. Format: Christian talk, inspirational music. ♦Richard Van Zandt, gen mgr; John McBride, stn mgr; Howard Fouks, opns mgr; Richard Beaman, gen sls mgr.

Pittsfield

WBBA-FM— Aug 1, 1966: 97.5 mhz; 10 kw. Ant 300 ft TL: N39 34 53 W90 47 52. Stereo. Hrs opn: 24 Box 312, 62363. Phone: (217) 285-5975. Fax: (217) 285-5977. Web Site: www.wbbaradio.com. Licensee: DJ Two Rivers Radio Inc. (acq 1-9-2004; $320,000 with WJBM(AM) Jerseyville). Population served: 225,000 Format: Country. Target aud: General. ♦David Fuhler, gen mgr.

***WIPA(FM)**— Jan 4 1993: 89.3 mhz; 50 kw. 492 ft TL: N39 43 25 W90 41 09. Stereo. Hrs opn: 24
Rebroadcasts WUIS(FM) Springfield 100%.
1 Universiity Plaza, WUIS 130, Springfield, 62703-5407. Phone: (217) 206-6516. Fax: (217) 206-6527. E-mail: wuis@uis.org Web Site: www.wuis.org. Licensee: University of Illinois at Springfield. Natl. Network: NPR, PRI. Rgnl. Network: Ill. Radio Net. Law Firm: Dow, Lohnes & Albertson. Format: News, class, jazz. News staff: 4; News: 45 hrs wkly. Target aud: 25-54. ♦Bill Wheelhouse, gen mgr; Lisa Clemmons-Stott, dev dir; Sinta Seiber, opns mgr & prom dir; Rick Bradley, news dir; Greg Manfroi, engrg mgr & chief of engrg.

Plano

WSPY-FM— Jan 19, 1974: 107.1 mhz; 1.5 kw. 466 ft TL: N41 39 55 W88 34 34. Stereo. Hrs opn: 24 One Broadcast Ctr., 60545. Phone: (630) 552-1000. Fax: (630) 552-9300. E-mail: wspy@nelsonmultimedia.net Licensee: Nelson Enterprises Inc. Population served: 750,000 Format: Full service. News: 12 hrs wkly. Target aud: 25-54. Spec prog: Farm 18 hrs wkly. ♦Larry Nelson, pres; Carol Barrows, CFO; Beth Pierre, gen mgr & gen sls mgr; Paul Morgan, opns mgr; Paul Morga, progmg dir; Lane Lindstrom, chief of engrg; Jeni Beckman, traf mgr.

Polo

WLLT(FM)— Dec 12, 1989: 107.7 mhz; 1.35 kw. 476 ft TL: N41 53 52 W89 36 20. Stereo. Hrs opn: 24 260 Illinois Rt. 2, Dixon, 61021. Phone: (815) 284-1077. Licensee: Sauk Valley Broadcasting Co. Population served: 75,000 Format: Soft adult contemp. ♦Bob Burns, gen mgr; Bob Thomas Burns, pres & gen mgr.

Pontiac

***WPJC(FM)**— 2003: 88.3 mhz; 500 w. Ant 207 ft TL: N40 56 42 W88 38 46. Hrs open: 150 Lincoln Way, Suite 2001, Valparaiso, IN, 46383. Phone: (219) 548-8956. Fax: (219) 548-5808. E-mail: wpjc@csnradio.com Licensee: CSN International (group owner; acq 12-31-2001; $25,000 for CP). Format: Christian talk. ♦Jim Motshagen, gen mgr.

WTRX-FM— July 1969: 93.7 mhz; 12 kw. Ant 472 ft TL: N40 45 27 W88 37 40. Stereo. Hrs opn: 24 315 N. Mill St., 61764. Phone: (815) 844-6101. Fax: (815) 844-7235. Web Site: www.wtrxoldieschannel.com. Licensee: Livingston County Broadcasters Inc. Group owner: Regent Communications Inc. (acq 5-12-2004; grpsl). Population served: 175,000 Format: Oldies. News staff: one. Target aud: 25-54. ♦Red Pitcher, gen mgr & gen sls mgr; Ron Ross, progmg dir.

Princeton

***WPRC(FM)**—Not on air, target date: unknown: 88.3 mhz; 150 w vert. Ant 315 ft TL: N41 16 53 W89 35 12. Hrs open: c/o WCIC, 3902 S. Baring Trace, Peoria, 61615. Phone: (309) 282-9191. Fax: (309) 282-9192. E-mail: wcic@wcicfm.org Licensee: Illinois Bible Institute Inc. (group owner). Format: Christian talk. ♦Dave Brooks, gen mgr; Tracey Moushon, prom dir.

WZOE(AM)— Oct 25, 1961: 1490 khz; 1 kw-U. TL: N41 21 08 W89 28 05. Hrs open: 24 Box 69, Broadcast Ctr., 61356. Secondary address: S. Main St. 61356. Phone: (815) 875-8014. Web Site: www.wzoeradio.com. Licensee: WZOE Inc. (group owner; acq 11-1-73). Population served: 63,000 Natl. Network: CBS. Law Firm: Shaw Pittman. Wire Svc: Metro Weather Service Inc. Format: News/talk, sports. News staff: 3; News: 84 hrs wkly. Spec prog: Farm 15 hrs wkly. ♦Steve Samet, pres & gen mgr; Paul Bomleny, opns dir; Chris Compton, sls dir; Tommy Rose, progmg dir; Scott Mighle, news dir; Nedda Simon, pub affrs dir & women's int ed; Greg Stephens, chief of engrg; Mary Harmon, traf mgr.

WZOE-FM— July 1, 1980: 98.1 mhz; 6 kw. Ant 300 ft TL: N41 21 49 W89 23 36. Stereo. 24 Web Site: www.wzoeradio.com.70,700 Natl. Network: CNN Radio. Format: Classic hits. News staff: 3; News: 8 hrs wkly. ♦Mary Harmon, traf mgr; Nedda Simon, women's int ed.

Quincy

KGRC(FM)—See Hannibal, MO

WCOY(FM)—Listing follows WTAD(AM).

***WGCA-FM**— Sept 20, 1987: 88.5 mhz; 40 kw. 449 ft TL: N39 58 18 W91 19 42. Stereo. Hrs opn: 535 Main, Suite 10, 62306. Phone: (217) 224-9422. Fax: (217) 228-0504. E-mail: themix@wgca.org Web Site: www.wgca.org. Licensee: Great Commission Broadcasting Corp. Natl. Network: USA. Format: Christian contemp. Target aud: 25-45. ♦Bruce Rice, gen mgr & progmg dir; Jim Taylor, progmg dir; Jim Wilson, chief of engrg.

WGEM(AM)— Jan 1, 1948: 1440 khz; 5 kw-D, 1 kw-N, DA-2. TL: N39 58 47 W91 19 27. Stereo. Hrs opn: Box 80, 62306. Secondary address: 513 Hampshire 62301. Phone: (217) 228-6600. Fax: (217) 228-6670. Fax: TWX: 910-246-3209. E-mail: jlawrence@wgem.com Web Site: wgem.com. Licensee: Quincy Broadcasting Co. Group owner: Quincy Newspapers Inc. Population served: 150,000 Natl. Network: ESPN Radio. Rgnl. Network: Tribune. Natl. Rep: Christal. Law Firm: Wilkinson, Barker & Knauer. Wire Svc: AP Format: Sports. News: 40 hrs wkly. Target aud: 25-54; general. ♦Thomas A. Oakley, CEO; Ralph M. Oakley, COO & opns dir; Thomas Oakley, pres; Tom Allen, gen mgr.

WGEM-FM— 1947: 105.1 mhz; 27.5 kw. 500 ft TL: N39 57 03 W91 19 54. Stereo. 24 Web Site: www.wgem.com.300,000 Natl. Rep: Christal. Law Firm: Wilkinson, Barker, Knauer & Quinn. Format: News/talk. News: 110 hrs wkly. Target aud: 25-54. Co-owned TV: WGEM-TV affil

WLIQ(AM)— Dec 13, 1966: 1530 khz; 1.4 kw-D, 290 w-CH. TL: N39 55 51 W91 25 46. Stereo. Hrs opn: Sunrise-sunset Box 711, Hannibal, MO, 63401. Phone: (573) 221-3450. Fax: (573) 221-5331. Licensee: Bick Broadcasting Co. (group owner; (acq 10-14-2003). Population served: 103,010 Rgnl. Network: Brownfield. Format: Talk. ♦Ed Foxall, gen mgr.

WQCY(FM)— May 8, 1989: 103.9 mhz; 1.8 kw. Ant 436 ft TL: N39 56 30 W91 35 03. Hrs open: 329 Maine St., 62301. Phone: (217) 224-4102. Fax: (217) 224-4133. E-mail: wqcy@staradio.com Web Site: www.1039thefox.com. Licensee: STARadio Corp. (group owner; acq 8-13-98; grpsl). Natl. Network: CBS. Format: Hot 80s. Target aud: 18-44; general. ♦Michael J. Moyers, gen mgr; Brenda Park, gen sls mgr; Sean Secrease, progmg dir; Mary Griffith, mus dir & news dir; Phillip Reilly, chief of engrg; Jerry Shoup, traf mgr & disc jockey.

***WQUB(FM)**— April 1974: 90.3 mhz; 28 kw. 417 ft TL: N39 57 22 W91 23 22. Stereo. Hrs opn: 24 1800 College Ave., 62301. Phone: (217) 228-5410. Fax: (217) 228-5616. E-mail: info@wqub.org Web Site: www.wqub.org. Licensee: Quincy University Corp. Population served: 240,000 Natl. Network: NPR, AP Radio. Law Firm: Wilkinson Barker Knauer. Format: Class, news, jazz. News staff: 2; News: 31 hrs wkly. Target aud: 25-64; male & female. Spec prog: Folk 2 hrs, blues 2 hrs, alternative rock 12 hrs, hip hop 2 hrs, oldies 2 hrs wkly. ♦Patrick Mays, dev dir; Jim Lenz, pub affrs dir; Jim Cate, engrg mgr & chief of engrg.

WTAD(AM)— July 25, 1925: 930 khz; 5 kw-D, 1 kw-N, DA-N. TL: N39 53 31 W91 25 25. Hrs open: 24 Lincoln-Douglas, 329 Maine St., 62306. Phone: (217) 224-4102. Fax: (217) 224-4133. Web Site: www.wtad.com. Licensee: STARadio Corp. (group owner; (acq 8-13-98; grpsl). Population served: 80,000 Natl. Network: CBS. Natl. Rep: McGavren Guild. Format: News/talk. News staff: one; News: 22 hrs wkly. Target aud: 35 plus; general. ♦Brenda Parks, gen mgr; Michael J. Moyers, gen mgr.

WCOY(FM)—Co-owned with WTAD(AM). 1948: 99.5 mhz; 100 kw. Ant 489 ft TL: N39 56 30 W91 35 03.24 80000 Natl. Network: CBS. Format: Hot country classics. ♦Mike Moyers, progmg dir.

Ramsey

***WJLY(FM)**— 1999: 88.3 mhz; 25 kw. Ant 502 ft TL: N39 08 06 W89 06 02. Hrs open: 24 Box 456, 62080. Secondary address: R.R. 2 Box 51A Phone: (618) 423-2082. Fax: (618) 423-2394. E-mail: wjly@frontiernet.net Licensee: Countryside Broadcasting. Population served: 73,184 Natl. Network: Moody. Format: Christian. News: 14 hrs wkly. Target aud: 35 plus. ♦Richard Wheeler, gen mgr; John Stanley, gen sls mgr; Dave Carruthers, progmg dir; Henry Voss, chief of engrg.

WTRH(FM)— Nov 21, 1990: 93.3 mhz; 3 kw. Ant 466 ft TL: N39 08 06 W89 06 02. Stereo. Hrs opn: 24 Box 456, 62080. Phone: (618) 423-2082. Fax: (618) 423-2394. E-mail: wtrh@frontiernet.net Licensee: Countryside Broadcasting Inc. Population served: 31,000 Format: Oldies, talk. News: 14 hrs wkly. Target aud: 35 plus; men & women who love old radio prgms & mus. ♦Richard Wheeler, gen mgr & opns mgr; Henry Voss, chief of engrg.

Rantoul

WKJR(AM)—Licensed to Rantoul. See Champaign

WMYE(FM)—Licensed to Rantoul. See Champaign

WQQB(FM)—Licensed to Rantoul. See Urbana

River Grove

***WRRG(FM)**— Mar 10, 1975: 88.9 mhz; 100 w. 128 ft TL: N41 54 56 W87 50 12. Stereo. Hrs opn: 9 AM-midnight (M-F); 10 AM-midnight (S, Su) 2000 N. 5th Ave., R113, 60171. Phone: (708) 583-3110. E-mail: info@wrrg.org Web Site: www.wrrg.org. Licensee: Triton College. Population served: 500,000 Format: CHR, alternative. Target aud: 14-40. Spec prog: Jazz 5 hrs, loc 4 hrs, metal 4 hrs, world mus 3 hrs, oldies 11 hrs, classic rock 2 hrs wkly. ♦Kelli A. Lynch, gen mgr, stn mgr & progmg mgr.

Robinson

WTAY(AM)— Jan 9, 1956: 1570 khz; 250 w-D. TL: N39 00 29 W87 46 41. Hrs open: 24 Box 245, Rt. 33 W., 62454. Phone: (618) 544-2191. Fax: (618) 544-3621. E-mail: wtaywtye@yahoo.com Licensee: Ann Broadcasting Corp. (acq 1994). Natl. Network: ABC. Format: Adult contemp. News staff: one; News: 15 hrs wkly. Spec prog: C&W 12 hrs, farm 3 hrs, polka 3 hrs, big band 10 hrs wkly. ♦Jerry F. Tye, pres & gen mgr; Roy Rice, gen sls mgr; Jerry Tye, progmg dir.

WTYE(FM)—Co-owned with WTAY(AM). Jan 4, 1963: 101.7 mhz; 1.45 kw. 449 ft TL: N39 00 29 W87 46 41. Stereo. 35,000

Rochelle

WRHL(AM)— Sept 16, 1966: 1060 khz; 250 w-D, DA. TL: N41 55 24 W89 03 30. Hrs open: 24 Box 177, 61068. Secondary address: 400 May Mart Dr. 61068. Phone: (815) 562-7001. Fax: (815) 562-7002. E-mail: wrhlamfm@rochelle.net Web Site: www.wrhl.net. Licensee: Rochelle Broadcasting Co. Inc. (acq 10-11-70). Population served: 45,000 Natl. Network: AP Network News. Rgnl. Network: Ill. Radio Net., Tribune. Wire Svc: AP Format: News/talk. News staff: 2; News: 140 hrs wkly. Target aud: 25-75. ♦David Van Drew, gen mgr; Penny Helm, gen sls mgr; Greg Saunders, progmg dir; Jeffrey Leon, news dir; Doug White, chief of engrg; Becky Leininger, traf mgr & women's int ed.

WRHL-FM— Oct 5, 1973: 102.3 mhz; 4.6 kw. 180 ft TL: N41 55 24 W89 03 30. Stereo. 24 E-mail: jb@wrhl.net Web Site: www.hitsandfavorites.com.45,000 Natl. Network: ABC. Format: Adult contemp. News: one hr wkly. Target aud: 25-54; female. ♦Becky Leininger, traf mgr & women's int ed.

Rock Falls

WSDR(AM)—See Sterling

Rock Island

WKBF(AM)— Feb 16, 1925: 1270 khz; 5 kw-U, DA-N. TL: N41 29 40 W90 28 00. Hrs open: 24 4020 N. 128th St., Brookfield, WI, 53005. Phone: (888) 321-1270. Web Site: www.truth1270.com. Licensee: Quad Cities Media LLC (group owner; (acq 12-4-2006; $150,000). Population served: 350,000 Format: Christian talk. ♦Randall R. Melchert, pres.

WLKU(FM)— October 1947: 98.9 mhz; 39 kw. Ant 900 ft TL: N41 19 40 W90 22 47. Stereo. Hrs opn: 24 2351 Sunset Blvd., Suite 170-218, Rocklin, CA, 95765. Phone: (916) 251-1600. Fax: (916) 251-1650. Web Site: www.klove.com. Licensee: Educational Media Foundation. (acq 2-3-2006; $3.5 million). Natl. Network: K-Love. Format: Contemp Christian. ♦Richard Jenkins, pres; Mike Novak, VP; Keith Whipple, dev dir; David Pierce, progmg dir; Ed Lenane, news dir; Sam Wallington, engrg dir; Karen Johnson, news rptr; Marya Morgan, news rptr; Richard Hunt, news rptr.

***WVIK(FM)**— Feb 25, 1963: 90.3 mhz; 31 kw. 1,096 ft TL: N41 32 52 W90 28 29. Stereo. Hrs opn: 24 Augustana College, 639 38th St., 61201. Phone: (309) 794-7500. Fax: (309) 794-1236. Web Site: www.wvik.org. Licensee: Augustana College. Population served: 780,000 Natl. Network: NPR. Law Firm: Dow, Lohnes & Albertson. Wire Svc: AP Format: Class, news. News staff: 2; News: 35 hrs wkly. Target aud: General. Spec prog: Jazz 9 hrs wkly. ♦Lowell Dorman, gen mgr; David Garner, opns dir; Sonita Oldfield-Carlson, dev dir; Mindy Heusel, gen sls mgr & mus dir; Herb Trix, news dir.

Rockford

***WFEN(FM)**— Aug 25, 1991: 88.3 mhz; 8.5 kw. 575 ft TL: N42 21 51 W89 08 15. Hrs open: 4701 S. Main St., 61102. Phone: (815) 964-9336. Fax: (815) 964-9318. E-mail: fred@wfen.org Web Site: www.wfen.org. Licensee: Faith Center (acq 10-2-91; FTR: 10-28-91). Format: Contemp Christian, praise & worship. Target aud: 35-54. Spec prog: Sp one hr wkly. ♦Fred Tscholl, gen mgr.

WGFB(FM)—See Rockton

WLUV(AM)—See Loves Park

WNIJ(FM)—See De Kalb

***WNIU(FM)**— Apr 28, 1991: 90.5 mhz; 50 kw. 367 ft TL: N42 00 55 W89 00 07. Stereo. Hrs opn: 24 NIU Broadcast Ctr., 801 N. First St., DeKalb, 60115. Phone: (815) 753-9000. Phone: (815) 961-8000. Fax: (815) 753-9938. E-mail: npr@niu.edu Web Site: www.northernpublicradio.org. Licensee: Northern Illinois University. Population served: 500,000 Natl. Network: PRI, NPR. Law Firm: Arter & Hadden. Format: Classical. News staff: 2; News: 52 hrs wkly. Target aud: General. Spec prog: New age 2 hrs, blues 4 hrs wkly. ♦Tim Emmons, gen mgr; Jan Kilgard, dev dir; Bill Drake, progmg dir; Susan Stephens, news dir; Jeff Glass, chief of engrg.

WNTA(AM)— Dec 24, 1953: 1330 khz; 1 kw-D, 91 w-N, DA-2. TL: N42 13 32 W89 02 47. Hrs open: 24 2830 Sandy Hollow Rd., 61109. Secondary address: 2830 Sandy Hollow Rd. 61109. Phone: (815) 874-7861. Fax: (815) 874-2202. E-mail: wnta@wnta.com Web Site: www.wnta.com. Licensee: Maverick Media of Rockford License LLC. Group owner: RadioWorks Inc. (acq 4-27-2005; grpsl). Population served: 280,000 Law Firm: Shaw Pittman. Format: MOR, news, talk. News staff: 2; News: 2 hrs wkly. Target aud: 35 plus. Spec prog: Gospel 20 hrs wkly. ♦Gary Rozynek, pres; Jay Chapman, gen mgr; Ken DeCoster, progmg dir, news dir & chief of engrg.

WQFL(FM)— May 2, 1974: 100.9 mhz; 2.7 kw. Ant 489 ft TL: N42 19 20 W89 00 41. Stereo. Hrs opn: 24 Box 2730, 61132-2730. Secondary address: 5375 Pebble Creek Tr., Loves Park 61111. Phone: (815) 654-1200. Fax: (815) 282-7779. E-mail: positive@101qfl.com Web Site: www.101qfl.com. Licensee: Quest for Life Inc. (acq 6-80; $590,000; FTR: 6-2-80). Population served: 200,000 Law Firm: Wilkinson, Barker, Knauer, L.L.P. Format: Christian pop. News staff: one; News: one hr wkly. Target aud: 25-44; female dominant, educated, upscale & middle class. ♦Ralph Trentadue, gen mgr; Rick Hall, progmg dir.

WROK(AM)— 1923: 1440 khz; 5 kw-D, 270 w-N, DA-D. TL: N42 16 50 W89 02 16. Stereo. Hrs opn: 3901 Brendenwood Rd., 61107. Phone: (815) 399-2233. Fax: (815) 399-8148. Web Site: www.cumulus.com. Licensee: Cumulus Licensing Corp. Group owner: Cumulus Media Inc. (acq 10-2-00; grpsl). Population served: 150,000 Natl. Rep: McGavren Guild. Law Firm: Wiley, Rein & Fielding. Format: News/talk. Target aud: 35 plus. ♦Mary Gerard, rgnl sls mgr; Erika Mohr, prom dir; Jesse Garcia, progmg dir; Kelly Dukes, chief of engrg; Jan Thorpe, traf mgr.

WZOK(FM)—Co-owned with WROK(AM). 1949: 97.5 mhz; 50 kw. 235 ft TL: N42 16 50 W89 02 16. (CP: Ant 429 ft.). Stereo. Web Site: www.97zok.com. Format: CHR. Target aud: 25-34. ♦J.J. Morgan, progmg dir.

WRTB(FM)—See Winnebago

WXRX(FM)—See Belvidere

Rockton

WGFB(FM)— March 1963: 103.1 mhz; 1.2 kw. 525 ft TL: N42 22 02 W89 05 13. Stereo. Hrs opn: 24 2830 Sandy Hollow Rd., Rockford, 61109. Phone: (815) 874-7861. Fax: (815) 874-2202. Web Site: www.B103fm.com. Licensee: Maverick Media of Rockford License LLC. Group owner: RadioWorks Inc. (acq 4-27-2005; grpsl). Population served: 400000 Natl. Rep: Katz Radio. Law Firm: Shaw Pittman. Format: Adult contemp. Target aud: 25-54; adult women. ♦Gary Rozynek, pres; Jay Chapman, gen mgr; Michelle Markhan, prom dir; Jim Stone, progmg dir; Ken DeCoster, news dir; Chuck Ingle, chief of engrg.

Rushville

WKXQ(FM)— May 1, 1985: 92.5 mhz; 6 kw. Ant 328 ft TL: N40 08 20 W90 39 26. Stereo. Hrs opn: 5 AM-midnight 119 W. Caroll, Macomb, 61455. Phone: (815) 758-8686. Fax: (309) 833-3460. E-mail: wkxq92@frontiernet.net Licensee: WPW Broadcasting Inc. (group owner; (acq 12-6-99; $550,000 with WPWQ(FM) Mount Sterling). Population served: 50,000 Natl. Network: CNN Radio. Law Firm: Reddy, Begley & McCormick. Format: Oldies. News staff: one; News: 12 hrs wkly. Target aud: 18-54. Spec prog: Relg 6 hrs, farm 6 hrs wkly. ♦Don Davis, pres; Vanessa Wetterling, gen mgr; Michael Weaver, progmg dir.

Saint Anne

WXNU(FM)— 2006: 106.5 mhz; 1.95 kw. Ant 462 ft TL: N41 00 20 W87 41 42. Hrs open: 70 Meadowview Ctr., Kankakee, 60901. Phone: (815) 935-9555. Fax: (815) 935-9593. Web Site: www.wxnu.com. Licensee: STARadio Corp. Format: Country. ♦Bob Kersmarki, gen mgr; Brendan Michaels, opns mgr; Larry Regnier, gen sls mgr; Phil Reilly, chief of engrg.

Saint Joseph

***WGNJ(FM)**— 1999: 89.3 mhz; 50 kw. Ant 351 ft TL: N40 05 16 W87 53 42. Stereo. Hrs opn: 24 Box 12345, Champaign, 61826. Secondary address: 2421 N. 1450 E. Rd., White Heath 61884. Phone: (217)897-6333. E-mail: staff@greatnewsradio.org Web Site: www.greatnewsradio.org. Licensee: Good News Radio Inc. Population served: 1,000,000 Natl. Network: Salem Radio Network. Format: Christian, religious, talk. Target aud: 35 plus; general. ♦David Herriott, chmn; Mark Burns, pres & gen mgr; Carrie Burns, opns dir.

Salem

WJBD(AM)— Dec 16, 1956: 1350 khz; 430 w-D, 60 w-N. TL: N38 37 56 W88 55 02. Hrs open: 24 Box 70, 310 W. McMackin St., 62881. Secondary address: 221 E. Broadway, Suite 107, Centralia 62801. Phone: (618) 548-2000. Phone: (618) 532-9600. Fax: (618) 548-2079. E-mail: wjbd@accessus.net Licensee: NRG License Sub. LLC. (group owner; (acq 10-31-2005; grpsl). Population served: 24,000 Format: Country. News staff: 3; News: 30 hrs wkly. Target aud: General. Spec prog: Farm 4 hrs, relg 6 hrs wkly. ♦Bruce Kropp, gen mgr, opns VP, sls VP, progmg dir & news dir.

WJBD-FM— June 1, 1972: 100.1 mhz; 1.5 kw. 450 ft TL: N38 33 45 W88 59 57. Stereo. 24 E-mail: wjbd@accessus.net Format: Adult contemp, news. ♦Matt Tackett, opns mgr.

***WSLE(FM)**— 2005: 91.3 mhz; 900 w. Ant 154 ft TL: N38 37 34 W88 56 41. Hrs open: Drawer 3206, Tupelo, MS, 38803. Phone: (662) 844-8888. Fax: (662) 842-6791. Licensee: American Family Association. Group owner: American Family Radio. Format: Christian. ♦Marvin Sanders, gen mgr; John Riley, progmg dir.

Sandwich

WAUR(AM)—Licensed to Sandwich. See Aurora

Savanna

WCCI(FM)— Nov 7, 1971: 100.3 mhz; 25 kw. 450 ft TL: N42 07 49 W90 08 24. Stereo. Hrs opn: 24 Box 310, 316 Main, 61074. Phone: (815) 273-7757. Fax: (815) 273-2760. E-mail: radio@wccilive.com Web Site: www.wcciradio.com. Licensee: Carroll County Communications Inc. (acq 9-1-76). Population served: 188,700 Natl. Network: Fox News Radio. Rgnl. Network: Brownfield. Law Firm: Lauren A. Colby. Format: New hit country, news. News staff: one; News: 35 hrs wkly. Target aud: 25-54. ♦John L. Miller, pres & gen mgr; Edward F. Bock, VP; Brian Reusch, stn mgr; Leslie Smith, progmg dir; Mark Schoening, news dir.

Seneca

WJDK-FM— 1993: 95.7 mhz; 3 kw. 328 ft TL: N41 13 12 W88 32 27. Stereo. Hrs opn: 219 W. Washington St., Morris, 60450. Phone: (815) 941-1000. Fax: (815) 941-9300. Licensee: Grundy County Broadcasters Inc. (acq 4-20-98). Population served: 49,000 Natl. Network: ABC. Format: Adult contemp. News staff: one; News: 14 hrs wkly. Target aud: 25-49. ♦Larry Nelson, pres; Jack Daly, gen mgr; Mike Williams, opns mgr.

Shelbyville

WEJT(FM)— Dec 31, 1969: 105.1 mhz; 13 kw. 459 ft TL: N39 35 39 W88 50 44. Stereo. Hrs opn: 410 N. Water, Suite C, Decatur, 62523. Phone: (217) 428-4487. Fax: (217) 428-4501. Web Site: www.wejt.com. Licensee: Cromwell Group Inc. of Illinois. Group owner: The Cromwell Group Inc. (acq 8-1-89; $320,000 with co-located AM; FTR: 7-31-89) Population served: 250,000 Law Firm: Pepper & Corazzini. Format: Classic hits. Target aud: 25-54; baby boomers. ♦Chris Bullock, gen mgr; Wayne Robbins, opns mgr; Jerry Scott, chief of engrg.

WINU(AM)— Nov 24, 1972: 870 khz; 500 w-D, DA. TL: N39 29 14 W88 57 31. Hrs open: 4:30 AM-7:30 PM 126 W. Main St., 62565. Phone: (217) 774-5277. Fax: (217) 774-5280. Web Site: www.hereshelpnet.org. Licensee: New Life Evangelistic Center Inc. (acq 7-31-98; $75,000). Population served: 60,000 Format: Southern gospel. Target aud: 25-54. ♦Hank Zeniewicz, gen mgr.

Sherman

WABZ(FM)— May 10, 1971: 93.9 mhz; 15 kw. Ant 430 ft TL: N39 59 25 W89 30 46. Hrs open: 3501 E. Sangamon Ave., Springfield, 62707. Phone: (217) 753-5400 (business). Fax: (217) 753-7902. Licensee: Saga Communications of Illinois LLC. Group owner: Saga Communications Inc. (acq 7-96; grpsl). Population served: 15,000 Format: Continuous soft favorities. Target aud: 25-54. ♦Leanne Arndt, gen mgr; Brandy Moore, prom dir; Bob Parrish, progmg dir; Michelle Eecles, news dir.

Skokie

WTMX(FM)— Aug 18, 1961: 101.9 mhz; 4.2 kw. 1,561 ft TL: N41 52 44 W87 38 10. Stereo. Hrs opn: 24 One Prudential Plaza, Suite 2700, Chicago, 60601. Phone: (312) 946-1019. Fax: (312) 946-4747. Web Site: www.wtmx.com. Licensee: Bonneville International Corp. (group owner; (acq 8-70). Population served: 600,000 Natl. Rep: Katz Radio. Format: Adult contemp. News staff: one. ♦Drew Horowitz, pres & gen mgr; Barry James, stn mgr; Jessy Ferdman, prom dir; Mary Ellen Kachinske, progmg VP & progmg dir; Barry Keefe, news dir & pub affrs dir; Kent Lewin, chief of engrg.

South Beloit

WTJK(AM)—Licensed to South Beloit. See Beloit WI

South Jacksonville

WJVO(FM)—Licensed to South Jacksonville. See Jacksonville

Sparta

WHCO(AM)— February 1955: 1230 khz; 1 kw-U. TL: N38 07 25 W89 43 20. Hrs open: 24 Box 255, 1230 W. Broadway, 62286. Secondary address: 47 W. Maine, Mascoutah 62258. Phone: (618) 443-2121. Fax: (618) 443-2800. Licensee: Hirsch Communication Engineering Co. Population served: 150,000 Natl. Network: CBS, Westwood One. Rgnl. Network: Brownfield. Format: News/talk, sports. News staff: 2; News: 10 hrs wkly. Target aud: 25-65. Spec prog: Pol 2 hrs, farm 20 hrs, relg 10 hrs wkly. ♦Jack L. Scheper Sr., pres & gen mgr; Mike Hoeft, news dir.

Spring Valley

WIVQ(FM)— December 1993: 103.3 mhz; 4.9 kw. 361 ft TL: N41 18 09 W89 14 11. Hrs open: 24
Rebroadcasts WSTQ 100%.
3905 Progress Blvd., Peru, 61354. Phone: (815) 224-2100. Fax: (815) 224-2066. E-mail: q@theradiogroup.net Web Site: qhitmusic.com. Licensee: Mendota Broadcasting Inc. Group owner: Studstill Broadcasting (acq 7-17-97; $700,000 with WBZG(FM) Peru). Population served: 130,000 Natl. Rep: Rgnl Reps. Law Firm: Booth, Freret, Imlay & Tepper. Wire Svc: AP Format: Top-40, adult contemp. Target aud: 18-44. ♦Lamar Studstill, chmn; Cole Charles Studstill, CFO; Owen L. Studstill, CEO, pres & gen mgr; Cole Studstill, stn mgr & opns mgr; Doris Studstill, opns VP.

***WSOG(FM)**— 12/2002: 88.1 mhz; 4 kw vert. Ant 262 ft TL: N41 17 32 W89 07 59. Hrs open: 24 Box 34, 61362. Phone: (815) 220-1929. Fax: (815) 220-1929. E-mail: wsog881@hotmail.com Licensee: Spirit Education Association Inc. ♦Louis J. Perona, pres.

Springfield

WCVS-FM—Virden, May 10, 1982: 96.7 mhz; 6 kw. Ant 328 ft TL: N39 38 26 W89 39 24. Stereo. Hrs opn: 24 3055 S. 4th St., 62703. Phone: (217) 528-3033. Fax: (217) 528-5348. E-mail: wcvs@wcvs.com Web Site: www.wcvsfm.com. Licensee: Neuhoff Family L.P. Group owner: Clear Channel Communications Inc. (acq 8-1-2007; grpsl). Population served: 400,000 Natl. Rep: Clear Channel. Law Firm: Wiley, Rein & Fielding LLP. Format: Classic rock. News staff: one. Target aud: 25-54. ♦Kevin O'Dea, VP, gen mgr & gen sls mgr; Danielle Outlaw, sls dir; Michelle Mitchell, prom dir; Jeremy Anderson, progmg dir & progmg dir; Jeff Hofmann, news dir; Ron Bluhm, chief of engrg.

WDBR(FM)—Listing follows WTAX(AM).

WFMB(AM)— 1922: 1450 khz; 1 kw-U. TL: N39 45 36 W89 39 05. Hrs open: 24 3055 S. 4th St., 62703. Phone: (217) 528-3033. Phone: (217) 544-9855. Fax: (217) 528-5348. E-mail: sportsradio1450@sportsradio1450.com Web Site: www.sportsradio.com. Licensee: Neuhoff Family L.P. Group owner: Clear Channel Communications Inc. (acq 8-1-2007; grpsl). Population served: 161,000 Natl. Network: ABC. Natl. Rep: Clear Channel. Format: Sports/personality. News staff: 2; News: 5 hrs wkly. Target aud: 25-54; upscale professionals. ♦Kevin O'Dea, gen mgr; D. Outlaw, sls dir.

WFMB-FM— July 1965: 104.5 mhz; 43 kw. 465 ft TL: N39 45 36 W89 39 05. Stereo. 24 E-mail: wfmb@wfmb.com Web Site: www.wfmb.com.300,000 Format: Country. Target aud: Professional adults. ♦Kevin O'Dea, VP.

***WLUJ(FM)**— May 24, 1995: 89.7 mhz; 20 kw. 328 ft TL: N39 48 30 W89 37 30. (CP: 10 kw). Stereo. Hrs opn: 600 W. Mason St., 62702. Phone: (217) 528-2300. Fax: (217) 528-2400. E-mail: wluj897@ameritech.net Web Site: www.wluj.org. Licensee: Cornerstone Community Radio Inc. Population served: 300,000 Natl. Network: Moody. Format: Christian talk. Target aud: General. ♦Arthur Gregg, sr VP; Dick Reed, VP; Richard Van Zandt, pres & gen mgr; John McBride, stn mgr; Howard Fouks, opns mgr; Richard Beaman, gen sls mgr.

WMAY(AM)— Oct 15, 1950: 970 khz; 1 kw-D, 500 w-N, DA-2. TL: N39 51 42 W89 32 32. Stereo. Hrs opn: Box 460, 62705. Secondary address: 1510 N. Third, Riverton 62561. Phone: (217) 629-7077. Fax: (217) 629-7952. Web Site: www.wmay.com. Licensee: Long Nine Inc. Group owner: The Mid-West Family Broadcast Group (acq 12-7-76). Population served: 350,000 Natl. Rep: D & R Radio. Law Firm: Fisher, Wayland, Cooper, Leader & Zaragoza. Format: News/talk. Target aud: 25-64. Spec prog: Big band 5 hrs wkly. ♦Kevan Kavanough, gen mgr; Dave Doetsch, sls VP; Amanda Johnson, prom dir; Robb Rose, progmg dir; Jim Leach, news dir; Greg Stephens, chief of engrg.

WNNS(FM)—Co-owned with WMAY(AM). Nov 1, 1980: 98.7 mhz; 50 kw. 500 ft TL: N39 41 59 W89 46 55. Stereo. Web Site: www.wnns.com.300,000 Format: Adult contemp. Spec prog: Jazz 6 hrs wkly. ♦Kavan Kavanough, gen mgr & progmg dir; Kellie Michaels, pub affrs dir; Greg Stephens, engrg dir.

***WQNA(FM)**— Aug 31, 1979: 88.3 mhz; 250 w. Ant 256 ft TL: N39 44 03 W89 38 18. Stereo. Hrs opn: 24 Capital Area Career Ctr., 2201 Toronto Rd., 62712. Phone: (217) 529-5431. Fax: (217) 529-7861. E-mail: info@wqna.org Web Site: www.wqna.org. Licensee: Capital Area Career Center. Population served: 300,000 Rgnl rep: Illinois Student News Network Law Firm: Shaw Pittman. Wire Svc: U.S. Newswire Format: Div. News staff: one; News: 6 hrs wkly. Target aud: 13-24; high school & college students. Spec prog: Varied. ♦Jim Grimes, gen mgr; Jim Pemberton, progmg dir; Ed Davis, news dir; Kerri Donovan, engr.

WQQL(FM)— Nov 15, 1993: 101.9 mhz; 50 kw. 300 ft TL: N39 42 39 W89 38 42. Stereo. Hrs opn: 24 3501 E. Sangamon Ave., 62707. Phone: (217) 753-5400. Fax: (217) 753-7902. Web Site: www.cool1019.com. Licensee: Saga Communications of Illinois LLC. Group owner: Saga Communications Inc. (acq 9-10-93; $1.44 million; FTR: 10-4-93). Population served: 2,000,000 Natl. Rep: Katz Radio. Law Firm: Smithwick & Belendiuk. Format: Oldies. News: 2 hrs wkly. Target aud: 25 plus; upscale educated adults. ♦Leanne Arndt, gen mgr; Kevin Anfield, gen sls mgr & prom dir.

***WSCT(FM)**— November 1993: 90.5 mhz; 3.8 kw. Ant 410 ft TL: N39 38 38 W89 30 51. Stereo. Hrs opn: 24
Rebroadcasts WIBI(FM) Carlinville 100%.
Box 140, Carlinville, 62626. Phone: (217) 854-4800. Fax: (217) 854-4810. E-mail: wibi@wibi.org Web Site: www.wibi.org. Licensee: Illinois Bible Institute. Population served: 150,000 Format: Christian. News: 5 hrs wkly. Target aud: 25-44; Christian & seeking non-Christians. ♦Reverend Larry Griswold, pres; Barry Copeland, gen mgr; Jeremiah Beck, stn mgr & opns dir; Jessica Barton, mktg dir & prom dir; Joe Buchanan, mus dir; Sally Braundmeier, chief of engrg & traf mgr.

WTAX(AM)— 1930: 1240 khz; 1 kw-U. TL: N39 47 36 W89 36 18. Hrs open: 3501 E. Sangamon Ave., 62707. Phone: (217) 753-5400. Fax: (217) 753-7902. Web Site: www.wtax.com. Licensee: Saga Communications of Illinois LLC. Group owner: Saga Communications Inc. (acq 1996). Population served: 91,753 Natl. Network: Moody, CBS. Natl. Rep: Christal. Wire Svc: UPI Format: News/talk, sports. News staff: 3; News: 20, hrs wkly. Target aud: 30 plus. Spec prog: Farm 16 hrs wkly. ♦Leanne Arndt, gen mgr; Michelle Eccles, news dir.

WDBR(FM)—Co-owned with WTAX(AM). April 1948: 103.7 mhz; 50 kw. 320 ft TL: N39 47 36 W89 36 18. (CP: 20 kw, ant 768 ft.). Web Site: www.wdbr.com.301,500 Format: CHR. Target aud: 18-49; general.

***WUIS(FM)**— Jan 3, 1975: 91.9 mhz; 50 kw. 524 ft TL: N39 47 00 W89 26 46. Stereo. Hrs opn: 24
Rebroadcasts WIPA(FM) Pittsfield 100%.
Box 19243, CBM-130, Univ. of Illinois at Springfield, 62794-9243. Secondary address: One University Plaza, MS CBM-130 62703. Phone: (217) 206-6516. Fax: (217) 206-6527. E-mail: wuis@uis.edu Web Site: www.wuis.org. Licensee: University of Illinois at Springfield. Population served: 149,600 Natl. Network: NPR, PRI. Rgnl. Network: Ill. Radio Net. Law Firm: Dow, Lohnes & Albertson. Format: News, class, jazz. News staff: 4; News: 45 hrs wkly. Target aud: 25-54. Spec prog: NPR entertainment 15 hrs, bluegrass 2 hrs, Singer/Songwriter 2 hrs, ambient 2 hrs wkly. ♦Bill Wheelhouse, gen mgr; Sinta Seiber, opns mgr; Lisa Clemmons-Stott, dev dir; Karl Scroggin, mus dir.

WYMG(FM)—Jacksonville, March 1948: 100.5 mhz; 50 kw. 500 ft TL: N39 39 40 W89 55 18. Stereo. Hrs opn: 3501 E. Sangamon Ave., 62707. Phone: (217) 753-5400. Fax: (217) 753-7902. E-mail: wymg@wymg.com Web Site: www.wymg.com. Licensee: Saga Communications of Illinois LLC. Group owner: Saga Communications Inc. (acq 10-1-86). Population served: 500,000 Format: Classic rock. Target aud: 18-49. Spec prog: Jazz 2 hrs, comedy one hr wkly. ♦Leanne Arndt, gen mgr; Jane Cochran, progmg.

Staunton

WAOX(FM)— December 1, 1999: 105.3 mhz; 6 kw. Ant 285 ft TL: N39 02 37 W89 44 56. Stereo. Hrs opn: 24 Box 10, Litchfield, 62056. Phone: (217) 532-2085. Fax: (217) 532-2431. E-mail: waox@theox1053.com Web Site: www.waox.com. Licensee: Talley Broadcasting Corp. Group owner: Talley Radio Stations Natl. Network: ABC. Format: Adult contemp. Target aud: 18-54. ♦Hayward L. Talley, pres & gen mgr; Brian Talley, opns VP; Brian C. Talley, opns mgr; Barbara Valentine, traf mgr.

Sterling

***WNIQ(FM)**— unknown: 91.5 mhz; 2.1 kw. 331 ft TL: N41 53 12 W89 35 43. Stereo. Hrs opn: 24
Rebroadcasts WNIJ(FM) De Kalb & WNIU(FM) Rockford 50%.
NIU Broadcast Ctr., 801 N. First St., DeKalb, 60115. Phone: (815) 753-9000. Fax: (815) 753-9938. E-mail: npr@niu.edu Web Site: www.northernpublicradio.org. Licensee: Northern Illinois University. Population served: 70,000 Natl. Network: PRI, NPR. Law Firm: Arter & Hadden. Format: News/talk, class. News: 2 hrs wkly. Target aud: General. ♦Tim Emmons, gen mgr; Jan Kilgard, dev dir; Bill Drake, progmg dir.

WSDR(AM)— Aug 21, 1949: 1240 khz; 500 w-D, 1 kw-N. TL: N41 48 59 W89 40 13. Hrs open: 24 3101 Freeport Rd., 61081. Phone: (815) 625-3400. Fax: (815) 625-6940. E-mail: wsdr1240@theramp.net Licensee: Withers Broadcasting Co. of Rock River LLC. Group owner: Withers Broadcasting Co. (acq 1-21-98; grpsl). Population served: 165,000 Natl. Network: CBS, ABC. Rgnl. Network: Ill. Radio Net. Natl. Rep: Christal. Format: News, sports, talk. News staff: 2; News: 10 hrs wkly. Target aud: 25 plus. Spec prog: Farm 16 hrs, Sp 4 hrs wkly. ♦Brian Zschiesche, gen mgr; Sherry Smith, gen sls mgr; Lisa Taylor, progmg dir.

WSSQ(FM)—Co-owned with WSDR(AM). August 1966: 94.3 mhz; 6 kw. 309 ft TL: N41 51 06 W89 42 38. Stereo. 190,000 Natl. Network: Westwood One, ABC. Natl. Rep: Christal. Format: Adult contemp. Target aud: 25-54; women.

Streator

WSPL(AM)— Sept 26, 1953: 1250 khz; 500 w-D, 100 w-N, DA-D. TL: N41 09 30 W88 50 13. Hrs open: 24 3905 Progress Blvd., Peru, 61354. Phone: (815) 672-2947. Fax: (815) 673-1833. E-mail: wspl@theradiogroup.net Web Site: am1250wspl.com. Licensee: Mendota Broadcasting Inc. Group owner: Studstill Broadcasting (acq 5-30-2000; grpsl). Population served: 110,000 Format: News/talk, sports. News staff: 3; News: 25 hrs wkly. Target aud: 35 plus. ♦Lamar Studstill, chmn; Owen L. Studstill, pres; Cole Studstill, VP & opns mgr; Lee Studstill, gen mgr; Cheryl Knirlberger, sls dir; Dave Noesen, news dir; Mark Baker, chief of engrg.

WSTQ(FM)—Co-owned with WSPL(AM). Sept 15, 1964: 97.7 mhz; 6 kw. Ant 328 ft TL: N41 10 49 W88 52 06. Stereo. 24 Phone: (815) 224-2100. Fax: (815) 224-2066. E-mail: q@theradiogroup.net Web Site: qhitmusic.com.130,000 Format: CHR. Target aud: 18-44. ♦Cole Studstill, progmg dir.

WYYS(FM)— 1995: 106.1 mhz; 6 kw. Ant 292 ft TL: N41 10 49 W88 52 06. (CP: 2.45 kw, ant 520 ft. TL: N41 16 30 W88 57 56). Hrs opn: 24 3905 Progess Blvd., Peru, 61354. Phone: (815) 224-2100. Fax: (815) 224-2066. E-mail: wyys@theradiogroup.net Licensee: Mendota Broadcasting Inc. Group owner: Studstill Broadcasting (acq 3-8-2000; grpsl). Population served: 100,000 Natl. Network: ABC. Natl. Rep: Rgnl Reps. Format: Oldies. Target aud: 35 plus. ♦Lamar Studstill, chmn; Cole Studstill, CFO; Lee Studstill, pres & gen mgr.

Sugar Grove

***WSRI(FM)**— 2005: 88.7 mhz; 250 w. Ant 129 ft TL: N41 47 01 W88 26 14. Hrs open: 24 2351 Sunset Blvd., Suite170-218, Rocklin, CA, 95765. Phone: (916) 251-1600. Fax: (916) 251-1650. E-mail: klove@klove.com Web Site: www.klove.com. Licensee: Educational Media Foundation. Group owner: EMF Broadcasting. Population served: 176,000 Natl. Network: K-Love. Law Firm: Shaw Pittman. Format: Contemp Christian. News staff: 3. Target aud: 25-44; female-Judeo/ Christian. ♦Richard Jenkins, pres; Mike Novak, VP; Lloyd Parker, gen mgr; Ed Lenane, opns dir & news dir; Keith Whipple, dev dir; David Pierce, progmg mgr; Sam Wallington, engrg dir; Karen Johnson, news rptr; Marya Morgan, news rptr; Richard Hunt, news rptr.

Sullivan

WZNX(FM)— April 1992: 106.7 mhz; 9.5 kw. 550 ft TL: N39 36 38 W88 41 32. Hrs open: 24 410 N. Water St., Suite C, Decatur, 62523. Phone: (217) 428-4487. Fax: (217) 428-4501. Web Site: www.wznx.com.

Licensee: WSHY Inc. Group owner: The Cromwell Group Inc. (acq 2-14-97; $750,000). Population served: 300000 Format: Classic rock. News staff: one. Target aud: 25-54; strong men. ♦Chris Bullock, gen mgr & opns dir; Jerry Scott, gen mgr & chief of engrg; Wayne Robbins, opns dir.

Summit

***WARG(FM)**— January 1976: 88.9 mhz; 500 w. 98 ft TL: N41 46 36 W87 48 17. Stereo. Hrs opn: 8 AM-10 PM 7329 W. 63rd St., 60501. Phone: (708) 728-8368. Fax: (708) 728-3155. Licensee: Community High School District No. 217. Format: Alternative, rock. Target aud: High School Students; alternative subculture.

Sycamore

WDEK(FM)—See De Kalb

WLBK(AM)—See De Kalb

WSQR(AM)— June 11, 1981: 1180 khz; 900 w-D, 1 w-N. TL: N42 00 24 W88 40 40. Hrs open: 1851 Coltonville Rd., 60178. Phone: (630) 552-1000. Fax: (630) 552-9300. E-mail: wspy-news@nelsonmultimedia.net Licensee: De kalb County Broadcasters Inc. (acq 9-94). Natl. Network: ABC. Format: News info, full service. Target aud: 35-55. Spec prog: Farm 6 hrs wkly. ♦Larry Nelson, pres; Beth Pierre, gen mgr & sls VP; Pam Nelson, CFO & gen mgr.

Taylorville

***WIHM(AM)**— 1952: 1410 khz; 1 kw-D, 63 w-N, DA-1. TL: N39 32 38 W89 16 36. Hrs open: 3515 Hampton Ave., St. Louis, MO, 63139. Phone: (314) 752-7000. E-mail: office@covenantnet.net Web Site: www.covenantnet.net. Licensee: Covenant Network (acq 7-31-98; $60,000). Population served: 30,000 Format: Christian; Religious; inspirtional. ♦Tony Holman, pres & gen mgr.

WQLZ(FM)— December 1967: 92.7 mhz; 11.5 kw. 482 ft TL: N39 38 38 W89 30 51. Stereo. Hrs opn: 24 Box 460, Springfield, 62705. Secondary address: 1510 N. Third, Riverton 62561. Phone: (217) 629-7077. Fax: (217) 629-7952. Web Site: www.wqlz.com. Licensee: Long Nine Inc. Group owner: Mid-West Family Stations (acq 2-3-93; $1 million; FTR: 2-22-93). Population served: 30,000 Law Firm: Fisher, Wayland, Cooper, Leader & Zaragoza. Format: AOR. News staff: 4; News: 3 hrs wkly. Target aud: 18-34. ♦Dave Duetsch, gen sls mgr; Valerie Knight, progmg dir; Jim Leach, news dir; Greg Stephens, chief of engrg; Quinn Fagg, traf mgr.

WTIM-FM— Nov 13, 1997: 97.3 mhz; 4.6 kw. Ant 374 ft TL: N39 27 08 W89 17 10. Hrs open: 24 Box 169, 62568-0169. Secondary address: 918 E. Park St. 62568. Phone: (217) 824-3395. Fax: (217) 824-3301. Web Site: www.randyradio.com. Licensee: Miller Communications Inc. Group owner: Miller Media Group Population served: 197,000 Natl. Network: Westwood One. Natl. Rep: Commercial Media Sales. Law Firm: Womble, Carlyle, Sandridge & Rice. Format: News/talk. News staff: one; News: 25 hrs wkly. Target aud: 25 plus. Spec prog: Farm 20 hrs, relg 4 hrs wkly. ♦Randal J. Miller, pres & gen mgr; Kami Payne, stn mgr; Brandon Fellows, mus dir; Darrin Wright, news rptr.

Teutopolis

WKJT(FM)— 1994: 102.3 mhz; 6 kw. 328 ft TL: N39 08 30 W88 33 36. Hrs open: 24 206 S. Willow, Effingham, 62401. Phone: (217) 347-5518. Fax: (217) 347-5519. E-mail: info@kjcountry.com Web Site: www.kjcountry.com. Licensee: Kirby Broadcasting Inc. Format: Country. ♦John W. Kirby, pres; Greg Sapp, stn mgr; Tonya Siner, opns VP; George Flexter, progmg dir.

Tower Hill

WRAN(FM)— Nov 25, 1997: 98.3 mhz; 3.7 kw. 420 ft TL: N39 16 48 W88 58 22. Stereo. Hrs opn: 24 918 E. Park, Box 169, Taylorville, 62568. Phone: (217) 824-3395. Fax: (217) 824-3301. Web Site: www.randyradio.com. Licensee: Kaskaskia Broadcasting Inc. Group owner: Miller Media Group Population served: 197,000 Natl. Network: CBS. Natl. Rep: Commercial Media Sales. Law Firm: Womble, Carlyle, Sandridge & Rice. Format: Soft adult contemp music. News staff: one; News: 25 hrs wkly. Target aud: 35—64. Spec prog: Farm 6 hrs, relg 3 hrs wkly. ♦Randal J. Miller, pres & gen mgr; Kami Payne, stn mgr; Brandon Fellows, mus dir; Darrin Wright, news rptr.

Tuscola

WEBX(FM)—Licensed to Tuscola. See Champaign

Urbana

WBCP(AM)— 1948: 1580 khz; 250 w-D, 10 w-N. TL: N40 07 32 W88 17 29. Hrs open: Unit D, 904 N. 4th St., Champaign, 61820. Phone: (217) 359-1580. Fax: (217) 359-1583. E-mail: wbcpradio@sbcglobal.net Licensee: WBCP Inc. (acq 12-89; $135,000; FTR: 12-4-89). Natl. Network: American Urban, ABC. Format: Gospel, rhythm and blues, smooth jazz. Spec prog: "". ♦Lonnie Clark, pres; J.W. Pirtle, VP & gen mgr; Sam Britten, opns VP; Lynn Randall, progmg dir; Steve Hamm, chief of engrg.

WCFF(FM)— Dec 4, 1967: 92.5 mhz; 11.5 kw. 485 ft TL: N40 01 29 W88 08 28. Stereo. Hrs opn: 24 2603 W, Bradley Ave., Champaign, 61821. Phone: (217) 352-4141. Fax: (217) 352-1256. E-mail: studio@wkio.com Web Site: www.925thechief.com. Licensee: Saga Communications of Illinois LLC. Group owner: Saga Communications Inc. (acq 2000; $7 million). Population served: 469,000 Natl. Rep: Katz Radio. Format: Oldies. Target aud: 35-64. ♦Ed Christian, CEO & chmn; Steve Goldstein, exec VP; Alan Beck, gen mgr; Jonathan Drake, opns mgr; Gary Saladino, mktg dir; Ryan Leskis, prom dir; Mike Cation, news dir; Mark Spalding, chief of engrg.

***WILL(AM)**— Mar 28, 1922: 580 khz; 5 kw-D, DA-D. TL: N40 04 53 W88 14 18. Hrs open: 24 Campbell Hall for Public Telecommunications, 300 N. Goodwin Ave., 61801-2316. Phone: (217) 333-0850. Fax: (217) 244-9586. Fax: (217) 333-7151. E-mail: willamfm@uiuc.edu Web Site: www.will.uiuc.edu. Licensee: University of Illinois Board of Trustees. Population served: 158,700 Natl. Network: NPR, PRI. Law Firm: Dow, Lohnes & Albertson. Wire Svc: AP Format: News/talk, div. News staff: 3; News: 115 hrs wkly. Target aud: 25-60; educated, upper middle income, professionals. Spec prog: Farm 7 hrs wkly. ♦Mark Leonard, gen mgr; Jay H. Pearce, stn mgr & progmg dir; Mike Pritchard, opns dir & dev dir; Kate Dobrovolny, prom dir; Rick Finnie, chief of engrg; Denise Perry, traf mgr; Tom Rogers, news dir & local news ed.

WILL-FM— Sept 1, 1941: 90.9 mhz; 105 kw. 850 ft TL: N40 06 52 W88 13 27. Stereo. 24 158,700 Format: Class, var. News: 2 hrs wkly. Target aud: 35-70. ♦Jake Schumacher, progmg dir. Co-owned TV: *WILL-TV affil

WLRW(FM)—See Champaign

WPGU(FM)— Apr 17, 1967: 107.1 mhz; 3 kw. 235 ft TL: N40 06 34 W88 14 06. Stereo. Hrs opn: 24 512 E. Green St., Champaign, 61820. Phone: (217) 337-3100. Fax: (217) 337-3162. E-mail: wpgu@wpgu.com Web Site: www.wpgu.com. Licensee: Illini Media Co. Population served: 233,800 Law Firm: Fisher, Wayland, Cooper, Leader & Zaragoza. Format: Alternative Rock. News: 7 hrs wkly. Target aud: 18-34. ♦Mary Cory, gen mgr; Beth Rehn, prom dir; Becky Brothman, progmg dir; Jon Hansen, news dir; Scott S. Downs, sls dir & chief of engrg; Melissa Pasco, traf mgr; Rachel Buenting, traf mgr.

WQQB(FM)—Rantoul, January 1993: 96.1 mhz; 3.8 kw. 403 ft TL: N40 12 27 W88 17 56. Stereo. Hrs opn: 24 4108 Fieldstone Rd., Suite C, Champaign, 61822. Phone: (217) 367-1195. Fax: (217) 367-3291. Web Site: www.wqqb.com. Licensee: RadioStar Inc. Group owner: AAA Entertainment LLC (acq 5-23-2006; grpsl). Population served: 175,000 Format: CHR. Target aud: 18-34; female. ♦Jim Glassman, CEO; John Ginzkey, gen mgr; Linda Bosch, gen sls mgr & rgnl sls mgr; Ken Cunningham, progmg dir; Tascha Turner, traf mgr.

Vandalia

WKRV(FM)—Listing follows WPMB(AM).

WPMB(AM)— Dec 9, 1963: 1500 khz; 250 w-D. TL: N38 57 30 W89 07 27. Hrs open: 6 AM-sunset Box 100, 62471. Secondary address: 232 S. 4th St. 662471. Phone: (618) 283-2325. Phone: (618) 283-2355. Fax: (618) 283-1503. E-mail: wkrv@sbcglobal.net Licensee: Two Petaz Inc. (acq 2-3-2005; $350,000 with co-located FM). Population served: 6,300 Rgnl. Network: Tribune, Ill. Radio Net. Wire Svc: Metro Weather Service Inc. Format: Big band, adult standards. News staff: 2; News: 8-10 hrs wkly. Target aud: General. Spec prog: Farm 4 hrs, gospel 3 hrs wkly. ♦Bayard H. Walters, pres; John D. Harris, gen mgr; John Harris, gen sls mgr; Todd Stapleton, opns mgr, progmg dir & news dir.

WKRV(FM)—Co-owned with WPMB(AM). May 28, 1974: 107.1 mhz; 6 kw. Ant 164 ft TL: N38 57 30 W89 07 27. Stereo. 24 Format: CHR, adult contemp. News staff: one; News: 10 hrs wkly. Target aud: 20-45. ♦Dan Michael, progmg dir.

***WVNL(FM)**— 10/7/2002: 91.7 mhz; 100 w. 164 ft Hrs opn: 24 hrs WIBI, Carlinville,IL, 100%.
Box 140, Carlinville, 62626. Phone: (217) 854-4800. Fax: (217) 854-4810. E-mail: wibi@wibi.org Web Site: www.wibi.org. Licensee: Illinois Bible Institute Inc. ♦Barry Copeland, gen mgr; Jeremiah Beck, stn mgr & opns dir; Jessica Barton, prom dir; Rob Regal, progmg dir; Joe Buchanan, mus dir; Sally Braundmeier, traf mgr.

Vernon Hills

WNVR(AM)— Mar 1, 1988: 1030 khz; 5 kw-D, 120 w-N, DA-N. TL: N42 15 10 W88 23 45. Hrs open: 24 3656 W. Belmont, Chicago, 60618. Phone: (773) 588-6300. Fax: (773) 267-4913. E-mail: polskieradio@polskieradio.com Web Site: www.polskieradio.com. Licensee: Polnet Communications Ltd. (group owner; acq 3-15-91; $495,000; FTR: 1-25-93). Population served: 10,000,000 Law Firm: Wiley, Rein and Fielding. Wire Svc: AP Format: Polish language. News staff: 7; News: 20 hrs wkly. Target aud: 18-54; Polish speaking audience. ♦Walter Kotaba, pres; Kamilla Dworska, gen mgr.

Virden

WCVS-FM—Licensed to Virden. See Springfield

Virginia

WVIL(FM)— February 1998: 101.3 mhz; 4 kw. 390 ft TL: N40 00 52 W90 19 55. Stereo. Hrs opn: 24 Box 101, Jacksonville, 62650. Phone: (217) 245-5700. Fax: (217) 245-5701. E-mail: lbostwick@mchsi.com Web Site: www.wvilfm.com. Licensee: Mark J. Langston. Population served: 65,000 Law Firm: Gammon & Grange. Format: All sports. News staff: one; News: one hr wkly. Target aud: 18-65; general. ♦Larry Bostwick, gen mgr.

Warsaw

***WIUW(FM)**— May 17, 1995: 89.5 mhz; 10 kw. 449 ft TL: N40 20 44 W91 24 11. Hrs open: 24
Rebroadcasts WIUM(FM) Macomb 100%.
515 Univ. Svcs. Bldg., Western Illinois Univ., Macomb, 61455. Phone: (309) 298-2424. Phone: (309) 298-1873. Fax: (309) 298-2133. E-mail: publicradio@wiu.edu Web Site: www.tristatesradio.com. Licensee: Western Illinois University. Population served: 57,326 Law Firm: Cohn & Marks. Format: Class, news. News staff: 2; News: 58 hrs wkly. Target aud: General. ♦Dorothy Vallillo, gen mgr; Ken Thermon, opns dir; Sharon Faust, dev dir; Rich Egger, news dir.

Watseka

WGFA-FM— Mar 2, 1962: 94.1 mhz; 50 kw. Ant 364 ft TL: N40 47 37 W87 45 17. Stereo. Hrs opn: 24 1973 E. 1950 North Rd., 60970. Phone: (815) 432-4955. Fax: (815) 432-4957. E-mail: 941fm@wgfaradio.com Web Site: www.wgfaradio.com. Licensee: Iroquois County Broadcasting Co. Population served: 750000 Law Firm: Borsari & Paxson. Format: Adult contemp. News staff: one; News: 12 hrs wkly. Target aud: 25-54. Spec prog: Farm 18 hrs, business 2 hrs, sports 18 hrs wkly. ♦Margaret Martin, gen mgr, dev dir & prom mgr; Justin Kaiser, opns dir, progmg dir & progmg dir; Stacey Ferguson, natl sls mgr; Milissa Long, rgnl sls mgr & pub affrs dir; Carl Gerdovich, news dir; Mark Spalding, chief of engrg; Stacey Smith, stn mgr, gen sls mgr & spec ev coord; Bill Yohnka, sports cmtr.

WGFA(AM)— Sept 1, 1961: 1360 khz; 1 kw-D, DA. TL: N40 47 48 W87 45 11.6 AM-6 PM Web Site: www.wgfaradio.com.60,000 Format: Big band, easy lstng. News staff: 6. Target aud: 30-65; wealthy, retiring, entrepreneurs. Spec prog: Talk 5 hrs wkly. ♦Margaret Martin, mktg dir, prom dir, adv dir, progmg VP & spec ev coord.

WMLF(FM)—Not on air, target date: unknown: 95.9 mhz; 6 kw. Ant 234 ft TL: N40 46 17 W87 46 13. Hrs open: 1717 Dixie Hwy., Suite 650, Fort Wright, KY, 41011. Phone: (859) 331-9100. Licensee: Radioactive LLC. ♦Benjamin L. Homel, pres.

Waukegan

WKRS(AM)— Sept 25, 1949: 1220 khz; 1 kw-D, DA. TL: N42 20 59 W87 52 53. (CP: 99 w-N). Hrs opn: 24 3250 Belvidere Rd., 60085. Phone: (847) 336-7900. Fax: (847) 336-1523. Web Site: www.wkrs.com. Licensee: NM Licensing LLC. Group owner: NextMedia Group LLC (acq 11-26-01; grpsl). Population served: 900,000 Format: News/talk. News staff: 4; News: 40 hrs wkly. Target aud: 25 plus. ♦Kira La Fond, gen mgr; Libby Collins, progmg dir.

WXLC(FM)—Co-owned with WKRS(AM). May 1963: 102.3 mhz; 3 kw. 322 ft TL: N42 20 59 W87 52 53. Stereo. Web Site: www.1023xlc.com. Format: Adult contemp. Target aud: 25-44. ♦Rory Fraley, gen mgr; Mike Peof, gen sls mgr; Haynes Johns, progmg dir.

West Frankfort

WFRX(AM)— May 2, 1951: 1300 khz; 1 kw-D. TL: N37 53 04 W88 55 44. Hrs open: Box 127, 1822 N. Court St., Marion, 62959. Phone: (618) 997-8123. Phone: (618) 932-8121. Fax: (618) 993-2319. Web Site: www.wfrx.com. Licensee: CC Licenses LLC. Group owner: Clear Channel Communications Inc. (acq 1-18-2001; grpsl). Population served: 8,836 Format: News, big band. News staff: one; News: 20 hrs wkly. Target aud: Adult. ♦Jerry Crouse, gen mgr; Steve Browning, gen sls mgr; Stavey Malick, prom dir; Paxton Guy, progmg dir & news dir; Tim Deterding, chief of engrg.

WQUL(FM)—Co-owned with WFRX(AM). Mar 14, 1972: 97.7 mhz; 3.5 kw. 433 ft TL: N37 45 15 W88 56 05.6 AM-10 PM Web Site: www.thebear977fm.com.8,836 Format: Pure classic rock. News: 2 hrs wkly. Target aud: 25-54; adult. ♦Matt Mellen, progmg dir.

Wheaton

***WETN(FM)**— Feb 27, 1962: 88.1 mhz; 250 w. 140 ft TL: N41 52 09 W88 05 56. Stereo. Hrs opn: 24 Wheaton College, 60187. Phone: (630) 752-5074. Fax: (630) 752-5286. E-mail: wetn@wheaton.edu Web Site: www.wetn.org. Licensee: Trustees of Wheaton College. Population served: 300,000 Format: Christian , classical. News: 2 hrs wkly. Target aud: 18-49. Spec prog: Live sports 5 hrs, live church svcs 3 hrs, live concerts 2 hrs wkly. ♦Dr. A. Duane Litfin, pres; John Rorvik, gen mgr; Mark Bartlebaugh, stn mgr.

Willow Springs

WAIT(AM)—Licensed to Willow Springs. See Chicago

Wilmington

WYKT(FM)— Sept 29, 1980: 105.5 mhz; 1.3 kw. 482 ft TL: N41 17 11 W88 14 23. Stereo. Hrs opn: 70 Meadowview Ctr., Kankakee, 60901. Phone: (815) 935-9555. Fax: (815) 935-9593. Web Site: www.1055thepickle.com. Licensee: STARadio Corp. (group owner; acq 7-6-98). Population served: 2,250,000 Format: Hits of the 60s & 70s. News staff: one; News: 4 hrs wkly. Target aud: 25-54. Spec prog: Gospel 4 hrs, pub service 4 hrs, sports 12 hrs wkly. ♦Brendan Michaels, opns mgr & progmg dir; Larry Regnier, gen sls mgr & chief of engrg; Robert Kersmarki, VP, gen mgr & mktg VP.

Winnebago

WRTB(FM)— 1971: 95.3 mhz; 1.25 kw. Ant 512 ft TL: N42 17 26 W89 09 51. Stereo. Hrs opn: 24 2830 Sandy Hollow Rd., Rockford, 61109. Phone: (815) 874-7861. Fax: (815) 874-2202. Web Site: www.953bobfm.com. Licensee: Maverick Media of Rockford License LLC. Group owner: RadioWorks Inc. (acq 4-27-2005; grpsl). Population served: 247,400 Natl. Rep: Katz Radio. Law Firm: Shaw Pittman. Format: Classic hits. Target aud: 25-54. ♦Gary Rozynek, pres; Jay Chapman, gen mgr; Tim Krull, progmg dir.

Winnetka

***WNTH(FM)**— Dec 10, 1960: 88.1 mhz; 100 w. 105 ft TL: N42 05 40 W87 43 07. Stereo. Hrs opn: 385 Winnetka Ave., 60093. Phone: (847) 784-2330. Fax: (847) 501-6400. Web Site: www.newtrier.kiz.il.us. Licensee: New Trier Township Board of Education. (acq 1960). Population served: 60,000 Format: Div. ♦Nina Lynn, stn mgr.

Wood River

KFNS(AM)— Oct 5, 1961: 590 khz; 1 kw-U, DA-2. TL: N38 55 43 W90 05 08. Hrs open:
KFNS-AM 590 simulcasts it's programs to 100.7FM (KFNS-FM) KFNS-FM, Troy/St. Louis, Missouri, 100%.
8045 Big Bend Blvd., Webster Groves, MO, 63119. Phone: (314) 962-0590. Fax: (314) 962-7576. Web Site: www.kfns.com. Licensee: Big Stick One LLC. Group owner: Big League Broadcasting LLC (acq 7-13-2004; grpsl). Natl. Rep: Interep. Format: Sports. Target aud: 25-54; men. ♦Evan Crocker, VP & gen mgr; Jim Goessling, gen sls mgr; Jason Komito, prom dir; Rob Weingarten, progmg dir; John Masters, chief of engrg; Annie Dailey, traf mgr.

Woodlawn

WDML(FM)— Nov 5, 1993: 106.9 mhz; 3 kw. 328 ft TL: N38 21 29 W89 05 56. Stereo. Hrs opn: 24 Box 1591, 3501 Broadway, Mount Vernon, 62864. Phone: (618) 242-3333. Fax: (618) 242-3334. E-mail: wdml@mvn.net Web Site: www.wdml.com. Licensee: Volunteer Broadcasting of Illinois Inc. Population served: 100,000 Natl. Network: Westwood One. Law Firm: Dennis Kelly. Format: Adult rock. News staff: one; News: everyday 5 3 minute segments. Target aud: 30 plus; male. Spec prog: Christian rock 3 hrs wkly, House of Blues Radio Hour. ♦David M. Lister, CEO, pres & gen mgr; Ryan Roddy, COO & stn mgr.

Woodstock

WZSR(FM)— May 24, 1974: 105.5 mhz; 3 kw. 429 ft TL: N42 15 30 W88 21 48. (CP: 1.95 kw, ant 567 ft. TL: N42 15 34 W88 21 45). Stereo. Hrs opn: 8800 Rt. 14, Crystal Lake, 60012. Phone: (815) 459-7000. Fax: (815) 459-7027. Web Site: www.star105.com. Licensee: NM Licensing LLC. Group owner: NextMedia Group L.L.C. (acq 11-26-01; grpsl). Population served: 200,000 Wire Svc: UPI Format: Adult contemp. Target aud: 25-54; female. Spec prog: Relg one hr wkly. ♦Kira Lefond, gen mgr; Don Oberbillig, gen sls mgr; Stew Cohen, news dir & chief of engrg.

Zion

WPJX(AM)— Sept 19, 1967: 1500 khz; 250 w-D, DA. TL: N42 27 18 W87 54 01. Hrs open:
Simulcast with WEEF(AM) Highland Park 100%.
4320 Dundee Rd., Northbrook, 60062. Phone: (847) 498-3350. Fax: (847) 498-5743. Licensee: Polnet Communications Ltd. (group owner; (acq 5-15-2006; $230,000). Format: Ethnic. ♦Sara Vargas, gen mgr; Fernando jaramillo, progmg dir.

WWDV(FM)— 1962: 96.9 mhz; 50 kw. 500 ft TL: N42 30 36 W87 53 11. Stereo. Hrs opn: 875 N. Michigan Ave., Suite 1510, Chicago, 60611. Phone: (312) 274-9710. Fax: (312) 274-1304. Web Site: www.wdrv.com. Licensee: Bonneville Holding Co. Group owner: Bonneville International Corp. (acq 1-25-01; $165 million with WDRV(FM) Chicago). Population served: 18,500 Format: Classic rock. ♦Jerry Schnacke, pres, VP & gen mgr; Greg Solk, opns VP.

Indiana

Alexandria

WHTI(FM)— Sept 3, 1980: 96.7 mhz; 2.5 kw. Ant 351 ft TL: N40 10 38 W85 40 23. Stereo. Hrs opn: 24 800 E. 29th St., Muncie, 47302. Phone: (765) 288-4403. Fax: (765) 378-2091. E-mail: maxstudio@maxrocks.net Web Site: www.maxrocks.net. Licensee: Indiana Sabrecom Inc. Group owner: Backyard Broadcasting LLC (acq 12-1-02; grpsl). Population served: 300,000 Natl. Network: Motor Racing Net. Format: Classic rock. News staff: one; News: 2 hrs wkly. Target aud: 25-54. ♦Brian Thomas, prom mgr & progmg dir; Sean Mattingly, chief of engrg.

Anderson

***WBSB(FM)**— December 1996: 89.5 mhz; 400 w. 364 ft TL: N40 10 38 W85 40 23. Hrs open:
Rebroadcasts WBST(FM) Muncie 100%.
c/o WBST(FM), Ball State Univ., Muncie, 47306-0550. Phone: (765) 285-5888. Fax: (765) 285-8937. E-mail: ipr@bsu.edu Web Site: www.bsu.edu/ipr. Licensee: Ball State University. Format: Class, news. News staff: one. ♦Marcus Jackman, gen mgr; Pam Coletti, gen sls mgr; Carol Trimmer, prom mgr; Steven Turpin, progmg dir; Robert Mittendorf, chief of engrg; Dorothy Marvell, traf mgr; Brian Beaver, news rptr.

***WGNR-FM**— Sept 11, 1973: 97.9 mhz; 50 kw. 489 ft TL: N40 03 43 W85 42 34. Stereo. Hrs opn: 24 2000 W. 53rd St., 46013. Phone: (765) 642-2750. Fax: (765) 642-4033. E-mail: wgnr@moody.edu Web Site: www.wgnr.org. Licensee: Moody Bible Institute of Chicago. Group owner: The Moody Bible Institute of Chicago (acq 12-17-97; $5.5 million with co-located AM). Population served: 70,787 Format: Inspirational, Christian. News staff: one. Target aud: 35-54. ♦Dr. Joe Stowell, pres; Ray Hashley, gen mgr & stn mgr; Tom Winn, progmg dir & progmg mgr; Sam Sundin, news dir; Jim Wagner, chief of engrg.

WGNR(AM)— 1946: 1470 khz; 1 kw-D, 35 w-N. TL: N40 03 43 W85 42 37.12 hrs E-mail: wgnr@moody.com Web Site: www.wgnr.org. Format: Christian talk. Target aud: 35-54. ♦Ray Hashley, gen mgr.

WHBU(AM)— April 1923: 1240 khz; 1 kw-U. TL: N40 06 17 W85 40 45. Hrs open: 24 800 E. 29th St., Muncie, 47302. Phone: (765) 288-4403. Fax: (765) 288-0429. Web Site: www.1240whbu.com. Licensee: Indiana Sabrecom Inc. Group owner: Backyard Broadcasting LLC (acq 12-1-02; grpsl). Population served: 164,000 Format: News/talk info. News staff: one; News: 40 hrs wkly. Target aud: 25-54. ♦Steve Lindell, VP, gen mgr & stn mgr; Brett Beshore, mktg VP.

WQME(FM)— Nov 29 1990: 98.7 mhz; 4.5 kw. 400 ft TL: N39 58 59 W85 42 41. Stereo. Hrs opn: 6 AM-Midnight 1100 E. 5th St., 46012-3495. Phone: (765) 641-4349. Fax: (765) 641-3825. E-mail: email@wqme.com Web Site: www.wqme.com. Licensee: Anderson University Inc. Population served: 2,368,000 Natl. Network: CNN Radio. Law Firm: Fletcher, Heald & Hildreth. Wire Svc: AP Format: Adult contemp, Christian. News: 8 hrs wkly. Target aud: 25-54. Spec prog: Relg 9 hrs wkly. ♦Donald Boggs, gen mgr; Gerald Longenbaugh, gen sls mgr; Matt Rust, progmg dir; Jerry Morton, chief of engrg & engr; Norma Armogum, traf mgr; Jill O'Malia, mktg.

Angola

***WEAX(FM)**— September 1979: 88.3 mhz; 920 w. Ant 151 ft TL: N41 37 53 W85 00 37. Stereo. Hrs opn: 24 1 University Ave, 46703-1750. Phone: (260) 665-4288. E-mail: weaxfm@tristate.edu Web Site: www.88xradio.com. Licensee: Tri-State University. Natl. Network: CNN Radio. Law Firm: Reddy, Begley & McCormick. Format: Alt/college. News: one hrs wkly. Target aud: 18-44. ♦Josh Hornbacker, gen mgr & opns mgr.

WLKI(FM)— July 15, 1974: 100.3 mhz; 4 kw. Ant 393 ft TL: N41 40 51 W85 00 05. Stereo. Hrs opn: 24 Box 999, 46703. Secondary address: 2655 State Rd. 127N 46703. Phone: (260) 665-9554. Fax: (260) 665-9064. E-mail: wlki@wlki.com Web Site: www.wlki.com. Licensee: Lake Cities Broadcasting Corp. Population served: 25,000 Format: Hot adult contemp. Target aud: 25-49; adults with youthful outlook, skews female. ♦Bill Kerner, VP; Thomas R. Andrews, pres & gen mgr; Andy St. John, progmg dir; Jim Measel, news dir; Greg Case, chief of engrg.

Attica

***WFWR(FM)**— 2002: 91.5 mhz; 160 w. Ant 171 ft TL: N40 16 47 W87 14 50. Hrs open: 909 S. McDonald St., 47918. Phone: (765) 764-1934. Web Site: www.atticaonline.com/wfwr.htm. Licensee: Fountain Warren Community Radio Corp. Format: Var. ♦Larry Grant, gen mgr.

WSHP(FM)— April 1990: 95.7 mhz; 3.1 kw. Ant 433 ft TL: N40 23 02 W87 07 55. Stereo. Hrs opn: 3824 S. 18th St., Lafayette, 47909. Phone: (765) 474-1410. Fax: (765) 474-3442. Licensee: Artistic Media Partners L.P. Group owner: Artistic Media Partners Inc. (acq 10-3-94; $410,000; FTR: 10-17-94). Natl. Rep: Christal. Law Firm: Rosenman & Colin. Format: Classic rock. Target aud: 25-54. ♦Arthur Angotti, pres, exec VP & gen mgr; Bob Henning, gen sls mgr & chief of engrg; Steve Clark, progmg dir.

Auburn

WGBJ(FM)— Apr 10, 1967: 102.3 mhz; 3 kw. Ant 300 ft TL: N41 20 01 W85 03 08. Stereo. Hrs opn: 24 2000 Lower Huntington Rd., Fort Wayne, 46819. Phone: (260) 747-1511. Fax: (260) 747-3999. E-mail: kfoate@summitcityradio.com Licensee: Three Amigo's Broadcasting Inc. Group owner: Summit City Radio Group (acq 11-30-2006; $1.35 million). Natl. Rep: McGavren Guild. Format: Sp. ♦Robert J. Britt, VP; Robert J. Britt, gen mgr.

WGLL(AM)— Sept 3, 1968: 1570 khz; 500 w-D, 151 w-N, DA-2. TL: N41 20 01 W85 03 08. Hrs open: 17
Rebroadcasts WGL(AM) Fort Wayne 100%.
5446 C. R. 29, Fort Wayne, 46706. Phone: (260) 925-4300. Fax: (260) 432-0986. Licensee: Kovas Communications of Indiana Inc. (acq 11-9-01; grpsl). Population served: 33,500 Natl. Network: CBS. Rgnl. Network: Network Indiana. Natl. Rep: Rgnl Reps. Law Firm: Lauren A. Colby. Format: Relg progmg. Target aud: 25-54. ♦Raymond Alexander, pres & gen mgr.

Aurora

WSCH(FM)— Oct 29, 1970: 99.3 mhz; 1.15 kw. 525 ft TL: N38 57 55 W84 56 51. Stereo. Hrs opn: 24 20 E. High St., Lawrenceburg, 47025-1820. Phone: (812) 438-2777. Fax: (812) 438-3495. E-mail: wsch@one.net Licensee: Columbus Radio Inc. (acq 12-2-02; with WXCH(FM) Versailles). Population served: 60,234 Natl. Network: ABC. Format: Young country. News staff: one; News: 10 hrs wkly. Target aud: 25-plus. Spec prog: Farm 3 hrs wkly. ♦Marty Pieratt, pres & gen mgr; Eric Boulander, progmg dir & mus dir; Bob Shannon, news dir; Robert Hawkins, chief of engrg.

Austin

WJAA(FM)— 1991: 96.3 mhz; 3 kw. 328 ft TL: N38 50 39 W85 49 26. Hrs open: 1531 W. Tipton St., Seymour, 47274. Phone: (812) 523-3343. Fax: (812) 523-5116. E-mail: coolbus@wjaa.net Web Site: www.wjaa.net. Licensee: Midland Media Inc. (acq 6-28-91; $15,000; FTR: 7-22-91). Natl. Network: ABC, Westwood One. Format: AOR, classic rock. News staff: News progmg 5 hrs wkly Target aud: 25-54; men & women. ♦Robert Becker, gen mgr; Tony Starkey, gen sls mgr; Shannon Pyle, progmg dir.

WXKU-FM— December 1993: 92.7 mhz; 2 kw. Ant 400 ft TL: N38 49 23 W85 47 24. Stereo. Hrs opn: 24 2470 N. Hwy. 7, North Vernon, 47265. Phone: (812) 346-1927. Fax: (812) 346-9722. Licensee: BK Media LLC (acq 7-31-2006; $850,000). Population served: 300,000 Natl. Network: USA. Format: Country. News staff: one; News: 21 hrs wkly. Target aud: 25-65. ♦Marty Pieratt, gen mgr.

Batesville

WRBI(FM)— May 14, 1977: 103.9 mhz; 1.95 kw. Ant 360 ft TL: N39 13 22 W85 15 28. Stereo. Hrs opn: 24 133 S. Main St., 47006. Phone: (812) 934-5111. Fax: (812) 934-2765. E-mail: wrbi@wrbiradio.com Web Site: www.wrbiradio.com. Licensee: White River Broadcasting Co. Inc. Group owner: The Findlay Publishing Co. (acq 7-31-97; grpsl). Population served: 102,000 Natl. Rep: Rgnl Reps. Format: Country. News staff: one; News: 10 hrs wkly. Target aud: General. Spec prog: Farm 5 hrs wkly. ♦David Glass, VP; Ronald E. Green, gen mgr; Caz Burdetter, progmg dir; Mary Mattingly, news dir.

Battle Ground

WASK-FM— Mar 11, 1993: 98.7 mhz; 4.4 kw. Ant 384 ft TL: N40 29 57 W86 52 25. Stereo. Hrs opn: 24 Box 7880, Lafayette, 47903-7880. Secondary address: 3575 McCarty Ln., Lafayette 47905. Phone: (765) 447-2186. Fax: (765) 448-4452. Web Site: www.wask.com. Licensee: WASK Inc. Group owner: Schurz Communications Inc. (acq 3-6-95; $860,000; FTR: 6-26-95). Population served: 238,000 Natl. Rep: Christal. Rgnl rep: Rgnl Reps. Law Firm: Hogan & Hartson. Wire Svc: AP Format: Oldies. News staff: 4; News: 20 hrs wkly. Target aud: 35 plus; general. ♦John A. Trent, pres, gen mgr & gen sls mgr; Mark Allen, opns mgr; Brian Green, gen sls mgr; Bryan McGarvey, progmg dir & progmg mgr; Steve Truex, chief of engrg.

Bedford

WBIW(AM)— October 1948: 1340 khz; 1 kw-U. TL: N38 52 23 W86 28 34. Hrs opn: 24 424 Heltonville Rd., 47421. Phone: (812) 275-7555. Fax: (812) 279-8046. E-mail: wbiw1340am@yahoo.com Web Site: www.wbiw.com. Licensee: Ad-Venture Media Inc. (acq 1-30-89; $1 million with co-located FM; FTR: 1-30-89) Population served: 13,087 Natl. Network: Westwood One, USA. Rgnl. Network: Network Indiana. Natl. Rep: Rgnl Reps. Law Firm: Reed, Smith, Shaw & McClay. Format: Sports, news, talk. News staff: one; News: 28 hrs wkly. Target aud: 25 plus; general. Spec prog: Sports, weather, farm 3 hrs wkly. ♦Dean Spencer, pres & gen mgr.

Beech Grove

WNTS(AM)— Dec 10, 1956: 1590 khz; 5 kw-D, 500 w-N, DA-3. TL: N39 44 21 W86 05 29. Hrs open: 24 Box 2368, Davidson, NC, 28036-5368. Phone: (317) 359-5591. Fax: (317) 359-3885. Web Site: www.lapoderosa1590am.com. Licensee: Davidson Media Station WNTS Licensee LLC. (acq 9-28-2005; $2 million). Population served: 744,624 Law Firm: Fletcher, Heald & Hildreth. Format: Rgnl Mexican. ♦Steve Stiegelmeyer, gen mgr; Mayra Elisa Arroyo, progmg dir.

Berne

WZBD(FM)— Aug 27, 1993: 92.7 mhz; 4.1 kw. 394 ft TL: N40 46 15 W85 56 05. Hrs open: 5 AM-10 PM 1891 W. State Rd. 97, Portland, 47371. Secondary address: 955 US 27 N. 46711. Phone: (260) 726-8729. Fax: (260) 726-4311. E-mail: wpgw@jayco.net Licensee: Adams County Radio Inc. (acq 3-11-99). Population served: 150,000 Format: Adult contemp, local news. Target aud: General. ♦Rob Weaver, pres & gen mgr; Tony Giltner, opns mgr.

Bicknell

WUZR(FM)—Licensed to Bicknell. See Vincennes

Bloomington

WBWB(FM)— July 17, 1978: 96.7 mhz; 1.65 kw. 439 ft TL: N39 09 46 W86 28 21. Stereo. Hrs opn: 24 Box 7797, 47407. Secondary address: 304 State Rd. 446 47401. Phone: (812) 336-8000. Fax: (812) 336-7000. E-mail: wbwb@wbwb.com Web Site: www.wbwb.com. Licensee: Artistic Media Partners L.P. Group owner: Artistic Media Partners Inc. (acq 1-89; grpsl; FTR: 1-23-89). Population served: 230,000 Natl. Rep: McGavren Guild. Rgnl rep: Rgnl Reps. Law Firm: Haley, Bader & Potts. Format: CHR. News staff: one. Target aud: 18-49. ♦Art Angotti, pres; Sandy Zehr, gen mgr; Dale Clark, gen sls mgr & progmg dir; Bob Henning, chief of engrg.

***WFHB(FM)**— December 1992: 91.3 mhz; 2.5 kw horiz, 2.45 kw vert. 266 ft TL: N39 01 55 W86 36 33. Hrs open: 24 Box 1973, 47402. Secondary address: 108 W. 4th St. 47404. Phone: (812) 323-1200. Fax: (812) 323-0320. E-mail: wfhb@wfhb.org Web Site: www.wfhb.org. Licensee: Bloomington Community Radio Inc. Format: Div, news, pub affrs. News staff: one; News: 5 hrs wkly. Target aud: General. Spec prog: Folk 10 hrs, Latin 3 hrs, Finnish 3 hrs wkly. ♦Markus Lowe, gen mgr & progmg dir; Jim Manion, mus dir; Chad Carrothers, news dir & pub affrs dir.

***WFIU(FM)**— Sept 30, 1950: 103.7 mhz; 29 kw. Ant 646 ft TL: N39 08 31 W86 29 43. Stereo. Hrs opn: 24 Radio-TV Ctr., Indiana Univ., 1229 E. 7th St., 47405. Phone: (812) 855-1357. Fax: (812) 855-5600. E-mail: wwfiu@indiana.edu Web Site: www.indiana.edu/-wfiu. Licensee: Trustees of Indiana University. Population served: 200,000 Natl. Network: NPR. Law Firm: Crowell & Moring. Format: Class, jazz, news. News staff: one; News: 7 hrs wkly. Target aud: General. ♦Christina Kuzmych, stn mgr & prom dir; Will Murphy, news dir; Bradley Howard, chief of engrg. Co-owned TV: *WTIU(TV) affil.

WGCL(AM)— Mar 11, 1949: 1370 khz; 5 kw-D, 500 w-N, DA-2. TL: N39 11 25 W86 38 02. Stereo. Hrs opn: 24 400 One City Ctr., 47404. Phone: (812) 332-3366. Fax: (812) 331-4570. Web Site: www.am1370wgll.com. Licensee: Sarkes Tarzian Inc. (group owner) Population served: 120,000 Natl. Network: ABC, ESPN Radio. Natl. Rep: Christal. Law Firm: Leventhal, Senter & Leman. Format: News/talk. News staff: 2; News: 5 hrs wkly. Target aud: 30 plus. ♦Ron Tarsi, gen mgr; Ducan Myers, gen sls mgr; Don Pratt, progmg dir; Marc Antonetti, chief of engrg.

WTTS(FM)—Co-owned with WGCL(AM). Jan 7, 1960: 92.3 mhz; 37 kw. 1,090 ft TL: N39 24 27 W86 08 52. Stereo. 24 10 S. New Jersey St., Indianapolis, 46204. Phone: (317) 972-9887. Fax: (317) 972-9886. Web Site: www.wttsfm.com.2,600,000 Natl. Rep: Christal. Format: AAA. News staff: one. Target aud: 25-54. Spec prog: Blues 2 hrs, acoustic show 4 hrs wkly. ♦Roger Ingram, gen sls mgr & prom dir; Ben Holte, progmg dir; Laura Duncan, news dir.

Bluffton

WNUY(FM)— Dec 10, 1963: 100.1 mhz; 6 kw. 298 ft TL: N40 52 10 W85 10 20. Stereo. Hrs opn: 24 Box 483, 46714. Phone: (260) 824-2804. Fax: (260) 824-2805. E-mail: wnuy@wnuy.com Web Site: www.wnuy.com. Licensee: IM IN Licenses LLC. (acq 3-21-2006; $1 million). Population served: 425,000 Natl. Rep: Rgnl Reps. Format: Adult contemp. News staff: one; News: 2 hrs wkly. Target aud: 25-54; professional women. Spec prog: Relg 4 hrs wkly. ♦Dayle Mentzer, gen mgr & gen sls mgr; Rob Caylor, progmg dir; Mike Peters, chief of engrg.

Boonville

WBNL(AM)— Sept 10, 1950: 1540 khz; 250 w-D. TL: N38 03 58 W87 16 27. Hrs open: 24 Box 270, 47601. Secondary address: 2177 N. Hwy. 61, 47601. Phone: (812) 897-2080. Fax: (812) 897-2130. E-mail: rtradio@sigecom.net Web Site: www.radio1540.net. Licensee: Turpen Communications LLC (acq 8-22-01). Population served: 71,000 Natl. Network: USA. Rgnl. Network: Network Indiana. Format: MOR, adult contemp. News staff: one; News: 14 hrs wkly. Target aud: 25-54; Women 25-54. Spec prog: Gospel 5 hrs wkly. ♦Ralph E. Turpen, pres & gen mgr.

WEJK(FM)— Dec 19, 1967: 107.1 mhz; 3 kw. 185 ft TL: N38 03 58 W87 16 27. Hrs open: 24 Box 3848, Evansville, 47736. Secondary address: 1162 Mt. Auburn Rd., Evansville 47720. Phone: (812) 424-8284. Fax: (812) 426-7928. E-mail: tim@sccradio.com Licensee: Boonville Broadcasting Co. Inc. Group owner: South Central Communications Corp. (acq 8-14-2000; $400,000 for stock with co-located AM). Population served: 200,000 Format: Jack FM. ♦John P. Engelbrecht, CEO; Tim Huelsing, VP & gen mgr; Rsuty James, progmg dir; Chris Myers, chief of engrg.

Brazil

WSDM-FM— Nov 13, 1973: 92.7 mhz; 6 kw. Ant 298 ft TL: N39 30 44 W87 08 18. Stereo. Hrs opn: 24 1301 Ohio St., Terre Haute, 47807. Phone: (812) 234-9770. Fax: (812) 238-1576. E-mail: info@crock927.com Web Site: www.crock927.com. Licensee: Crossroads Investments LLC. (acq 8-1-90; with co-located AM). Population served: 300,000 Natl. Rep: Roslin. Law Firm: Booth, Freret, Imlay & Tepper. Format: Country, rock. News staff: one; News: 2 hrs wkly. ♦Michael Petersen, pres & gen mgr.

WSDX(AM)—Co-owned with WSDM-FM. 1959: 1130 khz; 500 w-D, 20 w-N. TL: N39 30 44 W87 08 18.6 AM-2 hrs past sunset
Rebroadcasts WBOW(AM) Terre Haute 95%.
E-mail: score@espnsportsradio.com Web Site: espnsportsradio.com. Group owner: Crossroads Communications Inc. 25,000 Natl. Rep: Roslin. Rgnl rep: Rgnl Reps. Law Firm: Booth, Freret, Imlay & Tepper. Format: Sports. News staff: one. Target aud: General; sports fans.

Bremen

WHPZ(FM)— Mar 1, 1993: 96.9 mhz; 2.99 kw. Ant 462 ft TL: N41 26 37 W86 01 18. Hrs open: 61300 S. Ironwood Rd., South Bend, 46614. Phone: (574) 291-8200. Fax: (574) 291-9043. E-mail: thale@lesea.com Web Site: www.pulsefm.com. Licensee: Le Sea Broadcasting Corp. Group owner: Le Sea Broadcasting Corp. (acq 1-4-2000; $280,296). Format: Contemp Christian. ♦Tony Hale, CFO; Anna Riblet, stn mgr & gen sls mgr; Wes Hylton, chief of engrg.

Brookston

WLFF(FM)— Apr 16, 1967: 95.3 mhz; 2.3 kw. Ant 505 ft TL: N40 32 48 W86 50 59. Stereo. Hrs opn: 24 3824 S. 18th Street, Lafayette, 47909. Phone: (765) 474-1410. Fax: (765) 474-3442 / (574) 273-9090. Web Site: www.wlff.com. Licensee: Artistic Media Partners Inc. (group owner; (acq 9-1-98; $1.8 million). Population served: 100,000 Law Firm: Rosenman & Colin. Format: Country. News staff: one; News: 14 hrs wkly. Target aud: 35 plus; affluent, educated & upscale. ♦Arthur Angotti III, pres; Arthur Angotti, gen mgr; Mike Warner, progmg dir; Bob Henning, chief of engrg.

Brownsburg

WKLU(FM)— Mar 23, 1992: 101.9 mhz; 6 kw. Ant 252 ft TL: N39 49 07 W86 22 40. Stereo. Hrs opn: 24 8120 Knue Rd., Indianapolis, 46250. Phone: (317) 841-1019. Fax: (317) 841-5167. E-mail: bart@wklu.net Web Site: www.wklu.net. Licensee: Indy Radio LLC (acq 7-30-2004; $6.2 million). Population served: 950,000 Rgnl. Network: Metronews Radio Net. Format: Classic hits. Target aud: 25-54. Spec prog: Beetles brunch. ♦Bart Johnson, gen mgr; Monica Lephart, prom dir; Libby Zabriskie-Farr, progmg dir; Aimee McGrath, traf mgr.

Cannelton

WLME(FM)— July 1990: 102.9 mhz; 12.5 kw. 466 ft TL: N37 46 57 W86 36 26. Stereo. Hrs opn: 24 1115 Tamarock Rd., Suite 500, Owensboro, KY, 42301. Phone: (270) 683-5200. Fax: (270) 688-0108. Licensee: WLME Inc. Group owner: The Cromwell Group Inc. Population served: 47,200 Natl. Network: Jones Radio Networks. Natl. Rep: Rgnl Reps. Law Firm: Pepper & Corazzini. Format: Oldies. News staff: one; News: 3 hrs wkly. Target aud: 25-54; general. Spec prog: Sports 6 hrs wkly. ♦Bayard H. Walters, pres; Kevin Riecke, gen mgr.

Carmel

***WHJE(FM)**— September 1963: 91.3 mhz; 400 w. 100 ft TL: N39 58 45 W86 07 10. Stereo. Hrs opn: 24 520 E. Main St., 46032. Phone: (317) 571-4055. Phone: (317) 846-7721. Fax: (317) 571-4066. Web Site: www.whje.com. Licensee: Carmel Clay Schools. Population served: 30,000 Rgnl. Network: Network Indiana. Wire Svc: UPI Format: Classic rock, alternative. Target aud: 12 plus. ♦Tom Schoeller, gen mgr.

Centerville

WHON(AM)—Licensed to Centerville. See Richmond

Chandler

WLFW(FM)— Apr 2, 1994: 93.5 mhz; 3.2 kw. Ant 446 ft TL: N38 01 27 W87 21 43. Hrs open: 24 Box 3848, Evansville, 47736. Secondary address: 1162 Mt. Auburn Rd., Evansville 47720. Phone: (812) 424-8284. Fax: (812) 426-7928. Web Site: www.935thewolf.com. Licensee: South Central Communications Corp. (group owner; (acq 1996; $860,000). Population served: 625,000 Natl. Network: Westwood One. Format: Country. News staff: 3. Target aud: 35-49. ♦John D. Englebrecht, CEO & chmn; Craig Jacobus, pres; Paul Brayfield, gen mgr & gen sls mgr; Tim Huelsing, gen mgr; James Ashley, prom dir; Rusty James, progmg dir.

Charlestown

WLRX(FM)— Apr 7, 1998: 104.3 mhz; 3 kw. Ant 328 ft TL: N38 28 55 W85 37 33. Stereo. Hrs opn: 24
Simulcast with WLRS(FM) Shepherdsville, KY 100%.
520 S. 4th St., Suite 200, Louisville, KY, 40202. Phone: (502) 625-1220. Fax: (502) 584-1051. Web Site: www.wlrs.com. Licensee: Blue Chip Broadcasting Licenses Ltd. Group owner: Radio One Inc. (acq 2-14-2003; $2 million). Format: New rock.

Chesterton

***WBEW(FM)**— 2001: 89.5 mhz; 7 kw. Ant 216 ft TL: N41 42 58 W86 51 47. (CP: 23 kw, ant 187 ft.). Hrs opn: 848 E. Grand Ave., Navy Pier, Chicago, IL, 60611-3462. Phone: (312) 948-4600. Fax: (312) 948-4837. E-mail: tmalatia@chicagopublicradio.org Web Site: www.chicagopublicradio.org. Licensee: The WBEZ Alliance Inc. (acq 10-4-02; $550,000). Format: News, talk, jazz. ♦Torey Malatia, pres & gen mgr; Greg Salustro, dev VP.

***WDSO(FM)**— November 1976: 88.3 mhz; 400 w. 135 ft TL: N41 36 29 W87 03 37. Stereo. Hrs opn: 6 AM- 6 PM (M-F) Chesterton High School, 2125 S. 11th St., 46304. Phone: (219) 983-3777. Fax: (219) 983-3773. Web Site: www.wdso.org. Licensee: Duneland School Corp. Population served: 149,200 Format: Rock. News: 6 hrs wkly. Target aud: General. Spec prog: Class one hr, specialty rock 8 hrs wkly. ♦Brent Barber, stn mgr & chief of engrg; Michele Stipanovich, opns mgr.

Churubusco

WNHT(FM)— August 1994: 96.3 mhz; 6.7 kw. 554 ft TL: N41 06 13 W85 10 44. Hrs open: 24 2000 Lower Huntington Rd., Fort Wayne, 46819. Phone: (260) 747-1511. Fax: (260) 747-3999. Web Site: wild963.com. Licensee: Summit City License Sub, LLC. Group owner: Summit City Radio Group (acq 11-6-2006; grpsl). Natl. Rep: McGavren Guild. Format: Rhythmic CHR. Target aud: 18-34; women. ♦Lloyd Roach, VP & gen mgr; Dave Wisniewski, gen sls mgr; Mojo Wilson, progmg dir.

Cicero

***WJCY(FM)**— 2005: 91.5 mhz; 475 w vert. Ant 193 ft TL: N40 11 53 W86 07 44. Hrs open: CSN International, 4002N. 3300E., Twin Falls, ID, 83301. Phone: (208) 734-6633. Fax: (208) 736-1958. Licensee: CSN International. (group owner). ♦Mike Kestler, pres.

Clarksville

WTFX-FM— 1998: 93.1 mhz; 2.15 kw. Ant 387 ft TL: N38 17 02 W85 54 17. Hrs open: 24
Simulcast with WJZO(FM) Shelbyville, KY 100%.
4000 #1 Radio Dr., Louisville, KY, 40218. Phone: (502) 479-2222. Fax: (502) 479-2223. Web Site: www.foxrocks.com. Licensee: CC Licenses LLC. Group owner: Clear Channel Communications Inc. Population served: 850,000 Format: Active rock. ♦Kevin Hughes, gen mgr.

Clinton

WAXI(FM)—See Rockville

***WPFR-FM**— 1998: 93.9 mhz; 2.3 kw. 531 ft TL: N39 33 18 W87 28 40. Stereo. Hrs opn: 24 18889 N. 2350th St., Dennison, IL, 62423. Phone: (217) 826-9673. E-mail: wpfr@joink.com Licensee: Word Power Inc. Population served: 146,605 Natl. Network: Moody. Format: Christian. News: 17 hrs wkly. Target aud: 12 plus. ♦Paul Dean Ford, pres & gen mgr; Mark S. Ford, opns dir; Dan Watson, chief of opns.

Cloverdale

***WSPM(FM)**— 2003: 89.1 mhz; 25 w horiz, 49 kw vert. Ant 298 ft TL: N39 41 19 W86 42 03. Hrs open: 24 3500 DePauw Blvd., Suite 2085, Indianapolis, 46268. Phone: (317) 870-8400 ext 21. Fax: (317) 870-8404. E-mail: jim@catholicradioindy.org Web Site: www.catholicradioindy.org. Licensee: Hoosier Broadcasting Corp. Format: Relg-Catholic. ♦Chuck Cunningham, gen mgr; Ed Roehling, gen sls mgr; Bill Shirk, progmg dir; Jim Ganley, news dir; Marty Hensley, chief of engrg.

Columbia City

***WJHS(FM)**— Aug 12, 1985: 91.5 mhz; 2.65 kw. 219 ft TL: N41 10 04 W85 29 41. Stereo. Hrs opn: 24 600 N. Whitley St., 46725. Phone: (260) 248-8915. Phone: (260) 244-6136. Fax: (260) 244-5610. E-mail: comments@wjhs915.org Web Site: www.wjhs915.org. Licensee: Whitley County Consolidated Schools Board of Control. Population served: 26,000 Format: Adult alternative. Target aud: Men 25-54; general. ♦Krystal Walker Zoltek, stn mgr; Laurie Walls, mus dir.

WVBB(FM)— Oct 13, 1968: 106.3 mhz; 5.6 kw. Ant 339 ft TL: N41 12 49 W85 12 04. Stereo. Hrs opn: 2100 Goshen Rd., Suite 332, Fort Wayne, 46808. Phone: (260) 482-9288. Fax: (260) 482-8655. Web Site: www.1063thevibe.com. Licensee: Artistic Media Partners Inc. (group owner; (acq 7-30-2004; $2.61 million). Law Firm: Tieerney & Swift. Format: Rhythm and blues oldies. ♦Roger Diehm, gen mgr; Phil Becker, progmg dir.

Columbus

WCSI(AM)— 1950: 1010 khz; 500 w-D, 19 w-N. TL: N39 11 05 W85 57 17. Stereo. Hrs opn: 24 Box 1789, 47202-1789. Secondary address: 3212 Washington St. 47203. Phone: (812) 372-4448. Fax: (812) 372-1061. Licensee: White River Broadcasting Co. Group owner: The Findlay Publishing Co. (acq 11-1-57). Population served: 100,000 Rgnl. Network: AgriAmerica. Rgnl rep: Rgnl Reps. Wire Svc: CBS Format: News/talk, weather, sports. News staff: 3. Target aud: 35 plus. ♦John Foster, opns mgr, progmg dir & pub affrs dir; Tasha Mann, gen mgr & gen sls mgr; Kevin Keith, news dir; Chuck Weber, chief of engrg.

WKKG(FM)—Co-owned with WCSI(AM). 1958: 101.5 mhz; 50 kw. 492 ft TL: N39 11 05 W85 57 17. Stereo. 24 E-mail: wkkg@wkkg.com Web Site: wkkg.com.300,000 Format: Country. Target aud: 25-54. ♦Scott Michaels, progmg dir; Judy Watkins, traf mgr; Sam Simmermaker, sports cmtr.

WINN(FM)— Jan 30, 1975: 104.9 mhz; 6 kw. Ant 300 ft TL: N39 11 09 W85 57 37. Stereo. Hrs open: 24 Box 1789, 47202-1789. Secondary address: 3212 Washington St. 47203. Phone: (812) 372-4448. E-mail: studio@1049theriver.fm Web Site: www.1049theriver.fm. Licensee: White River Broadcasting Co. Inc. Group owner: The Findlay Publishing Co. (acq 1-8-2002). Population served: 100,000 Format: Classic hits. Target aud: 35-54. ♦Kurt Kah, pres; David Glass, VP; Tasha Mann, gen mgr; John Foster, opns mgr; Rich Anthony, progmg dir; Kevin Keith, news dir; Chuck Weber, engr.

***WKJD(FM)**—Not on air, target date: unknown: 90.3 mhz; 1.5 kw vert. Ant 138 ft TL: N39 12 27 W86 01 27. Hrs open: 1680 Hwy. 62 N.E., Corydon, 47112. Phone: (812) 738-3482. Fax: (812) 375-2555. Licensee: Good Samaritan Educational Radio Inc. ♦Keith Reising, CEO.

WRZQ-FM—See Greensburg

WYGS(FM)— Feb 27, 2003: 91.1 mhz; 380 w vert. Ant 328 ft TL: N39 13 35 W85 44 47. Hrs open: 24 Box 2626, 47202. Secondary address: 825 Washington St. 47201. Phone: (812) 738-3482. Fax: (812) 375-2555. Web Site: www.wygs.org. Licensee: Good Shepherd Radio Inc. Format: Southern gospel. ♦Keith Reising, CEO.

Connersville

WIFE(AM)— Apr 5, 1948: 1580 khz; 250 w-D, 5 w-N. TL: N39 38 18 W85 08 54. Hrs open: 24 Box 619, 47331. Secondary address: 406 Central Ave. Phone: (765) 825-6411. Phone: (765) 825-8561. Fax: (765) 825-2411. Web Site: www.wifefm.com. Licensee: Rodgers Broadcasting Corp. (group owner; (acq 8-88; grpsl; FTR: 8-29-88). Population served: 17,604 Rgnl rep: Rgnl Reps. Format: Oldies. News staff: 2; News: 3 hrs wkly. Target aud: 25-54; affluent, middle-aged country listeners. Spec prog: Relg 12 hrs wkly. ♦David A. Rodgers, pres; John Trine, gen mgr; Jeri L. Pruet, stn mgr; Bob Wills, progmg dir; Mike Reese, progmg dir; Barry Welsh, mus dir; Brett Briscoe, news dir; Kristin Deiwert, news dir; Mike Peacock, engrg mgr; Bob Hawkins, chief of engrg; Becky Hymer, traf mgr; Kristin Dewert, local news ed.

WMOJ-FM— Feb 27, 1948: 100.3 mhz; 28 kw. Ant 215 ft TL: N39 38 15 W85 08 45. (CP: COL Norwood, OH. 3.6 kw, ant 426 ft. TL: N39 07 19 W84 32 52). Stereo. Hrs opn: 24 1 Centennial Plaza, 705 Central Ave., Suite 200, Cincinnati, OH, 45202. Phone: (513) 679-6000. Fax: (513) 679-6014. Licensee: Blue Chip Broadcasting Licenses Ltd. (acq 9-21-2006; $18 million). Population served: 85,000 Rgnl. Network: Network Indiana. Format: Urban adult contemp. ♦Lisa Thal, gen mgr.

Corydon

WOCC(AM)— May 22, 1964: 1550 khz; 250 w-D. TL: N38 11 26 W86 08 00. Hrs open: 24 Box 838, 47112. Phone: (812) 738-9622. Fax: (812) 738-1676. E-mail: wocc1550@cs.com Web Site: www.woccam1550.com. Licensee: Richard Lee Brabandt. (acq 4-15-97). Natl. Network: USA. Natl. Rep: Rgnl Reps. Format: Classic oldies. News staff: one; News: 18 hrs wkly. Target aud: 34-55; baby boomers. ♦Richard Lee Brabandt, pres; MaryAnn Brabandt, gen mgr & progmg dir; Dave Riddle, chief of engrg.

WSFR(FM)— 1994: 107.7 mhz; 8.2 kw. 567 ft TL: N38 10 25 W85 54 50. (CP: 36 kw). Stereo. Hrs opn: 24 612 4th Ave., Suite 100, Louisville, KY, 40202. Phone: (502) 589-4800. Fax: (502) 583-4820. Web Site: www.1077sfr.com. Licensee: Cox Radio Inc. Group owner: Cox Broadcasting (acq 5-99). Population served: 976,800 Law Firm: Dow, Lohnes & Albertson. Format: Classic rock. Target aud: 25-54. ♦Todd Schumacher, VP; Don Nordin, gen mgr & progmg dir; Amy Torres, gen sls mgr.

Covington

***WFOF(FM)**— June 17, 1984: 90.3 mhz; 19 kw. 265 ft TL: N40 09 08 W87 27 58. Stereo. Hrs opn: 24 1920 W. 53rd St., Anderson, 46013.

Phone: (765) 642-2750. Fax: (765) 642-4033. E-mail: wfof@wfof.org Web Site: www.wfof.org. Licensee: Doxa Inc. Format: Relg. ♦Ray McDaniel, pres; Ray Hashley, gen mgr.

WKZS(FM)— June 1, 1982: 103.1 mhz; 3 kw. 300 ft TL: N40 08 46 W87 27 15. Stereo. Hrs opn: 24 Box 67, Danville, IL, 61834. Secondary address: 820 Railroad St. 47932. Phone: (217) 443-4004. Phone: (765) 793-4823. Fax: (765) 793-4644. E-mail: info@kisscountryradio.com Web Site: www.kisscountryradio.com. Licensee: Benton-Weatherford Broadcasting Inc. of Indiana. (acq 7-12-85; FTR: 6-3-85). Population served: 100,000 Natl. Network: Jones Radio Networks. Law Firm: Borsari & Paxson. Format: Country. News: 10 hrs wkly. Target aud: 18-49. ♦Larry Weatherford, pres & opns dir; Rhea Benton-Weatherford, gen mgr; Greg Green, stn mgr; Tara Duncan, prom mgr.

Crawfordsville

WCDQ(FM)— Aug 13, 1953: 106.3 mhz; 3.4 kw. 440 ft TL: N40 03 19 W86 55 57. Stereo. Hrs opn: 24 Box 603, 47933. Secondary address: 1800 N. 175 W. 47933. Phone: (765) 362-8200. Phone: (765) 364-1063. Fax: (765) 364-1550. E-mail: cd1063@keybroadcasting.net Licensee: C.V.L. Broadcasting Inc. Group owner: Key Broadcasting Inc. (acq 12-13-99; $400,000). Population served: 50,000 Rgnl. Network: Ohio Radio Net. Law Firm: Arter & Hadden. Format: Hot adult contemp. News staff: 2; News: 48 hrs wkly. Target aud: 25-49; middle & upper class, educated, socially aware. ♦Sherry Moodie, gen mgr.

WCVL(AM)— Dec 12, 1964: 1550 khz; 250 w-U, DA-N. TL: N40 03 54 W86 56 00. Hrs open: Box 603, 47933-0603. Phone: (765) 362-8200. Fax: (765) 364-1550. Web Site: www.crawfordsvilleradio.com. Licensee: C.V.L. Broadcasting Inc. Group owner: Key Broadcasting Inc. (acq 1986). Population served: 50,000 Natl. Rep: Rgnl Reps. Format: Music of Your Life. Target aud: 45 plus. Spec prog: Farm 5 hrs wkly. ♦Sherry Moodie, gen mgr; Bill Boy, progmg dir.

WIMC(FM)—Co-owned with WCVL(AM). June 1, 1974: 103.9 mhz; 1.35 kw. 500 ft TL: N40 08 05 W86 54 12. Stereo. 50,000 Format: Classic hits. Target aud: 25-49.

***WNDY(FM)**— 1997: 91.3 mhz; 2.2 kw. 194 ft TL: N40 03 19 W86 55 57. Hrs open: Box 352, 47933. Secondary address: 301 W. Wabash 47933. Phone: (765) 361-6240. Phone: (765) 361-6038. Fax: (765) 361-6437. Web Site: www.wabash.edu. Licensee: Wabash College Radio Inc. Format: College eclectic. ♦Kevin Kilgore, gen mgr.

Crothersville

***WOJC(FM)**— 2005: 89.9 mhz; 300 w vert. Ant 244 ft TL: N38 50 39 W85 49 26. Hrs open: CSN International, 4002N. 3300E., Twin Falls, ID, 83301. Phone: (208) 734-6633. Licensee: CSN International. (group owner). ♦Mike Kestler, pres.

Crown Point

WXRD(FM)— Nov 10, 1972: 103.9 mhz; 3 kw. 330 ft TL: N41 19 24 W87 21 22. Stereo. Hrs opn: 24 2755 Sager Rd., Valpraiso, 46383. Phone: (219) 462-6111. Fax: (219) 462-4880. Web Site: www.xrock1039.com. Licensee: Porter County Broadcasting Holding Corporation, LLC. Group owner: Porter County Broadcasting Corp. (acq 2-6-2004; $4.9 million with WZVN(FM) Lowell). Population served: 200,000 Natl. Network: ABC. Rgnl. Network: Network Indiana. Law Firm: Reddy, Begley & McCormick. Format: Classic rock. News staff: one; News: 12 hrs wkly. Target aud: 25-54; women. ♦Leigh Ellis, pres & gen mgr.

Danville

***WDVL(FM)**—Not on air, target date: unknown: 88.1 mhz; 10 kw vert. Ant 410 ft TL: N39 47 44 W86 48 04. Hrs open: 5331 Mount Alifan Dr., San Diego, CA, 92111-2622. Phone: (858) 277-4991. Fax: (858) 277-1365. Licensee: Horizon Christian Fellowship. ♦Michael MacIntosh, pres.

WEDJ(FM)— Jan 10, 1975: 107.1 mhz; 1.8 kw. Ant 604 ft TL: N39 48 06 W86 34 24. Stereo. Hrs opn: 24 1800 N. Meridian, Suite 605, Indianapolis, 46202. Phone: (317) 924-1071. Fax: (317) 924-7766. Web Site: www.wedjfm.com. Licensee: Continental Broadcast Group Inc. (acq 12-30-93; grpsl; FTR: 1-24-94). Population served: 744600 Natl. Rep: Univision Radio National Sales. Law Firm: Pillsbury, Winthrop & Shaw Pittman. Format: Hispanic. Target aud: 18-54; hispanic adults. ♦Russ Dodge, gen mgr; Stephanie Tatay-Myers, mktg mgr; Manuel Sepulveda, progmg dir; Phil Alexander, chief of engrg.

Decatur

WADM(AM)— May 22, 1964: 1540 khz; 250 w-D. TL: N40 49 14 W84 55 12. Hrs open: 6:00 am-8:45 pm Box 530, 46733-0530. Secondary address: 16007 Kestrel Drive, Huntertown 46748. Phone: (260) 724-7161. Fax: (260) 490-9614. E-mail: rick@wadm.com Web Site: wadm.com. Licensee: Wells County Radio Corp. (acq 8-25-94; $27,500; FTR: 9-5-94). Population served: 33,625 Rgnl. Network: AgriAmerica. Rgnl rep: Indiana Broadcasters Assn Law Firm: Pillsbury, Winthrop, Shaw, Pittman. Format: Country classics. Target aud: 30 plus. ♦Rick Holden, gen mgr; John Dube, chief of opns.

WQHK-FM— Nov 8, 1966: 105.1 mhz; 2 kw. 397 ft TL: N40 49 14 W84 55 12. (CP: 13.4 kw, ant 449 ft.). Stereo. Hrs opn: 2915 Maples Rd., Fort Wayne, 46816. Phone: (260) 447-5511. Fax: (260) 447-7546. Web Site: www.k105fm.com. Licensee: Jam Communications Inc. Group owner: Federated Media. Population served: 100,000 Format: Country. ♦Mark DePrez, gen mgr; Rob Kelley, opns mgr & progmg dir; Mogan David, chief of engrg.

Delphi

WXXB(FM)— May 24, 1989: 102.9 mhz; 2.2 kw. 420 ft TL: N40 34 57 W86 38 26. Stereo. Hrs opn: 24 Box 7093, Lafayette, 47903. Secondary address: 3575 McCarty Ln., Lafayette 47905. Phone: (765) 448-1566. Fax: (765) 448-1348. E-mail: trent@wask.com Web Site: www.b1029.com. Licensee: WASK, Inc. Group owner: RadioWorks Inc. (acq 10-00; $1 million). Population served: 292,700 Format: CHR/Top 40. News staff: one. Target aud: 18-49. ♦Robert Rhea, pres; Ernie Caldemone, VP; John Trent, gen mgr; John Schurz, rgnl sls mgr; Anthony Bannon, progmg dir.

Earl Park

WIBN(FM)— Oct 15, 1983: 98.1 mhz; 25 kw. 328 ft TL: N40 34 22 W87 27 42. Stereo. Hrs opn: Box 25, Oxford, 47971. Phone: (765) 385-2373. Fax: (765) 385-2374. E-mail: wibn@981wibn.com Web Site: www.981wibn.com. Licensee: Brothers Broadcasting Corp. (group owner; acq 8-95; $100,000). Format: Oldies. ♦John Balvich, pres, gen mgr & gen sls mgr; Dan McKay, progmg dir; Ken Stapleton, news dir; Don Kerawac, chief of engrg.

Edinburgh

WYGB(FM)— Aug 24, 2000: 100.3 mhz; 6 kw. Ant 318 ft TL: N39 11 10 W85 57 29. Hrs open: 24 Edinburgh Radio, 825 Washington St., Columbus, 47201. Phone: (812) 348-1029. Fax: (812) 375-2555. E-mail: korncountry@korncountry.com Web Site: www.korncountry.com. Licensee: Edinburgh Radio. Format: Country. Target aud: 25-54; Bartholomew & Johnson county folks. ♦Keith Reising Jr., CEO.

Elkhart

WAUS(FM)—See South Bend

WBYT(FM)—Listing follows WTRC(AM).

WFRN-FM— June 10, 1963: 104.7 mhz; 50 kw. 488 ft TL: N41 37 18 W85 57 37. Stereo. Hrs opn: 24 Box 307, 46517. Secondary address: 25802 County Rd. 26 46515. Phone: (574) 875-5166. Fax: (574) 875-6662. E-mail: moore@wfrn.com Web Site: www.wfrn.com. Licensee: Progressive Broadcasting System Inc. (group owner). Population served: 1,150,000 Natl. Network: USA. Rgnl. Network: Network Indiana. Law Firm: Reddy, Begley & McCormick. Format: Contemp Christian. News staff: one. Target aud: 25-54; families-primarily women. ♦Edwin Moore, pres & gen mgr; Joanne Matthews, prom dir & traf mgr; Don Wagner, news rptr.

WFRN(AM)— Mar 16, 1956: 1270 khz; 5 kw-D, 1 kw-N, DA-2. TL: N41 37 16 W85 57 40.24 Web Site: www.wfrn.com.200,000 Format: Christian relg, talk. Target aud: 30 plus. ♦Doug Moore, progmg dir & news rptr.

WOZW(FM)—See Goshen

WTRC(AM)— Nov 18, 1931: 1340 khz; 1 kw-U. TL: N41 40 28 W85 56 51. Hrs open: 24 421 S. 2nd St., Suite 100, 46516-3230. Phone: (574) 389-5100. Fax: (574) 389-5101. Web Site: www.am1340.com. Licensee: Pathfinder Communications Corp. Group owner: Federated Media Population served: 139,000 Natl. Network: ABC, Jones Radio Networks. Natl. Rep: Christal. Format: News/talk. News staff: 2; News: 21 hrs wkly. Target aud: 35-64; Elkhart County residents. ♦Kathy Uebler, gen mgr & gen sls mgr; Allan Strike, progmg dir; Gary Sieber, news dir.

WBYT(FM)—Co-owned with WTRC(AM). Apr 1, 1947: 100.7 mhz; 15 kw. 910 ft TL: N41 36 58 W86 11 38. Stereo. 237 Edison Rd., Mishawaka, 46545. Phone: (574) 258-5483. Fax: (574) 258-0930. Web Site: www.b100.com.41,305 Natl. Rep: Christal. Format: Country. Target aud: 25-54. Spec prog: Relg 2 hrs wkly. ♦Brad Williams, gen mgr; Barb Deniston, gen sls mgr; Clint Marsh, progmg dir; Greg Trobridge, chief of engrg.

***WVPE(FM)**— May 1972: 88.1 mhz; 10 kw. 400 ft TL: N41 36 20 W86 12 46. (CP: 10.5 w, ant 554 ft. TL: N41 36 59 W86 11 43). Stereo. Hrs opn: 24 EACC, 2424 California Rd., 46514. Phone: (574) 262-5660. Phone: (574) 674-9873. Fax: (574) 262-5520. E-mail: wvpe@wvpe.org Web Site: www.wvpe.org. Licensee: Elkhart Community Schools Corp. Population served: 43,152 Natl. Network: PRI, NPR. Wire Svc: UPI Format: Jazz, news/talk. News staff: one; News: 13 hrs wkly. Target aud: 25-55. Spec prog: Blues 15 hrs, folk 9 hrs wkly. ♦Anthony Hupp, gen mgr & stn mgr.

Ellettsville

WHCC(FM)— 1992: 105.1 mhz; 6 kw. 328 ft TL: N39 11 32 W86 41 46. Hrs open: 24 Box 7797, Bloomington, 47407. Secondary address: 304 State Rd. 446, Bloomington 47401. Phone: (812) 336-8000. Phone: (812) 335-1051. Fax: (812) 336-7000. Web Site: www.whcc105.com. Licensee: Artistic Media Partners L.P. Group owner: Artistic Media Partners Inc. (acq 7-96; $675,000). Population served: 230,000 Natl. Network: Jones Radio Networks. Rgnl rep: Russ Dodge Law Firm: Haley, Bader & Potts. Format: Country. News staff: 2. Target aud: 25-54. ♦Art Angotti, pres; Sandy Zehr, stn mgr; Rick Evans, opns dir; Deborah Green, gen sls mgr; Bob Henning, chief of engrg.

Elwood

WURK(FM)— July 1964: 101.7 mhz; 6 kw. Ant 328 ft TL: N40 16 33 W85 51 44. Hrs open: 24
Simulcasts WERK(FM) Muncie 80%.
800 E. 29th St., Muncie, 47302. Secondary address: 9821 S. 800 W., Daleville 47334. Phone: (765) 288-4403. Fax: (765) 378-2091. Web Site: www.werkradio.com. Licensee: Indiana Sabrecom Inc. Group owner: Backyard Broadcasting LLC (acq 12-1-02; grpsl). Population served: 37,100 Natl. Network: ABC. Rgnl. Network: Network Indiana, AgriAmerica, ABC. Format: Oldies. News staff: one; News: 2 hrs wkly. Target aud: 25-54; adult buying public. Spec prog: Gospel 4 hrs wkly. ♦Steve Lindell, opns dir; Jay Garrison, progmg dir; Sean Mattingly, chief of engrg.

Evansville

WABX(FM)— 1997: 107.5 mhz; 2.35 kw. 518 ft TL: N37 59 21 W87 35 48. Hrs open: 24 Box 3848, 47736. Secondary address: 1162 Mount Auburn 47720. Phone: (812) 424-8284. Fax: (812) 426-7928. Web Site: www.wabx.net. Licensee: South Central Communications Corp. (group owner) Natl. Rep: Katz Radio. Format: Classic rock. Target aud: 25-49; men. ♦John P. Engelbrecht, CEO; Craig Jacobus, pres; Tim Huelsing, gen mgr; Krista Seaton, prom dir; Jason Mack, progmg dir; Randy Wheeler, news dir.

WEOA(AM)— 1935: 1400 khz; 1 kw-U. TL: N37 56 17 W87 31 51. Stereo. Hrs opn: 24 1100 W. Lloyd Expwy., Suite 419, 47708. Phone: (812) 424-8864. Fax: (812) 424-9946. Licensee: South Central Communications Corp. (group owner; acq 11-81). Population served: 138,764 Law Firm: Bryan Cave. Format: Urban adult contemp. News staff: 3; News: 2 hrs wkly. Target aud: 25-49; general. ♦Ed Lander, pres, gen sls mgr & progmg dir; Regina Lander, gen sls mgr; Ron Lyles, news dir.

WIKY-FM—Co-owned with WEOA(AM). Aug 28, 1948: 104.1 mhz; 39 kw. 571 ft TL: N37 59 21 W87 35 48. Stereo. 1162 Mt. Auburn Rd., 47720. Phone: (812) 424-8284. Fax: (812) 426-7928. Web Site: www.wiky.com. (Acq 1948). Format: Adult contemp. Target aud: 25-54; females, workplace. Spec prog: Farm 17 hrs wkly.John P. Engelbrecht, CEO; Tim Huelsing, VP & gen mgr; Paul Broyfield, gen sls mgr; Nora Mitz, mktg dir; Stephanie Todich, prom mgr; Mark Baker, progmg dir & disc jockey; Randy Wheeler, news dir; Chris Myers, chief of engrg; Erin Johnson, traf mgr; Dave Lyons, news rptr; John Gibson, news rptr; Charles Blake, farm dir; Joe Blair, disc jockey; Toni Roberts, disc jockey; Trish Mathews, disc jockey

WGBF(AM)— Nov 22, 1923: 1280 khz; 5 kw-D, 1 kw-N, DA-N. TL: N37 59 53 W87 28 33. Stereo. Hrs opn: 24 Secondary address: 1133 Lincoln Ave. 47714. Phone: (812) 425-4226. Fax: (812) 421-0005. Web Site: iuhoosiers.com/iuradionetwork.html. Licensee: Regent Broadcasting of Evansville/Owensboro Inc. Group owner: Regent Communications Inc. (acq 12-3-2003; grpsl). Population served: 565,400 Natl. Network: CNN Radio, Westwood One. Natl. Rep: Katz Radio. Format: News/talk. News staff: one; News: 10 hrs wkly. Target aud: 25-54; affluent, mature. ♦Mark Thomas, gen mgr.

WGBF-FM—See Henderson, KY

WJLT(FM)— Dec 22, 1964: 105.3 mhz; 50 kw. Ant 480 ft TL: N38 04 47 W87 36 36. Stereo. Hrs opn: 24 1133 Lincoln Ave., 47714. Phone: (812) 425-4226. Fax: (812) 421-0005. Web Site: www.lite1053.com. Licensee: Regent Broadcasting of Evansville/Owensboro Inc. Group owner: Regent Communications Inc. (acq 12-3-2003; grpsl). Population served: 138,764 Natl. Network: ABC. Format: Oldies. Target aud: 25-54. ♦Mike Sanders, CFO & opns mgr; Mark Thomas, gen mgr; Kris Mattingly, prom dir & news dir; Cindy Patrick, progmg dir; Rick Crago, gen mgr & chief of engrg.

***WNIN-FM**— Feb 1, 1982: 88.3 mhz; 17 kw. Ant 840 ft TL: N37 59 01 W87 16 13. Stereo. Hrs opn: 24 405 Carpenter St., 47708. Phone: (812) 423-2973. Fax: (812) 428-7548. E-mail: wnin@wnin.org Web Site: www.wnin.org. Licensee: Tri-State Public Teleplex Inc. Population served: 750,000 Natl. Network: PRI, NPR. Law Firm: Dow, Lohnes & Albertson. Format: Class, news. News: 20 hrs wkly. Target aud: General. ♦David L. Dial, pres & gen mgr; Jean Noyes, stn mgr; Daniel Moore, progmg dir.

***WPSR(FM)**— September 1957: 90.7 mhz; 14 kw. 130 ft TL: N38 01 45 W87 34 42. Stereo. Hrs opn: 6:45 AM-2:45 PM 5400 First Ave., 47710. Phone: (812) 435-8241. Fax: (812) 435-8241. E-mail: wpsr@907wpsr.com Licensee: Evansville Vanderburg School Corp. (acq 9-57). Population served: 150,000 Format: Div, educ, var music. News: 3 hrs wkly. Target aud: General. ♦Michael H. Reininga, gen mgr & dev dir.

***WSWI(AM)**— Aug 6, 1947: 820 khz; 250 w-D. TL: N37 57 53 W87 40 06. Hrs opn: Liberal Arts Bldg., 8600 University Blvd., 47712. Phone: (812) 465-1665. Fax: (812) 461-5261. E-mail: wswi@usi.edu Web Site: www.usi.edu/wswi. Licensee: University of Southern Indiana. (acq 11-3-81). Population served: 150,000 Format: College alternative. News staff: News progmg 3 hrs wkly Target aud: 18-54; students, faculty & community members. ♦John Morris, gen mgr.

***WUEV(FM)**— Apr 1, 1951: 91.5 mhz; 6.1 kw. 150 ft TL: N37 58 24 W87 31 48. Stereo. Hrs opn: 24 1800 Lincoln Ave., 47722. Phone: (812) 479-2022. Fax: (812) 479-2320. E-mail: wuev@evansville.edu Web Site: wuev.evansville.edu. Licensee: University of Evansville. Population served: 250,000 Rgnl. Network: Network Indiana. Wire Svc: UPI Format: Div, jazz. News staff: 3; News: 10 hrs wkly. Spec prog: Children 5 hrs, American Indian one hr wkly. ♦Mike Crowley, gen mgr & opns mgr; Phil Bailey, chief of engrg.

WVHI(AM)— Oct 31, 1948: 1330 khz; 5 kw-D, 1 kw-N, DA-N. TL: N38 03 12 W87 35 40. Hrs open: Box 3636, 47735. Phone: (812) 425-2221. Fax: (812) 425-2078. Web Site: www.wvhi.com. Licensee: Word Broadcasting Network. (acq 3-17-99). Population served: 138,764 Format: Relg, adult contemp. Target aud: General. ♦Stan Hoffman, gen mgr & gen sls mgr.

WYNG(FM)—See Mount Carmel, IL

Ferdinand

WQKZ(FM)— Nov 1, 1997: 98.5 mhz; 6 kw. 328 ft TL: N38 10 02 W86 49 49. Hrs open: Box 167, Jasper, 47547. Phone: (812) 482-2131. Fax: (812) 482-9609. E-mail: wqkz@psci.net Licensee: Gem Communications L.L.P. (acq 4-17-98). Natl. Network: Jones Radio Networks. Format: Country. ♦G. Earl Metzger, pres; Gene Kuntz, gen mgr; Gary Hoffman, progmg dir & progmg mgr; Chris James, news dir; Frank Hertel, chief of engrg.

Fishers

WWFT(FM)— February 1993: 93.9 mhz; 2.95 kw. Ant 476 ft TL: N39 49 39 W85 58 51. Hrs open: 24 6810 N. Shadeland Ave., Indianapolis, 46220. Phone: (317) 842-9550. Fax: (317) 577-3361. Licensee: Indy Lico Inc. Group owner: Susquehanna Radio Corp. (acq 5-5-2006; grpsl). Population served: 1,500,000 Law Firm: McFadden, Evans & Sill. Format: Talk. Spec prog: Hymns of praise 3 hrs, family values one hr wkly. ♦Drew Medland, gen sls mgr; David Wood, progmg dir; Jeff Goode, chief of engrg.

Fort Branch

***WBGW(FM)**— July 20, 1990: 101.5 mhz; 1 kw. 561 ft TL: N38 10 45 W87 29 13. Stereo. Hrs opn: 24 Box 4164, Evansville, 47724. Secondary address: Box 4463 E. 1200 F, R.R. 2, County Rd.1200 S., Haubstadt 47629. Phone: (812) 386-3342. Fax: (812) 768-5552. E-mail: goodnews@evansville.net Web Site: www.thyword.org. Licensee: Music Ministries Inc. Natl. Network: Moody, USA. Format: Relg. News: 12 hrs wkly. Target aud: 35-54. ♦Floyd E. Turner, gen mgr; Susan Turner, prom dir.

Fort Wayne

WAJI(FM)— August 1959: 95.1 mhz; 39 kw. 680 ft TL: N41 06 13 W85 11 28. Stereo. Hrs opn: 24 347 W. Berry, Suite 417, 46802. Phone: (260) 423-3676. Fax: (260) 422-5266. Licensee: Sarkes Tarzian Inc. (group owner) Population served: 416,800 Natl. Rep: Katz Radio. Law Firm: Leventhal, Senter & Lerman. Wire Svc: AP Format: Adult contemp. News staff: one. Target aud: 25-54; women. ♦Thomas Tarzian, CEO; R. Geoffrey Vargo, pres; Robert Davis, CFO; Candace A. Wendling, gen mgr; Lee Tobin, opns mgr; Daryl McIntire, gen sls mgr; Barb Richards, progmg dir; Amy Collins, news dir & traf mgr; Geary Morrill, chief of engrg.

***WBCL(FM)**— Jan 8, 1976: 90.3 mhz; 26 kw. Ant 692 ft TL: N41 06 13 W85 11 46. Stereo. Hrs opn: 24 1025 W. Rudisill Blvd., 46807. Phone: (260) 745-0576. Fax: (260) 456-2913. E-mail: wbcl@wbcl.org Web Site: www.wbcl.org. Licensee: Taylor University Broadcasting Inc. (acq 6-24-92; FTR: 7-20-92). Population served: 800,000 Format: Contemp Christian. ♦Marsha Bunker, gen mgr; Scott Tsuleff, progmg dir; Craig Albrecht, chief of engrg.

***WBOI(FM)**— June 15, 1978: 89.1 mhz; 34 kw. Ant 604 ft TL: N41 06 13 W85 10 44. Stereo. Hrs opn: 24 Box 8459, 46898-8459. Secondary address: 3204 Clairmont Ct. 46808. Phone: (260) 452-1189. Fax: (260) 452-1188. E-mail: bhaines@nipr.fm Web Site: www.nipr.fm. Licensee: Northeast Indiana Public Radio Inc. (acq 1-15-82). Population served: 500,000 Natl. Network: PRI, NPR, AP Radio. Rgnl. Network: Network Indiana. Law Firm: Dow, Lohnes & Albertson. Format: News/talk, jazz. Target aud: 25 plus. ♦Bruce Haines, gen mgr; Colleen Condron, opns dir, opns mgr & progmg dir; Karen Fraser, dev dir & prom dir; Jeanette Dillon, progmg dir & news dir.

WBTU(FM)—Kendallville, Dec 16, 1964: 93.3 mhz; 50 kw. Ant 450 ft TL: N41 23 55 W85 15 08. Stereo. Hrs opn: 24 2100 Goshen Rd., Suite 232, 46808. Phone: (260) 482-9288. Fax: (260) 482-8655. E-mail: pd@wbtu.fm Web Site: www.us933.us. Licensee: Artistic Media Partners Inc. (acq 1-16-97). Population served: 900,000 Format: Country. Target aud: 18-54; upscale, young audience. ♦Roger Diehm, gen mgr; Dave Turpchinoff, gen sls mgr; Scott Roddy, progmg dir; Shelley Hall, traf mgr; Tami Gatchell, traf mgr & disc jockey.

WFCV(AM)— June 17, 1968: 1090 khz; 2.5 kw-D, 1 kw-CH, DA-2. TL: N41 05 01 W85 04 32. Hrs open: Sunrise-sunset 3737 Lake Ave., 46805-5554. Phone: (260) 423-2337. Fax: (260) 423-6355. Web Site: www.bottradionetwork.com. Licensee: Bott Broadcasting. (group owner; acq 5-1-80). Population served: 850,000 Natl. Network: USA. Format: Christian info. Target aud: 25-54; family oriented. ♦Richard P. Bott, pres; Richard Bott II, VP; Dale Gerke, stn mgr; Kathy McClish, opns mgr.

WFWI(FM)— Mar 4, 1993: 92.3 mhz; 2.2 kw. Ant 544 ft TL: N41 06 39 W85 11 44. Hrs open: 24 1005 Production Rd., 46808. Phone: (260) 471-5100. Fax: (260) 471-5224. Web Site: www.923thefort.com. Licensee: Pathfinder Communications Corp. Group owner: Federated Media (acq 3-1-97). Population served: 380,000 Natl. Network: ABC. Format: Classic rock. News staff: one; News: 6 hrs wkly. Target aud: 25-54; men and adults. ♦John Dille, pres; Jim Allgeier, gen mgr; Sonya Maldeney, mktg dir; Billy Elvis, progmg dir; Jack Didier, chief of engrg.

WGL(AM)— Jan 24, 1924: 1250 khz; 2.5 kw-D, 1.4 kw-N, DA-2. TL: N41 01 16 W85 09 46. Hrs open: 24 2000 Lower Huntington Rd., 46819. Phone: (260) 747-1511. Fax: (260) 747-3999. Web Site: 1250theriver.com. Licensee: Summit City License Sub, LLC. Group owner: Summit City Radio Group (acq 11-6-2006;. grpsl). Population served: 177,671 Natl. Network: CBS. Format: Adult standards. News: 2 hrs wkly. ♦J.J. Fabini, progmg dir.

WKJG(AM)— November 1947: 1380 khz; 5 kw-U, DA-2. TL: N41 00 15 W85 05 57. Stereo. Hrs opn: 2915 Maples Rd., 46816. Phone: (260) 447-5511. Fax: (260) 447-7546. Web Site: www.wise33.com. Licensee: Pathfinder Communications Corp. Group owner: Federated Media. Population served: 102,500 Natl. Network: ABC. Natl. Rep: Christal. Format: All sports. ♦Mark DePrez, gen mgr & gen sls mgr; Tony Richards, COO & gen mgr; Jim Tighe, sls dir; Jon Zimney, progmg dir, news dir & news dir; Jack Didion, engrg VP; Mogan David, chief of engrg; Eileen Strickland, traf mgr.

WMEE(FM)—Co-owned with WKJG(AM). Feb 5, 1965: 97.3 mhz; 26 kw. 689 ft TL: N41 06 42 W85 11 43. Stereo. 24 2915 Maples Rd., 46816. Phone: (260) 447-5511. Fax: (260) 447-7546. E-mail: info@wmee.com Web Site: www.wmee.com.218,400 Format: Hot adult contemp. News: one hr wkly. Target aud: 25-54. ♦John Dille, pres; Bob Watson, CFO; Mark Evans, opns mgr; Joel Pyle, gen sls mgr; Rob Kiley, progmg dir; Chris Cage, mus dir; Jack Didier, engrg dir.

***WLAB(FM)**— Aug 23, 1976: 88.3 mhz; 7 kw. 341 ft TL: N41 05 58 W85 08 43. Stereo. Hrs opn: 24 6600 N. Clinton St., 46825. Phone: (260) 483-8236. Fax: (260) 482-7707. E-mail: melissa@star883.org Web Site: www.star883.org. Licensee: The Indiana District Lutheran Church-Missouri Synod Inc. (acq 4-24-87). Population served: 416,800 Natl. Network: Salem Radio Network. Law Firm: Shaw Pittman. Wire Svc: Standard Broadcast News Format: Adult contemp Christian. Target aud: 25-44; Christian. ♦Melissa Montana, gen mgr; Don Buettner, progmg dir.

WLDE(FM)— Aug 24, 1970: 101.7 mhz; 3 kw. 328 ft TL: N41 04 58 W85 04 22. Stereo. Hrs opn: 24 347 W. Berry, Suite 417, 46802. Phone: (260) 423-3676. Fax: (260) 422-5266. Licensee: Sarkes Tarzian Inc. (group owner; acq 2-16-93; FTR: 3-8-93). Population served: 380,900 Natl. Rep: Katz Radio. Law Firm: Leventhal,Senter & Lerman. Wire Svc: AP Format: 60's &70's. Target aud: 35-64. ♦Thomas Tarzian, CEO; R. Geoffrey Vargo, pres; Robert Davis, CFO; Candace A. Wendling, gen mgr; Shelly Steckler, gen sls mgr; Lee Tobin, progmg dir; Geary Morrill, chief of engrg.

WLYV(AM)— Mar 28, 1948: 1450 khz; 1 kw-U. TL: N41 04 14 W85 07 10. Stereo. Hrs opn: 24 4705 Illinois Rd., Suite 104, 46804. Phone: (260) 436-9598. Fax: (260) 432-6179. E-mail: Info@redeemerradio.com Licensee: Fort Wayne Catholic Radio Group Inc. (group owner; (acq 12-1-2005; $700,000). Population served: 177,671 Natl. Network: USA. Format: Christian, religious. Target aud: 25-54; Christian adults. Spec prog: Spanish one hr wkly. ♦Chris Langford, CEO & pres; Jason Garrett, gen mgr; Patty Becker, progmg dir.

WOWO(AM)— Mar 31, 1925: 1190 khz; 50 kw-U, DA-N. TL: N40 59 47 W85 21 06. (CP: 9.8 kw-N). Stereo. Hrs opn: 24 2915 Maples Rd., 46816. Phone: (260) 447-5511. Fax: (260) 447-7546. E-mail: info@wowo.com Web Site: www.wowo.com. Licensee: Pathfinder Communications Corp. Group owner: Federated Media Natl. Network: CBS. Natl. Rep: Christal. Format: News/talk, sports. News staff: 5. Target aud: 25-54. ♦Tony Richards, COO; John Dille, pres; Bob Watson, CFO; Mark DePrez, gen mgr; Jon Zimney, opns mgr, prom mgr & progmg mgr; Jim Tighe, gen sls mgr & progmg dir; Andy Ober, news dir; Mogan David, chief of engrg.

WXKE(FM)— May 6, 1976: 103.9 mhz; 3 kw. Ant 380 ft TL: N41 06 31 W85 09 56. Stereo. Hrs opn: 24
Simulcast with WCKZ(FM) Roanoke 100%.
2000 Lower Huntington Rd., 46819. Phone: (260) 747-1511. Fax: (260) 747-3999. Web Site: www.rock104radio.com. Licensee: Summit City License Sub, LLC. Group owner: Summit City Radio Group (acq 11-6-2006; grpsl). Population served: 417,300 Natl. Rep: D & R Radio. Format: Classic rock. News staff: one. ♦Dave Wisniewski, sls dir; Doc West, progmg dir.

Frankfort

WILO(AM)— Nov 23, 1953: 1570 khz; 250 w-U. TL: N40 16 40 W86 29 07. Hrs open: 5 AM-11 PM Box 545, 46041. Secondary address: 1401 Barner St. 46041-1506. Phone: (765) 659-3338. Fax: (765) 659-3338. Web Site: www.wilo.net. Licensee: Kaspar Broadcasting Co. Inc. Group owner: Kaspar Broadcasting Group (acq 10-1-59). Population served: 14,956 Natl. Rep: Rgnl Reps. Format: Community svc, nostalgia. Spec prog: Farm 12 hrs wkly. ♦Russ Kaspar, gen mgr & stn mgr; Vernon Kaspar, pres & gen mgr; Randy Lawson, progmg dir.

WSHW(FM)—Co-owned with WILO(AM). Sept 14, 1962: 99.7 mhz; 50 kw. 460 ft TL: N40 25 14 W86 24 47. Stereo. Phone: (800) 447-4463. Fax: (765) 452-0299. Fax: (765) 452-0399. Web Site: www.shine99.com. Format: Adult contemp. Spec prog: Farm 8 hrs wkly. ♦Russ Kaspar, gen sls mgr; Randy Lawson, progmg dir.

Franklin

***WFCI(FM)**— Oct 15, 1960: 89.5 mhz; 1.15 kw. Ant 140 ft TL: N39 24 29 W86 08 52. Stereo. Hrs opn: 8 AM-2 AM Franklin College, 101 Branigan Blvd., 46131. Phone: (317) 738-8205. Phone: (317) 738-8204. Fax: (317) 738-8233. Web Site: www.franklincollege.edu. Licensee: Franklin College of Indiana. Population served: 14,956 Format: CHR. Target aud: 12-24; college & high school students.

WIAU(FM)— Dec 15, 1961: 95.9 mhz; 3 kw. Ant 300 ft TL: N39 30 49 W86 04 07. Stereo. Hrs opn: 645 Industrial Dr., 46131. Phone: (317) 736-4040. Fax: (317) 736-4781. Web Site: www.gold959.com. Licensee: Pilgrim Communications LLC (acq 7-2-99). Population served: 110,000 Natl. Rep: Rgnl Reps. Format: Oldies. ♦Dr. Gene Hood, CEO; Randy Tipmore, gen mgr, opns dir & progmg dir.

French Lick

WFLQ(FM)— Apr 12, 1983: 100.1 mhz; 6 kw. 300 ft TL: N38 35 41 W86 36 48. Stereo. Hrs opn: 24 Box 100, 47432. Phone: (812) 936-9100. Fax: (812) 936-9495. E-mail: wflqfm@smithville.net Web Site: www.wflq.com. Licensee: W.G. Willis dba Willtronics Broadcasting. Population served: 30,000 Natl. Network: ABC. Rgnl. Network: Brownfield; AgriAmerica. Format: Modern country. News staff: one; News: 11 hrs wkly. Target aud: 25 plus. Spec prog: Farm 3 hrs, relg 6 hrs, Gospel 4 hrs wkly. ♦Col. W.G. Willis, CEO & gen mgr; Bill Willis, gen sls mgr & chief of engrg; Randall Hamm, progmg dir & mus dir; Joe Randolph, news dir.

Gary

***WGVE(FM)**— January 1954: 88.7 mhz; 2.1 kw. 91 ft TL: N41 33 15 W87 19 05. Hrs open: 1800 E. 35th Ave., 46409. Phone: (219) 962-9483. Fax: (219) 962-3726. E-mail: wgve887fm@yahoo.com Licensee: Gary Community School Corp. Population served: 336,695 Format: Educ, pub affrs. ♦Sarita Stevens, gen mgr; Eric Johnson, progmg dir.

WLTH(AM)— Nov 5, 1950: 1370 khz; 1 kw-D, 500 w-N, DA-N. TL: N41 34 17 W87 19 02. Hrs open: 1563 E. 85th Ave., Merrillville, 46410. Phone: (219) 794-1370. Fax: (219) 794-1377. Licensee: WLTH Radio Inc. (acq 4-21-98; $750,000). Population served: 350,000 Natl. Network: CNN Radio. Format: News/talk, sports. ♦Pluria Marshall Jr., gen mgr.

WWCA(AM)— Dec 7, 1949: 1270 khz; 1 kw-U, DA-1. TL: N41 31 38 W87 22 36. Hrs open: Box10745, Meriville, 46411-0745. Phone: (219) 309-9327. Web Site: www.ca@revelantradio.com. Licensee: Starboard Media Foundation Inc. Group owner: Relevant Radio (acq 7-1-2004; $1.5 million). Population served: 155,700 Format: Talk radio. ♦Armand Ciabattari, stn mgr.

Goshen

***WGCS(FM)**— Oct 2, 1958: 91.1 mhz; 6 kw. 220 ft TL: N41 33 29 W85 51 06. Stereo. Hrs opn: 24 1700 S. Main, 46526. Phone: (574) 535-7488. Phone: (574) 535-7688. Fax: (574) 535-7293. E-mail: globe@goshen.edu Web Site: globeradio.org. Licensee: Goshen College Broadcasting Corp. Natl. Network: PRI. Law Firm: Reddy, Bagley, McCormick. Format: Folk. News: 8 hrs wkly. Target aud: Adults 25-49. Spec prog: Sp 8 hrs, news 8 hrs, sports 10 hrs wkly. ♦Jason Samuel, gen mgr.

WKAM(AM)— 1954: 1460 khz; 2.5 kw-D, 500 w-N, DA-N. TL: N41 35 24 W85 48 56. Hrs open: 24 930 E. Lincoln Ave., 46528. Phone: (574) 533-1460. Fax: (574) 534-3698. Web Site: www.wkam1460.com. Licensee: Fulmer Communications LLC (acq 5-16-02; $100,000). Population served: 156,198 Natl. Network: USA. Rgnl. Network: Network Indiana. Natl. Rep: Katz Radio, Rgnl Reps. Format: Adult contemp, Latin. News staff: one; News: 20 hrs wkly. Target aud: 30-65; mature, family oriented, goal oriented. Spec prog: Southern gospel 6 hrs wkly. ♦Kent Fulmer, gen mgr.

WOZW(FM)— Jan 17, 1977: 97.7 mhz; 3 kw. 482 ft TL: N41 36 04 W85 55 41. Stereo. Hrs opn: 24 3371 Cleveland Rd., Suite 300, South Bend, 46628. Phone: (574) 273-9300. Fax: (574) 273-9090. E-mail: michael@wzow.com Web Site: www.wzow.com. Licensee: Artistic Media Partners Inc. (group owner; (acq 4-1-2002; $925,000). Format: Classic rock. Target aud: 25-54. ♦Jack Swart, gen mgr; Carrie Jones, natl sls mgr; Teresa Holden, prom dir; Chili Walker, progmg dir; Bob Henning, chief of engrg; Rita Kinzie, traf mgr.

Granger

WRBR-FM—See South Bend

Greencastle

***WGRE(FM)**— Apr 25, 1949: 91.5 mhz; 1 kw. 160 ft TL: N39 39 16 W86 51 40. Stereo. Hrs opn: 24 Ctr. for Contemporary Media, 609 S. Locust, 46135. Phone: (765) 658-4642. Phone: (765) 658-4637. Fax: (765) 658-4693. E-mail: newton@depauw.edu Web Site: www.wgre.org. Licensee: DePauw University. Population served: 45,000 Natl. Network: AP Radio. Wire Svc: AP Format: Alternative. News: 12 hrs wkly. Target aud: 18-25; college campus & loc community. Spec prog: Jazz 3 hrs, Intl 2 hrs, regl 2 hrs wkly. ♦Jeff McCall, pres & gen mgr; Chris Newton, opns mgr; Greg Stephan, chief of engrg.

***WIKL(FM)**—Not on air, target date: unknown: 90.5 mhz; 22 kw vert. Ant 253 ft TL: N39 41 19 W86 42 03. Hrs open: 3500 DePauw Blvd., Suite 2085, Indianapolis, 46268. Phone: (317) 870-8400. Fax: (317) 870-8404. Licensee: Hoosier Broadcasting Corp. Format: Christian. ♦Chuck Cunningham, gen mgr.

WREB(FM)— May 16, 1966: 94.3 mhz; 3 kw. 165 ft TL: N39 39 38 W86 53 34. Stereo. Hrs opn: 24 2468 W. County Rd. 25 N., 46135. Phone: (765) 653-9717. Fax: (765) 653-6677. Web Site: www.wrebfm.com. Licensee: The Original Co. Group owner: The Original Co. Inc. (acq 6-22-94; $200,000; FTR: 7-11-94). Population served: 30,000 Rgnl. Network: Network Indiana, Brownfield. Format: Country, loc news, sports. News staff: one; News: 37 hrs wkly. Target aud: General. Spec prog: Farm 5 hrs wkly. ♦Mark Lange, gen mgr; Tonya Sanders, progmg dir.

Greenfield

***WRGF(FM)**— 2001: 89.7 mhz; 750 w horiz, 2 kw vert. Ant 164 ft TL: N39 44 55 W85 40 50. Hrs open: 110 W. North St., Greenfield Central Comm. School Corp., 46140. Secondary address: 810 N. Broadway 46140. Phone: (317) 462-9211. Fax: (317) 467-6755. E-mail: wrgf@insight66.com Web Site: gcsc.k12.in.us. Licensee: Greenfield Central Community School Corp. Format: Old & new rock. ♦Tim Renshaw, gen mgr.

WZPL(FM)— June 1, 1962: 99.5 mhz; 12.5 kw. 991 ft TL: N39 46 03 W86 00 12. Stereo. Hrs opn: 9245 N. Meridian, Suite 300, Indianapolis, 46260. Phone: (317) 816-4000. Fax: (317) 816-4080. Web Site: www.wzpl.com. Licensee: Entercom Indianapolis License LLC. (acq 8-26-2004; grpsl). Population served: 443,600 Natl. Rep: McGavren Guild. Format: Modern adult contemp. Target aud: 18-49; women. ♦Phil Hoover, CFO, VP & gen mgr; Steve Hartley, gen sls mgr; Toni Williams, prom dir; Scott Sands, progmg dir; Gary Hunvnel, news dir; Mike Rabey, chief of engrg.

Greensburg

***WAUZ(FM)**— Sept 1, 1998: 89.1 mhz; 1.2 kw vert. 420 ft TL: N39 14 13 W85 34 00. Hrs open: 24 c/o WYGS(FM), 825 Washington St., Columbus, 47201. Secondary address: Box 487 47201. Phone: (812) 738-3482. Fax: (812) 375-2555. E-mail: ygs@wygs.org Web Site: www.wygs.org. Licensee: Good Shepherd Radio Inc. Population served: 70,000 Format: Christian/ southern Gospel. Target aud: 25-40. ♦Keith Reising, CEO & pres.

WRZQ-FM—Listing follows WTRE(AM).

WTRE(AM)— July 1, 1968: 1330 khz; 500 w-D, 41 w-N, DA-2. TL: N39 19 41 W85 30 06. Hrs open: 18 Box 487, 1217 W. Park Rd., 47240. Phone: (812) 663-3000. Fax: (812) 663-8355. E-mail: wtre@hsonline.net Web Site: www.treecountry.com. Licensee: WTRE Inc. (acq 8-13-99; with co-located FM). Population served: 10,000 Rgnl. Network: Network Indiana. Format: Country, div, news/talk. News staff: one; News: 24 hrs wkly. Target aud: 25 plus. Spec prog: Farm 10 hrs, relg 3 hrs wkly. ♦Keith Reising Jr., pres; Dave Peach, sls dir; Robert Hawkins, chief of engrg; Mark Gravely, disc jockey; Sandy Biddinger, stn mgr, gen sls mgr, news dir & disc jockey.

WRZQ-FM—Co-owned with WTRE(AM). December 1962: 107.3 mhz; 41.8 kw. 531 ft TL: N39 14 13 W85 34 00. Stereo. 24 Radio Bldg., 825 Washington St., Columbus, 47201. Phone: (812) 379-1077. Fax: (812) 375-2555. E-mail: qmix@qmix.com Web Site: www.qmix.com.445,000 Format: Adult contemp. News staff: one; News: 4 hrs wkly. Target aud: 18-49. ♦Keith Reising Jr., pres; Mike King, gen mgr; Dave Wineland, opns dir; Dale Marks, sls dir; Sara Beth Clark, mktg dir; Matt Joyce, prom dir; C.J. Miller, progmg dir; Keith Maddox, news dir; Jim Burgan, engrg dir; Mark Gravely, disc jockey; Sandy Biddinger, disc jockey.

Greenwood

WTLC-FM— 1994: 106.7 mhz; 3 kw. 328 ft TL: N39 42 42 W86 08 45. Hrs open: 21 E. Saint Joseph St., Indianapolis, 46204. Phone: (317) 266-9600. Fax: (317) 328-3870. Web Site: www.wtlc.com. Licensee: Radio One of Indiana LLC. Group owner: Radio One Inc. (acq 2-15-01; grpsl). Format: Oldies, rhythm and blues. ♦Charles T. Williams, VP & gen mgr; Kay Feenye-Caito, prom dir; Brian Wallace, progmg dir; Terri Durrett, news dir; Don Payne, chief of engrg.

Hagerstown

***WBSH(FM)**— December 1996: 91.1 mhz; 300 w. 216 ft TL: N39 56 31 W85 11 41. Hrs open:
Rebroadcasts WBST(FM) Muncie 100%.
c/o WBST(FM), Ball State Univ., Muncie, 47306. Phone: (765) 285-5888. Fax: (765) 285-8937. E-mail: ipr@bsu.edu Web Site: www.bsu.edu/ipr. Licensee: Ball State University. Format: Class, news. News staff: one; News: 33 hrs wkly. Target aud: General. ♦Marcus Jackman, gen mgr; Pam Coletti, gen sls mgr; Carol Trimmer, prom mgr; Steven Turpin, progmg dir & mus dir; Robert Mittendorf, chief of engrg; Dorothy Marvell, traf mgr; Brian Beaver, news rptr.

Hammond

WJOB(AM)— 1928: 1230 khz; 1 kw-U. TL: N41 35 46 W87 28 42. Hrs open: 24 6405 Olcott, 46320. Phone: (219) 989-8502. Fax: (219) 844-6190. Web Site: www.heyregion.com. Licensee: Vazquez Development LLC (acq 12-4-03; $1.2 million with WIMS(AM) Michigan City). Population served: 450,000 Law Firm: Martin & McCormick. Format: News/talk. News staff: 10; News: 17 hrs wkly. Target aud: 25-49; general. Spec prog: Sports, Pol 2 hrs, relg 2 hrs, Greek one hr wkly. ♦Jim Dedelow, gen mgr; Michael Stewart, progmg dir; Ron Perzo, news dir.

WPWX(FM)— Sept 14, 1959: 92.3 mhz; 50 kw horiz, 44 kw vert. 492 ft TL: N41 37 50 W87 31 40. Stereo. Hrs opn: 24 6336 Calumet Ave., 46324. Phone: (773) 734-4455. Fax: (219) 933-0323. E-mail: wpwxinfo@crawfordbroadcasting.com Web Site: www.power92chicago.com. Licensee: Dontron Inc. Group owner: Crawford Broadcasting Co. (acq 9-14-59). Population served: 250,000 Format: Urban hip hop. Target aud: 18-34; urban. ♦Donald Crawford, pres; Taft Harris, gen mgr & stn mgr; Jay Allen, progmg dir.

Hanna

WHLP(FM)— 2001: 89.9 mhz; 8 kw. Ant 505 ft TL: N41 26 09 W86 50 48. Hrs open: 150 Lincoln Way, Suite 2001, Valparaiso, 46383. Phone: (219) 548-8956. Fax: (219) 548-5808. E-mail: whlp@csnradio.com Web Site: csnradio.com. Licensee: CSN International (group owner). Format: Relg. ♦Jim Motshagen, gen mgr; Kathy Motshagen, progmg dir.

Hardinsburg

WKLO(FM)— 2002: 96.9 mhz; 3.5 kw. Ant 433 ft TL: N38 28 21 W86 24 39. Hrs open: 514 N. JFK Ave., Loogootee, 47553. Phone: (812) 295-9480. Fax: (812) 295-4455. Licensee: Hembree Communications Inc. (acq 12-17-2003; $350,000). Format: Hot adult contemp. ♦Larry Hembree, gen mgr.

Hartford City

***WHCI(FM)**— 2003: 88.1 mhz; 100 w. 72 ft Hrs opn: 7:30 AM-4 PM 2392 N. State Rd. 3, Blacklford County School Corp., 47348. Phone: (765) 348-7560. Fax: (765) 348-7568. Web Site: www.bcs.k12.in.us. Licensee: Blackford County School Corp. Format: Var. ♦Harry Anderson, gen mgr.

WHTY(FM)— Feb 26, 1965: 93.5 mhz; 3.04 kw. 456 ft TL: N40 25 16 W85 25 40. Hrs open: 24 800 E. 29th St., Muncie, 47302. Phone: (765) 288-4403. Fax: (765) 378-2091. E-mail: maxstudio@maxrocks.com Web Site: www.maxrocks.net. Licensee: Indiana Sabrecom Inc. Group owner: Backyard Broadcasting LLC (acq 12-1-02; grpsl). Population served: 150,000 Natl. Network: Westwood One. Natl. Rep: Rgnl Reps. Wire Svc: UPI Format: Classic rock. News staff: one; News: one hr wkly. Target aud: 25-54. ♦Steve Lindell, gen mgr & opns mgr.

Howe

***WHWE(FM)**— May 1, 1970: 89.7 mhz; 100 w. 68 ft TL: N41 43 32 W85 25 30. Stereo. Hrs opn: 7 AM-4 PM Box 240, Howe Military School, 46746. Phone: (219) 562-2131. Fax: (219) 562-3678. Licensee: Howe Military School. Population served: 10,000 Format: Educ, CHR, div. Target aud: 7-20; high school. ♦Steve Clark, stn mgr.

***WQKO(FM)**— 1994: 91.9 mhz; 3 kw. Ant 298 ft TL: N41 38 59 W85 21 12. Hrs open: 150 Lincolnway, Suite 2001, Valparaiso, 46383. Phone: (260) 562-2242. E-mail: wqko@earthlink.net Web Site: www.jesusfanatic.com. Licensee: CSN International (group owner; (acq 7-13-98; $80,000). Format: Christian praise & worship, teaching. ♦Banner Kidd, gen mgr; Phil Jennings, progmg dir.

Huntingburg

WBDC(FM)—Licensed to Huntingburg. See Jasper

Huntington

WBZQ(AM)— May 25, 1957: 1300 khz; 500 w-D, DA. TL: N40 52 31 W85 28 27. Hrs open: 24 Box 5570, Fort Wayne, 46895. Phone: (260) 482-8500. Licensee: Larko Communications Inc. (acq 8-31-00; $16,500). Population served: 50,000 Format: Oldies. Target aud: 25-65. ♦Chris Larko, gen mgr.

WGL-FM— Sept 1, 1965: 102.9 mhz; 4.7 kw. Ant 298 ft TL: N40 55 33 W85 23 15. Stereo. Hrs opn: 24 2000 Lower Hunting Rd., Ft. Wayne, 46819. Phone: (260) 747-1511. Fax: (260) 747-3999. Web Site: www.1029mikefm.com. Licensee: Summit City License Sub, LLC. Group owner: Summit City Radio Group (acq 11-6-2006; grpsl). Population served: 417,300 Natl. Network: CBS. Natl. Rep: D & R Radio. Format: Var/div (anything). News staff: one. Target aud: 25-54. ♦J.J. Fabina, opns mgr; Rob Livergood, sls dir; J.J Fabina, progmg dir.

***WVSH(FM)**— Jan 1, 1950: 91.9 mhz; 920 w. 110 ft TL: N40 53 32 W85 30 38. Stereo. Hrs opn: 450 MacGahan St., 46750. Phone: (260) 356-2019. Fax: (260) 358-2210. E-mail: bwalker@hccsc.k12in.us Licensee: Huntington County Community School Corp. Population served: 18,000 Rgnl. Network: Network Indiana. Format: CHR. ♦Bill Walker, gen mgr; George Castle, engr.

Indianapolis

***WBDG(FM)**— Sept 13, 1965: 90.9 mhz; 400 w. 78 ft TL: N39 47 05 W86 17 27. Stereo. Hrs opn: 24 1200 N. Girls School Rd., 46214. Phone: (317) 244-9234. Fax: (317) 243-5506. E-mail: jon.easter @wayne.k12.in.us Web Site: www.wayne.k12.in.us/bdwbdg. Licensee: Metropolitan School District of Wayne Township. Population served: 3,000 Wire Svc: Reuters Format: Contemp hit, rock/AOR, urban contemp. News: 10 hrs wkly. Target aud: 12-49. ♦Jon Easter, gen mgr; Matt Reedy, progmg dir; Rob Woock, progmg dir; Paul McDonald, news dir; Kevin Van Wyk, chief of engrg & chief of engrg.

WBRI(AM)— Mar 10, 1964: 1500 khz; 5 kw-D, DA. TL: N39 52 14 W86 05 17. Hrs open: 6 AM-7 PM 4802 E. 62nd St., 46220. Phone: (317) 255-5484. Fax: (317) 255-8592. E-mail: wbri@wilkinsradio.com Licensee: Heritage Christian Radio Inc. (acq 7-1-2003; $1.5 million). Population served: 1,800,000 Natl. Network: Salem Radio Network. Law Firm: Womble, Carlyle, Sandridge & Rice. Format: Christian teaching/talk. Target aud: 35 plus. ♦Bob Wilkins, pres; LuAnn Wilkins, exec VP; Mitchell Mathis, VP; Keith Smiley, stn mgr; Greg Garrett, opns mgr & progmg mgr; Phil Alexander, engr.

***WEDM(FM)**— Sept 14, 1970: 91.1 mhz; 180 w vert. Ant 216 ft TL: N39 47 29 W85 59 53. Stereo. Hrs opn: 24 c/o Walker Career Ctr., 9651 E. 21st St., 46229. Phone: (317) 532-6301. Fax: (317) 532-6199. Licensee: Metropolitan School District of Warren Township. Population served: 30,000 Rgnl. Network: Network Indiana. Format: CHR. News: 3 hrs wkly. Target aud: General; Warren Township residents. ♦Daniel J. Henn, stn mgr; Mike Rabey, chief of engrg.

WFBQ(FM)—Listing follows WNDE(AM).

WFMS(FM)— Mar 17, 1957: 95.5 mhz; 13 kw. 1,000 ft TL: N39 46 03 W86 00 12. Stereo. Hrs opn: 24 6810 N. Shadeland Ave., 46220. Phone: (317) 842-9550. Fax: (317) 577-3361. Web Site: www.wfms.com. Licensee: WFMS Lico Inc. Group owner: Susquehanna Radio Corp. (acq 11-20-72). Population served: 275,000 Format: Modern country. ♦Charlie Morgan, gen mgr; Todd Schumacher, gen sls mgr; Lisa Jullerst, mktg dir; Bob Richards, progmg dir; Mike Orr, news dir; Jeff Goode, chief of engrg.

***WFYI-FM**— Oct 1, 1954: 90.1 mhz; 10 kw. Ant 560 ft TL: N39 53 59 W86 12 01. Stereo. Hrs opn: 24 1401 N. Meridian St., 46202-2389. Phone: (317) 636-2020. Fax: (317) 397-2976. E-mail: webmaster@wfyi.org Web Site: www.wfyi.org. Licensee: Metropolitan Indianapolis Public Broadcasting Inc. (acq 12-1-86). Population served: 1,314,960 Natl. Network: PRI, NPR. Rgnl rep: Indiana Public Broadcasting Stations Wire Svc: AP Format: Class, news/talk. News: 41 hrs wkly. Target aud: 25-64; general. Spec prog: Black 5 hrs, blues 4 hrs wkly. ♦Lloyd Wright, pres; Anthony Lorenz, CFO; Alan Cloe, exec VP & sr VP; Jeanelle Adamak, exec VP & VP; Theresa Tetrault, dev dir; Rena Barraclough, prom VP; Lori Plummer, prom mgr; Michael Toulouse, mus dir; Steve Jensen, engrg VP. Co-owned TV: *WFYI-TV affil.

WHHH(FM)— Oct 28, 1991: 96.3 mhz; 3.3 kw. 285 ft TL: N39 46 32 W86 09 10. Stereo. Hrs opn: 24 21 E. St. Joseph, 46204. Phone: (317) 266-9600. Fax: (317) 328-3870. Web Site: www.hot963.com. Licensee: Radio One of Indiana LLC. Group owner: Radio One Inc. (acq 11-8-01; grpsl). Natl. Network: CNN Radio. Format: Urban hip hop, rhythm and blues. Target aud: 18-49. ♦Alfred Liggins, CEO & pres; Charles Williams, VP & gen sls mgr; Charles T. Williams, gen mgr; Anna Fraser, prom dir; Brian Wallace, progmg dir; Don Payne, chief of engrg.

WIBC(AM)— 1938: 1070 khz; 50 kw-D, 10 kw-N, DA-2. TL: N39 57 21 W86 21 30. Stereo. Hrs opn: 24 40 Monument Cir., Suite 400, One Emmis Plaza, 46204. Phone: (317) 266-9422. Fax: (317) 684-2022. Web Site: www.wibc.com. Licensee: Emmis Radio License LLC. Group owner: Emmis Communications Corp. (acq 6-9-94; $26 million with co-located FM). Population served: 1,175,000 Rgnl. Network: Agri-America. Natl. Rep: D & R Radio. Format: News/talk, sports. News staff: 12. Target aud: 25-54; white collar, above average income & education. ♦Jeff Smulyan, CEO; Tom Severino, VP & gen mgr; Jon Quick, opns dir, progmg dir & farm dir; Jay Chapman, sls dir; Patty England, gen sls mgr & natl sls mgr; Jessica Butcher, prom mgr; Sherry Fisher, news dir; Jeff Dinsmore, chief of engrg; Trish Boone, traf mgr.

WNOU(FM)—Co-owned with WIBC(AM). Dec 5, 1960: 93.1 mhz; 13.5 kw. Ant 991 ft TL: N39 46 03 W86 00 12. Stereo. Phone: (317) 236-9300. Fax: (317) 971-6469.744624 Format: CHR. Target aud: 18-34; adults. ♦Jeff Simulyan, pres & stn mgr; David Edgar, opns dir & progmg dir; Susan Wells, prom mgr; David Hood, chief of engrg; Shelly Grimes, traf mgr.

***WICR(FM)**— Aug 20, 1962: 88.7 mhz; 5 kw. 1,000 ft TL: N39 53 59 W86 12 02. Stereo. Hrs opn: 24 1400 E. Hanna Ave., 46227. Phone: (317) 788-3280. Fax: (317) 788-3490. E-mail: wicr@uindy.edu Web Site: wicr.uindy.edu. Licensee: University of Indianapolis. Population served: 1,500,000 Natl. Network: PRI. Rgnl. Network: Network Indiana. Law Firm: John D. Pellegrin. Format: Class, jazz. News: 5 hrs wkly. Target aud: 35 plus; Educated, Affluent, Older. ♦Beverley Pitts, pres; Scott Uecker, gen mgr; Russell Maloney, chief of engrg.

***WJEL(FM)**— Sept 3, 1975: 89.3 mhz; 1 kw. 115 ft TL: N39 54 34 W86 07 39. Stereo. Hrs opn: 24 1901 E. 86th St., 46240. Phone: (317) 259-5278. Fax: (317) 259-5298. Web Site: www.geocities.com/wjelpower. Licensee: Metropolitan School District of Washington Township. Population served: 1,100,000 Format: Var. ♦John R. King, gen mgr; Robert L. Hendrix, progmg dir; Tyler Hindman, pub affrs dir; Mike Rabey, chief of engrg.

WJJK(FM)—Noblesville, Sep 25, 1950: 104.5 mhz; 50 kw. Ant 492 ft TL: N39 50 25 W86 10 34. Stereo. Hrs opn: 24 6810 N. Shadeland Ave., 46220. Phone: (317) 842-9550. Fax: (317) 577-3361. Web Site: www.gold1045.com. Licensee: Indy Lico Inc. Group owner: Susquehanna Radio Corp. (acq 10-7-93; $7.15 million; FTR: 11-8-93). Population served: 1,133,200 Law Firm: Haley, Bader & Potts. Format: Oldies. News staff: one. Target aud: 25-54. ♦Jenny Skjodt, gen mgr; Eric Lunnenberg, gen sls mgr; Sheri Aquisto, mktg dir; Steve Cannon, progmg dir; Mike Orr, news dir; Jeff Goode, chief of engrg.

WNDE(AM)— Oct 23, 1924: 1260 khz; 5 kw-U, DA-N. TL: N39 51 54 W86 03 43. Hrs open: 24 6161 Fall Creek Rd., 46220. Phone: (317) 257-7565. Fax: (317) 253-6501. Web Site: www.wnde.com. Licensee: Capstar TX L.P. Group owner: Clear Channel Communications Inc. (acq 8-30-00; grpsl). Population served: 744,624 Natl. Network: AP Radio, ESPN Radio. Natl. Rep: Clear Channel. Format: Sports, talk. Target aud: 25-54. Spec prog: 0. ♦Rick Green, gen mgr; Marty Bender, opns mgr; Lee Anne Brooks, gen sls mgr; Dan Anderson, prom dir; Drew Carey, progmg dir; Dan Mettler, engrg mgr; Scott Fenstermaker, chief of engrg; Debbie Tunny, traf mgr; Mark Patrick, sports cmtr.

WFBQ(FM)—Co-owned with WNDE(AM). Nov 26, 1959: 94.7 mhz; 58 kw. Ant 804 ft TL: N39 53 43 W86 12 04.24 Web Site: www.wfbq.com. Natl. Network: AP Radio. Format: Classic rock. ♦Jim Kendall, prom dir; Drew Carey, progmg dir.

WNTR(FM)— Oct 15, 1984: 107.9 mhz; 22 kw. Ant 762 ft TL: N39 53 43 W86 12 04. Stereo. Hrs opn: 24 9245 N. Meridian St., Suite 300, 46260. Phone: (317) 816-4000. Fax: (317) 816-4050. E-mail: akeddie@mystar.com Web Site: www.1079thetrack.com. Licensee: Entercom Indianapolis License LLC. (group owner; (acq 8-26-2004; grpsl). Population served: 1,500,000 Natl. Rep: Christal. Law Firm: Fletcher, Heald & Heldreth. Format: Adult contemp. News staff: 3. Target aud: 25-54. Spec prog: Jazz 6 hrs wkly. ♦Phil Hoover, VP & gen mgr; Alex Keddie, opns VP; Amy Dillon, gen sls mgr & pub affrs dir; Toni Williams, prom dir; Gary Hummel, progmg VP & news dir; Tom Watson, progmg dir & asst music dir.

WXNT(AM)—Co-owned with WNTR(FM). 1923: 1430 khz; 5 kw-U, DA-N. TL: N39 50 17 W86 11 53. Stereo. 24 Web Site: newstalk1430.com. Format: News/talk. News staff: one; News: 18 hrs wkly. Target aud: 35-64. Spec prog: Big band 2 hrs wkly. ♦Steve Hartley, sls dir & gen sls mgr; Gary Havens, progmg dir.

***WRFT(FM)**— June 6, 1978: 91.5 mhz; 130 w. 180 ft TL: N39 40 39 W86 00 58. Stereo. Hrs opn: 6215 S. Franklin Rd., 46259. Phone: (317) 803-5552. Fax: (317) 862-7262. Licensee: Franklin Township Community School Corp. Format: Educ, div. Target aud: General. ♦Steve George, gen mgr; Abby Wheeling, progmg dir.

WRZX(FM)— May 15, 1964: 103.3 mhz; 18 kw. 850 ft TL: N39 53 43 W86 12 04. Stereo. Hrs opn: 6161 Fall Creek Rd., 46220. Phone: (317) 257-7565. Fax: (317) 254-9619. Web Site: www.x103.com. Licensee: Capstar TX L.P. Group owner: Clear Channel Communications Inc. (acq 8-30-00; grpsl). Format: Modern rock. ♦Rick Green, VP & gen mgr; Marty Bender, opns mgr; Lee Anne Brooks, gen sls mgr; Scott Jameson, prom mgr & progmg dir; Scott Fenstermaker, chief of engrg.

WSYW(AM)— May 15, 1963: 810 khz; 250 w-D. TL: N39 43 32 W86 11 08. Hrs open: Sunrise-sunset 1800 N. Meridian St., Suite 603, 46202-1433. Phone: (317) 924-1071. Fax: (317) 924-7766. Licensee: Continental Broadcast Group Inc. (acq 12-30-93; grpsl; FTR: 1-24-94). Population served: 1,000,000 Natl. Rep: Univision Radio National Sales. Law Firm: Pillsbury, Winthrop, Shaw, Pittman. Format: Hispanic. Target aud: 18+; Hispanic Adults. ♦Russ Dodge, gen mgr; Stephanie Tatay-Myers, mktg mgr; Manuel Sepulveda, progmg dir; Phil Alexander, chief of engrg.

WTLC(AM)— July 27, 1941: 1310 khz; 5 kw-D, 1 kw-N, DA-N. TL: N39 43 08 W86 10 33. Stereo. Hrs opn: 24 21 E. Saint Joseph St., 46204. Phone: (317) 266-9600. Fax: (317) 261-4664. Web Site: www.1310thelight.com. Licensee: Radio One of Indiana LLC. Group owner: Radio One Inc. (acq 11-8-01; grpsl). Population served: 744,624 Natl. Network: ABC. Rgnl. Network: Network Indiana. Natl. Rep: Katz Radio. Format: Gospel, talk, Ammos Brown Show. News: one hr wkly. Target aud: 35 plus; black females. ♦Charles Williams, VP & gen mgr; Ian Banks, gen sls mgr; Paul Robinson, progmg dir; Don Payne, chief of engrg.

WXLW(AM)— August 1948: 950 khz; 5 kw-D, 117 w-N, DA-2. TL: N39 51 05 W86 14 39. Hrs open: 24 Box 47307, 46247. Phone: (317) 655-9999. Fax: (317) 655-9995. E-mail: david@espn950.com Web Site: www.espn950.com. Licensee: Pilgrim Communications L.L.C. (acq 10-1-95). Natl. Network: ABC, ESPN Radio. Format: Sports. News: 7 hrs wkly. Target aud: 25-54; men, secondary women. Spec prog: Gospel 5 hrs wkly. ♦Randy Tipmore, gen mgr & news dir; Charles Sears, chief of engrg.

WYJZ(FM)—Speedway, May 28, 1967: 100.9 mhz; 6 kw. Ant 328 ft TL: N39 48 01 W86 04 39. Hrs open: 24 21 E. St. Joseph, 46204. Phone: (317) 266-9600. Fax: (317) 328-3870. Web Site: www.wyjzradio.com. Licensee: Radio One of Indiana LLC. Group owner: Radio One Inc. (acq 11-8-2001; grpsl). Population served: 151,000 Natl. Network: ABC. Rgnl. Network: Agri-Net, Brownfield. Natl. Rep: Katz Radio. Format: Smooth jazz, new adult contemp. News staff: 2; News: 12 hrs wkly. Target aud: 25-54. Spec prog: F. ♦Alfred Liggins, pres; Brian Harrington, gen sls mgr; Carl Frye, progmg dir; Charles Williams, VP, gen mgr & progmg dir.

WYXB(FM)— Jan 22, 1968: 105.7 mhz; 50 kw. 445 ft TL: N39 48 01 W86 04 39. Stereo. Hrs opn: 24 One Emmis Plaza, 40 Monument Cir., 46204. Phone: (317) 684-1057. Fax: (317) 684-2021. E-mail: music@b1057.com Web Site: www.b1057.com. Licensee: Emmis Radio License LLC. Group owner: Emmis Communications Corp. (acq 9-8-97; with co-located AM). Natl. Rep: D & R Radio. Format: Soft Rock. News staff: one; News: 3 hrs wkly. Target aud: 25-34; adults, (secondary is 25-54). Spec prog: Gospel 10 hrs wkly. ♦Tom Severino, gen mgr; David Edgar, opns mgr; Mike Cortese, sls dir; Jenni Gray, rgnl sls mgr; Mary Young, mktg dir; Scott Wheeler, progmg dir; Dave Hood, chief of engrg.

WZPL(FM)—See Greenfield

Jasper

WBDC(FM)—Huntingburg, Dec 22, 1975: 100.9 mhz; 11 kw. 500 ft TL: N38 12 31 W86 54 00. Stereo. Hrs opn: 24 Box 1009, 511 Newton St., 2nd Fl., 47547-1009. Secondary address: Box 330, 501 Old State Rd., Huntingburg 47542-0330. Phone: (812) 683-4144. Phone: (812) 634-9232. Fax: (812) 683-5891. E-mail: wbdc@psci.net Web Site: www.dcbroadcasting.com.JRN Licensee: Dubois County Broadcasting Inc. Group owner: DCBroadcasting Inc. Population served: 125,000 Natl. Network: CNN Radio, Jones Radio Networks. Rgnl. Network: Brownfield Law Firm: Miller & Miller. Wire Svc: AP Format: Country. News staff: 2; News: 10 hrs wkly. Target aud: 18-54. Spec prog: Farm 5 hrs, relg 5 hrs, sports 6 hrs, Sp .5 hr wkly. ♦Paul Knies, pres; Bill Potter, gen mgr; Ron Spaulding, sls dir & gen sls mgr; Pat Oxley, progmg dir; Dave Ferguson, chief of engrg; Janice Potter, traf mgr.

WITZ-FM— Nov 1, 1954: 104.7 mhz; 50 kw. 490 ft TL: N38 21 02 W86 56 26. Stereo. Hrs opn: Box 167, 1978 S. WITZ Rd., 47546. Phone: (812) 482-2131. Fax: (812) 482-9609. E-mail: witzamfm@psci.net Web Site: www.witzamfm.com. Licensee: Jasper On The Air Inc. Population served: 37,000 Rgnl. Network: AgriAmerica, Network Indiana. Natl. Rep: Rgnl Reps. Format: Adult contemp. Target aud: 18-54. Spec prog: Paul Harvey 3 hrs wkly. ♦Earl Metzger, gen mgr; Bob Boyles, gen sls mgr; Walt Ferber, progmg dir; Reed Parker, news dir; Jeri Weisheit, chief of engrg & traf mgr.

WITZ(AM)— July 4, 1948: 990 khz; 1 kw-D. TL: N38 21 02 W86 56 26. Stereo. Web Site: www.witzamfm.com.37,000 ♦Jeri Weisheit, traf mgr.

***WJPR(FM)**— 2006: 91.7 mhz; 2.6 kw. Ant 276 ft TL: N38 25 23 W86 49 47. Hrs open: 514 N. JFK Ave., Loogootee, 47553. Phone: (812) 295-9480. Fax: (812) 295-3295. Licensee: Jasper Public Radio Inc. (acq 12-4-2006). Format: Oldies. ♦Larry Hembree, gen mgr.

Jeffersonville

WAVG(AM)— June 26, 1961: 1450 khz; 1 kw-U. TL: N38 17 41 W85 45 07. Hrs open: 24 Box 726, 47131-0726. Phone: (812) 283-3577. Fax: (812) 285-5060. E-mail: wavg1450@win.net Web Site: www.wavg1450.com. Licensee: Sunnyside Communications Inc. Group owner: Susquehanna Radio Corp. (acq 3-30-01; grpsl). Population served: 50,034 Natl. Network: Jones Radio Networks. Rgnl rep: Rgnl Reps. Law Firm: Dow, Lohnes & Albertson. Format: Classic hit country. News staff: one. Target aud: 35 plus; general. ♦Blair W. Trask, pres & stn mgr; Kelly Anderson, mktg dir & rsch dir; Kelly Trask, gen mgr & prom dir; Blair Trask, progmg dir & traf mgr; Gil Daugherty, news dir & news rptr; Bob Hawkins, chief of engrg.

WQMF(FM)— Apr 25, 1974: 95.7 mhz; 28.5 kw. 643 ft TL: N38 08 06 W85 56 05. Stereo. Hrs opn: 4000 Radio Rd., Suite 1, Louisville, KY, 40218. Phone: (502) 479-2222. Fax: (502) 479-2227. Web Site: www.wqmf.com. Licensee: CC Licenses LLC. Group owner: Clear Channel Communications Inc. (acq 1-23-97; $13.5 million). Population served: 361,472 Format: Classic Rock. Target aud: 35-54; men. ♦Kevin Hughes, VP, stn mgr & mktg mgr; Roz Jones, gen sls mgr; Damon Hildreth, rgnl sls mgr.

Kendallville

WAWK(AM)— Nov 9, 1955: 1140 khz; 250 w-D. TL: N41 27 16 W85 15 48. Hrs open: 931 East Ave., 46755. Phone: (260) 347-2400. Phone: (260) 347-2401. Fax: (260) 347-2524. E-mail: wawk@locl.net Web Site: www.wawk.com. Licensee: Northeast Indiana Broadcasting Inc. Population served: 10,000 Natl. Network: USA. Rgnl. Network: Brownfield. Law Firm: Irwin, Campbell. Format: Var, hits of the 50s to present. News: 7 hrs wkly. Target aud: 25-54. Spec prog: Big Band 2 hrs, bluegrass 2 hrs, Farm one hr wkly. ♦Don Moore, pres, VP & gen mgr; Karen White, gen sls mgr; Scott Paul, progmg dir; Mike Shultz, news dir; Greg Case, chief of engrg.

WBTU(FM)—Licensed to Kendallville. See Fort Wayne

Kentland

WIVR(FM)— 2000: 101.7 mhz; 3.2 kw. Ant 453 ft TL: N40 51 52 W87 35 14. Stereo. Hrs opn: 24 202 E. Walnut, Watseka, IL, 60970. Phone: (815) 432-0700. Phone: (815) 933-9287. Fax: (815) 432-6112. E-mail: wvvliradio@comcast.net Licensee: Milner Broadcasting Enterprises LLC (acq 12-11-2000). Population served: 48,000 Natl. Network: AP Network News. Law Firm: Womble, Carlyle, Sandridge & Rice. Wire Svc: AP Format: Traditional country. News staff: 2; News: 168 bcsts wkly. Target aud: 12 plus; anthology country with current & recurrent hits. Spec prog: Chicago Bears Football. ♦Jim Brandt, gen mgr & opns mgr; Chris Swain, gen sls mgr; Mickey Milner, progmg dir; Ken Zyre, news dir; Don Kerouac, chief of engrg; Jody Woloszyn, traf mgr.

Knightstown

***WKPW(FM)**— Sept 7, 1993: 90.7 mhz; 4.4 kw. TL: N39 46 08 W85 31 05. Hrs open: 10892 N. State Rd., 140, 46148. Phone: (765) 345-9070. Fax: (765) 345-7039. E-mail: wkpw@knightstown.net Web Site: www.wkpw.net. Licensee: IN Soldiers' & Sailors' Childrens' HME. Law Firm: Booth, Freret, Imlay & Tepper. Wire Svc: AP Format: Country. Target aud: General. Spec prog: Gospel. ♦Dr. John Wittkamper, pres; Mike York, gen mgr & progmg dir; Bob Hawkins, chief of engrg.

Knox

WKVI(AM)— June 30, 1969: 1520 khz; 250 w-D. TL: N41 19 20 W86 36 17. (CP: 1.8 kw-D). Hrs opn: Box 10, 400 W. Culver Rd., 46534. Phone: (574) 772-6241. Fax: (574) 772-5920. Web Site: www.wkvi.com. Licensee: Kankakee Valley Broadcasting Co. Inc. Population served: 10,000 Format: Adult contemp. ♦Ted Hayes, gen mgr; Lo Ann McDaniel, progmg dir; Anita Goodan, news dir.

WKVI-FM— July 21, 1969: 99.3 mhz; 3 kw. 303 ft TL: N41 19 20 W86 36 17. Stereo. Web Site: www.wkvi.com.10,000

Kokomo

WIOU(AM)— July 16, 1948: 1350 khz; 5 kw-D, 1 kw-N, DA-2. TL: N40 25 00 W86 06 49. Hrs open: 24 Box 2208, 46904-2208. Secondary address: 671 E. 400 S. 46902. Phone: (765) 453-1212. Fax: (765) 455-3882. E-mail: newsroom.wzwz.wiou@sbcglobal.net Licensee: Mid-America Radio Group Inc. (group owner; acq 3-24-93; $1.21 million with co-located FM; FTR: 4-12-93). Population served: 100,000 Natl. Network: CBS. Format: Sports, news/talk. News staff: 2; News: 9 hrs wkly. Target aud: 25-54. ♦Steve Lamar, gen mgr; Lora Lacy, gen sls mgr; Mike Turner, prom dir; Allan James, progmg dir.

WZWZ(FM)—Co-owned with WIOU(AM). Nov 20, 1964: 92.5 mhz; 6 kw. 324 ft TL: N40 28 18 W86 09 52. Stereo. 24 E-mail: wzwz@comteck.com Licensee: Mid-America Radio Group of Kokomo Inc.80,000 Format: Adult contemp. News: 4 hrs wkly. Target aud: 18-49.

***WIWC(FM)**— September 1993: 91.7 mhz; 2.1 kw. 299 ft TL: N40 36 00 W86 18 08. Hrs open: c/o WGNR, 2000 W. 53rd St., Anderson, 46013. Phone: (765) 642-2750. Fax: (765) 642-4033. E-mail: wiwc@moody.edu Web Site: wiwc.mbn.org. Licensee: The Moody Bible Institute of Chicago. (group owner) Format: Relg. Target aud: 34-55; general. ♦Ray Hashley, gen mgr; Tom Winn, progmg dir; Sam Sundin, news dir; Jim Wagner, chief of engrg.

WWKI(FM)— Oct 21, 1962: 100.5 mhz; 50 kw. 480 ft TL: N40 27 04 W86 02 12. Stereo. Hrs opn: 24 519 N. Main St., 46901-4661. Phone: (765) 459-4191. Fax: (765) 456-1111. Fax: (765) 456-1112. Licensee: Citadel Broadcasting Co. Group owner: Citadel Broadcasting Corp. (acq 6-30-99; grpsl). Population served: 132,000 Natl. Network: AP Radio. Rgnl. Network: AgriAmerica. Natl. Rep: Katz Radio. Rgnl rep: Rgnl Reps. Law Firm: Leventhal, Senter & Lehrman. Wire Svc: AP Format: Country. News staff: 2; News: 13 hrs wkly. Target aud: 25-54. ♦Mike Christopher, gen mgr; James Stonecipher, gen sls mgr; Dave Broman, progmg dir; Robert Longshore, chief of engrg.

La Porte

WCOE(FM)—Listing follows WLOI(AM).

WLOI(AM)— 1948: 1540 khz; 250 w-D. TL: N41 37 55 W86 45 43. Hrs open: Sunrise-sunset 1700 Lincolnway Pl., Suite 8, 46350. Phone: (219) 362-6144. Phone: (219) 872-8986. Fax: (219) 324-7418. E-mail: wcoe@csinet.net Licensee: La Porte County Broadcasting Company Inc. (acq 1955). Population served: 115,000 Natl. Network: ABC, Westwood One. Rgnl. Network: Network Indiana, Tribune. Natl. Rep: Rgnl Reps. Law Firm: Wiley, Rein & Fielding. Format: Adult standards, MOR. News staff: 2; News: 26 hrs wkly. Target aud: 35 plus. Spec prog: Farm 8 hrs wkly. ♦Kenneth S. Coe, pres, gen mgr & gen sls mgr; Norma Sabie, sls dir & prom mgr; Dennis Siddall, progmg dir & disc jockey; Kate O'Malley, disc jockey.

WCOE(FM)—Co-owned with WLOI(AM). Jan 23, 1964: 96.7 mhz; 3 kw. 265 ft TL: N40 37 55 W86 45 43. Stereo. 24 Phone: (219) 362-5290. Web Site: www.eaglewcoe.com.228,300 Format: Hot country. News staff: 2; News: 19 hrs wkly. Target aud: 35-54; upper income, middle aged. ♦Norma Sabie, adv dir; Kenneth S. Coe, adv mgr; Dennis Sidall, mus dir; Donna Eichelberg, pub affrs dir; Carl Fletcher, engrg dir; Bob Costigan, local news ed & news rptr; Chip Jones, sports cmtr; Bobby Rivers, disc jockey; Dennis Siddall, disc jockey; Kate O'Malley, disc jockey.

Ladoga

***WJCJ(FM)**—Not on air, target date: unknown: 88.9 mhz; 50 kw. Ant 207 ft TL: N40 31 19 W87 30 58. Hrs open: CSN International, 4002N. 3300E., Twin Falls, ID, 83301. Phone: (208) 734-6633. Fax: (208) 736-1958. Web Site: www.csnradio.com. Licensee: CSN International (group owner; ♦Mike Kestler, pres.

Lafayette

WASK(AM)— 1942: 1450 khz; 1 kw-U. TL: N40 24 08 W86 50 59. Hrs open: 24 Box 7880, 47903. Secondary address: 3575 McCarty Ln. 47903. Phone: (765) 447-2186. Fax: (765) 448-4452. Web Site: www.wask.com. Licensee: WASK Inc. Group owner: Schurz Communications Inc. (acq 1-28-91; $8.25 million with co-located FM; FTR: 1-28-91). Population served: 44,955 Natl. Network: ESPN Radio. Natl. Rep: Christal. Law Firm: Hogan & Hartson. Format: Sports. News staff: 6; News: 15 hrs wkly. Spec prog: Farm 2 hrs wkly. ♦John Trent, pres & gen mgr; Brian Green, gen sls mgr; Randy Jones, progmg dir; Steve Truex, news dir & chief of engrg; Bryan McGarvey, disc jockey.

WKOA(FM)—Co-owned with WASK(AM). Sept 28, 1964: 105.3 mhz; 50 kw. 375 ft TL: N40 24 08 W86 50 59. Stereo. 24 Web Site: www.wkoa.com. Group owner: Lafayette Broadcasting Inc. Format: Country. News staff: 3; News: 12 hrs wkly. Target aud: 25-54. Spec prog: Farm 3 hrs wkly. ♦Lindsay Reinert, mktg dir & prom dir; Mark Allen, progmg dir; Skip Davis, farm dir; Bob Vizza, disc jockey; Christine Davis, disc jockey.

WAZY-FM— March 1965: 96.5 mhz; 50 kw. 500 ft TL: N40 23 02 W87 07 55. Stereo. Hrs opn: 24 3824 S. 18th St., 47909. Phone: (765) 474-1410. Fax: (765) 474-3442. Web Site: www.wazy.com. Licensee: Artistic Media Partners L.P. Group owner: Artistic Media Partners Inc. (acq 10-86; $2 million; FTR: 9-22-86). Population served: 301,100 Natl. Network: ABC. Rgnl. Network: Network Indiana. Natl. Rep: Christal. Law Firm: Rosenman & Colin L.L.P. Format: Top-40, CHR. News staff: one; News: 4 hrs wkly. Target aud: 18-34. ♦Arthur Angotti, pres & gen mgr; Jack Swart, gen mgr; Kit Osborne, pres & sls dir; Chris Green, gen sls mgr, prom dir & progmg dir; Bob Henning, chief of engrg; Bitsy Matatall, traf mgr.

WSHY(AM)—Co-owned with WAZY-FM. Nov 28, 1959: 1410 khz; 1 kw-D, 65 w-N, DA-1. TL: N40 21 38 W86 52 38.24 Licensee: Artistic Media Partners Inc. (acq 9-30-98; $275,000). Format: Country. News: 3 hrs wkly. Target aud: 25-49.

***WJEF(FM)**— Feb 7, 1972: 91.9 mhz; 250 w. 100 ft TL: N40 23 52 W86 52 26. Stereo. Hrs open: 24 1801 S. 18th St., 47905. Secondary address: 2300 Cason St. 47904. Phone: (765) 772-4700. E-mail: rbrist@lsc.k12.in.us Web Site: www.jeff92.org. Licensee: Lafayette School Corp. Population served: 149,000 Format: Oldies. News: 6 hrs wkly. Target aud: General. Spec prog: Jefferson High School sports. ♦Randall J. Brist, gen mgr.

WKHY(FM)— Jan 1, 1970: 93.5 mhz; 6 kw. 311 ft TL: N40 23 13 W86 58 10. Stereo. Hrs opn: 24 Box 7093, 47903. Secondary address: 711 N. Earl Ave. 47904. Phone: (765) 448-1566. Fax: (765) 448-1348. Web Site: www.wkhy.com. Licensee: Stay Tuned Broadcasting Corp. Group owner: RadioWorks Inc. (acq 5-12-99; grpsl). Population served: 292,700 Natl. Network: AP Radio. Natl. Rep: Katz Radio. Law Firm: Shaw Pittman. Format: Classic rock/AOR. News staff: one. Target aud: 25-54; adults that are active, mobile & moderately affluent. ♦John Trent, gen mgr; John Schurz, gen sls mgr; Liz Hahn, prom dir; Jeff Strange, progmg dir; Eric Burch, news dir; Steve Truex, chief of engrg.

***WQSG(FM)**— 2005: 90.7 mhz; 17 kw vert. Ant 328 ft TL: N40 22 13 W86 30 06. Hrs open:
Rebroadcasts WAFR(FM) Tupelo, MS 100%.
Drawer 2440, Tupelo, MS, 38803. Phone: (662) 844-8888. Fax: (662) 842-6791. Web Site: www.afr.net. Licensee: American Family Association. Group owner: American Family Radio (acq 7-29-2003). Format: Christian. ♦Marvin Sanders, gen mgr.

Lafayette Township

***WCYT(FM)**— 1995: 91.1 mhz; 200 w. 213 ft TL: N40 58 58 W85 17 42. Stereo. Hrs opn: 24 Homestead High School, 4310 Homestead Rd., Fort Wayne, 46814. Phone: (260) 431-2299. Phone: (260) 431-2911. Fax: (260) 431-2330. Web Site: www.wcyt.org. Licensee: Southwest Allen County Schools. Format: Modern rock, alternative, contemporary hit. Target aud: General. Spec prog: Oldies 2 hrs, blues 2 hrs wkly. ♦Adam Schenkel, stn mgr; Joe Asher, progmg dir; Julian Shine, mus dir.

Lagrange

WTHD(FM)— Sept 2, 1994: 105.5 mhz; 2.4 kw. Ant 522 ft TL: N41 37 24 W85 20 49. Hrs open: 24 206 S. High St., 46761. Phone: (260) 463-8500. Phone: (800) 856-1055. Fax: (260) 463-8580. E-mail: wthd@wthd.net Web Site: www.wthd.net. Licensee: Lake Cities Broadcasting Corp. (acq 7-14-93; FTR: 8-9-93). Natl. Network: ABC. Format: Country. News staff: one; News: 2 hrs wkly. Target aud: 25-54. ♦Penny Mitchell, opns mgr; Tim Murray, gen mgr & news dir; Thomas Andrews, chief of engrg.

Lanesville

WGZB-FM— June 20, 1988: 96.5 mhz; 1.6 kw. Ant 638 ft TL: N38 10 25 W85 54 50. Stereo. Hrs opn: 520 S. 4th St., Suite 200, Louisville, KY, 40202. Phone: (502) 625-1220. Fax: (502) 625-1257. Web Site: www.b96jams.com. Licensee: Blue Chip Broadcasting Licenses Ltd. Group owner: Radio One Inc. (acq 4-30-2001; grpsl). Format: Urban contemp. ♦Dale Schafer, gen mgr.

Lebanon

***WIRE(FM)**— 2001: 91.1 mhz; 3.2 kw vert. Ant 220 ft TL: N40 03 48 W86 26 36. Stereo. Hrs opn: 24 3500 DePauw Blvd., Suite 2085, Indianapolis, 46268-6103. Phone: (317) 870-8400. Fax: (317) 870-8404. Web Site: www.radiomom.fm. Licensee: Hoosier Broadcasting Corp. Format: Adult contemp. Target aud: 25-54; adults. ♦William Shirk Poorman, pres; Annie Martin, CFO; Chuck Cunningham, VP & gen mgr.

***WWDL(FM)**—Not on air, target date: unknown: 91.5 mhz; 130 w vert. Ant 144 ft TL: N39 46 30 W86 25 44. Hrs open: 5331 Mt. Alifan Dr., San Diego, CA, 92111. Phone: (858) 277-4991. Fax: (858) 277-1365. Licensee: Horizon Christian Fellowship. (acq 3-6-2006). ♦Mike MacIntosh, pres.

Ligonier

WLEG(FM)—Licensed to Ligonier. See Warsaw

Linton

***KXJH(FM)**—Not on air, target date: unknown: 90.1 mhz; 400 w. Ant 141 ft TL: N39 02 22 W87 07 33. Hrs open: American Family Radio, Box 3206, Tupelo, MS, 38803. Phone: (662) 844-8888. Fax: (662) 842-6791. Web Site: www.afr.net. Licensee: American Family Association. Group owner: American Family Radio. Format: Inspirational Christian. ♦Marvin Sanders, gen mgr.

WBTO(AM)— Oct 10, 1953: 1600 khz; 500 w-D, 32 w-N. TL: N39 03 57 W87 11 19. Hrs open: 6 AM-6 PM
Rebroadcasts WQTY(FM) Linton.
Box 242, Vincennes, 47591. Phone: (812) 254-4300. Fax: (812) 254-4361. Web Site: www.wqtyfm.com. Licensee: The Original Co. Inc. (group owner; (acq 6-11-99; $350,000 with co-located FM). Population served: 33,000 Natl. Network: Moody. Rgnl. Network: Tribune. Natl. Rep: Rgnl Reps. Format: Country. ♦Mark Lange, pres & gen mgr; Kim Boothe, opns mgr; Michelle York, sls dir.

WQTY(FM)—Co-owned with WBTO(AM). Sept 14, 1970: 93.3 mhz; 12 kw. Ant 475 ft TL: N39 00 46 W87 22 23. Stereo. Web Site: www.wqtyfm.com.230,000 Format: Oldies.

***WYTJ(FM)**—Not on air, target date: Aug 2003: 89.3 mhz; 1 kw. Ant 292 ft TL: N39 05 59 W87 10 59. Hrs open: R.R. 3, Box 1034, 47441. Phone: (812) 847-7222. Licensee: Bethel Baptist Church. Format: Relg. ♦Doug Cassel, gen mgr; Harold Smith, stn mgr.

Logansport

WLHM(FM)—Listing follows WSAL(AM).

WSAL(AM)— Feb 24, 1949: 1230 khz; 1 kw-U. TL: N40 45 16 W86 18 40. Hrs open: 24 Box 719, 46947. Phone: (574) 722-4000. Fax: (574) 722-4010. Web Site: www.wsal.com. Licensee: Logansport Radio Corp. (acq 12-11-85; $850,000; FTR: 11-4-85). Population served: 150,000 Rgnl. Network: AgriAmerica. Natl. Rep: Rgnl Reps. Format: Adult contemp, news/talk. News staff: 2; News: 14 hrs wkly. Target aud: General. Spec prog: Farm 10 hrs wkly. ♦John P. Jenkins, CEO & pres; Andy Eubank, gen mgr; Lynne Ness, gen sls mgr; Eric Pfeiffer, news dir; Jeff Smith, chief of engrg.

WLHM(FM)—Co-owned with WSAL(AM). May 11, 1965: 102.3 mhz; 3 kw. 300 ft TL: N40 45 16 W86 18 40. Stereo. 24 Web Site: www.102fm.com. Format: CHR. News staff: 2; News: 5 hrs wkly. Target aud: 18-49. ♦Andy Eubank, progmg dir.

***WWTS(FM)**—Not on air, target date: unknown: 89.5 mhz; 5 w horiz, 24 kw vert. Ant 400 ft TL: N40 40 08 W86 41 44. Hrs open: CSN International, 4002N. 3300E., Twin Falls, ID, 83301. Phone: (208) 734-6633. Fax: (208) 736-1958. Licensee: CSN International (group owner). ♦Mike Kestler, pres.

Loogootee

***WBHW(FM)**— September 1995: 88.7 mhz; 1.7 kw. 761 ft TL: N38 38 30 W86 59 57. Hrs open:
Rebroadcasts WBGW(FM) Fort Branch 100%.
Box 4164, Evansville, 47724. Phone: (812) 386-3342. Fax: (812) 768-5552. E-mail: goodnews@evansville.net Web Site: www.thyword.org. Licensee: Music Ministries Inc. Format: Relg. ♦Floyd E. Turner, gen mgr.

WRZR(FM)— Dec 6, 1984: 94.5 mhz; 1.8 kw. Ant 426 ft TL: N38 37 09 W86 58 27. Stereo. Hrs opn: 24 Box 1009, Jasper, 47547. Secondary address: 514 JFK Ave. 47553. Phone: (812) 634-9232. Fax: (812) 482-3696. E-mail: wrzr@pscl.com Web Site: dcbroadcasting.com. Licensee: Hembree Communications Inc. Group owner: DCBroadcasting Inc. (acq 8-97). Natl. Network: Jones Radio Networks. Rgnl. Network: Network Indiana Rgnl rep: Ron Spaulding Law Firm: Miller & Miller. Wire Svc: AP Format: Classic rock. News staff: one; News: 2 hrs wkly. Target aud: 24-45. Spec prog: Farm 2 hrs wkly. ♦Paul Knies, pres; Bill Potter, gen mgr & opns mgr; Ron Spoulding, gen mgr & gen sls mgr; Alan Williams, progmg dir; Chris Eckstein, news dir; David Ferguson, chief of engrg.

Lowell

***WTMK(FM)**— 2005: 88.5 mhz; 1.5 kw. Ant 167 ft TL: N41 04 59 W87 10 47. Hrs open: 150 Lincolnway, Suite 2001, Valparaiso, 46383-5556. Phone: (219) 548-5800. Fax: (219) 548-5808. Web Site: www.csnmidwest.com. Licensee: CSN International (group owner). Format: Relg. ♦Jim Motshagen, gen mgr; Kathy Motshagen, progmg dir.

***WWLO(FM)**— 2006: 89.1 mhz; 2.4 kw. Ant 253 ft TL: N41 19 24 W87 21 22. Hrs open:
Rebroadcasts WAFR(FM) Tupelo, MS 100%.
Drawer 2440, Tupelo, MS, 38801-2440. Phone: (662) 844-8888. Fax: (662) 842-6791. Web Site: www.afr.net. Licensee: American Family Association. Format: Christian. ♦Marvin Sanders, gen mgr.

WZVN(FM)— Nov 24, 1972: 107.1 mhz; 1.29 kw. 499 ft TL: N41 21 09 W87 24 12. Hrs open: 2755 Sager Rd., Valparaiso, 46383. Phone: (219) 462-6111. Fax: (219) 462-4880. Web Site: www.z1071.com. Licensee: Porter County Broadcasting Holding Corp. LLC. Group owner: Porter County Broadcasting Corp. (acq 2-6-2004; $4.9 million with WXRD(FM) Crown Point). Population served: 650,000 Natl. Network: ABC. Law Firm: Wilmer, Cutler & Pickering. Format: Adult contemp. Target aud: 25-54. ♦Leigh Ellis, pres, pres & gen mgr; O.J. Jackson, gen sls mgr; Scott Wagner, progmg dir; Laura Waluszko, news dir; Carl Fletcher, chief of engrg.

Madison

WIKI(FM)—See Carrollton, KY

WORX-FM— March 1950: 96.7 mhz; 1.05 kw. Ant 551 ft TL: N38 44 32 W85 21 43. Stereo. Hrs opn: 24 Box 95, 47250. Secondary address: 1224 E. Telegraph Hill Rd. 47250. Phone: (812) 265-3322. Fax: (812) 273-5509. E-mail: manager@worxradio.com Web Site: www.worxradio.com. Licensee: Dubois County Broadcasting Inc. Population served: 50,000 Format: Adult contemp. News staff: one; News: 20 hrs wkly. Spec prog: Agriculture business 3 hrs wkly. ♦Paul Knies, pres; Bill Potter, gen mgr.

WXGO(AM)—Co-owned with WORX-FM. March 1956: 1270 khz; 1 kw-D, 58 w-N, DA-2. TL: N38 44 28 W85 21 41. (CP: COL Aurora. 330 w-D, DA. TL: N39 02 30 W84 56 28). 24 Web Site: www.worxradio.com. Group owner: DCBroadcasting Inc. 35,000 Natl. Network: USA. Natl. Rep: Rgnl Reps. Format: News, oldies. Target aud: General. Spec prog: Farm 3 hrs, relg 6 hrs wkly.

Marengo

***WBRO(FM)**— 2000: 89.9 mhz; 1 kw. Ant 279 ft TL: N38 21 49 W86 25 13. Hrs open: Box 181, 47140. Phone: (812) 365-9276. Fax: (812) 365-2127. E-mail: wbrofm@aol.com Web Site: www.wbro.org. Licensee: Crawford County Community Radio Inc. (acq 6-25-01). Format: Var. ♦Shawn Scott, gen mgr.

Marion

WBAT(AM)— June 7, 1947: 1400 khz; 1 kw-U. TL: N40 33 40 W85 41 30. Hrs open: 24 Box 839, 46952. Secondary address: 820 Pennsylvania St. 46953. Phone: (765) 664-6239. Fax: (765) 662-0730. E-mail: wbat@comteck.com Web Site: www.wbat.com. Licensee: Mid-America Radio Group. Group owner: Mid-America Radio Group Inc. (acq 12-88; grpsl; FTR: 12-19-88). Population served: 100,000 Natl. Network: CBS, ESPN Radio. Rgnl rep: Regional Reps Format: Sports, oldies. News staff: one; News: 7 hrs wkly. Target aud: 25-54. ♦David Keister, pres; David Poehler, exec VP; Carolyn Bush, gen mgr; James F. Brunner, gen sls mgr; Tim George, progmg dir; Mike Jenkins, news dir; Warren Arnett, chief of engrg.

***WBSW(FM)**— 1997: 90.9 mhz; 1 kw horiz, 2.4 kw vert. 308 ft TL: N40 40 01 W85 37 50. Hrs open:
Rebroadcasts WBST(FM) Muncie 100%.
c/o WBST(FM), Ball State Univ., Muncie, 47306-0550. Phone: (765) 285-5888. Fax: (765) 285-8937. E-mail: ipr@bsu.edu Web Site: www.bsu.edu/ipr. Licensee: Ball State University. Format: Class, news. News staff: one; News: 33 hrs wkly. Target aud: General. ♦Marcus Jackman, gen mgr; Pam Coletti, gen sls mgr; Carol Trimmer, prom

mgr; Steven Turpin, progmg dir & mus dir; Robert Mittendorf, chief of engrg; Dorothy Marvell, traf mgr; Brian Beaver, news rptr.

WCJC(FM)—Van Buren, Aug 28, 1989: 99.3 mhz; 3 kw. 328 ft TL: N40 40 01 W85 37 50. Stereo. Hrs opn: 24 Box 839, 820 S. Pennsylvania, 46952. Phone: (765) 664-6239. Fax: (765) 662-0730. E-mail: wcjc@comteck.com Web Site: www.wcjc.com. Licensee: Mid-America Radio Group Inc. (group owner; acq 12-19-88; grpsl; FTR: 12-19-88). Natl. Network: ABC. Rgnl rep: Regional Reps Format: Country. News staff: 2; News: 25 hrs wkly. Target aud: 25-54; consumer-oriented modern country fans. Spec prog: Relg 3 hrs wkly. ♦David Keister, pres; David Poehler, VP & stn mgr; Carolyn Bush, gen mgr; Tim George, opns mgr & progmg dir; James F. Brunner, gen sls mgr; Warren Arnett, chief of engrg.

WMRI(AM)— May 11, 1955: 860 khz; 1 kw-D, 500 w-N, DA-2. TL: N40 33 12 W85 38 45. Hrs open: 24 Box 1538, 46952. Secondary address: 820 S. Pennsylvania St. 46953. Phone: (765) 664-7396. Phone: (765) 664-9466. Fax: (765) 668-6767. Web Site: wmri.com. Licensee: Mid-America Radio of Indiana Inc. Group owner: Mid-America Radio Group Inc. (acq 5-12-2003; with co-located FM). Population served: 40,253 Natl. Network: Music of Your Life. Rgnl. Network: Network Indiana. Format: Nostalgia. Target aud: 35-70. ♦David Poehler, pres & VP; Carolyn Bush, gen mgr & chief of engrg; Vanessa Miller, opns mgr; Gloria Millspaugh, gen sls mgr; Mike Jenkins, news dir & pub affrs dir.

WXXC(FM)—Co-owned with WMRI(AM). Dec 19, 1948: 106.9 mhz; 50 kw. Ant 499 ft TL: N40 35 52 W85 39 21. Stereo. 24 E-mail: studio@1069wxxc.com Web Site: www.1069wxxc.com.500,000 Natl. Network: CNN Radio. Format: Classic hits. Target aud: 25-54.

Martinsville

WCBK-FM—Listing follows WMYJ(AM).

WMYJ(AM)— Apr 18, 1967: 1540 khz; 500 w-D. TL: N39 24 31 W86 25 10. Hrs open: Box 1577, 46151. Secondary address: 1639 Burton Ln. 46151-3004. Phone: (765) 342-3394. Fax: (765) 342-5020. Licensee: Mid-America Radio Group Inc. (group owner; (acq 8-4-97; with co-located FM). Population served: 100,000 Natl. Network: USA. Rgnl. Network: Network Indiana, Tribune. Format: Gospel. Target aud: 25-54. Spec prog: Farm one hr wkly. ♦David Keister, gen mgr.

WCBK-FM—Co-owned with WMYJ(AM). Oct 15, 1968: 102.3 mhz; 6 kw. 308 ft TL: N39 26 18 W86 27 58. Stereo. Web Site: www.wcbk.com. ♦Ruth Ann Arney, gen sls mgr; John Taylor, progmg dir.

Michigan City

WEFM(FM)— Sept 15, 1966: 95.9 mhz; 3 kw. Ant 230 ft TL: N41 42 58 W86 51 47. Stereo. Hrs opn: 24 1903 Springland Ave., 46360. Phone: (219) 879-8201. Fax: (219) 879-8202. E-mail: wefm@yahoo.com Licensee: Michigan City FM Broadcasters Inc. Natl. Network: USA. Rgnl. Network: Network Indiana. Format: Adult contemp, oldies. Target aud: General. Spec prog: Farm, relg 4 hrs wkly. ♦Thomas Burns, pres; Ronald Miller, stn mgr; Jim Spevak, gen sls mgr; Tod Allen, progmg dir & progmg mgr; Tim Volckmann, chief of engrg.

WIMS(AM)— Aug 10, 1947: 1420 khz; 5 kw-U, DA-2. TL: N41 40 26 W86 55 58. Hrs open: 24 6405 Olcott Ave., Hammond, 46320. Phone: (219) 989-8502. Fax: (219) 989-8516. Web Site: www.wimsradio.com. Licensee: Vazquez Development LLC (acq 12-4-2003; $1.2 million with WJOB(AM) Hammond). Population served: 39,369 Natl. Rep: Rgnl Reps. Format: Talk. Target aud: 30 plus; general. Spec prog: Pol 3 hrs wkly. ♦Jim Dedelow, gen mgr; Debbie Wargo, gen sls mgr; Michael Stewart, progmg dir; Ron Perzo, news dir.

Mitchell

***WMBL(FM)**—Not on air, target date: unknown: 88.1 mhz; 1 kw. Ant 400 ft TL: N38 45 50 W86 31 15. Hrs open: 820 N. LaSalle Blvd., Chicago, IL, 60610. Phone: (312) 329-4438. Phone: (800) 246-0691. Web Site: www.mbn.org. Licensee: The Moody Bible Institute of Chicago. Format: Relg. ♦Allen Henderson, gen mgr.

WPHZ(FM)— Aug 17, 1991: 102.5 mhz; 6 kw. Ant 282 ft TL: N38 38 16 W86 27 11. Hrs open: 24 Box 1307, Bedford, 47421. Phone: (812) 275-7555. Fax: (812) 279-8046. Web Site: www.wphz.com. Licensee: Mitchell Community Broadcast Co. (acq 2-26-92; $8,000 for CP; FTR: 3-16-92). Natl. Network: ABC, Jones Radio Networks. Natl. Rep: Rgnl Reps. Law Firm: Reed, Smith, Shaw & McClay. Format: Adult Contemp. News staff: one; News: 2 hrs wkly. Target aud: General. ♦Holly Lindsey, gen mgr.

Monticello

WMRS(FM)— March 1989: 107.7 mhz; 4.4 kw. 500 ft TL: N40 45 03 W86 48 17. Stereo. Hrs opn: 24 132 N. Main, 47960. Phone: (574) 583-8121. Phone: (574) 583-8933. Fax: (574) 583-8933. E-mail: kevinp@wmrsradio.com Web Site: www.wmrsradio.com. Licensee: Monticello Community Radio Inc. (acq 1-15-91; FTR: 2-11-91). Natl. Network: USA, Jones Radio Networks. Format: Adult contemp, div, talk. News staff: 2; News: 20 hrs wkly. Target aud: 25-60; motivated, intelligent, diverse. Spec prog: Gospel 5 hrs, bluegrass 2hrs wkly. ♦Kevin Page, gen mgr.

WXXB(FM)—See Delphi

Montpelier

***WJCO(FM)**—Not on air, target date: unknown: 91.3 mhz; 350 w vert. Ant 196 ft TL: N40 33 21 W85 17 39. Hrs open: CSN International, 4002N. 3300E., Twin Falls, ID, 83301. Phone: (208) 734-6633. Fax: (208) 736-1958. Licensee: CSN International (group owner). ♦Mike Kestler, pres.

Morgantown

***WCJL(FM)**— 2005: 90.9 mhz; 1 kw horiz, 13.5 kw vert. Ant 213 ft TL: N39 19 17 W86 31 08. Hrs open: 4002 N. 3300 E., Twin Falls, ID, 83301. Fax: (208) 734-6633. Fax: (208) 736-1958. Web Site: www.csnradio.com. Licensee: CSN International. (group owner). ♦Mike Kestler, pres.

Morristown

***WJCF(FM)**— 2000: 88.1 mhz; 2.7 kw vert. Ant 151 ft TL: N39 45 01 W85 33 19. Hrs open: Box 846, Greenfield, 46140. Secondary address: 15 Wood St., Greenfield 46140-2162. Phone: (317) 467-1064. Fax: (317) 467-1065. E-mail: wjcfradio@aol.com Web Site: www.wjcfradio.com. Licensee: Indiana Community Radio Corp. Format: Community radio. ♦Jennifer Cox-Hensley, pres; Marty Hensley, progmg dir; Pat Diemer, news dir.

Mount Vernon

WRCY(AM)— Aug 21, 1955: 1590 khz; 500 w-D, 35 w-N. TL: N37 56 03 W87 55 42. Hrs open: 7109 Upton Rd., 47620-9483. Phone: (812) 838-4484. Fax: (812) 838-6434. Web Site: www.wyfx.com. Licensee: The Original Co. Inc. (group owner; (acq 1999; $360,000 with co-located FM). Population served: 7200 Format: Real country. Target aud: 25 plus; loc county. Spec prog: Farm 5 hrs wkly. ♦Mark Lange, pres; Sean Dulaney, gen mgr; Frank Hertel, chief of engrg.

WYFX(FM)—Co-owned with WRCY(AM). August 1992: 106.7 mhz; 3 kw. Ant 295 ft TL: N37 56 03 W87 55 35. Natl. Network: ESPN Radio. Format: Sports/talk.

Muncie

***WBST(FM)**— Sept 12, 1960: 92.1 mhz; 3 kw. 300 ft TL: N40 12 48 W85 27 36. Stereo. Hrs opn: 24 Ball State Univ., 47306-0550. Phone: (765) 285-5888. Fax: (765) 285-8937. E-mail: ipr@bsu.edu Web Site: www.bsu.edu/ipr. Licensee: Ball State University. Population served: 1,500,000 Natl. Network: PRI, NPR. Format: Classical, news. News staff: one; News: 33 hrs wkly. Target aud: General. ♦Marcus Jackman, gen mgr; Pam Coletti, sls dir & gen sls mgr; Carol Trimmer, prom mgr; Steven Turpin, progmg dir & mus dir; Robert Mittendorf, chief of engrg; Dorothy Marvell, traf mgr; Brian Beaver, news rptr.

WERK(FM)— Jan 16, 1986: 104.9 mhz; 3 kw. 328 ft TL: N40 09 19 W85 25 48. Stereo. Hrs opn: 24
Simulcasts WURK(FM) Elwood 80%.
800 E. 29th St., 47302. Phone: (765) 288-4403. Fax: (765) 378-2091. E-mail: werkstudio@werkradio.com Web Site: www.werkradio.com. Licensee: Indiana Sabrecom Inc. Group owner: Backyard Broadcasting LLC (acq 12-1-02; grpsl). Population served: 300,000 Format: Oldies. ♦Steve Lindell, gen mgr; Jay Garrison, progmg dir & news dir.

***WKMV(FM)**—Not on air, target date: unknown: 88.3 mhz; 200 w vert. Ant 295 ft TL: N40 05 06 W85 23 52. Hrs open: 5700 West Oaks Blvd., Rocklin, CA, 95765. Phone: (916) 251-1600. Fax: (916) 251-1650. Licensee: Educational Media Foundation. (acq 3-23-2007; grpsl). ♦Richard Jenkins, pres.

WLBC-FM—Listing follows WXFN(AM).

WLHN(AM)— Feb 14, 1965: 990 khz; 250 w-D, 2 w-N, DA-1. TL: N40 06 54 W85 22 02. Stereo. Hrs opn: 3611 S. Post Rd., 47302. Phone: (765) 747-6970. Fax: (765) 747-5054. E-mail: wlhn990@yahoo.com Web Site: wlhnradio.com. Licensee: Electronic Applications Radio Service Inc. (acq 3-16-99). Population served: 300,000 Natl. Rep: Roslin, Rgnl Reps. Law Firm: Harris, Beach & Wilcox. Format: Southern gospel. Target aud: 25-54; upscale adults & families, professional & blue collar. ♦Steven Dugger, gen mgr.

WMDH-FM—See New Castle

***WWDS(FM)**— 1978: 90.5 mhz; 100 w. Ant 174 ft TL: N40 16 42 W85 20 52. Stereo. Hrs opn: 3400 E. State Rd. 28, 47303. Phone: (765) 288-5597. Fax: (765) 288-8498. E-mail: fclark@delcomschools.org Licensee: Delaware Community School Corp. Format: Adult contemp. ♦Ford Clark, stn mgr.

***WWHI(FM)**— 1950: 91.3 mhz; 310 w. 79 ft TL: N40 09 45 W85 22 45. Hrs open: 1601 E. 26th St., 47302. Phone: (765) 747-5339. Fax: (765) 747-5325. Licensee: Ball State University (acq 2-13-2004). Format: Educ. ♦Ken Wickliffe, gen mgr.

WXFN(AM)— November 1926: 1340 khz; 1 kw-U. TL: N40 09 42 W85 22 41. Hrs open: 800 E. 29th St., 47302. Phone: (765) 288-4403. Fax: (765) 288-0429. Licensee: Indiana Sabrecom Inc. Group owner: Backyard Broadcasting LLC (acq 12-1-2002; grpsl). Population served: 80,093 Natl. Network: ABC, ESPN Radio, Sporting News Radio Network. Wire Svc: AP Format: Sports. News staff: 2; News: 12 hrs wkly. Target aud: 25-54. Spec prog: Black 3 hrs wkly. ♦Barry Drake, CEO; Robin Smith, CFO; Sean Mattingly, chief of engrg; Joanna Black, traf mgr; Steve Lindell, VP, gen mgr & mus critic; Jay Garreson, sports cmtr.

WLBC-FM—Co-owned with WXFN(AM). October 1947: 104.1 mhz; 50 kw. 420 ft TL: N40 09 38 W85 22 42. Stereo. E-mail: steve@wlbc.com Web Site: www.wlbc.com.80,093 Format: News. News staff: one; News: 19 hrs wkly. Target aud: 18-49; female. ♦Joanna Black, traf mgr; Dave Stout, news rptr; Steve Lindell, progmg VP, farm dir & mus critic.

Nappanee

WYPW(FM)— Dec 16, 1991: 95.7 mhz; 1.4 kw. 500 ft TL: N41 24 43 W86 01 51. Stereo. Hrs opn: 24 237 W. Edison Rd., Mishawaka, 46545. Phone: (574) 258-5483. Phone: (888) 737-6244. Web Site: www.power957.com. Licensee: Talking Stick Communications LLC. (group owner; (acq 8-25-2000). Population served: 600,000 Natl. Network: ABC, CBS. Format: Adult contemp. News: 10 hrs wkly. Target aud: 35-70; secretaries, bankers. ♦Emily Wideman, exec VP & gen sls mgr; Abe Thompson, gen mgr; Alec Dille, stn mgr; Gene Walker, opns mgr; Chuck Wright, progmg dir; Greg Trobridge, chief of engrg.

Nashville

WVNI(FM)— August 1997: 95.1 mhz; 1.6 kw. 636 ft TL: N39 13 39 W86 25 05. (CP: 2.3 kw, ant 472 ft.). Hrs opn: 24 Box 1628, Bloomington, 47402. Secondary address: 4317 E.3rd St., Bloomington 47401. Phone: (812) 335-9500. Fax: (812) 335-8880. E-mail: spitit95@spirit95fm.com Web Site: www.sprint95fm.com. Licensee: Brown County Broadcasters Inc. Group owner: Mid-America Radio Group Inc. (acq 10-29-97; $20,000 for 51% of stock). Natl. Network: Salem Radio Network. Format: Contemp Christian. Target aud: 25-54. ♦Diana Nuchols, gen mgr; Denise Ray, opns dir.

New Albany

WFIA-FM— Jan 1, 1996: 94.7 mhz; 6 kw. 328 ft TL: N38 17 02 W85 54 17. Stereo. Hrs opn: 24 9960 Corporate Campus Dr., Suite 3600, Louisville, KY, 40223. Phone: (502) 339-9470. Fax: (502) 423-3139. Web Site: www.salemradiogroup.com. Licensee: Salem Media of Kentucky Inc. Group owner: Salem Communications Corp. (acq 1999; $5 million with WRVI(FM) Valley Station, KY). Format: Talk, teaching, Southern gospel, Christian. Target aud: 30 plus. ♦Gordon Marcy, gen mgr.

***WNAS(FM)**— May 28, 1949: 88.1 mhz; 2.85 kw. 3 ft TL: N38 17 56 W85 48 45. Hrs open: 1020 Vincennes St., 47150. Phone: (812) 949-4272. Fax: (812) 949-6926. Web Site: www.wnas.org. Licensee: New Albany-Floyd County Consolidated School Corp. Population served: 1,000,000 Format: Educ, Top-40. ♦Lee Kelly, gen mgr.

WWSZ(AM)— June 15, 1949: 1570 khz; 1.5 kw-D, 233 w-N. TL: N38 19 40 W85 46 56. Hrs open: 24 304 W Liberty St., Suite 110, Louisville, KY, 40202. Phone: (502) 583-1212. Fax: (502) 583-1414.

Web Site: www.wszradio.com. Licensee: New Albany Broadcasting Co. Inc. Group owner: Mortenson Broadcasting Co. (acq 1-27-2005; $1 million). Population served: 1,000,000 Natl. Network: CBS Radio, ESPN Radio. Format: Sports. Target aud: 35-54; blues and true jazz listeners. ♦Scott Thompson, gen mgr.

New Carlisle

WZOW(FM)— July 2, 1991: 102.3 mhz; 2 kw. 397 ft TL: N41 43 38 W86 24 30. Stereo. Hrs opn: 24 3371 Cleveland Rd., Suite 300, South Bend, 46628. Phone: (574) 273-9300. Fax: (574) 273-9090. E-mail: michael@wzow.com Web Site: www.wzow.com. Licensee: Artistic Media Partners Inc. (group owner; (acq 3-22-2002; $1.5 million). Format: Classic rock. ♦Jack Swart, gen mgr; Carrie Jones, natl sls mgr; Teresa Holden, prom dir; Chili Walker, progmg dir; Bob Henning, chief of engrg; Rita Kinzie, traf mgr.

New Castle

WMDH(AM)— Nov 14, 1960: 1550 khz; 250 w-U, DA-2. TL: N39 55 59 W85 24 26. Hrs open: Box 690, 1134 W. State Rd. 38, 47362. Phone: (765) 529-2600. Fax: (765) 529-1688. Web Site: www.wmdh.com. Licensee: Citadel Broadcasting Co. Group owner: Citadel Broadcasting Corp. (acq 7-1-99; grpsl). Population served: 24,000 Rgnl. Network: Network Indiana. Natl. Rep: Katz Radio. Law Firm: Leventhal Senter & Lerman. Format: Adult standards. Target aud: 49 plus. Spec prog: Farm 2 hrs, relg 2 hrs wkly. ♦Pam Price, gen sls mgr; Paulette Lees, gen mgr & gen sls mgr; Jon Sipes, progmg dir.

WMDH-FM— Aug 6, 1976: 102.5 mhz; 50 kw. 500 ft TL: N40 03 18 W85 23 05. Stereo. 24 Web Site: www.wmdh.com.240,000 Format: Country. News staff: one; News: 2 hrs wkly. Target aud: 25-54. Spec prog: Farm one hr wkly.

New Haven

WJFX(FM)— April 1990: 107.9 mhz; 3.2 kw. Ant 453 ft TL: N41 01 26 W85 03 51. Stereo. Hrs opn: 24 5936 E. State Blvd., Fort Wayne, 46815. Phone: (260) 493-9539. Fax: (260) 749-5151. Web Site: www.hot1079online.com. Licensee: Fort Wayne Radio Corp. (acq 12-1-98; $1.3 million). Population served: 424,900 Natl. Rep: Interep. Law Firm: Wiley, Rein & Fielding. Format: CHR. Target aud: 18-49; adults. ♦Russ Oasis, pres; Roger Diehm, VP & gen mgr; Beth Thornton, gen sls mgr; Phil Becker, progmg dir.

New Paris

WVXR(FM)—See Richmond

New Washington

***WARA(FM)**— 1994: 88.3 mhz; 950 w. Ant 300 ft TL: N38 35 40 W85 28 06. Stereo. Hrs opn: 24 2351 Sunset Blvd., Suite 170-218, Rocklin, CA, 95765. Phone: (916) 251-1600. Fax: (916) 251-1650. E-mail: info@air1.com Web Site: www.air1.com. Licensee: Educational Media Foundation. Group owner: EMF Broadcasting (acq 10-2-2003; grpsl). Natl. Network: Air 1. Law Firm: Shaw Pittman. Format: Contemp Christian. News staff: 3. Target aud: 18-35; Judeo Christian, female. ♦Richard Jenkins, pres; Mike Novak, VP; Keith Whipple, dev dir; Eric Allen, natl sls mgr; David Pierce, progmg dir; Ed Lenane, news dir; Sam Wallington, engrg dir; Karen Johnson, news rptr; Marya Morgan, news rptr; Richard Hunt, news rptr.

New Whiteland

***WWDN(FM)**—Not on air, target date: unknown: 88.3 mhz; 1 w horiz, 4.6 kw vert. Ant 495 ft TL: N39 24 27 W86 08 52. Hrs open: 5331 Mt. Alifan Dr., San Diego, CA, 92111. Phone: (858) 277-4991. Fax: (858) 277-1365. Licensee: Horizon Christian Fellowship. (acq 3-6-2006). ♦Mike MacIntosh, pres.

Newburgh

WDKS(FM)— Feb 11, 1991: 106.1 mhz; 6 kw. 328 ft TL: N37 57 16 W87 25 07. Stereo. Hrs opn: 24 1133 Lincoln Ave., Evansville, 47714. Phone: (812) 425-4226. Fax: (812) 428-5895. E-mail: mthomas@regentcomm.com Web Site: www.kissevansville.com. Licensee: Regent Broadcasting of Evansville/Owensboro Inc. Group owner: Regent Communications Inc. (acq 12-3-2003; grpsl). Natl. Network: ABC, Westwood One. Format: Top 40. News staff: one. Target aud: 18-34; women. ♦Mark Thomas, gen mgr; Max Powers, prom dir; Cat Michaels, progmg dir; Gene Stewart, news dir; Rick Crazo, chief of engrg.

WGAB(AM)— Mar 5, 1984: 1180 khz; 670 w-D. TL: N37 57 16 W87 25 07. Hrs open: 24 hrs 2601 South Boeke Rd., Evansville, 47714. Phone: (812) 853-9422. Fax: (812) 474-4483. E-mail: faithmusicbb@aol.com Licensee: Faith Broadcasting LLC (acq 12-1-2004; $300,000). Population served: 232,000 Natl. Network: ABC, Jones Radio Networks, Salem Radio Network, Westwood One. Format: Christian progmg. Target aud: 18-54; men & women. ♦Gayle Russ, gen mgr, opns VP & opns mgr.

Noblesville

WJJK(FM)—Licensed to Noblesville. See Indianapolis

North Judson

***WTMW(FM)**—Not on air, target date: unknown: 91.3 mhz; 50 kw. Ant 269 ft TL: N41 02 21 W86 30 55. Hrs open: 125 S. Main St., Bishop, CA, 93514-3414. Phone: (760) 954-6655. Fax: (760) 872-4155. Licensee: Living Proof Inc. Law Firm: Fletcher, Heald & Hildreth. ♦Daniel McClenaghan, pres & gen mgr.

North Manchester

***WBKE-FM**— May 1967: 89.5 mhz; 3 kw. 80 ft TL: N41 00 40 W85 45 45. Stereo. Hrs opn: 24
Rebroadcasts WBNI-FM Ft. Wayne 50%.
Box 19, Manchester College, 604 E. College Ave., 46962. Phone: (260) 982-5424. Fax: (260) 982-5043. E-mail: wbke@manchester.edu Web Site: www.wbke.manchester.edu. Licensee: Manchester College. Population served: 6,800 Format: Div, free form. News staff: one; News: 10 hrs wkly. Target aud: General. Spec prog: Class 10 hrs, Sp one hr, classic rock 6 hrs, AOR 6 hrs, Top 40/rap 10 hrs, alternative 7 hrs wkly. ♦Sunday Isang, gen mgr; Logan Condon, stn mgr.

North Vernon

WJCP(AM)— Jan 8, 1955: 1460 khz; 1 kw-D, 92 w-N. TL: N38 59 46 W85 39 02. Stereo. Hrs opn: 2470 N. State Hwy. 7, 47265-7184. Phone: (812) 346-1927. Fax: (812) 346-9722. Licensee: Columbus Radio Inc. (acq 11-20-2001; swap for WWWY(FM) Columbus plus $1.2 million). Population served: 25,000 Format: Sports. ♦Marty Pieratt, gen mgr.

WWWY(FM)— Mar 19, 1963: 106.1 mhz; 50 kw. 486 ft TL: N39 04 02 W85 42 10. Stereo. Hrs opn: 24 Box 1789, Columbus, 47202-1789. Secondary address: 3212 Washington St., Columbus 47203. Phone: (812) 372-4448. E-mail: rockme@y106.com Web Site: www.y106.com. Licensee: White River Broadcasting Co. Inc. Group owner: The Findlay Publishing Co. (acq 8-1-97; grpsl). Format: Rock. Target aud: 25-44. ♦Kurt Kah, pres; David Glass, VP; Tasha Mann, gen mgr; John Foster, opns mgr; Scott Michaels, progmg dir; Kevin Keith, news dir; Chuck Weber, engr.

Notre Dame

***WSND-FM**— Sept 17, 1962: 88.9 mhz; 3.4 kw. Ant 361 ft TL: N41 36 20 W86 12 46. Stereo. Hrs opn: 315 LaFortune Student Ctr., 46556. Phone: (574) 631-7342 / (574) 631-9059. Fax: (574) 631-3653. E-mail: mcfadden.20@nd.cdn Web Site: www.nd.edu/~wsnd. Licensee: Voice of the Fighting Irish Inc. Population served: 7,700 Format: Class. ♦Laurie McFadden, gen mgr; Patrick Longenbaker, gen mgr & stn mgr; Ed Jaroszewski, progmg dir.

Oolitic

***WMYJ-FM**— 2005: 88.9 mhz; 5.2 kw vert. Ant 256 ft TL: N38 59 14 W86 27 31. Hrs open: Box 1970, Martinsville, 46151. Phone: (765) 349-1485. Fax: (765) 342-3569. Licensee: Spirit Educational Radio Inc. (acq 6-30-2005; $45,000 for CP). Population served: 136,000 Format: Southern gospel. ♦David Keister, chmn; Diana Nuchols, gen mgr.

Orland

***WCKZ(FM)**— Feb 2, 2002: 91.3 mhz; 2 kw. Ant 298 ft TL: N41 44 36 W85 05 48. Stereo. Hrs opn: 24 Box 8459, Fort Wayne, 46898. Phone: (260) 452-1189. Fax: (260) 452-1188. E-mail: ccondron@nipr.fm Web Site: www.nipr.fm. Licensee: Northeast Indiana Public Radio Inc. Natl. Network: NPR. Law Firm: Dow, Lohnes & Albertson, PLLC. Format: Classical. ♦Bruce Haines, gen mgr & prom dir; Colleen Condron, opns dir & progmg dir; Karen Fraser, dev dir; Janice Furtner, mus dir; Jeanette Dillon, news dir; Jackie Didier, traf mgr.

Paoli

WSEZ(AM)— Nov 7, 1963: 1560 khz; 250 w-D. TL: N38 32 25 W86 28 42. Hrs open: 6 AM-6 PM Box 26, 192 S. Court St., 47454. Phone: (812) 723-4484. Fax: (812) 723-4966. E-mail: wume@blueriver.net Web Site: hitsandfavorites.com. Licensee: Ironic Broadcasting Inc. (acq 3-21-97; with co-located FM). Population served: 20,000 Rgnl. Network: Network Indiana. Natl. Rep: Rgnl Reps. Law Firm: Haley, Bader & Potts. Format: Oldies. News staff: one; News: 7 hrs wkly. Target aud: General. Spec prog: Farm 7 hrs wkly. ♦Jerry Wall, gen mgr & gen sls mgr; Jason Archer, progmg mgr & pub affrs dir; Dave Dedrick, news dir; Todd Edwards, chief of engrg.

WUME-FM—Co-owned with WSEZ(AM). September 1972: 95.3 mhz; 3 kw. 300 ft TL: N38 32 25 W86 28 42. Stereo. 24 Web Site: hitsandfavorites.com.20,000 Natl. Network: ABC. Rgnl rep: Rgnl Reps Format: Comtemp hit. News staff: one; News: 9 hrs wkly. Target aud: General. ♦Jason Archer, traf mgr; Dave Dedrick, news rptr, farm dir & sports cmtr.

Pendleton

***WEEM-FM**— Nov 1, 1971: 91.7 mhz; 1.2 kw. Ant 154 ft TL: N39 59 52 W85 44 07. Stereo. Hrs opn: One Arabian Dr., 46064. Phone: (765) 778-2161, EXT. 236. Fax: (765) 778-0605. Licensee: South Madison Community School Corp. Population served: 250,000 Law Firm: Pillsbury Winthrop Shaw Pittman LLP. Format: Triple A. Target aud: Adults 18-45, students 13-18. Spec prog: High school sports 10 hrs, country 4 hrs, educ progmg 10 hrs wkly. ♦Jeff Dupont, gen mgr; Josh Brown, prom dir; Rodney Conner, prom dir; Amanda Downham, progmg dir; Leigh Montano, mus dir; Steve Longenecker, chief of engrg.

Peru

WARU(AM)— Sept 12, 1954: 1600 khz; 1 kw-D. TL: N40 45 53 W86 02 26. Hrs open: 24 Box 1010, 46970. Secondary address: 1711 E. Wabash Rd. 46970. Phone: (765) 473-4448. Fax: (765) 473-4449. E-mail: waru@sbcglobal.net Web Site: www.warufm.com. Licensee: Miami County Broadcasting Inc. Group owner: Mid-America Radio Group Inc. Population served: 14,139 Natl. Network: AP Radio. Rgnl. Network: Tribune, Agri-Net. Natl. Rep: Rgnl Reps. Format: Real country. News staff: one; News: 20 hrs wkly. Target aud: 25-54. ♦David Keister, pres; David Poehler, VP; Dan Keister, gen mgr & stn mgr; Steve Morris, progmg dir.

WARU-FM— 2001: 101.9 mhz; 3.6 kw. Ant 423 ft TL: N40 48 30 W85 56 07. Stereo. 24 1711 E. Wabash Rd., 46970. Web Site: www.warufm.com. Licensee: Mid-America Radio Group Inc.30,000 News staff: one; News: 4 hrs wkly.

WMYK(FM)— Apr 5, 1965: 98.5 mhz; 6 kw. Ant 328 ft TL: N40 37 46 W86 02 28. Stereo. Hrs opn: 24 Box 2208, Kokomo, 46904-2208. Secondary address: 671 E. 400 S., Kokomo 46902. Phone: (765)455-9850. Fax: (765) 455-3882. E-mail: classichits@radio.fm Web Site: www.classichits.org. Licensee: Miami County Broadcasting Inc. Group owner: Mid-America Radio Group Inc. (acq 1996; $360,000 with WARU(AM) Peru). Population served: 14,139 Format: Classic rock. News staff: one. ♦Steve LaMar, gen mgr; Bryan Michaels, opns mgr & progmg dir; Amber Stearns, news dir; Steve Ross, chief of engrg.

Petersburg

WBTO-FM— Oct 8, 1984: 102.3 mhz; 3 kw. 321 ft TL: N38 30 33 W87 17 28. Stereo. Hrs opn: 24 Box 616, Washington, 47501. Secondary address: Box 242, Vincennes 47591. Phone: (812) 254-4300. Phone: (812) 882-6060. Fax: (812) 254-4361. Fax: (812) 885-2604. Web Site: www.wbtofm.com. Licensee: The Original Co. Inc. (group owner; acq 11-24-99; $400,000). Natl. Rep: Rgnl Reps. Format: Classic rock. News: 9 hrs wkly. Target aud: General. ♦Mark Lange, pres.

Plainfield

WRDZ-FM— July 1, 2003: 98.3 mhz; 3 kw. 300 ft TL: N39 45 33 W86 22 30. Stereo. Hrs opn: 24 630 W. Carmel Dr., Indianapolis, 46032. Phone: (317) 574-2000. Fax: (317) 581-1985. E-mail: jim.mcconville@abc.comc Web Site: www.radiodisney.com. Licensee: Radio Disney Group LLC. Group owner: ABC Inc. (acq 7-1-03; $5.6 million). Population served: 1,750,000 Natl. Network: Radio Disney. Natl. Rep: Interep. Format: Pop/CHR. News staff: one; News: 20 hrs wkly. Target aud: 25-44. ♦Jim McCondville, gen mgr; Laura Sanchez, prom mgr & chief of engrg.

Plymouth

***WIKV(FM)**— 2005: 89.3 mhz; 400 w. Ant 249 ft TL: N41 20 51 W86 20 23. Hrs open: 5700 West Oaks Blvd., Rocklin, CA, 95765. Phone: (916) 251-1600. Fax: (916) 251-1650. Web Site: www.klove.com. Licensee: Educational Media Foundation. Group owner: American Family Radio. (acq 3-23-2007; grpsl). Natl. Network: K-Love. Format: Contemp Christian. ♦Richard Jenkins, pres.

WTCA(AM)— Aug 18, 1964: 1050 khz; 250 w-U, DA-2. TL: N41 19 06 W86 18 41. Hrs open: 112 W. Washington St., 46563. Phone: (574) 936-4096. Fax: (574) 936-6776. Licensee: Community Service Broadcasters Inc. (acq 11-27-98). Population served: 100,000 Law Firm: Reddy, Begley & McCormick. Format: Oldies. Target aud: 25-65. Spec prog: Farm 11 hrs wkly. ♦Kathryn E. Bottorff, stn mgr; Andrew Banas, progmg dir; Tony Ross, news dir; James Kunze, chief of engrg.

WZOC(FM)— July 20, 1966: 94.3 mhz; 11.5 kw. 492 ft TL: N41 31 41 W86 15 53. Stereo. Hrs opn: 112 W. Washington St., 46563. Phone: (574) 936-4096. Fax: (574) 936-6776. Licensee: Plymouth Broadcasting Inc. (acq 9-12-96; $575,000). Format: Oldies. ♦James Kunze, stn mgr.

Portage

WNDZ(AM)— May 13, 1987: 750 khz; 5 kw-D, DA. TL: N41 33 49 W87 09 18. (CP: 15 kw-D, DA). Hrs opn: 6012 S. Pulaski Rd., Chicago, IL, 60629. Phone: (773) 767-1000. Fax: (773) 767-1100. Licensee: WNDZ Inc. Group owner: Newsweb Corp. (acq 3-16-2004; $24 million with WRZA(FM) Park Forest, IL). Format: Relg, ethnic. Target aud: General. Spec prog: Ger 2 hrs, gospel 2 hrs, Serbian 2 hrs, Lithuanian 4 hrs, Bosnian one hr wkly. ♦Harvey Wells, VP; Mark Pinski, gen mgr; Mike McCarthy, engrg dir.

Portland

***WBSJ(FM)**— December 1996: 91.7 mhz; 2.1 kw. 210 ft TL: N40 24 26 W85 02 15. Hrs open:
Rebroadcasts WBST(FM) Muncie 100%.
c/o WBST(FM), Ball State Univ., Muncie, 47306-0550. Phone: (765) 285-5888. Fax: (765) 285-8937. E-mail: ipr@bsu.edu Web Site: www.bsu.edu/ipr. Licensee: Ball State University. Format: Class, news. News staff: one; News: 33 hrs wkly. Target aud: General. ♦Marcus Jackman, gen mgr; Pam Coletti, gen sls mgr; Carol Trimmer, prom mgr; Steven Turpin, progmg dir; Robert Mittendorf, chief of engrg; Dorothy Marvell, traf mgr; Brian Beaver, news rptr.

WPGW(AM)— Jan 14, 1951: 1440 khz; 500 w-D, 35 w-N, DA-1. TL: N40 26 10 W85 00 56. Hrs open: 1891 W. State Rd 67, 47371. Phone: (260) 726-8729. Fax: (260) 726-4311. E-mail: wpgw@jayco.net Licensee: WPGW Inc. (acq 8-1-74). Population served: 25,000 Rgnl. Network: AgriAmerica. Natl. Rep: Rgnl Reps. Format: Adult contemp. Target aud: General. ♦Robert A. Weaver, pres & gen mgr.

WPGW-FM— May 19, 1975: 100.9 mhz; 4.6 kw. 180 ft TL: N40 26 10 W85 00 54. Stereo. 25,000 Format: Country. Target aud: General. ♦Jeff Overholser, disc jockey; Laurette Horn, traf mgr & disc jockey.

Princeton

WRAY(AM)— Dec 16, 1950: 1250 khz; 1 kw-D, 59 w-N. TL: N38 21 25 W87 35 25. Stereo. Hrs opn: 24 Box 8, 1900 W. Broadway, 47670-0008. Phone: (812) 386-1250. Fax: (812) 386-6249. E-mail: wray@wrayradio.com Web Site: www.wrayradio.com. Licensee: Princeton Broadcasting Co. Inc. Population served: 250,000 Format: News, talk. News staff: 3. Target aud: 25-54. ♦Richard Langford, pres; Lynn Roach, gen sls mgr, prom mgr & spec ev coord; Stephen R. Langford, gen mgr, opns mgr & progmg dir; Cliff Ingram, news dir; Floyd Turner, chief of engrg; Dave Kunkel, disc jockey; Paul Viton, disc jockey.

WRAY-FM— May 15, 1960: 98.1 mhz; 50 kw. 420 ft TL: N38 21 25 W87 35 25. Stereo. 24 Web Site: www.wrayradio.com. Format: Country. ♦Dave Kunkel, progmg dir & disc jockey; Charlene K. Garrison, traf mgr; Paul Viton, disc jockey.

WSJD(FM)— Oct 1, 1994: 100.5 mhz; 6 kw horiz, 5.5 kw vert. Ant 328 ft TL: N38 23 24 W87 34 23. Stereo. Hrs opn: 24 606 Market St., Mount Carmel, IL, 62863. Phone: (618) 262-4102. Fax: (618) 262-4103. E-mail: wsjd@midwest.net Licensee: WSJD Inc. Group owner: Southern Wabash Communications Corp. (acq 8-3-01). Population served: 650,000 Format: Oldies. News staff: one; News: 20 hrs wkly. Target aud: 25 plus. ♦Randolph V. Bell, pres; Sally Dorgan Potts, exec VP; Kevin Madden, gen mgr.

Rensselaer

WLQI(FM)—Listing follows WRIN(AM).

***WPUM(FM)**— Sept 6, 1977: 90.5 mhz; 10 w. 190 ft TL: N40 55 12 W87 09 27. Stereo. Hrs opn: 24 Box 651, St. Joseph's College, 47978. Phone: (219) 866-6000. E-mail: wpum@saintjoe.edu Licensee: St. Joseph's College. (acq 8-1-76). Population served: 5,000 Natl. Network: Superadio. Format: Rock. News staff: one; News: 10 hrs wkly. Target aud: 18-34; general. Spec prog: Country 3 hrs, classical 3 hrs, blues 3 hrs, talk 1 hr wkly. ♦Sally Nesselroad, stn mgr.

WRIN(AM)— Sept 14, 1963: 1560 khz; 1 kw-D, 500 w-CH. TL: N40 57 41 W87 09 07. Hrs open: Box D, 47978. Secondary address: 560 W. Amster Rd. Phone: (219) 866-5105. Phone: (219) 866-4104. Fax: (219) 866-5106. Web Site: www.1560wrin.com. Licensee: Brothers Broadcasting Corp. (acq 6-18-86). Population served: 50,000 Rgnl. Network: AgriAmerica. Natl. Rep: Rgnl Reps. Law Firm: Leventhal, Senter & Lerman. Format: Adult standards. News staff: one; News: 10 hrs wkly. Target aud: 30+. Spec prog: Farm 12 hrs, gospel 2 hrs, relg 10 hrs wkly. ♦John Balvich, pres & gen mgr; Connie Graham Luthi, sls dir; Bob Burt, progmg dir; Bob Kurtz, news dir.

WLQI(FM)—Co-owned with WRIN(AM). 1973: 97.7 mhz; 3.3 kw. 300 ft TL: N40 58 14 W87 09 10. Stereo. 24 Web Site: www.1560wrin.com.100,000 Natl. Network: Jones Radio Networks. Law Firm: Cohn & Marks. Format: Classic hits. News staff: one; News: ndws progmg 10 hrs wkly. Target aud: 25 plus.

Richmond

***WECI(FM)**— September 1964: 91.5 mhz; 400 w. 106 ft TL: N39 49 22 W84 54 39. Stereo. Hrs opn: 6 AM-3 AM 801 National Rd. W., Drawer 45, 47374. Phone: (765) 983-1246. Fax: (765) 983-1641. E-mail: hennja@earlham.edu Web Site: www.earlham.edu/~weci. Licensee: Earlham College. Population served: 70,000 Format: Class, country, var/div. Spec prog: Bluegrass/folk 19 hrs, classic rock 16 hrs, progressive 18 hrs wkly. ♦Alice Edgerton, stn mgr; Kate Galligan, progmg dir; Sam Robinson, news dir.

WFMG(FM)—Listing follows WKBV(AM).

WHON(AM)—Centerville, Feb 17, 1964: 930 khz; 500 w-D, 114 w-N, DA-2. TL: N39 53 33 W84 56 09. Hrs open: 24 Box 1647, 47375. Phone: (765) 962-1595. Fax: (765) 966-4824. Web Site: www.whon930.com. Licensee: Brewer Broadcasting Corp. (group owner; (acq 11-20-97). Population served: 43,999 Natl. Rep: Rgnl Reps. Format: News/talk. News staff: one. Target aud: 35 plus. ♦Dave Strycker, gen mgr; Troy Derengowski, progmg dir.

WQLK(FM)—Co-owned with WHON(AM). Oct 15, 1973: 96.1 mhz; 50 kw. 350 ft TL: N39 53 33 W84 56 09. Stereo. 24 Web Site: www.kicks96.com. Format: Hot country. News staff: one. Target aud: 25-54. ♦Steve Baker, progmg dir.

WKBV(AM)— Sept 27, 1926: 1490 khz; 1 kw-U. TL: N39 49 30 W84 55 50. (CP: N39 49 41 W84 55 57). Hrs opn: 24 Box 1646, 2301 W. Main St., 47374. Phone: (765) 962-6533. Fax: (765) 966-1499. Licensee: Rodgers Broadcasting Corp. (group owner; (acq 8-4-97; with co-located FM). Population served: 500,000 Natl. Network: ABC, ESPN Radio. Format: News/talk. News staff: 2; News: 10 hrs wkly. Target aud: 25-54. Spec prog: Farm 4 hrs wkly. ♦David Rodgers, pres; Steve Frey, stn mgr & sls dir; Rick Duncan, progmg dir & local news ed; Bob Phillips, local news ed.

WFMG(FM)—Co-owned with WKBV(AM). Dec 17, 1960: 101.3 mhz; 50 kw. 280 ft TL: N39 49 30 W84 55 50. Stereo. 24 2301 W. Main St., 47374. Web Site: www.g1013.com.500,000 Format: Hot adult contemp. News staff: 2; News: 2 hrs wkly. Target aud: 18-44. Spec prog: Miami University Sports 8 hrs wkly. ♦Rick Duncan, opns mgr.

***WVXR(FM)**— Dec 24, 1988: 89.3 mhz; 4.2 kw. Ant 187 ft TL: N39 52 08 W84 47 47. Stereo. Hrs opn: 24 Box 793, New Albany, OH, 43054. Phone: (614) 855-9171. Fax: (614) 855-9280. Licensee: Christian Voice of Central Ohio Inc. (acq 5-15-2007; grpsl). Rgnl. Network: Ohio Radio Net. Format: Contemp Christian. ♦Dan Baughman, gen mgr.

Rising Sun

WSCH(FM)—See Aurora

Roann

WARU-FM—Licensed to Roann. See Peru

Roanoke

***WBNI-FM**— 1991: 94.1 mhz; 6 kw. Ant 328 ft TL: N40 58 51 W85 16 48. Hrs open: 24
Simulcast with WCKZ(FM) Orland 100%.
Box 8459, Fort Wayne, 46898-8459. Phone: (260) 452-1189. Fax: (260) 452-1188. E-mail: ccondron@nipr.fm Web Site: www.nipr.fm. Licensee: Northeast Indiana Public Radio Inc. Group owner: Summit City Radio Group (acq 2-7-2007; $1.75 million). Natl. Network: NPR. Law Firm: Dow, Lohnes & Albertson, PLLC. Format: Classical. ♦Bruce Haines, gen mgr; Karen Fraser, dev dir; Colleen Condron, progmg dir; Jeanette Dillon, news dir.

Rochester

***WQKV(FM)**— 2006: 88.5 mhz; 250 w. Ant 171 ft TL: N41 03 14 W86 16 12. Hrs open: 5700 West Oaks Blvd., Rocklin, CA, 95765. Phone: (916) 251-1600. Fax: (916) 251-1650. Licensee: Educational Media Foundation. Group owner: American Family Radio. (acq 3-23-2007; grpsl). Format: Christian. ♦Richard Jenkins, pres.

WROI(FM)— Aug 29, 1971: 92.1 mhz; 4.2 kw. 240 ft TL: N41 03 02 W86 15 39. Stereo. Hrs opn: 24 110 E. 8th St., 46975. Phone: (574) 223-6059. Fax: (574) 223-2238. E-mail: wroi@rtcol.com Web Site: www.wroifm.com. Licensee: Bair Communications Inc. (acq 10-21-92; FTR: 10-19-92). Population served: 65,000 Rgnl. Network: Brownfield, Network Indiana Natl. Rep: Rgnl Reps. Format: Oldies. News staff: one; News: 20 hrs wkly. Target aud: General. Spec prog: Farm 10 hrs, relg 6 hrs wkly. ♦Tom Bair, pres & gen mgr; Sue Bair, gen sls mgr; Matt Bair, progmg dir; Baron Imhoof, news dir.

Rockville

WAXI(FM)— August 1977: 104.9 mhz; 3 kw. Ant 400 ft TL: N39 43 44 W87 17 56. Stereo. Hrs opn: 24 1301 Ohio St., Terre Haute, 47807. Phone: (812) 234-9770. Fax: (812) 238-1576. E-mail: waxi@waxi.com Web Site: www.waxifm.com. Licensee: Crossroads Investments LLC. Group owner: Crossroads Communications Inc. (acq 4-20-98; $485,000). Population served: 100,000 Natl. Network: ABC. Rgnl. Network: AgriAmerica, Network Indiana. Natl. Rep: Rgnl Reps. Law Firm: Booth, Freret, Imlay & Tepper. Format: Oldies. News staff: one; News: 10 hrs wkly. Target aud: 35 plus; local to Parke, Vermillion counties, affluent boomers & seniors in Terre haute market. Spec prog: Gospel 2 hrs, Cubs baseball & sports. ♦Mike Petersen, pres; Doug Edge, gen mgr & prom dir; Brad Simon, prom dir; John Sigman, progmg dir; Tom Mulvihill, chief of engrg.

Royal Center

WHZR(FM)— Oct 16, 1989: 103.7 mhz; 6 kw. 328 ft TL: N40 48 43 W86 21 56. Stereo. Hrs opn: 24 Box 103, Logansport, 46947. Phone: (574) 732-1037. Fax: (574) 739-1037. E-mail: whza@verizon.net Licensee: Mid-America Radio Group of Logansport-Peru Inc. Group owner: Mid-America Radio Group Inc. (acq 5-1-95; $450,000; FTR: 6-5-95). Format: Country. News staff: one; News: 6 hrs wkly. Target aud: 18-49; mass appeal. ♦David Keister, pres; Dan Keister, gen mgr & news dir.

Rushville

WIFE-FM— Aug 5, 1971: 94.3 mhz; 1.05 kw. 561 ft TL: N39 42 22 W85 29 41. Stereo. Hrs opn: 102 N. Perkins St., 46173. Phone: (765) 932-3983. Phone: (765) 932-3409. Fax: (765) 938-1916. Licensee: Rodgers Broadcasting Corp. (acq 7-5-2007; $1.5 million). Population served: 50,000 Rgnl. Network: Brownfield. Natl. Rep: Christal, Rgnl Reps. Format: Country. Target aud: 35 plus. Spec prog: Farm 18 hrs wkly. ♦David Rodgers, pres; Scott Huber, gen mgr; Kevin Stone, gen sls mgr; Doug Raab, progmg dir, engrg mgr & chief of engrg; Martha Swain, chief of engrg & traf mgr.

Salem

WSLM(AM)— Feb 14, 1953: 1220 khz; 5 kw-D, 384 w-N, DA-2. TL: N38 36 55 W86 05 10. Hrs open: 16 Box 385, 47167. Phone: (812) 883-5750. Fax: (812) 883-2797. Licensee: Don H. Martin. Population served: 24,000 Rgnl. Network: AgriAmerica, Network Indiana, Tribune. Natl. Rep: Rgnl Reps. Law Firm: Baraff, Koerner & Olender. Format: Farm, C&W, gospel. News staff: 5; News: 12 hrs wkly. Target aud: 21-70. ♦Don H. Martin, pres, gen mgr & gen sls mgr; Becky L. White, adv dir; David Stuart, mus dir; J.R. Martin, stn mgr, prom mgr & chief of engrg.

WSLM-FM— 1992: 97.9 mhz; 3 kw. 220 ft TL: N38 38 07 W86 10 37. Stereo. 18 Licensee: Rebecca L. White. Format: News, talk shows. Target aud: 18-65. ♦David Stuart, progmg dir; Don H. Martin, adv mgr & progmg mgr.

WZKF(FM)— 1962: 98.9 mhz; 50 kw. 300 ft TL: N38 35 59 W86 05 17. (CP: Ant 492 ft. TL: N38 21 56 W85 58 55). Stereo. Hrs opn: 24 4000 #1 Radio Dr., Louisville, KY, 40218-4568. Phone: (502) 479-2222. Fax: (502) 479-2223. Web Site: www.kisslouisville.com. Licensee: CC Licenses LLC. Group owner: Clear Channel Communications Inc. (acq 12-31-96). Population served: 7,000 Format: CHR. Target aud: 18-54; country music listeners. ♦Kevin Hughes, gen mgr.

Santa Claus

WAXL(FM)— July 30, 1996: 103.3 mhz; 6 kw. 462 ft TL: N38 12 31 W86 54 00. Stereo. Hrs opn: 24 Box 1009, Jasper, 47547. Secondary address: 501 Old State Rd., Huntingburg 47547. Phone: (812) 683-1215. Phone: (800) 522-1033. Fax: (812) 683-5891. E-mail: waxl@psci.net Web Site: www.dcbroadcasting.com.ABC Licensee: Dubois County Broadcasting Inc. Group owner: DCBroadcasting Inc. (acq 7-25-97). Natl. Network: ABC. Rgnl. Network: Network Indiana, Brownfield Law Firm: Miller & Miller, P.C. Wire Svc: AP Format: Adult contemp. News staff: one; News: 3 hrs wkly. Target aud: 24-49. Spec prog: Agriculture 3 hrs wkly. ♦Paul Knies, pres; Bill Potter, gen mgr; Ron Spaulding, sls dir & gen sls mgr; Pat Oxley, progmg VP & progmg dir.

Scottsburg

WMPI(FM)— Dec 16, 1966: 105.3 mhz; 2.2 kw. 511 ft TL: N38 37 12 W85 45 15. Stereo. Hrs opn: 24 Box 270, 22 E. McClain Ave., 47170. Phone: (812) 752-5612. Fax: (812) 752-2345. E-mail: Ray@scottsburg.com Web Site: www.i1053.com. Licensee: D. R. Rice Broadcasting Inc. (acq. 1987). Population served: 50,000 Natl. Rep: Rgnl Reps. Format: C&W. News staff: one; News: 5 hrs wkly. Target aud: 25-54. ♦Donald R. Rice, pres; Raymond Rice, gen mgr; Tom Cull, stn mgr; John Ross, progmg dir & news dir; Steve Woodruff, chief of engrg.

Seelyville

WWSY(FM)— Sept 12, 1996: 95.9 mhz; 4.1 kw. Ant 397 ft TL: N39 34 29 W87 24 06. Hrs open: 824 S. 3rd St., Terre Haute, 47807. Phone: (812) 232-4161. Fax: (812) 234-9999. Web Site: www.y959thevalley.com. Licensee: Midwest Communications Inc. (acq 6-13-2005; $3.39 million with WMGI(FM) Terre Haute). Natl. Network: Jones Radio Networks. Rgnl. Network: Jones Satellite Audio. Natl. Rep: Christal. Format: Var rock. Target aud: 35-54. ♦Kathleen Walker, sls VP & gen sls mgr; Karl Wertzler, mktg mgr; Chad Edwards, progmg dir; Jerry Arnold, chief of engrg.

Seymour

***WJLR(FM)**— August 1995: 91.5 mhz; 5.6 kw. Ant 351 ft TL: N38 49 23 W85 47 24. Stereo. Hrs opn: 24 2351 Sunset Blvd., Suite 170-218, Rocklin, CA, 95765. Phone: (916) 251-1600. Fax: (916) 251-1650. Licensee: Educational Media Foundation. (acq 11-9-2004; $150,000). Natl. Network: K-Love. Law Firm: Shaw Pittman LLP. Format: Christian info & educ. Target aud: 30-80; family oriented. ♦Richard Jenkins, pres; Keith Whipple, dev dir; Eric Allen, natl sls mgr; David Pierce, progmg mgr; Ed Lenane, news dir; Sam Wallington, engrg dir; Karen Johnson, news rptr; Marya Morgan, news rptr; Richard Hunt, news rptr.

WQKC(FM)—Listing follows WZZB(AM).

WZZB(AM)— Nov 4, 1949: 1390 khz; 1 kw-D, 74 w-N. TL: N38 58 23 W85 53 20. Hrs open: 24 Box 806, 47274. Secondary address: 1534 Ewing St. 47274. Phone: (812) 522-1390. Fax: (812) 522-9541. E-mail: wzzb@comcast.net Licensee: SCI Broadcasting Inc. Group owner: Susquehanna Radio Corp. (acq 5-25-01; grpsl). Population served: 97,000 Natl. Network: USA, Jones Radio Networks. Rgnl. Network: AgriAmerica, Network Indiana. Rgnl rep: Rgnl Reps. Law Firm: Dow, Lohnes & Albertson. Format: Full service, news, sports, adult contemp, oldies. News staff: 2; News: 17 hrs wkly. Target aud: 25 plus; community oriented. Spec prog: Farm 2 hrs, relg 6 hrs wkly. ♦Blair W. Trask, pres, gen mgr & gen sls mgr; Bud Shippee, opns dir, progmg dir & news dir; Bob Hawkins, chief of engrg.

WQKC(FM)—Co-owned with WZZB(AM). Feb 23, 1961: 93.7 mhz; 25 kw horiz, 24.5 kw vert. Ant 699 ft TL: N38 58 22 W86 10 03. (CP: COL Sellersburg. 93.9 mhz; 2.67 kw, ant 499 ft. TL: N38 15 21.7 W85 45 29.1). Stereo. 24 500,000 Natl. Network: Jones Radio Networks. Format: Country. News staff: one; News: 7 hrs wkly. Target aud: General. ♦Bud Shippee, pub affrs dir & local news ed.

Shelbyville

WLHK(FM)— Nov 6, 1964: 97.1 mhz; 23 kw. 739 ft TL: N39 40 02 W86 01 51. Stereo. Hrs opn: 24 One Emmis Plaza, 40 Monument Cir., Suite 600, Indianapolis, 46204. Phone: (317) 266-9700. Fax: (317) 684-2021. Web Site: www.real971.com. Licensee: Emmis Radio License LLC. Group owner: Emmis Communications Corp. (acq 6-81). Population served: 1,100,000 Natl. Rep: D & R Radio. Law Firm: Gardner, Carton & Douglas. Format: Country. News staff: one. Target aud: 25-54; female. ♦Tom Severino, gen mgr; David Edgar, opns dir, progmg dir & news dir; Mike Cortese, gen sls mgr; Dave Hood, chief of engrg; Shelly Grimes, traf mgr.

WSVX(AM)— Jan 14, 1961: 1520 khz; 1 kw-D, 250 w-N, DA-2. TL: N39 33 25 W85 46 18. Hrs open: 2356 N. Morristown Rd., 46176. Phone: (317) 398-2200. Fax: (317) 392-3292. E-mail: info@wsvx.com Web Site: www.wsvx.com. Licensee: RSE Broadcasting LLC. (acq 11-4-99). Population served: 300,000 Format: Top-40. ♦Scott A. Huber, gen mgr; Kevin Stone, gen sls mgr & news dir; Doug Raab, progmg dir; Doug Raad, chief of engrg; Martha Swain, traf mgr.

South Bend

***WAUS(FM)**—Berrien Springs, MI) 1971: 90.7 mhz; 50 kw. 492 ft TL: N41 57 42 W86 21 02. Stereo. Hrs opn: 24 Waus Berrien Spring, Berrien Springs, MI, 49104-0240. Phone: (269) 471-3400. Fax: (269) 471-3804. E-mail: waus@andrews.edu Web Site: www.waus.org. Licensee: Andrews Broadcasting Corp. Population served: 200,000 Natl. Network: PRI. Law Firm: Donald E. Martin. Format: Class. News: 3 hrs wkly. Target aud: 35 plus; listeners with interest in classical music. Spec prog: Relg 10 hrs wkly. ♦Niels-Erik Andreasen, chmn; Sharon Dudgeon, gen mgr; Bill Brent, opns dir.

WBYT(FM)—See Elkhart

WDND(AM)— 1944: 1490 khz; 1 kw-U. TL: N41 41 38 W86 13 50. Hrs open: 24 3371 Cleveland Rd., Suite 310, 46628. Phone: (574) 273-9300. Fax: (574) 273-9090. Web Site: www.artisticradio.com/wdnd.htm. Licensee: Artistic Media Partners Inc. (group owner; (acq 10-22-98; $6,123,180 with co-located FM). Population served: 238,000 Natl. Network: ESPN Radio. Natl. Rep: McGavren Guild. Format: Sports. Target aud: 25-54. ♦Arthur A. Angotti, pres; Jack Swart, gen mgr; Mike Sullivan, stn mgr; Joe Cook, gen sls mgr; Sean Stires, progmg dir.

WNDV-FM—Co-owned with WDND(AM). 1962: 92.9 mhz; 12.5 kw. 800 ft TL: N41 36 20 W86 12 45. Stereo. 24 Web Site: www.u93.com.255,800 Format: CHR. Target aud: 25-44; women. Spec prog: Notre Dame football & basketball, Rick Dees 4 hrs wkly. ♦Karen Rite, progmg dir.

***WETL(FM)**— Nov 17, 1958: 91.7 mhz; 3 kw. 200 ft TL: N41 37 24 W86 14 15. Stereo. Hrs opn: 1902 S. Fellows, 46613. Phone: (574) 283-8432. Fax: (574) 283-8405. E-mail: jovermyer@sbcsc.k12.in.us Licensee: South Bend Community School Corp. (acq 11-17-58). Population served: 135,000 Format: Educ, instructional. Target aud: General; student in the South Bend community school and community. ♦Anita Brown, gen mgr; John Overmyer, progmg dir; Allen Wujcik, chief of engrg.

WHLY(AM)— Dec 22, 1947: 1580 khz; 10 kw-D, 500 w-N, DA-N. TL: N41 41 09 W86 09 53. Hrs open: 24 Box 1322, Elkhart, 46515. Phone: (574) 361-4618. Licensee: Times Communications Inc. Population served: 250,000 Natl. Network: EWTN Radio. Format: Catholic.

WHME(FM)— January 1968: 103.1 mhz; 3 kw. 300 ft TL: N41 36 11 W86 12 51. Hrs open: 24 61300 Ironwood Rd., 46614. Phone: (574) 291-8200. Fax: (574) 291-9043. E-mail: thale@lesea.com Web Site: www.lesea.com. Licensee: Le Sea Broadcasting Corp. Group owner: Le Sea Broadcasting Population served: 302,000 Law Firm: Gardner, Carton & Douglas. Format: Adult contemp Christian. News: 3 hrs wkly. Target aud: 24-36; general. ♦Tony Hale, CFO; Anna Riblet, stn mgr, gen sls mgr & natl sls mgr; Wes Hylton, chief of engrg & disc jockey. Co-owned TV: WHME-TV affil.

WNSN(FM)—Listing follows WSBT(AM).

WOZW(FM)—See Goshen

WRBR-FM— 1965: 103.9 mhz; 3 kw. 328 ft TL: N41 41 53 W86 09 20. Stereo. Hrs opn: 237 W. Edison Rd., Suite 200, Mishawaka, 46545. Phone: (574) 258-5483. Fax: (574) 258-0930. Web Site: www.wrbr.com. Licensee: Hicks Broadcasting of Indiana LLC (acq 6-5-2002; $840,879). Natl. Rep: Christal. Format: Hard rock. Target aud: 25-54; affluent, older people. ♦Abe Thompson, gen mgr; Alec Dillie, gen mgr & stn mgr; Gene Walker, opns mgr; Emily Wideman, gen sls mgr; Chuck Wright, progmg dir; George Trobridge, chief of engrg.

WSBT(AM)— April 1922: 960 khz; 5 kw-U, DA-2. TL: N41 37 00 W86 13 01. Hrs open: 24 300 W. Jefferson Blvd., 46601. Phone: (574) 233-3141. Fax: (574) 289-7382. Web Site: www.wsbtradio.com. Licensee: WSBT Inc. Group owner: Schurz Communications Inc. Population served: 211,500 Natl. Network: Fox News Radio. Natl. Rep: Katz Radio. Format: News/talk, sports. News staff: 3; News: 10 hrs wkly. Target aud: 25-54. Spec prog: Relg 2 hrs wkly. ♦Sally Brown, VP & gen mgr. Co-owned TV: WSBT-TV affil

WNSN(FM)—Co-owned with WSBT(AM). Aug 1, 1962: 101.5 mhz; 13 kw. 970 ft TL: N41 37 00 W86 13 01. Stereo. Web Site: www.sunny1015.com.816,000 Format: Adult contemp. News staff: one; News: 2 hrs wkly. Target aud: 25-54; adults. Co-owned TV: WSBT-TV affil

***WUBS(FM)**— 1993: 89.7 mhz; 1.5 kw. 79 ft TL: N41 40 51 W86 15 34. Hrs open: Box 3931, 46619. Phone: (574) 287-4700. Fax: (574) 287-2478. E-mail: broshane@wubs.org Licensee: Interfaith Christian Union Inc. Format: Inspirational. ♦Rev. Sylvester Williams Jr., gen mgr; Shane R. Williams, progmg dir; Brian Hoover, chief of engrg.

WUBU(FM)— October 1992: 106.3 mhz; 3 kw. Ant 292 ft TL: N41 44 11 W86 17 19. Hrs open: 24 237 Edison Rd., Suite 200, Mishawaka, 46545. Phone: (574) 258-5483. Fax: (574) 258-0930. Licensee: Partnership Radio LLC (acq 7-15-99). Natl. Network: Jones Radio Networks. Natl. Rep: Interep, McGavren Guild. Law Firm: Wiley, Rein & Fielding. Format: Smooth jazz. Target aud: 35-64; adults. ♦Abe Thompson, gen mgr; Alec Dille, stn mgr; Gene Walker, opns mgr; Emily Wideman, gen sls mgr; Chuck Wright, progmg dir; Greg Trobridge, chief of engrg.

WWLV(AM)— Nov 6, 1998: 1620 khz; 10 kw-D, 1 kw-N. TL: N41 38 11 W86 17 06. Hrs open: 3371 Cleveland Rd., Suite 310, 46628. Phone: (574) 273-9300. Fax: (574) 273-9090. E-mail: u93@u93.com Licensee: Artistic Media Partners Inc. (acq 3-31-2000; with WHLY(AM) South Bend). Format: Sports. Target aud: 25-54. ♦Jack Swart, gen mgr; Greg DeRue, stn mgr; Carrie Jones, natl sls mgr; Teresa Holden, prom dir; Chili Walker, progmg dir; Bob Henning, chief of engrg; Rita Kinzie, traf mgr.

South Whitley

WLZQ(FM)— Dec 2, 1992: 101.1 mhz; 6 kw. Ant 328 ft TL: N41 04 42 W85 31 20. Stereo. Hrs opn: 24 Box 5570, Ft. Wayne, 46895. Phone: (260) 482-8500. E-mail: q101@wlzq.com Licensee: Larko Communications. Population served: 130,000 Natl. Network: ABC. Format: Hot adult contemp. Target aud: 25-44. ♦Chris Larko, CEO & gen mgr.

Speedway

WYJZ(FM)—Licensed to Speedway. See Indianapolis

Spencer

WCLS(FM)— Sept 15, 1983: 97.7 mhz; 6 kw. Ant 328 ft TL: N39 13 22 W86 38 40. Stereo. Hrs opn: 201 N. VanDalia Ave., 47460. Phone: (812) 829-9393. Fax: (812) 829-9747. Licensee: Mid-America Radio of Indiana Inc. Group owner: Mid-America Radio Group Inc. (acq 11-13-2002; $321,100). Natl. Network: Westwood One. Format: Mainstream country. Spec prog: Relg 6 hrs wkly. ♦Ruth Ann Arney, stn mgr; Tony Kale, opns mgr; Johnnie Robbins, gen sls mgr & mktg; Monica Witt, progmg dir & traf mgr; Steve Vail, news dir; Steve Ross, chief of engrg.

Sullivan

WNDI(AM)— Oct 7, 1963: 1550 khz; 250 w-D. TL: N39 04 32 W87 23 57. Hrs open: 556 E. State Rd. 54, 47882. Phone: (812) 268-6322. Fax: (812) 268-6652. Licensee: JTM Broadcasting Corp. (acq 7-13-94; $237,000 with co-located FM; FTR: 8-1-94) Population served: 200,000 Format: Country. Target aud: 24-54. Spec prog: Farm 6 hrs wkly. ♦John Montgomery, gen mgr.

WNDI-FM— Aug 10, 1982: 95.3 mhz; 6 kw. Ant 328 ft TL: N39 09 36 W87 32 32.4,683

Syracuse

WAWC(FM)— May 31, 1991: 103.5 mhz; 3 kw. 328 ft TL: N41 22 57 W85 41 35. Stereo. Hrs opn: 24 216 W. Market St., Suite 1, Warsaw, 46580. Phone: (574) 457-8181. Phone: (800) 779-1094. Fax: (574) 457-4488. E-mail: bill@hoosier1035.com Licensee: Talking Stick Communications LLC. (acq 11-1-2006; $600,000). Population served: 120,000 Natl. Network: CBS. Rgnl. Network: Network Indiana. Format: Adult contemp. News staff: one; News: 7 hrs wkly. Target aud: 25-54; people in Kosciusko, Elkhart & Noble counties. Spec prog: Relg 4 hrs wkly. ♦Patrick Brown, gen mgr; Jay Michaels, progmg dir; Bill Dixon, news dir; Stacey Page, news dir; Brent Randall, pub affrs dir & sports cmtr; Greg Stoddard, chief of engrg; Jack Didier, chief of engrg.

Tell City

WTCJ(AM)— Feb 1, 1948: 1230 khz; 1 kw-U. TL: N37 56 16 W86 45 28. Hrs open: 24 1115 Tamarack Rd., Suite 500, Owensboro, KY, 42301. Phone: (270) 683-5200. Fax: (270) 688-0108. Web Site: www.tellcityradio.com. Licensee: Hancock Communications Inc. Group owner: The Cromwell Group Inc. (acq 12-20-99; $25,000). Population served: 47,200 Natl. Network: ABC. Natl. Rep: Rgnl Reps. Format: Timeless classics. News staff: one. Target aud: 25-54; community-oriented listeners. Spec prog: Gospel 6 hrs wkly. ♦Bayard Walters, pres; Kevin Riecke, gen mgr & gen sls mgr; Jeff Morgan, progmg dir & news dir.

WTCJ-FM— May 2001: 105.7 mhz; 4.8 kw. Ant 364 ft TL: N37 55 33 W86 43 19.24 47,200 Format: Classic rock.

Terre Haute

WBOW(AM)— May 23, 1958: 1300 khz; 500 w-D, 75 w-N. TL: N39 28 01 W87 25 34. Hrs open: 24
Rebroadcasts WSDX(AM) Brazil.
1301 Ohio St., 47807. Phone: (812) 234-9770. Fax: (812) 238-1576. E-mail: score@espnsportsradio.com Web Site: www.espnsportsradio.com. Licensee: Crossroads Investments LLC. Group owner: Crossroads Communications Inc. (acq 9-10-97; $57,500 assumption of debt). Population served: 70,286 Natl. Network: ESPN Radio. Natl. Rep: Roslin. Rgnl rep: Rgnl Reps. Law Firm: Booth, Freret, Imlay & Tepper. Format: All sports. News: 5 hrs wkly. Target aud: 25-64; Sports Fans. ♦Mike Petersen, pres; Doug Edge, gen mgr; Brad Simon, prom dir; John Sherman, progmg dir & news dir; Tom Mulvihill, chief of engrg.

WBOW-FM— Sept 11, 1962: 102.7 mhz; 28 kw. 659 ft TL: N39 20 13 W87 28 00. Stereo. Hrs opn: 24 1301 Ohio St., 47807. Phone: (812) 234-9770. Fax: (812) 238-1576. E-mail: wbow@literock1027.com Web Site: www.literock1027.com. Licensee: Crossroads Investments LLC (acq 3-14-03; $2.09 million). Natl. Rep: Roslin. Rgnl rep: Rgnl Reps Law Firm: Booth, Freret, Imlay & Tepper. Format: Soft adult contemp. News: 2 hrs wkly. Target aud: Adults 25-54. ♦Mike Petersen, pres; Doug Edge, gen mgr; Brad Simon, prom dir; Chris Carter, progmg dir; Tom Mulvihill, chief of engrg.

***WCRT-FM**— January 1992: 88.5 mhz; 550 w. 308 ft TL: N39 30 14 W87 26 37. Hrs open: 2108 W. Springfield, Champaign, IL, 61821. Phone: (217) 359-8232. Fax: (217) 359-7374. E-mail: wbgl@wbgl.org Web Site: www.wbgl.org. Licensee: Illinois Bible Institute. Format: Adult contemp Christian. ♦Jeff Scott, stn mgr; Jennifer Briski, prom dir; Ryan Springer, progmg dir; Joe Buchanan, mus dir.

***WHOJ(FM)**— 1997: 91.9 mhz; 1 kw. Ant 95 ft TL: N39 28 06 W87 23 56. Hrs open: Covenant Network, 4424 Hampton Ave., St. Louis, MO, 63109. Phone: (314) 752-7000. Web Site: www.covenantnet.net. Licensee: Covenant Network. (acq 3-30-2004; $112,500 with KBKC(FM) Moberly, MO). Format: Christian, relg, talk. ♦Tony Holman, gen mgr.

***WISU(FM)**— Sept 13, 1964: 89.7 mhz; 13.5 kw. 512 ft TL: N39 30 26 W87 31 50. Stereo. Hrs opn: 11 AM-2 AM Rm. 217, 217 N. 6th St., 47809. Phone: (812) 237-3248. Phone: (812) 237-3252. Fax: (812) 237-8970. Fax: (812) 237-3241. E-mail: cmwisufm@ruby.indstate.edu Web Site: wisu.indstate.edu. Licensee: Indiana State University Board of Trustees. Population served: 200,000 Law Firm: Crowell & Moring. Format: Urban contemp, AOR. News: 4 hrs wkly. Target aud: 18-25; young professionals, students. ♦Joe Tenerelli, gen mgr; David Sabaini, progmg dir; Dan Watson, chief of engrg.

WMGI(FM)— June 13, 1960: 100.7 mhz; 50 kw. 500 ft TL: N30 27 22 W87 28 50. Stereo. Hrs opn: 824 S. 3rd St., 47807. Phone: (812) 232-4161. Fax: (812) 234-9999. E-mail: chad@1007mixfm.com Web Site: www.1007mixfm.com. Licensee: Midwest Communications Inc. (acq 6-13-2005; $3.39 million with WWSY(FM) Seelyville). Population served: 70,286 Natl. Network: Westwood One. Natl. Rep: Christal. Format: CHR. Target aud: 18-34. ♦Karl Wertzler, gen mgr & mktg mgr; Kathleen Walker, gen sls mgr; Chad Edwards, progmg dir; Jerry Arnold, chief of engrg.

***WMHD-FM**— 1981: 90.7 mhz; 160 w. 79 ft TL: N39 28 57 W87 19 33. Stereo. Hrs opn: 8 AM-2 AM 5500 Wabash Ave., 47803. Phone: (812) 872-6923. Fax: (812) 872-6926. E-mail: wmhd@wmhd.rose-hulman.edu Web Site: wmhd.rose-hulman.edu. Licensee: Rose Hulman Institute of Technology. Population served: 2,000 Format: Educ, AOR, progsv. News: 2 hrs wkly. Target aud: General; loc & college audience. Spec prog: Classical 4 hrs, bluegrass one hr, Jazz 2 hrs, contemp Christian 2 hrs wkly. ♦Brandon Inzego, gen mgr; Brooks Borchers, opns dir; Ben Braun, progmg dir.

WPFR(AM)— Jan 6, 1948: 1480 khz; 5 kw-D, 1 kw-N, DA-2. TL: N39 30 02 W87 23 10. Hrs open: 24 18889 N. 23 50th St., Dennison, IL, 62423. Phone: (217) 826-9673. E-mail: wpfr@joink.com Licensee: Word Power Inc. (acq 1-1-00; $350,000 donation). Population served: 70286 Natl. Network: Moody. Format: Christian. Target aud: 12+. ♦Paul Dean Ford, gen mgr; Mark S. Ford, opns VP; Dan Watson, chief of opns.

WTHI-FM— October 1948: 99.9 mhz; 50 kw. 494 ft TL: N39 27 57 W87 24 12. Hrs open: 24 Box 1486, 47808. Secondary address: 918 Ohio St. 47808. Phone: (812) 232-9481. Fax: (812) 234-0089. E-mail: jconner@wthi.emmis.com Web Site: www.hi99.com. Licensee: Emmis Radio License LLC. Group owner: Emmis Communications Corp. (acq 1998 grpsl). Population served: 390,000 Natl. Network: ABC. Natl. Rep: Interep, D & R Radio. Format: Country. ♦James Conner, stn mgr; Barry Kent, opns mgr; Robert Rhodes, sls dir & gen sls mgr; Chris Perrot, prom dir & progmg mgr.

WWVR(FM)—See West Terre Haute

Union City

***WJYW(FM)**— 1997: 88.9 mhz; 4.1 kw. 285 ft TL: N40 11 32 W84 47 58. Hrs open: Box 445, 47390. Secondary address: 505 S. Division St., OH 45390. Phone: (937) 968-5633. Fax: (937) 968-3320. E-mail: office@899joyfm.com Web Site: www.889joyfm.com. Licensee: Positive Alternative Radio Inc. Natl. Network: Salem Radio Network. Law Firm: Booth, Freret, Imlay & Tepper. Format: Contemp Christian music. ♦Vernon H. Baker, CEO; Dan Franks, gen mgr.

Upland

***WTUR(FM)**— Sept 4, 1995: 89.7 mhz; 150 w. 112 ft TL: N40 25 02 W85 29 31. Hrs open: 236 W. Reade Ave., 46989-1001. Phone: (765) 998-5263. Phone: (765) 998-2751. Fax: (765) 998-4810. Web Site: www.tayloru.edu. Licensee: Taylor University. Format: Contemp Christian music. Target aud: College age. ♦Sonja Paul, stn mgr; Tim Walters, progmg dir.

Valparaiso

WAKE(AM)— Nov 4, 1964: 1500 khz; 1 kw-D, 25 w-N, DA-2. TL: N41 26 36 W87 02 54. Hrs open: 24 2755 Sager Rd., 46383. Phone: (219) 462-6111. Fax: (219) 462-4880. Licensee: Porter County Broadcasting Holding Corp. LLC. Group owner: Porter County Broadcasting Corp. Population served: 500,000 Law Firm: Miller & Fields, P.C. Format: Adult pop standards. News staff: 2; News: 21 hrs wkly. Target aud: 30 plus; community oriented, middle to middle-upper class. ♦Leigh Ellis, chmn, pres & gen mgr; O.J. Jackson, gen sls mgr; Don Clark, progmg dir; Laura Waluszko, news dir; Carl Fletcher, chief of engrg; Jennifer Malmquist, traf mgr.

WLJE(FM)—Co-owned with WAKE(AM). Oct 6, 1967: 105.5 mhz; 1.25 kw. 513 ft TL: N41 31 28 W87 01 08. Stereo. Phone: (219) 462-8125. Web Site: www.indiana105.com.800,000 Format: Country. Target aud: 25-55; family, middle income.

***WVUR-FM**— Sept 25, 1966: 95.1 mhz; 36 w. 125 ft TL: N41 27 57 W87 02 29. Stereo. Hrs opn: 24 1809 Chapel Dr., 46383. Phone: (219) 464-5383. Fax: (219) 464-6742. E-mail: wvur@valpo.edu Web Site: www.valpo.edu/wvur. Licensee: The Lutheran University Association Inc. Population served: 350,000 Format: Free-form. News staff: 2; News: 8 hrs wkly. Target aud: 18-34. Spec prog: Class 3 hrs, jazz 3 hrs, urban contemp 3 hrs, metal 3 hrs, classic rock 3 hrs wkly. ♦Ken LaVicka, gen mgr; Lauren LaVicka, gen sls mgr; Rachel Cooper, progmg dir; Ben Hampton, news dir; Rich Robertson, chief of engrg.

Van Buren

WCJC(FM)—Licensed to Van Buren. See Marion

Veedersburg

WSKL(FM)— July 15, 1999: 92.9 mhz; 4.5 kw. 269 ft TL: N40 08 46 W87 27 15. Hrs open: 24 Box 67, Danville, IL, 61834. Phone: (765) 793-4823. Fax: (765) 793-4644. E-mail: fmkool929@aol.com Web Site: www.koololdies.net. Licensee: Zona Communications Inc. (acq 10-29-99). Population served: 110,000 Natl. Network: AP Radio, Jones Radio Networks. Format: Oldies. News staff: one; News: 5 hrs wkly. Target aud: 35-65. ♦Rhea Benton-Weatherford, pres; Rhea Bonton-Weatherford, gen mgr; Greg Green, stn mgr; J.J. McKay, progmg dir; Tara Duncan, traf mgr.

Versailles

***WKRY(FM)**— Apr 11, 2003: 88.1 mhz; 600 w. Ant 233 ft TL: N39 03 55 W85 19 00. Hrs open: 24 825 Washington St., Columbus, 47201. Phone: (812) 738-3482. Fax: (812) 375-2555. Licensee: Good Shepherd Radio Inc. Format: Relg. ♦Keith Reising, CEO & gen mgr.

WXCH(FM)— Nov 15, 1984: 103.1 mhz; 3 kw. 328 ft TL: N39 10 38 W85 17 00. Stereo. Hrs opn:
Rebroadcasts WSCH(FM) Aurora 98%.
20 E. High St., Lawrenceburg, 47025-1820. Phone: (812) 438-2777. Fax: (812) 438-3495. E-mail: wsch@one.net Licensee: Columbus Radio Inc. (acq 12-2-02; with WSCH(FM) Aurora). Population served: 24,360 Format: Country. ♦Dennis Drees, opns VP; Bob Shannon, news dir.

Vevay

WKID(FM)— Sept 6, 1974: 95.9 mhz; 2.7 kw. 308 ft TL: N38 50 12 W85 01 48. Stereo. Hrs opn: 24 118 W. Main St., 47043. Phone: (812) 427-9590. Fax: (812) 427-2492. Web Site: www.k959froggy.com. Licensee: Dial Broadcasting Inc. (acq 1996). Natl. Network: Jones Radio Networks. Rgnl rep: Regl Reps Law Firm: Kaye, Scholer, Fierman, Hays & Handler. Wire Svc: AP Format: Country. News: 7 hrs wkly. Target aud: 25-49; middle-income families. ♦Ken Trimble, gen mgr.

Vincennes

WAOV(AM)— Oct 22, 1940: 1450 khz; 1 kw-U. TL: N38 42 26 W87 29 42. Hrs open: 24 Box 242, 47591-0242. Phone: (812) 882-6060. Fax: (812) 885-2604. E-mail: waov@originalcompany.com Web Site: www.waovam.com. Licensee: Old Northwest Broadcasting Inc. Group owner: The Original Co. Inc. (acq 9-28-93; $250,000 with WWBL(FM) Washington; FTR: 10-18-93) Population served: 40,000 Natl. Rep: Rgnl Reps. Format: News/talk, sports. News staff: 2; News: 56 hrs wkly. Target aud: 25 plus. ♦Mark R. Lange, pres, gen mgr & progmg dir; Jim Evans, news dir & chief of engrg.

***WATI(FM)**— 2002: 89.9 mhz; 500 w. Ant 157 ft TL: N38 41 47 W87 26 27. Hrs open: Box 3206, American Family Radio, Tupelo, MS, 38803. Phone: (662) 844-8888. Fax: (662) 842-6791. Web Site: www.afr.net. Licensee: American Family Association. Group owner: American Family Radio Format: Relg (Christian), inspirational. ♦Marvin Sanders, gen mgr.

WFML(FM)— May 16, 1965: 96.7 mhz; 3 kw. 377 ft TL: N38 42 26 W87 29 42. Stereo. Hrs opn: 24 1002 N. First St., 47591. Phone: (812) 888-5830. Fax: (812) 882-2237. E-mail: kdoades@hot96wfml.com Web Site: hot96wfml.com. Licensee: The Vincennes University Foundation (acq 8-29-86). Population served: 113,000 Natl. Network: Fox News Radio. Rgnl. Network: Network Indiana, AgriAmerica. Rgnl rep: Rgnl Reps Wire Svc: AP Format: Contemp country. News staff: 2; News: 2 hrs wkly. Target aud: 18-54. ♦Phil Smith, stn mgr; Kim Donaldson, sls dir; Keith Doades, gen sls mgr; Dave Folly, prom dir; Kevin Watson, progmg dir; John Szink, news dir; Steve McClure, chief of engrg.

WUZR(FM)—Bicknell, June 4, 1991: 105.7 mhz; 1.8 kw. 426 ft TL: N38 43 47 W87 24 44. Stereo. Hrs opn: 24 Box 242, Historic Brevoort House, 522 Busseron St., 47591. Phone: (812) 882-6060. Fax: (812) 885-2604. E-mail: wuzr@originalcompany.com Web Site: www.wuzr.com. Licensee: The Original Co. Inc. (group owner; acq 4-20-98; $682,000). Format: Good time rock and roll oldies. News staff: one; News: 7 hrs wkly. Target aud: 25-54. Spec prog: Loc news, high school sports, Univ. of Evansville basketball. ♦Mark Lange, pres & gen mgr; Brad Deetz, opns dir; Michelle York, gen sls mgr; Dave Young, progmg dir.

***WVUB(FM)**— Dec 7, 1970: 91.1 mhz; 50 kw. Ant 500 ft TL: N38 39 06 W87 28 37. Stereo. Hrs opn: 1002 N. First St., 47591. Phone: (812) 888-5830. Phone: (812) 888-5354. Fax: (812) 882-2237. E-mail: blazerwvub@hotmail.com Licensee: Board of Trustees for Vincennes University. Natl. Network: PRI. Rgnl. Network: Network Indiana. Format: Hot adult contemp/CHR. Spec prog: Class 6 hrs wkly. ♦Phil Smith, stn mgr & gen sls mgr; Michael Woods, progmg dir; John Szink, news dir; Michael Murphy, chief of engrg. Co-owned TV: WVUT(TV) affil

WZDM(FM)— September 1988: 92.1 mhz; 4.1 kw. 400 ft TL: N38 43 18 W87 33 37. Stereo. Hrs opn: 24 Box 242, Historic Brevoort House, 522 Busseron St., 47591. Phone: (812) 882-6060. Fax: (812) 885-2604. E-mail: wzdm@originalcompany.com Web Site: www.wzdm.com. Licensee: The Original Co. Inc. (group owner) Format: Adult contemp. News staff: 2; News: 10 hrs wkly. Target aud: 25-54; upscale. ♦Mark R. Lange, pres & gen mgr; Michelle York, gen sls mgr; Dave Young, progmg dir.

Wabash

WJOT-FM— July 1, 1993: 105.9 mhz; 3 kw. 318 ft TL: N40 47 11 W85 49 19. Hrs open: 1360 S. Wabash St., 46992. Phone: (260) 563-1161. Fax: (260) 563-0883. E-mail: wjot@comtek.com Licensee: Mid-America Radio of Wabash Inc. Group owner: Mid-America Radio Group Inc. (acq 7-1-98; $190,000 with co-located AM). Natl. Network: Westwood One. Rgnl rep: Rgnl Reps. Law Firm: Fletcher, Heald & Hildreth. Format: Oldies. Target aud: 25-64. ♦Bill Barrows, opns dir, opns mgr & news dir; Wade Weaver, gen mgr, gen sls mgr & progmg dir; Deb Dale, pub affrs dir & traf mgr; Jack Elmore, chief of engrg.

WJOT(AM)— November 1971: 1510 khz; 250 w-D. TL: N40 47 11 W85 49 19.13,379 Natl. Network: Westwood One. ♦Wade Weaver, prom mgr & progmg dir.

WKUZ(FM)— Apr 1, 1965: 95.9 mhz; 4.2 kw. 394 ft TL: N40 41 54 W85 45 03. Stereo. Hrs opn: 24 Box 342, 1864 S. Wabash St., 46992. Phone: (260) 563-4111. Fax: (260) 563-4425. E-mail: wkuz@kconline.com Web Site: www.wkuz.com. Licensee: Upper Wabash Broadcasting Corp. Population served: 170,200 Natl. Network: USA. Rgnl. Network: Brownfield. Format: Country. News staff: one; News: 10 hrs wkly. Target aud: General. Spec prog: Farm 5 hrs wkly. ♦Toni Adams, pres, CFO & gen mgr; Dawn Hughes, stn mgr; Amber Childers, opns mgr, disc jockey & sls; Jack Elmore, mktg dir; John Morgan, progmg dir; Grant Miller, news dir; Paul Adams, chief of engrg.

Wadesville

***WRFM(FM)**— 2005: 90.1 mhz; 6 kw vert. Ant 285 ft TL: N37 56 03 W87 55 35. Hrs open: Box 846, Greenfield, 46140. Secondary address: 15 Wood St., Greenfield 46140-2162. Phone: (317) 467-1064. Fax: (317) 467-1065. E-mail: hensleym31@aol.com Licensee: Indiana Community Radio Corp. Format: Adult contemp, Christian. ♦Jennifer Cox-Hensley, pres & gen mgr.

Wakarusa

***WYBV(FM)**— 2006: 89.9 mhz; 1.75 kw. Ant 328 ft TL: N41 27 50 W85 49 22. Hrs open: Bible Broadcasting Network Inc., 11530 Carmel Commons Blvd., Charlotte, NC, 28226-3976. Phone: (704) 523-5555. Fax: (704) 522-1967. Web Site: www.bbnradio.org. Licensee: Bible Broadcasting Network Inc.

Walton

WFRR(FM)— 1995: 93.7 mhz; 6 kw. 328 ft TL: N40 43 30 W86 10 30. Stereo. Hrs opn: 24
Rebroadcasts WFRN-FM Elkhart 85%.
c/o WFRN-FM Box 307, Elkhart, 46515. Secondary address: 25802 CR 26, Elkhart 46517. Phone: (574) 875-5166. Phone: (574) 674-6626. Fax: (574) 875-6662. E-mail: moore@wfrn.com Web Site: www.wfrn.com. Licensee: Christian Friends Broadcasting Inc. Population served: 185,000 Natl. Network: USA. Natl. Rep: Salem. Law Firm: Reddy, Begley & McCormick. Format: Contemp Christian. News staff: one. Target aud: 25-54; general. ♦Edwin Moore, pres & gen mgr; James Carter, progmg dir; Joe Guadagnoli, gen sls mgr & mus dir; Don Wagner, news dir.

Warsaw

WLEG(FM)—Ligonier, June 10, 1991: 102.7 mhz; 2 kw. Ant 394 ft TL: N41 27 52 W85 44 40. Stereo. Hrs opn: 24 Box 699, Elkhart, 46515. Secondary address: 421 S. 2nd St., Elkhart 46516. Phone: (574) 389-5100. Fax: (574) 389-5101. E-mail: bwilliams@federatedmedia.com Web Site: www.ilovemyfroggy.com. Licensee: Pathfinder Communications Corp. Group owner: Federated Media (acq 9-26-2002; $550,000). Format: Hot adult contemp. ♦Kathy Uebler, gen mgr; Jeff Deweese, progmg dir; George Trobridge, chief of engrg.

WRSW(AM)— 1951: 1480 khz; 1 kw-D, 500 w-N. TL: N41 13 21 W85 50 17. Stereo. Hrs opn: 24 216 W. Market St., 46580. Phone: (574) 372-3064. Phone: (574) 267-3111. Fax: (574) 267-2230. Web Site: www.wrsw.net. Licensee: Talking Stick Communications LLC. (group owner; (acq 12-19-2003; $1.2 million with co-located FM). Population served: 60,000 Natl. Network: Westwood One. Rgnl. Network: Tribune, Network Indiana. Format: Sports. News staff: one; News: 18 hrs wkly. Target aud: General; 60% male, 40% female. Spec prog: Farm 11 hrs, Sp 2 hrs wkly. ♦Patrick Brown, gen mgr & progmg dir.

WRSW-FM— 1948: 107.3 mhz; 50 kw. 293 ft TL: N41 13 21 W85 50 17. Stereo. 90,000 Natl. Network: Westwood One. Natl. Rep: Rgnl Reps. Rgnl rep: Rgnl reps. Format: Oldies, Sp, classic rock. Target aud: General; affluent adults. Spec prog: Sp 2 hrs, farm 2 hrs wkly.

Washington

WAMW(AM)— January 1955: 1580 khz; 500 w-D, DA-D. TL: N38 39 04 W87 09 55. Hrs open: 800 W. National Hwy., 47501. Phone: (812) 254-6761. Fax: (812) 254-3940. E-mail: wamw@rtccom.net Web Site: www.wamwamfm.com. Licensee: Greene Electronics. Population served: 100,000 Natl. Network: ABC. Wire Svc: AP Format: Adult contemp. News staff: one. Target aud: 45 plus. ♦Dave Crooks, gen mgr; Andy Morrison, opns mgr & news dir; Macy Kalb, gen sls mgr; Taylor Brown, news dir.

WAMW-FM— Nov 20, 1989: 107.9 mhz; 3 kw. 328 ft TL: N38 38 47 W87 16 47. Stereo. 24 Web Site: www.wamwamfm.com.100,000 Natl. Network: ABC. Format: Soft adult contemp. News staff: one. Target aud: 25+; soft adult contemporary.

WWBL(FM)— February 1948: 106.5 mhz; 50 kw. 340 ft TL: N38 39 04 W87 09 55. Stereo. Hrs opn: Box 616, 47501-0616. Secondary address: Box 242, Vincennes 47591-0242. Phone: (812) 254-4300. Phone: (812) 882-6060. Fax: (812) 254-4361. Fax: (812) 885-2604. Web Site: www.wwbl.com. Licensee: Old Northwest Broadcasting Inc. Group owner: The Original Co. Inc. (acq 10-93; $250,000 with WAOV(AM) Vincennes; FTR: 10-18-93) Population served: 200,000 Natl. Network: ABC. Natl. Rep: Rgnl Reps. Format: Country. News staff: one; News: 15 hrs wkly. Target aud: 18 plus. Spec prog: Farm 15 hrs wkly. ♦Mark Lange, pres; Ken Booth, opns mgr.

West Lafayette

***WBAA(AM)**— Apr 4, 1922: 920 khz; 5 kw-D, 1 kw-N, DA-N. TL: N40 20 29 W86 53 01. Hrs open: Purdue University, 712 3rd St., 47907. Phone: (765) 494-5920. Fax: (765) 496-1542. E-mail: wbaa@wbaa.org Web Site: www.wbaa.org. Licensee: Purdue University. Natl. Network: NPR, PRI. Law Firm: Wiley & Rein. Format: Jazz, news/talk. News staff: 3; News: 20 hrs wkly. ♦Tim Singleton, gen mgr; Bette Carson, opns mgr; Laura Loy, dev dir; David Bunte, progmg dir; Maurie Mogridge, chief of engrg.

WBAA-FM— February 1993: 101.3 mhz; 5 kw. Ant 358 ft TL: N40 17 50 W86 54 05. Stereo. Natl. Network: NPR. Format: Classical, news. News staff: 3; News: 20 rs wkly.

WGLM(FM)— June 15, 1992: 106.7 mhz; 6 kw. 328 ft TL: N40 31 20 W86 58 57. Hrs open: 24 2700-A Kent Ave., 47906. Phone: (765) 497-9456. Fax: (765) 497-3299. E-mail: ron@wglm.fm Web Site: www.wglm.fm. Licensee: KVB Broadcasting. Natl. Network: CNN Radio. Format: Adult contemp. News staff: one; News: 8 hrs wkly. Target aud: 25-54. ♦Kelly Busch, CEO & gen mgr; Ron Schuessler, stn mgr, sls dir & gen sls mgr.

***WHPL(FM)**— Sept 10, 1993: 89.9 mhz; 2 kw. 328 ft TL: N40 17 50 W86 54 05. Hrs open: 24 1920 W. 53rd St., Anderson, 46013. Phone: (765) 449-0899. Fax: (765) 449-3025. E-mail: whpl@moody.edu Licensee: The Moody Bible Institute of Chicago. (group owner; acq 6-20-97). Population served: 100,000 Natl. Network: Moody. Law Firm: Southmayd & Miller. Format: Relg. News: 14 hrs wkly. Target aud: 35 plus; relg. ♦Ray Hashley, gen mgr.

WLFF(FM)—See Brookston

West Terre Haute

WWVR(FM)— Jan 20, 1967: 105.5 mhz; 3.3 kw. 314 ft TL: N39 27 15 W87 28 18. Hrs open: 6 AM-2 AM Box 1486, Terre Haute, 47808. Secondary address: 918 Ohio St., Terre Haute 47808. Phone: (812) 232-9481. Fax: (812) 234-0089. E-mail: jconner@wthi.emmis.com Web Site: www.1055theriver.net. Licensee: Emmis Radio License LLC. Group owner: Emmis Communications Corp. Population served: 500,000 Format: Classic rock. News: 6 hrs wkly. Target aud: 35-64. Spec prog: Gospel, news/talk, Black 6 hrs wkly. ♦James Conner, stn mgr & gen sls mgr; Chris Perrot, prom dir & prom mgr; Barry Kent, progmg dir; Jeff Tucker, chief of engrg.

Winamac

WFRI(FM)— 1998: 100.1 mhz; 6 kw. 328 ft TL: N41 02 21 W86 30 55. Stereo. Hrs opn: 24
Rebroadcasts WFRN-FM Elkhart 80%.
Box 307, Elkhart, 46515. Phone: (800) 522-9376. Fax: (219) 875-6662. E-mail: comments@wfrn.com Web Site: www.wfrn.com. Licensee: Progressive Broadcasting System Inc. (group owner). Natl. Network: USA. Format: Contemp Christian. Target aud: 25-54. ♦Edwin Moore, pres & gen mgr.

Winchester

WZZY(FM)— May 1967: 98.3 mhz; 3 kw. 300 ft TL: N40 05 23 W84 56 13. Stereo. Hrs opn: 24 2301 W. Main St., Richmond, 47374. Phone: (765) 962-6533. Fax: (765) 966-1499. E-mail: promotions @todaysmusicmix.com Web Site: www.todaysmusicmix.com. Licensee: Rodgers Broadcasting Corp. (group owner; acq 1-1-00). Population served: 300,000 Rgnl rep: Rgnl Reps. Format: Full service, adult contemp. News staff: 2; News: 10 hrs wkly. Target aud: 25-54; general. ♦David Rodgers, pres; Steve Frey, gen mgr; Rick Duncan, opns dir & progmg dir; Bob Phillips, news dir; Keith Wade, disc jockey.

Zionsville

***WITT(FM)—**Not on air, target date: unknown: 91.9 mhz; 1 kw. Ant 351 ft TL: N40 00 13 W86 28 14. Hrs open: 6218 Kingsley Dr., Indianapolis, 46220. Phone: (317) 251-3851. Licensee: Kids First Inc. ♦James E. Walsh, pres.

Iowa

Adel

***KIHS(FM)—** 2004: 88.9 mhz; 10 kw. Ant 154 ft TL: N41 36 12 W94 02 53. Hrs open: CSN International, 4002N. 3300E., Twin Falls, ID, 83301. Phone: (208) 734-6633. Fax: (208) 736-1958. E-mail: csn@csnradio.com Web Site: www.csnradio.com. Licensee: CSN International (group owner). Format: Christian. ♦Ray Garney, gen mgr.

Albia

KLBA-FM— June 15, 1995: 96.7 mhz; 10 kw. Ant 508 ft TL: N41 01 47 W92 47 12. Stereo. Hrs opn: 10 N. Clinton, 52531. Phone: (641) 932-2112. Fax: (641) 932-2113. Licensee: H&H Broadcasting Corp. Format: Country, hits of the 50s & 60s. ♦Harold Mick, pres; Larry Mikesell, gen mgr & gen sls mgr; Sue Mikesell, prom dir & progmg dir; Joe Milledge, chief of engrg.

Algona

KLGA(AM)— 1956: 1600 khz; 5 kw-D, 500 w-N, DA-2. TL: N43 03 52 W94 18 13. Hrs open: 24 Box 160, 50511. Secondary address: 2102 80th Ave. 50511. Phone: (515) 295-2475. Fax: (515) 295-3851. Web Site: www.waittmedia.com. Licensee: NRG License Sub. LLC. Group owner: Waitt Broadcasting Inc. (acq 10-31-2005; grpsl). Natl. Network: ABC. Rgnl. Network: Radio Iowa. Law Firm: Bryan Cave. Format: Adult contemp. News staff: one; News: 44 hrs wkly. Target aud: 25-54. Spec prog: Farm, news, weather. ♦Bob Ketchum, gen mgr; Dana Myee, progmg dir.

KLGA-FM— Aug 17, 1970: 92.7 mhz; 3.5 kw. 449 ft TL: N43 04 05 W94 12 08. Stereo. 6 AM-10:30 PM Web Site: www.waittmedia.com.

Alta

KBVU-FM— 1999: 97.5 mhz; 6 kw. Ant 315 ft TL: N42 38 05 W95 10 10. Hrs open: Buena Vista University, 610 W. 4th St., Storm Lake, 50588. Phone: (712) 749-1234. Fax: (712) 749-1211. Web Site: edge.bvu.edu. Licensee: Buena Vista University. Format: Alternative rock. Target aud: 18-25; college students. ♦Bruce Ellingson, gen mgr.

Ames

KASI(AM)— 1948: 1430 khz; 1 kw-D, 32 w-N. TL: N42 02 15 W93 41 21. Hrs open: 5 AM-midnight 415 Main St., 50010. Phone: (515) 232-1430. Fax: (515) 232-1439. Web Site: www.1430kasi.com. Licensee: Citicasters Licenses L.P. Group owner: Clear Channel Communications Inc. (acq 8-24-99; with co-located FM). Population served: 100,000 Natl. Network: ABC. Format: Oldies, news/talk. News staff: 2; News: 25 hrs wkly. Target aud: 25 plus. ♦Joel McCrea, gen mgr; Tony Calumet, gen sls mgr; Linda Thede, progmg dir & traf mgr; Trent Rice, news dir; Mike Stover, chief of engrg; B.J. Schaben, sports cmtr.

KCCQ(FM)—Co-owned with KASI(AM). June 20, 1968: 105.1 mhz; 25 kw. 328 ft TL: N42 04 33 W93 38 54. Stereo. 24 Web Site: 1051channelq.com.100,000 Format: CHR. Target aud: 18-40. ♦Linda Thede, opns dir & traf mgr; B.J. Schaben, sports cmtr.

KLTI-FM— June 2, 1967: 104.1 mhz; 100 kw. Ant 1,010 ft TL: N41 54 09 W93 54 15. Stereo. Hrs opn: 24 1416 Locust St., Des Moines, 50309. Phone: (515) 280-1350. Fax: (515) 280-3011. E-mail: jms720@aol.com Web Site: www.lite1041.com. Licensee: Saga Communications of Iowa LLC. Group owner: Saga Communications Inc. (acq 1-1-97; $3.2 million). Population served: 600,000 Natl. Rep: Katz Radio. Law Firm: Smithwick & Belendiuk. Format: Soft adult contemp. News staff: one. Target aud: Women; 25-54. ♦Ed Christian, CEO; Bill Wells, gen mgr; Jim Schaefer, opns dir; Celia Rodine, natl sls mgr; Tiffany Tauscheck, prom dir.

***KURE(FM)—** Apr 17, 1970: 88.5 mhz; 250 w. 100 ft TL: N42 01 24 W93 39 00. Stereo. Hrs opn: 24 1199 Friley Hall, Iowa State Univ., 50012. Phone: (515) 294-4332. Phone: (515) 294-9292. Fax: (515) 294-8093. E-mail: generalmanager@kure885.org Web Site: www.kure885.org. Licensee: Residence Associations Broadcasting Service Inc. Population served: 60,000 Format: Var/div. News: 10 hrs wkly. Target aud: 18-25; Iowa State Univ students & Ames community. ♦Rob McMahon, gen mgr; Rezza Rahmoni, opns dir; Katherine Beaver, mktg dir; James Bishop, progmg dir.

***WOI(AM)—** 1922: 640 khz; 5 kw-D, 1 kw-N, DA-N. TL: N41 59 34 W93 41 27. Hrs open: 24 2022 Communicatons Bldg., Iowa State Univ., 50011. Phone: (515) 294-2025. Fax: (515) 294-1544. Web Site: www.woi.org. Licensee: Iowa State University. Population served: 3,000,000 Natl. Network: NPR, PRI. Law Firm: Dow, Lohnes & Albertson. Format: News/talk. News staff: 3; News: 40 hrs wkly. Target aud: General. Spec prog: Jazz 7 hrs, blues 3 hrs wkly. ♦Gregory Geoffroy, pres; Donald Wirth, gen mgr; Dave Becker, progmg dir; David Knippel, chief of engrg.

WOI-FM— July 1, 1949: 90.1 mhz; 100 kw. 1,490 ft TL: N41 48 33 W93 36 53. Stereo. 24 Web Site: www.woi.org.340,000 Format: Class, jazz. News staff: 3; News: 12 hrs wkly. Target aud: General.

Anamosa

KKSY(FM)—Not on air, target date: unknown: 95.7 mhz; 6 kw. Ant 328 ft TL: N42 08 19 W91 27 38. Hrs open: 600 Old Marion Rd. N.E., Cedar Rapids, 52402-2152. Phone: (319) 395-0530. Fax: (319) 393-9600. Licensee: Citicasters Licenses L.P. ♦John Laton, gen mgr.

Ankeny

KPTL(FM)— July 1, 1991: 106.3 mhz; 6 kw. Ant 328 ft TL: N41 40 45 W93 35 46. Hrs open: 2141 Grand Ave., Des Moines, 50312. Phone: (515) 245-8900. Web Site: bus1063.com. Licensee: Citicasters Licenses L.P. Group owner: Clear Channel Communications Inc. (acq 5-4-99; grpsl). Population served: 350,000 Format: AAA. Target aud: 35 plus. ♦Joel McCrea, gen mgr; Matt Gillon, sls dir; Andy Roat, gen sls mgr; Cathy Erickson, prom dir; Molly Pins, prom mgr; John McKeighan, progmg dir; Jared Goldberg, mus dir; Jim Boyd, news dir; Raleigh Rubenking, chief of engrg.

Asbury

WJOD(FM)— Mar 31, 1994: 103.3 mhz; 6.6 kw. 643 ft TL: N42 34 19 W90 30 55. Stereo. Hrs opn: 24 5490 Saratoga Rd., Dubuque, 52002-2593. Phone: (563) 557-1040. Fax: (563) 583-4535. Web Site: www.103wjod.com. Licensee: Cumulus Licensing Corp. Group owner: Cumulus Media Inc. (acq 2-6-98). Population served: 120,000 Natl. Network: Jones Radio Networks. Law Firm: Wiley, Rein & Fielding. Format: Country. Target aud: 18-49. ♦Scott Lindahl, gen mgr; Ken Peiffer, opns mgr & progmg VP.

Atlantic

KJAN(AM)— September 1950: 1220 khz; 250 w-D, 86 w-N. TL: N41 25 02 W95 00 15. Stereo. Hrs opn: Box 389, N. Olive St., 50022. Phone: (712) 243-3920. Fax: (712) 243-3937. E-mail: kjan@metc.net Web Site: www.kjan.com. Licensee: Wireless Communications Corp. (acq 1-13-88; FTR: 11-16-87). Population served: 40,000 Natl. Network: Fox News Radio. Rgnl. Network: Radio Iowa, Brownfield Format: Adult contemp, MOR, news. News staff: 1; News: 40 hrs wkly. Target aud: 25 plus; general. Spec prog: Farm 12 hrs wkly. ♦J.C. Van Ginkel, chmn; Merlyn Christensen, pres; James M. Field, gen mgr.

KSWI(FM)— July 2000: 95.7 mhz; 20 kw. Ant 358 ft. TL: N41 26 07 W94 50 00. Stereo. Hrs opn: 24 413 Chestnut St., 50022. Phone: (712) 243-6885. Fax: (712) 243-1691. E-mail: ksom@mchsi.com Web Site: www.iowasuperstation.com. Licensee: Meredith Communications L.C. Population served: 30,000 Natl. Network: ABC. Rgnl. Network: Iowa Radio Net. Format: CHR. ♦Stephen O. Meredith, pres; Bill Saluk, gen mgr; Jill Christensen, progmg dir.

Audubon

KSOM(FM)— August 1995: 96.5 mhz; 100 kw. 528 ft TL: N41 26 07 W94 50 00. Stereo. Hrs opn: 24 413 Chestnut St., Atlantic, 50022. Phone: (712) 243-6885. Fax: (712) 243-1691. E-mail: ksom@mchsi.com Web Site: www.iowasuperstation.com. Licensee: Meredith Communications L.C. Population served: 180,000 Natl. Network: ABC, Motor Racing Net, Premiere Radio Networks. Rgnl. Network: Iowa Radio Net. Format: Country. News staff: 2; News: 6 hrs wkly. Target aud: General; upscale & farmers. ♦Bill Saluk, gen mgr; Jill Christensen, opns mgr & progmg dir.

Belle Plaine

KZAT-FM— May 30, 1997: 95.5 mhz; 4.4 kw. 384 ft TL: N41 56 35 W92 23 51. Stereo. Hrs opn: 24 303 McClellan St., 205 W. 3rd St., Tama, 52339. Phone: (641) 484-5958. Fax: (641) 484-5962. E-mail: kzat@kzat.com Web Site: www.kzat.com. Licensee: Camrory Broadcasting Inc. (acq 8-27-2004). Natl. Network: CBS, Westwood One, ABC. Law Firm: Katten Muchin Zavis Rosenman. Wire Svc: AP Format: Classic hits. News staff: one; News: 6 hrs wkly. Target aud: 25-54; listeners who are professionals, laborers, commuters, tourists & truckers. Spec prog: Polka 2 hrs, Sp 2 hrs wkly. ♦Catherine A. Campbell Currier, pres & gen mgr.

Bettendorf

KQCS(FM)— July 7, 1984: 93.5 mhz; 6 kw. 300 ft TL: N41 35 59 W90 24 33. Stereo. Hrs opn: 24 1229 Brady St., Davenport, 52803. Phone: (563) 326-2541. Fax: (563) 326-0844. Web Site: www.93rock.net. Licensee: Cumulus Licensing Corp. Group owner: Cumulus Media Inc. (acq 3-15-00; grpsl). Population served: 295,000 Law Firm: Putbrese, Hunsaker & Trent. Format: Active rock. Target aud: 18-34. ♦Jack Swart, gen mgr; Julie Derrer, gen sls mgr; Jeff James, progmg dir; Andy Andresen, chief of engrg; Tracy Hall, traf mgr.

***KWNJ(FM)—**Not on air, target date: unknown: 91.1 mhz; 20 kw vert. Ant 617 ft TL: N41 18 44 W90 22 46. Hrs open: University of Northern Iowa, 324 Communications Arts Center, Cedar Falls, 50614-0359. Phone: (319) 273-6400. Fax: (319) 273-2682. Licensee: University of Northern Iowa. ♦Wayne Jarvis, gen mgr.

Bloomfield

KDMU(FM)— June 26, 1982: 106.9 mhz; 14 kw. Ant 367 ft TL: N40 46 39 W92 23 54. Stereo. Hrs opn: 24 Box 186, 52537. Secondary address: 22620 195th St. 52537. Phone: (641) 664-3721. Fax: (641) 664-3738. E-mail: mmcvey@kmemfm.com Web Site: www.kdmufm.com. Licensee: Bloomfield Broadcasting Co. Inc. (acq 1-16-2006; $460,000). Natl. Network: AP Radio, Jones Radio Networks. Rgnl. Network: Brownfield. Rgnl rep: Judy Shepherd Law Firm: Miller & Neely. Format: Classic Hits. News: 14 hrs wkly. Target aud: 25-54 yrs; 50/50 male/female split. ♦Lana Norfleet, gen mgr & traf mgr; Mark McVey, pres & gen mgr; Mark Denney, progmg dir.

Boone

KFFF(AM)— 1927: 1260 khz; 5 kw-D, 33 w-N, DA-D. TL: N42 02 55 W93 53 54. Hrs open: 900 8th St., 50036. Phone: (515) 432-5014. Fax: (515) 432-2092. Licensee: Boone Biblical Ministries Inc. Format: Relg. ♦Robert Stumbo, pres; Jamie Johnson, gen mgr & progmg dir; Bob Pink, chief of engrg.

KFFF-FM— 1950: 99.3 mhz; 5.2 kw. Ant 351 ft TL: N42 02 55 W93 53 54. Format: Relg. ♦Jamie Johnson, progmg dir.

KWBG(AM)— Jan 15, 1950: 1590 khz; 1 kw-D, 500 w-N, DA-N. TL: N42 01 22 W93 52 36. Hrs open: 6 AM-11 PM 724 Story St., 50036. Phone: (515) 432-2046. Fax: (515) 432-1448. E-mail: ckuster@nrgmedia.com Web Site: www.kwbg.com. Licensee: NRG License Sub. LLC. Group owner: Waitt Broadcasting Inc. (acq 10-31-2005; grpsl). Population served: 30,000 Natl. Network: ABC, ESPN Radio. Format: News/talk. News staff: one; News: 36 hrs wkly. Target aud: 35 plus; Boone County, Iowa residents. Spec prog: Farm 15 hrs wkly. ♦Carol Kuster, gen mgr; Jim Turbes, news dir; Ben Parsons, sls.

KWQW(FM)— May 15, 1975: 98.3 mhz; 41 kw. 541 ft TL: N41 49 51 W93 43 54. Stereo. Hrs opn: 24 4143 109th St., Urbandale, 50322. Phone: (515) 331-9200. Fax: (515) 331-9292. Web Site: www.983wowfm.com. Licensee: Citadel Broadcasting Co. Group owner: Citadel Broadcasting Corp. (acq 8-29-2003; grpsl). Natl. Rep: Christal. Format: Talk. News: 15 hrs wkly. Target aud: 25-54. ♦Jack O'Brien, opns mgr; Doug Wood, sls dir; Terry Peters, gen mgr & mktg VP.

Britt

KHAM(FM)—Not on air, target date: unknown: 99.5 mhz; 6 kw. Ant 253 ft TL: N43 06 05 W93 55 10. Hrs open: 1296 Marian Ln., Green Bay, WI, 54304. Phone: (920) 494-6310. Licensee: Steven A. Roy, Personal Representative, Estate of Lyle Evans (acq 6-27-2006; with WRMO(FM) Milbridge, ME). ♦Lyle R. Evans, gen mgr.

Brooklyn

KSKB(FM)— Mar 1, 1988: 99.1 mhz; 50 kw. Ant 175 ft TL: N41 42 36 W92 27 54. Stereo. Hrs opn: Box 440, 52211. Secondary address: 505 Josephine St., Titusville, FL 32796. Phone: (641) 522-7202. Fax: (641) 522-7239. E-mail: wpio@gate.net Web Site: noncomradio.net. Licensee: Florida Public Radio Inc. (acq 1-8-90). Format: Adult contemp Christian mus. Target aud: General. Spec prog: Ger one hr, Pol one hr wkly. ♦Bill Korns, gen mgr.

Burlington

***KAYP(FM)**— Nov 1, 2000: 89.9 mhz; 9 kw vert. Ant 440 ft TL: N40 47 59 W91 32 35. Hrs open: 14267 Washington Rd., West Burlington, 52655. Phone: (319) 758-6911. Fax: (319) 758-6922. E-mail: kayp@mchsi.com Web Site: www.afr.net. Licensee: American Family Association. Group owner: American Family Radio Format: Christian adult contemp. ♦Marvin Sanders, gen mgr.

KBUR(AM)— July 1941: 1490 khz; 1 kw-U. TL: N40 49 26 W91 08 33. Stereo. Hrs opn: 24 Box 70, 1411 N. Roosevelt Ave., 52601. Phone: (319) 752-2701. Fax: (319) 752-5287. Web Site: www.kbur.com. Licensee: Citicasters Licenses L.P. Group owner: Clear Channel Communications Inc. (acq 5-4-99; grpsl). Population served: 275,000 Rgnl. Network: Radio Iowa. Natl. Rep: Clear Channel. Format: Adult contemp, MOR, news/talk. News staff: 3; News: 28 hrs wkly. Target aud: 25 plus; general. Spec prog: Farm 19 hrs wkly. ♦Steve Staebell, gen mgr & natl sls mgr; Steve Hexom, progmg dir & disc jockey; Carl Lensgraf, mus dir & disc jockey; J.K. Martin, news dir & local news ed; Brad Bostrom, engrg mgr; Mark Hempen, traf mgr; John Weir, farm dir; Don Brandt, sports cmtr; Matt Frisbec, disc jockey; Patrick Noon, disc jockey; Rod Cary, disc jockey.

KGRS(FM)—Co-owned with KBUR(AM). Nov 27, 1968: 107.3 mhz; 100 kw. 429 ft TL: N40 49 26 W91 08 33. Stereo. 24 Web Site: www.kgrsfm.com.256,500 Format: Adult contemp. News staff: 3; News: 15 hrs wkly. Target aud: 25-45. ♦Cosmo Leone, progmg dir & disc jockey; Mark Hempen, traf mgr; J.K. Martin, local news ed; Tim Brown, disc jockey.

KCPS(AM)— July 30, 1965: 1150 khz; 500 w-D, 67 w-N, DA-1. TL: N40 51 11 W91 08 10. Hrs open: 24 Box 946, 208 Jefferson, 52601. Phone: (319) 754-6698. Fax: (319) 754-8899. E-mail: kcps@aol.com Web Site: www.kcps.com. Licensee: John Giannettino. (acq 1-88). Population served: 100,000 Natl. Network: CBS, Westwood One, ABC. Natl. Rep: Katz Radio. Law Firm: Shaw Pittman. Format: Talk. News staff: one; News: 7 hrs wkly. Target aud: 25-54; middle-aged, upscale & well-informed adults. Spec prog: Agriculture-business 10 hrs, pro sports 10 hrs wkly. ♦John Giannettino, gen mgr.

KDMG(FM)— July 19, 1993: 103.1 mhz; 12 kw. 445 ft TL: N40 44 04 W91 15 16. Stereo. Hrs opn: 24 Box #112, 2850 Mt. Pleasant St., 52601. Phone: (319) 752-5402. Fax: (319) 752-4715. E-mail: johnp@burlingtonradio.com Licensee: Pritchard Broadcasting Co. Population served: 109,600 Format: Country. News staff: one; News: 3 hrs wkly. Target aud: 25-54. ♦John T. Pritchard, pres & gen mgr; Kathy Jolly Vance, opns mgr; Chet Young, gen sls mgr; Kathy Vance, progmg dir.

KKMI(FM)— Oct 22, 1981: 93.5 mhz; 6.0 kw. 305 ft TL: N40 49 11 W91 07 02. Stereo. Hrs opn: 24 Suite 112, 2850 Mt. Pleasant St., 52601. Phone: (319) 752-5402. Fax: (319) 752-4715. E-mail: johnp@burlingtonradio.com Licensee: Pritchard Broadcasting Co. (acq 8-5-91). Population served: 90,000 Format: Adult contemp. News staff: one; News: 3 hrs wkly. Target aud: 25-55; upscale. Spec prog: Pu. ♦John T. Pritchard, pres & gen mgr; Kathy Vance, opns mgr; Chet Young, gen sls mgr.

Carroll

KCIM(AM)— June 8, 1950: 1380 khz; 1 kw-U, DA-2. TL: N42 02 29 W94 53 06. Hrs open: 1119 E. Plaza Dr., 51401. Phone: (712) 792-4321. Fax: (712) 792-6667. E-mail: kcimkkrl@win-4-u.net Web Site: carrollbroadcasting.com. Licensee: Carroll Broadcasting Co. (group owner; acq 8-1-85; $1.5 million with co-located FM; FTR: 5-20-85) Population served: 200,000 Natl. Network: CBS. Natl. Rep: Katz Radio. Law Firm: Womble, Carlyle, Sandridge & Rice. Format: Classic hits of the 50s, 60s & 70s. ♦Mary Collison, CEO; Kim Hackett, gen mgr & natl sls mgr; Lynda Dukes Franey, gen mgr & gen sls mgr; John Ryan, progmg mgr; Bob Grote, chief of engrg.

KKRL(FM)—Co-owned with KCIM(AM). Jan 18, 1967: 93.7 mhz; 100 kw. 300 ft TL: N42 03 14 W94 53 06. Stereo. 24 Web Site: carrollbroadcasting.com. Natl. Rep: Katz Radio. Law Firm: Womble, Carlyle, Sandridge & Rice. Format: Hot AC 70s, 80s, 90s & today. News staff: one. Target aud: 18 plus. ♦John Ryan, progmg dir.

***KWOI(FM)**— 2004: 90.7 mhz; 10 kw. Ant 289 ft TL: N42 07 14 W94 48 49. Stereo. Hrs opn: 24
WOI-AM.
2022 Communications Bldg. ISU, WOI Radio Group, Ames, 50011-3241. Phone: (515) 294-2025. Fax: (515) 294-1544. E-mail: woi@iastate.edu Web Site: www.woi.org. Licensee: Iowa State University of Science and Technology. Natl. Network: NPR, PRI. Format: Class, news. News: 40 hrs wkly. Target aud: General; Educated. ♦Dave Becker, progmg dir.

Castana

***KILV(FM)**— 2001: 107.5 mhz; 25 kw. Ant 328 ft TL: N42 12 26 W96 07 26. Stereo. Hrs opn: 24 2351 Sunset Blvd., Suite 170-218, Rocklin, CA, 95765. Phone: (916) 251-1600. Fax: (916) 251-1650. E-mail: klove@klove.com Web Site: www.klove.com. Licensee: Educational Media Foundation. Group owner: EMF Broadcasting (acq 10-26-01). Population served: 5,500 Natl. Network: K-Love. Law Firm: Shaw Pittman. Format: Contemp Christian. News staff: 3. Target aud: 25-44; female-Judeo/Christian. ♦Richard Jenkins, pres; Mike Novak, VP & progmg dir; Keith Whipple, dev dir; David Pierce, progmg mgr; Sam Wallington, engrg dir; Karen Johnson, news rptr; Marya Morgan, news rptr; Richard Hunt, news rptr.

Cedar Falls

KCNZ(AM)— September 1998: 1650 khz; 10 kw-D, 1 kw-N. TL: N42 24 47 W92 26 15. Stereo. Hrs opn: 24 Box 248, 50613. Phone: (319) 277-1918. Fax: (319) 277-5202. E-mail: kcnz@kcnzam.com Web Site: www.1650thefan.com. Licensee: Fife Communications Co. LLC. Population served: 120,000 Natl. Network: CBS. Format: Sports/talk. News staff: 2; News: 25 hrs wkly. Target aud: 25-54; eastern Iowa adults. Spec prog: Farm 6 hrs. ♦John Coloff, pres; Jim Coloff, gen mgr; Doug Petersen, progmg dir.

KDNZ(AM)— Feb 2, 1958: 1250 khz; 500 w-U, DA-2. TL: N42 32 41 W92 29 16. Hrs opn: 24 Box 248, 721 Shirley St., 50613. Phone: (319) 277-1918. Fax: (319) 277-5202. E-mail: radio@radio1250.com Licensee: Fife Communications L.C. (acq 1995; $90,000). Population served: 120,000 Format: Sp. ♦Jim Coloff, pres & gen mgr; Tony Coloff, VP; Jeff Ryant, stn mgr; Sue Coloff, opns VP.

***KHKE(FM)**— Apr 1, 1974: 89.5 mhz; 10 kw. 410 ft TL: N42 23 58 W92 19 15. (CP: Ant 417 ft. TL: N42 23 55 W92 19 34). Stereo. Hrs opn: 6 AM-2 AM 324 Communications Arts Center, Univ. of Northern Iowa, 50614-0359. Phone: (319) 273-6400. Fax: (319) 273-7911. E-mail: kuni@uni.edu Web Site: www.khke.org. Licensee: University of Northern Iowa. Population served: 29,517 Format: Class. News staff: 3; News: 5 hrs wkly. Target aud: General. ♦Scott Vezdos, mktg mgr; Wayne Jarvis, gen mgr, gen mgr & progmg dir; Al Schares, mus dir; Greg Shanley, news dir; Steve Schoon, chief of engrg.

KOEL-FM— Jan 7, 1994: 98.5 mhz; 25 kw. 328 ft TL: N42 28 09 W92 29 05. Hrs open: 24 501 Sycamore St., Suite 300, Blacks Bldg., Waterloo, 50703. Phone: (319) 833-4800. Phone: (319) 833-4985 (Contest line). Fax: (319) 833-4866. Web Site: www.k985.com. Licensee: Cumulus Licensing Corp. Group owner: Cumulus Media Inc. (acq 3-15-00; grpsl). Population served: 230,000 Format: Country. News staff: one; News: one hr wkly. Target aud: 18-49; general. ♦Lew Dickey, pres; William Hathaway, gen mgr.

***KUNI(FM)**— Sept 15, 1960: 90.9 mhz; 100 kw. 1,782 ft TL: N42 18 59 W91 51 31. Stereo. Hrs opn: 24 324 Communications Art Center, Univ. of Northern Iowa, 50614-0359. Phone: (319) 273-6400. Fax: (319) 273-7911. E-mail: kuni@uni.edu Web Site: www.kuniradio.org. Licensee: University of Northern Iowa. Natl. Network: NPR, PRI. Format: News and Information, Triple A. News staff: 3; News: 77 hrs wkly. Target aud: General. Spec prog: Folk 4 hrs, rhythm and blues 5 hrs wkly. ♦Scott Vezdos, mktg mgr; Wayne Jarvis, gen mgr, prom dir & progmg dir; Al Shares, mus dir; Greg Shanley, news dir; Steve Schoon, chief of engrg.

Cedar Rapids

***KCCK-FM**— Sept 5, 1972: 88.3 mhz; 10 kw. 420 ft TL: N41 54 33 W91 39 17. Stereo. Hrs opn: Box 2068, 214 Linn Hall, 6301 Kirkwood Blvd. S.W., 52406. Phone: (319) 398-5446. Fax: (319) 398-5492. E-mail: studio@kcck.org Web Site: www.kcck.org. Licensee: Kirkwood Community College. Population served: 109,642 Natl. Network: PRI, AP Radio. Law Firm: Wilkinson, Barker & Knauer. Format: Jazz. News staff: one; News: 8 hrs wkly. Target aud: 25-54; educated, affluent, active in community. Spec prog: New age 7 hrs wkly. ♦Cheryle Mitvalsky, exec VP; Dennis Green, gen mgr; Debra Umbdenstock, dev dir; Bob Stewart, progmg dir; George Dorman, opns dir & news dir; Dave Maley, chief of engrg.

KDAT(FM)— May 1971: 104.5 mhz; 100 kw. 500 ft TL: N42 04 51 W91 41 45. Stereo. Hrs opn: 24 4th Fl., 425 Second St. S.E., 52401. Phone: (319) 365-9431. Fax: (319) 363-8062. E-mail: kdat@kdat.com Web Site: www.kdat.com. Licensee: Cumulus Licensing Corp. Group owner: Cumulus Media Inc. (acq 8-7-00; grpsl). Law Firm: Latham & Watkins. Format: Light rock. Target aud: 25-54. ♦Jim Worthington, gen mgr; Dick Stadler, progmg dir.

KFMW(FM)—See Waterloo

KGYM(AM)— 1947: 1600 khz; 5 kw-U, DA-N. TL: N41 58 15 W91 32 01. Stereo. Hrs opn: 24 1110 26th Ave. S.W., 52404-3430. Phone: (319) 363-2061. Fax: (319) 363-2948. E-mail: info@1600espn.com Web Site: www.1600espn.com. Licensee: KZIA Inc. (acq 10-30-2006; $775,000). Population served: 315,000 Natl. Network: ESPN Radio. Rgnl. Network: Radio Iowa Natl. Rep: D & R Radio. Law Firm: Dow, Lohnes & Albertson. Wire Svc: AP Format: Sports. News: one hr wkly. Target aud: 18-49. ♦Eliot Keller, pres & gen mgr; Robert Norton Jr., exec VP & opns mgr; Julie Hein, sls dir; Kellie Lala, gen sls mgr; Jamie Burgin, prom dir; Dorothy Roach, traf mgr; Scott Unash, progmg dir & sports cmtr.

KHAK(FM)— July 1, 1961: 98.1 mhz; 100 kw. 485 ft TL: N41 55 28 W91 36 55. Stereo. Hrs opn: 425 Second St. S.E., 4th Fl., 52401. Phone: (319) 365-9431. Phone: (319) 365-3698. Fax: (319) 363-8062. E-mail: khak@khak.com Web Site: www.khak.com. Licensee: Cumulus Licensing Corp. Group owner: Cumulus Media Inc. (acq 8-7-00; grpsl). Population served: 406,000 Rgnl. Network: Brownfield. Law Firm: Latham & Watkins. Format: Modern country. Target aud: 25-54; general. ♦Jim Worthington, gen mgr; Bob James, progmg dir.

KMJM(AM)— July 1, 1961: 1360 khz; 1 kw-D, 124 w-N, DA-1. TL: N41 55 28 W91 36 55. Hrs open: 24 600 Old Marion Rd. N.E., 52402. Phone: (319) 395-0530. Fax: (319) 393-9600. Licensee: Capstar TX L.P. Group owner: Clear Channel Communications Inc. (acq 8-30-2000; grpsl). Population served: 142,000 Format: Sports. ♦John Laton, gen mgr.

KMRY(AM)— August 1949: 1450 khz; 1 kw-U. TL: N42 00 25 W91 42 29. (In-band On-channel). Hrs opn: 24 1957 Blairsferry Rd. N.E., 52402. Phone: (319) 393-1450. Fax: (319) 393-1407. E-mail: kmry@kmryradio.com Web Site: www.kmryradio.com. Licensee: Sellers Broadcasting Inc. (acq 3-5-98; $475,000). Population served: 200,000 Natl. Network: CBS. Law Firm: Katten, Muchin, Rosenman, LLP. Wire Svc: AP Format: Adult standards. News staff: one; News: 20 hrs wkly. Target aud: 40 plus; affluent, upscale adults with large disposable income. Spec prog: 50's oldies 3 hrs, polka 3 hrs, big band 2 hrs weekly. ♦Rick Sellers, pres; Kevin Alexander, VP; Rick Sampson, opns mgr; Eric Christopher, progmg dir & progmg mgr; Jim Davies, chief of engrg.

KZIA(FM)— Apr 29, 1975: 102.9 mhz; 100 kw. Ant 853 ft TL: N42 03 25 W91 41 42. Stereo. Hrs opn: 24 1110 26th Ave. S.W., 52404-3430. Phone: (319) 363-2061. Fax: (319) 363-2948. E-mail: kzia@kzia.com Web Site: www.kzia.com. Licensee: KZIA Inc. (acq 5-13-94; $2 million; FTR: 2-25-85). Population served: 315,000 Natl. Rep: D & R Radio. Law Firm: Dow, Lohnes & Albertson. Wire Svc: AP Format: CHR. News staff: one; News: one hr wkly. Target aud: 18-49. ♦Eliot Keller, pres & gen mgr; Robert Norton Jr., exec VP & opns mgr; Julie Hein, gen sls mgr; Kellie Lala, gen sls mgr; Jamie Burgin, prom dir; Greg Runyon, progmg dir; Ric Swann, mus dir; Scott Schulte, news dir; Dorothy Roach, traf mgr.

WMT(AM)— 1922: 600 khz; 5 kw-U, DA-N. TL: N42 03 40 W91 32 44. Stereo. Hrs opn: 24 600 Old Marion Rd. N.E., 52402-2152. Phone: (319) 395-0530. Fax: (319) 393-0918. Web Site: www.wmtradio.com. Licensee: Citicasters Licenses L.P. Group owner: Clear Channel Communications Inc. (acq 5-4-99; grpsl). Population served: 170,000 Natl. Network: CBS. Format: Full svc, news/talk. Target aud: 35 plus. Spec prog: Farm 19 hrs wkly. ♦John Laton, gen mgr; Andy Roat, gen sls mgr; Lisa Pucelik, mktg dir; Teisha Welsh, mktg dir; Randy Lee, progmg mgr; Jeff Schmidt, news dir; Tom Spaight, chief of engrg.

WMTD-FM— Oct 1, 1985: 102.3 mhz; 160 w. 1,008 ft TL: N37 42 56 W80 56 55. (CP: 368 w, ant 1,273 ft.). Stereo. 24 38,000 News staff: one; News: 1 hr wkly. Target aud: Adults 18-49. ♦Rhonda Pritt, traf mgr.

WMTM-FM— Nov 17, 1964: 93.9 mhz; 100 kw. 555 ft TL: N31 12 54 W83 47 13. Stereo. 6 AM-midnight Format: Oldies. ♦Jim Turner, opns mgr, mktg mgr & traf mgr.

WMTX(FM)—Co-owned with WHNZ(AM). November 1947: 100.7 mhz; 100 kw. 1,358 ft TL: N28 02 21 W82 39 21. (CP: TL: N27 50 32 W82 15 46). Stereo. 24 Web Site: www.wmtx.com. Format: Top 40, adult contemp. ♦Dan DiLoreto, gen mgr; Tony Florentino, progmg dir; Wilson Welch, chief of engrg; Diana Roselle, traf mgr.

WMTC-FM— Jan 1, 1991: 99.9 mhz; 6 kw. Ant 328 ft TL: N37 36 23 W83 26 48. Stereo. 24 Phone: (606) 666-5006. E-mail: wmtc@asburyusa.net 2,000,000 Format: Christian, relg. Spec prog: Farm one hr wkly. ♦Seldon Short, VP; Seldon Short, gen mgr, sls dir, edit dir & farm dir; Jennifer Cox, progmg dir; Gordon Sampsel, mus dir; Theresa Kerley, disc jockey.

WMT-FM— Feb 16, 1963: 96.5 mhz; 100 kw. 540 ft TL: N42 01 43 W91 38 27. Stereo. Web Site: www.mix965.com. Natl. Network: CBS. Format: Adult contemp. Target aud: 25-49. ♦Randy Lee, progmg dir.

Centerville

KCOG(AM)— Mar 1, 1949: 1400 khz; 500 w-D, 1 kw-N. TL: N40 44 40 W92 54 32. Hrs opn: 5 AM-midnight 402 N. 12th St., 52544. Phone: (641) 437-4242. E-mail: kcogam@lisco.net Web Site: www.kmgo.com. Licensee: KCOG Inc. (acq 6-1-84; $406,000; FTR: 4-16-84). Population served: 6,531 Natl. Network: USA. Rgnl. Network: Brownfield. Format: Adult contemp. ♦Fred Jenkins, gen mgr & progmg dir.

KMGO(FM)—Co-owned with KCOG(AM). Oct 1, 1974: 98.7 mhz; 100 kw. 500 ft TL: N40 47 34 W92 52 47. Stereo. 24 Phone: (641) 856-3996. Fax: (641) 856-3337. E-mail: kmgofm@lisco.net Web Site: www.kmgo.com. Licensee: KMGO Inc. (acq 6-5-85). Natl. Network: USA. Format: Country. ♦Larry Stout, progmg dir.

Chariton

KELR-FM— Nov 15, 1979: 105.5 mhz; 50 kw. 390 ft TL: N41 00 50 W93 17 23. Stereo. Hrs opn: 24 215 N. Main St., 50049. Phone: (641) 774-8494. Fax: (641) 774-8495. E-mail: KELR@lisco.com Licensee: FMC Broadcasting Inc. (acq 6-15-99). Population served: 70,000 Natl. Network: Westwood One. Format: Adult contemp. News staff: one; News: 30 hrs wkly. Target aud: 28 plus. Spec prog: Gosp 5 hrs wkly. ♦Thomas A. Palen, CEO & pres; Cindy Spidle, adv mgr; Nick Hoffman, progmg VP & asst music dir; Fred Jenkins, engrg VP; Jill Schull, traf mgr & local news ed; John Johnston, traf mgr & local news ed.

Charles City

KCHA(AM)— November 1949: 1580 khz; 500 w-D, 10 w-N. TL: N43 03 05 W92 40 00. Hrs open: 24 207 N. Main St., 50616. Phone: (641) 228-1000. Fax: (641) 228-1200. E-mail: kcha@clearchannel.com Web Site: www.kchafm.com. Licensee: Clear Channel Broadcasting Licenses Inc. Group owner: Clear Channel Communications Inc. (acq 9-25-00; grpsl). Population served: 200,000 Natl. Rep: Farmakis. Format: Adult standards. News staff: news progmg 7 hrs wkly News: one;. Target aud: 35 plus. ♦Hal Hoffman, gen mgr; Mike Watson, progmg dir.

KCHA-FM— October 1971: 95.9 mhz; 6 kw. 300 ft TL: N43 03 05 W92 40 00. Stereo. 24 Web Site: www.kchafm.com.245,000 Format: Adult contemp. News staff: one; News: 7 hrs wkly. Target aud: 25-54; adults.

Cherokee

KCHE(AM)— January 1953: 1440 khz; 500 w-D. TL: N42 47 21 W95 33 06. Hrs open: Box 1440, 201 S. 5th, 51012. Phone: (712) 225-2511. Fax: (712) 225-2513. E-mail: kche1@nen.net Web Site: www.kcheradio.com. Licensee: J & J Broadcasting Corp. (acq 11-14-03; $600,000 with co-located FM). Population served: 15,000 Natl. Network: ABC. Format: Oldies. News staff: 2; News: 14 hrs wkly. Target aud: 45-80; general. Spec prog: Farm 12 hrs, Sp one hr wkly. ♦Jeff Fuller, pres, gen mgr & gen sls mgr; Curt Carlson, VP, gen mgr & sls VP; Dick Keane, chief of opns & chief of engrg; Bill Beroni, progmg dir; Greg Slotsky, news dir; Lynn Dittmer, prom dir & pub affrs dir; Hallie Dessell, traf mgr.

KCHE-FM— Dec 9, 1976: 92.1 mhz; 6 kw. 210 ft TL: N42 47 21 W95 33 08. Stereo. 24 Web Site: www.kcheradio.com.13,500 Natl. Network: ABC. Natl. Rep: Farmakis. Format: Adult contemp. News staff: 2; News: 28 hrs wkly. ♦Hallie Dessell, traf mgr.

Clarinda

KKBZ(FM)— Sept 25, 1990: 99.3 mhz; 50 kw. 492 ft TL: N40 33 12 W95 07 18. Stereo. Hrs opn: 24 Box 960, 209 N. Elm, Shenandoah, 51601. Phone: (712) 246-5270. Fax: (712) 246-5275. Web Site: www.kmakkbz.com. Licensee: May Broadcasting Co. Natl. Network: Westwood One, CNN Radio. Law Firm: Duane Morris. Wire Svc: AP Format: Adult Contemp. News staff: 2; News: 5 hrs wkly. Target aud: 25-44. ♦Edward W. May, pres; Don Hansen, stn mgr; Chuck Morris, opns mgr.

Clarion

KIAQ(FM)— May 18, 1964: 96.9 mhz; 100 kw. 578 ft TL: N42 40 18 W94 09 11. Stereo. Hrs open: 200 North 10th St., Fort Dodge, 50501. Phone: (515) 555-5656. Fax: (515) 555-5844. E-mail: bdennis@kuel.threeeagles.com Web Site: www.kiaqfm.com. Licensee: Three Eagles of Ft. Dodge Inc. Group owner: Three Eagles Communications (acq 4-22-97; $1,244,117). Population served: 304,000 Format: Country. Target aud: 25-54. ♦Gary Buchanan, pres & traf mgr; Patrick Kolar, gen mgr; Gregg Ellendson, opns mgr; Travis Reeves, sls VP; Michael Moody, news dir; Barb Dennis, traf mgr.

Clear Lake

KLKK(FM)— Feb 16, 1978: 103.7 mhz; 25 kw. Ant 328 ft TL: N43 07 15 W93 11 36. Stereo. Hrs opn: 24 341 Yorktown Pike, Mason City, 50401. Phone: (641) 423-1300. Fax: (641) 423-2906. Web Site: www.klkkfm.com. Licensee: Clear Channel Broadcasting Licenses Inc. Group owner: Clear Channel Communications Inc. (acq 9-25-2000; grpsl). Population served: 100,000 Natl. Rep: Clear Channel. Format: Classic rock. News staff: one; News: 42 hrs wkly. Target aud: 25-54. ♦Hal Hofman, gen mgr & gen sls mgr; Drew Kelly, progmg dir; Laurie Gansen, traf mgr.

Clinton

KCLN(AM)— Dec 21, 1956: 1390 khz; 1 kw-D, 91 w-N, DA-2. TL: N41 54 32 W90 13 16. Hrs open: 24 1853 442nd Ave., 52732. Phone: (563) 243-1390. Fax: (563) 242-4567. E-mail: kcln@kcln.com Web Site: www.kcln.com. Licensee: WPW Broadcasting Inc. (group owner; (acq 4-29-99; $800,000 with co-located FM). Population served: 150,000 Rgnl. Network: Tribune. Law Firm: Miller & Fields. Format: Music of the 40s, 50s & 60s, big band. News staff: one; News: 2 hrs wkly. Target aud: 40 plus. Spec prog: Farm 10 hrs wkly. ♦Don Davis, pres; Larry Timpe, gen mgr; Chris Streets, progmg dir; Brad Seward, news dir; Aaron Winski, chief of engrg; Tracie Morgan, traf mgr.

KMCN(FM)—Co-owned with KCLN(AM). Dec 7, 1970: 94.7 mhz; 3 kw. Ant 300 ft TL: N41 54 32 W90 13 20. Stereo. 24 1853 442nd Ave., 52732. Phone: (563) 243-5256. Web Site: kcln.com.150,000 Format: Adult hits. News staff: one; News: one hr wkly. Target aud: 25-54.

KMXG(FM)— July 1974: 96.1 mhz; 100 kw. 980 ft TL: N41 37 58 W90 24 38. Stereo. Hrs opn: 24 3535 E. Kimberly Rd., Davenport, 52807. Phone: (563) 344-7000. Fax: (563) 344-7006. Web Site: www.kmxg.com. Licensee: Citicasters Licenses L.P. Group owner: Clear Channel Communications Inc. (acq 11-15-00; grpsl). Population served: 350,000 Natl. Rep: Christal. Format: Hot adult contemp. News staff: one; News: 3 hrs wkly. Target aud: 25-54; yuppies, baby boomers, upscale professional females. Spec prog: Jazz 3 hrs wkly. ♦Larry Rosmilso, VP & gen mgr; Jim O'Hara, opns mgr & progmg dir; Kevin Allensworth, chief of engrg.

KROS(AM)— Sept 28, 1941: 1340 khz; 1 kw-U. TL: N41 51 36 W90 12 18. Stereo. Hrs opn: 5:30 AM-midnight Box 0518, William Scott Broadcast Ctr., 870 13th Ave. N., 52733-0518. Phone: (563) 242-1252. Fax: (563) 242-4825. E-mail: kros@clinton.net Web Site: www.krosradio.com. Licensee: KROS Broadcasting Inc. (acq 7-28-98; $23,000 for 28). Population served: 34,719 Natl. Network: CNN Radio. Rgnl. Network: Radio Iowa. Format: Full service. News staff: one; News: 38 hrs wkly. Target aud: General; loc audience. Spec prog: Folk 2 hrs, jazz one hr, blues one hr, gospel one hr, women 5 hrs. ♦Brad Parker, pres; Dave Vickers, gen mgr.

Council Bluffs

***KIWR(FM)—** Nov 23, 1981: 89.7 mhz; 100 kw. 1,100 ft TL: N41 18 40 W96 01 37. Stereo. Hrs open: 24 2700 College Rd., 51503. Phone: (712) 325-3254. Fax: (712) 325-3391. E-mail: sjohn@iwcc.edu Web Site: www.897theriver.com. Licensee: Iowa Western Community College. Population served: 1,000,000 Format: Progsv. News: 5 hrs wkly. Target aud: 18-34; well-educated, upper & middle-upper income. Spec prog: Var/div 16 hrs wkly. ♦Dan Kinney, pres; Tom Johnson, CFO; Sophia John, gen mgr.

KLNG(AM)— 1947: 1560 khz; 1 kw-D. TL: N41 12 28 W95 54 04. Hrs open: 6 AM-sunset 120 S. 35th St., Suite 2, 51501. Phone: (712) 323-0100. Fax: (712) 323-0022. E-mail: klgn@wilkinsradio.com Web Site: www.wilkinsradio.com. Licensee: Wilkins Communications Network Inc. (group owner; (acq 4-89; $250,000). Population served: 1,500,000 Natl. Network: Salem Radio Network. Law Firm: Womble, Carlyle, Sandridge & Rice. Format: Christian teaching/talk. Target aud: 35 plus. Spec prog: Sp 10 hrs, Black 6 hrs wkly. ♦Bob Wilkins, pres; LuAnn Wilkins, exec VP; Mitchell Mathis, VP; Charles Yates, stn mgr; Greg Garrett, opns mgr; John Bible, engr.

KOTK(AM)—See Omaha, NE

KQKQ-FM— 1969: 98.5 mhz; 100 kw. Ant 1,102 ft TL: N41 18 25 W96 01 37. Stereo. Hrs opn: 5011 Capitol Ave., Omaha, NE, 68132. Phone: (402) 342-2000. Fax: (402) 346-5748. Web Site: www.q985fm.com. Licensee: Waitt Omaha LLC. (group owner; (acq 1-7-2002; grpsl). Population served: 134,800 Format: Modern adult contemp. Target aud: 18-44. ♦Mary Quass, CEO, pres & sr VP; Chuck DuCoty, COO & mus dir; Jim McKernan, gen mgr & chief of engrg; Rhonda Gerrard, sls dir; Sam Coughlin, gen sls mgr; Brandon Pappas, prom dir; Mark Todd, prom dir & progmg VP; Nevin Dane, progmg dir; Lori Storz, traf mgr.

KSRZ(FM)—See Omaha, NE

Cresco

KCZQ(FM)— Apr 1, 1991: 102.3 mhz; 3 kw. Ant 328 ft TL: N43 25 47 W92 09 49. Stereo. Hrs opn: 116 First Ave. W., 52136-1514. Phone: (563) 547-1000. Phone: (563) 547-3366. Fax: (563) 547-2200. E-mail: superc@iowatelecom.net Licensee: Mega Media Ltd. Population served: 250,000 Natl. Rep: Farmakis. Format: Adult contemp. Target aud: General. Spec prog: Farm 12 hrs wkly. ♦James B. Hebel, pres, gen mgr & gen sls mgr; Debra Lowe, opns mgr; Jim Bernard, progmg dir; Stan McHenry, mus dir.

Creston

***KLOX(FM)—** 2005: 90.9 mhz; 100 kw vert. Ant 335 ft TL: N41 04 29 W94 22 35. Hrs open: 505 Josephine St., Titusville, FL, 32796. Phone: (321) 267-3000. Fax: (321) 264-9370. E-mail: wpio@gate.net Web Site: noncomradio.net. Licensee: Florida Public Radio Inc. Format: Adult contemp Christian. ♦Archie Shetler, exec VP; Randy Henry, pres & gen mgr.

KSIB(AM)— Dec 7, 1946: 1520 khz; 1 kw-D. TL: N41 02 16 W94 23 38. Hrs open: Box 426, 50801. Phone: (641) 782-2155. Fax: (641) 782-6963. Licensee: G.O. Radio Ltd. (acq 2-82; grpsl; FTR: 2-22-82). Population served: 68,234 Natl. Network: ABC. Rgnl. Network: Iowa Radio Net. Format: C&W. Target aud: General. ♦Dave Rieck, pres & gen mgr; Chad Riek, gen sls mgr; Ben Walter, progmg dir; Mark Saylor, news dir; Charlie Maley, chief of engrg.

KSIB-FM— March 1966: 101.3 mhz; 19 kw. 364 ft TL: N41 05 41 W94 22 30. Stereo. 24 68,234 Format: C&W. Target aud: General.

Davenport

***KALA(FM)**— Nov 4, 1967: 88.5 mhz; 100 w. 110 ft TL: N41 32 28 W90 34 57. Stereo. Hrs opn: 24 518 W. Locust St., 52803. Phone: (563) 333-6219. Fax: (563) 333-6218. E-mail: kala@sau.edu Web Site: www.sau.edu/kala. Licensee: St. Ambrose University. (acq 11-4-67). Population served: 100,000 Format: Jazz, progsv, urban contemp. News staff: one; News: 34.5 hrs wkly. Target aud: General. Spec prog: Sp 15 hrs, gospel 13 hrs wkly. ♦David Baker, gen mgr & opns mgr.

KBEA-FM—Muscatine, February 1949: 99.7 mhz; 100 kw. 895 ft TL: N41 26 43 W91 04 36. Stereo. Hrs opn: 24 1229 Brady St., 52803. Phone: (563) 326-2541. Fax: (563) 326-1819. Licensee: Cumulus Licensing Corp. Group owner: Cumulus Media Inc. (acq 3-15-00; grpsl). Population served: 600000 Law Firm: Putbrese, Hunsaker & Trent. Format: Top 40. News staff: one. Target aud: 25-54. ♦Jack Swart, pres & gen mgr; Julie Derrer, gen sls mgr; Steve Fuller, progmg dir; Andy Andresen, chief of engrg; Tracy Hall, traf mgr.

KCQQ(FM)— Sept 1, 1996: 106.5 mhz; 100 kw. 210 ft TL: N41 32 14 W90 34 30. Stereo. Hrs opn: 24 3535 E. Kimberly Rd., 52807. Phone: (563) 344-7000. Fax: (563) 359-8524. E-mail: jimhunter@clearchannel.com Web Site: www.kcqq106.com. Licensee: Citicasters Licenses L.P. Group owner: Clear Channel Communications Inc. (acq 11-15-00; grpsl). Population served: 200,000 Format: Classic rock. News staff: one; News: 2 hrs wkly. Target aud: 25-54. Spec prog: Relg one hr wkly. ♦Larry R. Rosmilso, gen mgr; Jim Hunter, stn mgr & progmg dir; Teri Van Dyke, gen sls mgr; Kevin Allensworth, chief of engrg.

KJOC(AM)— 1947: 1170 khz; 1 kw-U, DA-2. TL: N41 23 22 W90 31 08. Hrs opn: 24 1229 Brady St., 52803. Phone: (563) 326-2541. Fax: (563) 326-1819. Web Site: www.kjoc.com. Licensee: Cumulus Licensing Corp. Group owner: Cumulus Media Inc. (acq 3-15-00; grpsl). Population served: 49,900 Natl. Network: CBS. Law Firm: Putbrese, Hunsaker & Trent, P. Format: Sports. Target aud: 18-49. ♦Jack Swart, gen mgr.

WFXN(AM)—See Moline, IL

WLLR-FM—Listing follows WOC(AM).

WOC(AM)— February 1922: 1420 khz; 5 kw-U, DA-2. TL: N41 33 00 W90 28 37. Hrs opn: 24 3535 E. Kimberly Rd., 52807. Phone: (563) 344-7000. Fax: (563) 344-7065. Web Site: www.woc1420.com. Licensee: Citicasters Licenses L.P. Group owner: Clear Channel Communications Inc. (acq 11-15-2000; grpsl). Population served: 50,000 Rgnl. Network: Ill. Radio Net, Radio Iowa. Natl. Rep: Christal. Law Firm: Baker & Hostetler. Format: News/talk, info. News staff: 5. Target aud: 35-64; info-oriented adults. Spec prog: Farm 10 hrs wkly. ♦Larry Rosmilso, gen mgr; Scott Bitting, gen sls mgr; Caressa Clearman, prom dir; Dan Kennedy, progmg dir; Kevin Allensworth, chief of engrg.

WLLR-FM—Co-owned with WOC(AM). October 1948: 103.7 mhz; 100 kw. 1,191 ft TL: N41 32 49 W90 28 35. Stereo. 24 Phone: (563) 359-9557. Fax: (563) 344-7016. E-mail: jimohara@clearchannel.com Web Site: www.wllr.com.200,000 Format: Country. News: 2 hrs wkly. Target aud: 25-54. ♦Mike Weindruch, gen sls mgr; Carrie Clearman, prom dir; Jim O'Hara, progmg dir; Kevin Allensworth, engrg dir; Lorraine Meier, traf mgr.

De Witt

KBOB-FM— Jan 12, 1977: 104.9 mhz; 12.5 kw. Ant 469 ft TL: N41 43 11 W90 34 13. Stereo. Hrs opn: 24 1229 Brady St., Davenport, 52803. Phone: (563) 326-2541. Fax: (563) 326-1819. Web Site: www.97rock.net. Licensee: Cumulus Licensing Corp. Group owner: Cumulus Media Inc. (acq 10-2-2000; grpsl). Format: Active rock. ♦Jack Swart, gen mgr; Julie Derrer, gen sls mgr; Ryan Chase, progmg dir; Andy Andresen, chief of engrg; Deanna Flynn, traf mgr.

Decorah

KDEC-FM— Sept 2, 1986: 100.5 mhz; 30 kw. Ant 420 ft TL: N43 19 26 W91 47 04. Stereo. Hrs opn: 24 Box 27, 52101. Secondary address: 110 Highland Dr. 52101. Phone: (563) 382-4251. Fax: (563) 382-9540. E-mail: kdec@kdecradio.com Web Site: www.kdecradio.com. Licensee: Decorah Broadcasting Inc. (acq 3-1-96; $696,500). Population served: 100,000 Law Firm: Reddy, Begley & McCormick. Format: Adult contemp, AAA. News staff: 2; News: 3 hrs wkly. Target aud: 18-54. ♦Bob Holtan, pres & gen mgr; Colleen Holtan, VP; Jennifer Grouws, stn mgr.

KDEC(AM)— May 1947: 1240 khz; 1 kw-U. TL: N43 19 26 W91 47 04.5 AM-10 PM (M-F) 40,000 Natl. Network: Westwood One. Natl. Rep: Farmakis. Format: MOR. News staff: 2; News: 12 hrs wkly. Target aud: 35 plus.

***KLCD(FM)**— July 15, 1977: 89.5 mhz; 100 w. 140 ft TL: N43 18 56 W91 47 18. Stereo. Hrs opn: 206 S. Broadway, Suite 735, Rochester, MN, 55904. Phone: (507) 282-0910. Fax: (507) 282-2107. E-mail: mail@mpr.org Web Site: www.mpr.org. Licensee: Minnesota Public Radio Inc. Natl. Network: NPR, PRI. Format: Class. News staff: one. ♦Chris Cross, gen mgr; Mary Stapek, dev dir; Sea Stachura, news rptr.

***KLNI(FM)**— 1993: 88.7 mhz; 100 w. -36 ft TL: N43 18 35 W91 48 30. Hrs open: 24 206 S. Broadway, Suite 735, Rochester, MN, 55904. Phone: (507) 282-0910. Fax: (507) 282-2107. E-mail: mail@mpr.org Web Site: www.mpr.org. Licensee: Minnesota Public Radio (group owner; (acq 6-10-92). Population served: 10,000 Natl. Network: NPR. Format: News & info. News staff: one. ♦Chris Cross, gen mgr; Mary Stapek, dev dir; Sea Stachura, news rptr.

***KWLC(AM)**— December 1926: 1240 khz; 1 kw-U. TL: N43 18 38 W91 48 41. Hrs open: 700 College Dr., 52101. Phone: (563) 387-1240. Fax: (563) 387-1489. Web Site: kwlc.luther.edu. Licensee: Luther College. Population served: 30,000 Format: Var/div, progsv. Target aud: General.

Denison

KDSN(AM)— Apr 11, 1956: 1530 khz; 500 w-D, 13 w-N. TL: N42 02 10 W95 19 44. Hrs open: 6 AM-10 PM Box 670, 51442. Secondary address: 1530 Ridge Rd. 51442. Phone: (712) 263-3141. Fax: (712) 263-2088. E-mail: info@kdsnradio.com Web Site: www.kdsnradio.com. Licensee: M & J Radio Corp. (acq 8-3-93; $450,000 with co-located FM; FTR: 8-23-93). Population served: 30,000 Rgnl. Network: Agri-Net, Radio Iowa. Natl. Rep: Farmakis. Format: Country, adult contemp, farm markets. News staff: one; News: 8 rs wkly. Target aud: General. Spec prog: Farm 12 hrs, polka 4 hrs, Sp 3 hrs wkly. ♦Michael Dudding, pres, exec VP & gen mgr; Phyllis Rohlin, exec VP & gen mgr.

KDSN-FM— Aug 1, 1968: 107.1 mhz; 6 kw. 300 ft TL: N42 02 11 W95 19 50. Stereo. 24 Web Site: www.kdsnradio.com.30,000 Format: Adult contemp. News staff: one; News: 8 hrs wkly. Target aud: 25-54. ♦Michael Dudding, adv dir; Dick Keane, engrg dir; Kathy Dudding, traf mgr; Brian Schmid, farm dir; Randy Grossman, sports cmtr; Joelle Cooper, women's int ed; Nicole Thunder, disc jockey.

Des Moines

KBGG(AM)— 1998: 1700 khz; 10 kw-D, 1 kw-N. TL: N41 35 30 W93 31 43. Hrs open: 24 4143 109th St., Urbandale, 50322. Phone: (515) 331-9200. Fax: (515) 331-9292. Fax: (515) 254-1037. Licensee: Citadel Broadcasting Co. Group owner: Citadel Broadcasting Corp. (acq 8-29-2003; grpsl). Natl. Rep: Christal. Wire Svc: UPI Format: Spanish. Target aud: 25 plus. ♦Jack O'Brien, opns mgr; Doug Wood, sls dir; Terry Peters, VP & mktg mgr.

***KDFR(FM)**— Mar 24, 1989: 91.3 mhz; 32 kw. 446 ft TL: N41 36 59 W93 31 36. Stereo. Hrs opn: 24 Box 57023, 50317. Secondary address: 2350 N.E. 44th Ct. 50317. Phone: (515) 262-0449. E-mail: kdfr@familyradio.org Web Site: www.familyradio.com. Licensee: Family Stations Inc. (group owner) Population served: 600,000 Format: Relg, inspirational. News staff: one; News: 6 hrs wkly. Target aud: 25 plus; general. Spec prog: Class 2 hrs wkly. ♦Harold Camping, pres; Larry Vavroch, opns mgr.

KDRB(FM)—Listing follows WHO(AM).

KGGO(FM)— May 31, 1964: 94.9 mhz; 100 kw. 1,066 ft TL: N41 37 55 W93 27 27. Stereo. Hrs opn: 24 4143 109th St., Urbandale, 50322. Phone: (515) 331-9200. Fax: (515) 312-9292. Web Site: www.kggo.com. Licensee: Citadel Broadcasting Co. Group owner: Citadel Broadcasting Corp. (acq 8-29-03; grpsl). Natl. Rep: Christal. Format: Classic Rock. ♦Jack O'Brien, gen mgr & opns mgr; Doug Wood, sls dir; Terry Peters, mktg VP.

KHKI(FM)— July 4, 1964: 97.3 mhz; 105 kw. Ant 469 ft TL: N41 39 46 W93 45 24. Stereo. Hrs opn: 24 4143 109th St., Urbandale, 50322. Phone: (515) 331-9200. Fax: (515) 331-9292. Web Site: 973thehawk.com. Licensee: Citadel Broadcasting Co. Group owner: Citadel Broadcasting Corp. (acq 8-29-03; grpsl). Population served: 400,000 Natl. Rep: Christal. Format: Country. Target aud: 18-49. ♦Jack O'Brien, opns VP & progmg dir; Doug Wood, sls dir; Terry Peters, gen mgr & mktg VP.

KIOA(FM)— Sept 18, 1964: 93.3 mhz; 100 kw. 1,063 ft TL: N41 37 54 W93 27 24. Stereo. Hrs opn: 24 1416 Locust St., 50309. Phone: (515) 280-1350. Fax: (515) 280-3011. Web Site: www.kioa.com. Licensee: Saga Communications of Iowa LLC. Group owner: Saga Communications Inc. (acq 4-19-93; $2.7 million with co-located AM; FTR: 5-3-93). Population served: 410,000 Law Firm: Smithwick & Belendiuk. Format: Oldies. News staff: one. Target aud: 25-54. ♦Bill Wells, gen mgr; Jeff Delvaux, sls dir; Jill Olsen, prom mgr; Don Tool, adv mgr; Tim Fox, progmg dir; Polly Carverkimm, news dir; Joe Farrington, chief of engrg; Lee Ann Rose, traf mgr.

KPSZ(AM)—Co-owned with KIOA(FM). April 1947: 940 khz; 10 kw-D, 5 kw-N, DA-2. TL: N41 28 35 W93 22 26.750,000 Format: Christian contemp mus, talk. ♦Mary Sayre, sls dir; Joe Acker, prom dir; Steve Gibbons, progmg mgr.

KJJY(FM)—West Des Moines, Feb 4, 1978: 92.5 mhz; 41 kw. Ant 541 ft TL: N41 39 53 W93 45 25. Stereo. Hrs opn: 24 4143 109th St., Urbandale, 50322. Phone: (515) 331-9200. Fax: (515) 331-9292. Web Site: www.kjjy.com. Licensee: Citadel Broadcasting Co. Group owner: Citadel Broadcasting Corp. (acq 8-29-03; grpsl). Population served: 758,000 Natl. Rep: Christal. Format: Country. Target aud: 25-54; general. ♦Jack O'Brien, opns mgr; Doug Wood, sls dir & mktg dir; Terry Peters, gen mgr & mktg VP.

***KJMC(FM)**— May 1999: 89.3 mhz; 7.1 kw. Ant 200 ft TL: N41 39 21 W93 35 51. Stereo. Hrs opn: 24 1169 25th St., 50311. Phone: (515) 279-1811. Fax: (515) 279-1802. Licensee: Minority Communications Inc. Population served: 400,000 Natl. Network: ABC. Format: Urban contemp, hits, oldies. ♦Larry Rollins, gen mgr; Larry Neville, opns VP; John Farington, chief of opns.

KKDM(FM)— Aug 22, 1995: 107.5 mhz; 100 kw. 705 ft TL: N41 38 36 W93 17 21. Hrs open: 24 2141 Grand Ave., 50312. Phone: (515) 245-8900. Fax: (515) 245-8906. Web Site: www.kkdm.com. Licensee: Clear Channel Broadcasting Licenses Inc. Group owner: Clear Channel Communications Inc. (acq 9-1-99; $7.35 million). Natl. Rep: Clear Channel. Format: CHR. Target aud: 18-49. ♦Joel McCrea, gen mgr; Matt Gillon, sls dir; Greg Chance, progmg dir; Sean Cage, mus dir.

KRNT(AM)—Listing follows KSTZ(FM).

KSTZ(FM)— 1970: 102.5 mhz; 100 kw. 1,248 ft TL: N41 48 01 W93 36 27. Stereo. Hrs opn: 24 1416 Locust St., 50309. Phone: (515) 280-1350. Fax: (515) 280-3011. E-mail: jms720@aol.com Web Site: www.star1025.com. Licensee: Saga Communications of Iowa LLC. Group owner: Saga Communications Inc. (acq 8-88; $3.2 million with co-located AM; FTR: 8-1-88) Population served: 456,400 Natl. Network: CNN Radio. Natl. Rep: Katz Radio. Law Firm: Smithwick & Belendiuk. Wire Svc: AP Format: Hot adult contemp. News staff: one. Target aud: 25-54; emphasis on upscale women. ♦Jeremy Dresen, gen mgr & sls.

KRNT(AM)—Co-owned with KSTZ(FM). Mar 17, 1935: 1350 khz; 5 kw-U, DA-N. TL: N41 33 34 W93 34 40.24 E-mail: krntpdsteve@hotmail.com Web Site: www.1350krnt.com. Natl. Network: CBS. Natl. Rep: Katz Radio. Law Firm: Smithwick & Belendiuk. Format: MOR. News staff: one. Target aud: 50 plus. ♦Jill Olsen, mktg dir, prom dir & adv dir; Jim Brown, progmg dir.

KWKY(AM)— Feb 2, 1948: 1150 khz; 1 kw-U, DA-2, 2.5 kw-N. TL: N41 27 07 W93 40 44. Hrs opn: 24 St. Gabriel Communications, Box 160, Norwalk, 50211. Phone: (515) 223-1150. Fax: (515) 981-0840. E-mail: info@kwky.com Web Site: www.kwky.com. Licensee: Putbrese Communications Ltd. (acq 10-3-2006; $2.04 million). Population served: 400,000 Natl. Network: EWTN Radio. Format: Catholic radio, talk, sports. ♦John Putbrese, pres; Charles E. Putbrese, gen mgr; Matthew Phelps, opns mgr; Dennis Ray, mus dir; Jon Farrington, chief of engrg.

KXNO(AM)— July 21, 1921: 1460 khz; 5 kw-U, DA-N. TL: N41 38 45 W93 32 12. Stereo. Hrs opn: 24 2141 Grand Ave., 50312. Phone: (515) 245-8900. Fax: (515) 245-8906. E-mail: kxno@clearchannel.com Web Site: www.kxno.com. Licensee: Capstar TX L.P. Group owner: Clear Channel Communications Inc. (acq 8-30-00; grpsl). Population served: 350,000 Natl. Network: Fox Sports. Format: Sports. Target aud: 35 plus; 25-54 Male. ♦Joel McCrea, gen mgr; Geoff Conn, opns mgr; Matt Gillon, sls dir & gen sls mgr; Van Harden, progmg dir; Jim Boyd, news dir; Raleigh Rubenking, chief of engrg; Julie Traver, traf mgr; Molly Pins, spec ev coord.

WHO(AM)— Apr 10, 1924: 1040 khz; 50 kw-U. TL: N41 39 10 W93 21 01. Hrs open: 24 2141 Grand Ave., 50312. Phone: (515) 245-8900. Web Site: www.whoradio.com. Licensee: Citicasters Licenses L.P. Group owner: Clear Channel Communications Inc. (acq 5-4-99; grpsl). Population served: 419,000 Natl. Network: Fox News Radio. Rgnl. Network: Iowa Radio Net. Natl. Rep: Christal. Format: News/talk. Target aud: General; info seeking adults. Spec prog: Farm 15 hrs wkly. ♦Cheryl Pannier, opns mgr; Cathy Erickson, prom dir & spec ev coord; Van Harden, progmg dir; Jim Boyd, news dir; Bonnie Lucas, pub affrs dir; Raleigh Ruben King, chief of engrg; Julie Traver, traf mgr; Geoff Conn, sports cmtr.

KDRB(FM)—Co-owned with WHO(AM). Feb 1, 1948: 100.3 mhz; 100 kw. Ant 1,700 ft TL: N41 48 33 W93 36 53. Stereo. 111,300 Natl. Network: Westwood One. Format: Hot adult contemp. Target aud: 25-54.

Dubuque

KATF(FM)—Listing follows KDTH(AM).

KDTH(AM)— May 4, 1941: 1370 khz; 5 kw-U, DA-N. TL: N42 29 06 W90 38 39. Hrs open: 24 Box 659, 52004-0659. Secondary address: 346 W. 8th St. 52001. Phone: (563) 690-0800. Fax: (563) 588-5688. E-mail: kdth@kdth.com Web Site: www.kdth.com. Licensee: Radio Dubuque Inc. (group owner; acq 7-1-00; $3.68 million with co-located FM). Population served: 129,800 Natl. Network: CBS. Natl. Rep: Katz Radio. Law Firm: Pepper & Corazzini. Wire Svc: NWS (National Weather Service) Format: Full service. News staff: 3; News: 25 hrs wkly. Target aud: 35 plus; responsible adults with established careers & households. Spec prog: Farm 17 hrs wkly. ♦Thomas Parsley, stn mgr; Perry Mason, gen sls mgr & natl sls mgr; Michael Kaye, progmg dir; Ed Anderson, news dir.

KATF(FM)—Co-owned with KDTH(AM). June 25, 1967: 92.9 mhz; 89.7 kw. Ant 1,014 ft TL: N42 31 44 W90 36 58. Stereo. 24 Phone: (563) 690-0800. Format: Adult contemp. News: 3 hrs wkly. Target aud: 25-54; adults establishing families, careers & households. ♦Thomas Parsley, gen mgr.

***KDUB(FM)**— 2005: 89.7 mhz; 530 w horiz, 2.6 kw vert. Ant 646 ft TL: N42 36 18 W90 47 57. Hrs open:
Rebroadcasts KUNI(FM) Cedar Falls 100%.
324 Communications Arts Center, Univ. of Northern Iowa, Cedar Falls, 50614. Phone: (319) 273-6400. Fax: (319) 273-2682. E-mail: kuni@uni.edu Web Site: www.kuniradio.org. Licensee: University of Northern Iowa. Format: News/talk. News staff: 3; News: 77 hours. Spec prog: Folk, Blues. ♦Scott Vezdos, mktg dir; Wayne Jarvis, gen mgr & progmg dir; Al Shares, mus dir; Greg Shanley, news dir; Steve Schoon, chief of engrg.

***KIAD(FM)**— 2006: 88.5 mhz; 750 w vert. Ant 518 ft TL: N42 24 16 W90 34 12. Hrs open:
Rebroadcasts WAFR(FM) Tupelo, MS 100%.
Drawer 2440, Tupelo, MS, 38803. Phone: (662) 844-8888. Fax: (662) 842-6791. Web Site: www.afr.net. Licensee: American Family Association. Format: Christian. ♦Marvin Sanders, gen mgr.

KLYV(FM)—Listing follows WDBQ(AM).

KXGE(FM)— Mar 8, 1980: 102.3 mhz; 2.4 kw. 410 ft TL: N42 32 28 W90 36 46. Stereo. Hrs opn: 24 5490 Saratoga Rd., 52002. Phone: (563) 557-1040. Fax: (563) 583-4535. Web Site: www.eagle102online.com. Licensee: Cumulus Licensing Corp. Group owner: Cumulus Media Inc. (acq 12-17-98; grpsl). Population served: 246,290 Natl. Network: ABC. Format: Classic rock. News staff: one; News: 2 hrs wkly. Target aud: 18-49; in high school or college in the 60s & 70s. ♦Dan Sullivan, gen mgr; Doris Garius, gen sls mgr; Scott Thomas, progmg dir; Tom Berryman, news dir.

WDBQ(AM)— Oct 30, 1933: 1490 khz; 1 kw-U. TL: N42 30 10 W90 42 24. Stereo. Hrs opn: 5490 Saratoga Rd., 52002. Phone: (319) 583-6471. Fax: (319) 583-4535. Web Site: www.cumulus.com. Licensee: Cumulus Licensing Corp. Group owner: Cumulus Media Inc. (acq 12-17-98; grpsl). Population served: 64,000 Natl. Network: ABC, Westwood One. Law Firm: C. Reynolds. Format: News, talk, sports. ♦Jack Kilcoyne, progmg dir, news dir & sports cmtr; Alan Williams, traf mgr; Mike Field, disc jockey.

KLYV(FM)—Co-owned with WDBQ(AM). Sept 1, 1965: 105.3 mhz; 50 kw. 330 ft TL: N42 30 10 W90 42 11. Stereo. Phone: (563) 557-1040. Web Site: www.y105online.com.130,000 Format: CHR. Target aud: 18-49. ♦Scott Thomas, progmg dir.

Dunkerton

KCOO(FM)—Not on air, target date: unknown: 103.9 mhz; 6 kw. Ant 312 ft TL: N42 42 23.9 W92 13 03.7. Hrs open: 2801 Via Fortuna Dr., Suite 675, Austin, TX, 78746. Phone: (713) 528-2517. Licensee: Ace Radio Corp. ♦Stephen Hackerman, pres.

Dyersville

KDST(FM)— Aug 25, 1985: 99.3 mhz; 3 kw. 298 ft TL: N42 25 43 W91 12 50. Stereo. Hrs opn: 24 1931 20th Ave. S.E., 52040. Phone: (563) 875-8193. Fax: (563) 875-6001. E-mail: kdst993@iowatelecom.net Web Site: www.realcountryonline.com. Licensee: Design Homes Inc. (acq 12-88; $22,079; FTR: 12-26-88). Natl. Network: ABC. Rgnl. Network: Brownfield. Law Firm: Miller & Miller. Format: Country. News staff: one. Target aud: 45-60. Spec prog: Farm. ♦Randy Weeks, CEO; Franklin Weeks, pres; Doug Langston, stn mgr & opns mgr.

Eagle Grove

***KJYL(FM)**— Feb 20, 1994: 100.7 mhz; 25 kw. 328 ft TL: N42 40 18 W94 09 11. Hrs open: 24 Box 325, 103 W. Broadway, 50533. Phone: (515) 448-4588. Fax: (515) 448-5267. Web Site: www.kjyl.org. Licensee: Minn-Iowa Christian Broadcasting Inc. (group owner). Population served: 200,000 Format: Christian. News staff: one. Target aud: 30-55. ♦Jay Rudolph, opns mgr; Chris Sloan, disc jockey.

Eddyville

KKSI(FM)— July 30, 1990: 101.5 mhz; 49 kw. 498 ft TL: N41 07 57 W92 42 12. Stereo. Hrs opn: 24 416 E. Main St., Ottumwa, 52501. Phone: (641) 684-5563. Fax: (641) 684-5832. E-mail: mail@ottumwaradio.com Web Site: www.ottumwaradio.com. Licensee: "O"-Town Communications Inc. (acq 12-10-99; $162,400). Law Firm: Miller & Neely, P.C. Format: Classic Rock. News staff: 2; News: 4 hrs wkly. Target aud: 25-54. ♦Greg H. List, pres; Bruce Linder, VP; Jeff Downing, opns dir; Mike Buchanan, news dir.

Eldon

KRKN(FM)— 1996: 104.3 mhz; 23.5 kw. Ant 341 ft TL: N40 52 06 W92 18 20. Stereo. Hrs opn: 24 416 E. Main St., Ottumwa, 52501. Phone: (641) 684-5563. Fax: (641) 684-5832. E-mail: mail@ottumwaradio.com Web Site: www.ottumwaradio.com. Licensee: O-Town Communications Inc. (acq 12-10-99; $162,400). Law Firm: Miller & Neely, P.C. Format: New country. News staff: 2; News: 4 hrs wkly. Target aud: 18-54. ♦Greg H. List, pres & stn mgr; Bruce Linder, VP; Jeff Downing, opns mgr; Mike Buchanan, news dir.

Eldora

KDAO-FM— June 1, 1992: 99.5 mhz; 3 kw. 328 ft TL: N42 15 49 W93 03 57. Stereo. Hrs opn: 24 Box 538, Marshalltown, 50158. Secondary address: 1930 N. Center St., Marshalltown 50158. Phone: (641) 752-4122. Fax: (641) 752-5121. Licensee: Eldora Broadcasting Co. Inc. (acq 12-18-91; $15,000 for CP; FTR: 1-13-92). Natl. Network: Fox News Radio. Format: Adult contemp. Target aud: 25-54. ♦Mark Osmundson, gen mgr.

Elkader

KADR(AM)— May 15, 1983: 1400 khz; 1 kw-U. TL: N42 50 57 W91 24 43. Hrs open: Box 990, 52043. Phone: (563) 245-1400. Fax: (563) 245-1402. Web Site: www.hitsandfavorites.com. Licensee: KADR-AM 14, div of Design Homes Inc. (acq 3-20-85). Natl. Rep: Farmakis. Format: Adult contemp. ♦Dan Berns, gen mgr; Troy Thein, chief of opns.

KCTN(FM)—See Garnavillo

Emmetsburg

KUYY(FM)— Jan 10, 1977: 100.1 mhz; 16 kw. Ant 300 ft TL: N43 01 20 W94 41 59. Stereo. Hrs opn: 24 2303 W. 18th St., Spencer, 51301. Phone: (712) 264-1074. Fax: (712) 264-1077. E-mail: mspies@nrgmedia.com Licensee: Jim Dandy Broadcasting Inc. (acq 1-13-2003; $2.5 million with KKIA(FM) Ida Grove). Population served: 12,000 Rgnl. Network: Iowa Radio Network. Format: Active adult contemp. Target aud: 25-54. ♦Marty Spies, gen mgr; Stan Calvert, opns mgr & progmg dir; Stephanie Haviland, gen sls mgr; Steve Heaton, chief of engrg.

Epworth

KGRR(FM)— Dec 10, 1994: 97.3 mhz; 19 kw. 380 ft TL: N42 26 13 W90 50 43. Hrs open: 24 Box 659, Dubuque, 52004. Secondary address: 346 W. 8th St., Dubuque 52004. Phone: (563) 690-0800. Fax: (563) 588-5688. E-mail: kgrr@kgrr.com Web Site: kgrr.com. Licensee: Radio Dubuque Inc. (group owner; acq 7-1-00; $1.5 million). Population served: 150,000 Natl. Rep: Katz Radio. Format: Classic hits, classic rock. News staff: one; News: 2 hrs wkly. Target aud: 25-54; families. ♦Don Rabbitt, CEO; Paul Hemmer, VP; Thomas Parsley, pres, gen mgr & progmg dir.

Estherville

KILR(AM)— Dec 23, 1967: 1070 khz; 250 w-D, 48 w-N, DA-2. TL: N43 25 45 W94 49 23. Hrs open: 6 AM-2 hrs past sunset Box 453, 3875 150th St., 51334. Secondary address: 3875 150th St. 51334. Phone: (712) 362-2644. Fax: (712) 362-5951. E-mail: ubcbroadcast@netins.net Licensee: Jacobson Broadcasting Co. Inc. (acq 7-1-82; $610,000 with co-located FM; FTR: 7-5-82) Population served: 40,000 Natl. Network: ABC. Natl. Rep: Farmakis. Law Firm: Lauren A. Colby. Format: News/talk. News staff: one; News: 24 hrs wkly. Target aud: 29-65; loc baby boomers. Spec prog: Farm 9 hrs, relg 11 hrs wkly. ♦Barbara J. Jacobson, CFO; Peggy Zahrt, opns mgr; Ed Funston, news dir; Roger J. Jacobson, pres, gen mgr, gen sls mgr, prom dir, progmg dir & chief of engrg.

KILR-FM— Oct 17, 1969: 95.9 mhz; 20 kw. Ant 325 ft TL: N43 25 45 W94 49 23. Stereo. 3875 150th St., 51334. E-mail: kilrprod@netins.net Natl. Rep: Salem. Format: Country, sports. Target aud: 25-54.

Fairfield

***KHOE(FM)**— 1994: 90.5 mhz; 100 w. 98 ft TL: N41 00 59 W91 58 09. Stereo. Hrs opn: 24 Box 1017, 1000 N. 4th St., 52557. Phone: (641) 469-5463. E-mail: khoe@mum.edu Licensee: Fairfield Educational Radio Station. Population served: 14,000 Format: World music, class, educ. News: 2 hrs wkly. Target aud: 18-35; Univ. audience, College. Spec prog: Children 3 hrs, folk 6 hrs, gospel 3 hrs, jazz 2 hrs, Sp 2 hrs wkly. ♦Bill Goldstein, CEO; Jeffrey Hedquist, pres; Stan Stansberry, gen mgr.

KIIK-FM—Listing follows KMCD(AM).

KMCD(AM)— Mar 3, 1958: 1570 khz; 250 w-D, 108 w-N. TL: N41 00 25 W92 00 50. Hrs open: 24 Box 648, 57 S. Court St., 52556. Phone: (641) 472-4191. Fax: (641) 472-2071. E-mail: jay.mitchell@radiovillage.com Web Site: www.radiovillage.com. Licensee: Fairfield Media Group Inc. (group owner; (acq 2-28-94; $200,000 with co-located FM; FTR: 5-2-94). Population served: 30,000 Format: News/talk. News staff: 2; News: 22 hrs wkly. Target aud: 30 plus; community leaders & Jefferson County. Spec prog: Farm 10 hrs wkly. ♦Jay Mitchell, pres & gen mgr; Bob Harvey, gen sls mgr; Steve Smith, progmg dir; Emily Humble, news dir; E. Marie Kiefer, traf mgr.

KIIK-FM—Co-owned with KMCD(AM). 1977: 95.9 mhz; 6 kw. 400 ft TL: N40 58 47 W92 05 45. (CP: 4.1 kw). Stereo. 24 Web Site: www.radiovillage.com.70,000 Natl. Network: ABC. Format: Adult contemp. News staff: one; News: 12 hrs wkly. Target aud: 25-54.

***KUNJ(FM)**—Not on air, target date: unknown: 88.1 mhz; 250 w. Ant 325 ft TL: N41 05 21 W91 58 05. Hrs open: University of Northern Iowa, 324 Communications Arts Center, Cedar Falls, 50614-0359. Phone: (319) 273-6400. Fax: (319) 273-2682. Licensee: University of Northern Iowa. ♦Wayne Jarvis, gen mgr.

Forest City

KIOW(FM)— Nov 8, 1978: 107.3 mhz; 25 kw. 328 ft TL: N43 17 02 W93 37 50. Stereo. Hrs opn: 24 Box 308, 50436. Secondary address: 18643 360th St. 50436. Phone: (641) 585-1073. Fax: (641) 585-2990. E-mail: kiow@kiow.com Web Site: www.kiow.com. Licensee: Pilot Knob Broadcasting Inc. Population served: 150,000 Natl. Network: CNN Radio. Rgnl. Network: Radio Iowa. Natl. Rep: Farmakis. Wire Svc: AP Format: Country, adult contemp, news. News staff: one; News: 15 hrs wkly. Target aud: General; Adults 25 +. Spec prog: Farm 15 hrs, contemp hits 19 hrs wkly. ♦Susan I. Coloff, CFO; Tony Coloff, pres & gen mgr.

Fort Dodge

***KEGR(FM)**— 2005: 89.5 mhz; 17 kw vert. Ant 364 ft TL: N42 40 18 W94 09 11. Hrs open: Box 286, Shenandoah, 51601. Phone: (515) 545-4841. Web Site: www.familyradio.com. Licensee: Family Stations Inc. (group owner). Natl. Network: Family Radio. Format: Relg.

***KICB(FM)**— September 1971: 88.1 mhz; 200 w. 130 ft TL: N42 29 27 W94 12 01. Stereo. Hrs opn: 330 Ave. M, 50501. Phone: (515) 576-6049. Fax: (515) 576-5656. Licensee: Iowa Central Community College. Population served: 31,263 Format: Alternative. Target aud: 13-34; young men & women with progsv tastes. ♦Robert Paxton, pres; Amy Simpson, gen mgr; Jeff Nelsen, mus dir & chief of engrg.

KKEZ(FM)—Listing follows KWMT(AM).

***KTPR(FM)**— Sept 15, 1980: 91.1 mhz; 100 kw. Ant 1,052 ft TL: N42 49 03 W94 24 41. Stereo. Hrs opn: 24
WOI-AM.
WOI Radio Group, 2022 Communications Bldg., Ames, 50011-3241. Phone: (515) 294-2025. E-mail: woi@iastate.edu Web Site: www.woi.org. Licensee: Iowa State University of Science and Technology. Population served: 17,000 Natl. Network: NPR, PRI. Format: Class, jazz, news. News staff: 2; News: 41 hrs wkly. Target aud: General; educated. Spec prog: New age 10 hrs wkly. ♦Don Wirth, opns mgr; Dave Becker, progmg dir.

KUEL(FM)—Listing follows KVFD(AM).

KVFD(AM)— Dec 24, 1939: 1400 khz; 1 kw-U. TL: N42 28 44 W94 12 10. Hrs open: 24 Box Y, 200 N. 10th St., 50501. Phone: (515) 955-1400. Fax: (515) 955-5844. Licensee: Three Eagles of Joliet Inc. Group owner: Three Eagles Communications (acq 6-18-2004; grpsl). Population served: 135,000 Natl. Network: ABC. Wire Svc: AP Format: Sports, news, oldies. ♦Gary Buchanan, pres; Dennis Martin, gen mgr; Jay Alexander, opns mgr; Mike Laughter, engrg VP.

KUEL(FM)—Co-owned with KVFD(AM). July 28, 1975: 92.1 mhz; 3 kw. 300 ft TL: N42 28 44 W94 12 10. (CP: Ant 321 ft.). Stereo. Phone: (515) 955-5656.58,000 Format: Adult contemp. ♦Rolland C. Johnson, chmn; Jay Alexander, progmg dir; Mike Laughter, engrg dir.

KWMT(AM)— April 1956: 540 khz; 5 kw-D, 200 w-N, DA-2. TL: N42 22 94 W94 12 27. Hrs open: 540 A St., 50501. Phone: (515) 576-7333. Fax: (515) 955-4250. Web Site: www.kwmt.com. Licensee: Clear Channel Broadcasting License Inc. Group owner: Clear Channel Communications Inc. (acq 7-19-99; $7.5 million with co-located FM). Population served: 68,000 Natl. Rep: McGavren Guild. Law Firm: Reddy, Begley & McCormick. Format: Country. Target aud: General. Spec prog: Farm. ♦Ron Revere, gen mgr.

KKEZ(FM)—Co-owned with KWMT(AM). 1966: 94.5 mhz; 100 kw. 640 ft TL: N42 29 43 W94 12 33. Stereo. 24 E-mail: kkez@clearchannel.com Web Site: www.kkez.com.31,623 Format: Adult contemp. News staff: 3; News: 5 hrs wkly. Target aud: 18-49.

Fort Madison

KBKB(AM)— Feb 6, 1948: 1360 khz; 1 kw-D, 35 w-N. TL: N40 39 30 W91 16 20. Hrs open: 18 Box 70, Burlington, 52601. Secondary address: 2060 Hwy. 61, Burlington 52601. Phone: (319) 372-5252. Fax: (319) 752-5287. Web Site: www.1360kbkb.com. Licensee: Citicasters Licenses L.P. Group owner: Clear Channel Communications Inc. (acq 5-4-99; grpsl). Population served: 125,000 Natl. Network: ABC. Natl. Rep: Farmakis. Law Firm: Fisher, Wayland, Cooper, Leader & Zaragoza L.L.P. Format: News/talk. News staff: 2; News: 30 hrs wkly. Target aud: 30-55. ♦Steve Staebell, gen mgr; Steve Hexom, progmg dir; J.K. Martin, news dir & local news ed; Brad Bostrom, chief of engrg; Naomi Martig, news rptr.

KBKB-FM— June 1, 1973: 101.7 mhz; 50 kw. 466 ft TL: N40 43 25 W91 13 49. Stereo. 24 Web Site: www.1017thebull.com.187,000 Natl. Network: ABC. Format: Country. News staff: 2; News: 7 hrs wkly. ♦Kosmo Leone, progmg dir; J.K. Martin, local news ed.

Garnavillo

KCTN(FM)— Dec 6, 1982: 100.1 mhz; 3 kw. 300 ft TL: N42 53 06 W91 19 11. Stereo. Hrs opn: 24 Box 990, Elkader, 52043. Phone: (563) 245-1400. Fax: (563) 245-1402. E-mail: kctn@alpinecom.net Web Site: kctn.com. Licensee: KCTN-FM 100 div of Design Homes Inc. Rgnl. Network: Brownfield. Natl. Rep: Farmakis. Format: Country. News staff: one. Target aud: 24-55; farmers & rural communities. Spec prog: Farm. ♦Randy Weeks, CEO; Dan Berns, gen mgr & opns mgr; Troy Thein, chief of opns.

Glenwood

KXKT(FM)— Apr 8, 1966: 103.7 mhz; 100 kw. Ant 1,086 ft TL: N41 18 32 W96 01 33. Stereo. Hrs open: 24 5010 Underwood Ave., Omaha, NE, 68132. Phone: (402) 561-2000. Fax: (402) 556-8937. E-mail: request@thekat.com Web Site: www.thekat.com. Licensee: Capstar TX L.P. Group owner: Clear Channel Communications Inc. (acq 8-30-2000; grpsl). Population served: 1250000 Law Firm: Haley, Bader & Potts. Format: Country. News staff: one. Target aud: 18-54; general. ♦Donna Baker, gen mgr; Mitch Baker, opns mgr.

Grinnell

KGRN(AM)— Nov 15, 1957: 1410 khz; 500 w-D, 47 w-N. TL: N41 44 44 W92 42 36. (CP: 300 w-D, 33 w-N. TL: N41 46 35 W92 38 56). Hrs opn: Box 660, 50112. Phone: (641) 236-6106. Fax: (641) 236-8896. E-mail: kgrn@iowatelecom.net Web Site: www.kgrn1410.com. Licensee: Crawford Broadcasting Co. (acq 12-5-97; $560,000 for stock). Population served: 55,000 Rgnl. Network: Iowa Radio Net. Natl. Rep: Farmakis. Format: Adult Contemp. Spec prog: Farm 12 hrs, C&W 12 hrs wkly. ♦Russ Crawford, pres & gen mgr.

KRTI(FM)— May 1993: 106.7 mhz; 50 kw. 492 ft TL: N41 48 16 W92 40 09. Hrs open: 24 Box 306, 50112. Secondary address: 1801 N. 13th Ave. E., Newton 50208. Phone: (641) 792-5262. Fax: (641) 792-8403. Web Site: www.energy1067.com. Licensee: Central Iowa Broadcasting Inc. (acq 12-6-93; $350,000; FTR: 1-3-94). Population served: 200,000 Format: CHR mainstream. ♦Frank Liebl, gen mgr; Tim Graves, opns dir.

Grundy Center

KCRR(FM)— Oct 8, 1983: 97.7 mhz; 16 kw. 407 ft TL: N42 23 28 W92 13 57. Stereo. Hrs opn: 24 501 Sycamore St., Suite 300 Black's Bldg., Waterloo, 50703. Phone: (319) 833-4800. Fax: (319) 833-4866. Licensee: Cumulus Licensing Corp. Group owner: Cumulus Media Inc. (acq 3-15-00; grpsl). Population served: 168,000 Format: Classic rock. News staff: 2; News: 3 hrs wkly. Target aud: 25-54. ♦Lew Dickey, CEO; William Hathaway, gen mgr; Dick Stadlen, opns mgr.

Hampton

KLMJ(FM)— May 16, 1983: 104.9 mhz; 6 kw. 255 ft TL: N42 49 45 W93 11 10. Stereo. Hrs opn: 24 Box 495, 50441. Secondary address: 1509 4th St. N.E. 50441. Phone: (641) 456-5656. Fax: (641) 456-5655. E-mail: klmj@klmj.com Web Site: www.klmj.com. Licensee: C.D. Broadcasting Inc. (acq 10-93; $60,000; FTR: 10-11-93). Population served: 100,000 Rgnl. Network: Radio Iowa, Brownfield. Natl. Rep: Farmakis. Law Firm: Fletcher, Heald & Hildreth, P.L.C. Format: Adult contemp,country, oldies. News staff: 2; News: 14 hrs wkly. Target aud: 25 plus; general. Spec prog: Iowa State & Univ. of Northern Iowa, farm 8 hrs wkly. ♦Craig Donnelly, gen mgr; Marlin Burrier, opns dir.

Harlan

KNOD(FM)— Nov 12, 1979: 105.3 mhz; 25 kw. 300 ft TL: N41 37 00 W95 16 10. Stereo. Hrs opn: 24 Box 723, 51537. Phone: (712) 755-3883. Fax: (712) 755-7511. E-mail: knodnews@harlannet.com Web Site: knodfm.com. Licensee: Wireless Broadcasting L.L.C. (acq 5-23-02). Population served: 22,500 Rgnl. Network: Brownfield. Format: Oldies. News staff: one; News: 5 hrs wkly. Target aud: 25-50. Spec prog: Farm 3 hrs, relg 2 hrs wkly. ♦Judy Storm, gen mgr & gen sls mgr; Richard Keane, chief of opns; Jason Dinesen, news dir.

Hiawatha

***KWOF-FM**— 2002: 89.1 mhz; 400 w vert. Ant 400 ft TL: N42 03 13 W91 44 35. Hrs open: 1450 Boyson Rd, Bldg C 3-2, 52233. Phone: (319) 378-8600. Phone: (319) 236-5700. Fax: (319) 236-8777. E-mail: studio@891thespirit.com Web Site: www.891thespirit.com. Licensee: Friendship Communications Inc. Format: Christian hit radio. ♦Michael James, stn mgr.

Hudson

KCVM(FM)— Aug 27, 1997: 96.1 mhz; 6 kw. 312 ft TL: N42 23 33 W92 30 44. Stereo. Hrs opn: 24 Box 248, 721 Shirley St., Cedar Falls, 50613. Phone: (319) 277-1918. Phone: (319) 266-6499. Fax: (319) 277-5202. E-mail: themix@mix96.net Web Site: www.mix96.net. Licensee: Fife Communications Co. L.C. Population served: 120,000 Format: Adult contemp. News staff: one; News: 2 hrs wkly. Target aud: 25-54; eastern Iowa adult females. ♦Jim Coloff, pres, gen mgr, opns VP & opns mgr; Tony Coloff, VP; Jay Rhymer, prom dir; Teri Lynn, progmg dir.

Humboldt

KHBT(FM)— Aug 5, 1970: 97.7 mhz; 5.8 kw. 275 ft TL: N42 43 57 W94 12 23. Stereo. Hrs opn: 24 Box 217, 50548. Secondary address: 2196 Montana Ave. 50548. Phone: (515) 332-4100. Fax: (515) 332-2723. E-mail: thebolt@waittradio.com Licensee: NRG License Sub. LLC. (group owner; (acq 10-31-2005; grpsl). Population served: 44,500 Law Firm: Pepper & Corazzini. Wire Svc: AP Format: Adult contemp. News staff: one; News: 30 hrs wkly. Target aud: 30-65; general. Spec prog: Farm 10 hrs wkly. ♦Bob Ketchum, gen mgr.

Ida Grove

KKIA(FM)— September 1981: 92.9 mhz; 25 kw. Ant 328 ft TL: N42 29 23 W95 17 40. Stereo. Hrs opn: 24 606 1/2 Lake Ave., Storm Lake, 50588. Secondary address: P.O. Box 108, Storm Lake 50588. Phone: (712) 732-3520. Fax: (712) 732-1746. Licensee: Jim Dandy Broadcasting Inc. (acq 1-13-2003; $2.5 million with KUYY(FM) Emmetsburg). Population served: 50,000 Natl. Network: Fox News Radio. Natl. Rep: Farmakis. Format: Hot country. News staff: one; News: 5 hrs wkly. Target aud: 18-54. Spec prog: Farm 10 hrs wkly. ♦Buzz Paterson, stn mgr; Matt Fisher, prom mgr.

Independence

KQMG(AM)— Dec 10, 1959: 1220 khz; 250 w-D, 166 w-N. TL: N42 28 34 W91 52 31. Hrs open: 1812 Third Ave. S.E., 50644. Phone: (319) 334-3300. Fax: (319) 334-6158. E-mail: lite953@indyte.com Web Site: www.lite953.com. Licensee: KM Radio of Independence L.L.C. Group owner: KM Communications Inc. (acq 10-9-03; $500,000 with co-located FM). Population served: 20,000 Natl. Network: ABC. Format: Adult contemp. ♦Noel Showers, mus dir; Rick Peters, chief of engrg.

KQMG-FM— Jan 1, 1972: 95.3 mhz; 2.9 kw. 410 ft TL: N42 28 34 W91 52 31. Stereo. 20,000 ♦Noel Showers, progmg dir.

Indianola

***KSTM(FM)**— Apr 15, 1994: 88.9 mhz; 100 w. 124 ft TL: N41 22 00 W93 33 57. Hrs open: Simpson College, 701 N. C St., 50125. Phone:

(515) 961-1747. Phone: (515) 961-1803. Fax: (515) 961-1674. E-mail: KSTM@storm.simpson.edu Licensee: Simpson College. Format: Alt. ♦Rich Ramos, gen mgr.

KXLQ(AM)— July 22, 1963: 1490 khz; 500 w-D, 1 kw-N. TL: N41 21 24 W93 35 16. Hrs open: 810 Main St., # 228, Pella, 50219. Phone: (515) 263-0000. Licensee: Birach Broadcasting Corp. (acq 6-4-2007; $800,000 with WCXN(AM) Claremont, NC). Rgnl. Network: Brownfield. Format: Sp. ♦Joe Milledge, chief of engrg.

Iowa City

KCJJ(AM)— Oct 14, 1998: 1630 khz; 10 kw-D, 1 kw-N. TL: N41 36 03 W91 30 04. Stereo. Hrs opn: 24 Box 2118, 52244-2118. Phone: (319) 354-1242. Fax: (319) 354-1921. E-mail: kcjjam@aol.com Web Site: www.1630kcjj.com. Licensee: River City Radio Inc. (acq 9-1-94; $650,000). Population served: 100,000 Natl. Network: ABC, CBS. Format: Hot talk, hot hits. News staff: 4. Target aud: 25-54. ♦Tom Suter, gen mgr.

KKRQ(FM)—Listing follows KXIC(AM).

KRNA(FM)— Oct 4, 1974: 94.1 mhz; 100 kw. 981 ft TL: N41 45 00 W91 50 16. Stereo. Hrs opn: 24 4th Floor, 425 2nd St. S.E., Cedar Rapids, 52401-1819. Phone: (319) 365-9431. Fax: (319) 363-8062. Web Site: www.krna.com. Licensee: Cumulus Licensing Corp. Group owner: Cumulus Media Inc. (acq 2000; grpsl). Population served: 371,000 Natl. Rep: D & R Radio. Law Firm: Dow, Lohnes & Albertson. Format: Classic Rock. News staff: 2; News: 2 hrs wkly. Target aud: 18-49. ♦Jim Worthington, gen mgr; Dick Stadlen, progmg dir.

***KRUI-FM**— Mar 28, 1984: 89.7 mhz; 100 w. 90 ft TL: N41 39 29 W91 32 40. Stereo. Hrs opn: 24 379 Iowa Memorial Union, 52242. Phone: (319) 335-9525. Fax: (319) 335-9526. E-mail: krui@uiowa.edu Web Site: www.uiowa.edu/~krui. Licensee: Student Broadcasters Inc. Population served: 60,000 Format: Div, educ, progsv. News: 7 hrs wkly. Target aud: 18-34; Univ. ♦Brian Anstey, prom dir; Nate George, progmg dir; Bill Penisten, news dir; Aaron Roemig, pub affrs dir; Adam Erickson, chief of engrg; Rick Oswavay, news rptr; Ryal Brier, sports cmtr.

KSUI(FM)—Listing follows WSUI(AM).

KXIC(AM)— June 7, 1948: 800 khz; 1 kw-D, 199 w-N, DA-2. TL: N41 41 15 W91 32 39. Hrs open: 3365 Dubuque St. N.E., 52240-7970. Phone: (319) 354-9500. Fax: (319) 354-9504. Web Site: www.kxic.com. Licensee: Citicasters Licenses L.P. Group owner: Clear Channel Communications Inc. (acq 5-4-99; grpsl). Population served: 81,000 Format: News & info. ♦John Laton, gen mgr; Roy Justis, news dir.

KKRQ(FM)—Co-owned with KXIC(AM). May 1, 1966: 100.7 mhz; 100 kw. 981 ft TL: N41 45 26 W91 31 31. Stereo. Web Site: www.thesox.com.185,000 Format: Classic rock.

***WSUI(AM)**— 1919: 910 khz; 5 kw-U, DA-N. TL: N41 39 45 W91 34 30. Hrs open: 710 S. Clinton St. Bldg., Univ. of Iowa, 52242-1030. Phone: (319) 335-5730. Fax: (319) 335-6116. E-mail: wsui@uiowa.edu Web Site: wsui.uiowa.edu. Licensee: The University of Iowa. Population served: 46,850 Natl. Network: NPR. Format: News/talk. ♦John Monick, gen mgr; Dennis Reese, progmg dir; Jim Davis, chief of engrg.

KSUI(FM)—Co-owned with WSUI(AM). 1948: 91.7 mhz; 100 kw. 1,292 ft TL: N41 43 15 W91 20 30. Web Site: ksui.uiowa.edu. Format: Fine arts, class. ♦Joan Kjaer, progmg dir; Jim Davies, engrg dir & disc jockey.

Iowa Falls

KIFG(AM)— July 22, 1962: 1510 khz; 1 kw-D, 500 w-CH. TL: N42 30 49 W93 12 57. Hrs open: 406 Stevens St., 50126. Phone: (641) 648-4281. Phone: (641) 648-4282. Fax: (641) 648-4606. E-mail: kifg@iafalls.com Web Site: www.kifgradio.com. Licensee: Times-Citizen Communications Inc. (acq 9-99; $320,000 with co-located FM). Population served: 40,000 Natl. Network: CNN Radio, Westwood One. Rgnl. Network: Westwood One, CNN. Natl. Rep: Keystone (unwired net). Law Firm: Reddy, Begley & McCormick. Format: Adult contemp. News staff: one. Target aud: 25 plus. Spec prog: Farm 5 hrs wkly. ♦T.J. Norman, gen mgr & progmg dir.

KIFG-FM— Oct 1, 1965: 95.3 mhz; 6 kw. 194 ft TL: N42 30 49 W93 12 57. Stereo. 24 Web Site: www.kifgradio.com. Format: Sports, news, weather. News staff: one. ♦Ann Denholm, prom dir & prom mgr; Pat Dunn, disc jockey.

Jefferson

KGRA(FM)— Oct 1, 1981: 98.9 mhz; 11 kw. Ant 499 ft TL: N42 00 59 W94 22 26. Stereo. Hrs opn: 24 2260 141st St., Perry, 50220-0022. Phone: (515) 465-5357. Fax: (515) 465-3952. E-mail: kg98@netins.net Licensee: Coon Valley Communications (acq 1-19-94; FTR: 3-28-94). Natl. Network: ABC. Format: Classic Rock. News staff: one; News: 9 hrs wkly. Target aud: 25-49; Adults. ♦Patrick Delaney, pres, CFO, sls dir & chief of engrg; Sue Thomsen, stn mgr; Linda Hass, opns mgr.

Keokuk

***KMDY(FM)**— 2001: 90.9 mhz; 7.7 kw. Ant 197 ft TL: N40 30 41 W91 19 50. Hrs open: 521 Main St, Carthage, IL, 62321. Phone: (217) 357-3000. Fax: (217) 357-3001. Licensee: Cornerstone Community Radio Inc. (group owner) (acq 2-23-2006). Natl. Network: Moody. Format: Christian. ♦Robert Neff, VP.

KOKX(AM)— Oct 19, 1947: 1310 khz; 1 kw-D, 500 w-N, DA-N. TL: N40 22 50 W91 21 09. Hrs open: 24 Box 427, 108 Washington St., 52632. Phone: (319) 524-5410. Fax: (319) 524-7275. E-mail: gmfolluo@interlinet Licensee: Withers Broadcasting of Iowa. Group owner: Withers Broadcasting Co. (acq 7-15-81; $900,000 with co-located FM; FTR: 7-13-81) Population served: 100,000 Format: Adult standards, news/talk, sports. News staff: 2; News: 25 hrs wkly. Target aud: 25-54. Spec prog: Farm 6 hrs wkly. ♦W. Russell Withers Jr., pres; Gary M. Folluo, gen mgr, opns mgr & adv mgr; Greg Ikerd, progmg dir; Judy Hall, traf mgr; Jim Worrell, news rptr; Robert Bertram, sports cmtr.

KOKX-FM— Jan 30, 1973: 95.3 mhz; 100 kw. Ant 804 ft TL: N40 24 01 W91 35 09. Stereo. 24 Box108, 52632. Secondary address: 108 Washington St. 52632.190,000 Natl. Network: ABC. Format: Var/div. News staff: 2; News: 4 hrs wkly. Target aud: 25-54; women 50% Men 50%. ♦Jim Worrell, news dir & news rptr; Judy Hall, traf mgr; Robert Bertram, sports cmtr.

KRNQ(FM)— 1999: 96.3 mhz; 19 kw. Ant 804 ft TL: N40 24 01 W91 35 09. Stereo. Hrs opn: 24 108 Washington St., 52632. Phone: (319) 524-1111. Fax: (319) 524-7275. E-mail: gmkokx@imchsi.com Licensee: David M. Lister. Gary M. Folluo, (319) 524-5410; gmkokx@mchsi.com Population served: 100,000 Format: Classic rock. Target aud: 18-45; 50/50 male, female. ♦Gary M. Folluo, gen mgr; Dan Workman, opns dir & progmg dir; Gary Folluo, gen sls mgr; Jim Worrell, news dir.

Knoxville

KNIA(AM)— Aug 30, 1960: 1320 khz; 500 w-D, 222 w-N. TL: N41 19 40 W93 06 34. Hrs open: 24 Box 31, 50138. Secondary address: 1610 N. Lincoln 50138. Phone: (641) 842-3161. Fax: (641) 842-5606. E-mail: kniaakrls@kniakrls.com Web Site: www.kniakrls.com. Licensee: M & H Broadcasting Inc. (acq 2-23-93; $768,000 with co-located FM; FTR: 3-15-93) Population served: 7,755 Format: Real country. News staff: 3; News: 25 hrs wkly. Target aud: 25-54; female. Spec prog: Relg 18 hrs wkly. ♦Jim Butler, gen mgr.

Lake City

KIKD(FM)— 1997: 106.7 mhz; 25 kw. 328 ft TL: N42 07 14 W94 48 49. Hrs open: 24 Box 886, Carroll, 51401-0886. Secondary address: 1119 East Plaza Dr., Carroll 51401. Phone: (712) 792-4321. Fax: (712) 792-6667. E-mail: kikd@carrollbroadcasting.com Web Site: carrollbroadcasting.com. Licensee: Carroll Broadcasting Co. (group owner; acq 1999; $975,000). Population served: 85,000 Natl. Rep: Katz Radio. Law Firm: Womble, Carlyle, Sandridge & Rice. Format: Country. News: 2. Target aud: 18-49; contemp country with strong families. Spec prog: Sports. ♦Mary Collison, pres; Kim Hackett, gen mgr; John Ryan, opns mgr; Lynda Dukes-Francy, gen sls mgr.

Lamoni

***KOWI(FM)**— 2000: 97.9 mhz; 50 kw. Ant 492 ft TL: N40 48 52 W93 50 15. Stereo. Hrs opn: 24
WOI-AM.
WOI Radio Group, 2022 Communications Bldg., Iowa State University, Ames, 50011-3241. Phone: (515) 294-2025. Fax: (515) 294-1544. E-mail: woi@iastate.edu Web Site: www.woi.org. Licensee: Iowa State University of Science and Technology (acq 7-30-2004; $450,000). Natl. Network: NPR, PRI. Format: Class, news. News: 40 hrs wkly. Target aud: General; Educated. ♦Dave Becker, progmg dir.

Le Mars

KKMA(FM)—Listing follows KLEM(AM).

KLEM(AM)— Oct 12, 1954: 1410 khz; 1 kw-D, 63 w-N. TL: N42 49 05 W96 10 00. Hrs open: 24 37 2nd Ave. N.W., 51031. Phone: (712) 546-4121. Fax: (712) 546-9672. E-mail: daveg@lemarscomm.net Web Site: www.klem1410.com. Licensee: Powell Broadcasting Co Inc. (acq 7-6-99; with co-located FM). Population served: 15,000 Format: Adult contemp, news, sports. News staff: 2. Spec prog: Farm 18 hrs wkly. ♦Tom Spies, pres; Dennis Bullock, gen mgr; Dave Grosenheider, stn mgr & gen sls mgr; Dave Ruden, progmg dir, women's int ed & disc jockey; Larry Schmitz, news dir & farm dir; Stan Culley, chief of engrg; Christi Rush, traf mgr; Joanne Glamm, reporter; Corey Pithan, disc jockey; Denny Callahan, disc jockey.

KKMA(FM)—Co-owned with KLEM(AM). Jan 1, 1967: 99.5 mhz; 100 kw. 790 ft TL: N42 28 56 W96 15 30. Stereo. 2000 Indian Hills Dr., Sioux City, 51104. Phone: (712) 239-2100. Fax: (712) 239-3346. Web Site: www.kool995.com.285,000 Format: Oldies, classic hits. ♦Dennis Bullock, gen mgr & rgnl sls mgr; Kelli Erickson, gen sls mgr; Scott McKenzie, progmg dir; Stan Culley, chief of engrg; Christi Rush, traf mgr; Justin Barker, disc jockey.

Madrid

KNWM(FM)— Aug. 21, 1997: 96.1 mhz; 2.5 kw. Ant 515 ft TL: N41 58 49 W93 44 23. Hrs open: 24 3737 Woodland Ave., Suite 111, Des Moines, 50366. Phone: (515) 327-1071. Fax: (515) 327-1073. Web Site: www.desmoines.fm. Licensee: Northwestern College. Group owner: Northwestern College & Radio (acq 12-30-2003; $1.8 million with KNWI(FM) Osceola). Population served: 257,000 Format: Christian. ♦Richard Whitworth, gen mgr; Dave St. John, progmg dir.

Manchester

KMCH(FM)— Dec 5, 1991: 94.7 mhz; 6 kw. 328 ft TL: N42 31 42 W91 22 53. Stereo. Hrs opn: 24 Box 497, 212 E. Main St., 52057. Phone: (563) 927-6249. Fax: (563) 927-4372. E-mail: kmchradio@iowatelecom.net Web Site: www.kmch.com. Licensee: Fife Communication Co. L.C. Population served: 65,580 Natl. Network: CBS. Format: Adult contemp, C&W. News staff: one; News: 20 hrs wkly. Target aud: 25-64; northeast Iowa adults & farm population. Spec prog: Farm 7 hrs, sports 7 hrs, relg 4 hrs wkly. ♦Anthony G. Coloff, pres; James A. Coloff, VP & gen mgr; Jackie Coates, stn mgr & opns mgr.

Manson

KXFT(FM)—Not on air, target date: unknown: 99.7 mhz; 25 kw. Ant 285 ft TL: N42 31 03 W94 20 43. Hrs open: 540 A St., Fort Dodge, 50501. Phone: (515) 576-7333. Fax: (515) 955-4250. Licensee: Clear Channel Broadcasting Licenses Inc. ♦Tracey Williams, gen mgr.

Maquoketa

KMAQ(AM)— Aug 26, 1958: 1320 khz; 500 w-U. TL: N42 05 26 W90 37 43. Hrs open: 6 AM-10 PM Box 940, 129 N. Main St., 52060. Phone: (563) 652-2426. Fax: (563) 652-6210. Licensee: Maquoketa Broadcasting Co. (acq 1965). Population served: 40,000 Natl. Network: USA. Rgnl. Network: Brownfield, Radio Iowa. Natl. Rep: Farmakis. Law Firm: Miller & Fields, P.C. Format: C&W. News staff: one; News: 28 hrs wkly. Target aud: General; adults, high percentage of farmers. Spec prog: Farm 10 hrs, polka 3 hrs wkly. ♦Dennis W. Voy, pres, gen mgr & progmg dir; Leighton Hepker, opns dir & sls dir; Tom Messerli, chief of engrg.

KMAQ-FM— Sept 1, 1967: 95.1 mhz; 6 kw. 328 ft TL: N42 05 26 W90 37 43. Stereo. 45,000 News staff: one; News: 28 hrs wkly. ♦Leighton Hepker, adv dir.

Marion

***KZNJ(FM)**—Not on air, target date: unknown: 89.9 mhz; 1.1 kw. Ant 216 ft TL: N42 06 24 W91 42 05. Hrs open: University of Northern Iowa, 324 Communications Arts Center, Cedar Falls, 50614-0359. Phone: (319) 273-6400. Fax: (319) 273-2682. Licensee: University of Northern Iowa. ♦Wayne Jarvis, gen mgr.

Marshalltown

KDAO(AM)— Dec 16, 1978: 1190 khz; 250 w-D, 20 w-N. TL: N42 04 17 W92 55 19. Hrs open: 24 P.O. Box 538, 1930 N. Center St., 50158. Phone: (641) 752-4122. Fax: (641) 752-5121. Web Site: www.kdao.com. Licensee: MTN Broadcasting Inc. Format: Adult standards. Target aud: 25-54. ♦Mark K. Osmundson, gen mgr. Co-owned TV: KDAO-TV affil

KFJB(AM)— June 1923: 1230 khz; 1 kw-U. TL: N42 04 01 W92 58 10. Hrs open: 24 Box 698, 123 W. Main St., 50158. Phone: (641) 753-3361. Fax: (641) 752-7201. E-mail: office@marshalltownbroadcasting.com Web Site: www.1230kfjb.com. Licensee: Marshalltown Broadcasting Inc. (acq 12-29-86). Population served: 120,000 Natl. Network: ABC. Rgnl. Network: Brownfield. Natl. Rep: Katz Radio. Wire Svc: AP Format: News/talk. News staff: 2; News: 12 hrs wkly. Target aud: 35-64. ♦David L. Nelson, pres; Clark L. Wideman, gen mgr; Kyle Martin, progmg dir.

KXIA(FM)—Co-owned with KFJB(AM). January 1968: 101.1 mhz; 100 kw. 649 ft TL: N42 00 19 W92 55 45. Stereo. 24 Web Site: www.kixweb.com.700,000 Natl. Network: ABC. Natl. Rep: Katz Radio. Format: Country. News staff: 2; News: 6 hrs wkly. Target aud: 25-54. ♦Todd Collins, progmg dir.

***KRFH(FM)**—Not on air, target date: unknown: 88.7 mhz; 8.3 kw. Ant 95 ft TL: N42 04 17 W92 55 19. Hrs open: Box 538, 50158. Phone: (641) 752-4122. Licensee: Marshalltown Education Plus Inc.

Mason City

***KBDC(FM)**— 2001: 88.5 mhz; 1.8 kw vert. Ant 230 ft TL: N43 03 35 W93 22 47. Hrs open: Box 3206, American Family Radio, Tupelo, MS, 38803. Phone: (662) 844-8888, EXT. 204. Fax: (662) 842-6791. Licensee: American Family Association. Group owner: American Family Radio Format: Adult contemp. ♦Marvin Sanders, gen mgr.

***KCMR(FM)**— May 3, 1979: 97.9 mhz; 6 kw. 300 ft TL: N43 07 18 W93 11 32. (CP: Ant 315 ft.). Stereo. Hrs opn: 24 Box 979, 50402-0979. Secondary address: 600 First St. N.W. 50401. Phone: (641) 424-9300. Fax: (641) 423-2221. Licensee: TLC Broadcasting Corp. (acq 5-24-2004). Population served: 35,000 Format: Easy lstng, inspirational. Target aud: Over 30. Spec prog: Class 5 hrs, nostalgia 10 hrs wkly. ♦Bill Schickel, gen mgr; Bob Miller, dev dir.

KGLO(AM)— Jan 17, 1937: 1300 khz; 5 kw-U, DA-2. TL: N43 03 15 W93 12 17. Hrs open: Box 1300, 341 Yorktown Pike, 50401. Phone: (641) 423-1300. Fax: (641) 423-2906. Web Site: www.rivercitysquare.com. Licensee: Clear Channel Broadcasting Licenses Inc. Group owner: Clear Channel Communications Inc. (acq 9-25-00; grpsl). Population served: 110,000 Natl. Network: CBS. Format: Talk. News staff: 3. Target aud: 25-35; adults. Spec prog: Farm 15 hrs wkly. ♦Charlie Thomas, gen mgr; Tim Fleming, stn mgr; Tim Fleming, opns dir & progmg dir; Hall Hofman, gen sls mgr; Andy Roat, natl sls mgr; Tami Ramon, mktg dir; Jamie Larson, prom dir; Tim Renshaw, news dir; Greg Gade, chief of engrg; Darcy Piper, traf mgr; Chris Frenz, farm dir.

KIAI(FM)—Co-owned with KGLO(AM). November 1985: 93.9 mhz; 100 kw. Ant 790 ft TL: N43 10 04 W93 06 05. Stereo. E-mail: jbrooks@clearchannel.com Web Site: kiaifm.com. Format: Country. News: 2 hrs wkly. Target aud: 24-54. ♦J. Brooks, progmg dir; Darcy Piper, traf mgr; Chris Frenz, farm dir.

KLSS-FM—Listing follows KRIB.

KRIB(AM)— April 1948: 1490 khz; 1 kw-U. TL: N43 08 05 W93 12 30. Stereo. Hrs opn: 24 402 19th St. S.W., 50401. Phone: (641) 423-8634. Fax: (641) 423-8206. E-mail: krib@kribradio.com Web Site: www.kribradio.com. Licensee: Three Eagles of Mason City Inc. Group owner: Three Eagles Communications (acq 5-2-97; $3.596 million with co-located FM). Population served: 150,000 Rgnl. Network: Iowa Radio Net. Natl. Rep: McGavren Guild. Format: Adult standards, oldies. News staff: 2; News: 25 hrs wkly. Target aud: 35 plus; married up-scale adults, financially secure with two incomes or retired. Spec prog: Relg 5 hrs wkly. ♦Gary Buchanan, pres; Dalena Barz, gen mgr & natl sls mgr; John Swinton, progmg dir; Bob Fisher, news dir; Christi Lyman, pub affrs dir; Ron Schacts, chief of engrg.

KLSS-FM—Co-owned with KRIB. Nov 1, 1967: 106.1 mhz; 100 kw. 315 ft TL: N43 08 31 W93 06 40. Stereo. E-mail: klss@klssradio.com Web Site: www.klssradio.com.240,000 Natl. Network: ABC. Format: Adult contemp. Target aud: 18-54. ♦John Swinton, opns mgr & disc jockey; Pam Dzick, gen sls mgr; Harry O'Neil, mus dir & disc jockey; Brenda McWhorter, traf mgr; Brian Wilson, news rptr; Colleen Devine, disc jockey.

***KRNI(AM)**— Mar 1, 1948: 1010 khz; 1 kw-D, 16 w-N. TL: N43 08 31 W93 06 40. Hrs open: Sunrise-sunset
Rebroadcasts KUNI(FM) Cedar Falls 100%.
c/o KUNI-FM, Univ. of Northern Iowa, Cedar Falls, 50614-0359. Phone: (319) 273-6400. Fax: (319) 273-2682. E-mail: kuni@uni.edu Web Site: www.kuniradio.org. Licensee: University of Northern Iowa. (acq 10-30-98; grpsl). Population served: 40,000 Natl. Network: PRI, NPR. Format: News & info, AAA. News staff: 3; News: 77 hrs wkly. Target aud: General. Spec prog: Folk 4 hrs, blues 5 hrs wkly. ♦Scott Vezdan, mktg dir; Wayne Jarvis, gen mgr & progmg dir; Al Schares, mus dir & disc jockey; Greg Shanley, news dir & news rptr; Steve Schoon, engrg dir; Tony Dehner, traf mgr; Jeneane Beck, local news ed; Pat Blank, reporter; Bob Dorr, disc jockey; Jacqueline Halbloom, disc jockey; Karen Impola, disc jockey; Mark Simmet, disc jockey.

Milford

KUQQ(FM)—Licensed to Milford. See Spirit Lake

Mitchellville

***KDMR(FM)**—Not on air, target date: unknown: 88.9 mhz; 1 kw. Ant 236 ft TL: N41 40 05 W93 19 43. Hrs open:
KUNI-FM,Cedar Falls, IA.
University of Northern Iowa, 324 Communications Arts Center, Cedar Falls, 50614. Phone: (319) 273-6400. Fax: (319) 273-2682. E-mail: kuni@uni.edu Web Site: www.kuniradio.org. Licensee: University of Northern Iowa. Format: News and Information, Triple A. News staff: 3; News: 77 hrs. Spec prog: Folk 4hrs, Blues 5 hrs. ♦Scott Vezdos, mktg mgr; Wayne Jarvis, gen mgr & progmg dir; Al Schares, mus dir; Greg Shanley, news dir; Steve Schoon, chief of engrg.

Mount Pleasant

KILJ(AM)— December 1974: 1130 khz; 250 w-D. TL: N40 57 32 W91 35 01. Hrs open: 24 2411 Radio Dr., 52641. Phone: (319) 385-8728. Fax: (319) 385-4517. E-mail: kilj@iowatelecom.net Web Site: www.kilj.com. Licensee: KILJ Inc. (acq 10-29-2003; $1.01 million with co-located FM). Population served: 60,000 Natl. Network: ABC. Format: Country. News staff: one. ♦John R. Kuhens, gen mgr, stn mgr, sls dir & progmg dir; Paul Dennison, gen sls mgr; Bob Maltocks, news dir; Theresa Rose, news dir; Leo Septen, chief of engrg; Lora Roth, traf mgr.

KILJ-FM— October 1970: 105.5 mhz; 24 kw. 338 ft TL: N40 56 32 W91 34 08. (CP: 3.3 kw). Stereo. 24 Web Site: www.kilj.com. Format: Smooth sounds. News staff: one. Target aud: 25-54. ♦John Kuhens, gen mgr & disc jockey; Lori Roth, traf mgr & disc jockey.

Mount Vernon

***KRNL-FM**— Apr 1, 1948: 89.7 mhz; 36 w. 167 ft TL: N41 55 24 W91 25 18. Stereo. Hrs opn: Midnight-noon 810 Commons Cir., 52314. Phone: (319) 895-4431. Phone: (319) 895-5765. E-mail: krnl@cornellcollege.edu Web Site: www.cornellcollege.edu/krnl. Licensee: Cornell College. Population served: 100,000 Natl. Network: USA. Format: Free-form, progsv. Target aud: 18-25; collegians & those seeking an alternative to coml radio. Spec prog: Folk 2 hrs, Ger 2 hrs, jazz 2 hrs, Sp one hr wkly. ♦Sarah Altmann, gen mgr; Robin Schwab, stn mgr.

Muscatine

KBEA-FM—Licensed to Muscatine. See Davenport

KMCS(FM)—Listing follows KWPC(AM).

KWPC(AM)— Jan 5, 1947: 860 khz; 250 w-D, 8 w-N. TL: N41 26 43 W91 04 36. Stereo. Hrs opn: 24 3218 Mulberry Ave., 52761. Phone: (563) 263-2442. Fax: (563) 263-9206. E-mail: mail@voiceofmuscatine.com Web Site: www.voiceofmuscatine.com. Licensee: WPW Broadcasting Inc. (group owner; (acq 11-5-99; $2.2 million with co-located FM). Population served: 63,405 Natl. Network: USA. Law Firm: Fletcher, Heald & Hildreth. Format: Oldies. News staff: 2; News: 20 hrs wkly. Target aud: 25-54. ♦Don Davis, pres & mus dir; Terri Forbes, CFO; DeWayne Hopkins, gen mgr.

KMCS(FM)—Co-owned with KWPC(AM). June 16, 1996: 93.1 mhz; 4.4 kw. Ant 384 ft TL: N41 26 34 W91 04 33. Stereo. 24 Natl. Network: USA, AP Radio. Format: Country. News: 6 hrs wkly. Target aud: 25-54.

New Hampton

KCZE(FM)— Dec 1, 1992: 95.1 mhz; 5.5 kw. 328 ft TL: N43 02 46 W92 18 09. Stereo. Hrs opn: 207 N. Main St., Charles City, 50616. Phone: (641) 228-1000. Fax: (641) 228-1200. E-mail: kcze@clearchannel.com Web Site: www.951thebull.com. Licensee: Clear Channel Broadcasting Licenses Inc. Group owner: Clear Channel Communications Inc. (acq 9-25-00; grpsl). Population served: 200,000 Natl. Rep: Farmakis. Format: Country. Target aud: General. Spec prog: Farm 12 hrs wkly. ♦Hal Hofman, gen mgr; J. Brooks, opns mgr & progmg dir; Tami Ramon, mktg dir; Patrick Gwin, chief of engrg.

New London

KHDK(FM)— Oct 5, 2001: 97.3 mhz; 3.8 kw. Ant 410 ft TL: N40 47 53 W91 26 22. Stereo. Hrs opn: 24 2850 Mt. Pleasant St., Burlington, 52601. Phone: (319) 752-5402. Fax: (319) 752-4715. E-mail: johnp@burlingtonradio.com Licensee: Pritchard Broadcasting Co. (acq 12-27-99; $25,000 for CP). Population served: 147,800 Format: Adult contemp. News staff: one; News: 3 hrs wkly. Target aud: 25-54. ♦John T. Pritchard, pres & gen mgr; Chet Young, gen sls mgr; Kathy Vance, opns mgr & progmg dir.

New Sharon

KCWN(FM)— Oct 16, 1995: 99.9 mhz; 25 kw. 297 ft Hrs opn: 6 AM-11 PM Box 999, Pella, 50219. Secondary address: 304 Oskaloosa St., Pella 50219. Phone: (641) 628-9999. Phone: (800) 506-4562. Fax: (641) 628-9229. E-mail: kcwnfm@lisco.com Web Site: www.kcwnfm.org. Licensee: Crown Broadcasting Co. Format: Adult contemp Christian. ♦Marion L. Vink, CEO, pres & gen mgr; Beverly DeVries, stn mgr.

Newell

KWDN(FM)—Not on air, target date: unknown: 100.9 mhz; 6 kw. Ant 292 ft TL: N42 38 43 W95 10 33. Hrs open: 5331 Mt. Alifan Dr., San Diego, CA, 92111. Phone: (858) 277-4991. Fax: (858) 277-1365. Licensee: Horizon Christian Fellowship. (acq 2-9-2006; grpsl). ♦Mike MacIntosh, pres.

Newton

KCOB(AM)— Sept 15, 1955: 1280 khz; 1 kw-D, 500 w-N. TL: N40 44 11 W93 01 12. Hrs open: Box 66, 1801 N. 13th Ave. E., 50208. Phone: (641) 792-5262. Fax: (641) 792-8403. Web Site: kcobradio.com. Licensee: Central Iowa Broadcasting Inc. Population served: 15,619 Format: Country, news. Target aud: 25-50. Spec prog: Farm 2 hrs wkly. ♦Frank Liebl, CEO, pres, gen mgr & gen sls mgr; Terry Walter, progmg dir; Randy Van, news dir; Phil Benjamin, chief of engrg.

KCOB-FM— Jan 3, 1969: 95.9 mhz; 5.1 kw. Ant 354 ft TL: N41 44 11 W93 01 12. Stereo. 18 Web Site: kcobradio.com.15,619

***KKLG(FM)**— 2005: 88.3 mhz; 400 w. Ant 218 ft TL: N41 41 33 W93 00 37. Hrs open:
Rebroadcasts KLVR(FM) Santa Rosa, CA 100%.
2351 Sunset Blvd., Suite 170-218, Rocklin, CA, 95765. Phone: (916) 251-1600. Fax: (916) 251-1650. Web Site: www.klove.com. Licensee: Educational Media Foundation. (acq 9-22-2005; $20,000 for CP). Natl. Network: K-Love. Format: Contemp Christian. ♦Richard Jenkins, pres; Mike Novak, VP; Keith Whipple, dev dir; David Pierce, progmg mgr; Ed Lenane, news dir; Sam Wallington, engrg dir; Karen Johnson, news rptr; Marya Morgan, news rptr; Richard Hunt, news rptr.

Northwood

KYTC(FM)— Oct 15, 1990: 102.7 mhz; 25 kw. 318 ft TL: N43 29 18 W93 14 12. Stereo. Hrs opn: 24 402 19th St. S.W., Mason City, 50401. Phone: (800) 598-2858. Phone: (641) 423-8634. Fax: (641) 423-8206. E-mail: steveg@klssradio.com Web Site: www.ky102.net. Licensee:

Three Eagles of Mason City Inc. Group owner: Three Eagles Communications (acq 5-21-99). Population served: 75,000 Rgnl. Network: Tribune. Format: Country. News staff: one; News: 6 hrs wkly. Target aud: 25-64; primary audience men & women 35+. Spec prog: Gospel one hr, relg 2 hrs wkly. ♦Rolland Johnson, CEO; Gary Buchanan, pres; Dalena Barz, gen mgr & stn mgr; Henry O'Neil, chief of opns.

Oelwein

KKHQ-FM—Listing follows KOEL(AM).

KOEL(AM)— July 23, 1950: 950 khz; 5 kw-D, 500 w-N, DA-2. TL: N42 39 26 W91 54 02. Hrs open: 24 2502 S. Frederick, 50662. Phone: (319) 283-1234. Fax: (319) 283-3615. Licensee: Cumulus Licensing Corp. Group owner: Cumulus Media Inc. (acq 3-15-00; grpsl). Population served: 180,000 Format: News/talk, sports. News staff: 2; News: 30 hrs wkly. Target aud: 35 plus. Spec prog: Farm 16 hrs wkly.Jeffrey D. Warshaw, pres; Jeff Dientz, VP; Rob Murthum, gen mgr & mktg mgr; Dick Stadlen, opns mgr; Craig Friedrich, gen sls mgr; Bob Fisher, natl sls mgr; April Walker, prom mgr; Rich Calvert, progmg dir, progmg mgr & disc jockey; Matt Kelly, mus dir; Roger King, news dir, pub affrs dir & disc jockey; Arnold Zaruba, chief of engrg; Traci Berry, traf mgr; Mark Barber, farm dir & disc jockey

KKHQ-FM—Co-owned with KOEL(AM). Dec 29, 1971: 92.3 mhz; 100 kw. 1,000 ft TL: N42 40 53 W91 52 52. (CP: 95.3 mhz, ant 991 ft.). Stereo. 24 Box 720, Blacks Bldg., 501 Sycamore St., Waterloo, 50703. Phone: (319) 833-4800. Phone: (800) 923-5635. Fax: (319) 833-4866.150,000 Format: Country. News staff: one; News: 2 hrs wkly. Target aud: 35 plus. ♦Mark Anderson, gen sls mgr; April Walker, prom dir & mus dir; Bill Knight, progmg mgr; Elwin Huffman, news dir; Wes Davis, chief of engrg.

Okoboji

***KOJI(FM)**— 2002: 90.7 mhz; 4.5 kw. Ant 371 ft TL: N43 09 53 W95 19 29. Hrs open: 24
Rebroadcasts KWIT(FM) Sioux City 100%.
4647 Stone Ave., Sioux City, 51106-1997. Phone: (712) 274-6406. Fax: (712) 274-6411. E-mail: gondekg@witcc.com Web Site: www.kwit-koji.org. Licensee: Western Iowa Tech Community College. Natl. Network: NPR, PRI. Wire Svc: AP Format: Class, news/talk, Sp. News staff: one; News: 36 hrs wkly. Spec prog: Triple A 18 hrs, blues 4 hrs, jazz 17 hrs wkly. ♦Gretchen Gondek, gen mgr; Steve Smith, opns mgr.

Onawa

KZSR(FM)— Nov 6, 1995: 102.3 mhz; 100 kw. 643 ft TL: N42 10 29 W96 23 13. Hrs open: 24 522 14th St., Sioux City, 51105. Phone: (712) 258-5655. Fax: (712) 258-1511. Web Site: www.bobfm1023.com. Licensee: Powell Broadcasting Co. Inc. (acq 5-1-2007; $4.2 million with KKYY(FM) Whiting). Population served: 200,000 Format: Var. News: one. Target aud: 25-54. ♦Jerry Haack, gen mgr; Kelli Erickson, stn mgr; Jeff Heyer, progmg dir; Tim Guentz, chief of engrg.

Osage

KSMA-FM— July 9, 1980: 98.7 mhz; 25 kw. Ant 328 ft TL: N43 21 53 W93 02 53. Stereo. Hrs opn: 341 Yorktown Pike, Mason City, 50401. Phone: (641) 423-1300. Fax: (641) 423-2906. E-mail: patrickgwin @clearchannel.com Web Site: www.kiss987.com. Licensee: Clear Channel Broadcasting Licenses Inc. Group owner: Clear Channel Communications Inc. (acq 9-25-00; grpsl). Population served: 190,000 Natl. Rep: Farmakis. Format: Adult contemp, CHR. Target aud: General; 12-25. ♦Hal Hofman, gen mgr & gen sls mgr; Tim Fleming, opns dir; Tami Ramon, mktg dir; Dan Maynard, progmg dir; Laurie Gansen, traf mgr.

Osceola

***KNWI(FM)**— Oct 4, 1982: 107.1 mhz; 27 kw. Ant 649 ft TL: N41 01 34 W93 51 43. Stereo. Hrs opn: 24 3737 Woodland Ave., Suite 111, West Des Moines, 50266. Phone: (515) 327-1071. Fax: (515) 327-1073. E-mail: knwi@desmoines.fm Web Site: www.desmoines.fm. Licensee: Northwestern College. Group owner: Northwestern College & Radio. (acq 12-30-2003; $1.8 million with KNWM(FM) Madrid). Format: Christian. Target aud: 18-44; women. ♦Richard Whitworth, gen mgr; Dave St. John, progmg dir.

Oskaloosa

KBOE(AM)— Nov 15, 1950: 740 khz; 250 w-D, 12 w-N. TL: N41 19 15 W92 38 44. Hrs open: 24 Box 380, Hwy. 63 N., 52577. Phone: (515) 673-3493. Fax: (515) 673-3495. E-mail: kboe@kboeradio.com Web Site: www.kboeradio.com. Licensee: Jomast Corp. Population served: 72,000 Rgnl. Network: Brownfield, Radio Iowa. Format: Country. News staff: one; News: 15 hrs wkly. Target aud: 25-50. Spec prog: Gospel 9 hrs wkly. ♦Brad Muhl, pres; Glenda Lind-Booy, gen mgr & gen sls mgr; Scott Dailey, news dir; Gary Wilson, chief of engrg; Bob Allan, disc jockey; John Jacobs, disc jockey; Steve Shelter, disc jockey.

KBOE-FM— Feb 7, 1964: 104.9 mhz; 50 kw. 492 ft TL: N41 19 15 W92 38 44. Stereo. 24 Web Site: www.kboeradio.com. Format: News/talk.

***KCNJ(FM)**—Not on air, target date: unknown: 89.5 mhz; 420 w. Ant 308 ft TL: N41 19 15 W92 38 44. Hrs open: University of Northern Iowa, 324 Communications Arts Center, Cedar Falls, 50614-0359. Phone: (319) 273-6400. Fax: (319) 273-2682. Licensee: University of Northern Iowa. ♦Wayne Jarvis, gen mgr.

***KIGC(FM)**— 1975: 88.7 mhz; 230 w. 93 ft TL: N41 18 37 W92 38 49. (CP: Ant 123 ft.). Stereo. Hrs opn: 24 William Penn University, 201 Trueblood Ave., 52577. Phone: (641) 673-1095. Fax: (641) 673-1396. Licensee: William Penn University Population served: 15,000 Format: Oldies, alternative, Black. News: one hr wkly. Target aud: 13-25. Spec prog: Jazz 12 hrs, gospel 12 hrs wkly. ♦James Roberts, progmg dir; Larz G. Roberts, gen mgr & mus critic.

***KOSK(FM)**—Not on air, target date: unknown: 90.5 mhz; 1.5 kw. Ant 292 ft TL: N41 19 15 W92 38 44. Hrs open: 204A Communications Bldg., Ames, 50011. Phone: (515) 294-9478. Fax: (515) 294-1544. Licensee: Iowa State University of Science and Technology. ♦Warren R. Madden, VP.

Ottumwa

KBIZ(AM)— 1941: 1240 khz; 1 kw-U. TL: N41 00 00 W92 23 23. Hrs open: Box 190, Broadcast Ctr., 209 S. Market, 52501. Phone: (515) 682-4535. Fax: (515) 684-5892. Licensee: O-Town Communications Inc. (group owner; (acq 10-20-2005; $890,000 with co-located FM). Population served: 301,700 Natl. Network: CBS. Format: Classic oldies. Target aud: 25-54. Spec prog: Farm 12 hrs, relg 4 hrs wkly. ♦Greg List, pres & opns mgr; Judy Bushong, gen mgr; Mike Buchanan, news dir & farm dir; Phil Benjamin, chief of engrg; Chris Tubbs, sports cmtr.

KTWA(FM)—Co-owned with KBIZ(AM). December 1984: 92.7 mhz; 50 kw. 318 ft TL: N41 01 29 W92 28 09.24 Format: Adult contemp. ♦Judy Bushong, gen mgr & stn mgr; Phil Benjamin, chief of engrg.

KLEE(AM)— Aug 1, 1954: 1480 khz; 500 w-D, 33 w-N. TL: N41 01 27 W92 28 56. Stereo. Hrs opn: 24 601 W. 2nd St., 52501. Phone: (641) 682-8711. Phone: (641) 682-8712. Fax: (641) 682-8482. Licensee: FMC Broadcasting Inc. (acq 1-16-92; $400,000 with co-located FM; FTR: 2-10-92) Population served: 203,000 Natl. Network: Westwood One. Rgnl. Network: Iowa Radio Net., Brownfield. Format: Country, news/talk. News staff: one; News: 28 hrs wkly. Target aud: General; people on the move. Spec prog: Gospel 6 hrs, polka one hr wkly. ♦Thomas A. Palen, pres, gen mgr & gen sls mgr; Marcia Wagner, prom dir; Dave Michaels, progmg dir; Mike Dixon, news dir; Fred Jenkins, chief of engrg.

KOTM-FM—Co-owned with KLEE(AM). Mar 22, 1976: 97.7 mhz; 6 kw. 200 ft TL: N41 01 27 W92 28 56. Stereo. Web Site: www.kotm.com. Natl. Network: Westwood One. Format: CHR. Target aud: Teens-50.

***KUNE(FM)**—Not on air, target date: unknown: 88.3 mhz; 500 w. Ant 350 ft TL: N40 57 41 W92 22 13. Hrs open: University of Northern Iowa, 324 Communications Arts Center, Cedar Falls, 50614-0359. Phone: (319) 273-6400. Fax: (319) 273-2682. Licensee: University of Northern Iowa. ♦Wayne Jarvis, gen mgr.

***KUNZ(FM)**—Not on air, target date: unknown: 91.1 mhz; 1.9 kw. Ant 400 ft TL: N40 57 40 W92 22 11. Hrs open: University of Northern Iowa, 324 Communications Arts Center, Cedar Falls, 50614-0359. Phone: (319) 273-6325. Fax: (319) 273-2682. Web Site: www.kuniradio.org. Licensee: University of Northern Iowa. ♦John Hess, gen mgr; Wayne Jarvis, gen mgr; Scott Verdos, mktg dir; Al Schares, mus dir; Greg Shanley, news dir; Steve Schoon, chief of engrg.

Parkersburg

KQCR-FM— Oct 18, 2000: 98.9 mhz; 6 kw. 328 ft TL: N42 33 48 W92 57 22. Stereo. Hrs opn: 24 Box 495, Hampton, 50441-0495. Secondary address: 1509 4th St NE, Hampton 50441-1106. Phone: (641) 456-5656. Fax: (641) 456-5655. E-mail: kqcr@kqcr.fm Web Site: www.kqcr.fm. Licensee: CD Broadcasting Inc. Population served: 100,000 Law Firm: Fletcher, Heald & Hildreth, P.L.C. Format: Adult contemp. News staff: 2; News: 12 hrs wkly. Target aud: 25-45; Light, Soft AC 70's, 80's, 90's. ♦Craig Donnelly, gen mgr; Marlin Burrier, opns dir.

Patterson

KZLN(FM)—Not on air, target date: unknown: 105.9 mhz; 6 kw. Ant 235 ft TL: N41 20 40 W93 57 58. Hrs open: Connoisseur Media LLC, 136 Main St., Suite 202, Westport, CT, 06880-3304. Phone: (203) 227-1978. Fax: (203) 227-2373. Web Site: www.connoisseurmedia.com. Licensee: Connoisseur Media LLC.

Pella

KAZR(FM)— Aug 1, 1976: 103.3 mhz; 100 kw. Ant 745 ft TL: N41 32 18 W93 17 58. Stereo. Hrs opn: 24 1416 Locust St., Des Moines, 50309. Phone: (515) 280-1350. Fax: (515) 280-3011. Web Site: www.lazer1033.com. Licensee: Saga Communications of Iowa LLC. Group owner: Saga Communications Inc. (acq 9-17-96; $2.7 million). Population served: 1,033,400 Natl. Rep: Katz Radio. Law Firm: Smithwick & Belendiuk. Wire Svc: AP Format: Active rock. News staff: one; News: 4 hrs wkly. Target aud: 25-44. ♦Bill Wells, gen mgr; Jim Schaefer, opns mgr; Celia Rodine, natl sls mgr; Scott Allen, mktg mgr; Ryan Paatrick, progmg dir.

KNIA(AM)—See Knoxville

Perry

KDLS(AM)— May 10, 1961: 1310 khz; 500 w-D, 300 w-N, DA-2. TL: N41 49 58 W94 02 15. Hrs open: 6 AM-10 PM Box 548, 50220. Secondary address: 2260 141st Dr. 50220. Phone: (515) 465-5357. Fax: (515) 465-3952. E-mail: kdls@prairieinet.net Licensee: Coon Valley Communications Inc. (acq 2-15-2006; $300,000). Population served: 7,073 Natl. Network: Westwood One, CNN Radio. Natl. Rep: Farmakis. Wire Svc: AP Format: Var. News staff: one; News: 25 hrs wkly. Target aud: General. ♦Patrick Delaney, pres; Tom Quinlan, VP & farm dir; Patrick Graney, gen mgr & gen sls mgr; John Patrick, opns dir, progmg dir, news dir & local news ed; Bob Pink, chief of engrg; Marcia Murphy, traf mgr; Jerry Roberts, disc jockey; Shawn Kenney, disc jockey; Toni Allen, disc jockey.

KDLS-FM— Feb 26, 1971: 105.5 mhz; 6 kw. Ant 305 ft TL: N41 50 03 W94 02 12. Stereo. Hrs opn: 6 AM-midnight 301 Ashworth Rd., West Des Moines, 50265. Phone: (515) 278-4117. Fax: (515) 254-1037. Licensee: Perry Broadcasting Co. (acq 2-1-2006; with co-located AM). Format: Sp. ♦Joel Garcia, gen mgr.

Red Oak

KCSI(FM)—Listing follows KOAK(AM).

KOAK(AM)— Aug 16, 1968: 1080 khz; 250 w-D. TL: N41 01 00 W95 12 46. Hrs open: Sunrise-sunset Box 465, 1991 Ironwood, 51566. Phone: (712) 623-2584. Fax: (712) 623-2583. E-mail: kcsi@kcsifm.com Web Site: kcsifm.com. Licensee: Hawkeye Communications Inc. (acq 7-1-94). Population served: 7,000 Natl. Network: ABC. Format: Contemp country. ♦Jerry V. Dietz, pres & gen mgr; Melanie L. West, gen sls mgr; Marilyn Dietz, traf mgr.

Rock Valley

KIHK(FM)— 1998: 106.9 mhz; 25 kw. 328 ft TL: N43 20 28 W96 19 03. Hrs open: 6 AM-6 AM Box 298, Sioux Center, 51250. Phone: (712) 722-1090. Fax: (712) 722-1102. E-mail: ksou@waittradio.com Web Site: www.ksoufm.com. Licensee: Sorenson Broadcasting Corp. (group owner; (acq 7-1-2004; grpsl). Format: Country. Spec prog: Gospel bluegrass 3 hrs wkly. ♦Craig Aukes, gen mgr; Dan Bonnema, gen sls mgr; Doug Brock, news dir.

Sageville

KIYX(FM)— 1999: 106.1 mhz; 4.1 kw. 397 ft TL: N42 41 27 W90 37 26. Stereo. Hrs opn: 24 51 Means Dr., Platteville, WI, 53818. Phone:

(608) 349-2000. Fax: (608) 349-2002. E-mail: superhits@queenbradio.com Licensee: Queen B Radio Wisconsin Inc. (acq 6-8-99). Natl. Network: Westwood One. Format: Top 40 hits of the 70s 80s. News staff: one. ♦Dan Sullivan, gen mgr.

Saint Ansgar

KJCY(FM)— September 2001: 95.5 mhz; 6 kw. Ant 328 ft TL: N43 21 12 W93 02 48. Hrs open: 24 Box 1069, Mason City, 50402. Phone: (641) 424-5529. Fax: (641) 424-5597. E-mail: kjcy@kjcy.com Web Site: www.kjcy.com. Licensee: Minn-Iowa Christian Broadcasting Inc. (group owner; acq 3-20-01; $200,000). Format: Christian. ♦Matt Dorfner, exec VP; Rick Boyd, stn mgr & opns mgr.

Sheldon

KIWA(AM)— Oct. 27, 1961: 1550 khz; 283 w-D, 6 w-N. TL: N43 10 53 W95 51 56. Hrs open: 24 411 9th St., 51201. Phone: (712) 324-2597. Fax: (712) 324-2340. E-mail: newtips@kiwaradio.com Web Site: www.kiwaradio.com. Licensee: Sheldon Broadcasting Co. Inc. (acq 10-27-61). Population served: 140,000 Natl. Network: ABC. Rgnl. Network: Radio Iowa. Natl. Rep: Farmakis. Format: News/talk. News staff: 2; News: 15 hrs wkly. Target aud: General; adult. ♦Tim Torkildson, gen mgr & news dir; Walt Pruiksma, gen mgr, gen sls mgr & prom mgr; Wayne Barahona, progmg dir & chief of engrg; Jessica DeBoer, traf mgr; Larry Ahrens, sports cmtr.

KIWA-FM— Oct 1, 1971: 105.3 mhz; 50 kw. 292 ft TL: N43 11 00 W95 52 05. Stereo. 24 E-mail: walt@kiwaradio.com Web Site: www.kiwaradio.com.140,000 Natl. Rep: Farmakis. Format: Classic hits. News staff: 2; News: 15 hrs wkly. Target aud: General.

Shenandoah

KMA(AM)— Aug 12, 1925: 960 khz; 5 kw-U, DA-N. TL: N40 46 48 W95 21 23. Hrs open: 24 Box 960, 209 N. Elm, 51601. Phone: (712) 246-5270. Fax: (712) 246-5275. E-mail: marke@kmakkbz.com Web Site: www.kma960.com. Licensee: May Broadcasting Co. Population served: 5,968 Natl. Network: ABC. Rgnl. Network: Radio Iowa. Law Firm: Duane Morris. Format: News/talk. News staff: 2; News: 15 hrs wkly. Target aud: 35-54. Spec prog: Farm. ♦Edward W. May, pres; Mark Eno, gen mgr; Don Hansen, stn mgr.

***KYFR(AM)**— 1977: 920 khz; 5 kw-D, 2.5 w-N, DA-2. TL: N40 37 22 W95 14 42. Hrs open: 24 Box 286, 51601. Secondary address: 700 W. Sheridan Ave. 51601. Phone: (712) 246-5151. E-mail: kyfr@familyradio.org Web Site: www.shenessex.heartland.net/kyfr. Licensee: Family Stations Inc. (group owner; (acq 1976). Population served: 3,200,000 Natl. Network: Family Radio. Format: Christian. ♦Harold Camping, pres; Mike DeStefano, stn mgr.

Sioux Center

***KDCR(FM)**— Aug 16, 1968: 88.5 mhz; 100 kw. 320 ft TL: N43 05 00 W96 09 50. Stereo. Hrs opn: Dordt College Campus, 498 4th Ave. N.E., 51250. Phone: (712) 722-0885. Fax: (712) 722-6244. E-mail: kdcr@dordt.edu Web Site: www.kdcrdordt.edu. Licensee: Dordt College Inc. (acq 1-19-90). Natl. Network: USA. Format: Relg. Spec prog: Farm 2 hrs, Dutch one hr wkly. ♦Dennis DeWaard, gen mgr; Jim Bolkema, mus dir; John Slezers, news dir; Ralph Goemaat, chief of engrg.

KSOU(AM)— Nov 17, 1969: 1090 khz; 500 w-D, DA. TL: N43 03 22 W96 10 17. Hrs open: Sunrise-sunset Box 298, 128 20th St. S.E., 51250. Phone: (712) 722-1090. Phone: (712) 722-1091. Fax: (712) 722-1102. Web Site: www.ksoufm.com. Licensee: Sorenson Broadcasting Corp. (group owner; (acq 7-1-2004; grpsl). Population served: 80,000 Natl. Rep: Farmakis. Format: Contempory Christian. News staff: one; News: 17 hrs wkly. Target aud: General. ♦Craig Aukes, gen mgr; Dan Bonnema, gen sls mgr; James DeBoer, progmg dir; Doug Broek, news dir, local news ed & sports cmtr; Steve Heaton, chief of engrg; Shirley Wierda, spec ev coord & women's int ed.

KSOU-FM— Oct 17, 1974: 93.9 mhz; 3 kw. 300 ft TL: N43 03 22 W96 10 17. (CP: 50 kw, ant 492 ft.). Stereo. 24 82,000 Natl. Network: ABC. Format: Adult contemp. ♦Scott France, progmg VP & progmg dir; Steve Heaton, chief of opns & engrg VP; Shirley Wierda, women's int ed.

Sioux City

KGLI(FM)—Listing follows KWSL(AM).

KKYY(FM)—See Whiting

KMNS(AM)— May 1, 1949: 620 khz; 1 kw-U, DA-2. TL: N42 22 15 W96 27 00. Hrs open: Box 3009, 51102. Secondary address: 1113 Nebraska St. 51102. Phone: (712) 258-0628. Fax: (712) 252-2430. Web Site: www.620kmns.com. Licensee: AMFM Radio Licenses LLC. Group owner: Clear Channel Communications Inc. (acq 10-1-2002; grpsl). Population served: 970,000 Rgnl. Network: Linder Farm. Natl. Rep: Christal. Format: News/talk, farm. Target aud: 35 plus; rural, farm & country. Spec prog: Farm 20 hrs wkly. ♦Curtis Anderson, gen mgr; Laura Schiltz, sls dir & gen sls mgr.

KSEZ(FM)—Co-owned with KMNS(AM). Feb 6, 1960: 97.9 mhz; 100 kw. 643 ft TL: N42 29 48 W96 18 55. Stereo. Phone: (712) 258-5595. Web Site: www.298rocks.com.84,000 Natl. Network: ABC, Westwood One. Law Firm: Leventhal, Senter & Lerman. Format: Classic rock. Target aud: 18-49. ♦Debbie Scott-Miller, gen mgr & chief of engrg.

***KMSC(FM)**— April 1978: 88.3 mhz; 10 w. 105 ft TL: N42 28 28 W96 21 34. Stereo. Hrs opn: Library Bldg., 1501 Morningside Ave., 51106. Phone: (712) 274-5665. Phone: (712) 274-5299. Fax: (712) 274-5664. E-mail: fusion@morningside.edu Web Site: webs.morningside.edu/kmsc. Licensee: Morningside College Board of Directors. Format: Alternative. News: 2 hrs wkly. High school, college students, young professionals. Spec prog: Womens mus 5 hrs, techno 4 hrs, rock/AOR 4 hrs, urban contemp 4 hrs wkly. ♦John Reinders, pres; Ron Jorgensen, CFO; Bill Deeds, exec VP; Dr. Mark J. Heistad, gen mgr.

KSCJ(AM)— 1927: 1360 khz; 5 kw-D, 1 kw-N, DA-N. TL: N42 33 24 W96 20 12. Hrs open: 24 2000 Indian Hills Dr., 51104. Phone: (712) 239-2100. Fax: (712) 239-3346. Web Site: www.kscj.com. Licensee: Powell Broadcasting Co. (acq 1996; $3.8 million with KSUX(FM) Winnebago, NE). Population served: 283,400 Natl. Network: ABC. Rgnl. Network: Iowa Radio Net. Format: News/talk, sports. News staff: 2; News: 40-42 hrs wkly. Target aud: 35-64; educated, higher income, issues-oriented. ♦Dennis J. Bullock, gen mgr; Dave Grossenherder, sls dir & gen sls mgr; Steve Arthur, progmg dir; Randy Renshaw, news dir.

KTFC(FM)— July 1, 1965: 103.3 mhz; 100 kw. 669 ft TL: N42 29 26 W96 18 21. Stereo. Hrs opn: 24 1534 Buchanan Ave., 51106. Phone: (712) 252-4621. Licensee: Donald A. Swanson. Natl. Network: USA. Format: Gospel, all Bible. Spec prog: Farm one hr, news 10 hrs, children 5 hrs wkly. ♦Donald A. Swanson, pres & gen mgr.

***KWIT(FM)**— Jan 31, 1978: 90.3 mhz; 100 kw. Ant 910 ft TL: N42 28 56 W96 15 30. Stereo. Hrs opn: 24 4647 Stone Ave., 51106-1997. Phone: (712) 274-6406. Fax: (712) 274-6411. E-mail: gondekg@witcc.com Web Site: www.kwit-koji.org. Licensee: Western Iowa Tech Community College. Population served: 325,000 Natl. Network: PRI, NPR. Format: Class, news/talk, Sp. News staff: one; News: 36 hrs wkly. Target aud: 25-54. Spec prog: Blues 2 hrs, Triple A 12 hrs, Sp 20 hrs wkly. ♦Gretchen Gondek, gen mgr; Steve Smith, opns mgr.

KWSL(AM)— April 1938: 1470 khz; 5 kw-U, DA-2. TL: N42 24 42 W96 25 30. Stereo. Hrs opn: Box 3009, 51102. Phone: (712) 255-1470. Fax: (712) 252-2430. Licensee: AMFM Radio Licenses LLC. Group owner: Clear Channel Communications Inc. (acq 10-1-2002; grpsl). Population served: 272,300 Natl. Rep: Christal. Law Firm: Leventhal, Senter & Lerman. Format: Sp. Target aud: 35 plus. ♦Curtis Anderson, progmg dir.

KGLI(FM)—Co-owned with KWSL(AM). Mar 11, 1974: 95.5 mhz; 100 kw. 900 ft TL: N42 30 53 W96 18 13. Stereo. Phone: (712) 258-5595. Fax: (712) 252-2430. Format: Hot adult contemp. Target aud: 18-49. ♦Ryan Reid, gen mgr; Rob Powers, progmg dir.

Sioux Rapids

KTFG(FM)— 1991: 102.9 mhz; 50 kw. 479 ft TL: N42 54 34 W95 09 35. Hrs open:
Rebroadcasts KTFC(FM) Sioux City 100%.
1534 Buchanan Ave., Sioux City, 51106. Phone: (712) 252-0327. Licensee: Donald A. Swanson. Natl. Network: USA. Format: Gospel. Spec prog: Children 5 hrs wkly. ♦Donald A. Swanson, gen mgr.

Spencer

KICD(AM)— December 1942: 1240 khz; 1 kw-U. TL: N43 10 00 W95 08 45. Hrs open: 24 Box 260, 51301. Secondary address: 2600 N. Hwy. Blvd. 51301. Phone: (712) 262-1240. Fax: (712) 262-2076. Web Site: www.kicdam.com. Licensee: Saga Communications of Iowa LLC. Group owner: Saga Communications Inc. (acq 11-22-99; grpsl). Population served: 124,000 Natl. Network: CBS. Format: Talk. News staff: one; News: 26 hrs wkly. Target aud: 35 plus. ♦David Putnam, gen mgr & gen sls mgr; Bill Campbell, opns mgr & progmg dir; Brent Palm, news dir; Dave Inqualson, pub affrs dir; Joseph Schloss, chief of engrg.

KICD-FM— Sept 17, 1965: 107.7 mhz; 100 kw. 310 ft TL: N43 10 00 W95 08 45. Stereo. 24 Web Site: www.cd1077fm.com. Natl. Network: CBS. Format: Country. News staff: one. Target aud: 25 plus. ♦David Putman, gen sls mgr; Rhoda Wedeking, progmg dir.

KLLT(FM)— February 1979: 104.9 mhz; 25 kw. 279 ft TL: N43 17 13 W95 08 34. Stereo. Hrs opn: 24 Box 260, 2600 N. Hwy. Blvd., 51301. Phone: (712) 262-1240. Fax: (712) 262-2076. Fax: (712) 262-5821. E-mail: lite1049@ncn.net Web Site: www.lite1049.com. Licensee: Saga Communications of Iowa LLC. Group owner: Saga Communications Inc. (acq 11-22-99; grpsl). Population served: 80000 Natl. Rep: Katz Radio. Format: Light Rock. News: one hr wkly. Target aud: 25-54. ♦Edward Christian, CEO & pres; Dave Putnam, gen mgr; Bill Campbell, opns dir & opns mgr; Darby Bishop, prom dir; Kevin Tlam, progmg dir; Brent Palm, news dir.

Spirit Lake

***KJIA(FM)**—Not on air, target date: unknown: 88.9 mhz; 50 kw. Ant 272 ft TL: N43 20 34 W95 12 24. Hrs open: Box 738, Okoboji, 51355. Secondary address: 7 S. Okoboji Grove Rd., Arnolds Park 51331. Phone: (712) 332-7184. Fax: (712) 332-2428. E-mail: kjia@kjiaradio.com Web Site: www.kjiaradio.com. Licensee: Minn-Iowa Christian Broadcasting Inc. (group owner). Format: Christian. ♦Matt Dorfner, gen mgr & progmg dir; Mark Groom, chief of engrg.

KUOO(FM)— Apr 1, 1985: 103.9 mhz; 50 kw. 492 ft TL: N43 20 34 W93 12 24. Stereo. Hrs opn: 24 Box 528, 1039 Radio Dr., 51360. Phone: (712) 336-5800. Fax: (712) 336-1634. Licensee: Sorenson Broadcasting Corp. (group owner; (acq 7-1-2004; grpsl). Natl. Network: Fox News Radio. Format: Adult contemp. News staff: 2; News: 16 hrs wkly. Target aud: 25-54. ♦Dean Sorenson, pres; Marty Spies, gen mgr; Dan Larkin, opns mgr.

KUQQ(FM)—Milford, Oct 1, 1996: 102.1 mhz; 50 kw. 420 ft TL: N43 20 34 W93 12 24. Hrs open: Box 528, 1039 Radio Dr., 51360. Phone: (712) 336-5877. Fax: (712) 336-1634. Licensee: Sorenson Broadcasting Corp. (group owner; (acq 7-1-2004; grpsl). Format: Classic rock. Target aud: 18-44. ♦Dean Sorenson, pres; Marty Spies, gen mgr; Dan Larkin, opns mgr.

State Center

***KTDV(FM)**—Not on air, target date: unknown: 91.9 mhz; 25 kw. Ant 279 ft TL: N42 15 49 W93 03 57. Hrs open: Box 538, Marshalltown, 50158. Phone: (641) 752-4122. Fax: (641) 752-5121. Licensee: Marshalltown Education Plus Inc. ♦Mark Osmundson, gen mgr.

Storm Lake

KAYL(AM)— November 1948: 990 khz; 250 w-D, 6 w-N. TL: N42 38 05 W95 10 10. Hrs open: Box 1037, 50588. Secondary address: 606 1/2 Lake Ave. 50588. Phone: (712) 732-3520. Fax: (712) 732-1746. E-mail: kayl@waittradio.com Licensee: Sorenson Broadcasting Corp.

(group owner; (acq 7-1-2004; grpsl). Population served: 14,200 Natl. Network: ESPN Radio. News staff: one; News: 1.5 hrs wkly. Target aud: 38-50; male. Spec prog: Special 3 hrs wkly. ♦Dean Sorenson, pres; Buzz Paterson, gen mgr.

KAYL-FM— February 1949: 101.7 mhz; 50 kw. 400 ft TL: N42 38 05 W95 10 10. Stereo. 23,400 Format: Adult contemp, news, sports. News staff: one; News: 15 hrs wkly. Target aud: 25-54; male & female.

Stuart

KKRF(FM)— Aug 11, 1993: 107.9 mhz; 12.kw. Ant 472 ft TL: N41 30 25 W94 18 06. Stereo. Hrs opn: 204 S. Division St., 50250. Secondary address: 204 S. Division St. 50250-5021. Phone: (515) 465-5357. Fax: (515) 465-3952. E-mail: kkrf1079@aol.com Web Site: RealCountryOnline.Com. Licensee: Coon Valley Communications Inc. Population served: 60,100 Natl. Network: ABC. Format: Country. News staff: one; News: 10 hrs wkly. Target aud: 25-64; general. Spec prog: Farm 5 hrs wkly. ♦Pat Delaney, pres & CFO; Sue Thomsen, gen mgr; John France, opns mgr & progmg dir.

Twin Lakes

KTLB(FM)— Oct 5, 1975: 105.9 mhz; 25 kw. 328 ft TL: N42 32 09 W94 40 48. Stereo. Hrs opn: 18 1014 Central Ave., Fort Dodge, 50501. Phone: (515) 573-5748. Fax: (515) 573-3376. Licensee: Three Eagles of Ft. Dodge Inc. Group owner: Three Eagles Communications (acq 4-22-97; $248,883). Population served: 25,000 Format: Oldies. News staff: one; News: 15 hrs wkly. Target aud: 35-54; baby boomers. Spec prog: Farm 15 hrs, gospel 2 hrs, relg 2 hrs wkly. ♦Gary Buchanan, pres; Pat Kolar, gen mgr & gen sls mgr; Greg Allenson, opns mgr & gen sls mgr; Travis Reeves, gen mgr & opns mgr.

Vinton

KRQN(FM)— 2005: 107.1 mhz; 4.7 kw. Ant 371 ft TL: N42 08 56 W91 52 50. Hrs open: 425 2nd St. S.E., 4th Fl., Cedar Rapids, 52401. Phone: (319) 365-9431. Fax: (319) 363-8062. E-mail: bob@khah.com Licensee: George S. Flinn Jr. Format: 60s and 70s rock and roll hits. ♦Jim Worthington, gen mgr & opns mgr; Terry Weinacht, gen sls mgr; Bill Hahn, progmg dir; Ryan Brainard, news dir; Scott Wilcox, chief of engrg; Mike Meilly, traf mgr.

Wapello

***KAIP(FM)—** 2005: 88.9 mhz; 1 w hoirz, 13.5 kw vert. Ant 494 ft vert TL: N41 04 59 W91 10 18. Hrs open: 2351 Sunset Blvd., Suite 170-218, Rocklin, CA, 95765. Phone: (916) 251-1600. Fax: (916) 251-1650. Web Site: www.air1.com. Licensee: Educational Media Foundation. Group owner: EMF Broadcasting. Population served: 104,000 Natl. Network: Air 1. Law Firm: Shaw Pittman. Format: Christian. News staff: 3. Target aud: 25-44; Judeo Christian, female. ♦Richard Jenkins, pres; Mike Novak, VP; Keith Whipple, dev dir; David Pierce, progmg mgr; Ed Lenane, news dir; Sam Wallington, engrg dir; Karen Johnson, news rptr; Marya Morgan, news rptr; Richard Hunt, news rptr.

Washington

KCII(AM)— Nov 12, 1961: 1380 khz; 500 w-D. TL: N41 18 18 W91 42 36. Hrs open: 5 AM-11 PM Box 524, 110 E. Main St., 52353. Phone: (319) 653-2113. Fax: (319) 653-3500. Licensee: Home Broadcasting Inc. (acq 9-3-96; $800,000 with co-located FM). Population served: 140,000 Natl. Network: AP Radio. Law Firm: Shaw Pittman. Format: News, adult contemp. News staff: one; News: 14 hrs wkly. Target aud: 25-54; females. ♦Jack Davison, gen mgr; Joe Nichols, gen sls mgr; Ben Tyler, progmg dir; Jeremy Aitken, news dir; Calista Stout, traf mgr.

KCII-FM— 1975: 95.3 mhz; 3 kw. 300 ft TL: N41 18 18 W91 42 36. Stereo. 5 AM-11 PM Format: News, oldies. ♦Ben Tyler, spec ev coord; B.J. Hansen, news rptr; Kris Ellston, sports cmtr.

Waterloo

***KBBG(FM)—** July 26, 1978: 88.1 mhz; 9.5 kw. 150 ft TL: N42 30 35 W92 19 35. Stereo. Hrs opn: 19 918 Newell St., 50703-2720. Phone: (319) 234-1441. Phone: (319) 235-1515. Fax: (319) 234-6182. E-mail: lou@kbbg.org Web Site: www.kbbgfm.org. Licensee: Afro-American Community Broadcasting Inc. Natl. Network: American Urban. Format: Educ, gospel, rhythmn and blues, jazz. Target aud: General. Spec prog: Gospel. ♦Jimmie Porter, CEO; Lou Porter, pres; Beverly Douglas, stn mgr; Lou Lou Porter, dev dir & news dir.

KFMW(FM)—Listing follows KWLO(AM).

***KNWS(AM)—** 1953: 1090 khz; 1 kw-D. TL: N42 26 38 W92 17 58. Hrs open: Sunrise-sunset 4880 Texas, 50702. Phone: (319) 296-1975. Fax: (319) 296-1977. E-mail: knws@knws.org Web Site: www.knws.org. Licensee: Northwestern College. Group owner: Northwestern College & Radio (acq 4-2-53). Rgnl. Network: Skylight. Format: Relg, Christian, talk. Target aud: 35 plus. ♦Paul Virts, exec VP; Doug Smith, gen mgr & stn mgr; Dan Raymond, progmg dir & asst music dir; David Dobes, chief of engrg.

KNWS-FM— 1965: 101.9 mhz; 100 kw. 1,571 ft TL: N42 24 02 W91 50 36. Stereo. 24 Web Site: www.knws.org. (acq 1965).350,000 Natl. Network: AP Radio. Format: Adult contemp Christian music, relg. Target aud: 30-50; women. ♦Brent Manion, disc jockey; Mike Lanser, disc jockey.

KOKZ(FM)—Listing follows KXEL(AM).

KWLO(AM)— November 1947: 1330 khz; 5 kw-U, DA-2. TL: N42 28 56 W92 16 16. Stereo. Hrs opn: 24 Box 1540, 50704. Secondary address: 514 Jefferson St. 50704. Phone: (319) 234-2200. Fax: (319) 234-0149. Web Site: www.star1330.com. Licensee: KXEL Broadcasting Co. Inc. Group owner: Bahakel Communications (acq 8-16-96; grpsl). Natl. Network: ABC. Natl. Rep: Katz Radio. Law Firm: Brooks, Pierce, McLendon, Humphrey & Leonard. Format: Btfl mus, big band. News staff: 3. Target aud: 35 plus. ♦Cy N. Bahakel, pres; Tim Mathews, gen mgr; Dennis Lowe, opns dir, progmg dir & news dir; Mark Schumacher, chief of engrg; Amy Mollus, traf mgr.

KFMW(FM)—Co-owned with KWLO(AM). November 1968: 107.9 mhz; 100 kw. 1,850 ft TL: N42 24 04 W91 50 43. Stereo. 24 Web Site: www.rock108.com. Format: Rock. Target aud: 18-34; men. ♦Michael Cross, opns mgr; Mark Chapman, progmg dir; Dolly Fortier, pub affrs dir.

KWOF(AM)— Oct 31, 1972: 850 khz; 500 w-D, DA. TL: N42 28 56 W92 16 16. Hrs open: 3232 Osage Rd., 50703. Phone: (319) 319-236-5700. Fax: (319) 236-8777. E-mail: theprodigal @theprodigalradio.com Web Site: www.theprodigalradio.com. Licensee: Friendship Communications Inc. (acq 2-23-95; FTR: 5-22-95). Format: Classic Christian rock. Target aud: 34-55; Christian, interdenominational. ♦Michael James, gen mgr.

KXEL(AM)— July 14, 1942: 1540 khz; 50 kw-U, DA-N. TL: N42 10 47 W92 18 38. Stereo. Hrs opn: 24 Box 1540, 514 Jefferson St., 50701. Phone: (319) 234-2200. Fax: (319) 234-0149. Web Site: www.kxel.com. Licensee: KXEL Broadcasting Co. Inc. Group owner: Bahakel Communications (acq 1-11-58). Natl. Network: ABC. Natl. Rep: Katz Radio. Format: News/talk. News staff: 3; News: 28 hrs wkly. Target aud: 45-65. Spec prog: Relg 20 hrs wkly. ♦Cy N. Bahakel, pres; Tim Mathews, gen mgr; Dennis Lowe, opns dir, progmg dir & news dir; Mark Schumacher, chief of engrg.

KOKZ(FM)—Co-owned with KXEL(AM). Nov 21, 1962: 105.7 mhz; 100 kw. 1,403 ft TL: N42 24 35 W92 05 10. Stereo. 24 Web Site: www.cool1057.com. Format: Oldies. Target aud: 25-54. ♦Dolly Fortier, pub affrs dir.

Waukon

KHPP(AM)—Listing follows KNEI-FM.

KNEI-FM— Sept 1, 1968: 103.5 mhz; 37 kw. Ant 574 ft TL: N43 18 28 W91 27 18. Stereo. Hrs opn: 14 W. Main St., 52172. Phone: (563) 568-3476. Phone: (563) 568-3477. Fax: (563) 568-3391. E-mail: knei@direcway.com Licensee: Wennes Communications Stations Inc. (acq 4-5-2002; grpsl). Population served: 3,639 Natl. Network: CBS Radio. Law Firm: Sam Miller. Format: Real country. News: one hr wkly. Target aud: 25-45. ♦Greg Wennes, CEO; Chuck Bloxham, gen mgr, opns dir & gen sls mgr.

KHPP(AM)—Co-owned with KNEI-FM. July 1, 1967: 1160 khz; 880 w-D, 26 w-N. TL: N43 17 13 W91 28 06. Format: Oldies.

Waverly

***KWAR(FM)—** Sept 15, 1951: 89.1 mhz; 40 w. 125 ft TL: N42 43 24 W92 28 05. Hrs open: 24 Wartburg College, 100 Wartburg Blvd., 50677. Phone: (319) 352-8209/352-8306. Fax: (319) 352-8610. E-mail: business@kwar.org Web Site: www.kwar.org. Licensee: Wartburg College. Population served: 13,000 Format: Educ, div. News staff: one; News: 3 to 5 hrs wkly. Target aud: General. Spec prog: Folk 3 hrs, classical 3 hrs, jazz 5 hrs, world music 3 hrs, news publ affrs 3 hrs wkly. ♦Steven Murray, stn mgr; Adam Van Briesen, progmg dir.

KWAY(AM)— May 6, 1958: 1470 khz; 1 kw-D, 61 w-N, DA-2. TL: N42 42 13 W92 28 21. Hrs open: Box 307, 50677. Phone: (319) 352-3550. Fax: (319) 352-3601. E-mail: kwayradio@kwayradio.com Web Site: www.kwayradio.com. Licensee: Al Suhr Enterprises. Population served: 25,000 Format: Classic country. ♦Al Suhr, pres, gen mgr & chief of engrg; Steven Hatter, opns mgr.

KWAY-FM— Dec 21, 1971: 99.3 mhz; 3 kw. 180 ft TL: N42 42 13 W92 28 21.135,000 Format: Adult contemp.

***KWER(FM)—**Not on air, target date: unknown: 89.9 mhz; 100 w. Ant 52 ft TL: N42 43 38 W92 28 51. Hrs open: Wartburg College, 100 Wartburg Blvd., 50677. Phone: (319) 352-8276. Licensee: Wartburg College.

***KWVI(FM)—** 2006: 88.9 mhz; 20 kw vert. Ant 274 ft TL: N42 47 21 W92 14 22. Hrs open:
Rebroadcasts WAFR(FM) Tupelo, MS 100%.
Drawer 2440, Tupelo, MS, 38801. Phone: (662) 844-8888. Fax: (662) 842-6791. Web Site: www.afr.net. Licensee: American Family Association. Format: Christian. ♦Marvin Sanders, gen mgr.

Webster City

KQWC(AM)— Feb 5, 1950: 1570 khz; 250 w-D, 132 w-N. TL: N42 27 45 W93 48 05. (CP: 147 w). Hrs opn: Box 550, 50595. Phone: (515) 832-1570. Fax: (515) 832-2079. Licensee: NRG License Sub. LLC. Group owner: Waitt Broadcasting Inc. (acq 10-31-2005; grpsl). Population served: 50,000 Natl. Rep: Farmakis. Format: Btfl mus, big band. News staff: one; News: 45 hrs wkly. Target aud: 50 plus; affluent with max spendable income. Spec prog: Farm 8 hrs wkly. ♦Mike Delick, CFO; Mary Harris, stn mgr; Eli Savoie, opns mgr; Tracey Williams, gen sls mgr; Pat Powers, news dir, women's int ed & disc jockey; Brent Balbinot, disc jockey; Glenn R. Olson, disc jockey; Jarrod Gammond, disc jockey; Larry Schultz, disc jockey.

KQWC-FM— 1969: 95.7 mhz; 25 kw. 328 ft TL: N42 28 04 W93 47 48. Stereo. 1020 E. 2nd St., 50595. Web Site: www.kqradio.com. Format: Adult Contemp. News staff: one; News: 35 hrs wkly.

West Des Moines

KJJY(FM)—Licensed to West Des Moines. See Des Moines

***KWDM(FM)—** March 1976: 88.7 mhz; 100 w. 170 ft TL: N41 35 25 W93 45 10. Stereo. Hrs opn: 14 1140 35th St., 50266. Phone: (515) 226-2660. Phone: (515) 226-2600. Fax: (515) 226-2609. E-mail: kwdmfm@hotmail.com Web Site: www.wdm.k12.ia.us/kwdm. Licensee: West Des Moines Community School District. Population served: 300,000 Law Firm: Reddy, Begley & McCormick. Format: Alternative. News: 3 hrs wkly. Target aud: 12-25; educ facility-var progmg. Spec prog: Sports 3 hrs wkly. ♦Mack Wzie Carey, stn mgr; Marianne Coppock, dev dir; Nicole Faust, mktg dir & progmg dir.

Whiting

KKYY(FM)— Dec 11, 1979: 101.3 mhz; 50 kw. Ant 492 ft TL: N42 21 25 W96 08 02. Stereo. Hrs opn: 24 522 14th St., Sioux City, 51105. Phone: (712) 258-5655. Fax: (712) 258-1511. Web Site: www.y1013.net. Licensee: Powell Broadcasting Co. Inc. Group owner: Waitt Broadcasting Inc. (acq 5-1-2007; $4.2 million with KZSR(FM) Onawa). Population served: 200,000 Natl. Network: Motor Racing Net, Westwood One, ABC. Format: Country. News staff: one; News: 20 hrs wkly. Target aud: 18-54; adult men & women. ♦Jerry Haack, gen mgr; Kelli Erickson, gen sls mgr; Tim Guentz, progmg dir; Pam Guntz, chief of engrg.

Winterset

KZZQ(FM)— March 1994: 99.5 mhz; 6 kw. 328 ft TL: N41 24 02 W93 54 58. Stereo. Hrs opn: 24 3317 335th St., Waukee, 50263. Phone: (515) 987-9995. Fax: (515) 987-9808. Web Site: www.kzzq.com. Licensee: Positive Impact Media Inc. (acq 1-11-94; $600,000; FTR: 1-31-94). Format: Contemp Christian mus. Target aud: Families. ♦David Nadler Jr., gen mgr; David St. John, progmg dir.

Kansas

Abilene

KABI(AM)— Apr 8, 1963: 1560 khz; 250 w-D, 58 w-N. TL: N38 55 46 W97 14 46. Hrs open: 24 200 N. Broadway, 67410. Phone: (785)

263-1560. Fax: (785) 263-0166. Web Site: www.ksallink.com. Licensee: MCC Radio LLC. Group owner: Morris Radio LLC (acq 1-30-2004; grpsl). Population served: 18,000 Natl. Network: ABC. Format: Adult standards. News staff: one; News: 2 hrs wkly. Target aud: 35 plus; loc residents of Dickinson County. Spec prog: Relg 4 hrs wkly. ♦Robert Protzman, VP & gen mgr; Clarke Sanders, opns mgr & prom mgr; Richie Allen, progmg dir; John Anderson, news dir; Mark Beaver, traf mgr.

KSAJ-FM—Co-owned with KABI(AM). Dec 10, 1968: 98.5 mhz; 100 kw. 443 ft TL: N38 47 50 W97 13 01. Stereo. Box 80, Salina, 67402. Secondary address: 131 N. Santa Fe, Salina 67401. Phone: (785) 823-1111. Fax: (785) 823-2034. Web Site: www.ksallink.com.311,000 Natl. Network: ABC. Format: Oldies. News staff: news progmg one hr wkly News: 3;. Target aud: 35-64; baby boomers. ♦Clarke Sanders, gen mgr & prom mgr; Robert Protzman, gen mgr; Richie Hall, progmg dir; Mark Beaer, engrg mgr & traf mgr.

Andover

KDGS(FM)— Nov 1, 1993: 93.9 mhz; 25 kw. 328 ft TL: N37 37 00 W97 20 11. Hrs open: 2120 N. Woodlawn, Suite 352, Wichita, 67206. Phone: (316) 685-2121. Fax: (316) 685-3408. E-mail: info@power939.com Web Site: www.power939.com. Licensee: Entercom Wichita License LLC. Group owner: Entercom Communications Corp. (acq 4-21-00; $3.15 million). Format: CHR rhythmic. Target aud: 18-34. ♦Jackie Wise, gen mgr; Greg Williams, progmg dir.

Arkansas City

KACY(FM)— February 1999: 102.5 mhz; 6 kw. 328 ft TL: N37 05 01 W96 55 46. Hrs open: 24 106 N. Summit, 67005. Phone: (620) 442-1102. Fax: (620) 442-8102. Licensee: Third Coast Broadcasting. Format: AAA, AOR. ♦Marshall Ice, gen mgr.

***KAXR(FM)**— 2001: 91.3 mhz; 13.5 kw. Ant 321 ft TL: N36 55 32 W97 01 34. Hrs open: Box 3206, American Family Radio, Tupelo, MS, 38803. Phone: (662) 844-8888. Fax: (662) 842-6791. E-mail: comments@afr.net Web Site: www.afr.net. Licensee: American Family Association. Group owner: American Family Radio. Format: Inspirational Christian. ♦Marvin Sanders, gen mgr.

KSOK(AM)— Jan 1, 1947: 1280 khz; 1 kw-D, 100 w-N. TL: N37 05 19 W97 01 56. Hrs open: 24 Box 1014, 334 East Radio Lane, 67005. Phone: (620) 221-1440. Fax: (620) 442-5401. E-mail: brianc@ksokfm.com Licensee: Cowley County Broadcasting Inc. (acq 9-3-02; with KSOK-FM Winfield). Population served: 40,000 Format: Classic Country. News staff: one; News: 10 hrs wkly. Target aud: 34 plus; agriculturally oriented listeners seeking info. Spec prog: Farm 10 hrs, relg 5 hrs wkly. ♦Marty Mutti, gen mgr; Brian Cunningham, opns mgr & chief of opns; Christy Bursack, prom dir; Jonathon Stevens, news dir.

KYQQ(FM)— Nov 1, 1979: 106.5 mhz; 100 kw. Ant 1,278 ft TL: N37 21 24 W96 57 55. Stereo. Hrs opn: 4200 N. Old Lawrence Rd., Wichita, 67219. Phone: (316) 838-9141. Fax: (316) 838-3607. Licensee: Journal Broadcast Corp. Group owner: Journal Broadcast Group Inc. (acq 6-11-99; grpsl). Population served: 720,000 Law Firm: Shaw Pittman. Format: Mexican. Target aud: 18-49; Hispanic. ♦Rob Burton, VP & gen mgr; Eric McCart, gen sls mgr; Manny Cowzinski, prom dir; Beverlee Brannigan, progmg dir.

Atchison

KAIR(AM)— July 28, 1939: 1470 khz; 1 kw-U, DA-1. TL: N39 37 09 W94 59 27. Hrs open: 5 AM-midnight Box G, 200 N. 5th St., 66002. Phone: (913) 367-1470. Fax: (913) 367-7021. E-mail: kair@lunworth.com Licensee: Mark V Media Group Inc. (group owner; (acq 12-27-2004; $1.55 million with KAIR-FM Horton). Population served: 256,500 Natl. Network: EWTN Radio. Format: Catholic. Target aud: General; 25-65. Spec prog: Farm 7 hrs wkly. ♦Mark Oppold, pres & gen mgr.

Augusta

KFXJ(FM)— Apr 1, 1992: 104.5 mhz; 45 kw. Ant 515 ft TL: N37 48 15 W97 15 56. Stereo. Hrs opn: 24 4200 N. Old Lawrence Rd., Wichita, 67219. Phone: (316) 838-9141. Fax: (316) 838-3607. Web Site: www.1045thefox.com. Licensee: Journal Broadcast Corp. Group owner: Journal Broadcast Group Inc. (acq 6-11-99; grpsl). Population served: 500,000 Format: Classic rock hits. Target aud: 25-54. ♦Rob Burton, VP & gen mgr; Ron Eric Taylor, opns mgr; Eric McCart, gen sls mgr; Jason Wituk, rgnl sls mgr; Manny Cowzinski, prom dir.

KIBB(FM)— 2006: 100.5 mhz; 25 kw. Ant 276 ft TL: N37 44 13 W97 09 25. Hrs open: 136 Main St., Suite 202, Westport, CT, 06880-3304. Phone: (203) 227-1978. Fax: (203) 227-2373. Web Site: www.connoisseurmedia.com. Licensee: Connoisseur Media LLC. Format: Adult hits. ♦Doug Downs, gen mgr; Ron Allen, progmg dir.

Baldwin City

***KNBU(FM)**— Nov 29, 1965: 89.7 mhz; 100 w. 118 ft TL: N38 46 45 W95 11 15. Hrs open: Box 65, 66006. Phone: (785) 594-6451, EXT. 300. Phone: (785) 594-8300. Fax: (785) 594-3570. Licensee: Baker University. Population served: 2,520 Spec prog: Jazz 15 hrs wkly. ♦Tom Hedrick, gen mgr.

Baxter Springs

KMOQ(FM)— Feb 1, 1980: 107.1 mhz; 6 kw. 300 ft TL: N37 07 34 W94 42 12. Stereo. Hrs open: 2510 W. 20th St., Joplin, MO, 64804. Phone: (417) 781-1313. Fax: (417) 781-1316. Licensee: FFD Holdings I Inc. (acq 12-20-2004; grpsl). Natl. Network: ABC. Format: CHR. Target aud: 25-54. ♦Dave Clemons, gen sls mgr; Chris Yeager, progmg dir; Kathleen Pike, gen mgr & traf mgr.

Belle Plaine

KANR(FM)— Mar 4, 1996: 92.7 mhz; 12 kw. Ant 469 ft TL: N37 20 15 W97 27 56. Hrs open: 2120 N. Woodlawn, Wichita, 67208. Phone: (316) 652-9275. Fax: (316) 683-0818. Licensee: Daniel D. Smith (acq 2-16-93; $10,700; FTR: 3-8-93). Population served: 180,000 Format: Rgnl Mexican. ♦Daniel D. Smith, pres; Daniel Smith, gen mgr; Joe Roach, progmg dir; Bruce Adamek, traf mgr.

Belleville

KREP(FM)— June 26, 1984: 92.1 mhz; 14.5 kw. 276 ft TL: N39 45 00 W97 36 48. Stereo. Hrs opn: 24 2307 US Hwy. 81, 66935. Phone: (785) 527-2266. Phone: (785) 527-2267. Fax: (785) 527-5919. E-mail: kr-92@nckcn.com Licensee: First Republic Broadcasting Corp. (acq 6-84; FTR: 1-84). Natl. Network: ABC. Format: Country. News staff: one; News: 20 hrs wkly. Target aud: 25-55. ♦Deborah Sasser, pres, gen mgr, progmg dir & news dir; Christine Strutt, gen sls mgr & mktg dir; Marvin Hoffman, chief of engrg; Eric Allgood, sports cmtr.

Beloit

KVSV(AM)— Nov 21, 1979: 1190 khz; 2.3 kw-D, 90 w-N, DA. TL: N39 26 53 W98 04 45. Hrs open: 6 AM-9 PM Box 7, E. Hwy. 24, 67420. Phone: (785) 738-2206. Web Site: www.kvsvradio.com. Licensee: McGrath Publishing Co. (acq 11-1-99; $500,000 with co-located FM). Population served: 51,200 Rgnl. Network: Kan. Agri. Format: Adult Top-40. News staff: one; News: 11 hrs wkly. Target aud: General. Spec prog: Farm 9 hrs wkly. ♦John Swanson, gen mgr & progmg dir; Sharon Fuller, gen sls mgr.

KVSV-FM— Nov 11, 1980: 105.5 mhz; 50 kw. 443 ft TL: N39 28 09 W98 05 37. Stereo. 24 Web Site: www.kvsvradio.com. Format: Btfl mus, easy lstng. ♦John Swanson, progmg mgr.

Bronson

***KBJQ(FM)**— 2002: 88.3 mhz; 99 kw. Ant 380 ft TL: N37 53 56 W95 00 09. Hrs open: Drawer 3206, Tupelo, MS, 38803. Phone: (662) 844-8888 ext. 204. Fax: (662) 842-6791. Licensee: American Family Association. Group owner: American Family Radio Population served: 127,972 Format: Relg. ♦Marvin Sanders, gen mgr.

Burlington

KSNP(FM)— June 14, 1990: 97.7 mhz; 6 kw. 349 ft TL: N38 10 08 W95 39 07. Hrs open: 6 AM-11 PM Box 233, 1910 S. 6th, 66839. Phone: (620) 364-8807. Fax: (620) 364-2047. E-mail: ksnp@kans.com Licensee: Southeast Kansas Broadcasting Co. (group owner; acq 1999; $230,000). Rgnl. Network: Mid-American Ag. Format: Hot country. News staff: one; News: 5 hrs wkly. Target aud: 25-45; industrial employees. Spec prog: Farm 8 hrs, relg 3 hrs wkly. ♦Peg Downard, stn mgr, sls dir & gen sls mgr; Mindy Ryan, progmg dir.

Caney

KEOJ(FM)— Oct 15, 1992: 101.1 mhz; 3 kw. 328 ft TL: N36 58 19 W95 53 47. Stereo. Hrs opn: 24 City Plex Towers, 2448 E. 81 St., Tulsa, OK, 74137. Phone: (918) 492-2660. Fax: (918) 492-8840. E-mail: kxoj@kxoj.com Web Site: www.kxoj.com. Licensee: KXOJ Inc. Group owner: Adonai Radio Group (acq 4-29-92; grpsl). Format: Contemp Christian. Target aud: 18-35; young married Christians. ♦Mike Stephens, pres; Joy Stephens, VP; David Stephens, gen mgr & stn mgr; Bob Thornton, sls dir & progmg dir; Joe Hancock, chief of engrg; Dorothy Kimble, traf mgr.

Cawker City

KZDY(FM)— 1999: 96.3 mhz; 13 kw. Ant 230 ft TL: N39 30 29 W98 18 57. Stereo. Hrs opn: 24 Box 88, Glen Elder, 67446. Phone: (785) 545-3220. Fax: (785) 545-3220. E-mail: kdnskzdy@nckcn.com Licensee: Dierking Communications Inc. Group owner: Hoeflicker Stns. (acq 7-27-2006). Natl. Network: Jones Radio Networks, AP Radio. Law Firm: Kenkel & Associates. Format: Adult contemp. News staff: one; News: 6 hrs wkly. Target aud: 18-60. ♦Wade Gerstner, gen mgr.

Chanute

KKOY(AM)— Nov 17, 1952: 1460 khz; 1 kw-D, 57 w-N. TL: N37 41 25 W95 28 08. Hrs open: 24 Box 788, 66720-0788. Secondary address: 702 N. Plummer Sts. 66720. Phone: (620) 431-3700. Fax: (620) 431-4643. Web Site: www.kkoy.com. Licensee: Southeast Kansas Broadcasting Co. Inc. (group owner; acq 5-21-97; $464,447 with co-located FM). Population served: 53,000 Rgnl. Network: Kan. Info., Kan. Agri, Mid-American Ag. Format: News/talk. News staff: one. Target aud: 25-54. ♦Rhonda Kane, gen mgr; Heather Lee, progmg dir; Rob Strand, news dir.

KKOY-FM— Jan 1, 1971: 105.5 mhz; 8 kw. Ant 584 ft TL: N37 35 59 W95 39 10. Stereo. 24 E-mail: sales@kkoy.com Web Site: www.kkoy.com. Natl. Network: ABC. Format: Hot adult contemp.

Clay Center

KCLY(FM)— Jan 6, 1978: 100.9 mhz; 35.5 kw. Ant 581 ft TL: N39 28 03 W97 03 45. Stereo. Hrs opn: 1815 Meadowlark Rd., 67432. Phone: (785) 632-5661. Fax: (785) 632-5662. Web Site: www.kclyradio.com. Licensee: Taylor Communications. (acq 8-94; FTR: 1-78). Population served: 90,000 Format: C&W, adult contemp, relg. Target aud: 24-55; general. ♦Kyle Bauer, gen mgr; Joyce Beck, gen sls mgr; Jamie Bloom, progmg dir; Rod Keen, engrg mgr; Joe Woodward, traf mgr; Rocky Downing, sports cmtr.

Clearwater

KFH-FM—Licensed to Clearwater. See Wichita

Coffeyville

KGGF(AM)— 1930: 690 khz; 10 kw-D, 5 kw-N, DA-2. TL: N37 08 58 W95 28 27. Hrs open: Box 1087, 306 W. 8th St., 67337. Phone: (620) 251-3800. Fax: (620) 251-9210. Licensee: KGGF-KUSN Inc. Group owner: Mahaffey Enterprises Inc. (acq 12-26-90; $750,000 with co-located FM; FTR: 1-14-91). Population served: 100,000 Natl. Network: ABC. Rgnl. Network: Kan. Agri. Format: News/talk. News staff: one; News: 5 hrs wkly. Target aud: 35 plus. ♦John Leonard, gen mgr.

KKRK(FM)—Co-owned with KGGF(AM). Sept 1, 1983: 98.9 mhz; 6 kw. 305 ft TL: N37 06 28 W95 43 22. Stereo. Licensee: KGGF-KUSN, Inc.65,000. Format: Classic rock. News staff: one. Target aud: 35-54.

Colby

KQLS(FM)—Listing follows KXXX(AM).

***KTCC(FM)**— May 1974: 91.9 mhz; 3 kw. 199 ft TL: N39 22 34 W101 03 08. Stereo. Hrs opn: 24 1255 S. Range, 67701. Phone: (785) 462-3984, EXT. 309. Fax: (785) 462-4600. Web Site: www.colbycc.edu. Licensee: Colby Community College. Population served: 25,000 Format: CHR. News staff: one; News: 14 hrs wkly. Target aud: 18-25; young adults. Spec prog: Sports 3 hrs, classic rock 3 hrs, hard rock 7 hrs, hip hop 4 hrs wkly. ♦Corey Sorenson, stn mgr.

KWGB(FM)— Sept 1, 1998: 97.9 mhz; 100 kw. Ant 712 ft TL: N39 23 19 W101 33 34. Hrs open: 24 3023 W. 31st St., Goodland, 67765. Phone: (785) 899-2309. Fax: (785) 899-3062. Licensee: Melia Communications Inc. (group owner) Population served: 50,000 Format: Country. ♦Martin K. Melia, pres & gen mgr; Martin Melia, gen sls mgr; Curtis Duncan, progmg dir.

KXXX(AM)— August 1947: 790 khz; 5 kw-D. TL: N39 23 35 W101 00 06. Hrs open: 24 1065 S. Range, 67701. Phone: (785) 462-3305. Fax: (785) 462-3307. E-mail: kxkq@colbyweb.com Licensee: NRG Media LLC. Group owner: Waitt Broadcasting Inc. (acq 10-31-2005; grpsl). Population served: 55,000 Natl. Network: Westwood One. Rgnl. Network: Mid-American Ag. Format: Contemp country, farm. News staff: one; News: 5 hrs wkly. Target aud: 35-55; male, female, city & rural. ♦Mike Fell, gen mgr & stn mgr; Joe Vyzourek, opns mgr, progmg dir & engrg mgr; Radonda Buford, traf mgr; Chad Wolf, farm dir; Clayton Jacobs, disc jockey.

KQLS(FM)—Co-owned with KXXX(AM). September 1971: 100.3 mhz; 100 kw. 610 ft TL: N39 28 50 W100 54 34. Stereo. 24 Phone: (785) 462-3306.55,000 Natl. Network: ABC. Format: Adult contemp. Target aud: 18-49; general. ♦Lacy Stroup, progmg dir & farm dir; Rick James, disc jockey.

Columbus

KJML(FM)— Dec 25, 1982: 105.3 mhz; 6.1 kw. 308 ft TL: N37 14 47 W94 44 52. (CP: 25 kw, ant 210 ft. TL: N37 01 57 W94 49 44). Stereo. Hrs opn: 24 2510 W. 20th St., Joplin, MO, 64804. Phone: (417) 781-1313. Fax: (417) 781-1316. Licensee: FFD Holdings I Inc. Group owner: Petracom Media L.L.C. (acq 12-20-2004; grpsl). Population served: 325,000 Natl. Network: ABC. Format: Modern rock. News staff: one; News: 2 hrs wkly. Target aud: 18-49; growing families with needs for a wide range of goods & svcs. ♦Dave Clemons, CFO & gen sls mgr; Kathleen Pike, VP & traf mgr.

Concordia

KCKS(FM)—Listing follows KNCK(AM).

KNCK(AM)— Feb 6, 1954: 1390 khz; 500 w-D, 54 w-N. TL: N39 33 58 W97 41 04. Hrs open: 24 Box 629, Rt. 1 W. 11th St., 66901. Phone: (785) 243-1414. Fax: (785) 243-1391. Licensee: KNCK Inc. (acq 10-18-89; $190,000 with co-located FM; FTR: 11-6-89) Population served: 196,000 Rgnl. Network: Kan. Agri., Kan. Info. Format: Country. News staff: news progmg 7 hrs wkly News: one;. Target aud: 45 plus. Spec prog: Public affrs 2 hrs wkly. ♦Joe Jindra, pres & gen mgr; Marvin Hoffman, chief of engrg.

***KVCO(FM)**— May 1, 1977: 88.3 mhz; 127 w. 77 ft TL: N39 33 17 W97 39 48. Stereo. Hrs opn: Box 1002, 2221 Campus Dr., 66901. Phone: (785) 243-1435. Phone: (785) 243-4444. Fax: (785) 243-1043. Licensee: Cloud County Community College. Population served: 8,000 Natl. Network: CNN Radio. Format: Div/var. Target aud: 16-30 plus; students & young adults. ♦John Chapin, gen mgr.

Copeland

***KHYM(FM)**— Dec 23, 1997: 103.9 mhz; 100 kw. 702 ft TL: N37 28 35 W100 35 59. Stereo. Hrs opn: 24 Box 991, 909 W. Carthage, Meade, 67864-0991. Phone: (620) 873-2991. Fax: (620) 873-2755. E-mail: khym@khym.org Web Site: www.khym.org. Licensee: Great Plains Christian Radio Inc. Format: Relg, Christian. News staff: one. ♦Don Hughes, CEO, pres & gen mgr; Glenn Hascoll, stn mgr; Glenn Hascall, opns dir & sls dir; Peggy Burdick, dev dir; Steve Larsen, chief of engrg.

***KJIL(FM)**— Sept 5, 1992: 99.1 mhz; 100 kw. 935 ft TL: N37 23 35 W100 35 59. Stereo. Hrs opn: 24 Box 991, 909 W. Carthage, Meade, 67864-0991. Phone: (620) 873-2991. Fax: (620) 873-2755. E-mail: kjil@kjil.com Web Site: www.kjil.com. Licensee: Great Plains Christian Radio Inc. Population served: 368,164 Natl. Network: Moody, USA. Format: Contemp Christian, relg. Target aud: 25-60; Evangelical Christians. ♦Don Hughes, pres & gen mgr; Michael Luskey, opns dir; Delvin Kinser, news dir; Steve Larson, chief of engrg; Polly Hughes, traf mgr.

KSKZ(FM)— May 1, 1994: 98.1 mhz; 100 kw. Ant 666 ft TL: N37 30 00 W100 40 00. Hrs open: 24 Box 759, 1402 E. Kansas Ave., Garden City, 67846. Phone: (620) 276-2366. Fax: (620) 276-3568. Web Site: www.wksradio.com. Licensee: Ingstad Broadcasting Inc. Group owner: Robert Ingstad Broadcast Properties (acq 1-27-95; FTR: 3-20-95). Population served: 135,000 Format: Hot adult contemp. Target aud: 25-54. ♦Gil Wohler, gen mgr & gen sls mgr; James Janda, progmg dir; Andrew Mahoney, news dir; Tom Dial, chief of engrg; Rachel Wheet, traf mgr.

Dearing

KUSN(FM)— Oct 1, 1999: 98.1 mhz; 9.7 kw. 495 ft TL: N37 06 28 W95 43 22. Stereo. Hrs opn: 24 hours daily Box 4584, Springfield, MO, 65808. Secondary address: 306 W. 8th St., Coffeyville 67337. Phone: (417) 883-9180. Fax: (417) 883-9096. Licensee: KGGF-KUSN Inc. Group owner: Mahaffey Enterprises Inc. Format: Country. Target aud: 18-49. ♦Robert B. Mahaffey, pres; John Leonard, gen mgr.

Derby

KZCH(FM)— 1978: 96.3 mhz; 50 kw. 492 ft TL: N37 37 03 W97 20 11. Stereo. Hrs opn: 24 9323 E. 37th St. N., Wichita, 67226. Phone: (316) 494-6600. Fax: (316) 494-6730. Web Site: www.channel963.com. Licensee: Clear Channel Broadcasting Licenses Inc. Group owner: Clear Channel Communications Inc. (acq 8-30-2000; grpsl). Population served: 276,554 Natl. Rep: Clear Channel. Format: Chr. Target aud: 18-34; women.

Dodge City

***KAIG(FM)**—Not on air, target date: unknown: 89.9 mhz; 45 kw vert. Ant 590 ft TL: N37 55 56 W100 19 02. Hrs open: 2351 Sunset Blvd., Suite 170-218, Rocklin, CA, 95765. Phone: (916) 251-1600. Fax: (916) 251-1650. Licensee: Educational Media Foundation. ♦Richard Jenkins, pres; Mike Novak, VP; Keith Whipple, dev dir; David Pierce, progmg mgr; Ed Lenane, news dir; Sam Wallington, engrg dir; Marya Morgan, news rptr; Richard Hunt, news rptr.

KDCC(AM)— 1992: 1550 khz; 1 kw-D, 90 w-N, DA-2. TL: N37 47 14 W100 01 55. Hrs open: 7 AM-10 PM 3004 N. 14th, 67801-2007. Phone: (620) 225-6783. Phone: (620) 225-6720. Fax: (620) 225-0918. Licensee: Dodge City Community College. (acq 6-7-92; $11,400; FTR: 7-27-92). Population served: 30,000 Format: Educ, news, sp. News: 35 hrs wkly. Target aud: 18 plus. Spec prog: Christian, Sp. ♦John Ewy, gen mgr & sls dir.

KONQ(FM)—Co-owned with KDCC(AM). Apr 26, 1978: 91.9 mhz; 2.6 kw. 123 ft TL: N37 46 33 W100 02 12. Fax: (620) 225-0918. Web Site: www.dodgecitycommunitycollege.com.25,000 Format: Var/div, educ, MOR. News: 7 hrs wkly. Spec prog: Black 10 hrs, sports 5 hrs wkly. ♦John Ewy, gen mgr.

KGNO(AM)— June 30, 1930: 1370 khz; 5 kw-D, 230 w-N. TL: N37 45 36 W100 05 53. Hrs open: 2601 Central Ave., Suite C, Village Plaza, 67801. Phone: (620) 225-8080. Fax: (620) 225-6655. E-mail: rockswks@sbcgobal.net Licensee: NRG Media LLC. Group owner: Waitt Broadcasting Inc. (acq 9-12-2005; grpsl). Population served: 85,000 Rgnl. Network: Mid-American Ag. Format: News, sports, talk radio. Target aud: 25-54. Spec prog: Farm 15 hrs wkly. ♦Kurt Lampe, gen mgr; Peggy Burdick, gen sls mgr; Brian Nugen, progmg dir.

KOLS(FM)—Co-owned with KGNO(AM). May 1966: 95.5 mhz; 100 kw. 570 ft TL: N37 38 28 W100 20 40. (CP: 24 kw, ant 59 ft.). Stereo. 24 130,000 Format: Adult contemp. News: 5 hrs wkly. Target aud: 25-49.

KZRD(FM)— December 1997: 93.9 mhz; 100 kw. 511 ft TL: N38 00 07 W101 14 45. Hrs open: 2601 Central Ave., Village Plaza Suite C, 67801. Phone: (620) 225-8080. Fax: (620) 225-6655. Licensee: NRG Media LLC. Group owner: Waitt Broadcasting Inc. (acq 9-12-2005; grpsl). Format: Classic rock. ♦Brian Nugen, opns mgr; Keith Tallent, progmg dir; Michelle Stoll, traf mgr.

Downs

KDNS(FM)— Apr 11, 1994: 94.1 mhz; 28 kw. Ant 292 ft TL: N39 30 29 W98 18 57. Stereo. Hrs opn: 5 AM-1 AM Box 88, West Hwy. 24, Glen Elder, 67446. Phone: (785) 545-3220. Licensee: Dierking Communications Inc. (acq 7-27-2006; $276,000). Natl. Network: Jones Radio Networks. Rgnl. Network: Brownfield. Format: Country. News staff: one; News: 4 hrs wkly. Target aud: 25-54. Spec prog: Farm 6 hrs, gospel 5 hrs wkly. ♦Wade Gerstner, gen mgr.

El Dorado

KAHS(AM)— Nov 16, 1953: 1360 khz; 1 kw-D. TL: N37 48 47 W96 48 44. Hrs open: 24 201 Industrial Park Rd., Excelsior Springs, MO, 64024. Phone: (316) 320-1360. Web Site: www.1360kahs.com. Licensee: Kansas City Catholic Network Inc. (group owner; (acq 11-23-2005; $525,000). Population served: 500,000 Natl. Network: CNN Radio, Westwood One. Format: Relg. Target aud: 35-69. Spec prog: . Jazz 2 hrs wkly ♦James E. O'Laughlin, pres.

***KBTL(FM)**— March 1998: 88.1 mhz; 400 w. 92 ft TL: N37 48 16 W96 53 02. Stereo. Hrs opn: Butler County Community College, 901 S. Haverhill Rd., 67042. Phone: (316) 321-2222. Licensee: Butler County Community College. Format: Div. ♦Lance D. Hayes, gen mgr.

***KTLI(FM)**— Feb 15, 1972: 99.1 mhz; 100 kw. Ant 617 ft TL: N37 56 22 W96 59 20. Stereo. Hrs opn: 125 N. Market, Suite 1900, Wichita, 67202. Phone: (316) 303-9999. Fax: (316) 303-9900. Web Site: k-love.com. Licensee: El Dorado Licenses Inc. Group owner: KXOJ Inc. (acq 12-3-2004; $2.95 million). Population served: 500,000 Format: Adult contemp, Christian. News: 25 hrs wkly. Target aud: 25-54; women. ♦Crystal Wojtecko, gen mgr & rgnl sls mgr.

Emporia

***KANH(FM)**— 2002: 89.7 mhz; 3 kw. Ant 262 ft TL: N38 21 45 W96 07 00. Hrs open: 1120 W. 11th St., Lawrence, 66044. Phone: (785) 864-4530. Fax: (785) 864-5278. E-mail: kpr@ku.edu Web Site: www.kpr.ku.edu. Licensee: University of Kansas. Natl. Network: NPR. Format: Class, jazz. ♦Janet Campbell, gen mgr.

KANS(FM)— April 1998: 96.1 mhz; 6 kw. 318 ft TL: N38 24 21 W96 14 13. Hrs open: 1811 W. 6th Ave., 66801. Phone: (620) 343-9393. Fax: (620) 342-7617. E-mail: kans@ksradio.com Licensee: C&C Consulting Inc. (acq 11-14-97; $10,000 for CP). Format: Soft hits. ♦Brook Reed, pres & traf mgr; Marty Hill, gen mgr, opns mgr & dev dir; Angie Boden, progmg dir.

KFFX(FM)—Listing follows KVOE(AM).

***KNGM(FM)**— Jan 11, 1987: 91.9 mhz; 3 kw. 263 ft TL: N38 24 35 W96 13 30. Stereo. Hrs opn: 24 Box 506, 815 Graham St., 66801. Phone: (620) 343-9292. E-mail: kngm@osprey.net Licensee: Christian Action Team Inc. (acq 8-86; FTR: 1-87). Population served: 60,000 Format: Contemp Christian. News staff: one. Target aud: Young families and adults. ♦Jeff Shirley, pres & sr VP; April Reitmann, VP; Steve Pearson, gen mgr.

***KPOR(FM)**— June 19, 2002: 90.7 mhz; 2 kw. Ant 328 ft TL: N38 26 50 W96 07 42. Hrs open: 24 Family Stations Inc., 4135 Northgate Blvd., Sacramento, CA, 95834. Phone: (916) 641-8191. Licensee: Family Stations Inc. (group owner) Format: Relg. ♦Scott Miller, opns mgr.

KVOE(AM)— Jan 21, 1939: 1400 khz; 1 kw-U. TL: N38 23 10 W96 10 36. Hrs open: Box 968, 1420 C of E Dr., 66801. Phone: (620) 342-1400. Fax: (620) 342-0804. E-mail: kvoe@kvoe.com Web Site: www.kvoe.com. Licensee: Emporia Radio Stations Inc. Group owner: Emporia's Radio Stations Inc. (acq 1-7-87). Population served: 25,327 Law Firm: Irwin, Campbell & Tannenwald. Format: Adult contemp, oldies, news. Target aud: 35-54. Spec prog: Sp 3 hrs wkly. ♦Lee Schroeder, gen mgr & gen sls mgr; Ron Thomas, progmg dir; Jeff O'Dell, news dir; Charlie Allen, chief of engrg.

KFFX(FM)—Co-owned with KVOE(AM). June 15, 1966: 104.9 mhz; 3 kw. Ant 279 ft TL: N38 23 10 W96 10 36. Stereo. 24 Format: Hot adult contemp. News: 4 hrs wkly. Target aud: 20-40.

KVOE-FM— Jan 16, 1985: 101.7 mhz; 3.2 kw. Ant 298 ft TL: N38 21 45 W96 07 00. Stereo. Hrs opn: 24 1420 C of E Dr., 66801. Phone: (620) 342-1400. Fax: (620) 342-0804. E-mail: kvoe@kvoe.com Licensee: Emporia Radio Stations Inc. Group owner: Emporia's Radio Stations Inc. (acq 1994). Rgnl. Network: Kan. Agri. Law Firm: Irwin, Campbell & Tannenwald. Format: Country. News staff: 2; News: 7 hrs wkly.

Target aud: 25-54. ♦Erren Harter, pres & prom mgr; Steve Sauder, CEO, pres & gen sls mgr; Jef O'Dell, news dir & chief of engrg; Greg Rahe, sports cmtr.

Enterprise

***KBMP(FM)**— Mar 6, 2002: 90.5 mhz; 19 kw. Ant 384 ft TL: N39 07 53.9 W97 19 58.8. Hrs open: 24
Rebroadcasts KSIV-FM Saint Louis, MO 100%.
209 N. Meridian Rd., Newton, 67114. Phone: (316) 283-4592. Fax: (316) 283-3177. E-mail: comments@bottradionetwork.com Web Site: www.bottradionetwork.com. Licensee: Community Broadcasting Inc. Group owner: American Family Radio (acq 1-26-2006; $30,000 with KARF(FM) Independence). Format: Christian teaching and talk. Target aud: 25-54; adults. ♦Dan Snell, gen mgr.

Eureka

KOTE(FM)— October 1988: 93.5 mhz; 3 kw. 321 ft TL: N37 47 29 W96 17 25. Stereo. Hrs opn: 24 Box 350, 67045. Secondary address: 1275 P. Rd. 67045. Phone: (620) 583-7414. Fax: (620) 583-7233. E-mail: info@kotefm.com Web Site: www.koteinfo.com. Licensee: Niemeyer Communications LLC (acq 8-15-2005; $125,000). Population served: 15,000 Rgnl. Network: Kan. Info. Format: Classic rock, country. News staff: one. Target aud: General. ♦Steve Niemeyer, gen mgr.

Fairway

KCNW(AM)— Apr 16, 1953: 1380 khz; 2.5 kw-D, 29 w-N. TL: N39 04 19 W94 40 58. Hrs open: 24 4535 Metropolitan Ave., Kansas City, 66106. Phone: (913) 384-1380. Fax: (913) 236-9470. E-mail: kcnw@wilkinsradio.com Web Site: www.wilkinsradio.com. Licensee: Kansas City Radio Inc. Group owner: Wilkins Communications Network Inc. (acq 1-17-2001; $725,000). Population served: 2,000,000 Natl. Network: Westwood One. Law Firm: Womble, Carlyle, Sandridge & Rice. Format: Christian teaching, talk. News staff: one. Target aud: 35 plus; adults involved in community & family. ♦Bob Wilkins, pres; LuAnn Wilkins, exec VP; Mitchell Mathis, VP; Anthony Norman, gen mgr & stn mgr; Greg Garrett, opns mgr; Ed Treese, engr.

Fort Scott

KMDO(AM)— Oct 8, 1954: 1600 khz; 770 w-D, 35 w-N. TL: N37 47 01 W94 42 00. Hrs open: Box 72, 2 N. National, 66701. Phone: (620) 223-4500. Phone: (620) 223-4501. Fax: (620) 223-5662. Licensee: Fort Scott Broadcasting Co. (acq 2-1-60). Population served: 25,000 Format: Oldies, rock. Target aud: General; 30 plus. ♦Jon Hart, prom dir; Tim McKenney, pres, gen mgr & outdoor ed.

KOMB(FM)—Co-owned with KMDO(AM). Jan 23, 1981: 103.9 mhz; 25 kw. Ant 328 ft TL: N37 54 30 W94 45 58. Stereo.

***KVCY(FM)**— November 1983: 104.7 mhz; 16 kw. 410 ft TL: N37 47 47 W94 42 20. Stereo. Hrs opn: 24 3434 W. Kilbourn Ave., Milwaukee, WI, 53208. Phone: (414) 935-3000. Fax: (414) 935-3015. E-mail: kvcy@vcyamerica.org Web Site: www.vcyamerica.org. Licensee: VCY America Inc. (group owner) Natl. Network: USA, Moody. Format: Relg, Christian. ♦Vic Eliason, VP & gen mgr; Jim Schneider, progmg dir.

Fredonia

KGGF-FM— July 14, 1997: 104.1 mhz; 6 kw. 328 ft TL: N37 31 36 W95 49 39. Hrs open: Box 1087, Coffeyville, 67337. Phone: (316) 251-3800. Fax: (918) 251-9210. Licensee: KGGF-KUSN Inc. Group owner: Mahaffey Enterprises Inc. Format: Adult contemp. ♦John Leonard, gen mgr.

Galena

KCAR-FM— June 29, 2000: 104.3 mhz; 6 kw. Ant 328 ft TL: N37 03 13 W94 42 12. Hrs open: 2510 W. 20th St., Joplin, MO, 64804. Phone: (417) 781-1313. Fax: (417) 781-1316. Licensee: FFD Holdings I Inc. Group owner: Petracom Media L.L.C. (acq 12-20-2004; grpsl). Format: Classic rock, oldies. ♦Dave Clemons, gen mgr & gen sls mgr.

Garden City

***KANZ(FM)**— June 29, 1980: 91.1 mhz; 100 kw. Ant 958 ft TL: N37 46 43 W100 53 43.4. Stereo. Hrs opn: 5 AM-midnight
Rebroadcasts KZNA(FM) Hill City 100%.
210 N. 7th St., 67846-5519. Phone: (620) 275-7444. Fax: (620) 275-7496. Web Site: www.hppr.org. Licensee: KANZA Society Inc. (acq 11-77; FTR: 7-80). Population served: 270,000 Natl. Network: PRI, NPR. Format: Div, educ, class. News: 39 hrs wkly. Target aud: General. Spec prog: Jazz 15 hrs, folk 6 hrs, Sp 6 hrs wkly. ♦Robert Kirby, progmg dir.

KBUF(AM)—Holcomb, 1948: 1030 khz; 2.5 kw-D, 1.2 kw-N, DA-N. TL: N38 00 01 W100 53 54. Hrs open: 24 Box 759, 1402 E. Kansas, 67846. Phone: (620) 276-2366. Fax: (620) 276-3568. Licensee: KBUF Partnership. Group owner: Robert Ingstad Broadcast Properties (acq 11-1-79). Population served: 30000 Rgnl. Network: Mid-American Ag. Law Firm: Fisher, Wayland, Cooper, Leader & Zaragoza L.L.P. Wire Svc: NWS (National Weather Service) Wire Svc: UPI Format: C&W, talk. News staff: one. Target aud: 25-54; people interested in class country & info progmg. Spec prog: Farm 15 hrs wkly. ♦Gill Wohler, gen mgr; James Janda, progmg mgr.

KKJQ(FM)—Co-owned with KBUF(AM). Nov 20, 1962: 97.3 mhz; 100 kw. 850 ft TL: N37 46 48 W100 27 36. Stereo. 24 75,000 Natl. Network: ABC. Format: Country, adult contemp. Target aud: 18-49.

KIUL(AM)— May 20, 1935: 1240 khz; 1 kw-U. TL: N37 59 52 W100 54 25. Hrs open: 609 E. Kansas Plaza, 67846. Phone: (620) 276-3251. Fax: (620) 276-3568. Licensee: Steckline Communications Inc. (group owner; (acq 11-28-2006; $550,000 with KYUL(AM) Scott City). Population served: 50,000 Natl. Network: CBS, Westwood One. Rgnl. Network: Mid-American Ag. Law Firm: Dow, Lohnes & Albertson. Format: Sports, news/talk. News staff: 2; News: 20 hrs wkly. Target aud: 45 plus; upscale adults. Spec prog: Farm 5 hrs wkly. ♦Gil Wohler, gen mgr.

KWKR(FM)—See Leoti

Girard

KSEK-FM— Sept 1, 1988: 99.1 mhz; 6 kw. Ant 325 ft TL: N37 29 02 W94 51 08. Hrs open: 202 E. Centenial Dr., Suite 2B, Pittsburg, 66762. Phone: (620) 232-9912. Fax: (620) 232-9915. Licensee: Southeast Kansas Independent Living Resource Center Inc. (group owner; (acq 11-30-2004; $700,000 with KSEK(AM) Pittsburg). Natl. Network: AP Network News. Format: Classic rock. News: 10 hrs wkly. Target aud: 25-54; resident adults & univ. students. ♦Lynn Meredith, pres & gen mgr.

Goodland

***KGCR(FM)**— Mar 1, 1988: 107.7 mhz; 100 kw. 446 ft TL: N39 22 03 W101 26 44. Stereo. Hrs opn: 24 Box 9, Brewster, 67732. Secondary address: 3410 Rd. 66, Brewster 67732. Phone: (785) 694-2877. Fax: (785) 694-2875. Licensee: The Praise Network Inc. (acq 7-2-98). Population served: 40,000 Natl. Network: Moody, USA. Format: Relg. News staff: one; News: 30 hrs wkly. Target aud: 25-54; Christian families. Spec prog: Farm 2 hrs wkly. ♦Lloyd Mintzmyer, CEO & pres; James Claasson, gen mgr.

KKCI(FM)—Listing follows KLOE(AM).

KLOE(AM)— 1947: 730 khz; 1 kw-D, 20 w-N. TL: N39 20 04 W101 45 28. Hrs open: 24 Box 569, 3023 W. 31st St., 67735-0569. Phone: (785) 899-2309. Fax: (785) 899-3062. Web Site: www.kloe.com. Licensee: Melia Communications Inc. (group owner; acq 1-26-96; $990,000 with co-located FM). Population served: 100,000 Natl. Network: CBS. Rgnl. Network: Kan. Agri. Format: News/talk, country, farm. Target aud: General. ♦Martin K. Melia, pres & gen mgr; Curtis Duncan, progmg dir.

KKCI(FM)—Co-owned with KLOE(AM). Sept 15, 1990: 102.5 mhz; 100 kw. 712 ft TL: N39 23 19 W101 33 34. Stereo. 24 (Acq 4-90; $40,000; FTR: 5-21-90). Natl. Network: Jones Radio Networks. Format: Adult contemp, sports, jazz. Target aud: 25-55.

Great Bend

***KBDA(FM)**— 1999: 89.7 mhz; 1.4 kw. Ant 112 ft TL: N38 20 16 W98 45 48. Hrs open: American Family Radio Assoc., Box 3206, Tupelo, MS, 38803. Phone: (662) 844-8888, Ext 204. Fax: (662) 842-6791. Web Site: www.afr.net. Licensee: American Family Association. Group owner: American Family Radio Format: Inspirational Christian. ♦Marvin Sanders, gen mgr.

***KHCT(FM)**— Aug 3, 1992: 90.9 mhz; 50 kw. 781 ft TL: N38 37 04 W98 56 32. Hrs open: 815 N. Walnut St., Suite 300, Hutchinson, 67501. Phone: (620) 662-6646. Licensee: Hutchinson Community College. Natl. Network: NPR. Format: Class, new age, news. ♦David Horning, gen mgr; Geralyn Smith, opns dir; Sharon Webb, dev dir.

KHOK(FM)—Hoisington, 1978: 100.7 mhz; 100 kw. 430 ft TL: N38 32 49 W98 45 59. Stereo. Hrs opn: 24 Box 609, 1200 Baker St., 67530. Phone: (620) 792-3647. Fax: (620) 792-3649. Web Site: www.eagleradio.net. Licensee: Eagle Communications Inc. Group owner: Eagle Communications Group (acq 9-1-86; grpsl; FTR: 4-8-91). Population served: 50,000 Format: Country. News staff: one. Target aud: 18-44. Spec prog: Relg 2 hrs wkly. ♦Gary Shorman, pres; Rick Nulton, gen mgr.

KVGB(AM)— Mar 10, 1937: 1590 khz; 5 kw-U, DA-N. TL: N38 18 50 W98 47 35. Hrs open: Box 609, 1200 Baker St., 67530. Phone: (620) 792-4637. Fax: (620) 792-3649. Web Site: eagleradio.net. Licensee: Eagle Communications Inc. Group owner: Eagle Communications Group (acq 4-95). Population served: 25,000 Natl. Network: ABC. Format: News, talk, sports. Target aud: 28 plus. Spec prog: Farm 7 hrs, relg 2 hrs wkly. ♦Rick Nulton, gen mgr; Randy Goering, sls dir.

KVGB-FM— Jan 17, 1977: 104.3 mhz; 96 kw. 810 ft TL: N38 25 54 W98 46 18. Stereo. Web Site: eagleradio.net. Natl. Network: ABC. Format: Classic rock. ♦Randy Goering, sls VP.

***KWBI(FM)**— Oct 10, 2001: 91.9 mhz; 1.8 kw. Ant 259 ft TL: N38 20 16 W98 45 48. Stereo. Hrs opn: 24 2351 Sunset Blvd., Suite 170-218, Rocklin, CA, 95765. Phone: (916) 251-1600. Fax: (916) 251-1650. E-mail: klove@klove.com Web Site: www.klove.com. Licensee: Educational Media Foundation. Group owner: EMF Broadcasting. Population served: 27,000 Natl. Network: K-Love. Law Firm: Shaw Pittman. Format: Contemp Christian. News staff: 3. Target aud: 25-44; Judeo Christian, female. ♦Richard Jenkins, pres; Mike Novak, VP; Keith Whipple, dev dir; Eric Allen, natl sls mgr; David Pierce, progmg mgr; Ed Lenane, news dir; Sam Wallington, engrg dir; Karen Johnson, news rptr; Marya Morgan, news rptr; Richard Hunt, news rptr.

KZLS(FM)— Feb 3, 1986: 107.9 mhz; 100 kw. 886 ft TL: N38 46 16 W98 44 17. Stereo. Hrs opn: 24 5501 10th St., Great W. Bend, 67530. Phone: (620) 792-7108. Fax: (620) 792-7051. E-mail: kzls@waittradio.com Licensee: NRG Media LLC. Group owner: Waitt Broadcasting Inc. (acq 9-12-2005; grpsl). Rgnl. Network: Mid-American Ag. Law Firm: Blooston, Mordkofsky, Jackson & Dickens. Format: Adult contemp. News staff: one; News: 5 hrs wkly. Target aud: 25-54. ♦Ken Schwamborn, gen mgr; Chris Elson, opns dir & progmg dir; Rod Rogers, chief of engrg.

Haven

KGGG(FM)— 1998: 97.1 mhz; 11.5 kw. Ant 1,024 ft TL: N37 47 47 W97 31 59. Hrs open: 106 N. Main St., Hutchinson, 67501-5219. Phone: (620) 665-5758. Fax: (620) 665-6655. Licensee: Ad Astra Per Aspera Broadcasting Inc. (group owner; (acq 9-16-98). Format: Contemp. Target aud: 12-49. ♦Cliff C. Shank, gen mgr; Mike Hill, stn mgr; Aaron West, opns mgr.

Hays

KAYS(AM)—Listing follows KHAZ(FM).

KHAZ(FM)— May 1, 1985: 99.5 mhz; 100 kw. 515 ft TL: N38 56 29 W99 21 22. Stereo. Hrs opn: 24 Box 6, 67601. Secondary address: 2300 Hall St. 67601. Phone: (785) 625-2578. Fax: (785) 625-3632. Licensee: Eagle Communications Inc. Group owner: Eagle Communications Group Natl. Network: ABC. Format: Country. News staff: 2; News: 4 hrs wkly. Target aud: 25-54. Spec prog: Farm 10 hrs, gospel 3 hrs wkly. ♦Todd Nelson, gen mgr; Dwayne Detter, gen sls mgr; Theresa Trapp, progmg dir; Callie Kolacney, news dir; Mark Goff, chief of engrg.

KAYS(AM)—Co-owned with KHAZ(FM). Oct 15, 1948: 1400 khz; 1 kw-U. TL: N38 53 29 W99 22 03. Stereo. 24 Box 817, 2300 Hall St., 67601. (Acq 3-20-91; grpsl; FTR: 4-8-91).15,396 Format: Oldies. News staff: one; News: 6 hrs wkly. Target aud: Adults. ♦Mike Cooper, progmg dir.

KHYS(FM)—Not on air, target date: unknown: 89.7 mhz; 250 w. Ant 292 ft TL: N38 51 16 W99 22 55. Hrs open: Drawer 2440, Tupelo, MS, 38803. Phone: (662) 844-8888. Fax: (662) 842-6791. Licensee: Salt & Light Communications Inc.

KJLS(FM)— June 27, 1974: 103.3 mhz; 100 kw. 994 ft TL: N39 01 15 W99 28 12. Stereo. Hrs opn: 24 Box 6, 2300 Hall St., 67601. Phone: (785) 625-2578. Fax: (785) 625-3632. Licensee: Eagle Communications Inc. Group owner: Eagle Communications Group (acq 9-12-00; with KKQY(FM) Hill City). Population served: 110,000 Format: Adult contemp. News staff: one; News: 8 hrs wkly. Target aud: 25-49; 60% female, 40% male. ♦Todd Nelson, VP & gen mgr.

***KPRD(FM)**— 1994: 88.9 mhz; 83 kw. 636 ft TL: N38 46 16 W98 44 17. Hrs open: 301 W. 13th St., Suite 409, 67601. Phone: (785) 628-6300. Fax: (785) 628-6389. E-mail: kprd@kprd.org Web Site: www.kprd.org. Licensee: The Praise Network Inc. Population served: 75,000 Format: Relg. Target aud: 20-48. ♦Lloyd Mintzmyer, CEO; David Breedon, stn mgr.

***KZAN(FM)**—Not on air, target date: unknown: 91.7 mhz; 1.25 kw. Ant 246 ft TL: N38 56 28 W99 21 20. Hrs open: 24
Rebroadcast KANZ(FM) Garden City 100%.
Kanza Society Inc., 210 N. 7th St., Garden City, 67846. Phone: (800) 678-7444. Fax: (620) 275-7496. E-mail: hppr@hppr.org Web Site: www.hppr.org. Licensee: Kanza Society Inc. Natl. Network: NPR, PRI, AP Radio. Format: Educ, class, news/talk. Target aud: 25-80; educated. ♦Richard Hicks, gen mgr; Diana Gonzales, dev dir; Debra Stout, prom dir; Bob Kirby, progmg dir; Mary Palmer, mus dir; Chuck Springer, chief of engrg.

Haysville

KFBZ(FM)— Aug 25, 1985: 105.3 mhz; 100 kw. 1,000 ft TL: N37 46 40 W97 30 37. Stereo. Hrs opn: 2120 N. Woodlawn, Suite 352, Wichita, 67208-1847. Phone: (316) 685-2121. Fax: (316) 685-3408. Web Site: www.1053thebuzz.com. Licensee: Entercom Wichita License LLC. Group owner: Entercom Communications Corp. (acq 2000; grpsl). Format: Hot AC. News: 2 hrs wkly. Target aud: 25-54. Spec prog: Relg 2 hrs wkly. ♦Jackie Wise, gen mgr.

Herington

***KJRL(FM)**— Sept 6, 1997: 105.7 mhz; 12.5 kw. 500 ft TL: N38 37 01 W96 59 09. Stereo. Hrs opn: 24 Box 389, 165 Trail Rd., 67449. Phone: (785) 258-2660. Fax: (785) 258-2777. E-mail: kjrl@kjrl.org Web Site: www.kjil.org. Licensee: Great Plains Christian Radio Inc. (acq 9-5-01). Population served: 320,000 Natl. Network: Salem Radio Network, Moody. Rgnl. Network: Kan. Info. Wire Svc: AP Format: Adult Contemp Christian. News staff: one; News: 16 hrs wkly. Target aud: 18-54; farmers, railroad & transportation workers, military & professional. ♦Don Hughes, CEO & gen mgr; Lee Issaac, chmn; Michael Luskey, stn mgr; Mark Erdman, opns dir; Noreen Stowe, dev dir; Delvin Kinser, news dir.

Hiawatha

KNZA(FM)— Aug 18, 1977: 103.9 mhz; 50 kw. 492 ft TL: N39 34 41 W95 33 46. Stereo. Hrs opn: 24 Box 104, 66434-0104. Secondary address: 1828 Hwy. 73 66434-0104. Phone: (785) 547-3461. Fax: (785) 547-9900. E-mail: knza@rainbowtel.net Web Site: knzafm.com. Licensee: KNZA Inc. (group owner; acq 6-83; FTR: 6-20-83). Population served: 84,000 Format: Country. News staff: one; News: 10 hrs wkly. Target aud: 18-54; general. Spec prog: Farm 14 hrs wkly. ♦Greg Buser, gen mgr; Robert Hilton, opns mgr.

Hill City

KKQY(FM)— Aug 29, 1997: 101.9 mhz; 97 kw. 994 ft TL: N39 01 15 W99 28 13. Stereo. Hrs opn: 24 Box 6, Hays, 67601. Phone: (785) 625-2578. Fax: (785) 625-3632. Licensee: Eagle Communications Inc. Group owner: Eagle Communications Group (acq 9-12-00; with KJLS(FM) Hays). Population served: 110,000 Format: Country. News: 5 hrs wkly. Target aud: 25-54. ♦Todd Nelson, gen mgr; Todd Lynn, gen sls mgr; Craig Taylor, progmg dir.

***KZNA(FM)**— 1986: 90.5 mhz; 100 kw. 600 ft TL: N39 15 56 W99 49 48. Stereo. Hrs opn: 24
Rebroadcasts KANZ(FM) Garden City 100%.
210 N. 7th St., Garden City, 67846-5519. Phone: (620) 275-7444. Fax: (620) 275-7496. E-mail: hppr@hppr.org Web Site: www.hppr.org. Licensee: Kanza Society Inc. (acq 11-77; FTR: 7-80). Population served: 200,000 Natl. Network: PRI, NPR, AP Radio. Format: Educ, class, news/talk. News: 39 hrs wkly. Target aud: 25-80; educated. Spec prog: Jazz 15 hrs, folk 6 hrs, Sp 6 hrs wkly. ♦Bob Kirby, opns dir, opns mgr & progmg dir; Diana Aguilar-Gonzales, dev dir; Debra Stout, prom dir; Mary Palmer, mus dir; Chuck Springer, chief of engrg.

Hoisington

KHOK(FM)—Licensed to Hoisington. See Great Bend

Holcomb

KBUF(AM)—Licensed to Holcomb. See Garden City

Horton

KAIR-FM— Jan 25, 1995: 93.7 mhz; 25 kw. Ant 328 ft TL: N39 37 43 W95 18 53. Stereo. Hrs opn: 24 Box G, Atchison, 66002. Secondary address: 200 N. 5th St., Atchinson 66002. Phone: (913) 367-1470. Fax: (913) 367-7021. E-mail: thewakeupcrew@hotmail.com Licensee: Mark V Media Group Inc. (group owner; (acq 12-27-2004; $1.55 million with KAIR(AM) Atchison). Natl. Network: AP Radio. Wire Svc: AP Format: Country. News staff: 4; News: 133 hrs wkly. Target aud: 25-54. ♦Mark Oppold, pres & gen mgr.

Hugoton

KFXX-FM— Sept 16, 1983: 106.7 mhz; 35 kw. Ant 259 ft TL: N37 19 03 W101 20 16. Stereo. Hrs opn: 24 2917 S. Colorado, Ulysses, 67880. Phone: (620) 276-2366. Fax: (620) 356-3635. Licensee: KBUF Partnership. Group owner: Robert Ingstad Broadcast Properties. Population served: 18,000 Rgnl. Network: Kan. Info., Kan. Agri. Law Firm: Baraff, Koerner & Olender. Format: Rgnl Mexican. ♦Bob Dale, gen mgr.

Humboldt

KINZ(FM)— September 1998: 95.3 mhz; 24 kw. Ant 335 ft TL: N37 44 52 W95 33 39. Stereo. Hrs opn: 24 117 S. Grant St., Chanute, 66720. Phone: (620) 431-1333. Fax: (620) 431-1943. E-mail: mike@kinz.biz Web Site: www.kinz.biz. Licensee: Sutcliffe Communications LLC. Natl. Network: CNN Radio. Format: Classic rock. Target aud: 25-55. Spec prog: Gospel 3 hrs wkly. ♦Mike Sutcliffe, CEO & gen mgr; Sheri Sutcliffe, VP.

Hutchinson

***KHCC-FM**— Sept 11, 1972: 90.1 mhz; 100 kw. 1,080 ft TL: N38 03 40 W97 45 49. Hrs open: 24 815 N. Walnut, Suite 300, 67501-6217. Phone: (620) 662-6646. E-mail: rfragoza@radiokansas.org Web Site: www.radiokansas.org. Licensee: Hutchinson Community College. Population served: 1,000,000 Natl. Network: NPR. Format: Class, new age, news. News: 27 hrs wkly. ♦David M. Horning, gen mgr; Geralyn Smith, opns dir; Sharon Webb, dev dir; Melody Fisher, prom dir; Ken Baker, progmg dir.

KHUT(FM)—Listing follows KWBW(AM).

KWBW(AM)— May 28, 1935: 1450 khz; 1 kw-U. TL: N38 04 02 W97 57 53. Hrs open: Box 1036, 67504-1036. Phone: (620) 662-4486. Fax: (620) 662-5357. Web Site: www.khutfm.com. Licensee: Eagle Communications Inc. Group owner: Eagle Communications Group (acq 11-4-91; with co-located FM). Population served: 65,000 Format: Talk, new/sports. Spec prog: Black 2 hrs, gospel 11 hrs wkly. ♦Dan Deming, gen mgr & progmg dir; John Brennan, gen sls mgr; Rod Zook, news dir.

KHUT(FM)—Co-owned with KWBW(AM). Mar 15, 1972: 102.9 mhz; 28.5 kw. 496 ft TL: N38 02 36 W98 00 53. (CP: TL: N38 02 39 W98 00 56). Web Site: www.khutfm.com. Format: C&W. ♦Terry Drowhard, gen sls mgr; Jason Younger, progmg dir; Fred Gough, news dir; Mark Trotman, financial ed.

KWHK(FM)— 2007: 95.9 mhz; 2.85 kw. Ant 489 ft TL: N38 02 57 W98 00 44. Hrs open: 24 106 N. Main St., 67501-5219. Phone: (620) 665-5758. Fax: (620) 665-6655. Licensee: Ad Astra Per Aspera Broadcasting Inc. Natl. Network: ABC. Format: Oldies. News staff: one; News: 2 hrs wkly. ♦Cliff Shank, pres; Cliff C. Shank, gen mgr; Aaron West, opns mgr; Mike Hill, gen sls mgr; Lucky Kidd, news dir.

KZSN(FM)— Oct 7, 1968: 102.1 mhz; 100 kw. Ant 1,027 ft TL: N37 46 40 W97 30 37. Stereo. Hrs opn: 24 9323 E. 37th St. N., Wichita, 67226. Phone: (316) 494-6600. Fax: (316) 494-6730. Web Site: www.kzsn.com. Licensee: Clear Channel Broadcasting Licenses, Inc. Group owner: Clear Channel Communications Inc. (acq 8-30-00; grpsl). Population served: 276,554 Natl. Rep: Clear Channel. Format: Country. News staff: one. Target aud: 25-54; Adults.

Independence

***KARF(FM)**— 1997: 91.9 mhz; 250 w. Ant 180 ft TL: N37 15 54 W95 39 26. Hrs open: 10550 Barkley St., Suite 100, Overland Park, 66212-1824. Phone: (913) 642-7770. Fax: (913) 642-1319. Web Site: www.bottradionetwork.com. Licensee: Community Broadcasting Inc. Group owner: American Family Radio (acq 1-26-2006; $30,000 with KBMP(FM) Enterprise). Format: Christian. ♦Richard P. Bott II, VP.

***KBQC(FM)**— 2002: 88.5 mhz; 20 kw vert. Ant 476 ft TL: N37 03 11 W96 06 07. Hrs open: Drawer 2440, Tupelo, MS, 38801. Phone: (662) 844-8888. Fax: (662) 842-6791. Licensee: American Family Association. Group owner: American Family Radio (acq 12-18-00; buyer paid construction & bcst costs of CP). Format: Christian. ♦Marvin Sanders, gen mgr.

KIND(AM)— Dec 8, 1947: 1010 khz; 250 w-D, 32 w-N. TL: N37 13 07 W95 43 30. Hrs open: 24 122 W. Myrtle, 67301. Phone: (620) 331-3000. Fax: (620) 331-8008. E-mail: mojo@rwa911.com Licensee: Tallgrass Broadcasting Inc. (acq 10-25-2006; $306,000 with co-located FM). Population served: 12,100 Natl. Network: Westwood One. Wire Svc: AP Format: Adult Standards. Target aud: 35-65; baby boomers. Spec prog: Big band 2 hrs, class 3 hrs wkly. ♦Joseph E. Walker, pres; Mark Wilson, gen mgr.

KIND-FM— May 10, 1969: 102.9 mhz; 25 kw. Ant 272 ft TL: N37 15 42 W95 45 59.24 65,000 Natl. Network: CNN Radio. Format: Hot adult contemp. Target aud: 22-44. Spec prog: Alternative 4 hrs, Christian hot AC 2 hrs wkly.

Ingalls

KSSA(FM)— July 1, 1999: 105.9 mhz; 100 kw. 666 ft TL: N37 46 48 W100 27 36. Hrs open: 24 1402 E. Kansas Ave., Garden City, 67846. Phone: (620) 276-3251. Fax: (620) 276-3568. E-mail: kssa@wksradio.com Licensee: KBUF Partnership. Group owner: Robert Ingstad Broadcast Properties (acq 1999; $250,000). Law Firm: Shaw Pittman. Format: Sp. ♦G.L. Wohler, gen mgr; Rachel Wheet, traf mgr.

KSSH(FM)— Jan 1, 2001: 96.3 mhz; 100 kw. 659 ft TL: N37 38 28 W100 20 40. Stereo. Hrs opn: 24 2601 Central Ave. Village, Plaza Suite C, Dodge City, 67801. Phone: (620) 225-8080. Fax: (620) 225-6655. Licensee: NRG Media LLC. (group owner; (acq 9-12-2005; grpsl). ♦Brian Nugen, opns mgr.

Iola

KALN(AM)— July 25, 1961: 1370 khz; 500 w-D, 62 w-N, DA. TL: N37 54 07 W95 24 26. Hrs open: 24 Box 710, S. Hwy. 169, 66749. Phone: (620) 365-3151. Fax: (620) 365-5431. Licensee: Iola Broadcasting Inc. (acq 9-1-73). Population served: 14,125 Format: Oldies. ♦Tom Norris, gen mgr, stn mgr & progmg dir.

KIKS-FM—Co-owned with KALN(AM). June 9, 1977: 101.5 mhz; 11.5 kw. Ant 289 ft TL: N37 54 04 W95 24 04. Stereo. 24 Format: Adult contemp.

Junction City

KJCK(AM)— May 15, 1949: 1420 khz; 1 kw-D, 500 w-N, DA-N. TL: N39 01 33 W96 48 36. Hrs open: 24 Box 789, W. Ash & Hwy. 77, 66441. Phone: (785) 762-5525. Fax: (785) 762-5387. E-mail:

platinum@kjck.com Web Site: www.kjck.com. Licensee: Platinum Broadcasting Inc. (group owner; acq 9-4-86). Population served: 25,000 Natl. Network: ABC. Wire Svc: AP Format: News/talk. News staff: 2; News: 75 hrs wkly. Target aud: 35-54. ♦Mark Ediger, gen mgr & opns dir; Ed Klimek, gen sls mgr; Jerry Brecheisen, progmg dir; Dewey Terrill, news dir; Randy Stewart, chief of engrg; Crystal Garrels, news rptr.

KJCK-FM— July 22, 1965: 97.5 mhz; 100 kw. Ant 630 ft TL: N39 00 53 W96 52 15. Stereo. Web Site: www.kjck.com.690,000 Format: Top-40. News staff: 2; News: one hr wkly. Target aud: 18-34; young adults. ♦Robert Elfman, progmg dir; Keri Parker, traf mgr; Matt McBain, disc jockey; Rodney Baker, disc jockey.

Kansas City

KDTD(AM)— 1925: 1340 khz; 1 kw-U. TL: N39 06 50 W94 40 05. Hrs open: 24 1701 S. 55th St., 66106-2241. Phone: (913) 287-1480. Fax: (913) 287-5881. Licensee: Davidson Media Station KCKN Licensee LLC. (acq 10-11-2005; $1.9 million). Population served: 1,500,000 Format: Rgnl Mexican. News staff: 2. ♦Dan Perez, gen mgr; Carlos Mercado, opns dir.

KFKF-FM— May 28, 1963: 94.1 mhz; 100 kw. 994 ft TL: N39 00 57 W94 30 24. Stereo. Hrs opn: 4717 Grand Ave., Suite 600, MO, 64112. Phone: (816) 753-4000. Fax: (816) 753-4045. Web Site: www.kfkf.com. Licensee: Wilks License Co.-Kansas City LLC. Group owner: Infinity Broadcasting Corp. (acq 1-10-2007; grpsl). Population served: 280,000 Format: Contemp country. ♦Herndon Hasty, gen mgr.

KKHK(AM)— 1926: 1250 khz; 25 kw-D, 3.7 kw-N, DA-N. TL: N39 11 06 W94 27 28. Hrs open: 6220 Kansas Ave., 66611. Phone: (913) 788-1255. Fax: (913) 788-1254. Web Site: www.lasuperx1250.com. Licensee: Entercom Kansas City License LLC. Group owner: Entercom Communications Corp. (acq 3-3-99; $2.75 million). Format: Mexican rgnl. ♦Debra Sandoval, gen mgr.

KUDL(FM)— Oct 9, 1959: 98.1 mhz; 100 kw. 994 ft TL: N39 04 23 W94 29 06. Stereo. Hrs opn: 24 4935 Belinder Rd., Westwood, 66205. Phone: (913) 677-8998. Fax: (913) 677-8061. Web Site: www.kudl.com. Licensee: Entercom Kansas City License L.L.C. Group owner: Entercom Communications Corp. (acq 10-17-97; grpsl). Population served: 1,500,000 Format: Adult contemp. News staff: one. Target aud: 25-44; women. ♦Cindy Schloss, gen mgr.

KXTR(AM)— 2001: 1660 khz; 10 kw-D, 1 kw-N. TL: N39 06 50 W94 40 45. Hrs open: 24 4935 Belinder Rd., Westwood, 66205. Phone: (913) 677-8998. Fax: (913) 677-8061. Web Site: www.kxtr.com. Licensee: Entercom Kansas City License LLC. Group owner: Entercom Communications Corp. Format: Classical. ♦Cindy Schloss, gen mgr; Patrick Nease, progmg dir; John Verlin, sls.

Kingman

KCVW(FM)— December 1997: 94.3 mhz; 50 kw. 492 ft TL: N37 48 03 W97 56 49. Hrs open: 24 Mezzanine, 100 N. Main, Hutchinson, 67501. Phone: (620) 663-0943. Fax: (620) 663-0913. E-mail: kcvw@bottradionetwork.com Web Site: www.bottradionetwork.com. Licensee: Bott Communications Inc. Group owner: Bott Radio Network Natl. Network: USA. Format: Christian talk. Target aud: 25-55. ♦Richard P. Bott II, exec VP; Jason Potoenik, traf mgr.

KTCM(FM)— Sept 15, 1989: 100.3 mhz; 48 kw. Ant 505 ft TL: N37 29 59 W98 10 24. (CP: 7.8 kw. TL: N37 29 59 W98 10 25). Stereo. Hrs opn: 315 W. D Ave., 67010. Phone: (620) 532-1190. Fax: (620) 264-0562. Licensee: Connoisseur Media of Wichita LLC. (acq 10-11-2005; $1.7 million). Format: Sp, salsa. ♦Maria Salazar, gen mgr.

Larned

KBGL(FM)— 2001: 106.9 mhz; 100 kw. Ant 485 ft TL: N38 27 06 W99 10 03. Hrs open: 1200 Baker St., Great Bend, 67530. Phone: (620) 792-3647. Fax: (620) 792-3649. Web Site: www.eagleradio.net. Licensee: Hull Broadcasting Inc. (acq 9-12-00; with KFIX(FM) Plainville). Format: Oldies. ♦Rick Nulton, gen mgr; Phil Grossardt, opns mgr; Mike Durler, sls dir.

KGTR(FM)—Listing follows KNNS(AM).

KNNS(AM)— Nov 4, 1963: 1510 khz; 1 kw-D. TL: N38 09 54 W99 06 05. Hrs open: 5501 W. 10 St., Great Bend, 67530. Phone: (620) 792-7108. Fax: (620) 792-7051. E-mail: kzls@nrgmedia.com Licensee: NRG Media LLC. Group owner: Waitt Broadcasting Inc. (acq 9-12-2005; grpsl). Population served: 30,000 Rgnl. Network: Kan. Info., Kan. Agri. Format: ESPN sports. Target aud: General; people looking for loc info. Spec prog: Farm 10 hrs, gospel 6 hrs, relg 5 hrs wkly. ♦Jen Schwamborn, gen mgr; Chris Elsen, progmg dir.

KGTR(FM)—Co-owned with KNNS(AM). Nov 1, 1965: 96.7 mhz; 3 kw. 290 ft TL: N38 09 54 W99 06 05. (CP: Ant 265 ft.). Stereo. Format: Oldies. Target aud: 30-60; baby boomers with disposable income. ♦Dan Cormack, local news ed.

Lawrence

***KANU(FM)**— Sept 15, 1952: 91.5 mhz; 100 kw. 698 ft TL: N38 57 18 W95 15 57. Stereo. Hrs opn: 24 1120 W. 11th St., 66044. Phone: (785) 864-4530. Fax: (785) 864-5278. E-mail: kpr@ku.edu Web Site: www.kpr.ku.edu. Licensee: University of Kansas. Population served: 85,000 Natl. Network: PRI, NPR. Rgnl. Network: Kan. Pub. Law Firm: Arter & Hadden. Format: Class, jazz. News staff: 3; News: 35 hrs wkly. Target aud: 25-49; upscale. Spec prog: Bluegrass 4 hrs, Celtic 2 hrs, blues 4 hrs wkly. ♦Janet Campbell, gen mgr.

***KJHK(FM)**— 1975: 90.7 mhz; 2.9 kw. Ant 163 ft TL: N38 57 30 W95 15 00. Stereo. Hrs opn: 24 2051 A. Dole Ctr., 66045. Phone: (785) 864-4745. Fax: (785) 864-5173. E-mail: kjhk@mail.ku.edu Web Site: www.kjhk.org. Licensee: University of Kansas. Population served: 75,000 Format: Rock, jazz. News: 15 hrs wkly. Target aud: 18-34; Univ & community population. Spec prog: Reggae 3 hrs, blues 2 hrs wkly. ♦Tom Johnson, gen mgr; Danielle Basci, stn mgr; Joe Noh, prom dir; Tom Kimmel, progmg dir.

KLWN(AM)— Feb 22, 1951: 1320 khz; 500 w-D, 250 w-N. TL: N38 56 05 W95 17 12. Hrs open: 24 3125 W. 6th St., 66049-3101. Phone: (785) 843-1320. Fax: (785) 841-5924. Fax: (785) 843-4585. E-mail: mail@lazer.com Web Site: www.klwn.com. Licensee: Great Plains Media Inc. Group owner: Zimmer Radio Group (acq 6-30-2006; with co-located FM). Population served: 100,000 Law Firm: Fletcher, Heald & Hildreth. Format: News/talk, sports. News staff: 2; News: 10 hrs wkly. Target aud: 25-59; adults. Spec prog: Relg 4 hrs wkly. ♦John Flood, progmg dir.

KLZR(FM)—Co-owned with KLWN(AM). Aug 20, 1963: 105.9 mhz; 100 kw. 771 ft TL: N39 02 21 W95 26 59. Stereo. 24 Web Site: www.lazer.com.100,000 Format: CHR, top 40. News staff: 2; News: one hr wkly. Target aud: 18-34; young adults.

Leavenworth

KKLO(AM)— 1946: 1410 khz; 5 kw-D, 500 w-N, DA-2. TL: N39 16 24 W94 54 27. Hrs open: Box 473, Saint Louis, MO, 63166. Secondary address: 481 Muncie Rd. 66048. Phone: (913) 351-1410. Fax: (913) 351-1410. Web Site: www.hereshelpnet.org. Licensee: New Life Evangelistic Center Inc. (acq 10-29-99). Population served: 41,100 Format: Christian. Target aud: 25-49; upscale, educated, loyal Christian listeners. ♦Larry Rice, CEO, chmn, pres, gen mgr & progmg dir.

KQRC-FM—Licensed to Leavenworth. See Kansas City MO

Leoti

KWKR(FM)— Nov 1, 1983: 99.9 mhz; 100 kw. Ant 395 ft TL: N38 16 39 W101 17 50. Stereo. Hrs opn: Box 759, Garden City, 67846. Secondary address: 1402 E. Kansas, Garden City 67846. Phone: (620) 276-3251. Fax: (620) 276-3568. Web Site: www.wksradio.com. Licensee: KBUF Partnership. Group owner: Robert Ingstad Broadcast Properties (acq 12-1-97; $841,170). Population served: 70,000 Natl. Network: Westwood One. Law Firm: Dow, Lohnes & Albertson. Format: Classic rock. Target aud: 25-44. Spec prog: Sp 3 hrs wkly. ♦Gil Wohler, gen mgr; James Janda, VP & progmg dir.

Liberal

KLDG(FM)— October 1994: 102.7 mhz; 100 kw. 466 ft TL: N37 02 45 W101 06 11. Hrs open: 24 1410 Northwestern Ave., 67901. Phone: (620) 624-3891. Fax: (620) 624-7885. Web Site: www.kscb.net. Licensee: Seward County Broadcasting Co. Inc. (group owner) Population served: 85,000 Natl. Network: Jones Radio Networks. Natl. Rep: Roslin. Law Firm: Shaw Pittman. Format: Country. News staff: 2; News: 2 hrs wkly. Target aud: 18-49; young, mobile & impulsive consumers. ♦John Landon, chmn; Don Ford, pres; Bob Larrabee, VP; Stuart Melchert, gen mgr; Terry Miller, opns mgr; John Mulhern, chief of engrg; Mikki Hofferber, traf mgr.

KSCB(AM)— July 25, 1948: 1270 khz; 1 kw-D, 30 w-N. TL: N37 03 15 W100 53 39. Hrs open: 24 1410 N. Western Ave., 67901. Phone: (620) 624-3891. Fax: (620) 624-9472. E-mail: kscb@kscb.net Web Site: www.kscb.net. Licensee: Seward County Broadcasting Co. Population served: 13,471 Natl. Network: ABC, Westwood One. Rgnl. Network: Kan. Info. Natl. Rep: Roslin. Law Firm: Wiley, Rein & Felding. Format: News/talk. News staff: 3. Target aud: 35 plus. Spec prog: Farm 6 hrs wkly. ♦Stuart Melchert, gen mgr; Cheryl Collins, gen sls mgr; Terry Miller, VP, opns mgr & progmg dir; Brock Kappelmann, news dir; Joe Denoyer, news dir; John Mulhurn, engrg dir & chief of engrg; Mikki Hofferber, traf mgr.

KSCB-FM— July 10, 1978: 107.5 mhz; 100 kw. 511 ft TL: N37 02 45 W101 06 11. Stereo. 24 Web Site: www.kscb.net.50,000 Natl. Network: Jones Radio Networks. Natl. Rep: Roslin. Format: Adult contemp. News staff: 2; News: 7 hrs wkly. Target aud: 25-49; young adults. ♦Mikki Hofferber, traf mgr.

KSLS(FM)—Listing follows KYUU(AM).

KYUU(AM)— Sept 15, 1960: 1470 khz; 1 kw-D, 125 w-N. TL: N37 03 17 W100 53 06. Hrs open: 150 Plaza Dr., 67901. Phone: (620) 624-8156. Fax: (620) 624-4606. Licensee: Waitt Omaha LLC. Group owner: Waitt Broadcasting Inc. (acq 5-3-2001; grpsl). Population served: 16,500 Rgnl. Network: Mid-American Ag. Natl. Rep: McGavren Guild. Format: Sp. News: one hr wkly. Target aud: 25-54; Spanish speaking. ♦Steve Schiffner, gen mgr; Matt Younkin, progmg dir.

KSLS(FM)—Co-owned with KYUU(AM). July 1978: 101.5 mhz; 100 kw. 550 ft TL: N37 03 20 W100 48 40. Stereo. Format: Adult contemp.

KZQD(FM)— October 1997: 105.1 mhz; 50 kw. 492 ft TL: N37 17 39 W100 51 38. (CP: Ant 387 ft. TL: N37 02 53 W100 54 34). Hrs opn: Box 2636, 67905. Phone: (620) 626-8282. Web Site: www.kzqdradiolibertad.com. Licensee: Mario Loredo. (acq 3-8-94; FTR: 5-9-94). Format: Christian. ♦Mario Loredo, gen mgr & progmg dir.

Lindsborg

KQNS-FM— Oct 8, 1985: 95.5 mhz; 15.5 kw. Ant 417 ft TL: N38 40 00 W97 41 30. Stereo. Hrs opn: 24 1825 S. Ohio, Salina, 67401-4573. Phone: (785) 827-2100. Fax: (785) 827-3503. Web Site: www.0295.com. Licensee: NRG Media LLC. Population served: 325,000 Format: Adult contemp. News: 2 hrs wkly. Target aud: 25-49. ♦Jerry Hinrikus, gen mgr.

Lyons

KXKU(FM)— Apr 10, 1970: 106.1 mhz; 100 kw. 659 ft TL: N38 16 33 W98 12 11. Stereo. Hrs opn: 24 106 N. Main St., Hutchinson, 67501-5219. Phone: (620) 665-5758. Fax: (620) 665-6655. Licensee: Ad Astra Per Aspera Broadcasting Inc. (group owner; acq 9-17-86; $366,816; FTR: 6-9-86). Population served: 150,000 Format: Country. News staff: one; News: 2 hrs wkly. Target aud: 25-64; listeners throughout central KS. ♦Cliff C. Shank, pres; Mike Hill, VP, stn mgr & sls VP; Cheryl Dinwiddie, gen sls mgr & rgnl sls mgr.

Manhattan

***KGLV(FM)**—Not on air, target date: unknown: 88.7 mhz; 25 kw. Ant 380 ft TL: N39 08 40 W96 31 01. Hrs open:
Rebroadcasts KLVR(FM) Santa Rosa, CA 100%.

2351 Sunset Blvd., Suite 170-218, Rocklin, CA, 95765. Phone: (916) 251-1600. Fax: (916) 251-1650. Web Site: www.klove.com. Licensee: Educational Media Foundation. (acq 10-27-2006; $325,000 for CP). Natl. Network: K-Love. Format: Contemp Christian. ♦Richard Jenkins, pres.

KJCK-FM—See Junction City

KMAN(AM)— June 1950: 1350 khz; 500 w-D, 40 w-N. TL: N39 13 00 W96 33 30. Stereo. Hrs opn: 24 Box 1350, 66502. Secondary address: 2414 Casement Rd. 66502. Phone: (785) 776-1350. Phone: (785) 776-4851. Fax: (785) 539-1000. Web Site: www.1350kman.com. Licensee: Manhattan Broadcasting Co. Group owner: Seaton Stations Population served: 110,000 Natl. Network: CBS, ESPN Radio, Westwood One. Rgnl. Network: Kan. Info. Law Firm: Shaw Pittman. Format: News/talk, sports. News staff: one; News: 60 hrs wkly. Target aud: 30 plus. ♦Richard Seaton, chmn; Richard T. Wartell, pres & gen mgr; Danielle Runnebaum, opns mgr & gen sls mgr; Daneille Runnebaum, sls dir; Matt Walters, progmg dir; Cathy Dawes, news dir; Kevin Block, chief of engrg; Krista Wilder, traf mgr; Tyler Siefkes, sports cmtr.

KMKF(FM)—Co-owned with KMAN(AM). Sept 1, 1972: 101.5 mhz; 39 kw. 577 ft TL: N39 15 55 W96 27 56. Stereo. 24 Phone: (785) 776-1015. Web Site: www.purerock.com.400,000 Format: Rock/AOR. News staff: one; News: 2 hrs wkly. Target aud: 18-35. ♦Corey Dean, progmg dir & disc jockey; Kevin Block, engrg dir; Krista Wilder, traf mgr.

KQLA(FM)—Ogden, Feb 14, 1986: 103.5 mhz; 41 kw. Ant 312 ft TL: N39 09 21 W96 36 44. Stereo. Hrs opn: 24 Box 104, 66505. Secondary address: 104 S. 4th St. 66502. Phone: (785) 587-0103. Phone: (785) 323-9797. Fax: (785) 776-0110. E-mail: platinum@kjck.com Web Site: www.kqla.com. Licensee: Platinum Broadcasting Co. (group owner; (acq 9-24-97; $650,000). Population served: 110,000 Natl. Network: ABC. Format: Adult contemp. News staff: one; News: 10 hrs wkly. Target aud: 18-44; mobile, educated persons with quality income. ♦Ed Klimek, exec VP; Mark Ediger, gen mgr.

***KSDB-FM**— 1949: 91.9 mhz; 1.4 kw. Ant 290 ft TL: N39 09 49 W96 31 54. Stereo. Hrs opn: 24 A.Q. Miller School of Journalism, Rm. 105, Kedzie Hall, 66506. Phone: (785) 532-2971. Fax: (785) 532-5484. E-mail: radio@ksu.edu Web Site: www.wildcatradio.ksu.edu. Licensee: Kansas State University. Population served: 150,000 Format: Progsv, rock, urban contemp. News staff: one; News: 12 hrs wkly. Target aud: 18-34; young adults. Spec prog: Black 4 hrs, gospel 3 hrs, jazz 3 hrs wkly. ♦Steve Smethers, chmn & pres; Aaron Leiker, gen mgr.

KXBZ(FM)— September 1994: 104.7 mhz; 20 kw. 502 ft TL: N39 15 55 W96 27 56. Hrs opn: 2414 Casement Rd., 66502. Phone: (785) 776-1350. Fax: (785) 539-1000. E-mail: dubs@purerock.com Web Site: www.b1047.com. Licensee: Manhattan Broadcasting Co. Inc. Group owner: Seaton Stations (acq 3-2-99). Natl. Network: Westwood One. Format: Hot country. Target aud: 18-35; men & women. ♦Richard T. Wartell, pres & gen mgr.

Marysville

KNDY(AM)— July 10, 1956: 1570 khz; 250 w-D. TL: N39 51 02 W96 38 52. Hrs opn: 24 937 Jayhawk Rd., 66508. Phone: (785) 562-2361. Fax: (785) 562-2188. Licensee: Dierking Communications Inc. (group owner; (acq 9-6-88). Population served: 40,000 Rgnl. Network: Mid-American Ag. Format: Farm, C&W. News: 24 hrs wkly. Target aud: General. ♦Bruce Dierking, pres, gen mgr, gen sls mgr & progmg dir.

KNDY-FM— July 23, 1974: 95.5 mhz; 25 kw. Ant 328 ft TL: N39 57 36 W96 44 05. Stereo. 24 Format: C&W. ♦Myron Nolind, chief of engrg; Larry Steckline, farm dir.

McPherson

KBBE(FM)—Listing follows KNGL(AM).

KNGL(AM)— Jan 4, 1949: 1540 khz; 250 w-D. TL: N38 20 30 W97 40 12. Hrs opn: Box 1069, 67460. Phone: (620) 241-1504. Fax: (620) 241-3196. E-mail: kbbelngl@earthlink.net Web Site: midkansasmedia.com. Licensee: Davies Communications Inc. (acq 10-1-85; $589,000 with co-located FM; FTR: 8-19-85) Population served: 140,000 Format: Oldies. News staff: 2; News: 25 hrs wkly. Target aud: 25-54. Spec prog: Relg 5 hrs wkly. ♦Jerry Davies, pres; Diane Davies, exec VP, prom dir & progmg dir; Joe Johnston, gen mgr; Nick Gosnell, news dir; Shawn White, engrg dir; Kelsey Walker, traf mgr; Chris Swick, local news ed; Claude Hughes, farm dir; Bob Hapgood, disc jockey; Misty Souder, disc jockey; Resa Larson, disc jockey; Val Buck, disc jockey.

KBBE(FM)—Co-owned with KNGL(AM). Jan 12, 1974: 96.7 mhz; 6 kw. 245 ft TL: N38 20 30 W97 40 12. Stereo. 24 1137 14th Ave., 67460. Web Site: midkansasmedia.com.50,000 Format: Adult contemp. News staff: one. ♦Gary Jordon, disc jockey.

Medicine Lodge

KREJ(FM)— January 1990: 101.7 mhz; 50 kw. 492 ft TL: N37 13 58 W98 39 43. Hrs opn: 24 301 S. Main St., 67104-1513. Phone: (620) 886-3537. Fax: (321) 264-9370. Licensee: Florida Public Radio Inc. (acq 5-90; FTR: 6-11-90). Natl. Network: Moody. Format: Relg. Target aud: General. ♦Mike Henry, gen mgr; Randy Henry, chief of engrg.

***KSNS(FM)**— April 1999: 91.5 mhz; 48 kw. Ant 462 ft TL: N37 14 02 W98 39 55. Hrs open: 301 S. Main, 67104-1513. Secondary address: 505 Josephine St., Titusville, FL 32796. Web Site: noncomradio.net. Licensee: Florida Public Radio Inc. Format: Contemp Christian mus. ♦Mike Henry, gen mgr.

Minneapolis

KILS(FM)— Feb 24, 1993: 92.7 mhz; 50 kw. 492 ft TL: N39 00 52 W97 37 42. Hrs open: 24 1825 S. Ohio, Salina, 67401. Phone: (785) 827-2100. Fax: (785) 827-3503. Web Site: www.927thezoo.com. Licensee: NRG Media LLC. Group owner: Waitt Broadcasting Inc. (acq 9-12-2005; grpsl). Population served: 150,000 Format: Classic rock. Target aud: 18-54. ♦Jerry Hinrikus, gen mgr.

Mission

KCNW(AM)—See Fairway

KCZZ(AM)— October 1957: 1480 khz; 1 kw-D, 500 w-N, DA-2. TL: N39 04 05 W94 42 09. Hrs open: 24 1701 S. 55th St., Kansas City, 66106. Phone: (913) 287-1480. Fax: (913) 287-5881. Licensee: Davidson Media Station KCZZ Licensee LLC. (acq 1-28-2005; $3.9 million with KAKS(FM) Huntsville, AR). Population served: 1,500,000 Natl. Rep: Lotus Entravision Reps LLC. Format: Sp. Target aud: 18-49; adults. ♦Dan Perez, gen mgr; Carlos Mercado, opns dir.

KRBZ(FM)—See Kansas City, MO

Newton

KFTI-FM— 1959: 92.3 mhz; 100 kw. Ant 640 ft TL: N38 01 09 W97 23 01. Stereo. Hrs opn: 24 4200 N. Old Lawrence Rd., Wichita, 67219. Phone: (316) 838-9141. Fax: (316) 838-3607. Web Site: www.classiccountry923.com. Licensee: Journal Broadcast Corp. Group owner: Journal Communications Inc. (acq 3-20-2000; $4.25 million). Population served: 720,000 Format: Classic country. ♦Rob Burton, VP & gen mgr; Eric McCart, gen sls mgr; Manny Cowzinski, prom dir; Ray Micheals, progmg dir.

KJRG(AM)— May 24, 1953: 950 khz; 500 w-D, 147 w-N. TL: N38 02 45 W97 22 24. Hrs open: 209 N. Meridian Rd., 67114. Phone: (620) 663-0943. Fax: (620) 663-0913. E-mail: kjrg@bottradionetwork.com Web Site: www.bottradionetwork.com. Licensee: Community Broadcasting Inc. (acq 6-20-2006; $650,000). Population served: 15,439 Format: Relg. ♦Richard P. Bott, gen mgr; Eben Fowler, opns dir; Candy Green, progmg dir.

North Fort Riley

KBLS(FM)— Jan 1, 1993: 102.5 mhz; 100 kw. 492 ft TL: N38 57 05 W96 47 45. Stereo. Hrs opn: 5008 Skyway Dr., Manhattan, 66503. Phone: (785) 537-3232. Fax: (785) 587-9495. Licensee: MCC Radio LLC. Group owner: Morris Radio LLC (acq 1-14-2004; grpsl). Population served: 125,000 Format: Adult contemp. News staff: 3; News: 2 hrs wkly. Target aud: 25-54; women. ♦William S. Morris IV, pres; Robert Protzman, gen mgr; Clarke Sanders, prom mgr; John Anderson, progmg dir; Mark Beaver, traf mgr.

North Newton

***KBCU(FM)**— Apr 6, 1989: 88.1 mhz; 149 w. 56 ft TL: N38 04 26 W97 20 35. Hrs open: 24 (T-Su); 8 AM-midnight (M) 300 E. 27th St., 67117. Phone: (316) 284-5228. Phone: (316) 284-5271. Fax: (316) 284-5286. Licensee: Bethel College. Population served: 16,000 Rgnl. Network: Kan. Info. Format: Var/div. News: 5 hrs wkly. Target aud: General; college students & Harvey County. Spec prog: Sp 2 hrs wkly. ♦Christine Crouse-Dick, gen mgr & stn mgr.

Norton

KQNK(AM)— Oct 30, 1963: 1530 khz; 1 kw-D. TL: N38 35 04 W95 15 57. Hrs open: 1530 KQNK Road, 67654. Phone: (785) 877-3378. Fax: (785) 877-3379. E-mail: kqnk@ruraltel.net Web Site: www.kqnk.com. Licensee: Dierking Communications Inc. (group owner; (acq 7-13-99; $165,000 with co-located FM). Population served: 3,976 Natl. Rep: Keystone (unwired net). Format: Soft adult contemp. ♦Bruce Dierking, pres; Marvin Matchett, gen mgr; Mandi Fick, gen sls mgr & disc jockey; Deena Wente, progmg dir, news dir, chief of engrg & disc jockey.

KQNK-FM— Mar 1, 1993: 106.7 mhz; 51 kw. 92 ft TL: N39 49 37 W99 52 08. Format: Adult contemp.

Oberlin

KFNF(FM)— July 1977: 101.1 mhz; 100 kw. 420 ft TL: N39 49 33 W100 39 09. Stereo. Hrs opn: 24 Box 116 C, R.R. 2, 6 Miles W. of Oberlin, 67749. Phone: (785) 475-2225. Phone: (785) 475-2226. Fax: (785) 475-2510. Licensee: Armada Media - McCook Inc. (group owner; (acq 1-17-2007; grpsl). Population served: 60,000 Natl. Network: ABC. Rgnl. Network: Mid-American Ag. Format: Country. Target aud: 25-65; farmers. Spec prog: Gospel 3 hrs wkly. ♦Bryan Loker, gen mgr; Adam Kadavy, chief of opns.

***KRLE(FM)**—Not on air, target date: unknown: 91.3 mhz; 250 w. Ant 118 ft TL: N39 47 14 W100 31 54. Hrs open: 5700 West Oaks Blvd., Rocklin, CA, 95765. Phone: (916) 251-1600. Fax: (916) 251-1650. Licensee: Educational Media Foundation. (acq 3-23-2007; grpsl). ♦Richard Jenkins, pres.

Ogden

KQLA(FM)—Licensed to Ogden. See Manhattan

Olathe

KCCV-FM— Dec 1, 1993: 92.3 mhz; 8.3 kw. 564 ft TL: N38 56 10 W94 50 41. Stereo. Hrs opn: 24 10550 Barkley, Suite 112, Overland Park, 66212. Phone: (913) 642-7600. Fax: (913) 642-2424. E-mail: kccv@bottradionetwork.com Web Site: www.bottradionetwork.com. Licensee: Bott Broadcasting Co. (group owner; acq 7-1-92; $537,500; FTR: 8-3-92). Population served: 3,000,000 Natl. Network: USA. Format: Christian talk. Target aud: 25-54; family oriented. ♦Trace Thurlby, COO; Richard P. Bott Sr., pres; Tom Holdeman, CFO; Richard P. Bott II, exec VP & VP; Pat Rulon, natl sls mgr; Rachel Moser, mktg mgr; Jason Potocnik, traf mgr.

Olpe

KEKS(FM)—Not on air, target date: unknown: 103.1 mhz; 2.45 kw. Ant 315 ft TL: N38 17 37 W96 13 03. Hrs open: 1711 N. Grand St., Pittsburg, 66762-3229. Phone: (620) 404-8108. Licensee: Andrew A. Wachter. ♦Andrew A. Wachter, gen mgr.

Olsburg

***KANV(FM)**— 2003: 91.3 mhz; 6 kw. Ant 328 ft TL: N39 00 55 W96 53 55. Hrs open: 1120 W. 11th St., Lawrence, 66044. Phone: (785) 864-4530. Fax: (785) 864-5278. E-mail: kpr@ku.edu Web Site: www.kpr.ku.edu. Licensee: The University of Kansas. Natl. Network: NPR. Format: Class, jazz. ♦Janet Campbell, gen mgr; Cordelia Brown, opns mgr.

Osage City

KMXN(FM)— July 26, 1982: 92.9 mhz; 7.9 kw. Ant 538 ft TL: N38 48 21 W95 42 58. (CP: 36 kw, ant 564 ft. TL: N38 31 47 W96 05 09). Stereo. Hrs opn: 24 3125 W. 6th St., Lawrence, 66049. Phone: (785) 843-1320. Fax: (785) 841-5924. Licensee: Great Plains Media Inc. (acq 6-30-2006). Rgnl. Network: Mid-American Ag. Format: Rhythmic CHR. ♦Ron Covert, gen mgr; Jon Thomas, progmg dir; Mike Mayfield, chief of engrg.

Ottawa

KCHZ(FM)— Mar 1, 1962: 95.7 mhz; 96 kw. Ant 981 ft TL: N39 00 45 W95 01 46. Hrs open: 4240 Blueridge Blvd., Suite 820, Kansas City, MO, 64133. Phone: (816) 356-2400. Fax: (816) 356-2479. Web Site: www.z957.net. Licensee: Kansas City Trust LLC, Trustee Group owner: Cumulus Media Inc. (acq 3-3-2006). Population served: 11,036 Format: Rhythmic top-40. ♦Maurice DeVoe, opns mgr.

KOFO(AM)— Sept 24, 1949: 1220 khz; 250 w-D, 40 w-N. TL: N38 35 04 W95 15 57. Hrs open: 24 320 E. Radio Rd., 66067. Phone: (785) 242-1220. Fax: (785) 242-1442. E-mail: kofo@kofo.com Web Site: www.kofo.com. Licensee: Brandy Communications Inc. Population served: 55,000 Natl. Network: ABC. Rgnl. Network: Kan. Agri. Format: C&W. News staff: 1; News: 7 hrs wkly. Target aud: 25-54; Male & Female. Spec prog: Farm 2 hrs wkly. ♦Brad Howard, pres & gen mgr.

***KRBW(FM)**— 1997: 90.5 mhz; 250 w. 187 ft TL: N38 35 04 W95 15 57. Hrs open: American family Radio, 320 E. Radio Rd., 66067. Phone: (785) 242-9050. Fax: (662) 842-6791. Web Site: www.afr.net. Licensee: American Family Association Group owner: American Family Radio (acq 1-24-97). Format: Comtemp Christian. ♦Marvin Sanders, gen mgr.

***KTJO-FM**— May 1951: 88.9 mhz; 145 w. 66 ft TL: N38 36 16 W95 15 49. Stereo. Hrs opn: 7 AM-midnight Ottawa Univ., 1001 S. Cedar St., 66067. Phone: (785) 242-5200. Fax: (785) 242-7429. Licensee: Ottawa University. Population served: 13,102 Format: Div, CHR, Contemp Christian. News: 5 hrs wkly. Target aud: General; Ottawa Univ community & City of Ottawa, KS. ♦Bradley A. Howard, CEO; Ben Weiss, engrg VP.

Overland Park

KCCV(AM)— 1962: 760 khz; 6 kw-D, 200 w-N, DA-2. TL: N39 02 26 W94 30 34. Hrs open: Sunrise-sunset 10550 Barkley, Suite 112, 66212. Phone: (913) 642-7600. Fax: (913) 642-2424. E-mail: kccv@bottradionetwork.com Web Site: www.bottradionetwork.com. Licensee: Bott Broadcasting Co. Group owner: Bott Radio Network (acq 1962). Population served: 3,000,000 Natl. Network: USA. Format: Christian talk, Christian, relg, news/talk. Target aud: 25-54; family-oriented. ♦Trace Thurlby, COO; Richard P Bolt, pres; Tom Holdeman, CFO; Richard P. Bott II, exec VP; Pat Rulon, natl sls mgr; Rachel Moser, mktg mgr; Jason Potocnik, traf mgr.

Parsons

KLKC(AM)— 1948: 1540 khz; 250 w-D. TL: N37 20 35 W95 13 55. Hrs open: Box 853, 1812 Main St., 67357-0853. Phone: (620) 421-6400. Fax: (620) 421-5570. E-mail: judynance@klkc.com Web Site: klkc.com. Licensee: Southeast Kansas Independent Living Resource Center Inc. (acq 11-23-2005; $334,932 with co-located FM). Population served: 113,015 Format: Talk, sports. News: 28 hrs wkly. Target aud: 12-60; general. Spec prog: Farm 2 hrs, relg 2 hrs, big band 3 hrs wkly. ♦Lynn Meredith, gen mgr; Colt Smith, gen sls mgr; Ed Hernandez, gen sls mgr & sls; Steve Lardy, progmg dir; Terry Blackburn, mus dir; Annette Tucker, news dir & chief of engrg.

KLKC-FM— October 1978: 93.5 mhz; 3 kw. Ant 267 ft TL: N37 20 35 W95 13 55.6 AM-11 PM Format: Oldies.

Phillipsburg

KKAN(AM)— Dec 31, 1959: 1490 khz; 1 kw-U. TL: N39 47 32 W99 19 55. Hrs open: Box 548, 67661. Phone: (785) 543-2151. Phone: (785) 543-6593. Fax: (785) 543-2152. Web Site: www.kkankqma.com. Licensee: Walter C. Seidel. (acq 3-29-88). Population served: 120,000 Rgnl. Network: Kan. Info. Format: Div, news. Target aud: General; rural population & small towns. Spec prog: Farm 10 hrs, gospel 12 hrs wkly. ♦Bob Yates, gen mgr, stn mgr & progmg dir; Tad Felts, news dir.

KQMA-FM—Co-owned with KKAN(AM). July 14, 1984: 92.5 mhz; 100 kw. 510 ft TL: N39 37 02 W99 17 55. Stereo. 205 F St., 67661. Web Site: www.kkankqma.com.80,000 Format: Var.

Pittsburg

KKOW(AM)— Oct 11, 1937: 860 khz; 10 kw-D, 5 kw-N, DA-N. TL: N37 24 46 W94 38 16. Hrs open: 24 1162 E. Hwy. 126, 66762. Phone: (620) 231-7200. Fax: (620) 231-3321. E-mail: kkow@kkowradio.com Web Site: www.kkowam.com. Licensee: American Media Investment Inc. (acq 6-89; $400,000 with co-located FM; FTR: 6-26-89) Population served: 342,000 Natl. Network: CBS. Natl. Rep: McGavren Guild. Format: Classic country. News staff: 2; News: 5 hrs wkly. Target aud: General. ♦Chris Kelly, gen mgr & prom dir.

KKOW-FM— Apr 20, 1975: 96.9 mhz; 100 kw. 278 ft TL: N37 23 44 W94 40 42. Stereo. Hrs opn: 1162 E. Hwy. 126, 66762. Phone: (620) 231-7200. Fax: (620) 231-3321. E-mail: kkow@kkowradio.com Web Site: www.kkowfm.com. Licensee: American Media Investments Inc. Format: Contemp country. ♦Chris Kelly, gen mgr.

***KRPS(FM)**— Apr 29, 1988: 89.9 mhz; 100 kw. 1,000 ft TL: N37 18 44 W94 48 58. Stereo. Hrs opn: 24 Box 899, 66762. Phone: (620) 235-4288. Fax: 620(235-4290). E-mail: krps@pittstate.edu Web Site: www.krps.org. Licensee: Pittsburg State University. Natl. Network: NPR, PRI. Format: Class, jazz, news. News: 39 hrs wkly. Spec prog: Folk 3 hrs. ♦Missi Kelly, gen mgr; Vicki Pritchett, dev dir.

KSEK(AM)— 1948: 1340 khz; 1 kw-U. TL: N37 23 44 W94 40 42. Hrs open: 202 E. Centenial Dr., Suite 2B, 66762. Phone: (620) 232-9912. Fax: (620) 232-9915. Licensee: Southeast Kansas Independent Living Resource Center Inc. (group owner; (acq 11-30-2004; $700,000 with KSEK-FM Girard). Natl. Network: ESPN Radio, AP Network News. Format: Sports. News: 10 hrs wkly. Target aud: 25 plus. Spec prog: High school basketball & football. ♦Lynn Meredith, gen mgr.

Plainville

KFIX(FM)— May 11, 1998: 96.9 mhz; 10.5 kw. 876 ft TL: N39 01 15 W99 28 13. Stereo. Hrs opn: 24 Box 6, Hays, 67601. Secondary address: 2300 Hall, Hays 67601. Phone: (785) 625-2578. Fax: (785) 625-3632. E-mail: studio@kfix.com Web Site: www.kfix.com. Licensee: Hull Broadcasting Inc. (acq 9-12-00; with KBGL(FM) Larned). Population served: 65,000 Format: AOR. Target aud: 25-64. ♦Richard C. Hull, pres; Nancy E. Baumrucker, gen mgr; Cameron Perry, progmg dir; Callie Kolacny, news dir; Kristy Pfeifer, traf mgr.

Pratt

KHMY(FM)— July 1, 1965: 93.1 mhz; 100 kw. Ant 1,007 ft TL: N37 55 50 W98 19 04. Stereo. Hrs opn: 24 Box 1036, Hutchinson, 67504-1030. Phone: (620) 662-5900. Fax: (620) 662-5797. Licensee: Eagle Communications Inc. Group owner: Eagle Communications Group (acq 3-17-03; swap for KSSH(FM) Ingalls). Population served: 100,000 Natl. Rep: McGavren Guild. Format: Adult contemp. ♦Mark Trotman, gen mgr & sls dir; Jason Younger, opns mgr; Terry Drouhard, gen sls mgr; Fred Gough, news dir.

KWLS(AM)— Sept 19, 1963: 1290 khz; 5 kw-D, 500 w-N, DA-2. TL: N37 38 34 W98 40 39. Hrs open: 24 Box 486, 30129 E. Hwy. 54, 67124. Phone: (620) 672-5581. Fax: (620) 672-5583. E-mail: kwls@socencom.net Licensee: NRG Media LLC. Group owner: Waitt Broadcasting Inc. (acq 9-12-2005; grpsl). Population served: 22,000 Rgnl. Network: Mid-American Ag. Natl. Rep: McGavren Guild. Law Firm: Fisher, Wayland, Cooper, Leader & Zaragoza L.L.P. Format: Oldies 60s & 70s. News staff: one; News: 3 hrs wkly. Target aud: 25 plus; rural. ♦Eric Strobel, gen mgr.

Riley

KACZ(FM)— Sept 16, 2003: 96.3 mhz; 11.5 kw. Ant 479 ft TL: N39 13 34 W96 37 00. Hrs open: Box 1350, 2414 Casement Rd., Manhattan, 66502. Phone: (785) 776-1350. Fax: (785) 539-1000. Web Site: www.z963.com. Licensee: Manhattan Broadcasting Co. Inc. Group owner: Seaton Stations. Wire Svc: AP Format: CHR. News staff: 3; News: 4 hrs wkly. Target aud: 18-59; woman. ♦Richard T. Wartell, pres & gen mgr.

Rozel

KKCV(FM)— 2007: 102.5 mhz; 100 kw. Ant 488 ft TL: N37 57 28 W99 25 45.2. Hrs open: 24 10550 Barkley, Suite 100, Overland Park, 66212. Phone: (913) 642-7770. Fax: (913) 642-1319. Web Site: www.bottradionetwork.com. Licensee: Bott Communications Inc. Format: Christian talk. Target aud: 25-55. ♦Trace Thurlby, COO; Richard P. Bott Sr., pres; Tom Holdeman, CFO; Richard P. Bott II, VP & gen mgr; Eben Fowler, opns dir; Pat Rulon, natl sls mgr; Rachel Moser, mktg mgr; Jason Potocnik, traf mgr.

Russell

KCAY(FM)—Listing follows KRSL(AM).

KRSL(AM)— Jan 11, 1956: 990 khz; 250 w-D, 30 w-N. TL: N38 54 22 W98 51 39. Hrs open: Box 666, 1984 N. Main St., 67665. Phone: (785) 483-3121. Fax: (785) 483-6511. E-mail: wayne@krsl.com Web Site: www.krsl.com. Licensee: West Central Radio Inc. (acq 10-24-89; $404,000 with co-located FM; FTR: 11-13-89). Population served: 5,371 Format: Adult contemp. Target aud: 24 plus; general. Spec prog: Polka 4 hrs, farm 2 hrs wkly. ♦Larry Calvery, gen mgr; Carol McKenna, news dir.

KCAY(FM)—Co-owned with KRSL(AM). July 1, 1965: 95.9 mhz; 1.35 kw. Ant 487 ft TL: N38 54 22 W98 51 39. Stereo. Web Site: www.krsl.com. Format: Adult contemp.

Saint Marys

KQTP(FM)— Dec 4, 1994: 102.9 mhz; 50 kw. 320 ft TL: N39 05 34 W95 47 05. Hrs open: 825 S. Kansas Ave., Topeka, 66603. Phone: (785) 272-2122. Fax: (785) 272-6219. E-mail: oldieskqtp@aol.com Web Site: www.cumulus.com. Licensee: Cumulus Licensing Corp. Group owner: Cumulus Media Inc. (acq 4-13-01; with KWIC(FM) Topeka). Format: Classic country. Target aud: 35-54. ♦Kevin Klein, gen mgr.

Salina

***KAKA(FM)**— 2002: 88.5 mhz; 46 kw. Ant 394 ft TL: N39 04 12 W97 51 14. Hrs open: American Family Radio, Box 2440, Tupelo, MS, 38803. Phone: (662) 844-8888. Fax: (662) 842-6791. Web Site: www.afr.net. Licensee: American Family Association. Group owner: American Family Radio Format: Relg. ♦Marvin Sanders, stn mgr.

***KCVS(FM)**— 1994: 91.7 mhz; 11.5 kw. Ant 748 ft TL: N38 39 58 W97 41 30. Stereo. Hrs opn: 24 3434 W. Kilbourn Ave., Milwaukee, WI, 53208. Phone: (414) 935-3000. Fax: (414) 935-3015. E-mail: kcvs@vcyamerica.org Web Site: www.vcyamerica.org. Licensee: VCY/America Inc. (group owner; (acq 7-2-97). Natl. Network: Moody, USA. Format: Christian. ♦Vic Eliason, VP & gen mgr; Jim Schneider, progmg dir.

KFRM(AM)— 1947: 550 khz; 5 kw-D, 110 w-N, DA-2. TL: N39 26 10 W97 39 40. Hrs open: sun up to sun down 1815 Meadowlark Rd., Clay Center, 67432. Phone: (785) 632-5661. Fax: (785) 632-5662. E-mail: kbauen@kfrm.com Web Site: kfrm.com. Licensee: Taylor Communications Inc. (acq 9-12-96; $500,000). Population served: 1,800,000 Format: Farm, talk. News staff: 2; News: 6 hrs wkly. Target aud: 25-55; agricultural. Spec prog: Farm 5 hrs, gospel 5 hrs wkly. ♦Kyle Bauer, gen mgr; Rod Keen, opns mgr; Joyce Beck, gen sls mgr; Jamie Bloom, progmg dir; Joe Woodward, traf mgr; Rocky Downing, sports cmtr.

***KHCD(FM)**— Jan 28, 1988: 89.5 mhz; 100 kw. 925 ft TL: N39 06 16 W97 23 15. Stereo. Hrs opn: 24
Rebroadcasts KHCC-FM Hutchinson 100%.
815 N. Walnut, Suite 300, Hutchinson, 67501. Phone: (620) 662-6646. Licensee: Hutchinson Community College. Natl. Network: NPR. Format: Class, new age, news. ♦David M. Horning, gen mgr; Geralyn Smith, opns dir; Sharon Webb, dev dir.

KINA(AM)—Listing follows KSKG(FM).

KSAL(AM)— May 18, 1937: 1150 khz; 5 kw-U, DA-N. TL: N38 53 08 W97 30 58. Hrs open: 24 Box 80, 67402. Secondary address: 131 N. Santa Fe 67401. Phone: (785) 823-1111. Fax: (785) 823-2034. Web Site: www.ksallink.com. Licensee: MCC Radio LLC. Group owner: Morris Radio LLC (acq 1-30-2004; grpsl). Population served: 300,000 Natl. Network: ABC. Format: News/talk. News staff: 5; News: 20 hrs wkly. Target aud: 35-64. Spec prog: Farm 3 hrs wkly. ♦Robert

Protzman, gen mgr; Bill Ray, opns mgr & gen sls mgr; Clarke Sanders, prom mgr; Rich Alexander, progmg dir; Todd Pittenger, news dir.

KYEZ(FM)—Co-owned with KSAL(AM). May 1, 1975: 93.7 mhz; 100 kw. 510 ft TL: N38 57 14 W97 36 29. Stereo. 24 Web Site: www.y937.com.150,000 Format: Country. News staff: 5; News: 2 hrs wkly. Target aud: 25-54.

KSAL-FM— October 1988: Stn currently dark. 104.9 mhz; 14 kw. 440 ft TL: N38 53 23 W97 38 46. Stereo. Hrs opn: Box 80, 67402-0080. Secondary address: 131 N.Santa Fe 67401. Phone: (785) 823-1111. Fax: (785) 823-2034. Web Site: www.ksallink.com. Licensee: MCC Radio LLC. Group owner: Morris Radio LLC (acq 1-14-2004; grpsl). Population served: 110,000 Format: Contemp classic hits. News staff: 2; News: 5 hrs wky. Target aud: 18-44. ♦Robert Protzman, gen mgr; Clarke Sanders, prom mgr; J.J. Hill, progmg dir; Mark Beaver, traf mgr.

KSKG(FM)— 1961: 99.9 mhz; 100 kw. 570 ft TL: N38 47 36 W97 31 33. Hrs open: 24 1825 S. Ohio St., 67401-0198. Phone: (785) 825-4631. Fax: (785) 825-4600. Web Site: www.999kskg.com. Licensee: Eagle Communications Inc. Group owner: Eagle Communications Group. Population served: 300,000 Format: Modern country. Target aud: 25-54; 51% female, 49% male (baby boomers). Spec prog: Gospel 3 hrs wkly. ♦Gary Shorman, pres; Jerry Hinrikus, gen mgr; Scott Carroll, opns mgr & progmg dir; Randy Picking, news dir; Mark Goff, chief of engrg; Cher Richards, traf mgr.

KINA(AM)—Co-owned with KSKG(FM). Apr 20, 1964: 910 khz; 500 w-D, 29 w-N, DA-2. TL: N38 45 52 W97 32 30.24 hrs (Acq 11-1-95; $235,000).142,000 Natl. Network: CNN Radio. Format: News/talk, sports. Target aud: 45 plus; middle to upper income adults.

Scott City

KSKL(FM)— Nov 9, 1964: 94.5 mhz; 100 kw. Ant 345 ft TL: N38 31 35 W100 34 42. Stereo. Hrs opn: Box 246, 67871. Phone: (620) 872-5345. Fax: (620) 872-5422. Web Site: www.wksradio.com. Licensee: Western Kansas Wireless Inc. Group owner: Robert Ingstad Broadcast Properties (acq 3-26-93; $175,000 with co-located AM; FTR: 4-12-93). Rgnl. Network: Kan. Info. Format: Oldies. ♦Gil Wohler, gen mgr.

KYUL(AM)— Oct 13, 1962: 1310 khz; 500 w-D, 147 w-N. TL: N38 31 35 W100 54 42. Hrs open: Box 246, 67871. Phone: (620) 872-5345. Fax: (620) 872-5422. Licensee: Steckline Communications Inc. (group owner; (acq 11-28-2006; $550,000 with KIUL(AM) Garden City). Population served: 4,100 Rgnl. Network: Kan. Agri., Kan. Info. Format: Relg. Spec prog: Farm 5 hrs wkly. ♦Gil Wohler, gen mgr.

Seneca

KMZA(FM)— Oct 15, 1992: 92.1 mhz; 4.5 kw. 377 ft TL: N39 49 50 W96 02 39. Stereo. Hrs opn: 24
Rebroadcasts KNZA(FM) Hiawatha 90%.
28 S. 4th St., 66538. Phone: (785) 336-6166. Fax: (785) 336-3600. Web Site: www.kmzaseneca.com. Licensee: KNZA Inc. (group owner) Format: Country, loc. News staff: one; News: 10 hrs wkly. Target aud: General. Spec prog: Farm 7 hrs wkly. ♦Greg Buser, pres & gen mgr; Robert Hilton, opns mgr.

Silver Lake

KCVT(FM)— 1996: 92.5 mhz; 6.7 kw. 387 ft TL: N39 08 42 W95 55 37. Hrs open: 534 S. Kansas, Suite 930, Topeka, 66603. Phone: (785) 233-9250. Fax: (785) 233-9260. E-mail: kcvt@bottradionetwork.com Web Site: www.bottradionetwork.com. Licensee: Richard P. Bott II. Group owner: Bott Radio Network Natl. Network: USA. Format: Christian talk. ♦Richard P. Bott II, VP & gen mgr.

Sterling

KSKU(FM)— June 12, 1995: 94.7 mhz; 50 kw. Ant 485 ft TL: N38 13 50 W98 18 53. Hrs open: 24 106 N. Main St., Hutchinson, 67501-5219. Phone: (620) 665-5758. Fax: (620) 665-6655. E-mail: ksku@ourtownusa.com Licensee: Ad Astra Per Aspera Broadcasting Inc. (group owner) Population served: 100,000 Format: CHR. News staff: one; News: 2 hrs wkly. Target aud: 18-54; general. ♦Cliff C. Shank, pres; Aaron West, opns mgr; Mike Hill, VP, stn mgr & sls VP; Cheryl Dinwiddie, gen sls mgr & natl sls mgr.

Topeka

***KBUZ(FM)**— 1994: 90.3 mhz; 11 kw. 840 ft TL: N39 00 19 W96 02 58. Hrs open: Box 2440, Tupelo, MS, 38803. Phone: (785) 272-6191. Fax: (785) 272-2132. Licensee: American Family Association. Group owner: American Family Radio (acq 12-30-94; FTR: 3-20-95). Format: Relg. ♦Bob Faulkner, stn mgr, sls VP & progmg VP; Jennie Crable, prom VP; George McGurk, chief of engrg.

KDVV(FM)—Listing follows KTOP(AM).

***KJTY(FM)**— Aug 31, 1985: 88.1 mhz; 100 kw. Ant 426 ft TL: N39 11 25 W95 39 29. Stereo. Hrs opn: 24 1005 S.W. 10th St., 66604. Phone: (785) 357-8888. Fax: (785) 357-0100. E-mail: joy88@joy88.org Licensee: Family Life Broadcasting Inc. (acq 3-27-2007; grpsl). Natl. Network: USA. Format: Relg. News: 15 hrs wkly. Target aud: 25-49. Spec prog: Children 5 hrs wkly. ♦Randy L. Carlson, pres; Tom P. Bush, gen mgr; Anthony Barber, progmg mgr.

KMAJ(AM)— July 1947: 1440 khz; 5 kw-D, 1 kw-N, DA-1. TL: N39 01 17 W95 34 15. Hrs open: 825 S. Kansas Ave., 66603. Phone: (785) 272-2122. Fax: (785) 272-6219. Web Site: www.kmaj.com. Licensee: Cumulus Licensing Corp. Group owner: Cumulus Media Inc. (acq 7-31-98; grpsl). Population served: 155,500 Format: News/talk, sports. Target aud: General. ♦Kevin Klein, gen mgr; Chris Rundel, progmg dir.

KMAJ-FM— July 1, 1971: 107.7 mhz; 100 kw. Ant 1,214 ft TL: N39 01 34 W95 54 58. Stereo. Hrs opn: 825 S. Kansas Ave., Suite 100, 66603. Phone: (785) 272-2122. Fax: (785) 272-6219. Web Site: www.kmaj.com. Licensee: Kansas City Trust LLC, Trustee (acq 3-3-2006). Population served: 200,000 Format: Adult contemp. ♦Kevin Klein, gen mgr; Rose Diehl, progmg dir.

KTOP(AM)— July 1947: 1490 khz; 1 kw-U. TL: N39 04 39 W95 40 46. Hrs open: 825 S. Kansas Ave., 66612. Phone: (785) 272-2122. Fax: (785) 272-6219. Licensee: Cumulus Licensing Corp. Group owner: Cumulus Media Inc. (acq 7-31-98; grpsl). Population served: 125,011 Natl. Network: CNN Radio, Jones Radio Networks. Format: Nostalgia. Target aud: 45 plus. ♦Kevin Klein, gen mgr.

KDVV(FM)—Co-owned with KTOP(AM). May 29, 1960: 100.3 mhz; 100 kw. 984 ft TL: N38 57 15 W95 54 43. Stereo. Web Site: www.v100rocks.com.125,011 Natl. Network: Westwood One. Format: AOR. Target aud: 18-54.

KTPK(FM)— Nov 25, 1974: 106.9 mhz; 100 kw. Ant 1,210 ft TL: N39 01 34 W95 55 01. Stereo. Hrs open: 24 2121 S.W. Chelsea, 66614. Phone: (785) 273-1069. Phone: (785) 297-1069. Fax: (785) 273-0123. E-mail: Jallangm@countrylegends1069.com Web Site: www.countrylegends1069.com. Licensee: JMJ Broadcasting Co. Inc. (acq 11-24-2004; $5.7 million). Population served: 512,000 Natl. Rep: Katz Radio. Format: Country. News staff: 2; News: 6 hrs wkly. Target aud: 25-64; mobile, family-oriented, high-income professional adults. ♦Herbert W. McCord, pres; Jim Allan, gen mgr; Bill Kentling, sls dir; Michael Newman, prom dir; Chris Fisher, adv VP & progmg dir; Trevor Kirkwood, mus dir; Megan Kirkwood, traf mgr; Roy Baum, engr.

KWIC(FM)— Oct 15, 1993: 99.3 mhz; 6 kw. 292 ft TL: N39 02 56 W95 40 32. Hrs open: 24 825 S. Kansas Ave., 66603. Phone: (785) 272-2122. Fax: (785)2726219. Web Site: www.eagle993.com. Licensee: Cumulus Licensing Corp. Group owner: Cumulus Media Inc. (acq 4-13-01; with KQTP(FM) Saint Marys). Format: Classic rock. ♦Kevin Klein, gen mgr.

WIBW(AM)— May 8, 1927: 580 khz; 5 kw-U, DA-N (ST-KKSU). TL: N39 05 05 W95 46 58. Hrs open: 24 1210 S.W. Executive Dr., 66615. Phone: (785) 272-3456. Fax: (785) 228-7282. Web Site: www.580radio.com. Licensee: Morris Communications Corp. Group owner: Morris Communications Inc. (acq 12-22-97; grpsl). Population served: 485,700 Natl. Network: ABC. Law Firm: Wiley, Rein & Fielding. Format: News/talk, sports. News staff: 4. Target aud: 25 plus. Spec prog: Relg 6 hrs wkly. ♦Michael Osterhaus, VP; Larry Riggins, gen mgr; Michelle Hay, gen sls mgr; Amber Rome, prom dir; Bruce Steinbrock, progmg dir; Liz Montano, news dir; Roy Baum, chief of engrg.

WIBW-FM— Sept 1, 1961: 94.5 mhz; 100 kw. Ant 1,161 ft TL: N39 01 34 W95 55 01. Stereo. 24 Web Site: www.94country.com. Format: Country. News staff: 4; News: 4 hrs wkly. Target aud: 25-54. ♦Keith Montgomery, progmg dir; Stephanie Lynn, mus dir.

Ulysses

KULY(AM)— Mar 1, 1965: 1420 khz; 1 kw-D, 500 w-N, DA-N. TL: N37 14 28 W101 21 49. Hrs open: 24 Box 1420, 67880. Phone: (620) 276-2366. Fax: (620) 356-3635. Licensee: KBUF Partnership. Group owner: Robert Ingstad Broadcast Properties. Population served: 5,779 Natl. Network: Westwood One. Rgnl. Network: Kan. Info., Kan. Agri. Law Firm: Fisher, Wayland, Cooper, Leader & Zaragoza. Format: Country. News staff: one; News: 24 hrs wkly. Target aud: 21-65; middle to upper class workers, farmers & housewives. Spec prog: Farm 12 hrs, Sp 3 hrs wkly. ♦Gil Wohler, gen mgr.

Wamego

KHCA(FM)— Mar 6, 1986: 95.3 mhz; 6 kw. 328 ft TL: N39 12 35 W96 21 05. Stereo. Hrs opn: 18 Box 1471, Manhattan, 66505. Secondary address: 103 N. 3rd, Manhattan 66502. Phone: (785) 537-9595. Fax: (785) 537-2955. E-mail: angel95@kansas.net Licensee: KHCA Inc. (acq 9-18-91; $126,000; FTR: 10-7-91). Population served: 90,000 Natl. Network: Salem Radio Network. Format: Christian, adult contemp, rock. ♦Jerry Hutchinson, pres & gen mgr.

Wellington

KLEY(AM)— Nov 19, 1966: 1130 khz; 250 w-D, 1 w-N. TL: N37 14 28 W97 24 04. Hrs open: 24 338 S. Kley Dr., 67152. Phone: (620) 326-3341. Fax: (620) 326-8512. E-mail: kley@sutv.com Licensee: Johnson Enterprises Inc. (acq 5-1-89; $575,000 with co-located FM; FTR: 3-27-89) Natl. Network: USA. Rgnl. Network: Kan. Info., Kan. Agri. Format: Talk, news. News staff: one; News: 20 hrs wkly. Target aud: General. Spec prog: Farm 10 hrs, relg 4 hrs wkly. ♦E. Gordon Johnson, pres, gen mgr & gen sls mgr; Travis Turner, opns mgr, prom mgr, progmg dir & news dir; Larry Waggoner, engrg dir & chief of engrg.

KWME(FM)—Co-owned with KLEY(AM). Aug 27, 1979: 93.5 mhz; 6 kw. 321 ft TL: N37 14 28 W97 24 04.24 Natl. Network: USA. Format: Oldies. News staff: one. Target aud: 35-64. ♦Travis Turner, progmg mgr.

Wichita

***KCFN(FM)**— Apr 23, 1978: 91.1 mhz; 100 kw. 345 ft TL: N37 48 01 W97 17 50. Stereo. Hrs opn: 24 4000 N. Old Lawrence Rd., 67219. Phone: (316) 831-9111. Fax: (316) 831-9119. E-mail: kcfn@afo.net Web Site: www.kcfn.net. Licensee: American Family Association Group owner: American Family Radio (acq 5-94). Population served: 650,000 Format: Christian. News: 2 hrs wkly. Target aud: 35-65; general. Spec prog: Relg, news/talk. ♦Don Wildmon, chmn; Tim Widmon, pres; Cindy Kreyer, gen mgr, stn mgr & opns dir.

KEYN-FM— October 1968: 103.7 mhz; 95 kw. 859 ft TL: N37 46 37 W97 31 01. Stereo. Hrs opn: 24 2120 N. Woodlawn, Suite 352, 67208. Phone: (316) 685-2121. Fax: (316) 685-3408. Web Site: www.keyn.com. Licensee: Entercom Wichita License LLC. Group owner: Entercom Communications Corp. (acq 2000; grpsl). Population served: 420,000 Natl. Network: ABC. Format: Oldies. News: 5 hrs wkly. Target aud: 25-54; baby boomers. Spec prog: Dr. Demento 2 hrs wkly. ♦Jackie Wise, gen mgr.

KFDI-FM—Listing follows KFTI(AM).

KFH(AM)— Oct 28, 1947: 1240 khz; 637 w-U. TL: N37 43 06 W97 19 05. Hrs open: 2120 N. Woodlawn, Suite 352, 67208. Phone: (316) 685-2121. Fax: (316) 685-3408. E-mail: letters@kfhradio.com Web Site: www.kfhradio.com. Licensee: Entercom Wichita License L.L.C. Group owner: Entercom Communications Corp. (acq 2000; grpsl). Natl. Network: CBS. Natl. Rep: D & R Radio. Format: Hot talk. Target aud: 35 plus; professionals. ♦Jackie Wise, gen mgr.

KFH-FM—Clearwater, July 4, 1995: 98.7 mhz; 50 kw. Ant 492 ft TL: N37 24 11 W97 35 22. Stereo. Hrs opn: 24 2120 N. Woodlawn, Suite 352, 67208. Phone: (316) 685-2121. Fax: (316) 685-3408. E-mail: letters@kfhradio.com Web Site: www.kfhradio.com. Licensee: Entercom Witicha License LLC. Group owner: Entercom Communications Corp. (acq 5-8-00; $2 million). Population served: 494,569 Format: Sports, talk. Target aud: 18-49; general. ♦Jackie Wise, gen mgr; Steve McIntosh, sls dir & news dir; Mark Yearout, gen sls mgr & mus dir; Tony Duesing, progmg VP & progmg dir; Jessie Yearout, pub affrs dir & traf mgr; Craig Maudlin, chief of engrg.

KFRM(AM)—See Salina

KFTI(AM)— September 1923: 1070 khz; 10 kw-D, 1 kw-N, DA-N. TL: N37 42 47 W97 19 59. Stereo. Hrs opn: 24 Box 1402, 67201. Secondary address: 4200 N. Old Lawrence Rd. 67219. Phone: (316) 838-9141. Fax: (316) 838-3607. Web Site: www.kfdi.com. Licensee: Journal Broadcast Corp. Group owner: Journal Communications Inc. (acq 6-11-99; grpsl). Population served: 411,000 Natl. Network: ABC. Law Firm: Dow, Lohnes & Albertson. Format: C&W, oldies. News staff: 7. Target aud: 25-54. ♦Beverlee Brannigan, progmg dir; Dan Dillon, news dir; Krysti Bradford, traf mgr; Dugg Collins, disc jockey; Johnny Western, disc jockey; Orin Friesen, disc jockey.

KFDI-FM—Co-owned with KFTI(AM). June 6, 1963: 101.3 mhz; 100 kw. Ant 1,139 ft TL: N37 47 47 W97 31 59. Stereo. 276,554 Format: Modern country. Target aud: 25-54; adults. ♦Dugg Collins, disc jockey; Johnny Western, disc jockey; Orin Friesen, disc jockey.

KGSO(AM)— 1950: 1410 khz; 5 kw-D, 1 kw-N, DA-2. TL: N37 44 05 W97 21 06. Hrs open: 24 1632 S. Maize Rd., 67209. Phone: (316) 721-4407. Fax: (316) 721-8276. Web Site: www.kgso.com. Licensee: Steckline Communications Inc. (acq 7-1-2005; $1.3 million). Population served: 400,000 Natl. Network: ESPN Radio, NBC Radio, USA. Format: Sports. Target aud: Men 25-54. ♦Todd Johnson, gen mgr.

KICT-FM— Apr 28, 1972: 95.1 mhz; 100 kw. 1,026 ft TL: N37 47 58 W97 31 58. Stereo. Hrs opn: 24 4200 N. Old Lawrence Rd., 67219. Phone: (316) 838-9141. Fax: (316) 838-3607. Web Site: www.t95.com. Licensee: Journal Broadcast Corp. Group owner: Journal Broadcast Group Inc. (acq 6-14-99; grpsl). Population served: 720,000 Format: Active rock. News staff: 2; News: 15 hrs wkly. Target aud: 18-44. ♦Rob Burton, gen mgr; Eric McCart, gen sls mgr; Jasen Wituk, rgnl sls mgr; Manny Cowzinski, prom dir; Ray Michaels, progmg dir.

***KMUW(FM)**— Apr 26, 1949: 89.1 mhz; 100 kw. 450 ft TL: N37 45 01 W97 18 12. Stereo. Hrs opn: 24 3317 E. 17th St., 67208. Phone: (316) 978-6789. Fax: (316) 978-3946. E-mail: info@kmuw.org Web Site: www.kmuw.org. Licensee: Wichita State University. Population served: 517,000 Natl. Network: NPR, PRI. Rgnl. Network: Kan. Info., Kan. Pub. Law Firm: Schwartz, Woods & Miller. Wire Svc: AP Format: AAA, jazz, news. News staff: 2; News: 121 hrs wkly. Target aud: General. Spec prog: Gospel 2 hrs, jazz 14 hrs, folk/world 5 hrs, AAA 14 hrs wkly. ♦Mark McCain, gen mgr; Lu Stephens, prom dir; Jon Cyphers, engr.

KNSS(AM)— May 26, 1922: 1330 khz; 5 kw-U, DA-N. TL: N37 42 47 W97 14 51. Stereo. Hrs opn: 24 2120 N. Woodlawn St., Suite 352, 67208-1847. Phone: (316) 685-2121. Fax: (316) 685-3408. Web Site: www.knssradio.com. Licensee: Entercom Wichita License LLC. Group owner: Entercom Communications Corp. (acq 2000; grpsl). Population served: 276,554 Natl. Network: CBS. Wire Svc: Weather Wire Format: News/talk. Target aud: 25-54. ♦Jackie Wise, gen mgr.

KQAM(AM)— 1936: 1480 khz; 5 kw-D, 1 kw-N, DA-2. TL: N37 44 21 W97 16 14. Hrs open: 5610 E. 29th St. N., 67220. Phone: (316) 686-5726. Fax: (316) 686-5728. Web Site: www.radiodisney.com/wichita. Licensee: Radio Disney Group LLC. Group owner: ABC Inc. (acq 7-29-02; $2 million). Population served: 400,000 Natl. Network: Radio Disney. Format: Children. ♦Bob Martin, gen mgr.

KRBB(FM)— Sept 19, 1948: 97.9 mhz; 100 kw. Ant 1,027 ft TL: N37 46 40 W97 30 37. Stereo. Hrs opn: 24 9323 E. 37th St. N., 67226. Phone: (316) 494-6600. Fax: (316) 494-6730. Web Site: www.698fm.com. Licensee: Capstar TX L.P. Group owner: Clear Channel Communications Inc. (acq 8-30-00; grpsl). Population served: 276,554 Natl. Rep: Clear Channel. Format: Adult contemp. News staff: one. Target aud: W 25-54; a 25-54; working & family oriented. Spec prog: Jazz 2 hrs, Sp 3 hrs, love songs 18 hrs wkly.

KSGL(AM)— August 1957: 900 khz; 250 w-D, 28 w-N, DA-2. TL: N37 41 33 W97 22 54. Hrs open: 3337 W. Central, 67203. Phone: (316) 942-3231. Fax: (316) 942-9314. E-mail: ksgl900@aol.com Licensee: Agape Communications Inc. (acq 1977). Population served: 500,000 Natl. Network: USA. Format: Relg, big band. ♦Don Clifford, pres; Norbert Atherton, sr VP; Terry Atherton, gen mgr.

KTHR(FM)— Apr 17, 1967: 107.3 mhz; 100 kw. 884 ft TL: N37 46 37 W97 31 01. Stereo. Hrs opn: 24 9323 E. 37th N., 67226. Phone: (316) 494-6600. Fax: (316) 494-6730. Web Site: www.1073theroad.com. Licensee: Clear Channel Broadcasting Licenses, Inc. Group owner: Clear Channel Communications Inc. (acq 8-30-2000; grpsl). Population served: 276,554 Natl. Rep: Clear Channel. Wire Svc: UPI Format: Classic rock. Target aud: 25-54; Males.

***KYFW(FM)**— Sept 24, 1988: 88.3 mhz; 17 kw. 141 ft TL: N37 40 22 W97 20 08. Stereo. Hrs opn: 24 11530 Carmel Commons Blvd., Charlotte, 28226-3976. Phone: (704) 523-5555. Fax: (316) 788-7883. E-mail: kyfw@bbnradio.org Web Site: www.bbnradio.org. Licensee: Bible Broadcasting Network. (group owner; acq 6-26-89). Format: Christian mus & progmg. Target aud: General. ♦Lowell Davey, pres; Matt Johnson, gen mgr.

KYQQ(FM)—See Arkansas City

***KYWA(FM)**— Mar 25, 1990: 90.7 mhz; 25 kw horiz, 23 kw vert. 335 ft TL: N37 21 53 W97 20 30. Hrs open: 24 110 S. Main St., Suite 1050, 67202-3732. Phone: (316) 436-1091. Fax: (316) 838-0691. E-mail: email@kzzd.org Web Site: www.kzzd.org. Licensee: WAY-FM Media Group Inc. (group owner; acq 4-12-2004; $485,000). Format: Sp, relg talk, CHR, modern rock. News: 10 hrs wkly. Target aud: 25-54; upscale females. ♦Robert D. Augsburg, pres; Dan Wemmer, stn mgr.

KZCH(FM)—See Derby

Winfield

***KBDD(FM)**— 2000: 91.9 mhz; 48 kw. Ant 492 ft TL: N37 22 56 W96 57 20. Hrs open: Box 262550, Baton Rouge, LA, 70826. Secondary address: 8919 World Ministry Ave., Baton Rouge, LA 70810. Phone: (225) 768-3688. Phone: (225) 768-8300. Fax: (225) 768-3724. E-mail: kawikfish@yahoo.com Web Site: www.jsm.org. Licensee: Family Worship Center Church Inc. (group owner; acq 6-10-2004; $1.15 million). Format: Christian. ♦David Whitelaw, COO; Jimmy Swaggart, pres; John Santiago, progmg dir.

KKLE(AM)— Aug 19, 1963: 1550 khz; 250 w-D, 52 w-N. TL: N37 14 21 W97 00 43. Hrs open:
Rebroadcasts KLEY(AM) Wellington 98%.
Box 249, Wellington, 67152. Phone: (620) 221-3341. Fax: (620) 326-8512. Licensee: Johnson Enterprises Inc. (group owner; acq 1990). Population served: 35000 Natl. Network: ESPN Radio, USA. Format: Talk, news/talk, sports. News staff: one; News: 36 hrs wkly. Target aud: General. ♦Gordon Johnson, pres & gen mgr.

KSJM(FM)— 1980: 107.9 mhz; 50 kw. 397 ft TL: N37 14 42 W96 54 19. Stereo. Hrs opn: 24 7701 E. Kellogg, Suite 107, Wichita, 67207. Phone: (316) 612-1079. Fax: (316) 612-1077. E-mail: ksjm@1079jamz.com Web Site: www.1079jamz.com. Licensee: Carter-Sherman Broadcast Group Inc. Group owner: Carter Broadcast Group Inc. (acq 5-17-2004; $900,000). . Population served: 228519 Natl. Rep: McGavren Guild. Law Firm: Bryan Cave. Format: Rhythm and blues, hip hop. News staff: one; News: 10 hrs wkly. Target aud: 18-54. ♦Michael Carter, CEO; Don Sherman, pres & gen mgr; Denise Sherman, exec VP & gen sls mgr; Chris Carter, VP; Jude "Jaz" Zeno, prom dir; Hozie Mack, progmg dir.

KSOK-FM— 1996: 95.9 mhz; 15.2 kw. Ant 420 ft TL: N37 04 32 W96 56 13. Hrs open: 24 334 E. Radio Ln., Arkansas City, 67005. Phone: (620) 442-5400. Fax: (620) 442-5401. E-mail: ksok@ksokradio.com Web Site: www.ksok.com. Licensee: Cowley County Broadcasting Inc. (acq 9-3-02; with KSOK(AM) Arkansas City). Format: Today's best country. Target aud: 22-55; blue collar, middle America, people who have children, are still working, & have mortgages. ♦Marty Mutti, gen mgr; Brian Cunningham, opns mgr; Christy Bursack, prom dir; Jonathan Stevens, news dir.

***KSWC(FM)**— November 1967: 100.3 mhz; 10 w. 70 ft TL: N37 14 42 W96 54 19. Hrs open: 100 College St., 67156. Phone: (620) 229-6351. Web Site: www.sckans.edu. Licensee: Southwestern College. Population served: 11,405 Format: College rock. Target aud: 21 & younger. ♦Tom Jacobs, gen mgr.

Kentucky

Albany

WANY(AM)— Oct 25, 1958: 1390 khz; 1 kw-D. TL: N36 41 54 W85 09 00. Hrs open: Box 400, 42602. Phone: (606) 387-5186. Fax: (606) 387-6595. E-mail: mix1063@hotmail.com Licensee: Pamela Allred dba Albany Broadcasting Co. (acq 11-13-01; with co-located FM). Population served: 1,891 Rgnl. Network: Ky. Agri. Natl. Rep: Keystone (unwired net). Format: Country. Spec prog: Farm 2 hrs, gospel 6 hrs wkly. ♦Randy Speck, gen mgr, prom mgr, progmg dir & news dir; Larry Nelson, chief of engrg.

WANY-FM— Apr 18, 1966: 106.3 mhz; 2.7 kw. 155 ft TL: N36 41 54 W85 09 00.1,891 Format: Country.

Allen

WMDJ-FM— Sept 1, 1984: 100.1 mhz; 1.3 kw. 492 ft TL: N37 35 12 W82 42 57. (CP: 2.6 kw). Stereo. Hrs opn:
Rebroadcasts WMDJ-FM Martin 100%.
Box 1530, Martin, 41649. Secondary address: Old Hwy. Rt. 80, Martin 41649. Phone: (606) 874-8005. Fax: (606) 874-0057. Licensee: Floyd County Broadcasting Co. Inc. (acq 12-84; grpsl; FTR: 12-31-84). Population served: 40,000 Natl. Rep: Katz Radio. Format: Country, oldies. Target aud: 25-65. Spec prog: Solid Gold Saturday nights, gospel. ♦Dale McKinney, pres & gen mgr; Mona Dingus, gen sls mgr.

Annville

WANV(FM)— 2006: 96.7 mhz; 1.85 kw. Ant 499 ft TL: N37 13 24 W84 02 01. Hrs open: Box 1227, Corbin, 40702-5656. Phone: (606) 528-8787. Fax: (606) 528-9824. Licensee: F.T.G. Broadcasting Inc. Format: Oldies. ♦Trevor Grigsby, gen mgr.

Ashland

WCMI(AM)— 1935: 1340 khz; 1 kw-U. TL: N38 28 02 W82 35 50. Hrs open: 24 401 11th St., Suite 200, Huntington, WV, 25701. Phone: (304) 523-8401. Web Site: www.wcmi.am. Licensee: Fifth Avenue Broadcasting Co. Inc. Group owner: Kindred Communications Inc. (acq 1-26-98; with WRVC-FM Catlettsburg). Population served: 24,000 Natl. Network: Air America. Rgnl. Network: Ky. Net. Format: Progressive talk. ♦Mike Kirtner, pres & gen mgr; Cameron Smith, opns VP.

WDGG(FM)— 1948: 93.7 mhz; 100 kw. 741 ft TL: N38 23 14 W82 39 45. Stereo. Hrs opn: 401 11th St., Suite 200, Huntington, WV, 25701. Phone: (304) 523-8401. Fax: (304) 523-4848. Web Site: www.wdgg.fm. Licensee: Fifth Avenue Broadcasting Co. Inc. Group owner: Kindred Communications Inc. (acq 1988). Population served: 30,700 Law Firm: Arent, Fox, Kintner, Plotkin & Kahn. Format: Country. Target aud: 25-49; male. ♦Mike Kirtner, pres & gen mgr.

***WKAO(FM)**—Not on air, target date: unknown: 91.1 mhz; 7 kw. Ant 354 ft TL: N38 25 11 W82 24 06. Hrs open: Box 889, Blacksburg, VA, 24063. Phone: (540) 552-4282. Fax: (540) 951-5282. Web Site: www.parfm.com. Licensee: Positive Alternative Radio Inc. ♦Edward A. Baker, pres.

WRVC-FM—See Catlettsburg

WTCR-FM—See Huntington, WV

Auburn

***WAYD(FM)**— 2005: 88.1 mhz; 1 kw. Ant 371 ft TL: N36 57 37 W86 32 49. Hrs open: WAY-FM, 1012 McEwen Dr., Franklin, TN, 37067. Phone: (615) 261-9293. Fax: (615) 261-3967. E-mail: waym@wayfm.com Web Site: www.wayfm.com. Licensee: WAY-FM Media Group Inc. (group owner). Format: Christian. ♦Matt Austin, gen mgr; Teresa White, dev dir; Jeff Brown, progmg dir.

WBVR-FM— May 1965: 96.7 mhz; 45 kw. Ant 423 ft TL: N36 50 35 W86 15 30. Stereo. Hrs opn: 24 1919 Scottsville Rd., Bowling Green, 42104. Phone: (270) 843-3333. Fax: (270) 843-0454. E-mail: mark@forevercomm.com Web Site: www.beaverfm.com. Licensee: Forever Communications Inc. (group owner); (acq 1984). Population served: 280,000 Natl. Rep: Christal. Law Firm: Kaye, Scholer, LLP. Format: Country. Target aud: 18-54. ♦Christine Hillard, pres & opns dir; Mark Mackey, gen mgr.

Barbourville

WKKQ(FM)—Listing follows WYWY(AM).

WYWY(AM)— Dec 13, 1955: 950 khz; 1 kw-D. TL: N36 50 26 W83 52 16. Hrs open: 222 Daniel Boone Dr., 40906. Phone: (606) 546-4128. Fax: (606) 546-4138. Licensee: Barbourville Community Broadcasting Co. (acq 11-66). Population served: 3,549 Format: Relg, southern gospel. ♦Mildred Engle, pres; Chad Engle, gen mgr; Pat Jordan, opns mgr & traf mgr; Orville Burnett, chief of engrg; Sherry Moore, sls.

WKKQ(FM)—Co-owned with WYWY(AM). Oct 2, 1974: 96.1 mhz; 25 kw. 300 ft TL: N36 51 55 W83 53 55. Stereo. 90,000 Format: Hot adult contemp. Target aud: 25-34. ♦Randy Brock, gen sls mgr; Sean Terrell, disc jockey.

Bardstown

WBRT(AM)— December 1954: 1320 khz; 1 kw-D. TL: N37 49 09 W85 29 10. Hrs open: 106 S. 3rd St., 40004. Phone: (502) 348-3943. Fax: (502) 348-4043. E-mail: wbrt@commonwealthbroadcasting.com Licensee: Central Kentucky Broadcasting Inc. Group owner: Commonwealth Broadcasting Corp. (acq 7-17-2006; $240,000). Population served: 5,800 Rgnl. Network: Ky. Net. Natl. Rep: Rgnl Reps. Format: C&W, info. Target aud: 20 plus. Spec prog: Farm 10 hrs wkly. ♦Kenny Fogle, gen mgr.

Beattyville

WLJC(FM)— May 12, 1965: 102.1 mhz; 1.2 kw. 520 ft TL: N37 36 23 W83 41 16. Hrs open: 219 Radio Stn. Loop, 41311. Phone: (606) 464-3600. Fax: (606) 464-5021. E-mail: wljc@wljc.com Web Site: www.wljc.com. Licensee: Hour of Harvest Inc. Population served: 260,000 Natl. Rep: Rgnl Reps. Format: Relg. ♦Margaret Drake, pres; Jonathan Drake, gen mgr; Kim Mitchell, gen sls mgr & progmg; Alan Mulford, chief of engrg.

Beaver Dam

WAIA(AM)—Licensed to Beaver Dam. See Hartford

WXMZ(FM)—See Hartford

Benton

***WAAJ(FM)**— 1996: 89.7 mhz; 6 kw vert. Ant 298 ft TL: N36 48 31 W88 13 26. Hrs open: Box 281, Hardin, 42048. Secondary address: 219 College St., Harding 42048. Phone: (270) 437-4095. Fax: (270) 437-4098. E-mail: info@hmiradio.com Web Site: www.hmiradio.com. Licensee: Heartland Ministries. Format: Gospel and bluegrass. ♦Darrell Gibson, pres.

WCBL(AM)— Dec 13, 1954: 1290 khz; 5 kw-D. TL: N36 51 30 W88 20 13. Hrs open: 24 Box 387, 1039 Eggners Ferry Rd., 42025. Phone: (270) 527-3102. Fax: (270) 527-5606. E-mail: wcbl@bellsouth.net Licensee: Jim W. Freeland. (acq 11-17-98; with co-located FM). Population served: 25,000 Rgnl. Network: Ky. Net. Format: Talk/sports. News staff: one; News: 7 hrs wkly. Target aud: General. ♦Chris Freeland, gen mgr & gen sls mgr; Sherry Rickman, opns mgr; Gregg Leath, progmg dir; Sam Rickmon, news dir; Shane Freeland, disc jockey.

WCBL-FM— Mar 3, 1966: 99.1 mhz; 3 kw. 298 ft TL: N36 51 30 W88 20 13. (CP: 3.3 kw). Format: Oldies. ♦Chad Winstead, disc jockey.

***WTRT(FM)**— December 1998: 88.1 mhz; 600 w. 253 ft TL: N36 47 53 W88 20 50. Stereo. Hrs opn: 24 Box 281, Hardin, 42048. Secondary address: 219 College St., Harding 42048. Phone: (270) 437-4095. Fax: (270) 437-4098. E-mail: info@hmiradio.com Web Site: www.hmiradio.com. Licensee: Heartland Ministries Format: Christian adult contemp. ♦Darrell Gibson, pres; Jeremy Johnson, stn mgr & progmg dir; Eddie Sheriden, mus dir.

***WVHM(FM)**— June 1989: 90.5 mhz; 8 kw. 351 ft TL: N36 48 31 W88 13 26. Hrs open: Box 281, Hardin, 42048. Secondary address: 219 College St., Hardin 42048. Phone: (270) 437-4095. Fax: (270) 437-4098. E-mail: info@hmiradio.com Web Site: www.hmiradio.com. Licensee: Heartland Ministries. Natl. Network: USA. Format: Southern gospel. Target aud: 18-49. ♦Darrell Gibson, pres; Jeremy Johnson, progmg dir; Eddie Sheridan, mus dir.

Berea

WKXO(AM)— July 18, 1971: 1500 khz; 250 w-D. TL: N37 35 12 W84 18 04. Hrs open: 6 AM-7 PM 128 Big Hill Ave., Richmond, 40475. Phone: (859) 623-1389. Fax: (859) 623-1341. Licensee: Wallingford Communications LLC. Group owner: Wallingford Broadcasting Co. (acq 1999; grpsl). Population served: 10,000 Natl. Network: Jones Radio Networks. Rgnl. Network: Ky. Net. Format: News/talk. News staff: one; News: 3 hrs wkly. Target aud: 21 plus. ♦Kelly Wallingford, gen mgr & stn mgr; Ray White, progmg dir & edit dir.

WLFX(FM)—Co-owned with WKXO(AM). Sept 27, 1990: 106.7 mhz; 1.95 kw. 584 ft TL: N37 30 15 W84 12 58. Stereo. 24 Format: Classic rock. Spec prog: Gospel 12 hrs wkly.

Bowling Green

WBGN(AM)— Nov 24, 1959: 1340 khz; 1 kw-U. TL: N37 00 34 W86 27 09. Hrs open: 24 1919 Scottsville Rd., 42101. Phone: (270) 843-3333. Fax: (270) 783-0454. E-mail: mark@forevercomm.com Web Site: www.1340wbgn.com. Licensee: Forever Communications Inc. (group owner; (acq 2001). Population served: 70,000 Natl. Network: ABC, ESPN Radio. Rgnl. Network: Ky. Net. Natl. Rep: Christal. Law Firm: Kaye, Scholer, LLP. Format: Sports. Target aud: 35-54. ♦Mark Mackey, gen mgr; Chris Idle, opns dir & progmg dir.

***WCVK(FM)**— Apr 22, 1986: 90.7 mhz; 14 kw. 448 ft TL: N37 00 18 W86 31 19. Stereo. Hrs opn: Box 539, 42102. Secondary address: 1407 Scottsville Rd. 42104. Phone: (270) 781-7326. Fax: (270) 781-8005. E-mail: mail@christianfamilyradio.com Web Site: www.christianfamilyradio.com. Licensee: Bowling Green Community Broadcasting Inc. Natl. Network: USA. Format: Relg, MOR, adult contemp Christian. Target aud: 25-54; Christian men & women. Spec prog: Black 2 hrs wkly. ♦Mike Wilson, gen mgr.

WDNS(FM)—Listing follows WKCT(AM).

WGGC(FM)— June 23, 1961: 95.1 mhz; 100 kw. Ant 987 ft TL: N36 54 43 W86 11 21. Hrs open: 24 Box 70163, 42101. Phone: (270) 783-8730. Fax: (270) 783-8665. E-mail: wggc@wggc.com Web Site: www.wggc.com. Licensee: Heritage Communications Inc. (acq 10-20-97; $400,000 for stock). Format: Country. ♦Bill Evans, gen mgr; Darrin Evans, stn mgr.

WKCT(AM)— Nov 1, 1947: 930 khz; 5 kw-D, 500 w-N, DA-N. TL: N37 01 53 W86 26 18. Hrs open: 24 Box 930, 42102. Secondary address: 804 College St. 42101. Phone: (270) 781-2121. Fax: (270) 842-0232. E-mail: alan@wdnsfm.com Web Site: www.93wkct.com. Licensee: Daily News Broadcasting Co. Population served: 221,000 Natl. Network: CBS. Format: News/talk, info. News staff: one; News: 25 hrs wkly. Target aud: 25 plus. ♦Alan Cooper, gen mgr; Chad Young, pres & progmg dir.

WDNS(FM)—Co-owned with WKCT(AM). Mar 12, 1973: 93.3 mhz; 12 kw. 472 ft TL: N36 56 39 W86 15 11. Stereo. 24 Web Site: www.wdnsfm.com. Format: Classic Rock. News staff: one; News: 7 hrs wkly. Target aud: 18-54.

***WKYU-FM**— November 1980: 88.9 mhz; 100 kw. 721 ft TL: N37 05 22 W86 38 05. Stereo. Hrs opn: 24 Western Kentucky Univ., 1906 College Heights Blvd., # 11035, 42101. Phone: (270) 745-5489. Phone: (800) 599-9598. Fax: (270) 745-6272. E-mail: wkyufm@wku.edu Web Site: www.wkyu.org. Licensee: Western Kentucky University. Natl. Network: NPR, PRI. Rgnl. Network: Ky. Net. Law Firm: Leventhal, Senter & Lerman. Format: Class, news. News staff: 3; News: 35 hrs wkly. Target aud: General. ♦Peter Bryant, gen mgr. Co-owned TV: *WKYU-TV affil

***WWHR(FM)**— Aug 18, 1988: 91.7 mhz; 1.3 kw. 10 ft TL: N36 59 00 W86 27 24. Stereo. Hrs opn: 24 1906 College Heights Blvd. #11070, 42101. Phone: (270) 745-5439. Phone: (270) 745-5350. Fax: (270) 745-5835. E-mail: gm@revolution.fm Web Site: www.revolution.fm. Licensee: Western Kentucky University. Population served: 50,000 Format: College progsv. News: 5 hrs wkly. Target aud: Adults 18-34. Spec prog: Black 2 hrs, punk 2 hrs, gothic 2 hrs, loc 2 hrs wkly. ♦Dr. Marjorie Yambor, gen mgr; Michael Dean, stn mgr.

Brandenburg

WMMG(AM)— July 1984: 1140 khz; 250 w-D. TL: N37 59 05 W86 09 24. Hrs open: Box 505, 40108. Secondary address: 1715 Bypass Rd. 40108. Phone: (270) 422-3961. Fax: (270) 422-3464. E-mail: wmmg935@bbtel.com Web Site: www.wmmgradio.com. Licensee: Meade County Communications Inc. Rgnl. Network: Ky. Net. Natl. Rep: Rgnl Reps. Format: Country. Spec prog: Relg 8 hrs wkly. ♦Gwen Blevins, gen mgr.

WMMG-FM— Aug 23, 1972: 93.5 mhz; 3.4 kw. 290 ft TL: N37 59 05 W86 09 24. Stereo. Web Site: www.wmmgradio.com.500,000 Natl. Rep: Rgnl Reps. Target aud: 18-65.

Brodhead

WPNS(FM)—Not on air, target date: unknown: 101.9 mhz; 6 kw. Ant 325 ft TL: N37 30 13.6 W84 19 40. Hrs open: 1717 Dixie Hwy., Suite 650, Fort Wright, 41011. Phone: (859) 331-9100. Licensee: Radioactive LLC. ♦Benjamin L. Homel, pres.

Brownsville

WKLX(FM)— 2000: 100.7 mhz; 8 kw. Ant 584 ft TL: N37 09 17 W86 19 33. Hrs open: Box 457, Glasgow, 42142. Phone: (270) 651-6050. Fax: (270) 651-7666. Licensee: Charles M. Anderson. Law Firm: Leventhal, Senter & Lerman. Format: Adult hits. ♦Darron Steenbergen, gen mgr.

Buffalo

WXAM(AM)— Nov 26, 1974: 1430 khz; 1 kw-D. TL: N37 31 49 W85 42 49. Hrs open: 611 W. Poplar St., Suite C2, Elizabethtown, 42701-2483. Phone: (270) 763-0800. Fax: (270) 769-6349. Licensee: Mark Goodman Productions Inc. (acq 2-1-89; $99,292; FTR: 2-13-89). Natl. Network: ESPN Radio. Format: ESPN radio. ♦Roth Stratton, gen mgr & gen sls mgr.

Burgin

WKYB(FM)—Not on air, target date: unknown: 105.9 mhz; 1.1 kw. Ant 472 ft TL: N37 47 18 W84 42 50. Hrs open: 2351 Sunset Blvd., Suite 170-218, Rocklin, CA, 95765. Phone: (916) 251-1600. Fax: (916) 251-1650. Licensee: Educational Media Foundation. ♦Mike Novak, sr VP.

Burkesville

WKYR-FM— October 1988: 107.9 mhz; 6 kw. Ant 312 ft TL: N36 47 26 W85 22 47. Stereo. Hrs opn: 24 Box 340, 42717. Secondary address: Hwy. 90 E. 42717. Phone: (270) 433-7191. Fax: (270) 433-7195. E-mail: wkyr@mchsi.com Licensee: Cumberland Broadcasting LLC (acq 7-31-2007; $153,375). Natl. Network: ABC, Jones Radio Networks. Rgnl. Network: Ky. News Net. Format: Country. ♦Jessie Crabtree, gen mgr & gen sls mgr.

Burnside

WSEK(FM)—Listing follows WSFE(AM).

WSFE(AM)— Feb 28, 1984: 910 khz; 430 w-D, 115 w-N. TL: N37 01 46 W84 36 28. Hrs open: 24 Box 740, Somerset, 42502. Secondary address: 101 First Radio Ln., Somerset 42503. Phone: (606) 678-5151. Fax: (606) 678-2026. Web Site: www.wsfeam.com. Licensee: Capstar TX L.P. Group owner: Clear Channel Communications Inc. (acq 12-8-2000; grpsl). Format: Talk. News staff: one; News: 5 hrs wkly. Target aud: 25-54. ♦Richard Dills, gen mgr & stn mgr; Jo-Ella Shelly, gen sls mgr; Rod Zimmerman, progmg dir; Jim Mercer, chief of engrg.

WSEK(FM)—Co-owned with WSFE(AM). Aug 17, 1985: 93.9 mhz; 50 kw. 492 ft TL: N37 09 15 W84 27 35. Stereo. E-mail: wsek@clearchannel.com Web Site: www.wsfeam.com. Format: Country.

Cadiz

WKDZ(AM)— Apr 8, 1966: 1110 khz; 1 kw-D. TL: N36 52 57 W87 50 44. Hrs open: 24 Box 1900, 42211-0316. Secondary address: 19 Wooldridge Ln. 42211-0316. Phone: (270) 522-3232. Fax: (270) 522-1110. E-mail: bmann@wkdzradio.com Web Site: www.oldies1480.com. Licensee: Ham Broadcasting Co. Inc. (acq 1-22-91; $200,000 with co-located FM; FTR: 2-11-91) Population served: 100,000 Natl. Network: ABC, CNN Radio, Fox News Radio. Format: Oldies. News staff: 3; News: 18 hrs wkly. Target aud: 25-54. Spec prog: Farm 2 hrs wkly. ♦D.J. Everett III, pres; Beth A. Mann, gen mgr; Alan Watts, news dir.

WKDZ-FM— May 18, 1972: 106.5 mhz; 13.4 kw. 449 ft TL: N36 48 29 W87 38 09. Stereo. 24 E-mail: wkdz@wkdzradio.com Web Site: wkdzradio.com.225,000 Format: Real country. News staff: 2; News: 20 hrs wkly. Target aud: 35-64.

Calvert City

WCCK(FM)— 1993: 95.7 mhz; 960 w. 505 ft TL: N37 04 21 W88 15 04. Hrs open: 24 Box 1116, 42029. Phone: (270) 395-5133. Fax: (270) 395-5231. E-mail: wcck@freelandbroadcasting.com Licensee: Jim Freeland DBA Freeland Broadcasting. Population served: 175,000 Natl. Network: AP Radio. Format: Classic Country. News: 10 hrs wkly. Target aud: 30 plus; professionals and retired. ♦Jim Freeland, CEO, gen mgr & gen sls mgr; Sherry Darnall, opns dir; Greg Leath, progmg dir; Loal D. Cole, chief of engrg.

Campbellsville

***WAPD(FM)**— 1996: 91.7 mhz; 2.323 kw vert. Ant 216 ft TL: N37 19 59 W85 19 53. Hrs open: St. Andrew United Methodist Church, 1001 S. Central Ave., 42718. Phone: (270) 465-7559. Licensee: American Family Association. Group owner: American Family Radio Format: Christian. ♦Linda Collins, gen mgr.

WCKQ(FM)— Dec 1, 1964: 104.1 mhz; 17 kw. 374 ft TL: N37 19 29 W85 18 36. Stereo. Hrs opn: 24 50 Friendship Pike, 42718. Secondary address: Box 1053 42719. Phone: (270) 789-2401. Fax: (270) 789-1450. E-mail: wckq@commonwealthbroadcasting.com Web Site: www.myq104.com. Licensee: CBC of Marion and Taylor Counties Inc. Group owner: Commonwealth Broadcasting Corp. (acq 6-30-97; $720,000 with co-located AM). Population served: 100,000 Natl. Network: ABC, Jones Radio Networks. Natl. Rep: Rgnl Reps. Law Firm: Womble, Carlyle, Sandridge & Rice, PLLC. Format: Hot adult contemp. News staff: one; News: 7 hrs wkly. Target aud: 18-44. ♦Steve Newberry, pres; Barb Smith, stn mgr; Marty Bagby, opns mgr & news dir; Greg Bowen, gen sls mgr; Rob Collins, prom mgr & progmg dir; Mike Graham, chief of engrg.

WTCO(AM)—Co-owned with WCKQ(FM). March 1948: 1450 khz; 1 kw-U. TL: N37 20 07 W85 22 33.24 Phone: (270) 469-9826.23,000 Natl. Network: ESPN Radio. Format: All sports. News staff: one. Target aud: 25-64.

WGRK-FM—See Greensburg

Campton

WCBJ(FM)— 1999: 103.7 mhz; 6 kw. 328 ft TL: N37 44 23 W83 33 59. Hrs open: 129 College St., West Liberty, 41472. Phone: (606) 743-3145. Fax: (606) 743-9557. Licensee: Morgan County Industries Inc. (group owner) Format: Rock. ♦Tina Moore, gen mgr; Sharon Williams, gen sls mgr; Paul Lyons, chief of engrg.

Cannonsburg

WOKT(AM)— December 1987: 1080 khz; 1.8 kw-D. TL: N38 23 10 W82 41 53. Hrs open: Sunrise-sunset 3027 Lester Ln., Ashland, 41102. Phone: (606) 928-3778. Fax: (606) 928-1659. E-mail: wokt@windstream.net Licensee: Big River Radio Inc. Group owner: Baker Family Stations (Positive Radio Group) Population served: 340,000 Format: Christian teaching. Target aud: General. ♦Jeremy Holbrook, stn mgr.

Carlisle

WBVX(FM)— December 1994: 92.1 mhz; 32 kw. Ant 610 ft TL: N38 11 19 W84 22 13. Stereo. Hrs opn: 24 401 W. Main, Suite 301, Lexington, 40507. Phone: (859) 233-1515. Fax: (859) 233-1517. Licensee: L.M. Communications of Kentucky LLC. Group owner: L.M. Communications Inc. (acq 8-17-01; $4.8 million). Population served: 41,000 Format: Classic hits. Target aud: General. ♦Lynn Martin, pres; James MacFarlane, gen mgr.

Carrollton

WIKI(FM)— Apr 12, 1968: 95.3 mhz; 3 kw. 423 ft TL: N38 39 58 W85 16 51. Stereo. Hrs opn: 24 2470 N. State Hwy. 7, North Vernon, IN, 47265. Phone: (812) 346-1927. Fax: (812) 346-9722. Licensee: Star Media Inc. (acq 1999; $550,000). Population served: 400,000 Natl. Network: Jones Radio Networks, CNN Radio. Natl. Rep: Rgnl Reps. Law Firm: Pepper & Corazzini. Format: Country. News: 10 hrs wkly. Target aud: 10-90; general. ♦Marty Pieratt, gen mgr.

Catlettsburg

WRVC-FM— Jan 19, 1972: 92.7 mhz; 3 kw. 298 ft TL: N38 27 58 W82 35 27. Stereo. Hrs opn: 20 401 11th St., Suite 200, Huntington, WV, 25701. Phone: (304) 523-8401. Fax: (304) 523-4848. Web Site: www.wrvc.fm. Licensee: Fifth Avenue Broadcasting Co. Inc. Group owner: Kindred Communications Inc. (acq 7-7-98; with WCMI(AM) Ashland). Format: Rock. ♦Mike Kirtner, pres & gen mgr.

Cave City

WPTQ(FM)— Sept 2, 1988: 103.7 mhz; 13.5 kw. 449 ft TL: N37 06 39 W85 58 41. Stereo. Hrs opn: 24 Box 475, Glasgow, 42142. Secondary address: 113 W. Public Sq., Suite 400, Glasgow 42141. Phone: (270) 651-6060. Fax: (270) 651-7666. E-mail: wptq@commonwealthbroadcasting.com Web Site: www.1037thepoint.net. Licensee: Commonwealth Broadcasting Corp. (acq 11-25-97). Population served: 150,000 Natl. Network: Westwood One. Law Firm: Pepper & Corazzini. Format: Classic rock & the best new rock. News staff: one; News: 7 hrs wkly. Target aud: 25-44. ♦Darren Steenbergen, gen mgr; Kellie Wood, opns mgr.

Central City

WMTA(AM)— Feb 19, 1955: 1380 khz; 500 w-D, 23 w-N. TL: N37 16 34 W87 08 39. Stereo. Hrs open: 24 Box 2463, Evansville, IN, 47728. Phone: (812) 479-5342. Web Site: www.faith1380.com. Licensee: WMTA LLC (acq 2-26-2004; $65,000). Population served: 30,000 Law Firm: Shaw Pittman. Format: Christian. News: 20 hrs wkly. Target aud: General. ♦Gayle Russ, CEO.

WNES(AM)— Jan 1, 1955: 1050 khz; 1 kw-D, 172 w-N. TL: N37 16 09 W87 08 32. Hrs open: Box 471, 42330. Phone: (270) 754-3000. Fax: (270) 754-9484. Licensee: Starlight Broadcasting. (acq 9-28-89). Population served: 33,400 Natl. Network: CBS. Rgnl. Network: Ky. Net. Format: Sports, talk. Spec prog: Farm 7 hrs wkly. ♦Andy Anderson, pres; Jowanna Bandy, gen mgr; Stan Barnett, progmg mgr.

WQXQ(FM)—Co-owned with WNES(AM). Dec 18, 1956: 101.9 mhz; 100 kw. 215 ft TL: N37 16 09 W87 08 32. (CP: Ant 676 ft.). E-mail: wqxq@ocdirect.net Web Site: www.q1019.com.200,000 Format: Hot adult contemp.

Clinton

WLLE(FM)— 1997: 102.1 mhz; 12.5 kw. Ant 476 ft TL: N36 44 22 W88 47 13. Hrs open: Box 2397, Paducah, 42002. Secondary address: 1176 State Rt. 45 N., Mayfield 42066. Phone: (270) 247-5122. Fax: (270) 554-5468. Licensee: Bristol Broadcasting Co. Inc. (group owner; (acq 3-15-2004; grpsl). Format: Country. ♦Gary Morse, gen mgr.

Coal Run

WPKE-FM— Sept 21, 1974: 103.1 mhz; 1.2 kw. Ant 741 ft TL: N37 27 57 W82 33 04. Stereo. Hrs opn: 24 Box 2200, Pikeville, 41502. Secondary address: 1240 Radio Dr., Pikeville 41501. Phone: (606) 437-4051. Fax: (606) 432-2809. E-mail: wdhr@wdhr.com Licensee: East Kentucky Broadcasting Corp. (group owner; acq 6-94; $480,000 with WBPA(AM) Elkhorn City). Population served: 56,000 Natl. Network: ABC. Rgnl rep: Rgnl Reps Law Firm: Womble, Carlyle, Sandridge & Rice. Format: Rock. News staff: one. Target aud: General. ♦Keith Casebolt, gen mgr.

Columbia

WAIN(AM)— Aug 1, 1951: 1270 khz; 1 kw-D, 68 w-N. TL: N37 06 26 W85 16 42. Hrs open: 24 Box 69, 42728. Secondary address: 1521 Liberty Rd. 42728. Phone: (270) 384-2134. Fax: (270) 384-6722. E-mail: wain@keybroadcasting.net Web Site: www.935wain.com. Licensee: Tri-County Radio Broadcasting Corp. Group owner: Key Broadcasting Inc. Population served: 240,000 Natl. Rep: Rgnl Reps. Rgnl rep: Rgnl Reps Format: Oldies 50s, 60s & 70s. News staff: one; News: 8 hrs wkly. Target aud: 16-65. Spec prog: Farm 2 hrs. ♦Gary Phelps, gen mgr; Louise Wooten, gen mgr & gen sls mgr.

WAIN-FM— Mar 1, 1968: 93.5 mhz; 5.2 kw. 220 ft TL: N37 06 36 W85 16 42. Stereo. 24 Web Site: 935wain.com.240,000 Natl. Network: ABC. Format: Country. News staff: one. Target aud: 16-65. ♦Delno Salmon, disc jockey; Lisa Fisher, news dir, farm dir & disc jockey.

WHVE(FM)—See Russell Springs

Corbin

WCTT(AM)— May 9, 1947: 680 khz; 1 kw-U, DA-N. TL: N36 54 09 W84 04 50. Hrs open: 5 AM-midnight Box 742, 40702-0742. Secondary address: 821 Adams Rd. 40701. Phone: (606) 528-4717. Fax: (606) 528-4487. Licensee: Encore Communications Inc. (acq 5-95; with co-located FM; FTR: 6-22-81). Population served: 7,988 Natl. Network: ABC. Rgnl. Network: Ky. Net. Natl. Rep: Rgnl Reps. Law Firm: Shaw Pittman. Format: Oldies, news/talk, MOR. News staff: one. ♦Stephanie Mullins, gen mgr.

WCTT-FM— June 1, 1967: 107.3 mhz; 50 kw. 492 ft TL: N36 54 09 W84 04 55. Stereo. 24 Phone: (606) 528-6617. Natl. Rep: Rgnl Reps. Format: Adult contemp. News: one hr wkly. Target aud: 18-54.

***WEKF(FM)**— June 24, 2003: 88.5 mhz; 21 kw vert. Ant 499 ft TL: N37 01 13 W84 23 41. Stereo. Hrs opn: 24
Rebroadcasts WEKU(FM) Richmond 100%.
102 Perkins Bldg., 521 Lancaster Ave., Richmond, 40475-3102. Phone: (859) 622-1660. Fax: (859) 622-6276. E-mail: wekunews@eku.edu Web Site: www.weku.fm. Licensee: Eastern Kentucky University. Natl. Network: NPR, PRI. Law Firm: Hardy, Carey & Chautin. Wire Svc: AP Format: News, classical. News staff: 3; News: 35 hrs wkly. Target aud: General. ♦Tim Singleton, gen mgr; Mary Ellyn Cain, opns mgr; Carol Siler, dev dir; Laura Allen, progmg dir; Charles Compton, news dir; Bill Browning, chief of engrg.

WKDP(AM)— Nov 23, 1961: 1330 khz; 5 kw-D, DA. TL: N36 56 20 W84 04 44. Hrs open: Box 742, 40702. Secondary address: 821 Adams Rd. 40701. Phone: (606) 528-6617. Fax: (606) 528-4487. E-mail: swaggoner@wkdp.com Licensee: Eubanks Broadcasting Inc. (acq 12-28-89). Population served: 30,000 Natl. Rep: Rgnl Reps. Format: Relg, news/talk. Target aud: 30-64. ♦Dallas R. Eubanks, pres; Stephanie Mullins, gen mgr; Derek Eubanks, chief of engrg.

WKDP-FM— 1967: 99.5 mhz; 50kw. 709 ft TL: N36 57 14 W84 58 41. Stereo. 250,000 Natl. Network: ABC. Law Firm: Fisher, Wayland, Cooper, Leader & Zaragoza. Format: Country. Target aud: General.

Covington

WCVG(AM)— Oct 29, 1965: 1320 khz; 500 w-D, 430 w-N, DA-2. TL: N39 02 44 W84 30 30. Hrs open: 24 Box 14425, Cincinnati, OH, 45250-0425. Phone: (859) 291-2255. Fax: (859) 655-4345. Licensee: Richard L. Plessinger Sr. Group owner: Plessinger Radio Group (R.L. Plessinger Holding Co.) (acq 1987). Population served: 1,500,000 Format: Sp. ♦Tracie Hunter, gen mgr; Jeff Eldred, opns mgr; Avery Corbin, prom dir & prom mgr; John Jones, mus dir.

Cumberland

WCPM(AM)— October 1951: 1280 khl; 1 kw-D. TL: N36 58 25 W82 59 15. Hrs open: 24 hrs a day 101 Keller St., 40823. Phone: (606) 589-4623. Web Site: www.wcpmradio.com. Licensee: Cumberland City Broadcasting Inc. (acq 8-22-2003). Population served: 25,000 Natl. Network: Jones Radio Networks, AP Radio. Format: Country; relg, news. News: 9 hrs wkly. Target aud: 18-49; general. Spec prog: Black one hr, farm one hr wkly. ♦Laura Hewitt, traf mgr; Susan Burton, gen mgr, sls & progmg.

WVEK-FM— December 1994: 102.7 mhz; 175 w. Ant 1,824 ft TL: N36 54 59 W82 54 02. Hrs open: 24 Rt. 2, Box 2114, Jonesville, VA, 24263. Phone: (276) 346-2000. Fax: (276) 346-2049. Licensee: JBL Broadcasting Inc. (acq 5-24-2005; $265,000). Population served:

150,000 Natl. Network: Jones Radio Networks. Format: Classic rock. News: 4 hrs wkly. Target aud: 18-49; young adults. ♦Regina Moore, gen mgr & adv mgr.

Cynthiana

WCYN(AM)— Sept 1, 1956: 1400 khz; 500 w-D, 1 kw-N. TL: N38 24 20 W84 17 32. Hrs open: 111 Court St., 41031. Phone: (859) 234-1400. Fax: (859) 234-1425. Web Site: www.wcyn.com. Licensee: WCYN Broadcasting Inc. (acq 12-29-2004; $122,000). Population served: 16,000 Rgnl. Network: Ky. Agri., Ky. Net. Natl. Rep: Keystone (unwired net), Rgnl Reps. Format: Oldies. Target aud: General. ♦Chris Winkle, gen mgr.

WCYN-FM— June 1, 1970: 102.3 mhz; 3.4 kw. Ant 400 ft TL: N38 24 39 W84 19 07. Hrs open:
Rebroadcasts WLXX(FM) Lexington 100%.
300 W, Vine St., Suite 3, Lexington, 40507. Phone: (859) 253-5900. Fax: (859) 253-5940. Web Site: www.wlxxthebear.com. Licensee: Cumulus Licensing LLC. (acq 11-26-2002). Population served: 16,000 Format: Country. ♦Dave Kabakoff, gen mgr.

Danville

***WDFB-FM**— Sept 1992: 88.1 mhz; 170 w. 328 ft TL: N37 35 46 W84 50 19. Hrs open: 24 Box 106, 40423-0106. Secondary address: 3596 Alum Springs Rd. 40422. Phone: (859) 236-9333. Fax: (859) 236-3348. E-mail: wdfb@searnet.com Web Site: www.wdfb.com. Licensee: Alum Springs Educational Corp. (acq 6-8-92). Natl. Network: USA. Format: Educ, Christian. Target aud: General. ♦Donald A. Drake, pres; Jim Gaskin, gen sls mgr; Mildred Drake, exec VP, gen mgr & progmg dir; Cindy Pike, traf mgr.

WHIR(AM)— Oct 27, 1947: 1230 khz; 1 kw-U. TL: N37 40 28 W84 46 06. Hrs open: 24 2063 Shakertown Rd., 40422. Phone: (859) 236-2711. Fax: (859) 236-1461. E-mail: hometownradio@bellsouth.net Web Site: www.hometownlive.net. Licensee: Hometown Broadcasting of Danville Inc. (acq 1995; $525,000 with co-located FM). Population served: 200,000 Natl. Network: Westwood One. Natl. Rep: Rgnl Reps. Format: News/talk. News staff: one; News: 2 hrs wkly. Target aud: 25-54; business owners, sports fans, housewives. ♦Bruce Leslie, pres; Robert Wagner, gen mgr; Jim Parman, opns dir & progmg dir; Vicki Hyde, news dir.

***WLAI(FM)**— Oct 27, 1969: 107.1 mhz; 4.9 kw. Ant 157 ft TL: N37 40 28 W84 46 06. (CP: 4.4 kw, ant 387 ft. TL: N37 45 40 W84 44 46). Stereo. Hrs opn: 24
Rebroadcasts KLRD(FM) Yucaipa, CA 100%.
2351 Sunset Blvd., Suite 170-218, Rocklin, CA, 95765. Phone: (916) 251-1600. Fax: (916) 251-1650. Web Site: www.air1.com. Licensee: Educational Media Foundation. (group owner; (acq 5-20-2005; $1 million). Population served: 200,000 Natl. Network: Air 1. Format: Alternative, Christian. ♦Richard Jenkins, pres; Mike Novak, VP; Keith Whipple, dev dir; David Pierce, progmg mgr; Ed Lenane, news dir; Sam Wallington, engrg dir; Karen Johnson, news rptr; Marya Morgan, news rptr; Richard Hunt, news rptr.

Drakesboro

WNTC(FM)— 2001: 103.9 mhz; 1.95 kw. Ant 407 ft TL: N37 06 50 W87 03 52. Hrs open: 2514 Eugenia Ave., Nashville, TN, 37211. Phone: (615) 844-1039. Phone: (615) 251-1222. Fax: (615) 313-9933. Licensee: Nashville's SportsRadio Inc. Group owner: Southern Wabash Communications Corp. (acq 10-17-01). Population served: 150,000 Format: Mexican. ♦Randolph V. Bell, pres; Wayne DeSylvia, gen mgr & stn mgr.

Eddyville

WWLK(AM)— May 2, 1981: Stn currently dark. 900 khz; 1 kw-D, 120 w-N, DA-2. TL: N37 04 26 W88 04 48. Hrs open: Box 90, 42038. Licensee: Tilent Inc. (acq 7-20-89; $65,000; FTR: 8-14-89). Format: Relg. ♦Jim Baggett, pres.

Edmonton

WHSX(FM)— Apr 5, 1990: 99.1 mhz; 3 kw. 328 ft TL: N37 01 33 W85 33 14. Stereo. Hrs opn: 24 Box 85, Horse Cave, 42749. Phone: (270) 786-1000. Phone: (270) 432-7991. Fax: (270) 786-4402. Fax: (866) 999-hoss. E-mail: 991@scrtc.com Web Site: www.thehoss.com. Licensee: Hart County Communications Inc. (acq 7-5-2001; $350,000). Population served: 50,000 Format: Country. News staff: one; News: 10 hrs wkly. Target aud: 25-54. Spec prog: Farm 15 hrs wkly. ♦Dewayne Forbis, gen mgr.

Elizabethtown

WIEL(AM)— Oct 1, 1950: 1400 khz; 1 kw-U. TL: N37 41 11 W85 52 19. Hrs open: 24 611 W. Poplar St., 42701. Phone: (270) 763-0800. Fax: (270) 769-6349. Licensee: Elizabethtown CBC Inc. Group owner: Commonwealth Broadcasting Corp. (acq 5-12-00; grpsl). Population served: 150,000 Natl. Network: ABC. Natl. Rep: Rgnl Reps. Wire Svc: AP Format: ESPN, sports. News staff: one; News: one hr wkly. Target aud: 24-54; upscale adult. ♦Roth Stratton, gen mgr; Holli Lee, traf mgr; Dan Michaels, opns.

WKMO(FM)—See Hodgenville

***WKUE(FM)**— Oct 15, 1990: 90.9 mhz; 5.2 kw. 633 ft TL: N37 44 46 W85 53 18. Stereo. Hrs opn: 24
Rebroadcasts WKYU-FM Bowling Green 100%.
Western Kentucky Univ., 1906 College Heights Blvd., Bowling Green, 42101. Phone: (270) 745-5489. Phone: (800) 599-9598. Fax: (270) 745-6272. E-mail: wkyufm@wku.edu Web Site: www.wkufm.org. Licensee: Western Kentucky University. Natl. Network: NPR, PRI. Rgnl. Network: Ky. Net. Law Firm: Leventhal, Senter & Lerman. Format: Class, news. News staff: 3; News: 30 hrs wkly. Target aud: General. Spec prog: Jazz 15 hrs, folk 5 hrs wkly. ♦Peter Bryant, gen mgr.

WQXE(FM)— Nov 24, 1969: 98.3 mhz; 8.5 kw. Ant 531 ft TL: N37 43 18 W86 02 10. Stereo. Hrs opn: 233 W. Dixie Ave., 42701. Phone: (270) 737-8000. Fax: (270) 737-7229. E-mail: bill@wqxe.com Web Site: www.wqxe.com. Licensee: Skytower Communications E'town Inc. Population served: 100,000 Natl. Network: Westwood One. Format: Hot adult contemp. Target aud: 25-54; upscale, dual income families. ♦Bill Evans, pres & gen mgr.

Elkhorn City

WEKB(AM)— Nov 24, 1979: 1460 khz; 5 kw-D, 114 w-N. TL: N37 18 25 W82 19 53. Hrs open: 24
Simulcast with WPKE(AM) Pikeville 100%.
Box 2200, Pikeville, 41502. Secondary address: 1240 Radio Dr., Pikeville 41501. Phone: (606) 437-4051. Fax: (606) 432-2809. E-mail: wdhr@wdhr.com Licensee: East Kentucky Broadcasting Corp. (group owner; (acq 6-94; $480,000 with co-located FM). Population served: 55,000 Natl. Network: ABC. Rgnl. Network: Ky. Net. Rgnl rep: Rgnl Reps Law Firm: Womble, Carlyle, Sanridge & Rice. Format: Oldies. Target aud: 25-49. ♦Keith Casebolt, gen mgr.

Elkton

WEKT(AM)— July 21, 1977: 1070 khz; 500 w-D. TL: N36 48 33 W87 09 38. Hrs open: Box 577, 42220. Phone: (270) 265-5636. E-mail: wektan1070@yahoo.com Licensee: M&R Broadcasting Inc. (acq 1-22-98; $55,000 for 50% of stock). Population served: 200,000 Natl. Network: USA. Format: Southern gospel.

Eminence

WTSZ(AM)— June 1, 1956: 1600 khz; 500 w-D, 48 w-N. TL: N38 21 02 W85 11 11. Hrs open: 304 W. Liberty St., Suite 110, Louisville, 40202. Phone: (502) 583-1212. Fax: (502) 583-1414. Web Site: www.wszradio.com. Licensee: Metro East CBC Inc. Group owner: Commonwealth Broadcasting Corp. (acq 4-13-00; $600,000 with WTSZ-FM Eminence). Population served: 16,000 Rgnl. Network: Ky. News Net. Format: Sports. Target aud: 25-54. ♦Scott Thompson, gen mgr.

WTUV-FM— July 4, 1988: 105.7 mhz; 3 kw. Ant 325 ft TL: N38 21 09 W85 11 09. Hrs open: 4109 Bardstown Rd., Suite 104, Louisville, 40218. Phone: (502) 583-6200. Fax: (502) 587-2979. E-mail: info@lacalienteradio.com Web Site: www.lacalienteradio.com. Licensee: Davidson Media Station WTSZ Licensee LLC. Group owner: Commonwealth Broadcasting Corp. (acq 5-3-2006; $500,000). Population served: 452,524 Format: Rgnl Mexican. ♦Paul Dendy, gen mgr; Dennis Mendez, progmg dir.

Erlanger

WIZF(FM)— Sept 22, 1965: 101.1 mhz; 2.5 kw. Ant 508 ft TL: N39 06 18 W84 33 25. Stereo. Hrs opn: 705 Central Ave., Suite 200, Cincinnati, OH, 45202. Phone: (513) 679-6000. Fax: (513) 679-6014. Web Site: www.wize.com. Licensee: Blue Chip Broadcasting Licenses Ltd. Group owner: Radio One Inc. (acq 4-30-2001; grpsl). Population served: 10,000 Format: Urban mainstream. Target aud: 18-54. ♦Alfred Wiggins, CEO; Lisa Thal, gen mgr.

Falmouth

WIOK(FM)— June 1981: 107.5 mhz; 6 kw. 695 ft TL: N38 43 15 W84 22 27. Stereo. Hrs opn: 24 Box 50, 41040. Phone: (859) 472-1075. Fax: (859) 472-2875. E-mail: wiok@fuse.net Web Site: www.wiok.com. Licensee: Hammond Broadcasting Inc. (acq 1993). Population served: 3,100,000 Natl. Network: USA. Natl. Rep: Rgnl Reps. Format: Southern gospel. News: 12 hrs wkly. Target aud: 25-64; women. ♦Jan Hammond, VP; Jamie Porter, gen sls mgr.

Flemingsburg

WFLE(AM)— November 1981: 1060 khz; 1 kw-D, DA. TL: N38 27 01 W83 44 06. Hrs open: 334 Recreation Park Rd., 41041. Phone: (606) 849-4433. Fax: (606) 845-9353. Licensee: DreamCatcher Communications Inc. (group owner; acq 5-23-02; $607,491 with co-located FM). Format: Gospel, country. Target aud: 25-54. ♦Don Bowles, pres; Carl Haight, gen mgr; Kim Hester, gen sls mgr; Eddie Plummer, prom mgr; Brent Mullikin, progmg dir.

WFLE-FM— February 1993: 95.1 mhz; 1.61 kw. 449 ft TL: N38 24 42 W83 34 41. Format: Country.

Florence

WDJO(AM)— September 1984: 1160 khz; 5 kw-D, 990 w-N, DA-2. TL: N38 58 09 W84 40 56. Hrs open: 24 635 W. 7th St., Suite 400, Cincinnati, OH, 45203. Phone: (513) 533-2500. Fax: (513) 533-2527. Web Site: www.oldies1160.com. Licensee: Christian Broadcasting System Ltd. Group owner: Salem Communications Corp. (acq 2-10-2006; swap of WDJO(AM) and WCVX(AM) Cincinnati, OH plus $6.75 million cash for WLQV(AM) Detroit, MI). Population served: 1,800,000 Format: Oldies. News: 6 hrs wkly. Target aud: 25-54; 35-64; adults. ♦Jon R. Yinger, pres; Brian Kauffman, gen mgr; Rodger Kay, opns mgr; Michael Gavin, gen sls mgr; Dusty Rhodes, progmg mgr.

Fort Campbell

WCVQ(FM)—Listing follows WJQI(AM).

WJQI(AM)— July 27, 1963: 1370 khz; 1 kw-D, 53 w-N. TL: N36 38 28 W87 26 04. Hrs open: 24 1640 Old Rossellville Pike, Clarksville, TN, 37043. Phone: (931) 431-4984. Fax: (931) 648-7769. E-mail: q108@q108.com Web Site: www.q108.com. Licensee: Saga Communications of Tuckessee L.L.C. Group owner: Saga Communications Inc. (acq 2-1-2001; grpsl). Population served: 180,000 Format: Adult contemp. News staff: one; News: 14 hrs wkly. Target aud: 18-34. Spec prog: Gospel 10 hrs, relg 4 hrs wkly. ♦Scott Farkas, pres; Susan Quesenberry, gen mgr; Lee Erwin, prom dir; J.C. Morrow, chief of engrg.

WCVQ(FM)—Co-owned with WJQI(AM). August 1969: 107.9 mhz; 100 kw. 950 ft TL: N36 32 23 W87 39 45. Stereo. 24 Phone: (931) 648-7720. E-mail: q108@q108.com Web Site: www.q108.com.198,000 News staff: one; News: 2 hrs wkly. Target aud: 25-40. ♦Lee Erwin, prom mgr.

Fort Knox

WLVK(FM)— Oct 1, 1967: 105.5 mhz; 6 kw. 299 ft TL: N37 46 57 W85 54 38. Stereo. Hrs opn: 24 Box 2087, Elizabethtown, 42702. Secondary address: 519 N. Miles St., Elizabethtown 42702. Phone: (270) 766-1035. Fax: (270) 769-1052. E-mail: rbell@wase.org Web Site: www.bigcat1055.com. Licensee: W & B Broadcasting Co. Inc. (acq 6-26-2000; $900,000). Population served: 120,000 Law Firm: Miller & Miller. Format: Country. News staff: 14; News: 4 hrs wkly. Target aud: 25-49; young & middle age country fans. Spec prog: Lou Helton Country Countdown. ♦Bill Walters, pres; Rene Bell, gen mgr; Cale Tharp, opns mgr & chief of engrg.

Fort Thomas

WYGY(FM)— Apr 18, 1994: 97.3 mhz; 2.55 kw. Ant 508 ft TL: N39 12 01 W84 31 22. Stereo. Hrs opn: 24 2060 Reading Rd., Cincinnati, OH, 45202. Phone: (513) 699-5959. Fax: (513) 699-5000. Web Site: www.wygy.com. Licensee: CBS Radio Stations Inc. Group owner: Infinity Broadcasting Corp. (acq 11-13-98; grpsl). Population served: 395,000 Law Firm: Tierney & Swift. Format: Country. News: 8 hrs wkly. ♦Mike Fredrick, gen mgr; Rob McClacken, gen sls mgr; Jeff Nagel, prom dir.

Frankfort

WFKY(AM)— February 1946: 1490 khz; 1 kw-U. TL: N38 12 46 W84 52 31. Hrs open: 24 115 W. Main St., 40601. Phone: (502) 875-1130. Fax: (502) 875-1225. Web Site: www.wfky.com. Licensee: CC Licenses LLC. Group owner: Clear Channel Communications Inc. (acq 8-27-2001; grpsl). Population served: 75,000 Natl. Network: Westwood One. Natl. Rep: Rgnl Reps. Format: Oldies. News staff: 2; News: 12 hrs wkly. Target aud: 25-55. ♦John Roberts, gen mgr.

WKYW(FM)—Co-owned with WFKY(AM). Jan 1, 1967: 104.9 mhz; 3 kw. 300 ft TL: N38 13 19 W84 54 55. Stereo. 24 Web Site: www.capitalcountry 1049.com.150,000 Format: Country. News staff: 2; News: 168 hrs wkly. Target aud: 25-54.

WKED-FM— Apr 15, 1991: 103.7 mhz; 2.5 kw. 350 ft TL: N38 13 17 W84 54 52. Stereo. Hrs opn: 24 115 W. Main St., 40601-2807. Phone: (502) 875-1130. Fax: (502) 875-1225. Web Site: www.star1037.com. Licensee: CIC Licenses LLC. Group owner: Clear Channel Communications Inc. (acq 8-27-2001; grpsl). Population served: 42,000 Natl. Network: Westwood One. Format: Adult contemp. News staff: 2; News: 4 hrs wkly. Target aud: 25-54. ♦John Roberts, gen mgr.

Franklin

WFKN(AM)— Apr 25, 1954: 1220 khz; 250 w-D, 90 w-N. TL: N36 44 20 W86 34 42. Hrs open: 24 103 N. High St., 42135-0390. Phone: (270) 586-4481. Fax: (270) 586-6031. E-mail: wfkn@franklinfavorite.com Licensee: WFKN LLC (acq 11-27-01). Population served: 25,000 Rgnl. Network: Ky. Net. Natl. Rep: Rgnl Reps. Format: Country. News staff: 2; News: 16 hrs wkly. Target aud: General. Spec prog: Relg, farm 6 hrs wkly. ♦Jamie Johnson, gen mgr; Shelly Jent, news dir.

Fredonia

WKEN(FM)—Not on air, target date: unknown: 92.1 mhz; 6 kw. Ant 328 ft TL: N37 15 22 W88 01 49. Hrs open: 194 Godfrey Rd., Edgewater, FL, 32141. Phone: (386) 690-2200. Licensee: Patricia Van Zandt. ♦Patricia Van Zandt, gen mgr.

Fulton

WFUL(AM)— July 8, 1951: 1270 khz; 1 kw-D, 54 w-N. TL: N36 30 54 W88 54 16. Stereo. Hrs opn: 8807 Middle Rd. State Rt. 166 E., 42041. Phone: (270) 472-1270. Fax: (270) 472-1189. Licensee: River County Broadcasting Inc. (acq 8-23-2004; $350,000). Population served: 3,250 Format: Country, gospel. Target aud: 40 plus. ♦Max McDade, gen mgr.

WWKF(FM)— September 1954: 99.3 mhz; 3.3 kw. 337 ft TL: N36 27 59 W88 56 47. Stereo. Hrs opn: 24 1729 Nailling Dr., Union City, TN, 38261. Phone: (731) 885-1240. Fax: (731) 885-3405. Web Site: www.kf99kq105.com. Licensee: WENK of Union City Inc. Group owner: WENK Broadcast Group Inc. (acq 10-1-82; $473,131; FTR: 10-18-82). Population served: 73,500 Natl. Rep: Rgnl Reps. Law Firm: Shainis & Peltzman. Format: CHR. News staff: one. Target aud: 18-34. ♦Terry L. Hailey, pres, gen mgr & prom mgr.

Garrison

WOKE(FM)— Sept 7, 1998: 98.3 mhz; 5.2 kw. 492 ft TL: N38 36 19 W83 03 37. Hrs open: 492 Main St., South Shore, 41175. Phone: (606) 932-2223. Fax: (606) 932-6132. E-mail: info@wokejoyfm.org Web Site: www.wokejoyfm.org. Licensee: Big River Radio Inc. (acq 10-25-94; FTR: 11-14-94). Law Firm: Booth, Freret, Imlay & Tepper. Format: Southern gospel. ♦Paul Hunt, gen mgr.

Georgetown

***WRVG(FM)**— Oct 1, 1963: 89.9 mhz; 50 kw. Ant 410 ft TL: N38 12 15 W84 32 51. Hrs open: 24 2351 Sunset Blvd., Suite 170-218, Rocklin, CA, 95765. Phone: (916) 251-1600. Fax: (916) 251-1650. Web Site: www.klove.com. Licensee: Educational Media Foundation. Group owner: EMF Broadcasting (acq 3-12-2004; $1.7 million). Population served: 260,000 Format: Christian. ♦Richard Jenkins, pres & progmg dir; Mike Novak, VP; Keith Whipple, dev dir; David Pierce, progmg mgr; Ed Lenane, news dir; Sam Wallington, engrg dir; Arthur Vassar, traf mgr; Karen Johnson, news rptr; Marya Morgan, news rptr; Richard Hunt, news rptr.

WXRA(AM)— Sept 6, 1957: 1580 khz; 10 kw-D, 45 w-N, DA-2. TL: N38 10 05 W84 35 37. Hrs open: 2601 Nicholasville Rd., Lexington, 40503. Phone: (859) 422-1000. Fax: (859) 422-1071. Web Site: www.sunny1580.com. Licensee: Citicasters Licenses L.P. Group owner: Clear Channel Communications Inc. (acq 6-30-97; grpsl). Population served: 300,000 Natl. Rep: Christal. Format: Sp. News: 2 hrs wkly. Target aud: 45-70. ♦Gene Guinn, gen mgr.

WXZZ(FM)— Sept 10, 1973: 103.3 mhz; 1.8 kw. Ant 607 ft TL: N38 02 07 W84 27 02. Stereo. Hrs opn: 300 W. Vine St., Suite 3, Lexington, 40507. Phone: (859) 253-5900. Fax: (859) 253-5940. Web Site: www.zrock103.com. Licensee: Cumulus Licensing Corp. Group owner: Cumulus Media Inc. (acq 7-22-99; grpsl). Format: Rock. News staff: 3. Target aud: 18-34. ♦Dave Kabakoff, gen mgr & prom dir.

Glasgow

WCDS(AM)—Listing follows WOVO(FM).

WCLU(AM)— Sept 25, 1946: 1490 khz; 1 kw-U. TL: N37 00 19 W85 54 42. Hrs open: Box 1628, 42142. Phone: (270) 651-9149. Fax: (270) 651-9222. Web Site: www.wcluradio.com. Licensee: Royse Radio Inc. Natl. Network: CBS. Format: Full service. Target aud: 30 plus; listeners with disposable income. ♦Henry Royse, pres & gen mgr.

WLYE-FM— 1997: 94.1 mhz; 4.5 kw. Ant 298 ft TL: N36 59 02 W85 52 20. Stereo. Hrs opn: 24 1919 Scottsville Rd., Bowling Green, 42104-3303. Phone: (270) 843-3333. Fax: (270) 843-0454. E-mail: mark@forevercom.com Licensee: Forever Communications Inc. (group owner; acq 9-3-03). Population served: 125,000 Natl. Rep: Christal. Law Firm: Kaye, Scholer, LLP. Format: Country. News staff: one; News: 6 hrs wkly. Target aud: 25-54; adults, serving southern central Kentucky. ♦Christine Hillard, pres; Mark Mackey, gen mgr.

WOVO(FM)— July 14, 1972: 105.3 mhz; 25 kw. Ant 318 ft TL: N36 54 50 W85 43 20. Stereo. Hrs opn: Box 457, 42142-0457. Secondary address: 113 W. Public Sq., Suite 400 42141. Phone: (270) 651-6050. Fax: (270) 651-7666. E-mail: wovo@cbcradio.net Licensee: Newberry Broadcasting Inc. Group owner: Commonwealth Broadcasting Corp. (acq 11-25-97; grpsl). Population served: 500,000 Format: Oldies. ♦Derron Steenbergen, gen mgr & gen sls mgr; Kelly McKay, progmg dir & disc jockey.

WCDS(AM)—Co-owned with WOVO(FM). Oct 1, 1962: 1440 khz; 5 kw-D. TL: N36 58 10 W85 56 24. E-mail: wcds@cbsradio.net Licensee: Newberry Broadcasting Inc.250,000 Format: Sports.

***WSGP(FM)**— 2002: 88.3 mhz; 13 kw. Ant 298 ft TL: N36 49 05 W85 41 30. Hrs open: Secondary address: 93 Rainbow Terr., Somerset 42503. Phone: (606) 679-6300. Fax: (606) 679-1342. Web Site: www.kingofkingsradio.net. Licensee: Somerset Educational Broadcasting Foundation. Format: Conservative, traditional relg & edu. ♦David Carr, gen mgr; Carolyn Jones, progmg dir; Marvin Whitaker, chief of engrg.

WWKU(AM)—Not on air, target date: 2007: Stn currently dark. 1230 khz; 750 w-U. TL: N37 00 17 W85 56 27. Hrs open: 1519 Euclid Ave., Bowling Green, 42103. Phone: (270) 782-0246. Licensee: Anderson Communications LLC. ♦Charles M. Anderson, gen mgr.

Grayson

WGOH(AM)— June 1, 1959: 1370 khz; 5 kw-D, 21 w-N. TL: N38 19 44 W82 58 33. Hrs open: 6 AM-2 hrs past sunset Box 487, 150 Radio Tower Dr., 41143. Phone: (606) 474-5144. Fax: (606) 474-7777. E-mail: mail @wgohwugo.com Web Site: www.wgohwugo.com. Licensee: Carter County Broadcasting Co. Natl. Network: CBS. Natl. Rep: Rgnl Reps. Law Firm: Booth, Freret, Imlay & Tepper. Wire Svc: AP Format: Classic country. News staff: one; News: 30 hrs wkly. Target aud: 35-65. ♦Francis M. Nash, gen mgr; Jeff Roe, opns dir & opns mgr; Melodie Carter, progmg dir & traf mgr; Mike Phillips, mus dir; Jim Phillips, news dir; William H. Craig, chief of engrg & engr.

WUGO(FM)—Co-owned with WGOH(AM). February 1967: 102.3 mhz; 4.8 kw. 360 ft TL: N38 19 44 W82 58 33. Stereo. 24 Web Site: www.wgohwugo.com. Format: Adult contemp. News: 30 hrs wkly. Target aud: 25-54.

Greensburg

WGRK(AM)— Mar 15, 1972: 1540 khz; 1 kw-D. TL: N37 15 34 W85 30 57. Hrs open: Box 1053, Campbellsville, 42719. Secondary address: 50 Friendship Pike, Campbellsville 42719. Phone: (270) 932-7401. Phone: (270) 789-1464. Fax: (270) 789-1450. E-mail: wgrk@commonwealthbroadcasting.com Web Site: www.103kcoutry.com. Licensee: Green County CBC Inc. (acq 10-30-97; $600,000 with co-located FM). Population served: 11,000 Natl. Network: ABC, Jones Radio Networks. Natl. Rep: Rgnl Reps. Format: Country. News staff: one; News: 7 hrs wkly. Target aud: 25-54. Spec prog: Farm 5 hrs wkly. ♦Steve Newberry, pres; Barb Smith, stn mgr; Marty Bagby, opns mgr; Greg Bowen, gen sls mgr; Trent Ford, progmg dir; Mike Graham, chief of engrg.

WGRK-FM— Dec 15, 1977: 103.1 mhz; 2.2 kw. 375 ft TL: N37 15 34 W85 30 57. (CP: 4.6 kw). Stereo. 24 20,000 Law Firm: Womble, Carlyle, Sandridge & Rice, PLLC. Format: Mainstream country hits. News staff: one.

Greenup

WLGC-FM— Sept 1, 1982: 105.7 mhz; 12.5 kw. Ant 466 ft TL: N38 35 44 W82 51 22. Stereo. Hrs opn: 1401 Winchester Ave., Ashland, 41101. Phone: (606) 920-9565. Fax: (606) 920-9523. E-mail: wlgc@inet99.net Licensee: Greenup County Broadcasting Inc. Rgnl. Network: Ky. Net. Wire Svc: AP Format: Country. News staff: one; News: 3 hrs wkly. Target aud: 25-54; middle income listeners. ♦Bob Hall, gen mgr; Scott Martin, gen sls mgr; Mark Justice, progmg dir.

WLGC(AM)— Apr 1, 1985: 1520 khz; 5 kw-D. TL: N38 35 44 W82 51 20. Stereo. Format: Sports/talk, gospel.

Greenville

WKYA(FM)— Dec 11, 1981: 105.5 mhz; 3 kw. 300 ft TL: N37 11 45 W87 12 38. Stereo. Hrs opn: 464 St. Rt. 189 S., 42345. Phone: (270) 338-6655. Fax: (270) 338-7388. Licensee: Starlight Broadcasting Co. (group owner; acq 1996; grpsl). Rgnl. Network: Ky. Net. Format: Good time oldies. Target aud: 18-40. ♦Andy Anderson, gen mgr.

Hardinsburg

WULF(FM)— July 9, 1970: 94.3 mhz; 43 kw. 290 ft TL: N37 45 40 W86 26 22. (CP: Ant 525 ft.). Stereo. Hrs opn: 233 W. Dixie Ave., Elizabethtown, 42701. Phone: (270) 765-0943. Fax: (270) 737-7229. E-mail: jodie@wqxe.com Licensee: Skytower Communications - 94.3 LLC (acq 12-12-01; $1.15 million). Population served: 100,000 Format: Country. ♦Bill Evans, pres & gen mgr.

WXBC(FM)— Aug 15, 1992: 104.3 mhz; 3 kw. 328 ft TL: N37 45 12 W86 26 08. Stereo. Hrs opn: 24 Box 104, 40143. Secondary address: 110 S. Main St. 40143. Phone: (270) 756-1043. Fax: (270) 756-1086. E-mail: wxbc@bbtel.com Licensee: Breckinridge Broadcasting Co. Inc. Population served: 50,000 Rgnl. Network: Ky. Net. Law Firm: Booth, Freret, Imlay & Tepper. Format: Classic hit country & today's hits. News staff: one; News: 15 hrs wkly. Target aud: 25-55. ♦Jo Ann Keenan, CEO, pres, CFO & gen mgr; Dennis Day, chief of opns.

Harlan

WFSR(AM)— April 1976: 970 khz; 5 kw-D, 94 w-N. TL: N36 50 59 W83 23 41. Hrs open: Box 818, 40831-0818. Secondary address: 125 S. Main 40831. Phone: (606) 573-1470. Fax: (606) 573-1473. E-mail: wtuk-wfsr@harlanonline.net Licensee: Eastern Broadcasting Co. (acq 5-26-98; $400,000 with co-located FM). Population served: 18,000 Rgnl. Network: Ky. Net. Format: Southern gospel. Target aud: 25-54; adult purchasers. ♦Jeff Capps, gen mgr, gen sls mgr, chief of engrg & local news ed.

WTUK(FM)—Co-owned with WFSR(AM). June 26, 1991: 105.1 mhz; 270 w. 1,037 ft TL: N36 54 09 W83 18 01. Format: Country.

WHLN(AM)— May 30, 1941: 1410 khz; 5 kw-D, 94 w-N. TL: N36 52 02 W83 19 36. Hrs open: Box 898, 40831. Secondary address: 100 Eversole St., Suite 1 40831. Phone: (606) 573-2540. Fax: (606) 573-7557. E-mail: whln@harlanonline.net Licensee: Radio Harlan Inc. (acq 6-1-56). Population served: 200,000 Natl. Network: ABC. Natl. Rep: Rgnl Reps. Wire Svc: AP Format: Adult Contemp. Target aud: 25-54. ♦James T. Morgan, pres; James O. Morgan, VP & gen mgr.

Harold

WXLR(FM)— January 1994: 104.9 mhz; 370 w. 922 ft TL: N37 31 59 W82 29 40. Hrs open: Box 1049, 41635. Phone: (606) 478-1200. Fax: (606) 478-4202. E-mail: prod@mikrotec.com Web Site: www.thedoublex.com. Licensee: Adam D. Gearheart. Format: New country. ♦Adam D. Gearheart, pres & gen mgr.

Harrodsburg

WHBN(AM)— June 25, 1955: 1420 khz; 1 kw-D, 46 w-N. TL: N37 44 03 W84 48 50. Hrs open: 24 2063 Shakertown Rd., Danville, 40422. Phone: (859) 236-2711. Fax: (859) 236-1461. E-mail: hometownradio@bellsouth.net Web Site: www.hometownlive.net. Licensee: Hometown Broadcasting of Harrodsburg Inc. (acq 1-4-01). Population served: 200,000 Natl. Network: Jones Radio Networks. Rgnl. Network: Ky. Net. Rgnl rep: Rgnl Reps. Format: Country, gospel. News staff: one; News: 21 hrs wkly. Target aud: General; residents of Mercer county. ♦Robert Wagner, gen mgr, sls dir & prom mgr; Jim Parman, opns mgr & progmg dir; Vicki Hyde, news dir.

Hartford

WAIA(AM)—Beaver Dam, June 21, 1969: 1600 khz; 1 kw-D. TL: N37 26 36 W86 53 57. Hrs open: 6 AM-sunset
Rebroadcasts WKYA(FM) Greenville 100%.
Box 106, 42347. Phone: (270) 298-3268. Phone: (270) 298-3269. Fax: (270) 298-9326. Licensee: Starlight Broadcasting Co. (group owner; acq 1996; grpsl). Population served: 20,000 Rgnl. Network: Ky. News Net. Natl. Rep: Rgnl Reps. Format: News/talk, sports. Target aud: General. ♦Andy Anderson, pres, gen mgr, opns mgr, gen sls mgr, progmg dir & chief of engrg.

WXMZ(FM)—Co-owned with WAIA(AM). May 18, 1972: 106.3 mhz; 3 kw. 280 ft TL: N37 26 36 W86 53 57. Stereo. 16
Rebroadcasts WKYA(FM) Greenville 100%.
Web Site: www.wxmzfm.com. Format: Oldies.

Hawesville

WKCM(AM)— Nov 7, 1972: 1160 khz; 2.5 kw-D, 1 kw-N, DA-N. TL: N37 54 20 W86 45 30. Stereo. Hrs opn: 24 1115 Tamarack Rd., Suite 500, Owensboro, 42301. Phone: (270) 683-5200. Fax: (270) 688-0108. E-mail: dpowers@cromwellradio.com Web Site: www.owensbororadio.com. Licensee: Hancock Communications Inc. Group owner: The Cromwell Group Inc. Population served: 1,262 Rgnl. Network: Ky. Agri. Format: Real country. News staff: one; News: 10 hrs wkly. Target aud: 25-54; general. Spec prog: Farm 3 hrs, sports 6 hrs wkly. ♦Bayard H. Walters, pres; Dale Powers, gen mgr; Jeff Morgan, progmg dir; Amy Spalding, traf mgr; Jeff Nalley, farm dir.

WXCM(FM)—Co-owned with WKCM(AM). May 1993: 97.1 mhz; 4 kw. Ant 403 ft TL: N37 41 50 W86 59 28. Web Site: www.owensbororadio.com. Licensee: The Cromwell Group Inc. of Kentucky. (acq 1993; $170,000; FTR: 9-6-93). Format: Rock. Target aud: Males in 30's. ♦Bayard H. Walters, CEO; Kevin Riecke, opns mgr; Amy Spalding, traf mgr; Jeff Nalley, farm dir.

Hazard

***WEKH(FM)**— February 1985: 90.9 mhz; 33 kw. 1,004 ft TL: N37 11 34 W83 11 16. Stereo. Hrs opn: 24
Rebroadcasts WEKU-FM Richmond 100%.
102 Perkins Bldg., 521 Lancaster Ave., Richmond, 40475-3102. Phone: (859) 622-1660. Fax: (859) 622-6276. E-mail: wekunews@eku.edu Web Site: www.weku.fm. Licensee: Board of Regents, Eastern Kentucky University. Natl. Network: NPR, PRI. Law Firm: Hardy, Carey & Chautin. Wire Svc: AP Format: Class, news magazine, info. News staff: 3; News: 35 hrs wkly. Target aud: General. ♦Tim Singleton, gen mgr & stn mgr; Mary Ellyn Cain, opns mgr; Carol Siler, dev dir; Laura Allen, progmg dir; Charles Compton, news dir; Bill Browning, chief of engrg.

WJMD(FM)— July 26, 1989: 104.7 mhz; 2.5 kw. 1,135 ft TL: N37 11 36 W83 11 04. Stereo. Hrs opn: 24 125 Main St., 41701. Phone: (606) 439-3358. Fax: (606) 439-3371. E-mail: wjmd@alltel.net Web Site: www.wjmdfm.com. Licensee: Hazard Broadcasting Services (acq 4-13-01; $250 for 25%). Natl. Network: Salem Radio Network. Format: Relg. News staff: 2; News: 7 hrs wkly. Target aud: General. ♦Michael R. Barnett, gen mgr.

WKIC(AM)— Nov 23, 1947: 1390 khz; 5 kw-D. TL: N37 14 19 W83 12 41. Stereo. Hrs opn: Box 7428, 41702. Secondary address: 516 Main St. 41701. Phone: (606) 436-2121. Fax: (606) 436-4172. Licensee: Mountain Broadcasting Service Inc. (acq 12-67). Population served: 5,459 Natl. Network: Westwood One. Rgnl. Network: Ky. Net. Natl. Rep: Rgnl Reps. Format: Adult standards. ♦Faron Sparkman, gen mgr, gen sls mgr & gen sls mgr.

WSGS(FM)—Co-owned with WKIC(AM). Feb 3, 1959: 101.1 mhz; 100 kw. 1,463 ft TL: N37 11 38 W83 10 52. Stereo. Web Site: www.wsgs.com.1,000,000 Natl. Network: ABC. Format: Country.

WQXY(AM)— Mar 1, 1988: 1560 khz; 1 kw-D, 500 w-CH, DA. TL: N37 16 27 W83 11 29. Stereo. Hrs opn: Box 864, Hindman, 41822. Phone: (606) 785-6129. Fax: (606) 785-0106. Licensee: Black Gold Broadcasting. (acq 8-90; $97,500; FTR: 9-24-90). Natl. Network: Jones Radio Networks, AP Radio, CNN Radio. Format: Oldies. Target aud: 25-54; educated, mobile, child-rearing couples in suburbs, blue collar workers. ♦Randy Thompson, gen mgr.

Henderson

WGBF-FM— Dec 1, 1971: 103.1 mhz; 6 kw. 460 ft TL: N37 46 54 W87 37 24. (CP: 3.16 kw, ant 453 ft.). Stereo. Hrs opn: 24 1133 Lincoln Ave., Evansville, IN, 47714. Phone: (812) 425-4226. Fax: (812) 421-0005. Web Site: www.103gbfrocks.com. Licensee: Regent Broadcasting of Evansville/Owensboro Inc. Group owner: Regent Communications Inc. (acq 12-3-2003; grpsl). Population served: 185,000 Natl. Network: ABC. Format: Rock/AOR. News staff: one. Target aud: 18-49. ♦Mark Thomas, gen mgr; Bobby Gates, prom dir; Mike Sanders, progmg dir.

WKDQ(FM)— 1947: 99.5 mhz; 100 kw. 944 ft TL: N37 49 36 W87 33 00. Stereo. Hrs opn: 24 1133 Lincoln N.E., Evansville, IN, 47714. Phone: (812) 425-4226. Web Site: www.wkdq.com. Licensee: Regent Broadcasting of Evansville/Owensboro Inc. Group owner: Regent Communications Inc. (acq 2-25-03; grpsl). Population served: 250,000 Natl. Rep: Christal. Format: Country. News staff: 2. Target aud: 25-54. ♦Lori Tevault, prom dir; Jon Prell, progmg dir.

***WKPB(FM)**— Apr 1, 1990: 89.5 mhz; 43 kw. 377 ft TL: N37 51 06 W87 19 43. Stereo. Hrs opn: 24
Rebroadcasts WKYU-FM Bowling Green 100%.
Western Kentucky Univ., 1906 College Heights Blvd., Bowling Green, 42101. Phone: (270) 745-5489. Phone: (800) 599-9598. Fax: (270) 745-6272. E-mail: wkyufm@wku.edu Web Site: www.wkyufm.org. Licensee: Western Kentucky University. Natl. Network: NPR, PRI. Rgnl. Network: Ky. Net. Law Firm: Leventhal, Senter & Lerman. Format: Class, news. News staff: 3; News: 35 hrs wkly. Target aud: General. ♦Peter Bryant, gen mgr.

WSON(AM)— Dec 17, 1941: 860 khz; 500 w-U, DA-N. TL: N37 51 11 W87 32 12. Stereo. Hrs opn: 24 Box 418, 42419-0418. Secondary address: 230 2nd St. 42420. Phone: (270) 826-3923. Fax: (270) 826-7572. Licensee: Henry G. Lackey (acq 7-31-79). Population served: 30,000 Rgnl. Network: Ky. Net. Natl. Rep: Rgnl Reps. Format: Adult standards. News staff: one; News: 8 hrs wkly. Target aud: 35+. Spec prog: Farm 2 hrs wkly. ♦Henry G. Lackey, pres & gen mgr; Bill Stephens, news dir.

Highland Heights

***WNKU(FM)**— Apr 29, 1985: 89.7 mhz; 12 kw. 318 ft TL: N39 02 21 W84 27 57. Stereo. Hrs opn: 21 (M-F); 20 (S); 19 (Su) 301 Landrum Academic, 41099. Phone: (859) 572-6500. Fax: (859) 572-6604. E-mail: wnku@nku.edu Web Site: www.wnku.org. Licensee: Northern Kentucky University. Natl. Network: PRI, NPR. Law Firm: Arter & Hadden. Format: AAA, news. News staff: 2; News: 40 hrs wkly. Target aud: 35-49. ♦Aaron Sharpe, dev dir; Grady Kirkpatrick, progmg dir.

Hindman

WKCB(AM)— Jan 26, 1971: 1340 khz; 6 kw-U. TL: N37 19 45 W83 00 17. Hrs open: Box 864, 41822. Secondary address: 1517 Hwy. 550 W. 41822. Phone: (606) 785-3129. Fax: (606) 785-0106. Web Site: www.wkcb.com. Licensee: Hindman Broadcasting Corp. (acq 9-15-89; $100,000 with co-located FM; FTR: 10-23-89) Population served: 18,200 Rgnl. Network: Ky. Net. Natl. Rep: Rgnl Reps. Format: Christian. ♦Randy Thompson, pres & gen mgr; Paul Hoskins, progmg mgr & news dir.

WKCB-FM— Dec 13, 1974: 107.1 mhz; 770 w. 650 ft TL: N37 19 56 W82 56 52. Stereo. Web Site: www.wkcb.com.65,000 Format: Heart of rock.

Hodgenville

WKMO(FM)— March 1974: 106.3 mhz; 3 kw. 400 ft TL: N37 40 21 W85 44 34. Stereo. Hrs opn: 611 W. Poplar St., Suite C2, Elizabethtown, 42701. Phone: (270) 763-0800. Fax: (270) 769-6349. Web Site: www.1063thebear.com. Licensee: Elizabethtown CBC Inc. Group owner: Commonwealth Broadcasting Corp. (acq 7-1-00; grpsl). Population served: 98,000 Natl. Rep: Keystone (unwired net), Rgnl Reps. Law Firm: Verner, Liipfert, Bernhard, McPherson & Hand. Format: Mainstream country. Target aud: 24-65. Spec prog: Farm 2 hrs wkly. ♦Steve Newberry, pres; Dale Thornhill, VP; Roth Stratton, gen mgr.

WXAM(AM)—See Buffalo

Hopkinsville

WHOP(AM)— Jan 8, 1940: 1230 khz; 830 w-U. TL: N36 52 54 W87 30 44. Hrs open: 24 Box 709, 220 Dink Embry's Buttermilk Rd., 42241-0709. Phone: (270) 885-5331. Fax: (270) 885-2688. E-mail: whopamfm@bellsouth.net Web Site: www.lite987whop.com. Licensee: Hop Broadcasting Inc. (acq 10-28-99; with co-located FM). Population served: 190,000 Natl. Network: CBS. Rgnl. Network: Ky. Net. Natl. Rep: Rgnl Reps. Format: News/talk. News staff: 2. Target aud: 30 plus. ♦Mike Chadwell, gen mgr.

WHOP-FM— May 1948: 98.7 mhz; 100 kw. Ant 620 ft TL: N36 55 41 W87 32 50. Stereo. 24 Web Site: www.lite987whop.com.300,000 Format: Adult contemp. News staff: 2; News: 15 hrs wkly. Target aud: 18-54.

WHVO(AM)— Sept 19, 1954: 1480 khz; 1 kw-D, 24 w-N. TL: N36 52 15 W87 30 43. Hrs open: 24 Oldies Radio, Box 1900, Cadiz, 42211-1900. Phone: (270) 886-1480. Fax: (270) 886-6286. E-mail: oldies@oldies1480.com Web Site: www.oldies1480.com. Licensee: Ham Broadcasting Inc. (acq 10-95). Population served: 100,000 Natl. Network: AP Network News, Jones Radio Networks, Fox News Radio. Natl. Rep: Rgnl Reps. Format: Oldies. News staff: two; News: 3 hrs wkly. Target aud: 35-54; Upscale Baby-boomers. Spec prog: Relg 6 hrs, gospel 3 hrs wkly. ♦D.J. Everett, pres; Beth Mann, gen mgr; Bill Booth, opns mgr; Amy Berry, gen sls mgr; Alan Watts, news dir.

***WNKJ(FM)**— Aug 3, 1981: 89.3 mhz; 12 kw. 330 ft TL: N36 48 34 W87 24 20. Stereo. Hrs opn: 24 Box 1029, 42241-1029. Secondary address: 1100 E. 18th St. 42240. Phone: (270) 886-9655. Fax: (270) 885-7210. E-mail: wnkj@wnkj.org Web Site: www.wnkj.org. Licensee: Pennyrile Christian Community Inc. Natl. Network: Moody. Format: Christian. News: 12 hrs wkly. Target aud: General. Spec prog: Black 5.5 hrs, Korean one hr, Sp one hr wkly. ♦Jim Dozier Adams, gen mgr.

WVVR(FM)— July 1, 1960: 100.3 mhz; 100 kw. 1,000 ft TL: N36 56 58 W87 40 18. Stereo. Hrs opn: 24 1640 Old Russellville Pike, Clarksville, TN, 37043. Phone: (931) 648-7720. Fax: (931) 648-7769. Web Site: www.thebeaver.com. Licensee: Saga Communications of Tuckessee LLC. Group owner: Saga Communications Inc. (acq 11-27-00; $7 million). Population served: 1,100,000 Rgnl rep: Rgnl Reps. Format: Country. News staff: one; News: 7 hrs wkly. Target aud: 18-54; working class. ♦Susan Quesenberry, gen mgr.

WZZP(FM)— Feb 28, 2001: 97.5 mhz; 6 kw. Ant 328 ft TL: N36 45 47 W87 26 59. Hrs open: 1640 Old Russellville Pike, Clarksville, TN, 37043. Phone: (931) 648-7720. Fax: (931) 648-7769. Web Site: www.z975.com. Licensee: Saga Communications of Tuckessee L.L.C. Group owner: Saga Communications Inc. (acq 2-1-01; grpsl). Format: Rock. ♦Susan Quesenberry, gen mgr.

Horse Cave

WHHT(FM)— Sept 19, 1994: 106.7 mhz; 2.9 kw. Ant 476 ft TL: N37 13 57 W85 52 06. Hrs open: Box 457, Glasgow, 42142-0457. Secondary address: 113 W. Public Sq., Suite 400, Glasgow 42141. Phone: (270) 651-6050. Fax: (270) 651-7666. Licensee: Commonwealth Broadcasting Corp. (acq 11-25-97; grpsl). Natl. Network: Westwood One. Format: Adult contemp. ♦Derron Steenbergen, gen mgr & stn mgr; Kellie Wood, opns mgr.

Hyden

WZQQ(FM)— Nov 7, 1988: 97.9 mhz; 1.75 kw. 1,207 ft TL: N37 11 36 W83 11 04. Stereo. Hrs opn: Box 7280, Hazard, 41702. Secondary address: 516 Main St., Hazard 41701. Phone: (606) 436-9898. Fax: (606) 436-4172. E-mail: wzqq@alltel.net Licensee: Leslie County Broadcasting Inc. (acq 4-3-01; $50 for 50%). Natl. Network: ABC. Format: Hot adult contemp, CHR. Target aud: General. ♦Stuart Shane Sparkman, CEO; Mike Reeves, gen mgr & gen sls mgr; Bob Hale, chief of engrg.

Inez

WBTH(AM)—See Williamson, WV

WXCC(FM)—See Williamson, WV

Irvine

WCYO(FM)—Listing follows WIRV(AM).

WIRV(AM)— July 2, 1960: 1550 khz; 1 kw-D. TL: N37 42 26 W83 58 15. (CP: TL: N37 42 57 W83 58 29). Hrs opn: 128 Big Hill Ave., Richmond, 40475. Phone: (859) 623-1386. Fax: (859) 623-1241. Licensee: Kentucky River Broadcasting Co Inc. Population served: 100,000 Rgnl. Network: Ky. Net. Natl. Rep: Rgnl Reps. Format: Oldies. ♦Kelly T. Wallingford, gen mgr; Ray White, gen sls mgr & progmg dir.

WCYO(FM)—Co-owned with WIRV(AM). August 1991: 100.7 mhz; 9.2 kw. 505 ft TL: N37 39 40 W84 08 55.24 200,000 Format: Country.

Jackson

WEKG(AM)— Mar 7, 1969: 810 khz; 5 kw-D. TL: N37 34 41 W83 24 19. Hrs opn: 1501 Hargas Ln., 41339. Phone: (606) 666-7531. Fax: (606) 666-4946. Licensee: Intermountain Broadcasting Co. Population served: 100,000 Rgnl. Network: Ky. Net. Format: Gospel. ♦Doug Neace, gen mgr & gen sls mgr; Gloria Hay, gen mgr.

WJSN-FM—Co-owned with WEKG(AM). Jan 1, 1979: 97.3 mhz; 19 kw. Ant 813 ft TL: N37 40 19 W83 24 21. Stereo. 50,000 Format: Country.

Jamestown

WJKY(AM)— Sept 3, 1967: 1060 khz; 1 kw-D. TL: N37 01 31 W85 04 23. Hrs open: Box 800, 42629. Secondary address: 2804 South US Hwy 127, Russell Springs 42642. Phone: (270) 866-3487. Phone: (270) 343-4444. Fax: (270) 866-2060. Web Site: www.lakercountry.com. Licensee: Lake Cumberland Broadcasters. (acq 7-1-70). Population served: 50,000 Format: Country. ♦Mae Hoover, gen mgr & gen sls mgr.

WJRS(FM)—Co-owned with WJKY(AM). Sept 3, 1966: 104.9 mhz; 2 kw. 360 ft TL: N37 01 31 W85 04 23.

Jeffersontown

WMJM(FM)— Dec 1, 1978: 101.3 mhz; 2 kw. 194 ft TL: N38 13 42 W85 38 22. Stereo. Hrs opn: 520 S. 4th St., Suite 200, Louisville, 40202. Phone: (502) 625-1220. Fax: (502) 625-1259. Web Site: www.1013online.com. Licensee: Blue Chip Broadcasting Licenses Ltd. Group owner: Radio One Inc. (acq 4-30-2001; grpsl). Format: Urban adult contemp. ♦Dale Schaefer, gen mgr.

Jenkins

WIFX-FM— May 10, 1975: 94.3 mhz; 4.2 kw. 1,565 ft TL: N37 06 38 W82 44 18. Hrs open: 24 Box 1049, Harold, 41635. Phone: (606) 478-1200. Fax: (606) 478-4202. E-mail: psod@foxy943.com Licensee: Letcher County Broadcasting Inc. (acq 7-1-93; $37,000; FTR: 7-26-93). Population served: 250,000. Law Firm: Henry Crawford. Format: Adult rock. News: 2 hrs wkly. Target aud: 25-45. ♦Barry Boyd, gen mgr.

WKVG(AM)— Feb 1, 1970: 1000 khz; 1 kw-D. TL: N37 09 59 W82 37 13. Hrs open: 7:30 AM- 6 PM Box 613, Pound, VA, 24279-0613. Secondary address: Box 1474 41537. Phone: (606) 832-4655. Fax: (606) 832-4656. Licensee: Martins and Assoc. Inc. (acq 6-15-92; $40,000; FTR: 6-7-92). Population served: 10,000 Format: Gospel, relg. News: 9 hrs wkly. Target aud: General. ♦Emma Jean Martin, gen mgr.

Junction City

WDFB(AM)— May 20, 1985: 1170 khz; 1 kw-D, DA. TL: N37 35 46 W84 50 19. Hrs open: Sunrise-sunset Box 106, Danville, 40423-0106. Secondary address: 3596 Alum Springs Rd., Danville 40422. Phone: (859) 236-9333. Fax: (859) 236-3348. E-mail: wdfb@searnet.com Web Site: www.wdfb.com. Licensee: Alum Springs Vision and Outreach Corp. Natl. Network: USA. Format: Relg. Target aud: General. ♦Donald A. Drake, pres; Mildred Drake, exec VP & gen mgr; Cindy Pike, traf mgr.

Keavy

***WVCT(FM)**— January 1984: 91.5 mhz; 100 w. 341 ft TL: N36 58 21 W84 07 28. Stereo. Hrs opn: 24 968 W. City Dam Rd., 40737. Phone: (606) 528-4671. Fax: (606) 526-0589. E-mail: csivley@bellsouth.net Licensee: Victory Training School Corp. Format: Gospel. ♦Charles Sivley, gen mgr.

Keene

WJMM-FM— Dec 9, 1969: 99.1 mhz; 4.5 kw. Ant 384 ft TL: N37 57 37 W84 32 42. Stereo. Hrs opn: 24 3270 Blazer Pkwy., Suite 101, Lexington, 40509. Phone: (859) 264-9700. Fax: (859) 264-9705. Web Site: www.wjmm.com. Licensee: Christian Broadcasting System Ltd. (acq 7-1-2006; grpsl). Format: Christian teaching/inspirational. ♦Ron Kight, gen mgr.

Lancaster

WRNZ(FM)— Oct 1, 1988: 105.1 mhz; 3 kw. 325 ft TL: N37 36 06 W84 34 27. Stereo. Hrs opn: 24 2063 Shakertown Rd., Danville, 40422. Phone: (859) 236-2711. Fax: (859) 236-1461. E-mail: hometownradio@bellsouth.net Web Site: www.hometownLIVE.net. Licensee: Hometown Broadcasting of Lancaster Inc. Population served: 200,000 Rgnl rep: Rgnl Reps Format: Hot adult contemp. News staff: one; News: 2 hrs wkly. Target aud: 25-54; upscale, white collar, baby boomers, business owners. Spec prog: Relg 4 hrs wkly. ♦Robert Wagner, gen mgr; Vicki Hyde, news dir.

Lawrenceburg

WKYL(FM)— May 11, 1993: 102.1 mhz; 6 kw. Ant 328 ft TL: N38 01 37 W84 52 59. Stereo. Hrs opn: 24 1010 Industry Rd., 40342. Phone: (502) 839-1021. Licensee: Davenport Broadcasting Inc. (acq 1-16-97; $525,000). Population served: 150,000 Natl. Network: Jones Radio Networks. Law Firm: Pepper & Corazzini. Format: Smooth jazz. Target aud: 30-50; higher income; especially at-work listeners. Spec prog: Relg 2 hrs wkly. ♦C. Michael Davenport, CEO & gen mgr.

Lebanon

WLBN(AM)— October 1954: 1590 khz; 1 kw-D, 74 w-N, DA-1. TL: N37 35 55 W85 14 47. Hrs open: 24 253 W. Main St., 40033. Phone: (270) 692-3126. Fax: (270) 692-6003. Web Site: www.1590wlbn.com. Licensee: CBC of Marion County Inc. Group owner: CBC of Marion and Taylor Counties Inc. (acq 7-3-97; $360,000 with co-located FM). Population served: 126,000 Natl. Network: Jones Radio Networks. Natl. Rep: Rgnl Reps. Law Firm: Leonard S. Joyce. Format: Oldies. News staff: one; News: 13 hrs wkly. Target aud: 35-64. Spec prog: Gospel 5 hrs, open mike 5 hrs wkly. ♦Lisa Kearnes, gen mgr, gen mgr, gen sls mgr, gen sls mgr, rgnl sls mgr, prom mgr & adv dir; Andy Colley, opns mgr, progmg mgr & mus dir; Patty Brown, news dir; Mike Graham, chief of engrg.

WLSK(FM)—Co-owned with WLBN(AM). Oct 1, 1979: 100.9 mhz; 16.5 kw. Ant 410 ft TL: N37 41 43 W85 19 06. Stereo. 24 E-mail: wlsk@commonwealthbroadcasting.com Web Site: www.lebanonmike.com.121,000 Format: Mike 80s & 90s—whatever. News staff: one; News: 9 hrs wkly. Target aud: 30-49. ♦Lisa Kearnes, mktg dir & traf mgr; Kevin Johnson, disc jockey.

Lebanon Junction

WTHX(FM)— October 1979: 107.3 mhz; 6 kw. Ant 321 ft TL: N37 44 26 W85 49 28. Stereo. Hrs opn: 611 W. Poplar St., Suite C2, Elizabethtown, 42701. Phone: (270) 763-0800. Fax: (270) 769-6349. Web Site: www.etownstar.com. Licensee: Elizabethtown CBC Inc. Group owner: Commonwealth Broadcasting Corp. (acq 12-23-02; $900,000). Natl. Rep: Rgnl Reps. Format: CHR. News staff: one; News: one hr wkly. Target aud: 18-44; females. ♦Steve Newberry, CEO & chmn; Dale Thornhill, sr VP; Roth Stratton, gen mgr; Holli Lee, traf mgr; Dan Michaels, opns & progmg.

Ledbetter

***WHMR(FM)**— 2004: 90.1 mhz; 1 kw vert. Ant 328 ft TL: N37 06 10 W88 24 15. Hrs open: Box 281, Hardin, 42048. Secondary address: 219 College St., Harding 42048. Phone: (270) 437-4095. Fax: (270) 437-4098. E-mail: info@hmiradio.com Web Site: www.hmiradio.com. Licensee: Heartland Ministries Inc. Format: Traditional christian. ♦Darrell Gibson, pres; Jeremy Johnson, progmg dir.

Leitchfield

WKHG(FM)—Listing follows WMTL(AM).

WMTL(AM)— Jan 17, 1959: 870 khz; 500 w-D. TL: N37 30 40 W86 17 15. Hrs open: 2160 Brandenburg Rd., 42754. Phone: (270) 259-3165. Fax: (270) 259-5693. Licensee: Heritage Media of Kentucky Inc. (acq 1-26-95; $350,000 with co-located FM; FTR: 3-20-95) Format: Country. ♦Mark Buckles, pres, gen mgr & gen sls mgr; Ed Thomas, chief of engrg.

WKHG(FM)—Co-owned with WMTL(AM). Oct 29, 1967: 104.9 mhz; 3.5 kw. 250 ft TL: N37 30 40 W86 17 15. Stereo. Phone: (270) 259-5692. Web Site: www.k105.com. Format: Adult contemp.

Lerose

***WOCS(FM)**— March 1999: 88.3 mhz; 1 kw. Ant 321 ft TL: N37 36 23 W83 41 16. Hrs open: 3 PM-9 PM (M-F) Owsley County High School, Hwy. 28/Shepherd Ln., Booneville, 41314. Phone: (606) 593-5185. Fax: (606) 593-6312. E-mail: tburns@owsley.k12.ky.us Web Site: www.owsley.k12.ky.us. Licensee: Board of Regents - Morehead State University (acq 4-13-01; $15,000). Population served: 15,000 Format: Div. Target aud: 12-35; poor & uneducated in need of information. ♦Diana Gross, chmn; Stephen F. Jackson, CEO & pres; Jerry McIntosh, CFO; Dan Conti, gen mgr; Bill Hodges, mus dir.

Lewisport

WKCM(AM)—See Hawesville

Lexington

WBUL-FM— July 15, 1969: 98.1 mhz; 100 kw. 561 ft TL: N38 02 07 W84 27 02. Stereo. Hrs opn: 2601 Nicholasville Rd., 40503. Phone: (859) 422-1000. Fax: (859) 422-1038. Web Site: www.wbul.com. Licensee: Citicasters Licenses L.P. Group owner: Clear Channel Communications Inc. (acq 5-4-99; grpsl). Population served: 120,000 Format: Country. Target aud: 25-49; baby boomers who grew up with Stones & Beatles. ♦Gene Guinn, gen mgr.

WGKS(FM)—Listing follows WLXG(AM).

WLAP(AM)—Listing follows WMXL(FM).

WLXG(AM)— 1946: 1300 khz; 2.5 kw-D, 1 kw-N, DA-N. TL: N38 05 50 W84 31 45. Hrs open: 401 W. Main St., Suite 301, 40507. Phone: (859) 233-1515. Fax: (859) 233-1517. E-mail: jmac@lmcomm.com Web Site: www.wlxg.com. Licensee: L.M. Communications Inc. (group owner; (acq 7-1-84). Population served: 190,000 Natl. Network: ESPN Radio. Format: Sports radio. Target aud: 25-54; adults. ♦Lynn Martin, pres; James E. MacFarlane, gen mgr.

WGKS(FM)—Co-owned with WLXG(AM). June 5, 1968: 96.9 mhz; 50 kw. 492 ft TL: N38 07 32 W84 21 12. Stereo. Web Site: www.wgks.com. Natl. Network: ABC. Format: Soft rock. News staff: one; News: 15 hrs weekly. ♦Skip Elliot, progmg dir.

WLXX(FM)—Listing follows WVLK(AM).

WMXL(FM)— 1940: 94.5 mhz; 100 kw. 640 ft TL: N38 07 25 W84 26 45. Stereo. Hrs opn: 24 2601 Nicholasville Rd., 40503. Phone: (859) 422-1000. Fax: (859) 422-1038. Web Site: www.wmxl.com. Licensee: Citicasters Licenses L.P. Group owner: Clear Channel Communications Inc. (acq 5-4-99; grpsl). Population served: 195,100 Format: Adult contemp. News: 3 hrs wkly. Target aud: 25-54; women. ♦Gene Guinn, gen mgr; Barry Fox, opns mgr; Dale O'Brien, progmg dir; Gerry Westerberg, mus dir & chief of engrg; Karyn Czar, news dir; Michael Jordan, mktg dir, traf mgr & disc jockey.

WLAP(AM)—Co-owned with WMXL(FM). September 1922: 630 khz; 5 kw-D, 1 kw-N, DA-2. TL: N38 07 25 W84 26 45.24 Web Site: www.wlap.com.1,951,000 Natl. Network: CBS. Natl. Rep: Christal. Format: News. News staff: one. Target aud: 18-49; men. ♦Kevin Bell, progmg dir.

***WRFL(FM)**— Mar 3, 1988: 88.1 mhz; 250 w. 289 ft TL: N38 02 19 W84 30 16. Hrs open: Box 777, University Stn., Music Director, 40506. Phone: (859) 257-4636. Phone: (859) 257-9735. Fax: (859) 323-1039. E-mail: gm@wrfl881.org Licensee: Radio Free Lexington Inc. Format: Var. ♦John Clark, gen mgr.

***WUKY(FM)**— Mar 13, 1941: 91.3 mhz; 95 kw. 1,004 ft TL: N37 47 18 W84 40 49. Stereo. Hrs opn: 24 Univ. of Kentucky, 340 McVey, 40506-0045. Phone: (859) 257-3221. Fax: (859) 257-6291. E-mail: wuky@uky.edu Web Site: www.wuky.org. Licensee: University of Kentucky. Population served: 50,000 Natl. Network: NPR, PRI. Law Firm: Sanchez Law Firm. Format: News, AAA music. News staff: 2; News: 62 hrs wkly. Target aud: 35-54. ♦Tom Godell, gen mgr; John Lumagui, opns mgr; Gail Bennett, mktg mgr & mus dir.

WVLK(AM)— October 1947: 590 khz; 5 kw-D, 1 kw-N, DA-2. TL: N38 06 42 W84 34 36. (CP: 1.6 kw-N). Stereo. Hrs opn: 24 300 W. Vine St., Suite 3, 40507. Phone: (859) 253-5900. Fax: (859) 253-5940. Web Site: www.wvlkam.com. Licensee: Cumulus Licensing Corp. Group owner: Cumulus Media Inc. (acq 7-22-99; grpsl). Population served: 275,400 Natl. Network: CBS. Rgnl. Network: Ky. Net. Law Firm: Latham & Watkins. Format: News/talk, sports. News staff: 5. Target aud: 25-54. ♦Ken Fearnow, gen mgr; Matt Hyland, gen sls mgr; Curtis Johnson, prom dir; Robert Lindsey, opns mgr & progmg dir.

WLXX(FM)—Co-owned with WVLK(AM). February 1962: 92.9 mhz; 100 kw. Ant 854 ft TL: N38 02 22 W84 24 11. Stereo. 24 Web Site: www.wlxxthebear.com. Format: Country. ♦Marshall Stewart, progmg dir.

Lexington-Fayette

WLKT(FM)— July 30, 1992: 104.5 mhz; 50 kw. 492 ft TL: N38 05 54 W84 18 38. Hrs open: 2601 Nicholasville Rd., Lexington, 40503. Phone: (859) 422-1000. Fax: (859) 422-1038. Web Site: www.wlkt.com. Licensee: Citicasters Licenses L.P. Group owner: Clear Channel Communications Inc. (acq 5-4-99; grpsl). Format: Contemp hit. ♦Gene Guinn, gen mgr.

Liberty

WKDO(AM)— November 1963: 1560 khz; 1 kw-D. TL: N37 18 22 W84 55 02. Hrs open: Box 990, 42539. Secondary address: 988 Dry Ridge Rd. 42539. Phone: (606) 787-7331. Fax: (606) 787-2166. Licensee: Radio Station WKDO. (acq 11-27-75). Population served: 2,500 Natl. Network: USA. Format: Country. Target aud: 18-49. ♦Carlos Wesley, pres, gen mgr & gen sls mgr; David Smith, chief of engrg.

WKDO-FM— January 1977: 98.7 mhz; 25 kw. 239 ft TL: N37 18 22 W84 55 02. Stereo. 16 15,500 News staff: 3; News: 21 hrs wkly. Target aud: 15-35.

London

WFTG(AM)— Sept 1, 1955: 1400 khz; 1 kw-U. TL: N37 08 28 W84 04 45. Hrs open: 24 Box 1988, 40743-0647. Secondary address: 534 Tobacco Rd. 40741. Phone: (606) 864-2148. Fax: (606) 864-0645. E-mail: trgrigsby@broadcasting.net Licensee: F.T.G. Broadcasting Inc. Group owner: Key Broadcasting Inc. (acq 8-5-92; $410,000; FTR: 8-24-92). Population served: 43,537 Rgnl. Network: Ky. News Net. Format: Talk. News staff: one. Target aud: 40 plus. ♦Mike Tarter, pres; Trever Grigsby, gen mgr; Travis Stevens, progmg dir; Phillip Fraley, chief of engrg.

WWEL(FM)—Co-owned with WFTG(AM). Sept 15, 1970: 103.9 mhz; 3 kw. 190 ft TL: N37 08 28 W84 04 45. Stereo. 24 Web Site: www.1039thewolf.com. Format: Country. News staff: one; News: 21 hrs wkly. Target aud: 18-50.

WGWM(AM)— Aug 8, 1981: 980 khz; 900 w-D, 109 w-N. TL: N37 10 16 W84 06 39. Hrs open: 24 948 Moriah Church Rd., 40741-7635. Phone: (606) 878-0980. Fax: (606) 878-0980. Licensee: WGWM Broadcasting Inc. (acq 1996; $35,000). Format: Southern gospel. News: 5 hrs wkly. Target aud: 25-54; male/female. ♦Elmer Oakley, gen mgr.

WYGE(FM)— 1994: 92.3 mhz; 23.5 kw. 722 ft TL: N37 09 01 W83 59 32. Hrs open: 24 201 E. 2nd St., 40741. Phone: (606) 877-1326. Fax: (606) 864-3702. E-mail: wygeradio@yahoo.com Web Site: www.good-news-outreach.org. Licensee: Ethel Huff Broadcasting LLC. Population served: 200,000 Natl. Network: Salem Radio Network, USA. Format: Relg. ♦Ethel Huff, chmn; Gene Huff, gen mgr.

Louisa

WBTH(AM)—See Williamson WV

WXCC(FM)—See Williamson WV

WZAQ(FM)— May 17, 1991: 92.3 mhz; 4.48 kw. 377 ft TL: N38 10 33 W82 37 39. Hrs open: 112 Madison St., 41230. Phone: (606) 638-9203. Fax: (606) 638-9210. Licensee: Louisa Communications Inc. Format: Country. ♦Harold Britton, pres; Marti Fairchild, gen mgr.

Louisville

WAMZ(FM)—Listing follows WHAS(AM).

WAVG(AM)—See Jeffersonville, IN

WDJX(FM)— Aug 1, 1963: 99.7 mhz; 24 kw. 720 ft TL: N38 21 53 W85 50 18. Stereo. Hrs opn: 520 S. 4th, Suite 200, 40202. Phone: (502) 625-1220. Fax: (502) 625-1256. Web Site: www.wdjx.com. Licensee: Blue Chip Broadcasting Licenses Ltd. Group owner: Radio One Inc. (acq 4-30-2001; grpsl). Format: Adult CHR.

WFIA(AM)— March 1947: 900 khz; 1 kw-U. TL: N38 16 12 W85 42 25. Stereo. Hrs opn: 24 9960 Corporate Campus Dr., Suite 3600, 40223. Phone: (502) 339-9470. Fax: (502) 423-3139. Web Site: www.salemradiogroup.com. Licensee: SCA License Corp. Group owner: Salem Communications Corp. (acq 1-24-01; $1.75 million). Population served: 1,000,000 Natl. Rep: Salem. Format: Christian teaching, talk. News: 2 hrs wkly. Target aud: 30 plus; general. ♦Gordon Marcy, gen mgr.

***WFPK(FM)**— Oct 4, 1954: 91.9 mhz; 100 kw. 236 ft TL: N38 14 40 W85 45 27. Stereo. Hrs opn: 24 619 S. 4th St., 40202. Phone: (502) 814-6500. Fax: (502) 814-6599. Web Site: www.wfpk.org. Licensee: Kentucky Public Radio Inc. Population served: 1,300,000 Natl. Network: PRI, NPR. Format: Alternative. Target aud: 25 plus. ♦Donovan Reynolds, gen mgr & gen sls mgr; Stacy Owen, progmg dir & mus dir.

***WFPL(FM)**— Feb 20, 1950: 89.3 mhz; 21 kw. 774 ft TL: N38 21 55 W85 50 24. Stereo. Hrs opn: 24 619 S. 4th St., 40202. Phone: (502) 814-6500. Fax: (502) 814-6599. Web Site: www.wfpl.org. Licensee: Kentucky Public Radio Inc. Population served: 1,300,000 Natl. Network: NPR, PRI. Format: News/talk. News staff: 3; News: 124 hrs wkly. Target aud: General. ♦Donovan Reynolds, gen mgr; Rick Howlett, progmg dir.

WGTK(AM)— Dec 30, 1933: 970 khz; 5 kw-U, DA-2. TL: N38 19 05 W85 44 39. Stereo. Hrs opn: 24 9960 Campus Dr., Suite 3600, 40223. Phone: (502) 339-9470. Fax: (502) 423-3139. Web Site: www.970wgtk.com. Licensee: Salem Media of Kentucky Inc. (group owner; (acq 10-4-2000). Population served: 900000 Law Firm: Brooks, Pierce, McLendon, Humphrey & Leonard. Format: News/talk. News staff: one; News: 21 hrs wkly. Target aud: 35 plus; people with most discretionary incomes. ♦Mark Thomas, stn mgr.

WHAS(AM)— July 18, 1922: 840 khz; 50 kw-U. TL: N38 15 40 W85 25 43. Stereo. Hrs opn: 24 4000 #1 Radio Dr., 40218. Phone: (502) 479-2222. Fax: (502) 479-2308. E-mail: info@whas.com Web Site: www.whas.com. Licensee: CC Licenses LLC. (acq 8-86; with co-located FM). Population served: 1,300,000 Natl. Rep: Clear Channel. Law Firm: Cohn & Marks. Format: News/talk. News staff: 12; News: 14 hrs wkly. Target aud: 25-54. ♦Kevin Hughes, gen mgr; Doug Wethington, stn mgr & gen sls mgr; Kelly Carls, progmg dir; Kirk Wesley, chief of engrg.

WAMZ(FM)—Co-owned with WHAS(AM). September 1966: 97.5 mhz; 100 kw. 672 ft TL: N38 03 49 W85 43 52. Stereo. 24 E-mail: info@wamz.com Web Site: www.wamz.com.1,300,000 Format: Country. ♦Coyote Calhoun, progmg dir.

WKJK(AM)— November 1948: 1080 khz; 10 kw-D, 1 kw-N, DA-2. TL: N38 18 29 W85 49 45. Hrs open: 24 4000 #1 Radio Dr., 40218. Phone: (502) 479-2222. Fax: (502) 479-2223. Web Site: www.talkradio1080.com. Licensee: CC Licenses LLC. Group owner: Clear Channel Communications Inc. (acq 9-13-96; $1 million with intellectual property of WSFR(FM) Corydon, IN) Population served: 1,000,000 Natl. Network: ABC. Natl. Rep: Clear Channel. Format: Talk. Target aud: 18-34; women. ♦Kevin Hughes, gen mgr.

WKRD(AM)—Listing follows WLUE(FM).

WLLV(AM)— June 1940: 1240 khz; 1 kw-U. TL: N38 14 49 W85 42 19. Hrs open: 2001 W. Broadway, 40203. Phone: (502) 776-1240. Fax: (502) 776-1250. E-mail: wlouwllv@aol.com Licensee: Davidson Media Station WLLV Licensee LLC. Group owner: Mortenson Broadcasting Co. (acq 4-12-2006; $2.65 million with WLOU(AM) Louisville). Population served: 361,472 Format: Black gospel. ♦Archie Dale, gen mgr.

WLOU(AM)— 1948: 1350 khz; 5 kw-U, DA-N. TL: N38 13 45 W85 46 47. (CP: 2.2 kw). Hrs opn: 2001 W. Broadway, 40203. Phone: (502) 776-1240. Fax: (502) 776-1250. E-mail: wlouwllv@aol.com Licensee: Davidson Media Station WLOU Licensee LLC. Group owner: Mortenson Broadcasting Co. (acq 4-12-2006; $2.65 million with WLLV(AM) Louisville). Population served: 361,472 Natl. Network: American Urban. Format: Gospel. Target aud: 25-54; mature adults. ♦Archie Dale, gen mgr.

WLUE(FM)— June 7, 1993: 100.5 mhz; 37.4 kw. Ant 554 ft TL: N38 03 49 W85 43 52. Hrs open: 24 4000 #1 Radio Dr., 40218. Phone: (502) 479-2222. Fax: (502) 479-2308. E-mail: louie@louieonline.com Web Site: www.louieonline.com. Licensee: Clear Channel Broadcasting Licenses Inc. Group owner: Clear Channel Communications Inc. (acq 9-13-96; $6.9 million with co-located AM). Natl. Rep: Clear Channel. Format: Adult hits. News: 3 hrs wkly. Target aud: 18-49; general. ♦Bill Gentry, gen mgr & gen sls mgr.

WKRD(AM)—Co-owned with WLUE(FM). 1936: 790 khz; 5 kw-D, 1 kw-N, DA-2. TL: N38 11 34 W85 31 14.24
Simulcast with WKRD-FM Shelbyville 100%.
E-mail: jimfenn@clearchannel.com Web Site: www.790wkrd.com. Licensee: CC Licenses LLC. (acq 1996). Natl. Network: Fox Sports, Premiere Radio Networks. Natl. Rep: Clear Channel. Format: Sports. News: 2 hrs wkly. Target aud: 25-54. ♦Kevin Hughes, stn mgr & gen sls mgr; Jim Fenn, progmg dir.

WMJM(FM)—See Jeffersontown

WPTI(FM)— 1974: 103.9 mhz; 1.35 kw. Ant 490 ft TL: N38 15 20 W85 45 28. Stereo. Hrs opn: 24 612 4th Ave., Suite 100, 40202. Phone: (502) 589-4800. Fax: (502) 587-0212. Web Site: newcountry1039.com. Licensee: Cox Radio Inc. Group owner: Cox Broadcasting (acq 8-26-99; $1.77 million). Population served: 1,000,000 Law Firm: Dow, Lohnes & Albertson. Format: Country. News: one hr wkly. Target aud: 25-54; emphasis on 25-44. ♦Tim Hartlage, gen sls mgr & natl sls mgr; Todd Schumacher, VP, gen mgr & mktg mgr; Matt Killion, progmg dir.

WQMF(FM)—See Jeffersonville, IN

WTUV(AM)— Aug 20, 1958: 620 khz; 500 w-U, DA-2. TL: N38 18 59 W85 42 08. Hrs open: 24 4109 Bardstrom Rd., Suite 104, 40218. Phone: (502) 583-6200. Fax: (502) 589-2979. E-mail: info@bcalienteradio.com Web Site: www.lacalienteradio.com. Licensee: Davidson Media Station WTMT Licensee LLC. (acq 6-30-2006; $1 million). Population served: 361,472 Format: Rgnl Mexican. ♦Paul Dendy, gen mgr; Dennis Mendez, progmg dir.

***WUOL(FM)**— Dec 20, 1976: 90.5 mhz; 21 kw. Ant 774 ft TL: N38 21 55 W85 50 24. Stereo. Hrs opn: 24 619 S. 4th St., 40202. Phone:

(502) 814-6500. Fax: (502) 814-6599. Web Site: www.wuol.org. Licensee: Kentucky Public Radio Inc. Population served: 2,000,000 Rgnl. Network: Ky. Pub. Format: Class. News: 2 hrs wkly. Target aud: General; those interested in quality music & info. ♦Donovan Reynolds, gen mgr; Alan Brandt, progmg dir.

WVEZ(FM)— Apr 1, 1967: 106.9 mhz; 24.5 kw. 670 ft TL: N38 22 20 W85 49 32. Stereo. Hrs opn: 612 4th Ave., Suite 100, 40202. Phone: (502) 589-4800. Fax: (502) 583-4820. Web Site: www.lite1069.com. Licensee: Cox Radio Inc. Group owner: Cox Broadcasting (acq 5-99). Population served: 976,800 Law Firm: Dow, Lohnes & Albertson. Format: Adult contemp. Target aud: 24-54; upper-scale, working women. Spec prog: Delilah. ♦Todd Schumacher, VP & gen mgr; Kitty Malone, gen sls mgr; Don Nordin, progmg dir.

WWSZ(AM)—See New Albany, IN

WXMA(FM)— October 1964: 102.3 mhz; 3 kw. 300 ft TL: N38 14 37 W85 45 34. Hrs opn: 24 520 S. 4th Ave., Suite 200, 40202-2532. Phone: (502) 625-1220. Fax: (502) 625-1258. Web Site: www.themaxfm.com. Licensee: Blue Chip Broadcasting Licenses Ltd. Group owner: Radio One Inc. (acq 8-10-2001; grpsl). Population served: 361,472 Format: Hot adult contemp. Target aud: 25-49; young adults who enjoy modern/alternative rock. ♦Dale Schaefer, gen mgr.

Madisonville

WFMW(AM)— January 1947: 730 khz; 500 w-D, 215 w-N. TL: N37 21 03 W87 29 25. Stereo. Hrs opn: 24 Box 338, 42431. Secondary address: 2380 N. Main St. 42431. Phone: (502) 821-4096. Fax: (502) 821-5954. E-mail: wfmw@wfmw.net Web Site: www.wfmw.net. Licensee: Sound Broadcasters Inc. Population served: 50,000 Natl. Network: CNN Radio. Natl. Rep: Rgnl Reps. Wire Svc: AP Format: C&W. News staff: one; News: 13 hrs wkly. Target aud: 18 plus. ♦Robert T. Kelley, pres & gen mgr; Danny Koeber, progmg dir, farm dir, women's int ed & disc jockey; Chris Gardener, news dir; Chris Meyers, chief of engrg; Erin Grant, disc jockey; Rick Stevens, disc jockey; Steven Strait, disc jockey.

WKTG(FM)—Co-owned with WFMW(AM). Apr 19, 1949: 93.9 mhz; 50 kw. 584 ft TL: N37 21 05 W87 29 25. Stereo. 24 Phone: (270) 821-1156. E-mail: wktg@wktg.com Web Site: www.wktg.com.500,000 Natl. Network: USA. Law Firm: Pepper & Corazzini. Format: Rock/AOR. News staff: one; News: 3 hrs wkly. Target aud: 20-45. ♦Robert T. Kelley, stn mgr; Bill McClone, progmg dir; Erin Grant, disc jockey; Kevin O'Connor, disc jockey.

***WSOF-FM**— February 1977: 89.9 mhz; 39.4 kw. 282 ft TL: N37 19 11 W87 30 57. Stereo. Hrs opn: 1415 Island Ford Rd., 42431. Phone: (270) 825-3004. Fax: (270) 825-3005. E-mail: comments@wsof.org Web Site: www.wsof.org. Licensee: Madisonville Christian School a division of Madisonville Baptist Temple Inc. Natl. Network: USA. Format: Christian educ. Target aud: General; Christian. ♦Gary Hall, gen mgr.

WTTL(AM)— Sept 16, 1956: 1310 khz; 1.5 kw-D, 500 w-N, DA-N. TL: N37 20 12 W87 32 41. Hrs open: 24 Box 1310, 42431. Secondary address: 265 S. Main St. 42431. Phone: (270) 821-1310. Fax: (270) 825-3260. Licensee: Madisonville CBC Inc. Group owner: Commonwealth Broadcasting Corp. (acq 2-8-2000; $1.31 million with co-located FM). Population served: 80,000 Format: News, talk, sports. Target aud: 25-54. ♦Lee Ann Oliver, gen sls mgr; Tom Rogers, gen mgr & progmg dir; Stephanie Vandygraiff, traf mgr.

WYMV(FM)—Co-owned with WTTL(AM). Sept 7, 1992: 106.9 mhz; 2 kw. 528 ft TL: N37 22 51 W87 28 04. Stereo. 24 Phone: (270) 825-1079. Natl. Network: ABC, Jones Radio Networks. Format: Adult contemp. Target aud: 25-34.

Manchester

WKLB(AM)— Sept 26, 1981: 1290 khz; 50 kw-U. TL: N37 09 29 W83 47 06. Stereo. Hrs opn: 24 Box 448, 40962. Secondary address: 106 Richmond Rd. 40962. Phone: (606) 598-2445. Fax: (606) 598-2653. E-mail: wklb1stchoice@yahoo.com Licensee: Barker Broadcasting Co. Group owner: Larry Barker (acq 1981). Population served: 500,000 Rgnl. Network: Ky. Net. Natl. Rep: Rgnl Reps. Rgnl rep: Barker Broadcasting Law Firm: Robert Olender. Format: Country. News staff: one; News: 8 hrs wkly. Target aud: 24-65; working people. ♦Larry Barker, pres & gen mgr.

WTBK(FM)— October 1989: 105.7 mhz; 7.5 kw. 462 ft TL: N37 08 57 W83 45 09. Stereo. Hrs opn: 19 Box 453, 40962. Secondary address: 107 Dickerson St. 40962. Phone: (606) 598-7588. Fax: (606) 598-7598. E-mail: wtbkradio@yahoo.com Licensee: Manchester Communications Inc. (acq 3-24-89). Natl. Network: Westwood One, ABC. Format: Classic rock. News staff: one; News: 10 hrs wkly. Target aud: General; 18 plus in the morning, 16-45 at night. Spec prog: Talk 8 hrs wkly. ♦Tim Finley, gen mgr.

WWLT(FM)— Aug 9, 1967: 103.1 mhz; 2.25 kw. Ant 538 ft TL: N37 04 30 W83 49 14. Stereo. Hrs opn: 24 8686 Michael Ln., Fairfield, OH, 45014. Phone: (513) 829-7700. Web Site: www.klove.com. Licensee: Wilderness Hills Broadcasting Co. Group owner: Vernon R. Baldwin Inc. (acq 1956). Population served: 100,000 Natl. Network: K-Love. Format: Contemp Christian. ♦Vernon R. Baldwin, pres.

WWXL(AM)— 1956: 1450 khz; 1 kw-U. TL: N37 09 04 W83 45 45. Hrs open: 24 Hours 24 Manchester Sq. Shopping Ctr, #205, 40962. Phone: (606) 598-9995. Fax: (606) 598-9995. Licensee: Juanita H. Nolan (acq 1-3-2004). Population served: 18,600 Format: Oldies. News staff: one; News: 3 hrs wkly. Target aud: 35-54; programmed for adults 35-54. ♦Joe Burchell, gen mgr.

Mannsville

WVLC(FM)— Dec 31, 1994: 99.9 mhz; 11 kw. 492 ft TL: N37 10 04 W85 11 26. Stereo. Hrs opn: 24 Box 4190, Campbellsville, 42719. Secondary address: 101 East Main St., Campbellsville 42719. Phone: (270) 789-4998. Fax: (270) 789-4584. E-mail: bigdawg@wvlc.com Web Site: www.wvlc.com. Licensee: Patricia Rodgers. Natl. Network: Jones Radio Networks, CNN Radio. Format: Country. News staff: one. Target aud: General. ♦Greg Gribbins, gen mgr.

Marion

WMJL(AM)— July 10, 1968: 1500 khz; 250 w-D. TL: N37 20 11 W88 04 12. Stereo. Hrs opn: 6 AM-sunset Box 68, 42064. Secondary address: 251 Club Dr. 42064. Phone: (270) 965-2271. Licensee: Joe Myers Production Inc. Population served: 20,000 Rgnl. Network: Ky. Net. Natl. Rep: Rgnl Reps. News staff: one; News: 12 hrs wkly. Target aud: General. ♦Joe Myers, pres, gen mgr, gen sls mgr, progmg dir & chief of engrg.

WMJL-FM— June 1993: 102.7 mhz; 6 kw. 328 ft TL: N37 20 16 W88 04 03. Stereo. 24 (Acq 3-12-91; FTR: 4-1-91). Format: Oldies.

Mayfield

WNGO(AM)— Jan 7, 1947: 1320 khz; 1 kw-D, 97 w-N. TL: N36 45 37 W88 38 20. Hrs open: 5 AM-10 PM
Rebroadcasts WNBS(AM) Murray 100%.
Box 2397, Paducah, 42002. Secondary address: 6000 Bristol Dr., Paducah 42003. Phone: (270) 554-8255. Fax: (270) 554-5468. Web Site: www.wkyx.com. Licensee: Bristol Broadcasting Co. Inc. (group owner; acq 2-20-2004; grpsl). Population served: 33,000 Rgnl. Network: Ky. Agri. Law Firm: Mullin, Rhyne, Emmons & Topel. Format: News/talk. News staff: one; News: 12 hrs wkly. Target aud: 24-54. ♦Gary Morse, gen mgr; Jamie Futrell, gen sls mgr & prom dir; Greg Dunker, progmg dir; Greg Walker, chief of engrg.

WQQR(FM)—Co-owned with WNGO(AM). Nov 2, 1955: 94.7 mhz; 32 kw. Ant 443 ft TL: N36 45 58 W88 38 50. (CP: COL Clinton. 50 kw, ant 472 ft. TL: N36 45 19 W88 39 36.6). Stereo. 24 Web Site: www.wqqr.com. Rgnl rep: Rgnl Reps. Format: Classic rock. Target aud: 25 plus. ♦Nick Black, progmg dir.

WYMC(AM)— Oct 18, 1976: 1430 khz; 1 kw-U, DA-N. TL: N36 47 12 W88 39 16. Hrs open: 24 Box V, 42066. Secondary address: 197 WYMC Rd. 42066. Phone: (270) 247-1430. Fax: (270) 247-1825. E-mail: radio@wymcradio.com Licensee: JDM Communications Inc. (acq 12-31-90; $277,649; FTR: 1-21-91). Population served: 38,000 Law Firm: Wiley, Rein & Fielding. Format: MOR. News staff: one. Target aud: 35-64; affluent, business oriented. ♦Jim Moore, gen mgr.

Maysville

WFTM(AM)— Jan 1, 1948: 1240 khz; 1 kw-U. TL: N38 38 10 W83 45 38. Hrs open: 6 AM-11 PM Box 100, 41056. Phone: (606) 564-3361. Fax: (606) 564-4291. E-mail: wftmnews@maysvilleky.net Licensee: Standard Tobacco Co. Population served: 100,000 Natl. Rep: Keystone (unwired net), Rgnl Reps. Format: Music of Your Life. News staff: one; News: 10 hrs wkly. Target aud: 50-70. Spec prog: Farm 6 hrs, gospel 5 hrs, relg 5 hrs wkly. ♦J.A. Finch, pres; Jeff Cracraft, VP; Doug McGill, gen mgr, gen sls mgr & chief of engrg; Dave Gray, news dir.

WFTM-FM— Oct 26, 1965: 95.9 mhz; 3 kw. 207 ft TL: N38 38 04 W83 46 48. Stereo. 24 E-mail: wftmsales@maysvilleky.net Web Site: soft96.com.165,000 Natl. Network: AP Radio. Format: Soft hits. News staff: one. Target aud: 18-55. ♦Danny Weddle, sls dir; Philip Hay, mus dir & sls.

McDaniels

***WBFI(FM)**— Sept 7, 1987: 91.5 mhz; 5 kw. Ant 328 ft TL: N37 36 06 W86 22 13. Stereo. Hrs opn: 24 Box 2, 40152. Phone: (270) 257-2689. Phone: (888) 333-9234. Fax: (270) 257-8344. Web Site: www.wbfiradio.com. Licensee: Bethel Fellowship Inc. Population served: 20,000 Law Firm: Reddy, Begley & McCormick. Format: Relg, educ, news/talk. News: 20 hrs wkly. Target aud: General; Christians. ♦Ronald W. Miller, pres; Roger Goostree, gen mgr & opns mgr; Daryl Cook, progmg dir; James Coates, mus dir & engrg mgr.

McKee

WWAG(FM)— Nov 1, 1990: 107.9 mhz; 3.9 kw. 400 ft TL: N37 23 39 W83 54 32. Hrs open: 24 1680 State Rd. 1071, Tyner, 40486-9543. Phone: (606) 287-9924. Licensee: Dandy Broadcasting Inc. (acq 1994). Population served: 2000 Natl. Network: ABC. Natl. Rep: Rgnl Reps. Law Firm: Lauren A. Colby. Format: Country. News staff: one; News: 10 hrs wkly. Target aud: General. Spec prog: Bluegrass 9 hrs wkly. ♦Dan Brockman, pres; Sherry Handy, gen mgr.

Middlesboro

WFXY(AM)— Mar 1, 1969: 1490 khz; 1 kw-U. TL: N36 36 47 W83 42 34. Stereo. Hrs opn: 24 Box 999, 40965. Secondary address: 2118 Cumberland Ave. 400965. Phone: (606) 337-2100. Fax: (606) 248-6397. Web Site: 1490wfxy.com. Licensee: Country-Wide Broadcasters Inc. (acq 5-1-01). Population served: 120,000 Natl. Network: Jones Radio Networks. Rgnl. Network: Ky. Net., Tenn. Radio Net. Natl. Rep: Rgnl Reps. Law Firm: Bechtel & Cole. Format: Hot adult contemp. News staff: 2; News: 20 hrs wkly. Target aud: 25-54; community-oriented. Spec prog: Black 2 hrs, gospel 3 hrs, relg 3 hrs wkly. ♦Brian O'Brien, opns mgr.

WMIK(AM)— Nov 15, 1948: 560 khz; 500 w-D, 88 w-N. TL: N36 37 38 W83 42 52. (CP: 2.5 kw-D). Hrs opn: Box 608, 40965. Secondary address: N. 19th St. 40965. Phone: (606) 248-5842. Fax: (606) 248-7660. Licensee: Gateway Broadcasting Inc. Population served: 11,844 Format: Southern gospel. Spec prog: Farm one hr, gospel 2 hrs wkly. ♦Roy Shotten, gen mgr & progmg dir; Chuck Owens, chief of engrg.

WMIK-FM— June 4, 1971: 92.7 mhz; 130 w. 1,438 ft TL: N36 35 50 W83 47 49. Stereo. 11,844 Format: Christian.

Midway

WBTF(FM)— 1998: 107.9 mhz; 6 kw. 328 ft TL: N38 11 41 W84 38 25. Hrs open: 401 W. Main, Suite 301, Lexington, 40507. Phone: (859) 233-1515. Fax: (859) 233-1517. Licensee: L.M. Communications of Kentucky L.L.C. Group owner: L.M. Communications Inc. (acq 4-10-01). Format: Urban contemp, CHR. ♦Lynn Martin, pres; James MacFarlane, gen mgr.

Monticello

WFLW(AM)— May 19, 1955: 1360 khz; 1 kw-D. TL: N36 49 30 W84 51 20. Hrs open: 6 AM-6 PM Box 696, 150 Worsham Ln., 42633.

Phone: (606) 348-8427. Phone: (606) 348-7083. Web Site: wflw.com. Licensee: Stephen Staples Jr. (acq 11-9-94; with co-located FM; FTR: 1-2-95). Population served: 20,000 Rgnl. Network: Ky. Net. Natl. Rep: Rgnl Reps. Format: Gospel. News staff: one. Target aud: General. Spec prog: Farm 5 hrs, news/talk 10 hrs wkly. ♦Stephen Staples Jr., gen mgr; Debbie Brown, mus dir; Bruce Correll, chief of engrg.

WKYM(FM)—Co-owned with WFLW(AM). Dec 19, 1965: 101.7 mhz; 1.75 kw. 617 ft TL: N36 48 08 W84 50 51. Stereo. 24 E-mail: wkymmail@wkym.com Web Site: www.wkym.com.80,000 Natl. Rep: Rgnl Reps. Format: Classic rock. Target aud: 18-50; baby boomers. ♦Stephen Staples Jr., progmg dir.

WMKZ(FM)— June 1, 1990: 93.1 mhz; 2.15 kw. 558 ft TL: N36 48 29 W84 50 46. Stereo. Hrs opn: 24 183 Old Hwy. 90, 42633. Phone: (606) 348-3393. Fax: (606) 348-3330. Web Site: www.wmkz.com. Licensee: Monticello-Wayne County Media Inc. Natl. Network: USA. Format: Country. News: 9 hrs wkly. Target aud: 24-55; general. ♦Joel Catron, gen mgr.

Morehead

***WBMK(FM)**— 2002: 88.5 mhz; 600 w. Ant 522 ft TL: N38 10 38 W83 24 24. Hrs open: Box 3206, Tupelo, MS, 38803. Phone: (662) 844-8888. Web Site: www.afr.net. Licensee: American Family Association. Group owner: American Family Radio (acq 11-26-99). Format: Christian. ♦Marvin Sanders, gen mgr.

WIVY(FM)— July 1, 1994: 96.3 mhz; 6 kw. Ant 328 ft TL: N38 10 56 W83 26 56. Stereo. Hrs opn: 24 Box 963, 40351. Secondary address: 123 E. First St. 40351. Phone: (606) 784-9966. Fax: (606) 674-6700. Licensee: Gateway Radio Works Inc. (group owner). Population served: 150,000 Natl. Network: ABC. Law Firm: William Silva. Wire Svc: NWS (National Weather Service) Format: Unforgettable Favorites. News staff: one. Target aud: 25 plus; affuent, well educ, mature adult, higher spendable income. ♦Hays McMakin, pres & gen mgr; Jeff Ray, stn mgr & progmg mgr.

WKCA(FM)—See Owingsville

***WMKY(FM)**— June 1965: 90.3 mhz; 37 kw. Ant 895 ft TL: N38 10 38 W83 24 18. Stereo. Hrs opn: 24 132 Breckinridge Hall, Morehead State Univ., 40351. Phone: (606) 783-2001. Phone: (606) 783-2334. Fax: (606) 783-2335. Web Site: www.msuradio.com. Licensee: Morehead State University. Population served: 155,317 Natl. Network: PRI, NPR. Rgnl. Network: Ky. News Net, Ky. Pub. Format: Var. News staff: 2; News: 53 hrs wkly. Target aud: 25-54; high education level (58% college or beyond). ♦Paul W. Hitchcock, gen mgr & stn mgr; Greg Jenkins, opns dir.

WMOR(AM)— Feb 18, 1955: 1330 khz; 570 w-D. TL: N38 10 56 W83 26 56. Hrs open: Box 338, 129 College St., West Liberty, 41472. Phone: (606) 743-3145. Fax: (606) 743-9557. Licensee: Morgan County Industries Inc. (group owner; acq 3-16-99; $300,000 with co-located FM). Population served: 15,000 Natl. Network: Moody. Format: Country. ♦C.C. Smith, pres, pres & gen mgr.

WMOR-FM— June 15, 1965: 106.1 mhz; 8 kw. 348 ft TL: N38 10 56 W83 26 56.20,000 Format: Adult contemp.

Morganfield

WEZG(FM)—Listing follows WMSK(AM).

WMSK(AM)— Nov 21, 1960: 1550 khz; 250 w-D. TL: N37 40 00 W87 55 40. Hrs open: 24 Box 369, 42437. Secondary address: 1339 US 60 W. 42437. Phone: (270) 389-1550. Fax: (270) 389-1553. E-mail: wmsk@bellsouth.net Jones U.S. Country Licensee: Union County Broadcasting Inc. Population served: 50000 Natl. Network: Jones Radio Networks. Natl. Rep: Rgnl Reps. Law Firm: Womble, Carlyle, Sandridge & Rice. Wire Svc: AP Format: Relg, country. News: 20 hrs wkly. Target aud: General; adults 25-64. ♦J.B. Crawley, pres; John Robinson, gen mgr, gen sls mgr, adv VP & progmg dir; Don Sheridan, sls VP & news dir; Rhonda Gibson, traf mgr; Bob Hite, edit mgr.

WEZG(FM)—Co-owned with WMSK(AM). Aug 8, 1967: 95.3 mhz; 25 kw. Ant 269 ft TL: N37 46 38 W87 37 26. 1339 Hwy. 60, 42437. Secondary address: Box 36 42437. Phone: (270) 389-1551.Jones 50,000 Natl. Rep: Rgnl Reps. Law Firm: Pepper & Corazzini. Target aud: 25-64; general. ♦J.B. Crawley, CEO; John Robinson, mktg dir, prom dir & adv dir; Bob Hite, local news ed; Don Sheridan, sports cmtr, disc jockey & disc jockey; Melissa Moore, women's int ed; John Gipson, disc jockey; Rhonda Gibson, opns mgr, traf mgr & disc jockey.

Morgantown

WLBQ(AM)— 1976: 1570 khz; 1 kw-D, 150 w-N. TL: N37 14 10 W86 42 29. (CP: TL: N37 13 09 W86 41 21). Hrs opn: 6 AM-10 PM Box 130, 42261. Phone: (270) 526-3321. Fax: (270) 526-5393. E-mail: info@wlbqam.com Licensee: Butler County Broadcasting Co. Population served: 11,000 Natl. Network: ABC. Rgnl. Network: Ky. Net. Format: C&W. News staff: one; News: 8 hrs wkly. Target aud: General; residents of Morgantown & Butler County, KY. ♦Charles Black, pres; Jan Embry, VP; Mary Alice Black, gen mgr; Howard Phelps, stn mgr.

WWKN(FM)—Not on air, target date: unknown: 99.1 mhz; 6 kw. Ant 184 ft TL: N37 17 38 W86 44 30. Hrs open: 8226 Douglas Ave., Suite 627, Dallas, TX, 75225. Phone: (469) 619-1001. Licensee: Independence Media Holdings LLC. ♦David Jacobs, CEO.

Mount Vernon

WANK(FM)—Not on air, target date: unknown: 102.9 mhz; 360 w. Ant 492 ft TL: N37 20 26.4 W84 22 57.9. Hrs open: 1717 Dixie Hwy., Suite 650, Fort Wright, 41011. Phone: (859) 331-9100. Licensee: Radioactive LLC. ♦Benjamin L. Homel, pres.

Mt. Sterling

***WAXG(FM)**— 1998: 88.1 mhz; 300 w. 174 ft TL: N38 03 39 W83 57 20. Hrs open: Box 2440, American Family Radio, Tupelo, MS, 38803. Phone: (662) 844-8888. Fax: (662) 842-6791. Web Site: www.afr.net. Licensee: American Family Association. Group owner: American Family Radio Format: Relg. ♦Marvin Sanders, gen mgr.

WKCA(FM)—See Owingsville

WMKJ(FM)— May 28, 1968: 105.5 mhz; 3 kw. 300 ft TL: N38 05 36 W83 56 39. Stereo. Hrs opn: 2601 Nicholasville Rd., Lexington, 40503-3307. Phone: (859) 422-1000. Fax: (859) 422-1038. Web Site: www.wmkj.com. Licensee: Citicasters Co. Group owner: Clear Channel Communications Inc. (acq 4-13-01). Population served: 28,000 Format: Oldies. ♦Gene Guinn, gen mgr.

WMST(AM)— Oct 17, 1957: 1150 khz; 2.5 kw-D, 53 w-N. TL: N38 02 41 W83 54 05. Hrs open: 24 22 West Main, 40353. Phone: (859) 498-1150. Fax: (859) 498-7930. Web Site: www.wmstradio.com. Licensee: Gateway Radio Works Inc. (group owner; acq 1-1-00). Population served: 165,000 Rgnl. Network: Ky. Net. Natl. Rep: Rgnl Reps. Law Firm: William Silva. Format: Timeless classics, news/talk. News staff: one; News: 37 hrs wkly. Target aud: 25 plus; affluent, mature adult, high spendalbe income, well educated. Spec prog: Farm 2 hrs wkly. ♦Hays McMakin, pres; Jeff Ray, VP & gen mgr; Vernice Taylor, stn mgr; Frances Denny, opns mgr; Tom Byron, prom dir; Dan Manley, news dir; Frank Folsom, chief of engrg.

Mt. Vernon

WRVK(AM)— April 30, 1957: 1460 khz; 500 w-D. TL: N37 23 49 W84 19 45. Hrs open: 6 AM-9 PM Box 7, Renfro Valley, 40473. Phone: (606) 256-2146. Fax: (606) 256-9146. E-mail: manager@wruk1460.com Web Site: www.wrvk1460.com. Licensee: Saylor Broadcasting Inc. (acq 2-1-02). Population served: 35,000 Format: Classic country, country gospel. Target aud: General. ♦Charles W. Saylor, pres; Charles Saylor, gen mgr; Charles Napier, sls VP & gen sls mgr.

Mt. Washington

WLCR(AM)— Oct 29, 1955: 1040 khz; 1.5 kw-D. TL: N38 00 11 W85 40 51. Stereo. Hrs opn: 3600 Goldsmith Ln., Louisville, 40220. Phone: (502) 451-9527. Fax: (502) 451-9527. Web Site: www.wlcr.net. Licensee: LCR Partners L.P. (acq 1999; $162,500). Population served: 1,000,000 Format: Southern gospel, relg, talk. Target aud: General; Those interested in the existance of God. ♦Vince Heuser, gen mgr.

Munfordville

WCLU-FM— Aug 1, 1964: 102.3 mhz; 3 kw. 99 ft TL: N37 16 30 W85 55 00. Hrs open: Box 1628, Glasgow, 42142. Phone: (270) 651-9149. Fax: (270) 651-9222. Web Site: www.wcluradio.com. Licensee: Royse Radio Inc. (acq 3-9-98; $225,000 with co-located AM). Format: Adult contemp. ♦Henry Royse, pres & gen mgr.

WLOC(AM)— February 1993: 1150 khz; 1 kw-D, 61 w-N. TL: N37 16 30 W85 55 00. Hrs open: 24 P.O. Box 98, Horse Cave, 42749. Secondary address: 1130 South Dixie, Horse Cave 42749. Phone: (270) 786-4400. Fax: (270) 786-4402. E-mail: wloc@scrtc.com Web Site: www.wloconline.com. Licensee: Forbis Communications Inc. (acq 12-5-2003; $120,000). Population served: 6,000 Format: Country, gospel. Target aud: 25 plus; serve entire area. ♦DeWayne Forbis, pres; Dewayne Forbis, gen mgr & gen sls mgr; Chris Jessie, progmg dir; Joe Berry, news dir.

Murray

WFGE(FM)—Listing follows WOFC(AM).

***WKMS-FM**— May 11, 1970: 91.3 mhz; 100 kw. 602 ft TL: N36 55 18 W88 05 50. Stereo. Hrs opn: 24 Box 2018, University Stn., 42071. Phone: (800) 599-4737. Phone: (270) 809-4359. Fax: (270) 809-4667. E-mail: wkms@murraystate.edu Web Site: www.wkms.org. Licensee: Board of Regents, Murray State University. Population served: 110,250 Natl. Network: NPR, PRI. Law Firm: Don Martin. Wire Svc: AP Format: news, diverse. News staff: 2; News: 82 hrs wkly. Target aud: 35 plus; life long learners. Spec prog: New age 3 hrs, folk & bluegrass 3 hrs, triple-A 3 hrs, urban contemp 3 hrs, blues 2 hrs, world music 1 hr, independent 1 hr wkly. ♦Kate Lochte, stn mgr; Tracy Ross, opns mgr; Rhonda Gibson, dev dir & dev mgr; Allen Fowler, prom mgr & chief of engrg; Mark Welch, progmg dir & mus dir.

WNBS(AM)— July 1948: 1340 khz; 1 kw-U. TL: N36 37 42 W88 18 04. Hrs open: 24 1500 Diuguid Dr., 42071-1669. Phone: (270) 753-2400. Fax: (270) 753-9434. Licensee: Forever Communications Inc. (group owner; acq 12-31-02; grpsl). Population served: 35,000 Natl. Network: ESPN Radio, CBS Radio. Rgnl rep: Rgnl Reps. Format: News, talk, sports. News staff: one; News: 10 hrs wkly. Target aud: 25-55. ♦Debbie Howard, gen mgr; Candi Freeland, news dir; Neal Bradley, progmg dir & sports cmtr.

WOFC(AM)— Sept 12, 1978: 1130 khz; 1.5 kw-D. TL: N36 38 08 W88 19 10. Hrs open: 24 1500 Diuguid Dr., 42071. Phone: (270) 753-2400. Fax: (270) 753-9434. Licensee: Forever Communications Inc. (group owner; (acq 12-31-2002; grpsl). Population served: 200,000 Natl. Network: CBS. Rgnl. Network: Ky. Net., Ky. Agri. Rgnl rep: Rgnl reps. Format: Adult contemp. News staff: 2; News: 20 hrs wkly. Target aud: 35 plus. ♦Scott Swalls, gen mgr; Mary Ellen Smith, gen sls mgr; Neil Bradley, progmg dir; Adam Bittel, chief of engrg.

WFGE(FM)—Co-owned with WOFC(AM). June 23, 1967: 103.7 mhz; 100 kw. 659 ft TL: N36 32 58 W88 19 52. Stereo. 24 Web Site: www.froggy103.com.800,000 Natl. Network: Westwood One. Format: Country. News staff: one; News: 4 hrs wkly. Target aud: 18-45. ♦Jay Crockett, stn mgr & progmg dir.

Neon

WVSG(AM)— Aug 31, 1956: 1480 khz; 5 kw-D. TL: N37 11 54 W82 42 42. Hrs open: 12 486 Lakeside Dr., Jenkins, 41537. Phone: (606) 634-9430. Licensee: Letcher County Broadcasting Inc. (acq 2-7-2007; $30,000). Population served: 150,000 Format: Relg. Target aud: 24-60; general. ♦Ernestine Kincer, pres; G.C. Kincer, gen mgr.

Newburg

WDRD(AM)— 1992: 680 khz; 1.3 kw-D, 450 w-N, DA-2. TL: N38 05 31 W85 40 56. Hrs open: 24 11700 Commonwealth Dr., Suite 800, Louisville, 40299. Phone: (502) 240-0602. Fax: (502) 240-0940. E-mail: john.salzman@abc.com Web Site: www.radiodisney.com. Licensee: Radio Disney Group LLC. Group owner: ABC Inc. (acq 2-14-02; $1.92 million). Format: Family hits. Target aud: Age 25-44 mothers of children; Mothers of children younger than 15. ♦John Salzman, gen mgr.

Newport

WNOP(AM)— Aug 21, 1948: 740 khz; 2.5 kw-D, 30 w-N, DA-2. TL: N39 05 41 W84 34 59. Stereo. Hrs opn: 24 5440 Moeller Ave., Norwood, OH, 45212. Phone: (513) 731-7740. Fax: (513) 731-6465. Licensee: Catholic Radio Foundation of Greater Cincinnati Inc. (acq 9-10-01). Format: Catholic relg. ♦Bill Levitt, stn mgr.

Nicholasville

WCGW(AM)— Sept 15, 1986: 770 khz; 1 kw-D. TL: N37 53 07 W84 31 46. Stereo. Hrs opn: 3270 Blazer Pkwy., Lexington, 40509. Phone: (859) 264-9700. Fax: (859) 264-9705. Web Site: www.wcgwam.com. Licensee: Christian Broadcasting System Ltd. (group owner) (acq 7-1-2006; grpsl). Population served: 988,000 Natl. Network: USA. Format: Southern gospel. Target aud: 25-54; above average in educ, family size, income. ♦Ron Knight, gen mgr.

WLTO(FM)— Aug 29, 1988: 102.5 mhz; 2 kw. Ant 400 ft TL: N37 49 52 W84 30 18. Stereo. Hrs opn: 24 300 W. Vine St., Suite 3, Lexington, 40507. Phone: (859) 253-5900. Fax: (859) 253-5940. Licensee: Cumulus Licensing Corp. Group owner: Cumulus Media Inc. (acq 7-22-99; grpsl). Format: Top-40. ♦Dave Kabakoff, gen mgr.

WVKY(AM)— December 1962: 1250 khz; 500 w-D, 59 w-N. TL: N37 54 18 W84 33 25. Hrs open: 24 3270 Blazer Pkwy., Lexington, 40509. Phone: (859) 264-9700. Fax: (859) 264-9705. Licensee: Christian Broadcasting System Ltd. Group owner: Mortenson Broadcasting Co. (acq 7-1-2006; grpsl). Population served: 350,000 Format: News/talk. ♦Jonathon R. Yinger, pres; Ron Kight, gen mgr.

Oak Grove

WEGI(FM)— Aug 31, 1964: 94.3 mhz; 6 kw. Ant 256 ft TL: N36 38 28 W87 26 01. Stereo. Hrs opn: 24 1640 Old Russellville Pike, Clarksburg, TN, 37043. Phone: (931) 648-7720. Fax: (931) 648-7769. Web Site: eagle943.com. Licensee: Saga Communications of Tuckessee LLC. Group owner: Saga Communications Inc. (acq 10-4-2002; $1.5 million with co-located AM). Population served: 85,000 Format: Classic hits. News staff: one; News: 5 hrs wkly. Target aud: 25-54. ♦Susan Quesenberry, gen mgr.

Okolona

***WJIE-FM**— Jan 1, 1988: 88.5 mhz; 24.5 kw. 623 ft TL: N38 01 59 W85 45 16. Stereo. Hrs opn: 24 Box 197309, Louisville, 40259. Secondary address: 5400 Minors Ln., Louisville 40219. Phone: (502) 968-1220. Fax: (502) 962-3143. Web Site: www.wjie.org. Licensee: Evangel Schools Inc. Population served: 900000 Natl. Network: Moody. Law Firm: Pepper & Corazzini. Format: Contemp christian music. News: 7 hrs wkly. Target aud: 25-49; Christian adults. ♦Jim Fraser, gen mgr.

Owensboro

WBKR(FM)—Listing follows WOMI(AM).

***WJVK(FM)**— 2004: 91.7 mhz; 100 w. Ant 174 ft TL: N37 44 48 W87 06 58. Hrs open: Box 539, Bowling Green, 42102. Secondary address: 1407 Scottsville Rd., Bowling Green 42104. Phone: (270) 781-7326. Fax: (270) 781-8005. E-mail: mail@christianfamilyradio.com Web Site: www.christianfamilyradio.com. Licensee: Bowling Green Community Broadcasting Inc. Format: Christian. ♦Mike Wilson, gen mgr; Dale McCubbins, progmg dir.

***WKWC(FM)**— Jan 21, 1983: 90.3 mhz; 5 kw. 100 ft TL: N37 44 37 W87 07 12. Stereo. Hrs opn: 8 AM-midnight 3000 Frederica St., 42301. Phone: (270) 852-3601. Fax: (270) 853-3597. Licensee: Kentucky Wesleyan College. Format: Relg. News: 2 hrs wkly. Target aud: 12 plus. ♦Pam Gray, gen mgr & progmg dir; Brandon Bartlett, news dir; Rick Graves, chief of engrg.

WOMI(AM)— Mar 7, 1938: 1490 khz; 830 w-U. TL: N37 44 29 W87 06 58. Hrs open: 24 3301 Frederica St., 42301.6082. Phone: (270) 683-1558. Fax: (270) 683-2128. E-mail: lcraig@wbkr.com Web Site: www.wbkr.com. Licensee: Regent Broadcasting of Evansville/Owensboro Inc. Group owner: Regent Communications Inc. (acq 2-25-03; with co-located FM). Population served: 54,000 Rgnl. Network: Ky. Net. Natl. Rep: Christal. Format: News/talk. News staff: 2. Target aud: 35-64. ♦Bill Stakelin, CEO; Mark Thomas, pres & gen mgr; Joe O'Neal, gen mgr & gen sls mgr; Chad Benefield, prom dir; Rick Crago, engrg dir.

WBKR(FM)—Co-owned with WOMI(AM). 1948: 92.5 mhz; 91 kw. 1,049 ft TL: N37 36 29 W87 03 15. (CP: 96 kw, ant 1,000 ft. TL: N37 46 20 W87 21 27). Stereo. 24 Web Site: wbkr.com.250,000 Format: Country. News staff: 2. Target aud: 25-54. Spec prog: Farm 2 hrs wkly. ♦Moon Mullins, progmg dir; Dave Spenser, mus dir; Cathy Carton, news dir; Rick Crago, engrg mgr; Michael Owns, traf mgr.

WSTO(FM)— June 7, 1948: 96.1 mhz; 100 kw. 1,000 ft TL: N37 46 20 W87 21 27. Stereo. Hrs opn: 24 Box 3848, Evansville, IN, 47736. Secondary address: 1162 Mt. Auburn Rd., Evansville 47720. Phone: (812) 421-9696. Fax: (812) 421-3273. Web Site: www.hot96.com. Licensee: South Central Communications Corp. (group owner; (acq 12-30-2003; $13 million). Population served: 187,600 Format: CHR. News staff: 4; News: one hr wkly. Target aud: 18-34. ♦Robert Shirel, CFO; Tim Huelsing, VP & gen mgr; Missy Bentley, prom dir; Jason Addams, progmg dir.

WVJS(AM)— Nov 26, 1947: 1420 khz; 5 kw-D, 1 kw-N, DA-2. TL: N37 46 32 W87 09 31. Stereo. Hrs opn: 24 1115 Tamarack Rd., Suite 500, 42301. Phone: (270) 683-5200. Fax: (270) 688-0108. Licensee: Cromwell Group Inc. of Kentucky. Group owner: The Cromwell Group Inc. (acq 11-20-02; $300,000). Population served: 150,000 Natl. Network: ABC. Natl. Rep: Rgnl Reps. Law Firm: Pepper & Corazzini. Format: Adult standards, memories. News staff: news progmg 3 hrs wkly News: one;. Target aud: 35-54. Spec prog: Farm one hr wkly. ♦Kevin Riecke, gen mgr.

Owingsville

WKCA(FM)— Dec 1, 1983: 107.7 mhz; 6 kw. 370 ft TL: N38 11 16 W83 46 34. Stereo. Hrs opn: 24 17 S. Court St., 40360. Phone: (606) 674-2266. Fax: (606) 674-6700. Licensee: Gateway Radio Works Inc. (group owner). Population served: 105,000 Natl. Network: ABC. Rgnl. Network: Ky. Agri., Ky. News Net. Law Firm: William Silva. Wire Svc: NWS (National Weather Service) Format: Real country. News staff: one; News: 10 hrs wkly. ♦Hays McMakin, pres; Jeff Ray, gen mgr; Becky Young-Black, opns mgr.

Paducah

WDDJ(FM)— Nov 26, 1946: 96.9 mhz; 100 kw. 340 ft TL: N37 05 55 W88 37 19. (CP: Ant 777 ft. TL: N37 02 56 W88 36 52). Stereo. Hrs opn: 24 Box 2397, 42002. Secondary address: 6000 Bristol Dr. 42003. Phone: (270) 534-9690. Fax: (270) 554-4613. E-mail: pd@electric969.com Web Site: www.electric969.com. Licensee: Bristol Broadcasting Co. Inc. Group owner: Nininger Stations (acq 6-24-97; $2.7 million with co-located AM). Population served: 250,000 Format: Adult top 40. News staff: 2; News: 2 hrs wkly. Target aud: 18-49; active, white & blue collar adults. ♦Gary Morse, gen mgr; Jamie Futrell, gen sls mgr & rgnl sls mgr; Mark Summer, progmg dir; Greg Walker, chief of engrg.

WPAD(AM)—Co-owned with WDDJ(FM). Aug 23, 1930: 1560 khz; 10 kw-D, 5 kw-N, DA-3. TL: N37 03 08 W88 36 03.24 Web Site: www.electric969.com.100,000 Natl. Network: Westwood One. Format: Sports. News staff: 2; News: 22 hrs wkly. Target aud: 35-64; upscale, white-collar.

WDXR(AM)— Dec 24, 1957: 1450 khz; 1 kw-U. TL: N37 05 55 W88 37 19. Stereo. Hrs opn: 24 Secondary address: 1176 Stat Rt. 45 N., Mayfield 42066. Phone: (270) 247-5122. Fax: (270) 554-5468. Licensee: Bristol Broadcasting Co. Inc. (group owner; (acq 3-15-2004; grpsl). Population served: 250,000 Natl. Network: ABC. Law Firm: Rosenman & Colin L.L.P. Format: Urban contemp. News staff: one; News: 5 hrs wkly. Target aud: 30-65. ♦Gary Morse, gen mgr & stn mgr.

***WGCF(FM)**— December 1996: 89.3 mhz; 12 kw. 492 ft TL: N37 11 31 W88 58 41. Hrs open: 24 1112 E. Kentucky Ave., Kevil, 42053. Phone: (270) 462-3020. Fax: (270) 462-3024. E-mail: info@wgcf.org Web Site: www.wgcf.org. Licensee: American Family Association. Group owner: American Family Radio (acq 11-25-2003; $200,000). Population served: 250,000 Format: Adult contemp, CHR, contemp Christian. ♦Bill Hughes, chmn & gen mgr.

WKYQ(FM)—Listing follows WKYX(AM).

WKYX(AM)— 1946: 570 khz; 1 kw-D, 500 w-N, DA-2. TL: N37 00 53 W88 36 46. Stereo. Hrs opn: 24
Simulcast with WKYX-FM Golconda, IL 100%.
Box 2397, 42002. Secondary address: 6000 Bristol Dr. 42003. Phone: (270) 554-8255. Fax: (270) 554-5468. Web Site: www.wkyx.com. Licensee: Bristol Broadcasting Co. Inc. Group owner: Nininger Stations (acq 11-23-71). Population served: 450,000 Natl. Rep: Christal. Law Firm: Fisher, Wayland, Cooper, Leader & Zaragoza L.L.P. Format: News/talk. News staff: 3; News: 20 hrs wkly. Target aud: 25-54; middle to upper income. ♦Gary Morse, gen mgr; Jamie Futrell, gen sls mgr.

WKYQ(FM)—Co-owned with WKYX(AM). 1947: 93.3 mhz; 100 kw. 915 ft TL: N37 00 53 W88 36 46. Stereo. 24 E-mail: production@wkyq.com Web Site: www.wkyq.com.450,000 Format: Country. News staff: 3. Target aud: 25-54.

WREZ(FM)—See Metropolis, IL

WRIK-FM—See Metropolis, IL

WZZL(FM)—See Reidland

Paintsville

WKLW-FM— June 18, 1993: 94.7 mhz; 4.9 kw. 731 ft TL: N37 42 42 W82 48 03. Stereo. Hrs opn: 24 Box 1407, 41240. Secondary address: 865 S. Mayo Tr. 41240. Phone: (606) 789-6664. Fax: (606) 789-6669. Web Site: www.wklw.com. Licensee: B & G Broadcasting Inc. Format: Hot adult contemp. ♦Alan Burton, gen mgr & stn mgr.

WKYH(AM)— Mar 18, 1985: 600 khz; 5 kw-D, 43 w-N. TL: N37 47 19 W82 47 07. Stereo. Hrs opn: 330 2nd St, 41240. Phone: (606) 789-3333. Fax: (859) 402-0260. Web Site: www.wkyh.com. Licensee: Highlands Broadcasting Corp. (acq 11-30-99). Natl. Network: Westwood One. Law Firm: Midlen & Guillot. Format: News/talk/sports info. Target aud: 25-49; middle to upper class adults. ♦Charles K. Belhasen, gen mgr, pres, opns mgr, gen sls mgr, progmg dir, news dir & chief of engrg.

WSIP(AM)— April 24, 1949: 1490 khz; 1 kw-U. TL: N37 48 21 W82 46 01. Hrs open: Box 597, 41240. Secondary address: 124 Main St. 41240. Phone: (606) 789-5311. Fax: (606) 789-7200. Licensee: S.I.P. Broadcasting Inc. Group owner: Key Broadcasting Inc. (acq 2-84). Population served: 66,300 Rgnl. Network: Ky. Net. Natl. Rep: Rgnl Reps. Format: Oldies. Target aud: General. ♦Spike Berkhimer, gen mgr, gen sls mgr & sports cmtr.

WSIP-FM— Jan 12, 1965: 98.9 mhz; 94 kw. Ant 600 ft TL: N37 47 45 W82 48 04. Stereo. 24 Web Site: www.wsipfm.com.579,000 Format: New country.

Paris

WGKS(FM)—Licensed to Paris. See Lexington

***WPTJ(FM)**— August 2003: 90.7 mhz; 10 kw. Ant 314 ft TL: N38 19 40 W84 07 44. Hrs open: 24 1811 Cynthiana Millersburg Rd., 40361. Secondary address: Lay Witness Broadcasting, Box 7 40362-0007. Phone: (859) 484-9691. E-mail: jsmith@wptj.org Web Site: www.wptj.org. Licensee: Lay Witness Outreach Inc. Population served: 225,000 Format: Relg. ♦John Smith, gen mgr; John Wesley Brett, progmg dir.

WYGH(AM)— January 1993: 1440 khz; 1 kw-D, 25 w-N. TL: N38 13 30 W84 14 59. Hrs open: 24 Box 50, Falmouth, 41040. Phone: (859) 472-1075. Fax: (859) 472-2875. E-mail: wygh@fuse.net Web Site: www.wygh.com. Licensee: Hammond Broadcasting Inc. Population served: 500,000 Format: Southern gospel. ♦Jan Hammond, gen mgr.

Philpot

WBIO(FM)— Nov 18, 1993: 94.7 mhz; 3 kw. 328 ft TL: N37 41 51 W86 59 26. Stereo. Hrs opn: 24 1115 Tamarack Rd., Suite 500, Owensboro, 42301. Phone: (270) 683-5200. Fax: (270) 688-0108. E-mail: wbioradio@adelphia.net Web Site: www.wbio.com. Licensee: Hancock Communications Inc. Group owner: The Cromwell Group Inc. (acq 6-17-93; $90,565; FTR: 7-5-93). Population served: 70,000 Natl. Network: ABC. Natl. Rep: Rgnl Reps. Law Firm: Pepper & Corazzini. Format: True country. News staff: one; News: 4 hrs wkly. Target aud: 25-54. ♦Bayard Walters, CEO; Kevin Riecke, gen mgr.

Pikeville

WBTH(AM)—Williamson WV

WDHR(FM)—Listing follows WPKE(AM).

***WJSO(FM)**— Apr 1989: 90.1 mhz; 3.8 kw. 455 ft TL: N37 27 52 W82 32 45. Hrs open: 24 Box 3237, 41502. Phone: (606) 432-0351. Phone: (312) 329-4300. Fax: (312) 329-8980. E-mail: wjso@moody.edu Web Site: wjso.mbn.org. Licensee: Moody Bible Institute of Chicago. (group owner; acq 12-18-91; donation; FTR: 1-13-92). Format: Relg. News: 15 hrs wkly. Target aud: 35-55. ♦Scott Keegan, gen mgr.

WLSI(AM)— Jan 20, 1949: 900 khz; 5 kw-D. TL: N37 29 06 W82 32 44. Hrs open: 6 AM-midnight
Simulcast with WPRT(AM) Prestonburg 100%.
Box 2200, 41502. Secondary address: 1240 Radio Dr. 41501. Phone: (606) 437-4051. Fax: (606) 432-2809. E-mail: wdhr@wdhr.com Web Site: www.900wlsi.com. Licensee: East Kentucky Broadcasting Corp. (acq 5-14-2003; $531,273 with WZLK(FM) Virgie). Population served: 50,000 Natl. Network: CNN Radio, NBC Radio, Sporting News Radio Network. Format: Talk. News staff: one; News: 21 hrs wkly. Target aud: 25-49. ♦Keith Casebolt, gen mgr.

WPKE(AM)— July 31, 1949: 1240 khz; 1 kw-U. TL: N37 28 53 W82 31 27. Stereo. Hrs opn: 24
Rebroadcasts WBPA(AM) Elkhorn City 100%.
Box 2200, 41502. Secondary address: 1240 Radio Dr. 41501. Phone: (606) 437-4051. Fax: (606) 432-2809. E-mail: wdhr@wdhr.com Web Site: www.wdhr.com. Licensee: East Kentucky Broadcasting Corp. (group owner; acq 1962). Population served: 35,000 Natl. Rep: Rgnl Reps. Rgnl rep: Rgnl Reps Law Firm: Womble, Carlylee, Sandridge & Rice. Format: Oldies. News staff: one; News: 5 hrs wkly. Target aud: General. ♦Keith Casebolt, gen mgr; Randy Jones, progmg dir; Walter Dingus, chief of engrg.

WDHR(FM)—Co-owned with WPKE(AM). Mar 25, 1966: 93.1 mhz; 22 kw. Ant 758 ft TL: N37 27 57 W82 33 04. Stereo. 24 Web Site: www.wdhr.com.600,000 Natl. Network: ABC. Format: Country. News staff: 2.

WXCC(FM)—Williamson WV

Pineville

WANO(AM)— Mar 16, 1957: 1230 khz; 1 kw-U. TL: N36 46 07 W83 42 59. Hrs open: Box 999, Middlesboro, 40965. Phone: (606) 337-2100. Fax: (606) 248-6397. Web Site: 1490wfxy.com. Licensee: Cumberland Media Group Inc. (acq 4-18-2006; grpsl). Population served: 12,000 Format: Oldies. ♦Brian O'Brien, opns mgr.

WRIL(FM)— Feb 24, 1973: 106.3 mhz; 350 w. 750 ft TL: N36 45 15 W83 42 23. (CP: 1.05 kw, ant 768 ft.). Hrs opn: Box 693, 40977. Secondary address: US 25 E. Log Mountain 40977. Phone: (606) 337-5200. Fax: (606) 337-8020. Licensee: Pine Hills Broadcasting Inc. (acq 2-22-84; $300,000; FTR: 3-5-84). Population served: 100,000 Format: Country. ♦Barry Himbre, gen mgr.

Pippa Passes

***WWJD(FM)**— Nov 1, 1986: 91.7 mhz; 7.3 kw. 544 ft TL: N37 19 45 W82 52 30. Stereo. Hrs opn: 24 Alice Lloyd College, 100 Purpose Rd., 41844. Phone: (606) 368-6131. Fax: (606) 368-6017. E-mail: wwjd@alc.com Licensee: Alice Lloyd College. Population served: 500,000 Format: Adult contemp, Christian. Target aud: 13-25. ♦Mike Sexton, gen mgr.

Prestonsburg

WDOC(AM)— November 1957: 1310 khz; 5 kw-D, 25 w-N. TL: N37 41 45 W82 45 24. Hrs open: Sunrise-sunset Box 345, 95 Jackson, 41653. Phone: (606) 886-2338. Phone: (606) 886-8409. Fax: (606) 886-1026. E-mail: q95prod@bellsouth.net Licensee: WDOC Inc. Population served: 135,000 Natl. Rep: Rgnl Reps. Format: Country, talk. Target aud: 25-64. ♦Gormon Collins Jr., pres & gen mgr; Samantha Osborne, gen sls mgr.

WQHY(FM)—Co-owned with WDOC(AM). Feb 11, 1968: 95.5 mhz; 100 kw. 1,000 ft TL: N37 41 45 W82 45 24. Stereo. 24 Web Site: www.q95fm.net.900,000 Natl. Network: ABC. Natl. Rep: Rgnl Reps. Format: Adult contemp. News staff: one; News: 2 hrs wkly. Target aud: 18-34. ♦Ben Sheperd, progmg dir; Ron Webb, mus dir; David Graff, news dir; Russ Lafforty, chief of engrg; Carla Hughes, traf mgr; Norm Marcum, local news ed.

WPRT(AM)— Dec 5, 1952: 960 khz; 5 kw-D. TL: N37 40 14 W82 45 14. Hrs open:
Simulcast with WLSI(AM) Pikeville 100%.
Box 2200, Pikeville, 41502. Secondary address: 1240 Radio Dr., Pikesville 41501. Phone: (606) 437-4051. Fax: (606) 432-2809. E-mail: wdhr@wdhrcom Web Site: www.900wlsi.com. Licensee: East Kentucky Radio Network Inc. (group owner; (acq 10-26-2001). Population served: 5,000 Natl. Network: CNN Radio, NBC Radio, Sporting News Radio Network. Format: Talk. ♦Keith Casebolt, gen mgr.

WXKZ-FM— Feb 10, 1967: 105.3 mhz; 4.7 kw. Ant 371 ft TL: N37 39 24 W82 45 58. Stereo. Hrs opn: Box 1049, Harold, 41635. Phone: (606) 478-1200. Fax: (606) 478-4202. E-mail: prod@mikrotec.com Web Site: www.thedoublex.com. Licensee: Adam Gearheart dba WXLR-FM (acq 1-17-97; with co-located AM). Population served: 5,000 Format: Oldies. ♦Barry Boyd, gen mgr & gen sls mgr; Mel Stevens, progmg dir.

Princeton

WAVJ(FM)—Listing follows WPKY(AM).

WPKY(AM)— Mar 15, 1950: 1580 khz; 250 w-D. TL: N37 07 14 W87 51 31. Hrs open: 24 Box 270, 42445. Secondary address: 108 W. Main St. 42445. Phone: (270) 365-2072. Fax: (270) 365-2073. E-mail: wavj@commonwealthbroadcasting.com Licensee: Caldwell County CBC Inc. (acq 6-25-98; $362,000 with co-located FM). Population served: 25,000 Natl. Network: ESPN Radio. Format: Sports. News staff: one; News: 11 hrs wkly. Target aud: General. ♦Tom Rogers, gen mgr; Caroline Garcia-Quinn, opns mgr & progmg dir; Shirley Gray, gen sls mgr & progmg dir; Ed Thomas, news dir & chief of engrg.

WAVJ(FM)—Co-owned with WPKY(AM). Apr 1, 1969: 104.9 mhz; 3 kw. 187 ft TL: N37 07 14 W83 51 31. Stereo. 24 40,000 Format: Lite rock AC.

Providence

WWKY(FM)— Apr 9, 1976: 97.7 mhz; 6 kw. 328 ft TL: N37 24 52 W87 34 23. Stereo. Hrs opn: 24 Box 1310, Madisonville, 42431. Secondary address: 265 S. Main St., Madisonville 42431. Phone: (270) 825-9779. Fax: (270) 825-3260. E-mail: wwky@commonwealthbroadcasting.com Licensee: Hopkins-Webster CBC Inc. Group owner: Commonwealth Broadcasting Corp. (acq 5-21-98; $425,000). Population served: 152,752 Natl. Network: CNN Radio. Rgnl. Network: Ky. Net. Law Firm: Leonard S. Joyce. Format: Oldies. News staff: 2; News: 21 hrs wkly. Target aud: 25-54. ♦Tom Rogers, gen mgr.

Radcliff

WAKY(FM)— July 25, 1995: 103.5 mhz; 3.5 kw. Ant 761 ft TL: N37 52 45 W85 43 03. Stereo. Hrs opn: 24 Box 2087, Elizabethtown, 42702. Secondary address: 519 N. Miles, Elizabethtown 42701. Phone: (270) 766-1035. Fax: (270) 769-1052. Web Site: www.waky1035.com. Licensee: W & B Broadcasting Inc. Population served: 513,878 Natl. Network: ABC. Law Firm: Miller & Miller. Format: Oldies. News staff: 14; News: 4 hrs wkly. Target aud: 25-54; baby boomers. ♦Bill Walters, pres; Rene Bell, gen mgr; Cale Tharp, opns mgr & chief of engrg.

Reidland

WZZL(FM)— October 1992: 106.7 mhz; 1.35 kw. 492 ft TL: N37 03 23 W88 27 22. Hrs open: 24 Box 7501, Paducah, 42002-8123. Phone: (270) 538-5251. Fax: (270) 415-0599. Web Site: www.wzzl.com. Licensee: Withers Broadcasting Co. of Paducah LLC. Group owner: Withers Broadcasting Co. (acq 9-11-97; grpsl). Population served: 162,000 Format: Rock/AOR. Target aud: 18-49. ♦Rick Lambert, gen mgr.

Richmond

WCBR(AM)— March 1969: 1110 khz; 250 w-D. TL: N37 44 09 W84 16 05. Hrs open: Sunrise-sunset Box 570, 40476-0570. Secondary address: 509 Leighway Dr. 40475. Phone: (859) 623-1235. Fax: (859) 623-7094. E-mail: wcbrradio@bellsouth.net Web Site: wcbr1110.com. Licensee: WCBR Inc. (acq 3-12-2004). Population served: 21,141 Natl. Network: USA. Format: Southern gospel. News: 5 hrs wkly. Target aud: 35 plus; older adult listener. Spec prog: Loc talk shows, news, sports & relg talk 35 hrs wkly. ♦Bill Robbins, pres; David L. Humes, exec VP & gen mgr; Malissa Blair, traf mgr.

***WEKU(FM)**— September 1968: 88.9 mhz; 50 kw. 720 ft TL: N37 52 45 W84 19 33. Stereo. Hrs opn: 24 102 Perkins Bldg., 521 Lancaster Ave., 40475-3102. Phone: (859) 622-1660. Fax: (859) 622-6276. E-mail: wekunews@eku.edu Web Site: www.weku.fm. Licensee: Board of Regents, Eastern Kentucky University. Population served: 700,000 Natl. Network: NPR, PRI. Law Firm: Hardy, Carey & Chautin. Wire Svc: AP Format: Class, news magazine, info. News staff: 3; News: 35 hrs wkly. Target aud: General. ♦Tim Singleton, gen mgr & stn mgr; Mary Ellyn Cain, opns mgr; Carol Siler, dev dir; Laura Allen, progmg dir; Charles Compton, news dir; Bill Browning, chief of engrg.

WEKY(AM)— Oct 17, 1953: 1340 khz; 1 kw-U. TL: N37 43 00 W84 18 25. Hrs open: 24 128 Big Hill Ave., 40475. Phone: (859) 623-1386. Fax: (859) 623-1341. E-mail: coyote@chpl.net Web Site: www.wcyo.com. Licensee: Wallingford Communications Inc. Group owner: Wallingford Broadcasting Co. (acq 1999; grpsl). Population served: 16,861 Rgnl. Network: Ky. News Net. Natl. Rep: Rgnl Reps. Format: News/talk. News staff: one; News: 3 hrs wkly. Target aud: 25-54. Spec prog: Black 12 hrs, relg 6 hrs wkly. ♦Kelly Wallingford, gen mgr; Ray White, opns mgr & progmg dir.

WVLK-FM— May 12, 1972: 101.5 mhz; 7.2 kw. Ant 541 ft TL: N37 52 45 W84 19 33. Stereo. Hrs opn: 24 300 W. Vine St., Suite 3, Lexington, 40507. Phone: (859) 253-5900. Fax: (859) 253-5940. Licensee: Cumulus Licensing Corp. Group owner: Cumulus Media Inc. (acq 10-5-99; grpsl). Format: Sports talk. News staff: one; News: 6 hrs wkly. ♦Dave Kabakoff, gen mgr.

Russell Springs

WHVE(FM)— 1993: 92.7 mhz; 6 kw. 328 ft TL: N37 00 31 W85 12 14. Hrs open: Box 927, Columbia, 42728. Secondary address: 7955 Russell Springs Rd. 42642. Phone: (270) 384-7979. Fax: (270) 384-6244. E-mail: thewave@ridingthewave.com Web Site: www.ridingthewave.com. Licensee: Shoreline Communications Inc. (group owner; (acq 5-2002; $525,000). Format: Adult contemp. ♦Alan W. Reed, gen mgr; Jan Royse, gen mgr; Don Salmon, opns mgr.

WIDS(AM)— Oct 14, 1982: 570 khz; 500 w-D. TL: N37 05 42 W85 05 05. Hrs open: 24 Box 50, Falmouth, 41040. Phone: (859) 472-1075. Fax: (859) 472-2875. E-mail: wids@fuse.net Web Site: www.tri-stategospel.org. Licensee: Hammond Broadcasting Inc. (acq 8-16-94; FTR: 8-29-94). Format: Southern gospel. ♦Jan Hammond, gen mgr.

WJKY(AM)—See Jamestown

WJRS(FM)—See Jamestown

Russellville

WRUS(AM)— Aug 28, 1953: 610 khz; 2.5 kw-D, 73 w-N. TL: N36 48 51 W86 52 50. (CP: 500 w-N, TL: N36 52 29 W86 52 56 (night)). Hrs opn: Sunrise-sunset Box 1740, 42276. Phone: (270) 726-2471. Fax: (270) 726-3095. Licensee: Logan Radio Inc. (acq 10-24-02). Population served: 300,000 Format: News/talk, classic country. ♦Chris McGinnis, gen mgr & gen sls mgr.

WUBT(FM)— Mar 28, 1965: 101.1 mhz; 47 kw. 1,289 ft TL: N36 31 36 W86 41 14. Stereo. Hrs opn: 24 55 Music Sq. W., Nashville, TN, 37203. Phone: (615) 664-2400. Fax: (615) 664-2457. Web Site: www.101thebeat.com. Licensee: Capstar TX L.P. Group owner: Clear Channel Communications Inc. (acq 8-30-00; grpsl). Format: Urban contemp. ♦David Alpert, gen mgr; Bill Reed, sls dir; Keith Kaufman, mktg dir.

Saint Matthews

WRKA(FM)— Oct 19, 1964: 103.1 mhz; 6 kw. 312 ft TL: N38 16 03 W85 41 53. Stereo. Hrs opn: 24 612 4th Ave., Suite 100, Louisville, 40202. Phone: (502) 589-4800. Fax: (502) 587-0212. Web Site: www.wrka.com. Licensee: Cox Radio Inc. Group owner: Cox Broadcasting Population served: 976,800 Natl. Network: ABC. Natl. Rep: Christal. Law Firm: Dow, Lohnes & Albertson. Format: Classic Hits. Target aud: 35-54. ♦Tim Hartlage, gen sls mgr; Todd Schumacher, mktg mgr; Matt Killion, progmg dir.

Salyersville

WRLV(AM)— September 1979: 1140 khz; 1 kw-D. TL: N37 44 58 W83 05 19. Hrs open: 12 Box 550, 41465. Secondary address: 225 S. Church St. 41465. Phone: (606) 349-6125. Phone: (606) 349-6126. Fax: (606) 349-6129. E-mail: coyote2@foothills.net Web Site: www.wrlvradio.com. Licensee: Morgan County Industires Inc. (acq 7-5-2007; $460,000 with co-located FM). Population served: 65,000 Natl. Rep: Rgnl Reps. Format: Country. News staff: 2; News: 5 hrs wkly. Target aud: 35-65; middle age to elderly. ♦C.C. Smith, pres; Kathy Puckett, gen mgr, gen sls mgr, mktg VP, prom mgr & adv VP; Bryan Russell, prom dir, news dir & mus critic; Teresa Witten, progmg dir, progmg dir & traf mgr; Sanford Baca, local news ed; Terry Lykins, relg ed; Scott Ratliff, disc jockey.

WRLV-FM— Aug 25, 1989: 106.5 mhz; 5.9 kw. Ant 331 ft TL: N37 45 27 W83 03 50. Stereo. 24 Web Site: www.wrlvradio.com.100,000 Natl. Rep: Rgnl Reps. Format: Country. News staff: 2; News: 5 hrs wkly. Target aud: 18-80; young to elderly. ♦Kathy Puckett, sls VP, adv dir & spec ev coord; Bryan Russell, mus dir & mus critic; Teresa Witten, traf mgr; Sanford Baca, local news ed; Hershall Wright, relg ed; Terry Lykins, relg ed; Scott Ratliff, sports cmtr.

Scottsville

WLCK(AM)— Feb 27, 1958: 1250 khz; 1 kw-D. TL: N36 44 24 W86 10 20. Hrs open: 6 AM-9 PM Box 158, 42164. Secondary address: 104 1/2 Public Sq. 42164. Phone: (270) 237-3148. Fax: (270) 237-3533. E-mail: wlckwvle@nctc.com Web Site: www.wvleradio.com. Licensee: Sherandan Broadcasting Co. (acq 7-3-85). Population served: 20,000 Natl. Network: USA. Law Firm: Hardy & Carey. Format: Relg. News staff: one; News: 12 hrs wkly. Target aud: General. ♦Darrin Evans, pres, gen sls mgr, progmg dir & mus dir; Chris Nelson, gen mgr & news dir; Max Murphy, chief of engrg.

Shelbyville

WCND(AM)— June 3, 1964: 940 khz; 250 w-D. TL: N38 13 00 W85 09 45. Hrs open: 115 W. Main St., Frankfort, 40601-2807. Phone: (502) 875-1130. Fax: (502) 875-1225. Web Site: www.hometown1490.com. Licensee: CC Licenses LLC. Group owner: Clear Channel Communications Inc. (acq 2-1-2002; with co-located FM). Population served: 401,112 Rgnl. Network: Ky. Net. Format: Oldies. News staff: one; News: one hr wkly. Target aud: 35-64; upscale, white collar. ♦John Roberts, gen mgr.

WKRD-FM— Sept 30, 1989: 101.7 mhz; 6 kw. Ant 328 ft TL: N38 12 48 W85 10 16. Stereo. Hrs opn: 24
Simulcasts WKRD(AM) Louisville 100%.
4000 #1 Radio Dr., Louisville, 40218. Phone: (502) 479-2222. Fax: (502) 479-2223. Licensee: CC Licenses LLC. (acq 2-1-2002; with co-located AM). Population served: 750,000 Natl. Network: Fox Sports. Format: Sports. ♦Kevin Hughes, gen mgr.

Shepherdsville

WLRS(FM)— 1993: 105.1 mhz; 2.2 kw. Ant 446 ft TL: N38 02 54 W85 46 04. Hrs open: 520 S. 4th Ave., Suite 200, Louisville, 40202. Phone: (502) 625-1220. Fax: (502) 584-1051. Web Site: www.wlrs.com. Licensee: Blue Chip Broadcasting Licenses Ltd. Group owner: Radio One Inc. (acq 4-30-2001; grpsl). Population served: 1000000 Format: New rock. ♦Dale Schaeffer, gen mgr.

Smiths Grove

WUHU(FM)— Dec 1, 1986: 107.1 mhz; 50 kw. Ant 492 ft TL: N36 50 35 W86 15 30. Stereo. Hrs opn: 24 1919 Scottsville Rd., Bowling Green, 42103. Phone: (270) 843-3333. Fax: (270) 843-0454. E-mail: mark@forevercomm.com Web Site: www.allhitwuhu107.com. Licensee: Forever Communications Inc. (group owner; (acq 2001). Population served: 280,000 Natl. Rep: Christal. Law Firm: Kaye, Scholer, LLP. Format: Hot adult contemp. Target aud: 18-49. ♦Christine Hillard, pres; Brooke Summers, opns dir & progmg dir.

Somerset

***WDCL-FM**— July 1985: 89.7 mhz; 100 kw. Ant 570 ft TL: N37 09 29 W85 09 50. Stereo. Hrs opn: 24
Rebroadcasts WKYU-FM Bowling Green 100%.
Western Kentucky Univ., 1906 College Heights Blvd., Bowling Green, 42101. Phone: (270) 745-5489. Phone: (800) 599-9598. Fax: (270) 745-6272. E-mail: wkyufm@wku.edu Web Site: www.wkyufm.org. Licensee: Western Kentucky University. Natl. Network: PRI, NPR. Rgnl. Network: Ky. Net. Law Firm: Leventhal, Senter & Lerman. Format: Class, news. News staff: 3; News: 35 hrs wkly. Target aud: General. Spec prog: Folk 5 hrs, jazz 15 hrs wkly. ♦Peter Bryant, gen mgr.

WKEQ(FM)—Listing follows WSFC(AM).

***WKVY(FM)**— 2004: 88.1 mhz; 4 kw vert. Ant 526 ft TL: N37 04 36 W84 48 39. Stereo. Hrs opn: 24 2351 Sunset Blvd., Suite 170-218, Rocklin, CA, 95765. Phone: (916) 251-1600. Fax: (916) 251-1650. E-mail: klove@klove.com Web Site: www.klove.com. Licensee: Educational Media Foundation. Group owner: EMF Broadcasting. Natl. Network: K-Love. Law Firm: Shaw Pittman. Format: Contemp Christian. News staff: 3. Target aud: 25-44; Judeo Christian, female. ♦Richard Jenkins, pres; Mike Novak, VP; Keith Whipple, dev dir; David Pierce, progmg mgr; Ed Lenane, news dir; Sam Wallington, engrg dir; Karen Johnson, news rptr; Marya Morgan, news rptr; Richard Hunt, news rptr.

WLLK-FM— Aug 14, 1989: 102.3 mhz; 6 kw. 328 ft TL: N37 04 41 W84 40 39. Hrs open: 24 Box 740, 42502. Secondary address: 101 First Radio Ln. 42503. Phone: (606) 678-5151. Fax: (606) 678-2026. E-mail: wsek@clearchannel.com Web Site: www.somersetradio.com. Licensee: Capstar TX L.P. Group owner: Clear Channel Communications Inc. (acq 12-8-2000; grpsl). Population served: 100,000 Format: Hot adult contemp. News staff: one; News: 6 hrs wkly. Target aud: 25-54. ♦Richard Dills, gen mgr.

WSFC(AM)— Dec 14, 1947: 1240 khz; 790 w-U. TL: N37 07 06 W84 36 44. Stereo. Hrs opn: 24 Box 740, 42502. Secondary address: 101 First Radio Ln. 42503. Phone: (606) 678-5151. Fax: (606) 678-2026. Web Site: www.wsfcam.com. Licensee: Capstar TX L.P. Group owner: Clear Channel Communications Inc. (acq 12-8-2000; grpsl). Population served: 50,000 Natl. Rep: Rgnl Reps. Law Firm: Latham & Watkins. Wire Svc: NOAA Weather Format: Talk. News staff: one; News: 15 hrs wkly. Target aud: General. ♦Richard Dills, gen mgr, stn mgr, sls dir & gen sls mgr; Rod Zimmerman, progmg dir & traf mgr.

WKEQ(FM)—Co-owned with WSFC(AM). Sept 1, 1964: 97.1 mhz; 27.5 kw. 659 ft TL: N36 57 40 W84 34 07. Stereo. Web Site: www.wsfcam.com.85,000 Format: Classic hits.

***WTHL(FM)**— July 16, 1987: 90.5 mhz; 50 kw. 590 ft TL: N37 07 52 W84 33 15. Stereo. Hrs opn: Box 1423, 42502. Secondary address: 93 Rainbow Terr. 42503. Phone: (606) 679-6300. Fax: (606) 679-1342. Web Site: www.kingofkingsradio.net. Licensee: Somerset Educational Broadcasting Foundation. Natl. Network: Moody. Format: Conservative, traditional relg, educ. Target aud: 40 plus; people with conservative, traditional & relg values & interests. ♦David Carr, gen mgr; Carolyn Jones, progmg dir; Marvin Whitaker, gen sls mgr & chief of engrg.

WTLO(AM)— Nov 1, 1958: 1480 khz; 1 kw-D. TL: N37 05 15 W84 38 14. Stereo. Hrs opn: Box 1480, 42502. Secondary address: 290 WTLO Rd. 42502. Phone: (606) 678-8151. Fax: (606) 678-8152. E-mail: wtlo@usa.com Web Site: www.wtloradio.com. Licensee: Cumberland Communications Inc. (acq 11-6-74; $255,000). Population served: 67,800 Natl. Rep: Keystone (unwired net). Format: Unforgettable oldies, rhythm, news/talk. Target aud: 45 plus; upscale & highly mobile. Spec prog: Farm one hr, relg 4 hrs wkly. ♦Brooke Cary, gen mgr & opns mgr; John Moggard, gen sls mgr.

Springfield

WYSB(FM)— Feb 17, 1989: 102.7 mhz; 4 kw. Ant 354 ft TL: N37 41 43 W85 19 06. Hrs open: 24 Box 190, Bardstown, 40004. Phone: (502) 350-4482. Fax: (502) 350-4483. E-mail: kfogle@commonwealthbroadcashieg.com Licensee: Washington County CBC Inc. Group owner: Commonwealth Broadcasting Corp. (acq 10-30-97; $350,000). Population served: 50,000 Natl. Network: ABC. Rgnl. Network: Ky. Net. Format: Adult contemp. News staff: 4; News: 20 hrs wkly. Target aud: 25-55. Spec prog: Sports 12 hrs, farm 10 hrs wkly. ♦Kenny Fogle, gen mgr, gen sls mgr & adv dir.

Stamping Ground

WLXO(FM)— Dec 15, 1994: 96.1 mhz; 6 kw. Ant 328 ft TL: N38 12 15 W84 32 51. Hrs open: 401 W. Main, Suite 301, Lexington, 40507. Phone: (859) 233-1515. Fax: (859) 233-1517. Web Site: www.supertalk961.com. Licensee: Clarity Communications Inc. (acq 8-7-01; $400,000). Format: Talk. ♦Lynn Martin, pres; James MacFarlane, gen mgr.

Stanford

WRSL(AM)— Nov 1, 1961: 1520 khz; 500 w-D. TL: N37 33 03 W84 38 45. (CP: COL Corbin. 1600 khz; 2 kw-D, 27 w-N. TL: N37 01 06 W84 05 58). Stereo. Hrs opn: Sunrise-sunset 100 N. La Salle St., Ste 1111, Chicago, IL, 60602-3537. Phone: (312) 345-1111. Fax: (606) 365-7979. Licensee: Lincoln-Garrard Broadcasting Co. Inc. (acq 2-11-2002; with co-located FM). Population served: 70,000 Format: Country. News staff: 2. Target aud: 25-54; young Christian families. Spec prog: Farm & community affrs 4 hrs wkly. ♦Johnathan Smith, pres; Renee Schoebel, gen mgr; David Smith, chief of engrg.

WXKY-FM— May 22, 1967: 96.3 mhz; 12.5 kw. Ant 472 ft TL: N37 31 27 W84 52 12. Hrs open: 2351 Sunset Blvd., Suite 170-218, Rocklin, CA, 95765. Phone: (916) 251-1600. Fax: (916) 251-1650. Web Site: www.klove.com. Licensee: Educational Media Foundation. (acq 10-20-2004; $800,000). Population served: 100,000 Natl. Network: K-Love. Law Firm: Shaw Pittman LLP. Format: Christian music. ♦Richard Jenkins, pres; Mike Novak, VP; Keith Whipple, dev dir; David Pierce, progmg mgr; Ed Lenane, news dir; Sam Wallington, engrg dir; Karen Johnson, news rptr; Marya Morgan, news rptr; Richard Hunt, news rptr.

Stanton

WBFC(AM)— June 21, 1975: 1470 khz; 2500 w-D, 25 w-N. TL: N37 53 14 W83 52 58. Hrs open: Box 577, 40380. Secondary address: 2401 Paint Creek Rd. 40380. Phone: (606) 663-6631. Fax: (606) 663-2267. E-mail: beverly@wbfcam.com Web Site: www.wbfcam.com. Licensee: Combs Broadcasting Inc. (acq 10-13-98; $70,000). Population served: 150,000 Format: Southern gospel. ♦James Harold Combs, pres; Beverly Combs, gen mgr.

WSKV(FM)— Aug 10, 1974: 104.9 mhz; 440 w. 680 ft TL: N37 45 43 W83 50 36. Stereo. Hrs opn: 24 Box 610, 40380. Secondary address: 28 W. Hall's Rd. 40380. Phone: (606) 663-2811. Fax: (606) 663-2895. Licensee: Moore Country 104 LLC (acq 12-16-2004; $650,000). Population served: 50,000 Rgnl. Network: Ky. Net. Format: Classic Country. Target aud: General. ♦A.C. Moore, gen mgr.

Sturgis

WMSK-FM— 11/09/2006: 101.3 mhz; 6 kw. Ant 276 ft TL: N37 40 00 W87 55 40. Stereo. Hrs opn: 24
WMSK-FM simulcasts with WMSK-AM licenced to Henson Media of Union County, LLC.
P O Box 369, Morganfield, 42437. Secondary address: 1339 U.S. Highway 60, Morganfield 42437. Phone: (270) 389-1550. Fax: (270) 389-1553. E-mail: wmsk@bellsouth.net Licensee: Henson Media Inc. Natl. Network: Jones Radio Networks. Rgnl rep: Regional Reps Wire Svc: AP Format: Country; local news; sports; information. News: 20 hrs wkly. Target aud: 25-64; adults. Spec prog: High School, College & professional sports; Religion. ♦Ed Henson, pres; Rhonda Gibson, opns mgr.

Tompkinsville

WKWY(FM)— 2003: 102.7 mhz; 6 kw. Ant 315 ft TL: N36 43 27 W85 40 53. Hrs open: 6 AM-10 PM 341 Radio Station Rd., 42167. Phone: (270) 487-6119. Fax: (270) 487-8462. Licensee: Elizabeth Bernice Whittimore (acq 8-10-2006). Format: Country. News staff: 2; News: 20 wkly. All age group. Spec prog: Bluegrass 6 hrs, gospel 7 hrs wkly. ♦Elizabeth Whittimore, gen mgr; Bernice Whittimore, gen sls mgr.

WTKY(AM)— May 28, 1960: 1370 khz; 2.1 kw-D. TL: N36 43 27 W85 40 53. Stereo. Hrs opn: 341 Radio Station Rd., 42167. Phone: (270) 487-6119. Fax: (270) 487-8462. E-mail: wtky@alltel.net Licensee: Whittimore Enterprises Inc. (acq 8-10-2006). Population served: 29,000 Natl. Network: CBS. Format: Country. News staff: 2; News: progmg 17 hrs wkly. Target aud: All age groups. ♦Bernice Whittimore, gen mgr, gen mgr, gen sls mgr & progmg dir.

WTKY-FM— Jan 20, 1972: 92.1 mhz; 6 kw. 328 ft TL: N36 43 27 W85 40 53. Stereo. 6-10 AM 22,000 News staff: 2; News: 20 hrs wkly.

Upton

***WJCR-FM**— February 1990: 90.1 mhz; 100 kw. 383 ft TL: N37 25 57 W86 01 50. Stereo. Hrs opn: 24 Box 91, 42784-0091. Secondary address: 13101 Raider Hollow Rd. 42784. Phone: (270) 369-8614. Fax: (270) 369-7402. E-mail: wjcrfm@yahoo.com Web Site: www.wjcr.org. Licensee: FM 90.1 Inc. Law Firm: Reddy, Begley & McCormick. Format: Southern gospel. Target aud: General. Spec prog: 5 hrs live prayline wkly. ♦Don Powell, pres & gen mgr; Lauree K. Powell, CFO & VP; Gary Richardson, progmg dir; Larry Baysinger, engrg dir & chief of engrg.

Valley Station

WRVI(FM)— 1982: 105.9 mhz; 1.9 kw. 413 ft TL: N38 08 16 W85 56 06. Hrs opn: 9960 Corporate Campus Dr., Suite 3600, Louisville, 40223. Phone: (502) 339-9470. Fax: (502) 423-3139. Web Site: www.salemradiogroup.com. Licensee: Salem of Kentucky Inc. Group owner: Salem Communications Corp. (acq 1999; $5 million with WFIA-FM New Albany, IN). Format: Christian contemp music. ♦Gordon Marcy, gen mgr.

Vanceburg

WKKS(AM)— June 1, 1958: 1570 khz; 1 kw-D. TL: N38 35 50 W83 20 50. Hrs opn: 1106 Fairlane Dr., 41179. Phone: (606) 796-3031. Fax: (606) 796-6186. Licensee: Brown Communications Inc. (acq 1984). Population served: 1,773 Format: Classic country. ♦Dennis Brown, pres, gen mgr, gen sls mgr & progmg dir.

WKKS-FM— 1983: 104.9 mhz; 3 kw. 298 ft TL: N38 36 19 W83 19 57. Format: Top-40.

Vancleve

WMTC(AM)— June 1948: 730 khz; 5 kw-D, DA. TL: N37 36 12 W83 26 39. Hrs opn: 6 AM-sunset 1036 Hwy. 541, Jackson, 41339. Phone: (606) 666-5006. Phone: (606) 666-9512. Fax: (606) 666-7534. E-mail: wmtc@asburyusa.net Web Site: www.mountaingospel.org. Licensee: Kentucky Mountain Holiness Assn. Population served: 2,000,000 Natl. Network: Salem Radio Network. Law Firm: Womble, Carlyle, Sandridge & Rice. Format: Relg, Christian. News: 14 hrs wkly. Spec prog: Farm one hr wkly. ♦Philip Speas, pres; Jennifer Cox, gen mgr & mus dir; Anna Marshall, progmg dir; Kenneth Amspaugh, chief of engrg.

WMTC-FM— Jan 1, 1991: 99.9 mhz; 6 kw. Ant 328 ft TL: N37 36 23 W83 26 48. Stereo. 24 Phone: (606) 666-5006. E-mail: wmtc@asburyusa.net 2,000,000 Format: Christian, relg. Spec prog: Farm one hr wkly. ♦Seldon Short, VP; Seldon Short, gen mgr, sls dir, edit dir & farm dir; Jennifer Cox, progmg dir; Gordon Sampsel, mus dir; Theresa Kerley, disc jockey.

Versailles

WCDA(FM)— July 16, 1973: 106.3 mhz; 3 kw. Ant 316 ft TL: N38 02 44 W84 39 29. Stereo. Hrs opn: 24 401 W. Main, Suite 301, Lexington, 40507. Phone: (859) 233-1515. Fax: (859) 233-1517. Web Site: www.cd1063.com. Licensee: L.M. Communications Inc. (group owner; acq 9-3-98). Population served: 400,000 Natl. Network: USA. Format: Hot adult contemp. News: 2 hrs wkly. Target aud: 25-49; female. ♦Lynn Martin, pres; James E. MacFarlane, gen mgr.

Vine Grove

WRZI(FM)— Oct 5, 1993: 101.5 mhz; 6 kw. 328 ft TL: N37 35 07 W85 50 20. Stereo. Hrs opn: 24 611 W. Poplar St., Suite C2, Elizabethtown, 42701-2483. Phone: (270) 763-0800. Fax: (270) 769-6349. Web Site: etownpoint.com. Licensee: Elizabethtown CBC Inc. Group owner: Commonwealth Broadcasting Corp. (acq 7-1-00; grpsl). Population served: 200,000 Rgnl rep: Rgnl Reps. Law Firm: Verner, Liipfert, Bernhard, McPherson & Hand. Format: Rock. News staff: one; News: 3 hrs wkly. Target aud: 25-45; males. ♦Steve Newberry, pres; Roth Stratton, gen mgr & sls VP; Dan Diaz, opns VP & opns dir; Misty Russell, progmg dir; Mike Graham, engrg dir; Holli Lee, traf mgr.

Virgie

WZLK(FM)— Nov 15, 1992: 107.5 mhz; 580 w. TL: N37 22 47 W82 34 11. (CP: 1.12 kw). Hrs opn: Box 2200, Pikesville, 41502. Secondary address: 1240 Radio Dr. 41502. Phone: (606) 437-4051. Fax: (606) 432-2809. E-mail: wdhr@wdhr.com Licensee: East Kentucky Broadcasting Corp. (acq 5-14-2003; $531,273 with WLSI(AM) Pikeville). Format: Top 40s. ♦Keith Casebolt, gen mgr.

Warsaw

WKID(FM)—See Vevay, IN

West Liberty

WLKS(AM)— July 25, 1965: 1450 khz; 1 kw-U. TL: N37 55 36 W83 16 41. Stereo. Hrs opn: 24 Box 338, 129 College St., 41472. Phone: (606) 743-3145. Fax: (606) 743-9557. Licensee: Morgan County Industries Inc. (group owner) Population served: 83,920 Natl. Rep: Rgnl Reps. Format: Oldies. News staff: one; News: 35 hrs wkly. Target aud: General. Spec prog: Farm 5 hrs wkly. ♦Paul Lyons, gen mgr.

WLKS-FM— Jan 1, 1994: 102.9 mhz; 6 kw. 328 ft TL: N37 55 36 W83 16 41.24 129 College St., 41472. Phone: (606) 743-1029. Format: Country.

Whitesburg

***WMMT(FM)**— Nov 1, 1985: 88.7 mhz; 1 kw horiz 15 kw vert. 1,469 ft TL: N37 06 38 W82 44 15. Stereo. Hrs opn: 24 91 Madison Ave., 41858. Phone: (606) 633-0108. Fax: (606) 633-1009. E-mail: wmmtfm@appalshop.org Web Site: www.appalshop.org. Licensee: Appalshop Inc. Rgnl. Network: Ky. Pub. Format: Bluegrass, country. Target aud: General. ♦Cheryl Marshall, gen mgr & dev dir.

WTCW(AM)— Feb 19, 1953: 920 khz; 4.2 kw-D, 43 w-N. TL: N37 08 46 W82 46 01. Hrs opn: Box 288, Mayking, 41837. Phone: (606) 633-4434. Phone: (606) 633-2711. Fax: (606) 633-4445. E-mail: wxkq@yahoo.com Web Site: www.1039thebulldog.com. Licensee: T.C.W. Broadcasting Co. Inc. Group owner: Key Broadcasting Inc. (acq 1-1-86; $765,000 with co-located FM; FTR: 10-7-85) Population served: 80,000 Natl. Network: CBS. Rgnl. Network: Ky. Net. Natl. Rep: Rgnl Reps. Format: Classic country. Target aud: 30 plus. ♦Kevin Day, gen mgr.

WXKQ-FM—Co-owned with WTCW(AM). Nov 25, 1964: 103.9 mhz; 280 w. Ant 1,500 ft TL: N37 06 38 W82 44 15. Stereo. Web Site: www.1039thebulldog.com. Format: Classic hits.

Whitesville

WXCM(FM)—Licensed to Whitesville. See Hawesville

Whitley City

WHAY(FM)— Dec 1, 1990: 98.3 mhz; 5.1 kw. Ant 354 ft TL: N36 39 40 W84 26 53. Hrs opn: 24 Box 69, 42653. Phone: (606) 376-2218. Fax: (606) 376-5146. E-mail: whayradio@highland.net Web Site: www.hay98.com. Licensee: Tim Lavender. Population served: 18,000 Format: Americana. News: 5 hrs wkly. Target aud: 30 plus. ♦Dave Shelley, gen mgr.

Wickliffe

WBCE(AM)— Jan 4, 1981: 1200 khz; 1 kw-D. TL: N36 58 54 W89 04 39. Hrs opn: Box 128, 42087. Secondary address: 1136 Barlow Rd. 42087. Phone: (270) 335-5171. Fax: (270) 335-5172. Licensee: WBCE Inc. Format: Relg. Target aud: General. ♦Faye Crews, gen mgr.

WGKY(FM)— January 1987: 95.9 mhz; 3 kw. Ant 759 ft TL: N36 56 24 W88 57 59. Stereo. Hrs opn: 24 930 Wickliffe Rd., 42087. Phone: (270) 335-3696. Fax: (270) 335-3698. E-mail: wgky@brtc.net Web Site: www.96classiccountry.com. Licensee: W. Russell Withers Jr. (acq 2-21-2006; $400,000). Population served: 250,000 Natl. Network: Jones Radio Networks. Rgnl. Network: Ky. News Net., Brownfield. Rgnl rep: Rgnl Reps. Law Firm: Miller & Miller. Format: Oldies. News: 5 hrs wkly. Target aud: 24-54; rural homeowners, farmers. ♦Larry Kelley, gen mgr & opns mgr.

Williamsburg

WEKC(AM)— Sept 21, 1981: 710 khz; 4.2 kw-D. TL: N36 46 28 W84 10 05. Stereo. Hrs opn: Box 419, 40769. Secondary address: 402 Main St. 40769. Phone: (606) 549-3000. Fax: (606) 539-0916. E-mail: wekc@wekc.net Web Site: www.wekc.net. Licensee: Gerald Parks (acq 6-21-00). Natl. Network: USA. Format: Relg teaching, gospel. News: 5 hrs wkly. Target aud: General; young adults, All ages, race & creed. ♦Kay Parks, gen mgr & stn mgr.

WEZJ(AM)— Mar 7, 1959: 1440 khz; 2.5 kw-D, 500 w-N, DA-1. TL: N36 43 48 W84 09 04. Hrs opn: 522 Main St., 40769. Phone: (606) 549-2285. Fax: (606) 549-5565. Licensee: Whitley Broadcasting Co. Inc. (group owner; acq 5-23-02; grpsl). Population served: 200,000 Rgnl. Network: Ky. Net. Natl. Rep: Rgnl Reps. Format: News, talk. Target aud: General. ♦David Estes, gen mgr; Rick Campbell, progmg mgr; Frank Folson, chief of engrg.

WEZJ-FM— November 1990: 104.3 mhz; 1.4 kw. 656 ft TL: N36 44 43 W84 11 24. Format: Country.

Williamstown

WNKR(FM)— Apr 1, 1992: 106.5 mhz; 1.41 kw. Ant 476 ft TL: N38 40 54 W84 39 33. (CP: 106.7 mhz; 1.8 kw, ant 607 ft. TL: N38 41 19 W84 35 07). Stereo. Hrs opn: 24 Box 182, 118 S. Main St., Dry Ridge, 41035. Phone: (859) 824-9106. Phone: (800) 925-1220. Fax: (859) 824-9835. E-mail: prod@wnkr.net Licensee: Grant County Broadcasters Inc. (acq 1992). Population served: 350,000 Rgnl. Network: Ky. Net. Natl. Rep: Rgnl Reps. Law Firm: Koerner & Olender. Format: Country. News: 6 hrs wkly. Target aud: 25-54; adults. ♦Robert Wallace, pres; Jeffrey K. Ziesmann, gen mgr; T.C. Sommers, opns mgr; Laura Ziesmann, sls dir & gen sls mgr; Katherine Marshall, pub affrs dir & traf mgr; Jim Stitt, chief of engrg.

Wilmore

WVRB(FM)— Sept 18, 1995: 95.3 mhz; 4.1 kw. Ant 397 ft TL: N37 57 37 W84 32 42. Hrs opn: 24 5700 W. Oaks Blvd., Rocklin, CA, 95765. Phone: (707) 528-9236. Web Site: www.air1.com. Licensee: Vernon R. Baldwin Inc. (group owner; acq 7-26-94; FTR: 8-8-94). Population served: 500,000 Natl. Network: Air 1. Format: Contemp Christian. Target aud: 20-45; baby boomers, Christians. ♦Keith Whipple, gen mgr & stn mgr.

Winchester

WKQQ(FM)— 1974: 100.1 mhz; 32 kw. 490 ft TL: N38 07 25 W84 26 45. Stereo. Hrs opn: 24 2601 Nicholasville Rd., Lexington, 40503. Phone: (859) 422-1000. Fax: (859) 422-1038. Web Site: www.wkqq.com. Licensee: Citicasters Licenses L.P. Group owner: Clear Channel Communications Inc. (acq 5-4-99; grpsl). Population served: 70,000 Format: Classic rock. News staff: one; News: 3 hrs wkly. Target aud: 18-49; women. ♦Gene Guinn, gen mgr & prom dir.

WMJR(AM)— Oct 19, 1954: 1380 khz; 2.5 kw-D, 40 w-N. TL: N38 00 46 W84 09 38. (CP: COL Nicholasville. 5 kw-D, 38 w-N. TL: N37 54 27 W84 28 42). Hrs opn: 24 195 Moore Dr., Lexington, 40503. Phone: (859) 278-0894. Fax: (859) 278-0426. E-mail: info@wmjr.net Web Site: www.wmjr.net. Licensee: Thy Kingdom Come Network Inc. (acq 1999; $583,000). Population served: 28,000 Rgnl. Network: Ky. Net. Law Firm: Pepper & Corazzini. Format: Christian. Target aud: 35-64. ♦Leo Brown, pres & progmg dir.

Louisiana

Abbeville

KPEL-FM—Listing follows KROF(AM).

KROF(AM)— July 9, 1948: 960 khz; 1 kw-D, 95 w-N. TL: N30 00 40 W92 07 21. Hrs opn: 24 1749 Bertrand Dr., Lafayette, 70506. Phone: (337) 233-6000. Fax: (337) 234-7360. Web Site: www.960thegator.com. Licensee: Regent Broadcasting of Lafayette LLC. Group owner: Regent Communications Inc. (acq 12-7-2001; grpsl). Format: Cajun. ♦Mike Grimsley, gen mgr; Chuck Wood, gen sls mgr; Camey Doucet, progmg dir; Kyle Vidrine, chief of engrg.

KPEL-FM—Co-owned with KROF(AM). June 1, 1974: 105.1 mhz; 25 kw. 300 ft TL: N30 00 40 W92 07 21. (CP: Ant 292 ft.). Stereo. 24 Web Site: www.kpel1051news.com. Format: News/talk. ♦Bernadette Lee, progmg dir.

Alexandria

***KAPM(FM)**— June 1998: 91.7 mhz; 1 kw. 128 ft TL: N31 16 04 W92 26 24. Hrs open: Box 3206, American Family Radio, Tupelo, MS, 38803. Phone: (662) 844-8888. Fax: (662) 842-6791. Web Site: www.afr.net. Licensee: American Family Association. Group owner: American Family Radio Format: Inspirational Christian. ♦Marvin Sanders, gen mgr.

KBCE(FM)—Boyce, Mar 29, 1982: 102.3 mhz; 21 kw. Ant 289 ft TL: N31 22 21 W92 38 09. Hrs open: 1605 Murray St., Suite 216, 71301. Phone: (318) 445-0800. Fax: (318) 445-1445. Licensee: Trinity Broadcasting Corp. (acq 7-2-98; $26,248). Natl. Network: American Urban. Natl. Rep: D & R Radio. Format: Urban contemp. Target aud: General. ♦Alison Randolf, pres; James Alexander, gen mgr; Casandra Scott, traf mgr.

KDBS(AM)— December 1953: 1410 khz; 1 kw-D, 30 w-N. TL: N31 16 25 W92 25 43. Hrs open: 24 1115 JTexas Ave., 71301. Phone: (318) 443-7454. Phone: (318) 445-1234. Fax: (318) 473-1960. E-mail: daveg@kswl.com Web Site: www.kdixie.com. Licensee: Cenla Broadcasting Licensing Co. LLC. Group owner: Clear Channel Communications Inc. (acq 11-13-2006; grpsl). Population served: 140,000 Law Firm: Pepper & Corazzini. Format: Oldies. News staff: 2; News: 4 hrs wkly. ♦Taylor Thompson, gen mgr; Charlie Sopraz, opns mgr; Tish Robertson, gen sls mgr; Dave Grachien, progmg dir; Linnie Dupree, chief of engrg; Sybil Ford, traf mgr.

KRRV-FM—Co-owned with KDBS(AM). May 11, 1969: 100.3 mhz; 100 kw. 1,055 ft TL: N31 01 59 W92 30 08. Stereo. 24 E-mail: hollywood@cenlabroadcasting.com Web Site: www.krrv-fm.com.210,000 Format: Country. ♦Hollywood Harrison, progmg dir.

KEDG(FM)— 2001: 106.9 mhz; 6 kw. Ant 328 ft TL: N31 25 35 W92 24 25. Hrs open: Box 7057, 71306. Secondary address: 1115 Texas Ave 71301. Phone: (318) 445-1234. Fax: (318) 473-1960. Web Site: www.kiss1069.com. Licensee: Flinn Broadcasting Corp. Format: Hip hop, rhythm and blues. ♦Taylor Thompson, gen mgr.

KEZP(FM)—Bunkie, 1993: 104.3 mhz; 18 kw. 384 ft TL: N31 05 14 W92 21 34. Hrs open: 24 92 W. Shamrock Ave., Pineville, 71360-6414. Phone: (318) 487-1035. Fax: (318) 487-1045. Web Site: www.kezp.com. Licensee: Opus Broadcasting Alexandria LLC. Group owner: Opus Media Partners LLC (acq 7-30-2004; $1.83 million). Population served: 250,000 Natl. Network: Westwood One. Format: Classic hits. News staff: one. Target aud: 35-64. ♦Mark Jones, gen mgr.

KJMJ(AM)— Sept 21, 1935: 580 khz; 5 kw-D, 1 kw-N, DA-N. TL: N31 18 25 W92 25 00. Hrs open: 24 601 Washington St., 71301. Phone: (318) 561-6145. Fax: (318) 449-9954. E-mail: info.usa@radiomaria.org Web Site: www.radiomaria.us. Licensee: Radio Maria Inc. (group owner; acq 9-20-99). Population served: 502010 Law Firm: Putbrese, Hunsaker & Trent, P.C. Format: Christian, relg, talk. News: 8 hrs wkly. Homebound, prisoners & sick. ♦Dale DePerrodill, gen sls mgr; Duane Stenzel, progmg dir & progmg dir; Danny Brou, chief of engrg.

KLAA-FM—Tioga, May 25, 1984: 103.5 mhz; 50 kw. 476 ft TL: N31 25 39 W92 24 18. Stereo. Hrs opn: 24 92 W. Shamrock, Pineville, 71360. Phone: (318) 487-1035. Fax: (318) 487-4419. Web Site: www.la103.com. Licensee: Opus Broadcasting Alexandria LLC. Group owner: Opus Media Partners LLC (acq 9-30-2004; $3.38 million with KBKK(FM) Ball). Population served: 200,000 Natl. Rep: McGavren Guild. Law Firm: Dow, Lohnes & Albertson. Format: Country. News staff: one; News: 4 hrs wkly. Target aud: 25-54; working people, upscale professionals. ♦Kim Jones, pres & gen mgr.

***KLSA(FM)**— 1987: 90.7 mhz; 100 kw. Ant 1,243 ft TL: N31 33 56 W92 32 50. Hrs open: 24
Rebroadcasts KDAQ(FM) Shreveport 100%.
Box 5250, Shreveport, 71135. Phone: (318) 797-5150. Phone: (800) 552-8502. Fax: (318) 797-5265. E-mail: listenermail@redriverradio.com Web Site: www.redriverradio.org. Licensee: Board of Supervisors Louisiana State University & Agricultural Mechanical College. Natl. Network: NPR, PRI. Format: Classical, news, jazz. News: Nws progmg 40 hrs wkly. ♦Kermit Poling, gen mgr; Rick Shelton, opns mgr.

***KLXA-FM**— November 1998: 89.9 mhz; 3 kw. Ant 328 ft TL: N31 22 40 W92 28 27. Hrs open: 24 2351 Sunset Blvd., Suite 170-218, Rocklin, CA, 95765. Phone: (916) 251-1600. Fax: (916) 251-1650. E-mail: klov@klove.com Web Site: www.klove.com. Licensee: Educational Media Foundation. Group owner: EMF Broadcasting (acq 12-1-03; $125,000). Natl. Network: K-Love. Law Firm: Shaw Pittman. Format: Contemp Christian. News staff: 3. Target aud: 25-44; Judeo Christian, female. ♦Richard Jenkins, pres; Mike Novak, VP; Lloyd Parker, gen mgr; Ed Lenane, opns dir & news dir; Keith Whipple, dev dir; David Pierce, progmg mgr; Sam Wallington, engrg dir; Arthur Vassar, traf mgr; Karen Johnson, news rptr; Marya Morgan, news rptr; Richard Hunt, news rptr.

KMXH(FM)— February 1993: 93.9 mhz; 6 kw. Ant 328 ft TL: N31 16 04 W92 26 24. Hrs open: 1605 Murray St., Suite 216, 71301. Phone: (318) 445-0800. Fax: (318) 445-1445. Licensee: FM Broadcasting Corp. (acq 7-1-2005; $1.2 million). Format: Rhythm and blues, Southern soul.

KQID(FM)—Listing follows KSYL(AM).

KSYL(AM)— Apr 1, 1947: 970 khz; 1 kw-U, DA-N. TL: N31 19 33 W92 29 17. Stereo. Hrs opn: Box 7057, 71306. Secondary address: 1115 Texas Ave. 71301. Phone: (318) 445-1234. Fax: (318) 473-1960. Web Site: www.ksyl.com. Licensee: Cenla Broadcasting Inc. (acq 8-1-80). Population served: 41,557 Format: Talk. ♦Taylor Thompson, pres & gen mgr.

KQID(FM)—Co-owned with KSYL(AM). Sept 17, 1978: 93.1 mhz; 100 kw. 1,700 ft TL: N31 38 20 W92 12 18. Stereo. Web Site: www.q93fm.com.750,000 Format: Top-40.

KTTP(AM)—Pineville, Sept 13, 1974: 1110 khz; 2 kw-D. TL: N31 21 52 W92 27 15. Stereo. Hrs opn: 3419 Hyson St., 71303. Phone: (318) 473-4388. Fax: (318) 449-1779. E-mail: kttpam1110@aol.com Licensee: Benjamin-Dane LLC (acq 4-7-2005; $175,000). Format: Gospel. Target aud: 25-70. ♦Ronald Reeves, pres; Carolyn Frazier, gen mgr & stn mgr; Dave Grayso, gen sls mgr & chief of engrg.

KWDF(AM)—Ball, 1986: 840 khz; 10 kw-D. TL: N31 22 41 W92 28 27. Hrs open: 3735 Rigolette Rd., Pineville, 71360. Phone: (318) 640-4373. Fax: (318) 640-5971. Licensee: NWLA Broadcasting L.L.C. Group owner: EMF Broadcasting (acq 4-5-2006). Natl. Network: USA. Format: Southern gospel. ♦Al Moore, pres; Sharon Thorne, gen mgr; Jimmy Bryant, opns dir; Jimmy Bryant, progmg mgr; Tommy Moore, chief of engrg.

KZMZ(FM)— 1947: 96.9 mhz; 98 kw. Ant 1,053 ft TL: N31 01 59 W92 30 08. Stereo. Hrs opn: 24 1115 Texas Ave, 71301. Phone: (318) 445-1234. Fax: (318) 445-7231. Licensee: Cenla Broadcasting Licensing Co. LLC. Group owner: Clear Channel Communications Inc. (acq 11-13-2006; grpsl). Population served: 475,000 Law Firm: Pepper & Corazzini. Format: Classic rock. News staff: one; News: 2 hrs wkly. Target aud: 18-49; baby boomers. ♦Taylor Thompson, gen mgr.

Amite

WABL(AM)— January 1956: 1570 khz; 500 w-D. TL: N30 42 31 W90 31 31. Hrs open: 12 Box 787, 70422. Secondary address: 12515 Bankston Rd. 70422. Phone: (985) 748-8385. Fax: (985) 748-3918. E-mail: wabl1570@hotmail.com Licensee: Spotlight Broadcasting LLC (group owner; acq 6-8-01; $70,000). Population served: 200,000 Format: News/talk, country. Target aud: 20-64. ♦Patrick Andras, gen mgr.

WTGG(FM)— Mar 3, 1997: 96.5 mhz; 6 kw. 328 ft TL: N30 41 39 W90 26 41. Hrs open: 24 200 E. Thomas St., Hammond, 70401. Phone: (985) 345-0060. Fax: (985) 542-9377. Licensee: Southwest Broadcasting Inc. (group owner; acq 4-8-98; $650,000). Population served: 80,000 Natl. Network: Westwood One. Rgnl. Network: La. Net. Format: 50s, 60s & 70s oldies. News staff: one; News: one hr wkly. Target aud: 25-54; women. ♦Charles Dowdy, CEO; Eloise Dowdy, gen mgr & gen sls mgr; Ben Bickham, chief of engrg.

Angola

***KLSP(FM)**— Aug 12, 1986: 91.7 mhz; 100 w. 90 ft TL: N30 57 17 W91 35 45. Hrs open: Louisiana State Penitentiary, Hwy. 66, 70712. Phone: (225) 655-2001. Web Site: www.corrections.state.la.us/lsp/klsp.htm. Licensee: Angola Educational Foundation Inc. Format: Div. Target aud: General. Spec prog: Black 10 hrs, C&W 6 hrs, jazz 7 hrs, poets corner one hr, legal wave 3 hrs wkly. ♦Burl Cain, gen mgr; Cheryl M. Ranatza, stn mgr.

Arcadia

***KHCL(FM)**— Jan 20, 2001: 92.5 mhz; 6 kw. Ant 328 ft TL: N32 27 27 W92 59 38. Hrs open: 24
Rebroadcasts KHCB-FM Houston, TX 100%.
Houston Christian Broadcasters Inc., 2424 South Blvd., Houston, TX, 77098. Phone: (713) 520-5200. Web Site: www.khcb.org. Licensee: Houston Christian Broadcasters Inc. (group owner) Format: Christian. ♦Bruce Munsterman, stn mgr; Bonnie BeMent, mus dir & news dir; Dan Wales, chief of engrg.

Atlanta

KCIJ(FM)— January 2002: 106.5 mhz; 25 kw. Ant 328 ft TL: N31 48 29 W92 48 22. Stereo. Hrs opn: 24 213 Renee St., Natchitoches, 71457. Phone: (318) 354-4000. Fax: (318) 352-9598. Licensee: North Face Broadcasting L.L.C. (acq 5-9-2003; $348,000 with KNOC(AM) Natchitoches). Population served: 150,000 Format: Classic Hits. Target aud: 25-54; general. ♦Theresa Wilkerson, stn mgr; John Brewer, opns dir; Shane Evath, news dir.

Baker

WTGE(FM)— June 16, 1994: 107.3 mhz; 4.6 kw. 328 ft TL: N30 37 24 W91 09 50. Stereo. Hrs opn: 24 Box 2231, Baton Rouge, 70821. Secondary address: 929-B Government St., Baton Rouge 70802. Phone: (225) 388-9898. Fax: (225) 344-3077. E-mail: owen.weber@gbcradio.com Web Site: www.countrylegends1073.com. Licensee: Guaranty Broadcasting Co. of Baton Rouge LLC. Group owner: Guaranty Broadcasting Co. (acq 2-5-97). Population served: 750,000 Natl. Rep: D & R Radio. Law Firm: Wiley, Rein & Fielding. Format: Classic country. Target aud: 25-54; general. ♦George A. Foster Jr., chmn; Bridger Eglin, pres & gen mgr; Owen Weber, VP & gen mgr; Dave Dunaway, opns mgr.

Ball

KBKK(FM)— September 1998: 105.5 mhz; 6 kw. 318 ft TL: N31 25 39 W92 24 18. Hrs open: 24 92 W. Shamrock, Pineville, 71360. Phone: (318) 487-1035. Fax: (318) 487-4419. Web Site: www.1055kbuck.com. Licensee: Opus Broadcasting Alexandria LLC. Group owner: Opus Media Partners LLC (acq 9-30-2004; $3.38 million with KLAA-FM Tioga). Population served: 150,000 Natl. Rep: McGavren Guild. Format: Classic country. ♦Kim Jones, pres & gen mgr.

KWDF(AM)—Licensed to Ball. See Alexandria

Basile

KQIS(FM)— May 4, 1990: 102.1 mhz; 3 kw. 328 ft TL: N30 28 52 W92 35 50. Hrs open: 24 Box 60571, Lafeyette, 70596. Phone: (337) 783-2521. Fax: (337) 783-5744. E-mail: info@kqis.com Web Site: www.kqis.com. Licensee: Third Partner Broadcasting Inc. (acq 3-21-94; $380,000; FTR: 5-30-94). Natl. Network: ABC. Format: Todays hits & yesterdays favorites. News: 3 hrs wkly. Target aud: 25-55; middle-income. ♦Phil Lizotte, gen mgr; Jimmy Cole, gen sls mgr; Hans Nelson, progmg dir.

Bastrop

***KAXV(FM)**— 2000: 91.9 mhz; 12 kw. Ant 456 ft TL: N32 49 22 W92 07 28. Hrs open: Box 3206, American Family Radio, Tupelo, MS, 38803. Phone: (662) 844-8888. Fax: (662) 842-6791. E-mail: comments@afr.net Web Site: www.afr.net. Licensee: American Family Radio. (group owner) Format: Inspirational Christian. ♦Marvin Sanders, gen mgr.

KJMG(FM)— 1996: 97.3 mhz; 6 kw. 328 ft TL: N32 45 46 W91 57 35. Hrs open: 24 1109 Hudson Ln., Monroe, 71201. Secondary address:

Box 4808 71211. Phone: (318) 388-2323. Fax: (318) 388-0569. E-mail: kjmg@bayou.com Web Site: www.majic97.com. Licensee: Holladay Broadcasting of Louisiana LLC (group owner; acq 9-30-98; $700,000). Natl. Rep: McGavren Guild. Law Firm: Latham & Watkins. Format: Urban adult contemp. Target aud: 25-54. Spec prog: Blues 12 hrs wkly. ◆Bob Holladay, pres & gen mgr.

KRVV(FM)— 1977: 100.1 mhz; 50 kw. 490 ft TL: N32 40 20 W91 55 06. Stereo. Hrs opn: 24 Box 4808, Monroe, 71211. Secondary address: 1109 Hudson Ln., Monroe 71201. Phone: (318) 388-2323. Fax: (318) 388-0569. E-mail: krvv@bayou.com Web Site: www.thebeat.net. Licensee: Holladay Broadcasting of Louisiana LLC (group owner; acq 10-15-91; $1 million; FTR: 11-4-91). Population served: 254,000 Natl. Network: ABC. Natl. Rep: McGavren Guild. Law Firm: Latham & Watkins. Format: Urban. Target aud: 18-49. Spec prog: Gospel 4 hrs wkly. ◆Bob Holladay, pres & gen mgr.

Baton Rouge

KBRH(AM)—Listing follows WBRH(FM).

***KLSU(FM)**— October 1981: 91.1 mhz; 5 kw. 159 ft TL: N30 24 37 W91 10 37. Stereo. Hrs opn: 24 B-39 Hodges Hall, Louisiana State Univ., 70803. Phone: (225) 578-8688. Fax: (225) 388-1698. Fax: (225) 578-0579. Web Site: www.klsu.fm. Licensee: Louisiana State University. Population served: 325,000 Natl. Rep: Rgnl Reps. Format: Var. News: one hr wkly. Target aud: 18-25; university students & college age listeners. ◆Peyton Juneau, stn mgr.

***WBRH(FM)**— September 1977: 90.3 mhz; 21 kw. Ant 197 ft TL: N30 26 42 W91 09 33. Stereo. Hrs opn: 24 2825 Government St., 70806. Phone: (225) 383-3243. Fax: (225) 379-7685. Licensee: East Baton Rouge Parish School Board. Population served: 460,000 Natl. Network: NPR. Format: Jazz. Target aud: 25-54; men. ◆Larry Davis, gen mgr & stn mgr; Lyn Kenyon, sls dir; Rob Payer, mus dir.

KBRH(AM)—Co-owned with WBRH(FM). 1953: 1260 khz; 5 kw-D, 127 w-N. TL: N30 27 38 W91 14 37.24 (Acq 7-7-93; FTR: 8-2-93). Format: Classic rhythm and blues. Target aud: 35+.

WCDV(FM)—See Hammond

WDGL(FM)— Oct 1, 1968: 98.1 mhz; 100 kw. 1,550 ft TL: N30 21 58 W91 12 47. Stereo. Hrs opn: 24 Box 2231, 70821. Secondary address: 929-B Government St. 70802. Phone: (225) 388-9898. Fax: (225) 344-3077. E-mail: owen.weber@gbcradio.com Web Site: www.eagle981.com. Licensee: Guaranty Broadcasting Co. of Baton Rouge LLC. Group owner: Guaranty Broadcasting Co. Population served: 433,600 Natl. Rep: McGavren Guild. Law Firm: Wiley, Rein & Fielding. Wire Svc: AP Format: Classic rock. Target aud: 25-54; general. ◆George A. Foster Jr., chmn; Bridger Eglin, pres & VP; Owen Weber, VP, gen mgr & gen mgr; Dave Dunaway, opns mgr.

WDVW(FM)—See La Place

WFMF(FM)—Listing follows WJBO(AM).

WIBR(AM)— July 18, 1948: 1300 khz; 5 kw-D, 1 kw-N, DA-2. TL: N30 28 25 W91 13 34. Hrs open: 650 Wooddale Blvd., 70806. Phone: (225) 926-1106. Fax: (225) 928-1606. Web Site: 1300espn.com. Licensee: Citadel Broadcasting Co. Group owner: Citadel Broadcasting Corp. (acq 1999; grpsl). Population served: 245,752 Natl. Network: ABC. Rgnl. Network: La. Net. Natl. Rep: McGavren Guild. Wire Svc: UPI Format: Sports. Target aud: 25-50. ◆Greg Benefield, gen mgr.

WJBO(AM)— Dec 11, 1934: 1150 khz; 5 kw-U, DA-1. TL: N30 27 47 W91 16 10. Hrs open: 5555 Hilton Ave., Suite 500, 70808. Phone: (225) 231-1860. Fax: (225) 231-1873. E-mail: info@wjbo.com Web Site: www.wjbo.com. Licensee: Capstar TX L.P. Group owner: Clear Channel Communications Inc. (acq 8-30-00; grpsl). Population served: 525,700 Natl. Network: CBS, Westwood One. Format: Talk, news, sports. Target aud: 20 plus. ◆Donnie Picou, VP & gen mgr.

WFMF(FM)—Co-owned with WJBO(AM). 1941: 102.5 mhz; 85 kw. 1,260 ft TL: N30 17 49 W91 11 40. Stereo. Web Site: www.wfmf.com. Format: CHR. Target aud: 18-34; female.

***WJFM(FM)**— June 1995: 88.5 mhz; 25.5 kw. Ant 269 ft TL: N30 23 06 W91 05 28. Stereo. Hrs opn: 24 Box 262550, 70826-2550. Secondary address: 8919 World Ministry Ave. 70810. Phone: (225) 768-3688. Phone: (225) 768-8300. Fax: (225) 768-3729. E-mail: kawikfish@yahoo.com Web Site: www.jsm.org. Licensee: Family Worship Center Church Inc. (group owner; (acq 12-15-99). Population served: 150,000 Format: Christian. News staff: one; News: 2 hrs wkly. Target aud: 25-54; full gospel Christians & anyone searching for hope. ◆David Whitelaw, COO; Jimmy Swaggart, pres & stn mgr; John Santiago, dev dir & progmg dir.

WPFC(AM)—Port Allen, 1963: 1550 khz; 5 kw-D. TL: N30 30 07 W91 12 39. Hrs open: Sunrise-sunset 6943 Titian Ave., 70806. Phone: (225) 926-1506. Fax: (225) 590-3238. Web Site: www.1550wpfc.com. Licensee: Victory and Power Ministries Inc. (acq 11-15-94; $450,000; FTR: 12-12-94). Population served: 500,000 Natl. Network: USA. Law Firm: Latham & Watkins. Format: Relg, gospel music. News staff: 2; News: 3 hrs wkly. Target aud: 35-59; middle-class female. ◆Pastor Ralph Moore, CEO & gen mgr; Keith Richard, stn mgr.

WPYR(AM)— 1956: 1380 khz; 5 kw-D, DA. TL: N30 27 39 W91 13 23. Hrs open: 5555 Hilton Ave., Suite 500, 70808. Phone: (225) 231-1860. Fax: (226) 231-1879. Web Site: www.talkradio1380.com. Licensee: 6 Johnson Road Licenses Inc. Group owner: Clear Channel Communications Inc. (acq 7-24-2006; grpsl). Population served: 550,000 Format: Talk. ◆Mark Kennedy, progmg dir.

***WRKF(FM)**— Jan 18, 1980: 89.3 mhz; 28 kw. 935 ft TL: N30 22 22 W91 12 16. Stereo. Hrs opn: 24 3050 Valley Creek, 70808. Phone: (225) 926-3050. Fax: (225) 926-3105. Licensee: Public Radio Inc. Natl. Network: NPR, PRI. Format: Class, news/talk. News staff: one; News: 35 hrs wkly. Target aud: General. ◆Blythe Earl, gen mgr & opns dir; Jennifer Berniard, pres & gen mgr.

WUBR(AM)— Nov 1, 1946: Stn currently dark. 910 khz; 1 kw-U, DA-1. TL: N30 34 48 W91 07 50. Hrs open: 1111 Michigan Ave., East Lansing, MI, 48823. Phone: (517) 351-3333. Licensee: Communications Capital Co. III LLC. (acq 2-17-2006; $75,000). Target aud: 18-59. ◆Michael H. Oesterle, CEO; Sandra Pate, gen mgr.

WXOK(AM)— February 1953: 1460 khz; 5 kw-D, 1 kw-N, DA-3. TL: N30 28 08 W91 12 24. Hrs open: 24 650 Wooddale Blvd., 70806. Phone: (225) 926-1106. Fax: (225) 928-1606. E-mail: wxok.am@citcomm.com Web Site: www.heaven1460.com. Licensee: Citadel Broadcasting Co. Group owner: Citadel Broadcasting Corp. (acq 1-14-99). Population served: 180,000 Natl. Network: ABC. Format: Urban gospel. Target aud: 18 plus. ◆Greg Benefield, gen mgr.

WYNK-FM— Dec 7, 1968: 101.5 mhz; 97 kw. Ant 1,499 ft TL: N30 19 34 W91 16 36. Stereo. Hrs opn: 24 5555 Hilton Ave, Suite 500, 70808. Phone: (225) 231-1860. Fax: (225) 231-1873. Web Site: www.wynk.com. Licensee: Capstar TX L.P. Group owner: Clear Channel Communications Inc. (acq 8-30-2000; grpsl). Format: Country. News staff: 2. Target aud: 18-54. ◆Donnie Picou, gen mgr; Bob Murphy, progmg dir.

WYPY(FM)— Sept 10, 1966: 100.7 mhz; 97 kw. 1,499 ft TL: N30 19 35 W91 16 36. Stereo. Hrs opn: 24 Box 2231, 70821. Secondary address: 929-B Government St. 70802. Phone: (225) 388-9898. Fax: (225) 344-3077. E-mail: owen.weber@gbcradio.com Web Site: www.newcountry1007.com. Licensee: Guaranty Broadcasting Co. of Baton Rouge LLC. Group owner: Guaranty Broadcasting Co. (acq 1996; $5.5 million). Population served: 299,300 Natl. Rep: D & R Radio. Law Firm: Wiley, Rein & Fielding. Format: Country. Target aud: 25-54; general. ◆George A. Foster Jr., chmn & VP; Bridger Eglin, pres; Owen Weber, VP & gen mgr; Dave Dunaway, opns mgr & progmg dir.

Bayou Vista

KQKI(FM)— Dec 31, 1976: 95.3 mhz; 16.5 kw. 400 ft TL: N29 29 38 W91 17 41. Hrs open: 128 Pluto St., 70381. Phone: (985) 395-2853. Fax: (985) 395-5094. Web Site: www.kqki.com. Licensee: Teche Broadcasting Corp. Population served: 59,000 Natl. Network: ABC. Rgnl. Network: La. Net. Format: Country. News staff: one; News: 17 hrs wkly. Target aud: 30 plus; general. ◆Paul J. Cook, pres & gen mgr; Ernest Dean Polk, stn mgr & news dir; Julie Boyne, gen sls mgr & progmg dir.

Belle Chasse

KMEZ(FM)— March 1990: 102.9 mhz; 4.7 kw. 604 ft TL: N29 57 14 W89 56 58. Hrs open: 201 St. Charles Ave., Suite 201, New Orleans, 70170. Phone: (504) 581-7002. Fax: (504) 566-4857. E-mail: lbj.kmez@citcomm.com Web Site: www.oldschool1029.com. Licensee: Citadel Broadcasting Co. Group owner: Citadel Broadcasting Corp. (acq 8-29-03; grpsl). Population served: 2,965,100 Law Firm: Dow, Lohnes & Albertson. Format: Rhythm & blues, old school. Target aud: 35-54; female. ◆Dave Siebert, gen mgr & mktg dir; LeBron Joseph, progmg dir.

Benton

KSYR(FM)— 1981: 92.1 mhz; 3 kw. 299 ft TL: N32 39 19 W93 41 38. Stereo. Hrs opn: 24 208 N. Thomas Dr., Shreveport, 71107. Phone: (318) 222-3122. Fax: (318) 459-1493. Licensee: Access. 1 Louisiana Holding Co. LLC. Group owner: Access.1 Communications Corp. (acq 5-5-00; grpsl). Natl. Network: ABC. Format: Regional Mexican. News staff: 1. Target aud: 35 plus; upper income, upwardly mobile. ◆Cary D. Camp, pres, gen mgr & stn mgr; Don Zimmerman, gen sls mgr.

Berwick

KBZE(FM)— July 4, 1990: 105.9 mhz; 3.2 kw. Ant 403 ft TL: N29 45 27 W91 10 25. Stereo. Hrs opn: Box 1560, Morgan City, 70381. Secondary address: 1320 Victor II Blvd., Morgan City 70380. Phone: (985) 385-6266. Fax: (985) 385-6268. E-mail: kbze@petronet.net Web Site: ww.kbze.com. Licensee: HubCast Broadcasting Inc. (acq 6-6-94; $105,500). Population served: 75,000 Natl. Network: ABC. Format: Urban adult contemp, relg, sports. News staff: one; News: 10 hrs wkly. Target aud: 24-54; middle to upper income. ◆Howard Castay Jr., pres & gen mgr; Darlene Castay, opns VP.

Blanchard

KDKS-FM— Oct 19, 1998: 102.1 mhz; 14 kw. 440 ft TL: N32 35 57 W93 54 01. Hrs open: 24 208 N. Thomas Dr., Shreveport, 71107. Phone: (318) 222-3122. Fax: (318) 459-1493. Web Site: www.kdks.fm. Licensee: Access. 1 Louisiana Holding Co. LLC. Group owner: Access.1 Communications Corp. (acq 6-30-00; $7.9 million with KLKL(FM) Minden). Format: Urban Adult contemp. News staff: 1. Target aud: 25-54. ◆Cary Camp, gen mgr; Quinn Echols, progmg dir.

***KFLO-FM**— 2006: 89.1 mhz; 20 kw vert. Ant 406 ft TL: N32 18 28 W93 58 34. Hrs open: Box 7277, Shreveport, 71137. Secondary address: 2097 N. Hearne Ave., Shreveport 71107. Phone: (318) 550-2000. Fax: 318-550-2002. E-mail: info@miracle891.org Web Site: miracle891.org. Licensee: Family Life Educational Foundation (acq 10-7-2005; $97,000 for CP). Format: Inspirational music. ◆A.T. Moore, pres; Donna Cole, gen mgr; Dan Perkins, opns mgr; Joe Miot, progmg dir.

Bogalusa

WBOX(AM)— Mar 1, 1954: 920 khz; 1 kw, DA-N. TL: N30 50 29 W89 50 06. Hrs open: Box 280, 70429. Secondary address: 22037 Hwy. 436 70427. Phone: (985) 732-4288. Phone: (985) 732-4288. Licensee: Best Country Broadcasting LLC (acq 9-6-2002; $150,000 with WBOX-FM Varnado). Population served: 18,412 Format: Country. ◆Ben R. Strickland, pres & gen mgr.

WBOX-FM—See Varnado

WIKC(AM)— May 15, 1947: 1490 khz; 1 kw-U. TL: N30 47 30 W89 52 47. Hrs open: 24 Box 638, 70429. Secondary address: 607 Rio Grande St. 70429. Phone: (985) 732-4190. Fax: (985) 732-7594. E-mail: timberlands@bellsouth.net Licensee: Timberlands Broadcasting Corp. (acq 6-29-82). Population served: 65,000 Natl. Network: Salem Radio Network. Format: Relg, gospel, news/talk. News: 30 hrs wkly. Target aud: General. ◆G.S. Adams Jr., pres; Gardner Adams, gen mgr.

Bossier

KRMD(AM)—See Shreveport

KRMD-FM—See Shreveport

Bossier City

KBCL(AM)— September 1957: 1070 khz; 250 w-D. TL: N32 32 14 W93 43 28. Hrs open: 316 Gregg St., Shreveport, 71104. Phone: (318) 861-1070. E-mail: kbcl_radio@bellsouth.net Web Site: www.praise1070.org. Licensee: Barnabas Center Ministries (acq 8-26-02; donation). Population served: 182,064 Format: Christian, talk shows. ◆Leon McKee, gen mgr; Jean McKee, progmg dir.

Boyce

KBCE(FM)—Licensed to Boyce. See Alexandria

Breaux Bridge

KFTE(FM)— May 1, 1993: 96.5 mhz; 22.5 kw. 328 ft TL: N30 06 09 W91 59 30. Stereo. Hrs opn: 24 1749 Bertrand Dr., Lafayette, 70506-2054. Phone: (337) 233-6000. Fax: (337) 234-7360. Web Site: www.planet965.com. Licensee: Regent Broadcasting of Lafayette LLC. Group owner: Regent Communications Inc. (acq 12-7-2001; grpsl). Law Firm: Wiley, Rein & Fielding. Format: Modern rock. ♦Mike Grimsley, gen mgr; Scott Pessin, progmg dir.

Broussard

***WHFG(FM)**—Not on air, target date: unknown: 91.3 mhz; 25 kw. Ant 230 ft TL: N29 57 15 W92 08 18. Hrs open: 1115 Honeysuckle Dr., Keene, TX, 76059. Phone: (817) 641-3495. Licensee: Mary V. Harris Foundation. ♦Linda De Romanett, pres.

Brusly

KRVE(FM)— Sept 9, 1989: 96.1 mhz; 43 kw. 449 ft TL: N30 29 34 W91 00 15. Stereo. Hrs opn: 24 5555 Hilton Ave., Suite 500, Baton Rouge, 70808. Phone: (225) 231-1860. Fax: (225) 231-1869. E-mail: info@murphysamandjodi.com Web Site: www.961theriver.com. Licensee: Capstar TX L.P. Group owner: Clear Channel Communications Inc. (acq 8-30-00; grpsl). Format: Lite adult contemp. News staff: 3. Target aud: 25-54; women. ♦Dick Lewis, gen mgr; Jill Stokold, prom dir & pub affrs dir; Bob Murphy, progmg dir; Jodie Carson, mus dir.

Bunkie

KEZP(FM)—Licensed to Bunkie. See Alexandria

***KITA(FM)**—Not on air, target date: unknown: 89.5 mhz; 15 kw. Ant 131 ft TL: N30 57 11 W92 10 59. Hrs open: 2351 Sunset Blvd., Suite 170-218, Rocklin, CA, 95765. Phone: (916) 251-1600. Fax: (916) 251-1650. Licensee: Educational Media Foundation. (acq 11-1-2006; grpsl). ♦Richard Jenkins, pres.

Buras

***KMRL(FM)**— Apr 22, 1995: Stn currently dark. 91.9 mhz; 3 kw. Ant 164 ft TL: N29 20 15 W89 28 46. Hrs open: 3600 Manhattan Blvd., Harvey, 70058. Phone: (504) 362-3379. Licensee: White Dove Fellowship Inc. (acq 8-12-02; $25,000). Population served: 25,000 Target aud: General.

Clinton

***WBKL(FM)**— Sept 23, 1981: 92.7 mhz; 32 kw. Ant 604 ft TL: N30 51 03 W91 04 31. Stereo. Hrs opn: 24
Rebroadcasts KLVR(FM) Santa Rosa, CA 100%.
2351 Sunset Blvd., Suite 170-218, Rocklin, CA, 95765. Phone: (916) 251-1600. Fax: (916) 251-1650. Web Site: www.klove.com. Licensee: Educational Media Foundation. (acq 7-1-2005; $3.2 million). Natl. Network: K-Love. Format: Contemp Christian. ♦Richard Jenkins, pres; Mike Novak, VP; Keith Whipple, dev dir; David Pierce, progmg mgr; Ed Lenane, news dir; Sam Wallington, engrg dir; Karen Johnson, news rptr; Marya Morgan, news rptr; Richard Hunt, news rptr.

Columbia

KQLQ(FM)— Jan 21, 1980: 103.1 mhz; 25 kw. 348 ft TL: N32 09 25 W92 10 58. Stereo. Hrs opn: 24 1200 N. 18th St., Suite D, Monroe, 71201. Phone: (318) 387-3922. Fax: (318) 322-4585. Web Site: 1031theparty.com. Licensee: Opus Broadcasting Monroe L.L.C. Group owner: Opus Media Partners LLC (acq 7-19-2004; grpsl). Format: Hip hop. ♦Chris Zimmerman, gen mgr.

Coushatta

KRRP(AM)— May 1981: 950 khz; 500 w-D, 209 w-N, DA-2. TL: N31 56 49 W93 21 13. Stereo. Hrs opn: 24 Box 197, Rt. 4 Jordan Ferry Rd., 71019. Phone: (318) 932-6704. Fax: (318) 932-9700. E-mail: krrp@cp-tel.net Licensee: Roberto Feliz (acq 12-1-2003; $350,000). Population served: 950000 Natl. Network: ESPN Radio. Law Firm: Cohn & Marks. Format: Sports. News staff: one; News: 20 hrs wkly. Target aud: 35 plus; mature, educated, affluent listeners. ♦Chris Boyd, gen mgr & gen sls mgr; George Moore, progmg dir; Robert Abrahams, chief of engrg.

KSBH(FM)— Nov 15, 1992: 94.9 mhz; 25 kw. 328 ft TL: N31 51 34 W93 13 00. Stereo. Hrs opn: 24 213 Renee St., Natchitoches, 71457. Phone: (318) 354-4000. Fax: (318) 352-9598. Licensee: KSBH L.L.C. (acq 6-3-98; $350,000). Population served: 150,000 Format: Country. Target aud: 18-54. ♦Theresa Wilkerson, stn mgr; John Brewer, opns mgr; Shane Evath, news dir.

Covington

WASO(AM)— November 1953: 730 khz; 250 w-D, 25 w-N. TL: N30 29 37 W90 08 37. Hrs open: 3313 Kingmen St., Metairie, 70006. Phone: (504) 888-8255. Fax: (504) 888-8329. E-mail: info@hottalkradio.com Web Site: www.hottalkradio.com. Licensee: America First Communications Inc. (acq 7-7-2005). Population served: 75,000 Format: News/talk. Target aud: 25 plus. ♦Robert Namer, pres & gen mgr.

Crowley

KAJN-FM— Oct 1, 1977: 102.9 mhz; 95 kw. 1,499 ft TL: N30 02 19 W92 22 15. Stereo. Hrs opn: 24 Box 1469, 70527-1469. Secondary address: 110 W. 3rd St. 70526. Phone: (337) 783-1560. Fax: (337) 783-1674. Web Site: www.kajn.com. Licensee: Rice Capital Broadcasting. Population served: 1,500,000 Natl. Network: USA. Law Firm: Shaw Pittman. Wire Svc: AP Format: Relg. News staff: one; News: 3 hrs wkly. Target aud: 25-44; female, family oriented. Spec prog: Black 2 hrs wkly. ♦Annette G. Thompson, VP; Barry D. Thompson, CEO, pres & gen mgr.

KSIG(AM)— May 1947: 1450 khz; 1 kw-U. TL: N30 13 50 W92 21 45. Hrs open: Box 228, 70527. Secondary address: 320 N. Parkerson Ave. 70527. Phone: (337) 783-2520. Fax: (337) 783-5744. Licensee: Acadia Broadcast Partners Inc. (acq 12-7-92; $350,000; FTR: 1-4-93). Population served: 16,104 Format: Oldies. Spec prog: Fr 18 hrs, farm 5 hrs wkly. ♦Phil Lizotte, pres, gen mgr & stn mgr; Jimmy Cole, gen sls mgr; Hans Nelson, progmg dir; Tony Evans, chief of engrg.

De Quincy

KTSR(FM)— Nov 1, 1985: 92.1 mhz; 13.5 kw. Ant 448 ft TL: N30 13 24 W93 18 36. Stereo. Hrs opn: 24 900 N. Lakeshore Dr., Lake Charles, 70601. Phone: (337) 433-1641. Fax: (337) 433-2999. Web Site: www.star921online.com. Licensee: Apex Broadcasting Inc. (group owner; (acq 11-6-2001; $400,000). Format: Adult top-40. ♦Sara Cormier, gen mgr; Shelley Doucett, gen sls mgr; Eric Scott, progmg dir; Dave Chimeno, chief of engrg.

De Ridder

***KBAN(FM)**— 2001: 91.5 mhz; 20.5 kw. Ant 361 ft TL: N30 38 10 W93 02 33. Hrs open: American Family Radio, Box 3206, Tupelo, MS, 38803. Secondary address: Quicken Ministries %AFR, 1411 Parish Rd., Lake Charles 70611. Phone: (662) 844-8888. Phone: (337) 217-0252. Fax: (662) 842-6791. Fax: (337) 217-0253. E-mail: comments@myafr.com Web Site: www.afr.net. Licensee: American Family Association. Group owner: American Family Radio Format: Inspirational Christian. ♦Marvin Sanders, gen mgr; Elizabeth Arrington, stn mgr.

KDLA(AM)— Nov 11, 1950: 1010 khz; 1 kw-D, 40 w-N. TL: N30 52 43 W93 17 25. Hrs open: Sunrise-sunset 1825 Pelican Rd., 70634. Phone: (337) 460-7657. Fax: (337) 460-9099. Web Site: www.kdla1010.com. Licensee: Christian Broadcasting of De Ridder Inc. (group owner; (acq 2-18-98; $150,000). Population served: 150,000 Natl. Network: Reach Satellite. Format: Gospel. Target aud: 18-54. ♦Samuel Williams, gen mgr.

KQLK(FM)— Sept 6, 1991: 97.9 mhz; 50 kw. Ant 492 ft TL: N30 36 57 W93 13 31. Stereo. Hrs opn: 24 425 Broad St., Lake Charles, 70601. Phone: (337) 439-3300. Fax: (337) 433-7278. Web Site: www.kqlk.com. Licensee: Cumulus Licensing LLC. (group owner; (acq 12-6-2004; $3 million with KAOK(AM) Lake Charles). Format: Adult contemp. ♦Crash Kelley, progmg dir; Richard Rhodes, chief of engrg.

Delhi

KGGM(FM)— September 1991: 93.5 mhz; 3 kw. 328 ft TL: N32 37 45 W91 33 13. Hrs open: 1707 Louisa St., Rayville, 71269. Phone: (318) 878-8255. Fax: (318) 343-0476. Licensee: Kenneth W. Diebel (acq 9-10-2003; $120,000). Format: Southern gospel. ♦Ken Diebel, pres & gen mgr.

Denham Springs

WSKR(AM)— Apr 15, 1959: 1210 khz; 10 kw-D, 1 kw-N, DA-N. TL: N30 31 20 W90 58 15. Stereo. Hrs opn: 18 5555 Hilton Ave., Suite 500, Baton Rouge, 70808. Phone: (225) 231-1860. Fax: (225) 231-1869. Web Site: www.thescore1210.com. Licensee: Capstar TX L.P. Group owner: Clear Channel Communications Inc. (acq 8-30-00; grpsl). Population served: 750,000 Natl. Network: Westwood One. Format: Sports. News staff: 2. Target aud: 25-54.

Donaldsonville

KNXX(FM)— 1972: 104.9 mhz; 6 kw. Ant 299 ft TL: N30 05 57 W91 00 13. Hrs open: 24
Rebroadcasts WNXX(FM) Jackson 100%.
Box 2231, Baton Rouge, 70821-2231. Secondary address: 929 B Government St., Baton Rouge 70802-6033. Phone: (225) 388-9898 ext 148. Fax: (225) 499-9800. E-mail: owen.weber@98radio.com Web Site: www.104thex.com. Licensee: Guaranty Broadcasting Co. of Baton Rouge LLC. Group owner: Guaranty Broadcasting Co. (acq 2-18-2000; $1.2 million). Population served: 150,000 Natl. Rep: McGavren Guild. Law Firm: Wiley, Rein & Fielding. Format: Alternative, new rock. News: 2 hrs wkly. Target aud: 18-34. ♦George Foster Jr., chmn; Bridger Eglin, pres; Owen Weber, VP & gen mgr; Dave Dunaway, opns mgr.

Dry Prong

***KVDP(FM)**— Aug 13, 1985: 89.1 mhz; 4.5 kw. 295 ft TL: N31 35 20 W92 30 59. Stereo. Hrs opn: Box 249, 71423. Secondary address: 160 Bud Walker Rd. 71423. Phone: (318) 899-5837. Fax: (318) 899-7624. Licensee: Dry Prong Educational Broadcasting Foundation Inc. (acq 11-24-92; FTR: 12-21-92). Natl. Network: USA. Format: Relg, educ, Christian. ♦Donna Clina, gen mgr; Darris Cline, mus dir & chief of engrg.

Dubach

KNBB(FM)— June 4, 1984: 97.7 mhz; 50 kw. Ant 464 ft TL: N32 40 09 W92 37 58. Stereo. Hrs opn: 5 AM-11 PM Box 430, Ruston, 71270. Secondary address: 500 N. Monroe St., Ruston 71270. Phone: (318) 255-5000. Fax: (318) 255-5084. E-mail: seanfox@espn977.com Web Site: www.espn977.com. Licensee: Communications Capital Co. II of Louisiana LLC. Group owner: Communications Capital Managers LLC (acq 5-27-2003; $1.5 million). Population served: 250,000 Natl. Network: ESPN Radio. Format: Sports talk. Target aud: 25-44; men. ♦Gary McKenney, gen mgr; Sean Fox, progmg dir.

Empire

KNOU(FM)— June 2001: 104.5 mhz; 7.8 kw. Ant 850 ft TL: N29 33 45 W89 49 46. Hrs open: Hartman Leito & Bolt LLP, 6100 Southwest Blvd., Suite 500, Fort Worth, TX, 76109. Phone: (817) 738-2400. Licensee: On Top Communications of Louisiana LLC, Debtor-in-Possession Group owner: On Top Communications Inc. (acq 1-13-2006). Format: Oldies. ♦Bryan C. Rice, gen mgr.

Erath

KRKA(FM)— April 1992: 107.9 mhz; 25 kw. 328 ft TL: N30 02 54 W91 59 49. (CP: 10 kw, ant 469 ft.). Stereo. Hrs opn: 24 1749 Bertrand Dr., Lafayette, 70506-2054. Phone: (337) 233-6000. Fax: (337) 234-7360.

Web Site: www.1079ishot.com. Licensee: Regent Broadcasting of Lafayette LLC. Group owner: Regent Communications Inc. (acq 12-7-2001; grpsl). Population served: 300000 Rgnl. Network: La. Net. Natl. Rep: Katz Radio. Format: Rhythmic CHR. News: 25 hrs wkly. Target aud: 12-34. ♦Mike Grimsley, gen mgr; Chris Logan, progmg dir.

Erwinville

***KPAE(FM)**— Sept 30, 1985: 91.5 mhz; 5 kw. 167 ft TL: N30 32 09 W91 24 52. Stereo. Hrs opn: 24
Rebroadcasts WPAE (FM) Centreville, MS 75%.
Box 1390, Centreville, MS, 39631. Secondary address: 122 E. Main St., Centreville, MS 39631. Phone: (601) 645-6515. Fax: (225) 627-4970. Fax: (601) 645-9122. E-mail: wpaefm@telepak.net Web Site: www.soundradio.org. Licensee: Port Allen Educational Broadcasting Foundation. Natl. Network: Moody. Format: Relg teaching. Target aud: General. ♦Willie F. Kennedy, pres.

Eunice

KEUN(AM)— October 1952: 1490 khz; 1 kw-U. TL: N30 28 17 W92 24 51. Hrs open: 24 1237 E. Ardion St., 70535. Phone: (337) 457-3041. Fax: (337) 457-3081. E-mail: spots@keunworldwide.com Web Site: www.keunworldwide.com. Licensee: Tri-Parish Broadcasting Co. Inc. (acq 5-12-2003; with co-located FM). Population served: 25,000 Format: Sports/talk. News staff: one; News: 5 hrs wkly. Target aud: 25 plus. ♦Roger W. Cavaness, CEO & pres; Rick Nesbitt, gen mgr, progmg dir & news dir; Tony Evans, chief of engrg.

KEUN-FM— Oct 22, 1981: 105.5 mhz; 1 kw. Ant 485 ft TL: N30 26 16 W92 26 49. Stereo. 24 Web Site: www.keunworldwide.com. Natl. Network: ABC. Format: Contemp country. News staff: one; News: 5 hrs wkly. Target aud: 25-54. Spec prog: Cajun music 12 hrs wkly.

Farmerville

KBYO-FM— Apr 19, 1979: 92.7 mhz; 6 kw. Ant 328 ft TL: N32 40 31 W92 19 10. Stereo. Hrs opn: 24 1109 Hudson Ln., Monroe, 71201. Phone: (318) 998-6320. Fax: (318) 388-0569. Web Site: www.bayou92.com. Licensee: Union Broadcasting Co. Inc. Rgnl. Network: La. Net. Format: Hot adult contemp. Target aud: 24-65; adult audience with incomes to buy. ♦Hanson Holladay, gen mgr.

Ferriday

KFNV-FM— October 1971: 107.1 mhz; 18.5 kw. 233 ft TL: N31 36 08 W91 32 27. Stereo. Hrs opn: 24 Box 1510, 71334. Secondary address: 917 S. EE Wallace Blvd. 71334. Phone: (318) 757-4200. Fax: (318) 757-7689. E-mail: kfnv@bellsouth.net Licensee: Desi-Ray Productions. Group owner: The Radio Group Population served: 30,000 Format: Classic hits. News: 2 hrs wkly. Target aud: 25-55; baby boomers. ♦Desiree Smith, gen mgr.

Folsom

WJSH(FM)— March 1996: 104.7 mhz; 6 kw. Ant 328 ft TL: N30 39 55 W90 04 49. Hrs open: 24 200 E. Thomas St., Hammond, 70401. Phone: (985) 542-0060. Fax: (985) 542-9377. Licensee: Southwest Broadcasting. (acq 11-17-2000). Population served: 250000 Format: Smooth jazz. ♦Charles Dowdy, gen mgr; Eloise Dowdy, gen mgr & gen sls mgr; Ben Bickham, chief of engrg.

Franklin

KDDK(FM)— May 9, 1975: 105.5 mhz; 3 kw. Ant 300 ft TL: N29 50 14 W91 32 22. (application for CP: COL Addis. 6 kw, ant 328 ft. TL: N30 19 25 W91 16 52). Stereo. Hrs opn:
Rebroadcasts KJCB(AM) Lafayette 100%.
5047 Hwy. 1148, Plaquemine, 70764-52227. Phone: (225) 687-2882. E-mail: kddk@bellsouth.net Web Site: www.kddkfm.com. Licensee: Radio & Investments Inc. Population served: 100,000 Law Firm: James Cooke. Format: Adult standards. ♦Ken Noble, gen mgr, gen sls mgr & progmg dir.

KFRA(AM)— June 4, 1961: 1390 khz; 500 w-D. TL: N29 50 14 W91 32 22. Stereo. Hrs opn: 5047 Hwy. 87, 70538. Phone: (337) 836-2526. Licensee: Radio & Investments Inc. (acq 4-15-97; with co-located FM). Population served: 150,000 ♦Kenneth R. Noble II, pres; Roger Robinson, gen mgr.

Franklinton

WOMN(AM)— Dec 5, 1966: 1110 khz; 1 kw-D. TL: N30 51 34 W90 09 57. Hrs open: Sunrise-sunset 3015 E. Causeway Approach, Mandeville, 70448. Phone: (985) 624-9452. Fax: (985) 624-9559. E-mail: mpittman@pittmanbroadcasting.com Licensee: Pittman Broadcasting Services LLC. (group owner; (acq 6-4-2002; with co-located FM). Population served: 10,000 Format: Country. Target aud: General. ♦Mike Mitchell, gen sls mgr; Tony Evans, chief of engrg.

WUUU(FM)—Co-owned with WOMN(AM). Mar 3, 1997: 98.9 mhz; 6 kw. 108 ft TL: N30 51 34 W90 09 57.24

Galliano

WTIX-FM—Licensed to Galliano. See Golden Meadow

Garyville

***WCKW(AM)**— Dec 22, 1970: 1010 khz; 500 w-D, 42 w-N. TL: N30 04 35 W90 37 17. Stereo. Hrs opn: 24 4424 Hampton Ave., St. Louis, MO, 63109. Phone: (314) 752-7000. Fax: (314) 752-7702. Web Site: www.covenantnet.net. Licensee: Covenant Network (acq 10-17-2006). Population served: 96,000 Format: Relg. Target aud: 18-54. ♦John Anthony Holman, pres.

Gibsland

KBEF(FM)— May 23, 2001: 104.5 mhz; 6 kw. Ant 328 ft TL: N32 31 59 W93 11 34. Hrs open: 410 Lakeshore Dr., Minden, 71055. Phone: (318) 377-1240. Fax: (318) 377-4619. E-mail: staff@kbef.com Web Site: www.kbef.com. Licensee: Amistad Communications Inc. (acq 7-12-2000; $375,000 for CP with KASO(AM) Minden). Format: Contemp Christian. ♦Mike Griffith, gen mgr.

Golden Meadow

KLEB(AM)— May 13, 1963: 1600 khz; 5 kw-D, 250 w-N. TL: N29 22 41 W90 15 50. Hrs open: 24 Drawer 1350, 11603 Hwy. 308, Larose, 70373. Phone: (985) 798-7792. Fax: (985) 798-7793. E-mail: klrz@mobiletel.com Web Site: www.klrzfm.com. Licensee: Coastal Broadcasting of Larose Inc. (acq 1999; $250,000). Format: C&W, Fr, Oldies. Target aud: 25 plus; general. ♦Jerry J. Gisclair, pres & gen mgr; Andrea Galjour, opns mgr.

WTIX-FM—Galliano, Nov 16, 1975: 94.3 mhz; 100 kw. 982 ft TL: N29 33 46 W89 49 46. Hrs open: 24 4539 I-10 Service Rd., 3rd Fl., Metairie, 70006. Phone: (504) 454-9000. Fax: (504) 454-9002. Web Site: wtixfm.com. Licensee: Fleur de Lis Broadcasting Inc. (acq 6-8-95; $600,000). Population served: 1250000 Natl. Network: ABC. Law Firm: Reddy, Begley & McCormick. Format: Oldies. News: 4 hrs wkly. Target aud: 25-54. ♦George Buck, pres; Michael Costello, gen mgr.

Grambling

***KGRM(FM)**— January 1974: 91.5 mhz; 50 kw. 492 ft TL: N32 30 56 W92 43 27. Stereo. Hrs opn: 6 AM-midnight Box 4254, Grambling State Univ., 71245. Secondary address: Washington Johnson Complex 2nd Fl., 403 Main St. 71245. Phone: (318) 274-6343. Fax: (318) 274-3245. E-mail: evansjb@gram.edu Web Site: www.gram.edu/kgrm. Licensee: Grambling State University. Format: Gospel, urban contemp. Target aud: Black community. ♦Joyce Evans, gen mgr.

Gretna

KGLA(AM)— Jan 6, 1969: 1540 khz; 1 kw-D. TL: N29 53 27 W90 05 05. Hrs open: Box 428, Marrero, 70072. Secondary address: 3521 Industry St., Harvey 70058. Phone: (504) 347-1540. Fax: (504) 340-4737. E-mail: sales@tropical1540.com Web Site: www.tropical1540.com. Licensee: Crocodile Broadcasting Corp. (acq 6-12-2007; $245,000). Population served: 1,500,000 Format: Sp contemp. ♦Ernesto Schweikert, gen mgr.

KKNO(AM)— Sept 10, 1989: 750 khz; 250 w-D, DA. TL: N29 53 15 W90 05 03. Hrs open: Sunrise-sunset 980 Avenue A, Marrero, 70072. Phone: (504) 347-7775. Fax: (504) 347-7440. E-mail: kkno750am@aol.com Licensee: Robert C. Blakes Enterprises Inc. (acq 6-24-93; $275,000; FTR: 7-12-93). Format: Christian gospel, relg. News: 10 hrs wkly. Target aud: General. ♦Robert C. Blakes Sr., pres; Lois R. Blakes, gen mgr; Stacey Blakes, opns mgr, gen sls mgr & progmg dir.

Hammond

***KSLU(FM)**— Nov 11, 1974: 90.9 mhz; 3 kw. 141 ft TL: N30 30 53 W90 27 59. Stereo. Hrs opn: 24 D. Vickers Hall Rm 112, SLU 10783, 70402. Phone: (985) 549-2330. Fax: (985) 549-3960. E-mail: kslu@selu.edu Web Site: www.kslu.org. Licensee: Southeastern Louisiana University. Population served: 150,000 Natl. Network: PRI. Format: Alternative. News staff: one; News: 30 hrs wkly. Target aud: General. ♦Todd Delaney, gen mgr; Steve Portier, chief of engrg.

WCDV(FM)— Apr 3, 1965: 103.3 mhz; 100 kw. Ant 1,004 ft TL: N30 24 06 W90 50 43. Stereo. Hrs opn: 24 650 Wooddale Blvd., Baton Rouge, 70806-2930. Phone: (225) 926-1106. Fax: (225) 928-1606. E-mail: wcdv.fm@citcomm.com Web Site: www.diva103.com. Licensee: Citadel Broadcasting Co. Group owner: Citadel Broadcasting Corp. (acq 1999; grpsl). Natl. Network: ABC. Rgnl. Network: La. Net. Natl. Rep: McGavren Guild. Format: Soft rock. News staff: one; News: 3 hrs wkly. Target aud: 25-54. ♦Greg Benfield, gen mgr & sls dir.

WDVW(FM)—La Place

WFPR(AM)— Nov 15, 1947: 1400 khz; 1 kw-U. TL: N30 30 31 W90 30 18. Hrs open: 24 200 E. Thomas, 70401. Phone: (985) 345-0060. Fax: (985) 542-9377. E-mail: swapshop@nsbradiobroadcasting.com Licensee: North Shore Broadcasting Co. Inc. (acq 12-4-2003; $1.85 million with co-located FM). Population served: 80,000 Natl. Network: CBS. Format: Country. News staff: one; News: 7 hrs wkly. Target aud: 35-64. Spec prog: Farm one hr, gospel 12 hrs wkly. ♦Wayne Dowdy, pres; Eloise Dowdy, gen mgr & gen sls mgr; Ben Bickham, progmg dir, chief of engrg & disc jockey.

WHMD(FM)—Co-owned with WFPR(AM). Aug 26, 1974: 107.1 mhz; 6 kw. Ant 328 ft TL: N30 25 32 W90 17 01. Stereo. Format: New country.

Haughton

KBTT(FM)— 1993: 103.7 mhz; 6 kw. 328 ft TL: N32 31 20 W93 30 05. Hrs open: 24 208 N. Thomas Dr., Shreveport, 71107. Phone: (318) 222-3122. Fax: (318) 459-1493. Web Site: www.1037thabeat.fm. Licensee: Access. 1 Louisiana Holding Co. LLC. Group owner: Access.1 Communications Corp. (acq 5-10-00; grpsl). Format: Urban contemp. News staff: 1. Target aud: 18-34. ♦Cary D. Camp, gen mgr.

Hodge

KRLQ(FM)—Not on air, target date: Aug 15, 2007: 94.1 mhz; 47 kw. Ant 507 ft TL: N32 19 53 W92 53 07. Hrs open: P.O. Box 2941, Ruston, 71273. Secondary address: 1319 N. Vienna, Ruston 71270. Phone: (318) 255-7941. Fax: (318) 255-8211. E-mail: billbrown@kslq941.com Web Site: krlq941.com. Licensee: William W. Brown. Format: full service. Spec prog: New Orleans Saints. ♦William W. Brown, gen mgr.

Homer

KYLA(FM)— March 1998: 106.7 mhz; 50 kw. Ant 492 ft TL: N32 37 03 W93 14 36. Hrs open: 2351 Sunset Blvd., Suite 170-218, Rocklin, CA, 95765. Phone: (916) 251-1600. Fax: (916) 251-1650. Web Site: www.klove.com. Licensee: Educational Media Foundation. (acq 5-30-2006). Natl. Network: K-Love. Format: Christian. ♦Richard Jenkins, pres; Mike Novak, VP; Keith Whipple, dev dir; David Pierce, progmg mgr; Ed Lenane, news dir; Sam Wallington, engrg dir; Karen Johnson, news rptr; Marya Morgan, news rptr; Richard Hunt, news rptr.

Houma

KCIL(FM)—Listing follows KJIN(AM).

KJIN(AM)— Apr 1, 1946: 1490 khz; 1 kw-U. TL: N29 34 14 W90 43 42. Hrs open: 24 Box 2068, 70361. Secondary address: 120 Prevost Dr. 70364. Phone: (985) 851-1020. Fax: (985) 872-4403. Licensee: Sunburst Media-Louisiana LLC. (acq 1-23-2007; grpsl). Population served: 125,000 Natl. Network: ABC. Natl. Rep: Roslin. Law Firm: Wiley, Rein & Fielding. Format: Sports. News staff: one; News: 3 hrs wkly. Target aud: 35 plus. ♦Danny Fletcher, gen mgr & gen sls mgr; John Delise, opns mgr; Cade Voison, progmg dir; Bo Hoover, chief of engrg.

KCIL(FM)—Co-owned with KJIN(AM). Dec 31, 1965: 107.5 mhz; 69 kw. Ant 650 ft TL: N29 26 48 W90 44 34. Stereo. Web Site: www.1075kcil.net.250,000 Format: Country. News staff: one; News: 2 hrs wkly. Target aud: Adults; 25-54.

KYRK(FM)— Nov 15, 1968: 104.1 mhz; 100 kw. Ant 1,945 ft TL: N29 57 13 W90 43 25. Stereo. Hrs opn: 24 929 Howard Ave., New Orleans, 70113. Phone: (504) 679-7300. Fax: (504) 679-7358. Web Site: www.kissneworleans.com. Licensee: Clear Channel Broadcasting Licenses Inc. Group owner: Clear Channel Communications Inc. (acq 2-97; $6.75 million). Population served: 2,500,000 Natl. Rep: Clear Channel. Format: New rock. News staff: one; News: 2 hrs wkly. Target aud: 25-54. ♦Dick Lewis, pres & gen mgr; Michael Scott, gen sls mgr; Mike Kramer, progmg dir; Tom Courtney, chief of engrg.

Jackson

WNXX(FM)— Oct 17, 2001: 104.5 mhz; 6 kw. 328 ft TL: N30 43 07 W91 14 26. Stereo. Hrs opn: 24
Rebroadcasts KNXX(FM) Donaldsonville 100%.
Box 2231, Baton Rouge, 70821. Secondary address: 929-B Government St., Baton Rouge 70802. Phone: (225) 388-9898. Fax: (225) 344-3077. E-mail: owen.weber@gbcradio.com Web Site: www.104thex.com. Licensee: Guaranty Broadcasting Co. of Baton Rouge LLC. Group owner: Guaranty Broadcasting Co. (acq 10-5-2000; $1.044 million). Natl. Rep: McGavren Guild. Law Firm: Wiley, Rein & Fielding. Format: Alternative new rock. Target aud: 18-34. ♦George A. Foster Jr., chmn; Bridger Eglin, pres & gen mgr; Owen Weber, VP & gen mgr; Dave Dunaway, opns mgr.

Jena

***KAYT(FM)**— Jan 1, 2001: 88.1 mhz; 15.5 kw horiz, 70 kw vert. 1,007 ft TL: N31 33 54 W92 33 00. Hrs open: 5003 Masonic Dr., Suite 113, Alexandria, 71301. Phone: (318) 484-2500. Fax: (318) 487-0909. Licensee: Black Media Works Inc. (group owner). Format: Relg. ♦Raymond Kassis, gen mgr; Jocelyn Jacob, stn mgr.

KJNA-FM— November 1976: 102.7 mhz; 6 kw. 298 ft TL: N31 41 51 W92 05 43. Stereo. Hrs opn: 24 Box 2750, 71342. Secondary address: 1791 N. 2nd St. 71342. Phone: (318) 992-4155. Fax: (318) 992-4479. Licensee: Little River Radio Co. Population served: 75,000 Format: Country. News staff: one; News: 20 hrs wkly. Target aud: 25-54. ♦Larry Evans, gen mgr.

Jennings

KHLA(FM)—Listing follows KJEF(AM).

KJEF(AM)— November 1950: 1290 khz; 1 kw-U. TL: N30 12 38 W92 39 55. Stereo. Hrs opn: 24 900 N. Lakeshore Dr., Lake Charles, 70601. Phone: (337) 433-1641. Fax: (337) 433-2999. Licensee: Apex Broadcasting Inc. (group owner; acq 9-19-2000; $864,800 with co-located FM). Population served: 11,783 Rgnl. Network: La. Net. Format: Cajun. Target aud: General. ♦Sara Cormier, gen mgr; Shelly Doucette, gen sls mgr; Mike Soileau, progmg dir; Dave Chimeno, chief of engrg.

KHLA(FM)—Co-owned with KJEF(AM). January 1963: 92.9 mhz; 33 kw. Ant 600 ft TL: N30 00 31 W92 46 47. Stereo. 24 Web Site: www.929thelake.com.500,000 Format: Classic hits. ♦Aaron Turner, gen sls mgr & min affrs dir; Gary Shannon, progmg dir; Sara Cormier, stn mgr & political ed.

Jonesboro

***KTOC-FM**— Oct 1, 1967: 104.9 mhz; 25 kw. Ant 236 ft TL: N32 13 28 W92 43 27. Stereo. Hrs opn: Box 262550, Baton Rouge, 70826. Secondary address: 8919 World Ministry Ave., Baton Rouge 70810. Phone: (225) 768-3688. Phone: (225) 768-8300. Fax: (225) 768-3729. E-mail: kawikfish@yahoo.com Web Site: www.jsm.org. Licensee: Family Worship Center Church Inc. (acq 9-25-2002; $200,000 with co-located AM). Population served: 8,000 Format: Christian. ♦David Whitelaw, COO; Jimmy Swaggart, pres; John Santiago, progmg dir.

Jonesville

KTGV(FM)— 2001: 105.1 mhz; 6 kw. Ant 315 ft TL: N31 36 21 W91 50 06. Hrs opn: Box 768, Natchez, MS, 39121. Secondary address: 2 Oferrall St., Natchez 39120-3000. Phone: (601) 442-4895. Fax: (601) 446-8260. Licensee: First Natchez Corp. Group owner: First Natchez Radio Group (acq 8-30-99; $150,000). Format: Urban contemp. ♦Margaret Perkins, gen mgr; Mickey Alexander, progmg dir; Keith Sanders, chief of engrg; Brenda Green, traf mgr.

Kaplan

***KCKR(FM)**—Not on air, target date: unknown: 91.9 mhz; 12.5 kw. Ant 464 ft TL: N30 17 05 W92 04 03. Hrs open: Box 563, Tanner, AL, 35671-0563. Phone: (256) 345-2478. Licensee: North Alabama Educational Foundation. ♦Richard W. Dabney, pres.

KMDL(FM)— Aug 1, 1981: 97.3 mhz; 42 kw. 535 ft TL: N30 02 54 W91 59 49. Stereo. Hrs opn: 24 1749 Bertrand Dr., Lafayette, 70506-2054. Phone: (337) 233-6000. Fax: (337) 234-7360. Web Site: www.973thedawg.com. Licensee: Regent Broadcasting of Lafayette LLC. Group owner: Regent Communications Inc. (acq 12-7-2001; grpsl). Natl. Network: AP Radio. Law Firm: Wiley, Rein & Fielding. Format: Classic country. News: 7 hrs wkly. Target aud: 25-54. ♦Mike Grimsley, gen mgr; Scott Bryant, progmg dir.

Kenner

WWL-FM— Sept 8, 1970: 105.3 mhz; 96 kw. Ant 1,004 ft TL: N29 58 57 W89 57 09. Stereo. Hrs opn: 1450 Poydras, Suite 500, New Orleans, 70112. Phone: (504) 593-6376. Fax: (504) 593-2285. Web Site: www.wwl.com. Licensee: Entercom New Orleans License LLC. Group owner: Entercom Communications Corp. (acq 12-13-99; grpsl). Format: News, talk. ♦Ken Beck, gen mgr; Patrick Galloway, sls dir; Mark Broudreaux, gen sls mgr; Diane Newman, progmg dir & chief of engrg; Joe Pollet, chief of engrg.

Kentwood

WEMX(FM)— Dec 14, 1967: 94.1 mhz; 100 kw. 981 ft TL: N30 51 18 W90 39 59. Stereo. Hrs opn: 24 650 Wooddale Blvd., Baton Rouge, 70806. Phone: (225) 926-1106. Fax: (225) 928-1606. E-mail: wemx.fm@citcomm.com Web Site: www.max94one.com. Licensee: Citadel Broadcasting Co. Group owner: Citadel Broadcasting Corp. (acq 1-14-99; grpsl). Population served: 500,000 Format: Hip hop, rhythm and blues. ♦Greg Benefield, gen mgr.

La Place

WDVW(FM)— Jan 10, 1966: 92.3 mhz; 100 kw. Ant 1,945 ft TL: N29 57 10 W90 43 26. Stereo. Hrs opn: 24 201 St. Charles Ave., Suite 201, New Orleans, 70170. Phone: (504) 581-7002. Fax: (504) 566-4857. Web Site: www.diva923.com. Licensee: Citadel Broadcasting Co. Group owner: Citadel Broadcasting Corp. (acq 1-30-2004; $14.25 million). Population served: 2,068,800 Natl. Rep: Christal. Rgnl rep: Christal Radio Format: Hot adult contemp. ♦Dave Siebert, gen mgr; John McQueen, progmg dir.

Lacombe

WYLK(FM)— March 1996: 94.7 mhz; 5.3 kw horiz, 5.2 kw vert. Ant 348 ft TL: N30 15 08 W89 45 46. Hrs open: 24 324 Lockwood St., Covington, 70433. Phone: (985) 867-5990. Fax: (985) 867-9530. Licensee: North Shore Broadcasting Inc. (group owner; (acq 7-6-2005; $4.5 million with WPRF(FM) Reserve). Population served: 256,000 Format: Hot CHR. News staff: one; News: 2 hrs wkly. Target aud: 25-54; general. ♦Vicki Hays, gen mgr & sls dir.

Lafayette

KFXZ(AM)— Nov 15, 1960: 1520 khz; 10 kw-D, 500 w-N, DA-N. TL: N30 16 51 W92 00 53. Stereo. Hrs opn: 24 3225 Ambassador Caffery Pkwy., 70506. Phone: (337) 993-5500. Fax: (337) 993-5510. Web Site: fsr1520.com. Licensee: Pittman Broadcasting Services LLC (group owner; (acq 1-28-2004; grpsl). Population served: 176,000 Format: Sports, talk. ♦Charles Sagona, gen mgr.

***KIKL(FM)**— Feb 7, 1988: 90.9 mhz; 6 kw. Ant 476 ft TL: N30 17 08 W92 04 03. Stereo. Hrs opn: 24 2351 Sunset Blvd., Suite 170-218, Rocklin, CA, 95765. Phone: (916) 251-1600. Fax: (916) 251-1650. Licensee: Educational Media Foundation. Group owner: EMF Broadcasting (acq 4-25-2005; $1.5 million). Population served: 350,000 Law Firm: Shaw Pittman LLP. Format: Christian. ♦Richard Jenkins, pres; Mike Novak, VP; Keith Whipple, dev dir; David Pierce, progmg mgr; Ed Lenane, news dir; Sam Wallington, engrg dir; Karen Johnson, news rptr; Marya Morgan, news rptr; Richard Hunt, news rptr.

KJCB(AM)— Apr 9, 1982: 770 khz; 1 kw-D, 500 w-N, DA-N. TL: N30 17 55 W91 59 30. Stereo. Hrs opn: 604 St. John St., 70501. Phone: (337) 233-4262. Fax: (337) 235-9681. Web Site: www.blackaction.net. Licensee: R & M Broadcasting Inc. (acq 11-16-92; $100,000; FTR: 12-14-92). Format: Urban contemp, gospel. Target aud: 25-54. Spec prog: . ♦Jenelle Schargios, gen mgr.

KPEL(AM)— Jan 2, 1950: 1420 khz; 1 kw-D, 750 w-N, DA-N. TL: N30 16 38 W92 03 51. Hrs open: KPEL AM FM, 1749 Bertrand Dr., 70506. Phone: (337) 233-6000. Fax: (337) 234-7360. Web Site: www.espn1420.com. Licensee: Regent Broadcasting of Lafayette LLC. Group owner: Regent Communications Inc. (acq 12-7-2001; grpsl). Population served: 250,000 Rgnl. Network: La. Net. Natl. Rep: Christal. Format: Sports. Target aud: 35-54; male. ♦Mike Grimsley, gen mgr; Chuck Wood, gen sls mgr; Kyle Vidrine, chief of engrg.

KTDY(FM)—Co-owned with KPEL(AM). Sept 15, 1966: 99.9 mhz; 100 kw. 984 ft TL: N30 12 04 W91 46 33. Stereo. Web Site: www.999ktdy.com.750,000 Format: Adult contemp. Target aud: 25-54; female. ♦C.J. Clements, gen sls mgr & progmg dir.

KRRQ(FM)— 1996: 95.5 mhz; 50 kw. 443 ft TL: N30 21 08 W92 10 51. Hrs open: 202 Galbert Rd., 70506. Phone: (337) 232-1311. Fax: (337) 233-3779. E-mail: krrq@krrq.com Web Site: www.krrq.com. Licensee: Citadel Broadcasting Co. Group owner: Citadel Broadcasting Corp. (acq 1-14-99; grpsl). Format: Urban contemp/hip hop. ♦Mary Galyean, gen mgr.

***KRVS(FM)**— 1962: 88.7 mhz; 100 kw. 449 ft TL: N30 15 25 W92 09 38. Stereo. Hrs opn: Box 42171, 70504. Secondary address: 231 Hebrard Blvd. 70503. Phone: (337) 482-5787. Fax: (337) 482-6101. E-mail: krvs@louisiana.edu Web Site: www.krvs.org. Licensee: University of Southwestern Louisiana. Population served: 300,000 Natl. Network: NPR. Format: Var, Eclectic (cajun, zydeco, blues, jazz). Target aud: General. ♦Dave Spizale, gen mgr; James Hebert, opns mgr; Judith Meriwether, dev dir & prom mgr; Kim Neustrom Richard, mktg dir; Karl Fontenot, chief of engrg.

KSMB(FM)— 1964: 94.5 mhz; 100 kw. Ant 1,079 ft TL: N30 21 44 W92 12 53. Stereo. Hrs opn: 202 Galbert Rd., 70506. Phone: (337) 232-1311. Fax: (337) 233-3779. Web Site: www.ksmb.com. Licensee: Citadel Broadcasting Co. Group owner: Citadel Broadcasting Corp. (acq 4-26-01; grpsl). Population served: 414,800 Format: CHR/Top-40. Target aud: 18-49; active on-the-go adults. ♦Mary Galyean, gen mgr.

KVOL(AM)— May 18, 1935: 1330 khz; 5 kw-D, 1 kw-N, DA-2. TL: N30 14 29 W92 03 31. Hrs open: 3225 Ambassador Caffery Pkwy., 70506. Phone: (337) 993-5500. Fax: (337) 993-5510. E-mail: kvol@pittmanbroadcasting.com Web Site: www.kvol1330.com. Licensee: Pittman Broadcasting Services LLC (group owner; acq 1-28-2004; grpsl). Natl. Network: Westwood One. Format: News talk. Target aud: 25-54; middle & upper income. ♦Charles Sagona, gen mgr.

KXKC(FM)—See New Iberia

Lake Arthur

KJMH(FM)— Aug 1, 1998: 107.5 mhz; 50 kw. Ant 462 ft TL: N30 12 07 W92 56 47. Hrs open: 24 900 N. Lakeshore Dr., Lake Charles, 70601. Phone: (337) 433-1641. Fax: (337) 433-2999. Web Site: www.107jamz.com. Licensee: Apex Broadcasting Inc. (group owner; acq 9-17-97; $74,300). Natl. Rep: Christal. Format: Hip Hop, rhythm and blues. News staff: one. Target aud: 25-54. ♦Sara Cormier, gen mgr; Aaron Turner, gen sls mgr; Erik Johnson, progmg dir; Dave Chimeno, chief of engrg.

Lake Charles

KAOK(AM)— May 10, 1947: 1400 khz; 1 kw-U. TL: N30 12 35 W93 12 43. Hrs open: 24 425 Broad St., 70601. Phone: (337) 439-3300.

Fax: (337) 433-7278. Web Site: www.kaok.com. Licensee: Cumulus Licensing LLC. Group owner: Pittman Broadcasting Services LLC (acq 12-6-2004; $3 million with KQLK(FM) De Ridder). Population served: 167,000 Natl. Network: CBS. Rgnl. Network: La. Net. Format: News/talk, info. News staff: one; News: 168 hrs wkly. Target aud: 24 plus; baby boomers. ♦Lewis W. Dickey Jr., pres; Larry LeBlanc, stn mgr & gen sls mgr; Eric Nielson, progmg dir; Richard Rhodes, chief of engrg.

KBIU(FM)— Dec 1, 1976: 103.3 mhz; 35 kw. Ant 479 ft TL: N30 14 41 W93 20 37. Stereo. Hrs opn: 24 425 Broad St., Lakes Charles, 70601. Phone: (337) 439-3300. Fax: (337) 439-7701. E-mail: jim.vidler@cumulus.com Web Site: www.kbiu.com. Licensee: Cumulus Licensing Corp. Group owner: Cumulus Media Inc. (acq 12-17-98; grpsl). Population served: 270,000 Natl. Rep: Katz Radio. Law Firm: Bryan Cave. Format: Adult contemp. Target aud: 25-54; adult. ♦Jimmie Cole, gen sls mgr; Jim Vidler, mktg mgr; Steve Alex, progmg dir; Larry Beck, news dir; Richard Rhodes, chief of engrg; Jim Nettles, traf mgr.

KXZZ(AM)—Co-owned with KBIU(FM). 1947: 1580 khz; 1 kw-U, DA-N. TL: N30 15 28 W93 11 55. Stereo. 24 Web Site: www.kxzz1580am.com.100,000 Natl. Network: American Urban. Natl. Rep: Katz Radio. Format: Classic soul. Target aud: 25-54; adult.

KEZM(AM)—See Sulphur

KKGB(FM)—See Sulphur

KLCL(AM)—Listing follows KNGT(FM).

KNGT(FM)— Nov 8, 1965: 99.5 mhz; 100 kw. 984 ft TL: N30 23 59 W93 00 10. Stereo. Hrs opn: 900 N. Lake Shore Dr., 70601. Phone: (337) 433-1641. Fax: (337) 433-2999. Web Site: www.gater995.com. Licensee: Apex Broadcasting Inc. (group owner; (acq 7-26-2000; grpsl). Population served: 175,000 Natl. Rep: Christal. Format: Country. Target aud: 25-54. ♦Sara Cormier, pres & gen mgr; Shelly Doucette, gen sls mgr; Gene Michaels, progmg dir; Dave Chimeno, chief of engrg.

KLCL(AM)—Co-owned with KNGT(FM). May 12, 1935: 1470 khz; 5 kw-D, 500 w-N. TL: N30 15 31 W93 16 07.175,000 Format: Cajun. Target aud: 18-64. ♦Mike Soileau, progmg dir.

***KOJO(FM)**— 1990: 91.1 mhz; 4 kw horiz, 14 kw vert. Ant 387 ft TL: N30 12 07 W92 56 47. Hrs open: 24
Rebroadcasts KJMJ(AM) Alexandria.
601 Washington St., Alexandria, 71301. Phone: (318) 561-6145. Fax: (318) 449-9954. E-mail: info.usa@radiomaria.org Web Site: www.radiomaria.us. Licensee: Radio Maria Inc. (group owner; acq 10-13-99). Format: Christian, relg, talk. Target aud: General; Christians seeking training & encouragement through Bible teaching programs. ♦Dale dePerrodell, gen sls mgr; Duane Stenzel, gen mgr & progmg dir; Danny Brou, chief of engrg.

KYKZ(FM)— January 1976: 96.1 mhz; 97 kw. 1,204 ft TL: N30 17 26 W93 34 35. Stereo. Hrs opn: 24 425 Broad St., 70601. Phone: (337) 439-3300. Fax: (337) 436-7278. Web Site: www.kykz.com. Licensee: Cumulus Licensing Corp. Group owner: Cumulus Media Inc. (acq 12-17-98; grpsl). Population served: 150,000 Format: Country. News staff: 3; News: 8 hrs wkly. Target aud: General. ♦Eric Nielson, progmg dir; Richard Rhodes, gen mgr & chief of engrg.

***KYLC(FM)**— 2001: 90.3 mhz; 80 kw vert. 469 ft TL: N30 38 10 W93 02 33. Hrs open: American Family Radio, Box 3206, Tupelo, MS, 38803. Phone: (662) 844-8888. Fax: (662) 842-6791. E-mail: comments@afr.net Web Site: www.afr.net. Licensee: American Family Association. Group owner: American Family Radio (acq 3-14-01). Format: Inspirational Christian. ♦Marvin Sanders, gen mgr.

Lake Providence

KLPL(AM)— June 27, 1957: Stn currently dark. 1050 khz; 250 w-D, 22 w-N. TL: N32 48 59 W91 12 22. Hrs open: Willis Broadcasting Corp., 645 Church St., Suite 400, Norfolk, VA, 23510. Phone: (757) 622-4600. Fax: (757) 624-6515. Licensee: Willis Broadcasting Corp. (group owner; acq 4-21-98; $120,000 with co-located FM). Population served: 10,300 Rgnl. Network: Prog Farm. Target aud: General.

KLPL-FM— Jan 28, 1975: Stn currently dark. 92.7 mhz; 3 kw. Ant 154 ft TL: N32 48 59 W91 12 22. Stereo. 10,300

Larose

KLRZ(FM)— Mar 29, 1993: 100.3 mhz; 89 kw. Ant 586 ft TL: N29 33 01 W90 21 04. Hrs open: 24 11603 Hwy. 308, Drawer 1350, 70373. Phone: (985) 798-7792. Fax: (985) 798-7793. E-mail: klrz@mobilete.com Web Site: www.klrzfm.com. Licensee: Coastal Broadcasting of Larose Inc. Natl. Network: Westwood One. Format: All Louisiana all the time. News staff: one; News: 22 hrs wkly. Target aud: 25-54; professionals. ♦Andrea Galjour, opns mgr; Jerry Gisclair, gen mgr & chief of engrg.

Leesville

KJAE(FM)—Listing follows KLLA(AM).

KLLA(AM)— September 1956: 1570 khz; 1 kw-D. TL: N31 06 24 W93 17 38. Hrs open: Box 1323, 71446. Secondary address: 101 Lees Ln. 71446. Phone: (337) 239-3402. Fax: (337) 238-9283. Web Site: www.kjae935.com. Licensee: Pene Broadcasting Co. (acq 12-1-76). Population served: 60,000 Format: Oldies. ♦Penny Scogin, pres & gen mgr; Peggy Merritt, gen sls mgr; Tony Evans, chief of engrg.

KJAE(FM)—Co-owned with KLLA(AM). October 1979: 93.5 mhz; 7.5 kw. 328 ft TL: N31 08 28 W93 17 44. Stereo. Web Site: www.kjae935.com. Format: Country.

KVVP(FM)— Jan 20, 1977: 105.7 mhz; 25 kw. 400 ft TL: N31 00 17 W93 16 40. Stereo. Hrs opn: 24 168 KVVP Dr., 71446. Phone: (337) 537-5887. Fax: (337) 537-4152. E-mail: kvvp@kvvp.com Web Site: www.kvvp.com. Licensee: Stannard Broadcasting Co. Inc. Population served: 95,000 Law Firm: James Popham. Format: New country. News staff: one; News: 15 hrs wkly. Target aud: 18-54; adults with spending power. Spec prog: Relg 9 hrs wkly. ♦Alan Taylor, CFO & political ed; Doug Stannard, pres, gen mgr & progmg dir.

Mamou

KBON(FM)— June 1997: 101.1 mhz; 25 kw. Ant 328 ft TL: N30 39 33 W92 19 00. Hrs open: 24 109 S. 2nd St., Eunice, 70535. Phone: (337) 546-0007. Fax: (337) 546-0097. E-mail: 101.1@kbon.com Web Site: www.kbon.com. Licensee: Rose Ann Marx. (acq 9-8-98; $70,000). Format: Var, country. ♦Paul Marx, gen mgr.

Mansfield

KJVC(FM)— September 1976: 92.7 mhz; 3 kw. 299 ft TL: N32 01 18 W93 44 18. Stereo. Hrs open: 24 Box 700, Logansport, 71049. Phone: (318) 697-4000. Fax: (318) 697-4004. Licensee: Metropolitan Radio Group Inc. (group owner; acq 10-97; $85,000). Population served: 320,000 Format: Country. ♦David Grahams, gen mgr.

***KMSL(FM)**— 2006: 91.7 mhz; 12 kw vert. Ant 339 ft TL: N32 10 39 W93 55 02. Hrs open:
Rebroadcasts WAFR(FM) Tupelo, MS 100%.
Box 2440, Tupelo, MS, 38803-2440. Phone: (662) 844-8888. Fax: (662) 842-6791. Web Site: www.afr.net. Licensee: American Family Association. (acq 5-13-2005; $10 for CP). Natl. Network: American Family Radio. Format: Christian. ♦Marvin Sanders, gen mgr.

KORI(FM)— May 1994: 104.7 mhz; 25 kw. 328 ft TL: N31 57 49 W93 53 58. Hrs open: 24 Box 700, Logansport, 71049. Secondary address: 222 Main St., Logansport 71049. Phone: (318) 697-4000. Fax: (318) 697-4004. E-mail: inc7008@bellsouth.net Web Site: www.korifm.com. Licensee: Metropolitan Radio Group Inc. (group owner; acq 1-30-98; $390,250). Population served: 250,000 Format: Country. ♦Carmen Beatriz, opns dir & progmg dir; David Graham, gen mgr & gen sls mgr.

Mansura

KZLG(FM)— July 2000: 95.9 mhz; 6 kw. 328 ft Stereo. Hrs opn: 24 Box 516, Moreauville, 71355. Secondary address: 10586 Hwy. 1, Moreauville 71355. Phone: (318) 985-3070. Fax: (318) 985-2995. E-mail: kzlg@kricket.net Licensee: Amy M. Coco. Population served: 50,000 Natl. Network: AP Radio. Format: Adult Contemporary. News: 3 hrs. Target aud: 25 plus. ♦Louis Coco, Jr., gen mgr.

Many

***KAVK(FM)**— June 1998: 89.7 mhz; 1 kw. 430 ft TL: N31 32 06 W93 25 21. Hrs open: Box 3206, American Family Radio, Tupelo, MS, 38803. Phone: (662) 844-8888. Fax: (662) 842-6791. E-mail: comments@afr.net Web Site: www.afr.net. Licensee: American Family Association. Group owner: American Family Radio Format: Inspirational Christian. ♦Marvin Sanders, gen mgr.

KWLA(AM)— August 1962: 1400 khz; 1 kw-U. TL: N31 34 30 W93 29 47. Stereo. Hrs opn: 24 605 San Antonio, 71449. Phone: (318) 256-5177. Fax: (318) 256-0950. Licensee: Baldridge-Dumas Communications Inc. (group owner; acq 1-10-00; with co-located FM). Population served: 7,500 Rgnl. Network: La. Net. Format: Talk. News: 5 hrs wkly. Target aud: General; 38 plus. ♦Cindy Ezemack, gen sls mgr & adv VP; Rhonda Benson-Leach, gen mgr, sls VP & progmg dir; Kenny Carter, chief of engrg.

KWLV(FM)—Co-owned with KWLA(AM). Nov 12, 1977: 107.1 mhz; 25 kw. 253 ft TL: N31 36 27 W93 24 05. Format: Country. Target aud: 20 plus.

Marksville

KAPB-FM— Aug 14, 1971: 97.7 mhz; 6 kw. 328 ft TL: N31 07 27 W92 04 40. Stereo. Hrs opn: Box 7, 71351. Secondary address: 520 Chester 71351. Phone: (318) 253-5272. Fax: (318) 253-5262. E-mail: kapbfm@yahoo.com Licensee: Three Rivers Radio Co. Group owner: The Radio Group Population served: 100,000 Rgnl. Network: Prog Farm. Format: Classic hit country. ♦Pamela Couvillion, gen mgr; Larry Young, news dir & chief of engrg.

Maurice

KKSJ(FM)— June 13, 1985: 106.3 mhz; 1.3 kw. Ant 495 ft TL: N30 04 16 W92 11 53. Stereo. Hrs opn: 3225 Ambassador Caffery Pkwy., Lafayette, 70506. Phone: (337) 993-5500. Fax: (337) 993-5510. E-mail: smoothjazz@pittmanbroadcasting.com Web Site: www.smoothjazz106.com. Licensee: Pittman Broadcasting Services LLC (group owner; acq 1-28-2004; grpsl). Format: Smooth jazz. ♦Charles Sagona, gen mgr.

Minden

KASO(AM)— Apr 1, 1952: 1240 khz; 1 kw-U. TL: N32 37 50 W93 16 56. Hrs open: 24 410 Lakeshore Dr., 71055. Phone: (318) 377-1240. Fax: (318) 377-4619. Web Site: www.kbef.com. Licensee: Amistad Communications Inc. (acq 10-2000; $375,000 with CP for KBEF(FM) Gibsland). Population served: 15,500 Natl. Network: Jones Radio Networks. Rgnl. Network: La. Net. Format: Adult standards. News: 102 hrs wkly. Target aud: 35-64; male & female. ♦Fred Caldwell Sr., pres; Mike Griffith, gen mgr; Mark Cheesne, opns mgr & progmg dir.

KLKL(FM)—Licensed to Minden. See Shreveport

Monroe

***KBMQ(FM)**— Aug. 15, 1999: 88.7 mhz; 25 kw horiz, 24.5 kw vert. Ant 458 ft TL: N32 24 15 W92 02 07. Stereo. Hrs opn: 24 Box 3265, 71201. Phone: (318) 387-1230. Fax: (318) 387-8856. E-mail: info@887fm.org Web Site: www.887fm.org. Licensee: Media Ministries Inc. Population served: 135,000 Format: Christian. News: one hr wkly. Target aud: 25-54; women. ♦Phillip Brooks, progmg dir.

***KEDM(FM)**— Apr 23, 1991: 90.3 mhz; 87.1 kw. 863 ft TL: N32 39 38 W91 59 28. Stereo. Hrs opn: 24 ULM, 225 Stubbs Hall, 71209-6805. Phone: (318) 342-5556. Fax: (318) 342-5570. E-mail: kedm@ulm.edu Web Site: www.kedm.org. Licensee: University of Louisiana at Monroe. Natl. Network: NPR, PRI. Rgnl. Network: La. Net. Law Firm: Dow, Lohnes & Albertson. Format: Var. News staff: one; News: 40 hrs wkly. Target aud: 35 plus; involved, upscale, educated, movers & shakers. ♦Ray Davidson, opns mgr.

KJLO-FM— July 1946: 104.1 mhz; 100 kw. Ant 1,017 ft TL: N32 39 36 W92 05 15. Stereo. Hrs opn: 24 Box 4808, 71211. Secondary address: 1109 Hudson Ln. 71201. Phone: (318) 388-2323. Fax: (318) 388-0569. Web Site: www.kjlo.com. Licensee: New South Communications Inc. (group owner; acq 12-15-86). Population served: 397,000 Natl. Rep: McGavren Guild. Law Firm: Latham & Watkins. Format: Country. News staff: one. Target aud: 25-54. Spec prog: Gospel 4 hrs wkly. ♦Robert H. Holladay, pres.

KLIC(AM)— 1950: 1230 khz; 1 kw-U. TL: N32 29 16 W92 05 25. Stereo. Hrs opn: 24 130 N 2nd St. Ste C, 71201. Phone: (318) 387-1230. Fax: (318) 387-8856. E-mail: phillip@887fm.org Web Site: www.am1230thesource.com. Licensee: Media Ministries Inc. (acq 10-28-92; $165,000; FTR: 11-23-92). Population served: 150,000 Natl. Network: Salem Radio Network. Format: Christian, talk. News: 14 hrs wkly. Target aud: 25-54; Adults 35 +. ♦Tony Davis, pres; Mike

Downhour, gen mgr & progmg dir; Diane Osborne, sls dir & mktg dir; Naomi Thompson, news dir & pub affrs dir; Ernie Sandidge, engrg dir; Mark Kemp, traf mgr.

KLIP(FM)— April 1993: 105.3 mhz; 50 kw. 433 ft TL: N32 33 08 W92 08 33. Stereo. Hrs opn: 24 Box 4808, 71211. Secondary address: 1109 Hudson Ln. 71201. Phone: (318) 388-2323. Fax: (318) 388-0569. E-mail: la105@bayou.com Web Site: www.la105.com. Licensee: Holladay Broadcasting of Louisiana LLC (group owner; acq 11-21-2003; grpsl). Natl. Network: ABC. Natl. Rep: McGavren Guild. Law Firm: Latham & Watkins. Format: Classic hits. News: 2 hrs wkly. Target aud: 25-54. ♦Bob Holladay, pres & gen mgr.

KMLB(AM)— July 1, 1930: 1440 khz; 5 kw-D, 1 kw-N, DA-N. TL: N32 33 10 W92 04 24. Hrs open: 24 Box 4808, 71211. Secondary address: 1109 Hudson Ln. 71201. Phone: (318) 388-2323. Fax: (318) 388-0569. E-mail: kmlb@bayou.com Web Site: www.kmlb.com. Licensee: Holladay Broadcasting of Louisiana LLC (group owner; acq 11-21-2003; grpsl). Population served: 125,000 Natl. Network: Fox News Radio. Natl. Rep: McGavren Guild. Law Firm: Latham & Watkins. Format: Talk/news. Target aud: 25 plus. ♦Bob Holladay, gen mgr; Cynthia Halladay, gen sls mgr; Cory Crow, progmg dir.

KNOE(AM)— Oct 4, 1944: 540 khz; 5 kw-D, 1 kw-N, DA-2. TL: N32 32 36 W92 10 45. Hrs open: Box 4067, 71211. Secondary address: 1400 Oliver Rd. 71201. Phone: (318) 388-8888. Fax: (318) 325-9466. Licensee: Noe Corp. L.L.C. Population served: 130,000 Format: News/talk, sports. News: 25 hrs wkly. ♦George Noe, pres; Roy Frostenson, gen mgr; Bobby Richards, opns mgr; John Matherne, gen sls mgr; Jerry Harkins, chief of engrg.

KNOE-FM— Jan 29, 1967: 101.9 mhz; 97 kw horiz, 96 kw vert. 1,670 ft TL: N32 11 45 W92 04 10. Stereo. Web Site: knoefm102.com.300,000 Format: CHR. ♦Bobby Richards, mus dir. Co-owned TV: KNOE-TV affil

KRJO(AM)— May 2001: 1680 khz; 10 kw-D, 1 kw-N. TL: N32 27 24 W92 01 06. Stereo. Hrs opn: 24 Box 4808, 71211. Secondary address: 1109 Hudson Ln. 71201. Phone: (318) 338-2323. Fax: (318) 388-0569. E-mail: rejoice@bayou.com Web Site: www.krjo.com. Licensee: Holladay Broadcasting of Louisiana LLC (group owner; acq 11-21-2003; grpsl). Natl. Network: ABC. Rgnl. Network: ABC. Natl. Rep: McGavren Guild. Law Firm: Latham & Watkins. Format: Gospel. ♦Bob Holladay, gen mgr.

KRVV(FM)—See Bastrop

KXRR(FM)— Nov 15, 1965: 106.1 mhz; 97 kw. Ant 1,017 ft TL: N32 39 36 W92 05 15. Stereo. Hrs opn: 1200 N. 18th St., Suite D, 71201. Phone: (318) 387-3922. Fax: (318) 322-4585. Web Site: rock106kxrr.com. Licensee: Opus Broadcasting Monroe L.L.C. Group owner: Opus Media Partners LLC (acq 7-19-2004; grpsl). Population served: 85,000 Natl. Rep: Christal. Format: Rock. Target aud: 25-49. ♦Chris Zimmerman, gen mgr & stn mgr.

***KXUL(FM)**— May 9, 1973: 91.1 mhz; 8.5 kw. 716 ft TL: N32 39 38 W91 59 28. Stereo. Hrs opn: 24 ULM, 130 Stubbs Hall, 71209-6805. Phone: (318) 342-5985. Phone: (318) 342-5986. Web Site: www.kxul.com. Licensee: University of Lousiana at Monroe. Population served: 215000 Law Firm: Dow, Lohnes & Albertson. Format: Alternative, rock. News: 2 hrs wkly. Target aud: 12-34. ♦Joel Willer, gen mgr.

***KYFL(FM)**— Oct 8, 1992: 89.5 mhz; 25 kw. Ant 377 ft TL: N32 33 08 W92 08 33. Stereo. Hrs opn: 24 11530 Carmel Commons Blvd., Charlotte, NC, 28226. Phone: (704) 523-5555. E-mail: kyfl@bbnradio.org Web Site: www.bbnradio.org. Licensee: Bible Broadcasting Network Inc. Group owner: Bible Broadcasting Network. Format: Conservative Christian. News: 3 hrs wkly. Target aud: General. ♦Michael Thomson, gen mgr.

Moreauville

KLIL(FM)— July 25, 1980: 92.1 mhz; 6 kw. 300 ft TL: N31 02 53 W91 59 47. Stereo. Hrs opn: Box 365, 71355. Secondary address: 10586 Hwy. 1 71355. Phone: (318) 985-2929. Fax: (318) 985-2995. E-mail: klil@kricket.net Licensee: Cajun Broadcasting Inc. Population served: 50,000 Natl. Network: AP Radio. Format: Oldies. Target aud: 20 plus; working adults. ♦Louis B. Coco Jr., pres & gen mgr.

Morgan City

KMRC(AM)— April 1954: 1430 khz; 500 w-D, 100 w-N. TL: N29 45 03 W91 10 24. Hrs open: 24 409 Duke St., 70381. Phone: (985) 384-1430. Fax: (985) 384-2351. E-mail: kmrc@kmrc1430.com Web Site: www.kmrc1430.com. Licensee: Spotlight Broadcasting L.L.C. (group owner; acq 2-1-00; $109,000). Population served: 60,000 Format: Swamp pop. News staff: one; News: 5 hrs wkly. Target aud: 25-54; middle to upper income. ♦John Stork, gen mgr & news dir.

KMYO-FM— Aug 1, 1967: 96.7 mhz; 12 kw. Ant 476 ft TL: N29 41 39 W90 59 58. Stereo. Hrs opn: 24 Box 2068, Houma, 70361. Secondary address: 120 Prevost Dr., Houma 70364. Phone: (985) 851-1020. Fax: (985) 872-4403. Web Site: www.mix967.net. Licensee: Sunburst Media-Louisiana LLC. (acq 1-23-2007; grpsl). Natl. Network: Jones Radio Networks. Format: Oldies. News: 2 hrs wkly. Target aud: 18-34; majority women. ♦Danny Fletcher, gen mgr; John Delise, opns mgr & progmg dir.

Moss Bluff

KZWA(FM)— Aug 12, 1994: 104.9 mhz; 25 kw. Ant 328 ft TL: N30 27 15 W93 08 20. Hrs open: 305 Enterprise Blvd., Lake Charles, 70601. Phone: (337) 491-9955. Fax: (337) 433-8097. Web Site: www.kzwa.com. Licensee: B & C Broadcasting Inc. Format: Urban contemp mainstream. Target aud: 18-34. ♦Faye Brown-Blackwell, CEO & gen mgr.

Natchitoches

***KBIO(FM)**— July 2, 2002: 89.7 mhz; 100 w. Ant 295 ft TL: N31 47 13 W93 07 52. Hrs open:
Rebroadcasts KJMJ(AM) Alexandria.
601 Washington St., Alexandria, 71301. Phone: (318) 561-6145. Fax: (318) 449-9954. E-mail: info.usa@radiomaria.org Web Site: www.radiomaria.us. Licensee: Radio Maria Inc. (group owner; (acq 9-6-2001). Format: Christian, relg, talk. ♦Daune Stenzel, gen mgr; Dale dePerrodil, gen sls mgr; Duane Stenzel, progmg dir; Danny Brou, chief of engrg.

KDBH(FM)— July 1, 1965: 97.3 mhz; 25 kw. Ant 220 ft TL: N31 48 17 W93 01 27. Stereo. Hrs opn: 605 San Antonio Ave., Many, 71449. Phone: (318) 352-9696. Fax: (318) 357-9595. Licensee: Baldridge-Dumas Communications Inc. (group owner; acq 5-14-01; $340,000 with co-located AM including two-year noncompete agreement). Population served: 80,000 Natl. Network: Jones Radio Networks. Rgnl. Network: La. Net. Law Firm: Kaye, Scholer, Fierman, Hays & Handler. Format: Country. ♦Rhonda Benson, gen mgr & stn mgr; Gordon Rivet, news dir; Kenny Carter, chief of engrg.

KNOC(AM)— May 1, 1947: 1450 khz; 1 kw-U. TL: N31 45 47 W93 03 47. Hrs open: 24 213 Renee St., 71457. Phone: (318) 354-4000. Fax: (318) 352-9598. Licensee: North Face Broadcasting L.L.C. (acq 5-9-2003; $348,000 with KCIJ(FM) Atlanta). Population served: 150,000 Format: News/talk. News staff: one; News: 20 hrs wkly. Target aud: 35+; upper-middle class. ♦Theresa Wilkenson, stn mgr; John Brewer, opns dir; Shane Erath, news dir.

***KNWD(FM)**— September 1975: 91.7 mhz; 255 w horiz. 164 ft TL: N31 44 51 W93 05 47. Stereo. Hrs opn: 24 109 Kyser Hall, 165 Sam Sibley Dr., 71497. Phone: (318) 357-4180. Fax: (318) 357-4398. E-mail: knwd917@yahoo.com Web Site: www.nsula.edu/thedemon. Licensee: Northwestern State University of Louisiana. Population served: 35,000 Format: Var. News staff: one; News: 3 hrs wkly. Target aud: 18-25. ♦Elliot Westphal, gen mgr & stn mgr.

KZBL(FM)— Oct 8, 1985: 100.7 mhz; 3 kw. 299 ft TL: N31 48 18 W93 01 29. Stereo. Hrs opn: 24 605 San Antonio Ave., Many, 71449. Phone: (318) 352-4363. Fax: (318) 357-9595. Licensee: Baldridge-Dumas Communications Inc. (group owner; acq 6-21-99; $400,000). Natl. Network: Jones Radio Networks. Format: Oldies. News: 10 hrs wkly. Target aud: 25-50. ♦Rhonda Benson, gen mgr & stn mgr.

New Iberia

KANE(AM)— August 1946: 1240 khz; 1 kw-U. TL: N30 01 03 W91 50 10. Stereo. Hrs opn: 24 145 B. West, 70560. Phone: (337) 365-3434. Fax: (337) 365-9117. Web Site: www.kane1240.com. Licensee: Coastal Broadcasting of Lafourche L.L.C. (acq 12-31-01). Population served: 250,000 Rgnl. Network: La. Net. Format: Cajun. News: 30 hrs wkly. Target aud: 25-54. ♦Jerry Gisclair, gen mgr.

KNIR(AM)— June 1, 1951: 1360 khz; 1 kw-D, 209 w-N. TL: N30 01 32 W91 49 20. Hrs open: 24 Radio Maria Inc., 601 Washington St., Alexandria, 71301. Phone: (318) 561-6145. Fax: (318) 449-9954. E-mail: info.usa@radiomaria.org Web Site: www.radiomaria.us. Licensee: Radio Maria Inc. (group owner; (acq 6-10-2003; $45,000). Population served: 50,000 Format: Christian. ♦Daune Stenzel, gen mgr.

KRDJ(FM)— 1991: 93.7 mhz; 100 kw. Ant 971 ft TL: N30 20 19 W91 31 23. Hrs open: 24 650 Wooddale Blvd., Baton Rouge, 70806. Phone: (225) 926-1106. Fax: (225) 928-1606. E-mail: krdj.fm@citcomm.com Web Site: www.red937.com. Licensee: Citadel Broadcasting Co. Group owner: Citadel Broadcasting Corp. (acq 10-8-99; $9.5 million). Population served: 500,000 Format: 70s, 80s & whatever. Target aud: 18-44; men. ♦Greg Benefield, gen mgr.

KXKC(FM)— January 1969: 99.1 mhz; 100 kw. 1,039 ft TL: N30 12 06 W91 46 37. Stereo. Hrs opn: 24 202 Galbert Rd., Lafayette, 70506. Phone: (337) 232-1311. E-mail: office@kxkc.com Web Site: www.kxkc.com. Licensee: Citadel Broadcasting Co. Group owner: Citadel Broadcasting Corp. (acq 12-5-2003; $7.6 million). Population served: 500,000 Format: New country. Target aud: 18-49. ♦Mary Galyean, gen mgr.

New Orleans

KGLA(AM)—See Gretna

WBOK(AM)— February 1951: 1230 khz; 1 kw-U. TL: N29 59 18 W90 02 45. Hrs open: Bakewell Media of Louisiana LLC, 3800 Crenshaw Blvd., Los Angeles, CA, 90008. Phone: (323) 291-6803. Fax: (291) 291-6804. Licensee: Bakewell Media of Louisiana LLC. Group owner: Willis Broadcasting Corp. (acq 2-9-2007; $550,000). Population served: 593,471 Format: Gospel. Target aud: 25 plus. ♦Danny J. Bakewell Sr., gen mgr.

***WBSN-FM**— Feb 5, 1979: 89.1 mhz; 8.5 kw. Ant 623 ft TL: N29 56 59 W89 57 27. Stereo. Hrs opn: 24 3939 Gentilly Blvd., 70126. Phone: (504) 816-8000. Fax: (504) 816-8580. E-mail: onair@lifesongs.com Web Site: www.lifesongs.com. Licensee: Providence Educational Foundation. Population served: 1,200,000 Format: Contemp Christian. Target aud: 25-49; active, Christian oriented families. ♦Stan Watts, gen mgr.

WBYU(AM)— 1950: 1450 khz; 1 kw-U. TL: N29 57 27 W90 09 47. Hrs open: 3330 W. Esplanade Ave., Suite 605, Metairie, 70002. Phone: (504) 841-2800. Fax: (504) 841-2805. Web Site: www.radiodisney.com. Licensee: Radio Disney Group LLC. Group owner: ABC Inc. (acq 2-5-2003; $1.5 million). Natl. Network: ABC. Natl. Rep: Roslin. Format: Children. ♦Steve Finney, stn mgr.

WDVW(FM)—See La Place

WEZB(FM)— Sept 1, 1945: 97.1 mhz; 100 kw. 984 ft TL: N29 55 11 W90 01 29. Stereo. Hrs opn: 400 Poydras St., Suite 30, 70130. Phone: (504) 593-6376. Fax: (504) 593-2205. Web Site: www.b97.com. Licensee: Entercom New Orleans License LLC. Group owner: Entercom Communications Corp. (acq 12-13-99; grpsl). Population served: 593,471 Format: CHR/top-40. Target aud: 18-34; females. ♦Ken Beck, gen mgr; Patrick Galloway, sls dir & gen sls mgr; Mike Kaplan, progmg dir; Joe Pollet, chief of engrg.

WGSO(AM)— Jan 27, 1946: 990 khz; 1 kw-D, 400 w-N. TL: N29 57 24 W90 04 34. Hrs open: 24 2250 Gause Blvd. E., Suite 205, Slidell, 70461-4235. Phone: (985) 639-3820. Fax: (985) 639-3869. E-mail: info@wgso.com Web Site: wgso.com. Licensee: Northshore Radio LLC (acq 5-30-2007; $1.01 million). Format: News/talk. ♦Mike Starr, gen mgr.

WIST(AM)— 1948: 690 khz; 10 kw-D, 5 kw-N, DA-2. TL: N29 57 53 W89 57 31. Hrs open: 24 Box 8386, Metairie, 70011. Secondary address: 4539 N. I-10 Service Rd., Suite 205, Metaire 70006. Phone: (504) 885-4690. Fax: (504) 885-4671. E-mail: feedback@wistradio.com Web Site: www.wistradio.com. Licensee: WTIX Inc. Group owner: GHB Radio Group (acq 2-12-92; $800,000; FTR: 3-16-92). Population served: 914,700 Law Firm: Cohn & Marks. Format: News/talk. News: 30 hrs wkly. Target aud: 25 plus; affluent, educated, professional. ♦Daniel Frazier, gen mgr.

WKBU(FM)— February 1953: 95.7 mhz; 100 kw. 984 ft TL: N29 55 11 W90 01 29. Stereo. Hrs opn: 1450 Poydras, Suite 500, 70112. Phone: (504) 593-6376. Fax: (504) 593-1850. E-mail: mail@entercom.com Web Site: bayou957.com. Licensee: Entercom New Orleans License LLC. Group owner: Entercom Communications Corp. (acq 12-13-99; grpsl). Law Firm: Latham & Watkins. Format: Classic Rock. Target aud: 25-54. ♦Patrick Galloway, gen sls mgr; Mike Kaplan, progmg dir; Dave Cohen, news dir; Joe Pollet, chief of engrg.

WLMG(FM)—Listing follows WWL(AM).

WLNO(AM)— 1925: 1060 khz; 50 kw-D, 5 kw-N, DA-2. TL: N29 52 46 W89 59 51. Stereo. Hrs opn: 24 401 Whitney Ave., #160, Gretna, 70056. Phone: (504) 362-9800. Fax: (504) 362-5541. E-mail: wlno@i-55.com Web Site: www.wlno.com. Licensee: Communicom Co. of Louisiana L.P. (acq 1-25-95; $700,000; FTR: 3-20-95). Population served: 1,100,000 Format: Christian, Religious. Target aud: General. ♦Carl DiMaria, CEO, gen mgr & opns dir.

WNOE-FM— Sept 15, 1968: 101.1 mhz; 100 kw. 1,004 ft TL: N29 58 57 W89 57 09. Stereo. Hrs opn: 24 929 Howard Ave., 70113. Phone: (504) 679-7300. Fax: (504) 679-7345. Web Site: www.wnoe.com. Licensee: Clear Channel Communications Inc. (group owner: Clear Channel Communiaction Inc. acq 1996; grpsl). Natl. Network: ABC, Westwood One. Natl. Rep: Clear Channel. Law Firm: Verner, Liipfert, Bernhard, McPherson & Hand. Format: Country. News staff: one. Target aud: 25-54. ♦Dick Lewis, gen mgr; Michael Scott, gen sls mgr; Ron Brooks, progmg dir; Tom Courtney, chief of engrg.

WODT(AM)— July 23, 1923: 1280 khz; 5 kw-U, DA-1. TL: N29 53 43 W90 00 16. Stereo. Hrs opn: 24 929 Howard Ave., 70113. Phone: (504) 679-7300. Fax: (504) 679-7345. Web Site: sportsradio1280.com. Licensee: Clear Channel Radio Licenses Inc. Group owner: Clear Channel Communications Inc. (acq 7-24-92). Population served: 1,000,000 Natl. Network: ESPN Radio. Law Firm: Riley & Fielding. Format: Sports. News: one hr wkly. Target aud: 35 plus; general. ♦Dick Lewis, gen mgr; Michael Scott, sls dir; Bruce Buchert, progmg dir; Richard Atwood, chief of engrg.

WQUE-FM—Co-owned with WODT(AM). Jan 1, 1949: 93.3 mhz; 93 kw. 459 ft TL: N29 57 24 W90 04 31. (CP: 100 kw, ant 984 ft. TL: N29 55 11 W90 01 29). Stereo. 24 Web Site: www.q93.com.2,000,000 Format: Urban contemp. ♦Nate Bell, gen sls mgr & progmg dir.

***WRBH(FM)**— 1980: 88.3 mhz; 54 kw. 600 ft TL: N29 57 01 W89 57 29. Hrs open: 24 3606 Magazine St., 70115. Phone: (504) 899-1144. Fax: (504) 899-1165. E-mail: wrbh883@gmail.com Web Site: www.wrbh.org. Licensee: Radio for the Blind and Print Handicapped Inc. Format: Radio reading svc, news. News: 28 hrs wkly. Target aud: Blind & print handicapped. ♦Natalia Gonzales, gen mgr; Jackie Bullock, progmg dir; Ernie Kain, chief of engrg.

WRNO-FM— Oct 17, 1967: 99.5 mhz; 100 kw. Ant 1,004 ft TL: N29 58 57 W89 57 09. Stereo. Hrs open: 929 Howard Ave., 70113. Phone: (504) 679-7300. Fax: (504) 679-7345. Web Site: www.thenew995fm.com. Licensee: Clear Channel Broadcasting Licenses Inc. Group owner: Clear Channel Communications Inc. (acq 8-8-2002; swap for KKND(FM) Port Sulphur). Population served: 593,471 Law Firm: Dow, Lohnes & Albertson. Format: News/talk. ♦Dick Lewis, gen mgr; Michael Scott, opns mgr & gen sls mgr; Bob Christopher, progmg dir; Tom Courtney, chief of engrg.

WSHO(AM)— 1926: 800 khz; 1 kw-D, 233 w-N, DA-1. TL: N29 50 42 W90 06 39. Hrs open: 365 Canal St., Suite 1175, 70130. Phone: (504) 527-0800. Fax: (504) 527-0881. E-mail: whso@wsho.com Web Site: www.wsho.com. Licensee: Shadowlands Communications L.L.C. (acq 1996). Population served: 593,471 Natl. Network: Salem Radio Network. Format: Christian music & talk. Target aud: 25-54. ♦William Ainsworth, pres, gen mgr & opns mgr; Mike Patton, chief of engrg.

WSLA(AM)—See Slidell

***WTUL(FM)**— Nov 14, 1974: 91.5 mhz; 1.5 kw. 161 ft TL: N29 56 18 W90 07 07. Stereo. Hrs opn: 24 Tulane Univ. Ctr., 70118. Phone: (504) 865-5887. E-mail: wtul@wtul.fm Web Site: www.wtul.fm. Licensee: Tulane Educational Fund. Format: Progsv. News: 3 hrs wkly. Target aud: General. ♦Li Yaffe, gen mgr.

WVOG(AM)— Apr 23, 1964: 600 khz; 1 kw-D. TL: N29 57 25 W90 09 33. Hrs open: 5:30 AM-8:30 PM 2730 Loumour Ave., Metairie, 70001. Phone: (504) 831-6941. E-mail: wvog@gellsouth.net Web Site: www.wwcr.com. Licensee: F.W. Robbert Broadcasting Co. Inc. (group owner; acq 6-28-74). Population served: 593,471 Format: Christian talk. News: 2 hrs wkly. Target aud: 30 plus. ♦Fred P. Westenberger, pres; Eric Westenberger, gen mgr.

WWL(AM)— Mar 31, 1922: 870 khz; 50 kw-U, DA-1. TL: N29 50 14 W90 07 55. Hrs open: 24 1450 Poydras, Suite 500, 70112. Phone: (504) 593-6376. Fax: (504) 593-1850. Web Site: www.wwl.com. Licensee: Entercom New Orleans License LLC. Group owner: Entercom Communications Corp. (acq 12-13-99; grpsl). Population served: 1,022,800 Natl. Network: CBS. Natl. Rep: D & R Radio. Format: News/talk, sports. ♦Ken Beck, gen mgr & opns mgr; Malcolm Pelham, sls dir & prom dir; Diane Newman, progmg dir & progmg mgr; Mark Broudreaux, gen sls mgr & news dir; Joe Pollet, chief of engrg.

WLMG(FM)—Co-owned with WWL(AM). Mar 15, 1970: 101.9 mhz; 100 kw. 984 ft TL: N29 55 11 W90 01 29. Stereo. 24 Web Site: www.magic1019.com.1,022,800 Format: Adult contemp. ♦Patrick Galloway, gen sls mgr; Andy Holt, progmg dir & mus dir.

***WWNO(FM)**— Feb 20, 1972: 89.9 mhz; 85 kw vert. Ant 748 ft TL: N29 55 11 W90 01 29. Stereo. Hrs opn: 24 Univ. of New Orleans, 2000 Lakeshore Dr., 70148. Phone: (504) 280-7000. Fax: (504) 280-6061. E-mail: info@wwno.org Web Site: www.wwno.org. Licensee: Louisiana State University. Population served: 1,300,000 Natl. Network: PRI, NPR. Wire Svc: AP Format: Class, news jazz. News: 39 hrs wkly. Target aud: 35 plus; well-educated professionals, mgrs, artists & art patrons. ♦Chuck Miller, gen mgr; Ron C. Curtis, opns dir; Karen Anklam, dev dir & prom mgr; Fred Kasten, progmg dir.

***WWOZ(FM)**— Dec 6, 1980: 90.7 mhz; 19 w. 279 ft TL: N29 57 01 W90 09 16. (CP: 4 kw, ant 508 ft. TL: N29 57 24 W90 04 31). Stereo. Hrs opn: 24 Box 51840, 70151. Secondary address: 1008 N. Peters 70116. Phone: (504) 568-1239. Phone: (504) 568-1238. Fax: (504) 558-9332. E-mail: wwoz@wwoz.org Web Site: www.wwoz.org. Licensee: Friends of WWOZ Inc. (acq 10-14-86). Law Firm: Garvey, Schubert & Barer. Format: Jazz, rhythm and blues. News: 5 hrs wkly. Target aud: 35-55; upscale & educated males. ♦David Freedman, gen mgr; Dwayne Breashears, progmg dir.

WWWL(AM)— Apr 21, 1925: 1350 khz; 5 kw-U, DA-2. TL: N29 55 27 W90 02 04. Hrs open: 24 1450 Poydras, Suite 500, 70112. Phone: (504) 593-6376. Fax: (504) 593-1850. Licensee: Entercom New Orleans License LLC. Group owner: Entercom Communications Corp. (acq 12-13-99; grpsl). Population served: 593,471 Format: Talk, sports. ♦Ken Beck, gen mgr; Malcolm Pelham, gen mgr & sls dir; Mark Broudreaux, gen sls mgr; Diane Newman, progmg dir; Joe Pollet, chief of engrg.

WYLD(AM)— 1949: 940 khz; 10 kw-D, 500 w-N, DA-2. TL: N29 54 00 W90 00 17. Hrs open: 24 929 Howard Ave., 70113. Phone: (504) 679-7300. Fax: (504) 679-7343. Web Site: www.am940.com. Licensee: Clear Channel Radio Licenses Inc. Group owner: Clear Channel Communications Inc. (acq 3-25-93; FTR: 3-20-95). Population served: 593,471 Natl. Network: ABC, Westwood One. Natl. Rep: Clear Channel. Law Firm: Wiley, Rein & Fielding. Format: Gospel. Target aud: 25-54. ♦Dick Lewis, gen mgr.

WYLD-FM— 1971: 98.5 mhz; 100 kw. 984 ft TL: N29 55 11 W90 01 29. (CP: Ant 902 ft.). Stereo. 24 Web Site: www.wyldfm.com.1,000,000 Format: Urban adult contemp. News staff: 2; News: 5 hrs wkly.

New Roads

KQXL-FM— Oct 1, 1979: 106.5 mhz; 50 kw. Ant 485 ft TL: N30 37 24 W91 09 50. Stereo. Hrs opn: 24 650 Wooddale Blvd., Baton Rouge, 70806. Phone: (225) 926-1106. Fax: (225) 928-1606. E-mail: kqxl.fm@citcomm.com Web Site: www.q106dot5.com. Licensee: Citadel Broadcasting Co. Group owner: Citadel Broadcasting Corp. (acq 1-14-99; grpsl). Population served: 600,000 Natl. Network: CBS. Format: Urban contemp. News staff: one. Target aud: 18-54; Black adults. ♦Greg Benefield, gen mgr.

Norco

WFNO(AM)— 1987: 830 khz; 5 kw-D, 750 w-N, DA-2. TL: N30 03 00 W90 22 41. Hrs open: 1111 Veterans Memorial Blvd., Suite 1810, Metairie, 70005. Phone: (504) 260-9366. Fax: (504) 830-7200. Licensee: Davidson Media Station WFNO Licensee LLC. (acq 1-11-2007; $2 million). Wire Svc: AP Format: Sp contemp. News staff: 2; News: 12 hrs wkly. Target aud: 18-44. ♦Yadira Hernandez, gen mgr.

***WNKV(FM)**—Not on air, target date: unknown: 91.1 mhz; 1 kw vert. Ant 56 ft TL: N29 59 50 W90 24 48. Hrs open:
Rebroadcasts KLVR(FM) Santa Rosa, CA 100%.
2351 Sunset Blvd., Suite 170-218, Rocklin, CA, 95765. Phone: (916) 251-1600. Fax: (916) 251-1650. Web Site: www.klove.com. Licensee: Educational Media Foundation. (acq 11-1-2006; grpsl). Natl. Network: K-Love. Format: Contemp Christian. ♦Richard Jenkins, pres.

North Fort Polk

KUMX(FM)— May 10, 1995: 106.7 mhz; 6 kw. 328 ft TL: N31 03 46 W93 16 11. Stereo. Hrs opn: 24 168 KWP Dr., Leesville, 71446. Phone: (337) 537-9000. Fax: (337) 537-4152. Web Site: www.kumx1067.com. Licensee: West Central Broadcasting Co. Inc. (acq 3-8-02; $208,000). Population served: 50,000 Natl. Network: ABC. Format: Christian. Target aud: 22-42. ♦Roscoe Burwell, gen mgr.

Oak Grove

KWCL-FM— Jan 30, 1973: 96.7 mhz; 23 kw. Ant 341 ft TL: N32 51 32 W91 21 22. Stereo. Hrs opn: 24 Box 260, 71263. Secondary address: 230 E. Main St. 71263. Phone: (318) 428-9670. Fax: (318) 428-2476. E-mail: kwcl@bellsouth.net Licensee: KWCL-FM Broadcasting Co. Inc. (acq 12-10-90). Population served: 2,500 Natl. Network: ABC, Jones Radio Networks. Law Firm: Miller & Miller. Format: Good time oldies. News staff: one; News: 17 hrs wkly. Target aud: General. ♦Irene Robinson, pres & gen mgr; Kelley Lovell, progmg dir & disc jockey.

Oakdale

KKST(FM)— 1972: 98.7 mhz; 48 kw. Ant 1,053 ft TL: N31 01 59 W92 30 08. Stereo. Hrs opn: 24 1515 Texas Ave., Alexandria, 71301. Phone: (318) 445-1234. Fax: (318) 445-7231. E-mail: chad@cenlabroadcasting.com Web Site: www.cenlabroadcasting.com. Licensee: Cenla Broadcasting Licensing Co. LLC. Group owner: Clear Channel Communications Inc. (acq 11-13-2006; grpsl). Population served: 150,000 Format: Urban hip hop. News staff: one; News: 20 hrs wkly. Target aud: 18-49; women. ♦Taylor Thompson, gen mgr.

Opelousas

KFXZ-FM— Aug 3, 1989: 105.9 mhz; 3.4 kw. Ant 433 ft TL: N30 27 53 W92 04 31. Hrs open: 3225 Ambassador Caffery Pkwy., Lafayette, 70506. Phone: (337) 993-5500. Fax: (337) 993-5510. E-mail: kfxz@pittmanbroadcasting.com Web Site: www.myspace.com/1059thefox. Licensee: Pittman Broadcasting Services LLC (group owner; (acq 1-28-2004; grpsl). Format: Country legends. ♦Charles Sagona, gen mgr.

KOGM(FM)—Listing follows KSLO(AM).

KSLO(AM)— September 1947: 1230 khz; 1 kw-U. TL: N30 31 31 W92 06 17. Hrs open: 5 AM-11 PM Box 1150, 70571-1150. Secondary address: 216 N. Court St. 70570. Phone: (337) 942-2633. Fax: (337) 942-2635. Licensee: KSLO Broadcasting Co. Inc. Population served: 516,000 Format: Country. ♦Penny Smith, pres; Chris Lamke, gen mgr & progmg dir; Walley LeBlanc, chief of engrg; Jay Miller, traf mgr.

KOGM(FM)—Co-owned with KSLO(AM). June 18, 1965: 107.1 mhz; 3 kw. Ant 203 ft TL: N30 31 31 W92 06 17. Stereo. 5 AM-11 PM Format: Hot adult contemp. Target aud: 25 plus.

Pineville

KTTP(AM)—Licensed to Pineville. See Alexandria

Plaquemine

***KPAQ(FM)**—Not on air, target date: unknown: 88.1 mhz; 2.9 kw vert. Ant 308 ft TL: N30 15 41 W91 18 40. Hrs open: Drawer 2440, Tupelo, MS, 38803. Phone: (662) 844-8888. Fax: (662) 842-6791. Licensee: American Family Association. ♦Marvin Sanders, gen mgr.

Port Allen

WPFC(AM)—Licensed to Port Allen. See Baton Rouge

Port Sulphur

KAGY(AM)— Aug 17, 1966: 1510 khz; 1 kw-D. TL: N29 29 03 W89 42 15. Hrs open: 6 AM-6 PM 409 Duke St., Morgan City, 70380. Phone: (985) 384-1430. Fax: (985) 384-2351. E-mail: kmrc@kmrc1430.com

Web Site: www.kmrcradio.com. Licensee: Spotlight Broadcasting of New Orleans LLC Group owner: Spotlight Broadcasting LLC (acq 12-30-2002; $250,000). Population served: 750,000 Format: Swamp pop. Target aud: 24-54; general. ♦John Stork, gen mgr.

KKND(FM)— July 4, 1989: 106.7 mhz; 100 kw. Ant 981 ft TL: N29 48 30 W89 45 42. Hrs open: 24 201 St. Charles Ave., Suite 201, New Orleans, 70170. Phone: (504) 581-7002. Fax: (504) 566-4857. E-mail: trapper.john@citcomm.com Web Site: www.country1067.com. Licensee: Citadel Broadcasting Co. Group owner: Citadel Broadcasting Corp. (acq 8-29-2003; grpsl). Natl. Rep: Clear Channel. Format: Country rock. News staff: one. Target aud: 25-54. ♦Dave Siebert, VP & gen mgr; Trapper John, progmg dir.

***KSUL(FM)**—Not on air, target date: unknown: 91.5 mhz; 100 w. Ant 184 ft TL: N29 30 52 W89 43 48. Hrs open: Drawer 2440, Tupelo, MS, 38803. Phone: (662) 844-8888. Fax: (662) 842-6791. Web Site: www.afr.com. Licensee: American Family Association. Group owner: American Family Radio. Format: Christian. ♦Marvin Sanders, chmn.

Rayne

KBEB-FM— 1993: 106.7 mhz; 3 kw. 328 ft TL: N30 18 17 W92 20 47. Hrs open: Box 228, Crowley, 70527. Secondary address: 320 N. Parkerson Ave., Crowley 70526. Phone: (337) 783-2520. Fax: (337) 783-5744. E-mail: info@b1067.com Web Site: www.b1067.com. Licensee: Broadcast Partners Inc. (acq 12-18-92; $60,000; FTR: 1-11-93). Format: Oldies 60s, 70s, 80s. ♦Phil Lizotte, pres & gen mgr; Jimmy Cole, gen sls mgr; Hans Nelson, progmg dir; Tony Evans, chief of engrg.

Rayville

KMYY(FM)— September 1984: 92.3 mhz; 26 kw. 492 ft TL: N32 27 51 W91 39 10. Stereo. Hrs open: 16 1200 N. 18th St., Suite D, Monroe, 71201. Phone: (318) 387-3922. Fax: (318) 322-4585. Web Site: www.realcountry923.com. Licensee: Opus Broadcasting Monroe L.L.C. Group owner: Opus Media Partners LLC (acq 7-19-2004; grpsl). Format: Real country. ♦Chris Zimmerman, gen mgr.

Reserve

WPRF(FM)— August 1991: 94.9 mhz; 50 kw. Ant 482 ft TL: N29 43 48 W90 43 37. Hrs open: 24 3500 N. Causeway Blvd., Suite 400, Metairie, 70002. Phone: (504) 834-7095. Fax: (504) 834-7096. E-mail: praise@praisefm949.com Licensee: Southeastern Broadcasting Inc. Group owner: Citadel Broadcasting Corp. (acq 7-6-2005; $4.5 million with WOPR(FM) Lacombe). Population served: 1,000,000 Format: Gospel. News: 2 hrs wkly. Target aud: 25-54; general. ♦Tom Wilson, gen mgr.

Richwood

KHLL(FM)— March 1995: 100.9 mhz; 6 kw. 328 ft TL: N32 24 25 W92 04 13. Hrs open: 704-C Trenton St., West Monroe, 71291. Phone: (318) 323-5994. Fax: (318) 323-6680. E-mail: hillradio@centurytel.net Web Site: www.hillradio.com. Licensee: Dan Gilliland. (acq 3-95). Format: Christian hit radio. ♦Rick Godley, gen mgr.

Ruston

***KAPI(FM)**— February 1998: 88.3 mhz; 300 w. Ant 197 ft TL: N32 33 08 W92 39 21. Hrs open: Box 3206, American Family Radio, Tupelo, MS, 38803. Phone: (662) 844-8888. Fax: (662) 842-6791. E-mail: comments@afr.net Web Site: www.afr.net. Licensee: American Family Association. Group owner: American Family Radio Format: Inspirational Christian. ♦Marvin Sanders, gen mgr.

***KLPI-FM**— 1973: 89.1 mhz; 4 kw. 285 ft TL: N32 31 09 W92 39 02. (CP: 20 kw). Stereo. Hrs opn: Box 8638, 71272. Secondary address: Union Bldg., 101 Wysteria St. 71272. Phone: (318) 257-4851. Fax: (318) 257-5073. E-mail: general@891klpi.org Web Site: www.891klpi.org. Licensee: Louisiana Tech University. Population served: 35,000 Format: Alternative. News staff: 2. Target aud: 18-24; college students. ♦Bill Hosli, gen mgr; Trey Thomas, progmg dir & spec ev coord.

KPCH(FM)— 1999: 99.3 mhz; 15.5 kw. Ant 328 ft TL: N32 28 53 W92 40 37. Hrs open: 24 Box 430, 71270. Secondary address: 500 N. Monroe St. 71270. Phone: (318) 255-6993. Fax: (318) 255-5084. E-mail: thepeach@bayou.com Licensee: Communications Capital Co. II of Louisiana LLC. Group owner: Communications Capital Managers LLC (acq 3-4-2002; grpsl). Format: Oldies. ♦Gary McKenney, gen mgr & progmg dir; Tommy Gray, chief of engrg.

KRUS(AM)— Nov 7, 1947: 1490 khz; 1 kw-U. TL: N32 30 48 W92 39 56. Hrs open: Box 430, 71273. Secondary address: 500 N. Monroe St. 71270. Phone: (318) 255-5000. E-mail: z1075fm@bayou.com Licensee: Communications Capital Co. II of Louisiana LLC. Group owner: Communications Capital Managers LLC (acq 3-4-2002; grpsl). Population served: 47,000 Format: Blues, Black gospel. Target aud: 25-55; Black. ♦Gary McKenney, gen mgr, stn mgr & gen sls mgr; James Cooper, opns dir & progmg dir; Tommy Gray, mus dir & chief of engrg.

KXKZ(FM)—Co-owned with KRUS(AM). June 29, 1966: 107.5 mhz; 98 kw. 1,066 ft TL: N32 26 38 W92 42 42. Stereo. 24 Web Site: www.z1075fm.com.650,000 Format: Country. News staff: one; News: 7 hrs wkly. Target aud: 25-54. ♦Mickey Alexander, progmg dir.

Saint Martinville

***KSJY(FM)**— 2005: 89.9 mhz; 30 kw. Ant 466 ft TL: N30 08 03 W91 51 46. Hrs open: Box 3206, American Family Radio, Tupelo, MS, 38803. Phone: (662) 844-8888. Fax: (662) 842-6791. E-mail: comments@afr.net Web Site: www.afr.net. Licensee: American Family Association. Group owner: American Family Radio. Format: Inspirational Christian. ♦Marvin Sanders, gen mgr.

Shreveport

***KDAQ(FM)**— Dec 21, 1984: 89.9 mhz; 100 kw. 932 ft TL: N32 40 41 W93 55 35. Stereo. Hrs opn: 24 Box 5250, 71135. Secondary address: One University Pl. 71115. Phone: (318) 797-5150. Phone: (800) 552-8502. Fax: (318) 797-5265. E-mail: listenermail@redriverradio.com Web Site: www.redriverradio.org. Licensee: Louisiana State University Board of Supervisors. Natl. Network: NPR, PRI. Format: Classical, news, jazz. News: 40 hrs wkly. Target aud: General. ♦Kermit Poling, gen mgr; Rick Shelton, opns mgr.

KEEL(AM)— 1922: 710 khz; 50 kw-D, 5 kw-N, DA-2. TL: N32 40 35 W93 51 35. Hrs open: 24 6341 Westport Ave., 71129. Phone: (318) 688-1130. Fax: (318) 687-8574. Web Site: www.710keel.com. Licensee: GAP Broadcasting Shreveport License LLC. Group owner: Clear Channel Communications Inc. (acq 8-3-2007; grpsl). Population served: 896,600 Rgnl. Network: La. Net. Natl. Rep: D & R Radio. Format: News, talk. News staff: 5; News: 6 hrs wkly. Target aud: 25-54; men. ♦Charlie Thomas, gen mgr & opns dir; Lisa Janes, gen sls mgr; Erin McCarty, progmg dir; Craig Westbrook, chief of engrg & local news ed.

KXKS-FM—Co-owned with KEEL(AM). May 17, 1968: 93.7 mhz; 95 kw. 1,010 ft TL: N32 40 39 W93 55 41. Stereo. Web Site: www.kisscountry937.com.896,600 Format: Country. Target aud: 25-54; 30 yr old female. ♦Chris Evans, progmg dir & disc jockey.

KIOU(AM)— 1950: 1480 khz; 1 kw-D. TL: N32 31 30 W93 48 30. Hrs open: 6 AM-6 PM 4149 George Rd., 71107. Phone: (318) 222-0272. Fax: (318) 222-0271. Licensee: Metropolitan Radio Group Inc. (group owner; acq 10-97; $70,500). Population served: 182,064 Format: Gospel. Target aud: General. ♦Ernest Pickens, gen mgr.

KLKL(FM)—Minden, July 1, 1978: 95.7 mhz; 50 kw. 469 ft TL: N32 33 24 W93 31 45. Stereo. Hrs opn: 24 208 N. Thomas Dr., 71107. Phone: (318) 222-3122. Fax: (318) 459-1493. Web Site: www.oldies957.fm. Licensee: Access. 1 Louisiana Holding Co. LLC. Group owner: Access.1 Communications Corp. (acq 6-30-00; $7.9 million with KDKS-FM Blanchard). Format: Oldies. News staff: 1. Target aud: 25-54. ♦Cary D. Camp, gen mgr.

KMJJ-FM— Dec 5, 1976: 99.7 mhz; 50 kw. 462 ft TL: N32 30 24 W93 45 13. Hrs open: Box 5459, Bossier City, 71171. Phone: (318) 549-8500. Fax: (318) 549-8505. Web Site: www.997kmjj.com. Licensee: Cumulus Licensing Corp. Group owner: Cumulus Media Inc. (acq 8-7-00; grpsl). Format: Urban contemp. News: one hr wkly. Target aud: 18-49; African American & general. ♦Phil Robkin, gen mgr; Paul Farnham, gen sls mgr; Gary Robinson, prom dir; Jay Tek, progmg dir; Jasen Bragg, chief of engrg.

KOKA(AM)— Aug 1, 1954: 980 khz; 5 kw-D. TL: N32 34 18 W93 44 39. Hrs open: 24 208 N. Thomas Dr., 71107. Phone: (318) 222-3122. Fax: (318) 459-1493. Web Site: www.koka.am. Licensee: Access. 1 Louisiana Holding Co. LLC. Group owner: Access.1 Communications Corp. (acq 12-20-02; grpsl). Natl. Rep: D & R Radio. Format: Gospel. News staff: 1. Target aud: 25-64; middle-aged, middle class, Black adults. ♦Cary D. Camp, gen mgr; Don Zimmerman, gen sls mgr; Eddie Giles, progmg dir.

KRMD-FM— August 1948: 101.1 mhz; 98 kw. Ant 1,119 ft TL: N32 41 08 W93 56 00. (CP: 100 kw, ant 1,627 ft). Stereo. Hrs opn: Box 5459, 270 Plaza Loop, Bossier City, 71111. Phone: (318) 549-8500. Fax: (318) 549-8505. Web Site: www.krmd.com. Licensee: Cumulus Licensing Corp. Group owner: Cumulus Media Inc. (acq 8-7-2000; grpsl). Population served: 369,800 Natl. Rep: Christal. Format: Contemp country. News staff: 2. Target aud: 25-54. ♦Phil Robkin, gen mgr; Chuck Redden, gen sls mgr & disc jockey; Margie Bueche, gen sls mgr & prom mgr; Gary Robinson, prom dir & disc jockey; James Anthony, mus dir & disc jockey; Rick Taylor, news dir; Jasen Bragg, engrg dir & disc jockey.

KRMD(AM)— June 1928: 1340 khz; 1 kw-U. TL: N32 29 36 W93 45 55. Stereo. 24 Web Site: www.supertalk1340.com. Format: Sports, news, talk. ♦John Sherman, progmg dir.

KRUF(FM)—Listing follows KWKH(AM).

***KSCL(FM)**— Mar 11, 1976: 91.3 mhz; 2.6 kw. Ant 184 ft TL: N32 28 51.4 W93 43 51.1. Stereo. Hrs opn: 24 2911 Centenary Blvd., 71104. Phone: (318) 869-5296. Fax: (318) 869-5294. E-mail: kscl@centenary.edu Web Site: www.centenary.edu. Licensee: Centenary College of Louisiana. Population served: 250,000 Format: Alternative music. News staff: one; News: 4 hrs wkly. Target aud: General; college students & adults interested in div music. ♦John Schleuss, stn mgr; Jon Schleuss, stn mgr; Alyson Escude, progmg dir & progmg dir; Tyler Davis, mus dir.

KSYB(AM)— July 10, 1975: 1300 khz; 5 kw-D. TL: N32 31 48 W93 48 16. Hrs open: 24 Box 7685, 71137. Secondary address: 1526 Corporate Dr. 71107. Phone: (318) 222-2744. Fax: (318) 425-7507. E-mail: ksyb@amistadradiogroup.com Licensee: Amistad Communications Inc. (acq 10-26-2000; $900,000). Population served: 182,064 Format: Christian, gospel. ♦Fred Caldwell, CEO & chief of engrg; Rhonda Sanders, stn mgr; Steve Anderson, mus dir.

KTUX(FM)—Carthage, TX) Apr 1, 1985: 98.9 mhz; 100 kw. 1,049 ft TL: N32 23 19 W94 01 10. Stereo. Hrs opn: 24 6341 Westport Ave., 71129. Phone: (318) 688-1130. Fax: (318) 688-9839. Web Site: www.therockstation99x.com. Licensee: GAP Broadcasting Shreveport License LLC. Group owner: Clear Channel Communications Inc. (acq 8-3-2007; grpsl). Format: New rock. Target aud: 18-49; super-active adults. ♦Charlie Thomas, gen mgr.

KVKI-FM— May 1959: 96.5 mhz; 95 kw. 797 ft TL: N32 35 38 W93 51 39. Stereo. Hrs open: 24 6341 Westport Ave., 71129. Phone: (318) 688-1130. Fax: (318) 688-9839. Web Site: www.965kvki.com. Licensee: GAP Broadcasting Shreveport License LLC. Group owner: Clear Channel Communications Inc. (acq 8-3-2007; grpsl). Population served: 328,000 Natl. Rep: D & R Radio. Format: Adult contemp. News staff: one. Target aud: 25-54; female. ♦Charlie Thomas, gen mgr.

KVMA-FM— 2001: 102.9 mhz; 42 kw. Ant 535 ft TL: N32 29 36 W93 45 55. Hrs open: 24 Box 5459, Bossier City, 71171. Phone: (318) 549-8500. Fax: (318) 549-8505. Web Site: magic1029fm.com. Licensee: Cumulus Licensing Corp. Group owner: Cumulus Media Inc. (acq 10-23-2000). Wire Svc: AP Format: Urban contemp. ♦Phil Robkin, gen mgr; Paul Farnharm, gen sls mgr; Rashon Vance, progmg dir; Jasen Bragg, chief of engrg.

KWKH(AM)— September 1925: 1130 khz; 50 kw-U, DA-N. TL: N32 42 15 W93 52 52. Hrs open: 24 6341 Westport Ave., 71129. Phone: (318) 688-1130. Fax: (318) 687-8574. Web Site: www.kwkhonline.com. Licensee: GAP Broadcasting Shreveport License LLC. Group owner: Clear Channel Communications Inc. (acq 8-3-2007; grpsl). Population served: 896,600 Natl. Rep: D & R Radio. Format: Country. News staff:

4; News: 18 hrs wkly. Target aud: Male 25-54. ♦Charlie Thomas, gen mgr; Lisa Janes, gen sls mgr; Barney Cannon, progmg dir; Craig Westbrook, chief of engrg & farm dir.

KRUF(FM)—Co-owned with KWKH(AM). Nov 5, 1948: 94.5 mhz; 100 kw. Ant 1,096 ft TL: N32 40 13 W93 55 59. (CP: Ant 1,666 ft. TL: N32 39 57 W93 55 58). Stereo. 24 Web Site: www.k945.com.896,600 Format: CHR. ♦Erin Bristol, progmg dir.

Simmesport

KCJN(FM)—Not on air, target date: unknown: 105.3 mhz; 2.8 kw. Ant 483 ft TL: N30 54 06 W91 56 16. Hrs open: 3501 Northwest Evangeline Thruway, Carencro, 70520. Phone: (337) 896-1600. Fax: (337) 896-2695. Licensee: Delta Media Corp. ♦Eddie Blanchard, gen mgr.

Slidell

WSLA(AM)— Dec 5, 1963: 1560 khz; 1 kw-U, DA-N. TL: N30 15 08 W89 45 46. Stereo. Hrs opn: Daytime Box 1175, 70459. Secondary address: 38230 Coast Blvd. 70458. Phone: (985) 643-1560. Fax: (985) 649-9822. E-mail: 1560@bellsouth.net Licensee: MAPA Broadcasting L.L.C. (acq 7-2-93; FTR: 8-2-93). Population served: 1,200,000 Natl. Network: USA. Rgnl. Network: La. Net. Format: ESPN & loc sports. Target aud: 25 plus; news intensive audience & sports fans. ♦George Mayoral, gen mgr; Jim Sommers, opns mgr & progmg dir.

South Fort Polk

KROK(FM)— Feb 22, 2003: 95.7 mhz; 6 kw. Ant 289 ft TL: N31 03 05 W93 16 41. Hrs open: 168 KVVP Dr., Leesville, 71446. Phone: (337) 537-5887. Fax: (337) 537-4152. E-mail: krok@krok.com Web Site: www.krok.com. Licensee: West Central Broadcasting Co. Inc. (acq 1-25-02). Format: Adult album alternative. Target aud: 18-54. ♦Alan Taylor, CFO; Doug Stannard, pres & gen mgr.

Springhill

KBSF(AM)—Listing follows KTKC(FM).

KTKC(FM)— Sept 5, 1975: 92.9 mhz; 40 kw. 548 ft TL: N33 00 30 W93 28 38. Stereo. Hrs opn: 24 541 S. Main St., 71075. Phone: (318) 539-4616. Fax: (318) 539-2356. Licensee: Metropolitan Radio Group Inc. (group owner; acq 6-97; with co-located AM). Population served: 100,000 Natl. Network: ABC. Format: Black gospel & relg. Target aud: 35-54. ♦Ernest Pickens, gen mgr, gen sls mgr & progmg dir; Rudy Johnson, chief of engrg.

KBSF(AM)—Co-owned with KTKC(FM). June 30, 1954: 1460 khz; 1 kw-D, 220 w-N. TL: N33 00 02 W93 28 43.Sunrise-sunset 25,000 Format: Urban adult contemp. Spec prog: Gospel 15 hrs wkly.

Sulphur

KEZM(AM)— 1955: 1310 khz; 500 w-D, 50 w-N, DA-1. TL: N30 13 27 W93 22 44. Hrs open: 24 113 E. Napoleon St., 70663-3313. Phone: (337) 527-3611. Fax: (337) 527-0213. Licensee: Merchant Broadcasting Inc. (acq 1-30-98; $75,000). Population served: 185,000 Natl. Network: Sporting News Radio Network. Law Firm: Cohn & Marks. Format: Sports. News staff: one; News: 5 hrs wkly. Target aud: 18-63; upscale baby-boomers. ♦Bruce L. Merchant, pres; Bruce Merchant, gen mgr; Kathy Soileau, sls dir & gen sls mgr.

KKGB(FM)— Dec 17, 1977: 101.3 mhz; 12 kw. Ant 479 ft TL: N30 14 41 W93 20 37. Stereo. Hrs opn: 24 425 Broad St., Lake Charles, 70601-4225. Phone: (337) 439-3300. Fax: (337) 436-7278. Web Site: www.kkgb.com. Licensee: Cumulus Licensing Corp. Group owner: Cumulus Media Inc. (acq 12-17-98; grpsl). Population served: 230,000 Natl. Rep: Christal. Law Firm: Kaye, Scholer, Fierman, Hays & Handler. Format: Classic rock. News staff: one. Target aud: General; baby boomers. ♦Eric Nielson, opns mgr; Jim Vidler, mktg mgr.

***KRLR(FM)**— 2007: 89.1 mhz; 1 w horiz, 16 kw vert. Ant 394 ft TL: N30 21 06 W93 23 49. Hrs open:
Rebroadcasts KLVR(FM) Santa Rosa, CA 100%.
2351 Sunset Blvd., Suite 170-218, Rocklin, CA, 95765. Phone: (916) 251-1600. Fax: (916) 251-1650. Web Site: www.klove.com. Licensee: Educational Media Foundation. (acq 11-1-2006; grpsl). Natl. Network: K-Love. Format: Contemp Christian. ♦Richard Jenkins, pres.

KYKZ(FM)—See Lake Charles

Tallulah

KBYO(AM)— Sept 4, 1954: 1360 khz; 500 w-D. TL: N32 25 37 W91 13 15. Hrs open: Box 4808, Monroe, 71211. Secondary address: 1109 Hudson Ln., Monroe 71201. Phone: (318) 388-2323. Fax: (318) 388-0569. Web Site: www.rejoice1680.com. Licensee: Holladay Broadcasting of Louisiana LLC (acq 1-10-2003; $450,000 with co-located FM). Population served: 9,634 Format: Black gospel. Target aud: 18-54. ♦Robert Holladay, pres; Calvin Murray, gen mgr & progmg dir; Russell Kendrick, chief of engrg.

KLSM(FM)—Co-owned with KBYO(AM). Apr 29, 1983: 104.5 mhz; 3 kw. Ant 320 ft TL: N29 45 35 W90 49 30. (CP: 104.5 mhz, 25 kw). Stereo. 1601 N. Frontage Rd., Suite E, Vicksburg, MS, 39180. Phone: (601) 636-2340. Fax: (601) 638-0869. Law Firm: Latham & Watkins. Format: Adult hits. Target aud: 25-54. ♦Bob Holladay, gen mgr.

KTJZ(FM)—Not on air, target date: unknown: 97.5 mhz; 6 kw. Ant 302 ft TL: N32 25 42 W91 18 47. Hrs open: 3313 Government St., Baton Rouge, 70806-5629. Phone: (225) 334-7490. Fax: (225) 334-7491. E-mail: lanaapc1@juno.com Licensee: Mid South Communications Co. Inc. Format: Gospel, Hip hop, rhythm and blues. ♦Ernest L. Johnson, chmn & pres.

Thibodaux

***KNSU(FM)**— Feb 15, 1972: 91.5 mhz; 250 w vert. 148 ft TL: N29 47 29 W90 48 07. (CP: 91.3 mhz, 3 kw, ant 285 ft. TL: N29 45 35 W90 49 30). Hrs opn: 10 AM-2 AM (M-F); noon-2 AM (S, Su) Box 2001, Nicholls State Univ., 70310. Phone: (985) 448-4586. Fax: (985) 449-7106. E-mail: knsu@nicholls.edu Web Site: www.nicholls.edu/knsu. Licensee: Board of Trustees, Nicholls State University. Population served: 14,925 Format: Alternative. News: 10 hrs wkly. Target aud: 18 plus. ♦Katie Kingdon, stn mgr; Jonathan DeSilvie, progmg dir.

KTIB(AM)— Dec 24, 1953: 640 khz; 5 kw-D, 1 kw-N, DA-2. TL: N29 50 05 W90 54 48. Stereo. Hrs opn: 24 5617 Boca Raton, Dallas, TX, 75230. Phone: (214) 405-1979. Licensee: Gap Broadcasting LLC (acq 2-5-2007; $650,000). Population served: 2,963,913 ♦George Laughlin, gen mgr.

***KTLN(FM)**— May 1995: 90.5 mhz; 200 w. 357 ft TL: N29 43 18 W90 46 33. Stereo. Hrs opn: 24
Rebroadcasts WWNO(FM) New Orleans 100%.
Univ. of New Orleans, New Orleans, 70148. Phone: (504) 280-7000. Fax: (504) 280-6061. E-mail: info@wwno.org Web Site: www.wwno.org. Licensee: Board of Supervisors of Louisiana State University and Agricultural and Mechanical College, University of New Orleans. Population served: 100,000 Natl. Network: NPR, PRI. Format: Class, news, jazz. News: 39 hrs wkly. Target aud: 35-70; well educated professionals, managers, artists & arts patrons. ♦Chuck Miller, gen mgr; Ronald C. Curtis, opns dir; Karen Anklam, dev dir; Fred Kasten, prom mgr & progmg dir.

KXOR-FM— May 1, 1966: 106.3 mhz; 25 kw. Ant 328 ft TL: N29 38 52 W90 41 34. Stereo. Hrs opn: 24 Box 2068, Houma, 70361. Secondary address: 120 Prevost Dr., Houma 70364. Phone: (985) 851-1020. Fax: (985) 872-4403. Web Site: www.rock1063.net. Licensee: Sunburst Media-Louisiana LLC. (acq 1-23-2007; grpsl). Population served: 75,000 Rgnl. Network: La. Agri-News. Format: Rock. News: 20 hrs wkly. Target aud: 18-54. ♦Danny Fletcher, gen mgr; John Delise, opns mgr & progmg dir.

Tioga

KLAA-FM—Licensed to Tioga. See Alexandria

Varnado

WBOX-FM— November 1985: 92.9 mhz; 3 kw. 321 ft TL: N30 54 10 W89 57 36. Stereo. Hrs opn: Box 280, Bogalusa, 70429. Secondary address: 22037 Hwy.436, Bogalusa 70427. Phone: (985) 732-4288. Fax: (985) 732-4288. Licensee: Best Country Broadcasting LLC (acq 9-6-2002; $150,000 with WBOX(AM) Bogalusa). Population served: 70,000 Format: Contemp country. ♦Ben R. Strickland, pres & gen mgr.

Vidalia

WQNZ(FM)—See Natchez, MS

Ville Platte

KVPI-FM— Feb 26, 1967: 92.5 mhz; 3.9 kw horiz. Ant 220 ft TL: N30 41 39 W92 18 46. Stereo. Hrs opn: 24 Box J, 70586. Secondary address: 809 W. LaSalle St. 70586. Phone: (337) 363-2124. Fax: (337) 363-3574. E-mail: kvpi@cebridge.net Web Site: www.oldies925.com. Licensee: Ville Platte Broadcasting Co. (acq 11-7-2005; with co-located AM). Population served: 72,940 Wire Svc: AP Format: Oldies. News staff: one; News: 12 hrs wkly. Target aud: 32-65. ♦Rhonda Pucheu, pres; Mark Layne, gen mgr; Danny Poullard, progmg dir; Cheryl DeBallion, traf mgr; Randy Guillory, sports cmtr.

KVPI(AM)— November 1953: 1050 khz; 250 w-D, 10 w-N. TL: N30 41 39 W92 18 46.6 AM-midnight E-mail: kvpi@cebridge.net 79,692 Natl. Network: ABC. Format: Classic country. News staff: one; News: 12 hrs wkly. Target aud: 32-65. Spec prog: Cajun 12 hrs wkly.

Vivian

KNCB(AM)— Apr 9, 1966: 1320 khz; 5 kw-D. TL: N32 54 08 W93 58 59. Hrs open: Sunrise-sunset Box 1072, 71082. Secondary address: 17525 Hwy. 1 N. 71082. Phone: (318) 375-3278. Fax: (318) 375-3329. E-mail: rjc1072@cs.com Licensee: North Caddo Broadcasting Co. (acq 4-9-66). Population served: 90,000 Format: Country, gospel, news/talk. Target aud: General. ♦Ruby J. Collins, gen mgr; Ruby Collins, gen sls mgr; Rudy Johnson, chief of engrg.

KNCB-FM— Sept 28, 1996: 105.3 mhz; 3.2 kw. 449 ft TL: N32 55 54 W93 54 22. Stereo. 24 (Acq 9-27-96.). Natl. Network: ABC. Format: Real country.

Washington

KNEK(AM)— Aug 18, 1980: 1190 khz; 250 w-D. TL: N30 35 09 W92 04 00. Hrs open: 202 Galbert Rd., Lafayette, 70506-1806. Phone: (337) 232-1311. Fax: (337) 233-3779. Web Site: www.knek.com. Licensee: Citadel Broadcasting Co. Group owner: Citadel Broadcasting Corp. (acq 1-14-99; grpsl). Rgnl. Network: La. Net. Format: Urban contemp. Target aud: 25-54. ♦Mary Galyean, gen mgr.

KNEK-FM— 1989: 104.7 mhz; 25 kw. Ant 328 ft TL: N30 25 17 W92 06 50. Stereo. Hrs opn: 24 202 Galbert Rd., Lafayette, 70506-1806. Phone: (337) 232-1311. Fax: (337) 233-3779. Web Site: www.knek.com. Licensee: The Last Bastion Station Trust LLC, as Trustee (acq 6-12-2007; grpsl). Format: Adult contemp. ♦Dave Kubicki, gen sls mgr; Deidre Williams, progmg dir; Doug Allen, chief of engrg.

West Monroe

KMBS(AM)— August 1956: 1310 khz; 5 kw-D, 49 w-N. TL: N32 29 02 W92 09 10. Hrs open: Box 547, 71294. Phone: (318) 397-2511. Licensee: Red Bear Broadcasting (acq 6-10-93; $200,000; FTR: 6-28-93). Population served: 260,000 Natl. Network: ABC. Format: Adult standards. ♦Chuck Redden, gen mgr.

KZRZ(FM)— Aug 1, 1967: 98.3 mhz; 50 kw. 492 ft TL: N32 39 38 W91 59 28. Stereo. Hrs opn: 24 1200 N. 18th St., Suite D, Monroe, 71201. Phone: (318) 387-3922. Fax: (318) 322-4585. E-mail: sunny983@comcast.net Web Site: www.sunny983.com. Licensee: Opus Broadcasting Monroe L.L.C. Group owner: Opus Media Partners LLC (acq 7-19-2004; grpsl). Population served: 30286 Format: Adult contemp. Target aud: 18-54; mid to upper income. ♦Chris Zimmerman, gen mgr; Mike Dawnhour, gen sls mgr.

White Castle

KKAY(AM)— November 1976: 1590 khz; 1 kw-D. TL: N30 11 01 W91 06 27. Hrs open: 24 706 Railroad Ave., Donaldsonville, 70346. Phone: (225) 473-6397. Fax: (225) 473-5764. E-mail: dave@kkay1590.com Web Site: www.kkay1590.com. Licensee: Cactus Communications LLC. Format: Full service. ♦David Dawson, gen mgr.

Winnfield

KVCL(AM)— Dec 17, 1955: 1270 khz; 820 w-D. TL: N31 56 54 W92 37 37. Stereo. Hrs opn: 24 304 KVCL Rd., 71483. Phone: (318) 628-5822. Fax: (318) 628-7355. Licensee: Harrison Broadcast Organization Inc. (acq 3-7-90; $475,000 FTR: 3-26-90). Population served: 150,000 Natl. Network: CNN Radio, Jones Radio Networks. Rgnl. Network: La. Net. Format: Southern gospel. News staff: 2; News: 21 hrs wkly. Target aud: General; financially able persons, blacks and professionals.

Spec prog: Black 3 hrs, relg 15 hrs, big band 4 hrs wkly. ♦Michael Parker, opns mgr & adv mgr; Kresi Parker, gen sls mgr; Rhonda Leach, gen mgr & natl sls mgr.

KVCL-FM— Nov 3, 1966: 92.1 mhz; 6 kw. 210 ft TL: N31 56 54 W92 37 37. Stereo. 24 (Acq 3-7-90; $595,000).275,000 Format: Country. News staff: 2; News: 22 hrs wkly. Target aud: 24-60; professional financially able persons, blacks & general. Spec prog: Black 20 hrs, gospel 20 hrs wkly. ♦George B. Harrison, spec ev coord, edit mgr, political ed & disc jockey; Patricia J. Harrison, rsch dir, traf mgr, min affrs dir, relg ed & women's int ed.

Winnsboro

KMAR-FM— August 1969: 95.9 mhz; 6 kw. Ant 178 ft TL: N32 11 02 W91 44 51. Stereo. Hrs opn: 24 Box 312, 71295. Secondary address: 1823 Hwy. 618 71295. Phone: (318) 435-5141. Fax: (318) 435-5749. E-mail: kmarfm@bellsouth.net Licensee: Boeuf River Broadcasting Co. Group owner: The Radio Group (acq 11-89; $200,000 with co-located AM; FTR: 11-6-89) Population served: 65,000 Natl. Network: ABC. Wire Svc: UPI Format: Country. News staff: one; News: 20 hrs wkly. Target aud: 30-60; adults. ♦Tom Gay, pres & gen mgr.

Zwolle

KTEZ(FM)— July 4, 2002: 99.9 mhz; 6 kw. Ant 328 ft TL: N31 39 17 W93 29 04. Hrs open: 605 San Antonio Ave., Many, 71449. Phone: (318) 256-5924. Fax: (318) 256-0950. Web Site: www.bdc-radio.com/ktez.htm. Licensee: Baldridge-Dumas Communications Inc. (group owner; (acq 2-25-2002). Format: Adult contemp. ♦Tedd W. Dumas, VP.

Maine

Auburn

WFNK(FM)—See Portland

WLAM(AM)—See Lewiston

WTHT(FM)— February 1977: 99.9 mhz; 50 kw. 492 ft TL: N43 57 07 W70 17 46. Stereo. Hrs opn: 24 477 Congress St., Suite 3 A, Portland, 04101. Phone: (207) 782-1800. Phone: (207) 797-0780. Fax: (207) 783-7371. Fax: (207) 253-1971. Licensee: Nassau Broadcasting III L.L.C. Group owner: Nassau Broadcasting Partners L.P. (acq 4-6-2004; grpsl). Population served: 750,800 Format: Country. Target aud: Women 18-34, women 25-54; Maine's kiss 99.9. ♦Stan Manning, opns mgr & progmg dir; Tim Gatz, gen mgr & gen sls mgr; Peter Magee, chief of engrg.

Augusta

WABK-FM—See Gardiner

WFAU(AM)—See Gardiner

WJZN(AM)— Feb 23, 1932: 1400 khz; 1 kw-U. TL: N44 17 30 W69 46 27. Hrs open: 24 52 Western Ave., 04330. Phone: (207) 623-4735. Fax: (207) 626-5948. E-mail: 92moose@midmaine.com Licensee: Citadel Broadcasting Co. Group owner: Citadel Broadcasting Corp. (acq 4-26-2001; grpsl). Population served: 100,000 Natl. Rep: D & R Radio. Format: Nostalgia. News staff: one. Target aud: 20-40; young adults. ♦Al Perry, gen mgr; Julie Crocker, gen sls mgr; Renee Nelson, news dir; Bob Perry, chief of engrg.

WMME-FM—Co-owned with WJZN(AM). Jan 14, 1981: 92.3 mhz; 50 kw. Ant 500 ft TL: N44 20 07 W69 41 01. Stereo. Web Site: www.92moose.fm.310,000 Format: CHR.

WKCG(FM)— July 1961: 101.3 mhz; 50 kw. Ant 321 ft TL: N44 18 51 W69 50 03. Hrs open: 150 Whitten Rd., 04330. Phone: (207) 623-9000. Fax: (207) 623-9007. E-mail: kellyslater@clearchannel.com Licensee: Capstar TX L.P. Group owner: Clear Channel Communications Inc. (acq 1-18-01; grpsl). Format: Adult contemp. ♦Kelly Slater, gen mgr; Steve Smith, opns dir; Rick Dougle, gen sls mgr.

***WMDR(AM)**— Oct 2, 1946: 1340 khz; 1 kw-U. TL: N44 19 43 W69 45 53. Hrs open: 24 160 Riverside Dr, 04330. Phone: (207) 622-1340. Fax: (207) 623-2874. E-mail: wmdr@adelphia.net Web Site: lightoflife.info /stations.htm. Licensee: Light of Life Ministries Inc. (acq 12-94; FTR: 2-13-95). Population served: 100,000 Format: kids programming. News: 5 hrs wkly. Target aud: General. Spec prog: stories. ♦Denise LaFountain, gen mgr; Randy Todd, progmg dir.

Bangor

WABI(AM)— 1924: 910 khz; 5 kw-U, DA-N. TL: N44 46 44 W68 44 22. Hrs open: 18 184 Target Industrial Circle, 04410. Phone: (207) 947-9100. Fax: (207) 942-8039. Licensee: CC Licenses LLC. Group owner: Clear Channel Communications Inc. (acq 6-28-2001; $3.75 million including five-year noncompete agreement with co-located FM). Population served: 33,168 Law Firm: Davis Wright Tremaine. Format: Local sports, nostalgia. News staff: one; News: 4 hrs wkly. Target aud: 35 plus. ♦George Hale, opns dir; Jim Herron, gen mgr & gen sls mgr.

WWBX(FM)—Co-owned with WABI(AM). Mar 15, 1961: 97.1 mhz; 5 kw. 1,230 ft TL: N44 42 13 W69 04 07. Stereo. 24 Web Site: www.b97hits.com. (Acq 7-30-2001).200,000 Format: Contemp hit. News staff: one; News: 2 hrs wkly. Target aud: General. ♦Michael W. Hale, opns mgr & progmg dir.

WBFB(FM)—See Belfast

WEZQ(FM)— June 9, 1976: 92.9 mhz; 20 kw. 787 ft TL: N44 45 35 W68 33 55. Stereo. Hrs opn: 24 49 Acme Rd., Brewer, 04412. Phone: (207) 989-5631. Fax: (207) 989-5685. E-mail: blacey@jpc.com Web Site: www.wezq-fm.com. Licensee: Cumulus Licensing Corp. Group owner: Cumulus Media Inc. (acq 3-1-99; grpsl). Population served: 50,000 Natl. Rep: D & R Radio. Law Firm: Fisher, Wayland, Cooper, Leader & Zaragoza. Format: Easy lstng. News staff: one. Target aud: 25-54. ♦Tom Preble, gen mgr; Paul Dupois, opns VP & opns mgr; Cindy Campbell, progmg dir.

***WHCF(FM)**— Aug 10, 1981: 88.5 mhz; 100 kw. 1,604 ft TL: N45 07 46 W68 21 28. (CP: 35 kw, ant 1,620 ft.). Stereo. Hrs opn: 24 Box 5000, 04402-5000. Secondary address: 1476 Broadway 04401. Phone: (207) 947-2751. Fax: (207) 947-0010. E-mail: whcf@whcf.cc Web Site: whcf.cc. Licensee: Bangor Baptist Church. Population served: 450,000 Natl. Network: Salem Radio Network. Law Firm: Fletcher, Heald & Hildreth. Format: Inspirational Christian, gospel. News: 7 hrs wkly. Target aud: 35-55; Adults. Spec prog: Childdren 5 1/2 hrs wkly. ♦Scott Stewart, chmn; Jerry Mick, pres; Pencil Boone, gen mgr; Tina Collins, opns mgr; Ed Paradis, chief of opns; Hal Welch, chief of engrg.

***WHSN(FM)**— September 1974: 89.3 mhz; 3 kw. Ant 85 ft TL: N44 49 46 W68 47 39. Stereo. Hrs open: 24 One College Cir., 04401. Phone: (207) 941-7116. Phone: (207) 973-1011. Fax: (207) 947-3987. E-mail: whsn@nescom.edu Web Site: www.whsn-fm.com. Licensee: Husson College Board of Trustees. Population served: 2,200 Wire Svc: AP Format: Alternative. News staff: one; News: 7 hrs wkly. Target aud: 12-25; high school & college students. ♦Ben Haskell, gen mgr; Mark Nason, progmg VP & progmg dir; Susan Patten, news dir & chief of engrg; David MacLaughlin, engr.

WKIT-FM—See Brewer

***WMEH(FM)**— Sept 14, 1970: 90.9 mhz; 13.5 kw. 850 ft TL: N44 45 36 W68 33 59. Stereo. Hrs opn: 24 65 Texas Ave., 04401. Secondary address: 1450 Lisbon St., Lewiston 04240. Phone: (207) 874-6570. Fax: (207) 942-2857. Fax: (207) 761-0318. Web Site: www.mainepublicradio.org. Licensee: Maine Public Broadcasting Corp. (acq 6-23-92; FTR: 7-13-92). Population served: 35,000 Natl. Network: NPR, PRI. Law Firm: Dow, Lohnes & Albertson. Format: Class, pub affrs, news. ♦Alexander G. Maxwell, Jr., COO & stn mgr; P. James Dowe, Jr., CEO & pres; Christopher F. Amann, CFO; Alexander G. Maxwell, sr VP; Mary Mayo, dev VP; Charles Beck, progmg VP.

WWMJ(FM)—See Ellsworth

WZON(AM)— December 1926: 620 khz; 5 kw-U, DA-N. TL: N44 49 44 W68 47 08. Hrs open: 24 Box 1929, 04402. Phone: (207) 990-2800. Fax: (207) 990-2444. E-mail: wzon@zoneradio.com Web Site: www.zoneradio.com. Licensee: The Zone Corp. (group owner; acq 9-1-93; $236,200; FTR: 9-27-93). Format: Sports, talk. News staff: one; News: 18 hrs wkly. Target aud: General; info & entertainment seekers. ♦Stephen King, pres; Bobby Russell, gen mgr & stn mgr; Ken Wood, sls dir & gen sls mgr; Scotty Moore, progmg dir.

Bar Harbor

WBQI(FM)— May 6, 1995: 107.7 mhz; 11.5 kw. 489 ft TL: N44 33 13 W68 05 40. Stereo. Hrs opn: 24 169 Port Rd., Kennebunk, 04043. Phone: (207) 797-0780. Fax: (207) 967-8671. Web Site: www.1077.com. Licensee: Nassau Broadcasting III L.L.C. Group owner: Nassau Broadcasting Partners L.P. (acq 4-6-2004; grpsl). Natl. Network: CBS, Westwood One. Format: Classic rock, jazz, sports. News staff: one; News: 6 hrs wkly. Target aud: 25-54; baby boomers. Spec prog: Blues 15 hrs wkly. ♦Pat Collins, gen mgr.

WLKE(FM)— June 1, 1992: 99.1 mhz; 45 kw. 400 ft TL: N44 32 53 W68 18 53. Stereo. Hrs opn: 24 184 Target Cir., Bangor, 04401-5718. Phone: (207) 667-7573. Fax: (207) 667-9494. E-mail: larryjulius @clearchannel.com Web Site: www.lucky99.net. Licensee: CC Licenses LLC. Group owner: Clear Channel Communications Inc. (acq 10-23-2000; grpsl). Population served: 75,000 Natl. Network: ABC. Natl. Rep: Christal. Law Firm: Dow, Lohnes & Albertson. Format: Country. News staff: one. Target aud: General. ♦Larry Julius, gen mgr; Jeffrey Pierce, opns mgr, gen sls mgr & mus dir; Josh Scroggins, gen sls mgr.

Bath

WBCI(FM)— June 1971: 105.9 mhz; 50 kw. 499 ft TL: N44 04 09 W69 55 28. Stereo. Hrs opn: Box 359, Topsham, 04086. Secondary address: 122 Main St., Topsham 04086. Phone: (207) 725-9224. Fax: (207) 725-2686. E-mail: wbci@gwi.net Web Site: www.wbci.net. Licensee: Blount Communications Inc. Group owner: Blount Communications Group (acq 4-20-95; $375,000). Population served: 750,000 Natl. Network: Salem Radio Network. Natl. Rep: Salem. Format: Talk, Christian. Target aud: 25-54; 60% men, 40% women. ♦Bill Blount, pres; Deborah Blount, exec VP; David Young, sr VP; Janice Murphy, stn mgr.

WCME(FM)—See Boothbay Harbor

WJTO(AM)— Sept 30, 1957: 730 khz; 1 kw-D, 29 w-N. TL: N43 52 39 W69 50 49. Hrs open: 24 Box 308, 04530. Phone: (207) 443-6671. Licensee: Blue Jey Broadcasting Co. Group owner: Bob Bittner Broadcasting Inc. (acq 2-28-97; $150,000). Population served: 880,000 Format: Adult Standards. News: 2 hrs wkly. Target aud: 35 plus; adults along the Maine coastline. ♦Bob Bittner, gen mgr.

Belfast

WBFB(FM)— Mar 7, 1986: 104.7 mhz; 10 kw. 1,099 ft TL: N44 34 51 W68 53 51. Stereo. Hrs opn: 24 184 Target Industrial Circle, Bangor, 04401. Phone: (207) 947-9100. Fax: (207) 942-8039. Web Site: www.1047the bear.com. Licensee: CC Licenses LLC. Group owner: Clear Channel Communications Inc. (acq 10-23-2000; grpsl). Population served: 300,000 Law Firm: Dow, Lohnes & Albertson. Format: Country. News staff: 2. Target aud: 18-49. ♦Larry Julius, gen mgr.

Biddeford

WCYY(FM)— August 1972: 94.3 mhz; 12 kw. 472 ft TL: N43 32 34 W70 24 12. Stereo. Hrs opn: 24 One City Ctr., Portland, 04101. Phone: (207) 774-6364. Fax: (207) 774-8707. E-mail: mike.sambrook@citcomm.com Web Site: www.wcyy.com. Licensee: Citadel Broadcasting Co. Group owner: Citadel Broadcasting Corp. (acq 7-7-99; grpsl). Population served: 137,900 Natl. Rep: Katz Radio. Format: Modern rock. News staff: one. Target aud: 25-44; educated, affluent. ♦Michael Sambrook, VP & gen mgr; Herbert Ivy, opns VP; Wendell Clough, prom dir; Brian James, mus dir; Celeste Nadeau, news dir.

WVAE(AM)— 1948: 1400 khz; 1 kw-U. TL: N43 28 52 W70 29 08. Hrs open: 24 420 Western Ave., South Portland, 04106. Phone: (207) 774-4561. Fax: (207) 774-3788. Web Site: www.ilovethebay.com. Licensee: Saga Communications of New England LLC. Group owner: Saga Communications Inc. (acq 11-17-03; $350,000). Population served: 177,976 Natl. Network: Jones Radio Networks. Format: Adult standards. News staff: one. Target aud: 35 plus; upscale professional. Spec prog: Relg 2 hrs wkly. ♦Harry Nelson, stn mgr.

Blue Hill

***WERU-FM**— June 1, 1988: 89.9 mhz; 15 kw. 899 ft TL: N44 26 04 W68 35 25. Stereo. Hrs opn: 6 AM-1 AM Box 170, East Orland, 04431-0170. Secondary address: 1186 Acadia Hwy., East Orland 04431. Phone: (207) 469-6600. Fax: (207) 469-8961. E-mail: info@weru.org Web Site: www.weru.org. Licensee: Salt Pond Community Broadcasting Co. Population served: 150,000 Format: Div, educ. Target aud: General. ♦Matt Murphy, gen mgr.

Boothbay Harbor

WCME(FM)— Apr 1, 1984: 96.7 mhz; 25 kw. 449 ft TL: N44 01 31 W69 34 17. Stereo. Hrs opn: 24 150 Whitten Rd., Agusta, GA, 04330. Phone: (207) 623-9000. Fax: (207) 623-9007. E-mail: donaldshieldsjr@clearchannel.com Web Site: www.newstalk962.com. Licensee: Capstar TX L.P. Group owner: Clear Channel Communications Inc. (acq 1-18-01; grpsl). Law Firm: Brown, Nietert & Kaufman. Format: News/talk. News staff: one; News: 12 hrs wkly. Target aud: 25-49. ♦Kelly Slater, gen mgr; Don Shields, opns dir; Rick Dougal, gen sls mgr & chief of engrg; Steve Smith, progmg dir.

Brewer

WKIT-FM— Feb 14, 1979: 100.3 mhz; 50 kw. 850 ft TL: N44 40 39 W68 45 15. Stereo. Hrs opn: 24 Box 1929, 861 Broadway, Bangor, 04402. Phone: (207) 990-2800. Fax: (207) 990-2444. E-mail: wkit@zoneradio.com Web Site: www.zoneradio.com. Licensee: The Zone Corp. (group owner; acq 9-95; $800,000 with co-located AM). Law Firm: Fisher, Wayland, Cooper, Leader & Zaragoza. News staff: 2. Target aud: 18-49. ♦Stephen King, CEO; Bobby Russell, gen mgr & progmg dir; Ken Wood, gen sls mgr.

WQCB(FM)— Jan 20, 1986: 106.5 mhz; 98 kw. 1,079 ft TL: N45 03 26 W69 11 27. Stereo. Hrs opn: 24 Box 100, 04412. Secondary address: 49 Acme Rd. 04412. Phone: (207) 989-5631. Fax: (207) 989-5685. E-mail: tom.preble@midmaine.com Web Site: www.wqcb-fm.com. Licensee: Cumulus Licensing Corp. Group owner: Cumulus Media LLC (acq 2-20-98; $6.4 million with WBZN(FM) Old Town). Natl. Rep: McGavren Guild. Format: Country. News staff: 2; News: 4 hrs wkly. Target aud: 25-54; general. ♦Tom Preble, gen mgr; Paul Dupois, opns mgr; Cindy Campbell, progmg dir.

Brunswick

WBCI(FM)—See Bath

***WBOR(FM)**— April 1957: 91.1 mhz; 300 w. 154 ft TL: N43 54 34 W69 57 43. Stereo. Hrs opn: 7 AM-2 AM WBOR 91.1 FM, 6200 College Stn., Bowdoin College, 04011-8462. Phone: (207) 725-3210. Phone: (207) 725-3250. Fax: (207) 725-3510. E-mail: wbor@bowdoin.edu Web Site: www.wbor.org/wbor. Licensee: Trustees of Bowdoin College. Population served: 35,000 Format: Div. Target aud: General. ♦Zach Tcheyan, stn mgr.

WCLZ(FM)— Apr 11, 1965: 98.9 mhz; 48 kw. Ant 400 ft TL: N43 55 40 W69 59 43. Stereo. Hrs opn: One City Ctr., Portland, 04101. Phone: (207) 774-6364. Fax: (207) 774-8707. E-mail: wclz@989wclz.com Web Site: www.989wclz.com. Licensee: The Last Bastion Station Trust LLC, as Trustee Group owner: Citadel Broadcasting Corp. (acq 6-12-2007; grpsl). Natl. Rep: Christal. Format: Progsv. ♦Michael Sambrook, gen mgr; Michelle Morel, adv.

WJJB(AM)— December 1955: 900 khz; 1 kw-D, 66 w-N. TL: N43 55 40 W69 59 43. Hrs open: 24
Simulcasts WJAE(AM) Westbrook.
Atlantic Coast Radio, 779 Warren Ave., Portland, 04103-1007. Phone: (207) 773-9695. Fax: (207) 761-4406. E-mail: morningjab@yahoo.com Web Site: www.thebigjab.com. Licensee: Atlantic Coast Radio L.L.C. (group owner; acq 9-99). Population served: 35,000 Natl. Rep: Christal. Format: Sports, talk. News staff: 2; News: 4 hrs wkly. Target aud: 25-54. ♦Jon VanHoogenstyn, gen mgr; David Shumacher, progmg dir.

Calais

***WMED(FM)**— November 1983: 89.7 mhz; 30 kw. 525 ft TL: N45 01 44 W67 19 25. Stereo. Hrs opn:
Rebroadcasts WMEH(FM) Bangor 100%.
65 Texas Ave., Bangor, 04401. Secondary address: 1450 Lisbon St., Lewiston 04240. Phone: (207) 874-6570. Fax: (207) 761-0318. Web Site: www.mainepublicradio.org. Licensee: Maine Public Broadcasting Corp. (acq 6-23-92). Natl. Network: NPR, PRI. Law Firm: Dow, Lohnes & Albertson. Format: Class, pub affrs, news. ♦Alexander G. Maxwell, Jr., COO, sr VP & stn mgr; Charles Beck, pres & progmg VP; P. James Dowe, Jr., CEO & pres; Christopher Amann, CFO; Mary Mayo, dev VP.

WQDY-FM— Jan 14, 1976: 92.7 mhz; 3 kw. Ant 299 ft TL: N45 10 02 W67 16 38. Stereo. Hrs opn: 24 Box 403, 04619. Secondary address: 637 Main St. 04619. Phone: (207) 454-7545. Fax: (207) 454-3062. E-mail: wqdy@wqdy.fm Web Site: www.wqdy.fm. Licensee: WQDY Inc. (acq 11-26-96; for stock). Natl. Network: Jones Radio Networks. Rgnl rep: Cyr Associates Law Firm: Fletcher, Heald & Hildreth. Wire Svc: AP Format: Adult contemp, classic rock. News staff: one. ♦Bill McVicar, pres; Roger Holst, chief of engrg.

Camden

***WMEP(FM)**— Oct 2002: 90.5 mhz; 2 kw vert. 1,178 ft TL: N44 12 40 W69 09 06. Hrs open:
Rebroadcasts WMEH(FM) Bangor 100%.
65 Texas Ave., Bangor, 04401. Secondary address: 1450 Lisbon St., Lewiston 04240. Phone: (207) 874-6570. Fax: (207) 761-0318. E-mail: cbeck@mpbn.net Web Site: www.mpbn.net. Licensee: Maine Public Broadcasting Corp. Natl. Network: NPR, PRI. Law Firm: Dow, Lohnes & Albertson. Format: Class, pub affrs, news/talk. ♦Alexander G. Maxwell, Jr., COO & sr VP; P. James Dowe, Jr., CEO & pres; Christopher Amann, CFO; Mary Mayo, dev VP; Charles Beck, progmg VP.

WQSS(FM)— May 1988: 102.5 mhz; 7.9 kw. 1,201 ft TL: N44 12 40 W69 09 06. Stereo. Hrs opn: 24 15 Payne Ave., Rockland, 04841-2117. Phone: (207) 594-9400. Fax: (207) 594-2234. E-mail: kellyslater@clearchannel.com Web Site: www.1025thepeak.com. Licensee: CC Licenses LLC. Group owner: Clear Channel Communications Inc. (acq 5-3-2002; $1.72 million). Natl. Network: ABC. Natl. Rep: McGavren Guild. Format: Classic rock. News staff: one; News: 10 hrs wkly. Target aud: 25-54. ♦Kelly Slater, gen mgr; Steve Smith, opns dir; Rick Dougal, gen sls mgr; Don Shields, progmg dir; Marc Fisher, chief of engrg.

Caribou

WBPW(FM)—See Presque Isle

WCXU(FM)— Nov 15, 1986: 97.7 mhz; 20 kw. Ant 328 ft TL: N46 47 26 W67 55 07. Stereo. Hrs opn: 24 152 E. Green Ridge Rd., 04736. Phone: (207) 473-7513. Fax: (207) 472-3221. E-mail: channelxradio@yahoo.com Web Site: www.channelxradio.com. Licensee: The Canxus Broadcasting Corp. (group owner) Population served: 150,000 Rgnl rep: Cyr Associates. Law Firm: Koteen & Naftalin. Format: Adult contemp, news, oldies. News staff: one; News: 21 hrs wkly. Target aud: 25-54; educated, div occupations, affluent. ♦Dennis H. Curley, CEO, chmn, pres & CFO; Richard Chandler, gen mgr & opns mgr; Mark Stewert, progmg dir.

***WFST(AM)**— July 15, 1956: 600 khz; 5 kw-D, 127 w-N. TL: N46 53 12 W68 02 44. (CP: TL: N46 45 52 W67 59 23). Hrs opn: Box 600, 04736-0600. Secondary address: 670 Sweeden St. 04736-0600. Phone: (207) 492-6000. Fax: (207) 493-3268. E-mail: wfst@mfx.net Web Site: www.wfst.net. Licensee: Northern Broadcast Ministries Inc. (acq 6-8-93; $54,000; FTR: 6-28-93). Population served: 75,000 Natl. Network: USA. Format: Christian, gospel, relg. Target aud: General. ♦Donald Flewelling, pres; Tom Hale, VP; John Stephenson, gen mgr.

Dennysville

WCRQ(FM)— May 1998: 102.9 mhz; 100 kw. 456 ft TL: N45 01 44 W67 19 25. Stereo. Hrs opn: 24 637 Main St., Calais, 04619. Phone: (207) 454-7545. Fax: (207) 454-3062. E-mail: wcrq@wqdy.fm Licensee: WQDY Inc. (acq 5-30-03; $195,000). Population served: 300,000 Law Firm: Fletcher, Heald & Hildreth. Format: Hot adult contemp. News: 6 hrs wkly. Target aud: 18-49; mass appeal. ♦Bill McVicar, pres; Bill Conley, mus dir.

Dexter

WGUY(FM)— 1993: 102.1 mhz; 50 kw. 672 ft TL: N45 02 40 W69 15 01. Stereo. Hrs opn: 24 184 Target Industrial Cir., Bangor, 04401. Phone: (207) 947-9100. Fax: (207) 942-8039. Licensee: CIC Licenses LLC. Group owner: Clear Channel Communications Inc. (acq 10-20-2003; $1.2 million). Population served: 165,000 Natl. Rep: CYR Associates. Format: Hits of the 50s, 60s, & 70s. News staff: one; News: 2 hrs wkly. Target aud: 25-54. ♦Larry Julius, gen mgr.

Dover Foxcroft

WDME-FM— November 1980: 103.1 mhz; 4.8 kw. 385 ft TL: N45 12 58 W69 14 34. Stereo. Hrs opn: 24 Box 1929, Bangor, 04402. Phone: (207) 990-2800. Fax: (207) 990-2444. E-mail: wdme@zoneradio.com Web Site: zoneradio.com. Licensee: The Zone Corp. (group owner; acq 2-16-01; $175,100). Population served: 100,000 Natl. Network: ABC. Law Firm: Verner, Liipfert, Bernhard, McPherson & Hand. Format: Adult contemp. News staff: one. Target aud: General. ♦Bobby Russell, VP, stn mgr & progmg dir; Howard Soule, chief of engrg; Ken Woods, pub affrs dir & sls.

Eastport

***WSHD(FM)**— April 1984: 91.7 mhz; 12 w. Ant 115 ft TL: N44 54 30 W66 59 24. Stereo. Hrs opn:
Rebroadcasts WQDY-FM Calais 100%.
Shead High School, 89 High St., 04631. Phone: (207) 853-6254. Fax: (207) 853-2919. Licensee: Shead High School. Population served: 3,500 Format: Classic hits. Target aud: General.

Ellsworth

WDEA(AM)— Dec 13, 1958: 1370 khz; 5 kw-U, DA-2. TL: N44 28 00 W68 28 11. Hrs open: Box 100, Brewer, 04412. Secondary address: 49 Acme Rd. 04412. Phone: (207) 989-5631. Fax: (207) 989-5685. Licensee: Cumulus Licensing Corp. Group owner: Cumulus Media Inc. (acq 1999; grpsl). Population served: 40,000 Natl. Network: CBS. Natl. Rep: D & R Radio. Law Firm: Shaw Pittman. Format: Nostalgia. Target aud: 35 plus. ♦Tom Preble, gen mgr & gen sls mgr; Michael O'Hara, prom dir; Fred Miller, progmg dir; Allison Bankston, news dir; Richard Hyatt, chief of engrg.

WWMJ(FM)—Co-owned with WDEA(AM). Dec 27, 1965: 95.7 mhz; 11.5 kw. Ant 1,029 ft TL: N44 39 31 W68 36 20. Stereo. E-mail: q1065@midmaine.com Web Site: www.wwmj-fm.com.300,000 Format: Classic hits. Target aud: 25-54.

WKSQ(FM)— May 27, 1982: 94.5 mhz; 11.5 kw. 1,027 ft TL: N44 39 31 W68 36 17. Stereo. Hrs opn: 24 184 Target Cir., Bangor, 04401-5718. Phone: (207) 667-7573. Fax: (207) 667-9494. E-mail: larryjulius@clearchannel.com Web Site: www.kiss945.com. Licensee: CC Licenses LLC. Group owner: Clear Channel Communications Inc. (acq 10-23-2000; grpsl). Population served: 250,000 Natl. Rep: Christal. Law Firm: Dow, Lohnes & Albertson. Format: Adult contemp. News staff: 3; News: 7 hrs wkly. Target aud: 25-54. ♦Larry Julius, gen mgr; Jeff Pierce, opns dir; Josh Scroggins, gen sls mgr & natl sls mgr.

Fairfield

WCTB(FM)— November 1993: 93.5 mhz; 10.5 kw. Ant 499 ft TL: N44 44 42 W69 41 32. Stereo. Hrs opn: Box 159, Skowhegan, 04976. Phone: (207) 474-5171. Fax: (207) 474-3299. Licensee: Mountain Wireless Inc. (group owner; acq 4-20-94; $60,000 FTR: 7-4-94). Law Firm: Schwartz, Woods & Miller. Format: Classic rock. ♦Jay Hanson, gen mgr.

Farmington

WKTJ-FM— Aug 21, 1973: 99.3 mhz; 1.5 kw. 400 ft TL: N44 39 22 W70 11 48. Hrs open: 18 Box 590, 04938. Phone: (207) 778-3400. Fax: (207) 778-3000. E-mail: wktj@wktj.com Web Site: www.wktj.com. Licensee: Franklin Broadcasting Corp. Population served: 125,000 Wire Svc: AP Format: Adult contemp. News: 12 hrs wkly. Target aud: 25-54. ♦Nelson Doak, gen mgr & mus dir; Steve Bull, gen sls mgr; Marc Fisher, chief of engrg.

***WUMF-FM**— February 1972: 100.1 mhz; 13 w. Ant -190 ft TL: N44 40 09 W70 09 00. Stereo. Hrs opn: 111 South St., 04938. Phone: (207) 778-7352. Fax: (207) 778-7113. E-mail: wumf@umf.maine.edu Web Site: http://wumf.umf.maine.edu. Licensee: University of Maine System. Population served: 3,096 Format: AOR, progsv, div. Target aud: 18-45; college students & local residents. Spec prog: Jazz 15 hrs, tech/industrial 15 hrs wkly. ♦Megan Littlefield, gen mgr & stn mgr; Will McArthur, progmg dir; Rob Graham, mus dir.

Fort Kent

***WMEF(FM)**— March 1994: 106.5 mhz; 25 kw. 302 ft TL: N47 15 30 W68 33 30. Hrs open:
Rebroadcasts WMEH(FM) Bangor 100%.
65 Texas Ave., Bangor, 04401. Secondary address: 1450 Lisbon St., Lewiston 04240. Phone: (207) 941-1010/783-9101. Fax: (207) 942-2857/783-5193. E-mail: cbeck@mpbn.net Web Site: www.mpbn.net. Licensee: Maine Public Broadcasting Corp. Natl. Network: PRI, NPR. Law Firm: Dow, Lohnes & Albertson. Format: Class, pub affrs, news/talk. ♦Alexander G. Maxwell Jr., COO & sr VP; P. James Dowe Jr., CEO & pres; Charles Beck, CFO & progmg VP; Christopher Amann, CFO; Mary Mayo, dev VP.

Freeport

***WMSJ(FM)**— Dec 1, 1997: 89.3 mhz; 7.5 kw vert. 394 ft TL: N43 45 45 W70 19 30. Hrs open: Box 287, 04032. Phone: (207) 865-3448. Fax: (207) 865-1763. E-mail: info@positive.fm Web Site: www.positive.fm. Licensee: The Positive Radio Network. Format: Contemp Christian music. Target aud: 25-48. ♦John A. Libby, chmn & pres; Chris Scotland, stn mgr & prom; Paula K, gen mgr & progmg dir.

Gardiner

WABK-FM—Listing follows WFAU(AM).

WFAU(AM)— Sept 23, 1968: 1280 khz; 5 kw-U, DA-N. TL: N44 14 53 W69 48 51. Stereo. Hrs opn: 24 150 Whitten Rd., Augusta, 04330. Phone: (207) 623-9000. Fax: (207) 623-9035. Licensee: Capstar TX L.P. Group owner: Clear Channel Communications Inc. (acq 1-18-01; grpsl). Population served: 200,000 Law Firm: Reddy, Begley & McCormick. Format: Adult contemp, news/talk. News staff: one; News: 34 hrs wkly. Target aud: 25-54. ♦Jim Herron, gen mgr.

WABK-FM—Co-owned with WFAU(AM). Apr 1, 1974: 104.3 mhz; 50 kw. 492 ft TL: N44 18 36 W69 49 51. Stereo. Format: Oldies. ♦Mark Jackson, progmg dir.

Gorham

WLVP(AM)— Mar 3, 1980: 870 khz; 10 kw-D, 1 kw-N. TL: N43 41 19 W70 30 34. Hrs open: 477 Congress St., Suite 3B, Portland, 04101. Phone: (207) 797-0780. Fax: (207) 797-0368. Licensee: Nassau Broadcasting III L.L.C. Group owner: Nassau Broadcasting Partners L.P. (acq 4-6-2004; grpsl). Natl. Network: ESPN Radio. Format: Sports talk. ♦Patrick Collins, gen mgr.

***WMPG(FM)**— Sept 1, 1973: 90.9 mhz; 110 w horiz, 1 kw vert. 233 ft TL: N43 40 50 W70 26 59. Stereo. Hrs opn: 24 Box 9300, 96 Falmouth St., Portland, 04104-9300. Phone: (207) 780-4943. Fax: (207) 780-4590. E-mail: stationmanager@wmpg.org Web Site: www.wmpg.org. Licensee: Trustees University of Maine. Population served: 10,000 Format: Div, community. News: 10 hrs wkly. Target aud: General; any group currently underserved by other loc stns. Spec prog: Sp 4 hrs, Balkan 2 hrs, Cambodian 2 hrs, African 4 hrs, Irish 2 hrs wkly, Vietnamese 2 hrs, Middle Eastern 2 hrs. ♦James Rand, stn mgr; Dave Bunker, progmg dir; Ron Raymond, mus dir; Brian Dyer, chief of engrg.

Hampden

WRME(AM)—Not on air, target date: unknown: 750 khz; 50 kw-D, 10 kw-N, DA-N. TL: N44 51 27 W68 49 36. Hrs open: 16 Doe Run, Pittstown, NJ, 08867. Phone: (908) 730-7959. Fax: (908) 730-7408. Licensee: Charles A. Hecht and Alfredo Alonso. ♦Charles A. Hecht, gen mgr.

Harpswell

***WYFP(FM)**— July 8, 1993: 91.9 mhz; 6 kw. 144 ft TL: N43 44 14 W69 59 39. Hrs open: 24 11530 Carmel Commons Blvd., Charlotte, NC, 28228-3976. Phone: (207) 729-9919. Fax: (207) 729-9919. E-mail: wyfp@bbnradio.org Web Site: www.bbnradio.org. Licensee: Bible Broadcasting Network Inc. (group owner; acq 9-30-97; $150,000). Population served: 350,000 Natl. Network: USA. Format: Bible teaching, religious. Target aud: 25-49. Spec prog: Christian rock 4 hrs, praise & worship 3 hrs wkly. ♦T. A. Smith, gen mgr; Dennis Gast, opns mgr; Teddi Wilson, disc jockey.

Houlton

WHOU-FM— Jan 13, 1976: 100.1 mhz; 9.6 kw. 525 ft TL: N46 08 35 W68 06 50. Hrs open: 24 Box 40, 04730. Secondary address: 39 Court St., Suite 215 04730. Phone: (207) 532-3600. Fax: (207) 521-0056. E-mail: sales@whoufm.com Web Site: www.whoufm.com. Licensee: County Communications Inc. (acq 4-96; $350,000). Population served: 240,000 Natl. Network: ABC. Law Firm: Crowell & Moring. Format: Adult contemp. News: 8 hrs wkly. Target aud: 25-54. Spec prog: Sacred one hr wkly. ♦David Moore, pres & gen mgr; Jacqueline Spencer, opns mgr; George Kelley, progmg dir; Barrett Quinn, chief of engrg.

Howland

WVOM(FM)— June 1993: 103.9 mhz; 54 kw. 1,509 ft TL: N45 07 46 W68 21 28. Hrs open:
Rebroadcasts WBYA(FM) Searsport 80%.
184 Target Industrial Cir., Bangor, 04401. Phone: (207) 942-3311. Fax: (207) 942-8039. E-mail: wvom@midmaine.com Web Site: www.thevoicemaine.com. Licensee: CC Licenses LLC. (acq 3-12-97). Natl. Network: CBS, Westwood One. Format: News/talk. Target aud: 25-64; upper income, professional, managerial. ♦Jeffery Pierce, CEO & opns dir; Jim Herron, gen mgr; Susan Patten, progmg dir & news dir.

Islesboro

WBYA(FM)— 1999: 105.5 mhz; 20 kw. Ant 305 ft TL: N44 18 58 W68 58 12. Hrs open: 1119 Tillson Ave., Rockland, 04841. Phone: (207) 594-9283. Fax: (207) 594-1620. E-mail: jlynch@nassaubroadcasting.com Web Site: www.frank1055fm.com. Licensee: Nassau Broadcasting III L.L.C. Group owner: Nassau Broadcasting Partners L.P. (acq 4-6-2004; grpsl). Format: Classic hits. News staff: one; News: 7 hrs wkly. Target aud: 50 plus. ♦Pat Collins, gen mgr; Stan Manning, opns mgr.

Kennebunk

WBQQ(FM)— November 1991: 99.3 mhz; 3 kw. 324 ft TL: N43 24 16 W70 26 15. Hrs open: 24 Unit 99.3, 169 Port Rd., 04043. Phone: (207) 797-0780. Fax: (207) 967-8671. E-mail: wbach@wbach.fm Web Site: www.wbachradio.com. Licensee: Nassau Broadcasting III L.L.C. Group owner: Nassau Broadcasting Partners L.P. (acq 4-6-2004; grpsl). Natl. Network: ABC. Format: Class. Target aud: 35-64; upscale. ♦Pat Collins, pres & gen mgr; Stan Manning, opns dir & opns mgr; Scott Hooper, progmg dir.

Kennebunkport

WHXQ(FM)— Dec 1, 1994: 104.7 mhz; 6 kw. 292 ft TL: N43 26 36 W70 26 38. Hrs open: 477 Congress St., Suite 3 A, Portland, 04101. Phone: (207) 797-0780. Fax: (207) 797-0368. E-mail: portlandspots@nassaubroadcasting.com Web Site: www.boneradio.com. Licensee: Nassau Broadcasting III L.L.C. Group owner: Nassau Broadcasting Partners L.P. (acq 4-6-2004; grpsl). Format: Classic rock. ♦Pat Collins, VP & gen mgr; Stan Manning, opns mgr & chief of engrg.

Kittery

WSHK(FM)— October 1992: 105.3 mhz; 2.2 kw. 371 ft TL: N43 10 28 W70 46 50. Stereo. Hrs opn: 24
Rebroadcasts WSAK(FM) Hampton, NH 100%.
Box 576, Dover, NH, 03821-0576. Secondary address: 292 Middle Rd., Dover, NH 03820-4901. Phone: (603) 749-9750. Fax: (603) 749-1459. E-mail: shark.mail@citicomm.com Web Site: www.shark1053.com. Licensee: Citadel Broadcasting Co. Group owner: Citadel Broadcasting Corp. (acq 7-7-99; grpsl). Population served: 500000 Natl. Network: CNN Radio. Natl. Rep: Christal. Law Firm: Wiley, Rein & Fielding. Wire Svc: AP Format: Classic rock. News staff: 2. Target aud: 25-54. ♦Farid Suleman, CEO; Judy Ellis, pres; Marty Lessard, gen mgr; Mark Ericson, opns mgr; Jonathan Smith, progmg dir.

Lewiston

WCYI(FM)— Feb 29, 1948: 93.9 mhz; 27.5 kw. Ant 633 ft TL: N44 08 40 W70 01 22. Hrs open: 24
Rebroadcasts WCLZ(FM) Brunswick 100%.
One City Ctr., Portland, 04101. Phone: (207) 774-6364. Fax: (207) 774-8707. E-mail: mike.sambrook@citcomm.com Web Site: www.wcyy.com. Licensee: The Last Bastion Station Trust LLC, as Trustee Group owner: Citadel Broadcasting Corp. (acq 6-12-2007; grpsl). Population served: 600,000 Natl. Rep: Christal. Format: Triple A. Target aud: General. ♦Mike Sambrook, gen mgr; Herbert Ivy, progmg dir; Michelle Morel, pub affrs dir & adv.

WEZR(AM)— Aug 21, 1938: 1240 khz; 1 kw-U. TL: N44 06 55 W70 14 56. Hrs open: 24 555 Center St., Auburn, 04210. Phone: (207) 784-5868. Fax: (207) 784-4700. E-mail: dick@gleasonmedia.com Web Site: www.gleasonmedia.com. Licensee: Mountain Valley Broadcasting Inc. Group owner: Gleason Radio Group (acq 11-28-90). Population served: 100,000 Natl. Rep: CYR Associates. Law Firm: Womble, Carlyle, Sandridge & Rice, PLLC. Format: EZ oldies. News staff: one; News: 17 hrs wkly. Target aud: General. ♦Richard D. Gleason, pres & gen mgr; Milton Simon, stn mgr; Scott Garrett, news dir.

WFNK(FM)—Licensed to Lewiston. See Portland

WLAM(AM)— Sept 4, 1947: 1470 khz; 5 kw-U, DA-1. TL: N44 03 47 W70 15 00. Hrs open: 477 Congress St., Annex, Suite 3B, Portland, 04101. Phone: (207) 797-0780. Fax: (207) 797-0368. Web Site: www.1470wlam.com. Licensee: Nassau Broadcasting III L.L.C. Group owner: Nassau Broadcasting Partners L.P. (acq 4-6-2004; grpsl). Population served: 75,000 Natl. Network: ESPN Radio. Natl. Rep: D & R Radio. Format: Sports. ♦Pat Collins, sr VP & gen mgr; Tim Gatz, stn mgr; Stan Manning, opns mgr.

***WRBC(FM)**— Oct 6, 1958: 91.5 mhz; 150 w. 16 ft TL: N44 06 18 W70 12 32. Stereo. Hrs opn: 24 31 Frye St., 04240. Phone: (207) 777-7532. Fax: (207) 795-8793. E-mail: mgraham3@bates.edu Web Site: www.bates.edu/wrbc. Licensee: President and Trustees of Bates College. Population served: 180,000 Format: Div, rock. Target aud: General; anyone searching for something different. Spec prog: Fr 2 hrs, jazz 4 hrs, metal 4 hrs, classic rock 11 hrs, folk 6 hrs wkly. ♦Molly Graham, gen mgr; Drew Faller, progmg dir; Bill Morse, mus dir.

Lincoln

***WHMX(FM)**— Apr 1, 1975: 105.7 mhz; 50 kw. 413 ft TL: N45 20 34 W68 30 25. Stereo. Hrs opn: 24 Box 5000, Bangor, 04402-5000. Secondary address: 1476 Broadway, Bangor 04401. Phone: (207) 262-1057. E-mail: contact@solutionfm.com Web Site: www.solutionfm.com. Licensee: Bangor Baptist Church. (acq 12-31-96; $80,000 with co-located AM). Population served: 200,000 Law Firm: Fletcher, Heald & Hildreth. Format: Contemp Christian. Target aud: 18-35. ♦Pencil Boone, gen mgr; Jolie Littlefield, prom dir & chief of engrg; Tim Collins, progmg dir; Morgan Smith, mus dir.

WSYY(AM)—See Millinocket

Machias

WALZ-FM— Nov 25, 1978: 95.3 mhz; 3 kw. 220 ft TL: N44 44 08 W67 30 11. Stereo. Hrs opn: 24
Rebroadcasts WQDY-FM Calais 50%.
637 MAin St., Calais, 04619. Phone: (207) 454-7545. Fax: (207) 454-3062. E-mail: wqdy@wqdy.fm Web Site: www.wqdy.fm. Licensee: William McVicar & Roger Holst, general partnership (acq 10-26-01). Natl. Network: ABC, Jones Radio Networks. Format: classic hits. ♦William G. Mcvicar, gen mgr; Roger Holst, opns dir.

Madawaska

WCXX(FM)— Jan 30, 1988: 102.3 mhz; 1.75 kw. 384 ft TL: N47 19 54 W68 20 31. Stereo. Hrs opn: 24
Rebroadcasts WCXU(FM) Caribou 50%.
152 E. Green Ridge Rd., Caribou, 04736. Phone: (207) 473-7513. Fax: (207) 472-3221. E-mail: channelxradio@yahoo.com Web Site: www.channelxradio.com. Licensee: Canxus Broadcasting Corp. Population served: 30,000 Natl. Network: CNN Radio. Rgnl rep: Cyr Associates.

Law Firm: Koteen & Naftalin. Format: Adult contemp, news. News staff: one; News: 16 hrs wkly. Target aud: 18-54. ♦Dennis H. Curley, pres; Richard Chandler, gen mgr; Mark Stewart, opns mgr & mus dir.

Madison

WIGY(FM)— 1995: 97.5 mhz; 6 kw. 328 ft TL: N44 47 32 W69 58 10. Stereo. Hrs opn: 24 150 Whitten Rd., Augusta, 04330. Phone: (207) 623-9000. Fax: (207) 623-9007. E-mail: donaldshields@clearchannel.com Web Site: www.foxsportsmaine.com. Licensee: Clear Channel Radio Licenses, Inc Group owner: Clear Channel Communications Inc. (acq 1-18-01; grpsl). Population served: 68,000 Rgnl rep: Cyr Associates. Law Firm: Fletcher, Heald & Hildreth. Format: Sports. Target aud: 25-54. ♦Kelly Slater, gen mgr & progmg mgr; Rick Dugal, gen sls mgr; Donald Shield, progmg dir.

Mexico

WTBM(FM)— Sept 15, 1988: 100.7 mhz; 850 w. Ant 1,273 ft TL: N44 34 56 W70 37 59. Stereo. Hrs opn: 24
Simulcast with WOXO-FM Norway 99%.
PO Box 72, 243 Main St., Norway, 04268. Phone: (207) 743-5911. Fax: (207) 743-5913. E-mail: info@woxo.com Web Site: www.woxo.com. Licensee: Mountain Valley Broadcasting Inc. Group owner: Gleason Radio Group (acq 12-90; FTR: 10-22-90). Population served: 150,000 Natl. Network: USA. Natl. Rep: CYR Associates. Law Firm: Womble, Carlyle, Sandridge & Rice ,PLLC. Format: Country, sports. News staff: 2; News: 12 hrs wkly. Target aud: General. ♦Richard Gleason, pres & gen mgr; Vic Hodgkins, stn mgr; Jeremy Rush, opns mgr & progmg dir.

Milbridge

WRMO(FM)— 2005: 93.7 mhz; 130 w. Ant 7 ft TL: N44 32 19 W67 52 58. Hrs open: Box 241, 04658. Phone: (207) 546-7510. Licensee: Steven A. Roy, Personal Representative, Estate of Lyle Evans (acq 6-27-2006; with KHAM(FM) Britt, IA). Format: Classic hits. ♦Mike McSorley, gen mgr.

Millinocket

WSYY(AM)— Dec 7, 1963: 1240 khz; 1 kw-U. TL: N45 40 24 W68 43 07. Hrs open: 24
Fox Sports Network.
Box 1240, 04462. Phone: (207) 723-9657. Fax: (207) 723-5900. E-mail: calendar@themountain949.com Web Site: www.themountain949.com. Licensee: Katahdin Communications Inc. (acq 12-29-86; $295,000 with co-located FM; FTR: 11-10-86) Population served: 20,000 Rgnl rep: Cyr Associates. Format: Sports. Target aud: General. ♦James Talbot, pres & gen mgr; Dave Keys, chief of engrg & traf mgr.

WSYY-FM— Apr 12, 1978: 94.9 mhz; 23.5 kw. 692 ft TL: N45 52 58 W68 47 54. Stereo. 24
Westwood One.
Web Site: www.themountain949.com.150,000 Format: Country. Target aud: 20-45.

Monticello

WCXH(AM)— Sept 2, 1981: 780 khz; 5 kw-D, 60 w-N. TL: N46 20 30 W67 49 04. Hrs open:
Simulcast with WCXU(FM) Caribou 100%.
152 E. Green Ridge Rd., Caribou, 04736. Phone: (207) 473-7513. Fax: (207) 472-3221. E-mail: channelxradio@yahoo.com Web Site: www.channelxradio.com. Licensee: Allan H. Weiner (acq 2-12-2002). Population served: 50,000 Format: Adult contemp. Target aud: 25-54; women 43%, men 57%. ♦Allan H. Weiner, pres; Richard Chandler, gen mgr.

North Windham

WHXR(FM)— 1996: 106.7 mhz; 810 w. 623 ft TL: N43 51 06 W70 19 40. Stereo. Hrs opn: 24
Rebroadcasts Wmtww(AM) Gorham 100%.
Box 8, Auburn, 04210. Secondary address: 99 Danville Corner Rd. 04210. Phone: (207) 782-1800. Fax: (207) 783-7371. E-mail: wmtw@wmtw.com Web Site: www.wmtw.com. Licensee: Nassau Broadcasting III L.L.C. Group owner: Nassau Broadcasting Partners L.P. (acq 4-6-2004; grpsl). Natl. Network: AP Radio. Format: News. Target aud: 35 plus; general. ♦David Kaufman, gen mgr; Bill Whitten, gen sls mgr; Scan Baker, progmg mgr; Jennifer Sullivan, news dir; Peter Magec, chief of engrg; Jennifer Bachelder, traf mgr.

Norway

WOXO-FM— Dec 12, 1970: 92.7 mhz; 2 kw. 360 ft TL: N44 12 24 W70 33 18. Stereo. Hrs opn: 24
Rebroadcasts WTBM(FM) Mexico 99%.
PO Box 72, 243 Main St., 04268. Phone: (207) 743-5911. Fax: (207) 743-5913. E-mail: info@woxo.com Web Site: www.woxo.com. Licensee: Mountain Valley Broadcasting Inc. Group owner: Gleason Radio Group (acq 12-12-75). Population served: 150,000 Natl. Network: USA. Natl. Rep: CYR Associates. Law Firm: Womble, Carlyle, Sandridge & Rice, PLLC. Format: Country, sports. News staff: 2; News: 12 hrs wkly. Target aud: General. ♦Richard D. Gleason, pres & gen mgr; Vic Hodgkins, stn mgr; Jeremy Rush, opns mgr; Jay Philips, progmg dir & chief of engrg.

Oakland

***WMDR-FM**— 2006: 88.9 mhz; 600 w vert. Ant 574 ft TL: N44 42 48 W69 43 39. Hrs open: 24 160 Bangor St., Augusta, 04330. Phone: (207) 622-1340. Fax: (207) 623-2874. E-mail: denise@lightoflife.info Web Site: www.worshipradionetwork.org. Licensee: Light of Life Ministries Inc. Population served: 300,000 Format: Christian country. News: 1 hr wkly. ♦Ray Bouchard, gen mgr.

Old Town

WBZN(FM)— Jan 1, 1995: 107.3 mhz; 50 kw. 308 ft TL: N45 02 06 W68 40 57. Hrs open: Box 100, Brewer, 04412. Secondary address: 49 Acme Rd., Brewer 04412. Phone: (207) 989-5631. Fax: (207) 989-5685. E-mail: z1073@midmaine.com Web Site: www.wbzn-fm.com. Licensee: Cumulus Licensing Corp. Group owner: Cumulus Media Inc. (acq 2-20-98; $6.4 million with WQCB(FM) Brewer). Format: CHR. ♦Tom Preble, gen mgr; Paul Dupois, opns dir; Dick Hyatt, chief of engrg.

Orono

WFGO(AM)—Not on air, target date: unknown: 1530 khz; 50 kw-D, 270 w-N, 9 kw-CH, DA-3. TL: N44 51 48 W68 40 06. Hrs open: 6930 Cahaba Valley Rd., Suite 202, Birmingham, AL, 35242. Phone: (205) 618-2020. Fax: (205) 618-2029. Licensee: Brantley Broadcast Assoiates LLC. (acq 5-8-2007; $100 for CP). ♦Paul Reynolds, gen mgr.

***WMEB-FM**— Apr 1, 1963: 91.9 mhz; 600 w. 150 ft TL: N44 55 08 W68 39 58. Stereo. Hrs opn: 24 5748 Memorial Union, 04469. Phone: (207) 581-4340. Phone: (207) 581-2333. Fax: (207) 581-4343. E-mail: wmeb919@hotmail.com Web Site: www.umaine.edu/wmeb. Licensee: Board of Trustees, University of Maine. Population served: 70,000 Format: Progsv, rock, div. Target aud: General. ♦Thomas Grucza, stn mgr.

Pittsfield

WJCX(FM)— December 1993: 99.5 mhz; 6 kw. 328 ft TL: N44 48 11 W69 10 06. Hrs open: 2881 Ohio St., Suite 8, Bangor, 04401. Phone: (207) 884-6052. Fax: (207) 884-6052. E-mail: wjcx@calvarychapel.com Licensee: CSN International. (acq 1996; $87,500). Format: News/talk, Christian. Target aud: 18-34; young adults. ♦Mike Archer, gen mgr, stn mgr, progmg dir & mus dir.

Portland

WBAE(AM)— March 1946: 1490 khz; 1 kw-U. TL: N43 39 48 W70 16 16. Hrs open: 24 420 Western Avenue, 04106. Phone: (207) 774-4561. Fax: (207) 774-3788. E-mail: feedback@ilovethebay.com Web Site: www.ilovethebay.com. Licensee: Saga Communications of New England LLC. Group owner: Saga Communications Inc. (acq 1996; $10 million with co-located FM). Population served: 150,000 Natl. Network: CNN Radio. Format: Music of Your Life. News staff: 5. Target aud: 25-54; general. ♦Cary Pahigian, pres & gen mgr.

WPOR(FM)—Co-owned with WBAE(AM). Oct 31, 1967: 101.9 mhz; 32.5 kw. 606 ft TL: N43 45 45 W70 19 30. Stereo. 24 E-mail: wpor@wpor.com Web Site: www.wpor.com. Format: Country.

WBLM(FM)— February 1966: 102.9 mhz; 100 kw. 1,460 ft TL: N43 55 28 W70 29 28. Stereo. Hrs opn: One City Ctr., 04101. Phone: (207) 774-6364. Fax: (207) 774-8707. Web Site: www.wblm.com. Licensee: Citadel Broadcasting Co. Group owner: Citadel Broadcasting Corp. (acq 7-7-99; grpsl). Population served: 220,000 Natl. Rep: Katz Radio. Format: AOR. Target aud: 25-54; active, involved, fun-loving. ♦Michael Sambrook, gen mgr; Michelle Morel, adv.

WFNK(FM)—Lewiston, Mar 1, 1973: 107.5 mhz; 100 kw. Ant 928 ft TL: N44 00 12 W70 25 24. Stereo. Hrs opn: 477 Congress St., Ste. 3B, 04101. Phone: (207) 797-0780. Fax: (207) 797-0368. E-mail: portlandspots@nassaubroadcasting.com Web Site: www.1075frankfm.com. Licensee: Nassau Broadcasting III L.L.C. Group owner: Nassau Broadcasting Partners L.P. (acq 4-6-2004; grpsl). Population served: 150,000 Natl. Rep: D & R Radio. Format: Classic rock. ♦Pat Collins, VP & gen mgr; Tim Gatz, gen mgr; Stan Manning, opns mgr.

WGAN(AM)— Aug 3, 1938: 560 khz; 5 kw-U, DA-2. TL: N43 41 22 W70 19 00. (CP: 4.8 kw-U, DA-1). Hrs opn: 420 Western Ave., South Portland, 04106. Phone: (207) 774-4561. Fax: (207) 774-3788. E-mail: wgan@560wgan.com Web Site: www.560wgan.com. Licensee: Saga Communications of New England LLC. Group owner: Saga Communications Inc. (acq 6-2-92; grpsl, including co-located FM). Natl. Network: CNN Radio. Format: News/talk. ♦Cary Pahigian, CEO, pres, VP & gen mgr.

WMGX(FM)—Co-owned with WGAN(AM). June 10, 1977: 93.1 mhz; 50 kw. Ant 443 ft TL: N43 41 17 W70 15 27. Stereo. Web Site: www.coast931.com. Format: Adult contemp.

WHOM(FM)—See Mt. Washington, NH

WJBQ(FM)— June 1, 1960: 97.9 mhz; 16 kw. 889 ft TL: N43 51 06 W70 19 40. (CP: 37.5 kw, ant 567 ft. TL: N43 45 32 W70 19 14). Stereo. Hrs opn: One City Center, 04101. Phone: (207) 774-6364. Fax: (207) 774-8087. Web Site: www.wjbq.com. Licensee: Citadel Broadcasting Co. Group owner: Citadel Broadcasting Corp. (acq 7-7-99; grpsl). Population served: 150,000 Natl. Rep: Christal. Format: CHR. Target aud: 25-44; female listeners. ♦Michael Sambrook, gen mgr; Tim Moore, opns mgr.

WLOB(AM)— Feb 2, 1957: 1310 khz; 5 kw-U, DA-2. TL: N43 41 22 W70 20 06. Hrs open: 24
Rebroadcasts WLOB-FM Rumford 100%.
779 Warren Ave., 04103-1007. Phone: (207) 773-9695. Fax: (207) 761-4406. E-mail: newstalkWLOB@yahoo.com Licensee: Atlantic Coast Radio L.L.C. (group owner; acq 9-8-00; grpsl). Population served: 250,000 Natl. Network: USA. Format: News/talk. News: 17 hrs wkly. Target aud: General. ♦J.J. Jeffrey, pres & opns mgr; Jon VanHoogenstyn, gen mgr; Jon Van Hoogenstyn, progmg dir.

***WMEA(FM)**— April 1974: 90.1 mhz; 49 kw. 1,919 ft TL: N43 51 33 W70 42 43. Stereo. Hrs opn: 24
Rebroadcasts WMEH(FM) Bangor 100%.
65 Texas Ave., Bangor, 04401. Secondary address: 1450 Libson St., Lewiston 04240. Phone: (207) 874-6570. Fax: (207) 761-0318. E-mail: cbeck@mpbn.net Web Site: www.mpbn.net. Licensee: Maine Public Broadcasting Corp. (acq 6-23-92). Population served: 60,000 Natl. Network: NPR, PRI. Law Firm: Dow, Lohnes & Albertson. Format: Class, pub affrs, news/talk. News staff: 8; News: 20 hrs wkly. Target aud: 25-64. ♦Alexander G. Maxwell, Jr., COO, sr VP, gen mgr & mktg dir; P. James Dowe, Jr., pres; Christopher Amann, CFO; Charles Beck, stn mgr & progmg VP; Mary Mayo, dev VP & dev dir.

WPKQ(FM)—See North Conway, NH

WTHT(FM)—See Auburn

WYNZ(FM)—See Westbrook

WZAN(AM)— July 13, 1925: 970 khz; 5 kw-U, DA-N. TL: N43 36 19 W70 19 18. Hrs open: 420 Western Ave., S. Portland, 04106. Phone: (207) 774-4561. Fax: (207) 774-3788. E-mail: feedback@970wzan.com Web Site: www.970wzan.com. Licensee: Saga Communications of New England LLC. Group owner: Saga Communications Inc. (acq 6-23-93; $350,000 with WYNZ-FM Westbrook; FTR: 7-12-93). Natl. Network: CBS, CNN Radio. Natl. Rep: Katz Radio. Format: Talk. Target aud: 25-54 Males. ♦Cary Pahigian, pres & gen mgr; Chris McGorrill, opns.

Presque Isle

WBPW(FM)— September 1973: 96.9 mhz; 100 kw. 440 ft TL: N46 45 52 W67 59 23. Hrs open: 24 551 Main St., 04769-2450. Phone: (207) 769-6600. Fax: (207) 764-5274. E-mail: wbpw.radio@citcomm.com Licensee: Citadel Broadcasting Co. Group owner: Citadel Broadcasting Corp. (acq 4-26-00; grpsl). Population served: 150,000 Natl. Rep: Katz Radio. Format: Hot country. News staff: 1; News: 2 hrs wkly. Target aud: 25-54. Spec prog: NASCAR, American Country Countdown. ♦Lisa Miles, gen mgr & gen sls mgr; Mark Shaw, news dir.

WEGP(AM)— June 24, 1960: 1390 khz; 25,000 W-U, DA-N. TL: N46 39 17 W68 03 01. Hrs open: 24
WREM.
Box 4088, 04769. Phone: (207) 762-6700. Fax: (207) 762-3319.

E-mail: wegp@mfx.net Web Site: www.wegp.net. Licensee: Decelles/Smith Media Inc. (acq 9-18-00). Population served: 100,000 Natl. Network: Fox News Radio, Premiere Radio Networks, Talk Radio Network, Westwood One. Format: Talk, news. Target aud: Adults; Mature listeners over age 29. ♦Paul Decelles, pres; Patrick Patterson, gen mgr, opns mgr & progmg dir; Bonnie Pack, disc jockey.

***WMEM(FM)**— 1975: 106.1 mhz; 99 kw. 1,079 ft TL: N46 33 06 W67 48 38. Hrs open:
Rebroadcasts WMEH(FM) Bangor 100%.
65 Texas Ave., Bangor, 04401. Secondary address: 1450 Lisbon St., Lewiston 04240. Phone: (207) 874-6570. Phone: (800) 884-1717. Fax: (207) 942-2857. E-mail: cbeck@mpbn.net Web Site: www.mpbn.net. Licensee: Maine Public Broadcasting Corp. (acq 6-23-92). Natl. Network: NPR, PRI. Law Firm: Dow, Lohnes & Albertson. Format: Class, pub affrs, news/talk. ♦Alexander G. Maxwell, Jr., COO & sr VP; P. James Dowe, Jr., CEO & pres; Christopher Amann, CFO; Charles Beck, stn mgr & progmg VP; Deb Turner, dev VP.

WOZI(FM)— Feb 2, 1981: 101.9 mhz; 7.9 kw. Ant 1,207 ft TL: N46 32 51 W67 48 35. Stereo. Hrs opn: 24 551 Main St., 04769-2450. Phone: (207) 769-6600. Fax: (207) 764-5274. E-mail: wozi.radio@citcomm.com Licensee: Citadel Broadcasting Co. Group owner: Citadel Broadcasting Corp. (acq 4-26-01; grpsl). Population served: 150,000 Natl. Network: Westwood One. Natl. Rep: Katz Radio. Format: Classic rock, Rock. News staff: one; News: 1 hr wkly. Target aud: 25-54. ♦Lisa Miles, gen mgr; Chris O' Brien, progmg dir.

WQHR(FM)— 1981: 96.1 mhz; 95 kw. 1,309 ft TL: N46 32 55 W67 48 35. Stereo. Hrs opn: 24 551 Main St., 04769-2450. Phone: (207) 769-6600. Fax: (207) 764-5274. E-mail: wqhr.radio@citcomm.com Licensee: Citadel Broadcasting Co. Group owner: Citadel Broadcasting Corp. (acq 4-26-01; grpsl). Population served: 250,000 Natl. Rep: Katz Radio. Format: Hot adult contemp. News staff: 1; News: 2 hrs wkly. Target aud: 18-49. Spec prog: Bob & Sheri; Rick Dees Weekly Top 40; Backtracks USA. ♦Lisa Miles, gen mgr; Mark Shaw, progmg dir.

***WUPI(FM)**— July 26, 1973: 92.1 mhz; 17 w. -39 ft TL: N46 40 15 W68 01 00. Hrs open: 15 Box 64 Normal Hall, 04769. Phone: (207) 768-9711. Phone: (207) 768-9615. Fax: (207) 768-9617. Licensee: University of Maine Trustees. Population served: 5,000 Natl. Network: Westwood One. Format: Div. News staff: 2; News: 3 hrs wkly. Target aud: 16-35. Spec prog: Black 2 hrs, Russian 2 hrs, Kenyan 2 hrs, Indian 2 hrs wkly. ♦Dr. Nancy Hensel, pres; Tim Cramer, stn mgr.

Rockland

WMCM(FM)—Listing follows WRKD(AM).

WRKD(AM)— Oct 1, 1952: 1450 khz; 1 kw-U. TL: N44 06 22 W69 06 31. Stereo. Hrs opn: 24 150 Whitten Rd., August, 04330. Phone: (207) 623-9000. Fax: (207) 594-2234. Licensee: Clear Channel Broadcasting Licenses Inc. Group owner: Clear Channel Communications Inc. (acq 2-12-01; $3.5 million with co-located FM including two-year, $5,000 noncompete agreement). Population served: 30,600 Natl. Network: Westwood One. Natl. Rep: CYR Associates. Law Firm: Wilmer, Cutler & Pickering. Format: Sports. News staff: one; News: 11 hrs wkly. Target aud: 35 plus. ♦Kelly Slater, gen mgr; Rick Dugal, gen sls mgr; Don Shields, progmg dir.

WMCM(FM)—Co-owned with WRKD(AM). Apr 16, 1968: 103.3 mhz; 20.5 kw. 771 ft TL: N44 07 35 W69 08 18. Stereo. 24 Web Site: www.realcountry1033.com.100,000 Format: Country. News staff: one. Target aud: General. ♦D.J. McCoy, progmg dir; Elaine Knowlton, traf mgr; Don Shields, local news ed, political ed & sports cmtr; Peter K. Orne, rsch dir & edit dir.

Rumford

WLOB-FM— Nov 15, 1975: 96.3 mhz; 100 kw. Ant 1,433 ft TL: N44 34 56 W70 37 59. (CP: 36 kw, ant 1,453 ft. N44 15 06 W70 25 24). Stereo. Hrs opn: 779 Warren Ave., Portland, 04103. Phone: (207) 773-9695. Fax: (207) 761-4406. E-mail: wlob@yahoo.com Licensee: Atlantic Coast Radio L.L.C. (group owner; (acq 9-8-00; grpsl). Population served: 950,000 Format: News/talk. ♦Jon Van Hoogenstyn, gen mgr.

WTME(AM)— Aug 21, 1953: 780 khz; 10 kw-D, 18 w-N. TL: N44 30 53 W70 31 01. Hrs open:
Simulcast with WKTQ(AM) South Paris 100%.
PO Box 72, 243 Main St., Norway, 04268. Phone: (207) 743-5911. Fax: (207) 743-5913. E-mail: info@woxo.com Web Site: www.wtme.com. Licensee: Mountain Valley Broadcasting Inc. Group owner: Gleason Radio Group (acq 11-2-2000; $50,000). Population served: 100,000 Natl. Network: CNN Radio, Westwood One. Natl. Rep: CYR Associates. Law Firm: Womble Carlyle. Format: Relg, news. News staff: one. Target aud: General. ♦Richard Gleason, pres & gen mgr; Jeremy Rush, opns mgr & progmg dir.

Saco

WCYY(FM)—See Biddeford

WRED(FM)— July 18, 1982: 95.9 mhz; 4.1 kw. Ant 397 ft TL: N43 32 33 W70 24 17. Stereo. Hrs opn: 24 779 Warren Ave., Portland, 04103. Phone: (207) 773-9695. Fax: (207) 761-4406. E-mail: info@redhot959.com Web Site: www.redhot959.com. Licensee: Atlantic Coast Radio L.L.C. (group owner; acq 7-12-99; $1.15 million). Natl. Network: Westwood One. Natl. Rep: McGavren Guild. Law Firm: Rick Hayes. Wire Svc: AP Format: Hip-Hop, rhythm and blues, CHR. Target aud: 18-34. ♦J.J. Jeffrey, pres; Lisa Menconi, CFO; John Van Hoogenstyn, gen mgr; Buzz Bradley, stn mgr & opns VP; Jon Van Hoogenstyn, sls dir; Gene Terwiliger, chief of engrg; Linda Petrin, chief of engrg & traf mgr.

WVAE(AM)—See Biddeford

Sanford

WPHX(AM)— Nov 9, 1957: 1220 khz; 1 kw-D, 234 w-N. TL: N43 25 53 W70 45 44. Hrs open: 24 One Washington St., Dover, NH, 03820. Phone: (603) 749-5900. Fax: (603) 749-0088. Web Site: fnxradio.com. Licensee: FNX Broadcasting LLC. Group owner: Phoenix Media Communications Group (acq 5-17-99; $1.025 million with co-located FM). Population served: 20000 Format: Sports. ♦Sam Pselfle, gen mgr.

WPHX-FM— Oct 10, 1975: 92.1 mhz; 1.2 kw. 525 ft TL: N43 35 24 W70 22 20. Stereo. 24 Web Site: www.fnxradio.com. Natl. Rep: McGavren Guild. Format: Alternative. News staff: one. Target aud: General. Spec prog: Jazz 6 hrs, talk 2 hrs wkly. ♦Gary Kurtz, gen mgr; Michael Snow, mktg dir; Keith Dakin, prom dir & progmg dir; Chris Hall, engrg mgr & chief of engrg.

***WSEW(FM)**— Mar 2, 1992: 88.5 mhz; 100 w. 387 ft TL: N43 25 11 W70 48 09. Stereo. Hrs opn: Box 398, New Durham, NH, 03855. Phone: (603) 859-9170. Fax: (603) 859-8172. E-mail: wsew@wsew.org Web Site: www.wsew.org. Licensee: Word-Radio Educational Foundation. Natl. Network: USA. Format: Relg. ♦Sharon Malone, gen mgr.

Scarborough

WBQW(FM)— 1960: 106.3 mhz; 3 kw. 299 ft TL: N43 35 24 W70 22 20. Stereo. Hrs opn:
Rebroadcasts WBQQ(FM) Kennebunk 90%.
169 Port Rd., Kennebunk, 04043. Phone: (207) 797-0780. Fax: (207) 967-8671. E-mail: wbach@wbach.fm Web Site: www.wbach.fm. Licensee: Nassau Broadcasting III L.L.C. Group owner: Nassau Broadcasting Partners L.P. (acq 4-6-2004; grpsl). Population served: 200,000 Natl. Network: AP Radio. Format: Class. Target aud: 25-54; upscale, affluent, management, professionals. Spec prog: Jazz 5 hrs wkly. ♦Pat Collins, pres, pres & gen mgr; Stan Manning, opns mgr; Scott Hooper, progmg dir.

Searsport

WFZX(FM)— Oct 10, 1994: 101.7 mhz; 2.65 kw. Ant 1,004 ft TL: N44 34 51 W68 53 47. Hrs open: 6 AM-midnight 184 Target Industrial Cir., Bangor, 04401. Phone: (207) 947-9100. Fax: (207) 942-8039. E-mail: susanfaloon@clearchannel.com Web Site: www.wfzxfm.com. Licensee: CC Licenses LLC. (group owner; (acq 10-8-98; $265,000). Format: Classic rock. ♦Larry Julius, gen mgr, gen mgr & mktg dir; Jeffrey Pierce, stn mgr & progmg dir; Josh Scroggins, gen sls mgr; Susan Faloon, news dir; Stacey Brann, chief of engrg; Amy Rowe, traf mgr; Lynn Nadeau, traf mgr.

Skowhegan

WFMX(FM)— September 1989: 107.9 mhz; 6 kw. Ant 666 ft TL: N44 42 46 W69 43 36. Stereo. Hrs opn: Box 159, 04976. Phone: (207) 474-5171. Fax: (207) 474-2399. E-mail: maine.radio@verizon.net Licensee: Mountain Wireless Inc. (group owner; (acq 11-20-97; $222,355). Format: Talk. Target aud: 25-54; baby boomers who grew up with Top-40 radio. ♦Jay Hansen, gen mgr & opns mgr.

WSKW(AM)— 1956: 1160 khz; 10 kw-D, 1 kw-N. TL: N44 44 43 W69 41 36. Hrs open: Box 159, 04976. Secondary address: 208 Middle Rd. 04976. Phone: (207) 474-5171. Fax: (207) 474-2399. E-mail: maine.radio@verizon.net Licensee: Mountain Wireless Inc. (group owner; (acq 1999; $1.6 million with WCTB(FM) Fairfield). Population served: 70,000 Natl. Network: ESPN Radio. Format: Sports. Target aud: 12 plus; loc sports fans. ♦Jay Hanson, gen mgr & opns mgr.

WTOS-FM— Nov 13, 1969: 105.1 mhz; 50 kw. 2,431 ft TL: N45 01 54 W70 18 50. Stereo. Hrs opn: 24 150 Whitten Rd., Augusta, 04330. Phone: (207) 623-9000. Fax: (207) 623-9035. E-mail: reverend@clearchannel.com Web Site: www.wtosfm.com. Licensee: Capstar TX L.P. Group owner: Clear Channel Communications Inc. (acq 1-18-01; grpsl). Format: Active Rock. News staff: one; News: 3 hrs wkly. Target aud: 18-49. ♦Kelly Slater, mktg mgr; Steve Smith, progmg dir.

South Paris

WKTQ(AM)— Oct 28, 1955: 1450 khz; 1 kw-U. TL: N44 13 16 W70 31 43. Hrs open: 24
Simulcast with WTME(AM) Rumford 100%.
PO Box 72, 243 Main St., Norway, 04268. Phone: (207) 743-5911. Fax: (207) 743-5913. E-mail: dick@gleasonmedia.com Web Site: www.wtme.com. Licensee: Mountain Valley Broadcasting Inc. Group owner: Gleason Radio Group (acq 7-27-76). Population served: 20,000 Natl. Network: CNN Radio, Westwood One. Natl. Rep: CYR Associates. Law Firm: Womble, Carlyle, Sandridge & Rice, PLLC. Format: Relg, news. News staff: one; News: 126 hrs wkly. Target aud: General. ♦Richard D. Gleason, pres & gen mgr; Victor Hodgkins, stn mgr; Jeremy Rush, opns dir.

WOXO-FM—See Norway

Standish

***WSJB-FM**— Apr 1, 1984: 91.5 mhz; 360 w. 85 ft TL: N43 49 32 W70 29 03. Stereo. Hrs opn: 6 AM-midnight St. Joseph's College, 278 White's Bridge Rd., 04084. Phone: (207) 892-6766. Fax: (207) 893-7873. Licensee: Trustees of St. Joseph's College. Population served: 10,000 Format: CHR. News: 3 hrs wkly. Target aud: College students. Spec prog: Class 2 hrs, C&W 3 hrs, jazz 3 hrs wkly. ♦Bill Yates, gen mgr.

Thomaston

WBQX(FM)— May 29, 1992: 106.9 mhz; 29.5 kw. 633 ft TL: N44 06 30 W69 09 28. Stereo. Hrs opn: 24
Rebroadcasts WBQQ(FM) Kennebunk 100%.
119 Tillson Ave., Suite 101A, Rockland, 04841. Phone: (207) 594-9283. Fax: (207) 594-1620. E-mail: jlynch@nassaubroadcasting.com Web Site: www.wbach.fm. Licensee: Nassau Broadcasting III L.L.C. Group owner: Nassau Broadcasting Partners L.P. (acq 4-6-2004; grpsl). Population served: 170,000 Rgnl rep: Kettell-Carter. Law Firm: Smithwick & Belendiuk. Format: Class. News: 4 hrs wkly. Target aud: 35 plus; affluent, upscale adults. Spec prog: Jazz 2 hrs, children one hr wkly. ♦Louis F. Mercatanti, pres; Pat Collins, VP; Joe Lynch, gen mgr; Scott Hooper, progmg dir & mus dir.

Topsham

WJJB-FM— 1993: 95.5 mhz; 6 kw. 456 ft TL: N43 54 12 W70 02 13. Hrs open: 24 779 Warren Ave., Portland, 04103. Phone: (207) 773-9695. Fax: (207) 761-4406. E-mail: morningjab@yahoo.com Web Site: www.bigjab.com. Licensee: Atlantic Coast Radio L.L.C. (group owner; acq 9-30-99). Population served: 700,000 Natl. Network: Westwood One. Law Firm: Arter & Hadden. Format: Sports, talk. News staff: one; News: 4 hrs wkly. Target aud: 25-49; middle class, active lifestyle with discretionary income. ♦Carla Thibodeau, chmn; J.J. Jeffrey, pres; Jon Van Hoogenstyn, gen mgr; Bruce Biett, stn mgr; David Schumacher, opns mgr & progmg mgr; Gene Terwilliger, engrg dir & chief of engrg.

Van Buren

WCXV(FM)—Not on air, target date: unknown: 98.1 mhz; 6 kw. Ant 89 ft TL: N47 10 04 W67 57 43. Hrs open: 152 E. Green Ridge Rd., Caribou, 04736-3737. Phone: (207) 473-7513. Fax: (207) 472-3221. Licensee: Canxus Broadcasting Corp. ♦Dennis Curley, pres.

Veazie

WNZS(AM)— August 2002: 1340 khz; 1 kw-D, 630 w-N. TL: N44 51 10 W68 40 44. Hrs open: 24 Box 8526, Bangor, 04402. Secondary address: 379 Riverside Dr, Eddington 04428. Phone: (207) 947-9697. Fax: (207) 989-5251. Licensee: Waterfront Communications Inc. Group owner: Daniel F. Priestley Stns. Population served: 150,000 Natl. Network: CNN Radio, Salem Radio Network, Talk Radio Network. Natl. Rep: Commercial Media Sales. Rgnl rep: Cyr Association Law Firm: Fletcher, Healld & Hildreth. Format: News. News staff: 2; News: 15 hrs wkly. Target aud: 25-54; 35-64; adults in metro Banger area. Spec prog: Maine news. ♦Jocelynn Priestley, stn mgr.

WWNZ(AM)— August 2004: 1400 khz; 1 kw-D, 810 w-N. TL: N44 50 50 W68 40 48. Hrs open: 24 Box 8526, Bangor, 04402-8526. Secondary address: 379 Riverside Dr., Eddington 04428. Phone: (207) 947-9697. Fax: (207) 989-5251. E-mail: wnzproduction@aol.com Licensee: Waterfront Communications Inc. Group owner: Daniel F. Priestley Stns. Population served: 150,000 Natl. Network: USA, Fox News Radio. Natl. Rep: Commercial Media Sales. Rgnl rep: Cyr Associates Law Firm: Fletcher, Heald & Hildreth. Format: News/talk. News: 50+. Target aud: 25-54; 35-64; Banger metro area adults 25 plus. ♦Daniel F. Priestley, pres; Jocelynn Priestley, stn mgr.

Waterville

WEBB(FM)—Listing follows WTVL(AM).

***WMEW(FM)**— November 1983: 91.3 mhz; 3 kw. 299 ft TL: N44 29 23 W69 39 05. Stereo. Hrs opn: 24
Rebroadcasts WMEH(FM) Bangor 100%.
65 Texas Ave., Bangor, 04401. Secondary address: 1450 Lisbon St., Lewiston 04240. Phone: (207) 783-9101. Phone: (207) 941-1010. Fax: (207) 942-2857. E-mail: cbeck@mpbn.net Web Site: www.mpbn.net. Licensee: Maine Public Broadcasting Corp. (acq 6-23-92; FTR: 7-13-92). Natl. Network: NPR, PRI. Law Firm: Dow, Lohnes & Albertson. Format: Class, pub affrs, news/talk. News staff: 8; News: 20 hrs wkly. Target aud: 25-64. ♦Alexander G. Macwell, Jr., COO; Christopher Amann, CFO & mktg dir; P. James Dowe, Jr., CEO, pres & CFO; Alexander G. Maxwell, Jr., sr VP; Mary Mayo, dev VP & dev dir; Charles Beck, progmg dir.

***WMHB(FM)**— Oct 1, 1974: 89.7 mhz; 110 w. 98 ft TL: N44 33 57 W69 39 49. Stereo. Hrs opn: 6 AM-2 AM Colby College, 4000 Mayflower Hill Dr., 04901-8840. Phone: (207) 859-5451. Phone: (207) 854-5450. Fax: (207) 859-5455. E-mail: wmhb@colby.edu Web Site: www.colby.edu/wmhb. Licensee: Mayflower Hill Broadcasting Corp. Population served: 40,000 Format: Div. News: 2.5 hrs wkly. Target aud: 5-100; we cater to everyone. Spec prog: Folk 12 hrs, Black 8 hrs, jazz 8 hrs, world 6 hrs wkly. ♦Alan Dzarowski, gen mgr; Miranda Geranlos, dev dir; Justine Ludwig, prom dir; Lena Barady, mus dir; Tim Williams, news dir.

WTVL(AM)— June 19, 1946: 1490 khz; 1 kw-U. TL: N44 33 52 W69 36 39. Hrs open: Box 5070, Augusta, 04330. Secondary address: 52 Western Ave., Augusta 04330. Phone: (207) 623-4735. Fax: (207) 626-5948. E-mail: b98.5@midmaine.com Web Site: www.b985.fm. Licensee: Citadel Broadcasting Co. Group owner: Citadel Broadcasting Corp. (acq 4-26-2001; grpsl). Population served: 18,192 Format: Country favorites. Target aud: 25-54. ♦Farid Suleman, CEO; Bob Proffitt, pres; Julie Crocker, gen sls mgr; Al Perry, rgnl sls mgr; Mac Dickson, prom dir; Andy Capwell, progmg dir; Renee Nelson, news dir & pub affrs dir; Bob Perry, chief of engrg.

WEBB(FM)—Co-owned with WTVL(AM). Mar 26, 1968: 98.5 mhz; 50 kw. 305 ft TL: N44 33 52 W69 36 39. Stereo. 24 Web Site: www.b985.fm.400,000 Law Firm: Fisher, Wayland, Cooper, Leader & Zaragoza. News staff: one.

Westbrook

WBQW(FM)—See Scarborough

WJAE(AM)— Nov 8, 1959: 1440 khz; 5 kw-D, 1 kw-N, DA-1. TL: N43 40 50 W70 22 47. Hrs open: 779 Warren Ave., Portland, 04103. Phone: (207) 773-9695. Fax: (207) 761-4406. E-mail: shoe@thebigjab.com Web Site: www.thebigjab.com. Licensee: Atlantic Coast Radio L.L.C. (group owner; (acq 9-99). Population served: 150,000 Natl. Network: Westwood One. Rgnl rep: Roslin Format: Sports, talk. News staff: News progmg 6 hrs wkly Target aud: General; male. ♦Bruce Biette, gen mgr; Dave Shoe, opns mgr.

WYNZ(FM)— February 1976: 100.9 mhz; 25 kw. Ant 305 ft TL: N43 41 26 W70 19 05. Stereo. Hrs opn: 420 Western Ave., South Portland, 04106. Phone: (207) 774-4561. Fax: (207) 774-3788. E-mail: bighits@y1009.com Web Site: www.y1009.com. Licensee: Saga Communications of New England LLC. Group owner: Saga Communications Inc. (acq 6-23-93; $350,000 with WYNZ(AM) Portland; FTR: 7-12-93) Natl. Network: CNN Radio. Natl. Rep: Katz Radio. Format: Oldies. Target aud: 25-54. ♦Cary Pahigian, pres & gen mgr; Chris McGorrill, opns mgr, mktg VP & prom mgr; Tina Segerstrom, sls VP; Tina Seuerstrom, gen sls mgr; Randi Kirshbaun, progmg dir; Jeffrey Wade, news dir; Andy Armstrong, engrg VP; Michael Bray, traf mgr.

Winslow

***WWWA(FM)**— Apr 23, 1999: 95.3 mhz; 12 kw. Ant 672 ft TL: N44 42 48 W69 43 39. Stereo. Hrs opn: 24 160 Bangor St., Augusta, 04330. Phone: (207) 622-1340. Fax: (207) 623-2874. E-mail: wwwa@adelphia.net Web Site: lightoflife.info/wwwa.htm. Licensee: Light of Life Ministries Inc. Population served: 130,000 Format: Christian. ♦Denise LaFountain, stn mgr; Ryan Gagne, gen mgr & progmg dir.

Winter Harbor

WNSX(FM)— 1999: 97.7 mhz; 50 kw. Ant 489 ft TL: N44 33 13 W68 05 40. Stereo. Hrs opn: Box 1171, Ellsworth, 04605. Phone: (207) 667-0002. Fax: (207) 667-0627. Licensee: Stony Creek Broadcasting LLC Group owner: Clear Channel Communications Inc. (acq 9-30-2005; $800,000). Wire Svc: AP Format: Soft classic rock. News: one hr wkly. Target aud: Adults; 25+. ♦Mark Osborne, gen mgr; Bill Da Butler, opns mgr; Irene Hafford, gen sls mgr; Bill Ducharme, chief of engrg; Natalie Knox, sls.

Yarmouth

***WYAR(FM)**— Nov 16, 1998: 88.3 mhz; 1 kw horiz. Ant 79 ft TL: N43 45 56 W70 08 27. Hrs open: 24 Box 219, Heritage Radio Society Inc., Cousins St., 04096. Phone: (207) 847-3169. E-mail: wyar@maine.rr.com Web Site: www.wyar.org. Licensee: Heritage Radio Society Inc. Population served: 250,000 Format: Btfl music, big band, oldies, class, educ. Target aud: General; senior citizens & young people. ♦Gary D. King Sr., CEO, pres, opns dir, pub affrs dir & chief of engrg; James Brown, VP & gen mgr.

York Center

WUBB(FM)— June 1987: 95.3 mhz; 1.4 kw. 682 ft TL: N43 13 24 W70 41 35. Stereo. Hrs opn: 24 815 Lafayette Rd., Portsmouth, NH, 03801. Phone: (603) 436-7300. Fax: (603) 430-9415. E-mail: ianhorne@clearchannel.com Web Site: www.wubbfm.com. Licensee: Clear Channel Communications Group owner: Clear Channel Communications Inc. (acq 8-30-00; grpsl). Natl. Rep: Katz Radio. Format: Country. News staff: one; News: 7 hrs wkly. Target aud: 25-54. ♦Ian Horne, progmg dir.

Maryland

Aberdeen

WAMD(AM)— May 1, 1957: 970 khz; 500 w-U, DA-2. TL: N39 30 35 W76 11 38. Hrs open: 24 400 Miob Ln., 21001. Phone: (410) 272-4400. Fax: (410) 575-6890. Licensee: First Broadcasting Investment Partners LLC (group owner; (acq 6-13-2005; grpsl). Population served: 30,000 Natl. Network: ABC. Format: Oldies. News: 24 hrs wkly. Target aud: 35+; primarily female. ♦Steve Clendenin, opns dir; Chuck McKay, gen sls mgr; Steve Clendenin, progmg dir; Carol Powel, traf mgr.

Annapolis

WBIS(AM)— Jan 10, 1947: 1190 khz; 10 kw-D, DA. TL: N38 56 32 W76 28 54. (CP: COL Garrison. 5 kw-D, 7 kw-N, DA-2. TL: N39 24 29 W76 46 32). Stereo. Hrs opn: 1610 West St., Suite 209, 21401. Phone: (410) 269-0700. E-mail: wbis@businessradio.net Web Site: www.wbis1190.com. Licensee: Nations Radio L.L.C. (acq 3-31-98; $400,000). Population served: 6,000,000 Law Firm: Reynolds & Manning. Format: Business talk. ♦Alan Pendelton, gen mgr.

***WFSI(FM)**— May 16, 1960: 107.9 mhz; 50 kw. 500 ft TL: N38 59 45 W76 39 27. Stereo. Hrs opn: 24 918 Chesapeake Ave., 21403. Phone: (410) 268-6200. Fax: (410) 268-0931. Licensee: Family Stations Inc. (group owner; acq 1-7-72). Natl. Network: Family Radio. Format: Relg, educ. ♦W.A. Sadlier, stn mgr.

WLZL(FM)— 1947: 99.1 mhz; 50 kw. Ant 459 ft TL: N38 59 46 W76 39 26. Stereo. Hrs opn: 24 4200 Parliament Pl., Suite 300, Lanham, 20706. Phone: (301) 306-0991. Fax: (301) 731-0431. Web Site: www.elzol991.com. Licensee: CBS Radio East Inc. Group owner: Infinity Broadcasting Corp. (acq 11-13-98; grpsl). Population served: 5,000,000 Natl. Rep: CBS Radio. Law Firm: Leventhal, Senter & Lerman. Format: Sp. News staff: one; News: 5 hrs wkly. Target aud: Adults 18-49; upscale professionals. ♦Michael Hughes, gen mgr; Areacely Rivera, progmg dir.

WNAV(AM)— 1949: 1430 khz; 5 kw-D, 1 kw-N, DA-N. TL: N38 59 00 W76 31 21. Hrs open: Box 6726, 21401. Phone: (410) 263-1430. Fax: (410) 268-5360. E-mail: stevehopp@wnav.com Web Site: www.wnav.com. Licensee: Sajak Broadcasting Corp. (acq 5-26-98; $2.2 million). Population served: 500,000 Natl. Network: CBS Radio, Westwood One. Format: Adult contemp, full service. News staff: 2. Target aud: 35 plus. Spec prog: Baltimore Orioles baseball, Naval Academy sports. ♦Patrick L. Sajak, pres; Steve Hopp, VP & gen mgr; Bill Lusby, mus dir.

WRNR-FM—Grasonville, Apr 1, 1980: 103.1 mhz; 6 kw. 328 ft TL: N38 56 37 W76 10 43. Stereo. Hrs opn: 112 Main St., 21401. Phone: (410) 626-0103. Fax: (410) 267-7634. E-mail: info@wrnr.com Web Site: www.wrnr.com. Licensee: Empire Broadcasting System Inc. (acq 6-17-97; $2.15 million). Population served: 2,000,000 Natl. Network: CBS Radio. Law Firm: Leventhal, Senter & Lerman. Format: Progressive /diversified. News staff: one; News: 2 hrs wkly. Target aud: 25-54; adults. ♦Don Cavaleri, chmn; Jon Peterson, opns mgr & mus dir; Steve Kingston, CEO, gen mgr & pub affrs dir; Judy Buddensick, sls.

Baltimore

WBAL(AM)— Nov 2, 1925: 1090 khz; 50 kw-U, DA-N. TL: N39 22 33 W76 46 21. Hrs open: 24 3800 Hooper Ave., 21211. Phone: (410) 467-3000. Fax: (410) 338-6483. Web Site: www.wbal.com. Licensee: WBAL Div., The Hearst Corp. (acq 1-14-35). Population served: 2,100,000 Natl. Network: CBS Radio. Natl. Rep: D & R Radio. Law Firm: Brooks, Pierce, McLendon, Humphrey & Leonard. Format: News/talk, sports. Target aud: 25-54. ♦Bob Cecil, VP & sls dir; Edward C. Kiernan, gen mgr; Jeffrey Beauchamp, stn mgr; Kerry Plackmayer, opns mgr; Steve Hartman, natl sls mgr; Arthur Hawkins, rgnl sls mgr; Alison Jessie, prom mgr; Mark Miller, news dir; Hank Volpe, chief of engrg; Donna Valentine, traf mgr.

WIYY(FM)—Co-owned with WBAL(AM). Dec 7, 1958: 97.9 mhz; 13.5 kw. 945 ft TL: N39 20 05 W76 39 03. Stereo. Web Site: www.98online.com. Format: AOR. Target aud: 18-49. ♦Hughes Jean, gen sls mgr; Steve Hartman, rgnl sls mgr; Lori Smyth, prom mgr; Dave Hill, progmg dir. Co-owned TV: WBAL-TV affil

WBGR(AM)— July 27, 1955: 860 khz; 2.5 kw-D, 66 w-N, DA-2. TL: N39 18 43 W76 29 26. Hrs open: 918 Chesapeake, 4th Flr., Annapolis, 21403. Phone: (410) 821-9000. Fax: (410) 268-0931. Licensee: Family Stations Inc. Group owner: Infinity Broadcasting Corp. (acq 3-2-2005; $7.5 million with WBMD(AM) Baltimore). Population served: 905,759 Format: Gospel, black. Target aud: 18-49. ♦Herold Camping, gen mgr.

WBIS(AM)—See Annapolis

***WBJC(FM)**— Apr 6, 1951: 91.5 mhz; 50 kw. 500 ft TL: N39 23 11 W76 43 52. Stereo. Hrs opn: 6776 Riestertown Rd., 21215. Phone: (410) 462-8444. Fax: (410) 333-7016. E-mail: info@wbjc.com Web Site: www.wbjc.com. Licensee: Baltimore City Community College.

(acq 4-22-91). Population served: 200,000 Natl. Network: PRI. Format: Class. ♦Cary Smith, gen mgr; Kati Kershaw, opns dir & opns mgr; Jim Ward, dev dir.

WBMD(AM)— Dec 7, 1947: 750 khz; 1 kw-D. TL: N39 19 26 W76 32 56. Hrs open: Sunrise-sunset 918 Chesapeake Ave., Annapolis, 21203. Phone: (410) 821-9000. Fax: (410) 268-0931. Licensee: Family Stations Inc. Group owner: Infinity Broadcasting Corp. (acq 3-2-2005; $7.5 million with WBGR(AM) Baltimore). Format: Relg. Target aud: 12 plus. Spec prog: Ger one hr, Pol 2 hrs, Greek 2 hrs, Lithuanian one hr wkly. ♦Herold Camping, gen mgr.

WCAO(AM)— May 8, 1922: 600 khz; 5 kw-U, DA-1. TL: N39 25 47 W76 45 42. Hrs open: 711 W. 40th St., Suite 350, 21211. Phone: (410) 366-7600. Fax: (410) 467-0011. E-mail: yourvoice@heaven600.com Web Site: www.heaven600.com. Licensee: Citicasters Licenses L.P. Group owner: Clear Channel Communications Inc. (acq 5-99; grpsl). Population served: 2,000,000 Format: Contemporary Black gospel. ♦Jim Dolan, VP, gen mgr & mktg mgr; Bill Hopkinson, sls dir & gen sls mgr; Lee Michaels, progmg dir; Danielle Brown, mus dir & news dir.

WCBM(AM)— 1924: 680 khz; 10 kw-D, 5 kw-N, DA-2. TL: N39 24 30 W76 46 34. Hrs open: 24 Hilton Plaza, 1726 Reisterstown Rd., Suite 117, 21208. Phone: (410) 580-6800. Fax: (410) 580-6810. E-mail: am680@wcbm.com Web Site: www.wcbm.com. Licensee: M-10 Broadcasting (acq 9-5-95). Population served: 1,500,000 Natl. Network: CBS. Law Firm: Fisher, Wayland, Cooper, Leader & Zaragoza L.L.P. Format: Talk. News staff: 3; News: 11 hrs wkly. Target aud: 25-54; informed adults with major purchasing power. ♦Nick Mangione Jr., sr VP; Bob Pettit, gen mgr; Marc Beavin, gen sls mgr; Niles Seaberg, progmg dir; Eddie Applefeild, prom.

***WEAA(FM)**— Jan 10, 1977: 88.9 mhz; 12.5 kw. Ant 220 ft TL: N39 20 31 W76 35 13. Stereo. Hrs opn: 24 1700 East Coldspring Ln., 21251. Phone: (443) 885-3564. Fax: (443) 885-8206. E-mail: weaa@moac.morgan.edu Web Site: www.weaa.org. Licensee: Morgan State University. Population served: 1,100,000 Natl. Network: NPR. Law Firm: Schwartz, Woods & Miller. Wire Svc: AP Format: Jazz, news/talk. News staff: one; News: 10 hrs wkly. Target aud: 25-49; 85% Black. Spec prog: Urban oldies 5 hrs, Caribbean 7 hrs, Africian world 4 hrs, gospel 13 hrs, hip hop 5 hrs wkly. ♦LaFontaine Oliver, gen mgr; Sandi Mallony, progmg dir & progmg mgr.

WERQ-FM— 1960: 92.3 mhz; 37 kw. 571 ft TL: N39 20 20 W76 40 02. Stereo. Hrs opn: 24 1705 Whitehead Rd., 21207. Phone: (410) 332-8200. Fax: (410) 944-7182. Web Site: www.92qjams.com. Licensee: Radio One Licenses LLC. Group owner: Radio One Inc. (acq 6-21-93; $9 million with co-located AM; FTR: 7-19-93). Population served: 325,000 Law Firm: Arent, Fox, Kintner, Plotkin & Kahn. Format: Urban contemp. Target aud: 18-34; young adults. ♦Alfred Liggins, CEO & prom dir; Mary Catherine Sneed, COO; Howard Mazer, gen mgr; Karl Goehring, chief of engrg; Victor Starr, traf mgr.

WOLB(AM)—Co-owned with WERQ-FM. Nov 25, 1947: 1010 khz; 1 kw-D, 27 w-N. TL: N39 16 38 W76 37 59.24 Phone: (410) 907-0401. Fax: (410) 944-1047.21,000 Format: News/talk. News: 5 hrs wkly. Target aud: 35 plus; African American adults. Spec prog: Relg 2 hrs, Sp 2 hrs wkly. ♦Dave Wilner, stn mgr & traf mgr.

WHFS(FM)—See Catonsville

WJFK(AM)— June 8, 1922: 1300 khz; 5 kw-U, DA-2. TL: N39 20 00 W76 46 13. Hrs open: 24
Rebroadcasts WJFK-FM Manassas, VA 77%.
600 Washington Ave., Suite 201, Towson, 21204. Phone: (410) 825-5400. Fax: (410) 821-5482. Web Site: www.1300wjfk.com. Licensee: CBS Radio WLIF-AM Inc. (group owner; Infinity Broadcasting Corp. (acq 5-29-89; $32 million with co-located FM; FTR: 4-24-89) Population served: 460,000 Natl. Network: Westwood One. Natl. Rep: CBS Radio. Law Firm: Leventhal, Senter & Lerman. Format: Talk, sports. Target aud: 18-49; men. ♦Mel Karmazin, pres; Bob Philips, opns mgr.

WLIF(FM)—Co-owned with WJFK(AM). Dec 24, 1970: 101.9 mhz; 13.5 kw. 960 ft TL: N39 25 02 W76 33 23. Stereo. Web Site: www.1019litefm.com.905,759 Format: Lite adult contemp. Spec prog: Jazz 8 hrs wkly.

WNST(AM)—See Towson

WPOC(FM)— 1959: 93.1 mhz; 16 kw. Ant 860 ft TL: N39 17 13 W76 45 16. Stereo. Hrs opn: 711 W. 40th St., Suite 350, 21211. Phone: (410) 366-7600. Fax: (410) 235-3899. E-mail: kboesen@wpoc.com Web Site: www.wpoc.com. Licensee: Citicasters Licenses L.P. Group owner: Clear Channel Communications Inc. (acq 4-29-99; grpsl). Population served: 2,075,600 Format: Country. ♦Jim Dolan, VP, gen mgr & mktg dir; Gary LaFrance, natl sls mgr; Lew Munza, mktg mgr; Steve Reus, prom dir & prom mgr; Ken Boesen, progmg dir.

WQSR(FM)— Dec 15, 1947: 102.7 mhz; 50 kw. 436 ft TL: N39 23 11 W76 43 52. Stereo. Hrs opn: 1423 Clarkview Rd., Suite 100, Towson, 21204-3913. Phone: (410) 825-1000. Fax: (410) 823-0816. Web Site: www.1027jackfm.com. Licensee: CBS Radio Inc. of Chesapeake. Group owner: Infinity Broadcasting Corp. (acq 11-13-98; grpsl). Format: Mixed/automated. ♦Jason Kidd, progmg dir.

WRBS(AM)— Mar 1, 1941: 1230 khz; 1 kw-U. TL: N39 18 58 W76 36 03. Stereo. Hrs opn: 24 3600 Georgetown Rd., 21227. Phone: (410) 247-4100. Fax: (410) 247-4533. E-mail: AM1230@wrbs.com Web Site: www.wrbsam.com. Licensee: WRBS-AM LLC Group owner: Salem Communications Corp. (acq 12-22-2006; $3.25 million). Population served: 8,000,000 Law Firm: Davis Wright Tremaine LLP. Format: Christian. ♦Joe Norris, opns mgr.

WRBS-FM— Aug 1, 1964: 95.1 mhz; 50 kw. 499 ft TL: N39 15 21 W76 40 29. Stereo. Hrs opn: 24 3600 Georgetown Rd., 21227. Phone: (410) 247-4100. Fax: (410) 247-4533. E-mail: info@wrbs.com Web Site: www.wrbs.com. Licensee: Peter and John Radio Fellowship Inc. (acq 9-64). Population served: 2,00,000 Format: Contemp Christian. News staff: one; News: 6 hrs wkly. ♦Steven D. Lawhon, gen mgr; David Paul, progmg dir.

WSMJ(FM)— 1949: 104.3 mhz; 50 kw. 420 ft TL: N39 25 46 W76 27 01. Stereo. Hrs opn: 711 W. 40th St., Suite 350, 21211. Phone: (410) 366-7600. Fax: (410) 235-3899. E-mail: wsmj@smoothjazz1043.com Web Site: www.smoothjazz1043.com. Licensee: Citicasters Licenses L.P. Group owner: Clear Channel Communications Inc. (acq 5-4-99; grpsl). Population served: 1,300,000 Natl. Rep: Katz Radio. Format: Smooth jazz. Target aud: 25-44. ♦Jim Dolan, VP, gen mgr & mktg mgr; Bill Hopkinson, sls dir; Chuck Allen, gen sls mgr; Lori Lewis, progmg dir.

WWIN(AM)— 1951: 1400 khz; 1 kw-U. TL: N39 19 21 W76 36 33. Stereo. Hrs opn: 24 1705 Whitehead Rd., 21207. Phone: (410) 907-0401. Fax: (410) 944-1047. Licensee: Radio One Licenses LLC. Group owner: Radio One Inc. (acq 1-23-92; $7.5 million with WWIN-FM Glen Burnie). Population served: 905,759 Natl. Rep: D & R Radio. Rgnl rep: Allied Radio Partners. Law Firm: Verner, Liipfert, Bernhard, McPherson & Hand. Format: Gospel. News staff: one; News: one hr wkly. Target aud: 25-54; Black, relg.

WWIN-FM—See Glen Burnie

WWMX(FM)— 1960: 106.5 mhz; 7.4 kw. 1,217 ft TL: N39 20 10 W76 38 59. Stereo. Hrs opn: 24 1423 Clarkview Rd., Suite 100, 21209. Phone: (410) 825-1065. Fax: (410) 321-4548. E-mail: dave.labrozzi@infinitybroadcasting.com Web Site: www.mix1065.fm. Licensee: CBS Radio Stations Inc. Group owner: Infinity Broadcasting Corp. Population served: 905,759 Format: Adult contemp. Target aud: 25-54. ♦Tracy Brandys, gen mgr; Dave Labrozzi, progmg dir. Co-owned TV: WJZ-TV

***WYPR(FM)**— May 23, 1979: 88.1 mhz; 10 kw. 360 ft TL: N39 19 53 W76 39 28. Stereo. Hrs opn: 2216 N. Charles St., 21218. Phone: (410) 235-1660. Fax: (410) 235-1161. E-mail: tbrandon@wypr.org Web Site: www.wypr.org. Licensee: WYPR License Holding LLC (acq 1-16-02). Natl. Network: NPR, PRI. Format: Jazz, news/talk. ♦Anthony Brandon, pres & gen mgr; Andy Bienstock, progmg dir.

Bel Air

***WHFC(FM)**— 1972: 91.1 mhz; 1.10 kw. 226 ft TL: N39 33 22 W76 16 48. Stereo. Hrs opn: 24 Harford Community College, 401 Thomas Run Rd., 21015-1698. Phone: (410) 836-4151. Phone: (410) 836-4305. Fax: (410) 836-4180. E-mail: whfc@harford.edu Web Site: www.whfc911.org. Licensee: Harford Community College. Population served: 250,000 Format: Var. Target aud: 24-42; upwardly mobile professionals. Spec prog: AAA 15 hrs, class 18 hr, jazz 18 hrs, Christian 6 hrs, Americain Indian 3 hrs wkly. ♦Gary Helton, gen mgr.

Berlin

WOCQ(FM)— June 25, 1981: 103.9 mhz; 3 kw. 328 ft TL: N38 22 58 W75 18 58. Stereo. Hrs opn: 24 20200 DuPont Blvd., Georgetown, 19947-3105. Phone: (410) 641-0001. Fax: (410) 641-1294. Web Site: www.oc104.com. Licensee: Great Scott Broadcasting. (group owner; acq 11-7-97; $2.775 million). Population served: 228,000 Format: Rap, hip hop. News staff: one; News: 4 hrs wkly. Target aud: 18-49. ♦Cathy Deighan, gen mgr & opns mgr.

Bethesda

WMMJ(FM)— Nov 12, 1961: 102.3 mhz; 2.9 kw. 480 ft TL: N38 56 09 W77 05 33. Stereo. Hrs opn: 5900 Princess Garden Pkwy. #800, Lanham, 20706. Phone: (301) 306-1111. Fax: (301) 306-9510. Licensee: Radio One Licenses LLC. Group owner: Radio One Inc. (acq 11-8-01; grpsl). Population served: 3,500,000 Format: Adult contemp, motown. ♦Alfred Liggins, CEO & pres; Catherine Hughes, chmn; Scott Royster, CFO; Michele Williams, gen mgr.

WTGB-FM— October 1959: 94.7 mhz; 20.5 kw. Ant 771 ft TL: N38 57 49 W77 06 18. Stereo. Hrs opn: 24 8403 Colesville Rd., Ste. 1500, Silver Spring, 20910-3474. Phone: (301) 683-0947. Fax: (301) 881-8746. Web Site: 947theglobe.com. Licensee: CBS Radio East Inc. Group owner: Infinity Broadcasting Corp. (acq 8-1-85; grpsl; FTR: 6-10-85). Natl. Network: CNN Radio. Natl. Rep: CBS Radio. Format: Classic rock. Target aud: 25-49. ♦Michael Hughes, gen mgr.

WTNT(AM)— Jan 2, 1946: 570 khz; 5 kw-D, 1 kw-N, DA-2. TL: N39 02 07 W77 10 11. Hrs open: 24 1801 Rockville Pike, Rockville, 20852. Phone: (301) 231-7798. Fax: (301) 881-8030. E-mail: contact@wtntam570.com Web Site: www.wtntam570.com. Licensee: AMFM Radio Licenses LLC. Group owner: Clear Channel Communications Inc. (acq 8-30-00; grpsl). Natl. Rep: Clear Channel. Format: Talk. Target aud: 25-54. ♦Kaiya Ramsey, gen sls mgr; Kevin Cannady, prom dir & prom mgr.

WTOP-FM—See Washington, DC

WUST(AM)—See Washington, DC

Braddock Heights

WTLP(FM)— Apr 8, 1972: 103.9 mhz; 350 w. Ant 958 ft TL: N39 27 50 W77 29 44. Stereo. Hrs opn: 24
Rebroadcasts WTOP-FM Washington, DC 100%.
3400 Idaho Ave. N.W., Washington, DC, 20016. Phone: (202) 895-5000. Fax: (202) 895-5103. Web Site: www.wtopnews.com. Licensee: Bonneville Holding Co. Group owner: Bonneville International Corp. (acq 1996; grpsl). Population served: 125,000 Natl. Network: CBS Radio. Natl. Rep: Katz Radio. Format: News. Target aud: General. ♦Joel Oxley, gen mgr.

Brunswick

WTRI(AM)— Oct 2, 1966: 1520 khz; 9.3 kw-D, 14 kw-CH. TL: N39 18 45 W77 36 31. Stereo. Hrs opn: 214 13th Ave., 21716. Phone: (301) 834-1991. Web Site: www.vegas-radio.com. Licensee: WTRI Holding LLC (acq 10-20-2004; $1.6 million). Population served: 500,000 Format: Sp, top 40, contemp. Target aud: 20-45. Spec prog: Talk 2 hrs, farm 2 hrs, sports 5 hrs wkly. ♦Martin F. Sheehan, pres; Alfred Hammond, gen mgr.

California

WKIK-FM— December 1994: 102.9 mhz; 3.7 kw. 407 ft TL: N38 20 53 W76 37 40. Stereo. Hrs opn: 24 Box 2908, La Plata, 20646. Secondary address: 28095 Three Notch Rd., Suite 2-B, Mechanicsville 20659. Phone: (301) 870-5550. Phone: (301) 884-5550. Fax: (301) 884-0280. E-mail: wsmdfm@aol.com Licensee: Somar Communications Inc. (group owner; acq 1993; $130,000; FTR: 5-24-93). Format: Country. Target aud: 25-54. ♦Roy Robertson, pres & gen mgr; Terrell Soellner, opns mgr; Sharon McGuire, gen sls mgr.

Cambridge

WCEM(AM)— 1947: 1240 khz; 1 kw-U. TL: N38 35 02 W76 04 56. Stereo. Hrs opn: 24 Box 237, 21613. Phone: (410) 228-4800. Fax: (410) 228-0130. E-mail: espn@intercom.net Web Site: www.mtslive.com. Licensee: MTS Broadcasting L.C. Group owner: MTS Broadcasting (acq 6-20-93; $1.8 million with co-located FM; FTR: 8-9-93) Population served: 11,595 Natl. Network: Westwood One, ESPN Radio. Format: Sports. News staff: one. Target aud: 25-54. Spec prog: Relg 5 hrs wkly. ♦Troy D. Hill, gen mgr; Shane Walker, opns mgr; Shan Shariff, progmg dir; Mike Detmer, news dir; Bryan Harz, engrg dir; Dwight Cromwell, reporter; Bob Kinnamon, sports cmtr.

WCEM-FM— Jan 29, 1968: 106.3 mhz; 6 kw. Ant 325 ft TL: N38 35 03 W76 04 54. Stereo. 2 Bay Street, 21613. Secondary address: Box 237 21613. E-mail: theheat@intercom.net Web Site: www.mtslive.com. (Acq 1993; FTR: 8-23-93).60,000 Format: Hot adult contemp. Target aud: 18-49. ♦Mike Detmer, news rptr; Shan Shariff, sports cmtr.

WINX-FM— 2000: 94.3 mhz; 4.6 kw. Ant 3617 ft TL: N38 37 49 W76 03 24. Stereo. Hrs opn: 112 Main St., 3rd Fl., Annapolis, 21401. Phone: (800) 353-9430. Fax: (410) 267-7634. E-mail: info@winxfm.com Web Site: www.winxfm.com. Licensee: CWA Broadcasting Inc. (acq 9-23-2005). Format: Country. ♦Roy Deutschman, gen mgr.

Catonsville

WHFS(FM)— Nov 22, 1963: 105.7 mhz; 50 kw. 492 ft TL: N39 19 26 W76 32 56. Stereo. Hrs opn: 24 600 Washington Ave., Suite 202, Towson, 21204. Phone: (410) 825-5400. Fax: (410) 825-5404. Web Site: www.1057freefm.com. Licensee: CBS Radio Stations Inc. Group owner: Infinity Broadcasting Corp. (acq 11-13-98; grpsl). Population served: 905,759 Natl. Rep: Christal. Format: Talk. News staff: one. Target aud: 25-54. ♦Robert Philips, VP, gen mgr & stn mgr.

Chestertown

WCTR(AM)— June 16, 1963: 1530 khz; 1 kw-D, 270 w-CH. TL: N39 13 35 W76 05 20. Hrs open: Box 700, 231 Flatland Rd., 21620. Phone: (410) 778-1530. Fax: (410) 778-4800. E-mail: wctr@wctr.com Web Site: www.wctr.com. Licensee: WCTR Broadcasting LLC (acq 5-12-2004; $340,000). Population served: 50,000 Natl. Network: ABC. Rgnl rep: Rgnl Reps Format: Adult contemp. News: 10 hrs wkly. Target aud: 35+. ♦Richard Gelfman, pres; Ken Collins, gen mgr; John Link, opns mgr; Scan Hall, progmg dir & news dir.

College Park

***WMUC-FM**— Sept 10, 1979: 88.1 mhz; 10 w. 3 ft TL: N38 58 59 W76 56 37. Stereo. Hrs opn: 24 Box 3130, South Campus Dining Hall, Univ. of Maryland, 20742-8431. Phone: (301) 314-7865. Phone: (301) 314 7868. Fax: (301) 314-7879. Web Site: wmuc.umd.edu. Licensee: University of Maryland. Population served: 50,000 Format: Var/div. News: 5 hrs wkly. Target aud: College students. ♦Mark Burdett, CFO; Anton Kropp, gen mgr; Nestor Diaz, opns mgr.

Crisfield

WBEY-FM— July 1995: 97.9 mhz; 4.3 kw. Ant 379 ft TL: N38 01 45 W75 45 05. Hrs open: 1637 Dunn Swamp Rd., Pocomoke, 21851. Phone: (410) 957-6081. Fax: (410) 957-6080. E-mail: wbey@direcway.com Web Site: www.easternshoreradio.com. Licensee: Bay Broadcasting. Format: Adult country. ♦Michael Powell, gen mgr, opns mgr & gen sls mgr; Adam Riggin, progmg dir.

Cumberland

WCBC(AM)— June 24, 1953: 1270 khz; 5 kw-D, 1 kw-N, DA-2. TL: N39 40 28 W78 46 48. Hrs open: Box 1290, 21501. Phone: (301) 724-5000. Fax: (301) 722-8336. E-mail: dnorman@wcbc1270am.com Web Site: www.wcbc1270am.com. Licensee: Cumberland Broadcasting Co. Inc. (acq 4-8-76). Population served: 30,000 Natl. Network: ABC, Westwood One. Format: News/talk. News staff: 2; News: 3 hrs wkly. Target aud: 25 plus. ♦David N. Aydelotte Sr., pres; Jim Robey, gen mgr & stn mgr; Mary Clites, gen mgr.

WCMD(AM)—Listing follows WROG(FM).

WKGO(FM)—Listing follows WTBO(AM).

WROG(FM)— 1948: 102.9 mhz; 32 kw. Ant 1,440 ft TL: N39 34 56 W78 53 53. Stereo. Hrs opn: 24 516 White Ave., 21502. Phone: (301) 777-5400. Fax: (301) 777-5404. E-mail: mail@hitcoutry1029.com Web Site: www.hitcountry1029.com. Licensee: Broadcast Communications Inc. (group owner; (acq 1-8-2004; $2 million with co-located AM). Population served: 350,000 Law Firm: Shaonis & Peltzman. Format: Country. News staff: 2; News: 20 hrs wkly. Target aud: 25-49. ♦Eva Geiger, gen mgr & sls; Grant Garland, progmg dir & news dir.

WCMD(AM)—Co-owned with WROG(FM). 1948: 1230 khz; 1 kw-U. TL: N39 38 36 W78 44 35.24 Web Site: www.hotcountry1029.com.65,000 Law Firm: Shaonis & Peltzman. Format: Oldies memories. News staff: 2; News: 20 hrs wkly. Target aud: 25-54.

WTBO(AM)— Dec 13, 1928: 1450 khz; 1 kw-U. TL: N39 38 43 W78 45 05. Stereo. Hrs opn: 24
Rebroadcasts WFRB(AM) Frostburg 75%.
Box 1644, 350 Byrd Ave., 21502. Phone: (301) 722-6666. Fax: (301) 722-0945. Licensee: WTBO-WKGO Corp. LLC. Group owner: Dix Communications (acq 11-1-77). Population served: 2,500,000 Natl. Network: CSN. Law Firm: Baker & Hostetler. Format: Nostalgia, adult standards. News staff: one; News: 20 hrs wkly. Target aud: 40 plus. ♦G. Charles Dix II, pres; Richard L. Cornwell, gen mgr & gen sls mgr; Tom Martin, prom dir & progmg dir; Jim Van, news dir; Mark Workman, chief of engrg; R. DiBuono, traf mgr; Linda Ward, spec ev coord.

WKGO(FM)—Co-owned with WTBO(AM). April 1962: 106.1 mhz; 4 kw. 1,400 ft TL: N39 34 54 W78 53 58. (CP: 5.4 kw, ant 1,410 ft.). Stereo. 24 2,000,000 Natl. Network: Westwood One. Format: Adult contemp. News staff: one; News: one hr wkly. Target aud: 25-54. ♦Richard Cornwell, adv mgr; Jim Van, local news ed; Mark St. John, disc jockey; Ray Wagner, disc jockey; Tim Martin, disc jockey.

Denton

WKDI(AM)— Dec 27, 1988: 840 khz; 1 kw-D, DA. TL: N38 53 53 W75 51 10. Stereo. Hrs opn: Sunrise-sunset Box 309, 21629. Secondary address: 24580 Station Rd. 21629. Phone: (410) 479-2288. Fax: (410) 479-5188. E-mail: wkdi@broadcast.net Licensee: Bayshore Communications Inc. Format: Christian/talk. News: 12 hrs wkly. Target aud: 25-49; middle-income Christians. ♦Edward Baker, CEO; Michael A. McCoy, gen mgr & progmg dir.

Easton

WCEI-FM—Listing follows WEMD(AM).

WEMD(AM)— Sept 29, 1960: 1460 khz; 1 kw-D, 500 w-N, DA-2. TL: N38 46 13 W76 04 55. Stereo. Hrs opn: 306 Port St., 21601. Phone: (410) 822-3301. Fax: (410) 822-0576. E-mail: sandy@wceiradio.com Web Site: www.wceiradio.com. Licensee: First Media Radio L.L.C. (group owner; (acq 11-30-99; $4 million with co-located FM). Population served: 70,000 Natl. Network: Jones Radio Networks. Law Firm: Dow, Lohnes & Albertson. Format: Music of Your Life. Target aud: 45 plus; mature adults. ♦Matt Spence, CEO & progmg dir.

WCEI-FM—Co-owned with WEMD(AM). May 14, 1975: 96.7 mhz; 25 kw. 245 ft TL: N38 46 13 W76 04 55. Stereo. 24 Web Site: www.wceiradio.com.120,000 Format: Adult contemp. News staff: one. Target aud: 25-54. ♦Alex Kolbiaski, CEO; Sandy Reeves, gen mgr & sls dir; Julie Johnson, traf mgr; Jodi Calvert, local news ed.

Elkton

***WOEL-FM**— September 1978: 89.9 mhz; 3 kw. 259 ft TL: N39 35 35 W75 51 49. Hrs open: Box 246, 21922. Phone: (410) 398-3764. Fax: (410) 392-3229. E-mail: adp.saved@juno.com Web Site: www.mbcmin.org. Licensee: Maryland Baptist Bible College. Population served: 200,000 Natl. Network: USA. Format: Relg-Educ. ♦Ray Linzy, gen mgr.

WSRY(AM)— Aug 22, 1963: 1550 khz; 1 kw-D, 10 w-N, DA-2. TL: N39 35 45 W75 47 50. Hrs open: Box 372, Wilmington, DE, 19899. Phone: (302) 731-7270. Fax: (302) 738-3090. Licensee: Priority Radio Inc. (group owner; (acq 12-10-99). Population served: 70,000 Natl. Network: ESPN Radio. Format: Sports, sports talk. Target aud: 25-64. Spec prog: Relg 3 hrs, farm one hr wkly. ♦Dan Edwards, gen mgr.

Emmittsburg

***WMTB-FM**— Oct 1, 1977: 89.9 mhz; 100 w. 144 ft TL: N39 41 02 W77 21 25. Hrs open: Noon-3 PM Mount Saint Mary's College, 16300 Old Emmittsburg Rd., 21727-7799. Phone: (301) 447-5239. Web Site: www.msmary.edu/wmtb. Licensee: Mount Saint Mary's College. Format: Classic rock, new age, alternative. News: 2 hrs wkly. Target aud: General; college & community. Spec prog: Folk one hr, gospel one hr, relg 4 hrs wkly.

Essex

WYRE(AM)— 1946: 810 khz; 240 w-D, DA. TL: N39 18 44 W76 29 24. Hrs open: Sunrise-Sunset 12216 Parklawn Dr., Suite 203, Rockville, 20852. Phone: (301) 424-9292. Fax: (301) 424-8266. Licensee: Bay Broadcasting Corp. (acq 2-28-2002). Population served: 400,000 Law Firm: Baraff, Koerner & Olender. Format: Sp. ♦Richard Dent, pres; Raul Lopez Bastidas, gen mgr.

Federalsburg

WTDK(FM)— Dec 2, 1978: 107.1 mhz; 3.9 kw. 408 ft TL: N38 46 02 W75 44 46. Hrs open: 24 Box 1495, Cambridge, 21613. Phone: (410) 288-4800. Fax: (410) 228-0130. Web Site: www.mtslive.com. Licensee: MTS Broadcasting. (group owner; acq 1-30-97). Natl. Network: USA, Westwood One. Format: Hits of the 50s 60s & 70s. News: 4 hrs wkly. Target aud: 25-54; affluent listeners. ♦Thomas Mulitz, pres; Troy Hill, gen mgr; Joel Scott, opns mgr.

Frederick

WFMD(AM)—Listing follows WFRE(FM).

WFRE(FM)— Feb 19, 1961: 99.9 mhz; 7.6 kw. Ant 1,164 ft TL: N39 30 00 W77 29 58. Stereo. Hrs opn: 24 5966 Grove Hill Rd., 21703. Phone: (301) 663-4337. Fax: (301) 682-8018. Web Site: www.wfre.com. Licensee: Capstar TX L.P. (acq 8-30-2000; grpsl). Population served: 500,000 Natl. Rep: Clear Channel. Wire Svc: AP Format: Country. News staff: 3; News: one hr wkly. Target aud: 25-54. ♦Doug Hillard, gen mgr; Blaine Young, sls dir.

WFMD(AM)—Co-owned with WFRE(FM). Jan 1, 1936: 930 khz; 5 kw-D, 2.5 kw-N, DA-2. TL: N39 24 55 W77 27 41.24 5966 Grove Hill Rd., 21703-6012. Phone: (301) 663-4181. Web Site: www.wfmd.com. Group owner: Clear Channel Communications Inc. 200,000 Natl. Network: ABC. Format: News/talk, sports. News staff: 3. Target aud: 35-64. ♦Frank Mitchell, opns mgr, prom dir & progmg dir; Renee Dutton-O'Hara, news dir.

WTWT(AM)— Dec 15, 1960: 820 khz; 4.3 kw-D, 430 w-N, DA-N. TL: N39 24 42 W77 28 20. Hrs open: 24
Rebroadcasts WTWP(AM) Washington, DC 100%.
6633 Mt. Phillip Rd., 21703. Phone: (202) 895-5000. Fax: (202) 895-5149. E-mail: wpr@washingtonpostradio.com Web Site: www.washingtonpostradio.com. Licensee: Bonneville Holding Co. Group owner: Bonneville International Corp. (acq 1996; grpsl). Population served: 125,000 Natl. Rep: Katz Radio. Format: In-depth news. Target aud: General. ♦Joel Oxley, gen mgr.

WWEG(FM)—See Hagerstown

***WYPF(FM)**— May 1991: 88.1 mhz; 4 kw. Ant 554 ft TL: N39 25 05 W77 30 03. Stereo. Hrs opn: 24 Box 319, 4707 E. Schley Ave., Braddock Heights, 21714. Phone: (301) 662-9090. Phone: (301) 371-8888. Fax: (301) 371-1777. E-mail: wjtm@wjtm.org Web Site: www.wjtm.org. Licensee: Your Public Radio Corp. (acq 11-24-2004; $1.2 million). Natl. Network: USA. Format: Christian, talk, music. Target aud: General. ♦Anthony S. Brandon, pres; Michael Payne, gen mgr.

Frostburg

WFRB(AM)— Dec 20, 1958: 560 khz; 5 kw-D. TL: N39 41 02 W78 57 57. Hrs open: 24
Rebroadcasts WTBO(AM) Cumberland 75%.
242 Finzel Rd., Frostberg, 21532. Phone: (301) 689-8871. Phone: (301) 722-6666. Fax: (301) 689-8880. E-mail: wfrb@wfrb.com Web Site: wtboam.com. Licensee: WTBO-WKGO Corp. L.L.C. Group owner: Dix Communications (acq 6-1-97; $3.5 million with co-located FM). Population served: 2,500,000 Format: Adult standards, relg, news/talk. News staff: one; News: 10 hrs wkly. Target aud: 40 plus; those gainfully employed in the market for goods & svcs. ♦G. Charles Dix II, pres; Richard Cornwell, gen mgr & gen sls mgr; Hannah Ford, prom dir; Carson Yoder, progmg dir; Tim Martin, progmg mgr; Jim Van, news dir, pub affrs dir & local news ed; Mark Workman, engrg mgr; Chris Bagley, traf mgr.

WFRB-FM— Oct 1, 1965: 105.3 mhz; 16.5 kw. 960 ft TL: N39 41 02 W78 57 57. Stereo. 24 Web Site: www.wfrb.com.2,000,000 Format: Country. News staff: 2; News: 5 hrs wkly. ♦Chris Bagley, traf mgr; Jim Frantz, local news ed.

***WFWM(FM)**— April 1986: 91.9 mhz; 255 w horiz, 1.3 kw vert. Ant 1,424 ft TL: N39 34 54 W78 53 53. Stereo. Hrs opn: 24 Stangle Bldg., Frostburg State Univ., 21532. Phone: (301) 687-4143. Fax: (301)

687-7040. E-mail: wfwm@frostburg.edu Web Site: www.wfwm.org. Licensee: Frostburg State University. Natl. Network: NPR. Wire Svc: AP Format: Classical, jazz, alternative, NPR. News staff: one; News: 2 hrs wkly. Target aud: General. Spec prog: Educ 11 hrs wkly. ♦Rene G. Atkinson, gen mgr; Chuck Dicken, progmg dir & progmg mgr.

***WLIC(FM)**— October 1989: 97.1 mhz; 3 kw. 1,401 ft TL: N39 34 56 W78 53 53. (CP: 150 w, ant 1,355 ft.). Stereo. Hrs opn: 24
Rebroadcasts WAIJ(FM) Grantsville 100%.
Box 540, Grantsville, 21536-0540. Secondary address: He's Alive Corp. Offices, 34 Springs Rd., Grantsville 21536. Phone: (301) 895-3292. Fax: (301) 895-3293. E-mail: hesalive@hesalive.net Web Site: www.hesalive.net. Licensee: He's Alive Inc. (group owner) Population served: 250,000 Natl. Network: USA. Format: Gospel, Christian, relg. Target aud: 18-35. ♦Dewayne Johnson, pres; Monte Palmer, stn mgr.

Fruitland

WKHI(FM)— 1972: 107.5 mhz; 18.5 kw. Ant 339 ft TL: N38 11 54 W75 40 50. Stereo. Hrs opn: 24 20200 Dupont Blvd, Georgetown, DE, 19947. Phone: (302) 856-2567. Fax: (302) 856-7633. E-mail: sue@greatscottbroadcasting.com Web Site: www.joeontheweb.net. Licensee: Great Scott Broadcasting. (group owner; acq 7-16-99; $700,000 with WKHW(FM) Pocomoke City, MD). Natl. Network: NBC. Law Firm: Cohn & Marks. Format: Joe. Target aud: 25-54. Spec prog: Black 5 hrs, gospel 5 hrs wkly. ♦Faye Scott, pres; Sue Timmons, gen mgr; Adam Davis, progmg dir; Tracy Baker, traf mgr.

Gaithersburg

WMET(AM)— Jan 31, 1983: 1160 khz; 50 kw-D, 1.5 kw-N, DA-2. TL: N39 11 16 W77 12 56. Hrs opn: 24 8121 Georgia Ave., Silver Spring, 20910. Phone: (202) 969-9884. Fax: (202) 969-9900. Web Site: www.wmet1160.com. Licensee: Beltway Acquisition Corp. (acq 7-24-2002; $7.03 million). Population served: 1,000,000 Natl. Network: ABC. Format: Variety. News staff: 2. Spec prog: Indian 4 hrs wkly. ♦Dennis R. Israel, gen mgr.

Glen Burnie

WFBR(AM)— May 15, 1963: 1590 khz; 1 kw-U, DA-2. TL: N39 10 36 W76 37 20. Hrs open: 159 8th Ave. N.W., 21061. Phone: (410) 761-1590. Fax: (410) 761-9220. Licensee: Way Broadcasting Licensee LLC (group owner; (acq 8-1-2005; exchange for WKDV(AM) Manassas, VA). Population served: 1,300,000 Format: Black gospel. ♦Arthur S. Liu, pres; Keith Baldwin, gen mgr.

WWIN-FM— Sept 15, 1964: 95.9 mhz; 3 kw horiz, 3 kw vert. 299 ft TL: N39 12 16 W76 34 07. Hrs open: 24 1705 Whitehead Rd., Baltimore, 21207. Phone: (410) 332-8200. Fax: (410) 944-1282. Web Site: www.magic959baltimore.com. Licensee: Radio One Licenses LLC. Group owner: Radio One Inc. (acq 1-23-92; $7.5 million with WWIN(AM) Baltimore). Population served: 915,800 Natl. Network: ABC. Natl. Rep: Christal. Law Firm: Verner, Liipfert, Bernhard, McPherson & Hand. Format: Urban contemp. News staff: one; News: one hr wkly. Target aud: 35-54; Black adult. ♦Howard Mazer, gen mgr & natl sls mgr.

Grantsville

***WAIJ(FM)**— October 1984: 90.3 mhz; 10kw. 561 ft TL: N39 42 14 W79 05 31. Stereo. Hrs opn: 19 Box 540, 21536-0540. Secondary address: He's Alive Corp. Offices, 34 Springs Rd. 21536. Phone: (301) 895-3292. Fax: (301) 895-3293. E-mail: hesalive@hesalive.net Web Site: www.hesalive.net. Licensee: He's Alive Inc. (group owner) Population served: 1,000,000 Natl. Network: USA. Format: Gospel, Christian, relg. Target aud: 18-35. ♦Dewayne Johnson, pres; Tim Eutin, progmg dir & relg ed.

Grasonville

WRNR-FM—Licensed to Grasonville. See Annapolis

Hagerstown

WARK(AM)— July 20, 1947: 1490 khz; 925 w-U. TL: N39 37 36 W77 42 40. Hrs open: 24 880 Commonwealth Ave., 21740. Phone: (301) 733-4500. Phone: (800) 222-9279. Fax: (301) 733-0040. E-mail: webmaster@wark.am Licensee: Nassau Broadcasting III L.L.C. (acq 2-25-2005; $18 million with co-located FM). Population served: 35,862 Law Firm: Shaw Pittman. Format: Talk, oldies. News staff: 2; News: 8 hrs wkly. Target aud: 25-54. Spec prog: Jazz 2 hrs wkly. ♦Eugene J. Manning, gen mgr; J. Frederick Manning, opns VP; Marcia Cason, gen sls mgr; Roger Lide, engrg dir; Caroline Henneberger, traf mgr.

WWEG(FM)—Co-owned with WARK(AM). March 1957: 106.9 mhz; 15.5 kw. Ant 853 ft TL: N39 29 57 W77 36 42. Stereo. 24 Web Site: www.warx.com.250,000 Natl. Network: Westwood One. Format: Classic hits. News staff: 2.

WAYZ(FM)— 1946: 104.7 mhz; 8.3 kw. 1,379 ft TL: N39 41 47 W77 30 47. Stereo. Hrs opn: 24 Box 788, Greencastle, PA, 17225. Phone: (717) 597-9200. Fax: (717) 597-9210. E-mail: info@wayz.com Web Site: www.wayz.com. Licensee: H.J.V. L.P. (acq 8-28-2000; $2.5 million and WWMD(FM) Waynesboro, PA). Population served: 200,000 Wire Svc: UPI Format: Country. Target aud: 25-54. Spec prog: Relg 3 hrs wkly. ♦Dottie Hedglin, gen mgr & pub affrs dir; Gary Kirtley, gen sls mgr; Chris Maestle, progmg dir & local news ed; Toni Anderson, mus dir.

WDLD(FM)—Listing follows WHAG(AM).

***WGMS(FM)**— June 15, 1993: 89.1 mhz; 900 w. Ant 1,338 ft TL: N39 41 39 W77 30 50. Hrs open: 24
Rebroadcasts WETA-FM Washington 100%.
2775 S. Quincy, Arlington, VA, 22206-2304. Phone: (703) 998-2790. Fax: (703) 824-7288. Web Site: www.weta.org. Licensee: Greater Washington Education Telecommunication Association. Population served: 250,000 Natl. Network: NPR, PRI. Format: News, pub affrs. Target aud: General; educated adults. ♦Joseph Bruns, COO & sr VP; Sharon Rockefeller, CEO & chmn; Polly Heath, CFO; Dan De Vany, VP & gen mgr; Cynthia Cotton, opns mgr; Adam Gronski, mktg mgr; DeLinda Mrowka, prom dir; Ingrid Lakey, progmg dir; David Ginder, mus dir; Mitra Keykhah, asst music dir; Ed Kennedy, engrg dir; Mike Byrnes, chief of engrg; Andrea Murray, reporter.

WHAG(AM)—Halfway, June 9, 1962: 1410 khz; 1 kw-D, 99 w-N, DA-2. TL: N39 37 03 W77 44 17. Hrs open: 6 AM-7 PM 1250 Maryland Ave., 21740. Phone: (301) 797-7300. Fax: (301) 797-2659. Licensee: MLB-Hagerstown-Chambersburg IV LLC. (group owner; (acq 7-20-2005; grpsl). Population served: 250,000 Format: News/talk. News staff: 2; News: 7 hrs wkly. Target aud: 25-64; news-talk information profile; older, upscale. ♦Rich Bateman, gen mgr.

WDLD(FM)—Co-owned with WHAG(AM). January 1965: 96.7 mhz; 4.8 kw. 164 ft TL: N39 37 03 W77 44 17.351,000 Natl. Network: ABC. Format: New & classic rock, AOR. Target aud: 18-54; adults, young families.

WICL(FM)—See Williamsport

WJEJ(AM)— October 1932: 1240 khz; 1 kw-U. TL: N39 40 00 W77 43 30. Hrs open: 24 1135 Haven Rd., 21742. Phone: (301) 739-2323. Fax: (301) 797-7408. E-mail: wjej@myactv.net Web Site: www.wjejradio.com. Licensee: Hagerstown Broadcasting Co. Inc. (acq 12-21-72). Population served: 35,862 Natl. Network: CBS. Rgnl. Network: Metro. Law Firm: Koerner & Olender, P.C. Wire Svc: Metro Weather Service Inc. Format: Easy lstng/pop standards. News staff: one; News: 21 hrs wkly. Target aud: 35 plus. Spec prog: Farm one hr, talk 8 hrs wkly. ♦John T. Staub, pres, gen mgr, gen sls mgr & rgnl sls mgr; Joanna C. Staub, opns dir; Louis J. Scally, progmg dir & chief of engrg; Tom Bradley, news dir; Jackie Hall, traf mgr.

***WZXH(FM)**—Not on air, target date: unknown: 91.7 mhz; 3 kw. Ant -13 ft TL: N39 33 06 W77 40 28. Hrs open: Box 186, Sellersville, PA, 18960. Phone: (215) 721-2141. Fax: (215) 721-9811. Web Site: www.wordfm.org. Licensee: Four Rivers Community Broadcasting Corp. ♦Charles W. Loughery, pres & gen mgr.

Halfway

WDLD(FM)—Licensed to Halfway. See Hagerstown

WHAG(AM)—Licensed to Halfway. See Hagerstown

Havre de Grace

WJSS(AM)— May 15, 1948: 1330 khz; 5 kw-D, 500 w-N, DA-N. TL: N39 33 55 W76 07 08. Hrs open: 24 1605 Level Rd., 21078. Phone: (410) 939-0800. Fax: (410) 939-2156. Web Site: www.wjss1330.net. Licensee: Benjamin-Dane LLC (acq 5-12-2004; $350,000). Population served: 2,000,000 Format: News/talk. Target aud: General. ♦Ronald Reeves, pres & gen mgr.

WXCY(FM)— June 19, 1960: 103.7 mhz; 50 kw. 341 ft TL: N39 33 55 W76 07 08. Stereo. Hrs opn: 24 Box 269, 707 Revolution St., 21078. Secondary address: 2727 Shipley Rd., Wilmington, DE 19803. Phone: (410) 939-1100. Fax: (888) 766-1037. E-mail: wxcy@wxcyfm.com Web Site: www.wxcyfm.com. Licensee: Delmarva Broadcasting Co. (group owner) Wire Svc: Metro Weather Service Inc. Format: Modern country. News staff: 2. Target aud: 25-54. Spec prog: Relg 2 hrs, NASCAR info updates on race day 6 hrs wkly. ♦Pete Booker, CEO & pres; Willis Schenk, chmn; Bob Bloom, gen mgr; Bob Mercer, opns dir.

Hurlock

WAAI(FM)— June 1, 1989: 100.9 mhz; 1.3 kw. 502 ft TL: N38 37 28 W75 53 20. Stereo. Hrs opn: 24 Box 1495, Cambridge, 21613. Secondary address: 2 Bay St., Cambridge 21613. Phone: (410) 228-4800. Fax: (410) 228-0130. E-mail: waai@intercom.net Web Site: www.mtslive.com. Licensee: MTS Broadcasting. (group owner; acq 1-30-97). Population served: 30,000 Natl. Network: USA. Format: Country. News staff: one; News: 6 hrs wkly. Target aud: 25-54; general. Spec prog: Gospel 3 hrs wkly. ♦Thomas Mulitz, pres; Troy Hill, gen mgr; Joel Scott, opns mgr & progmg dir; Thomas Latimer, sls dir; Chris Singleton, chief of engrg.

Indian Head

WWGB(AM)— June 1986: 1030 khz; 50 kw-D, DA. TL: N38 33 53 W76 49 01. Stereo. Hrs opn: Sunrise-sunset 5210 Auth Rd., Suite 500, Suitland, 20746. Phone: (301) 899-1444. Fax: (301) 899-7244. E-mail: info@wwgb.com Web Site: www.wwgb.com. Licensee: Good Body Media LLC. (acq 7-15-2002). Population served: 100,000 Law Firm: Roy F. Perkins. Format: Sp, Christian. ♦Ruth Salmeron, stn mgr.

La Plata

WKIK(AM)— October 1965: 1560 khz; 1 kw-D. TL: N38 32 36 W76 59 37. Hrs open: 12 Box 2908, 20646. Phone: (301) 884-5550/870-5550. Fax: (301) 884-0280. E-mail: wsmdfm@aol.com Licensee: Somar Communications Inc. (group owner; acq 4-12-91; $65,000; FTR: 5-6-91). Natl. Network: ABC. Format: Country. News: 5 hrs wkly. Target aud: 25-54. Spec prog: Local news, Baltimore Ravens football. ♦Jimmy R. Pyle, gen mgr; Roy Robertson, gen mgr; Terrell Soellner, opns mgr.

Laurel

WILC(AM)— Dec 23, 1965: 900 khz; 1.9 kw-D, 500 w-N. TL: N39 04 57 W76 50 19. Hrs open: 5 AM-4 AM 13499 Baltimore Ave., Suite 200, 20707. Phone: (301) 419-2122. Fax: (301) 419-2409. E-mail: viva900@tvcontacto.net Web Site: www.radiovivc900.com. Licensee: ZGS Radio Inc. (acq 2-11-02; $5.5 million). Population served: 200,000 Natl. Network: CNN Radio. Natl. Rep: Univision Radio National Sales. Format: Sp, adult contemp. News: 10 hrs wkly. Target aud: General. ♦Patricia Omana, gen mgr; Anebel Marcano, prom mgr; Sergio Uriola, mus dir.

Lexington Park

WPTX(AM)— July 1998: 1690 khz; 10 kw-D, 1 kw-N. TL: N38 16 57 W76 33 35. Hrs open: Box 2908, La Plata, 20646. Phone: (301)

884-5550. Phone: (301) 870-5550. E-mail: wsmdfm@aol.com Licensee: Somar Communications Inc. (group owner; acq 2-12-2001; $2.25 million with WYRX(FM) Lexington Park including three-year, $100,000 noncompete agreement). Format: News/talk. ♦Roy Robertson, gen mgr & progmg dir; Terrell Soellner, opns mgr; Sharon Maguire, sls.

WYRX(FM)— Dec 16, 1976: 97.7 mhz; 3 kw. Ant 273 ft TL: N38 16 57 W76 33 35. Stereo. Hrs opn: Box 2908, La Plata, 20646. Phone: (301) 884-5550. Fax: (301) 884-0280. E-mail: wmdmfm@aol.com Licensee: Somar Communications Inc. (group owner; acq 2-12-2001; $2.25 million with WPTX(AM) Lexington Park including three-year, $100,000 noncompete agreement). Population served: 100,000 Format: Modern rock. Target aud: 25-54. ♦Roy Robertson, gen mgr & progmg dir; Sharon Robertson, sr VP & gen sls mgr; Patrick Wood, traf mgr.

Mechanicsville

WSMD-FM— Sept 1, 1988: 98.3 mhz; 3 kw. 328 ft TL: N38 24 49 W76 46 31. Stereo. Hrs opn: 24 Box 2908, La Plata, 20646. Secondary address: 28095 Three Notch Rd., Suite 2-B 20659. Phone: (301) 870-5550. Phone: (301) 884-5550. Fax: (301) 884-0280. E-mail: wsmdfm@aol.com Licensee: Somar Communications Inc. (group owner). Natl. Network: ABC. Format: Classic rock. News staff: one; News: 6 hrs wkly. Target aud: 25-54. Spec prog: Local news. ♦Roy Robertson, pres, gen mgr & progmg dir; Terrell Soellner, opns mgr; Sharon Robertson, gen sls mgr.

Middletown

WAFY(FM)— May 7, 1990: 103.1 mhz; 1 kw. Ant 571 ft TL: N39 25 05 W77 30 03. Stereo. Hrs opn: 24 5742 Industry Ln., Frederick, 21704. Phone: (301) 620-7700. Phone: (301) 620-1031. Fax: (301) 696-0509. Web Site: www.wafy.com. Licensee: Nassau Broadcasting III L.L.C. (acq 2-14-2005; $15.7 million). Law Firm: Dickstein Shapiro Morin & Oshinsky. Format: Adult contemp. News staff: 2. Target aud: 25-54; upscale, well-educated, great radio commitment. ♦Brian Unger, sls dir; Rob Marmet, gen mgr & mktg dir; Caroline Wood, news dir & news rptr; Fred Klimes, chief of engrg; Marci Whyle, traf mgr.

Midland

WVMD(FM)—Not on air, target date: unknown: 99.5 mhz; 920 w. Ant 833 ft TL: N39 40 43 W78 57 34. Hrs open: 1251 Earl L. Core Rd., Morgantown, WV, 26505-5896. Phone: (304) 296-0029. Fax: (304) 296-3876. Licensee: West Virginia Radio Corp. of the Alleghenies. (acq 1-18-2007; $375,000 for CP). ♦Dale B. Miller, pres.

Morningside

WPGC(AM)— October 1954: 1580 khz; 50 kw-D, 250 w-N, DA. TL: N38 52 07 W76 53 48. Hrs open: 24 4200 Parliament Place, Suite 300, Lanham, 20706. Phone: (301) 918-0955. Fax: (301) 459-9509. Web Site: www.heaven1580am.com. Licensee: CBS Radio WPGC(AM) Inc. (group owner; (acq 1994; with co-located FM). Population served: 756,510 Law Firm: Leventhal, Senter & Lerman. Format: Gospel. News staff: one; News: 3 hrs wkly. Target aud: 25-54. ♦Sam Rogers, gen mgr.

WPGC-FM— February 1959: 95.5 mhz; 50 kw. 500 ft TL: N38 51 48 W76 54 38. Stereo. 24 Web Site: www.wpgc955.com. Licensee: CBS Radio Inc. of Maryland. Format: CHR. Target aud: 18-54.

Mountain Lake Park

WKHJ(FM)— July 9, 1990: 104.5 mhz; 1.5 kw. 663 ft TL: N39 24 37 W79 17 15. Stereo. Hrs opn: 24 Box 2337, 21550. Phone: (301) 344-4272. Phone: (301) 334-2086. Fax: (301) 334-2152. E-mail: wkhj@verizon.net Licensee: Southern Highlands Inc. (acq 11-14-2005). Population served: 100,000 Natl. Network: CNN Radio. Format: Adult contemp. News staff: one; News: 12 hrs wkly. Target aud: 18-49. ♦Pam Trickett, prom mgr; Terry King, gen mgr, opns mgr, gen sls mgr & mus dir; James Shaffer, news dir; Roger L. Ruff, chief of engrg.

Oakland

WKHJ(FM)—See Mountain Lake Park

WMSG(AM)— May 19, 1963: 1050 khz; 1 kw-D, 75 w-N. TL: N39 25 15 W79 25 00. Hrs open: Box 449, 21550. Phone: (301) 334-3800. Fax: (301) 334-2152. Licensee: Oakland Media Group Inc. (acq 11-14-2005; with co-located FM). Population served: 1,786 Natl. Network: CBS. Format: Adult standard. Target aud: General. ♦Paul Mullan, gen mgr & opns mgr.

WWHC(FM)—Co-owned with WMSG(AM). 1966: 92.3 mhz; 1.4 w. 689 ft TL: N39 26 41 W79 31 42. Stereo. 2,500 Natl. Network: ABC. Format: Country.

Ocean City

WKHZ(AM)— July 1, 1960: 1590 khz; 1 kw-D, 500 w-N, DA-2. TL: N38 24 16 W75 07 37. Hrs open: 24 11500 Coastal Hwy., Sea Watch Suite #1, 21842. Secondary address: 12216 Parklawn Dr., Suite 203, Rockville 20852. Phone: (410) 723-9100. Fax: (410) 723-6561. E-mail: wkhz1590@aol.com Web Site: khzradio.com. Licensee: Radio Broadcast Communications Inc. (acq 2-2001). Population served: 550,000 Format: News. News: 2 hrs wkly. Target aud: 25-54; active, thinking, responsive, affluent adults, with high disposable income. ♦Bill Parris, pres; Richard Marcus, gen mgr.

WOCQ(FM)—See Berlin

WOSC(FM)—See Bethany Beach, DE

***WRAU(FM)**—Not on air, target date: unknown: 88.3 mhz; 50 kw. Ant 492 kw TL: N38 23 12 W75 17 27. Hrs open: c/o WAMU(FM) - Brandywine Bldg., 4400 Massachusetts Ave. N.W., Washington, DC, 20016-8082. Phone: (202) 885-1200. Fax: (202) 885-1269. Licensee: Exec. Comm. of Bd. of Trustees of American University. ♦Caryn Mathes, gen mgr.

WRXS(FM)— Apr 15, 1994: 106.9 mhz; 6 kw. 303 ft TL: N38 20 57 W75 11 07. Hrs open: 12010 Industrial Park Rd., Suite 6, Bishopville, 21813. Phone: (410) 352-0001. Fax: (410) 352-0005. E-mail: skip@x1069.com Web Site: www.x1069.com. Licensee: Atlantic Radio Broadcasting L.L.C. (acq 4-98). Format: Alternative. Target aud: 18-34. ♦Crystal Layton, VP & stn mgr; Ronald J. Gillenardo, gen mgr; Skip Dixxon, opns mgr & progmg dir; Amy Mulford, traf mgr.

***WSDL(FM)**— Feb 13, 1998: 90.7 mhz; 15 kw. Ant 331 ft TL: N38 30 06 W75 10 07. Stereo. Hrs opn: 24 Box 2596, Salisbury, 21802. Phone: (410) 543-6895. Web Site: www.wscl.org. Licensee: Salisbury State University Foundation Inc. Natl. Network: NPR, PRI. Format: News/talk. News staff: one; News: 24 hrs wkly. Target aud: General. ♦Fred Marino, gen mgr.

WWFG(FM)— June 30, 1978: 99.9 mhz; 38 kw. Ant 469 ft TL: N38 25 20 W75 08 23. Hrs open: Gateway Crossing, 351 Tilghman Rd., Salisbury, 21804. Phone: (410) 742-1923. Fax: (410) 742-2329. E-mail: froggyemail@yahoo.com Web Site: www.froggy999 .com. Licensee: Capstar TX L.P. Group owner: Clear Channel Communications Inc. (acq 8-7-00; grpsl). Natl. Rep: Clear Channel. Format: Country. Target aud: 25-54; affluent, upwardly mobile. ♦Frank Hamilton, gen mgr; Brian Cleary, opns mgr; Dixie Penner, prom dir.

Ocean City-Salisbury

WQHQ(FM)—Licensed to Ocean City-Salisbury. See Salisbury

Ocean Pines

WQJZ(FM)— March 1994: 97.1 mhz; 4.6 kw. 374 ft TL: N38 22 75 W75 10 32. Stereo. Hrs opn: 24 Box 909, Salisbury, 21803. Secondary address: 919 Ellegood St., Salisbury 21801. Phone: (410) 219-3500. Fax: (410) 548-1543. E-mail: wqjz@radiocenter.com Web Site: www.wqjz.com. Licensee: Delmarva Broadcasting Co. (group owner; acq 6-26-97; grpsl). Natl. Rep: Katz Radio. Law Firm: Hogan & Hartson. Wire Svc: AP Format: Smooth jazz. News staff: 3. Target aud: 30-60. ♦Michael Reath, gen mgr; Joe Edwards, opns mgr; Joe Beail, gen sls mgr.

Owings Mills

WCBM(AM)—See Baltimore

Pikesville

WVIE(AM)— Apr 5, 1955: 1370 khz; 50 kw-D, 7.7 kw-N, DA-2. TL: N39 26 23 W76 21 20. Hrs open: 1726 Reisterstown, Suite 117, Baltimore, 21208. Phone: (410) 580-6800. Fax: (410) 580-6810. E-mail: bpettit680@yahoo.com Web Site: www.wcbm.com. Licensee: M-10 Broadcasting Inc. (acq 6-12-98; $1.1 million with WJSS(AM) Havre de Grace). Population served: 2,500,000 Format: Female-oriented talk. ♦Nick Mangione Jr., sr VP & opns mgr; Bob Pettit, gen mgr; Marc Beavin, gen sls mgr; Niles Seaberg, progmg dir; Eddie Applefeild, prom.

Pocomoke City

WBEY(AM)—Not on air, target date: unknown: 1070 khz; 500 w-D, 250 w-N, DA-2. TL: N38 04 45 W75 34 34. Hrs open: 1637 Dunn Swamp Rd., 21851-3300. Phone: (410) 957-6081. Fax: (410) 957-6080. Licensee: Bay Broadcasting Inc. ♦Michael Powell, gen mgr.

WGOP(AM)— Aug 1, 1955: 540 khz; 500 w-D, 243 w-N. TL: N38 03 11 W75 34 11. (CP: COL: Damascus, 1 kw-U, DA-2. TL: N39 17 46 W77 13 12). Hrs opn: 1637 Dunn Swamp Rd., 21851. Phone: (410) 957-6081. Fax: (410) 957-6080. E-mail: wbey@direcway.com Licensee: Birach Broadcasting Corp. (group owner; acq 11-25-92; $127,500; FTR: 12-14-92). Population served: 80,000 Law Firm: Pepper & Corazzini. Format: Adult standards. ♦Michael Powell, gen mgr; Chuppy Layton, opns dir, mus dir & farm dir.

WKHW(FM)— May 1, 1992: 106.5 mhz; 1.8 kw. Ant 341 ft TL: N37 58 38 W75 32 36. Hrs open: Box 69, Crisfield, 21851. Phone: (410) 957-4300. Fax: (410) 957-6080. E-mail: sue@greatscottbroadcasting.com Licensee: Great Scott Broadcasting. (group owner; (acq 7-16-99; $700,000 with WKHI(FM) Fruitland). Law Firm: Cohn & Marks. Format: Classic rock. Spec prog: Gospel 5 hrs, bluegrass 4 hrs wkly. ♦Michael Powell, gen mgr.

WXMD(FM)— October 2000: 92.5 mhz; 2.95 kw. Ant 472 ft TL: N38 08 35 W75 39 53. Hrs open: Box 909, Salisbury, 21803. Phone: (410) 219-3500. Fax: (410) 548-1543. E-mail: max925@radiocenter.com Web Site: www.max925.com. Licensee: Delmarva Broadcasting Co. (group owner; acq 7-10-00; $425,000). Natl. Rep: Katz Radio. Law Firm: Hogan & Hartson. Wire Svc: AP Format: Adult contemp, rock. News staff: one; News: 2 hrs wkly. Target aud: Adults 35-54; baby boomers. ♦Michael Reath, gen mgr; Joe Edwards, opns mgr & progmg mgr; Bill Reddish, news dir; Jeff Twilley, chief of engrg.

Potomac-Cabin John

WCTN(AM)— 1965: 950 khz; 2.5 kw-D, 47 w-N, DA-2. TL: N39 02 12 W77 12 09. Hrs open: 24 100-25 Queens Blvd., Suite 1CC, Forest Hills, NY, 11375. Phone: (301) 424-9292. Fax: (301) 424-8266. Licensee: Win Radio Broadcasting Corp. (acq 1-15-2004; $2.2 million). Population served: 2,500,000 Law Firm: Fisher, Wayland, Cooper, Leader & Zaragoza L.L.P. Format: Sp. ♦Richard S. Yoon, pres; Raul Lopez Bastidas, gen mgr.

Prince Frederick

WWXT(FM)— August 1971: 92.7 mhz; 2.85 kw. Ant 476 ft TL: N38 40 26 W76 35 40. Stereo. Hrs opn: 24 8121 Georgia Ave., Suite 1050, Silver Spring, 20910. Phone: (301) 562-5800. Fax: (301) 589-9772. Web Site: www.triplexespnradio.com. Licensee: Red Zebra Broadcasting Licensee LLC. Group owner: Mega Communications Inc. (acq 5-9-2006; grpsl). Population served: 500,000 Natl. Network: ESPN Radio. Format: Sports. ♦Bruce Gilbert, CEO.

Princess Anne

***WESM(FM)**— Mar 29, 1987: 91.3 mhz; 50 kw. 347 ft TL: N38 12 37 W75 40 56. Stereo. Hrs opn: 24 Univ. of Maryland Eastern Shore, Backbone Rd., 21853. Phone: (410) 651-8001. Fax: (410) 651-8005. E-mail: wesm913@umes.edu Web Site: www.umes.edu/wesm. Licensee: University of Maryland Eastern Shore. Population served: 150,000 Natl. Network: NPR, PRI. Format: Jazz, blues, NPR. News staff: one; News: 15 hrs wkly. Target aud: General. Spec prog: Blues 5 hrs, reggae 2 hrs, big band 10 hrs, gospel 20 hrs wkly. ♦Dr. Thelma B. Thompson, pres; Marva Copeland, gen mgr & mktg dir; Angel Resto Jr., opns VP; Brian Daniels, mktg mgr & progmg; Yancy Carrigan, mus dir.

WOLC(FM)— Dec 24, 1976: 102.5 mhz; 50 kw. Ant 500 ft TL: N38 06 43 W75 39 14. Stereo. Hrs opn: 24 Box 130, 11890 Crisfield Ln., 21853. Phone: (410) 543-9652. Fax: (410) 651-9652. E-mail: wolc@wolc.org Web Site: www.wolc.org. Licensee: Maranatha Inc. Population served: 200,000 Law Firm: Shainis - Peltzman. Format: Relg. Target aud: 25-64. ♦Robert L. Shores, pres; Deborah G. Byrd, gen mgr; Jack Tucker, gen sls mgr; Rodney Baylous, progmg dir; Mark Bohnett, chief of engrg.

Rockville

WLXE(AM)— November 1951: 1600 khz; 1 kw-D, 500 w-N, DA-N. TL: N39 05 51 W77 09 07. Hrs open: 24 Radio Ctr., 12216 Parklawn Dr., Suite 203, 20852. Phone: (301) 424-9292. Fax: (301) 424-8266. Licensee: Multicultural Radio Broadcasting Licensee LLC. Group owner: Multicultural Radio Broadcasting Inc. (acq 7-31-01; $800,000). Population served: 4,000,000 Format: Sp. ♦Bill Parris, gen mgr.

Salisbury

***WDIH(FM)**— June 1990: 90.3 mhz; 378 w. 180 ft TL: N38 24 28 W75 36 16. Hrs open: Box 186, 21801. Phone: (410) 860-5000. Fax: (410) 546-7772. E-mail: biscope@acninc.net Licensee: Salisbury Educational Broadcasting Foundation. Format: Christian preaching & mus, info progmg. ♦Bishop Dr. George Copeland, gen mgr.

WDKZ(FM)— July 25, 1982: 105.5 mhz; 2.1 kw. Ant 384 ft TL: N38 24 26 W75 35 57. Stereo. Hrs opn: Gateway Crossing, 351 Tilghman Rd., 21804. Phone: (410) 742-1923. Fax: (410) 742-2329. E-mail: kissfm@kiss1055.com Web Site: www.kiss1055.com. Licensee: Capstar TX L.P. Group owner: Clear Channel Communications Inc. (acq 8-25-2000; grpsl). Natl. Rep: Clear Channel. Format: Top-40. News staff: one. Target aud: 18-44. ♦Frank Hamilton, gen mgr; Brian Cleary, opns mgr; Dixie Penner, prom dir.

WGOP(AM)—See Pocomoke City

WICO(AM)— September 1957: 1320 khz; 1 kw-D, 36 w-N. TL: N38 21 39 W75 37 00. Hrs open: 24 Box 909, 21803. Secondary address: 919 Ellegood St. 21801. Phone: (410) 219-3500. Fax: (410) 548-1543. E-mail: wico@radiocenter.com Web Site: www.wicoam.com. Licensee: Delmarva Broadcasting Co. (group owner; (acq 6-26-97; grpsl). Population served: 100,000 Natl. Network: ABC. Natl. Rep: Katz Radio. Law Firm: Hogan & Hartson. Wire Svc: ABC Wire Svc: AP Format: News/talk, sports. News staff: 2; News: 18 hrs wkly. Target aud: 35-64. Spec prog: Farm one hr wkly. ♦Michael Reath, gen mgr; Joe Edwards, opns mgr & progmg dir; Joe Beail, gen sls mgr; Bill Reddish, news dir & pub affrs dir; Jeff Twilley, chief of engrg.

WICO-FM— Sept 3, 1969: 97.5 mhz; 4.5 kw. 299 ft TL: N38 21 39 W75 37 00. Stereo. 24
Rebroadcasts WXJN(FM) Lewes, DE 100%.
E-mail: catcountry@radiocenter.com Web Site: www.catcountryradio.com.250,000 Format: Country. News staff: one; News: 3 hrs wkly. Target aud: 25-54. ♦Joe Edwards, chief of opns; E.J. Foxx, mus dir; Brian K. Hall, disc jockey; Dixie Kelly, disc jockey; Joe Alan, disc jockey; Mike Chaney, disc jockey.

WJDY(AM)—Listing follows WSBY-FM.

WQHQ(FM)—Listing follows WTGM(AM).

WSBY-FM— Dec 13, 1989: 98.9 mhz; 6 kw. 328 ft TL: N38 18 00 W75 37 41. Hrs open: 24 Gateway Crossing, 351 Tilghman Rd., 21804. Phone: (410) 742-1923. Fax: (410) 742-2329. E-mail: kennylove@clearchannel.com Web Site: www.wsby.com. Licensee: Capstar TX L.P. Group owner: Clear Channel Communications Inc. (acq 8-7-00; grpsl). Population served: 250,000 Law Firm: Mullin, Rhyne, Emmons & Topel. Format: Urban contemp. News staff: one. Target aud: 25-54. ♦Frank Hamilton, gen mgr; Brian Cleary, opns mgr, progmg dir & news dir; Ed Sennessy, gen sls mgr; Chris Kelly, chief of engrg; Marie Merrill, traf mgr.

WJDY(AM)—Co-owned with WSBY-FM. Mar 14, 1958: 1470 khz; 5 kw-D, 500 w-N, DA-D. TL: N38 23 30 W75 38 48.6 AM-midnight Web Site: www.wjdy,com. Format: Gospel. Target aud: 04-12. ♦John Thomas-Mason, traf mgr.

***WSCL(FM)**— May 29, 1987: 89.5 mhz; 33 kw. 600 ft TL: N38 39 15 W75 36 42. Stereo. Hrs opn: 24 Box 2596, 21802. Secondary address: S. Salisbury Blvd. 21802. Phone: (410) 543-6895. Fax: (410) 548-3000. E-mail: prd@salisbury.edu Web Site: www.wscl.org. Licensee: Salisbury State University Foundation Inc. Population served: 415,800 Natl. Network: NPR, PRI, AP Radio. Format: Class, news. News staff: one; News: 29 hrs wkly. Target aud: General. ♦Fred Marino, gen mgr; Pamela Andrews, progmg dir; Bill Bukowski, opns.

WTGM(AM)— Sept 13, 1940: 960 khz; 5 kw-U, DA-2. TL: N38 25 44 W75 37 26. Stereo. Hrs opn: 24 351 Tilghman Rd., 21804-1891. Phone: (410) 742-1923. Fax: (410) 742-2329. Web Site: www.delmarvaradio.com. Licensee: Capstar TX L.P. Group owner: Clear Channel Communications Inc. (acq 8-7-00; grpsl). Population served: 250,000 Format: Sports. News staff: one. Target aud: 25-64. ♦Doug Hillard, gen mgr.

WQHQ(FM)—Co-owned with WTGM(AM). July 31, 1965: 104.7 mhz; 33 kw. 610 ft TL: N38 23 15 W75 17 30. Stereo. Law Firm: Arent, Fox, Kintner, Plotkin & Kahn. Format: Adult contemp. Target aud: 25-54.

WWFG(FM)—See Ocean City

Silver Spring

WFED(AM)—Licensed to Silver Spring. See Washington DC

WIHT(FM)—See Washington, DC

WWRC(AM)—See Washington, DC

Snow Hill

WQMR(FM)— February 2004: 101.1 mhz; 1.2 kw. Ant 489 ft TL: N38 12 57 W75 19 21. Stereo. Hrs opn: 24 Snow Hill Broadcasting L.L.C., 7200 Coastal Hwy., Suite 101, Ocean City, 21843. Phone: (410) 524-6862. Fax: (410) 524-6808. E-mail: kevin@wqmr.com Web Site: www.wqmr.com. Licensee: Snow Hill Broadcasting L.L.C. (acq 5-21-2004; $200,000). Population served: 296,100 Format: News/talk, sports. Target aud: 18-64 Persons. Spec prog: Power Talk(local) Travel Show/Car Doc/Garison Show/ Tasting Room/ Satellite Sisters. ♦Jack Gillen, pres & gen mgr; Kevin Brenahan, VP; Corey Duices, progmg dir; Heather Renee Shingleton, news dir.

Takoma Park

***WGTS(FM)**— May 8, 1957: 91.9 mhz; 23.5 kw. Ant 610 ft TL: N38 53 30 W77 07 55. Stereo. Hrs opn: 24 7600 Flower Ave., 20912. Phone: (301) 891-4200. Fax: (301) 270-9191. E-mail: wgts@wgts919.org Web Site: www.wgts.org. Licensee: Columbia Union College Broadcasting Inc. Format: Educ, relg. Target aud: General. Spec prog: Inspirational. ♦Gerry Fuller, chmn; John Konrad, gen mgr; Becky Wilson Ali Gray, progmg dir.

Towson

WLIF(FM)—See Baltimore

WNST(AM)— Oct 27, 1955: 1570 khz; 5 kw-D, 236 w-N. TL: N39 25 04 W76 33 23. Hrs open: 1550 Hart Rd., Baltimore, 21286. Phone: (410) 821-9678. Fax: (410) 828-4698. E-mail: nasty@wnst.net Web Site: www.wnst.net. Licensee: Nasty 1570 Sports LLC. (acq 1-11-01; $1 million). Population served: 200,000 Format: Sports. ♦Paul Kopleke, gen mgr; Steve Hennessey, gen sls mgr.

***WTMD(FM)**— Feb 12, 1976: 89.7 mhz; 10.16 kw. 236 ft TL: N39 23 45 W76 36 29. Stereo. Hrs opn: 24 8000 York Rd., 21252. Phone: (410) 704-8938. Fax: (410) 704-2609. E-mail: wtmd@towson.edu Web Site: www.wtmd.org. Licensee: Towson University. Population served: 2,088,400 Format: AAA. News: 6 hrs wkly. Target aud: 25-64. ♦Stephen Yasko, gen mgr; Jeri Jenkins, dev dir; Dan Reed, progmg dir; Mike Matthews, mus dir; T. Scott Dunbar, chief of engrg.

Waldorf

WPRS-FM— February 1965: 104.1 mhz; 20 kw. Ant 800 ft TL: N38 37 07 W76 50 42. Stereo. Hrs opn: 24 5900 Princess Garden Pkwy., 7th Fl., Lanhan, 20706. Phone: (301) 306-1111. Fax: (301) 306-9540. Licensee: Bonneville Holding Co. Group owner: Bonneville International Corp. (acq 1996; grpsl). Population served: 2,800,000 Format: Black gospel. ♦Michelle Williams, gen mgr.

Walkersville

WDMV(AM)— December 1994: 700 khz; 5 kw-D, DA. TL: N39 27 27 W77 19 27. Hrs open: Birach Broadcasting Corp., 21700 Northwestern Hwy., Tower 14, Suite 1190, Southfield, MI, 48075. Phone: (703) 934-6300. Web Site: www.dcradio700.com. Licensee: Birach Broadcasting Corp. (group owner; (acq 9-95). Format: Talk. ♦Sima Birach Jr., pres & gen mgr.

Westernport

WWPN(FM)— Oct 1, 1993: 101.1 mhz; 6 kw. -541 ft TL: N39 29 14 W79 03 13. Stereo. Hrs opn: Box 3382, Lavale, 21502. Phone: (301) 463-5100. Licensee: Ernest F. Santmyire. Population served: 2,000,000 Format: Relg, contemp Christian. Target aud: 18-45; working class. ♦Ernest F. Santmyire, CEO & gen mgr.

Westminster

WTTR(AM)— July 1953: 1470 khz; 1 kw-U, DA-N. TL: N39 34 37 W77 01 21. Hrs open: 24 101 WTTR Ln., 21158. Phone: (410) 876-1515. Fax: (410) 876-5095. E-mail: wttr@toad.net Web Site: www.wttr.com. Licensee: Sajak Broadcasting Corp. (acq 12-20-2004; $540,000). Population served: 160,000 Format: Oldies. News staff: one; News: 12 hrs wkly. Target aud: 35-64. Spec prog: Farm 4 hrs wkly. ♦Steve Hopp, gen mgr; Dwight Dingle, stn mgr & mus dir; Mark Woodworth, news dir & pub affrs dir.

WZBA(FM)— Nov 1, 1959: 100.7 mhz; 27 kw. Ant 660 ft TL: N39 27 01 W76 46 37. Stereo. Hrs opn: 24 11350 McCormick Rd., Executive Plaza 3, Suite 701, Hunt Valley, 21031. Phone: (410) 771-8484. Fax: (410) 771-1616. E-mail: jlaird@thebayonline.com Web Site: www.wzbathebay.com. Licensee: Shamrock Communications Inc. (group owner; (acq 4-7-81; $1.74 million with co-located AM; FTR: 5-4-81). Population served: 1000000 Format: Classic rock. News: one hr wkly. Target aud: 25-49; men & women active in the country life group. ♦Jeff Laird, gen mgr; Mark Sheely, gen sls mgr; Jon McGann, progmg dir; Fred Klims, chief of engrg; Kelly Laymen, traf mgr.

Wheaton

WACA(AM)— 1954: 1540 khz; 5 kw-D. TL: N39 00 50 W77 01 46. Hrs open: 11141 Georgia Ave., Suite 310, 20902. Phone: (301) 942-3500. Fax: (301) 942-7798. E-mail: news@radioamerican.net Web Site: www.radioamerican.net. Licensee: Entravision Holdings LLC. Group owner: Entravision Communications Corp. (acq 3-14-00; grpsl). Law Firm: Leventhal, Senter & Lerman. Format: Sp. Target aud: General; Hispanic, Central & Latin American, Caribbean listeners. ♦Alejandro Carrasco, gen mgr.

WASH(FM)—See Washington, DC

Williamsport

***WCRH(FM)**— July 24, 1976: 90.5 mhz; 10 kw. Ant 884 ft TL: N39 39 34 W77 57 56. Stereo. Hrs opn: 24 Box 439, 21795. Secondary address: 12146 Cedar Ridge Rd. 21798. Phone: (301) 582-0285. Fax: (301) 582-2707. E-mail: wcrh@wcrh.org Web Site: www.wcrh.org. Licensee: Cedar Ridge Children's Home and School Inc. Population served: 1,000,000 Natl. Network: Moody. Law Firm: Hardy, Carey & Chautin, L.L.P. Format: Relg. News staff: one; News: 9 hrs wkly. Target aud: 25-45. ♦David Swacina, CEO; Jeff Ward, opns mgr.

WICL(FM)— Nov 15, 1972: 95.9 mhz; 3 kw. 300 ft TL: N39 36 17 W77 46 49. Stereo. Hrs opn: 24 1606 W. King St., Martinsburg, WV, 25401. Phone: (304) 263-8868. Fax: (304) 263-8906. Web Site: www.cool959.com. Licensee: Prettyman Broadcasting Co. (group owner; (acq 3-10-98; $1.05 million). Population served: 500,000 Law Firm: Dow, Lohnes, & Albertson, PLLC. Format: True Oldies. News: 15 hrs wkly. Target aud: 35-64. ♦Norm Slemenda, gen mgr.

Worton

***WKHS(FM)**— Mar 28, 1974: 90.5 mhz; 17.5 kw. 215 ft TL: N39 16 55 W76 05 26. Stereo. Hrs opn: Box 905, 21678. Secondary address: Rts. 297 & 298 21678. Phone: (410) 778-4249. Fax: (410) 778-3802. E-mail: wkhs@kent.k12.md.us Licensee: Board of Education of Kent County. Population served: 20,000 Format: Div. Target aud: 12 plus. Spec prog: Oldies 6 hrs, children 5 hrs, country 2 hrs, big band 2 hrs, rhythm and blues 2 hrs, jazz 2 hrs wkly. ♦Steve Kramarck, gen mgr.

Massachusetts

Acton

***WHAB(FM)**— Aug 1, 1979: 89.1 mhz; 9.1 w. 53 ft TL: N42 28 48 W71 27 28. Hrs open: 10 AM-5:30 PM (M-F) Acton Boxboro Regional High School, 36 Charter Rd., 01720. Phone: (978) 264-4700, EXT. 3470. Web Site: www.quadphonic.com. Licensee: Acton-Boxborough Regional School District. Format: Div, news. ♦Dan Drinkwater, gen mgr.

Allston

WGBH(FM)—See Boston

Amherst

***WAMH(FM)**— 1955: 89.3 mhz; 150 w. 718 ft TL: N42 21 51 W72 25 24. Stereo. Hrs opn: 24 AC# 1907Campus Center, 2171 Amherst College, 01002-5000. Phone: (413) 542-2224. Phone: (413) 542-2288. E-mail: wamh@amherst.edu Web Site: www.amherst.edu/~wamh. Licensee: Trustees of Amherst College. Population served: 17,926 Format: Var/div. News: 2 hrs wkly. Target aud: 16-32; youth of today. ♦Claire Kiechel, gen mgr.

***WFCR(FM)**— May 6, 1961: 88.5 mhz; 13 kw. 895 ft TL: N42 21 49 W72 25 24. Stereo. Hrs opn: 24 131 County Circle, Hampshire House, Univ. of Mass., 01003-9257. Phone: (413) 545-0100. Fax: (413) 545-2546. E-mail: radio@wfcr.org Web Site: www.wfcr.org. Licensee: University of Massachusetts. Population served: 1,183,119 Natl. Network: PRI, NPR. Law Firm: Wiley, Rein & Fielding. Wire Svc: AP Format: News, Classical, Jazz. News staff: 6; News: 40 hrs wkly. Target aud: General. Spec prog: Sp 4 hrs, folk 4 hrs wkly. ♦Martin Miller, gen mgr.

***WMUA(FM)**— 1949: 91.1 mhz; 1 kw. 26 ft TL: N42 23 31 W72 31 13. Stereo. Hrs opn: 24 Univ. of Massachusetts, 105 Campus Ctr., 01003. Phone: (413) 545-2876. Fax: (413) 545-0682. E-mail: adviser@wmua.org Web Site: www.wmua.org. Licensee: Board of Trustees of University of Massachusetts. Population served: 1,000,000 Format: Var/div. News: 3 hrs wkly. Target aud: General; Univ. ♦Zach Claudio, gen mgr; Leila Denna, progmg dir; Corey Charron, mus dir; Dan Ferreira, chief of engrg.

WPNI(AM)— Apr 2, 1963: 1430 khz; 5 kw-D, DA. TL: N42 21 25 W72 29 13. Hrs open: 6 AM-midnight 98 Lower Westfield Rd., 3rd Fl., Holyoke, 01040-2712. Phone: (413) 536-1105. Licensee: 6 Johnson Road Licenses Inc. Group owner: Pamal Broadcasting Ltd. (acq 5-29-2003; $8 million with co-located FM). Population served: 17926 Law Firm: Ginsburg, Feldman & Bress.

WRNX(FM)— Nov 12, 1990: 100.9 mhz; 1.35 kw. Ant 692 ft TL: N42 18 24 W72 31 59. Hrs open: 1331 Main St., 4th Fl., Springfield, 01103. Phone: (413) 781-1011. Fax: (413) 734-4434. Web Site: www.wrnx.com. Licensee: CC Licenses LLC. (acq 4-1-2007; grpsl). Format: AAA. ♦Sean Davey, gen mgr.

Ashland

WSRO(AM)— May 1967: 650 khz; 250 w-D. TL: N42 17 17 W71 25 53. Hrs open: 100 Mt. Wayte Ave., Framingham, 01702. Phone: (508) 424-2568. Fax: (508) 820-2473. E-mail: wsroam650@yahoo.com Web Site: www.wsro.com. Licensee: Langer Broadcasting Group L.L.C. (group owner; acq 1996; $10,000). Format: Talk, relg. Target aud: 20-80; general. ♦Carl Abrams, gen mgr.

Athol

WJOE(AM)—See Orange-Athol

WNYN-FM— Dec 4, 1989: 99.9 mhz; 1.85 kw. Ant 407 ft TL: N42 35 39 W72 12 02. Stereo. Hrs opn: 24 362 Green St., Gardner, 01440. Phone: (978) 630-3473. Fax: (978) 630-3011. E-mail: info@theeagle.com Web Site: www.theeagle999.com. Licensee: County Broadcasting Co. LLC. Group owner: Northeast Broadcasting Company Inc. (acq 10-6-2003; $650,000 with WJOE(AM) Orange-Athol). Population served: 50,000 Natl. Network: ABC. Format: Classic rock. ♦Glenn Cardinal, gen mgr; Spencer Marshall, progmg dir.

Attleboro

WARL(AM)— Oct 8, 1950: 1320 khz; 5 kw-U, DA-2. TL: N41 57 33 W71 19 37. Hrs open: 127 Dorrance St., 5th Fl., Providence, RI, 02903. Phone: (508) 989-5013. Fax: (401) 521-5878. E-mail: scott@spojo.com Web Site: www.1320thedrive.com. Licensee: The ADD Radio Group Inc. (acq 6-1-98; $600,000). Population served: 800,000 Law Firm: Arent, Fox, Kintner, Plotkin & Kahn. Format: News/talk, sports radio. Target aud: 18 plus; middle to upper middle class. ♦Scott MacPherson, gen mgr.

Barnstable

WQRC(FM)— July 20, 1970: 99.9 mhz; 50 kw. 378 ft TL: N41 41 19 W70 20 49. Stereo. Hrs opn: 24 737 W. Main St., Hyannis, 02601. Phone: (508) 771-1224. Fax: (508) 775-2605. E-mail: wqrc@cape.com Web Site: www.wqrc.com. Licensee: Sandab Communications Limited Partnership II (group owner; acq 4-16-92; grpsl; FTR: 1-13-92). Population served: 225,000 Natl. Network: AP Radio. Natl. Rep: Clear Channel. Law Firm: Covington & Burling. Wire Svc: AP Format: Adult contemp. News staff: 4; News: 32 hrs wkly. Target aud: Adults/women; 25-54. ♦Gregory D. Bone, gen mgr; Wayne White, opns mgr; Stephen Colella, sls dir; Michelle Dodd, prom mgr; Donna Credit, traf mgr.

Beverly

WNSH(AM)— Dec 23, 1963: 1570 khz; 1 kw-U, DA-2. TL: N42 33 08 W70 55 43. Hrs open: 24 31 Woodbury St., South Hamilton, 01982. Secondary address: 31 Woodbury St., South Hamilton 01982. Phone: (978) 921-1570. Fax: (978) 468-1954. E-mail: jackwhite@wnsh.com Web Site: www.wnsh.com. Licensee: Willow Farm Inc. (acq 9-24-97; $50,000). Population served: 150,000 Format: Women's talk. News staff: 2; News: 15 hrs wkly. Target aud: 35 plus. ♦Jack White, gen mgr.

Boston

WBCN(FM)— May 1968: 104.1 mhz; 20.9 kw. 771 ft TL: N42 20 50 W71 04 59. Stereo. Hrs opn: 83 Leo M Birmingham Pkwy., Brighton, 02135. Phone: (617) 746-1400. Fax: (617) 746-1408. Web Site: www.wbcn.com. Licensee: Hemisphere Broadcasting Corp. Group owner: Infinity Broadcasting Corp. (acq 11-13-98; grpsl). Population served: 650,000 Natl. Network: CBS. Format: Alternative. Target aud: 18-34; men. ♦Mel Karmazin, pres; Tony Berardini, gen mgr.

WBMX(FM)— 1948: 98.5 mhz; 9 kw. 1,145 ft TL: N42 18 27 W71 13 27. Stereo. Hrs opn: 24 1200 Soldiers Field Rd., 02134. Phone: (617) 779-2000. Fax: (617) 779-2002. Web Site: www.mix985.com. Licensee: CBS Radio Stations Inc. Group owner: Infinity Broadcasting Corp. (acq 6-5-98; grpsl). Natl. Network: CBS. Natl. Rep: Christal. Format: Modern adult contemp, var. ♦Barbara Jean Scannell, gen mgr; Jerry McKenna, progmg dir.

WBOS(FM)—See Brookline

***WBUR-FM**— March 1950: 90.9 mhz; 7.2 kw. Ant 1,046 ft TL: N42 18 27 W71 13 27. Stereo. Hrs opn: 24 890 Commonwealth Ave., 3rd Fl., 02215. Phone: (617) 353-0909. Fax: (617) 353-4747. Web Site: www.wbur.org. Licensee: The Executive Committee of Trustees of The Boston University. Population served: 3,100,000 Natl. Network: NPR, PRI. Format: News, talk. Target aud: 25-54; intelligent adults interested in news natl, internatl & local. ♦Paul LaCamera, gen mgr & progmg dir; Corey Lewis, stn mgr; Sam Fleming, progmg dir; John Davidow, news dir; Jeff Hutton, engrg dir.

WBZ(AM)— Sept 19, 1921: 1030 khz; 50 kw-U, DA-1. TL: N42 16 44 W70 52 34. Stereo. Hrs opn: 24 1170 Soldiers Field Rd., 02134. Phone: (617) 787-7000. Fax: (617) 787-5969. Web Site: www.wbz1030.com. Licensee: CBS Radio East Inc. Group owner: Infinity Broadcasting Corp. Population served: 641,071 Natl. Network: ABC, CBS. Natl. Rep: CBS Radio. Format: News/talk. Target aud: 25-54. ♦Ted Jordan, gen mgr; Peter Casey, progmg dir & news dir.

WCRB(FM)—See Lowell

WEEI(AM)— Dec 1, 1926: 850 khz; 50 kw-U, DA-2. TL: N42 16 41 W71 16 02. Hrs open: 24 20 Guest St., 3rd Flr., Brighton, 02135-2040. Phone: (617) 779-3500. Fax: (617) 779-3557. Web Site: www.weei.com. Licensee: Entercom Boston License L.L.C. Group owner: Entercom Communications Corp. (acq 10-15-98; $82 million with WRKO(AM) Boston). Population served: 3,300,000 Natl. Network: CBS. Format: Sports talk. News staff: 6. Target aud: 25-54. ♦Jason Wolfe, progmg dir & progmg; Julie Kahn, gen mgr & engrg dir; Jim Rushton, chief of engrg & adv.

***WERS(FM)**— Nov 14, 1949: 88.9 mhz; 4 kw. 614 ft TL: N42 21 08 W71 03 25. Stereo. Hrs opn: 24 c/o Emerson College, 120 Boylston St., 02116. Phone: (617) 824-8891. Fax: (617) 824-8804. E-mail: aldenfertig@emerson.edu Web Site: www.wers.org. Licensee: Emerson College. Population served: 3,000,000 Format: Var. News: 3.5 hrs wkly. Spec prog: Black 15 hrs, blues 15 hrs, jazz 15 hrs, Broadway 4 hrs, relg one hr wkly. ♦Jack Casey, gen mgr; Alden Fertig, opns mgr.

WEZE(AM)— Sept 29, 1924: 590 khz; 5 kw-U, DA-1. TL: N42 24 24 W71 05 14. Hrs open: 24 308 Victory Rd., North Quincy Phone: (617) 328-0880. Fax: (617) 328-0375. E-mail: mnillas@wezeradio.com Web Site: www.wezeradio.com. Licensee: Pennsylvania Media Associates Inc. Group owner: Salem Communications Corp. (acq 1-31-97; $6 million). Population served: 5,000,000 Natl. Network: ABC. Law Firm: Irwin, Campbell, Crowe & Tannenwald. Format: Relg, talk. News: 5 hrs wkly. Target aud: 25 plus. ♦Edward G. Atsinger III, pres; Alex Canavan, gen mgr.

***WGBH(FM)**— Oct 6, 1951: 89.7 mhz; 98 kw. 650 ft TL: N42 12 42 W71 06 51. Stereo. Hrs opn: 24 1 Gwert St., Brighton, 02135. Phone: (617) 300-2000. Fax: (617) 300-1026. E-mail: wgbh@wgbh.org Web Site: www.wgbh.org. Licensee: WGBH Educational Foundation. Population served: 330,000 Natl. Network: PRI, NPR. Format: Class, jazz, news. News: 22 hrs wkly. Target aud: General. Spec prog: Folk 10 hrs, blues 8 hrs, Irish 2 hrs, cultural 3 hrs wkly. ♦John Abbott, pres; Marita Rivero, gen mgr. Co-owned TV: *WGBH-TV, *WGBX-TV affils

WHRB(FM)—See Cambridge

WILD(AM)— 1946: 1090 khz; 5 kw-D. TL: N42 24 40 W71 04 28. Hrs open: 500 Victory Rd., Quincy, 02179. Phone: (617) 472-9447. Fax: (617) 472-9474. Web Site: wild1090.com/home.asp. Licensee: Radio One of Boston Licenses LLC. Group owner: Radio One Inc. (acq 12-20-2000; $5 million in cash & stock merger). Population served: 2,805,911 Natl. Network: ABC. Natl. Rep: Roslin. Format: News/talk. Target aud: General. ♦Frank Kelley, pres & gen mgr.

WJIB(AM)—Cambridge, 1948: 740 khz; 250 w-D, 5 w-N. TL: N42 23 13 W71 08 21. Stereo. Hrs opn: 24 443 Concord Ave., Cambridge, 02138. Phone: (617) 868-7400. Licensee: Bob Bittner Broadcasting Inc. (group owner; acq 9-12-91). Population served: 3,200,000 Format: Adult Standards. News: 3 hrs wkly. Target aud: 40-75; locally-programmed for those enjoying good adult mus. Spec prog: French 10 hrs, gospel 4 hrs wkly. ♦Bob Bittner, pres & gen mgr.

WJMN(FM)— Mar 31, 1948: 94.5 mhz; 11.5 kw. 1,053 ft TL: N42 18 27 W71 13 27. Stereo. Hrs opn: 10 Cabot Rd., Suite 302, Medford, 02155. Phone: (781) 663-2500. Fax: (781) 290-0722. E-mail: management@jamn945.com Web Site: www.jamn945.com. Licensee: AMFM Radio Licenses L.L.C. Group owner: Clear Channel Communications Inc. (acq 8-30-00; grpsl). Population served: 3,672,000 Natl. Rep: Katz Radio. Law Firm: Latham & Watkins. Format: Top-40. Target aud: 12-44. ♦Tom McConnell, gen mgr.

WKLB-FM—See Waltham

WMJX(FM)— Jan 6, 1982: 106.7 mhz; 21.5 kw. 750 ft TL: N42 20 50 W71 04 59. Stereo. Hrs opn: 55 Morrissey Blvd., 02125. Phone: (617) 822-9600. Fax: (617) 822-6571. E-mail: dkelley@magic1067.com Licensee: Greater Boston Radio Inc. Group owner: Greater Media Inc. (acq 2-85). Population served: 3,300,000 Natl. Rep: Katz Radio. Format: Adult contemp, soft rock. ♦Phil Redo, gen mgr; Jackie Laudry, gen sls mgr; Don Kelley, progmg dir.

WMKI(AM)— 1922: 1260 khz; 5 kw-U, DA-N. TL: N42 16 30 W71 02 31. Hrs open: 226 Lincoln St., Allston, 02134. Phone: (617) 787-0146. Fax: (617) 787-1236. Web Site: www.disney.com. Licensee: Radio Disney Group LLC. Group owner: ABC Inc. (acq 8-22-00; grpsl). Population served: 3000000 Natl. Network: USA. Format: Family. Target aud: 6-14. ♦Michael Kellogg, gen mgr & stn mgr.

WMKK(FM)—Lawrence, April 1960: 93.7 mhz; 50 kw. 430 ft TL: N42 40 26 W71 11 26. (CP: 29.5 kw, ant 640 ft. TL: N42 35 42 W71 02 18). Stereo. Hrs opn: 24 Entercom Boston, 20 Guest St., 3rd Fl., Brighton, 02135. Phone: (617) 779-5300. Fax: (617) 931-7827. Web Site: www.937mikefm.com. Licensee: Entercom Boston II License L.L.C. Group owner: Entercom Communications Corp. (acq 10-15-98; grpsl). Population served: 300,000 Format: Var. Target aud: 25-54. ♦Julie Kahn, gen mgr; Christina Anders, prom dir.

WNTN(AM)—See Newton

WODS(FM)— 1948: 103.3 mhz; 16.5 kw. 938 ft TL: N42 18 27 W71 13 27. Stereo. Hrs opn: 24 83 Leo Birmingham Pkwy., 02135. Phone: (617) 787-7500. Fax: (617) 787-7523. E-mail: murley.tina@cbsradio.com Web Site: www.oldies1033.com. Licensee: CBS Radio East Inc. Group owner: Infinity Broadcasting Corp. (acq 11-13-98; grpsl). Natl.

Rep: CBS Radio. Format: Oldies. ♦Ted Jordan, gen mgr; Tina Murley, gen sls mgr; Tanya Frazier, mktg dir; Courtney Conners, prom mgr; Pete Falconi, progmg dir; Don Albanese, chief of engrg.

***WRBB(FM)**— October 1970: 104.9 mhz; 10.9 w. 89 ft TL: N42 20 19 W71 05 28. Hrs open: 24 #174 Curry Student Ctr., 360 Huntington Ave., 02115. Phone: (617) 373-4338. Fax: (617) 373-5095. Web Site: www.wrbbradio.org. Licensee: Northeastern University. Population served: 900,000 Format: Var/div. News staff: one; News: 2 hrs wkly. Target aud: 12-35; college, urban. ♦Emily Rodrigues, gen mgr; Michelle Bablo, progmg; Ryan Scianino, engr.

WRKO(AM)— 1922: 680 khz; 50 kw-U, DA-2. TL: N42 29 25 W71 13 05. Hrs open: 20 Guest St., 3rd Fl., Brighton, 02135. Phone: (617) 779-3400. Fax: (617) 779-3467. Web Site: www.wrko.com. Licensee: Entercom Boston License L.L.C. Group owner: Entercom Communications Corp. (acq 10-15-98; $82 million with WEEI(AM) Boston). Population served: 641,071 Natl. Network: ABC. Format: Talk. Target aud: 25-54. ♦Julie Kahn, gen mgr; Christina Andres, prom mgr; Jim Rushton, adv.

WROL(AM)— Oct 8, 1950: 950 khz; 5 kw-D, 500 w-N. TL: N42 26 15 W70 59 40. Hrs open: 24 308 Victory Rd., N. Quincy, 02171. Phone: (617) 328-0880. Fax: (617) 328-0375. Web Site: www.wrolradio.com. Licensee: SCA License Corp. Group owner: Salem Communications Corp. (acq 3-2-01; $11 million). Population served: 3,000,000 Format: Christian. Target aud: Adults. ♦Alex Canavan, gen mgr.

WROR-FM—See Framingham

WTKK(FM)— 1945: 96.9 mhz; 12.5 kw. 1,010 ft TL: N42 18 12 W71 13 08. (CP: 22.5 kw, ant 735 ft.). Stereo. Hrs opn: 55 Morrissey Blvd., 02125. Phone: (617) 822-9600. Fax: (617) 822-6871. Web Site: www.wtkk.com. Licensee: Greater Boston Radio Inc. Group owner: Greater Media Inc. (acq 3-31-93; $11.65 million; FTR: 4-19-93). Population served: 641071 Natl. Rep: Katz Radio. Format: Talk. ♦Chris Paquin, gen mgr & gen sls mgr; Phil Redo, gen mgr; Paula O'Connor, progmg dir.

WTTT(AM)— Jan 1, 1979: 1150 khz; 5 kw-U, DA-2. TL: N42 24 48 W71 12 40. Stereo. Hrs opn: 308 Victory Rd/, N. Quincy, 02171. Phone: (617) 328-0880. Fax: (617) 328-0375. Web Site: www.talk1150.com. Licensee: Pennsylvania Media Associates Inc. Group owner: Salem Communications Corp. (acq 10-31-2003; $8.6 million). Population served: 1,000,000 Format: Talk. ♦Alex Canavan, gen mgr.

***WUMB-FM**— Sept 19, 1982: 91.9 mhz; 320 w. 207 ft TL: N42 15 27 W71 01 44. Stereo. Hrs opn: 24 Univ. of Massachusetts Boston, 100 Morrissey Blvd., 02125-3393. Phone: (617) 287-6900. Fax: (617) 287-6916. E-mail: wumb@umb.edu Web Site: www.wumb.org. Licensee: The University of Massachusetts. (group owner) Population served: 2,500,000 Natl. Network: PRI, NPR. Format: Folk. News: 5 hrs wkly. Target aud: 25-40. ♦Patricia A. Monteith, gen mgr; Brian Quinn, progmg dir.

WUNR(AM)—Brookline, 1947: 1600 khz; 5 kw-U, DA-1. TL: N42 17 20 W71 11 22. Hrs open: 160 N. Washington St., 02114. Phone: (617) 367-9003. Fax: (617) 367-2265. Web Site: wunr.com. Licensee: Champion Broadcasting System Inc. Population served: 641,071 Format: Ethnic, Sp. ♦Steve Lalli, gen mgr; Velma May, progmg dir & pub affrs dir.

WWZN(AM)— 1934: 1510 khz; 50 kw-U, DA-2. TL: N42 23 10 W71 12 01. Stereo. Hrs opn: 1 Van De Graaff Dr., Suite 300, Burlington, 01803-5171. Phone: (781) 221-7878. Fax: (781) 221-7877. E-mail: wwzn@1510thezone.com Web Site: www.1510thezone.com. Licensee: Rose City Radio Corp. (group owner; acq 3-23-01; grpsl). Population served: 535,000 Natl. Network: Sporting News Radio Network. Law Firm: Haley, Bader & Potts. Format: Sports. Target aud: 25-54. ♦Anthony Pepe, gen mgr; Jon Anik, progmg dir; Brad Parsons, chief of engrg.

WXKS-FM—See Medford

WZLX(FM)— Jan 1, 1979: 100.7 mhz; 21.5 kw. Ant 777 ft TL: N42 20 50 W71 04 59. Stereo. Hrs opn: 83 Leo M. Birmingham Pkwy., Brighton, 02135. Phone: (617) 746-5100. Fax: (617) 746-5105. Web Site: www.wzlx.com. Licensee: CBS Radio Inc. of Boston. Group owner: Infinity Broadcasting Corp. (acq 11-13-98; grpsl). Population served: 641,071 Natl. Network: CBS. Natl. Rep: CBS Radio. Format: Classic rock. Target aud: 25-54; males. ♦Mark Hannon, gen mgr; Joe Soucise, chief of engrg.

Boxford

***WBMT(FM)**— Jan 30, 1978: 88.3 mhz; 710 w. 17 ft TL: N42 37 39 W70 58 21. Hrs open: 2:30 PM-9 PM(M-F); 10 AM-6 PM(S,Su) 20 Endicott Rd., Topsfield, 01983. Phone: (978) 887-8830. Fax: (978) 887-7243. E-mail: gwalker@masconomet.org Licensee: Masconomet Regional High School System. Population served: 35,000 Format: AOR. ♦Glenn Walker, gen mgr.

Brewster

WZAI(FM)— 2006: 94.3 mhz; 4.7 kw. Ant 372 ft TL: N41 46 36 W70 00 40. Hrs open:
Rebroadcasts WNAN-FM Nantucket and WCAI-FM Woods Hole 100%.
Box 82, Woods Hole, 02543. Phone: (508) 548-9600. Fax: (508) 548-5517. E-mail: cainan@wgbh.org Web Site: www.cainan.org. Licensee: WGBH Educational Foundation. Natl. Network: NPR. Format: News/talk. ♦Eric A. Brass, gen mgr.

Bridgewater

***WBIM-FM**— November 1972: 91.5 mhz; 180 w. Ant 71 ft TL: N41 59 15 W70 58 21. Stereo. Hrs opn: 24 Campus Ctr. 109, Bridgewater State College, 02325. Phone: (508) 531-1303. Phone: (508) 531-1366. Fax: (508) 531-1786. E-mail: wbim@bridgew.edu Web Site: www.bridgew.edu/wbim. Licensee: Bridgewater State College. Population served: 20,000 Format: Var. News staff: one; News: 14 hrs wkly. Target aud: 18-35; college students & loc residents. ♦Mark C. Lilly, gen mgr & stn mgr; Kevin Kennedy, progmg dir.

Brockton

WKAF(FM)— July 21, 1948: 97.7 mhz; 1.7 kw. Ant 567 ft TL: N42 12 42 W71 06 51. Hrs open: 500 Victory Rd., Quincy, 02171. Phone: (617) 427-9447. Fax: (617) 472-9581. Web Site: www.radio-one.com. Licensee: Entercom Boston License L.L.C. Group owner: Radio One Inc. (acq 12-27-2006; $30 million). Population served: 1,500,000 Format: Hip-Hop, rhythm and blues. ♦Frank Kelley, gen mgr.

WMSX(AM)— July 17, 1961: 1410 khz; 1 kw-D, DA. TL: N42 03 30 W71 02 40. Hrs open: 288 Linwood St., 02301. Phone: (508) 587-5454. Fax: (508) 587-1950. E-mail: molinahbone@aol.com Licensee: Hispanic Broadcasters Inc. (acq 2-12-2004; $1.43 million). Population served: 400,000 Format: Sp. ♦Antonio Molina, gen mgr.

WXBR(AM)— Nov 27, 1946: 1460 khz; 5 kw-D, 1 kw-N, DA-N. TL: N42 04 23 W71 02 39. Hrs open: 60 Main St., 02301. Phone: (508) 587-2400. Fax: (508) 587-4786. Web Site: www.1460wxbr.com. Licensee: BTR Boston Inc. (acq 11-13-2006; $1 million). Population served: 250,000 Format: News/talk, sports. Target aud: 35 plus; general. ♦Richard Muserlian, gen mgr.

Brookline

WBOS(FM)— 1955: 92.9 mhz; 8.8 kw. 1,100 ft TL: N42 18 27 W71 13 27. Stereo. Hrs opn: 24 55 Morrissey Blvd., Boston, 02125. Phone: (617) 822-9600. Fax: (617) 822-6771. Web Site: www.wbos.com. Licensee: Greater Los Angeles Radio Inc. Group owner: Greater Media Inc. (acq 7-23-97). Population served: 400,000 Natl. Rep: Katz Radio. Format: AAA. News staff: one; News: 3 hrs wkly. Target aud: 25-49; baby boomers seeking diverse quality music. ♦Phil Redo, gen mgr; David Straws, gen sls mgr; David Ginsburg, progmg dir.

WUNR(AM)—Licensed to Brookline. See Boston

Cambridge

WHRB(FM)— May 1957: 95.3 mhz; 3 kw. 110 ft TL: N42 22 20 W71 07 09. (CP: 1.55 kw, ant 508 ft.). Stereo. Hrs opn: 24 389 Harvard St., 02138. Phone: (617) 495-8138. Fax: (617) 496-3990. E-mail: mail@whrb.org Web Site: www.whrb.org. Licensee: Harvard Radio Broadcasting Co. Inc. Population served: 3,000,000 Format: Class, jazz, AOR. News: 4 hrs wkly. ♦Stanley Chang, pres & gen mgr.

WJIB(AM)—Licensed to Cambridge. See Boston

***WMBR(FM)**— Apr 10, 1961: 88.1 mhz; 360 w. 285 ft TL: N42 21 42 W71 05 03. Stereo. Hrs opn: 6 AM-2 AM 3 Ames St., 02142. Phone: (617) 253-4000. Fax: (617) 232-1384. Web Site: www.wmbr.mit.edu. Licensee: Technology Broadcasting Corp. Population served: 600,000 Format: Var/div. Target aud: General. ♦Dugan Hayes, gen mgr; Gloria Apolinario, gen mgr; Christopher Bobko, stn mgr.

WTKK(FM)—See Boston

Charlton

***WYCM(FM)**— 1976: 90.1 mhz; 100 w. Ant 390 ft TL: N42 08 01 W71 57 26. Stereo. Hrs opn: 24 Box 573, 01507. Phone: (508) 248-0049. Fax: (508) 248-4518. E-mail: stationmanager@wycm.com Web Site: www.wycm.com. Licensee: Christian Mix Radio Inc. (acq 3-17-2004; $200,000). Population served: 750,000 Format: Christian music. ♦Stephen Binley, stn mgr; Judy Pelletier, prom.

Chatham

WFCC-FM— Mar 24, 1987: 107.5 mhz; 50 kw. 341 ft TL: N41 44 14 W70 00 40. Stereo. Hrs opn: 24
Rebroadcasts WCRB(FM) Waltham 100%.
582 Rt. 28, W. Yarmouth, 02673. Phone: (508) 790-3772. Fax: (508) 790-3773. E-mail: mail@wfcc.com Web Site: www.wfcc.com. Licensee: Cape Cod Broadcasting License I LLC (acq 7-10-2007; $7.5 million with WOCN-FM Orleans). Population served: 200,000 Format: Class. News: 2 hrs wkly. Target aud: 25 plus; upscale, affluent, educated adults. Spec prog: Children one hr wkly. ♦Christopher Jones, pres; Jan D'Antuono, gen mgr.

Chicopee

WACE(AM)— Dec 1, 1946: 730 khz; 5 kw-U. TL: N42 10 01 W72 37 31. Hrs open: 24 Box 1, Springfield, 01101. Secondary address: 326 Chicopee St. 01101. Phone: (413) 594-6654. Licensee: Carter Broadcasting Corp. (acq 12-24-86). Population served: 3,000,000 Format: Relg, talk. Spec prog: Pol one hr, Irish 2 hrs, Por one hr wkly. ♦Ken Carter, pres & gen mgr; Michael Durocher, opns mgr.

Concord

WBNW(AM)— Aug 28, 1989: 1120 khz; 5 kw-D, 1 kw-N, DA-2. TL: N42 26 54 W71 25 39. Hrs open: 5 AM-10 PM 144 Gould St. Suite 155, Needham, 02494. Phone: (781) 433-0001. Fax: (781) 433-0002. Web Site: www.moneymattersradio.net. Licensee: Money Matters Radio Inc. (acq 6-17-98; $550,000). Population served: 8,000 Rgnl rep: New England. Format: Business, personal finance. News: 17 hrs wkly. Target aud: 35 plus; upscale, suburban families. ♦Barry Armstrong, pres; Scott Cooper, gen mgr.

***WIQH(FM)**— December 1971: 88.3 mhz; 100 w. 30 ft TL: N42 26 48 W71 20 49. Stereo. Hrs opn: 1-9:30 PM (M-F); 10 AM-10 PM (S) 500 Walden St., 01742. Phone: (978) 318-1400 ext. 7185. Phone: (978) 369-2440. E-mail: wigh@colonial.net Web Site: www.wiqh.org. Licensee: Concord-Carlisle Regional School District. Population served: 17,500 Format: AOR, progsv. Target aud: 12-21; teenagers. ♦Ned Roos, gen mgr & stn mgr.

Danvers

WNSH(AM)—See Beverly

Dedham

WAMG(AM)— June 2005: 890 khz; 25 kw-D, 3.4 kw-N, DA-2. TL: N42 14 49 W71 25 30. Hrs open: 24 529 Main St., Suite 200, Charlestown, 02129-1119. Phone: (617) 830-1000. E-mail: laura.wareck@espnboston.com Web Site: www.890espn.com. Licensee: J Sports Licensee LLC Group owner: Mega Communications Inc. (acq 6-22-2005; $9 million with WLLH(AM) Lowell). Population served: 2,900,000 Natl. Network: ESPN Radio. Natl. Rep: McGavren Guild. Law Firm: Davis, Wright, Tremaine. Format: Sports. Target aud: Males 18+. ♦Jessamy Tang, gen mgr; Neil Kelleter, sls dir; Kara Lachance, prom mgr; Len Weiner, progmg dir.

Deerfield

***WGAJ(FM)**— May 1982: 91.7 mhz; 100 w. 314 ft TL: N42 32 05 W72 35 32. Stereo. Hrs opn: 7 AM-8 PM (M-F); 4 AM-11 PM (S); 7 AM-11 PM (Su) Deerfield Academy, 01342. Phone: (413) 774-1539. Fax: (413) 772-1100. Licensee: Trustees of Deerfield Academy. Population served: 150,000 Format: Var. Target aud: 10-20; teens, young adults, pre-teens. ♦Christopher Stacy, gen mgr & stn mgr.

Dudley

***WXRB(FM)**— 1975: 95.1 mhz; 14 w. Ant 125 ft TL: N42 02 40 W71 55 52. Stereo. Hrs opn: 24 48 Homeland Dr., Whitman, 02382. Phone: (508) 213-2138. E-mail: wxrbfm@yahoo.com Web Site: www.wxrbfm.com. Licensee: WXRB-FM Educational Broadcasting Inc. (acq 5-17-2006; $1,000). Population served: 5,000 Format: Golden oldies, educ. Target aud: 18 plus. ♦Andrea Becker, stn mgr.

East Longmeadow

WHNP(AM)— 1947: 1600 khz; 5 kw-D, 2.5 kw-N, DA-2. TL: N42 04 30 W72 31 40. Hrs open: 24 Box 268, Northampton, 01060. Secondary address: 15 Hampton Ave., Northampton 01060. Phone: (413) 586-7400. Web Site: www.whmp.com. Licensee: Saga Communications of New England LLC. Group owner: Saga Communications Inc. (acq 6-2-92; grpsl). Format: News/talk. Target aud: 18-49; upscale young adults. ♦Sean O'Mealy, gen mgr; Dave Musante, gen sls mgr; Chris Collins, progmg dir.

Easthampton

WVEI-FM— October 1967: 105.5 mhz; 720 w horiz, 706 w vert. Ant 918 ft TL: N42 14 29 W72 38 57. Stereo. Hrs opn: 24
Simulcast with WEEI(AM) Boston 100%.
326 Chicopee St., Chicopee, 01013. Phone: (413) 594-6585. Fax: (413) 592-1891. Web Site: www.wveifm.com. Licensee: Great Northern Radio LLC. (acq 9-13-2002; grpsl). Population served: 139,000 Format: Sports. ♦Julie Kahn, gen mgr.

Easton

***WSHL-FM**— Jan 1, 1973: 91.3 mhz; 100 w. 66 ft TL: N42 03 27 W71 04 47. Stereo. Hrs opn: 24 Stonehill College, 320 Washington St., North Easton, 02357. Phone: (508) 565-1000. Phone: (508) 565-1919. Fax: (508) 565-1974. E-mail: wshl@stonehill.edu Web Site: www.stonehill.edu/wshl. Licensee: Stonehill College. Population served: 10,000 Wire Svc: UPI Format: Div. Target aud: 19-30. ♦Ryan Delenunt, gen sls mgr; Megan Manshield, rgnl sls mgr.

Everett

WXKS(AM)—Licensed to Everett. See Medford

Fairhaven

WFHN(FM)— Mar 1, 1989: 107.1 mhz; 6 kw. 325 ft TL: N41 37 43 W71 00 24. Stereo. Hrs opn: 24 22 Sconticutneck Rd., 02719. Phone: (508) 999-6690. Fax: (508) 999-1420. E-mail: petebraley@wbsm.com Web Site: www.fun107.com. Licensee: Citadel Broadcasting Co. Group owner: Citadel Broadcasting Corp. (acq 2-23-00; grpsl). Population served: 800,000 Natl. Rep: McGavren Guild. Format: CHR. Target aud: 18-49. ♦Wayne Leland, exec VP; Gail Leblanc, gen mgr.

Fall River

WCTK(FM)—See Providence, RI

WHTB(AM)— May 13, 1948: 1400 khz; 1 kw-U. TL: N41 41 23 W71 08 43. Hrs open: 5 AM-11 PM 1 Home St., Somerset, 02725. Phone: (508) 678-9727. Fax: (508) 673-0310. Licensee: SNE Broadcasting Ltd. (acq 5-8-89; $650,000; FTR: 5-29-89). Natl. Rep: McGavren Guild. Format: Ethnic talk. Target aud: 25-64; Portuguese (ethnic). Spec prog: English 10 hrs, Pol one hr, Cambodian one hr, Fr one hr wkly. ♦Robert S. Karam, pres; Hector Gauthier, stn mgr & chief of opns.

WSAR(AM)— 1921: 1480 khz; 5 kw-U, DA-1. TL: N41 43 26 W71 11 21. Hrs open: One Home St., Somerset, 02725. Phone: (508) 678-9727. Fax: (508) 673-0310. E-mail: paul@wsar.com Web Site: www.wsar.com. Licensee: Bristol County Broadcasting Inc. (acq 1992; FTR: 11-23-92). Population served: 96,898 Format: News/talk, sports. Target aud: 25 plus. Spec prog: Por 3 hrs wkly. ♦Hector A. Gauthier Jr., gen mgr; Paul Giammarco, opns mgr & progmg dir.

Falmouth

WCIB(FM)— 1970: 101.9 mhz; 50 kw. 479 ft TL: N41 33 31 W70 35 46. Stereo. Hrs opn: 154 Barnstable Rd., Hyannis, 02601. Phone: (508) 778-2888. Fax: (508) 778-9651. E-mail: info@cool102.com Web Site: www.cool102.com. Licensee: Qantum of Cape Cod License Co. LLC. Group owner: Qantum Communications Corp. (acq 6-11-03; grpsl). Population served: 200,000 Natl. Rep: Eastman Radio. Format: Classic hits. Target aud: 25-54. ♦Allison Makkay, gen mgr; Steve McVie Solomon, opns dir.

***WFPB-FM**— 1996: 91.9 mhz; 300 w. 177 ft TL: N41 36 50 W70 35 56. Hrs open:
Rebroadcasts WUMB-FM Boston 100%.
Univ. of Massachusetts, 100 Morrissey Blvd., Boston, 02125-3393. Phone: (617) 287-6900. Fax: (617) 287-6916. E-mail: wumb@umb.edu Web Site: www.wumb.org. Licensee: University of Massachusetts. Natl. Network: NPR. Format: Folk. ♦Patricia A. Monteith, gen mgr; Danielle Knight, dev dir; Brian Quinn, progmg dir.

Fitchburg

WEIM(AM)— Oct 6, 1941: 1280 khz; 5 kw-D, 1 kw-N, DA-1. TL: N42 35 40 W71 50 12. Hrs open: 24 762 Water St., 01420. Phone: (978) 343-3766. Fax: (978) 345-6397. E-mail: radio@weim.com Web Site: www.am1280theblend.com. Licensee: Central Broadcasting Co. LLC (acq 11-1-2005; $795,000). Population served: 500,000 Natl. Network: ABC. Wire Svc: AP Format: news/talk, sports(Red Sox & Patriots—High School & College). News staff: 2; News: 10 hrs wkly. Target aud: 28 plus; loc listeners in the heart of New England. ♦William J. Macek, gen mgr; Ben Parker, stn mgr; Anne Bisbee, opns mgr.

WFGL(AM)— February 1950: 960 khz; 2.5 kw-D, 1 kw-N, DA-2. TL: N42 35 24 W71 49 41. Hrs open: 24
Rebroadcasts KAWZ(FM) Twin Falls, ID 50%.
356 Broad St., 01420-3030. Phone: (978) 342-5025. E-mail: mail@wfgl.org Web Site: www.wfgl.org. Licensee: CSN International. (group owner; acq 1993). Population served: 250,000 Format: Christian. Target aud: 25-54; college & career age, young families. ♦Jim Mottshager, gen mgr; Pete Cesnoia, stn mgr.

WXLO(FM)—Licensed to Fitchburg. See Worcester

***WXPL(FM)**— August 1985: 91.3 mhz; 100 w. 134 ft TL: N42 35 18 W71 47 26. Stereo. Hrs opn: Fitchburg State College, 160 Pearl St., 01420. Phone: (978) 665-3163. Fax: (978) 665-3693. E-mail: wxpl@fsc.edu Web Site: falcon.fsc.edu/~wxpl. Licensee: Fitchburg State College. Format: Var. Target aud: 16-25. ♦Sherry Horeanopoulos, gen mgr & stn mgr.

Framingham

WBIX(AM)—See Natick

***WDJM-FM**— 1973: 91.3 mhz; 100 w. 89 ft TL: N42 17 44 W71 26 18. Stereo. Hrs opn: Framingham State College, 100 State St., 01701. Phone: (508) 626-4622. Fax: (508) 626-4939. Licensee: Framingham State College. Format: Alternative. Target aud: 15-35; college, surrounding community & commuters.

WKOX(AM)— April 1947: 1200 khz; 10 kw-D, 1 kw-N, DA-N. TL: N42 17 17 W71 25 53. (CP: COL Newton. 50 kw-U, DA-2. TL: N42 17 20 W71 11 21). Hrs opn: 24 99 Revase Beach Pkwy., Medford, 02155. Phone: (781) 396-1430. Fax: (781) 391-3064. Web Site: www.1200rumba.com. Licensee: Capstar TX L.P. Group owner: Clear Channel Communications Inc. (acq 2-15-2001; $10 million). Population served: 1,700,000 Law Firm: Haley, Bader & Potts. Format: Sp. ♦Tom McConnell, VP & gen mgr.

WROR-FM— 1959: 105.7 mhz; 23 kw. Ant 735 ft TL: N42 20 50 W71 04 59. Stereo. Hrs opn: 55 Morrissey Blvd., Boston, 02125. Phone: (617) 822-9600. Fax: (617) 822-6471. Web Site: www.wror.com. Licensee: Greater Boston Radio Inc. Group owner: Greater Media Inc. (acq 10-11-96). Population served: 3,200,000 Format: Classic hits. Target aud: 25-54. ♦Phil Redo, gen mgr; Chris Paquin, gen sls mgr & news dir.

Franklin

***WGAO(FM)**— 1975: 88.3 mhz; 125 w. 174 ft TL: N42 05 08 W71 23 54. Stereo. Hrs opn: Dean College, 99 Main St., 02038. Phone: (508) 528-4210. Web Site: www.dean.edu. Licensee: Dean College. Population served: 100,000 Natl. Network: AP Radio. Format: Classic rock, CHR. Target aud: 15-25. Spec prog: Relg 8 hrs wkly. ♦Vic Michaels, gen mgr, opns dir & progmg dir.

Gardner

WGAW(AM)— 1946: 1340 khz; 1 kw-U. TL: N42 35 33 W71 59 20. Hrs open: 362 Green St., NH, 01440. Phone: (978) 630-8700. Fax: (978) 632-1332. Fax: (603) 577-8682. E-mail: wotw900am@hotmail.com Licensee: County Broadcasting Co. LLC. Group owner: Northeast Broadcasting Company Inc. (acq 12-2-2003; $235,000). Population served: 30,000 Natl. Network: ABC. Format: News/talk. ♦Chris Thompson, gen mgr; William B. Curtis, opns dir; Chuck Wright, progmg dir & chief of engrg; Kevin Kistler, traf mgr.

***WJWT(FM)**— 2006: 91.7 mhz; 850 w. Ant 276 ft TL: N42 33 29 W72 03 06. Hrs open: CSN International, 3232 W. MacArthur Blvd., Santa Ana, CA, 92704. Phone: (714) 825-9663. Fax: (714) 825-9661. Licensee: CSN International (group owner). Format: Relg. ♦Patrick Lannoye, gen mgr & opns mgr.

Gloucester

WBOQ(FM)— Sept 14, 1964: 104.9 mhz; 3.2 kw. 446 ft TL: N42 35 36 W70 43 28. Stereo. Hrs opn: 24 8 Enon St., North Beverly, 01915. Phone: (978) 927-1049. Fax: (978) 921-2635. Web Site: www.wboq.com. Licensee: Westport Communications L.P. Population served: 650,000 Law Firm: Akin, Gump, Strauss, Hauer & Feld. Format: Classic hits. News: 3 hrs wkly. Target aud: 25-54; mass appeal classical favorites. ♦Todd Tanger, gen mgr; Sam Koffman, opns mgr.

Great Barrington

***WAMQ(FM)**— November 1988: 105.1 mhz; 730 w. 918 ft TL: N42 09 36 W73 28 48. Stereo. Hrs opn: 24
Rebroadcasts WAMC-FM Albany, N.Y. 100%.
318 Central Ave., Box 66600, Albany, NY, 12206-6600. Phone: (518) 465-5233. Fax: (518) 432-6974. E-mail: mail@wamc.org Web Site: www.wamc.org. Licensee: WAMC. Group owner: WAMC/Northeast Public Radio (acq 3-5-93; $325,000; FTR: 3-29-93). Natl. Network: NPR, PRI. Law Firm: Dow, Lohnes & Albertson. Wire Svc: AP Format: News/talk. Target aud: General. Spec prog: Folk 7 hrs, jazz 13 wkly. ♦Alan Chartock, CEO; David Galletly, VP & progmg dir; Selma Kaplan, VP; Dona Frank, dev dir.

WSBS(AM)— December 1956: 860 khz; 2.7 kw-D. TL: N42 12 52 W73 20 45. Hrs open: 24 425 Stockbridge Rd., 01230. Phone: (413) 528-0860. Fax: (413) 528-2162. E-mail: fun@wsbs.com Web Site: www.wsbs.com. Licensee: Vox Communications Group LLC. Group owner: Vox Radio Group L.P. (acq 5-10-2004; grpsl). Population served: 188,594 Natl. Network: AP Radio. Natl. Rep: D & R Radio. Law Firm: Wilkinson Barker Knauer. Wire Svc: AP Format: Adult contemp,. News staff: 3; News: 10 hrs wkly. Target aud: 25 plus; general. ♦David Isby, gen mgr.

Greenfield

WHAI(FM)—Listing follows WHMQ(AM).

WHMQ(AM)— May 15, 1938: 1240 khz; 1 kw-U. TL: N42 35 21 W72 37 08. Hrs open: 24
Rebroadcasts WHMP(AM) Northampton 98%.
15 Hampton Ave., Northampton, 01060. Phone: (413) 586-7400. Fax: (413) 585-0927. Web Site: www.whmp.com. Licensee: Saga Communications of New England LLC. Group owner: Saga Communications Inc. (acq 4-1-01; $2.2 million with co-located FM). Population served: 65,000 Natl. Network: CBS. Law Firm: Wiley, Rein & Fielding. Format: News info. News staff: 2; News: 20 hrs wkly. Target aud: 35 plus. ♦Sean O'Mealy, gen mgr; Dave Musante, gen sls mgr & news dir; Chris Collins, progmg dir & news dir; Barbara Kuschka, traf mgr.

WHAI(FM)—Co-owned with WHMQ(AM). May 15, 1948: 98.3 mhz; 2 kw. 403 ft TL: N42 34 15 W72 38 42. Stereo. 24 81 Woodard Rd., 01301. Phone: (413) 774-4301. Fax: (413) 773-5637. E-mail: info@whai.com Web Site: whai.com.100,000 Format: Adult contemp. News staff: one. Target aud: 25-54. ♦Dan Guin, gen mgr; Nick Danjer, progmg dir & disc jockey.

WINQ(FM)—See Winchester NH

WIZZ(AM)— Aug 26, 1980: 1520 khz; 10 kw-D, DA. TL: N42 36 12 W72 36 21. Hrs open: Box 983, 01302. Secondary address: 369 S. Shelburne Rd. 01370. Phone: (413) 774-5757. Fax: (413) 625-8274. E-mail: info@wizzradio.com Web Site: www.wizzradio.com. Licensee: P. & M. Radio LLC (acq 1-31-03; $150,000). Natl. Network: AP Network News. Format: Nostalgia. ♦Phillip G. Drumheller, pres & gen mgr.

WPVQ(FM)— July 26, 1981: 95.3 mhz; 320 w. 780 ft TL: N42 41 50 W72 36 20. Stereo. Hrs opn: 24 81 Woodard Rd., 01301. Phone: (413) 774-4301. Fax: (413) 773-5637. E-mail: info@bear953.com Web Site: www.bear953.com. Licensee: Saga Communications of New England LLC. Group owner: Saga Communications Inc. (acq 2-13-2004; grpsl). Format: Country. Target aud: 18-45. ♦Dan Guin, gen mgr & gen sls mgr.

Harwich

***WCCT-FM**— May 1988: 90.3 mhz; 160 w horiz, 640 w vert. 125 ft TL: N41 42 40 W70 04 34. Hrs open: 24
Rebroadcasts WBUR(FM) Boston 90%.
Cape Cod Tech., 351 Pleasant Lake Ave., 02645. Phone: (508) 432-4500. Fax: (508) 432-7916. E-mail: jganss@capetech.us Web Site: www.wbur.org. Licensee: Cape Cod Regional Technical High School. (acq 11-11-87). Natl. Network: NPR. Format: Music & talk. News: 30 hrs wkly. Target aud: 20-65; educated adults. ♦John Ganss, gen mgr.

Harwich Port

WFQR(FM)— May 11, 1989: 93.5 mhz; 3 kw. Ant 328 ft TL: N41 44 19 W70 00 40. Stereo. Hrs opn: 24
Rebroadcasts WFRQ(FM) Mashpee.
278 S. Sea Ave., W. Yarmouth, 02673. Phone: (508) 760-5252. Fax: (508) 862-6329. Licensee: Nassau Broadcasting III L.L.C. Group owner: Boch Broadcasting (acq 11-7-2005; grpsl). Natl. Network: Westwood One. Natl. Rep: Katz Radio. Format: Adult hits. Target aud: 35-64. ♦Jake Demmin, gen mgr.

Haverhill

WCEC(AM)— 1947: 1490 khz; 1 kw-U. TL: N42 46 22 W71 06 01. Hrs open: 24 462 Merrimac St., Methuen, 01844. Phone: (978) 683-7171. Phone: (978) 686-9966. Fax: (978) 687-1180. Web Site: www.1490wcec.com. Licensee: Costa-Eagle Radio Ventures L.P. (group owner; (acq 1998). Population served: 249,700 Natl. Rep: Roslin. Law Firm: Bryan Cave. Format: Sp talk and info. Target aud: Sp speaking. ♦Patrick J. Costa, gen mgr; Luis Reyes, opns mgr.

WXRV(FM)— June 1959: 92.5 mhz; 25 kw. 710 ft TL: N42 46 23 W71 06 01. Stereo. Hrs opn: 24 30 How St., 01830. Phone: (978) 374-4733. Fax: (978) 373-8023. E-mail: info@wxrv.com Web Site: www.wxrv.com. Licensee: Beanpot License Corp. Group owner: Northeast Broadcasting Co. Inc. (acq 1981). Population served: 2,000,000 Format: AAA, adult contemp. Target aud: 25-54. ♦Terry Lieberman, gen mgr; Steve Young, gen sls mgr & chief of engrg; Ron Bowen, progmg dir & mus dir.

Holliston

***WHHB(FM)**— Apr 17, 1979: 99.9 mhz; 10 w. 52 ft TL: N42 12 29 W71 26 19. (CP: 170 w, ant 203 ft.). Hrs opn: Holliston High School, 370 Hollis St., 01746. Phone: (508) 429-0677. Fax: (508) 429-8225. E-mail: requests@whhbfm.com Licensee: Holliston High School. Format: Var. ♦Christopher Murphy, gen mgr & stn mgr.

Holyoke

***WCCH(FM)**— 1977: 103.5 mhz; 10 w. 258 ft TL: N42 11 55 W72 38 27. Hrs open: 6 AM-11 PM Holoyoke Community College, 303 Homestead Ave., 01040. Phone: (413) 552-2488. Licensee: Holyoke Community College. Format: Var. ♦Joanne Kostides, gen mgr.

Hyannis

WCOD-FM— June 2, 1967: 106.1 mhz; 50 kw. 450 ft TL: N41 43 46 W70 10 01. Stereo. Hrs opn: 24 154 Barnstable Rd., 02601-2930. Phone: (508) 778-2888. Fax: (508) 778-9651. Web Site: www.106wcod.com. Licensee: Qantum of Cape Cod License Co. LLC. Group owner: Boch Broadcasting (acq 4-11-2005; grpsl). Natl. Rep: Eastman Radio. Format: Hot adult contemp. News staff: 2. Target aud: 25-54. ♦Allison Makkay, gen mgr; Kevin Matthews, progmg dir.

WPXC(FM)— Jan 9, 1987: 102.9 mhz; 6 kw. 325 ft TL: N41 41 19 W70 20 49. Stereo. Hrs opn: 24 278 S. Sea Ave., West Yarmouth, 02673. Phone: (508) 775-5678. Fax: (508) 862-6329. E-mail: info@pixy103.com Web Site: www.pixy103.com. Licensee: Nassau Broadcasting III L.L.C. Group owner: Qantum Communications Corp. (acq 11-8-2005; grpsl). Population served: 200,000 Natl. Rep: McGavren Guild. Format: Rock. News staff: 3; News: 4 hrs wkly. Target aud: General. ♦Jake Demmin, gen mgr; Suzanne Tonaire, progmg dir.

WQRC(FM)—See Barnstable

Lawrence

WLLH(AM)—See Lowell

WMKK(FM)—Licensed to Lawrence. See Boston

WNNW(AM)— August 1947: 800 khz; 1 kw-D, 250 w-N. TL: N42 40 26 W71 11 26. Hrs open: 24 462 Merrimack St., Methuen, 01844. Phone: (978) 686-9966. Fax: (978) 687-1180. Web Site: www.power800am.com. Licensee: Costa-Eagle Radio Ventures L.P. (group owner; acq 3-27-98; $405,000). Population served: 525,000 Natl. Network: CNN Radio. Rgnl. Network: Metronews Radio Net. Natl. Rep: Lotus Entravision Reps LLC. Law Firm: Bryan Cave. Wire Svc: AP Format: Sp. News staff: 2; News: 10 hrs wkly. Target aud: 35-64; general. ♦Patrick Costa, gen mgr; Johnny McKenzie, opns mgr & progmg dir.

Leicester

WVNE(AM)— June 19, 1991: 760 khz; 25 kw-D. TL: N42 14 57 W72 04 41. Hrs open: Sunrise-sunset 70 James St., Suite 201, Worcester, 01603. Phone: (508) 831-9863. Fax: (508) 831-7964. E-mail: info@wvne.net Web Site: www.wvne.net. Licensee: Blount Masscom Inc. Group owner: Blount Communications Group (acq 5-15-90; FTR: 6-4-90). Population served: 3,000,000 Natl. Network: Salem Radio Network. Natl. Rep: Salem. Format: Relg. Target aud: 25-54. ♦William A. Blount, pres; Deborah C. Blount, exec VP; David O. Young, VP & gen mgr; Steve Tuzeneu, stn mgr & opns mgr.

Leominster

WCMX(AM)— Nov 13, 1967: 1000 khz; 1 kw-D. TL: N42 31 25 W71 44 07. Hrs open: 6 AM-2 hrs past sundown 194 Electric Ave., Lunenburg, 01462. Phone: (978) 582-4901. Fax: (978) 582-4978. E-mail: nate@hope1000.com Web Site: hope1000.net. Licensee: Twin City Baptist Temple Inc. (acq 1-95). Population served: 1,500,000 Natl. Network: Salem Radio Network. Format: Christian music, praise & worship. News: 7 hrs wkly. Target aud: 35-54 women. ♦Pastor Erven Burke, gen mgr; Nathan Burke, stn mgr, stn mgr & progmg dir.

WEIM(AM)—See Fitchburg

Lowell

WCAP(AM)— June 10, 1951: 980 khz; 5 kw-U, DA-2. TL: N42 39 16 W01 21 43. Hrs open: 24 243 Central St., 01852. Phone: (978) 454-0404. Fax: (978) 458-9124. E-mail: info@wcap.net Web Site: www.wcap.net. Licensee: Northeast Radio Inc. Population served: 2,000,000 Natl. Network: Westwood One, Premiere Radio Networks. Law Firm: Richard J. Hayes Jr. Wire Svc: AP Format: Talk, news, sports, nostalgia, adult standards. News: 20 hrs wkly. Target aud: 25 plus; Business people, professionals, factory workers, housewives. ♦Maurice Cohen, pres & gen mgr; Ryan Johnston, progmg dir.

WCRB(FM)— 1947: 99.5 mhz; 27 kw. Ant 653 ft TL: N42 39 14 W71 13 02. Stereo. Hrs opn: 750 South St., Waltham, 02453-1496. Phone: (781) 893-7080. Fax: (781) 893-0038. E-mail: wcrb@wcrb.com Web Site: www.wcrb.com. Licensee: Nassau Broadcasting II L.L.C. Group owner: Greater Media Inc. (acq 11-15-2006; exchange for WJJZ(FM) Burlington, NJ). Population served: 535,000 Format: Classical. Target aud: 30-64; adults. ♦Paul Kelley, gen mgr; Tim Neill, mktg dir & prom dir; Mark Edwards, progmg dir.

WLLH(AM)— June 2005: 1400 khz; 1 kw-U. TL: N42 39 29 W71 19 04. Hrs open: 24 529 Main St., Suite 200, Charlestown, 02129-1119. Phone: (617) 830-1000. E-mail: laura.wareck@espnboston.com Web Site: www.890espn.com. Licensee: J Sports Licensee LLC Group owner: Mega Communications Inc. (acq 6-22-2005; $9 million with WAMG(AM) Dedham). Population served: 94,239 Natl. Network: ESPN Radio. Natl. Rep: McGavren Guild. Law Firm: Davis, Wright, Tremaine. Format: Sports. Target aud: Males 18+. ♦Jessamy Tang, gen mgr; Neil Kelleher, sls dir; Kara Lachance, progmg dir; Len Weiner, progmg dir.

***WUML(FM)**— Nov 6, 1967: 91.5 mhz; 1.4 kw. 207 ft TL: N42 39 07 W71 19 15. Stereo. Hrs opn: 18 One University Ave., 01854. Phone: (978) 934-4975. Fax: (978) 934-3031. E-mail: wuml@wuml.org Web Site: www.wuml.org. Licensee: University of Massachusetts-Lowell Board of Trustees. Population served: 3,000,000 Format: Div, progsv rock. Target aud: 16-25. ♦Nate Osit, gen mgr; Joe Keefe, progmg dir.

Lynn

WFNX(FM)— Aug 5, 1963: 101.7 mhz; 1.7 kw. 626 ft TL: N42 21 08 W71 03 25. Stereo. Hrs opn: 24 25 Exchange St., 01901. Phone: (781) 595-6200. Fax: (781) 595-3810. E-mail: fnxradio@fnxradio.com Web Site: www.fnxradio.com. Licensee: MCC Broadcasting Inc. Group owner: Phoenix Media Communications Group (acq 11-10-82; FTR: 11-29-82). Population served: 3,880,000 Natl. Rep: McGavren Guild. Law Firm: Rubin, Winston, Diercks, Harris & Cooke. Wire Svc: AP Format: Alternative rock. News staff: one; News: 2 hrs wkly. Target aud: 18-49; well-educated, affluent & socially active trend setters. Spec prog: Jazz 8 hrs, loc music 2 hrs wkly. ♦Stephen Mindich, CEO; Brad Mindich, chmn; Rick Gallagher, CFO; Gary Kurtz, gen mgr & news dir; Jordi Chapdelaine, gen sls mgr; Keith Dakin, progmg dir & chief of engrg; Christopher Hall, chief of engrg.

WLYN(AM)— November 1947: 1360 khz; 700 w-D, 76 w-N. TL: N42 27 17 W70 58 44. Stereo. Hrs opn: 24 500 W. Cummings Park, Suite 2600, Woburn, 01801. Phone: (781) 938-0869. Fax: (781) 938-0933. E-mail: jeffk@mrbi.net Web Site: www.mrbi.net. Licensee: Multicultural Radio Broadcasting Licensee LLC. Group owner: Multicultural Radio Broadcasting Inc. (acq 8-7-2002; $1.78 million). Population served: 3,000,000 Format: Ethnic - leased time. Hispanic (Spanish & Portuguese). Spec prog: Greek 2 hrs wkly. ♦Jeff Kline, gen mgr.

Marion

***WWTA(FM)**— 1996: 88.5 mhz; 19 w horiz, 100 w vert. 53 ft TL: N41 42 32 W70 45 57. Hrs open: 66 Spring St., 02738. Phone: (508) 748-2000. Fax: (508) 291-6666. E-mail: kkistler@taboracademy.org Web Site: www.taboracademy.org. Licensee: Tabor Academy. Format: Var. ♦Karl Kistler, gen mgr.

Marshfield

WATD-FM— Dec 5, 1977: 95.9 mhz; 2.8 kw. 350 ft TL: N42 06 40 W70 42 14. Stereo. Hrs opn: 24 130 Enterprise Dr., 02050. Phone: (781) 837-1166. Fax: (781) 837-1978. E-mail: news@959watd.com Web Site: www.959watd.com. Licensee: Marshfield Broadcasting Co. Population served: 500,000 Format: Adult contemp, blues, oldies. News staff: 2; News: 10 hrs wkly. Target aud: 25-64; South Shore residents. ♦Edward Perry, Jr., pres & gen mgr.

Mashpee

WFRQ(FM)— Feb 12, 1987: 101.1 mhz; 3.7 kw. Ant 253 ft TL: N41 36 50 W70 35 56. Stereo. Hrs opn: 24
Rebroadcasts WFQR(FM) Harwich Port 100%.
278 South Sea Ave., West Yarmouth, 02673. Phone: (508) 775-5678. Fax: (508) 862-6329. Licensee: Nassau Broadcasting III L.L.C. Group owner: Boch Broadcasting (acq 11-7-2005; grpsl). Natl. Network: Westwood One. Natl. Rep: Katz Radio. Format: Adult hits. News staff: one. Target aud: 35-64. ♦Jake Demmin, gen mgr.

Maynard

***WAVM(FM)**— April 1973: 91.7 mhz; 16 w. -7 ft TL: N42 25 18 W71 27 02. Stereo. Hrs opn: 24 Maynard High School, One Tiger Dr., 01754. Phone: (978) 897-5179. Fax: (978) 897-6089. E-mail: studio@wavm.org Web Site: www.wavm.org. Licensee: Maynard Public Schools. Format: Var. Target aud: General. ♦Mark Minasian, gen mgr & stn mgr.

Medford

***WMFO(FM)**— March 1971: 91.5 mhz; 125 w. 135 ft TL: N42 24 27 W71 07 15. Stereo. Hrs opn: Box 65, 02155. Secondary address: 474 Boston Ave. 02155. Phone: (617) 625-0800. Fax: (617) 625-6072. E-mail: wmfo@wmfo.org Web Site: www.wmfo.org. Licensee: Tufts University. Population served: 74,397 Format: Var/div. ♦Annie Ross, gen mgr.

WXKS(AM)—Everett, Jan 20, 1952: 1430 khz; 5 kw-D, 1 kw-N, DA-N. TL: N42 24 11 W71 04 29. Hrs open: 24 99 Revere Beach Pkwy., 02155. Phone: (781) 396-1430. Fax: (781) 391-3064. Web Site: www.1200rumba.com. Licensee: AMFM Radio Licenses L.L.C. Group owner: Clear Channel Communications Inc. (acq 8-30-2000; grpsl). Population served: 743,900 Format: Sp. ♦Jake Karger, gen mgr.

WXKS-FM— Sept 1, 1960: 107.9 mhz; 20.5 kw. 771 ft TL: N42 20 50 W71 04 59. Stereo. 24 Web Site: www.kiss108.com. Format: Adult CHR. Target aud: 18-34.

Middleborough Center

WVBF(AM)— 1993: 1530 khz; 2.2 kw-D, 2 w-N, 940 w-CH. TL: N41 55 26 W70 56 07. Hrs open: 24 Box 329, 02346. Secondary address: 130 Enterprise Dr., Marshfield 02050. Phone: (781) 834-4400. Phone: (800) 696-9505. Fax: (781) 834-7716. Licensee: Steven J. Callahan. Format: Info. ♦Steven Callahan, gen mgr.

Milford

WMRC(AM)— Oct 6, 1956: 1490 khz; 1 kw-U. TL: N42 08 12 W71 30 50. Hrs open: 24 Box 421, 01757. Secondary address: 258 Main St. 01757. Phone: (508) 473-1490. Fax: (508) 478-2200. Licensee: First Class Radio Corp. (acq 12-4-2006). Population served: 250,000 Format: Adult contemp, var radio. News staff: 2; News: 32 hrs wkly. Target aud: 25-54. ♦Thomas M. McAuliffe Sr., pres; Thomas M. McAuliffe II, gen mgr.

Milton

***WMLN-FM**— Apr 1, 1975: 91.5 mhz; 170 w. 98 ft TL: N42 14 27 W71 06 52. Hrs open: 24 1071 Blue Hill Ave., 02186. Phone: (617) 333-0311. E-mail: afrank@curry.edu Web Site: www.curry.edu. Licensee: Curry College. Population served: 200,000 Natl. Network: CNN Radio. Wire Svc: AP Format: News/talk, adult contemp, div. News: 15 hrs wkly. Target aud: General. ♦Alan H. Frank, gen mgr.

Nantucket

***WNAN(FM)**— Mar. 15, 2000: 91.1 mhz; 2 kw. 72 ft TL: N41 18 22 W70 00 28. Hrs open: Box 82, Woods Hole, 02543. Secondary address: 3 Water St., Woods Hole 02543. Phone: (617) 300-2300. Fax: (508) 548-5517. Web Site: www.cainan.org. Licensee: WGBH Educational Foundation. (acq 12-31-97; $25,000 with WCAI(FM) Woods Hole). Natl. Network: NPR. Format: News/talk. ♦Henry Becton, gen mgr.

***WNCK(FM)**— June 28, 2002: 89.5 mhz; 78 w horiz, 500 w vert. Ant 118 ft TL: N41 17 06 W70 08 39. Hrs open: 24
Rebroadcasts WGBH(FM) Boston 100%.
Nantucket Public Radio Inc., Box 2185, 02534. Phone: (508) 825-8951. Fax: (508) 325-0030. Licensee: Nantucket Public Radio Inc. Format: Var. Target aud: 40 plus. ♦Robert Shapiro, gen mgr.

WRZE(FM)— June 15, 1981: 96.3 mhz; 50 kw. Ant 430 ft TL: N41 16 50 W70 10 10. Stereo. Hrs opn: 24 154 Barnstable Rd., Hyannis, 02601. Phone: (508) 778-2888. Fax: (508) 778-9651. E-mail: info@therose.net Web Site: www.therose.net. Licensee: Qantum of Cape Cod License Co. LLC. Group owner: Qantum Communications Corp. (acq 6-11-03; grpsl). Natl. Rep: Eastman Radio. Format: Top-40. Target aud: 18-49. ♦Allison Makkay, gen mgr; David Duran, gen mgr & progmg dir.

Natick

WBIX(AM)— November 1972: 1060 khz; 40 kw-D, 2.5 kw-N, 22 kw-CH, DA-3. TL: N42 17 17 W71 25 55. Hrs open: 24 100 Mount Wayte St., 100 Summer St., Framingham, 01702. Phone: (508) 820-2430. E-mail: alex@wbix.com Web Site: www.wbix.com. Licensee: WBIX Corp. Group owner: Langer Broadcasting Group L.L.C. (acq 11-29-2005). Population served: 2,900,000 Format: News/talk/business. ♦Alex Langer, gen mgr; Jim Harris, gen sls mgr.

New Bedford

WBSM(AM)— July 17, 1949: 1420 khz; 5 kw-D, 1 kw-N, DA-2. TL: N41 39 02 W70 54 58. Hrs open: 22 Sconticut Neck Rd., Fairhaven, 02719. Phone: (508) 993-1767. Fax: (508) 999-1420. E-mail: petebraley@wbsm.com Web Site: www.wbsm.com. Licensee: Citadel Broadcasting Co. Group owner: Citadel Broadcasting Corp. (acq 4-26-01; grpsl). Population served: 800,000 Natl. Rep: Christal. Format: News/talk, sports. ♦Gail Le Blanc, gen mgr; Deborah Aguiar, prom dir; Pete Braley, progmg dir.

WCTK(FM)—Licensed to New Bedford. See Providence RI

***WFHL(FM)**— 2003: 88.1 mhz; 300 w vert. Ant 134 ft TL: N41 38 15 W70 52 19. Stereo. Hrs opn: Box 3025, 02741. Secondary address: 71 William 02740. Phone: (508) 991-7600. E-mail: radio@radiowfhl.com Web Site: www.radiowfhl.com. Licensee: New Bedford Christian Radio Inc. Format: Sp, English, Portugese. ♦Manuel Pereira, gen mgr.

WJFD-FM— Feb 22, 1949: 97.3 mhz; 50 kw. 500 ft TL: N41 38 20 W70 52 27. Stereo. Hrs opn: 24 270 Union St., 02740. Phone: (508) 997-2929. Fax: (508) 990-3893. E-mail: jorge@wjfd.com Web Site: www.wjfd.com. Licensee: Edmund Dinis, trustee (acq 12-18-2001). Population served: 250,000 Format: Ethnic. Target aud: General; Portuguese-speaking community. ♦Edmund Dinis, pres.

WNBH(AM)—Licensed to New Bedford. See Providence RI

Newburyport

WNBP(AM)— Mar 10, 1957: 1450 khz; 1 kw-U. TL: N42 49 23 W70 51 42. Hrs open: 24 Box 1450, 44 Merrimac St., 01950. Phone: (978) 462-1450. Fax: (978) 462-0333. E-mail: wnbp.radio@verizon.net Web Site: www.wnbp.com. Licensee: Westport Communications L.P. (acq 11-23-2004; $500,000). Population served: 42,000 Natl. Network: AP Radio. Wire Svc: AP Format: Adults Standards. News staff: one; News: 5 hrs wkly. Target aud: 25-54. Spec prog: Irish 4 hrs wkly.Al Mozier, gen mgr; Matt Stevens, opns mgr, prom dir, progmg mgr, pub affrs dir, spec ev coord, local news ed & disc jockey; Bill Wayland, gen sls mgr; William Fuller, mus dir; Dan Guy, engrg dir & traf mgr; Bryan Keohane, disc jockey; Frank Messina, disc jockey; Peter Dearborn, disc jockey; Tom Healy, disc jockey; Win Damon, sls dir, news dir, sports cmtr & disc jockey

***WNEF(FM)**— 91.7 mhz; 400 w vert. 354 ft TL: N42 51 56 W70 56 18. Hrs open:
Rebroadcasts WUMB-FM Boston 100%.
Univ. of Massachusetts, 100 Morrissey Blvd., Boston, 02125-3393. Phone: (617) 287-6900. Fax: (617) 287-6916. E-mail: wumb@umb.edu Web Site: www.wumb.org. Licensee: University of Massachusetts. Format: Folk. ♦Patricia Monteith, gen mgr; Danielle Knight, dev dir; Brian Quinn, progmg dir.

Newton

WNTN(AM)— Apr 1, 1968: 1550 khz; 10 kw-D. TL: N42 21 27 W71 14 30. Hrs open: 143 Rumford Ave., 02466. Phone: (617) 969-1550. Web Site: www.wntn.com. Licensee: Colt Communications LLC. (acq 12-23-98; $602,800). Population served: 2,500,000 Format: Var, Greek, Haitian. Target aud: 40 plus. Spec prog: Irish 6 hrs, Indian 2 hrs wkly. ♦Rob Rudnick, gen mgr; Paul Roberts, opns dir; Leo Sullivan, chief of engrg.

***WZBC(FM)**— April 1974: 90.3 mhz; 1 kw. 220 ft TL: N42 20 05 W71 10 31. Stereo. Hrs opn: 24 Boston College, McElroy Commons 107, Chestnut Hill, 02467. Phone: (617) 552-3511. Fax: (617) 552-1738. Web Site: www.wzbc.org. Licensee: Trustees of Boston College. Population served: 7,500 Format: Underground rock. Target aud: 18-34. ♦Candace Savino, gen mgr; Court Hillman, opns dir; Nick Feeley, opns dir & progmg dir.

Norfolk

WDIS(AM)— Mar 20, 1978: 1170 khz; 1 kw-D, DA. TL: N42 05 32 W71 18 13. Hrs open: Day Time 100 Pond St., 02056. Phone: (508) 384-8255. Fax: (508) 384-1530. E-mail: wdismgmt@aol.com Web Site: www.wdis.com. Licensee: Discussion Radio Inc. (acq 8-12-92; $65,000; FTR: 9-7-92). Population served: 250,000 Natl. Network: Salem Radio Network. Format: News/talk. News staff: one; News: 6 hrs wkly. Target aud: 35-64. ♦Corine Slade, gen mgr; Dan Collier, progmg dir.

North Adams

***WJJW(FM)**— Sept 5, 1973: 91.1 mhz; 423 w. -830 ft TL: N42 41 27 W73 06 16. Stereo. Hrs opn: Mass. College of Liberal Arts, Murdock Hall, 375 Church St., 01247. Phone: (413) 662-5405. Fax: (413) 662-5010. E-mail: webmaster@mcla.edu Web Site: www.mcla.edu. Licensee: Massachusetts College of Liberal Arts. Population served: 20,000 Format: Progsv. ♦Alisha Cropper, gen mgr; Nick Strassel, progmg dir; Paul Wiley, chief of engrg.

WNAW(AM)— Nov 23, 1947: 1230 khz; 1 kw-U. TL: N42 41 03 W73 06 23. Hrs open: Box 707, 466 Curran Hwy., 01247-0707. Phone: (413) 663-6567. Fax: (413) 662-2143. E-mail: wnaw@wnaw.com Web Site: www.wnaw.com. Licensee: Vox Communications Group LLC. Group owner: Vox Radio Group L.P. (acq 5-10-2004; grpsl). Population served: 45,000 Natl. Rep: McGavren Guild. Law Firm: Wilkinson Barker Knauer. Format: Full service, adult contemp. News staff: 2. Target aud: Adults. ♦Earl Ingalls, gen mgr & gen sls mgr; Peter Barry, mktg mgr & progmg mgr; Ken Jones, chief of engrg.

WUPE-FM—Co-owned with WNAW(AM). July 12, 1964: 100.1 mhz; 1.3 kw. Ant 502 ft TL: N42 41 51 W73 03 52. Stereo. Web Site: www.wupe.com.150,000 Format: Oldies. Target aud: 35 plus. ♦Dick Savage, gen sls mgr.

North Dartmouth

***WTKL(FM)**— September 1973: 91.1 mhz; 1.2 kw. Ant 300 ft TL: N41 37 43 W71 00 24. Stereo. Hrs open: 2351 Sunset Blvd., Suite 170-218, Rocklin, CA, 95765. Phone: (916) 251-1600. Fax: (916) 251-1650. Web Site: www.klove.com. Licensee: Educational Media Foundation. (acq 6-30-2006; $725,000). Population served: 200,000 Natl. Network: K-Love. Format: Contemp Christian. ♦Richard Jenkins, pres.

***WUMD(FM)**— June 10, 2006: 89.3 mhz; 96 w horiz, 9.6 kw vert. Ant 305 ft TL: N41 37 43 W71 00 24. Hrs open: 285 Old Westport Rd., 02747. Phone: (508) 999-8149. Fax: (508) 999-8173. E-mail: wumd@umassd.edu Licensee: University of Massachusetts. Format: Progsv, alternative rock, techno. Target aud: 13-60; general, high school, college, community. ♦Jennifer Mulcare-Sullivan, stn mgr.

Northampton

WEIB(FM)— 2001: 106.3 mhz; 3 kw. Ant 289 ft TL: N42 22 25 W72 40 26. Stereo. Hrs opn: 24 8 North King St., 01060. Phone: (413) 585-1112. Fax: (413) 585-9138. E-mail: weibfm@aol.com Web Site: www.weibfm.com. Licensee: Cutting Edge Broadcasting Inc. Format: Smooth jazz. ♦Carol Moore Cutting, pres & gen mgr; Drew Dawson, progmg dir.

WHMP(AM)— December 1950: 1400 khz; 1 kw-U. TL: N42 19 36 W72 39 28. Hrs open: 15 Hampton Ave., 01061. Phone: (413) 586-7400. Fax: (413) 585-0927. Web Site: www.whmp.com. Licensee: Saga Communications of New England LLC. Group owner: Saga Communications Inc. (acq 2000; $12 million with co-located FM). Population served: 100,000 Natl. Network: CBS Radio, CNN Radio. Natl. Rep: Katz Radio. Wire Svc: AP Format: Sports, news/talk. News staff: 4; News: 40 hrs wkly. Target aud: 35 plus; upscale, well educated. Spec prog: Pol 3 hrs wkly. ♦Glenn Cardinal, gen mgr.

WLZX(FM)— Nov 1, 1956: 99.3 mhz; 3 kw. 321 ft TL: N42 22 29 W72 40 24. (CP: 6 kw). Stereo. Hrs opn: 24 45 Fisher Ave, East Longmeadow, 01028. Phone: (413) 525-4141. Fax: (413) 525-4334. Web Site: www.lazer993.com. Licensee: Saga Communications of New England LLC. Group owner: Saga Communications Inc. (acq 2000; $12 million with co-located AM). Population served: 200,000

Format: Rock/Active. News staff: 2. Target aud: 18-34; male. ♦Gary Zenobi, gen mgr; Bill Buller, gen sls mgr; Courtney Quinn, progmg dir; Kristin McCauley, traf mgr; Tina Shotwell, traf mgr.

***WOZQ(FM)**— 1981: 91.9 mhz; 200 w. 115 ft TL: N42 19 13 W72 38 14. Stereo. Hrs opn: 6 AM-2 AM Smith College, Davis Ctr., 01063. Phone: (413) 585-4956. Phone: (413) 585-4977. Fax: (413) 585-2075. E-mail: wozq@email.smith.edu Web Site: www.wozq.org. Licensee: Trustees of Smith College. Format: Educ, var/div, urban contemp. Target aud: 15 plus; college students & area businesses. ♦Jackie Batten, stn mgr; Lares Feliciano, mus dir; Ramona Tompkins, mus dir; Matthea Doughtry, news dir; Kasia Kornecki, chief of engrg.

Northfield

***WNMH(FM)**— Sept 10, 1984: 91.5 mhz; 235 w. 308 ft TL: N42 42 52 W72 26 38. Stereo. Hrs opn: Box WNMH, Northfield Mt. Hermon School, 206 Main St., 01360. Phone: (413) 498-3603. Fax: (413) 498-3664. Licensee: Northfield Mount Hermon School. Format: Div. Target aud: Student body & surrounding communities. ♦Bill Hattendorf, opns VP & opns mgr.

Orange

WJDF(FM)— 1995: 97.3 mhz; 3 kw. 328 ft TL: N42 37 19 W72 21 58. (CP: 5.8 kw, ant 82 ft.). Stereo. Hrs opn: Box 973, 01364. Phone: (978) 544-5335. Phone: (978) 544-0957. Fax: (978) 544-2131. E-mail: info@wjdf.com Web Site: www.wjdf.com. Licensee: Deane Brothers Broadcasting Corp. Format: Adult contemp. ♦Donn Deane, gen mgr; Chad Songer, gen sls mgr; Jay Deane, progmg dir.

Orange-Athol

WJOE(AM)— May 13, 1956: 700 khz; 2.5 kw-D. TL: N42 35 06 W72 16 56. Hrs open: Sunrise-sunset 362 Green St., Gardner, 01440. Secondary address: 660 E. Main St., Orange 01364. Phone: (978) 544-2321. Fax: (978) 544-6977. Web Site: www.eagle99.com. Licensee: County Broadcasting Co. LLC. Group owner: Northeast Broadcasting Company Inc. (acq 10-6-2003; $650,000 with WNYN-FM Athol). Population served: 50,000 Format: Oldies. Target aud: 45 plus. ♦Chris Thompson, gen mgr; Glenn Cardinal, gen mgr; Billy Curtis, progmg dir; Spencer Marshall, progmg dir.

Orleans

WFPB(AM)— Apr 10, 1970: 1170 khz; 1 kw-D, DA. TL: N41 46 48 W70 00 36. Hrs open: 24
Rebroadcasts WUMB-FM Boston 100%.
Univ. of Massachusetts, 100 Morrissey Blvd., Boston, 02125-3393. Phone: (617) 287-6900. Fax: (617) 287-6916. E-mail: wumb@umb.edu Web Site: www.wumb.org. Licensee: University of Massachusetts. (acq 10-30-98). Population served: 185,000 Format: Folk. ♦Patricia Monteith, gen mgr; Danielle Knight, dev dir; Brian Quinn, progmg dir.

WOCN-FM— July 25, 1974: 104.7 mhz; 50 kw. Ant 504 ft TL: N41 46 48 W70 00 36. Stereo. Hrs opn: 24 737 W. Main St., Hyannis, 02601. Phone: (508) 778-6200. Phone: (508) 771-1224. Fax: (508) 775-2605. E-mail: wocn@ocean1047.com Web Site: www.ocean1047.com. Licensee: Cape Cod Broadcasting License II LLC. (acq 7-10-2007; $7.5 million with WFCC-FM Chatham). Population served: 200,000 Natl. Network: AP Radio. Natl. Rep: Clear Channel. Law Firm: Covington & Burling. Wire Svc: AP Format: Full service, soft adult contemp. News staff: 4; News: 32 hrs wkly. Target aud: Adults 25-64. ♦Gregory D. Bone, gen mgr; Wayne White, opns mgr; Stephen Colella, sls dir; Michelle Dodd, prom mgr; Donna Credit, traf mgr.

Pittsfield

WBEC(AM)— March 1947: 1420 khz; 1 kw-U, DA-N. TL: N42 26 40 W73 16 43. Hrs open: 24 211 Jason St., 01201-5907. Phone: (413) 499-3333. Fax: (413) 442-1590. Licensee: Vox Communications Group LLC. Group owner: Vox Radio Group L.P. (acq 9-13-2002; grpsl). Population served: 80,000 Law Firm: Smithwick & Belendiuk. Format: News/talk, sports. News staff: 2; News: 15 hrs wkly. Target aud: 25-55; 60% male, 40% female. Spec prog: Relg 3 hrs wkly. ♦Laura Freed, gen mgr.

WBEC-FM—Listing follows WUPE(AM).

WBRK(AM)— Feb 20, 1938: 1340 khz; 1 kw-U. TL: N42 27 00 W73 12 55. Hrs open: 24 100 North St., 01201. Phone: (413) 442-1553. Fax: (413) 445-5294. E-mail: wbrk1340@aol.com Web Site: www.wbrk.com. Licensee: WBRK Inc. (acq 6-30-84). Population served: 57,020 Natl. Network: CBS, Westwood One. Natl. Rep: D & R Radio. Law Firm: Drinker, Biddle & Reath. Format: Full service. News staff: 2. Target aud: 35 plus. Spec prog: Pol 2 hrs, Irish one hr, relg 2 hrs wkly. ♦Willard H. Hodgkins III, CEO & pres; John Campoli, exec VP; Daniel Salzarulo, stn mgr & chief of engrg; Michael J. Bunn, opns VP & opns mgr; Cheryl Tripp, prom dir; Rick Beltaire, progmg VP & progmg dir.

WBRK-FM— Oct 10, 1970: 101.7 mhz; 3 kw. 145 ft TL: N42 28 31 W73 16 07. Stereo. 24 Web Site: www.wbrk.com.65,000 Natl. Network: ABC. Format: Adult contemp. Target aud: 25-54.

WUPE(AM)— Sept 9, 1971: 1110 khz; 5 kw-D, DA. TL: N42 26 22 W73 17 30. Hrs open: 24 211 Jason St., ., 01201. Phone: (413) 499-3333. Fax: (413) 442-1590. E-mail: wupe@wupe.com Web Site: www.wupe.com. Licensee: Vox Communications Group LLC. (acq 12-8-2003; $2.83 million with co-located FM). Population served: 250,000 Natl. Network: ABC. Natl. Rep: D & R Radio. Law Firm: Pepper & Corazzini. Format: Oldies. News staff: one; News: 14 hrs wkly. Target aud: 25-54; baby boomers. ♦Peter Barry, gen mgr; Mike Patrick, opns mgr; Dick Savage, sls dir; Larry Kratka, news dir & news rptr; Ken Jones, chief of engrg; Bob Heck, traf mgr.

WBEC-FM—Co-owned with WUPE(AM). 1975: 95.9 mhz; 1 kw. Ant 560 ft TL: N42 24 44 W73 17 05. Stereo. 24 Web Site: live959.com.150,000 Natl. Network: Westwood One. Format: Hot adult contemp. News staff: one; News: 28 hrs wkly. Target aud: 25-54; young adults with families. ♦Tom Conklin, news dir.

Plymouth

WPLM(AM)— Aug 8, 1955: 1390 khz; 5 kw-U, DA-2. TL: N41 58 05 W70 42 06. Hrs open: 17 Columbus Rd., 02360. Phone: (508) 746-1390. Fax: (508) 830-1128. Licensee: Plymouth Rock Broadcasting Co. Inc. Population served: 85,000 Natl. Rep: Roslin. Law Firm: Arent, Fox, Kintner, Plotkin & Kahn. Format: Soft adult contemp, business talk. ♦Dr. Laurie Campbell, pres; Alan Anderson, gen mgr & gen sls mgr; Pat Carroll, chief of opns & pub svc dir; Sean Casey, prom dir & progmg dir; Chip Morgan, chief of engrg.

WPLM-FM— June 25, 1961: 99.1 mhz; 50 kw. 430 ft TL: N41 58 02 W70 42 04. Format: Adult contemp.

Provincetown

***WOMR(FM)**— Mar 21, 1982: 92.1 mhz; 6 kw. Ant 161 ft TL: N42 03 54 W70 09 31. Stereo. Hrs opn: 24 Box 975, 494 Commercial St., 02657. Phone: (508) 487-2106. Fax: (508) 487-5524. E-mail: info@womr,org Web Site: www.womr.org. Licensee: Lower Cape Communications Inc. Law Firm: Garvey, Schubert, Baker. Format: Eclectic. News: 5.5 hrs wkly. Target aud: General; div. Spec prog: Black 6 hrs, class 16 hrs, educ 10 hrs, folk 19 hrs, oldies 9 hrs wkly. ♦Tina Lynde, pres; Dave Willard, VP; John Braden, opns mgr.

Quincy

WJDA(AM)— Sept 13, 1947: 1300 khz; 1 kw-D, 72 w-N. TL: N42 15 35 W70 58 36. Hrs open: 24 Box 690626, 02269-0626. Secondary address: 29 Brackett St. Phone: (617) 479-1300. Fax: (617) 479-0622. E-mail: info@wjda1300.com Web Site: www.wjda1300.com. Licensee: South Shore Broadcasting Co. Population served: 125,000 Natl. Network: ABC. Format: ABC " Memories ". News staff: 2; News: 10 hrs wkly. Target aud: 35 plus. Spec prog: Cantonese 3 hrs wkly. ♦Joe Catalano, progmg dir; Mike Logan, news dir.

Rockland

***WRPS(FM)**— Feb 8, 1974: 88.3 mhz; 100 w. 120 ft TL: N42 07 43 W70 55 01. Stereo. Hrs opn: 24 34 MacKinlay Way, 02370. Phone: (781) 871-0724. Fax: (781) 982-1483. E-mail: wrps883@yahoo.com Licensee: Rockland Public Schools. Population served: 50,000 Format: Public service, educ, adult contemp. Target aud: General. ♦David J. Cable-Murphy, gen mgr; Robert Mulligan, chief of engrg.

Salem

WESX(AM)— Jan 1, 1939: 1230 khz; 1 kw-U. TL: N42 31 06 W70 51 41. Hrs open: Box 710, 01970. Secondary address: 27 Naugus Ave., Marblehead 01945. Phone: (978) 744-1230. Fax: (978) 744-1853. E-mail: info@wesx1230.com Web Site: www.wesx1230.com. Licensee: North Shore Broadcasting Corp. (acq 4-1-50). Population served: 550,000 Natl. Network: ABC, Westwood One. Wire Svc: AP Format: MOR, news/talk. News staff: 2; News: 25 hrs wkly. Target aud: 35 plus; general. Spec prog: Auto repair 2 hrs, gardening 2 hrs, home improvement 2 hrs, restaurant/dining 2 hrs, Pol 2 hrs wkly. ♦James D. Asher, pres & gen mgr; Christopher Culkeen, opns mgr, progmg dir & progmg mgr; Bill Cooksly, news dir.

***WMWM(FM)**— 1976: 91.7 mhz; 130 w. Ant 132 ft TL: N42 30 14 W70 53 26. Stereo. Hrs opn: 7 AM-midnight Campus Ctr., 352 Lafayette St., 01970-5353. Phone: (978) 745-9401. Fax: (978) 542-8127. E-mail: eboard@wmwm.com Web Site: www.wmwmsalem.com. Licensee: Salem State College. Population served: 40,000 Format: Diversified. Target aud: General. ♦Justin Symington, gen mgr; Paul Collins, chief of engrg; Richard Tucker, progmg dir & disc jockey.

Sandwich

***WSDH(FM)**— 1976: 91.5 mhz; 310 w. 150 ft TL: N41 44 06 W70 27 35. Hrs open: 10 AM-4 PM (M-F) Sandwich High School, 365 Quaker Meetinghouse Rd., East Sandwich, 02537. Phone: (508) 888-0420. Fax: (508) 833-8392. Licensee: Sandwich Public Schools. Format: CHR, classic rock, educ. News: 4 hrs wkly. Target aud: 12-40. ♦Chip Hill, gen mgr.

Scituate

***WSMA(FM)**— May 2006: 90.5 mhz; 5 w horiz, 7.7 kw vert. Ant 492 ft TL: N41 56 02 W70 35 10. Hrs open: 400 Franklin St., Braintree, 02185. Phone: (978) 855-1782. Fax: (978) 343-0665. E-mail: doug@blesseddesigns.com Licensee: CSN International. (group owner). Format: Relg.

Sheffield

***WBSL-FM**— September 1973: 91.7 mhz; 250 w. 50 ft TL: N42 06 57 W73 25 00. Hrs open: Berkshire School, 245 N. Undermountain Rd., 01257-9672. Phone: (413) 229-8511. Fax: (413) 229-1014. E-mail: jharris@berkshireschool.org Web Site: www.berkshireschool.org. Licensee: Berkshire School Inc. Population served: 10,000 Format: Div. Target aud: General. Spec prog: Jazz 15 hrs, Black 2 hrs, folk 2 hrs, Sp 2 hrs, Pol one hr wkly. ♦James Harris, gen mgr; John Clinton, stn mgr; Thomas Jaworski, engr.

South Hadley

***WMHC(FM)**— May 14, 1957: 91.5 mhz; 100 w. Ant -18 ft TL: N42 15 12 W72 34 40. Stereo. Hrs opn: Box 9010, Mt. Holyoke College, 01075. Secondary address: Mt. Holyoke College, Blanchard Campus Center 01075. Phone: (413) 538-2044. Phone: (413) 538-2019. Fax: (413) 538-2431. E-mail: amlewis@mtholyoke.edu Web Site: www.mtholyoke.edu/org/wmhc. Licensee: President & Trustees of Mount Holyoke College. Population served: 50,000 Natl. Network: AP Radio. Format: Rock, urban contemp, var/div. Target aud: General; Mount Holyoke College Community. ♦Catherine Moldonado, gen mgr; Hilary Spring, progmg dir.

South Yarmouth

WKPE-FM— August 1994: 103.9 mhz; 5.5 kw. Ant 341 ft TL: N41 41 30 W70 08 43. Hrs open: 24 582 Main St., West Yarmouth, 02673. Phone: (508) 790-3772. Fax: (508) 790-3773. E-mail: cat@rocket1047.com Web Site: www.1047therocket.com. Licensee: Sandab Communications L.P. II. (group owner; acq 6-19-98; $1.2 million). Population served:

225,000 Law Firm: Garvey, Schubert & Barer. Format: Classic rock. Target aud: 25-54; males. ♦Jan D'Antuono, gen mgr.

Southbridge

WESO(AM)— Mar 20, 1955: 970 khz; 1 kw-D, 21 w-N. TL: N42 03 59 W71 59 28. Hrs open: 24 100 Foster St., 01550. Phone: (508) 909-0970. Fax: (508) 764-2682. Web Site: thespirit970.com. Licensee: Money Matters Inc. (acq 4-11-01; $250,000). Population served: 60,000 Format: Pop country, local news/talk, sports. News staff: 2; News: 30 hrs wkly. Target aud: 34-59. Spec prog: Pol 3 hrs. ♦Dick Vaughan, COO & pres; Lia Zaido, opns mgr; J.P. Ellery, news dir.

WWFX(FM)— Nov 1, 1968: 100.1 mhz; 2.85 kw. 295 ft (CP: 1.74 kw, ant 590 ft.). Hrs opn: 24 250 Commercial St., Worcester, 01608. Secondary address: WBA Inc., 295 Bridle Trail Rd, Needham 02192. Phone: (508) 752-1045. Fax: (508) 770-9964. Web Site: www.thefoxfm.com. Licensee: Citadel Broadcasting Co. Group owner: Citadel Broadcasting Corp. (acq 4-26-01; grpsl). Population served: 400,000 Natl. Network: Jones Radio Networks. Natl. Rep: D & R Radio. Format: Main Stream Rock. News staff: 2; News: 10 hrs wkly. Target aud: 25-54. ♦Joe Flynn, gen mgr; JayBeau Jones, opns mgr; Alex Byrne, prom dir.

Springfield

WACE(AM)—See Chicopee

WACM(AM)—See West Springfield

***WAIC(FM)**— February 1967: 91.9 mhz; 230 w. 66 ft TL: N42 06 44 W72 33 29. Stereo. Hrs opn: 1000 State St., 01109. Phone: (413) 205-3941. Fax: (413) 205-3943. E-mail: techsupport@waic.com Web Site: www.waic.com. Licensee: American International College. Population served: 163,905 Format: Div. Target aud: 16-40. Spec prog: Gospel. ♦Will Hughes, CEO, chmn & pres; Doc Holiday, gen mgr; Richard Innes, opns dir.

WAQY(FM)— Dec 17, 1966: 102.1 mhz; 50 kw. 780 ft TL: N42 05 00 W72 42 16. Stereo. Hrs opn: 24 45 Fisher Ave., East Longmeadow, 01028. Phone: (413) 525-4141. Fax: (413) 525-4334. E-mail: gzenobi@springfieldrock.com Web Site: www.rock102.com. Licensee: Saga Communications of New England LLC. Group owner: Saga Communications Inc. (acq 6-2-92; grpsl). Natl. Rep: Katz Radio. Law Firm: Smithwick & Belendiuk. Format: Classic rock. Target aud: General; upscale young adults with high income. ♦Gary Zenobi, gen mgr; Dave Cooper, gen sls mgr & progmg dir; Kristin McCauley, traf mgr.

WHYN(AM)— 1941: 560 khz; 5 kw-D, 1 kw-N, DA-2. TL: N42 11 37 W72 41 02. Stereo. Hrs opn: 24 1331 Main St., 01103-. Phone: (413) 781-1011. Fax: (413) 734-4434. Web Site: www.whynam560.com. Licensee: CC Licenses LLC. Group owner: Clear Channel Communications Inc. (acq 1996; grpsl). Population served: 163,905 Law Firm: Haley, Bader & Potts. Format: News/talk. News staff: 6. Target aud: General. ♦Debbie Wagner, gen mgr.

WHYN-FM— 1946: 93.1 mhz; 8.9 kw. 1,000 ft TL: N42 14 28 W72 38 56. Stereo. E-mail: fm@mix931.com Web Site: www.mix931.com. Format: Adult contemp. Target aud: 25-54.

WMAS(AM)— Sept 1, 1932: 1450 khz; 1 kw-U. TL: N42 06 32 W72 36 44. Stereo. Hrs opn: Box 9500, 01102. Secondary address: 101 West St. 01104. Phone: (413) 737-1414. Fax: (413) 737-1488. E-mail: susanwmas@aol.com Web Site: www.947wmas.com. Licensee: Citadel Broadcasting Co. Group owner: Citadel Broadcasting Corp. (acq 6-3-2004; $22 million with co-located FM). Population served: 493,000 Natl. Network: ABC. Format: Adult standards. Spec prog: Black one hr, relg 2 hrs wkly. ♦Susan Murray, VP, gen mgr & gen sls mgr; Dina Cox, prom dir & prom mgr; Paul Cannon, progmg dir & progmg mgr; Rob Anthony, news dir; Randy Place, chief of engrg.

WMAS-FM— Dec 1, 1947: 94.7 mhz; 50 kw. 194 ft TL: N42 06 32 W72 36 44. Stereo. E-mail: wmas@947wmas.com Web Site: www.947wmas.com.163,905 Format: Adult contemp. ♦Randy Place, engrg VP.

***WNEK-FM**— Feb 17, 1976: 105.1 mhz; 13 w. -23 ft TL: N42 06 55 W72 31 05. Stereo. Hrs opn: Western New England College, 1215 Wilbraham Rd., 01119-2684. Phone: (413) 782-1582. Fax: (413) 796-2111. Licensee: Trustees of Western New England College. Population served: 250,000 Format: Div. Target aud: 15-35; college community, greater Springfield area. ♦Ian Martin, opns dir.

WNNZ(AM)—Westfield, July 8, 1987: 640 khz; 50 kw-D, 1 kw-N, DA-2. TL: N42 10 46 W72 45 05. Stereo. Hrs opn: 24 1331 Main St., Suite 5, 01103. Phone: (413) 781-1011. Fax: (413) 734-4434. Web Site: www.wnnz.com. Licensee: CC Licenses LLC. Group owner: Clear Channel Communications Inc. (acq 11-24-98; $1.2 million). Population served: 1,500,000 Natl. Network: NPR. Law Firm: Akin, Gump, Strauss, Hauer & Feld. Format: News, public radio. Target aud: 25-54; upscale adults.

***WSCB(FM)**— Mar 1, 1958: 89.9 mhz; 100 w. 35 ft TL: N42 05 59 W72 33 30. Hrs open: 263 Alden St., 01109. Phone: (413) 748-3722. Phone: (413) 748-3712. Fax: (413) 748-3153. Licensee: President & Trustees of Springfield College. Population served: 200,000 Format: Div. ♦Shazz Wilson, stn mgr; Hunter Golden, progmg dir; Greg Antonelli, news dir.

WSPR(AM)— June 1936: 1270 khz; 5 kw-D, 1 kw-N, DA-2. TL: N42 05 24 W72 36 11. Hrs open: 34 Sylvan St., West Springfield, 01089. Phone: (413) 827-8484. Fax: (413) 734-2240. E-mail: pgois@aol.com Web Site: wspr1270.com. Licensee: Davidson Media Station WSPR Licensee LLC. (acq 5-16-2005; $6.8 million with WACM(AM) West Springfield). Format: Sp, tropical. ♦Antonio Gois, pres; Paul Gois, exec VP, gen mgr & opns dir.

***WTCC(FM)**— Aug 19, 1971: 90.7 mhz; 4 kw. 92 ft TL: N42 06 32 W72 34 45. Stereo. Hrs opn: Box 9000, 01103. Phone: (413) 746-9822. Fax: (413) 781-3747. E-mail: managerwtcc@stcc.edu Web Site: www.wtccfm.org. Licensee: Springfield Technical Community College. Population served: 200,000 Format: Div, multi-cultural. Target aud: General. ♦Denise Stewart, gen mgr & stn mgr; Mark Leak, progmg dir; Beverly Showell, news dir; Fred Krampito, chief of engrg.

Sudbury

***WYAJ(FM)**— September 1980: 97.7 mhz; 4 w. 220 ft TL: N42 22 30 W71 24 28. Hrs open: 390 Lincoln Rd., 01776. Phone: (978) 443-9961. Fax: (978) 443-8824. E-mail: paul_sarapas@lsrhs.net Web Site: www.lsrhs.net. Licensee: Lincoln-Sudbury Regional School District. Population served: 2,000 Format: CHR, new age, classic rock. Target aud: General. Spec prog: Black 6 hrs, class 3 hrs, jazz 5 hrs, loc rock artists 3 hrs wkly.

Taunton

WPEP(AM)— Dec 22, 1949: 1570 khz; 1 kw-D, 227 w-N. TL: N41 53 00 W71 03 50. Hrs open: 24 41 Taunton Green, 02780-3233. Phone: (508) 822-1570. Fax: (508) 822-6473. E-mail: info@wpep1570.com Web Site: www.wpep1570.com. Licensee: Anastos Media Group Inc. (group owner; acq 7-6-01; grpsl). Population served: 100,000 Natl. Network: USA, Westwood One. Law Firm: Leventhal, Senter & Lerman. Format: News/talk. News staff: 2; News: 20 hrs wkly. Target aud: 35 plus; male & female. Spec prog: Por 10 hrs, gospel 12 hrs wkly. ♦A.J. Nicholson, gen mgr.

WSNE-FM— Jan 26, 1966: 93.3 mhz; 30 kw. 620 ft TL: N41 51 56 W71 17 22. Stereo. Hrs opn: 75 Oxford St., Suite 302, Providence, 02905. Phone: (401) 781-9979. Phone: (401) 224-1933. Fax: (401) 781-9329. E-mail: feedback@coast933.com Web Site: www.wsne.com. Licensee: Capstar TX L.P. Group owner: Clear Channel Communications Inc. (acq 8-30-00; grpsl). Population served: 2,100,200 Natl. Network: AP Radio, Premiere Radio Networks. Natl. Rep: Clear Channel. Format: Adult contemp. Target aud: 25-54; mostly women. Spec prog: Pub affrs 4 hrs wkly. ♦James Corwin, gen mgr; Mark Coffey, sls dir; Michelle Maker, prom mgr; Rick Everett, progmg VP & progmg dir.

Tisbury

WMVY(FM)— June 1, 1981: 92.7 mhz; 3 kw. 300 ft TL: N41 26 17 W70 36 47. (CP: Ant 328 ft.). Stereo. Hrs opn: 21 Box 1148, Vineyard Haven, 02568. Secondary address: 57 Carrolls Way, Vineyard Haven 02568. Phone: (508) 693-5000. Fax: (508) 693-8211. E-mail: pj@mvyradio.com Web Site: www.mvyradio.com. Licensee: Aritaur Communications Inc. (acq 6-17-98; $1 million). Population served: 180,000 Natl. Network: Moody, AP Radio. Natl. Rep: McGavren Guild. Format: Album-Oriented Rock. News staff: one. Target aud: 25-49; upper income, active consumer group. Spec prog: Class 4 hrs, jazz 4 hrs wkly. ♦Greg Orcutt, gen mgr; Nick Ward, gen sls mgr & prom dir; P.J. Finn, progmg dir.

Truro

WGTX(FM)— 2000: 102.3 mhz; 340 w. Ant 98 ft TL: N42 01 03 W70 04 23. Hrs open: 300 Western Ave., Allston, 02134. Phone: (617) 254-6333. Fax: (617) 254-2234. E-mail: karl@karlnurse.com Licensee: Dunes 102FM LLC (acq 6-5-2007; $550,000). Format: News/talk. ♦Karl Nurse, gen mgr.

Turners Falls

WRSI(FM)— July 1994: 93.9 mhz; 3 kw. Ant 328 ft TL: N42 32 01 W72 35 34. Stereo. Hrs opn: 24 Box 268, Northampton, 01061. Phone: (413) 585-9555. Fax: (413) 585-0927. E-mail: dj@wrsi.com Web Site: www.wrsi.com. Licensee: Saga Communications of New England LLC. Group owner: Saga Communications Inc. (acq 2-13-2004; grpsl). Population served: 180,000 Natl. Network: ABC. Format: AAA. News: 9 hrs wkly. Target aud: 18-54; young, educated, spend money. ♦Sean O'Mealy, gen mgr.

Waltham

***WBRS(FM)**— Feb 5, 1968: 100.1 mhz; 25 w. 151 ft TL: N42 22 09 W71 15 28. Stereo. Hrs opn: 24 Brandeis Univ., 415 South St., 02453-2728. Phone: (781) 736-5277. E-mail: info@wbrs.org Web Site: www.wbrs.org. Licensee: Brandeis University. Population served: 300,000 Wire Svc: UPI Format: Div. News: 5 hrs wkly. Target aud: General. ♦Hadar Sayfan, gen mgr; Mayank Puri, progmg dir; Leor Galil, news dir.

WKLB-FM— 1948: 102.5 mhz; 8.1 kw. Ant 1,151 ft TL: N42 18 27 W71 13 27. Stereo. Hrs opn: 24 55 William T Morrissey Blvd., Dorchester, 02125-3315. Phone: (617) 822-9600. Fax: (617) 822-6671. Web Site: www.wklb.com. Licensee: Charles River Broadcasting WCRB License Corp. (acq 11-15-2006). Population served: 535,000 Natl. Rep: Katz Radio. Format: Country. Target aud: 25-54. ♦Phil Redo, gen mgr & sls VP; Cathy Cram, gen sls mgr; Mike Brophey, progmg dir.

WRCA(AM)— 1948: 1330 khz; 5 kw-U, DA-2. TL: N42 21 16 W71 15 44. (CP: COL Watertown. 25 kw-D, 17 kw-N, DA-2). Stereo. Hrs opn: 24 552 Massachusetts Ave., Suite 201, Cambridge, 02139. Phone: (617) 492-3330. Fax: (617) 492-2800. Web Site: 1330wrca.com. Licensee: WAEC License LP. Group owner: Beasley Broadcast Group Inc. (acq 5-2000; $6 million). Format: Sp/ethnic. News: 10 hrs wkly. Target aud: General. ♦Stu Fink, gen mgr, opns mgr & progmg dir.

Ware

WARE(AM)— July 11, 1948: 1250 khz; 5 kw-D, 2.5 kw-N, DA-2. TL: N42 14 41 W72 12 30. Hrs open:
Rebroadcasts WQVR(FM) Southbridge.
3 Converse St., Palmer, 01069. Phone: (413) 289-2300. Fax: (413) 289-2323. E-mail: info@realoldies1250.com Web Site: www.realoldies1250.com. Licensee: Success Signal Broadcasting Inc. (acq 12-3-02; $250,000). Population served: 85,000 Natl. Rep: D & R Radio. Law Firm: Cohn & Marks. Wire Svc: AP Format: Oldies. Target aud: 30 plus. Spec prog: Pol 4 hrs wkly. ♦John P. Slosek Jr., pres; Marshall Sanft, gen mgr; Joe Grivalsky, chief of opns.

Watertown

WAZN(AM)— January 1958: 1470 khz; 1.4 kw-D, 3.4 kw-N, DA-2. TL: N42 24 49 W71 12 40. Hrs open: 24
Rebroadcasts WLYN(AM) Lynn (partial schedule).
500 W. Cummings Park, Suite 2600, Woburn, 01801. Phone: (781) 938-0869. Fax: (781) 938-0933. E-mail: Jeffk@mrbi.net Web Site: www.mrbi.net. Licensee: Multicultural Radio Broadcasting Licensee LLC. Group owner: Multicultural Radio Broadcasting Inc. (acq 12-11-2002; $1.8 million). Population served: 2,000,000 Format: Multicultural - leased time. News staff: 3. Hispanic (Spanish & Portuguese), Russian. Spec prog: Russian 10 hrs wkly. ♦Jeff Kline, gen mgr.

Webster

WGFP(AM)— Apr 1, 1980: 940 khz; 1 kw-D. TL: N42 03 17 W71 50 00. Hrs open: 24 27 Douglas Rd., 01570. Phone: (508) 943-9400. Fax: (508) 943-0405. Web Site: www.coolcountry940.com. Licensee: Just Because Inc. (acq 6-12-2003). Population served: 250,000 Format: Country. News: 25 hrs wkly. Target aud: 25-54. Spec prog: Pol one hr, Greek one hr wkly. ♦Barry Sims, CEO.

WORC-FM— Apr 8, 1994: 98.9 mhz; 3 kw. 410 ft TL: N42 02 30 W71 59 18. Stereo. Hrs opn: 24 250 Commercial St., Suite 530, 01608. Phone: (508) 752-1045. Fax: (508) 793-0824. E-mail: jaybeau.jones@citcomm.com Web Site: www.oldies989.com. Licensee: Citadel Broadcasting Co. Group owner: Citadel Broadcasting Corp. (acq 6-8-99; $3.5 million). Population served: 500,000 Natl. Network: Westwood One, ABC. Format: Oldies. News: 8 hrs wkly. Target aud: 25-49. Spec prog: Sp one hr wkly. ♦Joe Flynn, gen mgr; Amy Wilfong, prom mgr; JayBeau Jones, progmg dir & progmg mgr.

Wellesley

***WZLY(FM)**— Sept 20, 1976: 91.5 mhz; 10 w. 164 ft TL: N42 17 35 W71 18 21. Stereo. Hrs opn: Schneider Ctr., 106 Central St., 02181-8201. Phone: (781) 283-2690. Fax: (781) 237-4433. Licensee: Wellesley College. Population served: 2,000 Natl. Network: AP Radio. Format: Div, progsv, AOR. Target aud: General; Wellesley town and college community. ♦Julia Luechtefeld, gen mgr.

Wellfleet

***WRYP(FM)**— 2006: 90.1 mhz; 2.5 kw vert. Ant 80 ft TL: N42 01 53 W70 05 26. Hrs open: 356 Broad St., Fitchburg, 01420-3030. Phone: (888) 310-7729. Web Site: www.renewfm.com. Licensee: Horizon Christian Fellowship (acq 3-24-2006; $150,000 for CP). Format: Christian. ♦George Small, gen mgr.

West Barnstable

***WKKL(FM)**— Sept 19, 1977: 90.7 mhz; 205 w. 71 ft TL: N41 41 31 W70 20 16. Hrs open: 24 Cape Cod Community College, Rt. 132, 02668. Phone: (508) 375-4030. Phone: (508) 362-2131, EXT. 4684. Fax: (508) 375-4020. E-mail: wkkl247@yahoo.com Web Site: www.geocities.com/wkkl247. Licensee: Board of Trustees Cape Cod Community Colleges. Population served: 40,000 Format: Alternative. ♦Lisa Zinsius, gen mgr.

West Springfield

WACM(AM)— Aug 28, 1949: 1490 khz; 1 kw-U. TL: N42 06 06 W72 37 22. Hrs open: 34 Sylvan St., 01089. Phone: (413) 781-5200. Fax: (413) 734-2240. E-mail: pgois@aol.com Web Site: www.wacm1490.com. Licensee: Davidson Media Station WACM Licensee LLC. (acq 5-16-2005; $6.8 million with WSPR(AM) Springfield). Population served: 1,200,000 Format: Sp. ♦Paul Gois, gen mgr; Antonio Gois, news dir & chief of engrg.

West Yarmouth

***WBUR(AM)**— October 1940: 1240 khz; 1 kw-U. TL: N41 38 07 W70 14 06. Hrs open: 24
Rebroadcasts WBUR-FM Boston 98%.
890 Commonwealth Ave., Boston, 02215. Phone: (617) 353-0909. Fax: (617) 353-4747. Web Site: www.wbur.org. Licensee: The Executive Committee of Trustees of The Boston University Group owner: WBUR Group (acq 12-17-96). Format: News/talk. News: 78 hrs wkly. Target aud: 25-54; intelligent adults interested in news & politics. Spec prog: Sp 5 hrs wkly. ♦Paul LaCamera, gen mgr; Corey Lewis, stn mgr; Sam Fleming, progmg dir; John Davidson, news dir; Jeffrey Hutton, engrg dir.

WXTK(FM)— Dec 30, 1948: 95.1 mhz; 50 kw. 246 ft TL: N41 38 08 W70 14 06. Stereo. Hrs opn: 154 Barnstable Rd., Hyannis, 02601-2930. Phone: (508) 778-2888. Fax: (508) 778-9651. Web Site: www.95wxtk.com. Licensee: Qantum of Cape Cod License Co. LLC. Group owner: Boch Broadcasting (acq 2005; grpsl). Natl. Network: ABC. Natl. Rep: Eastman Radio. Format: News/talk, sports. Target aud: 25 plus. ♦Allison Makkay, VP & gen mgr; Steve McVie Solomon, progmg dir.

Westborough

WAAF(FM)—Licensed to Westborough. See Worcester

Westfield

WNNZ(AM)—Licensed to Westfield. See Springfield

***WSKB(FM)**— October 1974: 89.5 mhz; 100 w. 130 ft TL: N42 07 55 W72 47 51. Stereo. Hrs opn: Ely Hall, 577 Western Ave., 01086. Phone: (413) 572-5579. Fax: (413) 572-5625. E-mail: wskb89.5fm@yahoo.com Web Site: wskb.nipod.com. Licensee: Westfield State College. Population served: 250,000 Format: Talk, alternative, div. Target aud: General. ♦Barbara Hand, opns mgr; Dave Kowalski, progmg dir.

Williamstown

***WCFM(FM)**— Sept 8, 1958: 91.9 mhz; 440 w. Ant -836 ft TL: N42 42 38 W73 12 06. Hrs open: Baxter Hall, Williams College, 01267. Phone: (413) 597-3265. Fax: (413) 597-2259. E-mail: wcfmbd@wso.williams.edu Web Site: wcfm.williams.edu. Licensee: The President & Trustees of Williams College. ♦Adam Ain, gen mgr.

Winchendon

***WKMY(FM)**—Not on air, target date: unknown: 91.1 mhz; 155 w. Ant 207 ft TL: N42 42 09 W72 02 18. Hrs open: 2351 Sunset Blvd., Suite 170-218, Rocklin, CA, 95765. Phone: (916) 251-1600. Fax: (916) 251-1650. Licensee: Educational Media Foundation. (acq 6-16-2005; $15,000 for CP). ♦Richard Jenkins, pres; Mike Novak, VP; Keith Whipple, dev dir; David Pierce, progmg mgr; Ed Lenane, news dir; Sam Wallington, engrg dir; Karen Johnson, news rptr; Marya Morgan, news rptr; Richard Hunt, news rptr.

WSNI(FM)— January 1983: 97.7 mhz; 6 kw. Ant 328 ft TL: N42 47 24 W72 09 06. Stereo. Hrs opn: 24 69 Stanhope Ave., Keene, NH, 03431. Phone: (603) 352-9230. Fax: (603) 357-3926. E-mail: winq@monadrockradiogroup.com Web Site: www.hitsandfavorites.com. Licensee: Saga Communications of New England LLC. Group owner: Saga Communications Inc. (acq 4-1-2003; $400,000). Population served: 250,000 Format: Adult contemp. News staff: 2; News: 7 hrs wkly. Target aud: 25-54; 60% female, 40% male. ♦Bruce Lyons, gen mgr; Steve Hamel, opns mgr; Vicki Lenanan, prom dir.

Woods Hole

***WCAI(FM)**— Sept. 25, 2000: 90.1 mhz; 6.5 kw. 298 ft TL: N41 25 26 W70 40 20. Hrs open: Box 82, 02543. Secondary address: 3 Water St. 02543. Phone: (508) 548-9600. Fax: (508) 548-5517. E-mail: cainan@wgbh.org Web Site: www.cainan.org. Licensee: WGBH Educational Foundation (acq 12-31-97; $25,000 with WNAN(FM) Nantucket). Natl. Network: NPR. Format: News/talk. ♦John Voci, gen mgr & stn mgr; Susan Larks, pres & dev dir; Steve Young, progmg dir.

Worcester

WAAF(FM)—Listing follows WVEI(AM).

***WBPR(FM)**— 1994: 91.9 mhz; 1 kw. 469 ft TL: N42 15 11 W71 57 41. Hrs open:
Rebroadcasts WUMB-FM Boston 100%.
Univ. of Massachutts, 100 Morrissey Blvd., Boston, 02125-3393. Phone: (617) 287-6900. Fax: (617) 287-6916. E-mail: wumb@umb.edu Web Site: www.wumb.org. Licensee: University of Massachusetts. Format: Folk. Target aud: 25-45. ♦Patricia A. Monteith, gen mgr; Danielle Knight, dev dir; Brian Quinn, progmg dir.

***WCHC(FM)**— Sept 12, 1977: 88.1 mhz; 100 w. -656 ft TL: N42 14 15 W71 48 31. Hrs open: 7 AM-2 AM Box G, Holy Cross College, One College St., 01610. Phone: (508) 793-2475. Fax: (508) 793-2471. Web Site: www.college.holycross.edu/wchc. Licensee: Trustees of the College of the Holy Cross. Population served: 5,000 Law Firm: Winston & Strawn. Format: Progsv, diversified. News: 5 hrs wkly. Target aud: 12-35; adventurous. Spec prog: Black 8 hrs, class 6 hrs, jazz 6 hrs, metal 6 hrs, funk 3 hrs wkly. ♦Michael Cunningham, gen mgr.

WCRN(AM)— Dec 5, 1994: 830 khz; 7 kw-D, 5 kw-N, DA-2. TL: N42 14 47 W71 55 51. Hrs open: 1049 Main St., 01603. Phone: (508) 792-5803. Fax: (508) 770-0659. E-mail: studio@wcrnradio.com Web Site: www.mikerobertswebdesign.com/wcrn/. Licensee: Carter Broadcasting Corp. (acq 1-16-90). Format: Talk. Target aud: 25-54. ♦Ken Carter, pres; Kurt Carberry, gen mgr; Art Dufault, stn mgr.

***WCUW(FM)**— Dec 4, 1973: 91.3 mhz; 630 w. 145 ft TL: N42 15 46 W71 47 59. Stereo. Hrs opn: 24 910 Main St., 01610. Phone: (508) 753-1012. E-mail: wcuw@wcuw.com Web Site: www.wcuw.com. Licensee: WCUW Inc. Population served: 500,000 Format: Div. News: one hr wkly. Target aud: General. Spec prog: Fr 2 hrs, Sp 19 hrs, Ger 2 hrs, Pol 6 hrs, ethnic 10 hrs wkly. ♦Joe Cutroni, gen mgr.

***WICN(FM)**— Nov 21, 1969: 90.5 mhz; 8.1 kw horiz, 7.2 kw vert. Ant 371 ft TL: N42 20 07 W71 42 54. (CP: N42 20 09 W71 42 57). Stereo. Hrs opn: 24 50 Portland St., 01608. Phone: (508) 752-0700. Fax: (508) 752-7518. E-mail: webmaster@winc.org Web Site: www.wicn.org. Licensee: WICN Public Radio Inc. Population served: 500,000 Natl. Network: NPR. Law Firm: Shaw Pittman. Format: Jazz, big band, folk. News: 12 hrs wkly. Target aud: 35 plus; high education, high income. ♦Mike Gorman, pres; Thomas Kenney, VP & mktg mgr; Brian Barlow, gen mgr; Kyle Warren, opns dir & opns mgr; Tyra Penn, dev dir.

WNEB(AM)— Dec 18, 1946: 1230 khz; 1 kw-U. TL: N42 16 23 W71 49 23. Hrs open: 24 70 James St., Suite 201, 01603. Phone: (508) 831-9863. Fax: (508) 831-7964. E-mail: info@wvne.net Web Site: www.wneb.net. Licensee: Blount Masscom Inc. Group owner: Blount Communications Group Population served: 500,000 Natl. Network: Salem Radio Network. Natl. Rep: Salem. Format: Relg. Target aud: 25-54. ♦William A. Blount, pres; Steve Tuzeneu, stn mgr.

WORC(AM)— February 1925: 1310 khz; 5 kw-D, 1 kw-N, DA-2. TL: N42 13 19 W71 49 02. Stereo. Hrs opn: 24 122 Green St., Ste 2R, 01604-4138. Phone: (508) 791-2111, x203. Fax: (508) 752-6897. E-mail: info@power1310.com Web Site: www.power1310.com. Licensee: Antonio F. Gois. (acq 1-7-2005; $950,000). Population served: 600,000 Natl. Network: Westwood One. Format: Tropical. News staff: one. Target aud: 29-54; Latinos. Spec prog: Sports 6 hrs, Pol 4 hrs wkly, Sp one hr wkly. ♦Ivon Gois, gen mgr.

WSRS(FM)—Listing follows WTAG(AM).

WTAG(AM)— May 1, 1924: 580 khz; 5 kw-U, DA-2. TL: N42 20 13 W71 49 15. Stereo. Hrs opn: 24 98 Stereo Ln., Paxton, 01612. Phone: (508) 795-0580. Phone: (508) 757-9696. Fax: (508) 757-1779. Web Site: www.wtag.com. Licensee: Capstar TX L.P. Group owner: Clear Channel Communications Inc. (acq 8-30-2000; grpsl). Population served: 620,000 Natl. Network: CBS. Format: News/talk. News staff: 6; News: 40 hrs wkly. Target aud: 25-54. Spec prog: Sports. ♦Michael Schaus, gen mgr; Susan Remkiewicz, natl sls mgr; Bruce Palmer, prom dir & prom mgr; George Brown, progmg dir & news dir; Greg Byrne, pub affrs dir; Dan Kelleher, chief of engrg.

WSRS(FM)—Co-owned with WTAG(AM). June 17, 1940: 96.1 mhz; 14 kw. 863 ft TL: N42 18 34 W71 54 10. Stereo. 24 Phone: (508) 757-9696. Web Site: www.wsrs.com.744,800 Natl. Network: ABC. Format: Adult contemp. News staff: one; News: 5 hrs wkly. Target aud: 25-54. ♦Bruce Palmer, prom dir; Tom Holt, progmg dir & mus dir; George Brown, news dir; Lanie Brown, traf mgr.

WVEI(AM)— 1926: 1440 khz; 5 kw-U, DA-N. TL: N42 20 13 W71 49 15. Hrs open: 24 181 Moreland St., 01609-1049. Phone: (508) 752-5611. Fax: (508) 752-1006. Licensee: Entercom Boston II License LLC. Group owner: Entercom Communications Corp. (acq 10-15-98; grpsl). Population served: 43600 Natl. Rep: CBS Radio. Format: Sports. ♦Tom Baker, gen mgr & stn mgr; Jim Rushton, sls dir & natl sls mgr; Jason Wolfe, progmg dir; Eric Fitch, chief of engrg.

WAAF(FM)—Co-owned with WVEI(AM). June 15, 1961: 107.3 mhz; 20 kw. Ant 784 ft TL: N42 18 11 W71 53 52. Stereo. 20 Guest St., 3rd Fl., Boston, 02135. Phone: (617) 779-5400. Web Site: waaf.com.542,000 Format: AOR. Target aud: 25-54. ♦Tom Baker, gen mgr; Julie Kahn, stn mgr; Jim Vereault, gen sls mgr; Bob Goodell, rgnl sls mgr; Keith Hasting, progmg dir; Mike Hsu, news dir; Eddie Webb, disc jockey; Greg "Hillman" Hill, disc jockey.

WVNE(AM)—See Leicester

WXLO(FM)—Fitchburg, August 1960: 104.5 mhz; 37 kw. 563 ft TL: N42 30 27 W71 49 37. Stereo. Hrs opn: 24 250 Commercial St., Suite 530, 01608. Phone: (508) 752-1045. Fax: (508) 793-0824. Web Site: www.wxlo.com. Licensee: Citadel Broadcasting Co. Group owner: Citadel Broadcasting Corp. Population served: 1,000,000 Natl. Rep: McGavren Guild. Law Firm: Kaye, Scholer, Fierman, Hays & Handler. Format: Hot A/C. News: 5 hrs wkly. Target aud: 25-54. Spec prog: 70s mus 5 hrs wkly. ♦Joe Flynn, VP & gen mgr; JayBeau Jones, progmg dir.

Michigan

Ada

WDSS(AM)— 1998: 1680 khz; 10 kw-D, 680 w-N. TL: N42 56 09 W85 27 26. Stereo. Hrs opn: 24 3777 44th St. S.E., Kentwood, 49512.

Phone: (616) 554-5958. Fax: (616) 656-9326. E-mail: ggoodrich@gqti.com Web Site: www.wdss1680.com. Licensee: Goodrich Radio L.L.C. Population served: 650,000 Natl. Network: Radio Disney. Format: Radio Disney. News staff: 1. Target aud: 6-16; 18-34; children-teen.young mothers ♦Robert Goodrich, chmn; Ross Pettinga, gen mgr & stn mgr.

Adrian

WABJ(AM)— Nov 13, 1946: 1490 khz; 1 kw-U. TL: N41 54 02 W84 00 51. Hrs open: 24 121 W. Maumee St., 49221. Phone: (517) 265-1500. Fax: (517) 263-4525. Licensee: Friends Communication of Michigan Inc. Group owner: Friends Communications Inc. (acq 10-1-90; grpsl; FTR: 10-29-90). Population served: 92,000 Rgnl. Network: Mich. Farm. Rgnl rep: Michigan. Law Firm: Fletcher, Heald & Hildreth. Format: News/talk. News staff: 2; News: 9 hrs wkly. Target aud: General. Spec prog: Farm 7 hrs, relg 3 hrs wkly. ♦Bob Elliot, chmn, gen mgr & gen mgr.

WQTE(FM)—Co-owned with WABJ(AM). Sept 1, 1976: 95.3 mhz; 3 kw. 299 ft TL: N41 48 15 W84 05 25. Stereo. 24 150,000 Format: Country. News: 2 hrs wkly. Target aud: 25-54.

WLEN(FM)— June 9, 1965: 103.9 mhz; 3 kw. 299 ft TL: N41 54 11 W83 59 13. Stereo. Hrs opn: 24 Box 687, 49221. Secondary address: 242 W. Maumee St. 49221. Phone: (517) 263-1039. Fax: (517) 265-5362. Web Site: www.wlen.com. Licensee: Lenawee Broadcasting Co. Natl. Network: CNN Radio. Format: Adult contemp. Target aud: 25-54. Spec prog: Sp 4 hrs wkly. ♦Julie M. Koehn, pres & gen mgr.

***WVAC-FM**— Feb 13, 1967: 107.9 mhz; 13 w horiz. 79 ft TL: N41 53 55 W84 03 33. Hrs open: Adrian College, 110 S. Madison St., 49221. Phone: (517) 265-5161, Ext 4540. Phone: (517) 264-3141. Fax: (517) 264-3331. Licensee: Adrian College Board of Trustees. Population served: 15,000 Format: Div. Target aud: 18-23; those affiliated to the college lifestyle. ♦Steven Shehan, gen mgr.

Albion

WRCC(FM)—Marshall, Oct 1, 1968: 104.9 mhz; 6 kw. Ant 328 ft TL: N42 18 47 W84 55 46. Hrs open: 390 Golden Ave., Battle Creek, 49015. Phone: (269) 963-5555. Fax: (269) 963-5185. Web Site: www.superrock1049.com. Licensee: Capstar TX L.P. Group owner: Clear Channel Communications Inc. Population served: 168,840 Format: Classic rock. ♦Jack McDevitt, gen mgr; Mike Klein, progmg dir.

***WUFN(FM)**— April 1971: 96.7 mhz; 3.2 kw. Ant 456 ft TL: N42 15 56 W84 38 43. Stereo. Hrs opn: 24 13799 Donovan Rd., 49224. Phone: (800) 776-1070. Fax: (517) 531-5009. E-mail: dphelps@flc.org Web Site: www.967flr.org. Licensee: Family Life Broadcasting System. (group owner) Population served: 150,000 Natl. Network: USA, AP Radio. Format: Inspirational, Christian. News: 1.5 hrs wkly. Target aud: 25-54; Christian families. ♦Randy Carlson, pres; Dave Phelps, gen mgr; Rod Robison, dev VP.

Allegan

WZUU(FM)— April 1991: 92.3 mhz; 860 w. 600 ft TL: N42 34 52 W85 45 17. Stereo. Hrs opn: 24 Box 80, 706 E. Allegan St., Otsego, 49078. Phone: (269) 343-1717. Fax: (269) 692-6861. E-mail: tflynn@wqxc.com Web Site: www.wzuu.com. Licensee: Forum Communications Inc. (acq 6-1-97). Population served: 500,000 Natl. Rep: Roslin. Rgnl rep: Michigan. Law Firm: Richard Hayes. Format: Classic rock. News staff: one; News: 2 hrs wkly. Target aud: 25-54; professionals. ♦Robert Brink, pres; Tom Bontrager, gen sls mgr; Tom Flynn, gen mgr & progmg dir.

Allendale

***WGVU-FM**— July 15, 1983: 88.5 mhz; 3 kw. 311 ft TL: N43 03 24 W85 57 31. Stereo. Hrs open: 24 Grand Valley State Univ., 301 W. Fulton, Grand Rapids, 49504-6492. Phone: (616) 331-6666. Fax: (616) 331-6625. E-mail: wgvu@gvsu.edu Web Site: www.wgvu.org. Licensee: Board of Control of Grand Valley State University. Population served: 700,000 Natl. Network: NPR, AP Radio. Rgnl. Network: Mich. Pub. Format: Jazz, news. News staff: 5; News: 26 hrs wkly. Target aud: 25 plus; mid to upper educ & income levels. ♦Michael T. Walenta, gen mgr; Ken Kolbe, opns mgr; Richard Nelson, gen sls mgr; Pamela Holtz, prom mgr; Fred Martino, news dir.

Alma

WFYC(AM)— Aug 17, 1948: 1280 khz; 1 kw-D, 45 w-N. TL: N43 22 08 W84 36 19. Hrs open: 24 Box 665, 48801. Phone: (989) 463-3175. Fax: (989) 463-6674. Licensee: Jacom Inc. (acq 1996). Population served: 100,000 Format: Sports. Target aud: 25-50. Spec prog: Farm 4 hrs wkly. ♦James Sommerville, pres, gen mgr & progmg dir; Susan Sommerville, prom mgr.

WQBX(FM)—Co-owned with WFYC(AM). November 1964: 104.9 mhz; 6 kw. Ant 328 ft TL: N43 22 08 W84 36 19. Stereo. 24 250,000 Natl. Network: ABC. Format: Adult contemp.

WMLM(AM)—See Saint Louis

***WQAC-FM**— Mar 27, 1993: 90.9 mhz; 100 w. 66 ft TL: N43 22 50 W84 40 14. Hrs open: 7 AM-2 AM (M-F); noon-2 AM (S, Su) Van Dusen Student Ctr., 614 W. Superior St., 48801. Phone: (989) 463-7095. Fax: (989) 463-7277. E-mail: wqaccharts@blazemail.com Web Site: students.alma.edu/organizations/wqac. Licensee: Alma College. Population served: 15,000 Format: Rock. News staff: one; News: 3 hrs wkly. Target aud: 13-24; high school & college students. Spec prog: Jazz 2 hrs, relg 2 hrs, hip hop/R&B 7 hrs, heavy metal 5 hrs, classic rock 2 hrs wkly. ♦Kat Lanphear, gen mgr; Kelly Gildersleeve, prom dir; Cailean Dinwoody, progmg dir; Mike Cruz, mus dir.

Alpena

WATZ(AM)— 1946: 1450 khz; 1 kw-U. TL: N45 03 58 W83 29 06. Hrs open: 20 Box 536, 49707. Secondary address: 123 Prentiss 49707. Phone: (989) 354-8400. Fax: (989) 354-3436. Web Site: www.watz.com. Licensee: WATZ Radio Inc. Group owner: Midwestern Broadcasting Co. Population served: 17,500 Rgnl. Network: Mich. Farm. Format: Talk. News staff: 2; News: 31 hrs wkly. Target aud: 35-64. Spec prog: Farm 3 hrs, Ger 2 hrs, Pol 2 hrs, relg 2 hrs wkly. ♦Mike Centala, gen mgr; Steve Wright, opns mgr; Bruce Johnson, news dir.

WATZ-FM— 1967: 99.3 mhz; 17 kw. Ant 843 ft TL: N44 51 25 W83 32 34. Stereo. 20 Web Site: www.watz.com.130,000 Format: Country. Target aud: 25-54. ♦Suzy Martin, mus dir.

***WCML-FM**— Apr 24, 1978: 91.7 mhz; 100 kw. 1,171 ft TL: N45 08 17 W84 09 44. Stereo. Hrs opn: 24
Rebroadcasts WCMU-FM Mount Pleasant 100%.
Public Broadcasting Ctr., Central Michigan Univ., Mount Pleasant, 48859. Phone: (989) 774-3105. Fax: (989) 774-4427. E-mail: cmuradio@cmich.edu Web Site: www.wcmu.org. Licensee: Central Michigan University. Population served: 30,000 Natl. Network: NPR, PRI. Rgnl. Network: Mich. Pub. Law Firm: Dow, Lohnes & Albertson. Wire Svc: AP Format: Jazz, class, news & info. News staff: 2; News: 45 hrs wkly. Target aud: General. ♦Ed Grant, gen mgr & rgnl sls mgr. Co-owned TV: *WCML-TV affil

WHSB(FM)— May 1965: 107.7 mhz; 99 kw. 760 ft TL: N45 03 40 W83 43 05. Stereo. Hrs opn: 1491 M-32 W., Apena, 49707. Phone: (989) 354-4611. Fax: (989) 354-4014. E-mail: thebay@1077thebay.com Web Site: www.1077thebay.com. Licensee: Edwards Communications LC. Group owner: Northern Radio Network (acq 12-21-2004; grpsl). Population served: 58,300 Natl. Rep: Michigan Spot Sales. Wire Svc: UPI Format: Adult contemp. Target aud: 25-54. ♦Jerry Edwards, pres; Darrel Kelly, progmg dir & chief of engrg.

WKJZ(FM)—Hillman, December 1993: 94.9 mhz; 50 kw. 492 ft TL: N45 01 33 W83 54 52. Hrs open:
Rebroadcasts WQLB(FM) Tawas City 85%.
Box 549, Tawas City, 48764. Phone: (989) 362-3417. Fax: (989) 362-4544. E-mail: wkjc@wkjc.com Web Site: www.wkjc.com. Licensee: Carroll Enterprises Inc. (group owner; acq 6-29-92; FTR: 7-27-92). Format: Classic rock. ♦John Carroll Jr., gen mgr.

Ann Arbor

WAAM(AM)— October 1947: 1600 khz; 5 kw-U, DA-2. TL: N42 11 32 W83 41 09. Stereo. Hrs opn: 24 4230 Packard Rd., 48108. Phone: (734) 971-1600. Fax: (734) 973-2916. E-mail: waamradio@aol.com Web Site: www.talkradio1600.com. Licensee: First Broadcasting Holdings LLC. (group owner; (acq 6-13-2005; grpsl). Population served: 250,000 Natl. Network: Westwood One. Law Firm: Bryan Cave. Format: MOR, news/talk. News staff: 4; News: 20 hrs wkly. Target aud: 35 plus; home owners & professionals. Spec prog: Relg 4 hrs, old time radio 6 hrs wkly. ♦Mabel Johnson, VP; Bob Hamilton, gen mgr; Dean Erskine, gen sls mgr.

***WCBN-FM**— Jan 23, 1972: 88.3 mhz; 200 w. 177 ft TL: N42 16 37 W83 44 07. Stereo. Hrs opn: 24 530 Student Activities Bldg., 48109-1285. Phone: (734) 647-4122. Phone: (734) 763-3535. E-mail: fm@wcbn.org Web Site: www.wcbn.org. Licensee: Regents of the University of Michigan. Population served: 99,797 Format: Div. News staff: 0. Target aud: 18-49. Spec prog: Jazz 17 hrs, Sp 3 hrs, folk 2 hrs, pub affrs 5 hrs, gospel one hr, Fr one hr wkly. ♦Sarah Herard, gen mgr; Malini Sridharan, dev dir.

WDEO(AM)—Ypsilanti, Nov 16, 1962: 990 khz; 9.2 kw-D, 250 w-N. TL: N42 15 53 W83 36 47 (D), N42 15 55 W83 36 42 (N). Stereo. Hrs opn: 24 Box 504, One Ave Maria Dr., 48106. Phone: (734) 930-5200. Fax: (734) 930-3179. E-mail: hroot@wdeo.net Web Site: www.wdeo.net. Licensee: Word Broadcasters Inc. 990 Investors LLC (acq 9-8-99; $2.5 million). Population served: 4,600,000 Law Firm: Dennis Kelly Law Offices. Format: Christian, talk. News: 9 hrs wkly. Target aud: 21 plus; adult Christian. ♦Al Kresta, CEO; Michael Jones, gen mgr; Steve Clarke, opns mgr.

WQKL(FM)—Listing follows WTKA(AM).

WSDS(AM)—Salem Township, 1962: 1480 khz; 750 w-D, 5 kw-N, DA-2. TL: N42 15 42 W83 37 10. Hrs open: 24 580 W. Clark Rd., Ypsilanti, 48198. Phone: (734) 484-1480. Fax: (734) 484-5313. E-mail: wsds@wsds1480.com Web Site: www.wsds1480.com. Licensee: Birach Broadcasting Corp. (acq 1-25-2005; $1.5 million). Population served: 2,050,000 Natl. Network: Westwood One. Format: Classic country, talk. News: 8 hrs wkly. Target aud: 25 plus. Spec prog: Bluegrass 6 hrs, Chinese 4 hrs, Greek 4 hrs wkly. ♦Sima Birach, pres; George Koch, gen mgr; Keith Jason, opns mgr, progmg mgr, mus dir & disc jockey; Hannah Koch, sls dir & prom mgr; John Petelka, pub affrs dir & disc jockey; Ralph Hines, chief of engrg; Andy Baron, disc jockey; Brian Barnum, disc jockey.

WTKA(AM)— Apr 26, 1945: 1050 khz; 10 kw-D, 500 w-N, DA-2. TL: N42 08 46 W83 39 36. Hrs open: 1100 Victors Way, Suite 100, 48108. Phone: (734) 302-8100. Fax: (734) 213-7508. Web Site: www.wtka.com. Licensee: Capstar TX L.P. Group owner: Clear Channel Communications Inc. (acq 8-7-00; grpsl). Population served: 99,797 Natl. Rep: Michigan Spot Sales. Format: Sports.

WQKL(FM)—Co-owned with WTKA(AM). Feb 14, 1967: 107.1 mhz; 3 kw. 289 ft TL: N42 16 41 W83 44 32. Web Site: annarbors1071.com. Format: Adult altenative.

***WUOM(FM)**— 1948: 91.7 mhz; 93 kw. 780 ft TL: N42 24 24 W83 54 54. Stereo. Hrs opn: 24 535 W. William St., 48103. Phone: (734) 764-9210. Fax: (734) 647-3488. E-mail: michigan.radio@umich.edu Web Site: www.michiganradio.org. Licensee: The Regents of University of Michigan. Population served: 130,000 Natl. Network: NPR, PRI. Rgnl. Network: Mich. Pub. Law Firm: Dow, Lohnes & Albertson. Wire Svc: AP Format: News/talk. News staff: 7; News: 140 hrs wkly. ♦Jon Hoban, stn mgr; Peggy J. Watson, opns mgr.

WWWW-FM— March 1962: 102.9 mhz; 49 kw horiz, 42 kw vert. 499 ft TL: N42 15 04 W83 48 28. Stereo. Hrs opn: 1100 Victors Way, Suite 100, 48108. Phone: (734) 302-8100. Fax: (734) 213-7508. E-mail: programming@w4country.com Web Site: www.w4country.com. Licensee: Capstar TX L.P. Group owner: Clear Channel Communications Inc. (acq 8-7-2000; grpsl). Population served: 200,000 Natl. Rep: McGavren Guild. Format: Country. ♦Bob Bolak, gen mgr; Shannon Brown, gen sls mgr.

Ashley

WJSZ(FM)— Mar 14, 1994: 92.5 mhz; 2 kw. 400 ft TL: N43 10 55 W84 26 58. Hrs open: 24 103 N. Washington, Owosso, 48867. Phone: (989) 725-1925. Fax: (989) 725-7925. E-mail: rodk@voyager.net Web Site: www.z925.com. Licensee: Krol Communications Inc. (acq 12-29-2005; $650,000). Law Firm: Miller & Neely. Format: Adult contemp. News: 90 mins wkly. Target aud: 25-54; general. ♦Rob Krol, pres & opns dir; Angie Bucsf, local news ed.

Atlanta

WFDX(FM)—Licensed to Atlanta. See Petoskey

Auburn Hills

***WAHS(FM)**— 1975: 89.5 mhz; 100 w. 141 ft TL: N42 37 42 W83 13 56. Hrs open: 2800 Waukegan St., 48326. Phone: (248) 852-9247. Fax: (248) 852-0595. Licensee: Avondale School District. Population served: 3,000 Format: CHR. ♦Rick Kreinbring, gen mgr.

WXOU(FM)—Licensed to Auburn Hills. See Rochester

Bad Axe

WLEW-FM— 1956: 102.1 mhz; 50 kw. Ant 492 ft TL: N43 53 28 W83 07 26. Stereo. Hrs opn: 24 935 S. Van Dyke Rd., 48413. Phone: (989) 269-9931. Fax: (989) 269-7702. Licensee: Thumb Broadcasting Inc. (acq 9-15-93; with co-located AM; FTR: 10-11-93). Population served: 620,000 Format: Adult contemp, classic rock. Target aud: 25-50. ♦Richard Aymen, CEO, VP, gen mgr & progmg dir; Matthew Aymen, VP & sls VP; Craig Routzahn, gen mgr, news dir & pub affrs dir; Jerry Stocker, sls VP & chief of engrg; Tina Hind, traf mgr & farm dir.

WLEW(AM)— 1950: 1340 khz; 1 kw-U, DA-D. TL: N43 47 56 W83 01 21.24 80,000 Format: Country. News staff: 2; News: 19 hrs wkly. Target aud: 18-50. ♦Richard Aymen, sls dir.

Baraga

***WVCN(FM)**— 1998: 104.3 mhz; 100 kw. Ant 859 ft TL: N46 39 50 W88 23 06. Stereo. Hrs opn: 24 3434 W. Kilbourn Ave., Milwaukee, WI, 53208. Phone: (414) 935-3000. Fax: (414) 935-3015. E-mail: wvcn@vcyamerica.org Web Site: www.vcyamerica.org. Licensee: Keweenaw Bay Broadcasting Inc. Group owner: VCY/America Inc. (acq 7-8-99). Format: Christian. ♦Dr. Randall Melchert, pres; Vic Eliason, VP & gen mgr; Jim Schneider, progmg dir & pub affrs dir; Tom Schlueter, mus dir; Gordon Morris, news dir; Andy Eliason, chief of engrg.

Battle Creek

WBCK(AM)— July 9, 1948: 930 khz; 5 kw-D, 1 kw-N, DA-2. TL: N42 17 40 W85 11 00. Hrs open: 390 Golden Ave., 49015. Phone: (269) 963-5555. Fax: (269) 963-5185. Web Site: www.am930wbck.com. Licensee: Capstar TX L.P. Group owner: Clear Channel Communications Inc. (acq 8-30-00; grpsl). Population served: 170,000 Format: News/talk. Target aud: 25-54. ♦Mike Klein, gen mgr & stn mgr; Tim Collins, prom mgr, progmg dir & progmg mgr; Walker Sisson, chief of engrg.

WBXX(FM)—Co-owned with WBCK(AM). Feb 28, 1975: 95.3 mhz; 3 kw. Ant 269 ft TL: N42 17 17 W85 09 54. Stereo. Web Site: www.softrock953.com.170,000 Format: Adult contemp. Target aud: 18-49.

WBFN(AM)— July 1, 1993: 1400 khz; 1 kw-U. TL: N42 18 15 W85 11 31. Hrs open: 390 Golden Ave., 49015. Phone: (269) 963-5555. Fax: (269) 963-5185. E-mail: bcnews@clearchannel.com Web Site: www.battlecreekradio.com. Licensee: Capstar TX Limited Partnership Group owner: Clear Channel Communications Inc. (acq 8-30-2000; grpsl). Population served: 168,840 Format: Sports. Target aud: 18-54; general. ♦Mike Klein, gen mgr; Tim Collins, progmg dir.

WKFR-FM— June 11, 1963: 103.3 mhz; 50 kw. 500 ft TL: N42 21 19 W85 20 28. Stereo. Hrs opn: 24 Box 50911, Kalamazoo, 49005-0911. Secondary address: 4154 Jennings Dr., Kalamazoo 49048. Phone: (269) 344-0111. Fax: (269) 344-4223. E-mail: radio@wkfr.com Web Site: www.wkfr.com. Licensee: Cumulus Licensing Corp. Group owner: Cumulus Media Inc. (acq 5-26-98; grpsl). Population served: 700,000 Format: CHR. News staff: one; News: 3 hrs wkly. Target aud: 25-54. ♦Lew Dickey, CEO & pres; John Pinch, COO; Martin Gausvik, CFO; Mike McKelly, opns mgr; Woody Houston, progmg dir.

WOLY(AM)— Nov 22, 1963: 1500 khz; 1 kw-D, DA. TL: N42 17 30 W85 10 08. Hrs open: 15074 6 1/2 Mile Rd., 49014. Phone: (269) 965-1515. Fax: (269) 965-1315. E-mail: wolyradio@sbcglobal.net Licensee: Christian Family Network. (acq 1-89; $100,000; FTR: 1-23-89). Population served: 100,700 Natl. Network: USA. Format: Traditional Christian mus & progmg. Target aud: General. ♦James Elsman, pres; Tim Smothers, gen mgr & progmg dir.

Bay City

***WCHW-FM**— Sept 1, 1973: 91.3 mhz; 110 w. 125 ft TL: N43 35 19 W83 52 28. Hrs open: 1624 Columbus Ave., 48708. Phone: (989) 892-1741. Phone: (989) 892-5533. Fax: (989) 892-7946. E-mail: wchwonline@fnmail.com Web Site: www.wchwonline.freewebspace.com. Licensee: School District Bay City. Population served: 49,449 Format: AOR. ♦Brian Bishop, gen mgr; Joe Bowker, progmg dir.

WHNN(FM)— 1947: 96.1 mhz; 100 kw. 1,020 ft TL: N43 33 10 W83 41 24. Stereo. Hrs opn: 24 1740 Champagne Dr. N., Saginaw, 48604. Phone: (989) 298-9466. Fax: (989) 754-9600. Web Site: www.whnn.com. Licensee: Citadel Broadcasting Co. Group owner: Citadel Broadcasting Corp. (acq 4-26-01; grpsl). Population served: 1,000,000 Format: Oldies. News staff: one; News: 6 hrs wkly. Target aud: 25-54. ♦Scott Meier, gen mgr.

WIOG(FM)— September 1969: 102.5 mhz; 86 kw. 860 ft TL: N43 28 24 W83 50 40. Stereo. Hrs opn: 24 1740 N. Champagne Dr., Saginaw, 48604. Phone: (989) 776-2100. Fax: (989) 754-5990. Web Site: www.wiog.com. Licensee: Citadel Broadcasting Co. Group owner: Citadel Broadcasting Corp. (acq 2-8-99; grpsl). Population served: 500,000 Natl. Network: ABC. Natl. Rep: McGavren Guild. Format: Adult contemp. News staff: one; News: 5 hrs wkly. Target aud: 25-54. ♦Scott Meier, VP & gen mgr; Bruce Johnston, sls dir; Tom Clark, gen sls mgr; Jerry Noble, progmg dir.

***WLKB(FM)**— July 25, 1993: 89.1 mhz; 50 kw vert. Ant 371 ft TL: N43 33 42 W83 58 52. Stereo. Hrs opn: 24
Rebroadcasts KLVR(FM) Santa Rosa, CA 100%.
2351 Sunset Blvd., Suite 170-218, Rocklin, CA, 95765. Phone: (916) 251-1600. Fax: (916) 251-1650. Web Site: www.klove.com. Licensee: Educational Media Foundation. (acq 8-8-2006; $800,000). Natl. Network: K-Love. Format: Contemp Christian. ♦Richard Jenkins, pres.

WMAX(AM)—Licensed to Bay City. See Saginaw

WSGW(AM)—See Saginaw

***WUCX-FM**— September 1989: 90.1 mhz; 30 kw. 479 ft TL: N43 33 10 W83 41 24. Stereo. Hrs open: 1961 Delta Rd., Univ. Ctr., 48710. Phone: (989) 686-9292. Fax: (989) 686-0155. E-mail: wucx@delta.edu Web Site: www.delta.edu/broadcasting. Licensee: Central Michigan University. Natl. Network: NPR, PRI. Rgnl. Network: Mich. Pub. Law Firm: Dow, Lohnes & Albertson. Format: Jazz, blues, news. Target aud: General. ♦Barry Baker, gen mgr; Ray Ford, progmg dir; Howard Sharper, progmg mgr; David Nicholas, news dir & chief of engrg; Tom Garnett, chief of engrg.

Bear Creek Township

***WTLI(FM)**— Sept 16, 1998: 89.3 mhz; 6 kw vert. 1,023 ft TL: N45 10 12 W84 45 04. Hrs open:
rebroadcast of WLGH(FM) Leroy Township.
Box 388, Williamston, 48895. Phone: (517) 381-0573. Fax: (877) 850-0881. E-mail: info@positivehits.com Web Site: www.positivehits.com. Licensee: Superior Communications. Format: Adult Christian hit. ♦Jenn Czelada, gen mgr.

Bear Lake

WCUZ(FM)— Nov 2, 1987: 100.1 mhz; 3 kw. Ant 328 ft TL: N44 25 18 W86 07 17. Hrs open:
Rebroadcasts WLDR-FM Traverse City 100%.
13999 S. West Bay Shore Dr., Traverse City, 49684-6206. Phone: (231) 947-3220. Fax: (231) 947-7201. Web Site: www.wldr.com. Licensee: Fort Bend Broadcasting Co. (group owner; (acq 9-27-2000; $590,000 with WOUF(FM) Beulah). Population served: 35,000 Law Firm: Baraff, Koerner & Olender. Format: Country. ♦Roy Henderson, pres & gen mgr.

Beaverton

WMRX-FM— Sept 15, 1980: 97.7 mhz; 4.1 kw. Ant 400 ft TL: N43 53 16 W84 31 45. Stereo. Hrs opn:
Rebroadcasts WMPX(FM) Midland 100%.
Box 1689, Midland, 48641-1689. Secondary address: 1510 Bayliss St., Midland 48640. Phone: (989) 631-1490. Fax: (989) 631-6357. E-mail: requests@wmpxwmrx.com Web Site: www.wmpxwmrx.com. Licensee: Steel Broadcasting Inc. (acq 1990). Natl. Network: ABC. Format: Adult standards, big band, classic. News: 8 hrs wkly. Target aud: General; 35+. Spec prog: Sounds of Sinatra 2 hrs wkly. ♦Thomas Steel, pres & gen mgr; Aaron Whiting, gen sls mgr.

Belding

***WSLI(FM)**—Not on air, target date: unknown: 90.9 mhz; 10 kw vert. Ant 98 ft TL: N43 05 27 W85 16 27. Hrs open: 3302 N. Van Dyke, Imlay City, 48444. Phone: (810) 724-2638. Fax: (877) 850-0881. Licensee: Superior Communications. ♦Edward Czelada, pres.

Benton Harbor

***WAYO-FM**—Not on air, target date: unknown: 89.9 mhz; 250 w. Ant 335 ft TL: N42 04 19 W86 22 14. Hrs open: 1159 E. Beltline Ave. N.E., Grand Rapids, 49525. Phone: (616) 942-1500. Fax: (616) 942-7078. Web Site: www.curadio.org. Licensee: Cornerstone University.

WCNF(FM)— June 15, 1998: 94.9 mhz; 2.2 kw. Ant 380 ft TL: N42 04 19 W86 22 14. Hrs open: 24 Box 107, St. Joseph, 49085. Secondary address: 580 E. Napier Ave. 49022. Phone: (269) 925-1111. Fax: (269) 925-1011. Web Site: www.thecoast.fm. Licensee: WSJM Inc. Group owner: The Mid-West Family Broadcast Group Rgnl rep: Michigan Spot Sales Format: Adult contemp, hits of the 80s & 90s. ♦Joe Daguanno, VP & sls dir; Gayle Olson, gen mgr; Jim Gifford, opns dir; Sue Patzen, prom dir.

WHFB-FM—Licensed to Benton Harbor. See Benton Harbor-St. Joseph

Benton Harbor-St. Joseph

WHFB(AM)— Sept 22, 1947: 1060 khz; 5 kw-D, 2.5 kw-CH. TL: N42 04 44 W86 28 00. Hrs open: 2100 Fairplain Ave., Benton Harbor, 49022. Phone: (269) 925-9300. Fax: (269) 925-0065. E-mail: whfbam@whfbam.com Web Site: www.whfbam.com. Licensee: WHFB Broadcast Associates L.P. Group owner: WinCom Communications Group Inc. (acq 8-85). Population served: 180,000 Rgnl. Network: Mich. Farm. Format: Great talk. News staff: one. Target aud: General. ♦Bill Stanley, gen mgr & progmg dir.

WHFB-FM— Oct 10, 1947: 99.9 mhz; 50 kw. 497 ft TL: N42 03 17 W86 27 31. Stereo. Web Site: www.catcountry999.com. Format: Country. Target aud: 25-54. ♦Mike Sullivan, gen sls mgr; Jim Roberts, progmg dir.

WIRX(FM)—Listing follows WSJM(AM).

WSJM(AM)—Saint Joseph, Nov 18, 1956: 1400 khz; 880 w-U. TL: N42 05 12 W86 26 40. Hrs open: 24 Box 107, St. Joseph, 49085. Secondary address: 580 E. Napier, Benton Harbor 49022. Phone: (269) 925-1111. Phone: (269) 925-9756. Fax: (269) 925-1011. Web Site: www.wsjm.com. Licensee: WSJM Inc. Group owner: The Mid-West Family Broadcast Group (acq 1-1-59). Population served: 170,000 Natl. Network: ABC. Law Firm: Davis Wright Tremaine. Wire Svc: AP Format: Full service, news/talk, sports. News staff: 4; News: 30 hrs wkly. Target aud: 35 plus. Spec prog: Black 5 hrs wkly. ♦Gayle Olson, pres & gen mgr; Jim Gifford, opns dir; Joe Daguanno, VP & sls dir; Annette Weston, news dir; Bob Bucholtz, sls.

WIRX(FM)—Co-owned with WSJM(AM). June 20, 1966: 107.1 mhz; 1.2 kw. 498 ft TL: N42 04 19 W86 22 14. Stereo. 24 Fax: (269) 925-1011. Web Site: wirx.com.293,000 Law Firm: Davis Wright Tremaine. Format: Rock. Target aud: 18-49. ♦Bob Bucholtz, gen sls mgr.

Berrien Springs

WAUS(FM)—Licensed to Berrien Springs. See South Bend IN

Beulah

WOUF(FM)— January 1998: 92.1 mhz; 1.6 kw. 600 ft TL: N44 30 54 W86 06 52. Stereo. Hrs opn: 13999 S.W. Bayshore Dr., 49684. Phone: (231) 947-3220. Fax: (231) 947-7201. Licensee: Fort Bend Broadcasting Co. (group owner; acq 9-27-00; $590,000 with WCUZ(FM) Bear Lake). Format: Alternative country. News staff: 3. Target aud: 18-54; male. ♦Ryan Henderson, pres & gen mgr.

Big Rapids

WBRN(AM)— Jan 6, 1953: 1460 khz; 5 kw-D, 2.5 kw-N, DA-N. TL: N43 39 57 W85 28 59. Stereo. Hrs opn: 24 18720 16 Mile Rd., 49307. Phone: (231) 796-7684. Fax: (231) 796-7951. E-mail: wbrnfm@wbrn.com Web Site: www.wbrn.com. Licensee: Mentor Partners Inc. (acq 6-21-2005; $850,000 with co-located FM). Population served: 110,900 Natl. Network: ESPN Radio. Natl. Rep: Michigan Spot Sales. Format: News/talk, sports. News staff: one; News: 12 hrs wkly. Target aud: 35 plus; adults. ♦Jeffrey Scarpelli, pres; Brian Goodenow, opns dir; George Keen, chief of engrg.

WWBR(FM)—Co-owned with WBRN(AM). September 1964: 100.9 mhz; 6 kw. 318 ft TL: N43 39 49 W85 28 54. Stereo. 24 Web Site: www.wwbrfm.com. Natl. Network: Jones Radio Networks, AP Network News. Format: Country. News staff: one; News: 10 hrs wkly. Target aud: 16-54.

WYBR(FM)— June 30, 1982: 102.3 mhz; 10.5 kw. 436 ft TL: N43 41 01 W85 34 56. Stereo. Hrs opn: 24 18720 16-Mile Rd., 49307. Phone: (231) 796-7000. Fax: (231) 796-7951. E-mail: wybr@tucker-usa.com Web Site: www.wybr.com. Licensee: Mentor Partners Inc. (acq 8-10-98). Population served: 175,000 Natl. Rep: Patt. Format: Hot adult contemp, CHR. News: 8 hrs wkly. Target aud: 25-54. Spec prog: Relg one hr wkly. ♦Jeffrey J. Scarpelli, pres & gen mgr.

Birmingham

WCSX(FM)— Mar 14, 1987: 94.7 mhz; 13.5 kw. 945 ft TL: N42 27 13 W83 09 50. Stereo. Hrs opn: 24 1 Radio Plaza, Ferndale, 48220. Phone: (248) 398-9470. Fax: (248) 541-9279. Web Site: www.wcsx.com. Licensee: Greater Michigan Radio Inc. Group owner: Greater Media Inc. (acq 7-3-73). Population served: 443,800 Natl. Rep: McGavren Guild. Format: Classic rock. Target aud: 25-54; males. ♦Tom Bender, VP & gen mgr.

Bloomfield Hills

***WBFH(FM)**— Oct 1, 1976: 88.1 mhz; 360 w. 100 ft TL: N42 34 42 W83 17 10. Stereo. Hrs opn: 24 4200 Andover Rd., 48302. Phone: (248) 341-5690. Fax: (248) 341-5679. E-mail: wbfh@bloomfield.org Web Site: www.wbfh.fm. Licensee: Board of Education of Bloomfield Hills School District. Population served: 250,000 Law Firm: Putbrese, Hunsaker & Trent, P.C. Format: Div, CHR, educ. Target aud: 12-34. Spec prog: Prep sports 6 hrs wkly. ♦Pete Bowers, gen mgr; Randy Carr, stn mgr; Cindi Hopkins, dev dir; Ron Wittebols, progmg.

Boyne City

WBCM(FM)— Apr 10, 1978: 93.5 mhz; 14.1 kw. 928 ft TL: N45 10 44 W85 05 42. Stereo. Hrs opn: 24
Simulcast of WTCM-FM Traverse City.
314 E. Front St., Traverse City, 49684. Phone: (231) 947-7675. Fax: (231) 929-3988. E-mail: country@wtcmradio.com Web Site: www.wtcmi.com. Licensee: Biederman Investments Inc. (acq 9-6-90; $250,000; FTR: 10-1-90). Natl. Rep: Katz Radio. Law Firm: Cordon & Kelly. Format: Modern country. News staff: 4; News: 25 hrs wkly. Target aud: 25-54. Spec prog: Farm one hr wkly. ♦Ross Biederman, pres; Chris Warren, gen mgr; Jack O'Malley, stn mgr & progmg dir; Joel Frank, news dir.

Bridgeport

WNEM(AM)— Nov 26, 1956: 1250 khz; 5 kw-D, 1.1 kw-N, DA-2. TL: N43 20 31 W83 53 57. Hrs open: 24 Box 531, Saginaw, 48606. Secondary address: 107 N. Franklin St., Saginaw 48607. Phone: (989) 755-8191. Fax: (989) 758-2110. E-mail: wnem@wnem.com Web Site: www.wnem.com. Licensee: Meredith Corp. Group owner: Meredith Broadcasting Group, Meredith Corp. (acq 5-18-2004; $1.1 million). Format: Local news & info. ♦Al Blinke, gen mgr.

Bridgman

WYTZ(FM)— March 1993: 97.5 mhz; 3.8 kw. 413 ft TL: N41 59 19 W86 31 46. Stereo. Hrs opn: 24 580 E. Napier, Benton Harbor, 48022. Phone: (269) 925-1111. Fax: (269) 925-1011. E-mail: robb@975country.com Licensee: WSJM Inc. Group owner: The Mid-West Family Broadcasting Group (acq 1996; grpsl). Natl. Rep: Christal. Law Firm: Shaw Pittman. Format: Country. News staff: 5. Target aud: 25-54. ♦Joe Daguanno, VP & chief of engrg; Gayle Olson, gen mgr; Jim Gifford, opns mgr; Bob Bucholtz, gen sls mgr; Sue Patzer, prom dir; Robb Rose, progmg dir.

Bronson

***WCVM(FM)**— 1998: 94.7 mhz; 4 kw. 403 ft TL: N41 44 32 W85 14 34. Hrs open: 24 150 W Lincolnway, Ste 2001, Valparaiso, IN, 46383. Phone: (219) 548-5800. Fax: (219) 548-5808. E-mail: wqko@wqko.com Licensee: CSN International. (group owner; acq 12-1-98; $80,000). Format: Cutting edge Christian music and teaching. Target aud: Teen-young adult. ♦Jim Motshagen, gen mgr.

Brooklyn

WKHM-FM— January 1994: 105.3 mhz; 2.2 kw. 377 ft TL: N42 06 29 W84 22 46. Hrs open: 24 1700 Glenshire Dr., Jackson, 49201. Phone: (517) 787-9546. Fax: (517) 787-7517. E-mail: bgoldsen@k1053.com Web Site: www.k1053.com. Licensee: Jackson Radio Works Inc. (group owner; (acq 12-8-97; grpsl). Population served: 175,000 Rgnl rep: Michigan Spot Sls Format: Adult contemp. News staff: 2. Target aud: 18-49. ♦Bruce I. Goldsen, pres & gen mgr; Sue Goldsen, VP; Jamie McKibbin, stn mgr & progmg dir; Michael Bradford, chief of engrg.

Buchanan

WSMK(FM)— 1991: 99.1 mhz; 3 kw. 328 ft TL: N41 52 51 W86 18 13. Hrs open: 24 925 N. 5th St., Niles, 49120. Phone: (269) 683-4343. Fax: (269) 683-7759. E-mail: sales@wsmkradio.com Web Site: www.wsmkradio.com. Licensee: Marion R. Williams. Population served: 800,000 Format: Rhythmic CHR. Target aud: 18-34; females. ♦Marion R. Williams, gen mgr.

Burton

***WTAC(FM)**— 2002: 89.7 mhz; 1 kw vert. Ant 187 ft TL: N43 05 07 W83 40 19. Hrs open: 3302 N. Van Dyke, Imlay City, 48444. Phone: (810) 724-2638. Fax: (877) 850-0881. Web Site: www.positivehits.com. Licensee: Superior Communications. Format: Adult, Christian hit radio. ♦Edward Czelada, pres.

Cadillac

WATT(AM)— September 1945: 1240 khz; 1 kw-U. TL: N44 13 27 W85 24 06. Hrs open: 24 Box 520, 49601. Secondary address: 7825 S. Mackinaw Trail 49601-0520. Phone: (231) 775-1263. Fax: (231) 779-2844. Web Site: www.lite96.com. Licensee: MacDonald Garber Broadcasting Inc. (group owner; (acq 11-17-98; grpsl). Population served: 11,500 Natl. Network: Westwood One. Natl. Rep: McGavren Guild. Rgnl rep: McGavren Guild. Format: News/talk. Target aud: General. ♦Trish MacDonald Garber, pres & gen mgr.

WLXV(FM)—Co-owned with WATT(AM). July 7, 1974: 96.7 mhz; 1.7 kw. 443 ft TL: N44 14 56 W85 18 48. Web Site: www.lite96.com.10,000 Format: Adult contemp. News staff: one; News: one hr wkly.

WCKC(FM)— Sept 15, 1985: 107.1 mhz; 2.75 kw. 482 ft TL: N44 10 16 W85 20 13. Stereo. Hrs opn: 1356 Mackinaw Ave., Cheboygan, 49721. Phone: (231) 627-2341. Fax: (231) 627-7000. Web Site: www.classicrockthebear.com. Licensee: Northern Star Broadcasting L.L.C. (group owner; (acq 9-11-98; grpsl). Format: Classic rock. Target aud: 25-54. ♦Palmer Pyle, pres; Chris Monk, VP & gen mgr.

WJZQ(FM)— Oct 15, 1961: 92.9 mhz; 100 kw. Ant 912 ft TL: N44 35 41 W85 11 53. Stereo. Hrs opn: 24 314 E. Front St., Traverse City, 49684. Phone: (231) 947-7675. Fax: (231) 929-3988. E-mail: wjzq@929thebreeze.com Web Site: www.929thebreeze.com. Licensee: WKJF Radio Inc. Group owner: Midwestern Broadcasting Co. (acq 10-29-2001; with co-located AM). Population served: 275,000 Natl. Rep: Katz Radio. Format: Smooth jazz. News staff: one; News: 2 hrs wkly. Target aud: 35-64; affluent adults. ♦John Dew, gen mgr; Steve Hibbard, opns mgr; Joel Franck, news dir; Eric Send, chief of engrg; Barbara Kanarek, traf mgr.

WLJW(AM)— Mar 15, 2004: 1370 khz; 5 kw-D, 1 kw-N, DA-N. TL: N44 13 54 W85 24 45. Stereo. Hrs opn: 24
Simulcast with WLJN(AM) Elmwood Township.
Box 1400, Traverse City, 49685. Secondary address: 7475 S. 41 Rd. 49601. Phone: (231) 946-1400. Fax: (231) 946-3959. E-mail: info@wljn.com Web Site: www.wljn.com. Licensee: Good News Media Inc. (group owner; acq 3-5-2004; $85,001). Population served: 45,000 Natl. Network: Moody, Salem Radio Network. Natl. Rep: Katz Radio. Law Firm: Southmayd & Miller. Wire Svc: AP Format: Christian talk, relg. Target aud: General. ♦Brian Harcey, pres & gen mgr; Pete Lathrop, progmg dir.

***WOLW(FM)**— May 26, 1988: 91.1 mhz; 50 kw horiz, 28 kw vert. 700 ft TL: N44 16 33 W85 42 49. Stereo. Hrs opn: 24
Rebroadcasts WPHN(FM) Gaylord 100%.
Box 695, Gaylord, 49734-0695. Secondary address: 1511 M-32 E., Gaylord 49735. Phone: (989) 732-6274. Fax: (989) 732-8171. E-mail: ncr@ncradio.org Web Site: www.ncradio.org. Licensee: Northern Christian Radio Inc. (group owner). Population served: 150,000 Natl. Network: Moody. Law Firm: Southmayd & Miller. Format: Relg. News: 10 hrs wkly. Target aud: 35-54. ♦George A. Lake Jr., CEO, gen mgr & chief of engrg; Joe Sereno, chmn.

Caro

WIDL(FM)—Listing follows WKYO(AM).

WKYO(AM)— May 19, 1962: 1360 khz; 1 kw-U, DA-2. TL: N43 27 32 W83 23 39. Hrs open: 1521 W. Caro Rd, 48723-9260. Phone: (989) 672-1360. Fax: (989) 673-0256. Licensee: Edwards Communications L.C. (group owner; (acq 2-25-98; with co-located FM). Population served: 500,000 Rgnl. Network: Mich. Farm. Format: Country. Spec prog: Farm 18 hrs wkly. ♦Jamie Lee McCoy, gen mgr & stn mgr.

WIDL(FM)—Co-owned with WKYO(AM). Oct 16, 1974: 92.1 mhz; 6 kw. Ant 318 ft TL: N43 28 51 W83 20 31. Stereo. 550,000 Format: Hot adult contemp.

Carrollton

WSGW-FM—Licensed to Carrollton. See Saginaw

Cassopolis

WGTO(AM)— August 1988: 910 khz; 1 kw-D, 35 w-N, DA-1. TL: N41 57 14 W86 00 59. Stereo. Hrs opn: 24 58176 O'Keefe Rd., 49031. Phone: (574) 258-8470. Web Site: www.wgtoradio.com. Licensee: Larry Langford Jr. Population served: 450,000 Law Firm: Lauren A. Colby. Format: Golden oldies. Target aud: 25-49; middle class Black adults. Spec prog: Blues 4 hrs, gospel 10 hrs wkly. ♦Larry Langford, pres & gen mgr.

Charlevoix

WCZW(FM)— Jan 31, 2003: 107.9 mhz; 5 kw. Ant 164 ft TL: N45 20 00 W85 14 47. Hrs open: 24
Rebroadcasts WCCW-FM Traverse City 100%.
Radio Centre, 300 E. Front St., Suite 450, Traverse City, 49684. Phone: (231) 946-6211. Fax: (231) 946-1914. Web Site: www.wccwi.com. Licensee: WCCW Radio Inc. Group owner: Midwestern Broadcasting Co. Natl. Network: ABC. Natl. Rep: Katz Radio. Format: Oldies. Target aud: 35 plus; baby boomers. ♦Hal Payne, gen mgr & sls dir; Brian Hale, opns mgr; Dave Gauthier, mus dir; Eric Send, chief of engrg; Wend Sobeck, traf mgr.

WKHQ-FM— May 16, 1980: 105.9 mhz; 100 kw. Ant 892 ft TL: N45 10 49 W85 05 50. Stereo. Hrs opn: 24 Box 286, Petoskey, 49770. Secondary address: 2095 U.S. 131 S., Petoskey 49770. Phone: (231) 347-8713. Fax: (231) 347-8782. Web Site: www.106khq.com. Licensee: MacDonald Garber Broadcasting Co. (group owner; acq 11-17-98; grpsl). Population served: 200,000 Natl. Network: ABC. Natl. Rep: McGavren Guild. Law Firm: Koteen & Naftalin. Format: Top 40. News staff: one. Target aud: 18-34. ♦Trish MacDonald-Garder, gen mgr; Tom Clemens, gen sls mgr; Mark Elliott, progmg dir; Lisa Knight, news dir; Brian Brachel, engrg dir & chief of engrg; Bob Sheen, traf mgr.

WMKT(AM)—Co-owned with WKHQ-FM. July 20, 1974: 1270 khz; 5 kw-U, DA-N. TL: N45 16 22 W85 15 08.24 Web Site: www.wmktthetalkstation.com.72,600 Format: News/talk. Target aud: 35 plus; listeners with spendable income. ♦Eric Michaels, progmg dir.

***WTCK(FM)**— 2006: 90.9 mhz; 600 w vert. Ant 659 ft TL: N45 10 49 W85 05 50. Hrs open: Box 1109, Indian River, 49749. Phone: (231) 238-0811. E-mail: baragabroadcasting@utmi.net Web Site: www.baragabroadcasting.com. Licensee: Baraga Broadcasting Inc. (acq 11-8-2006; $130,000). Format: Catholic. ♦Harry Speckman, opns mgr.

Charlotte

WJZL(FM)— Dec 29, 1965: 92.7 mhz; 1.5 kw. Ant 466 ft TL: N42 38 31 W84 47 55. Stereo. Hrs opn: 24 2495 N. Cedar, Suite 106, Holt, 48842. Phone: (517) 699-0111. Fax: (517) 699-1880. E-mail: djohnson@mmrglansing.com Licensee: Rubber City Radio Group. Group owner: Rubber City Radio Group Inc. (acq 2-15-2001; $600,000).

Population served: 397,000 Natl. Network: Jones Radio Networks. Law Firm: Baraff, Koerner & Olender. Wire Svc: AP Format: Smooth jazz. News: 3 hrs wkly. ♦Dave Johnson, pres & gen mgr; Paul Cashin, opns dir & progmg mgr.

WLCM(AM)— Aug 25, 1956: 1390 khz; 5 kw-D, 70 w-N, DA-2. TL: N42 34 02 W84 51 58. (CP: COL Holt. 5 kw-D, 4.5 kw-N, DA-2). Stereo. Hrs opn: 6 AM-sunset Box 338, 1613 W. Lawrence, 48813. Phone: (517) 543-8200. Fax: (517) 543-7779. E-mail: wlcm@sbcglobal.net Licensee: Christian Broadcasting System Ltd. (group owner; acq 1-5-93; assumption of land contract; FTR: 1-25-93). Population served: 700,000 Format: Relg, Christian progmg. News: 2 hrs wkly. Target aud: 25-55; general. Spec prog: Gospel 3 hrs wkly. ♦Jon R. Yinger, CEO & pres; Evelyn Shaw, VP; Jeff Frank, gen mgr, stn mgr & opns dir.

Cheboygan

WCBY(AM)— Oct 28, 1954: 1240 khz; 1 kw-U. TL: N45 39 38 W84 29 26. Hrs open: 1356 Mackinaw Ave., 49721. Phone: (231) 627-2341. Fax: (231) 627-7000. Licensee: Northern Star Broadcasting L.L.C. (group owner; (acq 11-1-2005; grpsl). Natl. Rep: Michigan Spot Sales. Format: Nostalgia, big band. Target aud: 35-64. ♦Palmer Pyle, pres; Chris Monk, gen mgr & gen sls mgr.

WGFM(FM)—Co-owned with WCBY(AM). Aug 15, 1968: 105.1 mhz; 100 kw. 610 ft TL: N45 26 50 W84 28 30. Stereo. Web Site: www.classicrockthebear.com. Format: Classic rock. ♦Chris Monk, VP.

Clare

WCFX(FM)— June 28, 1967: 95.3 mhz; 6 kw. 328 ft TL: N43 44 41 W84 48 09. Stereo. Hrs opn: 24 5847 Venture Way, Mount Pleasant, 48858. Phone: (989) 772-4173. Fax: (989) 773-1236. E-mail: kent@wcfx.com Web Site: www.wcfx.com. Licensee: Grenax Broadcasting LLC (acq 2-5-2004). Population served: 165,000 Format: Adult/CHR. Target aud: 18-49. ♦Greg Dinetz, pres; Jim Spangenberg, gen mgr; Kent Bergstrom, opns mgr; Rob Ryan, prom dir & progmg dir.

Clyde Township

***WXPZ(FM)**—Not on air, target date: unknown: 90.1 mhz; 1.5 kw. Ant 243 ft TL: N42 33 57 W86 12 26. Hrs open: 6808 Hanna Lake S.E., Caledonia, 49316. Phone: (616) 698-1831. Licensee: Larlen Communications Inc.

Coldwater

***WCWB(FM)**—Not on air, target date: unknown: 90.1 mhz; 2.5 kw vert. Ant 328 ft TL: N42 03 28 W84 59 50. Hrs open: 21781 Pearl Beach Dr., 49036. Phone: (517) 238-2502. Licensee: Michiana Christian Broadcasters Inc. ♦Wayne S. Reese, pres.

WNWN-FM—Licensed to Coldwater. See Kalamazoo

WTVB(AM)— Aug 7, 1949: 1590 khz; 5 kw-D, 1 kw-N, DA-N. TL: N41 54 34 W85 00 21. Hrs open: 24 182 N. Angola Rd., 49036. Phone: (517) 279-1590. Fax: (517) 279-4695. E-mail: wtvb@wtvbam.com Licensee: Midwest Communications Inc. (group owner; acq 6-1-95; grpsl). Population served: 44,000 Natl. Rep: Christal. Wire Svc: NOAA Weather Format: Oldies, full service. News staff: 2; News: 10 hrs wkly. Spec prog: Farm 6 hrs wkly. ♦D.E. Wright, pres; Peter Tanz, gen mgr; Ken Delaney, stn mgr.

Coleman

***WPRJ(FM)**— Dec 7, 1992: 101.7 mhz; 4.6 kw. Ant 374 ft TL: N43 48 39 W84 27 50. Stereo. Hrs opn: 24 Box 236, 227 Jackson St., 48618. Phone: (989) 465-9775. Fax: (989) 465-1060. E-mail: wprj@wprj.org Web Site: www.wprj.org. Licensee: Come Together Ministries Inc. (acq 11-9-89; $8,000; FTR: 11-27-89). Law Firm: Reddy, Begley & McCormick. Format: Full-time Christian adult contemp, CHR. Target aud: 18 plus; youth, young singles & married. ♦Gary H. Bugh, pres & gen mgr; Connie Wieber, stn mgr & opns mgr.

Coopersville

WHTS(FM)— Sept 14, 1983: 105.3 mhz; 20 kw. Ant 794 ft TL: N43 18 35 W85 54 45. Stereo. Hrs opn: 3rd Fl. Broadcast Ctr., 60 Monroe Ctr. N.W., Grand Rapids, 49503. Phone: (616) 774-8461. Fax: (616) 774-2491. Web Site: 1053hotfm.com. Licensee: Citadel Broadcasting Co. (acq 11-29-2005; $4.1 million). Format: Adult contemp. Target aud: 25-54; adult women. ♦Matt Hanlon, gen mgr; Kate Conley, stn mgr & gen sls mgr; Brent Alberts, opns mgr.

Crystal Falls

WOBE(FM)— June 2000: 100.7 mhz; 100 kw. Ant 653 ft TL: N45 49 15 W88 02 38. Stereo. Hrs opn: 24 212 W. J St., Iron Mountain, 49801. Phone: (906) 774-5731. Fax: (906) 774-4542. E-mail: trisha@frogcountry.com Licensee: Results Broadcasting of Iron Mountain Inc. Group owner: Results Broadcasting (acq 11-1-2001; $800,000). Natl. Network: ABC. Format: Classic hits. News staff: one. Target aud: 25-65. ♦Bruce Grassman, pres; Trisha Peterson, VP & gen mgr; Keith Huotari, opns mgr.

WUPG(FM)—Not on air, target date: unknown: 94.9 mhz; 6 kw. Ant 666 ft TL: N46 06 03 W88 32 25. Hrs open: 1717 Dixie Hwy., Suite 650, Fort Wright, KY, 41011. Phone: (859) 331-9100. Licensee: Radioactive LLC. ♦Benjamin L. Homel, pres.

Dearborn

WDTW(AM)— Dec 29, 1946: 1310 khz; 5 kw-U, DA-2. TL: N42 15 50 W83 15 14. Hrs open: 24 27675 Halsted, Farmington Hills, 48331. Phone: (248) 324-5800. Fax: (248) 848-0313. E-mail: elliotlerner@clearchannel.com Licensee: AMFM Radio Licenses L.L.C. Group owner: Clear Channel Communications Inc. (acq 8-30-2000; grpsl). Population served: 125,000 Natl. Network: Westwood One. Format: Talk, sports. News staff: 4; News: 27 hrs wkly. Target aud: 25-54; men 25-54. ♦Til Levsque, gen mgr; Dom Theodore, opns mgr.

***WHFR(FM)**— Dec 20, 1985: 89.3 mhz; 270 w. 98 ft TL: N42 19 26 W83 14 09. Stereo. Hrs opn: 24 Henry Ford Community College, 5101 Evergreen Rd., 48128. Phone: (313) 845-9676. Phone: (313) 845-9842. Fax: (313) 317-4034. E-mail: whfr@hfcc.edu Web Site: www.whfr.fm. Licensee: Henry Ford Community College. Population served: 800,000 Natl. Network: PRI. Format: Var, alternative. News: one hr wkly. Target aud: General. Spec prog: Jazz 16 hrs, world mus 2 hrs, big band 6 hrs, blues 12 hrs wkly. ♦Susan McGraw, gen mgr; Lara Hrycaj, opns mgr.

WNIC(FM)—Licensed to Dearborn. See Detroit

Dearborn Heights

WNZK(AM)—Licensed to Dearborn Heights. See Detroit

Detroit

WCAR(AM)—See Livonia

***WDET-FM**— Feb 13, 1949: 101.9 mhz; 48 kw. Ant 554 ft TL: N42 21 06 W83 03 48. Stereo. Hrs opn: 24 4600 Cass Ave., 48201. Phone: (313) 577-4146. Fax: (313) 577-1300. E-mail: wdetfm@wdetfm.org Web Site: www.wdetfm.org. Licensee: Wayne State University. (acq 5-52). Population served: 200,000 Natl. Network: NPR. Rgnl. Network: Mich. Pub. Law Firm: Paul, Weiss, Rifkind, Wharton & Garrison. Format: News, jazz, adult alternative acoustic. News staff: 6; News: 44 hrs wkly. Target aud: 35-54; sophisticated, varied mus tastes & news consumers. Spec prog: Folk 3 hrs, bluegrass 3 hrs, blues 10 hrs, reggae 2 hrs wkly. ♦Allen Mazeruk, gen mgr & progmg dir; Ken Munson, dev dir; Kevin Piotrowski, prom dir & prom mgr; Martin Bandyke, mus dir; Jerome Vaughn, news dir; Malloy Farley, chief of engrg; Matt Trevethan, traf mgr.

WDFN(AM)— Dec 17, 1939: 1130 khz; 50 kw-D, 10 kw-N, DA-2. TL: N42 06 39 W83 11 52. Hrs open: 27675 Halsted Rd., Farmington, 48331. Phone: (248) 324-5800. Fax: (248) 848-0313. Web Site: www.wdfn.com. Licensee: AMFM Radio Licenses L.L.C. Group owner: Clear Channel Communications Inc. (acq 8-30-2000; grpsl). Population served: 3,900,000 Natl. Network: Westwood One. Format: Sports talk. Target aud: 25-54. ♦Til Levsque, gen mgr; David Crumb, gen sls mgr.

WDTW-FM—Co-owned with WDFN(AM). Oct 16, 1960: 106.7 mhz; 61 kw. Ant 508 ft TL: N42 19 55 W83 02 42. Web Site: www.1067thedrive.com. Format: Country.

WDMK(FM)— May 26, 1960: 105.9 mhz; 20 kw. 725 ft TL: N42 28 16 W83 12 03. Stereo. Hrs opn: 3250 Franklin St., 48207. Phone: (313) 259-2000. Fax: (313) 259-7011. E-mail: kyoung@radio-one.com Web Site: www.kissdetroit.com. Licensee: Radio One of Detroit LLC. Group owner: Radio One Inc. (acq 11-8-2001; grpsl). Population served: 4,000,000 Format: Hip hop, Rhythm and Blues. Target aud: 25-49. Spec prog: Sports 2 hrs, entertainment guide 2 hrs wkly. ♦Alfred Liggins, pres; Carol Lawrence-Dobrusin, gen mgr; Benita Gray, mus dir.

WDRQ(FM)— July 9, 1947: 93.1 mhz; 26.5 kw. 669 ft TL: N42 28 16 W83 12 03. Stereo. Hrs opn: 24 3011 W. Grand Blvd. Fisher Building, Suite 800, 311 West Grand Blvd., 48202-9816. Phone: (313) 871-9300. Fax: (313) 872-0190. Web Site: www.drgradio.com. Licensee: Radio License Holding I LLC. Group owner: ABC Inc. (acq 6-12-2007; grpsl). Population served: 500,000 Natl. Rep: ABC Radio Sales. Target aud: 25-54. ♦Steve Kosbau, pres & gen mgr; Chris Arnaut, chief of engrg.

WDTK(AM)— 1926: 1400 khz; 1 kw-U. TL: N42 24 22 W83 06 44. Hrs open: Two Radio Plaza, Ferndale, 48220. Phone: (248) 581-1234. Fax: (248) 581-1231. E-mail: zaron@wdtkam.com Web Site: www.wdtkam.com. Licensee: Pennsylvania Media Associates Inc. Group owner: Salem Communications Corp. (acq 9-30-2004; $4.75 million). Population served: 151,148 Format: News/talk. ♦Christian D. MacCourtney, gen mgr; Zaron Frumin, opns mgr.

WDTW(AM)—See Dearborn

WDVD(FM)—Listing follows WJR(AM).

WGPR(FM)— 1961: 107.5 mhz; 50 kw. Ant 405 ft TL: N42 21 28 W83 03 55. Stereo. Hrs opn: 3146 Jefferson E., 48207. Phone: (313) 259-8862. Fax: (313) 259-6662. E-mail: wgprradioincs@aol.com Web Site: www.wgprdetroit.com. Licensee: WGPR Inc. (acq 7-64). Population served: 151,148 Natl. Network: American Urban. Law Firm: Hogan & Hartson. Format: "New AC Jazz". Target aud: 18-49. ♦George Mathews, CEO, pres & gen mgr; James O. Dogan, VP & stn mgr; Chaya Jordan, gen sls mgr; Carolyn James, progmg dir. Co-owned TV: WGPR-TV affil

WJLB(FM)— 1926: 97.9 mhz; 50 kw. 489 ft TL: N42 24 22 W83 06 44. Stereo. Hrs opn: 645 Griswold, Suite 633, 48226. Phone: (313) 965-2000. Fax: (313) 965-3965. E-mail: deansnyder@clearchannel.com Web Site: www.fm98wjlb.com. Licensee: AMFM Radio Licenses L.L.C. Group owner: Clear Channel Communications Inc. (acq 8-30-00; grpsl). Population served: 580,000 Format: Urban contemp. Target aud: 18-49; Black adults. ♦Dave Pugh, gen mgr; Dean Snyder, gen sls mgr; K. J. Holiday, progmg dir & traf mgr; Kris Kelley, mus dir; Charles Pugh, news dir; Thomas Christie, engrg mgr & chief of engrg.

WJR(AM)— May 4, 1922: 760 khz; 50 kw-U. TL: N42 10 07 W83 13 00. Hrs open: 3011 W. Grand Blvd., Suite 800, 48202. Phone: (313) 875-4440. Fax: (313) 875-9022. Web Site: www.wjr.net. Licensee: Radio License Holding I LLC. Group owner: ABC Inc. (acq 6-12-2007; grpsl). Population served: 300,000 Natl. Rep: ABC Radio Sales. Format: News. Target aud: 12 plus. ♦Mike Fezzey, pres & gen mgr; Tom O'Brien, progmg dir; Dick Haefner, news dir.

WDVD(FM)—Co-owned with WJR(AM). June 1, 1948: 96.3 mhz; 20 kw. 787 ft TL: N42 27 13 W83 09 50. Stereo. Phone: (313) 871-3030. Fax: (313) 875-9636. Web Site: www.963wdvd.com.280,000 Format: AC/modern AC.

WKQI(FM)— Feb 12, 1949: 95.5 mhz; 100 kw. 437 ft TL: N42 28 22 W83 11 59. Stereo. Hrs opn: 24 27675 Halsted, Farmington Hill, 48331. Phone: (248) 324-5800. Fax: (248) 848-0272. E-mail: programing@channel955.com Web Site: www.channel955.com. Licensee: AMFM Radio Licenses L.L.C. Group owner: Clear Channel Communications Inc. (acq 8-30-00; grpsl). Population served: 3,692,300 Natl. Network: Premiere Radio Networks. Law Firm: Latham & Watkins. Format: Top 40. Target aud: 18-49; active, upscale women. ♦Til Levesque, VP &

gen mgr; Rebecca Falk, mktg dir & prom dir; Beau Daniels, prom dir & mus dir; Dom Theodore, progmg VP & progmg dir.

WKRK-FM—Listing follows WWJ(AM).

WLQV(AM)— 1925: 1500 khz; 50 kw-D, 5 kw-N, DA-2. TL: N42 13 51 W83 11 55. Stereo. Hrs opn: 24 5210 S. Saginaw Rd., Flint, 48507. Secondary address: Two Radio Plaza, Ferndale 48220. Phone: (248) 581-1234. Fax: (248) 581-1231. E-mail: victory1500@wlqv.net Licensee: Caron Broadcasting Inc. (group owner; (acq 2-10-2006; swap for WDJO(AM) Florence, KY and WCVX(AM) Cincinnati, OH plus $6.75 million cash) Natl. Network: Salem Radio Network. Format: Relg. Target aud: 25-65 plus; middle class. Spec prog: Black 10 hrs wkly. ♦Chris MacCourtney, VP & gen mgr.

WMGC-FM— Mar 6, 1960: 105.1 mhz; 20 kw. 784 ft TL: N42 28 16 W83 12 03. Stereo. Hrs opn: One Radio Plaza, 48220. Phone: (248) 414-5600. Fax: (248) 542-7700. Web Site: www.detroitmagic.com. Licensee: Greater Boston Radio Inc. Group owner: Greater Media Inc. (acq 12-5-96). Population served: 4,300,000 Format: Adult contemp. Target aud: General; professional, upscale, educated. ♦Peter Smyth, pres; Tom Bender, gen mgr.

WMUZ(FM)— Nov 11, 1958: 103.5 mhz; 50 kw. 500 ft TL: N42 22 40 W83 14 32. Hrs open: 12300 Radio Pl., 48228. Phone: (313) 272-3434. E-mail: wmuzinfo@crawfordbroadcasting.com Web Site: www.crawfordbroadcasting.com. Licensee: WMUZ Radio Inc. Group owner: Crawford Broadcasting Co. Population served: 300,000 Format: Contemp Christian Music. ♦Donald B. Crawford, pres; Frank Franciosi, gen mgr.

WMXD(FM)— Dec 8, 1964: 92.3 mhz; 50 kw. 479 ft TL: N42 19 55 W83 02 42. Stereo. Hrs opn: 645 Griswold, Suite 633, 48226. Phone: (313) 965-2000. Fax: (313) 965-3965. Web Site: www.mix923fm.com. Licensee: AMFM Radio Licenses L.L.C. Group owner: Clear Channel Communications Inc. (acq 8-30-00; grpsl). Format: Urban adult contemp. ♦Til Levesque, gen mgr; Jeff Luckoff, gen sls mgr; Jamillah Muhammad, progmg dir; Randy Auerbach, engrg mgr & chief of engrg.

WNIC(FM)—Dearborn, December 1946: 100.3 mhz; 32 kw. 600 ft TL: N42 23 22 W83 08 53. Stereo. Hrs opn: 24 27675 Halstead, Farmington Hills, 48331. Phone: (248) 324-5800. Fax: (248) 848-0396. Web Site: www.wnic.com. Licensee: AMFM Radio Licenses L.L.C. Group owner: Clear Channel Communications Inc. (acq 8-30-00; grpsl). Population served: 498,700 Law Firm: Latham & Watkins. Format: Adult contemp. News staff: one; News: 2 hrs wkly. Target aud: 25-54; female. ♦Della Pizzati, gen sls mgr; Rebecca Falk, mktg dir; Don Gosselin, progmg dir.

WNZK(AM)—Dearborn Heights, Oct 12, 1985: 690 khz-D; 2.5 kw-U, DA-2. TL: N42 05 55 W83 19 48. (Note: Stn operates on 680 khz-N). Hrs opn: 21700 Northwestern Hwy., Suite 1190, Southfield, 48075. Phone: (248) 557-3500. Fax: (248) 557-2950. E-mail: sima@birach.com Web Site: www.birach.com/wnzk.html. Licensee: Birach Broadcasting Corp. (acq 1984). Format: Talk, news, ethnic. Target aud: General. ♦Sima Birach, gen mgr; Jim Henderson, opns mgr.

WOMC(FM)— Mar 5, 1948: 104.3 mhz; 190 kw. 361 ft TL: N42 28 25 W83 06 56. Stereo. Hrs opn: 24 2201 Woodward Heights Blvd., Ferndale, 48220. Phone: (248) 546-9600. Fax: (248) 546-5446. E-mail: kpmurphy@cbs.com Web Site: www.womc.com. Licensee: CBS Radio Inc. of Michigan. Group owner: Infinity Broadcasting Corp. (acq 4-28-88). Population served: 3,826,700 Natl. Network: Westwood One. Format: Oldies. News staff: one; News: 4 hrs wkly. Target aud: 25-54; upscale. ♦Joel Hollander, pres; Jacques Tortoroli, CFO; Scott Herman, exec VP; Stephen Schram, sr VP; Kevin Murphy, gen mgr; Lynn Montemayor, gen sls mgr.

***WRCJ-FM**— Feb 5, 1948: 90.9 mhz; 42 kw horiz, 38 kw vert. 437 ft TL: N42 22 25 W83 06 50. Stereo. Hrs opn: 8:30 AM-8:30 PM 9345 Lawton Ave., 48206. Phone: (313) 596-3507. Fax: (313) 596-3517. Web Site: wdtr.com. Licensee: Board of Education, City of Detroit. Population served: 1,000,000 Format: Educ, var/div. Target aud: General; intergenerational-urban/suburban. ♦Kathy Young-Welch, gen mgr, opns dir & gen sls mgr; Donald Walker, mktg dir & progmg dir; Aaron Alfano, prom mgr; Stephanie Davis, mus dir, news dir & pub affrs dir; Steve Johnson, chief of engrg.

WRIF(FM)— Jan 1, 1948: 101.1 mhz; 27.2 kw. 879 ft TL: N42 28 15 W83 15 00. Stereo. Hrs opn: 24 One Radio Plaza St., 48220-2140. Phone: (248) 547-0101. Fax: (248) 542-8800. Web Site: wrif.com. Licensee: Greater Media Inc. (group owner; acq 12-15-87). Population served: 3,860,000 Natl. Rep: Katz Radio. Format: Active rock. Target aud: 18-49; men. ♦Tom Bender, gen mgr.

WVMV(FM)— 1961: 98.7 mhz; 50 kw. 462 ft TL: N42 23 42 W83 08 58. Stereo. Hrs opn: 31555 W. Fourteen Mile Rd., Suite 102, Farmington Hills, 48334. Phone: (248) 855-5100. Fax: (248) 855-1302. Web Site: www.wvmv.com. Licensee: CBS Radio East Inc. Group owner: Infinity Broadcasting Corp. (acq 12-89; grpsl; FTR: 12-11-89). Population served: 3,500,000 Format: Smooth Jazz. ♦Debbie Kenyon, VP & gen mgr; Sheryl Coyne, gen sls mgr; Tom Sleeker, opns mgr & progmg dir.

WWJ(AM)— Aug 20, 1920: 950 khz; 5 kw-U, DA-N. TL: N42 26 47 W83 10 23. Hrs open: 24 26495 American Dr., Southfield, 48034. Phone: (248) 455-7200. Fax: (248) 304-4970. Web Site: www.wwj.com. Licensee: CBS Radio East Inc. Group owner: Infinity Broadcasting Corp. (acq 3-9-89; FTR: 2-27-89). Population served: 3,660,200 Natl. Network: CBS. Format: News. News staff: 32. Target aud: General. ♦Rich Homberg, VP & gen mgr; Georgeann Herbert, opns mgr; Pete Kowalski, gen sls mgr; Debbie Spatafora, prom mgr; Bob Mundie, news dir & local news ed; Ralph Hunt, chief of engrg; Rob Davidek, local news ed; Florence Walton, news rptr; Jeff Gilbert, news rptr; Vickie Thomas, news rptr; Tim Skubick, political ed; Larry Henry, sports cmtr.

WKRK-FM—Co-owned with WWJ(AM). May 9, 1941: 97.1 mhz; 15 kw. 890 ft TL: N42 28 59 W83 12 20. Stereo. 15600 W. 12 Mile Rd., Southfield, 48076. Phone: (248) 395-9797. Fax: (248) 423-7725. Web Site: www.wwj.com. Format: FM talk. ♦Scott Buie, gen sls mgr. Co-owned TV: WWJ-TV affil

WXYT(AM)— Oct 10, 1925: 1270 khz; 5 kw-U, DA-N. TL: N42 27 58 W83 15 00. Hrs open: 24 26495 American Dr., Southfield, 48076. Phone: (248) 455-7350. Fax: (248) 455-7369. E-mail: wxyt@wxyt.com Web Site: www.1270sports.com. Licensee: CBS Radio Inc. of Detroit. Group owner: Infinity Broadcasting Corp. (acq 11-13-98; grpsl). Population served: 3,660,200 Natl. Network: Westwood One. Law Firm: Covington & Burling. Format: Sports talk. News staff: 3. Target aud: 25-54. ♦Rich Homberg, VP & gen mgr; Georgeann Herbert, opns mgr.

WYCD(FM)— May 4, 1960: 99.5 mhz; 21 kw horiz, 19 kw vert. 755 ft TL: N42 28 16 W83 12 03. Stereo. Hrs opn: 26555 Evergreen, Suite 675, Southfield, 48076. Phone: (248) 799-0600. Fax: (248) 358-9216. E-mail: stephen.schram@infinitybroadcasting.com Web Site: www.wycd.com. Licensee: CBS Radio Inc. of Michigan. Group owner: CBS Radio (acq 1-96; grpsl). Population served: 3,500,000 Law Firm: Shaw Pittman. Format: Country. Target aud: 12-34. ♦Debbie Kenyon, gen mgr; Jay Jennings, gen sls mgr & natl sls mgr.

Dewitt

WQHH(FM)— 1991: 96.5 mhz; 3 kw. Ant 328 ft TL: N42 51 06 W84 40 06. Hrs open: Box 25008, Lansing, 48909. Secondary address: 600 W. Cavanaugh Rd., Lansing 48910. Phone: (517) 393-1320. Fax: (517) 393-0882. Licensee: The MacDonald Broadcasting Co. (acq 10-3-2006; $3.65 million with WXLA(AM) Dimondale). Format: Urban contemp. Target aud: 18-49. ♦Kenneth H. MacDonald Jr., CEO; Rick Sarata, gen mgr.

Dimondale

WXLA(AM)— Sept 20, 1982: 1180 khz; 1 kw-D, DA. TL: N42 39 01 W84 34 49. Hrs open: 12 Box 25008, Lansing, 48909. Secondary address: 600 W. Cavanaugh Rd., Lansing 28910. Phone: (517) 393-1320. Fax: (517) 393-0882. Licensee: The MacDonald Broadcasting Co. (acq 10-3-2006; $3.65 million). Format: Timeless classics. ♦Kenneth H. MacDonald Jr., CEO; Rick Sarata, gen mgr.

Dowagiac

WAUS(FM)—South Bend IN

WDOW(AM)—Listing follows WHPD(FM).

WHPD(FM)— January 1971: 92.1 mhz; 3.3 kw. Ant 299 ft TL: N41 59 52 W86 03 14. Stereo. Hrs opn: 24
Simulcast with WHPZ(FM) Bremen, IN.
Box 150, 26914 Marcellus Hwy., 49047. Phone: (269) 782-5106. Fax: (269) 782-5107. E-mail: q92radio@yahoo.com Web Site: www.pulsefm.com. Licensee: LeSea Broadcasting Corp. (acq 4-12-2005; $950,000 with co-located AM). Population served: 250,000 Natl. Rep: Michigan Spot Sales. Law Firm: Keorner & Olender. Format: Contemp Christian. Target aud: 25-64. ♦Tom Scott, gen mgr.

WDOW(AM)—Co-owned with WHPD(FM). September 1960: 1440 khz; 1 kw-D, 89 w-N. TL: N41 59 35 W86 05 10.24 150,000 Format: Sports. News staff: one. Target aud: 25-49.

Eagle

***WJOM(FM)**— 2006: 88.5 mhz; 4.3 kw vert. Ant 131 ft TL: N42 48 25 W84 47 18. Hrs open: Michigan Community Radio, 3302 N. Van Dyke, Imlay City, 48444. Phone: (810) 721-0891. Licensee: Michigan Community Radio. ♦Ed Czelada, pres.

East Jordan

***WICV(FM)**— June 25, 1989: 100.9 mhz; 2.8 kw. 489 ft TL: N45 10 40 W85 05 57. Stereo. Hrs opn: 24
Rebroadcasts WIAA(FM) Interlochen 100%.
Box 199, One Lyon St., Interlochen, 49643. Phone: (231) 276-4400. Fax: (231) 276-4417. Licensee: Interlochen Center for the Arts (acq 5-23-90). Population served: 60,000 Natl. Network: NPR, PRI, ABC. Rgnl. Network: Mich. Pub. Law Firm: Haley, Bader & Potts. Format: Class. Target aud: 35-80; upper income, arts-oriented, civic-minded professionals. ♦Thom Paulson, VP & gen mgr.

East Lansing

***WDBM(FM)**— Feb 24, 1989: 88.9 mhz; 2 kw. 279 ft TL: N42 42 20 W84 28 30. Stereo. Hrs opn: 24 G-4 Holden Hall, Michigan State Univ. Campus, 48825-1206. Phone: (517) 353-4414. Fax: (517) 355-6552. Licensee: Board of Trustees of Michigan State University. Format: Alternative rock. News: 10 hrs wkly. Target aud: 18-34; students of MSU. Spec prog: Blues 4 hrs, jazz 5 hrs, heavy metal 4 hrs, progsv country 4 hrs, Christian rock 4 hrs wkly. ♦Gary Reed, gen mgr.

WFMK(FM)— July 16, 1959: 99.1 mhz; 28 kw. 600 ft TL: N42 40 33 W84 30 00. Stereo. Hrs opn: 24 Secondary address: 3420 Pine Tree Rd., Lansing 48911. Phone: (517) 394-3999. Fax: (517) 394-3391. E-mail: wfmk@acd.net Web Site: www.99wfmk.com. Licensee: Citadel Broadcasting Co. Group owner: Citadel Broadcasting Corp. (acq 2000; grpsl). Population served: 379,500 Natl. Rep: Christal. Law Firm: Leventhal, Senter & Lerman. Wire Svc: AP Format: Adult contemp. News staff: one; News: 2 hrs wkly. Target aud: 25-54. ♦Rob Striker, gen mgr; Brent Alberts, opns mgr; Chris Reynolds, mus dir.

WJZL(FM)—See Charlotte

***WKAR(AM)**— Aug 18, 1922: 870 khz; 10 kw-D, DA. TL: N42 42 19 W84 28 30. Hrs open: 283 Communication Arts Bldg., Michigan State Univ., 48824-1212. Phone: (517) 432-9527. Fax: (517) 353-7124. E-mail: mail@wkar.org Web Site: wkar.org. Licensee: Board of Trustees of Michigan State University. Population served: 410,000 Natl. Network: NPR, PRI. Rgnl. Network: Mich. Pub. Law Firm: Schwartz, Woods & Miller. Format: News/talk. News staff: 5; News: 25 hrs wkly. Spec prog: Sp 3 hrs wkly. ♦DeAnne Hamilton, gen mgr; Gene Purdom, opns mgr; Cindy Herfindahl, dev dir; Diane Hutchens, prom dir; Curt Gilleo, progmg dir & news dir; Gary Blievernicht, engrg dir & engrg mgr.

WKAR-FM— Oct 10, 1948: 90.5 mhz; 86 kw horiz, 57 kw vert. 895 ft TL: N42 42 08 W84 24 51. Web Site: wkar.org.400,000 Format: Class, news. Spec prog: Jazz 7 hrs wkly. Co-owned TV: *WKAR-TV affil

WMMQ(FM)—Listing follows WVFN(AM).

WVFN(AM)— September 1964: 730 khz; 500 w-D, 17.5 w-N, DA-2. TL: N42 38 45 W84 33 39. Hrs open: 24 3420 Pine Tree Rd., Lansing, 48911. Phone: (517) 394-7272. Fax: (517) 394-3391. Web Site: www.730amthefan.com. Licensee: Citadel Broadcasting Co. Group owner: Citadel Broadcasting Corp. (acq 2000; grpsl). Population served: 379,500 Natl. Network: ESPN Radio. Natl. Rep: Christal. Law Firm: Leventhal, Senter & Leman. Format: All sports, talk. Target aud: 25-54. ♦Farid Suleman, CEO & chmn; Rob Striker, gen mgr; Ray Marshall, opns mgr; Linda Karl, traf mgr; Tim Staudt, sports cmtr.

WMMQ(FM)—Co-owned with WVFN(AM). Nov 16, 1963: 94.9 mhz; 49 kw. 499 ft TL: N42 38 44 W84 33 38. Stereo. E-mail: wmmq@voyager.net Web Site: www.wmmq.com.379,500 Natl. Network: CNN Radio. Format: Classic rock. News staff: one. Target aud: 25-54; baby boomers who grew up listening to the Beatles, the Who & the Stones. ♦Farid Suleman, chmn; Ray Marshall, opns dir; Dan Kelley, progmg dir; Deb Hart, news dir.

East Tawas

***WZHN(FM)**—Not on air, target date: unknown: 91.3 mhz; 5.4 kw vert. Ant 328 ft TL: N44 20 25 W83 37 18. Hrs open: Box 695, Gaylord, 49734-0695. Phone: (989) 732-6274. Fax: (989) 732-8171. E-mail: ncr@ncradio.org Web Site: www.ncradio.org. Licensee: Northern Christian Radio Inc. ♦George A. Lake Jr., gen mgr.

Elkton

***WJCE(FM)**—Not on air, target date: unknown: 88.9 mhz; 50 kw. Ant 262 ft TL: N43 16 25 W82 35 16. Hrs open: CSN International, 4002 N. 3300 E., Twin Falls, ID, 83301. Phone: (208) 734-6633. Fax: (208) 736-1958. Web Site: www.csnradio.com. Licensee: CSN International. ♦Mike Kestler, pres.

Elmwood Township

***WLJN(AM)**— Dec 23, 1982: 1400 khz; 640 w-U. TL: N44 46 36 W85 39 43. Hrs open: 24
Simulcast with WLJW(AM) Cadillac.
Box 1400, Traverse City, 49685. Secondary address: 1101 Cass St., Traverse City 49684. Phone: (231) 946-1400. Fax: (231) 946-3959. E-mail: info@wljn.com Web Site: www.wljn.com. Licensee: Good News Media Inc. (group owner). Format: Relg, contemp, talk. News: 4 hrs wkly. Target aud: General. ♦Brian Harcey, gen mgr.

Elsie

WOES(FM)—See Ovid-Elsie

Escanaba

WCHT(AM)— Dec 1, 1958: 600 khz; 1 kw-D, 191 w-N, DA-2. TL: N42 40 28 W87 08 41. Hrs open: 524 Ludington St., Suite 300, 49829. Phone: (906) 228-9700. Fax: (906) 789-9700. E-mail: wglq@yahoo.com Web Site: www.radioresultsnetwork.com. Licensee: Lakes Radio Inc. (group owner) Population served: 38,000 Natl. Rep: Christal. Format: News/talk. Target aud: 25-54. Spec prog: Farm one hr, forestry one hr wkly. ♦Rick Duerson, gen mgr & opns dir.

WGLQ(FM)—Co-owned with WCHT(AM). Sept 11, 1976: 97.1 mhz; 100 kw. 1,070 ft TL: N46 08 04 W85 56 02. Stereo. Web Site: www.radioresultsnetwork.com.50,000 Format: Adult contemp.

WDBC(AM)— Sept 4, 1941: 680 khz; 10 kw-D, 1 kw-N, DA-2. TL: N45 45 53 W87 05 48. Hrs open: 24 604 Ludington St., 49829. Phone: (906) 786-6144. Fax: (906) 789-9959. E-mail: wdbcam@chartermi.net Licensee: KMB Broadcasting Co. Inc. (acq 12-31-88). Population served: 90,000 Natl. Network: ABC. Natl. Rep: Katz Radio. Wire Svc: AP Format: Full service, nostalgia. News staff: one; News: 12 hrs wkly. Target aud: 25-54. Spec prog: Relg 4 hrs, children one hr wkly. ♦Betsy Cooke, pres; Alice Sabuco, gen mgr; Kim Rabitoy, gen sls mgr; Barry Zieglar, news dir.

WYKX(FM)—Co-owned with WDBC(AM). Dec 22, 1977: 104.7 mhz; 100 kw. 351 ft TL: N45 55 41 W87 16 00. (CP: Ant 1,000 ft. TL: N45 52 40 W87 28 00). Stereo. 24 Phone: (906) 786-3800. Web Site: kxcountry.net.110,000 Natl. Network: AP Radio. Format: Country. News staff: one; News: 4 hrs wkly. Target aud: General; 25-54. ♦Tommy Kareckas, progmg dir; Wayne Nault, mus dir.

Essexville

WMJO(FM)— Jan 1, 1992: 97.3 mhz; 3 kw. 328 ft TL: N43 36 48 W83 45 51. Hrs open: 24 Box 1776, Saginaw, 48605. Secondary address: 2000 Whittier St., Saginaw 48601. Phone: (989) 752-8161. Fax: (989) 752-8102. E-mail: mikeskot@macdonaldbroadcasting.com Web Site: www.973joefm.com. Licensee: The MacDonald Broadcasting Co. Group owner: MacDonald Broadcasting Co. (acq 12-20-2001; grpsl). Population served: 340,800 Rgnl rep: Interep. Law Firm: Fisher, Wayland, Cooper, Leader & Zaragoza. Format: Adult Hits. News staff: one; News: 3 hrs wkly. Target aud: 25-54; Women with families. ♦Kenneth H. MacDonald Jr., pres; Duane Alverson, stn mgr; Mike Skot, opns mgr.

Farmington Hills

WFDF(AM)— May 25, 1922: 910 khz; 50 kw-D, 19 kw-N, DA-2. TL: N42 03 57 W83 23 39. Stereo. Hrs opn: 24 3011 W. Grand Blvd., Detroit, 48202. Phone: (248) 304-4381. Fax: (248) 304-4391. Web Site: www.radiodisney.com/detroit. Licensee: Radio Disney Group LLC. Group owner: ABC Inc. (acq 8-15-2002; $3 million). Population served: 4,474,614 Natl. Network: Radio Disney. Natl. Rep: Interep. Format: Family hits. ♦Rich Padgen, gen mgr; Brian Christy, prom mgr; Elise Bennett, prom mgr.

Fenton

WCXI(AM)— Nov 15, 1985: 1160 khz; 1 kw-U, DA-1. TL: N42 38 30 W83 43 50. Hrs open: 15130 North Rd., 48430. Phone: (810) 750-1911. Phone: (248) 557-3500. Fax: (810) 750-9028. E-mail: sima@birach.com Web Site: www.wnzk.com. Licensee: Birach Broadcasting Corp. (group owner; acq 9-13-99; $708,000). Natl. Network: American Urban. Format: Traditional country. Target aud: General; average age 35, primarily female, average income $35,000. ♦Sima Birach, CEO; Brenda Charette, opns mgr; John Morris, progmg dir & mus dir.

Flint

***WAKL(FM)**— September 1997: 88.9 mhz; 380 w. Ant 263 ft TL: N42 58 49 W83 34 40. Stereo. Hrs opn: 24 2351 Sunset Blvd., Suite 170-218, Rocklin, CA, 95765. Phone: (916) 251-1600. Fax: (916) 251-1650. E-mail: klove@klove.com Web Site: www.klove.com. Licensee: Educational Media Foundation. Group owner: EMF Broadcasting (acq 11-19-01; $450,000). Population served: 279,000 Natl. Network: K-Love. Law Firm: Shaw Pittman. Format: Contemp Christian. News staff: 3. Target aud: 25-44; female-Judeo Christian. ♦Richard Jenkins, pres; Mike Novak, VP; Keith Whipple, dev dir; David Pierce, progmg mgr; Ed Lenane, news dir; Sam Wallington, engrg dir; Karen Johnson, news rptr; Marya Morgan, news rptr; Richard Hunt, news rptr.

WCRZ(FM)—Listing follows WFNT(AM).

WDZZ-FM— Sept 29, 1979: 92.7 mhz; 3 kw. Ant 260 ft TL: N43 00 57 W83 41 24. Stereo. Hrs opn: 6317 Taylor Dr., 48507. Phone: (810) 238-7300. Fax: (810) 743-2500. E-mail: jeff.wade@cumulus.com Web Site: www.wdzz.com. Licensee: Cumulus Licensing Corp. Group owner: Cumulus Media Inc. (acq 3-15-00; grpsl). Population served: 424,902 Format: Urban adult. Target aud: Adults. Spec prog: Gospel 8 hrs, teen talk one hr, concerned pastors one hr wkly. ♦Joe Mule, gen mgr; Jeff Wade, opns mgr & progmg dir.

WFBE(FM)— Oct 5, 1953: 95.1 mhz; 50 kw. 243 ft TL: N43 01 13 W83 40 40. Stereo. Hrs opn: 5 AM-1 AM G 4511 Miller Rd., 48507. Phone: (810) 720-9510. Fax: (810) 720-9513. E-mail: james.glendening@citcomm.com Web Site: www.b95.fm. Licensee: Citadel Broadcasting Co. Group owner: Citadel Broadcasting Corp. (acq 4-26-01; grpsl). Population served: 15,000 Natl. Network: PRI. Format: Country. News: 3 hrs wkly. Target aud: General; country music listeners. ♦Scott Meier, gen mgr; Jim Glendening, sls dir.

WFLT(AM)— Dec 5, 1955: 1420 khz; 500 w-D, 142 w-N, DA-2. TL: N43 01 19 W83 38 35. Hrs open: 317 S. Averill, 48506. Phone: (810) 239-5733. Fax: (810) 239-7134. E-mail: wflt1420am@aol.com Licensee: Metropolitan Missionary Baptist Church (acq 7-2-90; $225,000; FTR: 7-23-90). Natl. Rep: Michigan Spot Sales. Format: Black gospel. ♦Sammie Jordan, gen mgr.

WFNT(AM)— Apr 10, 1953: 1470 khz; 5 kw-D, 1 kw-N, DA-2. TL: N42 58 22 W83 38 24. Hrs open: 24 3338 E. Bristol Rd., Burton, 48529. Phone: (810) 742-1470. Fax: (810) 742-5170. E-mail: wfnt@aol.com Web Site: www.wfnt.com. Licensee: Regent Broadcasting of Flint Inc. Group owner: Regent Communications Inc. (acq 8-13-98; grpsl). Population served: 430,000 Law Firm: Haley, Bader & Potts. Format: Nostalgia. News staff: 3. ♦Zoe Burdine-Fly, gen mgr; Maggie McColman, prom dir; Chris Pavelich, news dir; Mike Hutchens, chief of engrg.

WCRZ(FM)—Co-owned with WFNT(AM). Nov 4, 1961: 107.9 mhz; 50 kw. 331 ft TL: N42 58 49 W83 34 40. Stereo. Phone: (810) 743-1080. E-mail: wcrz@aol.com Web Site: www.wcrz.com. Format: Adult contemp. ♦Kelly Quinn, gen sls mgr.

***WFUM-FM**— Aug 23, 1985: 91.1 mhz; 18 kw. 489 ft TL: N42 53 57 W83 27 42. Stereo. Hrs opn: 24
Rebroadcasts WUOM(FM) Ann Arbor 100%.
535 W. William St., Suite 110, Ann Arbor, 48103. Phone: (734) 764-9210. Fax: (734) 647-3488. E-mail: michigan.radio@umich.edu Web Site: michiganradio.org. Licensee: Regents of the University of Michigan. Natl. Network: NPR. Format: News/talk. ♦Jon Hoban, gen mgr & stn mgr; Peggy Watson, opns mgr; Michael Leland, dev dir & news dir. Co-owned TV: *WFUM-TV affil

WSNL(AM)— Apr 26, 1946: 600 khz; 440 w-D, 250 w-N, DA-2. TL: N42 54 27 W83 50 07. Hrs open: 24 5210 S. Saginaw St., 48507. Phone: (810) 694-4146. Fax: (810) 694-0661. Web Site: www.cbsl.biz. Licensee: Christian Broadcasting System Ltd. (group owner; (acq 1-22-93; $400,000; FTR: 2-8-93). Population served: 500,000 Format: Christian, teaching, talk. Target aud: 25-54; 35+. ♦Jon Yinger, pres; Evelyn Shaw, VP & gen mgr; Graham Parker, opns mgr & rgnl sls mgr.

WTRX(AM)— Oct 1, 1947: 1330 khz; 5 kw-D, 1 kw-N, DA-2. TL: N42 58 24 W83 39 02. Stereo. Hrs opn: G 4511 Miller Rd., 48507. Phone: (810) 720-9510. Fax: (810) 720-9513. Web Site: www.wtrxsports.com. Licensee: Citadel Broadcasting Co. Group owner: Citadel Broadcasting Corp. (acq 10-6-00; $180,000). Format: Sports, talk. 18-49 Males. ♦Ronnie Glover, gen mgr; Doug Fisher, progmg dir.

WWCK(AM)— Nov 11, 1946: 1570 khz; 1 kw-D, 238 w-N. TL: N43 00 38 W83 39 09. Hrs open: 6317 Taylor Dr., 48507. Phone: (810) 238-7300. Fax: (810) 238-7310. Web Site: www.wwck.com. Licensee: Cumulus Licensing Corp. Group owner: Cumulus Media Inc. (acq 3-15-00; grpsl). Population served: 193,317 Natl. Network: Westwood One. Format: CHR. Target aud: 18-34. ♦Joe Mule, gen mgr; Les Root, news dir; Dan Greer, chief of engrg.

WWCK-FM— September 1964: 105.5 mhz; 25 kw. 328 ft TL: N43 00 39 W83 39 04. Stereo. Web Site: www.wwck.com.193,317

Frankenmuth

WRCL(FM)— 2001: 93.7 mhz; 3.5 kw. 436 ft TL: N43 18 16 W83 33 07. Stereo. Hrs opn: 24 3338 E. Bristol Rd., Burton, 48529. Phone: (810) 742-1470. Fax: (810) 742-5170. E-mail: clover@regentflint.com Web Site: www.club937.com. Licensee: Regent Broadcasting of Flint Inc. Group owner: Regent Communications Inc. (acq 11-9-01; $7 million with WFGR(FM) Grand Rapids). Natl. Network: CNN Radio, Westwood One. Format: Rhythmic CHR. Target aud: 12-34; children & adults. ♦Burdine Fly, gen mgr; J. Patrick, opns mgr; Clay Church, mus dir; Nathan Reed, mus dir.

Frankfort

WBNZ(FM)— Oct 2, 1978: 99.3 mhz; 50 kw. 410 ft TL: N44 36 38 W86 09 38. Stereo. Hrs opn: 24 1532 Forrester Rd., 49635. Phone: (231) 352-9603. Fax: (231) 352-7877. E-mail: marc@wbnz.com Web Site: www.wbnz.com. Licensee: Fort Bend Broadcasting Co. (group owner; acq 8-7-01). Population served: 25,000 Natl. Rep: Patt. Format: Soft rock. News staff: one; News: 3 hrs wkly. Target aud: 25-54. Spec prog: Folk 2 hrs, big band 2 hrs wkly. ♦Roy Henderson, gen mgr.

Freeland

***WTRK(FM)**— 2005: 90.9 mhz; 430 w. Ant 324 ft TL: N43 33 42 W83 58 52. Hrs open:
Rebroadcasts KLVR(FM) Santa Rosa, CA 100%.
2351 Sunset Blvd., Suite 170-218, Rocklin, CA, 95765. Phone: (916) 251-1600. Fax: (916) 251-1650. Web Site: www.klove.com. Licensee: Educational Media Foundation. Group owner: American Family Radio. (acq 6-29-2005; $75,000). Natl. Network: K-Love. Format: Contemp Christian. ♦Richard Jenkins, pres; Mike Novak, VP; Keith Whipple, dev dir; David Pierce, progmg mgr; Ed Lenane, news dir; Sam Wallington, engrg dir; Arthur Vassar, traf mgr; Karen Johnson, news rptr; Marya Morgan, news rptr; Richard Hunt, news rptr.

Fremont

WSHN(AM)— May 23, 1961: 1550 khz; 1 kw-D. TL: N43 28 15 W85 56 25. Hrs open: Box 190, 49412. Phone: (231) 924-4700. Fax: (231) 924-9746. Licensee: WSHN Inc. (acq 3-11-97; grpsl). Population served: 200,000 Natl. Rep: Patt. Format: News, talk. ♦John Russell, news dir.

Gagetown

***WCTP(FM)**— 2006: 88.5 mhz; 6 kw. Ant 328 ft TL: N43 45 36 W83 05 45. Stereo. Hrs opn: 24 4330 Farver Rd., 48735. Phone: (989) 872-3525. Fax: (989) 872-3525. E-mail: dplonta@hotmail.com Licensee: Plonta Broadcasting Inc. Format: Christian. News staff: 3. ♦Duane Plonta, pres.

Gaylord

***WBLW(FM)**— 2000: 88.1 mhz; 3 kw vert. Ant 128 ft TL: N45 01 28 W84 43 44. Hrs open: Box 177, 49734. Secondary address: 232 S. Townline Rd. 49735. Phone: (989) 732-5676. Fax: (989) 731-1122. E-mail: info@wblwradio.com Web Site: www.gracebaptistministries.com. Licensee: Gaylord Baptist Christian School. Format: Christian.

WMJZ-FM— 1984: 101.5 mhz; 50 kw. Ant 492 ft TL: N45 01 10 W84 24 28. Stereo. Hrs opn: Box 1766, 49734. Secondary address: 3687 Old US Hwy. 27 S. 49735. Phone: (989) 732-2341. Fax: (989) 732-6202. Licensee: Darby Advertising Inc. (acq 1-1-98; with co-located AM). Population served: 45,000 Natl. Network: Motor Racing Net. Law Firm: Irwin Campbell & Tanneweld. Format: News, sports, adult hits. Target aud: 25-54. ♦Kent D. Smith, pres & gen mgr; Mike Reling, opns mgr; Rob Weaver, progmg dir.

***WPHN(FM)**— Apr 7, 1985: 90.5 mhz; 100 kw. 1,000 ft TL: N45 08 17 W84 09 44. Stereo. Hrs opn: 24 Box 695, 49734-0695. Secondary address: 1511 M-32 E. 49735. Phone: (989) 732-6274. Fax: (989) 732-8171. E-mail: ncr@ncradio.org Web Site: www.ncradio.org. Licensee: Northern Christian Radio Inc. (group owner). Population served: 150,000 Natl. Network: Moody, USA. Law Firm: Southmayd & Miller. Format: Relg. News: 10 hrs wkly. Target aud: 25-55. ♦George A. Lake Jr., CEO & gen mgr; Joe Sereno, pres.

WSRT(FM)— Nov 18, 1972: 106.7 mhz; 100 kw. 580 ft TL: N45 02 42 W84 50 44. Stereo. Hrs opn: 24 1020 Hastings, Traverse City, 49686. Phone: (231) 947-0003. Fax: (231) 546-4490. E-mail: wkpk@wkpk.com Web Site: www.wkpk.com. Licensee: Northern Radio of Gaylord Inc. (acq 9-23-96; $1.4 million with WMLQ(FM) Rogers City). Population served: 3,012 Natl. Rep: Christal. Wire Svc: UPI Format: Hot adult contemp. Target aud: 18-49; rgnl orientation including Traverse City, Petoskey, Cheboygan-active life style. ♦Charlie Ferguson, gen mgr; T.J. Clark, gen sls mgr; Todd Martin, progmg dir; Dennis Murray, chief of engrg.

Gladstone

WGKL(FM)— Feb 15, 1999: 105.5 mhz; 4.6 kw. 377 ft TL: N45 48 17 W87 10 15. Hrs open: 524 Ludington, Suite 300, Escanaba, 49829. Phone: (906) 228-9700. Fax: (906) 789-9700. E-mail: kool105fm@chartermi.net Web Site: www.radioresultsnetwork.com. Licensee: Lakes Radio Inc. (group owner) Format: Oldies. ♦Rick Duerson, gen mgr.

Gladwin

WGDN(AM)— Dec 7, 1974: 1350 khz; 1 kw-D, DA. TL: N43 57 03 W84 30 34. Hrs open: 3601 W. Woods Rd., 48624. Phone: (989) 426-1031. Licensee: Apple Broadcasting Co. Inc. (acq 3-87; with co-located FM; FTR: 12-22-86). Population served: 56,000 Format: Relg. Target aud: 35 plus. ♦Steve Coston, gen mgr.

WGDN-FM— Feb 7, 1978: 103.1 mhz; 11.5 kw. 453 ft TL: N43 57 03 W84 30 34. Stereo. 185,000 Natl. Network: Westwood One. Format: Country.

Glen Arbor

WGFN(FM)— February 1991: 98.1 mhz; 21 kw. Ant 738 ft TL: N44 49 16 W85 59 47. Stereo. Hrs opn: 24 1356 Mackinaw Ave., Cheboygan, 49721. Phone: (231) 627-2341. Fax: (231) 627-7000. Web Site: www.classicrockthebear.com. Licensee: Northern Star Broadcasting L.L.C. (group owner; (acq 11-1-2005; grpsl). Format: Classic rock. News staff: one. ♦Palmer Pyle, pres; Chris Monk, gen mgr.

WJZJ(FM)— Sept 1, 1997: 95.5 mhz; 21 kw. 738 ft TL: N44 49 16 W85 59 47. Stereo. Hrs opn: 24 1356 Mackinaw Ave., Cheboygan, 49721. Phone: (231) 627-2341. Fax: (231) 627-7000. Web Site: www.modernrockthezone.com. Licensee: Northern Star Broadcasting L.L.C. (group owner; (acq 11-1-2005; grpsl). Format: Modern rock. ♦Palmer Pyle, pres; Chris Monk, gen mgr.

Good Hart

***WJOG(FM)**— 2006: 91.3 mhz; 600 w vert. Ant 623 ft TL: N45 30 33 W85 02 11. Hrs open: Michigan Community Radio, 3302 N. Van Dyke, Imlay City, 48444. Phone: (810) 721-0891. Licensee: Michigan Community Radio. ♦Ed Czelada, pres.

Goodland Township

***WHYT(FM)**— 2004: 88.1 mhz; 400 w vert. Ant 581 ft TL: N43 10 30 W83 04 02. Hrs open: Box 388, Williamston, 48895. Phone: (517) 381-0573. Fax: (877) 850-0881. E-mail: info@positivehits.com Web Site: www.positivehits.com. Licensee: Superior Communications. ♦Jenn Czelada, gen mgr.

Grand Haven

WGHN(AM)— July 16, 1956: 1370 khz; 500 w-D, 22 w-N. TL: N43 02 17 W86 13 46. Hrs open: 24 Box 330, One S. Harbor, 49417. Phone: (616) 842-8110. Fax: (616) 842-4350. Web Site: www.wghn.com. Licensee: WGHN Inc. (acq 6-11-2007; $1.65 million with co-located FM). Population served: 120,000 Natl. Network: CBS. Rgnl. Network: Mich. Farm. Natl. Rep: Patt. Law Firm: Rothman, Gordon, Foreman & Groudine. Format: Adult contemp. News staff: 2; News: 30 hrs wkly. Target aud: 25-54. Spec prog: Agriculture & farm 5 hrs wkly. ♦Will Tieman, pres; William Struyk, gen mgr & gen sls mgr.

WGHN-FM— Jan 28, 1969: 92.1 mhz; 3 kw. 246 ft TL: N43 03 23 W86 14 27. Stereo. 24 Web Site: www.wghn.com.

Grand Rapids

***WAYG(FM)**— May 18, 1978: 89.9 mhz; 4.9 w. Ant 207 ft TL: N42 58 40 W85 35 44. Stereo. Hrs opn: 24
Rebroadcasts WAYK(FM) Kalamazoo 75%.
1159 E. Beltline Ave. N.E., 49525. Phone: (616) 942-1500. Fax: (616) 942-7078. Web Site: www.way.fm. Licensee: Cornerstone University (acq 1-21-98; $200,000). Format: Christian hits. News: 10 hrs wkly. Target aud: General. ♦Dr. Rex Rogers, pres; Lee Geysbeek, VP; Chris Lemke, gen mgr; Rich Anderson, stn mgr.

WBBL(AM)—Listing follows WLAV-FM.

WBCT(FM)— October 1951: 93.7 mhz; 320 kw. 781 ft TL: N42 37 56 W85 32 16. Stereo. Hrs opn: 24 77 Monroe Center, Suite 1000, 49503. Phone: (616) 459-1919. Fax: (616) 242-9373. Web Site: www.b93.com. Licensee: CC Licenses LLC. Group owner: Clear Channel Communications Inc. (acq 1996; grpsl). Natl. Rep: Clear Channel. Format: Country. News staff: one; News: 3 hrs wkly. Target aud: 25-49. ♦Skip Essick, VP & gen mgr; Rich Berry, gen sls mgr; Kani Kennedy, natl sls mgr.

WBFX(FM)—Listing follows WTKG(AM).

***WBLU-FM**— Aug 18, 1979: 88.9 mhz; 650 w. 400 ft TL: N42 59 15 W85 37 26. Stereo. Hrs opn:
Rebroadcasts WBLV(FM) Twin Lake 100%.
Blue Lake Fine Arts Camp, Twin Lake, 49457. Phone: (231) 894-2616. Phone: (231) 458-9258. Fax: (231) 893-2457. Web Site: www.bluelake.org. Licensee: Blue Lake Fine Arts Camp. (acq 3-1-93; $200,000; FTR: 3-15-93). Natl. Network: PRI, NPR. Rgnl. Network: Mich. Pub. Format: Class, jazz, news. Target aud: Adults. Spec prog: Folk 5 hrs wkly. ♦William F. Stansell, pres; Dave Myers, gen mgr; Gordon Christensen, opns dir; Steve Albert, progmg dir; Bonnie Bierma, mus dir; Don Hoogeboom, chief of engrg.

***WCSG(FM)**— June 9, 1973: 91.3 mhz; 37 kw. 570 ft TL: N42 47 46 W85 38 58. Stereo. Hrs opn: 1159 E. Beltline Ave. N.E., 49525. Phone: (616) 942-1500. Fax: (616) 942-7078. E-mail: wcsg@wcsg.org Web Site: www.wcsg.org. Licensee: Cornerstone University. Population served: 1,200,000 Natl. Network: AP Radio. Format: Christian. News staff: 2; News: 5 hrs wkly. Target aud: 35-49. ♦Dr. Rex Rogers, pres; Lee Geysbeek, VP; Chris Lemke, gen mgr & progmg dir.

WFGR(FM)— Aug 9, 1992: 98.7 mhz; 2.75 kw. Ant 492 ft TL: N43 01 57 W85 41 47. Hrs open: 50 Monroe N.W., Suite 500, 49503. Phone: (616) 451-4855. Fax: (616) 451-0113. Web Site: www.wfgr.com. Licensee: Haith Broadcasting Corp. Group owner: Regent Communications Inc. (acq 9-25-01; $3.9 million for stock). Format: Oldies. Target aud: 25 plus; affluent, well-educated professionals. ♦Terry Jacobs, pres; Phil Catlett, gen mgr; Rick Sarata, sls dir.

WFUR(AM)— November 1947: 1570 khz; 1 kw-D, 306 w-N. TL: N42 57 14 W85 41 52. Hrs open: 24 Box 1808, 49501. Secondary address: 399 Garfield Ave. S.W. 49501. Phone: (616) 451-9387. Fax: (616) 451-8460. Licensee: Furniture City Broadcasting Corp. Group owner: Kuiper Stations (acq 3-10-50). Population served: 500,000 Natl. Network: USA. Format: Relg. News: 5 hrs wkly. Target aud: 35 plus; 60% female, 30% male. ♦William E. Kuiper Sr., pres & gen mgr; Steven Kuiper, opns mgr & mus dir; Roger Peular, gen sls mgr; Dave Kuiper, news dir, local news ed, farm dir, relg ed & disc jockey; Pat Deja, pub affrs dir; Bill Kuiper Jr., chief of engrg.

WFUR-FM— September 1960: 102.9 mhz; 50 kw. 492 ft TL: N42 57 13 W85 41 55. Stereo. 24 1,000,000 Natl. Network: USA. Format: Relg music. News: 5 hrs wkly. Target aud: 35-64; homeowners, Christian families, Christian mus listeners. ♦Doug Wentworth, sports cmtr; Dave Kuiper, disc jockey; Karen Baker, disc jockey; Pat Deja, disc jockey.

WGRD-FM— Aug 1, 1962: 97.9 mhz; 13 kw. 590 ft TL: N42 47 46 W85 38 58. Stereo. Hrs opn: 24 50 Monroe N.W., Suite 500, Grands Rapids, 49503. Phone: (616) 451-4855. Fax: (616) 451-0113. Web Site: www.wgrd.com. Licensee: Regent Broadcasting of Grand Rapids Inc. Group owner: Regent Communications Inc. (acq 8-7-00; grpsl). Format: Modern rock, alternative. News staff: one; News: 3 hrs wkly. Target aud: 18-49. ♦Terry Jacobs, pres; Phil Catlett, gen mgr.

***WGVU(AM)**—Kentwood, Dec 25, 1954: 1480 khz; 2 kw-D, 5 kw-N. TL: N42 57 13 W85 41 36. Hrs open: 24 301 W. Fulton, 49504. Phone: (616) 331-6666. Fax: (616) 331-6625. E-mail: wgvu@gvsu.edu Web Site: www.wgvu.org. Licensee: Grand Valley State Univ. (acq 4-7-92; $240,000 part sale & part gift; FTR: 4-27-92) Population served: 400,000 Natl. Network: NPR, PRI. Format: News, info. News staff: 3; News: 89 hrs wkly. Target aud: 35-44; college-educated men with average income. ♦Michael T. Walenta, gen mgr; Ken Kolbe, opns mgr; Jayne Marsh, dev mgr; Richard Nelson, sls dir; Pamela Holtz, mktg mgr & prom mgr; Fred Martino, news dir; Bob Lumbert, chief of engrg; Ed Spier, traf mgr; Gary Kesler, spec ev coord.

WGVU-FM—See Allendale

WJNZ(AM)—See Kentwood

WKLQ(FM)—Greenville, October 1989: 107.3 mhz; 50 kw. Ant 492 ft TL: N43 01 10 W85 20 58. Stereo. Hrs opn: 24 60 Monroe Ctr. N.W., 3rd Fl., 49503. Phone: (616) 774-8461. Fax: (616) 774-2491. Web Site: www.wklq.com. Licensee: Citadel Broadcasting Co. Group owner: Citadel Broadcasting Corp. (acq 5-30-2000; grpsl). Population served: 600,000 Natl. Network: Westwood One. Natl. Rep: D & R Radio. Format: Active rock. News staff: one. Target aud: 25-54. ♦Matt Hanlon, gen mgr.

WLAV-FM— January 1947: 96.9 mhz; 50 kw. 499 ft TL: N43 02 01 W85 31 15. Stereo. Hrs opn: 24 60 Monroe Ctr. N.W., 3rd Fl., 49502. Phone: (616) 774-8461. Fax: (616) 774-2491. Web Site: www.wlav.com. Licensee: Citadel Broadcasting Co. Group owner: Citadel Broadcasting Corp. (acq 5-30-00; grpsl). Population served: 194,649 Natl. Network: ABC. Law Firm: Reddy, Begley & McCormick. Format: Classic rock. Target aud: 25-49. ♦Matthew R. Hanlon, VP; Matt Hanlon, gen mgr; Brent Alberts, opns dir & adv dir; Kat Conley, gen sls mgr; Rob Brant, news dir; Don Allen, chief of engrg; Melissa Martinez, traf mgr.

WBBL(AM)—Co-owned with WLAV-FM. Sept 18, 1940: 1340 khz; 1 kw-U. TL: N42 57 02 W85 41 55.24 Web Site: www.wbbl.com. Natl. Network: ABC. Format: Sports. Target aud: 18-49; men. ♦Bret Bakita, progmg dir.

WLHT-FM—Listing follows WNWZ(AM).

WMJH(AM)—See Rockford

WNWZ(AM)— Nov 1, 1947: 1410 khz; 1 kw-D, 48 w-N. TL: N42 59 14 W85 37 76. Stereo. Hrs opn: 50 Monroe N.W., Suite 500, 49503. Phone: (616) 459-4111. Fax: (616) 451-4225. Web Site: www.1401amaquina.com. Licensee: Regent Broadcasting of Grand Rapids Inc. Group owner: Regent Communications Inc. (acq 8-7-00; grpsl). Population served: 197,649 Natl. Network: Jones Radio Networks. Format: Latin mix /Sp language. Target aud: 35 plus; professionals. ♦Phil Catlett, gen mgr.

WLHT-FM—Co-owned with WNWZ(AM). Feb 28, 1962: 95.7 mhz; 40 kw. 551 ft TL: N43 01 57 W85 41 47. Stereo. Phone: (616) 451-4800. Fax: (616) 451-0113. Web Site: www.wlht.com.584,400 Law Firm: Dow, Lohnes & Albertson. Format: Adult contemp. Target aud: 25-54. ♦Terry Jacobs, pres & gen sls mgr.

WOOD(AM)— 1924: 1300 khz; 20 kw-U, DA-2. TL: N42 45 22 W85 39 24. Stereo. Hrs opn: 24 77 Monroe Ctr., Suite 1000, 49503. Phone: (616) 459-1919. Fax: (616) 242-6599. E-mail: web@woodradio.com Web Site: www.woodradio.com. Licensee: CC Licenses LLC. Group

owner: Clear Channel Communications Inc. (acq 5-10-96; grpsl). Population served: 680,000 Natl. Network: ABC. Natl. Rep: Clear Channel. Format: News/talk. News staff: 6; News: 24 hrs wkly. Target aud: 35-54. ♦Skip Essick, VP & gen mgr; Phil Tower, opns dir & progmg dir; Henry Capogna, gen sls mgr & rgnl sls mgr; Kami Alger, natl sls mgr; Glenn Del Vecchio, prom dir; Rich Jones, news dir; Don Missad, chief of engrg; Kay Jaarsma, traf mgr.

WOOD-FM— 1962: 105.7 mhz; 265 kw. Ant 810 ft TL: N42 41 13 W85 30 35. Stereo. 24 E-mail: web@star1057online.com Web Site: www.star1057online.com. Format: Soft adult contemp. News: 10 hrs wkly. ♦John Patrick, opns dir & progmg dir; Glenn Del Vecchio, prom mgr; Kay Jaarsma, traf mgr.

WTKG(AM)— February 1945: 1230 khz; 1 kw-U. TL: N42 59 42 W85 40 36. Stereo. Hrs opn: 24 77 Monroe Ctr., Suite 1000, 49503. Phone: (616) 459-1919. Fax: (616) 242-6599. E-mail: web@wtkg.com Web Site: www.wtkg.com. Licensee: CC Licenses LLC. Group owner: Clear Channel Communications Inc. (acq 1996; grpsl). Population served: 224,700 Natl. Rep: Clear Channel. Format: Talk, sports. News staff: 2; News: 7 hrs wkly. Target aud: 25-54; conservative. ♦Skip Essick, VP & gen mgr; Phil Tower, opns dir & progmg dir; Henry Capogna, gen sls mgr; Kami Kennedy, natl sls mgr; Glenn Delvecchio, prom mgr; Rich Jones, news dir; Don Missad, chief of engrg; Gail Warn, traf mgr.

WBFX(FM)—Co-owned with WTKG(AM). 1965: 101.3 mhz; 50 kw. Ant 420 ft TL: N43 02 28 W85 21 28. Stereo. 24 E-mail: web@101thefoxrocks.com Web Site: www.101thefoxrocks.com.197,649 Format: Classic rock. ♦Rich Berry, gen sls mgr; Glenn Delvecchio, prom dir; Doug Montgomery, progmg dir; Aris Hampers, mus dir.

WTNR(FM)—See Holland

***WVGR(FM)**— Dec 7, 1961: 104.1 mhz; 108 kw. 600 ft TL: N42 41 13 W85 30 35. Stereo. Hrs opn: 24
Rebroadcasts WUOM(FM) Ann Arbor 100%.
535 W. William St., Suite 10, Ann Arbor, 48103. Phone: (734) 764-9210. Fax: (734) 647-3488. E-mail: michigan.radio@umich.edu Web Site: www.michiganradio.org. Licensee: Regents of the University of Michigan. Population served: 720,000 Natl. Network: NPR, PRI. Rgnl. Network: Mich. Pub. Law Firm: Dow, Lohnes & Albertson. Format: News/talk. News staff: 7; News: 140 hrs wkly. ♦Bob Skon, stn mgr & chief of engrg; Jon Hoban, stn mgr; Peggy Watson, opns mgr & dev dir; Micheal Leland, news dir.

Grayling

WGRY(AM)— Aug 1, 1970: 1230 khz; 750 w-U. TL: N44 39 05 W84 44 18. Hrs open: 24 6514 Old Lake Rd., 49738. Phone: (989) 348-6171. Fax: (989) 348-6181. E-mail: radio@i2k.net Web Site: www.gannonbroadcasting.com. Licensee: Gannon Broadcasting. Population served: 50,000 Natl. Rep: Michigan Spot Sales. Format: Music of Your Life. News staff: one; News: 16 hrs wkly. Target aud: 25 plus. ♦William S. Gannon, pres & gen mgr; Pete Michaels, opns mgr.

WGRY-FM— June 16, 1977: 100.3 mhz; 50 kw. 436 ft TL: N44 36 50 W84 41 05. Stereo. Hrs opn: 6514 Old Lake Rd., 49738. Phone: (989) 348-6171. Fax: (989) 348-6181. E-mail: radio@i2k.net Web Site: www.gannonbroadcasting.com. Licensee: Gannon Broadcasting Systems Inc. Population served: 23,000 Natl. Network: ABC. Natl. Rep: Patt. Format: Country. Target aud: 25-54. ♦William Gannon, pres & gen mgr; Pete Michaels, opns mgr.

Greenville

***WDPW(FM)**—Not on air, target date: unknown: 91.9 mhz; 6 kw vert. Ant 157 ft TL: N43 05 27 W85 16 27. Hrs open: 6808 Hanna Lake S.E., Caledonia, 49316. Phone: (616) 698-1831. Licensee: Larlen Communications Inc.

WKLQ(FM)—Licensed to Greenville. See Grand Rapids

WSCG(AM)— May 19, 1960: 1380 khz; 1 kw-D, 500 w-N, DA-N. TL: N43 09 18 W85 15 25. Hrs open: Box 578, 9181 S. Greenville Rd., 48838. Phone: (616) 754-3656. Fax: (616) 754-2390. E-mail: wscgradio@chartermi.net Licensee: Stafford Broadcasting L.L.C. (acq 10-19-2004; with WSCG-FM Lakeview). Population served: 9,500 Format: News, talk, sports. ♦Chris Loiselle, CFO; Bruce Bentley, gen mgr, opns mgr & progmg dir; John P. Clark, gen mgr & gen sls mgr; Ralph Hayne, chief of engrg.

Gulliver

WCMM(FM)— 1982: 102.5 mhz; 100 kw. 813 ft TL: N45 58 01 W86 29 18. Stereo. Hrs opn: 524 Ludington, Suite 300, Escanaba, 49829. Phone: (906) 228-9700. Fax: (906) 789-9700. E-mail: countrymoose@chartermi.net Web Site: www.radioresultsnetwork.com/wcmm. Licensee: Lakes Radio Inc. (group owner; acq 11-30-99; grpsl). Population served: 200,000 Natl. Network: ABC. Format: Country. Target aud: 18-54; younger, contemp, mobile adult workers. ♦Rick Duerson, gen mgr.

Gwinn

WUPF(FM)—Not on air, target date: unknown: 100.3 mhz; 100 kw. Ant 852 ft TL: N46 34 13 W87 33 57. Hrs open: 1717 Dixie Hwy., Suite 650, Fort Wright, KY, 41011. Phone: (859) 331-9100. Licensee: Radioactive LLC. ♦Benjamin L. Homel, pres.

Hancock

WGLI(FM)— Feb 11, 2003: 98.7 mhz; 100 kw. Ant 522 ft TL: N47 06 13 W88 34 04. Hrs open: 24 805-B U.S. 41 S., Baraga, 49908. Phone: (906) 353-7625. Fax: (906) 353-9200. E-mail: rock985@up.net Web Site: www.wglifm.com. Licensee: Keweenaw Bay Indian Community (acq 2-13-03). Natl. Network: Jones Radio Networks. Format: Rock/AOR. ♦Ed Janisse, gen mgr; Deborah Hilscher, traf mgr; John Preston, sls.

WKMJ-FM—Listing follows WMPL(AM).

WMPL(AM)— Mar 2, 1957: 920 khz; 1 kw-D, 206 w-N. TL: N47 06 05 W88 35 26. Hrs open: Box 547, 49930. Phone: (906) 482-3700. Fax: (906) 482-1540. Licensee: Victor Broadcasting Corp. (acq 5-22-01; $237,500 with co-located FM). Population served: 4,820 Natl. Network: USA. Natl. Rep: Michigan Spot Sales. Format: Talk/news, info, sports. News staff: one; News: 15 hrs wkly. Target aud: General. ♦John Vertin, pres; Kathy Vertin, VP; Matt Vertin, gen mgr; Marianne Schulze, gen sls mgr; Brian Keranen, progmg dir, chief of engrg & traf mgr; Mitchell Lake, news dir; Phil Halonen, sports cmtr.

WKMJ-FM—Co-owned with WMPL(AM). 1968: 93.5 mhz; 3 kw. 249 ft TL: N47 06 05 W88 35 26. (CP: 13.5 kw, ant 456 ft.). 4,820 Natl. Network: Jones Radio Networks. Law Firm: Booth, Freret, Imlay & Tepper. Format: Adult contemp, CHR. News staff: one. Target aud: 18-45. ♦Brian Keranen, traf mgr; Phil Halonen, sports cmtr.

Harbor Beach

WCZE(FM)— 2005: 103.7 mhz; 50 kw. Ant 440 ft TL: N43 41 25 W82 56 27. Hrs open: Box 388, Williamston, 48895. Phone: (810) 721-0891. Web Site: www.smile.fm. Licensee: Jennifer & Edward Czelada. Format: Christian. ♦Jenn Czelada, gen mgr.

Harbor Springs

***WCMW-FM**— Aug 15, 1988: 103.9 mhz; 28 kw. 663 ft TL: N45 29 02 W84 58 00. Stereo. Hrs opn: 24
Rebroadcasts WCMU-FM Mount Pleasant 100%.
Public Broadcasting Ctr., Central Michigan Univ., Mount Pleasant, 48859. Phone: (989) 774-3105. Fax: (989) 774-4427. E-mail: cmuradio@cmich.edu Web Site: www.wcmu.org. Licensee: Central Michigan University. (acq 7-21-93; $325,000; FTR: 8-23-93). Natl. Network: NPR, PRI. Rgnl. Network: Mich. Pub. Law Firm: Dow, Lohnes & Albertson. Wire Svc: AP Format: Class jazz, news, info. News staff: 2; News: 45 hrs wkly. ♦Ed Grant, gen mgr & rgnl sls mgr.

Harrietta

WKAD(FM)— 2003: 93.7 mhz; 4.3 kw. Ant 390 ft TL: N44 16 41 W85 35 28. Hrs open: Box 520, Cadillac, 49601. Phone: (231) 775-1263. Fax: (231) 779-2844. Licensee: Cadillac Broadcasting LLC (acq 1-4-02). Format: Oldies. ♦Trish Garber, CEO & gen mgr.

Harrison

WTWS(FM)— Mar 26, 1975: 92.1 mhz; 6 kw. Ant 298 ft TL: N43 59 38 W84 50 13. Hrs open: 24 Box 468, Prudenville, 48651. Phone: (989) 366-5364. Fax: (989) 366-6200. Web Site: www.921thetwister.com. Licensee: Coltrace Communications Inc. (acq 6-7-2006; $200,000). Population served: 300,000 Format: Country. ♦Michael Jay, gen mgr.

Harrisville

***WJOJ(FM)**— December 2001: 89.7 mhz; 31 kw. Ant 469 ft TL: N44 42 12 W83 31 27. Hrs open: Box 388, Williamston, 48895. Phone: (810) 721-0891. E-mail: info@joyfm.net Web Site: www.joyfm.net. Licensee: Northland Community Broadcasters. Format: Contemporary Christian. ♦Jenn Czelada, gen mgr.

Hart

WWKR(FM)— July 1, 1995: 94.1 mhz; 13 kw. Ant 462 ft TL: N43 51 33 W86 18 25. Stereo. Hrs opn: 24 PO Box 855, Ludington, 49431. Phone: (231) 843-0941. Fax: (231) 843-9411. Web Site: www.94k-rock.com. Licensee: Synergy Media Inc. (acq 4-10-98; $250,000). Population served: 350,000 Natl. Network: AP Network News. Rgnl. Network: Michigan Sport Sales. Natl. Rep: Michigan Spot Sales. Rgnl rep: Michigan Spot Sales Law Firm: Drinker, Biddle & Reath. Format: Classic rock. News staff: one; News: 2 hrs wkly. Target aud: 25-54; baby boomers. Spec prog: Relg 3 hrs wkly. ♦Todd A. Mohr, pres & gen mgr; Barry Smith, opns mgr & progmg dir; Mary Mohr, opns mgr; Melissa Reed, gen sls mgr; Tom Green, chief of engrg.

Hartford

WHIT-FM— March 1996: 103.7 mhz; 3 kw. Ant 328 ft TL: N42 18 02 W86 15 03. Stereo. Hrs opn: 24 Box 107, St. Joseph, 49085. Secondary address: 580 E. Napier Ave., Benton Harbor 49022. Phone: (269) 925-1111. Fax: (269) 925-1011. Licensee: WSJM Inc. Group owner: The Mid-West Family Broadcast Group (acq 4-96; grpsl). Natl. Network: ABC. Natl. Rep: Rgnl Reps. Law Firm: Davis Wright Tremaine. Wire Svc: AP Format: Super hits 60s & 70s. Target aud: 35. ♦Gayle Olson, gen mgr; Jim Gifford, opns mgr; Joe Daguanno, sls dir; Sue Patzen, prom dir; Joe Jason, progmg dir; Annette Weston, news dir; Terry Green, engrg dir & chief of engrg.

Hastings

WBCH(AM)— November 1957: 1220 khz; 250 w-D, 48 w-N. TL: N42 37 36 W85 16 39. Hrs open: 24 Box 88, 49058. Secondary address: 119 W. State St. 49058. Phone: (269) 945-3414. Fax: (269) 945-3470. E-mail: wbch@wbch.com Web Site: www.wbch.com. Licensee: Barry Broadcasting Co. (acq 8-17-58). Population served: 50,000 Rgnl. Network: Mich. Farm. Rgnl rep: Patt. Wire Svc: NOAA Weather Format: Country, news/talk. News staff: one; News: 16 hrs wkly. Target aud: 25-54. ♦Kenneth Radant, gen mgr; Steven K. Radant, stn mgr.

WBCH-FM— December 1967: 100.1 mhz; 3 kw. 295 ft TL: N42 37 36 W85 16 39. Stereo. 24 Web Site: www.wbch.com.806,500 Natl. Network: ABC. Format: Hit country. ♦Dave McIntyre, local news ed, news rptr & farm dir; Sue Radant, women's int ed.

Hemlock

WCEN-FM— Aug 8, 1963: 94.5 mhz; 100 kw. Ant 981 ft TL: N43 43 36 W84 36 16. Stereo. Hrs opn: 1795 Tittabawassee Rd., Saginaw, 48604-9431. Phone: (989) 752-3456. Fax: (989) 754-5046. Web Site: www.945themoose.com. Licensee: NM Licensing LLC. Group owner: NextMedia Group L.L.C. (acq 12-30-02; grpsl). Population served: 1,000,000 Law Firm: Shaw Pittman. Wire Svc: Metro Weather Service Inc. Format: Hot country. Target aud: 25-54; medium income, rural & urban. ♦Floyd Evans, gen mgr; Dave Maurer, opns mgr.

Highland Park

***WHPR(FM)**— May 21, 1954: 88.1 mhz; 11 w. Ant 105 ft TL: N42 24 50 W83 05 48. Hrs open: 24 15851 Woodward, 48203. Phone: (313) 868-6612. Fax: (313) 868-8725. E-mail: tv68whpr@aol.com Licensee: R.J.s Late Night Entertainment Corp. Population served: 900,000 Format: Talk, CHR, oldies. Target aud: 21 & over; African Americans 40 plus politically aware & motivated. ♦Henry Tyler, VP & stn mgr; R. J. Watkins, sr VP & gen mgr.

Hillman

WKJZ(FM)—Licensed to Hillman. See Alpena

Hillsdale

WCSR(AM)— May 21, 1959: 1340 khz; 500 w-D, 1 kw-N. TL: N41 55 41 W84 38 10. Hrs open: Box 273, 49242. Secondary address: 170 N. West St. 49242. Phone: (517) 437-4444. Fax: (517) 437-7461. E-mail: wcsr@qcnet.net Web Site: www.radiohillsdale.com. Licensee: WCSR Inc. (acq 11-15-61). Rgnl. Network: Mich. Farm. Format: Adult contemp. Target aud: 25 plus; county-wide. Spec prog: Farm 3 hrs, relg 10 hrs wkly. ♦Anthony Flynn, pres; Michael Flynn, gen mgr.

WCSR-FM— May 19, 1973: 92.1 mhz; 6 kw. 243 ft TL: N41 55 41 W84 38 10. Stereo. Web Site: www.radiohillsdale.com.50,000

Holland

WHTC(AM)— July 31,1948: 1450 khz; 1 kw-U. TL: N42 47 41 W86 06 22. Hrs open: 24 87 Central Ave., 49423. Phone: (616) 392-3121. Fax: (616) 392-8066. E-mail: whtc@whtc.com Web Site: www.whtc.com. Licensee: Midwest Communications Inc. (group owner; (acq 8-1-00; grpsl). Population served: 120,000 Natl. Network: CBS. Natl. Rep: Michigan Spot Sales. Rgnl rep: Christal Radio Wire Svc: AP Format: News/talk, full service. News staff: one; News: 15 hrs wkly. Target aud: 25 plus. Spec prog: Sp 3 hrs wkly. ♦Duke Wright, pres; Peter Tanz, gen mgr; Kevin Oswald, gen sls mgr; Brent Alan, progmg mgr; Gary Stevens, news dir; Bridgett Bowmen, traf mgr.

WJQK(FM)—Zeeland, Aug 23, 1971: 99.3 mhz; 4.7 kw. 371 ft TL: N42 48 59 W85 57 24. Stereo. Hrs opn: 24 425 Centerstone Ct., Zeeland, 49464. Phone: (616) 931-9930. Phone: (888) 993-1260. Fax: (616) 931-1280. Web Site: www.jq99.com. Licensee: Lanser Broadcasting Corp. (acq 1-1-87). Population served: 1,200,000 Natl. Network: Fox News Radio. Natl. Rep: Salem. Rgnl rep: Mich Spot Sales Wire Svc: Metro Weather Service Inc. Format: Contemp Christian. News: 7 hrs wkly. Target aud: 25-49. ♦Les Lanser, pres; Brad Lanser, VP & gen mgr; Troy West, stn mgr.

WMAX-FM— September 1962: 96.1 mhz; 50 kw horiz, 45 kw vert. Ant 492 ft TL: N42 49 10 W85 52 09. Stereo. Hrs opn: 24 77 Monroe Ctr., Suite 1000, Grand Rapids, 49503. Phone: (616) 459-1919. Fax: (616) 235-9600. Web Site: 961maxfm.com. Licensee: CC Licenses LLC. Group owner: Clear Channel Communications Inc. (acq 2-27-97; $4.1 million). Natl. Rep: Clear Channel. Format: Contemp hit. Target aud: 25-34; women. ♦Skip Essick, VP & gen mgr.

***WTHS(FM)**— Oct 15, 1984: 89.9 mhz; 1 kw. 154 ft TL: N42 47 16 W86 06 02. Stereo. Hrs opn: 24 Box 9000, Hope College, 49423. Phone: (616) 395-7878. Phone: (616) 395-7880. Fax: (616) 395-7958. E-mail: wths@hope.edu Web Site: http://wths.hope.edu. Licensee: Hope College Board of Trustees. Law Firm: Lauren A. Colby. Format: Alternative. News: 7 hrs wkly. Target aud: 15-30; students & adults. Spec prog: Jazz 6 hrs, relg 14 hrs, Sp 8 hrs wkly. ♦Jason Cash, gen mgr; Gerry Ruffino, progmg dir.

WTNR(FM)— Mar 21, 1961: 94.5 mhz; 50 kw. Ant 499 ft TL: N42 51 20 W85 57 45. Stereo. Hrs opn: 24 60 Monroe Ctr. N.W., 3rd Fl., Grand Rapids, 49503. Phone: (616) 774-8461. Fax: (616) 774-2491. Licensee: Citadel Broadcasting Co. Group owner: Citadel Broadcasting Corp. (acq 4-26-2001; grpsl). Population served: 194,649 Natl. Network: ABC. Natl. Rep: Katz Radio. Law Firm: Fletcher, Heald & Hildreth. Format: Country. Target aud: 18-34; men. ♦Matt Hanlon, gen mgr; Jeff Morton, rgnl sls mgr.

Holton

WVIB(FM)— 1971: 100.1 mhz; 3 kw. Ant 302 ft TL: N43 28 17 W85 56 19. Stereo. Hrs opn: 3375 Merriam St., Muskegon, 49444. Phone: (231) 830-0176. Fax: (231) 830-0194. Licensee: Citadel Broadcasting Co. (acq 9-28-2005). Format: Urban contemp. ♦Jeff Morton, gen mgr.

Honor

WSRJ(FM)— 2002: 100.7 mhz; 4.7 kw. Ant 367 ft TL: N44 39 41 W85 48 53. Hrs open:
Rebroadcasts WKPK(FM) Gaylord 100%.
28 Old Colony Rd., Gaylord, 49735. Phone: (231) 947-0003. Fax: (213) 546-4490. Licensee: Northern Radio of Michigan Inc. Format: Hot adult contemp. ♦Charlie Ferguson, gen mgr.

Houghton

WCCY(AM)— 1929: 1400 khz; 1 kw-U. TL: N47 08 06 W88 33 53. Hrs open: 24 313 Montezuma Ave., 49931. Phone: (906) 482-7700. Fax: (906) 482-7751. Web Site: www.wccy.com. Licensee: Heartland Comm. Houghton License LLC. (acq 1-6-2005; grpsl). Population served: 20,000 Natl. Rep: Patt. Format: Easy lstng. News staff: one; News: 18 hrs wkly. Target aud: 25-65. Spec prog: Relg one hr, pub affrs one hr wkly. ♦Dallas Bond, gen mgr & gen sls mgr; Tim Murphy, opns mgr & progmg dir.

WOLV(FM)—Co-owned with WCCY(AM). Mar 7, 1980: 97.7 mhz; 875 w. 508 ft TL: N47 08 27 W88 32 26. Stereo. Web Site: www.thewolf.com.43,000 Format: Classic rock. Target aud: 18-45.

***WGGL-FM**— February 1982: 91.1 mhz; 100 kw. 809 ft TL: N47 02 08 W88 41 43. Stereo. Hrs opn: 24 45 E. 7th St., St. Paul, MN, 55101. Phone: (651) 290-1500. Fax: (651) 290-1224. Web Site: www.mpr.org. Licensee: Minnesota Public Radio Inc. Natl. Network: PRI, NPR. Rgnl. Network: Minn. Pub. Format: Class, news. News staff: one. Target aud: General. ♦William H. Kling, pres; Erik Nycklemoe, gen mgr; Ralph Hornberger, chief of engrg.

WHKB(FM)— Sept 1, 1989: 102.3 mhz; 1.05 kw. 554 ft TL: N47 06 13 W88 34 04. (CP: 35.5 kw, ant 492 ft.). Stereo. Hrs opn: 313 E. Montezuma Ave., 49931. Phone: (906) 482-7700. Fax: (906) 482-7751. Licensee: Heartland Comm. Houghton License LLC. (acq 1-6-2005; grpsl). Population served: 50,000 Natl. Network: ABC. Format: Country. Target aud: 18-60. Spec prog: Oldies 16 hrs wkly. ♦Dallas Bond, gen mgr; Darlene Basto, opns dir & news dir; Tim Murphy, progmg VP & progmg dir; Christy Tereschulk, chief of engrg & traf mgr.

***WMTU-FM**— Jan 26, 1994: 91.9 mhz; 4.4 kw vert. Ant 479 ft TL: N47 08 27 W88 32 26. Hrs open: West Wadsworth Hall, MTUniversity G03, 49931. Phone: (906) 487-2333. Fax: (906) 487-3016. E-mail: wmtu@mtu.edu Web Site: www.wmtu.mtu.edu. Licensee: Michigan Technological University. Format: Var. ♦Lindsay Worden, gen mgr.

Houghton Lake

WUPS(FM)— July 1, 1961: 98.5 mhz; 100 kw. Ant 981 ft TL: N44 17 18 W84 44 30. Stereo. Hrs opn: 24 Box 468, Prudenville, 48651. Phone: (989) 366-5364. Fax: (989) 366-6200. E-mail: wupsfm@yahoo.com Web Site: www.wups.com. Licensee: Coltrace Communications Inc. (acq 3-15-88; $900,000). Population served: 200,000 Natl. Network: ABC. Natl. Rep: Rgnl Reps. Law Firm: Drinker, Biddle & Reath, LLP. Format: Classic hits. News staff: one; News: 6 hrs wkly. Target aud: 25-54; general. ♦John M. Salov, pres & opns mgr; Sindy Winkler, sr VP.

Howell

WHMI-FM— Sept 1, 1977: 93.5 mhz; 5.2 kw. 354 ft TL: N42 39 47 W83 56 23. Stereo. Hrs opn: 24 Box 935, 48844. Secondary address: 1277 Parkway Dr. 48843. Phone: (517) 546-0860. Fax: (517) 546-1758. E-mail: whmi@whmi.com Web Site: www.whmi.com. Licensee: The Livingston Radio Co. (acq 3-3-89). Population served: 400,000 Law Firm: Irwin, Campbell & Tannenwald. Format: Classic hits. News staff: 3; News: 10 hrs wkly. Target aud: 25-64. ♦Greg Jablonski, pres & gen mgr; Reed Kittredge, opns mgr.

Hubbard Lake

***WKHN(FM)**—Not on air, target date: unknown: 88.1 mhz; 1.1 kw vert. Ant 367 ft TL: N44 54 20 W83 32 10. Hrs open: 901 Elizabeth Ct., Mount Pleasant, 48858. Phone: (989) 779-9178. Fax: (989) 779-1558. Licensee: Great Lakes Community Broadcasting Inc. Format: Oldies. ♦James J. McCluskey, gen mgr.

Hudson

WBZV(FM)— Mar 1, 1995: 102.5 mhz; 6 kw. 328 ft TL: N41 53 03 W84 31 24. Hrs open: 24 121 W. Maumee St., Adrian, 49221-2019. Phone: (517) 448-8988. Fax: (517) 263-4525. E-mail: friends@tc3net.com Licensee: Friends Communications of Hudson Inc. Group owner: Friends Communications Inc. Population served: 150,000 Natl. Network: ABC. Natl. Rep: Michigan Spot Sales. Law Firm: Fletcher, Heald & Hildreth. Format: Hot AC. News staff: one; News: 6 hrs wkly. Target aud: 25-54. ♦Bob Elliot, chmn, pres & gen mgr.

Hudsonville

WHIT(AM)— 1961: 940 khz; 300 w-D, DA. TL: N42 47 38 W85 52 24. Hrs open: 510 Williams St., South Haven, 49090. Phone: (269) 637-6397. Fax: (269) 637-2675. Licensee: WSJM Inc. (group owner; (acq 10-95; with WCSY-FM South Haven). Population served: 300,000 Rgnl. Network: Mich. Farm. Format: Nostalgia. Target aud: 35-45; females. Spec prog: Farm 3 hrs wkly. ♦Gayle Olson, pres & gen mgr.

Imlay City

***WWKM(FM)**— December 2000: 89.1 mhz; 1.5 kw. Ant 171 ft TL: N43 03 42 W83 05 44. Hrs open: Michigan Community Radio, Box 388, Williamston, 48896. Phone: (810) 721-0891. Fax: (413) 410-9708. E-mail: info@joyfm.net Web Site: www.positivehits.com. Licensee: Michigan Community Radio. Format: Contemporary Christian. ♦Jenn Czelada, gen mgr.

Inkster

WDMK(FM)—See Detroit

WDRJ(AM)— November 1956: 1440 khz; 1 kw-U, DA-2. TL: N42 15 22 W83 21 48. Hrs open: 24 2994 E. Grand Blvd., Detroit, 48202. Phone: (313) 871-1440. Fax: (313) 871-6088. Licensee: Davidson Media Station WMKM Licensee LLC. Group owner: Davidson Media Group LLC (acq 5-28-2004; $5.75 million). Format: Black gospel. Target aud: 35 plus; adult Black church audience. ♦Peter Davidson, pres; Crystal D. Sampson, gen mgr.

Interlochen

***WIAA(FM)**— July 22, 1963: 88.7 mhz; 100 kw. Ant 1,033 ft TL: N44 16 33 W85 42 49. Stereo. Hrs opn: 24 Box 199, Interlochen Ctr. for the Arts, 49643. Secondary address: One Lyon St. 49643. Phone: (231) 276-4400. Fax: (231) 276-4417. E-mail: ipr@interlochen.org Web Site: www.interlochen.org/ipr. Licensee: Interlochen Center for the Arts. Population served: 290,100 Natl. Network: NPR, PRI. Rgnl. Network: Mich. Pub. Format: Classical, news. Target aud: 35-80; professional, arts-oriented, upper-income. ♦Thom Paulson, CFO, VP & stn mgr.

Ionia

WION(AM)— Feb 1, 1953: 1430 khz; 5 kw-D, 330 w-N, DA-2. TL: N43 00 16 W85 05 09. pending for non directional signal. Hrs opn: 24 1150 Haynor Rd., 48846-8532. Phone: (616) 527-9466. Fax: (616) 775-5908. E-mail: office@i1430.com Web Site: www.i1430.com. Licensee: Packer Radio WION LLC (acq 12-14-2004; $127,000). Population served: 77,294 Natl. Network: Fox News Radio, Sporting News Radio Network. Rgnl. Network: Mich. Farm. Law Firm: Dykema Gossett PLLC. Format: Full service. News staff: one. ♦J.C. Angus, pres; Jim Aaron, VP & mus dir; Jim Carlyle, gen mgr; Tom Millard, gen sls mgr.

Iron Mountain

WHTO(FM)— 2003: 106.7 mhz; 6.1 kw. Ant 676 ft TL: N45 49 16 W88 02 34. Hrs open: 212 W. J St., 49801. Phone: (906) 774-5731. Fax: (906) 774-4542. Web Site: www.whtofm.com. Licensee: Results Broadcasting of Iron Mountain Inc. (acq 6-29-2005; $650,000). Format: Oldies. ♦Trisha Peterson, gen mgr.

WIMK(FM)— Dec 27, 1981: 93.1 mhz; 100 kw. 590 ft TL: N45 49 16 W88 02 28. Stereo. Hrs opn: 24
Rebroadcasts WUPK-FM Marquette 100%.
101 E. Kent St., 49801. Phone: (906) 774-4321. Fax: (906) 774-7799. E-mail: thebear@uplogon.com Web Site: rockthebear.com. Licensee: Northern Star Broadcasting L.L.C. (group owner; acq 11-5-01; grpsl). Population served: 300,000 Law Firm: Reddy, Begley & McCormick. Format: Classic rock, AOR. News staff: one. Target aud: 25-54. ♦Veronica Roberts, gen mgr; Steve Ponchaud, opns mgr; Tom Hill, news dir; Coral Howe, chief of engrg; Michelle Ellsworth, traf mgr.

WMIQ(AM)—Co-owned with WIMK(FM). January 1947: 1450 khz; 1 kw-U. TL: N45 49 16 W88 03 16.24 E-mail: talk1450wmiq@uplogon.com Web Site: talk1450.tripod.com.25,600 Natl. Network: USA. Natl. Rep:

Patt. Format: News/talk, sports. News staff: one; News: 24 hrs wkly. Target aud: 35-64; educated, middle to upper income listeners. ♦Kevin Richtig, progmg dir.

WJNR-FM— Aug 17, 1972: 101.5 mhz; 100 w. 620 ft TL: N45 49 15 W88 02 38. (CP: 100 kw, ant 613 ft.). Stereo. Hrs opn: 212 W. J St., 49801. Phone: (906) 774-5731. Fax: (906) 774-4542. E-mail: wjnr@chartermi.net Web Site: www.resultsbroadcasting.com. Licensee: Results Broadcasting of Michigan Inc. Group owner: Results Broadcasting (acq 6-5-97). Population served: 14,450 Format: Hot new country. Target aud: 25-54. ♦Bruce Grassman, pres; Trisha Peterson, gen mgr & gen sls mgr; Keith Huotari, opns mgr; Aaron Harper, news dir; Walt Baldwin, engrg VP & chief of engrg; John Kohler, disc jockey.

***WVCM(FM)**—Not on air, target date: unknown: 91.5 mhz; 500 w. 600 ft TL: N45 49 15 W88 02 25. Stereo. Hrs opn: 24 3434 W. Kilbourn Ave., Milwaukee, WI, 53208. Phone: (414) 935-3000. Fax: (414) 935-3015. E-mail: wvcm@vcyamerica.org Web Site: www.vcyamerica.org. Licensee: VCY America Inc. Group owner: VCY/America Inc. Format: Christian. ♦Randall Melchert, pres; Vic Eliason, VP & gen mgr; Jim Schneider, progmg dir.

Iron River

WIKB(AM)— Nov 18, 1949: 1230 khz; 1 kw-U. TL: N46 03 55 W88 38 17. Hrs open: 5 AM-11 PM Box AC, 809 W. Genesee St., 49935. Phone: (906) 265-5104. Fax: (906) 265-3486. E-mail: wikb@up.net Licensee: Heartland Communications License LLC. (group owner; (acq 5-10-2004; $1.25 million with co-located FM). Population served: 20,000 Natl. Rep: Roslin. Format: Oldies. News staff: one; News: 15 hrs wkly. Target aud: General. ♦Jay Barry, gen mgr, gen sls mgr & chief of engrg.

WIKB-FM— Sept 25, 1981: 99.1 mhz; 50 kw. Ant 492 ft TL: N46 06 03 W88 32 23. Stereo. 5 AM-11 PM

Ironwood

WIMI(FM)—Listing follows WJMS(AM).

WJMS(AM)— Nov 3, 1931: 590 khz; 5 kw-D, 1 kw-N, DA-N. TL: N46 25 25 W90 12 30. Hrs open: 24 222 S. Lawrence St., 49938. Phone: (906) 932-2411. Fax: (906) 932-2485. E-mail: wimi@broadcast.net Web Site: www.wjmsam.com. Licensee: Armada Media - Menominee Inc. Group owner: Badger Communications L.L.C. (acq 8-30-2006; grpsl). Natl. Network: CBS. Natl. Rep: D & R Radio. Format: Country, talk. News staff: one. Target aud: 25 plus. ♦Jim Coursolle, pres; Freddy Falls, progmg dir; Chuck Gennaro, engrg mgr & chief of engrg.

WIMI(FM)—Co-owned with WJMS(AM). March 1976: 99.7 mhz; 100 kw. 561 ft TL: N46 25 25 W90 14 53. Stereo. Web Site: www.wimifm.com. Format: Adult contemp.

***WLVM(FM)**—Not on air, target date: unknown: 88.3 mhz; 10 kw vert. Ant 515 ft TL: N46 26 28 W90 11 26. Hrs open: 2351 Sunset Blvd., Suite 170-218, Rocklin, CA, 95765. Phone: (916) 251-1600. Fax: (916) 251-1650. Licensee: Educational Media Foundation. (acq 7-23-2007; grpsl). ♦Mike Novak, sr VP.

WUPM(FM)— Oct 17, 1977: 106.9 mhz; 53 kw. 495 ft TL: N46 28 18 W90 00 43. Hrs open: Box 107, 49938. Secondary address: 209 Harrison 49938. Phone: (906) 932-5234. Fax: (906) 932-1548. E-mail: wupm@wupm-whry.com Web Site: www.wupm-whry.com. Licensee: Big G Little O Inc. Population served: 25,000 Natl. Network: ABC. Format: Adult contemp, CHR. ♦Charles H. Gervasio, pres & gen mgr; Laura Keller, progmg VP & progmg dir.

Ishpeming

WIAN(AM)— 1947: 1240 khz; 1 kw-U. TL: N46 30 16 W87 40 46. Hrs open: 1009 W. Ridge St., Marquette, 49855. Phone: (906) 225-1313. Fax: (906) 225-1324. Licensee: Northern Star Broadcasting L.L.C. (group owner; acq 11-5-01; grpsl). Format: News/talk. ♦Tammy Johnsen, gen mgr & rgnl sls mgr; Coral Howe, chief of engrg.

WJPD(FM)—Co-owned with WIAN(AM). May 15, 1975: 92.3 mhz; 100 kw. Ant 508 ft TL: N46 30 51 W87 28 58. Stereo. Web Site: www.wjpd.com.300,000 Format: Country.

WMQT(FM)—Listing follows WZAM(AM).

WZAM(AM)— June 26, 1959: 970 khz; 5 kw-D, 62 w-N. TL: N46 30 20 W87 32 24. Stereo. Hrs opn: 24 121 N. Front St., Suite A, Marquette, 49855. Phone: (906) 225-9100. Fax: (906) 225-5577. E-mail: tom@wmqt.com Web Site: www.espn970.com. Licensee: Taconite Broadcasting Inc. (acq 7-25-2005; $827,300 with co-located FM). Population served: 100,000 Format: Sports. Target aud: 25-54. ♦Tom Mogush, gen mgr & sls dir; Dennis Whitley, news dir.

WMQT(FM)—Co-owned with WZAM(AM). Jan 26, 1974: 107.7 mhz; 98 kw. Ant 639 ft TL: N46 30 08 W87 38 52. Stereo. 24 E-mail: tom@wmqt.com Web Site: www.wmqt.com.250,000 Format: Hot adult contemp. News staff: 2. Target aud: 18-49. ♦Jim Koski, progmg dir & rsch dir; Carol Keast, traf mgr; Tom Mogush, opns dir, sls VP, gen sls mgr & sports cmtr.

Jackson

WIBM(AM)— 1925: 1450 khz; 1 kw-U. TL: N42 13 16 W84 26 03. Hrs open: 24 1700 Glenshire Dr., 49201. Phone: (517) 787-9546. Fax: (517) 787-7517. Web Site: www.espnradio1450.com. Licensee: Jackson Radio Works Inc. (group owner; acq 11-14-97; grpsl). Population served: 175,000 Natl. Network: ESPN Radio. Rgnl rep: Michigan Spot Sales Law Firm: Davis, Wright , Tremain, LLP. Format: Sports. News staff: 2; News: one hr wkly. Target aud: 18-49; sports enthusiasts. Spec prog: Polish, Spanish. ♦Bruce I. Goldsen, pres; Jamie McKibbin, stn mgr; Sue Goldsen, VP, sls VP & mktg VP; Marc Daly, progmg dir & progmg mgr; Michael Bradford, chief of engrg.

***WJCQ(FM)**— 2004: 89.7 mhz; 500 w. Ant 121 ft TL: N42 17 05 W84 18 40. Hrs open: 901 Elizabeth Ct., Mount Pleasant, 48858. Phone: (989) 779-9178. Fax: (989) 779-1558. Licensee: Great Lakes Community Broadcasting Inc. Format: Oldies. ♦James J. McCluskey, gen mgr.

***WJKN(AM)**— January 1962: Stn currently dark. 1510 khz; 5 kw-D, DA. TL: N42 11 10 W84 22 39. (CP: TL: N42 10 08 W84 23 30). Stereo. Hrs opn: Spring Arbor University, 106 E. Main St., Spring Arbor, 49283. Phone: (517) 750-9723. Fax: (517) 750-6619. E-mail: wsae@arbor.edu Licensee: Spring Arbor University (acq 1-12-01). Population served: 45,484 Natl. Network: CBS, Westwood One. Natl. Rep: Patt. Format: Inspirational. ♦Carl Fletcher, gen mgr; Dave Benson, chief of engrg.

***WJKQ(FM)**— 2004: 88.5 mhz; 100 w vert. Ant 112 ft TL: N42 16 22 W84 21 27. Hrs open: 901 Elizabeth Ct., Mount Pleasant, 48858. Phone: (989) 779-9178. Fax: (989) 779-1558. Licensee: Great Lakes Community Broadcasting Inc. Format: Oldies. ♦James McCluskey, gen mgr.

WJXQ(FM)— May 30, 1976: 106.1 mhz; 50 kw. 489 ft TL: N42 23 28 W84 37 22. Stereo. Hrs opn: 24 2495 N. Cedar, Suite 106, Holt, 48842. Phone: (517) 699-0111. Fax: (517) 699-1880. Web Site: www.q106fm.com. Licensee: Rubber City Radio Group. Group owner: Rubber City Radio Group Inc. (acq 7-12-00; grpsl). Population served: 500,000 Natl. Network: Jones Radio Networks. Natl. Rep: Katz Radio. Wire Svc: AP Format: AOR. News staff: one; News: one hr wkly. Target aud: 25-44; baby boomers with an inclination for rock and roll. ♦Dave Johnson, gen mgr; Paul Cashin, opns mgr & mus dir; Scott Truman, gen sls mgr; Sheri Vegas, progmg dir.

WKHM(AM)— Dec. 7, 1951: 970 khz; 1 kw-U, DA-2. TL: N42 11 39 W84 25 50. Hrs open: 24 1700 Glenshire Dr., 49201. Phone: (517) 787-9546. Phone: (517) 787-3397. Fax: (517) 787-7517. E-mail: bgoldsen@k1053.com Web Site: www.wkhm.com. Licensee: Jackson Radio Works Inc. (group owner; acq 12-8-97; grpsl). Population served: 175,000 Natl. Network: ABC. Rgnl. Network: Mich. Farm. Rgnl rep: Mich. Spot Sales Law Firm: Davis, Wright, Tremaine, LLP. Format: News/talk. News staff: one; News: 15 hrs wkly. Target aud: 25-64. ♦Bruce I. Goldsen, pres; Jamie McKibbin, stn mgr; Deanna Stocker, sls dir; Sue Goldsen, VP, sls VP & mktg VP; Marc Daly, progmg dir; Michael Bradford, chief of engrg; Kathy Beauchamp, traf mgr & local news ed.

WVIC(FM)— 1955: 94.1 mhz; 40 kw. 551 ft TL: N42 23 32 W84 40 00. Stereo. Hrs opn: 2495 N. Cedar, Holt, 48842. Phone: (517) 699-0111. Fax: (517) 699-1880. Web Site: www.wvic.net. Licensee: Rubber City Radio Group. Group owner: Rubber City Radio Group Inc. (acq 7-12-00; grpsl). Population served: 864,300 Natl. Rep: Katz Radio. Wire Svc: AP Format: Classic hits. News staff: one; News: 17.5 hrs wkly. Target aud: 25-54; Female 25-49. ♦Dave Johnson, gen mgr; Paul Cashin, opns mgr & progmg dir; Scott Truman, gen sls mgr.

Kalamazoo

***WAYK(FM)**— Feb 3, 1997: 88.3 mhz; 10 kw. Ant 397 ft TL: N42 18 23 W85 39 25. Hrs open: 24 161 E. Michigan Ave., Suite 600, 49007. Phone: (269) 383-3600. Fax: (269) 381-0239. Web Site: www.way.fm. Licensee: Cornerstone University. Population served: 300,000 Format: Christian hit radio. ♦Rich Anderson, gen mgr; Tom Bos, rgnl sls mgr; Terri Brogan, prom dir; Mike Couchman, progmg dir.

***WIDR(FM)**— July 7, 1975: 89.1 mhz; 100 w. 158 ft TL: N42 16 55 W85 37 05. Stereo. Hrs opn: 24 Western Michigan Univ., 1501 Faunce Student Ser. Bldg., 49008-6301. Phone: (269) 387-6301. Phone: (269) 387-6303. Fax: (269) 387-2839. Web Site: www.widr.org. Licensee: Western Michigan University Board of Trustees. Population served: 120,000 Format: Var/div, educ, progsv. News: 7 hrs wkly. Target aud: 18-25; college students. ♦Andrew Grabowski, gen mgr; Allie Gruner, progmg dir; Ben Jones, news dir.

***WKDS(FM)**— October 1982: 89.9 mhz; 100 w. 150 ft TL: N42 14 36 W85 34 19. Stereo. Hrs opn: 8 AM-9 PM (M-F) 606 E. Kilgore Rd., 49001. Phone: (269) 337-0200. Fax: (269) 337-0251. Licensee: Kalamazoo Board of Education. Law Firm: Arent, Fox, Kintner, Plotkin & Kahn. Format: Div, educ. News: 3 hrs wkly. Target aud: High school & college students.

WKFR-FM—See Battle Creek

WKMI(AM)— August 1947: 1360 khz; 5 kw-D, 1 kw-N, DA-2. TL: N42 19 36 W85 31 39. Hrs open: 24 Box 50911, 49005-0911. Secondary address: 4154 Jennings Dr. 49048. Phone: (269) 344-0111. Fax: (269) 344-4223. E-mail: radio@wkmi.com Web Site: www.wkmi.com. Licensee: Cumulus Licensing Corp. Group owner: Cumulus Media Inc. (acq 5-26-98; grpsl). Population served: 400,000 Format: News/talk. News staff: one; News: 30 hrs wkly. Target aud: 25 plus. Spec prog: Sports. ♦Lew Dickey, CEO & pres; Jon Pinch, COO; Martin Gausvik, CFO; John Sterling, gen mgr; Mike McKelly, opns mgr; Andy Stone, progmg dir.

WKPR(AM)— Oct 20, 1960: 1420 khz; 1 kw-D, DA. TL: N42 18 46 W85 37 06. Hrs open: 6 AM-sunset 2244 Ravine Rd., 49004. Phone: (269) 381-1420. Licensee: Kalamazoo Broadcasting Co. Group owner: Kuiper Stations Population served: 150,000 Natl. Network: USA. Format: Relg. Target aud: 25 plus. ♦William E. Kuiper Sr., pres; Stan Gebben, stn mgr; William E. Kuiper Jr., chief of engrg.

WKZO(AM)— Sept 10, 1931: 590 khz; 5 kw-U, DA-N. TL: N42 21 00 W85 33 43. Hrs open: 24 4200 W. Main St., 49006. Phone: (269) 345-7121. Fax: (269) 345-1436. Web Site: www.wkzo.com. Licensee: Midwest Communications Inc. (acq 5-1-2006; grpsl). Natl. Network: CBS. Rgnl. Network: Mich. Farm. Format: News/talk. News staff: 5. Target aud: 25 plus; upscale, 60% male, 40% female. Spec prog: Farm 10 hrs, relg 5 hrs wkly. ♦Duey Wright, pres; Peter Tanz, gen mgr; Dennis Martin, opns dir & sls VP; Walker Sisson, chief of engrg.

***WMUK(FM)**— Jan 8, 1951: 102.1 mhz; 50 kw. 490 ft TL: N42 25 03 W85 31 55. Stereo. Hrs opn: 24 1903 W. Michigan Ave., 49008-5351. Phone: (269) 387-5715. Fax: (269) 387-4630. E-mail: wmukfm@wmich.edu Web Site: www.wmuk.org. Licensee: Western Michigan University Board of Trustees. Population served: 85,555 Natl. Network: NPR, PRI. Rgnl. Network: Mich. Pub. Law Firm: Gammon & Grange. Format: Class, jazz, news. News staff: 3; News: 38 hrs wkly. Target aud: General; educated adults. Spec prog: Bluegrass 4 hrs wkly. ♦Floyd Pientka, gen mgr; Gordon Bolar, dev dir; Michael Hahn, adv mgr; Klayton Woodworth, progmg dir; Andy Robins, news dir; Mark Tomlonson, chief of engrg.

WNWN(AM)—See Portage

WNWN-FM—Coldwater, Nov 11, 1950: 98.5 mhz; 50 kw. 500 ft TL: N42 03 28 W84 59 51. Stereo. Hrs opn: 24 25 W. Michigan Ave., Battle Creek, 49017. Phone: (269) 968-1991. Fax: (269) 968-1881. Web Site: www.wincountry.com. Licensee: Midwest Communications Inc. (group owner; acq 6-1-95; grpsl). Population served: 300,000 Natl. Rep: Christal. Wire Svc: NOAA Weather Format: Contemp country. News staff: 3; News: 5 hrs wkly. Target aud: 25-54. ♦D.E. Wright, pres; Peter Tanz, gen mgr; Michael Klein, gen sls mgr; P.J. Lacey, progmg dir & progmg; Bridgett Bowman, traf mgr.

WQLR(AM)— Sept 24, 1998: 1660 khz; 10 kw-D, 1 kw-N. TL: N42 14 11 W85 34 37. Hrs open: 4200 W. Main St., 49006. Phone: (269) 345-7121. Fax: (269) 345-1436. Web Site: www.wqsn.com. Licensee: Midwest Communications Inc. (acq 5-1-2006; grpsl). Format: Sports. ♦Duey Wright, pres; Dennis Martin, exec VP & gen sls mgr; Peter Tanz, gen mgr; Geary Morrill, opns dir & chief of engrg.

WRKR(FM)—Portage, Oct 13, 1988: 107.7 mhz; 50 kw. 500 ft TL: N42 07 43 W85 20 16. Stereo. Hrs opn: 24 Box 50911, 49005-0911. Secondary address: 4154 Jennings Dr. 49001-1087. Phone: (269) 344-0111. Fax: (269) 344-4223. E-mail: radio@wrkr.com Web Site: www.wrkr.com. Licensee: Cumulus Licensing Corp. Group owner: Cumulus Media Inc. (acq 5-26-98; grpsl). Population served: 700,000 Format: Classic rock, AOR. News staff: 2; News: 4 hrs wkly. Target aud: 25-54. Spec prog: Blues 5 hrs, jazz 4 hrs wkly. ♦Lew Dickey, CEO & pres; John Pinch, COO; Martin Gausvik, CFO; Mike McKelly, opns mgr; Jay Deacon, progmg dir & disc jockey; Dale Schiesser, chief of engrg.

WVFM(FM)— June 19, 1964: 106.5 mhz; 33 kw. Ant 600 ft TL: N42 28 32 W85 29 22. Stereo. Hrs opn: 4200 W. Main St., 49006. Phone: (269) 345-7121. Fax: (269) 345-1436. Web Site: www.wqlr.com. Licensee: Midwest Communications Inc. (acq 5-1-2006; grpsl). Format: Adult contemp. Target aud: 25-54; women. ♦Duey Wright, pres; Peter Tanz, gen mgr; Dennis Martin, sls VP; Ken Lanphear, progmg dir; Walker Sisson, chief of engrg.

Kalkaska

WKLT(FM)— Apr 8, 1979: 97.5 mhz; 32 kw. 670 ft TL: N44 47 29 W85 14 20. Stereo. Hrs opn: 24 1020 Hastings St., Traverse City, 49686. Phone: (231) 947-0003. Fax: (231) 947-7002. Web Site: www.wklt.com. Licensee: Northern Radio of Michigan (acq 1-82; $320,000; FTR: 1-18-82). Natl. Rep: Christal. Law Firm: Fletcher, Heald & Hildreth. Format: Classic AOR. News: 2 hrs wkly. Target aud: 25-54; baby boomers & young adults. Spec prog: Sunday night classics, blues 2 hrs wkly. ♦Charlie Ferguson, pres & gen mgr; Jackie Gordon, natl sls mgr; Terri Ray, progmg dir; Dennis Murray, chief of engrg.

Kentwood

WGVU(AM)—Licensed to Kentwood. See Grand Rapids

WJNZ(AM)— Sept 18, 1978: 1140 khz; 5 kw-D, DA. TL: N42 56 13 W85 27 20. Hrs open: 15 hrs 1919 Eastern Ave. S.E., Grand Rapids, 49507. Phone: (616) 475-4299. Fax: (616) 475-4335. E-mail: mjs@wjnz.com Web Site: www.wjnz.com. Licensee: WJNZ Radio L.L.C. (acq 9-22-2003; $360,000). Population served: 703,400 Natl. Network: ABC, Premiere Radio Networks. Law Firm: Koerner, & Olender. Format: Urban, rhythm and blues. News: Top of the hour 6a-7p. Target aud: 25-54; Baby Boomers. Spec prog: Jazz 6 hrs, gospel 4 hrs. ♦Mike St. Cyr, pres & gen mgr.

Kingsford

***WEUL(FM)**— Feb 11, 1990: 98.1 mhz; 240 w. 482 ft TL: N45 49 58 W88 04 57. Hrs open:
Simulcast with WHWL(FM) Marquette, WHWG(FM) Trout Lake.
130 Carmen Dr., Marquette, 49855. Phone: (906) 249-1423. Fax: (906) 249-4042. E-mail: gospelop@chartermi.net Web Site: www.gospelopportunities.com. Licensee: Gospel Opportunities Inc. Population served: 50,000 Format: Relg. News staff: 3. ♦W. Curtis Marker, gen mgr & progmg dir.

Kingsley

WJNL(AM)— Apr 17, 1947: 1210 khz; 50 kw-D, 2.5 kw-CH. TL: N44 33 34 W85 35 37. Hrs open: 2175 Click Rd., Petoskey, 49770-8818. Secondary address: 310 West Front St., Traverse City 49684. Phone: (231) 947-1210. Web Site: www.wjml.com. Licensee: Stone Communications Inc. (group owner; (acq 4-25-2007; . swap with WLDR(AM) Petoskey). Population served: 510,000 Natl. Network: USA. Format: Talk, sports, news. News: 72 hrs wkly. Target aud: 25 plus. ♦Philip Clever, stn mgr; Richard Stone, pres, gen mgr & gen sls mgr.

Lake City

***WAIR(FM)**—Not on air, target date: Fall 2003: 104.9 mhz; 1.6 kw. Ant 489 ft TL: N44 14 56 W85 18 48. Hrs open: Box 388, Williamston, 48895. Phone: (517) 381-0573. Fax: (877) 850-0881. E-mail: info@positivehits.com Web Site: www.positivehits.com. Licensee: Superior Communications. ♦Jenn Czelada, gen mgr.

Lakeview

WSCG-FM— November 1989: 106.3 mhz; 3 kw. 328 ft TL: N43 24 33 W85 15 53. Stereo. Hrs opn: Box 578, Greenville, 48838. Secondary address: 9181 S. Greenville Rd., Greenville 48838. Phone: (616) 754-3656. Fax: (616) 754-2390. Licensee: Stafford Broadcasting L.L.C. (acq 10-19-2004; with WSCG(AM) Greenville). Format: Country. ♦Bruce Bentley, gen mgr & opns mgr; John Clark, gen mgr.

L'Anse

WCUP(FM)— Jan 1, 1998: 105.7 mhz; 50 kw. 492 ft TL: N46 46 48 W88 32 06. Stereo. Hrs opn: 24 805 B US 41, Baraga, 49908. Phone: (906) 353-7625. Fax: (906) 353-9200. E-mail: wcupprod@up.net Web Site: www.wcupfm.com. Licensee: Keweenaw Bay Indian Community (acq 5-31-01; $176,000 for debt for 70%). Natl. Network: ABC. Format: Country. News: one hr wkly. Target aud: 18 plus. Spec prog: American Indian 2 hrs wkly. ♦Ed Janisse, gen mgr; John Preston, sls dir.

Lansing

WHZZ(FM)—Listing follows WILS(AM).

WILS(AM)— July 17, 1947: 1320 khz; 5 kw-D, 1 kw-N, DA-2. TL: N42 41 30 W84 33 38. Hrs open: Box 25008, 48909. Secondary address: 600 W. Cavanaugh Rd. 48910. Phone: (517) 393-1320. Fax: (517) 393-0882. Web Site: www.1320wils.com. Licensee: MacDonald Broadcasting Co. (group owner; (acq 12-20-2001; grpsl). Population served: 131,456 Natl. Rep: D & R Radio. Format: Talk. ♦Rick Sarata, gen mgr & gen sls mgr.

WHZZ(FM)—Co-owned with WILS(AM). January 1967: 101.7 mhz; 4 kw. 400 ft TL: N42 43 42 W84 30 54. Stereo. 24 Web Site: www.1017fm.com.350,000 Format: Adult hits. Target aud: 25-54; upscale adults. ♦Gary Harding, engrg dir & chief of engrg; Shane Pitman, traf mgr.

WITL-FM— Apr 15, 1964: 100.7 mhz; 26.5 kw. 640 ft TL: N42 40 33 W84 30 00. Stereo. Hrs opn: 3200 Pine Tree Rd., 48911. Phone: (517) 393-1010. Fax: (517) 394-3391. E-mail: witl@acd.net Web Site: www.witl.com. Licensee: Citadel Broadcasting Co. Group owner: Citadel Broadcasting Corp. (acq 2000; grpsl). Population served: 379,500 Natl. Rep: Christal. Law Firm: Leventhal,Senter& Lerman. Format: Country. News staff: one. Target aud: 25-54. ♦Farid Suleman, chmn; Rob Striker, gen mgr; Brent Alberts, opns mgr; Kelly Norton, sls dir; Chris Potter, rgnl sls mgr; Jordan Lee, prom mgr; Jay J. McCrae, progmg dir; Rick Housley, chief of engrg; Steve Goupil, traf mgr.

WJIM(AM)— 1934: 1240 khz; 1 kw-U. TL: N42 44 22 W84 30 39. (CP: 890 w). Hrs opn: 24 3420 Pine Tree Rd., 48911. Phone: (517) 394-7272. Fax: (517) 394-3391. E-mail: wjim@acd.net Licensee: Citadel Broadcasting Co. Group owner: Citadel Broadcasting Corp. (acq 2000; grpsl). Population served: 379,500 Natl. Network: Westwood One, ABC. Natl. Rep: Christal. Law Firm: Leventhal, Senter & Lerman. Format: News, info, talk. News staff: one; News: 4 hrs wkly. Target aud: 25-54. ♦Farid Suleman, CEO & chmn; Rob Striker, gen mgr; Ray Marshal, opns mgr; Deb Hart, news dir; Linda Karl, traf mgr.

WJIM-FM— June 1960: 97.5 mhz; 45 kw. Ant 512 ft TL: N42 40 33 W84 30 00. Stereo. 24 Web Site: www.new975.com.500,000 Natl. Network: CNN Radio. Format: CHR. News staff: one; News: 2 hrs wkly.

WJXQ(FM)—See Jackson

WJZL(FM)—See Charlotte

***WLNZ(FM)**— Feb 11, 1994: 89.7 mhz; 100 w. 98 ft TL: N42 44 16 W84 33 09. Hrs open: Lansing Community College, 400 N. Capitol Ave., Suite 001, 48933. Phone: (517) 483-1710. Phone: (517) 483-9897. Fax: (517) 483-1894. E-mail: wlnz@lansing.cc.mi.us Web Site: www.lcc.edu/wlnz. Licensee: Lansing Community College. Population served: 500,000 Natl. Network: PRI, NPR. Format: Jazz, blues, AAA. Spec prog: Reggae 4 hrs, big band 3 hrs, folk 3 hrs, Sp 4 hrs wkly. ♦Dave Downing, gen mgr; Lyn Peraino, progmg dir; Dae Lowry, mus dir; Lyle Layin, chief of engrg.

WMMQ(FM)—See East Lansing

WQTX(FM)—See Saint Johns

WVFN(AM)—See East Lansing

WWSJ(AM)—See Saint Johns

Lapeer

WLCO(AM)— Nov 16, 1962: 1530 khz; 5 kw-D, DA-D. TL: N43 01 35 W83 17 12. Hrs open: 3338 E. Bristol Rd., Burton, 48529. Phone: (810) 743-1080. Fax: (810) 742-7170. Licensee: Regent Broadcasting of Flint Inc. Group owner: Regent Communications Inc. (acq 7-18-2002; $1.3 million with co-located FM). Population served: 10,000 Natl. Network: ABC. Natl. Rep: Patt. Law Firm: Earl Stanley. Format: Country. Target aud: 35 plus. ♦Zoe Burdine-Fly, gen mgr.

WQUS(FM)—Co-owned with WLCO(AM). Feb 6, 1968: 103.1 mhz; 3 kw. 299 ft TL: N43 04 49 W83 11 30. Stereo. Web Site: www.us103.com. Format: Adult rock. Target aud: 25-40; college educated men & women. Spec prog: AOR, gospel, blues. ♦David Corley, sls dir; Mr. Brian Beddow, progmg dir; Tony LaBrie, mus dir.

***WMPC(AM)**— Dec 6, 1926: 1230 khz; 1 kw-U. TL: N43 04 46 W83 18 35. Hrs open: 24 Box 104, 1800 N. Lapeer Rd., 48446. Phone: (810) 664-6211. Fax: (810) 664-5361. E-mail: wmpc@chartermi.net Web Site: http://lapeer.org/ServiceOrg/WMPC. Licensee: The Calvary Bible Church of Lapeer Inc. Population served: 200,000 Natl. Network: Moody. Wire Svc: AP Format: Relg. News staff: one; News: 24 hrs wkly. Target aud: General. ♦Bob Baldwin, gen mgr & opns dir.

Leland

WFCX(FM)— Aug 9, 1991: 94.3 mhz; 3.6 kw. 426 ft TL: N44 54 48 W85 49 18. (CP: 14.88 kw). Stereo. Hrs opn: 24
Rebroadcasts WFDX(FM) Atlanta 100%.
1020 Hastings, Traverse City, 49686. Phone: (231) 947-0003. Fax: (231) 947-7002. Licensee: Northern Michigan Radio Inc. (acq 12-23-93; $1.1 million with WFDX(FM) Atlanta, MI; FTR: 1-17-94). Natl. Rep: Christal. Law Firm: Fletcher, Heald & Hildreth. Format: Classic Hits. News staff: one; News: 7 hrs wkly. Target aud: 25-54. Spec prog: Relg one hr wkly. ♦Charlie Ferguson, gen mgr.

Leroy Township

***WLGH(FM)**— December 1996: 88.1 mhz; 2.5 kw. Ant 328 ft TL: N42 42 20 W84 21 25. Hrs open: Box 388, Williamston, 48895. Phone: (517) 381-0573. Fax: (877) 850-0881. E-mail: info@positivehits.com Web Site: www.positivehits.com. Licensee: Superior Communications. Format: Adult, Christian hit radio. ♦Jenn Czelada, gen mgr.

Lexington

WBTI(FM)—Licensed to Lexington. See Port Huron

Linwood

WSAG(FM)— Nov 1, 2002: 104.1 mhz; 4.6 kw. Ant 325 ft TL: N43 43 30 W83 56 50. Hrs open: 2000 Whittier St., Saginaw, 48601. Phone: (989) 752-8161. Fax: 989-752-8102. Web Site: thebay104fm.com. Licensee: MacDonald Boadcasting Co. (acq 6-30-2005). Format: Soft rock. ♦Duane Alverson, pres & gen mgr; Mike Skot, progmg dir; Gary Harding, chief of engrg.

Livonia

WCAR(AM)— Oct 23, 1963: 1090 khz; 250 w-D, 500 w-N, DA-2. TL: N42 19 46 W83 21 43. Hrs open: 32500 Park Ln., Garden City, 48135. Phone: (734) 525-1111. Fax: (734) 525-3608. Licensee: 1090 Investments L.L.C. (acq 7-6-98; $2 million). Population served: 500,000 Format: Catholic. Target aud: 25 plus. Spec prog: Ethnic. ♦John F.X. Browne, pres.

Ludington

WKLA(AM)— Oct 9, 1944: 1450 khz; 1 kw-U. TL: N43 57 05 W86 25 28. Hrs open: 24 5941 W. U.S. 10, 49431. Phone: (231) 843-3438. Fax: (231) 843-1886. Web Site: www.wkla.com. Licensee: Lake Michigan Broadcasting Inc. (group owner; (acq 9-20-96; grpsl).

Population served: 28,000 Format: News/talk. News staff: one; News: 4 hrs wkly. Target aud: 40 plus; mature adults. ♦Lynn Baerwolf, gen mgr & prom mgr; Jason Wilder, opns mgr & mus dir; Alan Neushwander, news dir & chief of engrg; Richard Young, sls dir & mus critic.

WKLA-FM— May 1971: 106.3 mhz; 6 kw. 400 ft TL: N43 03 30 W86 24 59. Stereo. Web Site: www.wkla.com.70,000 Natl. Network: ABC. Rgnl rep: Patt Media Law Firm: Shaw Pittman. Format: Adult contemp. News staff: 2; News: 3 hrs wkly. Target aud: 25-50. ♦Richard Young, mktg dir & local news ed; Rod Beckman, sports cmtr.

WKZC(FM)—See Scottville

Luna Pier

WTWR-FM— July 16, 1967: 98.3 mhz; 3.4 kw. Ant 443 ft TL: N41 40 05 W83 27 11. Stereo. Hrs opn: 24 14930 Laplaisance Rd. Suite 113, Monroe, 48161. Phone: (734) 242-6600. Fax: (734) 242-6599. Web Site: www.tower983.com. Licensee: Cumulus Licensing Corp Group owner: Cumulus Media L.L.C. (acq 7-98; $2.8 million). Population served: 300,000 Natl. Rep: Michigan Spot Sales. Law Firm: Crowell & Moring. Format: CHR. News staff: one; News: 3 hrs wkly. Target aud: 25-54. Spec prog: Relg 3 hrs wkly. ♦Bill Bailey, gen mgr.

Mackinaw City

***WIAB(FM)**— Oct 1, 2000: 88.5 mhz; 20 kw. Ant 430 ft TL: N45 40 00 W84 38 05. Hrs open:
Simulcast with WIAA(FM) Interlochen 100%.
Box 199, Interlochen, 49643. Phone: (231) 276-4400. Fax: (231) 276-4417. Web Site: www.interlochen.org/pr. Licensee: Interlochen Center for the Arts (acq 3-14-2005; $580,000). Format: Classical, news. ♦Thom Paulson, stn mgr.

WLJZ(FM)— Sept 6, 1989: 94.5 mhz; 40 kw. Ant 380 ft TL: N45 40 00 W84 38 05. Stereo. Hrs opn:
Simulcasts with WJZJ(FM) Glen Arbor.
1356 Mackinaw Ave., Cheboygan, 49721. Phone: (231) 627-2341. Fax: (231) 627-7000. Web Site: www.modernrockthezone.com. Licensee: Northern Star Broadcasting L.L.C. (group owner; (acq 11-1-2005; grpsl). Format: Modern rock. ♦Palmer Pyle, pres; Chris Monk, VP; Chris Monk, gen mgr.

Manistee

WMLQ(FM)— Aug 1, 1970: 97.7 mhz; 2.5 kw. Ant 515 ft TL: N44 12 40 W86 17 53. Stereo. Hrs opn: 24 PO Box 855, Ludington, 49431. Phone: (231) 723-9462. Fax: (231) 843-9411. Web Site: www.977coastfm.com. Licensee: Synergy Media Inc. (acq 6-1-2006; $380,000). Population served: 40,000 Rgnl. Network: Ohio Radio Net. Natl. Rep: Michigan Spot Sales. Rgnl rep: Michigan Spot Sales Law Firm: Drinker, Biddle & Reath. Format: Soft adult contemp/easy lstng. Target aud: 35-64; Adults. ♦Todd Mohr, gen mgr; Mary Mohr, opns mgr; Melissa Reed, gen sls mgr; Tom Green, chief of engrg.

WMTE(AM)— June 7, 1951: 1340 khz; 1 kw-U. TL: N44 14 07 W86 19 05. Hrs open: 24 52 Greenbush St., 49660. Phone: (231) 723-9906. Fax: (231) 723-9908. Licensee: Lake Michigan Broadcasting Inc. (group owner; (acq 9-20-96; grpsl). Population served: 40,000 Natl. Network: ABC. Natl. Rep: Michigan Spot Sales. Format: News/talk. News staff: one. Target aud: General. ♦Judith Ouvry, stn mgr.

WMTE-FM— June 22, 1994: 101.5 mhz; 3 kw. 115 ft TL: N44 12 18 W86 17 22. Stereo. Hrs opn: 24 52 Greenbush, 49660. Phone: (231) 723-9906. Fax: (231) 723-9908. E-mail: judy@wkla.com Web Site: www.oldies1015.com. Licensee: Lake Michigan Broadcasting Inc. (group owner; (acq 2000; $300,000). Population served: 14056 Rgnl rep: Patt Media Format: Oldies. News staff: one. Spec prog: Pol 6 hrs, relg 2 hrs wkly. ♦Judith Ouvry, stn mgr, opns dir, sls dir & progmg dir; Mike Baerwolf, progmg dir.

Manistique

WPIQ(FM)— 2005: 99.9 mhz; 6 kw. Ant 151 ft TL: N45 58 13 W86 11 06. Hrs open: 2025 U.S. Hwy. 41 W., Marquette, 49855. Phone: (201) 221-4251. Fax: (906) 228-8128. E-mail: toddn@greatlakesradio.org Web Site: www.wpiqradio.com. Licensee: Todd Stuart Noordyk. Group owner: Great Lakes Radio Inc. Natl. Network: Premiere Radio Networks, ABC. Format: News/talk. News staff: 2. ♦Todd Noordyk, gen mgr; Devin Lawrence, stn mgr.

WTIQ(AM)— Feb 11, 1968: 1490 khz; 1 kw-U. TL: N45 57 51 W86 16 37. Stereo. Hrs opn: 7876W County Rd. 442, 49854-9000. Phone: (906) 341-8444. Fax: (906) 341-6222. E-mail: wtiq@chartermi.net Web Site: www.radioresultsnetwork.com. Licensee: Lakes Radio Inc. (group owner; acq 11-30-99; grpsl). Population served: 38,000 Rgnl. Network: MNN. Law Firm: Meyer, Faller, Weisman & Rosenberg, P. Format: Oldies. Target aud: 25-54; blue & white collar. ♦Rick Duerson, gen mgr; L. David Vaughan, stn mgr.

Marine City

WHLX(AM)—Licensed to Marine City. See Port Huron

Marlette

WBGV(FM)— July 25, 1999: 92.5 mhz; 3 kw. TL: N43 17 10 W82 58 17. Stereo. Hrs opn: 24 1260 Yosemite Blvd., Birmingham, 48009. Secondary address: 19 S. Elk St., Sandusky 48422. Phone: (810) 648-2700. Phone: (248) 540-3380. Fax: (248) 540-3379. E-mail: gebv@aol.com Licensee: GB Broadcasting Co. (acq 6-5-92). Population served: 35,000 Natl. Network: ABC. Format: Country. News staff: one; News: 2 hrs wkly. Target aud: General. ♦George Benko V, pres; Robert Armstrong, gen mgr & opns mgr; George Benko, reporter.

***WMSQ(FM)**— 2004: 89.3 mhz; 100 w. Ant 98 ft TL: N43 22 06 W83 07 00. Hrs open: 901 Elizabeth Ct., Mount Pleasant, 48858. Phone: (989) 779-9178. Licensee: Great Lakes Community Broadcasting Inc. Format: Oldies. ♦James McClusky, gen mgr.

Marquette

WDMJ(AM)— July 1, 1931: 1320 khz; 5 kw-D, 1 kw-N, DA-N. TL: N46 32 40 W87 26 42. Hrs open: 1009 W. Ridge St., Suite A, 49855-3963. Phone: (906) 225-1313. Fax: (906) 225-1324. Web Site: www.wjpd.com. Licensee: Northern Star Broadcasting L.L.C. (group owner; acq 11-5-01; grpsl). Population served: 75,000 Natl. Rep: Michigan Spot Sales. Format: News/talk. Target aud: 25-54. ♦Tammy Johnson, gen mgr.

WFXD(FM)— Apr 6, 1974: 103.3 mhz; 100 kw. Ant 938 ft TL: N46 36 14 W87 37 15. Stereo. Hrs opn: 24 2025 US 41 W., 49855. Phone: (906) 228-6800. Fax: (906) 228-8128. E-mail: toddn@greatlakesradio.org Web Site: www.wfxd.com. Licensee: Great Lakes Radio Inc. (group owner; (acq 11-30-99; grpsl). Format: Hot country hits. Target aud: 25-54. ♦Todd Noordyk, gen mgr.

***WHWL(FM)**— Dec 16, 1965: 95.7 mhz; 100 kw. Ant 531 ft TL: N46 29 52 W87 24 59. Stereo. Hrs opn: 130 Carmen Dr., 49855. Phone: (906) 249-1423. Fax: (906) 249-4042. E-mail: whwl@whwl.net Web Site: www.whwl.net. Licensee: Gospel Opportunities Inc. (acq 4-19-76). Population served: 150,000 Format: Relg. ♦W. Curtis Marker, gen mgr & progmg dir.

WIAN(AM)—See Ishpeming

***WNMU-FM**— August 1963: 90.1 mhz; 100 kw. 930 ft TL: N46 21 09 W87 51 32. Stereo. Hrs opn: 24 Learning Resources Ctr., Northern Michigan Univ., 1401 Presque Isle Ave., 49855. Phone: (906) 227-2600. Fax: (906) 227-2905. Web Site: www.nmu.edu/wnmufm. Licensee: Board of Trustees of Northern Michigan University. Population served: 250,000 Natl. Network: NPR, PRI. Rgnl. Network: Mich. Pub. Law Firm: Cohn & Marks. Wire Svc: AP Format: Class, jazz, news. News: 31 hrs wkly. Spec prog: Educ. ♦Eric Smith, gen mgr; Evelyn Massaro, stn mgr. Co-owned TV: *WNMU-TV affil

WUPK(FM)— May 1, 1992: 94.1 mhz; 4.4 kw. Ant 380 ft TL: N46 30 51 W87 28 58. Hrs open: 24
Rebroadcasts WIMK(FM) Iron Mountain 100%.
1009 W. Ridge St., Suite A, 49855-3963. Phone: (906) 225-1313. Fax: (906) 225-1324. Licensee: Northern Star Broadcasting L.L.C. (group owner; acq 11-5-01; grpsl). Natl. Rep: Patt. Law Firm: Reddy, Begley & McCormick. Format: Classic rock, AOR. News staff: one; News: 3 hrs wkly. Target aud: 25-54; baby boomers. ♦Chris Monk, VP; Tammy Johnson, gen mgr.

***WUPX(FM)**— 1994: 91.5 mhz; 200 w. 138 ft TL: N46 34 44 W87 23 42. Hrs open: 24 Northern Michigan Univ., 1204 University Ctr., 49855. Phone: (906) 227-2348. Phone: (906) 227-1844. Fax: (906) 227-2344. E-mail: wupx@nmu.edu Web Site: www.wupx.com. Licensee: Board of Control of Northern Michigan University. Format: Alternative. Target aud: College. Spec prog: Black 6 hrs, jazz 2 hrs wkly. ♦Troy Hanson, gen mgr. Co-owned TV: *WNMU-TV affil

Marshall

WRCC(FM)—Licensed to Marshall. See Albion

Mason

***WUNN(AM)**— May 11, 1967: 1110 khz; 1 kw-D, DA. TL: N42 33 04 W84 24 15. Hrs open: Sunrise-sunset 13799 Donovan Rd., Albion, 49224. Phone: (800) 776-1070. Fax: (517) 531-5009. E-mail: wunn@flc.org Web Site: www.967flr.org. Licensee: Family Life Broadcasting System. (group owner; (acq 1-1-69). Population served: 250,000 Format: Southern gospel. News: 1.5 hrs wkly. Target aud: 25-54; Christian families. ♦Randy Carlson, pres; Dave Phelps, gen mgr & dev mgr; Rod Robison, dev dir; Dick Lindley, chief of engrg.

McMillan

WMJT(FM)— 2006: 96.7 mhz; 50 kw. Ant 413 ft TL: N46 32 02 W85 35 24. Hrs open: Box 486, Newberry, 49868-0486. Secondary address: 210 W. John St., Newberry 49868-1125. Phone: (906) 293-1400. Fax: (906) 293-5161. E-mail: kent@radioeagle.com Licensee: David L. Smith. Format: Adult hits. ♦Kent Smith, gen mgr.

Menominee

WAGN(AM)— Nov 14, 1952: 1340 khz; 1 kw-U. TL: N45 06 27 W87 36 25. Hrs open: 20 N 2880 Roosevelt Rd., Marinette, WI, 54143. Phone: (906) 863-5551. Fax: (906) 863-5679. Licensee: Armada Media - Menominee Inc. (group owner; (acq 12-19-2006; grpsl). Population served: 55,000 Natl. Network: Moody. Law Firm: McCabe & Allen. Format: Oldies. News staff: 2; News: 15 hrs wkly. Target aud: 30 plus; older, affluent adults. Spec prog: CBS radio sports 6 hrs, loc sports hrs wkly. ♦Jeff Wagner, gen mgr & stn mgr.

WHYB(FM)—Co-owned with WAGN(AM). Oct 24, 1984: 103.7 mhz; 3 kw. 300 ft TL: N45 04 00 W87 39 55. (CP: 7.01 kw). Stereo. 24 Format: Country. News staff: 2; News: 5 hrs wkly. Target aud: 35-64. ♦Don Kitokowski, gen mgr.

WMAM(AM)—See Marinette, WI

Michigamme

***WKPK(FM)**—Not on air, target date: unknown: 88.1 mhz; 100 kw vert. Ant 410 ft TL: N46 28 30 W88 14 06. Hrs open: 3302 N. Van Dyke Rd., Imlay City, 48444. Phone: (810) 724-2638. Web Site: www.joyfm.net. Licensee: Northland Community Broadcasters. ♦Edward T. Czelada, pres; Jennifer Czelada, gen mgr.

Midland

WKQZ(FM)— Dec 14, 1976: 93.3 mhz; 39.2 kw. 554 ft TL: N43 50 46 W84 05 32. Stereo. Hrs opn: 1740 Champagne Dr., Saginaw, 48604. Phone: (989) 776-2100. Fax: (989) 754-5990. Web Site: www.z93kqz.fm. Licensee: Citadel Broadcasting Co. Group owner: Citadel Broadcasting Corp. (acq 2-8-99; grpsl). Natl. Rep: McGavren Guild. Law Firm: Reddy, Begley & McCormick. Format: Rock. Target aud: 25-44; males. ♦Scott Meier, pres & gen mgr; Bruce Johnston, sls dir; Stan Parman, progmg dir; Jay Randall, mus dir; Hal Maas, news dir; Bob Friedle, chief of engrg.

WMPX(AM)— Sept 11, 1948: 1490 khz; 1 kw-U, DA-2. TL: N43 36 48 W84 13 17. Hrs open: 24 Box 1689, 48641. Secondary address: 1510 Bayliss St. 48640. Phone: (989) 631-1490. Phone: (989) 631-2220. Fax: (989) 631-6357. E-mail: request@wmpxwmrx.com Web Site: www.wmpxwmrx.com. Licensee: Steel Broadcasting Inc. (acq 8-19-81; $900,000; FTR: 8-24-81). Population served: 38,500 Natl. Network: ABC. Natl. Rep: Patt. Format: Adult standards, big band. News staff: one; News: 9 hrs wkly. Target aud: General. Spec prog: Sounds of Sinatra 2 hrs wkly, relg 3 hrs wkly. ♦Thomas Steel, pres & gen mgr; Aaron Whiting, gen sls mgr.

***WUGN(FM)**— Dec 2, 1973: 99.7 mhz; 100 kw. 997 ft TL: N43 30 56 W84 32 49. Stereo. Hrs opn: 510 E. Isabella Rd., 48640. Phone: (989) 631-7060. Fax: (989) 631-4825. E-mail: 997@997.org Web Site: www.997.org. Licensee: Family Life Communications System. (acq 1996). Population served: 1,000,000 Natl. Network: Salem Radio Network. Format: Contemp Christian. News: 8 hrs wkly. Target aud: 35-54; female with young children. ♦Peter Brooks, gen mgr.

Mio

WAVC(FM)— Oct 1, 1994: 93.9 mhz; 50 kw. 433 ft TL: N44 43 40 W84 21 35. Stereo. Hrs opn:
Rebroadcasts WMKC(FM) Saint Ignace.
1356 Mackinaw Ave., Cheboygan, 49721. Phone: (231) 627-2341. Fax: (231) 627-7000. Web Site: www.bigcountry1029.com. Licensee: Northern Star Broadcasting L.L.C. (group owner; (acq 9-11-98; grpsl). Format: Country. ♦Palmer Pyle, pres; Chris Monk, VP & gen mgr.

Monroe

WCSX(FM)—See Birmingham

***WDTR(FM)**— 2003: 88.1 mhz; 910 w. Ant 144 ft TL: N41 55 08 W83 22 34. Hrs open:
rebroadcast of WHYT(FM) Imlay city 100%.
Box 388, Williamston, 48895. Phone: (810) 721-0891. Fax: (413) 410-9708. E-mail: info@joyfm.net Web Site: www.joyfm.net. Licensee: Northland Community Broadcasters. Format: Contemporary Christian. ♦Jenn Czelada, gen mgr.

WRDT(AM)— July 12, 1956: 560 khz; 500 w-D, 27 w-N. TL: N41 53 28 W83 25 39. (CP: 14 w-N). Hrs opn: 24 12300 Radio Pl., Detroit, 48228. Phone: (313) 272-3434. Web Site: wmuz.com. Licensee: WMUZ Radio Inc. Group owner: Crawford Broadcasting Co. (acq 6-16-97; $3.15 million). Population served: 100,000 Natl. Rep: McGavren Guild. Format: Bible teaching. ♦Frank Franciosi, gen mgr & gen sls mgr.

***WYDM(FM)**— November 1978: 97.5 mhz; 8 w. Ant 135 ft TL: N41 55 07 W83 26 12. Hrs open: Monroe High School, 901 Herr Rd., 48161. Phone: (734) 265-3550. Licensee: Monroe Public Schools (acq 8-77). Format: Var/div. Target aud: 15-24.

Mount Clemens

WHTD(FM)— Nov 6, 1960: 102.7 mhz; 50 kw. 499 ft TL: N42 32 39 W82 54 09. Stereo. Hrs opn: 24 3250 Franklin, Detroit, 48207. Phone: (313) 259-2000. Fax: (313) 259-7011. Web Site: www.kissdetroit.com. Licensee: Radio One of Detroit LLC. Group owner: Radio One Inc. (acq 1999; $27 million). Natl. Network: ABC. Natl. Rep: D & R Radio. Format: Urban / AC. Target aud: 25-49; rock and rollers of all ages. ♦Carol Lawrence-Dobrusin, gen mgr.

Mount Pleasant

WCFX(FM)—See Clare

***WCMU-FM**— Apr 6, 1964: 89.5 mhz; 100 kw. 423 ft TL: N43 34 24 W84 46 21. Stereo. Hrs opn: 24 Public Broadcasting Ctr., 1999 E. Campus Dr., 48859. Phone: (989) 774-3105. Fax: (989) 774-4427. E-mail: cmuradio@radio.cmich.edu Web Site: www.wcmu.org. Licensee: Central Michigan University. Population served: 45,000 Natl. Network: NPR, PRI. Rgnl. Network: Mich. Pub. Law Firm: Dow, Lohnes & Albertson. Format: Class, jazz, news & info. News staff: 2; News: 30 hrs wkly. Target aud: General. ♦Ed Grant, gen mgr. Co-owned TV: *WCMU-TV affil

WCZY-FM— Aug 20, 1991: 104.3 mhz; 3 kw. 328 ft TL: N43 35 39 W84 49 26. Stereo. Hrs opn: 24 4065 E. Wing Rd., 48858. Phone: (989) 772-9664. Fax: (989) 773-5000. E-mail: wczy@wczy.net Web Site: www.wczy.net. Licensee: Central Michigan Communications Inc. Natl. Network: Jones Radio Networks. Natl. Rep: Michigan Spot Sales. Rgnl rep: Patt. Law Firm: Reddy, Begley & McCormick. Format: Easy lstng, adult contemp. News staff: one; News: 6 hrs wkly. Target aud: 25-54. ♦Mike Carey, pres & gen mgr; Bob Peters, gen sls mgr; Bryan Carr, progmg dir & news dir; Lisa Johnson, traf mgr.

***WMHW-FM**— Nov 20, 1972: 91.5 mhz; 307 w. 112 ft TL: N43 35 12 W84 46 24. 13.0 KW. Stereo. Hrs opn: 24 180 Moore Hall, Central Michigan Univ., 48859. Phone: (989) 774-7287. Phone: (989) 774-3851. Fax: (989) 774-2426. E-mail: wmhw@mail.cmich.edu Web Site: www.bca.cmich.edu. Licensee: Board of Trustees, Central Michigan University. Population served: 188,800 Law Firm: Irwin, Campbell & Tannenwald. Wire Svc: AP Format: New age, progsv. News: 10 wkly. Target aud: 12-34. ♦Peter B. Orlik, gen mgr; Jerry Henderson, opns mgr; Randy Kapenga, chief of engrg.

Munising

WQXO(AM)— Sept 20, 1955: 1400 khz; 1 kw-U. TL: N46 24 30 W86 38 22. Hrs open: 24 2025 US Hwy. 41 W., Marquette, 49855. Phone: (906) 228-6800. Fax: (906) 228-8128. E-mail: toddn@greatlakesradio.org Web Site: wqxo.com. Licensee: Great Lakes Radio Inc. (group owner; (acq 11-30-99; grpsl). Population served: 3,677 Natl. Rep: Patt. Law Firm: Haley, Bader & Potts. Format: Good time oldies. Target aud: 25-54. ♦Todd Noordyk, gen mgr.

WRUP(FM)—Co-owned with WQXO(AM). June 21, 1974: 98.3 mhz; 32 kw. Ant 357 ft TL: N46 24 53 W86 40 27. Stereo. 24 Web Site: wrup.com.67,000 Natl. Network: Westwood One. Natl. Rep: Patt. Format: Classic rock. Target aud: 18-54.

Muskegon

WGVS(AM)— 1926: 850 khz; 1 kw-U, DA-1. TL: N43 08 05 W86 15 14. Hrs open: 24
Rebroadcasts WGVU(AM) Kentwood 100%.
c/o WGVU, 301 W. Fulton, Grand Rapids, 49504-6492. Phone: (616) 331-6666. Fax: (616) 331-6625. E-mail: wgvu@gvsu.edu Web Site: www.wgvu.org. Licensee: Grand Valley State University. (acq 4-9-99; with WGVS-FM Whitehall). Population served: 100,000 Natl. Network: NPR, PRI. Law Firm: Cohn & Marks. Format: News, info. ♦Michael T. Walenta, gen mgr; Ken Kolbe, opns mgr; Gary Hunt, gen sls mgr; Pamela Holtz, mktg mgr & prom mgr; Scott Vander Werf, mus dir; Fred Martino, news dir & pub affrs dir; Bob Lumbert, engrg dir; Ed Spier, traf mgr.

WGVS-FM— 1975: 95.3 mhz; 2 kw. 360 ft TL: N43 21 14 W86 19 38. Stereo.
Rebroadcasts WGVU-FM Allendale 100%.
Web Site: www.wgvu.edu. Format: Jazz, news. ♦Ed Spier, traf mgr.

WKBZ(AM)— June 15, 1947: 1090 khz; 1 kw-D. TL: N43 16 35 W86 15 10. Hrs open: 6 AM-2 hrs past sunset 3565 Green St., 49444. Phone: (231) 733-2600. Fax: (231) 733-7461. E-mail: info@talkmuskegon.com Web Site: www.talkmuskegon.com. Licensee: CC Licenses LLC. Group owner: Clear Channel Communications Inc. (acq 1-17-2001; grpsl). Population served: 44,631 Natl. Rep: D & R Radio. Law Firm: John Garziglia. Format: Talk, news. News staff: one. Target aud: 25-54 primary; 35-64 secondary. ♦Bart Brandmiller, gen mgr.

WMUS(FM)—Co-owned with WKBZ(AM). 1962: 106.9 mhz; 50 kw. Ant 479 ft TL: N43 13 48 W86 05 03. Stereo. 24 Web Site: www.107mus.com.160,000

***WMCQ(FM)**— Mar 31, 2005: 91.7 mhz; 6 kw. Ant 328 ft TL: N43 18 37 W85 54 44. Hrs open: Drawer 2440, Tupelo, MS, 38801. Phone: (662) 844-8888. Fax: (662) 842-6791. Licensee: American Family Association. (acq 12-20-2002). Format: Chirstian. ♦Marvin Sanders, gen mgr.

WMHG(AM)— 1949: 1600 khz; 5 kw-U, DA-N. TL: N43 11 50 W86 13 22. Hrs open: 24 3565 Green St., 49441. Phone: (231) 733-2126. Fax: (231) 739-9037. Licensee: Cumulus Licensing Corp. Group owner: Cumulus Media Inc. (acq 3-15-00; grpsl). Population served: 50,000 Natl. Network: ABC. Natl. Rep: D & R Radio. Format: MOR. News: 2 hrs wkly. Target aud: 35 plus. ♦Bart Brandmiller, gen mgr; Caroljean Lindquist, opns mgr & prom VP; Greg Mckitrick, gen sls mgr; Don Beno, progmg dir & pub affrs dir; Jason Stephenitch, engrg dir.

***WPQZ(FM)**— 2003: 88.1 mhz; 580 w. Ant 344 ft TL: N43 16 38 W86 20 05. Hrs open: 901 Elizabeth Ct., Mount Pleasant, 48858. Phone: (989) 779-9178. Fax: (989) 779-1558. Licensee: Great Lakes Broadcast Academy Inc. (acq 1-13-2003). Format: Oldies. ♦James McCluskey, pres & gen mgr.

WSHZ(FM)— February 1990: 107.9 mhz; 15 kw. 348 ft TL: N43 17 41 W86 13 12. Hrs open: 24 3565 Green St., 49444. Phone: (231) 733-2600. Fax: (231) 739-9037. Fax: (213) 733-7461. Web Site: www.star108.com. Licensee: CC Licenses LLC. Group owner: Clear Channel Communications Inc. (acq 1-17-2001; grpsl). Population served: 350,000 Format: Adult contemp. Target aud: 18-49. ♦Bart Brandmiller, gen mgr; Don Beno, progmg dir; Ron Steenwyk, chief of engrg.

WSNX-FM— Nov 18, 1971: 104.5 mhz; 50 kw. 361 ft TL: N43 12 13 W86 01 49. (CP: 32 kw, ant 620 ft. TL: N43 12 16 W86 01 35). Stereo. Hrs opn: 24 77 Monroe Ctr., Suite 1000, Grand Rapids, 49503. Phone: (616) 459-1919. Fax: (616) 235-9104. E-mail: web@wsnx.com Web Site: www.wsnx.com. Licensee: CC Licenses LLC. Group owner: Clear Channel Communications Inc. (acq 9-30-99). Population served: 125,000 Natl. Network: ABC. Natl. Rep: Clear Channel. Format: CHR, urban contemp. News staff: one; News: news progrmg one hr wkly. Target aud: 18-34; women. ♦Skip Essick, VP & gen mgr.

Muskegon Heights

WMRR(FM)— Mar 29, 1974: 101.7 mhz; 15 kw. 305 ft TL: N43 16 38 W86 20 05. Stereo. Hrs opn: 24 3565 Green St., Muskegon, 49444. Phone: (231) 733-2600. Fax: (231) 739-9037. Web Site: www.wmrr.com. Licensee: CC Licenses LLC. Group owner: Clear Channel Communications Inc. (acq 1-17-2001; grpsl). Population served: 350,000 Natl. Network: Westwood One. Wire Svc: UPI Format: Classic rock. Target aud: 25-54; male. ♦Bart Brandmiller, gen mgr.

Negaunee

WKQS-FM— Jan 5, 1998: 101.9 mhz; 12 kw. Ant 1,007 ft TL: N46 36 14 W87 37 15. Stereo. Hrs opn: 2025 U.S. 41 W., Marquette, 49855. Phone: (906) 228-6800. Fax: (906) 228-8128. E-mail: toddn@greatlakesradio.org Web Site: www.wkqsfm.com. Licensee: Great Lakes Radio Inc. (group owner) Law Firm: Booth, Freret, Imlay & Tepper. Format: Today's hits, yesterday's favorites. ♦Todd Noordyk, gen mgr.

WNGE(FM)— 2001: 99.5 mhz; 3.6 kw. Ant 430 ft TL: N46 30 51 W87 28 58. Hrs open: 1009 W. Ridge St., Suite A, Marquette, 49855-3963. Phone: (906) 225-1313. Fax: (906) 225-1324. Licensee: Northern Star Broadcasting L.L.C. (group owner; acq 11-5-01; grpsl). Format: Oldies. ♦Tammy Johnson, gen mgr.

Newaygo

WLAW(FM)— Aug 15, 2005: 92.5 mhz; 2.25 kw. Ant 543 ft TL: N43 18 37 W85 54 44. Hrs open: 60 Monroe Ctr. N.W., 3rd Fl., Grand Rapids, 49503. Phone: (616) 774-8461. Fax: (616) 774-2491. Licensee: Citadel Broadcasting Co. (acq 7-8-2005). Format: Classic country. ♦Matt Hanlon, gen mgr.

Newberry

WIHC(FM)— Apr 24, 1989: 97.9 mhz; 50 kw. 352 ft TL: N46 18 53 W85 33 45. Stereo. Hrs opn: 24 1356 Mackinaw Ave., Cheboygan, 49721. Phone: (231) 627-2341. Fax: (231) 627-7000. Web Site: www.classicrockthebear.com. Licensee: Northern Star Broadcasting L.L.C. (group owner; (acq 9-11-98; grpsl). Population served: 183,000 Format: Classic rock. Target aud: 18-44. ♦Palmer Pyle, pres; Chris Monk, VP & gen mgr.

WNBY(AM)— May 16, 1966: 1450 khz; 1 kw-U. TL: N46 18 48 W85 30 38. Hrs open: Box 501, 49868. Secondary address: Hwy. S. M-123 49868. Phone: (906) 293-3221. Fax: (906) 293-8275. E-mail: wnby@up.net Licensee: Sovereign Communications LLC (acq 8-21-2003; $400,000 with co-located FM). Population served: 20,000 Natl. Rep: Patt. Format: Country Gold. Target aud: 35 plus. Spec prog: Polka 2 hrs wkly. ♦Travis Freeman, gen mgr.

WNBY-FM— 1977: 93.9 mhz; 50 kw. Ant 443 ft TL: N46 26 58 W85 06 04. Stereo. 25,000 Format: Oldies. Target aud: 25-45.

Niles

WAOR(FM)—Listing follows WNIL(AM).

WAUS(FM)—See South Bend, IN

WNIL(AM)— Dec 6, 1956: 1290 khz; 500 w-D, 44 w-N. TL: N41 49 22 W86 17 03. Hrs open: 24 Box 370, 237 Edison Rd., Mishawaka, IN,

46545. Phone: (269) 683-6123. Fax: (269) 683-2758. Web Site: www.1290wnil.com. Licensee: Pathfinder Communications Corp. Group owner: Federated Media (acq 7-21-99; $2 million with co-located FM). Population served: 50,000 Natl. Network: Jones Radio Networks. Law Firm: Wilkinson Barker Knauer. Format: Oldies. News: 5 hrs wkly. Target aud: 35-54 women; pro-active, community-involved people. Spec prog: Relg 6 hrs wkly. ♦J. Eric Plym, pres; Patrick Redd, gen mgr; Barb Deniston, gen sls mgr; Pam Reed, gen sls mgr; Clint Marsh, progmg dir; Heather Richards, news dir; Bob Henning, chief of engrg; Sara Charisse, news dir & news rptr; Ric Clingaman, disc jockey.

WAOR(FM)—Co-owned with WNIL(AM). Sept 13, 1968: 95.3 mhz; 3.3 kw. 298 ft TL: N41 49 22 W86 17 03. Stereo. 24 E-mail: waor@waor.com Web Site: www.waor.com.500,000 Format: Classic rock, rock/AOR. News: 5 hrs wkly. Target aud: 25-54; predominantly male, socially active, economically secure. ♦Stephanie Michel, gen sls mgr; Mike Ragozino, progmg dir & disc jockey.

North Muskegon

WLCS(FM)— November 1983: 98.3 mhz; 1.6 kw. Ant 456 ft TL: N43 18 50 W86 09 17. Stereo. Hrs opn: 24 3375 Merriam St., Muskegon, 49444. Phone: (231) 830-0176. Fax: (231) 830-0194. Licensee: Citadel Broadcasting Co. (group owner; (acq 9-28-2005; grpsl). Rgnl. Network: Patt. Format: Oldies. News: one hr wkly. Target aud: 35-54; general. ♦Jeff Morton, gen mgr.

Norway

WZNL(FM)— Mar 15, 1990: 94.3 mhz; 2.4 kw. 649 ft TL: N45 49 15 W88 02 25. Hrs opn: 101 E. Kent St., Iron Mountain, 49801-8110. Phone: (906) 774-4321. Fax: (906) 774-7799. E-mail: star943@uplogon.com Web Site: www.wznl.tripod.com. Licensee: Northern Star Broadcasting L.L.C. (group owner; acq 11-5-01; grpsl). Natl. Network: Westwood One. Format: Adult contemp. Target aud: 18-54. ♦Veronica Roberts, gen mgr; Tom Hill, opns mgr.

Novi

***WOVI(FM)**— Sept 4, 1978: 89.5 mhz; 100 w. 67 ft TL: N42 27 49 W83 29 28. Stereo. Hrs opn: 24 Novi High School, 24062 Taft Rd., 48375. Phone: (248) 449-1526. Fax: (248) 449-1519. Licensee: Board of Education Novi School District. (acq 3-8-76). Population served: 77,000 Format: Alternative, classic rock. Target aud: General. ♦Dave Legg, gen mgr.

Olivet

***WOCR(FM)**— Apr 22, 1975: 89.7 mhz; 110 w. 75 ft TL: N42 26 31 W84 55 30. Stereo. Hrs opn: 16 Kirk Ctr., Olivet College, 320 S. Main St., 49076. Phone: (269) 749-7598. Fax: (269) 749-7695. E-mail: wocr@olivetcollege.edu Licensee: Olivet College. Population served: 25,000 Format: CHR. News: one hr wkly. Target aud: College; high school and college-age, variety. ♦Jim Collins, gen mgr; Karolyn Batt, stn mgr; Garth Sims, opns mgr & chief of engrg.

Onsted

***WAQQ(FM)**— 2001: 88.3 mhz; 250 w vert. Ant 77 ft TL: N42 03 33 W84 12 54. Hrs opn: 901 Elizabeth Ct., Mount Pleasant, 48858. Phone: (989) 779-9178. Fax: (989) 779-1558. Licensee: Great Lakes Community Broadcasting Inc. Format: Oldies. ♦James J. McCluskey, gen mgr.

Ontonagon

***WOAS(FM)**— Nov 15, 1978: 88.5 mhz; 10 w. 124 ft TL: N46 52 30 W89 18 00. Hrs opn: 8 AM-10 PM (M-F) 701 Parker, 49953. Phone: (906) 884-4433. Fax: (906) 884-2742. E-mail: ken@oasd.k12.mi.us Web Site: www.woas-fm.org. Licensee: Ontonagon Area School District. Population served: 1,000 Format: Var/div. Target aud: General; local residents of the area. ♦Ken Raisanen, gen mgr.

WUPY(FM)— 1987: 101.1 mhz; 30 kw. 620 ft TL: N46 44 49 W89 11 27. Stereo. Hrs opn: 24 Box 265, 49953. Phone: (906) 884-9668. Fax: (906) 884-4985. E-mail: wupy@jamadots.com Web Site: www.wupy101.com. Licensee: SNRN Broadcasting Inc. (acq 7-90). Natl. Network: ABC. Format: Country. News staff: 2; News: 13 hrs wkly. Target aud: 25 plus. Spec prog: relg 3 hrs, polka 1 hr wkly. ♦Ken Waldrop, gen mgr; Jackie Dobbins, opns mgr & progmg dir; Jay Nix, gen sls mgr.

Orchard Lake

***WBLD(FM)**— May 28, 1974: 89.3 mhz; 10 w. 110 ft TL: N42 33 56 W83 21 32. Stereo. Hrs opn: 2 4925 Orchard Lake Rd., West Bloomfield, 48323. Phone: (248) 865-6754. Fax: (248) 865-6756. E-mail: wbld@hotmail.com Web Site: wbld893.tripod.com. Licensee: West Bloomfield Board of Education. Format: Div, rock. ♦Paul S. Townley, stn mgr; Randy G. Long, chief of engrg.

Oscoda

***WCMB-FM**— June 1998: 95.7 mhz; 25 kw. 699 ft TL: N44 40 30 W83 31 06. Hrs open: 24
Rebroadcasts WCMU-FM Mount Pleasant 100%.
Central Michigan Univ., 1999 E. Campus Dr., Mount Pleasant, 48859. Phone: (989) 774-3105. Fax: (989) 774-4427. E-mail: cmuradio@radio.cmich.edu Web Site: www.wcmu.org. Licensee: Central Michigan University. Natl. Network: NPR, PRI. Wire Svc: AP Format: Class, jazz, News. News staff: 2; News: 45 hrs wkly. ♦Ed Grant, gen mgr; Ann Blatte, dev dir & prom mgr; Art Curtis, gen sls mgr & rgnl sls mgr; Ray Ford, progmg dir; Randy Kapenga, chief of engrg.

WWTH(FM)— Aug 29, 1992: 100.7 mhz; 20.5 kw. 360 ft TL: N44 34 42 W83 22 40. Stereo. Hrs opn: 24 4429 N. US Hwy. 23, 48750. Phone: (989) 345-4611. E-mail: sunny100.7@kwcom.net Licensee: Edwards Communications LC. Group owner: Northern Radio Network (acq 12-21-2004; grpsl). Population served: 150,000 Natl. Rep: Roslin. Rgnl rep: Michigan. Law Firm: Carter, Ledyard & Milburn. Format: Soft adult contemp. News staff: one; News: 4 hrs wkly. Target aud: 25-54. ♦Lucky Osborn, gen mgr.

Otsego

WAKV(AM)— 1958: 980 khz; 1 kw-D. TL: N42 27 33 W85 43 58. Hrs open: 213 Gilkey St., Plainwell, 49080. Phone: (269) 685-2438. E-mail: 980am@net-link.net Web Site: www.980.com. Licensee: Vintage Radio Enterprises L.L.C. (acq 7-17-98; $17,500). Population served: 300,000 Format: Adult standards. ♦Jim Higgs, gen mgr.

WQXC-FM— Apr 17, 1981: 100.9 mhz; 3 kw. 299 ft TL: N42 30 31 W85 46 08. Stereo. Hrs opn: Box 80, 49078. Secondary address: 706 E. Allegan St. 49078. Phone: (269) 692-6851. Fax: (269) 692-6861. E-mail: tflynn@wqxc.com Web Site: www.wqxc.com. Licensee: Forum Communications Inc. Population served: 300,000 Format: Oldies. ♦Robert Brink, pres; Deb Whiteman, CFO; Tom Flynn, gen mgr, gen sls mgr & progmg dir; Todd Overhuel, mus dir; Jim McKinney, news dir & pub affrs dir.

Ovid-Elsie

***WOES(FM)**— Mar 21, 1978: 91.3 mhz; 553 w. 140 ft TL: N43 02 44 W84 23 14. Stereo. Hrs opn: 24 8989 Colony Rd., Elsie, 48831. Phone: (989) 834-2271. Fax: (989) 862-4463. Licensee: Ovid-Elsie Area Schools. (acq 1978). Population served: 3,000 Format: Polka. Spec prog: Class one hr, Pol one hr, Czech one hr, stage & screen one hr wkly. ♦George Bishop, gen mgr; Kevin Somers, opns mgr.

Owosso

WOAP(AM)— Jan 1, 1948: 1080 khz; 1 kw-D. TL: N43 01 51 W84 10 41. (CP: COL Waverly. 50 kw-D, DA. TL: N42 37 10 W84 34 31). Hrs opn: 32500 Parklane St., Garden City, 48135-1527. Phone: (989) 725-8196. Fax: (734) 525-3608. Web Site: www.michigancatholicradio.org. Licensee: 1090 Investments L.L.C. (acq 2-18-2000). Population served: 73,000 Rgnl. Network: Mich. Farm, Mich. Pub. Natl. Rep: Patt. Format: Catholic radio.

WRSR(FM)— Dec 2, 1965: 103.9 mhz; 2.85 kw. 482 ft TL: N42 59 44 W83 59 33. Stereo. Hrs opn: 6317 Taylor Dr., Flint, 48507. Phone: (810) 238-7300. Fax: (810) 238-7310. Web Site: www.classicfox.com. Licensee: Cumulus Licensing Corp. Group owner: Cumulus Media Inc. (acq 3-15-00; grpsl). Population served: 75,000 Format: Classic Rock. Target aud: 25-54. Spec prog: Class one hr, relg one hr, sports 4 hrs wkly. ♦Joe Mule, gen mgr; Jeff Wade, progmg dir; Les Root, news dir; Dan Greer, chief of engrg.

Pentwater

WMOM(FM)— Sept 26, 1999: 102.7 mhz; 6 kw. Ant 328 ft TL: N43 52 10 W86 21 32. Hrs open: 206 E. Ludington Ave., Ludington, 49431. Phone: (231) 845-9666. Fax: (231) 845-9322. Web Site: www.wmom.fm. Licensee: Bay View Broadcasting Inc. Format: Hot adult contemp. ♦Jana Rogers, opns mgr; Pat Martin, progmg dir; Pat Marin, news dir.

Petoskey

WFDX(FM)—Atlanta, Oct 20, 1988: 92.5 mhz; 100 kw. 868 ft TL: N45 01 00 W84 21 10. Stereo. Hrs opn: 24 1020 Hastings St., Traverse City, 49686. Phone: (231) 947-0003. Fax: (231) 947-7002. E-mail: markelliot@classichitsthefox.com Licensee: Northern Michigan Radio Inc. (acq 12-23-93; $1.165 million with WFCX(FM) Leland; FTR: 1-17-94). Population served: 374,600 Natl. Rep: Christal. Format: Classic hits. News staff: one. Target aud: 25-54. ♦Charlie Ferguson, gen mgr; T.J. Clark, gen sls mgr; Mark Elliot, progmg dir.

WJML(AM)— Dec 6, 1966: 1110 khz; 10 kw-D, DA. TL: N45 20 05 W84 55 34. Hrs open: 24 2175 Click Rd., 49770. Phone: (231) 348-5000. Web Site: www.wjml.com. Licensee: Stone Communications Inc. (acq 10-8-91; $24,000; FTR: 1-6-92). Population served: 415,627 Natl. Network: CBS. Law Firm: Reddy, Begley & McCormick. Format: News/talk. News: 72 hrs wkly. Target aud: 25 plus. Spec prog: Loc professional and college sports, relg 8 hrs wkly. ♦Richard D. Stone, pres & gen mgr; Philip Clever, stn mgr.

WKHQ-FM—See Charlevoix

WKLZ-FM— Dec 7, 1965: 98.9 mhz; 50 kw. 800 ft TL: N45 28 40 W84 57 04. Stereo. Hrs opn: 24
Rebroadcasts WKLT(FM) Kalkaska 85%.
1020 Hastings St., Traverse City, 49686. Phone: (231 947-0003. Fax: (231) 947-7002. Web Site: www.wklt.com. Licensee: Northern Radio of Petoskey Inc. (acq 8-15-91; $800,000). Population served: 415,627 Natl. Network: ABC. Natl. Rep: Christal. Law Firm: Fletcher, Heald & Hildreth. Format: Classic, AOR. News: 2 hrs wkly. Target aud: 18-49; baby boomers. ♦Charlie Ferguson, gen mgr; Jackie Gordon, natl sls mgr; Terri Ray, progmg dir; Dennis Murray, chief of engrg; Kara Cooper, traf mgr.

WLDR(AM)— June 16, 2000: 750 khz; 1 kw-D, 330 w-N, DA-2. TL: N45 20 05 W84 55 34. Hrs open: 24 2175 Click Rd., 49770. Phone: (231) 348-5000. Licensee: Fort Bend Media Broadcasting Co. (acq 4-25-2007; swap with WJNL(AM) Kingsley). Population served: 198,000 Natl. Network: USA. Format: Country. Target aud: 25 plus. ♦Richard D. Stone, gen mgr; Philip Clever, stn mgr.

WLXT(FM)—Listing follows WMBN(AM).

WMBN(AM)— May 1946: 1340 khz; 1 kw-U. TL: N45 20 50 W84 58 01. Hrs open: 24 Box 286, 49770. Secondary address: 2095 U.S. 131 S. 49770. Phone: (231) 347-8713. Fax: (231) 347-9920. Licensee: MacDonald Garber Broadcasting Inc. (group owner; acq 11-17-98; grpsl). Population served: 6,500 Natl. Rep: D & R Radio. Format: Easy lstng. News staff: 2; News: 2 hrs wkly. Target aud: 25 plus. ♦Trish MacDonald-Garber, gen mgr; Brian Brachel, chief of engrg.

WLXT(FM)—Co-owned with WMBN(AM). Jan 1, 1967: 96.3 mhz; 100 kw. 981 ft TL: N45 19 17 W84 52 33. Stereo. Web Site: www.lite96.com.1,000,000 Format: Oldies.

Pickford

WMKD(FM)— Dec 22, 2000: 105.5 mhz; 55 kw. Ant 108 ft TL: N46 17 24 W84 18 53. Hrs open: 3183 Logan Valley Rd., Traverse City,

49684. Phone: (800) 968-0981. Web Site: www.nsbroadcasting.com. Licensee: Northern Star Broadcasting LLC. (acq 10-21-2005; $900,000). Format: Relg. ♦Palmer Pyle, gen mgr.

Pinconning

WYLZ(FM)— Nov 15, 1983: 100.9 mhz; 2.6 kw. Ant 495 ft TL: N43 50 46 W84 05 32. Stereo. Hrs opn: 24 1740 N. Champagne Dr., Saginaw, 48604. Phone: (989) 776-2100. Fax: (989) 754-5990. Licensee: The Last Bastion Station Trust LLC, as Trustee Group owner: Citadel Broadcasting Corp. (acq 6-12-2007; grpsl). Natl. Rep: McGavren Guild. Format: Country. Target aud: 25-54; general. ♦Scott Meier, gen mgr; Bruce Johnston, sls dir; Stan Parman, progmg dir; Bob Friedle, chief of engrg.

Pittsford

***WPCJ(FM)**— Oct 23, 1985: 91.1 mhz; 270 w. Ant 184 ft TL: N41 53 04 W84 28 15. Hrs open: 16 9400 Beecher Rd., 49271. Phone: (517) 523-3427. Fax: (517) 523-3427. E-mail: wpcj@freedomfarm.info Web Site: www.freedomfarm.info. Licensee: Pittsford Educational Broadcasting Foundation. Natl. Network: Moody. Format: Educ, relg, Christian. News: 10 hrs wkly. Target aud: General; rural. ♦Tim Neinas, stn mgr; Ed Trombley, chief of engrg.

Plymouth

***WSDP(FM)**— Feb 14, 1972: 88.1 mhz; 200 w. 110 ft TL: N42 20 50 W83 29 51. Stereo. Hrs opn: 24 46181 Joy Rd., Canton, 48187. Phone: (734) 416-7732. Phone: (734) 416-7745. Fax: (734) 416-7732. E-mail: keithb@pccs.klz.mi.us Web Site: www.881theescape.com. Licensee: Plymouth Canton Community Schools. Population served: 1,000,000 Format: New music. News: 2 hrs wkly. Target aud: General. ♦Bill Keith, gen mgr & stn mgr.

Port Huron

WBTI(FM)—Listing follows WPHM(AM).

WGRT(FM)— December 1991: 102.3 mhz; 3 kw. 318 ft TL: N43 04 08 W82 28 48. Hrs open: 24 624 Grand River Ave., 48060. Phone: (810) 987-3200. Fax: (810) 987-3325. E-mail: wgrtoffice@sbcglobal.net Web Site: www.wgrt.com. Licensee: Port Huron Family Radio Inc. (acq 11-10-2004; $100,000). Population served: 200,000 Natl. Network: ABC. Law Firm: David Oxenford. Format: Adult contemp. News staff: one. Target aud: General. ♦Martin Doorn, gen mgr; Cathie Martin, news dir.

WHLS(AM)— Aug 8, 1938: 1450 khz; 1 kw-U. TL: N42 58 37 W82 27 52. Hrs open: 24 Box 807, 48061-0807. Secondary address: 808 Huron Ave. 48060. Phone: (810) 982-9000. Fax: (810) 987-9380. Web Site: www.whls.net. Licensee: Liggett Communications L.L.C. (acq 1-1-56). Population served: 250,000 Natl. Rep: Michigan Spot Sales. Format: Oldies. Target aud: 18-50; middle class. Spec prog: Black one hr, Sp one hr wkly. ♦Robert Liggett, chmn; James A. Jensen, pres; Lawrence C. Smith, gen mgr; Kristine Sikkema, sls dir & gen sls mgr; Jim McKenzie, progmg dir; Bill Gilmer, news dir; Craig Bowman, chief of engrg; Staci O'Brien, traf mgr; Dennis Stuckey, sports cmtr.

WSAQ(FM)—Co-owned with WHLS(AM). Aug 7, 1964: 107.1 mhz; 6 kw. 298 ft TL: N42 58 37 W82 27 52. Stereo. 24 Web Site: www.wsaq.net.550,000 Format: Country. Target aud: 25-55.

WHLX(AM)—Marine City, Dec 10, 1951: 1590 khz; 1 kw-D, 102 w-N, DA-1. TL: N42 43 42 W82 31 15. Hrs open: 24
Rebroadcasts WHLS (AM) Port Huron 100%.
808 Huron Ave., 48060. Phone: (810) 982-9000. Fax: (810) 987-9380. Licensee: Liggett Communications LLC. Group owner: Liqqctt Communications (acq 5-1-2000). Population served: 300,000 Natl. Network: ABC. Format: Adult contemp, oldies. Target aud: 25-54. ♦Robert Liggett, chmn; James A. Jenson, pres; Lawrence Smith, VP; Lawrence C. Smith, gen mgr.

***WNFA(FM)**— May 15, 1986: 88.3 mhz; 1.3 kw. 227 ft TL: N42 59 36 W82 28 06. Stereo. Hrs opn: 24
Rebroadcasts WNFR(FM) Sandusky 100%.
2865 Maywood Dr., 48060. Phone: (810) 985-3260. Fax: (810) 985-7712. Web Site: www.wnradio.com. Licensee: Ross Bible Church. Population served: 150,000 Natl. Network: Moody, USA. Law Firm: Southmayd & Miller. Format: Relg, inspirational. News: 14 hrs wkly. Target aud: 25-44; females. ♦Lori McNaughton, opns mgr & pub affrs dir.

***WORW(FM)**— May 31, 1973: 91.9 mhz; 188 w. 78 ft TL: N43 01 30 W82 26 10. (CP: 180 w). Hrs opn: 1799 Krafft Rd., 48060. Phone: (810) 984-2675, Ext. 363. Fax: (810) 984-2747. Licensee: Port Huron Area School District. Population served: 200 Format: Top-40. ♦Connie Meggs, gen mgr.

WPHM(AM)— Dec 6, 1947: 1380 khz; 5 kw-U, DA-2. TL: N42 51 50 W82 29 40. Hrs open: 24 808 Huron St., 48060. Phone: (810) 982-9000. Fax: (810) 987-9380. Web Site: www.wphm.net. Licensee: Liggett Communicaions L.L.C. (group owner; acq 5-1-2000; grpsl). Population served: 300,000 Natl. Rep: Michigan Spot Sales. Wire Svc: NWS (National Weather Service) Format: News/talk. News: 40 hrs wkly. Target aud: 25-54. ♦Robert Liggett, chmn; James A. Jewsen, pres; Lawrence C. Smith, gen mgr; Kristine Sikkema, sls dir & gen sls mgr; Paul Miller, progmg dir; Craig Bowman, engrg VP.

WBTI(FM)—Co-owned with WPHM(AM). July 13, 1991: 96.9 mhz; 3 kw. 380 ft TL: N43 12 34 W82 32 10. Stereo. 24 Box 807, 48061-0807. Web Site: www.wbti.net.200,000 Natl. Rep: Michigan Spot Sales. Format: Adult contemp. Target aud: 18-34. ♦Ben Coburn, progmg mgr.

***WSGR-FM**— October 1971: 91.3 mhz; 100 w. 87 ft TL: N42 58 43 W82 25 45. Stereo. Hrs opn: Box 5015, 323 Erie St., 48061-5015. Phone: (810) 989-5564. Fax: (810) 984-8991. Web Site: www.stclair.cc.mi.us. Licensee: St. Clair County Community College. Population served: 180,000 Format: Pop jazz. Target aud: General; all age groups. Spec prog: Metal-hard rock 12 hrs, urban 6 hrs wkly. ♦John Hill, gen mgr.

Portage

WFAT(FM)—Listing follows WNWN(AM).

WNWN(AM)— July 25, 1986: 1560 khz; 4.1 kw-D, DA. TL: N42 10 59 W85 35 30. Hrs opn: Sunrise-sunset 6021 S. Westnedge Ave., Kalamazoo, 49002. Phone: (269) 327-7600. Fax: (269) 327-0726. Web Site: www.1560radio.com. Licensee: Midwest Communications Inc. (group owner; acq 1995; grpsl). Population served: 300,000 Natl. Rep: Christal. Format: Urban contemp. News staff: 2; News: 5 hrs wkly. Target aud: 25-54; emphasis on 35-50 age group. Spec prog: Blues 3 hrs wkly. ♦D.E. Wright, pres; Dennis Martin, gen mgr; Janine Seals, gen sls mgr; John McNeill, news dir; Walker Sisson, chief of engrg.

WFAT(FM)—Co-owned with WNWN(AM). June 1992: 96.5 mhz; 3 kw. 321 ft TL: N42 12 55 W85 36 37. Web Site: www.wfat.com. Format: Classic hits. Target aud: 25-49. ♦Brian Hayes, progmg dir; Walker Sisson, engrg dir.

WRKR(FM)—Licensed to Portage. See Kalamazoo

Powers

WXPT(FM)—Not on air, target date: unknown: 107.3 mhz; 11.14 kw. Ant 490 ft TL: N45 38 36 W87 22 37. Hrs open: 1717 Dixie Hwy., Suite 650, Fort Wright, KY, 41011. Phone: (859) 331-9100. Licensee: Radioactive LLC. ♦Benjamin L. Homel, pres.

Raco

***WJOH(FM)**— 2006: 91.5 mhz; 5.5 kw. Ant 328 ft TL: N46 23 28 W84 27 52. Hrs open: Michigan Community Radio, 3302 N. Van Dyke, Imlay City, 48444. Phone: (810) 721-0891. Licensee: Michigan Community Radio. ♦Ed Czelada, pres.

Reed City

WDEE-FM— Aug 16, 1997: 97.3 mhz; 2.85 kw. Ant 479 ft TL: N43 46 53 W85 36 58. Hrs open: Box 722, Big Rapids, 49307. Phone: (231) 832-1500. Fax: (231) 832-1600. Web Site: www.sunny973.com. Licensee: Steven V. Beilfuss. Format: Classic hits, oldies. Target aud: 35 plus; anyone who likes oldies. ♦Steven V. Beilfuss, gen mgr.

Republic

WUPZ(FM)—Not on air, target date: unknown: 96.7 mhz; 16.11 kw. Ant 410 ft TL: N46 28 30 W88 14 06. Hrs open: 1717 Dixie Hwy., Suite 650, Fort Wright, KY, 41011. Phone: (859) 331-9100. Licensee: Radioactive LLC. ♦Benjamin L. Homel, pres.

Richland

***WTNP(FM)**—Not on air, target date: unknown: 91.9 mhz; 4.5 kw vert. Ant 328 ft TL: N42 29 01 W85 22 49. Hrs open: 5331 Mt. Alifan Dr., San Diego, CA, 92111. Phone: (858) 277-4991. Fax: (858) 277-1365. Licensee: Horizon Christian Fellowship. (acq 2-3-2006; $250,000 for CP). ♦Mike MacIntosh, pres.

Rochester

***WXOU(FM)**—Auburn Hills, August 1995: 88.3 mhz; 110 w. 256 ft TL: N42 42 35 W83 13 50. Stereo. Hrs opn: 18 69 Oakland Ctr., 48309. Phone: (248) 370-4273. Fax: (248) 370-2846. E-mail: wxougm@oakland.edu Web Site: www.wxou.org. Licensee: Oakland University. Format: Free-form. News: 13 hrs wkly. Target aud: General; univ & loc community not serviced by commercial media. ♦Adam Panchenko, gen mgr.

Rockford

WMJH(AM)— 1965: 810 khz; 3.6 kw-D. TL: N43 07 03 W85 34 06. Stereo. Hrs opn: 2422 Burton St. S.E., Grand Rapids, 49546. Phone: (616) 949-8585. Fax: (616) 949-9262. E-mail: jeremy@cookmediagr.com Licensee: Birach Broadcasting Corp. (group owner; (acq 11-6-2001; $1.9 million with WMFN(AM) Zeeland). Population served: 1,000,000 Natl. Network: CBS, Westwood One. Format: Nostalgia, adult standards. Target aud: 30 plus. ♦John Shepard, CEO; Tyrone Bynum, gen mgr.

Rogers City

WHAK-FM— April 1994: 99.9 mhz; 50 kw. 476 ft TL: N45 23 53 W83 55 19. Hrs open: 1491 M-32 W., Alpena, 49707. Phone: (989) 354-4611. Fax: (989) 354-4014. E-mail: nsn@chartermi.net Web Site: www.999thewave,com. Licensee: Edwards Communications LC. Group owner: Northern Radio Network (acq 12-21-2004; grpsl). Format: Oldies. ♦Tony Calumet, gen mgr & gen sls mgr; Danny Stann, progmg dir; Phil Heimerl, news dir; Darrel Kelly, chief of engrg; Mary Garrow, traf mgr.

WHAK(AM)— May 1949: 960 khz; 5 kw-D. TL: N45 23 53 W83 55 19. Natl. Rep: Michigan Spot Sales. Format: Talk. Target aud: 18-75. ♦Darrell Kelly, progmg dir; Mary Garrow, natl sls mgr & traf mgr.

WRGZ(FM)— June 16, 1984: 96.7 mhz; 42 kw. Ant 531 ft TL: N45 21 02 W83 46 59. Stereo. Hrs opn:
Simulcast with WATZ-FM Alpena 100%.
Box 536, Alpena, 49707. Secondary address: 123 Prentiss St., Alpena 49707. Phone: (989) 354-8400. Fax: (989) 354-3436. Licensee: WATZ Radio Inc. (acq 5-26-2006; $411,000). Format: Country. ♦Mike Centala, gen mgr.

Rogers Heights

***WAAQ(FM)**— 2004: 88.1 mhz; 100 w. Ant 85 ft TL: N43 37 05 W85 32 40. Hrs open: 901 Elizabeth Ct., Mount Pleasant, 48858. Phone: (989) 779-9178. Fax: (989) 779-1558. E-mail: great@broadcastingschool.com Licensee: Great Lakes Community Broadcasting Inc. Format: Oldies. ♦James J. McCluskey, pres & gen mgr.

Roscommon

WQON(FM)— March 1990: 101.1 mhz; 3.4 kw. 444 ft TL: N44 34 31 W84 42 19. Stereo. Hrs opn: 24 6514 Old Lake Rd., Grayling, 49738. Phone: (989) 348-6171. Fax: (989) 348-6181. E-mail: radio@i2k.net Web Site: www.gannonbroadcasting.com. Licensee: Gannon Broadcasting Systems Inc. (acq 9-88). Population served: 50,000 Natl. Rep: Michigan Spot Sales. Format: Adult contemp. News staff: one; News: 16 hrs wkly. Target aud: 25 plus. ♦William Gannon, gen mgr; Pete Michaels, opns mgr.

Rose Township

***WMSD(FM)**— Aug 11, 2000: 90.9 mhz; 5 kw vert. Ant 69 ft TL: N44 25 58 W84 00 33. Stereo. Hrs opn: 24 Box 42, Lupton, 48635. Phone: (989) 473-4616. E-mail: wmsd@m33access.com Web Site: www.bbc.northern-michigan.net. Licensee: Bible Baptist Church. Population served: 90,000 Format: Relg. ♦Paul E. Heaton, pres & stn mgr; Paul Heaton, progmg mgr.

Royal Oak

WEXL(AM)— October 1923: 1340 khz; 1 kw-U, DA-D. TL: N42 28 25 W83 06 56. Hrs open: 24 12300 Radio Pl., Detroit, 48228. Phone: (313) 272-3434. Licensee: WMUZ Radio Inc. Group owner: Crawford Broadcasting Co. (acq 4-18-97; $3.5 million). Population served: 3,500,000 Format: Gospel. Target aud: General. ♦Frank Franciosi, gen mgr.

Rust Township

***WKKM(FM)**— 2006: 88.5 mhz; 480 w. Ant 472 ft TL: N45 03 50 W83 42 57. Hrs open: Michigan Community Radio, 3302 N. Van Dyke, Imlay City, 48444. Phone: (810) 721-0891. Web Site: www.joyfm.net. Licensee: Michigan Community Radio. Format: Contemp Christian. ♦Edward Czelada, pres.

Saginaw

WGER(FM)—Listing follows WSGW(AM).

WHNN(FM)—See Bay City

WILZ(FM)— 1992: 104.5 mhz; 2.9 kw. Ant 413 ft TL: N43 23 34 W83 55 37. Hrs open: 1740 Champagne Dr. N., 48604. Phone: (989) 776-2100. Fax: (989) 754-5990. Web Site: www.wheelz.fm. Licensee: Citadel Broadcasting Co. Group owner: Citadel Broadcasting Corp. (acq 2-8-99). Natl. Rep: McGavren Guild. Law Firm: Reddy, Begley & McCormick. Format: Classic rock. Target aud: 35-54; adults. ♦Scott Meier, VP & gen mgr; Bob Kolen, sls dir; Stan Parman, progmg dir; Hal Maas, mus dir; Bob Friedle, chief of engrg.

WIOG(FM)—See Bay City

WKCQ(FM)—Listing follows WSAM(AM).

WMAX(AM)—Bay City, June 5, 1925: 1440 khz; 5 kw-D, 2.5 kw-N, DA-2. TL: N43 31 27 W83 57 58. Hrs open: 24
WDEO(AM) Ypsilanti 97%.
24 Frank Lloyd Wright Dr., Ann Arbor, 48105. Phone: (734) 930-5200. Fax: (989) 930-3179. E-mail: hroot@avemariaradio.net Web Site: www.avemariaradio.net. Licensee: 990 Investors L.L.C. (acq 6-01; $650,000). Population served: 1,500,000 Natl. Network: USA. Format: Christian, talk. News staff: 2; News: 14 hrs wkly. Target aud: 25-54; adult Christian. ♦Al Kresta, CEO; Michael P. Jones, exec VP & gen mgr.

WSAM(AM)— 1940: 1400 khz; 1 kw-U. TL: N43 25 00 W83 55 05. Hrs open: Box 1776, 2000 Whittier, 48601. Phone: (989) 752-8161. Fax: (989) 752-8102. Licensee: MacDonald Broadcasting Co. (group owner; acq 12-20-2001; grpsl). Population served: 400,000 Natl. Rep: D & R Radio. Law Firm: Fletcher, Heald & Hildreth. Format: MOR, full service. Target aud: 25 plus. ♦Kenneth MacDonald Jr., pres & gen mgr; Duane Alverson, sls VP & gen sls mgr; Barb Sheltraw, prom mgr; Rick Walker, VP, opns mgr & progmg dir; Gary Harding, chief of engrg.

WKCQ(FM)—Co-owned with WSAM(AM). 1947: 98.1 mhz; 50 kw. 500 ft TL: N43 25 04 W83 58 06. Stereo. Web Site: www.98fmkcq.com.91,849 Format: Country. Spec prog: Ger 3 hrs wkly. ♦Duane Alverson, adv dir; Gary Looper, traf mgr.

WSGW(AM)— Aug 11, 1950: 790 khz; 5 kw-D, 1 kw-N, DA-2. TL: N43 27 40 W83 48 48. Hrs open: 24 1795 Tittabawassee, 48604. Phone: (989) 752-3456. Fax: (989) 754-5046. Web Site: www.wsgw.com. Licensee: NM Licensing LLC. Group owner: NextMedia Group L.L.C. (acq 12-30-2002; grpsl). Population served: 91,849 Natl. Network: CBS. Rgnl. Network: Mich. Farm. Natl. Rep: Interep. Format: News/talk. News staff: 5; News: 40 hrs wkly. Target aud: 35-54; general. Spec prog: Farm 10 hrs wkly. ♦Floyd Evans, gen mgr; Dave Maurer, opns dir, progmg dir & news dir; Terry Henne, farm dir; Art Lewis, women's int ed.

WGER(FM)—Co-owned with WSGW(AM). Feb 19, 1969: 106.3 mhz; 4.4 kw. Ant 380 ft TL: N43 28 36 W83 57 06. Stereo. Web Site: www.magic1063.com.750,000 Format: Soft rock. Target aud: 25-54; upscale, mid/high level income.

WSGW-FM—Carrollton, Mar 11, 1991: 100.5 mhz; 3 kw. Ant 328 ft TL: N43 33 43 W85 58 54. Stereo. Hrs opn: 24 1795 Tittabauassee Rd., 48604. Phone: (989) 752-3456. Fax: (989) 754-5046. Licensee: NM Licensing LLC. Group owner: NextMedia Group L.L.C. (acq 12-30-2002; grpsl). Population served: 1,047,321 Wire Svc: Metro Weather Service Inc. Format: News/talk. News staff: one. Target aud: 18-49; women & teens. ♦Floyd Evans, gen mgr; David Mauer, opns mgr.

WTLZ(FM)— Nov 15, 1968: 107.1 mhz; 4.9 kw. 400 ft TL: N43 21 14 W83 55 06. Stereo. Hrs opn: 24 1795 Tittabawassee Rd., 48604. Phone: (989) 752-3456. Fax: (989) 754-5046. E-mail: brownerb@lflint.com Web Site: www.hotwtlz.com. Licensee: NM Licensing LLC. Group owner: NextMedia Group L.L.C. (acq 12-30-02). Population served: 234,000 Natl. Network: American Urban. Law Firm: Pepper & Corazzini. Format: Rhythm and blues. News: 6 hrs wkly. Target aud: 18-49; upscale, Blacks, women. Spec prog: Gospel 6 hrs wkly. ♦Floyd Evans, gen mgr.

Saint Ignace

WIDG(AM)— June 7, 1966: 940 khz; 5 kw-D. TL: N45 52 04 W84 47 09. Hrs open: Sunrise-sunset 1356 Mackinaw Ave., Cheboygan, 49721. Phone: (231) 627-2341. Fax: (231) 627-7000. Licensee: Northern Star Broadcasting L.L.C. (group owner; (acq 12-12-2005; grpsl). Population served: 3,130 Natl. Network: Sporting News Radio Network. Format: All sports. ♦Palmer Pyle, pres; Chris Monk, gen mgr & gen sls mgr; Nate Rose, progmg dir.

Saint Johns

WQTX(FM)— July 15, 1972: 92.1 mhz; 6 kw. Ant 400 ft TL: N42 53 29 W84 34 27. Stereo. Hrs opn: 24 2495 N. Cedar, Holt, 48842. Phone: (517) 699-0111. Fax: (517) 699-1880. Web Site: www.wqtx.net. Licensee: Rubber City Radio Group. Group owner: Rubber City Radio Group Inc. (acq 7-12-2000; grpsl). Population served: 397,000 Natl. Network: Jones Radio Networks. Natl. Rep: Katz Radio. Wire Svc: AP Format: Good Times Oldies. News staff: one; News: 8 hrs wkly. Target aud: 35 plus. ♦Thomas Mandel, pres; Mark Biviano, exec VP; Nick Anthony, exec VP; Dave Johnson, gen mgr; Paul Cashin, progmg dir.

WWSJ(AM)— Sept 23, 1959: 1580 khz; 1 kw-D, DA. TL: N42 58 14 W84 32 59. Hrs open: 24 Box 451, 1363 W. Parks Rd., St. John, 48879. Phone: (989) 224-7911. Fax: (989) 224-4683. Web Site: www.wwsj.com. Licensee: L. Harp, H. Harp, W. Hill, Elmira Hill. (acq 1-97; $160,000; FTR: 3-20-95). Natl. Network: American Urban. Format: Urban gospel. ♦Larry Harp, pres; Danielle Beckley, prom VP; Helen Harp, progmg dir; Dione Harp, mus dir; Ed Czelada, chief of engrg.

Saint Joseph

WHFB(AM)—See Benton Harbor-St. Joseph

WHFB-FM—See Benton Harbor-St. Joseph

WIRX(FM)—Licensed to Saint Joseph. See Benton Harbor-St. Joseph

WSJM(AM)—Licensed to Saint Joseph. See Benton Harbor-St. Joseph

Saint Louis

WFYC(AM)—See Alma

WMLM(AM)— Dec 15, 1977: 1520 khz; 1 kw-U, DA-2. TL: N43 21 08 W84 36 15. Hrs open: 24 Box 17, 48880. Secondary address: 4170 N. State Rd., Alma 48801. Phone: (989) 463-4013. Fax: (989) 463-4014. E-mail: wmlm@cmsinter.net Web Site: www.wmlm.com. Licensee: Siefker Broadcasting Corp. Population served: 150,000 Natl. Network: ABC. Format: Country. Target aud: 35 plus. Spec prog: Farm 5 hrs, gospel 2 hrs wkly. ♦Gregory Siefker, pres & gen mgr.

WQBX(FM)—See Alma

Salem Township

WSDS(AM)—Licensed to Salem Township. See Ann Arbor

Saline

WLBY(AM)— 1958: 1290 khz; 500 w-D, DA. TL: N42 12 17 W83 47 19. Hrs open: Sunrise-sunset 1100 Victors Way, Suite 100, Ann Arbor, 48108. Phone: (734) 302-8100. Fax: (734) 213-7508. E-mail: wcas@clearchannel.com Licensee: Capstar TX L.P. Group owner: Clear Channel Communications Inc. (acq 8-7-2000; grpsl). Population served: 231,600 Format: Talk/sports. ♦Bob Bolak, gen mgr; Shannon Brown, gen sls mgr.

Sandusky

WMIC(AM)— June 27, 1968: 660 khz; 1 kw-D, DA. TL: N43 23 34 W82 49 57. Hrs open: 19 S. Elk St., 48471. Phone: (810) 648-2700. E-mail: wmic@avc1.net Licensee: Sanilac Broadcasting Co. Population served: 1,000,000 Rgnl. Network: Mich. Farm. Format: Country, news/talk. News staff: 2; News: 20 hrs wkly. Target aud: 25 plus; general. Spec prog: Farm 12 hrs, Pol 5 wkly. ♦George E. Benko, pres; Robert Benko, VP; Bob Armstrong, gen mgr & gen sls mgr; Stan Grabitz, mus dir; Renae Davis, news dir; Kevin Larke, chief of engrg.

WTGV-FM—Co-owned with WMIC. Aug 16, 1971: 97.7 mhz; 3 kw. 325 ft TL: N43 23 33 W82 49 56. Stereo. 24 50,000 Format: Adult contemp. ♦Barbara Benko, mus critic.

***WNFR(FM)**— Feb 14, 1994: 90.7 mhz; 18 kw. 328 ft TL: N43 27 32 W82 57 42. Stereo. Hrs opn:
Rebroadcasts WNFA(FM) Port Huron 100%.
Wonderful News Radio, 2865 Maywood Dr., Port Huron, 48060. Phone: (810) 985-3260. Fax: (810) 985-7712. Web Site: www.wnradio.com. Licensee: Ross Bible Church. Population served: 100,000 Natl. Network: Moody, USA. Law Firm: Southmayd & Miller. Format: Relg, inspirational. Target aud: 25-44; females. ♦Lori McNaughton, opns mgr; Ellyn Davey, mus dir & news dir; Ed Czelada, chief of engrg.

Saugatuck

WYVN(FM)— July 4, 1987: 92.7 mhz; 2.15 kw. 387 ft TL: N42 41 10 W86 10 05. Stereo. Hrs opn: 24 87 Central Ave., Holland, 49423. Phone: (616) 392-3121. Phone: (616) 396-0927. Fax: (616) 392-8066. E-mail: advertising@thevan.fm Licensee: Midwest Communications Inc. (group owner; acq 9-5-01; $1.5 million). Rgnl rep: Christel Radio Wire Svc: AP Format: Classic hits. News staff: one; News: one hr wkly. Target aud: 25-54. ♦Duke Wright, CEO; Peter Tanz, gen mgr; Kevin Oswald, gen sls mgr; Phil Thompson, progmg mgr; Gary Stevens, news dir; Bridgett Bowman, traf mgr.

Sault Ste. Marie

***WCMZ-FM**— July 13, 1990: 98.3 mhz; 25 kw. 328 ft TL: N46 29 10 W84 13 49. Stereo. Hrs opn: 24
Rebroadcasts WCMU-FM Mount Pleasant 100%.
Central Michigan Univ., 1999 E. Campus Dr., Mount Pleasant, 48859. Phone: (989) 774-3105. Fax: (989) 774-4427. E-mail: cmuradio@radio.cmich.edu Web Site: www.wcmu.org. Licensee: Central Michigan University. Natl. Network: NPR, PRI. Rgnl. Network: Mich. Pub. Law Firm: Dow, Lohnes & Albertson. Wire Svc: AP Format: Class, jazz, news. News staff: 2; News: 45 hrs wkly. Target aud: General. ♦Ed Grant, gen mgr; Ann Blatte, dev dir & prom mgr; Art Curtis, gen sls mgr & rgnl sls mgr; Ray Ford, progmg dir; Randy Kapenga, chief of engrg.

WKNW(AM)—Listing follows WYSS(FM).

***WLSO(FM)**— 1995: 90.1 mhz; 100 w. 98 ft TL: N46 29 31 W84 21 48. Hrs open: 680 W. Easterday Ave., 49783. Phone: (906) 635-2107. Fax: (906) 635-2111. E-mail: wlso@gw.lssu.edu Web Site: www.lssu.edu/wlso. Licensee: Lake Superior State University. Format: Var. ♦Scott Korb, stn mgr.

WSOO(AM)— June 1, 1940: 1230 khz; 1 kw-U. TL: N46 26 16 W84 22 42. Hrs open: Box 1230, 49783. Phone: (906) 632-2231. Fax: (906) 632-4411. Licensee: Sovereign Communications LLC (acq 12-18-03; $2.6 million with co-located FM). Population served: 15,136 Natl. Rep: Michigan Spot Sales. Format: Adult contemp. ♦Tom Ewing, gen mgr.

WSUE(FM)—Co-owned with WSOO(AM). 1978: 101.3 mhz; 100 kw. 978 ft TL: N46 26 16 W84 22 42. Format: Classic rock.

***WTHN(FM)**— Jan 29, 2005: 102.3 mhz; 22.5 kw. Ant 344 ft TL: N46 29 08 W84 13 49. Hrs open: 24 Box 695, Gaylord, 49734. Phone: (989) 732-6274. Fax: (989) 732-8171. E-mail: ncr@ncradio.org Web Site: www.ncradio.org. Licensee: Northern Christian Radio Inc. (group owner). Format: Inspirational. ♦George A. Lake Jr., CEO & gen mgr.

WYSS(FM)— July 12, 1972: 99.5 mhz; 26.5 kw. 275 ft TL: N46 23 48 W84 23 52. (CP: 100 kw). Stereo. Hrs opn: 1402 Ashmun St., 49783. Phone: (906) 635-0995. Fax: (906) 635-1216. Web Site: www.995yesfm.com. Licensee: Northern Star Broadcasting L.L.C. (group owner; acq 11-5-01; grpsl). Population served: 13,000 Natl. Network: Westwood One. Law Firm: Cohn & Marks. Format: CHR. Target aud: 18-49. ♦Keith Neve, gen mgr, gen sls mgr & gen sls mgr; Tim Ellis, progmg dir & disc jockey; Candace Seward, traf mgr & disc jockey; Will Karr, news dir & disc jockey.

WKNW(AM)—Co-owned with WYSS(FM). Aug 25, 1990: 1400 khz; 250 w-U. TL: N46 29 18 W84 19 45. Stereo. 24 Web Site: www.talkradio1400.com. Format: Sports, news/talk. ♦Will Karr, progmg dir.

Schoolcraft

***WOFR(FM)**— May 2003: 89.5 mhz; 10 kw. Ant 138 ft TL: N42 06 38 W85 37 57. Hrs open: 24 Family Stations Inc., 4135 Northgate Blvd., Suite 1, Sacramento, CA, 95834. Phone: (916) 641-8191. Fax: (916) 641-8238. Web Site: www.familyradio.com. Licensee: Family Stations Inc. (group owner). Format: Relg. ♦Harold Camping, gen mgr; Craig Hulsebos, progmg dir; Rick Prime, chief of engrg.

Scottville

WKZC(FM)— Feb 16, 1983: 94.9 mhz; 17 kw. 400 ft TL: N44 03 27 W86 24 58. Stereo. Hrs opn: 24 5941 W. U.S. 10, Ludington, 49431. Phone: (231) 843-3438. Fax: (231) 843-1886. E-mail: mike@wkla.com.com Licensee: Lake Michigan Broadcasting Inc. (group owner; (acq 9-20-96; grpsl). Population served: 70,000 Natl. Network: ABC. Natl. Rep: Patt. Law Firm: Shaw Pittman. Format: Country. News staff: one; News: 5 hrs wkly. Target aud: 25-54. ♦Lynn Baerwolf, pres & gen mgr; Jason Wilder, opns mgr & prom VP.

Shepherd

WMMI(AM)— Feb 2, 1987: 830 khz; 1 kw-D. TL: N43 33 42 W84 45 00. Stereo. Hrs opn: Daytime 4065 E. Wing Rd., Mount Pleasant, 48858. Phone: (989) 772-9664. Fax: (989) 773-5000. E-mail: wczy@wczy.net Web Site: www.wczy.net. Licensee: Central Michigan Communications Inc. (acq 8-15-88). Natl. Rep: Michigan Spot Sales. Rgnl rep: Patt. Law Firm: Reddy, Begley & McCormick. Wire Svc: AP Format: Talk. News staff: 1; News: 6. Target aud: 25-54; general. ♦Mike Carey, pres; John Sebastian, progmg dir; Bryan Carr, news dir; Lisa Johnson, traf mgr.

South Haven

WCSY-FM— Oct 31, 1981: 98.3 mhz; 1.9 kw. Ant 403 ft TL: N42 18 02 W86 15 03. Stereo. Hrs opn: 510 Williams St., 49090. Phone: (269) 637-6397. Fax: (269) 637-2675. Web Site: www.wcsy.com. Licensee: WSJM Inc. (acq 10-95; with WHIT(AM) South Haven). Format: Middle of the road. ♦Paul Layendecker, stn mgr; Joe Daguanno, gen sls mgr.

Southfield

***WSHJ(FM)**— Feb 28, 1967: 88.3 mhz; 105 w. Ant 69 ft TL: N42 28 12 W83 15 51. Stereo. Hrs opn: 7:30 AM-10 PM Southfield High School, 24675 Lasher Rd., 48034. Phone: (248) 746-8630. Phone: (248) 746-8631. Web Site: www.southfield.l12.mi.us/itc. Licensee: Board of Education Southfield Public Schools. Population served: 69,285 Format: Oldies, hip-hop. Target aud: General; students & families. ♦Jamie Rudolph, gen mgr.

Spring Arbor

***KTGG(AM)**— Aug 15, 1985: 1540 khz; 450 w-D. TL: N42 09 13 W84 32 58. Hrs open: Sunrise-sunset Spring Arbor Univ., 106 Main St., 49283. Phone: (517) 750-6540. Fax: (517) 750-6619. Licensee: Spring Arbor University. Law Firm: Lauren A. Colby. Format: Inspirational, relg. Target aud: 18-49; rural to urban. ♦Hal Munn, pres; Carl Fletcher, gen mgr & stn mgr; Rachel Buchanan, progmg dir & pub affrs dir.

WSAE(FM)—Co-owned with KTGG(AM). Oct 2, 1963: 106.9 mhz; 3.9 kw. Ant 349 ft TL: N42 09 13 W84 32 58. Stereo. 24 E-mail: info@home.fm Web Site: www.home.fm.140,000 Law Firm: Lauren A. Colby. Format: Christian contemp. Target aud: 18-44.

***WJKN-FM**— 2005: 89.3 mhz; 2.5 kw vert. Ant 272 ft TL: N42 09 13 W84 32 57. Hrs open: Spring Arbor Univ., 106 E. Main St., 49283. Phone: (517) 750-6540. Fax: (517) 750-6619. E-mail: powerpraise@arbor.edu Web Site: www.arbor.edu. Licensee: Spring Arbor University. Format: Christian. ♦Carl Fletcher, gen mgr; Rachel Ryder, mus dir; Dave Benson, chief of engrg.

Springfield

***WCFG(FM)**—Not on air, target date: unknown: 90.9 mhz; 700 w vert. Ant 351 ft TL: N42 21 20 W85 20 28. Hrs open: 1159 E. Beltline Ave. N.E., Grand Rapids, 49525-5805. Phone: (616) 942-1500. Fax: (616) 942-7078. Licensee: Cornerstone University. ♦Chris Lemke, gen mgr.

Standish

WWCM(FM)— January 1990: 96.9 mhz; 3 kw. Ant 328 ft TL: N44 02 08 W84 00 31. Stereo. Hrs opn: 24
Rebroadcasts WCMU-FM Mount Pleasant 100%.
Public Broadcasting Ctr., 1999 E. Campus Dr., Mount Pleasant, 48859. Phone: (989) 774-3105. Fax: (989) 774-4427. E-mail: cmuradio@radio.cmich.edu Web Site: www.wcmu.org. Licensee: Central Michigan University (acq 9-14-00). Format: Class, jazz, news & info. Target aud: General. ♦Ed Grant, gen mgr.

Stephenson

WMXG(FM)— 1999: 106.3 mhz; 50 kw. Ant 492 ft TL: N45 38 36 W87 22 37. Hrs open: 1101 A Ludington St., Escanaba, 49829. Phone: (906) 786-0060. Fax: (906) 786-2990. E-mail: mix106@chartermi.net Web Site: www.wmxg.com. Licensee: Pacer Radio of the Near-North. Format: Hot adult contemp. ♦Mike DuBord, gen mgr.

Sterling Heights

***WUFL(AM)**— Oct 26, 1988: 1030 khz; 5 kw-D, DA. TL: N42 36 19 W82 54 37. Hrs open: Daytime Box 1030, 48311. Secondary address: 42669 Garfield Rd., Suite 328, Clinton Township 48038. Phone: (586) 263-1030. Fax: (586) 228-1030. E-mail: wufl@flc.org Web Site: wufl.org. Licensee: Family Life Broadcasting System. (group owner; acq 10-25-88). Natl. Network: USA. Format: Relg, Christian. News: 45 min wkly. Target aud: 35-54; Women. ♦Donald D. Aupperle, stn mgr.

Sturgis

WMSH(AM)— 1951: 1230 khz; 1 kw-U, DA-1. TL: N41 46 11 W85 25 09. (CP: 2.16 kw). Hrs opn: Box 7080, 49091. Secondary address: 70808 S. Nottawa Rd. 49091. Phone: (269) 651-2383. Phone: (269) 651-2384. Fax: (269) 659-1111. E-mail: wmsh@wmshradio.com Web Site: www.wmshradio.com. Licensee: Lake Cities Broadcasting Corp. (group owner; acq 1-12-98; $600,000 with co-located FM). Population served: 100,000 Natl. Network: ABC, Jones Radio Networks. Natl. Rep: Michigan Spot Sales. Format: News, sports. Target aud: 25-54. ♦Carter Snider, gen mgr, sls VP & gen sls mgr.

WMSH-FM— 1951: 99.3 mhz; 2.15 kw. 390 ft TL: N41 46 11 W85 25 09. Stereo. Web Site: www.wmshradio.com. Format: Oldies.

Tawas City

WHST(FM)— Nov 1, 1972: 106.1 mhz; 6 kw. 280 ft TL: N44 16 27 W83 39 42. Stereo. Hrs opn: 24
Rebroadcasts WPHN(FM) Gaylord 100%.
Box 695, Gaylord, 49734-0695. Secondary address: 1511 M-32 E., Gaylord 49735. Phone: (989) 732-6274. Fax: (989) 732-8171. E-mail: ncr@ncradio.org Web Site: www.ncradio.org. Licensee: Northern Christian Radio Inc. (group owner; acq 7-19-01). Population served: 75,000 Format: Relg. Target aud: 25-54. ♦George A. Lake Jr., gen mgr.

WIOS(AM)— Sept 27, 1958: 1480 khz; 1 kw-D, DA. TL: N44 15 48 W83 32 42. Hrs open: Box 549, 48764. Secondary address: 523 Meadow Rd. 48763. Phone: (989) 362-3417. Fax: (989) 362-4544. Licensee: Carroll Enterprises Inc. (group owner; acq 5-1-69). Population served: 50,000 Rgnl. Network: Mich. Farm. Natl. Rep: Michigan Spot Sales. Law Firm: Booth, Freret, Imlay & Tepper P. Format: Easy lstng, talk. Target aud: 25 plus; general. Spec prog: Big band 6 hrs wkly. ♦John Carroll Jr., CEO, pres & gen mgr; John Carroll Sr., chmn; Tim Carroll, gen sls mgr.

WKJC(FM)—Co-owned with WIOS(AM). October 1979: 104.7 mhz; 50 kw. 492 ft TL: N44 24 43 W83 37 17. Stereo. E-mail: wkjc@wkjc.com Web Site: www.wkjc.com.750,000 Format: Modern C&W.

WQLB(FM)— July 1997: 103.3 mhz; 25 kw. 328 ft TL: N44 21 06 W83 31 39. Hrs open: Box 549, 48764. Phone: (989) 362-3417. Fax: (989) 362-4544. E-mail: wkjc@wkjc.com Web Site: www.wkjc.com. Licensee: Carroll Broadcasting Inc. (acq 12-1-97). Law Firm: Booth,Freret, Imlay & Tepper. Format: Classic rock. ♦John Carroll Jr., gen mgr; Tim Carroll, gen sls mgr; Deb Michaels, progmg dir; Marvin Walther, chief of engrg; Mary Hill, traf mgr.

Taylor

WCHB(AM)— 1990: 1200 khz; 25 kw-D, 15 kw-N, DA-2. TL: N42 09 24 W83 19 56. Hrs open: 3250 Franklin St., Detroit, 48207. Phone: (313) 259-2000. Fax: (313) 259-7011. Web Site: www.wchb1200.com. Licensee: Radio One of Detroit LLC. Group owner: Radio One Inc. (acq 6-19-98; $34.2 million with WDTJ(FM) Detroit). Population served: 151,148 Format: News/Talk. Target aud: Adults 25-54. ♦Alfred Liggins, pres; Dr. Wendell Cox, VP; Carol Lawrence-Dobrusin, gen mgr.

Three Rivers

WLKM(AM)— May 3, 1962: 1520 khz; 430 w-D, 8 w-N. TL: N41 55 43 W85 38 15. Hrs open: 24 59750 Constantine Rd., 49093-9394. Phone: (269) 278-1815. Fax: (269) 273-7975. E-mail: info@wlkm.com Web Site: www.wlkm.com. Licensee: Impact Radio LLC (group owner; (acq 8-1-2002; grpsl). Population served: 30,000 Natl. Network: Westwood One, AP Radio. Rgnl. Network: Mich. Farm. Rgnl rep: Patt Media Sales Law Firm: Irwin, Campbell & Tannenwald. Format: Classic hit country. News staff: one; News: 14 hrs wkly. Target aud: General. ♦Dennis Rumsey, pres & gen mgr; Kathy Loker, gen sls mgr; Walker Sisson, chief of engrg.

WLKM-FM— March 1975: 95.9 mhz; 3 kw. Ant 315 ft TL: N41 55 43 W85 38 15. Stereo. 24 Web Site: www.wlkm.com.50,000 Natl. Network: NBC Radio. Format: Classic hits. News staff: one; News: 2 hrs wkly. Target aud: 25-54.

Traverse City

WCCW(AM)— July 15, 1960: 1310 khz; 15 kw-D, 7.5 kw-N, DA-2. TL: N44 40 38 W85 39 56. Hrs open: 300 E. Front, Suite 450, 49684. Phone: (231) 946-6211. Fax: (231) 946-1914. Web Site: www.wccwi.com. Licensee: Midwestern Broadcasting Co. (group owner; (acq 9-16-96; $2.2 million with co-located FM). Population served: 18,048 Natl. Rep: Michigan Spot Sales, Katz Radio. Format: Sports.

WCCW-FM— Nov 8, 1967: 107.5 mhz; 50 kw. 518 ft TL: N44 46 11 W85 41 22. (CP: 660 w, ant 702 ft. TL: N44 46 02 W85 41 26). Stereo. Web Site: www.wccwi.com. Format: Oldies.

***WICA(FM)**— Sept 13, 2000: 91.5 mhz; 4 kw. Ant 748 ft TL: N44 45 22 W85 40 42. Stereo. Hrs open: 24 Box 199, Interlochen, 49643. Phone: (231) 276-4400. Fax: (231) 276-4417. E-mail: ipr@interlochen.org Licensee: Interlochen Center for the Arts. Population served: 81,000 Wire Svc: AP Format: News, talk. News staff: 4; News: 168 hrs wkly. Target aud: 25-80. ♦Thom Paulson, VP & stn mgr.

WLDR-FM— July 17, 1966: 101.9 mhz; 100 kw. 538 ft TL: N44 46 13 W85 41 43. (CP: Ant 630 ft.). Stereo. Hrs opn: 24 13999 S. West Bay Shore Dr., 49684-6206. Phone: (231) 947-3220. Fax: (231) 947-7201. E-mail: wldr@wldr.com Web Site: www.wldr.com. Licensee: Great Northern Broadcasting System Inc. Group owner: Fort Bend Broadcasting Co. (acq 4-10-2001; $3.6 million for stock). Population served: 250,000 Format: Adult contemp. News staff: one. Target aud: 25-54. ♦Roy Henderson, CEO & gen mgr; Steve Smith, CFO; Dave Maxson, opns dir, gen sls mgr, prom dir & news dir; Trevor Tkach, gen sls mgr & chief of engrg.

***WLJN-FM—** Oct 1, 1989: 89.9 mhz; 39 kw vert. Ant 554 ft TL: N44 46 36 W85 39 43. Stereo. Hrs opn: 24 Box 1400, 49685. Secondary address: 13930 S. Morgan Hill Rd. 49684. Phone: (231) 946-1400. Fax: (231) 946-3959. E-mail: info@wljn.com Web Site: www.wljn.com. Licensee: Good News Media Inc. (group owner). Format: Relg, Christian music. News: 4 hrs wkly. Target aud: General. ♦Brian Harcey, gen mgr.

WLXT(FM)—See Petoskey

***WNMC-FM—** October 1967: 90.7 mhz; 600 w. 538 ft TL: N44 46 36 W85 41 02. Stereo. Hrs opn: 8 AM-2 AM 1701 E. Front St., 49686. Phone: (231) 995-2562. Phone: (231) 995-1090. Fax: (231) 922-8963. Web Site: www.wnmc.org. Licensee: Northwestern Michigan College. Population served: 25,000 Format: Div, jazz, rock,blues. Spec prog: American Indian one hr, Black 20 hrs, folk 11 hrs, Sp 2 hrs wkly. ♦Eric Hines, gen mgr.

WTCM(AM)— 1941: 580 khz; 50 kw-D, 1.1 kw-N, DA-2. TL: N44 43 18 W85 42 18. Stereo. Hrs opn: 24 314 E. Front St., 49684. Phone: (231) 947-7675. Fax: (231) 929-3988. Web Site: www.wtcmi.com. Licensee: WTCM Radio Inc. Group owner: Midwestern Broadcasting Co. Population served: 500,000 Law Firm: Cordon & Kelly. Format: News/talk. News staff: 4; News: 12 hrs wkly. Target aud: 25-54. Spec prog: Farm 5 hrs wkly. ♦Ross Biederman, pres; Chris Warren, gen mgr.

WTCM-FM— Dec 13, 1965: 103.5 mhz; 100 kw. 989 ft TL: N44 27 31 W85 42 02. Stereo. 24 Web Site: www.wtcmi.com.500,000 Format: Country. Target aud: 25-54.

Trout Lake

***WHWG(FM)—** 1999: 89.9 mhz; 500 w. Ant 390 ft TL: N46 11 17 W84 56 46. Hrs open:
Rebroadcasts WHWL(FM) Marquette 100%.
130 Carmen Dr., Marquette, 49855. Phone: (906) 249-1423. Fax: (906) 249-4042. E-mail: gospelop@chartermi.net Web Site: www.gospelopportunities.com. Licensee: Gospel Opportunities Inc. Format: Relg. ♦W. Curtis Marker, gen mgr & progmg dir.

Tuscola

WWBN(FM)— Sept 14, 1987: 101.5 mhz; 1.8 kw. Ant 489 ft TL: N43 12 00 W83 33 30. Stereo. Hrs opn: 24 G-3338 E. Bristol Rd., Burton, 48529. Phone: (810) 742-1470. Fax: (810) 742-5170. E-mail: banana1015@regentflint .com Web Site: www.banana1015.com. Licensee: Regent Broadcasting of Flint Inc. Group owner: Regent Communications Inc. (acq 12-19-97; grpsl). Natl. Rep: Katz Radio. Format: Rock/AOR. News staff: one; News: 20 hrs wkly. Target aud: 18-49; men. ♦Terry Jacobs, CEO; Bill Stakelin, pres; Fred Murr, sr VP; Mark Thomas, VP & gen mgr; J. Patrick, opns VP; Brian Beddow, progmg dir.

Twin Lake

***WBLV(FM)—** July 3, 1982: 90.3 mhz; 100 kw. Ant 649 ft TL: N43 33 00 W86 02 34. Stereo. Hrs opn: 24
Rebroadcasts WBLV-FM Grand Rapids 100%.
Blue Lake Fine Arts Camp, 300 East Crystal Lake Rd., 49457-9592. Phone: (231) 894-2616. Fax: (231) 893-2457. Web Site: www.bluelake.org/radio.html. Licensee: Blue Lake Fine Arts Camp. Natl. Network: PRI, NPR. Rgnl. Network: Mich. Pub. Law Firm: Booth, Freret, Imlay & Tepper. Format: Class, jazz, news. News: 20 hrs wkly. Target aud: Adult. Spec prog: Folk 5 hrs wkly. ♦Dave Myers, gen mgr; Steve Albert, progmg dir.

Vassar

WOWE(FM)— July 1, 1990: 98.9 mhz; 3 kw. 328 ft TL: N43 17 56 W83 30 34. Hrs open: 126 W. Kearsley St., Flint, 48502. Phone: (810) 234-4335. Fax: (810) 234-7286. E-mail: wowe98.9@netzero.com Licensee: Praestantia Broadcasting Inc. Format: Urban contemp. ♦Michael Shumpert, pres, gen mgr, progmg dir & chief of engrg.

Walker

WTRV(FM)— June 15, 1993: 100.5 mhz; 3 kw. Ant 328 ft TL: N43 00 59 W85 44 24. Hrs open: 24 50 Monroe N.W., Suite 500, Grand Rapids, 49503. Phone: (616) 451-4855. Fax: (616) 451-0113. Web Site: www.theriver-fm.com. Licensee: Regent Broadcasting of Grand Rapids Inc. Group owner: Regent Communications Inc. (acq 8-7-00; grpsl). Format: Soft adult contemp. Target aud: 35-64. ♦Terry Jacobs, pres; Phil Catlett, gen mgr; Rick Sarata, sls dir; Rob Wagley, gen sls mgr; Nikki Havner, prom dir; Bill Bailey, progmg dir; Gene Parker, mus dir; Chuck Latour, news dir; Mike Mackjewski, engrg dir; Kathie Bogerd, traf mgr.

Walled Lake

WPON(AM)— December 1954: 1460 khz; 1 kw-D, 760 w-N, DA-2. TL: N42 32 38 W83 29 58. Hrs open: 24 21700 Northwestern Hwy., Suite 1190, Southfield, 48075. Phone: (248) 557-3500. Fax: (248) 557-4321. E-mail: wpon@wpon.com Web Site: www.wpon.com. Licensee: Birach Broadcasting Corp. (group owner; acq 5-25-2004; $800,000). Population served: 3,750,000 Format: Oldies, talk. News staff: . Target aud: 35-65; 35 and above. ♦Sima Birach, gen mgr; Jimmie James, stn mgr & opns mgr.

Warren

***WPHS(FM)—** Mar 20, 1964: 89.1 mhz; 100 w. Ant 98 ft TL: N42 31 00 W83 00 36. Stereo. Hrs opn: 6:30 AM-8:30 PM P.K. Cousino High School, 30333 Hoover Rd., 48093. Secondary address: Warren Consolidated Schools, 31300 Anita 48093. Phone: (586) 698-4501. E-mail: wphs@wphs.com Web Site: www.wphs.com. Licensee: Warren Consolidated Schools. (acq 1963). Population served: 2,000,000 Format: Techno. News staff: 2; News: 10 hrs wkly. Target aud: 12-27; males. Spec prog: Blues 3 hrs, Pol 2 hrs, news 5 hrs, country 4 hrs, Christian rock 4 hrs wkly. ♦Jenny S. Stanczyk, gen mgr.

West Branch

WBMI(FM)— Nov 7, 1977: 105.5 mhz; 3 kw. Ant 312 ft TL: N44 17 57 W84 15 59. Stereo. Hrs opn: Box 807, 40001. Phone: (989) 345-3996. Fax: (989) 345-3996. Licensee: Kevin and Alana Beamish (acq 9-9-2005; $300,000). Population served: 11,000 Natl. Network: Jones Radio Networks. Format: Classic Country. Target aud: 25-54. Spec prog: Pol 6 hrs wkly. ♦Charlie Cobb, gen mgr & opns mgr; Mike McCall, progmg dir.

White Star

***WEJC(FM)—** July 2001: 88.3 mhz; 30 kw vert. 472 ft TL: N43 57 18 W84 32 57. Hrs open: Superior Communications, Box 388, Williamson, 48895. Phone: (517) 381-0573. Fax: (877) 850-0881. E-mail: info@positivehits.com Web Site: www.positivehits.com. Licensee: Superior Communications. Format: Contemporary Christian. ♦Jenn Czelada, gen mgr.

Whitehall

WEFG-FM— Apr 1, 1991: 97.5 mhz; 1.7 kw. Ant 426 ft TL: N43 23 04 W86 19 30. Stereo. Hrs open: 24 Box 190, Fremont, 49412-0190. Phone: (231) 924-4700. Fax: (231) 759-3410. Licensee: Unity Broadcasting Inc. (acq 12-16-2002; grpsl). Natl. Network: Westwood One. Format: Black, oldies. News: one hr wkly. Target aud: 25-49; male and female. ♦Bob Bolton, exec VP; Don Noordyk, gen mgr; Ernest Herrera, engrg VP; Rene Munoz, traf mgr.

WGVS-FM—Licensed to Whitehall. See Muskegon

WODJ(AM)— Oct 21, 1959: 1490 khz; 1 kw-U. TL: N43 23 04 W86 19 30. Stereo. Hrs opn: 24 3375 Merriam St., Muskegon, 49444. Phone: (231) 830-0176. Fax: (231) 830-0194. Licensee: Citadel Broadcasting Co. (group owner; (acq 9-28-2005; grpsl). Population served: 300,000 Format: Talk. ♦Jeff Morton, gen mgr.

Wyoming

***WYCE(FM)—** Nov 1, 1983: 88.1 mhz; 7 kw. 167 ft TL: N42 54 43 W85 41 00. Stereo. Hrs opn: 24 711 Bridge St., Grand Rapids, 49504. Phone: (616) 459-4788. Fax: (616) 742-0599. E-mail: comment@wyce.org Web Site: www.wyce.org. Licensee: Grand Rapids Cable Access Center Inc. (acq 5-31-89; $30,616; FTR: 6-19-89). Population served: 600,000 Format: Alternative, eclectic. Target aud: 25-54; general. Spec prog: Folk one hr, Sp 10 hrs. ♦Tim Goodwin, stn mgr; Pete Bruinsma, mus dir.

WYGR(AM)— Nov 14, 1964: 1530 khz; 500 w-D. TL: N42 55 38 W85 44 50. Hrs open: Box 9591, 49509. Phone: (616) 452-8589. Phone: (616) 248-9947. Fax: (616) 248-0176. E-mail: wygr@compuserv.com Licensee: WYGR Broadcasting. (acq 3-18-89). Natl. Rep: Patt. Law Firm: Meyer, Faller, Weisman & Rosenberg. Format: Sp, var/div. Target aud: Sp speaking Hispanics. Spec prog: Polka 3 hrs wkly. ♦Roland Rusticus, gen mgr, stn mgr, gen sls mgr, prom mgr & spanish dir; Scott Richards, progmg mgr, news dir & pub affrs dir; Robert Van Prooyen, chief of engrg.

Ypsilanti

WDEO(AM)—Licensed to Ypsilanti. See Ann Arbor

***WEMU(FM)—** Dec 8, 1965: 89.1 mhz; 15.5 kw. 289 ft TL: N42 15 48 W83 37 34. Stereo. Hrs opn: 24 Box 980350, 48198-0350. Secondary address: Eastern Michigan Univ., 426 King Hall 48197. Phone: (734) 487-2229. Phone: (734) 487-8936. Fax: (734) 487-1015. E-mail: wemu@emich.edu Web Site: www.wemu.org. Licensee: Eastern Michigan University. Population served: 295,000 Natl. Network: NPR. Rgnl. Network: Mich. Pub. Law Firm: Cohn & Marks. Wire Svc: AP Format: News, jazz, blues. News staff: 4; News: 39 hrs wkly. Target aud: General. ♦Arthur Timko, gen mgr; Michael Jewett, opns mgr; Mary Motherwell, dev dir; Clark Smith, progmg dir, progmg mgr & news dir; Linda Yohn, mus dir; Ray Cryderman, chief of engrg.

Zeeland

***WGNB(FM)—** Jan 21, 1989: 89.3 mhz; 30 kw. 499 ft TL: N42 50 14 W85 59 17. Stereo. Hrs opn: 24 Box 40, 3764 84th Ave., 49464. Phone: (616) 772-7300. Fax: (616) 772-9663. E-mail: wgnb@moody.edu Web Site: www.wgnb.org. Licensee: The Moody Bible Institute of Chicago Inc. (group owner; acq 2-5-91; FTR: 2-25-91). Natl. Network: Salem Radio Network, Moody. Law Firm: Southmayd & Miller. Format: Relg. News: 15 hrs wkly. Target aud: 35-54; Evangelical Christians. ♦Dr. Joseph Stowell, pres; Larry Mercer, sr VP; Robert Neff, VP; Scott Curtis, opns dir & mus dir; Dave Senzig, chief of engrg.

WJQK(FM)—Licensed to Zeeland. See Holland

WMFN(AM)— February 1990: 640 khz; 1 kw-D, 250 w-N. TL: N42 48 59 W85 57 24. Stereo. Hrs opn: 24 2422 Burton S.E., Grand Rapids, 49546. Phone: (616) 949-8585. Fax: (616) 949-9262. E-mail: jerenny@cookmediagr.com Licensee: Birach Broadcasting Corp. (group owner; acq 11-6-01; $1.9 million with WMJH(AM) Rockford). Population served: 500,000 Natl. Network: CBS. Format: News/talk, financial. Target aud: 25-54. ♦John Shepard, CEO & pres; Tyrone Bynum, gen mgr & opns mgr.

WPNW(AM)— Nov 2, 1956: 1260 khz; 10 kw-D, 1 kw-N, DA-2. TL: N42 43 56 W86 06 06. Hrs open: 24 425 Centerstone Ct., Suite 1, 49464. Phone: (616) 931-6620. Fax: (616) 931-1280. Web Site: www.1260thepledge.com. Licensee: Lanser Broadcasting Corp. (acq 11-1-83; $950,000; FTR: 10-3-83). Population served: 750,000 Natl. Network: CNN Radio. Natl. Rep: Salem. Law Firm: Reddy, Begley & McCormick. Format: News, talk. News staff: one; News: 9 hrs wkly. Target aud: 35 plus; mature adults. Spec prog: Sp 2 hrs, farm one hr wkly. ♦Leslie J. Lanser, pres; Bradley Lanser, exec VP; Troy West, stn mgr; Chad Millard, gen sls mgr & rgnl sls mgr; Jason Cramer, progmg dir.

Minnesota

Ada

KRJB(FM)— Sept 1, 1985: 106.3 mhz; 3 kw. 276 ft TL: N47 18 41 W96 31 13. Stereo. Hrs opn: 312 W. Main St., 56510. Phone: (218)

784-2844. Fax: (218) 784-3749. Web Site: www.krjbradio.com. Licensee: R & J Broadcasting. (acq 10-1-87). Population served: 326,000 Rgnl. Network: MNN. Format: Country. ♦Jim Birkemeyer, gen mgr & mktg VP.

Aitkin

KKIN(AM)— June 1, 1961: 930 khz; 2.5 kw-D, 400 w-N. TL: N46 32 26 W93 39 22. Stereo. Hrs opn: 24 Box 140, 56431. Secondary address: 37208 U.S. Hwy. 169 56431. Phone: (218) 927-2344. Fax: (218) 927-4090. E-mail: kkin@mlecmn.net Web Site: www.kkinradio.com. Licensee: Red Rock Radio Corp. (group owner; (acq 9-1-2006; grpsl). Population served: 120,000 Natl. Network: Jones Radio Networks. Law Firm: Timothy K. Brady. Format: Music of Your Life. News staff: one. Target aud: General. ♦Ro Grignon, pres; Terry Dee, gen mgr.

KKIN-FM— Jan 3, 1972: 94.3 mhz; 14 kw. Ant 436 ft TL: N46 41 18 W93 35 58. Stereo. 24 Web Site: www.kkinradio.com.120,000 Format: Classic country. News staff: one; News: 5 hrs wkly. Target aud: 35 plus.

Albany

KASM(AM)— Nov 20, 1950: 1150 khz; 2.5 kw-D, 23 w-N. TL: N45 37 59 W94 36 00. (CP: 2.1 kw. TL: N45 37 53 W94 36 00). Hrs opn: Box 390, 56307. Secondary address: 35223 238th Ave. Phone: (320) 845-2184. Fax: (320) 845-2187. Licensee: Starcom LLC (acq 1997; $1.25 million with co-located FM). Population served: 223,900 Rgnl. Network: MAGNET. Rgnl rep: Hyett/Ramsland. Law Firm: Baker & Hostetler. Format: Country, news, polka. News staff: one; News: 2 hrs wkly. Target aud: 36 plus. Spec prog: Oldies, Ger mus 2 hrs, farm 6 hrs wkly. ♦Randy Rothstein, gen mgr; Mark Sprint, progmg dir.

KDDG(FM)—Co-owned with KASM(AM). October 1993: 105.5 mhz; 6 kw. Ant 328 ft TL: N45 37 53 W94 36 00. Format: Adult contemp.

Albert Lea

KATE(AM)— October 1937: 1450 khz; 1 kw-U. TL: N43 38 00 W93 22 15. Stereo. Hrs opn: 24 305 S. 1st Ave., 56007. Phone: (507) 373-2338. Fax: (507) 373-4736. E-mail: copy@albertlearadio.com Licensee: Three Eagles of Luverne Inc. Group owner: Three Eagles Communications (acq 5-21-99; with co-located FM). Population served: 100,000 Natl. Rep: McGavren Guild. Rgnl rep: Midwest Radio. Law Firm: Reynolds & Manning. Wire Svc: AP Format: News/talk, MOR. News staff: 3; News: 30 hrs wkly. Target aud: 12 plus; general. Spec prog: Farm 18 hrs, Sp 2 hrs wkly. ♦Gary Buchanan, pres; Jackie Myran, gen mgr & opns mgr; Courtnay Doyle, gen sls mgr & farm dir; Steve Oman, news dir.

KQPR-FM— Aug 14, 1990: 96.1 mhz; 25 kw. 328 ft TL: N43 34 54 W93 23 42. Stereo. Hrs opn: 24 Box 1106, 56007-1106. Phone: (507) 373-9600. Fax: (507) 373-9045. E-mail: power96@chartermi.net Web Site: www.power96rocker.com. Licensee: Hometown Broadcasting Inc. (acq 11-21-01; grpsl). Natl. Network: Jones Radio Networks. Natl. Rep: Hyett/Ramsland. Format: Classic rock. News staff: one. Target aud: General. ♦Greg Jensen, CEO & CFO; Anna Rahn, gen mgr; Anna Rahn, stn mgr.

Alexandria

***KBHG(FM)**—Not on air, target date: unknown: 89.5 mhz; 7.2 kw. 321 ft TL: N45 55 55 W95 26 41. Hrs opn: 515 E. Pike St., 56360. Phone: (320) 859-3000. Fax: (320) 859-3010. E-mail: david@praisefm.org Licensee: Christian Heritage Broadcasting Inc. Format: Christian. ♦David McIver, gen mgr.

KULO(FM)— 1976: 94.3 mhz; 12 kw. Ant 466 ft TL: N45 56 25 W95 28 03. Stereo. Hrs opn: 24 Box 1024, 56308-1024. Secondary address: 604 Third Ave. W. 56308. Phone: (320) 762-2154. Fax: (320) 762-2156. E-mail: 100.7@kikvfm.com Web Site: cool943.com. Licensee: BDI Broadcasting Inc. Group owner: Omni Broadcasting Co. (acq 12-31-2001; $700,000). Population served: 198,000 Natl. Network: ABC. Rgnl. Network: MNN. Law Firm: Garvey, Schubert & Barer. Format: Oldies. News staff: one; News: 12 hrs wkly. Target aud: 35-64; adults. ♦Lou Buron, CEO & pres; Mary Campbell, CFO & VP; Dave Vagle, gen mgr; Trudy Blanshan, gen sls mgr & prom dir; Johnny Rocket, progmg dir & mus dir; Jim Rohn, news dir; Paul Sorum, pub affrs dir; Dave Cox, engrg dir & chief of engrg.

KXRA(AM)— July 27, 1949: 1490 khz; 1 kw-U. TL: N45 52 05 W95 21 47. Hrs open: 24 Box 69, 1312 Broadway, 56308. Phone: (320) 763-3131. Fax: (320) 763-5641. E-mail: thefolks@kxra.com Web Site: www.kxra.com. Licensee: Paradis Broadcasting of Alexandria Inc. (group owner; (acq 10-1-88). Population served: 40,000 Natl. Network: CNN Radio. Rgnl. Network: MNN. Rgnl rep: Midwest Radio. Law Firm: Fletcher, Heald & Hildreth. Format: News/talk. News staff: one; News: 25 hrs wkly. Target aud: 35-64. Spec prog: Farm 5 hrs, relg 2 hrs wkly. ♦Mel Paradis, chmn; Brett Paradis, pres & gen mgr.

KXRA-FM— May 1, 1968: 92.3 mhz; 13.5 kw. 446 ft TL: N45 52 30 W95 21 30. Stereo. 24 Web Site: www.kxra.com.50,000 Format: Classic rock. News staff: one; News: 5 hrs wkly. Target aud: 25-45; young adults, dual income households.

KXRZ(FM)— Apr 2, 1984: 99.3 mhz; 6 kw. Ant 285 ft TL: N45 52 47 W95 18 35. Stereo. Hrs opn: 24 1312 Broadway, 56308. Secondary address: Box 69 56308. Phone: (320) 763-3131. Fax: (320) 763-5641. E-mail: thefolks@kxra.com Web Site: www.z99radio.com. Licensee: Paradis Broadcasting of Alexandria Inc. (group owner; acq 5-1-00; $900,000). Population served: 40,000 Natl. Network: Jones Radio Networks. Rgnl rep: Midwest Radio. Law Firm: Fletcher, Heald & Hildreth. Format: Hits of the 80s, 90s & today. News staff: one; News: 5 hrs wkly. Target aud: 18-40; young adults. Spec prog: Relg 3 hrs wkly. ♦Brett Paradis, pres & gen mgr.

Anoka

KQQL(FM)— Aug 1, 1968: 107.9 mhz; 96 kw. Ant 1,092 ft TL: N45 20 20 W93 23 27. Stereo. Hrs opn: 1600 Utica Ave. S., Suite 400, Minneapolis, 55416. Phone: (952) 417-3000. Fax: (952) 417-3001. Web Site: www.kqql.com. Licensee: AMFM Broadcasting Licenses LLC. Group owner: Clear Channel Communications Inc. (acq 8-30-2000; grpsl). Population served: 2,600,000 Natl. Rep: Clear Channel. Law Firm: Wiley, Rein & Fielding. Format: Super hits. Target aud: 25-54.

KRJJ(AM)—Brooklyn Park, Apr 15, 1956: 1470 khz; 5 kw-U, DA-2. TL: N45 05 17 W93 22 59. Hrs open: 4640 W. 77th St., Suite 317, Edina, 55435. Phone: (952) 820-8520. Fax: (952) 820-8502. Licensee: Davidson Media Station KLBP Licensee LLC. (acq 9-7-2005; $5.2 million with KMNV(AM) Saint Paul). Population served: 100,000 Natl. Network: Westwood One. Rgnl. Network: MNN. Format: Gospel. ♦Tim Dennis, gen mgr.

Appleton

***KNCM(FM)**— February 1997: 88.5 mhz; 100 kw. 984 ft TL: N45 10 03 W96 00 02. Hrs open: Saint Johns University, Box 7711, Collegeville, 56321. Phone: (320) 363-7702. Fax: (320) 363-4948. E-mail: kncm@mpr.org Licensee: Minnesota Public Radio. Format: News & info. ♦William H. Kling, CEO & pres; Mark Alfuth, CFO; Thomas Kigin, exec VP; Mike Olson, gen mgr.

***KRSU(FM)**— Oct 25, 1989: 91.3 mhz; 75 kw. 1,158 ft TL: N45 10 03 W96 00 02. Stereo. Hrs opn: 24
Rebroadcast of KSJN(FM) Minneapolis-St. Paul.
45 E. 7th St., St. Paul, 55101. Phone: (800) 228-7123. Phone: (651) 290-1500. Fax: (651) 290-1224. Web Site: www.mpr.org. Licensee: Minnesota Public Radio Inc. Natl. Network: NPR, PRI. Rgnl. Network: Minn. Pub. Format: Class. ♦William H. Kling, pres & gen mgr; Ralph Hornberger, engrg dir & chief of engrg.

Atwater

KKLN(FM)—Licensed to Atwater. See Litchfield

Austin

KAUS(AM)— May 30, 1948: 1480 khz; 1 kw-U, DA-2. TL: N43 37 20 W92 59 26. Hrs opn: 24 18431 State Hwy. 105, 55912. Phone: (507) 437-7666. Phone: (507) 437-1480. Fax: (507) 437-7669. E-mail: kaus@kaus.com Web Site: www.kaus.com. Licensee: Three Eagles of Luverne Inc. Group owner: Three Eagles Communications (acq 4-1-00; grpsl). Population served: 65,000 Natl. Network: NBC. Rgnl. Network: MNN. Law Firm: Wiley, Rein & Fielding. Format: Oldies, adult contemp, news/talk. News staff: 2; News: 20 hrs wkly. Target aud: 25-54. ♦Rolland Johnson, chmn; Gary Buchanan, pres; Bob Mithuen, gen mgr; Gregory Soderberg, gen sls mgr; John Schramek, news dir; Ron Schat, engrg VP.

KAUS-FM— 1963: 99.9 mhz; 100 kw. 928 ft TL: N43 37 42 W93 09 12. Stereo. 24 Web Site: www.kaus.com.400,000 Format: Country. News staff: 2; News: 12 hrs wkly. Target aud: 25-54. ♦Tim Allen, prom mgr & progmg mgr.

***KMSK(FM)**— Jan 12, 1981: 91.3 mhz; 135 w. 221 ft TL: N43 40 39 W93 00 04. Stereo. Hrs opn: 24
Rebroadcasts KMSU(FM) Mankato 100%.
205 AFC, Mankato State University, 1536 Warren St., Mankato, 56001. Phone: (507) 389-5678. Fax: (507) 389-1705. Web Site: www.kmsu.org. Licensee: Mankato State University. (acq 12-23-91). Population served: 25,000 Natl. Network: NPR. Law Firm: Cohn & Marks. Format: Pub affrs, educ, mus. News staff: one; News: 50 hrs wkly. Target aud: General; upscale, educated. Spec prog: Drama 3 hrs, folk/ethnic 5 hrs, new age 5 hrs wkly. ♦Jim Gullickson, gen mgr; Karen Wright, opns dir.

KNFX(AM)— Apr 16, 1960: 970 khz; 5 kw-D, 500 w-N, DA-2. TL: N43 42 27 W92 56 45. Hrs open: 5 AM-midnight Radio Station KMFX(FM), 1530 Greenview Dr. S.W., Rochester, 55902. Phone: (507) 288-3888. Fax: (507) 288-7815. Licensee: CC Licenses LLC. Group owner: Clear Channel Communications Inc. (acq 9-25-2000; grpsl). Population served: 210,000 Rgnl. Network: Linder Farm, Tribune. Wire Svc: UPI Format: Sports/talk. News staff: one; News: 10 hrs wkly. Target aud: 25-54. ♦Bob Fox, gen mgr.

***KNSE(FM)**— 90.1 mhz; 6 kw. Ant 318 ft TL: N43 38 27 W93 08 51. Hrs open: summer 2003 Minnesota Public Radio, 45 E. 7th St., Saint Paul, 55101. Phone: (651) 290-1500. Fax: (651) 290-1224. E-mail: mail@mpr.org Web Site: www.mpr.org. Licensee: Minnesota Public Radio. Format: News. ♦William H. Kling, pres & gen mgr.

Babbitt

KAOD(FM)— 1999: 106.7 mhz; 19.8 kw. 790 ft TL: N47 41 18 W91 54 15. Hrs open:
Rebroadcast of KQDS-FM Duluth 100%.
501 Lake Ave. S., Suite 200A, Duluth, 55802. Phone: (218) 722-0921. Fax: (218) 723-1499. Licensee: Red Rock Radio Corp. (group owner; acq 1-10-00; grpsl). Format: Classic new rock. ♦Shawn Skramstad, gen mgr.

Bagley

KKCQ-FM— October 1997: 96.7 mhz; 25 kw. 328 ft TL: N47 36 08 W95 32 18. Hrs open: 24 Box 606, Fosston, 56542. Phone: (218) 435-1071. Fax: (218) 435-1480. E-mail: info@q107fm.com Web Site: www.q107fm.com. Licensee: Pine to Prairie Broadcasting Inc. (acq 6-16-97; $5,553 for CP). Population served: 40,000 Natl. Network: ABC. Format: Country. News staff: one. Spec prog: Farm 10 hrs, relg 9 hrs wkly. ♦Phil Ehlke, gen mgr; Karen Bingham, progmg dir & traf mgr.

Barnesville

KBVB(FM)— Jan 2, 1976: 95.1 mhz; 98 kw. Ant 991 ft TL: N46 40 29 W96 13 40. Stereo. Hrs opn: 24 1020 25th St. S., Fargo, ND, 58103. Phone: (701) 237-5346. Fax: (701) 235-4042. Web Site: www.bob95fm.com. Licensee: Radio Fargo-Moorhead Inc. Group owner: Clear Channel Communications Inc. (acq 1-19-2007; grpsl). Format: Country. ♦Jeff Hobert, gen mgr.

Baxter

WWWI(AM)— Aug 29, 1987: 1270 khz; 5 kw-U, DA-N. TL: N46 17 55 W94 16 42. Hrs open: 24 Box 783, 305 W. Washington St., Brainerd, 56401. Phone: (218) 828-9994. Fax: (218) 828-8327. E-mail: wwwi@brainerd.net Web Site: www.3wiradio.com. Licensee: Tower Broadcasting Corp. Natl. Network: CBS. Format: News/talk. Target aud: 25 plus. ♦James R. Pryor, pres, gen mgr, gen sls mgr & chief of engrg; Mary Pryor, VP & pub affrs dir.

Bemidji

KBHP(FM)—Listing follows KBUN(AM).

***KBSB(FM)**— Jan 19, 1970: 89.7 mhz; 115 w. 126 ft TL: N47 29 00 W94 52 27. Hrs open: 24 FM 90 KBSB Bemidji, State University Deputy Hall # 215, 1500 Birehmontn Dr. N.E., 56601. Phone: (218) 755-4120. Fax: (218) 755-4119. Web Site: www.fm90.org. Licensee: Bemidji State University. Population served: 11,490 Format: CHR. Target aud: 12-28; teens to young adults. Spec prog: American Indian 3 hrs, folk 3 hrs wkly. ♦Josh Harvey, stn mgr.

KBUN(AM)— 1946: 1450 khz; 1 kw-U, DA-1. TL: N47 27 56 W94 54 37. Hrs open: 24 Box 1656, 56619-1656. Secondary address: 502 Beltrami Ave. N.W. 56601. Phone: (218) 444-1500. Fax: (218) 759-0345. E-mail: phanson@pbbroadcasting.com Licensee: Paul Bunyan Broadcasting Co. Group owner: Omni Broadcasting Co. (acq 6-22-89; FTR: 6-26-89). Population served: 45,000 Natl. Network: ESPN Radio, Westwood One. Rgnl. Network: MNN. Law Firm: Garvey, Schubert & Barer. Format: Sports talk. News staff: one; News: 12 hrs wkly. Target aud: 18-54. ♦Lou Buron, CEO, pres & gen mgr; Mary Campbell, CFO

& VP; Kevin Jackson, opns dir, opns dir, opns mgr & progmg dir; Peggy Hanson, gen sls mgr; Brian Fisher, prom dir; Mardy Karger, news dir & pub affrs dir; Mark Anderson, engrg dir & chief of engrg.

KBHP(FM)—Co-owned with KBUN(AM). Aug 3, 1972: 101.1 mhz; 100 kw. Ant 522 ft TL: N47 22 12 W94 52 54. Stereo. 24 75,000 Natl. Network: ABC. Format: Country, mainstream. News staff: one; News: 12 hrs wkly. Target aud: 25-54. ♦Todd Haugen, opns dir, opns mgr, progmg dir & mus dir.

***KCRB-FM**— Dec 22, 1982: 88.5 mhz; 95 kw. 994 ft TL: N47 42 03 W94 29 15. Hrs open: 24
Rebroadcast of KSJN(FM) Minneapolis-St. Paul.
45 E. 7th St., St. Paul, 55101. Phone: (651) 290-1500. Fax: (651) 290-1224. Web Site: www.mpr.org. Licensee: Minnesota Public Radio Inc. Natl. Network: PRI, NPR. Rgnl. Network: Minn. Pub. Format: Class. News staff: one; News: 25 hrs wkly. ♦William H. Kling, gen mgr & stn mgr.

KKBJ(AM)— Oct 31, 1977: 1360 khz; 5 kw-D, 2.5 kw-N, DA-N. TL: N47 26 32 W94 55 07. Hrs open: 24 2115 Washington Ave. S., 56601. Phone: (218) 751-7777. Fax: (218) 759-0658. Web Site: www.kkbj.com. Licensee: R.P. Broadcasting Corp. (acq 4-1-95). Population served: 70,000 Natl. Network: AP Radio. Law Firm: Bechtel & Cole. Wire Svc: AP Format: Talk. News staff: one; News: 15 hrs wkly. Target aud: 25-54. ♦Daniel J. Voss, gen mgr & gen sls mgr; Chuck Sebastian, progmg dir; Roger Paskvan, pres & engrg VP; Rocky Coffin, relg ed.

KKBJ-FM— Aug 8, 1983: 103.7 mhz; 100 kw. 460 ft TL: N47 33 19 W94 47 59. Stereo. 24 75,000 Format: Hot adult contemp. News staff: one; News: 20 hrs wkly. Target aud: 18-49; 40% male & 60% female. ♦Tracy Bailey, prom dir; Daniel Voss, adv mgr.

KKZY(FM)— May 7, 1999: 95.5 mhz; 100 kw. Ant 423 ft TL: N47 22 12 W94 52 54. Stereo. Hrs opn: 24 Box 1656, 56619-1656. Secondary address: 502 Beltrami Ave. N.W. 56601. Phone: (218) 444-1500. Fax: (218) 759-0345. E-mail: phanson@pbbroadcasting.com Licensee: BG Broadcasting Inc. Group owner: Omni Broadcasting Co. (acq 6-22-98). Population served: 75,000 Natl. Network: ABC. Law Firm: Garvey, Schubert & Barer. Format: Adult contemp. News staff: one; News: 12 hrs wkly. Target aud: 25-54; adults. ♦Lou Buron, CEO, pres & gen mgr; Mary Campbell, CFO & VP; Peggy Hanson, gen sls mgr; Brian Fisher, prom dir; Kevin Jackson, progmg dir; Mardy Karger, news dir & pub affrs dir; Todd Haugen, opns dir, opns mgr, mus dir & news dir; Mark Anderson, engrg dir & chief of engrg.

***KNBJ(FM)**— July 1, 1994: 91.3 mhz; 60 kw. 974 ft Hrs opn: 24
Rebroadcasts KNOW-FM Minneapolis-St. Paul 90%.
45 E. 7th St., St. Paul, 55101. Phone: (651) 290-1500. Fax: (651) 290-1224. Web Site: www.mpr.org. Licensee: Minnesota Public Radio. Natl. Network: PRI, NPR. Rgnl. Network: Minn. Pub. Format: News. ♦William H. Kling, gen mgr & stn mgr.

Benson

KBMO(AM)— December 1956: 1290 khz; 500 w-D. TL: N45 19 06 W95 33 48. Hrs open: 24 105 13th St. N., 56215. Phone: (320) 843-3290. Fax: (320) 843-3955. E-mail: kscr@info-link.net Licensee: Quest Broadcasting Inc. (acq 5-2-94; $390,000 with co-located FM). Population served: 60687 Natl. Network: Jones Radio Networks. Rgnl. Network: MNN. Law Firm: Shainis & Peltzman. Format: Music of your life. News staff: one; News: 25 hrs wkly. Target aud: 40 plus. Spec prog: Farm 10 hrs wkly. ♦Paul Estenson, pres & gen mgr; Allison McGeary, gen sls mgr; Jason Brandt, progmg dir & news dir; Maynard Meyer, chief of engrg; Jeremy Goucet, traf mgr.

KSCR-FM—Co-owned with KBMO(AM). Apr 26, 1968: 93.5 mhz; 25 kw. Ant 328 ft TL: N45 19 06 W95 33 48. Stereo. 24 113,291 Format: Classic hits. Target aud: 18-54.

Blackduck

WBJI(FM)— 1991: 98.3 mhz; 50 kw. 456 ft TL: N47 33 19 W94 47 59. Stereo. Hrs opn: 24 2115 Washington Ave. S.E., Bemidji, 56601-8942. Phone: (218) 751-7777. Fax: (218) 759-0658. E-mail: wbji@paulbunyon.net Web Site: www.wbji.com. Licensee: R.P. Broadcasting Inc. Population served: 50,000 Natl. Network: ABC. Law Firm: Bechtel & Cole. Format: Real country. News staff: one; News: 8 hrs wkly. Target aud: 35-64; adults with above average income. Spec prog: NASCAR 4 hrs wkly. ♦Roger Paskvan, CEO; Marla Weckman, exec VP; Dan Voss, gen mgr.

WMIS-FM—Not on air, target date: unknown: 92.1 mhz; 36 kw. Ant 577 ft TL: N47 33 26 W94 48 04. Hrs open: 1410 30th St. N.W., Suite 115, Bemidji, 56601. Phone: (218) 766-7970. Licensee: Paskvan Media Inc. ♦Troy Paskvan, pres & gen mgr.

Blooming Prairie

KOWZ-FM— September 1995: 100.9 mhz; 100 kw. 620 ft TL: N44 02 46 W93 23 03. Hrs open: 255 Cedardale Dr. S.E., Owatonna, 55060. Phone: (507) 444-9224. Fax: (507) 444-9080. Licensee: Blooming Prairie Farm Radio Inc. Group owner: Linder Broadcasting Group. Format: Adult contemp. ♦Jeff Seaton, gen mgr.

Blue Earth

KBEW(AM)— Aug 29, 1963: 1560 khz; 1 kw-D. TL: N43 38 48 W95 33 48. Hrs open: Box 278, 56013. Phone: (507) 526-2181. Fax: (507) 526-7468. E-mail: kbew@bevcomm.net Licensee: KBEW Radio Inc. Group owner: Result Radio Group (acq 2-1-81). Population served: 50,000 Natl. Network: ABC. Rgnl. Network: Linder Farm. Natl. Rep: Katz Radio. Wire Svc: AP Format: Oldies, news/talk. News staff: one; News: 17 hrs wkly. Target aud: Farming community. Spec prog: Farm 15 hrs wkly. ♦Jerry Papenfuss, pres; Kevin Benson, stn mgr.

KBEW-FM— 1993: 98.1 mhz; 25 kw. 328 ft TL: N43 38 44 W94 05 33. Stereo. 24 , 56013. 50,000 Format: Country. News staff: one; News: 3 hrs wkly. Target aud: 18-54. ♦Kevin Benson, gen mgr & mktg mgr.

KJLY(FM)— Nov 1, 1983: 104.5 mhz; 50 kw. 453 ft TL: N43 39 41 W94 06 29. Stereo. Hrs open: Box 72, 56013. Secondary address: 12089 380th Ave. 56013. Phone: (507) 526-3233. Fax: (507) 526-3235. E-mail: kjly@kjly.com Web Site: www.kjly.com. Licensee: Minn-Iowa Christian Broadcasting Inc. (group owner). Natl. Network: Moody, Salem Radio Network. Format: Inspirational relg. News: 21 hrs wkly. Target aud: 45-65. Spec prog: Farm 5 hrs, children 4 hrs wkly. ♦Maurice Schwen, pres; Eugene Stallkang, VP; Doug Johnson, progmg dir; Mark Croom, chief of engrg.

Brainerd

***KBPN(FM)**— July 2003: 88.3 mhz; 5 kw. Ant 669 ft TL: N46 25 21 W94 27 41. Hrs open: 24 Box 578, Bemidji, 56619. Secondary address: Minnesota Public Radio, 45 E. 7th St., Saint Paul 55101. Phone: (218) 829-1072. Fax: (218) 751-8640. Web Site: www.mpr.org. Licensee: Minnesota Public Radio. Format: News. ♦Kristi Booth, stn mgr; Tim Post, local news ed; Barb Treat, sls.

KBPR(FM)—Licensed to Brainerd. See Collegeville

KLIZ(AM)— Aug 6, 1946: 1380 khz; 5 kw-U, DA-N. TL: N46 19 56 W94 10 26. Hrs open: 24 Box 746, 56401-0746. Secondary address: 13225 Dogwood Dr., Baxter 56425-8613. Phone: (218) 828-1244. Fax: (218) 828-1119. E-mail: production@brainerd.net Web Site: brainerdradio.net. Licensee: BL Broadcasting Inc. Group owner: Omni Broadcasting Co. (acq 4-1-2004; grpsl). Population served: 70,000 Natl. Network: Sporting News Radio Network, USA, Westwood One. Rgnl. Network: MNN. Law Firm: Garvey, Schubert & Barer. Format: Sports talk. News staff: one; News: 12 hrs wkly. Target aud: 25-64; adults. ♦Lou Buron, CEO & pres; Mary Campbell, CFO & VP; G. Michael Boen, gen mgr; Danny Wild, opns dir, opns mgr, progmg dir, mus dir & pub affrs dir; Jeff Hilborn, gen sls mgr; Tess Taylor, news dir; Dave Cox, engrg dir & chief of engrg.

KLIZ-FM— May 23, 1960: 107.5 mhz; 100 kw. Ant 350 ft TL: N46 19 56 W94 10 26. Stereo. 200,000 Natl. Network: ABC. Format: Classic rock. News staff: one; News: 12 hrs wkly. Target aud: 18-54; adults.

KUAL-FM—Listing follows KVBR(AM).

KVBR(AM)— May 16, 1964: 1340 khz; 1 kw-U. TL: N46 20 51 W94 10 52. Hrs open: 24 Box 746, 56401-0746. Secondary address: 13225 Dogwood Dr., Baxter 56425-8613. Phone: (218) 828-1244. Fax: (218) 828-1119. E-mail: production@brainerd.net Web Site: brainerdradio.net. Licensee: BL Broadcasting Inc. Group owner: Omni Broadcasting Co. (acq 4-1-2004; grpsl). Population served: 44,000 Natl. Network: ABC, USA, Westwood One. Rgnl. Network: MNN. Law Firm: Garvey, Schubert & Barer. Format: All sports. News staff: one; News: 12 hrs wkly. Target aud: 25-54; adults. ♦Lou Buron, CEO & pres; Mary Campbell, CFO & VP; G. Michael Boen, gen mgr; Danny Wild, opns dir, opns mgr, progmg dir, mus dir & pub affrs dir; Jeff Hilborn, gen sls mgr; Tess Taylor, news dir; Dave Cox, engrg dir & chief of engrg.

KUAL-FM—Co-owned with KVBR(AM). June 3, 1994: 103.5 mhz; 20 kw. Ant 279 ft TL: N46 20 55 W94 13 29. Natl. Network: ABC. Format: Oldies. News staff: one; News: 12 hrs wkly. Target aud: 25-54; adults. ♦Billy Holiday, progmg dir & mus dir; Tess Taylor, pub affrs dir.

WJJY-FM— July 21, 1978: 106.7 mhz; 100 kw. Ant 448 ft TL: N46 26 36 W94 22 58. Stereo. Hrs opn: 24 Box 746, 56401-0746. Secondary address: 13225 Dogwood Dr., Baxter 56425-8613. Phone: (218) 828-1244. Fax: (218) 828-1119. E-mail: production@brainerd.net Web Site: brainerdradio.net. Licensee: BL Broadcasting Inc. Group owner: Omni Broadcasting Co. (acq 3-2-94; $900,000; FTR: 5-2-94). Population served: 200,000 Natl. Network: ABC. Law Firm: Garvey, Schubert & Barer. Format: Full service adult contemp. News staff: one; News: 20 hrs wkly. Target aud: 25-54; adults. ♦Lou Buron, CEO & pres; Mary Campbell, CFO & VP; G. Michael Boen, gen mgr; Mark Hegstrom, opns mgr & progmg dir; Jeff Hillborn, gen sls mgr; Tess Taylor, news dir & pub affrs dir; David Cox, chief of engrg.

Breckenridge

KBMW(AM)—Licensed to Breckenridge. See Wahpeton ND

KLTA(FM)— Feb 17, 1970: 105.1 mhz; 100 kw. 713 ft TL: N46 32 41 W96 37 33. Stereo. Hrs open: 24 Box 9919, Fargo, ND, 58106. Secondary address: 2720 7th Ave. S., Fargo, ND 58103. Phone: (701) 237-4500. Fax: (701) 235-9082. E-mail: studio@fm1051.net Web Site: www.fm1051.net. Licensee: Monterey Licenses LLC. Group owner: Triad Broadcasting Co. LLC (acq 10-99; grpsl). Population served: 194,800 Natl. Rep: Christal. Law Firm: Shaw Pittman. Wire Svc: AP Format: Adult contemp. Target aud: 25-54; skews female. ♦Tom Douglas, CEO; David Benjamin, pres; Nancy Odney, gen mgr.

Breezy Point

KLKS(FM)— June 14, 1984: 104.3 mhz; 50 kw. 492 ft TL: N46 36 13 W94 15 04. Stereo. Hrs opn: 24 Box 300, 56472. Secondary address: 7170 Ski Chatet Dr. 56472. Phone: (218) 562-4884. Phone: (218) 829-2997. Fax: (218) 562-4058. Fax: (218) 829-9341. E-mail: klakes@uslink.net Web Site: www.klks.com. Licensee: Lakes Broadcasting Group Inc. Law Firm: Hogan & Hartson. Format: Adult standards, btfl mus, big band. News staff: 2; News: 25 hrs wkly. Target aud: 40 plus. ♦Bob Bundgaard, CEO & pres; Allen Gray, chmn; Diane Anderson, CFO & opns VP; Marj Bundgaard, gen mgr; Thomas Kenow, sls VP; David Pundt, news dir; Carol Bundgaard, traf mgr.

Brooklyn Park

KRJJ(AM)—Licensed to Brooklyn Park. See Anoka

Browerville

KXDL(FM)—Licensed to Browerville. See Long Prairie

Buffalo

KRWC(AM)— Nov 16, 1971: 1360 khz; 500 w-D. TL: N45 10 00 W93 55 11. Hrs open: 24 Box 267, 55313. Secondary address: 1472 10th St. N.W. 55313. Phone: (763) 682-4444. Fax: (763) 682-3542. E-mail: info@krwc1360.com Web Site: www.krwc1360.com. Licensee: Donnell Inc. (acq 6-15-98; $460,000). Population served: 150,000 Natl. Network: CNN Radio. Rgnl. Network: MNN. Format: Country, oldies, adult contemp, news/talk. News staff: one; News: 16 hrs wkly. Target aud: 25 plus. ♦Joe Carlson, pres, gen mgr & gen sls mgr; Tim Matthews, opns dir, progmg dir & news dir; John George, chief of engrg.

Buhl

***WIRN(FM)**— 1997: 92.5 mhz; 39 kw. 558 ft TL: N47 29 46 W92 47 05. Hrs open: Minnesota Public Radio, 224 Holiday Ctr., Duluth, 55802. Phone: (218) 722-9411. Fax: (218) 720-4900. Web Site: www.mpr.org. Licensee: Minnesota Public Radio. Format: News & info. ♦William Kling, pres; Kat Eldred, gen mgr; Bob Kelleher, news dir & local news ed; Doug Thompson, engrg dir; Cynthia Johnson, traf mgr.

Caledonia

KCLH(FM)— Nov 14, 1994: 94.7 mhz; 1.9 kw. Ant 584 ft TL: N43 41 24 W91 30 09. Stereo. Hrs opn: 24 201 State St., La Crosse, WI, 54602. Phone: (608) 782-1230. Fax: (608) 782-1170. Web Site: www.classichits947.com. Licensee: Family Radio Inc. Group owner: The Mid-West Family Broadcast Group (acq 7-19-01; grpsl). Population served: 300,000 Law Firm: Shaw Pittman. Wire Svc: AP Format: Classic hits. News staff: 4; News: one hr wkly. Target aud: General; 25-54. ♦Dick Record, pres; Brian Michaels, opns mgr & progmg dir; Dave Roberts, progmg dir; Kris Kody, prom.

Cambridge

WGVY(FM)— May 5, 1973: 105.3 mhz; 25 kw. Ant 298 ft TL: N45 31 17 W93 10 27. Stereo. Hrs opn: 24
Rebroadcasts WGVX(FM) Lakeville.
2000 S.E. Elm St., Minneapolis, 55414. Phone: (612) 617-4000. Fax: (612) 676-8292. Web Site: www.love105.fm. Licensee: Radio License Holding III LLC. Group owner: ABC Inc. (acq 6-12-2007; grpsl). Population served: 500,000 Natl. Rep: Interep. Law Firm: Haley, Bader & Potts. Format: Love songs of the 60s, 70s and 80s. Spec prog: Farm 8 hrs wkly. ♦Marc Kalman, pres & stn mgr; Dave Hamilton, opns mgr; Pete Frisch, sls dir; Susan Larkin, gen sls mgr; Leslie Heinemann, natl sls mgr; Brook Johnson, mktg dir; Shelley Miller, prom dir & prom mgr; Chris Rahn, progmg dir; Ben Gnam, mus dir; Christopher Taykalo, pub affrs dir; Dave Szaflarski, chief of engrg.

Cloquet

WKLK(AM)— Jan 31, 1950: 1230 khz; 1 kw-U, DA-1. TL: N46 44 58 W92 25 17. Hrs open: 24 1104 Cloquet Ave., 55720-1613. Phone: (218) 879-4534. Fax: (218) 879-1962. Web Site: www.wklkradio.com. Licensee: QB Broadcasting Ltd. (acq 5-12-92; $200,000 with co-located FM; FTR: 6-1-92). Population served: 150,000 Format: Music of Your Life. News staff: one; News: 16 hrs wkly. Target aud: Community oriented. ♦Al Quarstorom, gen mgr.

WKLK-FM— Apr 30, 1992: 96.5 mhz; 6 kw. 315 ft TL: N46 44 58 W92 25 17.24 Web Site: www.wklkradio.com. Format: Adult hit.

***WSCN(FM)**— Nov 17, 1975: 100.5 mhz; 100 kw. 875 ft TL: N46 47 21 W92 06 51. Stereo. Hrs opn: 24 224 Holiday Ctr., Duluth, 55802. Phone: (218) 722-9411. Fax: (218) 720-4900. Web Site: www.minnesotapublicradio.org. Licensee: Minnesota Public Radio. (acq 12-88; $200,000; FTR: 12-19-88). Natl. Network: PRI, NPR. Rgnl. Network: Minn. Pub. Format: News. ♦William H. Kling, pres; Kat Eldred, gen mgr.

Cold Spring

KMXK(FM)— Aug 30, 1968: 94.9 mhz; 50 kw. 492 ft TL: N45 23 53 W94 25 15. Stereo. Hrs opn: 24 640 Lincoln Ave. S.E., St. Cloud, 56304. Phone: (320) 251-4422. Fax: (320) 251-1855. Web Site: www.mix949.com. Licensee: Regent of St. Cloud Inc. Group owner: Regent Communications Inc. (acq 5-1-99; grpsl). Format: Hot adult contemp. Target aud: 35-54. ♦Terry Jacobs, CEO; Bill Stakelin, pres; Fred Murr, sr VP; David Engberg, gen mgr & sls dir; T.J. Randall, progmg dir; Lee Voss, news dir; Mark Young, chief of engrg.

Coleraine

KGPZ(FM)— July 1, 1995: 96.1 mhz; 100 kw. 577 ft TL: N47 19 31 W93 16 18. Hrs open: 24 Box 447, Grand Rapids, 55744-0447. Phone: (218) 327-3339. Fax: (218) 327-3425. E-mail: kgpz@paulbunyan.net Web Site: www.kgpzfm.com. Licensee: Latto Northland Broadcasting Inc. Group owner: Lew Latto Group of Northland Radio Stations Natl. Network: ABC. Law Firm: Pepper & Corazzini. Format: Real country. Target aud: 35-64. ♦Lew Latto, pres; Dory Butala, gen mgr; Dennis Yourczek, opns dir.

Collegeville

***KBPR(FM)**—Brainerd, February 1988: 90.7 mhz; 34.2 kw. 679 ft TL: N46 25 21 W94 27 41. Stereo. Hrs opn: 24
Rebroadcast of KSJN(FM) Minneapolis-St. Paul.
45 E. 7th St., St. Paul, 55101. Phone: (651) 290-1500. Fax: (651) 290-1224. Web Site: www.mpr.org. Licensee: Minnesota Public Radio. Natl. Network: PRI, NPR. Rgnl. Network: Minn. Pub. Format: Class. News staff: 2. Target aud: General. ♦William H. Kling, pres & gen mgr; Erik Nycklemoe, opns mgr & dev VP.

***KNSR(FM)**— Aug 29, 1988: 88.9 mhz; 100 kw. 728 ft TL: N45 29 52 W94 32 14. Stereo. Hrs opn: 24 Box 7711, St. John's Univ., 56321. Phone: (320) 363-7702. Fax: (320) 363-4948. E-mail: knsr@mpr.org Web Site: www.minnesotapublicradio.org. Licensee: Minnesota Public Radio. Natl. Network: PRI, NPR. Rgnl. Network: Minn. Pub. Format: News. News staff: 2. Target aud: General. ♦William H. Kling, pres; Mike Olson, gen mgr.

***KSJR-FM**— Jan 21, 1967: 90.1 mhz; 100 kw. 700 ft TL: N45 29 52 W94 32 14. Stereo. Hrs opn: 24
Rebroadcast of KSJN(FM) Minneapolis.
45 E. 7th St., St. Paul, 55101. Phone: (651) 290-1500. Fax: (651) 290-1224. Web Site: www.mpr.org. Licensee: Minnesota Public Radio. Population served: 20,000 Natl. Network: PRI. Rgnl. Network: Minn. Pub. Format: Class. News staff: 2. Target aud: General. ♦William H. Kling, pres, gen mgr & stn mgr.

Coon Rapids

WFMP(FM)—Licensed to Coon Rapids. See Minneapolis-St. Paul

Crookston

KQHT(FM)—Licensed to Crookston. See Grand Forks ND

KROX(AM)— April 1948: 1260 khz; 1 kw-D, 500 w-N, DA-N. TL: N47 47 20 W96 35 40. Hrs open: 208 S. Main St., 56716-0620. Phone: (218) 281-1140. Fax: (218) 281-5036. E-mail: kroxam@hotmail.com Web Site: www.kroxam.com. Licensee: Gopher Communications Co. (acq 5-11-87). Natl. Network: CNN Radio. Rgnl. Network: MNN Wire Svc: AP Format: Soft adult contemp, country, MOR, talk. Target aud: 35 plus; general. Spec prog: Farm 10 hrs wkly. ♦Frank Fee, pres & gen mgr; Jeanette Fee, VP.

KYCK(FM)— Mar 4, 1980: 97.1 mhz; 100 kw. 360 ft TL: N47 49 17 W96 49 03. Stereo. Hrs opn: 24 Box 13638, Grand Forks, ND, 58208. Phone: (701) 775-4611. Fax: (701) 772-0540. E-mail: morningkyck@97kyck.com Web Site: www.97kyck.com. Licensee: Leighton Enterprises Inc. (group owner). Population served: 175,000 News staff: 2. ♦Jarrod Thomas, pres & opns mgr; Jack Hansen, VP & gen mgr; Phil O'Reilley, progmg dir.

Crosby

KFGI(FM)— Oct 10, 1990: 101.5 mhz; 25 kw. Ant 328 ft TL: N46 33 52 W93 57 03. Hrs open: Box 140, Aitkin, 56431. Phone: (218) 927-2100. Fax: (218) 927-4090. Web Site: www.kkinradio.com. Licensee: Red Rock Radio Corp. (group owner; (acq 9-1-2006; grpsl). Format: Classic rock. ♦Terry Dee, gen mgr.

Dassel

KARP-FM—Licensed to Dassel. See Hutchinson

Deer River

KBAJ(FM)— 2000: 105.5 mhz; 100 kw. 508 ft TL: N47 20 22 W93 23 48. Hrs open:
Rebroadcasts KQDS-FM Duluth 100%.
501 Lake Ave. S., Suite 200, Duluth, 55802. Phone: (218) 722-0921. Fax: (218) 723-1499. Licensee: Red Rock Radio Corp. (group owner; acq 1-10-00; grpsl). Format: Classic new rock. ♦Shawn Skramstad, gen mgr.

Detroit Lakes

KDLM(AM)— October 1951: 1340 khz; 1 kw-U. TL: N46 50 14 W95 50 17. Hrs open: 24 Box 746, 56502-0746. Phone: (218) 847-5624. Fax: (218) 847-7657. E-mail: kdlmkbot@lakesnet.net Web Site: www.1340kdlm.com. Licensee: Leighton Enterprises Inc. (group owner) Population served: 30,000 Natl. Network: CBS. Rgnl. Network: MNN. Wire Svc: AP Format: Sports, news/talk, adult contemp. News staff: one; News: 10 hrs wkly. Target aud: 30 plus. Spec prog: Farm one hr, relg 8 hrs wkly. ♦Alver Leighton, chmn; John Sowada, pres; Denny Niess, VP; Jeff Leighton, gen mgr; Andy Lia, opns mgr.

KRCQ(FM)— July 4, 1994: 102.3 mhz; 50 kw. 492 ft TL: N46 48 24 W95 46 23. Stereo. Hrs opn: 24 Box 556, 1119 Jackson Ave., 56502. Phone: (218) 847-2001. Fax: (218) 847-2271. E-mail: krcq@lakesnet.net Licensee: Detroit Lakes Broadcasting Co. Inc. (acq 7-8-97; $1.2 million). Population served: 300,000 Law Firm: Miller & Miller. Wire Svc: AP Format: Real country. News staff: one; News: 10 hrs wkly. Target aud: General. ♦Robert D. Spilman, gen mgr.

Dilworth

WZFN(AM)—Not on air, target date: unknown: 1100 khz; 50 kw-D, 1 kw-N, 4.4 kw-CH, DA-N. TL: N46 45 54 W96 40 05. Hrs open: 415 N. College St., Brantley Broadcast Associates LLC, Greenville, AL, 36037. Phone: (205) 618-2020. Licensee: Brantley Broadcast Associates LLC. ♦Joan K. Reynolds, gen mgr.

Duluth

KDAL(AM)— Nov 26, 1936: 610 khz; 5 kw-U, DA-N. TL: N46 43 13 W92 10 34. Hrs open: 715 E. Central Entrance, 55811. Phone: (218) 722-4321. Fax: (218) 722-5423. Licensee: Midwest Communications Inc. (group owner; acq 8-1-01; grpsl). Population served: 100,578 Natl. Network: CBS. Rgnl. Network: MNN, Midwest Radio. Rgnl rep: Hyett/Ramsland. Law Firm: Rosenman & Colin. Format: Div, news/talk. Target aud: 35-64. ♦Duke Wright, CEO & pres; Gary Tesch, CFO & exec VP; Mark Fleischer, opns dir & opns mgr; Dave Strandberg, mus dir; John Talcott, engrg mgr; Pat Cadigan, disc jockey.

KDAL-FM— July 1985: 95.7 mhz; 100 kw. Ant 725 ft TL: N46 47 15 W92 07 21. Stereo. E-mail: tr@957thebridge.com Format: Adult contemp, gold. Target aud: 25-49.

KDNI(FM)—Licensed to Duluth. See Roseville

***KDNW(FM)**— December 1993: 97.3 mhz; 40 kw. 548 ft TL: N46 47 20 W92 07 04. Stereo. Hrs opn: 24 1101 E. Central Entrance, 55811. Phone: (218) 722-6700. Fax: (218) 722-1092. E-mail: kdnw@kdnw.fm Web Site: www.kdnw.fm. Licensee: Northwestern College. Group owner: Northwestern College & Radio (acq 12-4-91; $20,000; FTR: 1-6-92). Natl. Network: AP Radio. Law Firm: Bryan Cave. Format: Contemp Christian music. Target aud: 25-54. ♦Paul Virts, sr VP; Paul Harkness, stn mgr.

KKCB(FM)—Listing follows WEBC(AM).

KLDJ(FM)— Jan 1, 1994: 101.7 mhz; 18.5 kw. Ant 823 ft TL: N46 47 13 W92 07 17. Hrs open: 24 14 E. Central Entrance, 55811. Phone: (218) 727-4500. Fax: (218) 727-9356. Web Site: www.kool1017.com. Licensee: CC Licenses LLC. Group owner: Clear Channel Communications Inc. (acq 5-2-2003; grpsl). Natl. Rep: Christal. Format: Oldies. Target aud: 25-54; general. ♦Ron Stone, gen mgr; Derek Moran, opns mgr; Ryan Barnholdt, prom dir; Scott Klohan, progmg dir; Brian Dennish, sports cmtr.

KQDS-FM— Apr 1, 1976: 94.9 mhz; 100 kw. Ant 730 ft TL: N46 47 41 W92 07 05. Stereo. Hrs opn: 24 501 Lake Ave S., Suite 200, 55802. Phone: (218) 728-9500. Fax: (218) 723-1499. E-mail: production @redrockradio.org Licensee: Red Rock Radio Corp. (group owner; (acq 1-10-2000; grpsl). Population served: 377,300 Natl. Rep: McGavren Guild. Rgnl rep: O'Malley. Format: Classic rock, AOR. Target aud: 25-54. ♦Shawn Skramstad, gen mgr; Jeff Anderson, gen sls mgr; Bill Jones, progmg dir & news dir; Carlene Burstad, traf mgr.

KQDS(AM)— Mar 11, 1963: 1490 khz; 1 kw-U. TL: N46 47 42 W92 07 08.24 Format: Oldies.

KTCO(FM)— June 14, 1972: 98.9 mhz; 100 kw. 600 ft TL: N46 47 30 W92 06 59. Stereo. Hrs opn: 24 715 E. Central Entrance, 55811. Phone: (218) 722-4321. Fax: (218) 722-5423. E-mail: david@ktco.fm.net Web Site: www.ktco.fm.com. Licensee: Midwest Communications Inc. (group owner; acq 8-1-01; grpsl). Population served: 200,000 Rgnl rep: Hyett/Ramsland. Law Firm: Rosenman & Colin. Format: Country hits. Target aud: 25-49. ♦Duke Wright, pres; Gary Tesch, exec VP; Don Snyder, sls VP; Dave Strandberg, news dir; John Talcott, chief of engrg; Carla McCullough, disc jockey; Tim Roubik, disc jockey.

***KUMD-FM**— May 26, 1971: 103.3 mhz; 95 kw. 820 ft TL: N46 47 31 W92 07 21. Stereo. Hrs opn: 5 AM-3 AM (M-F); 6 AM-11 PM (Su) 130 Humanities Bldg., University of Minnesota, 55812. Phone: (218)

726-7181. Fax: (218) 726-6571. E-mail: kumd@kumd.org Web Site: www.kumd.org. Licensee: Board of Regents of University of Minnesota. (acq 8-75). Population served: 100,600 Natl. Network: PRI. Law Firm: Dow, Lohnes & Albertson. Format: Triple A. News: 12 hrs wkly. Target aud: 25-45. ♦Michael Dean, gen mgr; Paul Damberg, dev dir; John Ziegler, progmg dir.

WDSM(AM)—See Superior, WI

WEBC(AM)— June 1924: 560 khz; 5 kw-U, DA-2. TL: N46 38 37 W91 59 09. Hrs open: 14 E. Central Entrance, 55811-5508. Phone: (218) 727-4500. Fax: (218) 727-9356. Web Site: www.560webc.com. Licensee: CC Licenses LLC. Group owner: Clear Channel Communications Inc. (acq 5-2-2003; grpsl). Population served: 422,000 Natl. Rep: Christal. Format: News/talk, sports. ♦Ron Stone, gen mgr.

KKCB(FM)—Co-owned with WEBC(AM). 1966: 105.1 mhz; 100 kw. 789 ft TL: N46 47 21 W92 06 51. Stereo. Web Site: www.kkcb.com. Format: Country. Target aud: General.

WGEE(AM)—See Superior, WI

***WIRR(FM)**—Virginia-Hibbing, December 1985: 90.9 mhz; 21 kw. 552 ft TL: N47 29 46 W92 47 05. Stereo. Hrs opn: 24 224 Holiday Ctr., 55802. Phone: (218) 722-9411. Fax: (218) 720-4900. Web Site: www.mpr.org. Licensee: Minnesota Public Radio Inc. Natl. Network: PRI, NPR. Rgnl. Network: Minn. Pub. Format: Class. News staff: 3. Target aud: General. ♦William H. Kling, pres; Kat Eldred, gen mgr.

***WJRF(FM)**— Nov 1, 1982: 89.5 mhz; 2.85 kw vert. Ant 512 ft TL: N46 47 21 W92 07 09. Stereo. Hrs opn: 24 4604 Airpark Blvd., 55811. Phone: (218) 722-3017. Fax: (218) 722-1650. Web Site: www.refugeradio.com. Licensee: Refuge Media Group. Population served: 180,000 Format: contemp Christian. News: 5 hrs wkly. Target aud: 18-34; female. ♦Brett M. Gibson, CEO & gen mgr; Paul Hitchcock, pres; Keith Johnson, VP.

***WSCD-FM**— 1975: 92.9 mhz; 70 kw. 614 ft TL: N46 47 20 W92 07 04. Stereo. Hrs opn: 224 Holiday Ctr., 55802. Phone: (218) 722-9411. Fax: (218) 720-4900. Web Site: www.mpr.org. Licensee: Minnesota Public Radio. Population served: 22,000 Format: Classical music. ♦William Kling, pres; Kat Eldred, gen mgr.

WWJC(AM)— Apr 26, 1963: 850 khz; 10 kw-D. TL: N46 39 19 W92 12 40. Hrs open: 1120 E. McCuen St., 55808. Phone: (218) 626-2738. Fax: (603) 907-7881. E-mail: radio@wwjc.com Web Site: www.wwjc.com. Licensee: WWJC Inc. Population served: 225,000 Natl. Network: USA. Format: Solid gospel / talk. ♦Ted Elm, gen mgr.

Eagan

KKMS(AM)—See Minneapolis-St. Paul

East Grand Forks

KCNN(AM)— Aug 14, 1959: 1590 khz; 5 kw-D, 1 kw-N, DA-2. TL: N47 52 41 W97 00 24. Hrs open: 24 Box 13638, Grand Forks, 58208-3638. Secondary address: Old Belmont Rd. S., Grand Forks, ND 58201. Phone: (701) 772-2204. Fax: (701) 772-0540. E-mail: general@kcnn.com Web Site: www.leightonbroadcasting.com. Licensee: Leighton Enterprises Inc. (group owner; (acq 11-14-03;. $2.5 million). Population served: 100,000 Natl. Network: CBS, CNN Radio. Wire Svc: AP Format: News/talk. News: 2. Target aud: 25-60. Spec prog: Farm 6 hrs wkly. ♦Jack Hansen, gen mgr; Jarrod Thomas, opns mgr; Linn Hodgson, gen sls mgr.

KZLT-FM—Co-owned with KCNN(AM). Apr 1, 1975: 104.3 mhz; 100 kw. 550 ft TL: N47 48 37 W96 55 46. E-mail: general@kcnn.com Web Site: www.1043moremusic.com.100,000 Format: Adult contemp.

KSNR(FM)—Thief River Falls, May 1976: 100.3 mhz; 100 kw. Ant 620 ft TL: N47 58 38 W96 36 42. Stereo. Hrs opn: 24 505 University Ave., Grand Forks, ND, 58203. Phone: (701) 746-1417. Fax: (701) 746-1410. E-mail: koolradio@hotmail.com Web Site: www.ksnrfm100.com. Licensee: Citicasters Licenses L.P. Group owner: Clear Channel Communications Inc. (acq 10-26-99; grpsl). Population served: 200,000 Law Firm: Haley, Bader & Potts. Format: Country. News staff: one; News: 10 hrs wkly. Target aud: 25-54; boomers & kids. Spec prog: Farm one hr wkly. ♦Pat McLean, gen mgr & gen sls mgr; Susie Johnson, prom mgr; David Andrews, progmg dir; Ken Morgan, pub affrs dir; Dave Schroeder, engrg mgr & chief of engrg; Shannon Stone, local news ed; Josh Jones, disc jockey.

Eden Prairie

WGVZ(FM)— March 1993: 105.7 mhz; 6 kw horiz, 5.8 kw vert. Ant 239 ft TL: N44 53 51 W93 24 22. Hrs open: 24
Rebroadcasts WGVX(FM) Lakeville.
2000 S. Elm St., Minneapolis, 55414. Phone: (612) 617-4000. Fax: (612) 676-8292. Web Site: www.love105.fm. Licensee: Radio License Holding III LLC. Group owner: ABC Inc. (acq 6-12-2007; grpsl). Population served: 500,000 Natl. Rep: Interep. Format: Love songs of the 60s, 70s and 80s. ♦Marc Kalman, pres & gen mgr; Dave Hamilton, opns mgr; Pete Frisch, sls dir; Susan Larkin, gen sls mgr; Leslie Heinemann, natl sls mgr; Julia Schertz, rgnl sls mgr; Brook Johnson, mktg dir; Joni Schmidt, prom mgr; Chris Rahn, progmg dir; Ben Gnam, mus dir & pub affrs dir; Christopher Taykalo, pub affrs dir; Dave Szaflarski, chief of engrg.

Elk River

KLCI(FM)—Licensed to Elk River. See Princeton

Ely

WELY(AM)— Oct 2, 1954: 1450 khz; 1 kw-U, DA-N. TL: N47 53 37 W91 51 59. (CP: TL: N47 53 40 W91 51 50). Hrs opn: 133 E. Chapman St., 55731-1229. Phone: (218) 365-4444. Fax: (218) 365-3657. E-mail: wely@spacestar.net Web Site: www.wely.com. Licensee: Bois Forte Tribal Council (acq 6-1-2005; $445,000 with co-located FM). Population served: 20,000 Law Firm: Rini & Coran. Format: Var/div. Target aud: General; senior citizens. ♦Bill Roloff, gen mgr.

WELY-FM— July 25, 1992: 94.5 mhz; 6 kw. Ant 328 ft TL: N47 53 40 W91 51 50. (CP: 14.5 kw, ant 338 ft). Stereo. E-mail: wely@spacestar.net Web Site: www.wely.com.

Eveleth

KRBT(AM)— December 1948: 1340 khz; 1 kw-U. TL: N47 28 40 W92 32 00. Hrs open: 24 Box 650, 906 Old Hwy. 53, 55734. Phone: (218) 741-5922. Fax: (218) 741-7302. E-mail: weve@spacestar.net Licensee: Iron Range Broadcasting Inc. Group owner: Lew Latto Group of Northland Radio Stations (acq 5-1-78). Population served: 30,000 Format: News/talk. Target aud: 25-54. Spec prog: Finnish one hr, polka 3 hrs wkly. ♦Nancy Grummett, gen mgr & gen sls mgr; Dennis Jerrold, opns mgr & progmg dir; Steve Carlson, news dir; Dave Houston, chief of engrg; Dawn Hoyt, traf mgr.

WEVE-FM—Co-owned with KRBT(AM). June 26, 1978: 97.9 mhz; 71 kw. 555 ft TL: N47 35 53 W92 13 26. Stereo. 24 75,000 Format: Adult contemp. News staff: one; News: one hr wkly. ♦Annie Wargowski, stn mgr & traf mgr.

Fairmont

KFMC(FM)—Listing follows KSUM(AM).

KSUM(AM)— Jan 1, 1949: 1370 khz; 1 kw-U, DA-2. TL: N43 37 45 W94 29 00. Stereo. Hrs opn: Box 491, 56031. Secondary address: 1371 W. Lair Rd. 56031. Phone: (507) 235-5595. Fax: (507) 235-5973. E-mail: ksum@bevcomm.net Web Site: www.ksum.com. Licensee: Woodward Broadcasting Inc. (acq 11-1-62). Population served: 20,200 Natl. Rep: Hyett/Ramsland. Law Firm: Booth, Freret, Imlay & Tepper. Format: News. sports, agriculture info. Target aud: General. ♦Charles Woodward, gen mgr.

KFMC(FM)—Co-owned with KSUM(AM). July 31, 1978: 106.5 mhz; 100 kw. 400 ft TL: N43 37 45 W94 29 00. Stereo. 24 Web Site: www.ksum.com.113,530 Format: Classic rock. Target aud: 25-54.

Faribault

KBGY(FM)— October 2001: 107.5 mhz; 48 kw. 394 ft TL: N44 12 42 W93 20 18. Stereo. Hrs opn: 24 14589 Grand Ave. S., Burnsville, 55306. Phone: (952) 435-5777. Fax: (952) 435-3181. E-mail: info@spirit.fm Web Site: www.spirit.fm. Licensee: Milestone Radio II LLC (acq 10-3-01; $2.2 million). Format: Christian music. News staff: one; News: 10 hrs wkly. Target aud: General; 25-54. ♦Tom Payne, gen mgr.

KDHL(AM)— Jan 10, 1948: 920 khz; 5 kw-U, DA-2. TL: N44 15 47 W93 16 29. Hrs open: 24 601 Central Ave., 55021. Phone: (507) 334-0061. Fax: (507) 334-7057. Web Site: www.kdhlradio.com. Licensee: Cumulus Licensing Corp. Group owner: Cumulus Media Inc. (acq 7-21-98; grpsl). Population served: 250,000 Rgnl. Network: MNN. Format: News, sports, farm. News staff: one; News: 8 hrs wkly. Target aud: 35 plus. ♦Gary Foss, gen mgr; Bob Buck, progmg dir; Gordon Kosfeld, news dir.

KQCL(FM)—Co-owned with KDHL(AM). Jan 10, 1968: 95.9 mhz; 3 kw. 328 ft TL: N44 21 25 W93 11 31. Stereo. Web Site: www.cumulus.com.200,000 Format: Class rock. Target aud: 18-49. ♦Mike Eiler, progmg dir.

Fergus Falls

KBRF(AM)— Oct 20, 1926: 1250 khz; 5 kw-D, 2.2 kw-N, DA-N. TL: N46 16 22 W96 02 41. Hrs open: 24 Box 495, 56538. Secondary address: 728 Western Ave. N. 56537. Phone: (218) 736-7596. Fax: (218) 736-2836. E-mail: kbrfkzcr@prtel.com Licensee: Result Radio Inc. Group owner: The Result Radio Group (acq 1-30-78). Population served: 400,000 Natl. Network: Westwood One. Rgnl. Network: MNN. Natl. Rep: Hyett/Ramsland. Wire Svc: AP Format: Country, news/talk. News staff: one; News: 20 hrs wkly. Target aud: Adults 35+. Spec prog: Farm 15 hrs, relg 7 hrs wkly. ♦Greg Brady, pres & opns mgr; Doug Gray, gen mgr & gen sls mgr; Jerry Papenfuss, CEO, pres & opns mgr; Brian Lokken, news dir.

KZCR(FM)—Co-owned with KBRF(AM). Jan 19, 1968: 103.3 mhz; 100 kw. Ant 649 ft TL: N46 28 06 W96 11 54. Stereo. 24 E-mail: kzcr@prtel.com 400,000 Format: Rock. News staff: one; News: 10 hrs wkly. Target aud: Adults 25-49. ♦Greg Brady, progmg dir; David Bishop, pub affrs dir.

***KCMF(FM)**— June 6, 2003: 89.7 mhz; 2.7 kw. Ant 216 ft TL: N46 19 12 W96 05 32. Hrs open:
Rebroadcasts KSJN(FM) Minneapolis.
Minnesota Public Radio, 45 E. 7th St., St. Paul, 55101. Phone: (651) 290-1500. Fax: (651) 290-1224. Web Site: www.mpr.org. Licensee: Minnesota Public Radio. Format: Class. ♦William H. Kling, gen mgr.

KJJK(AM)— Dec 1, 1986: 1020 khz; 2 kw-D, 370 kw-N. TL: N46 14 43 W95 58 46. Hrs open: 24 Box 495, 728 Western Ave. N., 56537. Phone: (218) 736-7596. Fax: (218) 736-2836. E-mail: kbrfkzcr@prtel.com Licensee: Result Radio Inc. Group owner: The Result Radio Group (acq 3-27-97; $1.1 million with co-located FM). Population served: 54,000 Natl. Network: ABC, Westwood One. Natl. Rep: Hyett/Ramsland. Law Firm: Pepper & Corazzini. Format: Oldies. News staff: one; News: one hr wkly. Target aud: 35 plus; family, home owners, execs, mgrs, dual house income. Spec prog: Relg 3 hrs wkly. ♦Jerry Papenfuss, CEO & pres; Doug Gray, gen mgr, gen sls mgr & gen sls mgr; Gerg Brady, opns mgr; Jeff Swedberg, progmg dir; Brian Lokken, news dir.

KJJK-FM— Oct 14, 1981: 96.5 mhz; 100 kw. 480 ft TL: N46 14 43 W95 58 46. Stereo. 24 400,000 Natl. Network: ABC. Natl. Rep: Hyett/Ramsland. Format: Country. News staff: one; News: 3 hrs wkly. Target aud: 21-54; today's country music fans.

***KNWF(FM)**— April 2003: 91.5 mhz; 100 w. 226 ft Hrs opn:
Rebroadcast of KNOW-FM Minneapolis-St. Paul.
45 E. 7th St., St. Paul, 55101. Phone: (651) 290-1500. Fax: (651) 290-1224. Web Site: www.mpr.org. Licensee: Minnesota Public Radio. Format: News. ♦William H. Kling, gen mgr.

Forest Lake

WLKX-FM— Oct 28, 1978: 95.9 mhz; 3 kw. 300 ft TL: N45 17 40 W93 04 22. Stereo. Hrs opn: 24 15226 W. Freeway Dr., 55025. Phone: (651) 464-6796. Fax: (651) 464-3638. Web Site: www.spirit.fm. Licensee: Lakes Broadcasting Co. Inc. Population served: 50,000 Format: Adult contemp Christian. News staff: one; News: 20 hrs wkly. Target aud: 25-54; general. Spec prog: Auction show 9 hrs, relg 6 hrs wkly. ♦Gary Kastner, gen mgr.

Fosston

KKCQ(AM)— Dec 12, 1966: 1480 khz; 5 kw-D, 90 w-N. TL: N47 33 51 W95 43 27. Hrs open: 24 Box 606, 56542. Secondary address: 35006 Hwy. 2 E. 56542. Phone: (218) 435-1919. Fax: (218) 435-1480. E-mail: info@q107fm.com Web Site: www.q107fm.com. Licensee: Pine to Prairie Broadcasting Inc. (acq 2-1-92; $335,000 with co-located FM; FTR: 2-10-92) Population served: 1,684 Rgnl. Network: MNN. Law Firm: Eugene T. Smith. Format: Talk, oldies. News staff: one; News: 4 hrs wkly. Target aud: 25-54; family-oriented adults. Spec prog: Farm 5 hrs, relg 4 hrs wkly. ♦Bob Overmoe, pres; Phil Ehlke, gen mgr & gen sls mgr; Tom Lano, progmg dir; Jamie Nesvold, mus dir; Karen Bingham, news dir & pub affrs dir; Jim Offerdahl, chief of engrg.

KKEQ(FM)—Co-owned with KKCQ(AM). June 13, 1969: 107.1 mhz; 50 kw. 482 ft TL: N47 33 44 W95 43 30. Stereo. 24 Web Site: www.q107fm.com.250,000 Format: Christian. News: 2 hrs wkly. Target aud: 25-45.

Glencoe

KTTB(FM)— Sept 23, 1993: 96.3 mhz; 100 kw. Ant 577 ft TL: N44 56 25 W93 55 43. Hrs open: 24 5300 Edina Industrial Blvd., Suite 200, Edina, 55439. Phone: (952) 842-7200. Fax: (952) 842-1048. Web Site: www.b96online.com. Licensee: Northern Lights Broadcasting LLC Group owner: Radio One Inc. (acq 8-15-2007; $28 million). Rgnl. Network: Linder Farm. Format: Rhythmic CHR. News staff: one. Target aud: 18-34; adults. ♦Steve Woodbury, VP, gen mgr & engrg VP.

Glenwood

KMGK(FM)— Mar 11, 1983: 107.1 mhz; 3 kw. 300 ft TL: N45 36 53 W95 23 28. Stereo. Hrs opn: 24 Box 241, 56334. Phone: (320) 634-5358. Fax: (320) 634-5359. E-mail: traffic@kmgk1071.com Web Site: www.kmgk1071.com. Licensee: Branstock Communications Inc. Format: Adult contemp. News staff: one; News: 13 wkly. Target aud: 25-54. Spec prog: Rock weekend 70s. ♦Steven R. Nestor, CEO; Jeff Thornton, gen mgr.

Golden Valley

KDIZ(AM)— May 13, 1948: 1440 khz; 5 kw-D, 500 w-N, DA-N. TL: N44 59 20 W93 21 06. Hrs open: 24 2000 Elm St. S.E., Minneapolis, 55414. Fax: (612) 676-8214. Web Site: www.disney.com. Licensee: RD Minneapolis Assets LLC. (Acq 6-30-86). Population served: 659,300 Natl. Rep: ABC Radio Sales. Format: Children. News staff: one; News: 5 hrs wkly. Target aud: 18-49; baby boomers/Generation X. ♦Kevin McCarthy, prom dir; Laura Olson, stn mgr & progmg dir.

KQRS-FM— Sept 1, 1963: 92.5 mhz; 100 kw. 900 ft TL: N44 59 20 W93 21 06. (CP: Ant 1,033 ft.). Stereo. Hrs opn: 24 2000 Elm St. S.E., Minneapolis, 55414. Phone: (612) 617-4000. Fax: (612) 676-8292. Web Site: www.92kqrs.com. Licensee: Radio License Holding III LLC. Group owner: ABC Inc. (acq 6-12-2007; grpsl). Population served: 2,400,000 Natl. Network: ABC. Natl. Rep: ABC Radio Sales. Format: Classic rock. ♦Marc Kalman, pres & gen mgr; Pete Frisch, sls dir; Dave Hamilton, progmg dir; Reed Endersbe, mus dir; David Szaflarski, chief of engrg; Carolyn Kuhieke, rsch dir & traf mgr.

KYCR(AM)—Licensed to Golden Valley. See Minneapolis-St. Paul

Grand Marais

***WLSN(FM)**— 2005: 89.7 mhz; 6 kw. Ant 636 ft TL: N47 46 04 W90 20 47. Hrs open: 224 Holiday Ctr., Duluth, 55802. Phone: (218) 722-9411. Fax: (218) 720-4900. Web Site: www.mpr.org. Licensee: Minnesota Public Radio. Format: News & info. ♦William H. Kling, pres; Kat Eldred, gen mgr.

***WMLS(FM)**—Not on air, target date: unknown: 88.7 mhz; 6 kw. 613 ft Hrs opn: 224 Holiday Ctr., Duluth, 55802. Phone: (218) 722-9411. Fax: (218) 720-4900. Web Site: www.mpr.org. Licensee: Minnesota Public Radio. Format: Classical. ♦William Kling, pres; Kat Eldred, gen mgr; Bob Kelleher, news dir & local news ed; Doug Thompson, engrg dir; Aaron White, chief of engrg; Cynthia Johnson, traf mgr.

***WTIP(FM)**— July 1, 1998: 90.7 mhz; 25 kw. 584 ft TL: N47 46 09 W90 20 49. Hrs open: 5 AM-3 AM
Rebroadcasts KUMD-FM Duluth 70%.
Box 1005, 55604. Secondary address: 55 W. 5th St. 55604. Phone: (218) 387-1070. Fax: (218) 387-1120. E-mail: wtip@boreal.org Web Site: www.wtip.org. Licensee: Cook County Community Radio Corp. Population served: 150,000 Format: AAA, var, adult contemp. News: 15 hrs wkly. Target aud: General. Spec prog: Blues 15 hrs, AOR 15 hrs, progsv rock 15 hrs wkly. ♦Ann Possis, pres; Mike Raymond, exec VP & VP; Deb Benedict, stn mgr; Jeanne Wright, dev dir; Cathy Quinn, progmg dir; Jeff Nemitz, engrg dir.

WXXZ(FM)— 1999: 95.3 mhz; 100 kw. 699 ft TL: N47 39 55 W90 42 22. Hrs open:
Rebroadcasts KQDS-FM Duluth 100%.
501 Lake Ave. S., Suite 200A, Duluth, 55802. Phone: (218) 722-0921. Fax: (218) 723-1499. Licensee: Red Rock Radio Corp. (group owner; acq 1-10-00; grpsl). Format: Classic rock. ♦Shawn Skramstad, gen mgr.

Grand Rapids

***KAXE(FM)**— Apr 23, 1976: 91.7 mhz; 100 kw. 460 ft TL: N47 15 17 W93 26 03. Stereo. Hrs opn: 260 NE 2nd st., 55744. Phone: (218) 326-1234. Fax: (218) 326-1235. E-mail: comments@kaxe.org Web Site: www.kaxe.org. Licensee: Northern Community Radio Inc. Population served: 220,000 Natl. Network: NPR, PRI. Format: Var/div. Target aud: General. ♦Maggie Montgomery, gen mgr; John Bauer, dev dir; Dan Houg, progmg dir & engrg dir; Mark Tarner, progmg dir.

KMFY(FM)—Listing follows KOZY(AM).

KOZY(AM)— Jan 29, 1948: 1320 khz; 5 kw-U, DA-2. TL: N47 10 22 W93 27 10. Hrs open: 24 Box 597, 55744. Secondary address: 507 11th St. S.E. 55744. Phone: (218) 999-5699. Fax: (218) 990-5609. E-mail: kozykmfy@mchsi.com Licensee: Itasca Broadcasting Inc. (acq 4-15-02; with co-located FM). Population served: 44,000 Natl. Network: ABC. Rgnl. Network: MNN. Format: Gold classics. News staff: one. ♦Mike Iaizzo, pres & gen mgr.

KMFY(FM)—Co-owned with KOZY(AM). Dec 5, 1975: 96.9 mhz; 100 kw. 450 ft TL: N47 15 17 W93 26 03. Stereo. 24 Phone: (218) 999-5639.43,000 Format: Adult contemp. News staff: one. Target aud: 35-54.

Granite Falls

KKRC(FM)— Oct 5, 1993: 93.9 mhz; 6 kw. 262 ft TL: N44 54 06 W95 32 53. Hrs open: 24 Box 513, Montevideo, 56001. Phone: (320) 269-8815. Fax: (320) 269-8449. E-mail: kdma@info-link.net Web Site: www.kdmakrngkkrc.com. Licensee: Iowa City Broadcasting Co. Group owner: Tom Ingstad Broadcasting Group Natl. Network: ABC. Natl. Rep: Katz Radio. Format: Oldies. ♦Deanna Hodge, gen mgr & stn mgr; Dwight Mulder, opns mgr.

KMGM(FM)—See Montevideo

Hastings

KDWA(AM)— Oct 24, 1963: 1460 khz; 1 kw-D, 45 w-N. TL: N44 42 49 W92 50 30. Hrs open: 24 514 Vermillion St., 55033. Phone: (651) 437-1460. Fax: (651) 438-3042. E-mail: dan@kdwa.com Licensee: K & M Broadcasting Inc. (acq 6-30-92; $161,000; FTR: 7-20-92). Population served: 250000 Natl. Network: CNN Radio, USA. Rgnl. Network: MNN. Format: Loc news, sports, talk. News staff: 2; News: 18 hrs wkly. Target aud: 25-65; general. ♦Dan Massman, gen mgr.

Hermantown

WWAX(FM)— June 17, 1996: 92.1 mhz; 780 w. 905 ft TL: N46 47 13 W92 07 17. Hrs open: 501 Lake Ave. S., Suite 200A, Duluth, 55802. Phone: (218) 722-0921. Fax: (218) 723-1499. Licensee: Red Rock Radio Corp. (group owner; acq 1-10-00; grpsl). Format: Adult contemp. Target aud: 18-35; general. ♦Shawn Skramstad, gen mgr.

Hibbing

***KADU(FM)**— July 18, 1994: 90.1 mhz; 100 w. Ant 220 ft TL: N47 23 59 W92 57 47. (CP: 2.2 kw horiz, 22 kw vert, ant 371 ft. TL: N47 24 34 W92 57 02). Hrs opn:
Simulcast with KBHW(FM) International Falls 100%.
3309 6th Ave. W., 55746. Phone: (218) 263-4420. Licensee: Heartland Christian Broadcasters Inc. (acq 4-22-2005; $30,000). Format: Christian. ♦Steven Drees, pres & gen mgr.

KMFY(FM)—See Grand Rapids

WIRR(FM)—See Duluth

WMFG(AM)— 1935: 1240 khz; 1 kw-U. TL: N47 24 30 W92 57 04. Hrs open: 24 807 W. 37th St., 55746. Phone: (218) 263-7531. Fax: (218) 263-6112. Licensee: Midwest Communications Inc. (group owner; (acq 5-10-2004; grpsl). Population served: 70,000 Rgnl. Network: MNN. Format: Sports, talk. News staff: 2; News: 13 hrs wkly. Target aud: 25-55; men. Spec prog: Folk, relg 4 hrs, polka 4 hrs wkly. ♦Kristi Garrity, gen mgr, gen sls mgr, mktg dir, prom dir & adv dir; Ben Johnson, progmg dir; Dan Klasmat, engrg dir.

WMFG-FM— 1971: 106.3 mhz; 25 kw. 253 ft TL: N47 24 30 W92 57 04. (CP: 25 kw, ant 259 ft.). Stereo. 24 Format: Oldies. Target aud: 25-54. ♦Dennis Martin, gen sls mgr; Bill Meyes, chief of engrg; Sharon Maki, traf mgr; Tom Tario, news rptr; Kris Jeronimus, sports cmtr.

WNMT(AM)—Nashwauk, June 2, 1975: 650 khz; 10 kw-D, 500 w-N, DA-N. TL: N47 22 31 W93 00 56. (CP: 10 kw-D, 1 kw-N). Hrs opn: 807 W. 37th St., 55746. Phone: (218) 263-7531. Fax: (218) 263-6112. Web Site: www.wnmtradio.com. Licensee: Midwest Communications Inc. (group owner; acq 5-10-2004; grpsl). Population served: 356,000 Format: News/talk. Target aud: 35 plus. Spec prog: Pol 2 hrs wkly. ♦Kristy Garrity, gen mgr; Jim Heitzman, gen sls mgr; Craig Holgate, news dir & farm dir; Danny Klaysmat, chief of engrg.

WTBX(FM)—Co-owned with WNMT(AM). Dec 31, 1980: 93.9 mhz; 100 kw. Ant 531 ft TL: N47 22 24 W93 00 48. Stereo. Web Site: www.wtbx.com. Format: CHR. Target aud: 18-40.

Hutchinson

KARP-FM—Listing follows KDUZ(AM).

KDUZ(AM)— Sept 16, 1953: 1260 khz; 1 kw-D, 64 w-N. TL: N44 54 24 W94 21 59. Hrs open: 24 20132 Hwy. 15 N., 55350-5643. Phone: (320) 587-2140. Fax: (320) 587-5158. E-mail: kduz@hutchtel.net Web Site: www.kduz.com. Licensee: Iowa City Broadcasting Co. Inc. Group owner: Tom Ingstad Broadcasting Group (acq 4-1-2000; grpsl). Population served: 50,000 Rgnl rep: Hyett/Ramsland. Wire Svc: AP Format: Oldies, news, sports. News staff: 2; News: 21 hrs wkly. Target aud: 30 plus; general. Spec prog: Farm 18 hrs, gospel 3 hrs, polka 8 hrs, Sp one hr wkly. ♦Tom Ingstag, chmn; Dale Koktan, gen mgr & gen sls mgr; John Mons, opns mgr & disc jockey; Jim Ohnstad, progmg dir, pub affrs dir & disc jockey; Mark Wodarczyk, news dir & farm dir; Duane Wawyrzniak, chief of engrg; Joel Niemeyer, sports cmtr; Joel Neimeyer, disc jockey.

KARP-FM—Co-owned with KDUZ(AM). June 6, 1968: 106.9 mhz; 7 kw. 554 ft TL: N45 02 43 W94 33 32. Stereo. 24 E-mail: info@karpradio.com Web Site: www.karpradio.com.80,400 Format: Country. News staff: 2; News: 3 hrs wkly. Target aud: 18 plus.

International Falls

***KBHW(FM)**— Jan 4, 1983: 99.5 mhz; 100 kw. Ant 580 ft TL: N48 33 45 W93 49 22. Stereo. Hrs opn: 24 Box 433, 56649. Secondary address: 4090 Hwy.11 56649. Phone: (218) 285-7398. Fax: (218) 285-7419. E-mail: studio@psalm995.org Web Site: www.psalm995.org. Licensee: Heartland Christian Broadcasters. (acq 7-23-99; $1 with KXBR(FM) International Falls). Natl. Network: USA, Moody. Format: Christian. News: 15 hrs wkly. Target aud: General. ♦Bruce Christopherson, gen mgr & chief of engrg.

KGHS(AM)— Sept 1, 1959: 1230 khz; 500 w-D, 250 w-N. TL: N48 35 29 W93 22 54. Hrs open: 24 519 3rd St., 56649. Phone: (218) 283-3481. Fax: (218) 283-3087. E-mail: kghsksdm@northwinds.net Web Site: www.ksdmradio.com. Licensee: Red Rock Radio Corp. (group owner; (acq 9-1-2006; grpsl). Population served: 8,500 Natl. Network: Jones Radio Networks. Rgnl. Network: MNN. Format: Oldies. News staff: one; News: 10 hrs wkly. ♦Ro Grignon, pres; Dennis Martin, gen mgr; Jerry Franzen, news dir; Bill Meys, chief of engrg; Troy Edwards, sports cmtr.

KSDM(FM)—Co-owned with KGHS(AM). Mar 17, 1979: 104.1 mhz; 8.5 kw. 200 ft TL: N48 35 39 W93 22 56. Stereo. 24 Phone: (218) 283-2622. Web Site: www.ksdmradio.com.30,000 Natl. Network: ABC. Format: Country. News staff: one; News: 20 hrs wkly.

***KXBR(FM)**— June, 2000: 91.9 mhz; 1.5 kw. 128 ft TL: N48 34 15 W93 26 19. Hrs open: Box 433, 4090 Hwy. 11, 56649. Phone: (218) 285-9190. Fax: (218) 285-7419. E-mail: dj@edge919.com Web Site: www.edge919.com. Licensee: Heartland Christian Broadcasters. (acq 7-23-99; $1 with KBHW(FM) International Falls). Format: Christian rock. ♦Bruce Christopherson, gen mgr & chief of engrg.

Jackson

KKOJ(AM)— July 10, 1980: 1190 khz; 5 kw-D, DA. TL: N43 31 45 W95 00 02. Hrs open: Sunrise-sunset Box 29, 56143. Secondary address: 71991 US Hwy. 71 56143. Phone: (507) 847-5400. Fax: (507) 847-5745. E-mail: kkoj@rconnect.com Web Site: www.kkoj.com. Licensee: Kleven Broadcasting Co. of Minnesota. Population served: 275,000 Rgnl. Network: Linder Farm. Wire Svc: AP Format: Modern country. News staff: one; News: 20 hrs wkly. Target aud: General. Spec prog: Farm 15 hrs wkly. ♦Doug Johnson, pres, gen mgr & gen sls mgr; Dave Maschoff, news dir; Jerrie Johnson, chief of engrg & traf mgr; Lee Larson, sports cmtr.

KRAQ(FM)— Apr 25, 1994: 105.7 mhz; 25 kw. 328 ft TL: N43 36 54 W94 57 48. Stereo. Hrs opn: 24 Box 29, 56143. Secondary address: 71991 US Hwy. 71 56143. Phone: (507) 847-5400. Fax: (507) 847-5745. E-mail: kkoj@rconnect.com Web Site: www.kkoj.com. Licensee: Kleven Broadcasting Co. of Minnesota. (acq 2-27-98). Population served: 150,000 Natl. Network: AP Radio. Law Firm: Baraff, Koerner & Olender. Wire Svc: AP Format: Oldies/Classic Rock. News staff: one; News: new progmg 18 hrs wkly. ♦Doug Johnson, pres, gen mgr & gen sls mgr; Dave Maschoff, news dir; Jerrie Johnson, traf mgr & sports cmtr.

La Crescent

KQEG(FM)—Licensed to La Crescent. See La Crosse WI

***KXLC(FM)**— Nov 24, 1991: 91.1 mhz; 230 w. 843 ft TL: N43 48 16 W91 22 18. Stereo. Hrs opn: 24 206 S. Broadway, Suite 735, Rochester, 55904. Phone: (507) 282-0910. Fax: (507) 282-2107. Web Site: www.mpr.org. Licensee: Minnesota Public Radio. (acq 4-9-90). Natl. Network: NPR, PRI. Rgnl. Network: Minn. Pub. Format: News. News staff: one; News: 24 hrs wkly. Target aud: General. ♦Chris Cross, gen mgr; Mary Stapek, dev dir; Sea Stachura, news rptr.

Lake City

KLCH(FM)— December 2001: 94.9 mhz; 6 kw. Ant 328 ft TL: N44 22 56 W92 22 05. Hrs open: 24 474 Guernsey Ln., Redwing, 55066. Phone: (651) 388-7151. Fax: (651) 388-7153. E-mail: preding@waittradio.com Licensee: NRG Media LLC. (group owner; (acq 10-31-2005; grpsl). Population served: 35,000 Format: Adult contemp. News staff: 2. Target aud: 25-54. ♦George Pelletier, sr VP; Don Kliewer, gen mgr; Tom Hughes, opns dir.

KMFX-FM— Feb 14, 1991: 102.5 mhz; 9.4 kw. 528 ft TL: N44 16 45 W92 23 38. Hrs open: 1530 Greenview Dr., Rochester, 55902. Phone: (507) 288-3888. Fax: (507) 288-7815. Web Site: www.rochestersquare.com. Licensee: CC Licenses LLC. Group owner: Clear Channel Communications Inc. (acq 10-2000; grpsl). Natl. Rep: D & R Radio. Format: Country. ♦Bob Fox, gen mgr.

Lake Crystal

KQYK(FM)— 2005: 95.7 mhz; 6 kw. Ant 328 ft TL: N44 03 06 W94 17 59. Hrs open: 54934 210th Ln., Mankato, 56001. Phone: (507) 345-4646. Fax: (507) 345-3299. Licensee: Three Eagles of Luverne Inc. (acq 9-12-2005; $620,000 for CP). ♦Ron Gates, gen mgr.

Lakeville

WGVX(FM)— February 1993: 105.1 mhz; 2.6 kw. Ant 499 ft TL: N44 42 05 W93 09 02. Hrs open: 24 2000 S. E. Elm Street, Minneapolis, 55414. Phone: (612) 617-4000. Fax: (612) 676-8292. Web Site: www.love105.fm. Licensee: Radio License Holdiing III LLC. Group owner: ABC Inc. (acq 6-12-2007; grpsl). Population served: 500,000 Natl. Rep: Interep. Format: Love songs of the 60s, 70s and 80s. ♦Marc Kalman, pres & stn mgr; Dave Hamilton, opns mgr; Pete Frisch, sls dir; Susan Larkin, gen sls mgr; Leslie Heinemann, natl sls mgr; Brook Johnson, mktg dir; Joni Schmidt, prom dir; Chris Rahn, progmg dir; Ben Gnam, mus dir; Christopher Taykalo, pub affrs dir; Dave Szaflarski, chief of engrg.

Litchfield

KKLN(FM)—Atwater, Nov 26, 1988: 94.1 mhz; 3 kw. Ant 328 ft TL: N45 04 24 W94 45 20. Stereo. Hrs opn: 24 Kandi Mall, 1605 S. 1st St., Willmar, 56201-4234. Phone: (320) 235-1194. Fax: (320) 235-6894. E-mail: info@kkln.com Web Site: www.kkln.com. Licensee: Flagship Broadcasting. (acq 1999). Format: Classic rock. Target aud: General. ♦Rick Anderson, pres; Justin Klinghagen, gen mgr; Nate Thomas, opns mgr.

KLFD(AM)— Jan 2, 1959: 1410 khz; 500 w-D, 47 w-N. TL: N45 07 02 W94 33 13. Hrs open: 24 234 N. Sibley Ave., 55355. Phone: (320) 693-3281. Fax: (320) 693-3283. E-mail: klfd@hutchtel.net Licensee: Mid-Minnesota Broadcasting Co. (acq 11-26-91; FTR: 12-16-91). Population served: 22,000 Law Firm: Leventhal, Senter & Lerman. Format: Full service. News staff: one; News: 10 hrs wkly. Target aud: 25-54. Spec prog: Farm 20 hrs, relg 3 hrs wkly. ♦Steve Gretsch, pres & opns VP.

Little Falls

KFML(FM)—Listing follows KLTF(AM).

KLTF(AM)— October 1950: 960 khz; 5 kw-D, 35 w-N. TL: N46 00 16 W94 19 42. Hrs open: 24 16405 Haven Rd., 56345. Phone: (320) 632-2992. Fax: (320) 632-2571. E-mail: ads@fallsradio.com Web Site: www.fallsradio.com. Licensee: Little Falls Radio Corp. (group owner; acq 6-28-2004; grpsl). Population served: 100,000 Natl. Network: Fox News Radio. Rgnl. Network: MNN. Format: News/talk. News staff: one; News: 24 hrs wkly. Target aud: 35-65; loc audience who listen for news & info. Spec prog: Farm 8 hrs, relg 4 hrs, polka 2 hrs, party line 5 hrs wkly. ♦Chris Grams, gen mgr & stn mgr; Melanie Lintner, gen sls mgr; Rod Grams, pres & progmg dir; Fred Colby, news dir; Jack Hansen, chief of engrg; Lacy Beeker, traf mgr; Al Windsperger, sports cmtr; Donna Hilmerson, women's int ed.

KFML(FM)—Co-owned with KLTF(AM). November 1988: 94.1 mhz; 6 kw. 275 ft TL: N46 00 16 W94 19 42. Stereo. 24 Web Site: www.fallsradio.com. Format: Adult contemp. News staff: one; News: 3 hrs wkly. Target aud: 25-54. Spec prog: Relg 6 hrs wkly. ♦Cory Fink, disc jockey.

WYRQ(FM)— May 19, 1980: 92.1 mhz; 3 kw. 299 ft TL: N45 56 57 W94 17 48. Stereo. Hrs opn: 24 16405 Haven Rd., 56345. Phone: (320) 632-2992. Fax: (320) 632-2571. Web Site: www.fallsradio.com. Licensee: Little Falls Radio Corp. (group owner; acq 6-28-2004; grpsl). Population served: 50,000 Natl. Network: CNN Radio. Rgnl. Network: MNN. Wire Svc: UPI Format: Agriculture, country, news/talk. News staff: one; News: 20 hrs wkly. Target aud: 25-54; farmers & working people. ♦Chris Grams, gen mgr; Rod Grams, pres & prom mgr; Al Windsperger, progmg dir; Fred Colby, news dir; Mark Persons, chief of engrg.

Long Prairie

KEYL(AM)— Sept 15, 1959: 1400 khz; 1 kw-U. TL: N45 57 45 W94 52 09. Hrs open: 24 Box 187, 221 Central Ave., 56347. Phone: (320) 732-2164. Fax: (320) 732-2284. E-mail: keyl@keylrealcountry.com Licensee: Prairie Broadcasting Co. (acq 10-28-98; $375,000 for stock with KXDL(FM) Browerville). Population served: 28,000 Natl. Network: ABC. Law Firm: Miller & Miller, P.C. Format: Country. News: 15 hrs wkly. Target aud: 25 plus. Spec prog: Farm 5 hrs, sports 10 hrs, relg 4 hrs wkly. ♦Gene Sullivan, pres & gen mgr.

KXDL(FM)—Browerville, May 15, 1992: 99.7 mhz; 6 kw. 328 ft TL: N46 03 15 W94 50 50. Hrs open: 24 Box 187, 56347. Phone: (320) 732-2164. Fax: (320) 732-2284. Licensee: Prairie Broadcasting Co. (acq 10-28-98; $375,000 for stock with KEYL(AM) Long Prairie). Population served: 27,000 Rgnl rep: O'Malley. Law Firm: Miller & Miller. Format: Adult contemp. News: 4 hrs wkly. Target aud: 18-44; female. Spec prog: Sports 2 hrs wkly. ♦Gene Sullivan, pres & gen mgr; Clif Cline, opns mgr.

Luverne

KLQL(FM)—Listing follows KQAD.

KQAD(AM)— Mar 1, 1971: 800 khz; 500 w-D, 80 w-N, DA-2. TL: N43 39 01 W96 10 19. Stereo. Hrs opn: 24 Box 599, County Rd. 4 E., 56156. Phone: (507) 283-4444. Fax: (507) 283-4445. Web Site: www.kqad.net. Licensee: Three Eagles Communications, Luverne. Group owner: Three Eagles Communications (acq 1996; grpsl). Population served: 30,000 Natl. Rep: Hyett/Ramsland. Format: Soft adult contemp. News staff: one; News: 5 hrs wkly. Target aud: 50 plus. Spec prog: Relg 5 hrs wkly. ♦Steve Graphenteen, gen mgr.

Madison

KLQP(FM)— Jan 31, 1983: 92.1 mhz; 25 kw. 300 ft TL: N45 01 37 W96 11 15. Stereo. Hrs opn: 24 Box 70, 623 W. 3rd St., 56256. Phone: (320) 598-7301. Fax: (320) 598-7955. E-mail: klqpfm@farmerstel.net Web Site: www.klqpfm.com. Licensee: Lac Qui Parle Broadcasting Co. Inc. Population served: 40,000 Natl. Network: CNN Radio. Format: Country, oldies. News: 18 hrs wkly. Target aud: General. Spec prog: Farm 5 hrs wkly. ♦Maynard R. Meyer, CEO, pres & gen mgr; Kris Kuechenmeister, opns mgr.

Mahnomen

KRJM(FM)— Aug 27, 2001: 101.5 mhz; 25 kw. 328 ft TL: N47 27 23 W96 07 57. Hrs open: 24 213 N. Main St., 56510. Phone: (218) 935-5355. Fax: (218) 935-9020. Web Site: www.krjmradio.com. Licensee: R & J Broadcasting. (acq. 9-1-01; Format: Oldies. ♦Jim Birkemeyer, gen mgr.

Mankato

KDOG(FM)—Listing follows KTOE(AM).

KEEZ-FM— Apr 1, 1968: 99.1 mhz; 100 kw. 864 ft TL: N43 56 14 W94 24 41. Stereo. Hrs opn: 24 54934 210 Ln., 56001. Phone: (507) 345-4646. Fax: (507) 345-3299. E-mail: zdesk@keez.com Web Site: www.keez.com. Licensee: Three Eagles of Luverne Inc. Group owner: Three Eagles Communications (acq 6-19-00; grpsl). Population served: 350,000 Natl. Network: Westwood One. Rgnl rep: O'Malley. Format: Hot adult contemp. News staff: one. Target aud: 25-54. ♦Ron Gates, gen mgr.

KGAC(FM)—See Saint Peter

***KMSU(FM)**— Jan 7, 1963: 89.7 mhz; 20 kw. 400 ft TL: N44 08 34 W94 00 08. Stereo. Hrs opn: 24 AF 205 Mankato State Univ., 1536 Warren St., 56001. Phone: (507) 389-5678. Fax: (507) 389-1705. Web Site: www.kmsu.org. Licensee: Mankato State University. Population served: 140,000 Natl. Network: PRI. Law Firm: Cohn & Marks. Format: Pub affrs, educ, music. News staff: one; News: 50 hrs wkly. Target aud: General; upscale, educated. Spec prog: Drama 3 hrs, folk/ethnic 5 hrs, new age 5 hrs wkly. ♦Jim Gullickson, gen mgr; Karen Wright, opns dir.

KNGA(FM)—See Saint Peter

KTOE(AM)— 1950: 1420 khz; 5 kw-U, DA-N. TL: N44 10 06 W93 54 37. Hrs open: 24 Box 1420, 56002. Phone: (507) 345-4537. Fax: (507) 345-5364. Web Site: www.ktoe.com. Licensee: Minnesota Valley Broadcasting Co. Group owner: Linder Broadcasting Group Population served: 200,000 Natl. Rep: Katz Radio. Format: Talk. News staff: news progmg 21 hrs wkly News: 3;. Target aud: 25-54. ♦John Linder, CEO; Mike Parry, gen mgr.

KDOG(FM)—Co-owned with KTOE(AM). Apr 1, 1985: 96.7 mhz; 18 kw. Ant 390 ft TL: N44 13 20 W94 07 03. Stereo. Web Site: www.katoinfo.com. Format: Adult contemp. News staff: one; News: 4 hrs wkly.

KXLP(FM)—New Ulm, Nov 21, 1966: 93.1 mhz; 100 kw. 489 ft TL: N44 07 44 W94 11 15. Stereo. Hrs opn: 1807 Lee Blvd., 56003. Phone: (507) 388-2900. Fax: (507) 345-4675. Web Site: www.kxlpradio.com. Licensee: Minnesota Valley Broadcasting Co. Group owner: Clear Channel Communications Inc. (acq 6-8-2007; $3.13 million). Population served: 100,000 Format: Classic rock & roll. Target aud: 25-54. Spec prog: Oldies 15 hrs wkly. ♦Jo Guck Bailey, gen mgr.

KYSM(AM)—Licensed to Mankato. See North Mankato

KYSM-FM—Licensed to Mankato. See North Mankato

Maplewood

WCTS(AM)—Licensed to Maplewood. See Minneapolis-St. Paul

Marshall

KARZ(FM)— July 7, 1985: 107.5 mhz; 25 kw. 300 ft TL: N44 24 37 W95 51 43. (CP: 15 kw, ant 430 ft. TL: N44 19 32 W95 52 19). Stereo. Hrs opn: Box 61, 56258. Secondary address: 1414 E. College Dr. 56258. Phone: (507) 537-0566. Fax: (507) 532-3739. Web Site: www.marshallradio.net. Licensee: KMHL Broadcasting Co. Inc. Group owner: Linder Broadcasting Group (acq 5-6-97; $450,000). Format: Classic rock. Target aud: 25-54. ♦Brad Strootman, gen mgr & sls VP; Keith Petermeier, chief of opns; Scott Schmeling, chief of engrg.

KKCK(FM)—Listing follows KMHL(AM).

KMHL(AM)— Nov 30, 1946: 1400 khz; 1 kw-U. TL: N44 26 55 W95 45 27. (CP: TL: N44 26 59 W95 45 43). Hrs opn: 24 Box 61, 1414 E. College Dr., 56258. Phone: (507) 532-2282. Fax: (507) 532-3739. Web Site: www.marshallradio.com. Licensee: KMHL Broadcasting Co. Group owner: Linder Broadcasting Group Population served: 155,000 Natl. Network: ABC. Rgnl. Network: Linder Farm. Natl. Rep: Katz Radio. Format: Farm, news/talk. News staff: one; News: 20 hrs wkly. Target aud: 27 plus. Spec prog: Relg 7 hrs wkly. ♦Donald Linder, pres & sports cmtr; John Linder, VP; Brad Strootman, gen mgr & sls dir; Keith Petermeier, chief of opns; Justin Thordsen, progmg dir; Aaron Ziemer, pub affrs dir; Scott Schmelling, chief of engrg; Val Braun, traf mgr; Lynn Kettelson, farm dir.

KKCK(FM)—Co-owned with KMHL(AM). Dec 13, 1967: 99.7 mhz; 100 kw. 925 ft TL: N44 26 55 W95 45 27. Stereo. 24 Web Site: marshallradio.net.65,000 Format: Hot adult contemp, rock. News staff: one; News: 8 hrs wkly. Target aud: 21-39. ♦Brad Stoutman, exec VP & adv VP; Keith Petermierer, prom mgr; Russ Berreth, mus dir & mus critic; Val Braun, traf mgr; Aaron Ziemer, local news ed.

Minneapolis

KBEM-FM—Licensed to Minneapolis. See Minneapolis-St. Paul

KFAI(FM)—Licensed to Minneapolis. See Minneapolis-St. Paul

KFAN(AM)—Licensed to Minneapolis. See Minneapolis-St. Paul

KFXN(AM)—Licensed to Minneapolis. See Minneapolis-St. Paul

KMOJ(FM)—Licensed to Minneapolis. See Minneapolis-St. Paul

KSJN(FM)—Licensed to Minneapolis. See Minneapolis-St. Paul

KTCZ-FM—Licensed to Minneapolis. See Minneapolis-St. Paul

KTIS(AM)—Licensed to Minneapolis. See Roseville

KTIS-FM—Licensed to Minneapolis. See Minneapolis-St. Paul

KTLK-FM—Licensed to Minneapolis. See Minneapolis-St. Paul

KUOM(AM)—Licensed to Minneapolis. See Minneapolis-St. Paul

KXXR(FM)—Licensed to Minneapolis. See Minneapolis-St. Paul

WCCO(AM)—Licensed to Minneapolis. See Minneapolis-St. Paul

WLOL(AM)—Licensed to Minneapolis. See Minneapolis-St. Paul

WLTE(FM)—Licensed to Minneapolis. See Minneapolis-St. Paul

WWTC(AM)—Licensed to Minneapolis. See Minneapolis-St. Paul

Minneapolis-St. Paul

***KBEM-FM**—Minneapolis, Oct 4, 1970: 88.5 mhz; 2.15 kw. 370 ft TL: N44 58 38 W93 15 55. Stereo. Hrs opn: 24 1555 James Ave. N., Minneapolis, 55411. Phone: (612) 668-1735. Phone: (612) 529-5236. Fax: (612) 668-1766. E-mail: kbem@mpls.k12.mn.us Web Site: www.jazz88fm.com. Licensee: Special School District No. 1, Board of Education. Population served: 2,000,000 Natl. Network: PRI. Format: Jazz. News staff: one; News: 14 hrs wkly. Target aud: 35 plus; jazz/progsv adults, club/audiophiles. Spec prog: Bluegrass 4 hrs, Sp 4 hrs wkly. ♦Michele McKenzie, gen mgr & stn mgr; Ted Allison, dev dir.

KDIZ(AM)—See Golden Valley

KDWB-FM—Richfield, 1969: 101.3 mhz; 100 kw. Ant 1,033 ft TL: N45 03 30 W93 07 27. Hrs open: 24 1600 Utica Ave. S., Suite 400, Minneapolis, 55416. Phone: (952) 417-3000. Fax: (952) 417-3001. Web Site: www.kdwb.com. Licensee: AMFM Radio Licenses LLC. Group owner: Clear Channel Communications Inc. (acq 8-30-00; grpsl). Population served: 3,127,100 Natl. Rep: Clear Channel. Law Firm: Wiley Rein LLP. Format: CHR. News staff: one. Target aud: 18-34; women. ♦Mick Anselmo, pres & gen mgr.

KEEY-FM—Listing follows KFAN(AM).

***KFAI(FM)**—Minneapolis, May 1, 1978: 90.3 mhz; 125 w. 440 ft TL: N44 58 29 W93 16 17. Stereo. Hrs opn: 24 1808 Riverside Ave., Minneapolis, 55454-1035. Phone: (612) 341-3144. Fax: (612) 341-4281. Web Site: www.kfai.org. Licensee: Fresh Air Inc. Population served: 200,000 Format: Div. News staff: one; News: 7 hrs wkly. Target aud: General; underserved, under-represented communities. Spec prog: Black 10 hrs, folk 6 hrs, Fr 2 hrs, jazz 12 hrs, Sp 8 hrs wkly. ♦Janis Lane-Ewart, gen mgr.

KFAN(AM)—Minneapolis, 1923: 1130 khz; 50 kw-D, 25 kw-N, DA-2. TL: N44 38 48 W93 23 31. Hrs open: 24 1600 Utica Ave. S., Suite 400, Minneapolis, 55416. Phone: (952) 417-3000. Fax: (612) 417-3001. E-mail: kfanwebmaster@clearchannel.com Web Site: www.kfan.com. Licensee: AMFM Broadcasting Licenses LLC. Group owner: Clear Channel Communications Inc. (acq 8-30-2000; grpsl). Natl. Rep: Clear Channel. Format: Sports talk. News staff: 40; News: 20 hrs wkly. Target aud: 25-54; males. ♦Mick Anselmo, VP; Todd Kalman, gen sls mgr; Jeff Framke, natl sls mgr; Matt Tell, mktg dir; Lisa Sanderson, prom dir; Chad Abbott, progmg dir & progmg mgr; John Jansen, pub affrs dir; Dan Motler, engrg VP; Jess Meyer, engrg dir & chief of engrg; Cathy Maness, traf mgr.

KEEY-FM—Co-owned with KFAN(AM). June 1, 1969: 102.1 mhz; 100 kw. 1,033 ft TL: N45 03 30 W93 07 27. Stereo. 24 E-mail: k102online@clearchannel.com Web Site: www.k102.com.309,866 Natl. Rep: Clear Channel. Format: Country. News staff: 40; News: 2 hrs wkly. Target aud: 25-54; women. ♦Mike Anselmo, pres; Rob Berrell, gen sls mgr; Matt Tell, prom dir; Gregg Swedberg, progmg dir; Mary Gallas, mus dir; Cathy Maness, traf mgr.

KFXN(AM)—Minneapolis, Apr 5, 1962: 690 khz; 500 w-D, DA. TL: N45 01 25 W93 22 58. (CP: 1.5 kw-D, 500 w-N. TL: N44 44 58 W92 59 35). Stereo. Hrs opn: 1600 Utica Ave. S., Suite 400, Minneapolis, 55416. Phone: (952) 417-3000. Fax: (952) 417-3001. Web Site: www.thescore690.com. Licensee: AMFM Broadcasting Licenses LLC. Group owner: Clear Channel Communications Inc. (acq 8-30-2000; grpsl). Population served: 434,400 Law Firm: Wiley, Rein & Fielding. Format: Syndicated sports, talk. Target aud: 25-54. ♦Mick Anselmo, pres, opns dir & mktg mgr; Todd Kalman, gen sls mgr; Chad Abbott, progmg dir; Jess Meyer, chief of engrg.

KTCZ-FM—Co-owned with KFXN(AM). 1956: 97.1 mhz; 100 kw. 1,033 ft TL: N45 03 30 W93 07 27. Stereo. 24 Web Site: cities97.com. Law Firm: Wiley, Rein & Fielding. Format: Adult contemp. ♦Mick Anselmo, pres; Erik Christopherson, gen sls mgr; Dave Sheets, prom dir; Lauren MacLeash, progmg dir.

KKMS(AM)—Richfield, Oct 18, 1949: 980 khz; 5 kw-U, DA-1. TL: N44 47 18 W93 12 54. Stereo. Hrs opn: 24 2110 Cliff Rd., Eagan, 55122. Phone: (651) 405-8800. Fax: (651) 405-8222. Web Site: www.kkms.com. Licensee: Common Ground Broadcasting Inc. Group owner: Salem Communications Corp. (acq 9-27-96; $3 million). Population served: 1,961,380 Format: Christian, talk. Target aud: 18-50. ♦John Hunt, gen mgr.

KMNV(AM)—Saint Paul, 1936: 1400 khz; 1 kw-U. TL: N44 57 28 W93 12 23. Hrs open: 24 4640 W. 77th St., Suite 317, Edina, 55435. Phone: (952) 820-8520. Fax: (952) 820-8502. Licensee: Davidson Media Station KLBB Licensee LLC. (acq 9-7-2005; $5.2 million with KRJJ(AM) Brooklyn Park). Population served: 309,866 Rgnl. Network: MNN. Format: Rgnl Mexican. ♦Tim Dennis, gen mgr & opns mgr.

***KMOJ(FM)**—Minneapolis, Sept 15, 1978: 89.9 mhz; 1 kw. 600 ft TL: N44 59 00 W93 17 22. Stereo. Hrs opn: 24 555 Girard Terr., Suite 130, Minneapolis, 55405. Phone: (612) 377-0594. Fax: (612) 377-3990. Licensee: Center for Communication & Development. (acq 1975). Population served: 1,200,000 Format: Urban contemp. News staff: 2; News: 4 hrs wkly. Target aud: General. ♦Kelvin Quarles, gen mgr.

KNOF(FM)—Saint Paul, Apr 10, 1960: 95.3 mhz; 3 kw. 200 ft TL: N44 56 48 W93 09 26. Stereo. Hrs opn: 6 AM-10 PM 1347 Selby Ave., St. Paul, 55104. Phone: (651) 645-8271. Fax: (651) 645-4593. Licensee: Selby Gospel Broadcasting Co. Natl. Network: Salem Radio Network. Format: Gospel. News: 2 hrs wkly. Target aud: All ages. Spec prog: Black 5 hrs, Sp one hr, Russian one hr wkly. ♦Grace Adam, pres & gen mgr; Phil Mullen, opns mgr.

***KNOW-FM**— July 1, 1967: 91.1 mhz; 100 kw. Ant 1,310 ft TL: N45 03 44 W93 08 21. Stereo. Hrs opn: 45 E. 7th St., St. Paul, 55101. Phone: (651) 290-1500. Fax: (651) 290-1224. Web Site: www.mpr.org. Licensee: Minnesota Public Radio. Wire Svc: Reuters Format: News. ♦William H. Kling, pres & gen mgr; Erik Nycklemoe, opns dir.

KQRS-FM—See Golden Valley

***KSJN(FM)**—Minneapolis, 1956: 99.5 mhz; 100 kw. 1,033 ft TL: N44 03 30 W93 07 27. Stereo. Hrs opn: 45 E. 7th St., St. Paul, 55101. Phone: (651) 290-1500. Fax: (651) 290-1224. Web Site: www.mpr.org. Licensee: Minnesota Public Radio Inc. Population served: 2,500,000 Wire Svc: Reuters Format: Class. ♦William H. Kling, pres & gen mgr; Erik Nycklemoe, opns dir.

KSTP(AM)—Saint Paul, April 1924: 1500 khz; 50 kw-U, DA-N. TL: N45 01 32 W93 03 06. Hrs open: 3415 University Ave., Minneapolis, 55414. Phone: (651) 647-1500. Fax: (651) 649-1515. Web Site: www.am1500.com. Licensee: KSTP-AM L.L.C., a Delaware L.L.C. Group owner: Hubbard Broadcasting Inc. Population served: 300,000 Natl. Rep: Christal. Law Firm: Holland & Knight. Wire Svc: AP Format: Talk. News staff: 3; News: 5 hrs wkly. Target aud: 25-54; adults. ♦Stanley S. Hubbard, CEO; Virginia H. Morris, pres; Todd Fisher, VP & gen mgr. Co-owned TV: KSTP-TV affil

KSTP-FM— Nov 1, 1965: 94.5 mhz; 100 kw. 1,225 ft TL: N45 03 45 W93 08 22. Stereo. Phone: (651) 642-4141. Fax: (651) 642-4239. Web Site: www.ks95.com. Licensee: KSTP-FM L.L.C. a Delaware L.L.C.500,000 Natl. Network: ABC. Natl. Rep: Christal. Format: Adult contemp. News staff: one. Target aud: 25-54; female. ♦Dave Bestler, VP & gen mgr. Co-owned TV: KSTP-TV affil

***KTIS-FM**—Minneapolis, May 1949: 98.5 mhz; 100 kw. 1,033 ft TL: N45 03 30 W93 07 27. Stereo. Hrs opn: 24 3003 Snelling Ave. N., St. Paul, 55113-1598. Phone: (651) 631-5000. Fax: (651) 631-5084. Web Site: www.ktis.fm. Licensee: Northwestern College. Group owner: Northwestern College & Radio. Population served: 2,000,000 Natl. Network: AP Network News. Law Firm: Bryan Cave. Format: Inspirational. News staff: one; News: 20 hrs wkly. Target aud: 25-45. ♦Dr. Paul Virts, exec VP; Harv Hendrickson, VP; David Fitts, stn mgr.

KTLK-FM—Minneapolis, June 26, 1965: 100.3 mhz; 97 kw. 905 ft TL: N45 20 12 W93 23 28. (CP: Ant 922 ft.). Stereo. Hrs opn: 24 1600 Utica Ave. S., Suite 400, Minneapolis, 55416. Phone: (952) 417-3000. Fax: (952) 417-3001. Web Site: www.kjzi.com. Licensee: AMFM Broadcasting Licenses LLC. Group owner: Clear Channel Communications Inc. (acq 8-30-2000; grpsl). Population served: 2,500,000 Natl. Network: Fox News Radio. Natl. Rep: Clear Channel. Law Firm: Wiley, Rein & Fielding. Format: Adult contemp, smooth jazz, news/talk. News staff: 3. Target aud: Adults 25-54; adults. ♦Mick Anselmo, pres, gen mgr & mktg mgr.

KTNF(AM)—See Saint Louis Park

***KUOM(AM)**—Minneapolis, Jan 13, 1922: 770 khz; 5 kw-D. TL: N44 59 54 W93 11 18. Hrs open: Sunrise-sunset Univ. of Minn., 330 21st Ave. S., Minneapolis, 55455-0415. Phone: (612) 625-3500. Fax: (612) 625-2112. E-mail: radiok@umn.edu Web Site: www.radiok.org. Licensee: University of Minnesota. Population served: 2,000,000 Law Firm: Dow, Lohnes & Albertson. Wire Svc: AP Format: Alternative. News: 5 hrs wkly. Target aud: 18-34. ♦Andrew Marlow, stn mgr; Stuart Sanders, dev dir; Shelly Miller, progmg dir; Larry Oberg, chief of engrg.

KXXR(FM)—Minneapolis, Jan 6, 1961: 93.7 mhz; 100 kw. 1,033 ft TL: N45 03 30 W93 07 27. Hrs open: 24 2000 S. E. Elm Street, Minneapolis, 55414. Phone: (612) 617-4000. Fax: (612) 676-8293. E-mail: mail@93x.com. Web Site: www.93x.com. Licensee: Radio License Holding III LLC. Group owner: ABC Inc. (acq 6-12-2007; grpsl). Population served: 2,500,000 Natl. Rep: Interep. Format: Active rock. Target aud: 18-54. ♦Marc Kalman, stn mgr; Pete Frisch, sls dir; Wendy Ellis, mktg mgr; Wade Linder, progmg dir.

KYCR(AM)—Golden Valley, Oct 27, 1961: 1570 khz; 2.5 kw-D, 237 w-N. TL: N44 57 39 W93 21 25. Hrs open: 24 2110 Cliff Rd., Eagan, 55122. Phone: (651) 405-8800. Fax: (651) 405-8222. E-mail: info@kycr.com Web Site: www.kycr.com. Licensee: Common Ground Broadcasting Co. Inc. Group owner: Salem Communications Corp. (acq 7-2-98; $2.7 million with KTEK(AM) Alvin, TX). Population served: 2,400,000 Law Firm: Putbrese, Hunsaker & Trent, P. Format: Religious, talk. News: 6 hrs wkly. Target aud: 25-49; 60% female, 40% male. Spec prog: Sp 14 hrs wkly. ♦John Hunt, gen mgr.

WCCO(AM)—Minneapolis, Oct 2, 1924: 830 khz; 50 kw-U. TL: N45 10 40 W93 20 55. (CP: 46 kw-N. TL: N45 05 06 W93 31 06). Hrs opn: 24 625 2nd Ave. S., Minneapolis, 55402. Phone: (612) 370-0611. Fax: (612) 370-0159. E-mail: admin@wccoradio.cbs.com Web Site: www.wccoradio.com. Licensee: Infinity Media Corp. Group owner: Infinity Broadcasting Corp. (acq 11-13-98; grpsl). Population served: 2,000,000 Natl. Network: CBS. Natl. Rep: Interep. Format: News/talk. News staff: 7; News: 25 hrs wkly. Target aud: General. ♦Mary Niemeyer, gen sls mgr; Wendy Paulson, opns mgr & progmg mgr.

WLTE(FM)—Co-owned with WCCO-TV. Aug 27, 1973: 102.9 mhz; 100 kw. 1,033 ft TL: N45 03 30 W93 07 27. Stereo. 24 Phone: (612) 339-1029. Phone: (612) 339-1083. Fax: (612) 339-5653. Web Site: www.wlte.com. Format: Soft adult contemp. News: 5 hrs wkly. Target aud: 25-54. ♦John McMonagle, natl sls mgr.

WCTS(AM)—Maplewood, August 1964: 1030 khz; 50 kw-D, 1 kw-N, DA-2. TL: N44 52 01 W92 54 02. Hrs open: 24 900 Forestview Ln. N., Plymouth, 55441-5934. Phone: (763) 417-8270. Fax: (763) 417-8278. Web Site: www.wctsradio.com. Licensee: Central Baptist Theological Seminary of Plymouth. (acq 1-30-93; $1.5 million; FTR: 11-23-92). Population served: 370,000 Natl. Network: Moody. Wire Svc: AP Format: Christian, relg. News: 3 hrs wkly. Target aud: 40 plus; Christian. ♦Kimball Cummings Jr., gen mgr; Stephen Davis, opns mgr.

WDGY(AM)—Hudson, WI) Sept 19, 1959: 630 khz; 1 kw-D, 2.5 kw-N, DA-2. TL: N44 52 01 W92 54 02. Hrs open: 2619 E. Lake St., Minneapolis, 55406. Phone: (612) 729-3776. Fax: (612) 724-0437. E-mail: radiorey630am@yahoo.com Web Site: www.radiorey630am.com. Licensee: 630 Radio Inc. Population served: 1,200,000 Format: Spanish music. ♦Guadalupe Gonzales, pres; Manuel Robles, gen mgr.

WFMP(FM)—Listing follows WIXK(AM).

WIXK(AM)—New Richmond, WI) Sept 29, 1960: 1590 khz; 5 kw-D. TL: N45 05 10 W92 34 19. Hrs open: Sunrise-sunset 125 E. 2rd St., New Richmond, WI, 54017. Secondary address: Box 8, New Richmond, WI 54017. Phone: (715) 246-2254. Fax: (715) 246-7090. E-mail: jpetersen@hbi.com Licensee: WIXK-AM LLC. Group owner: Hubbard Broadcasting Inc. (acq 5-18-2000; with co-located FM). Population served: 323,000 Natl. Network: ABC. Law Firm: Miller & Miller. Format: Country. News staff: one; News: 9 hrs wkly. Target aud: 25-54. ♦Stanley S. Hubbard, CEO; Virginia H. Morris, pres; Todd Fisher, gen mgr.

WFMP(FM)—Co-owned with WIXK(AM). Sept 1, 1968: 107.1 mhz; 22 kw. Ant 587 ft TL: N45 03 45 W93 08 21. Stereo. 24 Phone: (651) 642-4107. Fax: (651) 647-2932. Web Site: www.fm107.fm. Licensee: WFMP-FM LLC. (acq 12-21-2000; $27 million). Natl. Rep: Christal. Format: Talk.

WLOL(AM)—Minneapolis, 1939: 1330 khz; 9.7 kw-D, 5.1 kw-N, DA-2. TL: N44 47 02 W93 20 38. Hrs open: 24 919 Lilac Dr. N., Golden Valley, 55422. Phone: (612)643-4119. Fax: (763)546-4444. E-mail: wlol@relevantradio.com Licensee: Starboard Media Foundation Inc. Group owner: Relevant Radio (acq 3-16-2004; $6.75 million). Population served: 2,300,000 Rgnl. Network: MNN. Format: Catholic. News: 3 hrs wkly. ♦Trish Leurck, CEO; Paul Sadek, stn mgr.

***WMCN(FM)**—Saint Paul, Sept 15, 1979: 91.7 mhz; 10 w. 1,004 ft TL: N44 36 22 W93 10 04. Stereo. Hrs opn: 8 AM-4 AM (M-F); 11 AM-4 AM (S, Su) 1600 Grand Ave., St. Paul, 55105. Phone: (651) 696-6082. Phone: (651) 696-6000. Fax: (651) 696-6689. E-mail: wmcn@macalester.edu Web Site: www.macalester.edu/~wmcn. Licensee: Macalester College. Population served: 2,000,000 Format: New rock, jazz. Target aud: All ages. Spec prog: Country 2 hrs, Latin 6 hrs, multicultural 10 hrs, class 4 hrs, folk 2 hrs, jazz 4 hrs, punk 4 hrs wkly. ♦Patrick McGrath, gen mgr; Ethan Torrey, chief of engrg.

WWTC(AM)—Minneapolis, Aug 10, 1925: 1280 khz; 5 kw-U, DA-N. TL: N44 57 41 W93 21 24. Hrs open: 2110 Cliff Rd., Egan, 55122. Phone: (651) 405-8800. Fax: (651) 405-8222. Web Site: www.am1280thepatriot.com. Licensee: SCA License Corp. Group owner: Salem Communications Corp. (acq 12-18-00; $7 million with WRRD(AM) Jackson, WI). Population served: 2,000,000 Format: News/talk. Target aud: 18-54. ♦John Hunt, gen mgr.

Montevideo

***KBPG(FM)**— July, 2002: 89.5 mhz; 500 w. 151 ft TL: N44 54 50 W95 44 10. Hrs open: American Family Radio, Box 3206, Tupelo, MS, 38803. Phone: (662) 844-8888. Fax: (662) 842-6791. Web Site: www.afr.net. Licensee: American Family Association. Group owner: American Family Radio (acq 11-26-99). Natl. Network: USA. Format: Inspirational. ♦Don Wildman, gen mgr.

KDMA(AM)— Dec 21, 1951: 1460 khz; 1 kw-U, DA-N. TL: N44 56 05 W95 44 50. Hrs open: 24 Box 513, 4454 Hwy. 212 W., 56265. Phone: (320) 269-8815. Phone: (320) 269-5131. Fax: (320) 269-8449. E-mail: deekdma@yahoo.com Web Site: www.kdmaradio.com. Licensee: Iowa City Broadcasting Co. Group owner: Tom Ingstad Broadcasting Group (acq 10-21-97; grpsl). Population served: 110,000 Rgnl. Network: Linder Farm. Rgnl rep: O'Malley. Wire Svc: Weather Wire Format: Country. News staff: one. Spec prog: Farm 7 hrs wkly. ♦Deanna Hodge, gen mgr; Dwight Mulder, opns dir, progmg dir & disc jockey; Roger Hill, gen sls mgr; Lynn Ketelson, farm dir.

KMGM(FM)—Co-owned with KDMA(AM). Oct 1, 1982: 105.5 mhz; 3 kw. 300 ft TL: N44 51 24 W95 37 46. Stereo. 24 Web Site: www.kdmaradio.com. Format: Soft hits. Target aud: 25-54. ♦Lynn Ketelson, farm dir; Dwight Mulder, disc jockey.

Moorhead

***KCCD(FM)**— June 1, 1992: 90.3 mhz; 100 kw. 495 ft TL: N46 45 35 W96 36 26. Hrs open: 24 901 S. 8th St., 56562. Phone: (218) 299-3666. Fax: (218) 299-3418. Web Site: www.mpr.org. Licensee: Minnesota Public Radio. Population served: 460,000 Natl. Network: NPR, PRI. Rgnl. Network: Minn. Pub. Format: News. News staff: 2. Target aud: General. ♦William Kling, pres; Vern Goodin, gen mgr; Julia Beaton, dev dir.

***KCCM-FM**— Oct 23, 1971: 91.1 mhz; 67 kw. 656 ft TL: N46 45 35 W96 36 26. Stereo. Hrs opn: 24 Concordia College, 901 8th St. S., 56562. Phone: (218) 299-3666. Fax: (218) 299-3418. Web Site: www.mpr.org. Licensee: Minnesota Public Radio Inc. Population served: 460,000 Natl. Network: NPR, PRI. Rgnl. Network: Minn. Pub. Format: Class mus, cultural progmg. News staff: 2; News: one hr wkly. Target aud: General. ♦William H. Kling, pres; Vern Goodin, gen mgr; Julia Beaton, dev dir.

KLTA(FM)—See Breckenridge

KQWB-FM—Licensed to Moorhead. See Fargo ND

KRWK(FM)—See Fargo, ND

KVOX(AM)— Nov 30, 1937: 1280 khz; 5 kw-D, 1 kw-N, DA-2. TL: N46 49 10 W96 45 56. Hrs open: 24 14378 40th St. S.E., Wheatland, ND, 58079-9732. Phone: (701) 866-2606. Licensee: Voice of Reason Radio Group owner: Clear Channel Communications Inc. (acq 7-31-2007). Population served: 70,000 Format: Catholic. ♦Robert A. Schumacher, pres.

KVOX-FM— Nov 30, 1966: 99.9 mhz; 100 kw. 444 ft TL: N46 49 09 W96 45 56. Stereo. Hrs opn: 24 Box 9919, Fargo, ND, 58106. Secondary address: 2720 7th Ave. St., Fargo, ND 58103. Phone: (701) 237-4500. Fax: (701) 235-9082. E-mail: studio@froggyweb.com Web Site: www.froggyweb.com. Licensee: Monterey Licenses L.L.C. Group owner: Triad Broadcasting Co. LLC. Population served: 194,800 Natl. Rep: Christal. Law Firm: Shaw Pittman. Wire Svc: AP Format: Country. News: 7 hrs wkly. Target aud: 25-54; female skew. ♦David Benjamin, pres; Nancy Odney, gen mgr.

Moose Lake

WMOZ(FM)— 2001: 106.9 mhz; 6 kw. Ant 118 ft TL: N46 30 20 W92 41 10. Hrs open: Agate Broadcasting Inc., 1104 Cloquet Ave., Cloquet, 55720. Phone: (218) 879-4534. Fax: (218) 879-1962. Licensee: QB Broadcasting Ltd. Format: Oldies. ♦Mark Senarighi, gen mgr.

Mora

KBEK(FM)— May 12, 1995: 95.5 mhz; 25 kw. 328 ft TL: N45 44 33 W93 22 48. Stereo. Hrs opn: 24 Box 136, 1947 Dennis Rd., 55051. Phone: (320) 679-6955. Phone: (320) 689-5500. Fax: (320) 679-2348. E-mail: kbek@besttimes.com Web Site: www.besttimes.com. Licensee: Colleen McKinney, personal representative (acq 11-5-2004). Format: Lite rock, golden oldies. News: 8 hrs wkly. Target aud: General. ♦Colleen McKinney, gen mgr, opns mgr & gen sls mgr; Robin Riley, progmg dir & mus dir; Ty Laugerman, sports cmtr.

Morris

KKOK-FM—Listing follows KMRS(AM).

KMRS(AM)— Sept 16, 1956: 1230 khz; 1 kw-U. TL: N45 36 11 W95 53 14. Hrs open: 19 Box 533, 56267. Phone: (320) 589-3131. Fax: (320) 589-2715. E-mail: kmrskkok@info-link.net Licensee: Iowa City Broadcasting Co. Group owner: Tom Ingstad Broadcasting Group (acq 1-11-2000; with co-located FM). Population served: 60,000 Natl. Rep: McGavren Guild. Format: News/talk, MOR. News staff: one; News: 80 hrs wkly. Target aud: 35-64; farmers & agri-business people. ♦Deb Mattheis, gen mgr.

KKOK-FM—Co-owned with KMRS(AM). Sept 16, 1976: 95.7 mhz; 100 kw. 474 ft TL: N45 36 11 W95 53 14. Stereo. Format: Country.

***KUMM(FM)**— Sept 17, 1970: 89.7 mhz; 225 w. Ant 56 ft TL: N45 35 20 W95 54 22. Stereo. Hrs opn: 24 KUMM, 600 E. 4th St., 56267. Phone: (320) 589-6076. Fax: (320) 589-6084. E-mail: kumm@kumm.org Web Site: www.kumm.org. Licensee: University of Minnesota. Population served: 7,000 Law Firm: Dow, Lohnes & Albertson. Wire Svc: AP Format: Adult alternative. News: 3 hrs wkly. Target aud: 18-30; primarily college students. ♦Mike Doucette, gen mgr.

Nashwauk

KMFG(FM)— October 1997: 102.9 mhz; 25 kw. 253 ft TL: N47 24 30 W92 57 05. Hrs open: 807 W. 37th St., Hibbing, 55746. Phone: (218) 263-7531. Fax: (218) 263-6112. Licensee: Midwest Communications Inc. (group owner; acq 5-10-2004; grpsl). Natl. Network: ABC. Format: Classic rock. Target aud: 24-55. ♦Kristi Garrity, gen mgr.

WMFG(AM)—See Hibbing

WNMT(AM)—Licensed to Nashwauk. See Hibbing

New Prague

KCHK(AM)— Sept 22, 1969: 1350 khz; 500 w-D, 70 w-N, DA-2. TL: N44 34 39 W93 30 16. Hrs open: 24 Box 251, 56071. Secondary address: 25821 Langford Ave. 56071. Phone: (952) 758-2571. Phone:

(952) 758-2572. Fax: (952) 758-3170. Licensee: Ingstad Brothers Broadcasting LLC (group owner; acq 5-1-2004; grpsl). Population served: 4,000 Natl. Network: ABC. Rgnl. Network: MNN. Law Firm: Rosenman & Colin L.L.P. Wire Svc: NWS (National Weather Service) Format: News/talk. News: 50 hrs wkly. Target aud: 35-59. Spec prog: Sp. 6 hrs, Pol, Czch, Ger 40 hrs wkly. ♦Ned Newberg, gen mgr; Dave Douglas, engrg mgr.

KRDS-FM—Co-owned with KCHK(AM). Dec 1, 1990: 95.5 mhz; 6 kw. 328 ft TL: N44 27 41 W93 35 21. Stereo. 24 E-mail: kchkamfm@beucomm.net 140,000 Format: Oldies. News staff: 3; News: 7 hrs wkly. Target aud: 25-64; general.

New Ulm

KNUJ(AM)— May 1949: 860 khz; 1 kw-U. TL: N44 17 10 W94 25 50. Hrs open: 24 Box 368, Grand Hotel Bldg., 56073. Phone: (507) 359-2921. Fax: (507) 359-4520. Web Site: www.knuj.net. Licensee: Ingstad Brothers Broadcasting LLC (group owner; acq 5-1-2004; grpsl). Population served: 43,000 Format: Country. News staff: 2; News: 30 hrs wkly. Target aud: 30 plus. Spec prog: Old-time 8 hrs wkly. ♦Jim Bartels, sr VP; Marj Frederickson, gen mgr.

KXLP(FM)—Licensed to New Ulm. See Mankato

Nisswa

KBLB(FM)— 2002: 93.3 mhz; 100 kw. Ant 558 ft TL: N46 26 34 W94 22 55. Stereo. Hrs opn: 24 Box 746, Brainerd, 56401-0746. Secondary address: 13225 Dogwood Dr., Baxter 56425-8613. Phone: (218) 828-1244. Fax: (218) 828-1119. E-mail: production@brainerd.net Web Site: brainerdradio.net. Licensee: BL Broadcasting Inc. Group owner: Omni Broadcasting Co. (acq 12-11-2000). Population served: 200,000 Natl. Network: ABC. Law Firm: Garvey, Schubert & Barer. Format: Country, mainstream. News staff: one; News: 12 hrs wkly. Target aud: 25-54; adults. ♦Lou Buron, CEO & pres; Mary Campbell, CFO & VP; G. Michael Boen, gen mgr; Al Davison, progmg dir; Tess Taylor, news dir & pub affrs dir; Dave Cox, chief of engrg.

North Branch

***KMKL(FM)**— Oct 6, 2001: 90.3 mhz; 195 w. Ant 364 ft TL: N45 24 42 W92 54 27. Stereo. Hrs opn: 24 2351 Sunset Blvd., Suite 170-218, Rocklin, CA, 95765. Phone: (916) 251-1600. Fax: (916) 251-1650. E-mail: klove@klove.com Web Site: www.klove.com. Licensee: Educational Media Foundation. Group owner: EMF Broadcasting. Population served: 52,400 Natl. Network: K-Love. Law Firm: Shaw Pittman. Format: Contemp Christian. News staff: 3. Target aud: 25-44; Judeo Christian, female. ♦Richard Jenkins, pres; Mike Novak, VP & progmg dir; Lloyd Parker, gen mgr; Ed Lenane, opns dir & news dir; Keith Whipple, dev dir; David Pierce, progmg mgr; Jon Rivers, mus dir; Sam Wallington, engrg dir; Arthur Vassar, traf mgr; Karen Johnson, news rptr; Marya Morgan, news rptr; Richard Hunt, news rptr.

North Mankato

KDOG(FM)—Licensed to North Mankato. See Mankato

KYSM-FM—Mankato, April 1948: 103.5 mhz; 100 kw. Ant 541 ft TL: N44 10 20 W94 02 23. Stereo. Hrs opn: 24 1807 Lee Blvd., 56003. Phone: (507) 388-2900. Fax: (507) 345-4675. E-mail: jobailey@clearchannel.com Web Site: country103.com. Licensee: Three Eagles Communications Inc. (acq 6-8-2007; grpsl). Natl. Rep: Katz Radio. Wire Svc: AP Format: New country. News staff: one; News: one hr wkly. Target aud: 25-54. Spec prog: Sp one hr wkly. ♦Jo Guck Bailey, gen mgr; Chris Painter, gen sls mgr; Kaaren Kohene, prom dir; Terry Cooley, progmg dir; Randall Harter, news dir.

KYSM(AM)— July 25, 1938: 1230 khz; 1 kw-U. TL: N44 10 20 W94 02 23. Stereo. 24 Web Site: www.kysmradio.com. Group owner: Clear Channel Communications Inc. 150,000 Natl. Network: CNN Radio, Fox Sports. Natl. Rep: Katz Radio. Format: News, sports talk. News staff: one; News: 20 hrs wkly. Target aud: 18 plus.

Northfield

***KCMP(FM)**— Apr 4, 1968: 89.3 mhz; 97.6 kw. Ant 768 ft TL: N44 41 21 W93 04 21. Stereo. Hrs opn: 24 480 Cedar St., Saint Paul, 55101. Phone: (651) 290-1500. Fax: (651) 290-1295. Web Site: www.mpr.org/thecurrent. Licensee: Minnesota Public Radio (acq 11-15-2004; $10.5 million with KMSE(FM) Rochester). Population served: 2,500,000 Format: AAA. ♦Chris Cross, gen mgr; Steve Nelson, progmg dir.

***KRLX(FM)**— Jan 25, 1975: 88.1 mhz; 100 w. 16 ft TL: N44 27 39 W93 09 21. Stereo. Hrs opn: Carleton College, One N. College St., 55057. Phone: (507) 646-4102. Web Site: krlxweb.carleton.edu. Licensee: Carleton College. Law Firm: Cohn & Marks. Format: Eclectic. Target aud: General; college-associated people and rural. ♦Jeremy Gantz, stn mgr.

KYMN(AM)— Sept 27, 1968: 1080 khz; 1 kw-D. TL: N44 29 12 W93 06 20. Stereo. Hrs opn: 24 Box 201, 55057. Phone: (507) 645-5695. Fax: (507) 645-9768. E-mail: kymn@clear.lakes.com Licensee: Ingstad Brothers Broadcasting LLC (group owner; acq 5-1-2004; grpsl). Population served: 100,000 Natl. Network: Motor Racing Net, Westwood One. Rgnl. Network: Linder Farm. Rgnl rep: O'Malley. Law Firm: Pepper & Corazzini. Format: Adult contemp. News staff: 1; News: 72 hrs wkly. Target aud: 35-54; parents with school-age children, well-educated. Spec prog: Big band 3 hrs, farm 6 hrs, relg 2 hrs, Latin/Hispanic 2 hrs wkly. ♦James Ingstad, CEO & pres; Ned Newberg, gen mgr.

Olivia

KOLV(FM)— June 27, 1983: 100.1 mhz; 6 kw. 285 ft TL: N44 45 51 W94 55 45. Stereo. Hrs opn: Box 6, 56277. Phone: (320) 523-1017. Fax: (320) 523-1018. Web Site: www.k100realcountry.com. Licensee: Bold Radio Inc. Group owner: Linder Broadcasting Group (acq 3-18-98; $335,000). Rgnl. Network: Linder Farm, AgriAmerica. Natl. Rep: Keystone (unwired net). Format: Country, farm, div. Spec prog: Big band, adult contemp, oldies, Top-40. ♦Steve Linder, pres; Doug Loy, gen mgr, stn mgr & gen sls mgr.

Ortonville

KCGN-FM—Licensed to Ortonville. See Milbank SD

KDIO(AM)— July 23, 1956: 1350 khz; 1 kw-D, 57 w-N. TL: N45 20 58 W96 27 10. Hrs open: 47 N.W. Second St., 56278. Phone: (320) 839-2581. Fax: (320) 839-2571. Licensee: Big Stone Broadcasting Inc. Group owner: Robert Ingstad Broadcast Properties (acq 10-15-99; grpsl). Population served: 115,000 Natl. Network: CBS. Rgnl. Network: Linder Farm. Law Firm: Richard Hayes. Format: Real country. Spec prog: Farm 18 hrs, relg 5 hrs wkly. ♦Jeff Kurtz, pres & gen mgr.

KPHR(FM)—Co-owned with KDIO(AM). 1996: 106.3 mhz; 6 kw. 328 ft TL: N45 20 59 W96 27 08. Format: Classic rock. ♦Jeff Kurtz, opns mgr & gen sls mgr.

Osakis

KBHL(FM)— Mar 11, 1985: 103.9 mhz; 3 kw. 341 ft TL: N45 50 24 W95 05 56. (CP: 6 kw, ant 328 ft.). Stereo. Hrs opn: Box 247, 56360. Secondary address: 515 E. Pike St. 56360. Phone: (320) 859-3000. Fax: (320) 859-3010. E-mail: mail@praisefm.org Web Site: www.praisefm.org. Licensee: Christian Heritage Broadcasting Inc. (acq 3-85; $14,127; FTR: 3-4-85). Natl. Network: Moody. Format: Christian. ♦David McIver, gen mgr.

Owatonna

KRFO(AM)— 1950: 1390 khz; 500 w-D, 100 w-N. TL: N44 04 29 W93 10 46. Hrs open: 5 AM-midnight 18th St., 55060. Phone: (507) 451-2250. Fax: (507) 451-8837. Licensee: Cumulus Licensing Corp. Group owner: Cumulus Media Inc. (acq 7-21-98; grpsl). Population served: 40,000 Format: Oldies. News staff: one; News: 18 hrs wkly. Target aud: 35 plus. Spec prog: Sp 2 hrs wkly. ♦Gary Foss, gen mgr.

KRFO-FM— Dec 29, 1966: 104.9 mhz; 4.7 kw. 200 ft TL: N44 04 29 W93 10 46. Stereo. 18 45,000 Format: Country. News staff: one; News: 12 hrs wkly. Target aud: 25-54.

Park Rapids

KDKK-FM—Listing follows KPRM(AM).

KPRM(AM)— Dec 1, 1962: 870 khz; 25 kw-D, 1 kw-N, DA-N. TL: N46 55 42 W95 00 22. Stereo. Hrs opn: Box 49, Hwy. 34 E., 56470. Phone: (218) 732-3306. Licensee: De La Hunt Broadcasting Corp. Population served: 150,000 Natl. Network: CBS. Format: Country. Target aud: 25 plus. ♦Cheryl Harle, pres; Ed DeLa Hunt, gen mgr.

KDKK-FM—Co-owned with KPRM(AM). December 1967: 97.5 mhz; 100 kw. 636 ft TL: N46 55 51 W95 00 27. Stereo. 100,000 Format: Music of Your Life. Target aud: 40 plus. ♦E.P. De La Hunt, gen mgr, progmg dir & farm dir; Bernie Schumacher, women's int ed.

KXKK(FM)— 1998: 92.5 mhz; 25 kw. 328 ft TL: N46 55 42 W95 00 22. Hrs open: Box 49, 56470. Phone: (218) 732-3306. Fax: (218) 732-3307. Licensee: Bernadine A. Schumacher. Format: Hot country. ♦Bernadine A. Schumacher, gen mgr.

Paynesville

KZPK(FM)— Dec 1, 1995: 98.9 mhz; 50 kw. 492 ft TL: N45 23 15 W94 24 47. Hrs open: 24 Box 1458, St. Cloud, 56302. Phone: (320) 251-1450. Fax: (320) 251-8952. Web Site: www.wildcountry989.com. Licensee: Leighton Enterprises Inc. (group owner; acq 4-15-97; $1 million). Format: Country. ♦Al Leighton, CEO; John Sowada, gen mgr; Denny Niess, sls VP; Denise Prozinski, gen sls mgr; Bri Zenzen, prom dir; Fred Colby, news dir; Dale Daley, chief of engrg; Cindy Niess, traf mgr; Kathy Carton, traf mgr.

Pelican Rapids

KBOT(FM)— June 1994: 104.1 mhz; 50 kw. Ant 492 ft TL: N46 29 57 W96 05 10. Stereo. Hrs opn: 24 Box 746, Detroit Lakes, 56502-0746. Secondary address: 128 Junius Ave. W., Fergus Falls 56537. Phone: (218) 847-5624. Fax: (218) 847-7657. E-mail: kdlmkbot@lakesnet.net Web Site: www.wild1041.com. Licensee: Leighton Enterprises Inc. (group owner; (acq 9-24-96; $700,000). Population served: 150,000 Rgnl rep: O'Malley. Format: Hot country. News: one hr wkly. Target aud: 25-54; Fargo-Moorhead, metro & TSA listeners. ♦Alver Leighton, CEO; John Sowada, pres; Denny Niess, VP; Jeff Leighton, gen mgr; Andy Lia, opns mgr; Kevin Flynn, progmg dir.

Pequot Lakes

KTIG(FM)— Apr 30, 1978: 102.7 mhz; 40 kw. 541 ft TL: N46 40 48 W94 25 02. Stereo. Hrs opn: 24 Box 409, 56472. Phone: (218) 568-4422. Fax: (218) 568-5950. E-mail: radio@ktig.org Web Site: www.ktig.org. Licensee: Minnesota Christian Broadcasters Inc. Population served: 100,000 Natl. Network: Moody, USA. Law Firm: Reddy, Begley & McCormick. Format: Christian inspirational, educ, adult contemp. News staff: one. Target aud: 35-55; general. ♦Jim Gammello, pres; Mike Heuberger, gen mgr; Joe Kimbler, progmg dir; Tom Bonar, news dir; Dwayne Walker, chief of engrg.

WZFJ(FM)— 2002: 100.1 mhz; 3.9 kw. Ant 407 ft TL: N46 40 48 W94 25 02. Hrs open: Box 409, 56472. Phone: (218) 568-4422. Fax: (218) 568-5950. Web Site: www.100thepulse.org. Licensee: Minnesota Christian Broadcasters Inc. Natl. Network: Moody, USA. Format: Christian rock. ♦Jim Gammello, pres; Mike Heuberger, gen mgr; Joe Kimbler, progmg dir; Tom Bonar, news dir; Dwayne Walker, chief of engrg.

Perham

KPRW(FM)— Aug 26, 1996: 99.5 mhz; 6 kw. 328 ft TL: N46 33 16 W95 27 12. Hrs open: Box 363, 56573. Phone: (218) 346-7596. Fax: (218) 346-7595. E-mail: kprw@eot.com Licensee: Jerry Papenfuss. Group owner: The Result Radio Group Format: Adult hit radio. Target aud: 25-54. ♦Doug Gray, gen mgr; David Howey, progmg dir; Cecelia Sundstrom, traf mgr.

Pillager

WWWI-FM— 2000: 95.9 mhz; 6 kw. Ant 239 ft TL: N46 15 03 W94 19 30. Hrs open: 305 W. Washington St., Brainard, 56401. Phone: (218) 828-9994. Fax: (218) 828-8327. Licensee: Tower Broadcasting Corp. (acq 5-28-2004; $360,000). Format: Hot country. ♦Jim Pryor, gen mgr.

Pine City

WCMP(AM)— June 13, 1957: 1350 khz; 1 kw-D. TL: N45 49 10 W92 59 45. Hrs open: 24 15429 Pokegama Lake Rd., 55063. Phone: (320) 629-7575. Fax: (320) 629-3933. E-mail: pinemill@ecenet.net Web Site: www.radiowcmp.net. Licensee: Quarnstrom Media Group LLC (group owner; acq 9-22-03). Population served: 127,500 Rgnl. Network: MNN. Law Firm: Blair, Joyce & Silva. Format: News, info, music of your life. News: 25 hrs wkly. Target aud: 30 plus; farmers, commuters, homemakers. Spec prog: Farm 6 hrs wkly. ♦Al Quarstrom, pres; Kim Cool, CFO; Don Welch, exec VP; Mike Hughes, gen mgr, opns mgr & progmg dir; Bill Mayes, chief of engrg; Paula Butterfield, traf mgr.

WCMP-FM— Oct 15, 1977: 100.9 mhz; 25 kw. Ant 300 ft TL: N45 54 07 W92 57 25. Stereo. Hrs opn: 24 15429 Pokegama Lake Rd., 55063. Phone: (320) 629-7575. Fax: (320) 629-3933. E-mail: pinemill@ecenet.com Web Site: www.radiowcmp.com. Licensee: Quamstrom Media Group LLC (group owner; acq 8-23-01; $1.2 million with co-located AM including five-year noncompete agreement). Population served: 127,500 Format: Contemp country, sports, news. Target aud: 18 plus; commuters, working adults with families. ♦Mike Hughes, gen mgr.

Pipestone

KISD(FM)—Listing follows KLOH(AM).

KLOH(AM)— June 1955: 1050 khz; 9 kw-D, 400 w-N. TL: N43 59 32 W96 20 37. Stereo. Hrs opn: 24 Box 456, 56164. Secondary address: 608 W. Hwy. 30 56164. Phone: (507) 825-4282. Fax: (507) 825-3364. E-mail: kloh@klohradio.com Web Site: www.klohradio.com. Licensee: Wallace Christensen. (acq 8-1-76). Population served: 750,000 Natl. Network: ABC. Rgnl. Network: Linder Farm. Wire Svc: AP Format: Talk, news, C&W, farm. Spec prog: Relg 5 hrs, Sp 2 hrs wkly. ♦Collin Christensen, gen mgr, natl sls mgr & chief of engrg; Carmen Christensen, gen sls mgr & farm dir; Mylan Ray, mus dir; Bernie Wieme, news dir; Diane Carlson, pub affrs dir & local news ed; Honee Lee Longstreet, traf mgr; Joel Herrig, sports cmtr.

KISD(FM)—Co-owned with KLOH(AM). Nov 20, 1968: 98.7 mhz; 100 kw. 1,014 ft TL: N43 53 52 W95 56 50. Stereo. 24 E-mail: kisd@kisdradio.com Web Site: kisdradio.com.600,000 Natl. Network: ABC. Format: Oldies. Target aud: 18-65. ♦Wally Christensen, gen mgr; Carmen Christensen, farm dir.

Preston

KFIL(AM)— May 21, 1966: 1060 khz; 1 kw-D. TL: N43 40 48 W92 08 27. Hrs open: Box 370, 300 St. Paul St. S.W., 55965. Phone: (507) 765-3856. Fax: (507) 765-2738. Licensee: KFIL Inc. Group owner: Cumulus Media Inc. (acq 3-30-2004; grpsl). Population served: 1,413 Rgnl. Network: MNN. Format: C&W. ♦Bruce Fishbaugher, gen mgr.

KFIL-FM— Sept 1, 1970: 103.1 mhz; 6 kw. 270 ft TL: N43 40 48 W92 08 27. Stereo. Format: Country.

Princeton

KLCI(FM)—Listing follows WQPM(AM).

***KPCS(FM)**—Not on air, target date: unknown: 89.7 mhz; 50 kw vert. Ant 105 ft TL: N45 35 54 W93 33 18. Hrs open: Box 18000, Pensacola, FL, 32523. Phone: (850) 479-6570. Fax: (850) 969-1638. Web Site: www.rejoice.org. Licensee: Pensacola Christian College Inc. ♦Arlin Horton, pres & gen mgr.

WQPM(AM)— Feb 1, 1967: 1300 khz; 1 kw-D, 83 w-N. TL: N45 32 58 W93 34 52. Hrs open: 24 Box 106, 55371. Secondary address: 32215 124th St. 55271. Phone: (763) 389-1300. Fax: (763) 389-1359. Licensee: Milestone Radio L.L.C. (acq 9-30-98; $1 million with co-located FM). Population served: 50,000 Natl. Network: ABC. Rgnl. Network: Minn. Pub. Law Firm: Mullin, Rhyne, Emmons & Topel. Format: Contemp country. News staff: one; News: 20 hrs wkly. Target aud: 25-54. Spec prog: Farm 2 hrs wkly. ♦Dennis Carpenter, pres; Neil Freeman, gen mgr.

KLCI(FM)—Co-owned with WQPM(AM). Dec 1, 1974: 106.1 mhz; 9.1 kw. Ant 538 ft TL: N45 14 20 W93 41 14. Stereo. 24 1,500,000 Format: Country. News: 7 hrs wkly.

Proctor

KBMX(FM)— 1994: 107.7 mhz; 7.7 kw. Ant 912 ft TL: N46 47 13 W92 07 17. Hrs open: 24 14 E. Central Entrance, Duluth, 55811. Phone: (218) 727-4500. Fax: (218) 727-9356. Web Site: www.mix108.com. Licensee: CC Licenses LLC. Group owner: Clear Channel Communications Inc. (acq 5-2-2003; grpsl). Population served: 375,000 Format: Adult Contemp. News: one hr wkly. Target aud: 25-54; working adults & families. ♦Ron Stone, gen mgr; Johnny Lee WAlker, opns mgr; Scott Klohan, progmg dir; Corey Carter, progmg mgr & disc jockey; Dave Huston, chief of engrg.

Red Wing

KCUE(AM)— Jan 29, 1949: 1250 khz; 1 kw-D, 110 w-N. TL: N44 32 20 W92 31 25. Hrs open: 24 474 Guernsey Ln., 55066. Phone: (651) 388-7151. Fax: (651) 388-7153. E-mail: preding@waittradio.com Web Site: www.1250kcue.com. Licensee: Sorenson Broadcasting Corp. (group owner; (acq 6-81; $1.1 million with co-located FM; FTR: 6-22-81) Population served: 30,000 Rgnl. Network: MNN. Format: News/talk, farm, relg. News staff: 2; News: 7 hrs wkly. Target aud: 35 plus; information consumer. ♦Don Kliewer, gen mgr & gen sls mgr; Tom Hughes, opns mgr; Jack Calwell, sports cmtr.

KWNG(FM)—Co-owned with KCUE(AM). Aug 26, 1965: 105.9 mhz; 20 kw. 300 ft TL: N44 29 15 W92 13 56. Stereo. 24 Web Site: www.1250kcue.com.39,000 Format: Rock classics of the 60s, 70s & 80s. News staff: 2; News: one hr wkly. Target aud: 25-44; family & yuppie.

Redwood Falls

KLGR(AM)— November 1954: 1490 khz; 1 kw-U. TL: N44 32 33 W95 07 57. (CP: 470 w. TL: N44 32 35 W95 07 57). Hrs opn: 639 W. Bridge, 56283. Phone: (507) 637-2989. Fax: (507) 637-5347. Web Site: www.klgram.com. Licensee: Three Eagles of Luverne Inc. Group owner: Three Eagles Communications (acq 12-13-99; with co-located FM). Population served: 50,000 Rgnl. Network: MNN. Format: Country. Target aud: General. ♦Mike Neudecker, gen mgr.

KLGR-FM— June 3, 1974: 97.7 mhz; 3 kw. 305 ft TL: N44 32 33 W95 07 57. (CP: 60 kw, ant 289 ft.). Stereo. Web Site: www.klgram.com. Format: Oldies.

Richfield

KDWB-FM—Licensed to Richfield. See Minneapolis-St. Paul

KKMS(AM)—Licensed to Richfield. See Minneapolis-St. Paul

Rochester

***KFSI(FM)**— Apr 28, 1981: 92.9 mhz; 6 kw. Ant 318 ft TL: N44 01 27 W92 32 36. Stereo. Hrs opn: 4016 28th St. S.E., 55904. Phone: (507) 289-8585. Fax: (507) 529-4017. E-mail: shine@kfsi.org Web Site: www.kfsi.org. Licensee: Faith Sound Inc. Natl. Network: Moody. Format: Adult contemp Christian. ♦Ray Logan, pres & gen mgr; Paul Logan, VP & progmg dir; Jim Royston, mus dir; Mike Anderson, asst music dir; Steve Schuh, engrg VP.

***KLSE-FM**— Dec 17, 1974: 91.7 mhz; 100 kw. Ant 953 ft TL: N44 02 26 W92 20 28. Stereo. Hrs opn: 24 206 S. Broadway, Suite 735, 55904. Phone: (507) 282-0910. Fax: (507) 282-2107. Web Site: www.mpr.org. Licensee: Minnesota Public Radio Inc. Population served: 463,000 Natl. Network: NPR, PRI. Rgnl. Network: Minn. Pub. Format: Class. News staff: one. ♦Chris Cross, stn mgr; Mary Stapek, dev dir; Sea Stachura, news rptr.

***KMSE(FM)**— Aug 1, 1998: 88.7 mhz; 250 w. Ant 531 ft TL: N44 02 32 W92 20 26. Hrs open: 24 206 S. Broadway, Suite 735, 55904. Phone: (507) 282-0910. Fax: (507) 282-2107. Web Site: www.mpr.org. Licensee: Minnesota Public Radio (acq 11-15-2004; $10.5 million with WCAL(FM) Northfield). Population served: 125,000 Format: AAA. News staff: one. ♦Chris Cross, gen mgr; Steve Nelson, progmg dir.

KNXR(FM)— Dec 24, 1965: 97.5 mhz; 100 kw. 1,040 ft TL: N44 02 28 W92 20 25. Stereo. Hrs opn: 1620 Greenview Dr. S.W., 55902-1034. Phone: (507) 288-7700. Fax: (507) 288-4531. Licensee: United Audio Corp. Population served: 225,000 Natl. Network: CBS, Wall Street. Natl. Rep: McGavren Guild. Law Firm: Miller & Neely. Wire Svc: AP Format: Adult traditional. Target aud: 35 plus. Spec prog: Class 4 hrs, talk 2 hrs wkly. ♦Thomas H. Jones, pres & gen mgr.

KOLM(AM)— November 1963: 1520 khz; 10 kw-D, 800 w-N. TL: N43 59 13 W92 25 05. Hrs open: 24 122 4th St. S.W., 55902. Phone: (507) 286-1010. Fax: (507) 286-9370. Web Site: www.1520theticket.com. Licensee: Cumulus Licensing LLC. Group owner: Cumulus Media Inc. (acq 3-31-2004; grpsl). Population served: 250,000 Natl. Network: Westwood One. Rgnl. Network: Linder Farm. Natl. Rep: McGavren Guild. Law Firm: Smithwick & Belendiuk. Format: Sports. News staff: 2; News: 3 hrs wkly. Target aud: 35-64; male 55%, female 45%. ♦Rosanne Rybak, gen mgr; Bretta Damson, gen sls mgr; Kim David, news dir; Bill Davis, chief of engrg.

KWWK(FM)—Co-owned with KOLM(AM). July 4, 1967: 96.5 mhz; 43 kw. 528 ft TL: N44 01 59 W92 36 10. Stereo. 24 Web Site: www.quickcountry.com. Format: Country. News: 2 hrs wkly. Target aud: 25-54; male & female 18-49, 25-54, 35-54.

KRCH(FM)—Listing follows KWEB(AM).

KROC(AM)— October 1935: 1340 khz; 1 kw-U. TL: N44 01 47 W92 29 31. Hrs open: 24 122 4th Ave. S.W., 55902. Phone: (507) 286-1010. Fax: (507) 280-0000. Web Site: www.kroc.com. Licensee: Cumulus Licensing LLC. Group owner: Cumulus Media Inc. (acq 3-29-2004; grpsl). Population served: 95,000 Natl. Rep: Hyett/Ramsland. Format: News/talk. News staff: 3; News: 42 hrs wkly. Target aud: 30-64. Spec prog: Farm 12 hrs wkly. ♦Rosanne Rybak, gen mgr & gen sls mgr; Joe O'Brien, progmg dir.

KROC-FM— July 1, 1965: 106.9 mhz; 100 kw. 1,110 ft TL: N43 34 15 W92 25 37. Web Site: www.kroc.com.300,000 Format: CHR. Target aud: 18-54.

***KRPR(FM)**— 1976: 89.9 mhz; 3.2 kw. Ant 590 ft TL: N44 02 28 W92 20 25. Stereo. Hrs opn: 24 Rochester Public Radio, 1620 Greenview Dr. S.W., 55902-1034. Phone: (507) 288-2376. Fax: (507) 288-4531. Licensee: Rochester Public Radio (acq 5-13-99). Population served: 100,000 Law Firm: Miller & Neely. Format: Classic rock. Target aud: General. ♦Thomas H. Jones, pres; Todd D. Brakke, gen mgr & stn mgr.

KWEB(AM)— Nov 27, 1957: 1270 khz; 5 kw-D, 1 kw-N, DA-2. TL: N43 58 47 W92 26 51. Hrs open: 1530 Greenview Dr. S.W., Suite 200, 55902. Phone: (507) 288-3888. Fax: (507) 288-7815. Licensee: CC Licenses LLC. Group owner: Clear Channel Communications Inc. (acq 9-25-2000; grpsl). Population served: 75,000 Natl. Network: CBS. Rgnl. Network: MNN. Natl. Rep: D & R Radio. Format: Sports, talk. Target aud: Men. ♦Bob Fox, gen mgr & opns mgr; Mary Anne Nonn, gen sls mgr; Mark Clark, prom dir & progmg dir; Craig Erpestad, chief of engrg.

KRCH(FM)—Co-owned with KWEB(AM). 1972: 101.7 mhz; 39.1 kw. 554 ft TL: N44 06 59 W92 41 22. Stereo. 24 Web Site: www.laser1017.net. Format: Solid rock. Target aud: 25-54.

***KZSE(FM)**— February 1989: 90.7 mhz; 1.38 kw. Ant 259 ft TL: N44 02 26 W92 20 28. Hrs open: 24 206 S. Broadway, Suite 735, 55904. Phone: (507) 282-0910. Fax: (507) 282-2107. Web Site: www.mpr.org. Licensee: Minnesota Public Radio Inc. Natl. Network: NPR, PRI. Rgnl. Network: Minn. Pub. Format: News & info. News staff: one. ♦Chris Cross, gen mgr; Mary Stapek, dev dir; Sea Stachura, news rptr.

Rockville

KYES(AM)—Not on air, target date: unknown: 1180 khz; 35 kw-D, 5 kw-N, 16 kw-CH, DA-3. TL: N45 21 42 W94 17 39. Hrs open: Box 547, Sauk Rapids, 56379-0547. Secondary address: 1310 2nd St. N., Sauk Rapids 56379-2532. Phone: (320) 251-1780. Licensee: Throw Fire Project. ♦Andrew W. Hilger, pres.

Roseau

KCAJ-FM— June 1996: 102.1 mhz; 50 kw. Ant 285 ft TL: N48 38 50 W95 44 10. Hrs open: 24 407 3rd St. N.W., 56751. Phone: (218) 463-3360. Fax: (218) 463-1977. Web Site: wild102fm.com. Licensee: Jack J. Swanson. Population served: 25,000 Natl. Network: CNN Radio. Format: Top-40. News staff: one; News: 10 hrs wkly. Target aud: General. ♦Jack Swanson, gen mgr & opns VP; Jack McDonald, gen sls mgr; Justin Gallo, prom dir.

KRWB(AM)— Apr 5, 1963: 1410 khz; 1 kw-U, DA-N. TL: N48 50 43 W95 43 34. Hrs open: 24 Box 69, Warroad, 56763. Phone: (218) 463-1410. Fax: (218) 463-3778. Web Site: www.1410krwb.com. Licensee: Border Broadcasting L.P. (acq 10-1-00; $62,000). Population served: 2,552 Natl. Network: ABC. Format: Classic Rock. News staff: one. Target aud: 25-54; general. ♦Mike Pederson, gen mgr.

***KRXW(FM)**—Not on air, target date: unknown: 103.5 mhz; 50 kw. Ant 422 ft TL: N48 51 10 W95 46 13. Hrs open: 480 Cedar St., St. Paul, 55101. Phone: (651) 290-1259. Fax: (651) 290-1243. Web Site: minnesota.publicradio.org. Licensee: Minnesota Public Radio. ♦Thomas J. Kigin, exec VP.

Roseville

***KDNI(FM)**—Duluth, Apr 16, 1983: 90.5 mhz; 2 kw. 728 ft TL: N46 47 21 W92 06 51. Stereo. Hrs opn: 24 1101 E. Central Entrance, Duluth, 55811. Secondary address: Northwestern College, 3003 N. Snelling Ave. N. 55811. Phone: (218) 722-6700. Fax: (218) 722-1092. E-mail: kdnw@kdnw.fm Web Site: www.kdnw.fm. Licensee: Northwestern College Radio Network. Group owner: Northwestern College & Radio (acq 12-18-92). Population served: 200,000 Natl. Network: AP Radio. Law Firm: Bryan Cave. Format: Talk, classic praise music. News: 5 hrs wkly. Target aud: 25-54; baby boomers. ♦Paul Virts, sr VP & VP; Paul Harkness, stn mgr.

***KTIS(AM)**—Minneapolis, Feb 7, 1949: 900 khz; 25 kw-D, 300 w-N, DA-2. TL: N44 59 51 W93 21 10. Hrs open: 24 3003 Snelling Ave. N., St. Paul, 55113. Phone: (651) 631-5000. Fax: (651) 631-5084. Web Site: www.ktis.fm. Licensee: Northwestern College. Group owner: Northwestern College & Radio. Population served: 2,000,000 Format: Relg, Christian, news. News staff: 2; News: 20 hrs wkly. Target aud: 35-45. ♦Paul Virts, sr VP; David Fitts, stn mgr; Marilyn Ryan, opns dir.

Rushford

KWNO-FM— Dec 18, 1991: 99.3 mhz; 11 kw. 495 ft TL: N43 56 32 W91 45 30. Hrs open: 24 Box 767, Winona, 55987. Secondary address: 752 Bluffview Cir., Winona 55987. Phone: (507) 452-4154. Fax: (507) 452-9494. E-mail: jpapenfuss@winonaradio.com Web Site: winonaradio.com. Licensee: KAGE Inc. Group owner: The Result Radio Group (acq 6-19-95; $1 million with KWNO(AM) Winona). Population served: 70,000 Law Firm: Wiley, Rein & Fielding. Wire Svc: AP Format: Hot country. News staff: one; News: one hr wkly. Target aud: 18-49; active young students & working persons. ♦Jerry Papenfuss, CEO, gen mgr & mktg VP; Les Guderian, sls dir; Pat Papenfuss, pres, exec VP, opns mgr & prom VP; Aaron Taylor, progmg dir; Darryl Smelser, news dir; Bob Sebo, pub affrs dir.

Saint Charles

KLCX(FM)— Apr 18, 1998: 107.7 mhz; 1.95 kw. 571 ft TL: N44 02 25 W92 13 05. Hrs open: 24 122 4th St. S.W., Rochester, 55902. Phone: (507) 286-1010. Fax: (507) 286-9370. Web Site: www.klcxfm.com. Licensee: Cumulus Licensing LLC. Group owner: Cumulus Media Inc. (acq 3-31-2004; grpsl). Population served: 200,000 Natl. Network: Westwood One. Natl. Rep: McGavren Guild. Law Firm: Smithwick & Belenduik. Wire Svc: AP Format: Classic hits. News staff: one; News: 2 hrs wkly. Target aud: 25-49. ♦Roseanne Rybak, gen mgr.

Saint Cloud

***KCFB(FM)**— Nov 17, 1986: 91.5 mhz; 15 kw. 348 ft TL: N45 30 02 W94 14 31. Stereo. Hrs opn: 24
Rebroadcasts KTIG(FM) Pequot Lakes 100%.
Box 409, Pequot Lakes, 56472. Phone: (218) 252-4422. Fax: (218) 568-5950. Web Site: www.kcfbradio.org. Licensee: Minnesota Christian Broadcasters Inc. (acq 9-4-97; $250,000). Natl. Network: Moody. Law Firm: Reddy, Begley & McCormick. Wire Svc: AP Format: Christian inspirational, educ, adult contemp. News staff: one; News: 14 hrs wkly. Target aud: General. ♦Mike Heuberger, gen mgr.

KCLD-FM—Listing follows KNSI(AM).

KKSR(FM)—Sartell, Aug 26, 1988: 96.7 mhz; 50 kw. 453 ft TL: N45 46 03 W94 08 04. Stereo. Hrs opn: 24 640 S.E. Lincoln Ave., St. Cloud, 56304. Phone: (320) 251-4422. Fax: (320) 251-1855. E-mail: studio@kiss96.com Web Site: www.kiss96.com. Licensee: Regent Licensee of St. Cloud Inc. Group owner: Regent Communications Inc. (acq 5-8-2001; grpsl). Law Firm: Wiley, Rein & Fielding. Format: Adult contemp. News staff: one. Target aud: 25-54. ♦Dave Engberg, gen mgr; Lee Voss, mus dir & news dir; Mark Young, chief of engrg.

KNSI(AM)— June 1938: 1450 khz; 1 kw-U. TL: N45 32 21 W94 10 05. Stereo. Hrs opn: Box 1458, 56302. Secondary address: 619 W. St. Germain St. 56302. Phone: (320) 251-1450. Fax: (320) 251-8952. Web Site: www.1450knsi.com. Licensee: Leighton Enterprises Inc. (group owner; acq 9-15-75). Rgnl rep: O'Malley. Format: News/talk. Target aud: 35 plus; males. ♦John J. Sowada, pres, VP & gen mgr; Denny Niess, sls VP; Gary Foss, sls dir; Bob Zensen, prom dir; Daniel Ochsner, progmg dir; Cory Kampschroer, news dir; Dale Daley, chief of engrg; Cindy Niess, traf mgr.

KCLD-FM—Co-owned with KNSI(AM). May 1, 1948: 104.7 mhz; 100 kw. 984 ft TL: N45 34 03 W94 30 43. Stereo. Web Site: www.1047kcld.com.200,000 Format: CHR. ♦J.J. Holiday, progmg dir.

***KVSC(FM)**— May 10, 1967: 88.1 mhz; 16.5 kw. Ant 446 ft TL: N45 31 00 W94 13 52. Stereo. Hrs opn: 24 St. Cloud State Univ., 27 Stewart Hall, 56301-4498. Phone: (320) 308-3066. Fax: (320) 308-5337. E-mail: info@kvsc.org Web Site: www.kvsc.org. Licensee: St. Cloud State University. Population served: 100,000 Wire Svc: Bay City News Service Format: Var, educ. Target aud: 17-60; educated, progsv. ♦Roya Majid, gen mgr; Jo McMullen-Boyer, stn mgr.

KXSS(AM)—See Waite Park

WJON(AM)— September 1950: 1240 khz; 1 kw-U. TL: N45 33 36 W94 08 20. Hrs open: 24 640 Southeast Lincoln Ave., 56304. Phone: (320) 251-4422. Web Site: www.wjon.com. Licensee: Regent Licensee of St. Cloud Inc. Group owner: Regent Communications Inc. (acq 5-1-99; grpsl). Population served: 100,000 Natl. Network: CBS. Rgnl. Network: MNN. Law Firm: Pepper & Corazzini. Format: News/talk, full service. News staff: 3; News: 30 hrs wkly. Target aud: 25 plus. ♦Dave Engberg, gen mgr & sls dir; Mike Dylan, opns mgr; Bob Hughes, progmg dir; Lee Voss, news dir; Mark Young, chief of engrg.

WWJO(FM)—Co-owned with WJON(AM). 1975: 98.1 mhz; 97 kw. 1,000 ft TL: N45 48 52 W94 01 38. Stereo. Web Site: www.98country.com. Format: Country. Target aud: 18 plus. ♦Terry Jacobs, CEO; Bill Stakelin, pres; Fred Murr, sr VP; Lynn Larson, mktg VP & chief of engrg; Bill Fink, progmg dir; Sandi Davis, asst music dir.

Saint James

KRRW(FM)— July 24, 1983: 101.5 mhz; 14 kw. Ant 446 ft TL: N43 52 29 W94 36 04. Stereo. Hrs opn: 24 Box 1420, Mankato, 56002. Phone: (507) 375-3386. Fax: (507) 375-5050. E-mail: krrw@linderradio.com Licensee: Minnesota Valley Broadcasting Co. Group owner: Linder Broadcasting Group (acq 1996; $800,000 with KXAC(FM) St). Rgnl. Network: Linder Farm. Law Firm: Miller & Miller. Format: Modern country. News staff: one; News: 16 hrs wkly. Target aud: 25-54. Spec prog: Sp one hr wkly. ♦Mike Parry, gen mgr; Dwayne Megaw, chief of opns.

KXAC(FM)— Nov. 1, 1992: 100.5 mhz; 50 kw. 433 ft TL: N43 52 29 W94 36 04. (CP: 34 kw, ant 590 ft. TL: N43 57 04 W94 23 27). Hrs opn: 24 Box 1420, Mankato, 56002. Phone: (507) 345-4537. Fax: (507) 345-5364. E-mail: kxac@linderradio.com Licensee: Minnesota Valley Broadcasting Co. Group owner: Linder Broadcasting Group (acq 1996; $800,000 with KXAX(FM) St). Rgnl. Network: MNN. Law Firm: Miller & Miller. Format: Oldies. News staff: one; News: 17 hrs wkly. Target aud: 25-58. ♦Mike Parry, gen mgr; Dwayne Megaw, chief of opns.

Saint Joseph

KCML(FM)— 1998: 99.9 mhz; 6 kw. 328 ft TL: N45 32 21 W94 10 05. Hrs open: Box 1458, St. Cloud, 56302. Secondary address: 619 W. Saint Germain, St. Cloud 56301. Phone: (320) 251-1450. Fax: (320) 251-8952. Web Site: www.lite999.com. Licensee: Leighton Enterprises Inc. (group owner) Format: Contemp lite. ♦Al Leighton, CEO; John Sowada, gen mgr; Denny Niess, sls VP; Denise Prozinski, gen sls mgr; Bri Zensen, prom dir; Ron Linder, progmg dir; Cory Kampschroer, news dir; Dale Daley, chief of engrg; Cindy Niess, traf mgr.

KKJM(FM)— May 7, 1996: 92.9 mhz; 25 kw. 328 ft TL: N45 38 19 W94 22 23. Stereo. Hrs opn: 24 1310 Second St. N., Sauk Rapids, 56379. Phone: (320) 251-1780. Fax: (320) 257-1624. E-mail: info@spirit929.com Web Site: www.spirit929.com. Licensee: Gabriel Communications Co., St. Cloud. (acq 11-30-99). Population served: 150,000 Natl. Network: Salem Radio Network. Wire Svc: AP Format: Christian adult contemp. News: 5 hrs wkly. Target aud: 25-54; females. ♦Deb Huschle, gen mgr.

Saint Louis Park

***KDXL(FM)**— Mar 17, 1977: 106.5 mhz; 10 w (ST: KUOM-FM). Ant 85 ft TL: N44 56 36 W93 21 39. Stereo. Hrs opn: 7:30 AM-10 PM (M-F) Senior High School, 6425 W. 33rd St., 55426. Phone: (952) 928-6149. Web Site: www.slpschools.org/sh/kdxl. Licensee: Independent School District 283. Format: AOR, classic rock, progsv. Target aud: 15-30; high school students & loc residents. ♦Charlie Fiss, stn mgr.

KTNF(AM)— May 13, 1958: 950 khz; 1 kw-U, DA-2. TL: N44 52 08 W93 25 11. Hrs open: 24 11320 Valley View Rd., Eden Prairie, 55344. Phone: (952) 946-8885. Fax: (952) 946-0888. Web Site: www.airamericaminnesota.com. Licensee: JR Broadcasting LLC Group owner: Infinity Broadcasting Corp. (acq 10-21-2004; $3 million). Population served: 6,938 Natl. Rep: McGavren Guild. Format: Talk. Target aud: 25-54. ♦Janet Robert, gen mgr.

***KUOM-FM**— Feb 17, 2003: 106.5 mhz; 8 w (ST: KDXL(FM)). Ant 253 ft TL: N44 56 46 W93 19 27. Stereo. Hrs opn: 4:30 PM-8 AM M-F; 24 S-S
Rebroadcasts KUOM(AM) Minneapolis 100%.
Univ. of Minnesota, 330 21st Ave. S., Minneapolis, 55455-0415. Phone: (612) 625-3500. Fax: (612) 625-2112. E-mail: radiok@umn.edu Web Site: www.radiok.org. Licensee: Regents of the University of Minnesota. Population served: 200,000 Law Firm: Dow, Lohnes & Albertson. Wire Svc: AP Format: Alternative. News: 5 hrs wkl;y. Target aud: 18-34. ♦Andrew Marlow, stn mgr; Larry Oberg, chief of opns; Stuart Sanders, dev dir; Shelley Miller, progmg dir.

KZJK(FM)— July 1, 1962: 104.1 mhz; 89 kw. Ant 1,033 ft TL: N45 03 30 W93 07 27. Stereo. Hrs opn: 625 2nd Ave., Ste 200, Minneapolis, 55402-1908. Phone: (952) 836-1041. Fax: (952) 915-6781. Web Site: www.mix1041fm.com. Licensee: The Audio House Inc. Format: Hits of the 80s. ♦Mary Niemeyer, gen mgr; John McMonagle, gen sls mgr.

Saint Paul

KEEY-FM—Licensed to Saint Paul. See Minneapolis-St. Paul

KMNV(AM)—Licensed to Saint Paul. See Minneapolis-St. Paul

KNOF(FM)—Licensed to Saint Paul. See Minneapolis-St. Paul

KSTP(AM)—Licensed to Saint Paul. See Minneapolis-St. Paul

KSTP-FM—Licensed to Saint Paul. See Minneapolis-St. Paul

WMCN(FM)—Licensed to Saint Paul. See Minneapolis-St. Paul

Saint Peter

***KGAC(FM)**— Mar 29, 1985: 90.5 mhz; 75 kw. 708 ft TL: N44 13 20 W94 07 03. Stereo. Hrs opn: 24 Minnesota Public Radio Inc., 45 E. 7th St., St. Paul, 55101. Phone: (651) 290-1500. Fax: (651) 290-1224. E-mail: mail@mpr.org Web Site: www.mpr.org. Licensee: Minnesota Public Radio Inc. Population served: 200,000 Natl. Network: PRI. Rgnl. Network: Minn. Pub. Format: Class, arts. News staff: 2. Target aud: General. Spec prog: Folk var 17 hrs wkly. ♦William H. Kling, pres & gen mgr.

***KNGA(FM)**— Mar 1, 1992: 91.5 mhz; 8.5 kw. 600 ft TL: N44 13 20 W94 07 03. Hrs open: 24
Rebroadcasts KNOW-FM Minneapolis-St. Paul.
Minnesota Public Radio, 45 E. 7th St., Saint Paul, 55101. Phone: (800) 228-7123. Fax: (507) 651-1295. E-mail: mail@mpr.org Web Site: www.mpr.org. Licensee: Minnesota Public Radio. Population served: 150,000 Natl. Network: NPR, PRI. Rgnl. Network: Minn. Pub. Format: News, info. News staff: 2. Target aud: General. ♦William H. Kling, gen mgr.

KRBI(AM)— Aug 5, 1957: 1310 khz; 1 kw-D, 343 w-N, DA-1. TL: N44 19 51 W93 58 19. Hrs open: 24 54934 210th Ln., Mankato, 56001. Phone: (507) 345-4646. Fax: (507) 345-3299. E-mail: krbi@krbi.com Web Site: www.krbi.com. Licensee: Three Eagles Communications LLC. Group owner: Three Eagles Communications (acq 7-16-2003; $3.2 million with co-located FM). Population served: 71,000 Rgnl. Network: Linder Farm. Rgnl rep: Midwest Radio Law Firm: Reddy, Begley & McCormick. Format: Country, news/talk, sports. News staff: 2; News: 3 hrs wkly. Target aud: 25 plus; people interested in loc & rgnl news, sports, weather & country music. ♦Ron Gates, gen mgr; Jen Jones, gen sls mgr; Jason Mediger, mus dir; Clay Kepner, sports cmtr.

KRBI-FM— Sept 1, 1966: 105.5 mhz; 25.kw. 200 ft TL: N44 19 41 W93 58 17. Stereo. 24 Web Site: www.river105.com.75,000 Format: Classic hits. News staff: one; News: 1 hr wkly. Target aud: 25-54; people liking classic hits - 70's, 80's, 90's & small amounts of news, weather, & sports.

Sartell

KKSR(FM)—Licensed to Sartell. See Saint Cloud

Sauk Centre

KIKV-FM— Dec 25, 1970: 100.7 mhz; 100 kw. Ant 790 ft TL: N45 41 10 W95 08 03. Stereo. Hrs opn: 24 Box 1024, Alexandria, 56308-1024. Secondary address: 604 Third Ave. W., Alexandria 56308. Phone: (320) 762-2154. Fax: (320) 762-2156. E-mail: 100.7@kikvfm.com Web Site: kikvradio.com. Licensee: BDI Broadcasting Inc. Group owner: Omni Broadcasting Co. (acq 9-25-89; $855,000; FTR: 10-16-89). Population served: 500,000 Natl. Network: ABC, AP Radio. Rgnl. Network: Linder Farm. Law Firm: Garvey, Schubert & Barer. Format: Country. News staff: one; News: 12 hrs wkly. Target aud: 25-54; adults. Spec prog: Farm 20 hrs wkly. ♦Lou Buron, CEO & pres; Mary Campbell, CFO & VP; Dave Vagle, gen mgr; Trudy Blanshan, gen sls mgr & prom dir; Rick Blanshan, progmg dir; Paul Sorum, mus dir & pub affrs dir; Jim Rohn, news dir; Dave Cox, engrg dir & chief of engrg.

Sauk Rapids

WBHR(AM)— Aug 3, 1963: 660 khz; 10 kw-D, 250 w-N, DA-2. TL: N45 36 18 W94 08 21. Hrs open: 24 1010 2nd St. N., 56379. Phone: (320) 252-6200. Fax: (320) 252-9367. Web Site: www.660wbhr.com. Licensee: Tri-County Broadcasting Inc. Population served: 1,752,103 Natl. Network: Radio Disney. Format: Sports. ♦Herb M. Hoppe, pres & gen mgr; Gary E. Hoppe, opns mgr & chief of engrg; Doug Kurtz, gen sls mgr.

WHMH-FM—Co-owned with WBHR(AM). Oct 31, 1975: 101.7 mhz; 50 kw. 423 ft TL: N45 35 48 W94 09 25. Stereo. 24 Web Site: www.rockin101.com.163,000 Format: Rock.

WPPI(AM)—Not on air, target date: unknown: 1010 khz; 1.7 kw-D, 240 w-N, DA-2. TL: N45 36 18 W94 08 21. Hrs open: Box 366, 56379-0366. Phone: (320) 252-6200. Fax: (320) 252-9367. Licensee: Herbert M. Hoppe. ♦Herb M. Hoppe, gen mgr.

WVAL(AM)— Mar 1, 1999: 800 khz; 2.6 kw-D, 850 w-N, DA-2. TL: N45 36 18 W94 08 21. Hrs open: Box 366, 56379. Secondary address: 1010 2nd St. N. 56379. Phone: (320) 252-6200. Fax: (320) 252-9367. E-mail: original@800wval.com Web Site: www.800wval.com. Licensee: Tri-County Broadcasting Inc. Format: Classical country. ♦Herb M. Hoppe, gen mgr.

Sebeka

***KOPJ(FM)**—Not on air, target date: unknown: 89.3 mhz; 100 kw. Ant 872 ft TL: N46 40 35.9 W94 43 02. Hrs open: Box 3006, Collegedale, TN, 37315-3006. E-mail: office@lifetalk.net Web Site: www.lifetalk.net. Licensee: LifeTalk Radio Inc. ♦James Gilley, pres.

Shakopee

KQSP(AM)— Oct 6, 1963: 1530 khz; 8.6 kw-D, 10 w-N, DA-2. TL: N44 48 26 W93 33 25. Hrs open: 24 2519 Osage Dr., Glenview, IL, 60026. Phone: (847) 687-6550. Licensee: Broadcast One Inc. Group owner: Relevant Radio (acq 9-8-2006; $1.2 million). Population served: 2,300,000 ♦Yong W. Kim, pres.

Slayton

KJOE(FM)— 1993: 106.1 mhz; 13 kw. 971 ft TL: N43 53 52 W95 56 50. Hrs open: 24 2660 Broadway Ave., 56172. Phone: (507) 836-6125. Phone: (507) 836-6126. Fax: (507) 836-6537. E-mail: kjoe@kjoeradio.com Web Site: www.kjoeradio.com. Licensee: Wallace Christensen. Format: Country. ♦Wallace Christensen, pres; Collin Christensen, gen mgr; Carmen Christensen, gen sls mgr; Bernard Wieme, prom dir; Mylan Ray, mus dir; Joel Herrig, news dir; Diane Marie, pub affrs dir; Honee Lee Longstreet, traf mgr.

Sleepy Eye

KNUJ-FM— June 1, 1995: 107.3 mhz; 1.9 kw. 400 ft TL: N44 19 38 W94 43 42. Stereo. Hrs opn: 24
Rebroadcasts KNUJ(AM) New Ulm 70%.
Box 368, New Ulm, 56073. Phone: (507) 359-2921. Fax: (507) 359-4520. E-mail: knuj@knuj.net Web Site: www.knuj.net. Licensee: Ingstad Brothers Broadcasting LLC (group owner; acq 5-1-2004; grpsl). Population served: 50,000 Format: Adult contemp. News staff: one; News: 20 hrs wkly. Target aud: 18-49; slightly more females than males. ♦Jim Bartels, VP; Marj Frederickson, gen mgr.

Spring Grove

KQYB(FM)— Aug 2, 1980: 98.3 mhz; 33 kw. 607 ft TL: N43 40 53 W91 45 28. Stereo. Hrs opn: 24 Box 308, Hwy. 44 W., 55974-0308. Phone: (507) 498-5720. Fax: (507) 498-5766. E-mail: email@kq98.com Web Site: www.kq98.com. Licensee: Family Radio Inc. Group owner: The Mid-West Family Broadcast Group (acq 7-19-01; grpsl). Population served: 209,000 Law Firm: Shaw Pittman. Wire Svc: AP Format: Hot country. News staff: one; News: one hr wkly. Target aud: General. ♦Dick Record, CEO, CEO & gen mgr.

Spring Valley

KVGO(FM)— 1993: 104.3 mhz; 2.8 kw. 472 ft TL: N43 33 46 W92 25 29. Hrs open: Box 370, Preston, 55965. Phone: (507) 765-3856. Fax: (507) 765-2738. Licensee: KVGO Inc. Group owner: Cumulus Media Inc. (acq 3-30-2004; grpsl). Natl. Rep: D & R Radio. Format: Oldies. ♦Bruce Fishbaugher, gen mgr & stn mgr.

Springfield

KNSG(FM)— 1995: 94.7 mhz; 50 kw. Ant 472 ft TL: N44 21 54 W95 19 27. Stereo. Hrs opn: 24 1414 E. College Dr., Marshall, 56258. Phone: (507) 532-2282. Fax: (507) 532-3739. Web Site: marshallradio.net. Licensee: Springfield Radio Inc. Group owner: Linder Broadcasting Group (acq 6-11-2007; $500,000). Natl. Network: Westwood One. Natl. Rep: Katz Radio. Wire Svc: AP Format: Adult contemp. Target aud: 30 plus; women. Spec prog: Farm 15 hrs, women 3 hrs wkly. ♦Brad Strootman, gen mgr; Heath Radke, chief of opns.

Staples

KNSP(AM)— June 3, 1982: 1430 khz; 1 kw-D, 199 w-N. TL: N46 21 34 W94 46 55. Hrs open: 24 Box 551, Wadena, 56482-0551. Secondary address: 201 1/2 Jefferson St. S., Wadena 56482. Phone: (218) 631-1803. Fax: (218) 631-4557. E-mail: kwadkkws@arvig.net Web Site: kwadknsp.net. Licensee: BL Broadcasting Inc. Group owner: Omni Broadcasting Co. (acq 4-1-2004; grpsl). Population served: 20,000 Natl. Network: ABC. Law Firm: Garvey, Schubert & Barer. Format: Country, mainstream. News staff: one; News: 20 hrs wkly. Target aud: 25-54; adults. Spec prog: Farm 10 hrs wkly. ♦Lou Buron, CEO & pres; Mary Campbell, CFO & VP; Rick Youngbauer, gen mgr & gen sls mgr; Sherry Linnes, prom dir; Kyle Gylsen, progmg dir & pub affrs dir; Chad Moyer, news dir; Dave Cox, engrg dir & chief of engrg.

KSKK(FM)— Aug 1, 1994: 94.7 mhz; 50 kw. 469 ft TL: N46 33 08 W94 39 03. Stereo. Hrs opn: 24 11 S.E. Bryant Ave., Wadena, 56482. Phone: (218) 631-3441. Fax: (218) 631-3414. E-mail: kskk@eot.com Licensee: NorMin Broadcasting Co. Natl. Network: CBS. Rgnl. Network: Linder Farm. Format: Soft hits. News: 20 hrs wkly. Target aud: 30 plus. ♦David J. De LaHunt, CEO, pres & CFO; Gene Marie Kanten, gen mgr; Joleen De LaHunt, VP & stn mgr; Heidi Hutson, gen sls mgr; Dave Lee, progmg VP.

Starbuck

KRVY-FM— 2001: 97.3 mhz; 50 kw. Ant 492 ft TL: N45 31 42 W95 32 52. Hrs open: 24 Box 380, Willmar, 56201. Secondary address: 730 N.E. Hwy. 71, Willmar 56201. Phone: (320) 231-1600. Fax: (320) 235-7010. Web Site: www.k-musicradio.com. Licensee: Iowa City Broadcasting Co. Group owner: Tom Ingstad Broadcasting Group (acq 7-19-99; $200,000 for stock). Wire Svc: UPI Format: Adult contemp, light rock. News staff: one; News: 8 hrs wkly. Target aud: 25-54; male & female. ♦Doug Hanson, gen mgr.

Stewartville

KYBA(FM)— Feb 1, 1993: 105.3 mhz; 50 kw. 492 ft TL: N43 40 23 W92 41 54. Hrs open: 122 4th St. S.W., Rochester, 55902-3320. Phone: (507) 286-1010. Fax: (507) 286-9370. Web Site: www.y105fm.com. Licensee: Cumulus Licensing LLC Group owner: Cumulus Media Inc. (acq 3-29-2004; grpsl). Natl. Network: ABC. Natl. Rep: Christal. Rgnl rep: Christal Radio Format: Adult contemp. Target aud: 25-54. ♦Roseanne Rybak, gen mgr.

Stillwater

KLBB(AM)— Mar 13, 1949: 1220 khz; 5 kw-D, 254 w-N. TL: N45 03 15 W92 49 42. Hrs open: 24 c/o Endurance Broadcasting LLC, 104 N. Main St., 55082. Phone: (651) 439-5006. Fax: (651) 439-5015. E-mail: dan@mighty1220.com Web Site: www.mighty1220.com. Licensee: Endurance Broadcasting LLC (acq 7-5-2001). Population served: 2,200,000 Natl. Network: ABC, Westwood One. Law Firm: Miller & Miller, P.C. Format: Adult standards. News staff: one. Target aud: 35-64. ♦Daniel Smith, CEO & pres; Gretchen Smith, VP; Scott Murray, gen mgr; Reed Hagen, progmg dir.

Sunburg

KLFN(FM)— 2003: 106.5 mhz; 2.3 kw. Ant 525 ft TL: N45 22 25 W95 08 23. Hrs open: Box 838, Willmar, 56201. Phone: (320) 235-3535. Fax: (320) 235-9111. Licensee: Lakeland Broadcasting Co. (acq 10-5-01). Format: Classic rock. ♦Doug Loy, gen mgr.

Thief River Falls

KKAQ(AM)— Nov 2, 1979: 1460 khz; 2.5 kw-U. TL: N48 07 21 W96 08 24. Hrs open: 24 Box 40, 56701. Secondary address: Hwy. 32 N., ThiefRiver Falls 56701. Phone: (218) 681-4900. Fax: (218) 681-3717. Web Site: www.trfradio.com. Licensee: Iowa City Broadcasting Co. Inc. Group owner: Tom Ingstad Broadcasting Group (acq 11-19-99; $620,000 with co-located FM). Population served: 30,000 Law Firm: Eugene T. Smith. Format: Country. News staff: one; News: 4 hrs wkly. Target aud: 25-54. Spec prog: Oldies 6 hrs wkly. ♦John Praska, gen mgr; Mark Stromstodt, progmg dir.

KKDQ(FM)—Co-owned with KKAQ(AM). Nov 1, 1989: 99.3 mhz; 6.5 kw. 167 ft TL: N48 07 25 W96 08 31. Stereo. 24 Web Site: www.trfradio.com.

***KNTN(FM)**— Dec 13, 1991: 102.7 mhz; 100 kw. 538 ft TL: N47 58 38 W96 36 32. Stereo. Hrs opn: 24 c/o KCCM, 901 S. 8th St., Moorhead, 56562. Phone: (218) 299-3666. Fax: (218) 299-3418. Web Site: www.mpr.org. Licensee: Minnesota Public Radio Inc. Natl. Network: NPR, PRI. Rgnl. Network: Minn. Pub. Format: News. News staff: 2. Target aud: General. ♦William H. Kling, pres; Vern Goodin, gen mgr; Julia Beaton, dev dir.

***KQMN(FM)**— Nov 26, 1990: 91.5 mhz; 100 kw. 449 ft TL: N47 58 38 W96 36 32. Stereo. Hrs opn: 24 901 S. 8th St., Concordia College, Moorhead, 56562. Phone: (218) 299-3666. Fax: (218) 299-3418. Web Site: www.mpr.org. Licensee: Minnesota Public Radio. Natl. Network: NPR, PRI. Format: Class music,cultural progmg. News staff: 2. ♦William H. Kling, pres; Vern Goodin, gen mgr.

KSNR(FM)—Licensed to Thief River Falls. See East Grand Forks

***KSRQ(FM)**— Nov 15, 1971: 90.1 mhz; 24 kw. 338 ft TL: N48 01 19 W96 22 12. Stereo. Hrs opn: 24 1101 Hwy. 1 E., 56701. Phone: (218) 681-0791. Phone: (800) 959-6282. Fax: (218) 681-0774. E-mail: travis.ryder@northlandcollege.edu Web Site: www.pioneer90.org. Licensee: Northland Community & Technical College. (acq 5-29-92). Population served: 140,000 Natl. Network: CNN Radio. Wire Svc: AP Format: Alternative, AAA. News: 10 hrs wkly. Target aud: 18-54; professionals. Spec prog: Sp one hr, adult standards 5 hrs wkly. ♦Anne Temte, pres; Mark Johnson, gen mgr & progmg dir; Travis Ryder, gen mgr & dev dir; Stan Mueller, chief of engrg.

KTRF(AM)— Jan 30, 1947: 1230 khz; 1 kw-U. TL: N48 07 47 W96 11 11. Hrs open: 24 Box 40, 56701. Phone: (218) 681-4900. Fax: (218) 681-3717. E-mail: ktrf@mncable.net Licensee: Iowa City Broadcasting Co. (acq 9-30-97; with KSNR(FM) Thief River Falls). Population served: 45,000 Natl. Network: CBS. Rgnl. Network: MNN. Law Firm: Haley, Bader & Potts. Format: MOR, news. News staff: 2; News: 35 hrs wkly. Target aud: General; 25 plus. Spec prog: Farm 12 hrs wkly. ♦Jon Praska, gen mgr.

Tracy

KARL(FM)— July 19, 1994: 105.1 mhz; 45 kw. 390 ft TL: N44 19 32 W95 52 19. Stereo. Hrs opn: 24 Box 61, KMHL Broadcasting Co., Marshall, 56258. Secondary address: 1414 E. College Dr., Marshall 56258. Phone: (507) 629-3355. Phone: (507) 532-2282. Fax: (507) 532-3739. E-mail: karl@marshallradio.net Web Site: karl.marshallradio.net. Licensee: KMHL Broadcasting Co. Group owner: Linder Broadcasting Group (acq 12-17-92; $22,100; FTR: 1-11-93). Natl. Network: ABC. Rgnl. Network: Linder Farm. Format: Hot country. News: 3 hrs wkly. Target aud: General. Spec prog: Farm 15 hrs wkly. ♦Donald Linder, pres; John Linder, VP; Brad Strootman, gen mgr; Justin Thordson, opns mgr.

Two Harbors

KZIO(FM)— September 1995: 104.3 mhz; 50 kw. 233 ft TL: N46 55 48 W91 53 01. Stereo. Hrs opn: 501 Lake Ave. S., Suite 200A, Duluth, 55802. Phone: (218) 722-0921. Fax: (218) 723-1499. Licensee: Red Rock Radio Corp. (group owner; acq 1-10-00; grpsl). Format: Active rock. ♦Sean Skramstad, gen mgr.

Verndale

KVKK(AM)— 2005: 1070 khz; 10 kw-D, 5 kw-N, DA-N. TL: N46 23 43 W94 57 54 (D), N46 23 45 W94 57 52 (N). Hrs open: Box 49, Park Rapids, 56470. Phone: (218) 732-3306. Fax: (218) 732-3307. Licensee: DJ Broadcasting Corp. Format: Country. ♦Edward P. DeLaHunt Sr., gen mgr.

Virginia

WUSZ(FM)— June 2, 1971: 99.9 mhz; 100 kw. Ant 567 ft TL: N47 22 52 W92 57 18. Stereo. Hrs opn: 24 807 W. 37th St., Hibbing, 55746. Phone: (218) 262-4545. Fax: (218) 263-6112. Web Site: www.radiousa.com. Licensee: Midwest Communications Inc. (group owner; acq 5-10-2004; grpsl). Population served: 330,000 Natl. Network: USA. Rgnl. Network: MNN. Format: Contemp country. News: one hr wkly. Target aud: 25-49; blue and white collar workers and families. ♦Kristi Garrity, gen mgr & sls VP.

Virginia-Hibbing

WIRR(FM)—Licensed to Virginia-Hibbing. See Duluth

Wabasha

KMFX(AM)— April 1976: 1190 khz; 1 kw-D. TL: N44 20 44 W91 58 28. Hrs open: 6 AM-sunset
Rebroadcasts KMFX-FM Lake City 90%.
1530 Greenview Dr. S.W., Rochester, 55902. Phone: (507) 288-3888. Fax: (507) 288-7815. Web Site: www.foxcountry.net. Licensee: CC Licenses LLC. Group owner: Clear Channel Communications Inc. (acq 9-25-2000; grpsl). Rgnl. Network: MNN. Format: Hot Country. Target aud: 18-54. ♦Bob Fox, gen mgr; Craig Erpestad, opns dir & progmg dir; Mary Anne Nons, gen sls mgr.

Wadena

KKWS(FM)—Listing follows KWAD(AM).

KWAD(AM)— Apr 24, 1948: 920 khz; 1 kw-U, DA-N. TL: N46 22 15 W95 08 58. Hrs open: 24 Box 551, 56482-0551. Secondary address: 201 1/2 Jefferson St. S. 56482. Phone: (218) 631-1803. Fax: (218) 631-4557. E-mail: kwadkkws@arvig.net Web Site: www.superstation106.com. Licensee: BL Broadcasting Inc. Group owner: Omni Broadcasting Co. (acq 4-1-2004; grpsl). Population served: 24,800 Natl. Network: ABC. Rgnl. Network: MN News Net., MN Farm Net. Law Firm: Garvey, Schubert & Barer. Format: C&W. News staff: one; News: 20 hrs wkly. Target aud: 25-54; adults. Spec prog: Farm 10 hrs wkly. ♦Lou Buron, CEO & pres; Mary Campbell, CFO & VP; Rick Youngbauer, gen mgr & gen sls mgr; Sherry Linnes, prom dir; Kyle Gylsen, progmg dir, mus dir & pub affrs dir; Chad Moyer, news dir; Dave Cox, engrg dir & chief of engrg.

KKWS(FM)—Co-owned with KWAD(AM). Sept 23, 1968: 105.9 mhz; 100 kw. Ant 564 ft TL: N46 36 00 W94 54 03. Stereo. 24 Web Site: www.superstationk106.com.65,000 Natl. Network: ABC. Format: Country, mainstream. News staff: one; News: 12 hrs wkly. Target aud: 25-54; adults. ♦Mike Danvers, progmg dir, mus dir & pub affrs dir.

Waite Park

KLZZ(FM)—Listing follows KXSS(AM).

KXSS(AM)— Jan 1, 1981: 1390 khz; 2.5 kw-D, 1 kw-N, DA-2. TL: N45 32 31 W94 15 41. Stereo. Hrs opn: 640 S.E. Lincoln Ave., St. Cloud, 56304. Phone: (320) 251-4422. Fax: (320) 251-1855. Web Site: www.1390thefan.com. Licensee: Regent Licensee of St. Cloud Inc. Group owner: Regent Communications Inc. (acq 5-8-2001; grpsl). Format: Sports. Target aud: 18-49. ♦William Stakelin, pres; David Engberg, gen mgr & gen sls mgr; Dick Nelson, progmg dir; Lee Voss, news dir; Mark Young, chief of engrg.

KLZZ(FM)—Co-owned with KXSS(AM). July 1989: 103.7 mhz; 3 kw. 328 ft TL: N45 32 35 W94 15 41. (CP: 25 kw. TL: N45 29 02 W94 08 12). Web Site: www.1037theloon.com. Format: Classic rock. ♦Don Monson, progmg dir & traf mgr.

Walker

KAKK(AM)— July 11, 1970: 1570 khz; 1 kw-D, 250 w-N. TL: N47 04 58 W94 35 21. Hrs open: 24 Box 1022, 56484. Phone: (218) 547-4000. Fax: (218) 547-4001. Licensee: Edward De La Hunt (acq 12-26-00). Population served: 25,000 Rgnl. Network: MNN. Format: Oldies. Target aud: General. ♦Brad Walhof, gen mgr & gen sls mgr.

KLLZ-FM— May 6, 1984: 99.1 mhz; 100 kw. Ant 505 ft TL: N47 12 52 W94 55 18. Stereo. Hrs opn: 24 Box 1656, Bemidji, 56619-1656. Secondary address: 502 Beltrami Ave. N.W., Bemidji 56601. Phone: (218) 444-1500. Fax: (218) 759-0345. E-mail: harry@pbbroadcasting.com Licensee: BG Broadcasting Inc. Group owner: Omni Broadcasting Co. (acq 10-24-2000). Population served: 75,000 Natl. Network: ABC. Law Firm: Garvey, Schubert & Barer. Format: Classic rock. News staff: one; News: 12 hrs wkly. Target aud: 25-54; adults. ♦Lou Buron, CEO, pres & gen mgr; Mary Campbell, CFO & VP; Jack Hicks, opns dir, opns mgr, progmg dir & mus dir; Harry Hastings, gen sls mgr; Brian Fisher, prom dir; Mardy Karger, news dir & pub affrs dir; Mark Anderson, engrg dir & chief of engrg.

KQKK(FM)— May 1, 1999: 101.9 mhz; 50 kw. 328 ft TL: N47 03 03 W94 28 12. Stereo. Hrs opn: 5:30 AM-Midday Box 1022, Hwy. 34 W., 56484. Phone: (218) 547-4000. Fax: (218) 547-4001. E-mail: kqkkkakk@eot.com Web Site: www.dbcradionet.com/kqkk. Licensee: CJ Broadcasting. Population served: 30,000 Natl. Network: CBS Radio. Wire Svc: AP Format: Adult contemp. ♦Bradley J. Walhof, gen mgr.

Warroad

KKWQ(FM)— August 1989: 92.5 mhz; 100 kw. 472 ft TL: N48 49 41 W92 23 16. Stereo. Hrs opn: 24 113A Lake St. Ctr., Box 69, 56763. Phone: (218) 386-3024. Fax: (218) 386-3090. Web Site: www.kq92.com. Licensee: Border Broadcasting LP (acq 1996). Natl. Network: ABC, Jones Radio Networks. Law Firm: Wombie, Carlyle. Format: Country. Target aud: 25-54. ♦Mike Pederson, pres & gen mgr.

Waseca

KOWZ(AM)— Dec 22, 1971: 1170 khz; 2.5 kw-D, 60 w-N, 1 kw-CH. TL: N44 02 45 W93 23 08 (D), N44 04 45 W93 30 24 (N). Hrs open: 24 255 Cedarale Dr. S.E., Owatonna, 55060. Phone: (507) 444-9224. Fax: (507) 444-9080. Licensee: Main Street Broadcasting Inc. Group owner: Linder Broadcasting Group (acq 12-26-01; with co-located FM). Population served: 65,000 Rgnl. Network: MNN, Midwest Radio. Law Firm: Miller & Miller. Format: News, talk. News staff: 2; News: 12 hrs wkly. Target aud: 30-65; general, farm. ♦Jeff Seaton, gen mgr.

KRUE(FM)—Co-owned with KOWZ(AM). June 1972: 92.1 mhz; 25 kw. 286 ft TL: N44 02 45 W93 23 08. Stereo. 24 Web Site: www.star92radio.com. Natl. Network: ABC. Format: Oldies. News staff: one; News: 4 hrs wkly. Target aud: 25-50. ♦Jay Kelly, progmg dir.

Watertown

KZGX(AM)— May 16, 1996: 1600 khz; 5 kw-U, DA-1. TL: N44 55 23 W93 46 56. Hrs open: 24 3300 Jaybird Rd., Bessemer, AL, 35020. Phone: (786) 253-5130. Fax: (205) 426-3178. Web Site: www.kwom1600.com. Licensee: WM Broadcasting Inc. (acq 2-9-2004; $600,000). Natl. Network: ABC. Format: Mexician rgnl. News staff: one. Target aud: 35 plus; general. ♦Mike Parry, gen mgr; Randy Asplund, opns mgr.

Willmar

***KBHZ(FM)**— Feb 16, 1996: 91.9 mhz; 25 kw. 328 ft TL: N45 00 40 W94 53 56. Hrs open: Box 247, Osakis, 56360. Secondary address: 106 Litchfield Ave. W. 56201. Phone: (320) 859-3000. Fax: (320) 859-3010. E-mail: mail@praisefm.org Web Site: www.praisefm.org. Licensee: Christian Heritage Broadcasting Inc. Format: Worship mus. ♦David McIver, gen mgr.

KDJS(AM)— Mar 2, 1981: 1590 khz; 1 kw-D, 89 w-N, DA-2. TL: N45 05 07 W95 00 19. Hrs open: 24 Box 380, 56201. Secondary address: 730 N.E. Hwy. 71 56201. Phone: (320) 231-1600. Fax: (320) 235-7010. Web Site: www.k-musicradio.com. Licensee: Iowa City Broadcasting Inc. (acq 4-1-2000; with co-located FM). Population served: 100,000 Format: Oldies. News staff: one. Target aud: 25-54. Spec prog: Farm 5 hrs wkly. ♦Doug Hanson, gen mgr & gen sls mgr.

KDJS-FM— May 17, 1993: 95.3 mhz; 50 kw. 436 ft TL: N45 01 23 W95 15 57. Stereo. 24 Web Site: www.k-musicradio.com.100,000 Rgnl rep: Hyett/Ramsland Format: Country. News staff: one. ♦Steve Youngberg, engrg mgr.

***KKLW(FM)**— Jan 29, 2004: 90.9 mhz; 400 w. Ant 423 ft TL: N45 11 52 W94 56 58. Stereo. Hrs opn: 24 2351 Sunset Blvd., Suite 170-218, Rocklin, CA, 95765. Phone: (916) 251-1600. Fax: (916) 251-1650. E-mail: klove@klove.com Web Site: www.klove.com. Licensee: Educational Media Foundation. Group owner: EMF Broadcasting. Natl. Network: K-Love. Law Firm: Shaw Pittman. Format: Contemp Christian. News staff: 3. Target aud: 25-44; Judeo Christian, female. ♦Richard Jenkins, pres; Mike Novak, VP; Keith Whipple, dev dir; David Pierce, progmg mgr; Ed Lenane, news dir; Sam Wallington, engrg dir; Marya Morgan, news rptr; Richard Hunt, news rptr.

KQIC(FM)—Listing follows KWLM(AM).

KWLM(AM)— 1940: 1340 khz; 1 kw-U. TL: N45 08 00 W95 02 35. Hrs open: 24 Box 838, 56201. Secondary address: 1340 N. 7th St. 56201. Phone: (320) 235-1340. Fax: (320) 235-9111. Web Site: www.kwlm.com. Licensee: Steven W. Linder. (acq 4-5-91; $691,937 with co-located FM; FTR: 4-22-91) Population served: 50,000 Rgnl. Network: Linder Farm. Format: News/talk. News staff: 3; News: 30 hrs wkly. Target aud: General. Spec prog: Farm 8 hrs wkly. ♦J.P. Cola, gen mgr & news dir; Doug Loy, gen sls mgr; Pete Hoagland, chief of engrg; Mary Overman, traf mgr.

KQIC(FM)—Co-owned with KWLM(AM). July 1, 1965: 102.5 mhz; 100 kw. 830 ft TL: N45 11 40 W95 05 01. Stereo. Phone: (320) 235-3535. Web Site: www.1025fm.com.120,000 Format: Adult contemp. Target aud: 18-49. ♦MaryElin Macht, mus dir.

Windom

KDOM(AM)— Dec 28, 1958: 1580 khz; 1 kw-D, 2 w-N, DA-2. TL: N43 51 41 W95 05 50. Hrs open: 24 Box 218, 56101. Phone: (507) 831-3908. Fax: (507) 831-3913. Web Site: www.kdomradio.com. Licensee: Windom Radio Inc. (acq 4-89; with co-located FM; FTR: 4-14-80). Population served: 165,500 Rgnl. Network: MNN. Format: Country, news/talk, sports. News staff: one; News: 21 hrs wkly. Target aud: General; farm audience, housewives, business owners & laborers. ♦Dave Cory, gen mgr; Dirk Abraham, news dir.

KDOM-FM— Dec 8, 1976: 94.3 mhz; 5.7 kw. 335 ft TL: N43 53 06 W95 10 53. Stereo. 6 AM-midnight Web Site: www.kdomradio.com.

KQRB(FM)— February 2003: 89.9 mhz; 250 w. Ant 171 ft TL: N43 51 15 W95 07 30. Hrs open: Box 3206, American Family Radio, Tupelo, MS, 38803. Phone: (662) 844-8888. Fax: (662) 842-6791. Web Site: www.afr.net. Licensee: American Family Association. Group owner: American Family Radio. Natl. Network: USA. Format: Christian. ♦Roy Willoff, gen mgr.

***KRLP(FM)**—Not on air, target date: unknown: 88.1 mhz; 250 w. Ant 171 ft TL: N43 51 15 W95 07 30. Hrs open: 5700 West Oaks Blvd.,

Rocklin, CA, 95765. Phone: (916) 251-1600. Fax: (916) 251-1650. Licensee: Educational Media Foundation. (acq 3-23-2007; grpsl). ♦Richard Jenkins, pres.

Winona

KAGE(AM)— Feb 17, 1957: 1380 khz; 4 kw-D. TL: N44 02 13 W91 37 09. Hrs open: Box 767, 752 Bluffview Cir., 55987-0767. Phone: (507) 452-4000. Phone: (507) 452-2867. Fax: (507) 452-9494. E-mail: jpapenfuss@winonradio.com Web Site: www.winonaradio.com. Licensee: KAGE Inc. Group owner: The Result Radio Group (acq 1-73). Population served: 90,420 Law Firm: Wiley, Rein & Fielding. Wire Svc: AP Format: C&W. News staff: one; News: 7 hrs wkly. Target aud: 35 plus; general. Spec prog: Farm, relg. ♦Pat Papenfuss, pres, opns mgr & progmg dir; Jerry Papenfuss, gen mgr; Les Guderian, sls dir & gen sls mgr; Darryl Smelser, news dir; Steve Schuh, chief of engrg; Paul Van Beck, sports cmtr.

KAGE-FM— Aug 14, 1971: 95.3 mhz; 11 kw. 495 ft TL: N44 02 31 W91 40 47. Stereo. 24 Web Site: www.winonaradio.com. Format: Adult contemp. News staff: one; News: 14 hrs wkly. Target aud: 25-54. ♦Jerry Papenfuss, CEO; Aaron Taylor, progmg dir.

KHME(FM)— June 4, 1992: 101.1 mhz; 25 kw. 741 ft TL: N44 04 26 W91 34 38. Stereo. Hrs opn: 24 Box 767, 55987. Secondary address: 752 Bluffview Cir. 55987. Phone: (507) 452-4000. Fax: (507) 452-9494. E-mail: khme@hbci.com Web Site: www.winonaradio.com/khme101. Licensee: KAGE Inc. Group owner: The Result Radio Group (acq 10-19-01; $1 million). Population served: 143,000 Format: Soft rock. News staff: 2; News: 21 hrs wkly. Target aud: 25-54; women. ♦Jerry Papenfuss, gen mgr; Les Guderian, gen sls mgr; Pat Papenfuss, opns mgr & chief of engrg.

***KQAL(FM)**— Dec 12, 1975: 89.5 mhz; 1.8 kw. 628 ft TL: N44 02 52 W91 38 40. Stereo. Hrs opn: 24 Box 5838, 55987-0838. Secondary address: 175 W. Mark St. 55987. Phone: (507) 453-2222. Fax: (507) 457-5226. Web Site: www.kqal.org. Licensee: Winona State University. Population served: 75,000 Natl. Network: AP Radio. Wire Svc: AP Format: Jazz, AOR. News: 5 hrs wkly. Target aud: General. Spec prog: Class 14 hrs, pub affrs 10 hrs wkly. ♦Ajit Daniel, gen mgr; Mike Martin, chief tech.

***KSMR(FM)**— Nov 1, 1978: 92.5 mhz; 4 w. -141 ft TL: N42 02 47 W91 41 43. Stereo. Hrs opn: St. Mary's Univ., #29, 700 Terrace Heights, 55987-1399. Phone: (507) 457-1613. Fax: (507) 457-1439. Web Site: www.smumn.edu. Licensee: St. Mary's University Population served: 35,000 Format: AOR, hip hop, varied. Target aud: 18-24. ♦Dean Beckman, gen mgr.

KWNO(AM)— January 1938: 1230 khz; 1 kw-U. TL: N44 01 52 W91 38 31. Hrs open: 24 752 Bluffview Cir., 55987-0767. Phone: (507) 452-4154. Fax: (507) 452-9494. E-mail: jrhyner@winonaradio.com Web Site: www.winonaradio.com. Licensee: KAGE Inc. Group owner: The Result Radio Group (acq 6-19-95; $1 million with KWNO-FM Rushford). Population served: 38,000 Rgnl. Network: MNN. Law Firm: Wiley, Rein & Fielding. Wire Svc: AP Format: Oldies, news/talk, sports. News staff: one; News: 21 hrs wkly. Target aud: 35 plus; Sports fans. Spec prog: Polka 5 hrs wkly. ♦Jerry Papenfuss, CEO & gen mgr; Pat Papenfuss, chmn, exec VP, opns VP & opns mgr.

Winthrop

KHRS(FM)—Not on air, target date: unknown: 105.9 mhz; 23 kw. Ant 344 ft TL: N44 28 25 W94 28 15. Hrs open: 120 W. Thompson Ave., Apt. 108, West St. Paul, 55118. Licensee: Matthew L. Ketelsen. ♦Matthew L. Ketelsen, gen mgr.

Worthington

***KBOJ(FM)**— 2002: 88.1 mhz; 250 w. Ant 144 ft TL: N43 35 53 W95 37 30. Hrs open: Box 3206, American Family Radio, Tupelo, MS, 38803. Phone: (662) 844-8888. Fax: (662) 842-6791. Web Site: www.afr.net. Licensee: American Family Association. Group owner: American Family Radio Natl. Network: USA. Format: Christian. ♦Marvin Sanders, gen mgr.

KITN(FM)— November 1994: 93.5 mhz; 50 kw. Ant 466 ft TL: N43 31 31 W95 24 47. Hrs open: 24 28779 County Hwy. 35, 56187. Phone: (507) 376-6165. Fax: (507) 376-5071. E-mail: contactus@935thebreeze.com Web Site: www.935thebreeze.com. Licensee: Three Eagles of Luverne Inc. Group owner: Three Eagles Communications (acq 12-23-99; grpsl). Population served: 121,300 Rgnl. Network: MNN. Format: Adult contemp. ♦Gary Buchanan, CFO & prom dir; Joel Koetke, gen mgr; Matt Widboom, opns mgr & farm dir.

***KRSW(FM)**— December 1973: 89.3 mhz; 100 kw. Ant 554 ft TL: N43 53 01 W95 55 44. Stereo. Hrs opn: 1450 College Way, 56187. Phone: (605) 335-6666. Fax: (605) 335-1259. Licensee: Minnesota Public Radio Inc. Population served: 20,000 Natl. Network: PRI, NPR. Format: Class. ♦William H. Kling, pres & gen mgr.

KWOA(AM)— Oct 11, 1947: 730 khz; 1 kw-U, 159 w-N. TL: N43 37 48 W95 40 32. Hrs open: 24 28779 County Hwy. 35, 56187. Phone: (507) 376-6165. Fax: (507) 376-5071. E-mail: contactus@kwoa.com Web Site: www.kwoa.com. Licensee: Three Eagles of Luverne Inc. Group owner: Three Eagles Communications (acq 12-23-99; grpsl). Population served: 100,000 Natl. Network: CBS. Natl. Rep: Hyett/Ramsland. Rgnl rep: Midwest Radio. Law Firm: Pepper & Corazzini. Format: News/talk. News staff: 2; News: 18 hrs wkly. Target aud: 35 plus. Spec prog: Farm. ♦Joel Koetke, gen mgr & gen sls mgr; Matt Widboom, progmg dir; Darrell Stitt, news dir.

KWOA-FM— May 3, 1961: 95.1 mhz; 100 kw. Ant 660 ft TL: N43 37 48 W95 40 32. Stereo. 20 E-mail: contactus@951theeagle.com Web Site: 951theeagle.com.103,500 Format: Classic hits. News staff: one; News: 10 hrs wkly. ♦Tony Winter, progmg dir & mus dir.

Worthington-Marshall

***KNSW(FM)**— 1979: 91.7 mhz; 99 kw. Ant 797 ft TL: N43 53 01 W95 55 44. Hrs open: Minnesota Public Radio, 45 7th St. E., Saint Paul, 55101-2202. Phone: (651) 290-1500. Fax: (651) 290-1224. Web Site: www.mpr.org. Licensee: Minnesota Public Radio. Format: News/talk. ♦William H. Kling, gen mgr.

Mississippi

Aberdeen

WACR-FM— June 1, 1975: 105.3 mhz; 50 kw. Ant 358 ft TL: N33 40 09 W88 40 08. Stereo. Hrs opn: 608 Yellow Jacket Dr., Starkville, 39759-3736. Phone: (662) 338-5424. Fax: (662) 338-5436. Web Site: www.wacrfm.com. Licensee: Urban Radio Licenses LLC. (acq 7-13-2005; $1.1 million). Population served: 485,700 Format: Blues-based urban adult contemp. ♦Pamela Hancock, gen mgr.

WWZQ(AM)— February 1952: 1240 khz; 1 kw-U. TL: N33 48 32 W88 32 33. Hrs open: Box 458, Amory, 38821. Phone: (662) 256-9726. Fax: (662) 256-9725. E-mail: fm95@fm95radio.com Web Site: www.fm95radio.com. Licensee: Stanford Communications Inc. (group owner; acq 12-2-99; $51,000). Population served: 11,600 Natl. Network: USA. Format: talk, news, sports. Spec prog: Gospel 8 hrs wkly. ♦Ed Stanford, gen mgr.

Ackerman

WFCA(FM)— 1986: 107.9 mhz; 100 kw. 1,007 ft TL: N33 25 25 W89 24 13. Stereo. Hrs opn: 24 R.R. 1 Box 12, 155 Mecklin, French Camp, 39745. Phone: (662) 547-6414. Fax: (662) 547-9451. E-mail: sales@wfcafm108.com Web Site: www.wfcafm108.com. Licensee: French Camp Radio Inc. Rgnl. Network: Miss. Net. Format: Southern gospel. Target aud: General. ♦Charles S. Carroll, stn mgr.

Amory

WACR-FM—See Aberdeen

WAFM(FM)—Listing follows WAMY(AM).

WAMY(AM)— Oct 23, 1955: 1580 khz; 1 kw-D. TL: N33 58 33 W88 29 29. Hrs open: 6 AM-10PM
Rebroadcasts WWZQ(AM) Aberdeen 75%.
Box 458, 38821. Secondary address: 521 Hwy.278 W. 38821. Phone: (662) 256-9726. Fax: (662) 256-9725. E-mail: fm95radio@fm95radio.com Web Site: fm95radio.com. Licensee: Stanford Communications Inc. (group owner; acq 9-21-92; $85,000 with co-located FM; FTR: 11-9-92) Population served: 45,000 Natl. Network: USA. Rgnl. Network: Miss. Net. Format: Sports, talk/news. News staff: one. Target aud: Genral. Spec prog: Relg 6 hrs, Gospel 6 hrs wkly. ♦Ed Stanford, CEO, gen mgr, sls VP & news dir; Teresa Stanford, VP; Ken Wardlaw, opns mgr, progmg mgr, mus dir & disc jockey; Olen Booth, chief of engrg; Clara Kennedy, traf mgr & women's int ed.

WAFM(FM)—Co-owned with WAMY(AM). 1974: 95.3 mhz; 6 kw. 272 ft TL: N33 58 33 W88 29 29. Stereo. 24 Web Site: www.fm95radio.com. Natl. Network: ABC. Format: Oldies. Spec prog: Relg 2 hrs wkly. ♦Ken Wardlaw, mus dir & disc jockey; Clara Kennedy, traf mgr; Olen Booth, engrg VP & women's int ed.

Artesia

WSMS(FM)— 1985: 99.9 mhz; 6 kw. Ant 328 ft TL: N33 39 14 W88 37 15. Hrs open: 24 200 6th St. N., Suite 205, Columbus, 39701. Phone: (662) 327-1183. Fax: (662) 328-1122. Web Site: www.999thefoxrocks.com. Licensee: Cumulus Licensing Corp. Group owner: Cumulus Media Inc. (acq 9-99; grpsl). Population served: 150,000 Format: Album Rock. Target aud: 18-49. ♦Cole Evans, gen sls mgr; C. S. Jones, mktg mgr.

Baldwyn

WESE(FM)— Oct 1, 1980: 92.5 mhz; 12 kw. Ant 472 ft TL: N34 21 46 W88 35 28. Hrs open: 24 Box 3300, Tupelo, 38803. Secondary address: 5026 Cliff Gookin Blvd., Tupelo 38801. Phone: (662) 842-1067. Fax: (662) 842-0725. Web Site: www.925jamz.com. Licensee: Clear Channel Broadcasting Licenses Inc. Group owner: Clear Channel Communications Inc. (acq 12-19-00; grpsl). Population served: 174,600 Natl. Network: ABC. Natl. Rep: Interep. Format: Urban contemp. Target aud: 18-54. Spec prog: Gospel 6 hrs, Blues 6 hrs wkly. ♦Mark Maharrey, gen mgr; Rick Stevens, opns VP.

Batesville

WBLE(FM)—Listing follows WJBI(AM).

WJBI(AM)— June 19, 1953: 1290 khz; 730 w-D, 91 w-N. TL: N31 18 13 W89 58 59. Hrs open: Box 1528, 38606. Phone: (662) 563-1290. Fax: (662) 563-9002. Licensee: Batesville Broadcasting Co. Inc. (acq 4-1-78). Population served: 24,700 Format: Nostalgia. Target aud: 30 plus. Spec prog: Gospel. ♦J. Boyd Ingram, pres; John P. Ingram, gen mgr.

WBLE(FM)—Co-owned with WJBI(AM). Aug 1, 1978: 100.5 mhz; 50 kw. 492 ft TL: N34 18 13 W89 58 59. Stereo. 225,000 Format: Country. Target aud: 25 plus.

Bay Springs

WIZK(AM)— 1570 khz; 3.2 kw-D. TL: N31 57 56 W89 18 03. Hrs open: 12 Box 548, 150 Bay Ave., 39422. Phone: (601) 764-9888. Fax: (601) 764-9887. E-mail: wizk@bayspringstel.net Licensee: M. Jerome Hughey (Acq 3-15-02.). Population served: 200,000 Format: Traditional Country, Southern gospel, oldies. Target aud: 25-54; baby boomers & older consumers. ♦Mitchell Jerome Hughey, CEO, pres & gen mgr; Mitchell ODell Hughey, opns VP; Tom Diaz, stn mgr & chief of engrg.

WKZW(FM)— July 7, 1975: 94.3 mhz; 3 kw. 328 ft TL: N31 59 00 W89 13 50. (CP: 50 kw, ant 531 ft.). Stereo. Hrs opn: Box 6408, Laurel, 39441. Secondary address: Box 16596, Hattiesburg 39404. Phone: (601) 649-0095. Fax: (601) 649-8199. E-mail: kz94@kz94.com Web Site: www.kz94.c0m. Licensee: Blakeney Communications Inc. (group owner; acq 3-25-98; $553,000 for stock). Format: Hot adult contemp. Target aud: 18-60; average working people. ♦Larry Blakeney, pres; Randy Blakeney, gen mgr & engrg dir; Stephen St. James, progmg dir.

Bay St. Louis

WBSL(AM)— March 1974: 1190 khz; 5 kw-D. TL: N30 19 25 W89 21 03. Hrs open: 13 1190 Casino Magic Dr., 39520. Phone: (228) 467-1190. Phone: (228) 467-7009. Fax: (228) 467-5295. Licensee: Hancock Broadcasting Corp. Population served: 15,000 Format: Blues/talk. News staff: one; News: 6 hrs wkly. Target aud: Men 25-54; sports enthusiasts. Spec prog: Gospel 13 hrs wkly. ♦Ira Hatchett, CEO; Benni Hatchett, pres; Barry Hatchett, exec VP; Delores Hatchett, VP.

WZKX(FM)—Licensed to Bay St. Louis. See Poplarville

Belzoni

WBYP(FM)— 1986: 107.1 mhz; 9.4 kw. Ant 531 ft TL: N33 03 04 W90 37 51. Stereo. Hrs opn: 24 Box 130, Yazoo City, 39194. Secondary address: 611 Center Park Ln., Yazoo City Phone: (662) 746-7676. Fax: (662) 746-1525. E-mail: power107@power107.org Web Site: www.power107.org. Licensee: Zoo-Bel Broadcasting LLC. Format: Country, southern gospel. News staff: 3; News: 18 hrs wkly. Target aud: 18-65; Adults. Spec prog: Black 7 hrs wkly. ♦Colon Johnston, gen mgr, sls dir, progmg dir & chief of engrg; Brenda Johnston, traf mgr; Dan Winstead, disc jockey.

WELZ(AM)— 1959: 1460 khz; 1 kw-D. TL: N33 10 24 W90 28 51. Hrs open: Box 130, Yazoo City, 39194. Phone: (662) 746-7676. Fax: (662) 746-1525. Web Site: www.power107.org. Licensee: Zoo-Bel Broadcasting LLC. (acq 6-9-98; $200,000). Population served: 10,000 Rgnl. Network: Miss. Net. Format: Southern gospel am, gospel pm. Target aud: 12 plus; general. Spec prog: Black 20 hrs wkly. ♦Colon Johnston, gen mgr.

Biloxi

***WMAH-FM**— December 1983: 90.3 mhz; 100 kw. 1,410 ft TL: N30 45 14 W88 56 44. Stereo. Hrs opn: 24 3825 Ridgewood Rd., Jackson, 39211. Phone: (601) 432-6565. Fax: (601) 432-6806. Web Site: www.mpbonline.org. Licensee: Mississippi Authority for Educational Television. Natl. Network: PRI, NPR. Law Firm: Schwartz, Woods & Miller. Wire Svc: AP Format: Class, news/talk, jazz. News staff: 5; News: 20 hrs wkly. Target aud: General. ♦Marie Antoon, chmn; Gene Edwards, gen mgr; Bob Holland, opns mgr & progmg mgr; Ty Warren, dev mgr; Tippy Garner, rgnl sls mgr; Jennifer Griffin, prom mgr; Greg Waxberg, mus dir; Dick Rizzo, news dir; Keith Martin, engrg dir; Betty Taylor, traf mgr.

WMJY(FM)— July 11, 1966: 93.7 mhz; 100 kw. 1,012 ft TL: N30 29 09 W88 42 53. Stereo. Hrs opn: 24 286 Debuys Rd., 39531. Phone: (228) 388-2323. Fax: (228) 388-2362. E-mail: reggiebates @clearchannel.com Web Site: www.magic937.com. Licensee: CC Licenses LLC. Group owner: Clear Channel Communications Inc. (acq 2-2-2004; grpsl). Population served: 350,000 Format: Adult contemp. News staff: one; News: 5 hrs wkly. Target aud: 25-54. ♦Reggie Bates, gen mgr.

WTNI(AM)— 2003: 1640 khz; 10 kw-D, 1 kw-N. TL: N30 28 27 W88 51 23. Hrs open: 1909 E. Pass Rd., Suite D-11, Gulfport, 39507. Phone: (228) 388-2001. Fax: (228) 896-9114. Web Site: www.1640wtni.com. Licensee: Monterey Licenses LLC. Group owner: Triad Broadcasting Co. LLC (acq 5-16-00). Format: News/talk, info. ♦Mary Bigelow, stn mgr.

WXBD(AM)— May 1948: 1490 khz; 1 kw-U. TL: N30 23 38 W88 59 58. Hrs open: 24 1909 E. Pass Rd., Suite D-11, Gulfport, 39507. Phone: (228) 388-2001. Fax: (228) 896-9736. Fax: (228) 896-9114. E-mail: wxbd@sportsradiowxbd.com Web Site: www.sportsradiowxbd.com. Licensee: Monterey Licenses LLC. Group owner: Triad Broadcasting Co. LLC (acq 6-30-99; grpsl). Law Firm: Shaw Pittman. Format: ESPN sports. ♦Jay Taylor, opns mgr.

Booneville

WBIP(AM)— Sept 1, 1950: 1400 khz; 1 kw-U. TL: N34 38 21 W88 34 33. Hrs open: 24 Box 356, 38829-0356. Secondary address: 1101 So. Second St. 38829-2572. Phone: (662) 728-0200. Fax: (662) 728-2572. E-mail: wbipam@avsia.com Licensee: Community Broadcasting Services of Mississippi Inc. (acq 8-31-98; $1,000 for 50% of stock with co-located FM). Population served: 50,000 Format: Classic country. News: 7 hrs wkly. Target aud: 24-54. ♦Larry Melton, pres; Jerry Thornton, VP; Larry Hill, gen mgr; Marty Williams, stn mgr, opns mgr, gen sls mgr & progmg dir.

***WMAE-FM**— December 1983: 89.5 mhz; 85 kw. Ant 660 ft TL: N34 40 00 W88 45 05. Stereo. Hrs opn: 24 3825 Ridgewood Rd., Jackson, 39211. Phone: (601) 432-6565. Fax: (601) 432-6806. Web Site: www.mpbonline.org. Licensee: Mississippi Authority for Educational Television. Natl. Network: PRI, NPR. Law Firm: Schwartz, Woods & Miller. Wire Svc: AP Format: Music, news, info. News staff: 5; News: 20 hrs wkly. Target aud: General. ♦Marie Antoon, chmn; Gene Edwards, gen mgr; Bob Holland, opns mgr.

Brandon

WRBJ-FM— Dec 1, 1974: 97.7 mhz; 6 kw. Ant 328 ft TL: N32 10 31 W89 56 10. Stereo. Hrs opn: 24 1985 Lakeland Dr., Suite 201, Jackson, 39216. Phone: (601) 713-0977. Fax: (601) 713-2977. Licensee: Roberts Radio Broadcasting LLC Group owner: On Top Communications Inc. (acq 9-29-2006;. $1.95 million). Format: Hip hop, rhythm & blues. Target aud: 18-34. ♦Bart Hocton, stn mgr.

WZQK(AM)— June 1967: 970 khz; 1 kw-D, DA. TL: N32 17 20 W89 59 50. Hrs open:
Simulcast with WQST(AM) Forest 100%.
Box 1040, Forest, 39074. Phone: (601) 469-1960. Fax: (601) 469-1366. Licensee: Jackson Radio LLC (acq 6-21-2005). Population served: 500,000 Format: Real country. ♦Ken Michaels, gen mgr.

Brookhaven

WBKN(FM)— July 29, 1976: 92.1 mhz; 5.2 kw. 351 ft TL: N31 36 00 W90 27 09. Stereo. Hrs opn: 24 Box 711, 39602. Secondary address: 911 Hwy. 550 39602. Phone: (601) 833-9210. Fax: (601) 833-6221. E-mail: wbkn92@vis-com.tv Licensee: Brookhaven Broadcasting Inc. (acq 2-27-2007; $1.4 million with WMJU(FM) Bude). Population served: 100,000 Rgnl. Network: Miss. Net. Law Firm: Fletcher, Heald & Hildreth. Format: Country. News: 5 hrs wkly. Target aud: 25-54. Spec prog: Gospel 3 hrs wkly. ♦C. Wayne Dowdy, pres; Ken Hollingsworth, gen mgr, traf mgr, local news ed, news rptr, edit dir & outdoor ed; Robbie Hamilton, sls dir & prom dir; Gaye Laird, progmg dir; Jamey Lambert, min affrs dir & farm dir; Tyler Bridge, spec ev coord & mus critic.

WCHJ(AM)— Aug 15, 1955: 1470 khz; 1 kw-D, 66 w-N. TL: N31 33 46 W90 26 51. Hrs open: 24 hrs Box 177, 39602. Secondary address: 983 Sawmill Ln. 39601. Phone: (601) 823-9006. Fax: (601) 823-0503. E-mail: wchjgospel@netsouth.com Licensee: Tillman Broadcasting Network Inc. (acq 4-9-99; $150,000). Population served: 100,000 Natl. Network: USA. Rgnl. Network: Miss. Net. Format: Black, gospel. News: one hr wkly. Target aud: 25-54. ♦Charles Tillman, CEO, gen mgr & opns mgr.

Brooksville

WAJV(FM)— August 1995: 98.9 mhz; 5.8 kw. 676 ft TL: N33 20 40 W88 32 47. Hrs open: 24 608 Yellow Jacket Dr., Starkville, 39759. Phone: (662) 338-5424. Fax: (662) 338-5436. Web Site: www.joy989.com. Licensee: Urban Radio Licenses LLC (acq 4-20-2001). Format: Urban contemp, gospel. News: 14 hrs wkly. Target aud: General. ♦Kevin Wagner, pres & gen mgr; James Alexander, opns mgr; Ron Davis, progmg dir.

Bruce

WCMR(FM)— 1995: 94.5 mhz; 5.1 kw. Ant 358 ft TL: N34 04 15 W89 13 29. Hrs open:
Rebroadcasts KSRD(FM) Saint Joseph, MO 100%.
KSRD Radio, 1212 Faraon St., St. Joseph, MO, 64501. Phone: (816) 233-5773. Fax: (816) 233-5777. E-mail: info@ksrdradio.com Web Site: www.ksrdradio.com. Licensee: Horizon Christian Fellowship. (acq 10-25-2006; $250,000). Format: Christian. ♦Michael MacIntosh, pres; Brian K.C. Jones, gen mgr.

Bude

***WMAU-FM**— December 1983: 88.9 mhz; 100 kw. 960 ft TL: N31 22 19 W90 45 05. Stereo. Hrs opn: 24 3825 Ridgewood Rd., Jackson, 39211. Phone: (601) 432-6565. Fax: (601) 432-6806. Web Site: www.mpbonline.org. Licensee: Mississippi Authority for Educational Television. Natl. Network: NPR, PRI. Law Firm: Schwartz, Woods & Miller. Wire Svc: AP Format: Class, news/talk, jazz. News staff: 5; News: 20 hrs wkly. Target aud: General. ♦Marie Antoon, chmn; Gene Edwards, gen mgr; Bob Holland, opns mgr.

WMJU(FM)— Aug 30, 1999: 104.3 mhz; 25 kw. Ant 328 ft TL: N31 33 33 W90 40 26. Stereo. Hrs opn: 24 Box 711, Brookhaven, 39602. Secondary address: 911 Hwy. 550, Brookhaven 39601. Phone: (601) 833-9210. Fax: (601) 833-6221. E-mail: majic104@yahoo.com Jonos Licensee: Brookhaven Broadcasting Inc. (group owner) (acq 2-27-2007; $1.4 million with WBKN(FM) Brookhaven). Population served: 100,000 Format: Adult contemp. News staff: one; News: 10 hrs wkly. Target aud: 25-49; adults who are middle income & above. ♦C. Wayne Dowdy, pres; Ken Hollingsworth, gen mgr; Robbie Hamilton, sls dir & prom dir; Gaye Laird, progmg dir.

Burnsville

***WOWL(FM)**— 2000: 91.9 mhz; 18 kw. Ant 548 ft TL: N34 55 47 W88 24 37. Stereo. Hrs opn: 121 Front St., Iuka, 38852. Phone: (662) 423-9919. Fax: (662) 423-9333. Licensee: Southern Community Services Inc. Population served: 500,000 Natl. Rep: Rgnl Reps. Law Firm: Garvey, Schubert & Barer. Format: Adult contemp. ♦Derrick Robinson, gen mgr.

Byhalia

***WKVF(FM)**— November 1994: 94.9 mhz; 6 kw. 403 ft TL: N34 55 30 W89 40 57. Stereo. Hrs opn: 24 2351 Sunset Blvd., Suite 170-218, Rocklin, CA, 95765. Phone: (916) 251-1600. Fax: (916) 251-1650. E-mail: klove@klove.com Web Site: www.klove.com. Licensee: Educational Media Foundation. Group owner: EMF Broadcasting (acq 2-1-00; $1.4 million). Population served: 750,000 Natl. Network: K-Love. Law Firm: Shaw Pittman. Format: Contemp, Christian. News staff: 3. Target aud: 25-44; Judeo Christian, female. ♦Richard Jenkins, pres; Mike Novak, VP & progmg dir; Lloyd Parker, gen mgr; Ed Lenane, opns dir; Keith Whipple, dev dir; Eric Allen, natl sls mgr; David Pierce, progmg mgr; Jon Rivers, mus dir; Sam Wallington, engrg dir; Arthur Vassar, traf mgr; Karen Johnson, news rptr; Marya Morgan, news rptr; Richard Hunt, news rptr.

Canton

WMGO(AM)— Dec 9, 1954: 1370 khz; 1 kw-D, 280 w-N. TL: N32 37 36 W90 01 47. Hrs open: 24 107 W. Peace St., 39046. Phone: (601) 859-2373. Phone: (601) 859-2374. Fax: (601) 859-2664. Licensee: WMGO Broadcasting Corp. Inc. (acq 5-3-93; $100,000; FTR: 5-24-93). Population served: 95,000 Rgnl. Network: Miss. Net. Format: Adult contemp, urban contemp, gospel. News staff: one; News: 12 hrs wkly. Target aud: 25-54; upscale & involved adults. ♦Jerry Lousteau, pres, gen mgr & progmg VP; John Woods, disc jockey.

WONG(AM)— April 1989: 1150 khz; 500 w-D. TL: N32 32 35 W90 03 36. Hrs open: 24 126 E. Sowell Rd., 39046. Phone: (601) 855-2035. Fax: (601) 855-2094. E-mail: wong1150am@cs.com Licensee: Marion R. Williams. (acq 7-26-99; $50,000). Natl. Network: American Urban. Format: Gospel, blues. Target aud: 25 plus. ♦Marion Williams, pres; Kaple Hill, gen mgr.

Carthage

WCKK(FM)— April 1979: 98.3 mhz; 20 kw. Ant 328 ft TL: N32 43 29 W89 32 44. Hrs open: Box 1700, Kosciusko, 39090. Secondary address: 1 Golf Course Rd., Kosciusko 39039. Phone: (662) 289-1340. Fax: (662) 289-7907. Licensee: Johnny Boswell Radio LLC (acq 8-15-2003; $450,000). Natl. Network: USA. Wire Svc: NOAA Weather Format: C&W. ♦Johnny Boswell, gen mgr; Ann Sheen, stn mgr; Eric Matthews, opns mgr.

Centreville

WAKK-FM— Nov 21, 1977: 104.9 mhz; 3 kw. Ant 298 ft TL: N31 06 07 W91 02 27. (CP: 6 kw, ant 328 ft. TL: N31 05 56 W91 02 27). Hrs opn: Box 1649, McComb, 39649. Phone: (601) 684-4116. Fax: (601) 684-4654. Licensee: Southwest Broadcasting. Format: Classic country. ♦Charles Dowdy, gen mgr.

***WPAE(FM)**— 1997: 89.7 mhz; 70 kw. 298 ft TL: N31 05 56 W91 02 27. Hrs open: 24
Rebroadcasts KPAE(FM) Erwinville, LA 100%.
Box 1390, 39631. Secondary address: 122 E. Main St. 39631. Phone: (601) 645-6515. Fax: (601) 645-9122. Licensee: Port Allen Educational Broadcasting Foundation. Population served: 1,000,000 Natl. Network: Moody. Format: Relg, educ. Spec prog: Children 5 hrs, Gospel 15 hrs wkly. ♦Willie F. Kennedy, gen mgr.

Charleston

WTGY(FM)— Apr 1, 1986: 95.7 mhz; 6 kw. Ant 328 ft TL: N33 53 28 W90 03 09. Stereo. Hrs opn: 18 Box 262550, Baton Rouge, LA, 70826. Phone: (225) 768-8300. Phone: (225) 768-3688. Fax: (225) 768-3729. E-mail: kawikfish@yahoo.com Web Site: jsm.org. Licensee: Family Worship Center Church Inc. (group owner; acq 7-15-02;

$300,000). Format: Relg. ♦David Whitelaw, COO; Jimmy Swaggart, pres; John Santiago, gen mgr & progmg dir.

Clarksdale

WAID(FM)— July 1, 1978: 106.5 mhz; 30 kw. 296 ft TL: N34 09 22 W90 37 52. Stereo. Hrs opn: 24 Box 668, 38614. Phone: (662) 627-2281. Fax: (662) 624-2900. Web Site: www.missradio.com. Licensee: Radio Cleveland Inc. (group owner; acq 8-2-83; $185,000; FTR: 8-1-83). Population served: 100,000 Natl. Network: USA. Format: Urban contemp. News staff: one; News: 2 hrs wkly. Target aud: General. ♦Clint Webster, gen mgr; Greg Shurden, gen sls mgr; Jim Thomas, progmg dir; Houston McDavid, chief of engrg.

WKDJ-FM— Nov 1, 1988: 96.5 mhz; 6 kw. 180 ft TL: N34 09 22 W90 37 52. Stereo. Hrs opn: Box 668, 38614. Phone: (662) 627-2281. Fax: (662) 624-2900. Web Site: www.missradio.com. Licensee: Clint Webster. (acq 12-3-93). Format: Country. Target aud: 25-55. ♦Clint Webster, gen mgr; Greg Shurden, stn mgr; Jim Thomas, progmg dir.

WKXY(FM)—Listing follows WROX(AM).

WROX(AM)— 1944: 1450 khz; 1 kw-U. TL: N34 12 40 W90 34 42. Hrs open: 24 330 Sunflower, 38614. Phone: (662) 627-1450. Fax: (662) 621-1176. Licensee: Delta Radio LLC. Group owner: Contemporary Communications (acq 12-18-98; $54,000 with WQMA(AM) Marks). Population served: 65,000 Rgnl. Network: Mississippi Network Law Firm: Wood, Maines & Brown. Format: Rhythm and Blues. News staff: one; News: 3 hrs wkly. Target aud: General. ♦George Hinds, gen mgr; Bill Perry Sr., stn mgr.

WKXY(FM)—Co-owned with WROX(AM). 2003: 92.1 mhz; 500 w. Ant 141 ft TL: N34 12 40 W90 34 42.

Cleveland

WCLD(AM)— 1949: 1490 khz; 1 kw-U. TL: N33 44 01 W90 42 50. Hrs open: 24 Hwy. 61 S., 38732. Secondary address: Drawer 780 Phone: (662) 843-4091. Fax: (662) 843-9805. E-mail: wcld@tecinfo.com Web Site: www.missradio.com. Licensee: Radio Cleveland Inc. (group owner; acq 1957). Population served: 175,000 Format: Black gospel. Target aud: 18 plus. ♦Clint L. Webster, gen mgr; Jim Thomas, opns mgr, progmg dir & news dir; Kevin Cox, gen sls mgr; Houston McDavitt, chief of engrg; Vicky Lowry, traf mgr.

WCLD-FM— 1972: 103.9 mhz; 24.5 kw. 300 ft TL: N33 44 01 W90 42 50. Stereo. 24 Web Site: www.missradio.com.200,000 Law Firm: Fletcher, Heald & Hildreth. Format: Urban contemp. Target aud: 18 plus. ♦Vicky Lowry, traf mgr.

WDFX(FM)— 1993: 98.3 mhz; 25 kw. 328 ft TL: N33 52 44 W90 43 04. Hrs open: Box 3206, American Family Radio, Tupelo, 38803. Phone: (662) 844-8888. Fax: (662) 842-6791. Web Site: www.afr.net. Licensee: American Family Association Inc. (acq 5-3-93; $6,150; FTR: 5-24-93). Natl. Network: USA. Format: Christian. ♦Don Wildman, gen mgr.

WDSK(AM)—Listing follows WDTL-FM.

WDTL-FM— May 22, 1970: 92.9 mhz; 50 kw. 492 ft TL: N33 44 17 W90 39 29. Stereo. Hrs opn: 24 Box 1438, 309 N. Chrisman Ave., 38732. Phone: (662) 846-0929. Fax: (662) 843-1410. E-mail: mrsviradio@aol.com Licensee: M.R.S. Ventures Inc. (group owner; acq 11-1-2003; grpsl). Population served: 130,000 Rgnl. Network: Miss. Net. Law Firm: Wood, Maines & Brown. Format: Country. News staff: one; News: 3 hrs wkly. Target aud: 25-54. ♦Wash Sellers Jr., opns VP, opns dir & progmg dir; Wendy Hodges, gen mgr, gen mgr, gen sls mgr, engrg dir & traf mgr.

WDSK(AM)—Co-owned with WDTL-FM. June 25, 1958: 1410 khz; 920 w-D, 23 w-N. TL: N33 45 56 W90 42 41. Stereo. 24 130,000 Natl. Network: CBS Radio. Format: News/talk. Target aud: 35 plus; male & female.

WMJW(FM)— 1993: 107.5 mhz; 25 kw. 328 ft TL: N33 43 36 W90 43 53. Stereo. Hrs opn: 24 Box 780, 38732. Phone: (662) 843-4091. Fax: (662) 843-9805. Web Site: www.missradio.com. Licensee: Radio Cleveland Inc. (group owner; acq 7-18-95). Law Firm: Fletcher, Heald & Hildreth. Format: Country. News: 8 hrs wkly. Target aud: 25-54; adults. ♦Clint L. Webster, pres & gen mgr; Kevin W. Cox, VP, sls dir, gen sls mgr & prom dir; Jim Thomas, opns mgr; Jim Gregory, pub affrs dir; Vickie Lowery, traf mgr.

WRKG(FM)—Drew, June 1, 1971: 95.3 mhz; 2.65 kw. Ant 492 ft TL: N33 44 17 W90 39 29. Stereo. Hrs opn: 24 Box 1438, 309 N. Chrisman Ave., 38732. Phone: (662) 846-0929. Fax: (662) 843-1410. E-mail: mrsviradio@aol.com Web Site: www.deltaradio.net. Licensee: M.R.S. Ventures Inc. (group owner; (acq 11-1-2003; grpsl). Population served: 85,000 Natl. Network: Jones Radio Networks. Law Firm: Wood, Maines & Brown. Format: Classic rock. News staff: one; News: 2 hrs wkly. Target aud: General. ♦Wendy Hodges, gen mgr.

WZYQ(FM)—Mound Bayou, Oct 10, 1997: 101.9 mhz; 6 kw. Ant 328 ft TL: N33 52 49 W90 42 24. Stereo. Hrs opn: 24 Box 1438, 309 N. Chrisman Ave., 38732. Phone: (662) 846-0929. Fax: (662) 843-1410. E-mail: mrsviradio@aol.com Licensee: M.R.S. Ventures Inc. (group owner; (acq 11-1-2003; grpsl). Population served: 85,000 Natl. Network: Jones Radio Networks. Law Firm: Wood, Maines & Brown. Format: Blues. News staff: one. Target aud: General. ♦Wendy Hodges, gen mgr.

Clinton

WHJT(FM)— 1974: 93.5 mhz; 6 kw. 328 ft TL: N32 20 15 W90 19 47. Stereo. Hrs opn: 24 Box 4048, 39058. Secondary address: 100 S. Jefferson 39058. Phone: (601) 925-3458. Fax: (601) 925-3337. Web Site: www.star93fm.com. Licensee: Mississippi College. Population served: 400,000 Law Firm: Smithwick & Belendiuk. Wire Svc: NOAA Weather Format: CHR Christian. News: 7 hrs wkly. Target aud: 18-54; upper & middle class Christian listeners. Spec prog: Relg 6 hrs wkly. ♦Billy Lytal, pres; Russ Robinson, gen mgr & stn mgr.

WTWZ(AM)— Oct 10, 1982: 1120 khz; 7.5 kw-D (2.5 kw-CH). TL: N32 21 03 W90 20 22. Hrs open: Sunrise-sunset 4611 Terry Rd., Suite C, Jackson, 39212. Phone: (601) 346-0074. Fax: (601) 346-0896. E-mail: wtwzam1120@bellsouth.net Licensee: Terry E. Wood. Population served: 500,000 Natl. Network: USA. Format: Bluegrass. News: 7 hrs wkly. Target aud: 18-50; 50% men & 50% women. ♦Terry Wood, pres & gen mgr.

Coldwater

WVIM-FM— 1976: 95.3 mhz; 3.6 kw. Ant 423 ft TL: N34 46 45 W89 58 01. Stereo. Hrs opn: 230 Goodman Rd., Bldg. 2, Suite 202, Southaven, 38671. Phone: (662) 349-9953. Fax: (662) 349-9255. E-mail: bgallagher@flash953.com Web Site: www.flash953.com. Licensee: First Broadcasting F Holdiings LLC. (group owner; (acq 6-24-2004; $2.1 million). Population served: 80,300 Format: Oldies. Target aud: 25-54. ♦Becky Gallagher, pres & stn mgr; Danny McGregor, opns mgr & progmg dir.

Collins

WKNZ(FM)— Aug 15, 1978: 107.1 mhz; 2.25 kw. Ant 541 ft TL: N31 31 49 W89 30 29. Stereo. Hrs opn: 24 Box 15935, Hattiesburg, 39404. Secondary address: 7501 Hwy. 49 N., Hattiesburg 39403. Phone: (601) 264-0443. Phone: (601) 268-1017. Fax: (601) 264-5733. Licensee: Educational Media Foundation. (acq 7-15-2005; $700,000). Population served: 200,000 Natl. Network: K-Love. Format: Christian. Target aud: 25-54; upwardly mobile. ♦Ted Tibbett, gen mgr.

Columbia

WCJU(AM)— Dec 20, 1946: 1450 khz; 1 kw-U. TL: N31 14 14 W89 50 24. Stereo. Hrs opn: 24 Box 472, 39429. Phone: (601) 736-2616. Fax: (601) 736-2617. Licensee: WCJU Inc. (acq 6-69). Population served: 45,000 Natl. Network: ABC. Natl. Rep: Keystone (unwired net). Format: News/talk, sports. News staff: 2; News: 30 hrs wkly. Target aud: 18-54. Spec prog: Gospel 4 hrs wkly. ♦Pam Ball, opns dir & rgnl sls mgr; T. McDaniel, pres & gen sls mgr; John Pittman Jr., mus dir.

WFFF(AM)— Apr 14, 1961: 1360 khz; 1 kw-D, 159 w-N. TL: N31 15 44 W89 50 41. Hrs open: 24 Box 550, 11 Gardner Shopping Ctr., 39429. Phone: (601) 736-1360. Fax: (601) 736-1361. E-mail: wfffradio@zzip.cc Licensee: Haddox Enterprises Inc. (acq 10-9-91; $250,000 with co-located FM; FTR: 11-4-91) Population served: 50,000 Natl. Network: ABC. Rgnl. Network: Miss. Net. Format: C&W, gospel. News staff: 4; News: 8 hrs wkly. Target aud: General. ♦Ronnie Geiger, pres, gen mgr, gen sls mgr, gen sls mgr, progmg dir, news dir, chief of engrg, local news ed & sports cmtr; Terri Geiger, VP & traf mgr.

WFFF-FM— October 1966: 96.7 mhz; 3 kw. 400 ft TL: N31 15 44 W89 50 41. Stereo. 24 60,000 Format: Adult contemp. News staff: 4; News: 8 hrs wkly. Target aud: 25-54. ♦Ronnie Geiger, opns VP, engrg dir, local news ed & sports cmtr; Terri Geiger, traf mgr.

***WPRG(FM)**—Not on air, target date: unknown: 89.5 mhz; 250 w. 177 ft TL: N31 16 50 W89 51 12. Hrs open: American Family Radio, Box 3206, Tupelo, 38803. Phone: (662) 844-8888. Fax: (662) 842-6791. Web Site: www.afr.net. Licensee: American Family Association. Group owner: American Family Radio (acq 10-1-01). Natl. Network: USA. Format: Christian. ♦Marvin Sanders, gen mgr.

Columbus

***WCSO(FM)**— 2006: 90.5 mhz; 10 kw vert. Ant 530 ft TL: N33 20 44 W88 14 06. Hrs open:
Rebroadcasts WAFR(FM) Tupelo 100%.
Drawer 3206, Tupelo, 38803. Phone: (662) 844-8888. Fax: (662) 842-6791. Web Site: www.afr.net. Licensee: American Family Association. Format: Christian. ♦Marvin Sanders, gen mgr.

WJWF(AM)— Nov 1, 1969: 1400 khz; 1 kw-U. TL: N33 29 30 W88 24 14. Hrs open: 601 2nd Ave. N., 39703. Phone: (662) 327-1183. Licensee: Cumulus Licensing Corp. Group owner: Cumulus Media Inc. (acq 2-14-02; with co-located FM). Format: Sports, news/talk. ♦C.J. Jones, VP, mktg mgr & progmg dir.

WMBC(FM)—Co-owned with WJWF(AM). Nov 1, 1969: 103.1 mhz; 22 kw. 754 ft TL: N33 29 30 W88 24 14. Stereo. 200 6th St. N., Court Square Towers, 39701. Phone: (662) 327-1183.130,000 Format: Hot adult contemp.

WKOR-FM— Dec 16, 1992: 94.9 mhz; 29.5 kw. 492 ft TL: N33 28 38 W88 16 25. Stereo. Hrs opn: 24 200 6th St. N., Suite 205, 39701. Phone: (662) 327-1183. Fax: (662) 328-1122. Web Site: www.k949.net. Licensee: Cumulus Licensing Corp. Group owner: Cumulus Media Inc. (acq 2-14-02; grpsl). Natl. Network: ABC. Format: Hot country. Target aud: 18-54. ♦Cole Evans, gen sls mgr; C. J. Jones, mktg mgr.

***WMUW(FM)**—Not on air, target date: unknown: 88.5 mhz; 980 w. Ant 89 ft TL: N33 29 23 W88 25 18. Stereo. Hrs opn: 6 AM-midnight Mississippi University for Women, 1100 College St. - MUW - 1619, 39701-5800. Phone: (662) 329-7255. Fax: (662) 329-7250. E-mail: wmuw@muw.edu Web Site: www.muw.edu/wmuw. Licensee: Mississippi University for Women. Population served: 30,000 Format: Eclectic (college radio). News: 5 hrs wkly. College students and those that like music. ♦Eric Harlan, gen mgr; Robert Pate, opns mgr; Dale Jones, mus dir; Jackie Doss, mus dir; Orlando Abrams, mus dir.

WTWG(AM)— 1950: 1050 khz; 1 kw-D, 48 w-N, DA. TL: N33 30 36 W88 24 46. Hrs open: Box 1078, 39703. Secondary address: 1910 14th Ave. N. 39703. Phone: (662) 328-1050. Fax: (662) 328-1054. Licensee: T & W Communications Inc. (acq 1997; $110,000 with co-located FM; FTR: 9-13-93). Population served: 25,795 Format: Gospel. Target aud: 25-54; general. ♦Edna Turner, gen mgr, gen sls mgr & traf mgr; J. Michael Bailey, progmg dir; Lloyd Mitchell, chief of engrg.

WWKZ(FM)— Dec 15, 1978: 103.9 mhz; 50 kw. Ant 492 ft TL: N33 24 27 W88 08 27. (CP: COL Okolona. Ant 394 ft. TL: N34 12 18 W88 41 49). Stereo. Hrs opn: 24 Box 3300, Tupelo, 38803. Phone: (662) 842-1067. Fax: (662) 842-0725. Licensee: Citicasters Licenses L.P. (acq 7-13-2005; $2.2 million). Population served: 182,000 Natl. Network: American Urban. Format: Urban contemp. News: 6 hrs wkly. ♦Mark Maharrey, gen mgr.

Como

WRBO(FM)— Sept 28, 1966: 103.5 mhz; 100 kw. Ant 587 ft TL: N34 51 44 W89 52 42. Stereo. Hrs opn: 24 5629 Murray Rd., Memphis, TN, 38119. Phone: (901) 682-1106. Fax: (901) 680-0457. Web Site: www.soulclassics.com. Licensee: Citadel Broadcasting Co. Group owner: Citadel Broadcasting Corp. (acq 3-23-2004; grpsl). Population served: 463,800 Natl. Network: ABC, Westwood One. Format: Soul classics. Target aud: 18-49. ♦Sherri Sawyer, gen mgr; Dan Baron, sls dir; Amy Goodman, gen sls mgr; Henry Nelson, progmg dir; Marvin Emilien, prom.

Corinth

WADI(FM)— Oct 26, 1968: 95.3 mhz; 2.6 kw. 472 ft TL: N34 55 47 W88 24 37. Stereo. Hrs opn: 121 Front St., Iuka, 38852. Phone: (662) 423-9533. Fax: (662) 423-9333. E-mail: biddleandsons@crossroadsisp.com Licensee: Power Valley Communications Inc. Population served: 30,000 Format: Country. ♦Frederick A. Biddle, pres & CFO; Rick Biddle, gen mgr; Brian Biddle, stn mgr; Mike Cannon, opns dir.

WKCU(AM)— Oct 24, 1965: 1350 khz; 900 w-D, 44 w-N. TL: N34 54 29 W88 30 06. Hrs open: 24 1608 S. Johns St., 38834. Phone: (662) 286-8451. Fax: (662) 286-8452. E-mail: wxrz@earthlink.net Licensee: TeleSouth Communications Inc. (group owner; acq 12-20-02; $350,000 with co-located FM). Population served: 100,000 Rgnl. Network: Miss. Net. Wire Svc: NWS (National Weather Service) Format: Today's Christian music. News: 12 hrs wkly. Target aud: 25-54; female. Spec prog: Black 2 hrs wkly. ♦James H. Anderson, gen mgr.

WXRZ(FM)—Co-owned with WKCU(AM). January 1967: 94.3 mhz; 25 kw. 328 ft TL: N34 48 36 W88 34 45. Stereo. 24 200,000 Format: Super talk Mississippi. ♦James H. Anderson, stn mgr.

WTKN(AM)— Mar 1, 1946: Stn currently dark. 1230 khz; 1 kw-U. TL: N34 52 07 W88 31 17. Hrs open: 121 Front St., Iuka, 38852. Phone: (662) 423-9533. Fax: (662) 423-9333. Licensee: Perihelion Global Inc. (acq 12-14-2004; $45,000). Population served: 30,000 ♦Rick Biddle, gen mgr.

Crenshaw

WHKL(FM)— Mar 1, 1997: 106.9 mhz; 6 kw. 328 ft TL: N34 26 51 W90 06 25. Hrs open: Box 1528, Batesville, 38606. Phone: (662) 563-4664. Fax: (662) 563-9008. E-mail: country101radio@yahoo.com Licensee: Batesville Broadcasting Co. Inc. Format: Oldies. ♦John Ingram, gen mgr & stn mgr.

De Kalb

WJXM(FM)— 1999: 105.7 mhz; 50 kw. Ant 384 ft TL: N32 38 37 W88 40 29. Hrs open: Box 5797, Meridian, 39302. Phone: (601) 693-2661. Fax: (601) 483-0826. E-mail: wjxm@wokk.com Licensee: Mississippi Broadcasters L.L.C. (group owner). Format: Urban contemp. ♦Clay Holladay, gen mgr; Scott Stevens, opns mgr.

Decatur

WZKR(FM)— 2001: 103.3 mhz; 4.8 kw. Ant 590 ft TL: N32 21 46 W88 54 48. Hrs open: 24 1106 18th Ave., Meridian, 39301. Phone: (601) 693-1103. Fax: (601) 693-9949. E-mail: 103zkp@comcast.net Web Site: theartofgreatmusic.com. Licensee: Ponytail Broadcasting LLC (acq 1-29-2003; $800,000). Population served: 300,000 Natl. Network: CNN Radio. Law Firm: Fletcher, Heald & Hildreth. Format: Adult contemp. Target aud: 18-54. ♦Al Brown, gen mgr.

D'Iberville

WCPR-FM— December 1992: 97.9 mhz; 50 kw. 466 ft TL: N30 36 59 W89 08 03. Hrs open: 24 1909 E. Pass Rd., Suite D-11, Gulfport, 39507. Phone: (228) 388-2001. Phone: (228) 388-2771 (request line). Fax: (228) 896-9736. E-mail: wcpr@wcprfm.com Web Site: www.wcprfm.com. Licensee: Monterey Licenses LLC. Group owner: Triad Broadcasting Co. LLC (acq 7-14-99; grpsl). Natl. Network: ABC. Format: Active Rock/alternative. ♦Buddy Burch, VP; Jay Taylor, opns dir.

Drew

WRKG(FM)—Licensed to Drew. See Cleveland

Duck Hill

***WAUM(FM)**— 1998: 91.9 mhz; 3 kw. Ant 466 ft TL: N33 38 34 W89 29 59. Hrs open: Box 3206, American Family Radio, Tupelo, 38803. Phone: (662) 844-8888. Fax: (662) 842-6791. Web Site: www.afr.net. Licensee: American Family Association. Group owner: American Family Radio Natl. Network: USA. Format: Classic gospel. ♦Marvin Sanders, gen mgr.

Durant

WLIN-FM— 1997: 101.1 mhz; 4.8 kw. 371 ft TL: N33 03 51 W89 36 12. Hrs open: Box1700, Kosciusko, 39090. Phone: (662) 289-1050. Fax: (662) 289-7907. E-mail: breezy@kopower.com Web Site: www.breezynews.com. Licensee: Boswell Radio LLC. Format: Adult contemp. ♦Johnny Boswell, gen mgr; Ann Steen, stn mgr; Jerry Price, gen sls mgr; Eric Matthews, progmg dir.

Ellisville

WJKX(FM)— Oct 5, 1973: 102.5 mhz; 50 kw. 492 ft TL: N31 46 05 W89 10 12. Stereo. Hrs opn: 24 2625 S. Memorial Dr., Suite A, Tulsa, 74129. Phone: (210) 822-2828. Licensee: CC Licenses LLC. Group owner: Clear Channel Communications Inc. (acq 12-19-2000; grpsl). Population served: 250,000 Format: Urban. ♦Urica Pleas, gen mgr.

Eupora

WLZA(FM)—Licensed to Eupora. See Starkville

Fayette

WTYJ(FM)— Oct 17, 1983: 97.7 mhz; 6 kw. 500 ft TL: N31 40 32 W91 06 18. (CP: 6 kw, 328 ft.). Stereo. Hrs opn: 20 E. Franklin St., Natchez, 39120. Phone: (601) 442-2522. Fax: (601) 446-9918. Licensee: Natchez Broadcasting Inc. (acq 4-86; $200,000; FTR: 4-14-86). Population served: 100,000 Format: Gospel, blues. Target aud: General. ♦Dianna Nutter, pres; James B. Nutter, VP & stn mgr; L. Weir, gen mgr.

Flora

WFMN(FM)— July 7, 1997: 97.3 mhz; 19.5 kw. 367 ft TL: N32 27 21 W90 15 32. Hrs open: 24 TeleSouth Communications Inc., 6311 Ridgewood Rd., Jackson, 39211. Phone: (601) 957-1700. Fax: (601) 956-5228. E-mail: pgallo@telesouth.com Web Site: www.supertalkms.com. Licensee: TeleSouth Communications Inc. (group owner; acq 9-8-97; $700,000). Population served: 500,000 Natl. Network: ABC. Format: Talk. ♦Steve Davenport, pres; Paul Gallo, gen mgr; John Winfield, opns mgr.

Flowood

WPBQ(AM)— January 1995: 1240 khz; 880 w-U. TL: N32 18 03 W90 08 12. Hrs open: 24 850 Brandon Ave., Jackson, 39209. Phone: (601) 982-3210. Fax: (601) 420-4114. Licensee: PDB Corp. (acq 2-10-92; $4,000; FTR: 3-2-92). Population served: 300,000 Law Firm: Haley, Bader & Potts. Format: News/talk, sports. News staff: 3; News: 10 hrs wkly. Target aud: 25-54. ♦Bill Fulgham, gen mgr, stn mgr & engrg dir.

Forest

***WMBU(FM)**— Oct 3, 1997: 89.1 mhz; 10 kw horiz, 100 kw vert. 640 ft TL: N32 18 54 W89 21 12. Stereo. Hrs opn: 24
Rebroadcasts WMBV(FM) Dixon's Mills, AL 95%.
Box 400, Lake, 39092. Secondary address: Box 91, Dixons Mills, AL 36736. Phone: (601) 775-3100. Fax: (601) 775-3400. E-mail: wmbu@moody.edu Web Site: www.wmbu.org. Licensee: The Moody Bible Institute of Chicago. (group owner) Population served: 600,000 Natl. Network: Moody. Law Firm: Southmayd & Miller. Wire Svc: AP Format: Christian. Target aud: 35-54. Spec prog: Children 2 hrs wkly. ♦Rob Moore, gen mgr; John Roger, progmg dir.

WQST(AM)— September 1955: 850 khz; 10 kw-D, DA. TL: N32 21 46 W89 25 09. Hrs open:
Simulcast with WZQK(AM) Brandon 100%.
Box 1040, 39074. Phone: (601) 469-1960. Fax: (601) 469-1366. Licensee: Ace Broadcasting Inc. (acq 1999; $45,000). Population served: 500,000 Law Firm: Garvey, Schubert & Barer. Format: Real country. ♦Ken Michaels, gen mgr.

WQST-FM— September 1962: 92.5 mhz; 100 kw. 1,040 ft TL: N32 21 48 W89 25 29. Stereo. Hrs opn: 1329 Deerfield Ln., Jackson, 39211. Phone: (601) 362-4277. Fax: (601) 362-1994. Web Site: www.afr.net. Licensee: American Family Association Inc. Group owner: American Family Radio Population served: 1,000,000 Natl. Network: USA. Format: Christian. ♦Jim Thorn, gen mgr & stn mgr.

***WQVI(FM)**— July 20, 2004: 90.5 mhz; 60 kw vert. Ant 430 ft TL: N32 42 51 W89 49 19. Hrs open: Box 3206, Tupelo, 38803. Phone: (662) 844-8888. Fax: (662) 842-6791. Web Site: www.afr.net. Licensee: American Family Association. Group owner: American Family Radio. Population served: 310,000 Natl. Network: USA. Format: Christian classics. ♦Marvin Sanders, gen mgr.

***WSQH(FM)**— 2005: 91.7 mhz; 15 kw. Ant 474 ft TL: N32 23 57 W89 05 02. Hrs open: Box 3206, Tupelo, 38803-3206. Phone: (662) 844-8888. Fax: (662) 842-6791. Web Site: www.afr.net. Licensee: Salt & Light Communications Inc. Natl. Network: USA. Format: Christian. ♦Marvin Sanders, gen mgr.

Friar's Point

WNEV(FM)—Not on air, target date: unknown: 98.7 mhz; 6 kw. Ant 328 ft TL: N34 21 56 W90 38 14. Hrs open: Box 2870, West Helena, AR, 72390. Phone: (870) 572-7000. Fax: (870) 572-1845. Licensee: L.T. Simes II & Raymond Simes. ♦Raymond Simes, gen mgr.

WWUN-FM— 1973: 101.5 mhz; 14 kw. Ant 395 ft TL: N34 34 02 W90 37 37. Stereo. Hrs opn: 301 S. State St., Clarksdale, 38614. Phone: (662) 624-5144. Fax: (662) 621-1833. E-mail: wwun@csnradio.com Web Site: www.csnradio.com. Licensee: CSN International. (group owner; (acq 8-27-2001). Population served: 750,000 Format: Relg, Christian. ♦Charles W. Smith, pres; Jeffrey W. Smith, VP & gen mgr; Clayton Collier, stn mgr.

Fulton

WFTA(FM)—Licensed to Fulton. See Tupelo

Gluckstadt

WYOY(FM)— Jan 7, 1976: 101.7 mhz; 50 kw. 300 ft TL: N32 30 03 W90 02 28. Stereo. Hrs opn: 24 265 High Point Dr., Ridgeland, 39157. Phone: (601) 956-0102. Fax: (601) 978-3980. Web Site: www.y101.com. Licensee: New South Radio Inc. Group owner: New South Communications Inc. (acq 11-10-94; $750,000 with WLRM(AM) Ridgeland; FTR: 12-12-94) Natl. Rep: McGavren Guild. Format: CHR. Target aud: 18-49. ♦Gwen Rakestraw, gen mgr.

Greenville

WBAD(FM)—See Leland

WBAQ(FM)— May 1, 1970: 97.9 mhz; 48 kw horiz. Ant 502 ft TL: N33 23 51 W91 00 35. Stereo. Hrs opn: 5:30 AM-midnight P.O. Box 1816, 38702-1816. Secondary address: 800 Hwy 1 South, Delta Plaza, Ste #39 38701. Phone: (662) 378-2617. Fax: (662) 378-8341. E-mail: JamesKarr@riverbroadcasting.com Licensee: Debut Broadcasting Corp. Inc. Group owner: The River Group (acq 6-19-2007; grpsl). Population served: 75,000 Natl. Network: ABC. Rgnl. Network: Prog Farm, Miss. Net. Format: Btfl music, easy lstng. News: 14 hrs wkly. Target aud: 25-54; quality-conscious adults with spendable income. Spec prog: Farm one hr, btfl sacred music 4 hrs wkly. ♦James P. Karr, gen mgr; Linda mcKee, opns mgr.

WDMS(FM)—Listing follows WGVM(AM).

WESY(AM)—See Leland

WGVM(AM)— 1948: 1260 khz; 5 kw-D, 32 w-N. TL: N33 25 20 W91 01 41. Hrs open: Box 1438, 38701. Secondary address: 1383 Pickett St. 38701. Phone: (662) 334-4550. Fax: (662) 332-1315. Licensee: WDMS Inc. (acq 11-9-2006; $780,000 with co-located FM). Population served: 100,000 Rgnl. Network: Miss. Net. Target aud: General.

WDMS(FM)—Co-owned with WGVM(AM). December 1967: 100.7 mhz; 100 kw. 449 ft TL: N33 24 20 W91 01 41. Stereo. 1383 Pickett St., 38701. Phone: (662) 334-4559. Web Site: www.wdmsradio.com.100,000 Format: Country. ♦Linda Tackett, traf mgr & disc jockey.

WIQQ(FM)—See Leland

WJIW(FM)—Not on air, target date: unknown: 104.7 mhz; 31 kw. Ant 620 ft TL: N33 28 10 W90 50 30. Hrs open: 830 Main St., 38701. Phone: (662) 332-5701. Fax: (870) 338-3166. Licensee: Mondy-Burke Broadcasting Network. Format: Gospel. ♦Elijah Mondy, gen mgr; April Mondy, mus dir.

***WLRK(FM)**— 2006: 91.5 mhz; 50 kw vert. Ant 321 ft TL: N33 32 25 W91 22 39. Hrs open: Broadcasting for the Challenged Inc., 188 S. Bellevue, Suite 222, Memphis, TN, 38104. Phone: (901) 516-8970. Licensee: Broadcasting for the Challenged Inc. Natl. Network: K-Love. Format: Contemp Chirstian. ♦George S. Flinn Jr., pres.

WNIX(AM)— August 1937: 1330 khz; 1 kw-D, 500 w-N, DA-N. TL: N33 24 36 W91 01 03. Hrs open: 24 Box 1816, 38702-1816. Secondary address: Unit 39 Delta Plaza Mall, 800 Hwy. 1 S. 38701. Phone: (662) 378-2617. Fax: (662) 378-8341. E-mail: lindamckee @debutbroadcasting.com Licensee: Debut Broadcasting Corp. Inc. Group owner: The River Group (acq 6-19-2007; grpsl). Population served: 250,000 Rgnl. Network: Ark. Radio Net. Law Firm: Stephen R. Ross. Format: Oldies. ♦James P. Karr Jr., gen mgr.

Greenwood

WABG(AM)— February 1950: 960 khz; 1 kw-D, 500 w-N, DA-N. TL: N33 33 45 W90 12 38. Hrs open: 6 AM-1 PM 2001 Garrard Ave., 38930. Secondary address: Box 408 38935-0408. Phone: (662) 453-7822. Fax: (662) 455-3311. Licensee: Greenwood Broadcasting Co. Inc. Group owner: Bahakel Communications Population served: 26,000 Format: Country, talk. News: 11 hrs wkly. Target aud: 35 plus. Spec prog: Black 4 hrs wkly. ♦Sherry Nelson, gen mgr.

WGNL(FM)— Dec 1, 1989: 104.3 mhz; 50 kw. 360 ft TL: N33 31 30 W90 09 52. (CP: TL: N33 21 56 W90 14 59). Stereo. Hrs opn: 24 Box 1801, 38930. Secondary address: 503 Ione St. 38930. Phone: (662) 453-1646. Fax: (662) 453-7002. Web Site: www.broadcasturban.net. Licensee: Team Broadcasting Co. Inc. Natl. Rep: Dora-Clayton. Law Firm: Mullin, Rhyne, Emmons & Topel. Format: Adult contemp. News staff: one; News: 12 hrs wkly. Target aud: 18 plus. Spec prog: Jazz 6 hrs wkly. ♦Maxine Hughes, opns mgr; Ruben C. Hughes, gen mgr & gen sls mgr.

WGRM(AM)— 1937: 1240 khz; 1 kw-U. TL: N33 31 55 W90 11 38. (CP: 730 w). Hrs opn: 1110 Wright St., 38930. Phone: (662) 453-1240. Fax: (662) 453-1241. Licensee: Christian Broadcasting of Greenwood Inc. (group owner; (acq 2-22-99; $500,000 with co-located FM). Population served: 180,000 Format: Gospel. Target aud: 25-45. Spec prog: Black 2 hrs wkly. ♦Gwen Riley, mus dir & news dir; Lee Hall, gen mgr, gen sls mgr, progmg dir & chief of engrg; Gwen Rilley, traf mgr.

WGRM-FM— July 17, 1989: 93.9 mhz; 3 kw. 328 ft TL: N33 32 02 W90 11 42. (CP: 25 kw). 24 180,000 ♦Gwen Riley, traf mgr.

WKXG(AM)— Jan 1, 1987: 1540 khz; 1 kw-D. TL: N33 31 12 W90 08 28. Hrs open: 6 AM-10 PM Box 1686, 38935-1686. Secondary address: 3192 Browning Rd. 38935. Phone: (662) 453-2174. Fax: (662) 455-5733. Licensee: TeleSouth Communications Inc. (group owner; acq 8-1-88). Rgnl. Network: Miss. Net. Natl. Rep: Rgnl Reps. Format: Gospel. News: 3 hrs wkly. Target aud: 18-44. ♦Charlotte Baglan, sls dir; Ellen Benish, progmg dir & news dir; Wes Sterling, gen mgr, gen sls mgr & chief of engrg; Rea Holm, traf mgr.

WYMX(FM)—Co-owned with WKXG(AM). June 15, 1965: 99.1 mhz; 100 kw. Ant 1,029 ft TL: N33 31 12 W90 08 28. Stereo. Phone: (662) 453-2174. Fax: (662) 455-5733.315,000 Format: Adult hits. ♦Herman Anderson, mus dir; Rea Holm, traf mgr.

***WMAO-FM**— December 1983: 90.9 mhz; 100 kw. 880 ft TL: N33 22 34 W90 32 32. Stereo. Hrs opn: 24 3825 Ridgewood Rd., Jackson, 39211. Phone: (601) 432-6565. Fax: (601) 432-6806. Web Site: www.mpbonline.org. Licensee: Mississippi Authority for Educational Television. Natl. Network: PRI, NPR. Law Firm: Schwartz, Woods & Miller. Wire Svc: AP Format: Class, news/talk, jazz. News staff: 5; News: 20 hrs wkly. Target aud: General. ♦Marie Antoon, chmn; Gene Edwards, gen mgr; Bob Holand, opns mgr; Ty Warren, dev mgr; Tippy Garner, rgnl sls mgr; Jennifer Griffin, prom dir; Bob Holland, progmg dir; Greg Waxberg, mus dir; Dick Rizzo, news dir; Keith Martin, engrg dir; Betty Taylor, traf mgr.

WTCD(FM)—Indianola, May 1990: 96.9 mhz; 12.5 kw. 469 ft TL: N33 35 35 W90 32 30. Stereo. Hrs opn: 24 Box 1686, 38935. Phone: (662) 453-2174. Fax: (662) 455-5733. E-mail: radio@wtcd.com Web Site: www.supertalkms.com. Licensee: TeleSouth Communications Inc. (group owner; acq 5-28-97; $325,000). Population served: 250,000 Natl. Network: USA. Rgnl. Network: Miss. Net. Law Firm: Smithwick & Belendiuk. Format: News/talk. News staff: one; News: 20 hrs wkly. Target aud: 35-64; strong family orientation, middle to upper incomes. Spec prog: Farm 5 hrs, relg 11 hrs, talk 10 hrs wkly. ♦Wes Sterling, gen mgr.

Grenada

WMUT(FM)— 2004: 101.3 mhz; 6 kw. Ant 328 ft TL: N33 49 20 W89 55 40. Hrs open: Box 2266, 38902. Secondary address: 157 Dowdle Rd. 38901. Phone: (662) 226-3133. Fax: (662) 226-3233. E-mail: rock101@cableone.net Licensee: George S. Flinn Jr. Natl. Network: CNN Radio, Westwood One. Format: Classic rock. Target aud: 18-54. ♦Will Stammerjohan, gen mgr; Connie Stammerjohan, gen sls mgr.

WOHT(FM)— 2003: 92.3 mhz; 4.1 kw. Ant 397 ft TL: N33 51 33 W89 55 13. Hrs open: Box 2266, 38902. Phone: (662) 226-3133. Fax: (662) 226-3233. E-mail: star92@star92fm.com Licensee: Century Broadcasting L.L.C. (acq 4-18-03). Format: Oldies. ♦Will Stammerjohan, gen mgr.

WQXB(FM)—Listing follows WYKC(AM).

WTGY(FM)—See Charleston

WYKC(AM)— February 1949: 1400 khz; 1 kw-U. TL: N33 46 48 W89 48 09. Hrs open: 1348 Sunset Dr., 38901. Phone: (662) 226-1400. Fax: (662) 226-1464. Licensee: Chatterbox Inc. (acq 1-16-81). Population served: 9,944 Rgnl. Network: Prog Farm. Format: Country. ♦Bob Evans Jr., pres & gen mgr.

WQXB(FM)—Co-owned with WYKC(AM). Oct 16, 1970: 100.1 mhz; 3 kw. 300 ft TL: N33 46 36 W89 49 23. Stereo. (Acq 2-24-78).9,944 Format: Hot country.

Gulfport

***WAOY(FM)**— 1999: 91.7 mhz; 78 kw. 1,089 ft TL: N30 42 29 W89 05 06. Hrs open:
Rebroadcasts WAFR(FM) Tupelo 80%.
Box 3206, American Family Radio, Tupelo, 38803. Phone: (601) 844-8888. Fax: (601) 842-6791. Web Site: www.afr.net. Licensee: American Family Association Inc. Group owner: American Family Radio Natl. Network: USA. Format: Christian. Target aud: General. ♦Marvin Sanders, pres & gen mgr.

WGCM(AM)— 1928: 1240 khz; 1 kw-U. TL: N30 22 38 W89 04 45. Hrs open: 24 10250 Lorrian, 39503. Phone: (228) 896-5500. Fax: (228) 896-0458. Licensee: JMD Inc. (acq 11-15-94; $950,000 with co-located FM; FTR: 12-12-94). Format: Country. Target aud: 35 plus. ♦Morgan Dowdy, pres & gen mgr; Buddy Baylor, opns mgr; Steve Spillman, gen sls mgr; Brian Rhodes, progmg dir; Gwen Wilson, news dir; Dave Melton, chief of engrg.

WGCM-FM— Nov 14, 1969: 102.3 mhz; 25 kw. 299 ft TL: N30 22 28 W89 04 45. (CP: 16 kw, ant 358 ft.). Stereo. 24 Format: Easy listening. Target aud: 25-54. ♦Buddy Baylor, gen sls mgr; Pat McGowan, progmg dir.

WQFX(AM)— May 7, 1975: 1130 khz; 500 w-D. TL: N30 23 21 W89 06 23. Hrs open: 336 Rodenberg Ave., Biloxi, 39531-3444. Phone: (228) 374-9739. Fax: (228) 374-9739. E-mail: wqfxradio@aol.com Web Site: www.wqfx.net. Licensee: Walking by Faith Ministries Inc. (acq 1994). Population served: 58,000 Format: Power gospel. ♦James Black, gen mgr.

WROA(AM)— Feb 27, 1955: 1390 khz; 5 kw-U, DA-2. TL: N30 27 30 W89 04 45. Hrs open: Box 2639, 39505. Phone: (228) 896-5500. Fax: (228) 896-0458. E-mail: www.morgan@kiker108.com Licensee: Dowdy & Dowdy Partnership. (acq 12-19-86). Population served: 100,000 Format: Music of your life. Spec prog: Farm one hr wkly. ♦Charles W. Dowdy, pres; Morgan Dowdy, gen mgr.

WUJM(FM)— July 13, 1977: 96.7 mhz; 3 kw. 245 ft TL: N30 23 21 W89 06 23. Stereo. Hrs opn: 24 1909 E. Pass Rd., Suite D11, 39507. Phone: (228) 388-2001. Fax: (228) 896-9736. E-mail: molly967@molly967.com Web Site: www.molly967.com. Licensee: Monterey Licenses LLC. Group owner: Triad Broadcasting Co. LLC (acq 7-14-99; grpsl). Population served: 200000 Format: Hot adult contemp. ♦Buddy Burch, VP; Jay Taylor, opns dir & opns mgr.

WXYK(FM)— 1964: 107.1 mhz; 1.85 kw. 394 ft TL: N30 27 32 W89 04 45. (CP: 2.8 kw, ant 400 ft.). Stereo. Hrs opn: 24 1909 E. Pass Rd., Suite D-11, 39507. Phone: (228) 388-2001. Fax: (228) 896-9736. E-mail: wxyk@monkeyradio.com Web Site: www.monkeyradio.com. Licensee: Monterey Licenses LLC. Group owner: Triad Broadcasting Co. LLC (acq 7-14-99; grpsl). Population served: 200,000 Natl. Network: ABC. Format: Top-40/CHR. ♦Jay Taylor, opns dir & opns mgr.

WZKX(FM)—See Poplarville

Guntown

WBVV(FM)— Jan 15, 1976: 99.3 mhz; 15.5 kw. Ant 420 ft TL: N34 21 46 W88 35 28. Hrs open: 24 Box 3300, Tupelo, 38803. Phone: (662) 728-5301. Fax: (662) 728-2572. Licensee: CC Licenses LLC. Group owner: Clear Channel Communications Inc. (acq 9-27-2001; $700,000 including 5-year noncompete agreement). Natl. Network: USA, Reach Satellite. Rgnl. Network: Miss. Net. Format: Contemp inspirational. ♦Mark Maharrey, gen mgr.

Hattiesburg

***WAII(FM)**— 1998: 89.3 mhz; 1 kw. 220 ft TL: N31 16 59 W89 21 01. Hrs open: American Family Radio, Box 3206, Tupelo, 38803. Phone: (662) 844-8888. Fax: (662) 842-6791. Web Site: www.afr.net. Licensee: American Family Association. Group owner: American Family Radio Natl. Network: USA. Format: Relg. ♦Marvin Sanders, gen mgr.

WFOR(AM)— May 1924: 1400 khz; 1 kw-U. TL: N31 20 03 W89 19 08. Hrs open: 24 One Commerce Dr., #106, 39402. Phone: (601) 544-1400. Phone: (601) 296-9800. Fax: (601) 582-5481. Licensee: CC Licenses LLC. Group owner: Clear Channel Communications Inc. (acq 12-19-2000; grpsl). Population served: 38,277 Format: Sports/talk. Target aud: 35 plus. Spec prog: Relg 8 hrs wkly. ♦Mike Comfort, gen mgr; Jack Walker, opns mgr & progmg dir; James Harris, gen sls mgr; Sherri Merringo, news dir; Glen Musgrove, chief of engrg; Terri Hudson, traf mgr.

WUSW(FM)—Co-owned with WFOR(AM). July 1, 1966: 103.7 mhz; 100 kw. 1,056 ft TL: N31 31 37 W89 08 07. Stereo. 24 Phone: (601) 544-1037. Fax: (601) 582-5481.500,000 Format: Rock. Target aud: 25-54. ♦Terri Hudson, traf mgr.

WGDQ(FM)— 2005: 93.1 mhz; 6 kw. Ant 328 ft TL: N31 22 58 W89 23 43. Hrs open: 704 River St., 39401. Phone: (601) 544-1941. Fax: (601) 544-1947. Licensee: Unity Broadcasters. ♦Victor Floyd, gen mgr.

WHSY(AM)— Sept 1, 1954: Stn currently dark. 950 khz; 5 kw-D, 64 w-N. TL: N31 22 33 W89 19 49. Hrs open: 24 63 Braswell Rd., 39401. Phone: (601) 582-7078. Fax: (601) 582-7122. E-mail: whsy950@yahoo.com Web Site: www.whsy950.com. Licensee: Southern Air Communications Inc. (acq 9-5-2004). Population served: 304,475 Natl. Network: CBS Radio. Format: News, talk, sport. ♦Charlie W. Holt, pres; Charlie Holt, gen mgr.

WJMG(FM)—Listing follows WORV(AM).

WORV(AM)— June 7, 1969: 1580 khz; 1 kw-D, 88 w-N. TL: N31 20 33 W89 17 53. Hrs open: 1204 Graveline St., 39401. Phone: (601) 544-1941. Fax: (601) 544-1947. Licensee: Vernon C. Floyd dba Circuit Broadcasting of Hattiesburg. Population served: 66,000 Natl. Network: American Urban. Natl. Rep: Dora-Clayton. Format: Gospel. ♦Vernon C. Floyd, pres & gen mgr.

WJMG(FM)—Co-owned with WORV(AM). May 10, 1982: 92.1 mhz; 6 kw. 300 ft TL: N31 20 33 W89 17 53. Stereo. 24 66,277 Format: Urban contemp.

***WUSM-FM**— May 10, 1973: 88.5 mhz; 3 kw. 282 ft TL: N31 21 02 W89 22 12. Stereo. Hrs opn: USM-118 College Dr., #10045, 39406-0045. Phone: (601) 266-4287. Phone: (601) 266-5649. Fax: (601) 266-4288. E-mail: wusmmik@yahoo.com Web Site: www.wusm.usm.edu. Licensee: University of Southern Mississippi. Population served: 250,000 Natl. Network: AP Radio. Format: Class, var/div, AAA. News staff: 4; News: 22 hrs wkly. Target aud: General; college students & upper income univ & community listeners. ♦Shelby Thames, pres; Dennis Webster, exec VP; Michael Davis, gen mgr.

WXRR(FM)— July 1, 1967: 104.5 mhz; 100 kw. 984 ft TL: N31 25 50 W89 08 51. (CP: TL: N31 25 52 W89 08 51). Stereo. Hrs opn: 24 Box 16596, 39404. Phone: (601) 544-0095. Fax: (601) 649-8199. E-mail: rock104fm@rock104fm.com Web Site: www.rock104fm.com. Licensee: Blakeney Communications Inc. (group owner; acq 8-30-94; $450,000 with co-located AM; FTR: 10-24-94) Population served: 100,000 Format: Classic rock. Target aud: General. ♦Larry Blakeney, pres & gen mgr.

WZLD(FM)—See Petal

Hazlehurst

WDXO(FM)—Listing follows WOEG(AM).

WOEG(AM)— June 1, 1953: 1220 khz; 250 w-D, 46 w-N. TL: N31 53 34 W90 24 08. Hrs open: 6 AM-6 PM Box 2016, Monticello, 39654. Phone: (601) 587-9363. Phone: (601) 587-7625. Fax: (601) 587-9401. Licensee: TeleSouth Communications Inc. (group owner; (acq 6-20-2006; grpsl). Population served: 36,000 Law Firm: Booth, Freret, Imlay & Tepper. Format: Urban gospel. Target aud: General; Black. ♦Heather Thurgood, opns mgr, progmg dir & traf mgr; Robert Byrd, gen sls mgr; Randy Bullock, prom dir; Rusty O'Neal, stn mgr, news dir & chief of engrg.

WDXO(FM)—Co-owned with WOEG(AM). Dec 24, 1970: 92.9 mhz; 6 kw. 295 ft TL: N31 53 34 W90 24 08. Stereo. 24 200,000 Natl. Network: ABC. Format: Oldies. News: 3 hrs wkly. Target aud: 18-50. ♦Rusty O'Neal, gen mgr.

Heidelberg

WHER(FM)— May 1, 1980: 99.3 mhz; 50 kw. 492 ft TL: N31 49 17 W89 18 37. Stereo. Hrs opn: 24 6555 Hwy. 98 W., Suite 8, Hattiesburg, 39402. Phone: (601) 296-9800. Fax: (601) 296-9838. E-mail: contact@eagle99.com Web Site: www.eagle99.com. Licensee: CC Licenses LLC. Group owner: Clear Channel Communications Inc. (acq 12-19-2000; grpsl). Population served: 350,000 Natl. Network: ABC, CNN Radio. Rgnl. Network: Miss. Net. Format: Classic country. Target aud: General. ♦Mike Comfort, gen mgr; Jackson Walker, opns mgr; Glenn Musgrove, chief of engrg.

Holly Springs

WKRA(AM)— Sept 2, 1966: 1110 khz; 1 kw-D. TL: N34 47 11 W89 25 00. Hrs open: Sunrise-sunset Box 398, 38635. Phone: (662) 252-1110. Fax: (662) 252-2739. E-mail: wkra@dixie-net.com Licensee: Bill Autrey. (acq 8-10-94; $250,000 with co-located FM; FTR: 9-5-94) Population served: 100,000 Rgnl. Network: Miss. Net. Format: Ethnic. News staff: one; News: 9 hrs wkly. Target aud: 25-55. Spec prog: Gospel 12 hrs wkly. ♦Pamela Rideout, gen mgr & stn mgr.

WKRA-FM— June 30, 1976: 92.7 mhz; 3 kw. 299 ft TL: N34 47 11 W89 25 00.24 News staff: one. Target aud: General; Black community.

***WURC(FM)**— Oct 14, 1988: 88.1 mhz; 3 kw. 328 ft TL: N34 46 53 W89 26 49. Hrs open: Brown Mass Communications, 150 Rust Ave., 38635. Phone: (662) 252-5881. Fax: (662) 252-8869. Licensee: Rust College Inc. Natl. Network: NPR. Format: Jazz, inspirational, info, news. Target aud: General; college students, alternative seekers, minority listeners. ♦David L. Beckley, pres; Wayne Fiddis, gen mgr.

Horn Lake

WHAL-FM— July 26, 1994: 95.7 mhz; 6 kw. Ant 289 ft TL: N35 08 09 W89 58 17. Hrs open: 24 2650 Thousand Oaks Blvd., Suite 4100, Memphis, TN, 38118. Phone: (901) 259-1300. Fax: (901) 259-6451. Web Site: www.hallelujahfm.com. Licensee: CC Licenses LLC. Group owner: Clear Channel Communications Inc. (acq 1996; grpsl). Natl. Rep: Clear Channel. Format: Gospel. News: 3 hrs wkly. Target aud: 35-54; baby boomers. ♦Tim Davies, gen mgr; Ralph Salierno, sls dir; Frank Gilbert, mktg dir; Eileen Collier, progmg dir.

Houston

WCPC(AM)— Oct 21, 1955: 940 khz; 50 kw-D, 250 w-N, DA-2. TL: N33 56 00 W89 00 33. Hrs open: 5 AM-9:15 PM 1189 N. Jackson St., 38851. Phone: (662) 456-3071. Fax: (662) 456-3072. Licensee: WCPC Broadcasting Co. Inc. Population served: 750,000 Natl. Network: USA. Format: Christian, country, gospel, Black. News staff: one; News: 14 hrs wkly. Adults. ♦Robin H. Mathis, pres & gen mgr; Melanie Mathis Munlin, opns dir, opns mgr & mus dir; Don Tallent, news dir.

WSYE(FM)— Sept 19, 1968: 93.3 mhz; 100 kw. 1,804 ft TL: N33 45 06 W88 52 40. Stereo. Hrs opn: 24 Box 410, Tupelo, 38802. Secondary address: 2214 S. Gloster, Tupelo 38802. Phone: (662) 842-7658. Fax: (662) 842-0197. Web Site: www.sunny93fm.net. Licensee: JMD Inc. (acq 9-28-99). Population served: 750,000 Natl. Rep: Christal. Format: Adult contemp. News staff: one; News: 2 hrs wkly. Target aud: 25-54. ♦Scott Bebout, gen mgr & mktg mgr; Brenda Bebout, stn mgr; Steve Drunam, opns mgr.

Indianola

WNLA(AM)— May 1953: 1380 khz; 500 w-D, 44 w-N. TL: N33 27 32 W90 37 45. Hrs open: 12 Box 667, Hwy. 448, 38751. Phone: (662) 887-1380. Fax: (662) 887-1396. E-mail: wnla@capital2.com Licensee: Debut Broadcasting Corp. Inc. (acq 6-7-2007; $300,000 with co-located FM). Population served: 36,000 Rgnl. Network: Miss. Net. Format: Black gospel. Target aud: 21-55; Black. ♦Robert Marquitz, pres; Erin Ely, gen mgr & progmg dir; Gerry Brophy, sls dir & engrg dir; Bob Taylor, chief of engrg.

WNLA-FM— Sept 1, 1969: 105.5 mhz; 4.4 kw. 200 ft TL: N33 28 41 W90 38 28. Stereo. 24 Format: Adult contemp. News: 21 hrs wkly. Target aud: 21-55.

WTCD(FM)—Licensed to Indianola. See Greenwood

***WYTF(FM)**— Aug 26, 2004: 88.7 mhz; 100 kw vert. Ant 636 ft TL: N33 35 03 W90 36 13. Hrs open: Drawer 3206, Tupelo, 38801. Phone: (662) 844-8888. Fax: (662) 842-6791. Web Site: www.afr.net. Licensee: American Family Association. Group owner: American Family Radio. Natl. Network: USA. Format: Christian. ♦Don Wildman, gen mgr.

Itta Bena

***WVSD(FM)**— June 23, 1991: 91.7 mhz; 3 kw. 292 ft TL: N33 31 05 W90 20 38. Hrs open: 6 AM-midnight 14000 Hwy. 82 W. MVSU, 38941. Phone: (662) 254-3612. Fax: (662) 254-3611. Licensee: Mississippi Valley State University. Format: Jazz, gospel, blues. Spec prog: Oldies 10 hrs, reggae/Latin 3 hrs, comedy 2 hrs wkly. ♦Dr. Lester Newman, pres; Larz G. Roberts, gen mgr & progmg dir; Debra Harmon, progmg mgr.

Iuka

WFXO(FM)— Nov 5, 1970: 104.9 mhz; 50 kw. 443 ft TL: N34 46 35 W88 23 40. Hrs open: 311 W. Eastport St., 38852. Phone: (662) 423-2369. Fax: (662) 423-6059. E-mail: fox@freedom2000net.com Licensee: Billy R. McLain. (acq 11-4-91; with co-located AM). Population served: 2,389 Format: Country, rock. ♦Billy McLain, gen mgr, gen sls mgr & progmg dir.

Jackson

WHLH(FM)— Nov 19, 1973: 95.5 mhz; 100 kw. Ant 1,115 ft TL: N32 14 26 W90 24 15. Stereo. Hrs opn: 1375 Beasley Rd., 39206. Phone: (601) 982-1062. Fax: (601) 362-1905. Web Site: www.hallelujah955.com. Licensee: Capstar TX L.P. Group owner: Clear Channel Communications Inc. (acq 8-30-00; grpsl). Natl. Network: CBS. Natl. Rep: D & R Radio. Law Firm: Cohn & Marks. Format: Gospel. Target aud: 18-34; female. ♦Jenell Roberts, gen mgr & progmg dir.

WJDX(AM)— 1929: 620 khz; 5 kw-D, 1 kw-N, DA-N. TL: N32 22 56 W90 11 26. Stereo. Hrs opn: 24 Box 31999, 39286. Secondary address: 1375 Beasley Rd. 39206. Phone: (601) 982-1062. Fax: (601) 362-1905. Web Site: www.wjdx.com. Licensee: Capstar TX L.P. Group owner: Clear Channel Communications Inc. (acq 8-30-00; grpsl). Population served: 153,968 Natl. Rep: McGavren Guild. Format: Sports/talk. News staff: one; News: 10 hrs wkly. Target aud: 25-54; middle to upper income contemp adults. Spec prog: Farm 2 hrs wkly. ♦Kenneth E. Windham, gen mgr; Mary Ann Kirby, gen sls mgr; Randy Bell, progmg dir & news dir; Jason Black, chief of engrg; Theresa Banks, traf mgr.

WMSI(FM)—Co-owned with WJDX(AM). 1948: 102.9 mhz; 100 kw. 1,800 ft TL: N32 12 46 W90 22 54. Stereo. 24 Web Site: www.miss103.com.153,968 Format: C&W. News: 4 hrs wkly. Target aud: 25 plus. Spec prog: Farm one hr wkly. ♦Sam McLeod, gen sls mgr; Rick Adams, progmg dir; Marshall Stewart, mus dir & news dir; Diana Bass, traf mgr.

WJMI(FM)—Listing follows WOAD(AM).

***WJSU(FM)**— August 1975: 88.5 mhz; 3 kw. 203 ft TL: N32 17 47 W90 12 23. Stereo. Hrs opn: Box 18450, Jackson State Univ., 39217. Phone: (601) 979-2140. Phone: (601) 979-2285. Fax: (601) 979-2878. Web Site: www.jsums.edu. Licensee: Jackson State University. Population served: 24,000 Natl. Network: NPR. Format: Jazz, news, world. Target aud: 25-54; middle-class multiracial who prefer jazz or alternative mus. Spec prog: Gospel 18 hrs, reggae 2 hrs, blues 2 hrs wkly. ♦Bobby Walker, gen mgr.

WKXI(AM)— 1947: 1400 khz; 1 kw-U. TL: N32 19 12 W90 11 25. Hrs open: 24 Box 9446, 39286. Secondary address: 731 S. Pear Orchard, Ridgeland 39157. Phone: (601) 957-1300. Fax: (601) 956-0516. Licensee: Urban Radio II L.L.C. Group owner: Inner City Broadcasting (acq 8-25-2000; grpsl). Population served: 300,000 Natl. Rep: D & R Radio. Format: Blues. News staff: one; News: 30 hrs wkly. Target aud: 25-54. ♦Kevin Webb, gen mgr.

***WMPN-FM**— November 1984: 91.3 mhz; 100 kw. 760 ft TL: N32 16 53 W90 17 41. Stereo. Hrs opn: 24 3825 Ridgewood Rd., 39211. Phone: (601) 432-6565. Fax: (601) 432-6806. Web Site: www.mpbonline.org. Licensee: Mississippi Authority for Educational Television. Natl. Network: PRI, NPR. Law Firm: Schwartz, Woods & Miller. Wire Svc: AP Format: Class, news, jazz. News staff: 5; News: 20 hrs wkly. Target aud: General. ♦Marie Antoon, chmn & pres; Gene Edwards, gen mgr; Bob Holland, opns mgr; Jennifer Griffin, prom dir.

***WMPR(FM)**— 1983: 90.1 mhz; 100 kw. 500 ft TL: N32 11 33 W90 05 28. Stereo. Hrs opn: Box 9782, 39286. Phone: (601) 948-5835. Fax: (601) 948-6162. E-mail: wmpr@wmpr901.com Web Site: www.wmpr901.com. Licensee: J.C. Maxwell Broadcasting Group Inc. Format: Blues, gospel, urban contemp. ♦Charles Evers, gen mgr.

WOAD(AM)— 1929: 1300 khz; 5 kw-D, 1 kw-N. TL: N32 23 12 W90 09 47. Hrs open: 731 S. Pear Orchard Rd., Suite 27, Ridgeland, 39157. Phone: (601) 957-1300. Fax: (601) 956-0516. Web Site: www.woad.com. Licensee: Urban Radio II L.L.C. Group owner: Inner City Broadcasting (acq 8-25-2000; grpsl). Population served: 330,000 Natl. Network: American Urban, ABC. Wire Svc: Weather Wire Format: Gospel. Target aud: 25-54. ♦Kevin Webb, gen mgr & gen sls mgr; Percy Davis, progmg dir; Emmett Rushing, chief of engrg; Kelly Greer, traf mgr.

WJMI(FM)—Co-owned with WOAD(AM). 1967: 99.7 mhz; 100 kw. 1,060 ft TL: N32 16 39 W90 17 41. Stereo. 24 Web Site: www.wjmi.com.700,000 Format: Urban hip-hop. News staff: one. Target aud: 18-49. ♦Stan Branson, progmg dir.

WSFZ(AM)— September 1938: 930 khz; 5 kw-U, DA-N. TL: N32 23 42 W90 09 14. Hrs open: 24 No. 5 Twelve Oaks Cir., Suite A, 39209. Phone: (601) 922-9307. Fax: (601) 922-5051. E-mail: espnradio930@espnradio930.com Web Site: www.supersport930.com. Licensee: Sportsrad Inc. (acq 12-20-01; $222,500). Population served: 153,968 Natl. Network: Westwood One. Natl. Rep: McGavren Guild. Format: Sports. ♦Bryan Eubank, gen mgr, stn mgr & opns dir.

WSTZ-FM—Vicksburg, June 1968: 106.7 mhz; 100 kw. 1,365 ft TL: N32 12 22 W90 24 50. (CP: Ant 1,059 ft. TL: N32 12 29 W90 24 50). Stereo. Hrs opn: Box 31999, 39286. Secondary address: 1375 Beasley Rd. 39206. Phone: (601) 982-1062. Fax: (601) 362-1905. E-mail: dougjones@clearchannel.com Web Site: www.z106.com. Licensee: Capstar TX L.P. Group owner: Clear Channel Communications Inc. (acq 8-30-00; grpsl). Population served: 25,478 Natl. Rep: D & R Radio. Format: Classic rock. Target aud: 25-54. ♦Kenneth Windham, gen mgr.

WWJK(FM)— Aug 10, 1971: 94.7 mhz; 100 kw. 1,168 ft TL: N32 16 53 W90 17 41. Stereo. Hrs opn: 24 222 Beasley Rd., 39206. Phone: (601) 957-3000. Fax: (601) 956-0370. E-mail: mail94@arrow94.com Web Site: www.arrow94.com. Licensee: Backyard Broadcasting Mississippi LLC Group owner: Backyard Broadcasting LLC (acq 5-31-2002; $4,830,000 with WRXW(FM) Pearl). Population served: 330,000 Natl. Rep: Christal. Law Firm: Fletcher, Heald & Hildreth. Format: Classic

rock and roll. News staff: one; News: one hr wkly. Target aud: 25-54. ♦Barry Drake, pres; Chris Butterich, VP & gen mgr; Bob Rall, opns dir.

WYOY(FM)—See Gluckstadt

WZQK(AM)—See Brandon

WZRX(AM)— Apr 8, 1965: 1590 khz; 5 kw-D, 1 kw-N, DA-N. TL: N32 22 01 W90 13 26. Stereo. Hrs opn: 24 Box 9734, 39286. Secondary address: 2980 Forest Ave. Ext. 39286. Phone: (601) 981-9080. Fax: (601) 981-9093. E-mail: radioair@bellsouth.net Licensee: Capstar MS L.P. Group owner: Clear Channel Communications Inc. (acq 5-29-98; grpsl). Natl. Network: American Urban. Format: Gospel. Target aud: 25 plus; general. ♦Carl Haynes, gen mgr & stn mgr; Emmitte Rushing, chief of engrg.

Kosciusko

***WJTA(FM)**— 1989: 91.7 mhz; 383 w. 171 ft TL: N33 05 54 W89 30 33. Stereo. Hrs opn: 24 Box 888, 319 N. Madison, 39090. Phone: (662) 289-5703. Fax: (662) 290-6080. Licensee: Kosciusko Educational Broadcasting Foundation. Population served: 125,000 Format: Gospel. News: 6 hrs wkly. Target aud: General. ♦Dr. William G. Suratt, pres & gen mgr.

WKOZ(AM)— Oct 31, 1947: 1340 khz; 1 kw-U. TL: N33 03 51 W89 36 12. Hrs open: 24 Box 1700, 39090-1700. Phone: (662) 289-1050. Fax: (662) 289-7907. E-mail: breezy@kopower.com Web Site: www.breezynews.com. Licensee: Boswell Radio LLC (acq 6-1-62). Population served: 10,000 Law Firm: William D. Silva. Format: News/talk. News staff: one. Target aud: 50 plus. ♦Johnny Boswell, gen mgr; Ann Steen, stn mgr; Eric Matthews, opns mgr.

WQJQ(FM)— June 25, 1965: 105.1 mhz; 100 kw. 981 ft TL: N32 41 25 W89 52 06. Stereo. Hrs opn: 24 Box 31999, Jackson, 39268. Phone: (601) 982-1062. Fax: (601) 362-8270. Web Site: www.q1051.com. Licensee: Capstar TX L.P. Group owner: Clear Channel Communications Inc. (acq 8-30-00; grpsl). Population served: 450,000 Format: Motown & Jammin Oldies. Target aud: 35-64. ♦Kenneth Windham, gen mgr; Steve Kelly, opns mgr.

Laurel

WAML(AM)— Oct 20, 1932: 1340 khz; 1 kw-U. TL: N31 40 01 W89 08 59. Hrs open: 24 Box 6226, 1425 Ellisville Blvd., 39440. Fax: (601) 425-0016. Licensee: Walking by Faith Ministries Inc. (acq 10-1-99). Population served: 24,145 Format: Gospel, relg. Target aud: General. ♦James Black, gen mgr.

***WATP(FM)**— 1998: 90.7 mhz; 350 w. 489 ft TL: N31 46 54 W89 09 31. Hrs open: American Family Radio, Box 3206, Tupelo, 38803. Phone: (662) 844-8888. Fax: (662) 842-6791. Web Site: www.afr.net. Licensee: American Family Association. Group owner: American Family Radio Natl. Network: USA. Format: Relg. ♦Marvin Sanders, gen mgr.

WEEZ(AM)— Feb 27, 1957: 890 khz; 10 kw-D. TL: N31 31 29 W89 14 31. Stereo. Hrs opn: Sunrise-sunset One Commerce Dr., Suite 106, Hattiesburg, 39402. Secondary address: 51 Victory Rd. 39443. Phone: (601) 296-9800. Fax: (601) 296-9838. Licensee: CC Licenses LLC. Group owner: Clear Channel Communications Inc. (acq 12-19-2000; grpsl). Population served: 143,000 Format: Blues. News staff: one; News: 4 hrs wkly. Target aud: General. ♦Mike Comfort, gen mgr & sls dir; Jackson Walker, opns mgr; James Harris, gen sls mgr; Denise Brooks, progmg dir; Glen Musgrove, chief of engrg; Terri Hudson, traf mgr.

WNSL(FM)—Co-owned with WEEZ(AM). Mar 10, 1959: 100.3 mhz; 100 kw. 1,050 ft TL: N31 31 37 W89 08 07. Stereo. 278,000 Natl. Network: ABC. Format: CHR. Target aud: 18-49. ♦Don King, progmg dir.

WIZK(AM)—See Bay Springs

WKZW(FM)—See Bay Springs

WMXI(FM)— April 1989: 98.1 mhz; 2.55 kw. Ant 512 ft TL: N31 33 22 W89 09 09. Hrs open: Box 15935, Hattiesburg, 39403. Secondary address: 7501 U.S. Hwy. 49, Hattiesburg 39403. Phone: (601) 261-0898. Fax: (601) 261-3798. E-mail: zoo107@bellsouth.net Web Site: www.wmxi.com. Licensee: Rainey Broadcasting Inc. (acq 10-21-96; $75,000). Population served: 200,000 Format: News/talk. ♦Ted Tibbett, gen mgr.

Leland

WBAD(FM)—Listing follows WESY(AM).

WESY(AM)— Apr 8, 1957: 1580 khz; 1 kw-D, 48 w-N. TL: N33 22 46 W90 55 47. (CP: 1 kw-N, DA-N). Hrs opn: 24 Box 5804, Greenville, 38704. Secondary address: 126 Seven Oaks Rd., Greenville 38701. Phone: (662) 335-9265. Fax: (662) 335-5538. E-mail: wbad@tecinfo.com Licensee: East Delta Communications Inc. (acq 1980). Population served: 400,000 Natl. Network: American Urban. Law Firm: David Tillotson. Format: Blues, relg, gospel. Target aud: 18-54. ♦Stanley S. Sherman, exec VP & gen sls mgr; William D. Jackson, pres & gen mgr.

WBAD(FM)—Co-owned with WESY(AM). 1973: 94.3 mhz; 50 kw. 300 ft TL: N33 24 55 W90 59 18. Stereo. 19 Licensee: Interchange Communications Inc. (acq 5-12-73). Natl. Network: American Urban. Format: Urban contemp.

WIQQ(FM)— Sept 1, 1985: 102.3 mhz; 1.65 kw. Ant 446 ft TL: N33 23 51 W91 00 35. Stereo. Hrs opn: 24 Box 1816, Greenville, 38702-1816. Secondary address: Unit 39, 800 Hwy. 1 S., Greenville 38702. Phone: (662) 378-2617. Fax: (662) 378-8341. E-mail: wiqq@sellsouth.net Licensee: Debut Broadcasting Corp. Inc. Group owner: The River Group. (acq 6-19-2007; grpsl). Population served: 200,000 Natl. Network: USA, Jones Radio Networks. Rgnl. Network: Ark. Radio Net. Natl. Rep: McGavren Guild. Law Firm: Baraff, Koerner & Olender. Format: Adult contemp. News staff: one; News: 4 hrs wkly. Target aud: 18-49; multi-paycheck & spendable income. Spec prog: Farm 6 hrs, relg 6 hrs wkly. ♦Robert Marquitz, pres; James P. Karr Jr., VP & gen mgr; Linda McKee, opns mgr; Percy Kuhn, chief of engrg.

Lexington

WAGR-FM— June 1, 1990: 102.5 mhz; 6 kw. 328 ft TL: N33 09 06 W90 07 45. (CP: 12.5 kw, ant 459 ft.). Stereo. Hrs opn: Box 369, 100 Radio Rd., 39095. Phone: (662) 834-1025. Phone: (662) 834-1254. Fax: (662) 834-1254. Licensee: Brad Maurice Cothran (acq 12-21-01). Format: Country, oldies. ♦Brad Maurice Cothran, gen mgr.

WXTN(AM)— Oct 23, 1959: 1000 khz; 5 kw-D. TL: N33 06 39 W90 02 21. Hrs open: Box 369, 39095. Phone: (662) 834-1025. Fax: (662) 834-1254. Licensee: Brad Maurice Cothran (acq 12-21-01). Population served: 500,000 Format: Gospel, Black. ♦Brad M. Cothran, gen mgr.

Liberty

WAZA(FM)— 1998: 107.7 mhz; 25 kw. Ant 328 ft TL: N31 17 12 W90 47 53. Hrs open: 215 E. Bay St., Magnolia, 39652. Phone: (601) 684-4116. Fax: (601) 684-4654. E-mail: sandow@telapak.net Licensee: Southwest Broadcasting Inc. Format: Oldies. Target aud: 18-30; adult contemporary, retro 80's shows. ♦Charles Dowdy, gen mgr.

Long Beach

WJZD(FM)— Mar 20, 1994: 94.5 mhz; 6 kw. 321 ft TL: N30 22 25 W89 06 38. Hrs open: 24 Box 6216, Gulfport, 39506. Secondary address: 10211 Southpark Dr., Gulfport 39503. Phone: (228) 896-5307. Fax: (228) 896-5703. E-mail: info@wjzd.com Web Site: www.wjzd.com. Licensee: WJZD Inc. Population served: 300,000 Natl. Network: ABC. Natl. Rep: Interep. Rgnl rep: Allied Radio Partners. Law Firm: Fletch, Heald & Hildreth. Format: Urban adult contemp, news/talk. News staff: 2; News: 2 hrs wkly. Spec prog: Gospel 20 hrs wkly. ♦Rip Daniels, CEO & gen mgr; Danielle Jewett, opns dir.

Lorman

***WPRL(FM)**— Oct 12, 1987: 91.7 mhz; 3 kw. 300 ft TL: N31 53 37 W91 08 54. Stereo. Hrs opn: 6 AM-2 AM (M-F); 6 AM-midnight (S, Su) Box 269, 39096. Secondary address: Alcorn State Univ., 1000 Alcorn Dr. 39096. Phone: (601) 877-6290. Phone: (601) 877-6613. Fax: (601) 877-2213. E-mail: lljunag@hotmail.com Web Site: alconstateuniv.edu. Licensee: Alcorn State University. Population served: 24,750 Natl. Network: PRI, NPR, AP Radio. Format: Var/div. News staff: one; News: 23 hrs wkly. Target aud: General; African-American, rural, University faculty & students. ♦Lijuana Weir, opns mgr.

Louisville

WLSM-FM— Apr 22, 1966: 107.1 mhz; 12.5 kw. Ant 466 ft TL: N33 07 20 W89 01 05. Stereo. Hrs opn: 24 Box 279, 39339. Secondary address: 2142 Hwy. 14 E. 39339. Phone: (662) 773-3481. Fax: (662) 773-3482. E-mail: wlsm@louisvillems.com Licensee: Harrison Communications Inc. Rgnl. Network: Miss. Net. Format: Adult contemp. Target aud: 18-54. ♦Phillip A. Harrison, pres, gen mgr & gen sls mgr; Stacy S. Harrison, stn mgr & progmg dir.

Lucedale

WRBE(AM)— Sept 3, 1960: 1440 khz; 5 kw-D. TL: N30 56 00 W88 36 20. (CP: TL: N30 55 58 W88 36 21). Hrs opn: Box 827, 39452. Secondary address: 3276 Hwy. 198 W. 39452. Phone: (601) 947-8151. Fax: (601) 947-8152. E-mail: jdl@datasync.com Licensee: JDL Corp. (acq 2-24-98; $220,000 with co-located FM). Population served: 35,000 Natl. Rep: Dora-Clayton, Keystone (unwired net). Format: Country, gospel. Target aud: General. ♦Larry Shirley, pres, gen mgr, progmg dir & news dir; Lillian Hodgel, gen sls mgr; Bob Bonnell, farm dir.

WRBE-FM— April 1993: 106.9 mhz; 6 kw. 258 ft TL: N30 55 58 W88 36 21. ♦Bob Bonnell, farm dir.

Lumberton

WZNF(FM)— Dec 10, 1983: 95.3 mhz; 50 kw. Ant 1,181 ft TL: N30 44 48 W89 03 30. Hrs open: 24 10250 Lorraine Rd., Gulfport, 39503. Phone: (228) 896-5500. Fax: (228) 896-0458. E-mail: patty@z95fm.com Web Site: www.z95fm.com. Licensee: JMD Inc. (acq 1-11-00; $5 million). Format: Classic rock. ♦Morgan Dowdy, CEO, chmn & pres; Buddy Baylor, gen mgr.

Madison

WUSJ(FM)— Sept 16, 1966: 96.3 mhz; 100 kw. Ant 1,284 ft TL: N32 11 29 W90 24 22. Stereo. Hrs opn: 265 High Point Dr., Ridgeland, 39157. Phone: (601) 956-0102. Fax: (601) 978-3890. Web Site: www.us963.com. Licensee: New South Communications Inc. (group owner; acq 8-24-99; $5 million). Format: Country. ♦Gwen Rakestraw, gen mgr.

Magee

WKXI-FM— Apr 11, 1970: 107.5 mhz; 98 kw. 952 ft TL: N32 15 28 W89 47 22. Stereo. Hrs opn: 731 S. Pear Orchard Rd., Suite 27, Ridgeland, 39157. Phone: (601) 957-1300. Fax: (601) 956-0516. Web Site: www.kixie107.com. Licensee: Urban Radio II L.L.C. Group owner: Inner City Broadcasting (acq 8-25-2000; grpsl). Format: Urban contemp. Target aud: 25-54. ♦Kevin Webb, VP & gen mgr; Stan Branson, opns mgr.

WSJC(AM)— July 5, 1957: 810 khz; 50 kw-D, 500 w-N, DA-N. TL: N31 52 00 W89 41 35. Hrs open: 130 Radio Station Dr., 39111. Phone: (601) 849-5838. Fax: (601) 849-5838. Licensee: Witko Broadcasting L.L.C. (acq 12-3-98). Format: Christian. ♦Norm Wick, gen mgr.

Marion

WJDQ(FM)— Mar 15, 1990: 95.1 mhz; 50 kw. 606 ft TL: N32 26 08 W88 36 24. Stereo. Hrs opn: 4307 Hwy. 39 N., Meridian, 39301. Phone: (601) 693-2381. Fax: (601) 485-2972. Licensee: CC Licenses LLC. Group owner: Clear Channel Communications Inc. (acq 3-16-2001; grpsl). Population served: 45,083 Format: Country. Target aud: 25-54. ♦Ron Harper, gen mgr.

McComb

WAKH(FM)—Listing follows WAPF(AM).

WAKK(AM)— Apr 18, 1948: 980 khz; 5 kw-D, 152 w-N. TL: N31 12 51 W90 27 42. Hrs open: 206 N. Front St., 39648. Secondary address: Drawer 1649 39648. Phone: (601) 684-7470. Fax: (601) 684-4654. Licensee: Southwest Broadcasting Inc. (group owner; (acq 9-86; $600,000 with co-located FM; FTR: 7-28-86) Population served: 22,000 Natl. Network: ABC. Format: News/talk, relg. Target aud: General. ♦Wayne Dowdy, pres; Charles Dowdy, gen mgr; David Hughes, progmg dir; Carl Lazenby, pub affrs dir; Susan Dowdy, local news ed.

WAPF(AM)— Apr 25, 1975: 1140 khz; 1 kw-D. TL: N31 14 51 W90 25 14. Hrs open: Box 1649, 39649. Secondary address: 206 N. Front 39648. Phone: (601) 684-4116. Fax: (601) 684-4654. E-mail: spats@k106.net Licensee: Southwest Broadcasting Inc. (group owner; (acq 8-5-93; $600,000; FTR: 8-23-93). Population served: 11,969 Rgnl. Network: Miss. Net. Format: Var/gospel. Target aud: General. ♦Charles Dowdy, gen mgr & gen sls mgr.

WAKH(FM)—Co-owned with WAPF(AM). Oct 15, 1978: 105.7 mhz; 100 kw. 489 ft TL: N31 16 50 W90 27 05. Stereo. 24 Format: Country. Target aud: General.

***WAQL(FM)**— 1999: 90.5 mhz; 3.75 kw. 331 ft TL: N31 16 40 W90 26 56. Hrs open: American Family Radio, Box 3206, Tupelo, 38803. Phone: (662) 844-8888. Fax: (662) 842-6791. Web Site: www.afr.net. Licensee: American Family Association. Group owner: American Family Radio Natl. Network: USA. Format: Relg. ♦Marvin Sanders, gen mgr.

WHNY(AM)— 1939: 1250 khz; 5 kw-D, 1 kw-N, DA-N. TL: N31 16 07 W90 26 03. Hrs open: 24 1114 Hwy. 570 E., 39648. Phone: (601) 250-1250. Fax: (601) 250-1254. E-mail: whny@eaglepc.net Licensee: C.W.H. Broadcasting Inc. (acq 1952; $43,000). Natl. Network: ABC. Format: News, talk, sports.

McLain

WXAB(FM)— January 1999: 96.9 mhz; 4 kw. Ant 400 ft TL: N31 06 56 W88 45 56. Stereo. Hrs opn: Box 723, Wiggins, 39577. Licensee: Tralyn Broadcasting Inc. (acq 2-19-97; grpsl). Population served: 45,000 ♦Mike Self, gen mgr.

Meridian

WALT(AM)— 1946: 910 khz; 5 kw-D, 1 kw-N. TL: N32 23 37 W88 40 08. Stereo. Hrs opn: 24 302 17th St., Suite C, 39302. Phone: (601) 693-3434. Fax: (601) 693-3439. E-mail: paul@waltnewstalk.com Licensee: New South Communications Inc. (group owner; acq 4-1-57). Population served: 200,000 Natl. Rep: McGavren Guild. Format: News/talk. Target aud: 35-64. ♦F.E. Holladay, pres; Paul Bucurel, gen mgr.

WOKK(FM)—Co-owned with WALT(AM). August 1967: 97.1 mhz; 100 kw. 600 ft TL: N32 19 45 W88 41 26. Stereo. Web Site: www.wokk.com.300,000 Format: Country. Target aud: 25-54.

***WMAW-FM**— December 1983: 88.1 mhz; 100 kw. 1,050 ft TL: N32 08 18 W89 05 36. Stereo. Hrs opn: 24 3825 Ridgewood Rd., Jackson, 39211. Phone: (601) 432-6565. Fax: (601) 432-6806. Web Site: www.mpbonline.org. Licensee: Mississippi Authority for Educational Television. Natl. Network: PRI, NPR. Law Firm: Schwartz, Woods & Miller. Wire Svc: AP Format: Music, news, info. News staff: 5; News: 20 hrs wkly. Target aud: General. ♦Marie Antoon, chmn; Gene Edwards, gen mgr; Bob Holland, opns mgr.

WMER(AM)— Oct 16, 1973: 1390 khz; 5 kw-D, 250 w-N. TL: N32 20 41 W88 41 32. Hrs open: 19 315 A St., 39301. Phone: (601) 693-9637. Fax: (601) 693-9637. Licensee: Michael H. Glass. (acq 1-9-98; $55,000). Population served: 279,000 Natl. Network: USA. Format: Gospel, Christian. Target aud: 25-54; upscale, young families, non-working mothers.

WMOX(AM)— Dec 1, 1945: 1010 khz; 10 kw-D, 1 kw-N, DA-2. TL: N32 23 42 W88 39 28. Hrs open: 24 Box 5184, 39302. Phone: (601) 693-1891. Phone: (601) 693-1010. Fax: (601) 483-1010. E-mail: wmox@wmox.net Web Site: www.wmox.net. Licensee: Magnolia State Broadcasting Inc. (acq 12-27-98; $125,000; FTR: 1-25-93). Population served: 45,083 Rgnl. Network: Miss. Net. Natl. Rep: Dora-Clayton. Format: Talk/news, sports. News staff: one. Target aud: 25 plus; College educated with 30k plus annual income. Spec prog: Relg 8 hrs wkly. ♦Eddie Smith, pres & gen mgr; William T. Smith, VP & opns dir.

WMSO(FM)— February 1968: 101.3 mhz; 99 kw. Ant 581 ft TL: N32 18 43 W88 41 33. Stereo. Hrs opn: 4307 Hwy. 39 N., 39301. Phone: (601) 693-2381. Fax: (601) 485-2972. Licensee: CC Licenses LLC. Group owner: Clear Channel Communications Inc. (acq 3-16-2001; grpsl). Population served: 101,300 Format: Adult contemp. ♦Ron Harper, gen mgr & sls dir; David Day, progmg dir.

WNBN(AM)— Nov 1, 1987: 1290 khz; 2.5 kw-D, 90 w-N. TL: N32 21 42 W88 37 26. Hrs open: 18 266 23rd St., 39301. Phone: (601) 483-3401. Fax: (601) 483-3411. Licensee: Frank Rackley Jr. Population served: 45,000 Format: Gospel, blues. News: 61 hrs wkly. Target aud: 18-54. Spec prog: Black, women's, business, inspirational. ♦Frank Rackley Jr., gen mgr & opns VP.

WUCL(FM)— 1994: 102.1 mhz; 800 w. Ant 610 ft TL: N32 21 51 W88 38 34. Hrs open: 3436 Hwy. 45 N., 39301. Phone: (601) 693-2661. Fax: (601) 483-0826. E-mail: wmmz@wokk.com Licensee: Mississippi Broadcasters L.L.C. (group owner; (acq 2-26-93; $243,500; FTR: 3-22-93). Format: Classic country. ♦Clay Holladay, pres & gen mgr; Scott Stevens, opns mgr; Karen Bostick, sls dir; Scott Shepperd, chief of engrg.

WYHL(AM)— December 1957: 1450 khz; 1 kw-U. TL: N32 23 09 W88 41 36. Hrs open: 4307 Hwy. 39 N., 39301. Phone: (601) 693-2381. Phone: (601) 693-2383. Fax: (601) 485-2972. Web Site: www.sports1450.com. Licensee: CC Licenses LLC. Group owner: Clear Channel Communications Inc. (acq 3-16-2001; grpsl). Population served: 45,083 ♦Ron Harper, gen mgr; Pam Gray, sls dir & gen sls mgr; Lee Taylor, progmg dir.

Mississippi State

***WMAB-FM**— December 1983: 89.9 mhz; 63 kw. 1,080 ft TL: N33 21 07 W89 08 56. Stereo. Hrs opn: 24 3825 Ridgewood Rd., Jackson, 39211. Phone: (601) 432-6565. Fax: (601) 432-6806. Web Site: www.mpbonline.org. Licensee: Mississippi Authority for Educational Television. Natl. Network: PRI, NPR. Law Firm: Schwartz, Woods & Miller. Wire Svc: AP Format: Music, news, info. News staff: 5; News: 20 hrs wkly. Target aud: General. ♦Marie Antoon, chmn; Gene Edwards, gen mgr; Bob Holland, opns mgr.

Monticello

WMLC(AM)— 1969: 1270 khz; 1 kw-D, 53 w-N. TL: N31 33 24 W90 08 06. Hrs open: 20 WMLC Rd., 39654. Phone: (601) 587-1270. Fax: (601) 587-2119. E-mail: espnradio1270@bellsouth.net Licensee: Walking by Faith Ministries Inc. (acq 6-28-2006; $50,000). Natl. Network: ESPN Radio. Format: Sports. ♦Will Watson, gen mgr.

WRQO(FM)— Nov 19, 1990: 102.1 mhz; 50 kw. Ant 500 ft TL: N31 36 13 W90 12 26. Stereo. Hrs opn: 24 Box 2016, Q102 Rd., 39654. Secondary address: Box 1084 39654. Phone: (601) 587-9363. Phone: (601) 587-7625. Fax: (601) 587-9401. Fax: (601) 835-5005. E-mail: country@wrqo-q102.com Licensee: TeleSouth Communications Inc. (group owner) (acq 6-20-2006; grpsl). Natl. Network: CBS. Rgnl. Network: Miss. Net. Law Firm: Booth, Freret, Imlay & Tepper. Format: Talk. News: 15 hrs wkly. Target aud: 25-54. Spec prog: Farm one hr, relg 10 hrs wkly. ♦Stephen C. Davenport, pres; Marcus Rusty O'Neal, gen mgr; Randy Bullock, opns mgr.

Morton

WQST(AM)—See Forest

Moss Point

WBUV(FM)—Licensed to Moss Point. See Pascagoula-Moss Point

Mound Bayou

WZYQ(FM)—Licensed to Mound Bayou. See Cleveland

Natchez

***WASM(FM)**— 2001: 91.1 mhz; 1 kw. Ant 482 ft TL: N31 29 10 W91 21 42. Hrs open: American Family Radio, Box 3206, Tupelo, 38803. Phone: (662) 844-8888. Fax: (662) 842-6791. Web Site: www.afr.net. Licensee: American Family Association. Group owner: American Family Radio Natl. Network: USA. Format: Relg. ♦Marvin Sanders, gen mgr.

WKSO(FM)— March 1993: 97.3 mhz; 1.45 kw. Ant 686 ft TL: N31 30 33 W91 24 19. Hrs open: Box 768, 39121. Secondary address: 2 O'Ferrall St. 39420. Phone: (601) 442-4895. Fax: (601) 446-8260. Licensee: Will Perk Broadcasting. Group owner: First Natchez Radio Group (acq 8-31-92; $36,000; FTR: 9-21-92). Natl. Network: ABC. Format: Adult contemp. ♦Margaret Perkins, gen mgr.

WMIS(AM)— May 18, 1941: 1240 khz; 1 kw-U. TL: N31 31 14 W91 23 09. Hrs open: 19 Box 1248, 39121. Secondary address: 20 E. Franklin St. 39120. Phone: (601) 442-2522. Fax: (601) 446-9918. E-mail: wmiswtyj@aol.com Licensee: Natchez Broadcasting Co. Population served: 100,000 Natl. Network: American Urban. Format: Black, gospel, blues. Target aud: General; Black. ♦Diana Ewing Nutter, pres; James B. Nutter, VP & gen mgr; Lijuna Weir, stn mgr.

WNAT(AM)— Dec 4, 1949: 1450 khz; 1 kw-U. TL: N31 33 24 W91 23 00. Stereo. Hrs opn: 24 Box 768, 39121. Secondary address: 2 O'Ferral St. 39121. Phone: (601) 442-4895. Fax: (601) 446-8260. Web Site: www.wnat1450am.com. Licensee: First Natchez Corp. Group owner: First Natchez Radio Group (acq 11-28-58). Population served: 48,000 Rgnl. Network: Miss. Net. Law Firm: Schwartz, Woods & Miller. Format: News/talk, sports. News staff: 2. Target aud: 25-54. Spec prog: Gospel 18 hrs wkly. ♦Marie Perkins, pres; Margaret Perkins, gen mgr, gen sls mgr & pub affrs dir; Mickey Alexander, progmg dir; Keith Sanders, news dir & chief of engrg; Brenda Green, traf mgr.

WQNZ(FM)—Co-owned with WNAT(AM). Mar 1, 1968: 95.1 mhz; 98 kw. 1,056 ft TL: N31 30 33 W91 24 19. (CP: ant 1,896 ft). Stereo. 24 Web Site: www.wnat1450.com.150,000 Format: Country. News staff: 2; News: 7 hrs wkly. Target aud: 25 plus.

WTYJ(FM)—See Fayette

New Albany

WNAU(AM)— Mar 27, 1955: 1470 khz; 500 w-U, DA-N. TL: N34 29 48 W89 00 52. Hrs open: Box 808, 38652. Phone: (662) 534-8133. Fax: (662) 538-4183. Licensee: MPM Investment Group (acq 11-9-2004). Population served: 10,000 Rgnl. Network: Miss. Net. Format: Oldies. Target aud: 25-54. Spec prog: Gospel. ♦Ricky McCollum, exec VP; Terry Cook, pres & gen mgr.

WTPO(FM)—Not on air, target date: unknown: 101.5 mhz; 6 kw. Ant 325 ft TL: N34 30 43 W89 03 02. Hrs open: 2801 Via Fortuna Dr., Suite 675, Austin, TX, 78746. Phone: (713) 528-2517. Licensee: Ace Radio Corp. ♦Stephen Hackerman, pres.

WWZD-FM— Mar 3, 1986: 106.7 mhz; 50 kw. 657 ft TL: N34 26 08 W88 57 35. Stereo. Hrs opn: 24 Box 3300, Tupelo, 38803. Secondary address: 5026 Cliff Gookin Blvd., Tupelo 38803. Phone: (662) 842-1067. Fax: (662) 842-0725. E-mail: rickstevens@clearchannel.com Licensee: Clear Channel Broadcasting Licenses Inc. Group owner: Clear Channel Communications Inc. (acq 12-19-00; grpsl). Population served: 174,600 Natl. Network: ABC. Natl. Rep: Interep. Format: Country. News staff: one. Target aud: 25-54. Spec prog: Southern gospel 4 hrs wkly. ♦Mark Maharrey, gen mgr; Rick Stevens, opns VP.

New Augusta

WZHL(FM)—Not on air, target date: unknown: 101.7 mhz; 5 kw. Ant 312 ft TL: N31 13 00.5 W89 10 56.8. Hrs open: 2801 Via Fortuna Dr., Suite 675, Austin, TX, 78746. Phone: (713) 528-2517. Licensee: Ace Radio Corp. ♦Stephen Hackerman, pres.

Newton

WHTU(FM)— Apr 17, 1975: 97.9 mhz; 11 kw. Ant 492 ft TL: N32 29 16 W89 01 23. Hrs open: 24 4307 Hwy. 39 N., Meridian, 39301-9704. Phone: (601) 693-2381. Fax: (601) 485-2972. Licensee: CC Licenses LLC. Group owner: Clear Channel Communications Inc. (acq 3-16-2001; grpsl). Format: Oldies. ♦Ron Harper, gen mgr.

Ocean Springs

WOSM(FM)— Feb 12, 1971: 103.1 mhz; 50 kw. Ant 459 ft TL: N30 24 34 W88 42 23. (CP: 100 kw, ant 679 ft. TL: N30 36 42 W88 39 17). Stereo. Hrs opn: 24 4720 Radio Rd., 39564. Phone: (228) 875-9031. Fax: (228) 875-6461. E-mail: wosm@wosmradio.com Licensee: Charles H. Cooper. Population served: 300,000 Natl. Network: AP Radio, Salem Radio Network. Wire Svc: AP Format: Southern gospel. News: 14 hrs wkly. Target aud: 18-54; family. ♦Charles H. Cooper, gen mgr; Phil Moss, opns dir & prom dir; Margaret Cooper, progmg dir.

WQYZ(FM)— Sept 1, 1992: 92.5 mhz; 6 kw. 197 ft TL: N30 23 40 W88 53 41. Hrs open: 24 286 DeBuys Rd., Biloxi, 39531. Phone: (228) 388-2323. Fax: (228) 388-2362. Web Site: www.kissfm925.com. Licensee: Capstar TX L.P. (acq 5-26-2005; $1,287,200). Format: CHR / Top 40. News: one hr wkly. Target aud: 28-42; adult families/singles. ♦Reggie Bates, gen mgr; Walter Brown, opns mgr.

Oxford

***WAVI(FM)**— 2002: 91.5 mhz; 8.13 kw. Ant 574 ft TL: N34 11 57 W89 49 09. Hrs open: American Family Radio, Box 3206, Tupelo, 38803. Phone: (662) 844-8888. Fax: (662) 842-6791. Web Site: www.afr.net. Licensee: American Family Association. Group owner: American Family Radio Natl. Network: USA. Format: Christian. ♦Marvin Sanders, gen mgr.

***WMAV-FM**— December 1983: 90.3 mhz; 100 kw. 1,240 ft TL: N34 17 26 W89 42 24. Stereo. Hrs opn: 24 3825 Ridgewood Rd., Jackson, 39211. Phone: (601) 432-6565. Fax: (601) 432-6806. Web Site: www.mpbonile.org. Licensee: Mississippi Authority for Educational Television. Natl. Network: PRI, NPR. Law Firm: Schwartz, Woods & Miller. Wire Svc: AP Format: Music, news, info. News staff: 5; News: news progmging 20 hrs wkly. Target aud: General. ♦Marie Antoon, chmn; Larry Miller, pres; Gene Edwards, gen mgr; Bob Holland, opns mgr; Merrill McKewen, dev dir.

WOXD(FM)— October 1988: 95.5 mhz; 6 kw. 328 ft TL: N34 18 10 W89 31 25. Stereo. Hrs opn: 24 302 Hwy 7 S., 38655-9799. Phone: (662) 533-4482. Phone: (662) 234-9631. Fax: (662) 236-5390. E-mail: info@bullseye955.com Web Site: www.bullseye955.com. Licensee: Taylor Communications. (acq 1996). Population served: 35,000 Natl. Rep: Rgnl Reps. Format: Classic Hits. Target aud: 25-54. Spec prog: Gospel 12 hrs wkly. ♦Jason T. Plunk, pres; Ron Cox, gen mgr.

WQLJ(FM)— Dec 31, 1984: 93.7 mhz; 25 kw. 328 ft TL: N34 20 05 W89 43 29. Stereo. Hrs opn: 24 Box 1077, 38655-1077. Secondary address: 461 Hwy. 6 W. 38655. Phone: (662) 236-0093. Fax: (662) 234-5155. E-mail: q937@exceedtech.net Web Site: www.wqlj.com. Licensee: TeleSouth Communications Inc. (group owner; acq 11-30-99; $1.4 million). Population served: 280,000 Rgnl. Network: Miss. Net. Law Firm: Fletcher, Heald & Hildreth. Format: Adult contemp. News staff: one; News: one hrs wkly. Target aud: 18-45. Spec prog: Contemp Christian 9 hrs wkly. ♦Steve Davenport, CEO & pres; Rick Mize, gen mgr; Jim Martin, opns dir; Judy McCormick, progmg dir & traf mgr; Bryan Hadley, news dir.

WWMS(FM)— Jan 1, 1969: 97.5 mhz; 100 kw. 1,000 ft TL: N34 10 05 W89 09 23. Stereo. Hrs opn: Box 410, 2214 S. Gloster St., Tupelo, 38801. Phone: (662) 842-7658. Fax: (662) 842-0197. Web Site: www.miss98.com. Licensee: San-Dow Broadcasting Inc. (acq 5-10-85). Population served: 176,000 Format: Country. Target aud: General. Spec prog: Farm 2 hrs wkly. ♦Bob Gipson, VP; Sam Cousley, stn mgr.

Pascagoula

WHGO(FM)—Licensed to Pascagoula. See Pascagoula-Moss Point

WKNN-FM— December 1964: 99.1 mhz; 100 kw. 1,012 ft TL: N30 29 09 W88 42 53. Stereo. Hrs opn: 24 286 Debuys, Biloxi, 39531. Secondary address: Box 4606, Biloxi 39535. Phone: (228) 388-2323. Fax: (228) 388-2362. Web Site: www.k99fm.com. Licensee: CC Licenses LLC. Group owner: Clear Channel Communications Inc. (acq 2-2-2004; grpsl). Population served: 110,000 Format: Country. News staff: one; News: 6 hrs wkly. Target aud: 25-54. ♦Reggie Bates, gen mgr.

***WPAS(FM)**— Mar 25, 2004: 89.1 mhz; 60 kw. Ant 574 ft TL: N30 33 03 W88 27 06. Hrs open: American Family Radio, Box 3206, Tupelo, 33880. Phone: (662) 844-8888. Fax: (662) 842-6791. Web Site: www.afr.net. Licensee: American Family Association. Group owner: American Family Radio. Population served: 585,000 Natl. Network: USA. Format: Christian. ♦Marvin Sanders, gen mgr.

Pascagoula-Moss Point

WBUV(FM)—Moss Point, June 1, 1964: 104.9 mhz; 33 kw. Ant 600 ft TL: N30 34 08 W88 22 48. Stereo. Hrs opn: 24 286 Debuys Rd., Biloxi, AL, 39531. Phone: (228) 450-0100. Phone: (228) 388-2323. Fax: (228) 388-2362. E-mail: kippgreggory@clearchannel.com Web Site: www.newsradio1049fm.com. Licensee: CC Licenses LLC. Group owner: Clear Channel Communications Inc. (acq 12-7-98; $1.4 million swap with WYOK(FM) Atmore, AL). Population served: 954,300 Natl. Network: Fox News Radio, Fox Sports, Premiere Radio Networks. Law Firm: Cohn & Marks. Format: News/talk. News staff: 4. Target aud: 25-54; men. ♦Reggie Bates, gen mgr.

WHGO(FM)—Pascagoula, June 1, 1976: 105.9 mhz; 25 kw. 312 ft TL: N30 22 05 W88 44 35. Stereo. Hrs opn: 24 1909 E. Pass Rd., Suite D-11, Gulfport, 39507. Phone: (228) 388-2001. Fax: (228) 896-9736. E-mail: gofm@cableone.net Web Site: www.1059gofm.com. Licensee: Monterey Licenses LLC. Group owner: Triad Broadcasting Co. LLC (acq 7-14-99; grpsl). Population served: 110,000 Natl. Network: ABC. Natl. Rep: Katz Radio. Format: Classic hits. Target aud: 25-54; male/female. ♦Buddy Birch, gen mgr; David Manning, gen mgr; Jay Taylor, opns dir; Kenny Vest, opns mgr; Wayne Watkins, progmg dir.

WPMP(AM)— September 1951: 1580 khz; 5 kw-D, 51 w-N, DA-2. TL: N30 23 01 W88 32 07. Hrs open: 24 5115 Telephone Rd., Pascagoula, 39567. Phone: (228) 762-5683. Fax: (228) 762-1222. Licensee: Flagship Radio Group Inc. (acq 5-26-2005; $88,000). Format: Talk.

Pearl

WJNT(AM)— Oct 28, 1980: 1180 khz; 50 kw-D, 500 w-N. TL: N32 17 43 W90 06 54. Hrs open: 24 Box 1248, Jackson, 39215-1248. Secondary address: 1985 Lakeland Dr., Suite 212, Jackson 39216. Phone: (601) 366-1150. Phone: (601) 366-1150. Fax: (601) 366-1627. E-mail: contactus@wjnt.com Web Site: www.wjnt.com. Licensee: Buchanan Broadcasting Co. Inc. Population served: 540,000 Natl. Network: CBS, ABC, Westwood One. Law Firm: Womble, Sandridge, Carlyle & Rice. Wire Svc: AP Format: News/talk. News staff: one; News: 28 hrs wkly. Target aud: 35 plus; high income, college educated, home owners. ♦Bob Buchanan, pres; Thena Gunn, gen mgr; Stan Carter, chief of opns.

WRXW(FM)— Nov 7, 1994: 93.9 mhz; 6 kw. 328 ft TL: N32 17 52 W89 59 56. Stereo. Hrs opn: 24 222 Beasley Rd., Jackson, 39206. Phone: (601) 957-3000. Fax: (601) 956-0370. E-mail: mail@wviv.com Web Site: www.rock939.com. Licensee: Backyard Broadcasting Mississippi LLC Group owner: Backyard Broadcasting LLC (acq 5-31-2002; $4,830,000 with WWJK(FM) Jackson). Population served: 250,000 Natl. Network: ABC, Westwood One. Rgnl rep: Christal. Law Firm: Fletcher, Heald & Hildreth. Format: Adult standards, rock/AOR. News staff: one; News: one hr wkly. Target aud: 35 plus. ♦Chris Butterich, gen mgr; Bob Rall, opns mgr; Janna Hughes, sls dir; Phil Conn, progmg mgr; Stan Carter, chief of engrg.

Petal

WZLD(FM)— January 1986: 106.3 mhz; 3 kw. 400 ft TL: N31 23 02 W89 10 44. Stereo. Hrs opn: 24 6555 Hwy. 98 W., Suite 8, Hattiesburg, 39402. Phone: (601) 296-9800. Fax: (601) 582-5481. E-mail: contact@wizldfm.com Web Site: www.wzld.com. Licensee: CC Licenses LLC. Group owner: Clear Channel Communications Inc. (acq 12-19-2000; grpsl). Natl. Network: CNN Radio. Format: Hip hop, rhythm and blues. News: 4 hrs wkly. Target aud: 25-54; upscale, educated & professional. Spec prog: Sports 3 hrs wkly. ♦Michael Comfort, gen mgr; Jackson Walker, opns mgr.

Philadelphia

WHOC(AM)— July 31, 1948: 1490 khz; 1 kw-U. TL: N32 45 52 W89 07 48. Hrs open: Box 26, 1016 W. Beacon St., 39350. Phone: (601) 656-1490. Fax: (601) 656-1491. E-mail: wwslfm@yahoo.com Licensee: WHOC Inc. (acq 1-31-89; $300,000; FTR: 2-20-89). Population served: 6,274 Rgnl. Network: Miss. Net. Format: Adult standard, talk. Target aud: General. Spec prog: Farm 2 hrs wkly. ♦Leah Jarrell, gen mgr; Joe Vines, opns mgr, gen sls mgr, progmg dir & traf mgr; Rex Smith, chief of engrg.

WWSL(FM)—Co-owned with WHOC(AM). Jan 1, 1981: 102.3 mhz; 4.9 kw. 364 ft TL: N32 43 35 W89 05 56. Stereo. Phone: (601) 656-7102. Licensee: H & GC Inc.40,000 Natl. Network: Westwood One. Format: Adult contemp.

Picayune

WMTI(FM)— November 1973: 106.1 mhz; 28 kw. Ant 659 ft TL: N30 31 17 W90 01 12. Stereo. Hrs opn: 24 201 St. Charles, Suite 201, New Orleans, LA, 70170. Phone: (504) 581-7002. Fax: (504)-566-4857. Web Site: www.martini1061.com. Licensee: Citadel Broadcasting Co. (acq 1-3-2006; $7 million). Population served: 1,000,000 Format: Soft adult standards. News staff: 2; News: 2 hrs wkly. Target aud: 18-45; young professionals. ♦Dave Siebert, gen mgr; Steven Kline, gen sls mgr; Jim Hanzo, progmg dir; Bill Major, chief of engrg.

WRJW(AM)— October 1949: 1320 khz; 5 kw-D, 75 kw-N. TL: N30 31 06 W89 38 41. Hrs open: 5a-10p Box 907, 39466. Secondary address: 2438 Hwy. 43 S. 39466. Phone: (601) 798-4835. Fax: (601) 798-9755. E-mail: wrjw@datasync.com Web Site: www.wrjw.com. Licensee: Pearl River Communications Inc. (acq 8-2-91). Population served: 36,000 Natl. Network: ABC. Rgnl. Network: Miss. Net., ABC. Format: Country/southern gospel. News staff: 2; News: 10 hrs wkly. Target aud: 18-54; contemp country listeners. Spec prog: Black 8 hrs, farm 6 hrs, relg 16 hrs, sports 4 hrs wkly. ♦Delores Wood, gen mgr; Denise Wilson, opns dir & prom mgr; Cleve Dawsey, news dir; Caroline Randolph, traf mgr.

Pickens

WOAD-FM— July 20, 1980: 105.9 mhz; 22.5 kw. 735 ft TL: N32 40 54 W90 06 06. (CP: Ant 745 ft.). Stereo. Hrs opn: 731 S. Pear Orchard, Suite 27, Ridgeland, 39157. Phone: (601) 957-1300. Fax: (601) 956-0516. Web Site: www.1059themaxx.com. Licensee: Urban Radio II L.L.C. Group owner: Inner City Broadcasting (acq 2000; grpsl). Format: Oldies. ♦Kevin Webb, gen mgr; Bill Wilson, prom dir.

Pontotoc

WSEL(AM)— Nov 30, 1962: 1440 khz; 890 w-D, DA. TL: N34 15 10 W88 57 36. Stereo. Hrs opn: Box 3788, Tupelo, 38803. Phone: (662) 489-0297. Fax: (662) 488-9735. Licensee: Ollie Collins Jr. (acq 5-5-92; $46,500 with co-located FM; FTR: 6-1-92) Format: Urban gospel. ♦Ollie Collins Jr., gen mgr, opns mgr & progmg dir; Jerry Campbell, chief of engrg.

WSEL-FM— Jan 1, 1966: 96.7 mhz; 3 kw. 299 ft TL: N34 15 10 W88 57 36. Stereo. Format: Urban gospel.

Poplarville

WRPM(AM)— 1963: 1530 khz; 10 kw-D, 1 kw. TL: N30 48 55 W89 30 24. Hrs open: 6 AM-6 PM Box 352, 39470. Phone: (601) 795-4900. Fax: (601) 795-0277. E-mail: wrpm@wrpm.com Licensee: Charles W. and J. Morgan Dowdy (acq 3-87; $2.25 million with co-located FM; FTR: 12-15-86). Population served: 500,000 Natl. Network: ABC. Format: Southern gospel. Target aud: 22-54. ♦Steve Spillman, gen sls mgr; Thomas Vaughn, gen mgr, progmg dir & chief of engrg.

WZKX(FM)—Co-owned with WRPM(AM). Feb 14, 1966: 107.9 mhz; 92 kw. 1,460 ft TL: N30 44 48 W89 03 30. Box 2639, Gulfport, 39505. Phone: (228) 896-5500. Fax: (228) 896-3724. Web Site: www.usasingles.com/wrpm.htm. Format: Country. ♦Morgan Dowdy, gen mgr.

Port Gibson

***WATU(FM)—** 1999: 89.3 mhz; 40 kw vert. Ant 384 ft TL: N32 07 56 W90 45 29. Hrs open: Box 3206, American Family Radio, Tupelo, 38803. Phone: (662) 844-8888. Fax: (662) 842-6791. Web Site: www.afr.net. Licensee: American Family Association. Group owner: American Family Radio Natl. Network: USA. Format: Relg. ♦Marvin Sanders, gen mgr.

WRTM-FM— July 16, 1999: 100.5 mhz; 44 kw. Ant 492 ft. TL: N31 59 38 W90 58 15. Stereo. Hrs opn: 24 Box 820583, Vicksburg, 39182. Secondary address: 1901 N. Frontage Rd. #8, Vicksburg 39180. Phone: (601) 636-7944. Fax: (601) 373-1343. E-mail: radioair@bellsouth.net Licensee: Commander Communications Corp. (acq 10-15-99). Format: Urban adult contemp. Target aud: 25-54. ♦Carl Haynes, gen mgr & gen sls mgr; Marty Hart, opns mgr.

Potts Camp

WCNA(FM)— Oct 1, 1995: 95.9 mhz; 14 kw. Ant 436 ft TL: N34 35 51 W89 06 12. Stereo. Hrs opn: 24 Box 2116, Radio Bldg., 1241 Cliff Gookin Blvd., Tupelo, 38803. Phone: (662) 842-7625. Fax: (662) 842-9568. Licensee: Olvie E. Sisk. Group owner: Air South Radio Inc. Format: Classic rock. News: 14 hrs wkly. ♦Gene Sisk, pres; Fred Blalock, stn mgr; Ivous Sisk, opns VP.

Prentiss

WCJU-FM— 2002: 104.9 mhz; 2.8 kw. Ant 436 ft TL: N31 31 56 W89 56 17. Hrs open: Box 472, Columbia, 39429. Phone: (601) 736-8889. Fax: (601) 736-2617. E-mail: wcju@zzip.cc Licensee: Sunbelt Broadcasting Corp. (group owner) Format: Oldies. ♦Tommy McDaniel, gen mgr; Glenn Beach, stn mgr.

WJDR(FM)— June 1, 1982: 98.3 mhz; 6 kw. 325 ft TL: N31 29 43 W89 53 33. Stereo. Hrs opn: 24 Box 351, 37 S. High School Ave., Columbia, 39429. Phone: (601) 731-2298. Phone: (601) 792-2056. Fax: (601) 792-2057. Licensee: Sunbelt Broadcasting Corp. (group owner; acq 12-1-85). Population served: 75,000 Natl. Network: ABC. Rgnl. Network: Miss. Net. Format: Hot country. News: 20 hrs wkly. Target aud: 25-54. Spec prog: Black 5 hrs wkly. ♦Thomas F. McDaniel, pres; Jody Fortenberry, stn mgr.

Quitman

***WLKO(FM)—** July 31, 1981: 98.9 mhz; 25 kw. Ant 315 ft TL: N32 03 51 W88 43 27. Stereo. Hrs opn: 24
Rebroadcasts KLVR(FM) Santa Rosa, CA 100%.
2351 Sunset Blvd., Suite 170-218, Rocklin, CA, 95765. Phone: (916) 251-1600. Fax: (916) 251-1650. Web Site: www.klove.com. Licensee: Educational Media Foundation. (acq 4-29-2005; $500,000). Natl. Network: K-Love. Format: Contemp Christian. ♦Richard Jenkins, pres; Mike Novak, VP; Keith Whipple, dev dir; David Pierce, progmg mgr; Ed Lenane, news dir; Sam Wallington, engrg dir; Karen Johnson, news rptr; Marya Morgan, news rptr; Richard Hunt, news rptr.

WQMS(AM)— Feb 2, 1968: Stn currently dark. 1500 khz; 1 kw-D. TL: N32 03 51 W88 43 27. Hrs open: 901 N.E. 173rd St., Miami, FL, 33162. Phone: (305) 770-1961. Licensee: Stephen C. Hellinger (acq 4-4-2006; $12,500). Population served: 2,702 Format: Sports. ♦Simcha Hellinger, gen mgr.

Redwood

WVBG-FM— 2005: 105.5 mhz; 1.95 kw. Ant 430 ft TL: N32 23 22 W90 48 34. Hrs open: 1102 Newitt Vick Dr., Vicksburg, 39183. Phone: (601) 883-0848. Licensee: Lendsi Radio LLC. Format: Oldies. ♦Lina H. Jones, gen mgr.

Richton

WXHB(FM)— 1995: 96.5 mhz; 6 kw. Ant 328 ft TL: N31 21 01 W88 59 11. Hrs open: Box 6408, Laurel, 39441. Phone: (601) 649-0095. Phone: (601) 544-0095. Fax: (601) 649-8199. Licensee: Blakeney Communications Inc. (group owner; acq 3-27-03; $650,000). Format: Solid gospel. ♦Larry Blakeney, gen mgr.

Ridgeland

WIIN(AM)— Dec 1, 1984: 780 khz; 5 kw-D. TL: N32 25 36 W90 12 19. Hrs open: Sunrise-sunset 265 Highpoint Dr., 39157. Phone: (601) 956-0102. Fax: (601) 978-3980. Licensee: New South Radio Inc. Group owner: New South Communications Inc. (acq 11-10-94; $750,000 with WLIN(FM) Gluckstadt; FTR: 12-12-94) Natl. Rep: McGavren Guild. Format: Nostalgia. Target aud: General; Adult professionals. ♦Gwen Rakestraw, gen mgr; Bill Rakestraw, gen sls mgr; Mark McCoy, opns mgr & progmg dir.

Ripley

WCSA(AM)— 1995: Stn currently dark. 1260 khz; 500 w-D, 38 w-N. TL: N34 43 15 W88 56 40. Hrs open: 4598 Appleville St., Memphis, TN, 38109. Phone: (601) 837-2816. Licensee: Keyboard Broadcasting Communication.

WKZU(FM)— June 1, 1979: 102.3 mhz; 3.5 kw. Ant 433 ft TL: N34 42 35 W88 50 36. Stereo. Hrs opn: 24 Box 572, 107 E. Spring St., 38663. Phone: (662) 837-1023. Phone: (662) 837-2990. Fax: (662) 837-2994. E-mail: scott@kudzu102.com Web Site: www.kudzu102.com. Licensee: Kudzu Communications Inc. (acq 7-23-98). Population served: 125,000 Rgnl rep: Miss. Net. Format: Classic country. News: 2 hrs wkly. Target aud: 25-54; 50% men & 50% women. Spec prog: Relg 2 hrs, gospel 6 hrs, bluegrass 2 hrs wkly. ♦Scott Peters, pres & gen mgr.

Saltillo

WWMR(FM)—Not on air, target date: unknown: 102.9 mhz; 12.5 kw. Ant 466 ft TL: N34 24 33 W88 32 24. Hrs open: 275 Goodwyn, Memphis, TN, 38111. Phone: (901) 375-9324. Web Site: mail@flinn.com. Licensee: George S. Flinn III. ♦George S. Flinn III, gen mgr.

Sardis

KBUD(FM)— 2005: 102.1 mhz; 4 kw. Ant 403 ft TL: N34 22 33 W89 45 52. Hrs open:
Simulcast with WHBQ-FM Germantown, TN 100%.
6080 Mt. Moriah Ext., Memphis, TN, 38115. Phone: (901) 375-9324. Fax: (901) 375-0041. Web Site: www.flinn.com. Licensee: George S. Flinn Jr. Format: Top-40. ♦Keith Parnell, gen mgr; Edrick Kearney, stn mgr.

Senatobia

***WMSB(FM)—** Jan 4, 1971: 88.9 mhz; 100 kw. Ant 380 ft TL: N34 37 39 W90 01 27. Stereo. Hrs opn: 24 Drawer 2440, Tupelo, 38803. Phone: (662) 844-5036. Fax: (662) 842-7798. Web Site: www.afr.net. Licensee: American Family Association. (acq 4-13-2007; $2 million). Population served: 600,000 Natl. Network: American Family Radio. Format: Christian. ♦Donald E. Wildmon, chmn.

WSAO(AM)— Aug 8, 1962: 1140 khz; 5 kw-D. TL: N34 36 56 W89 56 09. Hrs open: 5 AM-5 PM Box 190, 38668-0190. Secondary address: 15763 Hwy. 4 E. 38668. Phone: (662) 562-4445. Fax: (662) 562-4445. Licensee: Jesse C. Ross and Earnestine A. Ross. (acq 3-95). Population served: 25,000 Rgnl. Network: Miss. Net. Format: Christian, gospel, spiritual music. News staff: one. Target aud: General. Spec prog: Gospel. ♦Jesse Ross, gen mgr & stn mgr.

Southaven

WAVN(AM)— June 4, 1990: 1240 khz; 580 w-U. TL: N34 58 57 W90 00 45. Hrs open: 24 1336 Brookhaven Dr., 38671. Phone: (662) 393-8056. Fax: (662) 393-8066. Licensee: Arlington Broadcasting Co. Inc. (acq 8-31-92; $115,000; FTR: 9-21-92). Rgnl. Network: Miss. Net. Format: Traditional gospel. News: 7 hrs wkly. Target aud: 20-50; general. ♦Walter Stevens, gen mgr & progmg dir.

Starkville

***WJZB(FM)—** 1999: 88.7 mhz; 430 w. Ant 243 ft TL: N33 27 47 W88 49 01. Hrs open: American Family Radio, Box 3206, Tupelo, 38803. Phone: (662) 844-8888. Fax: (662) 842-6791. Web Site: www.afr.net. Licensee: American Family Association Inc. Group owner: American Family Radio (acq 10-8-97). Natl. Network: USA. Format: Relg. ♦Marvin Sanders, gen mgr.

WKOR(AM)— July 5, 1968: 980 khz; 1 kw-D. TL: N33 28 44 W88 44 40. Hrs open: 200 6th St. N., Suite 200, Columbus, 39701. Phone: (662) 327-1183. Fax: (662) 328-1122. Web Site: k949.net. Licensee: Cumulus Licensing Corp. Group owner: Cumulus Media Inc. (acq 2-14-02). Population served: 45,500 Format: ESPN/The Team. Target aud: General; business professionals. ♦Cole Evans, gen mgr & gen sls mgr; CJ Jones, mktg mgr.

WLZA(FM)—Eupora, Sept 1, 1978: 96.1 mhz; 50 kw. 500 ft TL: N33 28 18 W89 13 36. Stereo. Hrs opn: 24 Box 884, 1105 A Stark Rd, 39760. Phone: (662) 324-9601. Fax: (662) 324-7400. E-mail: wlza@bellsouth.net Licensee: Metro Radio. Group owner: Air South Radio Inc. Format: Adult contemp. Target aud: General. ♦Olvie E. Sisk, pres; Carolyn Jackson, gen mgr; David Ever, progmg dir.

WMSU(FM)— Sept 13, 1979: 92.1 mhz; 1.1 kw. 500 ft TL: N33 25 49 W88 45 17. Stereo. Hrs opn: 608 Yellow Jacket Dr., 39759. Phone: (662) 338-5424. Fax: (662) 338-5436. Web Site: www.power92fm.net. Licensee: Urban Radio Licenses LLC. (acq 12-7-2000). Format: Mainstream urban. Target aud: 25-54. ♦James Alexander, opns mgr; Kevin Wagner, pres & progmg mgr.

***WMSV(FM)—** March 1994: 91.1 mhz; 14.1 kw. 449 ft TL: N33 25 49 W88 45 17. Hrs open: Box 6210, Mississippi State Univ., 39762. Phone: (662) 325-8064. Phone: (662) 325-8034. Fax: (662) 325-8037. E-mail: wmsv@msstate.edu Web Site: www.wmsv.msstate.edu. Licensee: Mississippi State University. Format: Alternative. ♦Steve Ellis, gen mgr.

WMXU(FM)—Listing follows WSSO(AM).

WSSO(AM)— Nov 8, 1948: 1230 khz; 1 kw-U. TL: N33 27 09 W88 49 15. Hrs open: 200 6th St. N., Suite 205, Columbus, 39701. Phone: (662) 327-1183. Fax: (662) 328-1122. Licensee: Cumulus Licensing Corp. Group owner: Cumulus Media Inc. (acq 1998; grpsl). Population served: 120,000 Rgnl. Network: Miss. Net. Natl. Rep: Keystone (unwired net). Rgnl rep: Allied Radio Partners. Format: Sports. Spec prog: Black 12 hrs wkly. ♦C.J. Jones, VP.

WMXU(FM)—Co-owned with WSSO(AM). July 15, 1968: 106.1 mhz; 40 kw. Ant 502 ft TL: N33 17 38 W88 39 27. Stereo. 24 Web Site: mix1061.com.120,000 Format: Adult urban. ♦Bobby Holiday, progmg dir.

State College

WQJB(FM)—Not on air, target date: unknown: 104.5 mhz; 25 kw. Ant 328 ft TL: N33 24 14 W88 55 21. Hrs open: 6080 Mt. Moriah, Memphis, TN, 38115. Phone: (901) 375-9324. Fax: (901) 375-0041. Web Site: www.flinn.com. Licensee: George S. Flinn Jr. Format: Classic country. ♦Melanie Henkin-Booth, gen mgr.

Stonewall

WKZB(FM)— 1998: 106.9 mhz; 2.55 kw. 508 ft TL: N32 10 48 W88 40 22. Hrs open: 24 Box 5797, Meridian, 39302. Phone: (601) 693-2661. Fax: (601) 483-0826. E-mail: wmlv@wokk.com Licensee: Mississippi Broadcasters L.L.C. (group owner). Law Firm: Latham & Watkins. Format: Adult contemp. Target aud: 25-54. ♦Clay Holladay, pres & gen mgr; Scott Stevens, opns mgr & progmg mgr; Karen Bostick, sls dir; Van Mac, news dir & pub affrs dir; Scott Shepperd, chief of engrg.

Sumrall

WFMM(FM)— 1998: 97.3 mhz; 1 kw. 200 ft TL: N31 21 18 W89 31 19. Hrs open:
Rebroadcasts WFMN(FM) Flora 100%.
6310 I-55 N., Jackson, 39211. Phone: (601) 957-1700. Fax: (601) 956-5228. Web Site: www.supertalkms.com. Licensee: TeleSouth Communications Inc. (group owner; acq 1999; $200,000). Format: News/talk info. ♦Paul Gallo, gen mgr.

Taylorsville

WBBN(FM)— Mar 20, 1985: 95.9 mhz; 100 kw. Ant 731 ft TL: N31 38 03 W89 28 35. Stereo. Hrs opn: Box 6408, Laurel, 39441. Phone: (601) 649-0095. Phone: (601) 544-0095. Fax: (601) 649-8199. E-mail: b95@b95country.com Web Site: www.b95country.com. Licensee: Blakeney Communications Inc. (group owner) Format: Country. Target aud: 25-54. ♦Larry Blakeney, CEO & pres; Randall A. Blakeney, VP; David Blakeney, gen mgr; Debbie Blakeney, gen sls mgr.

Tchula

WGNG(FM)— 2001: 106.3 mhz; 7.1 kw. Ant 499 ft TL: N33 18 06 W90 07 31. Hrs open: 503 Ione St., Greenwood, 38930. Phone: (662) 453-1646. Fax: (662) 453-7002. E-mail: rhuge4@bellsouth.net Licensee: Team Broadcasting Co. Inc. Format: Urban contemporary. ♦Reuben C. Hughes, gen mgr.

Tunica

WIVG(FM)— 1998: 96.1 mhz; 25 kw. Ant 328 ft TL: N34 43 36 W90 09 43. Hrs open: Y96.1, Flinn Broadcasting, 6080 Mt. Moriah, Memphis, TN, 38115. Phone: (901) 375-9324. Fax: (901) 375-0041. Licensee: Flinn Broadcasting Corp. (acq 10-20-99). Format: Hispanic. ♦Lloyd Hekzer, gen mgr.

Tupelo

***WAFR(FM)**— Aug 31, 1991: 88.3 mhz; 75 kw. Ant 492 ft TL: N34 28 28 W88 43 41. Stereo. Hrs opn: 24 American Family Radio, Box 3206, 38803. Phone: (662) 844-8888. Phone: (662) 844-8893. Fax: (662) 842-6791. Web Site: www.afr.net. Licensee: American Family Association. Population served: 250,000 Natl. Network: American Family Radio. Format: Christian. News staff: one; News: 3 hrs wkly. Target aud: 30-60; conservative Christian. ♦Don Wildmon, gen mgr.

***WAJS(FM)**— 1996: 91.7 mhz; 23 kw. Ant 505 ft TL: N33 55 35 W88 39 46. Hrs open: Box 3206, American Family Radio, 38803. Phone: (662) 844-8888. Fax: (662) 842-6791. Web Site: www.afr.net. Licensee: American Family Association. Group owner: American Family Radio Natl. Network: USA. Format: Christian. ♦Marvin Sanders, gen mgr.

***WAQB(FM)**— 1997: 90.9 mhz; 9.5 kw. Ant 426 ft TL: N34 28 28 W88 43 41. Hrs open: American Family Radio, Box 3206, 38803. Phone: (662) 844-8888. Fax: (662) 842-6791. Web Site: www.afr.net. Licensee: American Family Association. Group owner: American Family Radio Natl. Network: USA. Format: Classic gospel. ♦Don Wildmon, gen mgr.

WELO(AM)— May 15, 1944: 580 khz; 1 kw-D, 500 w-N, DA-2. TL: N34 18 10 W88 42 17. Hrs open: 24 Box 410, 2214 S. Gloster Ave., 38801. Phone: (662) 842-7658. Fax: (662) 842-0197. Licensee: JMD Inc. Population served: 25,471 Format: Music of Your Life, big band. Target aud: 45 plus. Spec prog: Farm one hr wkly. ♦Bob Green, gen mgr & gen sls mgr; Dave Dunaway, opns mgr; Scott Kelly, progmg dir; Cathy Williams, news dir.

WZLQ(FM)—Co-owned with WELO(AM). September 1968: 98.5 mhz; 100 kw. 951 ft TL: N34 10 05 W89 09 23. Stereo. 150,000 Format: Adult contemp. Target aud: 25-54. ♦Steve Drumm, progmg dir.

WFTA(FM)—Fulton, Aug 19, 1976: 101.9 mhz; 100 kw. 560 ft TL: N34 15 46 W88 32 24. Stereo. Hrs opn: 24 Box 2116, 38803. Secondary address: 1241 Cliff Gookin Blvd., Radio Bldg. 38801. Phone: (662) 842-7625. Fax: (662) 842-9568. Licensee: Air South Radio Inc. (group owner) Population served: 325,000 Format: Adult contemp. News: 4 hrs wkly. Target aud: 14-44. ♦Gene Sisk, pres; Olvie E. Sisk, gen mgr; Fred Blalock, stn mgr & gen sls mgr.

WKMQ(AM)— Aug 25, 1972: 1060 khz; 1 kw-D, 33 w-N. TL: N34 15 19 W88 41 46. Hrs open: 24 Box 3300, 38803. Secondary address: 5026 Cliff Gookin Blvd. 38801. Phone: (662) 842-1067. Fax: (662) 842-0725. E-mail: rickstevens@clearchannel.com Licensee: Capstar TX L.P. Group owner: Clear Channel Communications Inc. (acq 12-19-00; grpsl). Population served: 62800 Format: Talk. Target aud: 35-54. ♦Mark Maharrey, gen mgr; Cynthia South, gen sls mgr; Rick Stevens, opns mgr & progmg dir; Jerry Mathis, chief of engrg.

WTUP(AM)— October 1953: 1490 khz; 1 kw-U. TL: N34 15 19 W88 41 46. Hrs open: 24 Box 3300, 38803. Secondary address: 5026 Cliff Gookin Blvd. 38801. Phone: (662) 842-1067. Fax: (662) 842-0725. E-mail: markmaharrey@clearchannel.com Web Site: www.wtup1490.com. Licensee: Capstar TX L.P. Group owner: Clear Channel Communications Inc. (acq 12-19-00; grpsl). Population served: 174,600 Rgnl. Network: Miss. Net. Natl. Rep: Interep. Law Firm: Gurman, Blask & Freedman. Format: Sports. News staff: one; News: 12 hrs wkly. Target aud: 25-54; men. ♦Mark Maharrey, gen mgr; Rick Stevens, opns dir.

Tylertown

WFCG(FM)— 2005: 107.3 mhz; 3.2 kw. Ant 457 ft TL: N31 04 39 W90 04 46. (CP: 2.2 kw, ant 550 ft). Hrs opn: Box 1649, McComb, 39648. Phone: (985) 839-3782. Fax: (985) 839-3783. Licensee: Southwest Broadcasting Inc. Format: Southern gospel. ♦C. Wayne Dowdy, pres; William Giles, gen mgr.

WTYL(AM)— Feb 8, 1969: 1290 khz; 1 kw-D. TL: N31 07 50 W90 08 13. Hrs open: 11 930 Union Rd., 39667. Phone: (601) 876-2105. Fax: (601) 876-9551. Licensee: Tylertown Broadcasting Co. Population served: 20,000 Format: Country. Spec prog: Farm 6 hrs wkly. ♦Carolyn Dillon, pres & gen mgr; Gail Ratcliff, progmg dir & traf mgr.

WTYL-FM— Apr 9, 1970: 97.7 mhz; 3 kw. 145 ft TL: N31 07 50 W90 08 13.24 ♦Gail Ratcliff, traf mgr.

Union

WZKS(FM)— October 1995: 104.1 mhz; 16 kw. 535 ft TL: N32 29 53 W88 53 20. (CP: 19 kw). Hrs opn: 4307 Hwy. 39 N., Meridian, 39301. Phone: (601) 693-2381. Fax: (601) 485-2972. Licensee: CC Licenses LLC. Group owner: Clear Channel Communications Inc. (acq 3-16-2001; grpsl). Format: Urban contemp. ♦Ron Harper, gen mgr.

University

WUMS(FM)— Apr 10, 1989: 92.1 mhz; 6 kw. 328 ft TL: N34 21 29 W89 32 30. Stereo. Hrs opn: 24 201 Bishop Hall, Oxford, 38677. Phone: (662) 915-5503. Fax: (662) 915-5703. Web Site: www.olemiss.edu. Licensee: Student Media Center of the University of Mississippi. Natl. Network: ABC, Premiere Radio Networks. Format: Alternative, hot adult contemp. News: 2. Target aud: 18-25. Spec prog: International 2 hrs, women one hr, the 80's 2 hrs wkly. ♦Melanie Store, pres; David Stil, stn mgr; Beth Vance, sls dir.

Utica

WJXN-FM— Aug 28, 1990: 100.9 mhz; 39 kw. Ant 551 ft TL: N32 03 13 W90 20 23. Stereo. Hrs opn: 24 1985 Lakeland Drive, Suite 201, Jackson, 39216. Phone: (601) 713-0977. Fax: (601) 713-2977. Licensee: Flinn Broadcasting Corp. (acq 12-31-97). Natl. Network: USA. Wire Svc: NWS (National Weather Service) Format: Urban adult contemp. Target aud: 25 plus. ♦George S. Flinn, pres; Karen Porter, stn mgr; Steve Poston, stn mgr & progmg dir.

Vicksburg

WBBV(FM)— Aug 21, 1989: 101.3 mhz; 13 kw. Ant 394 ft TL: N32 20 42 W90 52 55. Stereo. Hrs opn: 1601 E. North Frontage Rd., 39180. Phone: (601) 638-0101. Phone: (601) 636-2340 (business). Fax: (601) 638-0869. E-mail: spots@river101.com Web Site: www.river.com. Licensee: Holladay Broadcasting of Louisiana, LLC Group owner: New South Communications Inc. (acq 7-11-2005; $400,000 for stock). Population served: 65,000 Format: Country. Target aud: 24-54. ♦Bob Holladay, pres; Betsy McEachern, gen mgr; Ron Anderson, opns mgr; Deloris Dorbeck, gen sls mgr.

WJKK(FM)— Mar 19, 1966: 98.7 mhz; 100 kw. 950 ft TL: N32 12 29 W90 24 50. Stereo. Hrs opn: 24 265 Highpoint Dr., Ridgeland, 39157. Phone: (601) 956-0102. Fax: (601) 978-3980. Web Site: www.mix987.com. Licensee: New South Radio Inc. Group owner: New South Communications Inc. (acq 1-89; $1.1 million; FTR: 1-23-89). Format: Soft adult contemp. News staff: one; News: one hr wkly. Target aud: 18-49; upper income & educ. ♦Gwen Rakestraw, gen mgr.

WQBC(AM)— 1931: 1420 khz; 5 kw-D, 500 w-N. TL: N32 19 56 W90 51 00. Hrs open: 24 3190 Porter's Chapel Rd., 39180. Secondary address: Box 820483 39182. Phone: (601) 636-1108. Fax: (601) 631-0087. E-mail: WQBC@WQBC.net Web Site: www.WQBC.net. Licensee: Grace Media International LLC (acq 3-20-2001; $100,000). Population served: 26,091 Natl. Network: USA, Salem Radio Network. Format: News/talk, Sports. News staff: 2; News: 4 hrs wkly. Target aud: 35-55; Male & Female. ♦Jerry Rushins, gen mgr & gen sls mgr; Mike Corley, pres & opns dir.

WSTZ-FM—Licensed to Vicksburg. See Jackson

WVBG(AM)— 1948: 1490 khz; 1 kw-U. TL: N32 21 27 W90 51 29. Hrs open: 24 1102 Newit Vick Dr., MA, 39183. Phone: (601) 883-0848. Licensee: Commander Communications Corp. (acq 5-11-99). Population served: 35,000 Format: News/talk. News: 168 hrs. wkly. Target aud: 25 plus. ♦Mark Jones, gen mgr.

Walnut

WLRC(AM)— June 21, 1982: 850 khz; 963 w-D. TL: N34 56 46 W88 52 44. Hrs open: Box 37, 38683. Secondary address: 7760 Hwy. 72 E. 38683. Phone: (662) 223-4071. Fax: (662) 223-4072. Web Site: www.wlrcradio.com. Licensee: B.R. & Martha S. Clayton. (acq 11-83; $100,000; FTR: 11-28-83). Rgnl. Network: Miss. Net. Format: Christian. News staff: News progmg 14 hrs. wkly, Incl local news live Spec prog: Southern & country gospel mus, preaching & children's programs.

Water Valley

WTNM(FM)— Aug 1, 1996: 105.5 mhz; 4.7 kw. Ant 371 ft TL: N34 12 45 W89 44 49. Stereo. Hrs opn: 24 Box 1077, Oxford, 38655. Secondary address: 461 Hwy. 6 W., Oxford 38655. Phone: (662) 236-0073. Fax: (662) 234-5155. E-mail: supertalk1055@exceedtech.net Web Site: www.supertalkms.com. Licensee: TeleSouth Communications Inc. (group owner; acq 3-17-00). Population served: 89500 Format: Talk. News staff: one; News: 3 hrs wkly. Target aud: 25 plus. Spec prog: Christian, contemp 4 hrs wkly. ♦Rick Mize, gen mgr; Jim Martin, opns mgr; Steve Davenport, CEO & engrg VP.

Waynesboro

WABO(AM)— Sept 11, 1954: 990 khz; 1 kw-D. TL: N31 40 48 W88 40 34. Hrs open: Box 507, 39367. Secondary address: 6746 Hwy. 184 W. 39367. Phone: (601) 735-4331. Fax: (601) 735-4332. E-mail: waboradio@c-gate.net Web Site: www.wabo105.com. Licensee: Martin Broadcasting Inc. (acq 12-18-61). Population served: 16,000 Format: Country, soul. ♦Jamie Heathcock, progmg dir; Lisa Singley, traf mgr; Nancy N. Martin, pres, gen mgr, gen sls mgr & women's int ed.

WABO-FM— June 13, 1973: 105.5 mhz; 3 kw. 145 ft TL: N31 40 48 W88 40 34. Stereo. E-mail: waboradio@c-gate.net Web Site: www.wabo105.com.35,000 Format: Hot country.

***WZKM(FM)**—Not on air, target date: unknown: 89.7 mhz; 67 kw. Ant 581 ft TL: N31 50 09 W88 52 21. Hrs open: American Family Radio, Box 3206, Tupelo, 38803. Phone: (662) 844-8888. Fax: (662) 842-6791. Web Site: www.afr.net. Licensee: American Family Association. Group owner: American Family Radio (acq 1-24-03). Natl. Network: USA. Format: Relg. ♦Don Wildmon, gen mgr.

West Point

WKBB(FM)—Listing follows WROB(AM).

WROB(AM)— September 1947: 1450 khz; 1 kw-U. TL: N33 36 30 W88 39 15. Hrs open: Box 1336, 413 N. Forest St., 39773. Phone: (662) 494-1450. Fax: (662) 494-9762. E-mail: wrobwkbb@ebicom.net Licensee: TeleSouth Communications Inc. (group owner; acq 12-5-2003; $900,000 with co-located FM). Population served: 20,000 Rgnl. Network: Miss. Net. Format: Urban gospel. Target aud: General. Spec prog: Gospel 2 hrs wkly. ♦Greg Benefield, gen mgr, progmg dir & news dir; Bob McRaney Jr., gen sls mgr; Peggy Goode, pres & traf mgr.

WKBB(FM)—Co-owned with WROB(AM). Apr 14, 1974: 100.9 mhz; 10 kw. Ant 515 ft TL: N33 40 43 W88 48 18. Stereo. 150,000 Format: News/talk, jazz. Target aud: 35-54.

Wiggins

WIGG(AM)— February 1968: 1420 khz; 5 kw-D. TL: N30 52 18 W89 09 00. Hrs open: 24 Box 723, 39577. Secondary address: 959 N. Magnolia Dr. 39577. Phone: (601) 928-7281. Fax: (601) 528-5011. Licensee: Tralyn Broadcasting Inc. (acq 7-1-97; grpsl). Population served: 6,000 Rgnl. Network: Miss. Net. Format: Country. News: 5 hrs wkly. Target aud: 25-54; general. Spec prog: Gospel, sports. ♦A.R. Byrd, gen mgr.

Winona

WONA(AM)— Oct 25, 1958: 1570 khz; 1 kw-D. TL: N33 27 52 W89 44 11. Hrs open: 1006 S. Applegate St., 38967. Phone: (662) 283-1570. Fax: (662) 283-1520. Licensee: Southern Electronics Co. Population served: 25,521 Format: Country. ♦Johnny Pettit, pres; Seth Kent, opns mgr, gen sls mgr, progmg dir & chief of engrg; Sharon Kent, VP, gen mgr, news dir & traf mgr.

WONA-FM— Jan 4, 1976: 95.1 mhz; 3 kw. 328 ft TL: N33 29 34 W89 45 17. Stereo. E-mail: hawg@cablesouthmedia.net

Yazoo City

WJNS-FM— Dec 13, 1968: 92.1 mhz; 20 kw. 300 ft TL: N32 50 48 W90 23 18. Stereo. Hrs opn: 1405 Enchanted Dr., 39194. Phone: (662) 746-5921. Fax: (662) 746-5996. Licensee: Family Worship Center Church Inc. (group owner; acq 6-16-2004; $350,000). Population served: 84,355 Format: Relg. Target aud: 25-54. Spec prog: Farm 16 hrs, weather 16 hrs wkly.

WYAB(FM)— August 1997: 93.1 mhz; 4.1 kw. 394 ft TL: N32 49 20 W90 16 46. Hrs open: 24 740 Hwy. 49, Suite R, Flora, 39071. Phone: (601) 879-0093. Fax: (601) 427-8800. E-mail: matt@wyab.com Web Site: www.wyab.com. Licensee: SSR Communications Inc. (acq 4-1-03; $207,500). Population served: 100,000 Format: Oldies. News staff: one. Target aud: 25-64; div blend, not your typical oldies stn. ♦Matthew Wesolowski, CEO & engrg dir.

***WYAZ(FM)**— 2005: 89.5 mhz; 25 kw vert. Ant 518 ft TL: N32 48 04 W89 56 32. Hrs open: American Family Radio, Drawer 2440, Tupelo, 38801. Phone: (662) 844-8888. Fax: (662) 842-6791. Licensee: American Family Association. Group owner: American Family Radio (acq 5-13-2004). ♦Marvin Sanders, gen mgr.

Missouri

Albany

KAAN-FM—See Bethany

Anderson

***KGSF(FM)**—Not on air, target date: unknown: 88.5 mhz; 350 w vert. Ant 278 ft TL: N36 20 59 W94 20 54. Hrs open: 4002 N. 3300 E., Twin Falls, ID, 83301. Phone: (208) 734-6633. Fax: (208) 736-1958. Web Site: www.csnradio.com. Licensee: CSN International. ♦Mike Kestler, pres.

Arcadia

KTNX(FM)— 2006: 103.9 mhz; 450 w. Ant 932 ft TL: N37 34 23 W90 41 35. Hrs open: 540 Maple Valley Dr., Farmington, 63640. Phone: (573) 701-9590. Fax: (573) 701-9696. Licensee: Dockins Communications Inc. Format: Classic rock. ♦Fred M. Dockins Sr., pres & gen mgr.

Arnold

***KGNA-FM**— Mar 26, 1987: 89.9 mhz; 150 w horiz, 84 w vert. Ant 131 ft TL: N38 26 14 W90 23 24. Hrs open: 24
Rebroadcasts KGNV(FM) Washington 100%.
Box 187, Washington, 63090. Phone: (636) 239-0400. Fax: (636) 239-4448. Web Site: www.goodnewsvoice.org. Licensee: Missouri River Christian Broadcasting Inc. (acq 10-5-99). Population served: 150,000 Natl. Network: Moody, Salem Radio Network. Format: News/talk, Southern gospel, Inspirational. News: 14 hrs wkly. Target aud: 20-70:; inquisitive, conservative, liberal, philosophical. ♦James C. Goggan, pres.

Asbury

KWXD(FM)— October 1993: 103.5 mhz; 16 kw. Ant 413 ft TL: N37 23 44 W94 40 42. Hrs open: Box 383, 412 Locust St., Pittsburg, KS, 66762. Phone: (620) 232-5993. Fax: (620) 232-5550. Licensee: Innovative Broadcasting Corp. (group owner) Law Firm: Lauren A. Colby. Format: Country. News staff: one; News: 20 hrs wkly. Target aud: 25-54. ♦Lance Sayler, pres & gen mgr.

Ash Grove

KSGF-FM— Mar 1, 1994: 104.1 mhz; 21.5 kw. Ant 354 ft TL: N37 15 22 W93 41 14. Hrs open: 2330 W. Grand St., Springfield, 65781. Phone: (417) 865-6614. Fax: (417) 865-9643. Web Site: www.ksgf.com. Licensee: Journal Broadcast Corp. Group owner: Journal Communications Inc. (acq 11-26-2003; $5 million with KZRQ-FM Mount Vernon). Format: News, talk. ♦Rex Hansen, gen mgr; Chris Cannon, opns mgr.

Ashland

KOQL(FM)— October 1993: 106.1 mhz; 69 kw. Ant 958 ft TL: N38 45 01 W92 33 31. Stereo. Hrs opn: 24 503 Old 63 N., Columbia, 65201. Phone: (573) 449-4141. Fax: (573) 449-7770. Web Site: www.q1061.com. Licensee: Cumulus Licensing LLC. Group owner: Cumulus Media Inc. (acq 4-26-2004; grpsl). Population served: 260,000 Natl. Rep: Katz Radio. Format: Top-40. ♦Lewis W. Dickey Jr., pres; Scott Boltz, VP & mktg mgr.

Aurora

KSWF(FM)— Feb 19, 1968: 100.5 mhz; 33 kw. 600 ft TL: N37 05 39 W93 31 05. Stereo. Hrs opn: 1856 S. Glenstone, Springfield, 65804. Phone: (417) 890-5555. Fax: (417) 890-5050. E-mail: mycountry@mycountry.com Web Site: www.1005thewolf.com. Licensee: Clear Channel Broadcasting Licenses Inc. Group owner: Clear Channel Communications Inc. (acq 10-10-2000; grpsl). Population served: 5,359 Format: Country. Target aud: 18-54; general. ♦Paul Windisch, gen mgr; Paul Kelley, opns mgr.

KSWM(AM)— Oct 19, 1961: 940 khz; 1 kw-D, 30 w-N. TL: N36 59 39 W93 42 58. Hrs open: 24 126 S. Jefferson, 65605. Phone: (417) 678-0416. Fax: (417) 678-4111. Web Site: www.talonbroadcasting.com. Licensee: Falcon Broadcasting Inc. Group owner: Community Service Radio Group (acq 9-1-2005; $417,500). Population served: 250,000 Natl. Network: CNN Radio, USA. Law Firm: Fletcher, Heald & Hildreth. Format: News/talk. News staff: 2; News: 168 hrs wkly. Target aud: General. ♦DeWayne Gandy, gen mgr & engrg mgr.

Ava

KKOZ-FM— 1990: 92.1 mhz; 4 kw. 380 ft TL: N36 55 48 W92 39 19. Hrs open: 6 AM-10 PM Box 386, 65608. Phone: (417) 683-4191. Web Site: www.kkoz.com. Licensee: Corum Industries Inc. Rgnl. Network: Missourinet. Format: News/talk, farm. News: 15 hrs wkly. Target aud: 45 plus; farm oriented. ♦Joe Corum, pres & gen mgr; Art Corum, opns mgr, prom mgr & progmg dir; Bob Moore, engrg mgr & chief of engrg; Vickie Corum, traf mgr.

KKOZ(AM)— 1968: 1430 khz; 500 w-U. TL: N36 55 48 W92 39 19. Web Site: www.kkoz.com. (Acq 7-97; $11,200).50,000

Ballwin

***KYMC(FM)**— February 1978: 89.7 mhz; 120 w. Ant 171 ft TL: N38 37 23 W90 32 01. Stereo. Hrs opn: 24 Box 4038, 16464 Burkhardt Pl., Chesterfield, 63006. Phone: (636) 532-6515. Phone: (636) 532-3100. Fax: (636) 530-7928. E-mail: nhall@ymcastlouis.org Web Site: ymcastlouis.org. Licensee: YMCA of Greater St. Louis-W. County Branch. Law Firm: Womble, Carlyle, Sandridge & Rice. Format: Var/div. News staff: 3; News: 6 hrs wkly. Target aud: 12-35; families. Spec prog: Jazz 6 hrs, teens one hr, Christian rock 4 hrs, talk 3 hrs wkly. ♦Natalie Hall, gen mgr & chief of opns.

Bethany

KAAN(AM)— Dec 3, 1983: 870 khz; 1 kw-D. TL: N40 15 23 W94 09 23. Hrs open: Sunrise-sunset Box 447, Hwy. 69 S., 64424. Phone: (660) 425-6380. Fax: (660) 425-8148. Web Site: www.regionalradio.com. Licensee: KAAN Inc. Group owner: Shepherd Group Rgnl. Network: Missourinet. Format: Country, news. News staff: 3; News: 10 hrs wkly. Target aud: 25 plus. Spec prog: Farm 10 hrs, relg one hr wkly. ♦Mike Mattson, gen mgr & stn mgr; Rodney Harris, gen mgr & sls dir; Denise Fritzel, gen sls mgr; Stuart Johnson, progmg dir; Stuart Johsnson, news dir; Gregg Richwine, chief of engrg; Sheila Gallagher, traf mgr.

KAAN-FM— Oct 27, 1978: 95.5 mhz; 50 kw. 360 ft TL: N40 15 23 W94 09 23.19 Web Site: www.regionalradio.com.

Birch Tree

KBMV-FM— 1983: 107.1 mhz; 25 kw. Ant 328 ft TL: N36 56 03 W91 43 07. Stereo. Hrs opn: 24 Box 107, West Plains, 65775. Phone: (417) 255-0427. Fax: (417) 255-2907. Web Site: www.todaysbesthits.com. Licensee: Mountain Lakes Broadcasting Corp. (acq 9-8-2003; $175,000). Format: Hot adult contemp. ♦Connie P. Feifer, gen mgr.

KBSP(AM)— Sept 14, 1981: 1310 khz; 1 kw-D, 60 w-N. TL: N36 59 07 W91 32 54. Hrs open: 932 County Rd. 448, Poplar Bluff, 63901. Phone: (573) 686-3700. Fax: (573) 686-1713. Licensee: Eagle Bluff Enterprises. (group owner; (acq 9-8-99; grpsl). ♦Steven Fuchs, gen mgr.

Bismarck

KHCR(FM)— 2005: 99.5 mhz; 4.2 kw. Ant 798 ft TL: N37 38 52 W90 37 33. Hrs open: 627 State Hwy. 47, Bonne Terre, 63628. Phone: (573) 358-7700. Licensee: Joseph W. & Donna M. Bollinger. ♦Joseph W. Bollinger, gen mgr.

Blue Springs

KCWJ(AM)— Feb 2, 1984: 1030 khz; 5 kw-D, 500 w-N, DA-2. TL: N39 02 44 W94 14 06. Hrs open: 24 4240 Blue Ridge Blvd., Suite 530, Kansas City, 64133. Phone: (816) 313-0049. Fax: (816) 313-1036. E-mail: info@kcwj.org Web Site: www.kcwj.org. Licensee: Christian Broadcasting Associates L.P. (acq 1-13-99; $750,000). Population served: 1,800.000 Natl. Network: Salem Radio Network. Natl. Rep: Salem. Rgnl rep: PioneerSports Sales Law Firm: Miller & Neely, P.C. Wire Svc: Metro Weather Service Inc. Format: Christian,traditional christian. News: 5 hrs wkly. Target aud: 18-49; family oriented Christian audience. ♦Ken Ball, gen mgr.

Bolivar

KYOO(AM)— November 1961: 1200 khz; 1 kw-D. TL: N37 41 50 W93 25 45. Hrs open: 205 N. Pike Ave., 65613-1550. Phone: (417) 326-5259. Phone: (417) 326-5257. Fax: (417) 326-5900. E-mail: kyooradio@aol.com Web Site: www.realcountryonline.com. Licensee: KYOO Communications KYOO Communications (acq 7-29-97; $52,000 assumption of note). Population served: 500,000 Natl. Network: ABC. Rgnl. Network: Brownfield Wire Svc: NWS (National Weather Service) Format: News, country. News staff: one; News: 12 hrs. wkly. Target aud: 10-72 yrs. Spec prog: Farm 3 hrs, gospel 2 hrs wkly. ♦Ann Paris, VP; Stephen Paris, pres & gen mgr.

Bonne Terre

KDBB(FM)— September 1989: 104.3 mhz; 790 w. 630 ft TL: N37 48 04 W90 33 44. Hrs open: 24 Box 36, Park Hills, 63601. Phone: (573) 431-1000. Fax: (573) 431-0850. E-mail: radio@b104fm.com Web Site: www.b104fm.com. Licensee: MKS Broadcasting Inc. (acq 9-6-94; $315,753; FTR: 10-17-94). Natl. Network: Westwood One. Format: Rock. News staff: one; News: 20 hrs wkly. Target aud: 25-55. ♦M.L. Steinmetz, pres; Larry D. Joseph, gen mgr; Kelly Valle, gen sls mgr; Greg Camp, progmg mgr; Gib Collins Jr., news dir & pub affrs dir.

Boonville

KCLR-FM— Oct 1, 1974: 99.3 mhz; 33.2 kw. 590 ft TL: N38 46 34 W92 32 45. Stereo. Hrs opn: 24 3215 Lemone Industrial Blvd., Suite 200, Columbia, 65201. Phone: (573) 875-1099. Fax: (573) 875-2439. Web Site: www.clear99.com. Licensee: Zimmer Broadcasting Co. Format: Country. News staff: 3; News: 3 hrs wkly. Target aud: 25-54. ♦Teresa Davis, gen mgr & progmg dir.

KWJK(FM)—Listing follows KWRT(AM).

KWRT(AM)— Aug 11, 1953: 1370 khz; 1 kw-D, 84 w-N. TL: N38 56 44 W92 34 30. Hrs open: 24 1600 Radio Hill Rd., 65233. Phone: (660) 882-6686. Fax: (660) 882-6688. E-mail: kwrt@undata.com Licensee: Big Country of Missouri Inc. Population served: 150000 Natl. Network: Jones Radio Networks. Rgnl. Network: Missourinet. Format: Country. News staff: one; News: 5 hrs wkly. Target aud: 35 plus; general. Spec prog: Farm 5 hrs wkly. ♦Dick Billings, pres; Matt Billings, gen mgr & gen sls mgr; Pat Billings, opns mgr; Ted Bleil, news dir; Mike Mcgowan, chief of engrg; Sharon Korte, sports cmtr.

KWJK(FM)—Co-owned with KWRT(AM). Sept 15, 1999: 93.1 mhz; 7.2 kw. Ant 413 ft TL: N38 56 31 W92 34 32.24 Phone: (573) 441-9310. E-mail: kwrt@classicnet.net Web Site: www.931jack.fm. Licensee: Bittersweet Broadcasting Inc.150,000 Format: Adult hits.

Bowling Green

KPVR(FM)— Aug 1, 1975: 94.1 mhz; 7.5 kw. Ant 592 ft TL: N39 15 45 W91 04 09. Stereo. Hrs opn: 24 13358 Manchester Rd., Suite 100, Des Peres, 63131. Phone: (314) 909-8569. Fax: (314) 835-9739. E-mail: info@joyfmonline.org Web Site: www.joyfmonline.org. Licensee: Four Him Enterprises L.L.C. (acq 5-15-2001; $725,000 with co-located AM). Population served: 121,582 Format: Contemp Christian. ♦Sandi Brown, gen mgr.

Branson

***KLFC(FM)**— July 1988: 88.1 mhz; 1.8 kw. Ant 390 ft TL: N36 33 06 W93 14 17. Stereo. Hrs opn: 24 205 W. Atlantic, 65616-0921. Phone: (417) 334-5532. Fax: (417) 335-2437. E-mail: 881fm@klfcradio.com Web Site: www.klfcradio.com. Licensee: Mountaintop Broadcasting Inc. (acq 6-1-01). Population served: 25,000 Natl. Network: USA. Format: Christian. News: 7 hrs wkly. Target aud: General; resort & tourist community. ♦Herb Smith, pres, gen mgr & stn mgr; Vicky Smith, opns dir; Darin Ahrends, news dir.

KOMC(AM)— Dec 21, 1956: 1220 khz; 1 kw-D, 53 w-N. TL: N36 37 12 W93 12 40. Hrs open: 24
Rebroadcasts KDMC-FM Kimberling City 95%.
202 Courtney St., 65616. Phone: (417) 334-6003. Phone: (417) 334-6012. Fax: (417) 334-7141. E-mail: krzk@krzk.com Web Site: www.hometownradioonline.com. Licensee: KOMC-KRZX LLC. Group owner: Orr & Earls Broadcasting Inc. (acq 11-21-86; $335,000). Population served: 42,000 Natl. Network: CBS. Rgnl. Network: Missourinet. Format: Christian, relg, southern gospel. News staff: 3; News: 10 hrs wkly. Target aud: 40 plus. ♦Charles C. Earls, pres; Scottie Earls, gen mgr, stn mgr & opns mgr; Steve Willoughby, stn mgr & mktg mgr; Scott McCaulley, progmg dir; Morris James, news dir; Greg Pyron, chief of engrg.

KRZK(FM)—Co-owned with KOMC(AM). Mar 1, 1971: 106.3 mhz; 5.7 kw. 672 ft TL: N36 43 52 W93 10 03. (CP: 100 kw, ant 564 ft.). Stereo. 24 Web Site: hometowndailynews.com. Natl. Network: ABC. Format: Country. News staff: 3; News: 7 hrs wkly. Target aud: 25-54; Branson & loc tourists. ♦Charles C. Earles, CEO; Steve Willoughby, mktg dir; Scott Earls, engrg mgr & chief of engrg.

***KOZO(FM)**— 1998: 89.7 mhz; 150 w horiz, 20 kw vert. Ant 426 ft TL: N36 33 04 W93 14 36. Stereo. Hrs opn: 24 Box 1924, Tulsa, OK, 74101-1924. Phone: (918) 455-5693. Phone: (417) 339-3388. Fax: (417) 339-3410. E-mail: mail@oasisnetwork.org Web Site: www.oasisnetwork.org. Licensee: Creative Educational Media Corp. Inc. Natl. Network: USA. Rgnl rep: Rgnl Reps Format: Relg. Target aud: General. ♦David Ingles, pres & gen mgr.

Brookfield

KFMZ(AM)— Feb 14, 1956: 1470 khz; 500 w-D, 20 w-N, DA. TL: N39 50 26 W93 04 52. Hrs open: 24
Rebroadcasts KZBK-FM Brookfield.
107 S. Main, 64628. Phone: (660) 258-3383. Fax: (660) 258-7307. E-mail: kzbk@kzbkradio.com Web Site: www.kzbkradio.com. Licensee: Best Broadcasting Inc. Group owner: Best Broadcast Group (acq 6-14-93; $70,000 with co-located FM; FTR: 6-28-93). Population served: 54,910 Natl. Network: ABC. Law Firm: Bryan Cave. Format: Hot adult contemp. News: 4 hrs wkly. Target aud: 18-49; men & women with spendable income. ♦Phillip A. Chirillo, pres; Dale A. Palmer, VP & gen mgr.

KZBK(FM)—Co-owned with KFMZ(AM). September 1981: 96.9 mhz; 50 kw. 492 ft TL: N39 54 32 W93 04 34. Stereo. 24 70,000

Brookline

KQRA(FM)— May 28, 2002: 102.1 mhz; 4.9 kw. Ant 361 ft TL: N37 12 39 W93 13 42. Stereo. Hrs opn: 24 319 B E. Battlefield, Springfield, 65807. Phone: (417) 886-5677. Fax: (417) 886-2155. E-mail: info@q1021.fm Web Site: www.q1021.fm. Licensee: MW SpringMo Inc. Group owner: The Mid-West Family Broadcast Group. Population served: 285,500 Natl. Rep: McGavren Guild. Law Firm: Davis Wright Tremaine, LLP. Format: Rock. Target aud: 18-49; active adults. ♦Rick McCoy, pres & gen mgr; Mary Fleenor, opns mgr; Malcolm Hukriede, gen sls mgr; Keith Abercrombie, rgnl sls mgr; Kristen Bergman, progmg dir; Chris Louzader, traf mgr.

Buffalo

KBFL-FM— 1965: 99.9 mhz; 3.1 kw. Ant 476 ft TL: N37 31 14 W93 06 14. Stereo. Hrs opn: 24 Box 1385, 65622. Secondary address: 304 S. Pine 65622. Phone: (417) 345-2412. Fax: (417) 345-2410. Web Site: www.radiospringfield.com. Licensee: Meyer-Baldridge Inc. Group owner: Meyer Communications Inc. (acq 6-1-2000; $550,000). Population served: 30,000 Natl. Network: Music of Your Life. Rgnl. Network: Missourinet. Law Firm: Fletcher, Heald & Hildreth. Format: Nostalgia, news/talk, sports. News staff: one; News: 15 hrs wkly. Target aud: 34-54; male-female adults. Spec prog: Gospel 3 hrs wkly. ♦Kenneth E. Meyer, pres; Rob Evans, gen mgr.

Bunker

KHZA(FM)—Not on air, target date: unknown: 106.3 mhz; 25 kw. Ant 328 ft TL: N37 34 48 W91 19 38. Hrs open: 5331 Mt. Alifan Dr., San Diego, CA, 92111. Phone: (858) 277-4991. Fax: (858) 277-1365. Licensee: Horizon Christian Fellowship. (acq 2-9-2006; grpsl). ♦Mike MacIntosh, pres.

Butler

KMAM(AM)— May 11, 1962: 1530 khz; 500 w-D. TL: N38 14 56 W94 19 18. Hrs open: 16 800 E. Nursery St., 64730. Phone: (660) 679-4191. Fax: (660) 679-4193. E-mail: news@fm92radio.com Web Site: www.921kmoe.com. Licensee: Bates County Broadcasting Co. Population served: 183,400 Rgnl. Network: Brownfield. Format: Country. News staff: one; News: 15 hrs wkly. Target aud: General; family. Spec prog: Farm 15 hrs wkly. ♦Melody A. Thornton, pres & gen mgr.

KMOE(FM)—Co-owned with KMAM(AM). Jan 15, 1975: 92.1 mhz; 4.7 kw. Ant 148 ft TL: N38 14 56 W94 19 18. Stereo. 16 Web Site: www.921knoe.com.

Cabool

***KFFW(FM)**— 2003: 89.9 mhz; 10.5 kw. Ant 495 ft TL: N37 05 32 W92 03 10. Stereo. Hrs opn: 24 First Free Will Baptist Church, 401 S. Main, Mountain Grove, 65711. Phone: (417) 926-5396. Fax: (417) 926-7911. E-mail: info@kffw.org Web Site: www.kffw.org. Licensee: First Free Will Baptist Church (acq 12-11-01). Natl. Network: Salem Radio Network, American Family Radio. Format: Christian. News: 12 hrs wkly. ♦Brian Hurst, gen mgr; Rick Jesse, opns mgr.

KOZX(FM)— May 1978: 98.1 mhz; 3 kw. 220 ft TL: N37 07 58 W92 08 04. Stereo. Hrs opn: 800 N. Hubbard, Mountain Grove, 65711. Phone: (417) 926-4650. Fax: (417) 926-7604. Licensee: Quorum Radio Partners Inc. (group owner; (acq 8-8-2002; grpsl). Population served: 87,000 Format: Classic hits. Spec prog: Farm 2 hrs wkly. ♦Rick Vermillion, gen mgr.

California

KATI(FM)— July 27, 1984: 94.3 mhz; 50 kw. 492 ft TL: N38 31 25 W92 24 25. Stereo. Hrs opn: 19 3109 S. Ten Mile Dr., Jefferson City, 65109. Phone: (573) 893-5696. Fax: (573) 893-4137. E-mail: kati@zrgmail.com Web Site: www.kat943.com. Licensee: Zimmer Radio of Mid-Missouri Inc. Group owner: Zimmer Radio Group (acq 11-19-99; grpsl). Format: Country. ♦Ron Covert, gen mgr.

KRLL(AM)— July 27, 1984: 1420 khz; 500 w-D, 225 w-N. TL: N38 38 12 W92 35 00. Hrs open: 18 100 A.E. Buchanan, 65018. Phone: (573) 796-3139. Fax: (573) 796-4131. E-mail: krll01@earthlink.net Licensee: Moniteau Communications Inc. (acq 3-30-95; $50,000; FTR: 6-19-95). Law Firm: Leibowitz & Spencer. Format: Country. News staff: one; News: 19 hrs wkly. Target aud: 20 plus. Spec prog: Farm 5 hrs, gospel 3 hrs wkly. ♦Jeffrey G. Shackleford, pres & gen mgr.

Camdenton

***KCVO-FM**— Sept 23, 1985: 91.7 mhz; 10 kw. 435 ft TL: N38 01 13 W92 45 27. Stereo. Hrs opn: 24 Box 800, Lake Rd. 5-92, 65020. Phone: (573) 346-3200. Fax: (573) 346-1010. E-mail: email@spiritfm.org Web Site: www.spiritfm.org. Licensee: Lake Area Educational Broadcasting Foundation. Population served: 250,000 Format: Div, Christian. News: 7 hrs wkly. Target aud: 25-45. ♦Alice McDermott, CFO; James J. McDermott, pres & gen mgr.

Cameron

KKWK(FM)—Listing follows KMRN(AM).

KMRN(AM)— February 1971: 1360 khz; 500 w-D, 25 w-N. TL: N39 41 05 W94 14 22. Hrs open: 5:30 AM-7 PM 607 E. Platt Clay Way, 64429. Phone: (816) 632-6661. Fax: (816) 632-1334. Web Site: www.regionalradio.com. Licensee: KAAN Inc. Group owner: Shepherd Group (acq 7-13-99; with co-located FM). Population served: 50,000 Rgnl. Network: Missourinet, Brownfield. Format: News/talk. News staff: one; News: 40 hrs wkly. Target aud: General. Spec prog: Farm 12 hrs, relg 4 hrs wkly. ♦Rodney Harris, gen mgr & gen sls mgr; Barry Piatt, progmg dir; Greg Richwine, chief of engrg; Mary Ann Leichti, traf mgr.

KKWK(FM)—Co-owned with KMRN(AM). Apr 5, 1995: 100.1 mhz; 50 kw. 492 ft TL: N39 57 28 W94 06 55. Web Site: www.regionalradio.com.250000 Natl. Network: ABC. Format: Adult contemp. Target aud: 25-49. ♦Barry Piatt, progmg dir; Chris Ward, news dir & sports cmtr.

Campbell

KFEB(FM)— October 1998: 107.5 mhz; 17.5 kw. 390 ft TL: N36 29 55 W89 51 16. Stereo. Hrs opn: 932 CR Box 448, Poplar Bluff, 63901. Phone: (573) 686-3700. Fax: (573) 686-1713. Licensee: Eagle Bluff Enterprises. (group owner) Population served: 220,000 Format: Modern rock. ♦Steven C. Fuchs, gen mgr.

Canton

KRRY(FM)— May 4, 1971: 100.9 mhz; 28 kw. 656 ft TL: N40 07 33 W91 31 42. Hrs open: 24 408 N. 24th St., Quincy, IL, 62301. Phone: (217) 223-5292. Fax: (217) 223-5299. E-mail: mail@y101radio.com Web Site: www.y101radio.com. Licensee: Bick Broadcasting. Population served: 150,000 Natl. Rep: McGavren Guild. Format: Hot adult contemp. ♦Terry Bond, pres; Jeff Dorsey, gen mgr; Cheri Robertson, gen sls mgr; Dennis Oliver, progmg dir; Gary Gleanser, chief of engrg.

Cape Girardeau

KAPE(AM)— 1951: 1550 khz; 5 kw-D, 50 w-N, DA-2. TL: N37 16 45 W89 33 28. Hrs open: Box 558, 63702. Secondary address: 901 S. Kings Hwy. 63703. Phone: (573) 339-7000. Fax: (573) 651-4100. Licensee: Withers Broadcasting Co. of Missouri LLC. Group owner: Withers Broadcasting Co. (acq 6-72). Population served: 275,000 Natl. Network: Westwood One, Fox Sports. Natl. Rep: Katz Radio. Wire Svc: AP Format: News, talk, sports. Target aud: 25-54; active, aware adults. ♦W. Russell Withers Jr., pres; Rick Lambert, gen mgr; Chris Cook, progmg dir; Connie Hanner, news dir; Smokey King, chief of engrg.

KGMO(FM)—Co-owned with KAPE(AM). Mar 17, 1969: 100.7 mhz; 100 kw. 987 ft TL: N37 22 16 W89 31 52. Stereo. 800,000 Natl. Rep: Katz Radio. Format: Classic rock. ♦Chris Cook, prom mgr.

KCGQ-FM—Listing follows KGIR(AM).

KEZS-FM— Dec 10, 1970: 102.9 mhz; 100 kw. 947 ft TL: N37 24 23 W89 33 44. Stereo. Hrs opn: 24 Box 1610, 324 Broadway, 63702. Phone: (573) 335-8291. Fax: (573) 335-4806. E-mail: k103@zrgmail.com Web Site: www.k103fm.com. Licensee: MRR License LLC. Group owner: MAX Media L.L.C. (acq 6-2-2004; grpsl). Law Firm: Fletcher, Heald & Hildreth. Format: Country. News staff: one. Target aud: 25-54. ♦Steve Stephenson, gen mgr; Whitney Thomas, opns dir.

KZIM(AM)—Co-owned with KEZS-FM. 1925: 960 khz; 5 kw-D, 500 w-N, DA-N. TL: N37 18 59 W89 29 06.24 E-mail: kzim@zrgmail.com Web Site: www.960kzim.com.150,000 Natl. Network: CBS. Format: News/talk. Target aud: 35-64.

KGIR(AM)— June 10, 1966: 1220 khz; 250 w-D, 140 w-N. TL: N37 18 03 W89 29 27. Stereo. Hrs opn: 24 324 Broadway, 63701. Phone: (573) 335-8291. Fax: (573) 335-4806. E-mail: kgir@kgir.com Web Site: www.espn1220.com. Licensee: MRR License LLC. Group owner: MAX Media L.L.C. (acq 6-2-2004; grpsl). Population served: 200,000 Natl. Network: ESPN Radio. Law Firm: Leventhal, Senter & Lerman. Format: Sports talk. Target aud: 18 plus; men. ♦Steve Stephenson, gen mgr & sls dir; Erik Sean, progmg dir.

KCGQ-FM—Co-owned with KGIR(AM). 1978: 99.3 mhz; 5 kw. Ant 358 ft TL: N37 21 34 W89 37 16. Stereo. 24 Phone: (573) 335-8291. Web Site: www.realrock993.com. Format: Real rock. Target aud: 18-49; general. ♦Scott Hartline, progmg dir.

***KRCU(FM)**— Mar 3, 1976: 90.9 mhz; 6 kw. Ant 259 ft TL: N37 18 37 W89 31 57. Stereo. Hrs opn: One University Plaza, 63701. Phone: (573) 651-5070. Fax: (573) 651-5071. E-mail: comments@krcu.org Web Site: www.southeastpublicradio.org. Licensee: Board of Regents of Southeast Missouri State University. Population served: 102,106 Natl. Network: NPR, PRI. Law Firm: Dow, Lohnes & Albertson. Format: Jazz, news, classical. Target aud: General. ♦Dan Woods, gen mgr & opns dir; Jason Brown, opns dir; Amanda Lincoln, dev dir; Allen Lane, chief of engrg.

KREZ(FM)—Chaffee, July 1, 1990: 104.7 mhz; 7.7 kw. Ant 585 ft TL: N37 09 46 W89 28 59. Stereo. Hrs opn: 901 S. Kings Hwy., 63702-0558. Phone: (573) 339-7000. Fax: (573) 651-4100. Licensee: Dana R. Withers (acq 4-12-90; $33,587; FTR: 5-7-90). Natl. Rep: Katz Radio. Format: Adult contemp. Target aud: 18-49. ♦Rick Lambert, gen mgr.

Carrollton

KAOL(AM)— Apr 18, 1959: 1430 khz; 500 w-U, 27 w-N. TL: N39 19 58 W93 32 15. Hrs open: 24 KMZU Bldg., 102 N. Mason, 64633. Phone: (660) 542-0404. Fax: (660) 542-0420. E-mail: kmzu@carolnet.com Web Site: www.kmzu.com. Licensee: Kanza Inc. (acq 11-1-81; $665,000 with co-located FM; FTR: 11-23-81) Population served: 35,700 Natl. Rep: McGavren Guild. Format: Country, farm. News staff: 2; News: 10 hrs wkly. Target aud: 25-54; farm families & those with agricultural backrounds. Spec prog: Sp 3 hrs wkly. ♦Miles Carter, gen mgr; Rick Barton, gen sls mgr; Scott Powell, progmg dir; Jim Woods, mus dir; Chastity Anderson, news dir; Larry Tannons, chief of engrg; Sue Lightfoot, traf mgr.

KMZU(FM)—Co-owned with KAOL(AM). July 13, 1962: 100.7 mhz; 98.6 kw. 990 ft TL: N39 22 05 W93 29 40. Stereo.
Rebroadcasts WHB(AM) Kansas City 95%.
Web Site: www.kmzu.com.800,000 ♦Larry Timmons, engrg dir; Lynn Hammond, farm dir, women's int ed & women's cmtr.

Carthage

KDMO(AM)— June 3, 1947: 1490 khz; 1 kw-U. TL: N37 10 58 W94 21 43. Hrs open: 24 Box 426, 221 E. 4th St., 64836. Phone: (417) 358-6054. Phone: (417) 358-2648. Fax: (417) 358-1278. Licensee: Ronald L. Petersen. (acq 1-23-90). Population served: 397,000 Natl. Network: CNN Radio. Rgnl. Network: Missourinet, Brownfield. Format: Adult standards. News staff: one; News: 10 hrs wkly. Target aud: 55 plus. Spec prog: Sp 6 hrs wkly. ♦Ronald L. Petersen, pres & gen mgr.

KMXL(FM)—Co-owned with KDMO(AM). Jan 10, 1972: 95.1 mhz; 50 kw. 472 ft TL: N37 10 58 W94 21 35. Stereo. 24 E-mail: traffic@cbciradio.com Web Site: www.951mikefm.com.775,000 Format: Adult hits. News: one hr wkly. Target aud: 25-54; young adults & baby-boomers.

Carthage-

KKLL(AM)— Mar 10, 1984: 1100 khz; 5 kw-D. TL: N37 06 23 W94 16 50. Hrs open: 831 Moffitt, Joplin, 64801. Phone: (417) 781-1100. Fax: (417) 781-1100. Licensee: New Life Evangelistic Center Inc. (acq 8-10-98; $730,000 with KWAS(AM) Joplin). Format: Christian. ♦Charlie Hale, opns dir.

Caruthersville

KCRV(AM)— Feb 22, 1950: 1370 khz; 1 kw-D, 63 w-N. TL: N36 12 50 W89 41 25. Hrs open: Box 509, Kennett, 63857-0509. Secondary address: 1303 Southwest Dr. 63857. Phone: (573) 888-4616. Fax: (573) 888-4991. Licensee: Pollack Broadcasting Co. (group owner; (acq 9-21-99; with co-located FM). Population served: 23,000 Natl. Network: Moody. Rgnl. Network: Prog Farm, Brownfield. Format: Country. Target aud: General; residents of Pemiscot county. Spec prog: Relg 20 hrs wkly. ♦Perry Jones, gen mgr & progmg dir; Bill Page, news dir; P.J. Johnson, chief of engrg; Evelyn Bailey, traf mgr.

KCRV-FM— Apr 28, 1975: 105.1 mhz; 3 kw. 200 ft TL: N36 12 50 W89 41 25. (CP: 6 kw, ant 328 ft.). Stereo. 23,000 Format: Oldies.

Cassville

KRMO(AM)—Licensed to Cassville. See Monett

Cedar Hill

***KNLH(FM)**— October 1998: 89.5 mhz; 68 w. 699 ft TL: N38 21 40 W90 32 54. Hrs open: New Life Evangelistic Center Inc., 1411 Locust St., St. Louis, 63103. Phone: (314) 436-2424. Fax: (314) 436-2434. E-mail: larryr@hereshelpnet.org Web Site: www.hereshelpnet.org. Licensee: New Life Evangelistic Center Inc. Format: Adult contemp, gospel, talk. ♦Victor Anderson, gen mgr.

Centralia

KMFC(FM)— Feb 3, 1986: 92.1 mhz; 1.85 kw. Ant 418 ft TL: N39 09 58 W92 09 52. Stereo. Hrs opn: 24 1249 E. Hwy. 22, 65240. Phone: (573) 682-5525. Fax: (573) 682-2744. E-mail: info@kmfc.com Web Site: www.kmfc.com. Licensee: Clair Broadcasting Co. Population served: 260,000 Natl. Network: USA. Format: Relg, Christian contemp. Target aud: 25-50. Spec prog: Black 3 hrs, gospel 2 hrs, Sp one hr wkly. ♦Jerry D. Clair, pres & gen mgr; Sharon Dollens, stn mgr.

Chaffee

KREZ(FM)—Licensed to Chaffee. See Cape Girardeau

Charleston

KCHR(AM)— 1953: 1350 khz; 1 kw-D, 79 w-N. TL: N36 55 30 W89 17 45. Hrs open: 24 205 E. Commercial St., 63834. Phone: (573) 683-6044. Licensee: South Missouri Broadcasting Co. Inc. Population served: 5,200 Format: C&W, talk. News: one hr wkly. Target aud: General. Spec prog: Gospel 10 hrs, easy lstng 5 hrs wkly. ♦James L. Byrd III, pres; Danny Adams, gen mgr.

KWKZ(FM)— February 1993: 106.1 mhz; 34 kw. 384 ft TL: N36 57 29 W89 23 38. Stereo. Hrs opn: 24 753 Enterprise, Cape Girardeau, 63703. Phone: (573) 334-7800. Fax: (573) 334-7440. Web Site: www.kwkz.com. Licensee: Anderson Broadcasting Co. Inc. (acq 7-30-92). Format: Country, oldies. News: 2 hrs wkly. Target aud: 18-44; 35-55 male, 30-45 female. Spec prog: Farm one hr, gospel 6 hrs wkly. ♦Bill Anderson, CEO & gen mgr; Ann Anderson, chmn; Palmer Johnson, engrg dir & edit dir; Susan Bell, progmg dir & rsch dir.

Chillicothe

KCHI(AM)— Mar 3, 1950: 1010 khz; 250 w-D, 37 w-N. TL: N39 45 51 W93 33 21. Hrs open: 421 Washington St., 64601. Phone: (660) 646-4173. Fax: (660) 646-2868. Web Site: www.kchi.com. Licensee: Livingston Broadcasting Inc. (acq 7-1-84). Population served: 9,519 Rgnl. Network: Missourinet. Format: Todays news & yesterdays music. Target aud: 35-49. ♦Dan Leatherman, gen mgr; Randy Dean, progmg dir; Tom Tingerthal, news dir.

KCHI-FM— October 1976: 103.9 mhz; 4.1 kw. Ant 400 ft TL: N39 48 52 W93 35 20. Stereo. Web Site: www.kchi.com.

***KRNW(FM)**— Aug 30, 1993: 88.9 mhz; 38 kw. 498 ft TL: N39 48 50 W93 35 20. Stereo. Hrs opn: 24 Wells Hall, 800 University Dr., Maryville, 64468. Phone: (660) 562-1163. Phone: (660) 562-1164. Fax: (660) 562-1832. E-mail: kxcv@mail.nwmissouri.edu Web Site: www.kxcv.org. Licensee: Northwest Missouri State University. Population served: 64,000 Format: News, class, jazz. News staff: 2; News: 39 hrs wkly. Target aud: General. ♦Dean L. Hubbard, pres; Patty Andrews Holley, gen mgr, stn mgr & opns mgr.

Clayton

KFUO-FM— Jan 1, 1948: 99.1 mhz; 100 kw. 1,026 ft TL: N38 39 08 W90 17 03. Stereo. Hrs opn: 24 85 Founders Ln., St. Louis, 63105. Phone: (314) 725-0099. Fax: (314) 725-3801. E-mail: classic99@classic99.com Web Site: www.classic99.com. Licensee: Lutheran Church-Missouri Synod. Natl. Network: Wall Street, CNN Radio. Natl. Rep: Interep. Law Firm: Shaw Pittman. Format: Class. Target aud: General; upscale, educated. ♦Dennis Stortz, opns dir & chief of engrg; Oliver Trittler, sls dir & gen sls mgr; Jim Connett, progmg dir.

KFUO(AM)— Dec 14, 1924: 850 khz; 5 kw-D. TL: N38 38 20 W90 18 57.Sunrise-sunset Phone: (314) 725-3030. Fax: (314) 725-2538. E-mail: webmaster@kfuo.org Web Site: www.kfuo.org.1,625,236 Format: Relg, talk. Target aud: General.

KSIV(AM)— 1320 khz; 4.6 kw-D, 270 w-N, DA-N. TL: N38 36 26 W90 21 14. Hrs open: 24 1750 S. Brentwood Blvd., Suite 811, St. Louis, 63144. Phone: (314) 961-1320. Fax: (314) 961-7562. Web Site: www.bottradionetwork.com. Licensee: Bott Broadcasting. (group owner; acq 2-25-82; FTR: 3-15-82). Population served: 3,000,000 Natl. Network: USA. Format: Christian info. Target aud: 25-54; family-oriented. ♦Richard P. Bott, pres; Richard Bott II, VP; Michael McHardy, gen mgr; Joy Elder, sls dir & mktg dir.

***KWUR(FM)**— July 4, 1976: 90.3 mhz; 10 w. 136 ft TL: N38 38 45 W80 19 07. Stereo. Hrs opn: 24 Washington Univ. Box 1205, One Brookings Dr., St. Louis, 63105. Phone: (314) 935-5952. Fax: (314) 935-8833. Web Site: www.kwur.com. Licensee: Washington University. Format: Progressive/diversified. Target aud: 18 plus; those seeking alternative radio. ♦John Klacsmann, gen mgr.

Cleveland

KCTO(AM)— Nov 8, 2007: 1160 khz; 500 w-D, 230 w-N, DA-2. TL: N38 40 26 W94 36 28. Hrs open: 310 S. La Frenz Rd., Liberty, 64068. Phone: (816) 792-1140. Fax: (816) 792-8258. Licensee: Alpine Broadcasting Corp. Format: Talk. ♦Peter E. Schartel, pres & gen mgr; Jonne Santoli, adv mgr.

Clinton

KDKD(AM)— 1951: 1280 khz; 1 kw-D, 58 w-N. TL: N38 23 55 W93 46 19. Hrs open: 24 Box 448, 64735. Secondary address: 2201 N. Antioch Rd. Phone: (660) 885-6141. Fax: (660) 885-4801. Web Site: www.kdkd.net. Licensee: Legend Communications of Missouri LLC. Group owner: Legend Communications L.L.C. (acq 10-7-03; with co-located FM). Population served: 230,000 Rgnl. Network: Missourinet. Format: Oldies. News staff: one; News: 10 hrs wkly. Target aud: 25-55. ♦Bob May, gen mgr.

KDKD-FM— 1975: 95.3 mhz; 14.5 kw. Ant 433 ft TL: N38 22 18 W93 55 06. Stereo. 24 Web Site: www.kdkd.net. Licensee: Legend Communications of Missouri LLC. (acq 5-28-1985).250,000 Natl. Network: ABC, Motor Racing Net. Format: Hot new country, news, sports. News staff: one. Target aud: 25-55.

***KLRQ(FM)**— Oct 5, 1990: 96.1 mhz; 100 kw. Ant 987 ft TL: N38 28 27 W93 30 28. Hrs open: 24
Rebroadcasts KLVR(FM) Santa Rosa, CA 100%.
2351 Sunset Blvd., Suite 170-218, Rocklin, CA, 95765. Phone: (916) 251-1600. Fax: (916) 251-1650. Web Site: www.klove.com. Licensee: Educational Media Foundation. Group owner: EMF Broadcasting (acq 12-23-2003; $1.9 million). Population served: 920,000 Natl. Network: K-Love. Format: Contemp Christian. ♦Richard Jenkins, pres; Mike Novak, VP & progmg dir; Lloyd Parker, gen mgr; Ed Lenane, opns dir & news dir; Keith Whipple, dev dir; Eric Allen, natl sls mgr; David Pierce, progmg mgr; Jon Rivers, mus dir; Sam Wallington, engrg dir; Arthur Vassar, traf mgr; Karen Johnson, news rptr; Marya Morgan, news rptr; Richard Hunt, news rptr.

Columbia

***KBIA(FM)**— 1972: 91.3 mhz; 100 kw. 610 ft TL: N38 53 16 W92 15 48. Stereo. Hrs opn: 24 Univ. of Missouri, 409 Jesse Hall, 65211. Phone: (573) 882-3431. Fax: (573) 882-2636. E-mail: dunnm@missouri.edu Web Site: www.kbia.org. Licensee: Board of Curators, University of Missouri. Group owner: The Curators of the University of Missouri Population served: 300,000 Natl. Network: NPR, PRI. Law Firm: Fisher, Wayland, Cooper, Leader & Zaragoza. Format: News, class. News staff: 2; News: 55 hrs wkly. Target aud: 25-64. ♦Michael Dunn, gen mgr; Roger Karwoski, stn mgr; John Bailey, dev dir & progmg dir.

KBXR(FM)— Nov 11, 1994: 102.3 mhz; 88 kw. 420 ft TL: N38 57 21 W92 16 24. Stereo. Hrs opn: 24 503 Old 63 N., 65201. Phone: (573) 449-4141. Fax: (573) 449-7770. E-mail: bxr@bxr.com Web Site: www.bxr.com. Licensee: Cumulus Licensing LLC. Group owner: Cumulus Media Inc. (acq 4-26-2004; grpsl). Population served: 110,000 Natl. Rep: Katz Radio. Format: Album adult alternative. News staff: 3; News: one hr wkly. Target aud: 29-59; educated professional/technical. ♦Lewis W. Dickey Jr., pres; Scott Boltz, gen mgr & mktg mgr.

KCMQ(FM)— Dec 3, 1967: 96.7 mhz; 18 kw. 344 ft TL: N38 41 30 W92 05 44. (CP: 98 kw, ant 912 ft. TL: N38 41 30 W92 05 44). Stereo. Hrs opn: 24 3215 LeMone Industrial, Suite 200, 65201. Phone: (573) 875-1099. Fax: (573) 875-2439. Web Site: www.kcmq.com. Licensee: Zimmer Radio of Mid-Missouri Inc. (acq 7-15-93; $625,000 with co-located AM; FTR: 8-9-93) Population served: 130,000 Format: AOR. News: one hr wkly. Target aud: 25-54; male. ♦John Zimmer, chmn; Dave Wisniewski, sls dir; Jeff Studley, engrg mgr & chief of engrg.

KTGR(AM)—Co-owned with KCMQ(FM). 1955: 1580 khz; 214 w-D, 8 w-N. TL: N38 57 45 W92 18 14. Web Site: ktgr.com.130,000 Natl. Network: ABC, ESPN Radio. Format: Sports. Target aud: 18-34; male.

***KCOU(FM)**— Oct 31, 1973: 88.1 mhz; 435 w. 110 ft TL: N38 56 23 W92 19 20. Stereo. Hrs opn: 24 Univ. of Missouri, 101-F Pershing Hall, 65201. Phone: (573) 882-7820. Fax: (573) 882-6262. E-mail: kcou@mu.org Web Site: www.kcou.mu.org. Licensee: The Curators of the University of Missouri. (acq 12-14-98; $80,000). Population served: 75,000 Format: Progsv, rock. Target aud: 18-22; students & community members. ♦Jon Wujcik, gen mgr.

KFRU(AM)— Oct 10, 1925: 1400 khz; 1 kw-U. TL: N38 57 52 W92 18 26. Hrs opn: 24 503 Old Hwy., 63 N., 65201. Phone: (573) 449-4141. Fax: (573) 449-7770. Fax: (573) 499-1414. E-mail: news@kfru.com Web Site: www.kfru.com. Licensee: Cumulus Licensing LLC. Group owner: Cumulus Media Inc. (acq 4-26-2004; grpsl). Population served: 110,000 Rgnl. Network: Missourinet. Format: News/talk. News staff: 9; News: 40 hrs wkly. Target aud: General. ♦Lewis W. Dickey Jr., pres; Scott Boltz, mktg mgr.

***KOPN(FM)**— Mar 1, 1973: 89.5 mhz; 36 kw. 236 ft TL: N38 59 53 W92 11 48. (CP: 36.4 kw). Stereo. Hrs opn: 24 915 E. Broadway, 65201-4857. Phone: (573) 874-1139. Fax: (573) 499-1662. E-mail: mail@kopn.org Web Site: www.kopn.org. Licensee: New Wave Corp. Population served: 200,000 Natl. Network: NPR, PRI. Law Firm: Haley, Bader & Potts. Format: News/talk, Americana. News staff: News progmg 40 hrs wkly Target aud: 25-54; well educated, upwardly mobile. Spec prog: Blues 13 hrs, Black 10 hrs, AAA 10 hrs, Grateful Dead 6 hrs, jazz 4 hrs, gospel 3 hrs, bluegrass 6 hrs, folk 2 hrs wkly. ♦David Owens, gen mgr; Julie Baka, dev dir; Steve Jerrett, mus dir; Rich Winkel, chief of engrg.

KPLA(FM)— Feb 23, 1983: 101.5 mhz; 100 kw. 1,062 ft TL: N39 00 52 W92 16 32. Stereo. Hrs opn: 24 503 Old 63 N., 65201. Phone: (573) 449-4141. Fax: (573) 449-7770. E-mail: studio@kpla.com Web Site: www.kpla.com. Licensee: Cumulus Licensing LLC. Group owner: Cumulus Media Inc. (acq 4-26-2004; grpsl). Population served: 500,000 Format: Adult contemp. News staff: 2; News: one hr wkly. Target aud: 25-54. ♦Lewis W. Dickey Jr., pres; Scott Boltz, mktg mgr.

***KWWC-FM**— Feb 2, 1965: 90.5 mhz; 1.25 kw. 131 ft TL: N38 57 12 W92 19 05. Stereo. Hrs opn: 24 Box 2114, Stephens College, 65215. Phone: (573) 876-7297. Phone: (573) 876-7272. Fax: (573) 876-2330. E-mail: jwise@stephens.edu Licensee: Stephens College. Population served: 655,000 Format: Jazz, 80s hits, oldies, eclectic. Target aud: 25-60; college educated, professional or retired with middle upper income. ♦John Blakemore, gen mgr & mktg dir; Jonna Wiseman, gen mgr; Mark Smith, gen mgr; Max Ornles, chief of engrg.

Concordia

***KYRV(FM)**— 1998: 88.1 mhz; 1 kw. Ant 213 ft TL: N38 52 10 W93 32 58. Hrs open: 712 Chaucer Ln., Warrensburg, 64093. Phone: (660) 747-4155. Fax: (660) 747-4155. Licensee: Full Smile Inc. Format: Southern gospel. ♦Jim McCollum, gen mgr.

Country Club

***KJCV(FM)**— 2005: 89.7 mhz; 3.9 kw. Ant 548 ft TL: N39 42 35 W95 02 33. Hrs open: Bott Radio Network, 10550 Barkley, Overland Park, KS, 66212. Phone: (913) 642-7600. Fax: (913) 642-1319. Web Site: www.bottradionetwork.com. Licensee: Community Broadcasting Inc. Group owner: Bott Radio Network. Format: Relg.

Crestwood

KSHE(FM)— Feb 11, 1961: 94.7 mhz; 100 kw. 1,019 ft TL: N38 34 24 W90 19 30. Stereo. Hrs opn: 24 The Powerhouse, 800 St. Louis Union Stn., St. Louis, 63103. Phone: (314) 621-0095. Fax: (314) 621-3428. Web Site: www.kshe95.com. Licensee: Emmis Radio License LLC. Group owner: Emmis Communications Corp. (acq 3-19-84; grpsl; FTR: 1-30-84). Natl. Rep: D & R Radio. Format: Classic rock, AOR. News staff: one; News: one hr wkly. Target aud: 18-40. ♦John R. Beck Jr., gen mgr.

Cuba

KESY(FM)— 2005: 107.3 mhz; 6.7 kw. Ant 626 ft TL: N37 55 17 W91 26 36. Hrs open: 3418 Douglas Rd., Florissant, 63034. Phone: (314) 921-9330. Licensee: Twenty-One Sound Communications Inc. (acq 11-8-2004; $400,000 for CP). Format: Country. ♦Ruth Choate, gen mgr.

***KGNN-FM**— Jan 26, 1997: 90.3 mhz; 6.3 kw. Ant 325 ft TL: N38 05 11 W91 18 30. Hrs open: 24
Rebroadcasts KGNV(FM) Washington 100%.
Box 187, Washington, 63090-0187. Phone: (636) 239-0400. Fax: (636) 239-4448. Web Site: goodnewsvoice.org. Licensee: Missouri River Christian Broadcasting Inc. Population served: 20,000 Natl. Network: Moody, Salem Radio Network. Format: News/talk, southern gospel, inspirational. News: 14 hrs wkly. Target aud: 20-70; inquisitive, conservative, philosophical, liberal. Spec prog: Children 7 hrs wkly. ♦James Goggan, pres.

***KNLQ(FM)**— 2004: 91.9 mhz; 5 kw. Ant 249 ft TL: N38 02 14 W91 23 04. Hrs open: New Life Evangelistic Center Inc., 1411 Locust St., St. Louis, 63103. Phone: (314) 421-3020. Fax: (314) 436-2434. Web Site: www.hereshelpnet.org. Licensee: New Life Evangelistic Center Inc. Format: Gospel. ♦Rick Jesse, stn mgr.

De Soto

KDJR(FM)— Jan 29, 1991: 100.1 mhz; 2 kw. Ant 348 ft TL: N38 01 25 W90 34 02. Stereo. Hrs opn: Box 262550, Baton Rouge, LA, 70826. Secondary address: 8919 World Ministry Ave., Baton Rouge, LA 70810. Phone: (225) 768-3688. Phone: (225) 766-8300. Fax: (225) 768-3729. E-mail: kawikfish@yahoo.com Web Site: www.jsm.org. Licensee: Family Worship Center Church Inc. (acq 9-27-2005; $1.25 million). Format: Relg. ♦David Whitelaw, COO & gen mgr; Jimmy Swaggart, pres; John Santiago, progmg dir.

KRFT(AM)— Nov 1, 1968: 1190 khz; 10 kw-D, 22 w-N, DA-2. TL: N38 42 25 W90 03 10. Hrs open: Sunrise-sunset 8045 Big Bend Blvd., St. Louis, 63119. Phone: (314) 962-0590. Fax: (314) 962-7576. Web Site: www.kfns.com. Licensee: Big Stick Three LLC. Group owner: Big League Broadcasting LLC (acq 7-13-2004; grpsl). Natl. Network: Motor Racing Net. Format: Sports. Target aud: 25-64; sports fans, men 25-54. ♦Mike Phares, gen mgr.

Deerfield

KBZI(FM)— May 18, 2000: 100.7 mhz; 17.5 kw. Ant 390 ft TL: N37 43 08 W94 40 06. Hrs open: American Media Investments Inc., 1162 E. Hwy 126, Pittsburg, KS, 66762. Phone: (620) 231-7200. Fax: (620) 231-3321. Web Site: www.kbzi.com. Licensee: American Media Investments Inc. Format: Adult contemp. ♦Chris Kelly, gen mgr.

Dexter

KDEX-FM— July 17, 1969: 102.3 mhz; 6 kw. 279 ft TL: N36 47 18 W89 54 22. Stereo. Hrs opn: 24 Box 249, 20487 State Hwy. 114, 63841. Phone: (573) 624-3545. Fax: (573) 624-9926. E-mail: kdex1@dexter.net Licensee: Dexter Broadcasting Inc. (acq 7-15-88). Rgnl. Network: Prog Farm, Yancey Action. Law Firm: Shaw Pittman. Format: Modern country. News staff: 2; News: 5 hrs wkly. Target aud: 25-54. ♦Tony James, opns dir & progmg dir; Joeli Barbour, natl sls mgr & traf mgr; Walt Turner, gen mgr, gen sls mgr & rgnl sls mgr; Dave Obergoenner, chief of engrg.

KDEX(AM)— Feb 1, 1956: 1590 khz; 620 w-D, 78 w-N. TL: N36 47 20 W89 54 28.5 AM-midnight Box 249, 63841. Secondary address: 20487 State Hwy 114 63841. Phone: (573) 624-3545. Fax: (573) 624-9926. E-mail: kdexl@dexter.net 75,000

Dixon

***KCVZ(FM)**— May 2003: 92.1 mhz; 6 kw. Ant 328 ft TL: N37 57 59 W92 10 03. Stereo. Hrs opn:
Rebroadcasts KCVO-FM Camdenton 100%.
Box 800, Camdenton, 65020. Phone: (573) 346-3200. Fax: (573) 346-1010. E-mail: spiritfm@spiritfm.org Web Site: www.spiritfm.org. Licensee: Lake Area Educational Broadcasting Foundation (acq 12-20-01). Population served: 80,000 Format: Div, Christian. ♦Alice McDermott, CFO; James McDermott, pres & gen mgr.

Doniphan

KDFN(AM)— Feb 4, 1963: 1500 khz; 2.5 kw-D, DA. TL: N36 36 53 W90 49 23. Hrs open: 932 Country Rd. 448, Poplar Bluff, 639301. Phone: (573) 686-3700. Fax: (573) 686-1713. Licensee: Eagle Bluff Enterprises. (group owner; acq 9-8-99; grpsl). Population served: 10,000 Rgnl. Network: Missourinet. Format: Oldies. Target aud: General. Spec prog: Farm 5 hrs wkly. ♦Steven Fuchs, gen mgr; Shelley Fuchs, progmg dir; Ken Hosler, news dir; Maria Tillman, traf mgr.

KOEA(FM)—Co-owned with KDFN(AM). Apr 11, 1975: 97.5 mhz; 50 kw. 577 ft TL: N36 35 20 W90 49 10. Stereo. 25,000 Format: Country. ♦Steven Fuchs, gen sls mgr; Tammy Jameson, mktg mgr; Skeet Collins, progmg dir.

Doolittle

KUMR(FM)—Not on air, target date: unknown: 104.5 mhz; 3.9 kw. Ant 407 ft TL: N37 56 21.3 W91 56 44.9. Hrs open: 1282 Smallwood Dr., Suite 372, Waldorf, MD, 20603. Phone: (202) 251-7589. Licensee: Alma Corp. ♦Dennis Wallace, pres.

East Prairie

KYMO(AM)— Nov 15, 1965: 1080 khz; 500 w-D. TL: N36 47 42 W89 21 17. (CP: TL: N36 47 49 W89 21 19). Hrs opn: Box 130, 63845. Secondary address: 390 S. Hwy. 102 63845. Phone: (573) 649-3597. Fax: (573) 649-3983. E-mail: kymo@bootheel.net Licensee: Usher Broadcasting Inc. (acq 6-1-69). Population served: 15,000 Format: Easy Listening. ♦Barney L. Webster, pres, gen mgr & gen sls mgr; Michael Bennett, opns mgr.

KYMO-FM— Aug 5, 1991: 105.3 mhz; 3 kw. 207 ft TL: N36 47 49 W89 21 19. Format: Oldies.

El Dorado Springs

KESM(AM)— July 18, 1961: 1580 khz; 500 w-D. TL: N37 51 51 W94 00 54. Stereo. Hrs opn: 200 Radio Ln., 64744. Phone: (417) 876-2741. Fax: (417) 876-2743. Licensee: Wildwood Communications Inc. (acq 12-17-85). Population served: 44,500 Format: C&W, oldies. Target aud: General. ♦Donald Kohn, pres & gen mgr; Jena Worthington, stn mgr.

KESM-FM— June 1, 1965: 105.5 mhz; 6.0 kw. 187 ft TL: N37 51 51 W94 00 54. Stereo. (Acq 12-17-85; $200,000; FTR: 11-11-85).

Eldon

KLOZ(FM)— July 1, 1979: 92.7 mhz; 50 kw. Ant 620 ft TL: N38 20 27 W92 35 33. Stereo. Hrs opn: 24 160 Hwy 42, Kaiser, 65047. Phone: (573) 348-1958. Fax: (573) 348-1923. E-mail: mike@mix927.com Web Site: www.todaysbesthits.com. Licensee: Benne Broadcasting Co. L.L.C. Population served: 500,000 Natl. Network: ABC. Format: Hot AC. Target aud: 25-54; 70% female/30% male with average or above income. ♦Denny Benne, gen mgr; Greg Sullens, stn mgr & news dir; Mike Clayton, progmg dir; Dan Yeager, engrg mgr.

KZWV(FM)— 2006: 101.9 mhz; 42.7 kw. Ant 528 ft TL: N38 16 46 W92 35 06. Stereo. Hrs opn: 24 1081 Osage Beach Rd., Osage Beach, 65065-2232. Phone: (573) 746-7873. E-mail: jcaran@1019thewave.com Web Site: www.1019thewave.com. Licensee: Randall C. Wright. Format: Adult contemp w/ smooth jazz. Target aud: 25 plus; affluent adults. ♦John Caran, gen mgr & sls dir; Steve Richards, progmg dir; Stacy Johnson, news dir; Jessica Brink, traf mgr.

Ellington

KAUL(FM)— 1999: 106.7 mhz; 3 kw. 298 ft TL: N37 13 58 W90 51 08. Hrs open: 1411 Locust St., St. Louis, 63103. Phone: (314) 421-3020. Fax: (314) 436-2434. E-mail: larryr@hereshelpnet.org Web Site: www.hereshelpnet.org. Licensee: New Life Evangelistic Center. Format: Adult contemp, gospel, pub affrs. ♦Larry Rice, gen mgr; Judy Redlich, sls dir.

Excelsior Springs

***KEXS(AM)**— August 1968: 1090 khz; 1 kw-D. TL: N39 20 25 W94 14 26. Hrs open: 201 N. Industrial Park Rd., 64024. Phone: (816) 630-1090. Web Site: www.kexs.com. Licensee: Kansas City Catholic Network Inc. (acq 5-17-2004; $825,000). Population served: 500,000 Natl. Network: USA. Format: Catholic radio. Target aud: 25-54. ♦James E. O'Laughlin, pres & gen mgr.

Farmington

KREI(AM)— Dec 7, 1947: 800 khz; 1 kw-D, 150 w-N. TL: N37 47 45 W90 24 30. Hrs open: 24 Box 461, 63640. Secondary address: 1401 KREI Blvd. 63640. Phone: (573) 756-6476. Fax: (573) 756-1110. Web Site: www.krei.com. Licensee: KREI Inc. Group owner: Shepherd Group (acq 7-1-82; $160,000 with co-located FM; FTR: 5-17-82) Population served: 500,000 Rgnl. Network: Missourinet. Format: News/talk. News staff: 12; News: 40 hrs wkly. Target aud: General. Spec prog: Farm 5 hrs wkly. ♦Richard Womack, gen mgr; Kimberly Long, stn mgr; Scott Kubala, progmg dir; Kevin Brooks, chief of engrg.

KTJJ(FM)—Co-owned with KREI(AM). June 5, 1977: 98.5 mhz; 100 kw. 1,040 ft TL: N37 43 07 W90 33 01. Stereo. 24 Web Site: www.j98.com.750,000 Format: Country. News: 16 hrs wkly. Target aud: General.

***KSEF(FM)**— Sept 14, 2006: 88.9 mhz; 9.5 kw. Ant 640 ft TL: N37 47 57 W90 33 43. Stereo. Hrs opn:
Rebroadcasts KRCU(FM) Cape Girardeau 100%.
Southeast Missouri State University, One University Plaza, Cape Girardeau, 63701. Phone: (573) 651-5070. Fax: (573) 651-5071. E-mail: comments@krcu.org Web Site: www.southeastpublicradio.org. Licensee: Board of Regents, Southeast Missouri State University. Population served: 109,918 Natl. Network: NPR, PRI. Law Firm: Dow, Lohres & Albertson. Format: Classical, jazz, news. ♦Dan Woods, gen mgr; Jason Brown, opns dir; Amanda Lincoln, dev dir; Allen Lane, chief of engrg.

Fayette

KSSZ(FM)— July 15, 1994: 93.9 mhz; 25 kw. 328 ft TL: N39 03 28 W92 28 49. Stereo. Hrs opn: 24 3215 Lemone Industrial Blvd., Suite 200, Columbia, 65201. Phone: (573) 875-1099. Fax: (573) 875-2439. E-mail: eagle939@zrgmail.com Web Site: www.939theeagle.com. Licensee: Zimmer Radio of Mid-Missouri Inc. (acq 9-27-96; $550,000). Population served: 250,000 Natl. Network: ABC, Jones Radio Networks, Westwood One. Wire Svc: AP Format: News/talk. News staff: 2; News: 4 hrs wkly. Target aud: 25-54; adults. ♦Derek Gilbert, gen mgr & progmg dir.

Ferguson

***KCFV(FM)**— Apr 17, 1972: 89.5 mhz; 100 w. 159 ft TL: N38 46 07 W90 17 16. Stereo. Hrs opn: 16 3400 Pershall Rd., St. Louis, 63135-1499. Phone: (314) 513-4472. Phone: (314) 513-4478. Fax: (314) 513-4217. E-mail: dkirby@stlcc.edu Web Site: www.stlcc.edu/fv/kcfv. Licensee: St. Louis Community College District. Population served: 100,000 Law Firm: Dow, Lohnes & Albertson. Format: Hot ac, rhythmic. News: 2 hrs wkly. Target aud: General. Spec prog: Jazz 8 hrs, Black 4 hrs, hard rock 4 hrs wkly. ♦Dianna L. Kirby, gen mgr; Tim Croskey, chief of engrg.

Festus

KJFF(AM)— May 10, 1951: 1400 khz; 1 kw-U. TL: N38 13 56 W90 23 50. Hrs open: 24 Box 368, 63028. Phone: (636) 937-7642. Fax: (636) 937-3636. Web Site: www.kjff.com. Licensee: KREI Inc. Group owner: Shepherd Group (acq 2-1-89; $230,000; FTR: 2-1-89). Population served: 250,000 Natl. Network: ABC. Rgnl. Network: Missourinet. Format: News/talk. News staff: 4; News: 30 hrs wkly. Target aud: General. ♦David Shepherd, pres; Dick Womack, gen mgr; Kirk Mooney, stn mgr & sls dir; Matt West, progmg dir; Kevin Brooks, chief of engrg.

***KTBJ(FM)**— 1998: 89.3 mhz; 25 kw. Ant 371 ft TL: N38 09 16 W90 02 07. Stereo. Hrs opn: 115 E. Main St., 63028. Phone: (314) 892-9893. Fax: (314) 892-9527. E-mail: ktbj@csnradio.com Web Site: www.csnradio.com. Licensee: CSN International (group owner; acq 6-17-98; $100,000). Population served: 42,000 Format: Christian talk/music. ♦Scott Parker, stn mgr.

Florissant

KFTK(FM)— Apr 15, 1977: 97.1 mhz; 100 kw. 560 ft TL: N38 46 45 W90 43 43. Stereo. Hrs opn: 24 800 St. Louis Union St., The Powerhouse, St. Louis, 63103. Phone: (314) 231-9710. Fax: (314) 621-3000. Licensee: Emmis Radio License LLC. Group owner: Emmis Communications Corp. (acq 9-26-2000; grpsl). Natl. Rep: McGavren Guild. Format: Talk. ♦John Beck, sr VP, gen mgr & gen mgr.

Fredericktown

KYLS(AM)— June 29, 1963: 1450 khz; 1 kw-U. TL: N37 35 00 W90 17 31. Hrs open: 24 540 Maple Valley Dr., Farmington, 63640-1481. Phone: (573) 701-9590. Fax: (573) 701-9696. Web Site: www.froggy96.com. Licensee: Dockins Communications Inc. (acq 3-11-97). Population served: 300,000 Format: Oldies. News: 8 hrs wkly. Target aud: 35-65. ♦Fred M. Dockins Sr., pres & gen mgr.

Fulton

KFAL(AM)— Nov 14, 1950: 900 khz; 1 kw-D, 121 w-N. TL: N38 51 58 W91 57 15. Hrs open: 19 1805 Westminster, Jefferson City, 65251. Phone: (573) 642-3341. Fax: (573) 642-3343. Licensee: Zimmer Radio of Mid-Missouri Inc. Group owner: Zimmer Radio Group (acq 11-19-99; grpsl). Population served: 439,000 Natl. Network: Motor Racing Net. Format: Traditional country. News staff: one; News: 3 hrs wkly. Target aud: 35 plus. Spec prog: Other 6 hrs wkly. ♦Jerry Zimmer, CEO; John Zimmer, chmn; Don Zimmer, pres; Bob Steinberg, CFO; Jeremiah Washington, gen mgr; Andy Tutin, gen sls mgr.

KKCA(FM)—Co-owned with KFAL(AM). 1970: 100.5 mhz; 6 kw. 300 ft TL: N38 51 58 W91 57 15. Stereo. 24 12,248 Natl. Network: ABC, Westwood One, Jones Radio Networks. Format: Oldies. News staff: one; News: 2 hrs wkly. Target aud: 25-54.

Gainesville

KMAC(FM)— Mar 17, 1994: 99.7 mhz; 50 kw. Ant 492 ft TL: N36 36 06 W92 25 48. Stereo. Hrs opn: 24 Box 6610, 100 Bluebird St., Harrison, AR, 72601. Phone: (870) 743-1157. Fax: (870) 743-1168. E-mail: kmac997@hotmail.com Licensee: Pearson Broadcasting of Gainesville Inc. Group owner: Pearson Broadcasting (acq 12-19-94; $150,000; FTR: 2-13-95). Format: Hot adult contemp, classic rock. Target aud: General. ♦Dave Fransen, gen mgr.

Gallatin

KGOZ(FM)— June 1994: 101.7 mhz; 15 kw. Ant 423 ft TL: N39 53 14 W93 43 24. Stereo. Hrs opn: 24 Box 217, 804 Main, Trenton, 64683. Phone: (660) 359-2727. Fax: (660) 359-4126. E-mail: kttnamfm@grm.net Web Site: www.parbroadcastgroup.com. Licensee: PAR Broadcasting Co. Inc. (acq 1-3-94; $11,571 for CP; FTR: 1-24-94). Population served: 50,000 Natl. Network: Jones Radio Networks. Natl. Rep: Rgnl Reps. Law Firm: Reddy, Begley & McCormick. Format: Hot country. News: 2 hrs wkly. Target aud: 14-50. ♦John Ausberger, pres; John Anthony, gen mgr.

Garden City

KCJK(FM)— January 2001: 105.1 mhz; 69 kw. 1,145 ft TL: N39 05 26 W94 28 18. Stereo. Hrs opn: 24 5800 Foxridge Dr., Suite 600, Mission, KS, 66202. Phone: (913) 514-3000. Fax: (913) 514-3002. Web Site: www.1051jackfm.com. Licensee: CMP Houston-KC LLC. Group owner: Susquehanna Radio Corp. (acq 5-3-2006; grpsl). Population served: 1,475,000 Natl. Rep: Katz Radio. Format: Adult hits. News: 10.5 hrs wkly. Target aud: 25-54. ♦Pat Gibbs, gen sls mgr; Bryan Truta, progmg dir.

Gideon

KGLU(FM)—Not on air, target date: unknown: 103.9 mhz; 6 kw. Ant 328 ft TL: N36 32 10 W89 49 18. Hrs open: 5525 Yates Cove, Memphis, TN, 38120. Phone: (901) 685-0882. Licensee: Pollack Steel Supply Inc. (acq 3-8-2007; $155,000 for CP). ♦Sydney Pollack, pres.

Gladstone

KGGN(AM)— Nov 18, 1996: 890 khz; 1 kw-D, DA. TL: N39 20 03 W94 34 01. (CP: 960 w-D, DA). Hrs opn: 1734 E. 63rd St., Suite 600, Kansas City, 64110. Phone: (816) 333-0092. Fax: (816) 363-8120. E-mail: kggnproduction@aol.com Web Site: www.kggnam.com. Licensee: Mortenson Broadcasting Co. (group owner; acq 12-24-96; $450,000). Format: Gospel. ♦Doris Newman, gen mgr.

Gordonville

KCGQ-FM—Licensed to Gordonville. See Cape Girardeau

Halfway

KYOO-FM— April 1995: 99.1 mhz; 25 kw. 328 ft TL: N37 45 41 W93 15 42. Hrs open: 205 N. Pike Ave., Bolivar, 65613-1550. Phone: (417) 326-5259. Phone: (417) 326-5257. Fax: (417) 326-5900. E-mail: kyooradio@aol.com Web Site: www.todayshits.com. Licensee: KYOO Communications. Population served: 500,000 Natl. Network: ABC. Format: Adult contemp. News staff: one; News: 5 hrs wkly. Target aud: 10-72. ♦Ann Paris, VP; Stephen Paris, pres & gen mgr.

Hannibal

KGRC(FM)— Nov 28, 1968: 92.9 mhz; 100 kw. 489 ft TL: N39 43 45 W91 24 15. (CP: Ant 502 ft. TL: N39 43 48 W91 24 19). Stereo. Hrs opn: 329 Maine St., Lincoln Douglas Bldg., Quincy, IL, 62301. Phone: (217) 224-4102. Fax: (217) 224-4133. E-mail: jbates@staradio.com Web Site: www.real929.com. Licensee: STARadio Corp. (group owner; acq 12-2-98; $2.1 million with KZZK(FM) New London). Population served: 250,000 Natl. Network: Westwood One. Natl. Rep: Katz Radio. Law Firm: Pepper & Corazzini. Format: Hot adult contemp. News staff: one; News: one hr wkly. Target aud: 18-49; women. ♦Howard Doss, pres; Michael J. Moyers, VP & gen mgr; Casey, progmg dir & chief of engrg.

KHMO(AM)— April 1941: 1070 khz; 5 kw-D, 1 kw-N, DA-2. TL: N39 37 46 W91 22 33. Hrs open: 24 Box 711, 63401. Phone: (573) 221-3450. Fax: (573) 221-5331. E-mail: kickfm@bicbroadcasting.com Licensee: Bick Broadcasting. (acq 8-1-85; $1.35 million; FTR: 6-17-85). Population served: 100,000 Natl. Rep: McGavren Guild. Law Firm: Eugene T. Smith. Wire Svc: Weather Wire Format: News/talk, sports. News staff: 2; News: 22 hrs wkly. ♦Ed Foxall, gen mgr & gen sls mgr.

***KJIR(FM)**— April 2000: 91.7 mhz; 5.1 kw. Ant 554 ft TL: N39 43 48 W91 24 19. Stereo. Hrs opn: 24 Believers Broadcasting Corp., 220 N. 6th St., Quincy, IL, 62301. Phone: (217) 221-9410. Fax: (217) 228-0966. E-mail: kjir@motion.net Licensee: Believers Broadcasting Corp. Population served: 200,000 Format: Southern gospel. News: 7.5 hrs wkly. Target aud: Christian; 30-70. ♦I. Carl Geisendorfer, gen mgr; Michael Wartman, progmg dir.

Harrisonville

KCFX(FM)— July 19, 1974: 101.1 mhz; 97 kw. Ant 1,099 ft TL: N39 01 20 W94 30 49. Stereo. Hrs opn: 6th Fl., 5800 Foxridge Dr., Mission, KS, 66202. Phone: (913) 514-3000. Fax: (913) 514-3001. Web Site: www.101thefox.net. Licensee: Susquehanna Kansas City Partnership. Group owner: Susquehanna Radio Corp. (acq 7-14-00; grpsl). Population served: 1,300,000 Natl. Rep: Katz Radio. Format: Classic rock. Target aud: 25-54; baby boomers. ♦Dave Alpert, VP & gen mgr; Chris Hoffman, opns mgr; Jeanna White, gen sls mgr.

Hayti

KCRV(AM)—See Caruthersville

KCRV-FM—See Caruthersville

High Point

***KMCV(FM)**— 2001: 89.9 mhz; 18 kw vert. Ant 325 ft TL: N38 35 48 W92 32 17. Hrs open: 10550 Barkley St., Overland Park, KS, 66212. Phone: (913) 642-7770. Fax: 913) 642-1319. E-mail: kmcv@bottradionetwork.com Web Site: www.bottradionetwork.com. Licensee: Community Broadcasting Inc. Group owner: Bott Radio Network (acq 2-7-01; $1.25 million with KSCV(FM) Springfield). Format: Relg. ♦Richard Bott II, exec VP.

Hollister

KBCV(AM)— 2004: 1570 khz; 5 kw-D, 3 kw-N, DA-2. TL: N36 36 52 W93 12 49 (D), N36 36 51 W93 12 50 (N). Hrs open: 10550 Barkley, Suite 100, Overland Park, KS, 66212. Phone: (913) 642-7600. Fax: (913) 642-1319. Licensee: Bott Communications Inc. Group owner: Bott Radio Network. Format: Relg. ♦Trace Thurlby, COO; Richard P Bott Sr., pres; Tom Holdeman, CFO; Richard P Bott, exec VP; Rachel Moser, mktg mgr; Jason Potocnik, traf mgr.

Houston

KBTC(AM)—Listing follows KUNQ(FM).

KUNQ(FM)— May 1965: 99.3 mhz; 30 kw. Ant 604 ft TL: N37 05 32 W92 03 10. Stereo. Hrs opn: 24 Box 230, 17647 Hwy. B, 65483. Phone: (417) 967-3353. Fax: (417) 967-2281. E-mail: kunq@kunq.net Licensee: Metropolitan Radio Group Inc. (group owner; (acq 6-22-2000; $150,000 with co-located AM). Population served: 900,000 Law Firm: Shaw Pittman. Format: Classic country. News staff: one; News: 12 hrs wkly. Target aud: 25-69; blue collar. Spec prog: Farm 2 hrs, gospel 10 hrs wkly. ♦Beatrice Hall, stn mgr; Marilou Candela, sls dir, gen sls mgr & mus dir.

KBTC(AM)—Co-owned with KUNQ(FM). June 28, 1962: 1250 khz; 1 kw-D, 51 w-N. TL: N37 19 45 W91 53 55.42,178

Independence

KCTE(AM)— 1947: 1510 khz; 10 kw-D, DA. TL: N39 04 14 W94 26 58. Hrs open: 6721 W. 121 St., Overland Park, 66209. Phone: (913) 344-1500. Fax: (913) 344-1599. Web Site: www.1510.com. Licensee: Union Broadcasting Inc. (acq 8-19-98; $925,000). Natl. Network: ABC. Format: Sports. ♦Chad Boeger, gen mgr.

Ironton

KYLS-FM— Jan 6, 1984: 95.9 mhz; 3.2 kw. 922 ft TL: N37 34 23 W90 41 35. Stereo. Hrs opn: 24 540 S. Maple Valley Dr., Farmington, 63640. Phone: (573) 701-9590. Fax: (573) 701-9696. E-mail: freddockins@froggy96.com Web Site: www.froggy96.com. Licensee: Dockins Communications Inc. (acq 1-29-97). Format: Country. News staff: 2; News: 8 hrs wkly. Target aud: 18-54. ♦Fred M. Dockins Sr., pres & opns mgr; Sheila Dockins, gen sls mgr; Shannon Cavness, prom dir & traf mgr; Tom Colvin, news dir.

Jackson

KUGT(AM)— March 1972: 1170 khz; 250 w-D, 5 w-N. TL: N37 22 55 W89 39 12. Hrs open: 24 907 S. Kings Hwy., Cape Giradeau, 63702. Phone: (573) 339-7000. Fax: (573) 651-4100. Licensee: W. Russell Withers Jr. (acq 6-28-2005; $150,000). Natl. Network: Fox News Radio. Format: Relg, adult standards. Target aud: General. Spec prog: Parenting and family talk. ♦Rick Lambert, gen mgr.

KYRX(FM)—Marble Hill, December 1999: 97.3 mhz; 3.6 kw. Ant 426 ft TL: N37 22 49 W90 04 49. Hrs open: 901 S. Kings Hwy., Cape Girardean, 63702. Phone: (573) 339-7000. Fax: (573) 651-4100. Licensee: Dana R. Withers. Natl. Rep: Katz Radio. Format: Oldies. ♦Rick Lambert, gen mgr.

Jefferson City

KBBM(FM)—Listing follows KLIK(AM).

***KJLU(FM)**— August 1973: 88.9 mhz; 29.5 kw. 510 ft TL: N38 27 29 W92 13 32. Stereo. Hrs opn: 6 AM-midnight 1004 E. Dunklin St., 65102-0029. Phone: (573) 681-5301. Phone: (573) 681-5296. Fax: (573) 681-5299. E-mail: info@kjlu.com Web Site: www.lincolnu.edu/~kjlu/. Licensee: Board of Curators of Lincoln University. Population served: 200,000 Format: Jazz. Target aud: 18-54. ♦Michael P. Downey, gen mgr.

KLIK(AM)— January 1937: 1240 khz; 1 kw-U. TL: N38 33 50 W92 11 21. Hrs open: 24 3605 Country Club Dr., 65109. Phone: (573) 893-5100. Fax: (573) 893-8330. Web Site: www.klik1240.com. Licensee: Cumulus Licensing LLC. Group owner: Cumulus Media Inc. (acq 4-26-2004; grpsl). Population served: 68000 Format: News/talk. News staff: 4; News: 39 hrs wkly. Target aud: 35 plus; mid to upper income-well informed. ♦Lew Dickey, pres; Brian Wilson, gen mgr & progmg dir.

KBBM(FM)—Co-owned with KLIK(AM). 1974: 100.1 mhz; 33 kw. 600 ft TL: N38 31 25 W92 24 25. Stereo. 24 E-mail: buzz@buzz.fm Web Site: www.buzz.fm.200,000 Format: Active rock. News staff: 2. Target aud: 18-34.

KTXY(FM)—Listing follows KWOS(AM).

KWOS(AM)— February 1954: 950 khz; 5 kw-D, 500 w-N, DA-N. TL: N38 31 13 W92 10 42. Hrs open: 3109 S. 10 Mile Dr., 65109. Phone: (573) 893-5696. Fax: (573) 893-4137. Web Site: www.kwos.com. Licensee: Zimmer Radio of Mid-Missouri Inc. Group owner: Zimmer Radio Group (acq 11-19-99; grpsl). Population served: 58000 Format: News/talk. Spec prog: Farm 12 hrs wkly. ♦John Zimmer, chmn; Carla Leible, gen mgr; Dave Wisniewski, gen sls mgr; John Marsh, news dir; Jeff Studley, chief of engrg.

KTXY(FM)—Co-owned with KWOS(AM). Dec 1, 1969: 106.9 mhz; 100 kw. 1,250 ft TL: N38 38 16 W92 29 34. Stereo. 3215 LeMone Industrial, Suite 200, Columbia, 65201. Phone: (573) 875-1099. Fax: (573) 875-2439. E-mail: y107@zrgmail.com Web Site: www.y107.com.250,000 Format: Hot adult contemp.

KZJF(FM)— 2000: 104.1 mhz; 6 kw. Ant 312 ft TL: N38 34 45 W92 14 02. Hrs open: 24 3605 Country Club Dr., 65109. Phone: (573) 893-5100. Fax: (573) 893-8330. Licensee: Cumulus Licensing LLC. Group owner: Cumulus Media Inc. (acq 4-26-2004; grpsl). Population served: 100,000 Natl. Rep: Katz Radio. Format: Country. ♦Lewis W. Dickey Jr., pres; Scott Boltz, mktg mgr.

Joplin

KIXQ(FM)— November 1974: 102.5 mhz; 100 kw. 410 ft TL: N37 04 43 W94 32 26. Stereo. Hrs opn: 24 2702 E. 32nd, 64804. Phone: (417) 624-1025. Fax: (417) 781-6842. Web Site: www.kix1025.com. Licensee: Zimco Inc. Group owner: Zimmer Radio Group (acq 6-30-97; grpsl). Format: Contemp country. News staff: one; News: 2 hrs wkly. Target aud: 18-49. Spec prog: Class 3 hrs wkly. ♦John Zimmer, exec VP; Larry Boyd, gen mgr; Jason Knight, opns mgr.

***KOBC(FM)**— Mar 17, 1969: 90.7 mhz; 60 kw. 500 ft TL: N37 03 11 W94 23 17. Stereo. Hrs opn: 18 1111 N. Main St., 64801. Secondary address: 2711 Peace Church 64801. Phone: (417) 781-6401. Fax: (417) 782-1841. E-mail: kobc@kobc.org Web Site: www.kobc.org. Licensee: Ozark Christian College. Population served: 250,000 Format: Adult Christian contemp. Target aud: 25-45. ♦Rob Kime, gen mgr; T.C. Andrews, sls dir & gen sls mgr; Lisa Davis, progmg dir & mus dir; Lisa Satterfield, news dir & traf mgr; Mitch Piercy, chief of engrg.

KQYX(AM)— 1927: 1450 khz; 1 kw-U. TL: N37 04 43 W94 32 26. Hrs open: 24 2510 W. 20th St., 64804-0216. Phone: (417) 781-1313. Fax: (417) 781-1316. Licensee: FFD Holdings I Inc. Group owner: Petracom Media L.L.C. (acq 12-20-2004; grpsl). Population served: 240,000 Law Firm: Borsari & Paxson. Format: News/talk. Target aud: 18-39. ♦Dave Clemons, gen sls mgr; Matt Kruger, progmg dir.

KSYN(FM)— Dec 19, 1960: 92.5 mhz; 100 kw. 430 ft TL: N37 04 10 W94 32 49. Stereo. Hrs opn: 2702 E. 32nd, 64804. Phone: (417) 624-1025. Fax: (417) 781-6842. Web Site: www.ksyn925.com. Licensee: Zimco Inc. Group owner: Zimmer Radio Group (acq 6-30-97; grpsl). Format: CHR. Target aud: 18-39. ♦Larry Boyd, gen mgr; Jason Knight, opns mgr.

***KXMS(FM)**— Apr 5, 1986: 88.7 mhz; 10 kw. 185 ft TL: N37 05 57 W94 27 46. Stereo. Hrs opn: Missouri Southern State Univ., 3950 E. Newman, 64801-1595. Phone: (417) 625-9356. Fax: (417) 625-9742. E-mail: kxms@mssc.edu Web Site: www.kxms.org. Licensee: Board of Governors— Missouri Southern State College Format: Joplin's fine art stn. Spec prog: Big band 2 hrs wkly. ♦Jeffrey Skibbe, gen mgr.

KZRG(AM)— Nov 21, 1948: 1310 khz; 5 kw-D, 1 kw-N, DA-2. TL: N37 07 03 W94 32 41. Hrs open: 2702 E. 32nd St., 64804. Phone: (417) 624-1025. Fax: (417) 781-6842. Licensee: Zimmer Radio Inc. (acq 11-15-2005; $350,100). Format: News/talk. ♦Larry Boyd, gen mgr; Rob Kime, progmg dir.

KZYM(AM)— June 1, 1946: 1230 khz; 1 kw-U. TL: N37 04 48 W94 33 10. Hrs open: 24
Rebroadcasts KNLG(FM) New Bloomfield 100%.
2702 E. 32nd St., 64804. Phone: (417) 624-1025. Fax: (417) 626-7111. Licensee: Zimmer Radio Inc. (acq 9-30-2005; $300,000). Population served: 115,000 Format: MOR, contemp Christian. News staff: one; News: 2 hrs wkly. Target aud: General. Spec prog: Webb City High School sports. ♦James Zimmer, pres.

WMBH(AM)— May 25, 1962: 1560 khz; 10 kw-D, DA. TL: N37 04 10 W94 32 49. Hrs open: 6 AM-9 PM 611 S. Main St., 64801. Phone: (417) 781-1313. Fax: (417) 781-1316. Licensee: Hardman Broadcasting Inc. Group owner: Petracom Media L.L.C. (acq 6-3-2005; $1). Population served: 360,000 Format: Hip hop, rhythm and blues hits. ♦Dave Clemons, gen sls mgr.

Kansas City

KBEQ-FM— November 1960: 104.3 mhz; 100 kw. 987 ft TL: N39 04 59 W94 28 49. Stereo. Hrs opn: 24 4717 Grand Ave., Suite 600, 64112. Phone: (816) 753-4000. Fax: (816) 753-1045. Web Site: www.youngcountryq104.com. Licensee: Wilks License Co.-Kansas City LLC. Group owner: Infinity Broadcasting Corp. (acq 1-10-2007; grpsl). Population served: 1,239,000 Law Firm: Koteen & Naftalin. Format: Country. News staff: one; News: 6 hrs wkly. Target aud: 18-54; women. ♦Herndon Hasty, gen mgr & mktg mgr.

KCCV(AM)—See Overland Park, KS

KCFX(FM)—See Harrisonville

KCKC(FM)— Mar 5, 1961: 102.1 mhz; 100 kw. Ant 1,118 ft TL: N39 05 26 W94 28 18. Hrs open: CBS Radio, 4717 Grand Ave., Suite 600, 64112. Phone: (816) 561-9102. Fax: (816) 531-6547. Web Site: www.star102.net. Licensee: Wilks License Co.-Kansas City LLC. Group owner: Infinity Broadcasting Corp. (acq 1-10-2007; grpsl). Population served: 402,600 Wire Svc: UPI Format: Hot adult contemp. ♦Christy Roberts, prom dir; Mike Kennedy, prom dir & progmg dir.

KCMO(AM)— March 1922: 710 khz; 10 kw-D, 5 kw-N, DA-2. TL: N39 19 08 W94 29 48. Stereo. Hrs opn: 24 5800 Foxridge Dr., Suite 600, Mission, KS, 66202. Phone: (913) 514-3000. Fax: (913) 514-3007. Web Site: www.710kcmo.com. Licensee: Susquehanna Kansas City Partnership. Group owner: Susquehanna Radio Corp. (acq 7-14-2000; grpsl). Population served: 1,349,300 Natl. Rep: Katz Radio. Law Firm: Cohn & Marks. Format: Talk. News staff: 3; News: 30 hrs wkly. Target aud: 35-64. Spec prog: Pub affrs one hrs wkly. ♦Dave Alpert, VP & mktg mgr; Jeanna White, gen sls mgr; Chris Hoffman, progmg dir.

KCMO-FM— May 4, 1948: 94.9 mhz; 100 kw. Ant 1,120 ft TL: N39 05 26 W94 28 18. Stereo. 24 Phone: (913) 514-3000. Fax: (913) 514-3003. Natl. Rep: Katz Radio. Format: Greatest Hits of the 60's & 70's. News staff: one; News: 2 hrs wkly. Target aud: 25-54. ♦Page Olson, gen sls mgr; Don Daniels, progmg dir.

KCNW(AM)—See Fairway, KS

KCSP(AM)— Feb 16, 1922: 610 khz; 5 kw-U. TL: N38 59 03 W94 37 40. Stereo. Hrs opn: 24 4935 Belinder Rd., Westwood, KS, 66205. Phone: (913) 677-8998. Fax: (913) 677-8061. Web Site: www.610sports.com. Licensee: Entercom Kansas City License L.L.C. Group owner: Entercom Communications Corp. (acq 10-17-97; grpsl). Population served: 1,435,800 Natl. Network: Sporting News Radio Network. Natl. Rep: D & R Radio. Format: Country. News staff: News progmg 8 hrs wkly Target aud: 25-54; general. ♦Cindy Schloss, CEO; Michael Keck, stn mgr; Bob Olson, opns dir; Neal Jones, progmg dir; Scott Parks, news dir; Susan Wilson, rsch dir & traf mgr.

KCTE(AM)—See Independence

***KCUR-FM**— October 1957: 89.3 mhz; 100 kw. Ant 820 ft TL: N39 04 59 W94 28 49. Stereo. Hrs opn: 24 4825 Troost, Suite 202, 64110. Phone: (816) 235-1551. Fax: (816) 235-2864. E-mail: kcur@umkc.edu Web Site: www.kcur.org. Licensee: Curators of the University of Missouri. Group owner: The Curators of the University of Missouri Population served: 1,500,000 Natl. Network: NPR, PRI. Format: Pub affrs, news. News staff: 3; News: 50 hrs wkly. Target aud: General; educated. Spec prog: Sp 2 hrs wkly. ♦Patricia Deal Cahill, gen mgr; Parker Van Hecke, dev dir; Bill Anderson, progmg VP; Robert Moore, mus dir; Frank Morris, news dir & news rptr; Robin Cross, engrg mgr & chief of engrg; Steve Bell, reporter.

KCZZ(AM)—See Mission, KS

KEXS(AM)—See Excelsior Springs

***KKFI(FM)**— Feb 28, 1988: 90.1 mhz; 100 kw. 503 ft TL: N39 05 05 W94 28 47. Stereo. Hrs opn: 24 Box 32250, 64171-2250. Secondary address: 900 1/2 Westport Rd. 64111. Phone: (816) 931-3122. Phone: (816) 931-5534. Fax: (816) 931-7870. E-mail: kkfi901@aol.com Web Site: www.kkfi.org. Licensee: Mid-Coast Radio Project Inc. Format: News/ talk. News: 10 hrs wkly. Target aud: General; women & minorities. Spec prog: Jazz 10 hrs, blues 9 hrs, Sp 16 hrs, folk 4 hrs, American Indian 2 hrs wkly. ♦Dorothy Hawkins, gen mgr.

***KLJC(FM)**— Aug 9, 1970: 88.5 mhz; 100 kw. 745 ft TL: N39 04 24 W94 29 06. Stereo. Hrs opn: 24 c/o Calvary Bible College, 15800 Calvary Rd., 64147-1341. Phone: (816) 331-8700. Fax: (816) 331-3497. E-mail: kljc@kljc.org Web Site: www.kljc.org. Licensee: Calvary Bible College. Population served: 1,507,087 Natl. Network: Salem Radio Network. Format: Christian contemp. News: 6 hrs wkly. Target aud: 25-54. ♦Dr. Elwood Chipchase, pres & gen mgr; Michael Griman, progmg dir; Glenn Williams, chief of engrg.

KMBZ(AM)— 1921: 980 khz; 5 kw-U, DA-N. TL: N39 02 17 W94 36 55. Hrs opn: 7000 Squibb Rd., Mission, KS, 66202. Phone: (913) 677-8998. Fax: (913) 677-8901. Web Site: www.kmbz.com. Licensee: Entercom Kansas City News License L.L.C. Group owner: Entercom Communications Corp. (acq 3-6-97; grpsl). Population served: 1,349,300 Format: News radio. ♦Rich Deutsch, gen sls mgr; Neil Larrimore, progmg dir; Nicole Teich, news dir; Mike Cooney, chief of engrg; Megan Wilson, traf mgr.

KYYS(FM)—Co-owned with KMBZ(AM). October 1962: 99.7 mhz; 100 kw. 1,010 ft TL: N39 05 01 W94 30 57. Stereo. Web Site: www.kyys.com. Licensee: Entercom Kansas City License L.L.C. Format: Rock. ♦Kevin Klein, gen sls mgr; Greg Bergen, progmg dir.

KMXV(FM)— Mar 3, 1958: 93.3 mhz; 100 kw. 1,066 ft TL: N39 00 57 W94 30 57. Stereo. Hrs opn: 24 4714 Grand Ave., Suite 600, 64112. Phone: (816) 756-5698. Fax: (816) 931-8540. Web Site: www.mix93.com. Licensee: Wilks License Co.-Kansas City LLC. Group owner: Infinity Broadcasting Corp. (acq 1-10-2007; grpsl). Population served: 1,500,000 Format: CHR. News staff: one. Target aud: 18-49; women. ♦Herndon Hasty, gen mgr.

KPHN(AM)— Sept 1, 1971: 1190 khz; 5 kw-D, 250 w-N, DA-2. TL: N39 03 49 W94 30 37. Stereo. Hrs opn: 24 1212 Baltimore, 64105. Phone: (816) 421-1900. Fax: (816) 471-1320. E-mail: mark.t.ballard@abc.com Web Site: www.radiodisney.com. Licensee: Radio Disney Group LLC. Group owner: ABC Inc. (acq 7-19-2002; $3.8 million). Population served: 2,165,000 Natl. Network: Radio Disney. Format: Children. ♦Bob Martin, gen mgr.

KPRS(FM)—Listing follows KPRT(AM).

KPRT(AM)— 1950: 1590 khz; 1 kw-D, 47 w-N. TL: N39 04 05 W94 32 10. Hrs open: 24 11131 Colorado Ave., 64137. Phone: (816) 763-2040. Fax: (816) 966-1055. Licensee: Carter Broadcast Group Inc. (group owner). Population served: 507,087 Natl. Rep: McGavren Guild. Law Firm: Bryan Cave. Format: Gospel. ♦Cheryl Douglas, chmn & traf mgr; Michael Carter, gen mgr; Audrey Herbert, natl sls mgr; Vic Dyson, sls dir & rgnl sls mgr; Rich McCauley, mktg dir, prom mgr & news dir; Fred Bell, progmg dir; Debbie Rutledge, mus dir; Brooke Callowich, pub affrs dir; Mark Leaver, chief of engrg.

KPRS(FM)—Co-owned with KPRT(AM). 1963: 103.3 mhz; 100 kw. 994 ft TL: N39 00 57 W94 30 24. Stereo. 24 E-mail: 103@kprs.com Web Site: www.kprs.com. Format: Urban contemp. News: one hr wkly. Target aud: 25-54; mid-upper income. ♦Andre Carson, opns mgr; Rich McCauley, prom dir; Myron Fears, progmg dir; Beth Baker, traf mgr.

KQRC-FM—Leavenworth, KS) 1962: 98.9 mhz; 100 kw. 990 ft TL: N39 04 14 W94 54 39. Stereo. Hrs opn: 4935 Belinder Rd., Westwood, KS, 66206. Phone: (913) 677-8998. Fax: (913) 677-8061. Web Site: www.989therock.com. Licensee: Entercom Kansas City License LLC. Group owner: Entercom Communications Corp. (acq 7-14-00; grpsl). Law Firm: Crowell & Moring. Format: AOR. Target aud: 18-34; above average education & income; upscale professionals. ♦Cindy Schloss, gen mgr.

KRBZ(FM)— Jan 1, 1959: 96.5 mhz; 99 kw. 984 ft TL: N39 00 57 W94 30 24. Stereo. Hrs opn: 24 4935 Belinder Rd., West Wood, 66205. Phone: (913) 677-8998. Fax: (913) 677-7520. Web Site: www.965thebuzz.com. Licensee: Entercom Kansas City License L.L.C. Group owner: Entercom Communications Corp. (acq 7-14-00; grpsl). Population served: 1500000 Law Firm: Wiley, Rein & Fielding. Format: Rock/AOR. Target aud: 25 plus; adults with above-average disposable income. ♦Cindy Schloss, gen mgr.

KUDL(FM)—See Kansas City, KS

WHB(AM)— June 10, 1936: 810 khz; 50 kw-D, 5 kw-N, DA-N. TL: N39 18 21 W94 34 30. Hrs open: 24 6721 W. 121st St., Overland Park, KS, 66209. Phone: (913) 344-1500. Fax: (913) 344-1599. Web Site: www.810whb.com. Licensee: Union Broadcasting Inc. (acq 11-23-99; $8 million). Format: Sports talk. News staff: 2; News: 19 hrs wkly. Target aud: 25 plus; farm families. Spec prog: Sp 2 hrs wkly. ♦Chad Boeger, gen mgr & progmg dir; Nick McCabe, opns dir; Gabe Boucher, prom dir; Ed Treese, chief of engrg; Melissa Hoover, traf mgr.

Kennett

***KAUF(FM)**— June 1998: 89.9 mhz; 1 kw. 164 ft TL: N36 14 32 W90 03 54. Hrs open: Box 3206, American Family Radio, Tupelo, MS, 38803. Phone: (662) 844-8888. Fax: (662) 842-6791. Web Site: www.afr.net. Licensee: American Family Association. Group owner: American Family Radio Format: Inspirational Christian. ♦Marvin Sanders, gen mgr.

KBOA(AM)— 1963: 1540 khz; 1 kw-D. TL: N36 15 11 W90 02 56. Hrs open: Box 509, 63857. Phone: (573) 888-4616. Fax: (573) 888-4890. Licensee: Pollack Broadcasting Co. (group owner; acq 9-25-98; $450,000 with KBOA-FM Piggott, AR). Population served: 50,000 Format: Music of Your Life. Spec prog: Farm 5 hrs wkly. ♦Perry Jones, gen mgr.

KOTC(AM)— July 19, 1947: 830 khz; 10 kw-D. TL: N36 13 29 W90 04 31. Hrs open: Sunrise-sunset Box 271, 63857. Secondary address: 700 N. Bypass 63857. Phone: (573) 686-3700. Fax: (573) 686-6116. E-mail: kotc@sheltonbbs.com Web Site: www.foxradionetwork.com. Licensee: Eagle Bluff Enterprises (acq 9-18-96; $190,000). Population served: 20,000 Format: Classic country. News staff: one. Target aud: 28-55. ♦Steven Fuchs, pres & gen mgr; Charles Isabell, news dir; P.J. Johnson, chief of engrg.

KXOQ(FM)—Co-owned with KOTC(AM). Dec 13, 1995: 104.3 mhz; 6 kw. 328 ft TL: N36 21 01 W90 02 43. Web Site: www.foxradionetwork.com. Format: Rock & oldies mix.

Kimberling City

KOMC-FM— 1992: 100.1 mhz; 36 kw. Ant 577 ft TL: N36 31 58 W93 19 43. Stereo. Hrs opn: 24 202 Courtney St., Branson, 65616. Phone: (417) 334-6003. Fax: (417) 334-7141. E-mail: krzk@krzk.com Web Site: www.komc.com. Licensee: KOMC-KRZK LLC. Group owner: Orr & Earls Broadcasting Inc. (acq 6-27-97; $1,064,919). Population served: 70,000 Natl. Network: ABC, CBS. Format: Adult standards, big band. News staff: 2; News: 11 hrs wkly. Target aud: 45 plus; Branson & local tourists. ♦Charles Earls, pres; Scottie Earls, gen mgr & chief of opns; Steve Willoughby, stn mgr.

Kirksville

***KHGN(FM)**— Oct 6, 1997: 90.7 mhz; 32.5 kw. 325 ft TL: N40 13 46 W92 32 38. Stereo. Hrs opn: 24 Box 500, 63501. Secondary address: RR5, Box 14AB 63501. Phone: (660) 665-0466. Fax: (660) 665-7304. E-mail: khgn@kvmo.net Web Site: www.khgn.org. Licensee: Care Broadcasting Inc. Natl. Network: Moody. Rgnl. Network: Moody. Format: Relg. Target aud: 30 plus; general. ♦Dennis Phelps, pres; Tom Lloyd, chief of engrg.

KIRX(AM)— Oct 17, 1947: 1450 khz; 1 kw-U. TL: N40 12 24 W92 34 31. Hrs open: 24 Box 130, 1308 N. Baltimore, 63501. Phone: (660) 665-3781. Fax: (660) 665-0711. E-mail: kirx@cableone.net Web Site: www.1450kirx.com. Licensee: KIRX Inc. (acq 10-1-85; $1.3 million with co-located FM; FTR: 8-12-85) Population served: 25,000 Rgnl. Network: Brownfield. Format: Oldies. News staff: 2; News: 40 hrs wkly. Target aud: 25-54; general. Spec prog: Farm 10 hrs wkly. ♦David L. Nelson, pres; Steven D. Lloyd, exec VP, gen mgr & gen sls mgr.

KRXL(FM)—Co-owned with KIRX(AM). September 1967: 94.5 mhz; 100 kw. 1,010 ft TL: N40 13 32 W92 00 54. (CP: 90.4 kw). Stereo. 24 Phone: (660) 665-9828. E-mail: krxl@cableone.net Web Site: www.945thex.com.488,000 Natl. Network: ABC. Format: Classic rock. News: 2 hrs wkly. Target aud: 25-54.

***KKTR(FM)**— 2002: 89.7 mhz; 1 kw. Ant 197 ft TL: N40 10 40 W92 34 40. Hrs open: 409 Jesse Hall, Columbia, 65211-1310. Phone: (573) 882-3431. Fax: (573) 882-2636. Web Site: www.kbia.org. Licensee: Truman State University. Format: News, class, talk. ♦Mike Dunn, gen mgr; Robert Wells, gen sls mgr; John Bailey, progmg dir.

KLTE(FM)— May 20, 1991: 107.9 mhz; 100 kw. 715 ft TL: N39 57 23 W92 58 29. Hrs open: 24 3 Crown Dr., Suite 100, 63501. Phone: (660) 627-5583. Fax: (660) 665-8900. E-mail: klte@bottradionetwork.com Web Site: www.bottradionetwork.com. Licensee: Bott Communications Inc. Group owner: Bott Radio Network Population served: 1,500,000 Natl. Network: USA. Natl. Rep: Salem. Format: Christian. News: 3 hrs wkly. Target aud: 35 plus. ♦Dick Bott Sr., CEO & pres; Trace Thurlby, COO; Richard Bott II, chmn; Tom Holdeman, CFO; Judy Lene, gen mgr; Paul Shipman, sls.

***KTRM(FM)**— Feb 10, 1998: 88.7 mhz; 1 kw. Ant 197 ft TL: N40 10 40 W92 34 40. Hrs open: 7 AM-2 AM SUB Truman State University, Div. Language & Literature, 63501. Phone: (660) 785-4000. E-mail: ktrmtheedge@hotmail.com Web Site: ktrm.truman.edu. Licensee: Truman State University. Format: Alternative. News: 3 hrs wkly. ♦Clair Maronack, system mgr.

KTUF(FM)— Feb 14, 1983: 93.7 mhz; 50 kw. 492 ft TL: N40 13 38 W92 36 35. Stereo. Hrs opn: 24 Box 130, 63501. Secondary address: 1308 N. Baltimore Rd. 63501. Phone: (660) 627-5883. Fax: (660) 665-0711. E-mail: ktuf@cableone.net Web Site: www.937ktuf.com. Licensee: KIRX Inc. Natl. Network: ABC. Format: Country. News staff: 2; News: 40 hrs wkly. Target aud: 18-44. ♦David L. Nelson, pres; Steven D. Lloyd, exec VP & gen mgr; Duncan Miller, opns mgr.

Knob Noster

***KCVQ(FM)**— July 1998: 89.7 mhz; 5 kw. 230 ft TL: N38 52 10 W93 32 58. Stereo. Hrs opn: 24
Rebroadcasts KCVO-FM Camdenton 100%.
Box 800, c/o Spirit FM Radio, Camdenton, 65020. Phone: (573) 346-3200. Fax: (573) 346-1010. E-mail: email@spiritfm.org Web Site: www.spiritfm.org. Licensee: Lake Area Educational Broadcasting Foundation. Population served: 120,000 Format: Christian, div. Target aud: 25-45. ♦Alice McDermott, CFO; James J. McDermott, pres & gen mgr.

KXKX(FM)— June 24, 1983: 105.7 mhz; 40 kw. 502 ft TL: N38 46 28 W93 37 34. Stereo. Hrs opn: 24 2209 S. Limit Ave., Sedalia, 65301. Phone: (660) 826-1050. Fax: (660) 827-5072. E-mail: info@kxkx.com Web Site: www.kxkx.com. Licensee: Bick Broadcasting Co. (acq 7-19-89; $185,000; FTR: 8-7-89). Format: Hot Country. News staff: one; News: 5 hrs wkly. Target aud: 25-54. ♦Dennis Polk, gen mgr.

La Monte

KPOW-FM— Nov 18, 1998: 97.7 mhz; 100 kw. Ant 981 ft TL: N39 03 10 W93 16 01. Stereo. Hrs opn: 24 301 S. Ohio Ave., Sedalia, 65301-4431. Phone: (660) 826-5005. Phone: (660) 829-9700. Fax: (660) 826-5557. Web Site: www.power97.net. Licensee: Sedalia Investment Group L.L.C. Population served: 232,000 Natl. Network: CNN Radio. Format: Classic rock. News staff: one; News: 2 hrs wkly. Target aud: 25-54. Spec prog: Blues 6 hrs wkly. ♦Stu Steinmetz, gen mgr.

Lake Ozark

KQUL(FM)— May 9, 1994: 102.7 mhz; 6 kw. 328 ft TL: N38 02 06 W92 34 31. Stereo. Hrs opn: 24 160 Hwy. 42, Kaiser, 65047. Phone: (573) 348-1958. Fax: (573) 348-1923. E-mail: mike@mix927.com Licensee: Benne Broadcasting of Lake Ozark Inc. (acq 5-20-98; $800,000). Format: Oldies. Target aud: 35-60. ♦Denny Benne, gen mgr; Greg Sullens, stn mgr; Mike Clayton, opns mgr & progmg dir.

Lamar

KHST(FM)— May 1, 1992: 101.7 mhz; 22 kw. Ant 328 ft TL: N37 25 27 W94 16 11. Stereo. Hrs opn: Box 383, Pittsburg, KS, 66762. Secondary address: 412 Locust St., Pittsburg, KS 66762. Phone: (620) 232-5993. Fax: (620) 232-5550. Licensee: Innovative Broadcasting Corp. (group owner; (acq 9-22-98; $330,000). Format: New rock. ♦Lance Sayler, pres; Lance Saylor, gen mgr.

Lebanon

KBNN(AM)— Oct 20, 1973: 750 khz; 5 kw-D. TL: N37 41 10 W92 41 39. Hrs open: 6 AM-sunset Box 1112, 18553 Gentry Rd., 65536. Phone: (417) 532-9111. Fax: (417) 588-4191. E-mail: kjel@regionalradio.com Web Site: www.regionalradio.com. Licensee: Ozark Broadcasting Inc. Group owner: Shepherd Group (acq 7-83; $450,000 with co-located FM; FTR: 7-4-83) Population served: 208,000 Rgnl. Network: Agri-Net, Brownfield, Missourinet. Format: Talk. News staff: 5; News: 35 hrs wkly. Target aud: 35-64; middle America. Spec prog: News, farm 8 hrs wkly. ♦Mike Edwards, gen mgr; Jay Burns, opns mgr; Jim Anthony, gen sls mgr; Scott Smith, news dir; Kelly Nelson, chief of engrg; Barbara Johnson, traf mgr.

KJEL(FM)—Co-owned with KBNN(AM). Oct 20, 1973: 103.7 mhz; 100 kw. 984 ft TL: N37 49 10 W92 44 51. Stereo. 24 Web Site: www.regionalradio.com.804,000 Format: Country. News staff: 5; News: 45 hrs wkly. Target aud: 25-65; affluent, business oriented.

KCLQ(FM)—Listing follows KLWT(AM).

KLWT(AM)— July 4, 1948: 1230 khz; 1 kw-U. TL: N37 40 40 W92 41 16. Hrs open: 24 18785 Finch Rd., 65536. Phone: (417) 532-2962. Fax: (417) 532-5184. E-mail: klwt@klwt1230.com Web Site: www.klwt1230.com. Licensee: Pearson Broadcasting of Lebanon Inc. Population served: 30,000 Format: Country, news/talk, sports. News staff: 3; News: 10 hrs wkly. Target aud: 30 plus; adults. ♦Max H. Pearson, pres; Dan Caldwell, gen mgr; Kit Caldwell, opns dir; Brian McClendon, rgnl sls mgr; Tonya Bell, news dir.

***KTTK(FM)**— 1992: 90.7 mhz; 11 kw. Ant 476 ft TL: N37 37 58 W92 45 22. Hrs open: 5 AM-midnight Box 1232, 65536. Phone: (417) 588-1435. Fax: (417) 532-3055. Licensee: Lebanon Educational Broadcasting Foundation. Natl. Network: USA. Format: Christian. Spec prog: Southern gospel 80 hrs, gospel 7 hrs wkly. ♦Max Rhoades, gen mgr.

Lee's Summit

KCXM(FM)— 1998: 97.3 mhz; 55 kw. Ant 1,171 ft TL: N39 05 26 W94 28 18. Hrs open: 6721 W. 121st, Leawood, KS, 66209. Phone: (913) 344-1500. Fax: (913) 344-1599. Web Site: www.810whb.com. Licensee: Union First Broadcasting LLC (acq 12-11-2003; $10 million). Natl. Network: ESPN Radio. Format: Sports. ♦Chad Boeger, gen mgr & progmg dir; Nick McCabe, opns dir; Gabe Boucher, prom dir; Ed Treese, chief of engrg; Melissa Hoover, traf mgr.

Lexington

KLEX(AM)— Apr 19, 1956: 1570 khz; 250 w-D, 58 w-N. TL: N39 11 14 W93 50 03. Hrs open: 24
KAYX (FM).
111 W. Main St., Richmond, 64085. Phone: (816) 470-9925. Fax: (816) 470-8925. Web Site: www.bottradionetwork.com. Licensee: Bott Communications Inc. Group owner: Bott Radio Network (acq 1994; with KAYX(FM) Richmond). Population served: 50,000 Natl. Network: USA. Format: Christian talk. Target aud: 25-54. ♦Trace Thurlby, COO; Richard P. Bott, pres; Tom Holdeman, CFO; Richard P. Bott II, exec VP; Eben Fowler, opns dir; Pat Rulon, natl sls mgr; Rachel Moser, mktg mgr; Jason Potochik, traf mgr.

KMJK(FM)— Sept 11, 1969: 107.3 mhz; 100 kw. 1,184 ft TL: N39 02 15 W93 55 48. Stereo. Hrs opn: 24 Blue Ridge Towers, 4240 Blue Ridge Blvd., Suite 820, Kansas City, 64133. Phone: (816) 353-7600. Fax: (816) 353-2300. Licensee: CMP KC Licensing LLC. Group owner: Cumulus Media Inc. (acq 12-19-2003; $25 million with KCHZ(FM) Ottawa, KS). Population served: 2,000,000 Format: Urban. News staff: one; News: 5 hrs wkly. Target aud: 25-54. ♦Lewis W. Dickey Jr., CEO; Dave Alpert, gen mgr.

Liberty

KCXL(AM)— Feb 14, 1967: 1140 khz; 500 w-D, 5 w-N. TL: N39 14 18 W94 23 59. Hrs open: 310 S. La Frenz Rd., 64068. Phone: (816) 792-1140. Fax: (816) 792-8258. E-mail: kcxl@kcxl.com Web Site: www.kcxl.com. Licensee: Alpine Broadcasting Corp. FTR: (4-2-84). Population served: 1,700,000 Natl. Network: Jones Radio Networks. Law Firm: Reddy, Begley & McCormick. Format: Talk, variety, MOR. Target aud: 25-54; baby boomers. Spec prog: News 4 hrs, Sp 5 hrs, relg 3 hrs, health 12 hrs wkly. ♦Peter E. Schartel, pres & progmg dir; Vern Windsor, opns mgr; Jonne Santoli, rgnl sls mgr; Ed Treese, chief of engrg.

***KWJC(FM)**— Apr 14, 1974: 91.9 mhz; 240 w. Ant 166 ft TL: N39 14 52 W94 24 47. Stereo. Hrs opn: 24 500 College Hill, Box 1063, 64068. Phone: (816) 415-7594. Fax: (816) 415-5027. E-mail: wirtht@william.jewell.edu Licensee: William Jewell College. Population served: 100,000 Format: CHR, modern rock, class. News: 5 hrs wkly. Target aud: 12-34; men & women. Spec prog: Class 10 hrs, Christian 10 hrs wkly. ♦Dr. Todd Wirth, gen mgr, stn mgr, chief of opns & progmg dir.

WDAF-FM— Nov 9, 1979: 106.5 mhz; 100 kw. Ant 981 ft TL: N39 04 23 W94 29 06. Stereo. Hrs opn: 4935 Belinder Rd., Westwood, KS, 66205. Phone: (913) 677-8998. Fax: (913) 677-8061. Web Site: www.1065thewolf.com. Licensee: Entercom Kansas City License LLC. Group owner: Entercom Communications Corp. (acq 7-14-00; grpsl). Population served: 100,000 Natl. Rep: D & R Radio. Format: Smooth jazz. Target aud: 18-34. ♦Cindy Schloss, gen mgr & opns mgr.

Linn

KJMO(FM)— 2006: 97.5 mhz; 6 kw. Ant 328 ft TL: N38 29 56.9 W91 53 00.4. Hrs open: 3605 Country Club Dr., Jefferson City, 65109. Phone: (573) 893-5100. Fax: (573) 893-8330. Web Site: www.kjmo.com. Licensee: Cumulus Licensing LLC. Format: Oldies. ♦Scott Boltz, gen mgr.

Louisiana

KJFM(FM)— Sept 4, 1984: 102.1 mhz; 1.85 kw. 387 ft TL: N39 26 29 W91 02 19. Stereo. Hrs opn: Box 438, 63353. Secondary address: 615 Georgia St. 63353. Phone: (573) 754-5102. Fax: (573) 754-5544. E-mail: kjfmradio@yahoo.com Licensee: Foxfire Communications Inc. Natl. Network: CBS Radio. Rgnl. Network: Missourinet. Format: Country. News: 27 hrs wkly. Target aud: 25-54. ♦Thom T. Sanders, pres; Gordon Sanders, opns.

Lutesville

KMHM(FM)— Aug 4, 1995: 104.1 mhz; 2.5 kw. 508 ft TL: N37 22 40 W89 56 04. Hrs open: 24 Box 266E, Hwy. B, Marble Hill, 63764. Phone: (573) 238-1041. Fax: (573) 238-0104. E-mail: kmhm1041@clas.net Web Site: www.kmhm.net. Licensee: Southern Gospetality LLC. Natl. Network: Salem Radio Network. Format: Southern gospel. News: 14 hrs wkly. Target aud: 30-55; Christians and family-oriented listeners. ♦Harold L. Lawder, CEO; Will Stephens, gen mgr; Glen Aulgur, stn mgr & gen sls mgr; Joy Duprey, progmg mgr; Sheila Kirkpatrick, mus dir; Tom Beattie, chief of engrg.

Macon

KIRK(FM)— 1998: 99.9 mhz; 12.5 kw. 462 ft TL: N39 36 02 W92 34 24. Hrs open: 24 Box 619, Moberly, 65270. Secondary address: 300 W. Reed St., Moberly 65270. Phone: (660) 263-6999. Fax: (660) 263-2300. Web Site: regionalradio.com. Licensee: KIRK L.L.C. Group owner: Shepherd Group Format: Adult contemp. ♦David Shepherd, gen mgr.

KLTI(AM)— Jan 30, 1966: 1560 khz; 1 kw-D. TL: N39 42 34 W92 27 50. (CP: COL Springfield. 660 khz; 1.5 kw-D, 44 w-N, DA-2. TL: N37 11 30 W93 32 22). Hrs opn: 24 32968 US Hwy. 63 S., 63552. Phone: (660) 385-1560. Fax: (660) 385-7090. E-mail: klti@kltiradio.com Web Site: www.kltiradio.com. Licensee: Chirillo Electronics Inc. Group owner: Best Broadcast Group. Format: Country. Target aud: 25-44. ♦Dale A. Palmer, gen mgr.

Madison

KCDG(FM)—Not on air, target date: unknown: 97.3 mhz; 25 kw. Ant 328 ft TL: N39 28 52 W92 10 13. Hrs open: 525 S. Flagler Dr. #21A, West Palm Beach, FL, 33401. Phone: (561) 832-8192. Licensee: Christine Radio LLC. ♦Christine Z. Goodman, pres & gen mgr.

WGNU(AM)—See Saint Louis

Malden

KLSC(FM)—Listing follows KMAL(AM).

KMAL(AM)— Sept 15, 1954: 1470 khz; 1 kw-D. TL: N36 33 08 W89 58 42. Hrs open: 6 AM-sunset
Simulcasts KSIM (Sikeston).
Box 69, Sikeston, 63801-0069. Secondary address: 519 Greer Ave., Sikeston 63801. Phone: (573) 471-1400. Fax: (573) 471-1402. Licensee: MRR License LLC. Group owner: MAX Media L.L.C. (acq 6-2-2004; grpsl). Population served: 178000 Format: News/talk. News staff: 3. Target aud: 35 plus. ♦Bill Powers, gen mgr; Brenda Woodall, gen sls mgr; Tyler Morrison, progmg dir & news dir; Charley Lampe, chief of engrg; Libby Wilson, traf mgr.

KLSC(FM)—Co-owned with KMAL(AM). Nov 23, 1979: 92.9 mhz; 23.5 kw. 174 ft TL: N36 33 08 W89 58 42. Stereo. 6 AM-midnight Format: Hot adult contemp. News staff: 2; News: 30 hrs wkly. Target aud: General. ♦Bill Powers, progmg dir; Libby Wilson, traf mgr.

Malta Bend

KRLI(FM)— Oct 28, 1996: 103.9 mhz; 3.4 kw. Ant 879 ft TL: N39 21 59 W93 24 12. Stereo. Hrs opn: 24 615 Cherokee, Marshall, 65340. Phone: (660) 831-1234. Fax: (660) 831-1290. E-mail: sb@global.net Web Site: krli.com. Licensee: Kanza Inc. Format: Jazz, big band, oldies. News staff: 2; News: 6 hrs wkly. Target aud: 45 plus; baby boomers. ♦Miles Carter, CEO, pres & gen mgr.

Mansfield

KTRI-FM— 1978: 95.9 mhz; 6 kw. Ant 312 ft TL: N37 02 18 W92 40 29. Stereo. Hrs opn: 24 1569 N. Central St., Monett, 65708. Phone: (417) 235-6041. Fax: (417) 235-6388. Web Site: www.buzz959.com. Licensee: Thirteen Forty Productions Inc. (group owner; (acq 1-8-2007; $200,000). Population served: 57,000 Format: Travel info. ♦Gary W. Snadon, pres.

Marble Hill

KYRX(FM)—Licensed to Marble Hill. See Jackson

Marshall

KMMO-FM— December 1968: 102.9 mhz; 100 kw. 380 ft TL: N39 08 03 W93 13 19. Stereo. Hrs opn: 24 Box 128, Hwy. 65 N., 65340. Phone: (660) 886-7422. Fax: (660) 886-6291. Licensee: Missouri

Valley Broadcasting Inc. (acq 11-19-84; with co-located AM; FTR: 12-10-84). Population served: 40,000 Natl. Network: CBS. Format: Country. Target aud: General. ♦Mike Phillips, pres & progmg dir; John Wilson, gen mgr; Peter Hollabaugh, gen sls mgr.

KMMO(AM)— May 29, 1949: 1300 khz; 1 kw-D, 68 w-N. TL: N39 08 03 W93 13 19.24

***KMVC(FM)**— Nov 1, 1968: 91.7 mhz; 100 w vert. 51 ft TL: N39 06 31 W93 11 29. (CP: 93.1 mhz, 16 w). Stereo. Hrs opn: 7 AM-11 PM (M-F); 9 AM-11 PM (S); noon-8 PM (Su) Missouri Valley College, 500 E. College St., 65340. Phone: (660) 831-4193. Fax: (660) 886-9818. E-mail: kmvc@moval.edu Licensee: Missouri Valley College. Format: Alternative, rhythm and blues. News: 3 hrs wkly. Target aud: 17-26; pre-, current & post-college age. Spec prog: Black 10 hrs, progsv 10 hrs, relg 16 hrs, classic rock 4 hrs, hip hop 10 hrs, urban 10 hrs wkly. ♦Brent Foster, gen mgr; Josh Branch, stn mgr.

Marshfield

KKLH(FM)— June 1982: 104.7 mhz; 34 kw. Ant 594 ft TL: N37 12 21 W92 54 20. Stereo. Hrs opn: 24 319 B-East Battlefield, Springfield, 65807. Phone: (417) 886-5677. Fax: (417) 886-2155. E-mail: info@1047thecave.com Web Site: www.1047thecave.com. Licensee: MW SpringMo Inc. Group owner: The Mid-West Family Broadcast Group (acq 1996; $1.8 million). Population served: 295,300 Natl. Rep: McGavren Guild. Law Firm: Davis Wright Tremaine, LLP. Format: Classic rock. Target aud: 35-54. ♦Rick McCoy, pres & gen mgr; Mary Fleenor, opns mgr; Malcolm Hurriede, gen sls mgr; Keith Abercrombie, rgnl sls mgr; John Kimmons, progmg VP & progmg mgr; Chris Louzander, system mgr.

KMRF(AM)— Nov 1, 1969: 1510 khz; 250 w-D. TL: N37 20 55 W92 54 28. (CP: 1 kw). Hrs opn: Sunrise-sunset 3208 State Hwy 00, 65706-2438. Secondary address: Box 693 65706. Phone: (417) 468-6188. Fax: (417) 859-2916. Licensee: New Life Evangelistic Center Inc. (acq 4-4-94). Population served: 50,000 Natl. Network: USA. Rgnl. Network: Missourinet, Brownfield. Format: Southern gospel. News staff: one; News: 6 hrs wkly. Target aud: General. ♦Ed Moore, opns mgr.

KNLM(FM)—Co-owned with KMRF(AM).Not on air, target date: unknown: 91.9 mhz; 3 kw. 210 ft TL: N37 19 09 W92 57 43. Rebroadcasts KNLG(FM) New Bloomfield 100%. Format: Contemp Christian. Target aud: General.

Maryville

KNIM(AM)— 1953: 1580 khz; 500 w-D, 7 w-N. TL: N40 23 31 W94 58 04. Hrs open: 24 Box 278, 64468. Secondary address: 1618 S. Main 64468. Phone: (660) 582-2151. Fax: (660) 582-3211. E-mail: knim@knimmaryville.com Web Site: knimmaryville.com. Licensee: Nodaway Broadcasting Corp. (acq 5-14-2003; $50,000 for 10% of stock with co-located FM). Population served: 30,000 Natl. Network: CNN Radio. Natl. Rep: Keystone (unwired net). Format: News, sports. News staff: one. Target aud: 25-54. Spec prog: Farm 5 hrs wkly. ♦Joyce Cronin, pres; Jim Cronin, exec VP & gen mgr.

KNIM-FM— September 1972: 97.1 mhz; 21.5 kw. Ant 354 ft TL: N40 23 31 W94 58 04. Stereo. 24 Web Site: www.knimmaryville.com.120,000 Law Firm: Wiley, Rein & Fielding. Format: Rock. News staff: one; News: 25 hrs wkly.

***KXCV(FM)**— 1971: 90.5 mhz; 100 kw. 500 ft TL: N40 21 36 W94 53 00. Stereo. Hrs opn: 24 Wells Hall, 800 University Dr., 64468. Phone: (660) 562-1163. Phone: (660) 562-1164. Fax: (660) 562-1832. Web Site: www.kxcv.org. Licensee: Northwest Missouri State University. Population served: 138,000 Natl. Network: NPR, PRI. Format: News, class, jazz. News staff: 2; News: 39 hrs wkly. Target aud: General. ♦Dean L. Hubbard, pres; Patty Andrews Holley, gen mgr & mus dir; Patty Holley, opns mgr; Gayle Hull, dev mgr, mktg dir & spec ev coord; Marcia Fish, progmg mgr & traf mgr; John Coffey, news dir; Kirk Wayman, news dir; Charles Maley, engrg dir.

Memphis

KMEM-FM— Mar 29, 1982: 100.5 mhz; 25 kw. 298 ft TL: N40 29 59 W92 09 58. Stereo. Hrs opn: 24 Box 121, 63555. Secondary address: 650 N. Clay 63555. Phone: (660) 465-7225/465-2715. Fax: (660) 465-2626. E-mail: mdenney@kmemfm.com Web Site: www.kmemfm.com. Licensee: Boyer Broadcasting Co. Inc. (acq 2-14-01; $202,000). Population served: 50,000 Rgnl. Network: Missourinet, Iowa Radio Net., Brownfield. Wire Svc: AP Format: Country. News staff: one; News: 15 hrs wkly. Target aud: General; adult audience 30+. Spec prog: Farm 8 hrs, relg 4 hrs wkly. ♦Mark McVey, pres; Karen McVey, VP & gen mgr; Mark Denney, gen mgr.

Mexico

***KJAB-FM**— Oct 9, 1985: 88.3 mhz; 4.8 kw. 272 ft TL: N39 06 13 W91 53 35. Stereo. Hrs opn: 24 621 W. Monroe, 65265. Phone: (573) 581-8606. Fax: (573) 581-9655. E-mail: kjab@kjab.com Web Site: www.kjab.com. Licensee: Mexico Educational Broadcasting Foundation. Population served: 250,000 Natl. Network: USA. Format: Southern gospel. News staff: one; News: 2 hrs wkly. Target aud: General. Spec prog: Gospel 20 hrs, relg 20 hrs wkly. ♦Kevin Weber, pres, gen mgr & opns mgr.

KWWR(FM)— Dec 14, 1966: 95.7 mhz; 100 kw. 1,181 ft TL: N39 15 39 W92 08 06. Stereo. Hrs opn: 24 Box 475, 65265-0475. Secondary address: 1705 E. Liberty St. 65265-0475. Phone: (573) 581-5500. Fax: (573) 581-1801. E-mail: kwwr@country96.com Licensee: KXEO Radio Inc. (acq 2-4-91; with co-located AM; FTR: 2-18-91). Population served: 638,605 Natl. Network: CNN Radio, Westwood One. Format: Country. Target aud: 24-54. Spec prog: Farm 4 hrs wkly. ♦Anne Johnson, pres; Gary Leonard, gen mgr; Greg Holman, progmg dir.

KXEO(AM)—Co-owned with KWWR(FM). Dec 3, 1948: 1340 khz; 1 kw-U. TL: N39 10 01 W91 51 44.24 Phone: (573) 581-2340.368,605 Natl. Network: CNN Radio, Westwood One. Format: Adult contemp. News: 4 hrs wkly. Target aud: 25-54.

Miner

KBHI(FM)—Licensed to Miner. See Sikeston

Moberly

***KBKC(FM)**— 2004: 90.1 mhz; 250 w. Ant 256 ft TL: N39 24 39 W92 26 46. Hrs open: 4424 Hampton Ave., Saint Louis, 63109. Phone: (314) 752-7000. Web Site: www.covenantnet.net. Licensee: Covenant Network. (acq 3-30-2004; $112,500 with WHOJ(FM) Terre Haute, IN). Format: Christian. ♦Tony Holman, gen mgr.

KRES(FM)—Listing follows KWIX(AM).

KWIX(AM)— June 1950: 1230 khz; 1 kw-U. TL: N39 24 11 W92 25 57. Hrs open: 24 Box 619, 65270. Secondary address: 300 West Reed 65270. Phone: (660) 263-1600. Fax: (660) 269-8811. Web Site: www.regionalradio.com. Licensee: KWIX Inc. Group owner: Shepherd Group Population served: 212,000 Natl. Network: CBS. Wire Svc: NOAA Weather Format: Talk. News staff: 4; News: 30 hrs wkly. Target aud: General. ♦David Shepherd, pres & gen mgr; Howard Miedler, gen sls mgr; Ken Kujawa, progmg dir; Stephanie Ross, mus dir; Brad Boyer, news dir; Lloyd Collins, chief of engrg.

KRES(FM)—Co-owned with KWIX(AM). October 1966: 104.7 mhz; 100 kw. 1,025 ft TL: N39 27 53 W92 42 07. Stereo. Web Site: www.regionalradio.com.928,800 Format: Country.

KZZT(FM)— Apr 10, 1987: 105.5 mhz; 25 kw. 328 ft TL: N39 24 54 W92 24 36. (CP: 50 kw). Stereo. Hrs opn: Box 128, 65270. Secondary address: 107 S. Main St., Brookfield 65270. Phone: (660) 263-9390. Fax: (660) 258-3707. E-mail: kzzt@bestbroadcastgroup.com Licensee: FM 105 Inc. Group owner: Best Broadcast Group (acq 7-9-97; $200,000 for 43%). Natl. Network: ABC. Law Firm: Bryan Cave. Format: Oldies. News: 5 hrs wkly. Target aud: 25-54; men & women with spendable income. ♦Phil Chirillo, pres; Dale A. Palmer, gen mgr.

Monett

KKBL(FM)—Listing follows KRMO(AM).

KRMO(AM)—Cassville, August 1950: 990 khz; 2.5 kw-D, 47 w-N. TL: N36 56 15 W93 55 30. Hrs open: 24 1569 N. Central, 65708. Phone: (417) 235-6041. Fax: (417) 235-6388. E-mail: kkbl@talonbroadcasting.com Web Site: www.krmo.com. Licensee: Eagle Broadcasting Inc. (acq 8-4-03; $650,000 with KKBL(FM) Monett). Population served: 6,500 Rgnl. Network: Brownfield, Missourinet. Format: Country. News staff: one; News: 10 hrs wkly. Target aud: 35 plus; business professionals, farmers, elderly. ♦Dale D. Gandy, pres; Duane Gandy, gen mgr & gen sls mgr; Janet Gandy, stn mgr.

KKBL(FM)—Co-owned with KRMO(AM). December 1977: 95.9 mhz; 6 kw. 269 ft TL: N36 56 15 W93 55 30. Stereo. 24 Web Site: www.buzz959.com.65,000 Format: CHR. News staff: one; News: 4 hrs wkly. Target aud: General; young, adults, families. Spec prog: Children 2 hrs wkly.

Monroe City

KWBZ(FM)— July 4, 1981: 107.5 mhz; 10 kw. Ant 328 ft TL: N39 35 12 W91 47 57. Stereo. Hrs opn: 3702 Palmayra Rd., Hannibal, 63401. Phone: (573) 221-4000. Fax: (573) 221 1142. E-mail: thebreeze@socket.net Licensee: WPW Broadcasting Inc. (acq 8-17-2000). Format: Soft adult classic rock. Spec prog: Gospel 4 hrs wkly. ♦Phil Alexander, gen mgr.

Montgomery City

KMCR(FM)— Aug 15, 1977: 103.9 mhz; 3 kw. 300 ft TL: N38 59 12 W91 30 48. Stereo. Hrs opn: 24 205 E. Norman St., 63361. Phone: (573) 564-2275. Fax: (573) 564-8036. E-mail: kmcr@socket.net Web Site: www.bestbroadcastgroup.com. Licensee: Chirillo Electronics Inc. Group owner: Best Broadcast Group (acq 1994). Population served: 50,000 Rgnl. Network: Missourinet. Format: Hot adult contemp. News staff: one; News: 3 hrs wkly. Target aud: 25-60; male/female. Spec prog: Farm 2 hrs, relg 2 hrs wkly. ♦Dale A. Palmer, VP & gen mgr.

Mount Vernon

KZRQ-FM— July 29, 1993: 106.7 mhz; 25 kw. 328 ft TL: N37 09 16 W93 36 58. Hrs open: 2330 W. Grand St., Springfield, 65781. Phone: (417) 865-6614. Fax: (417) 865-9643. Web Site: www.2rocks.com. Licensee: Journal Broadcast Corp. Group owner: Journal Communications Inc. (acq 11-26-2003; $5 million with KSGF-FM Ash Grove). Format: Rock. ♦Rex Hansen, gen mgr; Chris Cannon, opns mgr; Janelle Carter, gen sls mgr.

Mountain Grove

KELE(AM)— Nov 16, 1954: 1360 khz; 1 kw-D, 60 w-N. TL: N37 08 07 W92 14 59. (CP: COL Ripley, OH. 1180 khz; 1 kw-D, 1 kw-CH. TL: N38 38 55 W84 00 42). Hrs open: 24 800 N. Hubbard, 65711. Phone: (417) 926-4650. Fax: (417) 926-7604. Licensee: Quorum Radio Partners Inc. (group owner; (acq 8-8-2002; grpsl). Population served: 50,000 Natl. Network: USA. Rgnl. Network: Brownfield. Format: Country. News staff: one; News: 10 hrs wkly. ♦Todd Fowler, CEO; Rick Vermillion, gen mgr; Tonya Shannon, traf mgr; Perry Dobson, local news ed & sports cmtr; Terry Dobson, sports cmtr.

KELE-FM— Jan 1, 1977: 92.5 mhz; 3 kw. Ant 299 ft TL: N37 08 07 W92 14 59. Stereo. 50,000 Format: Real country. News staff: one; News: 8 hrs wkly. Target aud: 24-59; general. Spec prog: Relg 3 hrs wkly.

Mountain View

KUPH(FM)— July 31, 1998: 96.9 mhz; 50 kw. 420 ft TL: N36 59 29 W91 47 41. Stereo. Hrs opn: 24 6962 U.S. Hwy. 60 W., 65548. Phone: (417) 934-1000. Phone: (417) 934-0969. Fax: (417) 934-2565. E-mail: ed@thefox969.com Web Site: www.thefox969.com. Licensee: Central Ozark Radio Network Inc. (acq 8-21-98; $196,500). Format: Hot adult contemp. News: 2 hrs wkly. Target aud: 25-54; upscale, mature individuals. ♦Tom Marhefka, CEO & pres; Bob Eckman, gen mgr & opns mgr; Gary Lee, opns mgr & progmg dir.

Naylor

KZMA(FM)— 2005: 99.9 mhz; 4.2 kw. Ant 387 ft TL: N36 39 43 W90 29 16. Hrs open: 1115 Nooney Dr., Poplar Bluff, 63901. Phone: (573) 778-1219. Fax: (573) 686-2377. Licensee: Daniel S. Stratemeyer (acq 6-1-2003; $30,000 for CP). Law Firm: Shaw Pittman LLP. Format: Adult contemp. ♦Jim Borders, gen mgr.

Neosho

KBTN-FM—Listing follows KQYS(AM).

***KNEO(FM)**— October 1986: 91.7 mhz; 4.6 kw. 374 ft TL: N36 52 49 W94 26 59. Hrs open: 24 10827 E. Hwy. 86, 64850. Phone: (417) 451-5636. Fax: (417) 451-1891. E-mail: cgr@kneo.org Web Site: www.kneo.org. Licensee: Sky High Broadcasting Corp. (acq 6-19-00). Population served: 250,000 Natl. Network: USA, Moody. Format: News/talk. News staff: 3; News: 10 hrs wkly. Target aud: 21-50; rural people & older shut-ins. ♦Mark Taylor, pres & gen mgr; Adam Winkler, opns dir; Roberta Foster, traf mgr.

KQYS(AM)— Feb 1, 1954: 1420 khz; 1 kw-D, 500 w-N, DA-N. TL: N36 50 52 W94 19 12. Hrs open: 5 AM-midnight Box 570, 64850. Secondary address: 216 W. Spring 64850. Phone: (417) 451-1420. Fax: (417) 451-2526. Licensee: FFD Holdings I Inc. Group owner:

Petracom Media L.L.C. (acq 12-20-2004; grpsl). Population served: 40,000 Rgnl. Network: Missourinet. Format: Talk. News staff: 2; News: 14 hrs wkly. Target aud: 18-54. Spec prog: Farm 6 hrs wkly. ♦Gail Johnson, gen mgr & gen sls mgr; David Horrath, news dir; Art Morris, chief of engrg; Monica Blain, progmg mgr, traf mgr & farm dir.

KBTN-FM—Co-owned with KQYS(AM). 1995: 99.7 mhz; 4.2 kw. Ant 393 ft TL: N36 46 05 W94 19 52.24 2510 W. 20th St, Joplin, 64804. Phone: (417) 781-1313. Fax: (417)781-1316. Format: Country. News staff: 2; News: 9 hrs wkly. Target aud: 18 plus. ♦Jennifer Isom, gen mgr.

Nevada

KNEM(AM)— 1949: 1240 khz; 500 w-U. TL: N37 51 37 W94 22 54. Hrs open: 24 Box 447, 414 E. Walnut., 64772. Phone: (417) 667-3113. Fax: (417) 667-9797. E-mail: mharbit@knemknmo.com Web Site: www.knemknmo.com. Licensee: Harbit Communications Inc. (acq 12-5-97; $475,000 with co-located FM). Population served: 30,000 Rgnl. Network: Brownfield, Missourinet. Format: Country. News staff: one; News: 30 hrs wkly. Target aud: General. Spec prog: Farm one hr, Christian 5 hrs wkly. ♦Mike Harbit, pres, gen mgr, gen sls mgr & progmg dir; Russ Warren, news dir; Daryl Nickolaus, chief of engrg.

KNMO(FM)—Co-owned with KNEM(AM). Sept 10, 1984: 97.5 mhz; 6 kw. Ant 281 ft TL: N37 52 45 W94 20 15. Stereo. 24 Web Site: www.knemknmo.com.30,000

New Bloomfield

***KNLG(FM)**— July 20, 1997: 90.3 mhz; 150 w. 216 ft TL: N38 42 16 W92 05 20. Stereo. Hrs opn: 24 c/o KNLJ(TV) Box 2525, 65603. Phone: (573) 896-5105. Fax: (573) 896-4376. Web Site: www.heresshelpnet.org. Licensee: New Life Evangelistic Center Inc. Population served: 2,000 Format: Southern gospel. News: 9 hrs wkly. Target aud: General. ♦Rev. Larry Rice, pres & gen mgr.

New London

KZZK(FM)— April 1996: 105.9 mhz; 10 kw. 515 ft TL: N39 43 45 W91 24 15. Stereo. Hrs opn: 24 329 Maine St., Quincy, IL, 62301. Phone: (217) 224-4102. Fax: (217) 224-4133. E-mail: kzzk@staradio.com Web Site: www.kzzk.com. Licensee: STARadio Corp. (group owner; acq 12-2-98; $2.1 million with KGRC(FM) Hannibal). Population served: 200,000 Format: Adult alternative, classic rock. News: 2 hrs wkly. Target aud: 18-49; skews male. ♦Howard Doss, pres; Michael J. Moyers, gen mgr; "Quaid", progmg dir.

New Madrid

KTMO(FM)—Licensed to New Madrid. See Portageville

Nixa

KGBX-FM—Licensed to Nixa. See Springfield

North Kansas City

WDAF-FM—See Liberty

Osage Beach

KMYK(FM)—Listing follows KRMS(AM).

KRMS(AM)— December 1952: 1150 khz; 1 kw-D, 55 w-N. TL: N38 07 29 W92 40 39. Hrs open: 24 Box 225, Hwy. 54, 65065. Phone: (573) 348-2772. Fax: (573) 348-2779. Licensee: Viper Communications Inc. Group owner: Viper Communications Broadcast Group (acq 11-97; $500,000 with co-located FM). Population served: 75,000 Natl. Network: CBS. Rgnl. Network: Missourinet. Format: News/talk. ♦Ken Kuenzie, pres & chief of engrg; Dennis Klautzer, VP, sls dir & progmg dir; Paul Hannigan, news dir; Tammy Pitts, traf mgr.

KMYK(FM)—Co-owned with KRMS(AM). Apr 12, 1964: 93.5 mhz; 39 kw. Ant 551 ft TL: N38 07 29 W92 40 39. Stereo. 85,000 Format: AOR, classic rock.

Osceola

***KCVJ(FM)**— June 29, 1990: 100.3 mhz; 6 kw. Ant 282 ft TL: N38 03 43 W93 33 24. Hrs open: 24
Rebroadcasts KCVO-FM Camdenton 100%.
Box 800, c/o Spirit FM Radio, Camdenton, 65020. Phone: (573) 346-3200. Fax: (573) 346-1010. E-mail: email@spiritfm.org Web Site: www.spiritfm.org. Licensee: Lake Area Educational Broadcasting Foundation (acq 1999; $70,000). Population served: 11,000 Format: Div, Christian. Target aud: 25-45. ♦Alice McDermott, CFO; James McDermott, pres & gen mgr.

Otterville

***KCVK(FM)**— 2001: 107.7 mhz; 2.7 kw. Ant 499 ft TL: N38 39 21 W92 54 27. Stereo. Hrs opn: 24
Rebroadcasts KCVO-FM Camdenton 100%.
Box 800, % Spirit FM Radio, Camdenton, 65020. Phone: (573) 346-3200. Fax: (573) 346 1010. E-mail: email@spirit.org Web Site: www.spiritfm.org. Licensee: Lake Area Educational Broadcasting Foundation (acq 8-15-01; at least $450,000 including two-year noncompete agreement). Population served: 100,000 Format: Christian, div. ♦Alice McDermott, CFO; James McDermott, pres & gen mgr.

Overland

***KRHS(FM)**— Nov 7, 1977: 90.1 mhz; 10 w. 60 ft TL: N38 42 38 W90 21 22. Hrs open: 9100 St. Charles Rock Rd., St. Louis, 63114. Phone: (314) 429-7111. Fax: (314) 429-6725. Licensee: Ritenour Consolidated School District. Format: Educ. Target aud: General. ♦Jane Bannester, gen mgr.

Owensville

KXMO-FM— Jan 1, 2001: 95.3 mhz; 37 kw. 564 ft TL: N38 08 06 W91 23 59. Hrs open: 24 Box 4584, Springfield, 65808. Phone: (417) 883-9180. Fax: (417) 883-9096. Licensee: KTTR-KZNN Inc. (acq 8-23-2001). Format: Oldies. Target aud: 35-64; male & female. ♦John B. Mahaffey, chmn; Robert B. Mahaffey, pres.

Ozark

KOMG(FM)— 1995: 92.9 mhz; 50 kw. Ant 492 ft TL: N36 58 26 W93 25 37. Stereo. Hrs opn: 24 319 E. Battlefield, Suite B, Springfield, 65807. Phone: (417) 886-5677. Fax: (417) 886-2155. E-mail: info@basscountry.fm Web Site: www.basscountry.fm. Licensee: MW Springmo Inc. Group owner: The Mid-West Family Broadcast Group (acq 12-15-99). Population served: 295,300 Natl. Rep: McGavren Guild. Law Firm: Davis Wright Tremaine, LLP. Format: Classic country. Target aud: 30-50. ♦Rick McCoy, pres & gen mgr; Mary Fleenor, opns mgr; Malcolm Hukriede, gen sls mgr; Keith Abercrombie, rgnl sls mgr; John Kimmons, progmg dir; Charlie Mason, prom.

Palmyra

KICK-FM— Sept 1, 1981: 97.9 mhz; 50 kw. 348 ft TL: N39 45 25 W91 29 57. Stereo. Hrs opn: 24 Box 711, Hannibal, 63401-0711. Phone: (573) 221-3450. Fax: (573) 221-5331. Web Site: www.979kickfm.com. Licensee: Bick Broadcasting Co. Population served: 100,000 Natl. Rep: McGavren Guild. Format: Country. News staff: 2. Target aud: 25-54; mainstream adults. ♦Ed Foxall, gen mgr.

Park Hills

***KBGM(FM)**— 2001: 91.1 mhz; 8 kw. Ant 620 ft TL: N37 48 04 W90 33 51. Hrs open: Box 3206, American Family Radio, Tupelo, MS, 38803. Phone: (662) 844-8888, EXT. 204. Fax: (662) 842-6791. Web Site: www.afr.net. Licensee: American Family Association. Group owner: American Family Radio Format: Inspirational Christian. ♦Marvin Sanders, gen mgr.

KFMO(AM)— July 1947: 1240 khz; 1 kw-U. TL: N37 51 10 W90 31 13. Hrs open: Box 36, 63601. Secondary address: 804 St. Joe Dr. 63601. Phone: (573) 431-2000. Fax: (573) 431-0850. E-mail: radio@b104fm.com Web Site: www.kfmo.com. Licensee: MKS Broadcasting Inc. (acq 3-16-92; FTR: 4-6-92). Population served: 74,000 Natl. Network: Westwood One. Format: Sports, news/talk loc information. Target aud: 25-54; females. ♦M.L. Steinmetz, pres; Larry D. Joseph, VP & gen mgr; Kelly Valle, gen sls mgr; Greg Camp, progmg dir; Gib Collins, news dir.

Parkville

***KGSP(FM)**— April 1972: 90.3 mhz; 100 w. 140 ft TL: N39 11 24 W94 40 49. Stereo. Hrs opn: 6 AM-midnight (M-F); 9 AM-midnight (S, Su) Box 2, 8700 N.W. River Park Dr., 64152. Phone: (816) 741-2000. Fax: (816) 741-4911. Web Site: www.park.com. Licensee: Board of Trustees of Park College. Population served: 57,000 Format: Alternative, var. News: 6 hrs wkly. Target aud: General; college students. Spec prog: Jazz 14 hrs, gospel 3 hrs, blues 12 hrs wkly. ♦Steve Youngblood, gen mgr.

Perryville

KBDZ(FM)— Jan 30, 1990: 93.1 mhz; 1.6 kw. ant 623 ft TL: N37 38 56 W89 56 21. Stereo. Hrs opn: 24 Box 344, 122 Perry Plaza, 63775. Secondary address: Box 428, Radio Hill, St. Genevieve 63670. Phone: (573) 883-2980. Phone: (573) 547-6780. Fax: (573) 883-2866. Fax: (573) 547-8005. E-mail: news@suntimenews.com Web Site: www.suntimesnews.com. Licensee: Donze Communications Inc. Format: Hot country, news. News staff: 3; News: 5 hrs wkly. Target aud: 25-54. Spec prog: Sports 5 hrs, relg 5 hrs, farm 2 hrs wkly. ♦Elmo L. Donze, pres, gen mgr & chief of engrg; Bob Scott, gen sls mgr & progmg dir; Don Pritchard, news dir, local news ed, edit dir, outdoor ed & women's cmtr; Brian Snider, reporter & mus critic; Susan White, relg ed; Michelle Hoog, women's int ed.

Piedmont

KPWB(AM)— May 16, 1966: 1140 khz; 1 kw-D. TL: N37 08 29 W90 42 11. Hrs open: 235 Business HH, 63957. Phone: (573) 223-4218. Fax: (573) 223-2351. Licensee: Dockins Communications Inc. (acq 7-28-2006; with co-located FM). Population served: 5,000 Natl. Network: USA. Rgnl. Network: Missourinet. Format: Gospel. Target aud: 18 plus; Christians. ♦Fred Dockins Sr., pres; Wanda Emert, gen mgr, mktg dir, progmg dir & news dir.

KPWB-FM— Sept 5, 1985: 104.9 mhz; 3 kw. 300 ft TL: N37 07 54 W90 41 28. Stereo. 12,350 Natl. Network: USA. Format: Country. Target aud: General. ♦Wanda Emert, opns mgr, gen sls mgr, mktg mgr, prom mgr & progmg mgr; Fred Dockins, chief of engrg.

Pleasant Hope

KTOZ-FM— May 1, 1993: 95.5 mhz; 50 kw. 497 ft TL: N37 25 32 W93 16 38. Hrs open: 1856 S. Glenstone, Springfield, 65804. Phone: (417) 890-5555. Fax: (417) 890-5050. E-mail: alice955@alice955.com Web Site: www.alice955.com. Licensee: Clear Channel Broadcasting Licenses Inc. Group owner: Clear Channel Communications Inc. (acq 10-10-00; grpsl). Format: Modern adult contemp. Target aud: 18-34; young adults, 60/40 female/male split. ♦Paul Windisch, gen mgr; Paul Kelley, opns dir.

Point Lookout

***KCOZ(FM)**— January 1995: 91.7 mhz; 200 w. 151 ft TL: N36 36 40 W93 14 29. Stereo. Hrs opn: College of the Ozarks, 65726. Phone: (417) 334-6411. Fax: (417) 335-2618. Licensee: College of the Ozarks. Population served: 500,000 Natl. Network: PRI, NPR. Format: News/talk, jazz, new age, blues. Target aud: Older & educated. Spec prog: Folk 10 hrs, new age 10 hrs wkly. ♦Ini Offong, opns VP, news dir & pub affrs dir; Courtney Hutton, progmg VP & mus dir; Jae Jones, gen mgr & chief of engrg.

***KSMS-FM**— Feb 12, 1962: 90.5 mhz; 8.5 kw. 768 ft TL: N36 33 44 W93 15 35. Hrs open:
Rebroadcasts KSMU(FM) Springfield 100%.

Missouri State Univ., 901 S. National Ave., Springfield, 65897. Phone: (417) 836-5878. Fax: (417) 836-5889. E-mail: ksmu@missouristate Web Site: www.ksmu.org. Licensee: Board of Governors, Southwest Missouri State University (acq 6-21-93; FTR: 7-19-93). Population served: 36,000 Law Firm: Dow, Lohnes & Albertson. Format: NPR News & Classical Music. Target aud: 25-54. ♦Tammy Wiley, gen mgr & stn mgr.

Poplar Bluff

KAHR(FM)— Mar 3, 1985: 96.7 mhz; 6 kw. Ant 328 ft TL: N36 45 59 W90 28 52. Stereo. Hrs opn: 932 County Rd. 448, 63901. Phone: (573) 686-3700. Fax: (573) 686-1713. Web Site: www.foxradionetwork.com. Licensee: Eagle Bluff Enterprises (acq 8-3-93; $350,000; FTR: 8-30-93). Format: Adult hits. Target aud: 18-54; listeners living in the middle-class strata. ♦Steven C. Fuchs, gen mgr.

KJEZ(FM)— Aug 20, 1977: 95.5 mhz; 100 kw. 860 ft TL: N36 50 50 W90 19 52. Stereo. Hrs opn: 24 1015 West Pine St., 63901. Phone: (573) 785-0881. Fax: (573) 785-0646. Web Site: www.kjez.com. Licensee: MRR License LLC. Group owner: MAX Media L.L.C. (acq 6-2-2004; grpsl). Population served: 125,000 Natl. Network: Westwood One. Format: Classic rock and roll. News: 10 hrs wkly. Target aud: 18-49; general. ♦John Rice, gen mgr.

KKLR(FM)—Listing follows KWOC(AM).

KLID(AM)— May 22, 1961: 1340 khz; 1 kw-U. TL: N36 46 03 W90 22 11. Hrs open: 24 KLID Bldg., 102 N. 11th St., 63901. Phone: (573) 686-1600. Fax: (573) 785-9844. Licensee: Browning Skidmore Broadcasting Inc. (acq 5-21-93; FTR: 6-14-93). Population served: 45,000 Format: Oldies, talk, sports. News: 15 hrs wkly. Target aud: 18-54; upper class, professionals. Spec prog: Relg 2 hrs, Black 2 hrs wkly. ♦Chris Browning, pres; Dolores Skidmore, gen mgr, dev mgr & progmg dir; Alverna Skidmore, sls dir, pub affrs dir & traf mgr; Palmer Johnson, chief of engrg; Nick Novak, sports cmtr; Dave Michaels, disc jockey; Paul White, disc jockey; Sunny Skidmore, disc jockey.

KLUE(FM)— Jan 1, 1995: 103.5 mhz; 50 kw. 492 ft TL: N36 53 56 W90 18 27. Hrs open: 24 6120 Waldo Church Rd., Metropolis, IL, 62960. Phone: (618) 564-9836. Fax: (618) 564-3202. Licensee: Benjamin Stratemeyer (acq 5-1-02; $800,000). Law Firm: Shaw Pittman. Format: Div. ♦Samuel Stratemeyer, gen mgr; Willie Kerns, opns mgr & progmg dir.

***KLUH(FM)**— Oct 8, 1988: 90.3 mhz; 25 kw. 300 ft TL: N36 43 07 W90 23 48. Hrs open: Box 1313, 63902-1313. Phone: (573) 686-1663. Fax: (573) 686-7703. Web Site: www.unity903.org. Licensee: Word of Victory Outreach Center Inc. (acq 4-5-95; FTR: 7-10-95). Format: Relg. Target aud: General. ♦David Craig, gen mgr.

***KOKS(FM)**— Oct 2, 1988: 89.5 mhz; 100 kw. 423 ft TL: N36 48 40 W90 27 50. Hrs open: 24 Box 280, 63902. Phone: (573) 686-5080. Fax: (573) 686-5544. E-mail: koksradio@mycitycable.com Web Site: www.koksradio.org. Licensee: Calvary Educational Broadcasting Foundation. Population served: 1,000,000 Format: Christian, gospel, relg. News: 14 hrs wkly. Target aud: General. ♦Don Stewart, gen mgr; Nina Stewart, stn mgr & progmg dir; Ben Stewart, mus dir; Charles Lampley, chief of engrg.

KPPL(FM)— 2003: 92.5 mhz; 25 kw. Ant 328 ft TL: N36 50 59 W90 22 20. Stereo. Hrs opn: 932 County Rd. 448, 63901. Phone: (573) 686-3700. Fax: (573) 686-1713. Licensee: George S. Flinn Jr. Population served: 220,000 Format: Country. ♦Steven Fuchs, gen mgr.

KWOC(AM)— May 10, 1938: 930 khz; 5 kw-D, 500 w-N, DA-N. TL: N36 43 15 W90 22 04. Hrs open: 24 Box 130, 63902. Phone: (573) 785-0881. Fax: (573) 785-0646. Web Site: www.kwoc.com. Licensee: MRR License LLC. Group owner: MAX Media L.L.C. (acq 6-2-2004; grpsl). Population served: 50,000 Rgnl. Network: Missourinet, Brownfield. Format: News/talk. News staff: 2; News: 5 hrs wkly. Target aud: 25-54; adults with middle to upper income. ♦John Rice, gen mgr; Katie Wylie, sls dir & gen sls mgr; Rick Carl, progmg dir & news dir; Charlie Lampe, chief of engrg; Pam Gray, traf mgr.

KKLR(FM)—Co-owned with KWOC(AM). 1952: 94.5 mhz; 100 kw. 807 ft TL: N36 43 18 W90 22 10. Stereo. Web Site: www.kklr.com.90,000 Format: Country. Target aud: 18-49. ♦Galen Stevens, progmg dir & mus dir.

Portageville

KMIS(AM)— Sept 1, 1960: 1050 khz; 1 kw-D, 87 w-N. TL: N36 25 30 W89 41 39. Hrs open: 24 Box 509, Kennett, 63857. Secondary address: 1303 Southwest Dr., Kennett 63857. Fax: (573) 888-4890. E-mail: ktme@il.net Licensee: Pollack Broadcasting Co. (group owner; (acq 5-7-2001; with KTMO(FM) New Madrid). Population served: 40,000 Rgnl. Network: missiourinet Format: News, sports. News staff: one. Target aud: General. Spec prog: Relg gospel 5 hrs wkly. ♦Bill Pollack, pres; Monte Lyons, opns mgr, prom dir & progmg dir; Perry Jones, gen mgr, sls dir & gen sls mgr; Charles Isabell, news dir; P.J. Johnson, chief of engrg; Evelyn Bailey, traf mgr & sports cmtr.

KTMO(FM)—Co-owned with KMIS(AM). Jan 31, 1976: 106.5 mhz; 50 kw. 469 ft TL: N36 25 30 W89 41 39. Stereo. 24 100,000 Natl. Network: ABC. Format: Country. News staff: one. Target aud: 18-49 adults; males.

Potosi

KHZR(FM)— Apr 17, 1997: 97.7 mhz; 9.4 kw. Ant 528 ft TL: N37 52 51 W90 47 01. Hrs open: 24 13358 Manchester Rd., Suite 100, Des Peres, 63131. Phone: (314) 909-8569. Fax: (314) 835-9739. E-mail: info@joyfmonline.org Web Site: www.joyfmonline.org. Licensee: Four Him Enterprises L.L.C. (acq 11-2-2000; $1.2 million). Format: Contemp Christian. ♦Sandi Brown, gen mgr.

***KNLP(FM)**— April 1998: 89.7 mhz; 2.3 kw. 262 ft TL: N37 55 42 W90 46 02. Hrs open: New Life Evangelistic Center, 1411 Locust St., St. Louis, 63103. Phone: (314) 436-2424. Phone: (573) 438-1473. Fax: (314) 436-2434. E-mail: larryr@hereshelpnet.org Web Site: www.hereshelpnet.org. Licensee: New Life Evangelistic Center Inc. Format: Adult contemp, gospel, talk. ♦Larry Rice, gen mgr.

KYRO(AM)— Feb 22, 1959: 1280 khz; 500 w-D. TL: N37 58 28 W90 45 44. Stereo. Hrs opn: 24 Box 176, Irondale, 63648. Phone: (573) 438-2136. Fax: (573) 438-3108. E-mail: news@kyro.com Web Site: www.kyro.com. Licensee: KYRO Inc. (acq 5-12-2005; $145,000). Population served: 30,000 Natl. Network: ABC. Rgnl. Network: Missourinet Format: New country, news. News staff: one; News: 12 hrs wkly. Target aud: 25 plus; general. ♦Debra S. Porter, VP; James T. Porter, pres & gen mgr.

Republic

KADI-FM— June 18, 1990: 99.5 mhz; 6 kw. 328 ft TL: N37 09 54 W93 23 44. Stereo. Hrs opn: 24 5431 W. Sunshine, Springfield, 65619. Phone: (417) 831-0995. Fax: (417) 831-4026. Web Site: www.kadi.com. Licensee: Vision Communications Inc. (acq 7-10-2000; $550,000). Format: Adult contemp Christian mus. News: 8 hrs wkly. Target aud: General; adults in their mid 30s. ♦R.C. Aner, gen mgr.

Richmond

KAYX(FM)— Aug 1, 1990: 92.5 mhz; 6 kw. 500 ft TL: N39 14 52 W93 58 16. Stereo. Hrs opn:
Rebroadcasts KCCV(AM) Overland Park, KS 85%.
111 W. Main St., 64085. Phone: (816) 470-9925. Fax: (816) 470-8925. E-mail: kayx@bottradionetwork.com Web Site: www.bottradionetwork.com. Licensee: Bott Communications, Inc. Group owner: Bott Radio Network (acq 1996). Format: Christian talk & info. ♦Richard P. Bott Sr., pres; Richard P. Bott II, VP.

Rolla

KDAA(FM)— Nov 20, 1964: 103.1 mhz; 2.05 kw. Ant 571 ft TL: N37 52 39 W91 44 45. Stereo. Hrs opn: 24 Box 727, 65402. Phone: (573) 364-2525. Fax: (573) 364-5161. Licensee: KDAA-KMOZ LLC. Group owner: Mahaffey Enterprises Inc. (acq 9-28-2001; $418,000 assumption of debt for 50% with co-located AM). Population served: 25,000 Format: Classic hits. Target aud: 18-44. ♦Joe Munsell, gen mgr.

***KMNR(FM)**— 1974: 89.7 mhz; 450 w. 230 ft TL: N37 57 12 W91 46 29. Hrs open: 24 Univ. of Missouri, 113 University Ctr. W., 65409-1440. Phone: (573) 341-4272. Phone: (573) 341-4273. Fax: (573) 341-6021. E-mail: kmnr@umr.edu Web Site: web.umr.edu/~kmnr/. Licensee: Curators of the University of Missouri. (group owner) Population served: 16,600 Natl. Network: AP Radio. Format: Div, educ. Target aud: 18-25; college community. Spec prog: Jazz 3 hrs wkly. ♦Fred Goss, gen mgr; Matt Rogers, stn mgr.

KMOZ(AM)— Aug 19, 1960: 1590 khz; 1 kw-D, 88 w-N. TL: N37 56 41 W91 48 40. Hrs open: Bott Radio Network, 10550 Barkley, Suite 108, Overland Park, KS, 66212. Phone: (913) 642-7770. Fax: (913) 642-1319. E-mail: comments@bottradionetwork.com Web Site: www.bottradionetwork.com. Licensee: Community Broadcasting Inc. (acq 5-5-2006; $40,000). Format: Country. Target aud: 50 plus; mature adults. ♦Trace Thurlby, COO; Tom Holdeman, CFO; Richard P Bott, VP; Richard P. Bott II, VP; Pat Rulon, natl sls mgr; Rachel Moser, mktg mgr; Jason Potocnik, traf mgr.

***KMST(FM)**— January 1964: 88.5 mhz; 100 kw. Ant 480 ft TL: N37 47 56 W91 43 28. Stereo. Hrs opn: 19 G-6 Library, 1870 Miner Cir., 65409-0130. Phone: (573) 341-4386. Fax: (573) 341-4889. E-mail: kumr@umr.edu Web Site: www.kmst.org. Licensee: The Curators of the University of Missouri. (group owner) Population served: 200,000 Natl. Network: NPR. Law Firm: Fisher, Wayland, Cooper, Leader & Zaragoza. Format: Class, div, news. News staff: one; News: 35 hrs wkly. Target aud: General. Spec prog: Bluegrass 5 hrs, jazz 3 hrs, folk 5 hrs wkly. ♦Jim Sigler, gen mgr; Tricia Crout, mktg mgr; John Francis, progmg dir; Charles Knapp, chief of engrg. Co-owned TV: *KOMU-TV affil.

KTTR(AM)— Sept 30, 1947: 1490 khz; 1 kw-U. TL: N37 56 42 W91 44 46. Hrs open: 24 Box 727, 65402. Secondary address: 1505 Soest Rd. 65401. Phone: (573) 364-2525. Fax: (573) 364-5161. Licensee: KTTR-KZNN Inc. Group owner: Mahaffey Enterprises Inc. (acq 6-1-84; with co-located FM; FTR: 4-16-84). Population served: 19,800 Rgnl. Network: Missourinet. Format: News/talk, sports. News staff: one. Target aud: General. ♦John Mahaffey, chmn & pres; Robert B. Mahaffey, pres & stn mgr; Joe Munsell, gen mgr; Sue Corey, gen sls mgr; Lee Buhr, news dir; Bob Moore, chief of engrg.

KZNN(FM)—Co-owned with KTTR(AM). Feb 12, 1973: 105.3 mhz; 100 kw. 631 ft TL: N37 52 39 W91 44 45. Stereo. 13,245 Natl. Network: ABC. Format: Modern country.

Saint Charles

***KCLC(FM)**— October 1968: 89.1 mhz; 35 kw. 257 ft TL: N38 47 12 W90 29 49. Stereo. Hrs opn: 24 209 S. Kings Hwy., 63301. Phone: (636) 949-4890. Phone: (636) 949-4891. Fax: (636) 949-4111. Web Site: www.lindenwood.edu/kclc. Licensee: Lindenwood University. Population served: 250,000 Format: Jazz, CHR. News: 12 hrs wkly. Target aud: 18-34; young adults. Spec prog: Bluegrass 12 hrs, progsv rock 12 hrs, gospel 9 hrs, contemp Christian 7 hrs wkly. ♦Dennis C. Spellmann, pres; Richard Reighard, opns mgr; Ralph Brancato, chief of engrg.

KFTK(FM)—See Florissant

KHOJ(AM)— Apr 13, 1958: 1460 khz; 5 kw-D, 85 w-N, DA-2. TL: N38 50 05 W90 28 08. Hrs open: 4424 Hampton Ave., Saint Louis, 63109. Phone: (314) 752-7000. E-mail: office@covenantnet.net Web Site: www.covenantnet.net. Licensee: Covenant Network. (acq 5-13-2005; $730,000). Population served: 2,500,000 ♦Tony Holman, gen mgr.

Saint James

KTTR-FM— 1994: 99.7 mhz; 12 kw. 472 ft TL: N37 56 41 W91 42 23. Stereo. Hrs opn: 24
Rebroadcasts KTTR(AM) Rolla 90%.
Box 4584, Springfield, MT, 65808. Secondary address: 1505 Soest Rd., Rolla 65808. Phone: (573) 364-2525. Fax: (573) 364-5161. Licensee: KTTR-KZNN Inc. Group owner: Mahaffey Enterprises Inc. Rgnl. Network: Missourinet. Format: News/talk. News staff: one; News: 20 hrs wkly. Target aud: 25-54. ♦John Mahaffey, chmn; Robert B. Mahaffey, pres.

Saint Joseph

KFEQ(AM)— Feb 16, 1926: 680 khz; 5 kw-U, DA-2. TL: N39 49 43 W94 48 20. Hrs open: 24 Box 8550, 64508. Secondary address: 4104 Country Ln. 64506. Phone: (816) 233-8881. Fax: (816) 279-8280. E-mail: garyexline@eagleradio.net Web Site: www.stjoeradio.com. Licensee: Eagle Communications Inc. Group owner: Eagle Communications Group (acq 3-20-69; grpsl; FTR: 4-8-91). Population served: 250,000 Natl. Network: ABC. Natl. Rep: Katz Radio. Law Firm: Wiley, Rein & Fielding. Wire Svc: AP Format: News/talk, Sports. News staff: 4; News: 50 hrs wkly. Target aud: 18 plus; adults. Spec prog: Farm 20 hrs wkly. ♦Gary Shorman, CEO; Gary Exline, VP & gen mgr; Kevin Wagner, opns dir.

KGNM(AM)— November 1985: 1270 khz; 1 kw-D, DA. TL: N39 44 39 W94 47 16. Hrs open: 24 2414 S. Leonard Rd., 64503. Phone: (816) 233-2577. Fax: (816) 233-2374. E-mail: kgnm@stjoelive.com Web Site: kgnmradio.com. Licensee: Orama Inc. (acq 6-80; $400,000; FTR: 6-30-80). Population served: 100,000 Natl. Network: USA. Format: Adult contemp, Christian music. Target aud: 30-55; conservative. ♦Rory Pullen, pres; Greg Glauser, VP; Chris Meikel, gen mgr; Marci Meikel, progmg dir.

KKJO(FM)—Listing follows KSFT(AM).

KSFT(AM)— June 1, 1946: 1550 khz; 5 kw-U, DA-N. TL: N39 42 23 W94 44 36. Hrs open: 24 Box 8550, 64508. Secondary address: 4104 Country Ln. 64506. Phone: (816) 233-8881. Fax: (816) 279-8280. Web Site: www.stjoeradio.com. Licensee: Eagle Communications Inc. Group owner: Eagle Communications Group (acq 3-1-99; $4 million with co-located FM). Population served: 150,000 Rgnl. Network: Missourinet. Law Firm: Wiley, Rein & Fielding. Format: Oldies. News: 5 hrs wkly. Target aud: 45-64. ♦Gary Shorman, CEO; Mark Vail, COO & gen mgr; Gary Exline, pres, gen mgr, sls dir & gen sls mgr; Kevin Wagner, opns dir & progmg dir; Teresa Hetz, prom mgr; Barry Birr, news dir; Ed Jurich, engrg dir & chief of engrg; Georgia Roades, traf mgr.

KKJO(FM)—Co-owned with KSFT(AM). Sept 1, 1962: 105.5 mhz; 100 kw. Ant 981 ft TL: N39 42 35 W95 02 33. Stereo. 24 Web Site: www.stjoeradio.com. Format: Adult contemp. News staff: 2; News: 2 hrs wkly. Target aud: 18-49. ♦Greg Lynn, progmg dir.

***KSRD(FM)**— 2004: 91.9 mhz; 10 kw. Ant 492 ft TL: N39 42 35 W95 02 33. Hrs open: 1212 Faraon St., 64501. Phone: (816) 233-5773. Fax: (816) 233-5777. E-mail: info@ksrdradio.com Web Site: www.ksrdradio.com. Licensee: Horizon Christian Fellowship. (acq 8-26-2004; $10,600). Format: Christian. ♦Brian KC Jones, gen mgr.

Saint Louis

KATZ(AM)— Jan 3, 1955: 1600 khz; 5 kw-U, DA-N. TL: N38 39 19 W90 07 53. Hrs open: 24 1001 Highlands Plaza Dr. W., Suite 100, 63110. Phone: (314) 333-8000. Fax: (314) 333-8300. Web Site: www.gospel1600.com. Licensee: Citicasters Licenses L.P. Group owner: Clear Channel Communications Inc. (acq 5-4-99; grpsl). Population served: 2,700,000 Natl. Network: American Urban. Format: Contemporary/Traditional Gospel. Target aud: 25-54; Adults. ♦Dennis Lamme, gen mgr; Tommy Austin, opns VP; Beth Davis, sls dir; Pierre Troupe, gen sls mgr & gen sls mgr; John Helmkamp, mktg dir.

***KDHX(FM)**— Oct 14, 1987: 88.1 mhz; 42.4 kw. 1,314 ft TL: N38 25 01 W90 25 59. Stereo. Hrs opn: 24 3504 Magnolia, 63118. Phone: (314) 664-3955. Fax: (314) 664-1020. Web Site: www.kdhx.org. Licensee: Double Helix Corp. Natl. Network: PRI. Format: Div. Target aud: General. ♦Beverly Hacker, gen mgr & stn mgr.

KEZK-FM— September 1968: 102.5 mhz; 100 kw. 400 ft TL: N38 36 47 W90 20 09. Stereo. Hrs opn: 24 3100 Market St., St. Louis, 63103. Phone: (314) 531-0000. Fax: (314) 969-7638. Web Site: www.kezk.com. Licensee: CBS Radio Holdings Inc. Group owner: Infinity Broadcasting Corp. (acq 11-13-98; grpsl). Population served: 2,112,400 Format: Soft adult contemp. Target aud: 25-54; high average household income. ♦Beth Davis, gen mgr.

KFNS(AM)—See Wood River, IL

KFUO(AM)—See Clayton

KHOJ(AM)—See Saint Charles

KIHT(FM)— Dec 22, 1959: 96.3 mhz; 100 kw. 650 ft TL: N38 36 47 W90 20 09. (CP: 80 kw, ant 1,027 ft.). Stereo. Hrs opn: 800 Saint Louis Union Stn., The Powerhouse, 63103. Phone: (314) 621-4106. Fax: (314) 621-3000. Web Site: www.k-hits.com. Licensee: Emmis Radio License LLC. Group owner: Emmis Communications Corp. (acq 9-26-2000; grpsl). Population served: 622,236 Format: Classic hits. Target aud: 25-54. ♦John Beck, sr VP.

KJFF(AM)—See Festus

KJSL(AM)— Sept 19, 1938: 630 khz; 5 kw-U, DA-2. TL: N38 40 18 W90 06 52. Hrs open: 24 KJSL(AM), 10845 Olive Blvd., Suite 160, 63141. Phone: (314) 878-3600. Fax: (314) 656-3608. E-mail: mflora@crawfordbroadcasting.com Web Site: www.kjslradio.net. Licensee: WMUZ Radio Inc. Group owner: Crawford Broadcasting Co. (acq 1994; $1.57 million). Population served: 500,000 Format: Christian, talk. Target aud: 30-60. ♦Don Crawford Jr., gen mgr; Micle Flora, stn mgr & gen sls mgr.

KLOU(FM)— November 1962: 103.3 mhz; 100 kw. Ant 920 ft TL: N38 31 47 W90 17 58. Stereo. Hrs opn: 24 10001 Highlands Dr., 63110. Phone: (314) 333-8000. Fax: (314) 333-8300. Web Site: www.playwhatiwant.com. Licensee: Citicasters Licenses L.P. Group owner: Clear Channel Communications Inc. (acq 5-4-99; grpsl). Population served: 2,700,000 Format: Oldies. News staff: one. Target aud: 25-54. ♦Dennis Lamme, gen mgr; Tommy Austin, opns mgr; Beth Davis, sls dir; Al Fox, gen sls mgr; John Helmkamp, mktg dir.

KMOX(AM)— Dec 24, 1925: 1120 khz; 50 kw-U. TL: N38 43 20 W90 03 16. Stereo. Hrs opn: 24 One Memorial Dr., St. Louis, 63102-2498. Phone: (314) 621-2345. Fax: (314) 444-1860 (SALES). E-mail: kmox@kmox.com Web Site: www.kmox.com. Licensee: CBS Radio East Inc. Group owner: Infinity Broadcasting Corp. (acq 11-13-98; grpsl). Population served: 2,098,500 Natl. Network: CBS. Law Firm: Leventhal, Senter & Lerman. Format: News/talk, info, sports. News staff: 16; News: 60 hrs wkly. Target aud: 25 plus. Spec prog: Jazz 4 hrs, relg one hr wkly.

KSD(FM)— November 1954: 93.7 mhz; 100 kw. 859 ft TL: N38 34 05 W90 19 55. Stereo. Hrs opn: 24 10001 Highlands Dr., 63110. Phone: (314) 333-8000. Fax: (314) 333-8300. Web Site: thebullrocks.com. Licensee: Citicasters Licenses L.P. Group owner: Clear Channel Communications Inc. (acq 5-4-99; grpsl). Population served: 2,700,000 Format: Country. News staff: one. Target aud: 18-34; adults. ♦Dennis Lamme, sr VP & gen mgr; Tommy Austin, opns mgr; Beth Davis, sls dir; Aaron Hyland, gen sls mgr; John Helmkamp, mktg dir.

KSHE(FM)—See Crestwood

KSIV(AM)—See Clayton

***KSIV-FM**— Apr 13, 1950: 91.5 mhz; 12.5 kw. Ant 400 ft TL: N38 37 10 W90 14 12. Hrs open: 1750 S. Brentwood Blvd., Suite 811, 63144. Phone: (314) 961-1320. Fax: (314) 961-7562. Web Site: www.bottradionetwork.com. Licensee: Community Broadcasting Inc. Group owner: Bott Radio Network (acq 1996; $1.625 million). Format: Christian info, relg. ♦Richard P. Bott, pres & VP; Michael McHardy, gen mgr; Joy Elder, sls dir & mktg dir.

KSLG(AM)— 1927: 1380 khz; 5 kw-D, 1 kw-N, DA-3. TL: N38 31 27 W90 14 17. Hrs open: 24 5261 Delmar, Suite 202, 63108. Phone: (314) 969-1380. Fax: (314) 367-8647. Web Site: www.1380espn.com. Licensee: Simmons Austin, LS LLC. Group owner: Simmons Media Group (acq 7-29-2004; $2.05 million). Population served: 504000 Natl. Network: ESPN Radio. Format: Sports. News: 168 hrs wkly. Target aud: 25-54; men and women. ♦Dave Green, gen mgr, gen sls mgr & progmg dir.

KSLZ(FM)— Sept 28, 1972: 107.7 mhz; 100 kw. 1,027 ft TL: N38 34 24 W90 19 30. Stereo. Hrs opn: 24 10001 Highlands Dr., 63110. Phone: (314) 333-8000. Fax: (314) 333-8300. Web Site: www.z1077.com. Licensee: Citicasters Licenses L.P. Group owner: Clear Channel Communications Inc. (acq 5-4-99; grpsl). Population served: 2,700,000 Format: Contemp hit. Target aud: 18-34; adults. ♦Dennis Lamme, gen mgr; Tommy Austin, opns mgr; Beth Davis, sls dir; Scott Adamec, gen sls mgr; John Helmkamp, mktg dir.

KSTL(AM)— 1948: 690 khz; 1 kw-D, 18 w-N. TL: N38 37 01 W90 10 17. Hrs open: 24 10845 Olive Blvd., Suite 160, Creve Coeur, 63141. Phone: (314) 878-3600. Phone: (618) 874-5785. Fax: (314) 656-3608. E-mail: dholmes@crawfordbroadcasting.com Licensee: WMUZ Radio Inc. Group owner: Crawford Broadcasting Co. (acq 1994). Population served: 3,000,000 Law Firm: Bryan Cave. Format: Gospel. ♦Donald Crawford, pres; Micle Flora, gen mgr; Deborah E. Holmes, stn mgr.

KTRS(AM)— Feb 14, 1922: 550 khz; 5 kw-U, DA-N. TL: N38 39 45 W90 07 43. Stereo. Hrs opn: 24 638 West Port Plaza, 63146. Phone: (314) 453-5500. Fax: (314) 453-9704. Web Site: www.ktrs.com. Licensee: KTRS-AM License L.L.C. (acq 1997). Population served: 622,236 Natl. Network: ABC. Natl. Rep: Katz Radio. Law Firm: Bryan Cave. Wire Svc: AP Format: News/talk, sports. News staff: 4. Target aud: 35-64. ♦Tim Dorsey, pres & gen mgr; Craig Unger, stn mgr & progmg dir; Geoff Witt, sls dir; Brian Kelly, news dir.

***KWMU(FM)**— June 2, 1972: 90.7 mhz; 97 kw. 981 ft TL: N38 34 50 W90 19 45. (CP: 100 kw, ant 1,000 ft.). Stereo. Hrs opn: 8001 Natural Bridge Rd., 63121. Phone: (314) 516-5968. Fax: (314) 516-5993. E-mail: kwmu@kwmu.org Web Site: www.kwmu.org. Licensee: The Curators of the University of Missouri. (group owner) Population served: 90,000 Natl. Network: NPR, PRI. Law Firm: Shaw Pittman. Wire Svc: AP Format: News info. News staff: 5; News: 40 hrs wkly. Target aud: 27-45; upscale. ♦Patricia Wente, gen mgr; Shelly Kerley, stn mgr; Shelley Kerley, dev dir; Mike Schrand, progmg dir.

KXEN(AM)— May 10, 1951: 1010 khz; 50 kw-D, 500 w-n, DA-2. TL: N38 45 46 W90 03 35. Hrs open: 24 Box 8085, Granite City, IL, 62040. Phone: (314) 436-6550. Fax: (618) 797-2293. Web Site: www.kxen1010.com. Licensee: BDJ Radio Enterprises LLC (acq 7-2-02). Population served: 3,000,000 Format: Relg. ♦Dirk L. Hallemeier, gen mgr; Jay Madas, progmg dir.

KYKY(FM)— 1960: 98.1 mhz; 90 kw. 1,027 ft TL: N38 34 24 W90 19 30. Stereo. Hrs opn: 24 3100 Market St., St. Louis, 63103. Phone: (314) 531-0000. Fax: (314) 531-9855. Web Site: www.y98.com. Licensee: CBS Radio Holdings Inc. Group owner: Infinity Broadcasting Corp. (acq 11-13-98; grpsl). Population served: 1,981,700 Natl. Network: Westwood One. Format: Hot adult contemp. Target aud: 25-54. ♦Beth Davis, gen mgr.

WARH(FM)—Granite City, IL) Nov 24, 1965: 106.5 mhz; 90 kw. Ant 1,027 ft TL: N38 34 24 W90 19 30. Stereo. Hrs opn: 11647 Olive Blvd., St. Louis, 63141. Phone: (314) 983-6000. Fax: (314) 994-9447. Web Site: www.wssm.com. Licensee: Bonneville Holding Co. Group owner: Bonneville International Corp. (acq 9-26-2000; grpsl). Format: Smooth jazz. ♦Bruce Reese, CEO & pres; Bob Johnson, CFO; John Kijowski, VP & gen mgr; Mike Jennewein, sls dir; Ben Granger, gen sls mgr & natl sls mgr; Amanda Koeppe, pub affrs dir; Marshall Rice, chief of engrg.

WEW(AM)— Apr 26, 1921: 770 khz; 1 kw-D. TL: N38 37 17 W90 04 36. Hrs open: 2 hrs past sunset (pssa) 2740 Hampton Ave., 63139. Secondary address: 21700 Northwestern Hwy, Tower 14, Ste 1190, Southfield, MI 48075. Phone: (314) 781-9397. Phone: (314) 969-7700. Fax: (314) 781-8545. E-mail: wewradio@oal.com Web Site: www.wewradio.com. Licensee: Birach Broadcasting Corp. (group owner; acq 1-6-2004; $1.35 million). Population served: 622,236 Natl. Network: CBS Radio, CNN Radio. Format: Ethnic. Target aud: 35-64; Mature audience/older. Spec prog: Ger 2 hrs, Pol 2 hrs wkly. ♦Sima Birach, CEO, pres & gen mgr; Rich Vannoy, opns mgr.

WGNU(AM)—Granite City, IL) Dec 1, 1961: 920 khz; 450 w-D, 500 w-N, DA-2. TL: N38 45 33 W90 03 00. Hrs open: 24 265 Union Blvd., Suite 1301, St. Louis, 63108-1262. Phone: (314) 454-6660. Fax: (314) 454-6609. E-mail: gm@wgnu.net Web Site: www.wgnu.net. Licensee: Radio Property Ventures LLC (acq 7-31-2007; $1.3 million). Population served: 2,300,000 Law Firm: Miller & Miller. Format: Talk. News staff: one; News: 2 hrs wkly. Target aud: 25-54; active, educated adults. Spec prog: German 2 hrs, Bosnian 2 hrs wkly. ♦Esther Wright, gen mgr; Joan Groceman, opns dir; Brian Boyland, gen sls mgr; Charles Geer, progmg dir, news dir & edit dir.

WHHL(FM)—See Jerseyville, IL

WIL(AM)— Feb 9, 1922: 1430 khz; 5 kw-U, DA-2. TL: N38 32 09 W90 11 26. Hrs open: 24 11647 Olive St., 63141. Phone: (314) 983-6000. Fax: (314) 994-9421. Web Site: www.legends1430.com. Licensee: Bonneville Holding Co. Group owner: Bonneville International Corp. (acq 9-26-2000; grpsl). Population served: 2,021,000 Natl. Network: Westwood One. Format: Country. News staff: one. Target aud: 35 plus; affluent, mature baby boomers. ♦John Kijowski, gen mgr; Keith Kraus, gen sls mgr; Greg Mozingo, progmg dir; Marshall Rice, chief of engrg; Tom Ennis, traf mgr.

WILI-FM— June 16, 1975: 98.3 mhz; 1.05 kw. 525 ft TL: N41 41 00 W72 13 01. Stereo. 24 Web Site: www.wili.com. Licensee: Nutmeg Broadcasting Co.125,000 Natl. Network: Superadio. Natl. Rep: D & R Radio. Format: CHR. News staff: one; News: 6 hrs wkly. Target aud: 22-44; college students, young married couples, young families.

WILL-FM— Sept 1, 1941: 90.9 mhz; 105 kw. 850 ft TL: N40 06 52 W88 13 27. Stereo. 24 158,700 Format: Class, var. News: 2 hrs wkly. Target aud: 35-70. ♦Jake Schumacher, progmg dir. Co-owned TV: *WILL-TV affil

WIL-FM— July 15, 1962: 92.3 mhz; 99 kw. 984 ft TL: N38 28 56 W90 23 53. Stereo. Web Site: www.wil92.com. Format: Contemp country.

Saint Robert

KFLW(FM)— Mar 22, 1994: 98.9 mhz; 6 kw. 328 ft TL: N37 52 41 W92 01 05. Hrs open: 24 250 Marshall Dr., St. Robert, 65584-8600. Phone: (573) 336-5359. Fax: (573) 336-7619. E-mail: toquinn@kflw99.com Web Site: kflw99.com. Licensee: Ozark Media (acq 2-19-02; $575,000). Format: Solid rock. News: 3 hrs wkly. Target aud: 25-55. ♦Dalton Wright, pres; Tracey O'Quinn, gen mgr.

Sainte Genevieve

KPNT(FM)— March 1967: 105.7 mhz; 100 kw. Ant 1,374 ft TL: N38 13 10 W90 35 44. Stereo. Hrs opn: 24 800 St. Louis Union Stn., The Power House, St. Louis, 63103. Phone: (314) 231-1057. Fax: (314) 621-3000. Web Site: www.1057thepoint.com. Licensee: Emmis Radio License LLC. Group owner: Emmis Communications Corp. Natl. Network: ABC. Format: New rock alternative. Target aud: 18-34. ♦John Beck, gen mgr; Tommy Mathern, natl sls mgr & progmg dir; Sam Caputa, chief of engrg.

KSGM(AM)—See Chester, IL

Salem

***KCVX(FM)**— February 2004: 91.7 mhz; 40kw. Ant 210 ft TL: N37 39 53 W91 32 00. Stereo. Hrs opn: 24
Rebroadcasts KCVO(FM) Campenton 100%.
Box 800, % Spirit FM Radio, Camdenton, 65020-0800. Phone: (573) 346-3200. Fax: (573) 346-1010. E-mail: email@spiritfm.org Web Site: www.spiritfm.org. Licensee: Lake Area Educational Broadcasting Foundation (acq 12-20-02; $35,000 for CP). Population served: 60,000 Format: Div, Christian. Target aud: 25-45; primarily females, married with children. ♦James J. McDermott, pres; Alice McDermott, CFO; James McDermott, gen mgr.

KKID(FM)— January 1971: 92.9 mhz; 21 kw. Ant 361 ft TL: N37 43 45 W91 28 23. Stereo. Hrs opn: 6 AM-midnight 1415 Forum Dr., Rolla, 65401-2508. Phone: (573) 364-4433. Fax: (573) 364-8385. Web Site: www.kkid929fm. Licensee: Ultra-Sonic Broadcast Stations Inc. Population served: 250,000 Natl. Network: USA. Format: Classic country. News staff: one; News: 20 hrs wkly. Target aud: 30-49. ♦David Wheeler, pres & gen mgr.

KSMO(AM)— November 1953: 1340 khz; 1 kw-U. TL: N37 37 36 W91 32 09. Hrs open: 24 800 S. Main, 65560. Phone: (573) 729-6117. Fax: (573) 729-7337. E-mail: ksl340@fidnet.com Web Site: www.ksmoradio.com. Licensee: KSMO Enterprises. (acq 11-84). Population served: 25,000 Natl. Network: AP Network News. Rgnl. Network: Missourinet, Brownfield. Law Firm: Booth, Freret, Imlay & Tepper. Wire Svc: NWS (National Weather Service) Format: Country, news/talk, sports. News staff: one; News: 40 hrs wkly. Target aud: General; middle class. Spec prog: Farm 18 hrs wkly. ♦Stanley M. Podorski, pres & gen mgr.

Savannah

KSJQ(FM)— September 1991: 92.7 mhz; 50 kw. 492 ft TL: N39 58 34 W94 58 37. Stereo. Hrs opn: 24 Box 8550, St. Joseph, 64508. Secondary address: 4104 Country Ln., St. Joseph 64506. Phone: (816) 233-8881. Fax: (816) 279-8280. Web Site: stjoeradio.com. Licensee: Eagle Communications Inc. Group owner: Eagle Communications Group (acq 1993; $450,000; FTR: 9-13-93). Population served: 200,000 Natl. Rep: Katz Radio. Law Firm: Wiley, Rein & Fielding. Wire Svc: AP Format: Country. News staff: 2; News: 3 hrs wkly. ♦Gary Shorman, CEO & pres; Mark Vail, COO & VP; Gary Exline, VP & gen mgr; Kevin Wagner, opns dir; Teresa Hetz, prom mgr; Brent Harmon, progmg dir; Barry Birr, news dir; Shannon Diggs, traf mgr; Tom Brand, farm dir.

Scott City

KGKS(FM)— 1998: 93.9 mhz; 5.4 kw. 344 ft TL: N37 22 07 W89 35 34. Hrs open: Box 1610, 324 Broadway, Cape Girardeau, 63701. Phone: (573) 335-8291. Fax: (573) 335-4806. E-mail: kiss@zrgmail.com Web Site: www.kiss939.com. Licensee: MRR License LLC. Group owner: MAX Media L.L.C. (acq 6-2-2004; grpsl). Format: Adult contemp. ♦Steve Stephensen, gen mgr; Whitney Thomas, progmg dir.

Sedalia

KDRO(AM)— Sept 13, 1939: 1490 khz; 1 kw-U. TL: N38 40 35 W93 15 18. Hrs open: 24 301 S. Ohio, 65301-4431. Phone: (660) 826-5005. Fax: (660) 826-5557. E-mail: 1490@kdro.com Web Site: www.kdro.com. Licensee: Mathewson Broadcasting Co. (acq 4-16-90; $300,000; FTR: 5-7-90). Population served: 231,830 Natl. Network: CBS. Rgnl. Network: Brownfield, Missourinet. Format: Country. News staff: 2; News: 11 hrs wkly. Target aud: General. Spec prog: Farm 6 hrs, Black one hr, relg 3 hrs wkly. ♦Stu Steinmetz, gen mgr.

KSDL(FM)—Listing follows KSIS(AM).

KSIS(AM)— Feb 18, 1954: 1050 khz; 1 kw-D, 86 w-N. TL: N38 43 52 W93 13 32. Stereo. Hrs opn: 24 2209 S. Limit, 65301. Phone: (660) 826-1050. Fax: (660) 827-5072. E-mail: ksis@bickbroadcasting.com Licensee: Bick Broadcasting Co. (acq 1-1-87). Population served: 33400 Wire Svc: U.S. Weather Service Format: News/talk. News staff: 2. Target aud: 25-54. ♦Dennis Polk, gen mgr.

KSDL(FM)—Co-owned with KSIS(AM). May 11, 1964: 92.1 mhz; 3 kw. Ant 280 ft TL: N38 43 52 W93 13 32. Stereo. 24 E-mail: radio92@ksdl.com Web Site: www.ksdl.com. Format: Adult contemp. News staff: 2. Target aud: 12-40; women.

Seligman

KIGL(FM)— Aug 1, 1986: 93.3 mhz; 100 kw. Ant 492 ft TL: N36 28 03 W94 10 25. Stereo. Hrs opn: 24 Box 8190, Fayetteville, AR, 72703. Secondary address: 4209 Frontage Rd., Fayetteville, AR 72703. Phone: (479) 973-9339. Fax: (479) 582-5302. Web Site: www.933theeagle.com. Licensee: Capstar TX L.P. Group owner: Clear Channel Communications Inc. (acq 8-30-00; grpsl). Population served: 250,000 Natl. Network: USA. Format: Classic rock. News staff: one; News: 7 hrs wkly. Target aud: 35 plus; mature, upscale professionals. ♦Tony Beriegee, gen mgr.

Shell Knob

KQMO(FM)— July 16, 1999: 97.7 mhz; 2.1 kw. Ant 558 ft TL: N36 44 54 W93 39 32. Stereo. Hrs opn: 24 126 S. Jefferson St., Aurora, 65605. Phone: (417) 678-0416. Fax: (417) 678-4111. Web Site: www.talonbroadcasting.com. Licensee: Falcon Broadcasting Inc. (acq 9-1-2005; $417,500). Population served: 150,000 Format: Sp. News staff: one; News: 21 hrs wkly. Target aud: Mexican-Hispanic. ♦Dewayn Gandy, gen mgr.

Sikeston

KBHI(FM)—Miner, 2001: 107.1 mhz; 3.7 kw. Ant 420 ft TL: N36 56 33 W89 41 47. Hrs open: 125 S. Kingshighway, 63801. Phone: (573) 471-2000. Fax: (573) 471-8525. Licensee: Dana R. Withers. Natl. Rep: Katz Radio. Format: Var. ♦Rick Lambert, gen mgr.

KBXB(FM)—Listing follows KRHW(AM).

KRHW(AM)— Mar 17, 1966: 1520 khz; 5 kw-D, 1.6 kw-N, DA-3. TL: N36 49 25 W89 35 45. Hrs open: Box 907, 125 S. Kings Hwy., 63801. Phone: (573) 471-2000. Fax: (573) 471-8525. Licensee: Withers Broadcasting Co. of Southeast Missouri LLC. Group owner: Withers Broadcasting Co. (acq 4-96; with co-located FM). Population served: 50,000 Format: Country, relg. Target aud: 45 plus. Spec prog: Farm 6 hrs wkly. ♦Rick Lambert, gen mgr; Joe Bill Davis, rgnl sls mgr; Kidd Manning, progmg dir; John Steeke, news dir; Smokey King, chief of engrg.

KBXB(FM)—Co-owned with KRHW(AM). Sept 12, 1968: 97.9 mhz; 50 kw. 469 ft TL: N36 59 52 W89 38 52. Stereo. 24 200,000 Natl. Rep: Katz Radio. Target aud: 18-49. ♦Hugh Robinson, farm dir.

KSIM(AM)— July 17, 1948: 1400 khz; 1 kw-U. TL: N36 52 12 W89 36 32. Hrs open: 324 Broadway, Cape Girardeau, 63701. Phone: (573) 335-8291. Fax: (573) 335-4806. E-mail: Ksim@riverradio.net Web Site: www.1400ksim.com. Licensee: MRR License LLC. Group owner: MAX Media L.L.C. (acq 6-2-2004; grpsl). Population served: 200,000 Natl. Rep: Christal. Format: News/talk. News staff: 4. Target aud: 25-54. Spec prog: Loc sports, news, Paul Harvey, various features. ♦Steve Stephenson, gen mgr; Whitney Thomas, opns mgr; Meg Davis, gen sls mgr; Faune Riggin, news dir.

South West City

KLTK(AM)— Mar 2, 1977: 1140 khz; 200 w-D. TL: N36 30 28 W94 36 35. (CP: COL Centerton, AR. 1 kw-D. TL: N36 18 10 W94 06 47). Hrs opn: 113 E. New Hope Rd., Rogers, AR, 72758. Phone: (479) 633-0790. Fax: (479) 631-9711. Licensee: KERM Inc. (group owner; (acq 4-4-2002; $350,000 with co-located FM). Population served: 100,000 Law Firm: Dow, Lohnes & Albertson. Format: Sports, talk. Target aud: 25-54; rural, agricultural. ♦Diane Womack, gen sls mgr & traf mgr; Kermit Womack, pres, pres, gen mgr & progmg dir; Eric Morris, chief of engrg.

KURM-FM—Co-owned with KLTK(AM). October 1989: 100.3 mhz; 3 kw. 328 ft TL: N36 30 28 W94 36 35. Web Site: www.kwmg.com. Natl. Network: Westwood One, CBS. Law Firm: Dow, Lohnes & Albertson. Format: News/talk. ♦Eric Morris, engrg VP.

Sparta

KSPW(FM)— Mar 1, 1989: 96.5 mhz; 3.2 kw. 453 ft TL: N37 05 17 W93 10 34. (CP: 50 kw, ant 492 ft. TL: N35 56 23 W93 17 15). Stereo. Hrs opn: 24 Box 2180, Springfield, 65801. Secondary address: 2330 W. Grand St., Springfield 65802. Phone: (417) 865-6614. Fax: (417) 865-9643. Web Site: www.power965jams.com. Licensee: Journal Broadcast Corp. Group owner: Journal Communications Inc. (acq 6-11-99; grpsl). Natl. Rep: Christal. Format: CHR. News staff: 6. Target aud: 18-34; young active adults. ♦Steven Smith, CEO & chmn; Doug Kiel, pres; Carl Gardner, exec VP; Rex Hansen, gen mgr & natl sls mgr; Chris Cannon, opns mgr & progmg dir; Janelle Carter, gen sls mgr.

Springfield

KADI(AM)— July 29, 1949: 1340 khz; 1 kw-U. TL: N37 12 30 W93 17 32. Hrs open: 5431 W. Sunshine St., Brookline Station, 65619. Phone: (417) 831-0995. Fax: (417) 831-4026. Web Site: www.kadi.com. Licensee: Vision Communications Inc. (acq 5-26-2005; $375,000). Population served: 250,000 Format: Talk. ♦R.C. Amer, gen mgr.

KBFL(AM)— 1972: 1060 khz; 500 w-D. TL: N37 11 29 W93 19 45. Stereo. Hrs opn: 6 AM-sunset + 2 hrs 610 W. College., 65806. Phone: (417) 832-1060. Fax: (417) 864-4111. E-mail: ktozam@pcis.net Web Site: www.ktozam.com. Licensee: Meyer-Baldridge Inc. (acq 2-27-2006; $275,000). Population served: 500,000 Format: Big band, nostalgia, MOR. Target aud: General. Spec prog: Jazz 12 hrs, blues 4 hrs, 50s mus 4 hrs wkly. ♦Kenneth E. Meyer, pres; James Cooper, VP; William H. Thomas, gen mgr; Jim Cooper, progmg dir.

KGBX-FM—Listing follows KGMY(AM).

KGMY(AM)— Oct 31, 1926: 1400 khz; 1 kw-U. TL: N37 11 46 W93 19 21. Stereo. Hrs opn: 1856 S. Glenstone, 65804. Phone: (417) 890-5555. Fax: (417) 890-5050. E-mail: studio@espn1400.com Web Site: www.espn1400.com. Licensee: Clear Channel Broadcasting Licenses Inc. Group owner: Clear Channel Communications Inc. (acq 10-10-00; grpsl). Population served: 220,000 Format: Sports. Target aud: 35 plus; affluent, educated white-collar skewing 35 plus year olds. ♦Paul Windisch, gen mgr; Paul Kelley, opns mgr; Mary Fleenor, sls VP; Kelli Presley, gen sls mgr; Sarah Green, prom dir; Lyn Hare, chief of engrg; Mary Brown, traf mgr.

KGBX-FM—Co-owned with KGMY(AM). December 1989: 105.9 mhz; 38 kw. 558 ft TL: N37 25 16 W93 24 06. Stereo. 24 E-mail: studio@kgbx.com Web Site: www.kgbx.com.220,000 Format: Adult contemp. News staff: one; News: 5 hrs wkly. Target aud: 25-54; educated, high income, women. ♦Brian Edwards, progmg dir; Paul Kelly, progmg dir.

KLFJ(AM)— Nov 1, 1974: 1550 khz; 5 kw-D, 28 w-N. TL: N37 11 45 W93 19 07. Hrs open: 24 610 W. College, 65806. Phone: (417) 831-1550. Licensee: 127 Inc. (acq 6-99; $432,500). Population served: 250,000 Format: Info, news. ♦Arlen Graves, gen mgr & stn mgr.

KLPW(AM)—See Union

***KSCV(FM)**— October 1995: 90.1 mhz; 9 kw. Ant 492 ft TL: N37 17 41 W93 09 10. (CP: 9 kw, ant 492 ft. TL: N37 17 41 W93 09 10). Stereo. Hrs opn: 24
Rebroadcasts KCCV(FM) Overland Park, KS 90%.
1111 S. Glenstone Ave., Suite 3-102, 65804. Phone: (417) 864-0901. Fax: (417) 862-7263. E-mail: pschneider@bottradionetwork.com Web Site: bottradionetwork.com. Licensee: Community Broadcasting Inc. Group owner: Bott Radio Network (acq 2-7-01; 1.25 million with KMCV(FM) High Point). Population served: 350,000 Natl. Network: USA. Format: Christian, talk. News: 4 hrs wkly. Target aud: 25-54 plus; women 60%, men 40%. ♦Paul Schneider, gen mgr.

KSGF(AM)— 1926: 1260 khz; 5 kw-U, DA-N. TL: N37 15 51 W93 19 04. Stereo. Hrs opn: 24
Rebroadcast KSGF-FM Ash Grove.
Box 2180, 65801. Secondary address: 2330 W. Grand 65802. Phone: (417) 865-6614. Fax: (417) 865-9643. Web Site: www.ksgf.com. Licensee: Journal Broadcast Corp. Group owner: Journal Communications

Inc. (acq 6-11-99; grpsl). Population served: 400,000 Rgnl. Network: Missourinet. Natl. Rep: Christal. Format: News/talk. News staff: 4; News: 30 hrs wkly. Target aud: 35-54.Steven Smith, CEO & chmn; Doug Kiel, pres; Carl Gardner, exec VP; Rex Hansen, VP, gen mgr & gen sls mgr; Chris Cannon, opns mgr; Karen Campbell, natl sls mgr; Kris Addison, prom dir & mus dir; David Hayes, progmg dir; Don Louzader, news dir & local news ed; David Rahmoeller, chief of engrg; Cristie Cummings, traf mgr & edit mgr; Jason Rima, reporter & outdoor ed; Nancy Simpson, reporter & mus critic

KTTS-FM—Co-owned with KSGF(AM). Aug1948: 94.7 mhz; 100 kw. 1,125 ft TL: N37 13 26 W93 14 33. Stereo. 24 2330 W. Grand, 65802. Web Site: www.ktts.com.675,800 Natl. Rep: Christal. Law Firm: Dow, Lohnes & Albertson. Format: Country. News staff: 4; News: 5 hrs wkly. Target aud: 25-54; adults. ♦Curly Clark, asst music dir & pub affrs dir; Jason Rima, reporter; Nancy Simpson, reporter.

***KSMU(FM)**— May 7, 1974: 91.1 mhz; 40 kw. Ant 403 ft TL: N37 10 14 W93 19 25. Stereo. Hrs opn: 24 Missouri State Univ., 901 S. National Ave., 65897. Phone: (417) 836-5878. Fax: (417) 836-5889. E-mail: ksmu@missouristate Web Site: www.ksmu.org. Licensee: Board of Governors, Southwest Missouri State University Population served: 360,000 Natl. Network: NPR. Law Firm: Dow, Lohnes & Albertson. Format: News & classical music. News staff: one; News: 54 hrs wkly. Target aud: 25-54. Spec prog: Jazz 10 hrs wkly. ♦Tammy Wiley, gen mgr.

KTXR(FM)— June 12, 1962: 101.3 mhz; 97.8 kw. Ant 1,488 ft TL: N37 11 41 W92 56 07. Stereo. Hrs opn: 24 3000 E. Chestnut Expwy., 65802. Secondary address: Box 3925 65802. Phone: (417) 862-3751. Fax: (417) 869-7675. E-mail: manager@radiospringfield.com Web Site: www.radiospringfield.com. Licensee: Stereo Broadcasting Inc. Group owner: Meyer Communications Inc. Population served: 1,000,000 Format: Contemp easy lstng. News staff: 2; News: 21. Target aud: 35 plus; female. Spec prog: MSU Bears sports, St. Louis Cardinals baseball. ♦Kenneth E. Meyer, pres & gen mgr; Bonnie Bell, gen sls mgr; Jamie Turner, progmg dir; Dale Blankenship, chief of engrg; Pat Willis, traf mgr.

***KWFC(FM)**— Apr 17, 1985: 89.1 mhz; 100 kw. 1,122 ft TL: N37 12 06 W92 56 33. Stereo. Hrs opn: 24 Box 8900, 65801-8900. Secondary address: 2316 N. Benton 65801. Phone: (417) 869-0891. Fax: (417) 866-7525. E-mail: info@kwfc.org Web Site: www.kwfc.org. Licensee: Baptist Bible College Inc. Population served: 151,000 Natl. Network: USA. Wire Svc: AP Format: Relg, Christian. News staff: one; News: 17 hrs wkly. Target aud: General; conservative, church-oriented. ♦Gary Longstaff, gen mgr; Kyle Dowden, progmg dir; Brady Shoemaker, news dir; Vickie Hawkins, traf mgr.

***KWND(FM)**— July 12, 1993: 88.3 mhz; 12 kw. 328 ft TL: N37 10 30 W93 02 35. Stereo. Hrs opn: 24 2550-100 S. Campbell, 65807. Phone: (417) 889-0883. Fax: (417) 886-8656. Web Site: www.88.3thewind.com. Licensee: The Radio Training Network. (acq 7-95). Format: Adult contemp, Christian. Target aud: 25-49. Spec prog: Gospel 3 hrs wkly. ♦Ben Birdsong, gen mgr.

KWTO(AM)— Dec 25, 1933: 560 khz; 5 kw-U, DA-N. TL: N37 08 08 W93 16 36. Hrs open: Box 3793, 65808. Secondary address: 3000 E. Chestnut Expwy. 65808. Phone: (417) 862-5600. Fax: (417) 869-7675. E-mail: manager@radiospringfield.com Web Site: www.radiospringfield.com. Licensee: KWTO Inc. Group owner: Meyer Communications Inc. (acq 3-20-95; $1.88 million with co-located FM; FTR: 6-19-95) Population served: 1,500,000 Format: Sports, news/talk. News: one hr wkly. Target aud: 25-55; male. Spec prog: Farm 20 hrs, relg one hr wkly. ♦Kenneth E. Meyer, pres, gen mgr & gen sls mgr; Bonnie Bell, dev mgr & spec ev coord; Dan Vaughn, progmg dir; Dale Blankenship, chief of engrg; Susie Proffitt, traf mgr; R.J. McAllister, news rptr; Lewis Miller, farm dir.

KWTO-FM— Nov 23, 1967: 98.7 mhz; 100 kw. 600 ft TL: N37 04 06 W93 18 31. Stereo. 24 Web Site: www.radiospringfield.com.200,000 Format: Sports, talk. Target aud: 25-45; male dominant middle class. ♦Susie Proffitt, traf mgr; Bonnie Bell, spec ev coord; R.J. McAllister, reporter; Lewis Miller, farm dir.

KXUS(FM)— Apr 17, 1969: 97.3 mhz; 100 kw. 479 ft TL: N37 14 23 W93 17 05. (CP: Ant 987 ft. TL: N37 11 10 W93 01 23). Stereo. Hrs opn: 24 1856 S. Glenstone Ave., 65804. Phone: (417) 890-5555. Fax: (417) 890-5050. E-mail: us97@us97.com Web Site: www.us97.com. Licensee: Clear Channel Broadcasting Licenses Inc. Group owner: Clear Channel Communications Inc. (acq 10-10-00; grpsl). Format: Classic rock. News staff: one; News: 5 hrs wkly. Target aud: 25-54; males -75%. ♦Paul Windisch, gen mgr; Paul Kelley, opns dir; Tony Metteo, progmg dir; Lyn Hare, chief of engrg; Mary Brown, traf mgr.

Steelville

KNSX(FM)— September 1985: 93.3 mhz; 8.5 kw. 1,168 ft TL: N38 06 16 W91 02 30. Hrs open: 3418 Douglas Rd., Florissant, 63034. Phone: (314) 921-9330. E-mail: 93x@knsx.com Web Site: www.knsx.com. Licensee: Broadcast Communications Inc. Format: Alternative. ♦Ruth Choate, gen mgr.

Stockton

KRWP(FM)— Jan 20, 1999: 107.7 mhz; 11.7 kw. Ant 479 ft TL: N37 31 24 W93 52 40. Hrs open: 24 Box 1020, 126 S. Jefferson, 65605. Secondary address: 1225 South Hwy. 39 65785. Phone: (417) 276-5253. Fax: (417) 276-2255. Licensee: Cumulus Licensing LLC. Group owner: Cumulus Media Inc. (acq 4-27-2004; $825,000). Population served: 79,000 Natl. Network: Jones Radio Networks. Rgnl. Network: Jones Satellite Audio, Missourinet. Law Firm: Fletcher, Heald & Hildreth. Format: Classic country. News: 16 hrs wkly. Target aud: 25-54; male and female. Spec prog: Local news, weather, Chief's football, farm 8 hrs wkly. ♦Lance Beamer, gen mgr.

Sullivan

KTUI(AM)— Feb 14, 1966: 1560 khz; 1 kw-D. TL: N38 11 42 W91 11 12. Hrs open: 6 AM-sunset Box 99, 63080-0099. Phone: (573) 468-5101. Fax: (573) 468-5884. Web Site: www.ktui.com. Licensee: Fidelity Broadcasting Inc. (acq 10-23-97; $497,000 with co-located FM). Population served: 100,000 Rgnl. Network: Missourinet. Format: News/talk. News staff: one. Target aud: General. ♦John C. Rice, gen mgr & gen sls mgr; Sam Scott, progmg dir & news dir; Wilma Scott, traf mgr.

KTUI-FM— 1981: 100.9 mhz; 3 kw. 276 ft TL: N38 11 42 W91 11 12. Stereo. Web Site: www.ktui.com.210,000 Format: Country, sports. Target aud: General. ♦John Rice, opns dir.

Sunrise Beach

***KCRL(FM)**— Sept 1, 1998: 90.3 mhz; 4.5 kw. 197 ft TL: N38 14 11 W92 46 03. Hrs open: 24 Community Broadcasting, 10550 Barkley St, Ste 100, Overland Park, KS, 66212. Secondary address: 30690 Gray Eagle Rd., Gravois Mills 65037. Phone: (573) 372-1903. Fax: (573) 372-3801. E-mail: kcrl@bottradionetwork.com Web Site: www.bottradionetwork.com. Licensee: Community Broadcasting Inc. Group owner: Bott Radio Network Natl. Network: USA. Format: Christian talk. Target aud: 25-55. ♦Trace Thurlby, COO; Richard P. Bott Sr., pres; Tom Holdeman, CFO; Richard P. Bott II, exec VP & gen mgr; Eben Fowler, opns dir & sls dir; Pat Rulon, natl sls mgr; Rachel Moser, mktg mgr; Jason Potocnik, traf mgr.

Tarkio

***KRSS(FM)**— Aug 22, 1977: 93.5 mhz; 11 kw. 489 ft TL: N40 31 11 W95 11 03. Hrs open: 24 23979 Hwy. 136, 64491. Phone: (660) 736-4321. Fax: (660) 736-5789. Web Site: www.calvarychapel.com/krss. Licensee: CSN International (group owner). Population served: 10,000 Format: Christian. Target aud: General. ♦Mick Miller, gen mgr.

Thayer

KALM(AM)— Dec 11, 1953: 1290 khz; 1 kw-D, 56 w-N. TL: N36 32 58 W91 33 05. Hrs open: 6 AM-sunset Box 15, N. Hwy. 63, 65791. Phone: (417) 264-7211. Phone: (417) 264-7063. Fax: (417) 264-7212. E-mail: kkountry@kkountry.com Web Site: www.kkountry.com. Licensee: Ozark Radio Network Inc. Population served: 40,000 Rgnl. Network: Brownfield, Missourinet. Law Firm: Greg Skall. Wire Svc: Weather Wire Format: News/talk. News staff: one; News: 70 hrs wkly. Target aud: 18 plus; farmers, ranchers, rural families. ♦Shawn N. Marhefka, pres; Robert Eckman, gen mgr; Jerry Elam, progmg dir.

KAMS(FM)—See Mammoth Spring, AR

KSAR(FM)— 92.3 mhz; 50 kw. 426 ft TL: N36 21 58 W91 28 35. Hrs open: 24 Box 458, Salem, AR, 72576. Secondary address: 352 Hwy. 62/412, Salem, AR 72576. Phone: (870) 895-2665. Fax: (870) 895-4088. E-mail: hometownradio@centurytel.net Web Site: www.yourhometownstations.com. Licensee: Bragg Broadcasting Corp. Population served: 75,000 Rgnl. Network: Ark. Radio Net. Format: Country, news, sports. Target aud: 25-54. Spec prog: Farm 4 hrs wkly. ♦James Bragg, gen mgr.

Trenton

KTTN-FM— Sept 15, 1978: 92.3 mhz; 18.5 kw. Ant 380 ft TL: N40 05 00 W93 33 30. Stereo. Hrs opn: 24 Box 307, 64683. Secondary address: 804 Main St. 64683. Phone: (660) 359-2261. Fax: (660) 359-4126. E-mail: kttnamfm@grm.net Web Site: www.kttn.com. Licensee: Luehrs Broadcasting Co. (acq 8-1-92). Population served: 40,000 Natl. Network: AP Radio. Rgnl rep: Rgnl Reps Law Firm: Reddy, Begley & McCormick. Format: Country, news, sports. News staff: 2; News: 15 hrs wkly. Target aud: General. Spec prog: Gospel 6 hrs wkly. ♦John Ausberger, pres; John Anthony, gen mgr.

KTTN(AM)— Apr 17, 1955: 1600 khz; 500 w-D, 35 w-N. TL: N40 05 00 W93 33 30. Stereo. 24 Web Site: www.kttn.com.10,000 Natl. Network: AP Radio, Jones Radio Networks. Law Firm: Reddy, Begley & McCormick. Format: Adult contemp. News: 8 hrs wkly. Target aud: 35 plus; general.

Troy

KFNS-FM— Nov 29, 1993: 100.7 mhz; 6 kw. 328 ft TL: N39 03 13 W90 59 47. Hrs open: 8045 Big Bend Blvd., Suite 200, St. Louis, 63119. Phone: (314) 962-0590 (main #). Fax: (314) 962-7576. E-mail: kfns@kfns.com Web Site: www.kfns.com. Licensee: Big Stick Two LLC. Group owner: Big League Broadcasting LLC (acq 7-13-2004; grpsl). Format: Sports radio. ♦Mike Phares, gen mgr.

Union

KLPW(AM)— Aug 18, 1954: 1220 khz; 1 kw-D, 126 w-N. TL: N38 28 57 W91 02 39. Hrs open: 24 Box 623, Washington, 63090. Phone: (636) 583-5155. Phone: (636) 239-3355. Fax: (636) 583-1644. E-mail: klpwam@klpw.com Web Site: www.klpwam.com. Licensee: Broadcast Properties Inc. (acq 7-5-2007; $200,000 for 50% of stock). Population served: 600,000 Format: All talk. News staff: 2; News: 40 hrs wkly. Target aud: 25-54; male. Spec prog: Relg 6 hrs wkly. ♦Tim McDonald, gen mgr & gen sls mgr; Ray Heller, opns dir & pub affrs dir; Dee Coppeans, sls dir; Diana Stanley, prom dir; Greg Marshall, progmg dir; John Covington, news dir, local news ed, news rptr & sports cmtr; Tom Lyons, chief of engrg; Marcy Frankenberg, traf mgr; Alex Pennock, reporter.

KLPW-FM— Aug 1, 1966: 101.7 mhz; 3.3 kw horiz. Ant 341 ft TL: N38 28 57 W91 02 39. Stereo. Hrs opn: 24 Box 623, Washington, 63090. Secondary address: 6531 Hwy. BB, Washington 63090. Phone: (636) 583-5155. Phone: (636) 239-3355. Fax: (636) 583-1644. Web Site: www.klpwfm.com. Licensee: Marathon Media Group L.L.C. (acq 1999; grpsl). Population served: 600,000 Format: Country. News staff: 2; News: 16.5 hrs wkly. Target aud: 18-49. ♦Tim McDonald, gen mgr; Steve Leslie, mus dir; Marcy Frankenberg, traf mgr; John Covington, local news ed, news rptr & sports cmtr.

Van Buren

***KBIY(FM)**— 2001: 91.3 mhz; 100 kw. Ant 492 ft TL: N37 06 25 W90 59 30. Hrs open: New Life Evangelistic Center Inc., 1411 Locust St., St. Louis, 63103. Phone: (314) 436-2424. Fax: (314) 436-2434. E-mail: larryr@hereshelpnet.org Web Site: www.hereshelpnet.org. Licensee: New Life Evangelistic Center Inc. Format: Adult contemp, gospel, loc pub affrs-news. ♦Larry Rice, pres & gen mgr.

Vandalia

KKAC(FM)—Not on air, target date: unknown: 104.3 mhz; 6 kw. 292 ft TL: N39 19 00 W91 28 22. Hrs open: 400 S. Lindell St., 63382. Phone: (573) 594-6000. Fax: (314) 594-2100. E-mail: kkacfm@vandaliamo.net Web Site: www.actioncountry.com. Licensee: Twenty-One Sound Communications Inc. Format: Country.

Versailles

KTKS(FM)— June 16, 1989: 95.1 mhz; 12.5 kw. Ant 462 ft TL: N38 24 32 W92 45 42. Stereo. Hrs opn: 24 Box 409, 65084. Secondary address: 16875 Hwy 52, Barnett 65011. Phone: (573) 378-5669. Fax: (573) 378-6640. E-mail: jay@lakeradio.net Web Site: lakeradio.com. Licensee: Twin Lakes Communications Inc. Population served: 100,000 Natl. Network: CNN Radio. Law Firm: Fletcher, Heald & Hildreth. Wire Svc: AP Format: Country. News staff: one; News: 23 hrs wkly. Target aud: 25-54; loc rural audience & transient tourist population. Spec prog: Farm 2 hrs, relg 3 hrs wkly. ♦Douglas A. Fisher, chmn; James D. Fisher, pres & gen mgr; Sheryl Lehman, gen sls mgr; J.T. Gerlt, progmg dir.

Vienna

***KNLN(FM)**—Not on air, target date: unknown: 90.9 mhz; 10 kw. 328 ft TL: N38 11 27 W92 07 22. Hrs open: New Life Evangelistic Center Inc., 1411 Locust St., St. Louis, 63103. Phone: (314) 421-3020. Fax: (314) 436-2434. E-mail: larryr@hereshelpnet.org Web Site: www.hereshelpnet.org. Licensee: New Life Evangelistic Center Inc. Format: Relg. ♦Larry Rice, gen mgr.

Warrensburg

KOKO(AM)— December 1953: 1450 khz; 1 kw-U. TL: N38 46 32 W93 43 12. Hrs open: 24 Box 398, 64093. Phone: (660) 747-9191. Fax: (660) 747-5611. Web Site: www.1450koko.com. Licensee: D & H Media L.L.C. (acq 10-3-01; $435,000). Population served: 42,514 Natl. Network: ABC. Rgnl. Network: Missourinet, Brownfield. Format: Oldies, Sports. News staff: one; News: 20 hrs wkly. Target aud: 25-54; educated-mainly female & sports enthusiasts. ♦Vance Delozier, pres; Greg Hassler, gen mgr & opns mgr.

***KTBG(FM)**— Apr 1, 1962: 90.9 mhz; 100 kw. 400 ft TL: N38 55 54 W93 49 06. Stereo. Hrs opn: Wood 11, 64093. Phone: (660) 543-4130. Fax: (660) 543-8863. Web Site: www.ktbg.fm. Licensee: Central Missouri State University Board of Regents. Population served: 1,000,000 Natl. Network: NPR, PRI. Format: AAA. Target aud: General. ♦Donald Peterson, gen mgr.

Warrenton

KFAV(FM)—Listing follows KWRE(AM).

KWRE(AM)— Mar 9, 1949: 730 khz; 1 kw-D, 120 w-N. TL: N38 49 20 W91 08 15. Hrs open: 5 AM-11 PM Box 220, 63383. Phone: (636) 456-3311. Fax: (636) 456-8767. E-mail: kwrekfav@socket.net Web Site: www.kwre.com. Licensee: Kaspar Broadcasting Co. Group owner: Kaspar Broadcasting Group. Population served: 500,000 Format: Traditional country. News staff: 3; News: 3 hrs wkly. Target aud: 35 plus. Spec prog: Farm one hr wkly. ♦Mike Thomas, opns dir, mus dir & news dir; Mark Becker, gen sls mgr; V.J. Kaspar, pres, gen mgr & chief of engrg.

KFAV(FM)—Co-owned with KWRE(AM). November 1991: 99.9 mhz; 10.5 kw. 512 ft TL: N38 50 20 W91 02 40.24 Fax: (636) 978-4710. Web Site: www.kfav.com. Format: Today's hot country. News staff: 3; News: 2 hrs wkly. Target aud: 20-49; general.

Warsaw

KAYQ(FM)— Mar 10, 1980: 97.1 mhz; 6 kw. Ant 239 ft TL: N38 17 19 W93 18 32. Stereo. Hrs opn: 24 Box 1420, 65355. Secondary address: Truman Hills Mall, Suite 6 Phone: (660) 438-7343. Fax: (660) 438-7159. E-mail: kayqtraffic@earthlink.net Web Site: www.971thelake.com. Licensee: Valkyrie Broadcasting Co. Inc. Population served: 30,000 Natl. Network: AP Radio. Rgnl. Network: Missourinet. Format: Classic country. News staff: one; News: 5 hrs wkly. ♦Jim McCollum, pres; Joey Anderson, gen mgr & chief of opns; Glenna Thrasher, prom mgr & traf mgr.

Washington

***KGNV(FM)**— Dec 25, 1990: 89.9 mhz; 1 kw. 213 ft TL: N38 35 49 W91 06 17. Hrs open: 24 Box 187, 63090. Phone: (636) 239-0400. Fax: (636) 293-4448. Web Site: goodnewsvoice.org. Licensee: Missouri River Christian Broadcasting Inc. Population served: 60,000 Natl. Network: Moody, Salem Radio Network. Format: News/talk, Southern gospel, inspirational. News: 14 hrs wkly. Target aud: 20-70; inquisitive, conservative, liberal, philosophical. Spec prog: Class 5 hrs, children 6 hrs, teen 5 hrs wkly. ♦James Goggan, pres & gen mgr; Charles Sachse, stn mgr.

KLPW-FM—See Union

KSLQ-FM—Listing follows KWMO(AM).

KWMO(AM)— Oct 19, 1985: 1350 khz; 500 w-D, 84 w-N, DA-1. TL: N38 34 44 W90 59 57. Hrs open: 511 W. 5th St., 63090. Phone: (636) 239-5432. Fax: (636) 239-0364. Licensee: Computraffic Inc. (acq 2-13-98; $200,000). Population served: 350,000 Natl. Network: USA. Rgnl. Network: Missourinet. Format: Oldies. Target aud: 35-54. ♦Waldo Zimarskie, gen mgr, opns mgr, gen sls mgr, chief of engrg & chief of engrg; Chris Dieckhause, traf mgr.

KSLQ-FM—Co-owned with KWMO(AM). Nov 21, 1989: 104.5 mhz; 3 kw. 328 ft TL: N38 36 03 W90 56 04. Stereo. 24 Licensee: Y2K Inc. (acq 6-24-98; $1.1 million). Natl. Network: USA. Format: Hot adult contemp. Target aud: 25-54.

Waynesville

KFBD-FM—Listing follows KOZQ(AM).

KJPW(AM)— Apr 3, 1962: 1390 khz; 5 kw-D, 67 w-N. TL: N37 49 09 W92 09 06. Hrs open: 19 Box D, 65583-0480. Secondary address: 313 Old Rte 66, St. Robert 65583-0480. Phone: (573) 336-4913. Phone: (573) 336-4450. Fax: (573) 336-2222. E-mail: kjcountry@webound.com Licensee: Ozark Broadcasting Inc. Group owner: Shepherd Group (acq 10-1-2003; $735,000 with co-located FM). Population served: 60,000 Natl. Network: NBC. Rgnl. Network: Missourinet. Law Firm: Cohn & Marks. Format: Talk radio. News staff: one; News: 14 hrs wkly. Target aud: General. Spec prog: Relg 3 hrs wkly. ♦David Shepherd, pres; Mike Edwards, exec VP, gen mgr & traf mgr; Gary Knehans, stn mgr & prom mgr; Jim Anthony, gen sls mgr; Warren McDonald, progmg dir; Warren Goforth, news dir; Bob Moore, chief of engrg.

KJPW-FM— May 2, 1968: 102.3 mhz; 2.65 kw. Ant 492 ft TL: N37 49 09 W92 09 06. Stereo. 19 60,000 Format: Adult contemp. News staff: one; News: 14 hrs wkly. Target aud: General.

KOZQ(AM)— May 9, 1968: 1270 khz; 500 w-D. TL: N37 49 42 W92 10 27. Hrs open: Box 4371, 65583. Phone: (573) 336-3133. Fax: (573) 336-1228. Licensee: Ozark Broadcasting Inc. (acq 10-12-2006; $500,000 with co-located FM). Population served: 100,000 Format: News/talk. Target aud: 40 plus. ♦John Rice, gen mgr; Alan Holcomb, opns mgr, gen sls mgr & asst music dir; Sam Scott, progmg dir; Wilma Scott, traf mgr.

KFBD-FM—Co-owned with KOZQ(AM). Dec 9, 1964: 97.9 mhz; 3 kw. Ant 259 ft TL: N37 49 42 W92 10 27. Stereo. 5 AM-midnight Format: Classic rock. ♦Woody Schuler, progmg dir.

Webb City

KJMK(FM)— Sept 10, 1985: 93.9 mhz; 48 kw. 505 ft TL: N37 14 34 W94 30 21. Stereo. Hrs opn: 2702 E. 32nd, Joplin, 64804. Phone: (417) 624-1025. Fax: (417) 781-6842. Web Site: www.magic939.com. Licensee: Zimco Inc. Group owner: Zimmer Radio Group (acq 6-17-97; grpsl). Format: Adult contemp. Target aud: 25-54. ♦Larry Boyd, gen mgr; Jason Knight, opns mgr.

KXDG(FM)— Sept 1, 1988: 97.9 mhz; 6 kw. 400 ft TL: N37 06 11 W94 24 11. Stereo. Hrs opn: 24 2702 E. 32nd St., Joplin, 64804. Phone: (417) 624-1025. Fax: (417) 781-6842. Web Site: www.bigdog979.com. Licensee: Zimco Inc. Group owner: Zimmer Radio Group (acq 6-17-97; grpsl). Natl. Network: USA. Format: Classic rock. Target aud: General. ♦Larry Boyd, gen mgr; Jason Knight, opns mgr.

West Plains

KKDY(FM)— Mar 31, 1984: 102.5 mhz; 50 kw. 485 ft TL: N36 41 22 W91 53 45. Stereo. Hrs opn: 24 983 E. Hwy. 160, 65775. Phone: (417) 256-1025. Fax: (417) 256-2208. E-mail: hotcountrykdy@kkdy.com Web Site: www.kkdy.com. Licensee: Central Ozark Radio Network Inc. (acq 8-1-94). Population served: 34,000 Natl. Network: CNN Radio. Law Firm: Haley, Bader & Potts. Format: Hot country. News staff: 2; News: 10 hrs wkly. Target aud: 18-49. Spec prog: Contemp Christian 3 hrs wkly. ♦Tom Marhefka, pres & gen mgr; Bob Eckman, opns VP; Chuck Boone, opns dir & progmg dir; Jonathan Bergman, sls dir & gen sls mgr; Bobby Helm, news dir & news rptr; Bill Martin, chief of engrg; Crystal Cook, traf mgr.

***KSMW(FM)**—Not on air, target date: unknown: 90.9 mhz; 350 w. Ant 387 ft TL: N36 45 00 W91 49 40. Hrs open: Missouri State Univ., 901 S. National Ave., Springfield, 65897. Phone: (417) 836-5878. Fax: (417) 836-5889. E-mail: ksmu@missouristate Web Site: www.ksmu.org. Licensee: Board of Governors, Southwest Missouri State University. Format: Class, news. ♦Tammy Wiley, gen mgr.

KSPQ(FM)—Listing follows KWPM(AM).

KWPM(AM)— 1947: 1450 khz; 1 kw-U. TL: N36 44 28 W91 50 01. Hrs open: 24 983 U.S. Hwy. 160 E., 65775. Phone: (417) 256-3131. Phone: (417) 256-5976. Fax: (417) 256-2208. Web Site: www.ozarkradionetwork.com. Licensee: Missouri Ozarks Radio Network. (acq 1996). Population served: 50,000 Rgnl. Network: Brownfield, Missourinet. Format: News/talk. News staff: 4. Target aud: 25-54. ♦Gerry Elan, gen mgr & progmg dir; Tom Marheska, opns mgr; Jonathan Bergman, gen sls mgr; Bobby Helm, news dir; Bill Martin, chief of engrg; Crystal Cook, traf mgr.

KSPQ(FM)—Co-owned with KWPM(AM). 1951: 93.9 mhz; 100 kw. Ant 650 ft TL: N37 00 12 W91 54 24. Stereo. Phone: (417) 256-2322.150,000 Format: Classic rock. Target aud: 45-65 plus. ♦Jonathan Bergman, natl sls mgr; Mike Crase, progmg dir.

Wheeling

KULH(FM)— May 3, 1999: 105.9 mhz; 6 kw. 328 ft TL: N39 54 25 W93 20 28. Hrs open: 24 802 Calhoun St., Chillicothe, 64601. Phone: (660) 646-2255. Fax: (660) 646-2242. E-mail: contactus@1059thewave.com Web Site: www.1059thewave.com. Licensee: Resources Management Unlimited, Inc. (acq 3-28-01; $350,000). Population served: 70,000 Natl. Network: USA. Law Firm: Reddy, Begley & McCormick. Format: Christian, adult contemp. Target aud: General. ♦Ean Leppin, gen mgr.

Willard

KOSP(FM)— Aug 15, 1992: 105.1 mhz; 50 kw. 492 ft TL: N37 01 01 W93 30 31. Stereo. Hrs opn: 24 319-B E. Battlefield, Springfield, 65807. Phone: (417) 886-5677. Fax: (417) 886-2155. E-mail: info@kosp.fm Web Site: www.kosp.fm. Licensee: MW SpringMo Inc. Group owner: The Mid-West Family Broadcast Group Population served: 295,300 Natl. Rep: McGavren Guild. Law Firm: Davis Wright Tremaine, LLP. Format: Oldies. Target aud: 35-64; baby boomers. ♦Rick McCoy, pres & gen mgr; Mary Fleenor, opns mgr; Malcolm Hukriede, gen sls mgr; Keith Abercrombie, rgnl sls mgr; Mike Edwards, prom mgr.

Willow Springs

KUKU(AM)— Oct 1957: 1330 khz; 1 kw-D, 52 w-N. TL: N36 58 47 W91 59 29. Hrs open: Sunrise-sunset
Rebroadcasts KWPM(AM) West Plains 100%.
6962 US Hwy. 60, Mountain View, 65548. Phone: (417) 469-2500. Fax: (417) 934-2565. E-mail: gto@kuku.com Web Site: kuku.com. Licensee: Missouri Ozarks Radio Network. (acq 1996). Population served: 20,000 Natl. Network: ABC. Format: News/talk.

KUKU-FM— June 15, 1985: 100.3 mhz; 50 kw. 492 ft TL: N37 03 49 W92 01 39. Stereo. 24 Web Site: kuku.com.100,000 Format: Oldies. News staff: 2; News: 27 hrs wkly. Target aud: 29 plus. ♦Gary Taylor, progmg dir & local news ed; Harlin Hutchinson, local news ed.

Windsor

KWKJ(FM)— Feb 21, 2002: 98.5 mhz; 2.3 kw. Ant 535 ft TL: N38 35 37 W93 31 26. Stereo. Hrs opn: 24 Box 398, Warrensburg, 64093. Phone: (660) 747-9191. Phone: (660) 747-3883. Fax: (660) 747-5611. Web Site: www.kwkj.com. Licensee: D & H Media LLC (acq 7-26-00; $47,500 for CP). Wire Svc: AP Format: CHR. News staff: one; News: 1 hr wkly. Target aud: Students; Central MO State Univ. Students and like age group. ♦Vance DeLozier, pres; Greg Hassler, VP & gen mgr.

Montana

Alberton

KRQS(FM)—Not on air, target date: unknown: 105.5 mhz; 1.1 kw. Ant 787 ft TL: N47 02 05 W114 41 11. Hrs open: Box 4106, Missoula, 59806. Phone: (406) 728-5000. Fax: (406) 721-3020. Licensee: CCR-Missoula IV LLC. (acq 10-31-2006; grpsl). ♦Chad Parrish, gen mgr.

Anaconda

KANA(AM)— Aug 1947: 580 khz; 1 kw-D, 197 w-N. TL: N46 07 50 W112 55 07. Hrs open: 105 Main St., 59711. Phone: (406) 563-8011. Fax: (406) 563-8259. Licensee: Jimmy Ray Carroll. Group owner: Jimmy Ray Carroll Stns (acq 10-9-01; grpsl). Format: Oldies. ♦Sissy Vigil, gen mgr.

KGLM-FM— Jan 18, 1974: 97.7 mhz; 210 w. Ant 941 ft TL: N46 06 07 W112 56 59. Stereo. Hrs opn: 24 105 Main St., 59711. Phone: (406) 563-8011. Fax: (406) 563-8259. E-mail: kkglm@montana.com Licensee: Jimmy Ray Carroll. Group owner: Jimmy Ray Carroll Stns (acq 10-9-01; grpsl). Population served: 30,000 Format: Hot adult contemp. News staff: one; News: 4 hrs wkly. Target aud: 18 plus. ♦Jim Carroll, pres; Sissy Vigil, gen mgr; Dave Michaels, progmg dir.

Arlee

***KJFT(FM)**— 2007: 90.3 mhz; 110 w. Ant 1,906 ft TL: N47 01 04 W114 00 49. Hrs open: 1601 S. Sixth St., Missoula, 59801. Web Site: www.csnradio.com. Licensee: CSN International (group owner). Format: Christian.

Baker

KATQ-FM—See Plentywood

KFLN(AM)— July 14, 1964: 960 khz; 5 kw-D, 91 w-N. TL: N46 22 31 W104 16 25. Hrs open: Box 790, 3600 Hwy. 7, 59313. Phone: (406) 778-3371. Fax: (406) 778-3373. E-mail: kfln@midrivers.com Licensee: Newell Broadcasting Corp. (acq 3-1-84; $870,000; FTR: 3-5-84). Population served: 2,584 Natl. Network: ABC. Wire Svc: AP Format: C&W. Spec prog: Farm 10 hrs wkly. ♦Russ Newell, pres, gen mgr & progmg dir; Devin Bannister, gen sls mgr; Darrin Nutt, news dir; Tony Cuesta, chief of engrg; Alysia Putnam, traf mgr.

KJJM(FM)—Co-owned with KFLN(AM). May 26, 2001: 100.5 mhz; 6 kw. Ant 108 ft TL: N46 22 31 W104 16 25. Stereo. 24 Format: Classic rock.

Belgrade

KCMM(FM)—Listing follows KGVW(AM).

KGVW(AM)— Feb 1, 1959: 640 khz; 10 kw-D, 1 kw-N, DA-2. TL: N45 46 15 W111 13 26. Hrs open: 24 2050 Amsterdam Rd., 59714. Phone: (406) 388-4281. Fax: (406) 388-1700. Licensee: Gallatin Valley Witness Inc. (acq 1996). Population served: 76,000 Natl. Network: USA. Format: Relg, news/talk. News staff: one; News: 16 hrs wkly. Target aud: 35-64; business people, farmers & housewives. ♦Bryan Brucks, sr VP, dev VP & adv dir; Mark Brashear, pres, gen mgr, opns mgr, gen sls mgr & adv VP; C.J. Swoboda, progmg dir; Dale Heidner, chief of engrg.

KCMM(FM)—Co-owned with KGVW(AM).Not on air, target date: unknown: 99.1 mhz; 6 kw. 200 ft

KISN(FM)— Nov 1, 1963: 96.7 mhz; 18.5 kw. Ant 813 ft TL: N45 40 24 W110 52 02. Hrs open: 125 W. Mendenhall, Suite 1, Bozeman, 59715. Phone: (406) 586-2343. Fax: (406) 587-2202. Web Site: www.bozemanskissfm.com. Licensee: Capstar TX L.P. Group owner: Clear Channel Communications Inc. (acq 2-21-01; grpsl). Population served: 3,500 Format: CHR. Target aud: 25-54; women. ♦Sammy Suarez, progmg dir; Sylvia Drain, gen mgr & progmg dir.

***KQLU(FM)**— 2006: 90.9 mhz; 5.5 kw vert. Ant 623 ft TL: N45 57 25 W111 22 11. Hrs open:
Rebroadcasts KLVR(FM) Santa Rosa, CA 100%.
2351 Sunset Blvd., Suite 170-218, Rocklin, CA, 95765. Phone: (916) 251-1600. Fax: (916) 251-1650. Web Site: www.klove.com. Licensee: Educational Media Foundation. Natl. Network: K-Love. Format: Contemp Christian. ♦Richard Jenkins, pres; Mike Novak, VP; Keith Whipple, dev dir; Eric Allen, natl sls mgr; David Pierce, progmg mgr; Ed Lenane, news dir; Sam Wallington, engrg dir; Karen Johnson, news rptr; Marya Morgan, news rptr; Richard Hunt, news rptr.

Belt

KZUS(FM)—Not on air, target date: unknown: 101.7 mhz; 13.1 kw. Ant 2,061 ft TL: N47 09 34 W111 00 39. Hrs open: 980 N. Michigan Ave., Suite 1880, Chicago, IL, 60611. Phone: (312) 204-9900. Licensee: College Creek Media LLC. ♦Neal J. Robinson, pres.

Big Sky

KBZM(FM)— July 31, 1998: 104.7 mhz; 5 kw. Ant 3,336 ft TL: N45 16 41 W111 26 57. Hrs open: 24 102 S. 19th Ave., Suite 5, Bozeman, 59718-6860. Phone: (406) 582-1045. Fax: (406) 582-0388. Web Site: www.kbzm.com. Licensee: Orion Media LLC (acq 10-29-2003; $400,000). Format: Classic hits. Target aud: Adults; 25-54. ♦Jeff Balding, gen mgr; Susan Balding, gen sls mgr; Colter Langan, progmg dir.

KSCY(FM)—Not on air, target date: unknown: 106.9 mhz; 690 w. Ant 3,326 ft TL: N45 16 41 W111 26 57. Hrs open: Box 161268, 59716. Phone: (406) 581-4818. Fax: (406) 995-3271. Licensee: Radick Construction Inc. ♦John Radick, VP.

Billings

KBBB(FM)— Dec 6, 1987: 103.7 mhz; 100 kw. 480 ft TL: N45 46 00 W108 27 27. Stereo. Hrs opn: 24 Box 1276, 59103. Phone: (406) 248-7827. Fax: (406) 252-9577. Web Site: www.bee104.com. Licensee: CC Licenses LLC. Group owner: Clear Channel Communications Inc. (acq 4-13-2001; grpsl). Natl. Rep: Tacher. Law Firm: Reddy, Begley & McCormick. Format: Adult contemp. Target aud: 25-54; general. ♦Dennis Koffman, gen mgr; Roy Brown, opns dir & prom VP.

KBLG(AM)—Listing follows KRKX(FM).

***KBLW(FM)**— August 2002: 90.1 mhz; 250 w vert. Ant 331 ft TL: N45 45 51 W108 27 18. Stereo. Hrs opn: 24
Rebroadcasts KXEI(FM) Havre 100%.
Box 2426, Havre, 59501. Secondary address: 317 First St., Havre 59501. Phone: (406) 265-5845. Fax: (406) 265-8860. E-mail: ynop@ynopradio.org Web Site: www.ynopradio.org. Licensee: Hi-Line Radio Fellowship Inc. (acq 8-7-02). Natl. Network: Moody, Salem Radio Network. Wire Svc: AP Format: Inspirational. Target aud: General; those looking for Christian inspirational music & progmg. ♦Brenda Boyum, stn mgr; Roger Lonnquist, dev dir; Brian Jackson, progmg dir.

KBUL(AM)— Mar 20, 1951: 970 khz; 5 kw-U, DA-N. TL: N45 44 35 W108 32 37. Stereo. Hrs opn: 24 Box 1276, 59103. Secondary address: 27 N. 27th St., 23rd Fl. 59103. Phone: (406) 248-7827. Fax: (406) 252-9577. E-mail: denniscoffman@clearchannel.com Licensee: CC Licenses LLC. Group owner: Clear Channel Communications Inc. (acq 4-13-2001; grpsl). Population served: 120,000 Natl. Rep: Christal. Format: News. News staff: one; News: 2 hrs wkly. Target aud: 25-54. ♦Dennis Coffman, pres & gen mgr; Roy Brown, sls dir & gen sls mgr; Tommy Braaten, progmg dir & mus dir; Dick Jones, chief of engrg; Stacy Ulstad, traf mgr.

KCTR-FM—Co-owned with KBUL(AM). Aug 14, 1979: 102.9 mhz; 100 kw. 500 ft TL: N45 45 59 W108 27 19. Stereo. Format: Country. ♦Erik Bowen, progmg dir.

***KEMC(FM)**— Apr 25, 1973: 91.7 mhz; 100 kw. Ant 520 ft TL: N45 39 51 W108 34 14. Stereo. Hrs opn: 24 1500 N. 30th St., 59101-0298. Phone: (406) 657-2941. Fax: (406) 657-2977. E-mail: mail@yellowstonepublicradio.org Web Site: www.yellowstonepublicradio.org. Licensee: Montana State University/Billings. Population served: 100,000 Natl. Network: NPR, AP Radio. Format: News, class, jazz. News staff: one; News: 24 hrs wkly. Target aud: General. Spec prog: Folk 5 hrs wkly. ♦Lois Bent, gen mgr.

KGHL(AM)— June 8, 1928: 790 khz; 5 kw-U, DA-N. TL: N45 43 34 W108 36 35. Stereo. Hrs opn: 24 222 N. 32nd St., 59101. Phone: (406) 238-1000. Fax: (406) 238-1038. Licensee: New Northwest Broadcasters LLC (group owner; acq 8-10-99; grpsl). Population served: 165,500 Natl. Network: CBS. Law Firm: Dow, Lohnes & Albertson. Format: Classic country. News staff: one; News: 4 hrs wkly. Target aud: 25-54. ♦Pete Benedetti, CEO; Tommy Ehrman, gen mgr; Dave Tester, sls dir; Jeff Howell, progmg dir; Mike Powers, chief of engrg; Tracey McCarthy, traf mgr & farm dir.

KGHL-FM— August 1978: 98.5 mhz; 85 kw. 370 ft TL: N45 45 51 W108 27 18. Stereo. 192800 Format: Continuous country. Target aud: 18-54. ♦Karen Gallagher, progmg dir.

KKBR(FM)— Dec 17, 1963: 97.1 mhz; 28 kw. 325 ft TL: N45 45 51 W108 27 18. Stereo. Hrs opn: Box 1276, 59103. Phone: (406) 248-7827. Fax: (406) 252-9577. Web Site: www.kbear.com. Licensee: CC Licenses LLC. Group owner: Clear Channel Communications Inc. (acq 4-13-2001; grpsl). Format: Oldies. ♦Dennis Koffman, gen mgr; Keith Todd, progmg dir.

***KLMT(FM)**— Dec 18, 2002: 89.3 mhz; 1 kw vert. Ant 335 ft TL: N45 45 41 W108 27 19. Hrs open: 24 Western Inspirational Broadcasters Inc., 6363 Hwy. 50 E., Carson City, NV, 89701. Phone: (775) 883-5647. Licensee: Western Inspirational Broadcasters Inc. Natl. Network: AP Radio. Wire Svc: AP Format: Contemp Christian. News: 16 hrs wkly. ♦Tom Hesse, gen mgr; Tim Weidemann, opns mgr; Bill Feltner, progmg dir; Janet Santana, mus dir; Paul Lierman, chief of engrg.

***KLRV(FM)**— 2005: 90.9 mhz; 7.5 kw vert. Ant 593 ft TL: N45 45 54 W108 27 19. Hrs open:
Rebroadcasts KLVR(FM) Santa Rosa, CA).
2351 Sunset Blvd., Suite 170-218, Rocklin, CA, 95765. Phone: (916) 251-1600. Fax: (916) 251-1650. Web Site: www.klove.com. Licensee: Educational Media Foundation. (acq 12-8-2004; $100,000 for CP with CP for KLWC(FM) Casper, WY). Natl. Network: K-Love. Format: Christian. ♦Richard Jenkins, pres; Mike Novak, VP; Keith Whipple, dev dir; David Pierce, progmg dir; Ed Lenane, news dir; Sam Wallington, engrg dir; Arthur Vassar, traf mgr; Karen Johnson, news rptr; Marya Morgan, news rptr; Richard Hunt, news rptr.

KMZK(AM)— Sept 8, 1946: 1240 khz; 1 kw-U. TL: N45 45 26 W108 32 08. Hrs open: 24 Box 31038, 636 Haugen St., 59102. Phone: (406) 245-3121. Fax: (406) 245-0822. E-mail: www.genmgr@kmzk.com Web Site: www.kmzk.com. Licensee: Elenbaas Media Inc. (acq 1-8-98; $115,000). Population served: 87,500 Natl. Network: Salem Radio Network. Wire Svc: AP Format: Today's Christian Music. Target aud: 18-44; young & energetic high school & college students & young adults. ♦Herm Elenbaas, pres; John Black, progmg dir; Holly Howard, pub affrs dir; Deb Padilla, traf mgr.

KQBL(FM)— December 1998: 105.1 mhz; 6 kw. 233 ft TL: N45 45 57 W108 27 17. Hrs open: 24 222 N. 32nd St., 10th Floor, 59101. Phone: (406) 238-1000. Fax: (406) 238-1038. Web Site: www.1051theend.com. Licensee: New Northwest Broadcasters LLC. (group owner; (acq 10-26-99; grpsl). Target aud: 25-54. ♦Pete Benedetti, CEO; Tommy Ehrman, gen mgr; Tom Oakes, opns mgr.

KRKX(FM)— July 1989: 94.1 mhz; 100 kw. 590 ft TL: N45 32 25 W108 38 31. Stereo. Hrs opn: 24 2075 Central Ave., 59102. Phone: (406) 652-8400. Fax: (406) 652-4899. Web Site: www.krkx.com. Licensee: CCR-Billings IV LLC. Group owner: Fisher Broadcasting Company (acq 10-31-2006; grpsl). Population served: 140,000 Format: Classic rock. Target aud: 25-54; affluent. ♦Debbie Sundberg, gen mgr; Terry Keys, opns mgr; Augie Aga, gen sls mgr.

KBLG(AM)—Co-owned with KRKX(FM). Sept 25, 1955: 910 khz; 1 kw-D, 63 w-N. TL: N45 45 10 W108 30 57.24 Web Site: www.kblg.com.120,000 Natl. Network: CBS. Format: News/talk, sports. News staff: one; News: 46 hrs wkly. Target aud: 35-64; upscale executives.

KRPM(FM)— 2001: 107.5 mhz;; 100 kw. Ant 984 ft TL: N45 44 29 W108 08 19. Hrs open: 222 N. 32nd St., 59101. Phone: (406) 238-1000. Fax: (406) 238-1038. Licensee: New Northwest Broadcasters LLC. (group owner; (acq 10-26-99). Format: Hot adult contemp. Target aud: 25-54. ♦Pete Benedetti, CEO; Dave Tester, gen mgr.

KRZN(FM)— 1998: 96.3 mhz; 100 kw. 695 ft TL: N45 45 37 W108 27 09. Hrs open: 2075 Central Ave., 59102. Phone: (406) 652-8400. Fax: (406) 652-4899. Web Site: www.thezone963.com. Licensee: CCR-Billings IV LLC. Group owner: Fisher Broadcasting Company. (acq 10-31-2006; grpsl). Format: New rock. ♦Dan Reese, gen mgr & stn mgr.

KURL(AM)— Oct 15, 1959: 730 khz; 5 kw-D, 236 w-N. TL: N45 45 29 W108 29 53. Hrs open: 24 Box 31038, 59107. Secondary address: 636 Haugen 59107. Phone: (406) 245-3121. Fax: (406) 245-0822. E-mail: genmgr@kurlradio.com Web Site: www.kurlradio.com. Licensee: Elenbaas Media Inc. (acq 11-14-94; $300,000; FTR: 1-2-95). Population served: 125,000 Natl. Network: USA, AP Radio. Format: Relg, syndicated talk. Target aud: 35-64. ♦Herm Elenbaas, pres & gen mgr.

KYYA-FM— Apr 5, 1969: 93.3 mhz; 100 kw. 700 ft TL: N45 45 37 W108 27 09. Stereo. Hrs opn: 24 2075 Central Ave., 59102. Phone: (406) 652-8400. Fax: (406) 652-4899. Web Site: www.y93.com. Licensee: CCR-Billings IV LLC. Group owner: Fisher Broadcasting Company (acq 10-31-2006; grpsl). Population served: 102,000 Natl. Network: ABC. Format: Hot adult contemp. News staff: one; News: 2 hrs wkly. Target aud: 18-49; women. ♦Debbie Sundberg, gen mgr, gen sls mgr, natl sls mgr & rgnl sls mgr; Ted Brown, progmg dir; Michael Lyon, news dir; Bruce Faulkner, chief of engrg.

Boulder

KMTZ(FM)—Not on air, target date: unknown: 107.7 mhz; 6 kw. Ant -39 ft TL: N46 15 34 W112 09 09. Hrs open: Box 309, Missoula, 59806-0309. Phone: (406) 542-1025. Fax: (406) 721-1036. Licensee: Sheila Callahan and Friends Inc. ♦M. Sheila Callahan Murphy, pres.

Bozeman

***KBMC(FM)**— October 1991: 102.1 mhz; 20.5 kw. 728 ft TL: N45 38 18 W111 16 05. Stereo. Hrs opn:
Rebroadcasts KEMC(FM) Billings 100%.
1500 University Dr., Billings, 59101-0298. Phone: (406) 657-2941. Fax: (406) 657-2977. Licensee: Montana State University/Billings. (acq 3-29-91; FTR: 4-15-91). Format: Class, jazz, news. ♦Lois Bent, gen mgr.

KBOZ(AM)— Dec 19, 1975: 1090 khz; 5 kw-U, DA-N. TL: N45 36 58 W111 05 16. Hrs open: Box 20, 59718. Secondary address: 5445 Johnson Rd. 59715. Phone: (406) 587-9999. Fax: (406) 587-5855. Licensee: Reier Broadcasting Co. Inc. (group owner; acq 10-18-96; grpsl). Population served: 60,000 Natl. Network: CBS, Jones Radio Networks. Format: Talk. Target aud: 25-64. ♦Bill Reier, gen mgr & opns mgr; Eric Reier, gen sls mgr; Brian Bennett, progmg dir; Les Clay, news dir; Dick Jones, chief of engrg; Diane Stovall, traf mgr.

KOBB-FM—Co-owned with KBOZ(AM). Nov 1, 1980: 93.7 mhz; 100 kw. 245 ft TL: N45 41 35 W110 58 50. Stereo. 45,000 Format: Oldies. Target aud: 25-54. ♦Tuck Reier, opns dir; Dave Visscher, progmg dir.

KBOZ-FM—Listing follows KOBB(AM).

***KGLT(FM)**— December 1963: 91.9 mhz; 2 kw. 365 ft TL: N45 41 35 W110 59 00. Stereo. Hrs opn: Montana State Univ., Rm. 324, 59717. Phone: (406) 994-3001. Fax: (406) 994-1987. Licensee: Montana State University. Population served: 61,581 Natl. Network: AP Radio. Format: Div, educ, alternative. Target aud: General. Spec prog: Black one hr, class 11 hrs, folk 12 hrs wkly. ♦Philip Charles, gen mgr; Jim Kehoe, mus dir; John Campbell, chief of engrg.

***KLBZ(FM)**—Not on air, target date: unknown: 89.3 mhz; 7 kw vert. Ant 679 ft TL: N45 57 25 W111 22 11. Hrs open: 2351 Sunset Blvd., Suite 170-218, Rocklin, CA, 95765. Phone: (916) 251-1600. Fax: (916) 251-1650. Licensee: Educational Media Foundation. ♦Richard Jenkins, pres.

KMMS(AM)— Oct 15, 1939: 1450 khz; 1 kw-U. TL: N45 39 33 W111 03 22. Hrs open: 125 W. Mendenhall, 59715. Phone: (406) 586-2343. Fax: (406) 587-2202. Web Site: www.kmmsam.com. Licensee: Capstar TX L.P. Group owner: Clear Channel Communications Inc. (acq 2-21-01; grpsl). Population served: 66,000 Natl. Network: ABC. Natl. Rep: Clear Channel. Law Firm: Hogan & Hartson. Format: News/talk, sports. News: one hr wkly. Target aud: 35-64. ♦Lowery Mays, chmn; John Hogan, pres; Sylvia Drain, gen mgr; Kay Ruh, natl sls mgr; Mary Atkins, prom dir & prom mgr; George Carter, progmg dir; John Russell, news dir; Dennis Mountford, chief of engrg; Lenny Jones, traf mgr.

KMMS-FM— Aug 14, 1986: 95.1 mhz; 94 kw. 781 ft TL: N45 40 24 W110 52 02.24 Web Site: www.mooseradio.com.66,000 Format: AAA. ♦Michelle Wolfe, progmg dir.

KOBB(AM)— May 22, 1950: 1230 khz; 1 kw-U, DA-2. TL: N45 42 02 W111 02 49. Hrs open: 24 Box 20, 59718. Secondary address: 5445 Johnson Rd. 59718. Phone: (406) 587-9999. Fax: (406) 587-5855. E-mail: reier@bigsky.net Licensee: Reier Broadcasting Co. Inc. (group owner; (acq 2-19-93; $125,000; FTR: 5-17-93). Population served: 50,000 Natl. Network: ABC. Rgnl. Network: Agri-Net. Rgnl rep: Tacher. Law Firm: Reddy, Begley & McCormick. Format: Adult Standards. News: 15 hrs wkly. Target aud: 30 plus; affluent adults. ♦William Reier Sr., pres & gen mgr; Eric Reier, gen sls mgr; Diane Stovall, progmg dir & traf mgr; Dick Jones, chief of engrg.

KBOZ-FM—Co-owned with KOBB(AM). 1983: 99.9 mhz; 100 kw. 338 ft TL: N45 41 34 W110 58 57. Format: Hot country. ♦Terry Michaels, progmg dir; Diane Stovall, chief of engrg & traf mgr.

KOZB(FM)—See Livingston

KZMY(FM)— 2004: 103.5 mhz; 100 kw. Ant 948 ft TL: N45 57 25 W111 22 11. Hrs open: 24 125 W. Mendenhall St., Suite 102, 59715. Phone: (406) 556-0123. Fax: (406) 587-2202. E-mail: kzmy@hotmail.com Web Site: my1035.com. Licensee: Capstar TX L.P. Group owner: Clear Channel Communications Inc. (acq 2-27-2004; $1.4 million for CP). Format: Adult contemp. ♦Nick Shannon, gen mgr & progmg dir.

Butte

KAAR(FM)— Nov 1, 1988: 92.5 mhz; 4.5 kw. 1,840 ft TL: N46 00 29 W112 26 30. Hrs open: 24 750 Dewey Blvd., Suite 1, 59701. Secondary address: Box 3788 59702. Phone: (406) 494-1030. Fax: (406) 494-6020. Licensee: CCR-Butte IV LLC. Group owner: Fisher Broadcasting Company (acq 10-31-2006; grpsl). Natl. Rep: McGavren Guild. Format: Country. News: 8 hrs wkly. Target aud: General. ♦Chris Ackerman, gen mgr; Jeff Gray, opns dir; Rene Wimberley, sls dir.

***KAPC(FM)**— 1999: 91.3 mhz; 880 w. 1,893 ft TL: N46 00 29 W112 26 30. Hrs open: c/o KUFM(FM), Univ. of Montana, Missoula, 59812. Phone: (406) 243-4931. Fax: (406) 243-3299. Web Site: www.kufm.org. Licensee: University of Montana. Format: Jazz, classical, news. News: 2 hrs wkly. ♦William Marcus, gen mgr.

KBOW(AM)— Feb 14, 1947: 550 khz; 5 kw-D, 1 kw-N, DA-N. TL: N45 58 30 W112 34 18. Hrs open: Box 3389, 59702. Secondary address: 660 Dewey Blvd. 59702. Phone: (406) 494-7777. Fax: (406) 494-5534. E-mail: bbi@inpch.com Licensee: Butte Broadcasting Inc. (acq 1-13-94; $550,000 with co-located FM; FTR: 1-31-94) Natl. Network: CBS. Law Firm: Haley, Bader & Potts. Format: Sports. Target aud: 25 plus; general. Spec prog: Farm 5 hrs, relg 2 hrs wkly. ♦Fran Workman, opns mgr & progmg dir; Ron Davis, pres, gen mgr & gen sls mgr; Mike Beckworth, prom dir; Paul Panisko, progmg dir; Pat Schulte, news dir; Chuck Beardslee, engrg VP; Araka Williams, traf mgr.

KOPR(FM)—Co-owned with KBOW(AM). Oct 26, 1972: 94.1 mhz; 100 kw. 1,840 ft TL: N46 00 23 W112 26 28. (CP: 58.4 kw). Stereo. 24 (Acq 4-1-94).70,000 Format: 80's & more. News staff: 2; News: 5 hrs wkly. Target aud: 25-45; women. ♦Fran Workman, pub affrs dir.

***KFRD(FM)**— 2006: 88.9 mhz; 2.8 kw vert. Ant 1,729 ft TL: N46 00 27 W112 26 30. Hrs open:
Rebroadcasts KUFR(FM) Salt Lake City, UT 100%.
c/o KUFR(FM), 136 E.S. Temple, Suite 1630, Salt Lake City, UT, 84111. Phone: (801) 359-3147. Fax: (801) 359-8112. Web Site: www.familyradio.com. Licensee: Family Stations Inc. Format: Christian relg. ♦Harold Camping, gen mgr.

***KFRT(FM)**— 2003: Stn currently dark. 88.1 mhz; 850 w vert. Ant 1,729 ft TL: N46 00 27 W112 26 30. Hrs open:
Rebroadcasts KUFR(FM) Salt Lake City, UT 100%.
c/o Radio Station KUFR(FM), 136 E.S. Temple, Suite 1630, Salt Lake City, UT, 84111. Phone: (801) 359-3147. Fax: (801) 359-8112. Web Site: www.familyradio.com. Licensee: Family Stations Inc. (group owner). Natl. Network: Family Radio. Format: Christian relg. ♦Harold Camping, gen mgr.

***KJLF(FM)**—Not on air, target date: unknown: 90.5 mhz; 200 w. Ant 1,896 ft TL: N46 00 24 W112 26 30. Hrs open: Box 2426, Havre, 59501. Phone: (406) 265-5845. Fax: (406) 265-8860. E-mail: info@ynopradio.org Web Site: www.ynopradio.org. Licensee: Hi-Line Radio Fellowship Inc. Natl. Network: Salem Radio Network. General; those looking for inspirational Christian music & programming. ♦Roger Lonnquist, gen mgr; Brenda Boyum, stn mgr; Brian Jackson, progmg mgr.

KMBR(FM)—Listing follows KXTL(AM).

***KMSM-FM**— 1975: 106.9 mhz; 500 w. 93 ft TL: N46 00 43 W112 33 23. Stereo. Hrs opn: 24 Student Union Bldg, Montana Tech., 59701. Phone: (406) 496-4601. E-mail: kmsm@mtech.edu Web Site: www.mtech.edu/kmsm. Licensee: Associated Students of Montanta Tech. Population served: 3,000 Law Firm: Pepper & Corazzini. Format: Educ, div, alternative. News: 2 hrs wkly. Target aud: General; very diversified group. Spec prog: Jazz 5 hrs, relg 3 hrs, class 2 hrs wkly. ♦Wendy Dyer, gen mgr; Ben Carter, stn mgr.

KXTL(AM)— 1927: 1370 khz; 5 kw-U. TL: N46 00 21 W112 37 54. Stereo. Hrs opn: 24 Box 3788, 59702. Secondary address: 750 Dewey Blvd., Suite 1 59702. Phone: (406) 494-4442. Fax: (406) 494-6020. Web Site: www.kxtl.com. Licensee: CCR-Butte IV LLC. Group owner: Fisher Broadcasting Company (acq 10-31-2006; grpsl). Population served: 40,000 Natl. Rep: McGavren Guild. Format: Hits of the 50s, 60s & 70s. News: 14 hrs wkly. Target aud: 25-54. Spec prog: Relg one hr wkly. ♦Chris Ackerman, gen mgr & gen sls mgr; Jeff Gray, opns mgr, progmg dir & news dir; Roger Bennett, chief of engrg; Tammy Gordon, traf mgr.

KMBR(FM)—Co-owned with KXTL(AM). Feb 7, 1980: 95.5 mhz; 50 kw. 1,820 ft TL: N46 00 29 W112 26 30. Stereo. Phone: (406) 494-5895. Web Site: www.955kmbr.com.50,000 Natl. Rep: McGavren Guild. Format: Classic rock.

Cascade

KIKF(FM)— January 2002: 104.9 mhz; 94 kw. Ant 2,037 ft TL: N47 09 34 W111 00 39. Hrs open: Box 3129, Great Falls, 59403. Phone: (406) 761-2800. Fax: (406) 727-7218. E-mail: tjlee@mykikfm.com Web Site: www.mykikfm.com. Licensee: Fisher Radio Regional Group Inc. Group owner: Fisher Broadcasting Company (acq 3-12-01). Natl. Rep: McGavren Guild. Format: Country. ♦Terry Strickland, gen mgr; Linda Cushingham, gen sls mgr; T.J. Lee, progmg dir.

Chinook

KRYK(FM)—Licensed to Chinook. See Havre

Choteau

KEAU(FM)—Not on air, target date: unknown: 102.3 mhz; 100 kw. Ant 557 ft TL: N47 49 13 W111 47 56. Hrs open: 980 N. Michigan Ave., Suite 1880, Chicago, IL, 60611. Phone: (312) 204-9900. Licensee: College Creek Media LLC. ♦Neal J. Robinson, pres.

Colstrip

KMCJ(FM)— August 2001: 99.5 mhz; 100 kw. Ant 800 ft TL: N46 10 32 W106 24 21. Stereo. Hrs opn: 24
Rebroadcasts KXEI(FM) Havre 100%.
Box 2426, Havre, 59501. Secondary address: 317 First St., Havre 59501. Phone: (406) 265-5845. Fax: (406) 265-8860. E-mail: ynop@ynopradio.org Web Site: www.ynopradio.org. Licensee: Hi-Line Radio Fellowship Inc. (acq 4-13-01; $52,000 for CP). Population served: 4,000 Natl. Network: Salem Radio Network. Wire Svc: AP Format: Christian Inspirational. Target aud: General:; those looking for inspirational Christian music & programming. ♦Roger Lonnquist, gen mgr; Brenda Boyum, stn mgr; Brian Jackson, progmg dir.

Columbia Falls

KHNK(FM)— Nov 17, 1998: 95.9 mhz; 55 kw horiz, 5.6 kw vert. Ant 2,286 ft TL: N48 30 42 W114 22 16. Hrs open: 24 2432 Hwy. 2 E., Kalispell, 59901. Phone: (406) 755-8700. Fax: (406) 755-8770. E-mail: kkmt@beebroadcasting.com Web Site: www.beebroadcasting.com. Licensee: Bee Broadcasting Inc. (group owner; (acq 12-31-97; $337,500). Format: Country. ♦Mark Wagner, gen mgr.

KRVO(FM)— 2006: 103.1 mhz; 8 kw. Ant 2,362 ft TL: N48 30 43 W114 22 13. Hrs open: Box 5409, Kalispell, 59903. Phone: (406) 755-8700. Fax: (406) 755-8770. Web Site: www.beebroadcasting.com. Licensee: Cathleen R. Bee. Law Firm: Smithwick & Belendiuk. ♦Cathleen R. Bee, gen mgr.

Conrad

KTZZ(FM)— July 1, 1997: 93.7 mhz; 100 kw. 558 ft TL: N47 49 13 W111 47 56. Hrs open: 24 Box 1239, Great Falls, 59403. Secondary address: 3313 15th St. N.E., Black Eagle 59414. Phone: (406) 761-1310. Fax: (406) 454-3775. Licensee: Jeannine M. Mason. Population served: 85118 Natl. Network: ABC. Format: Classic rock. News staff: one; News: 5 hrs wkly. Target aud: 25-54; general. ♦Steven Dow, pres; Laurie Vosberg, adv dir.

Darby

KHDV(FM)—Not on air, target date: unknown: 107.9 mhz; 30 kw. Ant -961 ft TL: N46 05 03 W114 11 14. Hrs open: Box 309, Missoula, 59806-0309. Phone: (406) 542-1025. Fax: (406) 721-1036. Licensee: Sheila Callahan and Friends Inc. ♦Sheila Callahan, gen mgr.

Deer Lodge

KBCK(AM)— 1963: 1400 khz; 1 kw-U. TL: N46 24 26 W112 43 08. Hrs open: 105 Main St., Anaconda, 59711. Phone: (406) 563-8011. Fax: (406) 563-8259. Licensee: Jimmy Ray Carroll. Group owner: Jimmy Ray Carroll Stns (acq 10-9-01; grpsl). Population served: 30,000 Format: Real country. Target aud: 18 plus. Spec prog: Farm 2 hrs wkly. ♦Jimmy Ray Carroll, pres; Sissy Virgil, gen mgr.

KQRV(FM)— July 4, 1997: 96.9 mhz; 20 kw. Ant 984 ft TL: N46 06 03 W112 57 00. Hrs open: 302 Missouri Ave., 59722. Phone: (406) 846-1100. Fax: (406) 846-1100. E-mail: river@3riversdbs.net Licensee: Robert Cummings Toole. Format: Country, full service. ♦Robert Cummings Toole, gen mgr; Karen Toole, gen sls mgr.

Dillon

KBEV-FM—Listing follows KDBM(AM).

KDBM(AM)— Jan 1, 1957: 1490 khz; 1 kw-U. TL: N45 14 13 W112 38 32. Hrs open: 610 N. Montana St., 59725. Phone: (406) 683-2800. Phone: (406) 683-6171. Fax: (406) 683-9480. Licensee: Dead-Air Broadcasting Co. Inc. (acq 3-6-98; $330,000 with co-located FM). Population served: 4,548 Format: Country. ♦Jo Ann Juliano, pres; Kathy Wise, gen mgr & traf mgr; Kasey Briggs, sls dir; John Schuyler, progmg dir & news dir; Ron Huckaby, chief of engrg.

KBEV-FM—Co-owned with KDBM(AM). August 1972: 98.3 mhz; 10.5 kw. 495 ft TL: N45 14 22 W112 40 03. Stereo. Format: Contemp hit.

KDIL(AM)—Not on air, target date: 10/07: 940 khz; 10 kw-D, 350 w-N, DA-2. TL: N45 13 26 W112 35 58. Hrs open: 110 Green Meadows, Abilene, TX, 79605. Phone: (325) 829-6850. Licensee: Scott Powell. Law Firm: Shanis & Peltzman. ♦Amy Meredith, pres; Scott Powell, gen mgr.

***KDWG(FM)**—Not on air, target date: unknown: 90.9 mhz; 850 w. -236 ft TL: N45 12 33 W112 38 14. Hrs open: 24 Univ. of Montana, Western, Campus Box 52, 710 S. Atlantic St., 59725. Phone: (406) 683-7156. Licensee: Western Montana College University of Montana. Format: Mainstream, educ, rock. ♦Cory Craden, mus dir.

Dutton

KVVR(FM)— Aug 7, 2001: 97.9 mhz; 100 kw. Ant 715 ft TL: N47 36 52 W111 20 51. Hrs open: Box 3309, Great Falls, 59403. Phone: (406) 761-7600. Fax: (406) 761-5511. Licensee: CCR-Great Falls IV LLC. Group owner: Cherry Creek Radio LLC (acq 12-19-2003; grpsl). Format: Adult contemp. ♦Ron Korb, gen mgr.

East Helena

KHKR-FM— Apr 13, 1989: 104.1 mhz; 5 kw. Ant 653 ft TL: N46 46 11 W112 01 25. Stereo. Hrs opn: 110 Broadway St., Helena, 59601. Phone: (406) 442-4490. Fax: (406) 442-7356. Web Site: www.khkr.com. Licensee: CCR-Helena IV LLC. Group owner: Cherry Creek Radio LLC (acq 2-3-2004; grpsl). Format: Hot country. Target aud: 25-54. ♦Dewey Bruce, gen mgr.

KKGR(AM)— May 26, 1988: 680 khz; 5 kw-D. TL: N46 33 58 W111 54 12. Hrs open: 1400 11th Ave., Helena, 59601. Phone: (406) 443-5237. Phone: (406) 442-7595. Licensee: KKGR Inc. (acq 3-16-99). Format: Oldies. ♦Jim Schaffer, gen mgr; Ron Davidson, gen mgr.

East Missoula

KLCY(AM)—Licensed to East Missoula. See Missoula

Eureka

KWDE(FM)—Not on air, target date: unknown: 93.5 mhz; 2.4 kw. Ant 1,817 ft TL: N48 38 35 W115 05 31. Hrs open: 5331 Mt. Alifan Dr., San Diego, CA, 92111. Phone: (858) 277-4991. Fax: (858) 277-1365. Licensee: Horizon Christian Fellowship. (acq 2-9-2006; grpsl). ♦Mike MacIntosh, pres.

Fairfield

KUUS(FM)—Not on air, target date: unknown: 103.7 mhz; 100 kw. Ant 557 ft TL: N47 49 13 W111 47 56. Hrs open: 980 N. Michigan Ave., Suite 1880, Chicago, IL, 60611. Phone: (312) 204-9900. Licensee: College Creek Media LLC. ♦Neal J. Robinson, pres.

Florence

KDTR(FM)— 2005: 103.3 mhz; 1.95 kw. Ant 2,083 ft TL: N46 48 06 W113 58 22. Hrs open: 2425 W. Central Ave., Suite 203, Missoula, 59801. Phone: (406) 721-6800. Fax: (406) 329-1850. Web Site: www.trail1033.com. Licensee: Spanish Peaks Broadcasting Inc. Natl. Rep: Tacher. Format: Triple A. Target aud: 25-54; adults. ♦Rod Harsell, gen mgr; Robert Chase, progmg dir.

Forsyth

KIKC-FM— September 1980: 101.3 mhz; 100 kw. Ant 1,010 ft TL: N46 10 32 W106 24 21. Stereo. Hrs opn: 24 Box 1140, 59327. Secondary address: 210 W. Front St. 59327. Phone: (406) 346-2711. Fax: (406) 346-2712. E-mail: kikc@rangeweb.net Web Site: klkcamfm.com. Licensee: Miles City, Forsyth Broadcasting Inc. (acq 1996; grpsl). Population served: 60,000 Natl. Network: CNN Radio. Natl. Rep: Interep. Rgnl rep: Allied Radio Partners Wire Svc: AP Format: Country. News: 4 hrs wkly. Target aud: 18 plus; general. ♦Steve Marks, CEO; Dick Haugen, VP & gen mgr; Patti Haugen, opns mgr; Grant West, progmg dir & mus dir.

KIKC(AM)— Oct 10, 1975: 1250 khz; 5 kw-D, 132 w-N. TL: N46 15 30 W106 41 21.24 Web Site: klkcamfm.com.60,000 Format: CHR, oldies. Target aud: 18-35; general. ♦Stephen Marks, pres.

Fort Belknap Agency

***KGVA(FM)**— October 1996: 88.1 mhz; 95 kw. 797 ft TL: N48 11 18 W108 42 36. Hrs open: Box 159, Harlem, 59526. Phone: (406) 353-4656. Fax: (406) 353-2898. Licensee: Fort Belknap College. Natl. Network: NPR. Format: Eclectic, news/talk. Target aud: General. Spec prog: American Indian 15 hrs wkly. ♦Will Gray Jr., gen mgr & spec ev coord.

Glasgow

KLAN(FM)— Mar 1, 1983: 93.5 mhz; 3 kw. 300 ft TL: N48 05 42 W106 37 08. Stereo. Hrs opn: Box 671, 59230. Phone: (406) 228-9336. Fax: (406) 228-9338. E-mail: kltz@kltz.com Web Site: www.kltz.com. Licensee: Glasgow Broadcasting Corp. Format: Adult contemp. ♦Shirley Trang, gen mgr.

KLTZ(AM)— Aug 14, 1954: 1240 khz; 1 kw-U. TL: N48 13 09 W106 38 54. Hrs open: Box 671, 59230. Phone: (406) 228-9336. Fax: (406) 228-9338. E-mail: kltz@kltz.com Web Site: www.kltz.com. Licensee: Glasgow Broadcasting Inc. Population served: 5,000 Rgnl. Network: Agri-Net. Format: C&W. Target aud: 25 plus. ♦Shirley Trang, gen mgr.

Glendive

KDZN(FM)— Dec 21, 1969: 96.5 mhz; 100 kw. 400 ft TL: N47 05 15 W104 48 04. Stereo. Hrs opn: 24 210 S. Douglas St., 59330. Phone: (406) 377-3377. Fax: (406) 365-2181. E-mail: kxgnkdzn@midrivers.com Web Site: www.kxgn.com. Licensee: Magic Air Communications Co. Natl. Network: CBS Radio, Westwood One. Law Firm: Davis Wright Tremaine, LLP. Wire Svc: AP Format: Country. News: 4 hrs wkly. Target aud: 25-54. ♦Steven Marks, pres; Paul Sturlaugson, gen mgr & sports cmtr; Marcy Copp, progmg dir; Ed Agre, news dir.

KGLE(AM)— Aug 22, 1962: 590 khz; 1 kw-D. TL: N47 05 50 W104 47 09. Hrs open: 24 Box 931, 86 Seven Mile Dr., 59330. Phone: (406) 377-3331. Fax: (406) 377-3332. E-mail: kgle@midrivers.com Licensee: Friends of Christian Radio Inc. (acq 1-12-93; $90,000; FTR: 2-1-93). Population served: 8,800 Natl. Network: Moody. Format: Relg, farm. Target aud: 35-64; general. ♦Tom Fatzinger, pres; Jim McBride, gen mgr.

KXGN(AM)— Sept 23, 1948: 1400 khz; 1 kw-U. TL: N47 05 40 W104 42 50. Hrs open: 24 210 S. Douglas, 59330. Phone: (406) 377-3377. Fax: (406) 365-2181. E-mail: kxgnkdzn@midrivers.com Web Site: www.glendivebroadcasting.com. Licensee: Glendive Broadcasting Corp. Population served: 6,305 Natl. Network: ABC. Law Firm: Davis Wright Tremaine. Wire Svc: NWS (National Weather Service) Format: Adult contemp, oldies. News staff: one; News: 6 hrs wkly. Spec prog: Derry Brownfield 5 hrs, farm 2 hrs wkly. ♦Stephen Marks, pres; Paul Strulaugson, exec VP; Paul Sturlaugson, gen mgr. Co-owned TV: KXGN-TV affil

Great Falls

KAAK(FM)— June 19, 1972: 98.9 mhz; 100 kw. Ant 500 ft TL: N47 32 08 W111 17 02. Stereo. Hrs opn: 24 Box 3309, 59403. Phone: (406) 761-7600. Fax: (406) 761-5511. Licensee: CCR-Great Falls IV LLC. (acq 5-31-2007; grpsl). Population served: 86,000 Format: Hot adult contemp. Target aud: 25-44. ♦Ron Korb, gen mgr.

***KAFH(FM)**— 2006: 91.5 mhz; 1 kw. Ant 297 ft TL: N47 31 57 W111 16 38. Hrs open:
Rebroadcasts WAFR(FM) Tupelo, MS 100%.
Drawer 2440, Tupelo, MS, 38803. Phone: (662) 844-8888. Fax: (662) 842-6791. Web Site: www.afr.net. Licensee: American Family Association. Format: Christian. ♦Marvin Sanders, gen mgr.

KEIN(AM)— July 1922: 1310 khz; 5 kw-D, 1 kw-N. TL: N47 31 20 W111 23 18. Hrs open: 24 Box 1239, 59403. Secondary address: 3313 15th St. N.E., Black Eagle 59414. Phone: (406) 761-1310. Fax: (406) 454-3775. Licensee: Munson Radio Inc. (acq 7-1-97). Population served: 135,000 Format: Adult standards. News staff: one; News: 5 hrs wkly. Target aud: 35 plus. ♦Steven Dow, pres.

***KFRW(FM)**— 2007: 91.9 mhz; 50 kw. Ant 466 ft TL: N47 49 13 W111 47 56. Hrs open:
Rebroadcasts KUFR(FM) Salt Lake City, UT 100%.
136 E.S. Temple, Suite 1630, Salt Lake City, UT, 84111. Phone: (801) 359-3147. Fax: (801) 359-8112. Web Site: www.familyradio.com. Licensee: Family Stations Inc. Format: Christian relg. ♦Harold Camping, gen mgr.

***KGFA(FM)**— 2006: 90.7 mhz; 1 kw. Ant 297 ft TL: N47 31 57 W111 16 38. Hrs open: 5700 West Oaks Blvd., Rocklin, CA, 95765. Phone: (916) 251-1600. Fax: (916) 251-1650. Web Site: www.air1.com. Licensee: Educational Media Foundation. (acq 3-23-2007; grpsl). Natl. Network: Air 1. Format: Christian. ♦Richard Jenkins, pres.

***KGFC(FM)**— 1996: 88.9 mhz; 6 kw. Ant 243 ft TL: N47 27 53 W111 21 24. Stereo. Hrs opn: 24
Rebroadcasts KXEI(FM) Havre 100%.
Box 2426, Havre, 59501. Phone: (406) 265-5845. Fax: (406) 265-8860. E-mail: ynop@ynopradio.org Web Site: www.ynopradio.org. Licensee: Hi-Line Radio Fellowship Inc. Wire Svc: AP Format: Christian

Inspirational. Target aud: General; those looking for inspirational Christian music & progmg. ♦Roger Lonnquist, gen mgr; Brenda Boyum, stn mgr; Brian Jackson, progmg dir.

***KGPR(FM)**— April 1984: 89.9 mhz; 9.5 kw. 295 ft TL: N47 32 23 W111 17 06. Stereo. Hrs opn: 24
Rebroadcasts KUFM(FM) Missoula 100%.
Box 3343, 59403. Secondary address: Box 6010, 2100 16th Ave. S. 59406-6010. Phone: (406) 268-3739. Fax: (406) 268-3736. E-mail: kgpr@msugf.edu Web Site: www.kgpr.msugf.edu. Licensee: Great Falls Public Radio Association. Population served: 100,000 Natl. Network: PRI, NPR. Format: Class, educ, news, world mus. News: 44 hrs wkly. Target aud: General. ♦Joseph Duffy, pres; Bill Tacke, VP; Tom Halverson, stn mgr; Carol Spahr, dev dir.

KINX(FM)— Feb 4, 2002: 107.3 mhz; 94 kw. Ant 2,037 ft TL: N47 09 34 W111 00 39. Hrs open: Box 3129, 59403. Phone: (406) 761-2800. Fax: (406) 727-7218. E-mail: mail@1073.com Web Site: www.sam1073.com. Licensee: Fisher Radio Regional Group Inc. Group owner: Fisher Broadcasting Company. Natl. Rep: McGavren Guild. Format: Var. Target aud: 25-54. ♦Terry Strickland, gen mgr; Gary Goodan, progmg dir.

KLFM(FM)— Feb 14, 1982: 92.9 mhz; 98 kw. 410 ft TL: N47 32 19 W111 15 41. Hrs open: 24 Box 3309, 59403. Secondary address: 20 3rd St. N. 59403. Phone: (406) 761-7600. Fax: (406) 761-5511. Licensee: CCR-Great Falls IV LLC. Group owner: Cherry Creek Radio LLC (acq 12-19-2003; grpsl). Population served: 100,000 Format: Good time oldies. Target aud: 25-54. ♦Ron Korb, gen mgr & gen sls mgr.

KLSK(FM)— 2003: 100.3 mhz; 100 kw. Ant 495 ft TL: N47 15 57 W111 08 39. Hrs open: 6080 Mt. Moriah, Memphis, TN, 38115. Phone: (901) 375-9324. Fax: (901) 375-0041. Licensee: Flinn Broadcasting Corp. Format: Hip-hop-rap. ♦Karen Wheatley, gen mgr.

KMON(AM)— May 30, 1947: 560 khz; 5 kw-U, DA-N. TL: N47 25 29 W111 17 20. Stereo. Hrs opn: 24 Box 3309, 20 3rd St. N., Suite 231, 59401. Phone: (406) 761-7600. Fax: (406) 761-5511. E-mail: 560@kmon.com Web Site: www.kmon.com. Licensee: CCR-Great Falls IV LLC. Group owner: Cherry Creek Radio LLC (acq 12-19-2003; grpsl). Population served: 210,000 Law Firm: Pepper & Corazzini. Format: Country, farm. News staff: one; News: 20 hrs wkly. Target aud: 35-64. Spec prog: Sports 5 hrs wkly. ♦Ron Korb, gen mgr; Melissa Horton, gen sls mgr; Skip Walters, progmg dir & news dir; Ken Eklund, chief of engrg; Angie Depping, rsch dir.

KMON-FM— Oct 1, 1972: 94.5 mhz; 98 kw. 495 ft TL: N47 32 19 W111 15 41. Stereo. 24 Web Site: www.kmonfm.com. Format: Hot country. News staff: one; News: 5 hrs wkly. Target aud: 25-54. ♦Ron Korb, sls dir, gen sls mgr & spec ev coord; Scott Hershey, progmg dir.

KQDI(AM)— 1955: 1450 khz; 1 kw-U. TL: N47 31 26 W111 18 04. Hrs open: 1300 Central Ave. W., 59404. Phone: (406) 761-2800. Fax: (406) 727-7218. Licensee: Fisher Radio Regional Group Inc. Group owner: Fisher Broadcasting Company (acq 9-27-95; with co-located FM). Natl. Rep: Christal. Format: News/talk. Target aud: 25-54. ♦Terry Strickland, gen mgr; Dave France, opns mgr & progmg dir; Anna Palagi, gen sls mgr; Pam Bennett, mus dir & traf mgr; Joe Bower, chief of engrg.

KQDI-FM— Dec 31, 1963: 106.1 mhz; 100 kw. 276 ft TL: N47 31 57 W111 16 41. Stereo. Box 312965,000 Format: Classic rock, AOR. Target aud: 18-49.

KXGF(AM)— 1987: 1400 khz; 1 kw-U. TL: N47 27 56 W111 20 22. Hrs open: Box 3129, 59403. Secondary address: 1300 Central Ave. W. 59403. Phone: (406) 761-2800. Fax: (406) 727-7218. Licensee: Fisher Radio Regional Group Inc. Group owner: Fisher Broadcasting Company (acq 12-28-94; grpsl, including co-located FM; FTR: 2-20-95). Population served: 86,000 Natl. Rep: McGavren Guild. Format: MOR. Target aud: 35-64. Spec prog: Farm 2 hrs wkly. ♦Larry Roberts, pres; Terry Strickland, gen mgr; Dave France, opns mgr & sls dir; Kim Landers, gen sls mgr; Tammie Toren, progmg dir; Joe Bower, chief of engrg.

Hamilton

KBAZ(FM)—Listing follows KLYQ(AM).

KLYQ(AM)— Feb 3, 1961: 1240 khz; 1 kw-U. TL: N46 15 22 W114 09 45. Hrs open: 5:30 AM-midnight Box 660, 217 N. 3rd St., Suite L, 59840. Phone: (406) 363-3010. Fax: (406) 363-6436. E-mail: contact@klyq.com Web Site: www.klyq.com. Licensee: Capstar TX L.P. Group owner: Clear Channel Communications Inc. (acq 2-21-01; grpsl). Population served: 35,250 Format: News/talk. News staff: one; News: 25 hrs wkly. Target aud: 25-54; adults. ♦Gene Peterson, gen mgr; Jim Coulter, sls dir; Steve Fullerton, opns dir, progmg dir & news dir; Mike Daniels, chief of engrg; Don Davis, sports cmtr.

KBAZ(FM)—Co-owned with KLYQ(AM). Feb 11, 1969: 96.3 mhz; 85 kw. Ant 2,066 ft TL: N46 48 08 W113 58 21. Stereo. 24 400 Ryman, Missoula, 59801. Web Site: www.kluq.com. Format: Alternative rock. ♦Denny Bedard, opns dir; Jim Coulter, gen sls mgr.

***KMZO(FM)**—Not on air, target date: unknown: 90.3 mhz; 725 w. Ant 331 ft TL: N46 13 46 W114 14 01. Stereo. Hrs opn: 24 Faith Communications Corp., 2201 S. 6th St., Las Vegas, NV, 89104. Phone: (702) 731-5452. Fax: (702) 731-1992. E-mail: info@sosradio.net Web Site: www.sosradio.net. Licensee: Faith Communications Corp. Format: Contemp Christian. News: 5 hrs wkly. ♦Jack French, CEO; Brad Staley, gen mgr; Chris Staley, progmg mgr.

***KUFN(FM)**— October 1998: 91.9 mhz; 850 w. 499 ft TL: N46 13 46 W114 14 01. Hrs open: c/o KUFM(FM), Univ. of Montana, Missoula, 59812. Phone: (406) 243-4931. Fax: (406) 243-3299. Web Site: www.kurn.org. Licensee: The University of Montana. Format: Jazz, classical, news, eclectic. News: 2 hrs wkly. ♦William Marcus, gen mgr; Daniel Plante, stn mgr.

KXDR(FM)— July 16, 1999: 98.7 mhz; 100 kw. 417 ft TL: N46 30 36 W113 58 45. Hrs open: 24 1600 North Ave. W., Suite 101, Missoula, 59801. Phone: (406) 728-5000. Fax: (406) 721-3020. Web Site: www.starfm.net. Licensee: CCR-Missoula IV LLC. Group owner: Fisher Broadcasting Company (acq 10-31-2006; grpsl). Format: Hot adult contemp. ♦Chad Parrish, gen mgr.

Hardin

KHDN(AM)— Dec 28, 1962: 1230 khz; 1 kw-U. TL: N45 42 55 W107 35 59. Hrs open: 24 Box 230, 59034. Phone: (406) 665-2828. Fax: (406) 665-2131. Web Site: www.bigskyradio.net. Licensee: Sun Mountain Inc. (acq 11-30-2000). Natl. Network: Jones Radio Networks. Rgnl. Network: Intermountain Farm/Ranch Network. Format: Adult standards/news. News: 2 hrs wkly. Target aud: 25-54. ♦Richard Solberg, pres & gen mgr.

KMHK(FM)— 1975: 95.5 mhz; 100 kw. 984 ft TL: N45 44 29 W108 08 19. Stereo. Hrs opn: Box 1276, Billings, 59103. Phone: (406) 248-7827. Fax: (406) 252-9577. Web Site: www.kmhk.com. Licensee: CC Licenses LLC. Group owner: Clear Channel Communications Inc. (acq 4-13-2001; grpsl). Format: Rock. Target aud: 18-34; general. ♦Dennis Koffman, gen mgr; Jay Branden, mktg dir & progmg dir.

Havre

***KNMC(FM)**— February 1979: 90.1 mhz; 10 w. 56 ft TL: N48 32 30 W109 41 06. (CP: 375 w, ant -112 ft. TL: N48 32 31 W109 41 17). Stereo. Hrs opn: c/o KEMC, 1500 University Dr., Billings, 59101-0298. Phone: (406) 657-2941. Fax: (406) 657-2977. Licensee: Montana State University-Northern. Format: Class, jazz. Spec prog: Class 10 hrs wkly. ♦Marvin Granger, gen mgr.

KOJM(AM)— Oct 31, 1947: 610 khz; 1 kw-U, DA-2. TL: N48 34 48 W109 38 54. Stereo. Hrs opn: 19 2210 31st St. N., 59501. Phone: (406) 265-7841. Fax: (406) 265-8855. E-mail: nmb@nmbi.com Web Site: www.kojm.com. Licensee: New Media Broadcasters Inc. (group owner; (acq 12-30-2002; grpsl). Population served: 80,000 Natl. Network: ABC. Law Firm: Cohn & Marks. Wire Svc: AP Format: Adult hits. News staff: 2; News: 20 wkly. Target aud: 30-64; boomer generation. Spec prog: Agriculture 4 hrs wkly. ♦C. David Leeds, pres & natl sls mgr; Kyle Leeds, prom dir; Geoff Cole, progmg dir; Justin Krezelak, news dir; Bruce Faulkner, chief of engrg.

KPQX(FM)— Mar 8, 1975: 92.5 mhz; 100 kw. Ant 1,788 ft TL: N48 10 55 W109 41 01. Stereo. Hrs opn: 19 2210 31st St. N., 59501. Phone: (406) 265-7841. Fax: (406) 265-8855. E-mail: nmb@nmbi.com Web Site: www.kpqx.com. Licensee: New Media Broadcasters Inc. (acq 12-30-2002; grpsl). Population served: 80,000 Natl. Network: ABC. Law Firm: Cohn & Marks. Wire Svc: AP Format: Country. News staff: 2; News: 20 hrs wkly. Target aud: 25-54. Spec prog: Farm 10 hrs wkly. ♦C. David Leeds, pres & natl sls mgr; Kyle Leeds, prom dir; Geoff Cole, progmg dir; Justin Krezelak, news dir; Bruce Faulkner, chief of engrg.

KRYK(FM)—Chinook, Nov 19, 1983: 101.3 mhz; 100 kw. Ant 688 ft TL: N48 23 29 W109 17 50. Stereo. Hrs opn: 5 AM-midnight 2210 31st St. N., 59501. Phone: (406) 265-7841. Fax: (406) 265-8855. E-mail: nmb@nmbi.com Web Site: www.kryk.com. Licensee: New Media Broadcasters Inc. (group owner; (acq 12-30-2002). Population served: 80,000 Natl. Network: ABC. Law Firm: Cohn & Marks. Format: Hot adult contemp. News staff: 2; News: 5 hrs wkly. Target aud: 18-49. ♦C. David Leeds, pres & natl sls mgr; Kyke Leeds, prom dir; Geoff Cole, progmg dir; Justin Krezelak, news dir; Bruce Faulkner, engrg dir.

***KXEI(FM)**— July 28, 1983: 95.1 mhz; 98 kw. Ant 1,699 ft TL: N48 10 42 W109 41 21. Stereo. Hrs opn: 24 Box 2426, 59501. Secondary address: 317 First St. 59501. Phone: (406) 265-5845. Fax: (406) 265-8860. E-mail: ynop@ynopradio.org Web Site: www.ynopradio.org. Licensee: Hi-Line Radio Fellowship Inc. Natl. Network: Moody, Salem Radio Network. Law Firm: Cohn & Marks. Wire Svc: AP Format: Christian Inspirational. Target aud: General; those looking for Christian inspirational music & progmg. Spec prog: C&W one hr, farm one hr wkly. ♦Roger Lonnquist, gen mgr; Brenda Boyum, stn mgr; Brian Jackson, progmg dir.

Helena

KBLL(AM)— September 1937: 1240 khz; 1 kw-U. TL: N46 35 24 W112 00 59. Hrs open: 110 Broadway St., 59601. Secondary address: Box 4111 59604. Phone: (406) 442-4490. Fax: (406) 442-6161. Licensee: CCR-Helena IV LLC. Group owner: Cherry Creek Radio LLC (acq 6-30-2004; $2.8 million with co-located FM). Population served: 40,000 Law Firm: Dow, Lohnes & Albertson. Format: News/talk. Target aud: 29-54; high buying power. ♦Dewey Bruce, gen mgr; Chris McCarthy, gen sls mgr; Stan Evans, progmg dir; Cato Butler, news dir; Ken Eklund, chief of engrg; Michele McAlister, traf mgr.

KBLL-FM— August 1979: 99.5 mhz; 30 kw. 790 ft TL: N46 46 12 W112 01 22. Stereo. 40,000 Format: Country. ♦Kurt Kittelson, progmg dir.

KCAP(AM)— October 1949: 1340 khz; 1 kw-U. TL: N46 36 43 W112 03 13. Hrs open: 110 Broadway St., 59601. Secondary address: Box 4111 59604. Phone: (406) 442-4490. Fax: (406) 442-7356. Web Site: www.kcap.com. Licensee: CCR-Helena IV LLC. Group owner: Cherry Creek Radio LLC (acq 2-3-2004; grpsl). Population served: 30,000 Natl. Network: CBS, Moody. Format: News/talk. Target aud: 25-54. ♦Dewey Bruce, gen mgr; Chris McCarthy, gen sls mgr; Stan Evans, progmg dir; Cato Butler, news dir; Ken Eklund, chief of engrg; Michele McAlister, traf mgr.

KZMT(FM)—Co-owned with KCAP(AM). 1975: 101.1 mhz; 95 kw. 1,899 ft TL: N46 44 52 W112 19 47. Stereo. Web Site: www.kzmt.com.130,000 Format: Classic rock. Target aud: 18-54; upscale, entrepreneurial, adults. ♦Michele McAlister, traf mgr.

***KHLV(FM)**— 2005: 90.1 mhz; 1 w horiz, 3.5 kw vert. Ant 662 ft vert TL: N46 46 07 W112 01 21. Stereo. Hrs opn: 24
Rebroadcasts KLVR(FM) Santa Rosa, CA 100%.
2351 Sunset Blvd., Suite 170-218, Rocklin, CA, 95765. Phone: (916) 251-1600. Fax: (916) 251-1650. E-mail: klove@klove.com Web Site: www.klove.com. Licensee: Educational Media Foundation. Group owner: EMF Broadcasting. Natl. Network: K-Love. Law Firm: Shaw Pittman. Format: Contemp Christian. News staff: 3. Target aud: 25-44; Judeo Christian, female. ♦Richard Jenkins, pres; Mike Novak, VP; Keith Whipple, dev dir; David Pierce, progmg mgr; Ed Lenane, news dir; Sam Walington, engrg dir; Karen Johnson, news rptr; Marya Morgan, news rptr; Richard Hunt, news rptr.

KMTX(AM)— Nov 1, 1976: 950 khz; 5 kw-U, DA-N. TL: N46 40 28 W112 01 05. Stereo. Hrs opn: Box 1183, 59624. Secondary address: 516 Fuller 59601. Phone: (406) 442-0400. Fax: (406) 442-0491. Licensee: KMTX LLC. Population served: 40,000 Natl. Network: AP Radio. Format: Oldies. ♦James O'Connell, pres; Kevin Skaalure, gen mgr.

KMTX-FM— Jan 19, 1985: 105.3 mhz; 86.9 kw. 1,878 ft TL: N46 44 52 W112 19 47. (CP: 100 kw, ant 1,954 ft.). Stereo. Phone: (406) 443-1053. Natl. Network: ABC. Format: Adult contemp.

KUFM(FM)—See Missoula

***KUHM(FM)**— 2000: 91.7 mhz; 910 w. Ant 761 ft TL: N46 46 11 W112 01 22. Hrs open: PARTV Bldg., Univ. of Montana, Missoula, 59812. Phone: (406) 243-4931. Phone: (800) 325-1565. Fax: (406) 243-3299. Licensee: The University of Montana. Format: Jazz, classical, news. News staff: 2. ♦William Marcus, gen mgr.

***KVCM(FM)**— Aug 2, 1993: 103.1 mhz; 30 kw. Ant 679 ft TL: N46 46 11 W112 01 25. Stereo. Hrs opn: 24
Rebroadcasts KXEI(FM) Havre 100%.
Box 2426, 317 First St., Havre, 59501. Phone: (406) 265-5845. Fax: (406) 265-8860. E-mail: ynop@ynopradio.org Web Site: www.ynopradio.org. Licensee: Hi-Line Radio Fellowship Inc. Natl. Network: Moody, Salem Radio Network. Wire Svc: AP Format: Christian Inspirational. Target aud: General:; those looking for inspirational Christian music & progmg. ♦Brenda Boyum, stn mgr; Roger Lonnquist, gen mgr & stn mgr; Brian Jackson, progmg dir.

Joliet

KPBR(FM)— Mar 15, 2006: 105.9 mhz; 100 kw. Ant 440 ft TL: N45 39 31 W108 34 14. Stereo. Hrs opn: 24 101 Grand Ave., Billings, 59101. Phone: (406) 248-7777. Fax: (406) 248-8577. E-mail: thebar@1059thebar.com Web Site: www.1059thebar.com. Licensee: Connoisseur Media LLC. Format: Country. ♦Michael Schutta, gen mgr.

Kalispell

KALS(FM)— November 1974: 97.1 mhz; 26 kw. 2,488 ft TL: N48 00 48 W114 21 55. Stereo. Hrs opn: Box 9710, 59904-2710. Phone: (406) 752-5257. Fax: (406) 752-3416. Web Site: www.kals.com. Licensee: Kalispell Christian Radio Fellowship Inc. (acq 11-26-01; $700,000). Population served: 14,000 Format: Christian, adult contemp. Spec prog: Class one hr wkly. ♦Brad Rauch, gen mgr.

KBBZ(FM)— Sept 12, 1983: 98.5 mhz; 58 kw. 2,378 ft TL: N48 30 42 W114 22 14. (CP: 60 kw, ant 2,313 ft.). Stereo. Hrs opn: 2432 Hwy. 2 E., 59901. Phone: (406) 755-8700. Fax: (406) 755-8770. E-mail: kbbz@beebroadcasting.com Web Site: www.beebroadcasting.com. Licensee: Bee Broadcasting Inc. (group owner; acq 6-12-83; $315,000; FTR: 9-26-83). Format: Adult classic, contemp rock. ♦Benny Bee, pres; Mark Wagner, gen mgr; Benny Bee Jr., opns mgr.

KDBR(FM)— November 1993: 106.3 mhz; 30 kw. 413 ft TL: N48 10 34 N114 20 53. Hrs opn: Box 5409, 59903. Phone: (406) 257-5327. Phone: (406) 755-8700. Fax: (406) 755-8770. E-mail: kdbr@beebroadcasting.com Web Site: www.beebroadcasting.com. Licensee: Bee Broadcasting Inc. (group owner) Format: Country. ♦Benny Bee, pres; Mark Wagner, gen mgr.

KGEZ(AM)— Mar 24, 1927: 600 khz; 5 kw-D, 1 kw-N, DA-2. TL: N48 09 40 W114 16 51. Hrs opn: Box 923, 59903. Phone: (406) 752-2600. Fax: (406) 257-0459. Web Site: www.z600.com. Licensee: Skyline Broadcasters Inc. (acq 12-2-99). Population served: 80,000 Natl. Network: CBS. Format: News/talk, sports. Target aud: 25-60. ♦John Stokes, gen mgr.

***KLKM(FM)**— 2006: 88.7 mhz; 8.5 kw vert. Ant 312 ft TL: N48 04 07 W114 02 20. Hrs open:
Rebroadcasts KLVR(FM) Santa Rosa, CA 100%.
2351 Sunset Blvd., Suite 170-218, Rocklin, CA, 95765. Phone: (916) 251-1600. Fax: (916) 251-1650. Web Site: www.klove.com. Licensee: Educational Media Foundation. (acq 1-11-2005; $95,000 for CP). Natl. Network: K-Love. Format: Contemp Christian. ♦Richard Jenkins, pres; Mike Novak, VP; Keith Whipple, dev dir; Eric Allen, natl sls mgr; David Pierce, progmg mgr; Ed Lenane, news dir; Sam Wallington, engrg dir; Karen Johnson, news rptr; Marya Morgan, news rptr; Richard Hunt, news rptr.

KOFI(AM)— Nov 11, 1955: 1180 khz; 50 kw-D, 10 kw-N, DA-N. TL: N48 11 52 W114 15 03. Stereo. Hrs opn: 24 Box 608, 59903. Secondary address: 317 First Ave. E. 59901. Phone: (406) 755-6690. Fax: (406) 752-5078. E-mail: kofi@kofi.radio.com Web Site: www.kofi.com. Licensee: KOFI Inc. (acq 9-11-90; $750,000 with co-located FM; FTR: 10-1-90). Population served: 105,000 Natl. Network: ABC, CNN Radio. Law Firm: Reddy, Begley & McCormick. Format: CHR, oldies, news/talk. News staff: 2; News: 35 hrs wkly. Target aud: 25-54. ♦Dave Rae, gen mgr.

KZMN(FM)—Co-owned with KOFI(AM). June 10, 1988: 103.9 mhz; 100 kw horiz, 55 kw vert. 571 ft TL: N48 05 39 W114 16 11. Stereo. 24 Web Site: www.kzmn.com. Natl. Network: CNN Radio. Format: Classic rock. News: 3 hrs wkly. Target aud: 18-49. ♦Dave Rae, pres; Mike Jorgensen, VP.

KQJZ(AM)—Not on air, target date: unknown: 1340 khz; 1 kw-D, 670 w-N. TL: N48 11 59 W114 19 03. Hrs open: Box 13396, Green Bay, WI, 54307-3396. Phone: (888) 774-7968. Fax: (920) 469-3023. Web Site: www.relevantradio.com. Licensee: Advance Acquisition Inc. ♦Mark C. Follett, pres.

KQRK(FM)—See Ronan

***KSPL(FM)**— February 1997: 90.9 mhz; 250 w. 2,529 ft TL: N48 30 22 W114 20 49. Hrs open:
Rebroadcasts KMBI-FM Spokane, WA 100%.
c/o KMBI-FM, 5408 S. Freya, Spokane, WA, 99223. Phone: (509) 448-2555. Fax: (509) 448-6855. E-mail: kmbi@moody.edu Web Site: www.moody.edu. Licensee: Moody Bible Institute of Chicago. Group owner: The Moody Bible Institute of Chicago Format: Relg. Target aud: 35-54; Christian men & women. ♦Richard Monteith, gen mgr & progmg mgr; Scott Richardson, chief of engrg; Bret Bremberg, disc jockey; Derek Cutlip, disc jockey; Shelly Hogeweide, disc jockey; Steve Stewart, disc jockey.

***KUKL(FM)**— October 1998: 89.9 mhz; 850 w. 443 ft TL: N48 10 34 W114 20 53. Hrs open: c/o KUFM(FM), Univ. of Montana, Missoula, 59812. Phone: (406) 243-4931. Fax: (406) 243-3299. Fax: (800) 325-1565. Web Site: mtpr.org. Licensee: University of Montana. Format: Jazz, classical, news, eclectic. News: 2 hrs wkly. ♦William Marcus, gen mgr.

Laurel

KBSR(AM)— September 1979: 1490 khz; 1 kw-U. TL: N45 39 11 W108 45 09. Hrs open: 24 Box 248, 59044. Phone: (406) 665-2828. Fax: (406) 665-2131. Web Site: www.bigskyradio.net. Licensee: Sun Mountain Inc. (acq 11-30-2000). Natl. Network: Jones Radio Networks. Format: Adult contemp, radio theatre. Target aud: 35 plus; professional, business people. ♦Richard Solberg, pres.

KRSQ(FM)— June 9, 1994: 101.9 mhz; 100 kw. Ant 367 ft TL: N45 45 48 W108 27 20. Hrs open: 222 North 32nd St., 10th Floor, Billings, 59101. Phone: (406) 238-1000. Fax: (406) 238-1038. Web Site: www.hot1019.com. Licensee: New Northwest Broadcasters LLC (group owner; acq 8-10-99; grpsl). Natl. Network: ABC. Law Firm: Dow, Lohnes & Albertson. Format: CHR. Target aud: 18-49. ♦Pete Benedetti, CEO; Tommy Ehrman, gen mgr; Tom Oakes, opns mgr.

Lewistown

KLCM(FM)—Listing follows KXLO(AM).

***KLEU(FM)**— Oct 21, 2003: 91.1 mhz; 4 kw. Ant 1,879 ft TL: N47 10 46 W109 32 05. Hrs open: 24
Rebroadcasts KXEI(FM) Havre 100%.
Box 2426, Havre, 59501-2426. Phone: (406) 265-5845. Fax: (406) 265-8860. Web Site: www.ynopradio.org. Licensee: Hi-Line Radio Fellowship Inc. acq 12-10-2003; $20,000 for CP). Natl. Network: Moody, Salem Radio Network. Wire Svc: AP Format: Christian Inspirational. ♦Roger Lonnquist, gen mgr; Brenda Boyum, stn mgr; Brian Jackson, progmg dir.

KXLO(AM)— 1947: 1230 khz; 1 kw-U. TL: N47 04 13 W109 24 26. Hrs open: 620 N.E. Main St., 95457. Phone: (406) 707-5275. Fax: (406) 538-3495. E-mail: kxlo@lewistown.net Web Site: www.kxlo-klcm.com. Licensee: KXLO Broadcast Inc. (acq 4-16-73). Population served: 7,500 Natl. Network: CBS. Rgnl. Network: Intermountain Farm/Ranch Network. Format: Country. Target aud: General. Spec prog: Farm. ♦Fred Lark, pres & gen mgr; Bethany Lark, progmg dir.

KLCM(FM)—Co-owned with KXLO(AM). April 1975: 95.9 mhz; 3 kw. 205 ft TL: N47 04 13 W109 24 26. Stereo. Web Site: www.kxlo-klcm.com. Licensee: Montana Broadcast Communications Inc. Format: Classic hits. Target aud: 18-54.

Libby

KLCB(AM)— Dec 23, 1950: 1230 khz; 1 kw-U. TL: N48 22 14 W115 32 19. Hrs open: 16 Box 730, 59923. Secondary address: 251 W. Cedar St. 59923. Phone: (406) 293-6234. Fax: (406) 293-6235. Licensee: Lincoln County Broadcasters Inc. (acq 12-66). Population served: 14,000 Natl. Network: ABC. Format: Country. News staff: one; News: 13 hrs wkly. Target aud: 25-54. ♦Duane J. Williams, VP & gen mgr.

KTNY(FM)—Co-owned with KLCB(AM). Apr 5, 1986: 101.7 mhz; 3 kw. -1,029 ft TL: N48 22 14 W115 32 19. Stereo. 16 14,000 Natl. Network: ABC. Format: MOR, adult contemp, oldies. News staff: one; News: 16 hrs wkly. Target aud: 35-54. ♦Duane J. Williams, CEO.

Livingston

KOZB(FM)— December 1977: 97.5 mhz; 100 kw. 265 ft TL: N45 39 26 W110 48 22. (CP: Ant 790 ft.). Stereo. Hrs opn: 24 Box 20, 5445 Johnson Rd., Bozeman, 59718. Phone: (406) 587-9999. Phone: (406) 586-5858. Fax: (406) 587-5855. E-mail: reier@bigsky.net Licensee: Reier Broadcasting Co. Inc. (group owner; acq 10-18-96; grpsl). Population served: 60,000 Format: Rock alternative. News staff: 2. Target aud: 18-44. ♦Bill Reier, gen mgr.

KPRK(AM)— Jan 10, 1947: 1340 khz; 1 kw-U. TL: N45 40 21 W110 32 21. Hrs open: 5:30 AM-midnight Box1340, Hwy. 10 E., 59047. Phone: (406) 222-2841. Phone: (406) 222-1340. Fax: (406) 222-1341. E-mail: kprkam@mooseradio.com Licensee: Capstar TX L.P. Group owner: Clear Channel Communications Inc. (acq 2-21-01; grpsl). Population served: 18,000 Natl. Network: AP Radio. Rgnl. Network: AP. Format: Classic hits. News: 15 hrs wkly. Target aud: 25-64; general. Spec prog: Oldies 5 hrs, big band 4 hrs wkly. ♦Dave Cowan, gen mgr; Courtney Lehman, stn mgr; Kaye Rugh, gen sls mgr; Gary Weiss, news dir; Ron Huckeby, chief of engrg.

KXLB(FM)—Co-owned with KPRK(AM).Not on air, target date: unknown: 100.7 mhz; 94 kw. 813 ft TL: N45 40 24 W110 52 02. Format: Country.

Lockwood

KPLN(FM)— Mar 1, 2006: 106.7 mhz; 100 kw. Ant 512 ft TL: N45 45 54 W108 27 19. Hrs open: 101 Grand Ave., Billings, 59101. Phone: (406) 248-7777. Licensee: Connoisseur Media LLC. Format: CHR. ♦Michael Schutta, gen mgr.

KYLW(AM)— 2005: 1450 khz; 1 kw-U. TL: N45 48 37 W108 25 38. Hrs open: 9045 Hobble Creek, Billings, 59101. Phone: (406) 665-2828. Fax: (406) 665-2131. Web Site: www.bigskyradio.net. Licensee: Sun Mountain Inc. (acq 7-7-2005; $26,000 for CP). ♦Richard Solberg, pres & gen mgr.

***KYWH(FM)**— 2006: 88.9 mhz; 1.9 kw vert. Ant 452 ft TL: N45 51 12 W108 45 50. Hrs open: 2121 South 48th St. West, Billings, 59106. Phone: (406) 254-1944. Fax: (406) 294-1946. Web Site: www.calvarychapel.com/billings. Licensee: CSN International (group owner). Format: Christian. ♦Wayne Hathaway, gen mgr.

Malta

KLTZ(AM)—See Glasgow

KMMR(FM)— Sept 9, 1980: 100.1 mhz; 2.25 kw. 377 ft TL: N48 15 17 W107 49 18. Stereo. Hrs opn: 6 AM-11 PM Box 1073, 140 S. 2nd Ave. E., 59538. Phone: (406) 654-2472. Fax: (406) 654-2506. Licensee: KMMR Radio Inc. (acq 5-95; $160,000). Population served: 13,000 Natl. Network: ABC. Format: Country, MOR. News staff: one; News: 3 hrs wkly. Target aud: 18-65; general, rural. ♦Gregory A. Kielb, pres, gen mgr & gen sls mgr; Claudette Kielb, opns VP; Joyce Robinson, opns dir; Valene Kielb, progmg dir.

Manhattan

KKQX(FM)— 2006: 105.7 mhz; 12.3 kw. Ant 682 ft TL: N45 38 16 W111 16 05. Hrs open: 102 S. 19th Ave., Suite 5, Bozeman, 59718-6860. Phone: (406) 582-1045. Fax: (406) 582-0388. Licensee: Radick Construction Inc. Format: Classic hits. ♦Jeffrey P. Radick, pres; Jeff Balding, gen mgr.

Miles City

KATL(AM)— Sept 4, 1941: 770 khz; 10 kw-D, 1 kw-N, DA-N. TL: N46 23 46 W105 46 44. Hrs open: 24 Box 700, 59301. Secondary address: 818 Main St. 59301. Phone: (406) 234-7700. Fax: (406) 234-7783. E-mail: katlradio@katlradio.com Web Site: www.katlradio.com. Licensee: Star Printing Co. Population served: 28,700 Natl. Network: Westwood One, ABC. Law Firm: Cohn & Marks. Format: Adult contemp. News staff: one; News: 17 hrs wkly. Target aud: 25-54; Adults. ♦John Sullivan, pres; Donald L. Richard, gen mgr, progmg dir & chief of engrg; Albert Homme, gen sls mgr.

KIKC-FM—See Forsyth

*KYPR(FM)— Nov 17, 1988: 90.7 mhz; 500 w. 502 ft TL: N46 23 22 W105 45 22. Stereo. Hrs opn: 24
Rebroadcasts KEMC(FM) Billings 100%.
M.S.U. Billings, 1500 University Dr., Billings, 59101-0298. Phone: (406) 657-2941. Fax: (406) 657-2977. Web Site: www.yellowstonepublicradio.org. Licensee: Montana State University-Billings. Natl. Network: NPR, PRI. Format: Div. News: 39 hrs wkly. Target aud: General. ♦Lois Bent, gen mgr.

KYUS-FM— Nov 8, 1984: 92.3 mhz; 100 kw. Ant 984 ft TL: N46 24 04 W105 39 06. Stereo. Hrs opn: 24 Box 1426, 59301. Secondary address: 508 Main St., Rm. 200 53901. Phone: (406) 234-5626. Fax: (406) 232-7000. E-mail: studio@hotcountry925.com Licensee: Custer County Community Broadcasting Corp. (acq 1-26-2007; $540,000 with co-located AM). Population served: 44,000 Format: Country. News staff: one; News: 3 hrs wkly. Target aud: 18-54; programmed for general audience appeal. Spec prog: Farm one hr wkly. ♦Kevin J. Senger, gen mgr; Kevin Senger, opns mgr & gen sls mgr; Karla Ellison, progmg dir & traf mgr; C.W. Wilcox, news dir & sports cmtr; Tony Questa, chief of engrg.

KMTA(AM)—Co-owned with KKRY(FM). October 1986: 1050 khz; 10 kw-D, 136 w-N. TL: N46 24 04 W105 39 06.24 45,000 Format: Classic rock. News staff: one; News: 6 hrs wkly. Target aud: 25-54. ♦Kevin J. Senger, chmn.

Missoula

*KBGA(FM)— Aug 24, 1996: 89.9 mhz; 1 kw. -262 ft TL: N46 52 56 W113 59 08. Hrs opn: Univ. Center, Univ. of Montana, 59812. Phone: (406) 243-6758. Fax: (406) 243-6428. E-mail: kbga@selway.umt.edu Web Site: www.kbga.org. Licensee: The University of Montana. Format: Alternative, rock and roll, educ. ♦Carly Dandrea, gen mgr.

KGGL(FM)—Listing follows KGRZ(AM).

KGRZ(AM)— 1947: 1450 khz; 1 kw-U. TL: N46 52 36 W114 00 47. (CP: TL: N46 52 39 W114 02 36). Hrs opn: 24 Box 4106, 59806. Secondary address: 1600 N. Ave. W. 59801. Phone: (406) 728-1450. Fax: (406) 721-3020. Licensee: CCR-Missoula IV LLC. Group owner: Fisher Broadcasting Company (acq 10-31-2006; grpsl). Population served: 70,000 Natl. Rep: McGavren Guild. Law Firm: Fisher, Wayland, Cooper, Leader & Zaragoza L.L.P. Format: Sports, talk. Target aud: 25-54; male, sports orientated. ♦Chad Parrish, gen mgr; Bill McPherson, rgnl sls mgr; Scott Richards, progmg dir & news dir; Vern Argo, chief of engrg; Lily Konda, traf mgr.

KGGL(FM)—Co-owned with KGRZ(AM). Apr 29, 1977: 93.3 mhz; 43 kw. 2,440 ft TL: N47 02 24 W113 59 00. Stereo. Phone: (406) 728-9399. Format: Country. Target aud: 25-54. ♦Lily Konda, traf mgr & disc jockey.

KGVO(AM)— Jan 18, 1931: 1290 khz; 5 kw-U, DA-N. TL: N46 49 47 W114 04 45. Hrs open: Box 5417, 59806. Secondary address: 3250 S. Reserve, Suite 200 59801. Phone: (406) 728-9300. Fax: (406) 542-2329. Web Site: www.kgvo1290.com. Licensee: Capstar TX L.P. Group owner: Clear Channel Communications Inc. (acq 4-12-01). Population served: 90,000 Natl. Network: CBS. Format: News/talk. Target aud: General. ♦Gene Peterson, CEO & gen mgr; Jim Coulter, gen sls mgr.

*KJCG(FM)—Not on air, target date: unknown: 88.3 mhz; 1 kw vert. Ant 2,086 ft TL: N46 48 09 W113 58 21. Hrs open: 820 N. LaSalle St., Chicago, IL, 60610-3214. Phone: (312) 329-4438. Fax: (312) 329-8980. Web Site: www.mbn.org. Licensee: The Moody Bible Institute of Chicago. ♦Robert C. Neff, VP.

KLCY(AM)—East Missoula, June 27, 1959: 930 khz; 5 kw-D, 1 kw-N, DA-N. TL: N46 51 57 W114 04 57. Hrs open: 24 3250 S00. Reserve, Suite 200, 59801. Phone: (406) 728-9300. Fax: (406) 542-2329. Web Site: www.klcy930.com. Licensee: Capstar TX L.P. Group owner: Clear Channel Communications Inc. (acq 2-21-01; grpsl). Population served: 65,000 Law Firm: Cohn & Marks. Format: Adult standards. News: 14 hrs wkly. Target aud: 35-54; adult spenders. ♦Gene Peterson, gen mgr; Jim Colter, gen sls mgr; Kirk Patrick, progmg dir; Pete Denault, news dir; Todd Clark, chief of engrg.

KYSS-FM—Co-owned with KLCY(AM). May 11, 1969: 94.9 mhz; 62 kw horiz, 12.5 kw vert. 2,381 ft TL: N47 01 57 W113 59 30. Stereo. 24 Web Site: www.kyssfm.com.120,000 Format: Hot country. News staff: News progmg 12 hrs wkly Target aud: 25-54; adults. ♦Chris Morgan, progmg dir.

KMSO(FM)— Feb 9, 1985: 102.5 mhz; 21 kw. Ant 1,748 ft TL: N46 48 30 W113 58 38. Stereo. Hrs opn: 24 Box 309, 59806-0309. Secondary address: 725 Strand Ave. 59801. Phone: (406) 542-1025. Fax: (406) 721-1036. E-mail: info@kmso.com Web Site: www.moclub.com. Licensee: Sheila Callahan & Friends Inc. Population served: 150,000 Natl. Network: AP Radio. Rgnl rep: Tacher, Portland Law Firm: Keller & Heckman. Wire Svc: AP Format: Hot adult contemp. News: 6 hrs wkly. Target aud: 25-54; upscale professional, well-educated mgmt level. Spec prog: Relg one hr wkly. ♦Sheila Callahan, gen mgr; Diana Helms, gen sls mgr; Dale Desmond, progmg dir; Kris Hardy, traf mgr.

*KMZL(FM)— 1998: 91.1 mhz; 1 kw. 2,040 ft TL: N46 48 09 W113 58 21. Stereo. Hrs opn: 24 2201 S. 6th St., Las Vegas, NV, 89104. Phone: (800) 804-5452. E-mail: info@sosradio.net Web Site: sosradio.net. Licensee: Faith Communications Corp. Law Firm: Cohn & Marks. Format: Adult Contemp Christian. News: 5 hrs wkly. Target aud: 25-44. ♦Jack French, CEO; Brad Staley, gen mgr & opns VP; Chris Staley, progmg mgr.

*KUFM(FM)— Jan 31, 1965: 89.1 mhz; 17 kw. 2,510 ft TL: N47 02 24 W113 59 00. (CP: 32 kw, ant 2,473 ft. TL: N47 01 58 W113 59 29). Hrs opn:
Rebroadcasts KUHM(FM) Helena 100%.
Univ. of Montana, 59812. Phone: (406) 243-4931. Fax: (406) 243-3299. Web Site: www.mtpr.org. Licensee: University of Montana. Population served: 320,000 Natl. Network: NPR. Format: Class, jazz, pub radio. News staff: 2. ♦William Marcus, gen mgr & stn mgr.

KYJK(FM)— July 2005: 105.9 mhz; 1.84 kw. Ant 2,083 ft TL: N46 48 06 W113 58 22. Hrs open: 2425 W. Central Ave., Suite 203, 59801. Phone: (406) 721-6800. Fax: (406) 329-1850. Licensee: Spanish Peaks Broadcasting Inc. Natl. Rep: Tacher. Format: Adult contemp. ♦Rod Harsell, gen mgr.

KYLT(AM)— July 15, 1955: 1340 khz; 1 kw-U. TL: N46 52 56 W113 59 08. Hrs open: 24 Box 4106, 1600 North Ave. W., 59806. Phone: (406) 728-5000. Fax: (406) 721-3020. Licensee: CCR-Missoula IV LLC. Group owner: Fisher Broadcasting Company (acq 10-31-2006; grpsl). Population served: 47,538 Format: Oldies. Target aud: 35-55. ♦Chad Parrish, gen mgr; Bill McPherson, gen sls mgr; Mark Morris, natl sls mgr & progmg dir; Scott Richards, news dir; Vern Argo, chief of engrg; Lily Konda, traf mgr.

KZOQ-FM—Co-owned with KYLT(AM). July 29, 1974: 100.1 mhz; 13.5 kw. Ant 2,102 ft TL: N46 48 09 W113 58 21. Stereo. 24 Format: Classic rock. Target aud: 25-54. ♦Lily Konda, progmg dir & traf mgr.

Pablo

KKMT(FM)— 2006: 99.7 mhz; 1.8 kw. Ant 2,112 ft TL: N47 46 25 W114 16 04. Hrs open: 581 N. Reservoir Rd., Polson, 59860. Phone: (406) 883-5255. Fax: (406) 883-4411. Licensee: Anderson Radio Broadcasting Inc. ♦Dennis L. Anderson, pres.

Park City

*KBIL(FM)— 2006: 89.7 mhz; 2.7 kw vert. Ant 525 ft TL: N45 51 12 W108 45 50. Hrs open: 24
Rebroadcasts KLRD(FM) Yucaipa, CA 100%.
2351 Sunset Blvd., Suite 170-218, Rocklin, CA, 95765. Phone: (916) 251-1600. Fax: (916) 251-1650. Web Site: www.air1.com. Licensee: Educational Media Foundation. Group owner: EMF Broadcasting (acq 10-2-2003; grpsl). Natl. Network: Air 1. Law Firm: Shaw Pittman. Format: Christian. News staff: 3. Target aud: 25-44; Judeo Christian female. ♦Richard Jenkins, pres; Mike Novak, VP; Ed Lenane, opns dir & news dir; Keith Whipple, dev dir; David Pierce, progmg mgr; Sam Wallington, engrg dir; Karen Johnson, news rptr; Marya Morgan, news rptr; Richard Hunt, news rptr.

KWMY(FM)— Feb. 15, 2006: 92.5 mhz; 100 kw. Ant 620 ft TL: N45 45 54 W108 27 19. Stereo. Hrs opn: 24 101 Grand Ave., Billings, 59101. Phone: (406) 248-7777. Fax: (406) 248-8577. Web Site: www.my925fm.com. Licensee: Chaparral Broadcasting Inc. (acq 11-30-92; $215,000 with KPOW(AM) Powell, WY; FTR: 12-21-92) Format: Classic hits. ♦Cam Maxwell, gen mgr.

Pinesdale

KBQQ(FM)— 2003: 106.7 mhz; 13 kw. Ant 2,089 ft TL: N46 48 09 W113 58 19. Hrs open: Fisher Radio Regional Group Inc., 1600 North Ave., Missoula, 59801. Phone: (406) 728-5000. Fax: (406) 721-3020. Licensee: CCR-Missoula IV LLC. Group owner: Fisher Broadcasting Company (acq 10-31-2006; grpsl). Format: Oldies. ♦Chad Parrish, gen mgr; Bill McPherson, gen sls mgr.

Plains

*KPLG(FM)— 1998: 91.5 mhz; 470 w. 4,041 ft TL: N47 22 21 W114 51 31. Stereo. Hrs opn: 24
Rebroadcasts KXEI(FM) Havre 100%.
Box 2426, Havre, 59501. Phone: (406) 265-5845. Fax: (406) 265-8860. E-mail: ynop@ynopradio.org Web Site: www.ynopradio.org. Licensee: Hi Line Radio Fellowship Inc. Natl. Network: Moody, Salem Radio Network. Wire Svc: AP Format: Religious. Target aud: General; those who are looking for Christian progmg. ♦Roger Lonnquist, gen mgr; Brenda Boyum, stn mgr; Brian Jackson, progmg dir.

Plentywood

KATQ(AM)— Sept 14, 1979: 1070 khz; 5 kw-D. TL: N48 46 03 W104 32 45. Hrs open: 6 AM-6 PM 112 E. 3rd Ave., 59254. Phone: (406) 765-1480. Fax: (406) 765-2357. E-mail: katq@nemont.net Licensee: Radio International-KATQ Broadcast Association Inc. (acq 1-13-92; $5,000 with co-located FM; FTR: 2-10-92) Population served: 400,000 Format: Country. Target aud: 18-54. Spec prog: Top-40, farm 5 hrs, relg 6 hrs wkly. ♦Myrna Kampen, pres; Casandra Syme, gen mgr, sls dir, gen sls mgr & traf mgr; Bruce Lapke, opns dir, mus dir, news dir & news rptr; Art Gehnert, chief of engrg.

KATQ-FM— June 1, 1962: 100.1 mhz; 3 kw. 34 ft TL: N48 47 06 W104 32 00. Stereo. 24 6,000 Natl. Network: ABC, AP Radio. ♦Grant Lindsey, sports cmtr.

Polson

KERR(AM)— Mar 22, 1976: 750 khz; 50 kw-D, 1 kw-N, DA-N. TL: N47 38 34 W114 07 25. Hrs open: 581 N. Reservoir Rd., 59860. Phone: (406) 883-5255. Fax: (406) 883-4441. Web Site: www.750kerr.com. Licensee: Anderson Radio Broadcasting Inc. (group owner; acq 9-22-2003; grpsl). Population served: 100,000 Format: Country. Target aud: General. ♦Dennis Anderson, pres & gen mgr.

Pryor

*KPGB(FM)—Not on air, target date: unknown: 88.3 mhz; Hrs opn: Box 24, 59066. Phone: (406) 255-0994. Licensee: Faith Baptist Church. Format: Gospel. ♦Ronnie Henderson, gen mgr.

Red Lodge

KMXE-FM— Jan 24, 1994: 99.3 mhz; 30 kw. 1,210 ft TL: N45 11 15 W109 14 46. Stereo. Hrs opn: 24 Box 1678, 59068. Phone: (406) 446-1199. Fax: (406) 446-9178. E-mail: fm99mtn@starband.net Licensee: Silver Rock Communications Inc. (acq 7-19-89; $30,000; FTR: 8-7-89). Population served: 150,000 Format: Class rock. News staff: one; News: one hr wkly. Target aud: 25-49; upwardly mobile. ♦Jeffrey S. Oliphant, exec VP, opns mgr & news dir; Leslie Brent-Oliphant, pres & gen mgr.

Ronan

KQRK(FM)— Oct 4, 1981: 92.3 mhz; 60 kw. 3,500 ft TL: N47 46 25 W114 16 04. Stereo. Hrs opn: 24 581 N. Reservoir Rd., Polson, 59860. Phone: (406) 883-5255. Fax: (406) 883-4441. Web Site: www.750kerr.com. Licensee: Anderson Radio Broadcasting Inc. (group owner; (acq 9-22-2003; grpsl). Population served: 40,000 Format: Adult contemp. Target aud: 25-49. ♦A.L. Anderson, pres; Dennis Anderson, gen mgr.

Saint Regis

KHZS(FM)—Not on air, target date: unknown: 99.1 mhz; 850 w. Ant 2,758 ft TL: N47 22 22 W114 51 34. Hrs open: 5331 Mt. Alifan Dr., San Diego, CA, 92111. Phone: (858) 277-4991. Fax: (858) 277-1365. Licensee: Horizon Christian Fellowship. (acq 2-9-2006; grpsl). ♦Mike MacIntosh, pres.

Scobey

KCGM(FM)— June 21, 1971: 95.7 mhz; 52 kw. Ant 660 ft TL: N48 48 03 W105 21 00. Stereo. Hrs opn: 16 Box 220, 20 Main St., 59263. Phone: (406) 487-2293. Fax: (406) 487-5922. Licensee: Prairie Communications Inc. Population served: 11,570 Natl. Network: USA. Rgnl rep: Taylor Brown Wire Svc: AP Format: Country. News staff: 2; News: 8 hrs wkly. Spec prog: Farm 6 hrs wkly. ♦Clifford Hagfeldt, CEO; Dixie Halverson, gen mgr.

Shelby

KSEN(AM)— Aug 11, 1947: 1150 khz; 5 kw-U, DA-2. TL: N48 28 54 W111 53 03. Hrs open: 19 830 Oilfield Ave., 59474. Phone: (406) 434-5241. Fax: (406) 434-2122. E-mail: ksen@shelby.mt.us Licensee: Capstar TX L.P. Group owner: Clear Channel Communications Inc. (acq 2-21-01; grpsl). Population served: 58,750 Format: Golden oldies. News staff: one; News: 15 hrs wkly. Target aud: 25-59. Spec prog: Farm 8 hrs wkly. ♦Lowrey Maya, CEO & pres; Julie Martin, gen mgr, gen sls mgr & progmg dir; Jim Sargent, opns dir, opns mgr & progmg dir; Mark Daniels, news dir; Anne James, pub affrs dir; Tony Mulligan, chief of engrg; Jim Sargent, traf mgr.

KZIN-FM—Co-owned with KSEN(AM). Dec 9, 1978: 96.7 mhz; 100 kw. Ant 551 ft TL: N48 19 42 W112 02 03. Stereo. 24 35,000 Format: C&W. News staff: one; News: 6 hrs wkly. Target aud: 18-49. ♦Anne Weins, progmg dir & traf mgr.

Sidney

KGCX(FM)— June 1, 2004: 93.1 mhz; 55 kw. Ant 499 ft TL: N47 45 02 W104 18 22. Stereo. Hrs opn: 24 213 2nd Ave. S.W., 59270. Phone: (406) 433-5429. Fax: (406) 433-5430. E-mail: kgcxeagle@midrivers.com Web Site: www.kgcx.net. Licensee: Sidney Community Broadcasting Corp. (acq 7-30-2002; $10,000 for CP). Natl. Network: Fox News Radio. Format: Classic rock. News: 15 hrs wkly. ♦Stephen A. Marks, pres; Mitch Miller, gen mgr; Melissa Quilling, gen sls mgr.

KTHC(FM)— December 1996: 95.1 mhz; 100 kw. 718 ft TL: N48 02 52 W103 59 01. Stereo. Hrs opn: 120 E. Main, 59270. Secondary address: Box 2048, Williston, ND 58802. Phone: (406) 433-5090. Phone: (701) 572-5371. Fax: (406) 433-5095. Fax: (701) 572-7511. E-mail: power95@midrivers.com Licensee: CCR-Williston IV LLC. Group owner: Cherry Creek Radio LLC (acq 12-19-2003; grpsl). Format: Adult contemp. Target aud: General. ♦Larry Timpe, VP & gen mgr.

Stevensville

KKVU(FM)— July 16, 2005: 104.5 mhz; 14.15 kw. Ant 2,083 ft TL: N46 48 06 W113 58 22. Hrs open: 2425 W. Central Ave., Suite 203, Missoula, 59801. Phone: (406) 721-6800. Fax: (406) 329-1850. Licensee: Spanish Peaks Broadcasting Inc. Natl. Rep: Tacher. Format: Adult contemp. ♦Rod Harsell, gen mgr.

Superior

KENR(FM)— October 1999: 107.5 mhz; 100 kw horiz. Ant 945 ft TL: N47 01 45 W114 41 18. Stereo. Hrs opn: 3250 Reserve St., Suite 200, Missoula, 59801. Phone: (406) 822-1075. Fax: (406) 542-2329. Web Site: www.kltcfm.com. Licensee: CC Licenses LLC. Group owner: Clear Channel Communications Inc. (acq 8-30-2002; $900,000). Format: Rhythmic adult contemp. ♦Gene Peterson, gen mgr.

Valier

KWDV(FM)—Not on air, target date: unknown: 105.7 mhz; 100 kw. Ant 640 ft TL: N48 19 44 W112 02 03. Hrs open: 5331 Mt. Alifan Dr., San Diego, CA, 92111. Phone: (858) 277-4991. Fax: (858) 277-1365. Licensee: Horizon Christian Fellowship. (acq 2-9-2006; grpsl). ♦Mike MacIntosh, pres.

Victor

KDXT(FM)—Not on air, target date: unknown: 97.9 mhz; 7.5 kw. Ant 338 ft TL: N46 30 34 W113 58 45. Hrs open: 725 Strand Ave., Missoula, 59801. Phone: (406) 542-1025. Fax: (406) 721-1036. Licensee: Sheila Callahan and Friends Inc. ♦Sheila Callahan, gen mgr.

West Yellowstone

KEZQ(FM)— June 1, 1996: Stn currently dark. 92.9 mhz; 46 kw. Ant 2,732 ft TL: N44 33 41 W111 26 32. Hrs open: 4350 N. Fairfax Dr., Suite 900, Arlington, VA, 22203. Secondary address: 14 Cockenoe Dr., Westport, CT 06880. Phone: (208) 535-0704. Fax: (208) 535-0761. Licensee: Chaparral Broadcasting Inc. Group owner: Chaparral Communications (acq 7-30-2004; grpsl). Format: Soft adult contemp. ♦Scott Parker, gen mgr.

KWYS(AM)—Co-owned with KEZQ(FM). Dec 20, 1967: 920 khz; 1 kw-D. TL: N44 38 56 W111 05 50. 603 N. Canyon, 59758. 1,500 Natl. Network: CNN Radio. Format: Oldies. ♦Kim Davis, stn mgr.

Whitefish

KJJR(AM)— Feb 14, 1979: 880 khz; 10 kw-D, 500 w-N. TL: N48 23 44 W114 19 11. Hrs open: Box 5409, Kalispell, 59903. Phone: (406) 755-8700. Fax: (406) 755-8770. E-mail: KJJR@beebroadcasting.com Web Site: www.beebroadcasting.com. Licensee: Bee Broadcasting Inc. (group owner) Population served: 60,000 Format: News/talk. ♦Benny Bee, pres; Mark Wagner, gen mgr.

KSAM(AM)— 2006: 1240 khz; 400 w-U. TL: N48 23 44 W114 19 40. Hrs open: 2432 US Highway 2 E., Kalispell, 59901. Phone: (406) 755-8700. Fax: (406) 755-8770. Web Site: www.beebroadcasting.com. Licensee: Bee Broadcasting Inc. ♦Benny Bee, pres; Mark Wagner, gen mgr.

KWOL-FM— 2005: 105.1 mhz; 62 kw. Ant 2,404 ft TL: N48 30 43 W114 22 13. Hrs open: Box 5409, Kalispell, 59903. Phone: (406) 755-8700. Fax: (406) 755-8770. E-mail: info@1051cool.com Web Site: www.1051cool.com. Licensee: Cathleen R. Bee dba Rose Communications. Format: Oldies. ♦Cassie Bee, gen mgr.

Whitehall

***KQLR(FM)**— 2007: 89.7 mhz; 1.45 kw vert. Ant 1,794 ft TL: N46 00 22 W112 26 33. Hrs open:
Rebroadcasts KLVR(FM) Santa Rosa, CA 100%.
2351 Sunset Blvd., Suite 170-218, Rocklin, CA, 95765. Phone: (916) 251-1600. Fax: (916) 251-1650. Web Site: www.klove.com. Licensee: Educational Media Foundation. (acq 12-2-2005; $28,450 for CP). Natl. Network: K-Love. Format: Contemp Christian. ♦Richard Jenkins, pres.

Wolf Point

KVCK(AM)— Sept 1, 1957: 1450 khz; 1 kw-U. TL: N48 05 18 W105 39 22. Hrs open: 24 324 Main St., 59201. Phone: (406) 653-1900. Fax: (406) 653-1909. E-mail: kvck@nemont.net Licensee: Wolf Town Wireless Inc. (acq 8-31-92; $120,000 with co-located FM; FTR: 11-16-92) Population served: 18,000 Natl. Network: ABC. Format: Oldies. News: 15 hrs wkly. Target aud: General. Spec prog: Farm 6 hrs wkly. ♦Larry Corns, progmg dir.

KVCK-FM— Sept 1, 1981: 92.7 mhz; 11.5 kw. Ant 499 ft TL: N48 11 09 W105 40 08. Stereo. 24 Format: Country. News: 15 hrs wkly. Spec prog: Farm 6 hrs wkly. ♦Susan Allmer, gen mgr.

Nebraska

Ainsworth

KBRB(AM)— Feb 6, 1968: 1400 khz; 1 kw-U. TL: N42 33 16 W99 49 52. Hrs open: 24 Box 285, 122 E. 2nd St., 69210. Phone: (402) 387-1400. Fax: (402) 387-2624. E-mail: kbrb@sscg.net Web Site: kbrbradio.com. Licensee: K.B.R. Broadcasting Co. Population served: 2,073 Natl. Network: ABC. Rgnl. Network: Brownfield. Law Firm: Bryan Cave. Wire Svc: AP Format: C&W, MOR. News: 30 hrs wkly. Target aud: General. ♦Lorris C. Rice, pres & gen mgr; Angie Von Heeder, prom VP; Cody Goochey, progmg dir; Randy Brudigan, chief of engrg; Renee Adkisson, traf mgr.

KBRB-FM— May 30, 1983: 92.7 mhz; 4.5 kw. 331 ft TL: N42 33 16 W99 49 52. Natl. Network: ABC. Format: Adult contemp mix. ♦Renee Adkisson, traf mgr.

Albion

KUSO(FM)— May 10, 2000: 92.7 mhz; 50 kw. 492 ft TL: N41 49 50 W97 41 12. Hrs open: 24 Box 747, Norfolk, 68702-0747. Secondary address: 214 N. 7th St., Norfolk 68701. Phone: (402) 371-0100. Fax: (402) 371-0050. E-mail: us92@us92.com Web Site: www.us92.com. Licensee: Flood Communications L.L.C. (acq 4-27-99; $50,000). Population served: 85,000 Law Firm: Reddy, Begley & McCormick. Format: Full service, country. News staff: one; News: 3 hrs wkly. Target aud: General. Spec prog: Farm 10 hrs wkly. ♦Michael J. Flood, pres & gen mgr; Dave Amick, opns dir; Angela Richard, gen sls mgr; Brian Masters, progmg dir; Tammy Partch, news dir; Ann Neilsen, traf mgr.

Alliance

KAAQ(FM)—Listing follows KCOW(AM).

KCOW(AM)— Feb 15, 1949: 1400 khz; 1 kw-U. TL: N42 06 26 W102 53 15. Hrs open: 24 (M-S) Box 600, 69301. Secondary address: 1210 W. 10th 69301. Phone: (308) 762-1400. Fax: (308) 762-7804. E-mail: kcow@bbc.net Web Site: www.doubleqcountry.com. Licensee: Eagle Communications Inc. Group owner: Eagle Communications Group (acq 1965). Population served: 15,000 Natl. Network: ABC. Natl. Rep: Interep. Wire Svc: NWS (National Weather Service) Wire Svc: AP Format: Oldies, news/talk. News staff: 2; News: 22 hrs wkly. Target aud: 25-54. Spec prog: Farm 18 hrs wkly. ♦Gary Shorman, pres; Mark Vail, VP; Mike Garwood, gen mgr; John Jones, rgnl sls mgr; Jason Wentworth, progmg dir; Jennifer Schmid, traf mgr; Kevin Horn, news rptr; Mike Glesinger, sports cmtr.

KAAQ(FM)—Co-owned with KCOW(AM). Sept 30, 1985: 105.9 mhz; 100 kw. Ant 705 ft TL: N41 50 29 W103 05 07. Stereo. 24 Web Site: www.doubleqcountry.com.65,000 Natl. Network: ABC. Natl. Rep: Interep. Format: Country. News staff: 2; News: 15 hrs wkly. Target aud: 18-54. Spec prog: Farm 4 hrs wkly. ♦Mark Vail, opns VP.

KPNY(FM)— 1978: 102.1 mhz; 100 kw. Ant 521 ft TL: N42 07 01 W103 07 09. Stereo. Hrs opn: 24 Box 30345, Lincoln, 68503-0345. Phone: (402) 845-6595. Web Site: www.missionnebraska.org. Licensee: Mission Nebraska Inc. (acq 2-5-2007; $360,000). Population served: 72,000 Format: Relg. Target aud: 18-35. ♦Stan Parker, gen mgr.

KQSK(FM)—Chadron, June 1, 1983: 97.5 mhz; 100 kw. Ant 840 ft TL: N42 38 06 W103 06 12. Stereo. Hrs opn: 24
Simulcast with KAAQ(FM) Alliance 95%.
Box 600, 69301. Secondary address: 1210 W. 10th 69301. Phone: (308) 762-1400. Fax: (308) 762-7804. E-mail: kcow@bbc.net Web Site: www.doubleqcountry.com. Licensee: Eagle Communications Inc. Group owner: Eagle Communications Group (acq 6-13-91; $125,000; FTR: 7-1-91). Population served: 65,000 Natl. Network: ABC. Natl. Rep: Interep. Wire Svc: AP Wire Svc: NWS (National Weather Service) Format: Country. News staff: 2; News: 15 hrs wkly. Target aud: 18-54. Spec prog: Farm 4 hrs wkly. ♦Mike Garwood, gen mgr; John Howard, stn mgr; Michael Glesinger, opns mgr; John Howard, rgnl sls mgr; John Axtell, news dir; Tony Cuesta, chief of engrg.

***KTNE-FM**— May 1990: 91.1 mhz; 92 kw. 1,325 ft TL: N41 50 24 W103 03 18. Stereo. Hrs opn: 24
Rebroadcasts KUCV(FM) Lincoln 100%.
Secondary address: 1800 N. 33rd St., Lincoln 68503. Phone: (402) 472-3611. Fax: (402) 472-2403. Fax: (402) 472-1785. E-mail: radio@netnebraska.org Web Site: www.netnebraska.org. Licensee: Nebraska Educational Telecommunications Commission. Natl. Network: PRI, NPR. Rgnl. Network: Neb. Pub. Law Firm: Dow, Lohnes & Albertson. Format: Class, news. Target aud: 35 plus; general. ♦Rod Bates, gen mgr; Nancy Finken, progmg dir; Martin Wells, news dir; Michael Winkle, mktg; Michael Beach, engr.

Auburn

KNCY-FM— Sept 18, 1981: 103.1 mhz; 14 kw. Ant 436 ft TL: N40 27 57 W95 45 38. Hrs open: 24 Box 278, 814 Central Ave., Nebraska City, 68410. Phone: (402) 873-3348. Fax: (402) 873-7882. Web Site: kncycountry.com. Licensee: Arbor Day Broadcasting Inc. (acq 2-3-2003; $600,000 with KNCY(AM) Nebraska City). Natl. Network: ABC. Rgnl. Network: Brownfield. Wire Svc: AP Format: Country, news. News staff: 2; News: 30 hrs wkly. ♦Scott Kooistra, VP & gen mgr; Chris Yates, gen sls mgr; Doug Jennings, progmg dir.

Aurora

KRGY(FM)— Mar 1, 1980: 97.3 mhz; 50 kw. Ant 348 ft TL: N40 52 44 W98 05 36. Stereo. Hrs opn: 24 3205 W. North Front St., 68802. Secondary address: Box 4907 68802. Phone: (308) 381-1430. Fax: (308) 382-6701. E-mail: info@gifamilyradio.com Web Site: www.gifamilyradio.com/STAR/star.htm. Licensee: Legacy Communications LLC (group owner; (acq 5-17-2004; grpsl). Natl. Network: ABC. Format: Hot adult contemp. News staff: 2; News: 2 hrs wkly. Target aud: 18-49. ♦Lyle Nelson, gen mgr; Jim Davis, opns mgr.

KROA(FM)—See Grand Island

Bassett

***KMNE-FM—** June 1991: 90.3 mhz; 92.3 kw. 1,292 ft TL: N42 20 05 W99 29 01. Stereo. Hrs opn: 24
Simulcasts KUCV(FM) Lincoln 100%.
1800 N. 33rd St., Lincoln, 68503-1409. Phone: (402) 472-3611. Fax: (402) 472-2403. E-mail: radio@netnebraska.org Web Site: netnebraska.org/radio. Licensee: Nebraska Educational Telecommunications Commission. Natl. Network: NPR, PRI. Format: Classical, news. News staff: 3; News: 38 hrs wkly. Target aud: General. ♦Ray Dilley, gen mgr; Nancy Finken, progmg dir; Bill Stiber, mus dir; Martin Wells, news dir; Jeff Smith, traf mgr.

Beatrice

***KNBE(FM)—**Not on air, target date: November 1: 88.9 mhz; 7.5 kw vert. Ant 459 ft TL: N40 33 03 W96 38 45. Hrs open: Box 262550, Baton Rouge, LA, 20826. Phone: (225) 768-3102. Fax: (225) 768-3729. E-mail: kawikfish@yahoo.com Web Site: www.jsm.org. Licensee: Family Worship Center Church Inc. (acq 10-12-2006; grpsl). Format: Relg, christian. ♦David Whitelaw, COO.

KTGL(FM)—Licensed to Beatrice. See Lincoln

KWBE(AM)— June 12, 1949: 1450 khz; 1000 W. TL: N40 15 49 W96 46 27. Hrs open: 24 Box 10, 200 Sherman St., 68310. Phone: (402) 228-5923. Fax: (402) 228-3704. E-mail: kwbe@broadcasthouse.com Web Site: www.kwbe.com. Licensee: Nebraska Broadcasting LLC. Group owner: Triad Broadcasting Co. LLC (acq 5-00; grpsl). Population served: 90,000 Natl. Network: CBS, Westwood One. Rgnl rep: Howard Anderson. Format: Adult contemp, news/talk. News staff: one; News: 25 hrs wkly. Target aud: 24-54; mature, affluent adults. Spec prog: Farm 14 hrs wkly. ♦David Benjamin, CEO; Charlie Brogan, gen mgr & gen sls mgr; Jay Stalder, gen mgr & progmg dir; Doug Kennedy, news dir; Dave Neidfeldt, pub affrs dir & news rptr.

Bellevue

KOZN(AM)— June 1999: 1620 khz; 10 kw-D, 1 kw-N. TL: N41 16 12 W95 47 10. Hrs open: 24 5011 Capitol Ave., Omaha, 68132. Phone: (402) 342-2000. Fax: (402) 346-5748. Web Site: www.1620thezone.com. Licensee: Waitt Omaha LLC. Group owner: Waitt Radio Inc. (acq 1-7-2002; grpsl). Natl. Network: ESPN Radio, Westwood One. Natl. Rep: Katz Radio. Law Firm: Pepper & Corazzini. Format: Sports. Target aud: 25-54; men. ♦Jim McKernan, gen mgr; Mark Todd, opns mgr; Rhonda Gerrard, sls dir; Brandon Pappas, prom dir; Neil Nelkin, progmg dir; Lori Storz, traf mgr.

KYDZ(AM)— Mar 19, 1987: Stn currently dark. 1180 khz; 25 kw-D, 1 kw-N, DA-2. TL: N41 16 12 W95 47 10. Hrs open: 24 5011 Capitol Ave., Omaha, 68132. Phone: (402) 342-2000. Fax: (402) 346-5748. Web Site: 1180labonita.com. Licensee: Waitt Omaha LLC. (group owner; (acq 1-7-2002; grpsl). Population served: 350,000 Format: Spanish. Target aud: 18+; General. ♦Jim McKernan, pres & gen mgr; Mark Todd, opns mgr; Rhonda Gerrard, sls dir; Neil Nelkin, progmg dir; Darwin Stinton, chief of engrg; Lyn Farhenbruch, traf mgr.

Bennington

KHUS(FM)— June 10, 1991: 93.3 mhz; 6 kw. 350 ft TL: N41 22 57 W96 07 57. Stereo. Hrs opn: 5010 Underwood Ave., Omaha, 68132. Phone: (402) 561-2000. Fax: (402) 556-8937. E-mail: michellematthews@clearchannel.com Web Site: www.krrk.com. Licensee: Capstar TX L.P. Group owner: Clear Channel Communications Inc. (acq 8-30-00; grpsl). Format: Hard rock. ♦Donna Baker, gen mgr & engrg mgr; Rhonda Gerrard, gen sls mgr; Michelle Matthews, progmg dir.

Blair

KBLR-FM— Sept 10, 2002: 97.3 mhz; 25 kw. Ant 302 ft TL: N41 38 21 W96 12 31. Hrs open: 118 E. 5th St., Fremont, 68025. Phone: (402) 721-1340. Fax: (402) 721-5023. Licensee: Waitt Omaha LLC. (group owner; (acq 1-7-2002; grpsl). Format: Country. ♦Del Meyer, gen mgr; Chris Walz, opns mgr.

***KDCV-FM—** Oct 1, 1972: 91.1 mhz; 10 w. 60 ft TL: N41 33 07 W96 09 20. Stereo. Hrs opn: 24 2848 College Dr., 68008. Phone: (402) 426-7322. Fax: (402) 426-7382. E-mail: kdcv@dana.edu Web Site: www.huntel.net/kdcv. Licensee: Dana College. Population served: 8,000 Format: Var/div. Target aud: 25-54; community, 18-34 college community. ♦Vern Wirka, gen mgr.

Bridgeport

KMOR(FM)— 2001: 101.3 mhz; 100 kw. Ant 1,112 ft TL: N41 50 23 W103 49 36. Hrs open: Box 532, Scottsbluff, 69361-0532. Phone: (308) 632-5667. Fax: (308) 436-7296. Licensee: Tracy Broadcasting Corp. (group owner). Format: Classic rock. ♦Michael Tracy, pres; Larry Swikard, gen mgr & gen sls mgr; Krista Sarchett, prom dir; Jeff McKenzie, progmg dir; Pat Leach, engrg dir & chief of engrg.

Broken Bow

KBBN-FM—Listing follows KCNI(AM).

KCNI(AM)— Sept 28, 1949: 1280 khz; 1 kw-D. TL: N41 24 31 W99 40 28. Hrs open: 6 AM-6 PM Box 409, W. Hwy., 2 Calaway Rd., 68822. Phone: (308) 872-5881. Fax: (308) 872-3284. Licensee: Custer County Broadcasting Co. Population served: 6,200 Format: C&W. News staff: one; News: 24 hrs wkly. Target aud: General; rural audience of all ages with focus on 25-65. Spec prog: Farm & country mus specials. ♦David Birnie, VP, gen mgr, opns dir & sls dir; Brent Apperson, progmg dir; Dale Sell, news dir & pub affrs dir; Val Lane, chief of engrg.

KBBN-FM—Co-owned with KCNI(AM). June 15, 1982: 98.3 mhz; 25 kw. Ant 312 ft TL: N41 23 49 W99 37 02. Stereo. 6 AM-11 PM Format: Classic rock. News staff: one; News: 10 hrs wkly. Target aud: 24-45; baby boomers & on either edge of age breakdown.

Central City

KZEN(FM)— July 22, 1985: 100.3 mhz; 100 kw. 1,854 ft TL: N41 32 28 W97 40 45. Stereo. Hrs opn: 24 1418 25th St., Columbus, 68601. Phone: (402) 564-2866. Fax: (402) 564-2867. Licensee: Three Eagles of Columbus Inc. Group owner: Three Eagles Communications (acq 9-5-97; grpsl). Population served: 186,000 Natl. Network: ABC, AP Radio. Natl. Rep: Interep. Wire Svc: AP Format: Country. News staff: 2; News: 18 hrs wkly. Target aud: 25-54; rgnl, rural & small town audience. Spec prog: Relg 5 hrs, farm 20 hrs wkly. ♦Rolland Johnson, CEO; Gary Buchanan, pres; Cindy Harris, CFO; Greg Wells, gen mgr; Dean Johnson, opns dir & mus dir.

Chadron

***KCNE-FM—** Aug 29, 1991: 91.9 mhz; 8.4 kw. 338 ft TL: N42 48 47 W103 00 22. Stereo. Hrs opn:
Rebroadcasts KUCV(FM) Lincoln 100%.
Secondary address: 1800 N. 33rd St., Lincoln 68503. Phone: (402) 472-3611. Fax: (402) 472-1785. E-mail: nprn@unl.edu Web Site: www.nprn.org. Licensee: Nebraska Educational Telecommunications Commission. Natl. Network: PRI, NPR. Format: Class, news. Target aud: General. ♦Ray Dilley, gen mgr.

KCSR(AM)— May 9, 1954: 610 khz; 1 kw-D, 118 w-N. TL: N42 49 56 W103 01 00. Stereo. Hrs opn: 24 226 Bordeaux, 69337. Phone: (308) 432-5545. Phone: (308) 432-2233. Fax: (308) 432-5601. E-mail: kcsr@chadrad.com Web Site: www.chadrad.com. Licensee: Chadrad Communications Inc. (acq 8-30-91; $150,000). Population served: 75,000 Natl. Network: AP Radio. Rgnl. Network: Mid-American Ag Law Firm: Fletcher, Heald & Hildreth. Wire Svc: AP Format: Country, farm. News staff: 2; News: 20 hrs wkly. Target aud: 25-54; people in the ranch, farm & agricultural industry. Spec prog: Farm 6 hrs wkly. ♦Dennis A. Brown, pres; J.J. Archer, progmg dir; Joe Lowery, mus dir; Brian Taylor, asst music dir & chief of engrg; Chris Faukhauser, news dir; Duanne Ekwall, pub affrs dir; Kathi Brown, gen sls mgr & traf mgr; Greg Mahaco, sports cmtr.

KQSK(FM)—Licensed to Chadron. See Alliance

Columbus

KJSK(AM)— Apr 28, 1948: 900 khz; 1 kw-D, 66 w-N. TL: N41 26 12 W97 23 47. Hrs open: 24 1418 25th St., 68601. Phone: (402) 564-9101. Fax: (402) 564-1999. E-mail: kjskkdjs@megavision.com Web Site: www.kjsk.com. Licensee: Three Eagles of Columbus Inc. Group owner: Three Eagles Communications (acq 8-23-01; $2.7 million with co-located FM including five-year noncompete agreement). Population served: 19,000 Rgnl. Network: Brownfield. Format: News/talk, farm. News staff: one. Target aud: General. Spec prog: Pol 4 hrs, Sp 6 hrs, relg 20 hrs wkly. ♦Dean Johnson, opns mgr & progmg dir; Greg Wells, gen mgr & gen sls mgr; Denise Kollath, prom dir; Bob Cook, engrg dir & chief of engrg; Bobbie Freeborn, traf mgr.

KLIR(FM)—Co-owned with KJSK(AM). August 1964: 101.1 mhz; 100 kw. 760 ft TL: N41 16 55 W97 24 30. Stereo. Phone: (402) 564-9101. E-mail: klirnet@megavision.com Web Site: www.klir.net.25,000 Format: Adult contemp. News staff: one. Target aud: General. Spec prog: Oldies 12 hrs wkly. ♦Dave Rinehart, opns mgr, prom dir & progmg dir; Cec Hottory, sls dir; Bobbie Freeborn, mus dir & traf mgr; Sheila Young, spec ev coord, local news ed, news rptr & reporter; Dave Reinhart, mus critic.

KKOT(FM)—Listing follows KTTT(AM).

***KTLX(FM)—** July 1974: 91.9 mhz; 100 w. 78 ft TL: N41 26 26 W97 21 14. Hrs open: c/o Trinity Lutheran Church, 2200 25th St., 68601. Phone: (402) 564-8548. Fax: (402) 562-6003. E-mail: ktlx@megavision.com Licensee: TLC Educational Corp. Format: Educ, relg. ♦Gary Spuit, pres; Russ Rote, gen mgr.

KTTT(AM)— Dec 2, 1962: 1510 khz; 500 w-D. TL: N41 27 14 W97 24 20. Hrs open: 1418 25th St., 68601. Phone: (402) 564-2866. Fax: (402) 564-2867. Licensee: Three Eagles Communications Inc. (group owner; acq 1996). Population served: 50,000 Natl. Rep: McGavren Guild. Format: Talk. News staff: one; News: 10 hrs wkly. Target aud: 25-65. Spec prog: Polka, farm 5 hrs, Ger 5 hrs, Pol 5 hrs wkly. ♦Rolland Johnson, CEO; Gary Buchanan, pres; Cindy Harris, CFO; Greg Wells, gen mgr; Dean Johnson, opns dir; Melissa Sanford, sls dir; Jim Dolezel, progmg dir; Bob Cook, chief of engrg; Bobbie Freeborn, traf mgr; Susan Littlefield, farm dir.

KKOT(FM)—Co-owned with KTTT(AM). Nov 25, 1969: 93.5 mhz; 100 kw. 981 ft TL: N41 32 28 W97 40 45. Stereo. 24 450,000 Natl. Rep: McGavren Guild. Format: Classic Rock. News staff: 2; News: 15 hrs wkly. Target aud: 18-49; young families. Spec prog: Farm 8 hrs wkly. ♦Gary Buchanan, COO & opns dir; Dean Johnson, progmg dir; Bobbie Freeborn, traf mgr; Susan Littlefield, farm dir.

Cozad

KAMI(AM)—Listing follows KCVN(FM).

***KCVN(FM)—** Aug 4, 1983: 104.5 mhz; 100 kw. Ant 360 ft TL: N40 46 35 W100 01 47. Stereo. Hrs opn: 24 233 S. 13th St., Suite 1520, Lincoln, 68508. Phone: (402) 465-8850. Fax: (402) 465-8852. E-mail: kcvn@bottradionetwork.com Web Site: www.bottradionetwork.com/station_cozad/cozad_home.asp. Licensee: Community Broadcasting Inc. Group owner: Bott Radio Network (acq 7-9-2004; $365,000 with co-located AM). Format: Christian teaching and talk. Target aud: 25-54; adults. ♦Tom Millett, gen mgr.

KAMI(AM)—Co-owned with KCVN(FM). November 1965: 1580 khz; 1 kw-D, 17 w-N. TL: N40 50 18 W99 56 20.8 AM-5 PM

Crete

***KDNE(FM)—** Aug 30, 1993: 91.9 mhz; 200 w. 66 ft TL: N40 37 16 W96 57 04. Stereo. Hrs opn: 1014 Boswell Ave., 68333. Phone: (402) 826-8611. Fax: (402) 826-8600. E-mail: kdne@doane.edu Licensee: Doane College Board of Trustees. Population served: 7,500 Format: Progsv. Target aud: General; males & females between the ages of 12 to 34. ♦Jonathan Brand, pres; Lee Thomas, gen mgr; John Thayer, stn mgr; Corey Rotschafer, progmg dir.

KIBZ(FM)— Aug 20, 1976: 104.1 mhz; 50 kw. 613 ft TL: N40 31 06 W96 46 07. Stereo. Hrs opn: 24 4630 Antelope Creek Rd., Suite 200, Lincoln, 68506. Phone: (402) 826-4393. Fax: (402) 483-9138. Web Site: www.kibz.com. Licensee: Capstar TX L.P. Group owner: Clear Channel Communications Inc. (acq 8-30-00; grpsl). Format: New rock. News staff: one. Target aud: 18-34. ♦Julie Gade, stn mgr; Charlie Thomas, progmg dir.

Crookston

***KINI(FM)**— January 1978: 96.1 mhz; 90 kw. Ant 499 ft TL: N43 07 50 W100 54 02. Stereo. Hrs opn: 24 Box 499, St. Francis, SD, 57572-0499. Phone: (605) 747-2291. Fax: (605) 747-5791. E-mail: kinifm@gwtc.net Web Site: www.gwtc.net/~kinifm. Licensee: Rosebud Educational Society Inc. (acq 1-78). Wire Svc: AP Format: Adult contemp, rock, native American. News staff: one; News: 12 hrs wkly. Target aud: General; Indian & white. Spec prog: American Indian 15 hrs, gospel 6 hrs, relg 6 hrs wkly. ♦Fr. John Hatcher, pres; Marcy VanWinkle, exec VP; Richard K. Iyotte Jr., gen mgr & stn mgr.

Dakota City

KTFJ(AM)— 1991: 1250 khz; 500 w-D, 700 w-N, DA-2. TL: N42 26 33 W96 15 41. Hrs open: 24
Rebroadcasts KTFC(FM) Sioux City, IA.
1521 Buchanan Ave., Sioux City, IA, 51106. Phone: (712) 252-4621. Web Site: www.worldwidebibleradio.com. Licensee: Donald A. Swanson. Natl. Network: USA. Format: Gospel. ♦Donald A. Swanson, pres & gen mgr.

Fairbury

KGMT(AM)— June 13, 1960: 1310 khz; 500 w-D, 97 w-N. TL: N40 06 58 W97 09 05. Hrs open: 6 AM-6 PM 414 4th St., 68352. Phone: (402) 729-3382. Fax: (402) 729-3446. E-mail: kutt@diodecom.net Licensee: Siebert Communications Inc. (acq 8-1-84). Population served: 5,265 Natl. Rep: Farmakis. Law Firm: Shaw Pittman. Format: Oldies, news. News: one. Target aud: 25-52. Spec prog: Farm 18 hrs wkly. ♦Rick Siebert, pres; Randy Bauer, gen mgr.

KUTT(FM)—Co-owned with KGMT(AM). December 1983: 99.5 mhz; 100 kw. 692 ft TL: N40 10 57 W96 58 33. Stereo. 24 Natl. Network: ABC. Format: Hot country. ♦Randy Bauer, stn mgr & progmg mgr.

Falls City

KLZA(FM)— July 7, 1998: 101.3 mhz; 6 kw. 328 ft TL: N40 06 54 W95 39 06. (CP: 25 kw). Hrs opn: Box 101, 68355. Phone: (402) 245-6010. Fax: (402) 245-6040. E-mail: sunny1013fm@hotmail.com Licensee: KNZA Inc. (group owner) Format: Soft rock. ♦Mike Gilmore, stn mgr; Robert Hilton, opns mgr; Mike Slocum, chief of engrg.

KTNC(AM)— Aug 3, 1957: 1230 khz; 500 w-D, 1 kw-N. TL: N40 03 57 W95 36 55. Hrs open: 24 1602 Stone St., 68355. Phone: (402) 245-2453. Fax: (402) 245-5862. E-mail: ktnc@sentco.net Web Site: www.ktncradio.com. Licensee: C.R. Communications Inc. (acq 8-1-81; $270,000 FTR: 7-13-81). Population served: 60,000 Natl. Network: ABC. Rgnl. Network: Brownfield. Format: Oldies. News staff: one; News: 23 hrs wkly. Target aud: 25 plus; farmers, businessmen, employees, retirees. Spec prog: Christian mus one hr wkly. ♦Charles A. Radatz, pres & gen mgr; Gerald Hopp, news dir; Jackie Johnson, traf mgr; Aaron Wisdom, pub svc dir; Darlene Tisdel, adv.

Firth

KOLB(FM)—Not on air, target date: unknown: 93.7 mhz; 6 kw. Ant 226 ft TL: N40 34 57.4 W96 37 15.2. Hrs open: 5829 N. 60th St., Omaha, 68104. Phone: (402) 571-0200. Fax: (402) 571-0833. Licensee: VSS Catholic Communications Inc. ♦Jim Carroll, gen mgr.

Fremont

KFMT-FM—Listing follows KHUB(AM).

KHUB(AM)— December 1939: 1340 khz; 500 w-D, 1 kw-N. TL: N41 25 58 W96 27 16. Hrs open: 5 AM-midnight 118 E. Fifth St., 68025. Phone: (402) 721-1340. Fax: (402) 721-5023. Licensee: NRG Media LLC. Group owner: Waitt Radio Inc. (acq 10-31-2005; grpsl). Population served: 165,000 Format: News/talk. News staff: one; News: 25 hrs wkly. Target aud: 35 plus; mature adults. Spec prog: Farm 6 hrs wkly. ♦Del Meyer, gen mgr, dev dir & progmg dir; Chris Walz, opns mgr, progmg dir, chief of engrg & traf mgr; Barry Reker, gen sls mgr; Jessica Meistrell, news dir.

KFMT-FM—Co-owned with KHUB(AM). July 1972: 105.5 mhz; 1.2 kw. 450 ft TL: N41 24 40 W96 31 53. Stereo. 5 AM-midnight Web Site: www.kfmt.com. Format: Oldies, classic rock. Target aud: 25-54. ♦Chris Walz, traf mgr.

Gering

KOZY-FM— August 1996: 103.9 mhz; 7 kw. 102 ft TL: N41 51 50 W103 42 20. Hrs open: Box 1263, Scottsbluff, 69363-1263. Phone: (308) 632-5667. Fax: (308) 635-1905. E-mail: kmor@tracybroadcasting.com Web Site: www.tracybroadcasting.com. Licensee: Tracy Broadcasting Corp. (group owner). Population served: 125,000 Format: Soft rock. News staff: 2. Target aud: 25-54; general. ♦Larry Swikard, gen mgr.

Gordon

KSDZ(FM)— May 19, 1979: 95.5 mhz; 60 kw. 310 ft TL: N42 47 56 W102 15 40. Stereo. Hrs opn: 24 Box 390, W. Hwy. 20, 69343. Phone: (308) 282-2500. Fax: (308) 282-0061. Licensee: DJ Broadcasting Inc. (acq 12-26-91; FTR: 1-13-92). Population served: 35,000 Natl. Network: ABC. Rgnl. Network: Mid-American Ag. Wire Svc: AP Format: C&W, oldies. Target aud: 25-54. ♦Jim Lambley, pres.

Grand Island

***KLNB(FM)**— 2005: 88.3 mhz; 1.7 kw. Ant 147 ft TL: N40 54 50 W98 23 52. Hrs open: 24 2351 Sunset Blvd., Suite 170-218, Rocklin, CA, 95765. Phone: (916) 251-1600. Fax: (916) 251-1650. E-mail: klove@klove.com Web Site: www.klove.com. Licensee: Educational Media Foundation. Group owner: EMF Broadcasting (acq 3-11-2003; grpsl). Natl. Network: K-Love. Law Firm: Shaw Pittman. Format: Contemp Christian. News staff: 3. Target aud: 25-44; Judeo Christian, female. ♦Richard Jenkins, pres; Mike Novak, VP; Keith Whipple, dev dir; David Pierce, progmg mgr; Ed Lenane, news dir; Sam Wallington, engrg dir; Karen Johnson, news rptr; Marya Morgan, news rptr; Richard Hunt, news rptr.

KMMJ(AM)— November 1925: 750 khz; 10 kw-U, DA-1. TL: N41 08 05 W97 59 38. Hrs open: Sunrise-sunset 3205 W. North Front St., 68802. Secondary address: Box 4907 68802. Phone: (308) 382-2800. Phone: (308) 381-1430. Fax: (308) 382-6701. Web Site: missionnebraska.org/thebridge/. Licensee: Mission Nebraska Inc. (group owner; (acq 5-1-2006; $825,000). Population served: 250,000 Format: Christian. ♦Alan Usher, VP & gen mgr; Jim Davis, progmg dir.

***KNFA(FM)**—Not on air, target date: Spring, 2008: 90.7 mhz; 250 w. Ant 161 ft TL: N40 54 50 W98 23 52. Hrs open: Box 262550, Baton Rouge, LA, 20826. Secondary address: 8919 World Ministry Ave., Baton Rouge 70810. Phone: (225) 768-3102. Fax: (225) 768-3729. E-mail: kawikfish@yahoo.com Web Site: www.jsm.org. Licensee: Family Worship Center Church Inc. (acq 10-12-2006; grpsl). Format: Relg, christian. ♦David Whitelaw, COO.

KRGI(AM)— Apr 1, 1953: 1430 khz; 5 kw-D, 1 kw-N, DA-N. TL: N40 62 26 W98 16 24. Hrs open: Box 4907, 68802-4907. Secondary address: 3205 W. N. Front St. 68803. Phone: (308) 381-1430. Fax: (308) 382-6701. E-mail: krgi@krgi.com Web Site: www.gifamilyradio.com/KRGI/krgi.htm. Licensee: Legacy Communications LLC. (group owner; (acq 5-17-2004; grpsl). Population served: 135,000 Natl. Rep: Christal. Law Firm: Fletcher, Heald & Hildreth. Format: Adult contemp, news/talk. Target aud: 25-54. ♦Alan Usher, gen mgr & gen sls mgr; Chris Loghry, opns dir & progmg dir; Rob Fossberg, news dir; Chuck Walker, chief of engrg; Alie Schlachter, traf mgr.

KRGI-FM— Oct 30, 1975: 96.5 mhz; 100 kw. 416 ft TL: N40 51 53 W98 23 47. Stereo. 24 Web Site: www.gifamilyradio.com/C96/C96.htm. Natl. Network: ABC. Format: Hot C&W. ♦Alie Schlachter, traf mgr.

***KROA(FM)**— Aug 11, 1967: 95.7 mhz; 100 kw. Ant 460 ft TL: N40 47 11 W98 22 00. Stereo. Hrs opn: 24 Box 495, Doniphan, 68832. Phone: (402) 845-6595. Fax: (402) 845-6597. E-mail: kroafm@kroa.org Web Site: www.missionnebraska.org/thebridge. Licensee: Mission Nebraska Inc. (acq 11-25-2003; $1.5 million). Population served: 250,000 Natl. Network: Moody. Format: Adult contemp Christian. ♦Dr. James Eckman, pres; Gordon Wheeler, stn mgr; Taryn Julane, mus dir.

KSYZ-FM— November 1982: 107.7 mhz; 100 kw. 899 ft TL: N40 51 53 W98 23 47. (CP: TL: N40 42 07 W98 35 20). Stereo. Hrs opn: 24 3532 W. Captial Ave., 68803. Secondary address: Box 5108 68802. Phone: (308) 381-1077. Fax: (308) 384-8900. E-mail: ksyzprod@nrgmedia.com Web Site: www.ksyz.com. Licensee: NRG Media LLC. (acq 10-31-2005; $5.28 million). Format: Adult contemp. Target aud: 25-49; adults. ♦Dan Zabka, gen mgr; Jim Cartwright, opns mgr.

Grand Isle

***KNHA(FM)**—Not on air, target date: unknown: 90.9 mhz; 500 w. Ant 148 ft TL: N40 38 56 W98 23 01. Hrs open: Box 262550, Baton Rouge, LA, 20826. Phone: (225) 768-3102. Fax: (225) 768-3729. E-mail: kawikfish@yahoo.com Web Site: www.jsm.org. Licensee: Family Worship Center Church Inc. (acq 10-12-2006; grpsl). Format: Relg, christian. ♦David Whitelaw, COO.

Hastings

***KCNT(FM)**— Feb 22, 1971: 88.1 mhz; 2 kw. 182 ft TL: N40 34 52 W98 19 58. Hrs open: 24 Box 1024, 68902. Phone: (402) 461-2580. Fax: (402) 461-2507. E-mail: jbrooks@cccneb.edu Web Site: www.cccneb.edu/programs/mart/kcnt/index.html. Licensee: Central Community College. Population served: 23,580 Format: CHR, educ. ♦John L. Brooks, gen mgr.

***KFKX(FM)**— Aug 30, 1997: 90.1 mhz; 1 kw vert. 292 ft TL: N40 38 56 W98 23 01. Stereo. Hrs opn: 4 PM-midnight 710 Turner Hastings College, 68901. Phone: (402) 461-7367. Fax: (402) 461-7442. E-mail: kfkx@hastings.edu Licensee: Hastings College. Population served: 29,625 Law Firm: Booth, Freret, Imlay & Tepper. Format: Div, AOR. News: 7 hrs wkly. Target aud: General. Spec prog: Jazz 3 hrs, relg 3 hrs, urban contemp 8 hrs wkly. ♦Phillip Dudley, pres; Sharon Behl Brooks, gen mgr; Meg Bernt, stn mgr; Bart Jones, opns dir.

KHAS(AM)— Sept 30, 1940: 1230 khz; 1 kw-U. TL: N40 34 40 W98 24 17. Hrs open: 24 Box 726, 68902. Secondary address: 500 East J St. 68901. Phone: (402) 462-5101. Fax: (402) 461-3866. E-mail: khas@gtmc.net Web Site: hastingslink.com. Licensee: Platte River Radio Inc. (acq 2-1-2006; $560,000 with KICS(AM) Hastings). Population served: 130,000 Natl. Network: CBS. Rgnl. Network: Brownfield. Rgnl rep: Howard Anderson Law Firm: Miller & Neely. Wire Svc: AP Format: Adult contemp. News staff: one; News: 20 hrs wkly. Target aud: 35 plus; general. Spec prog: Farm 2 hrs, class 2 hrs wkly. ♦David Oldfather, pres; Wayne Specht, gen mgr; Jim Stevens, stn mgr, prom VP, progmg VP & progmg dir; Mike Smithson, news dir; Gwen Sheppard, traf mgr.

***KHNE-FM**— June 1990: 89.1 mhz; 64.3 kw. 328 ft TL: N40 46 17 W98 05 22. Stereo. Hrs opn: 24
Rebroadcasts KUCV(FM) Lincoln 100%.
1800 N. 33rd St., Lincoln, 68503-1409. Phone: (402) 472-3611. Fax: (402) 472-2403. E-mail: radio@netnebraska.org Web Site: netnebraska.org/radio. Licensee: Nebraska Educational Telecommunications Commission. Natl. Network: PRI, NPR. Rgnl. Network: Neb. Pub. Format: Classical, news. News staff: 3; News: 38 hrs wkly. Target aud: General. ♦Ray Dilley, gen mgr; Nancy Finken, progmg dir; Bill Stiber, mus dir; Martin Wells, news dir; Jeff Smith, traf mgr.

KICS(AM)— Apr 15, 1964: 1550 khz; 500 w-D. TL: N40 34 09 W98 21 57. Hrs open: 24 500 E. J St., 68901-7113. Phone: (402) 462-5101. Fax: (402) 461-3866. E-mail: khas@gtmc.not Web Site: www.khasradio.com. Licensee: Platte River Radio Inc. (acq 2-1-2006; $560,000 with KHAS(AM) Hastings). Population served: 23,580 Natl. Network: ESPN Radio. Rgnl rep: Howard Anderson Law Firm: Miller & Miller. Format:

Sports. News staff: one; News: 12 hrs wkly. Target aud: 18-54; males. ♦David Oldfather, pres; Wayne Specht, gen mgr; Jim Stevens, stn mgr.

KLIQ(FM)— 2001: 94.5 mhz; 97.7 kw. Ant 948 ft TL: N40 36 08 W98 50 21. Hrs open: 24 Box 276, 68902. Phone: (402) 461-4922. Fax: (402) 461-4950. E-mail: dbrock@waittradio.com Web Site: www.kliqfm.com. Licensee: Platte River Radio Inc. Group owner: Waitt Radio Inc. (acq 2-1-2006; $700,000). Population served: 30000 Law Firm: Pepper & Corazzini. Format: Adult contemp. Target aud: 25-54. Spec prog: News & special community interest. ♦David Brock, gen mgr & stn mgr; Theresa Parr, gen sls mgr; Cody Hanson, progmg dir; Deb Fountaine, traf mgr.

***KNHS(FM)—**Not on air, target date: unknown: 91.7 mhz; 1.3 kw. Ant 148 ft TL: N40 31 13 W98 22 10. Hrs open: Drawer 2440, Tupelo, MS, 38801. Phone: (662) 844-8888. Fax: (662) 842-6791. Web Site: www.afr.net. Licensee: American Family Association. ♦Marvin Sanders, gen mgr.

KROR(FM)— February 1965: 101.5 mhz; 100 kw. Ant 1,004 ft TL: N40 39 28 W98 52 04. Stereo. Hrs opn: 24 3532 W. Capital Ave., Grand Island, 68803. Phone: (308) 381-1077. Fax: (308) 384-8900. E-mail: ksyzprod@nrgmedia.com Web Site: www.rock1015.com. Licensee: NRG License Sub LLC. (acq 1-31-2006; swap for KLIQ(FM) Hastings). Population served: 250,000 Format: Classic rock. Target aud: 25-54. ♦Dan Zabka, gen mgr; Jim Cartwright, opns mgr.

Holdrege

KMTY(FM)—Listing follows KUVR(AM).

KUVR(AM)— Oct 20, 1956: 1380 khz; 500 w-D, 62 w-N. TL: N40 26 26 W99 24 00. Hrs open: Box 465, 68949. Secondary address: 613 4th Ave. 68949. Phone: (308) 995-4020. Fax: (308) 995-2202. Licensee: NRG License Sub. LLC. (group owner; (acq 10-31-2005; grpsl). Population served: 5,635 Law Firm: Borsari & Paxson. Format: Oldies. Target aud: 35-54. Spec prog: Farm 5 hrs, big band 5 hrs, contemp gospel 5 hrs wkly. ♦John McDonald, gen mgr, progmg dir & news dir; Jim Conner, gen sls mgr; Randy Issler, mus dir & traf mgr; Val Lane, chief of engrg.

KMTY(FM)—Co-owned with KUVR(AM). October 1970: 97.7 mhz; 55 kw. Ant 253 ft TL: N40 26 26 W99 23 59. Stereo. 24 150,000 Natl. Network: ABC. Format: Hot adult contemp. Target aud: 20-45. ♦Randy Issler, traf mgr.

Hubbard

***KAYA(FM)—** 1998: 91.3 mhz; 5.1 kw. Ant 377 ft TL: N42 21 10 W96 31 32. Hrs open: Box 3206, Sioux City, MS, 38803. Secondary address: 1211 Tri-View, Sioux City, IA 51103. Phone: (662) 844-8888. Phone: (712) 255-9191. Fax: (662) 842-6791. Fax: (712) 255-3177. E-mail: comments@afr.net Web Site: www.afr.net. Licensee: American Family Association. Group owner: American Family Radio Format: Christian inspirational, relg. ♦Marvin Sanders, gen mgr.

Hyannis

KNPE(FM)—Not on air, target date: unknown: 97.9 mhz; 100 kw. Ant 610 ft TL: N42 01 28 W102 00 22. Hrs open: Box 717, Pickerington, OH, 43147-0717. Phone: (239) 877-4605. Licensee: In Phase Broadcasting Inc. ♦Peter L. Cea, pres.

Imperial

KADL(FM)— 2003: 102.9 mhz; 300 w. Ant 223 ft TL: N40 30 45 W101 38 39. Hrs open: Box 333, McCook, 69001. Phone: (308) 345-5400. Fax: (308) 345-4720. Licensee: Armada Media - McCook Inc. (acq 1-17-2007; grpsl). Format: Oldies. ♦David M. Stout, gen mgr.

Kearney

KGFW(AM)— 1927: 1340 khz; 1 kw-U. TL: N40 40 05 W99 04 52. Hrs open: 24 Box 666, 68848. Secondary address: 2223 Central Ave. 68847. Phone: (308) 237-2131. Fax: (308) 237-0312. E-mail: mail@kgfw.com Web Site: www.kgfw.com. Licensee: NRG License Sub. LLC. Group owner: Waitt Radio Inc. (acq 10-31-2005; grpsl). Population served: 50,000 Natl. Network: Westwood One. Rgnl. Network: Brownfield. Natl. Rep: Christal. Wire Svc: AP Format: News/talk. News staff: 2; News: 25 hrs wkly. Target aud: 25 plus; adults in central Nebraska. Spec prog: Farm 8 hrs, sports 10 hrs wkly. ♦Norman Waitt Jr, chmn; Mary Quass, pres; John McDonald, gen mgr.

KQKY(FM)—Co-owned with KGFW(AM). October 1979: 105.9 mhz; 97.6 kw. Ant 1,204 ft TL: N40 36 08 W98 50 21. Stereo. 24 Phone: (308) 236-6464. Web Site: www.kqky.com.200,000 Natl. Rep: Christal. Format: Top-40. Target aud: 18-49.

KKPR-FM—Listing follows KXPN(AM).

***KLPR(FM)—** Mar 8, 1968: 91.3 mhz; 1 kw. 100 ft TL: N40 42 30 W99 05 45. Stereo. Hrs opn: 6 AM-midnight Univ. of Nebraska at Kearney, 68849. Phone: (308) 865-8217. Phone: (308) 865-8216. Fax: (308) 865-8217. Licensee: University of Nebraska at Kearney. Population served: 30,000 Format: Jazz, new age, AOR. Spec prog: Class 18 hrs wkly. ♦Roy Hyatte, gen mgr.

KRNY(FM)— 1987: 102.3 mhz; 77.1 kw. Ant 1,086 ft TL: N40 36 08 W98 50 21. Stereo. Hrs opn: 24 Box 669, 68848. Secondary address: 2223 Central Ave. 68847. Phone: (308) 237-2131. Phone: (308) 236-8600. Fax: (308) 237-0312. E-mail: mail@krny.com Web Site: www.krny.com. Licensee: NRG License Sub. LLC. Group owner: Waitt Radio Inc. (acq 10-31-2005; grpsl). Population served: 200,000 Law Firm: Haley, Bader & Potts. Format: Hot country. News: 2 hrs wkly. Target aud: 25 plus. ♦John McDonald, gen mgr; Dirk Christensen, opns mgr.

KXPN(AM)— Dec 5, 1956: 1460 khz; 5 kw-D, 56 w-N. TL: N40 42 45 W99 10 15. Hrs open: 24 Box 130, 68848. Secondary address: 403 E. 25th St. 68848. Phone: (308) 236-9900. Fax: (308) 234-6781. E-mail: espn1460@charter.net Web Site: www.kkpr.com. Licensee: Platte River Radio Inc. (acq 1-1-94; $750,000 with co-located FM; FTR: 12-13-93). Natl. Network: ESPN Radio. Format: Sports. News staff: one; News: 2 hrs wkly. Target aud: 25-54; men. ♦David Oldfather, pres; Craig Eckert, exec VP & gen mgr; Dan Beck, opns mgr; Johnnie McCann, sls dir; Mike Cahill, news dir.

KKPR-FM—Co-owned with KXPN(AM). Nov 1, 1962: 98.9 mhz; 100 kw. Ant 700 ft TL: N40 48 53 W98 46 12. Stereo. 24 E-mail: generalmanager@kkpr.com Web Site: www.kkpr.com. Format: Oldies. News staff: one; News: 3 hrs wkly. Target aud: 35-64. ♦Dan Beck, chief of opns; Johnnie McCann, gen sls mgr.

Kimball

KIMB(AM)— 1958: 1260 khz; 1 kw-D, 112 w-N. TL: N41 15 42 W103 40 06. (CP: COL Ogallala. 50 kw-D, 110 w-N. TL: N41 04 30 W101 45 24). Hrs opn: 3205 W. North Front St., Grand Island, 68803. Phone: (308) 381-1430. Fax: (308) 382-6701. E-mail: info@gifamilyradio.com Licensee: Kimball Radio LLC. (group owner; (acq 4-2-2007; $300,000 with KYOY(FM) Kimball). Population served: 30,000 Format: News/talk. Target aud: General. ♦Alan Usher, VP & gen mgr.

KYOY(FM)— 1999: 100.1 mhz; 6 kw. Ant 328 ft TL: N41 11 36 W103 31 45. (CP: 26 kw horiz, ant 223 ft). Stereo. Hrs opn: 24 Box 532, Scottsbluff, 69363-0532. Phone: (308) 632-5667. Fax: (308) 635-1905. Licensee: Kimball Radio LLC. (group owner). (acq 4-2-2007; $300,000 with KIMB(AM) Kimball). Population served: 125,000 Natl. Network: AP Radio. Format: Oldies. News staff: 2. ♦Larry Swikard, gen mgr.

Lexington

***KLNE-FM—** May 4, 1990: 88.7 mhz; 43.8 kw. 938 ft TL: N40 23 05 W99 27 30. Stereo. Hrs opn: 24
Rebroadcasts KUCV(FM) Lincoln 100%.
1800 N. 33rd St., Lincoln, 68503-1409. Phone: (402) 472-3611. Fax: (402) 472-2403. E-mail: radio@netnebraska.org Web Site: netnebraska.org/radio. Licensee: Nebraska Educational Telecommunications Commission. Natl. Network: PRI, NPR. Rgnl. Network: Neb. Pub. Format: Classical, news & info. News staff: 3; News: 38 hrs wkly. General. ♦Ray Dilley, gen mgr; Nancy Finken, progmg dir; Bill Stibor, mus dir; Martin Wells, news dir; Jeff Smith, traf mgr.

KRVN(AM)— Feb 1, 1951: 880 khz; 50 kw-U, DA-N. TL: N40 31 03 W99 23 20. Hrs open: 24 Box 880, 1007 Plum Creek Pkwy., 68850-0880. Phone: (308) 324-2371. Fax: (308) 324-5786. E-mail: krvnam@krvn.com Web Site: www.krvn.com. Licensee: Nebraska Rural Radio Assn. (group owner) (acq 2-1-51). Natl. Rep: Katz Radio. Law Firm: Garvey Schubert & Barer. Wire Svc: AP Format: Country, news, farm. News staff: 4; News: 30 hrs wkly. Target aud: General; Nebraska farm/ranch families & consumers. Spec prog: Relg 12 hrs wkly. ♦Eric Brown, gen mgr; Ed Bennett, opns mgr; Dennis Waddle, gen sls mgr; Pam Snyder, prom dir & spec ev coord; Stafford Thompson, progmg dir; Frank Snyder, news dir; Vern Killion, engrg dir; Mike LePorte, farm dir.

KRVN-FM— Nov 1, 1962: 93.1 mhz; 100 kw. 890 ft TL: N40 41 48 W99 47 18. Stereo. 24 Web Site: www.krvnfm.com.125,000 Natl. Network: Westwood One. Natl. Rep: Katz Radio. Format: News, new country. News staff: one; News: 20 wkly. Contemp young adults. Spec prog: Farm 10 hrs wkly.

Lincoln

KBBK(FM)—Listing follows KLIN(AM).

KFOR(AM)—Listing follows KFRX(FM).

KFRX(FM)— May 2, 1965: 102.7 mhz; 100 kw. 500 ft TL: N40 49 12 W96 39 29. (CP: TL: N45 46 51 W96 22 52). Stereo. Hrs opn: 3800 Cornhusker Hwy., 68504-1533. Phone: (402) 466-1234. Fax: (402) 467-4095. E-mail: kfrx@threeeagles.com Web Site: www.kfrxfm.com. Licensee: Three Eagles Communications Inc. Group owner: Three Eagles Communications (acq 1996; $5.3 million with co-located AM). Population served: 239,000 Natl. Rep: McGavren Guild. Law Firm: Tierney & Swift. Format: CHR. Target aud: 18-49; women. ♦Roland Johnson, CEO; Gary Buchanan, COO; Cindy Harris, CFO; James Keck, gen mgr; Matt McKay, stn mgr & progmg dir; Mark Taylor, opns mgr; Janessa Wiegand, gen sls mgr; Vicki Marker, prom dir; Adam Michaels, mus dir; Bob Cook, chief of engrg; Terri Hutchinson, traf mgr.

***KLCV(FM)—** 1996: 88.5 mhz; 4.7 kw vert. Ant 315 ft TL: N40 55 49 W96 32 42. Hrs open: 8 AM-5 PM (M-F) 233 S. 13th St., Suite 1520, 68508. Phone: (402) 465-8850. Fax: (402) 465-8852. E-mail: klcv1@juno.com Licensee: Community Broadcasting Inc. Population served: 700,000 Format: Relg, Christian talk. Target aud: 25 plus; Christian families. ♦Richard Bott Sr., pres; Tom Millett, gen mgr.

KLIN(AM)— August 1947: 1400 khz; 1 kw-U. TL: N40 50 54 W96 40 29. Hrs open: 24 4343 O St., 68510. Phone: (402) 475-4567. Fax: (402) 479-1411. Web Site: www.klin.com. Licensee: Monterey Licenses LLC. Group owner: Triad Broadcasting Co. LLC (acq 3-31-2000; grpsl). Population served: 200,000 Natl. Network: Fox News Radio. Format: News/talk. News staff: 4; News: 80 hrs wkly. Target aud: 35-64; upper income, business owner, educated with high disposable income. ♦David Benjamin, pres; Mark Halverson, gen mgr; John Bishop, progmg dir; Greg Jackson, news dir & pub affrs dir; Bill Frost, chief of engrg.

KBBK(FM)—Co-owned with KLIN(AM). Sept 1, 1968: 107.3 mhz; 100 kw. 551 ft TL: N40 43 38 W96 36 49. Stereo. 24 Web Site: www.b1073.com.330000 Format: Adult contemp. Target aud: 25-54; middle-to-upper income, households, in-office & in-store lstng. ♦T. Pat Miller, progmg dir.

KLMS(AM)—Listing follows KRKR(FM).

KLMY(FM)— Feb 23, 1973: 106.3 mhz; 3 kw. 213 ft TL: N40 48 48 W96 42 25. (CP: 50 kw, ant 190 ft.). Stereo. Hrs opn: 24 4630 Antelope Creek Rd., Suite 200, 68506. Phone: (402) 484-8000. Fax: (402) 483-9138. E-mail: sonny@my1063.com Web Site: www.my1063.com. Licensee: Capstar TX L.P. Group owner: Clear Channel Communications Inc. (acq 8-30-00; grpsl). Population served: 172,400 Format: AOR/current rock. News: one hr wkly. Target aud: 35-64. ♦Julie Gade, gen mgr; Michelle Hay, gen sls mgr; Sonny Valentine, progmg dir; Tim Cawley, news dir; Dave Agnew, chief of engrg; Angela Zoucha, traf mgr.

KLNC(FM)— 1992: 105.3 mhz; 3 kw. Ant 328 ft TL: N40 49 12 W96 39 29. Hrs open: 4343 O St., 68510. Phone: (402) 475-4567. Fax: (402) 479-1411. Licensee: Monterey Licenses LLC. Group owner: Triad Broadcasting Co. LLC (acq 3-31-2000; grpsl). Format: Oldies. ♦David Benjamin, CEO; Mark Halverson, VP & gen mgr; J. Pat Miller, opns mgr; Ami Graham, gen sls mgr; E.J. Marshall, progmg dir; Steve Looney, chief of engrg.

KLTQ(FM)— June 22, 1958: 101.9 mhz; 100 kw. Ant 1,132 ft TL: N40 47 09 W96 23 07. Stereo. Hrs opn: 24 5011 Capitol Ave., Omaha, 68132. Phone: (402) 342-2000. Fax: (402) 346-5748. Web Site: www.lite1019.com. Licensee: Waitt Omaha LLC. (group owner; (acq 1-7-2002; grpsl). Population served: 2,120,000 Natl. Rep: Katz Radio. Format: Adult contemp. Target aud: 25-54; general. ♦Jim McKernan, gen mgr; Mark Todd, opns mgr; Rhonda Gerrard, sls dir; Sam Coughlin, gen sls mgr; Chris Pflaum, prom dir; Billy Shears, progmg dir; Cynthia Wallace, traf mgr.

KRKR(FM)— Mar 6, 1975: 95.1 mhz; 50 kw. 287 ft TL: N40 58 49 W96 41 45. Stereo. Hrs opn: 24 3800 Cornhusker Hwy., 68504. Phone: (402) 466-1234. Fax: (402) 467-4095. Web Site: www.95rocklincoln.com. Licensee: Three Eagles Communications Co. Group owner: Three Eagles Communications (acq 1996; grpsl). Population served: 222,900 Law Firm: Tierney & Swift. Format: Classic rock. News staff: one; News: 2 hrs wkly. Target aud: 25-54. ♦Roland Johnson, CEO; Gary Buchanan, COO; Cindy Harris, CFO;

Mark Taylor, opns mgr; Vicki Marker, prom dir; Scott Kaye, progmg dir; Dale Johnson, news dir; Bob Cook, chief of engrg; Teri Hutchinson, traf mgr.

KLMS(AM)—Co-owned with KRKR(FM). October 1949: 1480 khz; 5 kw-D, 1 kw-N, DA-2. TL: N40 47 47 W96 34 56. Stereo. Web Site: www.espn1480.com.222,900 Format: Sports. ♦Bill Doleman, progmg dir; Terri Hutchinson, traf mgr; Dale Johnson, local news ed.

***KRNU(FM)**— Feb 23, 1970: 90.3 mhz; 100 w. 180 ft TL: N40 49 11 W96 42 11. Stereo. Hrs opn: 24 147 Anderson Hall, Univ. of Nebraska, 68588-0466. Phone: (402) 472-3054. Fax: (402) 472-8403. E-mail: krnu@unl.edu Web Site: www.krnu.unl.edu. Licensee: University of Nebraska. Population served: 200,000 Natl. Network: ABC. Law Firm: Dow, Lohnes & Albertson. Format: Alternative. News: 10 hrs wkly. Spec prog: Black 2 hrs, blues 2 hrs, folk 2 hrs, gospel 2hrs, Jazz 2 hrs, Sp 2 hrs, sports 2 hrs wkly. ♦Rick Alloway, gen mgr, opns mgr, gen sls mgr & progmg dir; Trina Creighton, news dir; Vance Payne, chief of engrg. Co-owned TV: *KUON-TV affil.

KTGL(FM)—Beatrice, Nov 26, 1962: 92.9 mhz; 100 kw. 809 ft TL: N40 31 06 W96 46 07. Stereo. Hrs opn: 24 4630 Antelope Creek Rd., Suite 200, 68506. Phone: (402) 484-8000. Phone: (402) 489-9291. Fax: (402) 489-9607. E-mail: juliegade@clearchannel.com Web Site: www.ktgl.com. Licensee: Capstar TX L.P. Group owner: Clear Channel Communications Inc. (acq 8-30-00; grpsl). Population served: 210,000 Format: Classic rock. News staff: one; News: 3 hrs wkly. Target aud: 18-49. ♦Julie Gade, gen mgr; Julie Broman, sls dir; Joe Skare, progmg dir; Tim Cawley, news dir; Eric Taylor, pub affrs dir; Mike Elliott, chief of engrg.

***KUCV(FM)**— Jan 1, 1968: 91.1 mhz; 19.5 kw horiz, 100 kw vert. Ant 689 ft TL: N40 31 06 W96 46 06. Stereo. Hrs opn: 24 1800 N. 33rd St., 68503-1409. Phone: (402) 472-3611. Fax: (402) 472-2403. E-mail: radio@netnebraska.org Web Site: netnebraska.org/radio. Licensee: Nebraska Educational Telecommunications Commission. (acq 8-88). Population served: 1,100,000 Natl. Network: NPR, PRI. Rgnl. Network: Neb. Pub. Format: Classical, news. News staff: 3; News: 38 hrs wkly. Target aud: General. ♦Ray Dilley, gen mgr; Nancy Finken, progmg dir; Bill Stibor, mus dir; Martin Wells, news dir; Jeff Smith, traf mgr.

KZKX(FM)—See Seward

***KZUM(FM)**— 1978: 89.3 mhz; 1.5 kw. 174 ft TL: N40 48 47 W96 42 24. Stereo. Hrs opn: 6 AM-2 AM 941 O St., Suite 1025, 68508-3608. Phone: (402) 474-5086. Fax: (402) 474-5091. E-mail: kzumradio@aol.com Web Site: www.kzum.org. Licensee: Sunrise Communications Inc. Format: Div, jazz, urban contemp. News: 11 hrs wkly. Target aud: General; the unserved & underserved population. Spec prog: Sp 4 hrs, rock/progsv 15 hrs, new age 8 hrs, blues 13 hrs, folk 8 hrs, gospel 3 hrs wkly. ♦Dennis Svoboda, dev dir; Rich Hoover, progmg dir.

McCook

KBRL(AM)— Sept 26, 1947: 1300 khz; 5 kw-D, DA. TL: N40 11 31 W100 39 06. Hrs open: Box 333, 69001. Secondary address: 1811 W. O St. 699001. Phone: (308) 345-5400. Fax: (308) 345-4720. E-mail: dave@kicx.net Web Site: www.kicx.net. Licensee: Armada Media - McCook Inc. (group owner; (acq 1-17-2007; grpsl). Population served: 40,000 Format: Oldies, full service. Target aud: 35-64. ♦Jim Coursolle, pres; David Stout, gen mgr; Connie Stout, gen sls mgr & adv dir; Rich Barnett, news dir; Ron Fritz, chief of engrg.

KICX-FM—Co-owned with KBRL(AM). Jan 31, 1979: 96.1 mhz; 55 kw. Ant 318 ft TL: N40 10 19 W100 41 05. Stereo. 24 Web Site: www.kicx.net.40,000 Format: Adult contemp. News staff: one. Target aud: 25-64. ♦Connie Stout, sls VP.

KIOD(FM)— May 1, 1981: 105.3 mhz; 100 kw. Ant 591 ft TL: N40 11 27 W100 48 29. Stereo. Hrs opn: 24 Box 939, 69001. Secondary address: 106 W. 8th St. 69001. Phone: (308) 345-1981. Fax: (308) 345-7202. Web Site: mccookfamilyradio.gifamilyradio.com /Coyote_Country/coyote.htm. Licensee: Legacy Communications LLC. (acq 10-13-2005; $1.3 million with KSWN(FM) McCook). Population served: 40,000 Rgnl. Network: Agri-Net. Law Firm: Larry D. Perry. Format: Hot country. News staff: one; News: 5 hrs wkly. Target aud: 25-54. Spec prog: Sports 10 hrs, farm 6 hrs wkly. ♦Alan Usher, gen mgr; Jesse Stevens, progmg dir & progmg mgr; Derek Beck, mus dir; Ann Doyle, traf mgr.

KNAX(AM)—Not on air, target date: unknown: 700 khz; 250 w-U, DA-N. TL: N40 16 00 W100 34 09. Hrs open: Box 333, 69001. Phone: (308) 345-5400. Fax: (308) 345-4720. Licensee: McCook Radio Group L.L.C. (group owner). ♦David Stout, gen mgr & opns mgr; Connie Stout, gen sls mgr; Rich Barnett, news dir.

***KNGN(AM)**— June 23, 1961: 1360 khz; 1 kw-D. TL: N40 11 45 W100 41 57. Hrs open: 6 AM-2 hrs past sunset 38005 Road 717, 69001-7217. Phone: (308) 345-2006. Fax: (308) 345-2052. E-mail: goodnews@mccooknet.com Web Site: www.kngn.org. Licensee: Kansas Nebraska Good News Broadcasting Corp. (acq 9-5-01). Format: Relg. News: 7 hrs wkly. Target aud: 35 plus; family oriented. ♦Greg Stuekwiseh, pres; Mike Nielsen, gen mgr.

KRKU(FM)— 2000: 98.5 mhz; 55 kw. Ant 361 ft TL: N40 10 19 W100 41 05. (CP: COL Maxwell. Ant 292 ft. TL: N41 07 51 W100 31 12). Hrs opn: 24 Box 333, 69001. Secondary address: 1811 W. O St. 69001. Phone: (308) 345-5400. Fax: (308) 345-4720. E-mail: dave@kicx.net Web Site: www.kicx.net. Licensee: Armada Media - McCook Inc. (group owner; (acq 1-17-2007; grpsl). Population served: 40,000 Format: Classic rock. Target aud: 18-54. ♦David Stout, gen mgr & opns mgr; Connie Stout, gen sls mgr; Rich Barnett, news dir.

KSWN(FM)— Sept 17, 1998: 93.9 mhz; 50 kw. Ant 492 ft TL: N40 11 27 W100 48 29. Stereo. Hrs opn: 24 Box 939, 106 W. 8th St., 69001-0218. Phone: (308) 345-1100. Fax: (308) 345-7202. E-mail: jay@coyote105.com Web Site: mccookfamilyradio.gifamilyradio.com/us939.htm. Licensee: Legacy Communications LLC. (acq 10-13-2005; $1.3 million with KIOD(FM) McCook). Population served: 72,000 Natl. Network: ESPN Radio. Law Firm: Larry D. Perry. Format: Adult contemp, sports. Target aud: 25-54. ♦Jay D. Austin, gen mgr; Eileen G. Austin, opns VP; Jay Austin, gen sls mgr; Jesse Stevens, progmg mgr.

KZMC(FM)— Oct 25, 2006: 102.1 mhz; 100 kw. Ant 590 ft TL: N40 11 27 W100 48 29. Hrs open: Box 939, 69001-0939. Secondary address: 106 W. 8th St. 69001-3508. Phone: (308) 345-1981. Fax: (308) 345-7202. E-mail: info@hometownfamilyradio.com Web Site: www.hometownfamilyradio.com/Z/index.php. Licensee: Legacy Communications LLC. Format: Rock. ♦Alan Usher, VP & gen mgr; Jesse Stevens, opns mgr.

Merriman

***KRNE-FM**— Aug 29, 1991: 91.5 mhz; 92 kw. 964 ft TL: N42 40 38 W101 42 36. Stereo. Hrs opn: 24
Rebroadcasts KUCV(FM) Lincoln 100%.
1800 N. 33rd St., Lincoln, 68503-1409. Phone: (402) 472-3611. Fax: (402) 472-2403. E-mail: radio@netnebraska.org Web Site: netnebraska.org/radio. Licensee: Nebraska Educational Telecommunications Commission. Natl. Network: NPR, PRI. Format: Classical, news. News staff: 3; News: 38 hrs wkly. Target aud: General. ♦Ray Dilley, gen mgr; Nancy Finken, progmg dir; Bill Stibor, mus dir; Martin Wells, news dir; Jeff Smith, traf mgr.

Milford

KFGE(FM)— 1996: 98.1 mhz; 100 kw. 981 ft TL: N40 51 52 W97 16 14. Hrs open: 4343 O St., Lincoln, 68510. Phone: (402) 475-4567. Fax: (402) 479-1411. Licensee: Monterey Licenses LLC. Group owner: Triad Broadcasting Co. LLC (acq 3-31-00; grpsl). Format: Country. ♦David Benjamin, CEO; Mark Halverson, VP & gen mgr; J. Pat Miller, opns mgr; Ami Graham, gen sls mgr; Steve Albertson, progmg dir; Steve Looney, chief of engrg.

Minatare

KHYY(FM)—Not on air, target date: unknown: 106.9 mhz; 50 kw. Ant 474 ft TL: N41 54 28 W103 28 34. Hrs open: College Creek Media LLC, 980 N. Michigan Ave., Suite 1880, Chicago, IL, 60611. Phone: (312) 204-9900. Licensee: College Creek Media LLC. ♦Neal J. Robinson, pres.

Mitchell

KETT(FM)—Not on air, target date: unknown: 99.3 mhz; 50 kw. Ant 474 ft TL: N41 54 28 W103 28 34. Hrs open: College Creek Media LLC, 980 N. Michigan Ave., Suite 1880, Chicago, IL, 60611. Phone: (312) 204-9900. Licensee: College Creek Media LLC. ♦Neal J. Robinson, pres.

Nebraska City

KBBX-FM— Feb 1, 1995: 97.7 mhz; 100 kw. 981 ft TL: N40 53 31 W96 09 10. Hrs open: 24 5030 N. 72nd St., Omaha, 68134. Phone: (402) 592-5300. Fax: (402) 592-6605. Licensee: Connoisseur Media of Omaha LLC. Group owner: Journal Broadcast Group Inc. (acq 9-25-2006; $7.5 million). Population served: 942,906 Natl. Network: ABC. Rgnl. Network: S.W. Agri-Radio. Law Firm: Rosenman & Colin. Format: Sp. Target aud: 25-54. ♦Steve Wexler, gen mgr; Tom Land, opns dir; Jim Timm, gen sls mgr; RosAnna Salcido, mktg mgr; Kurt Owens, progmg dir; Bill Jensen, news dir.

KNCY(AM)— June 29, 1959: 1600 khz; 500 w-D, 31 w-N, DA-2. TL: N40 40 27 W95 53 08. Hrs open: 24 Box 278, 814 Central Ave., 68410. Phone: (402) 873-3348. Fax: (402) 873-7882. E-mail: kncy@kncycountry.com Web Site: www.kncycountry.com. Licensee: Arbor Day Broadcasting Inc. (acq 2-3-2003; $600,000 with KNCY-FM Auburn). Population served: 400,000 Natl. Network: ABC. Format: Variety. News staff: 2; News: 21 hrs wkly. Target aud: 18-80; local residents, farmers, business owners, workers, students. Spec prog: Farm 3 hrs, sports 6 hrs wkly. ♦Chris Yates, gen sls mgr; Scott Kooistra, gen mgr & news dir.

Norfolk

KEXL(FM)—Listing follows WJAG(AM).

KNEN(FM)— Apr 6, 1979: 94.7 mhz; 100 kw. Ant 539 ft TL: N41 55 28 W97 36 22. Stereo. Hrs opn: Box 937, 68702. Secondary address: 300 Madison Ave. 68701. Phone: (402) 379-3300. Fax: (402) 379-3008. E-mail: knen@waittradio.com Licensee: NRG License Sub, LLC. (acq 4-30-2006; $1.8 million). Population served: 150,000 Format: Adult contemp. News staff: 2; News: 15 hrs wkly. Target aud: 25-54; young to middle-aged. Spec prog: Farm 10 hrs wkly. ♦Neil Lipetzky, gen mgr; Steve Farlee, gen sls mgr; Kevin Rahfelat, progmg dir; William Seifert, news dir.

***KPNO(FM)**— Sept 23, 1992: 90.9 mhz; 50 kw. 351 ft TL: N42 06 16 W97 20 11. Stereo. Hrs opn: 24 109 S. 2nd St., 68701-5327. Phone: (402) 379-3677. Fax: (402) 379-3662. E-mail: kpno@conpoint.com Web Site: www.kpno.org. Licensee: The Praise Network Inc. Population served: 200,000 Natl. Network: Moody, USA. Format: Relg, inspirational. News: 14 hrs wkly. Target aud: 25-54; family-oriented adults. ♦Herb Roszhart Jr., CEO; Jon Shipman, gen mgr.

***KXNE-FM**— May 29, 1990: 89.3 mhz; 42.3 kw. 984 ft TL: N42 14 15 W97 16 41. Hrs open: 24
Rebroadcasts KUCV(FM) Lincoln 100%.
1800 N. 33rd St., Lincoln, 68503-1409. Phone: (402) 472-3611. Fax: (402) 472-2403. E-mail: radio@netnebraska.org Web Site: netnebraska.org/radio. Licensee: Nebraska Educational Telecommunications Commission. Natl. Network: PRI, NPR. Format: Classical, news. News staff: 3; News: 38 hrs wkly. Target aud: General. ♦Ray Dilley, gen mgr; Nancy Finken, progmg dir; Bill Stibor, mus dir; Martin Wells, news dir; Jeff Smith, traf mgr.

WJAG(AM)— July 27, 1922: 780 khz; 1 kw-U (L-WBBM). TL: N42 01 54 W97 29 47. Stereo. Hrs opn: Sunrise-sunset Box 789, 68702. Secondary address: 309 Braasch Ave. 68701. Phone: (402) 371-0780. Fax: (402) 371-6303. E-mail: wjagkexl@wjag.com Web Site: www.wjag.com. Licensee: WJAG Inc. Population served: 23,435 Natl. Network: Fox News Radio. Law Firm: Fletcher, Heald & Hildreth. Wire Svc: AP Format: News/talk. News staff: 2; News: 10 hrs wkly. Target aud: 35-64; info-oriented. ♦Jerry Huse, pres; Robert G. Thomas, VP & gen mgr; Jeffrey Steffen, opns mgr & progmg dir; Brad Hughes, sls dir, gen sls mgr, mktg dir & prom dir; Todd Michaels, mus dir; Jim Curry, news dir; J. Alan Johnson, pub affrs dir; Susan Risinger, farm dir; Joe Tjaden, sports cmtr.

KEXL(FM)—Co-owned with WJAG(AM). Aug 1, 1971: 106.7 mhz; 100 kw. 1,027 ft TL: N41 55 59 W97 40 49. Stereo. 24 E-mail: wjagkexl@kexl.com Web Site: www.kexl.com.75,000 Format: Adult contemp. News staff: 2; News: 6 hrs wkly. Target aud: 18-49; full service FM adults. ♦Todd Michaels, mus dir; Denise Keikofski, traf mgr; Susan Risinger, farm dir; Joe Tjaden, sports cmtr.

North Platte

KELN(FM)—Listing follows KOOQ(AM).

***KJLT(AM)**— July 1, 1957: 970 khz; 5 kw-D, 55 w-N. TL: N41 09 30 W100 52 36. (CP: TL: N41 09 36 W100 52 42). Hrs opn: Sunrise-sunset Box 709, 69103. Secondary address: 201 S. Bailey Ave. 69101. Phone: (308) 532-5515. E-mail: kjlt@kjlt.org Web Site: www.kjlt.org. Licensee: Tri-State Broadcasting Assn. Inc. (acq 7-1-57). Population served: 100,000 Natl. Network: Moody, Salem Radio Network. Wire Svc: NOAA Weather Wire Svc: AP Format: Christian educ. Target aud: General; families. Spec prog: Sp one hr wkly. ♦John L. Townsend, pres, gen mgr & progmg dir; Gary Hofer, chief of engrg.

KJLT-FM— Sept 24, 1979: 94.9 mhz; 100 kw. 652 ft TL: N40 59 49 W100 52 47. Stereo. 24 Web Site: www.kjlt.org. (Acq 3-22-90; $85,000; FTR: 4-16-90). Natl. Network: Moody, Salem Radio Network. Format: Inspirational, gospel, adult contemp. Target aud: General; young adults.

KODY(AM)— July 5, 1930: 1240 khz; 1 kw-U. TL: N41 09 14 W100 46 23. Hrs open: 305 E. Fourth St., 69101. Phone: (308) 532-3344. Fax: (308) 534-6651. Web Site: www.kodyradio.com. Licensee: NRG License Sub. LLC. Group owner: Waitt Radio Inc. (acq 10-31-2005; grpsl). Population served: 38,000 Natl. Network: CBS, Moody, Westwood One. Natl. Rep: Katz Radio. Format: News/talk. Target aud: 25 plus; middle to upper income. ♦Rob Mandeville, gen mgr; Rob Mandeville, gen sls mgr; George Keltz, news dir; Tony Lama, opns mgr, progmg dir & chief of engrg; Lisa Arent, traf mgr.

KXNP(FM)—Co-owned with KODY(AM). June 7, 1982: 103.5 mhz; 100 kw. 479 ft TL: N41 12 49 W100 43 48. Stereo. 24 Web Site: www.kx104.com.125,000 Natl. Network: Jones Radio Networks. Format: Contemp country. ♦Lisa Arent, traf mgr.

KOOQ(AM)— January 1966: 1410 khz; 5 kw-D, 1 kw-N, DA-N. TL: N41 10 30 W100 45 07. Hrs open: 24 Box 248, 69103. Secondary address: 1301 E. 4th St. 69103. Phone: (308) 532-1120. Fax: (308) 532-0458. E-mail: chuck.schwartz@eagleradio.net Web Site: oldiesradio.com. Licensee: Eagle Communications Inc. Group owner: Eagle Communications Group. Population served: 25,500 Wire Svc: AP Format: Oldies. News staff: one; News: 8 hrs wkly. Target aud: 30-55; baby boomers. ♦Gary Shorman, pres & gen sls mgr; Chuck Schwartz, gen mgr; Jerome Gilg, gen sls mgr; Dianne Morales, prom dir; David Fudge, progmg dir & chief of engrg; Heather James, news cmtr.

KELN(FM)—Co-owned with KOOQ(AM). February 1979: 97.1 mhz; 100 kw. 458 ft TL: N41 14 20 W100 41 43. Stereo. Web Site: mix97one.com.40,000 Natl. Network: Westwood One. Format: Adult contemp. News staff: one; News: one hr wkly. Target aud: 21-44; Young adults.

***KPNE-FM**— July 1, 1991: 91.7 mhz; 16.5 kw horiz, 81 kw vert. 843 ft TL: N41 01 21 W101 09 13. Stereo. Hrs opn: 24
Rebroadcasts KUCV(FM) Lincoln 100%.
Box 83111, Lincoln, 68501. Secondary address: 1800 N. 33rd St., Lincoln 68583. Phone: (402) 472-3611. Fax: (402) 472-2403. E-mail: radio@netnebraska.org Web Site: netnebraska.org. Licensee: Nebraska Educational Telecommunications Commission. Natl. Network: PRI, NPR. Format: Class, news. ♦Rod Bates, gen mgr; Nancy Finken, progmg dir; Martin Wells, news dir; Michael Winkle, mktg.

Ogallala

KMCX(FM)— 1975: 106.5 mhz; 100 kw. 300 ft TL: N41 08 02 W101 41 42. Stereo. Hrs opn: 24 Box 509, 113 W. 4th St., 69153. Phone: (308) 284-3633. Fax: (308) 284-3517. Web Site: www.4koga.com. Licensee: Capstar TX L.P. Group owner: Clear Channel Communications Inc. (acq 8-30-2000; grpsl). Population served: 40,000 Format: Country. News staff: one; News: 10 hrs wkly. Target aud: 25-54; general. Spec prog: Farm 2 hrs wkly. ♦Katrina Twomey, gen mgr; Corey Andersen, opns dir; John Brandt, gen sls mgr; Dave Geho, chief of engrg; Susan Jones, traf mgr.

KOGA(AM)— Jan 23, 1955: 930 khz; 5 kw-U, DA-2. TL: N41 08 32 W101 42 48. Stereo. Hrs opn: Box 509, 69153. Secondary address: 113 W. 4th St. 69153. Phone: (308) 284-3633. Fax: (308) 284-3517. E-mail: thelake@lakemac.net Web Site: www.4koga.com. Licensee: Capstar TX L.P. Group owner: Clear Channel Communications Inc. (acq 8-30-00; grpsl). Population served: 85,000 Law Firm: Fletcher, Heald & Hildreth. Format: Oldies. Spec prog: Farm 10 hrs wkly. ♦Katrina Twomby, gen mgr; Corey Anderson, opns mgr; John Brandt, sls dir; Tracey Knapp, progmg dir; Greg Holl, news dir; Dave Geho, chief of engrg.

KOGA-FM— November 1978: 99.7 mhz; 100 kw. 805 ft TL: N41 03 50 W101 20 16. Stereo. 24 Web Site: 997thelake.com.100,000 Format: Adult rock. ♦Greg Hill, local news ed.

Omaha

KCRO(AM)— March 1922: 660 khz; 1 kw-D, 54 w-N. TL: N41 18 47 W96 00 36. Hrs open: 11717 Burt St, Suite 202, 68154. Phone: (402) 422-1600. Phone: (402) 422-1601. Fax: (402) 422-1602. E-mail: kcro@kcro.com Web Site: www.kcro.com. Licensee: Salem Media of Illinois LLC. (acq 9-1-2005; $3.1 million). Population served: 1,101,000 Natl. Network: Salem Radio Network. Format: Christian talk. Target aud: 25-54. ♦Johnny Andrews, gen mgr; Mike Shane, opns mgr; Greg Vogt, gen sls mgr; Jim Leedham, chief of engrg.

KEZO-FM— May 15, 1961: 92.3 mhz; 100 kw. Ant 1,250 ft TL: N41 18 40 W96 01 37. Stereo. Hrs opn: 5030 N. 72nd St., 68134-2363. Phone: (402) 592-5300. Fax: (402) 592-6605. Web Site: www.z92.com. Licensee: Journal Broadcast Corp. (acq 11-29-94; $9 million with co-located AM; FTR: 1-16-95) Population served: 347,328 Format: Rock/AOR. ♦Steve Wexler, gen mgr; James Barton, prom dir; Lester St. James, progmg dir; Susie Copenhaver, traf mgr.

KFAB(AM)— 1924: 1110 khz; 50 kw-U, DA-N. TL: N41 07 11 W96 00 06. Stereo. Hrs opn: 24 5010 Underwood Ave., 68132. Phone: (402) 561-2000. Fax: (402) 556-8937. Web Site: www.kfab.org. Licensee: Capstar TX L.P. Group owner: Clear Channel Communications Inc. (acq 8-30-00; grpsl). Population served: 347,328 Natl. Network: ABC. Natl. Rep: Christal. Format: News/talk. News staff: 3; News: 6 hrs wkly. Target aud: 35-64. Spec prog: Farm 5 hrs wkly. ♦Donna Baker, gen mgr; Jim Steel, opns mgr; Rhonda Gerrard, gen sls mgr; Kevin Simonson, prom dir; Gary Sadlemyer, progmg dir; Tom Stanton, news dir; Steve George, chief of engrg; Sarah McCabe, traf mgr; Barbara Hill, news rptr; Jim Rose, sports cmtr.

KGOR(FM)—Co-owned with KFAB(AM). 1959: 99.9 mhz; 110 kw. Ant 1,214 ft TL: N41 18 29 W96 01 36. Stereo. 24 Web Site: www.kgor.com. Format: Oldies. Target aud: 35-54. ♦Todd McCarty, sls dir; Drew Bentley, progmg dir; Sara McCabe, traf mgr.

***KGBI-FM**— May 17, 1966: 100.7 mhz; 100 kw. Ant 1,014 ft TL: N41 18 40 W96 01 37. Stereo. Hrs opn: 24 11717 Burt St., Suite 202, 68154-1500. Phone: (402) 422-1600. Fax: (402) 422-1602. E-mail: kgbi@kgbifm.com Web Site: www.kgbifm.com. Licensee: Pennsylvania Media Associates Inc. (acq 1-31-2005; $8 million). Population served: 1,480,000 Format: Contemp Christian. News staff: one; News: 28 hrs wkly. Target aud: 25-54; conservative, Evangelical. ♦Johnny Andrews, gen mgr; Greg Vogt, gen sls mgr; Melody Miller, progmg dir; Jim Leedham, chief of engrg.

***KIOS-FM**— September 1969: 91.5 mhz; 55 kw. 554 ft TL: N41 17 15 W95 59 37. Stereo. Hrs opn: 3230 Burt St., 68131. Phone: (402) 557-2777. Fax: (402) 557-2559. E-mail: edward.mcgrath@ops.org Web Site: www.kios.org. Licensee: Douglas County School District 001. Population served: 347,328 Natl. Network: NPR, PRI. Law Firm: Cohn & Marks. Format: Class, educ, news & jazz. Spec prog: Black 2 hrs, jazz 10 hrs wkly. ♦Wilson W. Perry, stn mgr; Edward McGrath, dev dir & sls dir.

KKAR(AM)— March 1925: 1290 khz; 5 kw-U, DA-N. TL: N41 11 20 W96 00 21. Hrs open: 24 5011 Capitol Ave., 68132. Phone: (402) 342-2000. Fax: (402) 346-5748. Web Site: www.1290kkar.com. Licensee: Waitt Omaha LLC. (group owner; (acq 1-7-2002; grpsl). Population served: 346,929 Natl. Network: ABC, Fox News Radio, Jones Radio Networks, Premiere Radio Networks, Talk Radio Network. Natl. Rep: Katz Radio. Format: News/talk. Target aud: General. ♦Jim McKernan, gen mgr; Mark Todd, opns mgr; Rhonda Gerrard, sls dir; Neil Nelkin, progmg dir; Terry Leahy, pres & news dir; Darwin Stinton, chief of engrg; Lori Storz, traf mgr.

KKCD(FM)— Aug 11, 1990: 105.9 mhz; 50 kw. 479 ft TL: N41 18 16 W96 01 41. Hrs open: 24 5030 N. 72nd St., 68134. Phone: (402) 592-5300. Fax: (402) 331-1348. Web Site: www.cd1059.com. Licensee: Journal Broadcast Corp. Group owner: Journal Broadcast Group Inc. (acq 2-95; $3.55 million; FTR: 3-13-95). Population served: 800,000 Natl. Network: AP Network News. Law Firm: Leventhal, Senter & Lerman. Format: Classic Rock. News staff: one; News: 10 hrs wkly. Target aud: 25-54. Spec prog: Jazz 4 hrs, blues one hr, reggae one hr wkly. ♦Steve F. Wexler, VP & gen mgr; Tom Land, opns dir; Mike Stodden, gen sls mgr; Kurt Owens, progmg dir; Bill Jensen, news dir & traf mgr; John Gaeta, chief of engrg.

KOMJ(AM)— March 1942: 1490 khz; 1 kw-U. TL: N41 14 06 W95 57 57. Stereo. Hrs opn: 24 5030 N. 72nd St., 68134. Phone: (402) 592-5300. Fax: (402) 592-9434. Licensee: Cochise Broadcasting LLC. Group owner: Journal Broadcast Group Inc. (acq 3-27-2007; $500,000). Format: Adult standards. News staff: one. Target aud: 18-49. ♦Tom Land, opns dir & opns mgr; Kathy Hedstrom, gen sls mgr; Heath Hedstrom, prom dir & prom mgr; Kurt Owens, progmg dir; Bill Jensen, news dir; John Gaeta, chief of engrg; Cheryl Brye, traf mgr.

KOTK(AM)— Mar 2, 1957: 1420 khz; 1 kw-D, 330 w-N, DA-2. TL: N41 11 59 W95 54 34. Hrs open: 11717 Burt St., Suite 202, 68154-1500. Phone: (402) 422-1600. Fax: (402) 422-1602. Licensee: Pennsylvania Media Associates Inc. Group owner: Journal Broadcast Group Inc. (acq 12-7-2005; $900,000). Population served: 347,328 Natl. Network: Salem Radio Network. Format: News/talk. ♦Johnny Andrews, gen mgr.

KQBW(FM)— Oct 21, 1983: 96.1 mhz; 100 kw. Ant 1,414 ft TL: N41 04 14 W96 13 33. Stereo. Hrs opn: 24 5010 Underwood Avenue, 68132. Phone: (402) 561-2000. Fax: (402) 558-3036. Web Site: www.961thebrew.com. Licensee: Clear Channel Broadcasting Licenses Inc. (acq 10-9-2003; $10.5 million). Population served: 537,700 Format: Classic rock. News staff: one; News: one hr wkly. ♦Donna Baker, gen mgr.

KQCH(FM)—Listing follows KXSP(AM).

KSRZ(FM)— May 12, 1972: 104.5 mhz; 100 kw. 1,040 ft TL: N41 18 25 W96 01 37. Stereo. Hrs opn: 5030 N. 72nd St., 68134. Phone: (402) 592-5300. Fax: (402) 592-6605. Web Site: www.104star.com. Licensee: Journal Broadcast Corp. (acq 1-98; $5.475 million with co-located AM). Population served: 500,000 Format: Hot adult contemp. ♦Steve Wexler, gen mgr; Jill Butler, stn mgr & gen sls mgr; Jim Timm, stn mgr & sls dir; Tom Land, opns dir; James Barton, prom dir; Darla Thomas, progmg dir; Dave Swan, mus dir; John Gaeta, chief of engrg; Kathi Knutson, traf mgr.

***KVNO(FM)**— Aug 27, 1972: 90.7 mhz; 8.9 kw. Ant 646 ft TL: N41 18 25 W96 01 37. Stereo. Hrs opn: 24 Engineering 200, 60th & Dodge, 68182. Phone: (402) 559-5866. Fax: (402) 554-2440. Web Site: www.kvno.org. Licensee: University of Nebraska Board of Regents. Population served: 750,000 Format: Class. News staff: one; News: 2.5 hrs wkly. Target aud: General. ♦Debra Aliano, gen mgr; Joe Toppi, opns mgr; Marzia Puccioni, mktg mgr; Mike Hagstrom, progmg dir; Kristi Enyart, mus dir; Cheril Lewis, news dir & pub affrs dir; Frank Vacek, engrg mgr & chief of engrg.

***KVSS(FM)**— Jan 9, 1999: 88.9 mhz; 1 kw. 380 ft TL: N41 18 40 W96 01 37. Hrs open: 24 Secondary address: 5829 N. 60th St. 68104. Phone: (402) 571-0200. Fax: (402) 571-0833. E-mail: kvss@kvss.com Web Site: www.kvss.com. Licensee: VSS Catholic Communications Inc. (acq 7-7-99). Format: Catholic, Christian. Target aud: General. ♦Jim Taphorn, pres & CFO; Mike Delich, VP; Jim Carroll, gen mgr; D. Ann Yeoman, opns mgr; Vicki Sempek, dev dir; Mary Beth Jorgensen, gen sls mgr; Bruce McGregor, progmg dir; T. Scott Marr, news dir; Chuck Ramold, chief of engrg.

KXSP(AM)— Apr 2, 1923: 590 khz; 5 kw-U. TL: N41 19 00 W95 59 52. Stereo. Hrs opn: 24 5030 N. 72nd St., 68134. Phone: (402) 592-5300. Fax: (402) 331-1348. Web Site: www.bigsports590.com. Licensee: Journal Broadcast Corp. Group owner: Journal Broadcast Group Inc. (acq 10-26-98 with co-located FM). Population served: 347,328 Format: Sports. News staff: 4. Target aud: General. ♦Steve Wexler, gen mgr; Jim Timm, gen sls mgr; Tom Land, opns mgr & progmg dir; Bill Jensen, news dir; John Gaeta, chief of engrg; Cheryl Brye, traf mgr.

KQCH(FM)—Co-owned with KXSP(AM). 1959: 94.1 mhz; 100 kw. 508 ft TL: N41 18 47 W96 00 36. Stereo. 24 574,100 Format: Rhythm-based contemp. Target aud: 25-54. ♦Larkin Cavanaugh, gen mgr & prom dir; Jill Butler, gen sls mgr; Erik Johnson, progmg dir; Kathi Knutson, traf mgr; Bill Jensen, news rptr; Peter Shinn, farm dir.

O'Neill

KBRX(AM)— November 1955: 1350 khz; 1 kw-D, 44 W-N. TL: N42 27 34 W98 39 23. Hrs open: 24 Box 150, 68763. Secondary address: 251 N. Jefferson Phone: (402) 336-1612. Fax: (402) 336-3585. E-mail: Gil@kbrx,com Web Site: www.kbrx.com. Licensee: Ranchland Broadcasting Co. Inc. (acq 7-1-61). Population served: 45,000 Rgnl. Network: Brownfield. Law Firm: Bryan Cave. Wire Svc: AP Format: Classic Rock. News staff: one; News: 25 hrs wkly. Target aud: 25-65. Spec prog: Farm 12 hrs, Ger 6 hrs wkly. ♦Gilbert L. Poese, pres; Scott Poese, gen mgr.

KBRX-FM— December 1973: 102.9 mhz; 100 kw. Ant 500 ft TL: N42 26 06 W98 33 39. Stereo. Hrs opn: 24 Box 150, 251 N. Jefferson,

68763. Phone: (402) 336-1612. Fax: (402) 336-3585. E-mail: gil@kbrx.com Web Site: www.kbrx.com. Licensee: Ranchland Broadcasting Co. Inc. Population served: 45,000 Rgnl. Network: Brownfield. Law Firm: Bryan Cave . Wire Svc: AP Format: Country. News staff: one. Target aud: 25-60. ♦Gil Poese, pres & local news ed; Pat Poese, sls VP; Scott Poese, gen mgr & farm dir.

Orchard

***KGRD(FM)**— June 14, 1987: 105.3 mhz; 100 kw. 502 ft TL: N42 20 45 W98 25 05. Stereo. Hrs opn: 24 128 S. 4th St., O'Neill, 68763-1814. Phone: (402) 336-3886. E-mail: kgrd@kgrd.org Web Site: www.kgrd.org. Licensee: The Praise Network Inc. (acq 10-16-91). Population served: 50,000 Natl. Network: Salem Radio Network. Wire Svc: AP Format: Christian, Inspirational. News: 10 hrs wkly. Target aud: 35-54. ♦Lloyd Mintzmeyer, pres; Todd Gunnarson, gen mgr & stn mgr; Bill Taylor, mus dir.

Ord

KNLV(AM)— July 15, 1965: 1060 khz; 1 kw-D. TL: N41 34 16 W98 55 29. Hrs open: 205 S. 16th St., 68862. Phone: (308) 728-3263. Fax: (308) 728-3264. E-mail: knlv@yahoo.com Licensee: Sandhills Advertising Corp. (acq 6-1-74). Population served: 2,642 Rgnl. Network: Brownfield. Target aud: 18-54. Spec prog: Farm 8 hrs, Pol/Czeck/Bohemian 5 hrs wkly. ♦Johnnie James, gen mgr & progmg dir; Jeannie Niehart, gen sls mgr; Johnnie James, news dir; Randy Brannigan, chief of engrg; Denise O'Neel, traf mgr.

KNLV-FM— July 10, 1981: 103.9 mhz; 3.85 kw. 379 ft TL: N41 34 16 W98 55 29. Stereo. 6 AM-10 PM 7,200 ♦Johnnie James, stn mgr & disc jockey.

Overton

KHZY(FM)— 2007: 99.3 mhz; 100 kw. Ant 751 ft TL: N40 41 49 W99 47 16. Hrs open:
Rebroadcasts KSRD(FM) Saint Joseph, MO 100%.
1212 Faraon St., Saint Joseph, MO, 64501. Phone: (816) 233-5773. Fax: (816) 233-5777. Web Site: www.ksrdradio.com. Licensee: Horizon Christian Fellowship. (acq 2-9-2006; grpsl). Format: Christian. ♦Mike MacIntosh, pres; Brian KC Jones, gen mgr.

Plattsmouth

KOIL(AM)— Oct 26, 1970: 1020 khz; 1 kw-D, 1400 kw-N. TL: N41 01 35 W95 54 00. Hrs open: Sunrise-sunset 5011 Capital Ave., Omaha, 68132. Phone: (402) 342-2000. Fax: (402) 346-5748. Web Site: www.radiodisney.com. Licensee: Waitt Omaha LLC. (group owner; (acq 1-17-2001; $750,000). Population served: 750,000 Natl. Network: Radio Disney. Rgnl. Network: Brownfield. Format: Children. ♦Jim Kernan, gen mgr; Mark Todd, opns mgr; Rhonda Gerrard, sls dir; Neil Nelkin, progmg dir; Terry Leahy, news dir; Darwin Stinton, chief of engrg; Cynthia Wallace, traf mgr.

KOPW(FM)— July 1993: 106.9 mhz; 25 kw. Ant 328 ft TL: N41 09 18 W95 45 42. Stereo. Hrs opn: 24 5011 Capitol Ave., Omaha, 68132. Phone: (402) 342-2000. Fax: (402) 346-5748. Web Site: power1069fm.com. Licensee: Platte Broadcasting Co. Inc. Population served: 450,000 Natl. Rep: Katz Radio. Format: Contemp rhythm. Target aud: 18-34; gen. ♦Jim McKernan, gen mgr; Mark Todd, opns mgr; Rhonda Gerrard, sls dir; Sam Coughlin, gen sls mgr; Marcey Gibson, prom dir; Bryant McCain, progmg dir; Cynthia Wallace, traf mgr.

Ponca

***KFHC(FM)**—Not on air, target date: unknown: 88.1 mhz; 14 kw vert. Ant 239 ft TL: N42 27 48 W96 37 29. Hrs open: Box 2140, Northside Station, Sioux City, IA, 51104-0140. Licensee: St. Gabriel Communications Ltd.

Ralston

***KMLV(FM)**— July 21, 2001: 88.1 mhz; 3.7 kw vert. Ant 794 ft TL: N41 18 40 W96 01 37. Stereo. Hrs open: 24 2351 Sunset Blvd., Suite 170-218, Rocklin, CA, 95765. Phone: (916) 251-1600. Fax: (916) 251-1650. E-mail: klove@klove.com Web Site: www.klove.com. Licensee: Educational Media Foundation. Group owner: EMF Broadcasting. Population served: 735,000 Natl. Network: K-Love. Law Firm: Shaw Pittman. Format: Contemp Christian music. News staff: 3. Target aud: 25-44; Judeo Christian, female. ♦Richard Jenkins, pres; Mike Novak, VP & progmg dir; Lloyd Parker, gen mgr; Ed Lenane, opns dir & news dir; Keith Whipple, dev dir; Eric Allen, natl sls mgr; David Pierce, progmg mgr; Jon Rivers, mus dir; Sam Wallington, engrg dir; Karen Johnson, news rptr; Marya Morgan, news rptr; Richard Hunt, news rptr.

Ravenna

KKJK(FM)— June 1, 2006: 103.1 mhz; 100 kw. Ant 640 ft TL: N40 48 57 W98 46 18. Hrs open: Box 5853, Grand Island, 68802. Secondary address: 3205 W. North Front St., Grand Island 68803. Phone: (308) 381-1430. Fax: (308) 382-6701. E-mail: info@familyradio.com Web Site: www.thunder1031.com. Licensee: Community Radio Inc. Format: Rock and roll. ♦Donald Wilks, pres; Alan Usher, gen mgr.

Sargent

KHZZ(FM)—Not on air, target date: unknown: 92.1 mhz; 100 kw. Ant 341 ft TL: N41 33 08 W99 18 09. Hrs open: 5331 Mt. Alifan Dr., San Diego, CA, 92111. Phone: (858) 277-4991. Fax: (858) 277-1365. Licensee: Horizon Christian Fellowship. (acq 2-9-2006; grpsl). ♦Mike MacIntosh, pres.

Scottsbluff

***KDAI(FM)**—Not on air, target date: unknown: 89.1 mhz; 100 kw. Ant 249 ft TL: N41 55 06 W103 51 49. Hrs open: 2351 Sunset Blvd., Suite 170-218, Rocklin, CA, 95765. Phone: (916) 251-1600. Fax: (916) 251-1650. Licensee: Educational Media Foundation. (acq 7-23-2007; grpsl). ♦Mike Novak, sr VP.

***KLJV(FM)**— Feb 20, 2003: 88.3 mhz; 390 w. Ant 259 ft TL: N41 56 24 W103 39 20. Stereo. Hrs opn: 24 2351 Sunset Blvd., Suite 170-218, Rocklin, CA, 95765. Phone: (916) 251-1600. Fax: (916) 251-1650. E-mail: klove@klove.com Web Site: www.klove.com. Licensee: Educational Media Foundation. Group owner: EMF Broadcasting. Population served: 32,600 Natl. Network: K-Love. Law Firm: Shaw Pittman. Format: Contemp Christian. News staff: 3. Target aud: 25-44. ♦Richard Jenkins, pres; Mike Novak, VP; Keith Whipple, dev dir; David Pierce, progmg mgr; Ed Lenane, news dir; Sam Wallington, engrg dir; Karen Johnson, news rptr; Marya Morgan, news rptr; Richard Hunt, news rptr.

KNEB(AM)— Jan 1, 1948: 960 khz; 5 kw-D, 350 w-N, DA-2. TL: N41 47 30 W103 38 29. Hrs open: 5 AM-1 AM Box 239, 1928 E. Portal Pl., 69363-0239. Phone: (308) 632-7121. Fax: (308) 635-1079. E-mail: kneb@actcom.net Web Site: www.kneb.com. Licensee: Nebraska Rural Radio Association. Group owner: Nebraska Rural Radio Assn. (acq 8-1-84). Population served: 46,750 Format: C&W, news/talk. News staff: 2. Target aud: 18 plus. Spec prog: Farm 18 hrs, Sp 5 hrs wkly. ♦Dale Hanson, pres; Barbara Martinson, gen sls mgr & rgnl sls mgr; Dennis Ernest, progmg dir; Marty Martinson, gen mgr, natl sls mgr & news dir.

KNEB-FM— Dec 25, 1960: 94.1 mhz; 100 kw. 680 ft TL: N41 42 04 W103 40 49. Stereo. 5 AM-1 AM 112,200 Format: Farm, C&W.

KOAQ(AM)—See Terrytown

KOLT(AM)— Feb 15, 1930: 1320 khz; 5 kw-D, 1 kw-N, DA-N. TL: N41 51 37 W103 41 53. Hrs open: 24 Box 660, 69363. Phone: (308) 635-1320. Fax: (308) 635-1905. Web Site: www.tracybroadcasting.com. Licensee: Tracy Corp. Group owner: Tracy Broadcasting Corp. (acq 2-25-92; $37,000; FTR: 3-23-92). Population served: 15,000 Natl. Network: ABC. Law Firm: Michael Glaser. Format: Talk. Target aud: 39 plus. Spec prog: Farm 10 hrs wkly. ♦Michael Tracy, pres; Larry Swikard, gen mgr, stn mgr & gen sls mgr; Jeff McKenzie, progmg dir.

Seward

KZKX(FM)— Nov 12, 1976: 96.9 mhz; 100 kw. 610 ft TL: N41 07 26 W96 50 03. Stereo. Hrs opn: 4630 Antelope Creek Rd., Suite 200, Lincoln, 68506. Phone: (402) 484-8000. Fax: (402) 489-9989. Web Site: www.kzkx.com. Licensee: Capstar TX L.P. Group owner: Clear Channel Communications Inc. (acq 8-30-00; grpsl). Natl. Rep: D & R Radio. Law Firm: Gammon & Grange. Format: Country. Target aud: 25-54. ♦Julie Gade, gen mgr; Julie Broman, gen sls mgr & news dir; Brian Jennings, progmg dir; Dave Agnew, chief of engrg; Angela Zoucha, traf mgr.

Sidney

KSID(AM)— June 2, 1952: 1340 khz; 1 kw-U. TL: N41 07 50 W102 58 15. Hrs open: 24 Box 37, Legion Park, 69162. Phone: (308) 254-5803. Fax: (308) 254-5901. Licensee: KSID Radio Inc. (acq 1962). Population served: 128,500 Format: Country. Target aud: General. Spec prog: Farm 5 hrs wkly. ♦Elizabeth Young, pres; Lana Butts, opns mgr & gen sls mgr; Marge Elliott, progmg dir; Jason Lockwood, news dir; Dennis Brothers, engrg mgr & chief of engrg; Jean Spruckmeyer, traf mgr; Susan Ernest, gen mgr & traf mgr.

KSID-FM— Sept 13, 1974: 98.7 mhz; 62 kw. 368 ft TL: N41 11 03 W103 11 37. Stereo. 134,493 Format: Adult contemp. ♦Susan Ernest, traf mgr.

South Sioux City

KSFT-FM— 1997: 107.1 mhz; 1.55 kw. 328 ft TL: N42 29 00 W96 35 34. Stereo. Hrs opn: 24 Box 3009, Sioux City, IA, 51102. Secondary address: 1113 Nebraska St., Sioux City, IA 51102. Phone: (712) 258-6740. Fax: (712) 252-2430. Web Site: www.soft107.com. Licensee: AMFM Radio Licenses LLC. Group owner: Clear Channel Communications Inc. (acq 10-1-2002; grpsl). Natl. Network: Westwood One. Natl. Rep: Christal. Format: Soft rock. News staff: one; News: one hr wkly. Target aud: 25-54. ♦Laura Schiltz, gen mgr & gen sls mgr; Rob Powers, opns mgr & progmg dir; Curtis Anderson, news dir; Stan Culley, chief of engrg; Monica Mattoon, traf mgr.

Superior

KRFS(AM)— Mar 17, 1959: 1600 khz; 500 w-D. TL: N40 01 30 W98 04 38. Hrs open: 6 AM-sunset Rte. 2, Box 149, 68978. Phone: (402) 879-4741. Fax: (402) 879-4741. E-mail: krfsfm@yahoo.com Licensee: CK Broadcasting Inc. (acq 1-4-02; $150,000 with co-located FM). Population served: 100,000 Rgnl. Network: Brownfield. Format: Adult standards. News: 11 hrs wkly. Target aud: 25-55; general. Spec prog: Farm 5 hrs, gospel 3 hrs, relg 3 hrs wkly. ♦Cory Kopsa, gen mgr, sls VP, gen sls mgr, prom mgr, progmg dir, news dir, traf mgr & sports cmtr; Marvin Hoffman, chief of engrg.

KRFS-FM— Feb 25, 1977: 103.9 mhz; 6 kw. 220 ft TL: N40 01 30 W98 04 38. (CP: TL: N40 06 20 W98 06 20). Stereo. 24 Format: Country. ♦Cory Kopsa, traf mgr & sports cmtr.

Terrytown

KCMI(FM)— Mar 1, 1981: 96.9 mhz; 100 kw. 692 ft TL: N41 42 08 W103 41 00. Stereo. Hrs opn: 24 Box 1888, 209 E. 15th, Scottsbluff, 69363-1888. Phone: (308) 632-5264. Fax: (308) 635-0104. E-mail: christianmedia@charterinternet.com Web Site: www.kcmi.cc. Licensee: Christian Media Inc. Population served: 85,000 Natl. Network: USA. Rgnl. Network: Mid-American Ag. Format: Relg. News: 12 hrs wkly. Target aud: 25 plus. Spec prog: Class 4 hrs wkly. ♦Glenn A. Hascall, gen mgr; Gary Almquist, gen sls mgr & progmg dir; Lorraine Brown, traf mgr.

KOAQ(AM)— June 15, 1961: 690 khz; 1 kw-D, 64 w-N, DA-1. TL: N41 50 02 W103 39 20. Stereo. Hrs opn: 24
Rebroadcasts KBFZ(FM), Kimball 100%.
Box 1263, Scottsbluff, 69361. Phone: (308) 635-1320. Fax: (308) 635-1905. Web Site: www.tracybroadcasting.com. Licensee: Tracy Broadcasting Corp. (group owner; acq 12-86; $164,000; FTR: 11-10-86). Population served: 45,000 Natl. Network: Jones Radio Networks. Law Firm: Michael Glaser. Format: Oldies. News staff: 2; News: 21 hrs

wkly. Target aud: 25-54; Baby Boomers. Spec prog: Farm 5 hrs wkly. ♦Michael Tracy, pres; Larry Swikard, gen mgr & gen sls mgr; Krista Sarchet, prom mgr; Jeff McKenzie, progmg dir; Bob Hinze, chief of engrg.

Valentine

KVSH(AM)— Mar 6, 1961: 940 khz; 5 kw-D, 20 w-N. TL: N42 51 54 W100 31 07. Hrs open: 16 126 W. 3rd St., 69201. Phone: (402) 376-2400. Fax: (402) 376-2402. Licensee: Heart City Radio Corp. (acq 6-90; $235,000; FTR: 6-4-90). Population served: 2,880 Law Firm: Fletcher, Heald & Hildreth. Wire Svc: AP Format: Country. News staff: one; News: 24 hrs wkly. Target aud: 35-60; general. ♦Dave Otradovsky, pres; Mike Burge, progmg dir, news dir & chief of engrg; Zach Dean, gen sls mgr & disc jockey.

Wayne

KCTY(FM)—Listing follows KTCH(AM).

KTCH(AM)— Mar 18, 1968: 1590 khz; 2.5 kw-D, 33.4 w-N, DA-2. TL: N42 14 03 W97 03 19. Hrs open: 24 Box 413, W. Hwy. 35, 68787. Phone: (402) 375-3700. Fax: (402) 375-5402. E-mail: ktch@ktch.com Web Site: www.ktch.com. Licensee: NRG License Sub. LLC. (group owner; (acq 10-31-2005; grpsl). Population served: 40,000 Rgnl. Network: Brownfield. Format: Country. News staff: one; News: 15 hrs wkly. Target aud: 30-64. Spec prog: Farm 20 hrs wkly. ♦Neil Lipetzky, gen mgr; Mick Kemp, gen sls mgr; Dan Baddorf, progmg dir; Tom McDermott, news dir; Tim Geuntz, chief of engrg.

KCTY(FM)—Co-owned with KTCH(AM). Oct 19, 1975: 104.9 mhz; 25,000 kw. Ant 300 ft TL: N42 14 03 W97 03 19. Stereo. 24 Web Site: www.ktch.com.40,000 Format: Oldies. News staff: one. Target aud: General.

***KWSC(FM)**— Oct 13, 1971: 91.9 mhz; 350 w. 96 ft TL: N42 14 30 W97 00 48. Hrs open: 1111 Main St., 68787. Phone: (402) 375-7536. Phone: (402) 375-7426. E-mail: k92radio@hotmail.com Web Site: www.wsc.edu/k92. Licensee: Wayne State College. Population served: 35,000 Natl. Network: Westwood One. Format: AOR, alternative. Target aud: 18 plus. Spec prog: Black 4 hrs, jazz 2 hrs, heavy metal 2 hrs, blues 2 hrs wkly.

West Point

KTIC(AM)— Mar 17, 1985: 840 khz; 5 kw-D. TL: N41 47 06 W96 40 39. Hrs open: Sunrise-sunset Box 84, 1011 N. Lincoln St., 68788-0084. Phone: (402) 372-5423. Fax: (402) 372-5425. E-mail: ktickwpn@kwpnfm.com Web Site: www.kticam.com. Licensee: Nebraska Rural Radio Association. Group owner: Nebraska Rural Radio Association (acq 8-1-97; $1.5 million with co-located FM). Population served: 1,921,459 Natl. Network: ABC. Natl. Rep: Katz Radio. Rgnl rep: Neb. Pub. Wire Svc: AP Format: Farm market news, country. News staff: one; News: 20 hrs wkly. Target aud: General; farmers, ranchers, stockmen and all involved in agri-business. ♦Charlie Brogan, gen mgr, stn mgr & progmg dir; Denny Waddle, sls dir & gen sls mgr; Judy Mauch, gen sls mgr; Richard Sterling, mus dir; Bob Flittie, news dir; Vern Killion, chief of engrg; Tammie Harrington, traf mgr; Randy Koenen, farm dir; Tom McMahon, sports cmtr.

KTIC-FM— Aug 1, 1988: 107.9 mhz; 50 kw. Ant 318 ft TL: N41 47 06 W96 40 39. Stereo. 18 (Acq 1997).888,400 Natl. Network: ABC. Format: Country. News staff: one; News: 12 hrs wkly. Target aud: 25-49; general audience, adults. Spec prog: Farm 10 hrs wkly. ♦Richard Sterling, progmg dir; Karen Benne, traf mgr; Randy Koenen, news rptr & farm dir; Tom McMahon, sports cmtr.

Wilber

***KFLV(FM)**— Mar 1, 2001: 89.9 mhz; 8.8 kw. Ant 351 ft TL: N40 30 33 W96 50 41. Stereo. Hrs opn: 24 2351 Sunset Blvd., Suite 170-218, Rocklin, CA, 95765. Phone: (916) 251-1600. Fax: (916) 251-1650. E-mail: klove@klove.com Web Site: www.klove.com. Licensee: Educational Media Foundation. Group owner: EMF Broadcasting. Population served: 246,000 Natl. Network: K-Love. Law Firm: Shaw Pittman. Format: Contemp Christian music. News staff: 3. Target aud: 25-44; Judeo-Christian, female. ♦Richard Jenkins, pres; Mike Novak, VP; Keith Whipple, dev dir; Eric Allen, natl sls mgr; David Pierce, progmg mgr & mus dir; Ed Lenane, news dir; Sam Wallington, engrg dir; Karen Johnson, news rptr; Marya Morgan, news rptr; Richard Hunt, news rptr.

Winnebago

KSUX(FM)— June 1, 1990: 105.7 mhz; 50 kw. 463 ft TL: N42 20 33 W96 31 13. Stereo. Hrs opn: 24 2000 Indian Hills Dr., Sioux City, IA, 51104. Phone: (712) 239-2100. Fax: (712) 239-3346. Web Site: www.ksux.com. Licensee: Powell Broadcasting Co. (acq 1996; $3.8 million with KSCJ(AM) Sioux City, IA). Population served: 283,400 Format: Country. News staff: 2. Target aud: 25-54; female average to above average income, secondary male. ♦Dennis Bullock, gen mgr; Dave Grosenheider, sls dir; Tony Michaels, progmg dir.

York

KAWL(AM)— September 1954: 1370 khz; 500 w-D, 176 w-N. TL: N40 50 30 W97 35 16. Hrs open: 24 1309 Rd. 11, 68467. Phone: (402) 362-4433. Fax: (402) 362-6501. E-mail: kawl@alltel.net Web Site: www.oldiesradioonline.com. Licensee: MWB Broadcasting LLC (acq 12-3-2004; $1 million with co-located FM). Population served: 115,000 Natl. Network: ABC. Rgnl. Network: Mid-American Ag. Natl. Rep: Interep. Wire Svc: AP Format: Oldies, talk. News staff: one; News: 10 hrs wkly. Target aud: 20 plus; general. Spec prog: Farm 7 hrs, women 3 hrs wkly. ♦Mark Jensen, gen mgr; Brenda Janzen, opns mgr & traf mgr; Bob Bedient, news dir; Linda Korbelik, chief of engrg & sls; Donna Panritz, sls.

KTMX(FM)—Co-owned with KAWL(AM). Sept 1, 1970: 104.9 mhz; 25 kw. 974 ft TL: N40 45 07 W97 27 04. Stereo. 24 Web Site: hitsandfavorites.com.390,000 Natl. Network: ABC. Natl. Rep: McGavren Guild. Format: Adult contemp. News staff: one; News: 3 hrs wkly. Target aud: 25-54; general.

Nevada

Amargosa Valley

KPKK(FM)— 2003: 101.1 mhz; 51 kw horiz. Ant -49 ft TL: N36 38 33 W116 23 53. Hrs open: Sky Media L.L.C., 980 N. Michigan Ave., Suite 1880, Chicago, IL, 60611. Phone: (312) 204-9900. Fax: (312) 587-9466. Licensee: Sky Media L.L.C. (acq 11-20-2002; $5.1 million for CP). ♦Bruce Buzil, gen mgr.

Beatty

KDAN(AM)—Not on air, target date: unknown: 1240 khz; 1 kw-U. TL: N36 54 59 W116 45 47. Hrs open: 24 Box 1450, St. George, UT, 84771-1450. Secondary address: 210 N. 1000 E., St. George, UT 84770-3155. Phone: (435) 628-1000. Fax: (435) 628-6636. Licensee: Radio 1240 LLC. (acq 5-3-2006; $20,000 for CP). Law Firm: Dan J. Alpert. ♦E. Morgan Skinner Jr., pres.

Boulder City

KSTJ(FM)—Licensed to Boulder City. See Las Vegas

Caliente

KMOA(FM)—Not on air, target date: unknown: 94.5 mhz; 100 kw. Ant 1,709 ft TL: N37 14 37 W114 36 01. Hrs open: 7251 W. Lake Mead Blvd., Suite 300, Las Vegas, 89128-8380. Phone: (702) 655-1249. Fax: (702) 655-6175. Licensee: Aurora Media LLC. ♦Scott G. Mahalick, gen mgr.

Cal-Nev-Ari

KVYL(FM)—Not on air, target date: unknown: 104.9 mhz; 500 w. Ant -298 ft TL: N35 18 42 W114 52 55. Hrs open: Number 10 Media Center Dr., Lake Havasu, AZ, 86403. Phone: (928) 855-1051. Fax: (928) 855-7996. Licensee: Smoke and Mirrors LLC. ♦Rick L. Murphy, gen mgr.

Carlin

KHIX(FM)— March 2001: 96.7 mhz; 33 kw. Ant 1,597 ft TL: N40 55 18 W115 50 58. Hrs open: 24 1250 Lamoille Hwy., # 944, Elko, 89801. Phone: (775) 777-1196. Fax: (775) 777-9587. E-mail: ken@mix96.fm Web Site: www.mix96.fm. Licensee: Ruby Radio Corp. (group owner; (acq 6-15-2003; $475,000 for CP). Population served: 25,000 Law Firm: David Tillotson. Format: Adult hit radio (contemp). ♦Alene Sutherland, VP; Ken Sutherland, stn mgr.

Carson City

KBUL-FM— Nov 30, 1984: 98.1 mhz; 72.5 kw. 2,273 ft TL: N39 15 32 W119 42 06. (CP: Ant 2,286 ft.). Stereo. Hrs opn: 24 595 E. Plumb Ln., Reno, 89502-3773. Phone: (775) 789-6700. Fax: (775) 789-6767. Web Site: www.kbul.com. Licensee: Citadel Broadcasting Co. Group owner: Citadel Broadcasting Corp. (acq 5-29-92). Population served: 275,000 Format: C&W. News staff: one; News: 4 hrs wkly. Target aud: 25-54. ♦Dana Johnson, gen mgr.

KCMY(AM)— May 14, 1955: 1300 khz; 5 kw-D, 500 w-N, DA-N. TL: N39 09 59 W119 43 37. Hrs open: 1960 Idaho, 89701. Phone: (775) 884-8000. Fax: (775) 882-3961. Licensee: The Evans Broadcast Co. Inc. (acq 5-17-2004; $700,000). Population served: 350,000 Natl. Network: Fox News Radio. Law Firm: Shainis & Peltzman. Format: Classic Country. News staff: 2; News: 16 hrs wkly. Target aud: 35-54. ♦Jerry Evans, gen mgr, progmg dir & progmg.

***KNIS(FM)**— Oct 15, 1989: 91.3 mhz; 67 kw. 2,165 ft TL: N39 15 30 W119 42 36. Stereo. Hrs opn: 24 Western Inspirational Broadcasters, Inc., 6363 Hwy. 50 E., 89701. Phone: (775) 883-5647. Licensee: Western Inspirational Broadcasters Inc. (acq 10-15-89). Population served: 340,000 Natl. Network: AP Radio. Wire Svc: AP Format: Contemp Christian, educ, talk. News: 16 hrs wkly. Target aud: 24-44. ♦Tom Hesse, gen mgr; Tim Weidemann, opns mgr; Bill Feltner, progmg dir; Janet Santana, mus dir; Paul Lierman, chief of engrg.

KZTQ(FM)—Licensed to Carson City. See Reno

Dayton

KTHX-FM—Licensed to Dayton. See Reno

Elko

KELK(AM)— Dec 7, 1948: 1240 khz; 1 kw-U. TL: N40 50 37 W115 44 58. Hrs open: 24 1800 Idaho St., 89802. Phone: (775) 738-1240. Fax: (775) 753-5556. Web Site: elkoradio.com. Licensee: Elko Broadcasting Co. (acq 11-1-74). Population served: 30,000 Format: Full service, adult contemp. News staff: one; News: 25 hrs wkly. Target aud: 25-54; upscale, family oriented, white collar workforce. ♦Paul G. Gardner, pres; Tyler Gunter, gen mgr.

KLKO(FM)—Co-owned with KELK(AM). May 1982: 93.7 mhz; 4.5 kw. Ant 1,538 ft TL: N40 55 20 W115 50 56. Stereo. Web Site: elkoradio.com.35,000 Format: Adult hits. News staff: one; News: 5 hrs wkly.

***KNCC(FM)**— 1992: 91.5 mhz; 50 w. 741 ft TL: N40 49 16 W115 42 04. Hrs open: 24
Rebroadcasts *KUNR(FM) Reno.
Great Basin College, 1500 College Pkwy., 89801. Phone: (775) 753-2181. Fax: (775) 738-8771. E-mail: carl@gbcnv.edu Licensee: Great Basin College. Population served: 50,000 Format: Class, big band, educ, jazz, news. Spec prog: Public radio. ♦Carl Diekhans, gen mgr.

KOYT(FM)— 2005: 94.5 mhz; 36 kw. Ant 1,519 ft TL: N40 55 18 W115 50 58. Hrs open: 1250 Lamoille Hwy., Suite 944, 89801. Phone: (775) 777-1196. Fax: (775) 777-9587. Licensee: Ruby Radio Corp. (acq 10-25-2005; exchange for KCLS(FM) Ruby). Format: Rock. ♦Ken Sutherland, pres & gen mgr.

KPHD(FM)—Not on air, target date: unknown: 97.5 mhz; 90 kw. Ant 1,604 ft TL: N40 55 43 W115 50 33. Hrs open: College Creek Media LLC, 980 N. Michigan Ave., Suite 1880, Chicago, IL, 60611. Phone: (312) 204-9900. Licensee: College Creek Media LLC. ♦Neal J. Robinson, pres.

KRJC(FM)— October 1981: 95.3 mhz; 25 kw. Ant 774 ft TL: N40 54 35 W115 49 05. Stereo. Hrs opn: 24 1250 Lamoille Hwy., Suite 1045, 89801-1626. Phone: (775) 738-9895. Fax: (775) 753-8085. E-mail: krjc@krjc.com Web Site: www.krjc.com. Licensee: Holiday Broadcasting of Elko. Group owner: Carlson Communications International. Population served: 40,000 Natl. Network: CNN Radio. Wire Svc: AP Format: Hot country. News staff: one; News: 4 hrs wkly. Target aud: 25-54; above median income & education averages. ♦Ralph J. Carlson, pres; Jennifer Ford, gen sls mgr; Tammy Stewart, gen mgr & rgnl sls mgr; John Hunt, progmg dir; Mike Allen, news dir.

KTSN(AM)— November 1996: 1340 khz; 1 kw-D. TL: N40 52 08 W115 43 09. Hrs open: 1250 Lamoille Hwy., Suite 1045, 89801. Phone: (775) 738-9895. Fax: (775) 753-8085. Web Site: www.ktsn1340.com. Licensee: Humboldt Broadcasting LLC. Group owner: Carlson Communications International. Format: Talk, sports, news. ♦Tammy Stewart, gen mgr; Jennifer Ford, gen sls mgr; Josh Ellison, progmg dir; Kenneth Meyer, chief of engrg; Amber Stump, traf mgr.

Ely

KCLS(FM)— Nov 1, 1986: 101.7 mhz; 480 w. Ant 804 ft TL: N39 14 46 W114 55 39. Stereo. Hrs opn: 24 College Creek Media LLC, 980 N. Michigan Ave., Suite 1880, Chicago, IL, 60611. Phone: (312) 204-9900. Licensee: College Creek Media LLC. (acq 10-25-2005; exchange for KOYT(FM) Elko). Format: Adult contemp. ♦Neal J. Robinson, pres.

KDSS(FM)— Dec 22, 1984: 92.7 mhz; 14 kw. Ant 941 ft TL: N39 14 46 W114 55 39. Stereo. Hrs opn: 24 501 Aultman, Suite 208, 89301. Phone: (775) 289-6474. Fax: (775) 289-6531. E-mail: kdss@wpis.net Licensee: Coates Broadcasting Inc. (acq 5-1-96; $180,000). Natl. Network: Jones Radio Networks. Law Firm: Irwin, Campbell & Tannenwald, P.C. Format: C & W. News: 10 hrs wkly. Target aud: 18-64; older demographics, new & classic C&W listeners. Spec prog: Nashville News, fishing, outdoor. ♦Pat Coates, pres; Samantha Coates, VP; James Lee, sls VP; Jim Liebsack, chief of engrg.

KELY(AM)— July 8, 1950: 1230 khz; 1 kw-U. TL: N39 15 45 W114 51 46. Hrs open: 24 Box 151465, 89301. Phone: (775) 289-2077. Fax: (775) 289-6997. Web Site: www.elyradio.com. Licensee: Ely Radio LLC. (group owner; (acq 3-21-2006; $140,000). Population served: 12,650 Format: Oldies. Target aud: 40 plus. Spec prog: Swap shop 3 hrs, radio bingo 5 hrs wkly. ♦Bob Bolton, gen mgr; Howard Tenke, news dir.

Fallon

KHWG(AM)— June 2005: 750 khz; 10 kw-D, 280 w-N, 10 kw-CH. TL: N39 28 57 W118 45 36. Hrs open: 1050 W. Williams Ave., 89406. Phone: (775) 428-1764. Fax: (775) 428-1765. Licensee: Media Enterprises Inc. (acq 1-8-2003). Format: Classic country. ♦Keily Miller, pres; Dee Gregory, gen mgr.

KVLV(AM)— May 9, 1957: 980 khz; 5 kw-D. TL: N39 29 47 W118 48 50. Hrs open: 6 AM-sunset 1155 Gummow Dr., 89406. Phone: (775) 423-2243. Phone: (775) 423-5858. Fax: (775) 423-8889. E-mail: kvlv@phonewave.net Licensee: Lahontan Valley Broadcasting LLC. Population served: 60,000 Natl. Network: ABC. Format: C&W. News: 10 hrs wkly. Target aud: 25 plus. ♦Mike McGinness, gen mgr.

KVLV-FM— Nov 26, 1966: 99.3 mhz; 3.7 kw. 250 ft TL: N39 29 47 W118 48 50. Stereo. 24 30,000 Natl. Network: AP Radio. Format: Adult contemp. News: 10 hrs wkly.

Fallon/ Reno

KRNG(FM)— July 4, 1997: 101.3 mhz; 1.65 kw. 2,207 ft TL: N39 42 30 W119 10 16. Stereo. Hrs opn: 24 Box 490, Wadsworth, 89442. Secondary address: 360 Pyramid St., Wadsworth 89442. Phone: (775) 575-7777. Fax: (775) 575-7737. E-mail: email@renegaderadio.org Web Site: www.renegaderadio.org. Licensee: Sierra Nevada Christian Music Association Inc. Population served: 200,000 Format: Christian rock. Target aud: 12-35; youth, young adults. Spec prog: Rap, alternative. ♦Rev. Karry Crites, pres & stn mgr; William E. Bauer PhD., opns VP.

Gardnerville-Minden

KKFT(FM)— Sept 19, 1985: 99.1 mhz; 410 w. Ant 2,006 ft TL: N39 15 34 W119 42 21. Stereo. Hrs open: 24 1960 Idaho St., Carson City, 89701. Phone: (775) 884-8000. Fax: (775) 882-3961. E-mail: jerry@991fmtalk.com Web Site: www.991fmtalk.com. Licensee: Jerry Evans (acq 11-28-2003; $850,000). Population served: 600,000 Natl. Network: Fox News Radio. Law Firm: Shainis & Peltzman. Format: News, talk. News staff: 2; News: 24 hrs wkly. Target aud: 25-54; btfl people. ♦Jerry Evans, CEO, gen mgr & stn mgr.

Hawthorne

KIFO(AM)—Not on air, target date: unknown: 1450 khz; 1 kw-U. TL: N38 30 46 W118 37 58. Hrs open: Box 1450, St. George, UT, 84771-1450. Secondary address: 210 N. 1000 E., St. George, UT 84770-3155. Phone: (435) 628-1000. Fax: (435) 628-6636. Licensee: Radio 1450 LLC. (acq 3-15-2006; $13,000 for CP). Law Firm: Dan J. Alpert. ♦E. Morgan Skinner Jr., pres.

***KQMC(FM)**—Not on air, target date: unknown: 90.1 mhz; 1.3 kw. Ant 3,677 ft TL: N38 47 04 W118 49 59. Hrs open: American Educational Broadcasting Inc., 3185 S. Highland Dr. #13, Las Vegas, 89109. Phone: (702) 731-5588. Fax: (702) 731-5851. Licensee: American Educational Broadcasting Inc.

Henderson

KDOX(AM)— May 1956: 1280 khz; 5 kw-D, 28 w-N. TL: N36 03 13 W114 58 30. Hrs open: 24 150 Spectrum Blvd., Las Vegas, 89101. Phone: (702) 258-0285. Fax: (702) 732-3060. Web Site: www.1280talk.com. Licensee: S & R Broadcasting Inc. (acq 9-28-2005; $2 million). Population served: 160,000 Natl. Rep: Lotus Entravision Reps LLC. Law Firm: Rosenman & Colin L.L.P. Format: Sp. Target aud: General; Hispanic, above-average income, high home ownership. ♦Paul Ruttan, pres; Scott Gentry, gen mgr; Ellen Walker, gen sls mgr; Roberto Ibarra, progmg dir; Warren Brown, chief of engrg.

KKJJ(FM)— Nov 28, 1982: 100.5 mhz; 100 kw. 1,105 ft TL: N36 00 28 W115 00 20. Stereo. Hrs open: 24 6655 W. Sahara Ave., Suite C 216, Las Vegas, 89146. Phone: (702) 889-5100. Fax: (702) 257-2936. E-mail: jack@jackbaby.com Web Site: www.jackbaby.com. Licensee: Infinity Radio Inc. Group owner: Infinity Broadcasting Corp. (acq 11-13-98; grpsl). Natl. Network: CBS Radio. Natl. Rep: Katz Radio. Law Firm: Steven Lerman. Format: Adult hits. News: 1 hr wkly. Target aud: 25-54. ♦Joel Hollander, CEO; John Sykes, pres; Tom Humm, VP, gen mgr & gen sls mgr; Lorene Malis, natl sls mgr; Sam Ballenger, mktg dir & prom dir; Craig Powers, progmg dir; Herb Perry, pub affrs dir; Tracy Teagarden, chief of engrg; Stephanie Lindelow, traf mgr.

KMXB(FM)— Feb 10, 1970: 94.1 mhz; 100 kw. 1,210 ft TL: N36 00 26 W115 00 24. Stereo. Hrs opn: 24 6655 W. Sahara Ave., Suite D-110, Las Vegas, 89146. Phone: (702) 889-5100. Fax: (702) 257-2936. E-mail: justin@mix941.fm Web Site: www.mix941.fm. Licensee: CBS Radio Stations Inc. Group owner: Infinity Broadcasting Corp. (acq 11-13-98; grpsl). Population served: 1,500,000 Natl. Network: CBS Radio. Law Firm: Leventhal, Senter & Lerman. Format: Modern adult contemp. News staff: one. Target aud: 18-49; female. ♦John Sykes, CEO & chmn; John Fullam, COO & pres; Jacques Tortoli, CFO; Tom Humm, VP & gen mgr; Lorene Malis, natl sls mgr; Lori Heeren, rgnl sls mgr; Jennifer DiFazio, mktg dir & progmg dir; Justin Chase, progmg dir; Tracy Teagarden, chief of engrg; Stephanie Lindelow, traf mgr.

KWNR(FM)— July 18, 1972: 95.5 mhz; 92 kw. 1,161 ft TL: N36 00 31 W115 00 22. Hrs open: 2880 Meade Ave., Suite 250, Las Vegas, 89102. Phone: (702) 238-7300. Fax: (702) 732-4890. Web Site: www.kwnr.com. Licensee: Citicasters Licenses L.P. Group owner: Clear Channel Communications Inc. (acq 1999; grpsl). Law Firm: Hogan & Hartson. Format: Country. Target aud: 18-54. ♦Sean Cassidy, gen mgr & gen sls mgr; Bill Lubitz, prom dir; Brooks O'Brien, progmg dir; Mitch Kelly, news dir; Miguel Randtree, engrg dir.

Incline Village

KRNO(FM)—Licensed to Incline Village. See Reno

Indian Springs

KRGT(FM)— Nov 22, 2002: 99.3 mhz; 31 kw. Ant 2,263 ft TL: N36 19 28 W115 33 58. Hrs open: 24 6767 W. Tropicana Ave., Suite 102, Las Vegas, 89103. Phone: (702) 284-6400. Fax: (702) 284-6403. Licensee: Univision Radio License Corp. Group owner: Univision Radio (acq 9-22-2003; grpsl). Format: Sp. ♦Dana Demerjian, VP & gen mgr; Cristina Valarezo, gen sls mgr; Joe Reynols, natl sls mgr; Brent Bergel, rgnl sls mgr; Zulenia Benjamin, prom mgr; Ratael Miranlontes, progmg dir; Nancy Garcia, chief of engrg; Glocia Salvador, traf mgr.

Jackpot

***KBSJ(FM)**—Not on air, target date: unknown: 91.3 mhz; 3.7 kw. 2,463 ft Hrs opn: Idaho State Board of Education, 1910 University Dr., Boise, ID, 83725. Phone: (208) 426-3663. Fax: (208) 344-6631. Licensee: Idaho State Board of Education. Format: Classical jazz. ♦John Hess, gen mgr; Erik Jones, opns mgr; Ele Ellis, progmg dir.

Las Vegas

KBAD(AM)— June 1953: 920 khz; 5 kw-D, 500 w-N, DA-2. TL: N36 11 25 W115 10 35. Hrs open: 8755 W. Flamingo Rd., 89147-8667. Phone: (702) 876-1460. Fax: (702) 876-6685. Web Site: www.komp.com. Licensee: Lotus Broadcasting Corp. (acq 11-4-92; $1.42 million with co-located FM; FTR: 11-23-92) Population served: 500,000 Format: Sports. Target aud: 18 plus; men. ♦Tony Bonnici, gen mgr; John Hanson, progmg dir.

KXPT(FM)—Co-owned with KBAD(AM). Nov 29, 1961: 97.1 mhz; 50 kw. 1,950 ft TL: N35 56 44 W115 02 31. (CP: 24 kw). Stereo. Web Site: www.point97.com.600,000 Format: Classic hits. Target aud: 35-49. ♦John Griffin, opns mgr & progmg dir.

***KCEP(FM)**— October 1973: 88.1 mhz; 10 kw. -39 ft TL: N36 10 51 W115 08 43. (CP: 10 kw, ant 1,079 ft.). Stereo. Hrs opn: 24 330 W. Washington St., 89106. Phone: (702) 648-4218. Phone: (702) 648-0104. Fax: (702) 647-0803. Web Site: www.power88lu.com. Licensee: Economic Opportunity Board of Clark County. Population served: 90,000 Format: Black, urban contemp, rhythm & blues. News: 5 hrs wkly. Target aud: 12-55; African-Americans. Spec prog: Gospel 19 hrs, Jazz 12 hrs wkly. ♦Lee Winston, gen mgr; William Thompson, progmg dir; Warren Brown, engrg VP.

***KCNV(FM)**— Mar 24, 1980: 89.5 mhz; 98 kw. Ant 1,532 ft TL: N35 56 50 W115 03 01. Stereo. Hrs opn: 24 1289 S. Torrey Pines, 89146. Phone: (702) 258-9895. Fax: (702) 258-5646. E-mail: info@knpr.org Web Site: www.knpr.org. Licensee: Nevada Public Radio Corp. Population served: 1,200,000 Law Firm: Dow, Lohnes & Albertson. Format: Classical. Target aud: 35-54. ♦Lamar Marchese, CEO, pres & gen mgr; Louis Castle, chmn; Kathleen Hechinger, CFO; Phil Burger, opns dir; Valerie Freshwater, dev dir; Florence Rogers, progmg dir; John Clare, mus dir; Jay Bartos, pub affrs dir; Warren Brown, chief of engrg.

KDOX(AM)—See Henderson

KDWN(AM)— Apr 7, 1975: 720 khz; 50 kw-U, DA-N. TL: N36 04 22 W114 58 20. Hrs open: 24 1455 E. Tropicana, .Suite 800, 89119. Phone: (702) 730-0300. Fax: (702) 736-8447. E-mail: kdwn@kdwn.com Web Site: www.kdwn.com. Licensee: KDWN License L.P. (acq 8-7-2006; $17 million). Population served: 1,000,000 Natl. Network: Fox News Radio. Format: News/talk. Target aud: 35-54. ♦Tom Davis, gen mgr; Mark Warlaumont, sls dir; Charlotte Burke, progmg dir. Co-owned TV: .

KEIP(AM)—Not on air, target date: unknown: 760 khz; 1 kw-D, 930 w-N, DA-2. TL: N36 03 30 W115 12 39. Hrs open: 3185 Highland Dr., Suite 13, 89109. Phone: (702) 731-5588. Fax: (702) 731-5851. Licensee: Las Vegas Broadcasters Partnership. Law Firm: Irwin, Campbell & Tannenwald. ♦Carl J. Auel, gen mgr.

KENO(AM)— 1940: 1460 khz; 10 kw-D, 620 w-N, DA-2. TL: N36 11 25 W115 10 35. Hrs open: 8755 W. Flamingo Rd., 89147-8667. Phone: (702) 876-1460. Fax: (702) 876-6685. Web Site: www.foxsportsradio1460.com. Licensee: Lotus Broadcasting Corp. (group owner; (acq 6-1-65). Population served: 850,000 Natl. Rep: Christal. Wire Svc: UPI Format: Sports. Target aud: 18 plus. ♦Tony Bonnici, gen mgr; John Hanson, progmg dir.

KOMP(FM)—Co-owned with KENO(AM). Sept 1, 1966: 92.3 mhz; 100 kw. 1,520 ft TL: N35 56 50 W115 03 01. (CP: 22.9 kw, ant 3,844 ft.). Stereo. 500,000 Format: AOR. ♦John Griffin, progmg dir.

KISF(FM)— March 1989: 103.5 mhz; 100 kw. Ant 1,158 ft TL: N36 00 29 W115 00 20. Hrs open: 6767 W. Tropicana Ave., Suite 102, 89103. Phone: (702) 284-6400. Fax: (702) 284-6403. Web Site: kisf.netmio.com. Licensee: HBC License Corp. Group owner: Univision Radio (acq 9-22-2003; grpsl). Format: Rgnl Mexican. ♦Dana Demerjian, VP &

gen mgr; Cristina Valarelo, gen sls mgr; Joe Reynolds, natl sls mgr; Brent Berger, rgnl sls mgr; Zulene Benjamin, prom mgr; Jose R. Braro, progmg dir; Marry Farua, chief of engrg; Gloria Salvador, traf mgr.

KKLZ(FM)— Jan 26, 1984: 96.3 mhz; 100 kw. Ant 1,170 ft TL: N36 00 29 W115 00 20. Stereo. Hrs opn: 24 1455 E. Tropican, Suite 800, 89119. Phone: (702) 739-9600. Fax: (702) 736-8447. Web Site: www.963kklz.com. Licensee: Beasley Broadcasting of NV LLC. Group owner: Beasley Broadcast Group Inc. (acq 2-1-2001; grpsl). Population served: 850,000 Format: Classic hits. News staff: one. Target aud: 25-44; baby boomers. ♦Harry Williams, gen mgr; Tom Davis, gen sls mgr; Don Hallett, progmg dir; Dan Lea, mus dir; Dennis Mitchell, news dir; Joe Sands, chief of engrg.

KKVV(AM)— May 1, 1990: 1060 khz; 5 kw-D, 43 w-N. TL: N36 09 22 W115 15 24. Hrs open: 24 3185 S. Highland Dr., Suite 13, 89109. Phone: (702) 731-5588. Phone: (702) 650-5588. Fax: (702) 731-5851. E-mail: kkvv@kkvv.com Web Site: www.kkvv.com. Licensee: Las Vegas Broadcasters Inc. (acq 11-8-93; $17,000; FTR: 11-29-93). Population served: 1.7,000,000 Natl. Network: Salem Radio Network. Format: Relg, talk, adult contemp, christian, Sp. News staff: 2; News: 2 hrs wkly min. Target aud: General; General. Spec prog: Sp christian 20 hrs wkly. ♦Carl J. Auel, pres; Jane A. Filler, VP; Fred Hodges, gen mgr.

KLAV(AM)— June 1947: 1230 khz; 1 kw-U. TL: N36 11 20 W115 08 40. Hrs open: 24 1130 East Desert Inn Rd., 89109. Phone: (702) 796-1230. Fax: (702) 853-2599. E-mail: klavradio@aol.com Web Site: www.klav1230am.com. Licensee: AIM Broadcasting-Las Vegas L.L.C. (acq 8-6-2004; $3.2 million). Population served: 1,500,000 Law Firm: Haley, Bader & Potts. Format: Talk, info, sports. Target aud: 25-54. Spec prog: Relg 3 hrs, Indian one hr, Hawaiian 4 hrs, Arabic 7 hrs, Filipino 5 hrs, Hebrew one hr wkly. ♦Patrice Donley, VP; Lisa Lupo, gen mgr; Peggy Merrill, stn mgr; Jon Lindquist, opns mgr.

KLSQ(AM)—See Whitney

KLUC-FM— 1956: 98.5 mhz; 100 kw. 1,191 ft TL: N36 00 29 W115 00 20. Stereo. Hrs opn: 24 6655 W. Sahara Ave., Suite D208, 89146. Phone: (702) 253-9800. Fax: (702) 889-7373. Web Site: www.kluc.com. Licensee: Infinity Radio Inc. Group owner: Infinity Broadcasting Corp. (acq 12-14-00; grpsl). Population served: 1,000,000 Format: CHR/top-40. News staff: one. Target aud: 18-34. ♦Marty Basch, gen mgr; Frank Feder, gen sls mgr; Cat Thomas, progmg dir; Tracy Teagarden, chief of engrg; Theresa Dunbar, traf mgr.

KSFN(AM)—Co-owned with KLUC-FM. 1956: 1140 khz; 10 kw-D, 2.5 kw-N, DA-N. TL: N36 16 05 W115 02 41.24 hrs Web Site: www.hottalk1140.com. Format: Talk. ♦Raina Weathers, gen sls mgr; Jave Patterson, prom dir & progmg dir; Steve Diamond, news dir; Mike Weaver, chief of engrg; Dawn Keeble, traf mgr.

***KNPR(FM)**— Oct 31, 2003: 88.9 mhz; 24.5 kw. Ant 3,680 ft TL: N35 58 02 W115 30 06. (CP: 22 kw, ant 3,903 ft. TL: N35 57 55 W115 29 58.77). Stereo. Hrs opn: 1289 S. Torrey Pines Dr., 89146. Phone: (702) 258-9895. Fax: (702) 258-5646. E-mail: info@knpr.org Web Site: www.knpr.org. Licensee: Nevada Public Radio. Population served: 1,800,000 Natl. Network: NPR, PRI. Format: All news and info. ♦Lamar Marchese, pres & gen mgr.

KNUU(AM)—Paradise, Feb 21, 1962: 970 khz; 5 kw-D, 500 w-N, DA-2. TL: N36 00 40 W115 14 28. Hrs open: 24 1455 E. Tropicana Ave., Suite 550, 89119. Phone: (702) 735-8644. Fax: (702) 734-4755. Web Site: www.knews970.com. Licensee: BTR West Inc. (acq 11-13-2006; $3.9 million). Population served: 1,600,000 Natl. Network: Wall Street, ABC, CNN Radio. Format: News/talk. News staff: 8; News: 154 hrs wkly. Target aud: 35 plus. ♦Michael Metter, pres; Jim Servino, gen mgr.

KPLV(FM)— Sept 1, 1977: 93.1 mhz; 24 kw. Ant 3,742 ft TL: N35 58 02 W115 30 06. Stereo. Hrs opn: 2880 Meade Ave., Suite 250, 89102. Phone: (702) 238-7300. Fax: (702) 792-4890. Fax: (702) 792-9018. Web Site: www.kqol.com. Licensee: Citicasters Licenses L.P. Group owner: Clear Channel Communications Inc. (acq 1999). Law Firm: Hogan & Hartson. Format: Rhythmic adult contemp. ♦Kelly Kibler, gen mgr; Marty Thompson, opns dir; Jason Courtemanche, natl sls mgr; David Himmel, mktg dir; Rik McNeil, progmg dir; Joe McCarthy, news dir; Tree Lee, chief of engrg; Tom Chase, farm dir.

KQRT(FM)— 1993: 105.1 mhz; 50 kw. 1,614 ft TL: N36 19 46 W115 21 49. Stereo. Hrs opn: 24 500 Pilot Rd., Suite D89119, 89119. Phone: (702) 597-3070. Fax: (702) 507-1081. Licensee: Entravision Holdings LLC. Group owner: Entravision Communications Corp. (acq 3-14-2000; grpsl). Natl. Rep: Lotus Entravision Reps LLC. Format: Sp, CHR. Target aud: Spanish; young. ♦Walter Ulloa, CEO & pres; Chris Roman, gen mgr; Kathy Koch, stn mgr & gen sls mgr; Kerina Barzena, prom dir; Nan Long, traf mgr. Co-owned TV: KINC(TV)

KRLV(AM)— 1947: 1340 khz; 1 kw-U. TL: N36 09 22 W115 15 24. Hrs open: 24 5010 S. Spencer St., 89119. Phone: (702) 736-3145. Fax: (702) 740-8196. E-mail: generalmanager@krlv.net Web Site: www.krlv.net. Licensee: Continental Radio Broadcasting Acquisition LLC (D.I.P.) (acq 7-13-2004). Law Firm: Cohn & Marks, LLP. Format: Sp, news, sports. News staff: 3; News: 12 hrs wkly. Target aud: 25-54. ♦Patrice Donley, gen mgr & sls; Rod Stowell, opns mgr.

KSHP(AM)—North Las Vegas, 1954: 1400 khz; 1 kw-U. TL: N36 12 52 W115 09 18. Stereo. Hrs opn: 24 2400 S. Jones, Suite 3, 89146. Phone: (702) 221-1200. Fax: (702) 221-2285. Licensee: Las Vegas Radio Co. Group owner: McNaughton-Jakle Stations (acq 9-20-96; $600,000). Population served: 2,400,000 Format: Radio shopping, sports. ♦K. Richard Jakle, pres; Brett Grant, VP, gen mgr, gen sls mgr & progmg dir; Joe Sands, chief of engrg.

KSNE-FM— Aug 18, 1987: 106.5 mhz; 100 kw. 1,155 ft TL: N36 00 30 W115 00 20. Stereo. Hrs opn: 24 2880 Meade Ave., Suite 250, 89102. Phone: (702) 238-7300. Fax: (702) 732-4597. Licensee: Citicasters Licenses L.P. Group owner: Clear Channel Communications Inc. (acq 1999; grpsl). Law Firm: Hogan & Hartson. Format: Soft adult contemp. News: 4 hrs wkly. Target aud: 25-54; emphasis on women. Spec prog: Relg one hr, pub affrs one hr wkly. ♦Tom Chase, opns mgr & progmg dir; Jason Courtemanche, natl sls mgr; Verena King, prom dir; John Berry, mus dir; Joe McCarthy, news dir; Tree Lee, engrg dir & chief of engrg.

***KSOS(FM)**— July 18, 1972: 90.5 mhz; 100 kw. 1,269 ft TL: N36 00 29 W115 00 20. Stereo. Hrs opn: 24 2201 S. 6th St., 89104. Phone: (702) 731-5452. Phone: (800) 804-5452. Fax: (702) 731-1992. E-mail: info@sosradio.net Web Site: www.sosradio.net. Licensee: Faith Communications Corp. (acq 12-31-71). Population served: 1,900,000 Law Firm: Cohn & Marks. Format: Christian, adult contemp. News: 5 hrs wkly. Target aud: 25-44; young families. ♦Jack French, CEO & VP; Brad Staley, gen mgr; Chris Staley, opns mgr & progmg dir.

KSTJ(FM)—Boulder City, Sept 1, 1982: 102.7 mhz; 96 kw. Ant 1,978 ft TL: N35 56 46 W115 02 34. Stereo. Hrs opn: 24 1455 E. Tropicana Ave., Suite 800, 89119. Phone: (702) 730-0300. Fax: (702) 736-8447. Web Site: www.star1027fm.com. Licensee: KJUL License LLC. Group owner: Beasley Broadcast Group Inc. (acq 9-6-2000; grpsl). Population served: 780,000 Format: Music of the 80s. News: one hr wkly. Target aud: 25-54. ♦Harry Williams, CEO; Allen Shaw, exec VP; Tom Davis, gen sls mgr; Don Hallett, progmg dir; Joe Sands, chief of engrg.

***KUNV(FM)**— Apr 21, 1981: 91.5 mhz; 15 kw. 1,100 ft TL: N36 00 28 W115 00 20. Stereo. Hrs opn: 24 Univ. of Nevada, 1515 E. Tropicana Ave., Suite 240, 89119. Phone: (702) 798-9169. Fax: (702) 736-0983. Web Site: www.kunv.org. Licensee: University of Nevada Board of Regents. Population served: 1,700,000 Format: Jazz, multi-cultural. Target aud: General. Spec prog: Sp 5 hrs, electronic 2 hrs, community affrs 4 hrs, Ger one hr wkly. ♦David Reese, gen mgr; Gig Brown, progmg dir; Joe Sands, chief of engrg.

***KVKL(FM)**— 2006: 91.1 mhz; 1.6 kw vert. Ant 994 ft TL: N35 37 37 W115 16 11. Hrs open: 3185 S. Highland Dr., Suite 13, 89108. Phone: (702) 731-5588. Licensee: Southern Nevada Educational Broadcasters. ♦Carl J. Auel, pres.

KWID(FM)— Mar 22, 1963: 101.9 mhz; 100 kw. Ant 1,181 ft TL: N36 00 28 W115 00 20. Stereo. Hrs opn: 2880-B Meade Ave., Suite 250, 89119. Phone: (702) 238-7300. Fax: (702) 792-9018. Licensee: Citicasters Licenses L.P. Group owner: Clear Channel Communications Inc. (acq 1999; grpsl). Population served: 750,000 Law Firm: Hogan & Hartson. Format: Sp adult hits. Target aud: 25-49. ♦Kelly Kibler, gen mgr.

KWWN(AM)—Not on air, target date: unknown: 1100 khz; 20 kw-D, 2 kw-N, DA-2. TL: N36 12 45 W115 09 45. Hrs open: 8755 W. Flamingo Rd., 89147-8667. Phone: (702) 876-1460. Fax: (702) 876-6685. Licensee: Lotus Broadcasting Corp. ♦Tony Bonnici, gen mgr.

Laughlin

KVGS(FM)— 1991: 107.9 mhz; 98 kw. Ant 1,984 ft TL: N35 39 07 W114 18 42. Hrs open: 24 2725 E. Desert Inn Rd, Suite 180, Las Vegas, 89121. Phone: (702) 784-4000. Phone: (928) 704-4540. Fax: (702) 784-4040. Web Site: www.v108fm.com. Licensee: RBG Las Vegas Licenses LLC (acq 10-3-2005; $38 million with KOAS(FM) Dolan Springs, AZ). Natl. Network: ABC. Format: Urban contemp. Target aud: 25-54; adults. ♦Frank Woodbeck, gen mgr; Jodie Dames, gen sls mgr; Craig Knight, prom dir; Tony Rankin, progmg dir; Amy Stone, mus dir & traf mgr; Al Kirsckner, engrg VP & chief of engrg; Joe Sands, chief of engrg; Rick Fulkerson, chief of engrg.

Logandale

KADD(FM)— September 1997: 93.5 mhz; 82 kw horiz. Ant 2,148 ft TL: N36 38 07 W114 07 18. Hrs open: Box 1866, Lake Havasu City, AZ, 86403. Phone: (928) 855-4560. Fax: (928) 855-7996. E-mail: epress@maddog.net Web Site: www.maddog.net. Licensee: M&M Broadcasting LLC (acq 5-11-2001; $150,000). Format: Hot adult contemp. ♦Chris Rolando, gen mgr.

Lund

***KWPR(FM)**— September 2000: 88.7 mhz; 3 kw. Ant 2,201 ft TL: N39 18 54 W115 05 19. Stereo. Hrs opn: 24
Rebroadcasts KNPR(FM) Las Vegas 100%.
Nevada Public Radio, 1289 S. Torrey Pines Dr., Las Vegas, 89146. Phone: (702) 258-9895. Fax: (702) 258-5646. Web Site: www.nevadapublicradio.org. Licensee: Nevada Public Radio. Population served: 4,000 Format: Div, news. ♦Lamar Marchese, gen mgr.

Mesquite

***KAIZ(FM)**— 2005: 91.1 mhz; 400 w. Ant 827 ft TL: N36 53 51 W114 17 10. Stereo. Hrs opn: 24
Rebroadcasts KLRD(FM) Yucaipa, CA 100%.
2351 Suinset Blvd., Suite 170-218, Rocklin, CA, 95765. Phone: (916) 251-1600. Fax: (916) 251-1650. E-mail: info@air1.com Web Site: www.air1.com. Licensee: Educational Media Foundation. Group owner: EMF Broadcasting. Natl. Network: Air 1. Law Firm: Shaw Pittman. Format: Contemp Christian. News staff: 3. Target aud: 18-35; Judeo-Christian, female. ♦Richard Jenkins, pres; Mike Novak, VP & progmg dir; Lloyd Parker, gen mgr; Keith Whipple, dev dir; Eric Allen, natl sls mgr; David Pierce, progmg mgr; Ed Lenane, news dir; Sam Wallington, engrg dir.

***KEKL(FM)**— 2005: 88.5 mhz; 20.5 kw. Ant 489 ft TL: N36 41 00 W114 30 48. Hrs open: Southern Nevada Educational Broadcasters, 3185 S. Highland Dr., Suite 13, Las Vegas, 89109-1029. Phone: (702) 731-5588. Licensee: Southern Nevada Educational Broadcasters. ♦Carl J. Auel, gen mgr.

KVEG(FM)— July 23, 2001: 97.5 mhz; 100 kw. Ant 981 ft TL: N36 34 52 W114 35 59. Hrs open: 24 3999 Las Vegas Blvd. S., Suite K, Las Vegas, 89119. Phone: (702) 736-6161. Fax: (702) 736-2986. E-mail: mail@kvegas.com Web Site: www.kvegas.com. Licensee: Kemp Broadcasting Inc. Population served: 1,400,000 Law Firm: Koerner & Olender, P.C. Format: CHR/rhythmic. Target aud: 25-39. ♦Gary Cox, gen mgr.

Moapa Valley

KJUL(FM)— July 1, 2001: 104.7 mhz; 100 kw. Ant 604 ft TL: N36 41 00 W114 30 48. Hrs open: 24 Summit American Inc., 5000 W. Oakey, Las Vegas, 89146. Phone: (702) 258-0039. Fax: (702) 258-6536. Licensee: Summit American Inc. Law Firm: KMZ Rosenman. Wire Svc: AP Format: Adult contemp, country. Target aud: 25-54. ♦Scott Gentry, gen mgr.

North Las Vegas

KCYE(FM)— April 1989: 104.3 mhz; 24.5 kw. Ant 3,700 ft TL: N35 58 02 W115 30 06. Stereo. Hrs opn: 24 1455 E. Tropicana Ave., Suite 800, Las Vegas, 89119. Phone: (702) 730-0300. Fax: (702) 736-8447. Web Site: www.kjul.com. Licensee: KJUL License LLC. Group owner: Beasley Broadcast Group Inc. (acq 1-31-2001; grpsl). Format: Country. News: one hr wkly. Target aud: 35-64. ♦Allen Shaw, CEO; Harry Williams, gen mgr; Tom Davis, sls dir & gen sls mgr; Patti Mills, natl sls mgr; Kevin Miskimmons, rgnl sls mgr; David Allen, progmg dir; Joe Sands, chief of engrg.

KSFN(AM)—Licensed to North Las Vegas. See Las Vegas

KSHP(AM)—Licensed to North Las Vegas. See Las Vegas

KXNT(AM)— 1986: 840 khz; 50 kw-D, 25 kw-N, DA-2. TL: N36 23 53 W114 54 57. Stereo. Hrs opn: 24 6655 W. Sahara Ave., Suite D208, Las Vegas, 89146. Phone: (702) 364-8400. Fax: (702) 889-7384. Web Site: www.kxnt.com. Licensee: Infinity Radio Inc. Group owner: Infinity Broadcasting Corp. (acq 11-13-98; grpsl). Natl. Network: CBS Radio. Law Firm: Baraff, Koerner & Olender. Format: News/talk. Target aud: 35-64; upscale adults. ♦Tom Humm, gen mgr; Dan Larson, gen sls mgr; Jack Landreth, progmg dir; Tracy Teagarden, engrg dir & chief of engrg.

Overton

KONV(FM)—Not on air, target date: unknown: 106.9 mhz; 92 kw. Ant 2,043 ft TL: N36 50 55 W114 28 23. Hrs open: Wells Fargo Tower, 17th Fl., 3800 Howard Hughes Pkwy., Las Vegas, 89109. Phone: (702) 385-6000. Fax: (702) 736-2986. Licensee: Kemp Communications Inc. ♦Will Kemp, pres.

Pahrump

KNYE(FM)— Nov 19, 2001: 95.1 mhz; 6 kw. Ant -92 ft TL: N36 11 52 W116 02 08. Hrs open: 24 1230 Dutch Ford Rd., 89048. Phone: (775) 751-6193. Fax: (775) 751-6193. E-mail: karen@knye.com Web Site: www.knye.com. Licensee: Pahrump Radio Inc. (acq 2-14-01). Format: Oldies. ♦Art Bell, pres; Karen Jackson, gen mgr; Joe Sands, chief of engrg.

KXTE(FM)— 1989: 107.5 mhz; 24.5 kw. 3,715 ft TL: N35 58 02 W115 30 06. Stereo. Hrs opn: 24 6655 W. Sahara Ave., Suite C-202, Las Vegas, 89146. Phone: (702) 257-1075. Fax: (702) 889-7575. Web Site: www.xtremeradio.fm. Licensee: Infinity Radio Inc. Group owner: Infinity Broadcasting Corp. (acq 11-13-98; grpsl). Format: Talk, alternative. Target aud: 18-49. ♦Marty Basch, gen mgr; Marc Isquith, gen sls mgr; Chris Ripley, mus dir.

Panaca

***KLNR(FM)**— May 1989: 91.7 mhz; 1 kw. Ant 3,424 ft TL: N37 53 38 W114 34 40. Stereo. Hrs opn: 24
Rebroadcasts KNPR(FM) Las Vegas 100%.
1289 S. Torrey Pines Dr., Las Vegas, 89146. Phone: (702) 258-9895. Fax: (702) 258-5646. Licensee: Nevada Public Radio. Population served: 2,000 Natl. Network: NPR. Law Firm: Dow, Lohnes & Albertson. Format: All news. ♦Lamar Marchese, gen mgr.

Paradise

KNUU(AM)—Licensed to Paradise. See Las Vegas

Pioche

KBZB(FM)— 2002: Stn currently dark. 98.9 mhz; 5 kw. Ant 3,375 ft TL: N37 53 44 W114 34 41. (CP: 100 kw, ant 1,958 ft. TL: N37 27 40 W114 27 55). Hrs opn: Box 692, 89043. Phone: (775) 962-5681. Licensee: Gla-Mar Broadcasting LLC (acq 5-16-2003). Format: Div, talk.

Reno

KBZZ(AM)—See Sparks

KDOT(FM)— Oct 12, 1966: 104.5 mhz; 25 kw. 2,929 ft TL: N39 18 48 W119 52 59. Stereo. Hrs opn: 24 Box 9870, 2900 Sutro St., 89512. Phone: (775) 329-9261. Fax: (775) 323-1450. E-mail: javet@kdot.com Web Site: www.kdot.com. Licensee: Lotus Radio Corp. Group owner: Lotus Communications Corp. (acq 3-30-93; $600,000 with KIRS(AM) Sun Valley; FTR: 4-19-93). Natl. Rep: D & R Radio. Format: Active rock. News staff: one; News: 3 hrs wkly. Target aud: 18-49; young active adults that like today's lifestyle. ♦Dane Wilt, gen mgr; Marc Isquith, gen sls mgr; Derek Sante, prom dir; Jack Landreth, progmg dir; Joanne Silvernail, traf mgr.

KPLY(AM)—Co-owned with KDOT(FM). Oct 25, 1928: 630 khz; 5 kw-D, 1 kw-N, DA-N. TL: N39 34 25 W119 50 48. E-mail: espnradio630@aol.com (Acq 1995; $325,000).200,000 Format: Sports. Target aud: 25-54. ♦Dane Wilt, VP; Ken Allen, progmg mgr.

KHIT(AM)— Jan 29, 1955: 1450 khz; 1 kw-U. TL: N39 33 26 W119 47 47. Hrs open: 24 Box 9870, 2900 Sutro St., 89512. Phone: (775) 329-9261. Fax: (775) 323-1450. E-mail: kena@kozzradio.com Licensee: Lotus Radio Corp. Group owner: Lotus Communications Corp. (acq 9-67). Population served: 225,000 Natl. Rep: D & R Radio. Format: Big band. News staff: one; News: 2 hrs wkly. Target aud: 25-49. ♦Dane Wilt, gen mgr; Jim McClain, opns mgr; Raina Weathers, gen sls mgr; Ken Allen, prom dir & progmg dir; Steve Diamond, news dir; Mike Weaver, chief of engrg; Dawn Keeble, traf mgr.

KOZZ-FM—Co-owned with KHIT(AM). September 1969: 105.7 mhz; 75 kw. 2,120 ft TL: N39 15 34 W119 42 21. Stereo. 24 Web Site: www.kozzradio.com. (Acq 1-1-78). Format: Classic rock. ♦Bill Shriftman, CFO; Dawn Keeble, gen sls mgr & traf mgr; Rick Carter, prom mgr & progmg dir.

***KIHM(AM)**— Jan 1, 1984: 920 khz; 4.6 kw-D, 850 w-N. TL: N39 30 41 W119 42 51. Hrs open: 3550 Berron Way, Suite 3B, 89511. Phone: (775) 828-4228. Fax: (775) 823-5444. Web Site: www.kihmradio.org. Licensee: IHR Educational Broadcasting (group owner; acq 8-24-2000). Format: Catholic. ♦Doug Pearson, stn mgr.

KJFK(AM)—Listing follows KRNO(FM).

KKOH(AM)— Oct 13, 1970: 780 khz; 50 kw-U, DA-N. TL: N39 40 41 W119 48 06. Hrs open: 24 595 E. Plumb Ln., 89502. Phone: (775) 789-6700. Fax: (775) 789-6767. E-mail: dan.mason@citcomm.com Web Site: www.kkoh.com. Licensee: Citadel Broadcasting Co. Group owner: Citadel Broadcasting Corp. (acq 5-18-92; $12.5 million; grpsl; FTR: 6-8-92). Population served: 1,500,000 Natl. Rep: McGavren Guild. Format: News/talk. News staff: 4; News: 28 hrs wkly. Target aud: 35-64. Spec prog: Sports, Sp 2 hrs wkly. ♦Farid Suleman, CEO & pres; Dana Johnson, gen mgr; Andrew Perini, gen sls mgr & chief of engrg; Dan Mason, progmg dir.

KLCA(FM)—Tahoe City, CA) Apr 5, 1985: 96.5 mhz; 4 kw. 2,965 ft TL: N39 18 47 W119 52 59. Stereo. Hrs opn: 24 961 Matley Ln., Suite 120, 89502. Phone: (775) 829-1964. Fax: (775) 825-3183. Web Site: www.alice965.com. Licensee: Americom Broadcasting. Group owner: Americom (acq 1996; $1.225 million). Format: Modern hits. News: 3 hrs wkly. Target aud: 18-34. Spec prog: Metal shop 2 hrs wkly. ♦Daniel Cook, gen mgr.

KNEV(FM)— Dec 25, 1953: 95.5 mhz; 60 kw. 2,280 ft TL: N39 15 34 W119 42 16. Stereo. Hrs opn: 595 E. Plumb, 89502. Phone: (775) 789-6700. Fax: (775) 789-6767. Web Site: www.magic95.com. Licensee: Citadel Broadcasting Co. Group owner: Citadel Broadcasting Corp. (acq 4-13-93; $500,000; FTR: 5-3-93). Population served: 450,000 Format: Hot adult contemp. Target aud: General. Spec prog: Jazz 2 hrs, relg one hr, pub affrs one hr wkly. ♦Dana Johnson, gen mgr.

KNIS(FM)—See Carson City

KODS(FM)—Carnelian Bay, CA) 1970: 103.7 mhz; 6.3 kw. 2,985 ft TL: N39 18 16 W119 53 00. Stereo. Hrs opn: 961 Matley Ln., Suite 120, 89502. Phone: (775) 829-1964. Fax: (775) 825-3183. Web Site: www.river1037.com. Licensee: Americom Broadcasting. (acq 1996). Population served: 442,800 Natl. Rep: CBS Radio. Format: Hits of the 60s & 70s. ♦Tom Quinn, pres; Daniel Cook, gen mgr; Heather Forcier, gen sls mgr; Bob Garrison, news dir & pub affrs dir; Steve Weber, chief of engrg; Kristen Connell, traf mgr & disc jockey.

KRNO(FM)—Incline Village, July 1974: 106.9 mhz; 35 kw. Ant 2,988 ft TL: N39 18 38 W119 53 01. Stereo. Hrs opn: 24 300 E. 2nd St., 14th Floor, 89501. Phone: (775) 829-1964. Fax: (775) 825-3183. Licensee: Americom Las Vegas L.P. Group owner: Americom (acq 4-16-98; grpsl). Population served: 332,457 Law Firm: Shaw Pittman. Format: Soft rock, soft adult contemp. News staff: one; News: 18 hrs wkly. Target aud: 25-54; women. ♦Daniel Cook, gen mgr; Eric Bowlin, gen sls mgr; Dan Fritz, progmg dir; Steve Weber, chief of engrg; Wanda Schiwart, news dir & traf mgr.

KJFK(AM)—Co-owned with KRNO(FM). Oct 30, 1963: 1230 khz; 1 kw-U. TL: N39 30 42 W119 42 48.24 Format: Progressive talk. Target aud: 35 plus. ♦Heather Forcier, gen sls mgr; Dan Fritz, progmg dir.

KRNV-FM— Aug 12, 1986: 102.1 mhz; 11 kw. Ant 492 ft TL: N39 35 03 W119 47 52. Hrs open: 24 300 S. Wells Ave., Suite 12, 89502. Phone: (775) 333-1017. Fax: (775) 333-9046. Web Site: www.entravision.com.Network progmg Licensee: Entravision Holdings LLC. Group owner: Entravision Communications Corp. (acq 3-14-00; grpsl). Format: Sp. Target aud: 18-49; adults. ♦Philip Wilkinson, COO; Walter F. Ulloa, chmn; Jeff Liberman, pres; John DeLorenzo, CFO.

KRZQ-FM—See Sparks

KTHX-FM—Dayton, June 10, 1983: 100.1 mhz; 12.2 kw. Ant 2,162 ft TL: N39 15 34 W119 42 21. Stereo. Hrs opn: 24 300 E Second St., 14th Fl., 89501. Phone: (775) 333-0123. Fax: (775) 322-7361. Web Site: www.kthxfm.com. Licensee: Wilks License Co.-Reno LLC. Group owner: NextMedia Group L.L.C. (acq 9-29-2005; grpsl). Natl. Network: ABC. Law Firm: Leventhal, Senter & Lerman. Format: AAA. News: 2 hrs wkly. Target aud: 18-49; upscale, high income & educated. ♦Craig B. Klosk, pres; April Clark, gen mgr.

***KUNR(FM)**— Oct 7, 1963: 88.7 mhz; 20 kw. 2,169 ft TL: N39 15 34 W119 42 16. Stereo. Hrs opn: Mail Stop 294, Univ. of Nevada, 89557. Phone: (775) 327-5867. Fax: (775) 784-1381. Web Site: www.kunr.org. Licensee: University of Nevada Board of Regents. Population served: 370,000 Natl. Network: NPR, PRI. Format: News, jazz, class music. Target aud: General. Spec prog: Folk 2 hrs, ethnic 9 hrs wkly. ♦Steven Zink, exec VP; Bobbi Lazzarone, stn mgr; Terry Joy, opns dir; Kate Grey, sls dir & news dir; James Shannon, chief of engrg.

KURK(FM)— November 1994: 92.9 mhz; 48 kw. Ant 502 ft TL: N39 35 03 W119 48 06. Hrs open: c/o KKOH(AM) and KNEV(FM), 595 East Plumb Ln., 89502. Phone: (775) 789-6700. Fax: (775) 789-6767. Licensee: Wilks License Co.-Reno LLC. Group owner: NextMedia Group L.L.C. (acq 9-29-2005; grpsl). Format: Classic rock and roll. ♦Dana Johnson, gen mgr.

KXEQ(AM)— July 1946: 1340 khz; 1 kw-U. TL: N39 32 22 W119 46 53. Hrs open: 225 Linden St., 89502. Phone: (775) 827-1111. Phone: (775) 827-1313. Fax: (775) 827-2082. Licensee: Azteca Broadcasting Corp. (group owner; acq 10-16-91; $30,000; FTR: 11-4-91). Population served: 250,000 Natl. Network: AP Radio. Format: Sp. ♦Juan Morales, gen mgr & progmg dir.

KXTO(AM)— 1991: 1550 khz; 2.5 kw-D, 94 w-N. TL: N39 34 39 W119 50 52. Hrs open: 24 1085 E. 2nd St., Suite 2, 89502. Phone: (775) 348-5850. Web Site: www.kxto.com. Licensee: First Broadcasting of Nevada Inc. Population served: 500,000 Format: Sp, relg. News: 10 hrs wkly. Target aud: Hispanics. ♦Yolanda Amaya, gen mgr.

KZTQ(FM)—Carson City, June 27, 1972: 97.3 mhz; 87 kw. 2,112 ft TL: N39 15 21 W119 42 37. Stereo. Hrs opn: 24 961 Matley Ln., Suite 120, 89502. Phone: (775) 829-1964. Fax: (775) 825-3183. Licensee: Americom Las Vegas L.P. Group owner: Americom (acq 4-27-98; grpsl). Population served: 285,000 Format: CHR. News staff: one; News: 2 hrs wkly. Target aud: 18-34; women. ♦Daniel Cook, gen mgr; Steve Webber, chief of engrg.

Smith

KSVL(FM)— 1999: 92.3 mhz; 490 w. Ant 2,073 ft TL: N38 41 06 W119 11 04. Hrs open: Box 123, 89430. Phone: (775) 465-2200. Licensee: Donegal Enterprises. Format: Class. ♦Wayne Donegal, gen mgr.

Sparks

KBDB(AM)— 2002: 1400 khz; 600 w-U. TL: N39 34 10 W119 45 03. Hrs open: 1085 E. 2nd St., Suite 1, Reno, 89502. Phone: (775) 348-5852. Fax: (775) 348-5865. E-mail: admin@mega1400.com Web Site: www.mega1400.com. Licensee: George S. Flinn Jr. Format: Sp. ♦Ruben Billalobos, gen mgr.

KBZZ(AM)— Aug 9, 1960: 1270 khz; 5 kw-U, DA-2. TL: N39 32 03 W119 39 44. Hrs open: 24 961 Matley Ln., Suite 120, Reno, 89502. Phone: (775) 829-1964. Fax: (775) 825-3183. Web Site: www.kbzz.com. Licensee: Americom Las Vegas L.P. Group owner: Americom (acq 1996; grpsl). Population served: 300,000 Natl. Network: CBS, Westwood One. Format: Sports & info, news/talk. Target aud: 25-54; primarily men who are interested in sports. ♦Tom Quinn, pres; Daniel Cook, gen mgr; Dan Fritz, gen sls mgr & progmg dir; Steve Webber, chief of engrg.

KJZS(FM)— 1993: 92.1 mhz; 8.9 kw. Ant 502 ft TL: N39 35 03 W119 48 06. Hrs open: 24 300 E. Second St. 14 th Fl., Reno, 89501. Phone: (775) 333-0123. Web Site: www.smoothjazzreno.com. Licensee: Wilks License Co.-Reno LLC. Group owner: NextMedia Group L.L.C. (acq 9-29-2005; grpsl). Population served: 300000 Law Firm: Latham & Watkins. Format: Jazz, adult Contempo. ♦Robert Dees, gen mgr & progmg dir.

***KLRH(FM)**—Not on air, target date: unknown: 88.3 mhz; 1.78 kw. Ant 2,886 ft TL: N39 45 38 W119 27 59. Hrs open: 2351 Sunset Blvd., Suite 170-218, Rocklin, CA, 95765. Phone: (916) 251-1600. Fax: (916) 251-1650. E-mail: klove@klove.com Web Site: www.klove.com. Licensee: Educational Media Foundation. Group owner: EMF Broadcasting. Natl. Network: K-Love. Law Firm: Shaw Pittman. Format: Contemp Christian. News staff: 3. Target aud: 25-44; Judeo Christian female. ♦Richard Jenkins, pres; Mike Novak, VP; Keith Whipple, dev dir; David Pierce, progmg mgr; Ed Lenane, news dir; Sam Wallington, engrg dir; Karen Johnson, news rptr; Marya Morgan, news rptr; Richard Hunt, news rptr.

KRZQ-FM— July 1, 1983: 100.9 mhz; 2.9 kw. 203 ft TL: N39 22 04 W119 47 07. Stereo. Hrs opn: 24 300E. 2nd St. 14th Floor, Reno, 89015. Phone: (775) 333-0123. Fax: (775) 333-0110. Licensee: Wilks License Co.-Reno LLC. Group owner: NextMedia Group L.L.C. (acq 9-29-2005; grpsl). Natl. Network: ABC. Law Firm: Leibowitz & Associates. Format: Alternative rock, talk. Target aud: 18-54; general. ♦Craig B. Klosk, pres; April Clark, gen mgr; Jeremy Smith, progmg dir; Matt Bates, mus dir; Robert Barfoot, chief of engrg; Lori Quinn, traf mgr.

Sun Valley

KQLO(AM)— 1946: 1590 khz; 5 kw-D, 67 w-N. TL: N39 24 57 W119 42 51. Hrs open: 101 Locust St., Reno, 89501-1012. Phone: (775) 322-0847. Fax: ((775) 322-0927. E-mail: business@kqlo.com Web Site: www.kqlo.com. Licensee: Universal Broadcasting Inc. (acq 12-22-2003; $140,000). Population served: 186,500 Natl. Rep: Interep. Format: Sp contemp. Target aud: General. ♦Lee Chavez, gen mgr & natl sls mgr; Lourdes Rincon, system mgr.

KUUB(FM)— 1999: 94.5 mhz; 12 kw. Ant 459 ft TL: N39 35 02 W119 47 53. (CP: 50 kw). Hrs opn: 2900 Sutro St., Reno, 89512. Phone: (775) 329-9261. Fax: (775) 323-1450. Web Site: www.945themountain.com. Licensee: Lotus Radio Corp. Group owner: Lotus Communications Corp. Format: Country. ♦Dane Wilt, gen mgr.

KWNZ(FM)— 2002: 93.7 mhz; 3.6 kw. Ant 423 ft TL: N39 35 02 W119 47 54. Hrs open: 595 E. Plumb Ln., Reno, 89502. Phone: (775) 789-6700. Fax: (775) 789-6767. E-mail: angel.garcia@citcomm.com Web Site: wild937.com. Licensee: Flinn Broadcasting Corp. Format: CHR. ♦Dana Johnson, gen mgr.

Tonopah

KHWK(FM)— July 29, 1982: Stn currently dark. 92.7 mhz; 290 w. Ant 971 ft TL: N38 04 22 W117 13 16. Stereo. Hrs opn: Box 1669, 89049. Phone: (775) 482-5724. Licensee: Donald W. Kaminiski Jr. (acq 3-16-92; $240,000; FTR: 4-6-92). ♦Don Kaminski, CEO & gen mgr.

KTNP(AM)—Not on air, target date: unknown: 1400 khz; 1 kw-D, 880 w-N. TL: N38 05 06 W117 13 18. Hrs open: Box 1254, Alameda, CA, 94501. Phone: (510) 769-5904. Licensee: Eastern Sierra Broadcasting. ♦Chris Kidd, pres.

***KTPH(FM)**— October 1988: 91.7 mhz; 100 w. Ant 1,433 ft TL: N38 03 07 W117 13 30. Stereo. Hrs opn: 24
Rebroadcasts KNPR(FM) Las Vegas 100%.
1289 S. Torrey Pines Dr., Las Vegas, 89146. Phone: (702) 258-9895. Fax: (702) 258-5646. Licensee: Nevada Public Radio. Population served: 4,000 Natl. Network: NPR. Format: Classical, news. ♦Lamar Marchese, pres & gen mgr.

Wendover

KVUW(FM)— 2006: 102.3 mhz; 3 kw. Ant 26 ft TL: N40 44 30 W114 02 10. Hrs open: 479 E. Wendover Blvd., 84083. Phone: (435) 665-0600. Fax: (435) 665-0600. Licensee: Murray Grey Broadcasting Inc. (group owner). (acq 12-21-2005; $750,000 with KUSZ(FM) Laramie, WY). ♦Steven A. Silberberg, pres.

Whitney

KLSQ(AM)— Aug 15, 1986: 870 khz; 5 kw-D, 430 w-N, DA-N. TL: N35 58 35 W114 57 03. Stereo. Hrs opn: 24 6767 W. Tropicana, Ste. 102, Las Vegas, 89103. Phone: (702) 284-6400. Fax: (702) 284-6403. Web Site: www.netmio.com. Licensee: HBC-Las Vegas Inc. Group owner: Univision Radio (acq 9-22-2003; grpsl). Format: Sp. Target aud: 35 plus. ♦Dana Demerjiah, VP; Dana Demerjian, gen mgr; Cristina Valarezo, gen sls mgr; Jose R. Bravo, progmg dir; Manny Garcia, chief of engrg; Gloria Salvador, traf mgr.

Winchester

KBET(AM)— May 22, 2006: Stn currently dark. 790 khz; 1 kw-D, 300 w-N, DA-2. TL: N36 05 27 W115 00 59. Hrs open: 1455 E. Tropicana, Suite 800, Las Vegas, 89119. Phone: (702) 730-0300. Licensee: WAEC License L.P. Group owner: Diamond Broadcasting Corp. (acq 3-28-2007; $2.5 million for CP). Format: Country legends. ♦Tom Davis, gen mgr; R.W. Smith, progmg dir.

Winnemucca

KWNA(AM)— Jan 28, 1955: 1400 khz; 1 kw-U. TL: N40 57 23 W117 42 48. Hrs open: 24 Box 1400, 89446. Secondary address: 5130 E. Weikel Dr. 89446. Phone: (775) 623-5203. Fax: (775) 625-1011. Web Site: www.kwnaradio.com. Licensee: Ely Radio LLC. (acq 11-6-2006; $500,000 with co-located FM). Population served: 9,500 Rgnl rep: Art Moore. Law Firm: Gardner, Carton & Douglas. Format: News/talk. News staff: 2; News: 12 hrs wkly. Target aud: General. Spec prog: Farm 2 hrs wkly. ♦Bob Bolton, gen mgr & gen sls mgr; Richard Smith, progmg dir; Eric Skye, news dir; Tim Carroll, chief of engrg; Ashley Matson, traf mgr.

KWNA-FM— Apr 3, 1982: 92.7 mhz; 60 w. 2,120 ft TL: N41 00 40 W117 45 59. (CP: 140 w). Stereo. 5 AM-midnight (M-F); 6 AM-midnight (S); 7 AM-midnight (Su) Format: C&W.

New Hampshire

Bedford

WMLL(FM)— June 1996: 96.5 mhz; 730 w. Ant 935 ft TL: N42 59 02 W71 35 22. Hrs open: 500 Commercial St., Manchester, 03101. Phone: (603) 669-7979. Fax: (603) 669-4641. Web Site: www.965themill.com. Licensee: Saga Communications of New England LLC. Group owner: Saga Communications Inc. (acq 9-29-97; $3.3 million). Population served: 300,000 Natl. Rep: Katz Radio. Format: Classic rock. Target aud: 35-54; baby boomers. ♦Edward Christian, CEO; Raymond R. Garon, pres; Samuel Bush, CFO; J.C. Haze, opns dir & progmg dir.

Belmont

WNHW(FM)— May 8, 1994: 93.3 mhz; 300 w. Ant 1,020 ft TL: N43 23 52 W71 33 03. Stereo. Hrs opn: 11 Kimball Dr., Ste 114, Hooksett, 03106. Phone: (603) 225-1160. Fax: (603) 225-5938. Web Site: www.933thewolf.com. Licensee: Nassau Broadcasting III L.L.C. Group owner: Nassau Broadcasting Partners L.P. (acq 10-1-2004; $8 million with WJYY(FM) Concord). Format: Country. ♦Brit Johnson, gen mgr; Dawn Parris, gen sls mgr; Matt Forrest, progmg dir; Steve Ordinetz, engrg mgr & chief of engrg.

Berlin

WMOU(AM)— 1947: 1230 khz; 1 kw-U. TL: N44 27 32 W71 10 16. Hrs open: 24 Box 489, 297 Pleasant St., 03570. Phone: (603) 752-1230. Fax: (603) 752-3117. E-mail: wmou@ncia.net Licensee: Barry P. Lunderville. (acq 11-11-2003; $75,000). Population served: 35,000 Natl. Network: Westwood One. Format: Adult standards. News staff: one; News: 6 hrs wkly. Target aud: 25-54; local residents of northern New Hampshire. Spec prog: Fr 3 hrs, talk 2 hrs, swap shop 3 hrs wkly. ♦Barry Lunderville, pres & progmg dir; Bob Barbin, opns dir; Randy Frank, gen sls mgr; Brian Lunderville, chief of engrg.

WRTN(AM)—Not on air, target date: unknow: 1490 khz; 1 kw-D, 930 w-N. TL: N44 28 58 W71 10 38. Hrs open: 297 Pleasant St., 03570. Phone: (603) 752-1230. Fax: (603) 752-3117. Licensee: Barry P. Lunderville. ♦Barry P. Lunderville, gen mgr.

Campton

WLKC(FM)— 1997: 105.7 mhz; 125 w. Ant 2,001 ft TL: N43 57 32 W71 33 23. Stereo. Hrs opn: 24
Rebroadcasts WXRU(FM) Wolfeboro 100%.
288 S. River Rd., Bedford, 03110. Phone: (603) 669-1250. Fax: (603) 528-1638. E-mail: nebco231@hotmail.com Licensee: Devon Broadcasting Co. Inc. (acq 3-5-99; $300,000). Format: AAA. Target aud: 25-54. ♦Steve Young, gen mgr; Dana Marshall, progmg dir; Lou Muise, chief of engrg; Stephanie Battaglia, traf mgr.

Claremont

WHDQ(FM)—Listing follows WTSV(AM).

WQTH(AM)—Not on air, target date: unknown: 720 khz; 50 kw-D, 670 w-N, 50 kw-CH, DA-3. TL: N43 19 35 W72 22 47. Hrs open: Box 2295, New London, 03257. Phone: (603) 448-0500. Fax: (603) 448-6601. Licensee: KOOR Communications Inc. (group owner) Population served: 376,428 ♦Robert L. Vinikoor, gen mgr.

WTSV(AM)— 1948: 1230 khz; 1 kw-U. TL: N43 22 15 W72 19 42. Hrs open: 106 N. Main St., West Lebanon, 03784. Phone: (603) 298-0332. Fax: (603) 727-0134. E-mail: espnthescore@aol.com Web Site: www.scoreradio.com. Licensee: Nassau Broadcasting III L.L.C. Group owner: Nassau Broadcasting Partners L.P. (acq 8-2-2004; grpsl). Population served: 39,500 Natl. Network: ABC, ESPN Radio. Rgnl rep: Roslin. Law Firm: Rini & Coran. Format: Sports. Target aud: 35 plus. ♦Jeffrey Shapiro, pres; Shirley Clark, gen mgr.

WHDQ(FM)—Co-owned with WTSV(AM). 1948: 106.1 mhz; 9.51 kw. 1,068 ft TL: N43 23 48 W72 18 01. Stereo. E-mail: info@q106rock.com Web Site: www.q106rock.com.150,000 Format: Classic rock.

Concord

***WEVO(FM)**— Aug 4, 1981: 89.1 mhz; 50 kw. 380 ft TL: N43 12 53 W71 34 28. Stereo. Hrs opn: 24 207 N. Main St., 03301. Phone: (603) 228-8910. Fax: (603) 224-6052. E-mail: admin@nhpr.org Web Site: www.nhpr.org. Licensee: New Hampshire Public Radio Inc. Natl. Network: NPR, PRI. Law Firm: Garvey, Schubert & Barer. Wire Svc: AP Format: News/talk. News staff: 9; News: 42 hrs wkly. Target aud: 25-54; well educated adults. Spec prog: Folk 3 hrs wkly. ♦Elizabeth Gardella, pres; Mark Bevis, news dir; John Huntley, engrg dir.

WJYY(FM)— Sept 15, 1983: 105.5 mhz; 1.55 kw. 456 ft TL: N43 16 46 W71 30 15. Stereo. Hrs opn: 24 11 Kimball Dr., Hooksett, 03106. Phone: (603) 225-1160. Fax: (603) 225-5938. Web Site: www.wjyy.com. Licensee: Nassau Broadcasting III L.L.C. Group owner: Nassau Broadcasting Partners L.P. (acq 10-1-2004; $8 million with WNHW(FM) Belmont). Format: CHR. News staff: one; News: 6 hrs wkly. Target aud: 25-54. ♦Brit Johnson, gen mgr; Dawn Parris, gen sls mgr; Joe Dukette, progmg dir.

WKXL(AM)— June 15, 1946: 1450 khz; 1 kw-U. TL: N43 11 39 W71 33 17. Hrs open: 24 37 Redington Rd., 03301. Phone: (603) 225-5521 Office. Phone: (603) 224-1450 Studio. Fax: (603) 224-6404. E-mail: info@wkxl1450.com Web Site: www.wkxl1450.com. Licensee: New Hampshire Family Radio LLC (acq 12-16-2004;. $800,000). Population served: 80,000 Natl. Network: AP Radio. Wire Svc: AP Format: News/talk. News staff: 2; News: 36 hrs wkly. Target aud: 35 plus; adults in Concord, Hillsboro, Manchester & contiguous towns. ♦Anthony Schilella, gen mgr, stn mgr, opns mgr & progmg dir.

WNNH(FM)—See Henniker

***WSPS(FM)**— 1974: 90.5 mhz; 200 w. 110 ft TL: N43 11 37 W71 34 29. Hrs open: 24 St. Paul's School, 325 Pleasant St., 03301. Phone: (603) 228-4810. Phone: (603) 229-4600. Fax: (603) 229-4891. E-mail: wsps@sps.edu Web Site: www.wsps.sps.edu. Licensee: St. Paul's School. Population served: 40,000 Law Firm: Drinker Biddle & Reath. Format: Div. Target aud: General. ♦David Harvey, gen mgr.

WTPL(FM)—Hillsboro, Oct 1, 1989: 107.7 mhz; 580 w. 738 ft TL: N43 09 00 W71 47 56. Stereo. Hrs opn: 24 501 South St., Bow, 03304. Phone: (603) 545-0777. Fax: (603) 545-0781. Web Site: www.wwtplfm.com. Licensee: Great Eastern Radio LLC (acq 6-3-2004; $1.5 million). Natl. Network: CBS Radio, ESPN Radio. Format: News/talk, sports. News staff: 2; News: 50 hrs wkly. Target aud: 35 plus; adult audience in Merrimack & Hillsborough counties. ♦Mike Johnson, gen mgr.

***WVNH(FM)**— Mar 7, 1999: 91.1 mhz; 650 w vert. 331 ft TL: N43 23 54 W71 25 24. Hrs open: Box 40, 03302. Phone: (603) 227-0911. E-mail: info@wvnh.org. Web Site: www.wvnh.org. Licensee: New Hampshire Gospel Radio Inc. Format: Christian. ♦Peter Stohrer, gen mgr; Cheryl Eggert, stn mgr.

WWHK(FM)— Mar 7, 1972: 102.3 mhz; 3 kw. Ant 285 ft TL: N43 13 00 W71 34 34. Stereo. Hrs opn: 24 11 Kimbell Dr., Suite 114, Hooksett, 03106. Phone: (603) 225-1160. Fax: (603) 225-8935. Licensee: Capitol Broadcasting Corp. Inc. Group owner: Vox Radio Group L.P. (acq 8-12-99). Population served: 100,000 Format: Classic rock. Target aud: 25-54; adult audience in Merrimack county - south central NH. ♦Brid Johnson, gen mgr.

Conway

WBNC(AM)—Listing follows WVMJ(FM).

WMWV(FM)— June 23, 1967: 93.5 mhz; 3 kw. 420 ft TL: N43 56 48 W71 08 24. Stereo. Hrs opn: Mt. Washington Radio, Box 2008, 03818. Secondary address: FedEx/UPS, A30 Settlers Green OVP, Rt 16, North Conway 03860. Phone: (603) 356-8870. Fax: (603) 356-8875. E-mail: office@wmwv.com Web Site: www.wmwv.com. Licensee: Mt. Washington Radio & Gramophone L.L.C. (group owner; acq 9-27-01; grpsl). Population served: 17,000 Natl. Rep: Roslin. Format: AAA. ♦Ronald Frizzell, gen mgr; Charles Osgood, opns VP & chief of engrg.

WVMJ(FM)— Oct 23, 1995: 104.5 mhz; 3 kw. 328 ft TL: N43 55 34 71 05 46. Stereo. Hrs opn: 24 Box 2008, 03818. Secondary address: Settlers' Green Rt. 16, N. Conway 03860. Phone: (603) 356-8870. Fax: (603) 356-8875. Web Site: www.conwaymagic.com. Licensee: Mt. Washington Radio & Gramophone L.L.C. (group owner; acq 10-15-01; grpsl). Format: Adult contemp. ♦Ron Frizzell, gen mgr; Charles Osgood, opns mgr & chief of engrg; Greg Frizzell, gen sls mgr; Cooper Fox, progmg dir; Dean Luttrell, news dir; Karen Stone, traf mgr.

WBNC(AM)—Co-owned with WVMJ(FM). Dec 21, 1955: 1050 khz; 1 kw-D, 63 w-N. TL: N43 58 48 W71 06 36. Web Site: www.conwaymagic.com. ♦Greg Frizzell, sls dir; Karen Stone, traf mgr.

Derry

WDER(AM)— October 1983: 1320 khz; 10 kw-D, 1 kw-N, DA-2. TL: N42 51 59 W71 17 14. Hrs open: 24 Box 465, 8 Lawrence Rd., 03038-6465. Phone: (603) 437-9337. Phone: (603) 434-9302. Fax: (603) 434-1035. E-mail: wderam1320@aol.com Web Site: www.lifechangingradio.com. Licensee: Blount Communications Inc. of NH. Group owner: Blount Communications Group (acq 9-5-00; $793,000). Natl. Network: Salem Radio Network. Format: Talk, relg. Target aud: Male & Female ages 25-54. ♦William Blount, pres; David Young, VP & gen mgr; Emanuel DaCunha, stn mgr; Steve Sobozenski, opns mgr.

Dover

WOKQ(FM)— August 1970: 97.5 mhz; 50 kw. 492 ft TL: N43 13 26 W70 58 18. Stereo. Hrs opn: 24 Box 576, 03821-0576. Secondary address: 292 Middle Rd. 03820-4901. Phone: (603) 749-9750. Fax: (603) 749-1459. E-mail: wokq.nh@citcomm.com Web Site: www.wokq.com. Licensee: Citadel Broadcasting Co. Group owner: Citadel Broadcasting Corp. (acq 9-1-99; grpsl). Population served: 1,000,000 Natl. Network: CNN Radio. Natl. Rep: Christal. Law Firm: Paul, Hastings, Janofsky & Walker. Wire Svc: AP Format: Country. News staff: 2. Target aud: 25-54; general. ♦Farid Suleman, CEO; Judy Ellis, pres; Martin Lessard, gen mgr; Mark Ericson, opns mgr; Mark Jennings, progmg dir.

WTSN(AM)— August 1956: 1270 khz; 5 kw-U, DA-2. TL: N43 11 01 W70 51 14. Hrs open: 24 Box 400, 101 Back Rd., 03821-0400. Phone: (603) 742-1270. Fax: (603) 742-0448. Licensee: Garrison City Broadcasting Inc. (acq 3-18-83). Population served: 250,000 Natl. Rep: Roslin. Format: News/talk, sports. News staff: 3. Target aud: 25-54; very affluent. ♦Bob Demers, CEO; Rick Bean, gen mgr.

Durham

***WUNH(FM)**— July 15, 1963: 91.3 mhz; 3 kw. 300 ft TL: N43 09 23 W70 56 26. Stereo. Hrs opn: 24 Memorial Union Bldg., Univ. of New Hampshire, 03824. Phone: (603) 862-2541. Phone: (603) 862-2087. Fax: (603) 862-2543. E-mail: gm@wunh.unh.edu Web Site: www.wunh.unh.edu. Licensee: University of New Hampshire. Population served: 500,000 Natl. Network: AP Radio. Format: Progsv. News: 7 hrs wkly. Target aud: Diverse. Spec prog: Black 4 hrs, blues 3 hrs, jazz 5 hrs, Pol 2 hrs, folk 4 hrs, celtic 2 hrs wkly. ♦Josh Cilley, gen mgr; Alexandra Buchalski, opns dir; Abbie Crocker, prom dir; Augie Ciotti, progmg dir; Greg Falla, mus dir; John Bosselman, news dir; Peter Geremia, chief of engrg.

Exeter

WERZ(FM)—Listing follows WGIP(AM).

WGIP(AM)— June 4, 1966: 1540 khz; 5 kw-D. TL: N42 59 23 W70 56 14. Hrs open: 815 Lafayette Rd., Portsmouth, 03801. Phone: (603) 436-7300. Fax: (603) 430-9415. Licensee: Capstar TX L.P. Group owner: Clear Channel Communications Inc. (acq 8-30-00; grpsl). Population served: 80,000 Natl. Rep: McGavren Guild. Format: News/talk, sports. ♦Robert Greer, gen mgr; Dan Pierce, opns mgr & progmg dir; Judy Figliulo, gen sls mgr; Jennifer McElreavy, prom dir; Kelly Brown, news dir; Roger Wood, news dir; Ken Neeman, chief of engrg; Beth LaRocque, traf mgr.

WERZ(FM)—Co-owned with WGIP(AM). Sept 21, 1972: 107.1 mhz; 5.2 kw. 351 ft TL: N43 01 38 W70 52 51. Web Site: www.werz.com.499,000 Format: Top-40. ♦Michael O'Donnell, progmg dir; Beth La Rocque, traf mgr.

***WPEA(FM)**— 1964: 90.5 mhz; 115 w. 170 ft TL: N42 58 44 W70 57 00. Stereo. Hrs opn: Phillips Exeter Academy, 20 Main St., 03833-2460. Phone: (603) 777-4414. Fax: (603) 777-4384. E-mail: WPEA@exeter.edu Licensee: Trustees of Phillips Exeter Academy. Population served: 25,000 Format: Var. Target aud: General; students.

Farmington

WMEX(FM)— July 9, 1999: 106.5 mhz; 2.9 kw. Ant 486 ft TL: N43 24 01 W71 09 27. Stereo. Hrs opn: 24 1 Wakefield St., Suite 302, Rochester, 03867-1913. Phone: (603) 335-6600. Fax: (603) 299-0325. E-mail: oldies1065@aol.com Web Site: wmexfm.com. Licensee: Wimmex LLC. Population served: 320,000 Law Firm: Cohn & Marks. Format: Oldies. Target aud: 25 plus. ♦Dennis Jackson, CEO; Gary James, VP & gen mgr; Gene Vallee, gen sls mgr; Ron Malone, chief of engrg.

Fitzwilliam Depot

WZNH(AM)—Not on air, target date: unknown: 870 khz; 780 w-D, 400 w-N, DA-N. TL: N42 45 59 W72 06 59. Hrs open: 17 Knightsbridge Ct., Nanuet, NY, 10954. Phone: (845) 356-9613. Licensee: Steven Wendell. ♦Steven Wendell, gen mgr.

Franklin

WFTN(AM)— Oct 30, 1966: 1240 khz; 1 kw-U. TL: N43 27 16 W71 38 33. Hrs open: Box 941, 110 Babbitt Rd., 03235. Phone: (603) 934-2500. Fax: (603) 934-2933. E-mail: onair@mix941fm.com Web Site: www.mix941fm.com. Licensee: Northeast Communications Corp. (group owner; acq 9-30-74). Population served: 30,000 Format: Mus of your life. ♦Jeff Fisher, pres, gen mgr & stn mgr; Fred Caruso, opns mgr & progmg dir; Jeff Levitan, gen sls mgr; Rick Ganley, prom dir; Gary Ford, mus dir; Amy Bates, news dir; Cathy Keyser, traf mgr.

WFTN-FM— Apr 10, 1987: 94.1 mhz; 6 kw. 328 ft TL: N43 28 23 W71 36 20. Stereo. Web Site: www.mix941fm.com.100,000 Format: Adult contemp.

Gorham

***WEVC(FM)**— May 1995: 107.1 mhz; 6 kw. 151 ft TL: N44 27 32 W71 10 16. Stereo. Hrs opn: 24
Rebroadcasts WEVO(FM) Concord 100%.
207 N. Main St., Concord, 03301. Phone: (603) 228-8910. Fax: (603) 224-6052. E-mail: admin@nhpr.org Web Site: www.nhpr.org. Licensee: New Hampshire Public Radio Inc. Natl. Network: NPR, PRI. Wire Svc: AP Format: News/talk. News staff: 9; News: 42 hrs wkly. Target aud: 25-54. Spec prog: Folk 3 hrs wkly. ♦Elizabeth Gardella, pres & gen mgr; Mark Bevis, news dir; John Huntley, engrg dir.

Groveton

WRNH(FM)—Not on air, target date: unknown: 101.5 mhz; 6 kw. Ant 318 ft TL: N44 37 50 W71 17 34. Hrs open: 339 North St., Medfield, MA, 02052. Phone: (508) 359-2700. Licensee: Liveair Communications Inc. ♦David M. Wang, pres.

WXBN(FM)—Not on air, target date: unknown: 93.7 mhz; 6 kw. Ant -24 ft TL: N44 33 55 W71 37 48. Hrs open: Box 896, Littleton, 03561. Phone: (603) 788-3636. Fax: (603) 788-3536. Licensee: Alexxon Corp. ♦Barry P. Lunderville, pres.

Hampton

WSAK(FM)— August 1992: 102.1 mhz; 3 kw. 328 ft TL: N42 53 51 W70 53 02. Stereo. Hrs opn: 24
Rebroadcasts WSHK(FM) Kittery, ME 100%.
Box 576, Dover, 03821-0576. Secondary address: 292 Middle Rd., Dover 03820-4901. Phone: (603) 749-9750. Fax: (603) 749-1459. E-mail: shark.mail@citcomm.com Web Site: www.shark1053.com. Licensee: Citadel Broadcasting Co. Group owner: Citadel Broadcasting Corp. (acq 7-7-99; grpsl). Natl. Network: CNN Radio. Natl. Rep: Christal. Law Firm: Wiley, Rein & Fielding. Wire Svc: AP Format: Classic rock. News staff: 2. Target aud: 25-49. ♦Farid Suleman, CEO; Judy Ellis, pres; Marty Lessard, gen mgr; Mark Ericson, opns mgr; Ken Hoffman, gen sls mgr; Jonathan Smith, progmg dir.

Hanover

WDCR(AM)— Mar 4, 1958: 1340 khz; 1 kw-U. TL: N43 41 59 W72 16 47. Stereo. Hrs opn: 24 Box 957, 03755. Secondary address: 3rd Fl., Robinson Hall, Dartmouth College 03826. Phone: (603) 646-3313. Phone: (603) 646-3826. Fax: (603) 643-7655. E-mail: heath.cole@dartmouth.edu Web Site: www.webdcr.com. Licensee: Trustees of Dartmouth College. Population served: 80,000 Format: Modern rock. News staff: 2; News: 4 hrs wkly. Target aud: General. ♦Alex Belser, gen mgr; Heath Cole, opns mgr; Julie Kaye, gen sls mgr & traf mgr; Rob Demick, progmg dir.

WFRD(FM)—Co-owned with WDCR(AM). Feb 19, 1976: 99.3 mhz; 6 kw. Ant 328 ft TL: N43 39 14 W72 17 44.2. Stereo. 24 Web Site: www.wfrd.com.150,000 Format: Modern rock. News staff: 2. Target aud: 18-45. ♦Pauel Sotskov, progmg dir.

***WEVH(FM)**— October 1993: 91.3 mhz; 150 w. 1,180 ft TL: N43 42 30 W72 09 16. Hrs open:
Rebroadcasts WEVO(FM) Concord 100%.
207 N. Main St., Concord, 03301. Phone: (603) 228-8910. Fax: (603) 224-6052. Web Site: www.nhpr.org. Licensee: New Hampshire Public Radio Inc. Natl. Network: NPR, PRI. Law Firm: Garvey, Schubert & Barer. Wire Svc: AP Format: News/talk. News staff: 9; News: 42 hrs wkly. Target aud: 25-54. Spec prog: Folk 3 hrs wkly. ♦Elizabeth Gardella, pres; Mark Bevis, gen mgr & news dir; John Huntley, engrg dir.

WGXL(FM)—Listing follows WTSL(AM).

WTSL(AM)— October 1950: 1400 khz; 1 kw-U. TL: N43 41 03 W72 17 46. Hrs open: 24 31 Hanover, Suite 4, Lebanon, 03766. Phone: (603) 448-1400. Fax: (603) 448-1755. Licensee: Capstar TX L.P. Group owner: Clear Channel Communications Inc. (acq 1-1-01; with co-located FM). Population served: 50,000 Natl. Network: CBS. Law Firm: David Tillotson. Format: News/talk, sports. News staff: 2; News: 25 hrs wkly. Target aud: 35 plus. ♦Christopher Olsen, gen mgr; Tim Plant, gen mgr; Michael Barrett, opns mgr, progmg dir & news dir; Gary Laperle, chief of engrg.

WGXL(FM)—Co-owned with WTSL(AM). Jan 12, 1987: 92.3 mhz; 6 kw. 326 ft TL: N43 39 17 W72 17 41. Stereo. Format: Hot adult contemp. Target aud: 25-49.

WXXK(FM)—See Lebanon

Haverhill

WYKR-FM— Feb 19, 1990: 101.3 mhz; 3 kw. 39 ft TL: N44 06 49 W71 58 54. Stereo. Hrs opn: 6 AM-10 PM Box 675, Rt. 302, Wells River, VT, 05081. Secondary address: Box 1013, Woodsville 03785. Phone: (802) 757-2773. Fax: (802) 757-2774. E-mail: wykr@kingcon.com Web Site: www.wykr.com. Licensee: Puffer Broadcasting Inc. Population served: 100,000 Natl. Network: Westwood One, NBC, Jones Radio Networks. Natl. Rep: Roslin. Law Firm: Fisher, Wayland, Cooper, Leader & Zaragoza. Format: Country. Target aud: 25 plus. ♦Stephen J. Puffer, pres, gen mgr, gen sls mgr, adv mgr & progmg dir; Teresa Puffer, opns mgr; Don Smith, chief of engrg.

Henniker

***WNEC-FM**— Feb 9, 1971: 91.7 mhz; 120 w. -210 ft TL: N43 10 34 W71 49 22. Hrs open: 17 New England College, 28 Bridge St., 03242. Phone: (603) 428-2278. Phone: (603) 428-6393. Fax: (603) 428-7230. E-mail: A.Metzegen@nec.edu Licensee: New England College. Population served: 10,000 Format: Progsv, adult contemp, Black, jazz, new age, urban contemp. News: one hr wkly. Target aud: 18-25; college students. Spec prog: Blues 4 hrs, country 3 hrs, American Indian 18 hrs, folk 18 hrs, farm 4 hrs wkly. ♦Ambrose Metzegen, CEO; Chris Collord, progmg VP; Kristen Westhoven, mus dir; Dale Carlow, engrg VP.

WNNH(FM)— Nov 17, 1989: 99.1 mhz; 6 kw. 712 ft TL: N43 09 17 W71 47 44. Stereo. Hrs opn: 24 11 Kimball Dr., Unit 114, Hooksett, 03106. Phone: (603) 225-1160. Fax: (603) 225-5938. E-mail: oldies99@wnnh.com Web Site: www.wnnh.com. Licensee: Nassau Broadcasting III L.L.C. Group owner: Nassau Broadcasting Partners L.P. (acq 3-16-2004; grpsl). Natl. Rep: McGavren Guild. Law Firm: Verner, Liipfert, Bernhard, McPherson & Hand. Format: Oldies. News staff: 2; News: 20 hrs wkly. Target aud: 25-54; mass appeal. ♦Scott Brady, gen mgr; Andy Mack, opns mgr, progmg dir & progmg mgr; Ken Cail, news dir.

Hillsboro

WTPL(FM)—Licensed to Hillsboro. See Concord

Hinsdale

WYRY(FM)—Licensed to Hinsdale. See Keene

Jackson

WEVJ(FM)— 8/02: Stn currently dark. 99.5 mhz; 4.7 kw. Ant 171 ft TL: N44 10 30 W71 10 07. Hrs open: 12-12
WEVO(FM) Concord 100%.
New Hampshire Public Radio, 207 N. Main St., Concord, 03301-5003. Phone: (603) 228-8910. Fax: (603) 224-6052. E-mail: admin@nhpr.org Web Site: www.nhpr.org. Licensee: New Hampshire Public Radio. Natl. Network: NPR, PRI. Law Firm: Garvey,Schubert & Barer. Wire Svc: AP Format: News/talk. News staff: 9. Target aud: 25-54. ♦Betsy Gardela, pres; Mark Bevis, news dir; John Huntley, engrg dir.

Jaffrey

WXNH(AM)—Not on air, target date: unknown: 540 khz; 250 w-D, 330 w-N, DA-2. TL: N42 50 55 W71 57 53. Hrs open: 17 Knightsbridge Ct., Nanuet, NY, 10954. Phone: (845) 356-9613. Licensee: Steven Wendell. ♦Steven Wendell, gen mgr.

Keene

***WEVN(FM)**— April 1994: 90.7 mhz; 1.5 kw. 938 ft Stereo. Hrs opn: 24
Rebroadcasts WEVO(FM) Concord 100%.
c/o Radio Stn WEVO(FM), 207 N. Main St., Concord, 03301. Phone: (603) 228-8910. Fax: (603) 224-6052. E-mail: admin@nhpr.org Web Site: www.nhpr.org. Licensee: New Hampshire Public Radio Inc. Natl. Network: NPR, PRI. Law Firm: Garvey, Schubert & Barer. Wire Svc: AP Format: News/talk. News staff: 9; News: 42 hrs wkly. Target aud: 25-54. Spec prog: Folk 3 hrs wkly. ♦Elizabeth Gardella, pres; Mark Bevis, news dir; John Huntley, engrg dir.

WINQ(FM)—Winchester

WKBK(AM)— June 2, 1927: 1290 khz; 5 kw-U, DA-1. TL: N42 56 56 W72 18 22. Hrs open: Box 466, 03431. Secondary address: 69 Stanhope Ave. 03431. Phone: (603) 352-9230. Fax: (603) 357-3926. Licensee: Saga Communications of New England LLC. Group owner: Saga Communications Inc. (acq 5-1-02; grpsl). Population served: 60,000 Natl. Network: CBS. Natl. Rep: McGavren Guild. Format: News, talk. Target aud: 25 plus. ♦Bruce Lyons, gen mgr; Stephen Hamel, opns dir; Vicky Lenahan, prom dir; Dan Mitchell, progmg dir; Paul Scheuring, news dir; Ira Wilner, chief of engrg; Jennifer Bond, traf mgr.

WKNE(FM)—Co-owned with WKBK(AM). May 1964: 103.7 mhz; 12.2 kw. 991 ft TL: N43 02 00 W72 22 04. Stereo. Format: Adultcontemp, top-40. Target aud: 18-49. ♦Jennifer Bond, traf mgr.

***WKNH(FM)**— November 1975: 91.3 mhz; 274 w. 79 ft TL: N42 55 29 W72 16 42. (CP: 91.7 mhz, 192 w, ant 363 ft.). Stereo. Hrs opn: 24 Keene State College, 229 Main St., 03435-2704. Phone: (603) 358-2420. Phone: (603) 358-2421. Fax: (603) 358-2417. E-mail: wknhinfo@aol.com Web Site: www.jumblue.com/wknh/. Licensee: Board of Trustees University System of New Hampshire. Population served: 30,000 Format: Progsv. News staff: 2; News: 3 hrs wkly. Target aud: General. Spec prog: Class 4 hrs, folk 6 hrs, jazz 3 hrs, blues 3 hrs, rap 4 hrs, new age 4 hrs, reggae 3 hrs, Christian 3 hrs, metal 4 hrs, experimental 3 hrs wkly. ♦James McCluskey, gen mgr.

WYRY(FM)—Hinsdale, June 30, 1987: 104.9 mhz; 1.55 kw. 456 ft TL: N42 46 33 W72 27 19. (CP: 725 w, ant 669 ft.). Stereo. Hrs opn: 24 30 Warwick Rd., Suite 10, Winchester, 03470. Phone: (603) 239-8200. Fax: (603) 239-6203. Web Site: www.wyry.com. Licensee: Tri-Valley Broadcasting Corp. (acq 8-86). Natl. Network: Jones Radio Networks. Law Firm: Reddy, Begley & McCormick. Format: Country. News staff: 3; News: 10 hrs wkly. Target aud: 25-49; upscale adults & business decision makers. ♦Brian McCormick, VP & gen mgr; Sean Patrick, sls VP & progmg dir; Dan Guy, chief of engrg.

WZBK(AM)— May 1959: 1220 khz; 1 kw-U. TL: N42 55 50 W72 17 56. Hrs open: 69 Stanhope Ave., 03431. Phone: (603) 352-9230. Fax: (603) 357-3926. Licensee: Saga Communications of New Hampshire LLC. Group owner: Saga Communications Inc. (acq 7-1-02; $2.63 million with WOQL(FM) Winchester). Population served: 60,000 Law Firm: Booth, Freret, Imlay & Tepper P. Format: Standards. Target aud: 25-54; general. ♦Bruce Lyons, stn mgr & prom VP; Susan Wells, gen sls mgr; Steve Hamill, progmg dir & chief of engrg; Paul Schering, news dir; Ira Wilner, chief of engrg; Jen Bond, traf mgr.

Laconia

WEMJ(AM)—Listing follows WLNH-FM.

WEZS(AM)— Aug 22, 1922: 1350 khz; 5 kw-D, 112 w-N. TL: N43 30 27 W71 31 00. Hrs open: 277 Union Ave., 03246. Phone: (603) 524-6288. Fax: (603) 528-1638. E-mail: info@wezs.com Web Site: www.wezs.com. Licensee: Gary W. Hammond. (acq 3-17-94; FTR: 6-6-94). Population served: 152,500 Natl. Network: USA. Format: Easy lstng, smooth jazz. Target aud: 45 plus. ♦Gary W. Hammond, gen mgr.

WLNH-FM— Nov 22, 1965: 98.3 mhz; 3.8 kw. 413 ft TL: N43 35 46 W71 29 55. Hrs open: 24 hrs Box 7326, Village West Bldg. 1, Gilford, 03247. Phone: (603) 524-1323. Fax: (603) 528-5185. E-mail: info@wlnh.com Web Site: www.wlnh.com. Licensee: Nassau Broadcasting III L.L.C. Group owner: Nassau Broadcasting Partners L.P. (acq 4-7-2004; grpsl). Population served: 40,000 Law Firm: Rosenman & Colin L.L.P. Format: Hot adult contemp. Target aud: 25-54. ♦Molly King, prom dir; Chris Ialuna, progmg dir.

WEMJ(AM)—Co-owned with WLNH-FM. Apr 9, 1961: 1490 khz; 1 kw-U. TL: N43 32 29 W71 27 45.14,888 Natl. Network: CBS. Natl. Rep: D & R Radio. Format: Talk.

Lancaster

WXXS(FM)— 1998: 102.3 mhz; 1.5 kw. 964 ft TL: N44 23 39 W71 39 20. Stereo. Hrs opn: 24 Box 896, Littleton, 03561. Secondary address: 195 Main St. 03584. Phone: (603) 444-4102. Fax: (603) 788-3536. E-mail: kiss102@together.net Licensee: Barry P. Lunderville. Natl. Network: CBS. Format: Contemp hit/top-40. ♦Barry P. Lunderville, gen mgr; Brian Lunderville, opns mgr; Barry Lunderville, progmg dir; Danielle Corbiel, traf mgr.

Lebanon

WGXL(FM)—See Hanover

WHDQ(FM)—See Claremont

WUVR(AM)— 2004: 1490 khz; 640 w-U. TL: N43 39 12 W72 14 16. Hrs open: Box 2295, New London, 03257. Phone: (603) 448-0500. Fax: (603) 448-6601. E-mail: bob@wntk.com Web Site: www.wntk.com. Licensee: KOOR Communications Inc. Format: News/talk. ♦Robert Vinikoor, gen mgr, gen sls mgr & progmg dir; Dave Shurtleff, news dir; Russ McCallister, chief of engrg.

***WVFA(FM)**— Feb 6, 2004: 90.5 mhz; 7 w. Ant 695 ft TL: N43 37 17 W72 10 30. Hrs open: 24 Box 126, Hartford, VT, 05047-0126. Secondary address: 48 Wescott Rd., Enfield 03748. Phone: (802) 295-9683. Fax: (802) 295-9683. E-mail: vtpreacher@aol.com Licensee: Green Mountain Educational Fellowship Inc. Population served: 80,000 Wire Svc: AP Format: Inspirational, educ, relg. News: 11 hrs wkly. Target aud: 25-49; primary. ♦William A. Wittik, pres, CFO & gen mgr; Betsy Murray, opns mgr; Elmer Murray, opns mgr.

WXXK(FM)— Dec 18, 1990: 100.5 mhz; 22 kw. Ant 325 ft TL: N43 37 17 W72 10 30. Hrs open: 24 31 Hanover St., Suite 4, 03766. Phone: (603) 448-1400. Fax: (603) 448-1755. Web Site: www.kixx.com. Licensee: Clear Channel Radio Licenses, Inc. Group owner: Clear Channel Communications Inc. (acq 11-27-00; grpsl). Population served: 150,000 Natl. Network: Westwood One, CNN Radio. Format: Country. News staff: 3; News: 20 hrs wkly. Target aud: 25-54. ♦Cheryl Frisch, CFO & dev VP; Robert Frisch, gen mgr; Kenny Michaels, opns mgr & prom dir; Matt Cross, sls dir & traf mgr; Michael Barrett, progmg dir & news dir.

Lisbon

WLTN-FM—Licensed to Lisbon. See Littleton

Littleton

WLTN(AM)— Oct 10, 1963: 1400 khz; 1 kw-U. TL: N44 18 47 W71 46 08. Hrs open: 24 15 Main St., 03561. Phone: (603) 444-3911. Fax: (603) 444-7186. E-mail: oldies1400@adelphia.net Licensee: Barry P. Lunderville L.L.C. (acq 6-30-2005; with WLTN-FM Lisbon). Population served: 150,000 Format: Oldies, Red Sox. News staff: one; News: 40 hrs wkly. Target aud: 21-65. ♦Barry Lunderville, gen mgr; Christina Brooks, gen sls mgr; Phil Rivera, progmg dir; Jim Clothey, news dir; Brian Lunderville, chief of engrg; Danielle Corbiel, traf mgr.

WLTN-FM— Sept 1, 1991: 96.7 mhz; 6 kw. 295 ft TL: N44 13 11 W71 52 07. Stereo. 24 E-mail: mix967@adelphia.net 200,000 Natl. Network: Westwood One. Format: Bright adult contemp. News staff: one; News: 7 hrs wkly. Target aud: 25-54. ♦Danielle Corbiel, traf mgr.

WMTK(FM)— Feb 23, 1985: 106.3 mhz; 390 w. 1,256 ft TL: N44 21 14 W71 44 23. Stereo. Hrs opn: 24 Box 106, 03561-0106. Phone: (603) 444-5106. Fax: (603) 444-1205. E-mail: thenotch@kington.net Licensee: Vermont Broadcast Associates Inc. (acq 8-2000; $250,000). Population served: 180,000 Law Firm: Bryan Cave. Format: Classic hits. News staff: one. Target aud: 30-50; slightly more males, active lifestyles. ♦Bruce James, gen mgr & progmg dir; Steve Nichols, gen sls mgr; Todd Wellington, news dir; Don Smith, chief of engrg.

Madbury

WWNH(AM)— May 20, 1989: 1340 khz; 1 kw-U. TL: N43 10 22 W70 55 00. Hrs open: 24 Box 69, Dover, 03821. Secondary address: 284 Rt. 155, Dover 03821. Phone: (603) 742-8575. Fax: (603) 743-6444. E-mail: info@loveradio.net Web Site: www.loveradio.net. Licensee: Harvest Broadcasting. Natl. Network: USA. Format: MOR. News: 3 hrs wkly. Target aud: General; 29 plus. Spec prog: Family 24 hrs wkly. ♦Patti Smith, CEO, gen mgr, stn mgr & gen sls mgr; Ernie Jenkins, chief of opns, progmg dir & news dir; Steve Donnell, engrg dir & chief of engrg.

Manchester

WFEA(AM)— Mar 8, 1932: 1370 khz; 5 kw-U, DA-2. TL: N42 54 26 W71 27 45. Hrs open: 500 Commercial St., 03101. Phone: (603) 669-5777. Fax: (603) 669-4641. Licensee: Saga Communications of New England LLC. Group owner: Saga Communications Inc. (acq 6-2-92; grpsl, including co-located FM). Population served: 250,000 Rgnl rep: Katz. Law Firm: Smithwick & Belendiuk. Format: Adult standards. Target aud: 50 plus; "Modern Maturity" market. Spec prog: Fr 3 hrs, Sp 2 hrs wkly. ♦Raymond R. Garron, gen mgr.

WZID(FM)—Co-owned with WFEA(AM). 1948: 95.7 mhz; 14.5 kw. 930 ft TL: N42 59 02 W71 35 22. E-mail: radionh@com Web Site: www.wzid.com.448,000 Format: Adult contemp. Target aud: 25-54.

WGIR(AM)— October 1941: 610 khz; 5 kw-D, 1 kw-N, DA-2. TL: N43 00 57 W71 28 48. Hrs open: 24 195 McGregor St., Suite 810, 03102. Phone: (603) 625-6915. Fax: (603) 669-0610. Web Site: www.wgiram.com. Licensee: Capstar TX L.P. Group owner: Clear Channel Communications Inc. (acq 8-30-00; grpsl). Population served: 200,000 Natl. Network: Fox News Radio, Westwood One. Format: News/talk, sports. News staff: 2; News: 38 hrs wkly. Target aud: 35-54. ♦Joseph Graham, gen mgr.

WGIR-FM— June 5, 1963: 101.1 mhz; 11.5 kw. Ant 1,027 ft TL: N42 58 54 W71 35 21. Stereo. 195 McGregor St., Suite 810, 03105. Web Site: www.rock101fm.com.500,000 Format: AOR.

WKBR(AM)— Oct 1, 1946: 1250 khz; 5 kw-U, DA-2. TL: N43 00 40 W71 30 19. Hrs open: 5 AM-11 PM 922 Elm St., Suite 301, 03101. Phone: (603) 669-1250. Fax: (603) 647-1260. E-mail: newstalk1250@aol.com Web Site: www.wkbr1250.com. Licensee: Devon Broadcasting Co. Inc. Group owner: Northeast Broadcasting Company Inc. (acq 7-22-97; $145,000). Population served: 400,000 Format: Sports. Target aud: 25 plus. Spec prog: Greek 8 hrs wkly. ♦Steve Young, gen mgr; Jerri Stanford, stn mgr; Charles Dent, gen sls mgr; Dana Marshall, progmg dir; Ron Travers, news dir; Lou Muse, chief of engrg; Stephanie Battaglia, traf mgr.

***WLMW(FM)**— September 1997: 90.7 mhz; 15 w. 869 ft TL: N42 58 59 W71 35 25. Stereo. Hrs opn: Box 366, Auburn, 03032. Secondary address: 134 Hollis Rd., Amherst 03031. Phone: (603) 483-8950. Fax: (603) 483-8908. E-mail: jim@nhfamilyradio.org Web Site: www.nhfamilyradio.org. Licensee: Knowledge For Life. Format: Christian family radio. ♦Jim Phelan, gen mgr.

WOKQ(FM)—See Dover

Meredith

WWHQ(FM)— Nov 16, 1988: 101.5 mhz; 6 kw. Ant 328 ft TL: N43 35 46 W71 29 55. Stereo. Hrs opn: Box 7326, Gilford, 03247. Phone: (603) 524-1323. Fax: (603) 528-5185. Web Site: www.big1015.com. Licensee: Nassau Broadcasting III L.L.C. Group owner: Nassau Broadcasting Partners L.P. (acq 4-7-2004; grpsl). Format: Classic rock. ♦Louis Mercatanti, pres; Dominic Biello, opns dir; Rob Fulmer, gen mgr & gen sls mgr.

Moultonborough

WSCY(FM)— May 31, 1993: 106.9 mhz; 130 w. 2,096 ft TL: N43 46 09 W71 18 52. Stereo. Hrs opn: Box 99, Franklin, 03235. Phone: (603) 253-8080. Fax: (603) 934-2933. Web Site: www.mix941fm.com. Licensee: Northeast Communications Corp. (group owner; acq 5-4-93; $399,072; FTR: 5-24-93). Format: Hot country. ♦Jeff Fisher, pres & stn mgr; Amy Bates, sls VP & news dir; Jeff Levitan, opns mgr & gen sls mgr; Gene Terwilliger, chief of engrg.

Mt. Washington

WHOM(FM)— July 9, 1958: 94.9 mhz; 48 kw. 3,760 ft TL: N44 16 13 W71 18 13. (CP: Ant 46 ft.). Stereo. Hrs opn: 1 City Center, Portland, ME, 04101. Phone: (207) 773-0200. Fax: (207) 774-8707. E-mail: whom@whom949.com Web Site: www.whom949.com. Licensee: Citadel Broadcasting Co. Group owner: Citadel Broadcasting Corp. (acq 7-7-99; grpsl). Population served: 200,000 Natl. Rep: Christal. Format: Light rock, adult contemp. Target aud: 35-64; professionals with active lifestyles. ♦Mike Sambrook, gen mgr; Tim Moore, progmg dir.

Nashua

***WEVS(FM)**— 2005: 88.3 mhz; 3.5 kw horiz, 5 kw vert. Ant 69 ft TL: N42 45 00 W71 28 47. Stereo. Hrs opn:
Rebroadcasts WEVO(FM) Concord 100%.
207 N. Main St., Concord, 03301-5003. Phone: (603) 228-8910. Fax: (603) 224-6052. Web Site: www.nhpr.org. Licensee: New Hampshire Public Radio Inc. Natl. Network: NPR, PRI. Format: News/talk. ♦Elizabeth Gardella, gen mgr; Mark Bevis, news dir; John Huntley, engrg dir.

WFNQ(FM)— Oct 19, 1987: 106.3 mhz; 3 kw. 100 ft TL: N42 44 07 W71 23 37. (CP: 950 w, ant 541 ft.). Stereo. Hrs opn: 24 11 Kimball Dr., Suite 114, Hooksett, 03106. Phone: (603) 889-1063. Fax: (603) 882-0688. Web Site: www.wjyy.com. Licensee: Nassau Broadcasting III L.L.C. Group owner: Nassau Broadcasting Partners L.P. (acq 3-16-2004; grpsl). Law Firm: Cole, Raywid & Braverman. Format: Hot adult contemp. News staff: one; News: 5 hrs wkly. Target aud: 18-49. ♦Louis F. Mercatanti, pres; Steve Garsh, gen mgr; Andy Mack, opns dir; Phyllis Knight, gen sls mgr & traf mgr; Sarah Sullivan, progmg dir; Dirk Nadon, chief of engrg.

WGAM(AM)— 1991: 900 khz; 910 w-D, 60 w-N. TL: N42 45 34 W71 28 37. Hrs open: 6 AM-6 PM (Oct-Apr); 6 AM-10 PM (May-Sept) One Indian Head Plaza, 5th Fl., 03060. Phone: (603) 880-9001. Phone: (603) 883-9900. Fax: (603) 577-8682. E-mail: info@wsnh900am.com Web Site: www.wgamradio.com. Licensee: Absolute Broadcasting LLC (acq 11-2-2005; $925,000). Format: Sports. Target aud: 35 plus. ♦Jerry DiGrezio, gen mgr; Marty Terrell, gen sls mgr; John Kosian, progmg dir & chief of engrg; Paul Hust, traf mgr.

WSMN(AM)— Mar 9, 1958: 1590 khz; 5 kw-U, DA-1. TL: N42 44 40 W71 29 52. Hrs open: 18 1 Indian Head Plaza, 03060. Phone: (603) 880-9001. Fax: (603) 577-8682. Web Site: www.wsmnradio.com. Licensee: Absolute Broadcasting LLC (acq 11-10-2005; $250,000). Population served: 90,000 Natl. Network: ESPN Radio. Format: News/talk, sports. ♦Jerry DiGrezio, gen mgr.

New London

WNTK-FM— Nov 30, 1992: 99.7 mhz; 620 w. 712 ft TL: N43 26 52 W72 02 04. Hrs open: 24
Rebroadcasts WNTK(AM) Newport 50%.
Box 2295, 25 Newport Rd., 03257. Secondary address: 103 Hanover St., Lebanon 03766. Phone: (603) 448-0500. Fax: (603) 448-6601. Web Site: www.wntk.com. Licensee: Koor Communications Inc. (group owner) Law Firm: Shaw Pittman. Format: Talk/news. News staff: 2. Target aud: 24-54. ♦Robert L. Vinikoor, CEO, gen mgr & progmg dir; Sheila E. Vinikoor, pres; Dave Shurtleff, news dir; Russ McCallister, chief of engrg.

***WSCS(FM)**— February 1996: 90.9 mhz; 63 w horiz, 250 w vert. 297 ft TL: N43 24 41 W71 58 33. Hrs open: Colby-Sawyer College, 100 Main St., 03257. Phone: (603) 526-3493. Fax: (603) 526-3452. E-mail: wscs@colby-sawyer.edu Web Site: www.colby-sawyer.edu/wscs. Licensee: Colby-Sawyer College. Format: Educ. ♦Sean Joncas, stn mgr; James Kovach, disc jockey.

Newport

WNTK(AM)— Aug 11, 1960: 1010 khz; 10 kw-D, 37 w-N. TL: N43 21 52 W72 10 47. Hrs open: 24 Box 2295, New London, 03257. Phone: (603) 448-0500. Fax: (603) 448-6601. E-mail: bob@wntk.com Web Site: www.wntk.com. Licensee: KOOR Communications. (group owner; acq 8-88; $250,000; FTR: 8-29-88). Law Firm: Shaw Pittman. Format: Classic country/Americanna. Target aud: 25-54; informed adults. ♦Robert L. Vinikoor, pres & gen mgr; Robert Vinikoor, progmg dir & chief of engrg; Sheila Vinikoor, traf mgr & disc jockey.

WVRR(FM)— 1971: 101.7 mhz; 260 w. Ant 1,115 ft TL: N43 23 45 W72 17 40. Stereo. Hrs opn: 24 31 Hanover St., Suite 4, Lebanon, 03766-1312. Phone: (603) 448-1400. Fax: (603) 448-5231. Web Site: www.wvrrfm.com. Licensee: Capstar TX L.P. Group owner: Clear Channel Communications Inc. (acq 11-28-00). Natl. Network: Westwood One. Format: Rock. News staff: one. Target aud: 25-54. ♦Tim Plante, gen mgr.

North Conway

WPKQ(FM)— October 1952: 103.7 mhz; 21.5 kw horiz, 16.5 kw vert. Ant 3,874 ft TL: N44 16 14 W71 18 15. Stereo. Hrs opn: 24
Rebroadcasts WOKQ(FM) Dover 80%.
P.O. Box 576, Dover, 03821-0576. Secondary address: 2617 White Mountain Hwy. 03860. Phone: (603) 749-9750. Fax: (603) 749-6589. E-mail: wpkq@nh@citcomm.com Web Site: www.wpkq.com. Licensee: Citadel Broadcasting Co. Group owner: Citadel Broadcasting Corp. (acq 7-7-99; grpsl). Population served: 3,000,000 Natl. Network: CNN Radio. Natl. Rep: Christal. Law Firm: Wiley, Rein & Fielding. Wire Svc: AP Format: Country. News staff: 2. Target aud: 25-54; New England residents. ♦Farid Suleman, CEO; Judy Ellis, pres; Mark Ericson, gen mgr & opns mgr; Martin Lessard, gen mgr; Ken Hoffman, gen sls mgr; Mark Jennings, progmg dir.

Peterborough

WFEX(FM)— June 1971: 92.1 mhz; 180 w. Ant 1,332 ft TL: N42 51 42 W71 52 46. Stereo. Hrs opn: 24
Rebroadcasts WFNX(FM) Lynn, MA 80%.
25 Exchange St., Lynn, MA, 01901. Secondary address: 32 Technology Way 2W8, Nashua 03060. Phone: (603) 882-9210. Fax: (603) 578-9210. E-mail: fnxradio@fnxradio.com Web Site: www.fnxradio.com. Licensee: FNX Broadcasting of New Hampshire LLC. Group owner: Phoenix Media Communications Group (acq 11-29-99). Population served: 511,000 Law Firm: Rubin, Winston, Dierks, Harris, & Cooke. Wire Svc: AP Format: Alternative rock. News staff: one; News: 2 hrs wkly. Target aud: Adults 18-44; young, educated white collar professionals with extremely active lifestyles. Spec prog: Gay talk 2 hrs, jazz 6 hrs wkly. ♦Stephen Mindich, CEO; Gary Kurtz, gen mgr; Peter Cawley, gen sls mgr; Keith Dakin, progmg dir; Chris Hall, chief of engrg.

Plymouth

***WPCR-FM**— Sept 29, 1974: 91.7 mhz; 215 w. 95 ft TL: N43 45 25 W71 38 59. Hrs open: 24 WPCR HUB, Plymouth State College, 17 High St., 03264-1594. Phone: (603) 535-2242 (office). Phone: (603) 536-5000 (univ. switchboard). Fax: (603) 535-2783. E-mail: genmgr@wpcr.plymouth.edu Web Site: wpcr.plymouth.edu. Licensee: Plymouth State College. Natl. Network: AP Radio. Format: AOR, progsv. Target aud: 15-35; college students & those interested in progressive alternative music. Spec prog: Class 3 hrs, jazz 3 hrs, reggae 3 hrs, blues 3 hrs, comedy 3 hrs wkly.

WPNH(AM)— Nov 10, 1965: 1300 khz; 5 kw-D, DA-D. TL: N43 46 32 W71 42 20. Hrs open: 24 Box 99, Franklin, 03235. Secondary address: 110 Babbitt Rd., Franklin 03235. Phone: (603) 536-2500. Phone: (603) 536-2501. Fax: (603) 934-2933. Licensee: Northeast Communications Corp. (group owner; acq 2-9-99; with co-located FM). Population served: 29,100 Law Firm: Reddy, Begley & McCormick. Format: Big band. Target aud: 35 plus. Spec prog: Breakfast with the bands 6 hrs wkly. ♦Jeff Fisher, gen mgr; Fred Caruso, opns mgr & progmg dir; Jess Levitan, gen sls mgr; Amy Bates, news dir; Cathy Keizer, traf mgr.

WPNH-FM— Oct 1, 1975: 100.1 mhz; 2.35 kw. 364 ft TL: N43 45 41 W71 38 59. (CP: 4.9 kw, ant 358 ft.). Stereo. Web Site: www.wpnhfm.com.64,000 Format: Alternative. ♦Rick Ganley, progmg dir & local news ed; Cathy Keizer, traf mgr; Bob Moulton, disc jockey; Derek Lavoy, disc jockey; Mark Decker, disc jockey; Mark Servente, disc jockey; Terry Hughes, disc jockey.

Portsmouth

WERZ(FM)—See Exeter

WGIP(AM)—See Exeter

WHEB(FM)— Jan 14, 1964: 100.3 mhz; 50 kw. 459 ft TL: N43 03 11 W70 46 04. (CP: Ant 446 ft. TL: N43 03 05 W70 46 09). Stereo. Hrs opn: 815 Lafayette Rd., 03801. Phone: (603) 436-7300. Fax: (603) 430-9415. Web Site: www.wheb.com. Licensee: Capstar TX L.P. Group owner: Clear Channel Communications Inc. (acq 8-30-00; grpsl). Population served: 79,700 Format: Rock and roll. Target aud: 18-49. ♦Robert Greer, gen mgr; Christopher Garrett, opns mgr, progmg dir & mus dir; Christine Sieks, gen sls mgr; Kelly Brown, news dir; Kenneth Neelan, chief of engrg; Sandy Nagle, traf mgr; Greg Kretschmar, disc jockey.

WMYF(AM)—Co-owned with WHEB(FM). Dec 5, 1960: 1380 khz; 1 kw-U, DA-N. TL: N43 03 48 W70 47 09.24 Format: Music of your life. Target aud: 25-54. ♦Judy Figliulo, gen sls mgr; Michael O'Donnell, progmg dir; Heather Salisbury, traf mgr.

Rochester

WGIN(AM)—Listing follows WQSO(FM).

WQSO(FM)— Oct 21, 1979: 96.7 mhz; 3 kw. 328 ft TL: N43 17 14 W70 56 49. (CP: 5.8 kw, ant 98 ft. TL: N43 23 40 W71 02 22). Stereo. Hrs opn: 24 815 Lafayette Rd., Portsmouth, 03801. Phone: (603) 436-7300. Fax: (603) 430-9415. Licensee: Capstar TX L.P. Group owner: Clear Channel Communications Inc. (acq 8-30-00; grpsl). Law Firm: Wiley, Rein & Fielding. Format: Oldies. News staff: 3; News: 12 hrs wkly. Target aud: 25-54. ♦Robert Greer, gen mgr.

WGIN(AM)—Co-owned with WQSO(FM). 1947: 930 khz; 5 kw-U, DA-N. TL: N43 17 13 W70 56 55.24 310,000 Format: News/talk, sports. Target aud: 25-64; decision-makers, heads of businesses, households. ♦Dan Pierce, progmg dir; Kelly Brown, news dir; Roger Wood, news dir; Beth LaRocque, traf mgr.

Salem

WCCM(AM)— Jan 10, 1977: 1110 khz; 5 kw-D, DA-D. TL: N42 45 42 W71 16 13. Hrs open: 462 Merrimack St., Methuen, MA, 01844. Phone: (978) 686-9966. Phone: (978) 683-7171. Fax: (978) 687-1180. Web Site: 1110wccm.com/wccm/. Licensee: Costa-Eagle Radio Ventures L.P. (group owner; (acq 1996). Population served: 90,000 Natl. Rep: Roslin. Law Firm: Bryan Cave. Format: News/talk. News staff: 2; News: 25 hrs wkly. Target aud: General. ♦Pat Costa, gen mgr; John Bassett, stn mgr; Bruce Arnold, progmg dir.

Somersworth

WBYY(FM)— Jan 25, 1995: 98.7 mhz; 6 kw. 315 ft TL: N43 14 12 W70 53 47. Hrs open: 24 Box 400, 101 Back Rd., Dover, 03820-0400. Phone: (603) 742-0987. Fax: (603) 742-0448. Licensee: Garrison City Broadcasting Inc. Format: Adult contemp. News staff: 2. Target aud: 25-54. ♦Bob Demers, CEO; Rick Bean, gen mgr; Mike Pomp, news dir; Mark Ward, chief of engrg; Melissa Knox, traf mgr.

Walpole

WPLY-FM— November 2000: 96.3 mhz; 320 w. Ant 407 ft TL: N43 08 14 W72 25 59. Hrs open:
Rebroadcasts WWOD(FM) Hartford, VT 100%.
106 N. Main St., West Lebanon, 03784. Phone: (603) 298-0332. Fax: (603) 298-7554. E-mail: info@bestoldies104.com Web Site: www.bestoldies104.com. Licensee: Nassau Broadcasting III L.L.C. Group owner: Nassau Broadcasting Partners L.P. (acq 8-13-2004; grpsl). Format: Oldies. ♦Camille Losapio, gen mgr.

Whitefield

WXRG(FM)— 2007: 99.1 mhz; 460 w. Ant 1,135 ft TL: N44 21 10 W71 44 15. Hrs open: 288 S. River Rd., Bedford, 03110. Phone: (603) 444-5204. Fax: (603) 668-6470. Licensee: White Park Broadcasting Inc. ♦Steven A. Silberberg, pres.

Winchester

WINQ(FM)— Oct 15, 1991: 98.7 mhz; 6 kw. Ant 328 ft TL: N42 49 56 W72 23 34. Stereo. Hrs opn: 24 69 Stanhope Ave., Keene, 03431. Phone: (603) 352-9230. Fax: (603) 357-3926. Web Site: www.cool987.com. Licensee: Saga Communications of New Hampshire LLC. Group owner: Saga Communications Inc. (acq 7-1-2002; $2.63 million with WZBK(AM) Keene). Population served: 100,000 Natl. Network: ABC, Jones Radio Networks. Law Firm: Booth, Freret, Imlay & Tepper. Format: Country. News staff: 4; News: 8 hrs wkly. Target aud: 25-54. ♦Bruce Lyons, gen mgr.

WZBK(AM)—See Keene

Wolfeboro

WASR(AM)— April 1970: 1420 khz; 5 kw-D, 137 w-N. TL: N43 35 31 W71 13 12. Hrs open: 5 AM-8 PM Box 900, 03894-0900. Secondary address: 73 Varney Rd. 03894-0900. Phone: (603) 569-1420. Fax: (603) 569-1900. E-mail: wkrp@metrocast.net Web Site: www.wasr.net. Licensee: Winnipesaukee Network Inc. (acq 3-31-2004; $350,000). Population served: 60,000 Format: Adult contemp, news. News staff: 4; News: 35 hrs wkly. Target aud: 25-54. ♦Grant P. Hatch, pres; Gary Hammond, engrg dir.

WLKZ(FM)— Feb 1, 1985: 104.9 mhz; 570 w. 1,053 ft TL: N43 32 44 W71 22 45. Stereo. Hrs opn: 24 21 Meadowbrook Ln., Suite 15, Gilford, 03249. Phone: (603) 524-0105. Fax: (603) 293-0699. E-mail: wlkz@metroczst.net Web Site: www.wlkz.com. Licensee: Nassau Broadcasting III L.L.C. Group owner: Nassau Broadcasting Partners L.P. (acq 3-16-2004; grpsl). Natl. Rep: McGavren Guild. Law Firm: Haley, Bader & Potts. Format: Oldies. News: 5 hrs wkly. Target aud: 25-54; baby boomers. Spec prog: Dick Clark/The Beatle Years/Lil Walters Time Machine. ♦Louis F. Mercatanti, pres; Jim Cande, sr VP; Rob Fulmer, gen mgr; Pat Kelly, opns dir & progmg dir; Ron Piro, gen mgr & gen sls mgr; Dirk Nadon, chief of engrg.

New Jersey

Andover

WOF(AM)— Oct 8, 1946: 1000 khz; 1 kw-D, 250 w-N, DA-D (L-KQSL). TL: N35 48 31 W121 43 28. (CP: 5 kw-U). Stereo. Hrs opn: 24 12 Coulter Pl., 07821. Fax: (908) 219-0182. Licensee: General Broadcasting Corp. (group owner; (acq 7-20-69; $255,000 with co-located FM; FTR: 2-12-83). Rgnl. Network: Mountain State Network. Format: MOR, C&W. Spec prog: Sp 3 hrs wkly. ♦Edgar Adcock, gen mgr.

Asbury Park

WADB(AM)— 1926: 1310 khz; 2.5 kw-D, 1 kw-N, DA-2. TL: N40 13 47 W74 05 27. Hrs open: 2401 Rt. 66, Ocean, 07712. Phone: (732) 897-8282. Fax: (732) 897-8283. Web Site: www.1310espndeportes.com. Licensee: Millennium Shore License Holdco LLC. Group owner: Millennium Radio Group LLC (acq 6-11-2002; grpsl). Population served: 170,000 Natl. Network: ESPN Deportes. Format: Sp sports. ♦Bill Saurer, gen mgr; Lou Russo, progmg dir; Jay Pierce, chief of engrg; John Surno, sls.

WJLK(FM)—Co-owned with WADB(AM). Nov 20, 1947: 94.3 mhz; 1.3 kw. Ant 499 ft TL: N40 13 45 W74 05 24. Stereo. Web Site: www.getthepoint.com.757,000 Format: Hot adult contemp. ♦Lou Russo, opns mgr & progmg dir; Debbie Mazzella, mus dir; Tara Hessline, traf mgr.

WHTG(AM)—See Eatontown

***WYGG(FM)—**Not on air, target date: unknown: 88.1 mhz; 100 w. 33 ft TL: N40 13 01 W74 00 35. Hrs open: 1488 New York Ave., Minority Business & Housing Development, Inc., Brooklyn, NY, 11210. Phone: (908) 775-0821. Web Site: www.radiobonnenouvelle.com. Licensee: Minority Business & Housing Development, Inc. Format: Relg.

Atlantic City

WAJM(FM)— 1997: 88.9 mhz; 150 w vert. 102 ft TL: N39 21 54 W74 28 31. Hrs open: Atlantic City High School, 1400 N. Albany Ave., 08401. Secondary address: 1809 Pacific Ave. 08402. Phone: (609) 343-7300. Fax: (609) 343-7347. E-mail: wajm@comcast.net Licensee: Atlantic City Board of Education. Format: Div. ♦Pamela Lewis, gen mgr; Albert Horner, stn mgr.

WAYV(FM)— April 1961: 95.1 mhz; 50 kw. Ant 331 ft TL: N39 22 51 W74 27 04. Stereo. Hrs opn: 24
Simulcast with WAIV(FM) Cape May 100%.
8025 Black Horse Pike, Suite 100-102, West Atlantic City, 08232. Phone: (609) 484-8444. Fax: (609) 646-6331. E-mail: gfequity@aol.com Web Site: www.wayv951fm@aol.com. Licensee: Equity Communications L.P. (group owner; (acq 6-21-96; $3.1 million). Population served: 400,000 Natl. Network: Westwood One. Natl. Rep: Katz Radio. Law Firm: Latham & Watkins. Format: Hot adult contemp. Target aud: 18-49; adults. ♦Gary Fisher, sr VP, VP & gen mgr; Paul Kelly, progmg dir.

WENJ(AM)— 1940: 1450 khz; 1 kw-U. TL: N39 22 42 W74 26 53. Hrs open: 24 950 Tilton Rd., Suite 200, Northfield, 08225. Phone: (609) 645-9797. Fax: (609) 272-9228. E-mail: harry.hurley@citcomm.com Web Site: www.literock969.com. Licensee: Millennium Atlantic City License Holdco LLC. Group owner: Millennium Radio Group LLC (acq 5-11-2001; grpsl). Population served: 300,000 Natl. Network: CBS. Format: Talk. News staff: one; News: 24 hrs wkly. Target aud: 25 plus; the population of South Jersey. ♦Dan Sullivan, gen mgr; Mike Ruble, gen sls mgr; Jennifer Doughton, prom dir; Eric Johnson, progmg dir; Tom McNally, chief of engrg.

WFPG(FM)—Co-owned with WENJ(AM). September 1962: 96.9 mhz; 50 kw. 400 ft TL: N39 22 42 W74 26 53. Stereo. 24 Web Site: www.literock969.com.500,000 Format: Light rock, soft adult contemp. News staff: one. Target aud: 25-54. ♦Gary Guida, progmg dir.

WJSE(FM)—Petersburg, August 1991: 102.7 mhz; 3.3 kw. Ant 295 ft TL: N39 12 18 W74 39 33. Stereo. Hrs opn: 24 1601 New Rd., Linwood, 08221. Phone: (609) 653-1400. Fax: (609) 601-0450. Web Site: www.theace1027.com. Licensee: Access 1 New Jersey License Co. LLC. (acq 3-22-2005; $4.75 million). Population served: 325,000 Format: Active rock. ♦Chesley Maddox-Dorsey, pres; Dick Irland, gen mgr.

WMGM(FM)— June 14, 1961: 103.7 mhz; 50 kw. 400 ft TL: N39 23 38 W74 30 34. Stereo. Hrs opn: 24 1601 New Rd., Linwood, 08221. Phone: (609) 653-1400. Fax: (609) 601-0450. E-mail: wmgm1037@aol.com Web Site: www.theshark1037.com. Licensee: Access.1 New Jersey License Co. Group owner: Access.1 Communications Corp. (acq 11-17-2003; grpsl). Population served: 545,000 Natl. Rep: McGavren Guild. Law Firm: Rubin, Winston, Diercks, Harris & Cooke. Format: Classic rock. Target aud: 25-54; men. ♦Chesley Maddox-Dorsey, pres; Dick Irland, gen mgr; Nick Giorno, progmg dir; Sydney L. Small, chmn & news dir; Dan Merlo, chief of engrg; Anne Pratt, traf mgr.

WMID(AM)— May 30, 1947: 1340 khz; 890 w-U. TL: N39 22 35 W74 27 08. Hrs open: 24 Equity Communications LP, 8025 Black Horse Pike, Suite 100, W. Atlantic City, 08232-2959. Phone: (609) 484-8444. Fax: (609) 646-6331. E-mail: gfequity@aol.com Web Site: classicoldieswmid.com. Licensee: Equity Communications L.P. (group owner; (acq 3-29-2002; grpsl). Population served: 250,000 Natl. Rep: Katz Radio. Law Firm: Latham & Watkins. Format: Classic oldies. Target aud: 35-64; adults. ♦Gary Fisher, VP & gen mgr; Keith Fader, sls dir; Rob Garcia, progmg dir.

***WNJN-FM—** September 1996: 89.7 mhz; 25 w horiz, 6 kw vert. Ant 272 ft TL: N39 27 40 W74 41 06. Hrs open: Box 777, Trenton, 08625-0777. Phone: (609) 777-5036. Fax: (609) 777-5217. E-mail: radio@njn.org Web Site: www.njn.net. Licensee: New Jersey Public Broadcasting Authority. Natl. Network: NPR, PRI. Law Firm: Schwartz, Woods & Miller. Format: News/talk. Target aud: General. ♦Elizabeth G. Christopherson, CEO; Bill Jobes, gen mgr; Pharoah Cranston, stn mgr & opns mgr.

WOND(AM)—See Pleasantville

WPUR(FM)— June 1998: 107.3 mhz; 13.5 kw. Ant 449 ft TL: N39 21 40 W74 25 05. Hrs open: 950 Tilton Rd., Suite 200, Northfield, 08225. Phone: (609) 645-9797. Fax: (609) 272-9224. Web Site: www.catcountry1073.com. Licensee: Millennium Atlantic City License Holdco LLC. Group owner: Millennium Radio Group LLC (acq 5-11-01; grpsl). Population served: 500,000 Natl. Rep: McGavren Guild. Format: Country. Target aud: 25-54. ♦Andy Santoro, gen mgr; Joe Kelly, opns dir & progmg dir; Mike Ruble, sls dir & gen sls mgr; Hank Weisbecher, news dir; Tom McNally, chief of engrg.

WZBZ(FM)—See Pleasantville

Avalon

WILW(FM)— Mar 29, 1976: 94.3 mhz; 3 kw. 300 ft TL: N39 07 48 W74 47 20. Stereo. Hrs opn: 24 3208 Pacific Ave., Wildwood, 08260. Phone: (609) 522-1987. Fax: (609) 522-3666. Web Site: www.oldies94wilw.com. Licensee: Coastal Broadcasting Systems Inc. (acq 3-5-98; $470,000). Population served: 300,000 Format: Oldies. ♦Bob Maschio, gen mgr; Rick Rock, progmg mgr.

Bass River Township

WBBO(FM)— Oct 1, 1972: 106.5 mhz; 1.45 kw. Ant 682 ft TL: N39 37 53 W74 21 12. Stereo. Hrs opn: 24 2355 W. Bangs Ave., Neptune, 07753. Phone: (609) 242-0223. Fax: (732) 774-4974. E-mail: grockradio@grockradio.com Web Site: www.grockradio.com. Licensee: Press Communications LLC. (acq 2-11-2005; $3.16 million). Population served: 500,000 Natl. Rep: McGavren Guild. Format: Alternitive rock, modern adult contemp. News staff: one. Target aud: 18-49. ♦Frank Calderaro, gen mgr; John Kaszuba, gen sls mgr; Mike Gravin, progmg dir; Mike Heilman, chief of engrg; Annemarie Cassone, traf mgr.

Beach Haven West

***WVBH(FM)—** 2003: 88.3 mhz; 1 w horiz, 100 w vert. Ant 426 ft TL: N39 42 56 W74 17 32. Hrs open:
Simulcast of WXHL (FM) Christiana 100%.
179 Stanton-Christiana Rd., Newark, DE, 19702. Phone: (302) 731-0690. Fax: (302) 738-3090. Web Site: www.thereachfm.com. Licensee: Priority Radio Inc. (group owner; (acq 11-14-2003; $400,000). Format: Adult contemp chrisitan mus. ♦Steve Hare, gen mgr; Dan Edwards, opns mgr.

Belvidere

WWYY(FM)— Oct 15, 1992: 107.1 mhz; 1.2 kw. Ant 718 ft TL: N40 56 53 W75 09 38. Hrs open: 22 S. 6th St., Stroudsburg, PA, 18360. Phone: (570) 421-2100. Fax: (570) 421-2040. Web Site: www.107thebone.fm. Licensee: Nassau Broadcasting Holdings Inc. Group owner: Nassau Broadcasting Partners L.P. (acq 2-25-03; grpsl). Natl. Network: ABC. Format: Rock. ♦Maureen Barth, gen mgr.

Berlin

***WNJS-FM**— Aug 21, 1992: 88.1 mhz; 1 w horiz, 20 w vert. Ant 781 ft TL: N39 43 41 W74 50 39. (CP: 1 w horiz, 250 w vert). Hrs opn: Box 777, Trenton, 08625-0777. Secondary address: 25 S. Stockton St., Trenton 08608. Phone: (609) 777-5000. Fax: (609) 777-5217. Web Site: www.njn.net. Licensee: New Jersey Public Broadcasting Authority (acq 3-6-91; FTR: 3-25-91). Natl. Network: NPR, PRI. Law Firm: Schwartz, Woods & Miller. Format: News/talk. Target aud: General. ♦Bill Jobes, gen mgr.

Blackwood

***WDBK(FM)**— June 7, 1979: 91.5 mhz; 100 w. Ant 87 ft TL: N39 47 06 W75 02 19. Hrs open: Box 200, 08012. Phone: (856) 374-4881. Fax: (856) 374-4969. Licensee: Camden County College. Population served: 75,000 Format: Alternative. ♦James Canonica, gen mgr; Greg Gaughan, stn mgr.

Blairstown

WHCY(FM)— Oct 21, 1973: 106.3 mhz; 340 w. 859 ft TL: N41 02 51 W74 58 22. Stereo. Hrs opn: 24 45 Mitchell Ave., Franklin, 07416. Phone: (973) 827-2525. Fax: (973) 827-2135. E-mail: robryan@max1063.com Web Site: www.max1063.com. Licensee: CC Licenses LLC. Group owner: Clear Channel Communications Inc. (acq 2-13-2001). Population served: 130,000 Natl. Rep: Katz Radio. Format: Hot adult contemp. Target aud: 25-54; baby boomers. Spec prog: Relg one hr wkly. ♦John Hogan, CEO, pres & CFO; Andrew Rosen, VP; Bob Dunphy, gen mgr; Vince Thomas, opns mgr; Laura Brockmann, sls dir; Lois Burmester, natl sls mgr; Elizabeth Toscano, prom dir; Rob Ryan, progmg dir.

Brick Township

***WBGD(FM)**— June 1975: Stn currently dark. 91.9 mhz; 195 w horiz. Ant 56 ft TL: N40 06 17 W74 07 34. Stereo. Hrs opn: Brick Memorial High School, 2001 Lanes Mill Rd., 08724. Phone: (732) 262-2500. Fax: (732) 836-9246. Licensee: Brick Township Board of Educ. Target aud: General. ♦Fran Bristol, gen mgr.

Bridgeton

***WNJB(FM)**— 1998: 89.3 mhz; 1 w horiz, 2.5 kw vert. Ant 220 ft TL: N39 27 35 W75 09 28. Hrs open: 5am - 12 mid Box 777, Trenton, 08625-0777. Phone: (609) 777-5036. Fax: (609) 777-5217. E-mail: raido@njn.org Web Site: www.njn.net. Licensee: New Jersey Public Broadcasting Authority. Natl. Network: NPR, PRI. Law Firm: Schwartz, Woods & Miller. Format: News/talk. Target aud: General. ♦Elizabeth G. Christopherson, CEO; Bill Jobes, gen mgr; Pharoah Cranston, stn mgr, opns mgr & progmg dir; Steve Prido, adv mgr; Bill Schorbus, engrg dir.

WSNJ(AM)— August 1937: 1240 khz; 1 kw-U. TL: N39 27 40 W75 12 21. Hrs open: 5:30 AM-midnight 1771 S. Burlington Rd., 08302. Phone: (856) 451-2930. Fax: (856) 453-9440. E-mail: information@wsnjam.com Web Site: wsnjam.com. Licensee: Quinn Broadcasting Inc. (acq 2-18-2004; $550,000). Population served: 20,435 Law Firm: Wiley, Rein & Fielding. Format: Var. News staff: one; News: 10 hrs wkly. Target aud: 25 plus. Spec prog: Big band, MOR, news/talk, farm 10 hrs wkly. ♦James F. Quinn, pres & exec VP; Toni Coogan, CFO; Greg Hennis, gen mgr; Fred Sharkey, gen sls mgr & progmg dir; John Casey, mus dir; Richard Arsenault, chief of engrg.

Bridgewater

WWTR(AM)— Dec 23, 1971: 1170 khz; 243 w-D, DA. TL: N40 33 30 W74 35 52. Hrs open: 2088 Highway 130 North, Monmouth Junction, 08852. Phone: (732) 821-6009. Fax: (732) 821-6003. E-mail: info@ebcmusic.com Web Site: www.ebcmusic.com. Licensee: The Sentinel Publishing Co. Group owner: Greater Media Inc. (acq 7-12-2001; grpsl). Format: South Asian ethnic. ♦Alka Agrawal, gen mgr; Kulraaj Anand, progmg VP & progmg dir; Neal Newman, chief of engrg.

Brigantine

***WWFP(FM)**— 2006: 90.5 mhz; 77 w vert. Ant 307 ft TL: N39 22 46 W74 25 45. Hrs open: CSN International, Box 391, Twin Falls, ID, 83303. Phone: (208) 734-6633. Fax: (208) 736-1958. Web Site: www.csnradio.com. Licensee: CSN International (group owner). ♦Mike Stocklin, gen mgr; Don Mills, progmg dir; Kelly Carlson, chief of engrg.

Burlington

WJJZ(FM)— Jan 19, 1949: 97.5 mhz; 50 kw horiz, 48 kw vert. Ant 430 ft TL: N40 14 05 W74 46 02. (CP: 26 kw, ant 682 ft. TL: N40 04 57 W75 10 53). Stereo. Hrs opn: One Bala Plaza, Mail Stop 429, Bala Cynwyd, PA, 19004-1428. Phone: (610) 771-9750. Web Site: www.975wjjz.com. Licensee: Greater Philadelphia Radio Inc. (acq 11-15-2006; exchange for WCRB(FM) Lowell, MA). Population served: 500,000 Law Firm: Leventhal, Senter & Lerman. Format: Smooth jazz. ♦Peter H. Smyth, pres; John Fullam, gen mgr; Jim Brown, stn mgr; Chrissy Sirianni, prom dir; Michael Tozzi, progmg dir; Margo Marano, mus dir.

Camden

WEMG(AM)— September 1925: 1310 khz; 1 kw-D, 250 w-N. TL: N39 57 28 W75 06 54. Hrs open: 1341 N. Delaware Ave., Suite 509, Philadelphia, PA, 19125. Phone: (215) 426-1900. Fax: (215) 426-1550. Licensee: Davidson Media Station WEMG Licensee LLC. Group owner: Mega Communications Inc. (acq 1-31-2006; $8.75 million). Population served: 5,500,000 Format: Tropical Hispanic. ♦Kevin Jones, gen mgr.

***WKDN-FM**— July 23, 1968: 106.9 mhz; 38 kw. 600 ft TL: N39 54 33 W75 06 00. Stereo. Hrs opn: 2906 Mt. Ephraim Ave., 08104. Phone: (215) 922-0282. Licensee: Family Stations Inc. (group owner; acq 7-23-68). Law Firm: Dow, Lohnes & Albertson. Format: Relg. Target aud: General; families. Spec prog: Class 2 hrs wkly. ♦Rich Archut, opns mgr.

WTMR(AM)— Nov 1, 1948: 800 khz; 5 kw-D, 500 w-N. TL: N39 54 33 W75 06 00. Hrs open: 2775 Mt. Ephraim Ave., 08104. Phone: (856) 962-8000. Fax: (856) 962-8004. E-mail: radioman@voicenet.com Licensee: KAAY License L.P. Group owner: Beasley Broadcast Group (acq 9-4-98; $8 million). Format: Relg Talk. ♦Louise Bessler, gen mgr & gen sls mgr; Mike Smith, progmg dir.

Canton

WJKS(FM)— Jan 15, 1972: 101.7 mhz; 3 kw. 263 ft TL: N39 25 51 W75 20 13. Hrs open: First Federal Plaza Bldg., 704 King St., Suite 604, Wilmington, DE, 19801. Phone: (302) 622-8895. Fax: (302) 622-8678. E-mail: tonyq@wjks1017.com Web Site: www.wjks1017.com. Licensee: QC Communication Inc. (acq 3-17-97; $1.8 million with WFAI(AM) Salem). Population served: 700,000 Format: Urban Contemporary. Target aud: 18-44. ♦Mel Brittingham, opns dir; Maria Sylvanus, gen sls mgr; Tony Quartarone, gen mgr & progmg dir; Jeff DePaulo, chief of engrg.

Cape May

WAIV(FM)— June 3, 1967: 102.3 mhz; 3.2 kw. Ant 292 ft TL: N39 00 33 W74 52 13. Stereo. Hrs opn:
Simulcast with WAYV(FM) Atlantic City 100%.
8025 Black Horse Pike, Suite 100-102, West Atlantic City, 08232. Phone: (609) 484-8444. Fax: (609) 646-6331. E-mail: gfequity@aol.com Web Site: 951wayv.com. Licensee: Equity Communications L.P. (group owner; (acq 3-29-2002; grpsl). Population served: 200,000 Natl. Network: Westwood One. Natl. Rep: Katz Radio. Law Firm: Latham & Watkins. Format: Hot adult contemp. Target aud: 18-49; adults. ♦Gary Fisher, sr VP, VP & gen mgr; Paul Kelly, progmg dir.

***WWCJ(FM)**— September 1999: 89.1 mhz; 15 kw. Ant 308 ft TL: N39 02 58 W74 51 14. Hrs open: 24
Rebroadcasts WWFM(FM) Trenton 100%.
Box B, Trenton, 08690. Phone: (609) 587-8989. Fax: (609) 570-3863. E-mail: wwfm@mccc.edu Web Site: www.wwfm.org. Licensee: Mercer County Community College. Format: Classical. ♦Jeffery R. Sekerka, gen mgr.

Cape May Court House

WGBZ(FM)— Sept 5, 1985: 105.5 mhz; 3.3 kw. Ant 295 ft TL: N39 07 32 W74 49 26. Stereo. Hrs opn: 24
Simulcast with WZBZ(FM) Atlantic City 100%.
8025 Black Horse Pike, Suite 100-102, West Atlantic City, 08232. Phone: (609) 484-8444. Fax: (609) 646-6331. E-mail: info@993thebuzz.com Web Site: 993thebuzz.com. Licensee: Equity Communications L.P. (group owner; acq 5-31-2002; grpsl). Population served: 400,000 Natl. Network: Westwood One. Natl. Rep: Katz Radio. Law Firm: Latham & Watkins. Format: CHR. Target aud: 18-49; adults. ♦Gary Fisher, sr VP, VP & gen mgr; Rob Garcia, progmg dir; Denise Carrington, traf mgr.

***WJPG(FM)**— 2004: 88.1 mhz; 550 w vert. Ant 213 ft TL: N39 07 32 W74 49 27. Hrs open: Joy Communications Inc., Box 603, Woodbine, 08270-0603. Phone: (609) 861-3700. Fax: (609) 861-3730. E-mail: letters@praise899.org Web Site: www.praise899.org. Licensee: Maranatha Ministries. Population served: 450,000 Format: Praise & worship. News: 10 hrs wkly. Target aud: 25-54; women. Spec prog: Gospel one hr wkly. ♦Kenneth Manri, gen mgr.

***WNJZ(FM)**— August 1999: 90.3 mhz; 6 kw. Ant 236 ft TL: N39 06 18 W74 48 06. Hrs open: Box 777, Trenton, 08625-0777. Phone: (609) 777-5036. Fax: (609) 777-5217. E-mail: radio@njn.org Web Site: www.njn.net. Licensee: New Jersey Public Broadcasting Authority. Natl. Network: NPR, PRI. Format: News/talk. ♦Elizabeth G. Christopherson, CEO; Bill Jobes, gen mgr; Pharoah Cranston, stn mgr.

Cherry Hill

***WKVP(FM)**— Jan 7, 1985: 89.5 mhz; 50 w horiz, 2 kw vert. Ant 180 ft TL: N39 51 33 W74 57 00. Hrs open: 24
Rebroadcasts KLVR(FM) Santa Rosa, CA 100%.
2351 Sunset Blvd., Suite 170-218, Rocklin, CA, 95765. Phone: (916) 251-1600. Fax: (916) 251-1650. Web Site: www.klove.com. Licensee: Educational Media Foundation. (acq 1-10-2007; $2.45 million). Population served: 800,000 Natl. Network: K-Love. Format: Contemp Chirstian. ♦Richard Jenkins, pres.

Delaware Township

***WDVR(FM)**— Feb 19, 1990: 89.7 mhz; 4.8 kw. Ant 302 ft TL: N40 30 37 W74 57 29. Stereo. Hrs opn: 24 Box 191, Rt. 604, Sergeantsville, 08557-0191. Phone: (609) 397-1620. Fax: (609) 397-5991. Web Site: www.wdvrfm.org. Licensee: Penn-Jersey Educational Radio Corp. Natl. Network: ABC. Law Firm: Schwartz, Woods & Miller. Format: Div. News: 2 hrs wkly. Target aud: 30 plus. Spec prog: Folk 6 hrs, relg 6 hrs, jazz 11 hrs, oldies 13 hrs, bluegrass 6 hrs, country classic, 12 hrs; Americana country 6 hrs wkly. ♦Frank W. Napurano, pres & gen mgr; Ginny Lee, prom dir; Frank Napurano, progmg dir; Carla Van Dyk, mus dir.

Dover

WDHA-FM— Feb 22, 1961: 105.5 mhz; 1 kw. Ant 574 ft TL: N40 51 19 W74 30 42. Stereo. Hrs opn: 24 55 Horsehill Rd., Cedar Knolls, 07927. Phone: (973) 538-1250. Phone: (973) 455-1055. Fax: (973) 538-3060. E-mail: rock@wdhafm.com Web Site: www.wdhafm.com. Licensee: The Sentinel Publishing Co. Group owner: Greater Media Inc. (acq 7-6-01; grpsl). Population served: 360,000 Natl. Rep: Katz

Radio. Law Firm: Pepper & Corazzini. Format: New & Classic Rock. Target aud: 18-49. ♦Nancy McKinley, stn mgr; Matt DeVoti, gen sls mgr; Pete Forester, natl sls mgr.

Dover Township

***WWNJ(FM)—** December 1991: 91.1 mhz; 50 w horiz, 50 kw vert. Ant 151 ft TL: N39 58 07 W74 04 19. Stereo. Hrs opn: 24
Rebroadcasts WWFM(FM) Trenton 100%.
Box B, Trenton, 08690. Phone: (609) 587-8989. Fax: (609) 570-3863. E-mail: wwfm@mccc.edu Web Site: www.wwfm.org. Licensee: Mercer County Community College Board of Trustees. (acq 11-4-91). Natl. Network: PRI. Format: Classical. Target aud: General. ♦Jeffery R. Sekerka, gen mgr & dev mgr.

East Orange

***WFMU(FM)—** 1958: 91.1 mhz; 1.25 kw. 360 ft TL: N40 47 15 W74 04 19. (CP: Ant 505 ft. TL: N40 47 19 W74 15 20). Stereo. Hrs opn: 24 Box 2011, Jersey City, 07303-2011. Phone: (201) 521-1416. Fax: (201) 521-1286. E-mail: wfmu@wfmu.org Web Site: www.wfmu.org. Licensee: Auricle Communications. Population served: 6,000,000 Format: Div, free-form. Target aud: General. Spec prog: International 15 hrs wkly. ♦Brian Turner, progmg dir & mus dir; John Fogarazzo, chief of engrg; Ken Freedman, gen mgr & spec ev coord.

Eatontown

WHTG(AM)— Nov 1, 1957: 1410 khz; 500 w-D, 126 w-N. TL: N40 16 10 W74 04 19. Hrs open: 2355 W. Bango Ave., Neptune, 07753. Phone: (732) 774-4755. Fax: (732) 774-4974. Licensee: Press Communications L.L.C. (group owner; acq 11-4-00; $15 million with co-located FM). Population served: 1,000,000 Natl. Rep: Christal. Format: Great gold. Target aud: 35 plus; general. Spec prog: Baseball 20 hrs, football 3 hrs, basketball 6 hrs wkly. ♦Robert McAllan, CEO & pres; Richard T. Morena, CFO; John Dziuba, gen mgr; Cindy Brennan, stn mgr; John Kaszuba, gen sls mgr; Mathew Schwenker, natl sls mgr & prom dir; Jack Aponte, progmg dir & mus dir; Mike Heilman, chief of engrg.

WHTG-FM— Oct 11, 1961: 106.3 mhz; 1.1 kw. Ant 528 ft TL: N40 16 41 W74 04 51. Stereo. E-mail: g1063@g1063.com Web Site: www.g1063.com. Law Firm: Leventhal, Senter & Lerman. Format: Alternative rock, modern adult contemp. Target aud: 18-34. ♦John Kaszuba, opns dir & rgnl sls mgr; Michael Gavin, progmg mgr; Brian Phillips, mus dir.

Egg Harbor City

WSJO(FM)— Sept 23, 1971: 104.9 mhz; 10 kw. Ant 508 ft TL: N39 32 49 W74 38 19. Stereo. Hrs opn: 109 Walters Ave., Trenton, 08638. Phone: (609) 771-8181. Fax: (609) 406-7956. Fax: (609) 771-0581. Web Site: www.sojo1049.com. Licensee: Millennium Egg Harbor License Holdco LLC. Group owner: Nassau Broadcasting Partners L.P. (acq 11-23-2004; $14 million). Population served: 65,000 Natl. Network: AP Radio. Format: Hot adult contemp. Target aud: 35 plus. Spec prog: Relg 4 hrs wkly. ♦Andy Santoro, pres & gen mgr.

Egg Harbor Township

***WXGN(FM)—** 2000: 90.5 mhz; 500 w vert. Ant 82 ft TL: N39 16 46 W74 34 34. Hrs open: 24 1512 Atkinson Ave., Somers Point, 08244-1119. Phone: (609) 926-5182. Fax: (609) 926-5185. Web Site: wxgn.com. Licensee: Joy Broadcasting Inc. Format: Contemp Christian. ♦Bob Green, gen mgr.

Elizabeth

WJDM(AM)— Mar 11, 1970: 1530 khz; 1 kw-D. TL: N40 38 56 W74 14 32. Hrs open: 407 N. Broad St., 07208. Phone: (908) 352-3400. Fax: (908) 352-4268. Web Site: www.puertadepaz.com. Licensee: Multicultural Radio Broadcasting Licensee LLC. Group owner: Multicultural Radio Broadcasting Inc. (acq 2-4-2004; grpsl). Population served: 2,000,000 Format: Sp. ♦Richard Dirocco, gen mgr; Didier Ugalde, stn mgr.

Ewing

WIMG(AM)—Licensed to Ewing. See Trenton

Flemington

WCHR(AM)— Jan 5, 1998: 1040 khz; 4.7 kw-D, 1 kw-N, DA-2. TL: N40 30 18 W74 58 37. (CP: 15 kw-D, 2.5 kw-N, 7.5 kw-CH, DA-3). Hrs opn: 24 119 Locktown Rd., 08822. Phone: (215) 493-4252. Fax: (215) 321-5583. E-mail: csimpson@nassaubroadcast.com Web Site: www.wchram.net. Licensee: Nassau Broadcasting II L.L.C. Group owner: Nassau Broadcasting Partners L.P. (acq 2-15-02; grpsl). Population served: 5,511,450 Format: Relg. Target aud: General; Religious adults. ♦John White, stn mgr; Cort Simpson, opns mgr; Chuck Zulker, sls dir.

***WCVH(FM)—** April 1974: 90.5 mhz; 78 w. 449 ft TL: N40 33 25 W74 54 18. Stereo. Hrs opn: 24/7 84 Rt. 31, 08822. Phone: (908) 782-9595. Fax: (908) 284-7109. E-mail: wcvh@hcrhs.k12.nj.us Web Site: www.hcrhs.k12.nj.us. Licensee: Hunterdon Central Board of Education. Population served: 100,000 Format: Div. Spec prog: 30 hrs. Student Programing. ♦David R. Kelber, gen mgr; Joanna Lynch, opns dir; John Anastasio, chief of engrg.

Florence

WIFI(AM)— 1985: 1460 khz; 5 kw-D, 500 w-N, DA-2. TL: N40 04 53 W74 47 41. Hrs open: 24 2025 Burlington-Columbus Rd., Burlington, 08016. Phone: (609) 499-4800. Fax: (609) 499-4905. Licensee: Real Life Broadcasting. Population served: 2,000,000 Natl. Network: USA. Format: Relg. ♦Ron Graban, gen mgr.

Franklin

WSUS(FM)— Feb 28, 1965: 102.3 mhz; 590 w. 745 ft TL: N41 08 37 W74 32 21. Stereo. Hrs opn: 45 Mitchell Ave., 07416. Phone: (973) 827-2525. Fax: (973) 827-2135. E-mail: wsus1023@aol.com Web Site: www.wsus1023.com. Licensee: CC Licenses LLC. Group owner: Clear Channel Communications Inc. (acq 2-15-2001; grpsl). Natl. Rep: Katz Radio. Wire Svc: AP Format: Adult contemp, news. News staff: 2; News: 10 hrs wkly. Target aud: 25-54; women. ♦John Hogan, CEO, pres & CFO; Andy Rosen, VP; Bob Dunphy, gen mgr; Vince Thomas, stn mgr & opns mgr; Laura Brockmann, sls dir; Lois Burmester, natl sls mgr; Alexandra Vallejo, news dir; Peter Draney, traf mgr.

Freehold Township

***WRDR(FM)—** Feb 20, 1997: 89.7 mhz; 10 w horiz, 2 kw vert. 170 ft TL: N40 11 19 W74 15 01. Stereo. Hrs opn: 24 6550 Rt. 9 S., Howell, 07731. Phone: (732) 901-9953. Fax: (732) 901-0356. Web Site: www.bridgefm.org. Licensee: Bridgelight LLC (acq 1-31-03; $875,000). Population served: 6,700,000 Format: Christian. News staff: . Target aud: 25-62. Spec prog: Relg 6 hrs wkly. ♦Chris McCarrick, gen mgr; Don Gates, progmg dir.

Glassboro

***WGLS-FM—** January 1964: 89.7 mhz; 750 w. 489 ft TL: N39 41 41 W75 17 55. Stereo. Hrs open: 24 Rowan Univ., 201 Mullica Hill Rd., 08028-1701. Phone: (856) 863-9457. Fax: (856) 256-4704. E-mail: wgls@rowan.edu Web Site: http://wgls.rowan.edu. Licensee: Rowan University. Population served: 1,300,000 Natl. Network: ABC. Law Firm: Booth, Freret, Imlay & Tepper, P.C. Format: Div, educ. News staff: one; News: one. Target aud: 18-45; general. Spec prog: Black 10 hrs wkly. ♦Frank Hogan, gen mgr; Joe Staudenmayer, opns mgr; Nichole Chiaravalloti, news dir; Julia Giacoboni, pub affrs dir; Colin Weir, traf mgr.

Hackensack

WWDJ(AM)— 1921: 970 khz; 5 kw-U, DA-2. TL: N40 54 40 W40 01 42. Hrs open: 777 Terrace Ave, 6th floor, Hasbrouck Heights, 07604-3100. Phone: (201) 298-9700. Fax: (201) 298-5797. E-mail: office@nycradio.com Web Site: www.wmca.com. Licensee: Salem Media Corp. Group owner: Salem Communications Corp. (acq 8-3-94). Population served: 13,000,000 Format: Christian/Talk. Target aud: 25-44. ♦Edward Atsinger, pres; Joe D. Davis, VP; M. Susan Lucchiesi, gen mgr; Tamela Kay Maxwell, gen sls mgr.

Hackettstown

***WNTI(FM)—** Dec 5, 1957: 91.9 mhz; 5.6 kw. 510 ft TL: N40 51 07 W74 52 35. Stereo. Hrs opn: 24 400 Jefferson St., 07840. Phone: (908) 852-4545. Phone: (908) 979-4355. Fax: (908) 852-8515. Web Site: www.wnti.org. Licensee: Centenary College. Population served: 400,000 Natl. Network: PRI. Format: Free form. News: 2 hrs wkly. Target aud: 15 plus. Spec prog: Big band 4 hrs, blues 11 hrs, heavy metal 6 hrs, reggae 3 hrs, oldies 3 hrs, jazz 9 hrs, relg 4 hrs wkly. ♦Paul Massen, gen mgr.

WRNJ(AM)— 1996: 1510 khz; 2.5 kw-ND, 230 w-N, DA-N. TL: N40 50 47 W74 48 16 (D), N40 48 55 W74 49 38 (N). Hrs open: Box 1000, 07840. Phone: (908) 850-1000. Fax: (908) 852-8000. E-mail: info@oldies1510.com Web Site: www.wrnj.com. Licensee: WRNJ Radio Inc. Population served: 35,000 Natl. Network: ABC. Law Firm: Shaw Pittman. Wire Svc: AP Format: News/talk, oldies. News staff: 3; News: 10 hrs wkly. Target aud: 25-50 plus; upwardly mobile. Spec prog: Talk 10 hrs wkly. ♦L.J. Tighe, pres; Norman Worth, gen mgr; Russ Long, opns dir; Dan Hollis, gen sls mgr & news dir; Chuck Reiger, progmg dir; Larry Tighe, chief of engrg; Pat Layton, traf mgr.

Hammonton

WGYM(AM)— May 11, 1961: 1580 khz; 1 kw-D, 7 w-N. TL: N39 37 33 W74 47 44. Hrs open: 24 hrs
Rebroadcasts WOND(AM) Atlantic City 100%.
1601 New Rd., Linwood, 08221. Phone: (609) 653-1400. Fax: (609) 601-0450. Web Site: www.wond1400am.com. Licensee: Access.1 New Jersey License Co. LLC. Group owner: Access.1 Communications Corp. (acq 11-17-2003; grpsl). Population served: 545,000 Natl. Network: Westwood One. Natl. Rep: McGavren Guild. Law Firm: Rubin, Winston, Diercks, Harris & Cooke. Format: News/talk. Target aud: Adults 35+. ♦Sydney L. Small, chmn; Chesley Maddox-Dorsey, pres; Dick Irland, gen mgr; John De Lucia, gen sls mgr; Stuart Abrams, progmg dir & traf mgr; Dan Merlo, chief of engrg.

Hazlet

***WDDM(FM)—** May 24, 1979: 89.3 mhz; 10 w. Ant 125 ft TL: N40 25 37 W74 11 40. Stereo. Hrs opn: Box 192, Fairless Hills, PA, 19030. Phone: (215) 269-4446. Fax: (215) 269-6848. E-mail: info@domesticchurchmedia.org Web Site: domesticchurchmedia.org. Licensee: Domestic Church Media Foundation (acq 1-3-2006; $500,000). Format: Catholic. ♦James Manfredonia, pres & gen mgr.

Jersey City

WSNR(AM)— December 1948: 620 khz; 3 kw-D, 7.6 kw-N, DA-2. TL: N40 47 53 W74 06 24. (CP: 8.5 kw-D, 5 kw-N, DA-2. TL: N40 50 52 W74 20 22). Stereo. Hrs opn: 475 Park Ave. S., New York, NY, 10016-6901. Phone: (847) 509-1661. Fax: (646) 424-2232. Fax: (847) 509-7750. Web Site: www.sportingnews.com. Licensee: Rose City Radio Corp. (group owner; acq 3-23-01; grpsl). Population served: 9,000 Natl. Network: CBS. Wire Svc: CBS Format: multicultural. Target aud: General. ♦Clancy Woods, pres, gen mgr & progmg dir; Colleen Mamzella, chief of engrg & traf mgr.

WWRU(AM)— Dec 8, 1995: 1660 khz; 10 kw-U, DA-2. TL: N40 49 13 W74 04 09 (D), N40 49 13 W74 04 09 (N). (CP: TL: N40 49 13 W74 04 04 (D), N40 49 13 W74 04 09 (N). Hrs opn: 449 Broadway, 5th Fl., New York, NY, 10013. Phone: (212) 966-8700. Fax: (212) 966-9580. Licensee: Multicultural Radio Broadcasting Licensee LLC. Group owner: Multicultural Radio Broadcasting Inc. (acq 2-4-2004; grpsl). Format: Korean. ♦Gene Heinemeyer, gen mgr.

Lakewood

WOBM(AM)— Nov 20, 1970: 1160 khz; 5 kw-D, 8.9 kw-N, DA-2. TL: N40 08 09 W74 13 48. Hrs open: 24 Box 927, Toms River, 08754. Phone: (732) 269-0927. Fax: (732) 269-9292. Web Site: www.wobmam.com. Licensee: Millennium Shore License Holdco LLC. Group owner: Millennium Radio Group LLC (acq 5-14-02; grpsl). Population served: 452,000 Natl. Network: AP Radio. Natl. Rep: Katz Radio. Law Firm: Blair, Joyce & Silva. Format: Oldies, Big Band, talk. News staff: 5; News: 14 hrs wkly. Target aud: 45 plus; educated. Spec prog: Talk 12 hrs wkly. ♦Bill Saurer, gen mgr; John Furno, gen sls mgr; Steve Ardolina, progmg dir; Tom Mongelli, news dir; Jay Pierce, chief of engrg; Nancy Cordiano, traf mgr.

WOBM-FM—See Toms River

Lawrenceville

***WRRC(FM)—** Sept 23, 1989: 107.7 mhz; 17 w. 36 ft TL: N40 16 44 W74 44 15. Stereo. Hrs opn: 16 Rider Univ., Bart Luedeke Ctr., 2083 Lawrenceville Rd., 08648. Phone: (609) 896-5369. Fax: (609) 219-4729. Licensee: Rider University Board of Trustees. Format: Var. News staff: 3; News: 4 hrs wkly. Target aud: 16-21; high school & college students. Spec prog: Black 10 hrs, heavy metal 10 hrs wkly.

Lincroft

***WBJB-FM**— Jan 13, 1975: 90.5 mhz; 11 kw. 135 ft TL: N40 19 19 W74 07 57. Stereo. Hrs opn: 24 Brookdale Community College, 765 Newman Springs Rd., 07738. Phone: (732) 224-2490. Phone: (732) 224-2252. Fax: (732) 224-2494. E-mail: comments@wbjb.org Web Site: www.90.5thenight.org. Licensee: Board of Trustees of Brookdale Community College. Population served: 1,000,000 Natl. Network: NPR. Law Firm: Erwin, Campbell & Tannenwald. Format: AAA, news. News: 16 hrs wkly. Target aud: 18-54; general. Spec prog: Haitian 3 hrs, pub affrs 5 hrs, bluegrass 3 hrs, Sp 4 hrs, blues 4 hrs wkly. ♦Tom Brennan, stn mgr; Jeff Raspe, mus dir; Michelle McBride, pub affrs dir; George Marshall, chief of engrg.

Lindenwold

WTTM(AM)— May 1999: 1680 khz; 10 kw-D, 1 kw-N. TL: N39 53 15 W75 00 05. Hrs open: Multicultural Radio Broadcasting Inc., 449 Broadway, New York, NY, 10013. Phone: (212) 966-1059. Fax: (212) 966-9580. Licensee: Multicultural Radio Broadcasting Licensee LLC. Group owner: Multicultural Radio Broadcasting Inc. (acq 5-24-2002; grpsl). ♦Troy Hall, gen mgr; C.W. Queen, stn mgr.

Long Branch

WWZY(FM)— June 1, 1960: 107.1 mhz; 4.7 kw. Ant 371 ft TL: N40 18 17 W73 59 08. Stereo. Hrs opn: 2355 W. Bangs Ave., Neptune, 07753. Phone: (732) 774-4755. Fax: (732) 774-4974. Web Site: www.breezeradio.com. Licensee: Press Communications LLC (group owner; acq 6-18-2003; $20 million). Population served: 1,000,000 Natl. Rep: Eastman Radio. Wire Svc: AP Format: Adult contemp. Target aud: Women; 25-54. ♦Frank Calderaro, gen mgr; John Kaszuba, gen sls mgr; Mike Fitzgerald, progmg dir; Mike Heilman, chief of engrg.

Madison

***WMNJ(FM)**— Sept 15, 1980: 88.9 mhz; 8 w. Ant 75 ft TL: N40 45 30 W74 25 48. Stereo. Hrs opn: 24 Drew Univ., 36 Madison Ave., 07940. Phone: (973) 408-4753. Phone: (973) 408-3000 (univ.). Fax: (973) 408-3939. Licensee: Drew University (acq 9-25-89). Format: Alternative, AOR. ♦Margaret E.L. Howard, VP; Alicia E. Lutes, progmg dir.

Mahwah

***WRPR(FM)**— July 15, 1980: 90.3 mhz; 100 w. 30 ft TL: N41 04 51 W74 10 34. Stereo. Hrs opn: 18 505 Ramapo Valley Rd., 07430. Phone: (201) 825-7449. Fax: (201) 327-9030. E-mail: wrpr@ramapo.edu Web Site: www.wrpr.org. Licensee: Ramapo College of New Jersey. Format: College contemp. News staff: 4; News: 10 hrs wkly. Target aud: 18-24; college students. Spec prog: Pub affrs 12 hrs wkly. ♦Evan Brown, gen mgr; Andrew Bernstein, progmg dir; Sarah Tucci, mus dir & news dir; Delores Smith, traf mgr.

Manahawkin

WCHR-FM— 2002: 105.7 mhz; 13 kw. Ant 459 ft TL: N39 42 56 W74 17 32. Hrs open: 2401 Rt. 66, Ocean, 07712. Phone: (732) 897-8282. Fax: (732) 897-8283. Web Site: www.1057thehawk.com. Licensee: Millennium Shore License Holdco LLC. Group owner: Millennium Radio Group LLC (acq 3-15-2004; $12 million). Format: Classic rock. ♦Bill Saurer, gen mgr; Phil LoCascio, opns mgr; John Furno, gen sls mgr; Tom Monzeli, news dir; Jay Pierce, engrg dir.

WJRZ-FM— July 4, 1976: 100.1 mhz; 1.7 kw. Ant 436 ft TL: N39 47 54 W74 12 10. Stereo. Hrs opn: 24 Box 1000, 1001 Beach Ave., 08050. Secondary address: Box 100, 22 W. Water St., Toms River 08754. Phone: (609) 597-1100. Phone: (732) 349-1100. Fax: (609) 597-4400. Fax: (732) 505-8700. Web Site: www.wjrz.com. Licensee: Jersey Shore Broadcasting Corp. Group owner: Greater Media Inc. (acq 7-19-02). Population served: 1,100,000 Natl. Network: AP Radio. Natl. Rep: Christal. Format: Oldies. News staff: one; News: 4 hrs wkly. Target aud: 18-54. ♦Dan Finn, exec VP & VP; Mike Kazala, gen mgr & stn mgr; Jeff Rafter, opns mgr & progmg dir; Andria Iridoy, gen sls mgr; Peter Iridoy, prom dir; Bill Clanton Sr., chief of engrg; Sharon Zarnowski, traf mgr.

***WNJM(FM)**— August 1999: 89.9 mhz; 1 w horiz, 200 w vert. Ant 259 ft TL: N39 41 57 W74 14 05. Hrs open: Box 777, Trenton, 08625-0777. Phone: (609) 777-5036. Fax: (609) 777-5217. E-mail: radio@njn.org Web Site: www.njn.net. Licensee: New Jersey Public Broadcasting Authority. Natl. Network: NPR, PRI. Format: News/talk. ♦Elizabeth G. Christopherson, CEO; Bill Jobes, gen mgr; Pharoah Cranston, stn mgr.

***WYRS(FM)**— Mar 27, 1995: 90.7 mhz; 1 w horiz, 15 kw vert. Ant 262 ft TL: N39 38 24 W74 17 32. Stereo. Hrs opn: 24 Box 730, 08050. Phone: (609) 978-1678. Fax: (609) 597-4146. E-mail: info@wyrs.org Web Site: www.wyrs.org. Licensee: WYRS Broadcasting (acq 2-18-2005; $1). Natl. Network: AP Radio. Format: Family & community. Target aud: General. ♦Bob Wick, CEO, gen mgr & chief of engrg.

Margate City

WTTH(FM)— Nov 19, 1991: 96.1 mhz; 2.8 kw. Ant 400 ft TL: N39 21 02 W74 26 55. Stereo. Hrs opn: 24
Simulcast with WDTH(FM) Wildwood Crest 100%.
8025 Black Horse Pike, Suite 100-102, West Atlantic City, 08232. Phone: (609) 484-8444. Fax: (609) 646-6331. E-mail: gfequity@aol.com Web Site: www.961wtth.com. Licensee: Equity Communications L.P. (group owner; (acq 5-30-2003; grpsl). Population served: 400,000 Natl. Network: Westwood One. Natl. Rep: Katz Radio. Law Firm: Laitham & Watkins. Format: Urban adult contemp. Target aud: 18-49; adults. Spec prog: Gospel 5 hrs, relg one hr wkly. ♦Gary Fisher, sr VP, VP & gen mgr; Keith Fader, sls dir; Rob Garcia, progmg dir.

Medford Lakes

***WVBV(FM)**— 2005: 90.5 mhz; 21 kw vert. Ant 453 ft TL: N39 33 20 W74 44 48. Stereo. Hrs opn: 24 x 7 Hope Christian Church of Marlton Inc., 55 E. Main St., Marlton, 08053. Phone: (856) 983-1662. Fax: (856) 983-1814. E-mail: info@ccmarlton.org Web Site: www.hopefm.net. Licensee: Hope Christian Church of Marlton Inc. Law Firm: Fletcher, Heald & Hildreth. Format: Christian/talk, praise, worship & music. ♦William C. Luebkemann Jr., pres & gen mgr.

Millville

WMVB(AM)— December 1953: 1440 khz; 1 kw-D, 65 w-N, DA-2. TL: N39 25 19 W75 01 14. Hrs open: 24
Simulcast with WSNJ(AM) Bridgeton.
415 N. High St., 08332. Phone: (609) 805-0047. Fax: (856) 327-0408. E-mail: greg@wmvb.net Web Site: www.wsmjam.com. Licensee: Quinn Broadcasting Inc (acq 4-26-2000; $500,000). Population served: 342,917 Format: Var. News staff: 4. Target aud: 25-54; general public. Spec prog: Sp 2 hrs, gospel 8 hrs, children 3 hrs wkly. ♦Greg Hennis, gen mgr.

WXKW(FM)— Feb 2, 1962: 97.3 mhz; 50 kw. Ant 466 ft TL: N39 19 15 W74 46 17. Stereo. Hrs opn:
Rebroadcasts WKXW-FM Trenton 100%.
950 Tilton Rd., Suite 200, Northfield, 08225. Phone: (609) 771-8181. Fax: (609) 926-5907. Web Site: www.nj1015.com. Licensee: Millennium Atlantic City II License Holdco LLC. Group owner: Millennium Radio Group LLC (acq 12-21-2001; grpsl). Population served: 139,000 Format: Talk. Target aud: 25-54. ♦Andy Santoro, gen mgr.

Morristown

***WJSV(FM)**— Feb 22, 1971: 90.5 mhz; 124 w. 17 ft TL: N40 50 10 W74 29 16. Stereo. Hrs opn: 8 AM-10 PM (M-F) WJSV c/o Morristown High School, 50 Early St., 07960. Phone: (973) 292-2168. Fax: (973) 539-5573. E-mail: norman.wallerstein@msdk12.net Licensee: Morris School District Board of Education. Population served: 750,000 Format: Free form AOR. News staff: one; News: 3 hrs wkly. Target aud: General. Spec prog: News/talk 3 hrs, sports 3 hrs wkly. ♦Norman Wallerstein, gen mgr; Dame Mallan, stn mgr; Lee Tyler, prom dir & adv mgr; Ryan Whitenack, prom dir; Michael O'Brien, progmg dir.

WMTR(AM)— Dec 12, 1948: 1250 khz; 5 kw-D, 7 kw-N, DA-2. TL: N40 48 45 W74 27 36. Stereo. Hrs opn: Box 1250, 07962-1250. Phone: (973) 538-1250. Fax: (973) 538-3060. Web Site: www.wmtram.com. Licensee: The Sentinel Publishing Co. Group owner: Greater Media Inc. (acq 7-6-01; grpsl). Population served: 1,000,000 Law Firm: Pepper & Corazzini. Format: Oldies. Target aud: 35 plus. Spec prog: Community connection 5 hrs wkly. ♦Dan Finn, exec VP; Chris Edwards, opns mgr, mktg dir & progmg mgr; Matt DeVoti, gen sls mgr; Nancy McKinley, gen mgr & natl sls mgr.

Mount Holly

WWJZ(AM)— November 1992: 640 khz; 50 kw-D, 950 w-N, DA-2. TL: N40 05 28 W74 50 30. Hrs open: 501 Office Center Dr., Ste 190, Fort Washington, PA, 19034-3268. Phone: (215) 591-0100. Fax: (215) 591-4527. Web Site: www.radiodisney.com. Licensee: Radio Disney Group LLC. Group owner: ABC Inc. (acq 12-30-99). Format: Top-40. Target aud: 45 plus. ♦Robert S. Minton, gen mgr.

New Brunswick

WCTC(AM)— Dec 12, 1946: 1450 khz; 1 kw-U. TL: N40 29 32 W74 25 11. Stereo. Hrs opn: 24 Box 100, Broadcast Ctr., 08903. Secondary address: 78 Veronica Ave., Somerset 08873. Phone: (732) 249-2600. Fax: (732) 249-9010. Licensee: The Sentinel Publishing Co. Group owner: Greater Media Inc. (acq 5-1-57). Population served: 2,000,000 Format: News/talk. News staff: 6; News: 15 hrs wkly. Target aud: 35-54. Spec prog: Rutgers Univ. & high school sports. ♦Dan Finn, VP; Frank Calderaro, gen mgr; John Ford, gen mgr & stn mgr; Bruce Johnson, opns mgr, progmg dir, news dir & sports cmtr; Jack Cahill, gen sls mgr; Dave Kirby, prom dir; Keith Smeal, chief of engrg; Susan Young, traf mgr.

WMGQ(FM)—Co-owned with WCTC(AM). 1947: 98.3 mhz; 1 kw. 525 ft TL: N40 28 33 W74 29 34. Stereo. 24 2,000,000 Format: Adult contemp. News: 3 hrs wkly. Target aud: 25-54. ♦Tim Tefft, progmg dir.

***WRSU-FM**— April 1974: 88.7 mhz; 1.4 kw. 150 ft TL: N40 28 00 W74 26 15. Stereo. Hrs opn: 24 126 College Ave., 08903. Phone: (732) 932-7800. Fax: (732) 932-1768. E-mail: wrsu@wrsu.rutgers.edu Web Site: www.wrsu.rutgers.org. Licensee: Board of Governors Rutgers University. Population served: 1,000,000 Wire Svc: AP Format: Var/div. News: 6 hrs wkly. Target aud: 15-30; college students, div group of young adults. ♦Eric Strain, gen mgr.

Newark

***WBGO(FM)**— Feb 7, 1948: 88.3 mhz; 10 kw. Ant 431 ft TL: N40 44 11 W74 10 15. Stereo. Hrs opn: 24 54 Park Pl., 07102. Phone: (973) 624-8880. Fax: (973) 824-8888. E-mail: jazz88@wbgo.org Web Site: www.wbgo.org. Licensee: Newark Public Radio Inc. (acq 12-77). Population served: 375,000 Natl. Network: NPR. Law Firm: Dow, Lohnes & Albertson. Wire Svc: AP Format: Jazz. News staff: 3; News: 5 hrs wkly. Target aud: General. ♦Tim Porter, chmn; Cephas Bowles, gen mgr; Arajua Backman, gen sls mgr & rgnl sls mgr; Grey Johnson, mktg mgr; Thurston Briscoe, progmg dir; Doug Doyle, news dir; Brian McCabe, chief of engrg; Liz Kent, traf mgr.

WCAA(FM)— August 1992: 105.9 mhz; 2.4 kw. Ant 722 ft TL: N40 45 04 W73 58 25. Hrs open: 485 Madison Ave., New York, NY, 10022. Phone: (212) 310-6000. Fax: (212) 888-3694. Web Site: www.univision.com. Licensee: WADO-AM License Corp. ("WADO"). Group owner: Univision Radio (acq 9-22-2003; grpsl). Population served: 18,000,000 Format: Sp, reggaton. ♦Joe Pagan, gen mgr.

***WFME(FM)**— 1959: 94.7 mhz; 38 kw. 570 ft TL: N40 47 18 W74 15 19. Stereo. Hrs opn: 24 289 Mt. Pleasant Ave., West Orange, 07052. Phone: (973) 736-3600. Fax: (973) 736-4832. E-mail: audiohq@aol.com Web Site: www.familyradio.com. Licensee: Family Stations Inc. (group owner; acq 3-10-66). Population served: 14,000,000 Natl. Network: Family Radio. Format: Christian educ. Target aud: General. ♦Harold Camping, pres & gen mgr; Charles Menut, stn mgr & chief of engrg; Jason Frentses, pub affrs dir.

WHTZ(FM)— June 1, 1961: 100.3 mhz; 7.8 kw. 1,220 ft TL: N40 44 54 W73 59 10. Hrs open: 36th Fl., 101 Hudson St., Jersey City, 07302. Phone: (212) 239-2300. Fax: (212) 239-2308. E-mail: z100radio@aol.com Web Site: www.z100.com. Licensee: AMFM Radio Licenses L.L.C. Group owner: Clear Channel Communications Inc. (acq 8-30-00; grpsl). Population served: 381,930 Natl. Rep: Christal. Format: Top 40. ♦Rob Williams, exec VP & gen mgr; Tom Poleman, opns mgr; Bob McCuin, gen sls mgr; Josh Hadden, engrg dir & chief of engrg.

WNSW(AM)— 1947: 1430 khz; 5 kw-U, DA-N. TL: N40 42 32 W74 14 31. Stereo. Hrs opn: 24 449 Broadway 2nd Fl., New York, NY, 10013. Phone: (212) 966-1059. Fax: (212) 966-9580. E-mail: geneh@mrbi.net Licensee: Multicultural Radio Broadcasting Licensee LLC. Group owner: Multicultural Radio Broadcasting Inc. (acq 1-30-98; grpsl). Population served: 13,000,000 Natl. Rep: Katz Radio. Law Firm: Hopkins & Sutter. Format: Sp contemp Christian. ♦Gene Heinemeyer, gen mgr & progmg dir; Harold Chou, chief of engrg.

Newton

WNNJ(AM)— Dec 15, 1953: 1360 khz; 2 kw-D, 320 w-N, DA-2. TL: N41 02 22 W74 44 19. Hrs open: 24 45 Mitchell Ave., Franklin, 07416. Phone: (973) 827-2525. Fax: (973) 827-2135. Web Site: www.oldies1360.com. Licensee: CC Licenses LLC. Group owner: Clear Channel Communications Inc. (acq 1-31-2001; grpsl). Population served: 130,000 Natl. Network: Westwood One. Natl. Rep: Katz Radio. Wire Svc: AP Format: Oldies. News staff: 2; News: 5 hrs wkly. Target aud: 35 plus; mature adults with high incomes. Spec prog: Relg one hr, pub affrs one hr wkly. ♦John Hogan, CEO, pres & CFO; Randy Michaels, chmn; Andy Rosen, VP; Bob Dunphy, gen mgr; Vince Thomas, stn mgr, opns dir & opns mgr; Laura Brockman, sls dir; Elizabeth Toscano, prom dir; Rob Ryan, progmg dir; Alexandra Vallejo, news dir.

WNNJ-FM— Oct 15, 1961: 103.7 mhz; 2.3 kw. 892 ft TL: N41 11 33 W74 45 13. Stereo. 24 Web Site: wnnj.com.250,000 Format: Classic rock. News staff: 2. Target aud: 18-49; upscale, young families with teenage children. ♦Ken O'Brien, progmg dir, mus dir & disc jockey; Judi Edwards, pub affrs dir & traf mgr; Jay Wulff, disc jockey; Paty Hunter, disc jockey; Rob Ryan, disc jockey.

North Cape May

WSJQ(FM)— 1993: 106.7 mhz; 3 kw. Ant 200 ft TL: N38 57 32 W74 55 23. Hrs open: 3208 Pacific Ave., Wildwood, 08260. Phone: (609) 522-1987. Phone: (609) 522-3666. Web Site: www.wsjq.com. Licensee: WBES LLC (acq 11-1-2004; $700,000). Natl. Network: USA, Moody. Format: Contemp hit radio. Spec prog: Class 3 hrs wkly. ♦Bob Maschio, VP; Mark Hunter, progmg dir.

Oakland

WVNJ(AM)— Dec 13, 1993: 1160 khz; 20 kw-D, 2.5 kw-N, DA-2. TL: N41 03 26 W74 15 00. Stereo. Hrs opn: 24 1086 Teaneck Rd., Suite 4F, Teaneck, 07666. Phone: (201) 837-0400. Fax: (201) 837-9664. E-mail: wvnj1160am@aol.com Web Site: www.wvnj.com. Licensee: Universal Broadcasting of New York Inc. (group owner; acq 3-24-94; $12,050,000. with WTHE(AM) Mineola, NY; FTR: 6-20-94) Population served: 8,250,000 Natl. Rep: Universal Broadcasting Inc. Rgnl rep: Universal Broadcasting Inc Law Firm: Cohn & Marks. Format: Adult standards, talk. News staff: one; News: 8 hrs wkly. Target aud: 35-64; upscale. Spec prog: Health related. ♦Miriam Warshaw, pres; Howard Warshaw, sr VP; Dr. Abe Warshaw, gen mgr; David Margalotti, opns dir, progmg dir & mus dir; Pete Bucky, prom dir, news dir & news dir.

Ocean Acres

WKMK(FM)— 1992: 98.5 mhz; 6 kw. Ant 328 ft TL: N39 45 06 W74 15 39. Hrs open: 24 703 Millcreek Rd., Manahawkin, 08050. Secondary address: Rt. 35 & 66, Ocean 08050. Phone: (609) 597-6700. Fax: (609) 597-0639. Licensee: Press Communications LLC. Group owner: Millennium Radio Group LLC (acq 8-9-2004; $17 million). Population served: 400,000 Format: Alternative rock, modern adult contemp. News staff: one; News: one hr wkly. Target aud: 25-55; women. ♦Frank Calderaro, gen mgr; John Kaszuba, gen sls mgr; Mike Gravin, progmg dir; Mike Heilman, chief of engrg; Annemarie Cassone, traf mgr.

Ocean City

WIBG(AM)— October 1992: 1020 khz; 1.9 kw-D, 680 w-CH. TL: N39 13 45 W74 40 54. Hrs open: Sunrise-sunset Traders Lane Professional Complex, 3328 Simpson Ave., 08226. Phone: (609) 398-1020. Fax: (609) 398-3736. E-mail: wibg@wibg.com Web Site: www.wibg.com. Licensee: Enrico S. Brancadora. (acq 12-1-92; $140,000; FTR: 12-21-92). Format: Adult contemp, Christian talk & programing, praise & worship. News staff: one; News: one hr wkly. Target aud: 25-45; young urban-suburban professional. ♦Nancy Manno, stn mgr, opns mgr, prom mgr & traf mgr; Nancy Zimmerman, sls dir & gen sls mgr.

***WRTQ(FM)**— Sept 27, 1994: 91.3 mhz; 82 w horiz, 10.5 kw vert. Ant 384 ft TL: N39 19 15 W74 46 17. Hrs open: 24
Rebroadcasts WRTI(FM) Philadelphia 100%.
1509 Cecil B Moore Ave, Philadelphia, PA, 19121. Phone: (215) 204-8405. Fax: (215) 204-7027. E-mail: comments@wrti.org Web Site: www.wrti.org. Licensee: Temple University of the Commonwealth System of Higher Education. Population served: 325,000 Natl. Network: NPR, AP Radio. Rgnl. Network: Radio Pa. Format: Jazz, class. News staff: one; News: 15 hrs wkly. Target aud: 30-65. ♦Dave Conant, gen mgr; Jack Moore, progmg dir; Jeff DePolo, chief of engrg; Lorna Dixon, traf mgr.

WTKU-FM— April 1983: 98.3 mhz; 6 kw. Ant 328 ft TL: N39 12 18 W74 39 33. Stereo. Hrs opn: 24 1601 New Rd., Linwood, 08221. Phone: (609) 653-1400. Fax: (609) 601-0450. Web Site: www.kool983.com. Licensee: Access.1 New Jersey License Co. Group owner: Access.1 Communications Corp. (acq 11-17-2003; grpsl). Population served: 545,000 Natl. Rep: McGavren Guild. Law Firm: Rubin, Winston, Diercks, Harris & Cooke. Format: Hits of the 60s & 70s. News: 5 hrs wkly. Target aud: 25-64. ♦Chesley Maddox-Dorsey, pres; Dick Irland, gen mgr; Nick Giorno, progmg dir; Sydney L. Small, chmn & news dir; Dan Merlo, chief of engrg; Anne Pratt, traf mgr.

Parsippany-Troy Hills

WXMC(AM)— Jan 13, 1973: 1310 khz; 1 kw-D, 88 w-N, DA-1. TL: N40 51 51 W74 21 06. Hrs open: 24 Box 160, TCB, West Orange, 07052. Phone: (973) 575-5561. Fax: (973) 575-5637. E-mail: hoyestudia@hotmail.com Licensee: James Chladek Chladek Broadcast Group (acq 1-15-93; $200,000; FTR: 2-8-93). Population served: 500,000 Law Firm: KMZ Roseman. Format: Tropical, romantic Sp mus, relg. News staff: one; News: 4 hrs wkly. Target aud: 25 plus; young upper middle class business professionals. ♦James Chladek, CEO & gen mgr; Edwin Blas, stn mgr & chief of engrg; Otto Gust, chief of opns.

Paterson

WPAT(AM)— May 3, 1941: 930 khz; 5 kw-U, DA-2. TL: N40 50 59 W74 10 59. Stereo. Hrs opn: 449 Broadway, New York, NY, 10013. Phone: (212) 966-1059. Fax: (212) 966-9580. Licensee: Multicultural Radio Broadcasting Licensee LLC. Group owner: Multicultural Radio Broadcasting Inc. (acq 7-22-98). Format: Multilingual, sports in Sp. ♦Gene Heinemeyer, gen mgr; Harold Chou, chief of engrg.

WPAT-FM— Mar 29, 1957: 93.1 mhz; 5.3 kw. 1,420 ft TL: N40 42 43 W74 00 49. (CP: 21.88 kw, ant 338 ft.). Stereo. Hrs opn: 26 W. 56th St., New York, NY, 10019. Phone: (212) 541-9200. Fax: (212) 246-9239. Licensee: WPAT Licensing Inc. Group owner: Spanish Broadcasting System Inc. (acq 1996; $83.5 million). Format: Sp adult contemp. ♦Raul Alarcon Jr., CEO & pres; Raul Alarcon Sr., chmn; Jose A. Garcia, CFO.

Pemberton

***WBZC(FM)**— Jan 24, 1995: 88.9 mhz; 470 w horiz, 10 kw vert. Ant 220 ft TL: N39 50 34 W74 32 40. Stereo. Hrs opn: 24 Burlington County College, 601 Pemberton Browns Mills Rd., 08068-1599. Phone: (609) 894-9311, EXT. 1189. Fax: (609) 894-9440. E-mail: bholcumb@bcc.edu Web Site: www.z889.org. Licensee: Burlington County College. Population served: 500,000 Law Firm: Bechtel & Cole. Format: Rhythm crossover. News: 4 hrs wkly. Target aud: 18-35. Spec prog: Folk 4 hrs, jazz 4 hrs, bluegrass 4 hrs, reggae 4 hrs,. ♦Brett Holcomb, opns mgr & progmg dir; Neil Shore, mus dir.

Pennsauken

WRNB(FM)— 1946: 107.9 mhz; 780 w. Ant 905 ft TL: N39 57 09 W75 10 05. Stereo. Hrs opn: 5:30 AM-midnight 1000 River Rd., Suite 400, Conshohocken, PA, 19428-2437. Phone: (610) 276-1100. Fax: (610) 279-1139. Licensee: Radio One Licenses LLC. Group owner: Radio One Inc. (acq 2-2-2004; $35 million). Population served: 100,000 Format: Urban contemp. ♦Chester Schofield, gen mgr.

Petersburg

WJSE(FM)—Licensed to Petersburg. See Atlantic City

Piscataway

***WVPH(FM)**— May 1976: 90.3 mhz; 200 w. 7 ft TL: N40 32 45 W74 28 25. Hrs open: 100 Behmer Rd., 08854-4173. Phone: (732) 981-0153. Fax: (732) 981-1985. E-mail: wvph@pway.org Licensee: Board of Education Piscataway High School. Format: Educ, progsv, talk. ♦Patricia Cardinal, gen mgr.

Pleasantville

WMGM(FM)—See Atlantic City

WOND(AM)— July 1950: 1400 khz; 1 kw-U. TL: N39 23 26 W74 30 47. Hrs open:
Rebroadcasts WGYM(AM) Hammonton 100%.
1601 New Rd., Linwood, 08221. Phone: (609) 653-1400. Fax: (609) 601-0450. Web Site: www.wond1400am.com. Licensee: Access.1 New Jersey License Co. Group owner: Access.1 Communications Corp. (acq 11-17-2003; grpsl). Population served: 545,000 Natl. Network: Westwood One. Natl. Rep: McGavren Guild. Law Firm: Rubin, Winston, Diercks, Harris & Cooke. Format: News/talk. Target aud: Adults 35 +. ♦Sydney L. Small, chmn; Chesley Maddox-Dorsey, pres; Dick Irland, gen mgr; Stuart Abrams, opns VP, natl sls mgr & traf mgr; John De Lucia, gen sls mgr; Dan Merlo, chief of engrg.

WTAA(AM)— Jan 1, 1955: 1490 khz; 400 w-U. TL: N39 23 24 W74 30 45. Hrs open: 24 hrs
Simulcast with WTKU-FM Ocean City 100%.
1601 New Rd., Linwood, 08221. Phone: (609) 653-1400. Fax: (609) 601-0450. Web Site: www.kool983.com. Licensee: Access.1 New Jersey License Co. Group owner: Access.1 Communications Corp. (acq 11-17-2003; grpsl). Population served: 27,000 Natl. Rep: McGavren Guild. Law Firm: Rubin, Winston,Diercks, harris & Cooke. Format: Oldies (60's & 70's). Target aud: Adults 25-64. ♦Sydney L. Small, chmn; Chesley Maddox-Dorsey, pres; Dick Irland, gen mgr; Nick Giorno, progmg dir.

WZBZ(FM)— 1974: 99.3 mhz; 3 kw. Ant 328 ft TL: N39 22 35 W74 27 08. Stereo. Hrs opn: 24
Simulcast with WGBZ(FM) Cape May Court House 100%.
Equity Communications L.P., 8025 Black Horse Pike, Suite 100-102, West Atlantic City, 08232. Phone: (609) 484-8444. Fax: (609) 646-6331. E-mail: info@993thebuzz.com Web Site: 993thebuzz.com. Licensee: Equity Communications L.P. (group owner; acq 5-31-2002; grpsl). Population served: 400,000 Natl. Network: Westwood One. Natl. Rep: Katz Radio. Law Firm: Lathmam & Watkins. Format: CHR. Target aud: 18-49; adults. ♦Gary Fisher, sr VP & gen mgr; Rob Garcia, progmg dir.

Point Pleasant

WRAT(FM)— Oct 4, 1968: 95.9 mhz; 4 kw. Ant 293 ft TL: N40 10 17 W74 01 39. Stereo. Hrs opn: 24 1731 Main St., 610 Main St., Belmar, 07719-3051. Phone: (732) 681-3800. Fax: (732) 681-5995. Web Site: www.wrat.com. Licensee: The Sentinel Publishing Co. Group owner: Greater Media Inc. (acq 7-6-01; grpsl). Population served: 1,021,400 Natl. Rep: Katz Radio. Format: Rock. Target aud: 21-44; men 25-54. ♦Dan Finn, VP & gen mgr; Mike Kazala, stn mgr; Carl Craft, opns mgr & progmg dir; Marge Guglielmo, gen sls mgr; William Clanton Sr., chief of engrg.

Pomona

***WLFR(FM)**— Oct 16, 1984: 91.7 mhz; 900 w. Ant 151 ft TL: N39 28 45 W74 32 23. Stereo. Hrs opn: 6 AM-2 AM Stockton State College, Jim Leeds Rd., G 204-205, 08240. Phone: (609) 652-4780. Web Site: www.wlfr.fm. Licensee: Stockton State College Population served: 186,000 Format: Alternative, var/rock, smooth jazz. News: one hr wkly. Target aud: General. Spec prog: Folk 3 hrs, class 4 hrs, jazz 10 hrs wkly.

Pompton Lakes

WGHT(AM)— Oct 3, 1964: 1500 khz; 1 kw-D, DA-D. TL: N40 58 51 W74 17 06. Hrs open: Sunrise-sunset Box 316, 1878 Lincoln Ave., 07442. Phone: (973) 839-1500. Fax: (973) 839-2400. E-mail: livestudio@ghtradio.com Web Site: www.ghtradio.com. Licensee: Mariana Broadcasting Inc. (acq 7-8-93; FTR: 8-23-93). Population served: 500,000 Natl. Network: AP Radio. Format: Oldies, talk. News staff: 3; News: 10 hrs wkly. Target aud: 25-54; general. Spec prog: Relg 2 hrs, polka one hr, loc sports 3 hrs wkly. ♦John Silliman, pres, gen mgr & stn mgr; Tom Niven, opns VP; Mary Hamilton, gen sls mgr; Jimmy Howes, prom VP & progmg dir; Debra Valentine, news dir.

Port Republic

***WXXY-FM**— 2003: 88.7 mhz; 760 w vert. Ant 131 ft TL: N39 35 34 W74 26 15. Hrs open: 465 Rt. 9 S., Little Egg Harbor, 08087. Phone: (609) 965-9100. Fax: (609) 965-9190. E-mail: comments@wxxy.fm Web Site: www.wxxy.fm. Licensee: In His Name Broadcasting Inc. (acq 10-23-01). Format: Gospel.

Princeton

WHWH(AM)— Sept 7, 1963: 1350 khz; 5 kw-U, DA-2. TL: N40 22 00 W74 44 38. Hrs open: 24 619 Alexander Rd., 08540. Phone: (609) 419-0300. Fax: (609) 419-0143. E-mail: epalladino @nassaubroadcasting.com Web Site: www.moneytalk1350.com. Licensee: Multicultural Radio Broadcasting Licensee LLC. Group owner: Multicultural Radio Broadcasting Inc. (acq 5-24-2002; grpsl). Population served: 358,000 Natl. Network: Wall Street. Natl. Rep: Katz Radio. Law Firm: Booth, Freret, Imlay & Tepper P. Format: News/talk. News staff: 4; News: 36 hrs wkly. Target aud: 35 plus. Spec prog: Princeton University football & basketball, Trenton Thunder baseball. ♦Lou Mercatani, pres; Peter D. Tonks, CFO; Michelle Stevens, exec VP & gen mgr; Josh Gertzog, VP, gen mgr, sls dir & pub affrs dir; Ed Palladino, opns dir.

WJJZ(FM)—See Burlington

WPRB(FM)— October 1955: 103.3 mhz; 14 kw. 731 ft TL: N40 17 00 W74 41 20. Hrs open: 24 30 Bloomberg Hall, 08544. Phone: (609) 258-3655. Fax: (609) 258-1806. E-mail: manager@wprb.com Web Site: www.wprb.com. Licensee: Princeton Broadcasting Service Inc. Format: Class, jazz, progsv. Target aud: 13-60. Spec prog: Asian Indian 6 hrs wkly. ♦Spencer Salazar, gen mgr.

Princeton Junction

***WWPH(FM)**— November 1975: 107.9 mhz; 10 w. 36 ft TL: N40 18 20 W74 37 16. Hrs open: West Windsor-Plainsboro High School, 346 Clarksville Rd., 08550-1518. Phone: (609) 716-5050. Fax: (609) 716-5092. E-mail: wwph107.9fmlogin@ww-p.org Web Site: www.wwph1079.com. Licensee: West Windsor Plainsboro Regional Board of Education. Format: Var. Target aud: 14-30; West Windsor & Plainsboro residents interested in their community. ♦Glenn Allison, gen mgr.

Salem

WFAI(AM)—Licensed to Salem. See Wilmington DE

WJKS(FM)—See Canton

South Belmar

WRAT(FM)—See Point Pleasant

South Orange

***WSOU(FM)**— Apr 14, 1948: 89.5 mhz; 2.4 kw. 370 ft TL: N40 44 44 W74 14 50. Stereo. Hrs opn: 24 400 S. Orange Ave., 07079. Phone: (973) 313-6110. Fax: (973) 275-2001. E-mail: wsou@shu.edu Web Site: www.wsou.net. Licensee: Seton Hall University. Population served: 8,000,000 Law Firm: Booth, Freret, Imlay & Tepper. Wire Svc: AP Format: Active rock. News: 7 hrs wkly. Target aud: 18-34. Spec prog: Pol 2 hrs, Ethnic 10 hrs, pub affrs 3 hrs, relg 5 hrs wkly. ♦Mark Maben, gen mgr; Frank Scafidi, chief of engrg.

Stirling

WKMB(AM)— February 1972: 1070 khz; 250 w-D. TL: N40 40 35 W74 28 36. Hrs open: Sunrise-sunset 120 W. 7th, Suite 201, Plainfield, 07060. Phone: (908) 822-1515. Fax: (908) 822-1927. E-mail: mprayer@harvestradio.com Web Site: www.harvestradio.net. Licensee: World Harvest Communications Inc. (acq 1-15-03). Population served: 1,000,000 Format: Christian / Talk. ♦Gary Kirkwood Sr., CEO & pres; Melissa Prayer, gen mgr; Robert Hunt, opns mgr.

Sussex

***WNJP(FM)**— 1998: 88.5 mhz; 450 w. Ant 636 ft TL: N41 08 37 W74 32 18. Hrs open: Box 777, Trenton, 08625-0777. Phone: (609) 777-5036. Fax: (609) 777-5217. Web Site: www.njn.net. Licensee: New Jersey Public Broadcasting Authority. Natl. Network: NPR, PRI. Law Firm: Schwartz, Woods & Miller. Format: News/talk. Target aud: General. ♦Elizabeth G. Christopherson, CEO; Bill Jobes, gen mgr; Pharoah Cranston, opns mgr.

Teaneck

***WFDU(FM)**— Aug 30, 1971: 89.1 mhz; 550 w. 550 ft TL: N40 57 39 W73 55 23. Stereo. Hrs opn: 1:15 AM-3:45 PM (M-F); 24 (S, Su) 1000 River Rd., 07666. Phone: (201) 692-2806. Fax: (201) 692-2807. E-mail: barrys@fdu.edu Web Site: www.wfdu.fm. Licensee: Fairleigh Dickinson University. Population served: 18,000,000 Law Firm: Schwartz, Woods & Miller. Wire Svc: AP Format: Var/div. News: 2 hrs wkly. Target aud: General. Spec prog: Sp 3 hrs hrs wkly. ♦Carl J. Kraus, gen mgr; Barry Sheffield, opns mgr.

Toms River

WOBM-FM— Mar 1, 1968: 92.7 mhz; 1.4 kw. 485 ft TL: N39 52 30 W74 09 52. Stereo. Hrs opn: 24 Box 927, 08754. Phone: (732) 269-0927. Fax: (732) 269-9292. E-mail: wobm@wobm.com Web Site: www.wobm.com. Licensee: Millennium Shore Holdco LLC. Group owner: Millennium Radio Group LLC (acq 5-14-02; grpsl). Population served: 452,000 Law Firm: William D. Silva. Format: Adult contemp. ♦William Saurer, gen mgr; John Furno, gen sls mgr; Teddy Maturo, prom dir & progmg dir; Tom Mongelli, news dir.

Trenton

WBUD(AM)— Jan 20, 1947: 1260 khz; 5 kw-D, 2.5 kw-N, DA-2. TL: N40 15 56 W74 45 27. Hrs open: Box 5698, 08638. Secondary address: 109 Walters Ave. 08638. Phone: (609) 771-8181. Phone: (800) 678-9599. Fax: (609) 406-7956. Web Site: www.timelessclassiconline.com. Licensee: Millennium Central New Jersey License Holdco LLC. Group owner: Millennium Radio Group LLC (acq 12-21-2001; grpsl). Population served: 104,638 Natl. Rep: Christal. Law Firm: Irwin, Campbell, Crowe & Tannenwald. Format: MOR. News: 20 hrs wkly. Target aud: 35 plus. ♦Jim Donahoe, CEO; Andy Santoro, VP & gen mgr; Ray Handel, mktg dir; Lorenzo Caldara, adv dir; Eric Johnson, progmg dir; Eric Scott, news dir; Ray Fodge, engrg dir; Laurie Roth, traf mgr.

WKXW(FM)—Co-owned with WBUD(AM). Aug 27, 1962: 101.5 mhz; 15.5 kw. Ant 902 ft TL: N40 16 58 W74 41 11. Stereo. Web Site: www.nj1015.com.400,000 Natl. Rep: Christal. Format: Talk. News staff: 15; News: 75 hrs wkly. Target aud: General; New Jersey residents. ♦Eric Johnson, progmg dir; Laurie Roth, traf mgr.

WIMG(AM)—Ewing, 1923: 1300 khz; 5 kw-D, 2.5 kw-N, DA-2. TL: N40 17 16 W74 52 23. Hrs open: 24
Westwood One.
Box 9078, 08650. Secondary address: 1842 S. Broad St. 08610. Phone: (609) 695-1300. Fax: (609) 278-1588. E-mail: wimg1300@aol.com Web Site: www.wimg1300.com. Licensee: Morris Broadcasting Co. of New Jersey Inc. (acq 12-3-93; FTR: 12-20-93). Population served: 580,000 Natl. Network: American Urban, NBC. Natl. Rep: Williams Radio Sales. Law Firm: Booth, Freret, Imlay & Tepper. Format: Urban adult contemp, gospel. News: 6. Target aud: 25-54. ♦Johnny Morris, CEO; Louise E. Morris, chmn; Michael Morris, pres; Maggie Guzzardo, exec VP & gen mgr; Felicia Brannon, opns VP; Pamela Pruitt, dev VP.

***WNJT-FM**— May 20, 1991: 88.1 mhz; 110 w. Ant 689 ft TL: N40 16 58 W74 41 11. Hrs open: Box 777, 08625-0777. Phone: (609) 777-5036. Fax: (609) 777-5217. E-mail: pcrast@njn.org Web Site: www.njn.net. Licensee: New Jersey Public Broadcasting Authority. Natl. Network: NPR, PRI. Law Firm: Schwartz, Woods & Miller. Format: News/talk. Target aud: General. ♦Elizabeth G. Christopherson, CEO; Bill Jobes, gen mgr; Pharoah Cranston, stn mgr; Andre Butts, progmg dir.

WPHY(AM)—Apr 11, 1941: 920 khz; 1.4 kw-D, 1 kw-N, DA-2. TL: N40 15 19 W74 51 44. Stereo. Hrs opn: 24 619 Alexander Rd., Princeton, 08540-6003. Phone: (609) 924-1515. Fax: (609) 419-0147. Web Site: www.920espn.com. Licensee: Nassau Broadcasting II L.L.C. (group owner; (acq 4-25-2002; with co-located FM). Natl. Network: ESPN Radio. Law Firm: Cohn & Marks. Format: Sports. ♦Josh Gertzog, sls dir; Tony Henry, prom dir; Tripp Rogers, gen mgr & progmg dir; Tim Anderson, news dir; Tony Gervasi, chief of engrg; Kerri Johnson, traf mgr.

WPST(FM)—Co-owned with WPHY(AM). Aug 7, 1965: 94.5 mhz; 50 kw. Ant 492 ft TL: N40 11 22 W74 50 47. Stereo. 619 Alexander Rd., Princeton, 08540-6003. Phone: (609) 924-1515. Fax: (609) 419-0143. Web Site: www.wpst.com. Format: CHR. ♦Jim Spector, progmg dir & disc jockey; Angela Hartman, traf mgr; Randy Ellis, mus dir & disc jockey.

***WTSR(FM)**— September 1966: 91.3 mhz; 1.5 kw. 35 ft TL: N40 16 17 W74 46 55. Stereo. Hrs opn: 24 The College of New Jersey, WTSR(FM), Box 7718, Ewing, 08628-7718. Phone: (609) 771-3200. Phone: (609) 771-2554. Fax: (609) 637-5113. Web Site: www.wtsr.org. Licensee: The College of New Jersey Radio System. Population served: 106,638 Format: Progsv, alternative. News: 15 hrs wkly. Target aud: 13-40; people who listen to div mus formats. Spec prog: Gospel 6 hrs, pub affrs 8 hrs, folk 4 hrs, jazz 4 hrs, oldies 6 hrs wkly. ♦Greg Miller, stn mgr.

***WWFM(FM)**— Sept 6, 1982: 89.1 mhz; 1.15 kw. Ant 292 ft TL: N40 15 30 W74 38 59. Stereo. Hrs opn: 24 Box B, 08690. Phone: (609) 587-8989. Fax: (609) 570-3863. E-mail: wwfm@mccc.edu Web Site: www.wwfm.org. Licensee: Mercer County Community College Board of Trustees. Natl. Network: PRI, NPR. Format: Classical. News: 3.5 hrs wkly. Target aud: General. ♦Jeffery R. Sekerka, gen mgr & dev mgr.

Tuckerton

WBHX(FM)— 1999: 99.7 mhz; 5.3 kw. Ant 108 ft TL: N39 33 41 W74 14 27. Hrs open: 2355 West Bangs Ave., Neptune, 07753. Phone: (732) 774-4755. Fax: (609) 597-0639. E-mail: thebreezeradio @thebreezeradio.com Web Site: www.the breezeradio.com. Licensee: Press Communications L.L.C. (group owner; acq 9-18-02; $1.15 million). Format: Adult Contemp. ♦Mike Fitzgerald, progmg dir.

Union Township

***WKNJ-FM**— January 1980: 90.3 mhz; 8.7 w. 88 ft TL: N40 40 35 W74 14 02. Stereo. Hrs opn: Kean Univ., CAS 401, 1000 Morris Ave., Union, 07083. Phone: (908) 737-0440. Fax: (908) 737-0445. E-mail: wknjfm@yahoo.com Web Site: www. kean.edu/~cahss/acad_dept/comm /wknj/index/html. Licensee: Kean University. Format: New mus. Spec prog: Black 2 hrs, jazz 8 hrs, new age 8 hrs wkly. ♦Scott McHugh, gen mgr; Cathleen Londino, stn mgr.

Upper Montclair

***WMSC(FM)**— Dec 9, 1974: 90.3 mhz; 10 w. 672 ft TL: N40 51 53 W74 12 03. Stereo. Hrs opn: 7 AM-1 AM Student Ctr. Annex, Montclair State University, 07043. Phone: (973) 655-4257. Phone: (973) 655-4256. Fax: (973) 655-7433. Web Site: www.montclair.edu/org/wmsc. Licensee: Montclair State University. Population served: 200,000 Format: Alternative. News: 5 hrs wkly. Target aud: Under 35. Spec prog: Black 4 hrs, gospel 2 hrs, jazz 2 hrs, Sp 2 hrs, sports 3 hrs wkly. ♦Walter Soto, gen mgr; Andrew Ward, opns dir; Dave Giumara, prom dir; Dan Maxwell, progmg dir; Lisa Hresko, mus dir.

Villas

WCZT(FM)— February 1992: 98.7 mhz; 3 kw. 292 ft TL: N39 00 33 W74 52 13. Hrs open: 24 3208 Pacific Ave., Wildwood, 08260. Phone: (609) 522-1987. Fax: (609) 522-3666. Web Site: www.987thecoast.com. Licensee: WZK LLC (acq 5-21-01; $1.4 million for stock). Format: Adult contemp. ♦Bob Maschio, gen mgr; Ed Rosenfeld, gen sls mgr; Scott Wahl, stn mgr & news dir; Ray Bradley, chief of engrg.

Vineland

WMIZ(AM)—Listing follows WVLT(FM).

WVLT(FM)— October 1968: 92.1 mhz; 3 kw. 328 ft TL: N39 29 53 W75 04 31. Stereo. Hrs opn: Box 689, 638 E. Landis Ave., 08360. Phone: (856) 692-8888. Fax: (856) 696-2568. Licensee: Clear Communications Inc. (acq 8-1-86; $400,000; FTR: 5-12-86). Natl. Network: ABC. Format: Adult contemp. Target aud: 25-54; baby boomers.

WMIZ(AM)—Co-owned with WVLT(FM). Aug 19, 1959: 1270 khz; 500 w-D, 350 w-N, DA-2. TL: N39 29 53 W75 04 31. (CP: 360 w-D, 210 w-N). 52,600 Format: Sp. Target aud: Hispanic.

Washington Township

WNJC(AM)— July 29, 1946: 1360 khz; 5 kw-D, 800 w-N, DA-2. TL: N39 47 23 W75 06 11. Hrs open: 6 AM-midnight 123 Egg Harbor Rd., Ste 302, Sewell, 08080. Phone: (856) 227-1360. Fax: (856) 232-9093. Web Site: www.wnjc1360.com. Licensee: Forsyth Broadcasting LLC (acq 1995; $161,000). Population served: 7,000,000 Law Firm: Bechtel & Cole. Format: Talk, progsv, gospel. News staff: 3; News: 20 hrs wkly. Target aud: 30 plus; 52% women, upper income. Spec prog: Relg 6 hrs wkly. ♦John Forsythe, pres; Al Jones, gen mgr & progmg dir.

Wayne

***WPSC-FM**— Nov 1, 1988: 88.7 mhz; 200 w. Ant 259 ft TL: N40 59 46 W74 16 51. Stereo. Hrs opn: 9 am-3 am M-F; 6 am-3 am S-S Hobart Hall, 300 Pompton Rd., 07470. Phone: (973) 720-3319. Fax: (973) 720-2454. E-mail: wpsc887fm@wpunj.edu Licensee: William Paterson University of New Jersey. (acq 7-90; $1; FTR: 7-16-90). Population served: 1,700,000 Format: Alternative. Target aud: 18-35; independent thinking. Spec prog: Punk 3 hrs, hip hop 18 hrs, metal 18 hrs, classic rock 12 hrs, jazz 12 hrs wkly. ♦Ron Stotyn, gen mgr.

West Long Branch

***WMCX(FM)**— May 2, 1974: 88.9 mhz; 1 kw. 118 ft TL: N40 16 44 W74 00 26. Stereo. Hrs opn: 24 Monmouth Univ., 400 Cedar Ave., 07764. Phone: (732) 571-3482. Fax: (732) 263-5145. E-mail: wmcxradio@monmouth.edu Web Site: www.wmcx.com. Licensee: Monmouth University. Population served: 200,000 Natl. Network: AP Radio. Format: Modern rock. News: 2.5 hrs wkly. Target aud: 18-25; college students, recent grads, young adults. Spec prog: Sports 11 hrs, jazz 3 hrs, changes w/semester. ♦Kahlil Thomas, gen mgr.

Wildwood

WCMC(AM)— Nov 25, 1951: 1230 khz; 1 kw-U. TL: N39 00 09 W74 48 46. Hrs open: 24 8025 Black Horse Pike, Suite 100-102, West Atlantic City, 08232. Secondary address: 3010 New Jersey Ave. 08260. Phone: (609) 484-8444 x317. Fax: (609) 646-6331. E-mail: gfequity@aol.com Licensee: Equity Communications L.P. (group owner; (acq 11-4-97; $7.1 million with co-located FM). Population served: 200,000 Natl. Network: ABC. Natl. Rep: Katz Radio. Law Firm: Laitham & Watkins. Format: Adult standards. Target aud: Adults 35-64. ♦Gary Fisher, pres, VP & gen mgr; Keith Fader, sls dir & gen sls mgr; Jim MacMillan, progmg dir.

WZXL(FM)—Co-owned with WCMC(AM). Dec 17, 1959: 100.7 mhz; 38 kw. Ant 331 ft TL: N39 07 28 W74 45 56. Stereo. 24 Fax: (609) 646-6331. Web Site: www.wzxl.com.400,000 Natl. Network: Westwood One. Natl. Rep: Katz Radio. Law Firm: Laitham & Watkins. Format: Classic rock. Target aud: Adults 18-49. ♦Steve Raymond, progmg dir.

Wildwood Crest

WEZW(FM)— Aug 15, 1993: 93.1 mhz; 4.2 kw. Ant 216 ft TL: N39 00 33 W74 52 13. Stereo. Hrs opn: 24
Simulcast with WTTH(FM) Margate City 100%.
8025 Black Horse Pike, Suite 100-102, West Atlantic City, 08232. Phone: (609) 484-8444. Fax: (609) 646-6331. E-mail: info@993thebuzz.com Web Site: www.961wtth.com. Licensee: Equity Communications L.P. (group owner; (acq 5-30-2003; grpsl). Population served: 200,000 Natl. Network: Westwood One. Natl. Rep: Katz Radio. Law Firm: Laitham & Watkins. Format: Urban adult contemp. Target aud: 18-49; adults. Spec prog: . ♦Gary Fisher, sr VP, VP & gen mgr; Rob Garcia, progmg dir; Denise Carrington, traf mgr.

Woodbine

***WJPH(FM)**— Feb 16, 1999: 89.9 mhz; 1 kw. Ant 105 ft TL: N39 16 51 W74 51 11. Stereo. Hrs opn: 24 Box 603, 08270-0603. Phone: (609) 861-3700. E-mail: letters@praise899.org Web Site: www.praise899.org. Licensee: Maranatha Ministries/Joy Communications Inc. Population served: 150,000 Law Firm: Booth, Freret, Imlay & Tepper. Format: Praise & worship. News: 10 hrs wkly. Target aud: 25-54; women. Spec prog: Gospel one hr wkly. ♦Kenneth Manri, pres.

Zarephath

WAWZ(FM)— Aug 22, 1954: 99.1 mhz; 28 kw. Ant 656 ft TL: N40 36 41 W74 34 12. Stereo. Hrs opn: 24 Box 9058, 08890. Phone: (732) 469-0991. Fax: (732) 469-2115. E-mail: info@star991fm.com Web Site: www.star991fm.com. Licensee: Pillar of Fire Inc. (group owner) Format: Adult contemp, Christian. Target aud: 25-54. ♦Rea Crawford, gen mgr; Scott Taylor, stn mgr; Allen Lewis Lewicki, opns dir; Stacey Stone, prom dir; Ed Abels, adv dir; Johnny Stone, progmg dir; Ron Habegger, engrg dir & chief of engrg.

New Mexico

Alamo Community

***KABR(AM)**— August 1983: 1500 khz; 1 kw-D. TL: N34 25 01 W107 30 04. Hrs open: Box 907, Magdalena, 87825. Phone: (505) 854-2632. Phone: (505) 854-2641, Ext 1600-01. Fax: (505) 854-2545. Web Site: www.alamo.bia.edu. Licensee: Alamo Navajo Community School. Format: Ethnic. Target aud: General; Native Americans, loc ranchers, tourists, teachers & health professionals. Spec prog: American Indian 10 hrs wkly. ♦Ann Kerr, pres; Sarah Apache, gen mgr & stn mgr.

Alamogordo

KINN(AM)— June 10, 1957: 1270 khz; 1 kw-D, 500 w-N. TL: N32 53 13 W105 57 04. Stereo. Hrs opn: 24 Box 1848, 88311. Phone: (505) 434-1414. Fax: (505) 434-2213. Licensee: Burt Broadcasting Inc. (group owner; acq 4-1-01; with co-located FM). Law Firm: Baraff, Koerner & Olender. Format: News/talk. News: 22 hrs wkly. Target aud: 24-50; military & civil service personnel employed in high-tech jobs. ♦William F. Burt, pres & gen mgr; Lori Swinford, gen sls mgr; James White, progmg dir & news dir; Ken Bass, engrg dir; Donnie Burt, traf mgr.

KZZX(FM)—Co-owned with KINN(AM). 1979: 105.5 mhz; 6 kw. 157 ft TL: N32 53 13 W105 57 04. (CP: 105.3 mhz, 2.23 kw, ant 524 ft.). Stereo. 24 Format: Country. News: 22 hrs wkly. Target aud: 20-55.

KNMZ(FM)— 1997: 103.7 mhz; 47 kw. 1,338 ft TL: N33 10 45 W105 53 53. Hrs open: 24 Box 2710, 88311. Secondary address: Cuba Ave. & Canyon Rd. 88310. Phone: (505) 437-1505. Fax: (505) 437-5566. E-mail: alamogordo@snmradio.com Web Site: www.snmradio.com. Licensee: WP Broadcasting LLC. Group owner: Runnels Broadcasting System L.L.C. (acq 5-19-2006; grpsl). Format: Classic rock. ♦Philip Runnels, gen mgr.

KQEL(FM)— 2006: 107.9 mhz; 3 kw. Ant -594 ft TL: N32 53 13 W105 57 04. Hrs open: 24 Box 1848, 88311. Phone: (505) 434-1414. Fax: (505) 434-2213. Licensee: Burt Broadcasting Inc. (group owner; (acq 10-29-2003; $93,000 for CP). Format: Late 50's - early 80's. ♦William F. Burt, gen mgr; Lori Swinford, gen sls mgr.

KRSY(AM)— June 28, 1950: 1230 khz; 1 kw-U. TL: N32 53 46 W105 56 42. Hrs open: 24 Box 2710, Cuba Ave. & Canyon Rd., 88311. Phone: (505) 437-1505. Phone: (505) 437-1230. Fax: (505) 437-5566. Web Site: www.snmradio.com. Licensee: WP Broadcasting LLC. Group owner: Runnels Broadcasting System L.L.C. (acq 5-19-2006; grpsl). Population served: 55300 Law Firm: Jones, Waldo, Holbrook & McDonough; Barry Wood. Format: Talk, sports. News staff: 2; News: 10 hrs wkly. Target aud: 25-54; active adults, community-oriented. Spec prog: Big band 6 hrs, farm one hr, gospel 8 hrs, relg 4 hrs wkly. ♦Phil Reynolds, gen mgr; Les Hanks, gen sls mgr; Kelly Lynch, progmg dir & traf mgr; Jim Huff, chief of engrg.

KRSY-FM— Jan 17, 1987: 92.7 mhz; 3 kw. 192 ft TL: N32 98 15 W105 59 21. (CP: 103.7 mhz, 50.2 kw, ant 1,338 ft. TL: N33 10 45 W105 53 53). Stereo. 24 Phone: (505) 437-1063. Web Site: www.snmradio.com. Format: Hot country. News staff: 2; News: 2 hrs wkly. Target aud: 18-34; active adults, young adults. ♦Les Henke, natl sls mgr; Kelly Lynch, progmg mgr; Jim Huff, engrg dir.

***KUPR(FM)**— Oct 14, 2000: 91.7 mhz; 100 w. Ant 1,679 ft TL: N32 49 47 W105 53 10. Stereo. Hrs opn: 24 3001 N. Florida Ave., 88310. Phone: (505) 437-0917. Fax: (505) 434-6060. Fax: (505) 437-9917. E-mail: kupr917@yahoo.com Licensee: Southern New Mexico Radio Foundation. Law Firm: Wood, Maines & Brown. Format: Country gospel loc. News: 4 hrs wkly. Target aud: 25-60; adults. ♦Bob Flotte, pres; Devere Johnson, opns dir.

***KYCM(FM)**— 2006: 89.9 mhz; 800 w. Ant 1,630 ft TL: N32 49 47 W105 53 13. Hrs open:
Rebroadcasts KYCC(FM) Stockton, CA 100%.
9019 West Ln., Stockton, CA, 95210-1401. Phone: (209) 477-3690. Fax: (209) 477-2762. E-mail: kycc@kycc.org Web Site: www.kycc.org. Licensee: Your Christian Companion Network Inc. Format: Gospel, inspirational, adult contemp. ♦Shirley Garner, gen mgr.

KYEE(FM)— July 21, 1980: 94.3 mhz; 3 kw. -492 ft TL: N32 56 42 W105 56 47. Stereo. Hrs opn: 24 Box 1848, 88311. Phone: (505) 434-1414. Fax: (505) 434-2213. E-mail: 94key@bbiradio.net Web Site: www.sos.state.nm.us/radio.htm. Licensee: Burt Broadcasting Inc. (group owner; acq 11-88; $230,000; FTR: 12-19-88). Population served: 60,000 Format: CHR. Target aud: 18-44; young adults. ♦Donnie L. Burt, VP; William F. Burt, pres & gen mgr; Lori Swinford, gen sls mgr.

Albuquerque

KABQ(AM)— 1947: 1350 khz; 5 kw-D, 500 w-N, DA-N. TL: N35 06 02 W106 40 34. Hrs open: 8am - 5:30pm 5411 Jefferson N.E., 87109. Phone: (505) 338-7400. Fax: (505) 830-6543. Web Site: www.abqtalk.com. Licensee: Clear Channel Broadcasting Licenses. Group owner: Clear Channel Communications Inc. (acq 3-01-00; grpsl). Population served: 243,751 Natl. Rep: Lotus Entravision Reps LLC. Format: Progressive talk. Target aud: General. ♦Chuck Hammond, pres & gen mgr; Bill May, progmg dir & disc jockey; Chris Williams, chief of engrg.

KALY(AM)—Los Ranchos de Albuquerque, 1982: 1240 khz; 1 kw-U. TL: N35 12 06 W106 35 56. Stereo. Hrs opn: 24 2505 6th St. N.W., 87102. Phone: (505) 244-1100. Fax: (505) 244-0612. Web Site: www.radiodisney.com. Licensee: Radio Disney Group LLC. Group owner: ABC Inc. (acq 2-21-03; $650,000). Format: Family Programing. Target aud: 25-49. ♦Lynn Southard, gen mgr.

***KANW(FM)**— October 1950: 89.1 mhz; 20 kw. 4,152 ft TL: N35 12 44 W106 26 57. Stereo. Hrs opn: 24 2020 Coal Ave. S.E., 87106. Phone: (505) 242-7163. Phone: (505) 242-7848. E-mail: brasher@aps.edu Web Site: www.kanw.com. Licensee: Board of Education of the City of Albuquerque. Population served: 850,000 Natl. Network: NPR, PRI. Format: Sp. News staff: News progmg 20 hrs wkly ♦Michael Brasher, gen mgr.

KBQI(FM)— Apr 27, 1979: 107.9 mhz; 22.5 kw. 4,130 ft TL: N35 12 43 W106 26 57. Stereo. Hrs opn: 24 5411 Jefferson St. N.E., Suite 100, 87109. Phone: (505) 830-6400. Fax: (505) 830-6543. Web Site: www.bigi1079.com. Licensee: Citicasters Licenses L.P. Group owner: Clear Channel Communications Inc. (acq 9-28-99; grpsl). Format: Country. Target aud: 18-49. ♦Chuck Hammond, gen mgr; Bill May, chief of opns.

KBZU(FM)— November 1954: 96.3 mhz; 20 kw. Ant 4,110 ft TL: N35 12 44 W106 26 58. Stereo. Hrs opn: 24 500 4th St. N.W., 5th Fl., 87102. Phone: (505) 767-6700. Fax: (505) 767-6767. Licensee: The Last Bastion Station Trust LLC, as Trustee Group owner: Citadel Broadcasting Corp. (acq 6-12-2007; grpsl). Population served: 523,000 Natl. Rep: McGavren Guild. Law Firm: Kaye, Scholer, Fierman, Hays & Handler. Format: Talk. News: 2 hrs wkly. Upscale affluent professionals. ♦Eddie Haskell, opns dir & progmg dir; Jeff Berry, gen sls mgr & natl sls mgr; Paul Bailey, mus dir; Art Ortega, pub affrs dir; Bill Harris, engrg dir.

KDAZ(AM)— 1969: 730 khz; 1 kw-D, 76 w-N, DA-2. TL: N35 00 31 W106 42 52. Hrs open: 24 Box 4338, 87196. Secondary address: 5010 4th St. N.W. 87107. Phone: (505) 345-7373. Fax: (505) 345-5669. E-mail: kdaz@kdaz.org Web Site: www.kdaz.org. Licensee: Pan American Broadcasting Inc. (acq 11-17-2003). Population served: 662,380 Natl. Network: USA. Law Firm: Gammon & Grange. Format: Variety. Target aud: 25-54. ♦Blackie Gonzalez, CEO, chmn & pres; Vickie Archiveque, CFO; Annette Garcia, VP & gen mgr; Jim Sandell, progmg dir.

KDEF(AM)— September 1953: 1150 khz; 5 kw-D, 500 w-N, DA-2. TL: N35 12 06 W106 35 54. Hrs open: 10424 Edith N.E., 87113. Phone: (505) 888-1150. Fax: (505) 899-1977. Licensee: RAMH Corp. (acq 7-95; $125,000). Population served: 500,000 Natl. Network: CNN Radio. Format: 80s & 90s music, youth sports, news. Target aud: General. ♦Henry Tafoya, gen mgr.

KDRF(FM)— Apr 20, 1988: 103.3 mhz; 22 kw. 4,069 ft TL: N35 12 50 W106 27 00. Stereo. Hrs opn: 24 Citadel Southwest, 500 4th St. N.W., 5th Fl., 87102. Phone: (505) 767-6700. Fax: (505) 767-6767. Web Site: www.ed.fm. Licensee: Citadel Broadcasting Co. Group owner: Citadel Broadcasting Corp. (acq 1996; $5 million). Natl. Rep: Christal. Format: Best of the Hits. Target aud: 18-49. ♦Linda Rosenberg, gen sls mgr.

***KFLQ(FM)**— Feb 20, 1983: 91.5 mhz; 20 kw. Ant 4,041 ft TL: N35 12 51 W106 27 02. Stereo. Hrs opn: 24 3801 Eubank N.E., 87111.

Phone: (505) 296-9100. Fax: (505) 296-6262. Web Site: 915flr.org. Licensee: Family Life Broadcasting System. Group owner: Family Life Communications Inc. (acq 1982). Population served: 500,000 Natl. Network: Moody, USA. Format: Inspirational, praise & worship. Target aud: 35-54; female. ♦Randy Carlson, pres; Dan Rosecrans, stn mgr.

KKIM(AM)— Apr 15, 1972: 1000 khz; 10 kw-D. TL: N35 10 14 W106 37 51. Hrs open: Box 30925, 87190-0925. Secondary address: 4125 Carlisle Blvd. N.E. 87107-4806. Phone: (505) 878-0980. Fax: (505) 878-0098. Web Site: www.kkimam1000.com. Licensee: AGM-Nevada L.L.C. Group owner: American General Media (acq 12-22-97; grpsl). Population served: 243,751 Format: Christian, talk. Target aud: 25-54. Spec prog: Black 2 hrs wkly. ♦Scott Hutton, gen mgr.

KKJY(AM)— Feb 22, 1971: 1550 khz; 5 kw-D, 20 w-N. TL: N35 06 02 W106 40 34. Hrs open: 24 1213 San Pedro N E, 87110. Phone: (505) 899-5029. E-mail: joyam@joyam Web Site: www.joyam.com. Licensee: Vanguard Media L.L.C. (acq 1-21-2000; $112,000). Population served: 650,000 Format: Soft adult contemp. News: 5 hrs wkly. Target aud: 35-64; Mature upscale adults. ♦Don Davis, CEO & pres; Craig Collins, gen mgr, opns mgr & mus dir; Josie Bunch, stn mgr; Crystal Felice, rgnl sls mgr & mktg mgr.

KKOB(AM)— Apr 5, 1922: 770 khz; 50 kw-U, DA-N. TL: N35 12 09 W106 36 41. Hrs open: 24 500 4th St. N.W., 87102. Phone: (505) 767-6700. Fax: (505) 767-6767. E-mail: kkobam@citcomm.com Web Site: www.770kkob.com. Licensee: Citadel Broadcasting Co. Group owner: Citadel Broadcasting Corp. (acq 3-15-94; $7.8 million with co-located FM; FTR: 4-11-94). Population served: 500,000 Natl. Rep: McGavren Guild. Law Firm: Haley, Bader & Potts. Format: News/talk. ♦Milt McConnell, gen mgr; Matt Woodcock, rgnl sls mgr; Pat Frisch, opns mgr & progmg dir; Alex Cuellar, news dir; Art Ortega, pub affrs dir; Bill Harris, chief of engrg.

KKOB-FM— Aug 1, 1967: 93.3 mhz; 21.5 kw. 4,150 ft TL: N35 12 42 W106 26 59. Stereo. 24 863,000 Format: Adult contemp.Milt McConnell, gen mgr; Kris Abrans, opns dir & opns mgr; Tim Gannon, sls VP & rgnl sls mgr; Mark Anderson, mktg dir; Kris Abrams, progmg dir; Linda Land, traf mgr; Carlos Duran, disc jockey; Greg Fite, disc jockey; Jakie Taylor, disc jockey; John Forsythe, disc jockey; Phil Sisneras, disc jockey; Randy Savage, disc jockey; Tony Manero, disc jockey

KKRG(FM)— October 1994: 101.3 mhz; 3.7 kw. Ant 420 ft TL: N35 04 06 W106 46 46. Hrs open: 24 8009 Marble Ave. N.E., 87110. Phone: (505) 262-1142. Fax: (505) 254-7106. Licensee: Univision Radio License Corp. Group owner: Univision Radio (acq 9-22-2003; grpsl). Format: Sp. Target aud: 35-54.

***KLYT(FM)**— Sept 11, 1976: 88.3 mhz; 4.1 kw. Ant 4,244 ft TL: N35 12 49 W106 27 01. Stereo. Hrs opn: 24 4001 Osunda Rd. N.E., 87109. Phone: (505) 344-9146. Fax: (505) 344-9193. Web Site: www.m88.org. Licensee: Calvary Chapel of Albuquerque, Inc. (acq 11-30-2000). Population served: 450,000 Format: Contemp Christian hits. ♦Chip Lusko, VP & gen mgr; Lynn Gilstrap, stn mgr; Darren Arnold, gen sls mgr.

KMGA(FM)—Listing follows KTBL(AM).

KNML(AM)— Mar 28, 1928: 610 khz; 5 kw-U, DA-N. TL: N35 01 56 W106 39 48. Hrs open: 24 500 4th St. N.W., 5th Flr., 87102. Phone: (505) 767-6700. Fax: (505) 767-6767. E-mail: knml@citcomm.com Web Site: www.610thesportsanimal.com. Licensee: Citadel Broadcasting Co. Group owner: Citadel Broadcasting Corp. (acq 3-23-00; swap with KSVA(AM) Albuquerque). Population served: 100,000 Format: All sports. Target aud: 25-54. ♦Milt McConnell, gen mgr; Pat Frisch, opns dir; Ian Martin, progmg dir.

KPEK(FM)— December 1974: 100.3 mhz; 22.5 kw. 4,110 ft TL: N35 12 51 W106 27 02. Stereo. Hrs opn: 8am - 5:30pm 5411 Jefferson N.E., Suite A100, 87109. Phone: (505) 830-6400. Fax: (505) 830-6543. Web Site: www.1003thepeak.com. Licensee: Citicasters Licenses L.P. Group owner: Clear Channel Communications Inc. (acq 9-28-99; grpsl). Population served: 369,900 Format: Modern adult contemp. Target aud: 25-54; adults, high income professional and technical. ♦Chuck Hammond, gen mgr; Bill May, chief of opns.

KQBT(FM)—See Rio Rancho

KRKE(AM)— May 14, 1956: 1600 khz; 10 kw-D, 128 w-N. TL: N35 10 14 W106 37 51. Hrs open: 24 307 Los Ranchos Rd. N.W., 87107. Phone: (505) 899-5029. Fax: (505) 899-6865. E-mail: realoldies1600@realoldies1600 Web Site: www.realoldies1600.com. Licensee: Vanguard Media LLC (acq 12-20-2004; $650,000). Population served: 600,000 Natl. Network: CNN Radio. Natl. Rep: Christal. Format: Oldies. Target aud: 30-50. ♦Don Davis, CEO, pres & gen mgr; Craig Collins, opns mgr, progmg dir & mus dir; Crystal Felice, mktg mgr; Doris Budris, prom dir.

KRST(FM)— Sept 15, 1965: 92.3 mhz; 22.5 kw. 4,110 ft TL: N35 12 55 W106 27 02. Stereo. Hrs opn: 24 5th Fl., 500 4th St. N.W., 87102. Phone: (505) 767-6700. Fax: (505) 767-6767. Web Site: www.923krst.com. Licensee: Citadel Broadcasting Co. Group owner: Citadel Broadcasting Corp. (acq 9-30-96; grpsl). Population served: 600,000 Format: Hot new country. ♦Milt McConnell, gen mgr, gen sls mgr & natl sls mgr; Eddie Haskell, opns mgr & progmg dir; Richard Piombino, prom dir; Paul Bailey, mus dir; Art Ortega, pub affrs dir; Bill Harris, engrg dir & chief of engrg.

KRZY(AM)— June 1956: 1450 khz; 1 kw-U. TL: N35 07 56 W106 37 18. Stereo. Hrs open: 24 2725 F. Broadbent Parkway N E, 87107. Phone: (505) 342-4141. Fax: (505) 344-8714. E-mail: mwilder@entravision.com Web Site: www.entravision.com. Licensee: Entravision Holdings LLC. Group owner: Entravision Communications Corp. (acq 3-14-2000; grpsl). Population served: 500,000 Format: Sp/rgnl Mexican. ♦Jeff Liberman, pres; Margarita Wilder, gen mgr & gen sls mgr; Juan Vavala, prom mgr.

KSVA(AM)— Mar 28, 1998: 920 khz; 1 kw-D, 130 w-N. TL: N35 07 56 W106 37 18. Hrs open: 24 Box 2378, Corrales, 87048. Phone: (505) 890-0800. Phone: (866) 578-2920. Fax: (505) 890-0808. E-mail: ksva@lobo.net Web Site: www.lifetalk.net. Licensee: Lifetalk Radio Inc. (acq 4-7-2000; swap with KNML(AM) Albuquerque). Population served: 1000000 Natl. Network: USA. Format: Inspirational Christian. News: 2 hrs wkly. Target aud: 35 plus; Christians. Spec prog: Sp 4 hrs wkly. ♦Phil Folett, CEO; Jep Choate, chmn; Ricardo Baratta, stn mgr; Robert Hardy, chief of opns; Clare Gallimore, dev dir; Jeremy Woodruff, progmg dir; Elvin Vence, chief of engrg.

KTBL(AM)—Los Ranchos de Albuquerque, Dec 16, 1987: 1050 khz; 1 kw-D, 500 w-N, DA-1. TL: N34 58 46 W106 44 13. Hrs open: 500 4th St. N.W., 87102. Phone: (505) 767-6700. Fax: (505) 767-6767. Web Site: www.1050kbull.com. Licensee: Citadel Broadcasting Co. Group owner: Citadel Broadcasting Corp. (acq 6-28-96; $5.725 million with KBZU(FM) Albuquerque). Population served: 100,000 Format: News/talk. Target aud: 25-54. ♦Milt McConnell, gen mgr; Blake Mendenhall, rgnl sls mgr; Glenn Herbert, mktg dir & prom dir; Pat Frisch, opns dir & progmg dir; Art Ortega, pub affrs dir; Bill Harris, chief of engrg; Lynda Ortega, traf mgr.

KMGA(FM)—Co-owned with KTBL(AM). Nov 11, 1963: 99.5 mhz; 19.5 kw. 4,134 ft TL: N35 12 44 W106 26 58. Stereo. E-mail: kmga@citconm.com Web Site: www.99.5magicfm.com. Natl. Rep: McGavren Guild. Format: Light contemp. Target aud: 25-54; upscale, business professionals & families. ♦Kim Gannon, gen sls mgr; Kris Abrams, progmg dir; Julia Zuniga, traf mgr; Alison Atwood, disc jockey; Phil Moore, disc jockey.

***KUNM(FM)**— Oct 17, 1966: 89.9 mhz; 13.6 kw. 4,070 ft TL: N35 12 44 W106 26 57. Stereo. Hrs open: 24 MSC06 3520, Univ. of New Mexico, 87131-0001. Phone: (505) 277-4806. Fax: (505) 277-8004. E-mail: kunm@kunm.org Web Site: www.kunm.org. Licensee: Regents of the University of New Mexico. Population served: 780,000 Natl. Network: NPR, PRI. Law Firm: Dow, Lohnes & Albertson. Format: Div, news/talk. News staff: 3; News: 50 hrs wkly. Target aud: 25-54; those who enjoy NPR and diverse community-produced programs. Spec prog: Class 12 hrs, Sp 9 hrs, Indian 9 hrs wkly. ♦Richard Towne, gen mgr; Mary Bokuniewicz, dev dir.

KXKS(AM)— Dec 16, 1969: 1190 khz; 10 kw-D, 24 w-N. TL: N35 03 04 W106 38 34. Hrs open: 24 Wilkins Communications Network Inc., Box 444, Spartanburg, SC, 29304. Phone: (864) 585-1885. Fax: (864) 597-0687. E-mail: info@wilkinsradio.com Web Site: www.wilkinsradio.com. Licensee: Wild West Radio Corp. (group owner; (acq 12-16-2004; $775,000). Population served: 886,000 Natl. Network: Salem Radio Network. Natl. Rep: Salem. Law Firm: Womble, Carlyle, Sandridge & Rice. Format: Christian preaching/talk. Target aud: 35+. ♦Robert L. Wilkins, pres; LuAnn Wilkins, exec VP; Mitchell Mathis, VP; Steven Stigall, stn mgr; Michael Stark, engr.

KZRR(FM)— June 25, 1961: 94.1 mhz; 100 kw. 4,130 ft TL: N35 12 44 W106 26 58. Stereo. Hrs opn: 24 5411 Jefferson St. N.E., Suite 100, 87109. Phone: (505) 830-6400. Fax: (505) 830-6543. Web Site: www.94rock.com. Licensee: Clear Channel Broadcasting Licenses Inc. Group owner: Clear Channel Communications Inc. (acq 9-28-99; grpsl). Natl. Network: Westwood One. Format: AOR. ♦Chuck Hammond, gen mgr.

Angel Fire

KKTC(FM)— Jan 15, 1990: 99.9 mhz; 1.75 kw. Ant 2,119 ft TL: N36 33 30 W105 11 38. Stereo. Hrs opn: 24 5542 NDCBU, Taos, 87571-6122. Secondary address: 125A Camino de la Merced, Taos 87571-5119. Phone: (505) 758-4491. Fax: (505) 758-4452. E-mail: production@kxmt.com Web Site: radiotaos.com. Licensee: DMC Broadcasting Inc. (acq 3-19-2003; $645,000 with KXMT(FM) Taos). Natl. Network: ABC. Law Firm: Haley, Bader & Potts. Format: Todays country. News staff: one; News: 7 hrs wkly. Target aud: Adults 25-54; middle-, upper-income residents & tourists. Spec prog: Jazz 4 hrs, relg 8 hrs wkly. ♦Jeff Singer, opns mgr.

Armijo, Albuquerque

KNKT(FM)— Dec 17, 1991: 107.1 mhz; 50 kw. 304 ft TL: N35 03 15 W106 51 31. (CP: 60 kw, 2,365 ft.). Stereo. Hrs opn: 24 4001 Osuna Rd. N.E., Albuquerque, 87109. Phone: (505) 344-9146. Fax: (505) 344-9193. Web Site: www.calvaryabq.org. Licensee: Calvary Chapel of Albuquerque Inc. (acq 9-29-94; $800,000 with KDEF(AM) Albuquerque; FTR: 11-21-94) Population served: 500,000 Format: Praise, worship, Bible teaching. News: 4 hrs wkly. Target aud: 25-54. ♦Chip Lusko, gen mgr; Lynn Gilstrap, stn mgr; Darren Arnold, gen sls mgr.

Artesia

KSVP(AM)— Nov 14, 1946: 990 khz; 1 kw-D, 250 w-N. TL: N32 49 29 W104 23 59. Hrs open: 24 317 W. Quay, 88210. Phone: (505) 746-2751. Fax: (505) 748-3748. E-mail: info@ksvpradio.com Web Site: www.ksvpradio.com. Licensee: Pecos Valley Broadcasting Co. (acq 1993; $150,000 with co-located FM; FTR: 9-13-93) Population served: 150,000 Natl. Network: CBS. Law Firm: Cohn & Marks. Format: Talk. News staff: one; News: 18 hrs wkly. Target aud: General. ♦Gene Dow, gen mgr, gen sls mgr & chief of engrg.

KTZA(FM)—Co-owned with KSVP(AM). May 9, 1969: 92.9 mhz; 100 kw. 1,089 ft TL: N32 47 39 W104 12 27. Stereo. 24 121 S. Canal St., Suite C, Carlsbad, 88220. E-mail: info@kz93.com Web Site: www.kz93.com.150,000 Natl. Network: ABC. Format: Country. News: 3 hrs wkly. Target aud: 25-54. ♦Gene Dow, VP.

Aztec

KCQL(AM)— Sept 4, 1959: 1340 khz; 1 kw-U. TL: N36 49 17 W107 59 58. Stereo. Hrs open: 24 200 E. Broadway, Farmington, 87401. Phone: (505) 325-1716. Fax: (505) 325-6797. Web Site: www.foxsports1340.com. Licensee: Capstar TX L.P. Group owner: Clear Channel Communications Inc. (acq 8-30-2000; grpsl). Population served: 85,000 Natl. Network: Fox Sports. Wire Svc: UPI Format: Sports. News: 14 hrs wkly. Target aud: 18-54. Spec prog: Sp 6 hrs wkly. ♦Bill Kruger, gen mgr; Steve Bortstein, progmg dir.

KWYK-FM— Jan 2, 1978: 94.9 mhz; 100 kw. Ant 433 ft TL: N32 47 39 W104 12 27. Stereo. Hrs opn: 24 1515 W. Main, Farmington, 87401. Phone: (505) 325-1996. Fax: (505) 327-2019. Licensee: Basin Broadcasting Co. Format: Adult contemp. News: 15 hrs wkly. Target aud: 25-54; mainstream population. ♦Kerwin Gober, gen mgr; Dana Childs, progmg dir; Jim Burk, chief of engrg.

Bayard

KNFT(AM)— July 4, 1968: 950 khz; 5 kw-D. TL: N32 46 51 W108 11 58. Hrs open: Box 1320, Silver City, 88062. Secondary address: 5 Racetrack Rd., Silver City 88061. Phone: (505) 388-1958. Fax: (505) 388-5000. Licensee: SkyWest Licenses New Mexico LLC. Group owner: Runnels Broadcasting System L.L.C. (acq 6-1-2006; grpsl).

Population served: 100,000 Format: Talk, sports. ♦Matthew Runnell, gen mgr, gen sls mgr & progmg dir; Rita Niccum, chief of engrg; Anna Gallegos, traf mgr.

KNFT-FM— June 15, 1981: 102.9 mhz; 3 kw. 135 ft TL: N32 50 40 W108 14 18. (CP: 29.14 kw, ant 491 ft.). Web Site: www.gilanet.com. Format: C&W.

Belen

KARS(AM)— Oct 7, 1961: 860 khz; 1.3 kw-D, 186 w-N. TL: N34 41 43 W106 46 13. Stereo. Hrs opn: 24 Box 860, 208 N. 2nd St., 87002. Phone: (505) 864-3024. Fax: (505) 864-2719. Licensee: AGM-Nevada L.L.C. Group owner: American General Media (acq 12-22-97; grpsl). Population served: 54,000 Natl. Rep: Lotus Entravision Reps LLC. Format: Country. News staff: one; News: 30 hrs wkly. Target aud: 25 plus. Spec prog: Relg. ♦Scott Hutton, gen mgr; Ron Ortega, stn mgr & gen sls mgr; Ron Travis, mus dir & news dir; Bob Picknell, chief of engrg; Russ Ortego, disc jockey.

KLVO(FM)—Co-owned with KARS(AM). 1982: 97.7 mhz; 100 kw. 859 ft TL: N34 47 55 W106 48 59. Stereo. 24 4125 Carlisle NE, Albuquerque, 87107. Phone: (505) 878-0980. Fax: (505) 878-0098. Web Site: www.agmradio.com. Licensee: AGM-Nevada L.L.C.600,000 Format: Sp. ♦Russ Ortego, disc jockey.

***KQRI(FM)**— 2004: 90.7 mhz; 1.6 kw. Ant 649 ft TL: N34 47 55 W106 48 59. Hrs open:
Rebroadcasts KLRD(FM) Yucaipa, CA 100%.
5700 W. Oaks Blvd., Rocklin, 95765. Phone: (916) 251-1600. Phone: (800) 525-5683. Fax: (916) 251-1650. Web Site: www.air1.com. Licensee: Educational Media Foundation. Group owner: EMF Broadcasting. Natl. Network: Air 1. Format: Christian. ♦Richard Jenkins, pres; Mike Novak, VP & progmg dir; Lloyd Parker, gen mgr; Ed Lenane, opns dir & news dir; Keith Whipple, dev dir; Eric Allen, natl sls mgr; David Pierce, progmg mgr; Jon Rivers, mus dir; Sam Wallington, engrg dir; Arthur Vassar, traf mgr; Karen Johnson, news rptr; Marya Morgan, news rptr; Richard Hunt, news rptr.

Bloomfield

KKFG(FM)— 1988: 104.5 mhz; 100 kw. Ant 1,086 ft TL: N36 38 33 W107 46 54. Stereo. Hrs opn: 200 E. Broadway, Farmington, 87401. Phone: (505) 325-1716. Fax: (505) 325-6797. Web Site: www.kool1045.com. Licensee: Capstar TX L.P. Group owner: Clear Channel Communications Inc. (acq 8-30-2000; grpsl). Format: Oldies. ♦Bill Kruger, gen mgr.

Bosque Farms

***KQLV(FM)**— Nov 16, 2001: 105.5 mhz; 22 kw. Ant 745 ft TL: N34 47 55 W106 48 59. Stereo. Hrs opn: 24 2351 Sunset Blvd., Suite 170-218, Rocklin, CA, 95765. Phone: (916) 251-1600. Fax: (916) 251-1650. E-mail: klove@klove.com Web Site: www.klove.com. Licensee: Educational Media Foundation. Group owner: EMF Broadcasting. Natl. Network: K-Love. Law Firm: Shaw Pittman. Format: Contemp Christian mus. News staff: 3. Target aud: 25-44; Judeo-Christian, female. ♦Richard Jenkins, pres; Mike Novak, VP; Keith Whipple, dev dir; Eric Allen, natl sls mgr; David Pierce, progmg mgr & mus dir; Ed Lenane, news dir; Sam Wallington, engrg dir; Karen Johnson, news rptr; Marya Morgan, news rptr; Richard Hunt, news rptr.

KTEG(FM)— July 1, 1987: 104.7 mhz; 100 kw. 1,822 ft TL: N34 47 55 W106 48 59. (CP: Ant 843 ft. TL: N34 46 12 W106 51 42). Stereo. Hrs opn: 5411 Jefferson N.E., Suite A100, Albuquerque, 87109. Phone: (505) 830-6400. Fax: (505) 830-6543. Web Site: www.1047edgeradio.com. Licensee: Clear Channel Broadcasting Licenses Inc. group owner: Clear Channel Communication Inc. Format: Alternative. ♦Bill May, pres & chief of opns; Chuck Hammond, gen mgr.

Cannon AFB

***KKCJ(FM)**— 2006: 90.7 mhz; 25 kw. Ant 194 ft TL: N34 26 58 W103 37 03. Hrs open:
Rebroadcasts KSGR(FM) Portland, TX 100%.
c/o KSGR(FM), 3001 Rodd Field Rd., Corpus Christi, TX, 78414-3987. Phone: (361) 814-7775. Fax: (361) 814-7779. E-mail: jim.shepherd@csnradio.com Web Site: www.ksgr.org. Licensee: CSN International (group owner). Natl. Network: CSN. Format: Contemp Christian. ♦Jim Shepherd, stn mgr.

Carlsbad

KAMQ(AM)— June 10, 1938: 1240 khz; 1 kw-U. TL: N32 23 43 W104 14 48. Hrs open: Box 1538, 88220. Secondary address: 1609 Radio Blvd. 88220. Phone: (505) 887-5323. Fax: (505) 887-7000. E-mail: info@kdoveradio.com Licensee: KAMQ Inc. (acq 1-21-76). Population served: 40,000 Format: Adult contemp, Christian. ♦Don Hughes, gen mgr & gen sls mgr; Reginald James, progmg dir; Frank Nymeyer, chief of engrg.

KCDY(FM)—Co-owned with KAMQ(AM). July 1989: 104.1 mhz; 100 kw. 676 ft TL: N32 34 22 W104 05 32. Box 1538, 88220. Secondary address: 1609 Radio Blvd. Phone: (505) 887-7563. Fax: (505) 887-7000. Format: Adult contemp. ♦Steve Sparks, progmg dir & progmg mgr.

KATK(AM)— May 17, 1950: 740 khz; 1 kw-D, 500 w-N. TL: N32 27 02 W104 12 47. Hrs open: 1609 Radio Blvd., 88220. Phone: (505) 887-7563. Fax: (505) 887-7000. E-mail: katk@pccnm.com Web Site: www.katkradio.com. Licensee: Stubbs Broadcasting Co. Inc. (acq 3-31-00; $475,000 with co-located FM). Law Firm: Reddy, Begley & McCormick. Format: Adult standards. News: 70 hrs wkly. Target aud: 50 plus; bilingual Hispanics. Spec prog: Gospel 2 hrs wkly. ♦Don Hughes, gen mgr & gen sls mgr; Reginald James, progmg dir; Frank Nymeyer, chief of engrg.

KATK-FM— Sept 15, 1966: 92.1 mhz; 3 kw. 190 ft TL: N32 27 02 W104 12 47. Stereo. Web Site: www.katkradio.com. Format: Country. News: 70 hrs wkly. Target aud: 18-54. Spec prog: Hispanic.

KCCC(AM)— July 1, 1966: 930 khz; 1 kw-D, 60 w-N. TL: N32 24 20 W104 11 21. Hrs open: 24 930 N. Canal, 88220. Phone: (505) 887-5521. Fax: (505) 885-5481. E-mail: kccc@carlsbadnm.com Licensee: Compass Enterprises Inc. (acq 8-14-95). Population served: 55,000 Format: Oldies. ♦Nick Jenkins, pres, gen sls mgr & prom dir; Michelle McCutcheon, opns mgr & progmg dir; Phil Tozier, news dir; Frank Nymeyer, engrg mgr & chief of engrg.

KPZE-FM— 2000: 106.1 mhz; 39 kw. Ant 558 ft TL: N32 34 22 W104 05 32. Hrs open: 24 317 W. Quay Ave., Artesia, 88210. Phone: (505) 746-2751. Fax: (505) 748-3748. Web Site: www.kpze.com. Licensee: Pecos Valley Broadcasting Co. Group owner: Runnels Broadcasting System L.L.C. (acq 2005; $475,000). Population served: 50,000 Law Firm: Cohn and Marks L.L.P. Format: Rgnl/Mexician. News staff: one; News: one hr wkly. ♦Gene Dow, gen mgr.

Chama

KZRM(FM)— Oct 8, 1999: 95.9 mhz; 1 kw. Ant 312 ft TL: N36 53 58 W106 36 07. Hrs open: 24 Box 307, 87520. Phone: (505) 756-1617. Fax: (505) 756-1317. Web Site: www.kzrmradio.com. Licensee: Lance Broadcasting LLC (acq 5-3-2004; $220,000). Format: Classic rock. ♦Scott Flury, gen mgr.

Clayton

KLMX(AM)— Nov 10, 1949: 1450 khz; 1 kw-U. TL: N36 26 39 W103 11 24. Hrs open: Box 547, Union County Fairgrounds, 88415. Phone: (505) 374-2555. Fax: (505) 374-2557. Licensee: Johnson County Broadcasters Inc. (acq 11-10-78). Population served: 3,500 Format: Country. Spec prog: Sp 2 hrs wkly. ♦Avis Green Tucker, pres; Jim McCollum, VP; Janet Dillon, gen mgr & gen sls mgr; Paula V. Maestas-Ballew, progmg dir & news dir; Henry Walker, chief of engrg.

Cloudcroft

***KHII(FM)**— 2004: 88.9 mhz; 100 w. Ant 1,187 ft TL: N32 59 48 W105 42 38. Stereo. Hrs opn: 24
Rebroadcasts KUPR(FM) Alamogordo 80%.
3001 N. Florida Ave., Alamogordo, 88310. Phone: (505) 437-0917. Fax: (505) 434-6060. Fax: (505) 437-9917. E-mail: khii889@yahoo.com Licensee: Southern New Mexico Radio Foundation. Format: All Gospel music formats. ♦Bob Flotte, pres; DeVere Johnson, progmg dir.

KNMB(FM)—Not on air, target date: unknown: 96.7 mhz; 25 kw. 2,880 ft Hrs opn: Box 2010, Ruidoso Downs, 88346. Phone: (505) 258-9922. Fax: (505) 258-2363. E-mail: w105@trailnet.com Licensee: MTD Inc. (group owner) Format: Classic Country. ♦Tim Keithley, gen mgr.

Clovis

***KAQF(FM)**— March 1998: 91.1 mhz; 1 kw. 174 ft TL: N34 24 05 W103 12 12. Hrs open: Box 3206, American Family Radio, Tupelo, MS, 38803. Phone: (662) 844-8888. Fax: (662) 842-6791. Licensee: American Family Association. Group owner: American Family Radio Format: Inspirational Christian. ♦Marvin Sanders, gen mgr.

KCLV-FM— Jan 8, 1970: 99.1 mhz; 74.2 kw. Ant 230 ft TL: N34 23 18 W103 11 07. Stereo. Hrs opn: 24 Box 1907, 88102-1907. Secondary address: 2112 Thornton St. 88101. Phone: (505) 763-4401. Fax: (505) 769-2564. E-mail: kclv@allsups.com Licensee: Zia Broadcasting Co. (acq 11-12-81). Population served: 70,000 Natl. Network: ABC. Format: Country. Target aud: 30 plus. ♦Rcik Keefer, gen sls mgr; Rick Keefer, gen mgr & progmg dir.

KCLV(AM)— February 1953: 1240 khz; 1 kw-U. TL: N34 22 40 W103 12 17.24 (Acq 7-1-71).50,000 Natl. Network: ESPN Radio. Format: Sports. ♦Lonnie D. Allsup, pres; Gary Jackson, chief of engrg; Lorraine Weingates, traf mgr.

***KELU(FM)**— 2006: 90.3 mhz; 14 kw. Ant 397 ft TL: N34 26 21 W103 12 22. Hrs open:
Rebroadcasts KLVR(FM) Santa Rosa, CA 100%.
2351 Sunset Blvd., Suite 170-218, Rocklin, CA, 95765. Phone: (916) 251-1600. Fax: (916) 251-1650. Web Site: www.klove.com. Licensee: Educational Media Foundation. (acq 9-22-2005; $40,000 for CP). Natl. Network: K-Love. Format: Contemp Christian. ♦Richard Jenkins, pres; Mike Novak, VP; Keith Whipple, dev dir; David Pierce, progmg mgr; Ed Lenane, news dir; Sam Wallington, engrg dir; Karen Johnson, news rptr; Marya Morgan, news rptr; Richard Hunt, news rptr.

KICA(AM)— 1933: 980 khz; 50 kw-D, 188 w-N, DA-D. TL: N34 20 55 W102 57 18. Hrs open: 24 1000 Sycamore St., 88101. Phone: (505) 762-6200. Fax: (505) 762-8800. Web Site: www.kkyckica@plateautel.net. Licensee: Tallgrass Broadcasting LLC. (group owner; (acq 4-2-2007; grpsl). Population served: 100,000 Natl. Network: USA. Law Firm: Shaw Pittman. Format: Talk. Target aud: 30 plus. Spec prog: Farm 5 hrs, high school sports 4 hrs wkly. ♦Dana Taylor, opns dir, progmg dir & news dir.

KKYC(FM)— 1993: 102.3 mhz; 25 kw horiz. Ant 177 ft TL: N34 24 31 W103 11 15. (CP: 100 kw, ant 485 ft. TL: N34 29 36 W103 23 46). Hrs opn: 24 1000 Sycamore St., 88101. Phone: (505) 762-6200. Fax: (505) 762-8800. E-mail: kica-kkyc@plateautel.com Licensee: Tallgrass Broadcasting LLC. (acq 4-2-2007; grpsl). Population served: 60,000 Format: Country. News staff: one; News: 2 hrs wkly. Target aud: 25-54. ♦Rick Keefer, chief of engrg; Shannon Phillips, traf mgr.

KRMQ-FM— 2003: 101.5 mhz; 100 kw. Ant 453 ft TL: N34 15 08 W103 14 21. Hrs open: Mount Rushmore Broadcasting Inc., 218 N. Wolcott St., Casper, WY, 82601. Phone: (307) 265-1984. Licensee: Mount Rushmore Broadcasting Inc. (group owner). ♦Jan Charles Gray, pres & gen mgr.

KSMX(FM)— November 1982: 107.5 mhz; 100 kw. 550 ft TL: N34 11 34 W103 16 44. Stereo. Hrs opn: 208 E. Grand Ave., 88101. Secondary address: 42437 U.S. 70, Portales 88130. Phone: (505) 763-4649. Fax: (505) 359-0724. E-mail: bettermix@bettermix.com Web Site: www.bettermix.com. Licensee: Rooney Moon Broadcasting Inc. (group owner; acq 7-15-02; grpsl). Format: Hot adult contemp. ♦Duffy Moon, opns dir & mktg VP; Steve Rooney, pres, gen mgr & progmg dir; Kevin Robbins, news dir; Jeff Burmeister, chief of engrg; Lisa Schmidt, traf mgr.

KTQM-FM—Listing follows KWKA(AM).

KWKA(AM)— 1971: 680 khz; 500 w-U, DA-1. TL: N34 21 48 W103 13 05. Hrs open: 24 Box 869, 88102. Secondary address: 710 Curry Rd. K 88101. Phone: (505) 762-4411. Fax: (505) 769-0197. E-mail: ktqm@plateautel.net Licensee: Curry County Broadcasting Inc. (acq 10-24-80; $650,000; with co-located FM; FTR: 11-10-80) Population served: 125,000 Natl. Network: Jones Radio Networks. Rgnl rep: Rgnl Reps. Law Firm: Fletcher, Heald & Hildreth. Format: Oldies. News: 3 hrs wkly. Target aud: 25 plus; baby boomers. ♦C. Hewel Jones, pres & gen mgr; Robert D. Coker, VP, stn mgr & natl sls mgr.

KTQM-FM—Co-owned with KWKA(AM). Mar 1, 1963: 99.9 mhz; 100 kw. 360 ft TL: N34 21 48 W103 13 05. Stereo. 24 125,000 Natl. Network: ABC, ESPN Radio. Format: Adult contemp. News: 3 hrs wkly. Target aud: 18-49; young affluent.

Corrales

KKNS(AM)— July 15, 1985: 1310 khz; 5 kw-D, 500 w-N, DA-N. TL: N35 12 00 W106 35 59. Stereo. Hrs opn: 24 1606 Central Ave., Suite104, Albuquerque, 87106. Phone: (505) 255-5015. Fax: (505) 262-4792. E-mail: vcamino@elcaminocomm.com Web Site: www.elcaminocomm.com. Licensee: El Camino Communications LLC Group owner: Simmons Media Group (acq 1-22-2007; $860,000). Format: Mexican rgnl. Target aud: 25-54. ♦Victor Camino, gen mgr.

KSYU(FM)— Apr 27, 1996: 95.1 mhz; 3 kw. -531 ft TL: N35 14 42 W106 36 18. Hrs open: 24 5411 Jefferson St. N.E., Suite 100,

Albuquerque, 87109. Phone: (505) 830-6400. Fax: (505) 830-6543. Web Site: www.hot951.com. Licensee: Clear Channel Broadcasting Licenses Inc. Group owner: Clear Channel Communications Inc. (acq 9-28-99). Format: Urban adult contemp. ♦Chuck Hammond, gen mgr.

Deming

KDEM(FM)—Listing follows KOTS(AM).

KOTS(AM)— Mar 10, 1954: 1230 khz; 1 kw-U. TL: N32 15 05 W107 45 28. Hrs open: 24 Box 470, 1700 S. Gold, 88031. Phone: (505) 546-9011. Fax: (505) 546-9342. E-mail: radio@demingradio.com Web Site: www.demingradio.com. Licensee: Luna County Broadcasting Co. (acq 3-14-90). Population served: 18,000 Natl. Network: Westwood One. Wire Svc: AP Format: Country. News staff: one; News: 13 hrs wkly. Target aud: General. Spec prog: Farm 5 hrs, Sp 8 hrs wkly. ♦Candie G. Sweetser, stn mgr.

***KZPI(FM)**— Mar 25, 1996: 91.7 mhz; 600 w. 62 ft TL: N32 15 31 W107 46 45. Hrs open: 24 Box 252, Paulino Bernal Evangelism, McAllen, TX, 78505. Phone: (956) 686-6382. Fax: (956) 686-2999. Licensee: Paulino Bernal Evangelism. (acq 2-13-98; $45,000). Format: Christian, relg, Sp. Target aud: General. ♦Paulino Bernal, pres.

Des Moines

KHOD(FM)— July 2007: 105.3 mhz; 82 kw. Ant 2,053 ft TL: N36 42 20 W103 52 36. Stereo. Hrs opn: 24 520 Monticello Dr., Las Vegas, NV, 89107-3616. Phone: (702) 878-0773. E-mail: khodson@earthlink.net Licensee: Hodson Broadcasting. Format: 70s & 80s Top-40. Target aud: 35-54; males/females. ♦Richard Hodson, CEO, chief of engrg, sls & progmg.

Dulce

***KCIE(FM)**— Dec 3, 1990: 90.5 mhz; 100 kw. 1,535 ft TL: N36 59 00 W106 58 12. Hrs open: Box 603, A.I.E. Bldg., Narrow Gauge Rd., 87528. Phone: (505) 759-3681. Phone: (505) 759-3023. Fax: (505) 759-9140. Licensee: Jicarilla Apache Tribe. Population served: 3,000 Format: Div. News staff: one; News: 2 hrs wkly. Target aud: General. ♦Lisa Vigil-Gomez, stn mgr; Romaine Wood, progmg dir; Darnell Muniz, mus dir; Jim Burt, chief of engrg.

Espanola

KDCE(AM)— 1963: 950 khz; 4.2 kw-D, 90 w-N. TL: N36 00 08 W106 03 59. Hrs open: 403 W. Pueblo Dr., 87532. Phone: (505) 753-2201. Fax: (505) 753-8685. E-mail: kdce@espanola.com Web Site: www.kdce.net. Licensee: Richard L. Garcia Broadcasting Inc. (acq 11-29-82; $625,000; FTR: 11-8-82). Population served: 200,000 Format: Sp. ♦Casey Gallegos, gen mgr; Richard Garcia, pres & gen sls mgr; Ray Casias, progmg dir; Ken Bass, chief of engrg; Ester Marquez, traf mgr.

KYBR(FM)—Co-owned with KDCE(AM). July 6, 1981: 92.9 mhz; 3 kw. -249 ft TL: N36 00 08 W106 03 59. (CP: 50 kw, ant 203 ft.). Web Site: www.kdce.net. Licensee: Rio Chama Broadcasting Co. (acq 1995; $50,000). Format: Rgnl Mexican. ♦Efrem Galindo, progmg dir.

Eunice

KEJL(FM)— 1996: 100.9 mhz; 50 kw. Ant 295 ft TL: N32 28 10 W103 09 36. Stereo. Hrs opn: 24 Box 5967, Hobbs, 88240. Secondary address: 1423 W. Bender, Hobbs 88240. Phone: (505) 393-6000. Fax: (505) 397-6088. E-mail: larryphilpot@basinbreadband.com Licensee: FiveStar Enterprises L.C. (acq 1999; $20,000). Population served: 47,000 Natl. Network: Jones Radio Networks. Format: Classic rock. News staff: one. Target aud: 18-49. ♦Larry Philpot, gen mgr; Al Lobeck, gen sls mgr.

Farmington

KDAG(FM)— Sept 1, 1969: 96.9 mhz; 100 kw. Ant 1,010 ft TL: N36 39 49 W108 12 55. Stereo. Hrs opn: 24 200 E. Broadway, 87401. Phone: (505) 325-1716. Fax: (505) 325-6797. Web Site: www.bigdog969.com. Licensee: Capstar TX L.P. Group owner: Clear Channel Communications Inc. (acq 8-30-2000; grpsl). Population served: 250,000 Law Firm: Fisher, Wayland, Cooper, Leader & Zaragoza. Format: Classic rock. Target aud: 18-49. ♦Bill Kruger, gen mgr.

KENN(AM)— November 1951: 1390 khz; 5 kw-D, 1.3 kw-N, DA-N. TL: N36 42 27 W108 08 50. Hrs open: 212 W. Apache, 87401. Phone: (505) 325-3541. Fax: (505) 327-5796. Licensee: Winton Road Broadcasting Co. LLC (group owner; acq 5-3-01; grpsl). Population served: 38,000 Law Firm: Pepper & Corazzini. Format: News/talk, sports. Target aud: 25-54; upper middle class. ♦Sara Olsen, gen mgr.

KRWN(FM)—Co-owned with KENN(AM). 1974: 92.9 mhz; 30 kw. 430 ft TL: N36 41 45 W108 13 23. (CP: 62 kw, ant 394 ft.). Stereo. 24 Phone: (505) 327-4449. Web Site: www.krwn.com. Format: Classic rock. Target aud: 18-49. ♦Leslie Granger, progmg dir.

KISZ-FM—See Cortez, CO

KNDN(AM)— Aug 1, 1957: 960 khz; 5 kw-D, 163 w-N. TL: N36 43 48 W108 13 47. Hrs open: 6 AM-10 PM 1515 W. Main, 87401. Phone: (505) 325-1996. Fax: (505) 327-2019. Licensee: Basin Broadcasting Co. Format: Navajo Indian. Target aud: General; Navajo Indian reservation; all Navajo language. ♦Kerwin Gober, gen mgr & opns mgr; George Werito, progmg dir & news dir; Jim Burt, chief of engrg.

***KNMI(FM)**— Mar 18, 1980: 88.9 mhz; 6.22 kw. 360 ft TL: N36 40 16 W108 13 54. Stereo. Hrs opn: 24 Box 1230, 2103 W. Main St., 87401. Phone: (505) 325-0255. Fax: (505) 325-9035. E-mail: email@verticalradio.org Web Site: www.verticalradio.org. Licensee: Navajo Ministries, Inc. Population served: 120,000 Natl. Network: USA. Format: Christian, talk, CHR hits. News: 9 hrs wkly. Target aud: 24-40; general. ♦Wilann Thomas, gen mgr.

KPCL(FM)— Dec 14, 1988: 95.7 mhz; 100 kw. 394 ft TL: N36 41 44 W108 13 11. Stereo. Hrs opn: 24 Box 232, 87499. Secondary address: 1105 W. Apache 87401. Phone: (505) 327-7202. Fax: (505) 327-2163. E-mail: kpcl@kpcl.org Web Site: www.kpcl.org. Licensee: Voice Ministries of Farmington Inc. Population served: 300,000 Natl. Network: Moody. Format: Christian contemp. News staff: one; News: 3 hrs wkly. Target aud: General; relg audience. Spec prog: Class one hr, Navajo 7 hrs wkly. ♦Fareed W. Ayoub, pres.

KRZE(AM)— July 1, 1958: 1280 khz; 5 kw-D. TL: N36 49 03 W108 05 47. Hrs open: 24 Radio Fiesta, 204 E. Broadway, 87401. Phone: (505) 327-5287. Fax: (505) 327-5289. Licensee: J. Thomas Development of New Mexico Inc. (acq 12-18-91; with co-located FM). Population served: 100,000 Law Firm: Shaw Pittman. Format: Sp var. ♦Jeff Thomas, pres; Rogelio Esparva, gen mgr.

***KSJE(FM)**— November 1990: 90.9 mhz; 15 kw. 390 ft TL: N36 41 52 W108 13 14. Stereo. Hrs opn: 24 4601 College Blvd., 87402. Phone: (505) 566-3517. Fax: (505) 566-3385. E-mail: gotsch@sjc.cc.mm.us Web Site: www.ksje.com. Licensee: San Juan College. Natl. Network: PRI. Format: Class. News: 15 hrs wkly. Target aud: 25-65. Spec prog: Jazz 15 hrs, folk 15 hrs wkly. ♦Carol Spenser, pres; Constance Gotsch, progmg dir; Jim Burt, chief of engrg.

KTRA-FM— Feb 19, 1987: 102.1 mhz; 100 kw. Ant 1,033 ft TL: N36 48 52 W107 53 32. Stereo. Hrs opn: 24 200 E. Broadway, 87401. Phone: (505) 325-1716. Fax: (505) 325-6797. Web Site: www.102ktra.com. Licensee: Clear Channel Radio Licenses Inc. Group owner: Clear Channel Communications Inc. (acq 8-30-2000; grpsl). Population served: 250,000 Natl. Network: ABC. Format: Classic country. Target aud: 25-54. ♦Dave Schaefer, progmg dir.

KUCU(AM)—Not on air, target date: unknown: 1060 khz; 10 kw-D, 250 w-N, DA-2. TL: N36 43 55 W108 06 38. Hrs open: 324 N. Vine #1, 87401. Phone: (505) 324-6434. Licensee: Western Broadcasters Inc. ♦E. Boyd Whitney, pres & gen mgr.

***KUUT(FM)**—Not on air, target date: unknown: 89.7 mhz; 500 w vert. Ant 600 ft TL: N36 40 16 W108 13 54. Hrs open:
Rebroadcasts KUTE (FM) Ignacio.
Box 737, Ignacio, CO, 81137-0737. Phone: (970) 563-0255. Fax: (970) 563-0399. Web Site: www.ksut.org. Licensee: KUTE Inc. Format: Triple A, americana. ♦Eddie Box Jr., pres; Beth Warren, gen mgr.

KWYK-FM—See Aztec

Flora Vista

***KUSW(FM)**—Not on air, target date: unknown: 88.1 mhz; 2.2 kw. Ant 577 ft TL: N36 40 14 W108 14 20. Hrs open:
Rebroadcasts KUTE (FM) Ignacio.
Box 737, Ignacio, CO, 81137-0737. Phone: (970) 563-0255. Fax: (970) 563-0399. Web Site: www.ksut.org. Licensee: KUTE Inc. (acq 6-7-2006). Format: Triple A, americana. ♦Eddie Box Jr., pres; Beth Warren, gen mgr.

Fruitland

***KTGW(FM)**—Not on air, target date: unknown: 91.7 mhz; 20 kw. 308 ft TL: N36 41 44 W108 13 11. Hrs open: 24 Attn: Fareed W. Ayoub, Box 232, Farmington, 87499. Secondary address: 1103 W. Apache St., Farmington 87401. Phone: (505) 327-7202. Fax: (505) 327-2163. E-mail: kpcl@kpcl.org Web Site: www.kpcl.org. Licensee: Native American Christian Voice Inc. Law Firm: Southmayd & MIller. Format: Christian-talk. ♦Fareed Ayoub, pres; Annette Ayoub, exec VP.

Gallup

KFMQ(FM)— 1996: 106.1 mhz; 26 kw. 185 ft TL: N35 32 27 W108 44 32. Hrs open: 1632 S. 2nd St., 87301. Phone: (505) 863-9391. Fax: (505) 863-9393. Licensee: Clear Channel Broadcasting Licenses Inc. (Group owner: Clear Channel Communications Inc. (acq 4-17-97). Format: Rock. ♦Mary Ann Armijo, gen mgr & gen sls mgr; Ted Foster, opns mgr; Blas Saucedo, progmg dir.

KGAK(AM)— Feb 9, 1945: 1330 khz; 5 kw-D, 1 kw-N, DA-N. TL: N35 32 34 W108 44 11. Hrs open: 5am-10pm 401 E. Coal Ave., 87301. Phone: (505) 863-4444. Fax: (505) 722-7381. E-mail: kgak@cia-g.com Licensee: KRJG Inc. (acq 6-30-98; $102,600). Population served: 13,779 Natl. Network: CBS. Format: Navajo, Indian. Target aud: 30-55. ♦Jim Gober, CEO & gen mgr; Jim Burt, gen sls mgr & chief of engrg; Leaudro Jodie, progmg dir.

***KGGA(FM)**—Not on air, target date: unknown: 88.1 mhz; 30 kw. Ant -13 ft TL: N35 30 01 W108 44 06. Hrs open: 2351 Sunset Blvd., Suite 170-218, Rocklin, CA, 95765. Phone: (916) 251-1600. Fax: (916) 251-1650. Licensee: Educational Media Foundation. (acq 7-23-2007; grpsl). ♦Mike Novak, sr VP.

***KGLP(FM)**— Sept 1, 1992: 91.7 mhz; 160 w. Ant 1,145 ft TL: N35 36 13 W108 40 45. Stereo. Hrs opn: 24
Rebroadcasts KSUT(FM) Ignacio, CO 50%.
Univ. of New Mexico, 200 College Rd., 87301. Phone: (505) 863-7626. Fax: (505) 863-7532. Fax: (505) 863-7633. E-mail: kglp@kglp.org Web Site: www.kglp.org. Licensee: Gallup Public Radio. Population served: 50,000 Natl. Network: NPR. Format: News, div, public stn. News: 40 hrs wkly. Target aud: General; adult professional, academic, business community. Spec prog: Native American,World, Reggae, Blues, Resonator, Unsigned Bands, Classical, Jazz, Jam Bands, ClassicRock Band, Soul and Hispanic. ♦David Pracy, opns mgr & progmg dir; Tom Funk, mus dir.

KGLX(FM)— Mar 1, 1989: 99.1 mhz; 51 kw. 1,249 ft TL: N35 36 18 W108 41 11. Stereo. Hrs opn: 24 1632 S. 2nd St., 87301. Phone: (505) 863-9391. Fax: (505) 863-9393. Licensee: Clear Channel Broadcasting Licenses Inc. Group owner: Clear Channel Communications Inc. (acq 8-18-00; grpsl). Population served: 80,000 Law Firm: Bechtel & Cole. Format: Country. News staff: 2; News: 14 hrs wkly. Target aud: 25-54. Spec prog: American Indian 3 hrs wkly. ♦Mary Ann Armijo, gen mgr & adv dir; Sylvester Paquin, sls dir; Ted Foster, opns mgr & progmg dir; Pat Jarvison, sports cmtr.

KKOR(FM)—Listing follows KYVA(AM).

***KLLU(FM)**—Not on air, target date: unknown: 88.9 mhz; 900 w. Ant 1,184 ft TL: N35 36 22 W108 41 26. Hrs open: 2351 Sunset Blvd.,

Suite 170-218, Rocklin, CA, 95765. Phone: (916) 251-1600. Fax: (916) 251-1650. Licensee: Educational Media Foundation. ♦Richard Jenkins, pres; Mike Novak, VP; Keith Whipple, dev dir; Eric Allen, natl sls mgr; David Pierce, progmg mgr; Ed Lenane, news dir; Sam Wallington, engrg dir; Karen Johnson, news rptr; Marya Morgan, news rptr; Richard Hunt, news rptr.

KQLP(FM)—Not on air, target date: unknown: 101.5 mhz; 47.36 kw. Ant 1,325 ft TL: N35 36 21.9 W108 41 25.7. Hrs open: 2801 Via Fortuna Dr., Suite 675, Austin, TX, 78746. Phone: (713) 528-2517. Licensee: Ace Radio Corp. ♦Stephen Hackerman, pres.

KXXI(FM)— Aug 15, 1975: 93.7 mhz; 62 kw. 161 ft TL: N35 36 22 W108 41 26. Stereo. Hrs opn: 24 Box 420, 87301. Secondary address: 300 W. Aztec, Suite 200 87301. Phone: (505) 863-6851. Fax: (505) 863-2429. E-mail: mm1@galluparadio.com Web Site: www.gallupradio.com. Licensee: Millennium Media Inc. (acq 6-7-94). Population served: 200,000 Natl. Network: ABC. Law Firm: Fletcher, Heald & Hildreth. Format: Classic rock. Target aud: 25-44. ♦George Malti, CEO; Sammy Chioda, pres & stn mgr; Thomas Devlin, sls dir; John McBreen, news dir; Keith Desautels, chief of engrg.

KYVA(AM)— July 15, 1959: 1230 khz; 1 kw-U. TL: N35 32 02 W108 42 22. Stereo. Hrs opn: 24 Box 420, 87305. Phone: (505) 863-6851. Fax: (505) 863-2429. Licensee: Millennium Media Inc. (acq 3-77). Population served: 180,000 Natl. Network: ABC. Format: Country. News staff: one; News: 12 hrs wkly. Target aud: 35-54; mature, with buying power. ♦George Malti, pres; Sammy Chioda, exec VP & gen mgr; Tom Devlin, sls dir; Brian Smith, prom dir & progmg dir; John McBreen, news dir; Keith DeSautels, chief of engrg.

KKOR(FM)—Co-owned with KYVA(AM). Oct 6, 1974: 94.5 mhz; 100 kw. 1,388 ft TL: N35 28 03 W108 14 25. Stereo. 24 Phone: (505) 863-5567. Web Site: www.gallupradio.com.180,000 Format: Adult contemp. News staff: one. Target aud: 25-44; young families with buying power.

Grants

KDSK(FM)— June 1, 1997: 92.7 mhz; 26 kw. Ant 171 ft TL: N35 07 09 W107 54 08. (CP: 45 kw, ant 1,351 ft. TL: N35 10 57.1 W107 36 12.7). Stereo. Hrs opn: 24 733 Roosevelt, 87020. Phone: (505) 285-5598. Fax: (505) 285-5575. Licensee: KD Radio Inc. (acq 11-16-00; with KMIN(AM) Grants). Population served: 25,000 Format: Oldies. Target aud: 30-50; earning boom. ♦Derek Underhill, pres.

***KIDS(FM)**—Not on air, target date: unknown: 88.1 mhz; 100 w. Ant 162 ft TL: N35 07 09 W107 54 02. Hrs open: 2020 Coal Ave. S.E., Albuquerque, 87106. Phone: (505) 242-7163. E-mail: brasher@aps.edu Web Site: www.kanw.com. Licensee: Board of Education of the City of Albuquerque, NM. Population served: 30,000 ♦Michael Brasher, gen mgr.

***KLGQ(FM)**— 2006: 90.3 mhz; 1 kw. Ant 2,713 ft TL: N35 15 08 W107 35 45. Hrs open:
Rebroadcasts KLVR(FM) Santa Rosa, CA 100%.
2351 Sunset Blvd., Suite 170-218, Rocklin, CA, 95765. Phone: (916) 251-1600. Fax: (916) 251-1650. Web Site: www.klove.com. Licensee: Educational Media Foundation. Natl. Network: K-Love. Format: Contemp Christian. ♦Richard Jenkins, pres; Mike Novak, VP & progmg dir; Keith Whipple, dev dir; David Pierce, progmg mgr; Ed Lenane, news dir; Sam Wallington, engrg dir; Karen Johnson, news rptr; Marya Morgan, news rptr; Richard Hunt, news rptr.

KMIN(AM)— Sept 1, 1956: 980 khz; 1 kw-D, 250 kw-N. TL: N35 09 05 W107 52 31. Hrs open: 24 733 Roosevelt, 87020. Phone: (505) 285-5598. Fax: (505) 285-5575. E-mail: info@kmin980.com Web Site: www.kmin980.com. Licensee: KD Radio Inc. (acq 1-2-01; $145,000 with KDSK(FM) Grants). Population served: 25,000 Format: Country. Target aud: 25-54; active, working adults. ♦Derek Underhill, pres & opns mgr.

KYVA-FM— Aug 7, 1997: 103.7 mhz; 100 kw. 1,299 ft TL: N35 28 03 W108 14 25. Hrs open: 24 Box 420, Gallup, 87305. Phone: (505) 863-6851. Fax: (505) 863-2429. E-mail: mm1@gallupradio.com Web Site: www.gallupradio.com. Licensee: Millennium Media Inc. (acq 6-98). Natl. Network: ABC. Format: Oldies. News staff: one; News: 12 hrs wkly. Target aud: 25+. 0. Spec prog: American Indian 10 hrs, Sp 4 hrs wkly. ♦George M. Malti, CEO & pub affrs dir; Sammy Chioda, pres; Thomas Devlin, sls dir; John McBreen, news dir; Keith Desautels, chief of engrg.

Hatch

KVLC(FM)— Apr 1, 1994: 101.1 mhz; 100 kw. 1,033 ft TL: N32 41 35 W107 04 06. Stereo. Hrs opn: 24 101 Perkins Dr., Las Cruces, 88005. Phone: (505) 527-1111. Fax: (505) 527-1100. E-mail: kvlc@101gold.com Web Site: www.101gold.com. Licensee: Bravo MIC Communications LLC. (acq 1-7-2005; $1.3 million). Population served: 1,100,000 Natl. Network: Jones Radio Networks. Format: Good time oldies. News staff: one; News: 3 hrs wkly. Target aud: 25-54. Spec prog: Bi-lingual Sp/English 6 hrs wkly. ♦Michael Smith, gen mgr; Allen Moore, opns mgr; K.C. Court, progmg dir.

Hobbs

KHOB(AM)— Aug 7, 1954: 1390 khz; 5 kw-D, 500 w-N, DA-N. TL: N32 44 21 W103 10 48. Hrs open: 24 3301, N. Bensing Rd., 88240. Phone: (505) 392-9292. Fax: (505) 392-7579. E-mail: khobam@aol.com Licensee: American Asset Management Inc. (acq 2-28-90; $255,000; FTR: 3-19-90). Population served: 35,000 Format: Oldies. News staff: 2; News: 48 hrs wkly. Target aud: General; adult. ♦Harmon Hann, gen mgr; Pat Hann, progmg dir.

KIXN(FM)— Feb 1, 1996: 102.9 mhz; 100 kw. Ant 518 ft TL: N32 43 26 W103 34 34. Stereo. Hrs opn: 24 619 N.Turner St., 88240. Phone: (505) 397-4969. Fax: (505) 393-4310. E-mail: paul@1radiosquare.com Web Site: www.1radiosquare.com. Licensee: Noalmark Broadcasting Corp. (group owner; acq 1995; $53,000 for CP). Population served: 100,000 Format: Country. News staff: one; News: 5 hrs wkly. Target aud: Adults 18-49. ♦William C. Nolan, CEO & pres; Edwin Alderson, exec VP; Paul J. Starr, VP & gen mgr; Harry Harlan, sls dir & adv dir; Dawn Morgan, news dir; Ken Bass, engrg dir; Cathy Cox, traf mgr.

***KLIT(FM)**—Not on air, target date: unknown: 90.9 mhz; 50 kw vert. Ant 223 ft TL: N32 28 10 W103 09 36. Hrs open: 188 S. Bellevue, Suite 222, Memphis, TN, 38104. Phone: (901) 726-8970. Licensee: Broadcasting for the Challenged Inc. ♦George S. Flinn Jr., pres.

KLMA(FM)— November 1993: 96.5 mhz; 13 kw. Ant 459 ft TL: N32 46 08 W103 07 00. Hrs open: 24 Box 457, 108 S. Willow, 88240. Phone: (505) 391-9650. Fax: (505) 397-9373. Web Site: www.klmaradio.com. Licensee: Ojeda Broadcasting Inc. Group owner: Ojeda Broadcasting Inc. Law Firm: Mullin, Rhyne, Emmons & Topel. Format: Sp. Target aud: Hispanic. ♦Hermilo Ojeda, CEO, gen mgr, gen sls mgr & mktg mgr; Pearl Ojeda, pres; Letiicia Ojeda, prom mgr.

***KOBH(FM)**—Not on air, target date: unknown: 91.7 mhz; 250 w. Ant 157 ft TL: N32 42 48 W103 05 28. Hrs open: Drawer 2440, Tupelo, MS, 38803. Phone: (662) 844-8888. Licensee: Abundant Life Broadcasting. ♦Tamara Durham, VP.

KPER(FM)— August 1965: 95.7 mhz; 36 kw. 255 ft TL: N32 43 28 W103 09 03. Stereo. Hrs opn: Box 5967, 88241-5967. Secondary address: 1423 W. Bender St. 88240. Phone: (505) 393-1551. Fax: (505) 397-6088. E-mail: kper@basinbroadband.com Web Site: www.hobbsradio.com. Licensee: Noalmark Broadcasting Corp. (group owner; acq 1-99). Population served: 100,000 Format: Country. Target aud: 25-54. ♦Al Lobeck, gen mgr & gen sls mgr; Brant Swisher, news dir; Ken Bass, chief of engrg.

KYKK(AM)— July 17, 1971: 1110 khz; 5 kw-D. TL: N32 48 59 W103 13 56. Hrs open: 6 AM-sunset Box 5967, 88241. Secondary address: 1423 W. Bender Blvd. 88240. Phone: (505) 393-1551. Fax: (505) 397-6088. E-mail: kykk@hobbsradio.com Web Site: www.hobbsradio.com. Licensee: Noalmark Broadcasting Corp. (group owner; acq 8-8-77). Population served: 140,000 Format: News/talk, sports. News staff: one; News: 10 hrs wkly. Target aud: 25-54; men & women. ♦William Nolan, pres; Al Lobeck, gen mgr & gen sls mgr; Dawn Morgan, news dir; Ken Bass, chief of engrg.

KZOR(FM)—Co-owned with KYKK(AM). March 1975: 94.1 mhz; 100 kw. 400 ft TL: N32 48 59 W103 13 56. Stereo. Web Site: www.1radiosquare.com.140,000 Format: Hot adult contemp. Target aud: 18-44; female. ♦William Nolan, CEO; Paul J. Starr, opns VP & mktg VP; Bill Rediker, prom VP.

Hurley

KSIL(FM)— 2002: 105.5 mhz; 23 kw. Ant 1,063 ft TL: N32 50 40 W108 14 19. Stereo. Hrs opn: 24 2700 Hwy. 180 E., Silver City, 88061. Phone: (505) 534-1055. Fax: (505) 534-1400. E-mail: info@ksilradio.com Web Site: www.ksilradio.com. Licensee: James S. Bumpous dba Yellow Dog Radio. Population served: 25,000 Format: Div. News: 5 hrs wkly. Target aud: 25-54. Spec prog: Su AM relg progmg. ♦Keven McCauley, gen mgr; Misty Menelos, progmg.

Jal

KPZA-FM— Nov 1, 1998: 103.7 mhz; 100 kw. Ant 371 ft TL: N32 25 53 W103 09 08. Hrs open: 24 KIXN(FM)KPZA-(FM)-KYKK(AM)-KZOR(FM), 619 N. Turner St., Hobbs, 88240. Phone: (505) 397-4969. Fax: (505) 393-4310. E-mail: paul@1radiosquare.com Web Site: www.1radiosquare.com. Licensee: Noalmark Broadcasting Corp. (group owner; (acq 5-29-98; $10,000 for CP). Population served: 100,000 Format: Sp. News staff: one; News: 5 hrs wkly. Target aud: Hispanic. ♦William C. Nolan, CEO & pres; Edwin Alderson, sr VP; Paul J. Starr, gen mgr & opns VP; Tony Guerrero, stn mgr & news rptr; Harry Harlan, sls dir & adv dir; Cathy Cox, traf mgr.

Kirtland

KAZX(FM)— 1999: 102.9 mhz; 100 kw. Ant 1,007 ft TL: N36 48 52 W107 53 32. Hrs open: 200 E. Broadway, Farmington, 87401. Phone: (505) 325-1716. Fax: (505) 325-6797. Web Site: www.star1029.com. Licensee: Capstar TX L.P. Group owner: Clear Channel Communications Inc. (acq 12-19-00; $1.26 million). Format: CHR. ♦Bill Kruger, gen mgr.

La Luz

KRSY-FM—Licensed to La Luz. See Alamogordo

Las Cruces

KGRT-FM— Sept 8, 1966: 103.9 mhz; 6 kw. 151 ft TL: N32 18 33 W106 49 24. Stereo. Hrs opn: 24 Box 968, 88004. Secondary address: 1355 E. California St. 88001. Phone: (505) 525-9298. Fax: (505) 525-9419. E-mail: radiolc@kgrt.com Web Site: www.kgrt.com. Licensee: Sunrise Broadcasting Inc. (acq 12-30-88; with co-located AM; FTR: 12-19-88). Population served: 200,000 Natl. Network: ABC. Format: Country. News staff: 2; News: 2 hrs wkly. Target aud: 25-54; adults. ♦Allen Lumeyer, VP & gen mgr; Tamara Blasser, natl sls mgr; Ernesto Garcia, progmg dir; Sheila Kirsch, mus dir.

KSNM(AM)—Co-owned with KGRT-FM. Dec 15, 1955: 570 khz; 5 kw-D, 155 w-N. TL: N32 18 33 W106 49 24.24 Web Site: www.ksnm570.am.1,000,000 Natl. Network: ABC. Format: Soft adult contemp. News staff: 2; News: 4 hrs wkly. Target aud: 25 plus; adults. Spec prog: Talk show.

KHQT(FM)— Dec 12, 1974: 103.1 mhz; 1 kw. Ant 551 ft TL: N32 24 18 W106 45 41. Stereo. Hrs opn: 24 Box 968, 88004. Secondary address: 1355 E. California 88001. Phone: (505) 525-9298. Fax: (505) 525-9419. E-mail: radiolc@kgrt.com Web Site: www.hot103.fm. Licensee: Richardson Commercial Corp. (acq 10-2-2001; $1,650,000 with KKVS(FM) Truth or Consequences). Population served: 200,000 Law Firm: Fletcher, Heald & Hildreth. Format: CHR. News staff: 2; News: one hr wkly. Target aud: 18-34; general. ♦Allen Lumeyer, VP & gen mgr; Ernesto Garcia, opns mgr; Tamara Blaeser, natl sls mgr.

KKVS(FM)—See Truth or Consequences

***KMBN(FM)**— 2000: 89.7 mhz; 500 w. Ant 171 ft TL: N32 16 41 W106 54 39. Hrs open: Box 16691, 88004. Phone: (505) 521-8053. Fax: (505) 521-8053 (505) 521-8053. E-mail: kmbn@moody.edu Web Site: www.kmbn.org. Licensee: Moody Bible Institute of Chicago. Group owner: The Moody Bible Institute of Chicago Format: Christian. ♦John Powell, gen mgr.

KMVR(FM)—See Mesilla Park

KOBE(AM)— April 1947: 1450 khz; 1 kw-U. TL: N32 18 07 W106 48 08. Hrs open: 24 Drawer 1838, 88004. Secondary address: 1832 W. Amador 88005. Phone: (505) 526-2496. Fax: (505) 523-3918. E-mail: kmvr-kobe@totacc.com Web Site: kobeam1450.com. Licensee: Bravo Mic Communications II LLC. (acq 1-24-2007; $1.9 million with KMVR(FM) Mesilla Park). Population served: 167,000 Natl. Network: CBS. Format: News/talk, sports. News staff: one; News: 25 hrs wkly. Target aud: 25 plus. ♦Larry Edwards, gen mgr; Amanda Riordan, opns VP, progmg dir, news dir & disc jockey; Keith Lamonica, chief of engrg & disc jockey.

***KRUC(FM)**— March 1998: 88.9 mhz; 500 w. 197 ft TL: N32 16 41 W106 54 39. Hrs open: 24 5120 Prince Edward Ave., El Paso, TX, 79924. Phone: (915) 544-9192. Web Site: www.wrn-rcm.org. Licensee: World Radio Network Inc. (group owner) Format: Sp relg.

***KRUX(FM)**— Sept 20, 1989: 91.5 mhz; 1 kw. -194 ft TL: N32 17 03 W106 45 00. Hrs open: 7 AM-2 AM Box 30004, Corbett Ctr., 88003. Phone: (505) 646-4640. Fax: (505) 646-5219. E-mail: infor@krux.fm Web Site: www.kruxradio.com. Licensee: Board of Regents New Mexico State University. Format: Var. News staff: one; News: 2 hrs wkly. Target aud: General. ♦Melissa Aguilera, gen mgr; Mathias Ortiz, prom dir; Adrian Perez, progmg dir; Bianca Villani, news dir; Art Fountain, chief of engrg.

***KRWG(FM)**— Oct 3, 1964: 90.7 mhz; 100 kw. 350 ft TL: N32 15 24 W106 58 34. Stereo. Hrs opn: 24 Box 3000, 88003. Phone: (505) 646-4525. Fax: (505) 646-1974. E-mail: krwgfm@nmsu.edu Web Site: krwgfm.org. Licensee: Regents of New Mexico State University. Population served: 250,000 Natl. Network: NPR, PRI. Law Firm: Wiley, Rein & Fielding. Wire Svc: AP Format: Class, jazz, news, SP. News staff: 2; News: 39 hrs wkly. Target aud: 18-60. Spec prog: Sp 10 hrs, bluegrass/folk 8 hrs wkly. ♦Colin Gromatzky, gen mgr; Carrie Hamblen, chief of opns; L. Ford Ballard, dev dir; Robert Nosbisch, news dir. Co-owned TV: *KRWG-TV affil

KXPZ(FM)— May 1994: 99.5 mhz; 100 kw. Ant 1,023 ft TL: N32 41 35 W107 04 06. Stereo. Hrs opn: 24 101 Perkins, 88005. Phone: (505) 527-1111. Fax: (505) 527-1100. E-mail: rocket@bravomic.com Web Site: www.rocket995.com. Licensee: Bravo Mic Communications LLC. (acq 4-18-2006; $1.4 million). Law Firm: Leventhal, Senter & Lerman. Format: Active rock. ♦Mike Smith, gen mgr; Edmundo Resendez, gen sls mgr; K.C. Counts, progmg dir; Glen Lessler, chief of engrg.

Las Vegas

KBAC(FM)— Nov 10, 1989: 98.1 mhz; 100 kw. Ant 1,036 ft TL: N35 22 20 W105 22 02. Stereo. Hrs opn: 24 2351 Sunset Blvd., Suite 170-218, Rocklin, CA, 95765. Phone: (916) 251-1600. Fax: (916) 251-1650. Licensee: Educational Media Foundation. Group owner: Clear Channel Communications Inc. (acq 7-6-2007; $1.5 million with KSFQ(FM) White Rock). ♦Mike Novak, sr VP.

***KEDP(FM)**— September 1968: Stn currently dark. 91.1 mhz; 72 w. Ant -215 ft TL: N35 35 46 W105 13 18. Hrs open: New Mexico Highlands Univ., Media Arts Dept., Studio 103, 87701. Phone: (505) 454-3238. Fax: (505) 454-3109. E-mail: martinezda@nmhu.edu Web Site: www.nmhu.edu. Licensee: Board of Regents, New Mexico Highlands University. Population served: 18,000 Format: Oldies rock mix. ♦Donna Martinez, gen mgr; David Chavez, mus dir; Doyle Hanschulz, engr.

KFUN(AM)— Dec 25, 1941: 1230 khz; 1 kw-U. TL: N35 35 48 W105 12 21. Hrs open: 24 Box 700, 87701. Phone: (505) 425-6766. Fax: (505) 425-6767. E-mail: jpbaca1946@yahoo.com Licensee: Meadows Media LLC. (acq 4-19-91; $400). Population served: 20,000 Format: Sp, C&W. Target aud: General. ♦Joseph Baca Jr., pres, gen mgr, gen sls mgr, news dir & sls; Loretta Baca, traf mgr.

KLVF(FM)—Co-owned with KFUN(AM). June 19, 1973: 100.7 mhz; 10 kw. -77 ft TL: N33 35 48 W105 12 21. Stereo. 24 Format: Adult contemp. Target aud: 17-40.

KMDZ(FM)— 2000: 96.7 mhz; 4.4 kw. Ant 380 ft TL: N35 36 16 W105 15 35. Hrs open: 24 Sangre de Cristo Broadcasting Co., 304 S. Grand, 87701. Phone: (505) 425-5669. Fax: (505) 425-3557. E-mail: mattmartinez@knmx.com Licensee: Sangre de Cristo Broadcasting Co. Format: Classic rock. ♦Matt Martinez, gen mgr.

KNMX(AM)— Oct 1, 1980: 540 khz; 5 kw-D, DA. TL: N35 34 25 W105 10 17. Hrs open: Sunrise-sunset 304 S. Grand Ave., 87701. Phone: (505) 425-3555. Fax: (505) 425-3557. Licensee: Sangre de Cristo Broadcasting Co. (acq 9-26-96; $235,000). Population served: 400,000 Format: Sp, news/talk. News staff: one; News: 15 hrs wkly. Target aud: 25-55; Hispanic, Anglo. ♦Matt Martinez, prom dir; Matt C. Martinez, pres, gen mgr & progmg dir; John Chichester, chief of engrg.

Lordsburg

KPSA-FM— July 4, 1986: 97.7 mhz; 250 w. Ant -134 ft TL: N32 20 57 W108 42 18. Stereo. Hrs opn: 24 Box 2577, Silver City, 88062. Phone: (505) 538-3396. Fax: (505) 388-5000. E-mail: kpsa@silvercityradio.com Web Site: www.silvercity.com. Licensee: SkyWest Licenses New Mexico LLC. Group owner: Runnels Broadcasting System L.L.C. (acq 6-1-2006; grpsl). Format: Classic rock. ♦Sabrina Park, gen mgr; Jim Meyers, stn mgr & progmg dir; Patrice Loretta, dev dir; Steve Mull, chief of engrg.

Los Alamos

KABG(FM)— June 1956: 98.5 mhz; 100 kw. 1,781 ft TL: N35 53 08 W106 23 14. (CP: 100 kw, ant 1,906 ft.). Stereo. Hrs opn: Box 30925, Albuquerque, 87190. Phone: (505) 878-0980. Fax: (505) 878-0098. Web Site: www.bigoldies.net. Licensee: AGM-Nevada L.L.C. Population served: 840,000 Format: Oldies. Target aud: 25-54; affluent, upscale male professionals. ♦Scott Hutton, gen mgr; Scott Sherwood, progmg dir.

KQBA(FM)— Mar 1, 1998: 107.5 mhz; 100 kw. Ant 298 ft TL: N36 01 34 W105 48 18. Hrs open: 2502C Camino Entrada, Santa Fe, 87505. Phone: (505) 471-1067. Fax: (505) 473-2667. Licensee: Hutton Broadcasting LLC (acq 12-4-2000; $1 million). Format: Country. Target aud: 18-49; male. ♦Scott Hutton, gen mgr.

KRSN(AM)— Dec 9, 1949: 1490 khz; 1 kw-U. TL: N35 53 46 W106 17 21. Hrs open: 24 145 Central Park Square, 87544. Phone: (505) 663-1490. Fax: (505) 663-0011. E-mail: info@krsnam1490.com Web Site: www.krsnam1490.com. Licensee: Gillian Sutton (acq 1997). Population served: 150,000 Natl. Network: CBS, Westwood One. Format: Community News, local sports & music. News staff: one; News: 25 hrs wkly. Target aud: 35 plus; well educated, affluent. Spec prog: Big Band 8 hrs, jazz 2 hrs wkly. ♦Gillian Sutton, CEO; David Sutton, chief of opns.

KZNM(FM)— Mar 19, 1987: 106.7 mhz; 15.5 kw. Ant 1,948 ft TL: N35 47 15 W106 31 35. Stereo. Hrs opn: 24 4125 Carlisle N.E., Albuqerque, 87107. Phone: (505) 878-0980. Fax: (505) 878-0098. Licensee: A.G.M.-Nevada L.L.C. Group owner: American General Media (acq 8-9-00; grpsl). Population served: 103,000 Format: Sp var. News: 7 hrs wkly. Target aud: 25-54; mainstream audience with all socio-economic cells represented. ♦Scott Hutton, gen mgr.

Los Lunas

KAGM(FM)— January 1995: 106.3 mhz; 100 kw. Ant 656 ft TL: N34 48 51 W106 50 29. Stereo. Hrs opn: 24 4125 Carlisle Blvd. N.E., Albuquerque, 87107-4806. Phone: (505) 878-0980. Fax: (505) 878-0098. Licensee: AGM-Nevada L.L.C. Group owner: American General Media (acq 12-22-97; grpsl). Format: Classic country. Target aud: 18-49. Spec prog: Club mix 18 hrs wkly. ♦Scott Hutton, gen mgr & mktg mgr; Matt Rader, prom dir.

KIOT(FM)— July 6, 1981: 102.5 mhz; 20 kw. Ant 4,159 ft TL: N35 12 55 W106 27 02. Stereo. Hrs opn: 24 8009 Marble Ave. N.E., Albuquerque, 87110-7901. Phone: (505) 262-1142. Fax: (505) 254-7106. Licensee: Univision Radio License Corp. Group owner: Univision Radio (acq 9-22-2003; grpsl). Population served: 50,000 Law Firm: Jones, Waldo, Holbrook & McDonough. Format: Classic rock. News staff: one. Target aud: 25-49; hip adults who like diversity & have disposable income. Spec prog: Gospel 4 hrs wkly. ♦Chuck Morgan, gen mgr.

Los Ranchos de Albuquerque

KALY(AM)—Licensed to Los Ranchos de Albuquerque. See Albuquerque

KTBL(AM)—Licensed to Los Ranchos de Albuquerque. See Albuquerque

Lovington

KLEA(AM)— Dec 25, 1952: 630 khz; 500 w-D, 69 w-N. TL: N32 56 30 W103 19 12. Hrs open: 24 Box 877, Country Club Rd., 88260. Phone: (505) 396-2244. Fax: (505) 396-3355. E-mail: klea@valornet.com Web Site: www.107oldies.com. Licensee: Lea County Broadcasting Co. Population served: 220,000 Format: Soft hits, Fox sports weekends. News staff: one; News: 12 hrs wlky. Target aud: 25-54. Spec prog: Relg 3 hrs wkly. ♦Keith Kelly, progmg dir; Susan Coe, gen mgr, gen sls mgr & news dir; Rita Niccum, chief of engrg; Annette Giese, traf mgr.

KLEA-FM— October 1965: 101.7 mhz; 25 kw. 280 ft TL: N32 56 30 W103 19 12. Stereo. 24 Web Site: www.107oldies.com.150,000 Format: Oldies. News staff: one; News: 10 hrs wkly. ♦Susan Coe, exec VP.

***KYCV(FM)**—Not on air, target date: unknown: 91.3 mhz; 6 kw. Ant 1,627 ft TL: N33 03 20 W103 49 12. Hrs open: 9019 N. West Ln., Stockton, CA, 95210. Phone: (209) 477-3690. Fax: (209) 477-2762. E-mail: kycc@kycc.org Web Site: www.kycc.org. Licensee: Your Christian Companion Network Inc. ♦Shirley Garner, gen mgr.

Magdalena

KANM(FM)—Not on air, target date: unknown: 95.9 mhz; 100 kw. Ant 981 ft TL: N34 30 39 W107 13 16. Hrs open: 5842 Westslope Dr., Austin, TX, 78731. Phone: (512) 467-0643. Licensee: Matinee Radio LLC. ♦Robert Walker, pres.

Maljamar

***KMTH(FM)**— Feb 14, 1985: 98.7 mhz; 100 kw. Ant 710 ft TL: N32 54 55 W103 46 31. Stereo. Hrs opn: 24
Rebroadcasts KENW-FM Portales 100%.
Eastern New Mexico Univ., 52 Broadcast Ctr., Portales, 88130. Phone: (505) 562-2112. Fax: (505) 562-2590. E-mail: kenwfm@enmu.edu Web Site: www.kenw.org. Licensee: Eastern New Mexico University. Population served: 350,000 Natl. Network: NPR, PRI. Law Firm: Dow, Lohnes & Albertson. Wire Svc: AP Format: Class, btfl mus, news. News staff: one; News: 41 hrs wkly. Target aud: General. ♦Steven G. Gamble, pres; Ronnie Birdsong, VP; Duane W. Ryan, gen mgr; Shannon Hearn, opns dir; Carla Howard, dev dir; Virginia McReynolds, mktg dir; James Lee, news dir; Jeff Burmeister, engrg dir; Bob Scott, engrg mgr.

KWMW(FM)— Jan 17, 1990: 105.1 mhz; 100 kw. 917 ft TL: N32 52 40 W103 41 13. Hrs open: Box 2010, Ruidoso Downs, 88346. Secondary address: 916 Ave. D, Levington 88260. Phone: (505) 396-0499. Fax: (505) 396-8349. E-mail: kruikwmw@trailnet.com Licensee: M.T.D. Inc. Group owner: MTD Inc. Format: Country. ♦Tim Keithley, gen mgr; Will Rooney, stn mgr.

Mentmore

***KPKJ(FM)**—Not on air, target date: unknown: 88.5 mhz; 1.3 kw. Ant 491 ft TL: N35 33 36 W109 06 30. Hrs open: 4002 N. 3300 E., Twin Falls, ID, 83301. Phone: (208) 734-6633. Fax: (208) 736-1958. Web Site: www.csnradio.com. Licensee: CSN International. ♦Mike Kestler, pres.

Mesilla Park

KMVR(FM)— June 1, 1974: 104.9 mhz; 3 kw. -32 ft TL: N32 18 07 W106 48 08. Stereo. Hrs opn: 24 Drawer 1838, Las Cruces, 88005. Secondary address: 1832 W. Amador, Las Cruces 88004. Phone: (505) 526-2496. Fax: (505) 523-3918. E-mail: kmvr-kobe@totacc.com Web Site: kmvrmagic105.com. Licensee: Bravo Mic Communications II LLC. (acq 1-24-2007; $1.9 million with KOBE(AM) Las Cruces). Population served: 167,000 Format: Hot adult contemp. Target aud: 18-54. ♦Larry Edwards, gen mgr, gen sls mgr & progmg dir; Keith Lamonica, chief of engrg.

Mesquite

***KELP-FM**— February 2004: 89.3 mhz; 680 w. Ant 66 ft TL: N32 09 42 W106 42 03. Stereo. Hrs opn: 24 6900 Commerce, El Paso, TX, 79915. Phone: (915) 779-0016. Fax: (915) 779-6641. E-mail: info@kelpradio.com Licensee: Sky High Broadcasting Inc. (acq 3-28-02). Population served: 174,267 Natl. Network: Salem Radio Network. Format: Christian. Target aud: 25-55 plus. ♦Arniold McClatchey, pres; Carmen Moody, traf mgr.

Milan

KQNM(AM)— Sept 1, 1989: 1100 khz; 250 w-D, 20 w-N. TL: N35 05 51 W107 52 19. Hrs open: 24 809 Wellesly N.E., Albuqueque, 87106. Phone: (505) 285-5598 Station. Phone: (505) 261-0130 Office. Licensee: Cibola Radio Co. (acq 7-12-99; $29,800). Population served: 22,000 Format: Active rock. Target aud: 25-54; upscale men. ♦Don Davis, pres.

***KXXQ(FM)**— June 1991: 100.7 mhz; 100 kw. Ant 1,361 ft TL: N35 28 07 W108 14 24. Stereo. Hrs opn: 24 Box 180, Tahoma, CA, 96142. Phone: (530) 584-5700. Fax: (530) 584-5705. Web Site: www.ihradio.org. Licensee: IHR Educational Broadcasting. (acq 5-31-2005; $450,000). Format: Relg, catholic. ♦Douglas M. Sherman, pres.

Pecos

KLBU(FM)— Aug 1, 2001: 102.9 mhz; 3.7 kw horiz. Ant 686 ft TL: N35 39 06 W105 33 15. Stereo. Hrs opn: 551C Cordova Rd., Santa Fe, 87505. Phone: (505) 984-1029. Fax: (505) 984-0880. Web Site: www.blu1029.com. Licensee: Blu Media LLC (acq 2-19-2004; $1.15 million). Law Firm: Wood, Maines & Brown, Chartered. Format: Modern adult contemp. ♦Courtney Jones, pres; Kerri Fama, gen mgr; Joann Orner, progmg dir.

KVSF-FM— 2004: 101.5 mhz; 25 kw. Ant 279 ft TL: N35 32 50 W105 45 54. Hrs open: Box 1863, Santa Fe, TX, 87504. Phone: (505) 438-7007. Fax: (505) 438-7007. Licensee: James S. Bumpous (acq 3-9-2004; grpsl). Format: Var. ♦James S. Bumpous, gen mgr.

Portales

***KENW-FM**— Oct 1, 1968: 89.5 mhz; 100 kw. Ant 590 ft TL: N34 15 08.11 W103 14 20.63. Stereo. Hrs opn: 24 Eastern New Mexico Univ., 52 Broadcast Ctr., 88130. Phone: (505) 562-2112. Fax: (505) 562-2590. E-mail: kenwfm@enmu.edu Web Site: www.kenw.org. Licensee: Eastern New Mexico University. Population served: 350,000 Natl. Network: NPR, PRI. Rgnl. Network: N.M. Pub. Format: Btfl mus, class, news/talk. News staff: one; News: 41 hrs wkly. Target aud: General. ♦Steven G. Gamble, pres; Ronnie Birdsong, VP; Duane W. Ryan, gen mgr; Shannon Hearn, opns dir; Carla Howard, dev dir; James Lee, news dir; Jeff Burmeister, engrg dir; Bob Scott, engrg mgr. Co-owned TV: *KENW-TV affil

KSEL(AM)— February 1950: 1450 khz; 1 kw-U. TL: N34 11 51 W103 19 24. Hrs open: 24 42437 US 70, 88130. Phone: (505) 359-4649. Fax: (505) 359-0724. Licensee: Rooney Moon Broadcasting Inc. (group owner; acq 7-15-02; grpsl). Population served: 100,000 Natl. Network: CNN Radio. Format: News/talk. News: 168 hrs wkly. Target aud: 35+. ♦Duffy Moon, opns mgr & prom dir; Steve Rooney, pres, gen mgr, sls dir & progmg dir; Jeff Burmeister, chief of engrg; Lisa Schmidt, traf mgr.

KSEL-FM— March 1980: 95.3 mhz; 6 kw. Ant 298 ft TL: N34 11 51 W103 19 24. Stereo. 150,000 Natl. Network: CNN Radio. Law Firm: Garvey, Schubert & Barer. Format: Country. Target aud: 18-54. Spec prog: Farm 4 hrs wkly. ♦Lisa Schmidt, traf mgr.

Questa

KLNN(FM)— 2006: 103.7 mhz; 51 kw. Ant -211 ft TL: N36 39 23 W105 37 57. Hrs open: Box 1844, Taos, 87571. Phone: (505) 758-5826. Fax: (505) 758-8430. Web Site: www.luna1037.com. Licensee: West Waves Inc. (acq 2-16-2006; $68,160 for CP). Format: Adult contemp. ♦David W. Rahn, pres; Dave Noll, gen mgr.

Ramah

***KTDB(FM)**— Apr 24, 1972: 89.7 mhz; 15 kw. 300 ft TL: N34 57 59 W108 25 31. (CP: Ant 288 ft.). Hrs opn: 5 AM-11 PM Box 40, B.I.A. Rt. 125, Pine Hill, 87357. Phone: (505) 775-3215. Fax: (505) 775-3551. Web Site: www.rnsbinc.com. Licensee: Ramah Navajo School Board Inc. Population served: 80,000 Natl. Network: NPR. Format: C&W, cultural info, educ. Target aud: General; Native American. Spec prog: Navajo. ♦Barbara Maria, gen mgr & prom mgr; Irene Beaver, progmg dir; Earl Ericcho, news dir; Bernard J. Bustos, chief of engrg.

Raton

KBKZ(FM)— Dec 20, 2001: 96.5 mhz; 5.4 kw. Ant 968 ft TL: N36 59 33 W104 28 24. Hrs open: 100 Fisher Dr., Trinidad, CO, 81082. Phone: (719) 846-3355. Fax: (719) 846-4711. E-mail: kcrt@comcast.net Licensee: Phillips Broadcasting Co. Inc. Group owner: Phillips Broadcasting Inc. Population served: 12,000 Format: Country. ♦Lory Phillips, gen mgr.

KRTN(AM)— 1948: 1490 khz; 1 kw-U. TL: N36 53 10 W104 26 35. Hrs open: 24 Box 638, 1128 State St., 87740. Phone: (505) 445-3652. Fax: (505) 445-2911. E-mail: krtn@raton.com Licensee: Enchanted Air Inc. (acq 5-31-2005; $750,000 with co-located FM). Population served: 25,000 Format: Adult contemp. Target aud: General. ♦Jim Roper, stn mgr & prom mgr; Bill Donati, progmg dir & mus dir; Flo Roper, progmg dir, traf mgr & women's int ed; Jim Veltri, chief of engrg; Billy Donati, disc jockey.

KRTN-FM— April 1982: 93.9 mhz; 26 kw. Ant 1,446 ft TL: N36 40 59 W104 24 50. Stereo. 24 Format: Oldies. ♦Billy Donoti, disc jockey; Mike Higgins, disc jockey; Robbie Ley, disc jockey.

Red River

***KRDR(FM)**— 2002: 90.1 mhz; 3.2 kw vert. Ant 718 ft TL: N36 41 25 W105 33 43. Hrs open: Box 788, Questa, 87556. Phone: (505) 586-1919. Fax: (505) 586-2332. E-mail: krdr@newmex.com Web Site: www.krdr.com. Licensee: Red River Radio Inc. Format: Classic rock, oldies. ♦Lynn Nolen, stn mgr & progmg dir; Mike Nolen, gen mgr & chief of engrg.

Reserve

KZXQ(FM)— 2005: 104.5 mhz; 500 w. Ant -751 ft TL: N33 42 35 W108 45 56. Hrs open: Box 41497, Mesa, AZ, 85274. Phone: (505) 533-6100. Fax: (505) 533-6103. Licensee: New Star Broadcasting LLC (acq 12-18-2002; $80,000). Format: Talk, news. ♦Karey Barbee, pres & gen mgr.

Rio Rancho

KQBT(FM)— Nov 2, 1984: 101.7 mhz; 3.2 kw. Ant 99 ft TL: N35 11 35 W106 28 15. Hrs open: 8009 Marble Ave. N.E., Albuquerque, 87110. Phone: (505) 262-1142. Fax: (505) 254-7106. Licensee: Univision Radio License Corp. Group owner: Univision Radio (acq 9-22-2003; grpsl). Population served: 90,000 Natl. Rep: Christal. Format: New adult contemp/smooth jazz. ♦Chuck Morgan, gen mgr.

Roswell

KBCQ(AM)— May 1947: 1230 khz; 1 kw-U. TL: N33 25 00 W104 30 40. Hrs open: 24 Box 670, 88201. Secondary address: 5206 W. 2nd St. 88201. Phone: (505) 622-0290. Fax: (505) 622-9041. E-mail: penny@roswellradio.org Web Site: www.roswellradio.org. Licensee: Roswell Radio Inc. Group owner: Roswell Radio Inc./Quay Broadcasters Inc. (acq 2-28-2003). Population served: 67,000 Natl. Network: CNN Radio. Format: Oldies. Target aud: 35-75. Spec prog: Talk 15 hrs wkly. ♦John M. Dunn, pres; J.R. Law, gen mgr; Penny Dunn, gen sls mgr.

KBCQ-FM— Oct 15, 1977: 97.1 mhz; 100 kw. 300 ft TL: N33 24 05 W104 22 45. Stereo. Hrs opn: 24 Box 670, 88202. Phone: (505) 622-6450. Fax: (505) 622-9041. E-mail: penny@roswellradio.org Web Site: www.roswellradio.org. Licensee: Roswell Radio Inc. Group owner: Roswell Radio Inc./Quay Broadcasters Inc. (acq 11-2000; grpsl). Population served: 200,000 Law Firm: Fletcher, Heald & Hildreth. Format: CHR. News staff: one. Target aud: 18-49. ♦John Dunn, CEO & exec VP; J.P. Law, stn mgr; Gabe Mendez, prom dir; Gary Babcock, engrg VP; Penny Dunn, sls.

KBIM(AM)— May 1953: 910 khz; 5 kw-D, 500 w-N, DA-N. TL: N33 26 26 W104 31 35. Hrs open: 24 Box 1953, 88202. Secondary address: 1301 N. Main 88201. Phone: (505) 623-9100. Fax: (505) 623-4775. E-mail: kbim@dfn.com Web Site: kbim-roswell.com. Licensee: King Broadcasting Co. Inc. Population served: 200,000 Natl. Network: ABC. Format: News/talk. News: 16 hrs wkly. Target aud: 25-54; upscale male & active working female. ♦Betty King, chmn & gen sls mgr; John King, pres, gen mgr, natl sls mgr, rgnl sls mgr, progmg dir & progmg mgr; Michael Liles, opns dir & chief of engrg.

KBIM-FM— June 1959: 94.9 mhz; 100 kw. 1,880 ft TL: N33 03 20 W103 49 12. Stereo. 24 Web Site: kbim-roswell.com.187,000 Law Firm: Dow, Lohnes & Albertson. Format: Adult contemp. News: 16 hrs wkly. Target aud: 25-54. ♦Betty King, sls dir & progmg dir; John King, CEO & gen sls mgr.

KCKN(AM)— Dec 20, 1965: 1020 khz; 50 kw-U, DA-2. TL: N33 27 53 W104 29 58. Hrs open: 24 Box 220, 88202. Phone: (505) 622-0658. Fax: (505) 622-0852. E-mail: kckn@swwmail.net Web Site: kckn1020.com. Licensee: JCE Licenses L.L.C. Group owner: James Crystal Inc. (acq 2000; $2.5 million). Population served: 500,000 Natl. Network: AP Network News, Jones Radio Networks. Natl. Rep: Salem. Law Firm: John Wells King. Format: Classic Country/Religious. News staff: 2; News: 5 hrs wkly. Target aud: 25-54; adult professionals. ♦Jim Hilliard Jr., pres; Jerry Kiefer, gen mgr, prom dir & progmg dir; Don Niccum, opns mgr; Bob Souza, gen sls mgr; Bob Williams, rgnl sls mgr & mus dir; Doug Michaels, news dir; Kathi Silvas, traf mgr.

KCRX(AM)— Mar 15, 1927: 1430 khz; 5 kw-D, 1 kw-N, DA-N. TL: N33 26 11 W104 36 18. Hrs open: 24 Box 2052, 88202-2052. Secondary address: 200 W. 1st St. 88203-2052. Phone: (505) 622-1432. Fax: (505) 622-1432. E-mail: kcrx@digicominc.net Web Site: oldiesradioonline.com. Licensee: Rosendo Casarez Jr. Population served: 300,000 Natl. Network: CBS, Westwood One. Format: Good time rock and roll oldies. News staff: one. Target aud: 35 plus. ♦Rosendo Casarez Jr., pres & opns dir.

***KELT(FM)**—Not on air, target date: unknown: 91.7 mhz; 55 kw vert. Ant 174 ft TL: N33 29 12 W104 29 48. Hrs open: 2351 Sunset Blvd., Suite 170-218, Rocklin, CA, 95765. Phone: (916) 251-1600. Fax: (916) 251-1650. Licensee: Educational Media Foundation. (acq 7-23-2007; grpsl). ♦Mike Novak, sr VP.

KEND(FM)— May 30, 1990: 106.5 mhz; 52 kw. 135 ft TL: N33 23 05 W104 43 22. Stereo. Hrs opn: 24 1405 W Second St., 88201. Phone: (505) 625-2098. Fax: (505) 622-3877. Web Site: www.themix1065.com. Licensee: Pecos Valley Broadcasting Co. (acq 4-1-2007; $500,000). Format: AOR. News: 6 hrs wkly. Target aud: 18-34; upscale adults. Spec prog: Morning Show, 10 hrs; Afternoon Attitude, 15 hrs; After Dark, 4 hrs; Flashback Program, 3 hrs; CD Feature of the Week, 1 hr. ♦Mike Winters, gen mgr; Sean McKellips, traf mgr.

KMOU(FM)— August 1992: 104.7 mhz; 100 kw. Ant 328 ft TL: N33 24 49 W104 22 49. Stereo. Hrs opn: 24 Box 670, 88202-0670. Secondary address: 5206 W. 2nd St. 88203. Phone: (505) 625-6450. Fax: (505) 622-9041. E-mail: penny@roswellradio.org Web Site: www.,roswellradio.org. Licensee: Roswell Radio Inc. Group owner: Roswell Radio Inc./Quay Broadcasters Inc. (acq 11-22-2000; $750,000). Population served: 200,000 Law Firm: Fletcher, Heald & Hildreth. Format: Country. News staff: 2; News: 12 hrs wkly. ♦John M. Dunn, CEO; Joe Fink, prom dir; Penny Dunn, sls.

***KQAI(FM)**— 2007: 89.1 mhz; 2 kw vert. Ant 207 ft TL: N33 23 36 W104 37 27. Hrs open:
Rebroadcasts KLRD(FM) Yucaipa, CA 100%.
2351 Sunset Blvd., Suite 170-218, Rocklin, CA, 95765. Phone: (916) 251-1600. Fax: (916) 251-1650. Web Site: www.air1.com. Licensee: Educational Media Foundation. (acq 9-22-2005; $40,000 for CP). Natl. Network: Air 1. Format: Alternative rock, div. ♦Richard Jenkins, pres.

KRDD(AM)— 1963: 1320 khz; 1 kw-D. TL: N33 24 14 W104 28 12. Hrs open: Box 1615, NC, 88201. Phone: (505) 623-8111. E-mail: krddam@yahoo.com Licensee: Media Mining Group LLC (acq 5-11-2004). Population served: 100,000 Format: Sp. ♦Carlos Espinoza, pres, gen mgr & prom mgr; Monica Cardeas, gen sls mgr; Ramiro Vasquez, progmg dir.

***KRLU(FM)**—Not on air, target date: unknown: 90.1 mhz; 2 kw vert. Ant 394 ft TL: N33 21 47 W104 38 11. Hrs open: 24 2351 Sunset Blvd., Suite 170-218, Rocklin, CA, 95765. Phone: (916) 251-1600. Fax: (916) 251-1650. E-mail: klove@klove.com Web Site: www.klove.com. Licensee: Educational Media Foundation. Group owner: EMF Broadcasting. Law Firm: Shaw Pittman. Format: Contemp Christian. News staff: 3. Target aud: 25-44; Judeo Christian, female. ♦Richard Jenkins, pres; Mike Novak, VP; Keith Whipple, dev dir; David Pierce, progmg mgr; Ed Lenane, news dir; Sam Wallington, engrg dir; Karen Johnson, news rptr; Marya Morgan, news rptr; Richard Hunt, news rptr.

KSFX(FM)— Mar 15, 1991: 100.5 mhz; 100 kw. 122 ft TL: N33 28 54 W104 39 12. Hrs open: 24 Box 670, 88201. Secondary address: 5206 W. 2nd St. 88201. Phone: (505) 622-6450. Fax: (505) 622-9041. E-mail: penny@roswellradio.org Web Site: roswellradio.org. Licensee: Roswell Radio Inc. Group owner: Roswell Radio Inc./Quay Broadcasters Inc. (acq 11-2000; grpsl). Population served: 200,000. Law Firm: Fletcher, Hearld & Hildreth. Format: Classic Hits. News staff: 2; News: 12 hrs wkly. Target aud: 25-49; mainstream upscale. ♦John Dunn, CEO; Gary Babock, exec VP; John M. Dunn, gen mgr; J.R. Law, stn mgr & sls dir; Tony Clayton, prom dir; Chris Johnson, news dir; Gary Babcock, engrg VP & chief of engrg; Penny Dunn, sls.

***KWFL(FM)**— Dec 21, 1989: 99.5 mhz; 6.1 kw. Ant 459 ft TL: N33 21 47 W104 38 11. Stereo. Hrs opn: 24 3801 Eubank NE, Albuquerque, 87111. Phone: (800) 776-1050. Fax: (505) 296-6262. E-mail: flrradio@flc.org Web Site: www.flc.org/flr/kwfl. Licensee: Family Life Broadcasting System. Group owner: Family Life Communications Inc. (acq 3-24-2004; $1). Natl. Network: Moody. Format: Relg. Target aud: Christian community. ♦Randy L. Carlson, pres; Dan Rosecrans, gen mgr.

Ruidoso

KBUY(AM)— November 1959: 1360 khz; 5 kw-D, 199 w-N. TL: N33 19 35 W105 40 02. Hrs open: 24 Box 39, 88355. Secondary address: 1096 Mechen Dr., Suite 230 88345. Phone: (505) 258-2222. Fax: (505) 258-2224. E-mail: kwesradio@kwes.net Web Site: www.kwes.net. Licensee: Walton Stations New Mexico Inc. Group owner: Walton Stations (acq 10-22-82; $475,000 with co-located FM; FTR: 11-15-82)

Natl. Network: Fox News Radio. Format: Oldies. News staff: one; News: 14 hrs wkly. Target aud: 38 plus; 25-54 females. Spec prog: Sp 4 hrs wkly. ♦Steve Hall, progmg dir; Steve Swayze, chief of engrg; Gary Herron, traf mgr.

KWES-FM—Co-owned with KBUY(AM). 1982: 93.5 mhz; 25 kw. 58 ft TL: N33 23 12 W105 40 14. Stereo. 24 Web Site: www.kwes.net. Natl. Network: Jones Radio Networks. Format: C&W. News staff: one; News: 17 hrs wkly. Target aud: 18-54. ♦Steve Swayze, mus dir; Gary Herron, traf mgr.

KIDX(FM)— 2000: 101.5 mhz; 920 w. Ant 2,850 ft TL: N33 24 14 W105 46 56. Hrs open: Box 2010, 88346. Phone: (505) 258-9922. Fax: (505) 258-2363. E-mail: w105@trailnet.com Licensee: MTD Inc. (group owner) Format: Classic rock. ♦Tim Keithley, gen mgr.

KWES(AM)—Not on air, target date: unknown: 1450 khz; 1 kw-U. TL: N33 20 23 W105 39 54. Hrs open: Box 39, 88355-0039. Phone: (505) 258-2222. Fax: (505) 258-2224. Licensee: Walton Stations New Mexico Inc. ♦John Walton, pres.

***KYCT(FM)**—Not on air, target date: unknown: 91.3 mhz; 210 w. Ant 2,867 ft TL: N33 24 14 W105 46 55. Hrs open: 9019 N. West Ln., Stockton, CA, 95210. Phone: (209) 477-3690. Fax: (209) 477-2762. E-mail: kycc@kycc.org Web Site: www.kycc.org. Licensee: Your Christian Companion Network Inc. ♦Shirley Garner, gen mgr.

Ruidoso Downs

KRUI(AM)— April 1984: 1490 khz; 1 kw-U. TL: N33 19 17 W105 35 24. Hrs open: 24 1086 Mechem Dr., Ruidoso, 88345. Phone: (505) 258-9922. Fax: (505) 258-2363. E-mail: w105@trailnet.com Licensee: MTD Inc. (group owner; acq 12-88; FTR: 12-19-88). Natl. Network: Westwood One. Format: News, talk, sports. News: 14 hrs wkly. ♦Tim Keithley, gen mgr.

Santa Clara

KNUW(FM)— 1996: 95.1 mhz; 7.7 kw. 1,548 ft TL: N32 51 47 W108 14 28. Hrs open: 24 106 S. Bullard St., Silver City, 88061. Phone: (505) 534-8700. Phone: (505) 534-8701. Fax: (505) 534-8702. E-mail: knuw@zianet.com Licensee: Duran-Hill, Inc. Format: Sp. Target aud: General; Hispanic. ♦George H. Mesa, pres, gen mgr, opns VP, sls VP & progmg VP; Ken Bass, engrg VP.

Santa Fe

KABQ-FM— Nov 24, 1983: 104.1 mhz; 100 kw. Ant 1,876 ft TL: N35 46 50 W106 31 35. Stereo. Hrs opn: 24 5411 Jefferson St. N.E., Suite 100, Albuquerque, 87109. Phone: (505) 830-6400. Fax: (505) 830-6599. Licensee: Citicasters Licenses L.P. Group owner: Clear Channel Communications Inc. (acq 9-28-99; grpsl). Law Firm: Daniel Brenner. Format: Smooth jazz. News staff: one. Target aud: 25-54. ♦Chuck Hammond, VP & gen mgr; Bill May, chief of opns.

KHFM(FM)— Aug 15, 1965: 95.5 mhz; 19 kw. 1,850 ft TL: N35 53 08 W106 23 14. Stereo. Hrs opn: 24 4125 Carlisle N. E., Albuquerque, 87107. Phone: (505) 878-0980. Fax: (505) 889-0617. E-mail: lgold@americangeneralmedia.com Web Site: www.classicalkhfm.com. Licensee: AGM-Nevada L.L.C. Format: Class. Target aud: 18-49. ♦Kip Allen, progmg dir.

KJFA(FM)— Sept 28, 1985: 105.1 mhz; 100 kw. Ant 1,937 ft TL: N35 47 15 W106 31 35. Stereo. Hrs opn: 8009 Marble Ave. N.E., Albuquerque, 87110-7901. Phone: (505) 262-1142. Fax: (505) 254-7106. Licensee: Univision Radio License Corp. Group owner: Univision Radio (acq 9-22-2003; grpsl). Format: Mexican rgnl. Target aud: 35-54. ♦Chuck Morgan, gen mgr.

KKIM-FM—Listing follows KTRC(AM).

KKOB Exp Stn— 1986: 770 khz; 230 w-U. TL: N35 40 56 W105 58 21. Hrs open:
Rebroadcasts KKOB(AM) Albuquerque 100%.
500 4th St. N.W., Suite 500, Albuquerque, 87102. Phone: (505) 767-6700. Fax: (505) 767-6767. Web Site: www.770kkob.com. Licensee: Citadel Broadcasting Co. Natl. Rep: McGavren Guild. Format: MOR. ♦Milt McConnell, gen mgr & stn mgr; Dennis Logsdon, rgnl sls mgr; Glen Hebert, mktg dir & prom dir; Pat Frisch, opns mgr & progmg dir; Art Ortega, pub affrs dir; Mike Langner, chief of engrg.

KKSS(FM)— March 1969: 97.3 mhz; 94 kw. Ant 1,876 ft TL: N35 46 50 W106 31 55. (CP: 240 w, ant 85 ft. TL: N35 04 41 W106 35 06). Stereo. Hrs opn: 24 8009 Marble Ave. N.E., Albuquerque, 87110. Phone: (505) 262-1142. Fax: (505) 254-7106. Licensee: Univision Radio License Corp. Group owner: Univision Radio (acq 9-22-2003; grpsl). Population served: 600,000 Natl. Rep: D & R Radio. Format: CHR. Target aud: 18-34; Hispanic females. ♦Chuck Morgan, gen mgr.

KRZY-FM— Nov 2, 1983: 105.9 mhz; 100 kw. Ant 1,919 ft TL: N35 46 49 W106 31 34. Stereo. Hrs opn: 24 2725 Broadbent Pkwy. N.E., Suite F, Albuquerque, 87107. Phone: (505) 342-4141. Fax: (505) 344-8714. Licensee: Entravision Holdings LLC. Group owner: Entravision Communications Corp. (acq 3-14-2000; grpsl). Law Firm: Leventhal, Senter & Lerman. Format: Sp. Target aud: 18-34. ♦Margarita Wilder, gen mgr.

***KSFR(FM)**— Mar 16, 1990: 90.7 mhz; 3 kw. 199 ft TL: N35 40 41 W105 59 29. Stereo. Hrs opn: 24 6401 So. Richards Ave., 87508. Phone: (505) 428-1527. Fax: (505) 424-8938. E-mail: info@ksfr Web Site: ksfr.org. Licensee: Santa Fe Community College. Population served: 100,000 Format: News/talk, jazz. News staff: one; News: 14 hrs wkly. Target aud: General. Spec prog: American Indian 4 hrs, Sp 4 hrs wkly. ♦Dallas Dearmin, stn mgr; Tim Pemberton, opns dir & chief of engrg; Christine Lord, sls VP; John P. Greenspan, progmg dir; William Dupuy, news dir.

KSWV(AM)— June 1966: 810 khz; 5 kw-D. TL: N35 39 17 W106 00 05. Hrs open: Box 1088, 102 Taos St., 87504. Phone: (505) 989-7441. Fax: (505) 989-7607. Licensee: La Voz Broadcasting Co. (acq 12-20-90; $150,000). Format: Sp loc info & mus. Target aud: 25-54. ♦Celina V. Gonzales, pres; Anthony Gonzales, gen mgr; John Chidester, progmg dir & chief of engrg.

KTRC(AM)— 1935: 1260 khz; 5 kw-D, 1 kw-N. TL: N35 40 36 W105 58 21. Hrs open: 6 AM-midnight 2502 C Camino Entrada, 87505. Phone: (505) 471-1067. Fax: (505) 473-2667. Licensee: A.G.M.-Nevada L.L.C. Group owner: American General Media (acq 8-9-2000; grpsl). Population served: 155,000 Law Firm: Jones, Waldo, Holbrook & McDonough. Format: Progressive talk. Target aud: 35-64; involved affluent adults. ♦Scott Hutton, gen mgr.

KKIM-FM—Co-owned with KTRC(AM). 2000: 94.7 mhz; 100 kw. Ant 797 ft TL: N36 05 21 W106 01 41. (Acq 1996; $96,250). Format: Christian talk. ♦Dewey Moede, stn mgr.

KVSF(AM)— Feb 20, 1947: 1400 khz; 1 kw-U. TL: N35 41 16 W105 56 04. Hrs open: 2502 Camino Entrada, Suite C, 87505. Phone: (505) 471-1067. Fax: (505) 473-2667. Licensee: Hutton Broadcasting LLC Group owner: American General Media (acq 12-27-2005; $350,000). . Natl. Network: ESPN Radio. Format: Sports. ♦Edward B. Hutton, pres; Scott Hutton, gen mgr; Susan Olivares, gen sls mgr; Eileen Munroe, prom dir; Dave Anderson, chief of engrg.

Santa Rosa

KIVA(FM)— 2001: 95.9 mhz; 1.5 kw. Ant 118 ft TL: N34 56 47 W104 39 10. Hrs open:
Rebroadcasts KSSR(AM) Santa Rosa 100%.
HC 69 Box 78, 88435. Secondary address: 2818 Historic Rt. 66 88435. Phone: (505) 472-5777. Phone: (505) 472-3752. Fax: (505) 472-5777. Licensee: KNXX Inc. Format: Adult contemp, country, Sp. ♦Joseph Esquibel, gen mgr.

***KNLK(FM)**— 2004: 91.9 mhz; 100 w. Ant -26 ft TL: N34 57 20 W104 40 53. Hrs open: 2020 Coal Ave. S.E., Albuquerque, 87106. Phone: (505) 242-7163. E-mail: brasher@aps.edu Web Site: www.kanw.com. Licensee: Board of Education of the City of Albuquerque, NM. Population served: 10,000 Format: Sp. ♦Michael Brasher, gen mgr.

KSSR(AM)— Nov 2, 1960: 1340 khz; 1 kw-U. TL: N34 56 40 W104 39 00. Hrs open: 24 HC 69 Box 78, 88435. Secondary address: 2818 Historic Rt. 66 88435. Phone: (505) 472-5777. Fax: (505) 472-5777. E-mail: kssrradio@yahoo.com Licensee: Joseph M. Esquibel. (acq 1989; $50,000). Natl. Network: Westwood One. Format: Adult contemp, country, Sp. News staff: one; News: 16 hrs wkly. Target aud: General; 85% Hispanic, plus largely transient motorists. ♦Gabriel Esquibel, gen mgr.

Silver City

KSCQ(FM)— Nov 28, 1989: 92.9 mhz; 11.5 kw. Ant 1,023 ft TL: N32 50 40 W108 14 18. Stereo. Hrs opn: 24 Box 2577, 88062. Secondary address: 1560 N. Corbin St. 88061. Phone: (505) 388-4116. Fax: (505) 388-1759. E-mail: events@silvercityradio.com Web Site: theq929.com. Licensee: Skywest Media LLC (acq 10-31-2005; $330,000). Population served: 62,000 Format: Hot adult contemp. News: 3 hrs wkly. Target aud: 25-55; baby boomers, generation X. ♦Michael Rowse, news dir; Sabrinaa Pack, pub affrs dir; Sky West, LLC., chief of engrg.

Socorro

KMXQ(FM)— Jan 22, 1995: 92.9 mhz; 6 kw. -177 ft TL: N34 02 43 W106 54 21. Hrs open: 24 Box 699, 87801. Secondary address: 834 Hwy. 60 W. 87801. Phone: (505) 835-1286. Fax: (505) 835-2015. E-mail: kmxq@sdc.org Licensee: Lakeshore Media L.L.C. (acq 10-29-2002; $450,000). Population served: 70,000 Format: Country. News staff: 4; News: 167 hrs wkly. Target aud: 12 plus. Spec prog: Farm 2 hrs, talk one hr wkly. ♦Virgil Vigil, gen mgr, gen sls mgr, mktg mgr & prom dir; John Gonzales, prom mgr, progmg VP & engrg dir.

***KNMA(FM)**—Not on air, target date: unknown: 88.1 mhz; 7 kw vert. Ant 1,900 ft TL: N32 49 49 W105 53 25. Hrs open: CSN International, 4002 N. 3300 E., Twin Falls, ID, 83301. Phone: (208) 734-6633. Fax: (208) 736-1958. Licensee: CSN International. ♦Mike Kestler, pres; Don Mills, stn mgr.

***KVLK(FM)**— 2007: 89.5 mhz; 10 kw horiz. Ant 161 ft TL: N34 23 44 W107 00 42. Hrs open: 24
Rebroadcasts KLVR(FM) Santa Rosa, CA 100%.
2351 Sunset Blvd., Suite 170-218, Rocklin, CA, 95765. Phone: (916) 251-1600. Fax: (916) 251-1650. Web Site: www.klove.com. Licensee: Educational Media Foundation. Group owner: EMF Broadcasting. Natl. Network: K-Love. Format: Contemp Christian.

***KXFR(FM)**—Not on air, target date: unknown: 91.9 mhz; 25 kw. Ant 243 ft TL: N34 23 44 W107 00 42. Hrs open: 11865 Moveno Ave, Lakeside, CA, 92040. Phone: (619) 390-3475. Licensee: Family Stations Inc. Format: Christian.

Taos

KKIT(FM)— 2005: 95.9 mhz; 4 kw. Ant -630 ft TL: N36 23 22 W105 35 09. Stereo. Hrs opn: 24 5542 NDCBU, 87571-6122. Secondary address: 125A Camino de la Merced 87571-5119. Phone: (505) 758-4491. Phone: (877) 737-KKIT. Fax: (505) 758-4452. E-mail: production@kxmt.com Web Site: www.radiotaos.com. Licensee: DMC Broadcasting Inc. Natl. Network: ABC. Format: Adult hits. ♦Jeff Singer, opns mgr.

KTAO(FM)— January 1978: 101.9 mhz; 1.05 kw. 2,824 ft TL: N36 14 48 W105 39 15. Stereo. Hrs opn: 6 AM-2 AM Box 1844, 87571. Phone: (505) 758-5826. Fax: (505) 758-8430. E-mail: ktaoo@newmex.com Web Site: www.ktao.com. Licensee: Taos Communications Corp. (acq 1-78). Population served: 110,000 Law Firm: Brown, Nietert & Kaufman. Format: AAA. News staff: 2; News: 7 hrs wkly. Target aud: 25-49; educated, responsive, upwardly mobile. Spec prog: Jazz 3 hrs, Roots & Wires, 5 hrs; maccasin wire, 3 hrs; world on tour, 2 hrs; celtic 4 hrs; Sonido del sol, 3 hrs. ♦Brad Hockmeyer, CEO, pres & progmg dir; Dave Noll, opns dir; Paddy Mac, mus dir.

KVOT(AM)— 2005: 1340 khz; 1 kw-U. TL: N36 23 22 W105 35 09. Hrs open: 24/7 5542 NDCBU, 87571-6122. Secondary address: 125A Camino de la Merced 87571-5119. Phone: (505) 758-4491. Fax: (505) 758-4452. Licensee: DMC Broadcasting Inc. (acq 12-20-2005). Natl. Network: ABC. Format: Talk/ Air America. ♦Jeff Singer, opns mgr.

KXMT(FM)— Dec 1, 2000: 99.1 mhz; 60 kw. Ant 2,135 ft TL: N36 51 32 W106 00 28. Hrs open: 5542 NDCBU, 87571-6122. Secondary address: 125A Camino de la Merced 87571-5119. Phone: (505)

758-4491. Fax: (505) 758-4452. Web Site: www.kxmt.com. Licensee: DMC Broadcasting Inc. (acq 3-19-2003; $645,000 with KKTC(FM) Angel Fire). Format: Mexican rgnl. ♦Darren Cordova, pres; Jeff Singer, opns mgr.

Tatum

KTUM(FM)— 2003: 107.1 mhz; 100 kw. Ant 918 ft TL: N32 52 50 W103 41 01. Hrs open: Box 2010, Ruidoso Downs, 88346. Phone: (505) 396-0499. Fax: (505) 396-8349. Licensee: MTD Inc. (group owner) Format: Classic rock. ♦Tim Keithley, gen mgr; Will Rooney, stn mgr.

Thoreau

KXTC(FM)— Oct 21, 1991: 99.9 mhz; 100 kw. Ant 1,210 ft TL: N35 36 13 W108 40 45. Stereo. Hrs opn: 24 1632 S. 2nd St., Gallup, 87301-5836. Phone: (505) 863-9391. Fax: (505) 863-9393. Web Site: www.999xtc.com. Licensee: Clear Channel Broadcasting Licenses Inc. (acq 9-7-2000). Law Firm: Borsari & Paxson. Format: CHR. News staff: 2; News: 2 hrs wkly. Target aud: 18-44; Women. Spec prog: American Indian one hr, Sp 8 hrs wkly. ♦MaryAnn Armijo, gen mgr.

Truth or Consequences

KCHS(AM)— September 1944: 1400 khz; 1 kw-U. TL: N33 08 26 W107 13 55. Hrs open: 6am-11pm 7days a wk Box 351, 87901. Phone: (505) 894-2400. Fax: (505) 894-3998. E-mail: kchs@gpkmedia.com Web Site: www.gpkmedia.com. Licensee: Myrna Baird-Kohs dba GPK Media LLC (acq 6-18-92). Population served: 10,000 Natl. Network: AP Radio. Format: Country, news, oldies. News staff: 3. Target aud: General; area residents & visitors at lake. ♦Myrna Kohs, gen mgr, gen sls mgr & mus dir; Patrick Kohs, pres, prom VP, progmg dir & news dir.

KKVS(FM)— Nov 1, 1984: 98.7 mhz; 49 kw. Ant 2,644 ft TL: N32 58 15 W107 13 26. Stereo. Hrs opn: 24 Box 968, Las Cruces, 88004. Secondary address: 1355 E. California St., Las Cruces 88001. Phone: (505) 525-9298. Fax: (505) 525-9419. E-mail: radiolc@kgrt.com Web Site: www.vista.fm. Licensee: Richardson Commercial Corp. (acq 10-2-2001; $1,650,000 with KHQT(FM) Las Cruces). Population served: 300,000 Law Firm: Fletcher, Heald & Hildreth. Format: Rgnl Mexican. News staff: 2; News: 2 hrs wkly. Target aud: 25-54; Hispanic. Spec prog: NMSU Aggies Sports-in Spanish. ♦Allen Lumeyer, VP & gen mgr; Ernesto Garcia, opns mgr; Tamara Blaeser, natl sls mgr.

Tse Bonito

KHAC(AM)— Mar 21, 1967: 880 khz; 10 kw-D, 430 w-N. TL: N35 38 41 W109 01 13. Hrs open: Box 9090, Western Indian Ministries, Window Rock, AZ, 86515. Phone: (505) 371-5587. Fax: (505) 371-5588. E-mail: 1harpor@westernindian.net Web Site: www.westernindian.org. Licensee: Western Indian Ministries. (group owner) Population served: 73,500 Format: Christian, CHR, Navajo. Indian. ♦Larry Harpor, gen mgr; Bruce Kinde, engrg mgr.

Tucumcari

KQAY-FM—Listing follows KTNM(AM).

KTNM(AM)— 1941: 1400 khz; 1 kw-U. TL: N35 10 15 W103 42 25. Hrs open: Box 668, 902 S. Date St., 88401. Phone: (505) 461-0522. Phone: (505) 461-1400. Fax: (505) 461-0092. E-mail: ktnmkqay@yahoo.com Web Site: www.tucumcari.ws. Licensee: Quay Broadcasters Inc. Group owner: Roswell Radio Inc./Quay Broadcasters Inc. (acq 1-10-2003; with co-located FM). Population served: 15,000 Natl. Network: ABC. Law Firm: Cohn & Marks. Format: C&W. Target aud: General. Spec prog: Sp 18 hrs wkly. ♦Diane Paris, gen mgr, gen sls mgr, prom mgr & news dir; Greg Carnefix, progmg dir.

KQAY-FM—Co-owned with KTNM(AM). Jan 19, 1968: 92.7 mhz; 3 kw. 64 ft TL: N35 10 15 W103 42 25. Stereo. Web Site: www.tucumari.was. Format: Adult contemp.

***KVLP(FM)—**Not on air, target date: unknown: 91.7 mhz; 250 w. Ant 105 ft TL: N35 10 41 W103 43 21. Hrs open: 5700 West Oaks Blvd., Rocklin, CA, 95765. Phone: (916) 251-1600. Fax: (916) 251-1650. Licensee: Educational Media Foundation. (acq 3-23-2007; grpsl). ♦Richard Jenkins, pres.

White Rock

KSFQ(FM)— 1991: 101.1 mhz; 2.5 kw. Ant 1,863 ft TL: N35 53 09 W106 23 16. Stereo. Hrs opn: 24 2351 Sunset Blvd., Suite 170-218, Rocklin, CA, 95765. Phone: (916) 251-1600. Fax: (916) 251-1650. Licensee: Educational Media Foundation. (acq 7-6-2007; $1.5 million with KBAC(FM) Las Vegas). Population served: 147,000 Law Firm: Davis Wright Tremaine LLP. Format: Contemp Christian. ♦Mike Novak, sr VP.

Zuni

***KSHI(FM)—** Apr 6, 1978: 90.9 mhz; 100 w. Ant -249 ft TL: N35 05 18 W108 47 22. Stereo. Hrs opn: 8am - 5pm Box 339, 87327. Phone: (505) 782-4144. Fax: (505) 782-5069. E-mail: zuniradio@gmail.com Licensee: Zuni Communications Authority. Population served: 10,000 Format: Educ. Target aud: 18-34; primarily Indian. Spec prog: Indian 20 hrs wkly. ♦Duane Chimoni, gen mgr.

New York

Albany

***WAMC(AM)—** 1934: 1400 khz; 1 kw-U. TL: N42 41 21 W73 47 37. Hrs open:
wamc(FM) 97%.
Box 66600, 12206. Secondary address: 318 Central Ave. 12206. Phone: (518) 465-5233. Fax: (518) 432-6974. E-mail: mail@wamc.org Web Site: www.wamc.org. Licensee: WAMC. Group owner: WAMC/Northeast Public Radio (acq 4-24-03; $500,000). Population served: 114,873 Natl. Network: NPR, PRI. Law Firm: Dow Lohnes & Albertson. Wire Svc: AP Format: News/talk. ♦David Galletly, VP; Selma Kaplan, VP.

***WAMC-FM—** Oct 1, 1958: 90.3 mhz; 10 kw. Ant 1,970 ft TL: N42 38 14 W73 10 07. Stereo. Hrs opn: 24 Box 66600, 318 Central Ave., 12206. Phone: (518) 465-5233. Phone: (800) 323-9262. Fax: (518) 432-6974. E-mail: mail@wamc.org Web Site: www.wamc.org. Licensee: WAMC. Group owner: WAMC/Northeast Public Radio (acq 7-1-82). Population served: 1,464,200 Natl. Network: NPR, PRI. Law Firm: Dow, Lohnes & Albertson. Wire Svc: AP Format: News, talk. News staff: 11; News: 48 hrs wkly. Target aud: General. Spec prog: Jazz 18 hrs, folk 7 hrs. ♦Alan Chartock, CEO & chmn; Alan S. Chartock, CEO; David Galletly, VP; Selma Kaplan, VP.

***WCDB(FM)—** Mar 1, 1978: 90.9 mhz; 100 w. 222 ft TL: N42 41 16 W73 49 19. Stereo. Hrs opn: 24 Campus Ctr. 316, 1400 Washington Ave., 12222. Phone: (518) 442-5234. Phone: (518) 442-5262. Fax: (518) 442-4366. Web Site: www.wcdb.albany.edu. Licensee: State University of New York. Population served: 1,000,000 Format: Div, AOR, urban contemp. News: 15 hrs wkly. Target aud: 15-55; students & surrounding community. Spec prog: Gospel 3 hrs, Sp 3 hrs, dance 3 hrs, jazz 10 hrs, metal 10 hrs wkly. ♦Ed Horn, mus dir; Jeff Harfield, chief of engrg.

WDCD(AM)— May 1948: 1540 khz; 50 kw-U, DA-1. TL: N42 44 01 W73 51 49. Hrs open: 24 4243 Albany St., 12212. Phone: (518) 862-1540. Fax: (518) 862-1545. Web Site: www.crawfordbroadcasting.com. Licensee: Kimtron Inc. Group owner: Crawford Broadcasting Co. (acq 1995; $700,000). Population served: 790,000 Target aud: 30-54; educated, conservative, committed religious. ♦Donald B. Crawford, pres; Robert Hammond, gen mgr; Mark Shuttleworth, prom dir & progmg dir.

WDDY(AM)— June 14, 1924: 1460 khz; 5 kw-U, DA-N. TL: N42 37 21 W73 48 09. Hrs open: 24 52 Corporate Cir., Suite K, 12203. Phone: (518) 464-1311. Fax: (518) 464-4185. Web Site: www.radiodisney.com. Licensee: Radio Disney Group LLC. Group owner: ABC Inc. (acq 2-12-02; $2 million). Population served: 750,000 Natl. Network: ABC. Natl. Rep: Katz Radio. Format: Family & children programs. News staff: one. Target aud: 25-54; general. ♦Rob Thomson, gen mgr; Diane Frank, prom mgr.

WGNA-FM— December 1973: 107.7 mhz; 12.5 kw. Ant 984 ft TL: N42 38 18 W73 59 51. Hrs open: 1241 Kings Rd., Schenectady, 12303. Phone: (518) 881-1515. Fax: (518) 881-1516. E-mail: wgna1077@aol.com Web Site: www.wgna.com. Licensee: Regent Licensee of Mansfield Inc. Group owner: Regent Communications Inc. (acq 8-24-2001; grpsl). Law Firm: Latham & Watkins. Format: Country. ♦Robert Ausfeld, gen mgr; John Hirsch, stn mgr.

WGY(AM)—See Schenectady

WHAZ(AM)—See Troy

WHRL(FM)— Sept 1, 1966: 103.1 mhz; 6 kw. 328 ft TL: N42 39 46 W73 40 37. Stereo. Hrs opn: 1203 Troy-Schenectady Rd., Latham, 12110. Phone: (518) 452-4800. Fax: (518) 452-4855. E-mail: feedback@channel1031.com Web Site: www.whrl.com. Licensee: CC Licenses LLC. Group owner: Clear Channel Communications Inc. (acq 8-5-98; grpsl). Population served: 114,873 Natl. Rep: Clear Channel. Format: Alternative. Target aud: 21-54; upscale arrivers. ♦Kristen Delaney, VP & gen mgr; John Cooper, stn mgr & progmg dir; Lisa Biello, opns dir.

WKLI-FM— 1972: 100.9 mhz; 6 kw. Ant 298 ft TL: N42 43 54 W73 52 56. Stereo. Hrs opn: 24 6 Johnson Rd., Latham, 12110. Phone: (518) 786-6600. Fax: (518) 786-6610. Web Site: www.albanymagic.com. Licensee: 6 Johnson Road Licenses Inc. Group owner: Pamal Broadcasting Ltd. (acq 10-9-2001). Population served: 103,200 Natl. Network: CBS Radio. Format: Adult standards. Target aud: 25-54; upscale women. ♦Dan Austin, gen mgr; Kevin Callahan, opns mgr; Suzette Anthony, sls dir; Jay Scott, progmg dir; Darrin Kibbey, prom.

WOFX(AM)—See Troy

WPYX(FM)— Sept 16, 1980: 106.5 mhz; 15.3 kw. 902 ft TL: N42 38 09 W74 00 05. Stereo. Hrs opn: 24 1203 Troy-Schenectady Rd., Latham, 12110. Phone: (518) 452-4800. Fax: (518) 452-4855. E-mail: feedback@pyx106.com Web Site: www.pyx106.com. Licensee: Capstar TX L.P. Group owner: Clear Channel Communications Inc. (acq 8-30-00; grpsl). Population served: 1,300,000 Format: Classic rock. Target aud: 25-54. ♦Kristen Delaney, VP & gen mgr; John Cooper, stn mgr; Nicholas Lombardi, gen sls mgr; Jill Manti, mktg dir; John Cooper, progmg dir.

WROW(AM)— Sept 30, 1947: 590 khz; 5 kw-D, 1 kw-N, DA-2. TL: N42 34 25 W73 47 12. Hrs open: 24 6 Johnson Rd., Lathan, 12110. Phone: (518) 786-6600. Fax: (518) 786-6610. Web Site: www.wrow.com. Licensee: 6 Johnson Road Licenses Inc. Group owner: Pamal Broadcasting Ltd. (acq 10-19-2001; grpsl). Population served: 62,600 Natl. Network: CBS. Natl. Rep: McGavren Guild. Wire Svc: AP Format: News/talk. News staff: 3; News: 8 hrs wkly. Target aud: 35 plus; affluent, educated, white collar, upwardly mobile, homeowners. Spec prog: Gospel 3 hrs wkly. ♦Jim Morrell, pres; Dan Austin, gen mgr; Kevin Callahan, opns VP & opns mgr; Suzette Anthony, gen sls mgr; Paul Vandenburg, progmg VP; Mike Carey, news dir.

WTMM(AM)—See Rensselaer

WYJB(FM)—Co-owned with WROW(AM). October 1966: 95.5 mhz; 12 kw. 1,020 ft TL: N42 38 11 W74 00 00. Stereo. Web Site: www.b95.com.1,179,500 Natl. Rep: McGavren Guild. Format: Soft adult contemp. News staff: 3. Target aud: 25-54. ♦Kevin Callahan, opns dir & opns mgr; Chad O'Hara, mktg dir & news rptr; Darrin Kibbey, prom dir, progmg VP, mktg & prom; Chuck Taylor, progmg dir & mus dir.

WTRY-FM—See Rotterdam

Albion

***WJCA(FM)—** Dec 27, 2001: 102.1 mhz; 3.7 kw. Ant 423 ft TL: N43 11 19 W78 08 53. Hrs open: Box 262550, Baton Rouge, LA, 70826. Secondary address: 8917 World Ministry Ave., Baton Rouge, LA 70810. Phone: (225) 768-3688. Phone: (225) 768-8300. Fax: (225) 768-3729. E-mail: kawikfish@yahoo.com Web Site: www.jsm.org. Licensee: Family Worship Center Church Inc. (group owner; (acq 1-30-2006; $950,000). Format: Christian. ♦David Whitelaw, COO; Jimmy Swaggart, pres; John Santiago, progmg mgr.

Alfred

***WALF(FM)—** 1971: 89.7 mhz; 200 w. 73 ft TL: N42 15 17 W77 47 13. Stereo. Hrs opn: 1 Saxon Dr, 14802. Phone: (607) 871-2287. Web Site: www.walfradio.org. Licensee: Alfred University. Population served: 25,000 Natl. Network: NPR. Format: Div. ♦Kelly Donohoe, stn mgr.

***WETD(FM)—** Mar 19, 1973: 90.7 mhz; 360 w. 282 ft TL: N42 15 37 W77 47 51. Stereo. Hrs opn: 24 WETD Studios, Alfred State College, 10 Upper Campus Dr., 14802. Phone: (607) 587-3694. E-mail: wetd@alfredstate.edu Web Site: web.alfredstate.edu/wetd. Licensee: State University of New York. Format: Rock/AOR. Target aud: 18-22; college students.

WZKZ(FM)— Feb 28, 1999: 101.9 mhz; 1.3 kw. 699 ft TL: N42 12 20 W77 48 46. Hrs open: 24 3012 Eastside Ave., Wellsville, 14895. Phone: (585) 593-9553; (607) 733-5626. Fax: (585) 593-9554. E-mail:

wzkz@wzkzradio.com Web Site: www.wzkzradio.com. Licensee: Pembrook Pines Inc. Group owner: Pembrook Pines Media Group. Natl. Network: Jones Radio Networks. Natl. Rep: Interep. Wire Svc: AP Format: Hot country. News staff: 2; News: 10 hrs wkly. Target aud: 18-54. ♦Robert Pfuntner, CEO; Rod Biehler, gen mgr; Bob Weigand, opns mgr & news dir; Jim Davison, progmg.

Altamont

WZMR(FM)— June 26, 1968: 104.9 mhz; 530 w. Ant 932 ft TL: N42 38 11 W74 00 02. Hrs open: 24 6 Johnson Rd., Lathan, 12110. Phone: (518) 786-6600. Fax: (518) 786-6610. Web Site: albanyedge.com. Licensee: 6 Johnson Road Licenses Inc. Group owner: Pamal Broadcasting Ltd. (acq 10-19-2001; grpsl). Population served: 400,000 Format: Active rock. Target aud: 18-44; men. ♦Dan Austin, gen mgr; Kevin Callahan, opns mgr; Suzette Anthony, sls dir; Terry O'Donnell, mktg dir.

Amherst

WUFO(AM)— 1948: 1080 khz; 1 kw-D. TL: N42 56 46 W78 49 43. Hrs open: Sunrise-sunset 89 LaSalle Ave., Buffalo, 14214. Phone: (716) 834-1080. Fax: (716) 837-1438. E-mail: wufo1080am@aol.com Web Site: www.wufoam.com. Licensee: McL/McM New York LLC. (acq 3-1-72). Population served: 125,000 Natl. Network: American Urban. Format: Gospel. News: 5 hrs wkly. Target aud: 25-54; Black adults, relg foundation, strong work ethics. Spec prog: Talk. ♦Ron Davenport, pres; Alan Lincoln, gen mgr; Carol M. Salter, stn mgr, opns mgr & progmg dir.

Amsterdam

WCSS(AM)— Apr 8, 1948: 1490 khz; 1 kw-U. TL: N42 57 40 W74 10 35. Hrs open: 24 Box 581 Riverfront Ctr., 12010. Phone: (518) 843-2500. Fax: (518) 843-2501. E-mail: wcss@verizon.net Licensee: IZ Communications Corp. (acq 9-13-99; $188,000). Population served: 800,000 Natl. Network: Jones Radio Networks, USA. Format: News/talk, Music of Your Life. News staff: one; News: 20 hrs wkly. Target aud: 25 plus; adults interested in loc, community info & mus. Spec prog: Local progmg & talk 12 hrs wkly. ♦Joseph Isabel, gen mgr.

WEXT(FM)— Aug 1, 1975: 97.7 mhz; 1.6 kw. Ant 623 ft TL: N42 59 05 W74 10 49. Stereo. Hrs opn: 4 Global View, Troy, 12180. Phone: (518) 880-3400. Fax: (518) 880-3409. Web Site: www.wmht.org. Licensee: WMHT Educational Telecommunications (acq 9-19-2005; $1.5 million). Format: Classical. ♦Deborah Onslow, pres, gen mgr & stn mgr.

WVTL(AM)— Aug 16, 1961: 1570 khz; 1 kw-D, 207 w-N. TL: N42 54 38 W74 13 04. Hrs open: 24 5816 State Hwy. 30, 12010. Phone: (518) 843-9284. Fax: (518) 843-5225. Web Site: www.1570wvtl.com. Licensee: Roser Communications Network Inc. (acq 10-21-94; $400,000 with WBUG-FM Fort Plain; FTR: 12-5-94). Format: News, talk, sports. News staff: 2; News: 10 hrs wkly. Target aud: 25 plus. ♦Ken Roser Jr., gen mgr; Roxanne Roser, stn mgr; Grant Roser, gen sls mgr.

Arcade

***WCOF(FM)**— 2005: 89.5 mhz; 1 kw. Ant 593 ft TL: N42 27 41 W78 18 26. Hrs open:
Rebroadcasts WCIK(FM) Bath 100%.
Box 506, Bath, 14819. Phone: (607) 776-4151. Fax: (607) 776-6929. E-mail: mail@fln.org Web Site: www.fln.org. Licensee: Family Life Ministries Inc. Group owner: Family Life Network. Natl. Network: Salem Radio Network. Law Firm: Hardy, Carey, Chautin & Balkin, LLP. Wire Svc: Metro Weather Service Inc. Format: Contemp Christian. News staff: 3; News: 14 hrs wkly. Target aud: 30-54; general. ♦Dick Snavely, CFO; Rick Snavely, pres & gen mgr; John Owens, progmg dir; Jim Travis, chief of engrg.

Argyle

***WNGN(FM)**— August 1994: 91.9 mhz; 240 w. 571 ft TL: N43 13 33 W73 26 34. Hrs open: 24 Box 36 King Rd., Buskirk, 12028. Phone: (518) 686-0975. Fax: (518) 686-0975. E-mail: wngn@wngn.net Licensee: Northeast Gospel Broadcasting Inc. (acq 3-17-93; FTR: 4-5-93). Population served: 400,000 Natl. Network: Moody. Format: Christian, inspirational. News: one hr wkly. Target aud: 35-54; general. ♦Brian Larson, pres.

Arlington

WRRB(FM)— December 1989: 96.9 mhz; 3 kw. 1,010 ft TL: N41 43 11 W73 59 45. (CP: 310 w). Stereo. Hrs opn: 24
Rebroadcasts WDST(FM) Woodstock 100%.
Box 416, Poughkeepsie, 12602. Secondary address: 2 Pendell Rd., Poughkeepsie 12602. Phone: (845) 471- 1500. Fax: (845) 454-1204. Web Site: www.cumulus.com. Licensee: Cumulus Licensing Corp. Group owner: Cumulus Media Inc. (acq 1-23-02; grpsl). Law Firm: Fisher, Wayland, Cooper, Leader & Zaragoza. Format: Progsv adult rock. News: 5 hrs wkly. Target aud: 25-54; upscale professionals. ♦Charles Benfer, gen mgr.

Attica

WLOF(FM)— Nov 9, 1977: 101.7 mhz; 3 kw. Ant 295 ft TL: N42 50 51 W78 21 01. Stereo. Hrs opn: 6325 Sheridan Dr., Williamsville, 14221. Phone: (716) 839-6117. Fax: (716) 839-0400. Web Site: www.wlof.net. Licensee: Holy Family Communications Inc. Group owner: Holy Family Communications acq 12-20-99; $655,000). Population served: 250000 Format: Relg. Target aud: 25-54; blue collar, housewives. ♦Jim Wright, gen mgr.

Auburn

WAUB(AM)— Dec 24, 1959: 1590 khz; 500 w-D, 1 kw-N, DA-2. TL: N42 54 34 W76 36 09. Hrs open: 5998 Experimental Blvd., Geneva, 13021. Phone: (315) 258-0937. Fax: (315) 258-9248. E-mail: tbaker@flradiogroup.com Web Site: www.fingerlakes1.com. Licensee: Auburn Broadcasting Inc. Group owner: Finger Lakes Radio Group (acq 7-2-97; $70,000 plus additonal consideration). Population served: 34,599 Natl. Network: CBS. Law Firm: James L. Oyster. Format: Talk radio. Target aud: General. ♦Alan Bishop, pres; Ted Baker, news dir.

***WDWN(FM)**— Oct 31, 1972: 89.1 mhz; 3 kw. 102 ft TL: N42 56 40 W76 32 33. Stereo. Hrs opn: 197 Franklin St., 13021. Phone: (315) 255-1743, EXT. 2282. Phone: (315) 253-0449. Fax: (315) 255-2690. Web Site: www.wdwn.fm. Licensee: Cayuga County Community College. (acq 10-72). Population served: 250,000 Format: AOR. Target aud: 18-25; high school & college students, young adults. ♦Philip Gover, pres; Steven Keeler, gen mgr; Douglas Brill, chief of engrg.

WPHR-FM— May 20, 1949: 106.9 mhz; 14 kw. 941 ft TL: N42 48 05 W76 26 14. Stereo. Hrs opn: 500 Plum St., Suite 100, Syracuse, 13243. Phone: (315) 472-9797. Fax: (315) 473-0049. Licensee: CC Licenses LLC. Group owner: Clear Channel Communications Inc. (acq 3-24-2000). Population served: 34,599 Format: Urban Contemp. Target aud: 25-54; educated, up-scale, high income, mobile, family oriented. ♦Joel Delmonico, gen mgr; Butch Charles, progmg dir; Kenny Dees, mus dir.

WWLF(AM)— Jan 26, 1927: 1340 khz; 1 kw-U. TL: N42 57 05 W76 35 05. Hrs open: 401 W. Kirkpatrick St., Syracuse, 13204. Phone: (315) 472-0222. Fax: (315) 478-7745. Web Site: www.radiodisney.com. Licensee: WOLF Radio Inc. (group owner; acq 6-26-98; $103,000). Population served: 34599 Natl. Network: Radio Disney. Format: Children. ♦Sam Furco, gen mgr; Becky Mullen, prom mgr.

Avon

WYSL(AM)— Jan 23, 1987: 1040 khz; 20 kw-D, 500 w-N, DA-2. TL: N42 51 16 W77 42 39. Stereo. Hrs opn: 24 Box 236, 14414-0236. Secondary address: 5620 S. Lima Rd. 14414. Phone: (585) 346-3000. Fax: (585) 346-0450. E-mail: info@wysl1040.com Web Site: www.wysl1040.com. Licensee: Radio Livingston Ltd. Population served: 1,300,000 Natl. Network: CNN Radio, ABC. Wire Svc: AP Format: News/sports. News staff: 3; News: 160 hrs wkly. Target aud: 35 plus; general. Spec prog: Relg 4 hrs wkly. ♦Robert Savage, CEO; Robert C. Savage, pres; Judith Day, CFO & exec VP; J.C. Delass, gen mgr & stn mgr; Bob D'Angelo, opns mgr.

Babylon

WBAB(FM)— Aug 27, 1958: 102.3 mhz; 6 kw. 269 ft TL: N40 47 58 W73 20 08. Stereo. Hrs opn: 24 555 Sunrise Hwy., West Babylon, 11704-6009. Phone: (631) 587-1023. Fax: (631) 587-1282. E-mail: wbab@wbab.com Web Site: wbab.com. Licensee: Cox Radio Inc. Group owner: Cox Broadcasting (acq 5-22-98; grpsl). Population served: 2,800,000 Natl. Rep: Christal. Format: AOR. News staff: one; News: 2 hrs wkly. Target aud: 25-54; men. ♦Kim Guthrie, gen mgr.

WNYG(AM)— Jan 1, 1958: 1440 khz; 1 kw-D, 38 w-N. TL: N40 42 32 W73 21 53. Hrs open: 24 404 Rt. 109, West Babylon, 11704. Phone: (631) 321-9640. Fax: (631) 422-5992. E-mail: spiritny@verizon.net Web Site: www.wnygspiritofny.com. Licensee: Multicultural Radio Broadcasting Licensee LLC. Group owner: Multicultural Radio Broadcasting Inc. (acq 6-14-00; $850,000). Population served: 2,500,000 Format: Contemp Christian. News staff: one; News: 3 hrs wkly. Target aud: 25-64. ♦Doug Edwards, pres & progmg dir; Phyllis Rose, gen mgr.

Baldwinsville

***WBXL(FM)**— Jan 29, 1975: 90.5 mhz; 175 w. 207 ft TL: N43 09 47 W76 18 47. Stereo. Hrs opn: 7 AM-11 PM Baker High School, 29 E. Oneida St., 13027. Phone: (315) 638-6010. Phone: (315) 638-6000. Licensee: Baldwinsville Central School District. Population served: 250,000 Format: CHR. Family of school district students. ♦Peter Hunn, gen mgr.

WSEN(AM)— Feb 25, 1959: 1050 khz; 2.5 kw-D, DA. TL: N43 10 46 W76 20 19. Stereo. Hrs opn: 24
Rebroadcasts WSEN-FM Baldwinsville.
Box 1050, 13027. Secondary address: 8456 Smoky Hollow Rd. 13027. Phone: (315) 635-3971. Fax: (315) 635-3490. E-mail: webmaster@wsenfm.com Web Site: www.wsenfm.com. Licensee: Buckley Broadcasting of New York LLC. Group owner: Buckley Broadcasting Corp. (acq 8-20-80; $700,000 with co-located FM; FTR: 8-11-80). Natl. Network: CBS. Natl. Rep: McGavren Guild. Format: Oldies. Target aud: 35 plus; well-educated professionals with disposable income. ♦Richard Buckley, pres; Judith Kelly, VP & gen mgr; Jody Frawley, gen sls mgr & rgnl sls mgr; Kristine Gladle, prom mgr; Jim Tate, progmg dir; Al Jenner, chief of engrg.

WSEN-FM— Nov 10, 1967: 92.1 mhz; 25 kw. 300 ft TL: N43 10 46 W76 20 19. Stereo. 24 E-mail: webmaster@wsenfm.com Web Site: www.wsenfm.com. Natl. Rep: McGavren Guild. Target aud: 35-64; well-educated professionals with disposable income.

Ballston Spa

WKKF(FM)— May 27, 1968: 102.3 mhz; 4.1 kw. 386 ft TL: N42 52 44 W73 51 47. Stereo. Hrs opn: 24 1203 Troy-Schenectady Rd., Latham, 12110. Phone: (518) 452-4800. Fax: (518) 452-4885. Web Site: www.1023kissfm.com. Licensee: CC Licenses LLC. Group owner: Clear Channel Communications Inc. (acq 3-6-97). Population served: 739900 Natl. Rep: Clear Channel. Law Firm: Wilmer, Cutler & Pickering. Format: Contemp hit. Target aud: 18-34; upscale, hip women. ♦John Cooper, stn mgr; Rob Dawes, opns dir; Randy McMartin, progmg dir.

Batavia

WBTA(AM)— Feb 6, 1941: 1490 khz; 1 kw-D, 71 kw-N. TL: N42 58 35 W78 11 12. Hrs open: 24 113 Main St., 14020. Phone: (585) 344-1490. Fax: (585) 344-1441. E-mail: debbie@wbta1490.com Web Site: www.wbta1490.com. Licensee: HPL Communications Inc. (acq 11-24-2003). Population served: 75,000 Natl. Rep: Rgnl Reps. Format: News/talk, soft rock. News staff: news progmg 10 hrs wkly News: 2;. Older, upscale audience. Spec prog: Sports 9 hrs wkly. ♦Daniel C. Fischer, pres; Daniel Fischer, gen mgr; Lorne Way, gen sls mgr.

***WGCC-FM**— Nov 13, 1985: 90.7 mhz; 880 w. 164 ft TL: N43 01 03 W78 08 18. Stereo. Hrs opn: 6 am-1 am One College Rd., 14020. Phone: (585) 343-0055, EXT. 6284. Phone: (585) 343-9422. Fax: (585) 345-6806. E-mail: cmplatt@genesee.edu Web Site: wgcc-fm.com. Licensee: Genesee Community College Board of Trustees. Population served: 100,000 Format: Serious rock, rock/AOR, modern rock, classic rock. News staff: one; News: 5 hrs wkly. Target aud: 13-30;

high school & college youth. Spec prog: .Heavy metal, alternative, oldies ♦Chuck Platt, pres & engrg dir; Jeremy Canute, stn mgr; Melody Nardone, stn mgr; Andrew Scutt, mus dir; Ashley Mouting, mus dir.

Bath

WABH(AM)— Nov 2, 1962: 1380 khz; 500 w-D. TL: N42 20 11 W77 17 34. (CP: 5 kw-D, 350 w-N). Hrs opn: 19 Box 72, E. Washington St. Ext., 14810. Phone: (607) 776-3326. Fax: (607) 776-6161. E-mail: wvinsales@stny.rr.com Web Site: www.wvinradio.com. Licensee: Pembrook Pines Mass Media Inc. Group owner: Pembrook Pines Media Group (acq 4-13-90; with co-located FM; FTR: 5-7-90). Population served: 150,000 Natl. Network: CBS. Format: Oldies. Target aud: General. Spec prog: Farm one hr wkly. ♦Bill Fleishman, gen mgr.

WVIN-FM—Co-owned with WABH(AM). Oct 10, 1971: 98.3 mhz; 3 kw. 351 ft TL: N42 19 06 W77 21 27. (CP: 2.75 kw, ant 341 ft.). Stereo. 19 Web Site: www.wvinradio.com.150,000 Format: Soft adult contemp. Spec prog: Jazz 2 hrs wkly.

***WCIK(FM)**— Aug 29, 1983: 103.1 mhz; 790 w. Ant 532 ft TL: N42 20 07 W77 27 27. Stereo. Hrs opn: 24 Box 506, 7634 Campbell Creek Rd., 14810. Phone: (607) 776-4151. Fax: (607) 776-6929. E-mail: mail@fln.org Web Site: www.fln.org. Licensee: Family Life Ministries Inc. Group owner: Family Life Network. Natl. Network: Salem Radio Network. Law Firm: Hardy, Carey, Chautin & Balkin, LLP. Wire Svc: Metro Weather Service Inc. Format: Christian contemp. News staff: 3; News: 14 hrs wkly. Target aud: 30-54; general. ♦Dick Snavely, CFO & VP; Rick Snavely, pres & gen mgr; John Owens, progmg dir; Jim Travis, chief of engrg.

Bay Shore

WBZO(FM)— February 1993: 103.1 mhz; 3 kw. 285 ft TL: N40 45 04 W73 12 52. Stereo. Hrs opn: 24 234 Airport Plaza Blvd., Farmingdale, 11735. Phone: (631) 770-4200. Fax: (631) 770-0101. Web Site: www.b103.com. Licensee: WCMB Broadcasting L.P. Group owner: Barnstable Broadcasting Inc. (acq 3-6-97; $12.45 million). Population served: 2,500,000 Natl. Rep: D & R Radio. Law Firm: Verner, Liipfert, Bernhard, McPherson & Hand. Format: Oldies. News staff: one; News: 25 hrs wkly. Target aud: General. ♦Dave Widmer, gen mgr; Bill Wise, progmg dir; Michael Glaser, chief of engrg; Keith Allen, disc jockey; Scotty Hart, disc jockey.

Beacon

WBNR(AM)— Dec 17, 1959: 1260 khz; 1 kw-D, 500 w-N, DA-2. TL: N41 29 32 W73 58 43. Stereo. Hrs opn: 24
Simulcast with WLNA(AM) Peekskill 100%.
Box 310, 12508. Secondary address: 715 Rt. 52 12508. Phone: (845) 831-8000. Fax: (845) 838-2109. Web Site: www.thesoundofthevalley.com. Licensee: 6 Johnson Road Licenses Inc. Group owner: Pamal Broadcasting Ltd. (acq 10-19-2001; grpsl). Population served: 200,000 Natl. Network: ABC. Format: Stardust. News staff: 2; News: 2 hrs wkly. Target aud: 35 plus. Spec prog: Relg 5 hrs wkly. ♦James Morrell, CEO; Fred Bennett, exec VP & stn mgr.

WGNY-FM—See Newburgh

WSPK(FM)—See Poughkeepsie

Beekman

***WBKW(FM)**—Not on air, target date: unknown: 88.3 mhz; 3 w. Ant 679 ft TL: N41 34 22 W73 41 01. Hrs open: 375 Monroe Tpke., Monroe, CT, 06468. Phone: (203) 268-9667. Web Site: www.wmnr.org. Licensee: Monroe Board of Education. ♦Kurt Anderson, gen mgr; Jane Stadler, opns mgr.

Big Flats

WENI-FM— April 1989: 97.7 mhz; 1.30 kw. Ant 482 ft TL: N42 09 43 W77 02 15. Hrs open: 24 Box 1047, Corning, 14830. Secondary address: 2309 Davis Rd., Corning 14830. Phone: (607) 962-4646. Fax: (607) 962-1138. Licensee: Group A Licensee LLC. Group owner: Route 81 Radio LLC (acq 10-31-2003; grpsl). Population served: 50,000 Natl. Network: ABC. Natl. Rep: Roslin. Format: Oldies. Target aud: 25-54. ♦Bob Eolin, gen mgr, natl sls mgr & progmg dir; Richard Burton, rgnl sls mgr; Dee Eolin, news dir & pub affrs dir; Jim Appleton, chief of engrg.

Binghamton

WAAL(FM)—Listing follows WYOS(AM).

WENE(AM)—See Endicott

***WHRW(FM)**— Mar 1, 1966: 90.5 mhz; 1.45 kw. -47 ft TL: N42 05 24 W75 58 05. Stereo. Hrs opn: 24 Box 2000, Univ. Union, Binghamton Univ., 13902-6000. Phone: (607) 777-2137. Fax: (607) 777-6501. E-mail: whrwfm@binghamton.edu Web Site: www.whrwfm.org. Licensee: State University of New York. Population served: 64,123 Wire Svc: UPI Format: Var/div. News: 5 hrs wkly. Target aud: General. Spec prog: It 3 hrs, Jazz 9 hrs, Pol 3 hrs, Relg 6 hrs, Sp 9 hrs wkly. ♦Sam Smith, gen mgr; Brian Napolitano, chief of engrg.

WHWK(FM)—Listing follows WNBF(AM).

***WIFF(FM)**— 1995: 90.1 mhz; 100 w. Ant 686 ft TL: N42 03 10 W75 42 07. Hrs open: 24 111 N. Main St., Elmira, 14901. Phone: (607) 732-2484. Fax: (607) 732-8704. E-mail: wiff@csnradio.com Web Site: www.csnradio.com. Licensee: CSN International (group owner; acq 5-30-2003; $67,000). Population served: 200,000 Law Firm: .Reddy, Begley & McCormick Format: Christian adult contemp. News staff: 2; News: 8 hrs wkly. Target aud: General. ♦Lorenzo Galletti, gen mgr; Gina Galletti, progmg mgr; Stacey Sosnoski, asst music dir.

WINR(AM)— 1946: 680 khz; 5 kw-D, 500 w-N, DA-2. TL: N42 06 53 W75 51 16. Hrs open: 24 320 N. Jensen Rd., Vestal, 13850-2111. Phone: (607) 584-5800. Fax: (607) 584-5900. Licensee: AMFM Radio Licenses LLC. Group owner: Clear Channel Communications Inc. (acq 2-13-2001; $1 million). Population served: 64,123 Natl. Network: CBS. Law Firm: Fisher, Wayland, Cooper, Leader & Zaragoza. Format: MOR, news/talk. News staff: one; News: 10 hrs wkly. Target aud: 35 plus. ♦Tom Barney, gen mgr; Doug Mosher, opns mgr.

WNBF(AM)— 1928: 1290 khz; 5 kw-U, DA-2. TL: N42 03 31 W75 57 14. Hrs open: Box 414, 13902. Phone: (607) 772-8400. Fax: (607) 772-9806. Web Site: www.wnbf.com. Licensee: Citadel Broadcasting Co. Group owner: Citadel Broadcasting Corp. (acq 6-9-99; grpsl). Population served: 202,000 Format: News/talk. Target aud: 35-64. ♦Roger Neal, progmg dir; Bernie Fionte, news dir; Larry Hodge, chief of engrg.

WHWK(FM)—Co-owned with WNBF(AM). September 1956: 98.1 mhz; 10 kw. 960 ft TL: N42 03 34 W75 57 06. Stereo. Web Site: www.whwk.com.223,000 Format: Country. Target aud: 25-54. ♦Ed Walker, progmg dir.

***WSKG-FM**— Oct 22, 1975: 89.3 mhz; 10.2 kw. 942 ft TL: N42 03 22 W75 56 39. Stereo. Hrs opn: 24 Box 3000, 13902. Phone: (607) 729-0100. Fax: (607) 729-7328. Fax: (607) 231-0996. E-mail: wskg_mail@wskg.pbs.org Web Site: www.wskg.com. Licensee: WSKG Public Telecommunications Council. Population served: 273,000 Natl. Network: NPR, PRI. Law Firm: Dow, Lohnes & Albertson. Format: Class, news. News staff: one; News: 33 hrs wkly. Target aud: General. Spec prog: Jazz, folk 5 hrs wkly. ♦Gary Reinbolt, CEO & gen mgr; Gregory Keeler, opns dir; Linda Cohen, prom dir; William Snyder, mus dir; Mike Pufky, engrg dir & chief of engrg; Stacy Coveny, traf mgr. Co-owned TV: *WSKG-TV affil.

***WSQX-FM**— Jan 17, 1995: 91.5 mhz; 3.5 kw. 380 ft TL: N42 07 54 W75 55 56. Stereo. Hrs opn: 24 Box 3000, 13902. Phone: (607) 729-0100. Fax: (607) 729-7328. Web Site: www.wskg.com. Licensee: WSKG Public Telecommunications Council. Population served: 219,800 Natl. Network: NPR, PRI. Law Firm: Dow, Lohnes & Albertson. Format: Jazz, news. Target aud: General. Spec prog: Talk 10 hrs wkly. ♦Gary Reinbolt, CEO & pres; Suzanne Miller-Cormier, sr VP; Gregory Keeler, opns VP; Mike Pufky, prom mgr. Co-owned TV: *WSKG-TV affil

WYOS(AM)— June 1947: 1360 khz; 5 kw-D, 500 w-N, DA-2. TL: N42 04 03 W75 54 20. Hrs open: 24 59 Court St., 13901. Secondary address: Box 414 13902. Phone: (607) 772-8850. Fax: (607) 772-9806. Web Site: www.wnbf.com. Licensee: Citadel Broadcasting Co. Group owner: Citadel Broadcasting Corp. (acq 6-9-99; grpsl). Population served: 64,123 Natl. Network: ESPN Radio. Natl. Rep: McGavren Guild. Format: Sports. ♦Roger Neal, gen sls mgr & progmg dir.

WAAL(FM)—Co-owned with WYOS(AM). March 1954: 99.1 mhz; 8.7 kw. Ant 954 ft TL: N42 03 31 W75 57 06. Stereo. 24 Web Site: www.waal.com.250,000 Format: Classic rock. News staff: one; News: 2 hrs wkly. Target aud: 18-49; CHR/rock listeners. ♦Don Morgan, progmg dir.

Blue Mountain Lake

***WXLH(FM)**— November 1992: 91.3 mhz; 78 w. 1,729 ft TL: N43 52 18 W74 24 02. Hrs open: 24
Rebroadcasts WSLU(FM) Canton 100%.
St. Lawrence Univ., North Country Public Radio, Canton, 13617. Phone: (315) 229-5356. Fax: (315) 229-5373. Web Site: www.ncpr.org. Licensee: St. Lawrence University. Law Firm: Donald E. Martin. Format: Eclectic public radio. News staff: 2; News: 35 hrs wkly. Target aud: General. Spec prog: Gospel, jazz, class, folk, pub affrs. ♦Ellen Rocco, gen mgr; Sandra Demarest, dev dir.

Boonville

WBRV(AM)— June 22, 1955: 900 khz; 1 kw-D, 52 w-N. TL: N43 30 47 W75 21 46. Hrs open: 24
Rebroadcasts WLLG(FM) Lowville 50%.
7606 State St., Lowville, 13367. Phone: (315) 942-4311. Phone: (315) 376-7500. Fax: (315) 376-8549. Web Site: www.themoose.net. Licensee: Flack Broadcasting Group L.L.C. Population served: 15,000 Natl. Network: USA. Law Firm: Shaw Pittman. Format: Country. News: 18 hrs wkly. Target aud: General. ♦Sara Flack, sr VP; William Flack, pres, gen mgr & progmg dir; Brian Best, news dir; Dana Cowles, disc jockey.

WBRV-FM— Jan 31, 1989: 101.3 mhz; 5.5 kw. 348 ft TL: N43 26 53 W75 20 48. Stereo. 24
Rebroadcasts WLLG (FM) Lowville 50%.
Web Site: www.themoose.net. News: 10 hrs wkly. Target aud: General. Spec prog: Farm 3 hrs, relg 3 hrs wkly.

Brentwood

***WXBA(FM)**— June 21, 1975: 88.1 mhz; 180 w. 90 ft TL: N40 46 19 W73 15 19. Stereo. Hrs opn: 24 Ross High School, First & 5th Aves., 11717. Phone: (631) 434-2581. Phone: (631) 434-2582. Fax: (631) 273-6572. E-mail: wxba@88x.net Web Site: www.88x.net. Licensee: Brentwood Public School District. Population served: 100,000 Format: Educ, CHR. News staff: 2; News: 3 hrs wkly. Target aud: 18-54; general. Spec prog: Black 5 hrs wkly. ♦Les Black, gen mgr; Jaimie Ottone, stn mgr; Charles Vollmer, progmg dir; Paul Bryant, news dir; Frank Lapple, chief of engrg; Pete Mandzych, sports cmtr.

Brewster

WPUT(AM)—Licensed to Brewster. See Patterson

Briarcliff Manor

WXPK(FM)— Apr 8, 1960: 107.1 mhz; 890 w. 590 ft TL: N41 04 49 W73 48 26. Stereo. Hrs opn: 24 56 Lafayette Ave., White Plains, 10603. Phone: (845) 838-6000. Fax: (848) 838-2109. Web Site: www.1071thepeak.com. Licensee: 6 Johnson Road Licenses Inc. Group owner: Nassau Broadcasting Partners L.P. (acq 11-5-2004; $18.4 million). Population served: 2,000,000 Law Firm: Hogan & Hartson. Format: AAA. News staff: one. Target aud: 18-44; upscale, young, suburban. ♦Darren DiPrima, stn mgr.

Bridgehampton

WBAZ(FM)— 1996: 102.5 mhz; 4.8 kw. 367 ft TL: N40 53 58 W72 23 06. Stereo. Hrs opn: 24 Box 7162, Amagansett, 11930. Secondary address: 249 Montauk Hwy., Amagansett 11930. Phone: (631) 267-7800. Fax: (631) 267-1018. E-mail: info@wbaz.com Web Site: www.wbaz.com. Licensee: AAA Licensing LLC. Group owner: AAA Entertainment L.L.C. (acq 8-22-2000; $2.75 million with WBEA(FM) Southold). Population served: 1,500,000 Law Firm: Pepper & Corazzini. Format: Soft adult contemp. Target aud: 25-44; adults with active lifestyles. Spec prog: News, sports. ♦Malcolm A. Kahn, pres & gen mgr; Laura Marie, prom dir & chief of engrg; John Lynch, mus dir & traf mgr.

Bridgeport

WTKW(FM)— Nov 9, 1992: 99.5 mhz; 5.7 kw. 338 ft TL: N43 09 07 W75 56 05. (CP: 2.85 kw). Stereo. Hrs opn: 24 235 Walton St., Syracuse, 13202. Phone: (315) 472-9111. Fax: (315) 472-1888. E-mail: geninfo@tk99.net Web Site: www.tk99.net. Licensee: Galaxy Syracuse Licensee LLC. (group owner; (acq 4-6-2000; grpsl). Law Firm: Leventhal, Senter & Lerman. Format: Classic rock. News staff: one; News: 3 hrs wkly. Target aud: 25-54 plus; stable, peak-earning adults. ♦Ed Levine, pres; Mike Lucarelli, CFO; Ed Levine, gen mgr; Lisa Morrow, sls VP; Mimi Grizwold, progmg VP.

Brighton

WZNE(FM)— November 1996: 94.1 mhz; 6 kw. 318 ft TL: N43 08 07 W77 35 07. (CP: 1.8 kw, ant 407 ft.). Stereo. Hrs opn: 24 1700 HSBC Plaza, Rochester, 14604. Phone: (585) 399-5700. Fax: (585) 399-5750. Web Site: www.thezone941.com. Licensee: Infinity Radio Inc. Group owner: Infinity Broadcasting Corp. (acq 6-5-98; grpsl). Natl. Network: CNN Radio. Format: Alternative. Target aud: Men 18-34; Affluent fans of Alternative Rock music. ♦Al Casazza, gen mgr.

Brockport

WASB(AM)— Feb 15, 1970: 1590 khz; 1 kw-U, DA-2. TL: N43 11 44 W77 57 05. Hrs open: 24 6675 4th Section Rd., 14420. Phone: (585) 637-7040. Web Site: www.fountain-of-truth.com. Licensee: David L. Wolfe (acq 12-21-2005; with WRSB(AM) Canandaigua). Population served: 16,000 Format: Christian. Target aud: All ages; rural audience, western Rochester & suburbs. ♦Daniel Wolfe, gen mgr & stn mgr.

***WBSU(FM)**— Jan 14, 1981: 89.1 mhz; 7.33 kw. 160 ft TL: N43 12 45 W77 57 17. Stereo. Hrs opn: 24 Seymour Union, 14420. Phone: (585) 395-2580. Fax: (585) 395-5334. E-mail: wkozires@brockport.edu Web Site: www.891thepoint.com. Licensee: State University of New York. Population served: 500,000 Natl. Network: AP Radio. Wire Svc: AP Format: CHR, AOR, alternative. News: 4 hrs wkly. Target aud: 17-34; college & young professional. Spec prog: Black one hr, pub affrs 8 hrs wkly. ♦Dr. John Halstead, pres; Warren Kozireski, gen mgr; Dean King, chief of engrg.

***WKDL(FM)**— 1999: 104.9 mhz; 6 kw. Ant 328 ft TL: N43 09 51 W77 47 02. Hrs open: 2351 Sunset Blvd., Suite 170-218, Rocklin, CA, 95765. Phone: (916) 251-1600. Fax: (916) 251-1650. Web Site: www.klove.com. Licensee: Brockport Licenses LLC. (acq 1-27-2006; $4 million). Natl. Network: K-Love. Format: Contemp Chirstian. ♦Richard Jenkins, pres.

Bronxville

WFAS-FM—Licensed to Bronxville. See White Plains

Brooklyn

WKRB(FM)—Licensed to Brooklyn. See New York

WNYE(FM)—See New York

Brookville

***WCWP(FM)**— April 1965: 88.1 mhz; 100 w. 190 ft TL: N40 49 00 W73 35 49. Stereo. Hrs opn: 24
Rebroadcasts WLIU(FM) Southampton 85%.
Long Island Univ., C.W. Post Campus., 11548. Phone: (516) 299-2683. Phone: (516) 299-2626. Fax: (516) 299-2767. E-mail: wcwp@cwpost.liu.edu Web Site: www.liu.edu/wcwp. Licensee: Long Island University. (acq 8-90; FTR: 8-13-90). Population served: 1,400,000 Natl. Network: NPR. Format: News, jazz, AOR. News staff: one; News: 15 hrs wkly. Target aud: General. Spec prog: Relg 4 hrs, Sp one hr, pub affrs 5 hrs, sports 3 hrs, classic rock 2 hrs wkly. ♦Dan Cox, gen mgr; Joe Manfredi, opns dir; Jan Robin, mktg mgr; Joe Sallo, progmg mgr; Lauren Shallash, mus dir.

Buffalo

WBBF(AM)—Listing follows WHTT-FM.

WBEN(AM)— Sept 8, 1930: 930 khz; 5 kw-U, DA-N. TL: N42 58 42 W78 57 27. Hrs open: 24 500 Corporate Pkwy., Suite 200, Amherst, 14226. Phone: (716) 843-0600. Fax: (716) 832-2872. Web Site: www.wben.com. Licensee: Entercom Buffalo License L.L.C. Group owner: Entercom Communications Corp. (acq 1999). Population served: 1,410,000 Natl. Network: CBS. Natl. Rep: D & R Radio. Format: News/talk, sports. News staff: 10; News: 20 hrs wkly. Target aud: 35-64; general. Spec prog: Buffalo Bills football. ♦L. Greene, gen mgr; Brian Meany, sls dir; Mike Krupa, natl sls mgr; Cheryl Klocke, prom mgr; Tim Wenger, progmg dir; John Zach, news dir; Kevin Keenan, pub affrs dir; Dennis Kavanaugh, engrg VP; Kevin Sylvester, sports cmtr.

WTSS(FM)—Co-owned with WBEN(AM). Nov 11, 1946: 102.5 mhz; 110 kw. 1,340 ft TL: N42 39 33 W78 37 33. Stereo. 24 Web Site: www.star1025.com. Natl. Rep: D & R Radio. Format: CHR, adult contemp. Target aud: 18-49. ♦Dave Gillen, progmg dir; Kevin Sylvester, sports cmtr.

***WBFO(FM)**— Jan 7, 1959: 88.7 mhz; 24 kw. 240 ft TL: N43 00 13 W78 45 54. (CP: 50 kw, ant 256 ft.). Stereo. Hrs opn: 24 3435 Main St., 205 Allen Hall, 14214. Phone: (716) 829-2880. Phone: (716) 829-6000. Fax: (716) 829-2277. E-mail: mail@wbfo.org Web Site: www.wbfo.org. Licensee: State University of New York. Population served: 462,768 Natl. Network: NPR. Format: Jazz, news. News staff: 2; News: 50 hrs wkly. Target aud: General; educated professionals. Spec prog: Blues 8 hrs, bluegrass 3 hrs, class one hr, Pol 3 hrs wkly. ♦Carole Smith Petro, VP & gen mgr; Matthew Katafiaz, gen mgr; Mark Wozniak, opns mgr; Joan Wilson, dev dir.

WBLK(FM)—See Depew

***WBNY(FM)**— 1982: 91.3 mhz; 100 w. TL: N42 55 59 W78 52 59. Hrs open: 24 Campbell Student Union, 1300 Elmwood, 14222. Phone: (716) 878-5104. Phone: (716) 878-3080. Fax: (716) 878-6600. E-mail: wbny@hotmail.com Web Site: www.wbny.org. Licensee: State University of New York. Population served: 10,000 Format: New mus, alternative rock. News: 6 hrs wkly. Target aud: 18-25; college student. Spec prog: Black 12 hrs, jazz 3 hrs, reggae 3 hrs, heavy metal 3 hrs, folk 3 hrs wkly. ♦Dan Organ, gen mgr & mus dir.

WBUF(FM)— 1947: 92.9 mhz; 93 kw. Ant 580 ft TL: N42 38 12 W78 42 58. Stereo. Hrs opn: 14 Lafayette Sq., Suite 1300, 14203. Phone: (716) 852-9292. Fax: (716) 852-9290. Web Site: www.wbuf.com. Licensee: Regent Broadcasting of Buffalo Inc. Group owner: Infinity Broadcasting Corp. (acq 12-15-2006; grpsl). Population served: 2000000 Natl. Network: CBS Radio, CNN Radio. Natl. Rep: Christal. Format: Talk. Target aud: 18-34; men. ♦Jeff Silver, sr VP & gen mgr; Scott McCandless, gen sls mgr; Mike Krupa, natl sls mgr.

WDCX(FM)— February 1963: 99.5 mhz; 115 kw. 640 ft TL: N42 38 07 W78 46 05. (CP: 17 kw, ant 430 ft.). Stereo. Hrs opn: 24 625 Delaware Ave., 14202. Phone: (716) 883-3010. Fax: (716) 883-3606. E-mail: wdcxinfo@crawfordbroadcasting.com Web Site: www.wdcxfm.com. Licensee: Kimtron Inc. Group owner: Crawford Broadcasting Co. Population served: 6,000,000 Format: Christian & relg talk. Target aud: General. ♦Donald B. Crawford, pres; Nevin W. Larson, gen mgr.

WECK(AM)—See Cheektowaga

WEDG(FM)— 1947: 103.3 mhz; 49 kw. 340 ft TL: N42 55 33 W78 50 28. Stereo. Hrs opn: 50 James E. Casey Dr., 14206. Phone: (716) 881-4555. Fax: (716) 884-2931. Web Site: www.wedg.com. Licensee: Citadel Broadcasting Co. Group owner: Citadel Broadcasting Corp. (acq 2-23-00; grpsl). Law Firm: Shaw Pittman. Format: Alternative rock. Target aud: 18-34; alternative rock listeners.

***WFBF(FM)**— 1989: 89.9 mhz; 16 kw. Ant 295 ft TL: N42 41 19 W78 45 15. Hrs open: 918 Chesapeake Ave., Annapolis, MD, 21403. Phone: (716) 312-0911. Fax: (410) 268-0931. Web Site: www.familyradio.com. Licensee: Family Stations Inc. (group owner) Format: Christian, educ. ♦Harold Camping, pres.

WGR(AM)— May 22, 1922: 550 khz; 5 kw-U, DA-2. TL: N42 46 04 W78 50 39. Stereo. Hrs opn: 24 500 Corporate Pkwy., Amherst, 14226. Phone: (716) 843-0600. Fax: (716) 832-3080. Fax: (803) 0550 (studio). E-mail: studio@wgr55.com Web Site: www.wgr55.com. Licensee: Entercom Buffalo License LLC. Group owner: Entercom Communications Corp. (acq 12-13-99; grpsl). Population served: 3,800,000 Law Firm: Fisher, Wayland, Cooper, Leader & Zaragoza L.L.P. Format: Sports. News staff: 15; News: 168 hrs wkly. Target aud: 25-54. ♦Greg Ried, VP & gen mgr; Andy Roth, progmg dir & chief of engrg.

WGRF(FM)— Sept 14, 1959: 96.9 mhz; 24 kw. 712 ft TL: N42 57 14 W78 52 37. Stereo. Hrs opn: 50 James E. Casey Dr., 14206. Phone: (716) 881-4555. Fax: (716) 882-ufax. Web Site: www.97rock.com. Licensee: Citadel Broadcasting Co. Group owner: Citadel Broadcasting Corp. (acq 2-23-00; grpsl). Law Firm: Fisher, Wayland, Cooper, Leader & Zaragoza. Format: Classic rock. Target aud: 25-49; classic rock listeners. ♦John Hager, opns mgr.

WHTT-FM— Oct 3, 1954: 104.1 mhz; 50 kw. 500 ft TL: N42 49 50 W78 47 54. Stereo. Hrs opn: 24 50 James E. Casey Dr., 14206. Phone: (716) 881-4555. Fax: (716) 884-2931. Web Site: www.whtt.com. Licensee: Citadel Broadcasting Co. Group owner: Citadel Broadcasting Corp. (acq 2-23-2000; grpsl). Population served: 641,071 Format: Oldies. ♦Kevin Legrett, gen mgr; Chet Osadchey, gen sls mgr; Joe Siragusa, progmg dir; Chris Klein, news dir; Al Marranca, pub affrs dir & chief of engrg; Cheryl Schutt, traf mgr.

WBBF(AM)—Co-owned with WHTT-FM. September 1947: 1120 khz; 1 kw-D. TL: N42 49 50 W78 47 54. 225 Delaware, Suite 1 A, 14202. Phone: (716) 848-1120. Fax: (716) 848-9518. E-mail: totallygospel@adelphia.net 462,768 Format: Gospel. ♦Michael Brummer, gen mgr; John Young, opns; Natalie Hutchen, opns.

WJYE(FM)— Nov 11, 1966: 96.1 mhz; 47.1 kw. 505 ft TL: N42 53 10 W78 52 25. Stereo. Hrs opn: 24 14 Lafayette Sq., Suite 1200, 14203. Phone: (716) 856-3550. Fax: (716) 852-0537. Web Site: www.wjye.com. Licensee: Regent Broadcasting of Buffalo Inc. Group owner: Infinity Broadcasting Corp. (acq 12-15-2006; grpsl). Population served: 220,800 Natl. Rep: Christal. Format: Adult contemp. News staff: one; News: 23 hrs wkly. Target aud: 25-54. Spec prog: Pub affrs 2 hrs wkly. ♦Jeff Silver, sr VP & VP; Joe Chille, opns mgr; Bob Courtney, gen sls mgr.

***WNED(AM)**— Oct 14, 1924: 970 khz; 5 kw-U, DA-1. TL: N42 44 41 W78 00 00. Hrs open: 24 Box 1263, 14240. Secondary address: 140 Lower Terr. 14202. Phone: (716) 845-7000. Fax: (716) 845-7043. Web Site: www.wned.org. Licensee: Western New York Public Broadcasting Assoc. (acq 8-14-76). Population served: 1,000,000 Natl. Network: PRI, NPR. Format: News/talk. News staff: 8. Target aud: 35 plus. Spec prog: Pub affrs. ♦Donald K. Boswell, pres; Richard J. Daly, stn mgr; Cynthia Dwyer, dev VP; Jim Dimino, sls VP; Gwen Mysiak, prom dir; Al Wallack, progmg dir; Jon Herrington, engrg VP; Monica Wilson, local news ed; Sam Anson, local news ed & sports cmtr.

WNED-FM— June 6, 1960: 94.5 mhz; 105 kw. 710 ft TL: N42 38 13 W78 46 05. Stereo. Web Site: www.wned.org.1,400,000 Natl. Network: PRI. Law Firm: Schwartz, Woods & Miller. Format: Class. ♦Peter Goldsmith, progmg dir. Co-owned TV: *WNED-TV affil

WWKB(AM)— 1925: 1520 khz; 50 kw-U, DA-1. TL: N42 46 10 W78 50 34. Hrs open: 24 500 Corporate Pkwy., 14226. Phone: (716) 843-0600. Fax: (716) 832-3323. Web Site: www.kb1520.com. Licensee: Entercom Buffalo License LLC. Group owner: Entercom Communications Corp. (acq 12-13-99; grpsl). Population served: 462,768 Natl. Rep: D & R Radio. Format: Progressive talk. News staff: 4; News: 20 hrs wkly. Target aud: 25-54. ♦Gregory Reed, gen mgr & prom mgr.

WWWS(AM)— 1934: 1400 khz; 1 kw-U. TL: N42 55 33 W78 50 28. Hrs open: 500 Corporate Pkwy., 14226. Phone: (716) 843-0600. Fax: (716) 843-3323. Licensee: Entercom Buffalo License LLC. Group owner: Entercom Communications Corp. (acq 12-13-99; grpsl). Population served: 992,100 Natl. Rep: Katz Radio. Format: Urban contemp. Target aud: 35-54.

WYRK(FM)— Nov 14, 1962: 106.5 mhz; 50 kw. 390 ft TL: N42 53 10 W78 52 25. Stereo. Hrs opn: 14 Lafayette Sq., Suite 1200, 14203. Phone: (716) 852-7444. Fax: (716) 852-5683. Web Site: www.wyrk.com. Licensee: Regent Broadcasting of Buffalo Inc. Group owner: Infinity Broadcasting Corp. (acq 12-15-2006; grpsl). Population served: 462,768 Natl. Rep: Katz Radio. Wire Svc: UPI Format: Country. Target aud: 25-54; adults. ♦Jeff Silver, sr VP & VP; Mark Plimpton, gen sls mgr; Mike Krupa, natl sls mgr; Sue Durwald, prom dir; John Paul, progmg dir.

Calverton-Roanoke

WDRE(FM)— 1998: 105.3 mhz; 1 kw. Ant 492 ft TL: N40 51 18 W72 46 12. (CP: 660 w, ant 607 ft). Hrs opn: 24 3075 Vets Memorial Hwy. #201, Ronkonkoma, 11779. Phone: (631) 648-2500. Fax: (516) 222-1391. Licensee: Jarad Broadcasting Co. of Calverton Inc. Group owner: The Morey Organization Inc. (acq 10-2-98). Natl. Rep: Christal. Format: Rhythmic dance top-40. ♦Beverly Fortune, gen mgr.

Canajoharie

***WCAN(FM)**— October 1988: 93.3 mhz; 6 kw. 268 ft TL: N42 53 46 W74 35 45. Stereo. Hrs opn: 24
Rebroadcasts WAMC-FM Albany 100%.
Box 66600, 318 Central Ave., Albany, 12206-6600. Phone: (518) 465-5233. Phone: (800) 323-9262. Fax: (518) 432-6974. E-mail: mail@wamc.org Web Site: www.wamc.org. Licensee: WAMC. Group owner: WAMC/Northeast Public Radio Natl. Network: PRI, NPR. Law Firm: Dow, Lohnes & Albertson. Wire Svc: AP Format: News, talk. News: 48 hrs wkly. Target aud: General. Spec prog: Jazz 17 hrs, folk 7 hrs wkly. ♦Alan Chartock, CEO; David Galletly, VP; Selma Kaplan, VP.

Canandaigua

WCGR(AM)— Apr 5, 1961: 1550 khz; 250 w-D. TL: N42 52 52 W77 15 02. Hrs open: 6 AM-6 PM 3568 Lenox Rd., Geneva, 14456. Phone: (315) 781-7000. Fax: (315) 781-7700. Licensee: Canandaigua Broadcasting Inc. (acq 12-93; with co-located FM; FTR: 1-3-94). Law Firm: James L. Oyster. Format: News/talk, middle of the road. News staff: one; News: 10 hrs wkly. Target aud: General. ♦George Kimble, pres; Alan Bishop, gen mgr.

***WCIY(FM)**— Dec 14, 1992: 88.9 mhz; 680 w. Ant 1,063 ft TL: N42 44 44 W77 25 34. Stereo. Hrs opn:
Rebroadcasts WCIK(FM) Bath 100%.
Box 506, Bath, 14810. Secondary address: 7634 Campbell Creek Rd., Bath Phone: (607) 776-4151. Fax: (607) 776-6929. E-mail: mail@fln.org Web Site: www.fln.org. Licensee: Family Life Ministries Inc. Group owner: Family Life Network Natl. Network: Salem Radio Network. Law Firm: Hardy, Carey, Chautin & Balkin, LLP. Wire Svc: Metro Weather Service Inc. Format: Contemp Christian. News staff: 3; News: 14 hrs wkly. Target aud: 30-54. ♦Dick Snavely, pres & CFO; Rick Snavely, pres, VP & gen mgr; John Owens, progmg dir; Jim Travis, chief of engrg.

WRSB(AM)— Apr 5, 1997: 1310 khz; 1 kw-U, DA-2. TL: N42 53 20 W77 19 09. Hrs open:
Rebroadcasts WASB(AM) Brockport 100%.
6675 Fourth Section Rd., Brockport, 14420. Phone: (585) 637-7040. Licensee: David Wolfe (acq 12-21-2005; with WASB(AM) Brockport). Natl. Network: ABC. Format: Christian. Target aud: Everyone; all ages. ♦Dr. David Wolfe, gen mgr.

WVOR(FM)— July 16, 1974: 102.3 mhz; 3.4 kw. 282 ft TL: N42 51 47 W77 19 22. Hrs open: 24 207 Midtown Plaza, Rochester, 14604. Phone: (585) 454-3942. Fax: (585) 454-5081. E-mail: wisy.wfxfproductions@clearchannel.com Web Site: www.radiosunny.com. Licensee: Citicasters Licenses L.P. Group owner: Clear Channel Communications Inc. (acq 5-4-99; grpsl). Population served: 40,000 Format: Soft adult contemp. News: 3 hrs wkly. Target aud: 25-54. ♦Karen Carey, gen mgr; David Lefrois, progmg dir.

Canton

WNCQ-FM— July 1984: 102.9 mhz; 23.5 kw. Ant 338 ft TL: N44 32 10 W75 05 46. Stereo. Hrs opn: 24 1 Bridge Plaza, Suite 204, Ogdensburg, 13669. Phone: (315) 393-1220. Fax: (315) 393-3974. E-mail: john@q1029.com Web Site: www.q1029.com. Licensee: Radio Power Inc. Group owner: Martz Communications Group (acq 9-28-99). Population served: 100,000 Law Firm: Eugene T. Smith. Format: Hot country. Target aud: 25-54; adults. ♦John Winter, gen mgr.

WRCD(FM)— Jan 1, 1997: 101.5 mhz; 2.4 kw. 364 ft TL: N44 32 01 W75 05 50. Stereo. Hrs opn: 24 Box 210, Massena, 13662. Phone: (315) 769-3333. Fax: (315) 769-3299. E-mail: studio@1015thefox.com Web Site: www.1015thefox.com. Licensee: Radio Power Inc. Group owner: Martz Communications Group (acq 6-16-99). Law Firm: Eugene T. Smith. Format: Rock. News staff: 2; News: 7 hrs wkly. Target aud: 30-50; country fans. ♦Michael Boldt, gen mgr; Drew Scott, progmg dir.

***WSLU(FM)**— December 1964: 89.5 mhz; 40.3 kw. 299 ft TL: N44 32 01 W75 05 50. Hrs open: 24 St. Lawrence Univ., 13617. Phone: (315) 229-5356. Fax: (315) 229-5373. Web Site: www.ncpr.org. Licensee: St. Lawrence University. Population served: 1,000,000 Natl. Network: NPR, PRI. Law Firm: Donald E. Martin. Format: Eclectic public radio. News staff: 2; News: 35 hrs wkly. Target aud: General. Spec prog: Gospel, jazz, class, folk, pub affrs, Black. ♦Ellen Rocco, gen mgr; Sandra Demarest, dev dir; Jacqueline Sauter, progmg dir; Martha Foley, news dir; Robert G. Sauter, chief of engrg.

Cape Vincent

WBDR(FM)— Apr 21, 1997: 102.7 mhz; 6 kw. 328 ft TL: N44 06 58 W76 20 21. Hrs open: 24
Rebroadcasts WBDI(FM) Copenhagen 100%.
199 Wealtha Ave., Watertown, 13601. Phone: (315) 782-0103. Fax: (315) 782-0312. Licensee: Border International Broadcasting Inc. Group owner: Clancy-Mance Communications (acq 10-15-98; $50,000). Population served: 150,000 Natl. Rep: Roslin. Format: CHR. ♦David Mance, CEO, pres & gen mgr; Todd Dalessandro, opns VP; Dick Whelan, rgnl sls mgr.

***WMHI(FM)**— Oct 1, 1990: 94.7 mhz; 6 kw. 284 ft TL: N44 02 42 W76 15 37. Stereo. Hrs opn: 24
Rebroadcasts WMHR(FM) Syracuse 98%.
4044 Makyes Rd., Syracuse, 13215. Phone: (315) 469-5051. E-mail: mhn@marshillnetwork.org Web Site: www.marshillnetwork.org. Licensee: Mars Hill Broadcasting Co. Inc. dba Mars Hill Network. (group owner) Natl. Network: Moody, Salem Radio Network. Law Firm: Wiley, Rein & Fielding. Wire Svc: AP Format: Relg, Christian. News: 6 hrs wkly. Target aud: General; Christian families. ♦Clayton Roberts, pres; Wayne Taylor, gen mgr; Chris Tetta, progmg dir; Jeremy Miller, news dir; Valerie Smith, traf mgr.

Carthage

WTOJ(FM)— Nov 1, 1984: 103.1 mhz; 6 kw. 500 ft TL: N43 57 16 W75 43 45. Stereo. Hrs opn: 24 199 Wealtha Ave., Watertown, 13601. Phone: (315) 782-1240. Fax: (315) 782-0312. Web Site: www.magic103.com. Licensee: Community Broadcasters LLC. (group owner; (acq 10-13-2006; grpsl). Population served: 120,000 Natl. Rep: Roslin. Format: Adult contemp. Target aud: 25-54. ♦James L. Leven, pres; David W. Mance, gen mgr; Joseph Brosk, stn mgr; Todd Dalesandro, opns dir.

Catskill

WCKL(AM)— Feb 6, 1970: 560 khz; 1 kw-D, DA. TL: N42 12 00 W73 50 07. Hrs open: 24 2271 Adam Clayton Powell Blvd., New York, 10030. Phone: (518) 828-5006. Fax: (518) 828-1080. Licensee: Black United Fund of New York Inc. (acq 6-10-2003; $100,000). Population served: 980,450 Format: Talk. ♦Kermit Eady, pres.

WCTW(FM)— September 1990: 98.5 mhz; 2.1 kw. Ant 393 ft TL: N42 12 00 W73 50 07. (CP: 4.7 kw). Stereo. Hrs opn: 24 20 Tucker Drive, Poughkepsie, 12534. Phone: (845) 471-2300. Fax: (845) 471-2683. Web Site: www.985litefm.com. Licensee: CC Licenses LLC. Group owner: Clear Channel Communications Inc. (acq 1-17-2002; grpsl). Natl. Network: Westwood One. Natl. Rep: Katz Radio. Format: Bright adult contemp. News staff: one; News: 2 hrs wkly. Target aud: 25-44; female. ♦Frank Curcio, gen mgr; Reggie Osterhoudt, opns mgr; Jeanette Relyea, natl sls mgr; Nick Smirnoff, prom dir; Michelle Taylor, progmg dir; Cameron Hendrix, news dir; Bill Draper, chief of engrg; Dawn Morvillo, traf mgr.

Cazenovia

***WITC(FM)**— April 1978: 88.9 mhz; 129 w. 33 ft TL: N42 55 53 W75 51 15. Hrs open: Noon-midnight Cazenovia College, 22 Sullivan St., 13035. Phone: (315) 655-7154. Licensee: Cazenovia College. Population served: 15,000 Format: Alternative. News: 3 hrs wkly. Target aud: 15-35; college & young area residents. Spec prog: News/talk 3 hrs, div 10 hrs wkly. ♦Roger Benn, gen mgr.

Center Moriches

WLVG(FM)— Mar 3, 1997: 96.1 mhz; 2.65 kw. Ant 499 ft TL: N40 51 08 W72 45 55. Hrs open: 24 3241 Rt. 112, Bldg. #7, Medford, 11763. Phone: (631) 451-1039. Fax: (631) 451-0891. Web Site: www.wrcn.com. Licensee: IW Limited Liability Co. Group owner: Barnstable Broadcasting Inc. acq 1-13-2004; $3.75 million). Format: Adult contemp. ♦Dave Widmer, gen mgr; Robin Schlakman, gen sls mgr; Wendy Summers, opns mgr & prom dir; Bob Anderson, chief of engrg.

Champlain

WCHP(AM)— Aug 20, 1985: 760 khz; 35 kw-D, DA. TL: N44 56 44 W73 25 48. Hrs open: Sunrise-sunset Box 888, 137 Rapids Rd., 12919. Phone: (518) 298-2800. Fax: (518) 298-2604. E-mail: wchp@wchp.com Web Site: www.wchp.com. Licensee: Champlain Radio Inc. (acq 1-31-91; FTR: 2-18-91). Population served: 4,000,000 Format: Relg, talk. Target aud: 25 plus. Spec prog: Fr, Sp. ♦Robert A. Jones, VP; Teri Billiter, gen mgr; Tonya Billiter, opns mgr; Brandi Lloyd, progmg dir.

Chateaugay

WYUL(FM)— Apr 15, 1997: 94.7 mhz; 1.9 kw. 2,081 ft TL: N44 41 43 W73 53 00. Stereo. Hrs opn: 24 86 Porter Rd., Malone, 12953. Phone: (518) 483-1100. Fax: (518) 483-1382. Licensee: Cartier Communications Inc. Group owner: Martz Communications Group Population served: 3,500,000 Law Firm: Smithwick & Belendiuk. Format: CHR. ♦Timothy D. Martz, CEO, pres & CFO; Michael T. Boldt, gen mgr; Kim Scott, sls dir; Drew Scott, progmg dir.

Cheektowaga

WECK(AM)— August 1956: 1230 khz; 1 kw-U. TL: N42 55 27 W78 46 41. Hrs open: 14 Lafayette Sq., Suite 1200, Buffalo, 14203. Phone: (716) 856-3550. Fax: (716) 852-0537. Web Site: www.weck1230.com. Licensee: Regent Broadcasting of Buffalo Inc. Group owner: Infinity Broadcasting Corp. (acq 12-15-2006; grpsl). Population served: 100,000 Natl. Network: Westwood One. Natl. Rep: Christal. Format: Country. Target aud: 35-64. Spec prog: Pol 2 hrs wkly. ♦Jeff Silver, sr VP; Joe Chille, opns mgr.

Chenango Bridge

WWYL(FM)— July 1, 1996: 104.1 mhz; 3.1 kw. 462 ft TL: N42 08 20 W75 52 24. Hrs open: Box 414, Binghamton, 13902. Phone: (607) 772-8400. Fax: (607) 772-9806. Web Site: www.wild104fm.com. Licensee: Citadel Broadcasting Co. Group owner: Citadel Broadcasting Corp. (acq 6-9-99; grpsl). Format: CHR. ♦Mary Beth Walsh, gen mgr; Tim Skinner, prom dir; K.J. Bryant, progmg dir.

Cherry Valley

WJIV(FM)— 1949: 101.9 mhz; 11.5 kw. 1,027 ft TL: N42 47 36 W74 41 41. Stereo. Hrs opn: 24 Box 507, 13320. Secondary address: 1668 Country Hwy. 50 13320. Phone: (607) 264-3062. Phone: (518) 437-1251. Fax: (607) 264-8277. Licensee: Christian Broadcasting System Ltd. (group owner; acq 6-5-00; $1.3 million). Population served: 4,000,000 Law Firm: Bechtel & Cole. Wire Svc: UPI Format: Relg, talk. Target aud: 25-54. ♦John Yinger, pres.

Clifton Park

WPTR(FM)— November 1985: 96.7 mhz; 4.7 kw. 328 ft TL: N42 52 44 W73 51 47. Stereo. Hrs opn: 24 4243 Albany St., Albany, 12212. Phone: (518) 862-1540. Fax: (518) 862-1545. Web Site: www.crawfordbroadcasting.com. Licensee: Kimtron Inc. Group owner: Crawford Broadcasting Co. (acq 1996; $820,000). Population served: 850,000 Format: Big band, oldies. Target aud: 30-64; financially capable. ♦Donald B. Crawford, pres; Robert Hammond, gen mgr & stn mgr; Mark Shuttleworth, progmg dir; David Groth, chief of engrg.

Clinton

***WHCL-FM**— Feb 18, 1963: 88.7 mhz; 270 w. 97 ft TL: N43 03 04 W75 24 24. Stereo. Hrs opn: 24 Hamilton College, 198 College Hill Rd., 13323. Phone: (315) 859-4200. E-mail: mngrwhcl@hamilton.edu Web Site: www.whcl.org. Licensee: The Trustees of Hamilton College. Population served: 170,000 Format: Div, progsv, AOR. Target aud: General. Spec prog: Class 9 hrs, jazz 9 hrs, relg 2 hrs. ♦Alan Clark, gen mgr.

Clyde

***WCOV-FM**— Dec 5, 1995: 93.7 mhz; 3.8 kw. Ant 328 ft TL: N42 59 38 W76 51 59. Hrs open:
Rebroadcasts WCIK(FM) Bath 100%.
Box 506, Bath, 14810. Secondary address: 7634 Campbell Creek Rd., Bath 14810. Phone: (607) 776-4151. Fax: (607) 776-6929. E-mail: mail@fln.org Web Site: www.fln.org. Licensee: Family Life Ministries Inc. Group owner: Family Life Network (acq 10-3-00). Natl. Network: Salem Radio Network. Law Firm: Hardy, Carey, Chautin & Balkin, LLP. Wire Svc: Metro Weather Service Inc. Format: Contemp Christian. News staff: 3; News: 14 hrs wkly. Target aud: 30-54. ♦Dick Snavely, CFO; Rick Snavely, pres & gen mgr; John Owens, progmg dir; Jim Travis, chief of engrg.

Cobleskill

WQBJ(FM)— Sept 1, 1986: 103.5 mhz; 50 kw. Ant 492 ft TL: N42 58 21 W74 29 30. Stereo. Hrs opn:
Rebroadcasts WQBK-FM Rensselaer 100%.
1241 Kings Rd., Schenectady, 12303. Phone: (518) 881-1515. Fax:

(518) 881-1516. Web Site: www.wqbk.com. Licensee: Regent Licensee of Mansfield Inc. Group owner: Regent Communications Inc. (acq 8-7-2000; grpsl). Law Firm: Haley, Bader & Potts. Format: Rock/AOR. Target aud: 18-49; general. ♦Robert Ausfeld, gen mgr; Bob O'Neal, chief of engrg.

WSDE(AM)— July 1, 1981: 1190 khz; 1 kw-D. TL: N42 41 26 W74 26 40. Hrs open: 6 AM-sunset Box 608, 12043. Phone: (518) 234-3400. Fax: (518) 234-4567. Web Site: www.wsde1190.com. Licensee: Viva Communications Group LLC (acq 4-26-2004; $120,000). Population served: 1,000,000 Format: Personality talk. Target aud: General. ♦Floyld Hamilton, opns mgr.

Conklin

WKGB-FM— Feb 11, 1989: 92.5 mhz; 1.45 kw. Ant 676 ft TL: N42 06 48 W75 51 09. Stereo. Hrs opn: 24 320 N. Jensen Rd., Vestal, 13850. Phone: (607) 584-5800. Fax: (607) 584-5900. Web Site: www.925kgb.com. Licensee: CC Licenses LLC. Group owner: Clear Channel Communications Inc. (acq 4-14-2000; grpsl). Natl. Rep: D & R Radio. Law Firm: Carr, Morris & Graeff. Format: AOR, classic rock. News: one hr wkly. Target aud: 25-49; baby boomers who grew up with rock and roll of the 60s & 70s. Spec prog: Jazz 2 hrs, farm one hr wkly. ♦Tom Burney, gen mgr; Jim Free, opns mgr & progmg dir; Michele Page, sls dir; Tom Barney, mktg mgr.

Copenhagen

WBDI(FM)— 1994: 106.7 mhz; 1.7 kw. 1,191 ft TL: N43 52 47 W75 43 11. Hrs open: 24 199 Wealthea Ave., Watertown, 13601. Phone: (315) 782-1240. Fax: (315) 782-0312. Licensee: Community Broadcasters LLC. Group owner: Clancy-Mance Communications (acq 10-13-2006; grpsl). Natl. Rep: Roslin. Format: CHR. ♦James L. Leven, pres; David W. Mance, gen mgr; Todd Dalesandro, opns mgr; Dick Whelan, rgnl sls mgr.

Corinth

WFFG-FM— June 26, 1967: 107.1 mhz; 2.85 kw. Ant 482 ft TL: N43 14 40 W73 46 18. Stereo. Hrs opn: 24 89 Everts Ave., Queensbury, 12804. Phone: (518) 793-7733. Fax: (518) 793-0838. Web Site: www.froggy107.com. Licensee: 6 Johnson Road Licenses Inc. Group owner: Pamal Broadcasting Ltd. (acq 4-1-2004; grpsl). Population served: 110,000 Format: Country. News staff: one; News: 2 hrs wkly. Target aud: 18-54. ♦Clay Ashworth, gen mgr.

Corning

WCBA(AM)— November 1948: 1350 khz; 2 kw-D. TL: N42 07 01 W77 02 25. Hrs open: 24 21 E. Market St., 14830. Phone: (607) 937-8181. Fax: (607) 962-1138. Licensee: Group A Licensee LLC. Group owner: Route 81 Radio LLC (acq 10-31-2003; grpsl). Population served: 250,000 Natl. Network: Westwood One. Natl. Rep: Roslin. Format: All sports. Target aud: 50 plus. ♦Jamie Evans, gen mgr & stn mgr.

WGMM(FM)—Co-owned with WCBA(AM). February 1989: 98.7 mhz; 2 kw. Ant 393 ft TL: N42 09 38 W77 02 19. Stereo. 24
Rebroadcasts WENY-FM Elmira 100%.
Natl. Network: ABC. Natl. Rep: McGavren Guild. Format: Adult contemp. ♦Paul Lyle, gen mgr.

***WCEB(FM)**— 1979: 91.9 mhz; 10 w. 1,784 ft TL: N42 07 10 W77 05 02. Hrs open: Corning Community College, One Academic Dr., 14830. Phone: (607) 962-9360. Fax: (607) 962-9456. E-mail: wceb919@hotmail.com Licensee: Corning Community College. Population served: 5,300 Format: AOR. Target aud: Students; CCC and surrounding community. Spec prog: Oldies 6 hrs wkly. ♦Talik Murphy, pres & gen mgr; Andrew Forbes, progmg dir.

WENI(AM)— November 1949: 1450 khz; 1 kw-D, 930 w-N. TL: N42 06 59 W77 02 24. Hrs open:
Rebroadcasts WENY(AM) Elmira 100%.
21 E. Market St., 14830. Phone: (607) 962-4646. Fax: (607) 962-1138. Licensee: Group A Licensee LLC. Group owner: Route 81 Radio LLC (acq 10-31-2003; grpsl). Population served: 40,000 Natl. Network: USA. Natl. Rep: McGavren Guild. Format: News/talk. Target aud: 25-64. ♦Paul Lyle, gen mgr.

WNKI(FM)— May 1947: 106.1 mhz; 40 kw. 532 ft TL: N42 09 43 W77 02 15. Stereo. Hrs opn: 2205 College Ave., Elmira, 14903. Phone: (607) 732-4400. Fax: (607) 732-7774. Web Site: www.wink106.com. Licensee: Chemung County Radio Inc. Group owner: Backyard Broadcasting LLC (acq 12-1-02; grpsl). Population served: 250,000 Natl. Rep: Christal. Format: Adult contemp, CHR. News staff: one; News: one hr wkly. Target aud: 25-54; women. ♦Kevin White, gen mgr; Ally Payne, news dir.

***WSQE(FM)**— 1995: 91.1 mhz; 3.6 kw. Ant 653 ft TL: N42 06 20 W76 52 17. Stereo. Hrs opn: 24
Rebroadcasts WSKG-FM Binghamton 100%.
Box 3000, Binghamton, 13902. Phone: (607) 729-0100. Web Site: www.wskg.com. Licensee: WSKG Public Telecommunications Council. Population served: 160,100 Natl. Network: NPR, PRI, AP Radio. Law Firm: Dow, Lohnes & Albertson. Format: Class, news. News staff: one; News: 33 hrs wkly. Target aud: General. Spec prog: Jazz, folk 5 hrs wkly. ♦Gary Reinbolt, CEO & gen mgr.

Cornwall

WWLE(AM)— Nov 22, 1969: 1170 khz; 1 kw-D, DA. TL: N41 26 24 W74 04 25. Hrs open: Box 2130, Newburgh, 12550. Phone: (845) 569-7010. Fax: (845) 562-1348. Licensee: 1170 Broadcast Radio Inc. (acq 1-1-00; $100,000). Natl. Network: USA. Law Firm: Gammon & Grange. Format: News/talk. Spec prog: Farm one hr wkly. ♦Charles Stewart, gen mgr.

Cortland

WIII(FM)—Listing follows WKRT(AM).

WKRT(AM)— Nov 15, 1947: 920 khz; 1 kw-D, 500 w-N, DA-N. TL: N42 33 22 W76 09 17. Hrs open: 24 277 Tompkins St., 13045. Phone: (607) 756-2828. Fax: (607) 756-2953. E-mail: wiii.wkrt@citcamm.com Web Site: www.wkrt.com. Licensee: Citadel Broadcasting Co. Group owner: Citadel Broadcasting Corp. (acq 2-23-00; grpsl). Population served: 45,000 Natl. Rep: Katz Radio. Law Firm: Leventhal, Senter & Lerman. Format: News/talk. News staff: one; News: 20 hrs wkly. Target aud: 30-60; general. ♦Todd Mallinson, gen mgr; Mark Vanness, opns mgr; Margaret Tollner, gen sls mgr; Dave Edwards, engrg mgr & chief of engrg; Sarah Owen, traf mgr; Ryan Dean, local news ed.

WIII(FM)—Co-owned with WKRT(AM). Nov 15, 1947: 99.9 mhz; 24 kw. 710 ft TL: N42 33 22 W76 09 17. Stereo. 24 E-mail: i100@wiii.com Web Site: www.wiii.com.200,000 Natl. Rep: Katz Radio. Law Firm: Leventhal, Senter & Lerman. Format: Classic rock. News staff: one; News: one hr wkly. Target aud: 25-54; men. ♦Tony DeFranco, opns mgr.

***WSUC-FM**— Nov 17, 1976: 90.5 mhz; 241 w. -110 ft TL: N42 35 53 W76 11 13. Stereo. Hrs opn: 24 State Univ. of New York, Brockway Hall, Graham Ave., 13045. Phone: (607) 753-2936. Fax: (607) 753-2807. Licensee: State University of New York. Population served: 32,000 Natl. Network: AP Radio. Format: Var/div, rock. Target aud: 12-50. ♦Peter Johams, gen mgr.

Dannemora

***WKVJ(FM)**— 2005: 89.7 mhz; 4.4 kw. Ant 1,096 ft TL: N44 34 24 W73 40 31. Stereo. Hrs opn: 24 American Educational Broadcasting Inc., 3185 S. Highland Dr., Suite 13, Las Vegas, NV, 89109. Secondary address: Box 888, Studio, Champlain 12919. Phone: (518) 298-8200. Fax: (518) 298-2604. Licensee: American Educational Broadcasting Inc. Law Firm: Fletcher, Heald & Hildreth. Format: Christian. ♦Carl J. Auel, pres; Teri Billiter, gen mgr.

Dansville

WDNY-FM— March 1990: 93.9 mhz; 570 w. 741 ft TL: N42 30 45 W77 38 07. Stereo. Hrs opn: 24 195 Main St., 14437. Phone: (585) 335-2273. Fax: (585) 335-9677. E-mail: wdny@frontiernet.net Licensee: Miller Media Inc. Population served: 50,000 Natl. Network: Jones Radio Networks, Westwood One. Format: Adult contemp. News staff: one; News: 8 hrs wkly. Target aud: 25-54. Spec prog: Relg one hr, big band 3 hrs, sports 4 hrs wkly. ♦Dorothy Hotchkiss, gen mgr.

WDNY(AM)— Oct 20, 1978: 1400 khz; 1 kw-U. TL: N42 32 19 W77 40 57.19 (Acq 4-13-92; $290,000; FTR: 5-4-92).20,000 Natl. Network: Jones Radio Networks, Westwood One. Format: Music of your Life. News staff: one; News: 8 hrs wkly. ♦Mark Miller, progmg dir.

Delhi

WDHI(FM)— Mar 16, 1992: 100.3 mhz; 770 w. 643 ft TL: N42 22 40 W74 50 23. Stereo. Hrs opn: 16 34 Chestnut St., Oneonta, 13820. Phone: (607) 432-1030. Fax: (607) 432-6909. Licensee: Double O Central New York Corp. (group owner; (acq 10-22-2004; grpsl). Natl. Network: USA. Format: CHR. ♦George Wells, gen mgr.

Depew

WBLK(FM)— December 1964: 93.7 mhz; 47 kw. 505 ft TL: N42 53 10 W78 52 25. Stereo. Hrs opn: Rand Bldg., 14 Lafayette Sq., Buffalo, 14203. Phone: (716) 852-9393. Fax: (716) 852-9390. Web Site: www.wblk.com. Licensee: Regent Broadcasting of Buffalo Inc. Group owner: Infinity Broadcasting Corp. (acq 12-15-2006; grpsl). Population served: 462,768 Natl. Network: CBS Radio. Natl. Rep: Katz Radio. Format: Urban contemp. Target aud: General. ♦Jeff Silver, sr VP & VP; Mike Krupa, gen sls mgr.

Deposit

WIYN(FM)— Jan 16, 1991: 94.7 mhz; 770 w. Ant 642 ft TL: N42 01 43 W75 28 25. Stereo. Hrs opn: 16 34 Chestnut St., Oneonta, 13820. Phone: (607) 432-1030. Fax: (607) 432-6909. Licensee: Double O Central New York Corp. (group owner; (acq 10-22-2004; grpsl). Population served: 40,000 Format: Oldies. Target aud: 28-55. ♦George Wells, gen mgr.

DeRuyter

WWDG(FM)— 1948: 105.1 mhz; 42 kw. Ant 541 ft TL: N42 46 58 W75 50 28. Stereo. Hrs opn: 500 Plum St., Suite 100, Syracuse, 13204. Phone: (315) 472-9797. Fax: (315) 472-0049. Licensee: CC Licenses LLC. Group owner: Clear Channel Communications Inc. (acq 3-12-2001; $5 million). Population served: 500,000 Format: Adult contemp. Target aud: 18-40; general. ♦Joel Delmonico, gen mgr.

Dewitt

WVOA(AM)—Not on air, target date: unknown: 720 khz; 2.5 kw-D, 390 w-N, DA-N. TL: N43 03 30 W76 10 01 (D), N42 56 02 W76 06 59 (N). Hrs open: 4853 Manor Hill Dr., Syracuse, 13215-1336. Phone: (315) 468-0908. Licensee: Cram Communications LLC. ♦Sam Furco, gen mgr.

Dundee

WFLR(AM)— Oct 1, 1956: 1570 khz; 5 kw-D, 442 w-N. TL: N42 32 40 W76 59 35. Hrs open: 24 30 Main St., 14837. Phone: (607) 243-7158. Phone: (607) 243-7070. Fax: (607) 243-7662. E-mail: wflr@linkny.com Web Site: www.linkny.com/wflr. Licensee: Finger Lakes Radio Group Inc. Group owner: Finger Lakes Radio Group (acq 12-11-03; $600,000 with co-located FM). Population served: 80,000 Natl. Network: Motor Racing Net. Format: Country, news/talk. News staff: one; News: 40 hrs wkly. Target aud: 25-55. Spec prog: Relg 5 hrs wkly. ♦Dick Evans, gen mgr; Mark Feiock, prom dir.

WFLR-FM— Aug 20, 1968: 95.9 mhz; 780 w. 600 ft TL: N42 32 40 W76 59 35. Stereo. 24 Format: Adult contemp, news/talk. News: 35 hrs wkly. Target aud: 21-55.

Dunkirk

WDOE(AM)— Dec 24, 1949: 1410 khz; 1 kw-D, 500 w-N, DA-N. TL: N42 27 51 W79 21 21. Stereo. Hrs opn: Box 209, Willow Rd., 14048. Phone: (716) 366-1410. Phone: (716) 366-8580. Fax: (716) 366-1416. Licensee: Chadwick Bay Broadcasting Corp. (acq 2-26-2001; with WBKX(FM) Fredonia). Population served: 75,000 Natl. Network: ABC. Format: Oldies, news/talk. News staff: News progmg 12 hrs wkly Target aud: 45-65. Spec prog: Pol 6 hrs, Sp 2 hrs wkly. ♦John Bulmer, pres; Chuck Telford, gen mgr.

East Aurora

WLKK(FM)—See Wethersfield Township

East Hampton

WEHN(FM)— Mar 1, 1993: 96.9 mhz; 4.3 kw. Ant 384 ft TL: N40 59 37 W72 10 19. Hrs open: 24
Simulcast with WEHM(FM) Southampton 100%.
Box 7162, Amagansett, 11930. Secondary address: 249 Montauk Hwy., Amagansett 11930. Phone: (631) 267-7800. Fax: (631) 267-1018. Web Site: www.wehm.com. Licensee: AAA Licensing LLC. Group owner: AAA Entertainment L.L.C. (acq 5-31-2000; grpsl). Population served: 110,000 Natl. Network: CNN Radio. Format: Progsv adult rock. Target aud: 24-54; upscale Hamptons residents and NYC second homeowners. ♦Brian Cosgrove, opns mgr.

East Patchogue

WALK(AM)—Licensed to East Patchogue. See Patchogue

East Syracuse

WSIV(AM)—Licensed to East Syracuse. See Syracuse

Ellenville

WELG(AM)— December 1964: 1370 khz; 5 kw-D. TL: N41 44 19 W74 23 48. Hrs open: 20 Tucker Dr., Poughkeepsie, 12603. Secondary address: 22 N. Main St. 12428. Phone: (845) 471-2300. Fax: (845) 471-2683. Web Site: www.1370welg.com. Licensee: CC Licenses LLC. Group owner: Clear Channel Communications Inc. (acq 7-14-2000; grpsl). Population served: 50,000 Natl. Network: Jones Radio Networks. Natl. Rep: Katz Radio. Format: Nostalgia. Target aud: 30-64. ♦Frank Curcio, gen mgr; Reggie Osterhoudt, opns mgr; Jim Brady, gen sls mgr; Jeanette Relyea, natl sls mgr; Nick Smirnoff, prom dir; Rick Knight, progmg dir; Cameron Hendrix, news dir.

WRWC(FM)—Co-owned with WELG(AM). August 1970: 99.3 mhz; 115 w. Ant 1,630 ft TL: N41 41 06 W74 21 23. Stereo. 24
Simulcast with WRWD-FM Highland 100%.
Web Site: www.wrwdfm.com.400,000 Natl. Network: Jones Radio Networks. Natl. Rep: Katz Radio. Format: Country. News staff: one.

Elmira

***WCIH(FM)**— July 31, 1989: 90.3 mhz; 4 kw. Ant 526 ft TL: N41 53 39 W76 51 32. Stereo. Hrs opn: 24
Rebroadcasts WCIK(FM) Bath 100%.
Box 506, Bath, 14810. Secondary address: 7634 Campbell Creek Rd., Bath 14810. Phone: (607) 776-4151. Fax: (607) 776-6929. E-mail: mail@fln.org Web Site: www.fln.org. Licensee: Family Life Ministries Inc. Group owner: Family Life Network Natl. Network: Salem Radio Network. Law Firm: Hardy, Carey, Chautin & Balkin, LLP. Wire Svc: Metro Weather Service Inc. Format: Contemp Christian. News staff: 3; News: 14 hrs wkly. Target aud: 30-54; general Christian public. ♦Dick Snavely, CFO & VP; Rick Snavely, pres & gen mgr; John Owens, progmg dir; Jim Travis, chief of engrg.

***WECW(FM)**— Jan 19, 1959: 107.7 mhz; 6 w. -312 ft TL: N42 05 52 W76 48 53. Stereo. Hrs opn: Elmira College, One Park Pl., 14901. Phone: (607) 735-1885. Phone: (607) 735-1815. E-mail: wecw@elmira.edu Licensee: Elmira College. Population served: 55,000 Format: Classic rock, Top-40. Target aud: 18-30.

WEHH(AM)—See Elmira Heights-Horseheads

WELM(AM)— April 1947: 1410 khz; 5 kw-D, 1 kw-N, DA-N. TL: N42 07 11 W76 48 37. Stereo. Hrs opn: 24 1705 Lake St., 14901. Phone: (607) 733-5626. Phone: (607) 732-1400. Fax: (607) 733-5627. E-mail: ppinesmedia1@stny.rr.com Licensee: Pembrook Pines Elmira Ltd. Group owner: Pembrook Pines Media Group (acq 10-1-77). Population served: 350,000 Natl. Network: CBS. Law Firm: Bechtel & Cole. Format: Sports. News staff: one; News: 10 hrs wkly. Target aud: 25-54. Spec prog: Relg one hr wkly. ♦Robert J. Pfuntner, CEO, pres & gen mgr; Gary Knight, opns dir; David Crum, gen sls mgr & adv mgr; Patrick Leiby, mktg dir; Bob Michaels, progmg dir & sports cmtr; Brian Stoll, mus dir; Mike Jacobs, news dir & local news ed; Nancy Nicastro, pub affrs dir.

WLVY(FM)—Co-owned with WELM(AM). Aug 1, 1966: 94.3 mhz; 800 w. Ant 745 ft TL: N42 07 51 W76 47 26. Stereo. E-mail: airstaff@wlvy94rock.com Web Site: www.wlvy94rock.com. Natl. Network: Westwood One. Rgnl rep: Pembrook Pines Format: CHR, hot adult contemp, Top 40s. News staff: one; News: 5 hrs wkly. Target aud: 18-36; young vibrant adults. ♦Dave Crum, sls dir & adv dir; Sue Schneck, mktg dir; Bob Smith, prom dir; Mike Strobel, progmg dir & disc jockey; Jim Reed, engrg dir; Mark Saia, engrg mgr; Donna Vande Bogart, traf mgr; Allison Barden, disc jockey; Brian Stoll, disc jockey; Gary Knight, disc jockey.

WENY(AM)— 1939: 1230 khz; 1 kw-U. TL: N42 04 30 W76 46 55. Hrs open: 24
Rebroadcasts WCLI(AM) Corning 100%.
21 E. Market St., Corning, 14830. Phone: (607) 937-8181. Fax: (607) 962-1138. Licensee: Group A Licensee LLC. Group owner: Route 81 Radio LLC (acq 12-31-2003; grpsl). Population served: 200,000 Natl. Rep: McGavren Guild. Format: News/talk. Target aud: 30 plus. ♦Paul Lyle, gen mgr.

WENY-FM— Aug 15, 1965: 92.7 mhz; 700 w. Ant 561 ft TL: N42 01 55 W76 47 02. Stereo. 24
Rebroadcasts WCBA-FM Corning 100%.
250,000 Natl. Network: ABC. Format: Adult contemp.

WNKI(FM)—See Corning

WPGI(FM)—See Horseheads

WWLZ(AM)—See Horseheads

Elmira Heights-Horseheads

WEHH(AM)— July 4, 1956: 1600 khz; 5 kw-D, 170 w-N, DA-2. TL: N42 07 11 W76 48 37. Hrs open: 24 1705 Lake St., Elmira, 14901. Phone: (607) 733-5626. Phone: (607) 732-1400. Fax: (607) 733-5627. E-mail: ppinesmedial@sty.rr.com Web Site: wehhradio.com. Licensee: Pembrook Pines Elmira Ltd. Group owner: Pembrook Pines Media (acq 5-19-99). Population served: 100,000 Format: Adult classics. News: 2 hrs wkly. Target aud: 45 plus; upscale adults. ♦Robert J. Pfuntner, CEO & gen mgr; Sue Schneck, opns VP.

Endicott

WENE(AM)— September 1947: 1430 khz; 5 kw-U, DA-N. TL: N42 04 56 W76 01 53. Stereo. Hrs open: 24 320 N. Jensen Rd., Vestal, 13850-2111. Phone: (607) 785-3351. Fax: (607) 584-5900. E-mail: info@1430theteam.com Web Site: www.1430theteam.com. Licensee: CC Licenses LLC. Group owner: Clear Channel Communications Inc. (acq 4-14-2000; grpsl). Population served: 250,000 Natl. Network: Westwood One. Natl. Rep: McGavren Guild. Law Firm: Proskauer, Rose, Goetz & Mendelsohn, L. Format: Talk, sports. News staff: 2; News: 28 hrs wkly. Target aud: 35 plus. ♦Tom Barney, gen mgr.

WMRV-FM—Co-owned with WENE(AM). 1969: 105.7 mhz; 35 kw. 570 ft TL: N42 08 20 W75 59 58. Stereo. 24 Web Site: www.1430theteam.com. Format: CHR. Target aud: 18-34.

Endwell

WBBI(FM)— 1998: 107.5 mhz; 1.1 kw. Ant 544 ft TL: N42 08 17 W75 59 59. Hrs open: 320 North Jensen Rd., Vestal, 13850. Phone: (607) 584-5800. Fax: (607) 584-5900. Web Site: www.1075thebear.com. Licensee: CC Licenses LLC. Group owner: Clear Channel Communications Inc. (acq 4-14-2000; grpsl). Format: Classic rock. ♦Joanna Alay, gen mgr.

Essex

WCPV(FM)— Oct 1, 1994: 101.3 mhz; 1 kw. 797 ft TL: N44 24 12 W73 26 02. Hrs open: 1500 Hegeman Ave., Colchester, VT, 05446. Phone: (802) 654-9300. Fax: (802) 655-0478. Web Site: www.champrocks.com. Licensee: Capstar TX L.P. Group owner: Clear Channel Communications Inc. (acq 8-30-00; grpsl). Format: Classic rock. ♦Karen Marshall, gen mgr; Steve Cormier, opns mgr; John Hill, sls dir.

Fairport

WFKL(FM)— 1993: 93.3 mhz; 4.4 kw. Ant 384 ft TL: N43 10 37 W77 28 39. Stereo. Hrs opn: 24 Entercom Rochester LLC, 70 Commercial St., Rochester, 14614-1010. Phone: (585) 423-2900. Fax: (585) 325-5139. Web Site: www.93bbf.com. Licensee: Entercom Rochester Inc. Group owner: Entercom Communications Corp. (acq 4-23-98; grpsl). Natl. Rep: Katz Radio. Law Firm: Akin, Gump, Strauss, Hauer & Feld. Format: Oldies. News staff: one. Target aud: 25-54; upscale. ♦Michael Doyle, gen mgr; Mike Rockwell, natl sls mgr; Mike Johnson, rgnl sls mgr; Christine Neenan, prom dir; Steve Hausmann, news dir; Joe Fleming, chief of engrg.

Fenner

***WXXE(FM)**— Dec 21, 1998: 90.5 mhz; 7 w. 413 ft TL: N42 58 12 W75 47 12. Hrs open: 826 Euclid Ave., Syracuse, 13210. Phone: (315) 426-0850. Fax: (315) 701-0303. E-mail: info@wxxe.org Web Site: www.wxxe.org. Licensee: Syracuse Community Radio Inc. Format: Var. ♦Paul Melnikow, gen mgr; Mark Hughson, mus dir.

Fort Plain

WBUG-FM— Mar 1, 1990: 101.1 mhz; 1.25 kw. 718 ft TL: N42 52 44 W74 47 07. Stereo. Hrs opn: 24 185 Genesee St., Suite 1601, Utica, 13501. Phone: (315) 734-9245. Fax: (315) 624-9245. Licensee: Roser Communications Network Inc. (acq 10-21-94; $400,000 with WVTL(AM) Amsterdam; FTR: 12-5-94). Natl. Network: ABC. Format: C&W. News staff: 2; News: 10 hrs wkly. Target aud: 25 plus. ♦Ken Roser, gen mgr; Roxanne Roser, stn mgr & gen sls mgr.

Frankfort

WKLL(FM)—Licensed to Frankfort. See Utica

Fredonia

WBKX(FM)— April 1989: 96.5 mhz; 1.4 kw. Ant 686 ft TL: N42 22 02 W79 23 12. Stereo. Hrs opn: 24 Box 209, 4561 Willow Rd., Dunkirk, 14048. Phone: (716) 366-8580. Phone: (716) 366-1410. Fax: (716) 366-1416. Licensee: Chadwick Bay Broadcasting Corp. (acq 2-26-2001; with WDOE(AM) Dunkirk). Population served: 75000 Natl. Network: ABC. Format: Adult Contemporary. News: 12 hrs wkly. Target aud: 25-54. ♦John Bulmer, pres; Chuck Telford, gen mgr; David Rowley, news dir.

***WCVF-FM**— July 6, 1978: 88.9 mhz; 130 w. Ant -115 ft TL: N42 27 08 W79 20 14. Stereo. Hrs opn: 24 115 McEwen Hall, State Univ. of New York, 14063. Phone: (716) 673-3420. Fax: (716) 673-3427. E-mail: fredoniaradio@gmail.com Web Site: www.fredoniaradio.com. Licensee: State University of New York. Population served: 40,000 Wire Svc: UPI Format: Div, progsv. News: 28 hrs wkly. All ages of campus & community of Fredonia. Spec prog: Reggae 4 hrs, folk 4 hrs, world 4 hrs, Sp 4 hrs, new age 4 hrs wkly.

Freeport

WGBB(AM)— August 1924: 1240 khz; 1 kw-U. TL: N40 38 44 W73 34 38. Hrs open: 1850 Lausdown Ave., 404 Rte 109, W. Babylon, 11704. Phone: (516) 623-1240. Fax: (516) 623-1240. E-mail: support@am1240wgbb.com Web Site: www.am1240wgbb.com. Licensee: WGBB-AM Inc. Group owner: Cox Broadbasting (acq 5-22-98; grpsl). Population served: 35,000 Format: Var. Target aud: 25-65. Spec prog: Relg 6 hrs, Sp 2 hrs wkly. ♦Jeff Lo, opns mgr.

Friendship

***WCID(FM)**— 1989: 89.1 mhz; 7 kw. Ant 492 ft TL: N42 07 07 W78 10 43. Stereo. Hrs opn: 24
Rebroadcasts WCIK(FM) Bath 100%.
Box 506, 7634 Campbell Creek Rd., Bath, 14810. Phone: (607) 776-4151. Fax: (607) 776-6929. E-mail: mail@fln.org Web Site: www.fln.org. Licensee: Family Life Ministries Inc. Group owner: Family Life Network Natl. Network: Salem Radio Network. Law Firm: Hardy, Carey, Chautin & Balkin, LLP. Wire Svc: Metro Weather Service Inc. Format: Christian, contemp. News staff: 3; News: 14 hrs wkly. Target aud: 30-54; general. ♦Dick Snavely, CFO; Rick Snavely, pres & gen mgr; John Owens, progmg dir; Jim Travis, chief of engrg.

Fulton

WAMF(AM)— Aug 19, 1949: 1300 khz; 1 kw-D. TL: N43 17 41 W76 26 35. Stereo. Hrs opn: 24 Box 1994, Cicero, 13039-1994. Phone: (315) 593-1300. Fax: (315) 598-5158. E-mail: wamf1300@alltel.net Web Site: realcountryonline.com. Licensee: Donald H. Derosa (acq 8-6-2002; $300,000). Population served: 300,000 Natl. Network: ABC. Format: Country. Target aud: 35 plus; hometown listeners, county coverage. Spec prog: It 2 hrs, Pol 5 hrs wkly. ♦Don Derosa, gen mgr.

WBBS(FM)— Aug 1, 1961: 104.7 mhz; 50 kw. 310 ft TL: N43 12 53 W76 23 44. (CP: Ant 479 ft.). Stereo. Hrs opn: 500 Plum St., Suite 100, Bridgewater Pl., Syracuse, 13204. Phone: (315) 448-1047. Fax: (315) 474-7879. Web Site: www.b1047.net. Licensee: Citicasters Licenses L.P. Group owner: Clear Channel Communications Inc. (acq 5-4-99; grpsl). Population served: 550,000 Format: Country. ♦Joel Delmonico, gen mgr.

Garden City

***WHPC(FM)**— Oct 12, 1972: 90.3 mhz; 500 w. Ant 213 ft TL: N40 43 47 W73 35 33. Stereo. Hrs opn: 24 Nassau Community College, One Education Dr., 11530-6793. Phone: (516) 572-7439. Fax: (516) 572-7831. Web Site: www.sunynassau.edu. Licensee: Nassau Community College Board of Trustees. Population served: 2,000,000 Format: Div, educ, adult contemp. News staff: 2; News: 5 hrs wkly. Target aud: 20-65; general. ♦Sean Fanelli, CEO; Jack Ostling, exec VP; Jim Green, opns mgr & progmg dir.

WQBU-FM— 1988: 92.7 mhz; 2 kw. Ant 521 ft TL: N40 45 26 W73 42 52. Stereo. Hrs opn: 485 Madison Ave., New York, 10022. Phone: (212) 310-6000. Fax: (212) 310-6095. Web Site: www.univision.com. Licensee: Univision Radio License Corp. Group owner: Univision Radio (acq 1-12-2004; $60 million). Format: Rgnl Mexican. ♦Joe Pagan, VP & gen mgr.

Geneseo

***WGSU(FM)**— Feb 18, 1963: 89.3 mhz; 1.8 kw. 11 ft TL: N42 47 51 W77 49 13. Stereo. Hrs opn: 24 Blake B 104, 1 College Cir., 14454. Phone: (585) 245-5486. Fax: (585) 245-5240. E-mail: pruszyns@geneseo.edu Web Site: www.geneseo.edu/~wgsu/. Licensee: State University of New York. Population served: 10,000 Format: News, alternative. News: 7 hrs wkly. Target aud: 12-55; college, immediate community. ♦Chris Pruszynski, gen mgr.

Geneva

***WEOS(FM)**— Mar 30, 1971: 89.7 mhz; 4 kw. 312 ft TL: N42 51 27 W76 59 21. Stereo. Hrs opn: 24 300 Pulteney St., 14456. Secondary address: 113 Hamilton St. 14456. Phone: (315) 781-3456. Phone: (315) 781-3897. Fax: (315) 781-3916. E-mail: weos@hws.edu Web Site: www.weos.org. Licensee: The Colleges of the Seneca. Population served: 300,000 Natl. Network: NPR, PRI. Wire Svc: AP Format: Jazz, progsv, news/talk. News staff: one; News: 40 hrs wkly. Target aud: 18-plus. Spec prog: AAA 12 hrs, world 10 hrs, metal 6 hrs, gospel 3 hrs, reggae 3 hrs wkly. ♦Mark Gearan, pres; Michael R. Black, gen mgr; Genoa Boswell, opns dir; Greg Cotterill, dev dir; Genooa Boswell, progmg dir; Jamie Agnello, mus dir; Liz Cost, news dir.

WFLK(FM)— 1974: 101.7 mhz; 5.4 kw. Ant 125 ft TL: N42 51 34 W77 00 29. Stereo. Hrs opn: 24 Box 1017, 14456. Phone: (315) 781-1101. Fax: (315) 781-6666. E-mail: k1017@fltg.net Web Site: www.k1017.com. Licensee: MB Communications Inc. (acq 1993). Population served: 300,000 Law Firm: Henry Crawford. Format: Super hit country. News staff: 2; News: 10 hrs wkly. Target aud: 25-49. ♦Russ Kimble, pres & stn mgr; John Thomas, opns mgr & pub affrs dir; Lori Rose, dev mgr; Deb Hunt, gen sls mgr; Matt Ripley, prom dir & progmg dir.

WGVA(AM)— 1947: 1240 khz; 1 kw-U. TL: N42 51 37 W77 00 59. Hrs open: 24 3568 Lenox Rd., 14456. Phone: (315) 781-1240. Fax: (315) 781-7700. Licensee: Geneva Broadcasting Inc. (acq 9-27-96). Population served: 325,000 Law Firm: James L. Oyster. Format: News/talk. News staff: one; News: 10 hrs wkly. Target aud: General. ♦George Kimble, pres; Alan Bishop, exec VP & gen mgr.

WNYR-FM—See Waterloo

Glens Falls

WCQL(FM)—Listing follows WWSC(AM).

WFFG-FM—See Corinth

***WGFR(FM)**— January 1977: 92.7 mhz; 13 w. 49 ft TL: N43 18 44 W73 38 58. Stereo. Hrs opn: Adirondack Community College, 640 Bay Rd., Queensbury, 12804-1498. Phone: (518) 743-2311. Fax: (518) 745-1433. Web Site: www.wgfr.org. Licensee: Board of Trustees of Adirondack Community College. Population served: 50,000 Format: Progsv, AAA, Indie rock. Target aud: 18 plus; adults. ♦Kevin Ankeny, gen mgr; Steve Tefft, stn mgr.

***WLJH(FM)**— 2001: 90.9 mhz; 360 w. Ant 663 ft TL: N43 19 55 W73 20 20. Hrs open:
Rebroadcasts WFGB(FM) Kingston 100%.
Box 777, Lake Katrine, 12449. Phone: (845) 336-6199. Fax: (845) 336-7205. Web Site: www.soundoflife.org. Licensee: Sound of Life Inc. Format: Christian. ♦Tom Michael Zahradnik, gen mgr.

WMML(AM)— May 28, 1959: 1230 khz; 1 kw-U. TL: N43 19 43 W73 38 58. Hrs open: 24 89 Everts Ave., Queensbury, 12804. Phone: (518) 793-7733. Fax: (518) 793-0838. Licensee: 6 Johnson Road Licenses Inc. Group owner: Pamal Broadcasting Ltd. (acq 4-1-2004; grpsl). Population served: 117,222 Natl. Network: ESPN Radio. Format: Sports. News staff: one; News: 6 hrs wkly. Target aud: 18-54 plus. Spec prog: Relg 3 hrs wkly. ♦Clay Ashworth, gen mgr.

WNYQ(FM)—See Hudson Falls

WWSC(AM)— Dec 18, 1946: 1450 khz; 1 kw-U. TL: N43 18 46 W73 35 57. Hrs open: 24 128 Glen St., 12801. Phone: (518) 761-9890. Fax: (518) 761-9893. Web Site: www.radiowins.com. Licensee: Entertronics Inc. Broadcasting. (acq 1999; with co-located FM). Population served: 104,000 Natl. Network: ABC, CNN Radio. Law Firm: Richard J. Hayes Jr. Format: News/talk, sports. News staff: one; News: 166 hrs wkly. Target aud: 12 plus; people who want full news radio & talk. ♦David Covey, pres & gen mgr; William H. Walker III, VP; Dan Miner, stn mgr; Robin Covey, opns dir; Robin Truax, opns dir; Paul Van Amburgh, gen sls mgr; Pete Cloutier, prom dir; Jim Scott, news dir; Kevin Smith, chief of engrg.

WCQL(FM)—Co-owned with WWSC(AM). September 1967: 95.9 mhz; 410 w. Ant 863 ft TL: N43 18 17 W73 45 07. Stereo. 24 Web Site: www.radiowins.com.200,000 Format: Hits of the 80s, 90s & now. News staff: one; News: 3 hrs wkly. Target aud: 18-49.

Gloversville

WENT(AM)— July 1, 1944: 1340 khz; 1 kw-U. TL: N43 01 30 W74 21 10. Hrs open: 5:30 AM-midnight Box 831, 138 Harrison St. Ext., 12078. Phone: (518) 725-7175. Fax: (518) 725-7177. E-mail: went@capital.net Web Site: www.am1340went.com. Licensee: Whitney Radio Broadcasting Inc. Population served: 102,000 Natl. Network: CNN Radio, ESPN Radio. Wire Svc: AP Format: Full service, adult contemp. News staff: 2; News: 14 hrs wkly. Target aud: 30 plus. Spec prog: Talk one hr wkly. ♦Jack Scott, pres & gen mgr; Jon W. Clark, VP; Shirley V. Clark, stn mgr.

WFNY(AM)— 3/2003: 1440 khz; 800 w-D, 500 w-N, DA-2. TL: N43 01 57 W74 21 02. Hrs open: 101 S. Main St., 12078. Phone: (518) 725-1108. Licensee: Michael A. Sleezer. Format: Hits past & present. ♦Michael A. Sleezer, gen mgr.

Gouverneur

WGIX-FM— Dec 5, 1967: 95.3 mhz; 6 kw. Ant 328 ft TL: N44 20 22 W75 24 00. Stereo. Hrs opn: 24 2315 Knox St., Ogdensburg, 13669. Phone: (315) 393-1100. Fax: (315) 393-6673. Web Site: www.cooloIdies.us. Licensee: Community Broadcasters LLC. Group owner: Clancy-Mance Communications (acq 10-13-2006; grpsl). Population served: 120,000 Format: Oldies. News staff: 3; News: 7 hrs wkly. Target aud: 35-54. ♦James L. Leven, pres & gen mgr; Tobi Newcombe, sls dir & progmg dir; Ken Ruhland, chief of engrg.

Grand Gorge

***WGKR(FM)**— November 1997: 105.3 mhz; 60 w. 1,342 ft TL: N42 23 58 W74 35 27. Hrs open:
Rebroadcasts WFGB(FM) Kingston 100%.
Box 777, Lake Katrine, 12449. Secondary address: 199 Tuytenbridge Rd., Lake Katrine 12449. Phone: (845) 336-6199. Fax: (845) 336-7205. E-mail: email@soundoflife.org Web Site: www.soundoflife.org. Licensee: Sound of Life Inc. Format: Contemp Christian. ♦Tom Michael Zahradnik, gen mgr.

Greece

***WGMC(FM)**— Nov 11, 1973: 90.1 mhz; 15 kw. Ant 138 ft TL: N43 14 40 W77 41 36. Stereo. Hrs opn: 24 Box 300, North Greece, 14515-0300. Secondary address: 1139 Maiden Ln., Rochester 14615. Phone: (585) 966-2660. Fax: (585) 581-8185. Web Site: www.jazz901.org. Licensee: Greece Central School District. Population served: 1,00,000 Law Firm: Dow, Lohnes & Albertson. Format: Jazz. Target aud: 25-50; upscale, educated, mus lovers. Spec prog: Pol 2 hrs, Sp 10 hrs, Lithuanian one hr, Turkish one hr, blues 3 hrs wkly. ♦Jack Mindy, opns mgr & disc jockey; Rob Linton, stn mgr & opns mgr.

Hamilton

***WRCU-FM**— Mar 22, 1970: 90.1 mhz; 1.9 kw. 155 ft TL: N42 48 38 W75 31 58. Stereo. Hrs opn: Colgate University, 13346. Phone: (315) 228-7104. Fax: (315) 228-7028. Licensee: Colgate University. Population served: 3,636 Format: Progsv, div, jazz. Spec prog: Jazz 12 hrs, class 4 hrs, Black 10 hrs wkly. ♦Matt Pysher, gen mgr; Andy Jackson, progmg mgr.

Hampton Bays

WLIR-FM— Nov 20, 1980: 107.1 mhz; 6 kw. Ant 279 ft TL: N40 52 10 W72 34 37. Stereo. Hrs opn: 1103 Stewart Ave., Garden City, 11530. Phone: (631) 648-2500. Fax: (516) 222-1391. E-mail: info@wlir.com Licensee: Jarad Broadcasting Co. of Hampton Bays LLC Group owner: The Morey Organization Inc. (acq 2-27-2004; $2 million). Population served: 100,000 Natl. Rep: Roslin. Format: Alternative. ♦Beverly Fortune, gen mgr.

Hempstead

WHLI(AM)— July 22, 1947: 1100 khz; 10 kw-D, DA. TL: N40 41 06 W73 36 38. Hrs open: Sunrise-sunset 234 Airport Plaza Blvd., #5, Farmingdale, 11735-3938. Phone: (631) 770-4200. Fax: (631) 770-0090. Web Site: www.whli.com. Licensee: Long Island Broadcasting Inc. Group owner: Barnstable Broadcasting Inc. (acq 12-15-84; $5 million with co-located FM; FTR: 9-24-84) Population served: 2,000,000 Natl. Rep: Katz Radio. Format: Adult standards. News staff: one; News: 2 hrs wkly. Target aud: 35-64; adults. Spec prog: Black one hr wkly. ♦Dave Widmer, pres, gen mgr & progmg VP; Mike Banks, sls dir & gen sls mgr; Cheryl Kampanis, prom dir; Frank Brinka, news dir; Antoinette Rodriguez, traf mgr; Joe Satta, disc jockey; Paul Richards, progmg dir & disc jockey.

WKJY(FM)—Co-owned with WHLI(AM). July 22, 1947: 98.3 mhz; 3 kw. 328 ft TL: N40 41 08 W73 36 37. Stereo. 24 Web Site: www.whli.com. Natl. Rep: Katz Radio. Format: Adult contemp. News staff: one. Target aud: 25-54. Spec prog: Black one hr wkly. ♦Alissa Marty, mktg dir; Bill George, progmg dir; Antoinette Rodriguez, traf mgr; Bill Edwards, disc jockey; Jim Douglas, disc jockey; Jodi Vale, disc jockey; Kim Berk, disc jockey; Mike Glaser, news rptr & engr.

***WRHU(FM)**— June 9, 1959: 88.7 mhz; 470 w. 200 ft TL: N40 43 03 W73 36 12. Stereo. Hrs opn: 24 Rm. 127, 111 Hofstra Univ., 11549-1110. Phone: (516) 463-5667. Fax: (516) 463-5668. E-mail: mail@wrhu.org Web Site: www.wrhu.org. Licensee: Hofstra University. Population served: 2,500,000 Law Firm: Dow, Lohnes & Albertson. Wire Svc: AP Format: Div. News: 11 hrs wkly. Target aud: General. ♦Bruce Avery, gen mgr; Jamie Morris, stn mgr; Joel Meyer, opns mgr; Meghan Attreed, prom dir; Dustin Gervais, progmg dir.

Henderson

WOTT(FM)— 1991: 100.7 mhz; 3 kw. 328 ft TL: N43 49 13 W76 05 29. (CP: 6 kw). Hrs opn: 199 Wealtha Ave., Watertown, 13601. Phone: (315) 782-1240. Fax: (315) 782-0132. Web Site: www.realrock1007.com. Licensee: Community Broadcasters LLC. (acq 10-13-2006; grpsl). Natl. Rep: Roslin. Format: Active rock. ♦James L. Leven, pres; Glenn Curry, gen mgr; Todd Dalesandro, opns mgr; Vickie Fenn, sls dir; Johnny Keegan, progmg dir.

Henrietta

***WITR(FM)**— Mar 7, 1975: 89.7 mhz; 910 w. 154 ft TL: N43 05 08 W77 40 05. Stereo. Hrs opn: 24 32 Lomb Memorial Dr., Rochester, 14623. Phone: (585) 475-2000. Fax: (585) 475-4988. Licensee: Rochester Institute of Technology. Population served: 750,000 Format: Modern music. Target aud: General. Spec prog: Reggae 5 hrs, jazz 8 hrs, contemp Christian rock 10 hrs, gospel 8 hrs, industrial 2 hrs, comedy 2 hrs, world beat 2 hrs wkly. ♦Mark Zuriga, gen mgr; Justin Ricci, prom dir; Steve Montario, progmg dir.

Herkimer

WNRS(AM)— October 1956: 1420 khz; 1 kw-D. TL: N43 03 40 W75 01 44. Hrs open: 24 Box 927, Ilion, 13357. Phone: (315) 866-9200. Fax: (315) 866-6906. E-mail: wxur@hotmail.com Licensee: Arjuna Broadcasting Corp. (acq 10-23-96). Population served: 120,000 Natl. Network: ESPN Radio. Natl. Rep: Roslin. Law Firm: Cohn & Marks. Format: Sports. News staff: one. Target aud: 18 plus; men. ♦Mindy Barstein, pres & gen mgr; Tim Barstein, sls VP; Tom Davenport, opns VP & progmg dir; Anthony Falvo, chief of engrg.

WXUR(FM)—Co-owned with WNRS(AM). Apr 28, 1979: 92.7 mhz; 6 kw. 299 ft TL: N43 03 50 W75 01 44. Stereo. 24 420,000 Natl. Network: Westwood One. Format: Oldies. News staff: one. Target aud: 35-64; adults. ♦Tim Barstein, dev VP; Jenna Davenport, mktg mgr; Chris Miller, progmg VP & local news ed; Tony Falvo, engrg VP; Max Davenport, traf mgr; Robert Huyck, sports cmtr; Bruce Melnick, disc jockey; Jim Leno, disc jockey.

***WVHC(FM)**— October 1993: 91.5 mhz; 350 w vert. -115 ft Stereo. Hrs opn: 24 Reservoir Rd., 13350. Phone: (315) 866-0300, EXT. 354. Fax: (315) 866-7253. Licensee: Herkimer County Community College. Format: Jazz. News: 5 hrs wkly. Target aud: General; residents of southern Herkimer county & college community. ♦Wade Lamb, gen mgr & stn mgr.

Highland

WJGK(AM)—Not on air, target date: unknown: 1200 khz; 4.7 kw-D, 1 kw-N, DA-2. TL: N41 44 07 W73 57 38. Hrs open: Box 2307, Newburgh, 12550-0451. Phone: (845) 561-2131. Fax: (845) 561-2138. Licensee: Sunrise Broadcasting Corp. ♦Joerg Klebe, pres.

WRWD-FM— Oct 3, 1989: 107.3 mhz; 330 w. Ant 968 ft TL: N41 41 58 W74 00 11. Stereo. Hrs opn: 20 Tucker Dr., Poughkeepsie, 12603. Phone: (845) 454-2800. Fax: (845) 471-0793. Web Site: www.wrwdfm.com. Licensee: AMFM Radio Licenses LLC. Group owner: Clear Channel Communications Inc. (acq 12-10-97; $7.5 million with WBWZ(FM) New Paltz). Population served: 660,000 Natl. Network: Jones Radio Networks. Natl. Rep: Katz Radio. Law Firm: Gammon & Grange. Format: C&W. Target aud: 18 plus. Spec prog: Farm one hr wkly. ♦Frank Curcio, gen mgr; Jim Brady, gen sls mgr; Jeanette Relyea, natl sls mgr; Nick Smirnoff, prom dir; Aaron McCord, progmg dir; Cameron Hendrix, news dir.

Homer

WXHC(FM)— 1991: 101.5 mhz; 1.3 kw. 489 ft TL: N42 41 12 W76 11 54. Stereo. Hrs opn: 24 Box 386, 12 S. Main St., 13077. Phone: (607) 749-9942. Fax: (607) 749-2374. E-mail: johneves@wxhc.com Web Site: www.wxhc.com. Licensee: John Eves. Population served: 75,000 Natl. Rep: Roslin. Law Firm: Cole, Raywid & Braverman. Format: Oldies. News staff: one; News: 10 hrs wkly. Target aud: 25-54. ♦John Eves, pres; Bruce Eves, exec VP; Patricia Eves, VP; Sonny King, opns VP.

Honeoye Falls

WFXF(FM)— 1948: 95.1 mhz; 50 kw. Ant 479 ft TL: N43 02 01 W77 25 18. Stereo. Hrs opn: 24 207 Midtown Plaza, Rochester, 14604. Phone: (585) 246-0440. Fax: (585) 454-5081. Licensee: Citicasters Licenses Inc. (NEW). Group owner: Clear Channel Communications Inc. (acq 1999; grpsl). Population served: 120,000 Format: Modern rock, AOR. News: one hr wkly. Target aud: 18-44. Spec prog: Rgnl news one hr wkly. ♦Karen Carey, gen mgr.

Hoosick Falls

WHAZ-FM— July 4, 1991: 97.5 mhz; 400 w. Ant 1,204 ft TL: N42 51 40 W73 13 59. (CP: 420 w, ant 1,184 ft. TL: N42 51 49 W73 13 59). Hrs opn: 30 Park Ave., Cohoes, 12047-3330. Phone: (518) 237-1330. Fax: (518) 235-4468. Web Site: www.whaz.com. Licensee: Capital Media Corp. Group owner: Vox Radio Group L.P. (acq 7-19-2005; $1.1 million). Format: Christian oldies. ♦Paul Lotters, gen mgr.

Hornell

WCKR(FM)—Listing follows WLEA(AM).

WHHO(AM)— 1949: 1320 khz; 5 kw-D. TL: N42 17 32 W77 40 27. Hrs open: 6 AM-6 PM Box 726, 1484 Beech St., 14843. Phone: (607) 324-2000. Phone: (607) 324-2002. Fax: (607) 324-2001. E-mail: devin@wkpq.com Web Site: www.wkpq.com. Licensee: Bilbat Radio Inc. (acq 6-10-83; $450,000 with co-located FM; FTR: 5-30-83) Population served: 300,000 Format: Talk. News staff: one; News: 14 hrs wkly. Target aud: 25-54. Spec prog: Farm. ♦William H. Berry, CFO; Bob Lee, gen sls mgr; Susan Macool, progmg dir; Mickey Slansburgs, mus dir; Johnathan Mark, news dir; Ralph Van, chief of engrg; Mike Allen, sports cmtr.

WKPQ(FM)—Co-owned with WHHO(AM). 1946: 105.3 mhz; 50 kw. 530 ft TL: N42 17 32 W77 40 27. Stereo. 24 E-mail: bilbat@wkpq.com Web Site: www.wkpq.com.750,000 Format: Hot adult comtemp. News staff: one; News: 7 hrs wkly. Target aud: 18-54; females.

WLEA(AM)— September 1951: 1480 khz; 2.5 kw-D. TL: N42 17 15 W77 38 47. Hrs open: 18 5942 Ashbaugh Hill Rd., 14843. Phone: (607) 324-1480. Fax: (607) 324-5415. Web Site: www.wckr.com. Licensee: PMJ Communications Inc. (acq 10-18-90; $538,000 with co-located FM; FTR: 11-19-90) Population served: 80,000 Format: Oldies, news/talk. News staff: 2; News: 16 hrs wkly. Target aud: 35 plus. ♦Tom Booth, gen mgr; Bill Dubensky, news dir.

WCKR(FM)—Co-owned with WLEA(AM). June 1981: 92.1 mhz; 1.25 kw. 512 ft TL: N42 20 38 W77 37 36. Stereo. 19 E-mail: radione@infoblvd.net Web Site: www.wckr.com.100,000 Natl. Network: USA. Format: Country. News: 11 hrs wkly. Target aud: 21 plus.

***WSQA(FM)**— 2000: 88.7 mhz; 4.5 kw. Ant 495 ft TL: N42 16 02 W77 37 55. Hrs open: Box 3000, Binghamton, 13902. Phone: (607) 729-0100. Fax: (607) 729-7328. Web Site: www.wskg.com. Licensee: WSKG Public Telecommunications Council. Format: Jazz, news. ♦Susan Miller-Cormien, sr VP; Gary Reinbolt, gen mgr.

Horseheads

WEHH(AM)—See Elmira Heights-Horseheads

WLNL(AM)— May 7, 1967: 1000 khz; 5 kw-D. TL: N42 09 14 W76 50 47. Hrs open: Sunrise-sunset 3134 Lake Rd., 14845. Phone: (607) 737-9208. Fax: (607) 737-9210. E-mail: inbox@wlnlradio.com Web Site: www.wlnlradio.com. Licensee: Trinity Media Ltd. (acq 1-21-92; $256,000; FTR: 11-11-91). Population served: 200,000 Natl. Network: USA, Salem Radio Network. Format: Relg. Target aud: 25-54; Christian families, women/mothers who work at home. Spec prog: Country/bluegrass one hr wkly. ♦Heather Clark, pres, gen mgr & mus dir; Heather Clark, traf mgr; John Earley, sls.

WPGI(FM)—Listing follows WWLZ(AM).

WWLZ(AM)— April 1966: 820 khz; 5 kw-D, 1 kw-N, DA-2. TL: N42 09 14 W76 50 47. Hrs open: 2205 College Ave., Elmira, 14903. Phone: (607) 732-4400. Fax: (607) 732-7774. Licensee: Chemung County Radio Inc. Group owner: Backyard Broadcasting LLC (acq 12-1-2002; grpsl). Population served: 30,500 Format: News/talk. Target aud: 25-54; baby boomers. ♦Kevin White, gen mgr; Jim Poteat, prom mgr & progmg dir.

WPGI(FM)—Co-owned with WWLZ(AM). July 4, 1970: 100.9 mhz; 3 kw. 245 ft TL: N42 12 00 W76 51 30. Stereo. 30,500 Format: Country. Target aud: General.

Houghton

***WJSL(FM)**— Jan 18, 1979: 90.3 mhz; 6 kw. 216 ft TL: N42 22 39 W78 10 45. Stereo. Hrs opn: 24
Rebroadcasts WMHR(FM) Syracuse 70%.
Box 30021, Rochester, 14603. Phone: (585) 325-7500. E-mail: newsroom@wxxi.org Web Site: www.wxxi.org. Licensee: WXXI Public Broadcasting Council. Population served: 150,000 Format: Class. Target aud: 18-36; college. Spec prog: Class 5 hrs wkly. ♦Norm Silverstein, CEO & pres; Sue Rogers, VP.

Hudson

WHUC(AM)— 1947: 1230 khz; 1 kw-U. TL: N42 15 13 W73 45 45. Hrs open: 24 20 Tucker Dr., Poughkeepsie, 12603. Secondary address: 5620 Rt. 96 12534. Phone: (518) 828-5006. Fax: (518) 828-1080. Licensee: CC Licenses LLC. Group owner: Clear Channel Communications Inc. (acq 1-17-2002; grpsl). Population served: 82,000 Natl. Network: Jones Radio Networks. Natl. Rep: Katz Radio. Format: Adult standards. News staff: one; News: 4 hrs wkly. Target aud: 35 plus; loc people in Columbia & Greene counties. ♦Frank Curcio, gen mgr; Reggie Osterhoudt, opns mgr; Jim Brady, gen sls mgr; Jeanette Relyea, natl sls mgr; Nick Smirnoff, prom dir; Bill Williams, progmg dir; Cameron Hendrix, news dir.

WZCR(FM)—Co-owned with WHUC(AM). Jan 20, 1969: 93.5 mhz; 3 kw. Ant -15 ft TL: N42 15 13 W73 45 45. Stereo. 24 Phone: (845) 471-2300. Fax: (845) 401-2683. Natl. Network: Westwood One. Natl. Rep: Katz Radio. Format: Oldies. News: 3 hrs wkly. Target aud: General; adults 35-54.

***WHVP(FM)**— May 1998: 91.1 mhz; 362 w vert. 991 ft TL: N42 17 52 W73 53 57. Stereo. Hrs opn:
Rebroadcasts WFGB(FM) Kingston 100%.
Box 777, Lake Katrine, 12449. Phone: (845) 336-6199. Fax: (845) 336-7205. E-mail: email@soundoflife.org Web Site: www.soundoflife.org. Licensee: Sound of Life Inc. Format: Christian. ♦Tom Michael Zahradink, gen mgr.

Hudson Falls

WNYQ(FM)— Sept 19, 1983: 101.7 mhz; 4.6 kw. Ant 180 ft TL: N43 22 40 W73 39 56. Stereo. Hrs opn: 24 89 Everts Ave., Queensbury, 12804. Phone: (518) 793-7733. Fax: (518) 793-0838. Licensee: 6 Johnson Road Licenses Inc. Group owner: Pamal Broadcasting Ltd. (acq 4-1-2004; grpsl). Format: Adult standards. News staff: one. Target aud: 35-64. ♦Clay Ashworth, gen mgr.

Huntington

WNYH(AM)— Sept 1, 1951: 740 khz; 25 kw-D, 43 w-N, DA-2. TL: N40 51 04 W73 26 16. (CP: 20 kw-D, 50 w-N, DA-2). Stereo. Hrs opn: 24 100-25 Queens Blvd., Suite 1CC, Forest Hills, 11375. Phone: (718) 335-3333. Licensee: Win Radio Broadcasting Corp. (acq 9-1-2005). Population served: 1,500,000 Format: Adult. ♦Richard S. Yoon, pres & gen mgr.

Hyde Park

WCZX(FM)— Aug 18, 1970: 97.7 mhz; 300 w. 1,030 ft TL: N41 43 11 W73 59 45. Stereo. Hrs opn: 24 Box 416, 2 Pendell Rd., Poughkeepsie, 12602. Phone: (845) 471-1500. E-mail: randyturner@mix97fm.com Web Site: www.mix97fm.com. Licensee: Cumulus Licensing Corp. Group owner: Cumulus Media Inc. (acq 1-23-02; grpsl). Population served: 300,000 Natl. Rep: Katz Radio. Format: Adult contemp. News staff: one; News: 10 hrs wkly. Target aud: 25-54. ♦John Dickie, CEO; Lew Dickie, pres; Charles Benfer, gen mgr.

WHVW(AM)— July 4, 1963: 950 khz; 500 w-D, 57 w-N. TL: N41 44 46 W73 54 46. Hrs open: 24 316 Main St., Poughkeepsie, 12601-3123. Phone: (845) 471-9500. Fax: (845) 452-8696. E-mail: whvw@hui.net Web Site: www.whvw.org. Licensee: Joseph-Paul Ferraro. (acq 3-9-92; $350,000; FTR: 3-30-92). Population served: 500,000 Format: Oldies. Target aud: 25-54. Spec prog: Ger one hr, It one hr, Irish one hr wkly. ♦J.P. Ferraro, pres & gen mgr.

Irondequoit

WKGS(FM)— March 1992: 106.7 mhz; 3.5 kw. 627 ft TL: N43 11 27 W77 37 11. Stereo. Hrs open: 24 207 Midtown Plaza, Rochester, 14604. Phone: (585) 232-8870. Fax: (585) 454-5081. Fax: (585) 262-2334. E-mail: wkgs@eznet.net Web Site: www.kiss1067.com. Licensee: Citicasters Licenses L.P. Group owner: Clear Channel Communications Inc. (acq 1999; grpsl). Population served: 830,000

Format: CHR. Target aud: 18-34. ♦Karen Carey, gen mgr; Jeff Oar, gen sls mgr; Erick Anderson, progmg dir; Marilee Terran, progmg dir.

Islip

WLIE(AM)— 1960: 540 khz; 2.5 kw-D, 220 w-N, DA-2. TL: N40 45 06 W73 12 50. Stereo. Hrs opn: 24 2137 Deer Park Ave., Deer Park, 11729. Phone: (631) 243-5400. Fax: (631) 243-5444. E-mail: Info@wlie.com Web Site: www.wlie.com. Licensee: Stuart Henry (acq 12-4-2003). . Population served: 2,500,000 Natl. Network: USA, Jones Radio Networks. Law Firm: Thompson, Hine L.L.P. Format: Talk radio. News: 10 hrs wkly. Target aud: 45 plus. Spec prog: Relg 3 hrs wkly. ♦Stuart Henry, pres & gen mgr.

Ithaca

WHCU(AM)— Jan 23, 1923: 870 khz; 5 kw-D, 1 kw-N, DA-N. TL: N42 21 49 W76 36 20. Hrs open: 24 1751 Hanshaw Rd., 14850. Phone: (607) 257-6400. Fax: (607) 257-6497. Licensee: Saga Communications of New England LLC. (acq 5-31-2005; grpsl). Population served: 680,000 Natl. Network: CBS, Westwood One, AP Radio. Natl. Rep: Christal. Law Firm: Richard Carr. Format: News/talk, sports. News staff: 3; News: 40 hrs wkly. Target aud: 25-64. ♦Edward K. Christian, pres; Susan Johnston, gen mgr; Chris Allinger, opns dir; Connie Fairfax-Ozmun, mktg dir; Geoff Dunn, progmg dir & news dir.

WYXL(FM)—Co-owned with WHCU(AM). Sept 1, 1947: 97.3 mhz; 26 kw. 879 ft TL: N42 27 54 W76 22 23. Stereo. Format: Adult contemp. Target aud: 25-54. ♦Kevin English, progmg dir.

***WICB(FM)**— Jan 14, 1947: 91.7 mhz; 4.1 kw. Ant 135 ft TL: N42 25 07 W76 29 39. Stereo. Hrs opn: 24 Ithaca College, 118 Park Hall, 14850. Phone: (607) 274-1040. E-mail: wicb@ithaca.edu Web Site: www.wicb.org. Licensee: Ithaca College. Population served: 250,000 Natl. Network: ABC. Format: Modern rock, urban contemp. News: 6 hrs wkly. Target aud: 18-34; young audience with taste for innovative mus. Spec prog: Jazz 13 hrs, folk 2 hrs, blues 2 hrs, reggae 2 hrs, world beat 2 hrs wkly. ♦Christopher Wheatley, gen mgr.

WIII(FM)—See Cortland

WNYY(AM)— April 1956: 1470 khz; 5 kw-D, 1 kw-N, DA-N. TL: N42 23 32 W76 28 29. Hrs open: 24 1751 Hanshaw Rd., 14850. Phone: (607) 257-6400. Fax: (607) 257-6497. Licensee: Saga Communications of New England LLC. (acq 5-31-2005; grpsl). Population served: 680,000 Natl. Network: Westwood One. Natl. Rep: Christal. Format: Progressive talk. News staff: 3. Target aud: 35-54. ♦Edward K. Christian, pres; Susan Johnston, gen mgr; Chris Allinger, opns dir; Connie Fairfax-Ozmun, mktg dir; Geoff Dunn, progmg dir & news dir.

WQNY(FM)—Co-owned with WNYY(AM). 1948: 103.7 mhz; 15.5 kw. Ant 879 ft TL: N42 23 13 W76 40 10. Stereo. 24 Law Firm: Richard Carr. Format: Country. News staff: 3. Target aud: 25-54. ♦Chris Allinger, progmg dir.

***WSQG-FM**— 1988: 90.9 mhz; 5 kw. 294 ft TL: N42 34 55 W76 33 22. Stereo. Hrs opn: 24
Rebroadcasts WSKG-FM Binghamton 100%.
Box 3000, Binghamton, 13902. Phone: (607) 729-0100. Fax: (607) 729-7328. Web Site: www.wskg.com. Licensee: WSKG Public Telecommunications Council. Population served: 83,200 Natl. Network: NPR, PRI. Law Firm: Dow, Lohnes & Albertson. Format: Class, news. News staff: one; News: 33 hrs wkly. Target aud: General. Spec prog: Jazz 7 hrs, folk/bluegrass 5 hrs wkly. ♦Gary Reinbolt, CEO & gen mgr.

WVBR-FM— June 7, 1958: 93.5 mhz; 3 kw. 250 ft TL: N42 25 42 W76 26 57. Stereo. Hrs opn: 24 957-B Mitchell St., 14850. Phone: (607) 273-4000. Fax: (607) 273-4069. E-mail: radio@wvbr.com Web Site: www.wvbr.com. Licensee: Cornell Radio Guild Inc. Population served: 100,000 Natl. Network: Westwood One. Natl. Rep: Eastman Radio, Katz Radio. Format: Full service, AOR. News: 10 hrs wkly. Target aud: 18-49; highly educated listeners. Spec prog: Oldies 5 hrs, heavy metal 6 hrs, folk 8 hrs,blues 5 hrs, Latin 5 hrs wkly. ♦Matt Todaro, gen mgr; Matthew Leftwich, opns VP; Jordan Gremli, dev VP; Jill Shaughnessy, prom dir; Laura Kehe, progmg dir.

Jamestown

***WCOT(FM)**— Dec 14, 1992: 90.9 mhz; 12 kw. Ant 653 ft TL: N42 00 06 W79 03 19. Stereo. Hrs opn:
Rebroadcasts WCIK(FM) Bath 100%.
Box 506, 7634 Campbell Creek Rd., Bath, 14810. Phone: (607) 776-4151. Fax: (607) 776-6929. E-mail: mail@fln.org Web Site: www.fln.org. Licensee: Family Life Ministries Inc. Group owner: Family Life Network Natl. Network: Salem Radio Network. Law Firm: Hardy, Carey, Chautin & Balkin, LLP. Wire Svc: Metro Weather Service Inc. Format: Contemp Christian. News staff: 3; News: 14 hrs wkly. Target aud: 30-54. ♦Dick Snavely, CFO; Rick Snavely, pres, VP & gen mgr; John Owens, progmg dir; Jim Travis, chief of engrg.

WHUG(FM)—Listing follows WKSN(AM).

WJTN(AM)— December 1924: 1240 khz; 500 w-D, 1 kw-N. TL: N42 06 18 W79 15 28. Hrs open: 24 Box 1139, 14702-1139. Secondary address: 2 Orchard Rd. W.E. 14701. Phone: (716) 487-1151. Fax: (716) 664-9326. E-mail: wjtn@wjtn.com Web Site: www.wjtn.com. Licensee: Media One Group LLC (acq 8-30-2002; $5.05 million with co-located FM). Population served: 140,000 Natl. Network: Westwood One. Natl. Rep: Rgnl Reps. Wire Svc: AP Format: News, talk, sports. News staff: 3; News: 16 hrs wkly. Target aud: 35 plus; adults seeking full service progmg. Spec prog: It one hr, Sp one hr, Swedish one hr, farm one hr wkly. ♦Merrill Rosen, gen mgr; Nick Keefe, opns mgr, prom dir & progmg dir; Larry Sazacki, sls dir; Wayne Goff, chief of engrg; Kathy Roselle, traf mgr; Terry Frank, local news ed; Jason Sample, news rptr; Dennis Webster, farm dir; Matt Krieg, sports cmtr; Jim Roselle, disc jockey.

WWSE(FM)—Co-owned with WJTN(AM). October 1947: 93.3 mhz; 26.5 kw. 643 ft TL: N42 05 06 W79 17 23. Stereo. 24 Phone: (716) 664-9393. E-mail: wwsefm@wwsefm.com Web Site: wwsefm.com.200,000 Format: Adult contemp. News staff: 3; News: 7 hrs wkly. Target aud: 12554; female. ♦Cheryl Akin, sls VP & news rptr; Nick Keefe, mus dir; Brian Papalia, asst music dir; Sammie Green, traf mgr; Matthew Hanley, news rptr; Andrew Hill, disc jockey; Bill Dossion, disc jockey; Lee John, disc jockey.

WKSN(AM)— Jan 26, 1948: 1340 khz; 500 w-D, 1 kw-N. TL: N42 05 46 W79 14 48. Hrs open: Box 1199, 202 Front St., 14701. Phone: (716) 664-2313. Fax: (716) 488-1471. E-mail: jadmin@wksn.com Web Site: www.wksn.com. Licensee: Media One Group II LLC. Group owner: Vox Radio Group L.P. (acq 5-31-2005; grpsl). Population served: 39,795 Format: Music of your life. News staff: 2; News: 3 hrs wkly. Spec prog: Relg 2 hrs, Swedish one hr wkly. ♦Daniel C. Fischer, VP & gen mgr; Guy Ditonto, gen sls mgr; Tom Marshall, progmg mgr; Joel Keefer, news dir; Burton O. Waterman, chief of engrg; Roseanne De Frisco, traf mgr.

WHUG(FM)—Co-owned with WKSN(AM). Feb 1, 1965: 101.9 mhz; 3.3 kw. 298 ft TL: N42 07 55 W79 13 09. Stereo. E-mail: jadmin@whug.com Web Site: www.whug.com. Format: Country. News staff: 2; News: 2 hrs wkly.

***WNJA(FM)**— 1991: 89.7 mhz; 6 kw. 754 ft TL: N42 02 48 W79 05 26. Hrs open: 24 Box 1263, Buffalo, 14240. Secondary address: 140 Lower Terr., Buffalo 14202-1263. Phone: (716) 845-7000. Fax: (716) 845-7043. Web Site: www.wned.org. Licensee: Western New York Public Broadcasting Association. Law Firm: Schwartz, Woods & Miller. Format: Class. Target aud: 35 plus. ♦Donald K. Boswell, CEO, pres & gen mgr; Michael Sutton, CFO; Richard Daly, sr VP; Peter Goldsmith, progmg dir.

***WUBJ(FM)**— July 11, 1994: 88.1 mhz; 265 w. Ant 558 ft TL: N42 05 06 W79 17 23. Hrs open: 24
Rebroadcasts WBFO(FM) Buffalo 100%.
c/o Radio Stn. WBFO(FM), 3435 Main St., 205 Allen Hall, Buffalo, 14214-3003. Phone: (716) 829-6000. Fax: (716) 829-2277. E-mail: mail@wbfo.org Web Site: www.wbfo.org. Licensee: State University of New York. Population served: 117,000 Natl. Network: NPR. Format: Jazz, news. News staff: 2; News: news progrmg 50 hrs wkly. Target aud: General; educated professional. Spec prog: Blues 8 hrs, bluegrass music 3 hrs, Pol 3 hrs wkly. ♦Carole Smith Petro, VP & gen mgr; Matthew Katafiaz, gen mgr; Mark Wozniak, opns mgr; Joan Wilson, dev dir.

Jeffersonville

WDNB(FM)— Nov 15, 1999: Stn currently dark. 102.1 mhz; 2.2 kw. Ant 535 ft TL: N41 44 30 W74 51 23. Stereo. Hrs opn: 24 267 N. Main St., Suite 3, Liberty, 12754. Phone: (570) 253-1616. Fax: (570) 253-6297. E-mail: vbebedetti@boldgoldmedia.com Web Site: www.boldgoldmedia.com. Licensee: Bold Gold Media Group L.P. (group owner; (acq 5-23-2005; grpsl). Law Firm: Schwartz, Woods & Miller. Format: Country. News staff: one; News: 5 hrs wkly. Target aud: 25 plus; male & female general high school education plus. ♦Vince Benedetto, CEO; Bob Vanderheyden, gen mgr; Brian Walker, gen sls mgr; Paul Ciliberto, prom dir; George Schmitt, progmg dir; Theresa Opeka, news dir.

***WJFF(FM)**— Feb 12, 1990: 90.5 mhz; 3.7 kw. 629 ft TL: N41 48 58 W74 47 15. Stereo. Hrs opn: 24 Box 546, 4765 State Rt. 52, 12748. Phone: (845) 482-4141. Fax: (845) 482-WJFF. E-mail: wjff@wjffradio.org Web Site: www.wjffradio.org. Licensee: Radio Catskill. Natl. Network: NPR, PRI. Law Firm: Haley, Bader & Potts. Format: News/talk/eclectic music. News: 75 hrs wkly. Target aud: 16-60; general. ♦Bill Duncan, pres; Christine Aherne, stn mgr.

WPDA(FM)— January 1993: 106.1 mhz; 1.6 kw. 627 ft TL: N41 48 57 W74 45 42. Stereo. Hrs opn:
Rebroadcasts WPDH(FM) Poughkeepsie 100%.
Box 416, Poughkeepsie, 12602. Phone: (845) 471-1500. Fax: (845) 454-1204. Web Site: www.wpda.com. Licensee: Cumulus Licensing Corp. Group owner: Cumulus Media Inc. (acq 1-23-02; grpsl). Natl. Rep: Katz Radio. Format: Main stream rock. ♦Charles Benfer, gen mgr.

Johnson City

WLTB(FM)— Sept 3, 1972: 101.7 mhz; 1.25 kw. Ant 699 ft TL: N42 03 45 W75 56 37. Stereo. Hrs opn: Box 7, Vestal, 13851. Secondary address: 1808 Vestal Pkwy. E., Vestal 13851. Phone: (607) 748-9131. Fax: (607) 748-0061. Web Site: www.magic1017fm.com. Licensee: GM Broadcasting Inc. (acq 9-2-97; $176,000 with co-located AM). Population served: 400,000 Format: Adult contemp. Target aud: 18-49; emphasis on females. ♦Steve Gilinsky, gen mgr.

Johnstown

WENT(AM)—See Gloversville

WIZR(AM)— 1964: 930 khz; 1 kw-D. TL: N42 59 54 W74 21 31. Stereo. Hrs opn: 24 135 Guy Park Ave., Amsterdam, 12010. Phone: (518) 762-4631. Fax: (518) 762-0105. Licensee: 6 Johnson Road Licenses Inc. Group owner: Pamal Broadcasting Ltd. (acq 10-19-2001; grpsl). Population served: 250,000 Law Firm: Shaw Pittman. Format: Adult Contemp. News staff: one; News: 6 hrs wkly. Target aud: 25-54. Spec prog: It one hr, Pol one hr, Sp one hr wkly. ♦Joey Caruso, gen mgr.

Kingston

***WAMK(FM)**— March 1988: 90.9 mhz; 940 w. 1,486 ft TL: N42 04 35 W74 06 26. Stereo. Hrs opn: 24
Rebroadcasts WAMC-FM Albany 100%.
Box 66600, 318 Central Ave., Albany, 12206-6600. Phone: (518) 465-5233. Phone: (800) 323-9262. Fax: (518) 432-6974. E-mail: mail@wamc.org Web Site: www.wamc.org. Licensee: WAMC. Group owner: WAMC/Northeast Public Radio Natl. Network: PRI, NPR. Law Firm: Dow, Lohnes & Albertson. Wire Svc: AP Format: News/talk. News: 77 hrs wkly. Target aud: General. Spec prog: Jazz 13 hrs, folk 7 hrs. ♦Alan Chartock, CEO, chmn & pres; David Galletly, VP & progmg dir; Selma Kaplan, VP & news dir.

WDST(FM)—See Woodstock

***WFGB(FM)**— January 1985: 89.7 mhz; 3.1 kw. 1,486 ft TL: N42 04 35 W74 06 26. Stereo. Hrs opn: 24 Box 777, Lake Katrine, 12449. Phone: (845) 336-6199. Fax: (845) 336-7205. E-mail: email@soundoflife.org Web Site: www.soundoflife.org. Licensee: Sound of Life Inc. Population served: 350,000 Format: Christian. News: 3 hrs wkly. Target aud: General. ♦Tom Michael Zahradnik, gen mgr.

***WFRH(FM)**— Sept 1993: 91.7 mhz; 950 w. 272 ft TL: N41 59 04 W74 02 56. Hrs open: 24 918 Chesapeake Ave., Annapolis, 21403. Phone: (845) 336-0234. Fax: (410) 268-0931. Web Site: www.familyradio.com. Licensee: Family Stations Inc. (group owner) Population served: 175,000 Format: Relg. ♦Harold Camping, pres & gen mgr; Dan Elmendorf, stn mgr.

WGHQ(AM)— Mar 4, 1956: 920 khz; 5 kw-D, 262 w-N, DA-1. TL: N41 53 09 W73 58 15. Hrs open: 24 715 Rt. 52, Beacon, 12508. Phone: (845) 838-6000. Fax: (845) 838-6088. Licensee: 6 Johnson Road Licenses Inc. Group owner: Clear Channel Communications Inc. (acq 4-1-2007; grpsl). Population served: 160,000 Format: News/talk. News staff: one. Target aud: 30 plus. Spec prog: Relg 3 hrs wkly. ♦Jason Finkelberg, gen mgr.

WKNY(AM)— Aug 1, 1939: 1490 khz; 1 kw-U. TL: N41 56 11 W74 00 30. Hrs open: 24 718 Broadway, 12401. Secondary address: Box 1398 12402. Phone: (845) 331-1490. Fax: (845) 331-9569. E-mail: wknynews@pendellrd.com Licensee: Cumulus Licensing Corp. Group owner: Cumulus Media Inc. (acq 1-23-02; grpsl). Population served: 200,000 Natl. Network: CBS. Natl. Rep: Katz Radio. Wire Svc: AP Format: Adult contemp. News staff: 2; News: 26 hrs wkly. Target aud: 25-54; 60% female. Spec prog: Ger one hr, Pol one hr, Irish one hr wkly. ♦Chuck Benfer, gen mgr; Warren Lawrence, progmg dir; Linda Rosner, news dir; Dan Gorham, news rptr; Dominic Fusco, sls & mktg.

WKXP(FM)— Dec 13, 1965: 94.3 mhz; 2.25 kw. Ant 544 ft TL: N41 53 44 W73 59 32. Stereo. Hrs opn: 24 Box 416, Poughkeepsie, 12602-0416. Secondary address: 2 Pendell Rd., Poughkeepsie 12602. Phone: (845) 471-1500. Fax: (845) 454-1204. E-mail: newsroom@pendelled.com Web Site: www.943thewolf.com. Licensee: Cumulus Licensing Corp. Group owner: Cumulus Media Inc. (acq 2-11-2004; $3.5 million). Population served: 200,000 Format: Country. Target aud: 18-49; women. ♦Charles Benfer, gen mgr.

Lake George

WCKM-FM— Apr 21, 1994: 98.5 mhz; 6 kw. 1,289 ft TL: N43 25 12 W73 45 37. Stereo. Hrs opn: 24 128 Glen St., Glens Falls, 12801-4432. Phone: (518) 761-9890. Fax: (518) 761-9893. E-mail: staffmail@radiowins.com Web Site: www.radiowins.com. Licensee: Entertronics Inc. Population served: 300,000 Natl. Network: ABC. Law Firm: Joseph E. Dunne III. Format: Hits of the 60s, 70s & 80s. News staff: one; News: 7 hrs wkly. Target aud: 25-54; upscale baby boomers. Spec prog: Interviews. ♦David L. Covey, CEO & gen mgr; William Walker, exec VP; Robin G. Covey, opns dir; Paul VanAmburgh, gen sls mgr.

Lake Luzerne

WBAR-FM— June 30, 1992: 94.7 mhz; 300 w. 892 ft TL: N43 17 22 W73 44 35. Stereo. Hrs opn: 24
Rebroadcasts WHAZ(AM) Troy 100%.
30 Park Ave., Cohoes, 12047-3330. Phone: (518) 237-1330. Fax: (518) 235-4468. E-mail: info@whaz.com Web Site: www.whaz.com. Licensee: Capital Media Corp. (group owner; acq 10-1-92; FTR: 11-9-92). Population served: 600,000 Format: Bible teaching & preaching. Target aud: 25-75. ♦Paul F. Lotters, pres, gen mgr & progmg mgr; Steven L. Klob, opns dir, dev dir, sls dir, prom dir & adv dir; Rex P. Gregory, mus dir, news dir, pub affrs dir & disc jockey; John W. Shafer, chief of engrg.

Lake Placid

WIRD(AM)— Nov 21, 1961: 920 khz; 5 kw-D, 250 w-N. TL: N44 15 36 W74 01 22. Hrs open: 24 Box 211, Saranac Lake, 12983. Phone: (518) 891-1544. Fax: (518) 891-1545. Licensee: Radio Lake Placid Inc. Group owner: Mountain Communications (acq 1-10-2005). Population served: 80,000 Natl. Network: CBS. Law Firm: Tierney & Swift. Format: ESPN radio. News staff: 2; News: 20 hrs wkly. Target aud: 25-54; working blue collar/college educated.

WLPW(FM)—Co-owned with WIRD(AM). October 1979: 105.5 mhz; 3 kw. -236 ft TL: N44 15 36 W74 01 22. Stereo. 24 80,000 Format: Classic rock.

Lake Ronkonkoma

***WSHR(FM)**— January 1966: 91.9 mhz; 2.8 kw. 141 ft TL: N40 50 00 W73 06 01. Stereo. Hrs opn: 24 Sachem North High School, 212 Smith Rd., 11779. Phone: (631) 471-1472. Phone: (631) 471-1400. Fax: (631) 471-1491. Licensee: Board of Education Sachem Central School District at Holbrook. (acq 1967). Population served: 1,500,000 Format: Var. Target aud: General. ♦Mark Laura, gen mgr; Isaic Ramaswamy, stn mgr.

Lake Success

WKTU(FM)—Licensed to Lake Success. See New York

Lakewood

WKZA(FM)— Mar 2001: 106.9 mhz; 5.2 kw. 715 ft TL: N41 57 31 W79 16 11. Hrs open: 106 W. T/hird St., Suite 106, Jamestown, 14701. Phone: (716) 487-1106. Fax: (716) 488-2169. Web Site: www.1069kissfm.com. Licensee: Cross Country Communications LLC. Format: Top-40 hits. ♦John Newman, gen mgr.

Lancaster

WXRL(AM)— 1964: 1300 khz; 5 kw-D, 2.5 kw-N, DA-2. TL: N42 52 58 W78 37 54. Hrs open: 24 Box 170, 5426 William St., 14086. Phone: (716) 681-1313. Fax: (716) 681-7172. Web Site: www.wxrl.com. Licensee: Dome Broadcasting Inc. (acq 11-1-70). Population served: 1,000,000 Natl. Network: CNN Radio. Law Firm: Smithwick & Belendiuk. Format: Country. Target aud: 35 plus; Mature men & women 35 and older. Spec prog: German one hr, Polish 19 hrs wkly. ♦Louis A. Schriver, pres & gen mgr; Joan C. Schriver, exec VP & progmg dir; Lori Arumygam, dev dir; Louis E. Schriver Jr., gen sls mgr; Linda Sukennik, prom dir, prom mgr & traf mgr; Lynn Carol Supparits, opns dir, opns mgr & mus dir.

Liberty

***WGWR(FM)**— November 1997: 88.1 mhz; 60 w. 561 ft TL: N41 48 55 W74 45 48. Hrs open: 24
Rebroadcasts WFGB(FM) Kingston 100%.
Box 777, Lake Katrine, 12449. Phone: (845) 336-6199. Fax: (845) 336-7205. E-mail: wmial@soundoflife.org Web Site: www.soundoflife.org. Licensee: Sound of Life Inc. Format: Christian. ♦Tom Michael Zahradnik, gen mgr.

WVOS(AM)— 1947: 1240 khz; 1 kw-U. TL: N41 46 54 W74 43 49. Hrs open: 5 AM-11 PM 198 Bridgeville Rd., Monticello, 12701. Phone: (845) 794-9898. Fax: (845) 794-0125. Licensee: Watermark Communications LLC (acq 11-2-2005; $1.7 million with co-located FM). Population served: 35,000 Law Firm: Barry Skidelsky. Format: Hot country. ♦Helena Manzione, gen mgr.

WVOS-FM— December 1964: 95.9 mhz; 6 kw. Ant 328 ft TL: N41 45 09 W74 43 01.24 70,000 Target aud: 25-54.

Little Falls

WIXT(AM)— June 10, 1952: 1230 khz; 1 kw-U. TL: N43 02 33 W74 51 31. Hrs open: 24 239 Genesee St., Suite 500, Utica, 13501-3407. Phone: (315) 797-0803. Fax: (315) 797-7813. Web Site: www.starsradionetwork.com. Licensee: Capstar TX L.P. Group owner: Clear Channel Communications Inc. (acq 4-16-2001; $500,000). Population served: 200,000 Law Firm: Richard Hayes. Format: Sports. News staff: one; News: 16 hrs wkly. Target aud: 25-54. Spec prog: Farm one hr, relg one hr wkly. ♦Brian Delaney, gen mgr.

WSKU(FM)— Jan 3, 1991: 105.5 mhz; 2.3 kw. 152 ft TL: N42 59 27 W74 55 06. Stereo. Hrs opn: 24 239 Genesee St., Suite 500, Utica, 13501. Phone: (315) 797-0803. Fax: (315) 797-7813. Web Site: www.cnykiss.com. Licensee: Capstar TX L.P. Group owner: Clear Channel Communications Inc. (acq 4-16-2001; $2.15 million with WSKS(FM) Whitesboro). Population served: 200,000 Natl. Rep: Roslin. Law Firm: Richard Hayes. Format: CHR, rhythmic. News staff: one; News: one hr wkly. Target aud: 25-54. ♦Brian Delaney, gen mgr; Stephen Lawrence, opns mgr.

Livingston Manor

WJZI(FM)—Not on air, target date: unknown: 107.1 mhz; 6 kw horiz. Ant 328 ft TL: N41 58 28 W74 55 05. Hrs open: 149 Penn Ave., Scranton, PA, 18503. Phone: (570) 348-9103. Fax: (570) 348-9109. Licensee: Shamrock Communications Inc. ♦William R. Lynett, pres.

Lockport

WLVL(AM)— May 8, 1947: 1340 khz; 1 kw-U. TL: N43 10 30 W78 42 39. Hrs open: Box 477, 14094. Secondary address: 320 Michigan St. 14094. Phone: (716) 433-5944. Fax: (716) 433-6588. E-mail: wlvl@wlvl.com Web Site: www.wlvl.com. Licensee: Culver Communications Inc. (acq 9-81; $600,000; FTR: 10-5-81). Population served: 45,000 Natl. Network: Westwood One. Format: News/talk, sports. Target aud: 25-64; adult Lockport area citizens. Spec prog: Farm one hr, lt 2 hrs, Pol one hr, relg 3 hrs wkly. ♦Richard C. Greene, pres & gen mgr.

Loudonville

***WVCR-FM**— Apr 26, 1963: 88.3 mhz; 2.8 kw. Ant 840 ft TL: N42 38 13 W74 00 05. Stereo. Hrs opn: 24 515 Loudon Rd., 12211-1462. Phone: (518) 782-6750. Fax: (518) 782-6498. Web Site: www.wvcr.com. Licensee: Siena College. Population served: 1,000,000 Format: Variety/"We Play Anything". Target aud: 12-34; female. Spec prog: Pol 3 hrs, Sp 3 hrs, gospel 3 hrs, Irish 3 hrs. ♦Darrin S. Kibbey, gen mgr; Joseph Doty, opns mgr; Dean Charette, progmg mgr.

Lowville

WLLG(FM)— Apr 1, 1987: 99.3 mhz; 1 kw. 561 ft TL: N43 45 12 W75 33 50. Stereo. Hrs opn: 24 7606 N. State St., 13367. Phone: (315) 376-7500. Fax: (315) 376-8549. E-mail: sales@themoose.net Web Site: www.themoose.net. Licensee: The Flack Broadcasting Group L.L.C. Natl. Network: USA. Law Firm: Shaw Pittman. Format: Country, news. News staff: one; News: 18 hrs wkly. Target aud: General. Spec prog: Farm 3 hrs, relg 2 hrs wkly. ♦William Flack, pres, gen mgr & progmg dir; Brian Best, news dir; Ken Ruhlend, chief of engrg.

Malone

WICY(AM)— Nov 4, 1946: 1490 khz; 1 kw-U. TL: N44 50 46 W74 16 07. Hrs open: 18 86 Porter Rd., 12953. Phone: (518) 483-1100. Fax: (518) 483-1382. Web Site: www.oldiesradioonline.com. Licensee: Cartier Communications Inc. Group owner: Martz Communications Group (acq 6-30-97; $761,000 with co-located FM). Population served: 51,000 Natl. Rep: Rgnl Reps. Law Firm: Arter & Hadden. Format: Oldies. News staff: 2; News: 15 hrs wkly. Target aud: 25-54. Spec prog: Farm one hr wkly. ♦Michael Boldt, gen mgr & gen sls mgr.

WVNV(FM)—Co-owned with WICY(AM). May 1, 1993: 96.5 mhz; 2.4 kw. 361 ft TL: N44 49 37 W74 22 46. Stereo. 20 Web Site: www.country965.com.100,000 Format: Country. News staff: 2; News: 2 hrs wkly. Target aud: 18-54. ♦Drew Scott, progmg dir.

***WMHQ(FM)**— Dec 3, 2003: 90.1 mhz; 3 kw. Ant 325 ft TL: N44 49 48 W74 22 35. Stereo. Hrs opn: 24
Rebroadcasts WMHR(FM) Syracuse 99%.
4044 Makyes Rd., Syracuse, 13215. Phone: (315) 469-5051. E-mail: mhn@marshillnetwork.org Web Site: www.marshillnetwork.org. Licensee: Mars Hill Broadcasting Co. Inc. Natl. Network: Moody, Salem Radio Network. Law Firm: Wiley, Rein & Fielding. Wire Svc: AP Format: Christian. News: 6 hrs wkly. Target aud: General; Christian families. ♦Clayton Roberts, pres; Wayne Taylor, gen mgr; Chris Tetta, progmg dir; Jeremy Miller, news dir; Valerie Smith, traf mgr.

***WSLO(FM)**— February 1989: 90.9 mhz; 200 w. 354 ft TL: N44 49 46 W74 22 31. Hrs open: 24
Rebroadcasts WSLU(FM) Canton 100%.
St. Lawrence Univ., Canton, 13617. Phone: (315) 229-5356. Fax: (315) 229-5373. E-mail: radio@ncpr.org Web Site: www.ncpr.org. Licensee: St. Lawrence University. Law Firm: Donald E. Martin. Format: Eclectic public radio. News staff: 2; News: 35 hrs wkly. Target aud: General. ♦Ellen Rocco, gen mgr; Shelly Pike, opns mgr & chief of opns; Sandra Demarest, dev dir.

Malta

WBZZ(FM)— October 1996: 105.7 mhz; 7.1 kw. Ant 613 ft TL: N42 47 09 W73 37 43. Hrs open: 1241 Kings Rd., Schenectady, 12303. Phone: (518) 881-1515. Fax: (518) 881-1516. E-mail: comments@buzz1057.com Web Site: www.1045thebuzz.com. Licensee: Regent Licensee of Mansfield Inc. Group owner: Vox Radio Group L.P. (acq 1-4-2007; $4.9 million). Population served: 275,000 Format: Hot adult contemp. Target aud: 18-54; general. ♦Robert Ausfeld, gen mgr.

Manlius

WAQX-FM— Aug 23, 1978: 95.7 mhz; 25 kw. 300 ft TL: N43 00 25 W76 05 38. Stereo. Hrs opn: 24 1064 James St., Syracuse, 13203. Phone: (315) 472-0200. Fax: (315) 472-1146. Web Site: www.95x.com. Licensee: Citadel Broadcasting Co. Group owner: Citadel Broadcasting Corp. (acq 4-26-01; grpsl). Natl. Network: ABC. Natl. Rep: D & R Radio. Law Firm: Shaw Pittman. Format: AOR. News: 3 hrs wkly. Target aud: 19-49; male. Spec prog: Pub service one hr wkly. ♦Tom Mitchell, opns dir & opns mgr; Dave Edwards, chief of engrg.

Massena

WMSA(AM)— Oct 12, 1945: 1340 khz; 1 kw-U. TL: N44 54 14 W74 53 01. Hrs open: 5:30 AM-10:15 PM Box 210, 2155 State Rt. 420, 13662. Phone: (315) 769-3594. Fax: (315) 769-3299. E-mail: info@1340wmsa.com Web Site: www.1340wmsa.com. Licensee: Seaway Broadcasting Inc. Group owner: Martz Communications Group (acq 5-6-99; $545,000). Population served: 14,100 Format: Adult contemp. News staff: one; News: 16 hrs wkly. Target aud: 18 plus. ♦Michael Boldt, gen mgr.

WYBG(AM)— Aug 18, 1958: 1050 khz; 1 kw-D, 500 w-N. TL: N44 53 42 W74 56 05. Hrs open: 6 AM-6 PM Box 298, 24 Andrews St., 13662. Phone: (315) 764-0554. Fax: (315) 764-0118. E-mail: wybgradio@slic.com Web Site: www.wybg1050.com. Licensee: Wade Communications Inc. (acq 8-15-88; $450,000; FTR: 8-15-88). Population served: 295,000 Natl. Network: USA. Format: News/talk. News staff: 2; News: 14 hrs wkly. Target aud: 25-65; baby boomers & seniors. Spec prog: American Indian, children, farm, folk. ♦Curran Wade, pres & gen mgr; Dorothy Wade, VP.

Mechanicville

WABY(AM)— Oct 19, 1981: 1160 khz; 5 kw-D, 570 w-N. TL: N42 55 12 W73 42 08. Hrs open: 24 100 Saratoga Village Blvd., Malta, 12020. Phone: (518) 899-3000. Fax: (518) 889-3057. E-mail: moonradioam@aol.com Licensee: The Anastos Media Group Inc. Group owner: Anastos Media Group Inc. (acq 12-7-00; $280,000). Natl. Network: ABC. Format: Big band. News staff: one; News: 50 hrs wkly. Target aud: 40 plus; adults, male & female. ♦Scott Collins, pres & gen mgr; John Meaney, opns mgr, progmg dir, progmg mgr & news dir.

WTMM-FM— Jan 4, 1993: 104.5 mhz; 5 kw. Ant 351 ft TL: N42 52 44 W73 51 47. Stereo. Hrs opn: 24 1241 Kings Rd., Schenectady, 12303. Phone: (518) 881-1515. Fax: (518) 881-1516. Web Site: www.wtmm.com. Licensee: Regent Licensee of Mansfield Inc. Group owner: Regent Communications Inc. (acq 8-24-2001; grpsl). Natl. Network: ESPN Radio. Format: Sports talk. Target aud: 25-54; serious sports fans. ♦John Hirsch, stn mgr; Buzz Brindle, progmg dir.

Mexico

WVOA-FM— 1997: 103.9 mhz; 3 kw. 292 ft TL: N43 28 36 W76 16 44. Hrs open: Renard Communications Corp., 401 W. Kirkpatrick St., Syracuse, 13204. Phone: (315) 472-0222. Fax: (315) 478-7745. E-mail: programming@WVOARadio.com Web Site: WVOARadio.com. Licensee: Renard Communications Corp. (acq 3-13-97; $3,000 for CP). Format: Relg. Spec prog: Ger 2 hrs, It 2 hrs, Pol 4 hrs , Sp 15 hrs wkly. ♦Sam Furco, gen mgr.

Middletown

WALL(AM)— Aug 6, 1942: 1340 khz; 1 kw-U. TL: N41 27 25 W74 26 24. Hrs open: 24 Box 416, Poughkeepsie, 12602. Secondary address: 2 Pendel Rd., Poughkeepsie 12602. Phone: (845) 471-1500. Fax: (845)-454-1204. E-mail: weoknews.@bestweb.net Web Site: www.cumulus.com. Licensee: Cumulus Licensing Corp. Group owner: Cumulus Media Inc. (acq 1-23-02; grpsl). Population served: 305,000 Law Firm: Hogan & Hartson. Format: Sp. News staff: 2; News: 30 hrs wkly. Target aud: 35-64; educated, upscale families. ♦Victor Goodman, gen sls mgr; Nick Robbins, progmg dir; Beth Christie, pub affrs dir; Bryan Jones, local news ed.

WRRV(FM)—Co-owned with WALL(AM). Nov 11, 1966: 92.7 mhz; 3 kw. 300 ft TL: N41 27 21 W74 26 22. Stereo. 24 Web Site: www.cumulus.com. Format: Alternative ROCK. News staff: one; News: 3 hrs wkly. Target aud: 18-44; younger, mobile, upscale families. Spec prog: New mus 2 hrs wkly. ♦Mike Harris, pres; Bill Palmeri, gen mgr; Greg O'Brien, progmg dir & disc jockey; Andrew Boris, mus dir & disc jockey; Chris Dorman, disc jockey; Jack George, disc jockey.

***WOSR(FM)**— Feb 3, 1992: 91.7 mhz; 1.8 kw. 630 ft TL: N41 36 04 W74 33 13. Stereo. Hrs opn: 24
Rebroadcasts WAMC-FM Albany 100%.
Box 66600, Albany, 12206-6600. Secondary address: 318 Central Ave. 12206-6600. Phone: (518) 465-5233. Phone: (800) 323-9262. Fax: (518) 432-6974. E-mail: mail@wamc.org Web Site: www.wamc.org. Licensee: WAMC. Group owner: WAMC/Northeast Public Radio Natl. Network: NPR, PRI. Law Firm: Dow, Lohnes & Albertson. Wire Svc: AP Format: News, talk. News: 77 hrs wkly. Target aud: General. Spec prog: Folk 7 hrs, jazz 13 hrs wkly. ♦Alan Chartock, CEO, chmn & pres; David Galletly, VP & progmg dir; Selma Kaplan, VP & news dir.

Mineola

WTHE(AM)— Jan 1, 1964: 1520 khz; 1 kw-D. TL: N40 44 45 W73 37 29. Hrs open: 260 E. Second St., 11501. Phone: (516) 742-1520. Fax: (516) 742-2878. E-mail: nygospelradio@aol.com Web Site: www.wthe1520am.com. Licensee: Universal Broadcasting of New York Inc. (group owner; acq 7-10-69; $235,000). Population served: 12,000,000 Law Firm: Cohn and Marks, LLP. Format: Relg, Black gospel. Target aud: General. ♦Howard Warshaw, CEO, exec VP & gen mgr; Miriam Warshaw, pres; Abe Warshaw, sr VP & VP; Howard Warshaw Sr., VP; Darren Greggs, opns VP; Clara Mark, mus dir.

Minetto

WKRH(FM)— October 1996: 106.5 mhz; 5.1 kw. 328 ft TL: N43 25 45 W76 32 14. Hrs open: 235 Walton St., Syracuse, 13202-1351. Phone: (315) 343-1440. Phone: (315) 472-9111. Fax: (315) 472-1888. E-mail: generalinfo@krock.com Web Site: www.krock.com. Licensee: Galaxy Syracuse Licensee LLC. (acq 8-31-2000). Law Firm: Leventhal, Senter & Lerman. Format: Alt rock. News staff: one; News: 2 hrs wkly. Target aud: 18-49; men. ♦Ed Levine, pres; Michael Lucarelli, CFO; Lisa Morrow, sr VP; Mimi Grisworld, progmg VP; Scott Petibone, progmg dir.

Monroe

***WLJP(FM)**— May 1991: 89.3 mhz; 200 w. 1,023 ft TL: N41 22 38 W74 07 55. (CP: Ant 1,038 ft.). Stereo. Hrs opn:
Rebroadcasts WFGB(FM) Kingston 100%.
Box 777, Lake Katrine, 12449. Phone: (845) 336-6199. Fax: (845) 336-7205. E-mail: email@soundoflife.org Web Site: www.soundoflife.org. Licensee: Sound of Life Inc. Population served: 350,000 Format: Contemp Christian. Target aud: General. ♦Tom Michael Zahradnik, gen mgr.

Montauk

WMOS(FM)— Feb 19, 1993: 104.7 mhz; 6 kw. 328 ft TL: N41 01 57 W71 58 31. Stereo. Hrs opn: 24 7 Governor Winthrop Blvd., New London, 06320. Phone: (866) 441-9653. Fax: (860) 444-7970. Licensee: Citadel Broadcasting Co. Group owner: Citadel Broadcasting Corp. (acq 4-3-03). Format: Classic rock. News staff: one. Target aud: 25-45; upscale Hamptons, at-work & New York City 2nd homeowners. ♦Kevin O'Connor, opns dir; Steve Ardolina, opns dir; Julie Johnson, progmg dir.

***WPKM(FM)**— 2004: 88.7 mhz; 8 w horiz, 2.7 kw vert. Ant 226 ft TL: N41 01 53 W71 58 32. Hrs open:
Rebroadcasts WPKN (FM) Bridgeport 100%.
244 University Ave., Bridgeport, CT, 06604. Phone: (203) 331-9756. E-mail: wpkn@wpkn.org Web Site: www.wpkn.org. Licensee: WPKN Inc. Format: Div. ♦Harry Minot, gen mgr.

Montgomery

***WGMY(FM)**—Not on air, target date: unknown: 88.1 mhz; 1.65 kw. Ant 115 ft TL: N41 28 37.4 W74 16 09.8. Hrs open: River Broadcasting Inc., 205 River Rd., Walden, 12586-2815. Phone: (845) 778-2400. Licensee: River Broadcasting Inc. ♦John H. Katonah, pres.

Monticello

WJUX(FM)— Nov 1, 1994: 99.7 mhz; 6 kw. 328 ft TL: N41 39 24 W74 43 40. Stereo. Hrs opn: 24 6550 Rt. 9 S., Howell, 07731. Phone: (732) 901-9953. Fax: (732) 901-0356. E-mail: info@bridgefm.org Web Site: www.bridgefm.org. Licensee: Bridgelight LLC (acq 11-13-03). Population served: 6,700,000 Law Firm: Koteen & Naftalin. Format: Relg. Target aud: 35-54. ♦Eugene Blabey, stn mgr.

WSUL(FM)— Apr 16, 1977: 98.3 mhz; 2.2 kw. 535 ft TL: N41 39 38 W74 41 14. Stereo. Hrs opn: 24 Box 98.3, 198 Bridgeville Rd., 12701. Phone: (845) 794-9898. Phone: (845) 794-0242. Fax: (845) 794-0125. E-mail: office@wsul.com Web Site: www.wsul.com. Licensee: Watermark Communications LLC (acq 3-17-2005; $2.5 million). Population served: 100,000 Law Firm: Wilkinson Barker Knauer. Format: Hot Adult Contemp. News staff: 2. Target aud: 25-54. ♦Helena Manzione, gen mgr; Shannon Marie Holland, opns mgr.

WVOS-FM—See Liberty

Montour Falls

WNGZ(FM)— June 1973: 104.9 mhz; 1 kw. 480 ft TL: N42 15 05 W76 52 53. Stereo. Hrs opn: 24 2205 College Ave., Elmira, 14903. Phone: (607) 732-4400. Fax: (607) 732-7774. Licensee: Chemung County Radio Inc. Group owner: Backyard Broadcasting LLC (acq 12-1-02; grpsl). Population served: 1,534 Format: Classic rock. News staff: one. Target aud: 20-49; baby boomers, young adults. ♦Kevin White, gen mgr & gen sls mgr; Vinnie Pagano, progmg dir.

Morristown

WYSX(FM)— Noverber 1998: 96.7 mhz; 17 kw. Ant 354 ft TL: N44 34 43 W75 30 51. Stereo. Hrs opn: 24 One Bridge Plaza, Suite 204, Ogdensburg, 13669. Phone: (315) 393-1220. Fax: (315) 393-3974. E-mail: john@yesfm.com Web Site: www.yesfm.com. Licensee: Waters Communications Inc. Group owner: Martz Communications Group (acq 3-5-99; $285,000 with WPAC(FM) Ogdensburg). Population served: 112,000 Format: CHR. Target aud: 18-34. ♦Tim Martz, pres; John Winter, gen mgr.

Mount Hope

***WXHD(FM)**— September 1994: 90.1 mhz; 1.1 kw. 600 ft TL: N41 25 36 W74 34 54. Stereo. Hrs opn: 24 Box 2011, Jersey City, NJ, 07303-2011. Secondary address: 4th Floor, 43 Montgomery St., Jersey City, NJ 07302. Phone: (201) 521-1416. Fax: (201) 521-1286. E-mail: wfmu@wfmu.org Web Site: www.wfmu.org. Licensee: Auricle Communications. (acq 6-97). Law Firm: Haley, Bader & Potts. Format: Div, free form. Target aud: General. ♦Ken Freedman, pres; Brian Turner, progmg dir & mus dir; John Fogarazzo, chief of engrg.

Mount Kisco

WFAF(FM)— Jan 15, 1964: 106.3 mhz; 1.4 kw. Ant 440 ft TL: N41 11 56 W73 41 37. Stereo. Hrs opn:
Rebroadcasts WPDH(FM) Poughkeepsie 100%.
Box 416, Poughkeepsie, 12602-0416. Secondary address: 2 Pendell Rd., Poughkeepsie 12602. Phone: (845) 471-1500. Fax: (845) 454-1204. E-mail: newsroom@pendelled.com Web Site: www.wpdh.com. Licensee: Cumulus Licensing Corp. Group owner: Cumulus Media Inc. (acq 1-23-2002; grpsl). Population served: 150000 Format: Classic rock. ♦Charles Benfer, gen mgr.

WRVP(AM)— Oct 27, 1957: 1310 khz; 5 kw-D, 33 w-N, DA-2. TL: N41 11 37 W73 44 22. Hrs open: 6 AM-6 PM Box 2908, Patterson, NJ, 07509. Phone: (973) 881-8700. Fax: (973) 881-8324. Licensee: Radio Vision Cristiana Management Corp. (acq 8-26-2002; $1.36 million). Population served: 100,000 Law Firm: Koteen & Naftalin. Format: Sp Christian. News staff: 3. Target aud: General. ♦Milton Donato, stn mgr.

Nanuet

WRCR(AM)—See Spring Valley

New City

WRKL(AM)— July 4, 1964: 910 khz; 1 kw-D, 800 w-N, DA-2. TL: N41 10 52 W74 02 53. Hrs open: 24 1551 Rt. 202, Pomona, 10970. Phone: (845) 354-2000. Fax: (845) 354-4796. E-mail: wrkl@polskieradio.com Web Site: www.polskieradio.com. Licensee: Polnet Communications Ltd. (group owner; acq 3-19-99). Population served: 300,000 Law

Firm: Wiley, Rein and Fielding. Format: Polish language. News staff: one; News: 50 hrs wkly. Target aud: 18-54; Polish language audience. ♦Kent D. Gustafson, CEO & gen mgr; Walter Kotaba, pres; Grzegorz Sliwecki, opns mgr.

New Paltz

WBWZ(FM)— Nov 19, 1992: 93.3 mhz; 350 w. 1,328 ft TL: N41 41 58 W74 00 11. Hrs open: 24 20 Tucker Dr., Poughkeepsie, 12603. Phone: (845) 471-2300. Fax: (845) 471-2683. Web Site: www.star933fm.com. Licensee: AMFM Radio Licenses LLC. Group owner: Clear Channel Communications Inc. (acq 12-22-2000; with WRWD-FM Highland). Population served: 660,000 Natl. Network: Jones Radio Networks. Natl. Rep: Katz Radio. Format: Adult contemp hits of the 70s, 80s & 90s. News staff: one; News: 7 hrs wkly. Target aud: 25-54; baby boomers. ♦Frank Curcio, gen mgr; Reggie Osterhoudt, opns mgr; Jim Brady, gen sls mgr; Jeanette Relyea, natl sls mgr; Nick Smirnoff, prom dir; Aaron McCord, progmg dir; Cameron Hendrix, news dir.

New Rochelle

WVIP(FM)—Listing follows WVOX(AM).

WVOX(AM)— 1950: 1460 khz; 500 w-D. TL: N40 55 42 W73 46 30. Hrs open: 24 One Broadcast Forum, 10801. Phone: (914) 636-1460. Fax: (914) 636-2900. Web Site: www.wvox.com. Licensee: Hudson-Westchester Radio Inc. (acq 5-1-68). Population served: 75,385 Natl. Network: Jones Radio Networks, AP Radio. Law Firm: Garvey, Schubert & Barer. Format: News/talk. News staff: 4; News: 50 hrs wkly. Target aud: 24 plus; community minded. Spec prog: Black one hr, gospel one hr, relg 3, Irish one, Jewish 2 hrs, gay-lesbian one hr wkly. ♦Cindy Gallagher, CFO & exec VP; Nancy Curry, VP; Don Stevens, opns mgr; Matthew O'Shaughnessy, dev dir; Judy Fremont, sls VP & gen sls mgr; David O'Shaughnessy, progmg VP; Richard Littlejohn, mus dir; William O'Shaughnessy, CEO, pres, edit dir & political ed.

WVIP(FM)—Co-owned with WVOX(AM). 1953: 93.5 mhz; 3 kw. 325 ft TL: N40 57 45 W73 50 32. Stereo. 24 E-mail: don@wvox.com 7,000,000 Law Firm: Garvey, Schubert & Barer. Format: Adult standards. News staff: 2; News: 14 hrs wkly. Target aud: 18 plus; adults. Spec prog: It 6 hrs, Jewish one hr wkly. ♦William O'Shaughnessy, chmn; Don Stevens, stn mgr; Richard LittleJohn, sls VP & progmg dir.

New York

WABC(AM)— Oct 7, 1921: 770 khz; 50 kw-U. TL: N40 52 50 W74 04 12. Hrs open: 17th Fl., 2 Penn Plaza, 10121. Phone: (212) 613-3800. Fax: (212) 613-3823. Web Site: www.wabcradio.com. Licensee: Radio License Holding X LLC. (acq 6-12-2007; grpsl). Population served: 789,556 Natl. Rep: Interep. Format: Talk. ♦Mitch Dolan, pres & gen mgr; Tim McCarthy, stn mgr & sls dir; Fred Bennett, gen sls mgr; Russ King, prom dir & pub affrs dir; Phil Boyce, progmg dir; Kevin Plumb, chief of engrg.

WPLJ(FM)—Co-owned with WABC(AM). Jan 18, 1960: 95.5 mhz; 6.7 kw. Ant 1,335 ft TL: N40 44 54 W73 59 10. Stereo. 24 17th Fl., 2 Penn Plaza, 10121. Phone: (212) 613-8900. Fax: (212) 613-8956. Fax: (212) 613-8950. E-mail: writeus@Plj.com Web Site: www.plj.com. Licensee: Radio License Holding IX LLC.8,000,000 Natl. Rep: Interep. Format: Hot adult contemp. News: 5 hrs wkly. Target aud: 18-54; females. ♦Steven W. Borneman, stn mgr & gen sls mgr; Tom Cuddy, opns VP; Theresa Angela, prom dir; Scott Shannon, progmg dir; Tony Mascaro, mus dir; Patty Steele, news dir; Kevin Plumb, engrg dir.

WADO(AM)— Mar 12, 1934: 1280 khz; 50 kw-D, 7.2 kw-N, DA-2. TL: N40 49 36 W74 04 32. Hrs open: 24 485 Madison Ave., 3rd Flr., 10022. Phone: (212) 310-6000. Fax: (212) 888-3694. Web Site: www.univision.com. Licensee: Wado-Am License Corp. Group owner: Univision Radio (acq 9-22-2003; grpsl). Population served: 3,000,000 Natl. Rep: Katz Radio. Format: Sp, news/talk, sports. News staff: 13; News: 50 hrs wkly. Target aud: 25-54; Hispanics in the NY metropolitan area. ♦Joe Pagan, gen mgr.

WAXQ(FM)— Dec 1, 1956: 104.3 mhz; 6 kw. 1,361 ft TL: N40 44 54 W73 59 10. Stereo. Hrs opn: 1180 Ave. of the Americas, 10036. Phone: (212) 575-1043. Fax: (212) 302-7814. Web Site: www.q1043.com. Licensee: AMFM Radio Licenses LLC. Group owner: Clear Channel Communications Inc. (acq 8-30-2000; grpsl). Population served: 1,000,000 Law Firm: Fleischman & Walsh. Format: Classic rock. News staff: one; News: 2 hrs wkly. ♦Rob Williams, gen mgr; Bob Buchmann, progmg dir; Eric Wellman, mus dir; Henry Behring, chief of engrg.

***WBAI(FM)**— January 1960: 99.5 mhz; 5.4 kw horiz, 3.9 kw vert. 1,220 ft TL: N40 44 54 W73 59 10. Stereo. Hrs opn: 24 120 Wall St., 10th Fl., 10005. Phone: (212) 209-2800. Phone: (212) 209-2800. Fax: (212) 747-1698. Web Site: www.wbai.org. Licensee: Pacifica Foundation. Group owner: Pacifica Foundation Inc. dba Pacifica Radio (acq 1-9-60). Population served: 18,000,000 Format: Div, educ, news/talk. News staff: 2; News: 5 hrs wkly. Target aud: General; NY metropolitan area. Spec prog: American Indian one hr, Black 10 hrs, class 5 hrs, folk 2 hrs, jazz 5 hrs, Sp 3 hrs wkly. ♦Indra Hardat, gen mgr.

WBBR(AM)— Feb 13, 1991: 1130 khz; 50 kw-U, DA-N. TL: N40 48 39 W74 02 24. Hrs open: 731 Lexington Ave., 10022. Phone: (212) 318-2000. Fax: (917) 369-5000. Web Site: www.bloomberg.com. Licensee: Bloomberg Communications Inc. (acq 11-4-92; $13.58 million; FTR: 11-23-92). Population served: 18,000,000 Format: Business news. ♦Ken Kohn, gen mgr.

WBLS(FM)—Listing follows WLIB(AM).

WCBS(AM)— 1924: 880 khz; 50 kw-U. TL: N40 51 35 W73 47 09. Hrs open: 24 524 W. 57th St., 9th Fl., 10019. Phone: (212) 975-4321. Fax: (212) 975-4674. Web Site: www.wcbs880.com. Licensee: CBS Radio East Inc. Group owner: Infinity Broadcasting Corp. (acq 11-13-98; grpsl). Natl. Network: CBS. Format: News. Target aud: 25-54. ♦Chad Brown, VP & gen mgr; Matt Timothy, gen sls mgr; Mary Butler, natl sls mgr; Manny Severin, mktg dir, prom dir & adv dir; Cry Quimby, progmg dir; Tim Scheld, news dir; Mark Olkowski, chief of engrg.

WCBS-FM— 1941: 101.1 mhz; 6.8 kw. Ant 1,353 ft TL: N40 44 54 W73 59 10. 1515 Broadway, 10036. Phone: (212) 258-6000. Fax: (212) 846-5188. Web Site: www.wcbsfm.com.1,000,000 Format: Oldies. ♦Chad Brown, VP & gen mgr; Ezio Torres, gen sls mgr; Joe McCoy, progmg dir.

WEPN(AM)— Aug 28, 1922: 1050 khz; 50 kw-U. TL: N40 48 26 W74 04 11. Stereo. Hrs opn: 24 2 Penn Plaza, 17th Fl., 10121. Phone: (212) 613-3800. Fax: (212) 613-3861. Web Site: www.1050espnradio.com. Licensee: New York AM Radio LLC. Group owner: ABC Inc. (acq 2-28-03; $78 million). Population served: 15,340,000 Natl. Network: ESPN Radio. Format: Sports, talk. News staff: 2. Target aud: Men 25-54. ♦Tim McCarthy, pres & gen mgr; Mike Thompson, progmg dir.

WFAN(AM)— 1930: 660 khz; 50 kw-U. TL: N40 51 35 W73 47 09. Stereo. Hrs opn: 24 Kaufman-Astoria Studios, 34-12 36th St., Astoria, 11106. Phone: (718) 706-7690. Fax: (718) 361-1059. Web Site: www.wfan.com. Licensee: Infinity Broadcasting East Inc. Group owner: Infinity Broadcasting Corp. (acq 2-25-92; $70 million; FTR: 4-92). Population served: 18,000,000 Natl. Network: CBS. Natl. Rep: CBS Radio. Wire Svc: SportsTicker Wire Svc: Sports Wire Format: Sports, talk. News staff: 34. Target aud: 25-54; sports fans. ♦Lee Davis, gen mgr; Mark Chernoff, progmg dir; Mike Fagan, gen sls mgr & news dir.

***WFUV(FM)**— July 1947: 90.7 mhz; 46 kw. Ant 508 ft TL: N40 52 48 W73 52 40. Stereo. Hrs opn: 24 Fordham University, Bronx, 10458. Phone: (718) 817-4550. Fax: (718) 365-9815. Fax: (718) 817-5595. E-mail: thefolks@wfuv.org Web Site: www.wfuv.org. Licensee: Fordham University, Executive Committee, Board of Trustees. Population served: 15,898,000 Natl. Network: NPR, PRI. Law Firm: Renouf & Polivy. Wire Svc: AP Format: AAA, div. News staff: 2; News: 8 hrs wkly. Target aud: 25 plus; intelligent & sophisticated mus listeners. Spec prog: Irish 10 hrs wkly. ♦Joseph McShane, pres; John Hollwitz, VP; Ralph M. Jennings, gen mgr; George Evans, opns dir; John Platt, mktg dir; Janeen Shalteman, prom dir; Chuck Singleton, progmg dir; Rita Houston, mus dir; Julianne Welby, news dir; George Bodarky, pub affrs dir.

***WHCR-FM**— February 1985: 90.3 mhz; 10 w. 266 ft TL: N40 49 09 W73 56 59. Hrs open: 24 138th & Convent Ave., Nac Building, Room 1515, 10031. Phone: (212) 650-7481. Fax: (212) 650-7480. E-mail: info@whcr.org Web Site: www.whcr.org. Licensee: City College of New York. Format: Jazz, Sp, Black. News: 70 hrs wkly. Target aud: Community of Harlem. ♦Angela Harden, gen mgr.

WHTZ(FM)—See Newark, NJ

WINS(AM)— 1924: 1010 khz; 50 kw-U, DA-1. TL: N40 48 16 W74 06 25. (CP: TL: N40 48 39 W74 02 24). Hrs opn: 24 888 7th Ave., 10106. Phone: (212) 315-7000. Fax: (212) 315-7015. Web Site: www.1010wins.com. Licensee: Infinity Broadcasting East Inc. Group owner: Infinity Broadcasting Corp. Population served: 17,272,000 Natl. Network: ABC, CNN Radio. Natl. Rep: CBS Radio. Law Firm: Leventhal. Senter & Lerman. Wire Svc: AP Format: News. News staff: 50; News: 168 hrs wkly. Target aud: General. ♦Joel Hollander, CEO; Greg Janoff, gen mgr; Mike Felicetti, gen sls mgr; Mark Mason, progmg dir; Ben Mevorach, news dir; Mark Olkowski, engrg dir & engrg mgr. Co-owned TV: WCBS-TV affil.

***WKCR-FM**— October 1941: 89.9 mhz; 1 kw. 849 ft TL: N40 42 43 W74 00 49. (CP: 630 w, ant 1,419 ft.). Stereo. Hrs opn: 24 2920 Broadway, Mailcode 2612, 10027. Phone: (212) 854-9920. Fax: (212) 854-9296. Web Site: www.wkcr.org. Licensee: Trustees of Columbia University. Population served: 18,000,000 Format: Var/div, class, jazz. News: 3 hrs wkly. Target aud: General. Spec prog: Country 6 hrs, news/sports 6 hrs, international 8 hrs, Sp 10 hrs, Black 12 hrs wkly. ♦Matt Herman, stn mgr.

WKDM(AM)— 1927: 1380 khz; 5 kw-U, DA-1. TL: N40 49 13 W74 04 09. (CP: 5 kw-D, 13 kw-N, DA-2). Hrs opn: 2nd Fl., 449 Broadway, 10013. Phone: (212) 966-1059. Fax: (212) 966-9580. Licensee: Multicultural Radio Broadcasting Licensee LLC. Group owner: Multicultural Radio Broadcasting Inc. (acq 6-30-2003; $37 million). Population served: 1,000,000 Format: Chinese Mandarin (M-F), Sp (weekends). ♦Arthur Liu, pres; Gene Heinemeyer, gen mgr.

***WKRB(FM)**—Brooklyn, May 28, 1978: 90.3 mhz; 10 w. Ant 133 ft TL: N40 34 36 W73 56 04. Stereo. Hrs opn: 24 Kingsborough Community College, 2001 Oriental Blvd., Brooklyn, 11235. Phone: (718) 368-5817. Fax: (718) 368-4776. Licensee: Kingsborough Community College. Population served: 250,000 Format: Div, CHR. Target aud: General; young adults. ♦Regina S. Peruggi, pres; Rob Herklotz, gen mgr; Joe DeRosa, chief of engrg.

WKTU(FM)—Lake Success, 1940: 103.5 mhz; 5.4 kw. 1,417 ft TL: N40 42 43 W74 00 49. Stereo. Hrs opn: 24 16th Fl., 525 Washington Blvd., Jersey City, NJ, 07310. Phone: (201) 420-3700. Fax: (201) 420-3770. E-mail: info@ktu.com Web Site: www.ktu.com. Licensee: AMFM Radio Licenses LLC. Group owner: Clear Channel Communications Inc. (acq 8-30-00; grpsl). Population served: 13700000 Natl. Rep: D & R Radio. Law Firm: Leventhal, Senter & Lerman. Format: CHR. News staff: one. Target aud: 18-54. ♦Rob Williams, gen mgr.

WLIB(AM)— 1942: 1190 khz; 10 kw-D, 30 kw-N, DA. TL: N40 47 48 W74 06 06. Hrs open: 24 3 Park Ave., 10016. Phone: (212) 447-1000. Fax: (212) 447-5193. E-mail: info@wlib.com Web Site: www.wlib.com. Licensee: Urban Radio I L.L.C. Group owner: Inner City Broadcasting (acq 7-72). Population served: 786,776 Natl. Rep: McGavren Guild. Format: Black gospel. ♦Pierre M. Sutton, chmn; Deon Levingston, VP & gen mgr; Leon Van Gelder, gen sls mgr; Gwen Kingsberry, prom dir & prom mgr; Vinny Brown, opns mgr & progmg dir.

WBLS(FM)—Co-owned with WLIB(AM). Sept 15, 1965: 107.5 mhz; 5.4 kw horiz, 3.8 kw vert. 1,220 ft TL: N40 44 54 W73 59 10.24 E-mail: info@wbls.com Web Site: www.wbls.com. Natl. Network: ABC. Format: Black, urban contemp. Target aud: 25-54; upscale, urban.Pierre M. Sutton, CEO; Kernie Anderson, gen mgr; Vinny Brown, progmg dir; Bill Stallman, engrg dir; Lucella Duncan, traf mgr; Larry Hardesty, sports cmtr; Charles Mitchell, disc jockey; Doug Banks, disc jockey; Hal Jackson, disc jockey; J.C. Jordan, disc jockey; Jeff Fox, disc jockey; Paul Porter, disc jockey; Raymond Anthony, disc jockey; Sergio Dean, disc jockey

WLTW(FM)— Jan 26, 1961: 106.7 mhz; 5.4 kw horiz, 7.8 kw vert. 1,220 ft TL: N40 44 54 W73 59 10. Stereo. Hrs opn: 1133 Ave. of the Americas, 34th Floor, 10036. Phone: (212) 603-4600. Fax: (212) 603-4602. Web Site: www.1067litefm.com. Licensee: AMFM Radio Licenses LLC. Group owner: Clear Channel Communications Inc. (acq 8-30-2000; grpsl). Natl. Network: AP Radio. Natl. Rep: Katz Radio. Law Firm: Latham & Watkins. Format: Adult contemp. ♦Andrew Rosen, gen mgr; Steve Chessare, gen sls mgr; Bridget Sullivan, prom dir; Jim Ryan, progmg dir.

WMCA(AM)— 1925: 570 khz; 5 kw-U, DA-1. TL: N40 45 10 W74 06 15. (CP: 50 kw-D, 30 kw-N). Hrs opn: 24 777 Terrace Ave., 6th floor, Hasbrouck Heights, NJ, 07604-3100. Phone: (201) 298-5700. Fax: (201) 298-5757. E-mail: office@nycradio.com Web Site: www.wmca.com. Licensee: Salem Media of New York LLC. Group owner: Salem Communications (acq 9-15-89; $13 million; FTR: 8-14-89). Population served: 20,000,000 Format: Christian/Talk. News: 5 hrs wkly. Target aud: General. Spec prog: Jewish 9 hrs wkly. ♦Edward G. Atsinger III, pres; Joe D. Davis, VP; M. Susan Lucchesi, gen mgr; Tamela Kay Maxwell, gen sls mgr.

***WNYC(AM)**— July 8, 1924: 820 khz; 10 kw-D, 1 kw-N, DA. TL: N40 45 10 W74 06 15. Hrs open: One Centre St., 10007. Phone: (212) 669-7800. Fax: (212) 669-8986. Web Site: www.wnyc.org. Licensee: WNYC Radio Broadcasting Foundation (acq 10-3-96; $20 million with co-located FM). Population served: 15,000,000 Natl. Network: PRI, NPR. Format: News/talk, info. Target aud: General. Spec prog: Big band 2 hrs, spoken word 3 hrs wkly. ♦Laura Walker, CEO & pres; Mitchell Heskel, CFO; Peter Wilderotter, dev VP; Ellen Reynolds, dev dir; Dean Cappello, progmg VP & progmg dir.

WNYC-FM— Sept 21, 1943: 93.9 mhz; 5.4 kw. 1,418 ft TL: N40 42 43 W74 00 49. Stereo. 24 Web Site: www.wnyc.com. Format: News, class. News: 35 hrs wkly. Spec prog: Drama & literature 5 hrs, jazz 4 hrs wkly.

***WNYE(FM)**— November 1938: 91.5 mhz; 18 kw. Ant 430 ft TL: N40 41 21 W73 58 37. Hrs open: 24 112 Tillary St., 11201. Phone: (718) 250-5800. Fax: (718) 855-8863. Licensee: New York City Dept. of Info Technology & Telecommunications. Population served: 12,000,000 Natl. Network: NPR, PRI. Rgnl. Network: NPR, PRI. Law Firm: Arnold & Porter. Format: Educ. Target aud: General. ♦Terence M. O'Driscoll, gen mgr; Chang Kim, chief of engrg. Co-owned TV: WNYE-TV affil

***WNYU-FM**— May 3, 1973: 89.1 mhz; 8.3 kw. 256 ft TL: N40 51 26 W73 54 48. Stereo. Hrs opn: 4 PM-1 AM (M-F) 194 Mercer St., 5th Fl., 10012. Phone: (212) 998-1660. Fax: (212) 998-1652. Web Site: www.wnyu.org. Licensee: New York University. Population served: 100,000 Natl. Network: ABC. Format: AOR. News: 4 hrs wkly. Spec prog: Dance mus 13 hrs, Black 5 hrs, reggae 2 hrs, Sp 2 hrs, oldies 3 hrs wkly. ♦Ryan Bennett, gen mgr.

WOR(AM)— Feb 22, 1922: 710 khz; 50 kw-U, DA-1. TL: N40 47 30 W74 05 38. Hrs open: 24 111 Broadway, 10008. Secondary address: 166 West Putnam Ave., Greenwich, CT 06830. Phone: (212) 642-4500. Fax: (212) 642-4486. Web Site: www.wor710.com. Licensee: Buckley Broadcasting/WOR LLC. Group owner: Buckley Broadcasting Corp. (acq 12-89; $25.1 million; FTR: 12-11-89). Population served: 18,000,000 Natl. Rep: McGavren Guild. Format: Info, news/talk. News staff: 5; News: 4 hrs wkly. Target aud: 35-64. Spec prog: Relg 4 hrs wkly. ♦Rick Buckley, pres; Joseph Bilotta, exec VP; Bob Bruno, VP & gen mgr; Eloise Maroney, opns dir & opns mgr.

WQCD(FM)— 1945: 101.9 mhz; 6.2 kw. Ant 1,355 ft TL: N40 44 54 W73 59 10. Stereo. Hrs opn: 24 7th Fl., 395 Hudson St., 10014. Phone: (212) 352-1019. Fax: (212) 929-8559. E-mail: cd1019@cd1019.com Web Site: www.cd1019.com. Licensee: Emmis Radio License LLC. Group owner: Emmis Communications Corp. (acq 3-26-98; grpsl). Population served: 1,245,000 Natl. Rep: D & R Radio. Format: Smooth jazz. News staff: one; News: 4 hrs wkly. Target aud: 25-54. ♦Alex Cameron, sls dir; Dan Halyburton, gen mgr & mktg VP; Brian D'Aurelio, mktg dir; Blake Lawrence, progmg dir.

WQEW(AM)—Listing follows WQXR-FM.

WQHT(FM)— 1940: 97.1 mhz; 6.7 kw. Ant 1,338 ft TL: N40 44 54 W73 59 10. Stereo. Hrs opn: 24 395 Hudson St., 7th Floor, 10014. Phone: (212) 229-9797. Fax: (212) 929-8559. Web Site: www.hot97.com. Licensee: Emmis License Corp. of New York. Group owner: Emmis Communications Corp. Population served: 2,000,000 Natl. Rep: D & R Radio. Format: CHR. Target aud: General. ♦Barry Mayo, sr VP & gen mgr.

WQXR-FM— Nov 8, 1939: 96.3 mhz; 6 kw. 1,361 ft TL: N40 44 54 W73 59 10. Stereo. Hrs opn: 24 122 5th Ave., 3rd Fl., 10011. Phone: (212) 633-7600. Phone: (212) 633-7650. Fax: (212) 633-7666. Licensee: The New York Times Electronic Media Co. Group owner: The New York Times Co. (acq 2-1-44). Wire Svc: Reuters Format: Class. ♦Tom Bartunek, pres & gen mgr; Hester Furman, mus dir.

WQEW(AM)—Co-owned with WQXR-FM. Dec 3, 1936: 1560 khz; 50 kw-U, DA-2. TL: N40 42 59 W73 55 04. c/o WABC(AM), 2 Penn Plaza, 17th Fl., 10121. Phone: (212) 760-1560. Fax: (212) 947-1340. Web Site: www.wabcradio.com.14,000,000 Law Firm: Koteen & Naftalin. Format: Children. Target aud: 25-64; educated & affluent. ♦Tim McCarthy, gen mgr.

WRKS(FM)— 1941: 98.7 mhz; 6 kw. 1, 361ft TL: N40 44 54 W73 59 10. Stereo. Hrs opn: 24 395 Hudson St., 7th Floor, 10014. Phone: (212) 242-9870. Fax: (212) 929-8559. E-mail: 987kissfm@987kissfm.com Web Site: www.987kissfm.com. Licensee: Emmis Radio License Corp. of New York. Group owner: Emmis Communications Corp. (acq 10-26-94; $68 million; FTR: 12-5-94). Population served: 2,500,000 Natl. Rep: D & R Radio. Format: Rhythm and blues, classic soul. News staff: one. ♦Barry Mayo, sr VP & gen mgr.

***WSIA(FM)**—Staten Island, Aug 31, 1981: 88.9 mhz; 10 w. 650 ft TL: N40 35 51 W74 06 53. Stereo. Hrs opn: 20 2800 Victory Blvd., Staten Island, 10314. Phone: (718) 982-3050. Fax: (718) 982-3052. E-mail: mailbox@wsia.fm Web Site: www.wsia.fm. Licensee: College of Staten Island. Wire Svc: Pacifica Network News Format: Alternative rock. Target aud: General. ♦John Ladley, prom dir & chief of engrg.

WSKQ-FM— 1950: 97.9 mhz; 7.6 kw horiz, 5.4 kw vert. 1,220 ft TL: N40 44 54 W73 59 10. Stereo. Hrs opn: 26 W. 56th St., 10019. Phone: (212) 541-9200. Fax: (212) 541-9408. Web Site: www.lamega.com. Licensee: WSKQ Licensing Inc. Group owner: Spanish Broadcasting System Inc. (acq 2-1-89; $55 million). Population served: 15,000,000 Natl. Rep: McGavren Guild. Format: Sp, tropical salsa. Target aud: 18-49; Hispanic. ♦Raul Alarcon Jr., CEO & pres; Raul Alarcon Sr., chmn; Jose Garcia, CFO.

WSNR(AM)—See Jersey City, NJ

WWFS(FM)— August 1958: 102.7 mhz; 6 kw. Ant 1,361 ft TL: N40 44 54 W73 59 10. Stereo. Hrs opn: 24 888 7th Ave., 9th Fl., 10106. Phone: (212) 489-1027. Fax: (212) 489-1263. Web Site: www.fresh1027.com. Licensee: CBS Radio East Inc. Group owner: Infinity Broadcasting Corp. (acq 12-89; grpsl; FTR: 12-11-89). Population served: 18,000,000 Natl. Network: Westwood One. Format: Soft rock. Target aud: 24-44; women. ♦Maire Mason, VP & gen mgr; Mark Olkowski, chief of engrg.

WWPR-FM— Dec 14, 1953: 105.1 mhz; 6 kw. 1,362 ft TL: N40 44 54 W73 59 10. Stereo. Hrs opn: 24 18th Fl., 1120 Ave. of Americas, 10036-6798. Phone: (212) 704-1051. Fax: (212) 398-3299. Web Site: www.power1051fm.com. Licensee: AMFM Radio Licenses LLC. Group owner: Clear Channel Communications Inc. (acq 8-30-00; grpsl). Law Firm: Wilkinson Barker Knauer. Format: Urban contemp. News staff: one. Target aud: 25-54. ♦Rob Williams, gen mgr.

WWRL(AM)— Aug 26, 1926: 1600 khz; 5 kw-U, DA-2. TL: N40 47 44 W74 03 18. Hrs open: 333 7th Ave. 14th, 10003. Phone: (212) 631-0800. Fax: (212) 239-7203. Web Site: www.wwrl1600.com. Licensee: Access. 1 New York License Co. LLC. Group owner: Access.1 Communications Corp. (acq 9-28-89; $1.98 million; FTR: 10-16-89). Natl. Network: Air America. Law Firm: Rubin, Winston, Diercks, Harris & Cooke. Format: Progressive talk. ♦Adriane Gaines, pres & gen mgr; Rennie Bishop, progmg dir.

***WWRV(AM)**— May 1, 1972: 1330 khz; 5 kw-U, DA-1. TL: N40 32 45 W74 12 11. Hrs open: 24 Box 2908, Paterson, NJ, 07509. Secondary address: 419 Broadway, Paterson, NJ 07501. Phone: (973) 881-8700. Fax: (973) 881-8324. Web Site: www.radiovision.net. Licensee: Radio Vision Christiana Management Corp. (acq 6-30-89; $13 million; FTR: 5-15-89). Population served: 18,000,000 Format: Relg, Sp, Christian. News staff: one. Target aud: General. ♦Rev. Milton Donato, pres & gen mgr; Julio Carbrera, stn mgr; Jose Lastra, chief of opns.

WXRK(FM)— 1951: 92.3 mhz; 6 kw. Ant 1,220 ft TL: N40 44 54 W73 59 10. Stereo. Hrs opn: 24 40 W. 57th St., 14th Fl., 10019. Phone: (212) 314-9230. Fax: (212) 314-9282. E-mail: wxrk923@aol.com Web Site: www.923krock.com. Licensee: CBS Radio East Inc. Group owner: Infinity Broadcasting Corp. Natl. Network: ABC. Natl. Rep: CBS Radio. Format: Rock. ♦Tom Chiusano, VP & gen mgr; Alan Leinwand, sls VP; Mike Peer, mus dir; Richard Herby, engrg mgr.

WZRC(AM)— 1925: 1480 khz; 5 kw-U, DA-2. TL: N40 50 42 W74 01 12. Hrs open: 449 Broadway, 2nd Fl., 10013. Phone: (212) 965-1480. Fax: (212) 965-8917. Licensee: Multicultural Radio Broadcasting Licensee LLC. Group owner: Multicultural Radio Broadcasting Inc. (acq 1-30-98; grpsl). Population served: 13,400,000 Format: Cantonese. Target aud: 12-34. ♦Sherman Ngan, gen mgr.

Newark

WACK(AM)— Oct 19, 1957: 1420 khz; 5 kw-D, 500 w-N, DA-2. TL: N43 01 08 W77 04 41. Hrs open: 24 Box 1420, 187 Vienna Rd., 14513. Phone: (315) 331-1420. Fax: (315) 331-7101. E-mail: 1420wack@rochester.rr.com Web Site: www.1420wack.com. Licensee: Waynco Radio Inc. Group owner: Pembrook Pines Media Group (acq 3-11-2005; $600,000). Population served: 250,000 Natl. Network: CNN Radio, Motor Racing Net, Westwood One. Law Firm: Fletcher, Heald & Hildreth. Format: News/talk, sports. News staff: one; News: 30 hrs wkly. Target aud: 25-54; active, affluent, upscale audience. Spec prog: Farm 5 hrs wkly. ♦John Tinkner, pres; John Tickner, gen mgr; John Derleth, rgnl sls mgr; Dennis Federico, progmg dir; Dick Reeves, sls; Ralph Vanderlinden, engr.

Newburgh

WBNR(AM)—See Beacon

WGNY(AM)— Feb 25, 1933: 1220 khz; 5 kw-D, DA. TL: N41 29 57 W74 03 54. (CP: 5 kw-D, 180 w-N, DA-1. TL: N41 31 53 W74 06 48). Hrs opn: 24 Box 2307, 12550. Secondary address: 661 Little Britain Rd., New Windsor 12553. Phone: (845) 561-2131. Phone: (845) 561-2132. Fax: (845) 561-2138. Web Site: www.wgny.net. Licensee: Sunrise Broadcasting Corp. Group owner: Sunrise Broadcasting Corp. (acq 8-90; $10,000 with co-located FM; FTR: 8-20-90). Natl. Rep: Katz Radio. Law Firm: Rosenman & Colin L.L.P. Format: Oldies. Target aud: General. Spec prog: Relg 4 hrs, Sp one hr wkly. ♦Joerg Klebe, pres; Robert A. DeFelice, gen mgr; Robert Maines, opns VP, opns dir & chief of engrg; Hank Gross, news dir; Tom Morel, pub affrs dir.

WGNY-FM— Oct 29, 1966: 103.1 mhz; 6 kw. 275 ft TL: N41 28 22 W74 08 22. Stereo. Web Site: www.wgnyfm.com. Format: Adult contemp.

Newport Village

WBGK(FM)— 2001: 99.7 mhz; 1.4 kw. Ant 676 ft TL: N43 08 28 W75 01 49. Hrs open: 185 Genesee St., Suite 1601, Utica, 13501. Phone: (315) 734-9245. Fax: (315) 624-9245. Web Site: www.bugcountry.com. Licensee: Roser Communications Network Inc. (acq 3-20-01; $575,000). Format: Country. ♦Ken Roser, gen mgr; Roxanne Roser, stn mgr.

Niagara Falls

WHLD(AM)— May 20, 1940: 1270 khz; 5 kw-D, 1 kw-N, DA-2. TL: N42 44 41 W78 53 13. Hrs open: 19 2495 Main St., Ste 355, Buffalo, 14214. Phone: (716) 855-1270. Fax: (716) 855-4681. E-mail: Rmarks@whldam1270.com Web Site: www.whld1270.com. Licensee: Citadel Broadcasting Co. Group owner: Citadel Broadcasting Corp. Population served: 1,500,000 Format: Gospel. ♦Brian Brown Cashdollar, CFO; Ray Marks, VP & gen mgr.

WJJL(AM)— Dec 21, 1947: 1440 khz; 1 kw-D, 55 w-N. TL: N43 04 43 W79 00 40. Hrs open: 24 920 Union Rd., West Seneca, 14224. Secondary address: 6929 Williams Rd. 14304. Phone: (716) 674-9555. Fax: (716) 674-0400. E-mail: wjjl@buffalo.com Web Site: www.wjjl.com. Licensee: M.J. Phillips Communications Inc. (acq 10-20-92; $600,000; FTR: 11-23-92). Population served: 1,900,000 Law Firm: Leonard S. Joyce. Format: Old time rock & roll. News staff: one; News: 3 hrs wkly. Target aud: 25-54; baby boomers. Spec prog: Black 2 hrs, It 4 hrs, news/talk 5 hrs wkly, gospel one hr wky, Pol 2 hrs wkly. ♦Earl Morgan, chmn; John Phillips, pres; Dennis Westberg, CFO; Mark Phillips, CEO & opns VP; M.J. Phillips, opns dir.

WJYE(FM)—See Buffalo

WKSE(FM)— Jan 1, 1946: 98.5 mhz; 46 kw. 420 ft TL: N43 00 18 W78 59 35. Stereo. Hrs opn: 24 500 Corporate Pkwy., Suite 200, Amherst, 14226. Phone: (716) 843-0600. Fax: (716) 843-0250. Fax: (716) 644-9fax. E-mail: Info@kiss985,com Web Site: www.kiss985.com. Licensee: Entercom Buffalo License LLC. Group owner: Entercom Communications Corp. (acq 12-13-99; grpsl). Natl. Network: ABC. Natl. Rep: D & R Radio. Law Firm: Mullin, Rhyne, Emmons & Topel. Format: CHR. News staff: 3. Target aud: 12-49. ♦Larry Robb, gen mgr; Jimmy Steele, progmg dir.

North Creek

***WXLG(FM)**— 1995: 89.9 mhz; 200 w. 1,994 ft TL: N43 40 22 W74 02 58. Hrs open: 24
Rebroadcasts WSLU(FM) Canton 100%.
St. Lawrence Univ., North Country Public Radio, Canton, 13617. Phone: (315) 229-5356. Fax: (315) 229-5373. Web Site: www.ncpr.org. Licensee: St. Lawrence University. Law Firm: Donald E. Martin. Format: Eclectic public radio. News staff: 2; News: 35 hrs wkly. Target aud: General. ♦Ellen Rocco, gen mgr; Shelly Pike, chief of opns; Sandra Demarest, dev dir; Jacqueline Sauter, progmg dir.

North Syracuse

WKRL-FM—Listing follows WTLA(AM).

WTLA(AM)— Aug 1, 1959: 1200 khz; 1 kw-U, DA-N. TL: N43 09 06 W76 07 58. Hrs open: 24 235 Walton, Syracuse, 13202. Phone: (315)

472-9111. Fax: (315) 472-1888. Licensee: Galaxy Syracuse Licensee LLC. Group owner: Galaxy Communications LP (acq 4-6-2000; grpsl). Natl. Network: Jones Radio Networks. Law Firm: James L. Oyster. Format: Music of your life. News staff: one; News: 2 hrs wkly. Target aud: 35-64; white collar executives. Spec prog: Ger 2 hrs, Pol 2 hrs, relg 2 hrs wkly. ♦Ed Levine, pres & gen mgr; Joy Putnam, CFO; Lisa Morrow, sls VP; Mimi Griswald, progmg VP; Tim Backer, chief of engrg; Lisa Steele, traf mgr.

WKRL-FM—Co-owned with WTLA(AM). March 1972: 100.9 mhz; 6 kw. 164 ft TL: N43 09 06 W76 07 58. Stereo. Web Site: www.krock.com.600,000 Natl. Network: ABC. Format: Modern rock. Target aud: 18-34; upscale, educated. ♦Scott Petibone, progmg dir.

Norwich

WBKT(FM)— June 1, 1997: 95.3 mhz; 470 w. Ant 841 ft TL: N42 26 08 W75 30 47. Hrs open: 24 34 Chestnut St., Oneonta, 13820. Phone: (607) 432-1030. Fax: (607) 432-6909. Licensee: Double O Central New York Corp. (group owner; (acq 10-22-2004; grpsl). Population served: 100,000 Natl. Network: ABC. Format: Country. News staff: 2. Target aud: 25-54; general. ♦George Wells, gen mgr; Bud Williamson, chief of engrg.

WCHN(AM)— January 1953: 970 khz; 1 kw-D. TL: N42 30 24 W75 29 29. Hrs open: 24 34 Chestnut St., Oneonta, 13820. Phone: (607) 334-2218. Fax: (607) 334-9867. Licensee: Double O Central New York Corp. (group owner; (acq 10-22-2004; grpsl). Population served: 50,000 Natl. Network: ABC. Format: Stardust memories. News: 20 hrs wkly. Target aud: 35-65; mature. ♦James V. Johnson, gen mgr; Skip Barlow, progmg dir; James Sargent, news dir & news rptr.

WKXZ(FM)—Co-owned with WCHN(AM). 1961: 93.9 mhz; 26 kw. 680 ft TL: N42 32 52 W75 27 07. Stereo. 5 AM-1 AM 500,000 Natl. Network: ABC. Format: Hot adult contemp. News: 10 hrs wkly. Target aud: 25-54; growing families.

Norwood

WVLF(FM)— 2001: 96.1 mhz; 25 kw. Ant 328 ft TL: N44 54 11 W74 53 02. Hrs open: Box 210, Massena, 13662. Phone: (315) 769-3333. Fax: (315) 769-3299. E-mail: frank@valley961.com Web Site: www.valley961.com. Licensee: Seaway Broadcasting Inc. Group owner: Martz Communications Group (acq 9-15-2000). Format: Yesterday's favorites, today's hits. ♦Michael Bolt, gen mgr.

Noyack

***WSUF(FM)**— Sept 15, 1996: 89.9 mhz; 12 kw. 357 ft TL: N41 06 35 W72 22 05. Stereo. Hrs opn: 24
Rebroadcasts WSHU(FM) Fairfield, CT 30%.
5151 Park Ave., Fairfield, CT, 06825. Phone: (203) 365-6604. Fax: (203) 371-7991. E-mail: lombardi@wshu.org Web Site: www.wshu.org. Licensee: Sacred Heart University Inc. Natl. Network: NPR, PRI. Law Firm: Mullin, Rhyne, Emmons & Topel. Format: News/talk. News staff: 4; News: 45 hrs wkly. Target aud: General. Spec prog: Folk 5 hrs, new age 6 hrs wkly. ♦George Lombardi, gen mgr; Barbara Bashar, opns mgr; Gillian Anderson, dev dir.

Nyack

***WNYK(FM)**— May 5, 1982: 88.7 mhz; 10 w. 55 ft TL: N41 04 59 W73 55 45. Stereo. Hrs opn: 13 One South Blvd., Nyack College, 10960. Phone: (845) 358-1828. Phone: (845) 358-1710. Fax: (845) 348-8838. E-mail: wnyk@nyack.edu Web Site: www.nyackonline.org. Licensee: Nyack College. Population served: 250,000 Format: Positive alternative, Christian. Target aud: 18-35; 60%/40% -F/M, well educated. Spec prog: Black 6 hrs, relg 2 hrs wkly.

Ogdensburg

WBDB(FM)—Listing follows WSLB(AM).

WPAC(FM)— June 1998: 98.7 mhz; 3 kw. Ant 92 ft TL: N44 43 41 W75 26 36. Hrs open: 24 1 Bridge Plaza, Suite 204, 13369. Phone: (315) 393-1220. Fax: (315) 393-3974. E-mail: john@q1029.com Web Site: pac987.com. Licensee: Waters Communications Inc. Group owner: Martz Communications Group (acq 3-5-99; $285,000 with WYSX(FM) Morristown). Population served: 112,000 Format: Super hits of the 60s & 70s. News staff: one. ♦Tim Martz, pres; John Winter, gen mgr.

WSLB(AM)— 1940: 1400 khz; 1 kw-U. TL: N44 42 21 W75 27 55. Hrs open: 24 2315 Knox St., 13669. Phone: (315) 393-1100. Fax: (315) 393-6673. E-mail: burgproduction@commbroadcasters.com Web Site: www.talk1400.com. Licensee: Community Broadcasters LLC. Group owner: Clancy-Mancy Communications (acq 10-13-2006; grpsl). Population served: 14,554 Natl. Network: CNN Radio. Natl. Rep: Eastman Radio, Katz Radio. Law Firm: Shaw Pittman L.L.P. Wire Svc: AP Format: Talk. News staff: News progmg 10 hrs wkly Target aud: 25-49; community connected, active, mature, responsible & responsive. ♦James L. Leven, pres & gen mgr; Bryan Mallette, gen mgr & gen sls mgr; John Astolfi, opns mgr.

WBDB(FM)—Co-owned with WSLB(AM). July 1981: 92.7 mhz; 3 kw. 310 ft TL: N44 42 21 W75 27 55. Stereo. E-mail: johnny@theborder.fm Web Site: www.theborder.fm.100,000 Natl. Network: CNN Radio. Natl. Rep: Eastman Radio, Katz Radio. Format: CHR. Target aud: 18-34; young, educated, active, oriented to recreation, travel & self development.

Old Forge

WZNY(FM)—Not on air, target date: unknown: 94.1 mhz; 500 w. Ant 446 ft TL: N43 41 37 W74 57 42. Hrs open: 339 North St., Medfield, MA, 02052. Phone: (508) 359-2700. Licensee: Liveair Communications Inc. ♦David M. Wang, pres.

Olean

WHDL(AM)— February 1929: 1450 khz; 1 kw-U. TL: N42 04 39 W78 28 32. Hrs open: 24 3163 New State Rt. 417, 14760. Phone: (716) 372-0161. Fax: (716) 372-0164. Web Site: www.whdlradio.com. Licensee: Arrow Communication of N.Y. Inc. Group owner: Backyard Broadcasting LLC (acq 12-1-2002; grpsl). Population served: 32,000 Rgnl rep: Rgnl Reps. Law Firm: Wiley, Rein & Fielding. Format: Oldies. News staff: 2; News: 6 hrs wkly. Target aud: 25-54. ♦John J. Morton, gen mgr; Mark Thomson, progmg dir.

WPIG(FM)—Co-owned with WHDL(AM). Feb 1, 1949: 95.7 mhz; 43 kw. 740 ft TL: N42 02 08 W78 26 47. Stereo. 24 Web Site: www.wpig.com.350,000 Natl. Network: ABC. Format: Contemp country.

WMXO(FM)—Listing follows WOEN(AM).

WOEN(AM)— May 20, 1957: 1360 khz; 1 kw-D, 30 w-N. TL: N42 06 24 W78 23 28. Hrs open: 24 231 N. Union St., 14760. Phone: (716) 375-1015. Fax: (716) 375-7705. E-mail: traffic@mix101.com Web Site: www.mix101.com. Licensee: Pembrook Pines Inc. Group owner: Vox Radio Group L.P. (acq 2-22-2005; $950,000 with co-located FM). Population served: 19169 Natl. Network: CBS, Westwood One. Natl. Rep: Dome. Format: MOR. News staff: one. Target aud: 45-65. ♦Robert J. Pfuntner, pres; John R. Sirianni, gen mgr; Michael McAdam, progmg dir; Ralph Vanderlinden, chief of engrg.

WMXO(FM)—Co-owned with WOEN(AM). Nov 1, 1978: 101.5 mhz; 1.55 kw. 405 ft TL: N42 06 24 W78 23 28. Stereo. Natl. Network: CBS, Westwood One. Format: Adult contemp. Target aud: 18-49.

***WOLN(FM)**— March 1993: 91.3 mhz; 115 w. Ant 656 ft TL: N42 02 08 W78 26 47. Hrs open: 24
Rebroadcasts WBFO(FM) Buffalo 100%.
c/o WBFO(FM), 3435 Main St., 205 Allen Hall, Buffalo, 14214. Phone: (716) 829-2880. Fax: (716) 829-2277. E-mail: mail@wbfo.org Web Site: www.wbfo.org. Licensee: State University of New York. Population served: 111,800 Natl. Network: NPR. Format: News, jazz. News staff: 2; News: 50 hrs wkly. Target aud: General; educated professionals. Spec prog: Blues 8 hrs, bluegrass music 3 hrs, Pol 3 hrs wkly. ♦Carole Smith Petro, VP; Carole Smith Petro, gen mgr; Mark Wozniak, opns mgr; Joan Wilson, dev dir.

Olivebridge

***WFSO(FM)**— Dec 27, 1996: 88.3 mhz; 100 w vert. Ant 69 ft TL: N41 54 30 W74 14 46. Hrs open: 314 Acorn Hill Rd., 12461. Phone: (845) 657-5723. Licensee: Redeemer Broadcasting Inc. Format: Relg. ♦Clarence Elmendorf, stn mgr.

Oneida

WMCR(AM)— Sept 26, 1956: 1600 khz; 1 kw-D, 20 w-N. TL: N43 05 04 W75 41 35. Hrs open: 16 237 Genesee St., 13421. Phone: (315) 363-6050. Fax: (315) 363-9149. Licensee: Warren Broadcasting Co. Inc. (acq 1-3-2006). Format: Adult contemp, current CD's, oldies. Target aud: General. ♦Joel Meltzer, gen mgr, opns mgr & progmg dir.

WMCR-FM— September 1972: 106.3 mhz; 1.25 kw. Ant 718 ft TL: N43 02 48 W75 39 58.

Oneonta

WDOS(AM)— Dec 1, 1947: 730 khz; 1 kw-D. TL: N42 27 29 W75 00 20. Hrs open: Box 649, 13820. Phone: (607) 432-1030. Fax: (607) 432-6909. Web Site: www.wdos.com. Licensee: Double O Central New York Corp. (acq 11-4-2005; $3.8 million with co-located FM). Population served: 23,200 Law Firm: Haley, Bader & Potts. Format: Country. Target aud: General; adult. Spec prog: Big band 7 hrs, nostalgia 2 hrs, relg 7 hrs wkly. ♦Lou Cerra, gen mgr; Janet Laytham, progmg dir.

WSRK(FM)—Co-owned with WDOS(AM). Jan 26, 1970: 103.9 mhz; 850 w. 520 ft TL: N42 25 33 W75 02 47. (CP: 2.05 kw). Stereo. 5 AM-midnight Web Site: www.wsrk.com.16,030 Format: Adult contemp. News: 5 hrs wkly. Target aud: 25-54; adult males & females. Spec prog: Class 2 hrs wkly.

***WONY(FM)**— 1975: 90.9 mhz; 177 w. -72 ft TL: N42 28 02 W75 03 40. Stereo. Hrs opn: 24 Alumni Hall, SUCO Campus, 13820. Phone: (607) 436-2712. Fax: (607) 436-2713. Licensee: State University of New York. Population served: 20,000 Format: Educ, div. Target aud: General.

***WRHO(FM)**— Jan 1, 1970: 89.7 mhz; 150 w. 150 ft TL: N42 27 24 W75 04 28. Stereo. Hrs opn: 18 Hartwick College Dewar Hall, Radio Stn. WRHO, 13820. Phone: (607) 431-4555. Phone: (607) 431-4556. Fax: (607) 431-4064. Licensee: Hartwick College. Natl. Network: AP Radio. Format: AOR, classic rock, progsv. News: 4 hrs wkly. Target aud: General; teenagers, college students & young adults. Spec prog: Folk 5 hrs, jazz 4 hrs, Sp 2 hrs, world beat 2 hrs, children's 2 hrs wkly. ♦Brian Knox, gen mgr.

***WSQC-FM**— 1992: 91.7 mhz; 570 w horiz, 2.3 kw vert. 528 ft TL: N42 25 27 W75 02 33. Hrs open: 24
Rebroadcasts WSKG-FM Binghamton 100%.
Box 3000, Binghamton, 13902. Phone: (607) 729-0100. Fax: (607) 729-7328. Web Site: www.wskg.com. Licensee: WSKG Public Telecommunications Council. Population served: 92,000 Natl. Network: NPR, PRI. Law Firm: Dow, Lohnes & Albertson. Format: Class, news. News staff: one; News: 33 hrs wkly. Target aud: General. Spec prog: Jazz, folk 5 hrs wkly. ♦Gary V. Reinbolt, CEO, pres & gen mgr; Linda Cohen, sls VP; Robert Armstrong, mktg VP & progmg dir; Bill Snyder, mus dir.

WZOZ(FM)— Nov 28, 1972: 103.1 mhz; 2 kw. 360 ft TL: N42 25 28 W75 04 36. Hrs open: 24 34 Chestnut St., 13820-2466. Phone: (607) 432-1030. Fax: (607) 432-6909. E-mail: banjoradio@stny.rr.com Licensee: Double O Central New York Corp. (group owner; (acq 10-22-2004; grpsl). Population served: 100,000 Format: Hits of the 80s. News staff: 2; News: 8 hrs wkly. Target aud: 25-54. Spec prog: Jazz 2 hrs, blues 2 hrs, oldies 2 hrs wkly. ♦George Wells, gen mgr.

Ontario

WMJQ(AM)—Not on air, target date: unknown: 1330 khz; 1 kw-D, 2 kw-N, DA-2. TL: N43 10 49 W77 18 15. Hrs open: 135 White Bridge Rd., Middletown, 10940. Phone: (845) 355-4001. Fax: (845) 355-4002. E-mail: bud@drf.cc Licensee: Charles Williamson. Law Firm: Rini Coran. ♦Bud Williamson, gen mgr.

Ossining

***WDFH(FM)**— July 15, 1995: 90.3 mhz; 200 w. -33 ft TL: N41 09 59 W73 51 22. Stereo. Hrs opn: 24 21 Brookside Ln., Dobbs Ferry, 10522. Phone: (914) 674-0900. Web Site: www.wdfh.org. Licensee: Westchester Council for Public Broadcasting Inc. Population served: 625,000 Law Firm: Carter, Ledyard & Milburn. Format: Alternative, rock/AOR, news. Target aud: 18-39. Spec prog: Pub affrs 20 hrs wkly. ♦Marc Sophos, chmn & pres.

***WOSS(FM)**— Feb 22, 1972: 91.1 mhz; 10 w. 100 ft TL: N41 09 36 W73 51 38. (CP: 91.9 mhz, 16.42 w, ant 69 ft.). Stereo. Hrs opn: 24 190 Croton Ave., 10562. Secondary address: 29 S. Highland Ave. 10562. Phone: (914) 762-5760 x370. Licensee: Board of Education Union Free School District 1. Population served: 21,659 Format: Top-40, educ, urban contemp. News staff: 3. Target aud: General. ♦Martin McDonald, stn mgr.

Oswego

WAMF(AM)—See Fulton

WBBS(FM)—See Fulton

***WNYO(FM)**— 1993: 88.9 mhz; 100 w. 10 ft TL: N43 27 07 W76 32 40. Hrs open: State Univ. of NY, 9B Hewitt Union, 13126. Phone: (315) 312-2101. Fax: (315) 312-3542. E-mail: wnyo@oswego.edu Web Site: www.oswego.edu/~wnyo. Licensee: State University of New York. Format: Div, rock, urban contemp. Target aud: 13-34. Spec prog: Sp 6 hrs, news/talk 4 hrs wkly. ♦Tom Turner, gen mgr & progmg dir.

WOLF-FM— July 1990: 96.7 mhz; 3 kw. Any 328 ft TL: N43 29 12 W76 23 10. Stereo. Hrs opn: 24 401 W. Kirkpatrick St., Syracuse, 13204. Phone: (315) 472-0222. Fax: (315) 478-7745. E-mail: programming@movin100.com Web Site: movin100.com. Licensee: WOLF Radio Inc. (group owner; (acq 8-4-97; $65,000). Population served: 140,000 Rgnl rep: Rgnl Reps Law Firm: James L. Oyster. Format: Rhythmic adult contemp. ♦Sam Furco, gen mgr.

***WRVO(FM)**— Jan 6, 1969: 89.9 mhz; 50 kw. Ant 440 ft TL: N43 25 14 W76 32 39. Stereo. Hrs opn: 24 7060 State Rt. 104, 13126. Phone: (315) 312-3690. Fax: (315) 312-3174. E-mail: wrvo@wrvo.fm Web Site: www.wrvo.fm. Licensee: State University of New York. Population served: 545,100 Natl. Network: NPR. Wire Svc: AP Format: News/talk, old time radio. News staff: 4; News: 140 hrs wkly. Target aud: 25-55. ♦John E. Krauss, gen mgr; Fred Vigeant, opns dir & progmg dir; Matt Seubert, dev dir; Chris Vlanowski, news dir; Jeff Windsor, chief of engrg.

WSGO(AM)— 1960: 1440 khz; 1 kw-D, 42 w-N. TL: N43 24 56 W76 28 00. Hrs open: 24
Rebroadcasts WTLA(AM) North Syracuse 100%.
235 Walton Street, Syracuse, 13202. Phone: (315) 472-9111. Fax: (315) 472-1888. Licensee: Galaxy Syracuse Licensee LLC. Group owner: Galaxy Communications LP (acq 4-6-2000; grpsl). Population served: 150,000 Natl. Network: Jones Radio Networks. Format: Big band, nostalgia. News: one hr wkly. Target aud: 40 plus; retired & mobile. Spec prog: Ger 2 hrs, Pol 2 hrs wkly. ♦Ed Levine, pres & gen mgr; Joy Putnam, CFO; Lisa Morrow, sls VP & gen sls mgr; Mimi Griswold, progmg VP; Tim Backer, chief of engrg.

WTKV(FM)—Co-owned with WSGO(AM). Mar 15, 1973: 105.5 mhz; 3 kw. 450 ft TL: N43 24 56 W76 27 54. Stereo. 24
Rebroadcasts WTKW(FM) Bridgeport 100%.
Web Site: www.classicrock.com.250,000 Format: Classic rock. News staff: one; News: 2 hrs wkly. Target aud: 25-54. Spec prog: Folk 3 hrs, blues one hr wkly. ♦Sheila Parkes, prom dir; Mimi Griswold, progmg dir; Lisa Steele, traf mgr.

Owego

WEBO(AM)— July 27, 1957: 1330 khz; 5 kw-D, 50 w-N. TL: N42 06 19 W76 16 22. Hrs open: 5 AM-10 PM 212 Main St., 13827. Phone: (607) 687-9605. Fax: (607) 687-4184. Licensee: Tioga Media Inc. (acq 6-30-00; $1). Population served: 150,000 Natl. Network: USA. Natl. Rep: D & R Radio. Law Firm: Baraff, Koerner & Olender. Format: News, talk. News staff: 2. Target aud: 35 plus. Spec prog: NASCAR racing 16 hrs, relg 6 hrs wkly. ♦Terry Coleman, gen mgr.

Palmyra

WZXV(FM)— May 1993: 99.7 mhz; 2.8 kw. 485 ft TL: N43 02 00 W77 25 17. Hrs open: 24 Box 25099, Farmington, 14425. Secondary address: 1777 Rt. 332, Farmington 14425. Phone: (315) 597-9574; (585) 398-3569. Fax: (585) 398-3250. E-mail: wzxv@ccfingerlake.org Web Site: www.wzxv.org. Licensee: Calvary Chapel of the Finger Lakes Inc. (acq 8-2-95; $70,000). Format: Christian worship, Bible teaching. ♦Jeff Gallatin, gen mgr.

Patchogue

WALK(AM)—East Patchogue, May 20, 1952: 1370 khz; 500 w-D, 102 w-N. TL: N40 45 14 W72 59 14. Hrs open: 24 Box 230, 11772. Phone: (631) 475-5200. Fax: (631) 475-9016. Web Site: www.1370walk.com. Licensee: AMFM Radio Licenses LLC. Group owner: Clear Channel Communications Inc. (acq 8-30-00; grpsl). Population served: 300,000 Format: Bib band, oldies. News: 15 hrs wkly. Target aud: 50 plus. ♦Andy Rosen, VP & gen mgr; Jim Condron, gen sls mgr; Linda Healy, mktg dir; Bill Terry, prom dir; Rob Miller, progmg dir; John Lorentz, chief of engrg.

WALK-FM— December 1952: 97.5 mhz; 39 kw. 544 ft TL: N40 50 41 W73 02 01. Stereo. Web Site: walkradio.com.2,600,000 Format: Adult contemp. Target aud: 25-54. Spec prog: Hits of the 70s, love songs. ♦Cindi Clifford, disc jockey; Freddie Colom, disc jockey; K.T. Mills, disc jockey; Mark Daniels, disc jockey.

WBLI(FM)— Dec 1, 1958: 106.1 mhz; 49 kw horiz, 47 kw vert. 499 ft TL: N40 50 32 W73 02 25. Stereo. Hrs opn: 24 555 Sunrise Hwy., W. Babylon, 11704. Phone: (631) 669-9254. Fax: (631) 376-0812. E-mail: wbli@wbli.com Web Site: www.wbli.com. Licensee: Cox Radio Inc. Group owner: Cox Broadcasting (acq 5-22-98; grpsl). Population served: 2,800,000 Natl. Network: AP Radio. Natl. Rep: Christal. Format: CHR. News staff: one; News: 5 hrs wkly. Target aud: 18-34; women. ♦Austin Vali, VP & gen mgr; Nancy Cambino, opns mgr; Suzanne Riccio, pub affrs dir.

WLIM(AM)— Dec 1, 1951: 1580 khz; 10 kw-D, 37 w-N, DA-N. TL: N40 47 45 W72 59 32. (CP: 1 kw-D, 500 w-N, DA-N). Stereo. Hrs opn: 24 41 Pennsylvania Ave., Medford, 11763. Secondary address: 41 Pennsylvania Ave, Medford 11763. Phone: (631) 475-1580. Phone: (845) 354-2000. Fax: (631) 475-1523. Fax: (845) 354-4796. E-mail: wlim@polskieradio.com Web Site: www.polskieradio.com. Licensee: Polnet Communications Ltd. (group owner; acq 6-1-01; $850,000 including five-year noncompete agreement). Population served: 1,500,000 Law Firm: Wiley, Rein and Fielding. Format: Pol language. News: 20 hrs wkly. Target aud: 18-54; Polish language audience. Spec prog: Class 7 hrs, It 4 hrs, gospel 2 hrs, jazz 4 hrs, Irish one hr, international one hr wkly. ♦Kent Gustafson, CEO; Walter Kotaba, pres; Tomasz Sliwicka, stn mgr & opns mgr; Grzegorz Sciwecki, opns mgr.

Patterson

WDBY(FM)— Jan 17, 1982: 105.5 mhz; 1.5 kw. 460 ft TL: N41 31 18 W73 38 06. Stereo. Hrs opn: 24 1004 Federal Rd., Brookfield, CT, 06804. Phone: (203) 775-1212. Fax: (203) 775-6452. Web Site: www.y105radio.com. Licensee: Cumulus Licensing Corp. Group owner: Cumulus Media Inc. (acq 1-23-2002; grpsl). Natl. Network: Westwood One. Law Firm: Haley, Bader & Potts. Format: Adult contemp. News staff: 2; News: 4 hrs wkly. Target aud: 25-54. ♦Brett Beshore, gen mgr; Tim Sheehan, opns mgr; Tom Principi, gen sls mgr; Tony Wise, progmg dir; Lisa Harris, news dir; Peter Partenio, chief of engrg; Sheila Alexson, traf mgr.

WPUT(AM)—Co-owned with WDBY(FM). July 3, 1958: 1510 khz; 1 kw-D. TL: N41 24 34 W73 37 29.200,000 Natl. Network: ESPN Radio. Format: Sports. ♦Matt Carey, progmg dir.

Pattersonville

***WPGL(FM)**— Aug 15, 1994: 90.7 mhz; 30 w. 653 ft TL: N42 51 00 W74 03 58. Stereo. Hrs opn: 24
Rebroadcasts WFGB(FM) Kingston 95%.
Box 777, Lake Katrine, 12449. Secondary address: 199 Tuytenbridge Rd., Lake Katrine 12449. Phone: (845) 336-6199. Fax: (845) 336-7205. E-mail: email@soundoflife.org Web Site: www.soundoflife.org. Licensee: Sound of Life Inc. Population served: 300,000 Format: Christian. News: 3 hrs wkly. Target aud: General. ♦Tom Michael Zahradnik, gen mgr.

Paul Smiths

***WPSA(FM)**— Jan 10, 1973: 98.3 mhz; 10 w. -7 ft TL: N44 26 04 W74 15 04. Hrs open: Co-ordinator of Student Activities, Paul Smiths College, Rts. 86 & 30, 12970. Phone: (518) 327-6401. Fax: (518) 327-6369. Licensee: Paul Smiths College of Arts & Sciences. Format: Educ, pub affrs, MOR.

Pawling

WDBY(FM)—See Patterson

Peekskill

WHUD(FM)—Listing follows WLNA(AM).

WLNA(AM)— 1948: 1420 khz; 5 kw-D, 1 kw-N, DA-2. TL: N41 18 31 W73 55 00. Hrs open:
Simulcast with WBRN(AM) Beacon 100%.
Box 310, Beacon, 12508. Secondary address: 715 Rt. 82, Beacon 12508. Phone: (914) 838-6000. Fax: (914) 838-2109. Web Site: www.hvnet.com. Licensee: 6 Johnson Road Licenses Inc. Group owner: Pamal Broadcasting Ltd. (acq 10-19-2001; grpsl). Population served: 2,000,000 Natl. Network: ABC. Format: Sports, Adult standards. News staff: 2; News: 40 hrs wkly. Target aud: 35 plus. ♦James Morrell, pres; Jake Russell, exec VP; Fred Bennett, gen mgr; Steve Petrone, opns VP; Bob Outer, gen sls mgr; Tom Michaels, progmg VP; Rich Flaaherty, news dir; Paul Thurst, chief of engrg.

WHUD(FM)—Co-owned with WLNA(AM). Oct 24, 1958: 100.7 mhz; 50 kw. 500 ft TL: N41 20 18 W73 53 41.320,000 Natl. Network: ABC. Format: Adult contemp. Target aud: Upscale adults. Spec prog: Oldies 5 hrs wkly. ♦Maggie Carbaugh, gen sls mgr; Jay Pugliese, natl sls mgr; Marie Martelli, prom mgr; Steve Petrone, opns mgr & progmg mgr.

Penn Yan

WYLF(AM)— 1988: 850 khz; 1 kw-D, 47 w-N. TL: N42 39 41 W77 07 14. Hrs open: 24 100 Main St., 14527. Phone: (315) 536-0850. Fax: (315) 536-3299. Fax: (315) 781-6666. E-mail: wylf@airxcess.net Licensee: M.B. Communications. (acq 10-88). Population served: 550,000 Law Firm: Henry E. Crawford. Format: Adult standards. News staff: 2. Target aud: 35 plus. ♦Russ Kimble, pres, gen mgr & stn mgr; Don Radigan, opns mgr; Mary Ann Hurlburt, gen sls mgr.

Peru

***WXLU(FM)**— 1991: 88.3 mhz; 200 w. 1,109 ft TL: N44 34 26 W73 40 29. Hrs open: 24
Rebroadcasts WSLU(FM) Canton 100%.
N. Country Public Radio, St. Lawrence Univ., Canton, 13617. Phone: (315) 229-5356. E-mail: radio@mcpr.org Web Site: www.ncpr.org. Licensee: St. Lawrence University. Law Firm: Donald E. Martin. Format: Eclectic public radio. News staff: 2; News: 35 hrs wkly. Target aud: General. ♦Ellen Rocco, gen mgr.

Phoenix

WZUN(FM)— May 22, 1995: 102.1 mhz; 6 kw. 220 ft TL: N43 06 03 W76 16 56. Stereo. Hrs opn: 24 235 Walton St., Syracuse, 13202. Phone: (315) 472-9111. Fax: (315) 472-1888. Web Site: www.thesunnyspot.com. Licensee: Galaxy Syracuse Licensee LLC. (group owner; (acq 12-15-2000; $3.75 million). Law Firm: Leventhal, Senter & Lerman. Format: Adult contemp. Target aud: 25-54; general. ♦Ed Levine, pres; Mike Lucarelli, CFO; Lisa Morrow, sls VP; Mimi Griswold, progmg VP; Ted Bradford, progmg dir.

Plainview

***WPOB-FM**— September 1973: 88.5 mhz; 125 w. 150 ft TL: N40 46 53 W73 27 36. (CP: Ant 259 ft. TL: N40 47 48 W73 27 44). Hrs opn: 50 Kennedy Dr., 11803-4098. Phone: (516) 937-6373. Fax: (516) 937-6384. Licensee: Plainview-Old Bethpage Central School District. Format: Educ, AOR. Target aud: General. ♦Adam Weinstock, gen mgr; Joel Genero, opns dir.

Plattsburgh

WBTZ(FM)— Feb 3, 1960: 99.9 mhz; 100 kw. 984 ft TL: N44 46 13 W73 36 47. Stereo. Hrs opn: 24 255 S. Champlain St., Burlington, VT, 05402. Phone: (802) 860-2465. Fax: (802) 860-1818. E-mail: mailbag@99thebuzz.com Web Site: www.99thebuzz.com. Licensee: Hall Communications Inc. (acq 7-31-2006; $2.5 million). Population served: 261,000 Format: Alternative rock. Target aud: 18-44. ♦Dan Dubonnet, gen mgr; Matt Grasso, progmg dir.

***WCEL(FM)**— Jan 14, 1991: 91.9 mhz; 380 w. 852 ft TL: N44 46 27 W73 36 48. Stereo. Hrs opn: 24
Rebroadcasts WAMC-FM Albany 100%.
Box 666000, Albany, 12206. Secondary address: 318 Central Ave., Albany 12206. Phone: (518) 465-5233. Phone: (800) 323-9262. Fax: (518) 432-6974. E-mail: mail@wamc.org Web Site: www.wamc.org. Licensee: WAMC. Group owner: WAMC/Northeast Public Radio (acq 1996; $160,000). Natl. Network: PRI, NPR. Law Firm: Dow, Lohnes & Albertson. Wire Svc: AP Format: News, talk. News: 77 hrs wkly. Target aud: General. Spec prog: Folk 7 hrs, jazz 13 hrs wkly. ♦Alan Chartock, CEO, chmn & pres; David Galletly, VP & progmg dir; Selma Kaplan, VP & reporter.

WEAV(AM)— Feb 3, 1935: 960 khz; 5 kw-U, DA-2. TL: N44 34 27 W73 26 54. Hrs open: 1500 Hegeman Ave., Colchester, VT, 05446. Phone: (802) 655-0093. Fax: (802) 655-0478. Web Site: www.wxzofm.com. Licensee: CC Licenses LLC. Group owner: Clear Channel Communications Inc. (acq 1-31-2002). Population served: 225,000 Format: Hot talk, sports. Target aud: General. ♦Karen Marshall, gen mgr; Steve Cormier, opns mgr.

WIRY(AM)— Jan 30, 1950: 1340 khz; 1 kw-U. TL: N44 41 49 W73 28 40. Hrs open: 5 AM-midnight 301 Cornelia St., 12901. Phone: (518) 563-1340. Fax: (518) 563-1343. E-mail: wiry@wiry.com Web Site: www.wiry.com. Licensee: Hometown Radio Inc. (acq 2-1-95; $175,000; FTR: 2-27-95). Population served: 25,000 Natl. Network: Westwood One. Natl. Rep: Roslin. Format: Adult contemp. News staff: 2; News: 11 hrs wkly. Target aud: 18 plus. ♦Dan Santa, exec VP; William D. Santa, pres & gen mgr.

WKOL(FM)— Aug 22, 1994: 105.1 mhz; 23.5 kw. 338 ft TL: N44 31 31 W73 31 07. Stereo. Hrs opn: 24 Box 4489, Burlington, VT, 05406. Secondary address: 70 Joy Dr., South Burlington, VT 05403. Phone: (802) 658-1230. Fax: (802) 862-0786. E-mail: kool105@hallradio.com Web Site: www.wkol.com. Licensee: Hall Communications Inc. (group owner; acq 6-13-95; $1.1 million). Natl. Rep: D & R Radio. Law Firm: Fletcher, Heald & Hildreth. Wire Svc: AP Format: Classic Hits. Target aud: 25-54. ♦Bonnie Rowbotham, chmn; Arthur Rowbotham, pres; Bill Baldwin, exec VP; Dan Dubonnet, gen mgr; Rod Hill, opns dir & progmg mgr.

***WQKE(FM)**— April 1979: 93.9 mhz; 10 w. 156 ft TL: N44 41 40 W73 28 00. Stereo. Hrs opn: 7 AM-2:30 PM 101 Broad St., Kehoe 202, State University of NY, 12901. Phone: (518) 564-2727. Fax: (518) 564-3994. Licensee: State University of N.Y. Population served: 30,000 Format: College alternative. Target aud: 18-24. Spec prog: Heavy metal 10 hrs, classic rock 12 hrs, Black 9 hrs, relg 3 hrs wkly. ♦Phil Czterwastek, opns dir; Andy Martinez, mus dir.

WTWK(AM)— January 1998: 1070 khz; 5 kw-D. TL: N44 36 14 W73 27 18. Hrs open: 2 N. Main St., Suite 401, St. Albans, VT, 05478. Phone: (802) 524-2133. Fax: (802) 527-1450. Web Site: www.eve1070.com. Licensee: Champlain Communications Corp. Group owner: Northeast Broadcasting Company Inc. (acq 1-11-2002; $150,000). Natl. Network: Air America. Format: Talk. Target aud: 25-54; women. ♦Richard C. DeLancey Sr., gen mgr; J.J. Prieve, progmg dir.

Port Henry

WVTK(FM)— Sept 5, 1982: 92.1 mhz; 18 kw. Ant 10 ft TL: N44 01 38 W73 28 54. Stereo. Hrs opn: 24 Box 1093, Burlington, VT, 05402. Phone: (802) 655-0093. Fax: (802) 655-0478. Web Site: www.trueoldieschannel.com. Licensee: Capstar TX L.P. Group owner: Clear Channel Communications Inc. (acq 12-22-2000; grpsl). Population served: 200,000 Law Firm: Mullin, Rhyne, Emmons & Topel. Format: Oldies. News: one hr wkly. Target aud: 25-54. ♦Karen Marshall, gen mgr.

Port Jervis

WDLC(AM)— July 4, 1953: 1490 khz; 1 kw-U. TL: N41 21 49 W74 40 41. Hrs open: 24 18 Neversink Dr., 12771. Phone: (845) 856-5185. Licensee: PJ Radio L.L.C. (acq 2-7-2005; $4 million with co-located FM). Population served: 50,000 Format: Oldies. ♦James Morley, gen mgr.

WTSX(FM)—Co-owned with WDLC(AM). Oct 30, 1970: 96.7 mhz; 3 kw. Ant 300 ft TL: N41 22 24 W74 43 49. Stereo. 24 Format: Country. News staff: 2. ♦Judi Edwards, traf mgr; Bob Oefinger, disc jockey; Rick Davis, disc jockey; Rob Ryan, disc jockey; Ryan Drean, disc jockey; Victoria Curtain, disc jockey.

***WRPJ(FM)**— October 1992: 88.9 mhz; 500 w. 590 ft TL: N41 25 36 W74 34 45. Stereo. Hrs opn:
Rebroadcasts WFGB(FM) Kingston 100%.
Box 777, Lake Katrine, 12449. Secondary address: 199 Tuytenbridge Rd., Lake Katrine 12449. Phone: (845) 336-6199. Phone: (800) 724-8518. Fax: (845) 336-7205. E-mail: email@soundoflife.org Web Site: www.soundoflife.org. Licensee: Sound of Life Inc. Population served: 300,000 Format: Christian. Target aud: General. ♦Tom Michael Zahradnik, gen mgr.

Potsdam

***WAIH(FM)**— Sept 10, 1998: 90.3 mhz; 100 w. -16 ft TL: N44 39 43 W74 58 26. Hrs open: Student Union, 9050 Barrington Dr., 13676. Phone: (315) 267-4888. Fax: (315) 267-2798. E-mail: waih@potsdam.com Web Site: www2.potsdam.edu/waih. Licensee: State University of New York. Format: Music, talk. ♦Jon Foote, gen mgr.

WPDM(AM)— Apr 30, 1955: 1470 khz; 1 kw-D. TL: N44 38 38 W75 03 32. Hrs open: Box 348, 13676. Phone: (315) 265-5510. Fax: (315) 265-4040. E-mail: hits@slic.com Licensee: St. Lawrence Radio Inc. Population served: 9,985 Law Firm: Cohn & Marks. Format: Adult contemp. News staff: one; News: 5 hrs wkly. Target aud: 25 plus; div audience. ♦Jane A. Kyle, pres; William Solomon, VP & gen mgr; Derry Loucks, gen sls mgr; Justin James, mus dir & disc jockey; Scott Dosztan, news dir; Dan Simmons, chief of engrg; Betty Bombarn, traf mgr; Andy Van Duyne, disc jockey; Dave Williams, disc jockey; Drew Bradley, disc jockey; Josh Henry, disc jockey.

WSNN(FM)—Co-owned with WPDM(AM). Oct 15, 1968: 99.3 mhz; 3 kw. Ant 155 ft TL: N44 38 38 W75 03 32.6 AM-midnight Web Site: www.99hits.com. Law Firm: Cohn & Marks. Format: Country. News staff: one; News: 5 hrs wkly. Target aud: 25-54. ♦Derry Loucks, stn mgr; Justin Gonyea, mus dir & traf mgr; Andy Van Duyne, disc jockey; Dave Williams, disc jockey; Drew Bradley, disc jockey; Josh Henry, disc jockey; Justin James, disc jockey.

***WTSC-FM**— Nov 3, 1963: 91.1 mhz; 700 w. 155 ft TL: N44 39 45 W75 00 07. Stereo. Hrs open: 24 hrs a day Clarkson University, Box 8743, 13699. Phone: (315) 268-7658. E-mail: radio@clarkson.edu Web Site: http://radio.clarkson.edu. Licensee: Clarkson University. Population served: 9,985 Format: Alternative. ♦James Heroux, gen mgr; Chaz Adams, stn mgr.

Poughkeepsie

WEOK(AM)— October 1949: 1390 khz; 5 kw-D, DA. TL: N41 43 14 W73 54 29. Hrs open: 24 Box 416, 12602-0416. Secondary address: 2 Penvell Rd. 12602-0416. Phone: (845) 471-1500. Fax: (845) 454-1204. Web Site: www.cumulus.com. Licensee: Cumulus Licensing Corp. Group owner: Cumulus Media Inc. (acq 1-23-2002). Population served: 245,000 Format: Sp. News staff: 3. Target aud: 35 plus. Spec prog: Farm 2 hrs, Pol one hr, relg 2 hrs, talk 5 hrs, Sinatra 2 hrs wkly. ♦Charles Benfer, pres; Nick Robbins, opns dir & prom dir; Victor Goodman, gen sls mgr.

WPDH(FM)—Co-owned with WEOK(AM). December 1962: 101.5 mhz; 4.5 kw. Ant 1,540 ft TL: N41 43 09 W73 59 47. Stereo. 24 Web Site: www.wpdh.com. Format: Classic rock. Spec prog: Blues deluxe, flashback 4 hrs wkly. ♦Nick Robbins, opns mgr; Greg O'Brien, progmg dir & mus dir.

WKIP(AM)— June 1940: 1450 khz; 1 kw-U, DA-D. TL: N41 42 18 W73 53 16. Hrs open: 20 Tucker Dr., 12603-1644. Phone: (845) 471-2300. Fax: (845) 471-2683. Licensee: CC Licenses LLC. Group owner: Clear Channel Communications Inc. (acq 7-12-2000; grpsl). Population served: 222,900 Natl. Network: Jones Radio Networks. Natl. Rep: Katz Radio. Format: Big band, MOR. News staff: one; News: 2 hrs wkly. Target aud: 35-64. ♦Frank Curcio, gen mgr; Reggie Osterhoudt, opns mgr; Jim Brady, gen sls mgr; Jeanette Relyea, natl sls mgr; Nick Smirnoff, prom dir; Joe Daily, progmg dir; Cameron Hendrix, news dir.

WRNQ(FM)—Co-owned with WKIP(AM). June 30, 1989: 92.1 mhz; 2.15 kw. Ant 384 ft TL: N41 40 36 W73 49 14. Stereo. 24 Web Site: www.921litefm.com.500,000 Natl. Network: Jones Radio Networks. Natl. Rep: Katz Radio. News staff: one; News: 20 hrs wkly. Target aud: 35-54; primary market is women. ♦Michelle Taylor, progmg dir & mus dir.

WPKF(FM)— 1996: 96.1 mhz; 3 kw. Ant 171 ft TL: N41 44 46 W73 54 46. Hrs open: 24 20 Tucker Dr., 12603. Phone: (845) 471-2300. Fax: (845) 471-2683. Web Site: www.kissfmjams.com. Licensee: CC Licenses LLC. Group owner: Clear Channel Communications Inc. (acq 7-14-2000; grpsl). Natl. Network: Jones Radio Networks. Natl. Rep: Clear Channel, Katz Radio. Format: Rhythmic CHR. ♦Frank Curcio, gen mgr; Reggie Osterhoudt, opns dir; Jim Brady, gen sls mgr; Jeanette Relyea, natl sls mgr; Nick Smirnoff, prom dir; Jimi Jamm, progmg dir; Cameron Hendrix, news dir; Bill Draper, chief of engrg.

***WRHV(FM)**— Sept 5, 1990: 88.7 mhz; 230 w. 1,289 ft TL: N41 43 09 W73 59 47. Stereo. Hrs opn:
Rebroadcasts Wmht-FM Schenectady 60%.
4 Global View, Troy, 12180. Phone: (518) 880-3400. Fax: (518) 880-3409. E-mail: email@wmht.org Web Site: www.wmht.org. Licensee: WMHT Educational Telecommunications. Population served: 90,000 Natl. Network: PRI. Law Firm: Schwartz, Woods & Miller. Format: Class. Target aud: 35-54; class mus lovers. Spec prog: Jazz 2 hrs, ethnic one hr wkly. ♦Deborah Onslow, pres; Dave Nicosia, chief of engrg.

WSPK(FM)— Dec 7, 1947: 104.7 mhz; 7.4 kw. 1,260 ft TL: N41 29 19 W73 56 52. Stereo. Hrs opn: 24 715 Rt. 52, 12508. Phone: (845) 838-8600. Fax: (845) 838-2109. Web Site: www.k104online.com. Licensee: 6 Johnson Road Licenses Inc. Group owner: Pamal Broadcasting Ltd. (acq 10-19-2001; grpsl). Population served: 500,000 Natl. Network: Westwood One, ABC. Natl. Rep: Katz Radio. Format: CHR. News staff: one. Target aud: 18-49. ♦James Morrell, CEO; Fred Bennett, VP & natl sls mgr; Paul Thurst, chief of engrg.

***WVKR-FM**— 1976: 91.3 mhz; 3.7 kw. 820 ft TL: N41 38 25 W74 01 16. Stereo. Hrs opn: 24 Box 726, Vassar College, 12604. Phone: (845) 437-5475. Phone: (845) 437-7010. Fax: (845) 437-7656. Web Site: www.wvkr.org. Licensee: Vassar College. Population served: 1,500,000 Format: Div. News staff: 4; News: 5 hrs wkly. Target aud: General. ♦Caitlin Buckley, stn mgr & news dir; Halimah Marcus, gen mgr & stn mgr.

Pulaski

***WGKV(FM)**— January 1987: 101.7 mhz; 5 kw. Ant 358 ft TL: N43 36 28 W75 58 23. Hrs open: 2351 Sunset Blvd., Suite 170-218, Rocklin, CA, 95765. Phone: (916) 251-1600. Fax: (916) 251-1650. Web Site: www.klove.com. Licensee: Educational Media Foundation. (group owner; (acq 7-6-2007; grpsl). Natl. Network: K-Love. Format: Contemp Christian. ♦Mike Novak, sr VP.

WSCP(AM)—See Sandy Creek-Pulaski

Ravena

***WYKV(FM)**— 1991: 94.5 mhz; 3 kw. Ant 328 ft TL: N42 33 23 W73 52 05. Hrs open: 24 2351 Sunset Blvd., Suite 170-218, Rocklin, CA, 95765. Phone: (916) 251-1600. Fax: (916) 251-1650. Web Site: www.klove.com. Licensee: Educational Media Foundation. Group owner: Galaxy Communications L.P. (acq 7-6-2007; grpsl). Natl. Network: K-Love. Format: Contemp Christian. ♦Mke Novak, sr VP.

Remsen

WADR(AM)— Dec 12, 1966: 1480 khz; 5 kw-D, 50 w-N. TL: N43 19 31 W75 10 29. Stereo. Hrs opn: 24
Rebroadcasts WRNY(AM) Rome 100%.
239 Genesee St., Suite 500, Utica, 13501. Phone: (315) 797-0803. Fax: (315) 797-7813. Web Site: www.starsradionetwork.com. Licensee: CC Licenses LLC. Group owner: Clear Channel Communications Inc. (acq 8-5-98; grpsl). Population served: 175,000 Natl. Network: Westwood One. Natl. Rep: Christal. Law Firm: Latham & Watkins. Format: Sports. News: 3 hrs wkly. Target aud: 35 plus; 60% female, 40% male. ♦Brian Delaney, gen mgr; Gene Conte, progmg dir; Joe Petro, chief of engrg.

WOKR(FM)—Co-owned with WADR(AM). Dec. 1, 1982: 93.5 mhz; 1.15 kw. Ant 748 ft TL: N43 20 44 W75 15 00. Stereo. 24 Web Site: www.warmfm.com.275,000 Natl. Network: ABC. Format: Lite adult contemp. News: one hr wkly. Target aud: 25-54. ♦Stew Schantz, opns mgr; Jack Moran, progmg mgr.

Rensselaer

WQBK-FM—Listing follows WTMM(AM).

WTMM(AM)— Dec 3, 1961: 1300 khz; 5 kw-U, DA-2. TL: N42 35 23 W73 44 37. Hrs open: 24 1241 Kings Rd., Schenectady, NY, 12303. Phone: (518) 881-1515. Fax: (518) 881-1516. E-mail: ESPN@wtmm.com Web Site: www.wtmm.com. Licensee: Regent Broadcasting of Albany Inc. Group owner: Regent Communications Inc. (acq 2000; grpsl). Population served: 800,000 Natl. Network: ESPN Radio. Wire Svc: UPI Format: Sports. Target aud: 25-54; serious sports fans. Spec prog: Auto repair one hr, money mgmt 3 hrs, real estate one hr wkly. ♦Bob Ausfeld, exec VP; John Hirsch, stn mgr; Bill Brindle, opns dir; Buzz Brindle, progmg dir; Bob O'Neal, chief of engrg.

WQBK-FM—Co-owned with WTMM(AM). Dec 1, 1972: 103.9 mhz; 6 kw. Ant 302 ft TL: N42 35 06 W73 46 29. Stereo. 24 Web Site: www.wqbk.com.114,873 Format: Rock. ♦Jim Clifford, sls dir; Shawn Murphy, progmg dir.

Riverhead

WFTU(AM)— Aug 8, 1963: 1570 khz; 1 kw-D, 500 w-N, DA-2. TL: N40 54 48 W72 39 16. Hrs open: 3241 Rt. 112, Bldg. 7, Medford, 11763. Phone: (631) 451-1039. Phone: (631) 424-7000. Fax: (631) 451-0891. Web Site: www.wftu.net. Licensee: Five Towns College (acq 5-24-01; $80,000). Population served: 100,000 Format: College radio. ♦David Cohen, stn mgr.

WRCN-FM— Aug 14, 1962: 103.9 mhz; 1.5 kw. 466 ft TL: N40 51 07 W72 45 55. Stereo. Hrs opn: 3241 Rt 112, Bldg. 7, Medford, 11763. Phone: (631) 451-1039. Fax: (631) 451-0891. Fax: (631) 451-0896. Web Site: www.wrcn.com. Licensee: IW L.L.C. Group owner: Barnstable Broadcasting Inc. (acq 10-1-97; grpsl). Population served: 750,000 Natl. Rep: Katz Radio. Law Firm: Haley, Bader & Potts. Format: Classic hits, rock. Target aud: 18-49. ♦Mike Kaneb, pres; Dave Widmer, stn mgr; Robin Schlakman, gen sls mgr; David Musser, prom dir.

WRIV(AM)— June 1955: 1390 khz; 1 kw-D, 64 w-N. TL: N40 55 22 W72 38 52. Hrs open: 6 AM-midnight Box 1390, 11901. Secondary address: 40 W. Main St. 11901. Phone: (631) 727-1390. Fax: (631) 369-WRIV (9748). Web Site: www.wrivonline.com. Licensee: Crystal Coast Communications. (acq 10-87). Population served: 104,000 Format: Adult standards. News staff: one; News: 14 hrs wkly. Target aud: 35-64. Spec prog: Farm 8 hrs, Pol 4 hrs wkly. ♦Bruce Tria, gen mgr.

Rochester

WBEE-FM— February 1961: 92.5 mhz; 50 kw. 500 ft TL: N43 10 37 W77 28 39. Stereo. Hrs opn: 24 70 Commercial St., 14614-1010. Phone: (585) 423-2900. Fax: (585) 325-5139. Fax: (585) 423-2947. Web Site: www.wbee.com. Licensee: Entercom Rochester License LLC. Group owner: Entercom Communications Corp. (acq 4-23-98; grpsl). Population served: 1,161,800 Natl. Network: Westwood One. Format: Country. News staff: one; News: 4 hrs wkly. Target aud: 25-54. ♦Michael Doyle, gen mgr; Dave Symonds, opns mgr & chief of engrg; Sue Munn, gen sls mgr; Billy Kidd, progmg dir; Steve Hausmann, news dir; Joe Fleming, chief of engrg; Courtney Nourse, traf mgr.

WROC(AM)—Co-owned with WBEE-FM. 1947: 950 khz; 1 kw-U, DA-2. TL: N43 06 25 W77 35 51.24 831,800 Format: News/talk. News staff: one; News: 3 hrs wkly. Target aud: 35 plus; high income empty-nesters. ♦Joe Fleming, gen sls mgr & chief of engrg; Jim White, progmg mgr; Steve Hausmann, VP & news dir; Alana Katz, traf mgr.

***WBER(FM)**— 1974: 90.5 mhz; 2.5 kw. 417 ft TL: N43 02 00 W77 25 11. (CP: 50 kw). Stereo. Hrs opn: 24 2596 Baird Rd., Penfield, 14526-2333. Phone: (585) 419-8190. Web Site: http://wber.monroe.edu. Licensee: Monroe B.O.C.E.S #1. Format: Alternative. Target aud: 25-34; male & female. ♦Joey Guisto, stn mgr & progmg dir.

WBZA(FM)— 1939: 98.9 mhz; 50 kw. 560 ft TL: N43 10 14 W77 40 23. Stereo. Hrs opn: 24 Entercom Rocherster LLC, 70 Commercial St., 14614-1010. Phone: (585) 423-2900. Fax: (585) 325-5139. Fax: (585) 423-2947. Web Site: www.rochesterbuzz.com. Licensee: Entercom Rochester Inc. Group owner: Entercom Communications Corp. (acq 4-23-98; grpsl). Population served: 1,161,800 Natl. Network: Westwood One. Natl. Rep: Katz Radio. Law Firm: Akin, Gump, Strauss, Hauer & Feld. Format: Classic Hits. News staff: one. Target aud: 25-54; upscale. ♦Mike Johnson, VP; Michael Doyle, gen mgr.

WCMF-FM— June 9, 1960: 96.5 mhz; 50 kw. 457 ft TL: N43 08 07 W77 35 02. Stereo. Hrs opn: 24 1700 HSBC Plaza, 14604. Phone: (585) 399-5700. Fax: (585) 399-5750. Web Site: www.wcmf.com. Licensee: Infinity Radio Inc. Group owner: Infinity Broadcasting Corp. (acq 11-13-98; grpsl). Population served: 296,233 Format: AOR. News staff: one; News: 10 hrs wkly. ♦Al Casazza, gen mgr.

WDKX(FM)— Apr 6, 1974: 103.9 mhz; 800 w. 540 ft TL: N43 09 17 W77 36 16. Stereo. Hrs opn: 24 683 E. Main St., 14605. Phone: (585) 262-2050. Fax: (585) 262-2626. Web Site: www.wdkx.com. Licensee: Monroe County Broadcasting Co. Ltd. (acq 9-19-02). Population served: 296,233 Format: Urban contemp. News staff: 2; News: 6 hrs wkly. Target aud: General. Spec prog: Jazz 4 hrs, gospel 7 hrs wkly. ♦Andrew A. Langston, CEO & gen mgr; Andrew Langston, pres; Marietta Avery, CFO; Camilla Maas, sr VP; Gloria M. Langston, stn mgr; Andre Langston, opns dir.

WDVI(FM)—Listing follows WHAM(AM).

WFXF(FM)—See Honeoye Falls

WHAM(AM)— July 11, 1922: 1180 khz; 50 kw-U. TL: N43 04 55 W77 43 30. Stereo. Hrs opn: 24 Box 40400, Euclid Bldg., 207 Midtown Plaza, 14604. Phone: (585) 454-4884. Fax: (585) 454-5081. Web Site: www.wham1180.com. Licensee: Citicasters Licenses L.P. Group owner: Clear Channel Communications Inc. (acq 3-7-97; grpsl). Population served: 296,233 Natl. Network: CBS. Format: News/talk. News staff: 6. Target aud: 35 plus. ♦Jeff Howlett, gen mgr & stn mgr.

WDVI(FM)—Co-owned with WHAM(AM). 1962: 100.5 mhz; 50 kw. Ant 480 ft TL: N43 02 00 W77 25 17. Stereo. 24 Phone: (585) 454-3942. Web Site: www.mydrivefm.com.219,100 Format: Adult contemp. News staff: one. Target aud: 25-54. ♦Ken Spitzer, stn mgr; Karen Kelly, gen sls mgr; Danyelle Dodge, prom dir; Dave LeFrois, progmg dir & mus dir; Randy Gorbman, news dir; Diane DeNiro, pub affrs dir & disc jockey; Alli Schill, traf mgr; Joe Bonacci, disc jockey; Marc Murphee, disc jockey; Michael Gately, disc jockey; Natalie Blake, disc jockey.

WHIC(AM)— Sept 11, 1925: 1460 khz; 5 kw-U, DA-N. TL: N43 06 34 W77 34 20. Hrs open: 24 Box 25433, 14625. Secondary address: 2 Cambridge Pl., 1840 Winton Rd. S. 14618. Phone: (585) 271-0530. Fax: (585) 271-0530. Licensee: Holy Family Communications (group owner; acq 7-1-2003; $300,000). Population served: 1,500,000 Format: Catholic. ♦James N. Wright, pres; Jack Palvino, gen mgr.

WHTK(AM)— Nov 22, 1947: 1280 khz; 5 kw-U, DA-N. TL: N43 05 54 W77 35 00. Hrs open: 24 207 Midtown Plaza, 14604. Phone: (585) 454-4884. Fax: (585) 262-2334. Web Site: www.whtk.com. Licensee: Citicasters Licenses L.P. Group owner: Clear Channel Communications Inc. (acq 5-4-99; grpsl). Population served: 235,000 Natl. Rep: McGavren Guild. Format: Talk. Target aud: 25-54; men. ♦Jeff Howlett, gen mgr & progmg dir; Scott Gordon, gen sls mgr.

***WIRQ(FM)**— January 1960: 104.7 mhz; 10 w. 485 ft TL: N43 12 59 W77 35 46. Stereo. Hrs opn: 1 PM-8 PM (M-F); Sept-June 260 Cooper Rd., 14617. Phone: (585) 336-3065. Fax: (589) 336-2929. Licensee: Board of Education West, Irondequoit Central School District. Population served: 600,000 Format: Progsv, alternative. News: one hr wkly. Target aud: 13-45. Spec prog: Top-35 countdown 3 hrs, Progressive Pioneers 3 hrs, techno 3 hrs wkly. ♦Hannah Jacobs, gen mgr.

WJZR(FM)— Jan 22, 1993: 105.9 mhz; 3 kw. 180 ft TL: N43 09 35 W77 34 44. Stereo. Hrs opn: 24 Fedder Industrial Park, 1237 E. Main St., 14609. Phone: (585) 288-5020. Licensee: North Coast Radio Inc. Population served: 650,000 Natl. Network: AP Radio. Law Firm: Cohn & Marks. Format: Blues, jazz. News: 14 hrs wkly. Target aud: 25 plus. Spec prog: News review one hr wkly. ♦Lee Rust, pres & stn mgr; Barry Vee, gen sls mgr.

WLGZ(AM)— February 1947: 990 khz; 5 kw-D, 2.5 kw-N, DA-2. TL: N43 13 54 W77 52 00. Hrs open: 24 2494 Browncroft Blvd., 14625. Phone: (585) 264-1027. Fax: (585) 264-1165. Web Site: www.crawfordbroadcasting.com. Licensee: Kimtron Inc. Group owner: Crawford Broadcasting Co. (acq 6-5-97; $650,000). Format: Adult standards. ♦Robert Hammond, gen mgr; Ben Martin, opns mgr.

WPXY-FM— Sept 14, 1959: 97.9 mhz; 50 kw. 456 ft TL: N43 08 08 W77 35 02. Stereo. Hrs opn: 1700 HSBC Plaza., 14604. Phone: (585) 399-5700. Fax: (585) 399-5750. Web Site: www.98pxy.com. Licensee: Infinity Radio Inc. Group owner: Infinity Broadcasting Corp. (acq 11-13-98; grpsl). Format: CHR. ♦Laurie Zsedely, gen mgr; Barbara Williams, gen sls mgr; Mike Danger, progmg dir.

WRMM-FM— Nov 14, 1966: 101.3 mhz; 27 kw. 640 ft TL: N43 10 14 W77 40 23. Stereo. Hrs opn: Phone: (585) 399-5700. Fax: (585) 399-5750. Web Site: www.warmradio.com. Licensee: Infinity Radio Inc. Group owner: Infinity Broadcasting Corp. (acq 11-13-98; grpsl). Population served: 149,000 Format: Adult contemp. Target aud: 25-54; baby boomers. ♦Al Casazza, gen mgr.

***WRUR-FM**— Mar 6, 1966: 88.5 mhz; 3 kw. 348 ft TL: N43 09 23 W77 36 31. Stereo. Hrs opn: 24 CPU Box 277356, Univ. of Rochester, 14627-7356. Phone: (585) 275-6400. Phone: (585) 275-7400. Fax: (585) 273-1357. E-mail: jlapin@wrur.org Web Site: www.wrur.org. Licensee: University of Rochester Broadcasting Corp. Population served: 1,500,000 Format: Div. News: 4 hrs wkly. Target aud: General. Spec prog: Jazz, gospel 3 hrs, relg one hr, world 10 hrs, folk 2 hrs, Sp 6 hrs, techno 5 hrs, death metal 8 hrs, industrial 3 hrs wkly. ♦Jared Lapin, gen mgr; Paul Szymanski, opns mgr; James Lamara, progmg dir.

***WXXI(AM)**— 1936: 1370 khz; 5 kw-U, DA-N. TL: N43 06 01 W77 34 23. Hrs open: 24 280 State St., 14614. Phone: (585) 325-7500. Fax: (585) 258-0339. E-mail: radio@wxxi.org Web Site: www.wxxi.org. Licensee: WXXI Public Broadcasting Council. Population served: 941,600 Natl. Network: NPR. Law Firm: Schwartz, Woods & Miller. Wire Svc: AP Format: News/talk. ♦Norm Silverstein, pres & gen mgr; Peter Iglinski, news dir; Bud Lowell, reporter.

WXXI-FM— December 1974: 91.5 mhz; 45 kw. 400 ft TL: N43 08 07 W77 35 03. Stereo. 24 Web Site: www.wxxi.org.941,600 Natl. Network: PRI. Format: Class. ♦Julia Figueras, mus dir & disc jockey; Morderai Lipshutz, disc jockey; Richard Gladwell, disc jockey; Simon Pontin, disc jockey. Co-owned TV: *WXXI-TV affil.

WYSL(AM)—See Avon

Rome

WFRG-FM—See Utica

WODZ-FM— August 1968: 96.1 mhz; 7.4 kw. 600 ft TL: N43 02 14 W75 26 40. Stereo. Hrs opn: 24 9418 State Rt.49, Marcy, 13403. Phone: (315) 768-9500. Fax: (315) 736-0720. Web Site: www.wodz.com. Licensee: Regent License of Utica/Rome Inc. Group owner: Regent Communications Inc. (acq 11-5-99; grpsl). Population served: 267,900 Format: Oldies. ♦Mary Jo Beach, gen mgr.

WRNY(AM)— Oct 12, 1959: 1350 khz; 500 w-D, 60 w-N. TL: N43 12 18 W75 29 08. Hrs open: 24 239 Genesee St., Suite 500, Utica, 13501. Phone: (315) 797-0803. Fax: (315) 797-7813. Licensee: CC Licenses LLC. Group owner: Clear Channel Communications Inc. (acq 8-5-98; grpsl). Population served: 44,000 Law Firm: Baraff, Koerner & Olender. Format: Sports. News staff: one. Target aud: 25 plus. Spec prog: Black 3 hrs wkly. ♦Brian Deleney, gen mgr; Chuck Hebbard, gen sls mgr; Gene Conte, progmg dir; Joe Petro, chief of engrg.

WUMX(FM)—Co-owned with WRNY(AM). May 1, 1983: 102.5 mhz; 27 kw. 649 ft TL: N43 02 14 W75 26 40. Stereo. 24 Web Site: www.1025kiss.com.280,000 Format: Country. News staff: one. Target aud: 18-49. Spec prog: Pub affrs one hr wkly. ♦Chris Spiwak, natl sls mgr & prom dir; Stew Schantz, progmg dir; Joe Petro, engrg mgr & chief of engrg; Ed O'Brien, sports cmtr.

WRUN(AM)—See Utica

WYFY(AM)— September 1946: 1450 khz; 1 kw-U. TL: N43 12 18 W75 28 48. Hrs open: 24 Bible Broadcasting Network, 11530 Carmel Commons Blvd., Charlotte, NC, 28226. Phone: (704) 523-5555. Fax: (704) 522-1967. Web Site: www.bbnradio.org. Licensee: Bible Broadcasting Network Inc. Group owner: Bible Broadcasting Network (acq 5-7-99; $50,000). Population served: 275,000 Format: Relg. ♦Jason Padgett, stn mgr.

Rosendale

***WFNP(FM)**— Sept 5, 1990: 88.7 mhz; 230 w. 1,289 ft TL: N41 43 09 W73 59 47. Stereo. Hrs opn: 7 PM-5 AM State Univ. of New York, SUB Rm. 413, New Paltz, 12561. Phone: (845) 257-3084. Fax: (845) 257-3099. Web Site: www.wfnp.org. Licensee: State University of New York, Albany. Population served: 330,000 Natl. Network: ABC, AP Radio. Law Firm: Dow, Lohnes & Albertson. Format: Progsv, urban contemp. News: 3 hrs wkly. Target aud: General; demographic-specific programs. Spec prog: Black 14 hrs, jazz 4 hrs, Sp 3 hrs, news/talk 5 hrs, metal 7 hrs wkly. ♦William Clark, opns dir & opns mgr.

Rotterdam

WTRY-FM— Dec 15, 1986: 98.3 mhz; 6 kw. 318 ft TL: N42 44 43 W74 04 10. Stereo. Hrs opn: 24 1203 Troy Schenectady Rd., Suite 201, Latham, 12110. Phone: (518) 452-4800. Fax: (518) 452-4855. Web Site: www.wtry.com. Licensee: Capstar TX L.P. Group owner: Clear Channel Communications Inc. (acq 8-30-00; grpsl). Population served: 1,300,000 Natl. Rep: Clear Channel. Format: Oldies. News staff: one; News: 20 hrs wkly. Target aud: 35-54. ♦Dennis Lamme, gen mgr; John Cooper, stn mgr; Kristen Delaney, mktg mgr.

Rouses Point

***WKYJ(FM)**— 2005: 88.7 mhz; 300 w vert. Ant 43 ft TL: N44 56 44 W73 25 41. Hrs open: 3185 S. Highland Dr., Suite 13, Las Vegas, NV, 89109. Phone: (702) 731-5588. Licensee: American Educational Broadcasting Inc. ♦Carl J. Auel, pres.

Sag Harbor

WLNG(FM)— Apr 13, 1969: 92.1 mhz; 5.3 kw. 350 ft TL: N40 58 19 W72 20 54. Hrs open: Box 2000, 11963. Phone: (631) 725-2300. Fax: (631) 725-5897. E-mail: info@wlng.com Web Site: www.wlng.com. Licensee: Mainstreet Broadcasting Co. Format: Oldies, Top-40. ♦Paul Sidney, pres & gen mgr.

Saint Bonaventure

***WSBU(FM)**— Apr 13, 1975: 88.3 mhz; 165 w. Ant -256 ft TL: N42 04 45 W78 29 07. Stereo. Hrs opn: 24 Box O, Saint Bonaventure University, Rm. 210, Reilly Ctr., 14778. Phone: (716) 375-2307. Fax: (716) 375-2583. E-mail: webmaster@wsbu.net Licensee: St. Bonaventure University. Population served: 50,000 Format: AOR. News: 10 hrs wkly. Target aud: General; primarily students. ♦Dr. Robert Wickenheiser, pres; Joe O'Neil, stn mgr; Todd Lewandowski, spec ev coord.

Salamanca

WGGO(AM)— June 18, 1957: 1590 khz; 5 kw-D. TL: N42 10 24 W78 41 07. Hrs open: Box 100, Killbuck, 14748. Secondary address: 4104 Killbuck Rd. 14779. Phone: (716) 945-1590. Fax: (716) 945-1515. E-mail: wgrt983@direcway.com Licensee: Pembrook Pines Inc. (acq 6-9-2006; $1.25 million with co-located FM). Population served: 7,877 Natl. Network: ESPN Radio. Format: Sports. Target aud: General. Spec prog: Country 5 hrs, Pol one hr wkly. ♦Robert J. Pfuntner, pres; Michael Washington, gen mgr & progmg dir; Sue Washington, gen sls mgr & mus dir; Scott Douglas, news dir; Russ Ehman, chief of engrg.

WQRS(FM)—Co-owned with WGGO(AM). Oct 15, 1988: 98.3 mhz; 1.6 kw. 430 ft TL: N42 06 32 W78 36 28. Stereo. E-mail: wqrt983@direcway.net Format: Classic rock.

Sandy Creek-Pulaski

WSCP(AM)— Aug 8, 1974: 1070 khz; 2.5 kw-D. TL: N43 36 19 W76 07 48. Hrs open: Box 640, 5090 U.S. Rt. 11, Pulaski, 13142. Phone: (315) 298-3185. Fax: (315) 298-6181. Web Site: www.wscp.net. Licensee: Galaxy Syracuse Licensee LLC. (group owner; (acq 7-17-2001; $400,000 with WSCP-FM Pulaski). Natl. Network: Jones Radio Networks. Format: Country. News staff: one; News: one hr wkly. Target aud: 35+. ♦Mimi Griswold, progmg dir.

Saranac Lake

WNBZ(AM)— Sept 11, 1927: 1240 khz; 1 kw-U. TL: N44 18 58 W74 07 08. Hrs open: 24 Box 211, 12983. Secondary address: Colony Ct. Ext. 12983. Phone: (518) 891-1544. Phone: (518) 891-3636. Fax: (518) 891-1545. E-mail: mail@wnbz.com Web Site: www.wnbz.com. Licensee: Saranac Lake Radio L.L.C. Group owner: Mountain Communications (acq 6-1-98; $397,500 with co-located FM). Population served: 19,134 Law Firm: Irwin, Campbell, Crowe & Tannenwald. Format: Adult contemp. News staff: one; News: 36 hrs wkly. Target aud: 35 plus; loc community. ♦Ted Morgan, pres & gen mgr; John Gagnon, opns mgr & progmg dir; James Williams, gen sls mgr; Chris Knight, news dir; Crystal Tatro, pub affrs dir; Chris Brescia, chief of engrg; Steve Borst, disc jockey.

WYZY(FM)—Co-owned with WNBZ(AM). July 12, 1989: 106.3 mhz; 50 kw. Ant 394 ft TL: N44 20 28 W74 07 43. Stereo. 24 Web Site: www.wnbz.com.30,000 Format: Adult contemp. News staff: 2; News: 10 hrs wkly. Target aud: 18-49; adults. ♦Crystal Tatro, opns mgr; Steve Borst, disc jockey.

***WSLL(FM)**— July 1, 1989: 89.5 mhz; 200 w. 355 ft TL: N44 20 28 W74 07 43. Hrs open:
Rebroadcasts WSLU(FM) Canton 100%.
St. Lawrence Univ., Canton, 13617. Phone: (315) 229-5356. Fax: (315) 229-5373. Web Site: www.ncpr.org. Licensee: St. Lawrence University. Format: Eclectic public radio. Target aud: General. ♦Ellen Rocco, stn mgr; Sandra Demarest, dev dir.

WSLP(FM)— 2007: 93.3 mhz; 11 kw. Ant -207 ft TL: N44 15 36 W74 01 22. Hrs open: Box 368, Lake Placid, 12946. Phone: (518) 523-4900. Fax: (518) 523-4290. Licensee: North Country Radio Inc. Format: Adult contemp. ♦Jon Lundin, gen mgr; Jim Williams, gen sls mgr.

Saratoga Springs

***WSPN(FM)**— Sept 9, 1974: 91.1 mhz; 253 w. 98 ft TL: N43 05 55 W73 47 10. Stereo. Hrs opn: 24 Skidmore College, 12866. Phone: (518) 580-5783. Licensee: Skidmore College. Population served: 30,000 Natl. Network: AP Radio. Format: College radio. News: 5 hrs wkly. Target aud: All ages. Spec prog: Folk 3 hrs, Pol 3 hrs, Sp 3 hrs, blues 9 hrs, world mus 3 hrs wkly. ♦Alissa DeVogel, gen mgr; Lily Gedney, mus dir.

***WSSK(FM)**— 2001: 89.7 mhz; 50 w. Ant 430 ft TL: N43 11 35 W73 45 25. Hrs open: Box 777, Lake Katrine, 12449. Phone: (845) 336-6199. Fax: (845) 336-7205. Web Site: www.soundoflife.org. Licensee: Sound of Life Inc. Format: Contemp Christian. ♦Tom Michael Zahradnik, gen mgr.

WUAM(AM)— Mar 23, 1964: 900 khz; 250 w-U. TL: N43 04 24 W73 48 07. Hrs open: 24 100 Saratoga Blvd., Sutie 21, Malta, 12020. Phone: (518) 899-3000. Fax: (518) 899-3057. E-mail: moonradioam@aol.com Web Site: www.sta1013.com. Licensee: Anastos Media Group Inc. (group owner; acq 9-99; $100,000). Population served: 170000 Natl. Network: USA. Law Firm: Gammon & Grange. Format: Big band. News staff: 2; News: 20 hrs wkly. Target aud: 25-54. Spec prog: Farm one hr wkly. ♦Scott Collins, gen mgr; John H. Meaney, stn mgr & opns dir.

Saugerties

WBPM(FM)— 1999: 92.9 mhz; 6 kw. Ant 289 ft TL: N41 59 20 W74 01 08. Hrs open: 24 715 Rt. 52, Beacon, 12508. Phone: (845) 838-6000. Fax: (845) 838-6088. Web Site: www.wbpmfm.com. Licensee: 6 Johnson Road Licenses Inc. Group owner: Clear Channel Communications Inc. (acq 4-1-2007; grpsl). Format: Classic hits. ♦Jason Finkelberg, gen mgr.

Schenectady

WGY(AM)— February 1922: 810 khz; 50 kw. TL: N42 47 37 W74 00 36. Hrs open: 24 1203 Troy-Schenectady Rd., Latham, 12110. Phone: (518) 452-4800. Fax: (518) 452-4855. Web Site: www.wgy.com. Licensee: CC License LLC. Group owner: Clear Channel Communications Inc. (acq 8-5-98; grpsl). Population served: 800,000 Natl. Network: Fox News Radio, Premiere Radio Networks. Natl. Rep: Clear Channel. Wire Svc: AP Format: News/talk. News staff: 12; News: 23 hrs wkly. Target aud: 25-54; college graduate, married, homeowner. ♦Kristen Delaney, VP; Greg Foster, opns dir, opns mgr & progmg dir; Mark Scott, traf mgr; Don Weeks, disc jockey; Ed Martin, disc jockey; Joe Gallagher, disc jockey.

WRVE(FM)—Co-owned with WGY(AM). April 1940: 99.5 mhz; 14.5 kw. 925 ft TL: N42 38 13 W73 59 48. Stereo. 24 Web Site: www.wrve.com. Natl. Network: Premiere Radio Networks. Format: Adult contemp. News staff: one; News: 3 hrs wkly. Target aud: 25-54. ♦Randy McCarten, progmg dir.

***WMHT-FM**— June 8, 1972: 89.1 mhz; 11 kw. Ant 930 ft TL: N42 38 13 W74 00 06. Stereo. Hrs opn: 4 Global View, Troy, 12180. Phone: (518) 880-3400. Fax: (518) 880-3409. E-mail: email@wmht.org Web Site: www.wmht.org. Licensee: WMHT Educational Telecommunications. Format: Class. Spec prog: Jazz one hr wkly. ♦Deborah Onslow, gen mgr; Dave Nicosia, chief of engrg. Co-owned TV: *WMHT(TV) affil.

WOFX(AM)—See Troy

***WRUC(FM)**— May 9, 1975: 89.7 mhz; 100 w. -88 ft TL: N42 49 04 W73 55 45. Stereo. Hrs opn: Union College, 12308. Phone: (518) 388-6151. Phone: (518) 388-6154. Fax: (518) 388-6790. Licensee: Trustees of Union College. Population served: 950,000 Natl. Network: AP Radio. Format: Alternative. Target aud: 18 plus; general. Spec prog: It one hr, Sp 3 hrs, jazz 15 hrs, sports 4 hrs wkly.

WTRY-FM—See Rotterdam

WVKZ(AM)— Apr 15, 1942: 1240 khz; 1 kw-U. TL: N42 48 37 W73 59 04. Hrs open: 24 100 Saratoga Village Blvd., Ste. 21, Malta, 12020. Phone: (518) 899-3000. Fax: (518) 899-3057. E-mail: talk1240@aol.com Licensee: The Anastos Media Group Inc. Group owner: Anastos Media Group Inc. (acq 4-10-2000; $137,500). Population served: 247,000 Format: Real oldies. News staff: one. Target aud: 25-54; men. ♦Scott Collins, pres & gen mgr; John H. Meaney, opns mgr.

Schoharie

WMYY(FM)— 1990: 97.3 mhz; 800 w. Ant 895 ft TL: N42 37 51 W74 16 01. Stereo. Hrs opn: 24
Rebroadcasts WHAZ(AM) Troy 100%.
30 Park Ave., Cohoes, 12047-3330. Phone: (518) 237-1330. Fax: (518) 235-4468. E-mail: info@whaz.com Web Site: www.whaz.com. Licensee: Capital Media Corp. (group owner; acq 2-14-92; FTR: 2-17-92). Population served: 1,000,000 Format: Adult Christian. Target aud: 25-75; young to old. Spec prog: Gospel, relg. ♦Paul F. Lotters, pres & gen mgr; Steven L. Klob, opns dir, dev dir, sls dir, mktg dir, prom dir & adv dir; Rex P. Gregory, progmg dir, news dir, pub affrs dir & disc jockey; John W. Shafer, chief of engrg.

Schuyler Falls

***WAVX(FM)**— 2004: 90.9 mhz; 2.7 kw. Ant 1,073 ft TL: N44 34 24 W73 40 31. Hrs open: Box 8310, Essex, VT, 05451-8310. Phone: (802) 878-8885. Fax: (802) 879-6835. E-mail: cmi.radio@verizon.net Licensee: Christian Ministries Inc. Format: Christian. ♦Mark Kinsley, pres; Richard McClary, gen mgr.

Scotia

***WYAI(FM)**— December 1981: 93.7 mhz; 1.25 kw. Ant 705 ft TL: N42 51 24 W74 04 03. Hrs open: 2351 Sunset Blvd., Suite 170-218, Rocklin, CA, 95765. Phone: (916) 251-1600. Fax: (916) 251-1650. Web Site: www.air1.com. Licensee: Educational Media Foundation. (group owner; (acq 7-6-2007; grpsl). Natl. Network: Air 1. Format: Christian rock. ♦Mike Novak, sr VP.

Seneca Falls

WLLW(FM)—Listing follows WSFW(AM).

WSFW(AM)— Oct 1, 1968: 1110 khz; 1 kw-D. TL: N42 54 55 W76 46 28. Hrs open: Sunrise-sunset 3568 Lenox Rd., Geneva, 14456. Phone: (315) 781-7000. Fax: (315) 781-7700. E-mail: wnyr@flare.net Licensee: Auburn Broadcasting Inc. Group owner: Finger Lakes Radio Group (acq 3-2-2001; with co-located FM). Population served: 350,000 Natl. Rep: Rgnl Reps. Law Firm: Borsari & Paxson. Format: Talk. News staff: one; News: 24 hrs wkly. Target aud: 25-54. Spec prog: Irish 2 hrs, It 2 hrs, jazz one hr, oldies 3 hrs, Pol 2 hrs wkly. ♦Allan Bishop, gen mgr.

WLLW(FM)—Co-owned with WSFW(AM). Nov 1, 1968: 99.3 mhz; 5 kw. Ant 358 ft TL: N42 59 38 W76 51 59. Stereo. 24 Format: Classic Rock. ♦Ken Paradise, progmg dir.

Sidney

WCDO(AM)— 1983: 1490 khz; 1 kw-U. TL: N42 19 24 W75 22 57. Hrs open: 75 Main St., 13838. Phone: (607) 563-3588. Phone: (607) 563-3589. Fax: (607) 563-7805. Licensee: CDO Broadcasting Inc. Group owner: Clancy-Mance Communications (acq 3-8-86; $180,000 with co-located FM; FTR: 1-13-86) Format: Adult contemp. Target aud: 25-54. ♦Craig Harris, gen mgr; Jim Tomeo, progmg dir.

WCDO-FM— May 1982: 100.9 mhz; 970 w. 577 ft TL: N42 17 33 W75 22 03. (CP: 1.88 kw). Format: Adult contemp, oldies. Target aud: 25-54. ♦Craig Stevens, gen mgr & gen sls mgr; Greg Davie, sls dir.

Smithtown

*WFRS(FM)— Oct 17, 1988: 88.9 mhz; 1.5 kw horiz, 1.45 kw vert. 453 ft TL: N40 48 27 W73 10 48. Stereo. Hrs opn: 24 3200 Expressway Dr. S., Islandia, 11749. Phone: (631) 234-4151. Web Site: www.familyradio.com. Licensee: Family Stations Inc. (group owner; acq 9-27-83). Population served: 1,000,000 Natl. Network: Family Radio. Format: Nondenominational Christian educ. News: 5 hrs wkly. Target aud: General.

WMJC(FM)— May 21, 1957: 94.3 mhz; 2.6 kw. Ant 315 ft TL: N40 48 08 W73 17 12. Stereo. Hrs opn: 24 234 Airport Plaza Blvd., Suite 5, Farmingdale, 11735. Phone: (631) 770-4200. Fax: (631) 770-0101. Web Site: www.island943.com. Licensee: IW L.L.C. Group owner: Barnstable Broadcasting Inc. (acq 10-1-97; grpsl). Population served: 2,000,000 Natl. Network: AP Radio. Rgnl. Network: Metronews Radio Net. Wire Svc: Standard Broadcast Wire Format: Hot adult contemp. News staff: one; News: 2 hrs wkly. Target aud: 25-54; men & women. ♦Al Kaneb, CEO & natl sls mgr; Dave Widmer, gen mgr.

Sodus

WUUF(FM)— 1991: 103.5 mhz; 6 kw. 243 ft TL: N43 16 05 W77 09 40. Hrs open: 24 Box 1420, Newark, 14513. Phone: (315) 331-9667. Fax: (315) 331-7101. E-mail: bigdogfm@rochester.rr.com Web Site: www.bigdog1035.com. Licensee: Waynco Radio (acq 8-90; $10,000; FTR: 8-13-90). Population served: 2,000,000 Natl. Network: Motor Racing Net, Westwood One. Law Firm: Fletcher, Heald & Hildreth . Format: Country. News staff: one; News: one hr wkly. Target aud: 25-54. ♦Robert Pfuntner, pres; John Tickner, VP & gen mgr; Jim Hill, opns dir.

South Bristol Township

WCRR(FM)— Jan 22, 1996: 107.3 mhz; 650 w. Ant 994 ft TL: N42 44 47 W77 25 35. Hrs open: 24 207 Midtown Plaza, Rochester, 14604. Phone: (585) 232-8870. Fax: (585) 454-5081. Web Site: www.mycountryfm.com. Licensee: Citicasters Licenses L.P. Group owner: Clear Channel Communications Inc. (acq 1999; grpsl). Format: Country. ♦Debbie Johnston, gen mgr.

South Glens Falls

WENU(AM)— September 1988: 1410 khz; 1 kw-D, 126 w-N. TL: N43 16 07 W73 40 14. Hrs open: 24 89 Everts Ave., Queensbury, 12804. Phone: (518) 793-7733. Fax: (518) 793-0838. Licensee: 6 Johnson Road Licenses Inc. Group owner: Pamal Broadcasting Ltd. (acq 4-1-2004; grpsl). Population served: 233,000 Natl. Network: Westwood One. Format: Contemporary. News staff: one. Target aud: 35 plus. ♦James Morrell, pres; Mike Morgan, opns mgr.

Southampton

WEHM(FM)— July 21, 2003: 92.9 mhz; 6 kw. Ant 276 ft TL: N40 52 10 W72 34 37. Hrs open: Box 7162, Amagansett, 11930. Secondary address: 249 Montauk Hwy., Amagansett 11930. Phone: (631) 267-7800. Fax: (631) 267-1018. E-mail: info@wehm.com Web Site: www.wehm.com. Licensee: AAA Licensing LLC. Group owner: AAA Entertainment L.L.C. (acq 4-30-2003). Format: Progsv adult rock.

WHFM(FM)— October 1971: 95.3 mhz; 5 kw. 354 ft TL: N40 56 05 W72 23 15. Stereo. Hrs opn: Box 674, Center Moriches, 11934. Phone: (631) 587-1023. Fax: (631) 283-9506. Web Site: www.wbab.com. Licensee: Cox Radio Inc. Group owner: Cox Broadcasting (acq 5-22-98; grpsl). Population served: 150,000 Natl. Rep: Christal. Format: Adult contemp, rock/AOR. Target aud: 25-49; upscale. ♦Kim Guthrie, VP; Tim Guthrie, gen mgr; Todd Dinetz, gen sls mgr; Vinny DiMarco, natl sls mgr; John Olsen, progmg dir; Matthew Connor, chief of engrg; Lori DeFilliis, traf mgr.

*WLIU(FM)— Mar 3, 1979: 88.3 mhz; 5.9 kw horiz, 25 kw vert. 217 ft TL: N40 53 17 W72 26 43. (CP: 16 kw, ant 748 ft. TL: N40 51 18 W72 46 12). Stereo. Hrs opn: 24
Rebroadcasts WCWP(FM) Brookville 60%.
PO Box 803, Southhampton, 11969-0803. Phone: (631) 591-7000. Fax: (631) 591-7080. E-mail: wally@wliu.org Web Site: www.wliu.org. Licensee: Long Island University. Population served: 115,000 Natl. Network: PRI, NPR. Law Firm: Lawrence Bernstein. Format: Jazz. News staff: 2; News: 34 hrs wkly. Target aud: 34-55; upscale, educ, public radio listeners. Spec prog: Pub affrs one hr wkly. ♦Dr. Wallace Smith, gen mgr; Jamie Berger, opns dir; Bonnie Grice, mus dir; Robert Anderson, chief of engrg.

*WRLI-FM— July 1999: 91.3 mhz; 10 kw. Ant 312 ft TL: N40 56 05 W72 23 15. Hrs open:
Rebroadcasts WPKT(FM) Meriden 100%.
1049 Asylum Ave., Hartford, CT, 06105. Phone: (860) 278-5310. Fax: (860) 244-9624. E-mail: info@wnpr.org Web Site: www.wnpr.org. Licensee: Connecticut Public Television & Radio. Natl. Network: NPR, PRI. Format: News/talk. ♦Jerry Franklin, CEO & pres; Kim Grehn, VP & gen mgr.

Southold

WBEA(FM)— July 3, 1985: 101.7 mhz; 6 kw. Ant 283 ft TL: N40 52 10 W72 34 37. Stereo. Hrs opn: 24 Box 7162, Amagansett, 11930. Phone: (631) 267-7800. Fax: (631) 267-1018. Web Site: www.1017blaze.com. Licensee: AAA Licensing LLC. Group owner: AAA Entertainment L.L.C. (acq 8-22-2000; $2.75 million with WBAZ(FM) Bridgehampton). Population served: 1,500,000 Natl. Network: Westwood One. Law Firm: Haley, Bader & Potts. Format: Hip hop. Target aud: 25-54; adults with active lifestyles. Spec prog: Health talk, financial news, CNN news. ♦Bonnie Gomes, gen mgr & gen sls mgr.

Southport

WOKN(FM)— Sept 15, 1993: 99.5 mhz; 1.25 kw. 485 ft TL: N42 07 49 W76 47 23. Hrs open: 24 1705 Lake St., Elmira, 14901. Phone: (607) 733-5626. Fax: (607) 733-4040. Fax: (607) 733-5627. Web Site: www.995wokn.com. Licensee: Pembrook Pines Elmira Ltd. Group owner: Pembrook Pines Media Group. Population served: 350,000 Natl. Network: Jones Radio Networks. Law Firm: Bechtel & Cole. Format: Country. News staff: one; News: 2 hrs wkly. Target aud: 18-49; female. ♦Robert J. Pfuntner, CEO & pres; Nancy E. Nicastro, gen mgr; Michael Williams, opns mgr.

Spencer

*WCII(FM)— Oct 1, 1989: 88.5 mhz; 17 kw. 590 ft TL: N42 00 50 W76 15 53. Stereo. Hrs opn: 24
Rebroadcasts WCIK(FM) Bath 100%.
Box 506, 7634 Campbell Creek Rd., Bath, 14810. Phone: (607) 776-4151. Fax: (607) 776-6929. E-mail: mail@fln.org Web Site: www.fln.org. Licensee: Family Life Ministries Inc. Group owner: Family Life Network Natl. Network: Salem Radio Network. Law Firm: Hardy, Carey, Chautin & Balkin, LLP . Wire Svc: Metro Weather Service Inc. Format: Christian, inspirational, educ. News staff: 3; News: 14 hrs wkly. Target aud: 30-54; general public. Spec prog: News 14 hrs wkly. ♦Dick Snavely, CFO; Rick Snavely, pres, VP & gen mgr; John Owens, progmg dir; Jim Travis, chief of engrg.

Spring Valley

WRCR(AM)— Sept 15, 1977: 1300 khz; 500 w-D, 83 w-N, DA-2. TL: N41 05 48 W74 00 18. Hrs open: 24 Nanuet Mall, 75 W. Rt. 59, Ste. 2126, Nanuet, 10954. Phone: (845) 624-1313. Fax: (845) 624-1639. E-mail: mail@wrcr.com Web Site: www.wrcr.com. Licensee: Alexander Broadcasting Inc. (acq 4-14-2000; $270,000). Population served: 400000 Natl. Network: USA. Format: Adult contemp, news. News staff: 2; News: 18 hrs wkly. Target aud: 25-54; upscale. ♦Alexander Medakovic, pres & gen mgr.

Springville

WSPQ(AM)— Apr 20, 1986: 1330 khz; 1 kw-U, DA-2. TL: N42 29 53 W78 41 10. Hrs open: 24 51 Franklin St., 14141. Phone: (716) 592-9500. Fax: (716) 592-9522. Licensee: Hawk Communications Ltd. (acq 1996). Population served: 400,000 Natl. Network: CNN Radio, ESPN Radio, Motor Racing Net. Format: Var/Diverse, adult contemp, country, sports. News staff: one; News: 10 hrs wkly. Target aud: 25-54. Spec prog: Farm 5 hrs, relg 2 hrs wkly. ♦Kevin Bower, gen mgr.

Staten Island

WSIA(FM)—Licensed to Staten Island. See New York

Stillwater

WQAR(FM)— Oct 3, 1988: 101.3 mhz; 2.9 kw. 470 ft TL: N43 00 42 W73 41 01. Stereo. Hrs opn: 24 100 Saratoga Blvd., Ste. 21, Malta, 12020. Phone: (518) 899-3000. Phone: (518) 899-1013. Fax: (518) 899-3057. E-mail: star1013fm@aol.com Web Site: www.star1013.com. Licensee: Anastos Media Group Inc. (group owner; acq 9-4-98; $900,000). Law Firm: Shaw Pittman. Format: Adult contemp. News staff: one; News: 6 hrs wkly. Target aud: 25-54; upscale. Spec prog: Saratoga Forum one hr. ♦Scott Collins, pres; J. Scott Collins, gen mgr; John Meaney, opns mgr.

Stony Brook

*WUSB(FM)— June 27, 1977: 90.1 mhz; 3.6 kw. 531 ft TL: N40 50 32 W73 02 23. Stereo. Hrs opn: 24 Union Building, University at Stony Brook, 11794-3263. Phone: (631) 632-6501. Fax: (631) 632-7182. E-mail: info@wusb.org Web Site: www.wusb.org. Licensee: State University of New York. Population served: 3,000,000 Law Firm: Dow, Lohnes & Albertson. Wire Svc: AP Format: News/talk, diversified, progsv. News: 20 hrs wkly. Target aud: 18-49; progsv & musically adventurous. Spec prog: Black 12 hrs, Pol one hr, Sp 3 hrs, Chinese one hr, Korean one hr, class 14 hrs, folk 15 hrs, jazz 20 hrs, blues 10 hrs wkly. ♦Norman L. Prusslin, gen mgr; Marko Srdanovic, opns dir.

Sylvan Beach

WWLF-FM— April 1999: 100.3 mhz; 6 kw. Ant 328 ft TL: N43 14 46 W75 46 25. Hrs open: 401 W. Kirkpatrick St., Syracuse, 13204. Phone: (315) 472-0222. Fax: (315) 478-7745. E-mail: programming@movin100.com Web Site: movin100.com. Licensee: WOLF Radio Inc. (group owner; (acq 2-28-2002; $350,000). Format: Rhythmic adult contemp. ♦Sam Furco, gen mgr.

Syosset

*WKWZ(FM)— July 24, 1973: 88.5 mhz; 125 w. 90 ft TL: N40 49 48 W73 28 57. (CP: Ant 259 ft.). Hrs opn: 70 Southwoods Rd., 11791. Phone: (516) 364-5745. Phone: (516) 364-5746. Fax: (516) 364-5737. E-mail: BigDave5@aol.com Web Site: www.wkwz.org. Licensee: Syosset Central School District. Population served: 400,000 Format: Div. Spec prog: C&W 6 hrs, class 6 hrs, jazz 12 hrs wkly. ♦David C. Favilla, gen mgr; Chris Hoffman, stn mgr; Roy Dippel, chief of engrg.

Syracuse

*WAER(FM)— Apr 1, 1947: 88.3 mhz; 50 kw. Ant 276 ft TL: N43 02 01 W76 07 53. Stereo. Hrs opn: 24 795 Olsrom Ave., 13244-2110. Phone: (315) 443-4021. Fax: (315) 443-2148. E-mail: waer@waer.org Web Site: www.waer.org. Licensee: Syracuse University. Population served: 700,000 Natl. Network: NPR. Law Firm: Arter & Hadden. Format: Jazz, sports, news. News staff: 3; News: 20 hrs wkly. Target aud: 25-49. Spec prog: Gospel 3 hrs, blues 3 hrs, world mus 4 hrs, new age 3 hrs wkly. ♦Joe Lee, gen mgr; Ron Ockert, progmg dir & progmg mgr; Eric Cohen, mus dir.

WAMF(AM)—See Fulton

WAQX-FM—See Manlius

WBBS(FM)—See Fulton

*WCNY-FM— Dec 4, 1971: 91.3 mhz; 18.6 kw. 740 ft TL: N42 56 42 W76 01 28. Stereo. Hrs opn: 5 AM-midnight Box 2400, 13220-2400. Secondary address: 506 Old Liverpool Rd., Liverpool 13088. Phone: (315) 453-2424. Fax: (315) 451-8824. E-mail: wcny-online@wcny.org Web Site: www.wcny.org. Licensee: Public Broadcasting Council of Central New York. Population served: 80,000 Natl. Network: NPR. Law Firm: Dow, Lohnes & Albertson. Format: Class. Target aud: General. Spec prog: Bluegrass 3 hrs, jazz 7 hrs wkly. ♦Colleen Edwards, CFO;

Peter Hirsch, mktg dir; Don Dolloff, progmg dir; John Duffy, chief of engrg. Co-owned TV: *WCNY-TV affil

WFBL(AM)— Feb 4, 1922: 1390 khz; 5 kw-U, DA-N. TL: N43 05 30 W76 05 19. Hrs open: Box 1050, Baldwinsville, 13027. Phone: (315) 635-3971. Fax: (315) 635-3490. Web Site: www.wfbl.com. Licensee: Buckley Broadcasting of New York LLC. Group owner: Buckley Broadcasting Corp. (acq 11-10-2003; $1.2 million). Format: Talk. ♦Judith C Kelly, VP, gen mgr & stn mgr; Bryan Richards, opns mgr.

WHEN(AM)— Apr 14, 1941: 620 khz; 5 kw-D, 1 kw-N, DA-N. TL: N43 05 35 W76 11 19. Stereo. Hrs opn: 24 500 Plum St., Suite 100, 13204. Phone: (315) 472-9797. Fax: (315) 472-1904. Web Site: www.sportsradio620.com. Licensee: CC Licenses LLC. Group owner: Clear Channel Communications Inc. (acq 1999). Population served: 179,800 Format: Sports. Spec prog: Syracuse Chiefs, Buffalo Bills, Syracuse Crunch. ♦Joel Delmonico, gen mgr.

WWHT(FM)—Co-owned with WHEN(AM). Sept 1, 1958: 107.9 mhz; 50 kw. 490 ft TL: N42 57 21 W76 06 36. Stereo. 24 Web Site: www.hot1079.com. Format: CHR.

***WJPZ-FM**— Jan 30, 1985: 89.1 mhz; 100 w. 120 ft TL: N43 02 01 W76 07 53. Stereo. Hrs opn: 24 316 Waverly Ave., 13210. Phone: (315) 443-4689. Phone: (315) 443-2106. Fax: (315) 443-4379. Web Site: www.z89.com. Licensee: WJPZ Radio Inc. Law Firm: Gardner, Carton & Douglas. Format: CHR. Target aud: 12-34; women & teenagers. Spec prog: Black 12 hrs, pub service 13 hrs wkly. ♦Geoff Herbert, gen mgr; Scott Purdy, opns VP; Louise Vazquez, dev VP; Joan Kump, prom dir; David McKinley, progmg dir.

WKRL-FM—See North Syracuse

WLTI(FM)— Apr 8, 1996: 105.9 mhz; 4 kw. 200 ft TL: N43 05 23 W76 09 10. Stereo. Hrs opn: 24 1064 James St., 13203. Phone: (315) 472-0200. Fax: (315) 478-5625. Web Site: www.lite1059.com. Licensee: Citadel Broadcasting Co. Group owner: Citadel Broadcasting Corp. (acq 2000; grpsl). Population served: 536,300 Natl. Network: CBS Radio. Format: Soft adult contemp. Target aud: 25-54; general. ♦Tom Mitchell, opns dir; Dave Allen, progmg dir & progmg mgr.

***WMHR(FM)**— Mar 9, 1969: 102.9 mhz; 20 kw. 784 ft TL: N42 58 00 W76 12 01. Stereo. Hrs opn: 24 4044 Makyes Rd., 13215. Phone: (315) 469-5051. E-mail: mhn@marshillnetwork.org Web Site: www.marshillnetwork.org. Licensee: Mars Hill Broadcasting Co. Inc. dba Mars Hill Network. (group owner) Population served: 1,546,800 Natl. Network: Moody, Salem Radio Network. Law Firm: Wiley, Rein & Fielding. Wire Svc: AP Format: Christian. News: 6 hrs wkly. Target aud: General; Christian families. Spec prog: Children 11 hrs wkly. ♦Clayton Roberts, pres; Chris Tetta, CFO & mus dir; Jeremy Miller, VP & news dir; Wayne Taylor, gen mgr; Valerie Smith, traf mgr.

WNSS(AM)—Listing follows WNTQ(FM).

WNTQ(FM)— 1956: 93.1 mhz; 97 kw. 659 ft TL: N42 56 47 W76 01 32. Stereo. Hrs opn: 24 1064 James St., 13203. Phone: (315) 472-0200. Fax: (315) 478-5625. Web Site: www.93q.com. Licensee: Citadel Broadcasting Co. Group owner: Citadel Broadcasting Corp. (acq 4-26-01; grpsl). Population served: 536,300 Natl. Rep: McGavren Guild. Format: CHR. Target aud: 25-54; women. ♦Darren Smith, gen mgr; Laura Serway, sls dir; Janice Cole, prom dir; Tom Mitchell, opns mgr & progmg dir; Phil Spevak, news dir; Dave Edwards, chief of engrg; Kelley Galuppo, traf mgr.

WNSS(AM)—Co-owned with WNTQ(FM). 1946: 1260 khz; 5 kw-U, DA-2. TL: N43 09 10 W76 11 35.24 Web Site: www.espnradio1260.com.536,300 Natl. Network: ESPN Radio. Format: Sports/Talk. Target aud: 25-54; men. ♦Jim Tully, natl sls mgr & progmg dir; Karolyn Bryanz, traf mgr.

WOLF(AM)— Apr 27, 1940: 1490 khz; 620 w-D, 750 w-N, DA-D. TL: N43 03 30 W76 10 00. (CP: 1510 khz. TL: N42 57 42 W76 06 13). Stereo. Hrs opn: 24 401 W. Kirkpatrick, 13204. Phone: (315) 472-0222. Fax: (315) 478-7745. E-mail: wolfam/fm@aol.com Web Site: www.radiodisney.com. Licensee: WOLF Radio Inc. (group owner; acq 10-5-82). Population served: 695,000 Natl. Network: Radio Disney. Law Firm: James L. Oyster. Format: Children. ♦Sam Furco, gen mgr & opns dir.

***WRVD(FM)**— June 1, 1999: 90.3 mhz; 280 w. Ant 43 ft TL: N43 02 27 W76 08 22. Stereo. Hrs opn: 24
Rebroadcasts WRVO(FM) Oswego 100%.
c/o WRVO(FM), Lanigan Hall, State Univ. College, Oswego, 13126. Phone: (315) 312-3690. Fax: (315) 312-3174. E-mail: wrvd@wrvo.fm Web Site: wrvo.fm. Licensee: State University of New York. Population served: 350,163 Natl. Network: NPR. Format: News/talk, old time radio. News staff: 4; News: 140 hrs wkly. Target aud: 35-54. ♦John E. Krauss, gen mgr; Matt Seuert, dev dir; Fred Vigeant, progmg dir; Jeff Windsor, chief of engrg.

WSEN(AM)—See Baldwinsville

WSIV(AM)—East Syracuse, Dec 6, 1955: 1540 khz; 1 kw-D. TL: N43 05 40 W76 02 00. (CP: 1.5 kw-D, 57 w-N). Hrs opn: 24 7095 Myers Rd., East Syracuse, 13057. Phone: (315) 656-2231. Phone: (315) 956-2250. Fax: (315) 656-2259. E-mail: wvoaradio@msn.com Licensee: CRAM Communications L.L.C. (acq 1-6-97; $900,000 with WVOA(FM) DeRuyter). Population served: 350,000 Format: Christian, relg, Black gospel, music. Spec prog: Black 20 hrs, Gospel music. ♦Sam Furco, CEO; James Wall, gen mgr; Suzanne Anderson, stn mgr; Allen Elson, opns mgr.

WSYR(AM)— 1922: 570 khz; 5 kw-U, DA-2. TL: N42 59 13 W76 09 09. Stereo. Hrs opn: 24 Bridgewater Pl., 500 Plum St., 13204. Phone: (315) 472-9797. Fax: (315) 472-1904. Web Site: www.sybercuse.com. Licensee: CC Licenses LLC. Group owner: Clear Channel Communications Inc. Population served: 600,000 Natl. Network: PRI. Format: Full service, news/talk. News: 35 hrs wkly. Target aud: 25-54. ♦Joel Delmonico, gen mgr.

WYYY(FM)—Co-owned with WSYR(AM). 1946: 94.5 mhz; 100 kw. 650 ft TL: N42 56 40 W76 07 08. Stereo. Web Site: www.sybercuse.com. Format: Adult contemp.

WTLA(AM)—See North Syracuse

WWDG(FM)—See DeRuyter

Ticonderoga

***WANC(FM)**— Sept 6, 1982: 103.9 mhz; 1.55 kw. 380 ft TL: N43 49 55 W73 24 28. Stereo. Hrs opn: 24
Rebroadcasts WAMC-FM Albany 100%.
Box 66600, 318 Central Ave., Albany, 12206-6600. Phone: (518) 465-5233. Phone: (800) 323-9262. Fax: (518) 432-6974. E-mail: mail@wamc.org Web Site: www.wamc.org. Licensee: WAMC. Group owner: WAMC/Northeast Public Radio (acq 8-90; $400,000; FTR: 8-13-90). Natl. Network: NPR, PRI. Law Firm: Dow, Lohnes & Albertson. Wire Svc: AP Format: News, Talk. Target aud: General. Spec prog: Folk 6 hrs, jazz 17 hrs wkly. ♦Alan Chartock, CEO, chmn & pres; David Galletly, VP; Selma Kaplan, VP.

WIPS(AM)— July 1955: 1250 khz; 1 kw-D. TL: N43 51 16 W73 23 24. Hrs open: 6 AM-sunset PO Box 600, Crown Point, 12928-0600. Phone: (518) 597-9477. Phone: (518) 597-3201. Fax: (518) 597-9479. E-mail: info@wipsradio.com Web Site: www.wipsradio.com. Licensee: BisiBlue L.L.C. (acq 3-17-2004; $93,000). Population served: 5,000 Format: Oldies. News staff: one; News: 24 hrs wkly. Target aud: 25-54. Spec prog: Farm 6 hrs wkly. ♦Gregg Trask, pres & gen mgr; Patricia Knapp, chmn & CFO.

Troy

WFLY(FM)— August 1948: 92.3 mhz; 17 kw. 850 ft TL: N42 38 16 W73 59 55. Hrs opn: 24 6 Johnson Rd., Latham, 12210. Phone: (518) 786-6600. Fax: (518) 786-6610. Licensee: 6 Johnson Road Licenses Inc. Group owner: Pamal Broadcasting Ltd. (acq 10-19-2001; grpsl). Population served: 220,000 Natl. Network: ABC. Natl. Rep: McGavren Guild. Format: CHR. News staff: one; News: 5 hrs wkly. Target aud: 18-49. ♦Dan Austin, gen mgr; Kevin Callahan, opns mgr; Suzette Anthony, sls dir; Stephen Roberts, gen sls mgr; Amanda Guldenstern, prom dir; John Foxx, progmg dir.

WGNA-FM—See Albany

WHAZ(AM)— August 1922: 1330 khz; 1 kw-U. TL: N42 46 35 W73 41 10. Hrs open: 24 30 Park Ave., Cohoes, 12047-3330. Phone: (518) 237-1330. Fax: (518) 235-4468. Web Site: www.whaz.com. Licensee: Capital Media Corp. (group owner; (acq 9-24-87). Population served: 1,000,000 Format: Adult Christian. Target aud: 25-75; young to old. Spec prog: Gospel, rel. ♦Paul F. Lotters, pres & gen mgr; Steven L. Klob, opns dir & dev dir.

WOFX(AM)— Apr 15, 1940: 980 khz; 5 kw-U, DA-N. TL: N42 46 56 W73 50 07. Hrs open: River Hill Ctr., 1203 Troy Schenectady Rd., Suite 201, Latham, 12110. Phone: (518) 452-4800. Fax: (518) 452-4832. Web Site: www.wofx.com. Licensee: Capstar TX L.P. Group owner: Clear Channel Communications Inc. (acq 8-30-00; grpsl). Population served: 1,300,000 Natl. Rep: Clear Channel. Format: Sports talk. Target aud: 18-49. ♦Dennis Lamme, exec VP, gen mgr & chief of engrg; John Cooper, stn mgr; Greg Foster, opns mgr.

WPYX(FM)—See Albany

***WRPI(FM)**— Nov 1, 1957: 91.5 mhz; 10 kw. 450 ft TL: N42 41 14 W73 42 22. Stereo. Hrs opn: 6 AM-2 AM One WRPI Plaza, 12180. Phone: (518) 276-6248. Fax: (518) 276-2360. Web Site: www.wrpi.org. Licensee: Rensselaer Polytechnic Institute. Population served: 1,500,725 Format: Div. News: 5 hrs wkly. Target aud: General; open minded, educated listeners. ♦John Corbett, pres; Colin Fredericks, gen mgr.

Trumansburg

WPIE(AM)— Jan 15, 1990: 1160 khz; 5 kw-D, 31 w-N, DA-2. TL: N42 32 42 W76 42 39. Hrs open: 24 1705 Lake St., Elmira, 14901. Phone: (607) 733-5626. Fax: (607) 733-5627. E-mail: ppinesmedia1@stny.rr.com Web Site: wpieradio.com. Licensee: Pembrook Pines Ithaca Ltd. Group owner: Pembrook Pines Media Group (acq 3-3-93; $150,000; FTR: 3-22-93). Population served: 250,000 Law Firm: Bechtel & Cole. Format: Sports. News staff: one; News: 18 hrs wkly. Target aud: 25-54; mature, upscale adults. ♦Bob Michaels, opns dir; Robert J. Pfuntner, pres, gen mgr & dev dir.

Tupper Lake

WRGR(FM)— Feb 29, 1980: 102.3 mhz; 150 w. 1,446 ft TL: N44 09 35 W74 28 34. Stereo. Hrs opn: 24
Rebroadcasts WLPW(FM) Lake Placid 100%.
Box 211, Saranac Lake, 12983-0211. Phone: (518) 891-1544. Fax: (518) 891-1545. E-mail: sales@wnbz.com Web Site: www.wnbz.com. Licensee: Radio Lake Placid Inc. Group owner: Mountain Communications (acq 2003; grpsl). Population served: 20,000 Natl. Network: ABC. Wire Svc: UPI Format: Classic rock, adult contemp. News staff: 2; News: 2 hrs wkly. Target aud: 25-54; men. Spec prog: Relg one hr, big band 2 hrs wkly. ♦Ted Morgan, pres & gen mgr.

Utica

WFRG-FM— Oct 10, 1948: 104.3 mhz; 100 kw. Ant 500 ft TL: N43 03 27 W75 25 04. Stereo. Hrs opn: 9418 River Rd., Marcy, 13403. Phone: (315) 768-9500. Fax: (315) 736-3311. Web Site: www.bigfrog104.com. Licensee: Regent Licensee of Utica/Rome Inc. (acq 11-5-99; grpsl). Population served: 264,900 Format: Country. Target aud: 25-54. ♦Mary Jo Beach, gen mgr.

WIBX(AM)— Dec 5, 1925: 950 khz; 5 kw-U, DA-1. TL: N43 06 16 W75 20 20. Hrs open: 24 9418 State, Rt. 49, Marcy, 13403. Phone: (315) 768-9500. Fax: (315) 736-0720. Web Site: www.wibx950.com. Licensee: Regent Licensee of Utica/Rome Inc. Group owner: Regent Communications Inc. (acq 2-11-00; grpsl). Population served: 91,611 Natl. Network: CBS. Format: News/talk, sports. News staff: 5. Target aud: 35-64; middle to upper income adults. Spec prog: Pol 3 hrs, farm 14 hrs wkly. ♦Tom Jacobson, gen mgr & opns dir.

WLZW(FM)—Co-owned with WIBX(AM). Jan 1, 1972: 98.7 mhz; 25 kw. 660 ft TL: N43 08 39 W75 10 45. Stereo. 24 Web Site: www.lite987.com. Format: Lite adult contemp. News staff: 6. Target aud: 25-54; middle to upper income & educ levels. ♦Peter Naughton, progmg dir.

WIXT(AM)—See Little Falls

WKLL(FM)—Frankfort, Feb 12, 1990: 94.9 mhz; 50 kw. 276 ft TL: N43 03 26 W75 07 24. (CP: 34 kw, ant 567 ft.). Hrs opn: 39 Kellogg Rd., New Hartford, 13413. Phone: (315) 797-1330. Fax: (315) 738-1073. Web Site: www.krock.com. Licensee: Galaxy Communications L.P. Group owner: Route 81 Radio LLC (acq 4-6-2000; grpsl). Natl. Network: ABC. Natl. Rep: D & R Radio. Format: Modern rock. ♦Mimi Griswald, gen mgr.

***WKVU(FM)**— July 11, 1994: 100.7 mhz; 1.2 w. 551 ft TL: N43 09 12 W75 09 32. Hrs open: 24 1017 Higby Rd., New Hartford, 13413. Phone: (315) 793-1007. Fax: (315) 793-1044. Web Site: www.klove.com. Licensee: Educational Media Foundation. Group owner: EMF Broadcasting (acq 6-7-01; $1.25 million). Natl. Network: K-Love. Format: Christian music. Target aud: 25-45. ♦Bob Cain, chief of opns.

WOUR(FM)—Listing follows WUTQ(AM).

***WPNR-FM**— November 1977: 90.7 mhz; 450 w. 30 ft TL: N43 05 35 W75 16 21. Stereo. Hrs opn: c/o Utica College, 1600 Burrstone Rd., 13502. Phone: (315) 792-3066. Fax: (315) 792-3292. Licensee: Utica College. (acq 9-12-96). Format: Div, urban contemp, AOR. Spec prog: Class 10 hrs, jazz 14 hrs, reggae 5 hrs wkly. ♦Todd Hutton, pres.

WRCK(FM)—Listing follows WTLB(AM).

WRNY(AM)—See Rome

***WRUN(AM)**— Apr 24, 1948: 1150 khz; 5 kw-D, 1 kw-N, DA-2. TL: N43 10 31 W75 21 03. Hrs open:
WAMC-FM, Albany,NY.
318 Central Ave., Albany, 12206. Phone: (518) 465-5233. Fax: (518) 432-6974. Web Site: www.wamc.org. Licensee: WAMC. Group owner: Regent Communications Inc. (acq 7-6-2005; $275,000). Population served: 91,611 Natl. Network: NPR, PRI. Law Firm: Dow, Lohnes, & Albertson. Wire Svc: AP Format: News/talk. ♦Alan S. Chartock, CEO; David Galletly, VP; Selma Kaplan, VP.

***WRVN(FM)**— Jun 4, 1986: 91.9 mhz; 1.9 kw. Ant -62 ft TL: N43 08 31 W75 13 36. Stereo. Hrs opn:
Rebroadcasts WRVO(FM) Oswego 100%.
7060 State Rt. 104, Oswego, 13126. Phone: (315) 312-3690. Fax: (315) 312-3174. E-mail: wrvo@wrvo.fm Web Site: www.wrvo.fm. Licensee: State University of New York. Population served: 216,000 Natl. Network: NPR, PRI. Wire Svc: AP Format: News/talk, old time radio. News staff: 3; News: 140 hrs wkly. ♦John E. Krauss, gen mgr; Thomas Herbert, dev dir; Fred Vigeant, progmg dir; Chris Ulanowski, news dir; Jeff Windsor, chief of engrg.

WTLB(AM)— 1946: 1310 khz; 5 kw-D, 500 w-N, DA-2. TL: N43 03 24 W75 16 42. Hrs open: 24 39 Kellegg Rd., New Hartford, 13413. Phone: (315) 797-1330. Fax: (315) 738-1073. Licensee: Galaxy Communications L.P. Group owner: Route 81 Radio LLC (acq 4-6-2000; grpsl). Population served: 25,000 Format: Btfl mus, MOR. News: one hr wkly. Target aud: 55 plus. ♦Ed Levine, pres; Jason Passante, gen mgr; Dave Doughty, chief of engrg.

WRCK(FM)—Co-owned with WTLB(AM). Apr 23, 1962: 107.3 mhz; 50 kw. 499 ft TL: N43 08 40 W75 10 32. Stereo. Web Site: www.wrck.com.8,500 Format: Classic rock.

WUMX(FM)—See Rome

***WUNY(FM)**— Oct 30, 1985: 89.5 mhz; 6.3 kw. 777 ft TL: N43 08 38 W75 10 40. Stereo. Hrs opn: 5 AM-midnight Box 2400, 506 Old Liverpool Rd., Syracuse, 13220-2400. Phone: (315) 453-2424. Fax: (315) 451-8824. E-mail: wcny—online@wcny.org Web Site: www.wcny.org. Licensee: Public Broadcasting Council of Central New York. Natl. Network: NPR. Law Firm: Haley, Bader & Potts. Format: Class. Target aud: General. Spec prog: Bluegrass 3 hrs, jazz 7 hrs wkly. ♦Colleen Edwards, CFO; Peter Hirsch, mktg dir.

WUTQ(AM)— Jan 29, 1962: 1550 khz; 1 kw-D. TL: N43 06 48 W75 15 25. Stereo. Hrs opn: 24
Rebroadcasts WRNY(AM) Rome 100%.
Mayro Bldg., 239 Genesee St., 13501. Phone: (315) 797-0803. Fax: (315) 797-7813. Licensee: CC Licenses LLC. Group owner: Clear Channel Communications Inc. (acq 8-5-98; grpsl). Population served: 91,611 Natl. Network: Westwood One. Natl. Rep: Christal. Law Firm: Latham & Watkins. Format: Sports. Target aud: 35 plus; 60% female, 40% male. Spec prog: It 2 hrs, Pol 4 hrs wkly. ♦Brian Delaney, gen mgr; Gene Conte, progmg dir & traf mgr; Jack Moran, news dir; Joe Petro, chief of engrg.

WOUR(FM)—Co-owned with WUTQ(AM). June 1967: 96.9 mhz; 16 kw. 790 ft TL: N43 08 46 W75 10 40. Stereo. Web Site: www.wour.com.91,611 Format: Rock/AOR. Target aud: 25-49; adults. ♦Jerry Kraus, prom mgr & disc jockey; Stew Schantz, progmg dir & mus dir; Alison Ryan, disc jockey; J.P. Hastings, disc jockey.

Valhalla

***WARY(FM)**— Oct 3, 1973: 88.1 mhz; 171 w. 403 ft TL: N41 04 13 W73 47 25. Hrs open: 10 AM-10 PM (M-F) 75 Grasslands Rd., 10595. Phone: (914) 606-6752. Phone: (914) 606-6753. Fax: (914) 606-6260. E-mail: radprime1@aol.com Licensee: Westchester Community College. Population served: 500,000 Law Firm: Garvey, Schubert & Barer. Format: AOR. Target aud: 12-24. Spec prog: Pub service 10 hrs wkly. ♦Radames Ocasio, gen mgr.

Vestal

WMXW(FM)— June 2, 1989: 103.3 mhz; 6 kw. 1,014 ft TL: N42 03 22 W75 56 39. (CP: 592 w). Stereo. Hrs opn: 24 320 N. Jensen Rd., 13850-2111. Phone: (607) 584-5800. Fax: (607) 584-5900. Licensee: CC Licenses LLC. Group owner: Clear Channel Communications Inc. (acq 9-2000; grpsl). Natl. Rep: Katz Radio. Format: Spectrum adult contemp. News staff: one. Target aud: 25-54. ♦Dave Lozzi, progmg dir & news dir.

Voorheesville

WAJZ(FM)— May 24, 1991: 96.3 mhz; 6 kw. 1,118 ft TL: N42 37 01 W74 00 46. Stereo. Hrs opn: 24 6 Johnson Rd., Latham, 12110. Phone: (518) 786-6600. Phone: (518) 786-6620. Fax: (518) 786-6610. Web Site: www.jamz963.com. Licensee: 6 Johnson Road Licenses Inc. Group owner: Pamal Broadcasting Ltd. (acq 10-19-2001; grpsl). Population served: 73,200 Natl. Rep: McGavren Guild. Format: Rhymthic CHR. News staff: 2; News: 3 hrs wkly. Target aud: 18-49. Spec prog: Relg one hr wkly. ♦Dan Austin, gen mgr; Suzette Anthony, sls VP; Stephen Roberts, gen sls mgr; J.D. Reoman, mktg dir; Ruben Pimentel, prom dir; Rob Torres, progmg dir.

Walton

WDLA(AM)— May 30, 1951: 1270 khz; 5 kw-D, 100 w-N. TL: N42 08 08 W75 04 52. Hrs open: Box 58, Rt. 206, 13856. Phone: (607) 865-4321. Licensee: Double O Central New York Corp. (group owner; (acq 10-22-2004; grpsl). Population served: 44,000 Natl. Network: Jones Radio Networks. Format: Music of your life. Target aud: 28-55; general. Spec prog: Farm 2 hrs wkly. ♦Jim Johnson, gen mgr; Skip Barlow, progmg dir.

WDLA-FM— Nov 16, 1973: 92.1 mhz; 690 w. 656 ft TL: N42 08 10 W75 04 48.

Warrensburg

WKBE(FM)— 1990: 100.3 mhz; 1.45 kw. 1,312 ft TL: N43 25 12 W73 45 39. Hrs open: 24 6 Johnson Rd., Latham, 12110. Phone: (518) 786-6600. Fax: (518) 786-6610. Licensee: 6 Johnson Road Licenses Inc. Group owner: Pamal Broadcasting Ltd. (acq 10-9-2001). Format: CHR. News staff: one; News: 10 hrs wkly. Target aud: 18-34; women. ♦Dan Austin, gen mgr.

Warsaw

WCJW(AM)— May 16, 1973: 1140 khz; 2.5 kw-D, DA. TL: N42 43 35 W78 06 47. Hrs open: Sunrise-sunset Box 251, 3258 Merchant Rd., 14569. Phone: (585) 786-8131. Fax: (585) 786-2241. E-mail: wcjw@wcjw.com Web Site: www.wcjw.com. Licensee: Lloyd Lane Inc. (acq 9-1-84). Population served: 750,000 Natl. Network: USA. Rgnl rep: Regional Reps Format: Country. News staff: one; News: 20 hrs wkly. Target aud: 25-54; adults. Spec prog: Farm 11 hrs wkly. ♦Lloyd Lane, pres & gen mgr; Lee Richey, progmg dir; Jenny Snow, news dir.

***WCOU(FM)**— Dec 14, 1992: 88.3 mhz; 11 kw. Ant 535 ft TL: N42 49 36 W78 12 25. Stereo. Hrs opn:
Rebroadcasts WCIK(FM) Bath 100%.
Box 506, Bath, 14810. Secondary address: 7634 Campbell Creek Rd., Bath 14810. Phone: (607) 776-4151. Fax: (607) 776-6929. E-mail: mail@fln.org Web Site: www.fln.org. Licensee: Family Life Ministries Inc. Group owner: Family Life Network Natl. Network: Salem Radio Network. Law Firm: Hardy, Carey, Chautin & Balkin, LLP. Wire Svc: Metro Weather Service Inc. Format: Contemp Christian. News staff: 3; News: 14 hrs wkly. Target aud: 30-54; general. ♦Dick Snavely, CFO; Rick Snavely, pres, VP & gen mgr; John Owens, progmg dir; Jim Travis, chief of engrg.

Warwick

WTBQ(AM)— July 24, 1969: 1110 khz; 500 w-D (non-directional). TL: N41 16 51 W74 21 46. Hrs open: Sunrise-sunset 62 N. Main St., Florida, 10921. Phone: (845) 651-1110. Fax: (845) 651-1025. E-mail: am1110@magiccarpet.com Web Site: www.wtbq.com. Licensee: FST Broadcasting Corp. (acq 7-94; $150,000). Population served: 500,000 Natl. Network: ABC, Jones Radio Networks. Law Firm: William D. Silva. Format: Oldies, talk. News staff: 2; News: 10 hrs wkly. Target aud: 24-55; affluent Orange County-New York City commuters. Spec prog: Folk, Pol, relg, farm, Spanish. ♦Frank Truatt, pres, gen mgr & opns mgr; Rob McLean, gen sls mgr; Logan Moscovitz, progmg dir; Rich Ball, progmg dir.

Waterloo

WNYR-FM— Apr 19, 1989: 98.5 mhz; 3.2 kw. 446 ft TL: N42 48 22 W76 50 47. Stereo. Hrs opn: 24 3568 Lenox Rd., Geneva, 14456. Phone: (315) 781-7000. Fax: (315) 781-7700. Web Site: www.fingerlakes.com. Licensee: Lake Country Broadcasting. Population served: 325,000 Law Firm: James L. Oyster. Format: Adult contemp. News staff: one; News: 5 hrs wkly. Target aud: 25-54. ♦George Kimble, pres; Alan Bishop, VP & gen mgr; Mike Smith, opns mgr.

Watertown

WATN(AM)— Feb 3, 1941: 1240 khz; 1 kw-U. TL: N43 58 49 W75 56 12. Hrs open: 24 199 Wealtha Ave., 13601. Phone: (315) 782-1240. Fax: (315) 782-0312. Licensee: Community Broadcasters LLC. (group owner; (acq 10-13-2006; grpsl). Population served: 120,000 Natl. Rep: Roslin. Format: Talk. ♦James L. Leven, pres; David W. Mance, gen mgr; Todd Dalesandro, opns dir.

WCIZ-FM—Listing follows WTNY(AM).

WFRY-FM—Listing follows WNER(AM).

***WJNY(FM)**— July 24, 1986: 90.9 mhz; 7.09 kw. 449 ft TL: N43 51 44 W75 43 40. Stereo. Hrs opn: 5 AM-midnight
Rebroadcasts WCNY-FM Syracuse.
Box 2400, Syracuse, 13220-2400. Secondary address: 506 Old Liverpool Pl., Syracuse 13220. Phone: (315) 453-2424. Fax: (315) 451-8824. E-mail: wcny-online@wcny.org Web Site: www.wcny.org. Licensee: Public Broadcasting Council of Central New York Inc. Natl. Network: NPR. Law Firm: Haley, Bader & Potts. Format: Class. Spec prog: Bluegrass 3 hrs, jazz 5 hrs wkly. ♦Coleen Edwards, CEO & CFO; Peter Hirsch, mktg dir.

***WKWV(FM)**— June 26, 2000: 90.1 mhz; 400 w. Ant 679 ft TL: N43 57 15 W75 43 45. Stereo. Hrs opn: 24
Rebroadcasts KLVR(FM) Santa Rosa, CA 100%.
2351 Sunset Blvd., Suite 170-218, Rocklin, CA, 95765. Phone: (916) 251-1600. Fax: (916) 251-1650. Web Site: www.klove.com. Licensee: Educational Media Foundation. (acq 1-13-2006; $300,000). Population served: 130,000 Natl. Network: K-Love. Format: Contemp Christian. ♦Richard Jenkins, pres; Mike Novak, VP; Keith Whipple, dev dir; David Pierce, progmg mgr; Ed Lenane, news dir; Sam Wallington, engrg dir; Karen Johnson, news rptr; Marya Morgan, news rptr; Richard Hunt, news rptr.

WNER(AM)— Nov 2, 1959: 1410 khz; 3.5 kw-D, 58 w-N. TL: N43 56 47 W75 56 52. Hrs open: 134 Mullin St., 13601. Phone: (315) 788-0790. Fax: (315) 788-4379. Licensee: Regent Licensee of Watertown Inc. Group owner: Regent Communications Inc. (acq 11-5-99; grpsl). Population served: 30,787 Natl. Network: ESPN Radio. Format: Sports. Target aud: 35-64. ♦Don Wagner, pres & CFO; Lance Thomas, progmg dir.

WFRY-FM—Co-owned with WNER(AM). Nov 22, 1968: 97.5 mhz; 100 kw. 285 ft TL: N43 57 23 W75 50 45. Stereo. Web Site: www.froggy97.com. Format: Country. ♦Matt Raisman, progmg dir; Annie Croakly, disc jockey; James Pond, disc jockey; Web Foote, disc jockey.

***WRVJ(FM)**— July 1, 1989: 91.7 mhz; 1.6 kw. Ant 443 ft TL: N43 51 44 W75 43 40. Stereo. Hrs opn: 24
Rebroadcasts WRVO(FM) Oswego 100%.
7060 State Rt. 104, Oswego, 13126. Phone: (315) 312-3690. Fax: (315) 312-3174. E-mail: wrvo@wrvo.fm Web Site: wrvd.fm. Licensee: State University of New York. Population served: 71,650 Natl. Network: NPR, PRI. Wire Svc: AP Format: News/talk, old time radio. News staff: 4; News: 140 hrs wkly. Target aud: 25-55. ♦John E. Krauss, gen mgr; Matt Seubert, dev dir; Fred Vigeant, progmg dir; Chris Ulanowski, news dir; Jeff Windsor, chief of engrg.

***WSLJ(FM)**— 1992: 88.9 mhz; 200 w. 454 ft TL: N43 57 23 W75 50 28. Hrs open: 24
Rebroadcasts WSLU(FM) Canton 100%.

St. Lawrence Univ., Canton, 13617. Phone: (315) 229-5356. Fax: (315) 229-5373. Web Site: www.ncpr.org. Licensee: St. Lawrence University. Law Firm: Donald E. Martin. Format: Eclectic public radio. News staff: 2; News: 35 hrs wkly. Target aud: General. ♦Ellen Rocco, gen mgr; Shelly Pike, chief of opns; Sandra Demarest, dev dir.

WTNY(AM)— Apr 29, 1941: 790 khz; 1 kw-U, DA-N. TL: N43 56 44 W75 56 54. Stereo. Hrs opn: 24 134 Mullin St., 13601. Phone: (315) 788-0790. Fax: (315) 788-4379. Web Site: www.production@790wtny.com. Licensee: Regent Licensee of Watertown Inc. Group owner: Regent Communications Inc. (acq 11-5-99; grpsl). Population served: 50,000 Natl. Network: CBS. Law Firm: Dow, Lohnes & Albertson. Format: News. News staff: 3; News: 20 hrs wkly. Target aud: 25 plus. Spec prog: Farm 3 hrs wkly. ♦Don Wagner, CFO & gen mgr; Lance Thomas, progmg dir.

WCIZ-FM—Co-owned with WTNY(AM). Aug 25, 1986: 93.3 mhz; 6 kw. 328 ft TL: N43 57 23 W75 50 45. Stereo. 24 Web Site: www.production@790wtny.com. Format: Classic hits. News staff: 3; News: 2 hrs wkly. Target aud: 35-54.

Watkins Glen

WNGZ(FM)—See Montour Falls

WTYX(AM)— June 22, 1968: 1490 khz; 400 w-U. TL: N42 21 11 W76 52 13. Hrs open: 2205 College Ave., Elmira, 14903. Secondary address: 1685 Four Mile Dr., Williamsport, PA 17701. Phone: (607) 732-4400. Fax: (607) 732-7774. Licensee: Chemung County Radio Inc. Group owner: Backyard Broadcasting LLC (acq 12-1-2002; grpsl). Population served: 16,700 Natl. Rep: D & R Radio. Format: Country. Target aud: 20-49; baby boomers. ♦Kevin White, gen mgr.

Waverly

WATS(AM)—See Sayre, PA

WAVR(FM)— October 1974: 102.1 mhz; 4.1 kw. Ant 400 ft TL: N42 03 48 W76 31 28. Stereo. Hrs opn:
Rebroadcasts WATS(AM) Sayre 100%.
204 Desmond St., Sayre, PA, 18840. Phone: (570) 888-7745. Fax: (570) 888-9005. Web Site: www.cyber-quest.com/watsvavr. Licensee: Wats Broadcasting Inc. (acq 10-28-86). Format: Adult contemp. Target aud: 25-54; upscale bedroom community. ♦Charles C. Carver, pres & gen mgr; Meade T. Murtland, stn mgr.

Webster

***WFRW(FM)**— October 1988: 88.1 mhz; 8.5 kw. 337 ft TL: N43 04 18 W77 05 35. Stereo. Hrs opn: 24 918 Chesapeake Ave., Annapolis, 21403. Phone: (315) 331-7482. Fax: (410) 268-0931. Web Site: www.familyradio.com. Licensee: Family Stations Inc. (group owner) Population served: 1,000,000 Format: Relg, educ. News: 4 hrs wkly. Target aud: General. Spec prog: Class 2 hrs wkly. ♦Harold Camping, pres & gen mgr.

***WMHN(FM)**— Feb 29, 1988: 89.3 mhz; 1 kw. 75 ft TL: N43 13 45 W77 26 52. Stereo. Hrs opn: 24
Rebroadcasts WMHR(FM) Syracuse 99%.
4044 Makyes Rd., Syracuse, 13215. Phone: (315) 469-5051. E-mail: mhn@marshillnetwork.org Web Site: www.marshillnetwork.org. Licensee: Mars Hill Broadcasting Co. Inc. (group owner) Population served: 1,000,000 Natl. Network: Moody, Salem Radio Network. Law Firm: Wiley, Rein & Fielding. Wire Svc: AP Format: Christian. News: 13 hrs wkly. Christian families. Spec prog: Children 11 hrs wkly. ♦Clayton Roberts, pres; Wayne Taylor, gen mgr & opns mgr; Chris Tetta, progmg dir; Jeremy Miller, news dir; Valerie Smith, traf mgr.

WRCI(FM)— Feb 15, 1993: 102.7 mhz; 6 kw. Ant 328 ft TL: N43 10 14 W77 40 23. Hrs open: 24 2494 Browncroft Blvd., Rochester, 14625. Phone: (585) 264-1027. Fax: (585) 264-1165. Web Site: www.wrcifm.com. Licensee: Kimtron Inc. Group owner: Crawford Broadcasting Co. (acq 11-25-92; $950,000; FTR: 12-21-92). Format: Christian, talk. ♦Robert Hammond, gen mgr; Ben Martin, opns mgr; Mark Shuttleworth, progmg dir; Brian Cunningham, chief of engrg.

Wellsville

WJQZ(FM)—Listing follows WLSV(AM).

WLSV(AM)— Oct 31, 1955: 790 khz; 1 kw-D, 41 w-N. TL: N42 04 37 W77 55 47. Hrs open: 82 Railroad Ave., 14895. Phone: (585) 593-6070. Fax: (585) 593-6212. E-mail: oldiesz103@yahoo.com Licensee: DBM Communications Inc. (acq 8-21-98; $850,000 with co-located FM). Population served: 336,600 Law Firm: Baraff, Koerner & Olender. Format: Country. Target aud: General. ♦Richard Mangels, pres; Bob Mangels, news dir.

WJQZ(FM)—Co-owned with WLSV(AM). Feb 3, 1986: 103.5 mhz; 3 kw. 466 ft TL: N42 09 26 W77 55 26. Stereo. Format: Oldies. Target aud: 25-54. ♦Robert Mangels, progmg dir.

WQRW(FM)— Feb 14, 2007: 93.5 mhz; 1.1 kw. Ant 768 ft TL: N42 11 25 W77 49 17. Hrs open: 74 Main St., Hornell, 14843. Phone: (607) 281-1935. Fax: (607) 281-1936. Licensee: Pembrook Pines Mass Media N.A. Corp. Format: Bright adult contemp. ♦Robert J. Pfuntner, pres.

Westhampton

WBZB(FM)— Nov 18, 1993: 98.5 mhz; 950 w. Ant 525 ft TL: N40 51 18 W72 46 11. Stereo. Hrs opn: 3075 Vets Hwy., Ronkonkoma, 11779. Phone: (631) 648-2500. Fax: (631) 648-2550. Web Site: www.businesstalkny.com. Licensee: Jarad Broadcasting Co. of Westhampton Inc. Group owner: The Morey Organization Inc. Natl. Rep: Christal. Format: Business talk. ♦John Carraciolo, pres & chief of engrg.

Westhampton Beach

WRCN-FM—See Riverhead

Westport

WCLX(FM)— January 1995: 102.9 mhz; 6 kw. Ant 312 ft TL: N44 13 15 W73 24 41. Stereo. Hrs opn: 24 Westport Broadcasting, 19 Boas Ln., Wilton, CT, 06897-1031. Phone: (203) 762-9425. Fax: (509) 752-4105. E-mail: dj@broadcast.net Web Site: www.wclxfm.com. Licensee: Westport Broadcasting. Population served: 150,000 Law Firm: Cohn & Marks. Format: Progressive rock. Target aud: 25-54. ♦Dennis Jackson, CEO; Russ Kinsley, stn mgr; Diane Desmond, progmg dir.

Wethersfield Township

WLKK(FM)— 1948: 107.7 mhz; 19.5 kw. Ant 800 ft TL: N42 37 23 W78 17 16. Stereo. Hrs opn: 24 500 Corporate Pkwy., Suite 200, Buffalo, 14226. Phone: (716) 843-0600. Fax: (716) 832-3323. Web Site: www.1077thelake.com. Licensee: Entercom Buffalo License LLC. Group owner: Entercom Communications Corp. (acq 5-5-2004; $9 million). Population served: 1,200,000 Natl. Network: Westwood One. Format: Progsv, classic rock. Target aud: 25-54; adults. ♦Greg Ried, gen mgr; Jeff Surdej, prom mgr; Hank Dole, progmg dir.

White Plains

WFAS(AM)— Aug 11, 1932: 1230 khz; 1 kw-U. TL: N41 01 32 W73 49 39. Hrs open: 365 Secor Rd., Hartsdale, 10530. Phone: (914) 693-2400. Fax: (914) 693-0000. Web Site: www.wfasam.com. Licensee: Cumulus Licensing Corp. Group owner: Cumulus Media Inc. (acq 1-23-2002; grpsl). Population served: 785,500 Natl. Network: AP Radio. Natl. Rep: McGavren Guild. Format: MOR. News staff: 2; News: 5 hrs wkly. Target aud: General. Spec prog: Sports progmg 8 hrs wkly. ♦Rod Colarco, gen mgr; Dave Ashton, opns mgr; Bob Barnum, progmg dir; Val Cichorek, traf mgr.

WFAS-FM— Sept 1, 1947: 103.9 mhz; 600 w. Ant 667 ft TL: N41 01 32 W73 49 39. Stereo. E-mail: music@wfasfm.com Web Site: www.wfasfm.com. Format: Adult contemp. News: one hr wkly. Target aud: 25-54. ♦Robert Bongiardino, gen sls mgr; Misty Wien, prom dir; Dave Ashton, progmg dir; Pam Puso, news dir; Joan Franzino, pub affrs dir; Valencia Cichorek, traf mgr; Jim Killfield, news rptr.

WXPK(FM)—See Briarcliff Manor

Whitehall

WNYV(FM)— July 14, 1990: 94.1 mhz; 3 kw. 328 ft TL: N43 28 37 W73 26 56. Stereo. Hrs opn: 5:30 AM-midnight Box 141, 12887. Secondary address: Box 568, East Poultney, VT 05741. Phone: (802) 287-9031. Licensee: Pine Tree Broadcasting. Natl. Rep: Commercial Media Sales. Format: Adult contemp, country, oldies. News staff: one; News: 3 hrs wkly. Target aud: 25-55; active community oriented, working professional & families. Spec prog: Big band 3 hrs, pub affrs 5 hrs, relg 2 hrs, Pol one hr wkly. ♦Michael Leech, pres; Judith E. Leech, VP & gen mgr.

Whitesboro

WSKS(FM)— 1994: 97.9 mhz; 1.5 kw. 669 ft TL: N43 02 14 W75 26 40. Hrs open:
Rebroadcasts WOWB(FM) Little Falls 100%.
239 Genesee St., Suite 500, Utica, 13501. Phone: (315) 797-0803. Fax: (315) 797-7813. Licensee: Clear Channel Broadcasting Licenses Inc. Group owner: Clear Channel Communications Inc. (acq 3-12-01; $2.15 million with WSKU(FM) Little Falls). Format: CHR, rhythmic. ♦Brian Delaney, gen mgr; Jack Moran, opns mgr; Ken Morrison, prom dir.

Willsboro

WXZO(FM)— 1997: 96.7 mhz; 1 kw. Ant 797 ft TL: N44 24 12 W73 26 02. Hrs open: Box 1093, Burlington, 05403. Phone: (866) 696-7967. Web Site: www.wxzofm.com. Licensee: Capstar TX L.P. Group owner: Clear Channel Communications Inc. (acq 8-30-2000; grpsl). Format: Talk. ♦Karen Marshall, gen mgr; Steve Cormier, opns dir.

Windham

WRIP(FM)— Aug 5, 1999: 97.9 mhz; 580 w. Ant 1,056 ft TL: N42 17 06 W74 15 52. Stereo. Hrs opn: 24 134 South St., P.O. Box 979, 12496-0979. Phone: (518) 734-4747. Fax: (413) 375-4711. E-mail: wrip@mhcable.com Web Site: www.wripfm.com. Licensee: Rip Radio LLC. Population served: 400,000 Natl. Network: AP Network News. Law Firm: Cohn & Marks. Format: Adult contemp, full service, Christian contemp. Target aud: 25 plus; mass appeal. Spec prog: Jazz 3 hrs, Christian contemp 2 hrs wkly. ♦Dennis Jackson, CEO; Guy Patrick Garraghan, VP & gen mgr; Jay Fink, gen sls mgr & mus dir.

Windsor

WRRQ(FM)— 2006: 106.7 mhz; 680 w. Ant 643 ft TL: N42 03 10 W75 42 07. Hrs open: 24 495 Court St., 2nd Fl., Binghamton, 13904. Phone: (607) 772-1005. Fax: (607) 772-2945. Web Site: myq107.com. Licensee: Equinox Broadcasting Corp. Natl. Rep: Katz Radio. Format: Hot A/C. Target aud: 25-40. ♦George Hawras, pres.

Woodside

WWRL(AM)—See New York

Woodstock

WDST(FM)— Apr 29, 1980: 100.1 mhz; 2.9 kw. 308 ft TL: N41 59 04 W74 02 56. Stereo. Hrs opn: 24 Box 367, 12498. Secondary address: 293 Tinker St. 12498. Phone: (845) 679-7266. Fax: (845) 679-5395. E-mail: live@wdst.com Web Site: www.wdst.com. Licensee: CHET-5 Broadcasting L.P. (acq 2-12-93; $1.65 million with WKNY(AM) Kingston; FTR: 3-8-93) Population served: 300,000 Natl. Network: CBS Radio. Natl. Rep: Christal. Law Firm: Shaw Pittman. Format: Progsv adult rock. News staff: one; News: 5 hrs wkly. Target aud: 24-55; upscale professionals. ♦Gary H. Chetkof, chmn & pres; Greg Gattine, opns dir & progmg dir; Stan Beinstein, gen sls mgr; Ike Phillips, natl sls mgr; Jimmy Buff, progmg dir.

Wurtsboro

WZAD(FM)— Sept 1, 1990: 97.3 mhz; 620 w. Ant 718 ft TL: N41 36 04 W74 33 17. Stereo. Hrs opn: 24 Box 416, 2 Pendell Rd., Poughkeepsie, 12602. Phone: (845) 471-1500. Fax: (845) 454-1204. Licensee: Cumulus Licensing Corp. Group owner: Cumulus Media Inc. (acq 1-23-2002; grpsl). Population served: 450,000 Law Firm: Akin, Gump, Strauss, Hauer & Feld. Format: Oldies. News staff: one; News: 5 hrs wkly. Target aud: 25-54; upscale, educated. ♦Lew Dickie, pres; Charles Benfer, gen mgr.

Yonkers

WVIP(FM)—See New Rochelle

Youngstown

WTOR(AM)— May 6, 1998: 770 khz; 9 kw-D. TL: N43 13 05 W78 56 53. Hrs open: 21700 Northwestern Hwy., TWR 14, Suite 1190, Southfield, MI, 48075. Phone: (716) 754-9514. Fax: (716) 754-9515. Licensee: Birach Broadcasting Corp. (group owner; acq 1996; $409,000

less land cost for CP). Format: International mus. Target aud: Ethnic; Serbian, Lithuanian, Sp, Pol, Macedonian. ♦Sima Birach, CEO, gen mgr & opns mgr.

North Carolina

Aberdeen

WEEB(AM)—See Southern Pines

WFVL(FM)—See Southern Pines

WQNX(AM)— January 1982: 1350 khz; 2.5 kw-D, 28 w-N, DA-2. TL: N35 07 20 W79 24 57. Hrs open: Box 1350, 28315. Phone: (910) 944-1350. Fax: (910) 944-8182. E-mail: qtalk@pinehurst.net Web Site: www.wqnxtalk.com. Licensee: Golf Capital Broadcasting Inc. (acq 1987; $128,000; FTR: 4-20-87). Format: News/talk. ♦T.O. Calcutt, gen mgr.

Ahoskie

***WBKU(FM)**— 2002: 91.7 mhz; 87 kw. Ant 430 ft TL: N36 05 45 W77 12 30. Hrs open: Drawer 3206, Tupelo, MS, 38803. Phone: (662) 844-8888. Fax: (662) 842-6791. Licensee: American Family Association. Group owner: American Family Radio Format: Christian. ♦Marvin Sanders, gen mgr.

WQDK(FM)— Sept 2, 1968: 99.3 mhz; 3 kw. 300 ft TL: N36 16 46 W77 01 59. Stereo. Hrs opn: 24 332 Hwy. 42 W., 27910. Phone: (252) 332-7993. Fax: (252) 332-6887. Licensee: Max Radio of the Carolinas Licenses LLC. Group owner: MAX Media L.L.C. (acq 11-12-2002; grpsl). Population served: 23,616 Rgnl. Network: Agri-Net, Ray Sports Radio Net. Format: Country. Spec prog: Farm 7 hrs wkly. ♦Don Upchurch, gen mgr & opns mgr.

WRCS(AM)— Apr 25, 1948: 970 khz; 1 kw-D. TL: N36 16 46 W77 01 59. Hrs open: 24 443 North Carolina Hwy 42 W, 27910. Phone: (252) 332-3101. Fax: (252) 332-3103. E-mail: wrcs@gate811.net Licensee: WRCS-AM 970 Inc. (acq 6-14-02). Population served: 23,616 Format: Gospel. Target aud: 12-70. ♦J. C. Watford, gen mgr & opns mgr.

Albemarle

WPZS(FM)— February 1958: 100.9 mhz; 3 kw. Ant 200 ft TL: N35 22 40 W80 11 38. (CP: COL Indian Trail. 6 kw, ant 328 ft. TL: N35 07 29 W80 43 30). Stereo. Hrs opn: 24 2303 W. Morehead St., Charlotte, 28208. Phone: (704) 358-0211. Fax: (704) 358-3752. Licensee: Radio One of North Carolina LLC. (group owner; (acq 11-12-2004; $11.5 million). Population served: 300,000 Format: Inspirational. ♦Debbie Kwei, gen mgr.

WSPC(AM)— July 1947: 1010 khz; 1 kw-D, 64 w-N. TL: N35 22 40 W80 11 38. Hrs open: 24 Box 549, 28002-0549. Secondary address: 1234 Magnolia St. 28001. Phone: (704) 983-1580. Fax: (704) 983-1436. E-mail: wspc@ctc.net Web Site: 1010wspc.com. Licensee: Stanly Communications Inc. (acq 2-5-2004; $600,000 with WZKY(AM) Albemarle). Population served: 150,000 Law Firm: Brooks, Pierce, McLendon, Humphrey & Leonard. Format: News/talk. News staff: one. Target aud: General. ♦Matt Smith, gen mgr & opns VP.

WZKY(AM)— July 9, 1956: 1580 khz; 1 kw-D, 12 w-N. TL: N35 21 38 W80 10 39. Stereo. Hrs opn: 24 Box 549, 28002-0549. Phone: (704) 983-1580. Fax: (704) 983-1436. E-mail: wspc@ctc.net Web Site: www.1010wspc.com. Licensee: Stanly Communications Inc. (acq 2-5-2004; $600,000 with WSPC(AM) Albemarle). Population served: 55,000 Rgnl. Network: N.C. News Net. Format: Oldies. Target aud: 30 plus. ♦Matt Smith, pres, sr VP, gen mgr & opns mgr; Sherri Smith, VP.

Asheboro

WKRR(FM)— November 1948: 92.3 mhz; 100 kw. 1,275 ft TL: N35 22 40 W80 11 38. Stereo. Hrs opn: 24 192 E. Lewis St., Greensboro, 27408. Phone: (336) 274-8042. Fax: (336) 274-1629. Web Site: www.rock92.com. Licensee: Dick Broadcasting Co. Inc. of Tennessee (acq 4-84). Format: Classic rock. Target aud: 18-49. ♦Bruce Wheeler, gen mgr; James Kerr, opns mgr.

WKXR(AM)— May 24, 1947: 1260 khz; 5 kw-D, 500 w-N, DA-2. TL: N35 43 26 W79 48 21. Hrs open: 24 1119 Eastview Dr., 27203. Phone: (336) 625-2187/ 625-1260. E-mail: wkxr@atomic.net Web Site: www.wkxr.com. Licensee: Randolph Broadcasting Inc. (acq 8-4-86; $500,000; FTR: 7-7-86). Population served: 100,000 Natl. Network: AP Network News, Jones Radio Networks. Rgnl. Network: N.C. News Net. Format: Country. News: 8 hrs wkly. Target aud: 18 plus. Spec prog: Farm one hr, gospel 10 hrs wkly. ♦Edward F. Swicegood II, pres & gen mgr; Ted Swicegood, opns mgr; Larry Reed, gen sls mgr & progmg dir; Larry Reid, mktg dir & chief of engrg.

***WTJY(FM)**— June 30, 1999: 89.5 mhz; 10 kw. 544 ft TL: N35 36 55 W79 53 28. Hrs open: 24
Rebroadcasts WXRI(FM) Winston-Salem 100%.
Box 25775, Winston-Salem, 27114. Phone: (336) 788-1155. Fax: (336) 788-7199. E-mail: joyfm@bellsouth.net Web Site: www.joyfm.org. Licensee: Positive Alternative Radio Inc. Group owner: Baker Family Stations (Positive Radio Group) Law Firm: Booth, Freret, Imlay & Tepper. Format: Southern gospel. ♦Rodney Baucom, gen mgr.

WZOO(AM)— May 3, 1971: 710 khz; 1 kw-D, DA. TL: N35 45 50 W79 50 04. Hrs open: Box 460, 27204. Phone: (336) 672-0944. Licensee: Faith Enterprises Inc. (acq 11-15-86). Format: Southern gospel. ♦Huey Turner, gen mgr.

Asheville

***WCQS(FM)**— 1975: 88.1 mhz; 1.6 kw. 1,168 ft TL: N35 35 23 W82 40 26. Stereo. Hrs opn: 24 73 Broadway, 28801. Phone: (828) 253-6875. Fax: (828) 253-6700. Web Site: www.wcqs.org. Licensee: Western N.C. Public Radio Inc. (acq 1984). Population served: 500,000 Natl. Network: NPR, PRI. Law Firm: Cohn & Marks. Format: Class, jazz, news. News staff: one; News: 35 hrs wkly. Target aud: 25 plus. Spec prog: Folk 9 hrs wkly. ♦Ed Subkis, gen mgr; Margaret Marchur, sls dir; Barbara Sayer, progmg dir; David Hurand, news dir; Richard J. Kowal, news dir; Tom Spaight, chief of engrg.

WFGW(AM)—See Black Mountain

WISE(AM)— 1939: 1310 khz; 5 kw-D, 1 kw-N, DA-N. TL: N35 37 09 W82 34 21. Stereo. Hrs opn: 24 1190 Patton Ave., 28806. Phone: (828) 253-1310. Fax: (828) 253-5619. Licensee: Saga Communications of North Carolina, LLC (acq 5-1-2002; $1.7 million). Population served: 500,800 Natl. Network: ESPN Radio. Rgnl. Network: N.C. News Net. Format: Sports. News staff: one; News: 15 hrs wkly. Target aud: 35 plus; mature upscale audience. ♦Randy Cable, VP & gen mgr.

WKSF(FM)—Listing follows WWNC(AM).

***WLFA(FM)**— 1975: 91.3 mhz; 440 w. 3,340 ft TL: N35 36 02 W82 39 07. Hrs open: 2420 Wade Hampton Blvd., Greenville, SC, 29615. Phone: (800) 849-8930. Phone: (828) 254-9532. Fax: (864) 292-8428. Web Site: www.hisradio.com. Licensee: Asheville Educational Association Inc. Format: Contemp Christian music. ♦Jim Campbell, pres; Alan Henderson, gen mgr; Ted McCall, chief of engrg.

WMIT(FM)—See Black Mountain

WMYI(FM)—See Greenville, SC

WSKY(AM)— Apr 11, 1947: 1230 khz; 1 kw-U. TL: N35 35 43 W82 33 57. Hrs open: 20 40 Westgate Pkwy., Suite F, 28806. Phone: (828) 251-2000. Fax: (828) 251-2135. E-mail: wsky@wilkinsradio.com Web Site: www.wilkinsradio.com. Licensee: Wilkins Communications Network Inc. (group owner; (acq 1996). Population served: 450,000 Natl. Network: CBS, Salem Radio Network. Law Firm: Womble, Carlyle, Sandridge & Rice. Format: Gospel, Christian teaching/talk. Target aud: 35 plus. ♦Bob Wilkins, pres; LuAnn Wilkins, exec VP; Mitchell Mathis, VP; Ruthie Spears, gen mgr; Greg Garrett, opns mgr; Tim Warner, engr.

WWNC(AM)— Feb 22, 1927: 570 khz; 5 kw-U, DA-N. TL: N35 35 49 W82 36 20. Hrs open: 24 13 Summerlin Rd., 28806. Phone: (828) 257-2700. Fax: (828) 255-7850. Web Site: www.wwnc.com. Licensee: Capstar TX L.P. Group owner: Clear Channel Communications Inc. (acq 8-30-00; grpsl). Population served: 500,000 Natl. Network: Motor Racing Net. Natl. Rep: McGavren Guild. Format: News/talk. News staff: 3; News: 30 hrs wkly. Target aud: 25-54. Spec prog: Farm one hr, gospel 3 hrs, relg 3 hrs wkly.

WKSF(FM)—Co-owned with WWNC(AM). August 1947: 99.9 mhz; 53 kw. 2,672 ft TL: N35 25 32 W82 45 25. Stereo. 24 Web Site: www.99kisscountry.com. Format: Country. News: 3 hrs wkly. Target aud: 25-44.

Atlantic

WTKF(FM)— May 1992: 107.3 mhz; 7 kw. Ant 607 ft TL: N34 53 01 W76 30 21. Hrs open: Box 70, Newport, 28570-0070. Secondary address: 5447 Hwy. 70, Morehead City 28557. Phone: (252) 247-6343. Phone: (800) 818-2255. Fax: (252) 247-7343. Web Site: www.wtkf107.com. Licensee: Atlantic Ridge Telecasters Inc. (acq 11-30-94; $430,000; FTR: 1-16-95). Natl. Network: Westwood One, Motor Racing Net, USA. Rgnl. Network: N.C. News Net. Format: News/talk, sports. Target aud: 25 plus; educated, informed. ♦Lockwood Phillips, CEO; Ben Ball, gen mgr; Shane Willis, opns mgr.

Atlantic Beach

***WBJD(FM)**— 1999: 91.5 mhz; 50 kw. Ant 384 ft TL: N34 45 34 W76 51 16. Hrs open: 24 c/o WTEB(FM), 800 College Ct., New Bern, 28562. Phone: (252) 638-3434. Fax: (252) 638-3538. Web Site: www.publicradioeast.org. Licensee: Craven Community College. Format: News & Ideas. ♦Kathleen Beal, gen mgr.

Aurora

WSTK(FM)—Not on air, target date: unknown: Stn currently dark. 104.5 mhz; 4.2 kw. Ant 393 ft TL: N35 18 09 W76 34 00. Hrs open: 702 Hartness Rd., Statesville, 28677. Phone: (704) 878-9004. Licensee: Media East LLC (acq 1-15-2003). ♦Ronald Benfield, pres & gen mgr.

***WZGO(FM)**— Aug 1, 2006: 91.1 mhz; 40 kw. Ant 351 ft TL: N35 18 09 W76 34 00. Stereo. Hrs opn: 24
Rebroadcasts WAGO(FM) Snow Hill.
Box 1895, Goldsboro, 27533. Phone: (252) 747-8887. Fax: (252) 747-7888. E-mail: wago@gomixradio.org Web Site: www.gomixradio.org. Licensee: Pathway Christian Academy Inc. Natl. Network: Moody, Salem Radio Network. Format: Christian. News staff: one; News: 14 hrs wkly. Target aud: General. ♦Dr. T.D. Worthington, pres & gen mgr; Ashley Worthington, prom dir; Keith Aycock, progmg dir; Tim Sutton, mus dir; Joe Patton, chief of engrg.

Banner Elk

WZJS(FM)— Aug 5, 1989: 100.7 mhz; 6 kw. Ant 758 ft TL: N36 10 34 W81 50 05. Hrs open: 24 738 Blowing Rock Rd., Boone, 28607. Phone: (828) 264-2411. Fax: (828) 264-2412. Web Site: www.1007macfm.com. Licensee: Aisling Broadcasting of Banner Elk LLC (group owner; (acq 12-1-2003; $2.2 million with WATA(AM) Boone). Population served: 50,000 Natl. Network: Motor Racing Net. Format: Eclectic for men 18-34. Target aud: 18-44. ♦Jonathan Hoffman, CEO & sr VP; Andy Glass, opns dir.

Bath

***WZPE(FM)**— 2005: 90.1 mhz; 675 w. Ant 128 ft TL: N35 28 32 W76 48 44. Stereo. Hrs opn: 24 Box 828, Wake Forest, 27588. Phone: (919) 556-5178. Fax: (919) 556-9273. Web Site: www.wcpe.org. Licensee: Educational Information Corp. Format: Classic. ♦Deborah S. Proctor, gen mgr; Rae Weaver, dev dir; Dick Storck, progmg dir; Will Woltz, mus dir; John Graham, engr.

Bayboro

WNBB(FM)— 2001: 97.9 mhz; 14.5 kw. Ant 433 ft TL: N35 00 02 W76 49 58. Stereo. Hrs opn: 24 233 Middle St., Suite 107B, New Bern, 28562. Phone: (252) 638-8500. Fax: (252) 638-8597. E-mail: mail@bear979.com Web Site: www.bear979.com. Licensee: Coastal Carolina Radio LLC (acq 11-25-2003; $800,000). Natl. Network: ABC. Wire Svc: Metro Weather Service Inc. Format: Classic country. Target aud: 35-64; adults. ♦Dann Miller, gen mgr.

Beaufort

***WXBE(FM)**— 2005: 88.5 mhz; 1 kw. Ant 180 ft TL: N34 43 26 W76 43 18. Hrs open: Drawer 2440, Tupelo, MS, 38801. Phone: (662) 844-8888. Fax: (662) 842-6791. Web Site: www.afr.net. Licensee: American Family Association. Group owner: American Family Radio. Format: Christian. ♦Marvin Sanders, gen mgr.

Beech Mountain

WECR-FM— 1996: 102.3 mhz; 130 w. 1,948 ft TL: N36 11 03 W81 52 48. Hrs open: 24 738 Blowing Rock Rd., Boone, 28607. Phone: (828) 264-2411. Fax: (828) 264-2412. E-mail: info@wecr1023.com Web Site: www.mix1023fm.com. Licensee: Aisling Broadcasting of Banner Elk LLC (group owner; acq 4-1-2004; grpsl). Population served: 150,000 Natl. Network: CBS Radio. Format: Light adult contemp. News staff: 2; News: 2 hrs wkly. Target aud: 25-54. Spec prog: North Carolina Univ. football, basketball, Great American Songbook, Mountainhome Bluegrass. ♦Jonathan Hoffman, CEO & gen mgr.

Belhaven

WQZL(FM)— Oct 15, 1980: 101.1 mhz; 31 kw. Ant 607 ft TL: N35 18 18 W76 45 45. Stereo. Hrs opn: 24
Rebroadcasts WQSL(FM) Jacksonville 100%.
1361 Colony Dr., New Bern, 28562. Phone: (252) 639-7900. Fax: (252) 639-7976. Web Site: www.thebeatnc.com. Licensee: NM Licensing LLC. Group owner: NextMedia Group L.L.C. (acq 11-26-2001; grpsl). Format: Rhythmic contemp hit radio. News: 3 hrs wkly. Target aud: 35-54; baby boomers. ♦Rolf Pepple, gen mgr; Jack Spade, progmg dir.

Belmont

WCGC(AM)— Dec 11, 1954: 1270 khz; 5 kw-D, 500 w-N, DA-2. TL: N35 15 05 W81 03 26. Stereo. Hrs opn: 24 Box 1360, 6021 W. Wilkinson Blvd., 28012. Phone: (704) 825-2812. Fax: (704) 825-2127. E-mail: wcgc1270am@yahoo.com Licensee: WHVN Inc. Group owner: GHB Radio Group (acq 4-17-98; $250,000). Population served: 12,000 Natl. Network: Westwood One. Format: Relg,talk. News staff: 2; News: 8 hrs wkly. Target aud: General. Spec prog: Sports. ♦Tom Gentry, pres & gen mgr.

Benson

WPYB(AM)— Sept 1, 1961: 1130 khz; 1 kw-D. TL: N35 21 40 W78 34 45. Hrs open: Box 215, 27504. Phone: (919) 894-1130. Fax: (919) 894-1530. E-mail: wpbyb@surrealnet.net Web Site: www.wpyb.org. Licensee: Benson-Dunn Broadcasting Inc. (acq 5-1-96; $250,000). Population served: 100,000 Format: Country, bluegrass, gospel. Target aud: General. ♦Jasper L. Tart, pres & gen mgr; Mable Sue Tart, exec VP.

Biltmore Forest

WOXL-FM— 2002: 96.5 mhz; 1.85 kw. Ant 1,171 ft TL: N35 35 23 W82 40 26. Hrs open: 1190 Patton Ave., Asheville, 28806. Phone: (828) 259-9695. Fax: (828) 253-5619. Web Site: www.965woxl.com. Licensee: Saga Communications of North Carolina LLC. (acq 7-7-2005; $8 million). Format: Oldies. ♦Larry Williams, gen mgr.

Black Mountain

WFGW(AM)— May 27, 1962: 1010 khz; 50 kw-D, 500 w-N, 19 kw-CH, DA-3. TL: N35 36 19 W82 21 00. Hrs open: 24 Box 159, 1330 U.S. Hwy. 70, 28711. Phone: (828) 669-8477. Fax: (828) 669-6983. E-mail: thankyou@brb.org Web Site: www.wfgw.org. Licensee: Blue Ridge Broadcasting Corp. Population served: 366,000 Natl. Network: Salem Radio Network. Law Firm: Pillsbury, Winthrop, Shaw Pittman. Format: Christian teaching, talk. News: 3 hrs wkly. Target aud: 35 plus. Spec prog: Black one hr wkly. ♦Billy Graham, chmn; Dr. David Bruce, pres; Jim Kirkland, gen mgr; Wayne Roper, dev mgr; Tom Greene, progmg dir; Keith Pittman, news dir; Paul Zettle, engr.

WMIT(FM)—Co-owned with WFGW(AM). June 1, 1942: 106.9 mhz; 36 kw. Ant 3,090 ft TL: N35 44 06 W82 17 10. Stereo. 24 Web Site: www.wmit.org. (Acq 1963).1,500,000 Natl. Network: Fox News Radio. Law Firm: Pillsbury, Winthrop, Shaw Pittman. Format: Contemp Christian music & teaching. Target aud: 35-54; women. ♦Matt Stockman, mus dir.

WZGM(AM)— Feb 26, 1966: 1350 khz; 1 kw-D, 74 w-N. TL: N35 37 19 W82 19 02. Hrs open: 24 Box 430, Lincolnton, 28093. Fax: (828) 669-6224. Fax: (704) 732-9567. Web Site: www.z1350.com. Licensee: HRN Broadcasting Inc. (acq 4-28-2005; $850,000). Population served: 22,000 Format: Oldies. Target aud: General. ♦D. Mark Boyd III, pres; Lanny Ford, gen mgr.

Blowing Rock

WXIT(AM)— 1983: 1200 khz; 10 kw-D, 7 kw-CH. TL: N36 09 17 W81 39 41. Hrs open: 738 Blowing Rock Rd., Boone, 28607. Phone: (828) 265-1023. Fax: (828) 264-8902. E-mail: wxit@newstalk1200.com Web Site: www.goblueridge.net. Licensee: Aisling Broadcasting of Banner Elk LLC (group owner; acq 2-10-2004; grpsl). Natl. Network: CBS. Format: News/talk. Target aud: 25-60; professionals. Spec prog: Relg 8 hrs, big band 4 hrs wkly. ♦Donna Hoffman, VP; Jonathan Hoffman, gen mgr; Andy Zlass, opns mgr & progmg dir.

Boiling Springs

***WGWG(FM)**— Jan 22, 1974: 88.3 mhz; 50 kw. 302 ft TL: N35 13 52 W81 42 57. Stereo. Hrs opn: 24 Box 876, 106 Emily Ln., 28017. Phone: (704) 406-3525. Fax: (704) 434-4338. E-mail: info@wgwg.org Web Site: www.wgwg.org. Licensee: Gardner-Webb University. Population served: 315,000 Format: Triple A. News: one hr wkly. Target aud: General. Spec prog: Gospel 15 hrs wkly. ♦Frank Campbell, pres; Dan McClellan, stn mgr; Matt Webber, gen mgr & opns mgr.

Boone

***WASU-FM**— May 18, 1972: 90.5 mhz; 220 w. 57 ft TL: N36 12 48 W81 41 10. Stereo. Hrs opn: Appalachian State Univ., Wey Hall, 28608. Phone: (828) 262-3170. Phone: (828) 262-2543. Licensee: Appalachian State University. Population served: 20,000 Format: Alternative. News: 2 hrs wkly. Target aud: 18-30; college students & area residents. Spec prog: Urban contemp 6 hrs, blues 2 hrs, Christian rock 3 hrs, country 8 hrs wkly. ♦Dan Hill, gen mgr.

WATA(AM)— September 1950: 1450 khz; 1 kw-U. TL: N36 12 59 W81 42 06. Hrs open: 5 AM-midnight (M-S); 6 AM-midnight (Su) 738 Blowing Rock Rd., 28607. Phone: (828) 264-2411. Fax: (828) 264-2412. E-mail: info@wataradio.com Web Site: www.wataradio.com. Licensee: Aisling Broadcasting of Banner Elk LLC (group owner; acq 12-1-2003; $2.2 million with WZJS(FM) Banner Elk). Population served: 49,000 Natl. Network: ABC. Format: Local newstalk. News staff: one; News: 4 hrs wkly. Target aud: 25-54. Spec prog: Gospel 5 hrs, Paul Harvey 2.5 hrs.,Watauga High sports wkly. ♦Jonathan Hoffman, CEO & gen mgr; Andy Glass, opns mgr.

Brevard

WGCR(AM)— Sept 16, 1985: 720 khz; 10 kw-D. TL: N35 15 10 W82 40 28. Stereo. Hrs opn: 3400New Hendersonville Hwy., Pisgah Forest, 28768. Phone: (828) 884-9427. Fax: (828) 883-9427. Web Site: www.wgcr.net. Licensee: Anchor Baptist Broadcasting Association. (acq 2-87). Natl. Network: USA. Rgnl. Network: N.C. News Net. Format: Relg, news. Target aud: General. Spec prog: Gospel. ♦Randy C. Barton, pres, gen mgr & gen sls mgr; Shanna Barton, prom mgr & progmg dir; Shamma Barton, news dir; Lamar Owen, chief of engrg.

WSQL(AM)— July 6, 1950: 1240 khz; 1 kw-U. TL: N35 13 23 W82 42 20. Hrs open: Box 1240, 28712. Secondary address: 1319 Wilson Rd., Pisgah Forest 28768. Phone: (828) 877-5252. Fax: (828) 877-5253. Licensee: A & L Broadcasting Inc. (acq 3-14-97; $110,000). Population served: 26,000 Natl. Network: CBS. Format: Talk, adult contemp, MOR. Target aud: General. Spec prog: Jazz 10 hrs, gospel 8 hrs, relg 4 hrs wkly. ♦Allen Reese, gen mgr, opns mgr & gen sls mgr; Leah Reese, progmg dir.

Bryson City

WBHN(AM)— Oct 1, 1967: 1590 khz; 500 w-D, 37 w-N. TL: N35 25 41 W83 26 18. Hrs open: Box 1309, 28713. Phone: (828) 488-2682. Fax: (828) 488-3594. E-mail: wbhn@dnet.net Licensee: Starcast South Inc. Group owner: Starcast Systems Inc. (acq 10-84; $355,000; FTR: 10-15-84). Population served: 10,000 Format: Oldies. ♦Jack Mullen Jr., pres; Rob Henline, gen mgr, gen sls mgr & mus dir; J.B. Jacobs, chief of engrg.

Buie's Creek

***WCCE(FM)**— Oct 7, 1974: 90.1 mhz; 3 kw. Ant 105 ft TL: N35 24 36 W78 44 21. Stereo. Hrs opn: 24 Box 1030, Science Bldg., Campbell Univ., 27506. Phone: (910) 893-1745. Fax: (910) 893-1746. E-mail: wcce@mailcenter.campbell.edu Web Site: www.campbell.edu/wcce/. Licensee: Campbell University. Population served: 50,000 Format: Smooth jazz, soft rock, relg. News: 10 hrs wkly. Target aud: General. Spec prog: Bluegrass 3 hrs, big band 4 hrs wkly. ♦Travis Autry, gen mgr; Carolyn Bowden, opns mgr.

Burgaw

WKXB(FM)— Dec 13, 1964: 99.9 mhz; 100 kw. Ant 774 ft TL: N34 14 37 W78 07 24. Stereo. Hrs opn: 24 25 N. Kerr Ave., Wilmington, 28405. Phone: (910) 791-3088. Fax: (910) 791-0112. E-mail: stanleyb@nextmediagroup.com Web Site: www.jammin999sm.com. Licensee: NM Licensing LLC. (group owner) (acq 2-1-2005; grpsl). Population served: 182,000 Natl. Rep: McGavren Guild. Format: Rhythmic gold. News staff: one; News: 3 hrs wkly. Target aud: 25-54. ♦Jeff Dinetz, pres; Barry Brown, VP; Barbara Raybourne, gen mgr; Gayle Brown, gen sls mgr; Missy Andrus, prom dir; Stanley B., progmg dir; Doug Carlisle, news dir.

WVBS(AM)— June 21, 1963: 1470 khz; 1 kw-D, 93 w-N. TL: N34 32 05 W77 54 31. (CP: TL: N34 31 22 W77 54 17). Hrs opn: Sunrise-sunset Box 914, Bible Baptist Church, 2190 Hwy. 117 S., 28425. Phone: (910) 259-5718. E-mail: oldtimer@intrstar.net Licensee: Grace Christian School. (acq 12-23-94; FTR: 2-27-95). Population served: 1,744 Format: Christian. ♦Carl Gibbs, gen mgr; Dick Jones, stn mgr.

Burlington

WKSL(FM)— 1946: 93.9 mhz; 100 kw horiz, 94 kw vert. Ant 1,263 ft TL: N35 52 15 W79 09 40. (CP: COL Cary. 100 kw, ant 1,486 ft. TL: N35 42 50 W78 49 04). Stereo. Hrs opn: 3100 Smoketree Ct., Suite 700, Raleigh, 27604-1052. Phone: (919) 877-0939. Fax: (919) 876-2929. Licensee: Capstar TX L.P. Group owner: Clear Channel Communications Inc. (acq 8-30-2000; grpsl). Population served: 661,000 Format: Adult contemp. Target aud: 25-49; men. ♦Jon Robbins, CFO & opns dir; Ken Spitzer, gen mgr; Tammy O'Dell, VP & sls dir; Jinnie Forsythe, gen sls mgr & rgnl sls mgr; Jessica Hayes, prom dir; Brian Taylor, progmg dir; Fred Pace, chief of engrg; Stephanie Vacendak, traf mgr.

WPCM(AM)— September 1941: 920 khz; 5 kw-D, 55 w-N. TL: N36 05 50 W79 29 03. Hrs open: Box 1119, 27215. Secondary address: 1109 Tower Dr. 27215. Phone: (336) 584-0126. Fax: (336) 584-6333. Web Site: www.920wpcm.com. Licensee: Carolina Radio Group Inc. Group owner: Curtis Media Group (acq 3-1-90). Population served: 100,000 Format: Oldies and beach. Target aud: 25 plus; upscale. ♦Bill Whitley, gen mgr.

WZTK(FM)—Co-owned with WPCM(AM). December 1946: 101.1 mhz; 100 kw. Ant 1,191 ft TL: N35 56 31 W79 26 33. Stereo. 24 Fax: (336) 854-1039. Web Site: www.fmtalk1011.com.1,887,200 Format: Talk. Target aud: 25-54. Spec prog: Bluegrass 3 hrs wkly. ♦Bryon Tucker, progmg dir.

Burlington-Graham

WBAG(AM)— 1946: 1150 khz; 1 kw-D, 48 w-N. TL: N36 06 48 W79 27 00. Hrs open: 24 Box 2450, 27216. Secondary address: 1745 Burch Bridge Rd. 27217. Phone: (336) 226-1150. Fax: (336) 226-1180. Licensee: Gray Broadcasting LLC (acq 10-27-98; $150,000). Population served: 108,000 Rgnl. Network: N.C. News Net. Format: Adult standards/talk. News staff: 2; News: 25 hrs wkly. Target aud: 25-54; general. Spec prog: Relg 5 hrs wkly. ♦Joe Gray, gen mgr; Harry Myers, opns mgr & progmg dir; Bill Huff, gen sls mgr & sports cmtr; Tim Walker, chief of engrg.

WSML(AM)—See Graham

Burnsville

WKYK(AM)— May 28, 1967: 940 khz; 5 kw-D, 250 w-N, DA-N. TL: N35 55 32 W82 16 20. Stereo. Hrs opn: 24 Box 744, Mark Group Bldg., 28714. Secondary address: 749 Sawmill Road 28714. Phone:

(828) 682-3510. Phone: (828) 682-3798. Fax: (828) 682-6227. Fax: (828) 682-0998. E-mail: 940@wkyk.com Web Site: www.wkyk.com. Licensee: Mark Media Inc. (acq 4-10-69). Population served: 463,000 Natl. Network: ABC. Rgnl. Network: N.C. News Net. Format: Real country. News staff: one; News: 10 hrs wkly. Target aud: 18-55. Spec prog: Gospel 12 hrs wkly. ♦J. Ardell Sink, CEO & pres; Remelle Sink, exec VP; Michael Sink, VP, gen mgr & chief of engrg; Holly S. Hall, opns mgr & mktg dir; Mary Marsh, prom dir; Steve Murphy, news dir & pub affrs dir.

Buxton

***WBUX(FM)**— 1999: 90.5 mhz; 5.9 kw. Ant 154 ft TL: N35 16 01 W75 32 38. Hrs open:
Rebroadcasts WUNC(FM) Chapel Hill 99.9%.
120 Friday Center Dr., Chapel Hill, 27517-9495. Phone: (919) 966-5454. Fax: (919) 966-5955. E-mail: wunc@wunc.org Web Site: www.wunc.org. Licensee: Board of Trustees/University of North Carolina at Chapel Hill. Population served: 3,434 Natl. Network: NPR, PRI, CBC Radio One. Format: News & Info. News staff: 7; News: 124 hrs wkly. Spec prog: Folk 20 hrs wkly.

WHDX(FM)—Not on air, target date: Apr 1, 2008: 99.9 mhz; 50 kw. Ant 164 ft TL: N35 14 44 W75 32 02. Hrs open: 1400 12th St. N., Suite 5, Arlington, VA, 22209-3666. Phone: (703) 527-1434. E-mail: radiobuxton@yahoo.com Licensee: David Wilson. ♦David Wilson, gen mgr.

WHDZ(FM)—Not on air, target date: Apr 1, 2006: 101.5 mhz; 50 kw. Ant 164 ft TL: N35 14 44 W75 32 02. Hrs open: 1400 12th St. N., Suite 5, Arlington, VA, 22209-3666. Phone: (703) 527-1434. E-mail: radiobuxton@yahoo.com Licensee: David Wilson. ♦David Wilson, gen mgr.

Calabash

WYNA(FM)— June 1964: 104.9 mhz; 15 kw. Ant 338 ft TL: N33 49 19 W78 46 18. Stereo. Hrs opn: 3926 Wesley St., Suite 301, Myrtle Beach, SC, 29578. Phone: (843) 903-9962. Fax: (843) 903-1797. Web Site: www.1049bobfm.com. Licensee: Coastline Communications of Carolina Inc. (acq 1-14-99; $1.1 million). Population served: 250,000 Format: Adult Hits. Target aud: 25-54; adults. ♦Jerome Bresson, pres; Will Isaacs, gen mgr.

Camp Lejeune

WSME(AM)— Sept 8, 1980: Stn currently dark. 1120 khz; 6 kw-D, 4.2 kw-CH. TL: N34 43 03 W77 16 57. Hrs open: 337 E. Centre St., Jacksonville, 28540. Secondary address: 333 Center St. 28546. Phone: (910) 355-9763. Fax: (910) 355-9763. Licensee: CTC Media Group Inc. (group owner). Format: Country. ♦Edwin Lee Afflerbach, VP.

Canton

WPTL(AM)— Aug 3, 1963: 920 khz; 500 w-D, 38 w-N. TL: N35 31 15 W82 48 24. Hrs open: 6 AM-6:30 PM Box 909, 133 Pisgah Dr., 28716. Phone: (828) 648-3576. Phone: (828) 648-3577. Fax: (828) 648-3577. E-mail: admin@wptlradio.com Web Site: www.wptlradio.com. Licensee: Skycountry Broadcasting Inc. (acq 3-1-78). Population served: 50,000 Natl. Network: AP Radio, Jones Radio Networks. Format: C&W, relg. News: 8 hrs wkly. Target aud: 25 plus; adult family. ♦Linda Reck, VP & stn mgr; William Reck, pres & gen mgr.

WYSE(AM)— July 12, 1954: 970 khz; 5 kw-D. TL: N35 31 58 W82 51 58. Hrs open: 90 Lookout, Asheville, 28804. Phone: (828) 259-9695. Fax: (828) 253-5619. Licensee: Saga Communications of North Carolina LLC. Group owner: Saga Communications Inc. (acq 3-11-2003). Population served: 50,000 Natl. Network: ESPN Radio. Law Firm: Miller & Fields, P.C. Format: Sports. ♦Ed Christian, pres; Randy Cable, gen mgr.

Carolina Beach

WMYT(AM)— July 1, 1989: 1180 khz; 10 kw-D, DA. TL: N34 09 03 W78 04 48. Hrs open: Box 957, Wilmington, 28402-0957. Phone: (910) 763-2452. Fax: (910) 763-6578. E-mail: life@life905.com Web Site: www.life905.com. Licensee: Carolina Christian Radio Inc. (group owner; (acq 3-2-2001; $100,000 with WDVV(FM) Wilmington). Format: Relg, Sp. Target aud: General - Spanish. ♦Jim Stephens, gen mgr; Roger Brace, engr.

WUIN(FM)— October 1996: 106.7 mhz; 5.6 kw. Ant 341 ft TL: N34 03 02 W77 57 20. Hrs open: 24
Simulcast with WPPG(FM) Fair Bluff.
122 Cinema Dr., Wilmington, 28403. Phone: (910) 772-6300. Web Site: www.carolinapenguin.com. Licensee: Ocean Broadcasting II LLC (acq 7-3-2003; $1.5 million with WMFD(AM) Wilmington). Format: Triple A. ♦Paul Knight, gen mgr; Beau Gunn, progmg dir.

Chadbourn

WVOE(AM)— Apr 23, 1962: 1590 khz; 1 kw-D. TL: N34 21 05 W78 50 38. Hrs open: 1528 Old 74 Hwy. W., 28431. Phone: (910) 654-5621. Fax: (910) 654-4385. E-mail: wvoe@weblink.net Licensee: Ebony Enterprises Inc. Population served: 500,000 Format: Rhythm and blues, jazz, gospel. Target aud: General; white & blue collar workers, housewives, students, sr citizens. ♦Willie J. Walls, pres; Willie J. Walls, gen mgr; Willie J. Walls, stn mgr.

Chapel Hill

WCHL(AM)— Jan 25, 1953: 1360 khz; 5 kw-D, 1 kw-N, DA-N. TL: N35 56 18 W79 01 36. Stereo. Hrs opn: 24 88 VilCom Cir., Suite 100, 27514. Phone: (919) 933-4165. Fax: (919) 968-3748. E-mail: cdixon@wchl1360.com Web Site: www.wchl1360.com. Licensee: Vilcom Interactive Media LLC (acq 8-5-2004; $775,000). Population served: 85,000 Natl. Network: ABC, CBS Radio, Jones Radio Networks. Wire Svc: AP Format: News/talk. News staff: 3; News: 25 hrs wkly. Target aud: 25-54; educated adults with high median incomes. ♦Christy Jones Taylor, VP & gen mgr; Christy Dixon, stn mgr; Ron Stutts, progmg dir.

WDCG(FM)—See Durham

WKSL(FM)—See Burlington

WLLQ(AM)— December 1973: 1530 khz; 10 kw-D, DA. TL: N35 58 07 W79 00 10. Hrs open: Sunrise-sunset
Rebroadcasts WRTP(AM) Chapel Hill 100%.
Estuardo Valdemar Rodriguez and Leonor Rodriguez Stns, 1010 Vermont Ave. N.W., Suite 100, Washington, DC, 20005. Phone: (202) 638-1959. Fax: (202) 393-7464. E-mail: valtravel@rcn.com Licensee: Estuardo Valdemar Rodriguez and Leonor Rodriguez. Group owner: WRTP Radio Network (acq 12-13-2004; grpsl). Population served: 95,438 Format: Mexican rgnl. Target aud: Spanish young adult. ♦Estuardo Valdemar Rodriguez, gen mgr.

***WUNC(FM)**— Nov 3, 1952: 91.5 mhz; 100 kw. Ant 1,361 ft TL: N35 51 59 W79 10 00. Hrs open: 24 120 Friday Center Dr., 27517-9495. Phone: (919) 966-5454. Fax: (919) 966-5955. E-mail: wunc@wunc.org Web Site: www.wunc.org. Licensee: University of North Carolina at Chapel Hill. Population served: 2,011,484 Natl. Network: NPR, PRI, CBC Radio One. Law Firm: Brooks, Pierce, McLendon, Humphrey & Leonard, LLP. Format: News & info. News staff: 7; News: 124 hrs wkly. Target aud: 25-54; highly educated, pro-active in the community, concerned about local issues. Spec prog: Folk 20 hrs wkly. ♦Joan Siefert Rose, gen mgr; Kevin Wolf, opns mgr & traf mgr; Regina Yeager, dev dir; George Boosey, progmg dir & progmg mgr; Connie Walker, news dir; John Francioni, engrg dir & chief of engrg.

***WXYC(FM)**— Mar 18, 1977: 89.3 mhz; 400 w. 280 ft TL: N35 54 15 W79 02 50. Stereo. Hrs opn: 24 Ch5210, Carolina Union, 27599. Phone: (919) 962-7768. Phone: (919) 962-8989 (request line). E-mail: wxyc@unc.edu Web Site: www.wxyc.org. Licensee: Student Educational Broadcasting Inc. Population served: 25,000 Format: Div. News: 3 hrs wkly. Target aud: General. ♦Jason Perlmutter, stn mgr.

Charlotte

WBT(AM)— Apr 10, 1922: 1110 khz; 50 kw-U, DA-N. TL: N35 07 56 W80 53 23. Stereo. Hrs opn: 24 One Julian Price Pl., 28208. Phone: (704) 374-3500. Fax: (704) 374-3889. Web Site: www.wbt.com. Licensee: Jefferson-Pilot Communications Co. of North Carolina. (group owner; (acq 9-45). Population served: 889,000 Natl. Network: CBS. Format: News/talk. News staff: 5; News: 20 hrs wkly. Target aud: 35-54; men. ♦David Stoneciper, CEO; Terry stone, pres; Rick Jackson, sr VP & gen mgr; Lisa Gergely, gen mgr; Terry Mace, dev VP; Larry Rideaux, natl sls mgr; Matt Dubois, prom mgr; Bill White, progmg dir; Marshall Adams, news dir; Jerry Dowd, chief of engrg; Nancy Haynes, rsch dir; Nancy Albright, traf mgr; Jim Barroll, news rptr; Pete Kaliner, reporter.

WLNK(FM)—Co-owned with WBT(AM). Aug 15, 1962: 107.9 mhz; 97 kw. 1,692 ft TL: N35 21 51 W81 11 13. Stereo. 24 Fax: (704) 338-3062. Web Site: www.1079thelink.com. Format: Hot adult contemp. ♦Neal Sharpe, progmg dir; Derek James, mus dir; Nancy Haynes, rsch dir; Nancy Albright, traf mgr; Jim Barroll, news rptr; Pete Kaliner, reporter. Co-owned TV: WBTV(TV) affil

WCGC(AM)—See Belmont

***WFAE(FM)**— June 29, 1981: 90.7 mhz; 100 kw. 760 ft TL: N35 15 06 W80 41 12. Stereo. Hrs opn: 24 8801 J.M. Keynes Dr., Suite 91, 28262-8485. Phone: (704) 549-9323. Fax: (704) 547-8851. E-mail: wfae@wfae.org Web Site: www.wfae.org. Licensee: University Radio Foundation Inc. (acq 4-12-93). Population served: 1,700,000 Natl. Network: NPR, PRI. Law Firm: Garvey, Schubert & Barer. Format: News/talk. News staff: 4; News: 42 hrs wkly. Target aud: 35-49; professionals. ♦Roger Sarow, gen mgr; Tena Simmons, opns dir; Barbara Vermeire, dev dir; Catherine Little, dev dir & sls dir; Renee Ballos, prom dir & pub affrs dir; Paul Stribling, progmg dir; Mark Rumsey, news dir; Jobie Sprinkle, engrg dir.

WFNA(AM)— Dec 1, 2003: 1660 khz; 10 kw-D, 1 kw-N. TL: N35 14 57 W80 51 41. Hrs open: 1520 South Blvd., Suite 300, 28203. Phone: (704) 342-2644. Fax: (704) 727-8985. Licensee: Infinity Radio Holdings Inc. Group owner: Infinity Broadcasting Corp. Natl. Network: Sporting News Radio Network. Natl. Rep: D & R Radio, Katz Radio. Law Firm: Leventhal, Senter & Leman. Format: Sports. ♦Bill Schoening, gen mgr; O.J. Stout, opns mgr.

WFNZ(AM)— 1941: 610 khz; 5 kw-D, 1 kw-N, DA-2. TL: N35 17 53 W80 53 40. Hrs open: 1520 South Blvd # 300, 28203. Phone: (704) 319-9369. Fax: (704) 319-3934. Web Site: www.wfnz.com. Licensee: CBS Radio Holdings Inc. Group owner: Infinity Broadcasting Corp. (acq 11-13-98; grpsl). Population served: 410,000 Format: Sports. Target aud: 18 plus; male, sports oriented. ♦Bill Schoening, gen mgr; D.J. Stout, opns mgr & progmg dir; Scott Vandivier, gen sls mgr; Chele Fassig, prom dir; Eric Lakey, chief of engrg.

WNKS(FM)—Co-owned with WFNZ(AM). July 21, 1962: 95.1 mhz; 100 kw. 1,542 ft TL: N35 21 44 W81 09 19. Stereo. 4015 Stuart Andrew Blvd., 28217. Phone: (704) 331-9510. Fax: (704) 344-8656. Web Site: www.kiss951.com.325,000 Format: CHR. ♦Keith Cornwell, gen mgr; John Renolds, opns mgr; Rob Whitehead, gen sls mgr; Natalie Kirby, mktg mgr; Chad Fitzsimmons, prom mgr.

WGFY(AM)— Jan 18, 1955: 1480 khz; 5 kw-U, DA-2. TL: N35 17 05 W80 52 34. Hrs open: 24 1100 S Troyn St., Suite 210, 28203. Phone: (704) 377-2223. Fax: (704) 373-2245. Web Site: www.radiodisney.com. Licensee: Radio Disney Group LLC. Group owner: ABC Inc. (acq 8-22-00; grpsl). Law Firm: Fisher, Wayland, Cooper, Leader & Zaragoza. Format: Children, Radio Disney. News staff: 6; News: 25 hrs wkly. Target aud: Under 12; kids, mothers, families. Spec prog: Children pop top 40. ♦Jon Pendleton, gen mgr & rgnl sls mgr; Kevin Campbell, pub affrs dir; Clay Steely, engrg VP.

WGSP(AM)— Aug 23, 1958: 1310 khz; 5 kw-D. TL: N35 15 23 W80 51 52. Hrs open: 4801 E. Independence Blvd., Suite 803, 28212. Phone: (704) 442-7277. Fax: (704) 442-9518. Licensee: Norsan Consulting and Management Inc. (acq 12-13-2004; $2 million). Population served: 100,000 Format: Hispanic music. ♦Norberto Sanchez, pres; Javier Placencia, progmg dir.

WHVN(AM)— 1958: 1240 khz; 1 kw-U, DA-1. TL: N35 12 00 W80 48 39. Hrs open: 5732 N. Tryon St., 28213. Phone: (704) 596-4900. Fax:

(704) 596-6939. E-mail: bboonewhvn@bellsouth.net Licensee: WHVN Inc. Group owner: GHB Radio Group (acq 7-11-83). Population served: 241,178 Law Firm: Reddy, Begley & McCormick. Format: Relg. Target aud: 35 plus; Christian. ♦George Buck, pres; Tom Gentry, gen mgr, sls VP, mktg VP, prom VP & adv VP; Buddy Boone, progmg dir; Brant Hart, pub affrs dir; Steward Albert, chief of engrg.

WKQC(FM)— 1972: 104.7 mhz; 96 kw. Ant 1,210 ft TL: N35 15 06 W80 41 12. Stereo. Hrs opn: 24 4015 Stuart Andrew Blvd., 28217. Phone: (704) 372-1104. Fax: (704) 523-1047. Web Site: www.star1047.com. Licensee: Infinity Radio Holdings Inc. Group owner: Infinity Broadcasting Corp. (acq 11-13-98; grpsl). Population served: 1,200,000 Law Firm: Leventhal, Senter & Lerman. Format: Hits of the 70s & 80s. News staff: 2. Target aud: 25-54. ♦Keith Cornwell, gen mgr; John Reynolds, prom mgr.

WOGR(AM)— May 7, 1964: 1540 khz; 2.5 kw-D, DA. TL: N35 13 45 W80 58 32. (CP: TL: N35 16 26 W80 51 50). Hrs opn: Sunrise-sunset Box 16408, 1501 N. 1-85 Service Rd., 28216. Secondary address: Box 16408 28297. Phone: (704) 393-1540. Phone: (704) 393-1588. Fax: (704) 393-1527. E-mail: cynthia@wordnet.org Web Site: wordnet.org. Licensee: Victory Christian Center Inc. (acq 7-27-88). Natl. Network: Salem Radio Network. Law Firm: Gardner, Carton & Douglas. Format: Relg. Target aud: General. ♦Robyn Gool, pres; Wayne Hammond, gen mgr; James Sims, opns mgr; Cynthia Neely, gen sls mgr, prom dir, pub affrs dir & spec ev coord; Eleasah Hammond, mus dir; Tamma Wylie, traf mgr; Andrea Watts, mus critic.

WPEG(FM)—See Concord

WSOC-FM— 1947: 103.7 mhz; 100 kw. 1,040 ft TL: N35 15 41 W80 43 38. (CP: Ant 1,059 ft.). Stereo. Hrs opn: 1520 South Blvd, Suite 300, 28203. Phone: (704) 522-1103. Fax: (704) 523-2104. Web Site: www.wsocfm.com. Licensee: Infinity Radio Holdings Inc. Group owner: Infinity Broadcasting Corp. (acq 11-13-98; grpsl). Population served: 1,100,000 Law Firm: Leventhal, Senter & Lerman. Format: Country. Target aud: 25-54. ♦Bill Schoening, VP & gen mgr; Barry Roach, natl sls mgr; Billy Grooms, rgnl sls mgr; Chele Fassig, prom dir; Jeff Roper, progmg dir; Rick McCracken, mus dir; Frank Laseter, pub affrs dir; Eric Lakey, chief of engrg; Terry Cunningham, traf mgr.

***WYFQ(AM)**— Oct 14, 1933: 930 khz; 5 kw-D, 1 kw-N, DA-N. TL: N35 16 00 W80 54 05. Hrs open: 24 11530 Carmel Commons Blvd., 28226. Secondary address: 2004 Walkup Ave., Monroe 28110. Phone: (704) 523-5555. Fax: (704) 291-7807. E-mail: wyfq@bbnradio.org. Licensee: Bible Broadcasting Network Inc. (group owner; acq 2-6-92; $475,000; FTR: 2-24-92). Population served: 325,000 Format: Traditional Christian. News: 3 hrs wkly. Target aud: General. ♦Dan Austin, gen mgr & stn mgr; John Woolery, progmg dir; Ron Muffley, engrg dir.

Cherry Point

WANG(AM)—See Havelock

WSSM(FM)—See Havelock

Cherryville

WCSL(AM)— June 28, 1967: 1590 khz; 1 kw-D, 42 w-N. TL: N35 22 28 W81 24 23. Hrs open: 24 Box 430, Lincolnton, 28093. Phone: (704) 735-8071. Fax: (704) 732-9567. E-mail: info@hrnb.com Web Site: www.hrnb.com. Licensee: HRN Broadcasting Inc. (acq 4-14-2004; $500,000 with WLON(AM) Lincolnton). Population served: 5,258 Natl. Network: Westwood One. Rgnl. Network: N.C. News Net. Format: Christian. Target aud: General. Spec prog: Loc sports 3 hrs wkly. ♦Mark Boyd, pres; Lanny Ford, gen mgr, adv dir & traf mgr; Calvin Hastings, sls dir & gen sls mgr; Milton Baker, progmg dir, sports cmtr & disc jockey; Wendy Stout, mus dir; Larry Seagle, news dir & pub affrs dir; Josh Pierce, chief of engrg; Richard Howell, sports cmtr; Tim Biggerstaff, disc jockey.

China Grove

WRNA(AM)— Nov 17, 1980: 1140 khz; 1 kw-D, 250 w-CH, DA-D. TL: N35 34 20 W80 35 21. Hrs open: 6 AM-2 hrs past sunset Box 8146, Kannapolis, 28083. Phone: (704) 857-1101. Fax: (704) 857-0680. E-mail: info@fordbroadcasting.com Web Site: www.fordbroadcasting.com. Licensee: South Rowan Broadcasting Co. Group owner: Ford Broadcasting Inc. Population served: 500,000 Natl. Network: USA. Format: Southern gospel. Target aud: General. ♦Carl Ford, pres, gen mgr, stn mgr, gen sls mgr & progmg mgr; Taylor Ford, exec VP; Angela Ford, sr VP.

Claremont

WCXN(AM)— Sept 5, 1985: 1170 khz; 10 kw-D. TL: N35 43 34 W81 08 52. Hrs open: 19 9th St. SW, Hickory, 28602. Phone: (828) 459-9803. Fax: (828) 459-9805. E-mail: metroradio@carolina.rr.com Licensee: Birach Broadcasting Corp. Group owner: Davidson Media Group LLC (acq 6-5-2007; $800,000 with KXLQ(AM) Indianola, IA). Natl. Network: USA. Format: Sp. Target aud: General. ♦Sima Birach, pres; Russ Jones, gen mgr & progmg dir; Larry Schropp, chief of engrg.

Clayton

WHPY(AM)— 1974: 1590 khz; 5 kw-D, DA. TL: N35 38 49 W78 30 21. Hrs open: Sunrise-sunset Box 535, Fellowship Baptist Church, 911 W. Main St., 27520. Phone: (919) 553-6774. Fax: (919) 359-0016. E-mail: whpy@dockpoint.net Licensee: Fellowship Baptist Church Inc. dba Fellowship Christian Academy. (acq 8-4-97). Format: Christian. ♦Charles Ennis, pres; Keith Holland, gen mgr & stn mgr.

Clemmons

WMKS(FM)— May 3, 1947: 105.7 mhz; 34 kw. Ant 1,453 ft TL: N36 22 28 W80 22 31. Stereo. Hrs opn: 24 2-B PAI Park, Greensboro, 27409. Phone: (336) 822-2000. Fax: (336) 887-0104. Web Site: www.1057kissfm.com. Licensee: Clear Channel Broadcasting Licenses Inc. (group owner; (acq 9-12-2006;. $15.65 million). Format: Urban adult contemp. ♦Cheryl Salomone, stn mgr; Tim Satterfield, opns mgr; Kevin Isaacs, sls dir.

Clinton

WCLN(AM)— Sept 27, 1975: 1170 khz; 5 kw-D. TL: N35 01 21 W78 20 58. Hrs open: Sunrise-sunset Box 28, 118 E. Main St., 28328. Phone: (910) 592-8949. Fax: (910) 592-3732. E-mail: grandpa@oldies1170.com Web Site: www.oldies1170.com. Licensee: Christian Listening Network Inc. Population served: 100,000 Natl. Network: ABC. Format: Oldies, beach. News staff: 4; News: 4.5 hrs. wkly. Spec prog: Community, gospel 8 hrs wkly. ♦George Wilson, pres; Pat Dixon, gen mgr; Nolan Wiggins, progmg dir; Don Smith, news dir; Debbie New, traf mgr.

WCLN-FM— June 11, 1967: 107.3 mhz; 13 kw. 453 ft TL: N35 02 14 W78 29 56. Stereo. Hrs opn: 24 996 Helen St., Fayetteville, 28303. Phone: (910) 864-5028. Fax: (910) 864-6270. E-mail: wcln@christian107.com Web Site: www.christian107.com. Licensee: Christian Listening Network Inc. (acq 7-94). Population served: 36,000 Format: Contemp Christian, inspirational. ♦George Wilson, pres; Linda Miller, gen mgr & sls dir; Dan DeBruler, prom dir; Steve Turley, progmg dir & progmg mgr; Van Clough, chief of engrg.

WRRZ(AM)— Apr 5, 1947: 880 khz; 1 kw-D. TL: N34 58 40 W78 18 15. Hrs open: Box 378, 28329. Phone: (910) 592-2165. Fax: (910) 592-8556. E-mail: wrrzradio@webtv.net Licensee: Sanchez Broadcasting Corp. (acq 10-14-2004). Population served: 50,000 Rgnl. Network: N.C. News Net. Law Firm: Brooks, Pierce, McLendon, Humphrey & Leonard. Format: Sp. Target aud: 25 plus. Spec prog: Black 5 hrs, relg 6 hrs, Sp 5 hrs wkly. ♦Victor Sanchez, pres; Martha Sanchez, gen mgr.

Columbia

WERX-FM— Mar 14, 1983: 102.5 mhz; 64 kw. 689 ft TL: N36 05 00 W76 36 00. Stereo. Hrs opn: 24 Box 1418, Nags Head, 27959. Secondary address: 2422 S. Wrightsville Ave., Nags Head 27959. Phone: (252) 449-8331. Fax: (252) 449-8354. Web Site: www.1025theshark.com. Licensee: East Carolina Radio of Elizabeth City Inc. Group owner: East Carolina Radio Group. Population served: 135,000 Format: Golden Oldies. Target aud: 18-49; moderate to high income, mobile professionals & families with children. Spec prog: Flashback, in concert, off the record, BBC classic tracks. ♦Rick Loesch, pres; R. Loesch, gen mgr; Tom Charity, opns mgr; John Maloney, gen sls mgr.

WRSF(FM)—Licensed to Columbia. See Elizabeth City

Concord

WEGO(AM)— Mar 5, 1943: 1410 khz; 1 kw-D, 182 w-N. TL: N35 24 29 W80 36 41. Hrs open: Box 126, 28026. Phone: (704) 788-9346. Fax: (704) 720-9346. Licensee: GHB of Waxhaw Inc. Group owner: GHB Radio Group (acq 10-28-2002; $450,000 with WSVM(AM) Valdese). Population served: 95,000 Rgnl. Network: N.C. News Net. Format: Timeless Classics. ♦Bob Brown, gen mgr.

WPEG(FM)— June 15, 1962: 97.9 mhz; 95 kw. 1,608 ft TL: N35 21 44 W81 09 19. Stereo. Hrs opn: 1520 South Blvd., Suite 300, Charlotte, 28203. Phone: (704) 342-2644. Fax: (704) 227-8979. Web Site: www.power98fm.com. Licensee: Infinity Radio Holdings Inc. Group owner: Infinity Broadcasting Corp. (acq 11-13-98; grpsl). Population served: 900,000 Natl. Network: Westwood One. Natl. Rep: Katz Radio. Format: Urban contemp. News staff: one; News: 20 hrs wkly. Target aud: 12 plus; Black. Spec prog: Gospel 6 hrs, mix show 8 hrs wkly. ♦Bill Schoening, gen mgr; Terri Avery, opns mgr.

Creedmoor

WCMC-FM— Feb 1, 1993: 99.9 mhz; 22 kw. Ant 292 ft TL: N36 04 52 W78 28 27. Stereo. Hrs opn: 24 Box 710, 27522-0710. Secondary address: Box 10100, Raleigh 27603. Phone: (919) 890-6299. Fax: (919) 890-6199. E-mail: webmaster@999genuinecountry.com Web Site: 999genuinecountry.com. Licensee: Capitol Broadcasting Co. Inc. Group owner: Joyner Radio Inc. (acq 4-22-2005; $7.25 million). Population served: 1,267,676 Natl. Rep: Katz Radio. Law Firm: Holland & Knight. Wire Svc: AP Format: Country. ♦Jim Goodmon, pres; Dan McGrath, CFO; Ardie Gregory, VP, gen mgr & prom VP; Joe Formicola, opns mgr; Karen Cates, gen sls mgr; Mark Tarak, natl sls mgr; Brandon Alexander, mktg mgr.

Cullowhee

***WWCU(FM)**— Jan 15, 1977: 90.5 mhz; 760 w. -771 ft TL: N35 18 40 W83 10 34. Stereo. Hrs opn: 24 Box 2728, Western Carolina Univ., 28723. Secondary address: WCU-Moore Bldg., Suite 18 28723. Phone: (828) 227-7454. Phone: (828) 227-7173 (request line). Fax: (828) 227-7099. E-mail: info@wwcufm.com Web Site: www.wwcufm.com. Licensee: Western Carolina University. Population served: 14,408 Natl. Network: ABC. Wire Svc: Reuters Format: Adult contemp; rock; sports;. Target aud: 25-54; univ students & faculty, general public. ♦Will Candler, gen mgr; Matt Sprinkler, opns dir.

Dallas

***WSGE(FM)**— Oct 27, 1980: 91.7 mhz; 6 kw. Ant 853 ft TL: N35 24 26 W81 07 48. Stereo. Hrs opn: 6 AM-midnight Ray Craig Classroom Bldg., 201 Hwy. 321 S., 28034-1499. Phone: (704) 922-4286. Phone: (704) 922-6552. Fax: (704) 922-2347. E-mail: hall.cathis@gaston.edu Web Site: www.wsge.org. Licensee: Gaston College Board of Trustees. Population served: 1,500,000 Format: AAA, eclectic. News: 5 hrs wkly. Target aud: General. ♦Pat Skinner, pres; Cathis Hall, gen mgr & stn mgr; Cliff Anderson, mus dir.

WZRH(AM)— Jan 1, 1963: 960 khz; 1 kw-D, 500 w-N, DA-N. TL: N35 18 03 W81 10 13. Hrs open: 6 AM-midnight Box 477, 28034. Secondary address: 407 Robinson Clemmer Rd. 28034. Phone: (704) 922-3411. Phone: (704) 922-5960. Fax: (704) 922-6998. E-mail: manager@wzrh.com Web Site: www.wzrh.com. Licensee: Truth Broadcasting Corp. (group owner; (acq 6-30-2004; $775,000). Population served: 1,500,000 Law Firm: Smithwick & Belendiuk. Format: Talk. News: 10 hrs wkly. Target aud: Male 25-59; college educ, income 50K. ♦Stuart Epperson, pres & natl sls mgr.

Davidson

***WDAV(FM)**— Sept 1, 1973: 89.9 mhz; 100 kw. 807 ft TL: N35 26 55 W80 50 24. Stereo. Hrs opn: 24 Box 7178, 28035-7178. Secondary address: 423 N. Main St. 28036. Phone: (704) 894-8900. Fax: (704) 894-2997. E-mail: wdav@davidson.edu Web Site: www.wdav.org. Licensee: Trustees of Davidson College. Population served: 2,000,000 Natl. Network: PRI, NPR. Law Firm: Fletcher, Heald & Hildrreth. Format: Class. News: 2 hrs wkly. Target aud: General. ♦F. Kim Hodgson, gen mgr; Luann Ritsema, mktg dir & prom dir; Frank Dominguez, progmg dir; Larry Schropp, chief of engrg; Liz Barr, traf mgr; Joe Brant, disc jockey; Liz Silverson Barr, disc jockey; Mike McKay, disc jockey; Ted Weiner, mus dir & disc jockey.

Dobson

WYZD(AM)— Oct 10, 1978: 1560 khz; 1 kw-D. TL: N36 23 36 W80 44 05. Hrs open: Box 797, 131 1/2 Atkin St., 27017. Phone: (336) 356-1560. Licensee: Gospel Broadcasting Inc. (acq 5-3-02). Rgnl. Network: N.C. News Net. Format: Gospel. ♦Ricky Cothren, gen mgr.

Dunn

WCKB(AM)— Dec 7, 1946: 780 khz; 7 kw-D, 1 w-N. TL: N35 17 00 W78 35 49. Hrs open: Sunrise-sunset Box 789, 28335. Secondary address: 17336 US 421 S. 28334. Phone: (910) 892-3133. Fax: (910) 892-3135. E-mail: wckb@wckb780.com Web Site: www.wckb780.com.

Licensee: N.C. Central Broadcasters Inc. (acq 9-15-89; $216,000; FTR: 10-2-89). Population served: 25,000 Rgnl. Network: N.C. News Net. Format: Southern gospel, religious. News: 6 hrs wkly. Target aud: 25 plus; Christian, family-oriented with regional interests. Spec prog: Buy-sell-trade 9 hrs wkly.Charles Fowler, pres; Ronald Tart, gen mgr, gen sls mgr, natl sls mgr, rgnl sls mgr, prom mgr & adv dir; Margie Hughes, progmg dir; Neal Wood, asst music dir; Lottie Squires, pub affrs dir, traf mgr & disc jockey; Bill Lambert, chief of engrg; Al Myatt, news rptr & sports cmtr; Graden Blackman, disc jockey; Margie Raynor, disc jockey; Mike West, disc jockey; Roger Layaou, disc jockey

WRCQ(FM)— May 17, 1971: 103.5 mhz; 48 kw. 502 ft TL: N35 03 09 W78 38 54. Stereo. Hrs opn: 24 1009 Drayton Rd., Fayetteville, 28303. Phone: (910) 864-5222. Fax: (910) 864-3065. Web Site: www.rock103rocks.com. Licensee: Cumulus Licensing Corp. Group owner: Cumulus Media Inc. (acq 3-12-01; grpsl). Population served: 500,000 Natl. Network: ABC. Law Firm: Borsari & Paxson. Format: Rock/AOR. News staff: one; News: one hr wkly. Target aud: 18-49. ♦Tom Haymond, gen mgr.

Durham

WDCG(FM)— Feb 28, 1948: 105.1 mhz; 100 kw. 1,141 ft TL: N35 52 20 W79 09 29. Stereo. Hrs opn: 3100 Smoketree Ct., Suite 700, Raleigh, 27604. Phone: (919) 871-1051. Fax: (919) 876-2929. Web Site: www.g105.com. Licensee: Capstar TX L.P. Group owner: Clear Channel Communications Inc. (acq 8-30-00; grpsl). Population served: 545,100 Natl. Network: ABC. Format: Contemporary Hit. Target aud: 18-49. ♦Ken Spitzer, gen mgr; Jon Robbins, opns dir; Tammy O'Dell, sls dir & disc jockey; Myron Bethea, gen sls mgr; Jessica Hayes, prom dir; Rick Schmidt, progmg dir; Dan McLeod, pub affrs dir; Fred Pace, chief of engrg; Tracy Leonard, traf mgr & farm dir.

WDNC(AM)— Apr 9, 1934: 620 khz; 5 kw-D, 1 kw-N, DA-2. TL: N36 02 10 W78 58 07. Hrs open: 24 4601 Six Forks Rd., Suite 520, Raleigh, 27609-5287. Phone: (919) 875-9100. Fax: (919) 510-6990. Web Site: www.620thebull.com. Licensee: WCHL-WDNC Inc. (Group owner: Curtis Media Group (acq 12-30-86). Natl. Network: Sporting News Radio Network. Natl. Rep: McGavren Guild. Wire Svc: AP Format: Sports. ♦Brian Maloney, gen mgr; Mike Stangl, prom dir; Adam Gold, progmg dir.

WDUR(AM)— 1947: 1490 khz; 1 kw-U. TL: N35 58 03 W78 53 18. Hrs open: 24 3100 Smoketree Ct., Suite 700, Raleigh, 27604. Phone: (919) 878-1500. Fax: (919) 876-8578. Web Site: www.wdur.com. Licensee: Triangle Sports Broadcasters LLC Group owner: Clear Channel Communications Inc. (acq 6-28-2005; $1.13 million). grpsl Population served: 95,438 Natl. Rep: Clear Channel. Law Firm: Wilmer, Cutler & Pickering. Format: Black Gospel. News: 0. Target aud: 35-54; 35-54. ♦Ken Spitzer, gen mgr; Jon Robbins, opns dir; Dan McLeod, traf mgr.

WFXC(FM)— May 15, 1971: 107.1 mhz; 2.6 kw. Ant 502 ft TL: N35 58 41 W78 48 59. Hrs open: 24 8001-101 Creedmoor Rd., Raleigh, 27613. Phone: (919) 848-9736. Fax: (919) 863-4859. Web Site: www.foxyhits.com. Licensee: Radio One Licenses LLC. Group owner: Radio One Inc. (acq 11-8-2001; grpsl). Population served: 95,438 Wire Svc: UPI Format: Urban contemp. News staff: one; News: 5 hrs wkly. Target aud: 25-54; African American. Spec prog: Gospel 4 hrs wkly. ♦Gary Weiss, gen mgr; Cy Young, opns VP.

WKSL(FM)—Burlington

***WNCU(FM)**— August 1995: 90.7 mhz; 50 kw. 433 ft TL: N36 03 33 W78 57 14. Stereo. Hrs opn: 24 1801 Fayetteville St., Box 19875, 27707. Phone: (919) 560-9628. Fax: (919) 560-5283. E-mail: ethorpe@wpo.nccu.edu Web Site: www.wncu.org. Licensee: North Carolina Central University. Population served: 600,000 Natl. Network: NPR, PRI. Format: Jazz, news/talk, info. News staff: one; News: 33 hrs wkly. Target aud: 25-54; middle class/middle age. Spec prog: Black, gospel. ♦Edith Thorpe, gen mgr, mktg VP, prom VP, adv VP & progmg VP; Chris Whitfield, opns mgr; Uchenna Johnson, dev dir & sls dir; B.H. Hudson, mus dir; Kimberley Pierce, news dir & pub affrs dir; Jim Davis, engrg mgr.

WRJD(AM)— Oct 14, 1954: 1410 khz; 5 kw-D, 290 w-N, DA-2. TL: N36 01 44 W78 51 00. Hrs open: 24 707 Leon St., 27704-4125. Phone: (919) 220-3226. Fax: (919) 220-0006. Licensee: Davidson Media Station WSRC Licensee LLC. Group owner: Willis Broadcasting Corp. (acq 3-20-2006; $1.2 million). Population served: 200,000 Format: Relg. News staff: one. Target aud: 35-54. ♦Chuck Harris, gen mgr.

WTIK(AM)— 1945: 1310 khz; 5 kw-D, 1 kw-N, DA-2. TL: N36 01 30 W78 54 08. Hrs open: 24 707 Leon St., 27704-4125. Phone: (919) 220-3226. Fax: (919) 220-0006. Licensee: Davidson Media Carolinas Stations LLC. Group owner: Davidson Media Group LLC (acq 5-10-2004; grpsl). Population served: 111,500 Format: Hispanic contemp. ♦Peter W. Davidson, pres; Chuck Harris, opns mgr.

***WXDU(FM)**— November 1983: 88.7 mhz; 1.18 kw. 103 ft TL: N36 02 08 W79 04 48. Stereo. Hrs opn: 24 Box 90689, Duke Station, 27708. Phone: (919) 684-2957. Fax: (919) 684-3260. E-mail: wxdu@duke.edu Web Site: www.wxdu.org. Licensee: Duke University. Format: Diversified. Target aud: General. Spec prog: Jazz 18 hrs, urban sound & hip hop 12 hrs wkly. ♦Jim Davis, chief of engrg.

Eden

WCLW(AM)— Aug 16, 1970: 1130 khz; 1 kw-D. TL: N36 31 21 W79 45 55. Hrs open: 116 S. Franklin St., Reidsville, 27320. Phone: (336) 634-1774. Web Site: www.carolinabaptistcollege.com/radio.html. Licensee: Dr. Jerry L. Carter dba Reidsville Baptist Church. (acq 6-26-98; $150,000). Format: Gospel. Target aud: 25-49. ♦Dean Lundy, gen mgr, opns mgr & progmg dir.

WGBT(FM)— Mar 20, 1949: 94.5 mhz; 100 kw. Ant 981 ft TL: N36 20 48 W79 54 30. Stereo. Hrs opn: 2-B Pai park, Greensboro, 27409. Secondary address: Box 3018, Winston Salem 27102. Phone: (336) 822-2000. Fax: (336) 887-0104. Licensee: Clear Channel Broadcasting Licenses Inc. Group owner: Clear Channel Communications Inc. (acq 1996; grpsl). Population served: 15,871 Natl. Rep: Clear Channel. Format: Sp. ♦Bill Dotson, gen mgr; Chris Rollins, progmg dir.

WLOE(AM)— Dec 20, 1946: 1490 khz; 1 kw-U. TL: N36 30 21 W79 46 18. Hrs open: 5 AM-10 PM
Rebroadcasts WMYN Mayodan NC 100%.
Box 279, Mayodan, 27027. Phone: (336) 427-9696; (336) 627-9563. Fax: (336) 548-4636. E-mail: info@wloewmyn.com Web Site: www.wloewmyn.com. Licensee: Mayo Broadcasting Corp. (acq 6-90; $100,000; FTR: 6-4-90). Population served: 200,000 Natl. Network: Salem Radio Network, USA. Format: Info, talk, relg. News staff: one; News: 30 hrs wkly. Target aud: 25 plus; general. ♦Richard D. Hall, pres; Mike Moore, gen mgr; Annette Moore, stn mgr.

Edenton

WBXB(FM)— June 18, 1976: 100.1 mhz; 50 kw. 302 ft TL: N36 07 11 W76 35 29. Stereo. Hrs opn: 24 Box 765, 27932. Secondary address: 1900 Paradise Rd. 27932. Phone: (252) 482-8680. Fax: (252) 482-4260. Licensee: Willis Family Broadcasting Inc. (group owner; acq 3-4-92; grpsl; FTR: 3-23-92). Population served: 7,000 Natl. Network: American Urban. Format: Gospel. News: 8 hrs wkly. Target aud: General. ♦Bishop L.E. Willis Sr., pres; Toina Willis, gen mgr.

WZBO(AM)— November 1955: 1260 khz; 1 kw-D, 34 kw-N. TL: N36 05 00 W76 36 00. Hrs open: Box 950, 27932. Phone: (252) 482-2104. Fax: (252) 482-5591. Web Site: www.ecri.net. Licensee: East Carolina Radio of Elizabeth City Inc. Group owner: East Carolina Radio Group (acq 3-12-90; $400,000 with co-located FM; FTR: 4-2-90) Population served: 115,000 Rgnl. Network: N.C. News Net. Law Firm: Tharrington, Smith & Hargrove. Format: Hot hits rgnl Mexican. ♦Rick Loesch, pres, gen mgr, stn mgr & opns mgr; Sam Walker, progmg dir.

Elizabeth City

WCNC(AM)— September 1939: 1240 khz; 1 kw-U. TL: N36 18 38 W76 13 56. Hrs open: 24 Box 1246, 27906-1246. Secondary address: 911 Parsonage St. Ext. 27909. Phone: (252) 335-4379. Fax: (252) 338-5275. E-mail: swalker@ecri.net Web Site: ecri.net. Licensee: East Carolina Radio of Elizabeth City Inc. Group owner: East Carolina Radio Group (acq 10-29-98; $230,000). Population served: 120,000 Format: Hot hits rgnl Mexican. ♦Rick Loesh, pres; Tom Charity, opns mgr & gen sls mgr; Sam Walker, progmg dir & progmg mgr.

WGAI(AM)— Nov 2, 1947: 560 khz; 1 kw-D, 500 w-N, DA-2. TL: N36 20 16 W76 14 49. Stereo. Hrs opn: 24 Box 1897, Kill Devil Hills, 27948. Phone: (252) 480-4655. Licensee: Max Radio of the Carolinas Licenses LLC. Group owner: MAX Media L.L.C. (acq 11-12-2002; grpsl). Population served: 250,000 Natl. Network: CNN Radio. Rgnl. Network: Agri-Net. Format: News/talk, sports. News staff: 3; News: 30 hrs wkly. Target aud: General. Spec prog: Relg 4 hrs, farm 7 hrs, Black 4 hrs, relg 4 hrs wkly. ♦Mike Smith, gen mgr.

***WGPS(FM)**— February 2003: 88.3 mhz; 50 kw. Ant 446 ft TL: N36 18 40 W76 17 34. Hrs open: 905 Halstead Blvd., Suite 29, 27909. Phone: (252) 334-1883. Fax: (252) 333-1459. E-mail: wgpsradio@earthlink.net Web Site: www.wgpsradio.com. Licensee: CSN International (group owner). Format: Christian teaching, praise & worship music. ♦Jeff Ozanne, gen mgr; Darla Ozanne, progmg dir.

WKJX(FM)— Aug 21, 1984: 96.7 mhz; 50 kw. Ant 407 ft TL: N36 12 10 W75 52 23. Stereo. Hrs opn: 24 Box 1246, 27906. Secondary address: 911 Parsonage St. E. 27906. Phone: (252) 338-0196. Fax: (252) 338-5275. E-mail: swalker@ecri.net Web Site: www.ecri.net. Licensee: East Carolina Radio of Elizabeth City Inc. Group owner: East Carolina Radio Group (acq 5-21-98; $475,000). Rgnl. Network: N.C. News Net. Rgnl rep: T-N. Format: Soft adult contemp. News staff: one; News: 1 hr wkly. Target aud: 18-55. ♦Rick Loesch, pres; Tom Charity, gen mgr & opns mgr; Sam Walker, progmg dir.

WRSF(FM)—Columbia, June 13, 1983: 105.7 mhz; 100 kw. 613 ft TL: N35 53 18 W76 13 50. (CP: Ant 987 ft.). Stereo. Hrs opn: 24 Box 1418, Nags Head, 27959. Secondary address: 2422 S. Wrightsville Ave., Nags Head 27959. Phone: (252) 449-8331. Fax: (252) 449-8354. Web Site: www.ecri.net. Licensee: East Carolina Radio of Elizabeth City Inc. Group owner: East Carolina Radio Group (acq 1996). Population served: 550,000 Format: Hot Country. News staff: one; News: 5 hrs wkly. Target aud: 18-54; young & mid-range adults. ♦John Maloney, gen sls mgr & natl sls mgr; Jerry Barco, traf mgr; Ray Hall, disc jockey; Tom Charity, opns mgr & disc jockey.

***WRVS-FM**— Mar 18, 1986: 89.9 mhz; 41 kw. 280 ft TL: N36 16 55 W76 12 44. Stereo. Hrs opn: 24 1704 Weeksville Rd., Campus Box 800, Williams Hall, 27909. Phone: (252) 335-3515. Fax: (252) 335-3745. Licensee: Elizabeth City State University. Natl. Network: NPR, PRI. Format: Urban contemp, var div, pub affrs. News staff: one; News: 5 hrs wkly. Target aud: 18-24; young adult, college. Spec prog: Jazz 6 hrs, Black 20 hrs, gospel 20. ♦Willie Gilchnist, CEO; Melbay Brown, gen mgr.

Elizabethtown

WBLA(AM)— Aug 3, 1956: 1440 khz; 5 kw-D, 197 w-N. TL: N34 37 32 W78 37 28. Hrs open: 24 Box 458, 512 Peanut Rd., 28337. Secondary address: Box 28, Clinton 28329. Phone: (910) 862-3184. Phone: (910) 862-2000. Fax: (910) 872-0100. E-mail: wgqr1057@carolina.net Web Site: www.wgqr1057.com. Licensee: Sound Business of Elizabethtown Inc. (acq 4-29-98; $525,000 with co-located FM). Population served: 92,000 Rgnl. Network: N.C. News Net. Format: Oldies, beach. News: 5 hrs wkly. Target aud: 25-54. Spec prog: Black gospel/relg 8 hrs wkly. ♦Lee Hauser, pres; Bruce Dickerson, VP; Patrick Dixon, gen mgr; Al Radlein, progmg dir, local news ed & disc jockey; Buddy Wommack, chief of engrg; Bill Monroe, disc jockey; Don Arnsan, disc jockey; Paul Reese, disc jockey.

WGQR(FM)—Co-owned with WBLA(AM). December 1989: 105.7 mhz; 7.7 kw. Ant 583 ft TL: N34 44 05 W78 47 25. Stereo. 24 Web Site: www.wgqr1057.com.200,000 Format: Southern gospel. News: 6 hrs wkly. ♦Al Radlein, local news ed & disc jockey; Buddy Edwards, disc jockey; Dan Arnsan, disc jockey; K.C Evers, disc jockey; Paul Brian, disc jockey.

Elkin

WIFM-FM— 1949: 100.9 mhz; 600 w. 709 ft TL: N36 11 33 W80 50 59. Hrs open: 24 Box 1038, 28621. Secondary address: 813 N. Bridge St. 28621. Phone: (336) 835-2511. Fax: (336) 835-5248. E-mail: wifm@wifmradio.com Web Site: www.wifmradio.com. Licensee: Yadkin Valley Broadcasting Corp. (acq 1-12-2004; $1.15 million). Natl. Network: ABC. Format: Classic hits/Today's hits, Adult contemp. News staff: one. Target aud: 25-45. ♦Gary York, pres; Paula Rice, gen mgr & gen sls mgr; Jerry Laws, progmg dir; Stony Owens, engrg mgr.

Elon

***WSOE(FM)**— November 1978: 89.3 mhz; 500 w. 104 ft TL: N36 06 25 W79 30 22. Stereo. Hrs opn: Campus Box 6000, 27244. Phone: (336) 584-9763. Phone: (336) 278-7510. Fax: (336) 278-7298. E-mail: wsoe@elon.edu Web Site: www.elon.edu/wsoe. Licensee: Elon University. Format: Diversified. ♦Jon Chuk, gen mgr; Greg Piel, progmg dir & progmg.

Enfield

WBOB-FM— 2007: 107.3 mhz; 4.1 kw. Ant 279 ft TL: N36 09 59 W77 46 46. Hrs open: 301 S. Church St., Suite 270, Rocky Mount, 27804. Phone: (252) 446-9262. Fax: (252) 446-9261. Licensee: Julie Epperson. Format: Gospel. ♦Noel Winslow, gen mgr.

Erwin

***WUAW(FM)**— May 11, 1990: 88.3 mhz; 3 kw. 191 ft TL: N35 20 15 W78 39 49. Stereo. Hrs opn: 24 Triton High School, 215 Maynard Lake Rd., 28339. Phone: (910) 897-8070. Fax: (910) 897-3148. E-mail: wuaw883fm@gaggle.net Web Site: www.wuaw.homestead.com. Licensee: Central Carolina Community College. Format: Variety/Diverse. ♦Matt Garrett, pres; Ron McLamb, gen mgr & progmg dir; Dr. Jim Davis, chief of engrg.

Fair Bluff

WODR(FM)— 2003: 105.3 mhz; 11 kw. Ant 492 ft TL: N34 17 01 W78 48 09. Hrs open: 629 S. Madison, Whiteville, 28472. Phone: (910) 640-1180. Licensee: The Padner Group LLC (acq 12-4-2003; $1.25 million). Format: Oldies. ♦Alan Goodman, pres & gen mgr; Michelle Hurley, mktg dir; Bill Shannon, progmg dir.

WSRC(AM)— July 1988: Stn currently dark. 1480 khz; 1 kw-D, 48 w-N. TL: N34 19 23 W79 00 07. Hrs open: 24 Box 158, 28439. Secondary address: 12045 Andrew Jackson Hwy. 28439. Phone: (910) 649-1480. Fax: (910) 649-7266. E-mail: tony1480@tds.net Licensee: Rama Radio of North Carolina Inc. (acq 4-1-2006; $120,000). Population served: 250,000 Format: Oldies. ♦Anthony Lee, stn mgr.

Fairmont

WFMO(AM)— July 13, 1953: 860 khz; 1 kw-D, 12 w-N. TL: N34 31 03 W79 06 19. (CP: COL Conway, SC. 50 kw-D, 740 w-N, DA-2. TL: N33 49 40 W79 10 20). Hrs opn: Box 668, Hwy 41 N., 28340. Phone: (910) 628-6781. Fax: (910) 628-6648. Licensee: Pro Media Inc. Group owner: Clark-Pittman Group (acq 12-31-86; $600,000 with co-located FM; FTR: 11-10-86) Population served: 150,000 Rgnl. Network: N.C. News Net. Format: Black, gospel, relg. Target aud: 25-54. Spec prog: Farm 5 hrs wkly. ♦James C. Clark, pres, gen mgr, stn mgr & gen sls mgr.

WSTS(FM)— August 1975: 100.9 mhz; 50 kw. Ant 489 ft TL: N34 16 17 W78 56 24. Stereo. Hrs opn: Box 668, 28340. Phone: (910) 628-6781. Fax: (910) 628-6648. E-mail: wstf@carolina.net Licensee: Davidson Media Station WSTS Licensee LLC. (acq 10-31-2005). Population served: 500,000 Rgnl. Network: N.C. News Net. Format: Southern gospel. ♦James Clark, gen mgr; Shanna Todd, disc jockey.

Fairview

WPEK(AM)— July 4, 1997: 880 khz; 1.1 kw-D, DA. TL: N35 32 52 W82 28 16. Hrs open: 13 Summerlin Rd., Asheville, 28806. Phone: (828) 255-1906. Fax: (828) 255-7850. Web Site: www880ThePEK. Licensee: Clear Channel Broadcasting Licenses Inc. Group owner: Clear Channel Communications Inc. (acq 3-21-01; grpsl). Population served: 225,000 Natl. Network: CBS. Format: Talk. Target aud: 25-64; generally upscale adults. ♦Ken Salyer, gen mgr.

Farmville

WGHB(AM)— Dec 12, 1959: 1250 khz; 5 kw-D, 2.5 kw-N, DA-2. TL: N35 36 17 W77 34 29. Hrs open: 24 Box 3333, Greenville, 27836. Phone: (252) 317-1250. Web Site: www.pirateradio1250.com. Licensee: Pirate Media Group LLC (acq 11-25-03; $650,000). Population served: 500,000 Natl. Network: USA. Format: Talk, sports. News: 8 hrs wkly. Target aud: 25-54. ♦Troy Dreyfus, gen mgr; Wesley Hines, disc jockey.

WWNK(FM)— Mar 24, 1974: 94.3 mhz; 3.9 kw. Ant 407 ft TL: N35 36 25 W77 28 05. Stereo. Hrs opn: 24 211 Commerce St., Suite C, Greenville, 27835. Phone: (252) 756-9898. Fax: (252) 355-2234. Licensee: Inner Banks Media LLC. Group owner: Archway Broadcasting Group (acq 3-12-2007; grpsl). Population served: 100,000 Rgnl. Network: N.C. News Net. Law Firm: Cole, Raywid & Braverman. Format: Country. News: 3 hrs wkly. Target aud: 25-54. ♦Tony Denton, gen mgr & chief of engrg.

Fayetteville

WAZZ(AM)— 1947: 1490 khz; 1 kw-U. TL: N35 03 45 W78 54 30. Hrs open: 24 508 Person St., 28301. Phone: (910) 486-2055. Phone: (910) 484-1490. Fax: (910) 323-5635. Licensee: WFLB License L.P. Group owner: Beasley Broadcast Group (acq 1996; $228,635). Population served: 200,000 Natl. Network: Westwood One. Rgnl. Network: Southern Farm. Natl. Rep: D & R Radio. Format: MOR. News: 14 hrs wkly. Target aud: 35-64. Spec prog: Atlanta Braves baseball, Charlotte Bobcats basketball, auto racing. ♦George G. Beasley, CEO & gen mgr; Mac Edwards, VP & mktg mgr; Curt Nunnery, opns mgr, sls dir & gen sls mgr; Bryan Kusilka, natl sls mgr; Van Clough, chief of engrg.

WFAY(AM)— 1947: 1230 khz; 1 kw-U. TL: N35 04 15 W78 52 45. Hrs open: 24 346 Wagoner Dr., 28303. Phone: (910) 222-3776. Fax: (910) 860-3329. Licensee: Norsan Consulting and Management Inc. (acq 5-25-2006; $850,000). Natl. Network: ESPN Radio. Rgnl. Network: N.C. News Net. Format: Sp. ♦Norberto Sanchez, CEO; Jeffrey M. Andrulonis, gen mgr; Cheryl Roberts, stn mgr.

WFLB(FM)—See Laurinburg

WFNC(AM)— 1940: 640 khz; 10 kw-D, 1 kw-N. TL: N35 04 46 W78 55 58. Hrs open: 24 1009 Drayton Rd., 28303. Phone: (910) 864-5222. Fax: (910) 864-6208. Web Site: www.cumulus.com. Licensee: Cumulus Licensing Corp. Group owner: Cumulus Media Inc. (acq 3-12-01; grpsl). Population served: 400,000 Natl. Network: CBS. Wire Svc: AP Format: News/talk. News staff: 4; News: 20 hrs wkly. Target aud: 35 plus. ♦Tom Haymond, gen mgr & stn mgr; Perry Stone, opns dir; Alan Buffalo, gen sls mgr; Flo Knight, prom dir; Laura Chais-Price, progmg dir, news dir & pub affrs dir; Gail Galbreath, engrg dir; Brandy Tew, traf mgr; Lindsay Ladd, news rptr; Steve Blackmon, news rptr.

WQSM(FM)—Co-owned with WFNC(AM). 1947: 98.1 mhz; 100 kw. 830 ft TL: N35 04 46 W78 55 58. Stereo. Web Site: www.cumulus.com. Format: CHR. Target aud: General. ♦Chris Chaos, progmg dir; Paul Michels, rsch dir; Robin Duff, traf mgr; Rick Jensen, disc jockey; Rick O'Shea, disc jockey.

***WFSS(FM)**— Dec 7, 1977: 91.9 mhz; 100 kw. 440 ft TL: N35 04 22 W78 53 27. Stereo. Hrs open: 24 1200 Murchison Rd., 28301. Phone: (910) 672-1381. Fax: (910) 672-1964. E-mail: wfss@uncfsu.edu Web Site: wfss.org. Licensee: Fayetteville State University Board of Trustees. Population served: 350,000 Natl. Network: NPR, PRI. Wire Svc: AP Format: Jazz, news. News staff: one; News: 38 hrs wkly. Target aud: 18 plus; general. Spec prog: Gospel 2 hrs, African rhythms 2 hrs, class 5 hrs, folk 3 hrs wkly. ♦Joseph C. Ross, gen mgr; Phyllis Washington, dev dir, mktg dir & prom dir; Janet G. Wright, progmg dir; Jimmy Miller, mus dir; Kathy Klaus, news dir; Eric Lanc, chief of engrg; Arvetra Jones Jr., relg ed; Susan Franzblau Ph.D., women's int ed.

WIDU(AM)— Jan 20, 1958: 1600 khz; 5 kw-D, 147 w-N, DA-2. TL: N35 02 58 W78 51 33. Hrs open: 24 Box 2247, 28302. Secondary address: 1338 Bragg Blvd. 28301. Phone: (910) 483-6111. Phone: (910) 486-9438. Fax: (910) 483-6601. E-mail: slofton3@aol.com Licensee: Charles W. Cookman (acq 1-17-89). Population served: 300,000 Format: Black, gospel, news/talk. ♦Wes Cookman, pres; Sandra Lofton, gen mgr; Robert Smith, opns mgr; Val Holiday, progmg dir.

***WYBH(FM)**—Not on air, target date: unknown: 91.1 mhz; 500 w vert. Ant 226 ft TL: N35 05 20 W78 54 37. Hrs open: 11530 Carmel Commons Blvd., Charlotte, 28226-3976. Phone: (704) 523-5555. Web Site: www.bbnradio.org. Licensee: Bible Broadcasting Network Inc. ♦Lowell Davey, pres.

WZFX(FM)—Whiteville, Feb 21, 1962: 99.1 mhz; 100 kw. 1,000 ft TL: N34 44 05 W78 47 25. Stereo. Hrs opn: Box 710, 28302. Phone: (910) 486-4991. Fax: (910) 486-6720. Web Site: www.fox99.com. Licensee: WDAS License L.P. Group owner: Beasley Broadcast Group (acq 5-8-97; $11.5 million). Natl. Rep: D & R Radio. Format: Urban contemp. Target aud: 18-49. ♦George G. Beasley, CEO; Daniel Highsmith, VP & gen mgr; Mac Edwards, opns VP; Tila Comstock, gen sls mgr; Bryan Kusilka, natl sls mgr; Jeff Anderson, progmg mgr; Van Clough, chief of engrg.

Fletcher

WQNQ(FM)— Feb 5, 1991: 104.3 mhz; 470 w. Ant 1,145 ft TL: N35 31 39 W82 29 49. Stereo. Hrs opn: 24 13 Summerlin Rd., Asheville, 28806. Phone: (828) 257-2700. Fax: (828) 281-3299. Licensee: Clear Channel Broadcasting Licenses Inc. Group owner: Clear Channel Communications Inc. (acq 3-21-2001; grpsl). Population served: 150,000 Law Firm: Reddy, Begley & McCormick. Format: 80s, 90s & now. News staff: 2; News: 10 hrs wkly. Target aud: 25-54. Spec prog: Relg 2 hrs, news/talk 5 hrs wkly. ♦Ken Falyer, gen mgr.

Forest City

WAGY(AM)— Oct 15, 1958: 1320 khz; 1 kw-D, 500 w-N, DA-N. TL: N35 21 19 W81 52 52. Hrs open: Box 280, 28043. Phone: (828) 245-9887. Fax: (828) 245-9880. Licensee: WAGY Inc. (acq 12-13-85; $310,000; FTR: 11-4-85). Format: Real country. ♦Malcolm Watson, gen mgr.

WTPT(FM)— Sept 10, 1947: 93.3 mhz; 93 kw. Ant 2,030 ft TL: N35 16 19 W82 14 00. Stereo. Hrs open: 25 Garlington Rd., Greenville, SC, 29615. Phone: (864) 271-9200. Fax: (864) 242-1567. Web Site: www.newrock933.com. Licensee: Entercom Greenville License LLC. Group owner: Barnstable Broadcasting Inc. (acq 10-7-2005; grpsl). Population served: 1,000,000 Format: Active rock. ♦David J. Field, pres; Sharon Day, gen mgr; Mark Hendrix, progmg dir; Paige Pirtle, news dir.

WWOL(AM)— Sept 10, 1947: 780 khz; 10 kw-D. TL: N35 21 02 W81 54 04. Hrs open: Sunrise-sunset 1381 W. Main St., 28043. Phone: (828) 245-0078. Fax: (828) 245-8528. E-mail: wwol@rfci.net Licensee: Holly Springs Baptist Church. (acq 3-1-90; $150,000; FTR: 3-19-90). Population served: 608,804 Natl. Network: USA. Format: Southern gospel, relg. News staff: 3. Target aud: General. Spec prog: Our community forum, NC family policy issues. ♦Wade H. Huntley, pres & stn mgr; Ray Davis, chief of opns, progmg dir & pub affrs dir; Terri Frashier, sls dir, gen sls mgr & mktg dir; Jean Bruce, traf mgr.

Franklin

***WFQS(FM)**— Mar 31, 1989: 91.3 mhz; 265 w. 2,304 ft TL: N35 10 24 W83 34 52. Stereo. Hrs opn: 24
Rebroadcasts WCQS(FM) Asheville 100%.
73 Broadway, Asheville, 28801. Phone: (828) 253-6875. Fax: (828) 253-6700. Web Site: www.wcqs.org. Licensee: Western N.C. Public Radio Inc. Population served: 500,000 Natl. Network: NPR, PRI. Law Firm: Cohn & Marks. Format: Class, jazz, news. News staff: one; News: 35 hrs wkly. Target aud: 25 plus. Spec prog: Folk 9 hrs wkly. ♦Edward Subkis, gen mgr; Margaret Marchuk, sls dir; Barbara Sayer, progmg dir; Richard J. Kowal, mus dir; David Hurand, news dir; Terry Spaight, chief of engrg.

WFSC(AM)— May 5, 1957: 1050 khz; 1 kw-D. TL: N35 12 42 W83 22 07. Hrs open: 24 Box 470, 28744. Secondary address: 180 Radio Hill Rd. 28734. Phone: (828) 524-4418. Phone: (828) 524-5395. Fax: (828) 524-2788. E-mail: jduke@gacaradio.com Web Site: www.1050wfsc.com. Licensee: Sutton Radiocasting Corp. Group owner: Georgia-Carolina Radiocasting Companies (acq 12-14-2001; grpsl). Population served: 75,000 Natl. Network: CBS, ABC. Rgnl. Network: N.C. News Net. Law Firm: Dan J. Alpert. Format: Oldies. News staff: one; News: 12 hrs wkly. Target aud: 35+. ♦Douglas M. Sutton Jr., pres; Jeremy Duke, VP & gen mgr; Chad Dorsette, news dir; Tim Stephens, chief of engrg.

WNCC-FM—Co-owned with WFSC(AM). Sept 1, 1965: 96.7 mhz; 6 kw. 204 ft TL: N35 12 42 W83 22 07. Stereo. 24 Web Site: www.967wncc.com.75,000 Natl. Network: ABC. Format: Country. News staff: one; News: 6 hrs wkly. Target aud: 25 plus.

WPFJ(AM)— May 24, 1979: 1480 khz; 5 kw-D, 13 w-N. TL: N35 10 58 W83 21 27. Stereo. Hrs opn: 24 185 Franklin Plaza, 28744. Phone: (828) 369-5033. Fax: (828) 369-3197. E-mail: thedove@wpfj.com Web Site: www.wpfj.com. Licensee: Drake Enterprises Ltd. (acq 2-16-94; $250,000; FTR: 3-28-94). Population served: 66,500 Natl. Network: Salem Radio Network. Format: Relg. News: 12 hrs wkly. Target aud: 25-54. ♦Jeremy Duke, stn mgr; Morris Stamey, rgnl sls mgr; Brenda Wooten, prom mgr; Randy Raby, progmg dir; Chad McConnell, disc jockey.

Fuquay-Varina

WNNL(FM)— Dec 1, 1980: 103.9 mhz; 7.9 kw. 577 ft TL: N35 35 47 W78 45 18. Stereo. Hrs open: 24 8001-101 Creedmoor Rd., Raleigh, 27613. Phone: (919) 848-9736. Fax: (919) 848-4724. E-mail: mmarinaro@radio-one.com Web Site: www.thelight1039.com. Licensee: Radio One Licenses LLC. Group owner: Radio One Inc. (acq 11-8-01;

grpsl). Population served: 870,000 Natl. Network: ABC. Natl. Rep: Christal. Format: Inspirational, gospel. News: 20 hrs wkly. Target aud: 25-54; educated, upper income, professionals. ♦Gary Weiss, gen mgr; Cy Young, opns dir; Kim Gattis, sls dir; Jodi Luke, natl sls mgr; Steven Walker, gen sls mgr & prom dir; Jerry Smith, progmg dir.

Garner

WRTG(AM)— Aug 11, 1969: 1000 khz; 1 kw-D. TL: N35 43 50 W78 36 12. Hrs open:
Rebroadcasts WRTP(AM) Chapel Hill 100%.
Estuardo Valdemar Rodriguez and Leonor Rodriguez Stns, 1010 Vermont Ave. N.W., Suite 100, Washington, DC, 20005. Phone: (202) 638-1959. Fax: (202) 393-7464. E-mail: valtravel@rcn.com Licensee: Estuardo Valdemar Rodriguez and Leonor Rodriguez. Group owner: WRTP Radio Network (acq 12-13-2004; grpsl). Population served: 121,577 Natl. Network: USA. Format: Mexican rgnl. ♦Estuardo Valdemar Rodriguez, gen mgr.

Gaston

WTRG(FM)— Nov 28, 1988: 97.9 mhz; 1.35 kw. Ant 488 ft TL: N36 27 38 W77 33 52. Stereo. Hrs opn: 24 Box 910, Roanoke Rapids, 27870. Phone: (252) 538-9790. Fax: (252) 538-0378. Licensee: First Media Radio LLC. (group owner; (acq 7-22-2003; grpsl). Rgnl. Network: N.C. News Net., Va. News Net. Format: Oldies. News: 2 hrs wkly. Target aud: 21-54. ♦Al Haskins, gen mgr; Les Atkins, opns dir.

Gastonia

WBAV-FM— September 1947: 101.9 mhz; 99 kw. Ant 987 ft TL: N35 13 56 W81 16 35. Stereo. Hrs opn: 1520 South Blvd., Suite 300, Charlotte, 28203. Phone: (704) 342-2644. Fax: (704) 227-8985. Web Site: www.v1019.com. Licensee: Infinity Radio Holdings Inc. Group owner: Infinity Broadcasting Corp. (acq 11-13-98; grpsl). Population served: 900,000 Natl. Rep: Christal. Law Firm: Leventhal, Senter & Leman. Format: Adult contemp. News staff: one; News: 20 hrs wkly. Target aud: 25-54; black adults. Spec prog: Gospel 6 hrs, mixed shows 8 hrs wkly. ♦Bill Schoening, gen mgr; Terri Avery, opns mgr.

WGAS(AM)—See South Gastonia

WGNC(AM)— March 1939: 1450 khz; 1 kw-U. TL: N35 16 32 W81 12 04. Hrs open: 24 1511 W. Dixon Blvd., Shelby, 28152. Phone: (704) 868-8222. Fax: (704) 482-4680. E-mail: netoldies@aol.com Web Site: www.theboss.us. Licensee: HRN Broadcasting Inc. (acq 8-31-2006; $1.5 million with WOHS(AM) Shelby). Population served: 47,142 Natl. Network: ABC, CBS Radio, Westwood One. Rgnl. Network: N.C. News Net. Format: Oldies, sports, beach. News staff: one; News: 5 hrs wkly. Target aud: 18-49. ♦D. Mark Boyd III, pres; Terresa Hastings, VP; Calvin R. Hastings, gen mgr & gen sls mgr; Harold Watson, sls dir & adv dir; Lori Deitz, prom mgr; Mike Slade, progmg dir; Andy Foster, mus dir; Anna McGinnis, news dir & pub affrs dir; Larry Schropp, chief of engrg.

WZRH(AM)—See Dallas

Goldsboro

WFMC(AM)— Nov 11, 1951: 730 khz; 1 kw-D, 98 w-N. TL: N35 22 25 W78 00 41. Stereo. Hrs opn: 24 2581 U.S. Hwy. 70 W., 27530. Phone: (919) 734-4211. Fax: (919) 736-3876. Web Site: www.730wfmc.com. Licensee: New Age Communications Inc. Group owner: Curtis Media Group (acq 6-95; $300,000). Population served: 110,000 Rgnl. Network: Southern Farm, N.C. News Net. Format: Black gospel, relg. News staff: one; News: 10 hrs wkly. Target aud: 18 plus. ♦Donald W. Curtis, pres; Bill Johnston, gen mgr; Mary Barnes, gen sls mgr & disc jockey; Thomas Vick, news dir & disc jockey.

WGBR(AM)— 1939: 1150 khz; 5 kw-D, 800 w-N, DA-2. TL: N35 22 26 W78 00 42. Stereo. Hrs opn: 24 2581 U.S. Hwy. 70 W., 27530. Phone: (919) 736-1150. Fax: (919) 736-3876. Licensee: New Age Communications L.P. Group owner: Curtis Media Group (acq 2-15-89; $2.2 million with co-located FM; FTR: 3-6-89) Population served: 107,000 Natl. Network: CNN Radio. Rgnl. Network: N.C. News Net. Wire Svc: ABC Format: News/talk. News staff: one; News: 11 hrs wkly. Target aud: 25 plus. ♦Donald W. Curtis, pres; Bill Johnston, gen mgr; Mary Barnes, sls dir & gen sls mgr; Wayne Alley, progmg dir; Thomas Vick, news dir; Kari DelaCruz, chief of engrg & opns.

WYMY(FM)—Co-owned with WGBR(AM). 1946: 96.9 mhz; 100 kw. 1,056 ft TL: N35 23 52 W78 08 07. Stereo. 24 3012 Highwoods Blvd., Raleigh, 27604. Phone: (919) 790-9392.1,500,000 Format: Mexican, Sp. News: 2 hrs wkly. Target aud: 25-54. ♦Jon Bloom, gen mgr.

WRSV(FM)—See Rocky Mount

WSSG(AM)— Oct 22, 1955: 1300 khz; 1 kw-D, 49 w-N. TL: N35 24 08 W78 01 20. Hrs open: 116 W. Mulberry St., 27530. Phone: (919) 734-1300. Licensee: Robert Swinson (acq 12-18-97; $75,000). Format: Christian music. ♦Reginald Swinson, gen mgr.

WWNF(FM)— Feb 2, 1972: 102.3 mhz; 2.1 kw. Ant 561 ft TL: N35 23 54 W78 00 38. Stereo. Hrs opn: 24
Simulcast with WWMY(FM) Raleigh 100%.
3012 Highwoods Blvd., Suite 201, Raleigh, 27604. Phone: (919) 790-9392. Fax: (919) 790-8369. E-mail: y1029@curtismedia.com Web Site: www.y1029.com. Licensee: New Age Communications Inc. Group owner: Curtis Media Group (acq 7-1-96; $550,000). Population served: 120,000 Format: Oldies 60s & 70s. News staff: 4. ♦Don Curtis, pres; Mike Hartel, gen mgr.

Graham

WBAG(AM)—See Burlington-Graham

WSML(AM)— Dec 2, 1967: 1200 khz; 10 kw-D, 1 kw-N, DA-N. TL: N36 08 01 W79 28 14. Hrs open: 24
Rebroadcasts WSJS-AM 600, Winston-Salem NC 90%.
875 W. 5th St., Winston Salem, 27101. Phone: (336) 227-4254. Fax: (336) 227-4254. Licensee: Crescent Media Group LLC. Group owner: Infinity Broadcasting Corp. (acq 2-14-2007; grpsl). Population served: 1,000,000 Law Firm: Miller & Fields,P.C. Format: News/talk. Spec prog: Black 18 hrs wkly. ♦Tom Hamilton, gen mgr & gen sls mgr; Larry Ingold, progmg dir; George Newman, chief of engrg.

Granite Falls

WYCV(AM)— Feb 22, 1963: 900 khz; 2.5 kw-D, 251 w-N. TL: N35 47 10 W81 25 00. Hrs open: 5 AM-11 PM (M-S); 6 AM-9 PM (Su) Box 486, 398 South Main St., 28630. Phone: (828) 396-3361. Phone: (828) 396-3362. Fax: (828) 396-9193. E-mail: wycvradio@charter.net Web Site: www.gospel9.com. Licensee: Freedom Broadcasting Corp. Group owner: Marvin L. Sizemore (acq 4-29-92). Population served: 684,000 Natl. Network: USA, AP Radio. Law Firm: Smithwick & Belendiuk. Format: Relg, southern gospel. News staff: one; News: 2 hrs wkly. Target aud: 3-100. Spec prog: Gospel. ♦Marvin Sizemore, pres; Buddy Sizemore, gen mgr & stn mgr; Clyde Smith, mus dir & sls; Ted Fuller, chief of engrg & engr; Teresa S. Sizemore, asst music dir & traf mgr.

Greensboro

WCOG(AM)— May 22, 1948: 1320 khz; 5 kw-U, DA-2. TL: N36 09 01 W79 54 48. Hrs open: 24 3404-H West Wendover Ave., 27407. Phone: (336) 294-0699. Fax: (336) 294-4988. E-mail: gerryjfranzen@disney.com Web Site: www.radiodisney.com. Licensee: Radio Disney Group, LLC (acq 4-29-99; $500,000). Population served: 144,076 Format: Family Hits. Target aud: General. ♦Gerry Franzer., pres & stn mgr; Chris Nowak, prom mgr.

WEAL(AM)—Listing follows WQMG-FM.

WJMH(FM)—See Reidsville

WKEW(AM)— Feb 16, 1942: 1400 khz; 1 kw-U. TL: N36 04 00 W79 47 49. Hrs open: 24 hrs 7 days 4405 Providence Ln., #A, Winston Salem, 27106-3226. Phone: (336) 759-0363. Fax: (336) 759-0366. Web Site: www.1340thelight.com. Licensee: Truth Broadcasting Corp. (group owner; acq 8-9-00; $800,000). Population served: 245,000 Format: Urban Gospel, Religious. Target aud: 35 plus; African-American. ♦Stuart Epperson, pres & gen mgr.

WMAG(FM)—See High Point

***WNAA(FM)**— 1979: 90.1 mhz; 10 kw. 467 ft TL: N36 04 58 W79 46 08. Stereo. Hrs opn: 24 North Carolina A&T State Univ., Price Hall, Suite 200, 27411-1135. Phone: (336) 334-7936. Fax: (336) 334-7960. E-mail: wnaafm@ncat.edu Web Site: www.aggienewsonline.com. Licensee: North Carolina Agricultural & Technical State University. Population served: 900,000 Format: Gospel, black, jazz. News: 7 hrs wkly. Target aud: 35-45; general. Spec prog: Blues 3 hrs, reggae 7 hrs, oldies 5 hrs wkly. ♦Tony Welborne, gen mgr, chief of opns & dev dir; Cherie Lofton, progmg dir; Mamie Johnson, pub affrs dir; Larry Allen, chief of engrg.

WPAW(FM)—Winston-Salem, April 1947: 93.1 mhz; 100 kw. Ant 1,050 ft TL: N36 16 33 W79 56 27. Stereo. Hrs opn: 24 7819 National Service Rd., Suite 401, 27409. Phone: (336) 605-5200. Fax: (336) 605-5221. Web Site: www.931wolfcountry.com. Licensee: Entercom Greensboro License LLC. Group owner: Entercom Communications Corp. (acq 12-13-99; grpsl). Population served: 940,600 Natl. Network: CBS. Natl. Rep: McGavren Guild. Format: Country. ♦Brent Millar, gen mgr; Lisa Powell, sls dir & gen sls mgr; Greg Carpenter, gen sls mgr; Randy Bliss, progmg dir; Larry Allen, chief of engrg; Brian McCall, traf mgr.

WPET(AM)— 1954: 950 khz; 500 w-D. TL: N36 02 16 W79 47 42. Hrs open: 7819 National Service Rd., Ste. 401, 27409. Phone: (336) 605-5200. Fax: (336) 387-7206. Web Site: www.wpetam950.com. Licensee: Entercom Greensboro License LLC. Group owner: Entercom Communications Corp. (acq 1-28-2002; $20.5 million with co-located FM). Population served: 13,200 Format: Southern gospel. Target aud: 25-54. ♦Brent Miller, gen mgr; Dave Compton, progmg dir.

WSMW(FM)—Co-owned with WPET(AM). Jan 9, 1958: 98.7 mhz; 100 kw. 1,000 ft TL: N36 02 16 W79 47 42. Stereo. 24 Web Site: www.987simon.com.354,400 Format: Adult contemporary. Target aud: 18-49. ♦Sean Sellers, progmg dir.

WPOL(AM)—Winston-Salem, Mar 25, 1937: 1340 khz; 1 kw-U. TL: N36 04 26 W80 15 19. Hrs open: 24 4405 Providence Ln., Winston-Salem, 27106. Phone: (336) 759-0363. Fax: (336) 759-0366. Web Site: www.1340thelight.com. Licensee: Truth Broadcasting Corp. (acq 5-10-2000). Format: Relg, Urban gospel. News: one hr wkly. Relg. ♦Stuart Epperson Jr., pres & gen mgr.

***WQFS(FM)**— January 1970: 90.9 mhz; 1.9 kw. 200 ft TL: N36 05 39 W79 53 21. Stereo. Hrs opn: 24 Box 17714, Founders Halls, 5800 W. Friendly Ave., 27410. Phone: (336) 316-2352. Phone: (336) 316-2444. Web Site: www.gilford.edu. Licensee: Guilford College Board of Trustees. Population served: 280,000 Format: Free-form. News staff: one; News: 4 hrs wkly. Target aud: 15-50. ♦Erin Kauffman, gen mgr.

WQMG-FM— July 8, 1962: 97.1 mhz; 100 kw. 1,289 ft TL: N36 05 09 W79 45 38. (CP: TL: N35 56 43 W79 51 44). Stereo. Hrs opn: 24 7819 National Service Rd., Suite 401, 27409. Phone: (336) 605-5200. Fax: (336) 605-0138. Web Site: www.wqmg.com. Licensee: Entercom Greensboro License LLC. Group owner: Entercom Communications Corp. (acq 12-13-99; grpsl). Population served: 1,000,000 Natl. Network: ABC. Natl. Rep: McGavren Guild. Format: Smooth rhythm and blues, classic soul. News staff: one. Target aud: 18-49; Black. ♦Brant Millar, gen mgr; Lisa Powell, sls dir; Joyce Staley, prom dir; Larry Allen, chief of engrg; Brian McCall, traf mgr.

WEAL(AM)—Co-owned with WQMG-FM. Oct 5, 1962: 1510 khz; 1 kw-D, 250 w-CH. TL: N36 03 42 W79 47 35.Sunrise-sunset Fax: (336) 605-0138.820,000 Natl. Rep: McGavren Guild. Format: Gospel. Target aud: 25-54; North Carolina A&T State Univ. ♦Joseph Level, progmg dir.

***WUAG(FM)**— July 20, 1964: 103.1 mhz; 18.1 w. 230 ft TL: N36 03 51 W79 48 37. (CP: Ant 259 ft.). Stereo. Hrs opn: 24 Taylor Bldg., Univ. of North Carolina at Greensboro, 27412. Phone: (336) 334-5450. E-mail: wuag@uncg.edu Web Site: www.wuag.net. Licensee: University of North Carolina at Greensboro Board of Trustees. Population served: 170,000 Format: Progsv rock. News staff: one; News: 2 hrs wkly. Target aud: 12-40; high school & college students. Spec prog: Hip hop 6 hrs, world music 2 hrs, bluegrass 2 hrs, and blues 4 hrs wkly. ♦Jack Bonney, gen mgr.

WWBG(AM)— 1998: 1470 khz; 3.5 kw-D, 5 kw-N, DA-2. TL: N36 12 46 W79 54 46. Hrs open: Box 12876, Winston-Salem, 27117. Phone: (336) 784-9004. Fax: (336) 784-8337. E-mail: quepasa@quepasamedia.com Web Site: www.quepasamedia.com. Licensee: Davidson Media Station WWBG Licensee LLC. (acq 3-25-2005; swap with WTOB(AM) Winston-Salem for WDRU(AM) Wake Forest). Natl. Rep: Salem. Format: Mexican rgnl. ♦Roger Martinez, gen mgr.

Greenville

WNCT(AM)— 1940: 1070 khz; 50 kw-D, 10 kw-N, DA-2. TL: N35 36 08 W77 25 35. Hrs open: 24 2929 Radio Station Rd., 27834. Phone: (252) 757-0011. Fax: (252) 757-0286. Web Site: www.talk1070.com. Licensee: WNCT License L.P. (acq 1996). Natl. Network: CBS, ABC. Format: Talk, sports. Target aud: General. ♦Brad Hood, gen mgr & stn mgr.

WNCT-FM— Dec 22, 1963: 107.9 mhz; 100 kw. 1,800 ft TL: N35 21 55 W77 23 38. Stereo. 24 Web Site: www.oldies1079.com.451,000 Format: Oldies. ♦Jerry Wayne, progmg dir. Co-owned TV: WNCT-TV affil.

WRSV(FM)—See Rocky Mount

***WZMB(FM)**— Feb 2, 1982: 91.3 mhz; 282 w. 134 ft TL: N35 36 01 W77 21 53. Stereo. Hrs opn: 8am - midnight Mendenhall Student Ctr., East Carolina Univ., Rm 110, 27858. Phone: (252) 328-4751. Phone: (252) 328-4752. Fax: (252) 328-4773. Web Site: www.wzmb.ecu.edu. Licensee: East Carolina University Media Board. (acq 2-82). Natl. Network: ABC. Format: Alternative rock. News: 12 hrs wkly. Target aud: 18-24; univ students.

Grifton

WXNR(FM)— Sept 11, 1989: 99.5 mhz; 16.5 kw. 830 ft TL: N35 12 07 W77 11 15. Stereo. Hrs opn: 24 207 Glenbernie Dr., New Bern, 28560. Phone: (252) 633-1500. Fax: (252) 633-6546. Web Site: www.995thex.com. Licensee: WXNR License L.P. Population served: 430,000 Rgnl. Network: N.C. News Net. Natl. Rep: D & R Radio. Format: New rock. Target aud: 18-34. ♦Bruce Simel, gen mgr; Joe Peters, sls dir; Wilbur Vitols, gen sls mgr; Jeff Sanders, progmg dir; Richard Banks, chief of engrg.

Hamlet

WJSG(FM)— Aug 25, 1991: 104.3 mhz; 2.5 kw. 489 ft TL: N34 48 44 W79 43 38. Hrs open: 180 Airport Rd., Rockingham, 28379. Phone: (910) 895-3787. Fax: (910) 895-8811. E-mail: g104fm@104fm.com Web Site: www.g104fm.com. Licensee: Jackson Broadcasting Co. Format: Christian country. ♦Sherrell Jackson, gen mgr & progmg dir; Jerry Stout, news dir.

WKDX(AM)— June 30, 1957: 1250 khz; 1 kw-D. TL: N34 53 06 W79 40 50. Hrs open: Box 827, 28345. Phone: (910) 582-1997. Fax: (910) 582-1920. Web Site: www.wkdx.net. Licensee: The McLaurin Group (acq 5-19-00). Format: Christian contemp gospel. ♦Howard McLaurin Jr., pres & gen mgr.

Harkers Island

WLGP(FM)— Aug 1, 1996: 100.3 mhz; 100 kw. Ant 485 ft TL: N34 48 17 W76 54 23. Stereo. Hrs opn: 24 2278 Wortham Ln., Grovetown, GA, 30813. Phone: (706) 309-9610. Fax: (706) 309-9669. E-mail: ctbarinowski@gnnradio.org Web Site: www.gnnradio.org. Licensee: Barinowski Investment Co. Group owner: Good News Network. Population served: 800,000 Format: Christian. ♦Clarence Barinowski, pres & gen mgr.

Harrisburg

WQNC(FM)— 1995: 92.7 mhz; 6 kw. 328 ft TL: N35 16 20 W80 45 54. Hrs open: 2303 W. Morehead St., Charlotte, 28208. Phone: (704) 358-0211. Fax: (704) 358-3752. Licensee: Radio One of North Carolina LLC. Group owner: Radio One Inc. (acq 11-8-01; grpsl). Natl. Rep: McGavren Guild. Format: R&B. Target aud: Adults 25-54. ♦Debbie Kwei, gen mgr; Michael Taylor, gen sls mgr; Latoya Whitt, mktg dir & prom dir; Phil Woods, chief of engrg; Gaynell Nichols, traf mgr.

Hatteras

WCMS-FM— May 1999: 94.5 mhz; 91 kw. 981 ft TL: N35 29 10 W75 59 58. Stereo. Hrs opn: 24 Box 1897, Kill Devil Hills, 27948. Phone: (252) 480-4655. Fax: (252) 441-8063. Web Site: www.wcms.com. Licensee: Max Radio of the Carolinas Licenses LLC. Group owner: MAX Media L.L.C. (acq 11-12-2002; grpsl). Natl. Network: Jones Radio Networks. Format: Hot country. News staff: 2. ♦Mike Smith, VP & gen mgr; Steve Batton, progmg dir.

WYND-FM— March 1995: 97.1 mhz; 48 kw. Ant 558 ft TL: N35 27 48 W76 02 07. Stereo. Hrs opn: 24 637 Harbor Rd., Wanchese, 27981. Phone: (252) 475-1888. Fax: (252) 475-1881. E-mail: hunt@capsanmedia.com Web Site: www.wilburandorville.com. Licensee: CapSan Media LLC. Group owner: Convergent Broadcasting LLC (acq 6-30-2006; grpsl). Natl. Network: Westwood One. Law Firm: Richard Hayes. Format: Country. ♦William Whitlow, pres & gen mgr; Hunt Thomas, opns dir & progmg dir.

Havelock

WANG(AM)— June 16, 1962: 1330 khz; 1 kw-D. TL: N34 55 24 W76 56 37. Hrs open: Sunrise-sunset
Rebroadcasts WANG(FM) Morehead City 100%.
1361 Colony Dr., New Bern, 28562. Phone: (252) 639-7900. Fax: (252) 639-7979. Licensee: NM Licensing LLC. Group owner: NextMedia Group L.L.C. (acq 11-26-2001; grpsl). Population served: 5,283 Rgnl. Network: N.C. News Net. Format: Adult hits. Target aud: 35 plus; active lifestyle, early militaary retirees, relocated retirees. ♦Rolf Pepple, gen mgr; Jack Spade, progmg dir & chief of engrg.

WSSM(FM)—Co-owned with WANG(AM). Nov 12, 1971: 105.1 mhz; 18.5 kw. Ant 384 ft TL: N34 45 07 W76 52 57. Stereo. 17,035

Henderson

WCBQ(AM)—See Oxford

WHNC(AM)— June 20, 1945: 890 khz; 1 kw-D. TL: N36 21 04 W78 22 35. Hrs open: Sunrise-sunset
Rebroadcasts WCBQ(AM) Oxford 100%.
PO Box 1005, One Broadcast Ctr., 601 Henderson St., Oxford, 27565. Phone: (919) 693-3540. Phone: (919) 693-1340. Fax: (919) 693-9054. Licensee: The Paradise Network (TPN) of North Carolina Inc. (acq 4-30-01; $650,000 with WCBQ(AM) Oxford). Population served: 37,000 Natl. Rep: Keystone (unwired net). Rgnl rep: T-N. Format: Black gospel. News staff: one; News: 7 hrs wkly. Target aud: 18 plus.Alvin Augustis Jones, chmn; Nathaniel Smith, stn mgr, relg ed & disc jockey; Ronald Smith, sls dir & chief of engrg; Jim Davis, engrg mgr; Jeff Rose, spec ev coord; Al Woodlief, news rptr & political ed; Aaron Woodlief, farm dir; Anita Woodlief, women's int ed & disc jockey; Adam Woodlief, disc jockey; Randel Chandler, disc jockey; Ronnie Davies, disc jockey

WIZS(AM)— May 1, 1955: 1450 khz; 1 kw-U. TL: N36 19 31 W78 24 36. Hrs open: 5 AM-midnight Box 1299, 27536. Phone: (252) 492-3001. Fax: (252) 492-3002. E-mail: wizs@vance.net Web Site: hendersonnews.com. Licensee: Rose Farm and Rentals Inc. (acq 6-1-89; $265,000; FTR: 6-19-89). Population served: 57,000 Rgnl. Network: N.C. News Net. Format: Country. News staff: one; News: 6 hrs wkly. Target aud: 25 plus; fans of top 60 country music. ♦John D. Rose III, pres, news dir & chief of engrg; George Rush, sls dir; Martha Pelaquin, opns mgr & mus dir; Angela Hendley, pub affrs dir.

WYFL(FM)— 1948: 92.5 mhz; 100 kw. Ant 1,020 ft TL: N36 13 23 W78 12 07. Stereo. Hrs opn: 203 Capcom Ave., Suite 102-D, Wake Forest, 27587. Secondary address: 120 E. Belle St. 27536. Phone: (919) 562-3198. E-mail: wyfl@bbnradio.org Web Site: www.bbnradio.org. Licensee: Bible Broadcasting Network Inc. (group owner; acq 10-3-81; $335,000; FTR: 9-14-81). Population served: 2,500,000 Format: Relg. Target aud: General. ♦Bryant Nelson, gen mgr & news dir.

Hendersonville

WHKP(AM)— Oct 24, 1946: 1450 khz; 1 kw-U. TL: N35 20 20 W82 27 20. Hrs open: 24/7 Box 2470, 1450 7th Ave. E., 28793. Phone: (828) 693-9061. Fax: (828) 696-9329. E-mail: 1450@whkp.com Web Site: www.whkp.com. Licensee: Radio Hendersonville Inc. (acq 6-4-86). Population served: 100,000 Law Firm: Brooks, Pierce, McLendon, Humphrey & Leonard. Format: Music of Your Life. News staff: 10; News: 10 hrs wkly. Target aud: 25 plus; middle to upper income. Spec prog: Var 18 hrs wkly. ♦Art Cooley, pres, gen mgr & prom VP; Richard Rhodes, sls VP; Larry Freeman, progmg dir & news dir; Dave Lyons, chief of engrg; Marge Duncan, women's int ed; Abby Ramsey, disc jockey; Chuck Hill, disc jockey; Tippy Creswell, disc jockey.

WMYI(FM)—Licensed to Hendersonville. See Greenville SC

WTZQ(AM)— Dec 25, 1964: 1600 khz; 1 kw-D, 12 w-N. TL: N35 18 53 W82 25 58. Hrs open: 24 418 Duncun Road, Flat Rock, 28731. Phone: (828) 692-1600. Phone: (828) 697-1506. Fax: (828) 697-1416. E-mail: 1600@wtzq.com Web Site: www.wtzq.com. Licensee: Houston Broadcasting Inc. (acq 2-11-02; $750,000). Population served: 125,000 Natl. Network: ABC. Rgnl. Network: N.C. News Net. Format: Adult standards. News staff: one; News: 15 hrs wkly. Target aud: 35 plus; mature upscale audiences. Spec prog: Gospel 4 hrs wkly. ♦Randy Houston, pres, gen mgr, gen sls mgr & disc jockey; Glenn Trent, opns dir, progmg dir, news dir & disc jockey; Mark Warwick, sls dir & adv dir; Susan Houston, VP & prom dir; Michael Sink, chief of engrg.

Hertford

WFMZ(FM)— December 1997: 104.9 mhz; 50 kw. Ant 492 ft TL: N36 10 45 W76 21 12. Hrs open: 24 637 Harbor Rd., Wancheese, 27981. Phone: (252) 475-1888. Fax: (252) 475-1881. E-mail: 1049@capsanmedia.com Web Site: classichits1049.com. Licensee: CapSan Media LLC. Group owner: Convergent Broadcasting LLC (acq 6-30-2006; grpsl). Population served: 250,000 Natl. Network: USA. Format: Classic hits. Target aud: 25-54. ♦William Whitlow, pres; William Whitlow, gen mgr; Hunt Thomas, opns dir & progmg dir.

Hickory

WAIZ(AM)— Dec 5, 1948: 630 khz; 1 kw-D, 57 w-N. TL: N35 43 07 W81 18 36. Hrs open: 24 hrs Box 938, 28603. Secondary address: Box 430, Newton 28658. Phone: (828) 322-9472. Fax: (828) 464-9662. E-mail: totalradio@aol.com Licensee: Newton-Conover Communications Inc. (acq 10-5-94; $225,000). Population served: 701,000 Natl. Network: ABC. Format: Oldiies. Target aud: 25 plus. ♦Dave Lingafelt, pres & gen mgr; Jim Turner, gen sls mgr; Karol Lowery, traf mgr.

***WFHE(FM)**— Aug 31, 1995: 90.3 mhz; 150 w. 804 ft TL: N35 39 27 W81 24 23. Hrs open: 24
Rebroadcasts WFAE(FM) Charlotte 100%.
c/o WFAE(FM), 8801 J.M. Keynes Dr., Suite 91, Charlotte, 28262-8485. Phone: (704) 549-9323. Fax: (704) 547-8851. E-mail: wfae@wfae.org Web Site: www.wfae.org. Licensee: University Radio Foundation Inc. Population served: 200,000 Format: NPR/PRI Affiliate. ♦Roger Sarow, pres & gen mgr; Debra Peterson, CFO; Barbara Vermeire, dev dir; Catherine Little, dev dir; Paul Stribling, mktg dir & progmg dir; Renee Rallos, prom dir & pub affrs dir; Tena Simmons, opns dir & mus dir; Mark Rumsey, news dir; Jobie Sprinkle, engrg dir.

WHKY(AM)— June 10, 1940: 1290 khz; 50 kw-D, 1 kw-N, DA-2. TL: N35 43 35 W81 18 02. Hrs open: 24 Box 1059, 526 Main Ave. S.E., 28603. Phone: (828) 322-1290. Fax: (828) 322-8256. E-mail: whky@whky.com Web Site: www.whky.com. Licensee: Long Communications LLC (acq 12-31-01; with WHKY-TV Hickory). Population served: 1,200,000. Natl. Network: ABC, ESPN Radio. Rgnl. Network: N.C. News Net. Natl. Rep: Rgnl Reps. Rgnl rep: Rgnl Reps Law Firm: Hardy & Carey. Format: News/talk. News staff: 4; News: 50 hrs wkly. Target aud: 35-54. ♦Thomas E. Long, gen mgr; Jeff Long, stn mgr; Patty Guthrie, gen sls mgr; Heather Isenhour, traf mgr. Co-owned TV: WHKY-TV affil.

WLYT(FM)— Jan 20, 1959: 102.9 mhz; 31 kw. 1,545 ft TL: N35 24 26 W81 07 47. Stereo. Hrs opn: 801 Wood Ridge Center Dr., Charlotte, 28217. Phone: (704) 714-9444. Fax: (704) 372-3208. Web Site: www.wlyt.com. Licensee: Capstar TX L.P. Group owner: Clear Channel Communications Inc. (acq 8-30-00; grpsl). Population served: 150,000 Format: Adult contemp. Target aud: 25-54. ♦Morgan Johannon, gen mgr; Nick Allen, opns mgr; Kim Kyle, sls dir; Tom Hunt, sls dir; Amanda Knepp, mktg dir; Anthony Testa, prom dir; Jeff Kent, progmg dir; Linda Silver, news dir; Alan Lane, chief of engrg; Ben Birnitzer, chief of engrg.

***WPIR(FM)**— Dec 3, 1985: 88.1 mhz; 10 kw. 300 ft TL: N35 43 34 W81 08 52. Stereo. Hrs opn: 24
Rebroadcasts WXRI(FM) East Bend 60%.
Box 909, Claremont, 28610. Secondary address: 3289 WCXN Radio Rd., Claremont 28610. Phone: (828) 459-2772. Fax: (828) 459-9805. Licensee: Positive Alternative Radio Inc. Group owner: Baker Family Stations Format: Southern gospel, educ. ♦Rodney Baucom, gen mgr.

***WRYN(FM)**—Not on air, target date: unknown: 89.1 mhz; 500 w. Ant 449 ft TL: N35 43 59 W81 19 51. Hrs opn: Drawer 2440, Tupelo, MS, 38803-2440. Phone: (662) 844-5036. Web Site: www.afr.net. Licensee: American Family Association. ♦Donald E. Wildmon, chmn.

WXRC(FM)— Dec 7, 1962: 95.7 mhz; 100 kw. 1,276 ft TL: N35 42 32 W81 31 32. Stereo. Hrs opn: 24 1515 Mocking Bird Lane., Suite 910, Charlotte, 28208 - 28209. Phone: (704) 527-0957. Fax: (704) 527-2720. E-mail: totalradio@aol.com Web Site: 957theride.com.

Licensee: Pacific Broadcasting Group Inc. (acq 10-5-94; $3.05 million; FTR: 10-17-94). Natl. Rep: McGavren Guild. Format: Classic hits. Target aud: 18-49. ♦Dave Lingafelt, pres & gen mgr; Jim Turner, natl sls mgr; Peggy Barrett, prom dir; Karol Lowery, pub affrs dir & traf mgr; Larry Schropp, chief of engrg.

WYCV(AM)—See Granite Falls

High Point

WGOS(AM)— July 1947: 1070 khz; 1 kw-D. TL: N35 54 58 W80 01 00. Hrs open: Sunrise-sunset 6223 Old Mendenhall Rd., 27263-7624. Phone: (336) 434-5024. Fax: (336) 434-6018. E-mail: wgosradio@triad.rr.com Web Site: www.wgos.net. Licensee: Ritchy Broadcasting Co. Inc. (acq 8-6-79). Natl. Network: USA. Format: Loc talk, Sp. News: 2 hrs wkly. Target aud: General. Spec prog: Loc college sports 10 hrs wkly. ♦Lynn Ritchy, gen mgr & gen sls mgr; Max Parrish, stn mgr & chief of engrg; Simon Ritchy, pres & progmg dir.

***WHPE-FM**— November 1947: 95.5 mhz; 100 kw. 440 ft TL: N35 55 10 W80 01 47. Stereo. Hrs opn: 24 1714 Tower Ave., 27260. Phone: (336) 889-9473. Fax: (336) 889-9773. E-mail: whpe@rrnradio.org Web Site: www.rrnradio.org. Licensee: Bible Broadcasting Network. (acq 10-74). Population served: 713,000 Natl. Network: USA. Format: Relg. ♦Lowell Davey, pres; Dan Austin, gen mgr.

WJMH(FM)—See Reidsville

WMAG(FM)— 1946: 99.5 mhz; 100 kw. 1,500 ft TL: N35 52 13 W79 50 25. Stereo. Hrs opn: 24 2 B PAI Park, Greensboro, 27409. Phone: (336) 822-2000. Fax: (336) 887-0104. Web Site: www.wmagradio.com. Licensee: Capstar TX L.P. Group owner: Clear Channel Communications Inc. (acq 8-30-00; grpsl). Population served: 120,000 Format: Adult contemp. Target aud: 25-54. ♦Cheryl Salamone, VP; Scott Keith, opns mgr; Shannon Sopina, prom dir; Ed Skurka, news dir; Bill Flynn, disc jockey; Ted Levin, disc jockey.

WMFR(AM)— Oct 15, 1935: 1230 khz; 1 kw-U. TL: N35 57 20 W80 00 22. Hrs open: 5:30 AM-1 AM 875 W. 5th St., Winston Salem, 27101-2505. Phone: (336) 777-3900. Fax: (336) 885-3299. Web Site: www.wmfr.com. Licensee: Crescent Media Group LLC. Group owner: Infinity Broadcasting Corp. (acq 2-14-2007; grpsl). Population served: 80,000 Format: News/talk. ♦Tom Hamilton, VP; Marty Holbrook, mktg dir & progmg dir; Bob Costner, news dir; George Newman, chief of engrg.

WVBZ(FM)— June 1953: 100.3 mhz; 100 kw. 1,049 ft TL: N35 58 09 W79 49 29. Stereo. Hrs opn: 24 2-B PAI Park, Greensboro, 27409. Phone: (336) 822-2000. Fax: (336) 887-0104. Web Site: www.buzzitrocks.com. Licensee: Capstar TX L.P. Group owner: Clear Channel Communications Inc. (acq 8-30-00; grpsl). Population served: 1,885,000 Format: Rock. Target aud: 25-54. ♦Sheryl Solomone, gen mgr; Kevin Isaacs, sls dir; Tim Fattafield, progmg dir; Travis Moore, disc jockey.

WYSR(AM)— June 1953: 1590 khz; 1.4 kw-D, 14 w-N. TL: N35 59 04 W80 04 08. Hrs open: 808 English Rd., Suite 101, 27262. Phone: (336) 883-8852. Fax: (336) 882-1594. E-mail: wysr@northstate.net Licensee: Latino Broadcasting LLC (group owner; (acq 2-23-2006; $780,000). Population served: 100,000 Format: Talk, sports. Target aud: General. ♦Jose A. Isasi, pres; Carrie Armstrong, gen mgr; L.A. Batchelor, opns dir & opns mgr.

Highlands

WHLC(FM)— July 1993: 104.5 mhz; 460 w. Ant 1,158 ft TL: N35 03 40 W83 11 05. Stereo. Hrs opn: 24 Box 1889, 28741. Secondary address: 2420 Hwy. 64 E. 28741. Phone: (828) 526-1045. Fax: (828) 526-4900. E-mail: info@whlc.com Web Site: www.whlc.com. Licensee: Charisma Radio Corp. Population served: 280,000 Format: Easy lstng. Target aud: 35 plus. ♦Charles B. Cooper, pres & gen mgr; Steve Day, opns mgr.

Hope Mills

WCCG(FM)— July 1997: 104.5 mhz; 6 kw. 305 ft TL: N34 56 34 W78 51 41. Hrs open: 115 Gillespie St., Fayetteville, 28301. Phone: (910) 484-4932. Fax: (910) 485-5192. Web Site: www.hot1045fm.com. Licensee: James E. Carson. Format: Oldies. ♦James Carson, gen mgr.

Jacksonville

WILT(FM)— Apr 28, 1965: 98.7 mhz; 100 kw. 1,015 ft TL: N34 29 38 W77 29 18. Stereo. Hrs opn: 24 1361 Colony Dr., New Bern, 28562. Phone: (910) 455-5300. Phone: (252) 639-7900. Fax: (910) 455-3112. Fax: (252) 639-7979. Licensee: NM Licensing LLC. Group owner: NextMedia Group L.L.C. (acq 11-26-2001; grpsl). Population served: 1,000,000 Law Firm: Wilmer, Cutler & Pickering. Format: Jack - Hip Hop. News staff: one; News: 3 hrs wkly. Target aud: 25-54; middle to upper-income adults. ♦Rolf Pepple, gen mgr; Tom Pierce, gen mgr & sls dir; Brian White, progmg dir; Kent Winrich, chief of engrg.

WJCV(AM)— Oct 10, 1968: 1290 khz; 5 kw-D, 47 w-N. TL: N34 45 58 W77 23 28. Hrs open: 24 Box 1216, 28541. Phone: (910) 347-6141. Fax: (910) 347-1290. Web Site: www.wjcv.com. Licensee: Down East Broadcasting Co. Inc. (acq 1996). Population served: 130,000 Natl. Network: USA. Natl. Rep: Salem. Format: Southern gospel, Christian. Target aud: 25-54; general. ♦Melvin Bland, stn mgr, opns mgr, progmg dir & disc jockey; Joe North, disc jockey; Michael Bland, pres, gen mgr, sls dir, adv mgr & disc jockey.

***WJKA(FM)**—Not on air, target date: unknown: 90.1 mhz; 20 kw vert. Ant 305 ft TL: N34 31 31 W77 35 27. Hrs open: Box 2440, Tupelo, MS, 38803-2440. Phone: (662) 844-8888. Fax: (662) 842-6791. Web Site: www.afr.net. Licensee: American Family Association. ♦Marvin Sanders, gen mgr.

WJNC(AM)— Oct 16, 1945: 1240 khz; 1 kw-U. TL: N34 44 56 W77 24 51. Hrs open: Box 70, Newport, 28570. Phone: (910) 455-7222. Fax: (252) 247-7343. Licensee: Heritage Broadcasting LLC (acq 8-10-01; $358,500). Population served: 25,000 Natl. Network: Westwood One. Law Firm: Hogan & Hartson. Format: News/talk, sports. Target aud: 18-45 plus. ♦Ben Ball, gen mgr; Dave Gremoske, chief of engrg.

WQSL(FM)— November 1993: 92.3 mhz; 22.5 kw. Ant 725 ft TL: N34 31 10 W77 26 52. Hrs open: 24 1361 Colony Dr., New Bern, 28562. Phone: (252) 639-7900. Fax: (252) 639-7979. Web Site: thebeatnc.com. Licensee: NM Licensing LLC. Group owner: NextMedia Group L.L.C. (acq 11-26-01; grpsl). Format: CHR rhythm. ♦Rolf Pepple, gen mgr.

WSRP(AM)— June 21, 1954: 910 khz; 5 kw-U, DA-N. TL: N34 47 45 W77 29 24. Stereo. Hrs opn: 24 3389 NC Highway 121, Farmville, 27828-9556. Phone: (910) 455-2202. Fax: (910) 355-2203. Licensee: Estuardo Valdemar Rodriguez & Leonor Rodriguez. (acq 6-30-2006; $475,000). Format: Hispanic bcstg. ♦Henry Gonzalez, stn mgr.

WXQR(FM)— Mar 14, 1966: 105.5 mhz; 19 kw. 794 ft TL: N34 31 10 W77 26 52. Stereo. Hrs opn: 24 1361 Colony Dr., New Bern, 28562. Phone: (252) 639-7900. Fax: (252) 639-7979. Web Site: www.carolinaspurerock.com. Licensee: NM Licensing LLC. Group owner: NextMedia Group L.L.C. (acq 11-26-01; grpsl). Population served: 142,500 Natl. Network: ABC. Format: AOR. ♦Rolf Pepple, gen mgr & stn mgr.

Jefferson

WMMY(FM)— October 1999: 106.1 mhz; 10.5 kw. Ant 508 ft TL: N36 19 53 W81 35 17. Hrs open: 738 Blowing Rock Rd., Boone, 28607. Phone: (828) 264-2411. Fax: (828) 264-2412. Web Site: www.highway106.com. Licensee: Aisling Broadcasting of Banner Elk LLC (group owner; acq 4-13-2004; $1.58 million). Format: Country. Target aud: 18-49; Adults. Spec prog: Southern Fried Friday Night; Mountainhome Music Live Bluegrass Saturday Nights. ♦Jonathan Hoffman, CEO.

Kannapolis

WRFX-FM— October 1964: 99.7 mhz; 100 kw. 1,044 ft TL: N35 33 45 W80 42 40. (CP: 84 kw, ant 1,056 ft.). Stereo. Hrs opn: 801 Wood Ridge Center Dr., Charlotte, 28217. Phone: (704) 714-9444. Fax: (704) 371-3238. Web Site: www.wrfx.com. Licensee: Capstar TX L.P. Group owner: Clear Channel Communications Inc. (acq 8-30-00; grpsl). Format: Classic rock, AOR. Target aud: 25-54; male. Spec prog: Talk 3 hrs wkly. ♦Morgan Bohannon, gen mgr; Nick Allen, opns mgr; Kim Kyle, sls dir; Amanda Knepp, mktg dir; Jeff Kent, progmg dir; Linda Silver, news dir; Ben Brinitzer, chief of engrg.

WRKB(AM)— Dec 11, 1960: 1460 khz; 500 w-D, 194 w-N. TL: N35 29 14 W80 36 18. Hrs open: 24
Rebroadcasts WRNA(AM) China Grove 90%.
Box 8146, 28083. Phone: (704) 938-1460. Phone: (704) 857-1101. Fax: (704) 857-0680. Web Site: www.fordbroadcasting.com. Licensee: Ford Broadcasting Inc. (group owner; (acq 1994). Population served: 200,000 Natl. Network: USA. Law Firm: Smithwick & Belendiuk. Format: Southern gospel. ♦Carl Ford, pres, gen mgr & opns mgr; Taylor Ford, exec VP; Angela Ford, sr VP.

Kernersville

WTRU(AM)— Aug 16, 1970: 830 khz; 50 kw-D, 10 kw-N, DA-2. TL: N36 11 58 W80 12 25. Hrs open: 24 4405 Providence Ln., Suite D, Winston-Salem, 27106. Phone: (336) 759-0363. Fax: (336) 759-0366. E-mail: info@wtru.com Web Site: www.wtru.com. Licensee: Truth Broadcasting Corp. (group owner; (acq 7-20-2000; $3.5 million with WGTK(AM) Louisville, KY). Natl. Network: Salem Radio Network. Format: News, Christian. News: 24 hrs wkly. Target aud: General. ♦Stuart Epperson, pres; Stuart Epperson Jr., gen mgr.

Kill Devil Hills

WCXL(FM)— January 1993: 104.1 mhz; 100 kw. 981 ft TL: N36 07 42 W75 49 39. Stereo. Hrs opn: 24 Box 1897, 27948. Phone: (252) 480-4655. Fax: (252) 441-4827. Web Site: www.beach104.com. Licensee: Max Radio of the Carolinas Licenses LLC. Group owner: MAX Media L.L.C. (acq 11-12-2002; grpsl). Format: Adult contemp. News staff: 2. Target aud: 18-54. Spec prog: Farm 2 hrs wkly. ♦Mike Smith, gen mgr; Bob Davis, gen sls mgr & rgnl sls mgr.

King

WKTE(AM)— Dec 4, 1963: 1090 khz; 1 kw-D. TL: N36 17 48 W80 22 18. Stereo. Hrs opn: Phone: (336) 983-3111. Fax: (336) 368-1090. Web Site: www.wktelogo.com. Licensee: Booth-Newsom Broadcasting Inc. (acq 3-1-86; $105,000; FTR: 1-6-86). Population served: 80,000 Rgnl. Network: Southern Farm. Format: Gospel, bluegrass, country. Spec prog: Farm 2 hrs wkly. ♦P.W. Booth, pres; Rodney Booth, gen mgr; Mike Bertaux, progmg dir; Dan Sykes, chief of engrg; Ron Wishon, disc jockey; Elizabeth Club, sls.

Kings Mountain

WDYT(AM)— Mar 12, 1953: 1220 khz; 1 kw-D, 106 w-N. TL: N35 15 59 W81 19 23. Stereo. Hrs opn: 5 AM-midnight 2100 Cleveland Ave., 28086. Phone: (704) 739-1220. Fax: (704) 739-4900. Licensee: CRN Communications LLC (acq 3-31-2006; $950,000). Population served: 8,465 Rgnl. Network: N.C. News Net. Format: Talk. ♦Daniel J. Fontana, gen mgr.

Kinston

WELS(AM)— September 1950: 1010 khz; 1 kw-D, 75 w-N. TL: N35 15 45 W77 37 35. (CP: TL: N35 17 02 W77 39 55). Hrs opn: 8 AM-5 PM Box 3384, 28502. Secondary address: 313 N. Queen St. 28501. Phone: (252) 523-5151. Fax: (252) 523-9357. E-mail: wels1029@yahoo.com Licensee: Willis Broadcasting. (acq 1-13-95; FTR: 2-27-95). Population served: 180,000 Rgnl. Network: N.C. News Net., Tobacco. Natl. Rep: Clayton-Davis. Format: Gospel. Target aud: 25-54; middle income. ♦Anthony Gonzales, gen mgr.

WELS-FM— Nov 21, 1990: 102.9 mhz; 3 kw. 295 ft TL: N35 17 03 W77 39 53.24 Natl. Network: ABC. Target aud: General; middle to upper income, married, working, college grads. Spec prog: East Carolina Univ. sports, Kinston Indians baseball. ♦Anthony Gonzales, disc jockey.

WKIX(FM)— Sept 15, 1976: 97.7 mhz; 2.65 kw. Ant 501 ft TL: N35 17 28 W77 49 25. Stereo. Hrs opn: 24 2581 US Hwy. 70 W., Goldsboro, 27530. Phone: (919) 736-1699. Fax: (919) 736-3876. Web Site: www.yourfavoritecountry.com. Licensee: New Age Communications Inc. Group owner: Curtis Media Group (acq 8-30-2004; $875,000). Population served: 200,000 Format: Country. ♦Bill Johnston, gen mgr.

***WKNS(FM)**— Mar 26, 1977: 90.3 mhz; 20 kw. 312 ft TL: N35 25 01 W77 48 57. Stereo. Hrs opn:
Rebroadcasting WTEB(FM) New Bern 100%.
c/o WTEB(FM), 800 College Ct., New Bern, 28562. Phone: (252) 638-3434. Fax: (252) 638-3538. Web Site: www.publicradioeast.org. Licensee: Craven Community College. (acq 1-18-95; FTR: 6-19-95). Format: Information. Spec prog: Jazz 6 hrs wkly. ♦Kathleen Beal, gen mgr; Charles Wethington, stn mgr & opns mgr; Jill McGuire, dev dir & dev mgr; Tomothy Kimble, progmg dir; Megan George, mus dir & news dir; J. Howard Jones, chief of engrg; George Olsen, reporter.

WLNR(AM)— May 1954: 1230 khz; 1 kw-U. TL: N35 15 31 W77 36 33. Hrs open: 24 1223 W. New Bern Rd., 28504-4713. Phone: (202) 638-1959. Fax: (252) 522-4501. Fax: (658) 558-7399. E-mail: estuardovaldemar@hotmail.com Licensee: Estuardo Valdemar Rodriguez & Leonor Rodriguez. Group owner: Estuardo Valdemar Rodriguez and Leonor Rodriguez Stns (acq 1-2004; $315,000). Population served: 35,000 Format: Sp. ♦Nancy Skyes, gen mgr.

WRNS-FM— Oct 12, 1968: 95.1 mhz; 100 kw. 1,499 ft TL: N35 06 18 W77 20 15. Stereo. Hrs opn: 24 1361 Colony Dr., New Bern, 28562. Phone: (252) 639-7900. Fax: (252) 639-7979. E-mail: mail@wrns.com Web Site: www.wrns.com. Licensee: NM Licensing LLC. Group owner: NextMedia Group L.L.C. (acq 11-26-01; grpsl). Population served: 420,000 Natl. Network: ABC. Law Firm: Wilmer, Cutler & Pickering. Format: Country. News staff: one. Target aud: 25-54. Spec prog: NASCAR racing 6 hrs wkly. ♦Rolf Pepple, gen mgr; Tim Knisley, gen sls mgr; Scott Green, progmg dir; Jeff Hackett, mus dir; Kent Winrich, chief of engrg.

WRNS(AM)— Feb 28, 1937: 960 khz; 5 kw-D, 1 kw-N, DA-N. TL: N35 16 59 W77 39 01.24 40,000 Format: Country. Target aud: 25-64.

Laurinburg

WEWO(AM)— Sept 1, 1947: 1460 khz; 5 kw-U, DA-2. TL: N34 47 00 W79 30 40. Hrs open: 24 Box 788, 28353. Phone: (910) 280-5209. Fax: (910) 276-9787. E-mail: wewo1460@aol.com Licensee: Service Media Inc. (acq 5-27-98; $150,000). Population served: 8,859 Format: Gospel. Target aud: 25-54. ♦Westley Johnson, gen mgr.

WFLB(FM)— May 1, 1951: 96.5 mhz; 100 kw. Ant 1,043 ft TL: N34 46 50 W79 02 45. Stereo. Hrs opn: 24 508 Person St., Fayetteville, 28301. Phone: (910) 486-4114. Fax: (910) 323-5635. Licensee: Beasley FM Acquisition Corp. Group owner: Beasley Broadcast Group (acq 7-31-96; $4.2 million with co-owned AM). Population served: 600,000 Natl. Rep: D & R Radio. Format: contemporary hits. Target aud: 25-54; affluent men & women in their peak earning years. Spec prog: University of North Carolina football & basketball. ♦George Beasley, pres; Mac Edwards, VP; Angela Godwin, gen sls mgr; Bryan Kusilka, natl sls mgr; Dave Stone, progmg dir; Van Clough, chief of engrg; Clara Glover, traf mgr.

WLNC(AM)— Jan 2, 1962: 1300 khz; 500 w-D. TL: N34 47 00 W79 26 22. Hrs open: sunrise-sunset Box 1748, 28353. Secondary address: 1300 Lila Dr. 28352. Phone: (910) 276-1300. E-mail: wlncradio@carolina.net Web Site: www.wlncradio.com. Licensee: Fox Broadcasting Inc. (acq 2-1-90; $325,000). Population served: 35,000 Format: Adult contemp. Target aud: General. Spec prog: Gospel 4 hrs wkly. ♦Fred Fox, gen mgr.

Leaksville

WGBT(FM)—See Eden

WLOE(AM)—See Eden

Leland

WAAV(AM)—Licensed to Leland. See Wilmington

WKXS-FM—Licensed to Leland. See Wilmington

Lenoir

WJRI(AM)— Mar 15, 1947: 1340 khz; 1 kw-U. TL: N35 53 47 W81 33 57. Hrs open: 24 827 Fairview Dr., 28645. Secondary address: Box 1678 28645. Phone: (828) 754-5361. Fax: (828) 757-3300. Web Site: www.foothillsradio.com. Licensee: Foothills Radio Group LLC (group owner; acq 11-28-01). Population served: 70,000 Rgnl. Network: N.C. News Net. Format: News/talk. News staff: one; News: 14 hrs wkly. Target aud: 20-45. ♦Al Bunch, pres; Patty Guthrie, gen mgr, sls VP & mktg dir; Davy Crockett, opns VP, progmg dir & pub affrs dir; Rob Eastwood, news dir, local news ed & sports cmtr; Stoney Owen, chief of engrg; Shannon Laws, traf mgr; Rocky Brooks, disc jockey.

WKGX(AM)— Feb 13, 1969: 1080 khz; 5 kw-D. TL: N35 54 38 W81 33 35. Stereo. Hrs opn: 12 827 Fairview Dr., ., 28645. Phone: (828) 758-1033. Fax: (828) 757-3300. E-mail: wxgx@twave.net Licensee: Foothills Radio Group LLC (group owner; acq 11-28-01). Population served: 75,000 Format: Country, bluegrass. News staff: one; News: 6 hrs wkly. Target aud: 24-55; older, mature wise spenders. Spec prog: Trading post show 16 hrs wkly. ♦Patty Guthrie, gen mgr.

WKVS(FM)— Sept 27, 1993: 103.3 mhz; 910 w. Ant 843 ft TL: N35 58 30 W81 33 07. Hrs open: Box 1678, 827 Fairview Dr., 28645. Phone: (828) 758-1033. Fax: (828) 757-3300. Licensee: Foothills Radio Group LLC. (group owner; (acq 11-28-2001). Format: Hot new country. Target aud: 18-54. ♦Al Bunch, pres; Patty Guthrie, gen mgr, sls dir & mktg dir; Davy Crockett, progmg dir & pub affrs dir; Rob Eastwood, news dir; Stonie Owen, engrg dir; Bill Bailey, disc jockey; Greg Ryan, disc jockey; Rick James, disc jockey; Rocky Brooks, disc jockey.

Lewisville

WSGH(AM)— 1986: 1040 khz; 10 kw-D, DA. TL: N36 08 06 W80 30 14. (CP: 10 kw-D, 182 w-N). Hrs opn: 24 4015 Brownsboro Rd., Winston Salem, 27106. Phone: (336) 759-0524. Fax: (336) 759-9327. Web Site: www.radiolamovidita.com. Licensee: Davidson Media Carolinas Stations LLC. Group owner: Davidson Media Group LLC (acq 5-10-2004; grpsl). Natl. Network: USA. Format: Sp. Target aud: 17-28. ♦Marco Antonio Saucedo, pres; Lucy Saucedo, VP; Samuel Saucedo, gen mgr.

Lexington

WLXN(AM)— Sept 22, 1946: 1440 khz; 5 kw-D, 1 kw-N, DA-N. TL: N35 50 22 W80 14 02. Hrs open: 24 200 Radio Dr., 27292. Phone: (336) 248-2716. Fax: (336) 248-2800. Web Site: www.wlxn.com. Licensee: Davidson County Broadcasting Co. Inc. Population served: 130,000 Rgnl. Network: N.C. News Net. Format: News/talk, sports. News staff: one; News: 30 hrs wkly. Target aud: 35 plus; those interested in news & sports. ♦Greeley N. Hilton Jr., pres; Tom Collins, opns VP & chief of opns; Bob Mahoney, news dir; Hal McGee, engrg dir; Hal V. McGee, disc jockey; Harold Bowen, disc jockey; Willie Edwards, prom dir & disc jockey.

WTHZ(FM)—Co-owned with WLXN(AM). Aug 24, 1949: 94.1 mhz; 100 kw. 1,014 ft TL: N35 55 02 W80 17 37. Stereo. 24 Web Site: www.hitz94.com.1,000,000 Rgnl rep: T-N. Format: 80's hits, 90's hits & now. News staff: one; News: 1 hr wkly. Target aud: General; 25-49. ♦Greeley N. Hilton Jr., exec VP; Bob Campbell, progmg dir; Hal McGee, chief of engrg.

Lillington

***WLLN(AM)**— Feb 12, 1979: 1370 khz; 5 kw-D, 49 w-N, DA-2. TL: N35 23 16 W78 48 22. Hrs open: Day time Box 969, 27546. Secondary address: 910 E. McNeil St. 27546. Phone: (910) 893-2811. Fax: (910) 893-2811. Licensee: Estuardo Valdemar Rodriguez (acq 9-21-99; $145,000). Population served: 300,000 Format: Spanish music. Target aud: General. ♦Estuardo Rodriguez, chmn & gen mgr; Leonor Rodriguez, pres; Helen Hernandez, stn mgr; Orlando Henao, mus dir.

Lincolnton

WLON(AM)— Aug 28, 1953: 1050 khz; 1 kw-D, 231 w-N. TL: N35 29 28 W81 16 03. Hrs open: 24
Rebroadcasts WCSL(AM) Cherryville 80%.
Box 430, 28093. Phone: (704) 735-8071. Fax: (704) 732-9567. E-mail: info@hrnb.com Web Site: www.hrnb.com. Licensee: HRN Broadcasting Inc. (acq 4-14-2004; $500,000 with WCSL(AM) Cherryville). Population served: 35,000 Natl. Network: Westwood One. Rgnl. Network: N.C. News Net. Format: Oldies, sports. Target aud: 25 plus. Spec prog: Gospel 5 hrs wkly. ♦Mark . Boyd, pres; Lanny Ford, gen mgr; Milton Baker, opns mgr.

Lockwoods Folly Town

***WGHW(FM)**— 2006: 88.1 mhz; 1.5 kw vert. Ant 311 ft TL: N34 03 48 W78 05 32. Hrs open: 1460 Old Ocean Hwy., Bolivia, 28422. Phone: (336) 577-6592. Web Site: www.kjbbfm.com. Licensee: Church Planters of America (acq 5-18-2005). Law Firm: Fletcher, Heald & Hildreth. Format: Christian. ♦Danny Hawkins, pres & gen mgr.

Louisburg

WKXU(FM)—Listing follows WYRN(AM).

WYRN(AM)— Sept 12, 1958: 1480 khz; 500 w-D. TL: N36 06 46 W78 16 50. Hrs open: 24 Box 463, 27549. Phone: (919) 496-3105. Fax: (919) 496-5864. Licensee: New Century Media Group LLC. Group owner: Curtis Media Group (acq 6-1-2003; $2.8 million with co-located FM). Population served: 1,200,000 Rgnl. Network: N.C. News Net. Format: Talk. News staff: one; News: 20 hrs wkly. Target aud: Adults 25-54. Spec prog: Black. ♦Randy Jordan, gen mgr & opns mgr; William M. McClatchey Jr., pres, dev mgr & gen sls mgr; Jackie Ayscue, traf mgr.

WKXU(FM)—Co-owned with WYRN(AM). Dec 5, 1989: 102.5 mhz; 6 kw. 328 ft TL: N36 07 12 W78 22 48. Stereo. 24 Format: Country. News: 6 hrs wkly. Spec prog: News, birthday celebration, country exchange, sports. ♦Jackie Ayscue, traf mgr.

Lumberton

WAGR(AM)— Nov 27, 1954: 1340 khz; 1 kw-U. TL: N34 35 58 W79 00 33. Hrs open: 24 145 Rowan St. A-3, Fayetteville, 28301. Secondary address: 1498 Alamac Rd. 28301. Phone: (910)486-9438. Fax: (910) 671-1812. E-mail: wcookman@aol.com Licensee: WAGR Broadcasting Inc. (acq 6-30-98; $50,000). License and equipment only Population served: 20,000 Rgnl rep: Williams Format: Gospel. News: 4 hrs wkly. Target aud: 25-54. ♦Charles W. Cookman, pres; Sandra Lofton, gen mgr; Val Halliday, opns mgr & progmg dir.

WFNC-FM— July 19, 1964: 102.3 mhz; 6 kw. Ant 267 ft TL: N34 35 58 W79 00 33. Stereo. Hrs opn: 24 1009 Drayton Rd., Fayetteville, 28303. Phone: (910) 864-5222. Fax: (910) 864-3065. Web Site: www.cumulus.com. Licensee: Cumulus Licensing Corp. Group owner: Cumulus Media Inc. (acq 3-12-2001; grpsl). Population served: 103,000 Format: Oldies. Target aud: 18-54. ♦Tom Haymond, gen mgr & chief of engrg.

WKML(FM)— Dec 1, 1960: 95.7 mhz; 100 kw. 1,064 ft TL: N34 46 56 W79 04 42. Stereo. Hrs opn: 24 Box 2563, Fayetteville, 28302. Secondary address: 508 Person St., Fayetteville 28301. Phone: (910) 483-9565. Fax: (910) 483-6008. Licensee: Beasley Broadcasting of Eastern North Carolina Inc. Group owner: Beasley Broadcast Group (acq 1981). Population served: 325,000 Natl. Rep: D & R Radio. Law Firm: Brooks, Pierce, McLendon, Humphrey & Leonard. Format: C&W. News staff: one; News: 5 hrs wkly. Target aud: 25-54. ♦George G. Beasley, pres; J. Daniel Highsmith, gen mgr; Mac Edwards, opns VP & disc jockey; Angela Godwin, gen sls mgr; Bryan Kusilka, natl sls mgr; Paul Johnson, progmg dir; Van Clough, chief of engrg; Don Chase, disc jockey; Larry K. Smith, disc jockey.

***WLPS-FM**—Not on air, target date: unknown: 89.5 mhz; 2 kw vert. Ant 440 ft TL: N34 42 02 W79 06 32. Hrs open: 3463 Oakgrove Church Rd., 28360-3181. Phone: (910) 521-3101. Licensee: Billy Ray Locklear Evangelistic Association. ♦Billy Ray Locklear, chmn.

Manteo

WOBX-FM— 2001: 98.1 mhz; 50 kw. Ant 295 ft TL: N35 51 52 W75 39 01. Hrs open: 24 Box 1418, Nags Head, 27959. Secondary address: 2422 S. Wrightsville Ave., Nags Head 27959. Phone: (252) 449-8331. Fax: (252) 449-8354. E-mail: wobx@ecri.net Web Site: www.wobx.net. Licensee: East Carolina Radio of Elizabeth City Inc. Format: Active rock. ♦R. Loesch, gen mgr; John Maloney, gen sls mgr.

***WUND-FM**— 2004: 88.9 mhz; 50 kw horiz, 47 kw vert. Ant 1,371 ft TL: N35 54 00 W76 20 45. Hrs open: 24
Rebroadcasts WUNC(FM) Chapel Hill 99.9%.
120 Friday Center Dr., Chapel Hill, 27517-9495. Phone: (919) 966-5454. Fax: (919) 966-5955. E-mail: wunc@wunc.org Web Site: www.wunc.org. Licensee: Board of Trustees of the University of North

Carolina at Chapel Hill. Population served: 129,691 Natl. Network: NPR, PRI. Format: News & info. News staff: 7; News: 124 hrs wkly. Spec prog: Folk 20 hrs wkly. ♦Joan Siefert Rose, gen mgr.

***WURI(FM)**— 1999: 90.9 mhz; 3.9 kw. Ant 187 ft TL: N35 54 28 W75 40 26. Hrs open: 24
Rebroadcasts WUNC(FM) Chapel Hill 99.9%.
120 Friday Center Dr., Chapel Hill, 27517-9495. Phone: (919) 966-5454. Fax: (919) 966-5955. E-mail: wunc@wunc.org Web Site: www.wunc.org. Licensee: Board of Trustees/University of North Carolina at Chapel Hill. Population served: 21,838 Natl. Network: CBC Radio One, NPR, PRI. Format: News & Info. Spec prog: Folk 20 hrs wkly.

WVOD(FM)— Mar 28, 1986: 99.1 mhz; 50 kw. 491 ft TL: N35 50 44 W75 38 50. Stereo. Hrs opn: 24 637 Harbor Rd., Wanchese, 27981. Phone: (252) 475-1888. Fax: (252) 475-1881. Web Site: www.991thesound.com. Licensee: CapSan Media LLC. Group owner: Convergent Broadcasting LLC (acq 6-30-2006; grpsl). Population served: 28,000 Format: AAA. News staff: one. Target aud: 25-49. Spec prog: Class 6 hrs, blues 2 hrs, reggae 2 hrs wkly. ♦William Whitlow, gen mgr; Hunt Thomas, opns dir; Matt Cooper, progmg dir; Tad Abbey, mus dir; Andy Booth, chief of engrg; Sharon Pro, traf mgr.

Marion

WBRM(AM)— May 9, 1949: 1250 khz; 5 kw-D, 62 w-N. TL: N35 40 59 W82 02 08. Hrs open: 24 147 N. Garden St., 28752. Phone: (828) 652-9500. E-mail: wbrm@charterinternet.com 6 pm-6 am Westwood One Mainstream Country Licensee: WBRM Inc. (acq 12-1-88; $450,000). Population served: 75,000 Format: Country. News staff: one; News: 9 hrs wkly. Target aud: 25-55; young adult to mature. Spec prog: Gospel 5 hrs, relg 7 hrs wkly. ♦Annette Bryant, CEO & pres; Kevin Estes, opns mgr.

Mars Hill

***WYQS(FM)**— 1974: 90.5 mhz; 250 w. 230 ft TL: N35 49 30 W82 33 00. Stereo. Hrs opn: 24 73 Broadway, Asheville, 28801. Phone: (828) 689-1407. Fax: (828) 689-1106. Licensee: Western North Carolina Public Radio Inc. (acq 8-26-2004; $177,000). Population served: 3,000 Format: Var. News staff: 3; News: 7 hrs wkly. Target aud: 14-30; college students & college community. Spec prog: Jazz 7 hrs, class 7 hrs, new age 6 hrs, Christian music 15 hrs wkly.

Marshall

WHBK(AM)— Sept 20, 1956: 1460 khz; 500 w-D, 139 w-N. TL: N35 48 01 W82 40 34. Hrs open: 24 1055 Skyway Dr., 28753. Phone: (828) 649-3914. Fax: (828) 649-2869. Licensee: Southern Broadcasting Inc. (acq 10-22-91; $145,000). Population served: 30,000 Format: Southern gospel. Spec prog: Farm 3 hrs wkly. ♦Bruce Philips, pres; Ricky Seay, gen mgr; Ricky West, disc jockey.

Mayodan

WMYN(AM)— July 15, 1957: 1420 khz; 1 kw-D, 70 w-N, DA-2. TL: N36 24 58 W79 59 29. Hrs open: 5 AM-10 PM
Rebroadcasts WLOE(AM) Eden 100%.
Box 279, 27027. Phone: (336) 427-9696. Fax: (336) 548-4636. E-mail: info@wloewmyn.com Web Site: www.wloewmyn.com. Licensee: Mayo Broadcasting Corp. (acq 1982; $110,000). Population served: 200,000 Natl. Network: Salem Radio Network, USA. Format: Info, talk, relg. News staff: one; News: 30 hrs wkly. Target aud: 25 plus; general. ♦Richard D. Hall, pres; Mike Moore, gen mgr; Annette Moore, stn mgr.

Mebane

WGSB(AM)— Dec 7, 1973: 1060 khz; 1 kw-D, DA. TL: N36 03 28 W79 16 36. Hrs open: Sunrise-sunset
Rebroadcasts WRTP(AM) Chapel Hill 100%.
Estuardo Valdemar Rodriguez and Leonor Rodriguez Stns, 1010 Vermont Ave. N.W., Suite 100, Washington, DC, 20005. Phone: (202) 638-1959. Fax: (202) 393-7464. E-mail: valtravel@rcn.com Licensee: Estuardo Valdemar Rodriguez and Leonor Rodriguez. Group owner: WRTP Radio Network (acq 12-13-2004; grpsl). Population served: 250,000 Format: Mexican rgnl. ♦Estuardo Valdemar Rodriguez, gen mgr.

Mint Hill

WNOW(AM)— Aug 1, 1987: 1030 khz; 10 kw-D, DA. TL: N35 08 30 W80 36 05. Hrs open: Sunrise-sunset 4201-J Stewart Andrew Blvd., Charlotte, 28217. Phone: (704) 665-9355. Fax: (208) 545-9888. Licensee: Davidson Media Carolinas Stations LLC. Group owner: Davidson Media Group LLC (acq 5-10-2004; grpsl). Population served: 400,000 Format: Mexican regional. ♦Peter W. Davidson, pres; Russ Douglass Jones, gen mgr, gen sls mgr & progmg dir; Aura Gavilan, prom mgr; Winston Hawkins, chief of engrg; Ann Freese, traf mgr.

Mocksville

WDSL(AM)— October 1964: 1520 khz; 5 kw-D, 1 kw-CH. TL: N35 52 50 W80 32 26. Hrs open: Box 1520, 27028. Secondary address: 125 W. Deport St. 27028. Phone: (336) 751-9375. E-mail: wdslradio@mailcity.com Licensee: Davie Broadcasting Inc. (acq 10-26-90; $52,000; FTR: 11-19-90). Population served: 750,000 Format: Country, bluegrass, gospel. Target aud: 25-80; general.

Monroe

WDEX(AM)— December 1983: 1430 khz; 2.5 kw-U, DA-2. TL: N34 59 04 W80 36 14. Hrs open: 24 Box 3272, Kannapolis, 28111. Secondary address: Weddington Rd. 28110. Phone: (704) 289-9339. Fax: (704) 283-1255. E-mail: wdex1430am@yahoo.com Licensee: New Life Community Temple of Faith Inc. (acq 11-15-99). Law Firm: Smithwick & Belendink. Format: Traditional, contemporary, gospel. Target aud: 25-55. ♦Ella Hood, CEO & gen mgr; Sharon Talford, pres & gen mgr.

WIXE(AM)— May 3, 1968: 1190 khz; 5 kw-D, 70 w-N. TL: N34 57 41 W80 32 40. Stereo. Hrs opn: 24 Box 1007, 28111. Secondary address: 1700 Buena Vista Dr. 28112. Phone: (704) 289-2525. Fax: (704) 289-1416. E-mail: wixeradio@carolina.rr.com Web Site: www.wixe.com. Licensee: Monroe Broadcasting Co. (acq 5-2-2000; $800,000). Population served: 1,300,000 Law Firm: Yelverton. Format: C&W, gospel, talk. News: 8 hrs wkly. Target aud: 18-55. Spec prog: Beach & oldies 5 hrs wkly. ♦Archie Morgan, pres & gen mgr.

WXNC(AM)— July 1947: 1060 khz; 4 kw-D. TL: N34 58 45 W80 30 48. Stereo. Hrs opn: 4801 E. Independence Blvd., Suite 803, Charlotte, 28212. Phone: (704) 537-9322. Fax: (704) 537-9735. E-mail: cheri@geddings.com Web Site: www.1060radio.com. Licensee: Norsan Consulting and Management Inc. (acq 8-3-2005; $1.15 million). Population served: 100,000 Natl. Network: CNN Radio. Format: Talk. News staff: one; News: 5 hrs wkly. ♦Norberto Sanchez, pres; Kris Phillips, CFO; Cheri Moore, gen mgr & traf mgr.

Mooresville

WHIP(AM)— 1950: 1350 khz; 1 kw-D, 670 w-N. TL: N35 36 04 W80 48 51. Hrs open: 6 AM-6:30 PM Box 600, 2432 Statesville Hwy., 28115. Phone: (704) 664-9447. Fax: (704) 664-5551. Licensee: Mooresville Media Inc. (acq 8-76). Population served: 225,000 Natl. Network: USA. Format: Oldies. News: 13 hrs wkly. Target aud: 25-45. Spec prog: Black 6 hrs, relg 6 hrs wkly. ♦Glenn Hamrick, pres; Martha Hamrick, VP, traf mgr & women's int ed; Norman Tindal, sls VP, gen sls mgr & disc jockey; Harrill Hamrick, chief of engrg; Gary Trexler, sports cmtr & disc jockey; Kevin Burchett, sports cmtr; Vivian Brandon, disc jockey.

Morehead City

***WOTJ(FM)**— Dec 12, 1988: 90.7 mhz; 24 kw. 466 ft TL: N34 46 41 W76 52 42. Stereo. Hrs opn: 24 520 Roberts Rd., Newport, 28570. Phone: (252) 223-4600/223-6088. Fax: (252) 223-2201. Web Site: www.fbnradio.com. Licensee: Grace Christian School. Natl. Network: USA. Format: Relg. News: 8 hrs wkly. Target aud: General; family. ♦Michael D. Ebron, gen mgr.

WRHT(FM)— Dec 20, 1972: 96.3 mhz; 100 kw. 492 ft TL: N34 44 18 W76 48 40. Stereo. Hrs opn: 24
Rebroadcasts WCBZ(FM) Williamston 100%.
1307 S. Glenburnie Rd., New Bern, 28562. Phone: (252) 672-5900. Fax: (252) 637-6872. E-mail: thehotfm@thehotfm.com Licensee: Inner Banks Media LLC. Group owner: Archway Broadcasting Group (acq 3-12-2007; grpsl). Population served: 450,000 Format: CHR. News staff: one; News: 7 hrs wkly. Target aud: 18-49; young active adults & military personnel. ♦Bill Bailey, gen mgr.

Morganton

WCIS(AM)— Mar 1, 1988: 760 khz; 3.5 kw-D. TL: N35 47 40 W81 43 12. Stereo. Hrs opn: Day station Box 1806, 28680-1806. Secondary address: 1399 Bost Rd. 28655. Phone: (828) 433-9247. Fax: (828) 433-1498. E-mail: powerhouse76@aol.com Licensee: W.F.M. Inc. (acq 10-1-93; $65,000). Population served: 90,000 Natl. Network: USA. Format: Southern gospel. ♦John L. Whisnant Sr., pres; Jeff Whisnant, VP; John L. Whisnant Jr., gen mgr; Jeff K. Whisnant, opns mgr.

WMNC(AM)— Sept 23, 1947: 1430 khz; 5 kw-D, 1 kw-N, DA-N. TL: N35 45 09 W81 43 03. Hrs open: 24 Box 969, 28680-0969. Secondary address: 1103 N. Green St. 28655. Phone: (828) 437-0521. Phone: (828) 437-0009. Fax: (828) 433-8855. E-mail: wmnc@bellsouth.net Web Site: www.wmnc.com. Licensee: Cooper Broadcasting Co. (acq 9-23-47). Population served: 17,200 Rgnl. Network: Southern Farm. Format: Classic country. ♦Joe Cooper, stn mgr; Cindy Byas, progmg dir; C.J. Stancil, news dir.

WMNC-FM— Aug 3, 1963: 92.1 mhz; 6 kw. 327 ft TL: N35 45 09 W81 43 19. Stereo. 24 Web Site: www.wmnc.com. Format: Hot new country.

WSVM(AM)—See Valdese

Mount Airy

WPAQ(AM)— February 1948: 740 khz; 10 kw-D, 1 kw-CH. TL: N36 32 04 W80 35 48. Hrs open: 6 AM-6:15 PM winter, loc sunset summer Box 907, 27030. Phone: (336) 786-6111. Fax: (336) 789-7792. E-mail: wpaq740am@earthlink.net Licensee: WPAQ Radio Inc. Population served: 25,000 Format: Bluegrass, old time, big band. News staff: one; News: 13 hrs wkly. Target aud: 25-64. Spec prog: Farm one hr, community affrs one hr, old time string mus 15 hrs wkly. ♦Kelly Epperson, gen mgr; Kelly D. Epperson, stn mgr; Kathy Edmonds, gen sls mgr; Susan Carroll, prom mgr; Bernie Phillips, news dir; John Mullins, chief of engrg.

WSYD(AM)— Oct 4, 1951: 1300 khz; 5 kw-D, 1 kw-N, DA-N. TL: N36 30 12 W80 35 35. Hrs open: 24 Box 1678, 27030. Phone: (336) 786-2147. Fax: (336) 789-9858. Licensee: Granite City Broadcasters Inc. (acq 1996). Population served: 60,000 Format: Gospel. News staff: one; News: 8 hrs wkly. Target aud: General. ♦Kelly D. Epperson, pres & gen mgr; Deborah Cochran, progmg dir; Bernie Phillips, news dir; John Mullins, chief of engrg.

Mount Olive

WDJS(AM)— Dec 27, 1961: 1430 khz; 1 kw-D. TL: N35 12 16 W78 03 06. Hrs open: Box 479, 28365. Secondary address: 990 N. Center St., Ext. 28365. Phone: (919) 658-9751. Fax: (919) 658-4894. Licensee: The Mount Olive Broadcasting Co. Population served: 250,000 Format: Relg, Christian. Spec prog: Black 5 hrs, gospel 5 hrs, Sp 5 hrs wkly. ♦Ann W. Mayo, CEO & gen mgr; Nancy West, progmg dir.

Moyock

WCDG(FM)— Oct 17, 1974: 92.1 mhz; 18 kw. Ant 384 ft TL: N36 41 39 W76 02 57. Stereo. Hrs opn: 1003 Norfolk Sq., Norfolk, VA, 23502-4948. Phone: (757) 466-0009. Fax: (757) 466-7043. Web Site: www.921thebeat.com. Licensee: CC Licenses LLC. Group owner: Clear Channel Communications Inc. Population served: 1,700,000 Format: Oldies. Target aud: 18-34; males 18-49. ♦Lowery Mays,

CEO; Reggie Jordan, gen mgr; Eric Mychaels, opns mgr; Terry Ratliff, gen sls mgr; Bob Rischitelli, natl sls mgr; Toni B. Jones, prom dir; Greg Gabriel, chief of engrg.

Murfreesboro

WDLZ(FM)—Listing follows WWDR(AM).

WWDR(AM)— Mar 20, 1965: 1080 khz; 930 w-D. TL: N36 26 24 W77 08 10. Hrs open: Box 38, 27855. Secondary address: 1714 W. Main St. 27855. Phone: (252) 398-4111. Fax: (252) 398-3581. Licensee: First Media Radio LLC. (group owner; (acq 1-7-2003; grpsl). Population served: 40,000 Natl. Network: Moody. Rgnl. Network: N.C. News Net. Format: Gospel. Spec prog: Farm 10 hrs wkly. ♦Earl Tellega, gen mgr; Neil Haskins, gen sls mgr; Frank White, prom mgr & chief of engrg; Bob Wood, progmg dir.

WDLZ(FM)—Co-owned with WWDR(AM). Oct 11, 1970: 98.3 mhz; 3 kw. Ant 328 ft TL: N36 26 24 W77 08 10. Stereo. 40,000 Format: Soft adult contemp.

Murphy

WCNG(FM)—Listing follows WCVP(AM).

WCVP(AM)— Oct 12, 1958: 600 khz; 1 kw-D, 20 w-N. TL: N35 04 00 W83 59 58. Hrs open: 5 AM-10 PM Box 280, 28906. Phone: (828) 837-2151. Phone: (828) 837-2152. Licensee: Cherokee Broadcasting Co. Population served: 220,000 Format: MOR, news, gospel. Target aud: All ages. Spec prog: Farm 3 hrs, class 20 hrs, C&W 12 hrs wkly. ♦Allan Blakemore, pres; Jane Blakemore, gen mgr, progmg dir & traf mgr; Dennis Blakemore, gen sls mgr, prom mgr, chief of engrg, rsch dir & sports cmtr; Skip Ballard, mus dir.

WCNG(FM)—Co-owned with WCVP(AM). Oct 23, 1990: 102.7 mhz; 3 kw. 426 ft TL: N35 04 00 W83 59 58. (CP: Ant 236 ft.). Stereo. 5 AM-midnight 175,000 Format: Soft rock. ♦Dennis Gene, gen mgr & disc jockey.

WKRK(AM)— Aug 8, 1958: 1320 khz; 5 kw-D, 62 w-N. TL: N35 06 42 W84 00 31. Hrs open: 24 427 Hill Street, 28906. Phone: (828) 837-1320. Fax: (828) 837-8610. Web Site: www.1320am.com. Licensee: Radford Communications Inc. (acq 1995; $250,000). Population served: 120,000 Natl. Network: Westwood One. Natl. Rep: Keystone (unwired net). Format: C&W, news/talk, relg. News: 7 hrs wkly. Target aud: 25-54. Spec prog: Pub affrs 3 hrs wkly. ♦Tim Radford, pres & gen mgr; Ab Radford, VP; Emma Ramsey, mktg mgr & adv mgr; Suzanne Crawford, prom dir; Bill Yonce, disc jockey; Jack Boxer, disc jockey; Jeff Young, disc jockey; Verna McKay, disc jockey.

Nags Head

WZPR(FM)— Apr 4, 1990: 92.3 mhz; 18.5 kw. Ant 384 ft TL: N35 50 49 W75 38 19. Stereo. Hrs opn: 24 637 Harbor Rd., Wanchese, 27981. Phone: (252) 475-1888. Fax: (252) 475-1881. E-mail: hunt@capsanmedia.com Web Site: www.wilburandorville.com. Licensee: CapSan Media LLC. Group owner: Convergent Broadcasting LLC (acq 6-30-2006; grpsl). Natl. Network: Motor Racing Net. Rgnl. Network: Capitol Radio Net., N.C. News Net. Format: Country. News staff: one. Target aud: 25-54. ♦William Whitlow, pres & gen mgr; Hunt Thomas, opns mgr.

Nashville

WZAX(FM)— February 1997: 99.3 mhz; 6 kw. Ant 328 ft TL: N35 57 01 W77 57 26. Hrs open: 12717 East N.C 97, Rocky Mount, 27803. Phone: (252) 442-8092. Fax: (252) 977-6664. Web Site: www.zaxradio.com. Licensee: First Media Radio LLC (group owner; (acq 7-22-2003; grpsl). Format: Adult contemp. ♦Alex Kolobielski, pres; Mike Binkley, gen mgr; David Perkins, opns mgr & progmg dir; Chris Hordy, sls dir & adv mgr; Karen Bullock, prom dir; Angie Webb, traf mgr.

New Bern

***WAAE(FM)**— 1997: 91.9 mhz; 1 kw. 164 ft TL: N35 09 17 W77 02 00. Hrs open:
Rebroadcasts WAFR(FM) Tupelo 100%.
Box 2440, Tupelo, MS, 38803. Phone: (662) 844-8888. Fax: (662) 840-3187. Web Site: www.afr.net. Licensee: American Family Association. Group owner: American Family Radio Format: Lite contemp, praise, talk. ♦Marvin Sanders, gen mgr; Joey Moody, chief of engrg.

WIKS(FM)— August 1977: 101.9 mhz; 100 kw. 1,020 ft TL: N35 12 07 W77 11 15. Stereo. Hrs opn: 24 207 Glenburnie Dr., 28560. Phone: (252) 633-1500. Fax: (252) 633-6546. Web Site: 1019online.com. Licensee: Beasley FM Acquisition Corp. Population served: 924,300 Natl. Rep: D & R Radio. Format: Urban adult contemp. Spec prog: Gospel 4 hrs, jazz 2 hrs wkly. ♦Bruce Beasley, pres; Bruce Simel, VP & gen mgr; J Dot, prom dir.

WNOS(AM)— Apr 23, 1942: 1450 khz; 1 kw-U. TL: N35 06 03 W77 04 33. Hrs open: 24 116 S. Business Plaza, 28562. Phone: (252) 638-8888. Fax: (252) 636-5848. E-mail: newbernamradio@earthlink.net Web Site: www.wnos1450.com. Licensee: CTC Media Group Inc. (group owner; acq 7-1-00; $65,000). Format: Big band, adult standards. Target aud: 45 plus; mature, retired & semi-retired adults. Spec prog: Jazz 12 hrs wkly. ♦Lee Afflerbach, pres; Mike Afflerbach, gen mgr.

WSFL-FM— July 20, 1968: 106.5 mhz; 100 kw. 915 ft TL: N35 02 27 W77 21 11. Stereo. Hrs opn: 24 207 Glenburnie Dr., 28560. Phone: (252) 633-1500. Fax: (252) 633-6546. Web Site: www.wsfl.com. Licensee: W & B Media Inc. Group owner: Beasley Broadcast Group (acq 7-10-91; $500,000 with co-located AM; FTR: 7-29-91) Natl. Rep: D & R Radio. Format: Classic rock, AOR. Target aud: 18-54. ♦Bruce Simel, gen mgr; Jeff Sanders, opns dir; Wendy Gatlin, prom dir; Richard Banks, engrg dir.

***WTEB(FM)**— June 4, 1984: 89.3 mhz; 100 kw. 522 ft TL: N35 06 32 W77 06 10. Stereo. Hrs opn: 24 800 College Ct., 28562. Phone: (252) 638-3434. Fax: (252) 638-3538. Web Site: www.publicradioeast.org. Licensee: Board of Trustees, Craven Community College. Natl. Network: NPR, PRI. Format: Class, news & info. News: 44 hrs wkly. Target aud: 35 plus; highly educated professionals. Spec prog: Jazz 4 hrs wkly. ♦Kathleen Beal, gen mgr; Charles Wethington, stn mgr & opns mgr; Jill McGuire, dev dir; Tim Kimble, progmg dir; Megan George, news dir; J. Howard Jones, chief of engrg; George Olsen, reporter.

WWNB(AM)— July 5, 1953: 1490 khz; 1 kw-U. TL: N35 07 59 W77 03 56. Hrs open: 24 114 S. Business Plaza, 28562. Phone: (252) 635-1490. Fax: (252) 636-5848. E-mail: wwnb1490@coastalnet.com Web Site: wwnb1490.com. Licensee: CTC Media Group Inc. (group owner; acq 11-15-90; $75,000). Natl. Network: USA. Law Firm: Jimmy Young. Format: Black gospel. Target aud: 25-54; general. ♦Mike Afflerbach, gen mgr; John Rawson, gen sls mgr.

***WZNB(FM)**— 2006: 88.5 mhz; 300 w. Ant 121 ft TL: N35 06 32 W77 06 10. Hrs open: Public Radio East, 800 College Ct., 28562. Phone: (252) 638-3434. Fax: (252) 638-3538. Web Site: www.publicradioeast.org. Licensee: Craven Community College. Format: Information. ♦Charles Wethington, stn mgr; Jill McGuire, dev dir; Tim Kimble, progmg dir; Megan George, news dir; J. Howard Jones, chief of engrg.

New Hope

WAUG(AM)— July 20, 1987: 750 khz; 500 w-D. TL: N35 47 28 W78 37 10. Stereo. Hrs opn: 1315 Oakwood Ave., Raleigh, 27610. Phone: (919) 516-4750. E-mail: waug@st-aug.edu Licensee: Saint Augustine's College. Natl. Network: American Urban. Format: Relg, news/talk, gospel. Target aud: 18 plus; Black adults. ♦Dr. Diane Suber, pres; Alan Riggs, gen mgr & stn mgr; Frank Butler, opns mgr; John Hardee, chief of engrg.

Newland

WECR(AM)— Aug 14, 1978: 1130 khz; 1 kw-D. TL: N36 04 39 W81 54 59. Hrs open: 1281 Newland Hwy., 28657. Phone: (828) 733-0188. Fax: (828) 733-0189. Licensee: Aisling Broadcasting of Banner Elk LLC (group owner; acq 2-10-2004; grpsl). Population served: 55,000 Format: Country. Target aud: 25-54; middle class, blue collar. Spec prog: Gospel 10 hrs, relg 5 hrs, bluegrass 2 hrs wkly. ♦Jonathan Hoffman, gen mgr.

Newport

WMGV(FM)— Sept 4, 1983: 103.3 mhz; 100 kw. 980 ft TL: N34 45 06 W76 52 57. Hrs open: 24 207 Glenburnie Dr., New Bern, 28560. Phone: (252) 633-1500. Fax: (252) 633-0718. Web Site: www.v1033.com. Licensee: WMGV License L.P. Group owner: Beasley Broadcast Group Inc. (acq 2-3-2000; grpsl). Natl. Rep: D & R Radio. Law Firm: Pepper & Corazzini. Format: Soft rock. Target aud: 18-54. ♦Bruce Simel, gen mgr; Colleen Jackson, progmg dir.

Newton

WNNC(AM)— June 18, 1948: 1230 khz; 1 kw-U. TL: N35 40 20 W81 14 12. Stereo. Hrs opn: 24 Box 430, 28658. Phone: (828) 464-4041. Fax: (828) 464-9662. E-mail: totalradio@aol.com Licensee: Newton-Conover Communications Inc. (acq 8-76). Population served: 202,000 Format: Adult contemp. News staff: one. Target aud: 25-49. Spec prog: Black 2 hrs, jazz 3 hrs wkly. ♦Dave Lingafelt, pres, gen mgr & chief of engrg; Jim Turner, gen sls mgr; Karol Lowery, traf mgr.

Norlina

***WJIJ(FM)**— January 2001: 94.3 mhz; 6 kw. Ant 328 ft TL: N36 29 46 W78 11 14. Hrs open: 24
Rebroadcasts WAJC(FM) Raleigh 100%.
5 W. Hargett St., Suite 801, Raleigh, 27607. Phone: (919) 899-6778. Fax: (919) 899-6779. Web Site: www.csnradio.com. Licensee: CSN International (group owner; acq 12-18-98). Format: Christian teaching/praise & worship music. ♦Jim Walker, gen mgr.

***WZRN(FM)**—Not on air, target date: unknown: 90.5 mhz; 2.3 kw. Ant 298 ft TL: N36 29 38 W78 11 23. Hrs open: 230-B Roanoke Ave., Roanoke Rapids, 27870. Phone: (252) 537-9999. Fax: (252) 537-3333. Web Site: www.wzru.org. Licensee: Roanoke Valley Communications Inc. Format: Talk, class, news. ♦George Campbell, pres; Bryan Lewis, gen mgr.

North Wilkesboro

WKBC(AM)— June 1947: 800 khz; 1 kw-D, 308 w-N. TL: N36 11 18 W81 08 07. Hrs open: 24 Box 938, 400 C St., 28659. Phone: (336) 667-2221. Fax: (336) 667-3677. E-mail: wkbctraffic@charter.net Licensee: Wilkes Broadcasting Co. Inc. (acq 1-24-03; with co-located FM). Population served: 60,000 Format: Country. ♦Robert Brown, pres & gen mgr; Ed Racey, news dir.

WKBC-FM— July 1962: 97.3 mhz; 100 kw. 1,350 ft TL: N36 04 34 W81 07 44. (CP: 100 kw horiz, 92 kw vert, ant 1,332 ft. TL: N36 04 34 W81 07 43). Stereo. 24 1,200,000 Format: CHR.

Oak Island

WSFM(FM)— July 2000: 98.3 mhz; 18.5 kw. 380 ft TL: N33 57 40 W78 01 37. Hrs open: 24 25 N. Kerr Ave., Suite C, Wilmington, 28405. Phone: (910) 791-3088. Fax: (910) 791-0112. E-mail: mud@surf983.com Web Site: www.surf983.com. Licensee: NM Licensing LLC. Group owner: NextMedia Group L.L.C. (acq 12-20-2004; grpsl). Format: Alternative. ♦Barbara Raybourne, gen mgr; Missy Andrus, prom dir; Mike Kennedy, mus dir; Walt Howard, engrg dir.

Ocean Isle Beach

WLQB(FM)— 1999: 93.5 mhz; 6 kw. Ant 328 ft TL: N33 55 37 W78 23 48. Hrs open:
Simulcast with WGTR(FM) Bucksport, SC 100%.
4841 Hwy. 17 Bypass S., Myrtle Beach, SC, 29577. Phone: (843) 293-0107. Fax: (843) 293-1717. Licensee: Qantum of Myrtle Beach License Co. LLC. Group owner: Qantum Communications Corp. (acq 7-2-2003; grpsl). Format: Country. ♦Michael Meeks, gen mgr; Serap Jackson, opns mgr & progmg mgr.

Oriental

WWHA(FM)— Mar 18, 1993: 94.1 mhz; 11 kw. 485 ft TL: N35 00 02 W76 49 58. Stereo. Hrs opn: 24
Rebroadcasts WZBR-FM Kinston 100%.
1307 S. Glenburnie Rd., New Bern, 28562. Phone: (252) 672-5900. Fax: (252) 637-6872. Licensee: Inner Banks Media LLC. Group owner: Archway Broadcasting Group (acq 3-12-2007; grpsl). Population served: 250,000 Rgnl. Network: Capitol Radio Net. Law Firm: Pepper & Corazzini. Format: Country. Target aud: 25 plus; adults with disposable incomes. ♦Bill Bailey, gen mgr.

Oxford

WCBQ(AM)— June 9, 1949: 1340 khz; 1 kw-U. TL: N36 18 27 W78 34 37. Stereo. Hrs opn: 18
Rebroadcasts WHNC(AM) Henderson 100%.
PO Box 1005, One Broadcast Ctr., 601 Henderson St., 27565. Phone: (919) 693-3540. Phone: (919) 693-1340. Fax: (919) 693-9054. E-mail: wcbgwhne@gloryroad.net Licensee: The Paradise Network (TPN) of North Carolina Inc. (acq 4-30-01; $650,000 with WHNC(AM) Henderson). Population served: 25,000 Rgnl. Network: N.C. News Net. Rgnl rep: T-N. Format: Black gospel. News: 10 hrs wkly. Target aud: General. Spec prog: Farm, professional & college sports, news/talk. ♦Dr. Alvin Augustis Jones, gen mgr & chief of engrg; Jim Davis, engrg mgr; Jeff Rose, spec ev coord; Nathaniel Smith, min affrs dir & relg ed.

Pinehurst

***WBFY(FM)**— September 2003: 90.3 mhz; 3.5 kw vert. Ant 328 ft TL: N35 09 13 W79 34 16. Hrs open: Drawer 2440, Tupelo, MS, 38803. Phone: (662) 844-8888. Fax: (662) 840-3187. Web Site: www.afr.net. Licensee: American Family Association. Group owner: American Family Radio Format: Christian. ♦Marvin Sanders, gen mgr.

WEEB(AM)—See Southern Pines

WFVL(FM)—See Southern Pines

WIOZ(AM)— Mar 25, 1980: 550 khz; 1 kw-D, 260 w-N, DA-2. TL: N35 09 04 W79 28 40. Hrs open: 24 200 Short Rd., Southern Pines, 28387. Phone: (910) 692-2107. Fax: (910) 692-6849. Licensee: Muirfield Broadcasting Inc. (group owner; acq 12-28-83). Population served: 62,000 Format: Adult standards. News staff: one; News: 5 hrs wkly. Target aud: General. ♦Walker Morris, pres; Tiffany Hewitt, gen mgr & gen sls mgr; Rich Rushforth, opns mgr.

Pinetops

WPWZ(FM)— Dec 2, 1996: 95.5 mhz; 12.5 kw. 459 ft TL: N35 56 45 W77 39 37. Stereo. Hrs opn: 24 12714 East NC 97, Rocky Mount, 27803. Phone: (252) 442-8092. Fax: (252) 977-6664. Licensee: First Media Radio LLC. (group owner; (acq 12-3-2003; grpsl). Format: Adult urban. News staff: one. Target aud: 24-54. ♦Alex Kolobielski, pres; Mike Binkley, gen mgr; David Perkins, opns mgr.

Pineville

WGIV(AM)— Mar 8, 1948: 1370 khz; 3 kw-D, 45 w-N. TL: N35 12 45 W80 52 06. (CP: COL Gastonia. 20 kw-D, 30 w-N. TL: N35 15 56 W81 09 01). Hrs opn:
Rebroadcasts WRNA(AM) China Grove 90%.
Box 11584, Rock Hill, SC, 29731. Phone: (803) 329-2760. Fax: (803) 329-3317. E-mail: fneely@rejoiceradio.com Web Site: www.RejoiceRadio.com. Licensee: Frank Neely. Group owner: Neely Enterprises (acq 4-1-98). Population served: 600,000 Natl. Network: USA. Law Firm: Smithwick & Belendiuk. Format: Christian. Target aud: 30 plus. ♦Emma Neely, VP; Frank Neely, gen mgr & stn mgr; Frankie Hemphill, stn mgr.

Plymouth

WJPI(AM)— Sept 11, 1959: 1470 khz; 5 kw-D. TL: N35 50 48 W76 45 22. Hrs open: Hwy. 64, 27962. Licensee: Free Temple Ministries Inc. (acq 12-14-98). ♦Terry Baylor, gen mgr.

WPNC-FM— December 1979: 95.9 mhz; 2.6 kw. 350 ft TL: N35 50 48 W76 45 22. Stereo. Hrs opn: 24 930 Hwy. 32 S., 27962. Phone: (252) 793-9995. Fax: (252) 793-4673. E-mail: magic959production@yahoo.com Web Site: www.gomagic959.com. Licensee: Durlyn Broadcasting Inc. (acq 1996). Population served: 100,000 Format: Adult contemp. ♦Bill Benjamin, CEO & gen mgr; W.B. Cox Jr., VP; Alex Rains, opns mgr.

Raeford

WMFA(AM)— Apr 25, 1963: 1400 khz; 1 kw-U. TL: N35 58 43 W79 12 32. Hrs open: 6 AM-10 PM 1085 E. Central Ave., 28376. Phone: (910) 875-6225. Phone: (910) 875-6477. Fax: (910) 875-3220. E-mail: wmfa1400@yahoo.com Licensee: W & V Broadcasting Enterprises Inc. (acq 6-2-93; $12,000; FTR: 6-21-93). Population served: 300,000 Rgnl. Network: Capitol Radio Net. Format: Gospel. Target aud: General. Spec prog: Sp 6 hrs wkly. ♦William Hollingsworth, CEO & pres; Vera Hollingsworth, CFO; Jeremy Hollingsworth, gen mgr & opns mgr.

***WRAE(FM)**— 2006: 88.7 mhz; 2.25 kw vert. Ant 466 ft TL: N34 54 57 W79 07 28. Hrs open:
Rebroadcasts WAFR(FM) Tupelo, MS 100%.
Drawer 2440, Tupelo, MS, 38801-2440. Phone: (662) 844-8888. Fax: (662) 842-6791. Web Site: www.afr.net. Licensee: American Family Association. Format: Christian. ♦Marvin Sanders, gen mgr.

Raleigh

WBBB(FM)— 1947: 96.1 mhz; 100 kw. 985 ft TL: N35 41 07 W78 43 14. Stereo. Hrs opn: 24 3012 High Woods Blvd., Suite 200, 27604. Phone: (919) 876-6464. Fax: (919) 790-8893. Web Site: www.96rockonline.com. Licensee: Carolina Media Group Inc. Group owner: Curtis Media Group (acq 1996; $16 million). Population served: 1,700,000 Natl. Rep: McGavren Guild. Format: Rock. Target aud: M 25-44. ♦Don Curtis, pres; Mike Hartel, gen mgr; Jay Naclis, progmg dir & progmg mgr; Allen Sherrill, chief of engrg; Ali Diatta, traf mgr; Shalon Lenfestey, prom dir & disc jockey.

WCLY(AM)— Aug 15, 1962: 1550 khz; 1 kw-D, 7 w-N. TL: N35 45 37 W78 39 27. Hrs open: 3012 Highwoods Blvd., Suite 200, 27604. Phone: (919) 954-1550. Fax: (919) 954-1556. Web Site: www.1550wcly.com. Licensee: Triangle Broadcast Associates LLC. (acq 4-5-99). Format: Relg. Target aud: 25-65; primarily Black. ♦Rick Heilmann, gen mgr.

***WCPE(FM)**— July 17, 1978: 89.7 mhz; 96.7 kw. Ant 1,178 ft TL: N35 56 25 W78 28 45. Stereo. Hrs opn: 24 Box 897, Wake Forest, 27588. Phone: (919) 556-5178. Fax: (919) 556-9273. E-mail: wcpe@wcpe.org Web Site: theclassicalstation.org.Galaxy 14, Tr 8, Vert, 6.30/6.48 mhz Licensee: Educational Information Corp. Population served: 981,000 Law Firm: Brooks, Pierce, McLendon, Humphrey & Leonard. Format: Classical. Target aud: 35 plus; class mus listeners. Spec prog: Sacred music, Opera. ♦Deborah S. Proctor, CEO, pres & gen mgr; Rae C. Weaver, dev dir; Dick Storck, progmg dir; Will Woltz, mus dir; John Graham, engr.

WDCG(FM)—See Durham

WDOX(AM)— Dec 1, 1981: 570 khz; 500 w-D, 54 w-N. TL: N35 45 37 W78 39 27. Hrs open: 24 3012 Highwoods Blvd., 27604. Phone: (919) 855-9383. Fax: (919) 790-6654. Web Site: www.570wdox.com. Licensee: Triangle Broadcast Associates LLC. Group owner: Curtis Media Group (acq 6-1-99). Population served: 600,000 Natl. Network: ABC. Natl. Rep: McGavren Guild. Format: News/talk. ♦Rick Heilmann, gen mgr; Peter Richon, progmg dir.

***WKNC-FM**— 1966: 88.1 mhz; 25 kw. 259 ft TL: N35 47 15 W78 40 14. Stereo. Hrs opn: 24 Box 8607, North Carolina State Univ. Mail Ctr., 27695-8607. Secondary address: 343 Witherspoon 27695. Phone: (919) 515-2401. Fax: (919) 513-2693. E-mail: gm@wknc.org Web Site: www.wknc.org. Licensee: North Carolina State University. Population served: 1,000,000 Format: Rock, alternative. News: 4 hrs wkly. Target aud: 18-59; adults & high school & college students of all demographics. Spec prog: Hip hop12 hrs, Metal 12 hrs, Eclectronia 16 hrs wkly. ♦Steve McCreery, gen mgr; Jamie Lynn Gilbert, stn mgr; Hannah Branigan, prom dir; kyle Robb, progmg dir; James Meyer, mus dir; Kelly Reid, mus dir; Rich Gurnsey, mus dir; Steve Salevan, mus dir; John Jeringan, chief of engrg; Will Patnaud, chief of engrg & engr.

WKSL(FM)—See Burlington

WPJL(AM)— March 1939: 1240 khz; 1 kw-U. TL: N35 46 25 W78 37 09. Stereo. Hrs opn: 5:30 AM-midnight Box 27946, 515 Bart St., 27611. Phone: (919) 834-6401. Licensee: WPJL Inc. (acq 7-86; $600,000; FTR: 4-21-86). Population served: 500,000 Natl. Network: USA. Format: Full-time Christian. News: 10 hrs wkly. Target aud: 25-54; Evangelical Christian community of greater Raleigh area. Spec prog: Black gospel. ♦William C. Suttles, pres & gen mgr; LaRue Porter, opns mgr; Jon Hardee, chief of engrg.

WPTF(AM)— Sept 22, 1924: 680 khz; 50 kw-U, DA-N. TL: N35 47 38 W78 45 41. Stereo. Hrs opn: 24 3012 Highwoods Blvd., Suite 200, 27604. Phone: (919) 876-0674. Fax: (919) 876-5291. Web Site: www.wptf.com. Licensee: First State Communications. Population served: 236,600 Natl. Network: CBS. Rgnl. Network: Southern Farm. Natl. Rep: McGavren Guild. Format: News/talk. News: 20 hrs wkly. Target aud: 35-64. Spec prog: Farm 10 hrs wkly. ♦David Stuckey, gen mgr, gen sls mgr & progmg dir.

WQDR(FM)—Co-owned with WPTF(AM). August 1949: 94.7 mhz; 96 kw. 1,679 ft TL: N35 40 35 W78 32 09. Stereo. 24 Phone: (919) 876-6464. Fax: (919) 790-8893. Web Site: www.wqdr.net. Licensee: Carolina Media Group Inc.1,000,000 Format: Modern country. News staff: one; News: 2 hrs wkly. Target aud: 25-54. Spec prog: NASCAR racing, bluegrass. ♦Trip Savery, gen mgr & gen sls mgr.

WQOK(FM)—South Boston, VA) Oct 1, 1960: 97.5 mhz; 100 kw. 981 ft TL: N36 20 52 W78 40 00. Stereo. Hrs opn: 24 8001-101 Creedmoor, Rd., 27613. Phone: (919) 848-9736. Fax: (919) 848-4724. E-mail: mmarinaro@radio-one.com Web Site: www.k975.com. Licensee: Radio One Licenses LLC. Group owner: Radio One Inc. (acq 11-8-01; grpsl). Population served: 870,000 Natl. Network: ABC. Natl. Rep: Christal. Format: Urban contemp. News staff: one; News: 20 hrs wkly. Target aud: 25-54; upwardly mobile with discretionary income. Spec prog: Gospel 9 hrs wkly. ♦Gary Weiss, gen mgr; Cy Young, opns dir & progmg dir; Saundra Lemaster, sls dir & gen sls mgr; Jodi Luke, natl sls mgr.

WRAL(FM)— 1947: 101.5 mhz; 96 kw. 1,820 ft TL: N35 40 35 W78 32 09. Stereo. Hrs opn: 24 Box 10100, 27605. Secondary address: 711 Hillsborough St. 27603. Phone: (919) 890-6101. Fax: (919) 890-6146. E-mail: mixonline@wralfm.com Web Site: www.wralfm.com. Licensee: Capitol Broadcasting Co. Inc. (group owner; acq 1946). Population served: 1,267,676 Rgnl. Network: N.C. News Net. Natl. Rep: Katz Radio. Law Firm: Holland & Knight. Wire Svc: AP Format: Adult contemp. News: 7 hrs wkly news progmg. Target aud: 25-54. Spec prog: Public Afffairs Block - 6:30-8:00am Sundays. ♦Jim Goodmon, pres; Dan McGrath, CFO; Ardie Gregory, VP & gen mgr; Joe Formicola, opns dir; Robert Wallace, gen sls mgr; Mark Turak, natl sls mgr; Brandon Alexander, mktg dir. Co-owned TV: WRAL-TV affil.

WRBZ(AM)— 1947: 850 khz; 10 kw-D, 5 kw-N, DA-N. TL: N35 48 04 W78 48 51. Stereo. Hrs opn: 24 4601 Six Forks Rd., Suite 520, Raliegh, 27609-5287. Phone: (919) 875-9100. Fax: (919) 510-6990. E-mail: brianm@850thebuzz.com Web Site: www.850thebuzz.com. Licensee: McClatchey Broadcasting Co. LLC (acq 2-9-2005). Population served: 1,490,000 Natl. Network: Westwood One, Fox Sports. Natl. Rep: McGavren Guild. Wire Svc: AP Format: All sports. News: 5 hrs wkly. Target aud: 25-54. ♦Brian Maloney, gen mgr; Mike Stangl, prom mgr; Adam Gold, progmg dir; Ted Sawyer, traf mgr.

WRDU(FM)—Wilson, Mar 1, 1961: 106.1 mhz; 100 kw. Ant 1,364 ft TL: N35 45 36 W78 11 04. Stereo. Hrs opn: 24 3100 Smoketree Ct., Suite 700, 27604. Phone: (919) 876-1061. Fax: (919) 876-2929. Web Site: www.1061rdu.com. Licensee: Clear Channel Communications Group owner: Clear Channel Communications Inc. (acq 8-30-2000; grpsl). Population served: 750,000 Law Firm: Fisher, Wayland, Cooper, Leader & Zaragoza. Format: Country. News staff: one; News: 3 hrs wkly. ♦Ken Spitzer, gen mgr.

WRTG(AM)—See Garner

WRVA-FM—Rocky Mount, November 1947: 100.7 mhz; 100 kw. Ant 1,968 ft TL: N35 49 53 W78 08 50. Stereo. Hrs opn: 3100 Smoketree Ct., Suite 700, 27604. Phone: (919) 878-1500. Fax: (919) 876-8578. Web Site: www.1007theriver.com. Licensee: Capstar TX L.P. Group owner: Clear Channel Communications Inc. (acq 8-30-2000; grpsl). Population served: 750,000 Natl. Network: ABC. Law Firm: Fisher, Wayland, Cooper, Leader & Zaragoza. Format: Classic rock. Target aud: 25-54; upscale adults. ♦Ken Spitzer, gen mgr; Jon Robbins, opns dir; Myron Bethea, sls dir & gen sls mgr; Tammy O'Dell, sls dir; Jessica Hayes, prom dir; Fred Pace, engrg VP & chief of engrg; Tracy Leonard, traf mgr.

***WSHA(FM)**— Nov 18, 1968: 88.9 mhz; 50 kw. Ant 456 ft TL: N35 45 05 W78 36 01. Stereo. Hrs opn: 24 118 E. South St., 27601. Phone: (919) 546-8432. Phone: (919) 546-8430. Fax: (919) 546-8315. E-mail: wsha@shawu.edu Web Site: www.wshafm.org. Licensee: Shaw University. Population served: 511,619 Natl. Network: NPR. Format: Jazz. News: 13.5 hrs wkly. Target aud: 25-55; high income, well educated. Spec prog: Sp 3 hrs, African 3 hrs, Caribbean 4 hrs, world mus 4 hrs, blues 8 hrs, gospel 20 hrs wkly. ♦Dr. Clarence G. Newsone, pres; Dr. Emeka Emekauwa, gen mgr; Michael Rochelle, dev dir; Rashad Muhaimin, progmg dir; Jim Davis, chief of engrg.

WWMY(FM)— 2000: 102.9 mhz; 1.7 kw. Ant 620 ft TL: N35 47 38 W78 45 41. Hrs open: 3012 Highwoods Blvd., Suite 201, 27604. Phone: (919) 790-6961. Fax: (919) 790-8369. Web Site: www.y1029.com. Licensee: WWND LLC. Group owner: Curtis Media Group (acq 10-2-98; $495,000 for stock). Population served: 1,100,000 Natl. Rep: McGavren Guild. Format: Oldies. ♦Mike Hartel, gen mgr; Shalon Lenfestry, prom dir; Bill Campbell, progmg dir; Allen Sherrill, chief of engrg; Ali Diatta, traf mgr.

Red Springs

WTEL(AM)— June 15, 1970: 1160 khz; 5 kw-D, 250 w-N. TL: N35 50 19 W79 10 36. Hrs open: 17 Box 711, 28377. Phone: (910) 843-5946.

Fax: (910) 843-8694. Licensee: WDAS License L.P. Group owner: Beasley Broadcast Group Inc. (acq 6-12-97; $1.2 million with WUKS(FM) Saint Pauls). Population served: 200,000 Rgnl. Network: Southern Farm. Format: Southern gospel, Black gospel. News staff: 2. Target aud: 24-54. Spec prog: Farm 5 hrs wkly. ♦Danny Highsmith, gen mgr; Towanna Locklear, gen sls mgr & disc jockey; Deanna Hodges, prom mgr; Garrette Davis, progmg dir; Gilbert Baez, news dir; Van Clough, chief of engrg; George McPhaul, local news ed; Montana Locklear, disc jockey.

Reidsville

WJMH(FM)— Sept 6, 1948: 102.1 mhz; 100 kw. 1,203 ft TL: N36 16 33 W79 56 27. Stereo. Hrs opn: 24 7819 National Service Rd., Suite 401, Greensboro, 27409. Phone: (336) 605-5200. Fax: (336) 605-5219. Web Site: www.102jamz.com. Licensee: Entercom Greensboro License LLC. Group owner: Entercom Communications Corp. (acq 12-13-99; grpsl). Population served: 940,600 Natl. Rep: McGavren Guild. Format: Urban, Hip-Hop. Target aud: 16-35; 65% Black, 35% white. ♦Brent Miller, gen mgr; Erin Casey, gen sls mgr; Brian McCall, prom dir & traf mgr; Brian Douglas, progmg dir; Larry Allen, chief of engrg.

WREV(AM)— 1948: 1220 khz; 1 kw-D. TL: N36 23 19 W79 38 51. Hrs open:
Rebroadcasts WRTP(AM) Chapel Hill 100%.
1010 Vermont Ave. N.W., Suite 100, Washington, DC, 20005. Phone: (202) 638-1959. Fax: (202) 393-7464. E-mail: valtravel@rcn.com Licensee: Estuardo Valdemar Rodriguez and Leonor Rodriguez. Group owner: Estuardo Valdemar Rodriguez and Leonor Rodriguez Stns (acq 8-5-2004; $125,000). Population served: 750,000 Format: Mexican reng. ♦Estuardo Valdemar Rodriguez, gen mgr.

Roanoke Rapids

WCBT(AM)— November 1940: 1230 khz; 1 kw-U. TL: N36 26 45 W77 39 51. Hrs open: 24 3 E. First St., Weldon, 27890. Phone: (252) 538-4184. Fax: (252) 538-0378. E-mail: haskinsal@yahoo.com Licensee: First Media Radio LLC. (group owner; (acq 7-22-2003; grpsl). Population served: 40,000 Natl. Network: ABC, ESPN Radio. Rgnl. Network: N.C. News Net. Format: Sports. ♦Al Haskin, gen mgr; John Green, opns mgr; Allen Garrett, progmg dir; Frank White, chief of engrg.

***WPGT(FM)**— January 2001: 91.1 mhz; 2 kw. Ant 69 ft TL: N36 28 08 W77 39 02. Hrs open:
Rebroadcasts WGPS(FM) Elizabeth City 100%.
Winchester Stn., 905 Halstead Blvd., Elizabeth City, 27909. Phone: (252) 334-1883. Fax: (252) 333-1459. E-mail: wpgt@csnradio.com Licensee: CSN International (group owner; (acq 5-5-2000; $20,000 for CP). Format: Christian. News: one hr wkly. ♦Jeff Ozanne, gen mgr; Darla Ozanne, progmg dir.

WPTM(FM)— 1973: 102.3 mhz; 6 kw. 300 ft TL: N36 30 12 W77 44 47. (CP: 5.4 kw, ant 344 ft.). Stereo. Hrs opn: 24 Box 910, 27870. Secondary address: 3 E. 4th St., Weldon 27890. Phone: (252) 536-3115. Fax: (252) 538-0378. E-mail: amyhmoran@yahoo.com Web Site: www.wptm1023.com. Licensee: First Media Radio LLC. (group owner; (acq 7-22-2003; grpsl). Population served: 112,000 Rgnl. Network: Southern Farm. Wire Svc: UPI Format: Country. News staff: 3; News: 14 hrs wkly. Target aud: 25-54; females with spendable income, decision-makers. Spec prog: Farm15 hrs, relg 3 hrs wkly. ♦Al Haskins, gen mgr.

***WRTP(FM)**— July 4, 1994: 88.5 mhz; 24 kw. Ant 479 ft TL: N36 17 44 W78 06 21. Stereo. Hrs opn: 24 7610 Falls of Neuse Rd., Suite 150, Raleigh, 27615. Phone: (919) 477-7222. Fax: (919) 477-4424. E-mail: wrtp@goodnews.org Web Site: www.hisradiowrtp.com. Licensee: Radio Training Network Inc. (acq 4-29-2005; swap for WZRU(FM) Roanoke Rapids). Population served: 96,000 Natl. Network: Salem Radio Network. Natl. Rep: Salem. Format: Contemp Christian music. News: 14 hrs wkly. Target aud: 25-54; Christian. ♦Mark G. Parker, CEO, gen mgr & progmg dir; James Campbell, pres; Randy Jordan, sls dir.

WTRG(FM)—See Gaston

***WZRU(FM)**— Dec 8, 1972: 90.1 mhz; 760 w. Ant 174 ft TL: N36 26 13 W77 38 12. (CP: 11 kw, ant 505 ft. TL: N36 14 39 W77 34 40). Stereo. Hrs opn: 24 232 Roanoke Ave., 27870-1916. Phone: (252) 308-0885. Fax: (252) 537-3333. E-mail: wzru@schoollink.net Web Site: www.wzru.org. Licensee: Roanoke Valley Communications Inc. (acq 5-6-2005; swap for WRTP(FM) Roanoke Rapids). Natl. Network: NPR. Law Firm: Arter & Hadden. Format: Adult contemp. News staff: 2; News: 30 hrs wkly. Target aud: 35 plus; community oriented, above-average education. Spec prog: Gospel 6 hrs, jazz 6 hrs, folk 5 hrs, oldies 4 hrs, new age 10 hrs, big band 4 hrs wkly. ♦Brian Lewis, gen mgr.

Robbins

WLHC(FM)— June 2, 2003: 103.1 mhz; 6 kw. Ant 388 ft TL: N35 26 33 W79 26 37. Hrs open: 24 Woolstone Corporation, Box 1087, Angier, 27501-1087. Secondary address: 102 S. Steele St., Suite 301, Sanford 27330. Phone: (919) 775-1031. Fax: (919) 775-1397. Web Site: www.life1031.com. Licensee: Woolstone Corporation Natl. Network: ABC. Format: Lifestyle. Spec prog: Folk one hr, jazz 2 hrs, Christian 5 hrs, bluegrass 5 hrs, news/talk one hr wkly. ♦Alan Button, pres; Al Mangum, stn mgr.

Robbinsville

WCVP-FM— 1987: 95.9 mhz; 60 w. 2,008 ft TL: N35 15 28 W83 47 44. Hrs open: 5:30 AM-10 PM (M-F); 6 AM-10 PM (S); 7 AM-10 PM (Su) Box 756, 129 N. By-Pass, 28771. Phone: (828) 479-8080. Phone: (828) 479-2296. Fax: (828) 479-2296. Licensee: Cherokee Broadcasting Co. Format: C&W. Target aud: General. ♦Dennis G. Blakemore, pres, gen mgr, gen sls mgr, prom mgr, progmg dir & chief of engrg; Penny Wade, pub affrs dir.

Rockingham

WAYN(AM)— September 1946: 900 khz; 1 kw-U, DA-2. TL: N34 55 30 W79 44 35. Hrs open: 6 AM-10 PM Box 519, 28380. Secondary address: 1223 Rockingham Rd. 28380. Phone: (910) 895-4041. Fax: (910) 895-4993. Licensee: WAYN Inc. (acq 5-10-01). Population served: 23,600 Rgnl. Network: N.C. News Net. Law Firm: Cohn & Marks. Format: Adult contemp, info. News: 20 hrs wkly. Target aud: 25-49; event-conscious adults. ♦William F. Futterer, pres, gen mgr & gen sls mgr; Jim Smith, progmg dir, news dir, farm dir, women's int ed & disc jockey; Mary Futterer Morgan, mus dir; Gene Shaw, chief of engrg; Brent Goodwin, disc jockey; Jay Parks, disc jockey.

WLWL(AM)— Oct 27, 1969: 770 khz; 5 kw-D. TL: N34 55 30 W79 47 11. Hrs open: Box 1536, 28380. Secondary address: 275 River Rd. 28379. Phone: (910) 997-2526. Fax: (910) 997-2527. E-mail: wlwl@77bibwaves.com Web Site: www.77bigwaves.com. Licensee: Sandhills Broadcasting Co. Inc. Population served: 75,000 Rgnl. Network: N.C. News Net. Format: Beach, oldies. Target aud: 25-60. ♦Keith Davis, gen mgr & opns mgr; Jeff Ballard, progmg dir.

***WRSH(FM)**— May 1973: 91.1 mhz; 10 w. 60 ft TL: N34 57 03 W79 42 56. (CP: 339.7 w, ant 161 ft.). Hrs opn: Box 1748, 28380. Secondary address: Richmond Sr. High School, 838 N. US Hwy. 1 28379. Phone: (910) 997-9812. Fax: (910) 997-9816. Licensee: Richmond County Board of Education. Format: Educ. ♦Kim Newton, gen mgr.

Rocky Mount

WDWG(FM)—Listing follows WRMT(AM).

WEED(AM)— Sept 10, 1933: 1390 khz; 5 kw-D, 30 w-N. TL: N35 57 43 W77 49 35. Hrs open: 24 Box 2666, 27802. Secondary address: 115 N. Church St. 27802. Phone: (252) 937-7400. Fax: (252) 443-5977. Licensee: Northstar Broadcasting Corp. (acq 7-22-03; with co-located FM). Population served: 45,000 Natl. Network: ABC. Format: Relg. News staff: one; News: 14 hrs wkly. Target aud: Males; 18+. ♦Charles Johnson II, VP, gen mgr & gen sls mgr; Milton Cannon, progmg dir.

WRSV(FM)—Co-owned with WEED(AM). 1949: 92.1 mhz; 2.35 kw. 531 ft TL: N35 48 40 W77 44 33. Stereo. 24 Phone: (252) 937-7400. Web Site: www.soul92jams.com.557,400 Format: Urban contemp. News: 3 hrs wkly. Target aud: General; African American consumers of all age groups.

WFXK(FM)—See Tarboro

WRMT(AM)— Dec 15, 1958: 1490 khz; 1 kw-U. TL: N35 55 57 W77 49 49. Stereo. Hrs opn: 24 12714 E. NC#97, 27803-0005. Phone: (252) 442-8092. Fax: (252) 977-6664. Licensee: First Media Radio LLC (group owner; (acq 1-7-2003; grpsl). Population served: 50,000 Rgnl. Network: N.C. News Net. Format: Sports. Target aud: 30 plus. ♦Alex Klasbiky, pres; Mike Binkley, gen mgr; Chris Hardie, sls dir & adv dir.

WDWG(FM)—Co-owned with WRMT(AM). Dec 18, 1989: 98.5 mhz; 16 kw. 417 ft TL: N35 54 43 W77 50 06.250,000 Format: Country. Target aud: 18 plus.

***WRQM(FM)**— April 1, 1996: 90.9 mhz; 6 kw. Ant 626 ft TL: N35 48 40 W77 44 33. Hrs open: 24
Rebroadcasts WUNC(FM) Chapel Hill 99.9%.
120 Friday Ctr Dr., CB-0915, Chapel Hill, 27517-9495. Phone: (919) 966-5454. Fax: (919) 966-5955. E-mail: wunc@unc.edu Web Site: www.wunc.org. Licensee: The Board of Trustees of the University of NC at Chapel Hill (acq 5-99). Population served: 250,665 Natl. Network: NPR, PRI, CBC Radio One. Format: News & Info. News staff: 7; News: 124 hrs wkly. Target aud: 35 plus; educated, successful, community active.

WRVA-FM—Licensed to Rocky Mount. See Raleigh

Rose Hill

WEGG(AM)— 1971: 710 khz; 250 w-D. TL: N34 51 48 W78 02 16. Hrs open: Sunrise-sunset Box 608, 28458. Secondary address: 3228 U.S. Hwy. 117 28458. Phone: (910) 289-2031. Fax: (910) 289-2032. Licensee: Conner Media Corp. Population served: 157,000 Rgnl. Network: Southern Farm. Natl. Rep: Keystone (unwired net). Format: Gospel/relg, Black. Spec prog: Farm 9 hrs, bluegrass gospel 10 hrs wkly. ♦Don Brown, chief of engrg; Suzanne Wilson, gen mgr, opns mgr, progmg dir, news dir & farm dir; C.D. Melvin, relg ed.

WZUP(FM)— January 1993: 104.7 mhz; 2.8 kw. Ant 256 ft TL: N34 55 41 W78 03 22. Hrs open: 3389 NC Highway 121, Farmville, 277828-9556. Phone: (910) 455-2202. Fax: (910) 355-2203. Licensee: Conner Media Corp. Format: Sp. ♦Rodney Rainey, gen mgr.

Roxboro

WKRX(FM)—Listing follows WRXO.

WRXO(AM)— 1949: 1430 khz; 1 kw-D, 65 w-N. TL: N36 22 04 W78 59 58. Hrs open: 6 AM-sunset
Simulcast with WKRX(FM) Roxboro.
Box 1176, 2070 Hurdle Mills Rd., 27573. Phone: (336) 599-0266. Fax: (336) 599-9411. E-mail: radiod@aol.com Licensee: Roxboro Broadcasting Co. (acq 5-8-92). Population served: 60000 Natl. Network: ABC. Rgnl. Network: N.C. News Net., Tobacco. Law Firm: Edmundson & Edmundson. Format: Country. News staff: one; News: 7 hrs wkly. Target aud: 18-49. Spec prog: Black 4 hrs, farm 5 hrs, Southern gospel 5 hrs wkly. ♦David Bradsher, pres, gen mgr, gen sls mgr, adv mgr & progmg dir; Wayne Tuck, news dir; Conrad Kimbrough, chief of engrg; Bill Lester, disc jockey; David Ramsey, disc jockey; Don Carroll, disc jockey.

WKRX(FM)—Co-owned with WRXO. 1958: 96.7 mhz; 3 kw. 300 ft TL: N36 22 04 W78 59 58. Stereo. 5:30 AM-11 PM Natl. Network: ABC. Law Firm: Edmundson & Edmundson. News staff: one; News: 7 hrs wkly. Target aud: 18-49. ♦David Bradsher, mktg dir, prom VP & adv VP; Bill Lester, disc jockey; David Ramsey, disc jockey; Don Carroll, progmg dir & disc jockey.

Rutherfordton

WCAB(AM)— Oct 19, 1966: 590 khz; 1 kw-D, 228 w-N. TL: N35 23 35 W81 55 23. Hrs open: 24 Box 511, 191 Whiteside Rd., 28139. Phone: (828) 287-3356. Fax: (828) 287-7182. E-mail: wcabam59@bellsouth.net Web Site: www.wcab59.com. Licensee: Isothermal Broadcasting Corp. (acq 8-1-84; FTR: 7-16-84). Population served: 125,000 Rgnl. Network: N.C. News Net. Format: Country, news/talk, sports. News: 25 hrs wkly. Target aud: 25 plus; adult consumers. ♦James H. Bishop, pres & gen mgr; Van Austin, progmg dir; Lou Gilliam, traf mgr.

Saint Pauls

WUKS(FM)— Oct 16, 1994: Stn currently dark. 107.7 mhz; 6 kw. 328 ft TL: N34 46 59 W79 07 11. Hrs open: Box 710, 508 Person St., Fayetteville, 28302. Phone: (910) 486-4114. Fax: (910) 486-2124. Web Site: www.kiss1077.com. Licensee: WDAS License L.P. Group owner: Beasley Broadcast Group (acq 6-12-97; $1.2 million with WTEL(AM) Red Springs). Population served: 600,000 Natl. Network: ABC. Natl. Rep: D & R Radio. Format: Urban adult contemp. Target aud: 25-54. ♦George Beasley, chmn; Bruce Beasley, pres; Caroline Beasley, CFO; Brian Beasley, exec VP; Mac Edwards, gen mgr & mktg mgr; Tila Comstock, gen sls mgr; Bryan Kusilka, natl sls mgr; Taylor Morgan, prom dir & mus dir; Jeff Anderson, progmg dir & progmg mgr; Val Jones, pub affrs dir; Dave Cooke, engrg dir; Van Clough, chief of engrg.

Salisbury

WEND(FM)— Mar 16, 1946: 106.5 mhz; 100 kw. 1,003 ft TL: N35 44 11 W80 38 52. (CP: 84 kw, ant 1,046 ft.). Stereo. Hrs opn: 24 801 Woodbridge Center Dr., Charlotte, 28217-1908. Phone: (704) 376-1065.

Fax: (704) 334-9525. Web Site: www.1065.com. Licensee: Capstar TX L.P. Group owner: Clear Channel Communications Inc. (acq 3-12-01). Population served: 1,810,000 Natl. Rep: McGavren Guild. Format: New rock, modern, alternative. Target aud: 18-34. ♦Jack Daniel, progmg dir & disc jockey; Liz Luke, pub affrs dir; Rob Caskey, chief of engrg; Chris Rozak, disc jockey; Jack Anthony, disc jockey; Kristen Pettus, disc jockey.

***WOGR-FM—** November 1996: 93.3 mhz; 10 w. 180 ft TL: N35 40 03 W80 28 13. Hrs open:
Rebroadcasts WOGR(AM) Charlotte 100%.
Box 16408, Charlotte, 28297. Phone: (704) 630-1075. Fax: (704) 393-1527. Web Site: www.wordnet.org. Licensee: Victory Christian Center Inc. Format: Christian contemp gospel. ♦Robyn Gool, pres; Cynthia Neely, gen mgr, sls dir, prom VP, pub affrs dir & disc jockey; Tamma Wylie, traf mgr; Kevin Blash, disc jockey; Lorenzo Peterson, disc jockey; Markus Moore, disc jockey.

WSAT(AM)— June 1947: 1280 khz; 1 kw-U, DA-N. TL: N35 40 30 W80 30 30. Hrs open: 24 1525 Jake Alexander Blvd., 28145. Phone: (704) 633-0621. Fax: (704) 636-2955. E-mail: buddy@WSAT1280.com Web Site: www.1280wsat.com. Licensee: Cap Communications Inc. (acq 6-28-02). Population served: 250,000 Natl. Network: Motor Racing Net. Format: Adult standards. Target aud: 25-64; people that can afford high ticket items. ♦Charles Poole, pres; Bubby Poole, mus dir; Ted Fuller, chief of engrg; Buddy Poole, disc jockey; Eddie Fuller, disc jockey; Lance Anderson, disc jockey.

WSTP(AM)— Jan 1, 1939: 1490 khz; 1 kw-U. TL: N35 41 12 W80 30 15. Stereo. Hrs opn: 24 Box 4157, 28145-4157. Secondary address: 1105 Statesville Blvd. 28144. Phone: (704) 636-3811. Fax: (704) 637-1490. E-mail: newsradio1490@yahoo.com Web Site: www.1490wstp.com. Licensee: Rowan Media INC. (acq 12-31-01). Population served: 180,000 Natl. Network: Fox News Radio, Talk Radio Network, Jones Radio Networks. Rgnl. Network: N.C. News Net. Rgnl rep: Capital Radio Format: News/talk 24/7. News: 24 hrs daily. Target aud: 25-59. ♦Timothy H. Coates, pres; Mike Mangan, VP, gen mgr, gen sls mgr, natl sls mgr & rgnl sls mgr; Mark Brown, news dir & pub svc dir; Hal McGee, chief of engrg.

Sanford

***WDCC(FM)—** 1971: 90.5 mhz; 3 kw. 148 ft TL: N35 28 19 W79 08 36. Hrs open: 1105 Kelly Dr., 27330. Phone: (919) 718-7257. Fax: (919) 718-7429. E-mail: wdcc@cccc.edu Web Site: www.wdccfm.com. Licensee: Central Carolina Community College. Population served: 30,000 Format: Rock, college, progsv, jazz, alternative. ♦Bill Freeman, gen mgr, progmg VP & mus dir.

WFJA(FM)—Listing follows WWGP.

WWGP(AM)— 1946: 1050 khz; 1 kw-D, 161 w-N. TL: N35 26 28 W79 12 54. Hrs open: 6 AM-1 AM Box 3457, 27331. Phone: (919) 775-3525. Fax: (919) 775-4503. E-mail: production@wfjaradio.com Licensee: Richard K. Feindel. (acq 1-13-94; $190,000 with co-located FM; FTR: 2-7-94) Population served: 70,000 Format: Country. News staff: one; News: 7 hrs wklyone. Target aud: 18-54. Spec prog: Farm 7 hrs wkly. ♦Richard K. Feindel, pres & gen mgr; Audrey R. Mason, progmg dir; Margaret Murchison, news dir; Jim Vest, chief of engrg.

WFJA(FM)—Co-owned with WWGP. 1950: 105.5 mhz; 2.25 kw. 377 ft TL: N35 26 28 W79 12 54. Stereo. 24 150,000 Format: Hits of the 50s, 60s, 70 & 80s. News staff: one. Target aud: 25-54.

WXKL(AM)— Oct 2, 1952: 1290 khz; 1 kw-D, 44 w-N. TL: N35 27 01 W79 09 30. Hrs open: 6 AM-8 PM 1516 Woodland Ave., 27330. Phone: (919) 774-1290. Phone: (919) 774-1080. Fax: (919) 774-1118. Licensee: Thomas Broadcasting Inc. (acq 7-8-2003). Population served: 50,000 Natl. Network: NBC. Format: Gospel. News: 6 hrs wkly. Target aud: 25 plus; general. ♦James Thomas, pres & gen mgr; Amos Marks, progmg dir & disc jockey; Danny Davis, disc jockey; Marilyn Cross, disc jockey; Tommy Mack, disc jockey.

Scotland Neck

WYAL(AM)— Apr 3, 1960: 1280 khz; 5 kw-D. TL: N36 08 03 W77 25 53. Hrs open: Box 425, 27874. Phone: (252) 826-3380. Web Site: 1280wyal.tripod.com. Licensee: Sky City Communications Inc. (acq 1-31-97; $100,000). Population served: 2,869 Rgnl. Network: N.C. News Net., Tobacco. Format: Gospel. Spec prog: Farm 2 hrs wkly. ♦Frank Knapper, gen mgr.

Scotts Hill

***WZDG(FM)—** March, 2007: 88.5 mhz; 8.9 kw vert. Ant 544 ft TL: N34 30 07 W78 04 58. Hrs open: Box 957, Wilmington, 28402-0957. Phone: (910) 763-2452. Fax: (910) 763-6578. E-mail: matt@edgeonover.com Web Site: www.edgeonover.com. Licensee: Carolina Christian Radio Inc. ♦Jim Stephens, pres.

Selma

WTSB(AM)— Aug 4, 1964: 1090 khz; 9 kw-D, 1.7 kw-CH. TL: N35 36 57 W78 24 33. Hrs open: Sunrise-sunset 3505 Durham Dr., Suite 111, Triangle Sports Broadcaster, Raleigh, 27603. Phone: (919) 329-9810. Fax: (919) 329-9803. Web Site: www.trianglesportstalk.com. Licensee: Triangle Sports Broadcasters LLC (acq 8-6-2004; $1.5 million). Population served: 90,000 Rgnl. Network: N.C. News Net. Format: Sports. Target aud: 18-45; general. Spec prog: Sp 5 hrs, Black 6 hrs wkly. ♦Brian Mishkin, stn mgr & progmg dir.

Semora

WKVE(FM)— Mar 1, 1996: 106.7 mhz; 6 kw. 328 ft TL: N36 29 24 W79 00 36. Hrs open: 2351 Sunset Blvd., Suite 170-218, Rocklin, CA, 95765. Phone: (843) 267-0036. Fax: (843) 399-9031. Web Site: www.klove.com. Licensee: Educational Media Foundation. Natl. Network: K-Love. Format: Christian Contemp. ♦Richard Jenkins, pres; Mike Novak, VP; Keith Whipple, dev dir; David Pierce, progmg dir; Ed Lenane, news dir; Sam Wallington, engrg dir; Karen Johnson, news rptr; Marya Morgan, news rptr; Richard Hunt, news rptr.

Shallotte

WBNE(FM)— Oct 31, 1977: 103.7 mhz; 25 kw. Ant 328 ft TL: N33 59 55 W78 22 25. (CP: COL Wrightsville Beach. 35 kw, ant 510 ft. TL: N34 03 02 W77 57 20). Hrs opn: 24 122 Cinema Dr., Wilmington, 28403. Phone: (910) 772-6300. Fax: (910) 772-6310. E-mail: newsroom@seacomm.com Web Site: www.937thebone.com. Licensee: Sea-Comm Inc. (group owner; (acq 12-30-2003; with WLTT(FM) Shallotte). Population served: 600,000 Natl. Network: Jones Radio Networks. Format: Classic rock. News staff: one; News: 7 hrs wkly. Target aud: 25-54; upscale female. ♦Paul Knight, gen mgr & stn mgr; Max Deutsch, gen sls mgr; Jonathan Knight, prom mgr; Zach McHugh, progmg dir.

WLTT(FM)— Sept 20, 1986: 106.3 mhz; 6 kw. 328 ft TL: N34 02 50 W78 16 12. Stereo. Hrs opn: 24 122 Cinema Dr., Wilmington, 28403. Phone: (910) 772-6300. Fax: (910) 772-6310. Web Site: www.thebigtalker1063fm.com. Licensee: Sea-Comm Inc. (group owner; (acq 12-30-2003; with WBNE(FM) Shallotte). Natl. Network: ABC, Jones Radio Networks. Format: News/talk info. Target aud: 25 plus; mature professionals. Spec prog: Beach mus 6 hrs wkly. ♦Paul Knight, gen mgr; Max Deutsch, stn mgr.

WVCB(AM)— June 11, 1964: 1410 khz; 500 w-D. TL: N33 58 20 W78 23 02. Hrs open: Box 314, 28459. Secondary address: 4640 Main St. 28459. Phone: (910) 754-4512. Fax: (910) 754-3461. E-mail: wvcb@atmc.net Licensee: John G. Worrell. (acq 3-1-84; $30,000; FTR: 1-30-84). Population served: 897 Rgnl. Network: N.C. News Net. Format: Relg, gospel. Target aud: General. ♦Rhonda Worrell, stn mgr & opns mgr.

Shelby

WADA(AM)— July 9, 1958: 1390 khz; 1 kw-D, 500 w-N, DA-N. TL: N35 19 28 W81 32 00. Stereo. Hrs opn: 24 205 S. Washington St., Suite 1, 28150. Secondary address: Box 2266 28151-2266. Phone: (704) 482-1390. Fax: (704) 481-9007. E-mail: wada.2@juno.com Web Site: www.us1390.com. Licensee: HRN Broadcasting Inc. (acq 2006; $350,000). Population served: 250,000 Natl. Network: ABC. Format: Country. Target aud: 25-54; middle and older. ♦D. Mark Boyd III, pres; Joe Martin, stn mgr & opns mgr; Andy Johnson, gen sls mgr.

WBT(AM)—See Charlotte

WIBT(FM)— 1948: 96.1 mhz; 100 kw. 1,738 ft TL: N35 21 44 W81 09 19. Stereo. Hrs opn: 24 801 Woodbridge Center Dr., Charlotte, 28217-1908. Phone: (704) 338-9600. Fax: (704) 334-9525. Web Site: www.magic96.com. Licensee: Clear Channel Broadcasting Licenses Inc. Group owner: Clear Channel Communications Inc. (acq 10-18-2000). Population served: 2,000,000 Natl. Rep: McGavren Guild. Format: Hits of the 60s & 70s. News staff: one; News: 4 hrs wkly. Target aud: 25-54. ♦Morgan Bohannon, gen mgr; Graves Upchurch, gen sls mgr; Nick Allen, progmg dir; Linda Silver, news dir; Ben Brinitzer, chief of engrg; Brenda Grubb, traf mgr; Bobby Lane, disc jockey; Harriet Coffey, disc jockey; Ron Harper, disc jockey.

WOHS(AM)— Aug 21, 1946: 730 khz; 1 kw-D, 168 w-N. TL: N35 17 27 W81 34 05. Stereo. Hrs opn: 24 1511 W. Dixon Blvd., 28152. Phone: (704) 482-4510. Phone: (704) 487-6313. Fax: (704) 482-4680. E-mail: thebossradio@bellsouth.net Web Site: www.theboss.us. Licensee: HRN Broadcasting Inc. (acq 8-31-2006; $1.5 million with WGNC(AM) Gastonia). Population served: 500,000 Natl. Network: ABC. Rgnl. Network: N.C. News Net. Rgnl rep: T-N. Format: Beach, Oldies, Sports. News staff: 2. Target aud: General. ♦D. Mark Boyd III, pres; Calvin R. Hastings, gen mgr.

Siler City

WNCA(AM)— Aug 19, 1952: 1570 khz; 5 kw-D, 290 w-N. TL: N35 43 40 W79 29 18. Hrs open: 6am - midnight Box 429, 17890 Hwy. 64 W., 27344. Phone: (919) 742-2135. Fax: (919) 663-2843. Licensee: Chatham Broadcasting Co. Inc. of Siler City. (acq 3-1-62). Population served: 50,000 Rgnl. Network: N.C. News Net. Law Firm: Smithwick & Belendiuk. Format: Loc prgmg, news/talk, christian, sports. News staff: 2; News: 15+ hrs wkly. Target aud: 25-55; rural, agri-oriented, blue-collar, growing spanish community. Spec prog: Gospel 5 hrs, relg 12 hrs, loc sports 6 hrs & Spanish 25 hrs wkly. ♦Barry Hayes, pres, gen mgr, news dir & engrg mgr; Dacia Hayes, dev VP; Renee Kennedy, stn mgr, opns dir, sls VP & sls dir; Debbie Applewhite, traf mgr; Jose Alvarado, spanish dir.

Smithfield

WMPM(AM)— 1950: 1270 khz; 5 kw-D. TL: N35 31 33 W78 20 01. Hrs open: Box 240, 27577. Phone: (919) 934-2434. Fax: (919) 989-6388. E-mail: Info@1270wmpm.com Web Site: www.1270wmpm.com. Licensee: Carolina Broadcasting Service Inc. (acq 12-1-58). Population served: 90,000 Natl. Network: CBS. Format: Country, bluegrass, gospel. Target aud: 30 plus; general. Spec prog: Farm 2 hrs, relg 8 hrs, news/talk 12 hrs wkly. ♦Carl E. Lamm, gen mgr, adv mgr & progmg dir; Larry D. Barnes, stn mgr, pub affrs dir & farm dir; Bill Lambert, chief of engrg; Mickey Lamm, news dir & local news ed.

WTSB(AM)—See Selma

Snow Hill

***WAGO(FM)—** July 1, 1998: 88.7 mhz; 17 kw. Ant 310 ft TL: N35 30 07 W77 36 22. Stereo. Hrs opn: 24 Box 1895, Goldsboro, 27533. Phone: (252) 747-8887. Fax: (252) 747-7888. E-mail: wago@gomixradio.org Web Site: www.gomixradio.org. Licensee: Pathway Christian Academy Inc. Natl. Network: Moody, Salem Radio Network. Law Firm: Steve Yelverton. Format: Christian. News staff: one; News: 14 hrs wkly. Target aud: General. ♦Dr. T.D. Worthington, pres & gen mgr; Ashley Worthington, prom dir; Keith Aycock, progmg dir; Tim Sutton, mus dir; Joe Patton, chief of engrg.

South Gastonia

***WGAS(AM)—** Aug 14, 1959: 1420 khz; 500 w-D. TL: N35 12 53 W81 10 31. Hrs open:
Rebroadcasts WOGR(AM) Charlotte.
Box 16408, Charlotte, 28297. Phone: (704) 865-9427. Fax: (704) 393-1527. E-mail: wayne@wordnet.org Web Site: www.wordnet.org. Licensee: Victory Christian Center Inc. Population served: 150,000 Format: Christian. Target aud: General. ♦Robyn Gool, pres; Wayne Hammond, stn mgr; James Sims, opns mgr; Cynthia Neely, sls dir, prom mgr & pub affrs dir; Eleasah Hammond, mus dir; Tamma Wylie, traf mgr.

Southern Pines

WEEB(AM)— Nov 15, 1947: 990 khz; 10 kw-D, 500 w-N. TL: N35 11 37 W79 24 42. Hrs open: 24 Box 1855, Midland Rd., 28388. Phone: (910) 692-7440. Fax: (910) 692-7372. E-mail: mdfweeb@yahoo.com Web Site: www.weeb990.com. Licensee: Pinehurst Broadcasting Corp. (acq 8-31-91; $275,000; FTR: 6-10-91). Population served: 120,000 Natl. Network: ABC, Fox News Radio, Salem Radio Network. Rgnl. Network: N.C. News Net. Law Firm: Maupin, Taylor, Ellis & Adams. Format: News/talk. News staff: 3; News: 26 hrs wkly. Target aud: 25 plus; business professionals, CEOs, retirees. Spec prog: High school & college sports, gospel 6 hrs wkly. ♦Rich McCarthy, opns mgr; Steve Adams, CFO, VP, gen mgr & progmg dir; Rose Sharp, relg ed.

WFVL(FM)— Aug 14, 1973: 106.9 mhz; 50 kw. Ant 492 ft TL: N35 09 04 W79 28 40. Stereo. Hrs open: 24 1009 Drayton Rd., Fayetteville, 28303. Phone: (910) 864-5222. Fax: (910) 864-3065. Web Site: kix1069fm.com. Licensee: Cumulus Licensing Corp. Group owner: Cumulus Media Inc. (acq 3-12-2001; $6.15 million). Population served: 600,000 Format: Oldies. News staff: 2; News: 18 hrs wkly. Target aud: 35 plus. ♦Tom Haymond, gen mgr; Perry Stone, opns mgr.

WIOZ-FM— 1995: 102.5 mhz; 3.4 kw. 436 ft TL: N35 09 04 W79 28 40. Hrs open: 200 Short Rd., 28387. Phone: (910) 692-2107. Fax: (910) 692-6849. Web Site: www.star1025fm.com. Licensee: Meridian Communications L.L.C. Group owner: Muirfield Broadcasting Inc. (acq 6-17-97; $316,500). Format: Adult contemp. ♦Walker Morris, pres; Tiffany Hewitt, gen mgr; Rich Rushforth, opns mgr.

Southern Shores

WFMI(FM)— 2003: 100.9 mhz; 39 kw. Ant 485 ft TL: N36 12 10 W75 52 23. Hrs open: 4801 Columbus St., Suite 202, Virginia Beach, VA, 23462. Phone: (757) 490-9364. Fax: (757) 490-2524. E-mail: rejoice@rejoice100point9.com Web Site: www.rejoice100point9.com. Licensee: Communications Systems Inc. Format: Gospel, talk. ♦Mike Chandler, gen mgr; Vickie Bright, stn mgr.

Southport

WAZO(FM)— Apr 15, 1978: 107.5 mhz; 32 kw. 594 ft TL: N34 03 02 W77 57 20. Stereo. Hrs opn: 24 25 N. Kerr Ave., Wilmington, 28405. Phone: (910) 791-3088. Fax: (910) 791-0112. E-mail: mark@z1075.com Web Site: www.z1075.com. Licensee: NM Licensing LLC. (group owner) (acq 2-1-2005; grpsl). Population served: 182,000 Natl. Rep: McGavren Guild. Format: CHR. News staff: one; News: one hr wkly. Target aud: 18-49; young, upwardly mobile professionals. ♦Jeff Dinetz, pres; Barry Brown, VP; Bea Raybourne, gen mgr; Gayle Brown, gen sls mgr; Mark Jacobs, progmg dir; Doug Carlisle, news dir; Walt Howard, chief of engrg.

Sparta

WCOK(AM)— April 1967: 1060 khz; 800 w-D. TL: N36 28 55 W81 05 35. Hrs open: Box 578, 28675. Phone: (336) 372-8231. Fax: (336) 372-5863. E-mail: luke@ls.net Licensee: Mountain Empire Broadcasting Inc. (acq 10-1-99). Population served: 25,000 Format: C&W, relg. Target aud: General. ♦Andy Wright, pres, gen mgr, traf mgr & disc jockey; Jos Reynoso, gen mgr; Christy Galyan, disc jockey; Johnathan Johnson, disc jockey; Michael Sexton, disc jockey.

Spindale

WGMA(AM)— October 1982: 1520 khz; 500 w-D. TL: N35 21 00 W81 56 18. Hrs open: Sunrise-sunset Box 805, 301 W. Main St., 28160. Phone: (828) 287-5151. Phone: (828) 287-5150. Fax: 1-828-287-0081. Licensee: Moonglow Broadcasting Inc. (acq 5-16-03). Population served: 35,000 Law Firm: Smithwick & Belendiuk. Format: Southern gospel. News staff: one; News: 20 hrs wkly. Target aud: 30-50; adults. ♦Barbara Martin, pres & exec VP; Kaye Cantrell, gen mgr; Andy Foster, opns dir; Neil Murray, mus dir; Jerrell Bedford, chief of engrg & engr.

***WNCW(FM)**— Oct 13, 1989: 88.7 mhz; 17 kw. 3,054 ft TL: N35 44 05 W82 17 10. Stereo. Hrs open: 24 Box 804, 28160. Phone: (828) 287-8000. Fax: (828) 287-8012. E-mail: info@wncw.org Web Site: www.wncw.org. Licensee: Isothermal Community College. Population served: 580,000 Natl. Network: PRI, NPR. Law Firm: Schwartz, Woods & Miller. Format: AAA, news. News staff: one; News: 31 hrs wkly. Target aud: 35-49; anyone interested in diverse info & culture. Spec prog: Blues 4 hrs, jazz 5 hrs, folk 12 hrs, drama 3 hrs, gospel 2 hrs wkly. ♦David Gordon, gen mgr; Kate Barkschat, dev mgr, gen sls mgr & prom mgr; Elle Ellis, progmg dir; Martin Anderson, mus dir; Dennis Jones, chief of engrg & news rptr; Ele Ellis, traf mgr.

Spring Lake

WCIE(AM)— May 22, 1963: 1450 khz; 950 w-U. TL: N35 11 11 W78 57 35. Hrs open: 24 5418 Yadkin Road, Fayetteville, 28303. Phone: (910) 222-1450. Fax: (910) 223-1451. E-mail: radiolatina1450@earthlink.net Web Site: www.radiolatina1450.com. Licensee: WCIE-AM Inc. (acq 4-20-2001). Population served: 257,000 Rgnl. Network: Capitol Radio Net. Format: Spanish/news. Target aud: 12-65; Male & Female/Hispanic orintated culture. ♦Teo Rodriguez, gen mgr.

***WZRI(FM)**— 2005: 89.3 mhz; 2 kw vert. Ant 179 ft TL: N35 10 14 W78 57 44. Stereo. Hrs opn: 24 2351 Sunset Blvd., Suite 170-218, Rocklin, CA, 95765. Phone: (916) 251-1600. Fax: (916) 251-1650. E-mail: info@air1.com Web Site: www.air1.com. Licensee: Educational Media Foundation. Group owner: EMF Broadcasting (acq 11-12-2002). Natl. Network: Air 1. Law Firm: Shaw Pittman. Format: Contemp Christian. News staff: 3. Target aud: 18-35; Judeo-Christian, female. ♦Richard Jenkins, pres; Mike Novak, VP; Keith Whipple, dev dir; Liz Morton, mus dir; Ed Lenane, news dir; Sam Wallington, engrg dir; Karen Johnson, news rptr; Marya Morgan, news rptr; Richard Hunt, news rptr.

Spruce Pine

WTOE(AM)— Dec 24, 1955: 1470 khz; 5 kw-D, 100 w-N. TL: N35 54 24 W82 06 21. Hrs open: 24 Box 607, Mark Group Bldg., 749 Sawmill Rd., Burnsville, 28714. Secondary address: 749 Sawmill Road, Burnsville 28714. Phone: (828) 765-7441. Phone: (828) 682-3798. Fax: (828) 682-6227. Fax: (828) 682-0998. E-mail: 1470@wtoe.com Web Site: www.wtoe.com. Licensee: Mountain Valley Media Inc. (acq 9-27-91; $140,000). Population served: 104,000 Natl. Network: ABC. Rgnl. Network: N.C. News Net. Format: Soft Oldies. News staff: one; News: 10 hrs wkly. Target aud: 25 plus. Spec prog: Relg 8 hrs wkly. ♦Remelle K. Sink, CEO & pres; J. Ardell Sink, exec VP; Michael Sink, VP, gen mgr & chief of engrg; Holly Hall, opns mgr & mktg dir; Mary Marsh, prom dir; Dennis Renfro, adv dir; Steve Murphy, news dir & pub affrs dir.

Statesville

WAME(AM)— Oct 7, 1957: 550 khz; 500 w-D. TL: N35 47 43 W80 51 17. Hrs open: 24 212 Signal Hill Dr., 28625. Phone: (704) 872-0550. Fax: (704) 872-0551. E-mail: wame@statesville.net Licensee: Statesville Family Radio Corp. Group owner: GHB Radio Group (acq 4-22-86; $210,000; FTR: 3-31-86). Population served: 150000 Natl. Network: USA. Format: Adult standards. News: 2 hrs wkly. Target aud: 35-64. Spec prog: Loc & pro sports, gospel 5 hrs wkly.

WKKT(FM)— Mar 16, 1961: 96.9 mhz; 100 kw. 1,550 ft TL: N35 31 57 W80 47 47. Stereo. Hrs opn: 24 801 Wood Ridge Center Dr., Charlotte, 28217. Phone: (704) 714-9444. Fax: (704) 332-8805. Web Site: www.wkktfm.com. Licensee: Capstar TX L.P. Group owner: Clear Channel Communications Inc. (acq 8-30-00; grpsl). Population served: 1,200,000 Format: Country. Target aud: 25-54; middle to upper income adults. ♦Morgan Bohannon, gen mgr; Bruce Logan, opns mgr; Robin Colfax, natl sls mgr & rgnl sls mgr; Valerie Gladden, prom dir; John Roberts, progmg dir & news dir; Linda Silver, news dir & traf mgr; Ben Brinitzer, chief of engrg.

WSIC(AM)— May 3, 1947: 1400 khz; 1 kw-U. TL: N35 48 09 W80 53 30. Stereo. Hrs opn: 24 1117 Radio Rd., 28677. Phone: (704) 872-6345. Fax: (704) 873-6921. Web Site: www.1400wsic.com. Licensee: Iredell Broadcasting Inc. Group owner: Clear Channel Communications Inc. (acq 8-25-2006; $700,000). Population served: 19,996 Natl. Rep: Rgnl Reps. Rgnl rep: T-N. Format: Sports, news. News staff: one; News: 21 hrs wkly. Target aud: 35 plus; upscale. ♦Mark Sanger, gen mgr; Billy Blevins, opns mgr & mus dir.

Swanquarter

***WHYC(FM)**— Mar 8, 1981: 88.5 mhz; 3 kw. 293 ft TL: N35 26 29 W76 13 09. Stereo. Hrs opn: Box 155-A, 204 72 US 264, Swan Quarter, 27885. Phone: (252) 926-7201. Fax: (252) 926-1557. Licensee: Hyde County Board of Education. Format: Var. Target aud: General; eastern North Carolina population.

Sylva

WRGC(AM)— Nov 8, 1957: 680 khz; 1 kw-D, 250 w-N, DA-N. TL: N35 23 35 W83 11 38. Hrs open: 24 Box 1044, 1846 Skyland Dr., 28779. Phone: (828) 586-2221. Fax: (828) 586-6834. E-mail: skinner@gacaradio.com Web Site: www.wrgc.com. Licensee: Georgia-Carolina Radiocasting Co. LLC. Group owner: Sutton Radiocasting Companies (acq 1-17-2002; $450,000). Population served: 50,000 Natl. Network: ABC. Law Firm: Putbrese, Hunsaker & Trent, P.C. Wire Svc: AP Format: Adult contemp. News staff: one; News: 14 hrs wkly. Target aud: General. ♦Douglas M. Sutton Jr., pres; David Skinner, VP & gen mgr; Will Chandler, opns mgr & progmg dir; Charlie Bauder, news dir; Tim Stephens, engrg dir & chief of engrg.

Tabor City

WTAB(AM)— July 1, 1954: 1370 khz; 5 kw-D, 109 w-N. TL: N34 09 00 W78 51 40. Hrs open: 6 AM-midnight Box 127, 28463. Secondary address: 210 Avon St. 28463. Phone: (910) 653-2131. Fax: (910) 653-5146. E-mail: wtab@wtabradio.com Web Site: www.wtabradio.com. Licensee: WTAB Inc. (acq 7-1-95; $175,000). Population served: 40,000 Rgnl. Network: N.C. News Net. Format: Southern gospel, country. News: 7 hrs wkly. Target aud: General. Spec prog: Swap shop, 12hrs wkly. ♦Jack Miller, pres, gen mgr & sls VP; Bonnie Miller, exec VP; Bobby Pait, mus dir & disc jockey; Bob Gause, chief of engrg & engr; Lloyd Gore, disc jockey.

Tarboro

WCPS(AM)— January 1947: 760 khz; 1 kw-D. TL: N35 55 40 W77 34 15. Hrs open: Sunrise-sunset Box 1202, 27886. Secondary address: 1406 St Andrew St. 27886. Phone: (252) 824-7878. Fax: (252) 824-7818. Web Site: ww.wcpsam760.com. Licensee: Johnson Broadcast Ventures Ltd. (acq 5-6-00; $100,000). Population served: 13,000 Rgnl. Network: N.C. News Net. Format: Gospel/Blues. Target aud: General. ♦Jimmy Johnson, pres & gen mgr; Stephanie Randolph, stn mgr.

WFXK(FM)— September 1952: 104.3 mhz; 100 kw. 987 ft TL: N35 48 40 W77 44 33. Hrs open: 24
100% Simulcast with WFXC. WFXC, Durham, NC, 100%.
8001-101 Creedmoor Rd., Raleigh, 27613. Phone: (919) 848-9736. Fax: (919) 848-4724. E-mail: mmarinaro@radio-one.com Web Site: www.foxyhits.com. Licensee: Radio One Licenses LLC. Group owner: Radio One Inc. (acq 11-8-01; grpsl). Population served: 870,000 Format: Adult contemp, urban. News staff: one; News: 3 hrs wkly. Target aud: 25-54. Spec prog: Gospel 3 hrs, jazz 4 hrs wkly. ♦Gary Weiss, gen mgr; Cy Young, opns dir & progmg dir; Kim Gattis, sls dir; Jodi Luke, natl sls mgr; Bruce Farmer, prom dir; Jodi Berri, mus dir; Jim Davis, chief of engrg; Regina Williams, traf mgr.

Taylorsville

WACB(AM)— May 2, 1964: 860 khz; 1 kw-D. TL: N35 55 57 W81 10 19. Hrs open: 24 133 E. Main Ave., 28681. Phone: (828) 632-4621. Fax: (828) 632-9081. Licensee: Apple City Broadcasting Co. Inc. (acq 9-24-93; $70,239; FTR: 10-11-93). Population served: 500,000 Format: Modern country, oldies. News staff: 2; News: 4 hrs wkly. Target aud: General. Spec prog: Gospel 12 hrs wkly.Norris Keever, pres; Mary Alice Brown, VP; Joyce Brown, gen sls mgr; Lisa McLain, prom dir & women's int ed; Lonnie Carrigan, mus dir & disc jockey; Pete Ray, asst music dir & disc jockey; Roger Brown, CEO, exec VP, gen mgr, opns dir, news dir & pub affrs dir; Jeff Watts, chief of engrg; Dean Bruce, traf mgr; Rick Gilbert, sports cmtr; Mark Daniels, disc jockey

WTLK(AM)— June 17, 1962: 1570 khz; 1 kw-D, 248 w-N. TL: N35 55 45 W81 09 44. Hrs open: 133 E. Main Ave., 28681. Phone: (828) 632-4621. Fax: (828) 632-9081. Licensee: Apple City Broadcasting Co. Inc. (acq 6-95; $225,000). Population served: 28,000 Rgnl. Network: N.C. News Net. Format: Southern gospel, bluegrass gospel. ♦Norris Keever, pres; Mary Alice Brown, exec VP & mus dir; Roger Brown, CEO, gen mgr & opns dir; Joyce Brown, gen sls mgr; Lisa McLain, prom dir; Jeff Watts, chief of engrg.

Thomasville

WBLO(AM)— September 1947: 790 khz; 2.5 kw-D, 26 w-N. TL: N35 57 41 W80 02 13. Hrs open: 24 hrs
Simulcasts with WIST-FM Thomasville.
Box 5663, High Point, 27262. Secondary address: 1607 Country Club Dr., High Point 27262. Phone: (336) 887-0983. Fax: (336) 887-3055. Web Site: www.790theball.com. Licensee: GHB Radio Inc. Group owner: GHB Radio Group (acq 4-3-2001; $350,000). Law Firm: Reddy, Begley & McCormick. Format: Sports/Talk. Target aud: 35+; older, mature audience. ♦George H. Buck Jr., pres; Susan Childress, gen mgr, gen sls mgr, prom VP & prom mgr; Wes Jones, opns mgr; Stan Thomas, prom VP & prom dir; Bill Timm, progmg dir; Ed Kasovic, chief of engrg.

WIST-FM— April 1949: 98.3 mhz; 1.68 kw. Ant 429 ft TL: N35 57 41 W80 02 13. Stereo. Hrs opn: 24 Box 5663, High Point, 27262. Secondary address: 1607 Country Club Dr., High Point 27262. Phone: (336) 887-0983. Fax: (336) 887-3055. Web Site: www.countrylegends983.com. Licensee: WEAM Quality Radio Corp. Group owner: GHB Radio Group (acq 1997; $925,000). Population served: 800,000 Format: Country. Target aud: 35 plus. ♦George H.

Buck Jr., pres; Susan Childress, gen mgr, gen sls mgr & prom dir; Wes Jones, opns mgr; Ed Kasovic, chief of engrg.

Topsail Beach

WWTB(FM)— Sept 12, 1993: 103.9 mhz; 21.5 kw. Ant 328 ft TL: N34 29 38 W77 29 18. (CP: COL Swansboro. 104.1 mhz; 3.9 kw, ant 407 ft. TL: N34 43 03 W77 16 57). Stereo. Hrs opn: 24 122 Cinema Dr., Wilmington, 28403. Phone: (910) 772-6300. Fax: (910) 772-6310. Web Site: www.thebigtalkerfm.com. Licensee: Sea-Comm Inc. (group owner; acq 1-15-2004; $2.3 million). Population served: 7,00.,000 Law Firm: Gardner, Carton & Douglas. Format: Talk, news. News: 6 hrs wkly. Target aud: 35 plus; baby boomers, upper income & affluent retirees. ♦Paul Knight, gen mgr.

Troy

WJRM(AM)— Dec 8, 1961: 1390 khz; 1 kw-D. TL: N35 21 43 W79 51 38. Hrs open: 5 AM-9 PM Box 706, 27371. Phone: (910) 576-1390. Fax: (910) 576-1393. E-mail: jeffrey@wjrm.com Web Site: wjrm.com. Licensee: Family Worship Ministries Inc. (acq 6-10-02; $115,000). Population served: 33,000 Format: Christian gospel. News staff: one. ♦Harold Pope, pres & gen mgr; Jeffrey Pope, opns mgr & gen sls mgr.

Tryon

WJFJ(AM)— Oct 1, 1954: 1160 khz; 10 kw-D, 500 w-N, DA-N. TL: N35 14 07 W82 14 27. Hrs open: 24 Box 279, Courthouse St., Columbus, 28772. Phone: (828) 894-5858. Fax: (828) 894-2957. E-mail: wjfjradio@wjfjradio.com Web Site: www.wjfjradio.com. Licensee: Columbus Broadcast Corp. Inc. (acq 1996; $265,000). Natl. Network: USA. Format: Christian. News staff: one; News: 25 hrs wkly. Target aud: 25 plus; middle-to-upper income. ♦John Owens, gen mgr & opns mgr.

Valdese

WSVM(AM)— Oct 6, 1961: 1490 khz; 1 kw-U. TL: N35 44 03 W81 34 04. Hrs open: 24 Box 99, 1117 S. Praley St., 28690. Phone: (828) 874-0000. Fax: (828) 874-2123. E-mail: radio@1490wsvm.com Web Site: www.1490wsvm.com. Licensee: GHB of Waxhaw Inc. Group owner: GHB Radio Group (acq 10-28-2002; $450,000 with WEGO(AM) Concord). Population served: 56,600 Natl. Network: ABC. Rgnl. Network: N.C. News Net. Format: Oldies. News: 4 hrs wkly. Target aud: 25-64. Spec prog: Gospel 4 hrs, sports 15 hrs wkly. ♦Jerry Clegg, gen mgr.

Wade

***WZRL(FM)**— 2007: 90.7 mhz; 3 kw. Ant 353 ft TL: N35 13 54 W78 22 11. Hrs open:
Rebroadcasts KLRD(FM) Yucaipa, CA 100%.
5700 West Oaks Blvd., Rocklin, CA, 95765. Phone: (916) 251-1600. Fax: (916) 251-1650. Web Site: www.air1.com. Licensee: Educational Media Foundation. (acq 10-26-2006). Natl. Network: Air 1. Format: Alternative rock, div. ♦Richard Jenkins, pres; Mike Novak, sr VP.

Wadesboro

WADE(AM)— July 23, 1947: 1340 khz; 1 kw-U. TL: N34 57 01 W80 03 23. Hrs open: 24 Box 416, Waxhaw, 28173. Phone: (704) 843-5418. Fax: (704) 695-1495. Licensee: Inspirational Deliverance Center Inc. (acq 6-8-93; $27,500; FTR: 6-28-93). Format: Adult contemp Christian. Spec prog: Farm one hr wkly. ♦Myra Davis, stn mgr.

***WYFQ-FM**— 1994: 93.5 mhz; 8.7 kw. 554 ft TL: N35 02 57 W80 18 38. Stereo. Hrs opn: 24 2004 Walkup Ave., Monroe, 28110. Phone: (704) 523-5555. Fax: (704) 522-1967. Web Site: www.bbnradio.net. Licensee: Bible Broadcasting Network Inc. (group owner; acq 1996; $2,425,000). Law Firm: Smithwick & Belendiuk. Format: Relg. Target aud: 18-55. ♦Lowell L. Davey, CEO & pres; Hank Crull, stn mgr; Richard Johnson, opns mgr.

Wake Forest

WDRU(AM)— Sept 1, 1989: 1030 khz; 50 kw-D, DA. TL: N36 10 43 W78 45 30. Stereo. Hrs opn: Daytime
Simulcast with WTRU(AM) Kernersville.
4405 Providence Lane, Ste D, Winston-Salem, 27106. Phone: (336) 759-0363. Fax: (336) 759-0366. E-mail: info@wtru.com Web Site: www.wtru.com. Licensee: Truth Broadcasting Corp. Group owner: Davidson Media Group LLC (acq 5-2-2005; swap for WWBG(AM) Greensboro and WTOB(AM) Winston-Salem). Natl. Network: Salem Radio Network. Format: Christian. Target aud: 25-54; middle-class families. ♦Bryan Brown, COO & gen sls mgr; Stuart W. Epperson Jr., pres; Stuart Epperson, Jr., gen mgr; Ed Park, progmg dir; Mandel Owens, chief of engrg.

Wallace

WZKB(FM)— July 20, 1972: 94.3 mhz; 3.3 kw. 300 ft TL: N34 45 29 W78 00 00. Stereo. Hrs opn: Box 28, Clinton, 28329-0026. Phone: 910-864-5028. Fax: 910-864-6270. E-mail: dan@wgqe1057.com Web Site: www.wgqr.com. Licensee: Christian Listening Network Inc. (acq 12-15-2003; $425,000). Population served: 120,000 Natl. Network: Salem Radio Network. Natl. Rep: Salem. Law Firm: Putbrese, Huntsaker & Trent. Format: Southern gospel. News: 11 hrs wkly. Target aud: 35-54; women. ♦George E. Wilson, pres; Dan DeBruler, gen mgr; Steve Turley, progmg dir.

Wanchese

WOBR-FM—Listing follows WOBX(AM).

WOBX(AM)— May 29, 1970: 1530 khz; 1 kw-D, DA. TL: N35 51 52 W75 39 01. Hrs open: Sunrise-sunset Box 340, 27981. Secondary address: 3855 Mill Landing Rd., Hwy. 345 27981. Phone: (252) 473-5402. Fax: (252) 473-5838. Licensee: East Carolina Radio Inc. Group owner: East Carolina Radio Group (acq 8-82; $110,000; FTR: 8-9-82). Population served: 25,000 Format: Relg, gospel. Target aud: Christian. ♦Elmo Daniels, pres & gen mgr; Jim Mills, chief of engrg.

WOBR-FM—Co-owned with WOBX(AM). June 1, 1973: 95.3 mhz; 25 kw. Ant 295 ft TL: N35 51 52 W75 39 01. Stereo. 24 Box 1419, Nags Head, 27959. Phone: (252) 441-1024. Fax: (252) 441-2109.22,000 Natl. Network: ABC. Format: Classic rock. News staff: one. Target aud: 25-54; upscale, affluent baby boomers. ♦Eddie James, pres & progmg dir; Rich Laesch, gen mgr & opns mgr.

Warrenton

WARR(AM)— 1970: 1520 khz; 5 kw-D, 1 kw-CH. TL: N36 24 18 W78 08 09. Hrs opn: sun up-sun down Box 611, 27589. Phone: (252) 257-5557/257-9277. Fax: (252) 257-5988. Web Site: www.warr1520am.com. Licensee: Quad Divisions Inc. dba Darensburg Broadcasting (acq 7-1-02). Rgnl. Network: N.C. News Net. Format: Soul gold gospel. News: 5 hrs wkly. Target aud: 25-56. ♦Logan Darensburg, pres & gen mgr; Ann Alston, stn mgr & prom dir.

Washington

WDLX(AM)— Mar 3, 1942: 930 khz; 5 kw-D, 1 kw-N, DA-N. TL: N35 31 34 W77 04 43. Hrs open: Box 3333, Greenville, 27836. Secondary address: 1813 US Hwy 17 S., Chocwinity 27817. Phone: (252) 317-1250. Fax: (252) 317-1255. Web Site: www.pirateradio930.com. Licensee: Pirate Media Group LLC Group owner: NextMedia Group L.L.C. (acq 6-27-2005; $400,000). Population served: 8,961 Rgnl. Network: N.C. News Net. Format: Talk. Target aud: 35 plus. ♦Paul Kingman, gen mgr.

WERO(FM)— Jan 20, 1961: 93.3 mhz; 100 kw. Ant 1,781 ft TL: N35 21 55 W77 23 38. Stereo. Hrs opn: 24 1361 Colony Dr., New Bern, 28560. Phone: (252) 639-7900. Fax: (252) 946-0330. Web Site: www.bob933.com. Licensee: NM Licensing LLC. (acq 11-26-2001; grpsl). Population served: 800,000 Format: Hot adult contemp. News staff: one; News: 12 hrs wkly. Target aud: 25-54. ♦Paul Kingman, gen mgr.

WLGT(FM)— December 1988: 98.3 mhz; 1.3 kw. Ant 490 ft TL: N35 29 14 W77 02 42. Stereo. Hrs opn: 8:30am - 5:30pm 211 Commerce St., Suite C, Greenville, 27858-5030. Phone: (252) 672-5900. Phone: (252) 830-0943. Fax: (252) 355-2234. Web Site: www.983litefm.com. Licensee: ABG North Carolina LLC. Group owner: Archway Broadcasting Group (acq 2-27-2003; $3 million with WWNK(FM) Farmville). Format: Adult contemp. Target aud: 25-54; upscale, affluent audience. ♦Tony Denton, gen mgr; Lee Cherry, opns mgr.

Waxhaw

WNMX-FM— Mar 1, 1995: 106.1 mhz; 32 kw. 365 ft TL: N34 53 01 W80 47 37. Stereo. Hrs opn: 24 5732 N. Tryon St., Charlotte, 28213. Phone: (704) 596-4900. Fax: (704) 599-1061. Web Site: www.mix106.net. Licensee: GHB of Waxhaw Inc. Group owner: GHB Radio Group (acq 6-95; $325,000). Natl. Network: ABC. Law Firm: Reddy, Begley & McCormick. Format: Adult Standards. News: 7 hrs wkly. Target aud: 45 plus; general. Spec prog: Southern Lawn and Garden Line; Charlotte Bobcats Basketball; Duke Football/Basketball. ♦George H. Buck Jr., pres; Tom Gentry, gen mgr; Brant Hart, opns dir & traf mgr; Bob Wood, gen sls mgr; Ken Conrad, pub affrs dir; Stu Albert, chief of engrg; Terri Miller, news rptr.

Waynesville

WMXF(AM)— August 1947: 1400 khz; 1 kw-U. TL: N35 30 14 W82 58 25. Hrs open: 24 13 Summerlin Rd., Asheville, 28806. Phone: (828) 257-2700. Fax: (828) 281-3299. Licensee: Clear Channel Broadcasting Licenses Inc. Group owner: Clear Channel Communications Inc. (acq 3-21-01; grpsl). Population served: 6488 Natl. Rep: Keystone (unwired net). Format: Adult standards. News staff: one. Target aud: General. Spec prog: Relg 3 hrs wkly.

WQNS(FM)—Co-owned with WMXF(AM). October 1979: 104.9 mhz; 245 w. 1,581 ft TL: N35 34 07 W82 54 27. Stereo. 24 1318-B Patton Ave., Asheville, 98806. Web Site: www.rock104rocks.com.75,000 Format: Classic rock. Target aud: General.

Weaverville

WTMT(FM)— October 1989: 105.9 mhz; 9.7 kw. Ant 1,102 ft TL: N35 42 18 W82 50 01. Stereo. Hrs opn: 24 1190 Patton Ave., Asheville, 28806. Phone: (828) 259-9695. Fax: (828) 253-3291. Licensee: Saga Communications of North Carolina LLC. (acq 8-7-2006; $650,000). Format: Rock. ♦Edward K. Christian, pres; Randy Cable, VP & gen mgr; Chris Hoffman, gen sls mgr.

Weldon

WSMY(AM)— 1957: 1400 khz; 1 kw-U. TL: N36 24 43 W77 37 06. Hrs open: 24 Box 910, Roanoke Rapids, 27870. Phone: (252) 536-0209. Fax: (252) 538-0378. E-mail: info@wsmy1400.com Web Site: www.wsmy1400.com. Licensee: First Media Radio LLC. (group owner; (acq 7-22-2003; grpsl). Population served: 112,000 Format: Urban, relg. Target aud: 18 plus; affluent adults. ♦Al Haskins, gen mgr; John Green, stn mgr; Allen Garrett, opns mgr.

Wendell-Zebulon

WETC(AM)— June 16, 1959: 540 khz; 5 kw-D, 500 w-N, DA-2. TL: N35 52 06 W78 25 56. (CP: 8 kw-D). Hrs opn: 24 2865 Amwiler, Ste 650, Doraville, GA, 30360. Phone: (770) 825-0095. Fax: (770) 246-0054. Web Site: www.prietobroadcasting.com. Licensee: Prieto Broadcasting Inc. (acq 4-13-2004; $1.8 million). Population served: 545,000 Format: Sp. ♦Everado Morales, stn mgr.

West Jefferson

WKSK(AM)— May 27, 1959: 580 khz; 5 kw-D, 34 w-N. TL: N36 24 39 W81 29 46. Stereo. Hrs opn: 6 AM-7 PM (M-S); 7 AM-7 PM (Su) Box 729, 28694. Phone: (336) 246-6001. E-mail: wksk@skybest.com Web Site: www.580wksk.com. Licensee: Caddell Broadcasting, Inc. (acq 8-1-78). Population served: 80,000 Format: C&W. News staff: one;

News: 16 hrs wkly. Target aud: General. Spec prog: Farm 3 hrs, gospel 5 hrs, Sp one hr wkly. ♦Jan Caddell, pres & gen mgr; Graham Caddell, opns mgr.

Whiteville

WENC(AM)— July 14, 1946: 1220 khz; 5 kw-D, 152 w-N. TL: N34 18 30 W78 43 00. Hrs open: 6 AM-10 PM 108 Radio Station Rd., 28472. Phone: (910) 642-2133. Fax: (910) 642-5981. Licensee: DHA Communications. (acq 1-5-94; $135,000; FTR: 1-17-94). Population served: 80,400 Rgnl. Network: N.C. News Net. Format: Urban contemp, gospel, blues. News staff: one; News: 10 hrs wkly. Target aud: 25-54; women. Spec prog: Farm 5 hrs, relg 4 hrs, talk 5 hrs wkly. ♦Jesse Lee Godwin, gen mgr.

WTXY(AM)— Jan 1, 1976: 1540 khz; 1 kw-D. TL: N34 19 23 W78 42 47. Hrs open: Sunrise-sunset Box 1038, 501 W. Virgil St., 28472. Phone: (910) 642-8214. Phone: (910) 642-8215. Fax: (910) 640-1540. E-mail: wtxy@earthlink.net Web Site: www.whitevillnc.com/wtxy. Licensee: Stanley Broadcasting System Inc. (acq 2-2-87; $80,000; FTR: 12-22-86). Population served: 55,000 Natl. Network: Westwood One, Motor Racing Net. Rgnl. Network: Tenn. Agri. Format: News/talk. News staff: 2; News: 84 hrs wkly. Target aud: General. Spec prog: Farm 2 hrs, relg 10 hrs wkly. ♦John H. Stanley, exec VP; Thomas V. Stanley Jr., pres & gen mgr; Linda Shaver, opns mgr.

WZFX(FM)—Licensed to Whiteville. See Fayetteville

Wilkesboro

***WSIF(FM)**— Apr 6, 1977: 90.9 mhz; 1 kw. Ant -171 ft TL: N36 08 12 W81 11 02. Stereo. Hrs opn: 24 Box 120, 28697. Secondary address: 1328 S. Collegiate Dr. 28697. Phone: (336) 838-6179. Phone: (336) 838-6222. Fax: (336) 838-6528. E-mail: al.delachica@wilkescc.edu Licensee: Wilkes Community College. Population served: 20,000 Format: Classic rock, div, progsv. Target aud: 18-44. ♦Dr. Gordon G. Burns Jr., pres; Al de Lachica, gen mgr.

WWWC(AM)— Jan 26, 1970: 1240 khz; 1 kw-U. TL: N36 09 00 W81 09 42. Hrs open: Box 580, 28697. Secondary address: 413 Wilkesboro Blvd. 28697. Phone: (336) 838-1241/838-9992. Fax: (336) 838-9040. E-mail: onair@12403wc.com Web Site: www.12403wc.com. Licensee: Foothills Media Inc. (acq 1994). Population served: 60000 Natl. Network: USA. Rgnl. Network: N.C. News Net. Format: Southern gospel. Target aud: General. ♦John Wishon, pres, gen mgr, prom dir, progmg dir & chief of engrg; Petrice Edwards, gen sls mgr.

Williamston

WIAM(AM)— March 1951: 900 khz; 1 kw-D, 258 w-N. TL: N35 51 27 W77 02 34. Hrs open: 24 Box 590, 27892. Phone: (252) 792-4161. Fax: (252) 809-0039. E-mail: bryant@opendoorradio.com Web Site: www.opendoorradio.com. Licensee: Lifeline Ministries Inc. (acq 6-18-90; FTR: 7-9-90). Population served: 30,000 Rgnl. Network: N.C. News Net. Format: Relg, gospel. Target aud: General. ♦Johnny Bryant, pres, gen mgr & progmg dir.

WRHD(FM)— Aug 1, 1962: 103.7 mhz; 100 kw. Ant 981 ft TL: N35 53 47 W76 58 58. Stereo. Hrs opn: 24
Rebroadcasts WRHT(FM) Morehead City 100%.
408 W. Arlington Blvd., Suite 101-C, Greenville, 27834. Phone: (252) 355-1037. Fax: (252) 355-2234. Licensee: Inner Banks Media LLC. Group owner: Archway Broadcasting Group (acq 3-12-2007; grpsl). Population served: 300,000 Law Firm: Davis Wright Tremaine P.C. Format: Contemp hit/Top-40. Target aud: 18-49; young adults. ♦Rodney Rainey, gen mgr; Lee Cherry, opns mgr; Tori Gray, gen sls mgr; Tony Denton, mktg dir; Fox 5 Feltman, progmg dir & news dir; Eddie Harrell, chief of engrg; Donna Spivey, traf mgr.

Wilmington

WAAV(AM)—Leland, Dec 20, 1957: 980 khz; 5 kw-U, DA-N. TL: N34 14 54 W78 00 09. Hrs open: 24 3233 Burnt Mill Rd., 28403-2654. Phone: (910) 763-9977. Fax: (910) 762-0456. Web Site: www.980waav.com. Licensee: Cumulus Licensing Corp. Group owner: Cumulus Media L.L.C. (acq 7-2-97; $1.6 million with co-located FM). Population served: 60,000 Natl. Rep: McGavren Guild. Wire Svc: AP Format: News/talk. Target aud: 35 plus. ♦Jim Principi, gen mgr; Perry Stone, opns mgr; Jennifer McLean, gen sls mgr; Jackie Jordon, mktg dir & prom dir; Mike Farow, progmg dir; Mark Ward, pub affrs dir; Tim Nelson, chief of engrg; Harvard Jennings, disc jockey.

WKXS-FM—Co-owned with WAAV(AM). Dec 10, 1994: 94.5 mhz; 3.8 kw. Ant 416 ft TL: N34 12 35 W77 56 53.24 Format: Urban contemp. ♦Lou Bennett, progmg dir.

***WDVV(FM)**— 1999: 89.7 mhz; 13.5 kw vert. Ant 348 ft TL: N34 10 52 W78 02 33. Stereo. Hrs opn: Box 957, 28402. Phone: (910) 763-2452. Fax: (910) 763-6578. E-mail: church@thedoveonline.org Web Site: www.thedoveonline.org. Licensee: Carolina Christian Radio Inc. (group owner; (acq 2-16-2001; $100,000 with WMYT(AM) Carolina Beach). Natl. Network: USA. Format: Christian praise & worship. ♦Jim Stephens, gen mgr; Roger Brace, chief of opns.

WGNI(FM)— Mar 1, 1970: 102.7 mhz; 100 kw. 981 ft TL: N34 03 00 W78 04 56. Stereo. Hrs opn: 24 3233 Burnt Mill Rd., 28403-2654. Phone: (910) 763-9977. Fax: (910) 762-0456. Web Site: www.cumulus.com. Licensee: Cumulus Licensing Corp. Group owner: Cumulus Media Inc. Population served: 200,000 Natl. Rep: McGavren Guild. Wire Svc: AP Format: Hot Adult Contemp. News staff: two. ♦Jim Principi, gen mgr; Perry Stone, opns dir; Jennifer McLean, gen sls mgr; David Carroll, prom dir & adv dir; Mike Farrow, progmg dir; Tim Nelson, chief of engrg.

***WHQR(FM)**— Apr 24, 1984: 91.3 mhz; 100 kw. Ant 1,141 ft TL: N34 07 53 W78 11 17. Stereo. Hrs opn: 24 254 N. Front St., Wilminton, 28401. Phone: (910) 343-1640. Fax: (910) 251-8693. E-mail: whqr@whqr.org Web Site: www.whqr.org. Licensee: Friends of Public Radio Inc. Population served: 320,000 Natl. Network: NPR, PRI. Rgnl rep: Megan Gorham Law Firm: Schwartz, Woods & Miller. Wire Svc: AP Format: Class, news. News staff: 3; News: 63 hrs wkly. Target aud: 35 plus. ♦Bob Klorkmon, pres & progmg dir; John Milligan, gen mgr; George Scheibner, opns mgr & chief of engrg; Ann Berry, prom mgr; Catherine Welch, traf mgr.

WKXB(FM)—See Burgaw

WLSG(AM)— Dec 24, 1946: 1340 khz; 1 kw-U. TL: N34 12 35 W77 56 53. Hrs open: 24 201 N. Front St., Suite 805, 28401. Secondary address: Box 957 28402. Phone: (910) 763-2452. Fax: (910) 763-6578. E-mail: life@life905.com Web Site: www.life905.com. Licensee: Carolina Christian Radio Inc. (group owner; (acq 6-30-2000; $75,000). Natl. Network: Salem Radio Network, USA. Format: Southern gospel, relg. Target aud: 30 plus. ♦Jim Stephens, gen mgr; Roger Brace, engr.

WMFD(AM)—Listing follows WRQR(FM).

WMNX(FM)— Feb 24, 1970: 97.3 mhz; 100 kw. 602 ft TL: N34 16 34 W78 09 09. (CP: Ant 977 ft. TL: N34 03 00 W78 04 56). Stereo. Hrs opn: 24 3233 Burnt Mill Dr., 28403-2654. Phone: (910) 763-9977. Fax: (910) 762-0456. Web Site: www.coast973.com. Licensee: Cumulus Licensing Corp. Group owner: Cumulus Media Inc. (acq 3-12-01; grpsl). Population served: 600,000 Natl. Rep: McGavren Guild. Wire Svc: AP Format: Urban contemp. News staff: one. Target aud: 18-49; general. ♦Jim Principi, gen mgr; Perry Stone, opns dir.

WRQR(FM)— February 1994: 104.5 mhz; 4.5 kw. 377 ft TL: N34 16 15 W77 57 23. Hrs open: 25 N. Kerr Ave., 28405. Phone: (910) 791-3088. Fax: (910) 791-0112. Web Site: www.rock1045.com. Licensee: NM Licensing LLC. (acq 12-20-2004; grpsl). Format: Classic rock. Target aud: 25-54. ♦Matt Carter, progmg dir & progmg mgr.

WMFD(AM)—Co-owned with WRQR(FM). Apr 15, 1935: 630 khz; 1 kw-U, DA-2. TL: N34 13 31 W77 59 17. Stereo. 24 Web Site: www.am630.net.65,000 Natl. Network: CBS. Format: ESPN sports. News staff: one; News: 3 hrs wkly. Target aud: 30 plus; upscale audience.

WVBS(AM)—See Burgaw

WWIL(AM)— Aug 25, 1963: 1490 khz; 1 kw-U. TL: N34 13 52 W77 57 18. Hrs open: 24 Box 957, 28402-0957. Phone: (910) 763-2452. Fax: (910) 763-6578. E-mail: life@life905.com Web Site: www.life905.com. Licensee: Carolina Christian Radio Inc. (group owner; (acq 10-28-92; $35,000; FTR: 11-23-92). Population served: 150,000 Natl. Network: USA. Format: Black gospel "the light". Target aud: 25-49. ♦Jim Stephens, gen mgr; Pastor James Utley, stn mgr & mktg mgr.

WWIL-FM— December 1995: 90.5 mhz; 1 kw horiz, 20 kw vert. 328 ft TL: N34 10 52 W78 02 33.24 Web Site: www.life905.com. (Acq 6-95; FTR: 1-9-95).300,000 Natl. Network: USA. Format: Adult contemp, Christian. Target aud: 25-54.

WWQQ-FM— Mar 31, 1969: 101.3 mhz; 50 kw. 525 ft TL: N34 13 31 W77 59 17. (CP: 40 kw, ant 544 ft.). Stereo. Hrs opn: 24 3233 Burnt Mill Rd., 28403-2654. Phone: (910) 763-9977. Fax: (910) 762-0456. Web Site: www.cumulus.com. Licensee: Cumulus Licensing Corp. Group owner: Cumulus Media L.L.C. (acq 7-3-97; Population served: 110,000 Natl. Rep: McGavren Guild. Wire Svc: AP Format: Today's Country. News staff: one. Target aud: 25-54. ♦Jim Principi, gen mgr; Perry Stone, opns mgr; Robin Batson, natl sls mgr; Dave Carroll, mktg dir & prom dir; Paul Johnson, progmg mgr; Tim Nelson, chief of engrg.

Wilson

***WAJC(FM)**— May 1990: 90.5 mhz; 3.8 kw. 100 ft TL: N35 47 48 W78 18 31. Hrs open: 24 5 W. Hargett St., Suite 801, Raleigh, 27601-1348. Phone: (919) 899-6778. Fax: (919) 899-6779. Licensee: CSN International (group owner; acq 6-23-2000; $150,000). Format: Christian teaching/praise & worship music. ♦Jim Walker, gen mgr.

WGTM(AM)— July 18, 1937: 590 khz; 5 kw-U, DA-2. TL: N35 43 04 W78 03 33. Hrs open: 8am - 5pm 4002 Hwy. 42 W., 27895. Phone: (252) 243-2188. Fax: (252) 237-8813. E-mail: wgtm590am@hotmail.com Licensee: Celestine L. Willis. Group owner: Willis Broadcasting Corp. (acq 12-15-89; $375,000; FTR: 3-3-86). Population served: 32,500 Format: Gospel. ♦Celestine L. Willis, gen mgr; Raymond Grant, progmg dir; Ray Taylor, disc jockey; Raymond Gant, disc jockey; Wayne Walker, disc jockey.

WLLY(AM)— 1961: 1350 khz; 1 kw-D, 79 w-N. TL: N35 43 24 W77 55 16. Hrs open: Daylight WLLY Radio Station, Box 637, 210 Beacon St. W., 27894-0637. Phone: (252) 237-5171. Fax: (252) 237-5172. Licensee: Estuardo Valdemar Rodriguez and Leonor Rodriguez, joint tenants. Group owner: Estuardo Valdemar Rodriguez and Leonor Rodriguez Stns (acq 10-11-2002; $255,000). Natl. Network: USA. Format: Southern Gospel. Target aud: General.

WRDU(FM)—Licensed to Wilson. See Raleigh

WVOT(AM)— June 1948: 1420 khz; 1 kw-D, 500 w-N, DA-N. TL: N35 44 08 W77 53 02. Hrs open: 24 103 N. Jackson St., 27893. Phone: (252) 243-5157. Phone: (252) 243-1420. Fax: (252) 291-5000. E-mail: wvot@earthlink.net Licensee: Kingdom Expansion Corp. (acq 1-18-01; $100,000). Population served: 65,000 Format: Christian, sports. News: 25 hrs wkly. Target aud: 25-55. ♦M.K. Smith, pres; Joyce Farmer, gen mgr & stn mgr; Noel Johnson, sports cmtr.

Windsor

WBTE(AM)— 1969: 990 khz; 1 kw-D, 25 w-N. TL: N35 58 00 W76 56 54. Hrs open: Box 1008, 27983. Phone: (252) 794-5590. Fax: (252) 794-5151. E-mail: wbte@earthlink.net Licensee: Dr. Tine Hicks & Associate Group owner: Willis Broadcasting Corp. (acq 12-2-2005; $70,000). Format: Gospel. ♦Arbutis Walston, gen mgr.

WGTI(FM)— 1980: 97.7 mhz; 3 kw. Ant 300 ft TL: N36 04 06 W76 58 35. Hrs open: 24 Box 590, Williamston, 27892. Phone: (252) 792-4161. Fax: (252) 809-0039. E-mail: bryant@opendoorradio.com Web Site: www.opendoorradio.com. Licensee: Lifeline Ministries Inc. (acq 12-13-2005; $300,000). Format: Gospel. ♦Johnny Bryant, pres & gen mgr.

WNBR-FM— Dec 5, 1988: 98.9 mhz; 6 kw. 350 ft TL: N35 54 25 W77 00 32. Hrs open: 24 Box 590, Williamston, 27892. Secondary address: 1012 East Blvd., Williamston 27892. Phone: (252) 792-4161. Fax: (252) 809-0039. E-mail: bryant@opendoorradio.com Web Site: www.opendoorradio.com. Licensee: Eure Communications Inc. (group owner; (acq 5-17-2004; $1.07 million). Format: Southern gospel. ♦Johnny Bryant, gen mgr, gen sls mgr & progmg dir; Doug Ferris, chief of engrg.

Wingate

***WRCM(FM)**— June 14, 1993: 91.9 mhz; 17.7 kw. 515 ft TL: N35 03 33 W80 40 14. Stereo. Hrs opn: 24 Box 17069, Charlotte, 28227. Secondary address: 1092 Radio Drive, Indian Trail 28079. Phone: (704) 821-9293. Phone: (704) 570-9200. Fax: (704) 821-9285. E-mail: newlife91.9@wrcm.org Web Site: www.wrcm.org. Licensee: Columbia Bible College Broadcasting Co. Population served: 2,000,000 Natl. Network: Salem Radio Network. Wire Svc: UPI Format: Adult contemp, Christian. Target aud: 25-44; female. ♦Joe Paulo, gen mgr; Elizabeth Poplin, prom dir & adv dir; Dwayne Harrison, progmg dir; Joyce Younts, pub affrs dir; Dave Morrison, chief of engrg; Steve McCranie, spec ev coord.

Winston-Salem

WBFJ(AM)— Oct 1, 1960: 1550 khz; 1 kw-D, DA. TL: N36 06 33 W80 14 47. Stereo. Hrs opn: Sunrise-sunset 1249 Trade St., 27101. Phone: (336) 721-1560. Fax: (336) 777-1032. Web Site: www.wbfj.org. Licensee: Word of Life Broadcasting Inc. (acq 6-29-83). Population served: 375,000 Natl. Network: USA. Format: Christian, talk, educational. Target aud: 29-54; general. ♦Philip T. Watson, pres & gen sls mgr; John Hill, progmg dir; Wally Decker, gen mgr & mus dir; Larry Schropp, chief of engrg.

***WBFJ-FM**— Sept 1, 1994: 89.3 mhz; 2.5 kw. 423 ft TL: N36 05 56 W80 15 00. Hrs open: 24 1249 Trade St., 27101. Phone: (336) 721-1560. Phone: (336) 777-1893. Fax: (336) 777-1032. E-mail: wbfj@wbfj.org Web Site: www.wbfj.fm. Licensee: Triad Family Network Inc. Population served: 750,000 Natl. Network: USA. Format: Contemp Christian mus. Target aud: 25-49. ♦Kurt Myers, prom mgr; Wally Decker, gen mgr, gen sls mgr & progmg dir; Verne Hill, news dir; Larry Shropp, chief of engrg.

***WFDD-FM**— Mar 13, 1961: 88.5 mhz; 60 kw. 345 ft TL: N35 58 12 W80 12 54. (CP: TL: N35 55 15 W80 17 37). Stereo. Hrs opn: 24 Box 8850, 27109. Secondary address: 56 Wake Forest Rd. 27109. Phone: (336) 758-8850. Fax: (336) 758-5193. E-mail: wfdd@wfu.edu Web Site: www.wfdd.org. Licensee: Trustees of Wake Forest University. Population served: 90,000 Natl. Network: PRI, NPR. Rgnl rep: Public Radio Adv. Alliance Law Firm: Fletcher, Heald & Hildreth. Format: Class, news. News staff: 3; News: 29 hrs wkly. Target aud: General; educated/public radio. Spec prog: Jazz 16 hrs wkly. ♦Jay Banks, gen mgr & stn mgr; Denise Franklin, news dir.

WKTE(AM)—See King

WKZL(FM)— 1972: 107.5 mhz; 100 kw. 994 ft TL: N36 16 33 W79 56 27. Stereo. Hrs opn: 192 E. Lewis St., Greensboro, 27406-1459. Phone: (336) 274-8042. Fax: (336) 274-1629. Web Site: www.1075kzl.com. Licensee: Dick Broadcasting Co. Inc. of Tennessee (acq 11-23-92; $6.5 million with WGFX(FM) Gallatin, TN; FTR: 12-14-92). Population served: 128,300 Law Firm: Kaye, Scholer, Fierman, Hays & Handler. Format: Contemp hit/Top-40. Target aud: 25-49; women. ♦Allen Dick, CEO, chmn & pres; David Henderlight, CFO; Bruce Wheeler, VP & gen mgr; James Kerr, opns mgr & natl sls mgr; Jennifer Hart, gen sls mgr; Jeff McHugh, progmg dir; Marcia Gan, mus dir; Tom Caldwell, chief of engrg.

WMAG(FM)—See High Point

WPAW(FM)—Licensed to Winston-Salem. See Greensboro

WPIP(AM)— June 1, 1995: 880 khz; 900 w-D. TL: N36 06 33 W80 14 47. Hrs open: Sunrise-sunset 4135 Thomasville Rd., 27107. Phone: (336) 785-0527. Fax: (336) 785-0529. Web Site: www.wpipbereanradio.org. Licensee: Berean Baptist Church. (acq 5-95; $80,000; FTR: 5-8-95). Natl. Network: USA. Format: Conservative Christian. ♦Dr. Ron Baity, gen mgr; Jeff Baity, chief of opns.

WPOL(AM)—Licensed to Winston-Salem. See Greensboro

WSJS(AM)— Apr 17, 1930: 600 khz; 5 kw-D, 5 kw-N, DA-2. TL: N36 07 00 W80 21 26. Stereo. Hrs opn: 875 W. 5th St., 27101. Phone: (336) 727-8826. Fax: (336) 777-3915. Web Site: www.wsjs.com. Licensee: Crescent Media Group LLC. Group owner: Infinity Broadcasting Corp. (acq 2-14-2007; grpsl). Population served: 1,450,000 Natl. Network: Wall Street. Natl. Rep: Clear Channel. Format: News/talk. Target aud: 25-64. ♦Tom Hamilton, gen mgr & gen sls mgr; Marty Holbrook, prom dir; Beth Ann McBride, progmg dir; Bob Costner, news dir; George Newman, chief of engrg.

WSMX(AM)— October 1964: 1500 khz; 1 kw-D, DA. TL: N36 06 34 W80 12 42. Hrs open: Sunrise-sunset 1225 E. 5th St., Suite 104, 27101. Phone: (336) 391-1497. Fax: (336) 724-6368. E-mail: wsmxradio@aol.com Licensee: Gospel Media Inc. (acq 6-82). Population served: 140,000 Format: Gospel, community affrs. News: 12 hrs wkly. Target aud: 30-50 plus; blue collar, minorities, church members. ♦Joe Watson, pres & gen mgr.

***WSNC(FM)**— 1982: 90.5 mhz; 125 w. 92 ft TL: N36 05 36 W80 13 53. (CP: 10 kw, ant 194 ft. TL: N36 05 24 W80 13 20). Stereo. Hrs opn: 24 601 Martin Luther King Dr., Hall-Patterson, Fst floor, 27110. Phone: (336) 750-2433. Web Site: www.wsncfm.org. Licensee: Winston-Salem State University. Format: Adult contemp, jazz, relg. Target aud: 12-70; African-Americans. Spec prog: Class 4 hrs wkly. ♦Dr. Brian Blount, CEO; Elvin Jenkins, gen mgr; Monica Melton, progmg dir & mus dir; Baxter Griffin, chief of engrg.

WTIX(AM)— Oct 28, 1950: 980 khz; 1 kw-D, 69.3 w-N. TL: N36 09 15 W80 16 34. Hrs open: Box 5663, High Point, 27262. Secondary address: 1607 Country Club Dr., High Point 27262. Phone: (336) 887-0983. Fax: (336) 887-3055. E-mail: wes@countryledgends983.com Licensee: GHB Radio Inc. (acq 3-6-2006; $235,000). Population served: 150,000 Format: Country. ♦George H. Buck Jr., pres; Susan Childress, gen mgr; Wes Jones, opns mgr; Ed Kasovic, chief of engrg.

WTOB(AM)— Apr 22, 1947: 1380 khz; 5 kw-D, 2.5 kw-N, DA-2. TL: N36 08 53 W80 19 11. Stereo. Hrs opn: 18 Box 12876, 27117. Phone: (336) 714-2774. Fax: (336) 714-8337. E-mail: quepasa@quepasamedia.com Web Site: www.quepasamedia.com. Licensee: Davidson Media Station WTOB Licensee LLC. (acq 3-25-2005; swap with WWBG(AM) Greensboro for WDRU(AM) Wake Forest). Population served: 134,676 Rgnl. Network: N.C. News Net. Law Firm: Smithwick & Belendiuk. Format: Sp. Target aud: 35 plus; affluent audience. ♦Roger Martinez, gen mgr.

WTQR(FM)— Dec 1, 1947: 104.1 mhz; 100 kw. 1,420 ft TL: N36 22 28 W80 22 31. Stereo. Hrs opn: 24 2-B PAI Park, Greensboro, 27409. Phone: (336) 822-2000. Fax: (336) 887-0104. Web Site: www.wtqr.com. Licensee: Clear Channel Radio Licenses Inc. Group owner: Clear Channel Communications Inc. (acq 1996; grpsl). Population served: 1,450,000 Format: Country. Target aud: 25-54. Spec prog: NASCAR, bluegrass 2 hrs wkly. ♦Sheryl Solomone, gen mgr.

WVBZ(FM)—See High Point

***WXRI(FM)**— May 17, 1997: 91.3 mhz; 50 kw. 216 ft TL: N36 08 06 W80 22 32. Hrs open: Box 25775, 27114. Phone: (336) 699-8036. Phone: (336) 788-1155. Fax: (336) 788-7199. E-mail: joyfm@bellsouth.net Web Site: www.joyfm.org. Licensee: Positive Alternative Radio Inc. Group owner: Baker Family Stations (Positive Radio Group) (acq 5-21-92). Population served: 115,000 Law Firm: Booth, Freret, Imlay & Tepper. Format: Southern gospel. ♦Vernon H. Baker, pres; Edward A. Baker, VP; Rodney Baucom, gen mgr & stn mgr; Sam Stutts, mktg mgr; Dean Lilly, chief of engrg.

Winterville

WECU(AM)— Feb 7, 2006: 1570 khz; 3.8 kw-D, 200 w-N. TL: N35 32 15 W77 25 06. Hrs open: 24 Box 1534, Greenville, 27835. Phone: (252) 902-9147. Fax: (252) 931-9328. E-mail: wwnbwecu@yahoo.com Web Site: www.wecu1570.com. Licensee: CTC Media Group. Group owner: CTC Media Group Inc. Format: Gospel. ♦Edwin Lee Afflerbach, pres; Mike Afflerbach, stn mgr.

Wrightsville Beach

WNTB(FM)— Nov 27, 2000: 93.7 mhz; 6 kw. Ant 328 ft TL: N34 18 04 W77 48 07. (CP: COL Topsail Beach). Hrs opn: 24 122 Cinema Dr., Wilmington, 28403. Phone: (910) 772-6300. Fax: (910) 772-6310. E-mail: newsroom@seacomm.com Web Site: www.937thebone.com. Licensee: Sea-Comm Inc. (group owner; (acq 6-30-2000; $1.2 million for CP). Format: Classic rock. ♦Paul Knight, gen mgr; Max Deutsch, gen sls mgr; Zach McHugh, progmg mgr; Jonathan Knight, news dir.

Yanceyville

WYNC(AM)— Nov 9, 1979: 1540 khz; 2.5 kw-D. TL: N36 24 52 W79 20 06. Hrs open: sun up-sun down Box 670, 27379. Secondary address: 545 Firetower Rd. 27379. Phone: (336) 694-7343. Fax: (336) 694-7514. E-mail: wync@earthlink.net Licensee: Semora Broadcasting Inc. (acq 12-9-91; $102,041; FTR: 1-6-92). Population served: 50,000 Natl. Network: Westwood One. Format: Gospel. Target aud: General; rural Caswell county & Danville, VA. Spec prog: Gospel 16 hrs wkly. ♦George Thaxton, gen mgr & stn mgr; Leroy Connally, chief of engrg.

North Dakota

Arthur

KVMI(FM)— April 1994: 103.9 mhz; 25 kw. Ant 328 ft TL: N47 07 20 W97 19 29. Hrs open: 4 Langer Ave., Casselton, 58012. Phone: (701) 866-0799. Fax: (218) 287-8274. Licensee: Vision Media Inc. (acq 2-25-2000). Natl. Network: Westwood One. Format: Country. ♦Mike McCain, gen mgr.

Belcourt

***KEYA(FM)**— October 1975: 88.5 mhz; 19 kw. 263 ft TL: N48 50 37 W99 45 02. Stereo. Hrs opn: 19 Box 190, Media Bldg., Hospital Rd., 58316. Phone: (701) 477-5686. Phone: (701) 477-3527. Fax: (701) 477-3252. E-mail: keya@utma.com Web Site: www.keya.utma.com. Licensee: KEYA Inc. Population served: 38,000 Natl. Network: NPR, PRI. Law Firm: Steptoe & Johnson. Wire Svc: AP Format: C&W, oldies, rock/AOR. News staff: 5; News: 2 hrs wkly. Target aud: General; members of the Turtle Mountain Band of Chippewa Indians. Spec prog: American Indian 6 hrs, relg 10 hrs, old-time fiddle mus 4 hrs, Chippewa 3 hrs wkly. ♦Kimberly Thomas, gen mgr; William Morin, adv dir; Jarle Kvale, progmg dir; Janice Keplin, chief of engrg.

Beulah

KDKT(AM)— Oct 5, 1978: 1410 khz; 1 kw-D, 180 w-N. TL: N47 17 15 W101 45 46. Hrs open: 24 Box 1064, 58523. Secondary address: 547 S. 7th St., Suite 166, Bismarck 58504. Phone: (701) 873-2215. Fax: (701) 873-2363. E-mail: info@dsnradio.com Web Site: www.foxsports1410.com. Licensee: Digital Syndicate Network LLC (acq 2-27-2006; $150,000). Population served: 20,000 Natl. Network: Fox Sports. Rgnl. Network: N.D. News Net., AgriAmerica. Format: Sports. News staff: one; News: 14 hrs wkly. Target aud: 24-65; general. ♦Dawson Austin, opns mgr; Dusty Rhodes, progmg dir.

KHRU(FM)—Not on air, target date: unknown: 97.9 mhz; 6 kw. Ant 315 ft TL: N47 18 23 W101 43 35. Hrs open: 5331 Mt. Alifan Dr., San Diego, CA, 92111. Phone: (858) 277-4991. Fax: (858) 277-1365. Web Site: www.horizonsd.org/radio.asp. Licensee: Horizon Christian Fellowship. ♦Mike MacIntosh, pres.

Bismarck

KACL(FM)— Apr 22, 1997: 98.7 mhz; 100 kw. Ant 837 ft TL: N46 35 24 W100 47 46. Hrs open: 24 Box 1377, 1830 N. 11th St., 58501. Phone: (701) 250-6602. Fax: (701) 250-6632. Web Site: www.cumulus.com. Licensee: Cumulus Licensing Corp. Group owner: Cumulus Media Inc. (acq 5-11-98; grpsl). Format: Oldies. ♦Syd Stewart, gen mgr; Debbie Boechler, gen sls mgr; Bob Beck, progmg dir; Matt Murphy, news dir; Dennis Wilson, chief of engrg; Brian Kocher, traf mgr.

***KBFR(FM)**— October 2003: 91.7 mhz; 780 w. Ant 348 ft TL: N46 49 38 W100 46 28. Hrs open: 24 Family Stations Inc., 4135 Northgate Blvd., Suite 1, Sacramento, CA, 95834. Phone: (916) 641-8191. Fax: (916) 641-8238. Licensee: Family Stations Inc. (group owner). Format: Relg. ♦Harold Camping, pres.

***KBMK(FM)**— 2006: 88.3 mhz; 5.5 kw vert. Ant 380 ft TL: N46 49 38 W100 46 28. Hrs open:
Rebroadcasts KLVR(FM) Santa Rosa, CA 100%.
2351 Sunset Blvd., Suite 170-218, Rocklin, CA, 95765. Phone: (916) 251-1600. Fax: (916) 251-1650. Web Site: www.klove.com. Licensee: Broadcasting for the Challenged Inc. Natl. Network: K-Love. Format: Contemp Christian. ♦Richard Jenkins, pres; Mike Novak, VP; Keith Whipple, dev dir; David Pierce, progmg mgr; Ed Lenane, news dir; Sam Wallington, engrg dir; Karen Johnson, news rptr; Marya Morgan, news rptr; Richard Hunt, news rptr.

KBMR(AM)— Aug 15, 1958: 1130 khz; 50 kw-D, DA. TL: N46 50 04 W100 31 19. (CP: 10 kw-D). Stereo. Hrs opn: 24 3500 E. Rosser Ave., 58501. Phone: (701) 255-1234. Fax: (701) 222-1131. E-mail: kbmr@clearchannel.com Web Site: www.kbmr.com. Licensee: CC Licenses LLC. Group owner: Clear Channel Communications Inc. (acq 2-13-2004;. grpsl). Population served: 71,000 Natl. Rep: McGavren Guild. Law Firm: Borsari & Paxson. Format: C&W. Target aud: 25 plus. Spec prog: Farm 6 hrs wkly. ♦Bob Denver, gen mgr; Neil Cary, gen sls mgr; Charlie Williams, progmg dir; Jeff Alexander, news dir; Elliott Davidson, chief of engrg; Clarissa Lynn, disc jockey.

KBYZ(FM)—Listing follows KLXX(AM).

***KCND(FM)**— Sept 1, 1981: 90.5 mhz; 50 kw. 1,216 ft TL: N46 35 23 W100 48 02. Stereo. Hrs open: 24 207 N 5th St., Fargo, 58102. Phone: (701) 224-1700. Fax: (701) 224-0555. E-mail: program@pol.org Web Site: www.prairiepublic.org. Licensee: Prairie Public Broadcasting Inc. Population served: 65,000 Natl. Network: PRI, NPR. Format: Class, jazz, news. News staff: 2; News: 40 hrs wkly. Target aud: General.

Spec prog: American Indian 2 hrs, folk 6 hrs, blues 2 hrs wkly. ♦Bill Thomas, gen mgr & stn mgr; Duane Lee, opns mgr; David Thompson, news dir. Co-owned TV: *KBME-TV affil.

KFYR(AM)— 1925: 550 khz; 5 kw-U, DA-N. TL: N46 51 12 W100 32 37. Stereo. Hrs opn: Box 2156, 58502. Secondary address: 3500 E. Rosser Ave. 58501. Phone: (701) 255-1234. Fax: (701) 222-1131. E-mail: kfyr@clearchannel.com Web Site: www.kfyr.com. Licensee: Citicasters Licenses L.P. Group owner: Clear Channel Communications Inc. (acq 5-4-99; grpsl). Population served: 192,200 Format: News/talk. ♦Bob Denver, gen mgr; Neil Cary, gen sls mgr.

KYYY(FM)—Co-owned with KFYR(AM). Aug 15, 1966: 92.9 mhz; 100 kw. 1,180 ft TL: N46 36 19 W100 48 30. Stereo. Web Site: www.y93.com. Format: Hot adult contemp. ♦Todd Mitchell, opns dir & progmg dir.

KKCT(FM)— 1994: 97.5 mhz; 100 kw. Ant 837 ft TL: N46 35 24 W100 47 46. Hrs open: Box 1377, 1830 N. 11th St., 58501. Phone: (701) 250-6602. Fax: (701) 250-6632. Web Site: www.cumulus.com. Licensee: Cumulus Licensing Corp. Group owner: Cumulus Media Inc. (acq 5-11-98; grpsl). Format: CHR. ♦Syd Stewart, gen mgr; Debbie Boechler, gen sls mgr; Chris Ryan, progmg dir; Matt Murphy, news dir; Dennis Wilson, chief of engrg; Brian Kocher, traf mgr.

KLXX(AM)—Bismarck-Mandan, 1925: 1270 khz; 1 kw-D, 250 w-N. TL: N46 48 37 W100 50 10. Hrs open: 24 Box 1377, 58502. Phone: (701) 663-6411. Phone: (701) 250-6602. Fax: (701) 663-8790. Fax: (701) 250-6632. E-mail: syd.steward@cumulus.com Web Site: supertalk1270.com. Licensee: Cumulus Licensing Corp. Group owner: Cumulus Media Inc. (acq 5-11-98; grpsl). Population served: 250,000 Natl. Network: CNN Radio. Natl. Rep: Katz Radio. Format: News/talk, sports. News staff: one; News: 90 hrs wkly. Target aud: 35 plus. Spec prog: Sports 6 hrs wkly. ♦Syd Stewart, gen mgr & gen sls mgr; Dean Mastel, progmg dir; Elliot Davidson, chief of engrg.

***KNRI(FM)**— 2006: 89.7 mhz; 250 w. Ant 157 ft TL: N46 51 00 W100 46 11. Hrs open: 24 2351 Sunset Blvd., Suite 170-218, Rocklin, CA, 95765. Phone: (916) 251-1600. Fax: (916) 251-1650. E-mail: info@air1.com Web Site: www.air1.com. Licensee: Educational Media Foundation. Group owner: EMF Broadcasting. Natl. Network: Air 1. Law Firm: Shaw Pittman. Format: Contemp Christian. News staff: 3. Target aud: 18-35; Judeo Christian female. ♦Richard Jenkins, pres; Mike Novak, VP; Keith Whipple, dev dir; David Pierce, progmg mgr; Ed Lenane, news dir; Sam Wallington, engrg dir; Karen Johnson, news rptr; Marya Morgan, news rptr; Richard Hunt, news rptr.

KQDY(FM)— Sept 13, 1968: 94.5 mhz; 100 kw. 1,117 ft TL: N46 51 31 W100 41 38. Stereo. Hrs opn: 24 3500 E. Rosser, 58501. Phone: (701) 255-1234. Fax: (701) 222-1131. Web Site: www.kqdy.com. Licensee: CC Licenses LLC. Group owner: Clear Channel Communications Inc. (acq 2-13-2004; grpsl). Population served: 71,900 Format: Contemp country. News: 4 hrs wkly. Target aud: 18-49. ♦Bob Denver, gen mgr.

KSSS(FM)— Aug 1, 1994: 101.5 mhz; 100 kw. 987 ft TL: N46 56 31 W100 41 38. Hrs open: 24 3500 E. Rosser Ave., 58501. Secondary address: Box 2156 58502. Phone: (701) 255-1234. Fax: (701) 222-1131. Web Site: www.1015.fm. Licensee: CC Licenses LLC. Group owner: Clear Channel Communications Inc. (acq 2-13-2004; grpsl). Format: Classic Rock. ♦Bob Denver, gen mgr; Terry Flack, gen sls mgr; Jeff Alexander, news dir; Cindy Lindsay, traf mgr.

KXMR(AM)— Mar 20, 1999: 710 khz; 50 kw-D, 4 kw-N, DA-3. TL: N46 50 04 W100 31 19 (D), N46 40 08 W100 46 33 (N). Hrs open: 24 3500 E. Rosser Ave., 58501. Secondary address: Box 2156 58502. Phone: (701) 255-1234. Fax: (701) 222-1131. Licensee: CC Licenses LLC. Group owner: Clear Channel Communications Inc. (acq 12-10-2003). Format: Sports. ♦Bob Denver, gen mgr.

Bismarck-Mandan

KLXX(AM)—Licensed to Bismarck-Mandan. See Bismarck

Bottineau

KBTO(FM)— Nov 9, 1980: 101.9 mhz; 94 kw. 492 ft TL: N48 51 10 W100 20 01. Stereo. Hrs opn: 24 1120 Highway 5 west, 58318. Phone: (701) 228-5151. Fax: (701) 228-2483. E-mail: sunnyradio@hotmail.com Licensee: Programmers Broadcasting Inc. (acq 1-2-2002; $595,000). Population served: 40,000 Natl. Network: ABC. Rgnl. Network: Midwest Radio. Law Firm: Fletcher, Hildreth & Herald. Format: Country. News: 15 hrs wkly. ♦John Kircher, pres; Jean Kircher, VP & gen mgr; Jean Schemmp, stn mgr; J. Davis, gen sls mgr; Jeff Bliss, chief of engrg.

Bowman

KPOK(AM)— Aug 9, 1980: 1340 khz; 1 kw-U. TL: N46 10 48 W103 22 12. Hrs open: 24 Box 829, 11 1/2 N. Main, 58623. Phone: (701) 523-3883. Fax: (701) 523-3885. E-mail: kpok@ndsupernet.com Web Site: www.kpokradio.com. Licensee: Tri-State Communications Inc. Population served: 50,000 Natl. Network: Westwood One. Format: Country. News staff: one; News: 14 hrs wkly. Target aud: 25-54. Spec prog: Farm 2 hrs wkly. ♦Larry Kemnitz, pres; Richard Peterson, VP; Brian Fischer, gen mgr.

Burlington

KWGO(FM)— 2005: 102.9 mhz; 100 kw. Ant 512 ft TL: N48 03 04 W101 20 23. Hrs open: 24 1408 20th Ave. S.W. #1, Minot, 58701. Phone: (701) 852-7449. Fax: (701) 837-6925. E-mail: pbiminot@srt.com Licensee: Programmers Broadcasting Inc. Population served: 100,000 Law Firm: Fletcher, Heald & Hildreth. Format: Hot adult contemp. News: 5 hrs wkly. Target aud: 18-49. ♦John Kircher, pres; Jean Kircher, VP; Jean Schemmp, stn mgr; J. Davis, gen sls mgr; Jeff Bliss, chief of engrg.

Carrington

KDAK(AM)— Oct 16, 1961: 1600 khz; 500 w-D, 90 w-N. TL: N47 25 43 W99 05 03. Hrs open: 24 Box 50, 58421. Secondary address: Box1170, Jamestown 58402. Phone: (701) 652-3151. Fax: (701) 652-2916. E-mail: kdakam@daktel.com Licensee: Two Rivers Broadcasting Inc. Group owner: Robert Ingstad Broadcast Properties (acq 7-1-94). Population served: 8,000 Rgnl. Network: N.D. News Net. Format: C&W. Target aud: 30 plus. ♦Robert J. Ingstad, pres; Scott Lane, gen mgr & stn mgr.

KXGT(FM)—Co-owned with KDAK(AM). 1997: 98.3 mhz; 100 kw. 866 ft TL: N47 05 38 W99 02 11. Stereo. Phone: (701) 252-1400. Fax: (701) 252-1402. Format: Oldies. ♦Dave Reed, gen mgr.

Cavalier

KAOC(FM)— Sept 29, 1998: 105.1 mhz; 44 kw. 512 ft TL: N48 37 44 W98 00 35. Hrs open: 24 1420 3rd St., Langdon, 58249. Phone: (701) 256-1067. Fax: (701) 256-1051. E-mail: kndk1080@utma.com Licensee: Simmons Broadcasting Inc. (group owner; (acq 9-7-2004; $1). Format: Hot country. ♦Bob Simmons, gen mgr; Jen Taylor, opns mgr.

Devils Lake

KDLR(AM)— Jan 25, 1925: 1240 khz; 1 kw-U. TL: N48 06 42 W98 50 43. Hrs open: 318 W. Walnut St., 58301. Phone: (701) 662-7563. Fax: (701) 662-7564. E-mail: kdlrkdvl@stellarnet.com Web Site: lrradioworks.com. Licensee: Double Z Broadcasting Inc. Group owner: Lake Region Radio Works (acq 1-1-2003; $820,000 with KDVL(FM) Devils Lake). Rgnl. Network: AgriAmerica, N.D. News Net. Format: Country, news. Target aud: 25 plus; general. Spec prog: Minnesota Twins baseball, Vikings football. ♦Curt Teigen, pres & gen mgr; Roger Mertens, sls dir; Eric Arndt, news dir.

***KDVI(FM)**— 2007: 89.9 mhz; 250 w. Ant 171 ft TL: N48 08 05 W98 46 20. Hrs open:
Rebroadcasts WAFR(FM) Tupelo, MS 100%.
Drawer 2440, Tupelo, MS, 38803. Phone: (662) 844-8888. Fax: (662) 842-6791. Web Site: www.afr.net. Licensee: American Family Association. (acq 6-9-2006). Natl. Network: American Family Radio. Format: Contemp Christian. ♦Donald E. Wildmon, chmn.

KDVL(FM)— Jan 1, 1967: 102.5 mhz; 100 kw. 471 ft TL: N47 59 16 W98 55 59. Stereo. Hrs open: 24 318 West Walnut St., Devil's Lake, 58301. Secondary address: 400 12 Ave. 58301. Phone: (701) 662-7563. Fax: (701) 662-7564. E-mail: kdlrkdvl@stellarnet.com Web Site: lrradioworks.com. Licensee: Double Z Broadcasting Inc. Group owner: Lake Region Radio Works (acq 1-1-2003; $820,000 with KDLR(AM) Devils Lake). Population served: 90,000 Format: Oldies. News staff: one. Target aud: 18-54. ♦Curt Teigen, pres & gen mgr; Roger Mertens, sls dir; Bob Gunderson, progmg dir & disc jockey; Eric Arndt, news dir; Paul Clementich, traf mgr; Mark Beighley, sports cmtr; Kara Danelle, disc jockey.

KQZZ(FM)— August 1996: 96.7 mhz; 45 kw. 512 ft TL: N47 58 46 W99 03 16. Hrs open: 24 318 W. Walnut St., 58301. Phone: (701) 662-7563. Fax: (701) 662-7564. Web Site: www.lrradioworks.com. Licensee: Two Rivers Broadcasting Inc. Group owner: Lake Region Radio Works (acq 3-11-99; $250,000). Law Firm: Shaw Pittman. Format: Current & Classic Rock. ♦Curt Teigen, gen mgr & stn mgr.

KZZY(FM)— March 1984: 103.5 mhz; 100 kw. 433 ft TL: N47 59 28 W98 56 57. Stereo. Hrs opn: 24 318 W. Walnut St., 58301. Phone: (701) 662-7563. Fax: (701) 662-7564. E-mail: kzzyfm@stellarnet.com Web Site: www.zzcountry.com. Licensee: Double Z Broadcasting Inc. Group owner: Lake Region Radio Works (acq 4-11-90). Population served: 13,000 Format: C&W. News staff: one. ♦Curt Teigen, gen mgr, opns mgr & chief of engrg; Roger Mertens, gen sls mgr; Rob Hendricks, progmg dir; Kaye Schwab, traf mgr.

Dickinson

KCAD(FM)— Nov 20, 1996: 99.1 mhz; 100 kw. Ant 794 ft TL: N46 56 09 W102 43 55. Hrs open: 24 11291 39th St. SW, 58601-9206. Phone: (701) 227-1876. Fax: (701) 483-1959. Web Site: www.roughridercountry.net. Licensee: CC Licenses LLC. Group owner: Clear Channel Communications Inc. (acq 9-1-2000; grpsl). Natl. Network: AP Radio, Jones Radio Networks. Format: Hot country. Target aud: 16-50; general. ♦Grant Giessinger, gen mgr & gen sls mgr; Bill Palenuk, progmg dir.

KLTC(AM)—Co-owned with KCAD(FM). July 4, 1978: 1460 khz; 5 kw-U, DA-N. TL: N46 50 54 W102 49 49.24 80,000 Natl. Rep: Hyett/Ramsland. Format: Classic country. Target aud: General.

KDIX(AM)— 1947: 1230 khz; 1 kw-U. TL: N46 53 44 W102 47 06. Hrs open: 24 119 Second Ave. W., 58601. Phone: (701) 225-5133. Phone: (800) 934-1230. Fax: (701) 225-4136. Licensee: Starrdak Inc. (acq 4-1-93). Population served: 47,000 Natl. Network: CBS. Format: Adult contemp, oldies. News staff: 8; News: 6 hrs wkly. Target aud: 35-60. Spec prog: College sports. ♦Lee Leiss, chmn & gen mgr; Rod Kleinjan, opns dir.

***KDPR(FM)**— Oct 12, 1987: 89.9 mhz; 12.5 kw. 488 ft TL: N46 43 34 W102 54 56. Stereo. Hrs opn: 24
Rebroadcast KCND(FM) Bismark 100%.
1814 N. 15th St., Bismarck, 58501. Phone: (701) 241-6900. Fax: (701) 239-7650. E-mail: program@prairiepublic.org Web Site: www.prairiepublic.org. Licensee: Prairie Public Broadcasting Inc. Population served: 20,000 Natl. Network: PRI, NPR. Format: Class, jazz, news. News staff: 2; News: 40 hrs wkly. Target aud: General. Spec prog: American Indian 2 hrs, folk 6 hrs wkly. ♦Bill Thomas, gen mgr; Duane Lee, opns mgr; Dave Thompson, news dir; Stephanie Chimeziri, spec ev coord.

KZRX(FM)— Aug 15, 1983: 92.1 mhz; 10.5 kw. Ant 492 ft TL: N46 56 09 W102 43 55. Stereo. Hrs opn: 5 AM-midnight 11291 39 St. SW, 58601. Phone: (701) 227-1876. Fax: (701) 483-1959. Web Site: www.z92fm.net. Licensee: CC Licenses LLC. Group owner: Clear Channel Communications Inc. (acq 9-1-2000; grpsl). Population served: 35,000 Natl. Rep: Hyett/Ramsland. Format: Rock. News staff: one. Target aud: 18-45. ♦George Smith, gen mgr; Don Reisenauer, gen sls mgr; Chad Barta, progmg dir; Brian Funk, chief of engrg; Kim Kramer, news dir & traf mgr.

Fargo

***KDSU(FM)**— Jan 17, 1966: 91.9 mhz; 100 kw. 991 ft TL: N47 00 48 W97 11 37. Stereo. Hrs opn: 24 Box 3240, 58108. Phone: (701) 241-6900. Fax: (701) 231-8899. Web Site: www.prairiepublic.org. Licensee: North Dakota State University. Population served: 500,000 Natl. Network: NPR, PRI. Wire Svc: AP Format: Var/div. News staff: 3; News: 45 hrs wkly. Target aud: 24 plus; general. ♦John Harris, CEO; Bill Thomas, gen mgr; Nancy Wood, dev dir.

***KFBN(FM)**— Dec 8, 1997: 88.7 mhz; 30 kw horiz, 100 kw vert. 869 ft TL: N47 00 48 W97 11 37. Hrs open: Box 107, 58107. Phone: (701) 298-8877. Licensee: Fargo Baptist Church. Format: Mus, world news/rgnl weather, bible instruction. ♦T.C. Scheving, pres & gen mgr.

KFGO(AM)— Mar 14, 1948: 790 khz; 5 kw-U, DA-N. TL: N46 04 05 W96 48 05. Stereo. Hrs opn: Box 2966, 58108. Secondary address: 1020 25th St. S. 58103. Phone: (701) 237-5346. Fax: (701) 235-4042. Web Site: www.kfgo.com. Licensee: Radio Fargo-Moorhead Inc. Group owner: Clear Channel Communications Inc. (acq 1-19-2007; grpsl). Population served: 500,000 Rgnl. Network: Minn. Pub. Natl. Rep: CBS Radio. Format: News/talk. Target aud: General. ♦Jeff Hoberg, gen mgr; Tank McNamara, progmg dir.

KRWK(FM)—Co-owned with KFGO(AM). Feb 23, 1984: 101.9 mhz; 93 kw. Ant 1,000 ft TL: N47 00 37 W97 11 40. Stereo. Web Site: www.thebox1019.com. Format: Classic rock.

***KFNW(AM)**—West Fargo, Oct 28, 1955: 1200 khz; 10 kw-D, 1 kw-N, DA-N. TL: N46 48 06 W96 52 57. Hrs open: 24 5702 52nd Ave. S., 58104. Phone: (701) 282-5910. Fax: (701) 282-5781. E-mail: kfnw@kfnw.org Web Site: www.kfnw.org. Licensee: Northwestern College. Group owner: Northwestern College & Radio. Format: Relg. News staff: one;

News: 6 hrs wkly. Target aud: 25-54. ♦Gary D. Herr, stn mgr; Gary Ellingson, chief of engrg; Phil Kvamme, progmg dir & local news ed.

KFNW-FM— Mar 12, 1965: 97.9 mhz; 100 kw. 1,000 ft TL: N46 48 07 W96 52 58. Stereo. 24 Format: Christian contemp. Target aud: 25-54.

KKAG(AM)— 2007: 740 khz; 50 kw-D, 940 w-N, 7.5 kw-CH, DA-3. TL: N46 58 29 W96 30 12. Hrs open: 1020 25th St. S., 58108-2966. Phone: (701) 237-5346. Fax: (701) 235-4042. Licensee: Radio Fargo-Moorhead Inc. (acq 7-31-2007). Natl. Network: Fox Sports. Natl. Rep: Hyett/Ramsland. Format: Sports. ♦Jeff Hoberg, gen mgr.

KPFX(FM)— Jan 4, 1993: 107.9 mhz; 100 kw. Ant 713 ft TL: N46 32 41 W96 37 33. Stereo. Hrs opn: 24 Box 9919, 58106. Secondary address: 2720 Seventh Ave. S. 58103. Phone: (701) 237-4500. Phone: (701) 237-4949. Fax: (701) 237-5400. E-mail: studio@1079thefox.com Web Site: www.1079thfox.com/. Licensee: Monterey Licenses LLC. Group owner: Triad Broadcasting Co. LLC (acq 8-18-99; grpsl). Population served: 194,800 Natl. Rep: Christal. Law Firm: Shaw Pittman. Format: Classic rock. Target aud: 25-54; skews male. ♦Nancy Odney, gen mgr; John Austin, opns mgr; Michael Brooks, gen sls mgr.

KQWB-FM—Moorhead, MN) November 1966: 98.7 mhz; 100 kw. 460 ft TL: N46 45 35 W96 36 26. Stereo. Hrs opn: Box 9919, 58106-9919. Secondary address: 2720 7th Ave. S. 58103. Phone: (701) 237-4500. Phone: (701) 234-9898. Fax: (701) 235-9082. E-mail: studio@q98.com Web Site: www.q98.com. Licensee: Monterey Licenses LLC. Group owner: Triad Broadcasting Co. LLC (acq 10-99; grpsl). Population served: 194,800 Natl. Rep: Christal. Law Firm: Shaw Pittman. Format: Active rock. Target aud: 18-49; men. ♦Nancy Odney, VP & gen mgr; John Austin, opns dir; Anne Phibian, opns mgr.

KVOX(AM)—See Moorhead, MN

KVOX-FM—See Moorhead, MN

WDAY(AM)— May 22, 1922: 970 khz; 5 kw-U, DA-N. TL: N46 52 43 W96 53 05. Stereo. Hrs opn: 24 Box 2466, 58108. Secondary address: 301 S. 8th St. 58103. Phone: (701) 237-6500. Fax: (701) 241-5373. E-mail: jsunday@wday.com Web Site: www.wday.com. Licensee: Forum Communications Co. Inc. (group owner) Population served: 500,000 Natl. Rep: Christal. Wire Svc: NWS (National Weather Service) Format: News/talk, farm, sports. News staff: 3; News: 35 hrs wkly. Target aud: 35-64. ♦William Marcil Sr., CEO & pres; Jack Sunday, progmg dir; Mike Tanner, news dir.

WDAY-FM— 1965: 93.7 mhz; 100 kw. 1,040 ft TL: N47 00 43 W97 11 58. Stereo. Hrs opn: 24 1020 25th St. S., 58103. Phone: (701) 237-5346. Fax: (701) 235-4042. E-mail: mikekapel@clearchannel.com Licensee: Radio Fargo-Moorhead Inc. Group owner: Clear Channel Communications Inc. (acq 1-19-2007; grpsl). Natl. Rep: Christal. Wire Svc: NWS (National Weather Service) Format: Top-40. Target aud: 18-49. ♦Jeff Hoberg, gen mgr.

Fort Totten

***KABU(FM)**— Mar 15, 1999: 90.7 mhz; 6 kw. 328 ft TL: N47 59 28 W98 56 57. Hrs open: Box 7, 58335. Secondary address: KABU Radio Station, 7889 Hwy. 57, St. Michaels 58370. Phone: (701) 766-1995. Fax: (701) 766-4774. E-mail: kabu@stellarnet.com Licensee: Dakota Circle Tipi Inc. Population served: 9,001 Format: Educ, community, mus. Target aud: General; community on Spirit Lake Nation & surrounding areas to reach all age groups. Spec prog: American Indian 19 hrs, children 12 hrs, gospel 7 hrs, community & school 5 hrs wkly. ♦Mark Blackcloud, gen mgr & stn mgr.

Four Bears

***KMHA(FM)**— March 1984: 91.3 mhz; 100 kw. 380 ft TL: N47 44 23 W102 43 24. Stereo. Hrs opn: 24 601 Lodge Rd., Newtown, 58763. Phone: (701) 627-3333/627-4306. Fax: (701) 627-3376. E-mail: kmha_fm@restel.net Licensee: Fort Berthold Communications Enterprise. Population served: 20,000 Format: Div. News staff: one. Target aud: Ranchers, farmers, Native Americans. Spec prog: American Indian-Mandan /Hidatsa/Arikara 4 hrs, country 8 hrs, farm one hr wkly. ♦Rose Crow Flies High, gen mgr; Clarence Sun, opns mgr.

Grafton

KAUJ(FM)—Listing follows KXPO(AM).

KXPO(AM)— July 12, 1958: 1340 khz; 1 kw-U. TL: N48 23 53 W97 26 56. Hrs open: 6 AM-midnight 856 12th St. W., 58237. Phone: (701) 352-0431. Fax: (701) 352-0436. Licensee: KGPC Co. (acq 12-12-72). Population served: 50,000 Natl. Network: ABC. Rgnl. Network: American Net. Law Firm: Sam Miller. Wire Svc: AP Format: Country. News staff: 2; News: 15 hrs wkly. Target aud: 30-70. Spec prog: Farm 18 hrs, gospel 5 hrs, relg 4 hrs wkly. ♦Del Nygard, pres; Brian James, gen mgr & progmg dir; Todd Ingstad, adv mgr; Don Brintmall, chief of engrg; Nicki Amico, traf mgr.

KAUJ(FM)—Co-owned with KXPO(AM). Sept 17, 1984: 100.9 mhz; 3 kw. 125 ft TL: N48 23 53 W97 26 56.24 E-mail: kzzpo@polarcomm.com 32,000 Format: Oldies radio. News staff: one; News: 2 hrs wkly. Target aud: 18-60. ♦Nick Amico, traf mgr.

Grand Forks

KCNN(AM)—See East Grand Forks, MN

***KFJM(FM)**— Mar 6, 1995: 90.7 mhz; 2.4 kw. 154 ft TL: N47 55 55 W97 04 26. Hrs open: 24 Box 8117, 58202-8117. Phone: (701) 777-2577. Licensee: University of North Dakota. Population served: 75,000 Natl. Network: NPR, PRI. Format: AAA, jazz. News staff: 2; News: 15 hrs wkly. Target aud: 25-44; well-educated. Spec prog: Blues 3 hrs wkly. ♦Michael Olson, gen mgr, opns mgr & dev dir.

KJKJ(FM)— Aug 1, 1985: 107.5 mhz; 100 kw. 500 ft TL: N48 07 24 W97 04 20. Stereo. Hrs opn: 24 Box 13598, 58206-3598. Secondary address: 505 University Ave. 58203. Phone: (701) 746-1417. Fax: (701) 746-1410. E-mail: patmclean@clearchannel.com Web Site: www.kjkj.com. Licensee: Citicasters Licenses L.P. Group owner: Clear Channel Communications Inc. (acq 10-26-99; grpsl). Population served: 100,000 Format: AOR. News: 5 hrs wkly. Target aud: 18-49. ♦Jeff Hoberg, gen mgr; Pat McLean, stn mgr; Laura Hammack, gen sls mgr.

KKXL(AM)— 1941: 1440 khz; 1 kw-D, 500 w-N. TL: N47 57 52 W97 01 46. Stereo. Hrs opn: 24 Box 13598, 58208-3598. Secondary address: 505 University Ave. 58203. Phone: (701) 746-1417. Phone: (701) 775-0575. Fax: (701) 746-1410. Web Site: www.1440kkxl.com. Licensee: Citicasters Licenses L.P. Group owner: Clear Channel Communications Inc. (acq 10-26-99; grpsl). Population served: 100,000 Natl. Network: Jones Radio Networks. Format: Adult standards, news, mus of the 40s, 50s & 60s. News staff: one; News: 10 hrs wkly. Target aud: 25-54; farm community. ♦Pat McLean, gen mgr.

KKXL-FM— March 1975: 92.9 mhz; 63 kw. 385 ft TL: N47 57 52 W97 01 46. Stereo. Web Site: www.xl93.com. Format: CHR. Target aud: 18-34.

KNOX(AM)— Sept 7, 1947: 1310 khz; 5 kw-U, DA-N. TL: N47 50 39 W97 01 30. Hrs open: 24 Box 13638, Old Belmont Rd. S., 58208-3638. Phone: (701) 775-4611. Fax: (701) 772-0540. Web Site: www.leightonbroadcasting.com. Licensee: Leighton Enterprises Inc. (group owner; acq 10-23-96; $1.1 million with co-located FM). Population served: 39,008 Natl. Network: ABC. Rgnl. Network: MNN. Wire Svc: AP Format: Farm, news/talk. News staff: 3; News: 80 hrs wkly. Target aud: 35 plus. ♦Jack Hansen, gen mgr; Jarrod Thomas, opns mgr; Lynn Hodgson, gen mgr & gen sls mgr; Doug Barrett, news dir.

KNOX-FM— Feb 4, 1967: 94.7 mhz; 100 kw. 325 ft TL: N48 00 20 W97 04 18. Stereo. 24 Web Site: www.leightonbroadcasting.com.43,765 Natl. Network: ABC. Format: Country classics. Target aud: 18-45.

KQHT(FM)—Crookston, MN) March 1986: 96.1 mhz; 100 kw. 413 ft TL: N47 50 43 W96 50 22. Stereo. Hrs opn: 505 University Ave., 58203. Phone: (701) 746-1417. Fax: (701) 746-1410. Web Site: www.961thefox.com. Licensee: Citicasters Licenses L.P. Group owner: Clear Channel Communications Inc. (acq 10-26-99; grpsl). Rgnl rep: Hyett/Ramsland. Format: World class rock. Target aud: 18-49. ♦Pat McLean, gen mgr & gen sls mgr; Dave Schroeder, chief of engrg.

***KUND-FM**— May 30, 1976: 89.3 mhz; 38 kw. 215 ft TL: N47 55 55 W97 04 26. Stereo. Hrs opn: Box 8117, 58202. Phone: (701) 777-2577. Licensee: University of North Dakota. Population served: 180,000 Format: Class. Spec prog: New age 5 hrs wkly. ♦Gary Olson, stn mgr.

***KWTL(AM)**— Oct 22, 1923: 1370 khz; 1 k-D, 250 w-N. TL: N47 55 55 W97 04 26. Hrs open: 24 Box 13703, 58208. Phone: (701) 795-5969. Web Site: www.realpresenceradio.com. Licensee: Real Presence Radio (acq 10-26-2004; $317,100). Population served: 200,000 Format: Catholic. ♦Steve W. Loegering, pres.

KYCK(FM)—See Crookston, MN

Harvey

KHND(AM)— July 21, 1981: 1470 khz; 1 kw-D, 160 w-N. TL: N47 45 23 W99 55 06. Hrs open: Box 6, 58341. Secondary address: 718 Lincoln Ave. 58341. Phone: (701) 324-4848. Fax: (701) 324-2043. E-mail: khndsj@gondtc.com Web Site: khnd1470.com. Licensee: Three Way Broadcasting Inc. (acq 1-16-03). Natl. Network: ABC. Rgnl. Network: AgriAmerica. Format: Adult contemp. Target aud: 12 -85. Spec prog: Big band 3 hrs, polka 3 hrs, talk show 8 hrs wkly. ♦Rick Jensen, VP; Sheila Jensen, pres, gen mgr, stn mgr & gen sls mgr.

Harwood

KKLQ(FM)— 2001: 100.7 mhz; 25 kw. Ant 328 ft TL: N47 08 43 W96 58 18. Hrs open: 2351 Sunset Blvd., Suite 170-218, Rocklin, CA, 95765. Phone: (916) 251-1600. Fax: (916) 251-1650. Licensee: Educational Media Foundation. Group owner: EMF Broadcasting (acq 1-8-2004; $750,000). Natl. Network: K-Love. Format: Christian. ♦Richard Jenkins, pres; Mike Novak, VP; Keith Whipple, dev dir; David Pierce, progmg mgr; Ed Lenane, news dir; Sam Wallington, engrg dir; Karen Johnson, news rptr; Marya Morgan, news rptr; Richard Hunt, news rptr.

Hazelton

KUSB(FM)— 2006: 103.3 mhz; 100 kw. Ant 964 ft TL: N46 35 23.8 W100 47 46.2. Hrs open: 1830 N. 11th St., Bismarck, 58501. Phone: (701) 250-6602. Fax: (701) 250-6632. Licensee: Cumulus Licensing LLC. ♦Syd Stewart, gen mgr.

Hettinger

KNDC(AM)— Mar 1, 1954: 1490 khz; 1 kw-U. TL: N46 01 11 W102 41 33. Hrs open: 5:45 AM-10 PM Box 151, 505 2nd Ave. S., 58639. Phone: (701) 567-2421. Phone: (701) 567-2889. Fax: (701) 567-4636. E-mail: kndc1490@ndsupernet.com Web Site: www.knde.com. Licensee: Schweitzer Media Inc. (acq 5-19-99). Population served: 17,230 Rgnl. Network: AgriAmerica, N.D. News Net. Format: C&W. News: 24 hrs wkly. Target aud: 24-52; rural residents. Spec prog: Farm. ♦Mike Schweitzer, pres; Nolan Dix, gen mgr, progmg dir & mus dir; Terry Spratta, sports cmtr.

Hope

KMXW(FM)— 2002: 104.7 mhz; 100 kw. Ant 702 ft TL: N47 03 15 W97 24 44. Hrs open: 24 1020 25th St. S., Fargo, 58103. Phone: (701) 237-5346. Fax: (701) 235-4042. Web Site: fargomix.com. Licensee: Radio Fargo-Moorhead Inc. Group owner: Clear Channel Communications Inc. (acq 1-19-2007; grpsl). Format: Adult contemp. ♦Jeff Hoberg, gen mgr; Mike Kapel, opns mgr; Paul Jurgens, news dir; Al Murray, chief of engrg; Joan John, traf mgr.

Jamestown

***KJTW(FM)**—Not on air, target date: unknown: 89.9 mhz; 400 w. Ant 154 ft TL: N46 53 30 W98 42 46. Hrs open: Drawer 2440, Tupelo, MS, 38803. Phone: (662) 844-8888. Licensee: Abundant Life Broadcasting.

***KLRX(FM)**—Not on air, target date: unknown: 88.9 mhz; 1.8 kw vert. Ant 144 ft TL: N46 55 27 W98 46 19. Hrs open: 2351Sunset Blvd.,

Suite 1270-218, Rocklin, CA, 95765. Phone: (916) 251-1600. Fax: (916) 251-1650. Licensee: Educational Media Foundation. ♦Richard Jenkins, pres; Mike Novak, VP; Keith Whipple, dev dir; David Pierce, progmg mgr; Ed Lenane, news dir; Sam Wallington, engrg dir; Karen Johnson, news rptr; Marya Morgan, news rptr; Richard Hunt, news rptr.

***KPRJ(FM)**— 1993: 91.5 mhz; 18.5 kw. 354 ft TL: N46 46 36 W98 31 20. Hrs open: 24
Rebroadcasts KCND(FM) Bismarck 100%.
207 N 5th St., Fargo, 58102. Phone: (701) 241-6900. Fax: (701) 224-0555. E-mail: bthomas@prairiepublic.org Web Site: www.prairiepublic.org. Licensee: Prairie Public Broadcasting Inc. Natl. Network: NPR, PRI. Format: Class, jazz, news. News staff: 3. Target aud: General. Spec prog: American Indian 2 hrs, folk 6 hrs wkly. ♦John Harris, pres; Bill Thomas, gen mgr & stn mgr; Duane Lee, opns mgr; Nancy Wood, dev dir; Marie Luceo, prom dir; David Thompson, news dir.

KQDJ(AM)— Aug 12, 1954: 1400 khz; 1 kw-U. TL: N46 53 37 W98 41 20. Hrs open: 24 Box 1170, 58401. Secondary address: 2625 8th Ave. S.W. 58402. Phone: (701) 252-1400. Fax: (701) 252-1402. E-mail: bigdog@daktel.com Licensee: Two Rivers Broadcasting Inc. Group owner: Robert Ingstad Broadcast Properties (acq 1994; $600,000). Population served: 25,000 Natl. Network: CBS. Format: News/talk, sports, ESPN sports radio. Spec prog: Farm 6 hrs wkly. ♦Dave Reed, gen mgr.

KYNU(FM)—Co-owned with KQDJ(AM). Aug 25, 1984: 95.5 mhz; 100 kw. 398 ft TL: N46 51 52 W98 40 11. Stereo. 24 50,000 Natl. Network: ABC. Format: Hot country. Target aud: 18-65.

KSJB(AM)— 1937: 600 khz; 5 kw-U, DA-1. TL: N46 49 03 W98 42 34. Hrs open: 24 Box 5180, 58402-1840. Secondary address: 2400 8th Ave. S. W. 58402-1840. Phone: (701) 252-3570. Fax: (701) 252-1277. Web Site: www.ksjbam.com. Licensee: Chesterman Communications Inc. (acq 8-20-90; $850,000 with co-located FM; FTR: 9-10-90) Population served: 16,500 Format: Classic country. News staff: one; News: 13 hrs wkly. Target aud: 25 plus. Spec prog: Farm 12 hrs wkly. ♦Patrick Pfieffer, gen mgr; Patrick Pfeiffer, gen sls mgr, mktg mgr & prom mgr.

KSJZ(FM)—Co-owned with KSJB(AM). 1968: 93.3 mhz; 57 kw. 256 ft TL: N46 49 03 W98 42 34. Stereo. 6 AM-midnight Web Site: www.ksjbam.com. Format: Adult hit radio. Target aud: 28-52; 60% male, 40% female.

Kindred

***KFNL(FM)**— June 6, 1986: 92.7 mhz; 25 kw. Ant 328 ft TL: N46 39 38 W96 43 01. Stereo. Hrs opn: 24 3003 Snelling Ave. N., St. Paul, MN, 55113-1598. Phone: (612) 631-5000. Fax: (612) 631-5086. Licensee: Northwestern College. Group owner: Clear Channel Communications Inc. (acq 1-19-2007; donation). Population served: 150,000 Law Firm: Bryan Cave L.L.P. Format: Contemp Christian. ♦Alan S. Cureton, pres.

Langdon

KNDK(AM)— June 27, 1967: 1080 khz; 1 kw-D. TL: N48 46 25 W98 21 50. Hrs open: 16 Box 9, Rt. 5, 58249. Phone: (701) 256-1080. Fax: (701) 256-1081. E-mail: kndk1080@utma.com Licensee: KNDK Inc. (group owner; (acq 12-1-87). Population served: 25,500 Natl. Network: CBS. Law Firm: Haley, Bader & Potts. Format: News/talk, country. News staff: 3; News: 42 hrs wkly. Target aud: 25 plus. Spec prog: Farm 12 hrs, relg 4 hrs wkly. ♦Bob Simmons, pres, gen mgr, gen sls mgr, progmg dir & chief of engrg.

KNDK-FM— Jan 15, 1992: 95.7 mhz; 6 kw. 328 ft TL: N48 45 18 W98 21 38.24 (Acq 11-20-91; $90,000; FTR: 12-16-91). Format: Hot adult contemp.

Lincoln

***KVLQ(FM)**— 2006: 89.1 mhz; 950 w. Ant 693 ft TL: N46 35 24 W100 47 47. Hrs open: 24
Rebroadcasts KLVR(FM) Santa Rosa, CA 100%.
2351 Sunset Blvd., Suite 170-218, Rocklin, CA, 95765. Phone: (916) 251-1600. Fax: (916) 251-1650. E-mail: klove@klove.com Web Site: www.klove.com. Licensee: Educational Media Foundation. Group owner: EMF Broadcasting. Natl. Network: K-Love. Law Firm: Shaw Pittman. Format: Contemp Christian. News staff: 3. Target aud: 25-44; Judeo Christian, female. ♦Richard Jenkins, pres; Mike Novak, VP; Keith Whipple, dev dir; David Pierce, progmg dir; Ed Lenane, news dir; Sam Wallington, engrg dir; Arthur Vassar, traf mgr; Karen Johnson, news rptr; Marya Morgan, news rptr; Richard Hunt, news rptr.

Lisbon

KQLX(AM)— November 1984: 890 khz; 1.8 kw-D. TL: N46 26 43 W97 39 07. Stereo. Hrs opn: Box 1008, 1206 S. Main, 58054. Phone: (701) 683-5287. Fax: (701) 683-9029. E-mail: kqlx@kqlx.com Web Site: www.kqlx.com. Licensee: Loomis Broadcasting Inc. Population served: 340,000 Rgnl. Network: AgriAmerica, Agri-Net. Format: News/talk. News staff: one; News: 5 hrs wkly. Target aud: 18-65; farmers. Spec prog: Farm 18 hrs, Gospel 6 hrs wkly. ♦Terry Loomis, pres; Rita Loomis, VP; Bruce Dougherty, gen mgr.

KQLX-FM— Oct 24, 1986: 106.1 mhz; 50 kw. 249 ft TL: N46 26 43 W97 39 07. Stereo. Web Site: www.kqlx.com. Licensee: Sheyenne Valley Broadcasting Inc.12,000 Natl. Network: CNN Radio. Format: Country music. News staff: news progmg 5 hrs wkly News: one;. Target aud: 18-54; general.

Mandan

KLXX(AM)—See Bismarck

KNDR(FM)— June 19, 1977: 104.7 mhz; 100 kw. 852 ft TL: N46 35 11 W100 48 20. Stereo. Hrs opn: 24 Box 516, 58554. Secondary address: 1400 NE 3rd Str. 58554. Phone: (701) 663-2345. Fax: (701) 663-2347. E-mail: kndr@midconetwork.com Web Site: www.kndr.fm. Licensee: Central Dakota Enterprise Inc. Population served: 85,000 Wire Svc: AP Format: Christian, Adult Contemp. Target aud: 35-54; women and famlies (with children). ♦La Rue Goetz, chmn; Paul Grojz, chmn; Brad Bales, gen mgr.

Mayville

KMAV(AM)— Oct 20, 1967: 1520 khz; 2.5 kw-D. TL: N47 29 45 W97 21 03. Hrs open: Sunrise-sunset Box 216, 58257. Phone: (701) 786-2335. Fax: (701) 786-2268. E-mail: sports@kmavradio.com Web Site: kmav.com. Licensee: R & J Broadcasting. (acq 7-7-93; $200,000 with co-located FM; FTR: 8-2-93) Population served: 150,000 Natl. Network: ABC. Rgnl. Network: N.D. News Net., AgriAmerica, Red River Farm Net. Format: Country. News staff: one; News: 15 hrs wkly. Target aud: 25-55. Spec prog: Area high school and Mayville State Univ. sports, Minnesota Twins baseball. ♦Jim Birkemeyer, pres & sls dir; Rich Haraldson, exec VP; Dan Keating, gen mgr, adv mgr, progmg dir & sports cmtr; Eric Michaels, prom dir; Mary Keating, news dir.

KMAV-FM— Jan 10, 1977: 105.5 mhz; 25 kw. 328 ft TL: N47 29 45 W97 21 03. Stereo. 24 Web Site: kmav.com.150,000

Medina

KCVG(FM)—Not on air, target date: unknown: 92.3 mhz; 100 kw. Ant 699 ft TL: N47 05 38 W99 02 11. Hrs open: 5331 Mt. Alifan Dr., San Diego, CA, 92111. Phone: (858) 277-4991. Fax: (858) 277-1365. Web Site: www.horizonsd.org/radio.asp. Licensee: Horizon Christian Fellowship. ♦Mike MacIntosh, pres.

Minot

KCJB(AM)— September 1950: 910 khz; 5 kw-D, 1 kw-N, DA-2. TL: N40 11 57 W101 17 37. Hrs open: Box 10, 3425 S. Broadway, 58702. Phone: (701) 852-0361. Fax: (701) 852-1953. Licensee: CC Licenses LLC. Group owner: Clear Channel Communications Inc. (acq 1-12-2000; grpsl). Population served: 156,000 Natl. Network: CBS. Law Firm: Fisher, Wayland, Cooper, Leader & Zaragoza L.L.P. Format: Full service, country. Target aud: 25 plus. Spec prog: Loc sports, various loc talk segments, farm 4 hrs wkly. ♦Rick Stensby, gen mgr.

KYYX(FM)—Co-owned with KCJB(AM). Nov 15, 1966: 97.1 mhz; 100 kw. 984 ft TL: N48 03 02 W101 20 29. Stereo. 24 News staff: one; News: 6 hrs wkly. Target aud: 18-49; young families. Co-owned TV: KXMC-TV affil.

KHRT(AM)— Nov 17, 1957: 1320 khz; 2.5 kw-D, 310 w-N. TL: N48 11 48 W101 14 30. Hrs open: 24 Box 1210, 58702. Secondary address: 3600 County Rd. 195 S. 58702. Phone: (701) 852-3789. Fax: (701) 852-8498. E-mail: khrt@srt.com Licensee: Faith Broadcasting Inc. (acq 9-1-82; $188,248; FTR: 8-30-82). Population served: 107,300 Format: Relg, news/talk, Southern gospel. News staff: one; News: 10 hrs wkly. Target aud: 25-54; large families, loyal, upper-income professionals. Spec prog: Farm one hr wkly. ♦Richard Leavitt, pres, gen mgr, gen sls mgr & progmg dir; Roy Leavitt, stn mgr; Johas Nelson, mus dir & disc jockey; John Kennedy, news dir; John McCann, chief of engrg & disc jockey; Marcia Leavitt, traf mgr.

KHRT-FM— 1992: 106.9 mhz; 26 kw. Ant 344 ft TL: N48 09 48 W101 17 55. Format: Contemp Christian. ♦Johas Nelson, progmg dir.

KIZZ(FM)— Sept 7, 1968: 93.7 mhz; 98 kw. 571 ft TL: N48 12 56 W101 19 05. Stereo. Hrs opn: 24 Box 10, 58702. Secondary address: 101 S. Main St. 58702. Phone: (701) 852-2494. Fax: (701) 852-1390. E-mail: kizz@srt.com Licensee: CC Licenses LLC. Group owner: Clear Channel Communications Inc. (acq 9-1-2000; grpsl). Population served: 43,900 Format: Adult contemp. News staff: one; News: 3 hrs wkly. Target aud: 25-54. ♦Rick Stensby, gen mgr & gen sls mgr; Allison Bostow, opns mgr & progmg dir; Don May, news dir; Brian Funk, chief of engrg; Rhonda Jensen, traf mgr; Bill Allen, disc jockey; Cat Collins, disc jockey; Shawn Scott, disc jockey; Tim Webster, disc jockey.

***KMPR(FM)**— Nov 23, 1983: 88.9 mhz; 100 kw. 930 ft TL: N48 03 03 W101 23 24. (CP: 50 kw). Stereo. Hrs opn: 24
Rebroadcasts KCND(FM) Bismarck 100%.
1814 N. 15th St., Bismarck, 58501. Phone: (701) 224-1700. Fax: (701) 224-0555. E-mail: ppr@prairiepublic.org Web Site: www.ndpr.org. Licensee: Prairie Public Broadcasting Inc. Population served: 20,000 Natl. Network: PRI, NPR. Format: Class, news, jazz. News staff: 2; News: 40 hrs wkly. Target aud: General. Spec prog: American Indian 2 hrs, folk 6 hrs wkly. ♦Bill Thomas, gen mgr; Duane Lee, opns mgr; Marie Lucero, prom dir; David Thompson, news dir. Co-owned TV: *KSRE(TV) affil.

KMXA-FM— April 1984: 99.9 mhz; 100 kw. 500 ft TL: N48 10 57 W101 31 57. Stereo. Hrs opn: 24 1000 20th Ave. S.W., 58701. Phone: (701) 852-2494. Fax: (701) 852-1390. E-mail: allisonbostow@clearchannel.com Web Site: www.mix999fm.com. Licensee: CC Licenses LLC. Group owner: Clear Channel Communications Inc. (acq 1-12-2000; grpsl). Population served: 157,000 Format: Hot adult contemp. News: 4 hrs wkly. Target aud: 25-54; adult upper middle class with teens at home. ♦Rick Stensby, gen mgr & stn mgr; Allison Bostow, opns mgr.

KRRZ(AM)— Oct 28, 1929: 1390 khz; 5 kw-D, 1 kw-N. TL: N48 12 45 W101 14 30. Stereo. Hrs opn: 1000 20th Ave. S.W., 58701. Phone: (701) 852-4646. Fax: (701) 852-1390. Web Site: www.oldies1390.com. Licensee: CC Licenses LLC. Group owner: Clear Channel Communications Inc. (acq 9-1-2000; grpsl). Population served: 32,290 Rgnl. Network: N.D. News Net. Natl. Rep: Roslin. Format: Oldies. Target aud: 25-54. Spec prog: Sports. ♦Rick Stensby, gen mgr; Allison Bostow, opns mgr.

KZPR(FM)—Co-owned with KRRZ(AM). July 8, 1985: 105.3 mhz; 100 kw. 579 ft TL: N48 03 13 W101 26 03. Stereo. 24 Format: Classic rock. News: 4 hrs wkly. Spec prog: Farm 4 hrs wkly. ♦Allison Bostow, adv dir & disc jockey; Rick Anthony, disc jockey; Rick Stensby, disc jockey.

New England

KCVD(FM)—Not on air, target date: unknown: 95.7 mhz; 100 kw. Ant 636 ft TL: N46 41 35 W102 37 07. Hrs open: 5331 Mt. Alifan Dr., San Diego, CA, 92111. Phone: (858) 277-4991. Fax: (858) 277-1365. Licensee: Horizon Christian Fellowship. (acq 2-9-2006; grpsl). ♦Mike MacIntosh, pres.

Oakes

KDDR(AM)— July 31, 1959: 1220 khz; 1 kw-D, 327 w-N. TL: N46 07 23 W98 05 21. Hrs open:
Rebroadcast KOVC(AM) Valley City.
412 Main Ave., 58474. Phone: (701) 742-2187. Fax: (701) 742-2009. E-mail: kddr@drtel.net Licensee: Sioux Valley Broadcasting Co. Group owner: Robert Ingstad Broadcast Properties (acq 2-1-93; $85,000; FTR: 2-22-93). Population served: 50,000 Rgnl. Network: N.D. News Net. Law Firm: Fletcher, Heald & Hildreth. Format: Country, news. Target aud: 28-59; farm/agriculture. Spec prog: Farm 15 hrs, relg 3 hrs wkly. ♦Tim Ost, gen mgr; Terry James, progmg dir.

Rugby

KZZJ(AM)— Aug 21, 1961: 1450 khz; 1 kw-U. TL: N48 21 14 W99 59 31. Hrs open: 24 230 Hwy. 2 S.E., 58368. Phone: (701) 776-5254. Fax: (701) 776-6154. E-mail: kzzj@kzzj.com Web Site: www.kzzj.com. Licensee: Rugby Broadcasters Inc. (acq 7-6-89; $10,000; FTR: 7-24-89). Population served: 38,000 Rgnl. Network: Midwest Radio Format: Modern country, farm. News staff: one. Target aud: 25-65. ♦Lila Brossart, gen mgr; Jay Schmalz, opns mgr & mus dir; Cheryl Holm, gen sls mgr & chief of engrg; Bruce Allen, news dir; Dee Dee Bishoff, traf mgr.

Sarles

KCVF(FM)—Not on air, target date: unknown: 105.9 mhz; 100 kw. Ant 459 ft TL: N48 37 58 W99 06 05. Hrs open: 5331 Mt. Alifan Dr., San Diego, CA, 92111. Phone: (858) 277-4991. Fax: (858) 277-1365. Licensee: Horizon Christian Fellowship. (acq 2-9-2006; grpsl). ♦Mike MacIntosh, pres.

Tioga

KTGO(AM)— Feb 27, 1966: 1090 khz; 1 kw-D. TL: N48 23 30 W102 56 12. Hrs open: Box 457, 58852. Secondary address: 301 S.E. 2nd St. 58852. Phone: (701) 664-3322. Phone: (701) 664-3432. Fax: (701) 664-3322. E-mail: ktgo@wccray.com Licensee: Tioga Broadcasting Corp. Population served: 14,000 Natl. Network: CBS. Format: Country. Target aud: 18-55. Spec prog: Gospel 11 hrs wkly. ♦David Guttormson, pres, gen mgr, gen sls mgr & progmg dir.

Valley City

KOVC(AM)— Oct 19, 1936: 1490 khz; 1 kw-U. TL: N46 54 48 W98 01 02. Hrs open: 19 136 Central Ave. N., 58072. Phone: (701) 845-1490. Fax: (701) 845-1245. Licensee: Sioux Valley Broadcasting Co. Group owner: Robert Ingstad Broadcast Properties. Population served: 68,500 Rgnl. Network: AgriAmerica. Format: Country, news, sports. News: 7 hrs wkly. Target aud: 25 plus. ♦Dave Reed, opns mgr; Ron Lee, prom dir, progmg dir, mus dir & disc jockey; Kerry Johnson, adv mgr; Ryan Cunningham, news dir; Don Brintnall, engrg mgr & chief of engrg; Terri Suhr, traf mgr; Rod Reel, edit dir & disc jockey; Dave Michaels, disc jockey; Dave O'Bannon, disc jockey; Tim Ost, gen mgr & disc jockey.

KQDJ-FM—Co-owned with KOVC(AM). Aug 1, 1983: 101.1 mhz; 12 kw. 1,000 ft TL: N46 54 24 W97 58 20.19 Format: news/talk, sports info, adult standards. Target aud: 25-54. ♦Dave Reed, progmg mgr.

Velva

KTZU(FM)— 2005: 94.9 mhz; 98 kw. Ant 512 ft TL: N48 03 04 W101 20 23. Hrs open: 24 1408 20th Ave. S.W. #1, Minot, 58701. Phone: (701) 852-7449. Fax: (701) 837-6925. Licensee: Programmers Broadcasting Inc. Population served: 100,000 Law Firm: Fletcher, Heald & Hildreth. Format: Classic rock. News: 5 hrs wkly. Target aud: 25-54. ♦John Kircher, pres; Jean Kircher, VP; Jean Schempp, stn mgr; J. Davis, gen sls mgr; Jeff Bliss, chief of engrg.

Wahpeton

KBMW(AM)—Breckenridge, MN) Aug 28, 1948: 1450 khz; 1 kw-U. TL: N46 16 41 W96 35 19. Hrs open: 605 Dakota Ave., 58075. Phone: (701) 642-8747. Fax: (701) 642-9501. E-mail: studio@kbmwam.com Web Site: www.kbmwam.com. Licensee: Monterey Licenses LLC. Group owner: Triad Broadcasting Co. LLC (acq 1-31-03; $1.2 million). Population served: 60,000 Format: Country. News staff: one; News: 18 hrs wkly. Target aud: 25-54; general. ♦Bill Dadlow, stn mgr.

KEGK(FM)— May 21, 1989: 106.9 mhz; 41 kw. Ant 538 ft TL: N46 32 46 W96 37 39. Stereo. Hrs opn: 24 Box 1115, 58074. Secondary address: 605 Dakota Ave. 58075. Phone: (701) 642-8747. Phone: (701) 237-4500. Fax: (701) 642-9501. E-mail: studio@eagle1069.com Web Site: www.eagle1069.com. Licensee: Guderian Broadcasting Inc. (acq 7-17-2002). Natl. Rep: Midwest Radio. Rgnl rep: Quest Marketing. Format: Oldies. News staff: one; News: 5 hrs wkly. Target aud: 25-54. ♦Nancy Odney, gen mgr; John Austin, opns mgr & gen sls mgr; Michael Brooks, sls dir; Tim Murphy, adv mgr & progmg dir.

Walhalla

KYTZ(FM)— Sept 1, 1998: 106.7 mhz; 16 kw. Ant 836 ft TL: N48 38 38 W97 58 46. Hrs open: 24 1420 3rd St., Langdon, 58249. Phone: (701) 256-1067. Fax: (701) 256-1051. E-mail: kndk1080@vtma.com Licensee: Simmons Broadcasting Inc. (acq 9-7-2004). Format: Hot adult contemp. ♦Bob Simmons, gen mgr; Jen Taylor, opns mgr.

West Fargo

KFNW(AM)—Licensed to West Fargo. See Fargo

KQWB(AM)— Sept 1, 2000: 1660 khz; 10 kw-D, 1 kw-N. TL: N46 58 33 W96 35 02. Stereo. Hrs opn: 2720 7th Ave. S., Fargo, 58103. Phone: (701) 237-4500. Fax: (701) 235-9082. Web Site: www.123fargo.com. Licensee: Monterey Licenses LLC. Group owner: Triad Broadcasting Co. LLC (acq 8-18-99; grpsl). Natl. Network: Westwood One. Natl. Rep: Christal. Law Firm: Fisher, Wayland, Cooper, Leader & Zaragoza. Format: Talk/personality. Target aud: 35 plus. ♦David Benjamin, pres; Tom Douglas, CFO; Nancy Odney, gen mgr; Anne Phibian, opns mgr; John Austin, progmg dir.

Williston

KDSR(FM)— Feb 28, 1985: 101.1 mhz; 98 kw. Ant 800 ft TL: N48 03 30 W104 00 00. Stereo. Hrs opn: 18 910 E. Broadway, 58801. Phone: (701) 572-4478. Fax: (701) 572-1419. E-mail: kdsr@dia.net Licensee: Williston Community Broadcasting Corp. dba KDSR(FM) (acq 6-28-2002). Population served: 90,000 Natl. Network: CNN Radio. Law Firm: Booth, Freret, Imlay & Tepper. Format: Rock. News: 7 hrs wkly. ♦Stephen A. Marks, pres; Ben Buckles, opns mgr & chief of engrg; Danita Bloom, progmg dir.

KEYZ(AM)— 1948: 660 khz; 5 kw-U, DA-2. TL: N48 14 20 W103 39 01. Stereo. Hrs opn: 24 Box 2048, 58802-2048. Secondary address: 410 E. 6th 58801. Phone: (701) 572-5371. Fax: (701) 572-7511. Licensee: CCR-Williston IV LLC. Group owner: Cherry Creek Radio LLC (acq 12-19-2003; grpsl). Population served: 175,000 Rgnl. Network: AgriAmerica. Format: Country, news/talk. News staff: one. Target aud: 25-54. Spec prog: Relg 5 hrs wkly. ♦Larry Timpe, gen mgr; Lyla Semenko, natl sls mgr; Jeff Nelson, progmg dir & news rptr; Scott Haugen, mus dir; Earl Gross, news dir, pub affrs dir, chief of engrg, spec ev coord, local news ed & news rptr; Tammy Otteson, traf mgr.

KYYZ(FM)—Co-owned with KEYZ(AM). Dec 1, 1979: 96.1 mhz; 100 kw. 873 ft TL: N48 02 52 W103 59 01. Stereo. 24 68,000 Format: Hot country. News staff: one. Target aud: 25-54. ♦Sean Archer, progmg dir; Tammy Otteson, traf mgr; Earl Gross, spec ev coord, local news ed & news rptr; Jeff Nelson, news rptr.

***KPPR(FM)**— Nov 20, 1986: 89.5 mhz; 10.5 kw. 492 ft TL: N48 08 30 W103 53 34. Stereo. Hrs opn: 24
Rebroadcasts KCND(FM) Bismarck 100%.
207 N 5th St., Fargo, 58102. Phone: (701) 224-1700. Phone: (800) 359-5566. Fax: (701) 224-0555. E-mail: program@pol.org Web Site: www.prairiepublic.org. Licensee: Prairie Public Broadcasting. Population served: 40,000 Natl. Network: PRI, NPR. Format: Jazz, class, news. News staff: 2; News: 40 hrs wkly. Target aud: General. Spec prog: American Indian 2 hrs, folk 6 hrs wkly. ♦John Harris, pres; Bill Thomas, stn mgr; Nancy Wood, dev dir; Marie Lucero, prom dir; Dave Thompson, news dir. Co-owned TV: KWSE(TV) affil

Wimbledon

KRVX(FM)— 2005: 103.1 mhz; 99 kw. Ant 472 ft TL: N46 56 21 W98 18 30. Hrs open: Secondary address: 2625 8th Ave. S.W., Jamestown 58402. Phone: (701) 252-1400. Fax: (701) 252-1402. Licensee: James River Broadcasting Inc. Natl. Network: NBC Radio. Format: Rock. Target aud: 18-54. ♦Dave Reed, gen mgr; Lynn Hermanson, gen sls mgr.

Ohio

Ada

***WONB(FM)**— Oct 18, 1991: 94.9 mhz; 3 kw. 328 ft TL: N40 45 58 W83 50 14. Stereo. Hrs opn: 24 Freed Ctr., 45810. Phone: (419) 772-1195. Phone: (419) 772-1195. Fax: (419) 772-2794. E-mail: wonb@onu.edu Web Site: www.wonb.net. Licensee: Ohio Northern University. Format: CHR. News: 8 hrs wkly. Target aud: 18-49; general. Spec prog: Relg one hr, gospel 3 hrs, smooth jazz 18 hrs wkly. ♦Dr. Kendall Baker, pres; John Green, CFO; G. Richard Gainey, gen mgr; Tara Suman, progmg mgr.

Akron

WAKR(AM)— Oct 16, 1940: 1590 khz; 5 kw-U, DA-N. TL: N41 01 14 W81 30 20. Hrs open: 24 1795 W. Market St., 44313. Phone: (330) 869-9800. Fax: (330) 864-6799. Fax: (330) 864-9750. Web Site: www.wakr.net. Licensee: Rubber City Radio Group Inc. (group owner; acq 10-6-93; $9.3 million with co-located FM; FTR: 10-25-93). Population served: 565,100 Natl. Rep: Christal. Law Firm: Verner, Liipfert, Bernhard, McPherson & Hand. Format: MOR/full service, news. News staff: 7; News: 40 hrs wkly. Target aud: 35 plus. ♦Henry Zelman, CFO; Mark Biviano, exec VP & sls VP; Nick Anthony, sr VP; Thomas Mandel, CEO, pres & gen mgr; Kevin Mason, opns dir; Dominic Rizzo, gen sls mgr; Joyce Lagios, mktg dir & mktg mgr; Ed Esposito, progmg dir & news dir; Al Hruska, chief of engrg.

WONE-FM—Co-owned with WAKR(AM). October 1947: 97.5 mhz; 12 kw. 900 ft TL: N41 03 57 W81 34 59. Stereo. Web Site: www.wone.net. Format: Rock. Target aud: 18-49. ♦Brett Russell, prom mgr; T.K. O'Grady, progmg dir; Dana Durban, news rptr; Bob Campbell, disc jockey; Sandra Miller, disc jockey; Tim Daugherty, traf mgr, mus critic, mus critic & disc jockey.

WAKS(FM)—Listing follows WTOU(AM).

***WAPS(FM)**— Oct 4, 1955: 91.3 mhz; 800 w. Ant 151 ft TL: N41 03 18 W81 31 35. Stereo. Hrs opn: 24 65 Steiner Ave., 44301. Phone: (330) 761-3099. Fax: (330) 761-3240. Web Site: www.913thesummit.com. Licensee: Board of Education, Akron City School District. Population served: 300,000 Format: AAA, div. Target aud: 25-54; college educated adults. Spec prog: Ger 2 hrs, It 2 hrs, Hungarian one hr, Slovenian 2 hrs, Latin 2 hrs wkly. ♦Tommy Bruno, gen mgr; Andrew James, opns dir; Ryan Humbert, gen sls mgr.

WARF(AM)— 1926: 1350 khz; 5 kw-U, DA-1. TL: N41 10 05 W81 30 45. Hrs open: 7755 Freedom Ave., North Canton, 44720. Phone: (330) 492-4700. Fax: (330) 836-5321. Web Site: www.sportsradio1350.com. Licensee: Capstar TX L.P. Group owner: Clear Channel Communications Inc. (acq 2000; grpsl). Population served: 275,425 Natl. Network: Sporting News Radio Network. Format: Sports. ♦Dan Lankford, VP & gen mgr.

WCUE(AM)—See Cuyahoga Falls

WHLO(AM)— October 1944: 640 khz; 5 kw-D, 500 w-N, DA-2. TL: N41 04 47 W81 38 45. Stereo. Hrs opn: 24 7755 Freedom Ave., N. Canton, 44720. Phone: (330) 836-4700. Fax: (330) 836-5321. Licensee: CC Licenses LLC. Group owner: Clear Channel Communications Inc. (acq 12-31-2001; $4.5 million). Population served: 275,425 Format: News/talk. ♦Don Lankford, gen mgr.

WJMP(AM)—See Kent

WNIR(FM)—See Kent

WQMX(FM)—Medina, 1960: 94.9 mhz; 16.2 kw. 880 ft TL: N40 04 58 W81 38 00. Stereo. Hrs opn: 24 1795 W. Market St., 44313. Phone: (330) 869-9800. Fax: (330) 864-6799. E-mail: thom@wakr.net Web Site: wqmx.com. Licensee: Rubber City Radio Group Inc. (group owner; acq 7-21-88). Population served: 565,100 Natl. Rep: Christal. Law Firm: Verner, Liipfert, Bernhard, McPherson & Hand. Format: Country. Target aud: 25-54; upwardly mobile adults. ♦Thomas Mandel, CEO; Mark Biviano, exec VP; Sue Wilson, progmg dir; Paul Christopher, sls.

***WZIP(FM)**— Dec 10, 1962: 88.1 mhz; 7.5 kw. 827 ft TL: N41 04 58 W81 38 00. Stereo. Hrs opn: 24 302 E. Buchtel Ave., Buchtel Mall, 44325-1004. Phone: (330) 972-7105. Fax: (330) 972-5521. E-mail: wzip@uakron.edu Web Site: www.wzip.fm. Licensee: University of Akron. Population served: 1,000,000 Natl. Network: AP Network News. Wire Svc: AP Format: CHR, AOR. Target aud: 18-34. Spec prog: Polka 4 hrs, pub affrs 11 hrs, sports talk 3 hrs wkly. ♦Thomas G. Beck, gen mgr; Blake Thompson, chief of engrg.

Alliance

WDPN(AM)—Listing follows WZKL(FM).

***WRMU(FM)**— Oct 17, 1970: 91.1 mhz; 2.8 kw. 190 ft TL: N40 54 16 W81 06 45. Stereo. Hrs opn: Mount Union College, 1972 Clark Ave., 44601. Phone: (330) 823-2414. Phone: (330) 823-3777. Fax: (330) 829-4913. E-mail: wrmu@muc.edu Web Site: www.muc.edu/wrmu. Licensee: Mount Union College. Population served: 26,547 Wire Svc: AP Format: Smooth jazz, Rock/AOR, oldies. Spec prog: Gospel 2 hrs wkly, news/talk 5 hrs wkly. ♦Dr. Jack Ewing, pres; Mark A. Bergmann, gen mgr; William Weisinger, chief of engrg.

WZKL(FM)— April 1947: 92.5 mhz; 50 kw. 500 ft TL: N40 47 24 W81 06 26. Stereo. Hrs opn: 24 Box 2356, 44601. Secondary address: 392 Smyth Ave. 44601. Phone: (330) 450-9250. Fax: (330) 821-0379. E-mail: radiosales@alliancelink.com Web Site: www.q92radio.com. Licensee: D.A. Peterson Inc. Population served: 628,400 Format: Hot adult contemp. News staff: 2; News: 4 hrs wkly. Target aud: 25-54. ♦Don Peterson III, gen mgr; Mark O'Brian, sls dir; Mark O'Brien, gen sls mgr; John Stewart, progmg VP & progmg dir; Clint M, news dir; Steve Hundt, engrg dir & chief of engrg; Dee Zink, traf mgr.

WDPN(AM)—Co-owned with WZKL(FM). Sept 2, 1953: 1310 khz; 1 kw-D, 500 w-N, DA-2. TL: N40 55 34 W81 07 41.24 50,000 Format: Unforgettable favorites. News staff: 2; News: 21 hrs wkly. Target aud: 35-64. Spec prog: Relg 4 hrs wkly. ♦Doug Lane, progmg dir; Rex Coombs, progmg dir & traf mgr.

Anna

***WHJM(FM)**—Not on air, target date: unknown: 88.7 mhz; 3 kw vert. Ant 272 ft TL: N40 18 01 W84 12 25. Hrs open: c/o Putbrese, Hunsaker & Trent, 202 S. Church St., Woodstock, VA, 22664. Phone: (540) 459-7646. Licensee: Friends of Radio Maria Inc. ♦Florinda M. Iannace, pres.

Archbold

***WBCY(FM)**— Dec 1, 1992: 89.5 mhz; 20 kw. Ant 315 ft TL: N41 28 59 W84 16 58. Hrs open: 24 c/o WBCL(FM), 1025 W. Rudisill Blvd., Fort Wayne, IN, 46807. Phone: (260) 745-0576. Fax: (260) 456-2913. Web Site: www.wbcl.org. Licensee: Taylor University Broadcasting Inc. (acq 6-24-92). Population served: 889,000 Wire Svc: UPI Format: Contemp Christian. ♦Marsha Bunker, gen mgr; Craig Albrecht, opns mgr; Scott Tusleff, progmg dir & progmg mgr.

WMTR-FM— Mar 1968: 96.1 mhz; 3.8 kw. Ant 400 ft TL: N41 33 29 W84 11 08. Stereo. Hrs opn: 24 303 1/2 N. Defiance St., 43502. Phone: (419) 445-9050. Fax: (419) 445-3531. E-mail: wmtr@adelphia.net Web Site: www.961wmtr.com. Licensee: Nobco Inc. Population served: 125,000 Natl. Network: Westwood One. Rgnl. Network: Agri Bcstg. Natl. Rep: Rgnl Reps. Law Firm: Hogan & Hartson. Format: Adult top-40. News staff: one; News: 8 hrs wkly. Target aud: 25-54. ♦Max E. Smith Sr., pres; Max E. Smith Jr., gen mgr; Mark Knapp, mus dir; Larry Christy, news dir.

Ashland

WNCO(AM)— 1949: 1340 khz; 1 kw-U. TL: N40 50 25 W82 21 18. Hrs open: 24 1197 US Hwy. 42, 44805. Phone: (419) 289-2605. Phone: (419) 526-5825. Fax: (419) 289-0304. Web Site: www.theohioradio.com. Licensee: Capstar TX L.P. Group owner: Clear Channel Communications Inc. (acq 2-12-01; grpsl). Population served: 53,900 Natl. Rep: Rgnl Reps. Law Firm: Arent, Fox, Kintner, Plotkin & Kahn. Format: Adult standards. News staff: 2; News: 20 hrs wkly. Target aud: 35 plus. Spec prog: Farm 3 hrs wkly. ♦Diana Coon, gen mgr.

WNCO-FM— May 1947: 101.3 mhz; 50 kw. 500 ft TL: N40 50 25 W82 21 18. Stereo. 24 230,000 Format: Country. News staff: 3; News: 14 hrs wkly. Target aud: 25 plus. ♦Dean Stampfli, opns dir & mktg dir; Martin Larsen, sls dir; Darla Stampfli, prom mgr & pub affrs dir; Lori Johnson, traf mgr; Gene Davis, local news ed; Steve Crabtree, sports cmtr.

***WRDL(FM)**— Aug 24, 1967: 88.9 mhz; 3 kw. 171 ft TL: N40 51 41 W82 19 11. Stereo. Hrs opn: 6 AM-1 AM 401 College Ave., 44805. Phone: (419) 289-5678. Phone: (419) 289-5311. Fax: (419) 289-5329. Licensee: Ashland University. Population served: 250,000 Format: Rock, educ. News: 7 hrs wkly. Target aud: 18-35; general. Spec prog: Christian contemp 7 hrs, jazz 5 hrs, oldies 4 hrs wkly. ♦Dr. G. William Benz, pres; Tom Griffiths, gen mgr & chief of engrg.

Ashtabula

WFUN(AM)— November 1937: 970 khz; 5 kw-D, 1 kw-N, DA-2. TL: N41 48 52 W80 46 45. Hrs open: 24 3226 Jefferson Rd., 44004. Phone: (440) 993-2126. Fax: (440) 992-2658. E-mail: danschulte@clearchannel.com Web Site: www.wfunam97.com. Licensee: CC Licenses LLC. Group owner: Clear Channel Communications Inc. (acq 7-11-2000; grpsl). Natl. Network: ABC. Format: News, talk, sports. News staff: 2; News: 20 hrs wkly. Target aud: General; socially conscious, community-oriented listeners. ♦Dana Schulte, VP; Dennis Brockman, pres & opns dir.

WREO-FM—Co-owned with WFUN(AM). 1949: 97.1 mhz; 50 kw. 500 ft TL: N41 48 58 W80 46 52. Stereo. E-mail: star97@star97.com Web Site: www.star97.com.1,200,000 Format: Adult contemp. Target aud: 25-54; professionals. ♦Dennis O'Brien, progmg dir.

WYBL(FM)— 2005: 98.3 mhz; 5.3 kw. Ant 344 ft TL: N41 50 23 W80 44 36. Hrs open: 3226 Jefferson Rd., 44004-9112. Phone: (440) 993-2126. Fax: (440) 992-2658. Web Site: www.983thebull.com. Licensee: CC Licenses LLC. Group owner: Clear Channel Communications Inc. (acq 8-30-2002; $525,000 for CP). Format: News. ♦Dana Schulte, gen mgr.

WZOO-FM—See Edgewood

Athens

WATH(AM)— Oct 25, 1950: 970 khz; 1 kw-D, 160 w-N. TL: N39 20 40 W82 06 21. Stereo. Hrs opn: 24 Box 210, 45701. Secondary address: 300 Columbus Rd. 45701. Phone: (740) 593-6651. Phone: (740) 593-7982 (News). Fax: (740) 594-3488. E-mail: palmerd@wxtq.com Web Site: www.970wath.com. Licensee: WATH Inc. (acq 9-5-73; with co-located FM). Population served: 155,000 Natl. Network: CBS. Rgnl rep: Rgnl Reps. Law Firm: Pepper & Corazzini. Wire Svc: AP Format: MOR, news/talk,sports. News staff: 3; News: 20 hrs wkly. Target aud: 45 plus. Spec prog: Big band 15 hrs wkly. ♦David W. Palmer, pres; Robert E. Lambert, VP; Kathy Malesick, gen mgr & gen sls mgr; Joel Witkowski, opns mgr; Thom Williams, progmg dir; Bob Beyette, news dir.

WXTQ(FM)—Co-owned with WATH(AM). Sept 16, 1964: 105.5 mhz; 6 kw. 312 ft TL: N39 21 18 W82 05 32. Stereo. 24 Web Site: www.wxtq.com.75,000 Law Firm: Pepper & Corazzini. Format: Hot adult contemp. News staff: 2; News: 5 hrs wkly. Target aud: 18-34.

WJKW(FM)— Sept 1, 1998: 95.9 mhz; 6 kw. 199 ft TL: N39 22 37 W81 57 46. Hrs open: 24 3809 Maple Ave., Castalia, 44824. Phone: (740) 592-9879. Fax: (740) 592-9952. E-mail: wjkw@cfbroadcast.net Licensee: Christian Faith Broadcast Inc. Group owner: Christian Faith Broadcasting Inc. Law Firm: Joseph E. Dunne III. Format: Adult contemp, Christian. Target aud: 25-44. ♦Rusty Yost, gen mgr; Kevin Ingle, stn mgr.

***WOUB(AM)**— Sept 14, 1957: 1340 khz; 500 w-D, 1 kw-N. TL: N39 19 45 W82 05 29. Stereo. Hrs opn: 24 9 S. College St., 45701. Phone: (740) 593-4554. Fax: (740) 593-0240. E-mail: woub@woub.org Web Site: www.woub.org. Licensee: Ohio University. Population served: 7,500 Natl. Network: NPR, PRI. Format: News/talk, progsv. News staff: 3. Spec prog: Black 8 hrs wkly. ♦Carolyn Lewis, gen mgr; David Wiseman, opns VP & engrg VP; Steve Skidmore, opns dir; Scott Martin, opns mgr; Doug Partusch, mktg dir; Bryan Gibson, progmg dir & progmg mgr; Tim Myers, progmg dir; Tim Sharp, news dir; Ted Ross, engrg dir.

WOUB-FM— Dec 13, 1949: 91.3 mhz; 50 kw. 500 ft TL: N39 18 50 W82 08 54. Stereo. 24 Web Site: www.woub.org.70,000 Format: Adult contemp. News staff: 3. ♦Rusty Smith, progmg mgr; Jan Sole, asst music dir; Mark Hellenberg, asst music dir. Co-owned TV: *WOUB-TV affil.

Bainbridge

***WKHR(FM)**— May 6, 1977: 91.5 mhz; 1.1 kw. 269 ft TL: N41 23 42 W81 18 25. (CP: 100 w). Stereo. Hrs opn: Kenston High School, 17425 Snyder Rd., Chagrin Falls, 44023. Phone: (440) 543-9646. Fax: (440) 543-9012. Web Site: www.wkhr.org. Licensee: Kenston Local School District. Population served: 1,500,000 Format: Big band. Target aud: 55 plus; well established, mature. ♦Chris Kofron, gen mgr, opns dir & dev dir.

Barnesville

WBNV(FM)— July 1, 1991: 93.5 mhz; 6 kw. 489 ft TL: N39 54 10 W81 12 37. Stereo. Hrs opn: 24 Box 338, 4988 Skyline Dr., Cambridge, 43725. Secondary address: Box 293, 175 E. Main St. 43713-0293. Phone: (740) 484-4430. Phone: (740) 425-9268. Fax: (740) 432-1991. Web Site: www.yourradioplace.com. Licensee: W. Grant Hafley. Population served: 100,000 Natl. Network: USA. Natl. Rep: Rgnl Reps. Format: Adult contemp. News: 15 hrs wkly. Target aud: 25-54. ♦W. Grant Hafley, gen mgr; David L. Wilson, opns mgr.

Batavia

***WOBO(FM)**— July 30, 1981: 88.7 mhz; 15.5 kw. 428 ft TL: N39 03 43 W84 05 50. Stereo. Hrs opn: 24 Box 338, Owensville, 45160. Phone: (513) 724-3939/724-2969. E-mail: df1littman@cs.com Web Site: www.wobofm.com. Licensee: Educational Community Radio Inc. Population served: 1,000,000 Format: Var/div. Target aud: 35 plus. Spec prog: Ger 5 hrs, Scottish one hr, Celtic 3 hrs, Pol 3 hrs wkly. ♦Mel Reifan, pres.

Beach City

***WOFN(FM)**— Sept 27, 2000: 88.7 mhz; 3.3 kw horiz, 21 kw vert. Ant 358 ft TL: N40 35 41 W81 34 39. Stereo. Hrs opn: 24 Box 1924, Tulsa, OK, 74101. Secondary address: 4916 Spruce Hill Dr., Suite 400, Canton 44617. Phone: (330) 244-9151. Phone: (918) 455-5693. Fax: (330) 244-0153. E-mail: mail@oasisnetwork.org Web Site: www.oasisnetwork.org. Licensee: Creative Educational Media Corp. Inc. Format: Relg. Target aud: General. ♦David Ingles, pres; Bobbie Cook, gen mgr.

Beavercreek

WXEG(FM)— June 18, 1972: 103.9 mhz; 1.15 kw. 522 ft TL: N39 44 12 W84 09 25. Stereo. Hrs opn: 24 101 Pine St., Dayton, 45402. Phone: (937) 224-1137. Fax: (937) 224-7655. Web Site: www.wxeg.com. Licensee: Citicasters Licenses L.P. Group owner: Clear Channel Communications Inc. (acq 1999; grpsl). Population served: 830,200 Format: Alternative rock. News staff: one. Target aud: 18-34. ♦Karrie Sudbrack, gen mgr.

Bellaire

WOMP(AM)— Dec 2, 1947: 1290 khz; 1 kw-D, 33 w-N. TL: N40 02 09 W80 46 16. Hrs open: 24 Box 448, Rt. 214, 56325 High Ridge Rd., 43906. Phone: (740) 676-5661. Fax: (740) 676-2742. Licensee: Keymarket Licenses LLC. Group owner: Keymarket Communications LLC (acq 1-15-93; $575,000 with co-located FM; FTR: 1-25-93). Population served: 156,400 Natl. Network: ESPN Radio. Natl. Rep: Rgnl Reps. Law Firm: Fleischman & Walsh L. Format: Sports. News staff: 2; News: 20 hrs wkly. Spec prog: Pol 2 hrs, Czech 2 hrs wkly. ♦Gerald Getz, pres; John Crawford, gen mgr.

WYJK-FM—Co-owned with WOMP(AM). 1947: 100.5 mhz; 48 kw. Ant 518 ft TL: N40 02 09 W80 46 16. Stereo. 24 Fax: (740) 671-4487. Web Site: www.wompfm.com. Format: CHR. News staff: one. Target aud: 18-44; educated adults.

Bellefontaine

WBLL(AM)— 1951: 1390 khz; 500 w-D, 81 w-N. TL: N40 22 05 W83 44 02. Hrs open: 24 1501 Rd. 235, 43311-9506. Phone: (937) 592-1045. Fax: (937) 592-3299. E-mail: cwilkinson@wbll.com Web Site: www.wbll.com. Licensee: V-Teck Communications Inc. (acq 12-2-87; grpsl; FTR: 10-19-87). Natl. Network: ABC. Rgnl. Network: Agri Bcstg. Natl. Rep: Rgnl Reps. Format: News/talk, sports. News staff: 2; News: 126 hrs wkly. Target aud: General. Spec prog: Relg 6 hrs wkly. ♦Lou Vito, pres; Amie Huffman, gen mgr, adv dir & traf mgr; Chad Wilkinson, opns mgr & pub affrs dir; Sheryl Godwin, dev dir; Pam Allen, gen sls mgr; Ken Keller, prom dir; Bill Tipple, news dir, news rptr & sports cmtr; Bill Bowin, chief of engrg; Jamie Ross, reporter.

WPKO-FM—Co-owned with WBLL(AM). July 15, 1969: 98.3 mhz; 1.75 kw. Ant 430 ft TL: N40 22 05 W83 44 02. Stereo. 24 E-mail: cwilkinson@wpko.com Web Site: www.wpko.com. Format: Adult contemp. News staff: 2. Target aud: 12 plus. ♦Pam Allen, mktg dir & mktg mgr; Chad Wilkinson, mus dir & disc jockey; Louie Vito, asst music dir; Bill Bowin, engrg dir; Ken Keller, disc jockey; Mark Brake, disc jockey; Matt Hull, disc jockey.

Bellevue

WOHF(FM)— Apr 4, 1973: 92.1 mhz; 6 kw. Ant 328 ft TL: N41 14 19 W82 50 16. Stereo. Hrs opn: 24 1281 North River Road, Fremont, 43420. Phone: (419) 332-8218. Fax: (419) 333-8226. Web Site:

www.classicrockexperience.com. Licensee: BAS Broadcasting Inc. (acq 10-1-2003; $550,000). Population served: 54,000 Natl. Network: CNN Radio. Rgnl. Network: Agri-Net. Rgnl rep: Rgnl Reps Format: Oldies. News staff: one; News: 5 hrs wkly. ♦Jim Lorenzen, pres.

Belpre

***WCVV(FM)**— 1986: 89.5 mhz; 4.4 kw. Ant 384 ft TL: N39 19 27 W81 37 33. Stereo. Hrs opn: 24 Box 405, 45714. Phone: (740) 423-5895. Fax: (740) 423-9951. Licensee: Belpre Educ. Broadcasting Foundation. Format: Christian, news. Target aud: General. ♦Clay Sloan, gen mgr & stn mgr; Ralph Matheny, chief of engrg.

***WLKP(FM)**— May 1991: 91.9 mhz; 4.5 kw. Ant 325 ft TL: N39 20 46 W81 29 55. Hrs open: 2351 Sunset Blvd., Suite 170- 218, Rocklin, CA, 95765. Phone: (916) 251-1600. Fax: (916) 251-1650. Web Site: www.klove.com. Licensee: Educational Media Foundation. (acq 3-31-2005; $700,000 with WLKV(FM) Ripley, WV). Natl. Network: K-Love. Format: Christian. ♦Richard Jenkins, pres; Mike Novak, VP & news dir; Keith Whipple, dev dir; David Pierce, progmg mgr; Sam Wallington, engrg dir; Karen Johnson, news rptr; Marya Morgan, news rptr; Richard Hunt, news rptr.

WNUS(FM)— Sept 12, 1981: 107.1 mhz; 4.7 kw. 370 ft TL: N39 18 36 W81 35 49. Stereo. Hrs opn: 24 Box 5559, Vienna, WV, 26105. Secondary address: 6006 Grand Central Ave., Vienna, WV 26105. Phone: (304) 295-9441. Phone: (740) 423-9687. Fax: (304) 295-4389. E-mail: roadcrew@wnus.com Web Site: www.wnus.com. Licensee: CC Licenses LLC. Group owner: Clear Channel Communications Inc. (acq 4-17-2001; grpsl). Rgnl. Network: Ohio Radio Net., Metronews Radio Net. Natl. Rep: Clear Channel. Format: Country. News staff: 2; News: 2 hrs wkly. Target aud: 18 plus. ♦Chuck Poet, gen mgr.

Berea

***WBWC(FM)**— Mar 2, 1958: 88.3 mhz; 5 kw. Ant 256 ft TL: N41 25 05 W81 54 03. Stereo. Hrs opn: 19 275 Eastland Rd., 44017. Phone: (440) 826-2145. Fax: (440) 826-3426. E-mail: jtaranto@bw.edu Web Site: www.wbwc.com. Licensee: Baldwin-Wallace College. Population served: 1,400,000 Natl. Network: AP Radio. Format: Modern rock. Target aud: 12-25; alternative mus listeners. ♦Allen Thompson, opns dir.

Bowling Green

***WBGU(FM)**— November 1951: 88.1 mhz; 450 w. 178 ft TL: N41 22 33 W83 38 34. Hrs open: 24 120 W. Hall Bowling Green Univ., 43403. Phone: (419) 372-8657. Fax: (419) 372-0202. Web Site: www.wbgufm.com. Licensee: Bowling Green State University. Population served: 110,000 Format: Div, jazz, Black. News: 7 hrs wkly. Target aud: General. Spec prog: Country 4 hrs, class 3 hrs, folk 4 hrs, Sp 4 hrs wkly. ♦Keely Miller, gen mgr; Jim Davis, chief of engrg. Co-owned TV: *WBGU-TV affil.

WJYM(AM)— December 1964: 730 khz; 1 kw-D, 359 w-N, DA-2. TL: N41 31 57 W83 33 55. Hrs open: 6 AM-midnight Box 262550, Baton Rouge, LA, 70826. Phone: (225) 768-3202. Fax: (225) 768-3729. E-mail: kawikfish@yahoo.com Web Site: www.jsm.org. Licensee: Family Worship Center Church Inc. (group owner; acq 12-15-99). Population served: 5,000,000 Natl. Network: USA. Format: Relg. News: 3 hrs wkly. Target aud: 25-49. ♦David Whitelaw, COO.

WRQN(FM)— June 1964: 93.5 mhz; 7 kw. Ant 397 ft TL: N41 27 28 W83 39 33. Hrs open: 3225 Arlington Ave., Toledo, 43614. Phone: (419) 725-5700. Fax: (419) 389-5172. Web Site: www.935wrqn.com. Licensee: Cumulus Licensing Corp. Group owner: Cumulus Media Inc. (acq 9-11-97; grpsl). Population served: 135,000 Format: Oldies. Target aud: 25-54. Spec prog: Pub service one hr wkly. ♦Brian Olson, gen mgr & stn mgr; Ron Finn, opns mgr & progmg dir.

Bryan

WBNO-FM— June 30, 1966: 100.9 mhz; 6 kw. Ant 299 ft TL: N41 28 44 W84 34 50. Hrs open: 24 12810 State Rd. 34, 43506-8809. Phone: (419) 636-3175. Fax: (419) 636-4570. E-mail: wbno@wbno-wqct.com Licensee: Impact Radio LLC. (acq 8-1-2002; grpsl). Population served: 120,000 Natl. Network: ABC. Rgnl rep: Rgnl Reps Law Firm: Irwin, Campbell & Tannenwald, P.C. Wire Svc: AP Format: Classic hits. News staff: one; News: 27 hrs wkly. Target aud: 25-54; loc oriented. ♦Dennis Rumsey, pres & gen mgr.

WQCT(AM)—Co-owned with WBNO-FM. December 1962: 1520 khz; 500 w-D, 5 w-N, 250 w-CH. TL: N41 28 43 W84 34 49.24 (group owner; .120,000 Rgnl rep: Rgnl Reps Law Firm: Irwin, Campbell & Tannenwald, P.C. Format: Oldies. News staff: one; News: 3 hrs wkly. Target aud: 35-65; general.

***WGBE(FM)**— 1996: 90.9 mhz; 850 w. 387 ft TL: N41 28 30 W84 35 14. Hrs open: 24
Rebroadcasts WGTE-FM Toledo 100%.
1270 S. Detroit Ave., Toledo, 43614. Phone: (419) 380-4600. Fax: (419) 380-4710. E-mail: firstname-lastname@wgte.pbs.org Web Site: www.wgte.org. Licensee: The Public Broadcasting Foundation of Northwest Ohio. Population served: 30,800 Natl. Network: NPR, PRI. Law Firm: Schwartz, Woods & Miller. Format: Class, pub affrs, news. News: 23 hrs wkly. Target aud: General. Spec prog: Jazz 16 hrs wkly. ♦Marlon P. Kiser, CEO, pres & gen mgr; George Jones, chmn; Chris Peiffer, opns mgr; Ross Pfeiffer, dev dir.

Buchtel

WAIS(AM)— Dec 3, 1984: 770 khz; 1 kw-D. TL: N39 25 56 W82 12 02. Hrs open: Sunrise-sunset 15751 U.S. Rt. 33 S., Nelsonville, 45764. Phone: (740) 753-4094. Fax: (740) 753-4965. E-mail: wseo33@sbcglobal.net Licensee: Nelsonville TV Cable Inc. Natl. Network: ABC. Law Firm: Frank Jazzo. Format: News, talk, classic country. News staff: 3; News: 21 hrs wkly. Target aud: 35 plus. Spec prog: Farm 5 hrs, gospel 3 hrs wkly. ♦Eugene Edwards, pres & gen mgr; Sharon Elliott, stn mgr.

Bucyrus

WBCO(AM)— Dec 22, 1962: 1540 khz; 500 w-D, DA. TL: N40 45 47 W82 56 05. Hrs open: Box 1140, 403 East Rensselaer St., 44820-1140. Phone: (419) 562-2222. Fax: (419) 562-0520. Licensee: Franklin Communications Inc. Group owner: Saga Communications Inc. (acq 12-1-2003; $2.2 million with co-located FM). Population served: 30,000 Natl. Network: CBS. Rgnl. Network: Agri Bcstg. Natl. Rep: Rgnl Reps. Format: Adult standards. News staff: one; News: 25 hrs wkly. Target aud: 30-60. Spec prog: Farm 4 hrs, relg 4 hrs wkly. ♦Debbi Gifford, gen mgr, stn mgr, sls dir & prom dir; Dave Jones, gen sls mgr; Jim Radke, progmg dir, news dir, pub affrs dir & news rptr; Bill Bowin, chief of engrg; Sharon Shealy, traf mgr & farm dir.

WQEL(FM)—Co-owned with WBCO(AM). Sept 5, 1964: 92.7 mhz; 3 kw. 300 ft TL: N40 45 45 W82 55 50.24 Natl. Network: CBS. Natl. Rep: Rgnl Reps. Format: Classic rock, sports. News staff: 2; News: 10 hrs wkly. Target aud: 25-54. ♦Jim Hahn, disc jockey; Will Beard, progmg dir, farm dir & disc jockey.

Byesville

WILE-FM— Oct 29, 1994: 97.7 mhz; 1.8 kw. 413 ft TL: N40 02 24 W81 38 50. Hrs open: 24 Box 338, Cambridge, 43725. Secondary address: 4988 Skyline Dr., Cambridge 43725. Phone: (740) 432-5605. Fax: (740) 432-1991. Web Site: www.yourradioplace.com. Licensee: AVC Communications Inc. (group owner; acq 7-13-00). Format: Adult standard. ♦Joel Losego, gen mgr; W. Grant Hafley, CEO & gen mgr; Dave Wilson, opns dir.

Cadiz

WCDK(FM)— Aug 28, 1985: 106.3 mhz; 6 kw. 360 ft TL: N40 15 14 W80 50 35. Stereo. Hrs opn: 24 2307 Pennsylvania Ave., Weirton, WV, 26062. Phone: (304) 723-1444. Fax: (304) 723-1688. E-mail: wcdk@weir.net Web Site: www.106.3theriver.com. Licensee: Priority Communications Ohio LLC. Group owner: Priority Communications (acq 12-98; $475,000 with WEIR(AM) Weirton, WV). Population served: 249,000 Natl. Network: Jones Radio Networks. Natl. Rep: Dome. Law Firm: Pepper & Corazzini. Format: Classic hits. News staff: one; News: 5 hrs wkly. Target aud: 25-54; general. Spec prog: OSU football, Cleveland Browns football, high school football. ♦Jay M. Philippone, pres & gen mgr; Jude Sheets, opns mgr; Judy Vavrek, stn mgr & sls dir.

Caldwell

WWKC(FM)— July 1, 1989: 104.9 mhz; 3 kw. 300 ft TL: N39 48 47 W81 56 38. Stereo. Hrs opn: 24 Box 338, Cambridge, 43725. Secondary address: Box 19 43724. Phone: (740) 432-5605. Phone: (740) 732-7555. Fax: (740) 432-1991. Web Site: www.yourradioplace.com. Licensee: W. Grant Hafley. (acq 8-23-89; $15,000; FTR: 9-18-89). Rgnl. Network: Agri Bcstg. Natl. Rep: Agri Reps. Format: Country. Target aud: 25-54. ♦W. Grant Hafley, gen mgr; David L. Wilson, opns mgr.

Cambridge

WCMJ(FM)—Listing follows WILE(AM).

WILE(AM)— Apr 9, 1948: 1270 khz; 1 kw-D. TL: N40 02 24 W81 38 50. Hrs open: Box 338, 4988 Skyline Dr., 43725. Phone: (740) 432-5605. Fax: (740) 432-1991. Web Site: www.yourradioplace.com. Licensee: AVC Communications Inc. (group owner; (acq 5-5-83). Population served: 80,000 Natl. Network: ESPN Radio. Rgnl rep: Rgnl Reps. Format: Sports. Target aud: 25-54. ♦Grant Hafley, gen mgr.

WCMJ(FM)—Co-owned with WILE(AM). October 1964: 96.7 mhz; 2.3 kw. 367 ft TL: N40 02 24 W81 38 50. Stereo. Web Site: www.yourradioplace.com. Format: Adult contemp. News: 2. Target aud: 18-49.

***WOUC-FM**— May 11, 1987: 89.1 mhz; 5 kw. 500 ft TL: N40 05 32 W81 17 19. Stereo. Hrs opn: 24
Rebroadcasts WOUB-FM Athens 100%.
9 S. College St., Athens, 45701. Phone: (740) 593-4554. Fax: (740) 593-0240. E-mail: woub@woub.org Web Site: www.woub.org. Licensee: Ohio University. Natl. Network: PRI, NPR. Format: News/talk. News staff: 3. ♦Carolyn Lewis, gen mgr; David Wiseman, opns VP; Steve Skidmore, opns dir; Scott Martin, opns mgr. Co-owned TV: *WOUC-TV affil

Campbell

WGFT(AM)—Licensed to Campbell. See Youngstown

WHOT-FM—See Youngstown

Canton

WCER(AM)— 1947: 900 khz; 500 w-D, 78 w-N. TL: N40 49 17 W81 25 34. Stereo. Hrs opn: 4537 22nd St. N.W., 44708. Phone: (330) 478-6655. Fax: (330) 478-6651. E-mail: wcerradio@neo.rr.com Web Site: www.wcer.us. Licensee: Melodynamic Broadcasting Corp. (acq 7-11-91; $85,000; FTR: 7-29-91). Format: Christian, news/talk, sports. Spec prog: Farm 6 hrs, gospel 11 hrs, relg 5 hrs wkly. ♦Jack Ambrozic, gen mgr; John Amrhein, progmg dir.

WDPN(AM)—See Alliance

WHBC(AM)— Mar 9, 1925: 1480 khz; 5 kw-U, DA-2. TL: N40 43 15 W81 26 28. Stereo. Hrs opn: 550 Market Ave. S., 44702. Phone: (330) 456-7166. Fax: (330) 456-7199. Web Site: www.whbc.com. Licensee: NM Licensing LLC. Group owner: NextMedia Group L.L.C. (acq 11-26-01; grpsl). Natl. Rep: Christal. Law Firm: Cohn & Marks. Format: Full service, oldies. Target aud: 25 plus. Spec prog: Farm one hr wkly. ♦Richard Bossler, gen mgr.

WHBC-FM— Feb 2, 1948: 94.1 mhz; 50 kw. 500 ft TL: N40 53 53 W81 19 07. Stereo. Web Site: www.mix941.com. Format: Adult contemp. Target aud: 25-54. ♦Terry Simmons, progmg dir.

WILB(AM)— Aug 11, 1946: 1060 khz; 5 kw-D, DA. TL: N40 50 04 W81 25 46. Hrs open: 4365 Fulton Dr. N.W., 44718. Phone: (330) 966-2903. Fax: (330) 966-3177. Web Site: www.livingbreadradio.com.

Licensee: Living Bread Radio Inc. (acq 7-1-2004; $300,000). Population served: 100,500 Format: Catholic talk. ♦Barbara Gaskell, pres; Kate Sell, stn mgr; Dan Clark, opns mgr.

WINW(AM)— Apr 14, 1966: 1520 khz; 1 kw-D, DA. TL: N40 50 41 W81 21 02. Hrs open: Sunrise-sunset 237 W. Tuscarawas, 44705. Phone: (330) 453-1520. Fax: (330) 454-3030. E-mail: christfirst@joy1520.com Web Site: www.joy1520am.com. Licensee: Pinebrook Corp. (acq 9-27-96; $75,000). Natl. Rep: Rgnl Reps. Format: Gospel. Target aud: 35-54; professionals. ♦Patrick Barb, pres; Curtis Perry, gen mgr.

WKDD(FM)— Nov 19, 1961: 98.1 mhz; 50 kw. Ant 344 ft TL: N40 57 10 W81 19 20. Stereo. Hrs opn: 24 7755 Freedom Ave. N.W., North Canton, 44720. Phone: (330) 492-4700. Fax: (330) 836-5321. Web Site: www.wkdd.com. Licensee: Citicasters Licenses L.P. Group owner: Clear Channel Communications Inc. (acq 12-22-2000; grpsl). Population served: 1,000,000 Format: Hot adult contemp. ♦Dan Lankford, gen mgr; Becky Clark, gen sls mgr; Keith Kennedy, progmg dir; Tom Duresky, news dir; Don Kreiger, chief of engrg; Teri Jones, traf mgr.

WRQK(FM)— Mar 1, 1961: 106.9 mhz; 27.5 kw. Ant 340 ft TL: N40 49 17 W81 25 34. Stereo. Hrs opn: 24 7755 Freedom Ave. N.W., North Canton, 44720. Phone: (330) 492-5630. Fax: (330) 492-5633. E-mail: wrqk@wrqk.com Web Site: www.wrqk.com. Licensee: Cumulus Licensing Corp. Group owner: Cumulus Media Inc. (acq 3-15-00; grpsl). Population served: 335,000 Format: Mainstream rock. News: one hr wkly. Target aud: 18-49; emphasis on men. ♦Dave Kelly, gen mgr & rgnl sls mgr; Larry Blum, opns mgr & progmg dir.

WZKL(FM)—See Alliance

Castalia

WGGN(FM)— January 1975: 97.7 mhz; 1.25 kw. 660 ft TL: N41 23 48 W82 47 31. Stereo. Hrs opn: 24 3809 Maple Ave., 44824. Secondary address: P. O. Box 247 44824. Phone: (419) 684-5311. Fax: (419) 684-5378. E-mail: fm977@cfbroadcast.net Licensee: Christian Faith Broadcasting Inc. (group owner). Population served: 750,000 Natl. Network: USA. Format: Adult contemp Christian. Target aud: 25-49; general. ♦Shelby Gillam, pres; Rusty Yost, gen mgr; Jeff Ferback, gen sls mgr; Dave Yost, prom dir & progmg dir.

Cedarville

***WCDR-FM**— Dec 1, 1962: 90.3 mhz; 30 kw. 354 ft TL: N39 45 46 W83 53 05. Stereo. Hrs opn: 24 Box 601, 45314-0601. Phone: (937) 766-7815. Fax: (937) 766-7927. E-mail: info@thepath.fm Web Site: www.thepath.fm. Licensee: The Cedarville University. Population served: 400,000 Natl. Network: AP Radio, CNN Radio, Moody. Format: Relg, full service. News staff: one; News: 16 hrs wkly. Target aud: 35-54. Spec prog: Black 2 hrs wkly. ♦William Brown, pres; Martin Clark, VP; Paul Gathany, gen mgr; Keith Hamer, opns mgr.

Celina

WCSM(AM)— Sept 11, 1963: 1350 khz; 500 w-D, 11 w-N, DA-1. TL: N40 32 17 W84 35 20. Hrs open: 24 Box 492, Meyers & Schunck Rds., 45822. Phone: (419) 586-5134. Fax: (419) 586-3814. E-mail: wcsm@bright.net Web Site: www.wcsmradio.com. Licensee: Hayco Broadcasting Inc. (acq 1977). Population served: 80,000 Rgnl. Network: Ohio News Network, Agri-Net. News staff: one; News: 20 hrs wkly. Target aud: 18-49. ♦John H. Coe, pres & gen mgr; Sue Heiser, gen sls mgr; Jim Hyatt, progmg dir & mus dir; Kevin Sandler, news dir & chief of engrg.

WCSM-FM— 1968: 96.7 mhz; 3 kw. 328 ft TL: N40 33 08 W84 30 46.24 100,000 Natl. Network: ABC, Jones Radio Networks. Format: Sports.

WKKI(FM)— Dec 18, 1960: 94.3 mhz; 2.2 kw. 448 ft TL: N40 33 08 W84 30 46. Hrs open: 24 126 W. Fayette St., 45822. Phone: (419) 586-7715. Fax: (419) 586-1074. E-mail: k94@bright.net Web Site: www.wkki.net. Licensee: The Sonshine Communications Corp. (acq 5-26-2004; $370,000 for stock). Population served: 250,000 Natl. Network: CNN Radio, Westwood One. Natl. Rep: Roslin, Rgnl Reps. Format: Adult contemp. Target aud: 25-54. Spec prog: Contemp Christian 2 hrs wkly. ♦Paul Schmitmeyer, pres, pres & gen mgr; Hilany Dickey, stn mgr; Dan Dietz, chief of opns; Brian Mathews, progmg dir.

Centerville

***WCWT-FM**— Sept 20, 1971: 101.5 mhz; 10 w. 110 ft TL: N39 37 39 W84 09 57. (CP: Ant 194 ft.). Hrs opn: 500 E. Franklin, 45459. Phone: (937) 439-3558. Phone: (937) 439-3557. Fax: (937) 439-3574. E-mail: wcwt@centerville.k12.oh.us Licensee: Centerville City Board of Education. Population served: 30000 Format: Classic rock. Target aud: General. ♦Bob Romond, gen mgr & progmg dir.

Chillicothe

WBEX(AM)— September 1947: 1490 khz; 1 kw-U. TL: N39 19 56 W82 59 50. Hrs open: 24 Box 94, 45601. Secondary address: 45 W.Main St. 45601. Phone: (740) 773-3000. Fax: (740) 774-4494. E-mail: newsroom@wkkj.com Web Site: www.wbex.com. Licensee: Citicasters Licenses L.P. Group owner: Clear Channel Communications Inc. (acq 1999; grpsl). Population served: 65,000 Natl. Network: CBS, Westwood One. Natl. Rep: Katz Radio. Format: News/talk, full service. News staff: 3; News: 10 hrs wkly. Target aud: 30-50. ♦Dan Latham, VP & gen mgr; Tracy Taylor, sls dir; Dan Ramey, progmg dir.

WCHI(AM)— Oct 1, 1951: 1350 khz; 1 kw-D, 28 w-N. TL: N39 19 13 W82 57 03. Hrs open: 24 Box 94, 45 W. Main St., 45601. Phone: (740) 775-1350. Phone: (740) 773-3000. Fax: (740) 774-4494. Web Site: www.wchiam.com. Licensee: CC Licenses LLC. Group owner: Clear Channel Communications Inc. (acq 10-19-99; $4 million with co-located FM). Population served: 24,842 Natl. Network: ABC. Natl. Rep: Katz Radio. Format: Nostalgia. News staff: 2. Target aud: 25-65. ♦Dan Latham, gen mgr; Bob Neal, opns mgr; Tracy Taylor, sls dir.

WKKJ(FM)—Co-owned with WCHI(AM). Dec 22, 1978: 94.3 mhz; 25 kw. 266 ft TL: N39 19 52 W82 59 49. Stereo. 24 Web Site: www.wkkj.com. Natl. Network: ABC. Format: Country. News staff: 3; News: 14 hrs wkly. ♦Dan Latham, VP; Mike Smith, news dir.

WLZT(FM)— July 1, 1961: 93.3 mhz; 50 kw. 335 ft TL: N39 19 52 W82 59 49. Stereo. Hrs opn: 24 2323 W. Fifth Ave., Suite 200, Columbus, 43204. Phone: (614) 486-6101. Fax: (614) 487-2537. E-mail: pegbuehrle@clearchannel.com Web Site: www.933litefm.com. Licensee: CC Licenses LLC. Group owner: Clear Channel Communications Inc. (acq 4-11-2003). Population served: 100,000 Natl. Network: ABC. Format: Adult contemp. News staff: 3. Target aud: 24-54. ♦Tom Thon, gen mgr; Peg Buehrle, gen sls mgr; Steve Cherry, progmg dir.

***WOHC(FM)**— May 1, 1992: 90.1 mhz; 2 kw. 393 ft TL: N39 20 45 W83 11 15. Stereo. Hrs opn: 24
Rebroadcasts WCDR(FM) Cedarville 100%.
Box 601, Cedarville, 45314. Phone: (937) 766-7815. Fax: (937) 766-7927. E-mail: info@thepath.fm Web Site: thepath.fm. Licensee: The Cedarville University. Natl. Network: AP Radio, CNN Radio, Moody. Law Firm: Cohen & Berfield. Format: Relg, full service. News staff: one; News: 16 hrs wkly. Target aud: 35-54; information-oriented Christians and/or church members. Spec prog: Black 2 hrs wkly. ♦William Brown, pres; Marvin D. Sparks, gen mgr; Keith Hamer, opns mgr; Chad Bresson, news dir; John Tocknell, chief of engrg.

***WOUH-FM**— October 1992: 91.9 mhz; 750 w. 649 ft TL: N39 19 46 W82 48 08. Stereo. Hrs opn: 24
Rebroadcasts WOUB-FM Athens 100%.
9 S. College St., Athens, 45701. Phone: (740) 593-4554. Fax: (740) 593-0240. E-mail: woub@woub.org Web Site: www.woub.org. Licensee: Ohio University. Format: News/talk. News staff: 3. Spec prog: Pub affrs. ♦Carolyn Lewis, gen mgr; David Wiseman, opns VP; Steve Skidmore, opns dir; Scott Martin, opns mgr; Tim Myers, progmg dir; Rusty Smith, progmg mgr; Tim Sharp, news dir; Kelly Martin, prom.

***WVXC(FM)**— Jan 15, 1988: 89.3 mhz; 2.5 kw. Ant 351 ft TL: N39 20 45 W83 11 15. Stereo. Hrs opn: 24 Box 793, New Albany, 43054. Phone: (614) 855-9171. Fax: (614) 855-9280. Licensee: Christian Voice of Central Ohio Inc. (acq 5-15-2007; grpsl). Rgnl. Network: Ohio Radio Net. Format: Contemp Christian. ♦Dan Baughman, gen mgr.

Cincinnati

WAKW(FM)— Nov 21, 1961: 93.3 mhz; 50 kw. 500 ft TL: N39 12 22 W84 33 23. Stereo. Hrs opn: 24 6275 Collegevue Pl., 45224. Phone: (513) 542-9393. Fax: (513) 542-9333. Web Site: www.wakw.com. Licensee: Pillar of Fire Inc. (group owner) Population served: 3,000,000 Natl. Network: Moody. Format: Contemp Christian radio. News staff: one. Target aud: General; families. ♦Gerald Croucher, gen mgr & stn mgr.

WCIN(AM)— October 1953: 1480 khz; 4.5 kw-D, 300 w-N, DA-2. TL: N39 12 43 W84 29 20. Hrs open: 3540 Reading Rd., 45229. Phone: (513) 281-7180. Fax: (513) 281-6125. E-mail: thepulseofthecity@hotmail.com Web Site: www.1480wcin.com. Licensee: J4 Broadcasting of Cincinnati Inc., debtor-in-possession. (acq 9-23-98). Population served: 452,524 Format: Urban oldies & black talk. Target aud: 25-64; affluent, upscale, $50,000 plus income. ♦John Thomas, CEO & progmg mgr; John C. Thomas, pres & gen mgr.

WCKY(AM)— Sept 16, 1929: 1530 khz; 50 kw-U, DA-N, (LSS-Sacramento, CA). TL: N39 03 55 W84 36 27. Hrs open: 8044 Montgomery Rd., Suite 650, 45236. Phone: (513) 686-8300. Fax: (513) 333-4269. Web Site: www.wcky.com. Licensee: Clear Channel Communications Group owner: Clear Channel Communications Inc. Population served: 300,000 Natl. Network: ESPN Radio, Fox Sports, Westwood One. Format: Sports. Target aud: 35 plus; special focus on ages 35-64. ♦Darryl Parks, opns dir.

WCVG(AM)—See Covington, KY

WCVX(AM)— 1947: 1050 khz; 1 kw-D, 278 w-N. TL: N39 04 50 W84 31 18. Hrs open: 24 635 W. 7th St., Suite 400, 45203. Phone: (513) 579-1050. Fax: (513) 421-0821. E-mail: 1050am@wtsj.com Licensee: Christian Broadcasting System Ltd. Group owner: Salem Communications Corp. (acq 2-10-2006; swap of WCVX(AM) and WDJO(AM) Florence, KY plus $6.75 million cash for WLQV(AM) Detroit, MI) Population served: 2,000,000 Natl. Network: Salem Radio Network. Format: Relg, talk. News: 10 hrs wkly. Target aud: 24-54; family oriented young adults. Spec prog: Gospel 15 hrs, Sp 3 hrs wkly. ♦Jon R. Yinger, pres; Errol Dengler, VP; Gaither Stephens, gen mgr.

WDBZ(AM)— 1927: 1230 khz; 1 kw-U. TL: N39 06 27 W84 30 09. Hrs open: 24 705 Central Ave., Suite 200, 45202. Phone: (513) 679-6000. Fax: (513) 948-1985. E-mail: rporter@radio-one.com Web Site: www.1230thebuzz.com. Licensee: Radio One Licenses LLC. (acq 6-19-2007; $2.69 million). Population served: 452524 Natl. Rep: Christal. Format: Talk. News staff: one. Target aud: 25-54. ♦Lisa Thal, gen mgr; Lincoln Ware, opns mgr & progmg dir; Josh Guttman, gen sls mgr & natl sls mgr.

WEBN(FM)—Listing follows WLW(AM).

WFTK(FM)—See Lebanon

WGRR(FM)—See Hamilton

***WGUC(FM)**— Sept 21, 1960: 90.9 mhz; 15 kw. Ant 960 ft TL: N39 07 27 W84 31 18. Stereo. Hrs opn: 24 1223 Central Pkwy., 45214. Phone: (513) 241-8282. Fax: (513) 241-8456. Web Site: www.wguc.org. Licensee: Cincinnati Public Radio, Inc. (acq 2-14-02). Population served: 1,300,000 Natl. Network: PRI. Format: Class. Target aud: 35 plus; well-educated. ♦Richard Eiswerth, CEO; Barry Weinstein, CFO; Chris Phelps, gen mgr & mktg mgr; Robin Gehl, opns dir & progmg VP; Sherri Mancini, dev VP & dev dir; Gordon Bayliss, sls VP; Don Danko, engrg VP.

WIZF(FM)—See Erlanger, KY

***WJVS(FM)**— Apr 7, 1976: 88.3 mhz; 175 w. 105 ft TL: N39 17 21 W84 24 52. Stereo. Hrs opn: 3254 E. Kemper Rd., 45241. Phone: (513) 771-8810. Fax: (513) 771-4928. Licensee: Great Oaks Institute of Technical and Career Development. (acq 1976). Format: Adult contemp. ♦Dave Angeline, gen mgr & progmg dir.

WKRC(AM)— 1922: 550 khz; 5 kw-D, 1 kw-N, DA-2. TL: N39 00 29 W84 26 39. Hrs open: 8040 Montgomery Rd., Suite 650, 45236. Phone: (513) 686-8300. Web Site: www.55krc.com. Licensee: Jacor Broadcasting Corp. Group owner: Clear Channel Communications Inc. (acq 5-4-99; grpsl). Population served: 1,800,000 Natl. Network: Westwood One, CBS. Law Firm: Hogan & Hartson. Format: News/talk info. Target aud: 35-64; adult, affluent, conservative. ♦Mike Kenney, VP; Karrie Sudbrack, gen mgr.

WOFX-FM—Co-owned with WKRC(AM). Aug 19, 1964: 92.5 mhz; 16 kw. Ant 866 ft TL: N39 06 59 W84 30 07. Fax: (513) 749-4925. Web Site: www.wofxcincinnati.com. Format: Classic rock. Target aud: 25-54. ♦Tony Tolliver, stn mgr & progmg dir. Co-owned TV: WKRC-TV affil.

WKRQ(FM)— 1947: 101.9 mhz; 16 kw. 876 ft TL: N39 06 58 W84 30 05. Stereo. Hrs opn: 2060 Reading Rd., Cincinatti, 45202. Phone: (513) 699-5102. Fax: (513) 699-5000. Web Site: www.wkrq.com. Licensee: Infinity Radio Inc. Group owner: Infinity Broadcasting Corp. (acq 11-13-98; grpsl). Population served: 400,000 Natl. Rep: Katz Radio. Format: Hot adult contemp. ♦Jim Bryant, VP & gen mgr; Bryson Lair, sls dir & prom dir; Patti Marshall, progmg dir; Brian Douglas, mus dir.

WLW(AM)— Mar 22, 1922: 700 khz; 50 kw-U. TL: N39 21 11 W84 19 30. Hrs open: 8044 Montgomery Rd., Suite 650, 45236. Phone: (513)

686-8300. Fax: (513) 665-9700. Web Site: www.700wlw.com. Licensee: Jacor Broadcasting Corp. Group owner: Clear Channel Communications Inc. (acq 4-99; grpsl). Population served: 750,000 Format: News/talk.

WEBN(FM)—Co-owned with WLW(AM). Aug 27, 1967: 102.7 mhz; 16.6 kw. 876 ft TL: N39 07 31 W84 29 57. Stereo. Fax: (513) 749-3299. Web Site: www.webn.com. (Acq 2-86; $8 million; FTR: 2-17-86). 1,340,000 Format: AOR.

WRRM(FM)— Oct 1, 1959: 98.5 mhz; 17.5 kw. 807 ft TL: N39 07 19 W84 32 52. Stereo. Hrs opn: 895 Centeral Ave., Suite 900, 45202. Phone: (513) 241-9898. Fax: (513) 749-3398. Fax: (513) 241-6689. Web Site: www.warm98.com. Licensee: WRRM Lico Inc. Group owner: Susquehanna Radio Corp. (acq 1-72). Population served: 1,150,000 Format: Adult contemp. ♦TJ Holland, opns dir.

WSAI(AM)— June 7, 1923: 1360 khz; 5 kw-U, DA-N. TL: N39 14 51 W84 31 52. Hrs open: 24 8044 Montgomery Rd., Suite 650, 45236. Phone: (513) 686-8300. Fax: (513) 665-9700. Web Site: www.1360espn.com. Licensee: Clear Channel Communication Inc. Group owner: Clear Channel Communications Inc. (acq 4-29-99; grpsl). Natl. Network: ESPN Radio. Law Firm: Wiley, Rein & Fielding. Format: Sports. ♦Mike Kenney, gen mgr; Darryl Parks, opns mgr & progmg VP; Mike Jamison, sls dir; Holly Nesser, mktg dir; Vince Marotta, prom dir; Ted Ryan, engrg dir.

WSWD(FM)—See Fairfield

WUBE-FM— July 20, 1949: 105.1 mhz; 14 kw. 920 ft TL: N39 07 31 W84 29 57. Stereo. Hrs opn: 2060 Reading Rd., 45202. Phone: (513) 699-5105. Fax: (513-) 699-5000. Web Site: www.b105.com. Licensee: Infinity Radio Inc. Group owner: Infinity Broadcasting Corp. (acq 2000; grpsl). Format: Country. ♦Jim Bryant, gen mgr; Christine Mello, gen sls mgr; Duke Hamilton, mus dir & disc jockey; Marty Thompson, progmg dir & mus dir.

WVMX(FM)— 1955: 94.1 mhz; 32 kw. 600 ft TL: N39 06 18 W84 33 24. Stereo. Hrs opn: 24 8044 Montomery Rd., Suite 650, 45236. Phone: (513) 686-8300. Fax: (513) 421-3299. Web Site: www.mixcincinnati.com. Licensee: Citicasters Licenses L.P. Group owner: Clear Channel Communications Inc. (acq 5-4-99; grpsl). Population served: 452,524 Law Firm: Koteen & Naftalin. Format: Adult contemp. News staff: one. Target aud: 25-54. ♦Bobby Dayer, gen mgr & progmg dir; Karrie Sudbrack, gen mgr.

***WVXU(FM)**— Oct 1, 1971: 91.7 mhz; 26.1 kw. Ant 683 ft TL: N39 07 31 W84 29 57. Stereo. Hrs opn: 1223 Central Pkwy., 45214. Phone: (513) 352-9170. Fax: (513) 241-8456. E-mail: wxu@cinradio.org Web Site: www.wvxu.org. Licensee: Cincinnati Public Radio, Inc. (acq 8-22-2005; grpsl). Population served: 1,300,000 Natl. Network: PRI, NPR. Rgnl. Network: Ohio Radio Net. Law Firm: Baker & Hostetler LLP. Format: News and info. News staff: 6; News: 107 hrs wkly. Target aud: 35 yrs plus; well educated. ♦Barry Weinstein, CFO; Richard Eiswerth, gen mgr; Sherri Mancini, dev VP; Chris Phelps, mktg VP; Robin Gehl, progmg VP; Maryanne Zeleznik, news dir & pub affrs dir; Don Danko, engrg VP.

Circleville

WNKK(FM)— Oct 1, 1965: 107.1 mhz; 3 kw. Ant 328 ft TL: N39 39 52 W82 51 04. Hrs open: 10th Floor, 280 Plaza N. High St., Columbus, 43215. Secondary address: 219 S. Court St. 43215. Phone: (614) 233-9208. Fax: (614) 677-0083. Web Site: www.thebigwazu.com. Licensee: Wilks License Co.-Columbus LLC. Group owner: Infinity Broadcasting Corp. (acq 1-10-2007; grpsl). Population served: 250,000 Natl. Network: Westwood One. Natl. Rep: Katz Radio. Law Firm: Leventhal, Senter & Lerman. Format: Country. ♦Valerie Brooks, gen mgr & stn mgr; Robert Golias, gen sls mgr; Jill McCarron, rgnl sls mgr.

Cleveland

***WCPN(FM)**— Sept 8, 1984: 90.3 mhz; 50 kw. 500 ft TL: N41 22 18 W81 42 48. Stereo. Hrs opn: 24 1375 Euclid Ave., 44115. Phone: (216) 916-6100. Fax: (216) 916-6299. Web Site: www.wcpn.org. Licensee: Ideastream (acq 2-27-01). Population served: 2,000,000 Natl. Network: PRI, NPR. Law Firm: Schwartz, Woods & Miller. Format: News, jazz. News staff: 16; News: 50 hrs wkly. Target aud: General. Spec prog: Ger one hr, Hungarian one hr, Lithuanian one hr, Pol one hr, Slovak one hr wkly. ♦Jerry Wareham, CEO; Keith Turner, opns dir & opns mgr; Maureen Paschke, dev VP & dev dir.

***WCRF(FM)**— Nov 23, 1958: 103.3 mhz; 25.5 kw. Ant 660 ft TL: N41 17 48 W81 39 27. Stereo. Hrs opn: 24 9756 Barr Rd., 44141. Phone: (440) 526-1111. Fax: (440) 526-1319. E-mail: wcrf@moody.edu Web Site: wcrfradio.org. Licensee: Moody Bible Institute of Chicago. (group owner) Population served: 4,000,000 Law Firm: Southmayd. Wire Svc: AP Format: Inspirational. Target aud: 25-55. ♦Dr. Michael Easley, pres; Richard Lee, stn mgr; Phil Villareal, progmg dir; Gary Bittner, mus dir; Doug Hainer, chief of engrg.

***WCSB(FM)**— May 10, 1976: 89.3 mhz; 1 kw. 190 ft TL: N41 30 12 W81 40 30. Stereo. Hrs opn: 24 Cleveland State Univ., Rhodes Tower, 44115. Phone: (216) 687-3523. Web Site: www.wcsb.org. Licensee: Cleveland State University. Population served: 2,000,000 Format: Alternative, commerical. News: 7 hrs wkly. Target aud: General. ♦Eric Schulte, gen mgr; Brian Detrow, dev dir.

WDOK(FM)— Apr 30, 1950: 102.1 mhz; 12 kw. 1,004 ft TL: N41 23 02 W81 42 06. Stereo. Hrs opn: One Radio Ln., 44114. Phone: (216) 696-0123. Fax: (216) 363-7189. Web Site: www.wdok.com. Licensee: Infinity Radio Inc. Group owner: Infinity Broadcasting Corp. (acq 2000; grpsl). Population served: 3,330,500 Format: Soft rock. Target aud: 25-54; general. ♦Chris Maduri, gen mgr.

WENZ(FM)— July 14, 1959: 107.9 mhz; 70 kw. 750 ft TL: N41 28 03 W81 17 25. Stereo. Hrs opn: 24 2510 Saint Clair Ave., 44114. Phone: (216) 579-1111. Fax: (216) 575-9141. E-mail: dbevins@radio-one.com Licensee: Radio One Licenses LLC. Group owner: Radio One Inc. (acq 11-8-01; grpsl). Population served: 175,000 Natl. Rep: Christal. Format: Main stream urban. News staff: one. Target aud: 18-34. ♦Chris Forgy, VP & gen mgr; Kim Johnson, opns mgr; Rick Bennett, chief of engrg.

WERE(AM)—See Cleveland Heights

WFHM-FM— Apr 1, 1960: 95.5 mhz; 31 kw. 620 ft TL: N41 26 32 W81 29 28. Stereo. Hrs opn: 24 4 Summit Park Dr., Suite 150, Indepedence, 44131. Phone: (216) 901-0921. Fax: (216) 901-1104. E-mail: office@whkradio.com Web Site: www.whkradio.com. Licensee: SCA License Corp. Group owner: Salem Communications Corp. (acq 12-22-00; grpsl). Population served: 1,850,000 Natl. Rep: Salem. Format: Christian, adult contemp. News: one hr wkly. Target aud: 25-54; female. ♦Edward Atsinger, pres; Errol Dengler, gen mgr & opns VP.

WGAR-FM— July 1948: 99.5 mhz; 50 kw. 500 ft TL: N41 22 18 W81 43 04. Stereo. Hrs opn: 24 6200 Oak Tree Blvd., 4th Floor, Independence, 44131-2510. Phone: (216) 520-2600. Fax: (216) 524-2600. Web Site: www.wgar.com. Licensee: Citicasters Licenses L.P. Group owner: Clear Channel Communications Inc. (acq 5-4-99; grpsl). Population served: 1,754,500 Natl. Network: AP Radio. Natl. Rep: Christal. Format: Contemp country. News staff: 2. Target aud: 25-54. ♦Mike Kenney, gen mgr; Bob Butts, gen sls mgr; Brian Jennings, progmg dir; Chuck Collier, mus dir.

WHK(AM)— July 28, 1921: 1420 khz; 5 kw-U, DA-N. TL: N41 21 30 W81 40 03. Hrs open: 4 Summit Park Dr., Suite 150, Independence, 44131. Phone: (216) 901-0921. Fax: (216) 901-5517. E-mail: office@whkradio.com Web Site: www.whkradio.com. Licensee: Caron Broadcasting Inc. (acq 9-1-2004; $10 million). Population served: 1,728,000 Natl. Network: Salem Radio Network. Law Firm: Hogan & Hartson. Wire Svc: AP Format: Conservative talk. Target aud: 25-54; male & female. ♦Michael Luczak, progmg dir.

WHKW(AM)— December 1930: 1220 khz; 50 kw-U, DA-1. TL: N41 18 26 W81 41 21. Stereo. Hrs opn: 4 Summit Park Dr., Suite 150, 44131. Phone: (216) 901-0921. Fax: (216) 901-5517. Web Site: www.whkwradio.com. Licensee: Caron Broadcasting Inc. Group owner: Salem Communications Corp. (acq 8-24-2000; grpsl). Format: Christian talk. ♦Errol Dengler, gen mgr.

WJMO(AM)— July 6, 1949: 1300 khz; 5 kw-U, DA-1. TL: N41 20 28 W81 44 29. Hrs open: 24 2510 St. Claire Ave. NE, 44114. Phone: (216) 579-1111. Fax: (216) 575-9141. E-mail: dbevins@radio-one.com Web Site: www.praise1300.com. Licensee: Radio One Licenses LLC. Group owner: Radio One Inc. (acq 11-8-2001; grpsl). Population served: 750,903 Natl. Rep: Christal. Format: Gospel. News staff: 2; News: 4 hrs wkly. Target aud: 25-54; Black adults. ♦Chris Forgy, VP & gen mgr; Kim Johnson, opns mgr, gen sls mgr, prom mgr & mus dir.

WKNR(AM)— 1926: 850 khz; 50 kw-D, 4.7 kw-N, DA-2. TL: N41 19 00 W81 43 51. Stereo. Hrs opn: 24 9446 Broadview Rd., 44147-2397. Phone: (440) 838-8585. Fax: (440) 838-1546. E-mail: slegerski@sportstalk850.com Web Site: www.espncleveland.com. Licensee: Good Karma Broadcasting LLC. Group owner: Salem Communications Corp. (acq 2-7-2007; $7 million). Population served: 2,200,000 Natl. Network: ESPN Radio. Format: Sports talk. Target aud: 25-54; male sports fans. ♦Craig Karmazin, CEO & pres; Sam Pines, stn mgr; Jason Gibbs, prom dir.

WKRI(FM)—See Cleveland Heights

WMJI(FM)— Dec 6, 1948: 105.7 mhz; 27 kw. 900 ft TL: N41 23 09 W81 41 23. (CP: 15.5 kw, ant 1,020 ft.). Stereo. Hrs opn: 24 6200 Oak Tree Blvd., 4th Fl., 44131-2510. Phone: (216) 520-2600. Fax: (216) 524-3200. Web Site: www.wmji.com. Licensee: Citicasters Licenses L.P. Group owner: Clear Channel Communications Inc. (acq 5-4-99; grpsl). Natl. Network: AP Radio. Natl. Rep: Christal. Wire Svc: UPI Format: Oldies. News staff: 4; News: 5 hrs wkly. Target aud: 25-54. ♦Mike Kenney, gen mgr; Kevin Metheny, opns dir & progmg dir; Roger Moorman, gen sls mgr.

WMMS(FM)— Nov 11, 1948: 100.7 mhz; 34 kw. 600 ft TL: N41 21 30 W81 40 03. Stereo. Hrs opn: 6200 Oak Tree Blvd., Fourth Floor, 44131. Phone: (216) 520-2600. Fax: (216) 901-8166. E-mail: buzzard@wmms.com Web Site: www.wmms.com. Licensee: Citicasters Licenses L.P. Group owner: Clear Channel Communications Inc. (acq 1999; grpsl). Population served: 1,500,000 Format: Rock. Target aud: 18-34. ♦Bo Matthews, opns mgr & mktg dir; Keith Hotchkiss, gen sls mgr & prom dir; Mike Kenney, gen mgr & gen sls mgr.

WMVX(FM)—Listing follows WTAM(AM).

WNCX(FM)— Oct 23, 1948: 98.5 mhz; 16 kw. 960 ft TL: N41 20 28 W81 44 29. Stereo. Hrs opn: 24 1041 Huron Rd., 44115. Phone: (216) 861-0100. Fax: (216) 696-0385. E-mail: wncx@wncx.com Web Site: www.wncx.com. Licensee: Infinity Radio License Inc. Group owner: Infinity Broadcasting Corp. Population served: 1,800,000 Natl. Network: ABC, Westwood One. Law Firm: Leventhal, Senter & Lerman, PLLC. Format: Classic rock. Target aud: 25-54; adults. ♦Tom Herschel, VP & gen mgr; Linda Rodriguez, gen sls mgr; George Cohn, natl sls mgr; Marshall Goudy, mktg dir, prom dir & progmg dir.

WQAL(FM)— 1948: 104.1 mhz; 11 kw. 1,060 ft TL: N41 22 45 W81 43 12. Stereo. Hrs opn: 24 1 Radio Lane, 44114. Phone: (216) 696-0123. Fax: (216) 363-7199. Web Site: www.q104.com. Licensee: Infinity Radio Inc. Group owner: Infinity Broadcasting Corp. (acq 12-14-00; grpsl). Population served: 750,903 Law Firm: Wiley, Rein & Fielding. Format: Hot adult contemp. News staff: one. Target aud: 25-49; women. ♦Chris Maduri, gen mgr.

***WRUW-FM**— Feb 26, 1967: 91.1 mhz; 15 kw. 292 ft TL: N41 31 14 W81 35 03. (CP: 14.5 kw, ant 276 ft.). Stereo. Hrs opn: 24 11220 Bellflower Rd., 44106. Phone: (216) 368-2207. Phone: (216) 368-2208. Fax: (216) 368-5414. E-mail: gm@wruw.org Web Site: www.wruw.org. Licensee: Case Western Reserve University. Population served: 500,000 Format: Free-form. Target aud: General; Cleveland & CWRU community. ♦Micah Waldstein, gen mgr; Peter McCall, stn mgr.

WTAM(AM)— 1923: 1100 khz; 50 kw-U. TL: N41 16 50 W81 37 22. Hrs open: 24 6200 Oak Tree Blvd., 4th Fl., 44131-2510. Phone: (216) 520-2600. Fax: (216) 901-8152 (progmg). Web Site: www.wtam.com. Licensee: Clear Channel Broadcasting Inc. Group owner: Clear Channel Communications Inc. (acq 5-4-99; grpsl). Population served: 750,903 Law Firm: Hogan & Hartson. Format: News/talk, sports. News staff: 14; News: 25 hrs wkly. Target aud: 25-54. ♦Jim Meltzer, VP & gen mgr; Kevin Metheny, opns mgr; Gary Mincer, sls dir; Dave Ianni, gen sls mgr; Gaye Ramstrom, natl sls mgr; Jeff Zukauckas, prom dir; Ray Davis, progmg dir; R.C. Bauer, news dir; Cheryl Zivich, pub affrs dir; Dave Szucs, engrg dir & chief of engrg; Dawn Lesiak, traf mgr; Mike Snyder, sports cmtr.

WMVX(FM)—Co-owned with WTAM(AM). May 4, 1960: 106.5 mhz; 11.3 kw. 1,036 ft TL: N41 22 45 W81 43 12. Stereo. 24 Fax: (216)

520-3008. Web Site: www.wmvx.com.2,200,000 Format: Adult contemp. ♦Mike Kenney, stn mgr; Dawn Lesiak, traf mgr; Dave Snyder, sports cmtr.

WWGK(AM)— 1947: 1540 khz; 1 kw-D. TL: N41 30 10 W81 37 57. Hrs open: 8000 Euclid Ave., 44103. Phone: (216) 229-7400. Fax: (216) 707-9025. E-mail: info@espncleveland.com Web Site: www.espncleveland.com. Licensee: Good Karma Broadcasting L.L.C. (acq 10-27-2006; $2.5 million). Population served: 2,100,000 Natl. Network: ESPN Radio, Fox Sports, Premiere Radio Networks. Format: Sports. Target aud: 25-54; males. ♦Craig Karmazin, CEO & pres; Sam Pines, stn mgr; Jason Gibbs, prom dir.

WWMK(AM)— Apr 3, 1950: 1260 khz; 10 kw-D, 5 kw-N. TL: N41 17 10 W81 38 34. Stereo. Hrs opn: 24 175 Kenmar Industrial Pkwy., Broadview Heights, 44147. Phone: (440) 746-1010. Fax: (440) 746-1720. Web Site: www.radiodisney.com. Licensee: Radio Disney Group LLC. Group owner: ABC Inc. (acq 8-26-98; $3.9 million). Natl. Network: USA. Law Firm: Haley, Bader & Potts. Format: Family Hits. Target aud: 3-12, women 21-44; children & families. ♦John Gucan, gen mgr; Jeniffer Hansen, mktg dir; Lindsey Keller, gen sls mgr & prom dir.

WZAK(FM)— May 26, 1963: 93.1 mhz; 27.5 kw. 620 ft TL: N41 16 50 W81 37 22. Stereo. Hrs opn: 24 2510 St. Clair Ave., 44114. Phone: (216) 579-1111. Fax: (216) 771-4164. E-mail: cforgy@radio-one.com Web Site: www.931wzak.com. Licensee: Radio One Licenses LLC. Group owner: Radio One Inc. (acq 11-8-01; grpsl). Population served: 750,903 Natl. Rep: Christal. Format: Urban adult contemp. News staff: one. Target aud: 18-49; Black adults. ♦Chris Forgg, gen mgr; Kim Johnson, opns mgr & mus dir; Larry Gawthrop, gen sls mgr.

Cleveland Heights

WERE(AM)— 1947: 1490 khz; 1 kw-U. TL: N41 30 48 W81 36 05. Hrs open: 24 2510 St. Clair Ave., Cleveland, 44114. Phone: (216) 579-1111. Fax: (216) 771-4164. E-mail: cforgy@radio-one.com Licensee: Radio One Licenses LLC. Group owner: Radio One Inc. (acq 8-7-2000; grpsl). Population served: 1,737,300 Natl. Rep: Christal. Format: News/talk. News staff: 4; News: 30 hrs wkly. ♦Cathy Hughes, CEO; Chris Forgy, VP & gen mgr; Kim Johnson, opns mgr.

WKRI(FM)— Nov 23, 1960: 92.3 mhz; 40 kw. Ant 548 ft TL: N41 30 01 W81 33 59. Stereo. Hrs opn: 24 1041 Huron Rd., Cleveland, 44115. Phone: (216) 861-0100. Fax: (216) 696-3710. Web Site: www.krockcleveland.com. Licensee: CBS Radio Stations Inc. Group owner: Infinity Broadcasting Corp. (acq 12-14-2000; grpsl). Population served: 1,800,000 Natl. Network: ABC. Law Firm: Leventhal, Senter & Lerman, PLLC. Format: Alternative rock. Target aud: 18-34; mass appeal, young adults. ♦Tom Herschel, VP & gen mgr; Jeff Miller, gen sls mgr; George Cohn, natl sls mgr & progmg dir; Marshall Goudy, mktg dir & prom dir.

Clyde

***WHVT(FM)**— December 1986: 90.5 mhz; 2.7 w. 154 ft TL: N41 17 45 W82 58 26. (CP: 2.6 kw, ant 154 ft.). Stereo. Hrs opn: Box 273, 43410. Phone: (419) 547-8254. Fax: (419) 547-7195. E-mail: radio@whvtfm.com Web Site: www.whvtfm.com. Licensee: Clyde Educ. Broadcasting Foundation. Format: Educ, relg. ♦James Lewis, pres & gen mgr.

WMJK(FM)— July 16, 1981: 100.9 mhz; 3 kw. 134 ft TL: N41 26 28 W82 41 14. Stereo. Hrs opn: 24 1640 Cleveland Rd., Sandusky, 44870-4357. Phone: (419) 625-3380. Fax: (419) 625-1348. E-mail: paulmize@clearchannel.com Web Site: www.coast1009.com. Licensee: Citicasters Licenses L.P. Group owner: Clear Channel Communications Inc. (acq 5-4-99; grpsl). Population served: 500000 Natl. Network: ABC. Natl. Rep: Katz Radio. Rgnl rep: Rgnl Reps. Format: Class. News staff: 2; News: 4 hrs wkly. Target aud: 25-54; Males. ♦Lisa J. Rich, gen mgr; Randy Hugg, opns mgr & progmg dir; Todd Lewis, gen sls mgr; Steve Shoffner, news dir; Gary Homza, chief of engrg.

WOHF(FM)—See Bellevue

Coal Grove

WBVB(FM)— Feb 1, 1990: 97.1 mhz; 3 kw. 472 ft TL: N38 25 27 W82 32 04. Hrs open: Box 2288, Huntington, WV, 25724. Secondary address: 134 4th Ave. 25701. Phone: (304) 525-7788. Fax: (304) 525-6281. Fax: (304) 525-7861 (Sales). E-mail: info@B97fm.com Web Site: www.B97fm.com. Licensee: Capstar TX L.P. Group owner: Clear Channel Communications Inc. (acq 8-30-00; grpsl). Format: Oldies. Target aud: 18-34. Spec prog: Winston Cup racing. ♦Judy Jennings, gen mgr; Gloria Ward, sls dir; Mark Wood, progmg dir.

Columbus

WBNS(AM)— 1922: 1460 khz; 5 kw-D, 1 kw-N, DA-N. TL: N39 57 06 W82 54 23. Hrs open: 24 605 S. Front St., 43215. Phone: (614) 460-3850. Fax: (614) 460-3757. Web Site: www.1460thefan.com. Licensee: RadiOhio Inc. Group owner: Dispatch Broadcast Group (acq 1933). Population served: 545,000 Natl. Network: ESPN Radio. Rgnl. Network: Ohio News Net. Natl. Rep: Christal. Wire Svc: AP Format: Sports. Target aud: Men 25-54. ♦Dave VanStone, gen mgr; Tom Bunyard, sls dir; Mike Kearney, gen sls mgr; Lorene Gillman, natl sls mgr; Jimmy Powers, progmg dir; Steve Clawson, chief of engrg.

WBNS-FM— June 1959: 97.1 mhz; 20.5 kw. Ant 781 ft TL: N39 58 16 W83 01 40. Stereo. Web Site: www.mix971.net. Natl. Rep: Christal. Format: Adult contemp. Target aud: Adults 25-54. ♦Dave VanStone, VP & gen sls mgr; Marc Herbert, gen sls mgr; Todd Reigle, prom mgr; Jay Taylor, progmg dir. Co-owned TV: WBNS-TV affil.

WBZX(FM)— Apr 26, 1962: 99.7 mhz; 20 kw. 784 ft TL: N39 58 16 W83 01 40. Stereo. Hrs opn: 24 1458 Dublin Rd., 43215. Phone: (614) 481-7800. Fax: (614) 481-8070. E-mail: mail@wbzx.com Web Site: www.wbzx.com. Licensee: North American Broadcasting Co. Inc. (group owner) Population served: 1,200,000 Natl. Rep: D & R Radio. Wire Svc: AP Format: Rock. News staff: 5. Target aud: 18-49. ♦Matthew Mnich, CEO & pres; Norma J. Mnich, chmn; Mark E. Jividen, VP & gen mgr; Jim Pontius, sls dir; Eric Feucht, gen sls mgr; Greg Moebius, prom dir; Hal Fish, progmg dir; Ronni Hunter, mus dir; Mark Nuce, news dir & pub affrs dir; Bill Bowin, engrg mgr.

WMNI(AM)—Co-owned with WBZX(FM). Apr 26, 1958: 920 khz; 1 kw-D, 500 w-N, DA-2. TL: N39 53 32 W83 02 51. Stereo. 24 E-mail: mail@wmni.com Web Site: www.wmni.com. Natl. Network: AP Network News. Format: Adult standards. News staff: 5; News: 20 hrs wkly. Target aud: 35 plus. Spec prog: Relg 4 hrs wkly.

***WCBE(FM)**— Sept 26, 1956: 90.5 mhz; 11 kw. 531 ft TL: N39 57 48 W83 00 17. Stereo. Hrs opn: 24 540 Jack Gibbs Blvd., 43215. Phone: (614) 365-5555. Fax: (614) 365-5060. E-mail: wcbe@wcbe.org Web Site: www.wcbe.org. Licensee: Board of Education, City School District of Columbus, Ohio. Natl. Network: NPR, PRI. Law Firm: Ernest Sanchez. Wire Svc: AP Format: Var, news. News staff: 3; News: 37 hrs wkly. Target aud: 35-54. Spec prog: Jazz 4 hrs, blues 3 hrs, Celtic 4 hrs wkly. ♦Dan Mushalko, gen mgr & opns dir; Wendy Craven, dev dir; Maggie Brennan, mus dir; Jim Letizia, news dir & pub affrs dir.

WCKX(FM)— February 1996: 107.5 mhz; 1.9 kw. 413 ft TL: N39 57 46 W82 59 46. Hrs opn: 24 350 E. 1st Ave., Suite 100, 43201. Phone: (614) 487-1444. Fax: (614) 487-5862. Web Site: www.power1075.com. Licensee: Blue Chip Broadcasting Licenses Ltd. Group owner: Radio One Inc. (acq 4-30-01; grpsl). Natl. Network: ABC. Natl. Rep: D & R Radio. Format: Adult urban contemp. News: one hr wkly. ♦Jeff Wilson, gen mgr.

WCOL-FM— 1947: 92.3 mhz; 22 kw. 754 ft TL: N39 58 16 W83 01 40. Stereo. Hrs opn: 24 2323 W. Fifth Ave., Suite 200, 43204. Phone: (614) 486-6101. Fax: (614) 487-2554. Web Site: www.wcol.com. Licensee: Citicasters Licenses L.P. Group owner: Clear Channel Communications Inc. (acq 5-4-99; grpsl). Format: Country. News staff: one. Target aud: 18-49. ♦Tom Thon, gen mgr.

WYTS(AM)—Co-owned with WCOL-FM. 1922: 1230 khz; 1 kw-U. TL: N39 56 31 W83 01 20. Stereo. Phone: (614) 487-2559. Web Site: www.progressive1230.com. Format: Talk. Target aud: 35 plus; general. ♦Jeff Rehl, gen sls mgr; Steve Konrad, progmg dir; Dave Isaacs, news dir; Sis Campbell, traf mgr.

***WHKC(FM)**— 2006: 91.5 mhz; 15 kw. Ant 689 ft TL: N39 56 14 W83 01 16. Hrs open: 1630 Strathshire Hall Pl., Powell, 43065. Phone: (614) 433-0433. Licensee: Christian Broadcasting Services Inc. ♦Robert G. Casagrande, pres & gen mgr.

WJYD(FM)—See London

WLVQ(FM)— Apr 1, 1959: 96.3 mhz; 40 kw. 550 ft TL: N39 58 16 W83 01 40. Stereo. Hrs opn: 24 10th Fl., 280 Plaza N. High St., 43215. Phone: (614) 227-9696. Fax: (614) 461-1059. Web Site: www.qfm96.com. Licensee: Wilks License Co.-Columbus LLC. Group owner:Infinity Broadcasting Corp. (acq 1-10-2007; grpsl). Population served: 1,500,000 Natl. Network: Westwood One. Natl. Rep: Katz Radio. Law Firm: Leventhal, Senter & Lerman. Format: Classic Rock. News staff: one; News: 2 hrs wkly. Target aud: 25-54. ♦Valerie Brooks, VP & gen mgr; David Cooper, opns mgr & prom mgr; Robert Golias, gen sls mgr; Jill McCarron, rgnl sls mgr; Dave Redelberger, mktg dir.

WNCI(FM)— July 1961: 97.9 mhz; 175 kw. 560 ft TL: N39 58 10 W83 00 10. Stereo. Hrs opn: 24 2323 W. 5th Ave., Suite 200, 43204. Phone: (614) 486-6101. Fax: (614) 487-3553. Web Site: www.wnci.com. Licensee: Citicasters Licenses L.P. Group owner: Clear Channel Communications Inc. (acq 1999; grpsl). Population served: 1,215,700 Law Firm: Holland & Knight. Format: Adult contemp. News staff: one; News: 4 hrs wkly. Target aud: 18-49. ♦Tom Thon, gen mgr.

***WOSU(AM)**— Apr 24, 1922: 820 khz; 5 kw-D, 790 w-N (L-WBAP Ft. Worth, Tex.). TL: N40 01 44 W82 03 22. (CP: 1 kw-N). Hrs opn: 24 2400 Olentangy River Rd., 43210. Phone: (614) 292-9678. Fax: (614) 292-0513. Web Site: www.wosu.org. Licensee: Ohio State University. Population served: 1,200,000 Natl. Network: NPR, PRI. Rgnl. Network: Ohio Radio Net. Law Firm: Dow, Lohnes & Albertson. Wire Svc: AP Format: Pub affrs, news/talk. News staff: 9; News: 114 hrs wkly. Target aud: 35 plus; general. Spec prog: Black one hr, bluegrass 12 hrs wkly. ♦Thomas Rieland, gen mgr; Tim Eby, stn mgr.

WOSU-FM— Dec 13, 1949: 89.7 mhz; 13.3 kw. 938 ft TL: N39 56 16 W83 01 16. Stereo. 24 Web Site: www.wosu.org.1,000,000 Natl. Network: PRI. Format: Class music. Target aud: 25 plus. Co-owned TV: *WOSU-TV affil.

WRFD(AM)—Columbus-Worthington, Sept 27, 1947: 880 khz; 23 kw-D. TL: N39 56 31 W83 01 20. Hrs open: Sunrise-sunset 8101 N. High St., Suite 360, 43235-1406. Phone: (614) 885-0880. Fax: (614) 885-6322. E-mail: mail@wrfd.com Web Site: www.wrfd.com. Licensee: Salem Media of Ohio Inc. Group owner: Salem Communications Corp. (acq 2-1-82; $1.8 million; FTR: 12-21-81). Population served: 6,461,184 Natl. Network: Salem Radio Network. Natl. Rep: Christal, Salem. Format: Relg, farm. News staff: 0; News: 1 hrs wkly. Target aud: 30-60; conservatives, Christians, farmers. ♦Edward Atsinger III, pres; David Ruleman, VP; Dan Craig, gen mgr; Ryan Moran, opns mgr & progmg dir; Tom Heyl, gen sls mgr; Greg Sauold, engrg dir.

WSNY(FM)— Aug 12, 1982: 94.7 mhz; 22 kw. 753 ft TL: N39 58 16 W83 01 40. Stereo. Hrs opn: 4401 Carriage Hill Ln., 43220. Phone: (614) 451-2191. Fax: (614) 451-1831. Web Site: www.sunny95.com. Licensee: Franklin Communications Inc. Group owner: Saga Communications Inc. (acq 9-86). Population served: 1,382,200 Natl. Rep: Christal, Katz Radio. Wire Svc: AP Format: Adult contemp. News staff: one; News: 3 hrs wkly. Target aud: 25-64; women, upscale families. ♦Alan Goodman, pres, VP & gen mgr; Chuck Knight, opns dir; Chris Forgy, sls dir & news dir; Katie Cyr, gen sls mgr & natl sls mgr; Michelle Hurley, mktg dir.

WTVN(AM)— 1924: 610 khz; 5 kw-U, DA-N. TL: N39 52 26 W82 58 36. Hrs open: 24 2323 W. 5th Ave., Suite 200, 43204. Phone: (614) 486-6101. Fax: (614) 487-2559. E-mail: tomthorn@clearchannel.com Web Site: www.610wtvn.com. Licensee: Citicasters Licenses L.P. Group owner: Clear Channel Communications Inc. (acq 1999; grpsl). Population served: 1,500,000 Law Firm: Koteen & Naftalin. Format: Talk. Target aud: 25-54; leaning male. ♦Tom Thon, VP; Jeff Rehl, gen sls mgr; Steve Konrad, progmg dir.

***WUFM(FM)**— Mar 22, 1996: 88.7 mhz; 5 kw. Ant 774 ft TL: N39 56 16 W83 01 16. Stereo. Hrs opn: 24 Box 1887, Westerville, 43086-1887. Secondary address: 116 County Line Rd., Westerville 43082. Phone: (614) 839-7100. Fax: (614) 839-1329. E-mail: radiou@radiou.com Web Site: www.radiou.com.GE-1 (ku) Licensee: Spirit Communications Inc. (acq 9-27-96; $95,000). Population served: 1,600,000 Law Firm: Gammon & Grange. Format: Contemporary hit/top-40, rock, progressive. Target aud: 12-24; Male. ♦John P. Shumate Sr., pres; Kathy Shumate, VP; Michael Buckingham, gen mgr; Cole Drake, prom dir; Nikki Cantu, progmg dir & mus dir.

WVKO(AM)— Nov 21, 1951: Stn currently dark. 1580 khz; 1 kw-D, 250 w-N, DA-2. TL: N40 02 50 W83 03 44. Hrs open: 24 74 S. 4th St., 43215. Phone: (614) 469-1930. Fax: (614) 573-8401. Web Site: www.1580wvko.com. Licensee: Stop 26 Riverbend Licenses LLC (acq 12-30-2002). Natl. Rep: McGavren Guild. Format: Gospel. ♦Bill Cusack, gen mgr.

WVKO-FM—Johnstown, June 16, 1975: 103.1 mhz; 1.6 kw. Ant 443 ft TL: N40 13 44 W82 39 35.8. Stereo. Hrs opn: 74 S. 4th St., 43215. Phone: (614) 821-1031. Fax: (614) 821-0002. Licensee: Stop 26 Riverbend Licenses LLC. Natl. Rep: D & R Radio. Format: Sp/Mexicana. Target aud: 18-54; upscale, professional; homeowners with disposable incomes. Spec prog: Jazz 10 hrs, relg 13 hrs, reggae 5 hrs wkly. ♦Bill Cusack, gen mgr; Scott Wooten, gen mgr & opns mgr.

Columbus Grove

WLWD(FM)— 2003: 93.9 mhz; 14 kw. Ant 436 ft TL: N40 57 21 W84 07 59. Hrs open: Box 1128, Lima, 45802-1128. Phone: (419) 223-2060. Fax: (419) 229-3888. E-mail: phil@wild939.com Web Site: www.wild939.com. Licensee: CC Licenses LLC. Group owner: Clear Channel Communications Inc. (acq 8-10-2000). Format: CHR. News: one hr wkly. ♦Bill Gentry, sr VP; Art Versnick, VP; Phil Austin, opns mgr; Aaron Matthews, progmg dir; Mark Gierhart, chief of engrg.

Columbus-Worthington

WRFD(AM)—Licensed to Columbus-Worthington. See Columbus

Conneaut

***WGOJ(FM)**— Apr 5, 1964: 105.5 mhz; 6 kw. 295 ft TL: N41 51 42 W80 31 01. Stereo. Hrs opn: 24 Box 725, 44030. Secondary address: 236 State St. 44030. Phone: (440) 599-7252. Phone: (440) 593-1127. Fax: (440) 593-4761. E-mail: wgoj@suite224.net Licensee: Developing Radio LLC (acq 1-28-2004; $750,000). Population served: 600,000 Natl. Network: Bible Bcstg Net. Format: Christian. Target aud: General. ♦Dr. Roger P. Hogle, gen mgr; Robert Jackson, progmg dir.

WWOW(AM)— Oct 25, 1959: 1360 khz; 5 kw-D, 35 w-N. TL: N41 55 32 W80 32 32. Hrs open: 24 229 Broad St., 44030. Phone: (440) 593-2233. Fax: (440) 593-6885. E-mail: mlandon@1360wwow.com Web Site: www.1360wwow.com. Licensee: Cause Plus Marketing LLC (acq 1-31-2007; $200,000). Population served: 14,552 Natl. Network: EWTN Radio, Fox News Radio. Format: Catholic, Talk. Spec prog: Live local morning show M-F. ♦John Marra, pres; Marty Landon, opns mgr; Gary Gersin, progmg dir; Pat Williams, news dir.

Cortland

WKTX(AM)— Apr 1, 1985: 830 khz; 1 kw-D. TL: N41 24 56 W80 43 49. Hrs open: 11906 Madison Ave., Lakewood, 44107. Phone: (216) 221-0330. Fax: (216) 221-3638. Licensee: Miklos Kossanyi, Maria Kossanyi (acq 10-91). Natl. Network: USA. Format: Variety, ethnic, polka. Target aud: 35 plus; homeowners. Spec prog: Slovenian 2 hrs, Greek 2 hrs, Pol one hr, German 5 hrs wkly. ♦Miklos Kossanyi, pres; Maria Kossanyi, VP; Jim Georgiades, opns dir & chief of opns; Jack Cory, progmg dir.

Coshocton

***WOSE(FM)**— 1996: 91.1 mhz; 6 kw. Ant 321 ft TL: N40 20 30 W81 57 56. Stereo. Hrs opn:
Rebroadcasts WOSU-FM Columbus 100%.
2400 Olentangy River Rd., Columbus, 43210. Phone: (614) 292-9678. Fax: (614) 292-0513. E-mail: radio@wosu.org Web Site: www.wosu.org. Licensee: The Ohio State University. Natl. Network: PRI, NPR. Law Firm: Dow, Lohnes & Albertson. Wire Svc: AP Format: Class, news/talk. ♦Thomas Rieland, gen mgr; Tim Elby, stn mgr; Mary Alice Akins, opns mgr.

WTNS(AM)— Nov 9, 1947: 1560 khz; 1 kw-D. TL: N40 16 30 W81 49 37. Hrs open: 114 N. 6th St., 43812. Phone: (740) 622-1560. Fax: (740) 622-7940. Licensee: Coshocton Broadcasting Co. (group owner; acq 9-86; $560,653; FTR: 9-22-86). Population served: 13,747 Format: Country. ♦Bruce Wallace, pres & gen mgr; Tom Thompson, gen sls mgr; Mike Bechtol, mus dir; Ken Smailes, news dir; Jay Drummond, chief of engrg.

WTNS-FM— Apr 25, 1968: 99.3 mhz; 1.2 kw. 440 ft TL: N40 16 30 W81 49 37.13,747 Format: Adult contemp. ♦Flo Murdock, women's int ed & disc jockey; Brad Haynes, disc jockey; Jim Parr, disc jockey; Tom Thompson, disc jockey.

Covington

WPTW(AM)—See Piqua

Crestline

WYKL(FM)—Licensed to Crestline. See Mansfield

Crooksville

WYBZ(FM)— Oct 26, 1990: 107.3 mhz; 3 kw. 328 ft TL: N39 47 23 W82 05 39. (CP: Ant 302 ft.). Stereo. Hrs opn: 24 Box 669, 2895 A Maysville Pike, Zanesville, 43702-0669. Phone: (740) 453-6004. Fax: (740) 453-5865. E-mail: wybz@rrohio.com Web Site: www.wybz.com. Licensee: Y Bridge Broadcasting Inc. (acq 12-26-90; $60,000; FTR: 1-14-91). Natl. Network: CNN Radio. Law Firm: Smithwick & Belendiuk. Format: Oldies. News staff: one; News: 9 hrs wkly. Target aud: 25-55. ♦Michael Jaye, opns mgr & prom dir; Monica Martinelli, sls VP; Rick Sabine, pres, gen mgr, gen sls mgr & progmg dir; Mark Hines, chief of engrg.

Cuyahoga Falls

WAKS(FM)—See Akron

***WCUE(AM)**— 1950: 1150 khz; 5 kw-U, DA-2. TL: N41 12 05 W81 31 25. Hrs open: 24 13 Fairlane Dr., Joliet, IL, 60435. Secondary address: 4075 Bellaire Ln., Peninsula 44264. Phone: (815) 725-1331. Web Site: www.familyradio.com. Licensee: Family Stations Inc. (group owner; acq 10-22-86). Population served: 275,425 Law Firm: Dow, Lohnes & Albertson. Format: Relg. News: 4 hrs wkly. Target aud: 25 plus; Christians. Spec prog: Class 2 hrs wkly. ♦Harold Camping, pres & gen mgr.

WQAL(FM)—See Cleveland

Dayton

WDAO(AM)— Mar 1, 1955: 1210 khz; 1 kw-D. TL: N39 43 36 W84 12 23. Hrs open: 1012 West 3rd St., 45407. Phone: (937) 222-9326. Fax: (937) 461-6100. E-mail: wdaoamizo@aol.com Licensee: Johnson Communications Inc. (acq 1-88; $725,000; FTR: 1-18-88). Population served: 2,436,010 Natl. Rep: Christal. Format: Rhythm and blues. ♦Jim Johnson, VP & gen mgr; Sophia Carr, gen sls mgr; Jim Johnston, mus dir.

WDPR(FM)—See West Carrollton

***WDPS(FM)**— 1976: 89.5 mhz; 6 kw. 198 ft TL: N39 45 28 W84 11 36. Hrs open: 9:15 AM-4:30 PM 441 River Corridor Dr., 45402. Phone: (937) 542-7182. Fax: (937) 542-6714. Licensee: Dayton Public Schools. (acq 1976). Population served: 100,000 Format: Jazz, AAA. ♦Michael Reisz, gen mgr, mus dir & news dir; Christopher Hartley, opns dir; Jennifer Bryant, asst music dir; Tom Nornhold, chief of engrg.

WFCJ(FM)—See Miamisburg

WGTZ(FM)—See Eaton

WHIO(AM)— Feb 9, 1935: 1290 khz; 5 kw-U, DA-N. TL: N39 40 41 W84 07 53. Hrs open:
Simulcast with WHIO-FM Piqua 100%.
Box 1206, 45401. Secondary address: 1414 Wilmington Ave. 45420. Phone: (937) 259-2111. Fax: (937) 259-2168. Fax: (937) 259-2024. Web Site: 1290whio.com. Licensee: Cox Radio Inc. Group owner: Cox Broadcasting Population served: 823,100 Natl. Rep: D & R Radio. Law Firm: Dow, Lohnes & Albertson. Format: Full service, news/talk. News staff: 4; News: 30 hrs wkly. Target aud: 35-54. Spec prog: Relg 2 hrs wkly. ♦Donna Hall, VP & gen mgr; Lisa Allan, gen sls mgr; Marc Herbst, natl sls mgr; Kathy Eagle-Norris, rgnl sls mgr; Vicky Forrest, mktg dir; Tracey Slife, prom mgr; Larry Hansgen, progmg dir; Jim Barrett, news dir & pub affrs dir; Ron Gaier, chief of engrg. Co-owned TV: WHIO-TV affil.

WHKO(FM)—Co-owned with WHIO(AM). 1946: 99.1 mhz; 50 kw. 1,066 ft TL: N39 44 02 W84 14 52. Stereo. 838,400 Natl. Rep: Christal. Format: Country. ♦Nick Roberts, opns mgr, progmg dir & mus dir. Co-owned TV: WHIO-TV affil.

WING(AM)— May 24, 1921: 1410 khz; 5 kw-U, DA-N. TL: N39 40 56 W84 09 33. Hrs open: 717 E. David Rd., 45429. Phone: (937) 294-5858. Fax: (937) 297-5233. Web Site: www.wingam.com. Licensee: Blue Chip Broadcasting Licenses Ltd. Group owner: Radio One Inc. (acq 8-7-01; grpsl). Population served: 4,100 Natl. Network: CBS, Westwood One. Natl. Rep: McGavren Guild. Format: News/talk, sports. Target aud: 25-54; well educated. ♦Don Griffin, VP & gen mgr.

WLQT(FM)—See Kettering

WMMX(FM)— September 1964: 107.7 mhz; 50 kw. 420 ft TL: N39 43 36 W84 12 23. Stereo. Hrs opn: 24 101 Pine St., 45402. Phone: (937) 224-1137. Fax: (937) 224-7655. Web Site: www.wmmx.com. Licensee: Citicasters Licenses L.P. Group owner: Clear Channel Communications Inc. (acq 1999; grpsl). Format: Adult contemp. Target aud: 25-54. ♦Karrie Sudbrack, gen mgr; Jeff Stevens, progmg dir.

WONE(AM)— Mar 20, 1949: 980 khz; 5 kw-U, DA-N. TL: N39 40 03 W84 10 01. Stereo. Hrs opn: 24 101 Pine St., 45402. Phone: (937) 224-1137. Fax: (937) 224-7665. Web Site: www.wone.com. Licensee: Citicasters Licenses L.P. Group owner: Clear Channel Communications Inc. (acq 5-4-99; grpsl). Population served: 115,300 Format: Sports. News staff: 3; News: 18 hrs wkly. Target aud: 35-64. ♦Rick Porter, VP; Mary Fleenor, opns mgr.

***WQRP(FM)**— 1976: 89.5 mhz; 6 kw. 270 ft TL: N39 45 26 W84 12 24. Stereo. Hrs opn: 4:30 pm-9:15 am 917 East Central Ave., West Carrollton, 45449. Phone: (937) 865-5900. Fax: (937) 865-0041. E-mail: radio@praise895.com Web Site: praise895.com. Licensee: WQRP Family Radio Inc. (acq 5-1-00). Format: Praise & worship. News staff: one. Target aud: 25-45. Spec prog: Ger 3 hrs, Hungarian 3 hrs wkly. ♦Joe Laber, gen mgr; Rex Wood, pres & dev dir.

WTUE(FM)— 1959: 104.7 mhz; 50 kw. 499 ft TL: N39 43 19 W84 12 36. Stereo. Hrs opn: 24 101 Pine St., 45402. Phone: (937) 224-1137. Fax: (937) 224-7655. E-mail: wtue@wtue.com Web Site: www.wtue.com. Licensee: Citicasters Licenses L.P. Group owner: Clear Channel Communications Inc. Population served: 263,000 Format: AOR. News staff: one; News: 2 hrs wkly. Target aud: 18-49. ♦Karrie Sudbrack, gen mgr; Tony Tilford, opns mgr; Mike Finney, gen sls mgr.

***WUDR(FM)**— 2003: 98.1 mhz; 13 w. Ant 90 ft TL: N39 47 14 W84 14 23. Hrs open: University of Dayton, 300 College Park, 45469-1679. Phone: (937) 229-3058. Web Site: flyer-radio.udayton.edu. Licensee: University of Dayton. Format: Var. ♦Greg Hansberry, gen mgr.

***WWSU(FM)**— Apr 4, 1977: 106.9 mhz; 10 w. 150 ft TL: N39 46 57 W84 03 43. Stereo. Hrs opn: 24 Wright State University, 45435. Phone: (937) 775-5554. Phone: (937) 775-5555. Fax: (937) 775-5553. Web Site: www.wright.edu. Licensee: Wright State University. Population served: 60,000 Format: Various/diverse. Target aud: 15-26; college & high school students. Spec prog: Black 12 hrs, relg 11 hrs, gospel 3 hrs, jazz 3 hrs, Sp 3 hrs wkly. ♦Rod Hissong, gen mgr; Matt Hughes, progmg dir; Annie Hall, news dir.

De Graff

***WDEQ-FM**— Sept 1, 1967: 103.3 mhz; 10 w. 23 ft TL: N40 18 48 W83 55 06. Hrs open: 2096 County Rd. 24 S., 43318. Phone: (937) 585-5981. Fax: (937) 585-4599. Web Site: www.riverside.k12.oh.us. Licensee: Riverside Local Board of Education. Format: Educ. ♦Jennifer Thompson, gen mgr.

Defiance

WDFM(FM)— June 25, 1985: 98.1 mhz; 50 kw. 500 ft TL: N41 17 28 W84 32 17. Stereo. Hrs opn: 24 118 Clinton St., 43512. Phone: (419) 782-9336. Fax: (419) 784-0306. Web Site: www.981mix.com. Licensee: Citicasters Licenses L.P. Group owner: Clear Channel Communications Inc. (acq 5-4-99; grpsl). Natl. Network: CNN Radio. Natl. Rep: Katz Radio. Law Firm: Fletcher, Heald & Hildreth. Format: Adult contemp. News staff: one; News: 7 hrs wkly. Target aud: 25-54. Spec prog: Relg 3 hrs wkly. ♦Rick Small, pres & opns dir; Bob McLimans, gen mgr; Russ Ryder, progmg dir.

***WGDE(FM)**— Mar 14, 1999: 91.9 mhz; 6 kw. 305 ft TL: N41 17 41 W84 23 24. Hrs open: 24
Rebroadcasts WGTE-FM Toledo 100%.
1270 S. Detroit, Toledo, 43614. Phone: (419) 380-4600. Fax: (419) 380-4710. Web Site: www.wgte.org. Licensee: Public Broadcast Foundation of NW Ohio. Law Firm: Schwartz, Woods & Miller. Format: Class, pub affrs, news. News: 23 hrs wkly. Target aud: General. Spec prog: Jazz 16 hrs, new age 4 hrs wkly. ♦George Jones, chmn; Marlon P. Kiser, CEO, pres & gen mgr.

WONW(AM)— 1949: 1280 khz; 1 kw-D, 500 w-N, DA-N. TL: N41 16 44 W84 23 50. Hrs open: 24 2110 Radio Dr., 43512. Secondary address: 709 N. Perry St., Napoleon 43545. Phone: (419) 782-8126.

Fax: (419) 784-4154. E-mail: bobmclimans@clearchannel.com Web Site: www.wonw1280.com. Licensee: CC Licenses LLC. Group owner: Clear Channel Communications Inc. (acq 11-5-99; grpsl). Natl. Network: CNN Radio. Rgnl. Network: Agri Bcstg. Natl. Rep: Katz Radio. Wire Svc: AP Format: News/talk, sports. News staff: one. Target aud: General. Spec prog: Rush Limbaugh. ♦Robert E. McLimans, VP & gen mgr; Rick Small, opns dir & opns mgr; John Schuette, gen sls mgr; Rusty Hoops, progmg dir.

WZOM(FM)— Aug 25, 1989: 105.7 mhz; 6 kw. 347 ft TL: N41 13 23 W84 22 36. Stereo. Hrs opn: 24 2110 Radio Dr., 43512. Secondary address: 709 N. Perry St., Napoleon 43512. Phone: (419) 782-8126. Fax: (419) 784-4154. E-mail: 1057thebull@clearchannel.com Web Site: www.1057thebull.com. Licensee: CC Licenses LLC. Group owner: Clear Channel Communications Inc. (acq 1-1-2000; grpsl). Natl. Rep: Katz Radio. Rgnl rep: Rgnl Reps. Format: Country. News staff: one; News: 4 hrs wkly. Target aud: 25-54. Spec prog: Relg 6 hrs wkly. ♦Robert E. McLimans, sr VP, VP & gen mgr; Rick Small, opns dir; Bill Murphy, progmg dir.

Delaware

WDLR(AM)— Jan 18, 1961: 1550 khz; 500 w-D, 29 w-N, DA-2. TL: N40 17 56 W83 02 46. Hrs open: 24 1630 Strathshire Hall, Powell, 43065. Phone: (740) 368-9357. Fax: (740) 369-9463. Licensee: The Fifteen Fifty Corp. (acq 3-21-2006). Population served: 73,000 Format: Sp. ♦Luis Orozco, gen mgr & stn mgr.

***WJJE(FM)**— 2005: 89.1 mhz; 6 kw vert. Ant 328 ft TL: N40 24 01 W82 46 43. Hrs open: Drawer 2440, Tupelo, MS, 38803. Phone: (662) 844-8888. Fax: (662) 842-6791. Licensee: American Family Association. Group owner: American Family Radio (acq 12-15-2003; $10 for CP). Format: Relg. ♦Marvin Sanders, gen mgr.

WODB(FM)— June 21, 1991: 107.9 mhz; 6 kw. Ant 285 ft TL: N40 17 57 W83 02 45. Stereo. Hrs opn: 24 4401 Carriage Hill Ln., Columbus, 43202. Phone: (614) 451-2191. Fax: (614) 451-1831. Web Site: www.b1079.com. Licensee: Franklin Communications Inc. Group owner: Saga Communications Inc. (acq 12-30-2002; $9 million). Population served: 350,000 Natl. Rep: Christal. Wire Svc: AP Format: Classic top-40 hits. News staff: one; News: 10 hrs wkly. Target aud: 25-44. ♦Alan Goodman, pres, gen mgr & gen mgr; Michelle Hurley, mktg dir; Bill Shannon, progmg dir.

***WSLN(FM)**— Apr 28, 1952: 98.7 mhz; 100 w. 105 ft TL: N40 17 46 W84 22 36. Stereo. Hrs opn: Ohio Wesleyan Univ., 61 S. Sandusky St., 43015. Phone: (740) 368-2918. Fax: (740) 368-3649. Web Site: www.wslnowu.edu. Licensee: The Trustees of Ohio Wesleyan University. Population served: 20,000 Format: Var, college. ♦Chris Andrus, stn mgr.

Delhi Hills

***WORI(FM)**— July 1998: 90.1 mhz; 15 kw. Ant 335 ft TL: N39 13 34 W84 42 59. Hrs open: 24 2351 Sunset Blvd., Suite 170-218, Rocklin, CA, 95765. Phone: (916) 251-1600. Fax: (916) 251-1650. E-mail: info@air1.com Web Site: www.air1.com. Licensee: Educational Media Foundation. Group owner: EMF Broadcasting (acq 10-2-2003; grpsl). Natl. Network: Air 1. Law Firm: Shaw Pittman. Format: Contemp Christian. Target aud: 18-35; Judeo Christian, female. ♦Richard Jenkins, pres; Mike Novak, VP; Keith Whipple, dev dir; David Pierce, progmg mgr; Ed Lenane, news dir; Sam Wallington, engrg dir; Karen Johnson, news rptr; Marya Morgan, news rptr; Richard Hunt, news rptr.

Delphos

***WBIE(FM)**— 2001: 91.5 mhz; 5.5 kw. Ant 321 ft TL: N40 56 48 W84 15 24. Hrs open: Drawer 2440, Tupelo, MS, 38803-2440. Phone: (662) 844-8888. Fax: (662) 842-7798. Web Site: www.afr.net. Licensee: American Family Association. Group owner: American Family Radio Format: Classic gospel. ♦Marvin Sanders, gen mgr.

WDOH(FM)— Dec 16, 1972: 107.1 mhz; 3.3 kw. Ant 298 ft TL: N40 49 55 W84 21 11. Stereo. Hrs opn: 1301 N. Cable Rd., Lima, 45805. Phone: (419) 331-1600. Fax: (419) 228-5085. Web Site: www.wdoh.com. Licensee: Maverick Media of Lima License LLC. (acq 11-15-2004; $1.15 million). Population served: 7,608 Natl. Network: CBS. Law Firm: Fletcher, Heald & Hildreth. Format: Lite rock. News staff: one; News: 7 hrs wkly. Target aud: 25 plus. Spec prog: Farm 8 hrs wkly. ♦Gary S. Rozynek, pres; David P. Roach, gen mgr; Deb Klaus, opns dir; Matt Childers, gen sls mgr; Justin Kage, prom dir & mus dir.

Delta

WRWK(FM)— September 1994: 106.5 mhz; 3 kw. 328 ft TL: N41 35 13 W83 54 11. Hrs open: 3225 Arlington Ave., Toledo, 43614. Phone: (419) 725-5700. Fax: (419) 389-5172. Licensee: Cumulus Licensing Corp. Group owner: Cumulus Media Inc. (acq 11-18-99; $4,925,000). Format: Alternative. Target aud: 18-34; male. ♦Kathy Stinehour, gen mgr; Tim Roberts, opns mgr; Larry Scott, gen sls mgr; Chris Ammel, progmg dir.

Dover-New Philadelphia

WJER(AM)— Feb 10, 1950: 1450 khz; 1 kw-U. TL: N40 30 46 W81 27 24. Hrs open: 646 Boulevard, Dover, 44622. Phone: (330) 343-7755. Fax: (330) 364-4538. E-mail: wjer@wjer.com Web Site: www.wjer.com. Licensee: WJER Radio LLC (acq 2-16-2007; $200,000). Natl. Rep: Rgnl Reps. Format: Oldies. ♦Gary Petricola, pres; Bob Scanlon, gen mgr & progmg dir; Dan Pitzo, gen sls mgr; Jennifer Clark, news dir.

East Liverpool

WOGF(FM)—Listing follows WOHI(AM).

WOHI(AM)— Dec 1, 1949: 1490 khz; 1 kw-U. TL: N40 37 47 W80 36 09. (CP: 660 w-U). Hrs opn: Box 2050, 15655 St. Rt. 170, 43920. Phone: (330) 385-1490. Fax: (330) 385-2339. Licensee: Keymarket Licenses LLC. Group owner: Keymarket Communications LLC (acq 2000; grpsl). Population served: 50,000 Natl. Network: Jones Radio Networks. Natl. Rep: Rgnl Reps. Format: MOR. Target aud: General. Spec prog: Big band 4 hrs, gospel one hr wkly. ♦Gerald Getz, pres; Bob Simpson, opns mgr; Ron Aughinbaugh, gen mgr & gen sls mgr; Jim Martin, prom mgr & progmg dir; Amy Atkins, traf mgr.

WOGF(FM)—Co-owned with WOHI(AM). Apr 15, 1959: 104.3 mhz; 50 kw. 330 ft TL: N40 37 48 W80 36 10. (CP: Ant 492 ft.). Stereo. 24 15655 State Rt. 170, 43920. 500000 Format: Country. Target aud: 25-54. ♦Frank Bell, prom mgr & progmg VP; Steve Kline, progmg dir; Bob Simpson, news dir; Amy Atkins, traf mgr; John Rambo, sports cmtr; Rob Pratte, sports cmtr.

Eaton

WEDI(AM)— January 1979: 1130 khz; 250 w-D, DA. TL: N39 44 55 W84 35 02. Hrs open: 486 W. Second St., Xenia, 45385. Phone: (937) 372-3531. Fax: (937) 372-3508. E-mail: jmullins@myclassiccountry.com Web Site: www.myclassiccountry.com. Licensee: Town and Country Broadcasting Inc. (acq 1-4-2005; $175,000). Population served: 825,000 Natl. Network: Fox News Radio. Rgnl. Network: Agri Bcstg. Natl. Rep: Rgnl Reps. Law Firm: Reddy, Begley & McCormick, L.L.P. Wire Svc: AP Format: Classic country. News staff: 2; News: 3 hrs wkly. Target aud: 35-64; adults. Spec prog: Gospel 5 hrs., Farm 2 hrs. wkly. ♦Joe Mullins, pres & gen mgr; Roy Hatfield, progmg dir; Darrin Johnston, news dir; Megan Brugger, traf mgr.

WGTZ(FM)— Nov 28, 1960: 92.9 mhz; 31.6 kw. 600 ft TL: N39 50 10 W84 24 16. Stereo. Hrs opn: 24 717 E. David Rd., Dayton, 45429. Phone: (937) 294-5858. Fax: (937) 297-5233. Web Site: www.wgtz93.com. Licensee: Blue Chip Broadcasting Licenses Ltd. Group owner: Radio One Inc. (acq 4-30-01; grpsl). Population served: 11,300 Format: CHR. Target aud: 18-49; contemp middle America. ♦Don Griffin, VP & gen mgr.

Edgewood

WZOO-FM— Jan 23, 1989: 102.5 mhz; 5.8 kw. Ant 328 ft TL: N41 49 44 W80 49 28. Stereo. Hrs opn: 24 3226 Jefferson Rd., Ashtabula, 44004-9112. Phone: (440) 993-2126. E-mail: danaschulte@clearchannel.com Web Site: www.102zoo.com. Licensee: CC Licenses LLC. Group owner: Clear Channel Communications Inc. (acq 7-11-2000; grpsl). Population served: 100,000 Law Firm: Miller & Miller. Format: Hot adult contemp. News staff: 3. Target aud: General. ♦Dana Schulte, VP & gen mgr; Dennis O'Brien, opns dir.

Elyria

WEOL(AM)— October 1948: 930 khz; 1 kw-U, DA-2. TL: N41 16 10 W82 00 21. Hrs open: 24 Box 4006, 4th Fl., 538 Broad St., 44036. Phone: (440) 322-3761. Fax: (440) 284-3189. Licensee: Elyria-Lorain Broadcasting Co. (group owner) Population served: 275,000 Natl. Network: ABC. Natl. Rep: McGavren Guild. Rgnl rep: Rgnl Reps. Law Firm: Putbrese, Hunsaker & Trent. Wire Svc: AP Format: News/talk, sports. News staff: 4. Target aud: 35 plus. Spec prog: Sp 2 hrs wkly. ♦Gary L. Kneisley, pres & gen mgr.

WNWV(FM)—Co-owned with WEOL(AM). October 1948: 107.3 mhz; 20 kw. Ant 781 ft TL: N41 16 10 W82 00 16. Stereo. Web Site: www.wnwv.com. Format: Smooth jazz. News: 4 hrs wkly. Target aud: 21 plus; upscale.

Englewood

WDKF(FM)— Dec 15, 1993: 94.5 mhz; 6 kw. 328 ft TL: N39 57 17 W84 18 25. Hrs open: 24 101 Pine St., Dayton, 45402. Phone: (937) 224-1137. Fax: (937) 224-7655. Web Site: www.945kissfm.com. Licensee: Citicasters Licenses L.P. Group owner: Clear Channel Communications Inc. (acq 5-4-99; grpsl). Format: Top 40. Target aud: 18-34. ♦Kerri Sudbrack, gen mgr.

Fairborn

WGNZ(AM)— Sept 1, 1968: 1110 khz; 2.5 kw-D, 1.7 kw-CH, DA. TL: N39 41 15 W83 57 55. Hrs open: Box 1100, Dayton, 45405-0879. Phone: (937) 454-9000. Fax: (937) 454-1980. E-mail: wgnz@wgnz.com Web Site: www.wgnz.com. Licensee: L & D Broadcasters Inc. (acq 11-16-01). Population served: 1,500,000 Natl. Network: Salem Radio Network. Law Firm: Miller & Miller, P.C. Format: Relg. Target aud: General; listeners who like family radio. ♦Tim Livingston, pres & gen mgr.

WXEG(FM)—See Beavercreek

Fairfield

WCNW(AM)— Feb 14, 1964: 1560 khz; 5 kw-D, DA. TL: N39 20 20 W84 31 30. Hrs open: 8686 Michael Ln., 45014. Phone: (513) 829-7700. Licensee: Vernon R. Baldwin Inc. (group owner; acq 6-11-84; $700,000; FTR: 3-19-84). Population served: 2,000,000 Format: Southern gospel. ♦Vernon R. Baldwin, pres, CFO & gen mgr; Mark Mitchell, stn mgr.

WSWD(FM)— 1925: 94.9 mhz; 10.5 kw. Ant 1,056 ft TL: N39 12 01 W84 31 22. Stereo. Hrs opn: 24 895 Central Ave., Ste. 900, Cincinnati, 45202. Phone: (513) 241-9898. Fax: (513) 241-6689. E-mail: kmitchell@cincyradio.com Web Site: www.949thesound.com. Licensee: WVAE LICO Inc. Group owner: Susquehanna Radio Corp. (acq 5-5-2006; grpsl). Law Firm: Akin, Gump, Strauss, Hauer & Feld. Format: Modern rock. News staff: one; News: 4 hrs wkly. ♦Gary Lewis, gen mgr; Kyle Simpson, gen sls mgr & chief of engrg; T.J. Holland, prom mgr & progmg dir.

Findlay

WBVI(FM)—See Fostoria

WFIN(AM)— Dec 15, 1941: 1330 khz; 1 kw-D, 79 w-N. TL: N41 00 36 W83 38 04. Stereo. Hrs opn: 24 Box 1507, 45840-1507. Secondary address: 551 Lake Cascades Pkwy. 45840. Phone: (419) 422-4545. Fax: (419) 422-6736. E-mail: wfin@wfin.com Web Site: www.wfin.com. Licensee: Blanchard River Broadcasting Co. Group owner: Findlay Publishing Co. (acq 1949). Population served: 45,000 Rgnl rep: Rgnl Reps. Format: Local news/talk. News staff: 2. Target aud: 45 plus. Spec prog: Farm 7 hrs, sports 12 hrs wkly. ♦Edwin L. Heminger, chmn; Kurt P. Kah, pres; Robert L. Gordon, CFO; David P. Glass, VP; Sandy Kozlevcar, gen mgr & gen sls mgr; Kurt F. Heminger, opns dir; John Marshall, progmg dir; Tom Sheldon, news dir; Dennis Rund, chief of engrg; Vaun Wickerham, farm dir; Chris Miller, sports cmtr.

WKXA-FM—Co-owned with WFIN(AM). 1948: 100.5 mhz; 20 kw. 440 ft TL: N40 55 00 W83 35 45. Stereo. 24 E-mail: wkxa@wkxa.com Web Site: www.wkxa.com. Format: Adult contemp, alternative. News staff: 2. Target aud: 25-54. ♦Meg Stevens, mus dir; Vaun Wickerham, farm dir; Chris Miller, sports cmtr.

***WLFC(FM)**— Nov 1, 1973: 88.3 mhz; 155 w. 66 ft TL: N41 03 11 W83 39 13. Stereo. Hrs opn: 7 AM-midnight 1000 N. Main St., 45840. Phone: (419) 424-6921. Fax: (419) 434-4822. Licensee: University of Findlay. Population served: 45,000 Format: Rock/AOR. News: 3 hrs wkly. Target aud: 18-40. Spec prog: Class 3 hrs, folk 3 hrs, relg 3 hrs, Sp 3 hrs wkly. ♦Nick Meyers, gen mgr.

WPFX-FM—See North Baltimore

***WTKC(FM)**— 2007: 89.7 mhz; 125 w. Ant 30 ft TL: N41 02 43 W83 39 02. Hrs open: Box 1212, 45839-1212. Phone: (419) 423-3285. Phone: (815) 935-5169. E-mail: wtkc89.7@sbcglobal.net Licensee: Church of the Living God Ministries. Format: Contemp Christian. ♦Juan Salinas, pres & gen mgr; Richard Lugo, progmg mgr.

Fort Shawnee

WZRX-FM— 1991: 107.5 mhz; 3 kw. 328 ft TL: N40 40 04 W84 01 41. Hrs open: 24 Box 1128, 667 W. Market St., Lima, 45802. Secondary address: 667 W. Market St., Lima 45801. Phone: (419) 223-2060. Fax: (419) 229-3888. E-mail: comments@wzrx.com Web Site: www.x1075fm.com. Licensee: Jacor Broadcasting Corp. Group owner: Clear Channel Communications Inc. (acq 5-4-99; grpsl). Population served: 150,000 Natl. Rep: Clear Channel. Format: Oldies, rock/AOR. News: one hr wkly. Target aud: 18-49; male dominated. ♦Art Versnick, pres & gen mgr; Phil Austin, opns mgr; Eric Michaels, progmg dir.

Fostoria

WBVI(FM)—Listing follows WFOB(AM).

WFOB(AM)— Dec 9, 1952: 1430 khz; 1 kw-U, DA-2. TL: N41 06 11 W83 24 00. (CP: TL: N41 06 06 W83 23 59). Stereo. Hrs opn: 5 AM-2 AM Box 1157, 44830. Secondary address: Box 1624, Findlay 45840. Phone: (419) 435-5666. Phone: (419) 422-9284. Fax: (419) 435-6611. E-mail: wfob1430@aol.com Web Site: www.wfob.com. Licensee: TCB Holdings Inc. c/o Roppe Corp. (acq 11-24-97; with co-located FM). Population served: 309,100 Natl. Network: CBS. Natl. Rep: Rgnl Reps. Law Firm: Baker & Hostetler. Format: Adult contemp, talk. Target aud: General. Spec prog: Sp 3 hrs wkly. ♦Greg Peiffer, pres, gen mgr & adv mgr.

WBVI(FM)—Co-owned with WFOB(AM). 1946: 96.7 mhz; 3 kw. 330 ft TL: N41 06 01 W83 28 41. (CP: Ant 298 ft. TL: N41 06 00 W83 28 32). Stereo. 24 Web Site: www.wbvi.com.660,000 Natl. Network: Westwood One. News staff: one. Target aud: 25-54.

Fredericktown

WXXR(FM)— Sept 14, 1987: 98.3 mhz; 1.8 kw. Ant 423 ft TL: N40 34 27 W82 30 27. Stereo. Hrs opn: 24
Rebroadcasts WFXN-FM Galion 100%.
1197 US Hwy. Rt. 42, Ashland, 44805. Phone: (419) 289-2605. Fax: (419) 289-0304. E-mail: jeffschendel@clearchannel.com Licensee: Capstar TX L.P. Group owner: Clear Channel Communications Inc. (acq 2-12-2001; grpsl). Natl. Network: Fox News Radio. Format: CHR. Target aud: 25-54; Male. Spec prog: Underground Garage, House of Hair. ♦Diana Coon, gen mgr; Joe Rinehart, stn mgr; Eric Hansen, opns mgr.

Fremont

WFRO-FM— Dec 15, 1946: 99.1 mhz; 11.5 kw. Ant 364 ft TL: N41 21 58 W83 05 20. Hrs open: 24 1281 N. River Rd., 43420. Phone: (419) 332-8218. Fax: (419) 333-8226. Web Site: www.hitsandfavorites.com. Licensee: BAS Broadcasting Inc. (acq 9-11-2002; $1.3 million). Population served: 400,000 Natl. Network: ABC. Rgnl rep: Rgnl Reps Law Firm: Erwin Krasnow. Format: Adult contemp. News staff: 2; News: 5 hrs wkly. Target aud: 25-54; adults. ♦Jim Lorensen, pres; Tom Klein, CEO & gen mgr; Dave Campbell, opns mgr.

Gahanna

***WCVO(FM)—** Oct 13, 1972: 104.9 mhz; 6 kw. 298 ft TL: N40 04 16 W82 48 35. Stereo. Hrs opn: 24 Box 783, 4400 Reynoldsburg-New Albany Rd., New Albany, 43054. Phone: (614) 855-9171. Fax: (614) 855-9280. E-mail: theriver@104theriver.org Web Site: www.1049theriver.com. Licensee: Christian Voice of Central Ohio Inc. Population served: 1,500,000 Natl. Network: USA. Format: Christian, Adult Contempo. News: 15 hrs wkly. Target aud: 25-54; Christian, politically aware, female, middle aged professionals. ♦Dan Baughman, pres, gen mgr, stn mgr & disc jockey; Mike Russell, mus dir; Tate Luck, opns mgr & disc jockey.

Galion

WFXN-FM— Nov 8, 1974: 102.3 mhz; 3.5 kw. Ant 430 ft TL: N40 45 26 W82 47 23. Hrs open: 24 1197 US Hwy. Rt 42, Ashland, 44805. Phone: (800) 529-1013. Fax: (419) 289-0304. E-mail: jeffschendel @clearchannel.com Web Site: www.foxclassicrock.com. Licensee: Capstar TX L.P. Group owner: Clear Channel Communications Inc. (acq 2-12-2001; grpsl). Population served: 13,123 Natl. Network: Fox News Radio. Format: Classic rock. Target aud: 25-54; Male. Spec prog: Underground Garage 2hrs, House of Hair 2hrs. ♦Diana Coon, gen mgr; Eric Hanson, opns mgr; Jeff Schendel, progmg dir.

Gallipolis

WJEH(AM)— June 19, 1950: 990 khz; 1 kw-D, 16 w-N, 250 w-CH. TL: N38 48 20 W82 13 23. Stereo. Hrs opn: 24 117 Portsmouth Rd., 45631. Phone: (740) 446-3543. Fax: (740) 446-3001. Licensee: Legend Communications of Ohio LLC. (acq 8-21-98; $1.45 million with co-located FM). Population served: 90,000 Natl. Rep: Rgnl Reps. Law Firm: Dean George Hill. Format: Mus memories. News staff: one; News: 10 hrs wkly. Target aud: 35 plus. ♦John Pelletier, gen mgr; Steve Reinhardt, progmg dir; Bob Triplett, news dir & chief of engrg.

WRYV(FM)— Dec 15, 1961: 101.5 mhz; 50 kw. Ant 500 ft TL: N38 48 19 W82 13 36. Stereo. Hrs opn: 24 Box 404, Huntington, WV, 25708. Secondary address: 919 Fifth Ave., Suite 210, Huntington, WV 25701. Phone: (304) 399-9603. Fax: (304) 399-9609. Web Site: www.1015theriver.net. Licensee: Connoisseur Media of WV-OH LLC. (acq 6-21-2006; $3.1 million). Population served: 500,000 Format: Classic rock. Target aud: 24-49. ♦B.J. Nielsen, gen mgr.

Gambier

***WKCO(FM)—** 1975: 91.9 mhz; 266 w. 190 ft TL: N40 22 25 W82 23 45. Stereo. Hrs opn: 19 Box 312, Kenyon College, 43022. Phone: (740) 427-5411. E-mail: wkco@kenyon.edu Web Site: www.wkco.kenyon.edu. Licensee: Kenyon College. Population served: 50,000 Format: Var. News: 8 hrs wkly. Target aud: 18-25; college population. ♦Phillip Thompson, gen mgr.

Geneva

WKKY(FM)— Nov 2, 1987: 104.7 mhz; 6 kw. 328 ft TL: N41 47 30 W81 05 31. Stereo. Hrs opn: 24 95 W. Main St., 44041. Phone: (440) 466-9559. Fax: (440) 466-3138. E-mail: wkky@wkky.com Web Site: www.wkky.com. Licensee: Music Express Broadcasting Corp. of Northeast Ohio. (acq 3-15-90; $441,965; FTR: 4-2-90). Population served: 128,000 Natl. Network: ABC. Natl. Rep: Rgnl Reps. Format: Country. News: 6 hrs wkly. Target aud: 25-54. Spec prog: Pub affrs 2 hrs wkly. ♦Warren Jones, pres; Gary Hayes, gen mgr; Cindy Steiner, traf mgr.

Georgetown

WAXZ(FM)— Apr 19, 1976: 97.7 mhz; 1.6 kw. 390 ft TL: N38 52 03 W83 48 44. Stereo. Hrs opn: 8354 Fryer Rd., 45121. Phone: (937) 378-6151. Fax: (937) 377-2200. E-mail: info@977waxz.com Web Site: www.977waxz.com. Licensee: Gateway Radio Works Inc. Group owner: First Broadcasting Investment Partners LLC (acq 4-26-2006; $60,294). Rgnl. Network: Ohio Radio Net., Agri Bcstg. Format: Modern country. Spec prog: Farm 10 hrs wkly. ♦Heather Frye, gen mgr.

Gibsonburg

WIMX(FM)— Jan 24, 1989: 95.7 mhz; 3.5 kw. 433 ft TL: N41 28 19 W83 25 05. Stereo. Hrs opn: 24 5744 Southwyck Blvd., # 200, Toledo, 43614. Phone: (419) 868-7914. Fax: (419) 868-8765. Web Site: www.mix957.fm. Licensee: Urban Radio Licenses LLC. (acq 5-13-2005; $2 million). Format: Urban contemp. News staff: one. Target aud: 20-40. Spec prog: Mexican 8 hrs wkly. ♦Jeffrey Hedgeman, gen mgr; Barbara Hubley, opns mgr.

Granville

***WDUB(FM)—** Feb 7, 1962: 91.1 mhz; 100 w. 171 ft TL: N40 04 16 W82 31 24. Stereo. Hrs opn: 24 Slayter Hall, Denison Univ., 43023. Phone: (740) 587-0810. Phone: (740) 587-6382. Fax: (740) 587-8364. Web Site: www.wdub.org. Licensee: Denison University. Natl. Network: USA. Wire Svc: UPI Format: Progsv, classic rock, free-form. News staff: 3. Target aud: General; college students, faculty & loc residents. Spec prog: Black 9 hrs, reggae 2 hrs, Sp 2 hrs, Swedish 2 hrs wkly. ♦Ben Berry, gen mgr & opns mgr.

Greenfield

WVNU(FM)— May 1, 1994: 97.5 mhz; 3.2 kw. 305 ft TL: N39 24 01 W83 26 48. Hrs open: 24 Box 329, 321 Jefferson St., 45123. Phone: (937) 981-5050. Fax: (937) 981-2107. E-mail: wvnu@bright.net Web Site: wvnu.com. Licensee: Southern Ohio Broadcasting Inc. (acq 1-12-94; $35,227; FTR: 2-7-94). Natl. Network: Jones Radio Networks, CNN Radio. Law Firm: Pepper & Corazzini. Format: Lite adult contemp. Target aud: 24-54. ♦Patrick Hays, pres & gen mgr; Tom Archibald, VP; Nelson Hunter, progmg dir.

Greenville

***WDPG(FM)—** February 1994: 89.9 mhz; 50 kw. 403 ft TL: N40 08 49 W84 36 36. Hrs open: 24
Rebroadcasts WDPR(FM) West Carrollton 100%.
126 N. Main St., Dayton, 45402. Phone: (937) 496-3850. Fax: (937) 496-3852. E-mail: dpr@dpr.org Web Site: wwww.dpr.org. Licensee: Dayton Public Radio Inc. Format: Class, fine arts. ♦Georgie Woessner, gen mgr; Charles Wendelken-Wilson, mus dir.

WDSJ(FM)— Oct 26, 1990: 106.5 mhz; 50 kw. 482 ft TL: N40 08 49 W84 36 36. Stereo. Hrs opn: 24 101 Pine St., Dayton, 45402-2925. Phone: (937) 224-1137. Fax: (937) 224-3667. Web Site: www.daytonjazz.com. Licensee: Citicasters Licenses L.P. Group owner: Clear Channel Communications Inc. (acq 5-4-99; grpsl). Population served: 544,000 Natl. Network: Jones Radio Networks. Law Firm: Reddy, Begley & McCormick. Format: Jazz. Target aud: 25-54. ♦Karrie Sudbrack, gen mgr.

Grove City

WWCD(FM)— Aug 21, 1990: 101.1 mhz; 6 kw. 328 ft TL: N39 48 50 W83 03 19. Stereo. Hrs opn: 24 503 S. Front St., Suite 101, Columbus, 43215. Phone: (614) 221-9923. Fax: (614) 227-0021. E-mail: wcbmaster@cd101.com Web Site: www.cd101.com. Licensee: Fun With Radio LLC (acq 8-16-01). Law Firm: Drinker Biddle & Reath. Format: Alternative rock. Target aud: 21-40; well educated, upscale professionals with discretionary income. Spec prog: Jazz 3 hrs, acoustic 2 hrs, mix show 2 hrs wkly. ♦Roger Vaughan, pres & gen mgr; Randy Malloy, opns dir.

***WWGV(FM)—**Not on air, target date: unknown: 88.1 mhz; 20 kw vert. Ant 262 ft TL: N39 43 17 W82 51 25. Hrs open: Drawer 2440, Tupelo, MS, 38803. Phone: (662) 844-5036. Fax: (662) 842-7798. Licensee: American Family Association. ♦Donald E. Wildmon, chmn.

Hamilton

WGRR(FM)— Apr 15, 1961: 103.5 mhz; 19.3 kw. 790 ft TL: N39 16 24 W84 31 37. Stereo. Hrs opn: 24 2060 Reading Rd., Cincinnati, 45202. Phone: (513) 699-5103. Fax: (513) 699-5000. Web Site: www.wgrr.com. Licensee: Infinity Radio Inc. Group owner: Infinity Broadcasting Corp. (acq 11-13-98; grpsl). Population served: 1,969,100 Format: Oldies. News staff: one. Target aud: 35-54. ♦Jim Bryant, gen mgr; Tim Closson, opns mgr; Stefan Schellhas, gen sls mgr & disc jockey.

***WHSS(FM)—** May 12, 1975: 89.5 mhz; 190 w. 282 ft TL: N39 25 51 W84 37 40. Hrs open: 7:30 AM-6 PM Hamilton High School, 1111 Eaton Rd., 45013. Phone: (513) 887-4818. Fax: (513) 887-4804. Licensee: Hamilton City Schools Board of Education. Format: Rock, alternative. News: 3 hrs wkly. Target aud: 12 plus; general. ♦David P. Spurrier, gen mgr & progmg dir.

WMOH(AM)— Aug 15, 1944: 1450 khz; 1 kw-U. TL: N39 24 10 W84 31 54. Stereo. Hrs opn: 24 2081 Fairgrove Ave., 45011. Phone: (513) 863-1111. Fax: (513) 863-6856. E-mail: christheiss@wmoh.com Web Site: www.wmoh.com. Licensee: Vernon R. Baldwin Inc. (group owner; acq 1-7-03; $950,000). Population served: 289,512 Natl. Network: ESPN Radio. Natl. Rep: Rgnl Reps. Format: Sports. News staff: 5; News: 25 hrs wkly. Target aud: 35-64; adults above medium income.

♦Chris Theiss, gen mgr, opns dir, opns mgr, gen sls mgr & sports cmtr; Bill Douglas, progmg dir; Steve Vaughn, news dir; Gail Moore, traf mgr; Jay Crawford, engr.

Harrison

WNLT(FM)— Sept 1, 1991: 104.3 mhz; 3 kw. 328 ft TL: N39 15 02 W84 50 10. Hrs open: 8686 Michael Ln., Fairfield, 45014. Phone: (513) 829-7700. Licensee: Vernon R. Baldwin Inc. (group owner) Population served: 2,000,000 Natl. Network: K-Love. Format: Adult contemp Christian mus. ♦Marci Baldwin, VP; Vernon R. Baldwin, pres & gen mgr; Mark Mitchell, stn mgr; Glenn Moore, progmg dir.

Heath

WHTH(AM)— Oct 16, 1970: 790 khz; 1 kw-D, DA-1. TL: N40 03 05 W82 28 08. Hrs open: 24 Box 1057, 1000 N. 40th St., Newark, 43058-1057. Phone: (740) 522-8171. Fax: (740) 522-8174. E-mail: sales@wnko.com Web Site: www.wnko.com. Licensee: Runnymede Corp. (acq 10-15-98; $100,000 for stock with WNKO(FM) Newark). Population served: 140,000 Natl. Network: CNN Radio. Wire Svc: AP Format: Talk radio. Target aud: 35-54. ♦Charles Franks, pres; J. Thomas Swank, gen mgr; John Franks, opns VP.

WNKO(FM)—See Newark

Hicksville

WFGA(FM)— 2002: 106.7 mhz; 2.85 kw. Ant 482 ft TL: N41 19 16 W84 43 12. Hrs open: 450 N. Grand Staff Dr., Auburn, IN, 46706. Phone: (260) 920-3602. Fax: (260) 920-3604. Web Site: ilovefroggy.com. Licensee: Fallen Timber Communications, LLC (acq 6-28-2000; $512,000). Format: Var. ♦Leann Didier, gen mgr.

Hilliard

WBWR(FM)— Feb 6, 1991: 105.7 mhz; 2.4 kw. 522 ft TL: N39 58 10 W83 00 10. Stereo. Hrs opn: 24 2323 W. 5th Ave., Suite 200, Columbus, 43229. Phone: (614) 486-6101. Fax: (614) 487-3575. Web Site: www.1057thefox.com. Licensee: Citicasters Licenses L.P. Group owner: Clear Channel Communications Inc. (acq 1999; grpsl). Natl. Network: ABC. Format: Classic Rock. News staff: one. Target aud: 18-34. ♦Tom Thon, gen mgr; Rob O'Boyle, gen sls mgr & prom dir; J.P. Hastings, progmg dir.

Hillsboro

WSRW(AM)— July 15, 1956: 1590 khz; 500 w-D. TL: N39 09 58 W83 36 25. Hrs open: Box 9, 5675 State Rt. 247, 45133. Phone: (937) 393-1590. Fax: (937) 393-1611. E-mail: wsrw@clearchannel.com Web Site: www.wsrwam.com. Licensee: CC Licenses LLC. Group owner: Clear Channel Communcations Inc. (acq 10-26-99; $2.5 million with WSRW-FM Hillsboro). Population served: 75,000 Natl. Network: AP Radio. Natl. Rep: Katz Radio. Format: Music of your life. News: 3 hrs wkly. Target aud: 35-54; general. ♦Dan Latham, sr VP & gen mgr; Joe Fisher, stn mgr; John Barney, sls dir & gen sls mgr; Damon Haught, progmg dir; Paul Levo, chief of engrg.

WSRW-FM— 1962: 106.7 mhz; 50 kw. 300 ft TL: N39 09 58 W83 36 25. Stereo. Hrs opn: 24 Box 9, 5675 St., Rt. 247, 45133. Phone: (937) 393-1590. Fax: (937) 393-1611. E-mail: wsrw@clearchannel.com Web Site: www.wsrw.com. Licensee: CC Licenses LLC. Group owner: Clear Channel Communications Inc. (acq 10-26-99; $2.5 million with WSRW(AM) Hillsboro). Population served: 3,000,000 Natl. Network: ABC. Natl. Rep: Katz Radio. Format: Country. News staff: one; News: 8 hrs wkly. Target aud: General. ♦Dan Latham, gen mgr; Kim Scaggs, opns mgr.

Holland

WPOS-FM— Sept 1, 1966: 102.3 mhz; 6 kw. 312 ft TL: N41 37 32 W83 42 41. Stereo. Hrs opn: 24 Box 457, 43528. Secondary address: 7112 Angola Rd. 43528. Phone: (419) 865-5551. Fax: (419) 862-0112. E-mail: radio@wposfm.com Web Site: www.wposfm.com. Licensee: Maumee Valley Broadcasting Association. (acq 8-1-65). Law Firm: Wiley, Rein & Fielding. Format: Christian. News: 10 hrs wkly. Spec prog: Gospel 20 hrs wkly. ♦Rick Waldron, gen mgr.

Hubbard

WRBP(FM)— Aug 16, 1993: 101.9 mhz; 3 kw. 328 ft TL: N41 05 29 W30 30 05. Hrs open: 20 Federal Plaza W., # T2, Youngstown, 44503. Phone: (330) 744-5115. Fax: (330) 744-2221. Licensee: Stop 26 Riverbend Licenses LLC. Format: Adult urban contemp, classic rock, oldies. Target aud: 25-54; general. Spec prog: Black, news/talk, jazz 12 hrs, relg 6 hrs, Sp 2 hrs wkly. ♦Percy Squire, CEO; Frank Halfacre, chmn; Bill Cusack, gen mgr; Linda Penny, gen mgr & stn mgr; Charles Rhodes, gen sls mgr; Madaline Halfacre, progmg dir; Kenneth King, news dir & min affrs dir; Savannah Thomas, pub affrs dir; Del King, engrg dir; Charlie Ring, chief of engrg; Davida Herron, traf mgr; Walter Halfacre, spec ev coord.

Huron

WKFM(FM)— Apr 1, 1996: 96.1 mhz; 3.4 kw. 436 ft TL: N41 18 05 W82 29 16. Stereo. Hrs opn: 24 10327 Milan Rd., US Rte. 250, Milan, 44846. Phone: (419) 609-5961. Fax: (419) 609-2679. E-mail: k96@wkfm.com Web Site: www.wkfm.com. Licensee: Elyria-Lorain Broadcasting Co. (group owner; acq 7-1-96; $450,000). Natl. Network: Westwood One. Law Firm: Putbrese, Hunsaker & Trent. Format: Country. News staff: one; News: 1 hr wjky. Target aud: General. ♦Gary Kneisley, pres & gen mgr; Tim Kelly, opns mgr; Shelly Luipold, gen sls mgr.

Ironton

WBKS(FM)— July 1, 1973: 107.1 mhz; 3 kw. 125 ft TL: N38 32 22 W82 40 17. (CP: Ant 285 ft.). Hrs opn: 18 Box 2288, Huntington, WV, 25724. Phone: (304) 525-7788. Fax: (304) 525-3299. E-mail: kiss107fm@clearchannel.com Web Site: www.1071kiss.com. Licensee: Capstar TX L.P. Group owner: Clear Channel Communications Inc. (acq 8-30-00; grpsl). Format: Hot CHR. News: 3 hrs wkly. Target aud: 35 plus; affluent, middle-aged. Spec prog: Relg 2 hrs wkly. ♦Judy Jennings, gen mgr; Matt Tweel, gen sls mgr; Jim Davis, progmg mgr; Gary Miller, mus dir; Bill Cornwell, news dir; Scott Hensley, chief of engrg.

WIRO(AM)— September 1951: 1230 khz; 1 kw-U. TL: N38 32 22 W82 40 17. Hrs open: 24 Box 2288, Huntington, WV, 25724. Phone: (304) 525-7788. Fax: (304) 525-6281. E-mail: paulswann@clearchannel.com Web Site: www.800wvhu.com. Licensee: Capstar TX Limited Partnership Group owner: Clear Channel Communications Inc. (acq 8-7-00; grpsl). Population served: 15,030 Natl. Rep: Keystone (unwired net), Rgnl Reps. Format: News/talk. News: 5 hrs wkly. Target aud: 21-49. Spec prog: Relg 7 hrs wkly. ♦Judy Jennings, gen mgr; Matt Tweet, gen sls mgr; Paul Swann, progmg dir & chief of engrg.

***WOUL-FM**— Oct 12, 1987: 89.1 mhz; 50 kw. 400 ft TL: N38 31 23 W82 39 20. Stereo. Hrs opn: 24
Rebroadcasts WOUB-FM Athens 100%.
9 S. College St., Athens, 45701. Phone: (740) 593-4554. Fax: (740) 593-0240. E-mail: woub@woub.org Web Site: www.woub.org. Licensee: Ohio University. Population served: 70,000 Natl. Network: PRI, NPR. Format: News/talk. News staff: 3. ♦David Wiseman, VP & opns VP; Carolyn Lewis, gen mgr; Steve Skidmore, opns dir; Scott Martin, opns mgr.

Jackson

WCJO(FM)— 1971: 97.7 mhz; 3 kw. 300 ft TL: N39 01 45 W82 35 51. Hrs open: 24 Box 667, 45640. Secondary address: 295 E. Main St. 45640. Phone: (740) 286-3023. Fax: (740) 286-6679. E-mail: rburtrand@jcbiradio.com Licensee: Jackson County Broadcasting Inc. (group owner; (acq 6-15-99; grpsl). Population served: 300,000 Natl. Network: Westwood One. Law Firm: Fletcher, Heald & Hildreth. Format: Hot country. News staff: one; News: 8 hrs wkly. Target aud: General; current-country music lovers. ♦Jerry Mossbarger, gen mgr; Bill Forthofer, gen sls mgr; John Pelletier, progmg dir; Paul Brown, disc jockey.

Jefferson

***WCVJ(FM)**— 1978: 90.9 mhz; 1.85 kw. Ant 643 ft TL: N41 37 50 W80 45 36. Stereo. Hrs opn: 2351 Suset Blvd., Suite 170-218, Rocklin, CA, 95765. Phone: (916) 251-1600. Fax: (916) 251-1650. Licensee: Educational Media Foundation. (acq 10-14-2005; $650,000). Population served: 20,000 Format: Christian. ♦Richard Jenkins, pres; Mike Novak, VP; Keith Whipple, dev dir; David Pierce, progmg mgr; Ed Lenane, news dir; Sam Wallington, engrg dir; Karen Johnson, news rptr; Marya Morgan, news rptr; Richard Hunt, news rptr.

Johnstown

WVKO-FM—Licensed to Johnstown. See Columbus

Kent

WJMP(AM)— March 1964: 1520 khz; 1 kw-D, DA. TL: N41 09 35 W81 18 19. Hrs open: Box 2170, Akron, 44309-2170. Secondary address: 2449 S.R. 59 44240. Phone: (330) 673-2323. Fax: (330) 673-0301. Licensee: Media-Com Inc. (acq 1971). Population served: 300,000 Format: Sports. Target aud: 18 plus. ♦Richard M. Klaus, pres; William Klaus, stn mgr; Robert Klaus, sls VP; Jim Midock, news dir; Bob Sassman, chief of engrg; Mary Stein, traf mgr.

WNIR(FM)—Co-owned with WJMP(AM). Feb 19, 1962: 100.1 mhz; 4.2 kw. Ant 394 ft TL: N41 06 28 W81 21 19.2,000,000 Law Firm: Wombel, Carlyle, Sandridge & Rice. Format: Talk. Target aud: General. ♦Mary Stein, traf mgr.

***WKSU-FM**— 1950: 89.7 mhz; 14.5 kw. Ant 909 ft TL: N41 04 58 W81 38 02. Stereo. Hrs opn: 24 Box 5190, 44242-0001. Secondary address: 1613 E. Summit St. 44242-0001. Phone: (330) 672-3114. Fax: (330) 672-4107. E-mail: letters@wksu.org Web Site: www.wksu.org. Licensee: Kent State University. Population served: 2,650,000 Natl. Network: PRI, NPR, AP Radio. Law Firm: Dow, Lohnes & Albertson. Format: Class, news. News staff: 5; News: 35 hrs wkly. Target aud: 35-65; college grad, professional & upper income. Spec prog: Folk 12 hrs wkly. ♦Allen E. Bartholet, gen mgr; Deborah Frazier, gen mgr; N. Vincent Duffy, opns dir & progmg dir; Abbe Turner, dev dir; Ronald Bartlebaugh, engrg dir.

Kenton

WKTN(FM)— June 20, 1963: 95.3 mhz; 3 kw. 270 ft TL: N40 38 41 W83 33 59. Stereo. Hrs opn: 5 AM-midnight 112 N. Detroit St., 43326. Phone: (419) 675-2355. Fax: (419) 673-1096. E-mail: wktn@kenton.com Web Site: www.wktn.com. Licensee: Radio General Ltd. (acq 5-12-77). Population served: 40,000 Rgnl. Network: Agri Bcstg. Law Firm: Arent, Fox, Kintner, Plotkin & Kahn. Format: Adult contemp. News staff: one; News: 10 hrs wkly. Target aud: 25-54. Spec prog: Farm 2 hrs wkly. ♦Keith P. Gensheimer, pres & gen mgr; Quentin White, gen sls mgr.

Kettering

***WKET(FM)**— May 5, 1975: 98.3 mhz; 10 w. 150 ft TL: N39 41 46 W84 09 43. Hrs open: 3301 Shroyer Rd., 45429. Phone: (937) 296-7669. Fax: (937) 297-7435. Licensee: Kettering City School District. Format: Educ, classic rock, AOR. Target aud: 13-18; high school students. ♦Karl Bremer, gen mgr & stn mgr.

WLQT(FM)— Feb 20, 1962: 99.9 mhz; 50 kw. 500 ft TL: N39 44 07 W84 10 10. Stereo. Hrs opn: 24 101 Pine St., Dayton, 45402. Phone: (937) 224-1137. Fax: (937) 222-5483. E-mail: theboss@wlqt.com Web Site: www.wlqt.com. Licensee: Citicasters Licenses L.P. Group owner: Clear Channel Communications Inc. (acq 1999; grpsl). Format: Adult contemp. Target aud: 35-64; persons 35-64. ♦Karrie Sudbrack, gen mgr; Sandy Collins, progmg dir.

WQRP(FM)—See Dayton

Lancaster

***WFCO(FM)**— August 1988: 90.9 mhz; 1.2 kw vert. Ant 256 ft TL: N39 40 49 W82 35 51. Stereo. Hrs opn: 24 201 S. Broad St., Studio 303, 43130. Phone: (740) 654-8556. Fax: (740) 654-8581. E-mail: wfco@wfcofm.com Web Site: wfcofm.com. Licensee: Lancaster Educational Broadcasting Foundation. Population served: 200,000 Natl. Network: Moody. Format: Btfl mus, Christian, news. News staff: one; News: one hr wkly. Target aud: 30 plus; Christian audience & those interested in community events. Spec prog: Live coverage of sports & community events. ♦Steve Rauch, gen mgr.

WHOK-FM— December 1958: 95.5 mhz; 50 kw. Ant 492 ft TL: N39 40 32 W82 40 34. Stereo. Hrs opn: 24 10th Fl., 280 Plaza N. High St., Columbus, 43215. Phone: (614) 225-9465. Fax: (614) 677-0116. Web Site: www.whok.com. Licensee: Wilks License Co.-Columbus LLC. Group owner: Infinity Broadcasting Corp. (acq 1-10-2007; grpsl). Population served: 1,500,000 Natl. Network: Westwood One. Natl. Rep: Katz Radio. Law Firm: Leventhal, Senter & Lerman. Format: Country legends. News staff: one; News: 2 hrs wkly. Target aud: 25-54. ♦Valerie Brooks, gen mgr; Dave Cooper, opns dir; Robert Golias, gen sls mgr; Jill McCarren, rgnl sls mgr; Chrystal Beins, prom dir; George Wolf, progmg dir.

WJZA(FM)— Oct 7, 1989: 103.5 mhz; 4 kw. 435 ft TL: N39 51 52 W82 38 19. Stereo. Hrs opn: 24 4401 Carriage Hill Ln., Columbus, 43220. Phone: (614) 451-2191. Fax: (614) 451-1831. Web Site: www.wjza.com. Licensee: Franklin Communications Inc. Group owner: Saga Communications Inc. (acq 10-1-2003; $13 million). Population served:

1,500,000 Natl. Rep: Christal. Wire Svc: AP Format: Smooth jazz. News: 2 hrs wkly. Target aud: 25-54. Spec prog: Various 15 hrs wkly. ♦Alan Goodman, pres & gen mgr; Chris Forgy, sls dir; Michelle Hurley, mktg dir; Bill Harman, progmg dir.

WLOH(AM)— October 1948: 1320 khz; 1 kw-D, 28 w-N. TL: N39 44 21 W82 37 48. Hrs open: 24 724 So. Columbus St., 43130. Phone: (740) 653-4373. Fax: (740) 653-0702. E-mail: wloh@greenapple.com Web Site: www.wloh.net. Licensee: Frontier Broadcasting LLC No. 3 (acq 2-13-01; $325,000). Population served: 108,000 Natl. Rep: D & R Radio. Law Firm: Covington & Burling. Format: Talk radio. News staff: 2; News: 24 hrs wkly. Target aud: General; Fairfield, Franklin & surrounding county residents. Spec prog: Cleveland Cavaliers, Cincinnati Reds, farm one hr wkly. ♦Bart Johnson, CEO; Mark Bonah, gen mgr; Mark Bonach, opns mgr; Michael O'Riley, gen sls mgr.

Lebanon

WFTK(FM)— May 26, 1958: 96.5 mhz; 19.5 kw. Ant 810 ft TL: N39 21 11 W84 19 30. Stereo. Hrs opn: 24 c/o Radio Cincinnati, 895 Central Ave., Suite 900, Cincinnati, 45202. Phone: (513) 241-9898. Fax: (513) 241-6689. Licensee: WVAE Lico Inc. Group owner: Susquehanna Radio Corp. (acq 5-5-2006; grpsl). Population served: 67,865 Natl. Rep: Christal. Format: Talk. ♦Gary Lewis, gen mgr & progmg dir.

Lexington

***WFOT(FM)**— February 2007: 89.5 mhz; 360 w vert. Ant 304 ft TL: N40 43 36 W82 36 59. Hrs open: Box 32, Mansfield, 44901. Phone: (937) 642-1270. Fax: (937) 644-1617. E-mail: info@stgabrielradio.com Web Site: www.stgabrielradio.com. Licensee: St. Gabriel Radio Inc. Natl. Network: EWTN Radio. Format: Catholic relg. ♦Christopher Gabrelcik, pres; Michael Barone, stn mgr.

Lima

WEGE(FM)— Nov 25, 1970: 104.9 mhz; 3 kw. 260 ft TL: N40 43 21 W84 05 04. (CP: Ant 286 ft.). Stereo. Hrs opn: 24 1301 N. Cable Rd., 45805. Phone: (419) 331-1600. Fax: (419) 222-3755. Licensee: Maverick Media of Lima License LLC. Group owner: Maverick Media LLC (acq 12-4-2003; grpsl). Population served: 150,000 Natl. Network: ABC. Natl. Rep: Christal. Format: Classic rock. Target aud: 25-54; affluent, community involved. ♦Dave Roach, gen mgr; Matt Childers, gen sls mgr; Bill Rice, progmg dir; Brandy Rader, traf mgr.

WFGF(FM)— 1985: 93.1 mhz; 3 kw. 328 ft TL: N40 45 47 W84 10 59. Hrs open: 24 1301 N. Cable Rd., 45805. Secondary address: Box 1487 45802. Phone: (419) 331-1600. Fax: (419) 222-3755. E-mail: polly@froggy93.com Web Site: www.froggy93.com. Licensee: Maverick Media of Lima License LLC. Group owner: Maverick Media LLC (acq 12-4-2003; grpsl). Population served: 150,000 Format: Country. Target aud: 25-54; young, affluent. Spec prog: Nascar Nextel Races. ♦Gary Rozynek, pres & prom dir; Dave Roach, gen mgr & mus dir; Bill McAdams, stn mgr; Stacy McAdams, prom dir; Brandy Rader, engrg dir & traf mgr.

***WGLE(FM)**— Dec 2, 1981: 90.7 mhz; 50 kw. 420 ft TL: N40 39 15 W84 06 36. Stereo. Hrs opn: 24
Rebroadcasts WGTE-FM Toledo 100%.
1270 S. Detroit Ave., Toledo, 43614. Phone: (419) 380-4600. Fax: (419) 380-4710. Web Site: www.wgte.org. Licensee: The Public Broadcasting Foundation of Northwest Ohio. Population served: 458,318 Natl. Network: PRI, NPR. Format: Class, NPR, News, Public Affairs. News: 23 hrs wkly. Target aud: General. Spec prog: Jazz 16 hrs, new age/eclectic 4 hrs wkly. ♦Marlon P. Kiser, CEO, pres & gen mgr; George Jones, chmn; Chris Peiffer, opns mgr; Ross Pfeiffer, dev dir. Co-owned TV: *WGTE-TV affil.

WIMA(AM)— Dec 5, 1948: 1150 khz; 1 kw-U, DA-2. TL: N40 40 47 W84 06 34. Hrs open: 24 667 W. Market St., 45801. Phone: (419) 223-2060. Fax: (419) 229-3888. E-mail: comments@1150wima.com Web Site: www.1150wima.com. Licensee: Jacor Broadcasting Corp. Group owner: Clear Channel Communications Inc. (acq 5-4-99; grpsl). Population served: 150,000 Natl. Network: Fox News Radio, Fox Sports. Natl. Rep: Clear Channel. Wire Svc: AP Format: News/talk, sports. News staff: one; News: 20 hrs wkly. Target aud: 35 plus. Spec prog: Farm 5 hrs wkly. ♦Art Versnick, VP & gen mgr; David Cook, stn mgr; Phil Austin, opns mgr; Jack Wheelbarger, natl sls mgr; Dave Woodward, progmg dir; Doug Jenkins, news dir; Mark Gierhart, chief of engrg.

***WTGN(FM)**— Sept 27, 1966: 97.7 mhz; 6 kw. 300 ft TL: N40 45 26 W84 08 12. Stereo. Hrs opn: 1600 Elida Rd., 45805. Phone: (419) 227-2525. Fax: (419) 222-5438. E-mail: info@wtgn.org Web Site: www.wtgn.org. Licensee: Associated Christian Broadcasters Inc. Population served: 53,734 Format: Christian. ♦Wesley Lytle, pres; Scott Young, gen mgr.

WWSR(FM)—See Wapakoneta

***WYSM(FM)**— 2001: 89.3 mhz; 3 kw. Ant 220 ft TL: N40 39 15 W84 06 36. Hrs open: 5115 Glendale Ave., Toledo, 43614. Phone: (419) 389-0893. Fax: (419) 381-0731. E-mail: yesfm@yeshome.com Web Site: www.yeshome.com. Licensee: Side by Side Inc. Natl. Network: Salem Radio Network. Format: Christian. ♦J. Todd Hostetler, gen mgr.

WZOQ(AM)— Aug 22, 1963: 940 khz; 250 w-D, DA-2. TL: N40 43 21 W84 05 04. Stereo. Hrs opn: 24 1301 N. Cable Rd., 45805. Phone: (419) 331-1600. Fax: (419) 222-5085. Web Site: www.940jamz.com. Licensee: Maverick Media of Lima License LLC. Group owner: Maverick Media LLC (acq 12-4-2003; grpsl). Population served: 220,000 Natl. Rep: Christal, Rgnl Reps. Format: Urban contemp. News staff: 2; News: 20 hrs wkly. Target aud: 35-54; affluent, community involved. Spec prog: Jazz 2 hrs, relg 11 hrs wkly. ♦Gary S. Rozynek, pres; Dave Roach, gen mgr; Mark Mackey, stn mgr; Bill McAdams, opns mgr.

Logan

WLGN(AM)— December 1967: 1510 khz; 1 kw-D, 250 w-CH. TL: N39 31 47 W82 23 10. Stereo. Hrs open: Sunrise-sunset Box 429, One Radio Ln., 43138. Phone: (740) 385-2151. Fax: (740) 385-4022. E-mail: wlgn@magicohio.com Licensee: WLGN LLC. (acq 1-4-2005; $675,000 with co-located FM). Population served: 25,000 Natl. Rep: Rgnl Reps. Format: Country. News staff: one; News: 10 hrs wkly. Target aud: 18-54. ♦Roger L. Hinerman, gen mgr; Judy Davis, gen sls mgr & traf mgr; Kevin Reed, mus dir; Steve Carmean, chief of engrg.

WLGN-FM— Dec 10, 1965: 98.3 mhz; 3 kw. 240 ft TL: N39 31 47 W82 23 10. Stereo. 24 250,000 News staff: one; News: 25 hrs wkly. ♦Roger L. Hinerman, CEO; Judy Davis, traf mgr.

London

WJYD(FM)— 1965: 106.3 mhz; 6 kw. 328 ft TL: N39 53 05 W83 25 23. Stereo. Hrs opn: 24 350 E. 1st Ave., Suite 100, Columbus, 43201. Phone: (614) 487-1444. Fax: (614) 487-5862. Web Site: www.joy106.com. Licensee: Blue Chip Broadcasting Licenses Ltd. Group owner: Radio One Inc. (acq 4-30-01; grpsl). Population served: 1603000 Natl. Network: ABC. Natl. Rep: D & R Radio. Format: Gospel. News staff: 3. Target aud: 18-34. Spec prog: Gospel 10 hrs wkly. ♦Jeff Wilson, gen mgr.

Lorain

WCLV(FM)— April 1961: 104.9 mhz; 6 kw. Ant 328 ft TL: N41 28 32 W81 59 24. Stereo. Hrs opn: 24 26501 Renaissance Pkwy., Cleveland, 44128. Phone: (216) 464-0900. Fax: (216) 464-2206. E-mail: wclv@wclv.com Web Site: www.wclv.com. Licensee: Radio Seaway Inc. (acq 11-1-01). Population served: 1,500,000 Natl. Rep: D & R Radio, Interep. Law Firm: Hogan & Hartson. Wire Svc: AP Format: Classical. News: 5 hrs wkly. Target aud: 35-64; high-income, college graduates & professionals. Spec prog: Jazz 5 hrs, financial news one hr wkly. ♦Robert D. Conrad, CEO & pres; Richard G. Marschner, CFO & exec VP; Jenny Northern, gen mgr & gen sls mgr; John Simna, opns mgr; Bill O'Connell, progmg mgr.

WDLW(AM)— December 1969: 1380 khz; 500 w-D, 67 w-N. TL: N41 25 48 W82 09 07. Hrs open: 24 45624 State Rt. 20, Oberlin, 44074. Phone: (440) 775-1380. Fax: (440) 774-1336. Licensee: WDLW Radio Inc. (acq 2-14-02; $250,000). Population served: 1,000,000 Wire Svc: AP News staff: 3; News: 10 hrs wkly. Target aud: 35-64. ♦Doug Wilber, pres & gen mgr; Lorie Wilber, VP; Terry Coffee, progmg dir.

***WNZN(FM)**— 1992: 89.1 mhz; 2.2 kw. 374 ft TL: N41 18 34 W82 26 31. Hrs open: 24 9712 State Rd. 113, Berlin Heights, 44814. Phone: (419) 588-3700. E-mail: tony10491@adelphia.net Licensee: Spanish Cultural Network. Format: Sp. ♦Milton Velazquez, gen mgr & opns mgr.

Loudonville

WXXF(FM)—Licensed to Loudonville. See Wooster

Manchester

WAGX(FM)— 1992: 101.3 mhz; 3 kw. 299 ft TL: N38 40 58 W83 39 45. Hrs open: 24 9503 Mason Lewis Rd., Maysville, KY, 41056. Phone: (606) 564-8474. Phone: (800) 845-9770. Fax: (606) 564-8383. Licensee: Jewell Schaeffer Broadcasting Inc. Population served: 100,000 Format: Oldies, adult contemp, classic rock. Target aud: 25-54; upscale adults. ♦James P. Wagner, CEO & pres; James P. Wagner, gen mgr.

Mansfield

WMAN(AM)— Dec 4, 1939: 1400 khz; 1 kw-U. TL: N40 46 13 W82 32 36. Hrs open: 24 1400 Radio Ln., 44906. Phone: (419) 529-2211. Fax: (419) 529-2516. Web Site: www.am1400.com. Licensee: Capstar TX L.P. Group owner: Clear Channel Communications Inc. (acq 8-7-00; grpsl). Population served: 45,047 Natl. Network: CBS, Westwood One. Format: News/talk. News staff: 3; News: 30 hrs wkly. Target aud: 35 plus; upscale, active mgmt/exec.

WYHT(FM)—Co-owned with WMAN(AM). Oct 18, 1962: 105.3 mhz; 50 kw. Ant 371 ft TL: N40 46 09 W82 32 23. Stereo. Web Site: www.wyht.com.55,047 Format: Hot adult contemp.

WNCO-FM—See Ashland

***WOSV(FM)**— June 27, 1989: 91.7 mhz; 750 w. 450 ft TL: N40 42 33 W82 29 11. Stereo. Hrs opn: 24
Rebroadcasts WOSU-FM Columbus 100%.
2400 Olentangy River Rd., Columbus, 43210. Phone: (614) 292-9678. Fax: (614) 292-0513. Web Site: www.wosu.org. Licensee: The Ohio State University. Population served: 121,000 Natl. Network: PRI, NPR. Law Firm: Dow, Lohnes & Albertson. Wire Svc: AP Format: Classical, NPR News. News: 28 hrs wkly. Target aud: 35 plus. ♦Thomas Rieland, gen mgr; Tim Eby, stn mgr; Kevin Petrilla, opns mgr.

WRGM(AM)—See Ontario

***WVMC-FM**— March 1979: 90.7 mhz; 170 w. 100 ft TL: N40 43 19 W82 31 52. Hrs open: 24 500 Logan Rd., 44907. Phone: (419) 756-5651 ext 227. Fax: (419) 756-7470. Web Site: wvmcfm.com. Licensee: Mansfield Christian School. Population served: 35,000 Format: Christian. Target aud: 18-34; middle income adults, mostly female. ♦Todd Stach, gen mgr.

WVNO-FM— Aug 11, 1962: 106.1 mhz; 40 kw. 545 ft TL: N40 45 50 W82 37 04. Stereo. Hrs opn: 2900 Park Ave. W., 44906. Phone: (419) 529-5900. Fax: (419) 529-2319. Web Site: www.wvno.com. Licensee: Johnny Appleseed Broadcasting Co. Population served: 350,000 Rgnl rep: Rgnl Reps Format: Adult contemp. News staff: 5; News: 10 hrs wkly. Target aud: 25-54; female. ♦Gunther S. Meisse, pres & gen mgr; Jim Holmes, opns mgr.

***WYKL(FM)**—Crestline, Dec 10, 1990: 98.7 mhz; 1.8 kw. Ant 418 ft TL: N40 46 08 W82 46 03. Stereo. Hrs opn: 24 2351 Sunset Blvd., Suite 170-218, Rocklin, CA, 95765. Phone: (916) 251-1600. Fax: (916) 251-1650. E-mail: klove@klove.com Web Site: www.klove.com. Licensee: Educational Media Foundation. Group owner: EMF Broadcasting (acq 12-18-03; $900,000). Population served: 150,000 Natl. Network: K-Love. Law Firm: Shaw Pittman. Format: Contemp Christian. Target aud: 25-44; Judeo Christian female. ♦Richard Jenkins, pres; Mike Novak, VP; Keith Whipple, dev dir; Eric Allen, natl sls mgr; David Pierce, progmg mgr; Ed Lenane, news dir; Sam Wallington, engrg dir; Karen Johnson, news rptr; Marya Morgan, news rptr; Richard Hunt, news rptr.

Mariemont

WKFS(FM)—See Milford

Marietta

***WCMO(FM)**— Oct 1, 1960: 98.5 mhz; 40 w. 105 ft TL: N39 25 07 W81 26 32. Hrs open: Marietta College, 215 5th St., 45750. Phone: (740) 376-4802. Phone: (740) 376-4800. Fax: (740) 376-4807. Web Site: www.wmrtfm.com. Licensee: Marietta College. Population served: 17,000 Format: AOR. ♦Marilee Morrow, gen mgr & progmg dir.

WLTP(AM)— May 8, 1996: 910 khz; 5 kw-D, 61 w-N, DA-2. TL: N39 26 07 W81 28 01. Stereo. Hrs opn: 24 6006 Grand Central Ave., Vienna, WV, 26105. Phone: (304) 295-6070. Fax: (304) 295-4389. Web Site: www.wltp.com. Licensee: CC Licenses LLC. Group owner: Clear Channel Communications Inc. (acq 9-4-2002). Population served: 16,861 Natl. Network: CBS, Westwood One, AP Radio. Format: New, talk. News staff: one; News: 8 hrs wkly. Target aud: 18-54; males. ♦Chuck Poet, gen mgr.

WMOA(AM)— Sept 8, 1946: 1490 khz; 1 kw-U. TL: N39 25 07 W81 28 34. Hrs open: 24
Rebroadcasts WJAW(FM) McConnelsville 70%.
925 Lancaster St., 45750. Phone: (740) 373-1490. Fax: (740) 373-1717. E-mail: dcastelli@wmoa1490.com Web Site: www.mariettaonline.com. Licensee: Jawco Inc. (acq 7-11-97; $659,000 with WJAW(FM) McConnelsville). Population served: 15,000 Law Firm: Pepper & Corazzini. Format: Adult contemp, news, sports. News staff: 2; News: 5 hrs wkly. Target aud: 35 plus; mature, middle-class to affluent. Spec prog: Farm one hr, relg one hr, sports 15 hrs wkly. ♦John A. Wharff III, pres, gen mgr & gen sls mgr; Dan Castelli, mus dir.

***WMRT(FM)**— Nov 13, 1975: 88.3 mhz; 9.2 kw. 205 ft TL: N39 25 07 W81 26 32. Stereo. Hrs opn: Marietta College, 45750. Phone: (740) 376-4802. Phone: (740) 376-4800. Fax: (740) 376-4807. Web Site: www.wmrtfm.com. Licensee: Marietta College. Population served: 16,861 Format: Class, news/talk, jazz. ♦Marilee Morrow, gen mgr.

WRVB(FM)— Dec 1, 1964: 102.1 mhz; 25 kw. 400 ft TL: N39 25 07 W81 28 34. Stereo. Hrs opn: 24 6006 Grand Central Ave., Box 5559, Vienna, WV, 26105. Phone: (304) 295-6070. Fax: (304) 295-4389. Web Site: www.102theriver.com. Licensee: CC Licenses LLC. Group owner: Clear Channel Communications Inc. (acq 4-17-2001; grpsl). Format: CHR. Target aud: 25-54; general. ♦Chuck Poet, gen mgr.

WXIL(FM)—See Parkersburg, WV

Marion

WDIF(FM)— Feb 27, 1975: 94.3 mhz; 3 kw. 300 ft TL: N40 36 27 W83 14 14. Stereo. Hrs opn: 24 1330 N. Main St., 43302. Phone: (740) 383-1131. Fax: (740) 387-6997. Web Site: www.wdif.com. Licensee: Citicasters Licenses L.P. Group owner: Clear Channel Communications Inc. (acq 1999; grpsl). Population served: 150,000 Natl. Rep: Rgnl Reps. Format: CHR. Target aud: 25-54; upscale adult females. ♦Diane Glassmeyer, gen mgr; Audra Meadows, prom dir; Mike Mitchell, adv dir; Scott Shawver, opns mgr & mus dir.

WMRN(AM)— Dec 23, 1940: 1490 khz; 1 kw-U. TL: N40 36 54 W83 07 54. Hrs open: 1330 N. Main St., 43302. Phone: (740) 383-1131. Fax: (740) 387-3697. Web Site: www.wmrn.com. Licensee: Citicasters Licenses L.P. Group owner: Clear Channel Communications Inc. (acq 1999; grpsl). Population served: 38,646 Format: Oldies, news/talk. Spec prog: Farm 5 hrs wkly. ♦Diane Glassmeyer, gen mgr.

WMRN-FM— April 1953: 106.9 mhz; 25 kw. 340 ft TL: N40 36 54 W83 07 54. (CP: Ant 358 ft. TL: N40 36 50 W83 07 47). Stereo. Web Site: www.buckeyecountry107.com.58,000 Natl. Network: Westwood One. Format: Buckeye country. ♦Kurt Kaniewski, traf mgr; Ben Failor, news rptr; Scott Shawver, disc jockey; Tim Kennedy, disc jockey.

***WOSB(FM)**— Apr 14, 1998: 91.1 mhz; 2.5 kw horiz, 6.8 kw vert. 285 ft TL: N40 41 06 W83 15 24. Hrs open: 24
Rebroadcasts WOSU-FM Columbus.
2400 Olentangy River Rd., Columbus, 43210. Phone: (614) 292-9678. Fax: (614) 292-0513. Web Site: www.wosu.org. Licensee: The Ohio State University. Natl. Network: NPR. Wire Svc: AP Format: Classical, News. ♦Thomas Rieland, gen mgr.

Marysville

WUCO(AM)— Dec 1, 1983: 1270 khz; 500 w-U, DA-2. TL: N40 14 46 W83 19 50. Hrs open: 24 107 N. Main St., 43040. Phone: (937) 642-1270. Fax: (937) 644-1617. E-mail: info@stgabrielradio.com Web Site: www.stgabrielradio.com. Licensee: St. Gabriel Radio Inc (acq 1-25-2006; $250,000). Population served: 40,000 Natl. Network: EWTN Radio. Rgnl. Network: Agri-Net, Ohio News Net. Format: Catholic talk. ♦Chris Gabrelcik, pres; Nick D'Orsi, gen mgr; Della Schoenecker, progmg mgr; Farris Wilhite, engrg mgr.

Massillon

WTIG(AM)— Aug 1, 1957: 990 khz; 250 w-D, 119-N, DA-2. TL: N40 49 56 W81 33 40. Hrs open: 24 Box 608, 44648. Secondary address: 3580 Karen Ave. N.W. 44647. Phone: (330) 837-9900. Fax: (330) 837-9844. E-mail: espn@espn990.com Web Site: www.espn990.com. Licensee: WTIG Inc. (acq 1991; FTR: 8-12-85). Population served: 32,539 Rgnl. Network: Ohio News Net. Format: Sports. Target aud: 25-54; male. Spec prog: Loc church svcs 6 hrs wkly. ♦Donovan Resh, VP, opns dir, chief of opns & progmg dir; Ray Jeske, pres & gen mgr.

Maumee

***WYSZ(FM)**— Nov 14, 1992: 89.3 mhz; 6.3 kw. 321 ft TL: N41 38 55 W83 42 22. Stereo. Hrs opn: 24 5115 Glendale Ave., Toledo, 43614. Phone: (419) 389-0893. Fax: (419) 381-0731. Licensee: Side By Side Inc. Population served: 500,000 Natl. Network: Salem Radio Network. Law Firm: Gammon & Grange. Format: Christian, CHR/Rock. Target aud: 15-25. ♦J. Todd Hostetler, gen mgr; Janet Yonke, dev dir & progmg dir.

McArthur

WYRO(FM)— 1994: 98.7 mhz; 6 kw. 328 ft TL: N39 08 59 W82 35 36. Hrs open: 24 Box 667, Jackson, 45640. Secondary address: 295 E. Main St., Jackson 45640. Phone: (740) 286-3023. Fax: (740) 286-6679. E-mail: jmossbarger@jcbiradio.com Licensee: Davis Broadcasting Media Inc. (acq 4-26-99). Natl. Network: Westwood One. Rgnl rep: Keystone (unwired net), Rgnl Reps. Format: Classic rock. News staff: one. Target aud: 18-65. ♦Jerry Mossbarger, gen mgr; Bill Forthoter, gen sls mgr; John Pelletier, progmg mgr.

McConnelsville

WJAW-FM— October 1992: 100.9 mhz; 930 w. Ant 577 ft TL: N39 33 24 W81 51 06. Stereo. Hrs open: 24 Box 708, Marietta, 45750. Phone: (740) 373-1490. Fax: (740) 373-1717. E-mail: swiles@wmoa1490.com Web Site: wmao1490.com. Licensee: Jawco Inc. (acq 6-25-97; $659,300 with WMOA(AM) Marietta). Population served: 50,000 Natl. Network: ABC. Rgnl. Network: Ohio Radio Net. Format: Sports. News staff: one. Target aud: 18-34; male. ♦John Wharff III, pres, sr VP & gen mgr.

Medina

WQMX(FM)—Licensed to Medina. See Akron

Miamisburg

WFCJ(FM)— Jan 7, 1961: 93.7 mhz; 50 kw. Ant 492 ft TL: N39 39 35 W84 18 53. Stereo. Hrs opn: 24 Box 93.7, Dayton, 45449-0999. Secondary address: 7333 Manning Rd. 45342. Phone: (937) 866-2471. Fax: (937) 866-2062. E-mail: inspiration@wfcj.com Web Site: www.wfcj.com. Licensee: Miami Valley Christian Broadcasting Association Inc. Population served: 2,600,000 Natl. Network: USA, Salem Radio Network. Natl. Rep: Salem. Law Firm: Miller & Neely. Wire Svc: AP Format: Relg, Christian. News: 10 hrs wkly. Target aud: 35-64; Evangelical Christians. Spec prog: Black 3 hrs, children 2 hrs wkly. ♦Bud Schindler, pres; Clair D. Miller, VP & gen mgr; Bill Nance, progmg dir; John Graham, chief of engrg.

Miamitown

***WMWX(FM)**— 08/05/06: 88.9 mhz; 4.6 kw. Ant 374 ft TL: N39 19 18 W84 57 33. Hrs open: 24 5114 Princeton-Glendale Rd., Hamilton, 45011. Phone: (513) 887-0590. E-mail: bspry@classxradio.com Web Site: classxradio.com. Licensee: Spryex Communications Inc. Format: Classic rock. Target aud: 30-58. ♦William J. Spry Jr., pres.

Middleport

WYVK(FM)—Licensed to Middleport. See Middleport-Pomeroy

Middleport-Pomeroy

WMPO(AM)— Aug 28, 1959: 1390 khz; 5 kw-D, 120 w-N. TL: N39 00 35 W82 04 14. Hrs open: Box 71, 39520 Bradbury Rd., Middleport, 45760. Phone: (740) 992-6485. Fax: (740) 992-6486. Licensee: Positive Radio Group Inc. of Ohio. Group owner: Baker Family Stations (acq 1999; $492,000 with WYVK(FM) Middleport). Population served: 27,304 Natl. Rep: Rgnl Reps. Format: Sports. Target aud: 35 plus. Spec prog: Relg 6 hrs, farm one hr, gospel 18 hrs wkly. ♦Kevin Nott, pres, gen mgr & progmg dir.

WYVK(FM)—Co-owned with WMPO(AM). Aug 27, 1973: 92.1 mhz; 4.7 kw. 113 ft TL: N39 03 30 W82 02 31. Stereo. 19 37845 Format: Top 40. News: 3 hrs wkly. Target aud: 25-54. ♦Kevin Nott, mus dir.

Middletown

WPFB(AM)— Sept 1, 1947: 910 khz; 1 kw-D, 100 w-N. TL: N39 30 57 W84 21 05. Hrs open: 24 4505 Central Ave., 45044. Phone: (513) 422-3625. Fax: (513) 424-9732. Web Site: www.wpfb.com. Licensee: Radio Stations WPAY/WPFB Inc. Group owner: WPAY/WPFB Inc. Population served: 2,000,000 Natl. Rep: Roslin. Format: Classic country. News staff: 2; News: 26 hrs wkly. Spec prog: Radio Movie of the Week 2 hrs wkly. ♦Douglas L. Braden, pres & gen mgr.

WPFB-FM— July 1, 1959: 105.9 mhz; 34 kw. 590 ft TL: N39 30 57 W84 21 05. Stereo. 24 Web Site: www.therebel1059.com.48,767 Format: Country. Target aud: 25-54.

Milford

WKFS(FM)— Aug 1, 1969: 107.1 mhz; 3 kw. 299 ft TL: N39 06 16 W84 20 10. (CP: 6 kw). Stereo. Hrs opn: 8044 Montgomery Rd., Suite 650, Cincinnati, 45236. Phone: (513) 686-8300. Web Site: www.kiss107fm.com. Licensee: Jacor Broadcasting Corp. Group owner: Clear Channel Communications Inc. (acq 5-4-99; grpsl). Law Firm: Verner, Liipfert, Bernhard, McPherson & Hand. Format: Top 40. ♦Mike Kenney, gen mgr; Chuck Fredrick, opns mgr & chief of opns.

Millersburg

WKLM(FM)— 1988: 95.3 mhz; 3 kw. 328 ft TL: N40 29 07 W81 50 40. Stereo. Hrs opn: 5:30 AM-midnight 7409 White Hill Ln., 44654. Phone: (330) 674-1953. Fax: (330) 674-9556. Licensee: Coshocton Broadcasting Co. (group owner; acq 7-10-90; $490,000; FTR: 8-6-90). Natl. Network: ABC. Format: Adult contemp. News staff: one; News: 12 hrs wkly. Target aud: General. Spec prog: Loc sports. ♦Bruce Wallace, pres & gen mgr; Tom Thompson, gen sls mgr; Mark Lonsinger, progmg dir.

***WVML(FM)**— June 2004: 90.5 mhz; 4.8 kw. Ant 367 ft TL: N40 36 08 W81 44 32. Stereo. Hrs opn: 24
Rebroadcasts WCRF(FM) Cleveland 100%.
WCRF Radio, 9756 Barr Rd., Cleveland, 44141. Phone: (440) 526-1111. Fax: (440) 526-1319. E-mail: wcrf@moody.edu Web Site: wcrfradio.org. Licensee: The Moody Bible Institute of Chicago (group owner). Law Firm: Southmayd. Wire Svc: AP Format: Inspirational. Target aud: 25-55; Adults. ♦Dr. Michael Easley, pres; Richard Lee, stn mgr; Phil Villareal, progmg dir; Doug Hainer, chief of engrg.

Montpelier

WLZZ(FM)— 1991: 104.5 mhz; 3 kw. Ant 328 ft TL: N41 30 54 W84 39 43. Hrs open: 24 209 W. Main St., 43543. Phone: (419) 485-5530. Phone: (800) 788-1045. Fax: (419) 485-5539. E-mail: wlzz@wlzzradio.com Licensee: Lake Cities Broadcasting Corp. Population served: 50,000 Format: Country. News staff: one; News: 12 hrs wkly. Target aud: 25-54. ♦Tom Andrews, CEO, chmn & pres; William Kerner, exec VP & gen mgr.

Morrow

***WLMH(FM)**— 1970: 89.1 mhz; 100 w. 200 ft TL: N39 20 51 W84 08 13. Hrs open: 3001 E. US. 22nd & 3rd, 45152. Phone: (513) 899-3884. Fax: (513) 899-4912. Licensee: Little Miami Local Schools. Format: Educ, oldies, classic rock. ♦Wayne Lyke, opns mgr.

Mount Gilead

WVXG(FM)— March 1994: 95.1 mhz; 6 kw. Ant 328 ft TL: N40 35 15 W82 48 20. Stereo. Hrs opn: 24 Box 102, Powell, 43065. Phone: (740) 549-7002. Licensee: ICS Holdings Sub 1 Inc. (acq 12-17-2003; $384,588). Population served: 76,000 ♦Mark Litton, gen mgr.

Mount Vernon

WMVO(AM)— Nov 26, 1953: 1300 khz; 500 w-D, DA. TL: N40 24 17 W82 26 23. Hrs open: 17421 Coshocton Rd., Box 348, 43050. Phone: (740) 397-1000. Fax: (740) 392-9300. Web Site: www.wmvo.com. Licensee: Capstar TX L.P. Group owner: Clear Channel Communications Inc. (acq 2-12-01; grpsl). Population served: 20,000 Natl. Network: ABC. Natl. Rep: Rgnl Reps. Wire Svc: AP Format: Var, news/talk. Spec prog: Relg 7 hrs wkly. ♦Diana Coon, gen mgr & mktg mgr; Michael Hayes, opns mgr & news dir.

WQIO(FM)—Co-owned with WMVO(AM). May 26, 1951: 93.7 mhz; 37 kw. Ant 565 ft TL: N40 24 18 W82 26 20. Stereo. 24 E-mail: info@ohioradio.com Web Site: www.wqiofm.com.13,373 Format: Adult contemp. News staff: one. Target aud: 35-54. Spec prog: Hit mus 4 hrs, gospel 2 hrs wkly.

***WNZR(FM)**— May 1, 1986: 90.9 mhz; 100 w. 193 ft TL: N40 22 14 W82 28 05. Stereo. Hrs opn: 24 800 Martinsburg Rd., 43050. Phone: (740) 392-9090. Fax: (740) 392-9155. E-mail: wnzr@mvnu.edu Web Site: www.wnzr.fm. Licensee: Mt. Vernon Nazarene University. Population served: 53,309 Natl. Network: AP Radio. Law Firm: Sciarrina & Associates. Wire Svc: AP Format: Christian adult contemp. News: 5 hrs wkly. Target aud: 25-54; Christian adults. ♦Mary Rinehart, stn mgr.

Napoleon

WNDH(FM)— June 1972: 103.1 mhz; 3.3 kw. 300 ft TL: N41 18 00 W84 09 22. Stereo. Hrs opn: 24 709 N. Perry St., 43545. Phone: (419) 592-8060. Fax: (419) 592-1085. E-mail: wndh@clearchannel.com Web Site: www.wndh1031.com. Licensee: CC Licenses LLC. Group owner: Clear Channel Communications Inc. (acq 1-1-2000; grpsl). Natl. Network: CBS. Rgnl. Network: Agri Bcstg, Ohio Radio Net. Rgnl rep: Rgnl Reps Wire Svc: AP Format: Adult contemp. News staff: one. Target aud: General. Spec prog: Ger Polka 2 hrs wkly. ♦Robert E. McLimans, sr VP, VP & gen mgr; Rick Small, opns dir; John Schuette, gen sls mgr.

Nelsonville

WSEO(FM)— September 1990: 107.7 mhz; 3 kw. 328 ft TL: N39 27 38 W82 13 09. Stereo. Hrs opn: 24 15751 U.S. Rt. 33 S., 45764. Phone: (740) 753-2154. Phone: (740) 753-4094. Fax: (740) 753-4965. E-mail: wseo33@sbcglobal.net Licensee: Nelsonville TV Cable Inc. Format: Contemp country. News staff: 3; News: 15 hrs wkly. Target aud: 25-49. Spec prog: Farm. ♦Eugene R. Edwards, pres & gen mgr; Nick Brooks, mus dir.

New Boston

WIOI(AM)— Sept 2, 1959: 1010 khz; 1 kw-D, 22 w-N. TL: N38 43 48 W82 57 10. Hrs open: Box 1233, Portsmouth, 45662. Phone: (606) 932-4796. Fax: (606) 932-4796. E-mail: chip@wioiradio.com Web Site: wioiradio.com. Licensee: Maillet Media Inc. (acq 1996). Population served: 27,633 Format: Adult standards. ♦Charles Maillet Jr., gen mgr.

New Concord

***WMCO(FM)**— Jan 28, 1961: 90.7 mhz; 1.3 kw. 84 ft TL: N39 59 46 W81 43 18. Stereo. Hrs opn: 6 AM-midnight Caldwell Hall, 163 Stormont St., 43762. Phone: (740) 826-8375. E-mail: wmco@muskingum.edu Web Site: www.muskingum.edu/~wmco. Licensee: Muskingum College. Population served: 2,500 Format: Div, educ, progsv. News: 10 hrs wkly. Target aud: General. Spec prog: Class 4 hrs, jazz 10 hrs, relg 2 hrs wkly. ♦Jeffrey D. Harman, gen mgr; Kim Fox, stn mgr; Matthew Hott, progmg dir.

New Lexington

WWJM(FM)— May 1, 1978: 105.9 mhz; 1.7 kw. 627 ft TL: N39 46 37 W82 09 54. Stereo. Hrs opn: 24 210 S. Jackson St., 43764. Secondary address: 247 Market St., Zanesville 43701. Phone: (740) 342-1988. Fax: (740) 342-1036. E-mail: wwjm@aol.com Web Site: wwjm.com. Licensee: Perry County Broadcasting Co. Population served: 100,000 Natl. Network: Westwood One. Format: Adult contemp. News staff: one; News: 2 hrs wkly. Target aud: 18-54; young to middle-aged. ♦Charles Edwards, chmn, pres & gen mgr.

New Philadelphia

WJER(AM)—See Dover-New Philadelphia

***WKRJ(FM)**— July 12, 1994: 91.5 mhz; 2 kw. 240 ft TL: N40 33 50 W81 31 05. Hrs open:
Rebroadcasts WKSU-FM Kent 100%.
c/o WKSU-FM, 1613 E. Summit St., Kent, 44242-0001. Phone: (330) 672-3114. Fax: (330) 672-4107. Web Site: www.wksu.org. Licensee: Kent State University. Population served: 76,525 Natl. Network: NPR, PRI. Law Firm: Dow, Lohnes & Albertson. Format: News, class. News staff: 5; News: 35 hrs wkly. Target aud: 35-65; college grad, professional & upper income. ♦Allen E. Bartholet, gen mgr; Patricia Gerber, dev dir; Robert Burford, mktg dir & prom dir; Eric Nuzum, progmg dir; David Roden, mus dir; Vincent Duffy, news dir; Ronald Bartlebaugh, engrg dir.

WNPQ(FM)— Feb 2, 1969: 95.9 mhz; 3 kw. 400 ft TL: N40 35 51 W81 29 32. Stereo. Hrs opn: 24 3969 Convenience Cir. N.W., Suite 205, Canton, 44718. Phone: (330) 492-9590. Fax: (330) 492-3702. Web Site: www.wnpqfm.com. Licensee: Tuscarawas Broadcasting Co. Population served: 300,000 Natl. Network: CBS Radio. Format: Christian contemp. Target aud: 18-49; family oriented. Spec prog: Black 4 hrs, southern gospel 4 hrs wkly. ♦James Natoli Jr., pres; Garry Meeks, gen mgr; Tom Bishop, gen sls mgr.

WTUZ(FM)—Uhrichsville, May 1, 1990: 99.9 mhz; 5.3 kw. 348 ft TL: N40 26 19 W81 26 01. Stereo. Hrs opn: 24 2424 E. High Ave., 44663. Phone: (330) 339-2222. Fax: (330) 339-5930. E-mail: info@wtuz.com Web Site: www.wtuz.com. Licensee: WTUZ Radio Inc. Population served: 120,000 Natl. Network: Fox News Radio. Law Firm: Smithwick & Belendiuk. Wire Svc: AP Format: Country. News staff: 2; News: 7 hrs wkly. Target aud: General. Spec prog: Farm 2 hr, relg 4 hr wkly. ♦Edward A. Schumacher, pres & gen mgr; Melanie Osborn, gen sls mgr; Pat Smith, prom dir; Greg Morrison, progmg dir; Brad Shupe, news dir; John Demuth, chief of engrg.

Newark

WCLT(AM)— Jan 4, 1949: 1430 khz; 500 w-D, 48 w-N. TL: N40 02 02 W82 24 08. Hrs open: 24 Box 5150, 43058-5150. Secondary address: 674 Jacksontown Rd. S.E., Heath 43056. Phone: (740) 345-4004. Fax: (740) 345-5775. E-mail: wclt@wclt.com Web Site: www.wclt.com. Licensee: WCLT Radio Inc. (acq 1-1-58). Population served: 145,000 Natl. Network: AP Radio, Fox News Radio. Wire Svc: AP Format: News/talk. News staff: 2; News: 12 hrs wkly. Target aud: General. ♦Robert H. Pricer, CEO; Douglas C. Pricer, pres & gen mgr.

WCLT-FM— Aug 7, 1947: 100.3 mhz; 50 kw. 390 ft TL: N40 02 02 W82 24 08. Stereo. 24 Natl. Network: Fox News Radio. Format: Country. Target aud: 25-54.

WHTH(AM)—See Heath

WNKO(FM)— Dec 8, 1972: 101.7 mhz; 3 kw. Ant 298 ft TL: N39 59 38 W82 30 13. Stereo. Hrs opn: 24 Box 1057, 1000 N. 40th St., 43058-1057. Phone: (740) 522-8171. Fax: (740) 522-8174. E-mail: sales@wnko.com Web Site: www.wnko.com. Licensee: Runnymede Corp. (acq 10-15-98; $100,000 for stock with WHTH(AM) Heath). Population served: 140,000 Natl. Network: CNN Radio. Wire Svc: AP Format: Classic hits. News staff: 2. Target aud: 25-54. ♦Charles Franks, pres; Tom Swank, gen mgr; John Franks, opns VP.

Niles

WBBG(FM)— May 15, 1988: 106.1 mhz; 3 kw. 328 ft TL: N41 15 52 W80 45 35. Hrs open: 7461 South Ave., Youngstown, 44512. Phone: (330) 965-0057. Fax: (330) 729-9991. E-mail: billkelly@clearchannel.com Web Site: www.wbbgfm.com. Licensee: Citicasters Licenses L.P. Group owner: Clear Channel Communications Inc. (acq 5-4-99; grpsl). Law Firm: Cohn & Marks. Format: Oldies. Target aud: 18-49. ♦Bill Kelly, gen mgr; Dan Rivers, opns mgr; Jeff Kelly, progmg dir; John Nagy, news dir; Jim Hartzler, chief of engrg.

WRTK(AM)— Nov 1, 1963: 1540 khz; 500 w-D, DA. TL: N41 07 56 W80 45 40. Hrs open: 124 N. Park Ave., Warren, 44482. Phone: (330) 394-7700. Web Site: www.wrtk.net. Licensee: Beacon Broadcasting Inc. (acq 9-14-2005; $400,000). Population served: 100000 Natl. Network: ABC. Format: Urban contemp, gospel. Spec prog: It one hr, Pol one hr wkly. ♦Dominic Baragona, gen mgr; Jay Curtis, opns mgr; Robert Hotchkiss, gen sls mgr & rgnl sls mgr; Chris Patrick, progmg dir.

North Baltimore

WPFX-FM— July 30, 1990: 107.7 mhz; 3 kw. 328 ft TL: N41 07 04 W83 32 38. Stereo. Hrs opn: 24 1624 Tiffin Ave., Findlay, 45840. Phone: (419) 425-1077. Fax: (419) 422-2954. Web Site: www.1077thefox.com. Licensee: Citicasters Licenses L.P. Group owner: Clear Channel Communications Inc. (acq 5-4-99; grpsl). Law Firm: Miller & Miller. Format: Classic rock. Target aud: General.

North Canton

WHOF(FM)— Aug 29, 1968: 101.7 mhz; 6 kw. Ant 266 ft TL: N40 49 22 W81 25 41. Stereo. Hrs opn: 24 7755 Freedom Ave. N.W., 44720. Phone: (330) 492-4700. Web Site: www.my1017.com. Licensee: CC Licenses LLC. Group owner: Clear Channel Communications Inc. (acq 1-30-2004; $4.3 million with WJER(AM) Dover-New Philadelphia). Population served: 11,516 Format: Adult contemp. Spec prog: Farm 2 hrs wkly.

North Kingsville

WFXJ-FM— Apr 8, 2002: 107.5 mhz; 6 kw. Ant 328 ft TL: N41 54 10 W80 39 36. Hrs open: 3226 Jefferson Rd., Ashtabula, 44004-9112. Phone: (440) 998-1075. Fax: (440) 992-2658. E-mail: danaaschulte @ckearchannel.com Web Site: www.theforx1075.com. Licensee: CC Licenses LLC. Group owner: Clear Channel Communications Inc. (acq 7-11-2000; grpsl). Format: Classic rock. Target aud: 18-54; males. ♦Dana Schulte, VP & gen mgr; Dennis O'Brien, opns dir; Michelle Baird, sls dir.

North Ridgeville

WJTB(AM)— Sept 16, 1984: 1040 khz; 5 kw-D. TL: N41 22 37 W82 00 27. Hrs open: 105 Lake Ave., Elyria, 44035. Phone: (440) 327-1844. Fax: (440) 322-8942. E-mail: wjtb1040am@aol.com Licensee: Taylor Broadcasting Co. Format: Urban contemp, gospel. ♦James Taylor, pres & gen mgr; Henry Dunn, opns mgr.

Norwalk

WLKR(AM)— Mar 18, 1968: 1510 khz; 500 w-D, DA. TL: N41 16 45 W82 39 23. Hrs open: Sunrise-sunset 10327 Milan Rd., U.S. Rt. 250, Milan, 44846. Phone: (419) 609-5961. Fax: (419) 609-2679. E-mail: wikr@acc.com Web Site: www.wlkrradio.com.yes Licensee: Elyria-Lorain Broadcasting Co. (group owner; acq 4-9-02; with co-located FM). Population served: 150,000 Natl. Network: Westwood One, ESPN Radio. Format: Oldies. News staff: one; News: 2 hrs wkly. Target aud: 40 plus. ♦Bill Hatheway, gen sls mgr; Shelly Luipold, rgnl sls mgr; Tim Kelly, stn mgr & progmg mgr; Scott Truxell, news dir & local news ed; Ken Wilde, chief of engrg; Mike Jeffries, traf mgr.

WLKR-FM— Sept 17, 1962: 95.3 mhz; 3 kw. 300 ft TL: N41 16 49 W82 39 27. Stereo. 24 Web Site: wlkrradio.com.yes 150,000 Format: Adult contemp. News staff: one; News: 3 hrs wkly. Target aud: General; residents of Huron & Erie counties. Spec prog: Farm 3 hrs wkly. ♦Tim Kelly, mus dir; Carol Walters, pub affrs dir; Mike Jeffries, traf mgr; Scott Truxel, local news ed.

***WNRK(FM)**— 2004: 90.7 mhz; 4 kw. Ant 407 ft TL: N41 10 50 W82 23 21. Hrs open: 24
Rebroadcasts WKSU-FM Kent 100%.
c/o WKSU-FM, Box 5190, Kent, 44242-0001. Secondary address: 1613 E. Summit St. 44242-0001. Phone: (330) 672-3114. Fax: (330) 672-4107. Web Site: www.wksu.org. Licensee: Kent State University. Natl. Network: AP Radio, NPR, PRI. Law Firm: Dow, Lohnes & Albertson. Format: Classical, news. ♦Allen E. Bartholet, gen mgr; Deborah Frazier, gen mgr; N. Vincent Duffy, opns dir & progmg dir; Abbe Turner, dev dir; Ronald Bartlebaugh, engrg dir.

Oak Harbor

WJZE(FM)— August 1993: 97.3 mhz; 4.3 kw. Ant 387 ft TL: N41 28 19 W83 25 05. Stereo. Hrs opn: 24 110 Ottawa St., Toledo, 43602. Phone: (419) 255-6600. Fax: (419) 255-6600. Licensee: Urban Radio Licenses LLC. (acq 6-30-2005; $2.6 million). Population served: 500,000 Law Firm: Smithwick & Belendiuk. Format: Active rock. News staff: one. Target aud: 25-64; general, upscale. Spec prog: Saturday nite Root Hoot blues show. ♦Clyde Roberts, gen mgr.

Oberlin

***WOBC-FM**— November 1951: 91.5 mhz; 440 w. 124 ft TL: N41 17 39 W82 13 26. (CP: 88.3 mhz, 3.5 kw). Stereo. Hrs opn: 24 Wilder Hall, 135 W. Lorain St., 44074. Phone: (440) 775-8107. Phone: (440) 775-8139. Fax: (440) 775-6678. Web Site: www.wobc.org. Licensee: Oberlin College Student Network Inc. Population served: 250,000 Format: Div, educ. News: 5 hrs wkly. Target aud: General. Spec prog: Folk 12 hrs, Fr one hr, jazz 15 hrs, electronic 20 hrs wkly. ♦Dean Bein, stn mgr; Evan Smith, opns mgr & dev dir.

WOBL(AM)— Dec 24, 1971: 1320 khz; 1 kw-U, DA-2. TL: N41 16 05 W82 12 40. Hrs open: 24 Box 277, 45624 Rt. 20 E., 44074. Phone: (440) 774-1320. Fax: (440) 774-1336. E-mail: woblwdlw@earthlink.net Licensee: WOBL Inc. Population served: 276,000 Rgnl. Network: Agri Bcstg. Law Firm: Reddy, Begley & McCormick, LLP. Wire Svc: AP Format: Country Gold. News staff: 3; News: 14 hrs wkly. Target aud: 35-55. Spec prog: Farm 5 hrs, relg one hr wkly. ♦Doug Wilber, gen mgr; Terry Coffee, progmg dir & progmg mgr.

Ontario

WRGM(AM)— July 17, 1987: 1440 khz; 1 kw-D, DA. TL: N40 46 05 W82 37 04. Stereo. Hrs opn: 24 2900 Park Ave. W., Mansfield, 44906. Phone: (419) 529-5900. Fax: (419) 529-2319. Web Site: www.wrgm.com. Licensee: GSM Media Corp. Population served: 155,000 Natl. Network: ESPN Radio. Natl. Rep: Rgnl Reps. Format: Sports. News staff: 5; News: 10 hrs wkly. Target aud: 25 plus. Spec prog: High school football & basketball, NASCAR races. ♦Gunther Meisse, pres & gen mgr.

Ottawa

WBUK(FM)— Feb 4, 1977: 106.3 mhz; 1.4 kw. 489 ft TL: N40 57 21 W83 54 42. Stereo. Hrs opn: 24 1624 Tiffin Ave., Findlay, 45840. Phone: (419) 425-1077. Fax: (419) 422-2954. Web Site: www.wbuk.com. Licensee: Citicasters Licenses L.P. Group owner: Clear Channel Communications Inc. (acq 1999; grpsl). Population served: 300,000 Rgnl. Network: Agri Bcstg. Format: Oldies. News staff: one; News: 3 hrs wkly. Target aud: 18-49; affluent, upscale adults. Spec prog: Farm 5 hrs, sports 2 hrs, MOR 4 hrs wkly. ♦Kim Fields, gen mgr.

Oxford

***WMUB(FM)**— 1950: 88.5 mhz; 24.5 kw. 499 ft TL: N39 33 26 W84 47 35. Stereo. Hrs opn: 24 Williams Hall, Miami Univ., 45056. Phone: (513) 529-5885. Fax: (513) 529-6048. E-mail: comment@wmub.org Web Site: www.wmub.org. Licensee: President & Trustees of Miami University. Population served: 716,726 Natl. Network: NPR, PRI. Law Firm: Baker & Hostetler. Wire Svc: AP Format: News/talk, jazz. News staff: 2; News: 42 hrs wkly. Target aud: General. Spec prog: Folk 5 hrs wkly. ♦Cleve Callison, gen mgr; John E. Hingsbergen, progmg dir; John Hingsbergen, mus dir; Gary Scott, news dir; Kim Keen, chief of engrg.

WOXY(FM)— Dec 24, 1959: 97.7 mhz; 3 kw. 255 ft TL: N39 28 44 W84 45 51. (CP: Ant 321 ft.). Stereo. Hrs opn: 24 5120 College Corner Pike, 45056. Phone: (937) 378-6151. Fax: (513) 377-2200. Licensee: First Broadcasting Capital Partners LLC. Group owner: First Broadcasting Investment Partners LLC (acq 3-17-2004; $5.64 million). Population served: 1,000,000 Law Firm: Fletcher, Heald & Hildreth. Wire Svc: AP Format: Alternative. Target aud: 18-34. ♦Heather Frye, gen mgr.

Painesville

WABQ(AM)— Apr 25, 1956: 1460 khz; 1 kw-D, 500 w-N, DA-2. TL: N41 44 20 W81 14 09. Hrs open: 24 One Radio Pl., 44077. Phone: (440) 951-1460. Fax: (440) 352-8194. Fax: (440) 357-7701. E-mail: wbkc@wbkc.com Web Site: www.wbkc.com. Licensee: Radio Advantage One LLC (acq 1-27-2005; $450,000). Population served: 30,000 Format: Gospel. ♦Dale Edwards, pres & gen mgr; Almira Byrd, exec VP; Danelle Caldwell, opns mgr.

Parma

WCCD(AM)— Jan 9, 1973: 1000 khz; 500 w-D, DA. TL: N41 19 11 W81 46 07. Hrs open: 4 Summit Park Dr., Suite 150, Independence, 44131. Phone: (216) 901-0921, EXT. 205. Fax: (216) 901-1104. Licensee: Caron Broadcasting, Inc. Group owner: Salem Communications Corp. (acq 4-20-2005; $2.1 million). Population served: 12,807 Natl. Network: Salem Radio Network. Format: Relg, talk. Target aud: 25-54. Spec prog: Black 3 hrs, Greek 2 hrs, Ukrainian one hr wkly. ♦Errol Dengler, gen mgr.

Paulding

WKSD(FM)— Aug 14, 1989: 99.7 mhz; 3 kw. 328 ft TL: N41 03 32 W84 35 30. Hrs open: 24 Box 487, Van Wert, 45891. Phone: (419) 238-1220. Fax: (419) 238-2578. Licensee: First Family Broadcasting Inc. (acq 1-3-95; $225,000; with WERT(AM) Van Wert; FTR: 3-6-95) Natl. Rep: Rgnl Reps. Format: Hot Adult Contemp. ♦Chris Roberts, pres, gen mgr, progmg dir, news dir & chief of engrg; Mona Kennedy, gen sls mgr & traf mgr.

Piketon

WXZQ(FM)— December 1997: 100.1 mhz; 6 kw. 328 ft TL: N39 05 53 W82 57 20. Stereo. Hrs opn: 24 Box 820, 45661. Phone: (740) 947-0059. Fax: (740) 947-4600. E-mail: wxiz@roadrunner.com Licensee: Piketon Communications. Population served: 30,000 Format: Current hit radio. Target aud: 18-49. ♦Gerald E. Davis, gen mgr; Brad Lambert, gen sls mgr.

Piqua

WHIO-FM— Nov 30, 1960: 95.7 mhz; 50 kw. Ant 476 ft TL: N40 13 02 W84 17 35. Stereo. Hrs opn: 24
Simulcast with WHIO(AM) Dayton 100%.
1414 Wilmington Ave., Dayton, 45420. Phone: (937) 259-2111. Fax: (937) 259-2328. Web Site: 1290whio.com. Licensee: Cox Radio Inc. Group owner: Cox Broadcasting (acq 1998; grpsl). Population served: 1,463,000 Format: News/talk. Target aud: 35-54. ♦Donna Hall, gen mgr; Nick Roberts, opns mgr; Marc Herbst, gen sls mgr.

WPTW(AM)— November 1947: 1570 khz; 250 w-U. TL: N40 08 14 W84 16 00. Stereo. Hrs opn: 24 1625 Covington Ave., 45356. Phone: (937) 773-3513. Fax: (937) 773-4345. E-mail: wptwnews@1570wptw.com Web Site: www.1570wptw.com. Licensee: Frontier Broadcasting L.L.C. #2 (acq 5-28-99; $75,000). Population served: 90,000 Natl. Network: CBS. Rgnl. Network: Ohio Radio Net. Natl. Rep: Rgnl Reps. Law Firm: Miller & Fields. Wire Svc: UPI Format: Oldies. News staff: one; News: 8 hrs wkly. Target aud: 35 plus. Spec prog: Farm 6 hrs, sports sports 8 hrs wkly. ♦Bart Johnson, pres; Joe Neues, gen mgr.

Pleasant City

WBIK(FM)— 2002: 92.1 mhz; 6 kw. Ant 169 ft TL: N40 01 37 W81 33 09. Hrs open: 24 4988 Skyline Dr., Box 338, Cambridge, 43725. Phone: (740) 432-5605. Fax: (740) 432-1991. Web Site: wbik.com. Licensee: David L. Wilson (acq 8-1-00). Rgnl rep: Rgnl Reps Format: Classic rock. ♦David L. Wilson, gen mgr.

Pomeroy

WMPO(AM)—See Middleport-Pomeroy

Port Clinton

WXKR(FM)— Oct 4, 1961: 94.5 mhz; 30 kw. 640 ft TL: N41 29 51 W83 16 12. Stereo. Hrs opn: 24 3225 Arlington Ave., Toledo, 43614. Secondary address: 2965 Pickle Rd., Oregon 43616. Phone: (419) 385-2507. Fax: (419) 385-2902. Web Site: www.wxkr.com. Licensee: Cumulus Licensing Corp. Group owner: Cumulus Media L.L.C. (acq 12-18-97; $5 million cash). Population served: 658,000 Natl. Network: ABC. Law Firm: Fletcher, Heald & Hildreth. Format: Classic Rock. News staff: one; News: one hr wkly. Target aud: 25-49. Spec prog: Sp one hr wkly.Kathy Stinehour, gen mgr; Tim Roberts, opns dir; Beth Gleason, gen sls mgr; Ryan Young, prom dir; Andi McKay, progmg dir; Dave Waters, news dir; Kevin Hawley, engrg dir; Joe Morgan, traf mgr; Julie Weitman, edit dir; Sal Olalde, spanish dir; DC Bash, disc jockey; Dusty Scott, disc jockey; Micki Morget, disc jockey; Mike McIntyre, disc jockey; Steve Duvall, disc jockey

Portsmouth

WNXT(AM)— Aug 30, 1951: 1260 khz; 5 kw-D, 1 kw-N, DA-2. TL: N38 48 38 W82 59 21. Hrs open: 24 Box 1228, Masonic Temple Bldg., 602 Chillicothe St., 45662. Phone: (740) 353-1161. Fax: (740) 353-8080. E-mail: wnxtradio@yahoo.com Web Site: www.wnxtradio.com. Licensee: Hometown Broadcasting of Portsmouth Inc. (acq 5-96; $477,500 with co-located FM). Population served: 83,000 Natl. Network: ABC, ESPN Radio. Natl. Rep: Rgnl Reps. Law Firm: Pepper & Corazzini. Format: ESPN sports radio/talk. News staff: one; News: 35 hrs wkly. Target aud: Adult males; 25-54. ♦Phillip Bruce Leslie, pres; Steve Hayes, exec VP & opns mgr; Rick Mayne, gen mgr & gen sls mgr; Chris Smith, progmg dir; Sam McKibbin, news dir; Tyrone Henry, chief of engrg.

WNXT-FM— Sept 15, 1965: 99.3 mhz; 2.55 kw. 512 ft TL: N38 43 20 W83 00 05. Stereo. 24 Fax: (740) 353-3191. Web Site: www.wnxtradio.com.83,000 Format: Adult contemp/gold. News staff: one; News: one hr wkly. Target aud: 25-54. ♦Rick Mayne, stn mgr; Chris Smith, progmg mgr.

***WOHP(FM)**— Feb 18, 1992: 88.3 mhz; 1 kw. 643 ft TL: N38 43 20 W83 00 05. Hrs open: 24
Rebroadcasts WCDR-FM Cedarville 100%.
Box 601, 251 N. Main St., Cedarville, 45314. Phone: (937) 766-7815. Fax: (937) 766-7927. Web Site: www.thepath.fm. Licensee: The Cedarville University. Population served: 100,000 Law Firm: Cohen & Berfield. Format: Relg, full-service. News staff: one; News: 16 hrs wkly. Target aud: 35-54; church oriented audience. Spec prog: Black 2 hrs wkly. ♦William Brown, pres; Marvin D. Sparks, gen mgr; Keith Hamer, opns mgr; Chad Bresson, news dir; John Tocknell, chief of engrg.

***WOSP(FM)**— May 25, 1993: 91.5 mhz; 110 w. 1,207 ft TL: N38 45 42 W83 03 41. Stereo. Hrs opn: 24
Rebroadcasts WOSU-FM Columbus 100%.
2400 Olentangy River Rd., Columbus, 43210. Phone: (614) 292-9678. Fax: (614) 292-0513. Web Site: www.wosu.org. Licensee: The Ohio State University. Population served: 75,000 Natl. Network: PRI, NPR, AP Radio. Law Firm: Dow, Lohnes & Albertson. Wire Svc: AP Format: Classical. Target aud: 35 plus. ♦Thomas Rieland, gen mgr; Tim Eby, stn mgr; Kevin Petrilla, opns mgr. Co-owned TV: *WPBO-TV affil.

WPAY(AM)— Apr 15, 1935: 1400 khz; 800 w-U. TL: N38 43 22 W83 00 05. Hrs open: 1009 Gallia St., 45662-4140. Phone: (740) 353-5176. Fax: (740) 353-1715. E-mail: comments@f1400wpay.com Web Site: www.1400wpay.com. Licensee: Radio Stations WPAY/WPFB Inc. Group owner: WPAY/WPFB Inc. (acq 2-1-57). Population served: 27,633 Natl. Network: CBS. Format: Talk. Target aud: 18-65; young, affluent & upwardly mobile adults. Spec prog: Gospel 6 hrs wkly. ♦Douglas Braden, pres; Frank Lewis, gen mgr; Lorenzo Bentley, news dir.

WPAY-FM— June 15, 1948: 104.1 mhz; 100 kw. Ant 1,486 ft TL: N38 41 00 W83 00 46. Stereo. Web Site: www.104wpay.com.88,500 Natl. Network: CBS. Format: Country.

WZZZ(FM)— January 2003: 107.5 mhz; 2.6 kw. Ant 495 ft TL: N38 43 22 W82 59 56. Stereo. Hrs opn: 24 Box 1228, 45662. Secondary address: 602 Chillicothe St. 45662. Phone: (740) 353-1161. Fax: (740) 353-3191. E-mail: classicrock1075thebreeze@yahoo.com Web Site: www.wzzz.com. Licensee: Hometown Broadcasting of Portsmouth 2 Inc. (acq 6-28-02). Rgnl rep: Rgnl Reps Law Firm: Smithwick & Bellendiuk, P.C. Format: Classic rock. News staff: one; News: .25 hrs wkly. Target aud: 35-54; working class & professional adults. ♦Rick Mayne, gen mgr; Steve Hayes, opns mgr; Bill Murphy, progmg dir; Mistie Cook, news dir; Tyrone Henry, chief of engrg.

Proctorville

***WMEJ(FM)**— Jan 25, 1986: 91.9 mhz; 3.5 kw. 220 ft TL: N38 27 14 W82 25 05. Stereo. Hrs opn: 24 Box 7575, Huntington, WV, 25777. Phone: (740) 867-5333. Licensee: Maranatha Broadcasting Inc.

Format: Easy lstng Christian mus. Target aud: General. ♦Paul S. Warren, pres; Tim Jenkins, gen mgr.

Racine

WNTO(FM)— July 15, 1996: 93.1 mhz; 4.1 kw. Ant 397 ft TL: N38 56 56 W82 03 02. Hrs open: Box 661, Gallipolis, 45631-0661. Phone: (740) 446-3643. Licensee: Legend Communications of West Virginia LLC. Group owner: Legend Communications L.L.C. (acq 11-14-2001). Format: Hot adult contemp. Target aud: 18-50; general. Spec prog: American Indian one hr, Black one hr, farm one hr, folk 2 hrs, gospel 7 hrs, relg 7 hrs wkly. ♦Dave Diddle, stn mgr.

Reading

***WMKV(FM)**— 1995: 89.3 mhz; 41 kw. 236 ft TL: N39 13 23 W84 25 56. Stereo. Hrs opn: 24 11100 Springfield Pike, Cincinnati, 45246. Phone: (513) 782-2427. Fax: (513) 782-2720. Web Site: www.wmkvfm.org. Licensee: Lifesphere. Format: Nostalgia. Target aud: 50 plus. ♦George Zahn, stn mgr.

Richwood

WJZK(FM)— Nov 30, 1995: 104.3 mhz; 3.4 kw. 436 ft TL: N40 21 52 W83 15 34. Stereo. Hrs opn: 24 4401 Carriage Hill Ln., Columbus, 43220. Phone: (614) 451-2191. Fax: (614) 451-1831. Web Site: wjza.com. Licensee: Franklin Communications Inc. Group owner: Saga Communications Inc. (acq 10-1-2003; $13 million with WJZA(FM) Lancaster). Population served: 1500000 Natl. Rep: Christal. Wire Svc: AP Format: Smooth jazz. News: 2 hrs wkly. Target aud: 25-54. Spec prog: Various 15 hrs wkly. ♦Alan Goodman, gen mgr; Chris Forgy, sls dir; Michelle Hurley, mktg dir; Bill Harman, progmg dir.

Ripley

WAOL(FM)— 1993: 99.5 mhz; 3 kw. 328 ft TL: N38 45 14 W83 50 24. Hrs open: 8354 Fryer Rd., Georgetown, 45121. Phone: (937) 378-6151. Fax: (937) 377-2200. Web Site: www.classiccountry995.com. Licensee: First Broadcasting Capital Partners LLC. Group owner: First Broadcasting Investment Partners LLC (acq 3-17-2004; $4.06 million with WAXZ(FM) Georgetown). Format: Classic country. ♦Heather Frye, gen mgr.

Rossford

WDMN(AM)— Nov 28, 1966: 1520 khz; 500 w-D, 400 w-N, DA-2. TL: N41 30 32 W83 33 07. Hrs open: 1510 Reynolds Rd., Maumee, 43537. Phone: (419) 725-9366. Fax: (419) 725-2600. Web Site: www.dominion1520am.com. Licensee: Cornerstone Church Inc. (acq 5-20-98; $200,000). Population served: 383,318 Natl. Network: American Urban. Natl. Rep: Roslin. Rgnl rep: Rgnl Reps. Format: Contemp Christian. Target aud: 25-59; older established audience. Spec prog: Relg (church ministry). ♦Robert Pitts, gen mgr.

Rushville

***WLRY(FM)**— December 1998: 88.9 mhz; 1.1 kw vert. Ant 298 ft TL: N39 46 41 W82 25 26. Hrs open: 24 Box 220, Arcangel Broadcasting Foundation, 43150. Phone: (740) 536-0885. Fax: (740) 536-1885. E-mail: wlry@wlry.org Web Site: www.wlry.org. Licensee: Arcangel Broadcasting Foundation. Population served: 120,000 Natl. Network: USA. Format: Relg, Christian, talk. News: 40 hrs wkly. Target aud: 15-55; youth & adult mentors. Spec prog: Issues talk 16 hrs wkly. ♦Richard Finke, gen mgr.

Saint Mary's

WKKI(FM)—See Celina

WMLX(FM)— 1998: 103.3 mhz; 950 w. 823 ft TL: N40 38 03 W84 12 29. Hrs open: 24 Box 1128, Lima, 45802. Secondary address: 667 W. Market St., Lima 45801. Phone: (419) 223-2060. Fax: (419) 229-3888. E-mail: comments@wmlx.com Web Site: www.wmlx.com. Licensee: Clear Channel Radio Licenses, Inc. Group owner: Clear Channel Communications Inc. (acq 5-4-99; grpsl). Population served: 150,000 Format: Hits of the 80s & 90s, adult contemp. News: 2 hrs wkly. Target aud: 18-49; women. ♦Bill Gentry, sr VP & opns mgr; Art Versnick, VP & gen mgr; Phil Austin, opns mgr; Jennifer Cartwright, sls dir; Jack Wheelbarger, natl sls mgr; Tom Francis, rgnl sls mgr; Kathy Hague, progmg dir; Mark Gierhart, engrg dir.

WWSR(FM)—See Wapakoneta

Salem

WQXK(FM)—Listing follows WSOM(AM).

WSOM(AM)— June 2, 1965: 600 khz; 1 kw-D, 45 w-N, DA-2. TL: N40 49 47 W80 55 54. Hrs open: 4040 Simon Rd, Youngstown, 44512. Phone: (330) 783-1000. Fax: (330) 783-0060. Web Site: www.600wsom.com. Licensee: Cumulus Licensing Corp. Group owner: Cumulus Media Inc. (acq 3-15-00; grpsl). Population served: 600,000 Law Firm: Kaye, Scholer, Fierman, Hays & Handler. Format: Nostalgia. Target aud: 35 plus. Spec prog: Farm 2 hrs wkly. ♦Lou Dickey, CEO; Donna Palowitz, gen sls mgr; Tim Roberts, progmg dir; Wes Boyd, chief of engrg.

WQXK(FM)—Co-owned with WSOM(AM). Nov 25, 1958: 105.1 mhz; 88 kw. Ant 430 ft TL: N40 53 06 W80 49 50. (CP: TL: N40 53 08 W80 49 55). Stereo. Web Site: www.q105country.com.1,400,000 Format: Country. Target aud: 25-54. ♦Dave Steele, progmg dir & traf mgr.

Sandusky

WCPZ(FM)— Aug 15, 1959: 102.7 mhz; 50 kw. Ant 141 ft TL: N41 26 29 W82 41 12. Stereo. Hrs opn: 1640 Cleveland Rd., 44870. Phone: (419) 625-1010. Fax: (419) 625-1348. Web Site: www.wcpz.com. Licensee: Citicasters Licenses L.P. Group owner: Clear Channel Communications Inc. (acq 5-4-99; grpsl). Population served: 1550000 Format: Hot adult contemp. News: 2 hrs wkly. Target aud: 18-54. ♦Paul Mize, gen mgr; Randy Hugg, progmg dir; Tammy Harrison, traf mgr.

WLEC(AM)—Co-owned with WCPZ(FM). Dec 7, 1947: 1450 khz; 1 kw-U. TL: N41 26 29 W82 41 10.24 Web Site: www.wlec.com.250,000 Rgnl rep: Rgnl Reps. Format: Sports. News: 2 hrs wkly.

WGGN(FM)—See Castalia

***WVMS(FM)**— December 1993: 89.5 mhz; 2.12 kw horiz, 5.36 kw vert. Ant 69 ft TL: N41 26 29 W82 48 20. Stereo. Hrs opn: 24 Rebroadcasts WCRF(FM) Cleveland 100%.
c/o Radio Stn WCRF(FM), 9756 Barr Rd., Cleveland, 44141. Phone: (440) 526-1111. Fax: (440) 526-1319. E-mail: wcrf@moody.edu Web Site: wcrfradio.org. Licensee: The Moody Bible Institute of Chicago. Law Firm: Southmayd. Wire Svc: AP Format: Inspirational, relg, Christian. Target aud: 25-55; Adults. ♦Michael Easley, pres; Richard Lee, stn mgr; Gary Bittner, mus dir; Doug Hainer, chief of engrg.

Shadyside

WVKF(FM)— Sept 1, 1990: 95.7 mhz; 6.8 kw horiz, 6.67 kw vert. Ant 626 ft TL: N40 03 41 W80 45 09. Hrs open: 24 Clear Channel Communications, 1015 Main St., Wheeling, WV, 26003-2709. Phone: (304) 232-1170. Fax: (304) 234-0067. Web Site: www.wvkffm.com. Licensee: Capstar TX L.P. Group owner: Clear Channel Communications Inc. (acq 2-26-2004; $930,000). Natl. Rep: Christal. Format: CHR. ♦Scott Miller, gen mgr; Jon Dickerson, gen sls mgr; Keith Mac, progmg dir.

Shelby

***WAUI(FM)**— November 1998: 88.3 mhz; 900 w. Ant 138 ft TL: N40 55 14 W82 38 51. Hrs open: Drawer 2440, Tupelo, MS, 38803-2440. Phone: (662) 844-8888. Fax: (662) 842-6791. Web Site: www.afr.net. Licensee: American Family Association. Group owner: American Family Radio Format: Inspirational Christian. ♦Marvin Sanders, gen mgr.

WSWR(FM)— Dec 1, 1981: 100.1 mhz; 3 kw. Ant 300 ft TL: N40 56 42 W82 39 42. Stereo. Hrs opn: 1400 Radio Ln., Mansfield, 44906. Phone: (419) 529-2211. Fax: (419) 529-2516. Licensee: Capstar TX L.P. Group owner: Regent Communications Inc. (acq 8-24-2000; grpsl). Format: Oldies. ♦Diana Coon, gen mgr.

Sidney

WMVR-FM— 1965: 105.5 mhz; 3 kw. Ant 155 ft TL: N40 18 04 W84 12 21. Stereo. Hrs opn: 24 2929 W. Russell Rd., 45365. Phone: (937) 498-1055. Fax: (937) 498-2277. E-mail: hits@hits1055.com Licensee: Dean Miller Broadcasting Corp. (acq 6-10-2004). Population served: 16,332 Natl. Rep: Rgnl Reps. Format: Adult contemp. ♦Dean Miller, pres; Brad Smith, gen mgr & mus dir; Debby McNeely, prom dir.

South Vienna

***WOAR(FM)**— 2006: 88.3 mhz; 1 kw vert. Ant 278 ft TL: N39 55 54 W83 36 36. Hrs open: 5700 West Oaks Blvd., Rocklin, CA, 95765. Phone: (916) 251-1600. Fax: (916) 251-1650. Web Site: www.air1.com. Licensee: Educational Media Foundation. (acq 3-23-2007; grpsl). Natl. Network: Air 1. Format: Christian. ♦Richard Jenkins, pres.

South Webster

***WEKV(FM)**— 1996: 94.9 mhz; 3 kw. Ant 328 ft TL: N38 45 39 W82 43 17. Hrs open: 24 2351 Sunset Blvd., Suite 170-218, Rocklin, CA, 95765. Phone: (916) 251-1600. Fax: (916) 251-1650. Web Site: www.klove.com. Licensee: Educational Media Foundation. (acq 12-30-2005; $450,000). Population served: 140,000 Natl. Network: K-Love. Format: Contemp Christian. ♦Richard Jenkins, pres.

South Zanesville

***WCVZ(FM)**— Jan 5, 1983: 92.7 mhz; 16 kw. 304 ft TL: N39 56 55 W81 57 48. (CP: Ant 407 ft.). Stereo. Hrs opn: 24 2477 E. Pike, 43701-4626. Secondary address: Box 3208, Zanesville 43701. Phone: (740) 455-3181. Fax: (740) 455-6195. E-mail: joyfm@927joyfm.com Web Site: www.927joyfm.com. Licensee: Christian Voice of Central Ohio Inc. Population served: 250,000 Natl. Network: USA. Format: Relg, talk, adult contemp. News: 10 hrs wkly. Target aud: General; young children 5-10 to senior citizens. ♦Dan Baughman, pres & gen mgr; Tate Luck, opns mgr; Michael James, progmg dir; Mike Russell, mus dir.

Spencerville

***WBCJ(FM)**— Sept 1, 1997: 88.1 mhz; 2.6 kw. 492 ft TL: N40 42 41 W84 23 01. Hrs open: 1025 W. Rudisill Blvd., Fort Wayne, IN, 46807. Phone: (260) 745-0576. Fax: (260) 745-2001. E-mail: wbcl@wbcl.org Web Site: www.wbcl.org. Licensee: Taylor University Broadcasting Inc. Format: Contemp Christian. ♦Marsha Bunker, gen mgr; Craig Albrecht, opns mgr & chief of engrg; Jeremy Lawrrence, prom dir.

Springfield

WDHT(FM)— August 1958: 102.9 mhz; 50 kw. 492 ft TL: N39 57 11 W83 52 07. Stereo. Hrs opn: 24 717 E. David Rd., Dayton, 45429. Phone: (937) 294-5858. Fax: (937) 297-5233. Web Site: www.hot1029.com. Licensee: Blue Chip Broadcasting Licenses Ltd. Group owner: Radio One Inc. (acq 4-30-01; grpsl). Population served: 81,926 Wire Svc: UPI Format: Mainstream Urban. Target aud: General. ♦Don Griffin, gen mgr; J.D. Kunes, opns mgr & progmg dir.

***WEEC(FM)**— Dec 15, 1961: 100.7 mhz; 50 kw. 469 ft TL: N39 57 42 W83 52 05. Stereo. Hrs opn: 24 2265 Troy Rd., 45504. Phone: (937) 399-7837. Fax: (937) 399-7802. E-mail: info@weec.org Web Site: www.weec.org. Licensee: World Evangelistic Enterprise Corp. Population served: 2,000,000 Natl. Network: USA, Moody, AP Radio. Law Firm: Miller & Neely. Wire Svc: AP Format: Relg, Christian. News staff: one; News: 16 hrs wkly. Target aud: 40 plus; general. Spec prog: Black one hr, farm one hr wkly. ♦Duane Helman, pres; Newell Moore, VP; Tracy Figley, CEO & gen mgr.

WIZE(AM)— Nov 1, 1940: 1340 khz; 1 kw-U. TL: N39 56 33 W83 47 15. Hrs open: 24 101 Pine St., Dayton, 45402. Phone: (937) 224-1137. Fax: (937) 224-7655. E-mail: tonytilford@clearchannel.com Web Site: www.wize.com. Licensee: Citicasters Licenses L.P. Group owner: Clear Channel Communications Inc. (acq 5-4-99; grpsl). Population served: 165,000 Rgnl rep: Rgnl Reps. Law Firm: Haley, Bader & Potts. Format: Sports. News staff: one; News: 14 hrs wkly. Target aud: 25 plus; upper income, businesses, offices. ♦Karrie Sudbrack, gen mgr; Tony Tilford, progmg dir; Jeff Bennett, chief of engrg.

WULM(AM)— 1947: 1600 khz; 1 kw-D, 34 w-N. TL: N39 57 11 W83 52 07. Hrs open: 24 Box 3132, 45501. Phone: (937) 390-1693. Fax: (937) 399-8767. E-mail: webmaster@1600wuim.net Web Site: www.1600wulm.net. Licensee: Urban Light Ministries Inc. (acq 4-9-02; $250,000). Population served: 165,000 Natl. Network: CNN Radio, Westwood One. Natl. Rep: Rgnl Reps. Law Firm: Haley, Bader & Potts. Format: Oldies. News staff: 2. Target aud: 34-54. Spec prog: Gospel. ♦Eli Williams, CEO & pres; Judy Williams, CFO; Robert Pitsch, gen mgr; Marco Simmons, stn mgr.

***WUSO(FM)**— Feb 20, 1966: 89.1 mhz; 100 w. Ant 85 ft TL: N39 56 09 W83 48 41. Hrs open: 24 Box 720, Wittenberg Univ., 45501. Phone: (937) 327-7026. Fax: (937) 327-6340. E-mail: wusoprogrock@yahoo.com Web Site: wuso.org. Licensee: Wittenberg University. Population served: 35,000 Format: Progsv, rock. Target aud: General; liberal arts students & residents of Springfield, OH. Spec prog: Jazz 6 hrs, class 3 hrs, blues 3 hrs, urban contemp 9 hrs wkly. ♦Brian Cataldi, gen mgr.

Steubenville

***WBJV(FM)**— 2002: 88.9 mhz; 125 w. Ant 256 ft TL: N40 21 56 W80 43 36. Hrs open: Drawer 3206, Tupelo, MS, 38803. Phone: (662) 844-8888. Fax: (662) 842-6791. Licensee: American Family Association. Group owner: American Family Radio. Format: Christian. ♦Marvin Sanders, gen mgr.

WDIG(AM)— Sept 25, 1973: 950 khz; 1 kw-D, DA. TL: N40 26 49 W80 34 06. Hrs open: 24 4039 Sunset Blvd., 43952. Phone: (740) 264-1760. Fax: (740) 264-5035. Web Site: www.wdigradio.com. Licensee: World Witness For Christ Ministries Inc. Population served: 30,771 Natl. Network: ABC. Law Firm: Dan J. Alpert. Format: Urban, oldies. Target aud: 25-54; general. Spec prog: Gospel. ♦Roy Dawkins, CEO; Del King, gen mgr & engrg VP.

WKWK-FM—See Wheeling, WV

WSTV(AM)— Nov 4, 1940: 1340 khz; 1 kw-U. TL: N40 26 49 W80 34 06. Hrs open: 24 Box 1340, 43952. Secondary address: 320 Market St. 43952. Phone: (740) 283-4747. Fax: (740) 283-3655. E-mail: wstv@wstv.com Web Site: www.wstv.com. Licensee: Keymarket Licences LLC. Group owner: Keymarket Communications LLC (acq 3-20-2000; grpsl). Population served: 141,000 Natl. Network: ESPN Radio. Natl. Rep: Rgnl Reps. Law Firm: Fleischman & Walsh L. Format: Sports. News staff: 2; News: 5 hrs wkly. Spec prog: Po 2 hrs, Czech 2 hrs wkly. ♦Gerald Getz, pres; Jim Seemiller, gen mgr; Joyce Nicholson, opns mgr, progmg mgr & pub affrs dir; Frank Bell, progmg VP; Marjie De Fede, news dir; Greg Harper, chief of engrg.

WOGH(FM)—Co-owned with WSTV(AM). May 1, 1947: 103.5 mhz; 19.5 kw. Ant 810 ft TL: N40 20 33 W80 37 14. Stereo. 24 Web Site: www.froggyland.com. Format: Country. Target aud: 25-54. ♦Stu Schroeder, prom mgr; Scott Feist, progmg dir & mus dir.

Streetsboro

***WSTB(FM)**— September 1973: 88.9 mhz; 1 kw. 125 ft TL: N41 14 02 W81 19 29. Hrs open: 7 AM-midnight 1900 Annalane Dr., 44241. Phone: (330) 626-4906. Fax: (330) 626-4906. E-mail: wstbfm@wstbfm.com Web Site: www.rock889.com. Licensee: Streetsboro City Schools. Population served: 25,000 Format: Modern rock. News staff: one; News: 4 hrs wkly. Target aud: 16-34. ♦Robert L. Long, gen mgr; Adam Oliver, opns mgr.

Struthers

***WKTL(FM)**— Sept 6, 1965: 90.7 mhz; 15 kw. 23 ft TL: N41 03 06 W80 35 56. Stereo. Hrs opn: Struthers High School, 111 Euclid Ave., 44471. Phone: (330) 755-1435. Fax: (330) 755-4525. Licensee: Struthers Board of Education. Population served: 15,343 Format: Adult contemp, classic rock. ♦Tom Krestal, gen mgr; Jim Hartzler, chief of engrg.

Swanton

WJUC(FM)—Licensed to Swanton. See Toledo

Sylvania

WWWM-FM— Nov 29, 1968: 105.5 mhz; 2.15 kw. 390 ft TL: N41 38 48 W83 36 22. (CP: 2.7 kw). Stereo. Hrs opn: 24 3225 Arlington Ave., Toledo, 43614. Secondary address: 2965 Pickle Rd., Oregon 43616. Phone: (419) 385-2507. Fax: (419) 385-2902. Web Site: www.star105toledo.com. Licensee: Cumulus Licensing Corp. Group owner: Cumulus Media L.L.C. (acq 9-11-97; $10 million with WLQR(AM) Toledo). Population served: 377,600 Natl. Rep: D & R Radio. Format: Hot Adult Contemp. Target aud: 25-54. ♦Clyde Roberts, exec VP; Kathy Stinehour, gen mgr; Carolyn Smithers, sls dir & adv dir; Lyn Casye, prom dir & news dir; Ryan Young, prom dir; Steve Mars, progmg dir; Kevin Howley, chief of engrg; Debbie Calevro, traf mgr.

Thompson

***WKSV(FM)**— June 1997: 89.1 mhz; 50 kw. Ant 472 ft TL: N41 41 34 W81 02 51. Stereo. Hrs opn: 24
Rebroadcasts WKSU-FM Kent 100%.
c/o WKSU-FM, Box 5190, 1613 E. Summit St., Kent, 44242-0001. Phone: (330) 672-3114. Fax: (330) 672-4107. Web Site: www.wksu.org. Licensee: Kent State University. Population served: 191,723 Natl. Network: NPR, PRI, AP Radio. Law Firm: Dow, Lohnes & Albertson. Format: Class, news. News staff: 5; News: 35 hrs wkly. Target aud: 35-65; college grad, professional & upper income. ♦Allen E. Bartholet, gen mgr; Deborah Frazier, gen mgr; N. Vincent Duffy, opns dir, prom dir, progmg dir & news dir; Abbe Turner, dev dir; Ronald Bartlebaugh, engrg dir.

Tiffin

WCKY-FM—Listing follows WTTF(AM).

WTTF(AM)— Dec 19, 1959: 1600 khz; 500 w-D, 20 w-N, DA-1. TL: N41 07 32 W83 13 45. Hrs open: 6 AM-10 PM 122 S. Washington St., 44883. Phone: (419) 447-2212. Fax: (419) 447-1709. Web Site: www.wttf.com. Licensee: Citicasters Licenses L.P. Group owner: Clear Channel Communications Inc. (acq 5-4-99; grpsl). Population served: 65,000 Rgnl. Network: Agri Bcstg. Law Firm: Miller & Miller, P.C. Format: Adult contemp, oldies. News staff: 2; News: 18 hrs wkly. Target aud: General. Spec prog: Farm 3 hrs wkly. ♦Kim Field, gen mgr; Jim Bickel, progmg dir; Greg Salbolda, chief of engrg & traf mgr.

WCKY-FM—Co-owned with WTTF(AM). July 11, 1963: 103.7 mhz; 50 kw. 492 ft TL: N41 08 20 W83 14 45. Stereo. Web Site: www.1037wcky.com.355,000 Format: Country.

Toledo

WCWA(AM)— Apr 10, 1938: 1230 khz; 1 kw-U. TL: N41 38 13 W83 33 52. Hrs open: 125 S. Superior, 43602. Phone: (419) 244-8321. Fax: (419) 244-7631. E-mail: wcwa@clearchannel.com Web Site: www.wcwa.com. Licensee: Jacor Broadcasting Corp. Group owner: Clear Channel Communications Inc. (acq 1999; grpsl). Population served: 383,318 Natl. Rep: Clear Channel. Law Firm: Hogan & Hartson. Format: News/talk. Target aud: 25-54; male. Spec prog: Ger one hr, Pol one hr, relg 3 hrs, sports 15 hrs wkly. ♦John Hogan, CEO; Andy Stuart, VP & gen mgr; Kellie Holeman, sls dir & mktg dir; Jack Jolly, gen sls mgr; Tom Riggs, prom mgr & progmg dir.

WIOT(FM)—Co-owned with WCWA(AM). October 1949: 104.7 mhz; 50 kw. 540 ft TL: N41 40 23 W83 25 31. Stereo. 24 Web Site: www.wiot.com. (Acq 1997).600,000 Format: Rock/AOR. Spec prog: Progsv rock 2 hrs, metal 2 hrs wkly. ♦Brian Kohler, gen sls mgr; Don Grosselin, progmg dir.

***WGTE-FM**— May 2, 1976: 91.3 mhz; 13.5 kw. 949 ft TL: N41 39 27 W83 25 55. Stereo. Hrs opn: 24 1270 S. Detroit Ave., 43614. Phone: (419) 380-4600. Fax: (419) 380-4710. Web Site: www.wgte.org. Licensee: The Public Broadcasting Foundation of Northwest Ohio. Population served: 1,101,300 Natl. Network: NPR, PRI. Law Firm: Schwartz, Woods & Miller. Format: Class, NPR, News, Public Affairs. News: 23 hrs wkly. Target aud: General. Spec prog: Jazz 16 hrs, new age 4 hrs wkly. ♦Marlon P. Kiser, CEO, pres & gen mgr; George Jones, chmn; Chris Peiffer, opns mgr; Ross Pfieffer, dev dir. Co-owned TV: *WGTE-TV affil.

WJUC(FM)—Swanton, Feb 27, 1997: 107.3 mhz; 3 kw. 328 ft TL: N41 38 30 W83 54 03. Hrs open: 24 Box 351450, 43635-1450. Secondary address: 5902 Southwyck Blvd. 43614. Phone: (419) 861-9582. Fax: (419) 861-2866. E-mail: wcharleswelch@aol.com Licensee: Welch Communications Inc. Population served: 792,000 Rgnl rep: Interep Law Firm: J. Richard Carr. Wire Svc: AP Format: Urban, rhythm & blues. Target aud: 18-54; African Americans 70%, others 30%. Spec prog: Blues, gospel. ♦W. Charles Welch, CEO, chmn, pres & gen mgr.

WJYM(AM)—See Bowling Green

WKKO(FM)—Listing follows WTOD(AM).

WLQR(AM)— October 1954: 1470 khz; 1 kw-U, DA-2. TL: N41 37 55 W83 28 45. Hrs open: 3225 Arlington Ave., 43614. Phone: (419) 385-2507. Fax: (419) 385-2902. Licensee: Cumulus Licensing Corp. Group owner: Cumulus Media L.L.C. (acq 9-11-97; $10 million with WWWM-FM Sylvania). Population served: 383,318 Format: Sports. Target aud: 25-54. ♦Kathy Stinehour, gen mgr.

***WOTL(FM)**— Mar 24, 1988: 90.3 mhz; 700 w. 377 ft TL: N41 38 48 W83 36 22. Stereo. Hrs opn: 13 Fairlane Dr., Joliet, IL, 60435. Secondary address: 716 N. Westwood Ave. 43607. Phone: (815) 725-1331. Web Site: www.familyradio.com. Licensee: Family Stations Inc. (group owner) Format: Relg. Target aud: General. ♦Harold Camping, pres; John Rorvik, gen mgr.

WRVF(FM)—Listing follows WSPD(AM).

WSPD(AM)— Apr 15, 1921: 1370 khz; 5 kw-U, DA-N. TL: N41 36 03 W83 32 11. Stereo. Hrs opn: 125 S. Superior St., 43602. Phone: (419) 244-8321. Fax: (419) 244-7631. Web Site: www.wspd.com. Licensee: Citicasters Licenses L.P. Group owner: Clear Channel Communications Inc. (acq 5-4-99; grpsl). Population served: 791,000 Rgnl. Network: Ohio Radio Network. Rgnl rep: Rgnl Reps. Law Firm: Hogan & Hartson. Format: News/talk. Target aud: 25-54; mostly males. Spec prog: Relg 5 hrs, farm 3 hrs wkly. ♦Andy Stuart, VP; Jack Jolly, gen sls mgr & rgnl sls mgr; Kellie Holeman, sls dir & natl sls mgr; A.T. Simen, prom dir & prom mgr; Al Brady Law, progmg dir.

WRVF(FM)—Co-owned with WSPD(AM). Aug 11, 1946: 101.5 mhz; 19.1 kw. 810 ft TL: N41 41 00 W83 24 29. Stereo. Web Site: www.wrvf.com.1,000,000 Natl. Rep: Clear Channel. Format: Adult contemp. Target aud: 25-54; mostly female. Spec prog: Jazz 6 hrs wkly. ♦Maureen DeTange, gen sls mgr; Don Gosselin, progmg dir.

WTOD(AM)— June 16, 1946: 1560 khz; 5 kw-D, DA. TL: N41 36 59 W83 37 22. Hrs open: 3225 Arlington Ave., 43614. Phone: (419) 725-5700. Fax: (419) 385-2902. Web Site: www.am1560wtod.com. Licensee: Cumulus Licensing Corp. Group owner: Cumulus Media Inc. (acq 9-11-97; grpsl). Population served: 383,318 Format: Contemp country. News staff: one; News: 3 hrs wkly. Target aud: 25-54; adults. Spec prog: Pol 4 hrs wkly. ♦Kathy Stinehour, gen mgr; Gary Shores, progmg dir; London Mitchell, news dir; Kevin Hawley, chief of engrg.

WKKO(FM)—Co-owned with WTOD(AM). Dec 7, 1956: 99.9 mhz; 50 kw. 499 ft TL: N41 40 05 W83 27 01. (CP: 6.8 kw, ant 180 ft. TL: N41 37 00 W83 37 19). Stereo. 24 Web Site: www.k100country.com.386,000 Natl. Network: ABC. ♦Gary Outlaw, mus dir.

WVKS(FM)— Oct 14, 1957: 92.5 mhz; 50 kw. 480 ft TL: N41 31 55 W83 35 37. Stereo. Hrs opn: 24 125 S. Superior, 43602. Phone: (419) 244-8321. Fax: (419) 244-7631. Web Site: www.925kissfm.com. Licensee: Citicasters Licenses L.P. Group owner: Clear Channel Communications Inc. (acq 5-15-99; grpsl). Population served: 3,000,000 Natl. Rep: Clear Channel. Law Firm: Hogan & Hartson. Wire Svc: AP Format: CHR. News staff: one; News: one hr wkly. Target aud: 18-49; educated, employed adults, mostly females. ♦Andrew Stuart, VP & gen mgr; Bill Michaels, opns dir & progmg dir; Kellie Holeman, sls dir; Amy Jo Simon, prom dir.

WWWM-FM—See Sylvania

***WXTS-FM**— February 1975: 88.3 mhz; 1 kw. 125 ft TL: N41 40 07 W83 33 15. Stereo. Hrs opn: 24 2400 Collingwood Blvd., 43620. Phone: (419) 244-6875. Fax: (419) 249-8248. Licensee: Toledo Board of Education. Population served: 250,000 Format: Jazz. Target aud: 28-55. Spec prog: Blues 5 hrs wkly. ♦John Kuschell, gen mgr.

***WXUT(FM)**— Nov 4, 1990: 88.3 mhz; 100 w horiz. 190 ft TL: N41 39 26 W83 36 57. Stereo. Hrs opn: 8 PM-2 AM (M-W); 8 PM-4 AM (Th, F); 9 AM-4 AM (S); 10 AM-2 AM (Su) Student Union, 2801 W. Bancroft St., 43606. Phone: (419) 530-4172. Phone: (419) 530-4455. Fax: (419) 530-2210. E-mail: wxut@wxut.com Web Site: www.wxut.com. Licensee: University of Toledo. Population served: 262,000 Format: Alternative. News: 4 hrs wkly. Target aud: General. Spec prog: Black 8 hrs, heavy metal 4 hrs, rhythm and blues 2 hrs, poetry one hr, women 2 hrs wkly. ♦Terrance Teagarden, gen mgr; Matt Kimura, stn mgr.

Troy

WHIO-FM—See Piqua

***WOKL(FM)**— 1991: 96.9 mhz; 3 kw. Ant 315 ft TL: N40 01 41 W84 11 28. Stereo. Hrs opn: 24
Rebroadcasts KLVR(FM) Santa Rosa 100%.
2351 Sunset Blvd., Suite 170-218, Rocklin, CA, 95765. Phone: (916) 251-1600. Fax: (916) 251-1650. E-mail: klove@klove.com Web Site: www.klove.com. Licensee: Educational Media Foundation. Group owner: EMF Broadcasting (acq 7-17-03; $1.2 million). Natl. Network: K-Love. Law Firm: Shaw Pittman. Format: Contemp Christian. News staff: 3. Target aud: 25-44; Judeo Christian, female. ♦Richard Jenkins, pres; Mike Novak, VP; Keith Whipple, dev dir; David Pierce, progmg mgr; Ed Lenane, news dir; Sam Wallington, engrg dir; Karen Johnson, news rptr; Marya Morgan, news rptr; Richard Hunt, news rptr.

Uhrichsville

WBTC(AM)— Dec 13, 1963: 1540 khz; 250 w-D, 5 w-N. TL: N40 25 26 W81 21 47. Hrs open: 125 Johnson Dr., 44683. Phone: (740) 922-2700. Fax: (740) 922-2702. E-mail: jim@wbtcam.com Licensee: Tuscarawas Broadcasting Co. Population served: 8,000 Natl. Network: CBS. Natl. Rep: Rgnl Reps. Format: News/talk, sports. Target aud: 30-55. Spec prog: Relg 2 hrs wkly. ♦James Natoli Jr., pres, stn mgr & gen sls mgr; J.R. Richards, progmg dir.

WTUZ(FM)—Licensed to Uhrichsville. See New Philadelphia

Union City

WTGR(FM)— Dec 31, 1994: Stn currently dark. 97.5 mhz; 6 kw. Ant 328 ft TL: N40 11 32 W84 47 58. Hrs open: 24 514 Martin St., Greenville, 45331. Phone: (937) 548-5085. Fax: (937) 548-5089. Web Site: www.wtgr.com. Licensee: Positive Radio Group Inc. of Ohio. Group owner: Baker Family Stations. Population served: 70,000 Natl. Network: CNN Radio. Format: Country. News staff: one; News: one hr wkly. Target aud: 25-54; 25-49 female. ♦Vernon H. Baker, CEO; Edward A. Baker, VP; Kevin Nott, stn mgr.

University Heights

***WJCU(FM)**— May 13, 1969: 88.7 mhz; 850 w. Ant 321 ft TL: N41 29 24 W81 31 54. (CP: 2.5 kw, ant 341 ft). Stereo. Hrs opn: 24 John Carroll Univ., 20700 N. Park Blvd., Cleveland, 44118. Phone: (216) 397-4437. Phone: (216) 397-4438. Fax: (216) 397-4439. E-mail: wjcu@jcu.edu Web Site: www.wjcu.org. Licensee: John Carroll University. Format: Div, modern, progsv. News: one hr wkly. Target aud: General. Spec prog: It 2 hrs, Chinese one hr, Pol 2 hrs, Lithuanian 2 hrs, Hungarian 3 hrs, Armenian 2 hrs, Celtic 2 hrs, Latino 4 hrs wkly. ♦Mark Krieger, gen mgr; Joe Madigan, stn mgr.

Upper Arlington

WXMG(FM)— May 25, 1989: 98.9 mhz; 3 kw. 328 ft TL: N39 58 16 W83 01 40. (CP: 2.6 kw, ant 505 ft.). Stereo. Hrs opn: 24 350 E 1st Ave., Suite 100, Columbus, 43201. Phone: (614) 487-1444. Fax: (614) 487-5862. Web Site: www.wxmg.fm.com. Licensee: Blue Chip Broadcasting Licenses Ltd. Group owner: Radio One Inc. (acq 4-30-01; grpsl). Natl. Rep: Christal. Law Firm: Shaw Pittman. Format: Soul/rhythm and blues. News staff: one. Target aud: 25-54; upscale, educated, active & responsive. ♦Jeff Wilson, gen mgr.

Upper Sandusky

***WXML(FM)**— Dec 26, 1992: 90.1 mhz; 3 kw. 328 ft TL: N40 50 10 W83 14 11. Stereo. Hrs opn: 24 Box 158, 1800 E. Wyandot Ave., 43351. Phone: (419) 294-2900. Fax: (419) 294-1786. E-mail: wxmlradio@udata.com Web Site: www.wxml.cc. Licensee: Kayser Broadcast Ministries Inc. Population served: 695,051 Law Firm: Garvey, Schubert, Barer. Format: Relg. News: 8 hrs wkly. Target aud: General. ♦Daniel L. Kayser, CEO, pres, CFO & gen mgr; Richard Johnson, VP; Jon Bowlus, progmg dir.

WYNT(FM)— Oct 1, 1986: 95.9 mhz; 3 kw. 299 ft TL: N40 49 30 W83 15 06. Stereo. Hrs opn: 19 1330 No. Main St., Marion, 43302. Phone: (877) 472-3464. Fax: (740) 387-3697. Licensee: CC Licenses LLC. Group owner: Clear Channel Communications Inc. (acq 4-2-2002; $825,000). Rgnl. Network: Agri Bcstg. Format: Adult contemp, oldies. Target aud: 27 plus; general. Spec prog: Farm 3 hrs wkly. ♦Forest Whitehead, pres; Diane Glass Meyer, gen mgr; Scott Shawver, progmg dir; Ben Failor, news dir.

Urbana

WKSW(FM)— Aug 1, 1965: 101.7 mhz; 3.2 kw. 407 ft TL: N40 02 57 W83 46 06. Stereo. Hrs opn: 24 2963 Derr Rd., Springfield, 45503. Phone: (937) 399-5300. Fax: (937) 399-3661. E-mail: email@kisscountry.com Web Site: www.kisscountry.com. Licensee: Blue Chip Broadcasting Licenses Ltd. Group owner: Radio One Inc. (acq 4-30-01; grpsl). Population served: 250,000 Format: Country. Target aud: 25-54; above-average income, blue-collar.Cathy Hughes, CEO; Mary Catherine Sneed, COO; Alfred Liggens, pres; Scott Royster, CFO; Don Griffin, gen mgr; Roger C. Mackall, stn mgr; J.D. Kunes, opns dir; Andy Lawrence, prom mgr; Lee Riley, progmg dir & disc jockey; Chris Daniels, news dir & disc jockey; Brian Kelly, pub affrs dir & disc jockey; Gene Simmons, chief of engrg; Mickie Cooper, traf mgr; Andy Lawarance, disc jockey

Van Wert

WBYR(FM)— Oct 1, 1962: 98.9 mhz; 50 kw. 450 ft TL: N40 53 33 W84 31 40. Stereo. Hrs opn: 1005 Production Rd., Fort Wayne, IN, 46808. Phone: (260) 471-5100. Fax: (260) 471-5224. Web Site: www.989thebear.com. Licensee: Pathfinder Communications Corp. Group owner: Federated Media (acq 1996; $5.85 million). Population served: 147,500 Format: Active rock. Target aud: 18-49; men. ♦Jim Allgeier, gen mgr.

WERT(AM)— Nov 27, 1958: 1220 khz; 250 w-U. TL: N40 52 19 W84 33 15. Hrs open: 24 Box 487, 45891. Phone: (419) 238-1220. Fax: (419) 238-2578. E-mail: wireless@wcoil.com Web Site: www.vanwert.com. Licensee: Richard Ford. (acq 1-03-95; $225,000 with WKSD-FM Paulding; FTR: 11-7-94) Population served: 30,464 Natl. Network: ABC. Rgnl. Network: Agri Bcstg. Natl. Rep: Rgnl Reps. Format: Adult standard. News staff: 2; News: 30 hrs wkly. Target aud: 35 plus; spendable income. Spec prog: Gospel 3 hrs, Sp one hr wkly. ♦Chris Roberts, pres & gen mgr.

Wapakoneta

WWSR(FM)— July 1, 1964: 92.1 mhz; 3 kw. 320 ft TL: N40 39 20 W84 06 54. Stereo. Hrs opn: 24 Secondary address: 1301 N. Cable Rd., Lima 45805. Phone: (419) 331-1600. Fax: (419) 222-3755. Web Site: starlima.com. Licensee: Maverick Media of Lima License LLC. Group owner: Maverick Media LLC (acq 12-4-2003; grpsl). Population served: 150,000 Natl. Rep: Christal. Rgnl rep: Rgnl Reps. Format: CHR. News staff: one; News: 6 hrs wkly. Target aud: 18-49; young, affluent women. ♦Gary S. Rozynek, pres; Dave Roach, gen mgr; Matt Childers, gen sls mgr; Dan Kennedy, progmg dir; Tiffany Binder, prom mgr & sls.

Warren

WANR(AM)— Apr 7, 1971: 1570 khz; 500 w-D, 116 w-N, DA-1. TL: N41 12 22 W80 50 29. Hrs open: 24 Box 1798, 44482-1798. Phone: (330) 394-7700. Fax: (330) 394-7701. Licensee: Beacon Broadcasting Inc. (group owner; (acq 9-14-2005; grpsl). Population served: 250,000 Natl. Network: Westwood One. Rgnl. Network: Ohio Radio Net. Format: News/talk info. News staff: one; News: 20 hrs wkly. Target aud: 25-49; adult men. ♦Bill Henry, CFO; Michael Arch, gen mgr.

WHKZ(AM)— Nov 11, 1941: 1440 khz; 5 kw-U, DA-2. TL: N41 09 52 W80 50 47. Hrs open:
Rebroadcasts WHKW(AM) Cleveland 100%.
4 Summit Park Dr., Suite 150, Independence, 44131. Phone: (216) 901-0921. Fax: (216) 901-5517. Web Site: www.whkwradio.com. Licensee: SCA License Corp. Group owner: Salem Communications Corp. (acq 1-19-2001; $675,000). Population served: 75,000 Natl. Rep: Salem. Format: Christian talk. ♦Errol Dengler, gen mgr.

Washington Court House

WCHO(AM)— February 1952: 1250 khz; 500 w-D. TL: N39 32 59 W83 27 10. Hrs open: 1535 N. North St., 43160. Phone: (740) 335-0941. Fax: (740) 335-6869. Web Site: www.wchoam.com. Licensee: Citicasters Licenses L.P. Group owner: Clear Channel Communications Inc. (acq 5-4-99; grpsl). Population served: 250,000 Natl. Network: ABC. Rgnl. Network: Agri Bcstg. Natl. Rep: Katz Radio. Format: MOR. News: 18 hrs wkly. Spec prog: Farm 5 hrs wkly. ♦Dan Latham, sr VP & gen mgr; Kim Skaggs, opns VP; Tracy Taylor, sls dir; John Barney, gen sls mgr; Carl Staffan, mus dir & news dir; Todd Jellison, engrg mgr.

WCHO-FM— December 1968: 105.5 mhz; 3 kw. 300 ft TL: N39 32 59 W83 27 10. Web Site: www.wcho.com.250,000 Natl. Network: ABC. Format: Country.

Wauseon

WMTR-FM—See Archbold

WNKL(FM)— 2003: 96.9 mhz; 5 kw. Ant 358 ft TL: N41 36 03 W83 54 27. Hrs open: Box 351630, Toledo, 43635. Secondary address: 1510 Reynolds Rd., Maunee 43537. Phone: (419) 725-9366. Fax: (419) 725-2600. E-mail: info@q969.fm Licensee: Cornerstone Church Inc. (acq 8-31-2000). Format: Rhythmic CHR. ♦Brandon Brandon, gen mgr & progmg dir; Alan Colwell, chief of engrg; Dawn Vaculik, traf mgr.

***WYSA(FM)**— 1996: 88.5 mhz; 25 kw. 292 ft TL: N41 33 29 W84 11 08. Hrs opn: 5115 Glendale Ave., Toledo, 43614. Phone: (419) 389-0893. Fax: (419) 381-0731. Web Site: www.yeshome.com. Licensee: Side by Side Inc. Natl. Network: Salem Radio Network. Format: Christian, CHR, rock/AOR. Target aud: 15-25. ♦Jim Oedy, pres; J. Todd Hostetler, gen mgr; Jeff Howe, opns dir.

Waverly

WXIC(AM)— 1954: 660 khz; 1 kw-D. TL: N39 07 50 W83 00 46. Hrs open: Sunrise-sunset Box 227, 6655 St. Rt. 220 W., 45690. Phone: (740) 947-2166. Fax: (740) 947-4600. E-mail: wxiz@roadrunner.com Web Site: am660wxic.com. Licensee: Crystal Communications Corp. (acq 7-1-79). Population served: 4,858 Natl. Rep: Keystone (unwired net). Format: Southern gospel mus. Gospel music lovers. ♦Gerald E. Davis, pres, gen mgr & chief of engrg; Rick Schweinburg, opns mgr; Brad Lambert, sls dir & adv mgr; Rick Schweinsburg, prom mgr; Rick Schweinburgh, progmg mgr.

WXIZ(FM)—Co-owned with WXIC(AM). March 1971: 100.9 mhz; 920 w. 500 ft TL: N39 13 17 W82 59 33. Stereo. 24 Web Site: www.wxiz.com.4,858 Format: Country. News staff: one; News: 10 hrs wkly. Target aud: 25-50. ♦Gerald E. Davis, stn mgr; Brad Lambert, gen sls mgr & prom mgr; Tim Hughes, mus dir; Roy Belt, news dir.

Wellston

WKOV-FM—Listing follows WYPC(AM).

WYPC(AM)— 1953: 1330 khz; 500 w-D, 50 w-N. TL: N39 06 22 W82 34 44. Hrs open: 24 Box 667, 295 E. Main, Jackson, 45640. Phone: (740) 286-3023. Fax: (740) 286-6679. E-mail: rburtrand@vcairadio.com Licensee: Jackson County Broadcasting Inc. (group owner; acq 9-14-70). Population served: 42,000 Natl. Network: Westwood One. Format: Adult standards. News staff: one; News: 8 hrs wkly. Target aud: 50 plus. ♦Jerry Mossbarger, gen mgr.

WKOV-FM—Co-owned with WYPC(AM). July 17, 1971: 96.7 mhz; 16.5 kw. 430 ft TL: N39 01 45 W82 35 51. Stereo. 24 585,000 Law Firm: Fletcher, Heald & Hildreth. Format: Hot adult contemp. News staff: one; News: 21 hrs wkly. Target aud: 20-55.

West Carrollton

***WDPR(FM)**— Apr 9, 1977: 88.1 mhz; 600 w. Ant 781 ft TL: N39 43 16 W84 15 00. Stereo. Hrs opn: 24 126 N. Main St., Dayton, 45402. Phone: (937) 496-3850. Fax: (937) 496-3852. E-mail: gmw@dpr.org Web Site: www.dpr.org. Licensee: Dayton Public Radio Inc. (acq 4-28-98). Population served: 500,000 Natl. Network: USA. Format: Class. Target aud: 24-50. ♦Georganne M. Woessner, gen mgr; Larry Coressel, opns dir & progmg dir.

WROU-FM— Nov 25, 1991: 92.1 mhz; 890 w. Ant 597 ft TL: N39 43 15 W84 15 39. Hrs open: 24 717 E. David Rd., Dayton, 45429. Phone: (937) 294-5858. Fax: (937) 297-5233. Web Site: www.wrou.com. Licensee: Radio One of Dayton Licenses LLC. Group owner: Radio One Inc. (acq 7-17-2003; $6.7 million). Natl. Network: ABC. Format: Urban contemp. Target aud: 25-54. ♦Don Griffin, gen mgr.

West Chester

***WLHS(FM)**— Sept 3, 1976: 89.9 mhz; 100 w. 338 ft TL: N39 19 10 W84 22 04. Hrs open: 9 AM-5 PM 6840 Lakota Ln., Liberty Township, 45044-9578. Phone: (513) 759-4163. Fax: (513) 759-4165. Licensee: Lakota School District. Population served: 500 Wire Svc: UPI Format: Educ, rock/AOR. Target aud: General; div, open minded crowd. ♦Mark Hattersley, stn mgr; R.C. Anderson, opns dir; Corey Wyatt, sls dir; Danny Hall, mus dir; Matt Townsley, mus dir; Brandon Enright, asst music dir.

West Union

WRAC(FM)— Dec 15, 1981: 103.1 mhz; 3.3 kw. Ant 426 ft TL: N38 51 28 W83 36 42. (CP: COL Georgetown. 6 kw, ant 328 ft. TL: N38 52 14 W83 45 55). Stereo. Hrs opn: 24 Box 103, 114 Manchester St., 45693. Phone: (937) 544-9722. Fax: (937) 544-5523. E-mail: c103country@yahoo.com Licensee: DreamCatcher Communications Inc. (group owner; acq 9-21-81; $4,820; FTR: 10-12-81). Natl. Rep: Rgnl Reps. Format: Country, gospel. Target aud: General. Spec prog: Farm 10 hrs wkly. ♦Donald Bowles, pres & gen mgr; Venita Bowles, VP; Ted Foster, stn mgr & progmg dir; Brad Rolfe, mus dir & news dir.

***WVXW(FM)**— 1990: 89.5 mhz; 3.2 kw. Ant 330 ft TL: N38 51 36 W83 36 42. Stereo. Hrs opn: 24 Box 793, New Albany, 43054. Phone: (614) 855-9171. Fax: (614) 855-9280. Licensee: Christian Voice of Central Ohio Inc. (acq 5-15-2007; grpsl). Rgnl. Network: Ohio Radio Net. Format: Contemp Christain. ♦Dan Baughman, gen mgr.

Westerville

***WOBN(FM)**— Oct 8, 1958: 101.5 mhz; 28 w. 40 ft TL: N40 07 28 W82 56 15. Stereo. Hrs opn: Otterbein College, Cowan Hall, 43081. Phone: (614) 823-1725. Phone: (614) 823-1557. Fax: (614) 823-3367. Web Site: www.wobn.net. Licensee: Otterbein College. Population served: 12,530 Format: Rock, alternative rock, progsv. News staff: News progmg one hr wkly Target aud: General; Westerville & Otterbein College community. Spec prog: Black 2 hrs, jazz one hr, relg 4 hrs, heavy metal 2 hrs wkly. ♦Brittani Stai, gen mgr.

WTDA(FM)— 1998: 103.9 mhz; 6 kw. Ant 328 ft TL: N40 09 33 W82 55 21. Stereo. Hrs opn: 24 1458 Dublin Rd., Columbus, 43215. Phone: (614) 481-7800. Fax: (614) 481-8070. E-mail: mail@1039tedfm.com Web Site: www.1039tedfm.com. Licensee: North American Broadcasting Co. Inc. (group owner; (acq 1999; $5 million). Population served: 1,200,000 Natl. Network: Fox Sports. Natl. Rep: D & R Radio. Law Firm: Hogan & Hartson. Wire Svc: AP Format: Talk. ♦Matthew Mnich, CEO & pres; Norma J. Mnich, chmn; Mark E. Jividen, VP & gen mgr.

Wilberforce

***WCSU-FM**— Dec 15, 1962: 88.9 mhz; 1 kw. 150 ft TL: N39 42 57 W83 52 27. Stereo. Hrs opn: Box 1004, 45384-1004. Phone: (937) 376-6371. Fax: (937) 376-6436. Licensee: Central State University. Population served: 150,000 Format: Gospel, urban contemp, jazz. Target aud: 12-49; African-Americans. ♦J.C. Logan, gen mgr; Tony Chappel, mus dir.

Willard

WLRD(FM)— January 2000: 96.9 mhz; 6 kw. Ant 328 ft TL: N40 57 36 W82 37 16. Hrs open: 3809 Maple Ave., Castalia, 44824. Phone: (419) 684-5311. Fax: (419) 684-5378. Licensee: Christian Faith Broadcast Inc. Group owner: Christian Faith Broadcasting Inc. Population served: 24 Format: Southern gospel. Target aud: 25-54. ♦Rusty Yost, gen mgr.

Willoughby-Eastlake

WELW(AM)— Jan 25, 1965: 1330 khz; 500 w-D. TL: N41 38 56 W81 25 19. Hrs open: 24 Box 1330, Willoughby, 44096. Phone: (440) 946-1330. Fax: (440) 953-0320. E-mail: email@welw.com Web Site: www.welw.com. Licensee: Spirit Broadcasting Corp. (acq 9-11-90; FTR: 10-1-90). Population served: 500,000 Format: Loc talk, sports. Target aud: 35 plus; community adults. Spec prog: Ger one hr; Croation 12 hrs, It one hr; Pol one hr; Polka 15 hrs wkly. ♦Ray Somich, pres & gen mgr; Tony Petkovsek, exec VP; Ron Somich, VP; Van Lane, gen sls mgr.

Wilmington

WKFI(AM)— 1963: 1090 khz; 1 kw-D, DA. TL: N39 26 12 W83 51 21. Hrs open: 486 W 2nd St., Xenia, 45385-3610. Phone: (937) 372-3531. Fax: (937) 372-3508. E-mail: jmullins@myclassiccountry.com Web Site: www.myclassiccountry.com. Licensee: Town and Country Broadcasting Inc. (group owner; (acq 12-16-2004; $300,000). Population served: 450,000 Natl. Network: Fox News Radio. Natl. Rep: Rgnl Reps. Law Firm: Reddy, Belley & McCormick, L. Wire Svc: AP Format: Classic country. News staff: one; News: 3 hrs wkly. Target aud: General. Spec prog: Big band 5 hrs wkly. ♦Rick Johnston, gen mgr; Roy Hatfield, progmg dir; Megan Brugger, traf mgr; Oarrin Johnson, news cmtr.

WKLN(FM)— 1974: 102.3 mhz; 3 kw. Ant 300 ft TL: N39 21 54 W83 46 08. Stereo. Hrs opn: 8686 Michael Ln., Fairfield, 45014-3015. Phone: (513) 829-7700. Web Site: www.klove.com. Licensee: Vernon R. Baldwin Inc. (acq 4-22-2003; $1.2 million with co-located AM). Population served: 600,000 Natl. Network: K-Love. Format: Contemp Christian.

Wooster

***WCWS(FM)**— April 1968: 90.9 mhz; 1.05 kw. 230 ft TL: N40 48 34 W81 56 18. Stereo. Hrs opn: 24 Box 3177, Wishart Hall, College of Wooster, 44691. Phone: (330) 263-2240. Fax: (330) 263-2690. E-mail: wcws@wooster.edu Web Site: www.woo91.wooster.edu. Licensee: The College of Wooster. Population served: 70,000 Law Firm: Booth, Freret, Imlay & Tepper P.C. Format: Div, music mix. News: 10 hrs wkly. Target aud: General; college students & people of the surrounding area. Spec prog: Edu 8 hrs wkly.

***WKRW(FM)**— Mar 29, 1993: 89.3 mhz; 2.1 kw. 318 ft TL: N40 46 28 W81 55 05. Stereo. Hrs opn: 24
Rebroadcasts WKSU-FM Kent 99%.
Box 5190, 1613 E. Summit St., Kent, 44242-0001. Phone: (330) 672-3114. Fax: (330) 672-4107. Web Site: www.wksu.org. Licensee: Kent State University. Population served: 91,826 Natl. Network: NPR, PRI. Law Firm: Dow, Lohnes & Albertson. Format: Class, in-depth news. News staff: 5; News: 35 hrs wkly. Target aud: 35-65; college grad, professional & upper income. Spec prog: Folk 12 hrs wkly. ♦Allen E. Perry, gen mgr; Patricia Gerber, dev dir.

WKVX(AM)—Listing follows WQKT(FM).

WQKT(FM)— 1947: 104.5 mhz; 52 kw. Ant 330 ft TL: N40 47 31 W81 54 17. Stereo. Hrs opn: 24 Box 39, 44691. Secondary address: 186 S. Hillcrest Dr. 44691. Phone: (330) 264-5122. Fax: (330) 264-3571. E-mail: wqkt@aol.com Web Site: www.wqkt.com. Licensee: WWST Corp. L.L.C. Population served: 85,000 Natl. Network: Westwood One. Rgnl rep: Rgnl Reps Law Firm: Baker & Hostetler. Wire Svc: AP Format: Country, sports. ♦Ken Nemeth, gen mgr; Craig Walton, gen sls mgr; Mike Breckenridge, progmg dir; Mike Breckenenridge, news dir.

WKVX(AM)—Co-owned with WQKT(FM). 1947: 960 khz; 1 kw-D, 32 w-N. TL: N40 47 31 W81 54 17.24 Web Site: www.wkvx.com. Group owner: Dix Communications 85,000 Natl. Network: Westwood One. Rgnl rep: Rgnl Reps Law Firm: Baker & Hostetler. Format: Oldies.

WXXF(FM)—Loudonville, March 1990: 107.7 mhz; 6 kw. Ant 328 ft TL: N40 36 58 W82 05 34. Stereo. Hrs opn: 24
Rebroadcasts WFXN-FM Galion 100%.
1197 US Hwy. Rt 42, Ashland, 44805. Phone: (419) 289-2605. Fax: (419) 289-0304. E-mail: jeffschendel@clearchannel.com Licensee: Capstar TX L.P. Group owner: Clear Channel Communications Inc. (acq 2-12-2001; grpsl). Population served: 250,000 Natl. Network: Fox News Radio. Format: Classic rock. Target aud: 25-54; Male. Spec prog: Underground Garage, House of Hair. ♦Diana Coon, gen mgr; Joe Rinehart, stn mgr; Eric Hansen, opns mgr; Jeff Schendel, progmg dir.

Xenia

WBZI(AM)— Nov 11, 1963: 1500 khz; 500 w-D. TL: N39 42 48 W83 54 48. Hrs open: Sunrise-sunset 486 W. Second St., 45385. Phone: (937) 372-3531. Fax: (937) 372-3508. E-mail: myclassiccountry@myclassiccountry.com Web Site: www.wbzi.com. Licensee: Town & Country Broadcasting Inc. (acq 10-4-95; $140,000). Population served: 825,000 Natl. Network: Fox News Radio. Rgnl. Network: Ohio News Net. Natl. Rep: Rgnl Reps. Law Firm: Reddy, Begley, & McCormick L.L.P. Wire Svc: AP Format: Classic country. News staff: one; News: 3 hrs wkly. Target aud: 35-64; upper income, married, homeowners. Spec prog: Gospel 5 hrs, farm 2 hrs wkly. ♦Joe Mullins, gen mgr; Roy Hatfield, prom dir; Oarrin Johnston, news dir; Megan Brugger, traf mgr.

WGNZ(AM)—See Fairborn

WZLR(FM)— Mar 3, 1967: 95.3 mhz; 6 kw. 300 ft TL: N39 37 54 W83 53 49. Stereo. Hrs opn: 24 1414 Wilmington Ave., Dayton, 45420. Phone: (937) 259-2111. Fax: (937) 259-2328. Web Site: daytonspoint.com. Licensee: Cox Radio Inc. Group owner: Cox Broadcasting (acq 1998; grpsl). Population served: 1,463,000 Natl. Rep: Christal. Law Firm: Dow, Lohnes & Albertson. Format: Classic rock. Target aud: 25-54. ♦Donna Hall, VP & gen mgr; Marc Herbst, gen sls mgr; Jason Michaels, progmg dir & mus dir.

Yellow Springs

***WYSO(FM)**— Feb 8, 1958: 91.3 mhz; 37 kw. 410 ft TL: N39 45 46 W83 53 05. Stereo. Hrs opn: 24 800 Livermore St., 45387. Phone: (937) 767-6420. Fax: (937) 769-1382. E-mail: wyso@wyso.org Web Site: www.wyso.org. Licensee: Antioch University. Population served: 1,400,000 Natl. Network: PRI, NPR. Rgnl. Network: Ohio Educ Bcstg. Law Firm: Garvey, Schubert & Barer. Format: News/talk, Americana, AAA. News staff: 2; News: 77 hrs wkly. Target aud: 25-54; college educated, professional, mid-upper income. Spec prog: Folk 2 hrs, jazz 12 hrs, blues 4 hrs, new age 4 hrs, Celtic/British Isles 3 hrs,bluegrass 6 hrs wkly. ♦Glenn Watts, CFO; Paul Maassen, gen mgr; Yana Davis, dev dir; Tim Tattan, progmg dir.

Youngstown

WAKZ(FM)—Sharpsville, PA) Dec 28, 1976: 95.9 mhz; 3 kw. Ant 328 ft TL: N41 13 05 W80 33 43. Stereo. Hrs opn: 7461 South Ave., Boardman, 44512. Phone: (330) 965-0057. Fax: (330) 729-9991. Web Site: www.959kiss.com. Licensee: Citicasters Licenses L.P. Group owner: Clear Channel Communications Inc. (acq 1-15-2004; grpsl). Format: CHR. ♦Bill Kelly, gen mgr; Cornell Bogdon, gen sls mgr & rgnl sls mgr; John Thomas, prom dir & prom mgr.

WASN(AM)— May 9, 1976: 1500 khz; 500 w-D, 250 w-CH, DA. TL: N41 06 26 W80 34 57. Hrs open: Sunrise-sunset 20 Federal Plaza W., # T2, 44503. Phone: (330) 744-5115. Fax: (330) 744-4020. Licensee: Stop 26 Riverbend Licenses LLC (acq 1995; $250,000). Format: Relg, Black, talk. News: 4 hrs wkly. Target aud: General; families. ♦Percy Squire, pres; Linda Penny, gen mgr; Charles Rhodes, gen sls mgr; Kenneth King, news dir.

WBBG(FM)—See Niles

WBBW(AM)— Feb 20, 1949: 1240 khz; 1 kw-U. TL: N41 04 50 W80 38 54. Hrs open: 4040 Simon Rd., 44512. Phone: (330) 783-1000. Fax: (330) 783-0060. Web Site: www.cumulus.com. Licensee: Cumulus Licensing Corp. Group owner: Cumulus Media Inc. (acq 3-15-00; grpsl). Population served: 150,000 Natl. Network: Westwood One. Format: Sports. Target aud: General. ♦Larry Weiss, gen mgr; Lee B. Jolly, prom mgr; Pat Mulrooney, progmg mgr; Wesley Boyd, chief of engrg; Angie Capaldi, traf mgr.

WHOT-FM—Co-owned with WBBW(AM). November 1959: 101.1 mhz; 24 kw. Ant 711 ft TL: N41 03 28 W80 38 24. (CP: 25 kw, ant 694 ft.). Stereo. Web Site: www.cumulus.com. Law Firm: Putbrese, Hunsaker & Trent. Format: CHR. Target aud: 18-54. ♦Kelley McGrath, gen sls mgr; Angie Capaldi, natl sls mgr & traf mgr; Lee B. Jolly, prom dir; Mike Thomas, mus dir; Pat Mulrooney, progmg dir & news dir.

WGFT(AM)—Campbell, Oct 16, 1955: 1330 khz; 500 w-D, 1 kw-N, DA-2. TL: N40 58 30 W80 35 15. Hrs open: 20 Federal Plaza W., # T2, Youngtown, 44503. Phone: (330) 744-5115. Fax: (330) 744-2221. Licensee: Stop 26 Riverbend Licenses LLC (acq 12-30-02). Population served: 500,000 Format: Gospel talk radio. Target aud: General; family. ♦Frank Halfacre, chmn; Percy Squire, CEO & pres; Linda Penny, gen mgr.

WHKZ(AM)—See Warren

WKBN(AM)— 1926: 570 khz; 5 kw-U, DA-N. TL: N40 59 07 W80 36 02. Hrs open: 24 Box 9248, 44513. Secondary address: 7461 South Ave., Boardman 44512. Phone: (330) 965-0057. Fax: (330) 965-8277. Licensee: Citicasters Licenses L.P. Group owner: Clear Channel Communications Inc. (acq 1-22-99; $11 million with co-located FM). Population served: 750,000 Natl. Network: ABC, CBS. Format: News/talk, sports. Spec prog: Polka 2 hrs, Croation 2 hrs wkly. ♦Bill Kelly, VP & gen sls mgr; Dan Rivers, progmg dir.

WMXY(FM)—Co-owned with WKBN(AM). Aug 26, 1947: 98.9 mhz; 4.5 kw. 1,370 ft TL: N41 03 24 W80 38 44. Stereo. Format: Adult contemp.

WLOA(AM)—Farrell, PA) Oct 3, 1954: 1470 khz; 1 kw-D, 500 w-N, DA-N. TL: N41 11 58 W80 31 22. Hrs open: Box 1798, Warren, 44482-1798. Phone: (330) 394-7700. Fax: (330) 394-7701. Licensee: Beacon Broadcasting Inc. (group owner; (acq 10-4-2005; $295,000). Population served: 11,022 ♦Harold Glunt, gen mgr.

WNCD(FM)—Listing follows WNIO(AM).

WNIO(AM)— Sept 7, 1939: 1390 khz; 9.5 kw-D, 4.8 kw-N, DA-N. TL: N41 07 17 W80 42 05 (day), N40 59 11 W80 35 54 (night). Stereo. Hrs opn: 24 7461 South Ave., 44512. Phone: (330) 965-0057. Fax: (330) 965-8277. Web Site: www.wnio.com. Licensee: Citicasters Licenses L.P. Group owner: Clear Channel Communications Inc. (acq 1-15-2004; grpsl). Population served: 535000 Format: Nostalgia. News staff: 2; News: 3 hrs wkly. Target aud: General. Spec prog: It 3 hrs wkly. ♦Bill Kelly, gen mgr & gen sls mgr; Dan Rivers, adv dir.

WNCD(FM)—Co-owned with WNIO(AM). June 1959: 93.3 mhz; 50 kw. Ant 280 ft TL: N41 04 50 W80 38 54. Stereo. Box 9248, 445123. Web Site: www.wncd.com. Format: Rock. Target aud: 25-54. ♦Thomas John, prom dir; Dan Rivers, progmg dir.

WRTK(AM)—See Niles

***WYSU(FM)**— September 1969: 88.5 mhz; 50 kw. 499 ft TL: N41 03 28 W80 38 42. Stereo. Hrs opn: 24 Youngstown State University, One University Plaza, 44555. Phone: (330) 941-3363. Fax: (330) 941-1501. E-mail: sexton@wysu.org Web Site: www.wysu.org. Licensee: Youngstown State University. Population served: 914,9000 Natl. Network: PRI, NPR. Law Firm: Bakeer & Hostetler. Format: Class, news. News: 48 hrs wkly. Target aud: General. Spec prog: Folk 3 hrs wkly. ♦Gary Sexton, gen mgr; David Linscher, opns mgr & progmg dir; Michele Grant, dev dir; William C. Panko, chief of engrg.

***WYTN(FM)**— May 1991: 91.7 mhz; 3 kw. 299 ft TL: N41 03 28 W80 38 42. Stereo. Hrs opn: 24 13 Fairlane Dr., Joliet, IL, 60435. Secondary address: 3930 Sunset Blvd. 60435. Phone: (815) 725-1331. Web Site: www.familyradio.com. Licensee: Family Stations Inc. (group owner) Law Firm: Dow, Lohnes & Albertson. Format: Relg. Target aud: 25 plus; Christians. Spec prog: Class 2 hrs wkly. ♦Harold Camping, pres; John Rorvik, gen mgr.

Zanesville

WHIZ(AM)— July 8, 1924: 1240 khz; 1 kw-U. TL: N39 55 42 W81 59 06. Hrs open: 24 629 Downard Rd., 43701. Phone: (740) 452-5431. Fax: (740) 452-6553. Web Site: www.whizamfmtv.com. Licensee: Southeastern Ohio Broadcasting System Inc. (acq 6-47). Population served: 33,045 Natl. Rep: Roslin, Rgnl Reps. Format: Adult standards, news/talk. News staff: 10; News: 30 hrs wkly. Target aud: 25-54; general. Spec prog: Farm progmg 2 hrs wkly. ♦N.J. Littick, chmn; Henry Littick, pres; Van Vannelli, VP; Jay Benson, stn mgr, sls dir & adv dir; Brian Wagner, opns dir, mktg dir, progmg dir, farm dir & disc jockey; George Hiotis, news dir & local news ed; Ken Cash, chief of engrg; Andy Jones, sports cmtr; Brenda Larrick, disc jockey; Jeff Ball, disc jockey.

WHIZ-FM— Dec 16, 1961: 102.5 mhz; 50 kw. 490 ft TL: N39 55 42 W81 59 06. Stereo. 24 Web Site: www.whizamfmtv.com.245,000 Format: Adult contemp. News: 12 hrs wkly. Target aud: 25 plus; general. Spec prog: Relg one hr, sports 3 hrs wkly. ♦George Hiotis, local news ed; Brian Wagner, farm dir & disc jockey; Andy Jones, sports cmtr; Jared Stewart, disc jockey; Jeff Ball, mus dir & disc jockey. Co-owned TV: WHIZ-TV affil.

***WJIC(FM)**— 2000: 91.7 mhz; 6 kw. 276 ft TL: N40 04 16 W82 11 30. Hrs open: 24 c/o VCY/America, 3434 W. Kilbourn Ave., Milwaukee, WI, 53208. Phone: (414) 935-3000. Fax: (414) 935-3015. E-mail: wjic@vcyamerica.org Web Site: www.wcyamerica.org. Licensee: VCY/America Inc. (group owner; Natl. Network: USA. Format: Relg, Christian. ♦Vic Eliason, gen mgr; Jim Schneider, progmg dir; Andy Eliason, chief of engrg.

***WOUZ(FM)**— Nov 1, 1993: 90.1 mhz; 3 kw. 279 ft TL: N39 48 50 W81 57 21. Stereo. Hrs opn: 24
Rebroadcasts WOUB-FM Athens 100%.
9 S. College St., Athens, 45701. Phone: (740) 593-4554. Fax: (740) 593-0240. E-mail: woub@woub.org Web Site: www.woub.org. Licensee: Ohio University. Format: News/talk. News staff: 3. ♦Carolyn Lewis, gen mgr.

WYBZ(FM)—See Crooksville

Oklahoma

Ada

KADA(AM)— September 1934: 1230 khz; 1 kw-U. TL: N34 47 06 W96 40 44. Hrs open: Box 609, 74821. Secondary address: 1019 N. Broadway 74820. Phone: (580) 332-1212. Fax: (580) 332-0128. Licensee: The Chickasaw Nation. Population served: 30,000 Rgnl. Network: Okla. Radio Net. Format: Sports. Target aud: 25-54. Spec prog: Gospel 5 hrs wkly. ♦Roger Harris, gen mgr, dev mgr, adv mgr & min affrs dir.

KADA-FM— 1979: 99.3 mhz; 5.5 kw. 299 ft TL: N34 42 31 W96 44 24. Stereo. (acq 7-88). Format: Country. ♦Roger Harris, mktg mgr & min affrs dir.

***KAJT(FM)**— 2006: 88.7 mhz; 31 kw. Ant 239 ft TL: N34 46 32 W96 35 15. Hrs open: 24 262550 Box, Baton Rouge, LA, 70826. Secondary address: 8919 World Ministry Ave., Baton Rouge, LA 70810. Phone: (225) 768-3688. Fax: (225) 768-3729. Fax: (225) 768-3729. E-mail: kawikfish@yahoo.com Web Site: www.jsm.org. Licensee: Family Worship Center Church Inc. Group owner: American Family Radio. (acq 10-7-2005; $500,000 with CP for KSSO(FM) Norman). Population served: 65,000 Format: Southern gospel. ♦David Whitelaw, COO; Jimmy Swaggart, pres; John Santiago, progmg dir.

***KAKO(FM)**— 2006: 91.3 mhz; 100 kw vert. Ant 442 ft TL: N35 13 36 W96 55 42. Hrs open: Drawer 2440, Tupelo, MS, 38803. Phone: (662) 844-8888. Fax: (662) 842-6791. Web Site: www.afr.net. Licensee: American Family Association. Natl. Network: American Family Radio. Format: Christian classics. ♦Marvin Sanders, gen mgr.

***KTGS(FM)**— January 1999: 89.5 mhz; 5.8 kw. Ant 581 ft TL: N34 41 01 W96 45 44. Hrs open: 24 Box 1343, 74821. Phone: (580) 332-0902. E-mail: email@thegospelstation.com Web Site: www.thegospelstation.com. Licensee: South Central Oklahoma Christian Broadcasting Inc. Format: Southern gospel. ♦Randall Christy, pres & gen mgr; Rick Cody, opns mgr; Danny Allen, chief of engrg.

Altus

KEYB(FM)— Dec 25, 1988: 107.9 mhz; 50 kw. Ant 492 ft TL: N34 46 15 W99 32 20. Stereo. Hrs opn: 24 Box 1077, 73522. Secondary address: 808 N. Main 73521. Phone: (580) 482-1555. Fax: (580) 482-8353. E-mail: keyb@keyb.net Web Site: www.keyb.net. Licensee: Altus FM Inc. (acq 2-5-91; FTR: 12-31-90). Population served: 76,321. Natl. Network: Jones Radio Networks, AP Network News. Law Firm: Shaw Pittman. Format: Country. News staff: one; News: 3 hrs wkly. Target aud: 25-54. Spec prog: Farm 2 hrs wkly. ♦Gayle Ledbetter, CEO; Jerry T. Butler, gen mgr, opns mgr & progmg VP; Larry Sisco, engrg dir; Tracie Tobitt, traf mgr.

***KKVO(FM)**— 1985: 90.9 mhz; 400 w. Ant 121 ft TL: N34 42 44 W99 19 03. Stereo. Hrs opn: 24 2351 Sunset Blvd., Suite 170-218, Rocklin, CA, 95765. Phone: (916) 251-1600. Fax: (916) 251-1650. Web Site: www.klove.com. Licensee: Educational Media Foundation. (acq 6-6-2005; $150,000). Population served: 25,000 Natl. Network: K-Love. Format: Relg, educ. ♦Richard Jenkins, pres; Mike Novak, VP & progmg dir; Lloyd Parker, gen mgr; Ed Lenane, opns dir & news dir; Keith Whipple, dev dir; Eric Allen, natl sls mgr; David Pierce, progmg mgr; Jon Rivers, mus dir; Sam Wallington, engrg dir; Arthur Vassar, traf mgr; Karen Johnson, news rptr; Marya Morgan, news rptr; Richard Hunt, news rptr.

***KOCU(FM)**— July 2002: 90.1 mhz; 5 kw. Ant 85 ft TL: N34 40 14 W99 20 13. Hrs open: 2800 W. Gore, Lawton, 73505. Phone: (580) 581-2472. Fax: (580) 581-5571. E-mail: kccu@cameron.edu Web Site: www.kccu.org. Licensee: Cameron University. Natl. Network: NPR, PRI. Format: News, classical. ♦Mark Norman, gen mgr; Michael V. Leal, progmg dir.

KRKZ(FM)—Listing follows KWHW(AM).

KWHW(AM)— Apr 2, 1947: 1450 khz; 1 kw-U. TL: N34 37 35 W99 20 10. Hrs open: Box 577, 73522. Secondary address: 212 W. Cypress 73522. Phone: (580) 482-1450. Fax: (580) 482-3420. E-mail: 1450@kwhw.com Web Site: www.kwhw.com. Licensee: Monarch Broadcasting Inc. (group owner; (acq 12-31-2003; grpsl). Population served: 50,000 Rgnl. Network: Okla. Radio Net. Format: C&W, news/talk, agriculture info. Spec prog: Sp 16 hrs wkly. ♦Jimmy Young, gen mgr & gen sls mgr.

KRKZ(FM)—Co-owned with KWHW(AM). Apr 1, 1974: 93.5 mhz; 45 kw. Ant 528 ft TL: N34 37 35 W99 20 10. Stereo. E-mail: 935@krkz.com 50,000 Format: Classic rock.

Alva

KALV(AM)— Oct 18, 1956: 1430 khz; 500 w-U, DA-2. TL: N36 49 06 W98 38 38. Stereo. Hrs opn: 6 AM-midnight Box 53, Rt. 1, 73717. Phone: (580) 327-1430. Fax: (580) 327-1433. Licensee: MM&K of Alva Inc. (acq 8-30-94; $165,000; FTR: 9-19-94). Population served: 7,440 Format: Oldies. News staff: one; News: 8 hrs wkly. Target aud: 45-70; loc residents. ♦Randy Mitchel, pres & gen mgr.

KNID(FM)—Licensed to Alva. See Enid

KPAK(FM)—Not on air, target date: unknown: 97.5 mhz; 50 kw. Ant 492 ft TL: N37 01 27 W98 41 22. Hrs open: 188 S. Bellevue Blvd., Suite 222, Memphis, TN, 38104. Phone: (901) 516-8970 (office). Phone: (901) 375-9324 (station). Licensee: George S. Flinn Jr. ♦George Flinn Jr., pres.

KTTL(FM)— 2001: 105.7 mhz; 50 kw. Ant 492 ft TL: N36 47 06 W98 33 01. Hrs open: 24 R.R. 1 Box 53, 73717. Phone: (580) 327-1430. Fax: (580) 327-1433. Licensee: Women, Handicapped Americans and Minorities for Better Broadcasting Inc. Population served: 25,000 Format: Adult contemp. ♦Randy Mitchell, pres & gen mgr; Craig Killman, progmg mgr.

Anadarko

KVSP(FM)— September 1981: 103.5 mhz; 100 kw. Ant 1,968 ft TL: N35 15 04 W98 36 53. Stereo. Hrs opn: 5 AM-midnight Box 1360, 73005. Secondary address: 115 W. Broadway 73005. Phone: (405) 247-6682. Fax: (405) 247-1051. Licensee: Perry Broadcasting of Southwest Oklahoma Inc. Group owner: Perry Publishing & Broadcasting Co. (acq 11-22-2002; grpsl). Natl. Network: ABC. Format: Urban. Target aud: 18 plus. ♦Kevin Perry, gen mgr & local news ed; Russell M. Perry, chief of opns.

Antlers

KDOE(FM)— 2006: 102.3 mhz; 3.3 kw. Ant 276 ft TL: N34 13 35 W95 37 20. Hrs open: 404 E. Jackson, Hugo, 74743. Phone: (580) 326-5541. Fax: (580) 326-5236. Licensee: Will Payne. ♦Will Payne, gen mgr.

Apache

KACO(FM)— Jan 1, 1989: 98.5 mhz; 18.5 kw. Ant 305 ft TL: N34 56 30 W98 22 33. Stereo. Hrs opn: 24 Box 1487, Ardmore, 73402. Secondary address: 115 W. Broadway, Suite 501, Ardmore 73401. Phone: (580) 226-9850. Fax: (580) 226-5113. E-mail: klcm@cableone.net Web Site: oldiesradio.com. Licensee: A.M. & P.M. Communications L.L.C. (acq 1-5-98; $475,000). Population served: 75,000 Natl. Network: ABC. Law Firm: Cordon & Kelly. Format: Oldies, rock and roll. Target aud: 25-54. ♦Bill Countrymen, VP; Steve Spain, gen mgr; Rob Carter, opns mgr & sls dir; Ron Ricord, progmg mgr.

Ardmore

KKAJ-FM— June 24, 1974: 95.7 mhz; 100 kw. Ant 449 ft TL: N34 05 53 W97 10 54. Stereo. Hrs opn: 24 1205 Northglen, 73401. Phone: (580) 226-0421. Fax: (580) 226-0464. E-mail: webmaster@kkaj.com Web Site: www.kkaj.com. Licensee: LKCM Radio Group L.P. (acq 2-26-2007; grpsl). Population served: 512,000 Rgnl. Network: Agri-net. Format: Country. News staff: one; News: 25 hrs wkly. Target aud: 18-54. ♦Michael Baer, gen mgr; Dave Hilton, opns mgr.

KVSO(AM)—Co-owned with KKAJ-FM. September 1935: 1240 khz; 1 kw-U. TL: N34 10 54 W97 08 48. Stereo. 24 E-mail: webmaster@kvso.com Web Site: www.kvso.com. Group owner: NextMedia Group L.L.C. 320,000 Natl. Network: ABC. Natl. Rep: Christal. Format: Oldies. News staff: one; News: 4 hrs wkly. Target aud: 25 plus.

***KLCU(FM)**— June 19, 1998: 90.3 mhz; 25 kw. 213 ft TL: N34 12 10 W97 09 12. Hrs open:
Rebroadcasts KCCU(FM) Lawton 98%.
c/o KCCU(FM), Admin. Bldg., 2800 W. Gore Blvd., Lawton, 73505. Phone: (580) 581-2425. Phone: (580) 581-2474. Fax: (580) 581-5571. E-mail: kccu@cameron.edu Web Site: www.kccu.org. Licensee: Cameron University. Population served: 100,000 Natl. Network: NPR, PRI. Format: Classical/National Public Radio. ♦Mark Norman, gen mgr; Terry Anderson, dev dir & dev dir; Michael V. Leal, progmg dir.

***KQPD(FM)**— 2003: 91.1 mhz; 250 w. Ant 167 ft TL: N34 11 01 W97 07 23. Hrs open: Drawer 2440, Tupelo, MS, 38803. Phone: (662) 844-8888. Fax: (662) 842-6791. Licensee: American Family Association. Group owner: American Family Radio. Format: Christian. ♦Marvin Sanders, gen mgr.

KYNZ(FM)—See Lone Grove

Atoka

KHKC-FM— June 15, 1984: 102.1 mhz; 3.3 kw. Ant 449 ft TL: N34 25 08 W96 11 24. Stereo. Hrs opn: Box 810, 74525. Secondary address: Hwy. 75 N. 74525. Phone: (580) 889-3392. Phone: (580) 889-6300. Fax: (580) 889-9308. Licensee: Keystone Broadcasting Corp. (acq 10-23-2001; with co-located AM). Format: Country. ♦Ricky Chase, gen mgr & progmg dir.

KKNG-FM—See Newcastle

Bartlesville

KRIG-FM—Nowata, 1965: 104.9 mhz; 8.3 kw. Ant 564 ft TL: N36 43 37 W95 46 18. Stereo. Hrs opn: Box 1100, 74005. Secondary address: 1200 S.E. Frank Phillips Blvd. 74003. Phone: (918) 336-1001. Phone: (918) 336-1400. Fax: (918) 336-6939. E-mail: radio@bartlesvilleradio.com Web Site: www.bartlesvilleradio.com. Licensee: KCD Enterprises Inc. (group owner; acq 6-26-98; $775,000). Natl. Network: ABC. Rgnl rep: Rgnl Reps Law Firm: Lauren A. Colby. Wire Svc: AP Format: Country. News: 20 hrs wkly. Target aud: 35-65; mature buyers. Spec prog: Gospel 4 hrs wkly. ♦Kevin Potter, pres & gen mgr; Charlie Taraboletti, opns mgr; Dorea Potter, prom mgr; Sharon Frahm, traf mgr.

KWON(AM)— April 1942: 1400 khz; 1 kw-U. TL: N36 45 53 W95 57 35. Stereo. Hrs opn: 24 Box 1100, 74005. Secondary address: 1200 S.E. Frank Phillips Blvd. 74003. Phone: (918) 336-1001. Phone: (918) 336-1400. Fax: (918) 336-6939. E-mail: radio@bartlesvilleradio.com Web Site: www.bartlesvilleradio.com. Licensee: KCD Enterprises Inc. (group owner; acq 2-1-97; $625,000 with co-located FM). Population served: 42,000 Natl. Network: CBS. Rgnl. Network: Okla. Radio Net. Rgnl rep: Rgnl Reps Law Firm: Lauren A. Colby. Wire Svc: AP Format: News/talk. News staff: 2; News: 25 hrs wkly. Target aud: 25-54; general. Spec prog: Relg 5 hrs wkly. ♦Charlie Taraboletti, opns mgr, opns mgr, progmg dir, news dir, engrg dir & engrg mgr; Kevin Potter, pres, stn mgr, sls dir & gen sls mgr; Dorea Potter, prom dir.

KYFM(FM)—Co-owned with KWON(AM). Nov 6, 1961: 100.1 mhz; 25 kw. Ant 695 ft TL: N36 37 42 W96 11 26. Stereo. 24 Web Site: www.bartlesvilleradio.com.72,000 Natl. Network: ABC. Rgnl rep: Rgnl Reps Law Firm: Lauren A. Colby. Format: Adult contemp. News: 15 hrs wkly. Target aud: 25-49. Spec prog: Gospel 4 hrs wkly.

***KWRI(FM)**— 2004: 89.1 mhz; 100 kw vert. Ant 626 ft TL: N36 42 13 W95 30 57. Hrs open: 24 2351 Sunset Blvd., Suite 170-218, Rocklin, CA, 95765. Phone: (916) 251-1600. Fax: (916) 251-1650. E-mail: info@air1.com Web Site: www.air1.com. Licensee: Educational Media Foundation. Group owner: EMF Broadcasting. Natl. Network: Air 1. Law Firm: Shaw Pittman. Format: Contemp Christian. News staff: 3. Target aud: 18-35; Judeo-Christian, female. ♦Richard Jenkins, pres; Mike Novak, VP; Keith Whipple, dev dir; David Pierce, progmg mgr; Ed Lenane, news dir; Sam Wallington, engrg dir; Karen Johnson, news rptr; Marya Morgan, news rptr; Richard Hunt, news rptr.

Bennington

KFYZ-FM— Nov 1, 1979: 98.1 mhz; 3.5 kw. Ant 210 ft TL: N34 02 40 W96 01 10. Stereo. Hrs opn: 5 AM-1 AM North Texas Radio Group L.P., 5946 Club Oaks Dr., Dallas, TX, 75248. Phone: (972) 931-6055. Fax: (972) 931-9141. Licensee: North Texas Radio Group L.P. (acq 11-2-98; $1.15 million with co-located AM). ♦Richard E. Witkovski, gen mgr.

Bethany

KKWD(FM)—Licensed to Bethany. See Oklahoma City

Bixby

KJMM(FM)— November 1994: 105.3 mhz; 10 kw. 879 ft TL: N35 51 41 W95 46 03. Hrs open: 24 7030 S. Yale, Suite 302, Tulsa, 74136. Phone: (405) 427-5877. Fax: (918) 494-9683. Web Site: www.1053kjamz.com. Licensee: KJMM Inc. Group owner: Perry Publishing & Broadcasting Co. (acq 1-95). Natl. Network: ABC, American Urban, Westwood One. Law Firm: Meyer, Faller, Weisman & Rosenberg. Format: Urban. News staff: one; News: 10 hrs wkly. Target aud: General. ♦Russell Perry, CEO; Kevin Perry, chmn; Bryan K. Robinson, gen mgr; Terry Monday, progmg VP.

Blackwell

KLOR-FM—See Ponca City

KOKB(AM)— October 1952: 1580 khz; 1 kw-D, 49 w-N. TL: N36 48 35 W97 15 50. Hrs open: 6 AM-9 PM
Rebroadcasts KOKP(AM) Perry 80%.
Box 2509, Ponca City, 74602. Secondary address: 122 N. Third St., Ponca City 74602. Phone: (580) 765-2485. Fax: (580) 767-1103. E-mail: kokb@eteamradio.com Web Site: www.eteamradio.com. Licensee: Team Radio LLC (group owner; acq 10-18-96; $90,000). Population served: 65,000 Format: Talk, sports. News: 30 hrs wkly. Target aud: 35-75; adult, upper-middle income. Spec prog: Gospel 11 hrs wkly. ♦Bill Coleman, pres, gen mgr & stn mgr.

Blanchard

KOJK(FM)— Aug 18, 1977: 97.3 mhz; 1 kw. Ant 800 ft TL: N35 10 38 W97 36 10. Hrs open: 5101 S. Shields Blvd., Oklahoma City, 73129. Phone: (405) 616-5500. Fax: (405) 616-5505. Web Site: www.jackokc.com. Licensee: Nick Radio LLC. (acq 1-31-2006; $1 million). Population served: 100,000 Format: Adult contemp. ♦Skip Stow, gen mgr.

Bristow

KREK(FM)— Nov 14, 1978: 104.9 mhz; 5 kw. 351 ft TL: N35 47 11 W96 27 35. Stereo. Hrs open: 24 Box 1280, 74010. Phone: (918) 367-5501. Licensee: Big Chief Broadcasting Co. of Bristow Inc. Population served: 75,000 Format: C&W. Target aud: 18-49. ♦Clifford W. Smith, pres & gen mgr.

Broken Arrow

***KNYD(FM)**— Aug 19, 1986: 90.5 mhz; 100 kw. 1,638 ft TL: N36 01 15 W95 40 32. Stereo. Hrs opn: 24 Box 1924, Tulsa, 74101. Secondary address: 11717 S. 129th East Ave. 74011. Phone: (918) 455-5693. Fax: (918) 455-0411. E-mail: mail@oasisnetwork.org Web Site: www.oasisnetwork.org. Licensee: Creative Educational Media Inc. (acq 1985). Format: Relg. Target aud: General. ♦David Ingles, pres & gen mgr.

KTBT(FM)— Dec 23, 1970: 92.1 mhz; 27 kw. 656 ft TL: N36 06 38 W96 01 57. Stereo. Hrs opn: 24 2625 S. Memorial, Tulsa, 74129. Phone: (918) 388-5100. Fax: (918) 665-0555. Web Site: www.921thebeat.com. Licensee: Clear Channel Broadcasting Licenses Inc. Group owner: Clear Channel Communications Inc. Natl. Rep: Clear Channel. Format: CHR. Target aud: 18-34; women. ♦Michael Oppenheimer, gen mgr; Don Cristi, opns mgr.

Broken Bow

KKBI(FM)— January 1983: 106.1 mhz; 50 kw. 817 ft TL: N34 14 45 W94 46 58. Stereo. Hrs opn: 24 Box 1016, 74728-1016. Phone: (580) 584-3388. Fax: (580) 584-3341. E-mail: kkbi@pine-net.com Web Site: www.kkbifm.com. Licensee: J.D.C. Radio Inc. (acq 1-27-98; $800,000). Population served: 34,000 Natl. Network: Jones Radio Networks. Law Firm: Putbrese, Hunsaker & Trent. Format: Country. News staff: one; News: 5 hrs wkly. Target aud: 24-55. Spec prog: Farm 5 hrs, gospel 4 hrs wkly. ♦Homer Coleman, pres; David Smulyan, exec VP & gen mgr; Rod Kennedy, progmg dir & mus dir.

Byng

KYKC(FM)— Sept 17, 1992: 100.1 mhz; 50 kw. 492 ft TL: N34 43 43 W96 42 45. (CP: 12.87 kw, ant 459 ft.). Hrs opn: 24 121 S. Constant, Ada, 74820. Phone: (580) 436-6351. Fax: (580) 436-3388. E-mail: kykc@cableone.net Web Site: www.kykc.net. Licensee: The Chickasaw Nation. (acq 1-14-2005; $900,000). Format: Country. Target aud: 12 plus; across the board. ♦Dexter Pruitt, exec VP; Mike Hall, gen mgr; Kay Hall, gen sls mgr; David Wayne, progmg mgr.

Cache

***KARU(FM)**— 2005: 88.9 mhz; 440 w vert. Ant 259 ft TL: N34 38 10 W98 41 32. Stereo. Hrs opn: 24 2351 Sunset Blvd., Suite 170-218, Rocklin, CA, 95765. Phone: (916) 251-1600. Fax: (916) 251-1650. E-mail: info@air1.com Web Site: www.air1.com. Licensee: Educational Media Foundation. Group owner: EMF Broadcasting. Natl. Network: Air 1. Law Firm: Shaw Pittman. Format: Contemp Christian. News staff: 3. Target aud: 18-35; Judeo-Christian, female. ♦Richard Jenkins, pres; Mike Novak, VP; Keith Whipple, dev dir; Eric Allen, natl sls mgr; David Pierce, progmg dir; Liz Morton, mus dir; Ed Leane, news dir; Sam Wallington, engrg dir; Karen Johnson, news rptr.

KJMZ(FM)—Licensed to Cache. See Lawton

Carnegie

***KJCC(FM)**— 2005: 89.5 mhz; 350 w vert. Ant 194 ft TL: N35 06 59 W98 28 26. Hrs open: CSN International, 3232 W. MacArthur Blvd., Santa Ana, CA, 92704. Secondary address: 300 Towakkonie Rd., Fort Cobb 73038. Phone: (405) 643-2635. Fax: (405) 643-2851. E-mail: leannafarmer99@yahoo.com Web Site: www.csnintl.com. Licensee: CSN International (group owner). Format: Christian praise & worship, Bible teaching. ♦Leanna Farmer, stn mgr.

Catoosa

KEOR(AM)— Jan 29, 1968: 1120 khz; 2 kw-D, DA. TL: N36 18 31 W95 58 25. Hrs open: Sunrise-sunset Box 810, Hwy. 75 N., Atoka, 74525. Phone: (580) 889-3392. Phone: (580) 889-6300. Fax: (580) 889-9308. E-mail: gospelradio@yahoo.com Licensee: M&M Broadcasters Ltd. Group owner: First Broadcasting Investment Partners LLC (acq 4-16-2006; grpsl). Format: Relg. Target aud: General. ♦Gary L. Moss, pres; Ricky Chase, gen mgr & stn mgr; John Clemmetsen, progmg dir; Chris Hoopes, mus dir & chief of engrg.

Chelsea

KTFR(FM)— March 1, 2001: 100.7 mhz; 6 kw. 328 ft TL: N36 30 12 W95 26 29. Hrs open: 24
Rebroadcasts KXOJ-FM Sapulpa.
2448 E. 81st St., Suite 4500, Tulsa, 74137. Phone: (918) 492-2660. Fax: (918) 492-8840. E-mail: kxoj@kxoj.com Web Site: www.kxoj.com. Licensee: Michael P. Stephens. Group owner: Adonai Radio Group (acq 2-17-95; FTR: 5-15-95). Format: Contemp Christian. ♦Mike Stephens, pres; David Stephens, gen mgr; Bob Thornton, progmg dir.

Chickasha

***KSSX(FM)**—Not on air, target date: unknown: 90.5 mhz; 1.3 kw. Ant 305 ft TL: N35 00 43 W97 56 07. Stereo. Hrs opn: 1101 N. 81 Hwy., Marlow, 73055. Phone: (580) 658-9292. Fax: (580) 658-2561. E-mail:

kfxi@cableone.net Licensee: Sister Sherry Lynn Foundation Inc. Format: Gospel/btfl mus. News staff: 3. ♦Sherry Austin, pres; Ken Austin, gen mgr; Lisa Hatchett, gen sls mgr; Marshall Slayden, progmg dir; Teddy Casteel, mus dir; James Wilson, engr.

KWCO-FM— Nov 4, 1966: 105.5 mhz; 3.3 kw. 443 ft TL: N35 00 38 W97 55 54. Stereo. Hrs opn: 24 627 West Chickasha Ave., Oklahoma City, 73129. Phone: (405) 224-9105. Fax: (405) 224-2890. Web Site: www.ktuz.com. Licensee: Kenny Communications Inc. (acq 1-1-2004; $114,400). Population served: 100,000 Natl. Network: Jones Radio Networks. Format: Classic rock. News: 10 hrs wkly. Target aud: 18-54; Spanish persons. ♦Matthew Mollman, gen mgr; Keith Michaels, progmg dir; George Plummer, news dir; Christopher Hoops, chief of engrg.

Claremore

***KRSC-FM**— Aug 4, 1980: 91.3 mhz; 2.2 kw. 364 ft TL: N36 19 06 W95 38 18. Stereo. Hrs opn: 6 AM-midnight Rogers State University., 1701 W. Will Rogers Blvd., 74017-3252. Phone: (918) 343-7670. Phone: (918) 343-7659. Fax: (918) 343-7952. E-mail: alambert@vsu.edu Web Site: www.rsu.edu. Licensee: Board of Regents of the University of Oklahoma. Population served: 300,000 Format: AAA, Alternative, new pop. Target aud: General; college & community, young & older adults. Spec prog: Folk 5 hrs, gospel 4 hrs, jazz 5 hrs, progsv 12 hrs, blues 6 hrs, country 5 hrs wkly. ♦Alan Lambert, gen mgr & stn mgr; Dr. David Nelson, opns mgr. Co-owned TV: *KRSC-TV affil.

KRVT(AM)— Jan 17, 1958: 1270 khz; 1 kw-U, DA-N. TL: N36 17 59 W95 37 16. Hrs open: 24 Box 702588, Tulsa, 74170. Phone: (918) 496-7700. Fax: (918) 746-7615. E-mail: river@krvt.net Web Site: www.krvt.com. Licensee: Reunion Broadcasting L.L.C. (group owner; (acq 2-25-2000). Natl. Network: CNN Radio. Law Firm: Hardy, Carey & Chautin. Format: Oldies. News: 4 hrs wkly. Target aud: 35 plus; upscale adults. Spec prog: St. Louis Cardinal Baseball. ♦D. Stanley Tacker, pres & gen mgr.

Clinton

KCLI(AM)— Apr 15, 1949: 1320 khz; 1 kw-D, 108 w-N. TL: N35 29 00 W98 58 54. Hrs open: 700 Frisco, 73601. Phone: (580) 323-5254. Fax: (580) 323-0717. E-mail: sales@wrightradio.com Licensee: Wright Broadcasting Systems Inc. Group owner: Wright Broadcasting Systems (acq 9-13-2000; $25,000). Rgnl. Network: Okla. Radio Net. Law Firm: Fletcher, Heald & Hildreth. Format: News/talk. Target aud: General. ♦Harold Wright, gen mgr.

KWEY-FM— Apr 9, 1978: 95.5 mhz; 40 kw. Ant 492 ft TL: N35 27 04 W98 58 19. Stereo. Hrs opn: 24 Box 587, Weatherford, 73096. Phone: (580) 772-5939. Fax: (580) 772-1590. E-mail: sales@wrightradio.com Web Site: www.kwey.com. Licensee: Wright Broadcasting Systems Inc. Group owner: Wright Broadcasting Systems (acq 1996; $300,000). Natl. Network: ABC. Law Firm: Putbrese, Hunsaker & Trent, P.C. Wire Svc: AP Format: Country. News: 8 hrs wkly. Target aud: 18-54; upwardly mobile adults. ♦Harold Wright, CEO, pres & gen mgr; Todd Brunner, opns mgr; Heston Wright, sls dir; Rob Grogan, progmg dir; Ray Bagby, engrg dir.

***KYCU(FM)**— September 2002: 89.1 mhz; 40 kw. Ant 633 ft TL: N35 26 40 W98 59 22. Hrs open: 2800 W. Gore Blvd., Lawton, 73505. Phone: (580) 581-2425. Fax: (580) 581-5571. Licensee: Cameron University. Natl. Network: NPR, PRI. Format: News, classical. ♦Mark Norman, gen mgr.

Coalgate

KKFC(FM)— Dec 7, 2001: 105.5 mhz; 20 kw. Ant 364 ft TL: N34 41 43 W96 23 17. Hrs open: 24 1188 North Hills Centre, Ada, 74820. Phone: (580) 332-2211. Fax: (580) 436-1629. E-mail: kkfc@kkfcradio.com Web Site: www.kkfcradio.com. Licensee: Woodstone Broadcasting Inc. (acq 10-9-01). Population served: 176,000 Natl. Network: AP Radio, Jones Radio Networks. Law Firm: Smithwick & Belendink. Format: Classic country. Target aud: 25+. ♦Rick Woodward, pres & gen mgr; Howard Stone, VP; Sam Jackson, progmg dir; Pam Pinnella, traf mgr.

Collinsville

KIZS(FM)— June 25, 1996: 101.5 mhz; 6.2 kw. Ant 656 ft TL: N36 20 02 W95 47 08. Hrs open: 2625 S. Memorial, Tulsa, 74129. Phone: (918) 388-5100. Fax: (918) 665-0555. Licensee: Clear Channel Broadcasting Licenses Inc. Group owner: Clear Channel Communications Inc. (acq 10-6-97; $1.9 million). Format: Sp. Target aud: 25-54; general. ♦Michael Oppenheimer, gen mgr & stn mgr; Don Cristi, opns mgr.

Comanche

KDDQ(FM)— Apr 1, 1982: 105.3 mhz; 6 kw. Ant 298 ft TL: N34 26 12 W97 54 47. Stereo. Hrs opn: 24 1701 W. Pine Ave., Duncan, 73533. Phone: (580) 255-1350. Fax: (580) 470-9993. Licensee: Perry Broadcasting of Southwest Oklahoma Inc. Group owner: Perry Publishing & Broadcasting Co. (acq 1-9-2003; grpsl). Natl. Network: ABC. Format: Classic rock. News staff: one; News: 4 hrs wkly. Target aud: 25-54; females with mid-level income. Spec prog: Gospel 2 hrs wkly. ♦Kevin Perry, gen mgr; Joy Chapman, sls dir & prom dir; Mike Hoffman, chief of engrg.

Cordell

KCDL(FM)— Sept 1, 1988: 99.3 mhz; 10.5 kw. 505 ft TL: N35 26 49 W98 59 17. Hrs open: 24 700 Frisco, Clinton, 73601. Phone: (580) 323-9900. Fax: (580) 323-0717. E-mail: sales@wrightradio.com Web Site: kcdl.com. Licensee: Wright Broadcasting Systems Inc. Group owner: Wright Broadcasting Systems (acq 8-30-99; $350,000). Natl. Network: ABC, CNN Radio. Rgnl. Network: Agri-Net. Format: Classic rock. Target aud: General. ♦Harold Wright, CEO & pres; Todd Brunner, gen mgr & opns dir; Rob Grogan, progmg dir.

Coweta

***KDIM(FM)**— February 2005: 88.1 mhz; 100 kw. Ant 551 ft TL: N35 42 24 W96 05 39. Stereo. Hrs opn: 24 Box 1924, Tulsa, 74101. Secondary address: 11717 S. 129th E. Ave., Broken Arrow 74011. Phone: (918) 455-5693. Fax: (918) 455-0411. E-mail: mail@oasisnetwork.org Web Site: www.oasisnetwork.org. Licensee: Creative Educational Media Corp. Inc. Format: Relg. Target aud: General. ♦David Ingles, pres & gen mgr.

Cushing

KUSH(AM)— 1953: 1600 khz; 1 kw-D, 70 w-N. TL: N35 39 11 W96 42 37. Hrs open: Box 791, 74023. Phone: (918) 225-0922. Fax: (918) 225-0925. Licensee: Cimarron Valley Broadcasters Inc. (acq 3-4-65). Format: News/talk, sports. ♦Sean Kelly, gen mgr.

Del City

KOCY(AM)— November 1946: 1560 khz; 1 kw-D, 250 w-N, DA-2. TL: N35 26 26 W97 29 24 (D), N35 26 27 W97 29 24 (N). Hrs open: 24 5101 S. Shields, Oklahoma City, 73129. Phone: (405) 616-5500. Fax: (405) 616-5551. Licensee: Oklahoma Land Co. L.L.C. Group owner: Tyler Media Broadcasting Corp. (acq 12-15-2003; $250,000). Population served: 45,000 Natl. Network: Radio Disney. Format: Children. News staff: one; News: 8 hrs wkly. Target aud: 18 plus. Spec prog: Loc sports. ♦Skip Stow, gen mgr.

Dickson

KTRX(FM)— June 2001: 92.7 mhz; 5.5 kw. Ant 341 ft TL: N34 06 56 W97 00 06. Stereo. Hrs opn: 24 1205 Northglen, Ardmore, 73402. Phone: (580) 226-0421. Fax: (580) 226-0464. E-mail: webmaster@texomarocks.com Web Site: www.texomarocks.com. Licensee: LKCM Radio Group L.P. Group owner: NextMedia Group L.L.C. (acq 2-26-2007; grpsl). Population served: 310,000 Natl. Network: Jones Radio Networks. Rgnl. Network: Agri-net. Natl. Rep: Christal. Format: Classic rock. News staff: 1; News: 25 hrs wkly. Target aud: 25-54; men. ♦Dave Smith, VP & gen mgr; Dave Hilton, opns mgr & progmg dir; David MacMullen, sls dir.

Duncan

KKEN(FM)—Listing follows KPNS(AM).

KPNS(AM)— Oct 31, 1947: 1350 khz; 250 w-D, 100 w-N. TL: N34 40 43 W97 58 05. Stereo. Hrs opn: 6 AM-midnight 1701 Pine St., W., 73533. Phone: (580) 255-1350. Fax: (580) 470-9993. Licensee: Perry Broadcasting of Southwest Oklahoma Inc. Group owner: Perry Publishing & Broadcasting Co. (acq 11-22-02; grpsl). Population served: 40,000 Format: Sports talk. News staff: one; News: 20 hrs wkly. Target aud: General. ♦Peggy Richardson, gen mgr; Kristin Rei, opns mgr; Joy Chatman, gen sls mgr & adv mgr; Terry Monday, progmg VP & progmg dir.

KKEN(FM)—Co-owned with KPNS(AM). Dec 31, 1975: 102.3 mhz; 3 kw. 207 ft TL: N34 40 43 W97 58 05. Stereo. 40000 Format: Country. Target aud: 25-54. ♦Pam Peck, rgnl sls mgr.

Durant

***KAYC(FM)**— 2000: 91.1 mhz; 403 w. Ant 210 ft TL: N34 01 17 W96 28 18. Hrs open: Box 3206, American Family Radio, Tupelo, MS, 38803. Phone: (662) 844-8888. Fax: (662) 842-6791. Web Site: www.afr.net. Licensee: American Family Association. Group owner: American Family Radio. Format: Inspirational Christian. ♦Marvin Sanders, gen mgr.

KLBC(FM)—Listing follows KSEO(AM).

KSEO(AM)— May 1947: 750 khz; 250 w-D. TL: N34 00 07 W96 25 19. Hrs open: 6 AM-7 PM Box 190, 74702. Phone: (580) 924-3100. Fax: (580) 920-1426. Web Site: www.klbc.com.yes Licensee: Texoma Broadcasting Inc. (acq 5-28-99; with co-located FM). Population served: 34,000 Format: Adult contemp, golden oldies. News staff: one; News: 3 hrs wkly. Target aud: Adults; 18-54. ♦Bob McKenzie, opns dir & progmg dir; Todd Tidwell, pres, gen mgr & gen sls mgr; J.B. Conley, prom dir; Ron Eudaly, pub affrs dir.

KLBC(FM)—Co-owned with KSEO(AM). November 1958: 107.1 mhz; 6 kw. 365 ft TL: N34 00 07 W96 25 19. Stereo. 24 1418 N. 1st Ave., 74701. E-mail: klbcradio@netcommander.com Web Site: www.klbcfm.com.yes Format: Country. News staff: one; News: 20 hrs wkly. Target aud: General.

***KSSU(FM)**— Feb 1, 1972: 91.9 mhz; 1.5 kw. 341 ft TL: N34 00 45 W96 19 45. Stereo. Hrs opn: 24 1405 N. 4th St., PMB 4129, 74701-0609. Phone: (580) 745-2000. E-mail: kssu@sosu.edu Web Site: www.sosu.edu/kssu. Licensee: Southeastern Oklahoma State University. Population served: 35,000 Format: CHR. News: 3 hrs wkly. Target aud: 18-25; college, high school students & area residents. ♦Dr. John Hendricks, exec VP & gen mgr.

Edmond

***KCSC(FM)**— April 1966: 90.1 mhz; 100 kw. Ant 840 ft TL: N35 34 24 W97 29 08. Stereo. Hrs opn: 24 Univ. of Central Okla., 100 N. University Dr., 73034-5209. Phone: (405) 974-3333. Fax: (405) 974-3844. E-mail: kcscfm@kcscfm.com Web Site: www.kcscfm.com. Licensee: University of Central Oklahoma. Population served: 1,006,629 Natl. Network: PRI. Format: Class. News: 5 hrs wkly. Target aud: 35 plus; educated, affluent. ♦Bradford Ferguson, gen mgr; Barbara Hendrickson, opns mgr; Susan Reger, dev dir.

***KOKF(FM)**— September 1977: 90.9 mhz; 100 kw. Ant 480 ft TL: N35 33 59 W97 28 28. Stereo. Hrs opn: 24 2351 Sunset Blvd., Suite 170-218, Rocklin, CA, 95765. Phone: (916) 251-1600. Fax: (916) 251-1650. Web Site: www.air1.com. Licensee: Educational Media Foundation. (acq 5-25-2006; $4 million). Population served: 1,700,000 Natl. Network: Air 1. Format: Christian. ♦Richard Jenkins, pres & gen sls mgr; Mike Novak, VP; Keith Whipple, dev dir; David Pierce, progmg mgr; Ed Lenane, news dir; Sam Wallington, engrg dir; Karen Johnson, news rptr; Marya Morgan, news rptr; Richaard Hunt, news rptr.

WWLS-FM—Licensed to Edmond. See Oklahoma City

El Reno

KZUE(AM)— Sept 9, 1962: 1460 khz; 500 w-D. TL: N35 30 30 W97 54 00. Hrs open: 2715 S. Radio Rd., 73036. Phone: (405) 262-1460. Fax: (405) 262-1886. Licensee: La Tremenda Inc. (acq 12-8-93; $40,000; FTR: 1-3-94). Population served: 25,000 Natl. Rep: Keystone (unwired net). Format: Sp. Spec prog: Farm 7 hrs, Indian one hr wkly. ♦Nancy Galvan, gen mgr; George Ochoa, disc jockey; Shannon Calderon, disc jockey.

Eldorado

KXOW(FM)—Not on air, target date: unknown: 96.9 mhz; 6 kw. Ant 328 ft TL: N34 27 00 W99 32 04. Hrs open: 2768 Pharmacy Rd., Rio Grande City, TX, 78582. Phone: (956) 487-5621. Licensee: James Falcon. ♦James Falcon, gen mgr.

Elk City

KADS(AM)— October 1932: 1240 khz; 1 kw-U. TL: N35 22 51 W99 24 25. Hrs open: 24 Box 945, 73648. Phone: (580) 225-9696. Fax: (580) 225-9699. E-mail: bbrewerkeco@cableone.net Web Site: kecofm.com. Licensee: Paragon Communications Inc. (group owner; acq 6-15-01; $15,000). Population served: 20,000 Natl. Network: ESPN Radio. Wire Svc: AP Format: Sports. News staff: one. ♦Blake Brewer, gen mgr.

KECO(FM)— July 20, 1982: 96.5 mhz; 100 kw. 500 ft TL: N35 24 22 W99 29 54. Stereo. Hrs opn: 24 Box 945, 73648. Secondary address: 220 S. Pioneer Rd. 73648. Phone: (580) 225-9696. Fax: (580) 225-9699. E-mail: bbrewerkeco@cableone.net Web Site: www.kecofm.com. Licensee: Paragon Communications Inc. (group owner; acq 4-22-98; $100,000 for 72% with KXOO(FM) Elk City). Population served: 60,000 Law Firm: Bryan Cave. Format: Country. News staff: one; News: 2.5 hrs wkly. Target aud: General. ♦Blake Brewer, pres, gen mgr & gen sls mgr; Connie Legrand, opns mgr.

KTIJ(FM)— July 15, 2000: 106.9 mhz; 100 kw. Ant 981 ft TL: N34 58 39 W99 24 35. Hrs open: Box 311, 1515 N. Broadway, Hobart, 73651. Phone: (580) 726-5656. Fax: (580) 726-2222. E-mail: thezone@itlnet.net Licensee: Fuchs Radio LLC. Format: CHR/pop. ♦Chad Fox, gen mgr; Shelly Fox, VP & opns mgr.

KXOO(FM)— April 1995: 94.3 mhz; 12 kw. 469 ft TL: N35 24 22 W99 29 54. Stereo. Hrs opn: 24 Box 945, 73648. Phone: (580) 225-9696. Fax: (580) 225-9699. Licensee: Paragon Communications Inc. (group owner; acq 4-22-98; $100,000 for 72% with KECO(FM) Elk City). Population served: 40,000 Format: Contemp christian. News staff: one; News: 2 hrs wkly. ♦Blake Brewer, pres & gen mgr; Guy Baker, chief of engrg.

Enid

KCRC(AM)— 1926: 1390 khz; 1 kw-U, DA-1. TL: N36 25 11 W97 52 28. Hrs open: 24 Box 952, 73702. Secondary address: 316 E. Willow 73701. Phone: (580) 237-1390. Fax: (580) 242-1390. E-mail: ctbradio@yahoo.com Licensee: Chisholm Trail Holding Co. Inc. (acq 6-1-83; $1.38 million; FTR: 6-20-83). Population served: 50,195 Natl. Network: Jones Radio Networks, ESPN Radio. Rgnl. Network: Okla. Radio Net. Format: News. News staff: one; News: 2 hrs wkly. Target aud: General. ♦Hiram Champlin, pres & gen mgr; Ricky Roggow, opns mgr & mus dir; Sandy Daniels, gen sls mgr; Suzi Lakin, prom mgr; Chad McKee, progmg dir; Rob Houston, news dir & pub affrs dir; G.B. Bonham, chief of engrg.

KQOB(FM)—Co-owned with KCRC(AM). May 1, 1967: 96.9 mhz; 97.5 kw horiz, 100 kw vert. Ant 1,450 ft TL: N35 58 50 W97 41 42. Stereo. Licensee: Champlin Broadcasting Inc.250,000 Format: Var hits. Target aud: 18-54.

KFXY(AM)— January 2004: 1640 khz; 10 kw-D, 1 kw-N. TL: N36 25 14 W97 52 28. Hrs open: 24 Box 952, 73702. Secondary address: 316 E. Willow 73701. Phone: (580) 237-1390. Fax: (580) 242-1390. E-mail: hchamplin@knid.com Licensee: Chisholm Trail Broadcasting Co. Format: Sports. ♦Hiram Champlin, gen mgr; Ricky Roggow, opns mgr; Sandy Daniels, gen sls mgr; Suzi Lakin, prom dir; Chad McKee, progmg dir.

KGWA(AM)— 1950: 960 khz; 1 kw-U, DA-1. TL: N36 26 13 W97 55 16. Hrs open: 24 Box 3128, 73702. Secondary address: 1710 W. Willow Rd., Suite 300 73703. Phone: (580) 234-4230. Fax: (580) 234-2971. E-mail: radio@kofm.com Web Site: www.kofm.com. Licensee: Williams Broadcasting LLC. (acq 9-10-99; with co-located FM). Population served: 68,000 Natl. Network: ABC. Law Firm: Putbrese, Hunsaker & Trent. Format: News/talk. News staff: 2; News: 15 hrs wkly. Target aud: 35 plus. Spec prog: Farm 3 hrs wkly. ♦Kyle Williams, pres & gen mgr; Cheryl Myatt, gen sls mgr; J. Curtis Huckleberry, progmg dir & news dir.

KOFM(FM)—Co-owned with KGWA(AM). March 1982: 103.1 mhz; 25 kw. 298 ft TL: N36 26 14 W97 55 15. Stereo. 24 Phone: (580) 234-6371. E-mail: kwilliams@kofm.com 68,000 Format: Country. News staff: one; News: one hr wkly.

***KKRD(FM)**— Oct 1, 1986: 91.1 mhz; 300 w. Ant 297 ft TL: N36 23 48 W97 52 38. Stereo. Hrs opn: 24 2351 Sunset Blvd., Suite 170-218, Rocklin, CA, 95765. Phone: (916) 251-1600. Fax: (916) 251-1650. E-mail: info@air1.com Web Site: www.air1.com. Licensee: Educational Media Foundation. (acq 10-15-2004; $102,500). Population served: 200,000 Natl. Network: Air 1. Format: Relg. ♦Richard Jenkins, pres; Mike Novak, VP; Keith Whipple, dev dir; David Pierce, progmg mgr; Ed Lenane, news dir; Sam Wallington, engrg dir; Arthur Vassar, traf mgr; Karen Johnson, news rptr; Marya Morgan, news rptr; Richard Hunt, news rptr.

KNID(FM)—Alva, Feb 1, 1981: 99.7 mhz; 100 kw. 850 ft TL: N36 35 41 W98 15 38. Stereo. Hrs opn: 24 Box 952, 73702. Secondary address: 316 E. Willow Rd. 73701. Phone: (580) 237-1390. Fax: (580) 242-1390. E-mail: hchamplin@knid.com Licensee: Chisholm Trail Holding Co. Inc. Population served: 132,000 Natl. Network: ABC. Rgnl. Network: Mid-American Ag. Format: Country. News staff: one; News: 2 hrs wkly. Target aud: 25-49. ♦Hiram Champlin, pres & gen mgr; Ricky Roggow, opns mgr & progmg dir; Sandy Daniels, gen sls mgr; Suzi Lakin, prom dir; G.B. Bonham, chief of engrg.

Eufaula

KTNT(FM)— June 15, 1967: 102.5 mhz; 25 kw. 150 ft TL: N35 22 25 W95 34 00. Hrs open: 24 Box 956, 74432. Phone: (918) 689-3663. Fax: (918) 689-5451. E-mail: mrogers@k955.com Web Site: www.kfoxradio.com. Licensee: K95.5 Inc. (group owner; acq 9-24-98; $400,000). Population served: 100,000 Rgnl. Network: Okla. Radio Net. Format: Today's country. Spec prog: Gospel 3 hrs wkly. ♦William H. Payne, pres; Mike Rogers, gen mgr & mus dir; Lisa Cotten, progmg dir & traf mgr.

Frederick

***KSYE(FM)**— July 1992: 91.5 mhz; 100 kw. 390 ft TL: N34 21 52 W98 50 04. Stereo. Hrs opn: 24 Box 582, 400 E. Gladstone Ave., 73542. Phone: (866) 355-5793. Fax: (580) 335-5900. E-mail: info@ksye.org Web Site: www.ksye.org. Licensee: Criswell College. (group owner) Population served: 360,000 Natl. Network: ABC. Format: Inspirational. Target aud: General. ♦Dr. Royce Laycock, chmn; Dr. Jerry Johnson, pres; Ronald L. Harris, gen mgr; Bill Bumpas, stn mgr & progmg dir; Doug Price, opns VP; James Nance, dev VP.

KTAT(AM)— 1948: 1570 khz; 250 w-D. TL: N34 23 30 W99 01 51. Hrs open: Box 1088, 73542. Secondary address: 207 W. Grand Ave. 73542. Phone: (580) 335-5923. Fax: (580) 335-7659. Licensee: Morey Broadcasting LLC (acq 12-18-90; $60,000 with co-located FM; FTR: 1-7-91) Population served: 6,800 Rgnl. Network: Okla. Radio Net. Format: Adult standards, Sp. Spec prog: Farm 2 hrs wkly. ♦Brent Morey, gen mgr & gen sls mgr.

KYBE(FM)— Aug 15, 1982: 95.9 mhz; 6 kw. Ant 249 ft TL: N34 23 30 W99 01 51. Stereo. Hrs opn: Box 1088, 73542. Phone: (580) 335-5923. Fax: (580) 335-7659. Web Site: www.coyotenews.com. Licensee: Fort Worth Media Group G.P. LLC. (acq 9-9-2005; $325,000). Format: Country.

Glenpool

KTSO(FM)— May 24, 1976: 94.1 mhz; 100 kw. Ant 691 ft TL: N36 07 52 W96 04 13. Hrs open: 24 5810 E. Skelly Dr., Suite 801, Tulsa, 74135. Phone: (918) 665-3131. Fax: (918) 663-6622. E-mail: production@softoldies.com Web Site: www.941THESOUND.com. Licensee: Shamrock Communications Inc. (group owner; acq 1996; $1.8 million). Natl. Rep: McGavren Guild. Format: Classic hits. News staff: one; News: 35 hrs wkly. Target aud: 35-54. ♦Chuck Browning, gen mgr; William Lynett, CEO & opns mgr; Tom Holiday, sls dir; Paul Kaiegler, progmg dir.

Goodwell

***KPSU(FM)**— September 1977: 91.7 mhz; 380 w. 121 ft TL: N36 35 41 W101 38 10. Stereo. Hrs opn: Noon-midnight Box 430, 73939. Phone: (580) 349-2414. Fax: (580) 349-2302. Web Site: www.opsu.edu. Licensee: Panhandle State University. Population served: 3,000 Format: Div. Target aud: College age. ♦Dr. David Bryant, pres; Russell Guthrie, gen mgr.

Grandfield

***KWKL(FM)**— Sept 19, 2003: 89.9 mhz; 11 kw vert. Ant 499 ft TL: N34 16 19 W98 25 30. Stereo. Hrs opn: 24 2351 Sunset Blvd., Suite 170-218, Rocklin, CA, 95765. Phone: (916) 251-1600. Fax: (916) 251-1650. E-mail: klove@klove.com Web Site: www.klove.com. Licensee: Educational Media Foundation. Group owner: EMF Broadcasting. Population served: 264,000 Natl. Network: K-Love. Law Firm: Shaw Pittman. Format: Contemp Christian. News staff: 3. Target aud: 25-44; Judeo Christian, female. ♦Richard Jenkins, pres; Mike Novak, VP; Keith Whipple, dev dir; David Pierce, progmg mgr; Ed Lenane, news dir; Sam Wallington, engrg dir; Karen Johnson, news rptr; Marya Morgan, news rptr; Richard Hunt, news rptr.

Grove

KGVE(FM)— Dec 12, 1980: 99.3 mhz; 15 kw. 312 ft TL: N36 36 49 W94 45 53. Stereo. Hrs opn: 24 Box 451749, One W. Third, 74345. Phone: (918) 786-2211. Fax: (918) 786-2284. Licensee: Caleb Corp. (acq 4-1-92; FTR: 3-16-92). Population served: 350,000 Natl. Network: ABC. Rgnl. Network: Okla. Radio Net. Format: C&W. News staff: one. Target aud: General. ♦Janell Hestand, VP, opns dir & progmg dir; Larry Hestand, pres, dev dir & prom dir.

***KWXC(FM)**—Not on air, target date: unknown: 88.9 mhz; 6 kw vert. Ant 266 ft TL: N36 35 42 W94 38 05. Hrs open: 69601 E. 290 Rd., 74344. Phone: (918) 854-3523. Licensee: Grove Broadcasting Inc.

Guthrie

KMFS(AM)— Nov 16, 1955: 1490 khz; 1 kw-U. TL: N35 52 56 W97 23 34. Hrs open: 24 Box 262550, Baton Rouge, LA, 70826. Secondary address: 8919 World Ministry Ave., Baton Rouge, LA 70810. Phone: (225) 768-3688/8300. Fax: (225) 768-3729. E-mail: kawikfish@yahoo.com Web Site: www.jsm.org. Licensee: Family Worship Center Church Inc. (group owner; acq 9-27-2002; $150,000). Population served: 18,000 Natl. Network: ABC, AP Radio. Format: Relg teaching. ♦David Whitelaw, COO; Jimmy Swaggart, pres; John Santiago, gen mgr & progmg dir.

Guymon

KGYN(AM)— Dec 12, 1948: 1210 khz; 10 kw-U, DA-N. TL: N36 40 34 W101 22 58. Hrs open: 24 Box 130, 73942. Phone: (580) 338-1210. Phone: (800) 227-1210. Fax: (580) 338-8255. E-mail: kgyn@ptsi.net Web Site: www.kgynam1210.com. Licensee: Telns Broadcasting Co. Inc. (acq 9-5-86; $400,000; FTR: 8-4-86). Population served: 450,000 Format: Country. News staff: one; News: 12 hrs wkly. Target aud: 25-65; broad based listenership. Spec prog: Relg 8 hrs, Sp 8 hrs wkly. ♦Ed Smith, pres; Jim Smith, gen mgr; Lindsey Smith, stn mgr; Richard Ryther, opns mgr; Curt Edenbourough, gen sls mgr & progmg dir.

KKBS(FM)— Dec 25, 1983: 92.7 mhz; 11.5 kw. 485 ft TL: N36 42 43 W101 27 27. Stereo. Hrs opn: 24 Box 1756, 2143 Hwy. 64 N., 73942. Phone: (580) 338-5493. Fax: (580) 338-0717. E-mail: kkbs@kkbs.com Web Site: www.kkbs.com. Licensee: MLS Communications Inc. (acq 8-7-90; 8-27-90). Population served: 100,000 Format: Adult rock. News staff: 2; News: 17 hrs wkly. Target aud: 24-66; working people, 2 income families, farmers. Spec prog: Financial markets 5 hrs wkly. ♦Marsha Strong, pres & gen mgr; Ramey Cozart, opns mgr.

KRBG(FM)—Not on air, target date: unknown: 99.5 mhz; 100 kw. Ant 417 ft TL: N36 45 52 W101 12 32. Hrs open: Box 7441, Amarillo, TX, 79114. Phone: (806) 353-1488. Fax: (806) 353-1542. Web Site: www.gracechurchamarillo.com. Licensee: Grace Community Church of Amarillo (acq 8-16-2006; $95,000 for CP). ♦William Gehm, pres.

Healdton

KICM(FM)— October 1978: 97.7 mhz; 25 kw. Ant 328 ft TL: N34 21 00 W97 27 35. Stereo. Hrs opn: Box 1487, Ardmore, 73402. Phone: (580) 226-5105. Fax: (580) 226-5113. Web Site: www.kicm.com. Licensee: Keystone Broadcasting Corp. (acq 5-31-2005; $1.2 million). Population served: 80,000 Format: Country. Target aud: 21-49. Spec prog: Relg 6 hrs wkly. ♦Bill Countrymen, gen mgr.

Heavener

KPRV-FM— Oct 1, 1989: 92.5 mhz; 1.55 kw. 640 ft TL: N34 53 54 W94 34 30. Stereo. Hrs opn: 24 Box 368, Poteau, 74953. Phone: (918) 647-3221. Fax: (918) 647-5092. E-mail: lbilly@clnk.com Web Site: www.kprvradio.com. Licensee: LeRoy Billy. Natl. Network: ABC. Rgnl. Network: Okla. Radio Net. Format: Country. News: 24 hrs wkly. Target aud: 24-54. Spec prog: Gospel 12 hrs wkly. ♦David Billy, progmg dir; LeRoy Billy, VP, gen mgr, gen sls mgr & mus dir; Larry Johnson, chief of engrg.

Henryetta

***KVAZ(FM)**— Dec 26, 1985: 91.5 mhz; 250 w. Ant 178 ft TL: N35 21 56 W96 00 34. Hrs open: 24 The Gospel Station Network, Box 1343, Ada, 74821. Phone: (580) 332-0902. E-mail: email@thegospelstation.com Web Site: www.thegospelstation.com. Licensee: South Central Oklahoma Broadcasting Inc. (acq 7-1-2005; $25,000). Format: Southern gospel. ♦Randall Christy, pres & gen mgr.

KXBL(FM)— Dec 20, 1966: 99.5 mhz; 100 kw. 984 ft TL: N35 50 02 W96 07 28. Stereo. Hrs opn: 24 4590 E. 29th, Tulsa, 74114. Phone: (918) 743-7814. Fax: (918) 743-7613. Web Site: bigcountry995.com. Licensee: Journal Broadcast Corp. Group owner: Journal Communications Inc. (acq 6-11-99; grpsl). Law Firm: Koteen & Naftalin. Format: Young country. Target aud: 18-34. ♦Carl Gardner, pres; Ron Kurtis, CFO; Randy Bush, gen mgr; Ric Hampton, opns mgr, mus dir & chief of engrg.

Hobart

KQTZ(FM)— May 28, 1979: 105.9 mhz; 100 kw. Ant 1,020 ft TL: N34 52 15 W99 17 36. Stereo. Hrs opn: 24 Box 577, Altus, 73522. Phone: (580) 482-1450. Fax: (580) 482-3420. Web Site: www.kwhw.com. Licensee: Monarch Broadcasting Inc. (group owner; (acq 12-31-2003; grpsl). Format: Hot adult contemp. Target aud: General; contemp adults during the day, rockers at night. ♦Matthew L. Ward, pres & gen mgr; Dean Minnick, chief of opns & sls dir; Dick Fontana, progmg mgr; Cameron Dole, mus dir.

KTJS(AM)— June 21, 1947: 1420 khz; 1 kw-D, 360 w-N. TL: N35 02 57 W99 05 48. Hrs open: 19 Box 311, 1515 N. Broadway, 73651. Phone: (580) 726-5656. Fax: (580) 726-2222. E-mail: thezone@itlnet.net Licensee: Fuchs Broadcasting Co. (acq 11-24-98; $182,000). Population served: 50,000 Format: Country, news/talk. News: 19 hrs wkly. Target aud: 30 plus; agri-related businessmen. Spec prog: Relg 15 hrs wkly. ♦Chad Fox, pres, gen mgr, progmg dir & mus dir.

Holdenville

KTLS-FM— Nov 30, 1991: 106.5 mhz; 25 kw. 328 ft TL: N34 54 50 W96 31 20. Hrs open: 24 1188 North Hills Centre, Ada, 74820. Phone: (580) 332-2211. Fax: (580) 436-1629. E-mail: ktls@ktlsradio.com Web Site: www.ktlsradio.com. Licensee: Woodstone Broadcasting Inc. (acq 10-9-01). Population served: 170,000 Natl. Network: Jones Radio Networks. Law Firm: Smithwick & Belendink. Format: Hot adult contemp. News staff: one. Target aud: 25-54. ♦Howard Stone, sr VP; Rick Woodward, pres, gen mgr & opns dir; Craig Stone, progmg dir; Pam Pinnella, traf mgr.

KWSH(AM)—See Wewoka

Hollis

KKRE(FM)— 2005: 92.5 mhz; 6 kw. Ant 328 ft TL: N34 36 34 W99 50 57. Hrs open: Box 1077, Altus, 73522. Phone: (580) 482-1555. Fax: (580) 482-8353. Licensee: Altus FM Inc. ♦Scott Wilmes, VP.

Hugo

KIHN(AM)— October 1948: 1340 khz; 1 kw-U. TL: N34 00 15 W95 29 20. Hrs open: 16 Box 430, Hwy. 70 E., 74743. Phone: (580) 326-6411. Fax: (580) 326-7921. E-mail: kihn@1starnet.com Licensee: Little Dixie Broadcasting Co. Population served: 25,000 Format: Var, news. News staff: one; News: 20 hrs wkly. Target aud: General. Spec prog: Gospel music 5 hrs, children one hr, farm one hr wkly. ♦Leeta M. Henson, pres & gen mgr.

Idabel

KBEL-FM— Oct 1, 1973: 96.7 mhz; 25 kw. Ant 300 ft TL: N33 52 54 W94 49 10. Stereo. Hrs opn: 24 Box 418, 74745. Secondary address: 813 Lincoln Rd. 74745. Phone: (580) 286-6642. Phone: (877) 329-8280. Fax: (580) 286-6643. E-mail: kbel967@yahoo.com Web Site: www.kbelcountry.com. Licensee: Box Broadcasting Corp. (acq 9-1-99; with co-located AM). Population served: 75,000 Rgnl. Network: Quinstar. Format: Country. Target aud: 25 plus; country audience. Spec prog: tball. ♦Donald R. Box, chmn; Paul W. Box, CEO, pres & gen mgr; Kelly P. Kirkland, opns dir.

KBEL(AM)— June 3, 1948: 1240 khz; 1 kw-U. TL: N33 52 54 W94 49 10.24 75,000 Format: Sports talk. News: 30 hrs wkly. Target aud: 18 plus; men.

KKBI(FM)—See Broken Bow

KQIB(FM)— Aug 1, 1999: 102.9 mhz; 6 kw. 318 ft TL: N33 59 57 W94 47 29. Stereo. Hrs opn: 24 617 S. Park Dr., Broken Bow, 74728. Phone: (580) 584-3388. Fax: (580) 584-3341. E-mail: kkbi@pine-net.com Licensee: JDC Radio Inc. (acq 9-2-98). Population served: 34,000 Natl. Network: ABC. Law Firm: Putbrese, Hunsaker & Trent. Format: Hot adult contemp. News staff: one; News: 5 hrs wkly. Target aud: 25-44. ♦Homer Coleman, pres; David Smulyan, gen mgr & gen sls mgr; Rod Kennedy, natl sls mgr & progmg dir.

***KXRT(FM)**— 2003: 90.9 mhz; 500 w. Ant 210 ft TL: N33 53 33 W94 49 26. Hrs open: Drawer 2440, Tupelo, MS, 38801. Phone: (662) 844-8888. Fax: (662) 842-6791. Licensee: American Family Association. Group owner: American Family Radio (acq 1-31-2001). Format: Christian. ♦Marvin Sanders, gen mgr.

Ketchum

***KOSN(FM)**— May 26, 1989: 107.5 mhz; 100 kw. Ant 981 ft TL: N36 46 13 W95 27 07. Stereo. Hrs opn: 24
Simulcast with KOSU-FM Stillwater 100%.
Oklahoma State University, 302 PM Bldg., Stillwater, 74078. Phone: (405) 744-6352. Fax: (405) 744-9970. Web Site: www.kosu.org. Licensee: PRC Tulsa I-LLC (acq 1-13-2005; $4 million). Population served: 1,000,000 Natl. Network: NPR. Rgnl. Network: Okla. Radio Net. Format: Class, educ, news. News staff: 2; News: 18 hrs wkly. Target aud: General. Spec prog: American Indian one hr wkly. ♦Craig Beeby, gen mgr & progmg dir; Don Crider, dev dir; Rachel Hubbard, news dir; Dan Schroeder, chief of engrg.

Kingfisher

KINB(FM)— 2000: 105.3 mhz; 800 w. Ant 840 ft TL: N35 43 38 W97 52 30. Hrs open: 4045 N.W. 64th St., Suite 600, Oklahoma City, 73116. Phone: (405) 848-0100. Fax: (405) 843-5288. Licensee: The Last Bastion Station Trust LLC, as Trustee Group owner: Citadel Broadcasting Corp. (acq 6-12-2007; grpsl). Format: Sp. ♦Larry Bastida, gen mgr; Steve English, stn mgr; Luis Medina, opns mgr.

Lahoma

KXLS(FM)— Nov 1, 1995: 95.7 mhz; 9.6 kw. Ant 502 ft TL: N36 25 14 W98 01 12. Hrs open: Box 952, Enid, 73702. Secondary address: 316 E. Willow Rd., Enid 73701. Phone: (580) 237-1390. Fax: (580) 242-1390. E-mail: hchamplin@knid.com Licensee: Chisholm Trail Broadcasting Co. (acq 11-1-99; $525,000). Natl. Network: ABC. Format: Adult contemp. News staff: one; News: 2 hrs wkly. Target aud: 30-60; female. ♦Hiram Champlin, gen mgr; Ricky Roggow, gen mgr, opns mgr & progmg dir; Sandy Daniels, gen sls mgr; Suzi Lakin, prom dir; Rob Houston, news dir; G.B. Bonham, chief of engrg.

Langston

***KALU(FM)**— Mar 3, 1975: 89.3 mhz; 150 w. Ant 200 ft TL: N35 56 36 W97 15 32. Hrs open: 24 c/o Gen. Mgr., Sanford Hall, Langston Univ., 73050. Phone: (405) 466-2924. Phone: (405) 466-3342. Fax: (405) 466-2921. E-mail: mfjackson@lunet.edu Web Site: www.lunet.edu. Licensee: Langston University. Format: Jazz, relg, urban contemp. ♦Michael Jay Jackson, gen mgr.

Lawton

KBZQ(FM)— May 1, 1992: 99.5 mhz; 16 kw. 338 ft TL: N34 35 31 W98 32 55. Stereo. Hrs opn: 24 Box 6888, 73506-6888. Secondary address: 1006 N.W. 47th St., Suite B, lawton 73505. Phone: (580) 357-9950. Fax: (580) 357-9995. E-mail: kbzq@sbcglobal.net Web Site: www.hitsandfavorites.com. Licensee: William R. Fritsch Jr. Population served: 113,000 Law Firm: Arter & Hadden. Format: Adult contemp. News: one hr wkly. Target aud: 25-54; baby boomers, upscale white collar workers. Spec prog: Jazz 4 hrs, Hits of the 70s 5 hrs, Sp 4 hrs wkly. ♦Chuck Pettigrew, opns dir, opns mgr & progmg dir; Rick Fritsch, gen mgr & sls dir; Lino Roldan, spanish dir.

***KCCU(FM)**— July 13, 1989: 89.3 mhz; 2 kw. Ant 463 ft TL: N34 37 26 W98 16 15. Stereo. Hrs opn: 24 2800 W. Gore Blvd., 73505. Phone: (580) 581-2425. Fax: (580) 581-5571. E-mail: markn@cameron.edu Web Site: www.kccu.org. Licensee: Cameron University. Population served: 100,000 Natl. Network: NPR, PRI. Format: News, classical. News staff: 5; News: 40 hrs wkly. Target aud: General. Spec prog: Jazz. ♦Mark Norman, gen mgr.

KJMZ(FM)—Listing follows KKRX(AM).

***KJRF(FM)**— Jan 1, 2001: 91.1 mhz; 10 kw. Ant 413 ft TL: N34 41 22 W98 07 34. Hrs open: 24 The Christian Center Inc., 2405 S.W. Lee Blvd., 73505. Phone: (580) 357-4498. Fax: (580) 357-1818. E-mail: kjrf911@yahoo.com Web Site: www.thechristian-center.org. Licensee: The Christian Center Inc. Format: Christian. ♦Reverend Paul Craig, pres; Alan Hampton, chief of opns; Randy Muirhead, stn mgr & progmg dir; Allan Hampton, engrg dir.

KKRX(AM)— May 27, 1956: 1050 khz; 250 w-D, DA. TL: N34 35 27 W98 21 10. Hrs open: 1525 S.E. Flowermound Rd., 73501. Phone: (580) 355-1050. Fax: (580) 355-1056. Web Site: www.kjmz.com. Licensee: Perry Broadcasting of Lawton Inc. Group owner: Perry Publishing & Broadcasting Co. (acq 1-31-97; $486,000 with co-located FM). Natl. Rep: D & R Radio. Format: Gospel. Spec prog: Ger one hr wkly. ♦Joy Chapman, gen mgr, opns mgr & gen sls mgr; Michelle Campbell, news dir; Shawn Bailey, chief of engrg.

KJMZ(FM)—Co-owned with KKRX(AM). Oct 23, 1970: 97.9 mhz; 6 kw. Ant 292 ft TL: N34 35 31 W98 32 55. Stereo. 24 Web Site: www.kjmz.com.800,000 Law Firm: Eugene T. Smith. Format: Urban contemp. ♦Day Lorrezz, progmg dir.

KLAW(FM)— Jan 1, 1965: 101.3 mhz; 100 kw. Ant 584 ft TL: N34 32 59 W98 32 21. Stereo. Hrs opn: 24 626 D Ave. S.W., 73501. Phone: (580) 581-3600. Fax: (580) 357-2880. E-mail: klaw@clearchannel.com Web Site: www.klaw.com. Licensee: GAP Broadcasting Lawton License LLC. Group owner: Clear Channel Communications Inc. (acq 8-3-2007; grpsl): Population served: 117,000 Law Firm: Wiley, Rein & Fielding. Format: Country. News staff: one; News: 4 hrs wkly. Target aud: 25-54; adults. ♦Kim Dodds, gen mgr; David Crawford, opns mgr; JoAnne Taylor, gen sls mgr.

KMGZ(FM)— Nov 1, 1982: 95.3 mhz; 14 kw. 312 ft TL: N34 34 36 W98 28 30. Stereo. Hrs opn: 24 1421 Great Plains Blvd., Suite C, 73505-2843. Phone: (580) 536-9530. Fax: (580) 536-3299. E-mail: gm@kmgz.com Web Site: www.kmgz.com. Licensee: Broadco of Texas Inc. (acq 3-9-92; trade and joint venture agreement for KMGZ(FM); FTR: 4-6-92) Format: Adult contemp. News: one hr wkly. Target aud: 18-49. ♦Chuck Morgan, pres, gen mgr & gen sls mgr; Albert Young, progmg mgr.

***KVRS(FM)**— December 1989: 90.3 mhz; 1 kw vert. Ant 187 ft TL: N34 37 32 W98 31 43. Stereo. Hrs opn: 24 1411 Parish Rd., Lake Charles, LA, 70611. Phone: (580) 536-8886. Fax: (580) 536-8891. E-mail: kvvs@kvvsfm.com Licensee: American Family Association. Group owner: American Family Radio (acq 4-29-2004; $10). Population served: 250,000 Natl. Network: American Family Radio. Format: Christian, relg. News: 14 hrs wkly. Target aud: General. ♦Dan Meir, stn mgr & engrg VP.

KVRW(FM)— Mar 13, 1992: 107.3 mhz; 50 kw. 492 ft TL: N34 36 27 W98 16 26. Stereo. Hrs opn: 24 626 S.W. D Avenue, 73501. Phone: (580) 581-3600. Fax: (580) 357-2880. E-mail: oldies@oldies107.com Web Site: www.oldies107.com. Licensee: GAP Broadcasting Lawton License LLC. (acq 8-3-2007; grpsl). Population served: 89,900 Format: Oldies. News: 2 hrs wkly. Target aud: 25-54. ♦Kim Dodds, CEO & gen mgr; Joanne Taylor, stn mgr; Eric Sharum, progmg dir.

KXCA(AM)— May 1, 1941: 1380 khz; 1 kw-U, DA-2. TL: N34 35 24 W98 21 44. Hrs open: 24 1525 SE Flower Mound, 73501. Phone: (580) 355-1050. Fax: (580) 355-1056. E-mail: trevor1380@hotmail.com Web Site: www.1380theticket.com. Licensee: Perry Broadcasting of Southwest Oklahoma Inc. Group owner: Perry Publishing & Broadcasting Co. (acq 11-22-2002; grpsl). Population served: 1,060,000 Natl. Rep: Roslin. Format: Sports, talk. Target aud: 35 plus. ♦Joy Chapman, gen mgr & gen sls mgr; Trevor Myers, progmg dir.

KZCD(FM)— June 8, 1987: 94.1 mhz; 18 kw. 524 ft TL: N34 34 24 W98 28 40. Stereo. Hrs opn: 24 626 S.W. D Ave., 73501. Phone: (580) 581-3600. Fax: (580) 357-2880. E-mail: z94@clearchannel.com Web Site: www.z94.com. Licensee: GAP Broadcasting Lawton License LLC. Group owner: Clear Channel Communications Inc. (acq 8-3-2007; grpsl). Population served: 117,000 Law Firm: Wiley, Rein & Fielding. Format: Rock. News staff: one; News: 2 hrs wkly. Target aud: 18-49; males. ♦Kim Dodds, gen mgr; JoAnne Taylor, gen sls mgr; Don Brown, progmg dir; Scott Maingi, engrg dir.

Lindsay

KBLP(FM)— Oct 1, 1988: 105.1 mhz; 850 w. 564 ft TL: N34 54 01 W97 33 56. Stereo. Hrs opn: 24 204 S. Main, 73052. Phone: (405) 756-4438. Fax: (405) 756-2040. Web Site: www.kbipsports.com. Licensee: South Central Oklahoma Broadcasting & Advertising Corp. Rgnl. Network: Okla. Radio Net. Format: Country. News staff: 2; News: 21 hrs wkly. Target aud: 21-65; working consumers. ♦Charlie Jones, pres & gen mgr.

Locust Grove

KEMX(FM)— Feb 14, 1991: 94.5 mhz; 2.3 kw. 367 ft TL: N36 15 05 W95 13 21. Stereo. Hrs opn: 24
Rebroadcasts KXOJ-FM Sapulpa 100%.
2448 E. 81st St., Suite 4500, Tulsa, 74137. Phone: (918) 492-2660. Fax: (918) 492-8840. E-mail: kxoj@kxoj.com Web Site: www.kxoj.com. Licensee: KXOJ Inc. Group owner: Adonai Radio Group (acq 4-29-92; grpsl). Format: Contemp Christian music. Target aud: 18-35; young married or single Christians. ♦Mike Stephens, pres; Joy Stephens, VP; David Stephens, gen mgr & adv dir; Bob Thornton, progmg dir.

Lone Grove

KYNZ(FM)— May 25, 1988: 107.1 mhz; 24.5 kw. Ant 335 ft TL: N34 17 52 W97 09 12. Stereo. Hrs opn: 24 1205 Northglen, Ardmore, 73401. Secondary address: Next Media, 1205 Northglen, Ardmore 73401. Phone: (580) 226-0421. Fax: (580) 226-0464. E-mail: webmaster@kynz.com Web Site: www.kynz.com. Licensee: LKCM Radio Group L.P. Group owner: NextMedia Group L.L.C. (acq 2-26-2007;. grpsl). Population served: 320,000 Rgnl. Network: Agri-Net. Natl. Rep: Christal. Format: Hot adult contemp. News staff: one; News: 25 hrs wkly. Target aud: 18-54. ♦David Smith, pres, VP, gen mgr & stn mgr; Dave Hilton, opns dir; David MacMullen, sls dir & natl sls mgr; Steve Bart, progmg dir.

Madill

KMAD(AM)— May 20, 1962: 1550 khz; 250 w-D. TL: N34 06 24 W96 46 30. Hrs open: Sunrise-sunset Box 576, 6/10 Mile N. on Hwy. 199, 73446. Phone: (580) 795-2345. Fax: (580) 795-5623. Licensee: Robert S. Sullins (acq 3-20-98). Natl. Network: Jones Radio Networks. Rgnl. Network: Agri-Net. Format: Classic hit country. News: 12 hrs wkly. Target aud: General. Spec prog: Farm 2 hrs wkly. ♦Jay Lindley, gen mgr.

Mangum

***KHIM(FM)**— 1998: 97.7 mhz; 1.5 kw. 138 ft TL: N34 52 27 W99 30 01. Hrs open: 509 N. Main, Altus, 73521. Phone: (580) 482-9797. Fax: (580) 379-9201. E-mail: info@97khim.com Licensee: Ray Broadcasting Inc. (acq 12-22-2003; $320,000 with KJCM(FM) Snyder). Format: Classic rock. ♦Terri Kamphaus, gen mgr.

Marlow

KFXI(FM)— August 1987: 92.1 mhz; 100 kw. Ant 390 ft TL: N34 42 35 W98 03 00. Stereo. Hrs opn: 24 1101 Hwy. 81 N., 73055. Phone: (580) 658-9292. Fax: (580) 658-2561. E-mail: kfxi@texhoma.net Licensee: DFWU Inc. (group owner). Population served: 400,000 Law Firm: Southmayd & Miller. Format: Country. Target aud: 25-55. Spec prog: Gospel 8 hrs wkly. ♦K.D. Austin, gen mgr; Amy Helton, opns mgr; Sherry Lynn, gen sls mgr; Bill Marshall, progmg dir; James Wilson, adv.

McAlester

***KBCW-FM**— 1999: 91.9 mhz; 700 w. Ant 446 ft TL: N34 59 13 W95 42 10. Stereo. Hrs opn: 24
Rebroadcasts KCSC(FM) Edmond 100%.
Univ. of Central Oklahoma, 100 N. University Dr., Edmond, 73034-5209. Phone: (405) 974-3333. Phone: (877) 359-3334. Fax: (405) 974-3844. E-mail: kcscfm@kcscfm.com Web Site: www.kcscfm.com. Licensee: The University of Central Oklahoma. Population served: 40,000 Natl. Network: PRI. Format: Class. News staff: one; News: 5 hrs wkly. Target aud: 35 plus; educ, affluent. ♦Bradford Ferguson, gen mgr; Barbara Hendrickson, opns dir & news dir; Preston Walker, chief of engrg.

KMCO(FM)—Listing follows KNED(AM).

KNED(AM)— Mar 14, 1950: 1150 khz; 1 kw-D, 500 w-N, DA-N. TL: N34 56 12 W95 43 59. Stereo. Hrs opn: 24 Box 1068, 74502-1068. Secondary address: 1801 E. Gene Stipe Blvd. 74501. Phone: (918) 423-1460. Fax: (918) 423-7119. E-mail: kmcokned@mcalesterradio.com Web Site: mcalesterradio.com. Licensee: Southeastern Oklahoma Radio LLC. (group owner; (acq 1-18-2005; $222,223). Population served: 175,000 Law Firm: Wiley, Rein & Fielding. Wire Svc: AP Format: C&W Classic. News: 10 hrs wkly. Target aud: 45 plus. ♦Lee Anderson, gen mgr & stn mgr; Sheila Turnbow, gen sls mgr; Megan Waters, progmg dir; John Yates, news dir.

KMCO(FM)—Co-owned with KNED(AM). November 1965: 101.3 mhz; 100 kw. 494 ft TL: N34 59 13 W95 42 10. Stereo. 24 Phone: (918) 426-1050. (Acq 1-18-2005; $766,666).250,000 Law Firm: Wiley, Rein & Fielding. News: 5 hrs wkly. Target aud: 18-45. ♦Rod Cook, progmg dir.

KTMC(AM)— Mar 3, 1946: 1400 khz; 1 kw-U. TL: N34 57 00 W95 45 00. Hrs open: 24 Box 1068, 74502. Secondary address: 1801 E. Electric Ave. 74502. Phone: (918) 426-1050. Fax: (918) 423-7119. E-mail: kmconed@mclesteradio.com Web Site: mcalesteradio.com. Licensee: Southeastern Oklahoma Radio LLC. (acq 1-18-2005; $444,445 with co-located FM). Population served: 60,000 Natl. Network: ABC, AP Radio. Law Firm: Wiley, Rein & Fielding. Format: Oldies. News: 2 hrs wkly. Target aud: 50 plus; older, middle-aged, mature & retired adults. Spec prog: Gospel 5 hrs wkly. ♦Lee Anderson, gen mgr & stn mgr; Sheila Turnbow, gen sls mgr; John Yates, news dir.

KTMC-FM— June 24, 1987: 105.1 mhz; 1.6 kw. 454 ft TL: N34 59 13 W95 42 10. Stereo. 24 100,000 Format: Classic rock. Target aud: 34-50.

Miami

KGLC(FM)—Listing follows KVIS(AM).

KVIS(AM)— February 1948: 910 khz; 1 kw-U, DA-1. TL: N36 53 27 W94 47 00. Hrs open: 24 Box 1555, 8400 S. Hwy. 137, 74355. Phone: (918) 542-1818. Phone: (918) 542-7175. Fax: (918) 542-1819. Licensee: Eagle Broadcasting Inc. (acq 2-15-91; with co-located FM; FTR: 3-11-91). Rgnl. Network: USA. Law Firm: Latham & Watkins. Format: Southern Gospel. News staff: one; News: 9 hrs wkly. Target aud: Christian/family. ♦Gordon Chirillo, pres; Robert Suman, gen mgr, gen sls mgr & progmg dir; Shanda Daugherty, prom dir & prom mgr; Kimberley Barnes, news dir; Rusty Wynn, chief of engrg.

KGLC(FM)—Co-owned with KVIS(AM). December 1975: 100.9 mhz; 3.6 kw. 273 ft TL: N36 53 27 W94 47 01. Stereo. 150,000 Natl. Network: USA. Format: Christian, inspirational. Target aud: 25-44.

Midwest City

KEBC(AM)—Licensed to Midwest City. See Oklahoma City

KTLV(AM)— April 1973: 1220 khz; 250 w-D, DA. TL: N35 23 50 W97 27 04. Hrs open: 6 am-7 pm 3336 S.E. 67th St., Oklahoma City, 73135. Phone: (405) 672-1220. Phone: (405) 672-3886. Fax: (405) 672-5858. E-mail: ktlv1220@aol.com Web Site: www.ktlv1220.com. Licensee: First Choice Broadcasting Inc. (acq 6-19-92). Population served: 650,000 Format: Gospel, Christian. Target aud: 24 plus. ♦Howard D. Williams, pres; Dale Williams, gen mgr & progmg dir.

Moore

***KMSI(FM)**— Mar 26, 1991: 88.1 mhz; 50 kw vert. 597 ft TL: N35 12 07 W97 35 18. Stereo. Hrs opn: 24 Box 1924, Tulsa, 74101. Secondary address: 120 S.W. 4th St. 73160. Phone: (405) 794-5674; (918) 455-5693. Fax: (405) 794-5112. E-mail: mail@oasisnetwork.org Web Site: www.oasisnetwork.org. Licensee: Creative Educational Media Corp. Inc. Format: Relg. Target aud: General. ♦David Ingles, pres; Cherri Willis, gen mgr; David Warren, progmg dir; Hal Smith, chief of engrg.

WWLS(AM)— Sept 26, 1922: 640 khz; 1 kw-U, DA-N. TL: N35 17 21 W97 30 08. Hrs open: 4045 N.W. 64th St., Suite 600, Oklahoma, 73116. Phone: (405) 848-0100. Fax: (405) 848-5288. Web Site: www.thesportsanimal.com. Licensee: Citadel Broadcasting Co. Group owner: Citadel Broadcasting Corp. (acq 10-28-99; grpsl). Population served: 1,000,000 Format: Sports. ♦Larry Bastida, gen mgr; Chris Baker, opns mgr.

Muskogee

KBIX(AM)— May 1, 1936: 1490 khz; 450 w-U. TL: N35 46 56 W95 22 37. Hrs open: 24 215 State, Suite 910, 74401. Phone: (918) 682-9700. Fax: (918) 682-6775. Licensee: KMMY Inc. Group owner: Adonai Radio Group (acq 12-11-2002; $1 million with KCXR(FM) Taft). Population served: 150,000 Format: Sports. ♦David Stephens, gen mgr.

KCXR(FM)—See Taft

KHTT(FM)— February 1972: 106.9 mhz; 100 kw. 1,005 ft TL: N35 51 41 W95 46 03. Stereo. Hrs opn: 24 7030 S. Yale Ave., Suite 711, Tulsa, 74136. Phone: (918) 492-2020. Fax: (918) 496-2681. E-mail: pbryson@rendabroadcasting.com Web Site: www.khits.com. Licensee: Renda Broadcasting Corp. Group owner: Renda Broadcasting Corp.-Renda Radio Inc. (acq 4-15-93; $1.6 million; FTR: 5-3-93). Population served: 42,500 Format: Hot contemp hits. News: 5 hrs wkly. Target aud: 18-34; young adults. ♦Tony Renda, pres; Tod Tucker, opns mgr & progmg dir; Pat Bryson, gen sls mgr; Frank Canoli, prom dir.

KYAL-FM— Jan 19, 1984: 97.1 mhz; 100 kw. Ant 1,274 ft TL: N35 17 05 W95 25 26. Stereo. Hrs opn: 24 2448 E. 81 St, Suite 5500, Tulsa, 74137. Phone: (918) 492-2660. Fax: (918) 492-8840. Web Site: www.thesportsanimal.com. Licensee: KMMY Inc. (acq 9-15-93; $500,000; FTR: 10-11-93). Format: Sports. Target aud: 21-49; middle, upper-middle class. Spec prog: Farm 5 hrs wkly. ♦David Stephens, gen mgr.

Newcastle

KKNG-FM— Apr 15, 1971: 93.3 mhz; 100 kw. 797 ft TL: N35 11 28 W97 35 49. Stereo. Hrs opn: 5101 S. Shields Blvd., Oklahoma City, 73129. Phone: (405) 616-5500. Fax: (405) 616-5505. E-mail: info@kkng.com Web Site: www.kkng.com. Licensee: Tyler Broadcasting Corp. Group owner: Tyler Media Broadcasting Corp. (acq 10-95; $441,000). Population served: 19,300 Natl. Network: ABC. Natl. Rep: Katz Radio. Format: Classic country. Target aud: 25-54. Spec prog: Oldies 12 hrs, gospel 8 hrs wkly. ♦Kevin Young, progmg dir; Skip Stow, gen mgr, opns mgr & news dir; Randy Mullinax, chief of engrg.

Norman

***KGOU(FM)**— Sept 25, 1970: 106.3 mhz; 3 kw. 300 ft TL: N35 17 22 W97 21 30. Stereo. Hrs opn: 5 AM-midnight Kaufman Hall, Rm. 339, University of Oklahoma, 73019. Phone: (405) 325-3388. Fax: (405) 325-7129. E-mail: manager@kgou.org Web Site: www.kgou.org. Licensee: University of Oklahoma. Population served: 800,000 Natl. Network: NPR. Law Firm: Dow, Lohnes & Albertson. Format: News/talk, jazz. News staff: one; News: 82 hrs wkly. Target aud: 25-54; general. Spec prog: Blues 8 hrs, new age 4 hrs wkly. ♦Karen Holp, gen mgr; Kurt Gwartney, opns mgr; Jolly Brown, dev dir; Jim Johnson, progmg dir; Scott Gurian, news dir; Patrick Roberts, chief of engrg.

KREF(AM)— November 1949: 1400 khz; 1 kw-U. TL: N35 13 04 W97 24 37. Hrs open: 24 2020 E. Alameda, 73071. Phone: (405) 321-1400. Fax: (405) 321-6820. E-mail: kref@telepath.com Web Site: www.kref.com. Licensee: Fox Broadcasting, Inc. (acq 1-9-98; $300,000). Population served: 78000 Rgnl. Network: Okla. Radio Net. Law Firm: Reed, Smith, Shaw & McClay. Format: Sports. News staff: one. Target aud:

25-54; middle, upper class adults. ♦John Fox, pres; Mike Holt, gen mgr & chief of opns; Deryk Ruth, gen sls mgr; T.J. Perry, progmg dir.

***KSSO(FM)**— 2007: 89.3 mhz; 2.3 kw. Ant 163 ft TL: N35 13 22 W97 26 21. Hrs open: Box 262550, Baton Rouge, LA, 70826. Secondary address: 8919 World Ministry Ave., Baton Rouge, LA 70810. Phone: (225) 768-3688. Phone: (225) 768-8300. Fax: (225) 768-3729. E-mail: kawikfish@yahoo.com Web Site: www.jsm.org. Licensee: Family Worship Center Church Inc. (acq 10-7-2005; $500,000 for CP with KQUJ(FM) Ada). ♦David Whitelaw, COO; Jimmy Swaggart, pres; John Santiago, progmg dir.

Nowata

KRIG-FM—Licensed to Nowata. See Bartlesville

Okarche

KTUZ-FM— September 1968: 106.7 mhz; 13 kw. Ant 958 ft TL: N35 36 49 W97 52 19. Stereo. Hrs opn: 5101 S. Shields Blvd., Oklahoma City, 73129. Phone: (405) 616-9900. Fax: (405) 616-0328. Web Site: www.ktuz.com. Licensee: Tyler Media Broadcasting Corp. Group owner: Tyler Media Broadcasting Corp. (acq 1-27-98; $100,000 with co-located AM). Population served: 145,000 Natl. Network: ABC. Format: Sp. News: 16 hrs wkly. Target aud: 18-65. ♦Skip Stow, gen mgr.

Oklahoma City

KATT-FM— Oct 17, 1960: 100.5 mhz; 97 kw. 1,191 ft TL: N35 35 22 W97 29 03. Stereo. Hrs opn: 4045 N.W. 64th, Suite 600, 73116. Phone: (405) 848-0100. Fax: (405) 843-5288. E-mail: todd.griffin@citcomm.com Web Site: www.katt.com. Licensee: Citadel Broadcasting Co. Group owner: Citadel Broadcasting Corp. (acq 10-28-99; grpsl). Population served: 632,300 Format: AOR. ♦Larry Bastida, gen mgr; Tricia York, gen sls mgr; Chad Lunsford, prom mgr; Chris Baker, progmg dir.

KEBC(AM)—Midwest City, 1922: 1340 khz; 1 kw-U. TL: N35 29 58 W97 30 33. Stereo. Hrs opn: 24 50 Penn Pl., Suite 1000, 73118. Phone: (405) 840-5271. Fax: (405) 840-5808. E-mail: derricknance@clearchannel.com Licensee: Clear Channel Broadcasting Licenses Inc. (acq 6-10-2002). Population served: 800,400 Natl. Network: Fox Sports. Natl. Rep: McGavren Guild. Law Firm: Kaye, Scholer, Fierman, Hays & Handler L.L.P. Format: Sports. ♦Bill Hurley, gen mgr; Derrick Nance, gen sls mgr; Ken Post, progmg dir.

KHBZ-FM— June 6, 1967: 94.7 mhz; 98 kw. 1,387 ft TL: N35 32 58 W97 29 50. Stereo. Hrs opn: 24 Box 1000, 73101. Phone: (405) 840-5271. Fax: (405) 842-1315. Licensee: Clear Channel Broadcasting Licenses, Inc. Group owner: Clear Channel Communications Inc. (acq 1-94; $7.5 million). Population served: 850,000 Format: Alt rock. News staff: one. Target aud: 25-54. ♦Jimmy Barreda, VP & progmg dir.

KJYO(FM)—Listing follows KTOK.

KKWD(FM)—Bethany, Oct 29, 1965: 104.9 mhz; 3 kw. Ant 299 ft TL: N35 29 58 W97 37 08. Hrs open: 24 4045 Northwest 64th St., Suite 600, 73116. Phone: (405) 848-0100. Fax: (405) 843-5288. Web Site: www.wild1049hd.com. Licensee: The Last Bastion Station Trust LLC, as Trustee Group owner: Citadel Broadcasting Corp. (acq 6-12-2007; grpsl). Format: Urban hits. Target aud: 35-49 & 25-34; young, professional. ♦Larry Bastida, gen mgr; Chris Baker, stn mgr.

KMGL(FM)— Nov 25, 1965: 104.1 mhz; 100 kw. 1,425 ft TL: N35 32 58 W97 29 18. Stereo. Hrs opn: 24 Box 14818, 73113. Secondary address: 400 E. Britton Rd. 73113. Phone: (405) 478-5104. Fax: (405) 478-0448. E-mail: vance@magic104.com Web Site: www.magic104.com. Licensee: Renda Broadcasting. (group owner; acq 4-88). Population served: 831,600 Format: Adult contemp. ♦Vance Harrison, gen mgr; Harold Patterson, gen sls mgr; Jeff Couch, progmg dir; Dennis Orcutt, chief of engrg.

KOKC(AM)— Dec 24, 1922: 1520 khz; 50 kw-U, DA-N. TL: N35 20 00 W97 30 16. Stereo. Hrs opn: Box 14818, 73113. Secondary address: 400 E. Britton Rd. 73113. Phone: (405) 478-5104. Fax: (405) 478-0448. E-mail: gmaier@rendabroadcasting.com Web Site: www.1520kokc.com. Licensee: Renda Broadcasting Corp. of Nevada. Group owner: Renda Broadcasting Corp. (acq 6-30-98; grpsl). Population served: 2,160,200 Natl. Network: CBS. Natl. Rep: CBS Radio. Format: News/talk. Target aud: 25-54. ♦Vance Harrison Jr., gen mgr; Garth Maier, progmg dir.

KOMA(FM)— 1964: 92.5 mhz; 98 kw. 984 ft TL: N35 32 52 W97 29 29. Stereo. Hrs opn: 24 400 E. Britton Rd., 73114. Phone: (405) 478-5104. Web Site: www.komaradio.com. Licensee: Renda Broadcasting Corp. of Nevada. Group owner: Renda Broadcasting Corp. (acq 6-30-98; grpsl). Format: Oldies. Target aud: 25-54. ♦Vance Harrison Jr., gen mgr.

KQCV(AM)— 1948: 800 khz; 2.5 kw-D, 500 w-N, DA. TL: N35 24 45 W97 40 26. Hrs open: 24 1919 N. Broadway Ave., 73103. Phone: (405) 521-0800. Phone: (405) 521-1414. Fax: (405) 521-1391. Web Site: www.bottradionetwork.com. Licensee: Bott Broadcasting Co. Group owner: Bott Radio Network (acq 1-76). Population served: 1,600,000 Format: Christian info, news. Target aud: 25-54; family oriented. ♦Richard P. Bott, pres; Richard Bott II, VP; Paul Sublett, gen mgr; Eben Fowler, opns dir.

KRMP(AM)— 1946: 1140 khz; 1 kw-D. TL: N35 23 14 W97 29 56. Hrs open: 1528 N.E. 23rd St., 73111. Phone: (405) 427-5877. Fax: (405) 424-6708. E-mail: bdmanday@kvsp.com Web Site: www.kvsp.com. Licensee: Perry Broadcasting Co. Inc. Group owner: Perry Publishing & Broadcasting Co. (acq 3-3-93; $375,000; FTR: 3-22-93). Population served: 632,300 Format: Urban contemp. ♦Russell Perry, CEO; Kevin Perry, gen sls mgr; Terry Monday, opns mgr & progmg dir.

KROU(FM)—See Spencer

KRXO(FM)— Aug 7, 1987: 107.7 mhz; 99 kw. 991 ft TL: N35 32 58 W97 29 18. Stereo. Hrs opn: Box 14818, 73113. Secondary address: 400 E. Britton Rd. 73113. Phone: (405) 478-5104. Fax: (405) 478-0448. E-mail: bwiley@krxo.com Web Site: www.krxo.com. Licensee: Renda Broadcasting Corp of Nevada. Group owner: Renda Broadcasting Corp. Population served: 2,160,200 Format: Classic rock. ♦Vance Harrison, gen mgr; Buddy Wiley, progmg dir & traf mgr; Steve Bennett, news dir.

KTLR(AM)— 1946: 890 khz; 1 kw-D. TL: N35 33 59 W97 28 28. Hrs open: 24 5101 S. Shields Blvd., 73129. Phone: (405) 616-5500. Fax: (405) 616-5505. Web Site: www.ktlr.com. Licensee: Tyler Broadcasting Corp. Group owner: Tyler Media Broadcasting Corp. (acq 1999; $40,000). Population served: 366481 Natl. Rep: Christal. Format: Community/talk. Target aud: 10-80. Spec prog: Sp 3 hrs wkly. ♦Skip Stow, gen mgr; Mike Miller, stn mgr, opns mgr, gen sls mgr & chief of engrg; Cody Sparks, progmg dir & disc jockey.

KTOK(AM)— Jan 29, 1927: 1000 khz; 5 kw-U, DA-2. TL: N35 21 29 W97 27 48. Hrs open: Box 1000, 73101. Secondary address: 10th Fl., 50 Penn Place 73101. Phone: (405) 840-5271. Fax: (405) 842-1315. Web Site: www.ktok.com. Licensee: Clear Channel Broadcasting Licences, Inc. (group owner; acq 8-5-92). Population served: 865,000 Rgnl. Network: Okla. Radio Net. Natl. Rep: Clear Channel. Format: News/talk. ♦Mike McCarville, CEO & progmg dir.

KJYO(FM)—Co-owned with KTOK. Apr 9, 1961: 102.7 mhz; 98 kw. 900 ft TL: N35 32 52 W97 29 29. (CP: Ant 984 ft.). Stereo. Web Site: www.kj103fm.com. Licensee: Clear Channel Broadcasting Licenses, Inc. Format: CHR. ♦Mike McCoy, progmg dir.

KTST(FM)— Mar 16, 1962: 101.9 mhz; 100 kw. 1,390 ft TL: N35 32 58 W97 29 50. Stereo. Hrs opn: 24 50 Penn Place / Ste 1000, 73118. Phone: (405) 858-1400. Fax: (405) 858-1106. E-mail: tomtravis@clearchannel.com Web Site: www.thetwister.com. Licensee: Clear Channel Broadcasting Licenses, Inc. (group owner; acq 1996; grpsl). Population served: 793,900 Format: Country. News staff: one. Target aud: 18-49. ♦Bill Hurley, gen mgr; Tom Travis, opns dir & progmg dir.

KXXY-FM— October 1964: 96.1 mhz; 100 kw. 1,167 ft TL: N35 32 58 W97 29 18. (CP: Ant 256 ft.). Stereo. Hrs opn: Box 1000, 73101. Phone: (405) 840-5271. Fax: (405) 842-1315. Licensee: Clear Channel Broadcasting Licenses Inc. Group owner: Clear Channel Communications Inc. (acq 1996; grpsl). Population served: 800,400 Format: Country. ♦Bill Reed, gen mgr & progmg dir.

KYIS(FM)— June 1969: 98.9 mhz; 100 kw. 1,108 ft TL: N35 33 36 W97 29 07. Stereo. Hrs opn: 24 4045 N.W. 64th St., Suite 600, 73116. Phone: (405) 848-0100. Fax: (405) 843-5288. E-mail: equest@kyis.com Web Site: www.kyis.com. Licensee: Citadel Broadcasting Co. Group owner: Citadel Broadcasting Corp. (acq 10-28-99; grpsl). Population served: 310,000 Natl. Network: AP Radio. Format: Hot adult contemp. News staff: one. Target aud: 25-54; female. ♦Larry Bastida, gen mgr; Tricia York, gen sls mgr; Don Sweeney, prom dir; Ray Kalusa, progmg dir.

***KYLV(FM)**— Nov 3, 1980: 88.9 mhz; 4.3 kw. 502 ft TL: N35 36 48 W97 28 29. Stereo. Hrs opn: 24 2351 Sunset Blvd., Suite 170-218, Rocklin, CA, 95765. Phone: (916) 251-1600. Fax: (916) 251-1650. E-mail: klove@klove.com Web Site: www.klove.com. Licensee: Educational Media Foundation. Group owner: EMF Broadcasting (acq 11-9-98; $1.2 million). Population served: 790,000 Natl. Network: K-Love. Law Firm: Shaw Pittman. Format: Contemp Christian. News staff: 3. Target aud: 25-44; Judeo-Christian female. Spec prog: Black 3 hrs, relg 2 hrs, gospel 4 hrs, pub affrs 4 hrs wkly. ♦Richard Jenkins, pres; Mike Novak, VP & progmg dir; Keith Whipple, dev dir; David Pierce, progmg mgr; Jon Rivers, mus dir; Ed Lenane, news dir; Sam Wallington, engrg dir; Karen Johnson, news rptr; Marya Morgan, news rptr; Richard Hunt, news rptr.

WKY(AM)— January 1920: 930 khz; 5 kw-U, DA-N. TL: N35 33 43 W97 30 27. Stereo. Hrs opn: 24 4045 NW 64th, Ste 600, Okahoma City, 73116. Phone: (405) 848-0100. Fax: (405) 843-5288. Licensee: Citadel Broadcasting Co. Group owner: Citadel Broadcasting Corp. (acq 1-31-2003; $7.7 million). Population served: 759,100 Law Firm: Wiley, Rein & Fielding. Format: Sports. ♦Larry Bastida, gen mgr & natl sls mgr.

WWLS-FM—Edmond, June 28, 1962: 97.9 mhz; 6 kw. Ant 315 ft TL: N35 34 11 W97 30 01. Stereo. Hrs opn: 24 4045 N.W. 64, Suite 600, 73116. Phone: (405) 848-0100. Fax: (405) 843-5288. Web Site: www.thesportsanimal.com. Licensee: Citadel Broadcasting Co. Group owner: Citadel Broadcasting Corp. (acq 10-28-99; grpsl). Law Firm: Birch, Horton, Bittner & Cherot. Format: Sports. Target aud: 25-54. ♦Larry Bastida, pres & gen mgr; Jay Davis, gen sls mgr; Chad Lunsford, prom dir; Dax Barry Jr., news dir.

Okmulgee

KOKL(AM)— October 1937: 1240 khz; 1 kw-U. TL: N35 36 31 W95 58 19. Hrs open: 24 100 E. 7th St., Suite 100, 74447. Phone: (918) 756-3646. Fax: (918) 756-1800. E-mail: kokl@aol.com Licensee: Regency Radio Inc. (acq 2-22-94; FTR: 3-21-94). Population served: 60,000 Natl. Network: ABC. Format: Oldies. News staff: one; News: 16 hrs wkly. Target aud: 25 plus; mid to upper income. Spec prog: Tulsa Univ. sports 10 hrs. ♦James R. Brewer, pres; Paul Brown, gen mgr.

Owasso

KQLL-FM— Oct 1, 1981: 106.1 mhz; 100 kw. Ant 1,315 ft TL: N36 31 36 W95 39 12. Stereo. Hrs opn: 24 2625 South Memorial, Tulsa, 74129. Phone: (918) 388-5100. Fax: (918) 388-5400. Web Site: www.kooltulsa.com. Licensee: Clear Channel Broadcasting Licenses Inc. Group owner: Clear Channel Communications Inc. (acq 1997; grpsl). Format: Classic top-40. Target aud: Adults 35-54. ♦Michael Oppenheimer, gen mgr; Don Cristi, opns mgr.

Pauls Valley

KVLH(AM)— December 1947: 1470 khz; 890 w-D, 35 w-N, DA-2. TL: N34 42 15 W97 15 19. Hrs open: 24 DFWU Inc., 1101 N. 81 Hwy., Marlow, 73055. Phone: (580) 658-9292. Fax: (580) 658-2561. Licensee: DFWU Inc. (group owner; acq 1-27-00; $25,000). Population served: 89,000 Rgnl. Network: . Law Firm: Southmayd & Miller. Format: Oldies. ♦Sherry Lynn, pres; K.D. Austin, gen mgr; Bill Marshall, progmg dir; Teddy Casteel, mus dir; James Wilson, engr.

Pawhuska

KOSG(FM)— 1997: 103.9 mhz; 3 kw. 328 ft TL: N36 44 56 W96 17 51. Hrs open: 319 S. Dewey Ave., Bartlesville, 74003. Phone: (918) 333-8550. Fax: (918) 333-8553. Licensee: Tallgrass Broadcasting Inc. (acq 10-25-2006; $294,000). Format: Top-40. ♦Jack Borgen, gen mgr.

KPGM(AM)— Oct 19, 1963: 1500 khz; 500 w-D. TL: N36 45 42 W96 11 58. Hrs open: 6 AM-sunset Box 1526, 74056. Secondary address: 129 W. Main 74056. Phone: (918) 287-1145. Fax: (918) 287-1473. E-mail: kpgm@bartlesvilleradio.com Web Site: www.bartlesvilleradio.com/kpgm. Licensee: Potter Radio LLC. (acq 7-1-2005; $100,000). Population served: 40,893 Natl. Network: Salem Radio Network. Law Firm: Lauren Colby. Format: News/Christian talk. News staff: one; News: 20 hrs wkly. ♦Kevin Potter, pres & gen mgr; Charlie Taraboletti, progmg dir.

Perry

KOKP(AM)— July 6, 1986: 1020 khz; 400 w-D, 250 w-N, DA-2. TL: N36 15 35 W97 13 01. Hrs open: 24 Box 2509, Ponca City, 74602. Phone: (580) 765-2485. Fax: (580) 767-1103. Licensee: Team Radio L.L.C. (group owner; (acq 7-14-98; $308,000 with co-located FM). Population served: 225,000 Law Firm: Booth, Freret, Imlay & Tepper P. Format: Sports. Target aud: 24 plus; agriculture-related country. ♦Bill Coleman, stn mgr & gen sls mgr; Danny Diamond, progmg dir; John Singer, pres, gen mgr & edit dir; Chris Henry, disc jockey.

KOSB(FM)—Co-owned with KOKP(AM). Nov 24, 1988: 105.1 mhz; 6 kw. Ant 328 ft TL: N36 14 15 W97 21 59. Stereo. 24 Natl. Network: Westwood One. Format: All 70s. News staff: one. Target aud: 25-55. ♦Sean Anderson, disc jockey.

Piedmont

***KZTH(FM)**—Not on air, target date: unknown: 88.5 mhz; 45 kw. Ant 512 ft TL: N35 34 31 W98 13 34. Hrs open: Box 14, Ponca City, 74602-0014. Phone: (580) 767-1400. Fax: (580) 765-1700. Web Site: www.thehousefm.com. Licensee: The Love Station Inc. ♦Doyle Brewer, gen mgr.

Pocola

***KKRI(FM)**— June 11, 2002: 88.1 mhz; 26 kw vert. Ant 420 ft TL: N35 09 02 W94 13 48. Stereo. Hrs opn: 24 2351 Sunset Blvd., Suite 170-218, Rocklin, CA, 95765. Phone: (916) 251-1600. Fax: (916) 251-1650. E-mail: info@air1.com Web Site: www.air1.com. Licensee: Educational Media Foundation. Group owner: EMF Broadcasting. Population served: 256,600 Natl. Network: Air 1. Law Firm: Shaw Pittman. Format: Contemp Christian. News staff: 3. Target aud: 18-35; Judeo-Christian, female. ♦Richard Jenkins, pres; Mike Novak, VP; Keith Whipple, dev dir; David Pierce, progmg mgr; Sam Wallington, engrg dir; Arthur Vassar, traf mgr; Karen Johnson, news rptr; Marya Morgan, news rptr; Richard Hunt, news rptr.

Ponca City

KIXR(FM)— June 1984: 104.7 mhz; 25 kw. Ant 292 ft TL: N36 47 21 W97 02 53. Stereo. Hrs opn: 24 Box 2631, 74602. Secondary address: 3924 Santa Fe Rd. 74602. Phone: (580) 765-5491. Fax: (580) 762-8329. E-mail: kixr@kixr.com Web Site: www.kixr.com. Licensee: Mur-Thom Broadcasting Inc. (acq 8-10-94; $80,000; FTR: 9-12-94). Population served: 101,000 Natl. Network: Westwood One. Format: Community radio. News: 4 hrs wkly. Target aud: 24-55; core audience of females between the ages of 24-45. Spec prog: Native American 3 hrs wkly. ♦Carol Murphy, pres; Gordon Thompson, gen mgr & progmg dir; Dave Foster, chief of engrg.

***KJTH(FM)**— 2004: 89.7 mhz; 100 kw. Ant 1,007 ft TL: N36 35 42 W97 34 38. Hrs open: Box 14, 74602. Secondary address: 6600 W. Hwy. 60 74601. Phone: (580) 767-1400. Fax: (580) 765-1700. E-mail: mail@klvv.com Web Site: www.thehousefm.com. Licensee: The Love Station Inc. Format: Christian music. Target aud: 25-45; young Christian adults. ♦Doyle Brewer, CEO, pres & gen mgr; Janelle Keith, prom dir; Shaun Michaels, progmg dir.

KLOR-FM— December 1965: 99.3 mhz; 3 kw. 300 ft TL: N36 46 59 W97 04 15. Stereo. Hrs opn: 24 122 N. 3rd St., 74601. Phone: (580) 762-9930. Fax: (580) 767-1103. E-mail: billc@eteamradio.com Web Site: www.eteamradio.com. Licensee: Team Radio L.L.C. (group owner; acq 3-18-99). Population served: 68,000 Natl. Network: Westwood One, Jones Radio Networks. Rgnl. Network: Okla. Radio Net. Format: Classic rock, oldies. News staff: one; News: 75 hrs wkly. Target aud: 18-55. ♦Bill Coleman, pres & gen mgr; Darrel Dye, gen sls mgr; Jerry Vaughn, progmg dir.

***KLVV(FM)**— December 1992: 88.7 mhz; 11.5 kw. 479 ft TL: N36 41 25 W97 10 20. Stereo. Hrs opn: 24 Box 14, 74602. Secondary address: 6600 W. Hwy. 60 74601. Phone: (580) 767-1400. Fax: (580) 765-1700. E-mail: mail@klvv.com Web Site: www.klvv.com; www.mychristianfm.com. Licensee: The Love Station Inc. Population served: 75,000 Format: Inspirational music, Christian teaching. Target aud: 25-45; young Christian adults. ♦Doyle Brewer, CEO, pres & gen mgr; Tony Weir, progmg.

KOKB(AM)—See Blackwell

KPNC-FM— June 5, 1979: 100.9 mhz; 6 kw. 285 ft TL: N36 39 56 W97 04 20. Stereo. Hrs opn: 24 Box 2509, 1000 Oakland Rd., 74602. Phone: (580) 765-2485. Phone: (580) 767-1101. Fax: (580) 767-1103. E-mail: billc@eteamradio.com Web Site: www.eteamradio.com. Licensee: Team Radio L.L.C. (group owner; acq 7-20-90). Population served: 50,000 Format: Country. News: 20 hrs wkly. Target aud: 25-54; working middle class. Spec prog: Farm 5 hrs wkly. ♦Bill Coleman, chmn, VP & gen mgr; Darrel Dye, sls dir & gen sls mgr; Jerry Vaughn, progmg dir.

WBBZ(AM)— 1927: 1230 khz; 1 kw-U. TL: N36 41 46 W97 03 07. Stereo. Hrs opn: 5 AM-midnight Box 588, 1601 E. Oklahoma, 74602. Phone: (580) 765-6607. Fax: (580) 765-6611. E-mail: wbbz@wbbz.com Web Site: www.wbbz.com. Licensee: Ponca City Publishing Co. (acq 1949). Population served: 35,000 Natl. Network: AP Radio. Format: Classic favorites. News staff: one; News: 12 hrs wkly. Target aud: 35 plus. Spec prog: Class 5 hrs, relg 5 hrs, big band 3 hrs wkly. ♦Tom Muchmore, CEO; Phil Turney, gen mgr, stn mgr & gen sls mgr.

Poteau

***KARG(FM)**— June 1998: 91.7 mhz; 3.25 kw. 1,866 ft TL: N35 04 17 W94 40 47. Hrs open: Box 3206, American Family Radio, Tupelo, MS, 38803. Phone: (662) 844-8888, EXT. 204. Fax: (662) 842-6791. Web Site: www.afr.net. Licensee: American Family Association. Group owner: American Family Radio Format: Inspirational Christian. ♦Marvin Sanders, gen mgr.

KOMS(FM)— Oct 18, 1969: 107.3 mhz; 100 kw. 1,810 ft TL: N34 57 50 W94 22 34. Stereo. Hrs opn: 24 4608 Radio Tower Rd., Van Buren, AR, 72956. Phone: (479) 474-3422. Fax: (479) 474-2649. E-mail: jpmorgan@hotmail.com Web Site: www.bigcountry1073.com. Licensee: Cumulus Licensing Corp. Group owner: Cumulus Media Inc. (acq 5-17-99; $950,000). Population served: 500,000 Natl. Network: CNN Radio. Wire Svc: AP Format: Class country. News staff: News progmg 60 hrs wkly Target aud: 25-54; but we gather in all from 12 plus. ♦J.P. Morgan, opns mgr & disc jockey; Smitty O'Loughlin, mktg mgr; Michael Hauser, progmg dir & disc jockey; Don Jones, engr.

KPRV(AM)— Nov 25, 1953: 1280 khz; 1 kw-D. TL: N35 01 08 W94 39 22. Hrs open: Sunrise-sunset Box 1331, Rt. 1, 74953. Phone: (918) 647-3221. Fax: (918) 647-5092. E-mail: billy@clnk.com Web Site: www.kprvradio.com. Licensee: LeRoy Billy. Natl. Network: ABC. Rgnl. Network: Okla. Radio Net. Law Firm: Robert Allen. Format: Classic country. News: 24 hrs wkly. Target aud: 24-54. Spec prog: Gospel 12 hrs wkly. ♦Joann Billy, gen mgr; LeRoy Billy, gen sls mgr.

KZBB(FM)— 1967: 97.9 mhz; 100 kw. 2,000 ft TL: N35 04 19 W94 40 46. Hrs open: 24 311 Lexington Ave., Fort Smith, AR, 72901. Phone: (479) 782-8888. Fax: (479) 782-0366. E-mail: b98@kzbb.com Web Site: www.kzbb.com. Licensee: Capstar TX L.P. Group owner: Clear Channel Communications Inc. (acq 8-30-00; grpsl). Population served: 500,000 Format: CHR. News: one hr wkly. Target aud: 18-49; upscale. Spec prog: Black 2 hrs, jazz 2 hrs, relg one hr wkly. ♦Paul Swint, gen mgr; Lee Matthews, opns mgr & progmg mgr; Daren Bobb, news dir & prom.

Pryor

KMUR(AM)— July 3, 1950: 1570 khz; 1 kw-D. TL: N36 18 04 W95 19 29. (CP: COL Catoosa. TL: N36 15 52 W95 42 34). Hrs opn: Box 702588, Tulsa, 74170. Phone: (918) 496-7700. Fax: (918) 746-7615. E-mail: KMUR@1570country.com Web Site: www.1570country.com. Licensee: Reunion Broadcasting L.L.C. (group owner; (acq 11-23-2003; $75,000). Population served: 825,031 Natl. Network: CNN Radio. Law Firm: Hardy, Carey & Chautin. Format: Oldies. News staff: one; News: 10 hrs wkly. Target aud: 25 plus. Spec prog: St. Louis Cardinal Baseball. ♦Stan Tacker, gen mgr; Terri Tacker, gen sls mgr.

KMYZ-FM— July 3, 1969: 104.5 mhz; 78 kw. 1,250 ft TL: N36 18 04 W95 19 29. Stereo. Hrs opn: 5810 E. Skelly Dr., Suite 801, Tulsa, 74135. Phone: (918) 665-3131. Fax: (918) 663-6622. Web Site: www.edgetulsa.com. Licensee: Shamrock Communications Inc. (group owner; acq 4-14-84). Format: Alternative rock. ♦William Lynett, CEO; J. Michael Demarco, gen mgr.

Roland

KREU(FM)— Dec 29, 1995: 92.3 mhz; 740 w. 932 ft TL: N35 31 22 W94 23 32. Hrs open: 601 N. Greenwood Ave., Fort Smith, AR, 72901-3511. Phone: (479) 785-2527. Fax: (501) 782-9127. Licensee: Star 92 Co. (acq 6-13-2003; $10,000). Format: Spanish. ♦Gary Keifer, gen mgr; Fred Baker Jr., opns mgr & gen sls mgr.

Sallisaw

KKBD(FM)— May 18, 1972: 95.9 mhz; 30 kw. 600 ft TL: N35 24 29 W94 41 13. Stereo. Hrs opn: 24 311 Lexington ve., Fort Smith, AR, 72901. Phone: (479) 782-8888. E-mail: info@bigdog959.com Web Site: www.bigdog959.com. Licensee: Clear Channel Radio Licenses, Inc. Group owner: Clear Channel Communiations Inc. (acq 8-30-00; grpsl). Population served: 250000 Format: Classic Rock. Target aud: 25-49; adults. ♦Paul Swint, gen mgr.

KKUZ(AM)— Sept 16, 1968: Stn currently dark. 1560 khz; 250 w-D. TL: N35 26 34 W94 46 33. Hrs open: 2582 Monroe St., Paducah, AR, 42001. Phone: (501) 782-9699. Licensee: Teddy Bear Communications Inc. Population served: 150,000 ♦Ted Hite Sr., gen mgr.

Sand Springs

KJMU(AM)— July 22, 1961: 1340 khz; 450 w-D, 900 w-N. TL: N36 07 58 W96 05 36. Hrs open: 24 8886 W. 21st St., 74063. Secondary address: 1701 S. 55th St., Kansas City, KS 66106. Phone: (918) 241-1971. Fax: (918) 241-1975. Licensee: Davidson Media Station KTFX Licensee LLC. (group owner; (acq 1-13-2006; $1.03 million). Population served: 750,000 Format: Rgnl Mexican. News: 3 hrs wkly. Target aud: 18-54; general. ♦Ivan Duin-Obregon, pres & gen mgr.

KKCM(FM)— June 1989: 102.3 mhz; 50 kw. Ant 492 ft TL: N36 12 39 W96 06 03. Hrs open: 24 7136 S. Yale, Suite 500, Tulsa, 74136. Phone: (918) 493-3434. Fax: (918) 493-2376. E-mail: chris.kelly@cox.com Web Site: spirit1023.com. Licensee: Cox Radio Inc. Group owner: Cox Broadcasting (acq 3-16-99; $3.5 million). Format: Christian music. ♦Dan Lawrie, gen mgr; Chris Kelly, progmg dir; Wayne Smith, opns mgr & chief of engrg.

Sapulpa

KXOJ-FM—Listing follows KYAL(AM).

KYAL(AM)— June 15, 1962: 1550 khz; 2.5 kw-D, 47 w-N, DA-1. TL: N36 01 08 W96 05 55. Stereo. Hrs opn: Cityplex Towers, 2448 E. 81st St., Tulsa, 74137-4272. Phone: (918) 492-2660. Fax: (918) 492-8840. Licensee: KXOJ Inc. Group owner: Adonai Radio Group (acq 5-2-73). Population served: 350,000 Format: Sports. Target aud: 35 plus. ♦David Stephens, pres, gen mgr, opns mgr, sls dir & mktg dir.

KXOJ-FM—Co-owned with KYAL(AM). Feb 22, 1977: 100.9 mhz; 5 kw. 360 ft TL: N36 03 38 W96 06 03. Stereo. 24 E-mail: mail@kxoj.com Web Site: www.kxoj.com.400,000 Format: Contemp Christian mus. ♦Mike Stephens, CEO; David Stephens, stn mgr, opns dir, dev dir & adv dir.

Seminole

KIRC(FM)— Nov 1, 1978: 105.9 mhz; 4.4 kw. 384 ft TL: N35 18 28 W96 45 18. Stereo. Hrs opn: 24 2 E. Main St., Shawnee, 74801-6906. Phone: (405) 878-1803. Phone: (405) 382-0105. Fax: (405) 878-0162. E-mail: kirc1059@aol.com Licensee: One Ten Broadcast Group Inc. (group owner; acq 9-6-91). Population served: 250,000 Format: Country. News staff: 9; News: 2 hrs wkly. Target aud: 12-55; general. Spec prog: Area tribes one hr wkly. ♦Linda Jones, pres & VP; Dennis Burton, gen mgr; David Beerley, gen sls mgr & progmg dir; Hal Smith, chief of engrg; Nichole Johnson, traf mgr.

KTLS-FM—See Holdenville

KWSH(AM)—See Wewoka

***KXTH(FM)**— October 2003: 89.1 mhz; 2.3 kw vert. Ant 387 ft TL: N35 12 53 W96 44 26. Hrs open: 24 Box 14, Ponca City, 74602-0014. Secondary address: 6600 W. Hwy. 60, Ponca City 74601. Phone: (580) 767-1400. Fax: (580) 765-1700. E-mail: mail@klvv.com Web Site: www.thehousefm.com. Licensee: The Love Station Inc. (acq 9-16-03). Format: Adult contemp, Christian. Target aud: 25-45; young Christian adults. ♦Doyle Brewer, CEO, pres & gen mgr; Janelle Keith, prom; Tony Weir, progmg.

Shawnee

KGFF(AM)— Dec 10, 1930: 1450 khz; 1 kw-U. TL: N35 21 39 W96 53 41. Hrs open: 24 Box 9, 74802. Secondary address: 1570 S. Gordon Cooper Drive 74801. Phone: (405) 273-4390. Fax: (405) 273-4530. E-mail: mike@kgff.com Web Site: www.kgff.com. Licensee: Citizen Band Potawatomi Indian Tribe of Oklahoma Inc. (acq 11-10-98; $155,000). Population served: 45,000 Natl. Network: ABC. Rgnl. Network: Okla. Radio Net., Texas State Net. Format: Adult standards. News staff: one; News: 20 hrs wkly. General. Spec prog: school, University of Oklahoma, Oklahoma Baptist University, St. Gregory's University, relg 4 hrs wkly. ♦Michael Askins, gen mgr & opns dir.

KQCV-FM— Apr 13, 1998: 95.1 mhz; 100 kw. 1,004 ft TL: N35 15 47 W96 22 43. Hrs open: 1919 N. Broadway, Oklahoma City, 73103. Phone: (405) 521-0800. Fax: (405) 521-1391. Web Site: bottradionetwork.com. Licensee: Community Broadcasting Inc. Group owner: Bott Radio Network Format: Christian info & teaching. ♦Paul Sublett, gen mgr; Jerry McCall, opns mgr.

Snyder

***KJCM(FM)**— 2000: 100.3 mhz; 18 kw. Ant 384 ft TL: N34 38 02 W99 05 03. Hrs open: 509 N. Main St., Altus, 73521. Phone: (580) 482-9797. Fax: (580) 397-9201. Licensee: Ray Broadcasting Inc. (acq 12-22-2003; $320,000 with KHIM(FM) Mangum). Format: Adult contemp. ♦Bat Masterson, gen mgr.

Soper

KMMY(FM)—Not on air, target date: unknown: 96.5 mhz; 6 kw. Ant 325 ft TL: N34 01 18 W95 41 49. Hrs open: 404 E. Jackson, Hugo, 74743. Phone: (580) 326-5541. Licensee: Will Payne. ♦Will Payne, gen mgr.

Spencer

***KROU(FM)**— Jan 28, 1993: 105.7 mhz; 4 kw. Ant 328 ft TL: N35 35 22 W97 29 03. Stereo. Hrs opn: 5 AM-midnight
Rebroadcasts KGOU(FM) Norman 100%.
780 Van Vleet Oval, Norman, 73019. Phone: (405) 325-3388. Fax: (405) 325-7129. E-mail: manager@kgou.org Web Site: www.kgou.org. Licensee: University of Oklahoma. Population served: 800,000 Natl. Network: NPR. Law Firm: Dow, Lohnes & Albertson. Format: News/talk, jazz. News staff: one; News: 82 hrs wkly. Target aud: 25-54; general. Spec prog: Blues 8 hrs, new age 4 hrs wkly. ♦Karen Holp, gen mgr; Jim Johnson, progmg dir.

Sperry

KMUS(AM)— July 3, 1948: 1380 khz; 7 kw-D, 250 w-N, DA-2. TL: N36 15 59 W95 58 15. Hrs open: 24 8321 E. 61st St., Tulsa, 74145. Phone: (918) 250-8484. Phone: (918) 686-9222. Fax: (918) 250-6464. Web Site: www.radiodisney.com. Licensee: Radio Disney Group LLC. Group owner: ABC Inc. (acq 2-28-2003; $1.5 million). Population served: 270,000 Format: Children radio. ♦Stan Tacker, gen mgr; Terri Tacker, gen sls mgr; Jason Walker, progmg dir.

Stigler

***KTKL(FM)**— 2003: 88.5 mhz; 1 w horiz, 22 kw vert. Ant 643 ft vert TL: N35 08 30 W95 21 20. Stereo. Hrs opn: 24 2351 Sunset Blvd., Suite 170-218, Rocklin, CA, 95765. Phone: (916) 251-1600. Fax: (916) 251-1650. E-mail: klove@klove.com Web Site: www.klove.com. Licensee: Educational Media Foundation. Group owner: EMF Broadcasting. Natl. Network: K-Love. Law Firm: Shaw Pittman. Format: Christian contemp. News staff: 3. Target aud: 25-44; Judeo Christian, female. ♦Richard Jenkins, pres; Mike Novak, VP; Keith Whipple, dev dir; David Pierce, progmg mgr; Ed Lenane, news dir; Sam Wallington, engrg dir; Karen Johnson, news rptr; Marya Morgan, news rptr; Richard Hunt, news rptr.

Stillwater

KGFY(FM)— Feb 6, 1967: 105.5 mhz; 4.9 kw. Ant 361 ft TL: N36 10 31 W97 00 51. Stereo. Hrs opn: 24 408 E. Thomas Rd., 74075. Phone: (405) 372-6000. Fax: (405) 372-6969. Licensee: Stillwater Broadcasting LLC. Group owner: Mahaffey Enterprises Inc. (acq 9-28-2001). Population served: 75,000 Law Firm: Fletcher, Heald & Hildreth. Format: Country. News staff: 2; News: 5 hrs wkly. Target aud: 25-54; young, college community & upscale educated people. Spec prog: Contemp Christian 4 hrs wkly. ♦David Webb, gen mgr.

***KOSU(FM)**— Dec 29, 1955: 91.7 mhz; 100 kw. Ant 1,010 ft TL: N36 06 33 W97 11 43. Stereo. Hrs opn: 24 Oklahoma State Univ., 303 P.M. Bldg., 74078. Phone: (405) 744-6352. Fax: (405) 744-9970. Web Site: www.kosu.org. Licensee: Oklahoma State University. Population served: 1,500,000 Natl. Network: NPR. Format: Class, educ, news. News staff: 2; News: 48 hrs wkly. Target aud: General. Spec prog: American Indian one hr wkly. ♦Craig Beeby, gen mgr & progmg dir; Don Crider, dev dir & dev mgr; Rachel Hubbard, news dir; Dan Schroeder, engr.

KSPI(AM)— June 1, 1947: 780 khz; 250 w-D. TL: N36 04 56 W97 03 13. Stereo. Hrs opn: 6 AM-6 PM Box 1269, 74076. Secondary address: 408 E. Thomas Rd. 74076. Phone: (405) 372-7800. Fax: (405) 372-6969. E-mail: stillwaterradio@provalue.net Licensee: Stillwater Broadcasting LLC. Group owner: Mahaffey Enterprises Inc. (acq 7-21-97; $650,000 with co-located FM). Population served: 150,000 Natl. Network: ESPN Radio. Format: News/talk, sports. News staff: one; News: 12 hrs wkly. Target aud: 30 plus. Spec prog: News, sports, features. ♦John Mahaffey, pres; David Webb, gen mgr & sls VP; Gil Stuart, progmg mgr; Bill Van Ness, news dir.

KSPI-FM— Nov 1, 1947: 93.7 mhz; 16 kw. 886 ft TL: N36 06 31 W97 11 46. Stereo. 18 50,000 Format: Adult contemp. News staff: 2; News: 11 hrs wkly. Target aud: 24 plus. ♦Diane Keenom, opns mgr & mus dir; David Webb, gen sls mgr; Gil Stuart, prom dir & progmg dir; Bill Vanness, news dir & pub affrs dir; Rex Holt, sports cmtr; Chris Greenert, disc jockey; Ferris O'Brian, disc jockey; Steven Jones, disc jockey.

KVRO(FM)— Apr 12, 1997: 98.1 mhz; 6 kw. Ant 328 ft TL: N36 13 10 W97 09 47. Stereo. Hrs opn: 24 Box 1269, 74076-1269. Secondary address: 408 E. Thomas 74075. Phone: (405) 372-6000. Fax: (405) 372-6969. E-mail: stillwaterradio@coxinet.net Licensee: Stillwater Broadcasting LLC. Group owner: Mahaffey Enterprises Inc. (acq 9-28-2001). Population served: 150,000 Format: Oldies. News staff: 2; News: 24 hrs wkly. Target aud: 25-54. ♦Steven Folks, gen mgr; Gil Stuart, opns mgr; Bill Van Ness, news dir.

Stuart

***KLRB(FM)**— July 25, 2003: 89.3 mhz; 3 kw vert. Ant 272 ft TL: N34 54 57 W96 08 10. Hrs open: 24 Box 145, 2.2 mi W of Stuart, 1/2 mi N, 74570. Phone: (918) 697-4019. Fax: (580) 892-3941. Licensee: Lighthouse of Prayer Inc. Format: Country. ♦Walter Kuhlman, pres; Stephen Burke, gen mgr & stn mgr.

Sulphur

***KFXT(FM)**— 2000: 90.7 mhz; 7 kw. Ant 298 ft TL: N34 32 57 W96 58 34. Hrs open: 24 Sister Sherry Lynn Foundation Inc., 1101 N. 81 Hwy., Marlow, 73055. Phone: (580) 658-9292. Fax: (580) 658-2561. Licensee: Sister Sherry Lynn Foundation Inc. Law Firm: Southmayd & Miller. Format: Gospel. ♦Ken Austin, gen mgr; Sherry Lynn, gen sls mgr; Bill Marshall, progmg dir; James Wilson, engr.

KIXO(FM)— Nov 11, 1979: 106.1 mhz; 2.65 kw. Ant 499 ft TL: N34 39 03 W96 59 24. Stereo. Hrs opn: 24 1101 Hwy. 81 N., Marlow, 73055. Phone: (580) 658-9292. Fax: (580) 658-2561. Licensee: DFWU Inc. (group owner; acq 10-1-90). Population served: 80,000 Law Firm: Southmayd & Miller. Format: Country. News: 5 hrs wkly. Target aud: 25-52. ♦Ken Austin, gen mgr; Sherry Lynn, gen sls mgr; Bill Marshall, progmg dir; Amy Helton, chief of engrg & traf mgr.

Taft

KCXR(FM)— Mar 20, 1990: 100.3 mhz; 6 kw. Ant 380 ft TL: N35 48 42 W95 34 12. Stereo. Hrs opn: 24 2448 E. 81st, Suite 5500, Tulsa, 74137. Phone: (918) 492-2660. Fax: (918) 492-8840. Licensee: KXOJ Inc. (group owner; (acq 12-11-2002; $1 million with KBIX(AM) Mukogee). Population served: 828,000 Format: Christian Rock. News staff: one; News: 14 hrs wkly. Target aud: 25-54. Spec prog: Farm one hr wkly. ♦Michael P. Stephens, pres; David Stevens, gen mgr.

Tahlequah

KEOK(FM)—Listing follows KTLQ(AM).

KTLQ(AM)— August 1957: 1350 khz; 1 kw-D, 61 w-N. TL: N35 53 43 W94 57 12. Stereo. Hrs opn: 24 Box 676, 74465. Phone: (918) 456-2511. Fax: (918) 456-3231. E-mail: okctry@fullnet.net Web Site: www.tahlequah.com/okcountry. Licensee: Payne 5 Communications LLC (acq 11-24-2003; $1.15 million with co-located FM). Population served: 100,000 Natl. Network: Westwood One. Format: Sports, southern gospel. News staff: one; News: 6 hrs wkly. Target aud: 25-54. ♦Ralph Lynch, gen mgr; Mitchell Johnson, progmg VP.

KEOK(FM)—Co-owned with KTLQ(AM). Aug 20, 1966: 101.7 mhz; 6.6 kw. 295 ft TL: N35 53 43 W94 57 12. Stereo. Web Site: www.tahlequah.com/okcountry. Format: Country. Target aud: 25-60. ♦Susan Shelton, traf mgr; C.H. Jackson, local news ed.

Tishomingo

***KAZC(FM)**— Sept 29, 1998: 88.3 mhz; 5.5 kw. Ant 922 ft TL: N34 21 34 W96 33 34. Hrs open: Box 1343, Ada, 74821. Phone: (580) 332-0902. E-mail: email@thegospelstation.com Web Site: www.thegospelstation.com. Licensee: South Central Oklahoma Christian Broadcasting Inc. Format: Southern gospel. ♦Randall Christy, pres; Rick Cody, gen mgr & opns mgr.

Tonkawa

***KAYE-FM**— June 1, 1976: 90.7 mhz; 1.2 kw. 67 ft TL: N36 40 42 W97 17 50. Stereo. Hrs opn: 7 AM-midnight (M-F) Central Hall 306, 1220 E. Grand, 74653. Phone: (580) 628-6446. Phone: (580) 628-6200. Fax: (580) 628-6209. E-mail: kaye@north-ok.edu Web Site: www.north-ok.edu. Licensee: Northern Oklahoma College. Population served: 60,000 Format: Top 40. News: 6 hrs wkly. Target aud: 13-25. ♦Dr. Joe Kinzer, pres.

Tulsa

KAKC(AM)— July 15, 1938: 1300 khz; 5 kw-D, 1 kw-N, DA-2. TL: N35 59 40 W95 51 27. Hrs open: 24 2625 S. Memorial, 74129. Phone: (918) 388-5100. Fax: (918) 388-5400. Web Site: 1300thebuzz.com. Licensee: Clear Channel Broadcasting Licenses Inc. (group owner; (acq 8-5-92). Natl. Network: ESPN Radio, Fox Sports. Natl. Rep: Clear Channel. Format: Sports. Target aud: 25-54; men. ♦M. Oppenheimer, gen mgr; Garry Weaver, prom dir.

KMOD-FM—Co-owned with KAKC(AM). Oct 10, 1959: 97.5 mhz; 100 kw. 1,800 ft TL: N36 11 46 W96 05 53. Stereo. Web Site: www.kmod.com. Format: AOR. Target aud: 25-49; men.

KBEZ(FM)— March 1964: 92.9 mhz; 100 kw. 1,319 ft TL: N36 11 26 W96 05 50. Stereo. Hrs opn: 7030 S. Yale Ave., Suite 711, 74136. Phone: (918) 496-9336. Fax: (918) 496-1937. E-mail: jobs@kbez.com Web Site: www.kbez.com. Licensee: Renda Broadcasting Corp. (group owner; acq 6-8-90; grpsl; FTR: 6-25-90). Population served: 648,300 Format: Adult contemp. Target aud: 25-54. Spec prog: Big band 5 hrs wkly. ♦Pat Bryson, VP & gen mgr; Keith Marlow, opns mgr & progmg dir; Richard Harley, chief of engrg; Sammy Carrillo, pub affrs dir & prom.

KCFO(AM)— 1946: 970 khz; 2.5 kw-D, 1 kw-N, DA-2. TL: N36 11 46 W96 02 22. Hrs open: 18 5800 E. Skelly Dr., Ste 150, 74135-6416. Phone: (918) 445-1186. Phone: (918) 622-0970. Fax: (918) 622-0985. Web Site: www.kcfo.com. Licensee: Friendship Broadcasting L.P. (acq 8-1-90; $953,000; FTR: 7-2-90). Population served: 750,000 Natl. Network: USA. Format: Relg, talk, sports. News: 3 hrs wkly. Target aud: 25-54; women. ♦Ray Clatworthy, pres; Kenneth Staley, gen mgr.

KFAQ(AM)— Jan 23, 1925: 1170 khz; 50 kw-U, DA-N. TL: N36 08 49 W95 48 27. Stereo. Hrs opn: 24 4590 E. 29th, 74114. Phone: (918) 743-7814. Fax: (918) 743-7613. Web Site: www.kfaq.com. Licensee: Journal Broadcast Corp. Group owner: Journal Broadcast Group Inc. (acq 6-11-99; grpsl). Population served: 237,000 Natl. Network: CBS. Natl. Rep: Clear Channel. Law Firm: Dow, Lohnes & Albertson. Format: News/talk. News staff: 3; News: 24 hrs wkly. Target aud: 35 plus. Spec prog: Farm 5 hrs, gospel 2 hrs wkly. ♦Carl Gardner, pres; Ron Kurtis, CFO; Jay Werth, gen mgr; Moon Mullins, opns mgr & mus dir; Randy Bush, gen sls mgr; Claudette Rogers, prom dir; Michael DelGiorno, progmg dir; Brian Gann, news dir; Ray Klotz, engrg dir; Bob Cooper, disc jockey; Bob O'Shea, disc jockey; Dave Williams, disc jockey; Tom Watson, disc jockey.

KVOO-FM—Co-owned with KFAQ(AM). Nov 16, 1973: 98.5 mhz; 100 kw. 1,229 ft TL: N36 11 26 W96 05 50.24 Web Site: www.kvoo.com. Format: Today's country. Target aud: 25-54. ♦Bob Cooper, disc jockey; Bob O'Shea, disc jockey; Dave Williams, disc jockey; Tom Watson, disc jockey.

KGTO(AM)— 1946: 1050 khz; 1 kw-D. TL: N36 09 40 W96 03 10. Hrs open: 6 AM-sunset 7030 S. Yale, Suite 302, 74136. Phone: (918) 494-9886. Fax: (918) 494-9683. Licensee: KJMM Inc. Group owner: Perry Publishing & Broadcasting Co. (acq 3-30-01; $455,000). Population served: 375,000 Natl. Network: Westwood One. Format: Urban Contemp. News staff: one; News: 3 hrs wkly. Target aud: 35-54. ♦Bryan K. Robinson, gen mgr.

KJMU(AM)—See Sand Springs

KJSR(FM)— Nov 1, 1966: 103.3 mhz; 100 kw. Ant 1,279 ft TL: N36 01 10 W95 39 24. Stereo. Hrs opn: 24 7136 S. Yale, Suite 500, 74136. Phone: (918) 493-3434. Fax: (918) 493-5383. Web Site: www.star103fm.com. Licensee: Cox Radio Inc. Group owner: Cox Broadcasting (acq 3-28-97; grpsl). Population served: 650,000 Format: Classic rock/classic hits. News staff: one. Target aud: 25-44. Spec prog: Pub affrs 2 hrs wkly. ♦Dan Lawrie, gen mgr; Steve Hunter, opns mgr.

KRAV(FM)— Nov 21, 1962: 96.5 mhz; 100 kw. Ant 137 ft TL: N36 11 46 W96 05 53. Stereo. Hrs opn: 7136 S. Yale, Suite 500, 74136. Phone: (918) 491-9696. Fax: (918) 493-5385. Web Site: www.mix96tulsa.com. Licensee: Cox Radio Inc. Group owner: Cox Broadcasting (acq 11-21-96; $5.5 million with co-located AM). Population served: 640,000 Format: Hot adult contemp. Target aud: 25-54; 30% men, 70% women. ♦Robert Neil, pres; Marc Morgan, exec VP; Dan Lawrie, VP; Dan Lawrie, gen mgr.

KRMG(AM)— Dec 31, 1949: 740 khz; 50 kw-D, 25 kw-N, DA-2. TL: N36 04 50 W96 17 09. Stereo. Hrs opn: 24 7136 S. Yale, 74136. Phone: (918) 493-7400. Fax: (918) 493-2376. Web Site: www.krmg.com. Licensee: Cox Radio Inc. Group owner: Cox Broadcasting (acq 3-28-97; grpsl). Population served: 633,300 Format: News/talk. News staff: 7. Target aud: 25-54; those interested in news, info & issue oriented talk. ♦Chuck Browning, gen mgr; Steve Laswell, gen sls mgr; Drew Anderssen, progmg dir.

KWEN(FM)—Co-owned with KRMG(AM). 1961: 95.5 mhz; 96 kw. 1,328 ft TL: N36 11 46 W95 05 53. Stereo. 24 Web Site: www.krmg.com.150,000 Format: Contemp country. News staff: one. Target aud: 25-54; country life group. ♦Jim Vidler, gen sls mgr; Gerry McCracken, progmg dir.

KRVT(AM)—See Claremore

KTBZ(AM)— Jan 22, 1934: 1430 khz; 5 kw-U, DA-N. TL: N36 14 10 W95 56 50. Hrs open: 2625 S. Memorial Dr., 74129-2600. Phone: (918) 388-5100. Fax: (918) 388-5400. Web Site: www.1430thebuzz.com. Licensee: Clear Channel Broadcasting Licenses Inc. Group owner: Clear Channel Communications Inc. (acq 1997; grpsl). Population served: 600,000 Format: Sports. Target aud: 25-49; men. ♦Michael Oppenheimer, gen mgr.

***KWGS(FM)**— Oct 19, 1947: 89.5 mhz; 50 kw. Ant 1,067 ft TL: N36 01 15 W95 40 32. Stereo. Hrs opn: 24 600 S. College, 74104. Phone: (918) 631-2577. Fax: (918) 631-3695. E-mail: answers@kwgs.org Web Site: www.kwgs.org. Licensee: The University of Tulsa. Population served: 753,163 Natl. Network: NPR, PRI. Law Firm: John D. Pellegrin. Format: News & info/public radio. News: 84 hrs wkly. Target aud: General. ♦Frank Christel, sr VP; Richard Fisher, gen mgr; P. Casey Morgan, dev dir; Brad Newman, chief of engrg; Michelle McKee, opns.

***KWTU(FM)**— Oct 15, 2004: 88.7 mhz; 5 kw. Ant 1,066 ft TL: N36 01 15 W95 40 32. Hrs open: 24 The University of Tulsa, 600 S. College, 74104. Phone: (918) 631-2577. Licensee: The University of Tulsa. Format: Classical. ♦Rich Fisher, gen mgr; Michelle McKee, opns mgr.

KYAL(AM)—See Sapulpa

Vinita

KGND(AM)—Listing follows KITO-FM.

KITO-FM— Apr 9, 1981: 96.1 mhz; 50 kw. 492 ft TL: N36 34 56 W95 01 35. Stereo. Hrs opn: 24 Box 961, 74301. Secondary address: 402 N. Wilson St. 74301. Phone: (918) 256-2255. Phone: (918) 542-9824. Fax: (918) 256-2633. Licensee: KXOJ Inc. (acq 8-1-2007; $1.8 million with co-located AM). Population served: 750,000 Format: Country. News: 28 hrs wkly. Target aud: General; traditional country music fans. ♦Leona Boyd, gen mgr & gen sls mgr; Dave Boyd, progmg dir; Troy Langham, chief of engrg.

KGND(AM)—Co-owned with KITO-FM. Dec 7, 1954: 1470 khz; 500 w-D, 88 w-N. TL: N36 38 44 W95 07 35.75,000 Format: Country. Target aud: 30-50. ♦David Boyd, gen mgr.

Wagoner

KXTD(AM)— Mar 1, 1966: 1530 khz; 5 kw-D, DA. TL: N35 58 30 W95 29 30. Stereo. Hrs opn: 5807 S. Garnett, Suite F, Tulsa, 74146. Phone: (918) 254-7556. Fax: (918) 252-0036. E-mail: maria@quebuenaok.com Web Site: www.quebuenaok.com. Licensee: Gaytan-Galvan Limited Liability Co. (acq 1-31-97). Format: Sp. ♦Maria Gaytan, gen mgr.

Warner

KTFX-FM— March 1995: 101.7 mhz; 25 kw. Ant 276 ft TL: N35 34 39 W95 12 36. Stereo. Hrs opn: 24 401 W. Broadway, Muskogee, 74401. Phone: (918) 683-1017. Fax: (918) 686-6159. E-mail: ktfx@k955.com Web Site: www.country1017.com. Licensee: K95.5 Inc. (group owner) Population served: 88,649 Law Firm: Womble, Carlyle, Sandridge & Rice. Format: Country. News staff: one; News: 21 hrs wkly. Target aud: General. ♦William H. Payne, CEO, pres & gen mgr; Katey Sherrick, gen sls mgr; Cliff Casteel, progmg dir; Erin McWilliams, news dir.

Watonga

KIMY(FM)— Dec 12, 1987: 93.9 mhz; 3 kw. Ant 328 ft TL: N35 54 17 W98 23 09. (CP: 4.2 kw, ant 394 ft. TL: N35 50 27 W98 19 09). Stereo. Hrs opn: 24 Box 1343, Ada, 74821. Phone: (580) 332-0902. E-mail: email@thegospelstation.com Web Site: www.thegospelstation.com. Licensee: South Central Oklahoma Broadcasting Inc. (acq 3-11-2004; $163,000). Population served: 10,000 Format: Southern gospel. Target aud: 25-54; general. ♦Randall Christy, pres; Ronald W. Gabe, gen mgr.

Weatherford

***KAYM(FM)**— 2000: 90.5 mhz; 2.7 kw. Ant 282 ft TL: N35 29 47 W98 44 10. Hrs open: Box 3206, American Family Radio, Tupelo, MS, 38803. Phone: (662) 844-8888, EXT. 204. Fax: (662) 842-6791. E-mail: comments@afr.net Web Site: www.afr.net. Licensee: American Family Association. Group owner: American Family Radio Format: Inspirational Christian. ♦Marvin Sanders, gen mgr.

KWEY(AM)— June 1, 1970: 1590 khz; 1 kw-D, DA. TL: N35 33 33 W98 43 11. Hrs open: Box 587, Hwy. 54 N., 73096. Phone: (580) 772-5939. Fax: (580) 772-1590. Web Site: www.kwey.com. Licensee: Wright Broadcasting Systems Inc. (acq 7-17-91; $407,435 with co-located FM; FTR: 8-5-91) Population served: 11,400 Natl. Network: ABC. Rgnl. Network: Agri-Net. Law Firm: Putbrese, Hunsaker & Trent. Format: C&W. News staff: one; News: 14 hrs wkly. Target aud: 25 plus; full service station. ♦G. Harold Wright, CEO; Heston Wright, sls dir; Todd Brunner, opns mgr & progmg dir; Ray Bagby', engrg dir; Chuck Edwards, sports cmtr.

Wewoka

KSLE(FM)—Listing follows KWSH(AM).

KWSH(AM)— July 1951: 1260 khz; 1 kw-U, DA-N. TL: N35 10 10 W96 32 30. Hrs open: 2 E. Main, Shawnee, 74801. Phone: (405) 382-1260. Phone: (405) 378-1803. Fax: (405) 257-2011. Fax: (405) 382-0128. E-mail: kwsh@earthlink.net Licensee: One Ten Broadcast Group Inc. (group owner; (acq 4-1-99; $400,000 with co-located FM). Population served: 126,400 Rgnl. Network: Okla. Radio Net. Format: Country. Target aud: 21-61. Spec prog: American Indian one hr wkly. ♦Dennis Burton, stn mgr; Garry Walker, opns mgr & progmg dir; Linda Jones, pres, gen mgr, sls dir, prom mgr & adv mgr; Hal Smith, chief of engrg.

KSLE(FM)—Co-owned with KWSH(AM). October 1997: 104.7 mhz; 6 kw. 328 ft TL: N35 04 51 W96 35 03.24 Format: Oldies. ♦Linda Jones, prom dir.

Wilburton

KOCD(FM)— Nov 1, 2002: 103.7 mhz; 100 kw. Ant 607 ft TL: N34 59 13 W95 42 10. Stereo. Hrs opn: 24 Box 1068, McAlester, 74502. Phone: (918) 423-1460. Fax: (918) 465-3622. Licensee: KESC Enterprises LLC (group owner) (acq 1-18-2005; $766,666). Natl. Network: AP Radio. Law Firm: Wiley, Rein & Fielding. Format: Adult contemp. Target aud: 25-54. ♦Lee Anderson, gen mgr.

Woodward

***KJOV(FM)**— 1998: 90.7 mhz; 4 kw. 400 ft TL: N36 24 08 W99 25 47. Hrs open: Box 991, Meade, KS, 67854. Secondary address: 922 Webster 73802. Phone: (620) 873-2991. Fax: (620) 873-2755. Licensee: Christian Community Radio. Format: Contemp Christian. ♦Don Hughes, pres & gen mgr; Michael Luskey, opns dir; Delvin Kinser, news dir; Steve Larson, chief of engrg; Polly Hughes, traf mgr.

KMZE(FM)— Oct 15, 1989: 92.1 mhz; 2.15 kw. 1,099 ft TL: N36 16 06 W99 26 56. Stereo. Hrs opn: 24 Box D, 2728 Williams Ave., 73801. Phone: (580) 256-3692. Fax: (580) 256-3825. Licensee: FM 92 Broadcasters Inc. (acq 8-17-89; FTR: 9-5-89). Natl. Network: Jones Radio Networks. Format: Adult contemp. News staff: one. Target aud: 25-54. ♦Mike Mitchel, CEO & chmn; Kevin Grice, gen mgr.

KOPA(FM)— 2001: 95.9 mhz; 6 kw. Ant 328 ft TL: N36 24 40 W99 21 05. Hrs open: 1222-10th St., Suite 111, TX, 73801. Phone: (580) 256-0959. Fax: (580) 256-4959. Licensee: Shaffer Communications Group. Format: Alternative. ♦Robert Fisher, gen mgr.

KSIW(AM)— September 1947: 1450 khz; 1 kw-U. TL: N36 25 42 W99 24 10. Hrs open: 18 Box 1600, 73802. Secondary address: 1922 22nd St. 73801. Phone: (580) 256-1450. Fax: (580) 254-9102. Web Site: www.thesportsanimal.com. Licensee: Classic Communications Inc. (acq 6-20-2005). Population served: 50,000 Format: Sports. ♦Sherre House, pres, gen mgr & gen sls mgr; Sam Piel, progmg dir.

KWFX(FM)—Co-owned with KSIW(AM). Nov 1, 1974: 93.5 mhz; 3 kw. Ant 151 ft TL: N36 25 42 W99 24 10. (CP: 106.3 mhz, 100 kw, ant 868 ft. TL: N36 22 31 W99 28 32). Stereo. 19 Phone: (580) 256-0935. (Acq 4-30-96). Format: Country, talk. Target aud: 25-45; affluent, predominantly females.

KWDQ(FM)— Jan 9, 1990: 102.3 mhz; 100 kw. Ant 868 ft TL: N36 22 31 W99 28 31. Hrs open: Box 1600, 73802. Phone: (580) 254-9103. Fax: (580) 254-9102. Licensee: Classic Communications Inc. (acq 3-20-92). Format: Classic rock. ♦Sherre House, CEO & gen mgr.

KWOX(FM)— Dec 16, 1983: 101.1 mhz; 100 kw. Ant 1,204 ft TL: N36 16 06 W99 26 56. Stereo. Hrs opn: K-101 Center, 2728 Williams Ave., 73801. Phone: (580) 256-4101. Phone: (580) 256-4101. Fax: (580) 256-3825. E-mail: k101@k101online.com Web Site: k101online.com. Licensee: Omni Communications Corp. Population served: 150,000 Natl. Network: Westwood One, CBS. Law Firm: Womble, Carlyle, Sandridge & Rice. Format: Country. News staff: 2. Target aud: General. ♦J. Douglas Williams, CEO, chmn, pres & gen mgr; Justin Stephenson, CFO; C. J. Montgomery, weather dir. Co-owned TV: KOMI-TV affil.

Oregon

Albany

KGAL(AM)—See Lebanon

KHPE(FM)— Jan 12, 1969: 107.9 mhz; 100 kw. 1,160 ft TL: N44 38 46 W123 16 11. Stereo. Hrs opn: 24 Box 278, 34545 Hwy. 20, 97321. Phone: (541) 926-2233. Fax: (541) 926-3925. Web Site: www.hope1079.com. Licensee: Extra Mile Media Inc. Population served: 1,750,000 Law Firm: Gammon & Grange. Format: Contemp Christian. Target aud: 25-54; female. ♦Bill Zipp, pres & sls dir; Randy Davison, gen mgr & gen sls mgr; Jeff McMahon, opns mgr & progmg dir; John Kenneke, chief of engrg; Vicki Webber, traf mgr.

KWIL(AM)—Co-owned with KHPE(FM). Jan 14, 1941: 790 khz; 1 kw-U, DA-2. TL: N44 37 54 W123 00 57.24 E-mail: randy@hope1079.com Web Site: www.kwil.com. (Acq 7-1-57).705,600 Format: Christian teaching.

KRKT-FM—Listing follows KTHH(AM).

KSHO(AM)—See Lebanon

KTHH(AM)— 1959: 990 khz; 250 w-D. TL: N44 35 43 W123 07 54. Hrs open: 24 2840 Marion St. S.E., 97322. Phone: (541) 926-8628. Fax: (541) 928-1261. Licensee: Bicoastal Willamette Valley LLC. Group owner: Clear Channel Communications Inc. (acq 7-2-2007; grpsl). Population served: 245,000 Natl. Rep: Tacher. Law Firm: Fisher, Wayland, Cooper, Leader & Zaragoza L.L.P. Format: Classic country. News staff: one; News: 10 hrs wkly. Target aud: 25-54. ♦Robert Dove, gen mgr & gen sls mgr; Scott Schuler, adv dir.

KRKT-FM—Co-owned with KTHH(AM). June 1978: 99.9 mhz; 100 kw. Ant 1,069 ft TL: N44 38 46 W123 16 11. Stereo. Web Site: www.krktcountry.com. Format: Country. ♦Scott Schuler, progmg dir.

Altamont

KRAT(FM)—Licensed to Altamont. See Klamath Falls

Ashland

KCMX-FM—Listing follows KTMT(AM).

KIFS(FM)— Nov 25, 1996: 107.5 mhz; 5.8 kw. Ant 1,374 ft TL: N42 17 54 W122 44 53. Hrs open: 3624 Avion Dr., Medford, 97504. Phone: (541) 858-5423. Fax: (541) 857-0326. Web Site: www.107kiss.com. Licensee: Bicoastal Rogue Valley LLC. Group owner: Clear Channel Communications Inc. (acq 7-2-2007; grpsl). Population served: 150,000 Natl. Rep: Tacher. Format: Contemp hit. Target aud: 18-49. ♦Bill Nielsen, gen mgr.

KSJK(AM)—See Talent

***KSMF(FM)**— Nov 7, 1987: 89.1 mhz; 2.3 kw. 1,340 ft TL: N42 17 54 W122 44 59. Stereo. Hrs opn: 5 AM-2 AM 1250 Siskiyou Blvd., 97520. Phone: (541) 552-6301. Fax: 9541) 552-8565. Web Site: www.ijpr.org. Licensee: The State of Oregon, acting by and through the State Board of Higher Education. Population served: 700,000 Natl. Network: NPR, PRI. Law Firm: Ernest Sanchez. Wire Svc: AP Format: Jazz, AAA, news. News staff: one; News: 45 hrs wkly. Target aud: General. Spec prog: Blues 6 hrs, folk 3 hrs, pub affrs 7 hrs wkly. ♦Ronald Kramer, CEO; Mitchell Christian, CFO; Bryon Lambert, opns dir; Jessica Robinson, engrg dir.

***KSOR(FM)**— April 1969: 90.1 mhz; 38 kw. 2,657 ft TL: N42 41 30 W123 13 44. Stereo. Hrs opn: 5 AM-2 AM Southern Oregon University, 1250 Siskiyou Blvd., 97520. Phone: (541) 552-6301. Fax: (541) 552-8565. Web Site: www.ijpr.org. Licensee: The State of Oregon, acting by and through the State Board of Higher Education. Population served: 700,000 Natl. Network: PRI, NPR. Law Firm: Ernest Sanchez. Wire Svc: AP Format: Class, news. News staff: one; News: 35 hrs wkly. Target aud: General. Spec prog: Pub affrs 7 hrs wkly. ♦Ronald Kramer, CEO.

***KSRG(FM)**— 1995: 88.3 mhz; 230 w. 410 ft TL: N42 17 52 W122 44 58. Hrs open: 5 AM- 2 AM Southern Oregon Univ., 1250 Siskiyou Blvd., 97520. Phone: (541) 552-6301. Fax: (541) 552-8565. Web Site: www.ijpr.org. Licensee: The State of Oregon, acting by and through the State Board of Higher Education, for the benefit of Southern Oregon State University. Natl. Network: NPR, PRI. Law Firm: Ernest Sanchez. Wire Svc: AP Format: Div, classical. News staff: one; News: 35 hrs wkly. Target aud: General. ♦Ronald Kramer, CEO & gen mgr.

KTMT(AM)— 1946: 580 khz; 1 kw-U, DA-N. TL: N42 09 46 W122 38 51. Hrs open: 1438 Rossanley Dr., Medford, 97501. Phone: (541) 779-1550. Fax: (541) 776-2360. Licensee: Mapleton Communications LLC (group owner; acq 10-26-01; grpsl). Population served: 150,000 Natl. Network: ABC. Law Firm: Dow, Lohnes & Albertson. Format: Rgnl Mexican. ♦Ron Hren, VP & gen mgr; Jamy Gilinsky, sls dir; Devin Harpole, mktg dir & prom dir; Joe Mussio, mktg mgr; Maria Chaney, traf mgr; Robert Probert, engr.

KCMX-FM—Co-owned with KTMT(AM). July 20, 1978: 101.9 mhz; 31.5 kw. 1,457 ft TL: N42 17 54 W122 44 59. (CP: 31.62 kw, ant 1,426 ft.). Stereo. 24 Web Site: www.lite102.com.250,000 Natl. Network: ABC. Format: Adult contemp. News staff: one. Target aud: 25-54. ♦Casey Baker, progmg dir & mus dir; Kelly Kline, disc jockey.

Astoria

KAST(AM)— 1922: 1370 khz; 1 kw-U, DA-N. TL: N46 10 30 W123 50 50. Hrs open: 5 AM-midnight 1006 W. Marine Dr., 97103. Phone: (503) 325-2911. Fax: (503) 325-5570. Licensee: New Northwest Broadcasters LLC (group owner; (acq 10-26-99; grpsl). Population served: 35,000 Format: News/talk, sports. News: 50 hrs wkly. Target aud: 35 plus. ♦Paul Mitchell, gen mgr.

KKEE(AM)— 1950: 1230 khz; 1 kw-U. TL: N46 11 15 W123 49 30. Hrs open: 24 1006 W. Marine Dr., 97103-5826. Phone: (503) 325-2911. Fax: (503) 325-5570. E-mail: kastam@newnw.com Web Site: www.yourcommunityradio.com. Licensee: New Northwest Broadcasters LLC (group owner; acq 8-24-99; grpsl). Population served: 75,000 Natl. Network: CBS. Format: ESPN. News staff: one; News: 7 hrs wkly. Target aud: 25-54; diverse. ♦Paul Mitchell, gen mgr.

***KLOY(FM)**— 2006: 88.7 mhz; 250 w. Ant 1,053 ft TL: N46 15 46 W123 53 09. Hrs open:
Rebroadcasts KLVR(FM) Santa Rosa, CA 100%.
2351 Sunset Blvd., Suite 170-218, Rocklin, CA, 95765. Phone: (916) 251-1600. Fax: (916) 251-1650. Web Site: www.klove.com. Licensee: Educational Media Foundation. Group owner: EMF Broadcasting (acq 2-2-2004). Natl. Network: K-Love. Format: Contemp Christian. ♦Richard Jenkins, pres; Mike Novak, VP & progmg dir; Lloyd Parker, gen mgr; Ed Lenane, opns dir & news dir; Keith Whipple, dev dir; Eric Allen, natl sls mgr; David Pierce, progmg dir; Jon Rivers, mus dir; Sam Wallington, engrg dir; Arthur Vassar, traf mgr; Karen Johnson, news rptr; Marya Morgan, news rptr; Richard Hunt, news rptr.

***KMUN(FM)**— Feb 2, 1982: 91.9 mhz; 3 kw. 1,060 ft TL: N46 15 46 W123 53 09. Stereo. Hrs opn: 5 AM-1 AM Box 269, 97103. Secondary address: 1445 Exchange St. 97103. Phone: (503) 325-0010. Fax: (503) 325-3956. E-mail: kmun@kmun.org Web Site: www.kmun.org. Licensee: Tillicum Foundation. Population served: 50,000 Natl. Network: NPR. Law Firm: Haley, Bader & Potts. Format: Eclectic. News staff: one; News: 12 hrs wkly. Target aud: General. Spec prog: Folk 18 hrs, children's 6 hrs, Sp 3 hrs, American Indian 2 hrs, Black 2 hrs wkly. ♦Ray Merritt, pres; David Hammock, gen mgr; Stephanie Stern, dev dir.

***KORM(FM)**— 2006: 90.5 mhz; 48 w vert. Ant 469 ft TL: N46 10 56 W123 48 09. Hrs open:
Rebroadcasts KRUC(FM) Las Cruces, NM 100%.
Box 3765, McAllen, TX, 78502. Phone: (956) 787-9788. Fax: (956) 787-9783. Licensee: World Radio Network Inc. (group owner). Format: Sp language/evangelical. ♦Dr. William Haney, gen mgr.

***KWYA(FM)**— 2001: 89.7 mhz; 200 w. Ant 1,027 ft TL: N46 15 46 W123 53 09. Stereo. Hrs opn: 24
KWYQ.
3609 Columbia Heights Rd., Longview, WA, 98632-9585. Phone: (360) 577-5433. E-mail: office@wayfm.com Web Site: kwyq.wayfm.com. Licensee: WAY-FM Media Group Inc. (group owner; acq 8-1-03; $135,000 with KWYQ(FM) Longview, WA). Format: CCM. ♦Danny Houle, gen mgr & stn mgr.

Baker City

***KANL(FM)**— 2005: 90.7 mhz; 250 w. Ant 653 ft TL: N44 45 58 W117 52 54. Hrs open: Box 3206, Tupelo, MS, 38803. Phone: (662) 844-8888. Web Site: www.afr.net. Licensee: American Family Association. Group owner: American Family Radio. Format: Christian. ♦Marvin Sanders, gen mgr.

KBKR(AM)— 1939: 1490 khz; 1 kw-U. TL: N44 47 18 W117 48 35. Hrs open: 24
Rebroadcasts KLBM(AM) La Grande 100%.
Box 907, 2510 E. Cove Ave., La Grande, 97850. Phone: (541) 963-4121. Phone: (541) 963-4122. Fax: (541) 963-3117. E-mail: supertalk@eoni.com Licensee: Pacific Empire Radio Corp. (group owner; acq 7-19-2004; grpsl). Population served: 40,000 Natl. Network: Westwood One. Natl. Rep: McGavren Guild. Format: News/talk. News staff: one; News: 25 hrs wkly. Target aud: 25-54. Spec prog: Farm 2 hrs wkly. ♦Mark Bolland, pres; Steve Ryner, gen mgr & stn mgr; Bobby Hollowwa, progmg dir.

KKBC-FM—Co-owned with KBKR(AM). Feb 1, 1981: 95.3 mhz; 6 kw. -200 ft TL: N44 47 18 W117 48 35. (CP: 25 kw). Stereo. 24 Phone: (541) 523-4431. E-mail: theboomer@eoni.com 18,000 Natl. Rep: McGavren Guild. Format: Oldies. News staff: one; News: 6 hrs wkly. Target aud: 25-54.

KCMB(FM)— June 26, 1988: 104.7 mhz; 100 kw. 1,747 ft TL: N45 07 26 W117 46 48. Stereo. Hrs opn: 1009-C Adams Ave., La Grande, 97850. Secondary address: 2950 Church St. 97814. Phone: (541) 523-3400. Fax: (541) 523-5481. Licensee: Oregon Trail Radio Inc. Group owner: Capps Broadcast Group. Natl. Network: ABC. Natl. Rep: Tacher. Format: Country. Target aud: 25-54. ♦Randy McKone, gen mgr.

***KDJC(FM)**— 2005: 88.1 mhz; 500 w vert. Ant 1,810 ft TL: N45 07 26 W117 46 48. Hrs open: Calvary Chapel La Grande, 1433 Jefferson St., La Grande, 97850. Phone: (541) 963-5884. E-mail: wade.twilegar@csnradio.com Web Site: www.csnradio.com. Licensee: CSN International. (group owner). Format: Relg. ♦Wade Twilegar, gen mgr.

***KOBK(FM)**— 2007: 88.9 mhz; 600 w. Ant 1,834 ft TL: N44 35 57 W117 46 58. Stereo. Hrs opn: 24
Rebroadcasts KOPB-FM Portland 100%.
Oregon Public Broadcasting, 7140 S.W. Macadam Ave., Portland, 97219-3099. Phone: (503) 244-9900. Fax: (503) 293-4877. Web Site: www.opb.org. Licensee: Oregon Public Broadcasting. Natl. Network: NPR. Law Firm: Swarz, Woods & Miller. Format: News/talk. ♦Steve Bass, CEO; Jeff Douglas, opns VP.

Bandon

KBDN(FM)— October 1996: 96.5 mhz; 1.5 kw. 1,296 ft TL: N42 57 27 W124 16 13. Hrs open: 320 Central Ave., Suite 519, Coos Bay, 97420. Phone: (541) 267-2121. Fax: (541) 267-5229. Licensee: Bicoastal CB LLC. Group owner: Bicoastal Media L.L.C. (acq 10-16-2003; grpsl). Natl. Rep: Tacher. Format: Classic rock and roll. Target aud: 35-54. ♦John Pundt, gen mgr; Mike O'Brien, opns mgr & chief of opns.

Banks

KVMX(FM)— June 1990: 107.5 mhz; 35 kw. Ant 1,443 ft TL: N45 30 58 W122 43 59. Hrs open: 24 2040 S.W. First Ave., Portland, 97201. Phone: (503) 497-1075. Fax: (503) 222-2047. Web Site: www.movin1075.com. Licensee: Infinity Radio Inc. Group owner: Infinity Broadcasting Corp. (acq 11-13-98; grpsl). Population served: 1500000 Natl. Network: Westwood One. Format: Adult contemp. Target aud: 25-49; adult. ♦Mark Walen, gen mgr.

Bay City

KIXT(FM)— 2005: 95.9 mhz; 450 w. Ant 1,181 ft TL: N45 27 59 W123 55 11. Hrs open: 1600 Gray Lynn Dr., Walla Walla, WA, 99362. Phone: (509) 527-1000. Fax: (509) 529-5534. Licensee: Alexandra Communications Inc. (acq 8-2-2005; $150,000 for CP). ♦Tom Hodgins, pres & gen mgr.

Beaverton

KKCW(FM)— February 1984: 103.3 mhz; 95 kw. Ant 1,542 ft TL: N45 31 21 W122 44 45. Stereo. Hrs opn: 24 4949 S.W. MacAdam Ave., Portland, 97239. Phone: (503) 222-5103. Fax: (503) 222-0030. Web Site: www.k103.com. Licensee: Citicasters Licenses L.P. Group owner: Clear Channel Communications Inc. (acq 5-4-99; grpsl). Natl. Rep: D & R Radio. Format: Adult contemp. News staff: 3. Target aud: 25-54. ♦Robert Dove, gen mgr; Tony Coles, opns mgr.

Bend

KBND(AM)— 1938: 1110 khz; 10 kw-D, 5 kw-N, DA-N. TL: N44 06 25 W121 14 39. Hrs open: 24 Box 5037, 97708. Secondary address: 711 N.E. Butler Market Rd. 97701. Phone: (541) 382-5263. Fax: (541) 388-0456. E-mail: kbnd@kbnd.com Web Site: www.kbnd.com. Licensee: Combined Communications. (group owner; acq 4-27-90). Population served: 100,000 Natl. Network: CBS. Natl. Rep: McGavren Guild. Law Firm: Dow, Lohnes & Albertson. Format: Sports. News staff: 2. Target aud: 35-64; upscale, professionals. ♦Mike Chaney, gen mgr & gen sls mgr; Frank Bonacquispi, progmg dir.

KLRR(FM)—Co-owned with KBND(AM). June 17, 1985: 101.7 mhz; 27.5 kw. 985 ft TL: N44 04 41 W121 19 57. Stereo. Web Site: www.kbnd.com.125,000 Format: Rock adult contemp. Target aud: 25-54; upscale, professional women & men. Spec prog: Jazz 5 hrs wkly. ♦Doug Danolo, progmg dir.

KICE(AM)— Feb 4, 1960: 940 khz; 10 kw-D, 60 w-N, DA-2. TL: N44 04 50 W121 16 51. Hrs open: 24 969 SW Colorado Ave., 97702. Phone: (541) 388-3300. Fax: (541) 388-3303. Licensee: GCC Bend LLC. (group owner; (acq 1999). Population served: 55000 Natl. Network: ABC. Format: Sports. News staff: one; News: 2 hrs wkly. Target aud: 35 plus. ♦Della Pizzati, gen mgr; Rob Walker, opns dir; Brian Canady, sls dir; Matt Green, engrg dir.

KXIX(FM)—Co-owned with KICE(AM). December 1974: 94.1 mhz; 86 kw. Ant 994 ft TL: N44 02 49 W121 31 50. Stereo. E-mail: clark@x94.com Web Site: www.power94.com. (Acq 2000.)240,000 Format: CHR. Target aud: 18-49. ♦John Gross, pres; Brian Canady, gen sls mgr; Jim Gross, natl sls mgr & chief of engrg; Mike Flanagan, progmg dir; R. L. Garrigus, news dir; Pam Hudspeth, traf mgr.

***KLBR(FM)**—Not on air, target date: unknown: 88.1 mhz; 5 kw. Ant 850 ft TL: N44 02 49 W121 31 50. Hrs open:
Rebroadcasts KLCC(FM) Eugene 100%.
4000 E. 30th Ave., Eugene, 97405-0640. Phone: (541) 463-6000. Fax: (541) 463-6046. Web Site: www.klcc.org. Licensee: Lane Community College. ♦Steve Barton, gen mgr.

KMGX(FM)— July 4, 1973: 100.7 mhz; 50 kw horiz, 20 kw vert. Ant 518 ft TL: N44 04 40 W121 19 49. Stereo. Hrs opn: 705 S.W. Bonnett Way, Suite 1100, 97702. Phone: (541) 388-3300. Fax: (541) 388-3303. E-mail: dhorner@bendradiogroup.com Web Site: www.magic100fm.com. Licensee: GCC Bend LLC. (group owner). Population served: 125000 Rgnl rep: Allied Radio Partners. Format: Adult contemp. Target aud: 25 plus; middle to upper income consumers. ♦John Gross, pres; Dana Horner, gen mgr; Marie McCallister, progmg dir.

KMTK(FM)— 2000: 99.7 mhz; 26 kw. Ant 682 ft TL: N44 04 39 W121 19 57. Hrs open: 24 711 N.E. Butler Market Rd., 97701. Phone: (541) 382-5263. Fax: (541) 388-0456. E-mail: country@mountain997.com Web Site: www.mountain997.com. Licensee: Combined Communications Inc. Group owner: Combined Communications Population served: 150,000 Format: Country. ♦Chuck Chackel, CEO & pres; Mike Cheney, gen mgr.

KNLR(FM)— Dec 31, 1984: 97.5 mhz; 97 kw. Ant 536 ft TL: N44 04 38 W121 19 49. Stereo. Hrs opn: 24 Box 7408, 97708. Phone: (541) 389-8873. Fax: (541) 389-5291. E-mail: info@knlr.com Web Site: www.knlr.com. Licensee: Terry A. Cowan. Natl. Network: USA. Format: Christian. ♦Terry A. Cowan, gen mgr.

***KOAB-FM**— 1994: 91.3 mhz; 25 kw. 604 ft TL: N44 04 41 W121 19 57. Hrs open: 7140 S.W. Macadam, Portland, 97219. Phone: (503) 293-1905. Fax: (503) 293-1919. Web Site: www.opb.org. Licensee: Oregon Public Broadcasting. (acq 9-20-93; grpsl; FTR: 10-11-93). Population served: 110,000 Natl. Network: NPR. Format: News, info music. News staff: news progmg 146 hrs wkly News: 5;. ♦Jack Galmiche, COO, exec VP & gen mgr. Co-owned TV: *KOAB-TV affil.

KQAK(FM)— Sept 5, 1986: 105.7 mhz; 40 kw. Ant 592 ft TL: N44 04 40 W121 19 48. Stereo. Hrs opn: 24 854 N.E. 4th St., 97701. Phone: (541) 383-3825. Fax: (541) 383-3403. Web Site: www.kqakfm.com. Licensee: Horizon Broadcasting Group L.L.C. (group owner; acq 2000; $3.45 million). Population served: 140,000 Natl. Rep: Tacher. Format: Rock and roll oldies. Target aud: 25-54. ♦Keith Shipman, pres & gen mgr; Larry Wilson, opns mgr; Brian Canady, gen sls mgr; Paul Valle, news dir; Ramona Hartmann, traf mgr.

KTWS(FM)— Dec 21, 1990: 98.3 mhz; 5.2 kw. Ant 731 ft TL: N44 04 39 W121 19 57. Stereo. Hrs opn: 24 Box 5037, 97701. Secondary address: 711 N.E. Butler Market Rd. 97701. Phone: (541) 382-5263. Fax: (541) 388-0456. E-mail: thetwins@thetwins.com Web Site: www.thetwins.com. Licensee: Combined Communications, Inc. (group owner; acq 9-1-96). Population served: 150,000 Natl. Rep: McGavren Guild. Format: Classic rock. Target aud: 25-54. ♦Chuck Chackel, pres; Mike Cheney, gen mgr.

***KVLB(FM)**— 2003: 90.5 mhz; 500 w. Ant 564 ft TL: N44 04 40 W121 19 48. Hrs open: 24 2351 Sunset Blvd., Suite 170-218, Rocklin, CA, 95765. Phone: (916) 251-1600. Fax: (916) 251-1650. E-mail: klove@klove.com Web Site: www.klove.com. Licensee: Educational Media Foundation. Group owner: EMF Broadcasting (acq 3-11-03; grpsl). Natl. Network: K-Love. Law Firm: Shaw Pittman. Format: Contemp Christian. News staff: 3. Target aud: 25-44; Judeo Christian, female. ♦Richard Jenkins, pres; Mike Novak, VP; Keith Whipple, dev dir; David Pierce, progmg mgr; Ed Lenane, news dir; Sam Wallington, engrg dir; Karen Vassar, news rptr; Marya Morgan, news rptr; Richard Hunt, news rptr.

Bonanza

KYSF(FM)— 1999: 102.9 mhz; 460 w. Ant 2,106 ft TL: N42 05 48 W121 37 57. Hrs open: Box 339, Klamath Falls, 97601. Phone: (541) 882-8833. Fax: (541) 882-8836. Web Site: www.kiss102fm.com. Licensee: New Northwest Broadcasters LLC (group owner; acq 10-20-98; grpsl). Format: Adult contemp. ♦Pete Benedetti, CEO & pres; Greg Dourian, gen mgr.

Brookings

***KMWR(FM)**— Oct 31, 2002: 90.7 mhz; 100 w. Ant 1,233 ft TL: N42 07 23 W124 17 56. Hrs open: 24
Rebroadcasts KVIP-FM Redding, CA 100%.
1139 Hartnell Ave., Redding, CA, 96002. Phone: (530) 222-4455. Fax: (530) 222-4484. E-mail: info@kvip.org Web Site: www.kvip.org. Licensee: Pacific Cascade Communications Corp. Format: Inspirational, Christian. News staff: 2. ♦David L. Morrow, VP; Steve Hafen, gen mgr & stn mgr; Ted Hering, news dir; Paul Brown, chief of engrg.

KURY(AM)— May 2, 1958: 910 khz; 1 kw-D, 37 w-N. TL: N42 04 32 W124 18 52. Hrs open: 24 Box 1029, 605 Railroad, 97415. Phone: (541) 469-2111. Phone: (541) 469-2112. Fax: (541) 469-6397. E-mail: kury@charbernet.com Licensee: Eureka Broadcasting Co. Inc. (acq 4-19-2005; $775,000 with co-located FM). Population served: 45,000 Natl. Network: Jones Radio Networks. Format: Oldies, country classic. Target aud: General. ♦Hugo Papstein, pres; Rick Wafler, gen sls mgr; Kevin Bane, progmg dir; Vern Garvin, gen mgr, engrg mgr & chief of engrg; Debby Phillips, traf mgr.

KURY-FM— May 1977: 95.3 mhz; 8.7 kw. 1,164 ft TL: N42 07 23 W124 17 56. Stereo. 24 E-mail: kury@charterinternet.com Format: Adult contemp. News staff: one; News: 11 hrs wkly. ♦Vern R. Garvin, stn mgr; Rick Waffler, gen sls mgr; Rich Moore, adv mgr & sports cmtr; Kevin Bane, progmg mgr & disc jockey; Debby Phillips, traf mgr; Bill Dwinell, local news ed; Angela Arneson, disc jockey; Kenny Horn, disc jockey; Paul Flemming, disc jockey; Steve Braun, disc jockey.

Brownsville

KEHK(FM)— Apr 1, 1991: 102.3 mhz; 100 kw horiz, 43 kw vert. 918 ft TL: N44 00 08 W123 06 50. Stereo. Hrs opn: 24 1200 Executive Pkwy., Suite 440, Eugene, 97401. Phone: (541) 485-5846. Fax: (541) 485-0969. Web Site: www.starfm1023.com. Licensee: Cumulus Licensing Corp. Group owner: Cumulus Media Inc. (acq 8-24-00; grpsl). Natl. Network: Jones Radio Networks. Natl. Rep: McGavren Guild. Format: Hot adult contemp. News staff: one; News: 2 hrs wkly. Target aud: 25-54. ♦Bill Bradley, pres.

Burns

KQHC(FM)—Listing follows KZZR(AM).

KZZR(AM)— Sept 28, 1957: 1230 khz; 1 kw-U. TL: N43 33 49 W119 03 22. Hrs open: 24 Box 877, Fairgrounds Rd., 97720. Phone: (541) 573-2055. Fax: (541) 573-5223. Licensee: Action Radio LLC (acq 4-18-2005; $72,500 with co-located FM). Population served: 7,500 Natl. Network: ABC. Natl. Rep: Tacher. Wire Svc: AP Format: Contemp country, news, talk. News: 30 hrs wkly. Target aud: 18-55. Spec prog: Farm 6 hrs wkly. ♦Stan Swol, gen mgr & progmg dir; Ryan Steineckert, mus dir; Toni Carson, traf mgr.

KQHC(FM)—Co-owned with KZZR(AM). Sept 1, 1997: 92.7 mhz; 750 w. 905 ft TL: N43 34 22 W119 07 50. Stereo. 24 E-mail: kzzr_amkqhc_fm@centurytel.net 10,000 Natl. Network: Jones Radio Networks. Natl. Rep: Tacher. Format: Classic hits.

Cannon Beach

KCBZ(FM)— 1997: 94.9 mhz; 7 kw horiz, 1.2 kw vert. Ant 302 ft TL: N45 57 08 W123 56 14. Hrs open: Calcomm Stations Oregon LLC, 615 Broadway, Seaside, 97138. Phone: (503) 738-8668. Fax: (503) 738-8778. E-mail: calbrady@pacbell.net Web Site: www.kcbzfm.com. Licensee: Calcomm Stations Oregon LLC (acq 12-7-2004; $240,000). Format: Hot adult contemp. ♦Cal Brady, gen mgr; Renee Hartford, gen sls mgr; John Chapman, progmg dir.

Canyon City

KJDY-FM— Dec 13, 1996: 94.5 mhz; 51 kw. Ant 1,364 ft TL: N44 17 50 W119 02 09. Stereo. Hrs opn: 24 Box 399, 413 N.W. Bridge St., John Day, 97845. Phone: (541) 575-1185. Fax: (541) 575-2313. Licensee: Blue Mountain Broadcasting Co. Inc. Population served: 8,000 Natl. Network: ABC. Format: Country. Target aud: 25-54. ♦Phil Gray, gen mgr.

Cave Junction

KCNA(FM)— Apr 30, 1985: 102.7 mhz; 100 kw. 1,976 ft TL: N42 15 30 W123 39 38. (CP: 50.7 kw). Hrs opn: 511 Rossanley Dr., Medford, 97501. Phone: (541) 772-0322. Fax: (541) 772-4233. E-mail: jim@opusradio.com Web Site: www.kcnafm.com. Licensee: Opus Broadcasting Systems Inc. (group owner; acq 12-94). Format: Classic hits. ♦Henry Flock, pres; Dean Flock, gen mgr.

Cherryville

***KLVP-FM**— 1997: 88.7 mhz; 3.7 kw. 1,568 ft TL: N45 19 57 W121 42 57. Stereo. Hrs opn: 24
Rebroadcast KLVP(AM) Tigard 100%.
2351 Sunset Blvd., Suite 170-218, Rocklin, CA, 95765. Phone: (916) 251-1600. Fax: (916) 251-1650. E-mail: klove@klove.com Web Site: www.klove.com. Licensee: Educational Media Foundation. Group owner: EMF Broadcasting. Population served: 1,184,000 Natl. Network: K-Love. Law Firm: Shaw Pittman. Format: Contemp Christian. News staff: 3. Target aud: 25-44; Judeo-Christian, female. ♦Richard Jenkins, pres; Mike Novak, VP; Keith Whipple, dev dir; David Pierce, progmg mgr; Ed Lenane, news dir; Sam Wallington, engrg dir; Karen Johnson, news rptr; Marya Morgan, news rptr; Richard Hunt, news rptr.

Condon

KHAL(FM)—Not on air, target date: unknown: 93.5 mhz; 100 kw. Ant 945 ft TL: N45 15 29 W120 17 11. Hrs open: First Broadcasting Operating Inc., 750 N. Saint Paul, 10th Fl., Dallas, TX, 75201. Phone: (214) 855-0002. Fax: (214) 855-5145. Licensee: First Broadcasting Investment Partners LLC. ♦Ronald Unkefer, gen mgr.

Coos Bay

KDCQ(FM)— May 24, 1995: 92.9 mhz; 4.5 kw. Ant 523 ft TL: N43 21 15 W124 14 34. Stereo. Hrs opn: 24 3505 S.E. Ocean Blvd., 97423. Phone: (541) 269-0935. Fax: (541) 267-9376. E-mail: oldies@kdcq.com Web Site: www.kdcq.com. Licensee: Bay Cities Building Co. Inc. Population served: 60,000 Natl. Network: ABC. Natl. Rep: Tacher. Wire Svc: AP Format: Oldies. News: 5 hrs wkly. Target aud: 35-54; baby boomers. Spec prog: Wolfman Jack. ♦Bruce Latta, pres; Stephanie Kilmer, gen mgr; Mikel Chavez, opns mgr.

KHSN(AM)— Mar 15, 1928: 1230 khz; 1 kw-U. TL: N43 22 11 W124 12 54. Hrs open: 24 320 Central, Suite 519, 97420. Phone: (541) 267-2121. Fax: (541) 267-5229. Licensee: W7 Broadcasting LLC (acq 8-7-03). Population served: 65,000 Rgnl rep: Allied Radio Partners. Format: Radio. News: 14 hrs wkly. Target aud: 35- plus. ♦Laura Peck, gen mgr; Mike O'Brien, opns mgr.

***KJCH(FM)**— 2006: 90.9 mhz; 25 kw. Ant 1,476 ft TL: N42 57 32 W124 16 23. Hrs open: 1190 Face Rock Dr., Bandon, 97411. Secondary address: CSN International, 3232 W. MacArthur Blvd., Santa Ana, CA 92704-6802. Phone: (541) 347-2709. Fax: (541) 347-1447. E-mail: joshuatanner@rocketmail.com Licensee: CSN International (group owner). Format: Relg. ♦Joshua Tanner, gen mgr.

KMHS(AM)— Dec 7, 1956: 1420 khz; 1 kw-D. TL: N43 21 45 W124 11 33. Hrs open: 10th & Ingersoll, 97420. Phone: (541) 267-1451. Phone: (541) 267-1420. Fax: (541) 269-0161. E-mail: stevew@coosbay.k12.or.us Web Site: www.marshfield.coos-bay.k12.or.us/kmhs/index.htm. Licensee: Coos Bay School District No. 9. (acq 7-22-97; $8,505 donation). Format: Var. ♦Steve Walker, gen mgr.

***KSBA(FM)**— Nov 4, 1988: 88.5 mhz; 2.2 kw. 532 ft TL: N43 23 26 W124 04 46. Stereo. Hrs opn: 5 AM-2 AM 1250 Siskiyou Blvd., Ashland, 97520. Phone: (541) 552-6301. Fax: (541) 552-8565. Web Site: www.ijpr.org. Licensee: The State of Oregon, acting by and through the State Board of Higher Education. Natl. Network: NPR, PRI. Law Firm: Ernest Sanchez. Wire Svc: AP Format: Jazz, news,

AAA. News staff: one; News: 45 hrs wkly. Target aud: General. Spec prog: Blues 6 hrs, folk 3 hrs, pub affrs 7 hrs wkly. ♦Ronald Kramer, CEO & gen mgr.

KTEE(FM)—See North Bend

KYSJ(FM)— Nov 1, 1979: 106.5 mhz; 4 kw. 544 ft TL: N43 21 15 W124 14 31. Stereo. Hrs opn: 580 Kingwood Ave., 97420. Phone: (541) 269-2022. Fax: (541) 267-0114. E-mail: kysj@lighthouseradio.com Web Site: www.lighthouseradio.com. Licensee: Lighthouse Radio Group. (acq 10-25-93; $64,400; FTR: 11-8-93). Population served: 844,401 Format: Christian. ♦Joshua Tanner, progmg dir; Rick Stevens, gen mgr & opns.

KYTT-FM— November 1978: 98.7 mhz; 31 kw. 551 ft TL: N43 23 26 W124 07 46. (CP: 12.8 kw, ant 962 ft.). Stereo. Hrs opn: 24 580 Kingwood, 97420. Phone: (541) 269-2022. Fax: (541) 267-0114. Web Site: www.lighthouseradio.com. Licensee: Lighthouse Radio Group. (acq 3-1-89). Format: Contemp Christian. News: 8 hrs wkly. ♦Rick Stevens, gen mgr & opns mgr.

Coquille

KSHR-FM—Listing follows KWRO(AM).

KWRO(AM)— Feb 1, 1949: 630 khz; 5 kw-D. TL: N43 10 17 W124 11 54. Hrs open: Box 180, Coos Bay, 97428. Secondary address: 1270 W. 13th 97423. Phone: (541) 396-2121. Fax: (541) 267-5229. E-mail: connie@crbradio.com Web Site: www.southcoastradio.com. Licensee: Bicoastal CB LLC. Group owner: Bicoastal Media L.L.C. (acq 10-16-2003; grpsl). Population served: 61,000 Natl. Rep: McGavren Guild. Format: News/talk. ♦Connie Williamson, gen mgr.

KSHR-FM—Co-owned with KWRO(AM). Nov 1, 1981: 97.3 mhz; 61 kw. 856 ft TL: N43 14 51 W124 06 46. Stereo. Web Site: www.kshr.com. Format: Hot country.

Corvallis

***KBVR(FM)**— Oct 26, 1965: 88.7 mhz; 340 w. -80 TL: N44 33 50 W123 16 30. Stereo. Hrs opn: 24 Oregon State Univ., M.U. East, Snell Hall, Rm. 210, 97331-1618. Phone: (541) 737-2008. Fax: (541) 737-4545. Licensee: State Board of Higher Education. Population served: 45,000 Format: Jazz, urban contemp, alternative rock. Target aud: 15-45; general. Spec prog: Class 4 hrs, Sp 4 hrs, folk 4 hrs wkly. ♦Ian Rose, stn mgr.

KEJO(AM)— August 1955: 1240 khz; 1 kw-U. TL: N44 35 44 W123 14 54. Hrs open: 2840 Marion St. S.E., Albany, 97321-3978. Phone: (541) 926-8628. Fax: (541) 928-1261. Web Site: www.kejoam.com. Licensee: Bicoastal Willamette Valley LLC. Group owner: Clear Channel Communications Inc. (acq 7-2-2007; grpsl). Population served: 76,600 Law Firm: Reddy, Begley & McCormick. Format: Talk. Target aud: 40 plus. ♦Gary Grossman, gen mgr; Glenn Nobel, opns mgr, opns mgr, prom mgr & mus dir; Robert Dove, gen sls mgr; Nicole Meltzer, sports cmtr.

KFLY(FM)—Co-owned with KEJO(AM). Oct 1, 1966: 101.5 mhz; 28 kw. 98 ft TL: N44 35 44 W123 14 54. Stereo. 1345 Olive St., Box1120, 97440. Phone: (541) 485-1120. Fax: (541) 484-5769. Web Site: www.kflyfm.com. Format: Hot adult contemp. Target aud: 25-49; general.

KGAL(AM)—See Lebanon

KLOO(AM)— Aug 23, 1947: 1340 khz; 1 kw-U. TL: N44 33 25 W123 16 22. Hrs open: 24 2840 Marion St. S.E., Albany, 97322. Phone: (541) 926-8628. Fax: (541) 928-1261. Web Site: www.news1340.com. Licensee: Bicoastal Willamette Valley LLC. Group owner: Clear Channel Communications Inc. (acq 7-2-2007; grpsl). Population served: 43,500 Format: News/talk, sports. News staff: one; News: 83 hrs wkly. Target aud: 35-54. ♦Robert Dove, gen mgr; Larry Rogers, gen sls mgr; Rick Rogers, news dir; Robin O'Kelley, chief of engrg.

KLOO-FM— January 1973: 106.1 mhz; 100 kw. 1,140 ft TL: N44 38 45 W123 16 13. Stereo. 24 Web Site: www.kloo.com.650,000 Format: Classic rock. News: 15 hrs wkly. Target aud: 18-54.

***KOAC(AM)**— Dec 7, 1922: 550 khz; 5 kw-U, DA-2. TL: N44 38 12 W123 11 33. Hrs open: 5 AM-midnight 7140 S. W. Macadam Ave., 97219. Phone: (503) 293-1905. Phone: (541) 737-5332. Fax: (503) 293-1919. Licensee: Oregon Public Broadcasting. (acq 9-20-93; grpsl; FTR: 10-11-93). Population served: 1,000,000 Natl. Network: PRI, NPR. Format: News/talk. News staff: 5; News: 40 hrs wkly. Target aud: 25-54; college educated with an interest in news & mus. Spec prog: Jazz 12 hrs wkly. ♦Lynne Clendenin, opns mgr; Roger Dominigues, chief of engrg. Co-owned TV: *KOAC-TV affil

KSHO(AM)—See Lebanon

Cottage Grove

KDPM(FM)— Mar 21, 1994: 100.5 mhz; 6 kw. Ant 115 ft TL: N43 44 41 W123 05 29. Hrs open: 24 321 Main St., 97424. Phone: (541) 942-2468. Fax: (541) 942-5797. Licensee: Diamond Peak Investments LLC (acq 8-31-2005; $350,000). Natl. Rep: Tacher. Format: Adult contemp, light rock, Mexican music. News: 6 hrs wkly. Target aud: 25-49. ♦Diane O'Renick, gen mgr; Paul Schwartzberg, news dir.

KNND(AM)— August 1953: 1400 khz; 950 w-U. TL: N43 45 43 W123 04 42. Hrs open: 24 321 Main St., 97424. Phone: (541) 942-2468. Fax: (541) 942-5797. E-mail: paul@knnd.com Licensee: Schwartzberg Communications Inc. (acq 5-2-2005; $300,000). Population served: 30,000 Natl. Network: AP Radio. Law Firm: Dow, Lohnes & Albertson. Format: Country, news/talk. News staff: one; News: 40 hrs wkly. Target aud: General. Spec prog: Relg 3 hrs wkly. ♦Clif Wilson, progmg dir & disc jockey; Paul Schwartzberg, pres, gen mgr & news dir.

Creswell

KUJZ(FM)— Sept 1, 1983: 95.3 mhz; 625 kw. 1,207 ft TL: N44 00 04 W123 06 45. Stereo. Hrs opn: 1200 Executive Pkwy., Suite 440, Eugene, 97401. Phone: (541) 484-8500. Fax: (541) 485-0969. Web Site: www.953themoose.com. Licensee: Cumulus Licensing Corp. Group owner: Cumulus Media Inc. (acq 2-29-00;; grpsl). Population served: 200,000 Natl. Rep: Christal. Format: Country. Target aud: 18-34. Spec prog: Blues 2 hrs, women 2 hrs, Grateful Dead 2 hrs, Veterans of Foreign Wars 1 hr, loc musicians 1 hr wkly. ♦B.J. O'Brien, gen mgr.

Dallas

KWIP(AM)—Licensed to Dallas. See Salem

Eagle Point

KZZE(FM)— March 1995: 106.3 mhz; 900 w. 1,591 ft TL: N42 21 13 W122 47 05. Hrs open: 3624 Avion Dr., Medford, 97504. Phone: (541) 857-0340. Fax: (541) 857-0326. Web Site: www.kzze.com. Licensee: Bicoastal Rogue Valley LLC. Group owner: Clear Channel Communications Inc. (acq 7-2-2007; grpsl). Population served: 175,000 Natl. Rep: Tacher. Format: Rock/AOR. News staff: one; News: one hr wkly. Target aud: 18-49; rock listeners. ♦Bill Nielsen, pres & gen mgr.

Elgin

KRJT(FM)— 2005: 105.9 mhz; 160 w. Ant 1,916 ft TL: N45 26 26 W117 53 31. Hrs open: 24 Box 907, La Grande, 97850. Secondary address: 2510 E. Cove Ave., La Grande 97850. Phone: (541) 963-4121. Fax: (541) 963-3117. Licensee: Pacific Empire Radio Corp. Natl. Rep: Interep, McGavren Guild. Format: Oldies. Target aud: 25-54. ♦Mark Bolland, pres; Linda Ashlock, gen mgr.

Enterprise

KWVR(AM)— June 1, 1960: 1340 khz; 1 kw-U. TL: N45 26 14 W117 17 30. Hrs open: 24 220 W. Main St., 97828-1244. Phone: (541) 426-4577. Fax: (541) 426-4578. E-mail: kwvramfm@eoni.com Web Site: kwvr.com. Licensee: Wallowa Valley Radio Broadcasting Corp. (acq 9-1-84; grpsl; FTR: 7-2-84). Natl. Network: ABC. Format: News/talk. News staff: news progmg 11 hrs wkly News: 2;. Target aud: General. Spec prog: Farm 4 hrs wkly. ♦Lee D. Perkins, pres, gen mgr & stn mgr; Carol-Lee Perkins, VP & opns mgr; Alyssa Werst, traf mgr; Patrick Channing, II, sls.

KWVR-FM— 1986: 92.1 mhz; 6 kw. Ant -689 ft TL: N45 19 19 W117 13 18. Stereo. 24 Natl. Network: ABC, Jones Radio Networks. Format: Country.

Eugene

KDUK-FM—Florence, Nov 21, 1983: 104.7 mhz; 63 kw. 2,326 ft TL: N44 17 35 W123 32 15. Stereo. Hrs opn: 24 Box 1120, 97401. Secondary address: 1345 Olive St. 977401. Phone: (541) 485-1120. Fax: (541) 484-5769. Web Site: www.kduk.com. Licensee: Bicoastal Willamette Valley LLC. Group owner: Clear Channel Communications Inc. (acq 7-2-2007; grpsl). Format: CHR/Top-40. ♦Robert Dove, gen mgr.

KKNU(FM)—Springfield-Eugene, Dec 18, 1958: 93.3 mhz; 100 kw horiz, 43 kw vert. Ant 1,296 ft TL: N44 00 04 W123 06 45. Stereo. Hrs opn: 24 925 Country Club Rd., Suite 200, 97401. Phone: (541) 484-9400. Fax: (541) 344-9424. Web Site: www.kknu.com. Licensee: McKenzie River Broadcasting Co. Inc. Group owner: McKenzie River Broadcasting Group (acq 11-17-92; $1.01 million with KEED(AM) Eugene; FTR: 12-14-92). Population served: 288,800 Natl. Rep: D & R Radio. Law Firm: Holland & Knight. Format: New country. Target aud: 25-49; country life group. ♦John Tilson, pres & gen mgr; Dave Wiles, gen sls mgr; Jim Davis, progmg dir.

KKNX(AM)— 1992: 840 khz; 1 kw-D, 220 w-N. TL: N44 05 48 W123 04 18. Stereo. Hrs opn: 24 945 Garfield St., 97402. Phone: (541) 342-1012. Fax: (541) 342-6201. E-mail: john@radio84.com Web Site: www.radio84.com. Licensee: John S. Mielke, Susan J. Mielke. (acq 7-18-96; $150,000). Population served: 275,000 Natl. Network: AP Radio. Rgnl rep: Tacher & Co. Wire Svc: AP Format: Oldies. News staff: one; News: 7 hrs wkly. Target aud: 25-64; general. Spec prog: Black 3 hrs wkly. ♦John S. Mielke, pres; John S. Mielke, gen mgr.

***KLCC(FM)**— Feb 17, 1967: 89.7 mhz; 81 kw horiz, 54 kw vert. Ant 1,161 ft TL: N44 00 05 W123 06 48. Stereo. Hrs opn: 24 4000 E. 30th Ave., 97405-0640. Phone: (541) 463-6000. Fax: (541) 463-6046. E-mail: klcc@lanecc.edu Web Site: www.klcc.org. Licensee: Lane Community College. Population served: 750,000 Natl. Network: NPR. Law Firm: Arter & Hadden. Format: News/talk, adult contemp. News staff: one; News: 60 hrs wkly. Target aud: 25-54. Spec prog: Sp 5 hrs, folk 12 hrs, Black 3 hrs, blues 4 hrs, world 3 hrs, electronic 6 hrs wkly. ♦Steve Barton, gen mgr; Paula Chan Carpenter, dev dir; Don Hein, progmg dir; Tripp Sommer, news dir.

KLZS(AM)— Sept 7, 1954: 1450 khz; 1 kw-U. TL: N44 04 54 W123 06 34. Hrs open: 24 925 Country Club Rd., Suite 200, 97401. Phone: (541) 343-4100. Fax: (541) 343-0448. Licensee: Churchill Communications LLC Group owner: McKenzie River Broadcasting Group (acq 11-3-2004; $87,500). Natl. Network: CNN Radio. Format: Progsv talk. News staff: 2; News: 30 hrs wky. Target aud: 25-54.

KMGE(FM)— Oct 10, 1965: 94.5 mhz; 49 kw horiz, 21 kw vert. Ant 1,299 ft TL: N44 00 04 W123 06 45. Stereo. Hrs opn: 925 Country Club Rd., Suite 200, 97401. Phone: (541) 484-9400. Fax: (541) 344-9424. Web Site: www.kmge.com. Licensee: McKenzie River Broadcasting Co. Inc. (acq 3-87; $950,000; FTR: 9-29-86). Population served: 750,000 Format: Adult contemp. Target aud: 18-49. ♦John Tilson, pres.

KNRQ-FM—Listing follows KUGN(AM).

KODZ(FM)—Listing follows KPNW(AM).

KOPT(AM)— Sept 19, 1947: 1600 khz; 5 kw-D, 1 kw-N, DA-N. TL: N44 03 05 W123 03 48. Hrs open: 895 Country Club Rd., Suite A200, 97401. Phone: (541) 343-4100. Fax: (541) 343-0448. Licensee: Churchill Communications I LLC. (acq 3-16-2005; $225,000). Format: Progressive talk.

KORE(AM)—See Springfield-Eugene

KPNW(AM)— July 22, 1968: 1120 khz; 50 kw-U, DA-1. TL: N43 57 24 W123 02 10. Hrs open: 24 Box 1120, 97440. Secondary address: 1345 Olive St. 97440. Phone: (541) 485-1120. Fax: (541) 484-5769. Web Site: www.kpnw.com. Licensee: Bicoastal Willamette Valley LLC. Group owner: Clear Channel Communications Inc. (acq 7-2-2007; grpsl). Population served: 250,000 Format: News/talk. News staff: 2. Target aud: 25 plus; upper income, conservative. Spec prog: Sports. ♦Robert Dove, gen mgr; Lee Chabre, progmg dir.

KODZ(FM)—Co-owned with KPNW(AM). November 1968: 99.1 mhz; 100 kw. 1,945 ft TL: N44 06 56 W122 59 56. Stereo. 250,000 Format: Oldies. Target aud: 25-54; working women. ♦Robert Dove, sls dir; Paul Walker, progmg dir; Scott Howard, traf mgr.

***KRVM(AM)**— Nov 9, 1949: 1280 khz; 5 kw-D, 1.5 kw-N, DA-N. TL: N44 06 03 W123 03 06. Hrs open: 24 P.M.B. 237, 1574 Cobug Rd., 97401. Phone: (541) 687-3370. Fax: (541) 687-3573. Web Site: www.krvm.org. Licensee: Lane County School District 4J. (acq 1-10-97). Natl. Network: NPR. Natl. Rep: McGavren Guild. Format: Talk. ♦Randy Larson, gen mgr.

KRVM-FM— Dec 8, 1947: 91.9 mhz; 1.1 kw. Ant 745 ft TL: N44 00 08 W123 06 50. Stereo. 24 Web Site: www.krvm.org.200,000 Natl. Network: NPR. Format: AAA, AOR, blues. Target aud: General. Spec prog: Black 2 hrs, country one hr, folk 3 hrs, Native American 2 hrs wkly.

KSCR(AM)— June 12, 1962: 1320 khz; 1 kw-D, 40 w-N. TL: N44 05 25 W123 06 43. Hrs open: 1200 Executive Pkwy., Suite 440, 97401. Phone: (541) 485-5846. Fax: (541) 485-0969. Licensee: Cumulus Licensing Corp. Group owner: Cumulus Media Inc. (acq 2-29-00; grpsl). Population served: 200,000 Natl. Network: ESPN Radio. Rgnl. Network: Christal. Natl. Rep: Christal. Format: Sports. Target aud: 18-44; general. ♦Steve Ries, gen mgr.

KUGN(AM)— July 4, 1946: 590 khz; 5 kw-D, 5 kw-N, DA-N. TL: N44 05 48 W123 04 18. Stereo. Hrs opn: 24 1200 Executive Pkwy., Suite 440, 97405. Phone: (541) 284-8500. Fax: (541) 284-8500. Web Site: www.kugn.com. Licensee: Cumulus Licensing Corp. Group owner: Cumulus Media Inc. (acq 6-15-00; grpsl). Population served: 220,000 Natl. Network: CBS. Law Firm: Dow, Lohnes & Albertson. Wire Svc: NWS (National Weather Service) Format: News/talk. News staff: 6; News: 28 hrs wkly. Target aud: 30-65; general. ♦Bill Bradley, gen mgr; Troy Murphy, gen sls mgr; Wendy Wintrode, prom dir; Jerry Allen, progmg dir; Rick Little, news dir; Cory Schruth, chief of engrg.

KNRQ-FM—Co-owned with KUGN(AM). Dec 26, 1958: 97.9 mhz; 100 kw. 1,230 ft TL: N44 00 08 W123 06 50. Stereo. Web Site: www.nrq.com. Format: Alternative. ♦Chris Crowley, progmg dir.

***KWAX(FM)**— Apr 4, 1951: 91.1 mhz; 21.5 kw horiz, 12.5 kw vert. Ant 1,214 ft TL: N44 00 04 W123 06 45. Stereo. Hrs opn: 24 Agate Hall, Univ. of Oregon, 97403. Phone: (541) 345-0800. Web Site: www.kwax.com. Licensee: State Board of Higher Education. Population served: 267,000 Law Firm: Akin, Gump, Strauss, Hauer & Feld. Format: Class. News: 7 hrs wkly. Target aud: 35 plus. ♦Paul C. Bjornstad, gen mgr; Rocky Lammana, opns mgr.

***KWVA(FM)**— May 27, 1993: 88.1 mhz; 500 w. -56 ft TL: N44 00 07 W123 06 53. Hrs open: 24 Box 3157, ERB Memoral Union, Univ. of Oregon, 97403. Secondary address: Univ. of Oregon, EMU, Suite M-112 97403. Phone: (541) 346-4091. Fax: (541) 346-0648. E-mail: kwva@gladstone.uoregon.edu Web Site: gladstone.uoregon.edu/~kwva. Licensee: Associated Students of University of Oregon. (acq 6-29-92; FTR: 7-20-92). Population served: 275,000 Format: Div, progsv, urban contemp. News: 12 hrs wkly. Target aud: 3-30; college, alternative, underrepresented, varying educ levels & music lover. Spec prog: Asian 4 hrs, Black 4 hrs, jazz 6 hrs, country 3 hrs, Japanese 2 hrs, Sp 6 hrs wkly. ♦Charlotte Nisser, gen mgr, opns dir & dev dir; Michael Zarkesh, prom dir; Anna Jensen, progmg dir.

KZEL-FM— Apr 22, 1962: 96.1 mhz; 100 kw. 1,093 ft TL: N44 00 04 W123 06 45. Stereo. Hrs opn: 1200 Executive Way, ste 440, 97401. Phone: (541) 284-8500. Fax: (541) 485-4070. Web Site: www.96kzel.com/main. Licensee: Cumulus Licensing Corp. Group owner: Cumulus Media Inc. (acq 2-29-00; grpsl). Population served: 221,000 Natl. Network: Westwood One. Natl. Rep: Christal. Format: Classic Rock. Target aud: 18-44. ♦Steve Ries, gen mgr; Russ Davidson, opns mgr.

Florence

KCST(AM)— May 5, 1985: 1250 khz; 1 kw-D, 68 w-N. TL: N44 00 38 W124 05 37. Hrs open: 24 Box 20000, 97439. Secondary address: Radio Center Bldg., 4480 Hwy. 101 N. 97439. Phone: (541) 997-9136. Fax: (541) 997-9165. E-mail: radioway@kcst.com Web Site: www.kcst.com. Licensee: Coast Broadcasting Co. Inc. (acq 12-18-97). Format: Music of your life. Target aud: 55+. ♦John Thompson, gen mgr.

KCST-FM— October 1992: 106.9 mhz; 2.3 kw. 508 ft TL: N43 57 19 124 04 26. Stereo. 24 Box 20,000, 97439. Secondary address: 4480 Hwy 101 N., Radio Centre Bldg. 97439. Phone: (541) 997-9136. Fax: (541) 997-9165. E-mail: radiowaves@kcst.com Web Site: www.kcst.com. Natl. Network: ABC. Rgnl rep: Tacher Company Format: Adult contemp, country, oldies. Target aud: 35+.

KDUK-FM—Licensed to Florence. See Eugene

***KLFO(FM)**— Aug 16, 1999: 88.1 mhz; 250 w. Ant 548 ft TL: N43 57 26 W124 04 26. Stereo. Hrs opn: 24
Rebroadcasts KLCC(FM) Eugene 100%.
Lane Community College, 4000 E. 30th Ave., Eugene, 97405. Phone: (541) 463-6000. Fax: (541) 463-6046. Web Site: www.klcc.org. Licensee: Lane Community College. Format: Var/div, jazz, news. ♦Steve Barton, gen mgr & stn mgr; Paula Chan Carpenter, dev dir; Don Hein, progmg dir; Chris Heck, chief of engrg.

***KWVZ(FM)**— 2001: 91.5 mhz; 150 w. Ant 548 ft TL: N43 57 26 W124 04 26. Hrs open:
Rebroadcasts KWAX(FM) Eugene 100%.
c/o KWAX(FM), 75 Centennial Loop, Eugene, 97401. Phone: (541) 345-0800. Licensee: Oregon State Board of Higher Education. Format: Classical. ♦Paul C. Bjornstad, gen mgr.

Garibaldi

KDEP(FM)— 2001: 105.5 mhz; 320 w. Ant 1,181 ft TL: N45 27 59 W123 55 11. Hrs open: 1000 Main Ave. N., Ste 5, Tillamook, 97141-9272. Secondary address: 1550 N. Main, Tillamook 97141. Phone: (503) 842-3888. Fax: (503) 842-5640. E-mail: coast105@earthlink.net Licensee: Alexandra Communications Inc. (acq 9-21-2005; $250,000). Format: Soft rock. ♦Chris Gilbreth, gen mgr.

Gladstone

KRYP(FM)— May 10, 1981: 93.1 mhz; 1.55 kw. Ant 1,269 ft TL: N45 29 20 W122 41 40. Stereo. Hrs opn: 6400 S.E. Lake Rd., Suite 350, Portland, 97222. Phone: (503) 786-0600. Fax: (503) 786-1551. Licensee: Salem Media of Oregon Inc. (acq 1-19-2005). Population served: 45,000 Format: Rgnl Mexican. ♦Dennis Hayes, gen mgr.

Gleneden Beach

***KOGL(FM)**—Not on air, target date: unknown: 89.3 mhz; 210 w. Ant 73 ft TL: N44 53 08 W124 00 51. Hrs open: 7140 S.W. Macadam Ave., Portland, 97219. Phone: (503) 293-1905. Fax: (503) 293-1919. Web Site: www.opb.org. Licensee: Oregon Public Broadcasting. ♦Jack Galmiche, COO; Maynard Orme, gen mgr.

***KQAC(FM)**—Not on air, target date: unknown: 88.5 mhz; 100 w. Ant 66 ft TL: N44 53 08 W124 00 51. Hrs open: KBPS Public Radio, 515 N.E. 15th Ave., Portland, 97232. Phone: (503) 943-5828. Fax: (503) 802-9456. Web Site: www.allclassical.org. Licensee: KBPS Public Radio Foundation. ♦James Draznin, pres.

KSHL(FM)— December 1992: 97.5 mhz; 17 kw. 843 ft TL: N44 45 22 W124 02 57. Stereo. Hrs opn: 24 Box 1180, 131 N.E. 15th St., Newport, 97365. Phone: (541) 265-6477. Fax: (541) 265-6478. E-mail: news@ksh.com Web Site: www.kshl.com. Licensee: Stephanie Linn. Rgnl rep: McGavren Guild Format: Modern country. News: 2 hrs wkly. Target aud: 25-55; general. ♦Dick Linn, gen mgr & opns mgr; Stephanie Linn, pres & gen sls mgr.

Gold Beach

KGBR(FM)— December 1984: 92.7 mhz; 265 w. 1,030 ft TL: N42 23 50 W124 21 50. (CP: 42.06 kw, ant 2,700 ft. TL: N42 23 44 W124 21 47). Hrs opn: 24 Box 787, BetGar Bldg., 29795 Ellensburg Ave., 97444. Phone: (541) 247-7211. Phone: (541) 247-7418. Fax: (541) 247-4155. E-mail: info@kgbr.com Web Site: www.ontheradio.net /radiostations/kgbrfm.aspx. Licensee: St. Marie Communications Inc. (acq 3-87; $60,000; FTR: 11-16-87). Population served: 20,000 Natl. Rep: Tacher. Law Firm: Fisher, Wayland, Cooper, Leader & Zaragoza. Format: Adult contemp. News staff: 2; News: 4 hrs wkly. Target aud: 25-54. ♦Dale L. St. Marie, gen mgr.

Gold Hill

KRWQ(FM)— Aug 11, 1980: 100.3 mhz; 30 kw. Ant 991 ft TL: N42 27 07 W123 03 20. Stereo. Hrs opn: 24 3624 Avion Dr., Medford, 97504. Phone: (541) 772-4170. Fax: (541) 858-5416. Web Site: www.krwq.com. Licensee: Bicoastal Rogue Valley LLC. Group owner: Clear Channel Communications Inc. (acq 7-2-2007; grpsl). Population served: 250,000 Natl. Rep: Tacher. Format: Contemp country. News staff: one. Target aud: 18-54. ♦Bill Nielsen, gen mgr.

Grants Pass

***KAGI(AM)**— Dec 16, 1939: 930 khz; 5 kw-D, 123 w-N. TL: N42 25 24 W123 20 04. Hrs open: 24 hrs Jefferson Public Radio, 1250 Siskiyou Blvd., Ashland, 97520. Phone: (541) 552-6301. Fax: (541) 552-8565. Web Site: www.ijpr.org. Licensee: The State of Oregon, acting by and through the State Board of Higher Education, for the benefit of Southern Oregon University. (acq 7-11-91; FTR: 7-29-91). Population served: 50,000 Natl. Network: PRI, NPR. Law Firm: Ernest Sanchez. Format: News, info. Spec prog: Sp 6 hrs wkly. ♦Mitchell Christian, CFO; Bryon Lambert, opns dir; Paul Westhelle, dev dir & mktg dir; Ronald Kramer, CEO, gen mgr & progmg dir; Eric Alan, mus dir; Darin Ransom, engrg dir.

KAJO(AM)— Aug 15, 1957: 1270 khz; 10 kw-D, 48 w-N. TL: N42 26 16 W123 21 27. Stereo. Hrs opn: 24 888 Rogue River Hwy., 97526. Phone: (541) 476-6608. Fax: (541) 476-4018. E-mail: kajo@kajo.com Web Site: www.kajo.com. Licensee: Grants Pass Broadcasting Corp. Population served: 70,000 Natl. Network: ABC, AP Radio. Natl. Rep: Tacher. Rgnl rep: Tacher Wire Svc: AP Format: Adult standards, news/talk. News staff: one; News: 22 hrs wkly. Target aud: 35 plus. Spec prog: Gospel one hr, relg 8 hrs wkly. ♦Matt Wilson, CEO & gen sls mgr; Carl Wilson, CFO; Marty Sether, prom dir & progmg dir; Joe Torsistano, chief of engrg; Will Schacher, mus dir & prom.

***KAPK(FM)**— April 1998: 91.1 mhz; 250 w. 13 ft TL: N42 27 44 W123 18 33. Hrs open: Box 3206, American Family Radio, Tupelo, MS, 38803. Phone: (662) 844-8888. Fax: (662) 842-6791. Web Site: www.afr.net. Licensee: American Family Association. Group owner: American Family Radio Format: Inspirational Christian. ♦Marvin Sanders, gen mgr.

KROG(FM)— Oct 2, 1981: 96.9 mhz; 25 kw. 2,058 ft TL: N42 22 56 W123 16 29. (CP: 74 kw). Stereo. Hrs opn: 24 511 Rossanley Dr., Medford, 97501. Phone: (541) 772-0322. Fax: (541) 772-4233. E-mail: krogstudio@yahoo.com Web Site: www.97therogue.com. Licensee: Opus Broadcasting Systems Inc. (group owner; acq 3-6-91; $63,634 with KRTA(AM) Medford; FTR: 7-29-91). Natl. Rep: Tacher. Format: Alternative. News staff: one. Target aud: 25-54; affluent middle America. ♦Dean Flock, gen mgr & stn mgr; Dave Hatton, opns dir.

Gresham

***KMHD(FM)**— January 1984: 89.1 mhz; 7.9 kw. Ant 1,433 ft TL: N45 30 58 W122 43 59. Stereo. Hrs opn: 24 26000 S.E. Stark St., 97030. Phone: (503) 661-8900. Phone: (503) 491-7233. Fax: (503) 491-6999. E-mail: station-manager@knhd.fm Web Site: www.kmhd.org. Licensee: Mt. Hood Community College. Population served: 1,850,000 Natl. Network: NPR. Law Firm: Garvey, Schubert & Barer. Format: Jazz/blues. News: 5 hrs wkly. Target aud: 35-65; music lovers. Spec prog: Blues 15 hrs, news 5 hrs wkly. ♦Doug Sweet, gen mgr; Dan Gurin, opns dir; Calvin Walker, dev dir; Greg Gomez, mus dir.

KMUZ(AM)—Licensed to Gresham. See Portland

Harbeck-Fruitdale

KLDR(FM)— May 3, 1991: 98.3 mhz; 185 w. 2,096 ft TL: N42 22 56 W123 16 29. Stereo. Hrs opn: 24 888 Rogue River Hwy., Grants Pass, 97527. Phone: (541) 474-7292. Fax: (541) 474-7300. E-mail: kldr@kldr.com Web Site: www.kldr.com. Licensee: Grants Pass Broadcasting Corp. Population served: 100,000 Format: Adult contemp. News staff: one; News: 10 hrs wkly. Target aud: 25-54; Middle Age demo- actually a wide range in listeners. ♦Jim Wilson, pres; Carl Wilson, gen mgr; Matt Wilson, gen sls mgr; Jason Allen, progmg dir.

Hermiston

KOHU(AM)— Feb 6, 1956: 1360 khz; 4.3 kw-D, 500 w-N, DA-N. TL: N45 51 57 W119 18 45. Hrs open: 24 Box 145, 80404 Cooney Ln., 97838. Phone: (541) 567-6500. Fax: (541) 567-6068. Licensee: Westend Radio L.L.C. (acq 4-16-97; with co-located FM). Population served: 30,000 Natl. Network: ABC. Natl. Rep: Farmakis. Rgnl rep: Target. Format: C&W. News staff: one; News: 10 hrs wkly. Target aud: General; two county loc audience. Spec prog: Sp 6 hrs wkly. ♦Angela

Pursel, gen mgr, stn mgr, sls VP, gen sls mgr, farm dir & farm dir; Ron Hughes, pres & gen mgr; Jeff Walker, opns dir, prom dir & progmg dir; Adam Russell, news dir; Richard Wilson, engrg dir; Pam Rebman, traf mgr.

KQFM(FM)—Co-owned with KOHU(AM). Sept 18, 1978: 100.5 mhz; 5.3 kw. Ant 298 ft TL: N45 51 57 W119 18 38. Stereo. 24 175,000 Natl. Network: ABC. Format: Pure gold, adult contemp. News staff: one; News: 5 hrs wkly. Target aud: 25-54. ♦Ron Hughes, gen mgr & prom VP; Jeff Walker, mus dir; Pam Rebman, traf mgr; Angela Pursel, farm dir.

Hillsboro

KUIK(AM)— 1954: 1360 khz; 5 kw-U, DA-N. TL: N45 29 13 W122 54 31. Hrs open: 24 Box 566, 97123. Secondary address: 3355 N.E. Cornell Rd. 97124. Phone: (503) 640-1360. Fax: (503) 640-6108. E-mail: dave@kuik.com Web Site: www.kuik.com. Licensee: Dolphin Communications Inc. (group owner; (acq 8-1-78). Population served: 455,000 Law Firm: David Tillotson. Wire Svc: AP Format: News/talk, sports, Sp. News staff: one; News: 24 hrs wkly. Target aud: 25-54; Seekers of locally produced unique programming. Spec prog: Relg 2 hrs, Sp 21 hrs wkly. ♦Don McCoun, pres & gen mgr; Donna McCoun, sr VP; Paul Warren, opns mgr.

Hood River

KCGB-FM—Listing follows KIHR(AM).

KIHR(AM)— Oct 17, 1950: 1340 khz; 1 kw-U. TL: N45 42 07 W121 32 10. Hrs open: 24 Box 360, 97031. Secondary address: 1190 22nd St. 97031. Phone: (541) 386-1511. Fax: (541) 386-7155. Web Site: www.kihrk105.com. Licensee: Columbia Gorge Broadcasters Inc. (group owner; (acq 4-1-67). Population served: 18,334 Law Firm: Fisher, Wayland, Cooper, Leader & Zaragoza L.L.P. Format: C&W. News staff: 3; News: 12 hrs wkly. Target aud: 25-54. ♦Gary Grossman, pres, gen mgr & stn mgr; Rick Cavagnaro, sls VP & sls dir; Jeff Skye, mus dir & disc jockey; Mark Bailey, news dir, local news ed, edit dir & sports cmtr; Jim Keightley, engrg dir & chief of engrg; Gwen Troutner, traf mgr; Ismael Pinedo, spanish dir.

KCGB-FM—Co-owned with KIHR(AM). Dec 4, 1978: 105.5 mhz; 3 kw. -460 ft TL: N45 42 07 W121 32 10.14,934 Natl. Network: ABC. Format: Hot adult contemp. Target aud: 18-49. ♦Jeff Skye, progmg dir & farm dir; Gwen Troutner, traf mgr; Mark Bailey, local news ed.

***KQHR(FM)**— January 2002: 90.1 mhz; 44 w. Ant 1,105 ft TL: N45 43 20 W121 26 16. Hrs open: 24
Rebroadcasts KBPS-FM Portland 100%.
515 N.E. 15th Ave., Portland, 97323. Phone: (503) 916-5828. Fax: (503) 916-2642. E-mail: musicinfo@allclassical.org Web Site: www.allclassical.org. Licensee: KBPS Public Radio Foundation. Population served: 30,000 Format: Class. News: 5 hrs wkly. ♦Suzanne White, gen mgr.

John Day

KJDY(AM)— Dec 13, 1963: 1400 khz; 1 kw-U. TL: N44 25 17 W118 57 09. Hrs open: 24
KJDY-FM.
413 N.W. Bridge St.-, ., 97845. Phone: (541) 575-1400. Fax: (541) 575-2313. E-mail: kjdy@centurytel.net Licensee: Blue Mountain Broadcasting Co. (acq 11-87; $150,000; FTR: 11-9-87). Population served: 9,500 Natl. Network: ABC. Rgnl rep: Tacher Law Firm: J. Dominic Monahan. Format: C&W. News staff: one. ♦Phil Gray, gen mgr; Patricia Webb, gen sls mgr; Kelly Workman, progmg dir; J. Kelly Carlson, engrg VP & chief of engrg.

Jordan Valley

***KIDH(FM)**— 2005: 90.9 mhz; 21.5 kw vert. Ant 2,161 ft TL: N43 00 26 W116 42 23. Stereo. Hrs opn: 24 2351 Sunset Blvd., Suite 170-218, Rocklin, CA, 95765. Phone: (916) 251-1600. Fax: (916) 251-1650. E-mail: info@air1.com Web Site: www.air1.com. Licensee: Educational Media Foundation. Group owner: EMF Broadcasting. Natl. Network: Air 1. Law Firm: Shaw Pittman. Format: Contemp Christian. News staff: 3. Target aud: 18-35; Judeo-Christian, female. ♦Richard Jenkins, pres; Mike Novak, VP; Keith Whipple, dev dir; David Pierce, progmg mgr; Ed Lenane, news dir; Sam Wallington, engrg dir; Karen Johnson, news rptr; Marya Morgan, news rptr; Richard Hunt, news rptr.

Junction City

***KPIJ(FM)**—Not on air, target date: unknown: 88.5 mhz; 630 w. Ant 2,312 ft TL: N44 16 48 W123 34 57. Hrs open: CSN International, 3000 W. MacArthur Blvd., Santa Ana, CA, 92704. Fax: (714) 825-9660. Licensee: CSN International.

KXOR(AM)— 1998: 660 khz; 10 kw-D, 75 w-N. TL: N44 12 36 W123 10 56. Hrs open: 24 895 Country Club Rd., Suite A200, Eugene, 97401. Phone: (541) 343-4100. Fax: (541) 343-0448. Web Site: www.lax660.com. Licensee: Churchill Communications LLC. Group owner: Pamplin Broadcasting (acq 1-14-2005; $550,000). Population served: 270,000 Natl. Rep: Univision Radio National Sales. Rgnl rep: Julie Schneidar & Paul Danitz Wire Svc: AP Format: Sp. News staff: 1; News: 6am-6pm on the hour. Target aud: 25-54; 18-49. ♦Paul Danitz, gen mgr, progmg dir & mus dir; Phil Polter, gen sls mgr.

Keizer

KYKN(AM)— 1951: 1430 khz; 5 kw-U. TL: N44 55 36 W122 57 19. Hrs open: 24 Box 1430, Salem, 97308. Secondary address: 4205 Cherry Ave. N.E. 97303. Phone: (503) 390-3014. Fax: (503) 390-3728. E-mail: mfrith@kykn.com Web Site: www.kykn.com. Licensee: Willamette Broadcasting Co. Inc. (acq 10-27-01). Population served: 500,000 Rgnl rep: Interep Format: News/talk. News staff: 3; News: 46 hrs wkly. Target aud: 25-64; $40-60K income, homeowners, white collar. Spec prog: Portland Trailblazers, University of Oregon, gospel 6 hrs wkly. ♦Michael Frith, pres, gen mgr & gen sls mgr.

Klamath Falls

KAGO(AM)— July 19, 1923: 1150 khz; 5 kw-D, 1 kw-N, DA-N. TL: N42 12 56 W121 47 51. Hrs open: 24 Box 339, 97601. Phone: (541) 882-2551. Fax: (541) 882-8836. Licensee: New Northwest Broadcasters LLC (group owner; acq 3-16-99; $1.6 million with co-located FM). Population served: 100,000 Natl. Network: CBS. Law Firm: Dan Alpert. Format: News/talk. News staff: 2; News: 25 hrs wkly. Target aud: 35-65; upscale, professional. Spec prog: Farm 3 hrs, Sp 5 hrs wkly. ♦Gregory J. Dourian, gen mgr & gen sls mgr.

KAGO-FM— Oct 15, 1973: 99.5 mhz; 60 kw. 360 ft TL: N42 12 56 W121 47 56. (CP: 100 kw, ant 994 ft. TL: N42 13 08 W121 48 56). Stereo. 24 Natl. Network: CBS. Format: Classic rock. ♦Rob Siems, prom dir; Vivian Dickey, pub affrs dir & traf mgr; Michael O'Shea, sports cmtr.

KFEG(FM)— 2002: 104.7 mhz; 51 kw. Ant 645 ft TL: N42 13 24 W121 49 02. Hrs open: 24 Box 938, 97601. Phone: (541) 850-5242. Fax: (541) 884-2845. E-mail: sales@theeagle1047.fm Web Site: www.theeagle1047.fm. Licensee: Cove Road Publishing LLC (acq 3-1-01). Format: Classic rock. Target aud: 25-54. ♦Bill Ifft, pres & gen mgr.

KFLS(AM)— 1946: 1450 khz; 1 kw-U. TL: N42 12 19 W121 46 04. Stereo. Hrs opn: 24 Box 1450, 1338 Oregon Ave., 97601. Phone: (541) 882-4656. Fax: (541) 884-2845. Web Site: www.klamathradio.com. Licensee: Wynne Enterprises LLC (group owner; acq 1-1-71). Population served: 50,000 Natl. Network: ABC. Rgnl rep: Tacher. Format: News/talk, sports. Target aud: 35 plus. ♦Robert Wynne, CEO, chmn, pres, gen mgr & gen sls mgr.

KKRB(FM)—Co-owned with KFLS(AM). Apr 1, 1983: 106.9 mhz; 100 kw. 1,200 ft TL: N42 13 26 W121 49 02. Stereo. Format: Adult contemp, top 40.

***KKLJ(FM)**— Mar 14, 2003: 88.9 mhz; 32 w. Ant 2,132 ft TL: N42 03 54 W121 58 14. Stereo. Hrs opn: 24 2351 Sunset Blvd., Suite 170-218, Rocklin, CA, 95765. Phone: (916) 251-1600. Fax: (916) 251-1650. E-mail: klove@klove.com Web Site: www.klove.com. Licensee: Educational Media Foundation. Group owner: EMF Broadcasting. Natl. Network: K-Love. Law Firm: Shaw Pittman. Format: Contemp Christian. News staff: 3. Target aud: 25-44; Judeo Christian, female. ♦Richard Jenkins, pres; Mike Novak, VP; Lloyd Parker, gen mgr; Keith Whipple, dev dir; Eric Allen, natl sls mgr; David Pierce, progmg mgr; Ed Lenane, news dir; Sam Wallington, engrg dir; Arthur Vassar, traf mgr.

KLAD(AM)— September 1955: 960 khz; 5 kw-U. TL: N42 09 42 W121 39 01. Hrs open: 24 Box 339, 97601. Secondary address: 4509 S. 6th St., Suite 201 97601. Phone: (541) 882-8833. Fax: (541) 882-8836. Licensee: New Northwest Broadcasters LLC (group owner; (acq 10-20-98; grpsl). Population served: 50,000 Format: Sports. News staff: one; News: 3 hrs wkly. Target aud: 25-54; mature with spendable income. Spec prog: Farm one hr wkly.Brent Phillipy, stn mgr; Rob Siems, opns mgr, progmg dir & disc jockey; Greg Dourian, sls dir & adv dir; Jamie Jackson, news dir, local news ed, news rptr & disc jockey; Scott Noland, pub affrs dir, sports cmtr & disc jockey; James Boyd, chief of engrg; Vivian Dicky, traf mgr; Brian Mobley, mus critic & disc jockey; Jamie Jacdson, women's int ed; Josh Thomas, disc jockey

KLAD-FM— July 19, 1974: 92.5 mhz; 63 kw. Ant 2,188 ft TL: N42 05 51 W121 37 58. E-mail: rodeoradio@aol.com Natl. Network: ABC. Rgnl rep: Allied Radio Partners. Law Firm: Dow, Lohnes & Albertson. Format: Country. ♦Michael O'Shea, CEO; Ivan Braiker, pres; Tricia Houston, CFO; Scott Allen, VP; Rob Siems, opns dir, mus dir, farm dir & disc jockey; Vivian Dickey, traf mgr; Brian Mobley, mus critic & disc jockey; Scott Noland, sports cmtr & disc jockey; Josh Thomas, disc jockey.

***KLMF(FM)**— 2002: 88.5 mhz; 95 w. Ant 2,162 ft TL: N42 05 50 W121 37 59. Hrs open: Jefferson Public Radio, 1250 Siskiyou Blvd., Ashland, 97520. Phone: (541) 552-6301. Phone: (541) 552-8565. Web Site: www.ijpr.org. Licensee: The State of Oregon, acting by and through the State Board of Higher Education, for the benefit of Southern Oregon University. Natl. Network: NPR, PRI. Law Firm: Ernest Sanchez. Wire Svc: AP Format: Class, news. News staff: one; News: 35 hrs wkly. ♦Mitchell Christian, CFO; Ronald Kramer, CEO & gen mgr; Bryon Lambert, opns dir; Paul Westhelle, dev dir.

KRAT(FM)—Altamont, 1991: 97.7 mhz; 22 kw. Ant 1,712 ft TL: N42 10 06 W122 09 06. Stereo. Hrs opn: Box 235, 97601. Phone: (541) 884-8167. Fax: (541) 884-8226. Licensee: George J. Wade. Format: Oldies.

***KSKF(FM)**— Nov 10, 1989: 90.9 mhz; 2 kw. 2,253 ft TL: N42 05 50 W121 37 59. Stereo. Hrs opn: 5 AM-2 AM 1250 Siskiyou Blvd., Ashland, 97520. Phone: (541) 552-6301. Fax: (541) 552-8565. Web Site: ww.ijpr.org. Licensee: The State of Oregon, acting by and through the State Board of Higher Education. Natl. Network: NPR, PRI. Law Firm: Ernest Sanchez. Format: AAA, jazz, news. News staff: one; News: 45 hrs wkly. Target aud: General. Spec prog: Blues 6 hrs, folk 3 hrs, pub affrs 7 hrs wkly. ♦Mitchell Christian, CFO; Ronald Kramer, CEO & gen mgr; Bryon Lambert, opns dir; Paul Westhelle, dev dir; Jessica Robinson, news dir & disc jockey.

***KTEC(FM)**— Dec 19, 1950: 89.5 mhz; 250 w. 184 ft TL: N42 12 59 W121 47 57. (CP: Ant 597 ft. TL: N42 13 26 W121 49 02). Stereo. Hrs opn: 9 AM-midnight Oregon Institute of Technology, Box 2009, 3201 Campus Dr., 97601. Phone: (541) 885-1840. Phone: (541) 885-1841. Fax: (541) 885-1857. E-mail: ktec@oit.edu Web Site: www.oit.edu/-ktec. Licensee: Oregon State Board of Higher Education. Population served: 20,000 Format: Freeform (diversified). News staff: one; News: 5 hrs wkly. Target aud: 15 plus; eclectic, free thinking, progsv individuals. Spec prog: American Indian one hr, Black 3 hrs, folk 3 hrs, Sp 3 hrs, world mus 6 hrs, electronic 9 hrs wkly. ♦Carola Roufs, gen mgr; Jake Byron, progmg mgr; Len Simpson, mus dir.

La Grande

***KEOL(FM)**— October 1973: 91.7 mhz; 310 w. -750 ft TL: N45 19 16 W118 05 26. Stereo. Hrs opn: 24 One Univ. Blvd., 97850. Phone: (541) 962-3698. Web Site: www.eou.edu/. Licensee: Oregon State Board of Higher Education. Population served: 20,000 Format: CHR, div, progsv. Target aud: 14-25; college students & loc youth. Spec prog: Black 12 hrs, class 4 hrs, jazz 6 hrs, reggae 7 hrs wkly. ♦Dave McDermot, stn mgr.

KLBM(AM)— 1938: 1450 khz; 1 kw-U. TL: N45 19 45 W118 04 00. Hrs open: 24

Rebroadcasts KBKR(AM) Baker City 100%.
Box 907, 97850. Secondary address: 2510 E. Cove Ave. 97850. Phone: (541) 963-4121. Phone: (541) 963-4122. Fax: (541) 963-3117. E-mail: supertalk@eoni.com Licensee: Pacific Empire Radio Corp. (group owner; acq 7-19-2004; grpsl). Population served: 40,000 Natl. Network: Westwood One, ABC. Rgnl rep: McGavren Guild Law Firm: Denise Moline, P.C. Format: News/talk. News staff: one; News: 25 hrs wkly. Target aud: 25-54. Spec prog: Farm 2 hrs wkly. ♦Mark Bolland, pres; Steve Ryner, gen mgr; Bobby Hollowwa, progmg dir.

KUBQ(FM)—Co-owned with KLBM(AM). Aug 15, 1977: 98.7 mhz; 2.25 kw. 1,942 ft TL: N45 26 26 W117 53 31. Stereo. 24 E-mail: q98@eoni.com 35,000 Natl. Rep: McGavren Guild. Format: Classic rock. News staff: one; News: 6 hrs wkly. Target aud: 25-54.

***KTVR-FM**— 2004: 90.3 mhz; 400 w. Ant 2,519 ft TL: N45 18 33 W117 43 54. Hrs open: 7140 S.W. Macadam Ave., Portland, 97219. Phone: (503) 293-1905. Fax: (503) 293-1919. Web Site: www.opb.org. Licensee: Oregon Public Broadcasting. Format: News, info music. News staff: 5; News: 146 hrs wkly. ♦Jack Galmiche, COO & exec VP.

KWRL(FM)— Sept 27, 1988: 99.9 mhz; 60 kw. 377 ft TL: N45 12 59 W118 00 00. Stereo. Hrs opn: 24 1009 Adams Ave., Suite C, 97850. Phone: (541) 963-7911. Fax: (541) 963-5090. E-mail: 999@eoni.com Licensee: KSRV Inc. Group owner: Capps Broadcast Group (acq 12-14-98; $800,000). Natl. Network: ABC, Jones Radio Networks. Natl. Rep: Tacher. Rgnl rep: Tacher. Format: Adult contemp. Target aud: 18-49; general. ♦Dave Capps, pres; Randy McKone, gen mgr.

La Pine

***KKLP(FM)**— 2005: 90.1 mhz; 2.5 kw vert. Ant 131 ft TL: N43 34 50 W121 34 13. Stereo. Hrs opn: 24
Rebroadcasts KLVR(FM) Santa Rosa, CA 100%.
2351 Sunset Blvd., Suite 170-218, Rocklin, CA, 95765. Phone: (916) 251-1600. Fax: (916) 2511650. E-mail: klove@klove.com Web Site: www.kove.com. Licensee: Educational Media Foundation. Group owner: EMF Broadcasting. Natl. Network: K-Love. Law Firm: Shaw Pittman. Format: Contemp Christian. News staff: 3. Target aud: 25-44; Judeo Christian, female. ♦Richard Jenkins, pres; Mike Novak, VP; Keith Whipple, dev dir; David Pierce, progmg mgr; Ed Lenane, news dir; Sam Wallington, engrg dir; Karen Johnson, news rptr; Marya Morgan, news rptr; Richard Hunt, news rptr.

Lake Oswego

KDZR(AM)—Licensed to Lake Oswego. See Portland

KLTH(FM)— Aug 1, 1977: 106.7 mhz; 100 kw. Ant 1,443 ft TL: N45 30 58 W122 43 59. Stereo. Hrs opn: 24 222 S.W. Columbia, Suite 350, Portland, 97201. Phone: (503) 223-0300. Fax: (503) 223-6542. Fax: (503) 223-6795. E-mail: reachus@literock1067.com Web Site: www.literock1067.com. Licensee: Infinity Radio of Portland Inc. Group owner: Infinity Broadcasting Corp. (acq 11-13-98; grpsl). Population served: 1,674,000 Law Firm: Leventhal, Senter & Lerman. Format: Oldies. News staff: one. Target aud: 35-54; 55% women, 45% men. ♦Michael Jordan, CEO; Mel Karmazin, chmn; Mark Walen, gen mgr; Maureen Pulicella, gen sls mgr; Rob Bertrand, prom dir; Chris Miller, progmg dir; Diana Jordan, news dir.

Lakeview

KLCR(FM)— 2003: 95.3 mhz; 780 w. Ant 1,378 ft TL: N42 12 40 W120 19 35. Hrs open: 24 613 S. G St., 97630-1829. Phone: (541) 947-3325. Licensee: Woodrow Michael Warren. Group owner: Woodrow Michael Warren Stns. Format: Classic rock. ♦Mike Warren, gen mgr.

***KOAP(FM)**— 2000: 88.7 mhz; 170 w. Ant -590 ft TL: N42 10 42 W120 21 19. Stereo. Hrs opn: Oregon Public Broadcasting, 7140 S.W. Macadam Ave., Portland, 97219. Phone: (503) 293-1905. Fax: (503) 293-4877. Web Site: www.opb.org. Licensee: Oregon Public Broadcasting. Law Firm: Swarz, Woods & Miller. Format: News, world beat. ♦Steve Bass, CEO; Jeff Douglas, opns VP.

KQIK(AM)— Dec 5, 1956: 1230 khz; 1 kw-U. TL: N42 12 30 W120 21 39. Stereo. Hrs opn: 24 629 Center St., 97630. Phone: (541) 947-3351. Fax: (541) 947-2309. Licensee: Crystal Clear Broadcasting Co. Inc. (acq 12-10-2003; $118,000 with co-located FM). Population served: 15,000 Natl. Network: ABC. Wire Svc: AP Format: Country. News staff: one; News: 2 hrs wkly. Target aud: General. Spec prog: Relg 2 hrs wkly. ♦Tommie S. Dodd, gen mgr.

KQIK-FM— 1987: 93.5 mhz; 1 kw. Ant 951 ft TL: N42 12 18 W120 19 39. Stereo. 24 30,000 Natl. Network: ABC. Format: Adult contemp. Target aud: General. ♦Walt Lawton, progmg dir.

Lebanon

KGAL(AM)— Aug 5, 1995: 1580 khz; 1 kw-U, DA-1. TL: N44 34 30 W122 55 15. Hrs open: 24 36991 KGAL Dr., 97355. Phone: (541) 451-5425. Fax: (541) 451-5429. E-mail: kgal@kgal.com Web Site: www.kgal.com. Licensee: EADS Broadcasting Corp. Population served: 350,000 Natl. Network: CBS, Westwood One, Salem Radio Network, Sporting News Radio Network, Talk Radio Network. Natl. Rep: McGavren Guild. Law Firm: Crowell & Moring. Wire Svc: AP Format: News, talk, sports. News staff: 5; News: 22 hrs wkly. Target aud: 25-54; active listeners. Spec prog: Local interview show 5 hrs wkly. ♦Richard B. Eads, pres; Florence R. Eads, CFO; Charlie Eads, exec VP & gen mgr; Jim Willhight, opns mgr; Shelly Garrett, gen sls mgr; Ted Jenne, progmg dir; Weldon Greig, news dir; Susie Dowding, traf mgr.

***KGRI(FM)**— 2005: 88.1 mhz; 1 w horiz, 170 w vert. Ant 2,473 ft vert TL: N44 28 59 W122 34 55. Stereo. Hrs opn: 24
Rebroadcasts KLRD(FM) Yucaipa, CA 100%.
2351 Sunset Blvd., Suite 170-218, Rocklin, CA, 95765. Phone: (916) 251-1600. Fax: (916) 251-1650. E-mail: info@air1.com Web Site: www.air1.com. Licensee: Educational Media Foundation. Group owner: EMF Broadcasting. Natl. Network: Air 1. Law Firm: Shaw Pittman. Format: Contemp Christian. News staff: 3. Target aud: 18-35; Judeo-Christian, female. ♦Richard Jenkins, pres; Mike Novak, VP & progmg dir; Lloyd Parker, gen mgr; Ed Lenane, opns dir & news dir; Keith Whipple, dev dir; Eric Allen, natl sls mgr; David Pierce, progmg mgr; Jon Rivers, mus dir; Sam Wallington, engrg dir; Karen Johnson, news rptr; Marya Morgan, news rptr; Richard Hunt, news rptr.

KSHO(AM)— 1950: 920 khz; 1 kw-U, DA-1. TL: N44 34 30 W122 55 15. Hrs open: 24 36991 KGAL Dr., 97355. Phone: (541) 451-5425. Fax: (541) 451-5429. E-mail: kgal@kgal.com Web Site: www.ksho.net. Licensee: Eads Broadcasting Corp. (acq 10-1-81; $425,000; FTR: 10-5-81). Population served: 350,000 Natl. Network: Jones Radio Networks, Music of Your Life. Natl. Rep: McGavren Guild. Law Firm: Crowell & Moring. Wire Svc: AP Format: MOR, adult standards. News staff: 5; News: 7 hrs wkly. Target aud: 35 plus; mature adults with money & leisure. ♦Richard B. Eads, pres; Florence R. Eads, CFO; Charlie Eads, exec VP & gen mgr; Jim Willhight, opns mgr; Shelly Garrett, gen sls mgr; Ted Jenne, progmg dir; Weldon Greig, news dir; Susie Dowdig, traf mgr.

KXPC(FM)— Apr 8, 1974: 103.7 mhz; 100 kw. Ant 1,099 ft TL: N44 30 17 W122 57 20. Stereo. Hrs open: 24 1207 9th Ave. S.E., Albany, 97322. Phone: (541) 928-1926. Fax: (541) 791-1054. E-mail: kxpc@kxpc.com Web Site: www.kxpc.com. Licensee: Portland Broadcasting L.L.C. (acq 4-11-2001; $4.1 million). Population served: 250,000 Natl. Rep: McGavren Guild. Format: Country. News: 2 hrs wkly. Target aud: 18-54. ♦Rich Coleman, gen mgr.

Lincoln City

KBCH(AM)—Listing follows KCRF-FM.

KCRF-FM— Nov 1, 1981: 96.7 mhz; 19.5 kw. Ant 872 ft TL: N44 45 22 W124 02 57. Stereo. Hrs opn: 24 Box 1430, Newport, 97365. Phone: (541) 265-2266. Fax: (541) 265-6397. E-mail: info@kcrffm.com Web Site: kcrffm.com. Licensee: Pacific West Broadcasting Inc. (group owner; acq 11-15-00; grpsl). Population served: 68,000 Rgnl rep: Tacher Format: Classic rock. News staff: 2; News: 2 hrs wkly. ♦David Miller, pres & gen mgr.

KBCH(AM)—Co-owned with KCRF-FM. May 27, 1955: 1400 khz; 1 kw-U. TL: N44 59 27 W123 58 45. E-mail: info@kbcham.com Web Site: kbcham.com.12,000 Rgnl rep: Tacher Format: MOR, full service. News staff: 2; News: 4 hrs wkly.

Malin

***KBUG(FM)**— 2000: 100.9 mhz; 750 w. Ant 899 ft TL: N42 05 48 W121 37 57. Stereo. Hrs opn: Box 111, Klamath Falls, 97601. Phone: (541) 883-6331. Fax: (541) 884-8226. Licensee: Malin Christian Church, Inc. (acq 7-1-99; $3,000). Format: Christian country.

McMinnville

KLYC(AM)— June 18, 1949: 1260 khz; 1 kw-U, DA-N. TL: N45 13 19 W123 10 21. Hrs open: 24 Box 1099, 97128. Secondary address: 1975 Colvin Ct. 97128. Phone: (503) 472-1260. Fax: (503) 472-3243. E-mail: klyc@viclink.com Licensee: Bohnsack Strategies Inc. (acq 10-2-90; $120,000; FTR: 10-22-90). Population served: 90,000 Natl. Network: CNN Radio. Format: Adult contemp, oldies. Target aud: 25-54. Spec prog: Sp 8 hrs wkly. ♦Larry Bohnsack, pres & gen mgr.

***KSLC(FM)**— Jan 17, 1972: 90.3 mhz; 320 w. -46 ft TL: N45 12 06 W123 11 52. Stereo. Hrs opn: 6 AM-noon Unit DD, 900 S.E. Baker St., 97128. Phone: (503) 434-2550. Phone: (503) 434-2666. Fax: (503) 434-2665. Web Site: www.linfield.edu/kslc. Licensee: Linfield College. Population served: 25,000 Format: Alternative rock. News: 3 hrs wkly. Target aud: 12-25; young people looking for new mus. Spec prog: Black 2 hrs, heavy metal 7 hrs wkly, relg 2 hrs wkly. ♦Julie Kanago, gen mgr; Dave McAdams, opns dir; Zach Bowden, dev VP.

Medford

KBOY-FM— February 1958: 95.7 mhz; 100 kw. 935 ft TL: N42 27 07 W23 03 20. (CP: 60 kw, ant 751 ft.). Stereo. Hrs opn: 24 1438 Rossanley Dr., 97501. Phone: (541) 779-1550. Fax: (541) 776-2360. E-mail: cbaker@radiomedford.com Web Site: www.957kboy.com. Licensee: Mapleton Communications LLC (group owner; acq 10-26-01; grpsl). Population served: 140,000 Format: Classic rock. News: One. ♦Ron Hren, VP & gen mgr; Joe Mussio, mktg mgr; Casey Baker, progmg dir; Maria Chaney, traf mgr; Robert Probert, engr.

KCMX(AM)—Phoenix, Apr 7, 1962: 880 khz; 1 kw-U. TL: N42 18 36 W122 48 41. Hrs open: 24 1438 Rossanley Dr., 97501. Phone: (541) 779-1550. Fax: (541) 776-2360. Web Site: www.kcmxam.com. Licensee: Mapleton Communications LLC (group owner; acq 10-26-01; grpsl). Population served: 150,000 Rgnl rep: Art Moore. Format: Talk, news. News staff: one. Target aud: 18 plus. ♦Ron Hren, VP & gen mgr; Jamy Gilinsky, sls dir; Devin Harpole, mktg dir & prom dir; Joe Mussio, mktg mgr & news dir; Robert Probert, chief of engrg & engr; Garth Harrington, disc jockey; Rosemary Harrington, disc jockey.

KTMT-FM—Co-owned with KCMX(AM). Oct 15, 1970: 93.7 mhz; 31 kw. 7,580 ft TL: N42 04 55 W122 43 07. Stereo. 24 Web Site: www.937mike.com.500,000 Law Firm: Dow, Lohnes & Albertson. Format: Adult hits. News staff: one; News: 3 hrs wkly. Target aud: 18-49. ♦Casey Baker, mktg dir, mus dir & disc jockey; Richard Tempelton, prom dir; Leslie Haze, disc jockey; Ron Scott, disc jockey.

KCMX-FM—See Ashland

KCNA(FM)—See Cave Junction

***KDOV(FM)**— Aug 1, 1995: 91.7 mhz; 26 kw. -364 ft TL: N42 20 13 W122 51 44. Hrs open: 24 1236 Disk Drive St. E., 97501. Phone: (541) 776-5368. Fax: (541) 776-0618. E-mail: kdov@kdov.net Web Site: www.kdov.net. Licensee: UCB USA Inc. (acq 1-16-2004; $750,000). Natl. Network: Salem Radio Network. Law Firm: Edmundson & Edmundson. Wire Svc: AP Format: Relg, news/talk. News staff: one; News: 5 hrs wkly. Target aud: 25-54; women. ♦Perry A. Atkinson, pres; Dallas Rhoden, VP; Perry Atkinson, gen mgr; Pat Daly, opns mgr.

***KEZX(AM)**— May 31, 1954: 730 khz; 1 kw-D, 74 w-N. TL: N42 18 36 W122 48 41. Stereo. Hrs opn: 24 511 Rossanley Dr., 97501. Phone: (541) 772-0322. Fax: (541) 772-4233. E-mail: talkradio730pd@yahoo.com Web Site: www.talkradio730.com. Licensee: Opus Broadcasting Systems Inc. (group owner; (acq 12-17-2003; $70,000). Natl. Network: Sporting News Radio Network. Format: Talk. News staff: 3. ♦Dean Flock, gen mgr.

KLDZ(FM)— Aug 19, 1991: 103.5 mhz; 100 kw. Ant 479 ft TL: N42 17 13 W123 00 15. Stereo. Hrs opn: 24 3624 Avion Dr., 97504. Phone: (541) 774-1324. Fax: (541) 857-0326. Web Site: www.kool103.net. Licensee: Bicoastal Rogue Valley LLC. Group owner: Clear Channel Communications Inc. (acq 7-2-2007; grpsl). Population served: 150,000 Natl. Rep: Tacher. Format: Super hits. Target aud: 25 plus. ♦Bill Nielsen, gen mgr.

KMED(AM)— 1922: 1440 khz; 5 kw-D, 1 kw-N. TL: N42 18 36 W122 48 41. Hrs open: 24 3624 Avion Dr., 97504. Phone: (541) 773-1440. Fax: (541) 857-0326. E-mail: news@kmed.com Web Site: www.kmed.com. Licensee: Bicoastal Rogue Valley LLC. Group owner: Clear Channel Communications Inc. (acq 7-2-2007; grpsl). Population served: 125,000 Natl. Network: Westwood One, CBS. Format: News/talk. News staff: one; News: 14 hrs wkly. Target aud: 35 plus. ♦Bill Nielsen, gen mgr; Bill Meyer, stn mgr.

KRTA(AM)— October 1947: 610 khz; 5 kw-U, DA-2. TL: N42 23 15 W122 46 11. Stereo. Hrs opn: 24 511 Rossanley Dr., 97501. Phone: (541) 772-0322. Fax: (541) 772-4233. E-mail: brian@opusradio.com Licensee: Opus Broadcasting Systems Inc. (group owner; acq 7-9-91; $63,634 with KROG(FM) Grants Pass; FTR: 7-29-91). Population served: 209,031 Law Firm: Leibowitz & Spencer. Format: Sp. News: 9 hrs wkly. Target aud: 12 plus; Hispanic. Spec prog: Southern Oregon A's, North Medford & Oregon State football & basketball, Portland Trailblazers. ♦Dean Flock, gen mgr; Oscar Bonilla, progmg dir.

KTMT(AM)—See Ashland

Merrill

KKKJ(FM)—Not on air, target date: unknown: 105.7 mhz; 110 w. Ant 2,184 ft TL: N42 05 48 W121 37 57. Hrs open: Box 218, Cheney, 99004. Phone: (509) 235-6184. Fax: (509) 235-2887. Licensee: Klamath Basin Broadcasting. ♦William Ifft, gen mgr.

Milton-Freewater

***KLRF(FM)**— Jan 1, 1999: 88.5 mhz; 5 kw. Ant 1,302 ft TL: N45 47 16 W118 10 31. Hrs open: 24 Box 3006, Collegedale, TN, 37315. Phone: (509) 524-0885. Fax: (509) 524-0884; (423) 884-2802. E-mail: office@lifetalk.net Web Site: www.lifetalk.net. Licensee: Lifetalk Broadcasting Association. Format: Relg, Christian. ♦Grant McPherson, gen mgr.

KZTB(FM)— Sept 10, 1992: 97.9 mhz; 100 kw. Ant 899 ft TL: N45 47 41 W118 10 06. Stereo. Hrs opn: 24 2730 West Lewis #8, Pasco, WA, 99301. Phone: (509) 543-3334. Fax: (509) 452-0541. Licensee: Bustos Media of Eastern Washington License LLC. Group owner: Clear Channel Communications Inc. (acq 2-22-2006; $900,000 plus swap for KUJJ(FM) Weston). Format: Sp. ♦Bob Berry, gen mgr.

Milwaukie

KSZN(AM)— February 1988: 1010 khz; 4.5 kw-D. TL: N45 29 03 W122 24 40. Stereo. Hrs opn: 5110 S.E. Stark St., Suite C, Portland, 97415. Phone: (503) 234-5550. Fax: (503) 234-5583. E-mail: rtatum@bustosmedia.com Web Site: www.bustosmedia.com. Licensee: Bustos Media of Oregon License LLC. Group owner: Bustos Media Holdings (acq 11-19-2003; $1 million). Format: Sp. ♦Ricky Tatum, gen mgr; Chitra Gade, opns mgr; Tom Oberg, gen sls mgr; Henry Cualio, progmg dir; James Boyd, chief of engrg.

Molalla

KRSK(FM)—Licensed to Molalla. See Portland

Monmouth

KSND(FM)— Mar 23, 1995: 95.1 mhz; 1 kw. Ant 3,307 ft TL: N44 53 19 W123 36 26. Stereo. Hrs opn: 24 285 Liberty St. N.E., #340, Salem, 97301. Phone: (503) 763-9951. Fax: (503) 763-2676. E-mail: ernie@ksnd.com Web Site: www.ksnd.com. Licensee: Radio Beam LLC (acq 6-7-2002; $400,000). Population served: 718,000 Law Firm: Wiley, Rein & Fielding. Wire Svc: AP Format: Adult contemp. News staff: one; News: 5 hrs wkly. Target aud: 25-54. ♦Ernie Hopseker, pres, gen mgr, gen mgr & gen sls mgr; Scott Forrest, progmg dir; Frank Rippey, pub affrs dir; Lyndi Miles, traf mgr.

Mount Angel

KTRP(AM)—Not on air, target date: unknown: 1130 khz; 25 kw-D, 490 w-N, DA-2. TL: N45 04 35 W122 48 27. Hrs open: Box 60991, Palo Alto, CA, 94306-0991. Phone: (650) 856-6823. Licensee: JNE Investments Inc. ♦Jeffrey N. Eustis, pres.

Myrtle Point

***KOOZ(FM)**— Aug 1, 1996: 94.1 mhz; 1 kw. Ant 1,456 ft TL: N42 57 32 W124 16 23. Hrs open: 5 AM-2 AM Jefferson Public Radio, 1250 Siskiyou Blvd., Ashland, 97520. Phone: (541) 552-6301. Phone: (541) 552-8565. Web Site: www.ijpr.org. Licensee: JPR Foundation Inc. (acq 4-19-02; $83,700 with KTBR(AM) Roseburg). Population served: 90,000 Natl. Network: NPR, PRI. Law Firm: Ernest Sanchez. Wire Svc: AP Format: Class, news. News staff: one; News: 35 hrs wkly. ♦Mitchell Christian, CFO; Ronald Kramer, CEO & gen mgr; Bryon Lambert, opns dir; Paul Westhelle, dev dir.

Newport

KCUP(AM)—See Toledo

***KLCO(FM)**— Sept 11, 1990: 90.5 mhz; 3.2 kw. 256 ft TL: N44 45 22 W124 02 57. Stereo. Hrs opn: 24
Rebroadcasts KLCC(FM) Eugene 100%.
4000 E. 30th Ave., Eugene, 97405-0640. Phone: (541) 463-6000. Fax: (541) 463-6046. E-mail: klcc@lanecc.edu Web Site: www.klcc.org. Licensee: Lane Community College. Population served: 50,000 Natl. Network: NPR. Format: Jazz, div, news. News staff: one; News: 60 hrs wkly. Target aud: 25-54. Spec prog: Sp 5 hrs, Black 3 hrs, folk 12 hrs, blues 4 hrs, world 3 hrs, electronic 6 hrs wkly. ♦Jerry Moskus, pres; Steve Barton, gen mgr.

KNCU(FM)— June 2000: 92.7 mhz; 3.8 kw. 840 ft TL: N44 45 22 W124 02 57. Stereo. Hrs opn: Box 1430, 97365. Phone: (541) 265-2266. Fax: (541) 265-6397. E-mail: info@u92fm.com Web Site: www.u92fm.com. Licensee: Pacific West Broadcasting Inc. (group owner; acq 10-27-00; grpsl). Population served: 44,000 Rgnl rep: Tacher. Format: Country. News staff: one; News: 2 hrs wkly. Target aud: 24-54; adults. ♦David J. Miller, pres & gen mgr.

KNPT(AM)— June 28, 1948: 1310 khz; 5 kw-D, 1 kw-N, DA-N. TL: N44 37 40 W123 59 15. Hrs open: 24 Box 1430, 906 S.W. Alder St., 97365. Phone: (541) 265-2266. Fax: (541) 265-6397. E-mail: info@knptam.com Web Site: knptam.com. Licensee: Yaquina Bay Communications Inc. (acq 1-96). Population served: 18,000 Rgnl rep: Tacher. Format: News/talk, sports. News staff: 2; News: 21 hrs wkly. Target aud: 34 plus. Spec prog: Relg 2 hrs wkly. ♦David J. Miller, pres & gen mgr; Vern Morris, gen sls mgr; Johnny Randolph, progmg dir; Bob Spangler, disc jockey.

KYTE(FM)—Co-owned with KNPT(AM). Oct 25, 1976: 102.7 mhz; 66 kw. 881 ft TL: N44 45 22 W124 02 57. Stereo. 24 E-mail: info@kytefm.com 120,000 Rgnl rep: Tacher. Format: Adult contemp. News staff: 2; News: 4 hrs wkly. Target aud: 24-49. ♦Dusty Baker, disc jockey; Johnny Randolph, disc jockey; Michael Blakeley, disc jockey; Tyler Romaro, disc jockey.

***KYOR(FM)**— 2006: 88.9 mhz; 35 w. Ant 899 ft TL: N44 45 23 W124 02 59. Hrs open:
Rebroadcasts KUFR(FM) Salt Lake City, UT 100%.
c/o Radio Station KUFR(FM), 136 E. S. Temple, Suite 1630, Salt Lake City, UT, 84111. Phone: (801) 359-3147. Fax: (801) 359-8112. Web Site: www.familyradio.com. Licensee: Family Stations Inc. Format: Christian relg. ♦Harold Camping, gen mgr.

North Bend

KBBR(AM)—Listing follows KOOS(FM).

KOOS(FM)— October 1990: 107.3 mhz; 51 kw. Ant 692 ft TL: N43 12 18 W124 18 07. Stereo. Hrs opn: 24 Box 180, Coos Bay, 97420. Secondary address: 320 Central Ave., Suite 519, Coos Bay 97420. Phone: (541) 267-2121. Fax: (541) 267-5229. Licensee: Bicoastal CB L.L.C. Group owner: Bicoastal Media L.L.C. (acq 10-16-2003; grpsl). Population served: 382,619 Rgnl rep: Tacher Format: Adult contemp. News: 14 hrs wkly. Target aud: 18-45. ♦Kenneth R. Dennis, pres; John Pundt, gen mgr.

KBBR(AM)—Co-owned with KOOS(FM). December 1950: 1340 khz; 1 kw-U. TL: N43 25 52 W124 12 23.24 70,000 Natl. Network: CBS Radio, Jones Radio Networks, Westwood One. Natl. Rep: Katz Radio, Tacher. Format: News/talk. News: 40 hrs wkly. Target aud: 25-54.

KTEE(FM)— Dec 10, 1979: 94.9 mhz; 89 kw. Ant 626 ft TL: N43 12 18 W124 18 07. Stereo. Hrs opn: 320 Central Ave., Suite 519, Coos Bay, 97420. Phone: (541) 267-2121. Fax: (541) 267-5229. Web Site: www.southcoastradio.com. Licensee: Bicoastal CB L.L.C. Group owner: Bicoastal Media L.L.C. (acq 10-16-2003; grpsl). Population served: 120,000 Natl. Network: ABC. Natl. Rep: Tacher. Format: Classic hits. News: 8 hrs wkly. Target aud: 25-54. ♦Kenneth R. Dennis, pres; John Pundt, gen mgr & rgnl sls mgr; Mike O.Brien, opns mgr.

North Powder

***KEFS(FM)**— 2006: 89.5 mhz; 120 w. Ant 1,787 ft TL: N45 07 26 W117 46 48. Hrs open: 1433 Jefferson Ave., La Grande, 97850-2643. Phone: (541) 963-5884. Web Site: www.csnradio.com. Licensee: CSN International. Format: Christian. ♦Mike Kestler, pres.

Nyssa

KARO(FM)— 1997: 98.7 mhz; 100 kw. Ant 968 ft TL: N43 24 09 W116 54 09. Hrs open: 24 Box 1600, Nampa, ID, 83653. Phone: (208) 463-1900. Licensee: Educational Media Foundation (acq 2-25-03; $1 million). Format: Christian contemp. ♦Steve Sumner, gen mgr.

Oakridge

***KAVE(FM)**— 2006: 88.5 mhz; 400 w. Ant -1,286 ft TL: N43 44 27 W122 26 50. Stereo. Hrs opn: 24 Lane County School District 4J, 200 N. Monroe St., Eugene, 97402. Phone: (541) 687-3123. Fax: (541) 687-3573. E-mail: randy@krvm.org Web Site: www.krvm.org. Licensee: Lane County School District 4J. Target aud: 18-35. ♦Ken Martin, progmg dir; Randy Larson, stn mgr & chief of engrg.

***KMKR(FM)**— Oct 22, 1990: 92.1 mhz; 580 w. Ant -817 ft TL: N43 44 34 W122 26 03. Hrs open: Oakridge High School, 47997 W. First St., 97463. Phone: (541) 782-2231. Fax: (541) 782-4692. E-mail: kave921@hotmail.com Web Site: www.geocites.com/kave921. Licensee: School District #76. Format: CHR. Target aud: General. ♦Debbie Gillespie, gen mgr & chief of opns; Abbie Pierce, prom dir; Aaron Stone, progmg dir.

Ontario

KSRV(AM)— Nov 23, 1946: 1380 khz; 5 kw-D, 1 kw-N, DA-N. TL: N44 02 45 W116 58 24. Hrs open: 24 Box 129, 1725 N. Oregon St., 97914. Phone: (541) 889-8651. Fax: (541) 889-8733. E-mail: ksrv@ksrv.com Licensee: FM Idaho Co. LLC (acq 9-9-2004; $2.5 million with co-located FM). Law Firm: Dow, Lohnes & Albertson. Format: Modern country. News staff: one; News: 15 hrs wkly. Target aud: 25-54 plus. Spec prog: Farm 15 hrs wkly. ♦Dale Jeffries, gen mgr; Gina Reid, gen sls mgr; Bryan Gregory, prom dir; Carl Follick, progmg dir & news dir.

KSRV-FM— July 4, 1977: 96.1 mhz; 47 kw. Ant 2,673 ft TL: N43 45 18 W116 05 51. Stereo. 24 E-mail: dale@ksrv.com Web Site: www.961bobfm.com. Format: Adult hits. News: 40 hrs wkly.

Oregon City

KGDD(AM)— July 4, 1947: 1520 khz; 50 kw-D, 10 kw-N, DA-2. TL: N45 24 44 W122 34 37. Hrs open: 24 5110 SE Stark St., Portland, 97215. Phone: (503) 234-5550. Fax: (503) 234-5583. Web Site: www.lagrand1520.com. Licensee: Bustos Media of Oregon License LLC. Group owner: Bustos Media Holdings (acq 11-17-2003; $2.8 million). Population served: 400,000 Natl. Rep: D & R Radio. Format: Sp. News staff: one. ♦Amador Bustos, pres; Ricky Tatum, gen mgr; Tom Oberg, gen sls mgr; Henry Cualio, prom dir; James Boyd, chief of engrg; Chitra Gade, traf mgr.

KGON(FM)—See Portland

Pendleton

***KRBM(FM)**— Apr 18, 1970: 90.9 mhz; 25 kw. 587 ft TL: N45 35 21 W118 59 53. Stereo. Hrs opn: 7140 S.W. Macadam Ave., Portland, 97219. Phone: (503) 293-1905. Phone: (1-888) 293-1982. Fax: (503) 293-1919. Web Site: www.opb.org. Licensee: Oregon Public Broadcasting. (acq 9-20-93; grpsl; FTR: 10-11-93). Population served: 25,000 Natl. Network: PRI, NPR. Rgnl. Network: Ore. Pub. Bcstg Radio Net. Format: News, info music. News staff: 5; News: 146 hrs wkly. Target aud: Teens to adults. ♦Jack Galmiche, COO, exec VP & VP.

KTIX(AM)— 1941: 1240 khz; 1 kw-U. TL: N45 39 49 W118 47 19. Stereo. Hrs opn: 24 2003 NW 56th Dr., 97801. Phone: (541) 278-2500. Fax: (541) 276-1480. Licensee: KSRV Inc. Group owner: Capps Broadcast Group (acq 5-14-98; $1.2 million with co-located FM). Population served: 15,200 Natl. Network: ESPN Radio. Natl. Rep: Tacher. Format: Sports. News staff: one. Target aud: 25-54; upscale adults. ♦Randy McKone, pres & stn mgr; J.J. Ford, opns mgr; John Thomas, prom mgr & progmg dir.

KWHT(FM)—Co-owned with KTIX(AM). May 1, 1984: 103.5 mhz; 100 kw. 720 ft TL: N45 47 51 W118 22 17. Stereo. 24 350,000 Natl.

Network: ABC. Format: Country. News: 2 hrs wkly. Target aud: 25-54; adults. ♦Randy McKone, CEO & gen mgr; Julie Thompson, gen sls mgr; J.J. Ford, engrg VP; Connie Shurtleff, engr.

KUMA(AM)— Aug 25, 1955: 1290 khz; 5 kw-U, DA-N. TL: N45 40 25 W118 44 48. Hrs open: 24 2003 N.W. 56th Dr., 97801. Phone: (541) 276-1511. Fax: (541) 276-1480. Licensee: Round-Up Radio Inc. Group owner: Capps Broadcast Group (acq 7-1-93; $340,000 with co-located FM; FTR: 7-26-93). Population served: 50,000 Rgnl rep: Tacher. Format: Talk, news. News staff: one; News: 20 hrs wkly. Target aud: 25 plus; adults. Spec prog: Farm 10 hrs wkly. ♦Dave Capps, pres; Randy McKone, VP & gen mgr; J.J. Ford, opns dir & opns mgr; Butch Thurman, news dir.

KUMA-FM— Oct 1, 1978: 107.7 mhz; 72 kw. Ant 1,115 ft TL: N45 35 27 W118 34 47. Stereo. 100,000 Natl. Network: ABC. Format: Adult contemp. Target aud: 18 plus. ♦J.J. Ford, progmg mgr.

Phoenix

KAKT(FM)— 1991: 105.1 mhz; 52 kw. 545 ft TL: N42 25 41 W123 00 04. Hrs open: 24 1438 Rossanley Dr., Medford, 97501. Phone: (541) 779-1550. Fax: (541) 776-2360. Web Site: www.kat105.com. Licensee: Mapleton Communications LLC (group owner; acq 10-26-01; grpsl). Population served: 150,000 Format: Country. News staff: one. Target aud: 25-49; female. ♦Ron Hren, VP & gen mgr; Casey Baker, opns mgr & prom dir; Joe Mussio, mktg mgr; Maria Chaney, traf mgr; Robert Probert, engr.

***KAPL(AM)**— Jan 2, 1977: 1300 khz; 20 kw-U, DA-N. TL: N42 17 44 W122 48 15. Hrs open: Box 1090, Jacksonville, 97530. Phone: (541) 899-5275. Fax: (541) 899-8068. E-mail: kapl@applegatefellowship.org Web Site: www.applegatefellowship.org. Licensee: Applegate Media Inc. (acq 10-24-91). Population served: 200,000 Format: News/talk, Christian. Target aud: 25-54. ♦Chris Thompson, gen mgr & opns mgr.

KCMX(AM)—Licensed to Phoenix. See Medford

Pilot Rock

KVAN-FM— 2006: 92.1 mhz; 6.9 kw. Ant 633 ft TL: N45 35 21 W118 59 54. Hrs open: 45 Campbell Rd., Walla Walla, WA, 99362. Phone: (509) 527-1000. Fax: (509) 529-5534. Licensee: Bruton Broadcasting LLC (acq 1-29-2007). Format: Oldies. ♦Aaron Bruton, gen mgr.

Pine Grove

***KPFR(FM)**— June 22, 2005: 89.5 mhz; 7 kw vert. Ant 1,673 ft TL: N45 19 58 W121 42 48. Hrs open: 24 Family Stations Inc., 4135 Northgate Blvd., Suite 1, Sacramento, CA, 95834. Phone: (916) 641-8191. Fax: (916) 641-8238. Licensee: Family Stations Inc. (group owner; (acq 9-9-2002). Format: Christian. ♦Harold Camping, pres.

Portland

KBMS(AM)—See Vancouver, WA

KBNP(AM)— 1949: 1410 khz; 5 kw-D, 250 w-N. TL: N45 28 24 W122 39 36. Hrs open: 24 278 S.W. Arthur St., 97201. Phone: (503) 223-6769. Fax: (503) 223-4305. E-mail: kbnp@kbnp.com Web Site: www.kbnp.com. Licensee: 2nd Amendment Foundation. (acq 8-8-90; $320,000; FTR: 8-27-90). Population served: 1,327,500 Law Firm: Davis Wright Tremaine. Format: Business news & info, financial. News staff: 2; News: 163 hrs wkly. Target aud: General; corporations & individuals concerned with how-to's of making & keeping money. Spec prog: People w/disabilities, computer shows, home improvement. ♦Keith P. Lyons, gen mgr.

***KBOO(FM)**— June 1968: 90.7 mhz; 25.5 kw. 1,266 ft TL: N45 29 20 W122 41 40. Stereo. Hrs opn: 24 20 S.E. 8th Ave., 97214. Phone: (503) 231-8032. Fax: (503) 231-7145. E-mail: program@kboo.org Web Site: www.kboo.fm. Licensee: KBOO Foundation. (acq 8-5-75). Population served: 1,614,744 Law Firm: Haley, Bader & Potts. Format: Div. News staff: 2; News: 5 hrs wkly. Target aud: General. Spec prog: Sp 10 hrs, Indian one hr, ethnic 4 hrs, African/reggae 10 hrs wkly. ♦Denise Kowalczyk, gen mgr & stn mgr; Gene Bradley, gen mgr & prom mgr; Justin Miller, adv mgr; Chris Merrick, progmg dir; John Mackey, mus dir & chief of engrg.

***KBPS(AM)**— Mar 23, 1923: 1450 khz; 1 kw-U. TL: N45 31 38 W122 29 03. Hrs open: 18 515 N.E. 15th Ave., 97232. Phone: (503) 916-5830. Fax: (503) 916-2642. E-mail: music.info@allclassical.org Web Site: www.allclassical.org. Licensee: School District No. 1 Multnomah County, OR. Population served: 1,500,000 Format: Children programs. News: one hr wkly. Spec prog: Sp one hr wkly. ♦Sally Lewis, dev dir & mktg dir.

***KBPS-FM**— August 1983: 89.9 mhz; 8.7 kw. Ant 964 ft TL: N45 30 58 W122 43 59. Stereo. Hrs opn: 24 515 N.E. 15th Ave., 97232. Phone: (503) 916-5828. Fax: (503) 916-2642. Web Site: www.allclassical.org. Licensee: KBPS Public Radio Foundation (acq 12-15-2003; $5.5 million). Population served: 1,750,000 Natl. Network: PRI. Law Firm: Garvey, Schubert & Barer. Format: Class. News: 5 hrs wkly. ♦Sally Lewis, dev dir; Larry Holtz, chief of engrg.

***KBVM(FM)**— Dec 8, 1989: 88.3 mhz; 3.5 kw. Ant 1,433 ft TL: N45 30 58 W122 43 59. Stereo. Hrs opn: 24 Box 5888, 97228-5888. Secondary address: 5000 N. Willamette, No. 44 97203. Phone: (503) 285-5200. Fax: (503) 285-3322. Web Site: www.kbvm.com. Licensee: Catholic Broadcasting NW Inc. Population served: 1,907,000 Format: Relg. Target aud: General; anyone desiring Christian mus, inspiration, Catholic prayer & evangelism. Spec prog: Sp 14 hrs wkly. ♦Steven J. Moffitt, CEO & gen mgr; Dina Marie Hale, progmg dir; Mark Andreas, progmg dir.

KCMD(AM)— Oct 18, 1925: 970 khz; 5 kw-U, DA-N. TL: N45 30 56 W122 43 56. Stereo. Hrs opn: 24 222 S.W. Columbia, 97201. Phone: (503) 223-0300. Fax: (503) 497-2314. Licensee: Infinity Radio Inc. Group owner: Infinity Broadcasting Corp. (acq 11-13-98; grpsl). Format: Country, class. Target aud: 35 plus. ♦Dave McDonald, gen mgr; Mark Whaler, gen sls mgr.

KUFO-FM—Co-owned with KCMD(AM). May 1, 1977: 101.1 mhz; 100 kw. 1,640 ft TL: N45 30 58 W122 43 59. Stereo. 24 20040 S.W. 1st Ave., 97201. Web Site: www.kufo.com.250,000 Format: AOR.

KDZR(AM)—Lake Oswego, 1996: 1640 khz; 10 kw-D, 1 kw-N. TL: N45 27 14 W122 32 47. Hrs open: 3030 S.W. Moody, Suite 210, 97201. Phone: (503) 228-4322. Fax: (503) 228-4325. Web Site: www.radiodisney.com. Licensee: Radio Disney Group LLC. Group owner: ABC Inc. (acq 2-03; $3.8 million with KKSL(AM) Lake Oswego). Natl. Network: Radio Disney. Format: Families. ♦Jean-Paul Colaco, pres & gen mgr; Pamela Herrold, stn mgr.

KEX(AM)— Dec 24, 1926: 1190 khz; 50 kw-U, DA-N. TL: N45 25 20 W122 33 57. Hrs open: 24 4949 S.W. Macadam Ave., 97201. Phone: (503) 225-1190. Fax: (503) 227-5873. Web Site: www.1190kex.com. Licensee: Citicasters Licenses L.P. Group owner: Clear Channel Communications Inc. Population served: 1,600,000 Law Firm: Hogan & Hartson. Format: News, talk. News: 25 hrs wkly. Target aud: 25-54; general. Spec prog: Portland Trailblazers basketball. ♦Ron Saito, pres & gen mgr; Mike Dirkx, opns dir & progmg dir; dave Milner, sls dir & gen sls mgr; Mike Lulich, natl sls mgr; Scott Thompson, mktg VP & mktg dir; Teri Rodrigues, prom mgr; Brad Ford, news dir; Shane Ruark, chief of engrg.

KKRZ(FM)—Co-owned with KEX(AM). May 1946: 100.3 mhz; 95 kw. 1,433 ft TL: N45 31 22 W122 45 07. Stereo. Phone: (503) 226-0100. Fax: (503) 295-9281. Web Site: www.2100portland.com.382,619 Format: Contemp hit. Target aud: 18-49. ♦Jen Dalton, prom mgr; Michael Hayes, progmg dir; Shane Raurk, news dir.

KFXX(AM)— Jan 17, 1925: 1080 khz; 50 kw-D, 10 kw-N, DA-2. TL: N45 33 26 W122 29 08. Stereo. Hrs open: 0700 S.W. Bancroft St., 97239. Phone: (503) 223-1441. Fax: (503) 223-6909. E-mail: comments@koth.com Web Site: www.1080thefan.com. Licensee: Entercom Portland License LLC. Group owner: Entercom Communications Corp. (acq 12-18-2003; $44 million with co-located FM). Population served: 1,598,900 Format: Sports. Target aud: 25-54. ♦Ron Carter, CFO & gen mgr.

KWJJ-FM—Co-owned with KFXX(AM). 1968: 99.5 mhz; 50 kw. 1,266 ft TL: N45 29 20 W122 41 40. Stereo. Web Site: www.thewolfonline.com. Format: Country. Target aud: 25-54.

KGON(FM)— December 1967: 92.3 mhz; 100 kw. 920 ft TL: N45 20 23 W122 41 47. Stereo. Hrs opn: 24 0700 S.W. Bancroft, 97239. Phone: (503) 223-1441. Fax: (503) 223-6909. E-mail: jhutchison@entercom.com Web Site: www.kgon.com. Licensee: Entercom Portland License LLC. Group owner: Entercom Communications Corp. (acq 8-1-95; grpsl). Population served: 400,000 Natl. Rep: D & R Radio. Format: Classic rock. News staff: 3; News: one hr wkly. ♦David Field, pres; Jack Hutchinson, sr VP & VP; Erin Hubert, gen mgr; Dick Loughney, gen sls mgr & chief of engrg; Keevin Wagner, gen mgr & prom dir; Clark Ryan, progmg dir.

KINK(FM)— Dec 24, 1968: 101.9 mhz; 97 kw. Ant 1,446 ft TL: N45 30 58 W122 43 59. Stereo. Hrs opn: 24 1501 S.W. Jefferson St., 97201. Phone: (503) 517-6000. Fax: (503) 517-6100. E-mail: lwarren@kink.fm Web Site: www.kink.fm. Licensee: Infinity Radio Inc. Group owner: Infinity Broadcasting Corp. (acq 11-13-98; grpsl). Population served: 1,972,900 Natl. Network: AP Radio. Format: AAA. News staff: 2; News: 3 hrs wkly. Target aud: 25-54; primary, secondary. ♦Stan Mak, gen mgr; Maureen Pulicella, gen sls mgr; Candace Gonzales, mktg dir; Dennis Constantine, progmg dir; Sheila Hamilton, news dir; Leana Warren, pub affrs dir.

KKCW(FM)—See Beaverton

KKPZ(AM)— Nov 12, 1923: 1330 khz; 5 kw-U, DA-1. TL: N45 27 13 W122 32 45. Hrs open: 5 am-12 am (M-F); 6 am-12 am (S, Su) 4700 S. W. Macadam Ave., Suite 102, 97239. Phone: (503) 242-1950. Fax: (503) 242-0155. E-mail: info@kkpz.com Web Site: www.kkpz.com. Licensee: KPHP Radio Inc. Group owner: Crawford Broadcasting Co. (acq 1995; $2 million). Population served: 2,200,000 Format: Christian, talk. Target aud: 34-54. ♦Donald Crawford Sr., pres; James Autry, stn mgr; Andrea Bolyard, opns mgr; John White, chief of engrg; James Autry, sls.

KLTH(FM)—See Lake Oswego

KMUZ(AM)—Gresham, Sept 28, 1956: 1230 khz; 1 kw-U. TL: N45 29 35 W122 24 40. Stereo. Hrs opn: 24 5110 S.E. Stark St., WA, 97215. Phone: (503) 234-5550. Fax: (503) 234-5583. E-mail: rtatum@bustosmedia.com Web Site: www.bustosmedia.com. Licensee: Bustos Media of Oregon License LLC. Group owner: Bustos Media Holdings (acq 7-15-2003; $1.13 million). Population served: 300,000 Format: Sp. News: 3 hrs wkly. Target aud: 12-54; lower to upper middle income. Spec prog: News 3 hrs, relg one hr wkly. ♦Amador S. Bustos, pres; Rick Tatum, gen mgr; Tom Oberg, gen sls mgr; Henry Cuallo, prom dir; James Boyd, chief of engrg.

KNRK(FM)—Camas, WA) Nov 1, 1992: 94.7 mhz; 6.3 kw. Ant 1,322 ft TL: N45 29 20 W122 41 40. Stereo. Hrs opn: 24 0700 S.W. Bancroft, 97239. Phone: (503) 223-1441. Fax: (503) 223-6909. Web Site: www.947.fm. Licensee: Entercom Portland License L.L.C. Group owner: Entercom Communications Corp. Format: Alternative rock. News staff: one. Target aud: 25-54. ♦David Field, pres; Jack Hutchison, exec VP; Jaime Cooley, progmg dir; Mark Hamilton, stn mgr & progmg dir.

***KOPB-FM**— 1962: 91.5 mhz; 70 kw horiz, 21 kw vert. Ant 1,558 ft TL: N45 31 22 W122 45 07. Stereo. Hrs opn: 7140 S.W. Macadam Ave., 97219. Phone: (503) 293-1905. Fax: (503) 293-1919. E-mail: opbnews@opb.org Web Site: www.opb.org. Licensee: Oregon Public Broadcasting. (acq 9-20-93; grpsl; FTR: 10-11-93). Population served: 1,000,000 Natl. Network: NPR, PRI. Format: News, info music. News staff: 5; News: 146 hrs wkly. Target aud: 34-54. ♦Jack Galmiche, COO, exec VP, dev VP & news dir; Virginia Breen, VP. Co-owned TV: *KOPB-TV affil.

KPDQ-FM— 1961: 93.9 mhz; 50 kw. Ant 1,269 ft TL: N45 29 20 W122 41 40. Stereo. Hrs opn: 24 6400 S.E. Lake Rd., 97222. Phone: (503) 786-0600. Fax: (503) 786-1551. Web Site: www.kpdq.com. Licensee: Salem Media of Oregon Inc. Group owner: Salem Communications Corp. (acq 8-86; grpsl; FTR: 7-28-86). Population served: 1,500,000 Format: Christian talk. Target aud: 25-54; listeners of Christian talk progmg. ♦Dennis Hayes, gen mgr; David Schulte, opns dir, opns mgr, progmg dir & progmg mgr; Mark Durkin, gen sls mgr; Georgene Rice, news dir; Don Perkins, chief of engrg; Melissa Shambaugh, traf mgr.

KPDQ(AM)— July 30, 1947: 800 khz; 1 kw-D, 500 w-N. TL: N45 28 45 W122 44 55.24 Web Site: www.kpdq.com. Format: Conservative talk, Christian. Target aud: 18-54; listeners of talk.

KPOJ(AM)— Mar 25, 1922: 620 khz; 25 kw-D, 10 kw-N, DA-2. TL: N45 25 20 W122 33 57. Hrs open: 24 4949 S.W. Macadam, 97201. Phone: (503) 323-6400. Fax: (503) 323-6664. E-mail: deaveosporne@clearchannel.com Web Site: www.620kpoj.com. Licensee: Citicasters Licenses L.P. Group owner: Clear Channel Communications Inc. (acq 1999; grpsl). Natl. Network: ABC. Format: News/talk. Target aud: 25-54. ♦Robert Dove, gen mgr; Mike Dirkx, opns mgr.

***KRRC(FM)**— May 1958: 97.9 mhz; 8 w. Ant 13 ft TL: N45 28 51 W122 37 50. Hrs open: Reed College, 3203 S.E. Woodstock, 97202. Phone: (503) 771-1112. Fax: (503) 777-7769. Licensee: The Reed Institute. (acq 1959). Population served: 385,000 Format: Div. Target aud: 17-21; Reed College student body. Spec prog: Black 10 hrs, class 4 hrs, country 2 hrs, Fr 2 hrs, jazz 10 hrs, Sp 2 hrs wkly. ♦Nicholas Wright, gen mgr; Kristin Holmberg, opns dir & mus dir.

KRSK(FM)—Molalla, July 3, 1970: 105.1 mhz; 21 kw. Ant 1,542 ft TL: N45 31 21 W122 44 45. Stereo. Hrs opn: 0700 S.W. Bancroft St., 97239. Phone: (503) 223-1441. Fax: (503) 223-6909. Web Site: www.1051thebuzz.com. Licensee: Entercom Portland License L.L.C. Group owner: Entercom Communications Corp. (acq 4-23-98; grpsl). Population served: 1,200,000 Format: Hot adult contemp. ♦David Field, pres; Erin Hubert, gen mgr; Brian Lee, gen sls mgr; Liz Kay, prom dir; Jeff McHugh, progmg dir; Sheryl Steward, mus dir.

KUPL-FM— 1948: 98.7 mhz; 37 kw. 1,443 ft TL: N45 30 58 W122 43 59. Stereo. Hrs opn: 24 222 S.W. Columbia, Suite 350, 97201. Phone: (503) 223-0300. Fax: (503) 223-6995. E-mail: laura.klein @infinitybroadcasting.com Web Site: www.kupl.com. Licensee: Radio Systems of Miami Inc. Group owner: Infinity Broadcasting Corp. (acq 11-13-98; grpsl). Population served: 1,674,000 Natl. Network: AP Radio. Law Firm: Leventhal, Senter & Lerman. Format: Country. News staff: one. Target aud: 25-54. ♦Mel Karmazin, chmn; Dan Mason, pres; Mark Walen, gen mgr; Lee Rogers, opns mgr; Tom Hunter, prom mgr; Cary Rolfe, progmg dir.

KXJM(FM)—Listing follows KXL(AM).

KXL(AM)— 1926: 750 khz; 50 kw-D, 20 kw-N, DA-2. TL: N45 24 05 W122 26 47. Hrs open: 0234 S.W. Bancroft, 97231. Phone: (503) 243-7595. Fax: (503) 417-7662. Web Site: www.kxl.com. Licensee: Rose City Radio Corp. (group owner; (acq 11-30-98; $55 million with co-located FM). Population served: 1,473,700 Natl. Network: CBS. Natl. Rep: McGavren Guild. Format: News/talk. Target aud: 25-54. ♦Rose City Radio, CFO; Tim McNamara, gen mgr; James Derby, opns mgr; Bill Ashenden, gen sls mgr.

KXJM(FM)—Co-owned with KXL(AM). June 18, 1965: 95.5 mhz; 100 kw. 990 ft TL: N45 29 23 W122 41 47. Stereo. 24 Web Site: www.jamminfm.com. Format: CHR. ♦Tim McNamara, gen sls mgr; Mark Adams, progmg dir; Christina Martinez, disc jockey; Jeff G., disc jockey; Louis Cruz, disc jockey; Mario DeVoe, disc jockey; T.J. Hooker, disc jockey.

KXMG(AM)— July 4, 1954: 1150 khz; 5 kw-D, 47 w-N, DA-1. TL: N45 38 34 W122 36 49. Hrs open: 5110 S.E. Stark, 97215. Phone: (503) 234-5550. Fax: (503) 234-5583. Web Site: www.bustosmedia.com. Licensee: Bustos Media of Oregon License LLC. Group owner: Bustos Media Holdings (acq 11-18-2003; $1.25 million). Population served: 2,000,000 Natl. Rep: Interep. Format: Sp contemp. Target aud: 25-54; general. Spec prog: Black one hr, Scandinavian one hr, URDU Hindi one hr, It one hr wkly. ♦Ricky Tatum, gen mgr; Tom Oberg, gen sls mgr; Henry Cualio, prom dir; Chitra Gade, chief of engrg & traf mgr; James Boyd, chief of engrg.

***KXPD(AM)**—Tigard, June 28, 1993: 1040 khz; 2.2 kw-D, 200 w-N. TL: N45 28 26 W122 39 33. Stereo. Hrs opn: 24 871 Country Club Rd., Eugene, 97401-6009. Phone: (541) 344-5500. Fax: (541) 485-2550. Licensee: Churchill Communications LLC. Group owner: EMF Broadcasting (acq 7-31-2006; $1.8 million). Format: Sp. ♦Suzanne Arlie, gen mgr.

KYCH-FM— Apr 1, 1980: 97.1 mhz; 100 kw. Amt 1,266 ft TL: N45 29 20 W122 41 40. Stereo. Hrs opn: 0700 S.W. Bancroft St., 97239. Phone: (503) 223-1441. Fax: (503) 223-6909. E-mail: cryan@entercom.com Web Site: www.charliefm.com. Licensee: Entercom Portland License L.L.C. Group owner: Entercom (acq 4-23-98; grpsl). Population served: 230,000 Natl. Rep: Christal. Format: Adult contemp hits. Target aud: 25-54. ♦David Field, pres; Jack Hutchinson, exec VP; Maureen Pulicella, gen sls mgr; Dan Persigehl, progmg dir; Shel Bailey, prom dir & news dir; Gary Hilliard, chief of engrg.

Prineville

KLTW-FM—Listing follows KRCO(AM).

KRCO(AM)— Feb 1, 1950: 690 khz; 1 kw-D, 77 w-N. TL: N44 20 30 W120 54 10. Hrs open: 24 Box 690, 97754. Secondary address: 854 N.E. 4th St., Bend 97701. Phone: (541) 447-6770. Fax: (541) 383-3403 (sales). Web Site: www.krcoam.com. Licensee: Horizon Broadcasting Group L.L.C (group owner; (acq 3-2-2000; grpsl). Natl. Network: ABC. Natl. Rep: Tacher. Format: Classic country. Target aud: 35-64. ♦Keith Shipman, pres & gen mgr; Brian Canady, stn mgr; Larry Wilson, opns mgr; Bryon Mengle, prom dir; Roy Lamela, progmg dir; Bill Baker, news dir; Jean Morgan, traf mgr; Ramona Hartmann, traf mgr.

KLTW-FM—Co-owned with KRCO(AM). Apr 8, 1981: 95.1 mhz; 100 kw. Ant 472 ft TL: N44 18 32 W120 55 47. Stereo. 24 Phone: (541) 383-3825. Web Site: www.lite951.com. Natl. Rep: Tacher. Format: Soft adult contemp. Target aud: 25-54. ♦Jeff Nelson, progmg dir.

KWDP(FM)—Not on air, target date: unknown: 98.9 mhz; 390 w. Ant 2,250 ft TL: N44 26 17 W120 57 12. Hrs open: 5331 Mt. Alifan Dr., San Diego, CA, 92111. Phone: (858) 277-4991. Fax: (858) 277-1365. Licensee: Horizon Christian Fellowship. (acq 2-9-2006; grpsl). ♦Mike MacIntosh, pres.

Rainier

KPPK(FM)—Not on air, target date: 11/05: 98.3 mhz; 1.6 kw. Ant 640 ft TL: N46 10 59 W122 57 29. Hrs open: 1130, 14th Avenue, Longview, WA, 98632. Phone: (360) 425-1500. Fax: (360) 423-1554. E-mail: grodman@biocoastalmedia.com Licensee: Bicoastal Longview LLC. Format: Adult contemp 70s, 80s, and 90s. ♦Kevin P. Mostyn, opns mgr.

Redmond

***KKJA(FM)**—Not on air, target date: unknown: 89.9 mhz; 750 w. Ant 2,217 ft TL: N44 26 17 W120 57 14. Hrs open: CSN International, 4002 N. 3300 E., Twin Falls, ID, 83301. Phone: (208) 734-6633. Fax: (208) 736-1958. Licensee: CSN International. ♦Mike Kestler, pres.

KLRR(FM)—Licensed to Redmond. See Bend

KRDM(AM)— June 2004: 1240 khz; 750 w-U. TL: N44 17 18 W121 11 02. Hrs open: 24 Box 1309, 97756. Secondary address: 1514 S.W. Highland Ave. 97756. Phone: (541) 548-7621. Fax: (541) 504-8145. E-mail: krdm@krdm.net Licensee: Sage-Com Inc. Population served: 20,000 Format: Sp. ♦Bud Hutchinson, pres & gen mgr; Sam KirKaldie, sr VP; Bob Smith, opns mgr.

KSJJ(FM)— Feb 4, 1981: 102.9 mhz; 100 kw. 885 ft TL: N44 10 25 W121 16 29. Stereo. Hrs opn: 24 969 S.W. Colorado, Bend, 97702. Secondary address: 1500 N.E. Butler Market Rd., Bend 97702. Phone: (541) 388-3300. Fax: (541)389-7885. Web Site: www.ksjj.com. Licensee: GCC Bend LLC. (group owner; (acq 1999; grpsl). Law Firm: Arent, Fox, Kintner, Plotkin & Kahn. Format: Country. News staff: one; News: 8 hrs wkly. Target aud: 25-54. ♦Dana Horner, stn mgr & sls dir.

***KWRX(FM)**— 2002: 88.5 mhz; 250 w. Ant 2,198 ft TL: N44 26 14 W120 57 12. (CP: 720 w). Hrs opn: Agate Hall, Univ. of Oregon, 97403. Phone: (541) 345-0800. E-mail: kwax@qwest.net Web Site: www.kwax.com. Licensee: State Board of Higher Education for the University of Oregon. Format: Classical. ♦Paul C. Bjornstad, gen mgr.

Reedsport

KDUN(AM)— June 2, 1961: 1030 khz; 50 kw-D, 630 w-N. TL: N43 44 17 W124 04 30. Hrs open: 24 136 N. 7th St., 97467. Phone: (541) 271-1030. Fax: (541) 271-2598. E-mail: traffic.kdun@gmail.com Web Site: www.kdun.com. Licensee: Pamplin Broadcasting-Oregon Inc. Group owner: Pamplin Broadcasting (acq 10-29-99). Population served: 50,000 Natl. Network: CBS Radio. Rgnl rep: McGavren-Guild Format: News, Talk. Target aud: 25 plus. ♦Bill Schweitzer, gen mgr; Joe Zelinski, opns mgr.

KJMX(FM)— 1993: 99.5 mhz; 11 kw. Ant 400 ft TL: N43 40 40 W124 06 36. Hrs open: 24 Box180, Coos Bay, 97467. Phone: (541) 267-2121. Fax: (541) 267-5229. Web Site: www.southcoastradio.com/themix.html. Licensee: Bicoastal CB LLC. Group owner: Bicoastal Media L.L.C. (acq 10-16-2003; grpsl). Format: Country. News staff: one; News: 2 hrs wkly. ♦John Pundt, gen mgr; Mike O'Brien, opns mgr.

***KLFR(FM)**— 1999: 89.1 mhz; 1 kw. 400 ft TL: N43 43 21 W124 05 40. Hrs open: 24 4000 E. 30th Ave., Eugene, 97405-0640. Phone: (541) 463-6000. Fax: (541) 463-6046. E-mail: klcc@lanecc.edu Web Site: www.klcc.org. Licensee: Lane Community College (acq 1-5-01; $32,500 for CP). Format: News/talk, adult contemp. ♦Steve Barton, gen mgr; Paula Chan Carpenter, dev dir; Gayle Chisholm, prom mgr; Don Heim, progmg dir.

***KSYD(FM)**— March 1990: 92.1 mhz; 3 kw. 358 ft TL: N43 39 26 W124 11 10. Hrs open: 24
KRVM FM 92.1.
1574 Coburg Rd. # 237, Eugene, 97401. Phone: (541) 687-3370. Fax: (541) 687-3573. Web Site: www.krvm.org. Licensee: School District 4J Lane County. Format: AAA. Target aud: 12-40. ♦Carl Sundberg, gen mgr & chief of engrg; Bobbie Cirel, dev dir & gen sls mgr; Ken Martin, progmg dir.

Rockaway Beach

***KLON(FM)**— 2005: 90.3 mhz; 1.8 kw vert. Ant 342 ft TL: N45 36 18 W123 55 30. Hrs open: 24
Rebroadcasts KLVR(FM) Santa Rosa, CA 100%.
2351 Sunset Blvd., Suite 170-218, Rocklin, CA, 95765. Phone: (916) 251-1600. Fax: (916) 251-1650. E-mail: klove@klove.com Web Site: www.klove.com. Licensee: Educational Media Foundation. Group owner: EMF Broadcasting. Natl. Network: K-Love. Law Firm: Shaw Pittman. Format: Contemp Christian. News staff: 3. Target aud: 25-44; Judeo Christian, female. ♦Richard Jenkins, pres; Mike Novak, VP; Ed Lenane, opns dir & news dir; Keith Whipple, dev dir; David Pierce, progmg mgr; Sam Wallington, engrg dir; Karen Johnson, news rptr; Marya Morgan, news rptr; Richard Hunt, news rptr.

Rogue River

KRRM(FM)— October 1994: 94.7 mhz; 130 w. 2,043 ft TL: N42 26 44 W123 12 56. Hrs open: 24 225 Rogue River Hwy., Grants Pass, 97527. Phone: (541) 479-6497. Fax: (541) 479-5726. E-mail: krrm@krrm.com Web Site: www.krrm.com. Licensee: Shirley M. Bell. Population served: 99,100 Format: Classic country. Target aud: 35 plus. ♦Herb Bell, gen mgr, opns dir & progmg mgr; Shirley Bell, opns dir & sls dir.

Roseburg

***KMPQ(FM)**— Nov 24, 2004: 88.1 mhz; 950 w. Ant 351 ft TL: N43 12 22 W123 21 50. Hrs open: 4000 E. 30th Ave., Eugene, 97405-0640. Phone: (541) 463-6000. Fax: (541) 463-6046. E-mail: klcc@lanecc.edu Web Site: www.klcc.org. Licensee: Lane Community College. Natl. Network: NPR. Format: News/talk, adult contemp. ♦Steve Barton, gen mgr; Paula Carpenter, dev dir; Don Hein, progmg dir.

KQEN(AM)— Sept 19, 1950: 1240 khz; 1 kw-U. TL: N43 11 44 W123 21 33. Hrs open: 24 Box 5180, 97470. Phone: (541) 672-6641. Fax: (541) 673-7598. Web Site: www.am1240kqen.com. Licensee: Brooke Communications Inc. (group owner; acq 5-1-86; $173,000). Population served: 90,000 Natl. Network: ESPN Radio. Natl. Rep: Tacher. Law Firm: Haley, Bader & Potts. Format: News/ talk, sports. News staff: 2; News: 4 hrs wkly. Target aud: 35 plus; general. Spec prog: Sports. ♦Patrick A. Markham, pres; Mike Carter, opns mgr; Brian Prawitz, news dir.

KRNR(AM)— August 1935: 1490 khz; 1 kw-U. TL: N43 11 35 W123 21 39. Hrs open: 24 1445 W. Harvard Ave., 97470. Phone: (541) 672-6641. Phone: (541) 673-5551. Fax: (541) 673-7598. Web Site: classic1490krnr.com. Licensee: Brooke Communications Inc. (acq 1-14-2005). Population served: 55,000 Natl. Network: CBS. Natl. Rep: Tacher. Law Firm: Garvey, Schubert & Barer. Format: Classic country. News staff: 2. Target aud: 25 plus. ♦Patrick A. Markham, pres; Mike Carter, gen mgr, progmg dir & mus dir; Pam Houck, gen sls mgr.

KRSB-FM— Oct 1, 1970: 103.1 mhz; 2.75 kw. 308 ft TL: N43 12 24 W123 21 47. (CP: 25.5 kw, ant 676 ft. TL: N43 13 59 W123 19 22). Stereo. Hrs opn: 24 1445 W. Harvard Ave., 97470. Phone: (541) 672-6641. Fax: (541) 673-7598. E-mail: country@bciradio.com Web Site: www.bestcountry103.com. Licensee: Brooke Communications Inc. (group owner; acq 4-30-89). Population served: 90,000 Natl. Network: ABC. Natl. Rep: Tacher. Law Firm: Garvey Schubert Barer. Format: Contemp country. News staff: 2; News: 15 hrs wkly. Target aud: 25-54. ♦Patrick A. Markham, pres & gen mgr; Mike Carter, chief of opns.

***KSRS(FM)**— December 1990: 91.5 mhz; 2 kw. 305 ft TL: N43 12 24 W123 21 47. Stereo. Hrs opn: 5 AM-2 AM 1250 Siskiyou Blvd., Ashland, 97520. Phone: (541) 552-6301. Fax: (541) 552-8565. Web Site: www.ijpr.org. Licensee: The State of Oregon, Acting By and Through the State Board of Higher Education, for the benefit of Southern Oregon University. Natl. Network: NPR, PRI. Law Firm: Ernest Sanchez. Wire Svc: AP Format: Class, news. News staff: one; News: 35 hrs wkly. Target aud: General. ♦Ronald Kramer, CEO & gen mgr; Bryon Lambert, opns dir; Paul Westhelle, dev dir.

***KTBR(AM)**— November 1955: 950 khz; 1 kw-D, 20 w-N. TL: N43 10 08 W123 22 28. Hrs open: 24 hrs Jefferson Public Radio, 1250 Siskiyou Blvd., Ashland, 97520. Phone: (541) 552-6301. Web Site: www.ijpr.org. Licensee: JPR Foundation Inc. (acq 4-19-02; $83,700 with KOOZ(FM) Myrtle Point). Population served: 75,000 Natl. Network: NPR, PRI. Law Firm: Ernest Sanchez. Format: News info. News staff: one. Target aud: General. ♦Mitchell Christian, CFO; Ronald Kramer, CEO & gen mgr; Bryon Lambert, opns dir; Paul Westhelle, dev dir.

Saint Helens

KOHI(AM)— Mar 2, 1960: 1600 khz; 1 kw-D, 12 w-N. TL: N45 51 15 W122 49 11. Hrs open: 24
Talkstar Radio Network.
36200 Pittsburg Rd., Ste C, 97051-1188. Phone: (503) 397-1600. Fax: (503) 397-1601. E-mail: kohiradio@gmail.com Web Site: www.am1600kohi.com. Licensee: Volcano Broadcasting. (acq 5-21-82; $150,000; FTR: 6-14-82). Population served: 26,000 Natl. Rep: Keystone (unwired net). Rgnl rep: Tacher Format: Talk, sports, news. News staff: one; News: 12 hrs wkly. Target aud: Adults 25-65. Spec prog: Sports Talk, Religion. ♦David Aldridge, pres; Marty Rowe, VP & gen mgr; Alex Rowe, progmg dir; Thad Houk, sports cmtr.

Salem

***KAJC(FM)**— 2006: 90.1 mhz; 560 w. Ant 128 ft TL: N44 45 33 W123 13 34. Hrs open: 1399 Monmouth St., Independence, 97351. Phone: (503) 838-2476. Fax: (503) 838-2476. E-mail: kajc@kajcfm.org Web Site: www.kajcfm.org. Licensee: CSN International (group owner). Format: Relg.

KBZY(AM)— May 1957: 1490 khz; 1 kw-U. TL: N44 57 03 W123 02 43. Hrs open: 24 2659 Commercial St. SE, Ste 204, 97302-4496. Phone: (503) 362-1490. Fax: (503) 362-6545. Web Site: www.kbzy.com. Licensee: Capital Broadcasting Inc. (acq 6-15-82; $365,000; FTR: 7-5-82). Population served: 415,000 Rgnl rep: Tacher Format: Local svc oldies. Target aud: 25-54. ♦Roy Dittman, pres & gen mgr; Terry Sol, progmg dir.

KGAL(AM)—Lebanon

KKSN(AM)— 1934: 1390 khz; 5 kw-D, 690 kw-N. TL: N44 59 43 W123 04 15. Hrs open:
Simulcasts KFXX(AM) Portland 100%.
0700 S.W. Bancroft St., Portland, 97239. Phone: (503) 223-1441. Fax: (503) 223-6909. Web Site: www.1080thefan.com. Licensee: Entercom Portland License LLC. Group owner: Entercom Communications Corp. (acq 10-22-98; $605,000). Population served: 300,000 Natl. Network: ESPN Radio. Rgnl rep: Allied Radio Partners. Format: Sports. ♦David Field, pres; Jack Hutchison, exec VP.

KPJC(AM)— Dec 12, 1961: 1220 khz; 1 kw-D, 171 w-N. TL: N44 58 57 W123 00 17. Hrs open: 24 Box 17008, 97305. Phone: (503) 316-1220. Fax: (503) 364-1022. Web Site: www.thejctown.com. Licensee: KCCS LLC (acq 4-15-2004; $500,000). Population served: 250,000 Natl. Network: USA. Law Firm: Reddy, Begley & McCormick. Format: Christian Family Radio. News: 14 hrs wkly. Target aud: 25-54; family. ♦Christina Evans, gen mgr; Phil Swearingin, opns mgr.

KSHO(AM)—Lebanon

***KWBX(FM)**— Apr 1, 2002: 90.3 mhz; 135 w vert. Ant 46 ft TL: N44 52 57 W122 57 34. Stereo. Hrs opn: 24 Corban College, 5000 Deer Park Dr. S.E., 97301. Phone: (503) 375-7591. Fax: (503) 585-4316. E-mail: kwbx@corban.edu Web Site: www.corban.edu/radio. Licensee: Corban College. Population served: 200,000 Law Firm: Reddy, Begley & McCormick. Format: Christian hit music, positive alternative. News staff: one; News: one hr wkly. Target aud: 25-44; young adults. ♦Dr. Reno Hoff, pres; Steve Hunt, gen mgr; Josh Bartlett, stn mgr; Steve Hendrix, chief of engrg.

KWIP(AM)—Dallas, Apr 15, 1955: 880 khz; 5 kw-D, 1 kw-N. TL: N44 55 45 W123 17 22. Stereo. Hrs opn: 1405 E. Ellendale, Dallas, 97338. Phone: (503) 623-0245. Fax: (503) 623-6733. Web Site: www.kwip.com. Licensee: Jupiter Communications Corp. (acq 6-10-91; $21,000; FTR: 7-1-91). Population served: 200,000 Format: Rgnl Mexican. Target aud: 18-54; families & blue collar workers. Spec prog: Talk 5 hrs wkly. ♦Diana Burns, gen mgr.

KYKN(AM)—See Keizer

Scappoose

KFIS(FM)— May 1986: 104.1 mhz; 6.9 kw. Ant 1,266 ft TL: N45 29 20 W122 41 40. Hrs open: 6400 S.E. Lake Rd., Suite 350, Portland, 97222. Phone: (503) 786-0600. Fax: (503) 786-1551. Web Site: www.1041thefish.com. Licensee: Caron Broadcasting Inc. Group owner: Salem Communications Corp. (acq 9-20-2001; $35.8 million). Population served: 25,000 Law Firm: Gardner, Carton & Douglas. Format: Christian. Target aud: 25-54; women. ♦Dennis Hayes, gen mgr; Leslie Pfau, mktg mgr; Dave Arthur, progmg dir.

Seaside

KCRX-FM— 1998: 102.3 mhz; 25 kw. Ant 328 ft TL: N45 57 08 W123 56 14. Hrs open: 1006 W. Marine St., Astoria, 97103. Phone: (503) 325-2911. Fax: (503) 325-5570. E-mail: kcrx@nnbradio.com Web Site: www.kcrx1023.com. Licensee: New Northwest Broadcasters LLC (group owner; acq 8-24-99; grpsl). Format: Classic rock. ♦Paul Mitchell, gen mgr; Tom Freel, opns mgr; Bob Castle, progmg dir.

KCYS(FM)— Nov 26, 1996: 98.1 mhz; 6 kw. 174 ft TL: N45 57 8 W123 56 14. Stereo. Hrs opn: 24 Box 1258, Astoria, 97103. Secondary address: 1324 N. Holladay Dr. 97138. Phone: (503) 717-9643. Fax: (503) 717-9578. Licensee: Dave's Broadcasting Co. (acq 6-23-2005; for 66.66% of stock). Population served: 40,000 Rgnl rep: Tacher. Format: Country. Target aud: 35-44; working moms with kids, some college. ♦Dave Heick, gen mgr.

KSWB(AM)— July 12, 1968: 840 khz; 1 kw-D, 500 w-N. TL: N45 58 55 W123 55 02. Hrs open: 1006 W. Marine Dr., Astoria, 97103. Phone: (503) 325-2911. Fax: (503) 325-5570. Licensee: Cannon Beach Radio (acq 3-10-2000). Format: Oldies. ♦Jim Servino, gen mgr.

Selma

***KJKL(FM)**— 2003: 88.7 mhz; 9 kw vert. Ant 1,916 ft TL: N42 15 29 W123 39 32. Hrs open: 2351 Sunset Blvd., Suite 170-218, Rocklin, CA, 95765. Phone: (916) 251-1600. Fax: (916) 251-1650. E-mail: klove@klove.com Web Site: www.klove.com. Licensee: Educational Media Foundation. Group owner: EMF Broadcasting. Natl. Network: K-Love. Law Firm: Shaw Pittman. Format: Contemp Christian. News staff: 3. Target aud: 25-44; Judeo Christian, female. ♦Richard Jenkins, pres; Mike Novak, VP; Keith Whipple, dev dir; David Pierce, progmg mgr; Ed Lenane, news dir; Sam Wallington, engrg dir; Karen Johnson, news rptr; Marya Morgan, news rptr; Richard Hunt, news rptr.

Sisters

***KVRA(FM)**— 2006: 89.3 mhz; 1.4 kw vert. Ant 633 ft TL: N44 04 40 W121 19 48. Hrs open:
Rebroadcasts KLRD(FM) Yucaipa, CA 100%.
2351 Sunset Blvd., Suite 170-218, Rocklin, CA, 95765. Phone: (916) 251-1600. Fax: (916) 251-1650. Web Site: www.air1.com. Licensee: Educational Media Foundation. Natl. Network: Air 1. Format: Alternative rock, div. ♦Richard Jenkins, pres; Mike Novak, VP; Keith Whipple, dev dir; David Pierce, progmg mgr; Ed Lenane, news dir; Sam Wallington, engrg dir; Karen Johnson, news rptr; Marya Morgan, news rptr; Richard Hunt, news rptr.

KWPK-FM— 2001: 104.1 mhz; 34 kw. Ant 590 ft TL: N44 04 40 W121 19 49. Hrs open: 24 854 N.E. 4th St., Bend, 97701. Phone: (541) 383-3825. Fax: (541) 383-3403. Web Site: www.thepeak1041.com. Licensee: Horizon Broadcasting Group LLC. (acq 3-31-2005; $475,000). Natl. Rep: Tacher. Format: Hot adult contemp. Target aud: 18-49. ♦Keith Shipman, pres; Brian Canady, gen sls mgr; Paul Valle, news dir; Ramona Hartmann, traf mgr.

Springfield

***KQFE(FM)**— Mar 7, 1989: 88.9 mhz; 2 kw. 418 ft TL: N44 02 01 W123 00 25. Hrs open: 24 4135 Northgate Blvd., Sacramento, CA, 95834. Phone: (800) 835-4810. Fax: (541) 726-9156. Web Site: www.familyradio.com. Licensee: Family Stations Inc. (group owner) Natl. Network: Family Radio. Format: Relg. Target aud: 30 plus; older relg. ♦Harold Camping, gen mgr; Carmen Brambora, opns mgr.

KSCR(AM)—See Eugene

Springfield-Eugene

KKNU(FM)—Licensed to Springfield-Eugene. See Eugene

KORE(AM)— September 1927: 1050 khz; 5 kw-D, 149 w-N. TL: N44 04 07 W123 01 45. Hrs open: 24 2080 Laura St., 97477-2197. Phone: (541) 747-5673. E-mail: kore@kore1050am.com Licensee: Support Christian Broadcasting Inc. (acq 8-87). Population served: 800,000 Natl. Network: USA. Format: Christian. Target aud: 18 plus. ♦Larry Knight, gen mgr.

KPNW(AM)—See Eugene

KSCR(AM)—See Eugene

KUGN(AM)—See Eugene

Stanfield

KLKY(FM)— 2005: 96.1 mhz; 8.5 kw. Ant 1,178 ft TL: N45 29 12 W119 25 52. Hrs open: 45 Campbell Rd., Walla Walla, WA, 99362. Phone: (509) 527-1000. Fax: (509) 529-5534. Licensee: Alexandra Communications Inc. Format: Top-40. ♦Tom Hodgins, gen mgr.

Stayton

KCKX(AM)— June 1, 1987: 1460 khz; 1 kw-D, 15 w-N. TL: N44 48 10 W122 44 03. Hrs open: 24 1665 James St., Woodburn, 97071. Phone: (503) 981-9400. Phone: (503) 769-1460. Fax: (503) 981-3561. E-mail: don@cowboycountryradio.net Web Site: www.cowboycountryradio.net. Licensee: Sanlee Broadcasting Corp. (acq 1-21-98; $130,000). Population served: 600,000 Natl. Network: ABC. Rgnl rep: Allied Radio Partners. Format: Classic country/western. News: 3 hrs wkly. Target aud: 25 plus; stable, mature adults with above average income. Spec prog: Portland Trailblazers basketball, Forest Dragons arena football, high school sports, farm 15 hrs wkly. ♦Donald Coss, pres & exec VP; Chris McCartney, gen mgr; Andy McGarrett, sls.

Sunriver

KRXF(FM)— 2006: 92.7 mhz; 18.5 kw. Ant 813 ft TL: N44 02 49 W121 31 50. Hrs open: 3718 Country Trails, Bonita, CA, 91902. Phone: (619) 475-7200. Web Site: www.927fmradio.com. Licensee: Fields Pond Group. Format: Modern rock. ♦Charles M. Wilkinson, gen mgr.

Sutherlin

KAVJ(FM)— 1999: 101.1 mhz; 3.6 kw. Ant 859 ft TL: N43 22 19 W123 21 15. Hrs open: 24 1445 W. Harvard Ave., Roseburg, 97470. Phone: (541) 672-6641. Fax: (541) 673-7598. E-mail: coolfm@bciradio.com Web Site: www.cool101fm.com. Licensee: Brooke Communications Inc. (group owner; (acq 12-16-2002). Population served: 90,000 Rgnl rep: Tacher Law Firm: Garvey Shubert Barer. Format: Oldies. News staff: 2; News: 24 hrs wkly. Target aud: 25-54; 60% female, 40% male. Spec prog: Black one hr, Jazz 2 hrs wkly. ♦Pat Markham, pres & gen mgr; Mike Carter, opns dir & progmg dir; Pam Houck, gen sls mgr.

Sweet Home

KFIR(AM)— Aug 7, 1968: 720 khz; 1 kw-D, 184 w-N. TL: N44 24 52 W122 44 22. Hrs open: Box 720, 28041 Pleasant Valley Rd., 97386. Phone: (541) 367-5115. Fax: (541) 367-5233. Licensee: Radio Fiesta Network LLC (acq 4-13-2007; $500,000). Population served: 250,000 Rgnl rep: Allied Broadcast Partners. Format: Traditional country. Target aud: 25 plus. ♦Bob Ratter, gen mgr.

***KLVU(FM)**— Sept 20, 1989: 107.1 mhz; 9 kw. 2,476 ft TL: N44 28 59 W122 34 55. Stereo. Hrs opn: 24 2351 Sunset Blvd., Suite 170-218, Rocklin, CA, 95765. Phone: (916) 251-1600. Fax: (916) 251-1650. E-mail: klove@klove.com Web Site: www.klove.com. Licensee: Educational Media Foundation. Group owner: EMF Broadcasting (acq 3-12-97; $4 million). Population served: 1,500,000 Natl. Network: K-Love. Law Firm: Shaw Pittman. Format: Contemp Christian. News staff: 3. Target aud: 25-44; Judeo-Christian female. ♦Richard Jenkins, pres; Mike Novak, VP; Keith Whipple, dev dir; David Pierce, progmg mgr; Ed Lenane, news dir; Sam Wallington, engrg dir; Karen Johnson, news rptr; Marya Morgan, news rptr; Richard Hunt, news rptr.

Talent

***KSJK(AM)**— October 1960: 1230 khz; 1 kw-U. TL: N42 13 27 W122 44 33. Hrs open: 24 hrs Southern Oregon State College, 1250 Siskiyou Blvd., Ashland, 97520. Phone: (541) 552-6301. Fax: (541) 552-8565. Web Site: www.ijpr.org. Licensee: The State of Oregon, acting by and through the State Board of Higher Education. for the benefit of Southern Oregon University. (acq 7-28-89). Population served: 200,000 Natl. Network: PRI, NPR. Law Firm: Ernest Sanchez. Format: News, info. Target aud: General. Spec prog: Talk 6 hrs wkly.

♦Mitchell Christian, CFO; Ronald Kramer, CEO & gen mgr; Bryon Lambert, opns dir; Paul Westhelle, dev dir.

The Dalles

KACI(AM)— June 1955: 1300 khz; 1 kw-D, 13 w-N. TL: N45 34 54 W121 07 53. Hrs open: 24 502 Washington St., 97058-8003. Phone: (541) 296-2211. Fax: (541) 296-2213. Licensee: Columbia Gorge Broadcasters Inc. (group owner; acq 4-1-98; $390,000 with co-located FM). Population served: 40,000 Natl. Network: Jones Radio Networks. Law Firm: Fisher, Wayland, Cooper, Leader & Zaragoza L.L.P. Format: News/talk. News staff: one; News: 14 hrs wkly. Target aud: 25-54. Spec prog: Relg one hr, home improvement 3 hrs, financial talk 6 hrs, computer talk 3 hrs, farm one hr, gardening one hr, pub affrs one hr wkly. ♦Gary M. Grossman, CEO, pres, gen mgr, stn mgr & gen sls mgr; Greg LeBlanc, opns dir; Rick Cavagnaro, sls dir; Greg LaBlanc, news dir; Paulette LaRoque, traf mgr.

KACI-FM— Feb 1, 1985: 97.7 mhz; 5 kw. 890 ft TL: N45 38 56 W121 16 20. Stereo. 24 65,000 Format: Oldies. ♦Brian Thompson, progmg mgr; Paulette LaRoque, chief of opns & traf mgr; Greg LeBlanc, local news ed.

KMCQ(FM)— Nov 28, 1968: 104.5 mhz; 100 kw. Ant 1,998 ft TL: N45 42 44 W121 06 50. (CP: COL Covington, WA. 25 kw, ant 318 ft. TL: N47 11 13 W121 54 11). Stereo. Hrs opn: 24 Box 104, 719 E. 2nd St., 97058. Phone: (541) 298-5116. Phone: (541) 298-5117. Fax: (541) 298-5119. E-mail: q104@q104radio.com Web Site: www.q104radio.com. Licensee: First Broadcasting Capital Partners LLC. (acq 3-22-2007; $5.1 million). Population served: 80,000 Natl. Network: CNN Radio. Natl. Rep: McGavren Guild. Law Firm: FCC Attorney Dominic Monahan, Luvaas Cobb, Eugene, Or. Format: Adult contemp. News staff: one; News: 7 hrs wkly. Target aud: Females 25-49; mothers & family friendly. Spec prog: Blues 4 hrs, teen show 3 hrs wkly. ♦Gary Lawrence, pres; John Huffman, VP & gen mgr; Linda Griswold, sls dir & gen sls mgr; Paula Fairclo, opns dir & progmg dir.

KMSW(FM)— Oct 1, 2002: 92.7 mhz; 3.4 kw. Ant 892 ft TL: N45 38 56 W121 16 20. Stereo. Hrs opn: 502 Washington St., 97058-8003. Phone: (541) 296-2211. Fax: (541) 296-2213. Web Site: www.gorgeradio.com. Licensee: M.S.W. Communications L.L.C. Rgnl rep: Tacher Format: Classic rock. Target aud: 25-54. ♦Gary Grossman, pres & gen mgr; Rick Cavagnaro, sls dir.

KODL(AM)— Oct 12, 1940: 1440 khz; 5 kw-D, 1 kw-N, DA-N. TL: N45 35 31 W121 11 57. Hrs open: Box 1488, 97058. Secondary address: 404 E. 2nd St. 97058. Phone: (541) 296-2101. Fax: (541) 296-3766. E-mail: web-master@kodl.net Web Site: www.kodl.com. Licensee: Larson-Wynn Inc. (acq 9-1-74). Population served: 55,000 Format: Adult standard. Spec prog: Farm 4 hrs, Sp 2 hrs wkly. ♦Al Wynn, pres & gen mgr; Marcia Wynn, opns dir.

Tigard

KXPD(AM)—Licensed to Tigard. See Portland

Tillamook

***KAIK(FM)**— 2006: 88.5 mhz; 60 w vert. Ant 1,276 ft TL: N45 27 59 W123 55 11. Hrs open:
Rebroadcasts KLRD(FM) Yucaipa, CA 100%.
2351 Sunset Blvd., Sute 170-218, Rocklin, CA, 95765. Phone: (916) 251-1600. Fax: (916) 251-1650. E-mail: info@air1.com Web Site: www.air1.com. Licensee: Educational Media Foundation. Group owner: EMF Broadcasting. Natl. Network: Air 1. Law Firm: Shaw Pittman. Format: Contemp Christian. News staff: 3. Target aud: 18-35; Judeo Christian female. ♦Richard Jenkins, pres; Mike Novak, VP; Keith Whipple, dev dir; David Pierce, progmg mgr; Ed Lenane, news dir; Sam Wallington, engrg dir; Karen Johnson, news rptr; Marya Morgan, news rptr; Richard Hunt, news rptr.

KMBD(AM)— August 1947: 1590 khz; 5 kw-D, 1 kw-N, DA-N. TL: N45 27 24 W123 52 36. Hrs open: Box 40, 97141. Secondary address: 170 W. 3rd St. 97141. Phone: (503) 842-4422. Fax: (503) 842-2755. E-mail: comments@ktil-kmbd.com Web Site: www.ktil-kmbd.com. Licensee: Oregon Eagle Inc. (acq 12-29-86; $250,000; grpsl; FTR: 10-26-86) Population served: 4,500 Format: News/talk, sports. ♦Van Moe, pres & gen mgr.

***KTCB(FM)**— Aug 25, 2004: 89.5 mhz; 380 w. Ant 1,151 ft TL: N45 27 59 W123 55 11. Stereo. Hrs opn: 24 Box 269, Astoria, 97103. Phone: (503) 325-0010. Fax: (503) 325-3956. E-mail: kmun@kmun.org Web Site: www.kmun.org. Licensee: Tillicum Foundation (acq 1-23-2003; swap for KTMK(FM) Tillamook). Natl. Network: NPR. Law Firm: Garvey, Schubert & Barer. Wire Svc: AP Format: Public/Eclectic. News: 35 hrs wkly. ♦David Hammock, gen mgr; Arlene Layton, dev dir; Elizabeth Grant, progmg dir; Joanne Rideout, news dir.

KTIL-FM— October 1998: 94.3 mhz; 1.8 kw. Ant 1,171 ft TL: N45 27 59 W123 55 11. Stereo. Hrs opn: 24 Box 40, 170 3rd St., 97141. Phone: (503) 842-4422. Fax: (503) 842-2755. E-mail: comments@ktil-kmbol.com Web Site: www.ktil-kmbd.com. Licensee: Oregon Eagle Inc. (acq 2-4-99). Format: MOR, Music of your life. Target aud: General. ♦Van Moe, pres & gen mgr.

***KTMK(FM)**— 2005: 91.1 mhz; 140 w. Ant 1,168 ft TL: N45 27 59 W123 55 11. Hrs open: Oregon Public Broadcasting, 7140 S.W. Macadam Ave., Portland, 97219-3099. Phone: (503) 244-9900. Phone: (503) 244-1905. Fax: (503) 293-1919. Web Site: www.opb.org. Licensee: Oregon Public Broadcasting (acq 1-16-2003; swap for KTCB(FM) Tillamook). ♦Maynard Orme, gen mgr.

Toledo

KCUP(AM)— Sept 26, 1960: 1230 khz; 1 kw-U. TL: N44 37 47 W123 56 35. Hrs open: Box 456, 145 N. Coast Hwy., Newport, 97365. Phone: (541) 265-5000. Fax: (541) 265-9576. Licensee: Agpal Broadcasting Inc. (acq 3-14-90; grpsl; FTR: 4-2-90). Population served: 34,000 Natl. Rep: McGavren Guild. Law Firm: Haley, Bader & Potts. Format: Oldies. Target aud: 25-54. ♦Cheryl Harle, gen mgr; Ed Kowas, gen sls mgr.

KPPT-FM—Co-owned with KCUP(AM). December 1980: 100.7 mhz; 3 kw. 430 ft TL: N48 38 40 W124 00 52. Stereo. 43,000 Format: Classic rock. ♦Cheryl Harle, VP.

Tri City

KKMX(FM)— June 1, 1993: 104.3 mhz; 5.6 kw. Ant 1,384 ft TL: N43 00 13 W123 21 26. Stereo. Hrs opn: 24 1445 W. Harvard Ave., Roseburg, 97470. Phone: (541) 672-6641. Fax: (541) 673-7598. E-mail: kissfm@bciradio.com Web Site: www.1045kiss.com. Licensee: Brooke Communications Inc. (group owner; acq 11-21-96). Population served: 90,000 Natl. Rep: Tacher. Law Firm: Garvey, Schubert & Barer. Format: Adult contemp. News staff: 2; News: 2 hrs wkly. Target aud: 25-54; general. ♦Pat Markham, CEO, pres & gen mgr; Mike Carter, opns dir.

Troutdale

KPAM(AM)— 1997: 860 khz; 50 kw-D, 5 w-N. TL: N45 33 24 W122 29 08. Hrs open: 24 6605 S.E. Lake Rd., Portland, 97222. Phone: (503) 223-4321. Fax: (503) 294-0074. E-mail: email@kpam.com Web Site: www.kpam.com. Licensee: Pamplin Broadcasting-Oregon Inc. Group owner: Pamplin Broadcasting (acq 12-29-97; $652,500 for 87% of stock). Natl. Network: ABC. Natl. Rep: Tacher. Rgnl rep: The Tacher Co., Inc. Wire Svc: AP Format: News/talk. News staff: 11; News: 35.4 hrs wkly. Target aud: 35-54; adults. Spec prog: Wall St. Journal. ♦Paul Clithero, gen mgr; Mark Ail, opns dir; Margaret Evans, sls dir & gen sls mgr; Jeanne Winters, natl sls mgr; Misty Osko, prom mgr; Bill Gallagher, progmg dir & news dir; Paul Duckworth, progmg dir; Dave Bischoff, chief of engrg; Paul Blaviding, traf mgr; Robert Eisinger, political ed; Phil Cassidy, sports cmtr.

Umatilla

KLWJ(AM)— June 1980: 1090 khz; 2.5 kw-D. TL: N45 52 46 W119 20 37. Hrs open: 6 AM-sunset 80898 Powerline Rd., 97882. Phone: (541) 567-2102. Fax: (541) 567-2103. E-mail: klwjradio@hotmail.com Licensee: Umatilla Broadcasting Inc. Population served: 250,000 Natl. Network: USA. Format: Relg, news/talk, contemp Christian mus. News staff: one. Target aud: General. Spec prog: Farm one hr, Sp one hr wkly. ♦Darrell Marlow, pres; John Marlow, VP, gen mgr & opns mgr.

Veneta

KEUG(FM)— 1998: 105.5 mhz; 2.8 kw. Ant 994 ft TL: N44 00 11 W123 06 48. Hrs open: 24 925 Country Club Rd., Suite 200, Eugene, 97401. Phone: (541) 484-9400. Fax: (541) 344-9424. Web Site: bob1055.com. Licensee: McKenzie River Broadcasting Co. Inc. Group owner: McKenzie River Broadcasting Group (acq 1-28-2004; $1.02 million). Natl. Rep: D & R Radio. Law Firm: Holland & Knight. Format: Adult contemp, classic hits. Target aud: 25-54; adults. ♦John Tilson, pres & gen mgr; Dave Wiles, gen sls mgr; Jeff Baird, progmg dir.

Waldport

KORC(AM)— July 1, 1988: 820 khz; 1000 w-D, 15 w-N. TL: N44 26 05 W124 01 20. Hrs open: Box 495, 97394. Phone: (541) 563-5100. Fax: (541) 563-5116. Licensee: Larry D. and Margaret E. Profitt, a General Partnership (acq 10-7-2003; $185,000). Format: Easy lstng. ♦Larry Profitt, gen mgr.

Warm Springs

KWLZ-FM— Jan 18, 1986: 96.5 mhz; 100 kw. Ant 1,092 ft TL: N44 50 24 W121 13 56. Stereo. Hrs opn: 24 854 N.E. 4th St., Bend, 97701. Phone: (541) 383-3825. Fax: (541) 383-3403. Web Site: www.lazer965.com. Licensee: Horizon Broadcasting Group L.L.C. (group owner; acq 3-2-2000; grpsl). Natl. Rep: Tacher. Format: Classic rock. Target aud: 25-64. ♦Keith Shipman, pres & gen mgr; Larry Wilson, opns mgr; Brian Canady, gen sls mgr; Paul Valle, news dir; Ramona Hartmann, traf mgr.

***KWSO(FM)**— Sept 22, 1986: 91.9 mhz; 3.3 kw. 203 ft TL: N44 50 24 W121 13 56. Stereo. Hrs opn: 18 Box 489, 97761. Secondary address: 97761 Kahneeta Hamlet Rd. 97761. Phone: (541) 553-1965. Fax: (541) 553-3348. E-mail: kwso@wstribes.org Web Site: www.ksmo.org. Licensee: Confederated Tribes of Warm Springs. Format: Native American, adult contemp. Target aud: General. ♦Sue Matters, gen mgr.

Warrenton

***KCPB-FM**— Apr 17, 2006: 90.9 mhz; 9 w. Ant 1,040 ft TL: N46 15 46 W123 53 09. Hrs open: Box 263, Astoria, 97103. Phone: (503) 325-0010. Fax: (503) 325-3956. Web Site: www.kmun.org. Licensee: Tillicum Foundation (acq 3-19-2004). Natl. Network: NPR. ♦Ray Merritt, pres.

Welches

***KZRI(FM)**— May 10, 2001: 90.3 mhz; 280 w. Ant 1,568 ft TL: N45 19 57 W121 42 57. Stereo. Hrs opn: 24 2351 Sunset Blvd., Suite 170-218, Rocklin, CA, 95765. Phone: (916) 251-1600. Fax: (916) 251-1650. E-mail: info@air1.com Web Site: www.air1.com. Licensee: Educational Media Foundation. Group owner: EMF Broadcasting. Population served: 38,700 Natl. Network: Air 1. Law Firm: Shaw Pittman. Format: Contemp Christian. News staff: 3. Target aud: 18-35; Judeo-Christian, female. ♦Richard Jenkins, pres; Mike Novak, VP; Keith Whipple, dev dir; David Pierce, progmg mgr; Ed Lenane, news dir; Sam Wallington, engrg dir; Arthur Vassar, traf mgr; Karen Johnson, news rptr; Marya Morgan, news rptr; Richard Hunt, news rptr.

West Klamath

KRAM(AM)— Dec 1, 1987: Stn currently dark. 1070 khz; 1 kw-D. TL: N42 10 38 W121 46 25. Hrs open: Sunrise-sunset Box 1270, Klamath Falls, 97601. Phone: (541) 884-8074. Fax: (541) 884-8226. Licensee: Scott D. MacArthur, Personal Rep., estate of Sandra A. Falk (acq 8-8-2007). ♦Scott D. MacArthur, gen mgr.

Weston

KUJJ(FM)— 1997: 101.9 mhz; 13.5 kw. Ant 958 ft TL: N45 47 41 W118 10 06. Stereo. Hrs opn: 24 45 Campbell Rd., Walla Walla, WA, 99362. Phone: (509) 527-1000. Fax: (509) 529-5534. Licensee: Alexandra Communications Inc. (group owner; (acq 2-22-2006; swap for KMMG(FM) Milton-Freewater). Format: Smooth jazz. ♦Tom Hodgins, gen mgr.

Winchester

***KLOV(FM)**— August 1997: 89.3 mhz; 3.8 kw. Ant 420 ft TL: N43 14 06 W123 19 20. Stereo. Hrs opn: 24 2351 Sunset Blvd., Suite 170-218, Rocklin, CA, 95765. Phone: (916) 251-1600. Fax: (916) 251-1650. E-mail: klove@klove.com Web Site: www.klove.com. Licensee: Educational Media Foundation. Group owner: EMF Broadcasting. Population served: 74,000 Natl. Network: K-Love. Law Firm: Shaw Pittman. Format: Contemp Christian mus. News staff: 3. Target aud: 25-44; Judeo-Christian, female. ♦Richard Jenkins, pres; Mike Novak, VP; Keith Whipple, dev dir; David Pierce, progmg mgr; Ed Lenane, news dir; Sam Wallington, engrg dir; Arthur Vassar, traf mgr; Karen Johnson, news rptr; Marya Morgan, news rptr; Richard Hunt, news rptr.

Winston

KGRV(AM)— Feb 12, 1984: 700 khz; 25 kw-D, 500 w-N. TL: N43 03 26 W123 23 48. Hrs opn: 24 Box 1598, 97496. Secondary address: 196 S.E. Main St. 97496. Phone: (541) 679-8185. Fax: (541) 679-6456. E-mail: info@kgru700.net Web Site: www.kgrv700.net. Licensee: Pacific Cascade Communications Corp. (acq 4-15-85). Natl. Network: Moody. Format: Christian, regl, inspirational music. News: 10 hrs wkly. Target aud: 25-54. Spec prog: Southern gospel 3 hrs wkly. ♦David Morrow, pres & exec VP; Phil Morrow, gen mgr.

Woodburn

KWBY(AM)— July 10, 1964: 940 khz; 10 k-D, 500 w-N. TL: N45 10 37 W122 50 58. Hrs opn: 24 1665 James St., 97071. Phone: (503) 981-9400. Fax: (503) 981-3561. E-mail: ddc@lapantera940.com Web Site: www.lapantera940.com. Licensee: Donald D. Coss (acq 10-18-91; $300,000; FTR: 11-4-91). Population served: 1,250,000 Natl. Network: CNN Radio. Natl. Rep: Lotus Entravision Reps LLC. Format: Sp/rgnl Mexican. News staff: one; News: 35 hrs wkly. Target aud: 18-49; younger-larger-than-gen mkt average Hispanic families. Spec prog: Relg 5 hrs, gospel 3 hrs wkly. ♦Donald Coss, pres; Dorecia Luse, gen mgr & opns dir; Natasha Holstein, mktg VP & prom VP; Gilberto Galvan, progmg mgr.

Pennsylvania

Allentown

WAEB(AM)— 1949: 790 khz; 1 kw-U, DA-2. TL: N40 37 05 W75 26 58 (day), N40 39 37 W75 30 50 (night). (CP: 3.8 kw-D, 1.5 kw-N, DA-2. TL (night): N40 39 33 W75 30 48). Hrs opn: 1541 Alta Dr., Suite 400, Whitehall, 18052. Phone: (610) 434-1742. Phone: (610) 434—3808 (News). Fax: (610) 434-6288. Web Site: www.waeb.com. Licensee: Capstar TX L.P. Group owner: Clear Channel Communications Inc. (acq 8-30-00; grpsl). Population served: 109,527 Natl. Network: CBS. Format: News/talk. Target aud: 35-64. ♦Chris Taylor, gen mgr; Pat Gremling, gen sls mgr; Leanne Costelli, rgnl sls mgr; Laura St. James, progmg dir.

WAEB-FM— June 30, 1961: 104.1 mhz; 50 kw. 500 ft TL: N40 43 13 W75 35 44. (CP: 19.4 kw, ant 164 ft.). Stereo. Web Site: www.b104.com. Format: CHR. Target aud: 18-44. ♦Diane Lee, gen sls mgr & prom dir; Craig Stevens, progmg dir.

WBYN(AM)—Lehighton, Apr 12, 1962: 1160 khz; 4 kw-D, 1 kw-N, DA-2. TL: N40 49 03 W75 41 31. Stereo. Hrs opn: 24
Rebroadcasts WBYN-FM Boyertown 100%.
107 Paxinosa Rd. W., Easton, 18040-1344. Phone: (610) 258-6155. Fax: (610) 253-3384. Licensee: Nassau Broadcasting II LLC (acq 4-25-2003; $375,000). Population served: 300,000 Natl. Network: ESPN Radio. Format: Christian. ♦Rick Musselman, gen mgr.

***WDIY(FM)**— Jan 8, 1995: 88.1 mhz; 100 w vert. 843 ft TL: N40 33 54 W75 26 26. Hrs opn: 5 AM-1 AM 301 Broadway, Bethlehem, 18015. Phone: (610) 694-8100. Fax: (610) 954-9474. E-mail: info@wdiyfm.org Web Site: www.wdiy.org. Licensee: Lehigh Valley Community Broadcasters Association Board of Directors Inc. Natl. Network: NPR. Law Firm: Schwartz, Woods & Miller. Format: News, class, pub affrs. News: 30 hrs wkly. Target aud: General. Spec prog: Folk 12 hrs, jazz 10 hrs, Sp 3 hrs, Asian-Indian one hr, Arabic one hr, Jewish one hr wkly. ♦Bill Dautremont-Smith, pres & sls dir; Rick Weaver, VP; Burr Beard, stn mgr; Sharon Ettinger, dev dir.

WHOL(AM)— Sept 12, 1948: 1600 khz; 500 w-D, 100 w-N, DA-2. TL: N40 35 33 W75 28 42. Hrs open: 24 1125 Colorado St., 18103. Phone: (610) 434-4801. Licensee: Matthew P. Braccili (acq 11-25-2003; $940,000). Population served: 700,000 Natl. Network: USA, Radio Unica. Natl. Rep: Salem. Format: Contemp Spanish topical top 40. Target aud: 18-65. ♦Matthew Braccili, stn mgr.

***WJCS(FM)**— Feb 29, 1996: 89.3 mhz; 125 w vert. 804 ft TL: N40 33 54 W75 26 26. Hrs open: 24 Box 8900, 18105-8900. Secondary address: 300 E. Rock Rd., Suite 205 18103. Phone: (610) 434-1742. Fax: (610) 797-6922. E-mail: wjcs@wjcs.org Web Site: www.wjcs.org. Licensee: Beacon Broadcasting Corp. Population served: 500,000 Natl. Network: Moody. Format: Educ, relg, news/talk, Christian. Target aud: General. ♦Frank Ginther, stn mgr; Kim Bretzik, gen sls mgr & chief of engrg; Paula Divello, prom dir; Craig Stevens, progmg dir.

WLEV(FM)— July 1947: 100.7 mhz; 11 kw. 1,073 ft TL: N40 33 54 W75 26 26. Stereo. Hrs opn: Box 25096, Lehigh Valley, 18002. Secondary address: 2158 Avenue C, Suite 100, Bethlehem 18017. Phone: (610) 266-7600. Fax: (610) 231-0400. Licensee: Citadel Broadcasting Co. Group owner: Citadel Broadcasting Corp. (acq 9-5-97; $23 million). Population served: 250,000 Natl. Rep: Christal, Katz Radio. Format: Adult contemp. ♦Wayne Leland, pres; John Fraunfelter, gen mgr; Shelly Easton, opns mgr & progmg dir; Elizabeth Pembleton, sls dir.

***WMUH(FM)**— Feb 6, 1966: 91.7 mhz; 500 w. -3 ft TL: N40 35 52 W75 30 38. Stereo. Hrs opn: 24 Muhlenberg College, 2400 Chew St., 18104. Phone: (484) 664-3456. Fax: (484) 664-3539. E-mail: wmuh@muhlenberg.edu Web Site: www.muhlenberg.edu/wmuh. Licensee: Muhlenberg College. Population served: 400,527 Natl. Network: NPR. Format: Div. Target aud: General. Spec prog: Sp 4 hrs, Arabic 2 hrs, Ger 2 hrs, It 2 hrs, Pol 2 hrs wkly. ♦Joe A. Swanson, gen mgr; Mike Calcagno, stn mgr; Rich Gensiak, progmg dir.

WSAN(AM)— May 24, 1923: 1470 khz; 5 kw-U, DA-N. TL: N40 38 10 W75 29 06. Stereo. Hrs opn: 1541 Alta Dr., Suite 400, Whitehall, 18052. Phone: (610) 434-1742. Fax: (610) 434-6288. Web Site: www.1470wyhm.com. Licensee: Capstar TX L.P. Group owner: Clear Channel Communications Inc. (acq 8-30-2000; grpsl). Population served: 109,527 Natl. Rep: D & R Radio. Format: Sports. ♦Tom Barney, gen sls mgr; Paula Divello, prom dir; Craig Stevens, progmg dir.

WTKZ(AM)— September 1948: 1320 khz; 5 kw-D, 1 kw-N, DA-2. TL: N40 37 40 W75 29 09. Hrs opn: 107 Paxinosa Rd. W., Easton, 18040. Phone: (610) 258-6155. Phone: (610) 829-5500. Fax: (610) 253-3384. Web Site: www.espnlv.com. Licensee: Nassau Broadcasting II L.L.C. Group owner: Mega Communications Inc. (acq 2-14-2005; $500,000). Population served: 100,800 Natl. Network: ESPN Radio. Format: Sports. ♦Pat Lincoln, gen mgr & stn mgr.

WZZO(FM)—See Bethlehem

Altoona

WALY(FM)—Bellwood, Mar 28, 1970: 103.9 mhz; 3 kw. 984 ft TL: N40 34 04 W79 26 26. Stereo. Hrs opn: 18 One Forever Dr., Hollidaysburg, 16648-3029. Phone: (814) 944-2221. Fax: (814) 943-2754. E-mail: rog@waly1039.com Web Site: www.waly1039.com. Licensee: Forever Broadcasting LLC. Group owner: Forever Broadcasting (acq 7-16-97; grpsl). Population served: 536,100 Natl. Network: AP Radio. Natl. Rep: Katz Radio. Rgnl rep: Dome. Format: Oldies. Target aud: 35-64; earlier boomers, socially & financially active. ♦Carol B. Logan, pres; Dave Davies, gen mgr; Bobbi Castelluci, gen sls mgr.

WFBG(AM)— Oct 30, 1924: 1290 khz; 5 kw-D, 1 kw-N, DA-N. TL: N40 27 20 W78 23 50. Stereo. Hrs opn: 24 One Forever Dr., Hollidaysburg, 16648. Phone: (814) 941-9800. Phone: (814) 944-1290. Fax: (814) 943-2754. Fax: (814) 941-7198. Web Site: www.wfbg.com. Licensee: Forever of PA L.L.C. Group owner: Forever Broadcasting (acq 12-24-90; $2.1 million with co-located FM; FTR: 1-14-91). Population served: 69,900 Natl. Rep: Christal. Format: Adult standards. News staff: 2; News: 2 hrs wkly. Target aud: 25-54. ♦Dave Davies, gen mgr.

WFGY(FM)—Co-owned with WFBG(AM). Oct 17, 1960: 98.1 mhz; 30 kw. Ant 941 ft TL: N40 34 01 W78 26 32. Stereo. 24 Web Site: www.froggyradio.com.102,500 Natl. Network: CBS. Format: Contemp country. News staff: one. Target aud: 25-64.

WRKY-FM—See Hollidaysburg

WRTA(AM)— June 12, 1946: 1240 khz; 1 kw-U. TL: N40 30 26 W78 25 15. Hrs open: 19 1417-19 12th Ave., 16603. Phone: (814) 943-6112. Fax: (814) 944-9782. E-mail: contactus@wrta.com Web Site: www.wrta.com. Licensee: Handsome Brothers Inc. (acq 1-9-2004; $500,000). Population served: 155,000 Natl. Network: Westwood One. Rgnl rep: Marv Roslin Law Firm: Pepper & Corazzini. Format: News/talk. News staff: 2; News: 15 hrs wkly. Target aud: 25 plus; middle/upper income, college educated, professional. Spec prog: Sports play-by-play/loc college & high schools. ♦David Barger, pres; David R. Wolf, gen mgr, gen sls mgr, edit dir, edit mgr & political ed; Dave Weaver, news dir, local news ed & news rptr; Bob Taylor, chief of engrg; Ken Maguda, stn mgr, opns mgr, rgnl sls mgr, progmg dir & traf mgr; Charlie Weston, sports cmtr.

WVAM(AM)— July 1948: 1430 khz; 5 kw-D, 1 kw-N, DA-N. TL: N40 29 42 W78 24 06. Hrs open: One Forever Dr., Hollidaysburg, 16648. Phone: (814) 941-9800. Fax: (814) 943-2754. Web Site: www.wvamam.com. Licensee: Forever Broadcasting LLC. Group owner: Forever Broadcasting (acq 12-12-2003; $2.1 million with co-located FM). Population served: 62,900 Natl. Network: ESPN Radio. Format: Sports. Target aud: 25 plus; white collar professionals. Spec prog: Pol one hr wkly. ♦Dave Davies, gen mgr; Rich DeLeo, progmg dir; Troy Barnhart, chief of engrg.

WWOT(FM)—Co-owned with WVAM(AM). July 1976: 100.1 mhz; 3 kw. Ant 954 ft TL: N40 34 11 W78 26 25. Stereo. Web Site: www.hot100radio.com.70,000 Format: CHR. ♦Jonathan Reed, progmg dir; Mark Haze, news dir.

Ambridge

WMBA(AM)— May 1957: 1460 khz; 500 w-U, DA-2. TL: N40 35 08 W80 12 11. Hrs open: 24 1316 7th Ave., Beaver Falls, 15010. Phone: (724) 846-4100. Fax: (724) 843-7771. E-mail: 1230@wbvp-wmba.com Web Site: www.wbvp-wmba.com. Licensee: Iorio Broadcasting Inc. (acq 5-23-2000; $325,000). Population served: 400,000 Natl. Network: Talk Radio Network. Wire Svc: AP Format: Talk, sports. News staff: one; News: 10 hrs wkly. Target aud: 35+. Spec prog: Polka review 2 hrs, oldies 3 hrs, Polish 2 hrs wkly. ♦Frank Iorio, pres; Mark Peterson, gen mgr; John Nuzzo, progmg dir; Pat Septak, local news ed & news rptr; Bob Barrickman, sports cmtr.

Annville-Cleona

WWSM(AM)— Aug 4, 1968: 1510 khz; 5 kw-D, DA. TL: N40 17 44 W76 27 46. Hrs open: 621 Cumberland St., Suite 4, Lebanon, 17042. Phone: (717) 272-1510. Fax: (717) 272-7074. Web Site: www.wwsm.us. Licensee: Patrick H. Sickafus. (acq 10-14-93; $1; FTR: 11-1-93). Population served: 28,572 Natl. Network: USA. Format: Classic western. Target aud: 34 plus. Spec prog: Polka 2 hrs, bluegrass 3 hrs, gospel music 3 hrs wkly. ♦Patrick H. Sickafus, pres; Gary Gruver, gen mgr.

Apollo

WAVL(AM)— Dec 13, 1947: 910 khz; 5 kw-D, 69 w-N, DA-2. TL: N40 35 01 W79 31 34. (CP: 1360 khz; 6.7 kw-D, 700 w-N, DA-2. TL: N40 27 42 W79 36 07). Hrs opn: 120 Beale Rd., Sarver, 16055. Phone: (724) 295-2000. Fax: (724) 295-9009. Web Site: www.praise910.com. Licensee: Evangel Heights Assembly of God (acq 5-16-01; $400,000). Population served: 3,000,000 Natl. Network: USA. Format: Contemp Christian. News: 2 hrs wkly. ♦John P. Kuert, pres; Paul Barton, gen mgr; Jeff Bogaczyk, opns dir.

Avis

WQBR(FM)— Aug 11, 1989: 99.9 mhz; 570 w. Ant 1,053 ft TL: N41 13 45 W77 22 02. Stereo. Hrs opn: 24 Box 999, McElhattan, 17748. Secondary address: 330 McElhattan Dr., McElhattan 17748. Phone: (570) 769-2327. Fax: (570) 769-7746. E-mail: bear@kcnet.org Web Site: www.bear999.com. Licensee: Maximum Impact Communications Inc. (acq 9-9-93; $270,000; FTR: 10-4-93). Population served: 300,000 Natl. Network: Jones Radio Networks. Natl. Rep: Dome. Format: Country/Americana. News staff: one; News: 2 hrs wkly. Target aud: 25-54. ♦Karyn O'Brien Stratton, pres, gen mgr, opns mgr & progmg mgr; Dave Stratton, gen sls mgr, mktg dir, prom mgr, adv mgr & news dir; Michael Ferriola, engrg mgr; Patti Knepp, traf mgr.

Avoca

WILK-FM— Apr 2, 1976: 103.1 mhz; 6 kw. Ant 72 ft TL: N41 18 20 W75 45 38. Stereo. Hrs opn: 24

Simulcast with WILK(AM) Wilkes-Barre 100%.
305 Hwy. 315, Pittstown, 18640. Phone: (570) 883-9850. Fax: (570) 883-9851. E-mail: jimr@102themountain.com Web Site: www.wilknetwork.com. Licensee: Entercom Wilkes-Barre Scranton LLC. Group owner: Entercom Communications Corp. (acq 12-13-99; grpsl). Population served: 35,000 Format: News/talk. Target aud: 18-54; general. ♦Jim Rising, opns mgr; Andy Zapotek, gen sls mgr.

Beaver Falls

WAMO-FM—Licensed to Beaver Falls. See Pittsburgh

WBVP(AM)— May 25, 1948: 1230 khz; 5 kw-U. TL: N40 44 16 W80 17 47. Hrs open: 24 Box 719, 1316 7th Ave., 15010. Phone: (724) 846-4100. Phone: (412) 761-6600. Fax: (724) 843-7771. E-mail: 1230@wbvp-wmba.com Web Site: www.wbvp-wmba.com. Licensee: Iorio Broadcasting Inc. (acq 1996). Population served: 170,000 Natl. Network: Talk Radio Network. Rgnl. Network: Radio Pa. Wire Svc: AP Format: Full service, News, Talk, Sports. News staff: 2. Target aud: 35 plus. Spec prog: Polka music 2 hrs, relg 2 hrs, gospel 2 hsr wkly. ♦Frank Iorio, pres & gen mgr; Mark Peterson, stn mgr; John Nuzzo, opns mgr.

***WGEV(FM)**— Nov 15, 1965: 88.3 mhz; 15 w. 240 ft TL: N40 46 21 W80 18 33. Hrs open: 18 Geneva College, 15010. Phone: (724) 847-6678. Fax: (724) 847-6675. Web Site: www.geneva.edu. Licensee: Geneva College Board of Trustees. Population served: 20,000 Format: Christian contemp hit. News: 2 hrs wkly. Target aud: 14-24; college plus surrounding community. ♦Pete Croisant, gen mgr; Todd Hughes, chief of opns.

***WITX(FM)**— June 19, 1986: 90.9 mhz; 100 w. 167 ft TL: N40 47 05 W80 20 36. Hrs open: 501 37th St., 15010. Phone: (724) 846-8738. Phone: (877) 477-5219. Licensee: Beaver Falls Educational Broadcasting Foundation. Format: Relg. ♦Rev. Kenneth Manypenny, pres.

Beaver Springs

WLZS(FM)— Feb 21, 1993: 106.1 mhz; 175 w. Ant 1,312 ft TL: N40 42 04 W77 12 50. Stereo. Hrs opn: 24 Box 209, Mexico, 17056. Secondary address: Box 146 17812. Phone: (717) 436-2135. Fax: (717) 436-8155. Licensee: Starview Media Inc. (acq 1996; $235,000). Format: Oldies. News staff: one; News: 2 hrs wkly. Target aud: 25-54. ♦Curt Dreibelbis, gen mgr.

Bedford

WAYC(FM)— Dec 22, 1966: 100.9 mhz; 190 w. ant 1,279 ft TL: N40 00 46 W78 33 12. Stereo. Hrs opn: 24 Box 1, 15522. Secondary address: 134 E. Pitt St., 2nd Fl. 15522. Phone: (814) 623-1000. Fax: (814) 623-9692. E-mail: cesscomm@earthlink.net Web Site: www.hitsandfavorites.com. Licensee: Cessna Communications Inc. (group owner; acq 3-22-93; $350,000 with WBFD(AM) Bedford; FTR: 3-29-93). Population served: 50,000 Natl. Network: ABC. Rgnl. Network: Radio Pa. Rgnl rep: Commercial Media Sales. Law Firm: G S B Law. Format: Adult contemp. News staff: one. Target aud: 18-49; general. ♦Jay B. Cessna, pres; John H. Cessna, VP & gen mgr; Chris Collins, opns mgr.

WBFD(AM)— July 2, 1955: 1310 khz; 2.5 kw-D, 85 w-N. TL: N40 02 37 W78 30 11. Hrs open: 24 Box one., 15522. Secondary address: 2nd Fl., 134 E. Pitt St. 15522. Phone: (814) 623-5131. Fax: (814) 623-9692. E-mail: cesscom@earthlink.net Licensee: Cessna Communications Inc. (group owner; acq 3-22-93; $350,000 with WAYC(FM) Bedford; FTR: 3-29-93) Population served: 50,000 Natl. Network: USA, Fox News Radio, Talk Radio Network, Premiere Radio Networks. Rgnl. Network: Radio Pa. Rgnl rep: Commercial Media Sales. Format: News/talk info. News staff: one. Target aud: 30 plus; general. ♦Jay B. Cessna, pres; John H. Cessna, VP & gen mgr.

WBVE(FM)—Co-owned with WBFD(AM). Aug 15, 1988: 107.5 mhz; 370 w. 1,309 ft TL: N40 00 46 W78 33 12. Stereo. 24 50,000 Natl. Network: Westwood One, Fox News Radio. Format: Classic rock. News staff: one. Target aud: 25-54.

WHJB(AM)— Aug 5, 1974: 1600 khz; 2.7 kw-D, 18 w-N. TL: N40 02 35 W78 30 13. Hrs open: 24 Box 672, 15522. Secondary address: 134 E. Pitt St., 2nd Fl. 15522. Phone: (814) 624-0016. Fax: (814) 623-9692. E-mail: johnwhjb@earthlink.net Licensee: John H. Cessna. (acq 1-1-99; $29,000). Population served: 35,000 Natl. Network: Salem Radio Network, Moody. Law Firm: G S B Law. Format: Relg. Target aud: 35 plus. ♦John H. Cessna, gen mgr.

***WUFR(FM)**—Not on air, target date: unknown: 91.1 mhz; 470 w vert. Ant 1,145 ft TL: N40 17 40 W78 34 25. Hrs open: 290 Hegenberger Rd., Oakland, CA, 94621. Phone: (510). Fax: (916) 641-8238. Licensee: Family Stations Inc. ♦Harold Camping, pres.

Bellefonte

WBLF(AM)— Aug 1, 1958: 970 khz; 1 kw-D, 61 w-N. TL: N40 54 12 W77 46 06. Hrs open: 315 S. Atherton St., State College, 16801. Phone: (814) 272-9700. Licensee: Magnum Broadcasting Inc. (group owner; (acq 8-31-2005; $150,000). . Population served: 110,000 Format: News/talk. ♦Michael M. Stapleford, pres; Diana Albright, gen mgr; Tor Michaels, news dir.

WTLR(FM)—See State College

WZWW(FM)—Licensed to Bellefonte. See State College

Bellwood

WALY(FM)—Licensed to Bellwood. See Altoona

Benton

WGGI(FM)— Oct 4, 1985: 95.9 mhz; 6 kw. 328 ft TL: N41 10 16 W76 24 37. Hrs open: 305 Hwy. 315, Box 729, Pittston, 18640. Phone: (570) 883-9850. Fax: (570) 883-9851. E-mail: feedback@froggy101.com Web Site: www.froggy101.com. Licensee: Entercom Scranton Wilkes-Barre License LLC. Group owner: Entercom Communications Corp. (acq 12-13-99; grpsl). Population served: 600,000 Format: Country. Target aud: 25-54. ♦John Burkavage, VP; Jim Rising, opns mgr.

Berwick

WFBS(AM)— Aug 1, 1957: 1280 khz; 1 kw-D, 175 w-N. TL: N41 04 36 W76 15 32. Hrs open: 114 N. Market St., 18603. Phone: (570) 752-8012. Fax: (570) 752-1131. E-mail: way750am@aol.com Licensee: Bold Gold Media Group L.P. (acq 1-11-2007; $10,000 plus assumption of debt). Population served: 50,000 Format: Oldies. ♦JoAnn Germers Hausen, VP; Kevin Fennessy, gen mgr.

WHLM-FM— Feb 14, 1992: 103.5 mhz; 4.1 kw. 387 ft TL: N41 05 11 W76 16 41. Hrs open: 124 E. Main St., Bloomsburg, 17815. Phone: (570) 784-1200. Fax: (570) 784-6060. Web Site: www.wkab.net. Licensee: Columbia FM Inc. (acq 2006; $800,000). Format: Classic hits. Target aud: 25-54; females at work, 18-39 men on weekends. ♦Joseph Reilly, gen mgr & stn mgr.

Bethlehem

WGPA(AM)— Feb 14, 1946: 1100 khz; 250 w-D. TL: N40 37 27 W75 21 19. Hrs open: Sunrise-sunset 528 N. New St., 18018. Phone: (610) 866-8074. Fax: (610) 866-9381. E-mail: joetimmer@jollyjoetimmer.com Web Site: www.regiononline.com/joetimmer. Licensee: Joseph Timmer dba Timmer Broadcasting Co. (acq 6-19-92; $100,000; FTR: 7-13-92). Population served: 310,000 Natl. Network: USA. Format: Var, news/talk. News staff: one; News: 2 hrs wkly. Target aud: General. Spec prog: Ger 2 hrs, polka 12 hrs, Sp 4 hrs wkly. ♦Joe Timmer, pres & gen mgr; Mark Staller, opns dir.

***WLVR(FM)**— May 3, 1973: 91.3 mhz; 185 w. 60 ft TL: N40 36 22 W75 22 42. Stereo. Hrs opn: 7 AM-4 AM Lehigh Univ., 29 Trembley Dr., 18015-3066. Phone: (610) 758-4187. Fax: (610) 758-4186. Licensee: Lehigh University. Population served: 500,000 Format: Var. News: 10 hrs wkly. Target aud: General. Spec prog: Black 12 hrs, class 8 hrs, reggae 6 hrs, jazz 12 hrs wkly. ♦Aimee Van House, gen mgr; Killian O'Conner, progmg dir.

WZZO(FM)— Feb 14, 1946: 95.1 mhz; 30 kw. 631 ft TL: N40 37 13 W75 17 37. Stereo. Hrs opn: 1541 Alta Dr., Suite 400, Whitehall, 18052. Phone: (610) 434-1742. Fax: (610) 434-6288. E-mail: webmaster@wzzo.com Web Site: www.wzzo.com. Licensee: Capstar TX L.P. Group owner: Clear Channel Communications Inc. (acq 8-30-00; grpsl). Population served: 841,000 Natl. Rep: Katz Radio. Format: AOR. ♦Tom Barney, gen mgr; Pat Gremling, gen sls mgr.

Blairsville

WLCY(FM)— Apr 15, 1985: 106.3 mhz; 2.4 kw. Ant 363 ft TL: N40 31 10 W79 13 26. Stereo. Hrs opn: 24 840 Philadephia St., Suite 100, Indiana, 15701. Phone: (724) 465-4700. Fax: (724) 349-6842. Licensee: The St. Pier Group LLC. Group owner: Renda Broadcasting Corp. (acq 8-1-2004; $900,000). Natl. Network: Jones Radio Networks. Rgnl rep: Dome. Format: Country. News staff: 2; News: 2 hrs wkly. Target aud: 25-54. ♦Mark Bertig, gen mgr.

Bloomsburg

***WBUQ(FM)**— Sept 16, 1986: 91.1 mhz; 600 w. 500 ft TL: N41 00 29 W76 26 51. Stereo. Hrs opn: 16 Bloomsburg Univ., 400 E. Second St., 1250 McCormick Center for Human Services, 17815. Phone: (570) 389-4686. Fax: (570) 389-5071. E-mail: wbuq@bloomu.edu Web Site: www.wbuqfm.com. Licensee: Bloomsburg University of Pennsylvania. Population served: 15,000 Format: Alternative, rock/AOR. News staff: one; News: 2 hrs wkly. Target aud: General; college & area high school students. Spec prog: Talk 10 hrs, urban 10 hrs, metal 15 hrs, indie 10 hrs wkly. ♦Ray Moskal, gen mgr; Chad Eddinger, progmg dir.

WFYY(FM)— September 1956: 106.5 mhz; 36.5 kw. 570 ft TL: N40 59 42 W76 29 51. Stereo. Hrs opn: Box 90, Selinsgrove, 17870. Secondary address: 246 W. Main St. 17815. Phone: (570) 374-5711. Fax: (570) 784-1004. E-mail: sales@wfyyradio.com Web Site: www.wfyyradio.com. Licensee: MMP License LLC. Group owner: MAX Media L.L.C. (acq 10-17-03; grpsl). Population served: 11,652 Natl. Network: Westwood One. Rgnl rep: Dome. Format: Adult contemp. Target aud: 18-49. ♦John A. Trinder, pres; Scott Richards, VP & gen mgr; Dawn Marie, opns dir; Greg Adair, gen sls mgr; Ted Koppen, chief of engrg.

WHLM(AM)— Sept 26, 1947: 930 khz; 1 kw-D, 23 w-N. TL: N41 01 00 W76 27 44. Hrs open: 24 Box One, 105 W. Main St., 2nd FL., 17815-3329. Phone: (570) 784-1200. Fax: (570) 784-6060. E-mail: whlmam@aol.com Web Site: www.whlm.com. Licensee: Columbia Broadcasting Co. (acq 9-5-01; $45,000). Population served: 63,500 Natl. Network: CBS Radio. Law Firm: Borsari & Paxson. Format: Var/div. News staff: one; News: 15 hrs wkly. Target aud: 25-54. Spec prog: Farm one hr, relg 2 hrs wkly. ♦Joseph Reilly, pres & gen mgr; Larry Hopper, gen sls mgr.

Boalsburg

WBUS(FM)— Apr 13, 1998: 93.7 mhz; 33 kw. Ant 1,361 ft TL: N40 45 08 W77 45 16. Stereo. Hrs opn: 2551 Park Center Blvd., State College, 16801. Phone: (814) 237-9800. Fax: (814) 237-2477. Web Site: www.thebus.net. Licensee: Forever Broadcasting LLC. (group owner; (acq 5-1-2005; $2.65 million with WRSC(AM) State College). Population served: 116,000 Format: Classic hits, classic rock. Target aud: 25-54. ♦Andy Sumereau, VP & gen mgr.

Boyertown

WFKB(FM)— Oct 31, 1960: 107.5 mhz; 30 kw. Ant 610 ft TL: N40 24 15 W75 39 09. Stereo. Hrs opn: 24 280 Mill St., 19512. Phone: (610) 369-7777. Fax: (610) 369-7780. E-mail: jwhite@nassaubroadcasting.com Web Site: frankfmonline.com. Licensee: WDAC Radio Co. (acq 11-5-91; $3 million; FTR: 12-2-91). Population served: 7,100,000 Natl. Rep: Katz Radio. Law Firm: Jones, Waldo, Holbrook & McDonough. Format: Classic Hits. Target aud: 25-54; families. ♦Richard Crawford, pres; John White, gen mgr.

Braddock

WLFP(AM)— June 1947: 1550 khz; 1 kw-D, 4 w-N. TL: N40 24 47 W79 51 14. (CP: COL Reserve Township. 2 kw-D, 12 w-N, DA-2. TL:

N40 29 27 W79 58 05). Hrs opn: 5 AM-midnight 4736 Penn Ave., Pittsburgh, 15224. Phone: (412) 942-0076. Web Site: www.theedge1550.com. Licensee: Urban Radio of Pennsylvania L.L.C. Group owner: Inner City Broadcasting (acq 5-24-2000; $1.5 million for 55% of both WHAT(AM) Philadelphia and WLFP(AM) Braddock) Natl. Network: American Urban. Format: Talk. News staff: one; News: 6 hrs wkly. Target aud: 25-54. ♦Coddy Anderson, chmn & dev dir; Chris Squire, gen mgr; Shelia Corley, stn mgr.

WRRK(FM)— June 1959: 96.9 mhz; 44.7 kw. Ant 530 ft TL: N40 24 42 W79 55 53. Stereo. Hrs opn: 24 650 Smithfield St., Suite 2200, Pittsburgh, 15222. Phone: (412) 316-3342. Fax: (412) 316-3388. E-mail: feedback@bobfm96.9.com Web Site: www.bobfm969.com. Licensee: WPNT Associates. Population served: 2,500,000 Natl. Network: ABC. Law Firm: Dow, Lohnes & Albertson. Format: Adult hits. ♦Greg Frischling, gen mgr; Chris Kohan, gen sls mgr; Vicki Wolfe, prom dir; John Robertson, progmg dir; Amy Crago, news dir; Paul Carroll, chief of engrg; Ed Lang, traf mgr.

Bradford

WBRR(FM)—Listing follows WESB(AM).

WESB(AM)— April 1947: 1490 khz; 1 kw-U. TL: N41 27 54 W78 37 01. Hrs open: 24 Box 545, 1490 St. Francis Dr., 16701. Phone: (814) 368-4141. Fax: (814) 368-3180. E-mail: 1490@wesb.com Web Site: www.wesb.com. Licensee: Radio Station WESB Inc. Population served: 25,000 Natl. Network: CNN Radio. Rgnl. Network: Radio Pa. Natl. Rep: Dome. Law Firm: Baraff, Koerner & Olender. Format: Adult contemp. News staff: one. Target aud: 25-54. ♦Donald J. Fredeen, pres & gen mgr; Frank Williams, opns dir & progmg dir; Peggy Austin, sls dir; Christine Brookins, prom dir; Anne Holliday, mus dir & news dir.

WBRR(FM)—Co-owned with WESB(AM). Dec 1, 1987: 100.1 mhz; 1.65 kw. 525 ft TL: N41 58 12 W78 42 03. Stereo. 24 Web Site: www.wbrrfm.com. (Acq 10-87; $21,000; FTR: 10-12-87). Format: 70s/80s hits. Target aud: 25-54.

Bristol

***WLBS(FM)**— April 1998: 91.7 mhz; 100 w vert. 68 ft TL: N40 09 33 W74 51 24. Hrs open: 24
Rebroadcasts WRDV(FM) Warminster 100%.
Box 2012, Warminster, 18974. Phone: (215) 674-8002. Fax: (215) 674-4586. E-mail: info@wrdv.org Web Site: www.wrdv.org. Licensee: Bux-Mont Educational Radio Association. Population served: 10,000 Wire Svc: AP Format: Variety. ♦Charles W. Loughary, chmn; Todd H. Allen, gen mgr.

Brookville

WKQL(FM)— Jan 17, 2000: 103.3 mhz; 10.5 kw. 508 ft TL: N41 04 04 W79 04 59. Hrs open: 24 904 N. Main St., Punxsutawney, 15767. Secondary address: Renda Radio Inc., Broadcast Plaza, Pittsburgh 15767. Phone: (814) 938-6000. Fax: (814) 938-4237. E-mail: rendaradio@adelphia.net Licensee: Renda Radio Inc. Group owner: Renda Broadcasting Corp. Population served: 200,000 Natl. Network: ABC. Law Firm: Latham & Watkins. Format: Oldies. News staff: one; News: 2 hrs wkly. Target aud: 35-54; adults. ♦Diana Albright, gen mgr.

WMKX(FM)— Aug 22, 1981: 105.5 mhz; 16 kw. Ant 418 ft TL: N41 07 21 W79 03 51. Stereo. Hrs opn: 24 51 Pickering St., 15825. Phone: (814) 849-8100. Fax: (814) 849-4585. E-mail: megarock@alltel.net Licensee: Strattan Broadcasting Inc. Format: Classic rock. News staff: one; News: 5 hrs wkly. Target aud: 25-54; general. Spec prog: Jazz 3 hrs, rock classics 6 hrs, oldies 8 hrs wkly. ♦Jim Farley, pres & gen mgr; Nathan Sharp, gen sls mgr.

Brownsville

WASP(AM)— Aug 3, 1968: 1130 khz; 5 kw-D, DA. TL: N40 02 33 W79 54 20. Hrs open: 123 Blaine Rd., 15417. Phone: (724) 938-2000. Fax: (724) 938-7824. E-mail: pickleonline@yahoo.com Licensee: Keymarket Licenses LLC. Group owner: Keymarket Communications LLC (acq 8-31-99; $2.875 million with WPKL(FM) Uniontown). Population served: 2,500,000 Natl. Network: USA. Format: Oldies. Target aud: 35-64. ♦Gerald Getz, pres; Andrew Powaski, gen mgr.

Burgettstown

WOGH(FM)—Listing follows WSTV(AM).

Burnham

WVNW(FM)— August 1994: 96.7 mhz; 450 w. 850 ft TL: N40 35 10 W77 41 40. Hrs open: 24 Box 911, One Juniata St., Lewistown, 17044. Phone: (717) 242-1493. Fax: (717) 242-3764. E-mail: traffic@star967.com Licensee: WVNW Inc. (acq 2-1-94; FTR: 4-25-94). Natl. Network: ABC, Fox News Radio. Format: Hot country. Target aud: 25-54. ♦Jed A. Donahue, VP; Jed A. Donahue, gen mgr; Tom Sheeder, progmg dir; Erik Lane, news dir; Dave Busman, sports cmtr.

Butler

WBUT(AM)— Mar 14, 1949: 1050 khz; 500 w-D, 65 w-N. TL: N40 53 51 W79 53 22. Hrs open: 24 112 Hollywood Dr., Suite 203, 16001. Phone: (724) 287-5778. Fax: (724) 282-9188. E-mail: frontdesk@bcnetwork.com Web Site: www.wbut.com. Licensee: Butler County Radio Network Inc. (group owner; (acq 5-19-98; grpsl). Population served: 174,000 Natl. Network: CNN Radio. Format: Country. News: 2. Target aud: General. ♦Bob Cupp, progmg dir; Victoria Hinterberger, gen mgr, progmg dir & traf mgr; Ron Willison, sports cmtr.

WLER-FM—Co-owned with WBUT(AM). Mar 14, 1949: 97.7 mhz; 4.6 kw. 374 ft TL: N40 53 51 W79 53 22. Stereo. 24 Web Site: www.wler.com.175,000 Natl. Network: Westwood One, CNN Radio. Format: All hits. News staff: 2. ♦Victoria Hinterberger, stn mgr; Jay Kline, mus dir & disc jockey.

WISR(AM)— Sept 26, 1941: 680 khz; 250 w-D, 50 w-N. TL: N40 52 39 W79 54 09. Hrs open: 24 112 Hollywood Dr., Suite 203, 16003-0151. Phone: (724) 283-1500. Fax: (724) 283-3005. E-mail: frontdesk@bcrnetwork.com Web Site: www.insidebutlercounty.com. Licensee: Butler County Radio Network Inc. (group owner; acq 5-19-98; grpsl). Population served: 130,000 Natl. Network: CBS. Format: News/talk, sports. News staff: 2; News: 28 hrs wkly. Target aud: 45 plus. Spec prog: Relg 6 hrs wlky. ♦Ron Wilson, CFO; Vicki Hinterberger, gen mgr; Dave Malazkey, progmg dir.

California

***WCAL(FM)**— September 1973: 91.9 mhz; 3 kw. 160 ft TL: N40 02 57 W79 54 01. Stereo. Hrs opn: 24 California Univ. of PA, 428 Hickory St., 15419. Phone: (724) 938-3000. Fax: (724) 938-5959. E-mail: wheeler@cup.edu Licensee: The Student Association Inc. (acq 6-3-78). Population served: 500,000 Natl. Network: Westwood One. Format: Active rock. News: 5 hrs wkly. Target aud: 18-25; students at California Univ. Spec prog: Contemp Christian 6 hrs, urban contemp 12 hrs, rap 4 hrs, alternative 6 hrs, metal 5 hrs wkly. ♦J.R. Willer, gen mgr & opns mgr.

Cambridge Springs

WXXO(FM)— July 14, 1997: 104.5 mhz; 2.65 kw. 502 ft TL: N41 42 17 W80 09 53. Stereo. Hrs open: 24 1411 Liberty St., Franklin, 16323. Phone: (814) 432-2188. Fax: (814) 437-9372. E-mail: radio@zoominternet.com Web Site: www.mykissfm.com. Licensee: Forever Broadcasting LLC. Group owner: Forever Broadcasting (acq 7-21-00; grpsl). Population served: 84,000 Format: Hot AC. Target aud: 25-54. ♦Carol Logan, pres; Terry Deitz, gen mgr; Joe Elan, gen sls mgr; Todd Adkins, opns mgr & progmg dir.

Canonsburg

WWCS(AM)— Nov 28, 1957: 540 khz; 7.5 kw-D, 500 w-N, DA-2. TL: N40 17 22 W80 11 07. (CP: 5 kw-D, 500 w-N, DA-2). Hrs opn: 24 38 Angerer Rd., 15317. Phone: (724) 745-5400. Fax: (724) 745-8790. Web Site: www.radiodisney.com. Licensee: Birach Broadcasting Corp. (acq 5-28-92; $475,000; FTR: 6-22-92). Population served: 5,000,000 Format: Children. Target aud: Educated adults, ethnic groups, open minded. ♦Sima Birach Sr., gen mgr.

Canton

WHGL-FM— Aug 30, 1978: 100.3 mhz; 3.9 kw. 846 ft TL: N41 44 32 W76 50 08. Stereo. Hrs opn: 24 Box 100, 170 Redington Ave., Troy, 16947. Phone: (570) 297-0100. Fax: (570) 297-3193. E-mail: wiggle100@sosbbs.com Web Site: www.wiggle100.com. Licensee: Cantroair Communications Inc. (acq 1-7-99; $560,000 for 85% of stock with WTZN(AM) Troy). Natl. Network: ABC. Format: Country. News staff: one. Target aud: 25-54. ♦Bob Gisler, VP; Mike Powers, pres & gen mgr; Georgia Pepper, prom mgr.

WTZN(AM)—See Troy

Carbondale

WCDL(AM)— January 1950: 1440 khz; 5 kw-D. TL: N41 33 28 W75 29 11. Hrs open: 1 N. Main St., Suite 1440, 18704. Phone: (570) 282-1440. Fax: (570) 282-1435. Licensee: Group B Licensee LLC. Group owner: Route 81 Radio LLC (acq 12-2-2003; grpsl). Population served: 500,000 ♦Ed Histed, gen mgr.

***WCIG(FM)**— 2005: 91.3 mhz; 70 w. Ant 735 ft TL: N41 28 41 W75 29 51. Hrs open:
Rebroadcasts WCIK(FM) Bath, NY 100%.
Box 506, Bath, NY, 14810. Phone: (607) 776-4151. Fax: (607) 776-6929. E-mail: mail@fln.org Web Site: www.fln.org. Licensee: Family Life Ministries Inc. Natl. Network: Salem Radio Network. Law Firm: Hardy, Carey, Chautin & Balkin, LLP. Wire Svc: Metro Weather Service Inc. Format: Contemp Christian. News staff: 3; News: 14 hrs wkly. Target aud: 30-54. ♦Dick Snavely, CFO; John Owens, progmg dir; Jim Travis, chief of engrg.

WLNP(FM)— 1965: 94.3 mhz; 1.1 kw. Ant 770 ft TL: N41 32 37 W75 27 44. Stereo. Hrs opn: 24 84 S. Prospect St., Nanticoke, 18634. Phone: (570) 735-0730. Fax: (570) 735-4844. Licensee: Group B Licensee LLC. Group owner: Route 81 Radio LLC (acq 12-2-2003; grpsl). Population served: 2,000,000 Natl. Network: CBS Radio. Format: MOR, btfl music, nostalgia. ♦Margie McQuillin, stn mgr.

Carlisle

WCAT-FM— 1959: 102.3 mhz; 3 kw. Ant 328 ft TL: N40 17 23 W77 08 10. Stereo. Hrs opn: 24 Box 450, Hershey, 17033. Secondary address: 1703 Walnut Bottom Rd. 17013. Phone: (610) 266-7600. Phone: (508) 752-1045. Fax: (717) 258-4638. Web Site: www.red1023.com. Licensee: Citadel Broadcasting Co. Group owner: Citadel Broadcasting Corp. (acq 1999; $4.5 million with WHYL(AM) Carlisle). Population served: 80,000 Natl. Network: CNN Radio. Format: Country. Target aud: 25-54. ♦Cindy Miller, gen mgr; John Fraunfelter, opns mgr; Jay Hunter, prom dir; Will Robinson, progmg dir.

***WDCV-FM**— January 1972: 88.3 mhz; 450 w. 150 ft TL: N40 12 09 W77 11 46. Stereo. Hrs opn: 7 AM-2 AM Box 1773, Dickinson College, Student Activities Office, 17013-2896. Phone: (717) 245-1444. Fax: (717) 245-1899. E-mail: webmaster@dickinson.edu Web Site: www.the-freq.com. Licensee: Board of Trustees Dickinson College. Format: Div. News staff: 5; News: 5 hrs wkly. Spec prog: Jazz 6 hrs, Ger one hr, Sp one hr, funk/rap 15 hrs, Russian one hr, politics one hr, blues 6 hrs wkly. ♦Nick Stamos, gen mgr.

WHYL(AM)— 1948: 960 khz; 5 kw-D, 22 w-N. TL: N40 11 34 W77 10 28. Hrs open: 24 1703 Walnut Bottom Rd., 17013. Phone: (717) 249-1717. Fax: (717) 258-4638. Web Site: www.whylradio.com. Licensee: Group B Licensee LLC. Group owner: Route 81 Radio LLC (acq 12-2-2003; grpsl). Population served: 350,000 Natl. Rep: McGavren Guild. Format: News/talk. News staff: one. Target aud: 45 plus; older, mature adults. Spec prog: Polka 2 hrs. ♦Lloyd Roach, CEO; Gavin Stief, COO; Bruce Collier, gen mgr & gen sls mgr; Cindy Miller, stn mgr; Ruth O'Brien, progmg dir; John Domen, news dir; Bonnie Parthemore, traf mgr; Ben Barber, disc jockey.

WIOO(AM)— July 8, 1965: 1000 khz; 1 kw-D. TL: N40 09 30 W77 11 49. Hrs open: Sunrise-sunset 180 York Rd., 17013. Phone: (717) 243-1200. Fax: (717) 243-1277. E-mail: wioo@pa.net Web Site: www.carlisle-pa.com/wioo. Licensee: Harold Swidler. Population served: 18,079 Rgnl. Network: Radio Pa. Format: Classic country. News staff: 2; News: 15 hrs wkly. Target aud: 21 plus. Spec prog: Relg 5 hrs wkly. ♦Harold Swidler, pres; Florence Fisher, gen mgr & opns mgr.

Carnegie

WZUM(AM)— July 1962: 1590 khz; 1 kw-D, 24 w-N, DA-2. TL: N40 25 28 W80 05 05. Hrs open: 6 AM-sunset Box 27, Monroeville, 15146. Phone: (412) 460-4000. Fax: (412) 460-0317. Licensee: Starboard Media Foundation Inc. (acq 9-23-2005; $435,000). Population served: 300,000 Law Firm: Lauren A. Colby. Format: Var, news, info. Target aud: General. Spec prog: Ethnic 2 hrs wkly. ♦Mike Horvath, gen mgr.

Cashtown

***WFKJ(AM)**— Dec 7, 1988: 890 khz; 1 kw-D. TL: N39 52 59 W77 20 43. Hrs open: 7 AM-6 PM Box 115, 17310. Secondary address: 3425 Chambersburg Rd., Biglerville 17307. Phone: (717) 337-1607. Phone: (717) 337-1635. Fax: (717) 334-8914. E-mail: jil@wordbroadcast.org Web Site: www.wordbroadcast.org. Licensee: Jesus is Lord Ministries International. Natl. Network: Moody. Format: Relg. Target aud: General. Spec prog: Country 4 hrs, children 12 hrs wkly. ♦Fred Bream, prom

mgr, prom mgr, progmg dir, mus dir & pub affrs dir; Rev. Michael H. Yeager, pres, gen mgr, gen sls mgr & news dir; Larry Angle, chief of engrg.

Central City

WCCL(FM)— Oct 19, 1972: 101.7 mhz; 725 w. Ant 643 ft TL: N40 06 42 W78 51 33. Stereo. Hrs opn: 24 970 Tripoli St., Johnstown, 15902. Phone: (814) 534-8975. Fax: (814) 534-8979. Web Site: www.cool101online.com. Licensee: 2510 Licenses LLC. (group owner; (acq 2-16-2005; grpsl). Format: Oldies. News staff: 4. Target aud: 25-54. Spec prog: Relg 4 hrs wkly. ♦Nick Ferrara, gen mgr.

Centre Hall

WLTS(FM)— May 24, 1989: 99.5 mhz; 850 w. Ant 1,368 ft TL: N40 45 09 W77 45 16. Stereo. Hrs opn: 24 2551 Park Center Blvd., State College, 16801. Phone: (814) 237-9800. Fax: (814) 237-2477. Licensee: Megahertz Licenses LLC. Group owner: Forever Broadcasting (acq 4-29-2002; $875,000 with WHUN(AM) Huntingdon). Population served: 37,000 Rgnl rep: Commercial Media Sales. Format: Lite, easy favorites. ♦Andy Sumereau, gen mgr.

Chambersburg

WCHA(AM)— Aug 11, 1946: 800 khz; 1 kw-D, 196 w-N. TL: N39 55 41 W77 41 44. Stereo. Hrs opn: 25 Penncraft Ave., 17201. Phone: (717) 263-0813. Fax: (717) 263-9649. E-mail: mix95@mix95.com Licensee: MLB-Hagerstown-Chambersburg IV LLC. (acq 7-20-2005; grpsl). Population served: 17,315 Natl. Network: ABC. Law Firm: Latham Watkins. Format: News/talk. News staff: one. Target aud: 25-54. Spec prog: Relg 8 hrs, gospel 2 hrs wkly. ♦Rich Bateman, gen mgr; Craig Stevens, opns mgr & gen sls mgr; Rick Alexander, chief of opns & progmg mgr; Tammy Heckman, prom mgr.

WIKZ(FM)—Co-owned with WCHA(AM). Apr 15, 1948: 95.1 mhz; 50 kw horiz, 42 kw vert. 449 ft TL: N39 55 41 W77 41 44. Stereo. Web Site: www.mix95.com.206,000 Format: Adult contemp. News staff: one. Target aud: 25-44. ♦Lisa Harding, prom mgr & disc jockey; J.P. McCartney, asst music dir; Jeff Baker, engrg mgr; Barbara Turkenton, traf mgr; Lisa Kline, local news ed & news rptr; Artie Shultz, disc jockey; Courtney Fox, disc jockey; J.P. McCartneyu, disc jockey; Rick Alexander, opns mgr, mus dir & disc jockey.

WHGT(AM)— 1956: 1590 khz; 5 kw-D, 1 kw-N, DA-N. TL: N39 54 15 W77 39 45. Hrs open: 24 Emmanuel Baptist Temple, 16221 National Pike, Hagerstown, MD, 21740. Phone: (301) 582-0378. Fax: (301) 582-1620. E-mail: aikens@emmanuelbaptisttemple.org Web Site: www.emmanuelbaptisttemple.org. Licensee: Emmanuel Baptist Temple (acq 2-15-2006; gift from M. Belmont Verstandig Inc.). Population served: 200,000 Law Firm: Reddy, Begley & McCormick, LLP. Format: Conservative Christian. News: 14 hrs wkly. Spec prog: Church worship svc one hr wkly. ♦Dr. Larry Aikens Jr., gen mgr.

***WZXQ(FM)**— 2005: 88.3 mhz; 110 w vert. Ant 1,155 ft TL: N39 57 40 W77 28 32. Hrs open:
Rebroadcasts WBYO(FM) Sellersville 100%.
Box 186, Sellersville, 18960. Phone: (215) 721-2141. Fax: (215) 721-9811. E-mail: wordfm@wordfm.org Web Site: www.wordfm.org. Licensee: Four Rivers Community Broadcasting Corp. Format: Adult contemp, Christian, relg. ♦David Baker, VP; Charles W. Loughery, gen mgr.

Charleroi

WFGI(AM)— 1947: 940 khz; 250 w-D, 5 w-N, U. TL: N40 07 24 W79 53 45. Hrs open: 24 123 Blaine Rd., Brownsville, 15417. Phone: (724) 938-2000. Fax: (724) 938-7824. Licensee: Keymarket Licenses LLC. Group owner: Keymarket Communications LLC (acq 12-15-99; $3.5 million with co-located FM). Population served: 200,000 Law Firm: Baraff, Koerner & Olender. Format: Country. News staff: one; News: 8 hrs wkly. Target aud: General; Mon Valley area. Spec prog: Croation one hr, Pol 2 hrs wkly. ♦David Zynkhamm, gen sls mgr; Terry Hunt, progmg dir.

WOGI(FM)—Co-owned with WFGI(AM). July 10, 1967: 98.3 mhz; 6 kw. 300 ft TL: N40 07 24 W79 53 45. Stereo. 24 100 Ryan Ct., Suite 98, Pittsburgh, 15205. Phone: (412) 279-5400. Fax: (412) 279-5500. Web Site: www.froggyland.com.398,000 Target aud: 18-49. ♦G. Getz, gen sls mgr; Mark Lindow, progmg dir; Scott Tavares, mus dir; Hollywood Haley, disc jockey; Larry Resick, disc jockey; Mike Valente, disc jockey.

Chester

***WDNR(FM)**— Apr 22, 1977: 89.5 mhz; 10 w. 117 ft TL: N39 51 42 W75 21 20. Stereo. Hrs opn: 9 Widener University, Box 1000, One University Pl., 19013. Phone: (610) 499-4439. Phone: (610) 499-4000. Fax: (610) 499-4531. E-mail: wdnr895@mail.widener.edu Web Site: www.wdnr.com. Licensee: Widener University. Population served: 50,000 Format: Free-form. Target aud: 16-30; high school & college age population. Spec prog: Jazz 2 hrs, blues 2 hrs, oldies 2 hrs, children 2 hrs wkly. ♦Art Kalemkarian, gen mgr; Sean Sheenan, opns dir; Drena Gwin, progmg dir & mus dir; John Blazek, chief of engrg.

WPWA(AM)— October 1947: 1590 khz; 3.2 kw-D, 1 kw-N, DA-N. TL: N39 52 39 W75 27 22. Stereo. Hrs opn: 24 12 Kent Rd., Aston, 19014. Phone: (610) 358-1400. Fax: (610) 358-1845. Web Site: www.wpwa.net. Licensee: Mount Ocean Media L.L.C. (acq 8-20-2001; $675,000). Population served: 1,700,000 Law Firm: Booth, Freret, Imlay & Tepper P. Format: Gospel, Christian, relg. News: 20 hrs wkly. Target aud: 25-64; affluent, mature adults. ♦Steve Skalish, gen mgr.

WVCH(AM)— Apr 4, 1948: 740 khz; 1 kw-D. TL: N39 52 38 W75 24 24. (CP: 50 kw-D, 450 w-N, DA-1). Hrs opn: 6 AM-6 PM Box 157, Blue Bell, 19477-0102. Secondary address: 308 Dutton Mill Rd., Brookhaven 19015. Phone: (610) 872-8861. Fax: (610) 279-9002. E-mail: wvch@juno.com Web Site: www.wvch.com. Licensee: WVCH Communications Inc. (acq 10-74). Population served: 4,300,000 Natl. Network: Moody, USA. Law Firm: Wiley, Rein & Fielding. Format: Relg, Christian. Target aud: General. ♦Thomas H. Moffit, pres; Tom Moffit Jr., sr VP; William Fenton, VP; Tom Harvey, gen mgr.

Clarendon

WKNB(FM)— Aug 31, 1995: 104.3 mhz; 4.7 kw. 371 ft TL: N41 47 21 W79 08 29. Stereo. Hrs opn: 24 Box 824, 310 Second Ave., Warren, 16365. Phone: (814) 723-1310. Fax: (814) 723-3356. E-mail: info@kibcoradio.com Web Site: www.kibcoradio.com. Licensee: Radio Partners LLC. (acq 9-19-2005; grpsl). Natl. Network: AP Radio. Rgnl rep: Commercial Media Sales Law Firm: Borsari & Paxson. Wire Svc: AP Format: Country. News staff: one. Target aud: 18-45. ♦W. LeRoy Schneck, gen mgr; David Whipple, gen sls mgr; Dale Bliss, prom dir & sls; Mark Silvis, progmg mgr; Robert Seiden, news dir; Dana Simmons, traf mgr.

Clarion

WCCR(FM)—Listing follows WWCH(AM).

***WCUC-FM**— April 1977: 91.7 mhz; 3.2 kw. Ant 318 ft TL: N41 12 35 W79 22 39. Stereo. Hrs opn: Clarion Univ. of Pa., G55 Becker Hall, 16214. Phone: (814) 393-2330. Phone: (814) 393-2514. Fax: (814) 393-2065. E-mail: wadams@clarion.edu Licensee: Clarion University of Pennsylvania. (acq 6-76). Population served: 43,500 Format: Top 40. Spec prog: Black 6 hrs, country 6 hrs, jazz 3 hrs, new wave 9 hrs wkly. ♦Bill Adams, gen mgr.

WWCH(AM)— June 12, 1960: 1300 khz; 1 kw-D, 36 w-N. TL: N41 10 34 W79 20 22. Hrs open: 24 Box 688, 16214. Secondary address: 1168 Greenville Pike 16214. Phone: (814) 226-4500. Fax: (814) 226-5898. E-mail: wccrwwch@apelphia.net Web Site: www.insideclarioncounty.com. Licensee: Clarion County Broadcasting Corp. Population served: 48,000 Natl. Network: CBS Radio. Rgnl. Network: Radio Pa. Natl. Rep: Dome. Law Firm: Frederick Polner. Format: Classic country/news/talk. News staff: one; News: 8 hrs wkly. Target aud: 25-54. Spec prog: Pub affrs, relg 8 hrs wkly. ♦William S. Hearst, pres & gen mgr.

WCCR(FM)—Co-owned with WWCH(AM). June 28, 1985: 92.7 mhz; 3 kw. 400 ft TL: N41 14 41 W79 15 42. Stereo. 24 Web Site: www.insideclarioncountry.com. Format: Adult contemp. News staff: one; News: 8 hrs wkly. Target aud: 25-54.

Clearfield

WCPA(AM)— 1947: 900 khz; 2.5 kw-D, 500 w-N, DA-2. TL: N41 02 32 W78 26 54. Hrs open: 110 Healy Ave., 16830. Phone: (814) 765-5541. Fax: (814) 765-6333. Web Site: www.besthitsbestvariety.com. Licensee: First Media Radio LLC. (acq 1-23-2007; $750,000 with co-located FM). Population served: 80,176 Natl. Rep: Dome. Format: Oldies. Target aud: 35 plus. ♦Bob Day, gen sls mgr & progmg dir.

WQYX(FM)—Co-owned with WCPA(AM). July 12, 1967: 93.1 mhz; 1.7 kw. Ant 941 ft TL: N41 04 05 W78 31 07. Stereo. 24 Web Site: www.besthitsbestvariety.com. Law Firm: Baraff, Koerner & Olender. Format: Hot adult contemp. Target aud: 18-44.

Coatesville

WCOJ(AM)— Nov 29, 1949: 1420 khz; 5 kw-U, DA-N. TL: N40 01 21 W75 48 53. Hrs open: 24 17 W. Gay St., Lower Atrium, West Chester, 19380-3090. Phone: (610) 701-9300. Fax: (610) 701-9412. E-mail: wcoj@wcoj.com Web Site: www.wcoj.com. Licensee: Group A Licensee LLC. Group owner: Route 81 Radio LLC (acq 2-10-2004). Population served: 485,000 Natl. Network: CBS. Law Firm: Booth, Freret & Imlay. Wire Svc: AP Format: News/talk. News staff: 3; News: 14 hrs wkly. Target aud: 35-64. Spec prog: Relg 8 hrs, children 2 hrs, farm 2 hrs, gospel 2 hrs wkly. ♦Lloyd B. Roach, pres & gen mgr; Lloyd Roach, opns dir; Michelle Witkowski, opns mgr; Steve Bryant, progmg dir; Robert Henson, news dir; Jeffrey Depolo, chief of engrg.

***WRTJ(FM)**—Not on air, target date: unknown: 89.3 mhz; 1 w horiz, 600 w vert. Ant 236 ft TL: N40 01 23 W75 48 56. Hrs open: 1509 Cecil B. Moore Ave., WRTI/3rd Fl., Philadelphia, 19121-3410. Phone: (215) 204-8405. Fax: (215) 204-4870. Web Site: www.wrti.org. Licensee: Temple University of The Commonwealth System of Higher Education. ♦David S. Conant, gen mgr.

Columbia

WVZN(AM)— 1957: Stn currently dark. 1580 khz; 500 w-D, 5 w-N. TL: N40 00 53 W76 28 13. Hrs open: 24 1927 Columbia Ave., Lancaster, 17603-4337. Phone: (717) 823-9300. Licensee: Esfuerzo de Union Cristiana (acq 2-1-2002; $165,000). Rgnl. Network: Metronews Radio Net, Radio PA. ♦Wilson Cortez, gen mgr.

Confluence

WKEL(FM)—Not on air, target date: unknown: 98.5 mhz; 1.1 kw. Ant 764 ft TL: N39 55 40 W79 25 06. Hrs open: 2351 Sunset Blvd., Suite 170-218, Rocklin, CA, 95765. Phone: (916) 251-1600. Fax: (916) 251-1650. Licensee: Educational Media Foundation. ♦Mike Novak, sr VP.

Connellsville

WYJK(AM)— Apr 23, 1947: 1340 khz; 1 kw-U. TL: N40 01 27 W79 36 35. Hrs open: 24 123 Blaine Rd., Brownsville, 15417. Phone: (724) 938-2000. Fax: (724) 938-7824. E-mail: pickieonline@yahoo.com Licensee: Keymarket Licenses LLC Group owner: Keymarket Communications LLC (acq 1-17-2001; $475,000 with WPKL(FM) Uniontown). Population served: 150,000 Format: Oldies. Target aud: 25-64. ♦Gerald Getz, pres; Andrew Ponaski, gen mgr & sports cmtr.

Cooperstown

WUUZ(FM)— 2002: 107.7 mhz; 4.5 kw. Ant 377 ft TL: N41 29 23 W79 44 07. Hrs open: 24 Forever Broadcasting LLC, One Forever Dr., Holidaysburg, 16648. Phone: (814) 941-9800. Licensee: Forever Broadcasting LLC. Group owner: Forever Broadcasting (acq 7-5-01; $342,000 for CP). Format: Classic hits. ♦Terry Deitz, gen mgr.

Corry

WWCB(AM)— Apr 2, 1955: 1370 khz; 1 kw-D, 500 w-N, DA-N. TL: N41 56 10 W79 39 20. Hrs open: 6 AM-11 PM (M-F); 7 AM-11 PM (S); 7 AM-10 PM (Su) Box 4, 16407. Secondary address: 418 N. Center 16407. Phone: (814) 664-8694. Fax: (814) 664-8695. Licensee: Corry Communications Corp. (acq 1-22-89; $140,000; FTR: 1-15-90). Population served: 10,000 Natl. Network: Motor Racing Net, Westwood One, CBS. Format: Adult contemp, classic rock, sports. Target aud: General. ♦William Hammond III, pres.

Coudersport

WFRM(AM)— May 1953: 600 khz; 1 kw-D, 46 w-N. TL: N41 45 11 W78 00 03. Hrs open: 9 S. Main St., 16915. Phone: (814) 274-8600. Phone: (814) 642-9396. Fax: (814) 274-0760. E-mail: gmiller@wfrm.net; radio@wfrm.net Web Site: www.wfrm.net. Licensee: Farm & Home Broadcasting Co. Group owner: Allegheny Mountain Network Stations Population served: 2,831 Natl. Network: ABC. Natl. Rep: Dome. Law Firm: Borsari & Paxson. Format: Hit country, strong loc news/talk. Target aud: General. Spec prog: Farm 2 hrs wkly. ♦Gerri Miller, gen sls mgr, prom mgr & progmg mgr.

WFRM-FM— Sept 18, 1985: 96.7 mhz; 1.45 kw. 666 ft TL: N41 45 11 W78 00 03. Stereo. 24 Web Site: www.wfrm.net. Format: Adult contemp. News: 9 hrs wkly.

Covington

WDKC(FM)— 1994: 101.5 mhz; 1.9 kw. Ant 594 ft TL: N41 43 25 W77 02 46. Stereo. Hrs opn: Box 101.5, Mansfield, 16933. Secondary address: 8767 Rt. 414, Liberty 16930. Phone: (570) 662-9000. Fax: (570) 324-1015. E-mail: kc101@sosbbs.com Licensee: Mid-Atlantic Broadcasting Inc. (acq 4-95; $105,000). Population served: 100,000 Format: Country. News staff: one; News: 3 hrs wkly. Target aud: 25-54; 70% female. ♦Kevin Thomas, CEO & pres; Thomas Gluszczak, chmn; Kevin Gluszczak, gen mgr.

Cresson

WBRX(FM)— November 1981: 94.7 mhz; 970 w. Ant 794 ft TL: N40 24 11 W78 31 35. Hrs open: 24 1417 12th Ave., Altoona, 16602. Phone: (814) 944-9320. Fax: (814) 944-9350. Web Site: www.wbrx.com. Licensee: Sounds Good Inc. Population served: 225,000 Natl. Network: Westwood One. Format: Classic rock. News staff: one; News: 7 hrs wkly. Target aud: 18-54; males. ♦Diane Boslet, gen mgr & stn mgr.

Curwensville

WOKW(FM)— Aug 1, 1989: 102.9 mhz; 350 w. Ant 945 ft TL: N41 04 29 W78 31 58. Stereo. Hrs opn: 24 Box 589, Clearfield, 16830. Secondary address: 712 River Rd., Clearfield 16830. Phone: (814) 765-4955. Fax: (814) 765-7038. E-mail: news@wokw.com Web Site: www.wokw.com. Licensee: Raymark Broadcasting Co. Inc. Population served: 250,000 Law Firm: Southmayd & Miller. Format: Adult contemp. News staff: one; News: 14 hrs wkly. Target aud: 21-54. Spec prog: Oldies 2 hrs wkly. ♦Mark E. Harley, pres; Yvonne Lehman, exec VP; Mark Harley, gen mgr.

Dallas

WSJR(FM)— May 29, 1989: 93.7 mhz; 750 w. 679 ft TL: N41 15 43 W75 58 04. (CP: 1.45 kw). Stereo. Hrs opn: 24
Rebroadcasts WCTP(FM) Carbondale 100%.
600 Baltimore Dr., 2nd Fl., Wilkes-Barre, 18702. Phone: (570) 824-9000. Fax: (570) 820-0520. Web Site: www.jr937.us. Licensee: Citadel Broadcasting Co. Group owner: Citadel Broadcasting Corp. (acq 2-4-98; grpsl). Population served: 2,000,000 Natl. Network: CBS. Natl. Rep: Roslin. Format: Country. News staff: one; News: 3 hrs wkly. Target aud: 25-54. Spec prog: Community affrs one hr wkly. ♦Taylor Walet, gen mgr; Jim Dorman, opns mgr.

Danville

***WPGM(AM)**— June 1963: 1570 khz; 2.5 kw-D. TL: N40 59 10 W76 37 37. Hrs open: 8 E. Market St., 17821. Phone: (570) 275-1570. Fax: (570) 275-4071. Web Site: www.wpgm.info. Licensee: Montrose Broadcasting Corp. (group owner; (acq 1-6-64). Format: Relg, btfl mus. Target aud: General; families. ♦George Vacca, gen sls mgr & progmg dir.

WPGM-FM— Sept 6, 1968: 96.7 mhz; 340 w. 760 ft TL: N40 59 16 W76 32 51. Stereo. Web Site: www.wpgm.info.300,000

Doylestown

WISP(AM)— 1948: 1570 khz; 5 kw-D, 900 w-N, DA-2. TL: N40 19 34 W75 09 40. (CP: 950 w-N). Hrs opn: 24 Box 798, 18901. Secondary address: 40 Rickerts Rd., Doyelstown 18901. Phone: (215) 345-1570. Fax: (215) 345-1946. E-mail: 1570am@holyspiritradio.org Web Site: www.holyspiritradio.org. Licensee: Holy Spirit Radio Foundation Inc. (acq 1999; $1,023,750). Population served: 475,000 Law Firm: Cohn and Marks. Format: Relg. News: 14 hrs wkly. Target aud: General. ♦Dale W. Meier, CEO & gen mgr.

DuBois

WCED(AM)— February 1941: 1420 khz; 5 kw-D, 500 w-N, DA-N. TL: N41 08 31 W78 48 07. Hrs opn: 24 12 W. Long Ave., Du Bois, 15801-2100. Phone: (814) 375-5260. Fax: (814) 375-5262. Web Site: www.1420wced.com. Licensee: WCED Radio LLC. Group owner: Priority Communications acq 11-28-2003; $150,000). Population served: 40,000 Natl. Network: ABC, ESPN Radio. Rgnl. Network: Radio Pa. Law Firm: Womble Carlyle. Format: News/Talk. News staff: one; News: 20 hrs wkly. ♦Jay Philippone, gen mgr; Lori Lewis, stn mgr; Lindsey Schoening, news dir; Polly Slie, traf mgr; Al Lockwood, engr.

WDBA(FM)— Nov 12, 1975: 107.3 mhz; 50 kw. 499 ft TL: N41 11 28 W78 41 27. Stereo. Hrs opn: 24 28 W. Scribner Ave., 15801. Phone: (814) 371-1330. Fax: (814) 375-5650. E-mail: sales@wdba.com Web Site: www.wdba.com. Licensee: DuBois Area Brdcst Co. Inc. (acq 10-6-93; $360,000; FTR: 10-25-93). Population served: 270,000 Natl. Network: Salem Radio Network. Natl. Rep: Salem. Wire Svc: AP Format: Inspirational, Christian. News: 6 hrs wkly. Target aud: 25-54; women. Spec prog: Children 2 hrs wkly, Southern Gospel 5 hrs wkly, Christian rock 5 hrs wkly. ♦Daniel Brownlee, pres; Dan Kennard, gen mgr & progmg mgr; Gerald Meloon, opns dir & opns mgr.

WOWQ(FM)— 1948: 102.1 mhz; 28 kw. Ant 663 ft TL: N41 02 43 W78 42 11. Stereo. Hrs opn: 24 801 E. DuBois Ave., 15801. Phone: (814) 371-6100. Fax: (814) 371-7724. E-mail: q102@adelphia.net Web Site: www.q102radio.fm. Licensee: First Media Radio LLC. (group owner; (acq 4-10-2002; $4.2 million with WCED(AM) DuBois). Population served: 250,000 Format: Country. News staff: one; News: 10 hrs wkly. Target aud: 18 plus. ♦Alex Kolobielski, CEO, chmn & pres; F. "Moose" Rosana, gen mgr.

Dunmore

WBHD(FM)—Olyphant, 1991: 95.7 mhz; 300 w. 1,010 ft TL: N41 26 10 W75 43 45. Hrs open: 24
Simulcast with WBHT(FM) Mountaintop.
600 Baltmore Dr., 2nd Fl., Wilkes-Barre, 18702. Phone: (570) 824-9000. Fax: (570) 820-0520. Web Site: www.97bht.com. Licensee: Citadel Broadcasting Co. Group owner: Citadel Broadcasting Corp. (acq 1999; $950,000). Format: CHR. Target aud: 18-49; men, sports fans. Spec prog: Talk. ♦Taylor Walet, gen mgr.

East Stroudsburg

***WESS(FM)**— Mar 10, 1971: 90.3 mhz; 1.37 kw. -165 ft TL: N40 59 55 W75 10 21. Stereo. Hrs opn: McGarry Communications Ctr., East Stroudsburg Univ., 18301. Phone: (570) 422-3512. Fax: (570) 422-3777. E-mail: wess@esu.edu Web Site: www.esu.edu/wess. Licensee: East Stroudsburg University Board of Trustees/Student Activities Association. (acq 3-79). Population served: 40,000 Format: Div, alternative, sports. Spec prog: Class 4 hrs, educ 7 hrs, jazz 6 hrs, news/talk 6 hrs,oldies 8 hrs wkly. ♦Jillian Kane, stn mgr & prom dir; Jennifer Haney, prom dir; Nicholas Frey, news dir.

Easton

WCTO(FM)— 1948: 96.1 mhz; 50 kw. 500 ft TL: N40 35 55 W75 25 12. Stereo. Hrs opn: 24 Box 25096, Lehigh Valley, 18002-5096. Secondary address: 2158 Avenue C, Bethlehem 18017. Phone: (610) 266-7600. Fax: (610) 231-0400. Web Site: www.catcountry96.fm. Licensee: Citadel Broadcasting Co. Group owner: Citadel Broadcasting Corp. Population served: 841,000 Natl. Rep: Christal, Katz Radio. Format: Country. ♦John Fraunfelter, gen mgr; Shelly Easton, opns mgr; Elizabeth Penbleton, sls dir.

WEEX(AM)— May 1956: 1230 khz; 840 w-D, 1 kw-N, DA-D. TL: N40 42 30 W75 13 00. Hrs open: 107 Paxinosa Rd. W., 18040-1344. Phone: (610) 258-6155. Fax: (610) 253-3384. Licensee: Nassau Broadcasting II LLC. Group owner: Nassau Broadcasting Partners L.P. (acq 1-31-01; grpsl). Population served: 60,000 Natl. Network: ESPN Radio. Format: Sports. ♦Rick Musselman, gen mgr & gen sls mgr; Tom Fallon, progmg dir.

WODE-FM—Co-owned with WEEX(AM). June 1950: 99.9 mhz; 50 kw. 449 ft TL: N40 42 30 W75 13 00. Stereo. 30,256 Format: Classic hits. ♦Bill Sheridan, progmg dir.

WEST(AM)— Feb 17, 1936: 1400 khz; 1 kw-U. TL: N40 40 23 W75 12 30. Stereo. Hrs opn: 24 436 Northampton St., 18042. Phone: (610) 258-9378. Fax: (610) 250-9675. E-mail: infor@am1400west.net Web Site: www.am1400west.net. Licensee: Maranatha Broadcasting Co. Inc. Population served: 557,200 Natl. Rep: McGavren Guild. Law Firm: Fleischman & Walsh L. Format: MOR, lt. News staff: 2; News: 20 hrs wkly. Target aud: General. ♦David Hinson, gen mgr; John Richetta, gen sls mgr; Terry Rich, progmg dir; Bob Kratz, chief of engrg.

***WJRH(FM)**— March 1953: 104.9 mhz; 8 w. 23 ft TL: N40 41 53 W75 12 30. (CP: 100 w). Stereo. Hrs opn: 15 Box 9473, Hogg Hall, Lafayette College, 18042. Phone: (610) 330-5316. Fax: (610) 250-5318. Web Site: www.lafayette.edu. Licensee: Lafayette College. Population served: 200,000 Format: Var. News staff: 3; News: 6 hrs wkly. Target aud: General; college students & community. Spec prog: Jazz 9 hrs, reggae 6 hrs, metal 6 hrs, Sp 6 hrs, classic rock 4 hrs wkly. ♦Sergey Tosninski, gen mgr; Brian Hertz, progmg dir; Fred Lott, chief of engrg.

Ebensburg

WRDD(AM)— May 25, 1961: 1580 khz; 1 kw-D. TL: N40 29 33 W78 42 54. Stereo. Hrs opn: Sunrise-sunset Box 1095, Northern Cambria, 15714. Secondary address: 104 S. Center St. 15931. Fax: (814) 471-0282. E-mail: whpa@verizon.net Licensee: Vernal Enterprises Inc. (group owner; acq 3-19-97; $20,000 with WNCC(AM) Northern Cambria). Population served: 250,000 Natl. Network: USA. Format: news/talk. News: 4 hrs wkly. Target aud: General; church goers. ♦Denny Pompa, pres; Larry Schrecongost, gen mgr.

WRKW(FM)— July 15, 1962: 99.1 mhz; 50 kw. Ant 499 ft TL: N40 24 41 W78 46 29. Stereo. Hrs opn: 109 Plaza Dr., Suite 2, Johnstown, 15905. Phone: (814) 255-4186. Fax: (814) 255-6145. Web Site: www.rocky99.com. Licensee: Forever Broadcasting LLC. (group owner; (acq 5-1-2005; $2.73 million with WJHT(FM) Johnstown). Population served: 42,476 Format: Rock. ♦Verla Price, gen mgr; Tina Perry, gen sls mgr; Mike Stevens, progmg dir; Rick Shepard, news dir; Jim Boxler, chief of engrg.

WWGE(AM)—See Loretto

Edinboro

***WFSE(FM)**— Apr 3, 1979: 88.9 mhz; 3 kw. 312 ft TL: N41 52 41 W80 10 40. Stereo. Hrs opn: 24 Edinboro Univ. of Pa., Faculty Annex 110, 16444. Phone: (814) 732-2641. Phone: (814) 732-2889 (request line). E-mail: dumbluck77@hotmail.com Licensee: Edinboro University. Population served: 230,000 Format: Alternative, modern rock. News: 18 hrs wkly. Target aud: 18-25; college students with community interest. Spec prog: Football & basketball, Black 15 hrs, relg 4 hrs, loc news 3 hrs, swing 2 hrs wkly. ♦Dr. Frank Pogue, CEO; Terrence Warburton, chmn; Chris Volack, gen mgr; Richard Smith, dev dir.

WXTA(FM)— Oct 15, 1988: 97.9 mhz; 10 kw. 505 ft TL: N41 57 59 W80 06 40. Stereo. Hrs opn: 24 471 Robison Rd., Erie, 16509. Phone: (814) 864-4835. Fax: (814) 868-1876. Web Site: www.country98wxta.com. Licensee: Citadel Broadcasting Co. Group owner: Citadel Broadcasting Corp. (acq 5-12-2004; grpsl). Natl. Rep: Katz Radio. Format: Country. Target aud: Adults; 25-64. ♦Farid Suleman, CEO; Jim Riley, gen mgr; Stephanie Lancaster, gen sls mgr; Fred Horton, progmg dir; Dave Benson, news dir.

Elizabethtown

WMHX(FM)—Hershey, Apr 30, 1964: 106.7 mhz; 14 kw. Ant 928 ft TL: N40 10 16 W76 35 50. Stereo. Hrs opn: 24 515 S. 32nd St., Camp Hill, 17011. Phone: (717) 367-7700. Fax: (717) 367-0239. Web Site: www.mix1067fm.com. Licensee: Citadel Broadcasting Co. Group owner: Citadel Broadcasting Corp. (acq 5-29-97; grpsl). Population served: 1,500,000 Law Firm: Fleischman & Walsh. Format: Adult contemp. ♦Bob Adams, gen mgr; Steve Gallagher, opns mgr; Jay Hunter, prom dir.

WPDC(AM)— May 1958: 1600 khz; 1 kw-D, 18 w-N. TL: N40 09 45 W76 34 36. Hrs open: 24 1051 Dairy Lane, 17022. Phone: (717) 367-1600. E-mail: teamespn@earthlink.net Licensee: JVJ Communications Inc. (acq 10-1-84; $125,000; FTR: 10-15-84). Population served: 20,000 Natl. Network: ESPN Radio. Rgnl. Network: Radio Pa. Format: Sports. News: 10 hrs wkly. Target aud: 25-54; men. ♦Vincent Grande, pres & gen mgr; Bill Wilson, opns VP; Sam Conrad, opns dir.

***WWEC(FM)**— Aug 25, 1990: 88.3 mhz; 100 w. 373 ft TL: N40 08 83 W76 35 38. Stereo. Hrs opn: 18 Elizabethtown College, One Alpha Dr., 17022-2298. Phone: (717) 361-1413. Phone: (717) 361-1589. Fax: (717) 361-1180. E-mail: wwec@etown.edu Web Site: www.etown.edu. Licensee: Elizabethtown College. Population served: 25,000 Format: Progressive, Alternative. News: 6 hrs wkly. Target aud: General; college students, high school, churches, community. ♦Dr. Randyll K. Yoder, gen mgr; John Treese, stn mgr; Sara Robinson, dev dir; Adam Steiner, mktg dir; Kate Norton, prom dir.

Elizabethville

WYGL-FM— Dec 7, 1989: 100.5 mhz; 1.2 kw. 515 ft TL: N40 37 24 W76 49 54. Stereo. Hrs opn: 24 Box 90, Selinsgrove, 17870. Phone: (570) 374-8819. Fax: (570) 374-7444. E-mail: bigcountryrequest@hotmail.com Web Site: www.bigcountrynow.com. Licensee: MMP License LLC. Group owner: MAX Media L.L.C. (acq 10-17-03; grpsl). Natl. Network: USA. Law Firm: Kaye, Scholer, Fierman, Hays & Handler. Format: Contemp country. News staff: one; News: 8 hrs wkly. Target aud: 25-54. ♦John A. Trinder, pres; Scott Richards, VP & gen mgr; Greg Adair, gen sls mgr; Ted Koppen, chief of engrg.

Ellwood City

WKPL(FM)— July 4, 1968: 92.1 mhz; 2.5 kw. Ant 512 ft TL: N40 46 09 W80 16 56. Stereo. Hrs opn: 131 Pleasant Dr., Ste 4, Aliquippa, 15001-1300. Phone: (724) 728-6955. Fax: (724) 728-6955. Licensee: Keymarket Licenses LLC. Group owner: Keymarket Communications LLC (acq 6-30-2004; grpsl). Population served: 18,458 Format: Oldies.

Emporium

WLEM(AM)— Mar 2, 1958: 1250 khz; 2.5 kw-D, 30 w-N. TL: N41 30 22 W78 13 26. Hrs open: 16 145 E. 4th, 15834. Phone: (814) 486-3712. Fax: (814) 486-1772. Licensee: Salter Communications Inc. (acq 11-9-2006; $700,000 with co-located FM). Population served: 25,000 Natl. Network: Westwood One. Natl. Rep: Commercial Media Sales. Law Firm: Pepper & Corazzini. Format: Country. News staff: one; News: 3 hrs wkly. Target aud: 25-65. ♦John M. Salter, pres; J. Philippone, gen mgr; Gary Mitchell, opns mgr & progmg dir.

WQKY(FM)—Co-owned with WLEM(AM). May 20, 1985: 98.9 mhz; 2 kw. 548 ft TL: N41 29 32 W78 15 19. Stereo. Format: Adult contemp.

Ephrata

WIOV-FM— Nov 9, 1962: 105.1 mhz; 25 kw. 702 ft TL: N40 10 30 W76 09 31. Stereo. Hrs opn: 44 Bethany Rd., 17522-2416. Phone: (717) 738-1191. Fax: (717) 738-1661. E-mail: dick.raymond@citcomm.com Web Site: www.wiov.com. Licensee: Citadel Broadcasting Co. Group owner: Citadel Broadcasting Corp. (acq 5-12-2004; grpsl). Population served: 1,000,000 Natl. Rep: McGavren Guild. Format: Country. ♦Mitch Carroll, gen sls mgr; Dick Raymond, progmg dir; Carrie Rey, traf mgr.

***WRTL(FM)**— 2000: 90.7 mhz; 1 w horiz, 850 w vert. Ant 869 ft TL: N40 19 22 W76 11 52. Hrs open: 24
Rebroadcasts WRTI(FM) Philadelphia 100%.
1509 Cecil B. Moore ave., Philadelphia, 19121-3410. Phone: (215) 204-8405. Fax: (215) 204-7027. Web Site: www.wrtl.org. Licensee: Temple University of The Commonwealth System of Higher Education. Format: Classical jazz. ♦Dave Conant, gen mgr.

Erie

***WEFR(FM)**— March 1992: 88.1 mhz; 630 w. 430 ft TL: N41 57 59 W80 06 40. Stereo. Hrs opn: 24 Family Stations Inc., 4135 Northgate Blvd., Sacramento, CA, 95834-1226. Phone: (916) 641-8191. Fax: (916) 641-8238. Licensee: Family Stations Inc. (group owner) Format: Relg. Target aud: General. ♦Harold Camping, pres; John Rorvik, opns mgr.

***WERG(FM)**— Dec 1, 1972: 90.5 mhz; 2.75 kw. Ant 374 ft TL: N42 02 34 W80 03 57. Stereo. Hrs opn: 20 Gannon Univ., University Sq., 16541. Phone: (814) 871-5841. Fax: (814) 871-7302. E-mail: comedian13@hotmail.com Web Site: www.wergfm.com. Licensee: Gannon University. Population served: 129,231 Format: Var/div. News staff: 4; News: 5 hrs wkly. Target aud: 12-30. Spec prog: It 3 hrs, Pol 3 hrs, Sp 3 hrs, gospel 3 hrs, reggae 4 hrs wkly. ♦Evan O'Polka, gen mgr; Maggie Bausin, prom dir; Lacey Johnson, progmg dir & chief of engrg.

WFNN(AM)— 1947: 1330 khz; 5 kw-U, DA-2. TL: N42 03 18 W80 02 24. Hrs open: One Boston Store Place, 16501. Phone: (814) 461-1000. Fax: (814) 874-0011. Fax: (814) 455-6000. Licensee: Connoisseur Media of Erie LLC. Group owner: NextMedia Group L.L.C. (acq 3-31-2006; grpsl). Population served: 129,231 Natl. Rep: Katz Radio. Format: Oldies. ♦Rick Rambaldo, gen mgr.

WJET(AM)— 1951: 1400 khz; 1 kw. TL: N42 07 28 W80 03 54. Hrs open: 24 1 Boston Store Pl., 16501. Phone: (814) 461-1000. Phone: (814) 874-0011. Fax: (814) 874-0011. E-mail: jet1400@jet1400.com Web Site: www.jetradio1400.com. Licensee: Connoisseur Media of Erie LLC. Group owner: NextMedia Group L.L.C. (acq 3-31-2006; grpsl). Population served: 236,700 Rgnl. Network: Radio Pa. Format: News/talk. News staff: one. Target aud: 35 plus; middle to upper middle income, business owners, upscale. ♦Rick Rambaldo, gen mgr & stn mgr.

***WMCE(FM)**— Feb 2, 1989: 88.5 mhz; 750 w. Ant 499 ft TL: N42 05 25 W79 56 37. Stereo. Hrs opn: 24 501 E. 38th St., 16546. Phone: (814) 824-2260. Phone: (814) 824-2261. Fax: (814) 824-2590. E-mail: wshannon@mercyhurst.edu Web Site: www.mercyhurst.edu. Licensee: Mercyhurst College. Population served: 280,000 Natl. Network: AP Radio. Format: Class. News: 6 hrs wkly. Target aud: General; Adults 45+. Spec prog: Pol 3 hrs, Ger 4 hrs, Sp 3 hrs, jazz 4 hrs wkly. ♦William T. Shannon, gen mgr.

WPSE(AM)— Apr 21, 1935: 1450 khz; 1 kw-U. TL: N42 08 11 W80 02 25. Hrs open: 24 Penn State-Behrend, Station Rd., 16563-1450. Phone: (814) 898-6495. Phone: (814) 898-6491. Licensee: Board of Trustees, Pennsylvania State University. (acq 12-23-89). Natl. Network: CBS, Westwood One. Format: Business news, sports. Target aud: General. ♦Ron Slomski, gen mgr.

WQHZ(FM)— Oct 15, 1951: 102.3 mhz; 1.7 kw. Ant 613 ft TL: N42 02 25 W80 04 08. Stereo. Hrs opn: 24 471 Robison Rd., 16509. Phone: (814) 868-5355. Fax: (814) 868-1876. Web Site: www.z1023online.com. Licensee: Citadel Broadcasting Co. Group owner: Citadel Broadcasting Corp. (acq 5-12-2004; grpsl). Natl. Rep: Katz Radio. Format: Classic rock. News staff: one. Target aud: 25-54; Adults. ♦Farid Suleman, CEO; Jim Riley, gen mgr & natl sls mgr; Stephanie Lancaster, gen sls mgr; Adam Reese, progmg dir; Dave Benson, news dir.

***WQLN-FM**— Jan 7, 1973: 91.3 mhz; 35 kw. 500 ft TL: N42 02 35 W80 03 59. Stereo. Hrs opn: 24 8425 Peach St., 16509. Phone: (814) 864-3001. Fax: (814) 864-4077. E-mail: dmiller@wqln.org Web Site: www.wqln.org. Licensee: Public Broadcasting of Northwest Pennsylvania Inc. Population served: 129,231 Natl. Network: NPR, PRI. Law Firm: Dow, Lohnes & Albertson. Format: Classical; news; jazz. News staff: one; News: 24 hrs wkly. Target aud: General. Spec prog: Sp one hr, pub affrs 5 hrs, new age 2 hrs, call-in show 3 hrs wkly. ♦Dwight Miller, pres & gen mgr; Tracy B. Ferrior, VP; Tom Pysz, opns mgr & progmg dir; Kim Young, news dir. Co-owned TV: *WQLN(TV) affil.

WRIE(AM)— 1941: 1260 khz; 5 kw-U, DA-2. TL: N42 03 18 W80 02 24. Hrs open: 24 471 Robison Rd. W., 16509. Phone: (814) 868-5355. Fax: (814) 868-1876. Licensee: Citadel Broadcasting Co. Group owner: Citadel Broadcasting Corp. (acq 5-12-2004; grpsl). Natl. Network: ESPN Radio. Natl. Rep: Katz Radio. Format: Sports talk. ♦Farid Suleman, CEO; Judy Ellis, COO; Gary Spurgeon, gen mgr & gen sls mgr; Marcia Diehl, opns mgr; Donna Palowitz, sls dir; Tina Achhammer, prom dir; Ron Arlen, progmg dir; Heather Rose, mus dir & traf mgr; Dave Benson, news dir; Rick Pogson, chief of engrg.

WXKC(FM)—Co-owned with WRIE(AM). 1949: 99.9 mhz; 50 kw. 492 ft TL: N42 05 24 W79 57 12. Stereo. 24 Web Site: www.classy100.com. Format: Adult contemp. News staff: one. Target aud: Adults; 35-64. ♦Heather Rose, traf mgr.

WRKT(FM)—See North East

WRTS(FM)— May 1, 1969: 103.7 mhz; 50 kw. 499 ft TL: N42 05 25 W79 56 37. Stereo. Hrs opn: 24 1 Boston Store Pl., 16501. Phone: (814) 461-1000. Fax: (814) 455-6000. E-mail: star104@star104.com Web Site: www.star104.com. Licensee: Connoisseur Media of Erie LLC. Group owner: NextMedia Group L.L.C. (acq 3-30-2006; grpsl). Population served: 226,600 Law Firm: Fletcher, Heald & Hildreth. Format: Top 40. News staff: one; News: one hr wkly. Target aud: 25-54. Spec prog: PSA one hr wkly. ♦Richard Rambaldo, gen mgr.

WXBB(FM)— Sept 1, 1993: 94.7 mhz; 1.7 kw. Ant 613 ft TL: N42 02 31 W80 03 57. Hrs open: 24 1 Boston Store Place, 16501. Phone: (814) 461-1000. Fax: (814) 874-0011. Web Site: www.947bobfm.com. Licensee: Connoisseur Media of Erie LLC. Group owner: NextMedia Group L.L.C. (acq 3-30-2006; grpsl). Format: Adult hits. ♦Rick Rambaldo, gen mgr.

WYNE(AM)—See North East

Everett

WSKE(FM)—Listing follows WZSK(AM).

WZSK(AM)— Mar 15, 1963: 1040 khz; 10 kw-D. TL: N40 00 26 W78 21 44. Hrs open: Box 133, 15537-0133. Phone: (814) 652-2600. Fax: (814) 652-9347. E-mail: wzsk@penn.com Licensee: New Millennium Communications Group Inc. (acq 9-27-01; with co-located FM). Natl. Network: Westwood One. Rgnl. Network: Radio Pa. Rgnl rep: Dome. Law Firm: Fletcher, Heald & Hildreth. Format: News/talk. News: 10 hrs wkly. Target aud: 25-54. ♦John C. Imler, gen mgr & progmg dir; Shane S. Imler, pres & adv dir; Bob Resconsin, chief of engrg; John Imler, disc jockey.

WSKE(FM)—Co-owned with WZSK(AM). Mar 15, 1988: 104.3 mhz; 820 w. Ant 886 ft TL: N40 00 11 W78 23 58. Stereo. 24 Format: Country. News: 7 hrs wkly. Target aud: 25-54. ♦John Imler, disc jockey.

Fairless Hills

WKXW(FM)—See Trenton, NJ

Fairview

WTWF(FM)— October 2001: 93.9 mhz; 3 kw. Ant 469 ft TL: N41 57 59 W80 06 40. Hrs opn: One Boston Store Pl., Erie, 16501. Phone: (814) 461-1000. Fax: (814) 874-0011. E-mail: us939@us939.com Web Site: www.us939.com. Licensee: Connoisseur Media of Erie LLC. Group owner: NextMedia Group L.L.C. (acq 3-30-2006; grpsl). Format: Country. ♦Richard Rambaldo, gen mgr.

Farrell

WAKZ(FM)—See Youngstown, OH

WLOA(AM)—Licensed to Farrell. See Youngstown OH

Folsom

***WRSD(FM)**— Jan 5, 1983: 94.9 mhz; 1.4 kw. 20 ft TL: N39 53 12 W75 20 01. Hrs open: Ridley School District Admin. Bldg., 901 Morton Ave., Ste. 100, 19033. Phone: (610) 534-1900. Fax: (610) 461-7083. Licensee: Ridley School District. Format: Div, adult contemp. ♦Kevin Hitchens, gen mgr.

Forest City

WQFN(FM)— 2000: 100.1 mhz; 750 w. Ant 935 ft TL: N41 35 35 W75 25 56. Hrs open:
Simulcast with WQFM(FM) Nanticoke.
149 Penn Ave., Scranton, 18503. Phone: (570) 346-6555. Fax: (570) 346-6038. E-mail: tbass@shamrocknepa.com Web Site: www.921qfm.com. Licensee: The Scranton Times L.P. Group owner: Shamrock

Communications Inc. (acq 3-23-2000). Format: Adult contemp. Target aud: 35-64; adults. ♦William R. Lynett, CEO; Jim Loftus, gen mgr.

Franklin

***WAWN(FM)—** 1998: 89.5 mhz; 1 kw. 315 ft TL: N41 23 39 W49 46 20. Hrs open: Box 3206, Tupelo, MS, 38803. Phone: (662) 844-8888. Fax: (662) 842-6791. E-mail: comments@afr.net Web Site: www.afr.net. Licensee: American Family Association. Group owner: American Family Radio Format: Christian, inspirational. ♦Marvin Sanders, gen mgr.

WFRA(AM)— Apr 13, 1958: 1450 khz; 1 kw-U. TL: N41 23 27 W79 48 43. Hrs open: 6 AM-midnight Box 908, 1411 Liberty St., 16323. Phone: (814) 432-2189. Fax: (814) 437-9372. Licensee: Forever Broadcasting LLC. Group owner: Forever Broadcasting (acq 7-20-00; grpsl). Population served: 64,000 Law Firm: Reddy, Begley & McCormick. Format: MOR, news, sports. News staff: one; News: 12 hrs wkly. Target aud: 45 plus. ♦Carol Logan, pres; Terry Deitz, gen mgr; Tim Snyder, progmg dir & news dir; Lynn Deppen, engrg VP & chief of engrg; Tim Shaw, sports cmtr.

WOXX(FM)—Co-owned with WFRA(AM). Mar 5, 1971: 99.3 mhz; 7.3 kw. 600 ft TL: N41 26 16 W79 55 29. Stereo. 6 AM-midnight 180,000 Format: Hot adult contemp. News: 4 hrs wkly. Target aud: 18-44. ♦Tim Snyder, prom dir; Tim Shaw, sports cmtr.

Galeton

***WCOG-FM—** 1996: 100.7 mhz; 7.7 kw. Ant 492 ft TL: N41 39 36 W77 38 02. Hrs open:
Rebroadcasts WCIK(FM) Bath, NY 100%.
Box 506, Bath, NY, 14810. Secondary address: 7634 Campbell Creek Rd., Bath, NY 14810. Phone: (607) 776-4151. Fax: (607) 776-6929. E-mail: mail@fln.org Web Site: www.fln.org. Licensee: Family Life Ministries Inc. Group owner: Family Life Network (acq 10-1-96; $20,130). Natl. Network: Salem Radio Network. Law Firm: Hardy, Carey, Chautin & Balkin, LLP. Wire Svc: Metro Weather Service Inc. Format: Contemp Christian. News staff: 3; News: 14 hrs wkly. Target aud: 30-54; general. ♦Dick Snavely, pres & CFO; Rick Snavely, pres, VP, gen mgr & stn mgr; John Owens, progmg dir; Jim Travis, chief of engrg.

Gallitzin

WHPA(FM)— 1999: 93.5 mhz; 1.25 kw. Ant 727 ft TL: N40 29 36 W78 32 31. Stereo. Hrs open: 24 Box 1095, Northern Cambria, 15714-3095. Phone: (814) 472-4060. Fax: (814) 472-9370. E-mail: whpa@verizon.net Licensee: Vernal Enterprises Inc. (group owner). Format: Oldies. News: 2 hrs wkly. Spec prog: Altoona Curve baseball/local high school. ♦Larry Schrengost, gen mgr.

Gettysburg

WGET(AM)— Aug 27, 1950: 1320 khz; 1 kw-D, 500 w-N, DA-2. TL: N39 50 30 W77 13 25. Hrs open: 24 Box 3179, 1560 Fairfield Rd., 17325. Phone: (717) 334-3101. Fax: (717) 334-5822. Web Site: www.wget.com. Licensee: Times and News Publishing Co. Population served: 150,000 Natl. Network: CBS. Rgnl. Network: Radio Pa. Law Firm: Hogan & Hartson. Format: Adult contemp, news, sports. News staff: 3; News: 40 hrs wkly. Target aud: 35-64; mainstream mature adults. ♦Philip Jones, CEO; Cindy Ford, pres; Dave Jackson, opns mgr & mus dir; John C. Martin, gen sls mgr; Kim Alexander, news dir; Scott Steffan, engrg mgr & chief of engrg; Lou Ann Milhimes, traf mgr; Larry Rhoten, spec ev coord.

WGTY(FM)—Co-owned with WGET(AM). July 5, 1962: 107.7 mhz; 16 kw. 829 ft TL: N39 51 23 W76 56 57. Stereo. 24 Web Site: www.wgty.com.400,000 Format: Country. News: 2 hrs wkly. Target aud: 25-54. ♦Cindy Ford, gen mgr, opns mgr & progmg dir; Casey Lee Summers, mktg mgr & pub affrs dir; Lisa Sneddin, prom mgr; Brad Austin, mus dir; John LeMay, chief of engrg; Lou Ann Milhimes, traf mgr; Larry Roten, spec ev coord.

***WZBT(FM)—** Oct 23, 1976: 91.1 mhz; 180 w. 380 ft TL: N39 50 15 W77 14 09. Hrs open: 8 AM-2 AM Gettysburg College, Box 435, 17325. Secondary address: 300 N. Washington St. 17325. Phone: (717) 337-6000. Fax: (717) 337-6666. E-mail: wzbtexec@gettysburg.edu Web Site: www.gettesburg.edu/~wzbt/. Licensee: Gettysburg College. Population served: 100,000 Format: Progsv. Target aud: General. Spec prog: Class 3 hrs, folk 6 hrs, jazz 4 hrs, Sp 4 hrs, gospel one hr wkly. ♦Ryan Gottschall, stn mgr; Laura Benincasa, progmg dir.

Glen Mills

***WZZE(FM)—** May 20, 1975: 97.3 mhz; 18 w. 180 ft TL: N39 55 15 W75 29 58. (CP: Ant 184 ft.). Hrs opn: Box 5001, Concordville, 19331. Secondary address: Glen Mills Schools, Glen Mills Rd. 19342. Phone: (610) 459-8100, ext: 307. Phone: (610) 459-4829. E-mail: msmith@glenmillerschools.org Licensee: Glen Mills Schools. (acq 2-28-84). Natl. Network: ABC. Format: CHR. ♦C.D. Ferrainola, pres; Mark Smith, opns mgr.

Grantham

***WVMM(FM)—** Sept 29, 1989: 90.7 mhz; 100 w. 300 ft TL: N40 09 34 W76 59 00. Stereo. Hrs opn: 7 AM-12 AM Messiah College, One College Ave., Box 3058, 17027. Phone: (717) 691-6081. Fax: (717) 796-5353. E-mail: earke@messiah.edu Web Site: www.messiah.edu/wvmm. Licensee: Messiah College. Natl. Network: PRI. Format: Christian rock, AAA, Indy. News: 22. Target aud: 13-25. Spec prog: Jazz 4, bluegrass 2 hrs, gospel 2 hrs, hip hop 4 hrs wkly. ♦Edward T. Arke, gen mgr; Sheryl Ezbiansky, prom dir & chief of engrg.

Greencastle

WQCM(FM)— May 6, 1967: 94.3 mhz; 3.5 kw. 430 ft TL: N39 47 29 W77 40 30. Stereo. Hrs opn: 24 25 Penncraft Ave., Chambersburg, 17201. Phone: (717) 263-0813. Fax: (717) 263-9649. E-mail: rbateman@damebroadcasting.net Web Site: www.wqcmfm.com. Licensee: MLB-Hagerstown-Chambersburg IV LLC. (group owner; (acq 7-20-2005; grpsl). Population served: 212,913 Law Firm: Latham & Watkins. Format: Classic rock. News staff: one. Target aud: 25-44. ♦Rich Bateman, gen mgr; Tammy Heckman, prom dir; Mike Holder, progmg dir.

Greensburg

WGSM(FM)— Feb 2, 2006: 107.1 mhz; 2.85 kw. Ant 482 ft TL: N40 15 54 W79 20 25. Stereo. Hrs opn: 24 960 Penn Ave., Suite 200, Pittsburgh, 15222. Phone: (412) 456-4064. Fax: (412) 391-3559. Web Site: www.wamo.com. Licensee: The St. Pier Group LLC. (group owner). (acq 11-7-2006; $2.2 million). Law Firm: Fletcher, Heald & Hildreth. Format: Var hits. News staff: one; News: one hr wkly.

Greenville

WEXC(FM)— July 1965: 107.1 mhz; 3 kw. Ant 328 ft TL: N41 22 50 W80 24 48. Stereo. Hrs opn: 24 44 McCracken Rd., 16125. Secondary address: 124 N. Park Avenue, Warren, OH 44481. Phone: (724) 588-8000. Fax: (724) 588-2470. Web Site: www.thefreq107.com. Licensee: Beacon Broadcasting Inc. (group owner; (acq 9-14-2005; grpsl). Population served: 9,960 Format: Positive rock, Christian. News staff: one. Target aud: 14-34. ♦Harold Glunt, gen mgr; Dana Schroyer, disc jockey; Matt Rhodes, disc jockey.

WGRP(AM)— Sept 19, 1959: 940 khz; 1 kw-D, 2 w-N, DA-2. TL: N41 23 10 W80 24 35. Hrs open: 24 Box 1798, Warren, OH, 44482-1798. Phone: (330) 394-7700. Fax: (330) 394-7701. Licensee: Beacon Broadcasting Inc. (acq 9-14-2005; grpsl). Population served: 9,960 Law Firm: Hogan & Hartson. ♦Harold Glunt, pres & gen mgr.

***WTGP(FM)—** Sept 3, 1971: 88.1 mhz; 1.1 kw. 6 ft TL: N41 24 51 W80 24 50. Stereo. Hrs opn: 9 AM-midnight Thiel College, 75 College Ave., 16125. Secondary address: 57 Irvine Dr. 16125. Phone: (724) 589-2210. Phone: (724) 589-2171. Fax: (724) 589-2010. Fax: (724) 589-2730. E-mail: dwest@pathway.net Web Site: www.thiel.edu/studentlife/student_org/wtgp. Licensee: Thiel College. Population served: 8,704 Format: Div, progsv. Target aud: 18-23; Thiel college students, faculty & staff. Spec prog: Relg one hr wkly. ♦Ang Baker, progmg dir.

Grove City

***WSAJ-FM—** September 1968: 91.1 mhz; 3 kw. 125 ft TL: N41 09 20 W80 04 47. Stereo. Hrs open: 24 Grove City College, 100 Campus Dr., 16127. Phone: (724) 458-2077. Fax: (724) 458-2329. E-mail: wsaj@gcc.edu Licensee: Grove City College. Population served: 100,000 Format: Class. News staff: one. Target aud: General; listeners who are generally unfamiliar with class mus & arts. ♦Darren Morton, gen mgr & stn mgr.

WWGY(FM)— Sept 10, 1962: 95.1 mhz; 19 kw. Ant 805 ft TL: N41 15 08 W80 21 28. Stereo. Hrs opn: 24 219 Savannah Gardner Rd., New Castle, 16101. Phone: (724) 346-5070. Fax: (724) 346-5075. E-mail: webmaster@foreverradio.com Web Site: fforeverradio.com. Licensee: Forever Broadcasting LLC. Group owner: Forever Broadcasting (acq 2-23-2004; $2.28 million). Format: Country. News staff: one; News: 2 hrs wkly. Target aud: 18-34. ♦Scott D. Cohagan, gen mgr; John Thomas, progmg dir.

Halifax

***WLVU(FM)—**Not on air, target date: unknown: 88.5 mhz; 1 kw. Ant 420 ft TL: N40 27 26 W76 54 13. Hrs open: 2351 Sunset Blvd., Suite 170-218, Rocklin, CA, 95765. Phone: (916) 251-1600. Fax: (916) 251-1650. Licensee: Educational Media Foundation. (acq 7-23-2007; grpsl). ♦Mike Novak, sr VP.

Hanover

WHVR(AM)— Jan 9, 1949: 1280 khz; 5 kw-D, 500 w-N, DA-2. TL: N39 49 11 W77 00 25. Hrs open: Box 234, 17331. Secondary address: 275 Radio Rd. Phone: (717) 637-3831. Fax: (717) 637-9006. Licensee: Radio Hanover Inc. Population served: 35,000 Format: Classic country. ♦Joan McAnall, gen mgr; Rick McCauslin, gen sls mgr; Deanna Forney, news dir; Daryll Harcock, chief of engrg.

WYCR(FM)—Co-owned with WHVR(AM). Dec 22, 1962: 98.5 mhz; 10.5 kw. Ant 928 ft TL: N39 51 30 W76 56 52. Stereo. E-mail: info@thepeak.com Web Site: www.thepeak985.com.93,200 Format: Classic hits. ♦Beth Mowren, traf mgr; Davy Crockett, disc jockey; Jeff Brown, disc jockey; Jim Cooke, disc jockey; Lee Sheldon, disc jockey; Paul Scott, disc jockey; Tom Jackson, disc jockey.

Harrisburg

WHKF(FM)—Listing follows WTKT(AM).

WHP(AM)— 1924: 580 khz; 5 kw-U, DA-N. TL: N40 18 11 W76 57 07. Stereo. Hrs opn: 24 600 Corporate Circle, 17110. Phone: (717) 540-8800. Fax: (717) 541-0094. Fax: (717) 540-9268. Web Site: www.whp580.com. Licensee: Clear Channel Radio License Inc. Group owner: Clear Channel Communications Inc. (acq 8-5-98; grpsl). Population served: 392,400 Natl. Network: Westwood One. Format: News/talk. News staff: 4. Target aud: 35-64. ♦Ron Roy, gen sls mgr & natl sls mgr.

WRVV(FM)—Co-owned with WHP(AM). 1946: 97.3 mhz; 17 kw. 840 ft TL: N40 20 44 W76 52 09. Stereo. 24 Web Site: www.wrvv.com. Format: Rock, adult contemp. Target aud: 25-54.

***WITF-FM—** Apr 1, 1971: 89.5 mhz; 5.9 kw. Ant 1,361 ft TL: N40 20 45 W76 52 06. Stereo. Hrs opn: 24 Box 2954, 17105. Secondary address: 1982 Locust Ln. 17109. Phone: (717) 236-6000. Fax: (717) 232-7612. E-mail: info@witf.org Web Site: www.witf.org. Licensee: WITF Inc. Population served: 12,500 Natl. Network: NPR, PRI. Law Firm: Dow, Lohnes & Albertson. Format: Class, news/talk. News staff: 3; News: 43 hrs wkly. ♦Kathleen Pavelko, pres; Mitzi Trostle, gen mgr & stn mgr. Co-owned TV: *WITF-TV affil.

WKBO(AM)— 1922: 1230 khz; 48 kw-U. TL: N40 16 52 W76 52 06. Hrs open: 24 600 Corporate Cir., 17110-9787. Phone: (717) 540-8800. Fax: (717) 540-8814. E-mail: fortress1230am@oneheartministries.com Web Site: www.oneheartministries.com. Licensee: Clear Channel Broadcasting Licenses Inc. Group owner: Clear Channel Communications Inc. (acq 8-5-98; grpsl). Population served: 68,061 Natl. Rep: Salem. Format: Contemp Christian. News staff: 6; News: 168 hrs wkly. Target aud: 35-54; well educated, upscale professionals. Spec prog: Pop standards, Music of Your Life. ♦Pete Hamel, gen mgr, stn mgr & gen sls mgr.

WNNK-FM—Listing follows WTCY(AM).

WRBT(FM)— Sept 30, 1962: 94.9 mhz; 25 kw. 699 ft TL: N40 18 57 W76 57 02. Stereo. Hrs opn: 24 600 Corporate Cir., 17110. Phone: (717) 671-9949. Fax: (717) 540-8814. Web Site: www.bobradio.com. Licensee: Clear Channel Radio License Inc. Group owner: Clear Channel Communications Inc. (acq 8-5-98; grpsl). Population served: 1,000,000 Natl. Rep: Christal. Law Firm: Latham & Watkins. Format: Hot country. News staff: one. Target aud: 25-54. ♦Ronald Roy, gen mgr. Co-owned TV: WHP-TV.

WSJW(FM)—See Starview

WTCY(AM)— May 28, 1945: 1400 khz; 1 kw-U. TL: N40 14 58 W76 52 03. Hrs open: 24 2300 Vartan Way, 17110-9720. Phone: (717) 238-1041. Fax: (717) 234-4842. Web Site: www.cumulus.com. Licensee: Cumulus Licensing Corp. Group owner: Cumulus Media Inc. (acq 11-28-00; grpsl). Population served: 510,000 Format: Urban adult contemp. News staff: 2; News: 2 hrs wkly. Target aud: 25-54. Spec prog: Gospel, sportscasting for Harrisburg Heat. ♦Ron Vioanviannell, mktg mgr.

WNNK-FM—Co-owned with WTCY(AM). 1962: 104.1 mhz; 22.5 kw. 725 ft TL: N40 18 59 W76 57 04. Stereo. 24 Web Site: www.cumulus.com. Format: CHR. News staff: 2; News: 15 hrs wkly.

WTKT(AM)— February 1948: 1460 khz; 5 kw-D, 4.2 kw-N, DA-N. TL: N40 18 32 W76 56 13. Hrs open: 24 600 Corporate Cir., 17110. Phone: (717) 540-8800. Fax: (717) 540-8814. Web Site: www.1460theticket.com. Licensee: Clear Channel Radio License Inc. Group owner: Clear Channel Communications Inc. (acq 8-5-98; grpsl). Population served: 550,000 Natl. Rep: Clear Channel. Law Firm: Fisher, Wayland, Cooper, Leader & Zaragoza L.L.P. Format: Oldies, sports. News staff: 4; News: 30 hrs wkly. Target aud: General. Spec prog: Gospel 2 hrs, pub service 2 hrs wkly. ♦Ken Austin, progmg dir.

WHKF(FM)—Co-owned with WTKT(AM). July 1965: 99.3 mhz; 6 kw. 328 ft TL: N40 15 44 W76 54 37. Stereo. Web Site: www.wwklfm.com.550,000 Target aud: 35-54. ♦Doug Baker, gen sls mgr; Kraig Nace, prom mgr; Peter MacArthur, news dir; Tom Presite, chief of engrg.

***WXPH(FM)**— 1995: 88.1 mhz; 540 w. Ant 105 ft TL: N40 15 44 W76 53 11. Hrs open: 24 3025 Walnut St., Philadelphia, 19104. Phone: (215) 898-6677. Fax: (215) 898-0707. E-mail: wxpndesk@xpn.org Web Site: www.xpn.org. Licensee: The Trustees of University of Pennsylvania. (acq 12-18-92; $5,000; FTR: 1-18-93). Population served: 400,000 Format: Adult alternative. News staff: one; News: one hr wkly. ♦Roger LaMay, gen mgr; Quyen Shanahan, dev VP.

Havertown

***WHHS(FM)**— Dec 6, 1949: 107.9 mhz; 14 w. Ant 161 ft TL: N39 58 59 W75 18 10. Stereo. Hrs opn: 2 PM-10 PM (M-F) 200 Mill Rd., 19083. Phone: (610) 446-7111. Fax: (610) 853-5952. E-mail: whhsnewsdirector@yahoo.com Web Site: www.whhs.org. Licensee: School District of Haverford Township. Population served: 50,000 Format: Div. Target aud: General. ♦Kevin Moran, gen mgr & opns dir.

Hawley

***WBYH(FM)**— December 2000: 89.1 mhz; 200 w. 525 ft TL: N41 24 43 W75 09 51. Hrs open: 24 Box 186, Sellersville, 18960. Phone: (215) 721-2141. Fax: (215) 721-9811. E-mail: wordfm@wordfm.org Web Site: www.wordfm.com. Licensee: Four Rivers Communications Broadcasting Co. Population served: 50,000 Law Firm: Schwartz, Woods & Miller. Format: Contemp Christian. ♦David Baker, gen mgr.

WYCY(FM)— Sept 13, 1993: 105.3 mhz; 2.9 kw. 479 ft TL: N41 35 01 W75 10 30. Stereo. Hrs opn: 24 575 Grove St., Honesdale, 18431. Phone: (570) 253-1616. Fax: (570) 253-6297. E-mail: vbenedetto @boldgoldmedia.com Web Site: www.boldgoldmedia.com. Licensee: Bold Gold Media Group L.P. (group owner; (acq 5-23-2005; grpsl). Natl. Network: ABC. Format: Oldies. News staff: one; News: 5 hrs wkly. Target aud: 25-55. ♦Vincent Benedetto, CEO; Bob Vanderheyden, gen mgr; Brian Wilken, gen sls mgr; Paul Ciliberto, prom dir; George Schmitt, progmg mgr; Theresa Opeka, news dir; Jessica Baglieri, traf mgr.

Hazleton

WAZL(AM)— Dec 19, 1932: 1490 khz; 1 kw-U. TL: N40 56 24 W75 58 04. Hrs open: 24 8 W. Broad St., 18201. Phone: (570) 455-1490. Fax: (570) 501-1112. E-mail: patwazl@nni.com Licensee: Group B Licensee LLC. Group owner: Route 81 Radio LLC (acq 12-2-2003; grpsl). Population served: 120,000 Wire Svc: Metro Weather Service Inc. Format: Oldies. News staff: 5; News: 1.5 hrs wkly. Target aud: 18-64. ♦Patrick Ward, gen mgr; Rich Savillo, gen sls mgr; Rocky Brown, progmg dir.

WBSX(FM)— 1949: 97.9 mhz; 6.3 kw. 1334 ft TL: N41 10 56 W75 52 22. Stereo. Hrs opn: 24 600 Baltimore Dr., 2nd Fl., Wilkes-Barre, 18702. Phone: (570) 824-9000. Fax: (570) 820-0520. Web Site: www.979x.com. Licensee: Citadel Broadcasting Co. Group owner: Citadel Broadcasting Corp. (acq 5-29-97; grpsl). Natl. Network: ABC, Moody. Law Firm: Pepper & Corazzini. Format: Active rock. Target aud: 18-34. ♦Taylor Walet, gen mgr; Jules Riley, opns mgr; Bill Palmeri, gen sls mgr & mktg mgr; Chris Lloyd, progmg dir; Phil Galasso, chief of engrg.

Hershey

WMHX(FM)—Licensed to Hershey. See Elizabethtown

Hollidaysburg

WKMC(AM)—See Roaring Spring

WRKY-FM— Dec 1, 1978: 104.9 mhz; 280 w. Ant 1,417 ft TL: N40 29 15 W78 21 09. Stereo. Hrs opn: One Forever Dr., Holidaysburg, 16648. Phone: (814) 941-9800. Fax: (814) 943-2754. E-mail: xman@rocky1049.com Web Site: www.rocky1049.com. Licensee: Forever of PA L.L.C. Group owner: Forever Broadcasting (acq 2-18-97; $2 million with WKMC(AM) Roaring Spring). Population served: 140,000 Natl. Rep: Roslin. Format: Rock, adult contemp. Target aud: 25-54; adults with significant income. ♦Carol B. Logan, pres; David Davies, gen mgr.

Homer City

WCCS(AM)— Oct 25, 1983: 1160 khz; 10 kw-D, 1 kw-N, DA-1. TL: N40 34 18 W79 10 12. Stereo. Hrs opn: 24 840 Philadelphia St., Suite 100, Indiana, 15701. Phone: (724) 479-1160. Phone: (724) 465-4700. E-mail: mbertig@rendabroadcasting.com Web Site: www.1160.com. Licensee: The St. Pier Group LLC. Group owner: Renda Broadcasting Corp. (acq 10-4-2002; $650,000). Population served: 95,000 Natl. Network: ABC. Rgnl rep: Dome & Associates Wire Svc: AP Format: Adult contemp. News staff: 2; News: 14 hrs wkly. Target aud: 25-49. Spec prog: Pol 3 hrs, oldies 9 hrs wkly. ♦Tony Renda Sr., CEO & pres; Mark A. Bertig, gen mgr; Alan Serena, opns VP; Jack Benedict, opns dir; Ron Nocco, news dir.

Honesdale

WDNH-FM—Listing follows WPSN(AM).

WPSN(AM)— September 1972: 1590 khz; 2.5 kw-D. TL: N41 33 13 W75 15 18. Hrs open: 575 Grove St., 18431. Phone: (570) 253-1616. Fax: (570) 253-6297. Web Site: www.infocow.net. Licensee: Bold Gold Media Group L.P. (group owner; (acq 5-23-2005; grpsl). Population served: 90000 Natl. Rep: Dome. Law Firm: Schwartz, Woods & Miller. Format: Sports. Target aud: General. ♦George Schmitt, progmg mgr; John Emerson, news dir.

WDNH-FM—Co-owned with WPSN(AM). Oct 12, 1981: 95.3 mhz; 3 kw. Ant 256 ft TL: N41 34 23 W75 11 30. Stereo. 24 Phone: (570) 253-9595. Web Site: www.wdnh.com.150,000 Natl. Network: USA. Format: Hot adult contemp. News staff: one; News: 6 hrs wkly. Target aud: 25-54. ♦George Schmitt, progmg dir.

***WZZH(FM)**—Not on air, target date: unknown: 90.9 mhz; 200 w. Ant 912 ft TL: N41 35 35 W75 25 56. Hrs open: Box 186, Sellersville, 18960-0186. Phone: (215) 721-2141. Fax: (215) 721-9811. E-mail: wordfm@wordfm.org Web Site: www.wordfm.org. Licensee: Four Rivers Community Broadcasting Corp. ♦Charles W. Loughery, pres.

Hughesville

WRKK(AM)— Aug 4, 1985: 1200 khz; 10 kw-D, 250 w-N, DA-2. TL: N41 12 43 W76 44 56. Hrs open: 24
Rebroadcasts WRAK(AM) Williamsport 100%.
1559 W. 4th St., Williamsport, 17701. Phone: (570) 327-1400. Fax: (570) 327-8156. Web Site: www.wrak.com. Licensee: Clear Channel Broadcasting License Inc. Group owner: Clear Channel Communications Inc. (acq 8-5-98; grpsl). Natl. Network: ABC, Westwood One. Format: News/talk. News staff: one. Target aud: 35 plus. ♦James Dabney, gen mgr; Ken Sawyer, opns dir & progmg dir.

Huntingdon

WHUN(AM)— Mar 2, 1947: 1150 khz; 5 kw-D, 36 w-N. TL: N40 27 18 W77 58 50. Hrs open: 24 RR 3 Box 225A, Huntington, 16652-8804. Phone: (814) 542-8648. Fax: (814) 643-9625. Licensee: Megahertz Licenses LLC. Group owner: Forever Broadcasting (acq 4-29-2002; $875,000 with WLTS(FM) Mount Union). Population served: 56,000 Natl. Network: Motor Racing Net. Rgnl rep: Commercial Media Sales Inc. Format: Country. News staff: one; News: 15 hrs wkly. Target aud: 25 plus; general. Spec prog: Relg 2 hrs wkly. ♦Kristin Cantrell, gen mgr.

***WKVR-FM**— March 1978: 92.3 mhz; 10 w. -376 ft TL: N40 30 00 W78 00 52. Stereo. Hrs open: 22 Juniata College, 16652. Phone: (814) 643-5031. Phone: (814) 641-3341. Fax: (814) 643-4477. Licensee: Juniata College Board of Trustees. Population served: 4,000 Format: Classic rock, progsv, AOR. News: 8 hrs wkly. Target aud: 18-25; college students. Spec prog: CHR 15 hrs, jazz 3 hrs, Black 10 hrs, contemp Christian 3 hrs, reggae 3 hrs wkly. ♦Chad Herzog, gen mgr; J. Andrew Scott, prom dir.

WLAK(FM)— Sept 12, 1967: 103.5 mhz; 160 w. 1,427 ft TL: N40 29 51 W78 08 00. Stereo. Hrs opn: 24
Rebroadcasts WMRF-FM Lewistown 95.7%.
Box 667, Lewistown, 17044. Secondary address: 12 East Market St., 2nd Floor, Lewistown 17044. Phone: (717) 248-6757. Fax: (717) 248-6759. E-mail: merfradio@acsworld.net Web Site: www.merfradio.com. Licensee: First Media Radio LLC. (group owner; (acq 3-28-2001; grpsl). Population served: 50,000 Format: Adult contemp. Target aud: 18-44. ♦Peter Herman, gen mgr; Jeff Stevens, opns dir, progmg dir & disc jockey; Mary Lee Shaffer, news dir.

Indiana

WCCS(AM)—See Homer City

WDAD(AM)— Nov 4, 1945: 1450 khz; 1 kw-U. TL: N40 37 01 W79 07 55. Hrs open: 24 840 Philadelphia St., Suite 100, 15701. Phone: (724) 465-4700. Fax: (724) 349-6842. Web Site: www.wdadradio.com. Licensee: The St. Pier Group. Group owner: Renda Broadcasting Corp. (acq 2-13-2004; $3.25 million). Population served: 16,100 Natl. Network: CBS. Natl. Rep: Dome. Law Firm: Pepper & Corazzini. Format: Good time oldies. News staff: one. Target aud: 35 plus. Spec prog: Relg 2 hrs wkly. ♦Mark Bertig, gen mgr; Tony Renda Sr., pres & gen mgr.

WQMU(FM)—Co-owned with WDAD(AM). Aug 14, 1968: 92.5 mhz; 3 kw. 108 ft TL: N40 38 17 W79 08 47. Stereo. 24 Web Site: www.wqmuradio.com.16,100 Rgnl rep: Dome Format: Adult hits. Target aud: 21-41.

***WIUP-FM**— October 1969: 90.1 mhz; 1.6 kw. 88 ft TL: N40 36 57 W79 09 40. Stereo. Hrs opn: 7 AM-2 AM Indiana Univ. of Pa., 121 Davis Hall, 15705. Phone: (724) 357-9487. Licensee: Indiana University of Pennsylvania. Population served: 92,000 Format: Div. News: 11 hrs wkly. Target aud: General. Spec prog: Black 14 hrs, class 15 hrs, folk 4 hrs, gospel one hr, jazz 15 hrs, new age 4 hrs, radio drama one hr wkly. ♦James Rogers, gen mgr. Co-owned TV: *WIUP-TV affil.

Irwin

WKHB(AM)—Licensed to Irwin. See Pittsburgh

Jackson Township

***WRTY(FM)**— Aug 23, 1991: 91.1 mhz; 3.5 kw. 862 ft TL: N41 02 40 W75 22 45. Stereo. Hrs opn: 24
Rebroadcasts WRTI(FM) Philadelphia 100%.
1509 Cecel B. Moore Ave., 3rd Fl., Philadelphia, 19121. Phone: (215) 204-8405. Fax: (215) 204-7027. E-mail: comments@wrti.org Web Site: www.wrti.org. Licensee: Temple University of The Commonwealth System of Higher Education. Population served: 1,000,000 Natl. Network: NPR. Rgnl. Network: Radio Pa. Format: Jazz, class. News staff: one; News: 15 hrs wkly. Target aud: 30-65. ♦Dave Conant, gen mgr.

Jeannette

WKFB(AM)— Jan 28, 1974: 770 khz; 750 w-D, 750 w-CH. TL: N40 17 20 W79 42 04. Hrs open: Sunrise-sunset Box 990, Greensburg, 15601-0990. Secondary address: 1918 Lincoln Hwy., North Versailles 15137. Phone: (412) 823-7000. Licensee: Broadcast Communications Inc. (group owner; (acq 4-98). Population served: 376,000 Format: Var. ♦Ashley R. Stevens, VP; Robert M. Stevens, pres & gen mgr.

Jenkintown

WPPZ-FM— Nov 1, 1960: 103.9 mhz; 270 w. Ant 1,109 ft TL: N40 02 29.6 W75 14 11.4. Hrs open: 24 1000 River Rd., Suite 400, Conshohocken, 19428. Fax: (215) 884-9400. Web Site: www.praise1039.com. Licensee: Radio One Licenses LLC. Group owner: Radio One Inc. (acq 11-8-2001; grpsl). Law Firm: Dickstein Shapiro Morin & Oshinsky. Format: Relg. Target aud: 18-34. ♦Chester Schofield, gen mgr; Daisy Davis, progmg dir.

Jersey Shore

WJSA(AM)— July 10, 1979: 1600 khz; 1 kw-D, 20 w-N. TL: N41 13 32 W77 16 01. Hrs open: 24 262 Allegheny St., Suite 4, 17740-1442. Phone: (570) 398-7200. Fax: (570) 398-7201. E-mail: am@wjsaradio.com Web Site: www.wjsaradio.com. Licensee: Covenant Broadcasting Co. Population served: 160,000 Natl. Network: Salem Radio Network. Law Firm: Gammon & Grange. Format: Relg. News staff: one; News: 14 hrs wkly. Target aud: General. Spec prog: Class one hr, southern gospel 2 hrs wkly. ♦John K. Hogg Jr., CEO, gen mgr & chief of engrg; Jerry G. Frear Jr., gen sls mgr; Justin S. Hogg, mus dir; Liz Brady, news dir.

WJSA-FM— Nov 1, 1984: 96.3 mhz; 4.4 kw. 777 ft TL: N41 13 28 W77 22 48. Stereo. 24 E-mail: fm@wjsaradio.com Web Site: www.wjsaradio.com.300,000 News staff: one; News: 14 hrs wkly.

Johnsonburg

WJNG(FM)— July 1998: 100.5 mhz; 1.3 kw. 666 ft TL: N41 23 11 W78 41 32. Hrs open: 24
Rebroadcasts WMKX(FM) Brookville 100%.
517 Market St., 15845. Phone: (814) 965-2921. Fax: (814) 965-2921. Licensee: Strattan Broadcasting Inc. Format: Classic rock. ♦James W. Farley, gen mgr; Kevin Heinrick, opns mgr & progmg dir; Nathan Sharpe, gen sls mgr; Cindy Perucci, traf mgr.

WKBI-FM—See Saint Marys

Johnstown

WCRO(AM)— September 1947: 1230 khz; 1 kw-U. TL: N40 19 55 W78 54 46. Hrs open: 24 222 Central Ave., 15904. Secondary address: 1089 Broad St 15906. Phone: (814) 533-5533. Fax: (814) 533-5698. Licensee: Greater Johnstown School District. (acq 1-11-99; $75,000). Population served: 100,000 Wire Svc: AP Format: Adult standards. News: 35 hrs wkly. Target aud: 45 - 64; Fastest growing and most financially secure demographically. Spec prog: University of Pittsburgh Football, basketball, NASCAR racing. ♦Ed Scherlock, stn mgr; Ed Sherlock, pres & opns mgr.

WFGI-FM—Listing follows WNTJ(AM).

***WFRJ(FM)**— June 6, 1986: 88.9 mhz; 900 w. 1,063 ft TL: N40 22 15 W78 59 02. Hrs open: 24 13 Fair Lane Dr., Suite 5, Jolette, IL, 60435. Phone: (814) 322-3144. Web Site: www.familyradio.com. Licensee: Family Stations Inc. (group owner) Population served: 600,000 Format: Conservative Christian. News: 6 hrs wkly. Target aud: General; every age group. ♦Harold Camping, pres; Gary Johnson, opns mgr.

WJHT(FM)— Sept 1, 1974: 92.1 mhz; 580 w. Ant 1,043 ft TL: N40 22 15 W78 59 02. Stereo. Hrs opn: 24 109 Plaza Dr., Suite 2, 15905. Phone: (814) 255-4186. Fax: (814) 255-6145. Web Site: www.hot92fm.net. Licensee: Forever Broadcasting LLC. (group owner; (acq 5-1-2005; $2.73 million with WRKW(FM) Ebensburg). Population served: 209,000 Format: Contemporary hit. ♦Terry Deitz, gen mgr; Tina Perry, gen sls mgr; Mitch Edwards, progmg dir; Rick Shepard, news dir.

WKYE(FM)— Aug 14, 1973: 96.5 mhz; 50 kw. Ant 489 ft TL: N40 19 45 W78 53 54. Stereo. Hrs opn: 24 109 Plaza Dr., 15905. Phone: (814) 255-4186. Fax: (814) 255-6145. Web Site: www.96key.com. Licensee: Forever Broadcasting LLC. (acq 1-30-2004; $9.13 million with co-located AM). Population served: 890,300 Format: Adult contemp. ♦Verla Price, gen mgr; Jack Michaels, progmg dir; Brian Wolfe, mus dir; Jim Boxler, chief of engrg.

WNTJ(AM)— April 1925: 850 khz; 10 kw-U, DA-1. TL: N40 10 54 W78 53 20. Hrs open: 109 Plaza Dr., 15905. Phone: (814) 255-4186. Fax: (814) 255-6145. Licensee: Forever Broadcasting LLC. Group owner: Forever Broadcasting (acq 9-9-97; grpsl). Population served: 200,000 Natl. Rep: McGavren Guild, Dome. Format: Country. Target aud: 25 plus. Spec prog: Sports 20 hrs, Pol 5 hrs wkly. ♦Carol Logan, pres; Verla Price, gen mgr & gen sls mgr; Mike Stevens, progmg dir; Rick Shepard, news dir; Jim Boxler, chief of engrg.

WFGI-FM—Co-owned with WNTJ(AM). Aug 1949: 95.5 mhz; 57 kw. Ant 1,060 ft TL: N40 22 18 W78 58 57. Stereo. 109 Plaza Dr., Suite 2, 15905. Web Site: www.key95.com. Format: Adult contemp. Target aud: 25-54. ♦Jack Michaels, progmg dir; Brian Wolfe, mus dir; Tegan Hayes, traf mgr.

WPRR(AM)— August 1946: 1490 khz; 1 kw-U. TL: N40 19 25 W78 53 49. Hrs open: 24 970 Tripoli St., 15902. Phone: (814) 534-8975. Fax: (814) 534-8979. Web Site: www.espn1490online.com. Licensee: 2510 Licenses LLC. Group owner: Forever Broadcasting (acq 5-1-2005; grpsl). Population served: 125,400 Natl. Network: ESPN Radio. Format: Sports. Target aud: 18-54. ♦Nick Ferrara, opns mgr; Zak McDowell, progmg dir.

***WQEJ(FM)**— 1997: 89.7 mhz; 3.3 kw. Ant 1,036 ft TL: N40 22 17 W78 58 58. Hrs open: 24
Rebroadcasts WQED-FM Pittsburgh 100%.
c/o WQED-FM, 4802 5th Ave., Pittsburgh, 15213. Phone: (412) 622-1436. Fax: (412) 622-7073. Fax: (412) 622-1488. E-mail: radio@wqej.org Web Site: www.wqed.org. Licensee: WQED Pittsburgh. Natl. Network: AP Radio, NPR, PRI. Law Firm: Schwartz, Woods & Miller. Format: Class. Target aud: 35-64; educated, influential, professional, community leaders, mid to high income. ♦George L. Miles Jr., CEO & pres; Deborah Acklin, gen mgr; Susan Johnson, stn mgr; Lilli Mosco, dev VP; Rick Vacarelli, sls VP; Rosemary Martinelli, mktg dir; Bryan Sejvar, progmg dir; Paul Byers, chief of engrg.

WRKW(FM)—See Ebensburg

Kane

WLMI(FM)— Sept 17, 1984: 103.9 mhz; 3 kw. 300 ft TL: N41 39 34 W78 48 42. Stereo. Hrs opn: 24 27 A Fraley St., 16735. Phone: (814) 837-9711. Fax: (814) 837-6154. E-mail: wlmifm@penn.com Web Site: www.wlmifm.com. Licensee: Colonial Radio Group Inc. (acq 7-28-2006; $390,000). Population served: 75,000 Natl. Network: ABC. Rgnl rep: Dome, Commercial Media Sales Law Firm: Garvey, Schubert & Barer. Format: Country. News staff: one; News: 12 hrs wkly. Target aud: 25-49; families. Spec prog: Polka one hr, bluegrass one hr wkly. ♦Jeffrey Andrulonis, pres; Ginny Crouse, VP; Charles W. Crouse, gen mgr.

***WPSX(FM)**— 1995: 90.1 mhz; 17 kw. 761 ft TL: N41 37 04 W78 48 14. Stereo. Hrs opn:
Rebroadcasts WPSU(FM) 100%.
WPSU-FM, 120 Outreach Bldg/, University Park, 16802. Phone: (814) 865-1877. Fax: (814) 865-4043. E-mail: wpsu@psu.edu Web Site: wpsu.org. Licensee: The Pennsylvania State University. Natl. Network: NPR, PRI. Law Firm: Paul, Hastings, Janufsky & Walker. Wire Svc: AP Format: Public radio. News staff: one; News: 35 hrs wkly. Upscale educated adults. Spec prog: Folk 10 hrs, jazz 3 hrs, blues 2 hrs wkly. ♦Ted Krichels, gen mgr; Greg Petersen, stn mgr; Steve Shipman, opns dir; Ashear Barr, gen sls mgr; Kris Allen, progmg mgr; Carl Fisher, chief of engrg; Leslie Dyer, traf mgr.

Kearsarge

WCXJ(AM)—Not on air, target date: unknown: 1590 khz; 500 w-D, 900 w-N, DA-2. TL: N42 01 47 W80 07 06. Hrs open: 4039 Sunset Blvd., Steubenville, OH, 43952. Phone: (412) 936-1500. Licensee: Eaton-Dietterich Partnership. ♦Randy Dietterich, gen mgr.

King of Prussia

WFYL(AM)— December 1976: 1180 khz; 420 w-D. TL: N40 08 06 W75 23 27. Hrs open: 2400 W. Main St., Jeffersonville, 19403-3071. Phone: (610) 539-5015. Licensee: Langer Broadcasting Group L.L.C. (group owner). Population served: 80,000 Format: Talk. ♦Helen Lenza, gen mgr.

Kittanning

WTYM(AM)— 1948: 1380 khz; 1 kw-D, 28 w-N. TL: N40 47 19 W79 32 05. Hrs open: Box 14A, R.D. 7, 16201. Phone: (724) 543-1380. Fax: (724) 543-1140. E-mail: wtym@alltel.net Web Site: www.wtym.8m.com. Licensee: Vernal Enterprises Inc. (group owner; acq 7-22-92; $85,000; FTR: 6-15-92). Population served: 150,000 Law Firm: Haley, Bader & Potts. Format: Oldies, sports. Target aud: 20-55. Spec prog: Relg 4 hrs wkly. ♦Larry L. Schrecongost, pres, gen mgr & opns mgr; Nancy W. Schrecongost, VP; John DeFeo, gen sls mgr & chief of engrg.

Lancaster

WDAC(FM)— Dec 13, 1959: 94.5 mhz; 19 kw. 810 ft TL: N39 53 46 W76 14 22. Stereo. Hrs opn: Box 3022, 17604. Secondary address: for UPS, Fed-Ex only:, 683 Lancaster Pike, New Providence 17560. Phone: (717) 284-4123. Fax: (717) 284-2300. Web Site: www.wdac.com. Licensee: WDAC Radio Co. Population served: 417,000 Natl. Network: Moody, Salem Radio Network. Law Firm: Wiley, Rein & Fielding. Format: Christian, talk. News staff: one; News: 8 hrs wkly. Target aud: 25-49; Evangelical Christians, families. Spec prog: Farm 4 hrs wkly. ♦Doug Myer, COO & gen mgr; Richard Crawford, pres; Mike Stike, opns mgr; Joe Hartman, sls dir; John E. Eby, progmg dir.

***WFNM(FM)**— May 1973: 89.1 mhz; 100 w. 150 ft TL: N40 02 43 W76 19 14. Stereo. Hrs opn: 20 Box 3220, Franklin and Marshall College, 17604-3003. Phone: (717) 291-4098. Fax: (717) 358-4437. Web Site: wfnm.fandm.edu. Licensee: Franklin and Marshall College. Population served: 300,000 Format: Var/div. News: 4 hrs wkly. Target aud: 13-35. Spec prog: Black 6 hrs, sports talk 2 hrs, class 2 hrs, jazz 8 hrs wkly.

***WJTL(FM)**— Aug 27, 1984: 90.3 mhz; 4.7 kw. 198 ft TL: N40 04 13 W76 17 19. (CP: 11.8 kw, ant 495 ft.). Stereo. Hrs opn: Box 1614, 17608. Phone: (717) 392-3690. Fax: (717) 390-2892. E-mail: contact@wjtl.com Web Site: www.wjtl.com. Licensee: Creative Ministries Inc. (acq 11-30-90; $500,000; FTR: 12-31-90). Natl. Network: USA. Law Firm: Fisher, Wayland, Cooper, Leader & Zaragoza. Format: Contemp Christian. ♦Fred McNaughton, stn mgr.

WLAN(AM)— Aug 9, 1946: 1390 khz; 5 kw-D, 1 kw-N, DA-2. TL: N40 03 12 W76 20 26. Hrs open: 24 1685 Crown Ave., Suite 100, 17601. Phone: (717) 295-9700. Fax: (717) 295-7329. E-mail: webmaster@1390wlan.com Web Site: www.1390wlan.com. Licensee: Clear Channel Radio Licenses Inc. Group owner: Clear Channel Communications Inc. (acq 1996; $7 million with co-located FM). Population served: 420,000 Natl. Network: ABC. Natl. Rep: Clear Channel. Wire Svc: AP Format: Adult standards. News staff: 3; News: 9 hrs wkly. Target aud: 35-64. ♦Dick Taylor, gen mgr & gen sls mgr.

WLAN-FM— January 1948: 96.9 mhz; 50 kw. 500 ft TL: N40 02 52 W76 27 25. Stereo. 24 E-mail: webmaster@fm97.com Web Site: www.fm97.com.420,000 Format: Adult contemp, Top-40. News staff: 3; News: 9 hrs wkly. Target aud: 18-49.

***WLCH(FM)**— Sept 14, 1987: 91.3 mhz; 160 w. 135 ft TL: N40 04 13 W76 17 19. Hrs open: 30 N. Ann St., 1st Fl., 17602. Phone: (717) 295-7996. Fax: (717) 295-7759. E-mail: radiocenter@aol.com Licensee: Spanish American Civic Association for Equality Inc. Format: Sp, educ, div. Target aud: General; Hispanics. ♦Mayra Guevar, CEO & pres; Carlos Groupera, exec VP & mktg mgr; Enid Vazquez, gen mgr.

WLPA(AM)— 1922: 1490 khz; 600 w-U. TL: N40 03 38 W76 18 59. Stereo. Hrs opn: 24 Box 4368, 17604. Secondary address: 1996 Auction Rd., Manheim 17545. Phone: (717) 653-0800. Phone: (800) 222-1013. Fax: (717) 653-0122. Web Site: www.wlpa.com. Licensee: Hall Communications Inc. (group owner; (acq 2-13-77). Population served: 63,000 Natl. Network: Fox Sports. Rgnl. Network: Radio Pa. Law Firm: Fletcher, Heald & Hildreth. Format: Sports. News: 8 hrs wkly. Target aud: 25-54; men. ♦Bonnie H.M. Rowbotham, chmn; Arthur J. Rowbotham, pres; William S. Baldwin, exec VP, sr VP & gen mgr.

WROZ(FM)—Co-owned with WLPA(AM). 1944: 101.3 mhz; 50 kw. 1,289 ft TL: N40 02 04 W76 37 08. Stereo. 24 E-mail: wroz@hallradio.com Web Site: www.roseradio.com.3,000,000 Format: Soft adult contemp. News staff: one. Target aud: 25-54; women. ♦Tom Shannon, progmg dir & disc jockey; Michael C. Anthony, mus dir & disc jockey; Valerie Baldwin, news dir; Justin Broka, disc jockey; Patsy Sympson, disc jockey; Val Baldwin, disc jockey.

Lansdale

WNPV(AM)— Oct 17, 1960: 1440 khz; 2.5 kw-D, 500 w-N, DA-2. TL: N40 14 18 W75 19 00. Hrs open: 24 Box 1440, 1210 Snyder Rd., 19446. Phone: (215) 855-8211. Fax: (215) 368-0180. Web Site: www.wnpv1440.com. Licensee: WNPV Inc. (acq 10-1-80). Population

served: 1,200,000 Natl. Network: Motor Racing Net, ESPN Radio. Rgnl. Network: Radio Pa. Law Firm: Pillsbury, Winthrop, Shaw & Pittman. Format: News/talk. News staff: 2; News: 20 hrs wkly. Target aud: 30 plus. Spec prog: Big band 3 hrs, relg 5 hrs, sports 6 hrs wkly. ♦Phillip Hunt, pres; Darryl Berger, VP & progmg dir; Phillip N. Hunt, gen mgr; Linda Muskal, gen sls mgr; Randy Brock, news dir; David McCrork, engr.

Lansford

WLSH(AM)— Dec 24, 1952: 1410 khz; 5 kw-D, DA. TL: N40 50 40 W75 50 37. Hrs open: 2147 Market St., 18232. Phone: (570) 645-3123. Fax: (570) 645-2159. E-mail: wmgh@ptdprolog.net Web Site: www.wmgh.com. Licensee: J-Systems Franchising Corp. (group owner; acq 1-89; $300,000; FTR: 1-16-89). Population served: 5,168 Natl. Network: Westwood One. Format: MOR. Target aud: 35-64; Mature adults. Spec prog: Big Band 4 hrs, Oldies 3 hrs wkly. ♦Harold G. Fulmer, III, CEO, chmn & pres; Christopher G. Fulmer, VP; Bill Lakatas, gen mgr.

Laporte

WCOZ(FM)— August 1998: 103.9 mhz; 6 kw. Ant 276 ft TL: N41 26 06 W76 28 28. Stereo. Hrs opn: 24 Box 230, 201 Bernice Rd., Suite 2, Dushore, 18614. Phone: (570) 928-7200. Fax: (570) 928-2100. E-mail: contact_us@cozy.com Web Site: www.cozyradio.com. Licensee: Smith and Fitzgerald, Partnership (acq 5-7-01). Population served: 20,000 Format: Adult contemp. News: 4 hrs wkly. Target aud: 25-54; adults. ♦Ben Smith, gen mgr & progmg VP; Cindi McCarty, adv mgr; Kevin Fitzgerald, engrg VP.

Latrobe

WCNS(AM)— Aug 11, 1956: 1480 khz; 500 w-D, 1 kw-N, DA-N. TL: N40 16 12 W79 23 13. Hrs open: 24 400 Unity St., Suite 200, 15650. Phone: (724) 537-3338. Fax: (724) 539-9798. E-mail: info@wcnsradio.com Web Site: www.1480wcns.com. Licensee: Longo Media Group. (acq 1-89). Population served: 147,444 Natl. Network: Westwood One. Format: full service. News staff: 3; News: 15 hrs wkly. Target aud: 25 +; general. Spec prog: Relg 2 hrs wkly. ♦John Longo, pres; Greg Zahornacky, stn mgr; Dow Carnahan, opns mgr & progmg dir.

WQTW(AM)— 1952: 1570 khz; 1 kw-D, 220 w-N. TL: N40 18 07 W79 21 26. Hrs open:
Rebroadcasts WLSW(FM) Scottdale.
Box 208, George St., 15650. Phone: (724) 532-1778. Fax: (724) 532-1779. Licensee: L. Stanley Wall. (acq 4-84; $66,000; FTR: 4-23-84). Population served: 11,749 Format: Hot adult contemp. ♦L. Stanley Wall, pres & gen mgr.

Lebanon

WADV(AM)— July 4, 1976: 940 khz; 1 kw-D, 5 w-N. TL: N40 22 22 W76 21 53. Stereo. Hrs opn: 19 720 E Kercher Ave., 17046. Phone: (717) 273-2611. Fax: (717) 273-7293. Licensee: WADV Radio Inc. (acq 12-4-01). Population served: 800,000 Natl. Network: Moody. Format: Southern & bluegrass gospel, country. News staff: one; News: 18 hrs wkly. Target aud: 25 plus; loyal, exclusive. ♦Jennifer Taylor Kochel, pres; Earl Kochel, gen mgr; Pearl Kochel, gen mgr; Julie Kochel, opns VP.

WLBR(AM)— Nov 13, 1946: 1270 khz; 5 kw-D, 1 kw-N, DA-2. TL: N40 21 35 W76 27 30. Hrs open: 5 AM-1 AM 440 Rebecca St., 17042. Phone: (717) 272-7651. Fax: (717) 274-0161. Licensee: Lebanon Broadcasting Co. (acq 4-16-2007; with co-located FM). Population served: 112,000 Rgnl. Network: Radio Pa. Natl. Rep: Roslin. Rgnl rep: Dome Law Firm: Shaw Pittman. Format: News/talk. News staff: 2. Target aud: 25-64. ♦Robert D. Etter, VP, gen mgr & progmg dir; Mickey Santora, gen sls mgr; Greg Lyons, mus dir & disc jockey; Gordon Weise, news dir; Glenn Waybright, chief of engrg; Gayle Reich, traf mgr; Laura Lebeau, news rptr; Dave Eisenhauer, disc jockey; Don Bowman, disc jockey; Scott Bradley, sports cmtr & disc jockey.

WQIC(FM)—Co-owned with WLBR(AM). January 1948: 100.1 mhz; 3 kw. 267 ft TL: N40 21 37 W76 27 31. Stereo. 5 AM-1 AM Format: Adult contemp. Target aud: 25-54. ♦Mike Ebersole, mus dir & disc jockey; Gayle Reich, traf mgr; Scott Bradley, sports cmtr; John Tuscano, disc jockey; Phil Liles, disc jockey; Steve Todd, progmg dir & disc jockey.

WWSM(AM)—See Annville-Cleona

Lehighton

WBYN(AM)—Licensed to Lehighton. See Allentown

Levittown-Fairless Hills

WBCB(AM)— Dec 8, 1957: 1490 khz; 1 kw-U. TL: N40 10 08 W74 50 08. Hrs open: 24 200 Magnolia Dr., Levittown, 19054. Phone: (215) 949-1490. Fax: (215) 949-3671. Web Site: www.wbcb1490.com. Licensee: Progressive Broadcasting Co. (acq 11-13-92; $550,000; FTR: 11-30-92). Population served: 750,000 Natl. Network: USA. Format: Community radio. Target aud: 18 plus; varied programming appeals to different age groups. Spec prog: Sports. ♦Pasquale T. Deon Sr., pres; Merrill Reese, VP & gen mgr; Erica Darragh, opns mgr; Lee Alexander, sls dir & gen sls mgr; Paul Baroli, progmg dir.

Lewisburg

WCXR(FM)— Oct 18, 1990: 103.7 mhz; 3 kw. 418 ft TL: N40 56 40 W76 52 45. (CP: 103.7 mhz, 1.35 kw, ant 715 ft.). Stereo. Hrs opn: 24
Rebroadcasts WZXR(FM) South Williamsport 100%.
1685 Four Mile Dr., Williamsport, 17740. Phone: (570) 323-8200. Fax: (570) 327-9138. E-mail: dfarr54@aol.com Licensee: South Williamsport SabreCom Inc. Group owner: Backyard Broadcasting LLC (acq 12-1-02; grpsl). Population served: 311,000 Natl. Network: ABC. Natl. Rep: Christal. Format: Classic rock. News staff: 3. Target aud: 25-54. ♦Barry Drake, pres; Robin Smith, CFO; Dan Farr, gen mgr.

***WGRC(FM)**— Apr 22, 1988: 91.3 mhz; 3 kw. Ant 321 ft TL: N40 56 40 W76 52 45. Stereo. Hrs opn: 24 101 Armory Blvd., 17837-9504. Phone: (570) 523-1190. Fax: (570) 523-1114. E-mail: email@wgrc.com Web Site: www.wgrc.com. Licensee: Salt and Light Media Ministries Inc. Population served: 592,000 Natl. Network: Salem Radio Network. Law Firm: Miller & Neely. Wire Svc: AP Format: Christian, adult contemp. News staff: 3; News: 16 hrs wkly. Target aud: 25-54; young to middle-aged adult. ♦Larry Weidman, gen mgr; Jim Diehl, progmg dir & news rptr; John Callahan, news dir & news rptr; Lamar Smith, chief of engrg; Linda Dantonio, traf mgr; Chris Miller, engr.

***WVBU-FM**— October 1965: 90.5 mhz; 500 w. -120 ft TL: N40 57 18 W76 52 46. (CP: 225 w, ant 66 ft.). Stereo. Hrs opn: 8 AM-2 AM Box C-3956, Bucknell Univ., 701 Moore Ave., 17837. Phone: (570) 577-2000. Phone: (570) 577-3824. Fax: (570) 577-1174. E-mail: wvbu@bucknell.edu Web Site: www.orgs.bucknell.edu/wvbu. Licensee: Bucknell University. Population served: 60,000 Format: Modern Rock. News: 7 hrs wkly. Target aud: 18-23; college students. Spec prog: Jazz 3 hrs, dance/club 6 hrs, prison request 2 hrs, modern/new age one hr wkly.

Lewistown

WCHX(FM)— July 1, 1987: 105.5 mhz; 3 kw. 817 ft TL: N40 39 43 W77 34 28. (CP: 465 w, ant 816 ft.). Stereo. Hrs opn: 24 Box 911, 17044. Secondary address: 114 N. Logan Blvd., Burnham 17009. Phone: (717) 242-1493. Fax: (717) 242-3764. Licensee: Mifflin County Communications Inc. Population served: 125,000 Natl. Network: Fox News Radio. Law Firm: Wilkinson Barker Knauer. Format: Classic rock. News staff: one; News: 10 hrs wkly. Target aud: 25-54; mature, affluent, middle & upper class adults. ♦Anna Hain, pres & stn mgr; Jed A. Donahue, VP; Jed Donahue, gen mgr; Ed Thompson, sls dir & gen sls mgr; Scott Shaw, progmg dir.

WIEZ(AM)— June 1, 1941: 670 khz; 5.4 kw-D. TL: N40 36 30 W77 34 45. Hrs open: Sunrise-sunset Box 667, 17044. Secondary address: 12 E. Market St. 2nd Floor 17044. Phone: (717) 248-6757. Fax: (717) 248-6759. E-mail: wiez@meradio.com Licensee: First Media Radio LLC. (group owner; (acq 3-28-2001; grpsl). Population served: 75,000 Natl. Rep: Dome. Format: News, info. News staff: 2; News: 12 hrs wkly. Target aud: 45 plus; adults who control the area's disposable income. ♦Pete Herman, gen mgr; Jeff Stevens, opns mgr; Mary Lee Schaeffer, news dir.

***WJRC(FM)**— July 1996: 90.9 mhz; 100 w. 1,128 ft TL: N40 34 58 W77 29 48. Stereo. Hrs opn: 24
Rebroadcasts WGRC(FM) Lewisburg 100%.
101 Armory Blvd., Lewisburg, 17837-0279. Phone: (570) 523-1190. Fax: (570) 523-1114. E-mail: email@wgrc.com Web Site: www.wgrc.com. Licensee: Salt and Light Media Ministries Inc. Population served: 76,000 Natl. Network: Salem Radio Network. Law Firm: Miller & Neely. Wire Svc: AP Format: Contemp Christian. News staff: 3; News: 9 hrs wkly. Target aud: 25-54. ♦Larry Weidman, gen mgr; John Callahan, news dir & news rptr; Lamar Smith, chief of engrg; Linda Dantonio, traf mgr; Jim Diehl, news rptr; Chris Miller, engr.

WKVA(AM)— Dec 4, 1949: 920 khz; 1 kw-D, 500 w-N, DA-N. TL: N40 34 45 W77 34 18. Hrs open: 5 AM-midnight Box 911, 17044. Secondary address: 114 N. Logan Blvd., Burnham 17009. Phone: (717) 242-1055. Phone: (717) 242-1495. Fax: (717) 242-3764. E-mail: wkva@oldies920.com Web Site: www.oldies920.com. Licensee: Mifflin County Communications Inc. (acq 2-18-98; $277,692). Population served: 75,000 Natl. Network: ABC, CBS Radio. Rgnl. Network: Radio Pa. Law Firm: Putbrese, Hunsaker & Trent, P.C. Format: Oldies, full service. News staff: 2; News: 31 hrs wkly. Target aud: 25-54; blue collar mix of agricultural & industrial adults. ♦Anna A. Hain, pres; Jed A. Donahue, VP; Erik Lane, news dir.

WMRF-FM— Oct 1, 1964: 95.7 mhz; 3.9 kw. 407 ft TL: N40 36 30 W77 34 45. Stereo. Hrs opn: 24 12 E. Market St., 2nd Floor, 17044. Phone: (717) 248-6757. Fax: (717) 248-6759. E-mail: merfradio@acsworld.net Web Site: www.merfradio.com. Licensee: First Media Radio LLC. (group owner; (acq 5-14-2001; grpsl). Population served: 60,000 Format: Hot adult contemp. News staff: 2; News: 8 hrs wkly. Target aud: 18-44. ♦Peter Herman, gen mgr; Jeff Stevens, opns dir, progmg dir & progmg mgr; Mary Lee Sheaffer, news dir.

Lincoln University

***WWLU(FM)**— Aug 1, 1975: 88.7 mhz; 10 w. 100 ft Hrs opn: Box 179, 1570 Baltimore Pike, 19352. Phone: (610) 932-8300. Fax: (610) 932-1095. Licensee: Lincoln University. Population served: 23,000 Format: Urban contemp, hip hop. ♦Whitney G. Walton, gen mgr.

Linesville

WMVL(FM)— May 4, 1970: 101.7 mhz; 1.4 kw. Ant 554 ft TL: N41 42 38 W80 16 29. Stereo. Hrs opn: 24 Box 846, Meadville, 16335. Secondary address: 16271Conneaut Lake Rd., Ste 102, Meadville 16335. Phone: (814) 337-8440. Fax: (814) 333-2562. E-mail: wmvl@zoominternet.net Licensee: Vilkie Communications Inc. (acq 4-1-2003; $330,000). Population served: 980,000 Natl. Network: ABC. Rgnl rep: Regional Reps, CLE, OH Law Firm: Hogan & Hartson. Format: Oldies. News: 8 hrs wkly. Target aud: 29 plus. ♦Eugene Vilkie, VP; Joseph M. Vilkie, pres & gen mgr; Chuck Stopp, progmg dir; Dave Hanahan, sls; Dave Voisin, sls; Jim Jewell, sls.

Lock Haven

WBPZ(AM)— Feb 20, 1947: 1230 khz; 1 kw-U. TL: N41 08 03 W77 28 09. Hrs open: 24 Box 420, 17745. Secondary address: 21 E. Main St. 17745. Phone: (570) 748-4038. Fax: (570) 748-0092. Licensee: Lipez Broadcasting Corp. (acq 3-27-86). Population served: 50,000 Natl. Rep: Keystone (unwired net), Dome. Format: Oldies. News staff: one; News: 10 hrs wkly. Target aud: General. Spec prog: Loc sports. ♦John Lipez, pres & gen mgr; John Lupez, gen sls mgr; Randy Dorey, progmg dir; Bill Daney, mus dir; Mark Sohmer, news dir; Dennis Sherman, chief of engrg; Michelle Grove, traf mgr.

WSNU(FM)—Co-owned with WBPZ(AM). September 1965: 92.1 mhz; 3 kw. 255 ft TL: N41 08 49 W77 29 16. (CP: Ant 328 ft.). Stereo. 24 70,000 Format: Adult contemp. News staff: one; News: 6 hrs wkly. Target aud: 21-48. ♦Michelle Grove, traf mgr.

Loretto

WWGE(AM)— Dec 7, 1963: 1400 khz; 1 kw-U. TL: N40 30 12 W78 38 10. Hrs open: 24 Box 88, Ebensburg, 15931. Secondary address: 104 S. Center St., Suite 401, Ebensburg 15931. Phone: (814) 255-9943. Fax: (814) 255-3343. Web Site: www.edge1400.com.

Licensee: Pennsylvania Radiowerks LLC (acq 10-15-98; $100,000). Population served: 100,000 Natl. Network: Jones Radio Networks. Format: News/talk, sports. ♦Rev. Michael H. Yeager, pres; Jennifer Strelnik, gen mgr.

Mansfield

WNBQ(FM)— June 1999: 92.3 mhz; 800 w. 643 ft TL: N41 53 53 W77 05 38. Hrs open: 24
Rebroadcasts WNBT-FM Wellsboro 100%.
Box 98, Wellsboro, 16901. Secondary address: R.R. 7, Rt. 6 Box 198-B, Wellsboro 16901. Phone: (570) 724-1490. Fax: (570) 724-6971. E-mail: wnbt@ptb.net Web Site: www.wnbt.net. Licensee: Farm & Home Broadcasting Co. Group owner: Allegheny Mountain Network Stations Natl. Network: Westwood One. Rgnl rep: Dome & Assoc. Law Firm: . Borsari & Paxson Format: Bright adult contemporary. ♦Cary Simpson, pres; Al Harer, gen mgr.

***WNTE(FM)**— Sept 15, 1968: 89.5 mhz; 115 w. -320 TL: N41 48 22 W77 04 27. Hrs open: Box 84, South Hall, Mansfield Univ., 16933. Phone: (570) 662-4653. Fax: (570) 662-4654. Web Site: mustuweb.mnsfld.edu. Licensee: Mansfield University (acq 9-15-78). Population served: 8,500 Format: AOR, Top-40. Target aud: 17-25; college students/community. Spec prog: Black 5 hrs, jazz 2 hrs wkly.

Markleysburg

***WLOG(FM)**— 2002: Stn currently dark. 89.1 mhz; 100 w vert. Ant 328 ft TL: N39 43 32 W79 28 53. Hrs open: Box 5459, Twin Falls, ID, 83301. Phone: (208) 733-3551. Fax: (208) 733-3548. Licensee: Edgewater Broadcasting Inc. (acq 5-27-2005; $10,000). ♦Clark Parrish, pres.

Martinsburg

WJSM(AM)— Feb 27, 1968: 1110 khz; 1 kw-D. TL: N40 18 14 W78 15 59. Hrs open: Box 87, Rt. 2, 16662. Phone: (814) 793-2188. Fax: (814) 793-9727. Licensee: Martinsburg Broadcasting Inc. (acq 10-1-89). Population served: 275,000 Law Firm: Harold McCombs. Format: Relg, news, talk. Target aud: General. Spec prog: Farm one hr wkly. ♦Deborah J. Walters, opns mgr; Bill Reed, news dir; Deborah Walters, news rptr & disc jockey; Cheryl A. Walters, women's int ed; Larry S. Walters, pres, gen mgr, progmg dir & disc jockey.

WJSM-FM— Apr 19, 1965: 92.7 mhz; 640 w. 964 ft TL: N40 17 37 W78 15 38.350,000 Natl. Network: USA. Format: Relg, news/talk, gospel. ♦Bill Reed, local news ed; Deborah J. Walters, news rptr; William Reed, reporter; Hap Ritchey, mus critic & disc jockey; Cheryl A. Walters, women's int ed; Deborah Walters, disc jockey; Larry S. Walters, disc jockey; Nicole Lewis, disc jockey.

WKMC(AM)—See Roaring Spring

Masontown

***WRIJ(FM)**— November 1990: 106.9 mhz; 3 kw. 328 ft TL: N39 47 15 W79 59 20. Stereo. Hrs opn: 19
Rebroadcasts WAIJ(FM) Grantsville, MD 100%.
Box 540, 34 Springs Rd., Grantsville, MD, 21536. Phone: (301) 895-3292. Fax: (301) 895-3293. E-mail: hesalive@hesalive.net Web Site: www.hesalive.net. Licensee: He's Alive Inc. (group owner) Population served: 500,000 Natl. Network: USA. Format: Gospel, Christian, relg, adult contemp. Target aud: 18-35. ♦Dewayne Johnson, pres.

***WYFU(FM)**— 2003: 88.5 mhz; 16 kw vert. Ant 328 ft TL: N39 47 15 W79 59 20. Hrs open: Bible Broadcasting Network, 11530 Carmel Commons Blvd., Charlotte, NC, 28226. Phone: (704) 523-5555. Fax: (704) 522-1967. E-mail: bbn@bbnradio.org Web Site: www.bbnradio.org. Licensee: Bible Broadcasting Network Inc. Group owner: Bible Broadcasting Network (acq 2-12-99; $250,000). Format: Christian. ♦Richard Johnson, gen mgr.

McConnellsburg

WEEO-FM— 1997: 103.7 mhz; 135 w. 1,555 ft TL: N39 55 25 W77 57 20. Hrs open: 37 South Main St., Suite 103, Chambersburg, 17201. Phone: (717) 709-0800. Phone: (717) 709-0801. Fax: (717) 709-0802. Web Site: revolution1037.net. Licensee: Allegheny Mountain Network. Group owner: Allegheny Mountain Network Stations (acq 10-95; $18,000). Rgnl rep: Dome Law Firm: Borsari & Paxson. Format: Modern Rock. ♦John F. Simpson, CEO & pres.

***WWCF(FM)**— 2005: 88.7 mhz; 9 w. Ant 1,194 ft TL: N39 54 58 W77 57 25. Hrs open: 611 Longview Rd., 17233-9740. Phone: (717) 485-5526. Licensee: Morris Broadcasting & Communications Inc. Format: Children. ♦Glenn Morris, pres & gen mgr.

McKean

WQHZ(FM)—See Erie

McKeesport

WEDO(AM)— 1947: 810 khz; 1 kw-D. TL: N40 21 52 W79 48 49. Hrs open: Sunrise-sunset 1985 Lincoln Way, White Oak, 15131. Phone: (412) 664-4431. Fax: (412) 664-1236. Web Site: www.am81wedo.com. Licensee: 810 Inc. (acq 5-72). Population served: 2,300,000 Format: Talk, health, var. Target aud: 35-65, 25-54, 65+.Relg one hr, Slovenian one hr, Slovak one hr, Greek one hr, Croation one hr, Serbian one hr, It one hr, Lithuanian one hr, Pol one hr, Sp one hr wkly. Health & Nutrition. Dr. M. Gallagher, Dr. J. Winer- Daily. ♦Judith Baron, pres; John James, VP & gen mgr; Jeremy Bosse, opns dir.

WPTT(AM)— April 1947: 1360 khz; 5 kw-D, 1 kw-N, DA-N. TL: N40 24 30 W79 55 40. (CP: COL Mount Lebanon. 910 khz; 7 kw-D, DA. TL: N40 16 05 W79 59 01). Hrs opn: 24 3rd Fl., 900 Parish St., Pittsburgh, 15220. Phone: (412) 875-9500. Fax: (412) 875-9474. Web Site: www.1360wptt.com. Licensee: Renda Broadcasting Corp. of Nevada. (acq 9-10-97). Natl. Network: ABC. Natl. Rep: McGavren Guild. Wire Svc: AP Wire Svc: Metro Weather Service Inc. Format: News/talk. News: 15 hrs wkly. Target aud: 25-54. Spec prog: Oldies 9 hrs, polka 2 hrs wkly. ♦Tony Renda Sr., CEO; Tony Renda Jr., gen mgr.

Meadville

***WARC(FM)**— Feb 3, 1963: 90.3 mhz; 150 w. 86 ft TL: N41 38 55 W80 08 45. (CP: 340 w, ant 75 ft.). Stereo. Hrs opn: Box C, Allegheny College, 520 N. Main St., 16335. Phone: (814) 332-5275 (studio). Phone: (814) 332-3376. Web Site: www.warcallegheny.org. Licensee: Allegheny College. Population served: 40,000 Format: Alternative, div. Spec prog: Black 8 hrs, class 10 hrs, jazz 4 hrs wkly. ♦Phil Denman, gen mgr.

WGYY(FM)—Listing follows WMGW(AM).

WMGW(AM)— 1947: 1490 khz; 1 kw-U. TL: N41 37 53 W80 10 37. Hrs open: Box 397, Downtown Mall, 16335. Phone: (814) 724-1111. Fax: (814) 333-9628. Web Site: www.radio@zoominternet.net. Licensee: Forever Broadcasting LLC. Group owner: Forever Broadcasting (acq 7-20-00; grpsl). Population served: 100,000 Format: News/talk, sports, info. Target aud: 25-54. ♦Terry Dietz, gen mgr & gen sls mgr; Dave Galentine, progmg dir.

WGYY(FM)—Co-owned with WMGW(AM). 1947: 100.3 mhz; 20 kw. 587 ft TL: N41 37 53 W80 10 37. Stereo. Web Site: www.radio@zoominternet.net.1,000,000 Format: Country.

***WVME(FM)**— 2002: 91.9 mhz; 4.4 kw. Ant 308 ft TL: N41 37 50 W80 10 38. Stereo. Hrs opn: 24
Rebroadcasts WCRF(FM) Cleveland, OH 100%.
WCRF Radio, 9756 Barr Rd., Cleveland, OH, 44141. Phone: (440) 526-1111. Fax: (440) 526-1319. E-mail: wcrf@moody.edu Web Site: www.wcrfradio.org. Licensee: The Moody Bible Institute of Chicago. (group owner) Law Firm: Southmayd. Wire Svc: AP Format: Inspirational. Target aud: 25-55; Adults. ♦Michael Easley, pres; Richard Lee, stn mgr; Phil Villareal, progmg dir; Gary Bittner, local news ed.

Mechanicsburg

WTPA(FM)— Nov 1, 1978: 93.5 mhz; 1.25 kw. 718 ft TL: N40 10 38 W76 52 38. Stereo. Hrs opn: Cumulus Media, 2300 Vartan Way, Harrisburg, 17110-9720. Phone: (717) 238-1041. Fax: (717) 234-4842. Web Site: www.935WTPA.com. Licensee: Cumulus Licensing Corp. Group owner: Cumulus Media Inc. (acq 2000; grpsl). Population served: 86,500 Law Firm: Wilkinson Barker Knauer. Format: Active rock. Target aud: 18-49. ♦Ron Giovaniello, stn mgr; John O'Dea, opns VP; Karen Richards, sls VP; John Butler, gen sls mgr; Diane Sohanuch, prom dir; Chris James, progmg dir; Dave Supplee, chief of engrg.

Media

WPHI-FM— November 1982: 100.3 mhz; 35 kw. Ant 600 ft TL: N39 58 29 W75 25 22. (CP: 17 kw, ant 863 ft. TL: N40 02 36 W75 14 33). Stereo. Hrs opn: 24 1000 River Rd., Suite 400, Conshohocken, 19428-2437. Phone: (610) 276-1100. Fax: (610) 276-1139. E-mail: cschofield@radio.one.com Web Site: www.1003thebeaatphilly.com. Licensee: Radio One Licenses LLC. Group owner: Radio One Inc. (acq 11-8-2001; grpsl). Population served: 4,000,000 Format: Rhythm and blues. News staff: one. Target aud: 18-44; savvy suburban educated professional. ♦Chester Schofield, gen mgr; Helen Little, opns mgr & chief of engrg.

Mercer

WLLF(FM)— January 1985: 96.7 mhz; 1.4 kw. Ant 485 ft TL: N41 18 43 W80 16 39. Stereo. Hrs opn: 24 4040 Simon Rd., Youngstown, OH, 44512. Phone: (724) 346-4113. Fax: (330) 783-0060. E-mail: jbilo@theradiocenter Web Site: www.cumulus.com. Licensee: Cumulus Licensing Corp. Group owner: Cumulus Media Inc. (acq 3-15-00; grpsl). Natl. Network: Jones Radio Networks. Format: Soft rock. ♦Clyde Bass, gen mgr; Joe Bilo, natl sls mgr & news dir; Bob Popa, progmg dir; Wes Boyd, chief of engrg.

WWIZ(FM)— October 1972: 103.9 mhz; 3 kw. 300 ft TL: N41 12 10 W80 21 30. Stereo. Hrs opn: 4040 Simon Rd., Youngstown, OH, 44512. Phone: (330) 783-1000. Fax: (330) 783-0060. Web Site: www.realrock104.com. Licensee: Cumulus Licensing Corp. Group owner: Cumulus Media Inc. (acq 3-15-00; grpsl). Population served: 200,000 Format: Active rock. Target aud: 25-54. ♦Clyde Bass, gen mgr.

Mercersburg

WPPT(FM)— Mar 23, 1976: 92.1 mhz; 3.3 kw. 295 ft TL: N39 48 34 W77 48 22. (CP: 2.7 kw, ant 465 ft.). Stereo. Hrs opn: 24 Box 788, 10960 John Wayne Dr., Greencastle, 17225. Phone: (717) 597-9200. Fax: (717) 597-9210. E-mail: webmaster@wayz.com Licensee: M. Belmont VerStandig Inc. Group owner: VerStandig Broadcasting (acq 10-1-93; $1.6 million with WCBG(AM) Chambersburg; FTR: 9-6-93). Population served: 225,000 Law Firm: Leventhal, Senter & Lerman. Format: Classic Country. Target aud: 25-54; double income households. ♦Yogi Yoder, gen mgr.

Mexico

WJUN(AM)— Sept 8, 1955: 1220 khz; 1 kw-D, 46 w-N. TL: N40 32 06 W77 20 26. Hrs open: 24 Box 209, Old Rt. 22 E., 17056. Phone: (717) 436-2135. Fax: (717) 436-8155. E-mail: wjun@nmax.net Licensee: Starview Media Inc. (acq 11-16-89; grpsl; FTR: 12-19-88). Population served: 61,500 Rgnl. Network: Radio Pa. Format: Sports. News staff: one. Target aud: 25-54. Spec prog: Relg 2 hrs wkly. ♦Douglas George, pres; Curt Dreibelbis, gen mgr & gen sls mgr; Dan Roland, news dir & farm dir; John Hess, chief of engrg; Betty Lou Rowles, outdoor ed.

WJUN-FM— July 4, 1989: 92.5 mhz; 440 w. Ant 1,181 ft TL: N40 34 58 W77 29 48. Stereo. 24 Format: Country. News staff: one; News: 5 hrs wkly. ♦Mel Thomas, progmg dir.

Meyersdale

WQZS(FM)— 1992: 93.3 mhz; 630 w. 964 ft TL: N39 47 49 W79 10 05. Hrs open: 128 Hunsrick Rd., 15552. Phone: (814) 634-9111. Fax: (814) 634-0882. E-mail: helenwahl27@hotmail.com Licensee: Roger Wahl. Format: Oldies. News staff: one. Target aud: 25-60; females 60%, males 40%. Spec prog: Gospel 5 hrs wkly. ♦Helen E. Wahl, gen mgr & progmg dir; Jessy Chabol, gen sls mgr.

Middletown

***WMSS(FM)**— Sept 7, 1978: 91.1 mhz; .56 w. 73 ft TL: N40 11 52 W76 43 30. Stereo. Hrs opn: 7 AM-9 PM 215 Oberlin Rd, 17057. Phone: (717) 948-9136. E-mail: sales@wmssfm.com Web Site: www.wmssfm.com. Licensee: Middletown Area School District. Population served: 250,000 Format: Adult contemp, progsv, educ. News: one hr wkly. Target aud: General. Spec prog: Sports 5 hrs, relg 8 hrs wkly. ♦John Wilsbach, gen mgr; Maureen Denis, opns dir; Steve Leedy, opns mgr.

***WZXM(FM)**— 2006: 88.7 mhz; 200 w. Ant 695 ft TL: N40 04 33 W76 48 23. Hrs open:
Rebroadcasts WBYO(FM) Sellersville 100%.
Box 186, Sellersville, 18960. Phone: (215) 721-2141. Fax: (215) 721-9811. E-mail: wordfm@wordfm.org Web Site: www.wordfm.org. Licensee: Four Rivers Community Broadcasting Corp. Format: Contemp Christian. ♦Charles Loughery, pres; Charlie Loughery, gen mgr; Dave Baker, VP & stn mgr.

Mifflinburg

WWBE(FM)— 1975: 98.3 mhz; 1.4 kw. 482 ft TL: N40 53 27 W76 59 54. Stereo. Hrs opn: 24
Rebroadcasts WUNS(FM) Lewisburg 100%.
Box 90, Rt. 204 & State School Rd., Selinsgrove, 17870-0090. Phone: (570) 374-8819. Fax: (570) 374-7444. Web Site: www.bigcountrynow.com. Licensee: MMP License LLC. Group owner: MAX Media L.L.C. (acq 10-17-03; grpsl). Population served: 225,000 Natl. Network: Westwood One, Jones Radio Networks. Natl. Rep: Dome. Format: Country. News staff: one; News: 2 hrs wkly. Target aud: 25-54. Spec prog: Gospel 2 hrs wkly. ♦Scott Richards, gen mgr; Greg Adair, stn mgr & gen sls mgr; Dawn Marie, progmg dir; Ted Koppen, chief of engrg.

Mifflintown

***WQJU(FM)**— Oct 15, 1985: 107.1 mhz; 370 w. Ant 1,302 ft TL: N40 34 20 W77 30 51. Hrs open: 24
Rebroadcasts WTLR(FM) State College 100%.
2020 Cato Ave., State College, 16801. Phone: (814) 237-9857. E-mail: mail@cpci.org Licensee: Central Pennsylvania Christian Institute. (acq 1-22-93; $132,500; FTR: 2-15-93). Population served: 164,464 Natl. Network: Moody, USA. Wire Svc: AP Format: Christian. News staff: one; News: 8 hrs wkly. Target aud: 30-55; adults, family oriented. ♦Mark Van Ouse, gen mgr & stn mgr.

Mill Hall

WVRT(FM)— Aug 20, 1979: 97.7 mhz; 6 kw. Ant 295 ft TL: N41 13 14 W77 16 39. Stereo. Hrs opn: 24 1559 W. 4th St., Williamsport, 17701. Phone: (570) 327-1400. Fax: (570) 327-8156. E-mail: kcote@clearchannel.com Web Site: www.variety977.com. Licensee: Clear Channel Radio Licenses, Inc. Group owner: Clear Channel Communications Inc. (acq 3-12-01; $1.5 million). Population served: 136,000 Natl. Rep: Christal. Format: Hot adult contemp. Target aud: 18-49. ♦Karen Cote, gen sls mgr & adv mgr; Tom Scott, progmg dir; Mike Myer, engrg dir.

Millersburg

WQLV(FM)— Feb 24, 1992: 98.9 mhz; 780 w. 895 ft TL: N40 30 18 W77 07 03. Stereo. Hrs opn: 24 234 Union St., 17061. Phone: (717) 692-2193. Fax: (717) 692-2080. E-mail: wqlv@love99.com Web Site: www.love99.com. Licensee: Hepco Communications Inc. Population served: 210,000 Natl. Network: Fox News Radio. Format: Soft adult contemp. News: 4 hrs wkly. Target aud: 25-54. Spec prog: High school sports 6 hrs wkly. ♦Ric Cooper, pres; James Hepler, gen mgr & dev VP; Jeff Fegley, opns; J.D. Cooper, adv.

Millersville

***WIXQ(FM)**— 1978: 91.7 mhz; 129 w. 69 ft TL: N39 59 53 W76 21 20. Hrs open: 7 AM-3 AM Box 1002, Millersville Univ., 17551. Phone: (717) 872-3518. Phone: (717) 871-2317. Fax: (717) 872-3383. E-mail: comments@wixq.com Web Site: www.wixq.com. Licensee: Millersville University. Format: Progsv, Black, Diversified. News: one hr wkly. Target aud: 18-24; college students. Spec prog: Jazz 2 hrs wkly. ♦Greg Park, stn mgr; Paul Galvin, prom dir; Steve Entrekin, progmg dir.

Millvale

WAMO(AM)—Licensed to Millvale. See Pittsburgh

Milton

WMLP(AM)— Oct 27, 1955: 1380 khz; 1 kw-D, 18 w-N. TL: N40 59 52 W76 52 17. Hrs open: 24 Box 1070, Sunbury, 17801. Secondary address: 1227 Countryline Rd., Selingsrove 17870. Phone: (570) 286-5838. Fax: (570) 743-7837. E-mail: valley@wvly.com Web Site: www.1380wmlp.com. Licensee: Sunbury Broadcasting Corp. (acq 2006; $3 million with co-located FM). Population served: 150,000 Natl. Network: CNN Radio. Natl. Rep: Dome, Roslin. Format: Talk. News staff: 3; News: 12 hrs wkly. Target aud: 25-54. ♦Roger Haddon Jr., pres & gen mgr; Kevin Herr, opns mgr & progmg VP; Nicole Shelley, gen sls mgr; Matt Paul, news dir; Harry Bingaman, chief of engrg.

WVLY-FM—Co-owned with WMLP(AM). Oct 1, 1967: 100.9 mhz; 1.3 kw. Ant 715 ft TL: N40 57 12 W76 45 05. Stereo. 24 Web Site: www.wvly.com.200,000 Natl. Rep: Dome, Roslin. Format: Adult contemp. News staff: 3; News: 5 hrs wkly. Target aud: 25-54; medium to upper income college grads. Spec prog: Smooth jazz 6 hrs wkly.

Monroeville

WPGR(AM)— Sept 27, 1964: 1510 khz; 5 kw-D, 1 w-N, 2.5 kw-CH, DA-2. TL: N40 28 13 W79 51 04. Hrs open: 24 960 Penn Ave., Suite 200, Pittsburgh, 15222. Secondary address: Sheridan Broadcasting Co., 960 Penn Ave., Pittsburgh 15222. Phone: (412) 456-4064. Fax: (412) 391-3559. E-mail: mdouglass@sbcol.com Web Site: www.wamo.com. Licensee: McL/McM Pennsylvania LLC. (group owner; (acq 9-28-2001; $625,000). Population served: 300000 Natl. Rep: McGavren Guild. Law Firm: Fletcher, Heald & Hildreth, P.L.C. Format: Urban gospel. Target aud: 25-54; middle class & higher income households. Spec prog: Relg 6 hrs wkly. ♦Ronald Davenport Jr., chmn & pres; Michael Douglass, gen mgr.

Montrose

***WPEL(AM)**— May 30, 1953: 1250 khz; 1 kw-D. TL: N41 51 16 W75 51 50. (CP: 800 khz; 1 kw-D, 135 w-N). Hrs opn: 6 AM-sunset Box 248, 9 Locust St., 18801. Phone: (570) 278-2811. Fax: (570) 278-1442. E-mail: mail@wpel.org Web Site: wpel.org. Licensee: Montrose Broadcasting Corp. (group owner) Population served: 40,000 Natl. Network: AP Network News. Rgnl. Network: Radio Pa. Law Firm: Gammon & Grange. Wire Svc: AP Format: Southern gospel music. News: 7 hrs wkly. Target aud: General; families. Spec prog: Class one hr, farm one hr wkly. ♦Larry Souder, pres & gen mgr; Lloyd Sheldon, opns mgr; LaVerne Sollick, prom mgr; Robert Brigham, chief of engrg.

WPEL-FM— June 5, 1961: 96.5 mhz; 57 kw. 459 ft TL: N41 51 16 W75 51 50. Stereo. 24 826,645 Natl. Network: Moody. Format: btfl mus, relg. News: 12 hrs wkly. Target aud: General.

Mount Carmel

WSPI(FM)— March 1993: 99.7 mhz; 790 w. 646 ft TL: N40 49 09 W76 27 45. Hrs open: 24 612 N. Shamokin St., Shamokin, 17872. Phone: (570) 644-0700. Fax: (570) 644-2232. E-mail: mail@litefm.us Web Site: www.litefm.us. Licensee: H & P Communications Ltd. (acq 12-28-92; $124,000; FTR: 1-25-93). Population served: 250,000 Format: Adult contemp. Target aud: 25-65. ♦Gene Picarella, CEO & gen mgr; Jesse James, stn mgr, gen sls mgr, rgnl sls mgr, mktg dir & chief of engrg; Rick Brody, progmg dir; Susie Good, traf mgr.

Mount Pocono

WPLY(AM)— Apr 8, 1981: 960 khz; 1 kw-D, 24 w-N, DA-2. TL: N41 04 41 W75 23 33. Hrs open: 24
Simulcast with WVPO(AM) Stroudsburg.
22 S. 6th St., Stroudsburg, 18360. Phone: (570) 421-2100. Fax: (570) 421-2040. Licensee: Nassau Broadcasting II LLC. Group owner: Nassau Broadcasting Partners L.P. (acq 6-28-2000). Format: Oldies. ♦Rick Musselman, gen mgr.

Mount Union

WFZY(FM)— Mar 30, 1992: 106.3 mhz; 120 w. Ant 1,440 ft TL: N40 24 53 W77 54 13. Stereo. Hrs opn: 24
Simulcast with WFGY(FM) Altoona.
R.R. 3 Box 225-A, Huntingdon, 16652. Phone: (814) 643-9620. Phone: (814) 643-1063. Fax: (814) 643-9625. Licensee: Megahertz Licenses LLC. Group owner: Forever Broadcasting (acq 3-13-2002; $620,000). Population served: 82,345 Natl. Rep: Dome. Law Firm: Mullin, Rhyne, Emmons & Topel. Format: Contemp country. News staff: one. Target aud: 25-54; women, above average educ, income & status. ♦Kristen Cantrell, gen mgr.

Mountain Top

WBHT(FM)— September 1992: 97.1 mhz; 500 w. 1,102 ft TL: N41 10 57 W75 52 19. Hrs open: 24 600 Baltimore Dr., 2nd Fl., Wilkes-Barre, 18702. Phone: (570) 824-9000. Fax: (570) 820-0520. Web Site: www.97bht.com. Licensee: Citadel Broadcasting Co. Group owner: Citadel Broadcasting Corp. (acq 10-23-98; grpsl). Law Firm: Cohn & Marks. Format: CHR. Target aud: 18-34. ♦Taylor Walet, gen mgr; Jules Riley, opns mgr; Mark McKay, progmg dir; Bill Palmeri, sls; Mike Jarvie, sls.

Muncy

WBZD-FM—Licensed to Muncy. See Williamsport

Murrysville

***WRWJ(FM)**— July 1994: 88.1 mhz; 250 w. 243 ft TL: N40 28 51 W79 43 26. Hrs opn: 19
Rebroadcasts WAIJ(FM) Grantsville, MD 100%.
Box 540, 34 Springs Rd., Grantsville, MD, 21536-0540. Phone: (301) 895-3292. Fax: (301) 895-3293. E-mail: hesalive@hesalive.net Web Site: www.hesalive.net. Licensee: He's Alive Inc. Natl. Network: USA. Format: Gospel, Christian, relg, adult contemp. Target aud: 18-35. ♦Dewayne Johnson, pres.

Nanticoke

WNAK(AM)— February 1947: 730 khz; 1 kw-D, 38 w-N. TL: N41 13 10 W75 59 28. Stereo. Hrs opn: 84 S. Prospect St., 18634. Phone: (570) 735-0730. Fax: (570) 735-4844. Licensee: Group B Licensee LLC. Group owner: Route 81 Radio LLC (acq 12-2-2003; $475,000). Population served: 16,632 Natl. Rep: Katz Radio. Format: MOR, btfl mus, nostalgia. Target aud: 35 plus. Spec prog: Relg 8 hrs wkly. ♦Lloyd B. Roach, pres; Margie McQuillin, stn mgr.

WQFM(FM)— Oct 31, 1973: 92.1 mhz; 280 w. Ant 1,056 ft TL: N41 10 59 W75 52 31. Stereo. Hrs opn: 24
Simulcast with WQFN(FM) Forest City.
149 Penn Ave., Scranton, 18503. Phone: (570) 346-6555. Fax: (570) 346-6038. E-mail: tbass@shamrocknepa.com Web Site: www.921qfm.com. Licensee: The Scranton Times L.P. Group owner: Shamrock Communications Inc. (acq 8-10-94). Natl. Rep: Roslin. Format: Adult contemp. News staff: one; News: 5 hrs wkly. Target aud: 25-54. Spec prog: Pol 3 hrs wkly. ♦William R. Lynett, CEO; Jim Loftus, gen mgr.

***WSFX(FM)**— Oct 25, 1987: 89.1 mhz; 100 w. 50 ft TL: N41 11 42 W75 59 28. Stereo. Hrs opn: Luzerne County Community College, Prospect St. & Middle Rd., 18634. Phone: (570) 740-0632. Fax: (570) 740-0605. Licensee: Luzerne County Community College. Format: Div. Target aud: 16-25; college age alternative mus audience. ♦Ron Reino, gen mgr.

Nanty Glo

***WPKV(FM)**— 2006: 90.7 mhz; 2.1 kw vert. Ant 482 ft TL: N40 30 20 W78 48 12. Hrs open: 5700 West Oaks Blvd., Rocklin, CA, 95765. Phone: (916) 251-1600. Fax: (916) 251-1650. Licensee: Educational Media Foundation. (acq 3-23-2007; grpsl). Format: Christian. ♦Richard Jenkins, pres.

New Berlin

***WBGM(FM)**— September 1996: 88.1 mhz; 550 w. 417 ft TL: N40 53 27 W76 59 54. Hrs open:
Rebroadcasts WPGM-FM Danville 100%.
8 E. Market St., Danville, 17821. Phone: (570) 275-1570. Fax: (570) 275-4071. E-mail: info@wpgm.org Licensee: Montrose Broadcasting Corp. (group owner) Format: Relg, Christian. ♦George Vacca, gen mgr; Deanna Force, mus dir.

New Castle

WJST(AM)— Oct 23, 1938: 1280 khz; 4.9 kw-D, 1 kw-N, DA-N. TL: N40 57 14 W80 19 05. Hrs open: 219 Savannah Gardner Rd., 16101-5546. Phone: (724) 346-5070. Fax: (724) 654-3101. Licensee: Forever Broadcasting LLC. Group owner: Forever Broadcasting (acq 6-30-2004; grpsl). Population served: 38,559 Natl. Network: ABC. Natl. Rep: Dome, Rgnl Reps. Format: Oldies. Target aud: 30 plus. Spec prog: Black one hr, class one hr wkly. ♦Scott D. Cohagan, gen mgr.

WKST(AM)— Aug 25, 1968: 1200 khz; 5 kw-D, 1 kw-N, DA-N. TL: N40 56 22 W80 23 38. Hrs open: 5:30 AM-1 AM 219 Savanah Gardner Rd., 16101. Phone: (724) 346-5070. Fax: (724) 654-3101. Web Site: www.wkst.com. Licensee: Forever Broadcasting LLC. Group owner: Forever Broadcasting (acq 6-30-2004; grpsl). Population served: 38559 Rgnl rep: Commercial Media Sales. Format: Nesw/talk, sports. News staff: 2; News: 7 hrs wkly. Target aud: 34 plus. Spec prog: Ger one hr, Pol one hr, Greek one hr wkly. ♦Scott D. Cohagan, gen mgr; Ken Hlebovy, progmg dir; Wade Sutton, news dir.

***WVMN(FM)**— Nov 22, 1995: 90.1 mhz; 2 kw. Ant 236 ft TL: N41 00 47 W80 17 36. Stereo. Hrs opn: 24
Rebroadcasts WCRF(FM) Cleveland, OH 100%.
c/o Radio Stn WCRF(FM), 9756 Barr Rd., Cleveland, OH, 44141. Phone: (440) 526-1111. Fax: (440) 526-1319. E-mail: wcrf@moody.edu Web Site: www.wcrfradio.org. Licensee: Moody Bible Institute of Chicago. (group owner) Law Firm: Southmayd. Wire Svc: AP Format: Inspirational. Target aud: 25-55. ♦Michael Easley, pres; Richard Lee, stn mgr; Gary Bittner, mus dir; Doug Hainer, chief of engrg.

New Kensington

WGBN(AM)— October 1940: 1150 khz; 1 kw-D, 70 w-N, DA-1. TL: N40 34 24 W79 46 58. Hrs open: 24 560 7th St., 15068. Phone: (724) 337-3588. Fax: (724) 337-1318. Licensee: Pentecostal Temple Development Corp. (acq 11-3-92; FTR: 11-23-92). Population served: 1,209,000 Natl. Network: USA. Natl. Rep: Dome. Law Firm: Fletcher, Heald & Hildreth. Format: Gospel. News staff: one; News: 19 hrs wkly. Target aud: 30 plus; older, upscale. Spec prog: Pol 3 hrs, It 2 hrs, Irish 2 hrs, relg 2 hrs wkly. ♦Lauren Mann, gen mgr; Stacy Taylor, asst music dir; Del King, chief of engrg.

WPGB(FM)—See Pittsburgh

WZPT(FM)— Aug 17, 1967: 100.7 mhz; 17 kw. Ant 850 ft TL: N40 29 43 W80 00 18. Stereo. Hrs opn: 24 651 Holiday Dr., Foster Plaza Five, Pittsburgh, 15220. Phone: (412) 920-9400. Fax: (412) 920-9444. Web Site: www.1007.com. Licensee: Infinity Radio Holdings Inc. Group owner: Infinity Broadcasting Corp. (acq 6-8-98; grpsl). Format: Hot adult contemp. Target aud: 18-49. ♦Joel Hollander, pres; Scott Herman, exec VP; Keith Clark, sr VP, opns VP, progmg VP & progmg dir; Don Oylear, VP & gen mgr; Keith Belden, sls dir & natl sls mgr; Ronda Zegarelli, gen sls mgr; Susie Barker, prom mgr; Jonny Hortwell, mus dir; Kerri Griffith, news dir & pub affrs dir; Chris Hudak, chief of engrg.

New Wilmington

***WWNW(FM)**— Jan 31, 1968: 88.9 mhz; 4 kw vert. Ant 128 ft TL: N41 06 41 W80 20 21. Stereo. Hrs opn: 24 Box 89, Westminster College, 16172. Phone: (724) 946-7242. Fax: (724) 946-7070. E-mail: barnerdl@westminster.edu Web Site: titanradio.net. Licensee: Westminster College Board of Trustees. Population served: 60,000 Natl. Network: ABC. Wire Svc: AP Format: Hot adult contemp. News: 3 hrs wkly. Target aud: 18-35; college students & staff. Spec prog: Relg 3 hrs wkly. ♦R. Thomas Williamson, pres; David L. Barner, gen mgr; Charles Chirozzi, chief of engrg.

Norristown

WNAP(AM)— Aug 6, 1946: 1110 khz; 4.8 kw-D, DA. TL: N40 08 05 W75 18 48. Hrs open: 2311 Old Arch Rd., 19401. Secondary address: Box 11, Philadelphia 19128. Phone: (610) 272-7600. Fax: (610) 272-5793. E-mail: wnap@nni.com Licensee: George H. Buck. Group owner: GHB Radio Group (acq 12-15-87; $725,000; FTR: 4-2-84). Population served: 3,500,000 Rgnl. Network: Metronews Radio Net. Format: Black gospel. Target aud: General. ♦Fred Blain, gen mgr & progmg dir; Orey Ferrell, gen sls mgr; Dave McCrork, chief of engrg.

North East

WRKT(FM)— Mar 29, 1970: 100.9 mhz; 4.2 kw. 252 ft TL: N42 11 51 W79 45 10. Stereo. Hrs opn: 24 Boston Store Pl., Erie, 16501. Phone: (814) 461-1000. Fax: (814) 461-1500. E-mail: rocket101@rocket101.com Licensee: Connoisseur Media of Erie LLC. Group owner: NextMedia Group L.L.C. (acq 3-30-2006; grpsl). Population served: 226,600 Law Firm: Fletcher, Heald & Hildreth. Format: Classic rock. News staff: one; News: one hr wkly. Target aud: 25-54. Spec prog: Loc bands one hr wkly. ♦Richard Rambaldo, gen mgr; Michael Malpiedi, gen sls mgr.

WYNE(AM)— Nov 24, 1966: 1530 khz; 1 kw-D, 250 w-CH. TL: N42 12 05 W79 51 43. Hrs open: Sunrise-set 501 E. 38th St., Erie, 16546. Phone: (814) 824-2261. Fax: (814) 824-2590. Web Site: www.mercyhurst.edu. Licensee: Mercyhurst College (acq 2-18-2005; $110,000). Population served: 12,000 Format: Oldies. Target aud: 25-54; men & women. ♦William T. Shannon, gen mgr.

Northern Cambria

WNCC(AM)— Oct 15, 1950: Stn currently dark. 950 khz; 500 w-D. TL: N40 40 47 W78 44 26. Hrs open: 24 Box 1095, 15714. Phone: (814) 472-4060. Fax: (814) 948-0950. E-mail: whpa@verizon.net Licensee: Vernal Enterprises Inc. (group owner; (acq 3-19-97; $20,000 with WRDD(AM) Ebensburg). Population served: 40000 Format: Adult contemp, oldies. News staff: one; News: 6 hrs wkly. Target aud: 35 plus; females & males in the 35 plus age range. ♦Larry Schrengost, gen mgr.

WPCL(FM)— Sept 30, 1991: 97.3 mhz; 6 kw. 610 ft TL: N40 38 26 W78 47 45. Stereo. Hrs opn: 19 Box 540, 34 Spring Rd., Grantsville, MD, 21536. Phone: (301) 895-3292. Fax: (301) 895-3293. E-mail: hesalive@hesalive.net Web Site: www.hesalive.com. Licensee: He's Alive Inc. (group owner; acq 3-18-97; $105,000). Natl. Network: USA. Natl. Rep: Commercial Media Sales. Format: Christian, relg. Target aud: 18-35. ♦Dewayne Johnson, pres; Monte Palmer, stn mgr.

Northumberland

WEGH(FM)— Aug 22, 1994: 107.3 mhz; 900 w. Ant 843 ft TL: N40 47 10 W76 41 49. Stereo. Hrs opn: 24 Box 1070, Sunbury, 17801. Secondary address: RD#2 County Line Rd., Selinsgrove 17870. Phone: (570) 286-5838. Phone: (570) 743-1841. Fax: (570) 743-7837. Fax: (570) 743-1605. E-mail: eagle107@eagle107.com Web Site: www.eagle107.com. Licensee: Sunbury Broadcasting Corp. Natl. Rep: Roslin. Rgnl rep: Dome. Format: Classic Hits. News staff: 4; News: one hr wkly. Target aud: 25-54. ♦Roger S Haddon Jr., CEO; Roger S. Haddon Jr., pres; Kevin Herr, opns mgr; Gayle Fedder, gen sls mgr & mktg mgr; Rob Senter, progmg dir & mus dir; Kelli Tyler, prom.

Oil City

WGYI(FM)—Listing follows WOYL(AM).

WKQW(AM)— Dec 1, 1986: 1120 khz; 1 kw-D. TL: N41 23 45 W79 39 53. Hrs open: 222 Seneca St., 16301. Phone: (814) 676-8254. Fax: (814) 677-4272. E-mail: traffic@wkqw.com Web Site: www.kqw.com. Licensee: Clarion County Broadcasting Corp. (acq 2-25-2005; $540,000 with co-located FM). Population served: 65,000 Natl. Network: CBS Radio. Natl. Rep: Dome. Format: Loc info, classic country. News staff: one; News: 4 hrs wkly. Target aud: 25-54. ♦Dave Lucus, prom mgr & progmg dir; Mary Heim, news dir; Amy Krizon, traf mgr; Mike Snyder, disc jockey; Ron Smith, disc jockey.

WKQW-FM— September 1992: 96.3 mhz; 6 kw. 328 ft TL: N41 23 45 W79 39 53. Stereo. 24 Web Site: www.kqw.com.120,000 Natl. Network: CBS Radio. Rgnl rep: Dome & Assoc. Law Firm: Polner Law Office. Format: Adult contemp, loc info. News staff: one; News: 4 hrs wkly. ♦William Hearst, pres.

WOYL(AM)— Feb 14, 1946: 1340 khz; 1 kw-U, DA-D. TL: N41 25 04 W79 42 53. Hrs open: 6 AM-midnight Box 908, 1411 Liberty St., Franklin, 16323. Phone: (814) 676-5744. Fax: (814) 437-9372. Licensee: Forever Broadcasting LLC. Group owner: Forever Broadcasting (acq 7-20-00; grpsl). Population served: 15,033 Natl. Rep: Dome, Keystone (unwired net). Format: MOR, news/talk. News staff: one. ♦Terry Deitz, gen mgr; Joe Elan, sls dir & gen sls mgr; Todd Adkins, progmg dir; Paul Joseph, news dir.

WGYI(FM)—Co-owned with WOYL(AM). May 1, 1957: 98.5 mhz; 20 kw. 299 ft TL: N41 25 04 W79 72 53. Stereo. 180,000 Format: America's Best Country.

Oliver

WOGG(FM)— June 11, 1993: 94.9 mhz; 1.65 kw. Ant 1,233 ft TL: N39 52 11 W79 38 22. Stereo. Hrs opn: 24 123 Blaine Rd., Brownsville, 15417. Phone: (724) 938-2000. Fax: (724) 938-7824. E-mail: jtrunzo@zoominternet.net Web Site: foggyland.com. Licensee: Keymarket Licenses LLC. Group owner: Keymarket Communications LLC (acq 8-31-99; $2.875 million with WASP(AM) Brownsville). Population served: 2,500,000 Format: Country. News staff: one. Target aud: 25-54. ♦Andrew Powaski, gen mgr, opns mgr & gen sls mgr; Jeffrey Trunzo, chief of engrg.

Olyphant

WBHD(FM)—Licensed to Olyphant. See Dunmore

WQOR(AM)— July 20, 1987: 750 khz; 1.6 kw-D. TL: N41 28 34 W75 29 41. Hrs open: Sunrise-sunset 6325 Sheridan Dr., Williamsville, NY, 14221. Phone: (716) 839-6117. Fax: (716) 839-0400. Web Site: www.holyfamily.ws. Licensee: Holy Family Communications (group owner; acq 3-24-2003; $170,000). Population served: 750,000 Format: Catholic radio. ♦James Wright, pres.

Palmyra

WWKL(FM)— Sept 22, 1959: 92.1 mhz; 3.3 kw. Ant 300 ft TL: N40 19 35 W76 36 33. Stereo. Hrs opn:
Rebroadcasts WTPA(FM) Mechanicsburg 100%.
Cumulus Media-WNNK FM-WTCY AM, 2300 Vartan Way, Harrisburg, 17110-9720. Phone: (717) 238-1041. Fax: (717) 234-4842. Web Site: www.hot92.com. Licensee: Cumulus Licensing Corp. Group owner: Cumulus Media Inc. (acq 11-28-2000; grpsl). Population served: 500,000 Rgnl. Network: Radio Pa. Law Firm: Wilkinson Barker Knauer. Format: CHR/ rhythmic. Target aud: 25-54. Spec prog: Relg 6 hrs, Sp 14 hrs, Hershey Bears hockey, Hershey Wildcats soccer wkly. ♦Ron Giovanniello, gen mgr; Karen Richards, sls dir; Todd Matthews, gen sls mgr; John O'Dea, progmg dir; Amy Warner, mus dir; Dave Supplee, chief of engrg.

Patton

WBXQ(FM)— 1991: 94.3 mhz; 2.1 kw. Ant 548 ft TL: N40 39 17 W78 40 34. Hrs open: 1417 12th Ave, Altoona, 16601. Phone: (814) 944-9344. Fax: (814) 944-9350. Web Site: www.wbxq.com. Licensee: Sherlock Broadcasting Inc. (acq 3-95; $450,000; FTR: 6-26-95). Format: Classic rock. Target aud: 35-65. ♦Diane Boslet, gen mgr.

Pen Argyl

***WWPJ(FM)**— 2001: 89.5 mhz; 1 w horiz, 40 w vert. Ant 1,125 ft TL: N40 53 03 W75 15 43. Hrs open: 24
Rebroadcasts WWFM(FM) Trenton, NJ 100%.
Mercer County Community College, 1200 Old Trenton Rd., Trenton, NJ, 08690. Phone: (609) 587-8989. Licensee: Mercer County Community College. Format: Classical. ♦Jeffery Sekerka, gen mgr.

Philadelphia

KYW(AM)— 1921: 1060 khz; 50 kw-U, DA-1. TL: N40 06 12 W75 14 56. Stereo. Hrs opn: 24 400 Market St, 10th floor, 19106. Phone: (215) 238-1060. Fax: (215) 238-4657. Web Site: www.kyw1060.com. Licensee: CBS Radio East Inc. Group owner: Infinity Broadcasting Corp. Population served: 4,500,000 Natl. Network: ABC, CBS, CNN Radio. Natl. Rep: CBS Radio. Law Firm: Leventhal Senter & Lerman. Wire Svc: AP Format: News. News staff: 34; News: 168 hrs wkly. Target aud: 25-54; adults. ♦David Yadgaroff, VP & gen mgr; Michael Berkowitz, natl sls mgr; Rich Iovanisci, rgnl sls mgr; Rob Kaloustian, rgnl sls mgr; Kyle Ruffin, mktg dir; Dee Patel, news dir; Steve Butler, progmg dir & news dir; Frank Sippel, chief of engrg. Co-owned TV: KYW-TV affil.

WBEB(FM)— May 13, 1963: 101.1 mhz; 14 kw. 941 ft TL: N40 02 21 W75 14 13. Stereo. Hrs opn: 24 10 Presidential Blvd., Bala Cynwyd, 19004. Phone: (610) 667-8400. Fax: (610) 667-6795. Web Site: www.b101radio.com. Licensee: WEAZ-FM Radio Inc. (acq 5-63). Population served: 500,000 Natl. Rep: McGavren Guild. Law Firm: Borsari & Paxson. Format: Adult contemp. Target aud: 25-54.Jerry Lee, CEO & pres; David L. Kurtz, chmn & CFO; Blaise Howard, VP & gen mgr; William F. Boone, stn mgr; Dave Giordano, sls dir, natl sls mgr & natl sls mgr; Agnes Fuller, rgnl sls mgr; Emily Elfenbein, mktg dir, pub affrs dir & women's int ed; Bonnie Hoffman, prom mgr; Chris Conley, progmg dir; Chris Sarris, engrg dir; Mari Beth Hoban, traf mgr; Chris McCoy, disc jockey; Jason Lee, disc jockey; Joan Jones, disc jockey; Juan Varleta, disc jockey; Mary Marlowe, disc jockey

WBEN-FM— Mar 1, 1949: 95.7 mhz; 50 kw. 500 ft TL: N40 03 33 W75 14 20. Stereo. Hrs opn: 24 One Bala Plaza, Suite 424, Bala Cynwyd, 19004. Phone: (610) 771-0933. Fax: (610) 771-9690. E-mail: gdefrancesco@greaterphila.com Web Site: www.mix957online.com. Licensee: Greater Philadelphia Radio Group. Group owner: Greater Media Inc. (acq 5-29-97; $41.8 million). Population served: 4,000,000 Law Firm: Shaw

Pittman. Format: Hot Adult Contemp. News: 3 hrs wkly. Target aud: 25-54; professional, upscale executives. ♦Rick Feinblatt, VP & gen mgr; Larry Paulauski, chief of engrg.

WDAS-FM—Listing follows WUBA(AM).

WFIL(AM)— 1922: 560 khz; 5 kw-U, DA-2. TL: N40 05 42 W75 16 38. Stereo. Hrs opn: 24 117 Ridge Pike, Lafayette Hill, 19444. Phone: (610) 941-9560. Fax: (610) 828-8879. E-mail: wfil@wfil.com Web Site: www.wfil.com. Licensee: Pennsylvania Media Associates Inc. Group owner: Salem Communications Corp. (acq 11-1-93; $4 million). Population served: 11,792,380 Natl. Network: Salem Radio Network. Natl. Rep: Salem. Law Firm: Borsari & Paxson. Format: Relg, Christian, talk. News: 2 hrs wkly. Target aud: 35-64; parents & grandparents. ♦Russ Whitnah, VP & gen mgr; Kevin Manna, opns dir; David Handler, gen sls mgr; Carol Healey, rgnl sls mgr; Mark Daniels, mktg mgr & progmg mgr; Fred Moore, chief of engrg; Rene Tetro, chief of engrg.

WHAT(AM)— 1925: 1340 khz; 1 kw-U. TL: N40 00 06 W75 12 35. Hrs open: 2471 N. 54 St., Suite 220, 19131. Phone: (215) 482-0956. Fax: (215) 581-5185. E-mail: gmwhat1340am@aol.com Licensee: Urban Radio I L.L.C. Group owner: Inner City Broadcasting (acq 2-2-99). Population served: 194,806 Format: Talk, urban. Target aud: 25-54. Spec prog: Gospel 12 hrs wkly. ♦Pierre Sutton, CEO; Charles Warfield, pres; Bill Cooper, CFO; Christopher Squire, VP & gen mgr.

***WHYY-FM**— 1954: 90.9 mhz; 13.5 kw. 920 ft TL: N40 02 30 W75 14 24. Stereo. Hrs opn: 24 150 N. 6th St., Independence Mall West, 19106. Phone: (215) 351-1200. Phone: (215) 351-9204. Fax: (215) 351-3352. E-mail: talkback@whyy.org Web Site: www.whyy.org. Licensee: WHYY Inc. Population served: 3,987,600 Natl. Network: NPR, PRI. Law Firm: Schwartz, Woods & Miller. Format: News, info. News staff: 7; News: 35 hrs wkly. Target aud: 35-49. Spec prog: Opera 4 hrs, folk 4 hrs, jazz 4 hrs wkly. ♦William J. Marrazzo, pres; Paul Gluck, gen mgr.

WIOQ(FM)— 1941: 102.1 mhz; 27 kw. 669 ft TL: N40 02 40 W75 14 30. Stereo. Hrs opn: 24 One Bala Plaza, Suite 243, Bala Cynwyd, 19004. Phone: (610) 667-8100. Fax: (610) 668-4657. Web Site: www.q102philly.com. Licensee: AMFM Radio Licenses LLC. Group owner: Clear Channel Communications Inc. (acq 8-30-00; grpsl). Population served: 4,065,300 Format: CHR. Target aud: 18-34; females & teens. Spec prog: Pub affrs. ♦Rich Lewis, VP & gen mgr; Cassandra Banko, gen sls mgr; Lisa Acchione, mktg dir; Jeff Jordan, prom dir; Brian Bridgman, progmg dir; Marian Newsome, mus dir; Wendy McClure, pub affrs dir; Michael Guidotti, chief of engrg; Chris Marino, disc jockey; Glenn Kalina, disc jockey.

WIP(AM)— Mar 16, 1922: 610 khz; 5 kw-U, DA-1. TL: N39 51 56 W75 06 43. Hrs open: 2 Bala Plaza, Bala Cynwyd, 19004. Phone: (610) 949-7800. Fax: (610) 949-7880. Web Site: www.610wip.com. Licensee: Infinity Broadcasting Corp. of Philadelphia. Group owner: Infinity Broadcasting Corp. (acq 8-12-93; FTR: 8-30-93). Population served: 6,000,000 Natl. Network: Westwood One. Natl. Rep: CBS Radio. Format: All sports. ♦Cecil R. Forster Jr., VP & gen mgr; Tom Bigby, progmg mgr.

WYSP(FM)—Co-owned with WIP(AM). August 1971: 94.1 mhz; 9.6 kw. Ant 1,109 ft TL: N40 02 29.6 W75 14 11.5. Stereo. 101 S. Independence Mall E., 19106. Phone: (215) 625-9460. Fax: (215) 625-6560. Web Site: www.94wysp.com. Licensee: Infinity Broadcasting East Inc. (acq 11-1-81; grpsl; FTR: 9-28-81). Natl. Network: Westwood One. Natl. Rep: CBS Radio. Format: Rock. ♦Peter Kleiner, gen mgr; Gil Edwards, progmg dir.

WISX(FM)— Nov 11, 1959: 106.1 mhz; 22 kw. Ant 740 ft TL: N40 04 58 W75 10 54. Stereo. Hrs opn: 111 Presidential Blvd., Suite 100, Bala Cynwyd, 19004. Phone: (610) 508-1200. Phone: (610) 784-3333. Fax: (610) 784-0501. Web Site: www.phillys1061.com. Licensee: AMFM Radio Licenses LLC. Group owner: Clear Channel Communications Inc. (acq 8-30-2000; grpsl). Population served: 750,000 Natl. Rep: Christal. Law Firm: Latham & Watkins. Format: Adult contemp. Target aud: 25-54.

***WKDU(FM)**— 1970: 91.7 mhz; 110 w. 155 ft TL: N39 57 36 W75 11 27. Stereo. Hrs opn: 24 3210 Chestnut St., 19104. Phone: (215) 895-5920. Phone: (215) 895-5917. Fax: (215) 895-1050. Web Site: www.wkdu.org. Licensee: Drexel University. Population served: 2,500,000 Format: Progsv, alternative rock, free format. Target aud: General. Spec prog: International 12 hrs, rhythm & blues 3 hrs, gospel 4 hrs, Israeli 3 hrs, new age 2 hrs, metal 4 hrs, Black 8 hrs wkly. ♦Evan Caposerri, gen mgr; Casey Ross, progmg dir; Ryan McIntyre, pub affrs dir; Jim Cavanaugh, chief of engrg.

WMGK(FM)—Listing follows WPEN(AM).

WMMR(FM)— Apr 20, 1942: 93.3 mhz; 18 kw. 827 ft TL: N39 57 09 W75 10 05. Stereo. Hrs opn: 24 One Bala Plaza, Suite 424, Bala Cynwyd, 19004. Phone: (610) 771-0933. Fax: (610) 771-9710. Web Site: www.wmmr.com. Licensee: Greater Philadelphia Radio Inc. Group owner: Greater Media Inc. (acq 7-23-97; grpsl). Natl. Network: Westwood One. Natl. Rep: McGavren Guild. Format: Main stream rock. News staff: one; News: 5 hrs wkly. Target aud: 25-54; suburban rockers. ♦Richard D. Feinblatt, sr VP; John Fullam, gen mgr; Paul Blake, gen sls mgr; Scott Segelbeum, mktg dir; Bill Weston, progmg dir; Ken Zipeto, mus dir; Larry Paulauski, chief of engrg; Queen Chandler, traf mgr.

WNTP(AM)— 1923: 990 khz; 50 kw-D, 10 kw-N, DA-2. TL: N40 05 43 W75 16 37. Hrs open: 24 117 Ridge Pike, Lafayette Hill, 19444. Phone: (610) 940-0990. Fax: (610) 828-8879. E-mail: wntp@wntp.com Web Site: www.wntp.com. Licensee: Pennsylvania Media Associates Inc. Group owner: Salem Communications Corp. (acq 1994; $3.5 million grpsl). Population served: 7,927,724 Natl. Network: Salem Radio Network. Law Firm: Borsari & Paxson. Format: News/talk. News: 4 times per hr. Target aud: 35-64; Adults. Spec prog: Sports, Sp. ♦Russ Whitnah, VP & gen mgr; Kevin Manna, opns mgr; David Handler, gen sls mgr; Carol Healey, rgnl sls mgr; Rolanda Myers, mktg dir; Mark Daniels, mktg mgr & progmg mgr; Rene Tetro, chief of engrg.

WNWR(AM)— July 11, 1947: 1540 khz; 50 kw-D, DA. TL: N40 02 46 W74 14 15. Hrs open: 200 Monument Rd., Suite 6, Bala Cynwyd, 19004. Phone: (610) 664-6780. Fax: (610) 664-8529. Web Site: www.wnwr.com. Licensee: Global Radio L.L.C. (acq 6-95; $1.4 million). Population served: 250,000 Natl. Rep: Roslin. Law Firm: Taylor, Thiemann & Aitken. Format: Var/div, ethnic multicultural. Target aud: 25-54. ♦Jim Weitzman, pres; Sam Speiser, gen mgr & stn mgr; Shawn Laughlin, opns mgr.

WOGL(FM)—Listing follows WPHT(AM).

***WPEB(FM)**— May 15, 1981: 88.1 mhz; 1 w. Ant 49 ft TL: N39 57 33 W75 12 13. Hrs open: 4134 Lancaster Ave., 3rd Fl., 19104. Phone: (215) 387-6155. E-mail: volta@phillyinc.org Web Site: www.radiovolta.org. Licensee: West Philadelphia Educational Broadcasting Foundation. Population served: 1,500,000 Format: Div. Target aud: General.

WPEN(AM)— April 1929: 950 khz; 5 kw-U, DA-N. TL: N39 58 28 W75 16 30. Stereo. Hrs opn: 24 One Bala Plaza, Mail Stop 429, Bala Cynwyd, 19004-1428. Phone: (610) 667-8500. Fax: (610) 771-9692. Web Site: www.sportstalk950.com. Licensee: Greater Philadelphia Radio Inc. Group owner: Greater Media Inc. (acq 1-6-75). Population served: 4,000,000 Format: Sports. ♦John Fullam, gen mgr; Paul Blake, sls dir; Marc Taub, gen sls mgr; Mike McMonagle, prom dir; Gregg Henson, progmg dir & progmg mgr.

WMGK(FM)—Co-owned with WPEN(AM). 1942: 102.9 mhz; 8.9 kw. Ant 1,148 ft TL: N40 02 21 W75 14 13. Stereo. 24 One Bala Plaza, Suite 339, Bala Cynwyd, 19004. Web Site: www.wmgk.com.1,800,000 Format: Classic rock. ♦Chris Kirchner, gen sls mgr; Ed Marshall, prom dir.

WPHE(AM)—See Phoenixville

WPHT(AM)— 1922: 1210 khz; 50 kw-U. TL: N39 58 46 W74 59 13. Hrs open: Two Plaza, Suite 800, Bala Cynwyd, 19004. Phone: (610) 668-5800. Fax: (610) 667-5886. E-mail: talkradio1210@cbs.com Web Site: www.thebigtalker1210.com. Licensee: CBS Radio East Inc. Group owner: Infinity Broadcasting Corp. (acq 8-58). Population served: 4,000,000 Natl. Network: CBS, Westwood One. Natl. Rep: CBS Radio. Law Firm: Leventhal, Senter & Lerman PLLC. Format: Talk. News: 2. Target aud: 25-64; adults. ♦David Yadgaroff, gen mgr; Mike Baldini, stn mgr; Grace Blazer, progmg dir; Dave Skalish, chief of engrg; Jennifer Miller, traf mgr.

WOGL(FM)—Co-owned with WPHT(AM). May 16, 1944: 98.1 mhz; 9.6 kw. Ant 1,109 ft TL: N40 02 29.6 W75 14 11.4. Stereo. 24 Phone: (610) 668-5900. Fax: (610) 668-5977. E-mail: questions@wogl.com Web Site: www.wogl.com.4,000,000 Law Firm: Leventhal, Senter & Lerman. Format: Hits of the 60s & 70s. News staff: one; News: 1.25 hrs wkly. Target aud: 25-54. ♦James F. Loftus, gen mgr; Kerry Mulvey, gen sls mgr; Diane Santilippo, natl sls mgr; Cindy Webster, mktg dir; Anne Gress, progmg dir; Tommy McCarthy, mus dir; Dave Skalish, chief of engrg; LeeAnn Smith, traf mgr.

WRDW-FM— 1957: 96.5 mhz; 17 kw. 866 ft TL: N40 02 21 W75 14 13. Stereo. Hrs opn: 24 555 City Line Ave., Ste. 330, Bala Cynwyd, 19004. Phone: (610) 667-9000. Fax: (610) 667-2972. Web Site: www.wired965.com. Licensee: WDAS License L.P. Group owner: Beasley Broadcast Group (acq 3-11-97). Population served: 7,218,400 Natl. Network: Wall Street. Natl. Rep: D & R Radio. Law Firm: Fisher, Wayland, Cooper, Leader & Zaragoza. Format: CHR. Target aud: 18 plus; general. ♦Bruce Beasley, CEO, chmn, pres & progmg dir; Lynn Bruder, gen mgr; Don Melnyk, chief of engrg.

WRFF(FM)— February 1965: 104.5 mhz; 12.5 kw. Ant 1,008 ft TL: N40 02 30 W75 14 24. Stereo. Hrs opn: 24 111 President Blvd., Suite 100, Bala Cynwyd, 19004. Phone: (610) 784-3333. Fax: (610) 784-2011. Web Site: www.radio1045.com. Licensee: AMFM Radio Licenses LLC. Group owner: Clear Channel Communications Inc. (acq 8-30-2000; grpsl). Population served: 4,114,800 Format: Rock. News staff: one. ♦L. Lowry Mays, CEO; Manuel Rodriguez, VP; Ron Decastro, sls dir; Becki West, gen sls mgr; Shelvia Williams, prom dir; Brian Check, progmg dir; Margo Marano, mus dir; Jennifer Ryan, news dir & pub affrs dir; Sandra Johnson, traf mgr.

***WRTI(FM)**— July 9, 1953: 90.1 mhz; 12.5 kw. 1,010 ft TL: N40 02 21 W75 14 13. Stereo. Hrs opn: 24 1509 Cecil B. Moore Blvd., 3rd Fl., 19121. Phone: (215) 204-8405. Fax: (215) 204-7027. E-mail: comments@wrti.org Web Site: www.wrti.org. Licensee: Temple University of The Commonwealth System of Higher Education. Population served: 7,500,000 Natl. Network: NPR, PRI. Rgnl. Network: Radio Pa. Wire Svc: AP Format: Class, jazz. News staff: one; News: 15 hrs wkly. Target aud: 30-65. ♦David S. Conant, CEO & gen mgr; Vic Scarpato, CFO; William P. Johnson, stn mgr; Tobias Poole, opns dir; Patricia Prevost, dev dir; Rick Torpey, natl sls mgr; Porsche Blakey, mktg mgr; Jack Moore, progmg dir; Windsor Johnston, news dir; Jeff DePolo, chief of engrg; Lorna Nixon, traf mgr; Lesley Valdes, mus critic.

WTMR(AM)—See Camden, NJ

WUBA(AM)— 1923: 1480 khz; 5 kw-D, 1 kw-N, DA-2. TL: N39 59 53 W75 12 43. Hrs open: 24 23 W. City Ave., Bala Cynwyd, 19004. Phone: (610) 617-8500. Fax: (610) 617-8501. Web Site: www.rumba1045.com. Licensee: AMFM Radio Licenses LLC. Group owner: Clear Channel Communications Inc. (acq 8-30-2000; grpsl). Population served: 560,000 Format: Sp. ♦Joseph Tamburro, gen mgr, gen sls mgr & progmg dir.

WDAS-FM—Co-owned with WUBA(AM). 1959: 105.3 mhz; 3.3 kw. 870 ft TL: N40 02 30 W75 14 24. (CP: 16.3 kw). Stereo. Web Site: www.wdas.fm.com. Format: Black adult contemp. Target aud: 25-54.

WURD(AM)— July 23, 1958: 900 khz; 1 kw-D, 105 w-N, DA-2. TL: N39 55 02 W75 13 18. Hrs open: 1341 N. Delaware Ave., Suite 300, 19125. Phone: (215) 425-7875. Fax: (215) 634-6003. Licensee: Levas Communications LLC Group owner: Levas Communications LLC (acq 4-30-2003;. $4.25 million). Population served: 4,334,000 Natl. Network: CNN Radio. Natl. Rep: McGavren Guild. Law Firm: Womble, Carlyle, Sandridge & Rice. Format: Urban talk. Spec prog: Gospel 9 hrs, It 3 hrs wkly. ♦Art Camiolo, pres; Cody Anderson, gen mgr; Steve Ballard, opns dir; Bill Anderson, progmg dir; Kia Long, traf mgr.

WUSL(FM)— 1961: 98.9 mhz; 18 kw. 830 ft TL: N40 02 31 W75 14 11. Stereo. Hrs opn: 24 111 Presidential Blvd., Suite 100, Bala Cynwyd, 19004. Phone: (610) 784-3333. Fax: (610) 784-0507. Web Site: www.power99.com. Licensee: Clear Channel Radio Licenses, Inc. Group owner: Clear Channel Communications Inc. (acq 8-30-00; grpsl). Population served: 4,729,000 Law Firm: Latham & Watkins. Format: Urban contemp. News staff: 2; News: 4 hrs wkly. Target aud: 18-49. Spec prog: Gospel 4 hrs wkly. ♦Richard Lewis, VP, gen sls mgr & progmg dir.

WWDB(AM)— 1925: 860 khz; 10 kw-D, DA. TL: N40 09 15 W75 22 10. (CP: 500 w-N, DA-2). Hrs open: Daytime 555 City Line Ave., Suite 330, Bala Cynwyd, 19004. Phone: (610) 822-1321. Phone: (610) 822-1320. Fax: (610) 667-5978. Web Site: www.wwdbam.com. Licensee: Beasley Broadcasting of Eastern Pennsylvania Inc. Group owner:

Beasley Broadcast Group (acq 9-9-86; $2.4 million; FTR: 8-11-86). Population served: 400,000 Format: Talk. Target aud: 18-49. ♦Bruce Gilbert, gen mgr; Tim Halloran, opns mgr.

***WXPN(FM)**— April 1957: 88.5 mhz; 5 kw. 918 ft TL: N40 02 36 W75 14 33. Stereo. Hrs opn: 24
Rebroadcasts WXPH(FM) Harrisburg 100%.
3025 Walnut St., 19104. Phone: (215) 898-6677. Fax: (215) 898-0707. E-mail: wxpndesk@xpnonline.net Web Site: www.xpn.org. Licensee: Trustees of the University of Pennsylvania. Population served: 4,315,028 Natl. Network: PRI, NPR. Law Firm: Levanthol Senter & Lerman. Format: Adult alternative. News: 3 hrs wkly. Target aud: 25-54; educated. Spec prog: Children 5 hrs, folk 5 hrs wkly.Quyen Shanahan, dev VP & mktg dir; Jay Ricci, sls VP; Tom Mara, gen sls mgr; Debby Seitz, prom mgr & spec ev coord; Bruce Warren, progmg dir & mus critic; Ann Reed, mus dir; Bob Bumbera, news dir, local news ed & sports cmtr; Jay Goldman, engrg dir; nda Myers, traf mgr; Philomena Rhodes, traf mgr; Deb D'Alessandro, women's int ed; Ali Costellini, disc jockey; David Dye, disc jockey; Helen Leicht, disc jockey; Jonny Meister, disc jockey; Kathy O'Connell, disc jockey; Michaela Majoun, disc jockey; Robert Drake, disc jockey

WXTU(FM)— September 1958: 92.5 mhz; 15 kw. Ant 915 ft TL: N40 02 19 W75 14 14. Stereo. Hrs opn: 24 555 City Line Ave., Suite 330, Bala Cynwyd, 19004. Phone: (610) 667-9000. Fax: (610) 667-1355. E-mail: comments@925xtu.com Web Site: www.925xtu.com. Licensee: Beasley Broadcasting of Eastern Pennsylvania Inc. Group owner: Beasley Broadcast Group (acq 7-83; $6 million; FTR: 7-11-83). Natl. Rep: D & R Radio. Format: Contemp country. Spec prog: Sundays 6am-6:30am-Philadelphia Focus. ♦Bruce Beasley, pres & VP; Natalie Conner, VP & gen mgr; Mark Vizza, mktg dir; Bob McKay, progmg dir; Don Melnyk, chief of engrg; Andie Summers, disc jockey; Kris Stevens, disc jockey; Leigh Richards, disc jockey; Scott Evans, disc jockey.

Philipsburg

WJOW(FM)—Listing follows WPHB(AM).

WPHB(AM)— June 1, 1956: 1260 khz; 5 kw-D, 34 w-N. TL: N40 53 39 W78 11 51. Hrs open: 24 1884 Port Matilda Hwy., Radio Park, 16866. Phone: (814) 342-2300. Fax: (814) 342-WPHB/9742. E-mail: wphb@wphbradio.com Web Site: www.wphbradio.com. Licensee: Magnum Broadcasting Inc. (acq 11-24-2004; $2,022,527 with co-located FM). Population served: 50,000 Natl. Network: CNN Radio. Rgnl. Network: Radio Pa. Rgnl rep: Dome & Assoc Law Firm: Haley, Bader, Potts. Format: Classic country & news/talk, sports. Target aud: Men & women; generally 25+. Spec prog: Bluegrass 4 hrs, polka 6 hrs, Gospel 6 hrs, big band 5 hrs wkly. ♦Michael M. Stapleford, pres; Laura Shore Mack, gen mgr; Cliff Mack, chief of opns; Marian Kovach, gen sls mgr; Jason Torrance, prom dir; C.J. Daniels, progmg dir; Sherry Flick, pub affrs dir; Joe Portelli, chief of engrg; Mary Beth Thompson, traf mgr; Tor Michaels, news rptr; Bud O'Brien, sports cmtr; Sheldon Sharpless, disc jockey.

WJOW(FM)—Co-owned with WPHB(AM). March 1989: 105.9 mhz; 710 w. Ant 951 ft TL: N40 47 34 W78 10 29. Stereo. 24 Phone: (814) 272-1320. E-mail: buzz@buzzfm.com Web Site: www.buzzfm.com.350,000 Format: Modern rock/alternative. News: 2.5 hrs wkly. Target aud: 18-49; men & women. ♦Austin Davis, progmg dir & disc jockey; Sherry Flick, local news ed; Tor Michaels, news rptr; Jill Gleeson, disc jockey; Kenny Marks, disc jockey; Laura Mack, disc jockey.

Phoenixville

WPHE(AM)—Aug 23, 1978: 690 khz; 1 kw-D, DA. TL: N40 08 08 W75 33 37. Hrs open: Box 46327, Philadelphia, 19160. Secondary address: 321 W. Sedgley Ave., Philadelphia 19140. Phone: (215) 291-7532. Fax: (215) 739-1337. E-mail: rs@radiosalvation.com Web Site: www.radiosalvation.com. Licensee: Salvation Broadcasting Co. (acq 12-1-88). Population served: 400,000 Format: Sp, relg, div. Spec prog: Por 3 hrs wkly. ♦Sarrial Salva, pres; Mr. Sarrial Salva, gen mgr; Isabel Salva, sls dir; Juan Izquierdo, progmg dir; Juan Pydeck, chief of engrg.

Pittsburgh

KDKA(AM)— Nov 2, 1920: 1020 khz; 50 kw-U. TL: N40 33 33 W79 57 11. Stereo. Hrs opn: One Gateway Ctr., 15222. Phone: (412) 575-2200. Fax: (412) 575-2845. Web Site: www.kdkaradio.com. Licensee: Infinity Broadcasting East Inc. Group owner: Infinity Broadcasting Corp. Population served: 640,000 Format: News/talk. News: 9 hrs wkly. Target aud: 25-54.Michael J. Young, VP & gen mgr; Scott Schutt, gen sls mgr; Gleyn Ward, natl sls mgr; Jeff Hathhorn, mktg dir; Greg Jena, prom dir; Steve Hansen, progmg dir & news dir; Vic Pasquarelli, engrg dir; Mary Muldowney, traf mgr; Dan Wonders, local news ed; Scott Rimmel, local news ed; Barbara Boylan, news rptr & edit dir; Bob Kopler, news rptr; Kyle Anthony, news rptr; Lisa Alexander, news rptr; Mike Whitely, news rptr; Rose Douglas, news rptr; Bob Logue, disc jockey; Fred Honsberger, disc jockey; Larry Richot, disc jockey; Mike Pintek, disc jockey; Paul Alexander, disc jockey; Rob Pratte, disc jockey

KQV(AM)— Nov 19, 1919: 1410 khz; 5 kw-U, DA-2. TL: N40 31 17 W80 00 34. Hrs open: Centre City Tower, 650 Smithfield St., 15222. Phone: (412) 562-5900. Phone: (412) 562-5960. Fax: (412) 562-5936. Fax: (412) 563-5603. Web Site: www.kqv.com. Licensee: Calvary Inc. (acq 12-1-82; $1.75 million; FTR: 1-3-83). Population served: 1,250,000 Natl. Network: Wall Street, AP Radio. Wire Svc: PR Newswire Format: News. News staff: 18; News: 168 hrs wkly. Target aud: 35 plus; affluent, info-oriented adults. Spec prog: NFL football (regular season, playoffs & Superbowl), Notre Dame football, Duquesne University men's basketball. ♦Robert W. Dickey Sr., pres & gen mgr; Erik F. Selby, chief of opns.

WAMO(AM)—Millvale, Aug 1, 1948: 860 khz; 1 kw-D, 830 w-N, DA-2. TL: N40 29 27 W79 58 55. Hrs open: 24 960 Penn Ave., Suite 200, 15222. Phone: (412) 456-4064. Fax: (412) 391-3559. Web Site: www.wamo.com. Licensee: McL/McM Pennsylvania LLC. (acq 3-1-73). Population served: 40,200 Natl. Network: American Urban. Natl. Rep: McGavren Guild. Law Firm: Fletcher, Heald & Hildreth, P.L.C. Format: Smooth rhythm and blues, classic soul.Ronald Davenport Sr., chmn; Ronald Davenport Jr., pres; Michael L. Davenport, gen mgr; Kathy Gersha, opns VP; Mickey Baker, natl sls mgr; Jon Plesser, rgnl sls mgr & mus dir; Laura Varner-Norman, mktg mgr; Tammy Sadler, prom dir; George Cook, progmg dir; Ron Atkins, progmg dir; Kode Wred, mus dir; Tene Croom, news dir & pub affrs dir; Bob Sharkey, chief of engrg

WAMO-FM—Beaver Falls, 1960: 106.7 mhz; 37 kw. Ant 554 ft TL: N40 37 11 W80 05 36. Stereo. Hrs opn: 24 960 Penn Ave., Suite 200, 15222. Phone: (412) 456-4064. Fax: (412) 391-3559. E-mail: mdouglass@sbcol.com Web Site: www.wamo.com. Licensee: McL/McM Pennsylvania LLC. (group owner). Population served: 235,800 Natl. Network: American Urban. Natl. Rep: McGavren Guild. Law Firm: Fletcher, Heald & Hildreth. Format: Urban & hip hop. News: 2 hrs wkly. ♦Ronald Davenport Jr., pres; Michael Douglass, gen mgr; Kathy Gersna, opns VP; Mickey Baker, natl sls mgr; Jon Plesser, rgnl sls mgr; Laura Varner-Norman, rgnl sls mgr; Tammy Sadler, prom dir; Ron Atkins, progmg dir; Tene Croom, news dir; Bob Sharkey, chief of engrg.

WBGG(AM)— 1932: 970 khz; 5 kw-U, DA-2. TL: N40 30 30 W80 00 30. Hrs open: 24 200 Fleet St., 15220. Phone: (412) 937-1441. Fax: (412) 937-0323. E-mail: feedback@970theburgh.com Web Site: www.970theburgh.com. Licensee: AMFM Radio Licenses L.L.C. Group owner: Clear Channel Communications Inc. (acq 8-30-00; grpsl). Population served: 550,000 Format: Talk, Sports. News staff: one. ♦John Rohn, gen mgr; Mark Turley, gen sls mgr; Milanna Miljenodic, natl sls mgr; Allison Hilliard, prom mgr; Greg Gillispie, progmg dir.

WWSW-FM—Co-owned with WBGG(AM). 1940: 94.5 mhz; 50 kw. 810 ft TL: N40 27 48 W80 00 18. Stereo. Web Site: www.970theburgh.com. ♦Missy Gawaldo, gen sls mgr.

WDSY-FM— September 1962: 107.9 mhz; 17.5 kw. Ant 827 ft TL: N40 28 20 W79 59 41. Stereo. Hrs opn: Foster Five, 651 Holiday Dr., 15220. Phone: (412) 920-9400. Fax: (412) 920-9449. Web Site: www.y108.com. Licensee: Infinity Radio Holdings Inc. Group owner: Infinity Broadcasting Corp. (acq 12-14-00; grpsl). Population served: 2,019,400 Natl. Network: Westwood One. Natl. Rep: Katz Radio. Law Firm: Leventhal, Senter & Lerman, P.L.L.C. Format: Country. Target aud: 25-54; general. ♦Joel Hollander, pres; Jacques Tortoroli, CFO; Scott Herman, exec VP; Don Oyleaar, VP; Don Oylear, gen mgr; Keith Clark, opns VP & opns dir; Christine Fallon-McKenna, gen sls mgr & news dir; Keith Belden, natl sls mgr; Norm Slemanda, rgnl sls mgr & traf mgr; Michael Young, mktg mgr & chief of engrg; Jane O'Malia, prom dir; Stoney Richards, mus dir.

***WDUQ(FM)**— Dec 15, 1949: 90.5 mhz; 25 kw. 480 ft TL: N40 25 52 W80 00 26. Stereo. Hrs opn: 24 600 Forbes Ave., 15282-0001. Phone: (412) 396-6030. Fax: (412) 396-5061. E-mail: info@wduq.org Web Site: www.wduq.org. Licensee: Duquesne University. Natl. Network: NPR, PRI. Natl. Rep: Interep. Law Firm: Cohn & Marks. Format: Jazz, news, pub affairs, NPR. News staff: 5; News: 47 hrs wkly. Educated, moderately affluent. ♦Scott Hanley, gen mgr; Helen Wigger, opns dir; Fred Serrino, dev dir; Mary Lloyd, sls dir; Cynthia Ference-Kelly, mktg dir & prom dir; Shaunna Morrison, mus dir; Kevin Gavin, news dir; Chuck Leavens, engrg dir.

WDVE(FM)— May 10, 1962: 102.5 mhz; 55 kw. 820 ft TL: N40 29 38 W80 01 09. Stereo. Hrs opn: 200 Fleet St., 4th Fl., 15220. Phone: (412) 937-1441. Fax: (412) 937-0323. Web Site: www.dve.com. Licensee: Capstar TX L.P. Group owner: Clear Channel Communications Inc. (acq 8-00; grpsl). Population served: 2,000,000 Natl. Rep: Christal. Format: News, all talk. News staff: one. Target aud: 25-54. ♦Missy Gawaldo, gen sls mgr.

WEAE(AM)— May 1922: 1250 khz; 5 kw-U, DA-N. TL: N40 23 50 W79 57 43. Hrs open: 24 400 Ardmore Blvd., 15221. Phone: (412) 731-1250. Fax: (412) 244-4596. Fax: (412) 244-4409. Web Site: www.ESPNradio1250.com. Licensee: Sports Radio Group LLC. Group owner: ABC Inc. (acq 4-26-99; $5 million). Population served: 800,000 Natl. Network: ESPN Radio. Natl. Rep: ABC Radio Sales. Format: Sports, talk. Target aud: General; men 25-54. Spec prog: Penn State University football & basketball. ♦Jessamy Tang, pres; Dennis Begley, gen mgr; David Waugaman, natl sls mgr; Bryan Engel, prom mgr; John Lund, progmg dir; Joe DeStio, news dir; Thad Mazur, engrg mgr; Michelle Freeman, traf mgr.

WEDO(AM)—See McKeesport

WJAS(AM)— Oct 19, 1921: 1320 khz; 5 kw-U, DA-N. TL: N40 25 11 W79 54 38. (CP: 7 kw-D, 3.3 kw-N, DA-2. TL: N40 28 46 W79 54 12). Hrs opn: 900 Parish St., 15220. Phone: (412) 875-4800. Phone: (412) 875-9500. Fax: (412) 875-9570. Web Site: www.1320wjas.com. Licensee: Renda Broadcasting Corp. (group owner; acq 7-16-85; $700,000; FTR: 12-17-84). Population served: 1,821,000 Format: Big band, nostalgia, MOR. Target aud: 35 plus; older, upscale. Spec prog: Big band jump, Frank Sinatra 2 hrs wkly. ♦Anthony F. Renda, pres; Lawrence Weiss, gen mgr; David Pavlic, gen sls mgr; Chris Shovlin, prom dir; Mike McGann, progmg dir; Phil Lenz, chief of engrg.

WSHH(FM)—Co-owned with WJAS(AM). Mar 8, 1948: 99.7 mhz; 10.5 kw. 928 ft TL: N40 27 47 W80 00 17. Stereo. Fax: (412) 875-9474. Web Site: www.wshh.com. (Acq 11-83; $2.7 million; FTR: 11-14-83). Format: Soft adult contemp. News staff: one. Target aud: 25-54; white collar, upscale office workers, professionals, managers. Spec prog: Pub affrs one hr wkly. ♦Susan Kelly, gen sls mgr; Ron Antill, progmg dir; Allan Freed, pub affrs dir.

WKHB(AM)—Irwin, Oct 28, 1934: 620 khz; 5.5 kw-D, 50 w-N. TL: N40 17 20 W79 42 04. Stereo. Hrs opn: 24 1918 Lincoln Hwy., North Versailles, 15137. Phone: (412) 823-7000. Licensee: Broadcast Communications Inc. (group owner; (acq 10-9-96; $300,000). Population served: 2,481,152 Rgnl. Network: Radio Pa. Format: Var. Target aud: Adults. ♦Ashley R. Stevens, VP; Robert M. Stevens, pres & gen mgr; Barry Banker, stn mgr; Clark Ingram, opns mgr.

WKST-FM— Aug 8, 1960: 96.1 mhz; 44 kw. 522 ft TL: N40 23 49 W79 57 43. Stereo. Hrs opn: 200 Fleet St., 4th Fl., 15220. Phone: (412) 937-1441. Fax: (412) 937-0323. Web Site: www.kissfm961.com. Licensee: Capstar TX L.P. Group owner: Clear Channel Communications Inc. (acq 8-30-00; grpsl). Format: CHR. ♦Missy Gawaldo, gen sls mgr.

WLFP(AM)—See Braddock

WLTJ(FM)— Apr 4, 1942: 92.9 mhz; 47 kw. 890 ft TL: N40 29 38 W80 01 09. Hrs open: 24 650 Smith Field St., Suite 2200, 15222. Phone: (412) 316-3342. Fax: (412) 316-3388. E-mail: info@wltj.com Web Site: www.wltj.com. Licensee: WPNT Inc. (acq 4-84; $3 million; FTR: 3-19-84). Population served: 2,002,000 Rgnl. Network: Metronews Radio Net. Natl. Rep: McGavren Guild. Format: Lite rock. Target aud: 25-54; affluent, professional, working public. ♦Saul Frischling, pres; Greg Frischling, gen mgr; Chris Kohan, gen sls mgr; Vicki Wolfe, prom dir; Chuck Stevens, progmg dir; Amy Crago, news dir; Paul Carroll, chief of engrg.

WORD-FM— 1948: 101.5 mhz; 48 kw. 505 ft TL: N40 29 02 W79 59 34. Stereo. Hrs opn: 24 Seven Parkway Ctr., Suite 625, 15220. Phone: (412) 937-1500. Fax: (412) 937-1576. E-mail: word@wordfm.com Web Site: www.wordfm.com. Licensee: Pennsylvania Media Associates Inc. Group owner: Salem Communications Corp. Population served: 520,117 Format: Christian. Target aud: 25-49. ♦Chuck Gratner, CEO & gen mgr; Randy Dietterich, chmn & chief of engrg; Kenny Woods, opns mgr & progmg dir; Smitty Boros, gen sls mgr; Shaun Pierce, news dir; Lisa Cook, traf mgr.

WPIT(AM)—Co-owned with WORD-FM. 1947: 730 khz; 5 kw-D. TL: N40 29 02 W79 59 34. Web Site: www.wpitam.com. (Acq 12-2-92; $6.5 million; FTR: 12-21-92). Format: Christian.

WPGB(FM)— Feb 4, 1963: 104.7 mhz; 13 kw. Ant 827 ft TL: N40 28 20 W79 59 41. Stereo. Hrs opn: 24 200 Fleet, 4th Fl., 15220. Phone: (412) 937-1441. Fax: (412) 937-0323. Web Site: www.wpgb.com. Licensee: Capstar TX L.P. Group owner: Clear Channel Communications Inc. (acq 8-30-00; grpsl). Population served: 2,500,000 Natl. Network: Fox News Radio. Wire Svc: AP Format: News, all talk, sports. News staff: 0; News: 5 hrs wkly. Target aud: 25-54; white collar workers. ♦John Rohm, VP; Fred Traynor, gen sls mgr & chief of engrg; Jay Bohannon, progmg dir.

***WPTS-FM**— Aug 26, 1984: 92.1 mhz; 16 w. 462 ft TL: N40 26 39 W79 57 12. Stereo. Hrs opn: 24 Univ. of Pittsburgh, 411 William Pitt Union, 15260. Phone: (412) 648-7990. Fax: (412) 648-7988. E-mail: wpts@pitt.edu Web Site: www.wpts.pitt.edu. Licensee: University of

Pittsburgh. (acq 8-26-84). Population served: 30,000 Format: Eclectic, contemp, progsv. News staff: 3; News: 10 hrs wkly. Target aud: General. ♦Gregory Weston, gen mgr.

***WQED-FM**— Jan 25, 1973: 89.3 mhz; 43 kw. Ant 500 ft TL: N40 26 46 W79 57 51. Stereo. Hrs opn: 24 4802 5th Ave., 15213. Phone: (412) 622-1300. Fax: (412) 622-1488. E-mail: radio@wqed.org Web Site: www.wqed.org. Licensee: WQED Multimedia. Population served: 2,500,000 Natl. Network: NPR, PRI. Law Firm: Schwartz, Woods & Miller. Format: Class. Target aud: 35-64; educated, influential, professional, community leaders, mid to high income.George L. Miles Jr., pres; B.J. Leber, sr VP, stn mgr & prom VP; Michelle Pagano Heck, gen mgr; Ted Sohier, opns mgr & disc jockey; Lilli Mosco, dev VP & dev mgr; Rick Vaccarielli, sls dir; Karen Colbert, mktg dir; Gigi Saladna, prom mgr & progmg mgr; George Hazimanois, adv dir; Paul Byers, chief of engrg; Bob Walsh, disc jockey; Jim Cunningham, disc jockey; Jim Sweenie, disc jockey; Judi Cannava, disc jockey; Paul Johnston, disc jockey; Stephen Baum, disc jockey; Susan Johnson, disc jockey; Tony Marino, disc jockey; Warren Andrews, disc jockey . Co-owned TV: *WQED(TV) affil.

***WRCT(FM)**— April 1974: 88.3 mhz; 1.75 kw. 53 ft TL: N40 26 39 W79 56 37. Stereo. Hrs opn: 24 One WRCT Plaza, 5000 Forbes Ave., 15213. Phone: (412) 621-0728. Phone: (412) 621-9728. Fax: (412) 268-6549. E-mail: info@wrct.org Web Site: www.wrct.org. Licensee: Carnegie Mellon Student Government Corp. Population served: l,500,000 Law Firm: Putbrese, Hunsaker & Trent. Format: Div, educ. News: 10 hrs wkly. Target aud: General. Spec prog: Black 12 hrs, class 3 hrs, country 3 hrs, folk 3 hrs, experimental 12 hrs, jazz 18 hrs wkly. ♦Matt Siko, gen mgr; Pauline Law, progmg dir.

WRRK(FM)—See Braddock

WTZN-FM— July 19, 1948: 93.7 mhz; 41 kw. Ant 550 ft TL: N40 26 28 W80 01 32. Stereo. Hrs opn: 651 Holiday Dr., Suite 310, Foster Plaza Bldg 5, 15220. Phone: (412) 920-9400. Fax: (412) 920-9444. Web Site: www.937thezone.com. Licensee: CBS Radio Holdings Inc. Group owner: Infinity Broadcasting Corp. (acq 11-13-98; grpsl). Population served: 625,000 Format: Talk. Target aud: 25-49; male. ♦Joel Hollander, pres; Jacques Tortoroli, CFO; Scott Herman, exec VP; Michael Young, VP & gen mgr; Keith Clark, opns VP; Norm Slemenda, gen sls mgr; Brandon Davis, prom dir; Shelley Duffy, news dir & pub affrs dir; Chris Hudak, chief of engrg.

WWCS(AM)—See Canonsburg

WWNL(AM)— 1947: 1080 khz; 50 kw-D, DA. TL: N40 36 17 W79 57 37. Hrs open: 5316 Rt. 8 William Flynn Hwy., Unit 3N, Gibsonia, 15044. Phone: (724) 443-4844. Fax: (724) 443-4847. Licensee: Steel City Radio Inc. Group owner: Wilkins Communications Network Inc. (acq 6-14-2001). Population served: 3,200,000 Law Firm: Womble, Carlyle, Sandridge & Rice. Format: Christian teaching, talk. Target aud: 35 plus. ♦Bob Wilkins, pres; LuAnn Wilkins, exec VP; Mitchell Mathis, VP; B.J. Forsyth, gen mgr & stn mgr; Greg Garrett, opns mgr; Cliff Bryson, engr.

WXDX-FM— 1960: 105.9 mhz; 72 kw. 440 ft TL: N40 29 27 W79 58 55. Stereo. Hrs opn: 200 Fleet St., 15220. Phone: (412) 937-1441. Fax: (412) 937-0323. Web Site: www.wxdx.com. Licensee: Capstar TX L.P. Group owner: Clear Channel Communications Inc. (acq 8-30-00; grpsl). Population served: 2,000,000 Natl. Rep: Christal. Format: Alternative. Target aud: 18-34. ♦Missy Gawaldo, gen sls mgr.

***WYEP-FM**— Apr 30, 1974: 91.3 mhz; 18.2 kw. 265 ft TL: N40 24 42 W79 55 33. Stereo. Hrs open: 24 2313 E. Carson St., 15203. Phone: (412) 381-9131. Phone: (412) 381-9900. Fax: (412) 381-9126. E-mail: info@wyep.org Web Site: www.wyep.org. Licensee: Pittsburgh Community Broadcasting Corp. Population served: 2,500,000 Natl. Network: PRI, NPR. Format: AAA. Target aud: 25-49; socially, politically & culturally aware & active; well-educated. Spec prog: Folk 9 hrs, blues 7 hrs, bluegrass 4 hrs, soul 3 hrs, celtic 2 hrs wkly. ♦Blaine Lucas, chmn; Sean Sebastian, pres; Lee Ferraro, gen mgr; Tony Pirollo, sls dir; Rosemary Welsch, progmg dir; Joe Resch, disc jockey.

WZUM(AM)—See Carnegie

Pittston

WDMT(FM)— November 1983: 102.3 mhz; 5.8 kw. Ant 72 ft TL: N41 18 20 W75 45 38. Stereo. Hrs opn: 305 Hwy. 315, 18640. Phone: (570) 883-9850. Fax: (570) 883-9851. Web Site: www.102themountain.com. Licensee: Entercom Wilkes-Barre Scranton LLC. Group owner: Entercom Communications Corp. (acq 12-13-99; grpsl). Natl. Network: Jones Radio Networks. Natl. Rep: D & R Radio. Format: AAA. Target aud: 35-64; female. Spec prog: Philadelphia Eagles, Penn State football. ♦John Burkavage, gen mgr; Jim Rising, stn mgr; Andy Zapotek, gen sls mgr; Michael Ignatz, prom dir; Jerry Padden, progmg dir; Elizabeth Masich, mus dir; Lamar Smith, chief of engrg.

WITK(AM)— June 21, 1953: 1550 khz; 10 kw-D, 500 w-N, DA-2. TL: N41 20 45 W75 47 08. Hrs open: Box 851, 18640. Phone: (570) 207-6515. Fax: (877) 711-8500. Web Site: www.holyfamily.ws. Licensee: Robert C. Cordaro. Group owner: Citadel Communications Corp. Population served: 22000 Format: Catholic radio. ♦James N. Wright, gen mgr.

Plains

WYCK(AM)— 1923: 1340 khz; 810 w-U. TL: N41 15 01 W75 49 32. Hrs open: 24
Rebroadcasts WICK(AM) Scranton 98%.
1049 N. Sekol Rd., Scranton, 18504. Phone: (570) 344-1221. Phone: (570) 655-6660. Fax: (570) 344-0996. Fax: (570) 300-1996. Web Site: www.wick-am.com. Licensee: Bold Gold Media WBS L.P. (acq 3-13-2006; grpsl). Population served: 350,000 Law Firm: Smithwick & Belendiuk. Wire Svc: Metro Weather Service Inc. Format: Sports. News staff: one; News: 8 hrs progmg wkly. Target aud: Adults 35-64; adults who love original hits of top 40 era. Spec prog: Relg 3 hrs wkly, Polish 3 hrs wkly. ♦Ed Kerber, opns mgr; Phil Bullwinkel, gen sls mgr.

Pleasant Gap

WSGY(FM)— 1997: 98.7 mhz; 2.2 kw. Ant 551 ft TL: N40 55 58 W77 45 40. Hrs open: 2551 Park Center Blvd., State College, 16801. Phone: (814) 237-9800. Fax: (814) 237-2477. Web Site: www.froggyradio.com. Licensee: Forever Broadcasting L.L.C. Group owner: Forever Broadcasting (acq 1-27-99). Format: Contemp country. ♦Carol Logan, pres; Andrew Sumereau, gen mgr; Glen TurnerFerrara, opns mgr; Bobbi Castellucci, gen sls mgr; Donna Dunkel Himes, progmg dir; Kellie Green, news dir; Robert Taylor, chief of engrg.

Pocono Pines

WPZX(FM)— 2000: 105.9 mhz; 6 kw. Ant 328 ft TL: N41 05 06 W75 38 09. Hrs open:
Simulcast with WEZX(FM) Scranton.
149 Penn Ave., Scranton, 18503. Phone: (570) 346-6555. Fax: (570) 346-6038. Web Site: www.rock107.com. Licensee: The Scranton Times L.P. Group owner: Shamrock Communications Inc. (acq 9-8-00). Format: Classic rock. ♦William R. Lynett, CEO; Jim Loftus, COO & gen mgr; Tim Durkin, sls dir; Jenny Arndt, natl sls mgr; Mark Hoover, prom dir; Eric Logan, mus dir; Ruth Miller, news dir; Kevin Fitzgerald, chief of engrg; Krista Saar, traf mgr.

Port Allegany

WHKS(FM)— 1990: 94.9 mhz; 1.15 w. Ant 758 ft TL: N41 48 36 W78 23 10. Stereo. Hrs opn: 24 42 N. Main St., 16743. Secondary address: 59 Lent Hollow Rd., Coudersport 16915. Phone: (814) 642-7004. Phone: (814) 274-5368. Fax: (814) 642-9491. E-mail: whks@verizon.net Web Site: whksradio.com. Licensee: L-Com Inc. Population served: 30,000 Natl. Network: Jones Radio Networks, AP Radio. Rgnl. Network: Radio Pa. Natl. Rep: Dome. Rgnl rep: Commercial Media Sales. Format: Adult contemp. News: 2 hrs wkly. Target aud: 25-54; general. Spec prog: Relg 2 hrs wkly. ♦David F. Lent, pres, gen mgr, opns dir, opns mgr, progmg mgr, news dir & traf mgr; Joe Taylor, gen sls mgr & mktg mgr.

Port Matilda

WKVB(FM)— Oct 17, 1994: 107.9 mhz; 450 w. Ant 1,174 ft TL: N40 55 11 W77 58 28. Hrs open: 24 160 W. Clearview Ave., State College, 16803. Phone: (814) 238-5085. Fax: (814) 238-8993. Web Site: www.klove.com. Licensee: 2510 Licenses LLC. (group owner; (acq 2-16-2005; grpsl). Natl. Network: K-Love. Format: Contemp Christian. ♦Nick Ferrara, gen mgr.

Portage

WLKJ(FM)— Nov 15, 1990: 105.7 mhz; 3 kw. Ant 321 ft TL: N40 22 59 W78 39 31. Stereo. Hrs opn: 24 970 Tripoli St., Johnstown, 15902. Phone: (814) 534-8975. Fax: (814) 534-8979. Web Site: www.klove.com. Licensee: 2510 Licenses LLC. Group owner: Forever Broadcasting. (acq 5-1-2005; grpsl). Natl. Network: K-Love. Format: Contemp Christian. Target aud: 18-54; middle to upper income. ♦Nick Ferrara, opns mgr.

Pottstown

WPAZ(AM)— Oct 1, 1951: 1370 khz; 1 kw-D. TL: N40 16 35 W75 37 44. Hrs open: 6 AM-7 PM 224 Maugers Mill Rd., 19464. Phone: (610) 326-4000. Phone: (610) 326-6832. Fax: (610) 326-7984. Web Site: www.1370wpaz.com. Licensee: Faye Scott. Group owner: Great Scott Broadcasting Population served: 125,000 Format: News/talk. News staff: 2; News: 8 hrs wkly. Target aud: 25 plus; most affluent people. Spec prog: Pol one hr, relg 12 hrs wkly. ♦Faye Scott, pres; Mike LiCata, gen mgr & sls VP; Jay Warren, progmg dir; Paul Fanelli, news dir; Terry Dalton, chief of engrg.

WRFY-FM—See Reading

Pottsville

WAVT-FM—Listing follows WPPA(AM).

WPAM(AM)— 1946: Stn currently dark. 1450 khz; 1 kw-U. TL: N40 41 27 W76 11 39. Hrs open: 24 145 Lawtons Hill, 17901. Secondary address: PO Box 732 17901-0732. Phone: (570) 622-1450. Fax: (570) 622-4690. E-mail: bob@phoenix1450.com Web Site: www.phoenix1450.com. Licensee: Curran Communications Inc. (acq 1-76). Population served: 155,000 Natl. Network: Jones Radio Networks. Rgnl. Network: Radio Pa. Law Firm: Blair, Joyce & Silva. Format: Classic rock. Target aud: 25-54; active, upwardly mobile adults. Spec prog: Gospel, Talk 6 hrs wkly. ♦Robert Murray, gen mgr.

WPPA(AM)— May 9, 1946: 1360 khz; 5 kw-D, 500 w-N, DA-2. TL: N40 41 56 W76 11 43. Hrs open: 24 Box 540, 212 S. Centre St., 17901. Phone: (570) 622-1360. Fax: (570) 622-2822. Web Site: www.wpparadio.com. Licensee: Pottsville Broadcasting Co. Inc. Population served: 155,000 Natl. Network: CBS. Format: Adult contemp. News staff: 2; News: 14 hrs wkly. Target aud: 25-54. ♦Argie D. Tidmore, pres, gen mgr & chief of engrg; Les Blankenhorn, opns mgr & disc jockey; William Tidmore, gen sls mgr & adv mgr; Al Kovy, progmg dir & progmg mgr; Jay Levan, news dir; Deb Daugherty, pub affrs dir, women's int ed & disc jockey; Dave Michaels, disc jockey.

WAVT-FM—Co-owned with WPPA(AM). Nov 20, 1948: 101.9 mhz; 50 kw. 540 ft TL: N40 49 50 W76 12 32. Stereo. 24 Web Site: www.t102radio.com.775,000 Format: CHR. ♦James A. Bowman, stn mgr, gen sls mgr & adv mgr; Chad Gerber, progmg dir & disc jockey; Deb Dougherty, women's int ed; Courtney Roberts, disc jockey; Dave Smith, disc jockey; Travis Sparks, disc jockey.

Punxsutawney

WECZ(AM)— Mar 18, 1953: 1540 khz; 5 kw-D, 1 kw-CH. TL: N40 57 36 W79 00 08. Hrs open: 12 904 N. Main St., 15767. Phone: (814) 938-6000. Fax: (814) 938-4237. Licensee: Renda Radio Inc. (group owner; acq 6-1-81; $512,000; FTR: 5-11-81). Population served: 48,000 Natl. Network: Westwood One. Law Firm: Latham & Watkins. Format: News, talk. News staff: 2; News: 10 hrs wkly. Target aud: 45 plus. Spec prog: Pol 3 hrs wkly. ♦Anthony F. Renda, pres; Doug Metherey, gen mgr & stn mgr; Jennifer Black, gen sls mgr; Marty Palmer, engrg dir; Jim Costanzo, disc jockey.

WPXZ-FM—Co-owned with WECZ(AM). Dec 12, 1973: 104.1 mhz; 3 kw. 300 ft TL: N40 57 36 W79 00 08. Stereo. 24 E-mail: rendaradio@adelphia.net 60,000 Natl. Network: ABC. Format: Adult contemp. Target aud: 35-64. ♦Larry McGuire, mus dir.

Radnor Township

***WYBF(FM)—** August 1991: 89.1 mhz; 700 w. 223 ft TL: N40 03 22 W75 22 30. Hrs open: 7 AM-2 AM (M, W, F); noon-2 AM (Su) Rebroadcasts WXVU(FM) Villanova.
Widener Ctr., 610 King of Prussia Rd., Radnor, 19087-3698. Phone: (610) 902-8457. Fax: (610) 902-8285. Licensee: Cabrini College. Format: Div, contemp hit, news/talk. ♦Dr. Jerry Zurek, chmn; Krista Mazzeo, gen mgr.

Reading

WEEU(AM)— 1931: 830 khz; 20 kw-D, 6 kw-N, DA-2. TL: N40 30 54 W76 07 24. Hrs open: 24 34 N. 4th St., 19601-3996. Phone: (610) 376-7335. Fax: (610) 376-7756. Web Site: www.weeu.com. Licensee: WEEU Broadcasting Co. (acq 12-46). Population served: 340,000 Natl. Rep: McGavren Guild. Law Firm: Cohn & Marks. Format: Full service, news/talk, sports. News staff: 2; News: 6 hrs wkly. Target aud: 30 plus; mature. Spec prog: Folk 3 hrs, Ger 2 hrs wkly. ♦Dave Kline, gen mgr & stn mgr.

WFKB(FM)—See Boyertown

WIOV(AM)— Sept 1, 1946: 1240 khz; 1 kw-U. TL: N40 19 28 W75 56 31. Hrs open: 44 Bethany Rd., Ephrata, 17522-2416. Phone: (717) 738-1191. Fax: (717) 738-1661. E-mail: widv@ptd.net Web Site: www.espn1240.com. Licensee: Citadel Broadcasting Co. Group owner: Citadel Broadcasting Corp. (acq 5-12-2004; grpsl). Population served: 315,000 Natl. Network: ESPN Radio. Format: Sports. Target aud: 35-54; Male. ♦Mitch Carroll, VP, gen mgr, gen sls mgr & natl sls mgr; Jim Rudley, opns mgr; Brenda Perkins, natl sls mgr; Crissy Wall, rgnl sls mgr & adv mgr; CJ Taylor, prom dir; Susie Summer, prom mgr; Bob Moody, progmg VP; Dick Raymond, progmg dir; Steve Haage, progmg mgr; Shanon Robinson, traf mgr.

WRAW(AM)— September 1922: 1340 khz; 1 kw-U. TL: N40 19 27 W75 55 10. Stereo. Hrs opn: 24 1265 Perkiomen Ave., 19602. Phone: (610) 376-7173. Phone: (610) 376-6671. Fax: (610) 376-1270. Web Site: www.1340praiseradio.com. Licensee: Clear Channel Radio Licenses Inc. Group owner: Clear Channel Communications Inc. (acq 1996; grpsl). Population served: 49,600 Law Firm: Arent, Fox, Kintner, Plotkin & Kahn. Format: Contemp Christian. ♦John Rizzuto, gen mgr; Brian Check, opns mgr; Nick Harris, prom dir; Al Burke, progmg dir; Steve McKenzie, chief of engrg.

WRFY-FM—Co-owned with WRAW(AM). Sept 23, 1962: 102.5 mhz; 19 kw. 807 ft TL: N40 19 19 W75 53 41. Stereo. Web Site: www.y102.com.250,000 Format: CHR. Target aud: 18-49. ♦Al Burke, progmg mgr.

***WXAC(FM)—** 1967: 91.3 mhz; 200 w. 33 ft TL: N40 22 08 W75 54 37. Stereo. Hrs opn: 1621 N. 13th St., 19612-5234. Phone: (610) 921-7545. Fax: (610) 921-7685. E-mail: wxac@albright.edu Licensee: Albright College. Population served: 300,000 Format: Progsv, Jazz, AOR, Sp. Target aud: General; Albright college community & Reading area. ♦Mindy Cohen, stn mgr.

Red Lion

WGLD(AM)— Oct 22, 1950: 1440 khz; 1 kw-D, 56 w-N. TL: N39 54 17 W76 34 49. Hrs open: 140 E. Market St., York, 17401. Phone: (717) 852-2305. Fax: (717) 246-1717. Licensee: Susquehanna License Co. LLC. (acq 5-11-2005; $280,000). Population served: 5,645 Format: Relg. ♦David E. Kennedy, pres; Bill Fenton, gen mgr; John Peeling, stn mgr; Lisa Dissinger, opns mgr.

WSOX(FM)— October 1960: 96.1 mhz; 13.5 kw. Ant 951 ft TL: N39 54 16 W76 34 48. Stereo. Hrs opn: Box 910, York, 17402. Phone: (717) 764-1155. Fax: (717) 755-3714. Web Site: www.oldies961.com. Licensee: Susquehanna License Co. LLC. Group owner: Susquehanna Radio Corp. (acq 8-1-2003; $23 million). Population served: 5,645 Format: Oldies. ♦Tom Ranker, VP & gen mgr; Dave Anthony, progmg dir; Scott Steffan, chief of engrg.

Renovo

WZYY(FM)— Sept 19, 1996: 106.9 mhz; 800 w. Ant 876 ft TL: N41 14 15 W77 45 02. Hrs open: Box 436, State College, 16804. Phone: (570) 923-9106. Fax: (570) 923-3291. Licensee: Magnum Broadcasting Inc. (acq 7-21-2004; $200,000). Format: Hot adult contemp. ♦Michael M. Stapleford, pres; Diana Albright, gen mgr.

Reynoldsville

WDSN(FM)— Feb 14, 1990: 106.5 mhz; 6 kw. 328 ft TL: N41 08 41 W78 52 41. Stereo. Hrs opn: 24 12 W. Long Ave., Du Bois, 15801-2100. Phone: (814) 375-5260. Fax: (814) 375-5262. Web Site: www.sunny1065.fm. Licensee: Priority Communications. (acq 11-6-90; $275,000; FTR: 11-26-90). Population served: 115,000 Natl. Network: Jones Radio Networks, Fox News Radio. Rgnl rep: Commerical Media Sales Law Firm: Womble, Carlyle, Sandridge & Rice, LLP. Wire Svc: AP Format: Adult contemp. News staff: one; News: 18 hrs wkly. Target aud: 25-54. ♦Jay M. Philippone, pres & gen mgr; Lori Lewis, stn mgr & opns mgr; Beth Walters, opns dir; Lindsay Schoening, news dir; Al Lockwood, pub affrs dir; Polly Slie, traf mgr.

Ridgebury

WREQ(FM)— 1991: 96.9 mhz; 3.6 kw. Ant 430 ft TL: N41 55 43 W76 46 58. Hrs open: 24 111 N. Main St., Elmira, NY, 14901. Phone: (607) 732-2484. Fax: (607) 732-8704. E-mail: wreq@csnradio.com Web Site: q969online.com. Licensee: CSN International. (group owner; (acq 6-14-2001; $300,000). Law Firm: Reddy, Begley & McCormick. Format: Edu, contemp Christian mus, modern praise & worship. News: 4 hrs wkly. Target aud: 25-44; women with families (small children), heads of households. ♦Chuck Smith, pres; Lorenzo Galletti, gen mgr & stn mgr; Mike Stocklin, gen mgr; Doug Wakil, progmg mgr & mus dir; Gina Galletti, mus dir; Nora Cooper, asst music dir.

Ridgway

WKBI(AM)—See Saint Marys

WKBI-FM—See Saint Marys

Riverside

WLGL(FM)— Oct 25, 1990: 92.3 mhz; 440 w. 833 ft TL: N40 57 30 W76 42 53. Stereo. Hrs opn: 24
Rebroadcasts WYGL(AM) Selinsgrove.
Box 90, Rt. 204, State School Rd, Selinsgrove, 17870. Phone: (570) 374-8819. Fax: (570) 374-7444. Licensee: MMP License LLC. Group owner: MAX Media L.L.C. (acq 10-17-03; grpsl). Population served: 150,000 Natl. Network: Jones Radio Networks, CNN Radio. Rgnl rep: Dome & Associates. Law Firm: Kaye, Scholer, Fierman, Hays & Handler. Format: Contemp hot country. News staff: one; News: 5 hrs wkly. Target aud: 25-54. ♦Scott Richards, gen mgr; Dawn Marie, opns dir & news dir; Greg Adair, gen sls mgr & mktg dir; Shelly Marx, prom dir & mus dir; Ted Koppen, chief of engrg.

Roaring Spring

WKMC(AM)— May 1, 1955: 1370 khz; 5 kw-D, 38 w-N, DA-2. TL: N40 19 26 W78 23 40. Stereo. Hrs opn: 1345 S. Main St., 16673. Phone: (814) 224-7501. Fax: (814) 224-7504. E-mail: wkmc@cove.net Web Site: www.wkmcam.com. Licensee: Handsome Brothers Inc. Group owner: Allegheny Mountain Network Stations. (acq 5-1-2005; $80,000). Population served: 140,000 Format: Adult standard. Target aud: 45 plus; mature, loyal listeners. ♦David Barger, pres; Mike Martin, gen mgr, mktg dir, prom dir, progmg dir & news dir; Robert Lynn, chief of engrg.

WRKY-FM—See Hollidaysburg

Russell

WQFX-FM— Nov 11, 1984: 103.1 mhz; 2.5 kw. 351 ft TL: N41 57 48 W79 09 42. Stereo. Hrs opn: Box 1199, Jamestown, NY, 14702-1199. Phone: (716) 664-2313. Fax: (716) 488-1471. E-mail: jadmin@wksn.com Licensee: Media One Group II LLC. (acq 5-31-2005; grpsl). Population served: 140000 Natl. Rep: Dome. Law Firm: Fisher, Wayland, Cooper, Leader & Zaragoza. Format: Classic rock. Target aud: 25-54. ♦Marcus Maloney, gen mgr.

Saegertown

WUZZ(FM)— Jan 19, 1979: 94.3 mhz; 3 kw. 298 ft TL: N41 42 23 W80 10 09. Stereo. Hrs opn: Box 397, Meadville, 16335. Phone: (814) 724-1111. Fax: (814) 333-9628. Licensee: Forever Broadcasting LLC. Group owner: Forever Broadcasting (acq 7-20-2000; grpsl). Format: Classic hits. ♦Terry Deitz, gen mgr.

Saint Marys

WDDH(FM)— Apr 22, 1986: 97.5 mhz; 23 kw. 705 ft TL: N41 37 04 W78 48 14. (CP: 19.5 kw, ant 800 ft.). Stereo. Hrs opn: 14902 Bootjack Rd., Ridgway, 15853. Phone: (814) 772-9700. Fax: (814) 772-9750. E-mail: brm@houndcountry.com Web Site: www.houndcountry.com. Licensee: Intrepid Broadcasting Inc. (acq 3-25-2004; $1.25 million). Population served: 620,000 Natl. Network: Westwood One, Jones Radio Networks, ABC. Natl. Rep: Rgnl Reps. Rgnl rep: Dome & AssociatesCommercial Media Sales Wire Svc: AP Format: Country. Target aud: 25-54. ♦Michael Stapleford, CEO; Bryan Mallette, gen mgr; Peter Butler, rgnl sls mgr; Tracie Cockrell, progmg dir.

WKBI(AM)— July 23, 1950: 1400 khz; 1 kw-U. TL: N41 24 56 W78 33 56. Hrs open: 24 Box 466, 137 Melody Rd., 15857. Phone: (814) 834-2821. Phone: (814) 834-2822. Fax: (814) 834-4319. E-mail: b94@wkbi.net Web Site: www.wkbi.net. Licensee: Elk-Cameron Broadcasting Co. Group owner: Allegheny Mountain Network Stations Population served: 53,200 Natl. Network: Westwood One. Rgnl. Network: Allegheny Mtn. Net. Natl. Rep: Dome. Law Firm: Borsari & Paxson. Format: Adult contemp, oldies, sports. News staff: one; News: 10 hrs wkly. Target aud: 35-55. ♦Cary H. Simpson, pres; Ted Simpson, gen mgr; Erik Lane, opns dir, pub affrs dir & disc jockey; Nancy Bowser, prom mgr; Phil Leslie, news dir; Robert Lynn, chief of engrg; Chris O'Donnell, disc jockey; Jason Lang, disc jockey; Nancy Kelly, disc jockey; Val Joseph, disc jockey.

WKBI-FM— August 1966: 93.9 mhz; 2.35 kw. 800 ft TL: N41 23 11 W78 41 32. Stereo. 24 Web Site: www.wkbi.net.6,022 Natl. Network: Westwood One, Jones Radio Networks. Target aud: General; young adults. Spec prog: Relg 2 hrs wkly. ♦Chris O'Donnell, disc jockey; Erik Lane, disc jockey; Jason Lang, disc jockey; Nancy Kelly, disc jockey; Val Joseph, disc jockey.

Salladasburg

WBYL(FM)— 1989: 95.5 mhz; 3.9 kw. 239 ft TL: N41 14 00 W77 12 09. Stereo. Hrs opn: 1559 W. 4th St., Williamsport, 17701. Phone: (570) 327-1400. Fax: (570) 327-8156. Web Site: www.billcountry.com. Licensee: Clear Channel Radio License Inc. Group owner: Clear Channel Communications Inc. (acq 8-5-98; grpsl). Population served: 250,000 Format: Country. News staff: one; News: 4 hrs wkly. Target aud: 35 plus. ♦James Dabney, gen mgr; Joe Daniels, gen sls mgr; Gary Chrisman, prom dir; Ken Sawyer, opns dir & progmg dir; Kathy Thomas, news dir; Jerry Maiolo, engrg dir.

Sayre

WATS(AM)— June 1950: 960 khz; 5 kw-D. TL: N41 59 48 W76 30 03. Hrs open:
Rebroadcasts WAVR-FM Waverly, NY 100%.
204 Desmond St., 18840. Phone: (570) 888-7745. Fax: (570) 888-9005. E-mail: wats.wavr@cqservices.com Licensee: WATS Broadcasting Inc. (acq 10-17-86). Population served: 100,000 Format: Adult contemp. Target aud: 25-54. Spec prog: Farm one hr wkly. ♦Charles C. Carver Jr., VP & gen mgr.

WAVR(FM)—See Waverly, NY

Schnecksville

***WXLV(FM)—** Sept 23, 1983: 90.3 mhz; 670 w. 177 ft TL: N40 39 52 W75 36 40. (CP: 420 w, ant 230 ft.). Stereo. Hrs opn: 24 4525 Education Park Dr., 18078. Phone: (610) 799-4141. Fax: (610) 799-1571. E-mail: wxlv@hotmail.com Web Site: www.wxlvfm.com. Licensee: Lehigh Carbon Community College. Population served: 800,000 Format: Div. Target aud: General. Spec prog: Class 13 hrs, jazz 6 hrs, Pol 6 hrs, blues 4 hrs, new wave 6 hrs wkly. ♦Tony Peiffer, gen mgr & chief of opns.

Scottdale

WLSW(FM)— Dec 21, 1971: 103.9 mhz; 325 w. 780 ft TL: N40 00 51 W79 31 01. Stereo. Hrs opn: 24 Box 763, Connellsville, 15425. Phone: (724) 628-2800. Fax: (724) 628-7380. Web Site: www.wlsw.com. Licensee: Wall Broadcasting. Population served: 70,000 Natl. Network: Westwood One. Format: Hot adult contemp, oldies. Target aud: 25-54; general. ♦L. Stanley Wall, pres; Chris Molton, gen mgr, gen sls mgr & chief of engrg; Debbie Larson, progmg dir & disc jockey; Connie LaPorte, traf mgr; Charlie Apple, disc jockey; Jamie Allen, disc jockey; Jeff Allen, disc jockey; Jerry Braveman, disc jockey.

Scranton

WARM(AM)— 1940: 590 khz; 5 kw-U, DA-2. TL: N41 28 44 W75 52 51. Hrs open: 24 600 Baltimore Dr., Wilkes Barre, 18702. Phone: (570) 824-9000. Fax: (570) 820-0520. E-mail: phil.galasso@citcomm.com Licensee: Citadel Broadcasting Co. Group owner: Citadel Broadcasting Corp. (acq 7-1-97; grpsl). Population served: 1,064,000 Natl. Network: ABC. Format: Oldies. News staff: 1; News: 1 hrs wkly. Target aud: 35 plus. Spec prog: Yankees baseball, Polka 4 hrs wkly. ♦John Crawford, sls dir; Bill Palmeri, mktg mgr; Erin Evans, prom mgr; Brian Hughes, news dir; Phil Galasso, progmg dir & chief of engrg; Lori Law, traf mgr.

WBAX(AM)—See Wilkes-Barre

WBZU(AM)— Jan 12, 1925: 910 khz; 1 kw-D, 500 w-N. TL: N41 22 56 W75 41 51. Hrs open: 24
Rebroadcasts WILK(AM) Wilkes Barre 100%.
305 Hwy. 315, Pittston, 18640. Phone: (570) 883-9850. Fax: (570) 883-0832. Web Site: www.wilknewsradio.com. Licensee: Entercom Scranton Wilkes-Barre License LLC. Group owner: Entercom Communications Corp. (acq 12-16-99; grpsl). Population served: 103,564 Natl. Rep: D & R Radio. Wire Svc: ABC Wire Svc: AP Wire Svc: Metro Weather Service Inc. Format: News/talk. News staff: 6; News: 25 hrs wkly. Target aud: 25-54; affluent, educated. Spec prog: Relg one hr wkly. ♦John Burkavage, gen mgr; Jim Rising, opns dir; Andy Zapotell, gen sls mgr; Bob DeMono, natl sls mgr; Liz Masich, prom mgr; Nancy Kman, progmg dir; Joe Thomas, news dir & sports cmtr; Lamar Smith, chief of engrg; Shannon Ball, traf mgr; Tom Ragan, news rptr; Shadol Steel, mus critic.

WGGY(FM)—Co-owned with WBZU(AM). Dec 25, 1948: 101.3 mhz; 7 kw. 1,110 ft TL: N41 25 38 W75 44 53. Stereo. 305 Hwy. 315, Pittston, 18640. Phone: (570) 883-1111. Fax: (570) 883-1360. Web Site: www.froggy101.com. Natl. Network: CBS. Format: Country.John Burkavage, VP & gen mgr; Jim Rising, opns mgr; Andy Zapotek, gen sls mgr; Bob De Mono, natl sls mgr; Elizabeth Masieh, mktg dir; Cheryl Willis, prom mgr & reporter; Mike Krinik, progmg dir; Jaymie Gordon, mus dir & news rptr; Kelly Green, mus dir; Laman Smith, chief of engrg; Shannon Ball, traf mgr; Shadol Steele, mus critic; Joe Thomas, sports cmtr

WEJL(AM)— Nov 29, 1922: 630 khz; 500 w-D, 32 w-N. TL: N41 24 35 W75 40 41. Hrs open:
Rebroadcasts WBAX(AM) Wilkes Barre 100%.
149 Penn Ave., 18503. Phone: (570) 346-6555. Fax: (570) 346-6038. Web Site: www.wejl-wbax.com. Licensee: The Scranton Times LP. Group owner: Shamrock Communications Inc. (acq 1922). Population served: 103,564 Format: Sports. ♦William R. Lynett, CEO; Jim Loftus, COO & gen mgr; Tim Durkin, sls dir; Jenny Arndt, natl sls mgr; Mark Hoover, mktg dir; Michael Neff, progmg dir; Ruth Miller, news dir; Kevin Fitzgerald, chief of engrg; Krista Saar, traf mgr.

WEZX(FM)—Co-owned with WEJL(AM). Nov 1, 1967: 106.9 mhz; 1.45 kw. 617 ft TL: N41 20 52 W75 39 03. Stereo. Fax: (570) 346-6038. (Acq 1967).103,564 Format: Classic rock. ♦Mark Hoover, prom mgr; Kevin Fritzgerald, engrg dir; Krista Saar, traf mgr.

WICK(AM)— Apr 17, 1954: 1400 khz; 1 kw-U. TL: N41 25 05 W75 39 43. Stereo. Hrs opn: 24 1049 N. Sekol Rd., 18504. Phone: (570) 344-1221. Fax: (570) 344-0996. Licensee: Bold Gold Media WBS L.P. (acq 3-13-2006; grpsl). Population served: 550,000 Law Firm: Smithwick & Belendiuk. Format: Sports. News staff: 2; News: 8 hrs wkly. Target aud: 35-64; adults who love original hits of Top 40 Era. Spec prog: Relg 3 hrs, Pol 3 hrs wkly. ♦Ed Kerber, opns mgr; Phil Bullwinkel, gen sls mgr.

WWRR(FM)—Co-owned with WICK(AM). Nov 26, 1964: 104.9 mhz; 270 w. Ant 1,092 ft TL: N41 26 06 W75 43 35. Stereo. 24 Phone: (570) 344-1221. Fax: (570) 344-0996.650,000 Format: Adult contemp. News staff: one; News: 8 hrs wkly. Target aud: 25-54; men & women. ♦Ed Kerber, prom dir, disc jockey & disc jockey; Bill Stewart, disc jockey; Don Murley, disc jockey; Doug Lane, disc jockey; Jim Riley, disc jockey; Scott Young, disc jockey.

WILK(AM)—See Wilkes-Barre

WITK(AM)—See Pittston

WKRZ(FM)—See Wilkes-Barre

WMGS(FM)—See Wilkes-Barre

***WUSR(FM)**— Feb 27, 1993: 99.5 mhz; 300 w. 1,014 ft TL: N41 26 09 W75 43 33. Stereo. Hrs opn: 11 AM-2 AM Univ. of Scranton, St. Thomas Hall, 800 Linden St., 18510. Phone: (570) 941-7648. Fax: (570) 941-4628. E-mail: wusr@scranton.edu Web Site: www.scranton.edu/wusr. Licensee: University of Scranton. Population served: 368,664 Wire Svc: Metro Weather Service Inc. Format: Alternative, jazz, rock/AOR, blues. Target aud: General. Spec prog: Class 5 hrs, relg 4 hrs, loud rock 8 hrs, urban contemp 6 hrs, Latin 10 hrs wkly. ♦Ken Sandrowicz, gen mgr; Margo Christiansen, stn mgr.

***WVIA-FM**— Apr 23, 1973: 89.9 mhz; 5 kw. 1,250 ft TL: N41 10 55 W75 52 17. Stereo. Hrs opn: 24 100 Wvia Way, Pittston, 18640-6197. Phone: (570) 655-2808. Phone: (570) 826-6144. Fax: (570) 655-1180. E-mail: webadmin@wvia.org Web Site: www.wvia.org. Licensee: N.E. Pa. Educational TV Association. Population served: 1,500,000 Natl. Network: NPR, PRI. Rgnl. Network: Pa. Pub. Format: Class, jazz, news. News: 31 hrs wkly. ♦A. William Kelly, CEO & pres; Chris Norton, VP; George Graham, mus dir; Joseph Glynn, engrg VP. Co-owned TV: *WVIA-TV affil.

***WVMW-FM**— September 1974: 91.7 mhz; 2 kw. Ant -285 ft TL: N41 25 57 W75 38 06. Stereo. Hrs opn: 14 Marywood University, 2300 Adams Ave., 18509. Phone: (570) 348-6202. Fax: (570) 961-4769. E-mail: mengoni@marywood.edu Web Site: www.vmfm917.com. Licensee: Marywood College. Population served: 40,000 Format: Alternative. News staff: 2; News: 7 hrs wkly. Target aud: 15-25; young adults. Spec prog: Black 2 hrs, class 7 hrs, jazz 10 hrs wkly. ♦Earnest Mengoni, stn mgr; George Graham, chief of engrg.

WYCK(AM)—See Plains

Selinsgrove

***WQSU(FM)**— September 1967: 88.9 mhz; 12 kw. 620 ft TL: N40 57 06 W75 45 03. Stereo. Hrs opn: 24 Susquehanna Univ., 514 University Ave., 17870. Phone: (570) 372-4030. Fax: (570) 372-2757. E-mail: augustin@susqu.edu Web Site: www.wqsu.com. Licensee: Susquehanna University. Population served: 225,000 Natl. Network: AP Radio. Wire Svc: AP Format: Modern rock. News: 7 hrs wkly. Target aud: 18-34. Spec prog: Classic country 6 hrs, sports 4 hrs, bluegrass 7 hrs wkly. ♦Larry D. Augustine, gen mgr; Harry Bingaman, chief of engrg; Patricia Wendt, traf mgr.

WYGL(AM)— Jan 16, 1967: 1240 khz; 1 kw-U. TL: N40 48 59 W76 52 13. Hrs open: 24 Box 90, Rt. 204 & State School Rd., 17870. Phone: (570) 374-1155. Phone: (570) 374-8819. Fax: (570) 374-7444. Web Site: www.bigcountrynow.com. Licensee: MMP License LLC. Group owner: MAX Media L.L.C. (acq 10-17-03; grpsl). Population served: 150,000 Natl. Network: USA. Law Firm: Kaye, Scholer, Fierman, Hays & Handler L.L.P. Format: Contemp hot country. News staff: one; News: 8 hrs wkly. Target aud: 18 plus. ♦John A. Trinder, pres; Scott Richards, VP & gen mgr; Dawn Marie, opns dir; Greg Adair, gen sls mgr; Shelly Marks, prom dir & asst music dir; Lisa Richards, mus dir; Nat O'Brien, news dir; Ted Koppen, chief of engrg.

Sellersville

***WBYO(FM)**— March 1991: 88.9 mhz; 900 w. Ant 436 ft TL: N40 23 02 W75 21 02. Stereo. Hrs opn: 24 Box 186, 18960. Phone: (215) 721-2141. Fax: (215) 721-9811. E-mail: wordfm@wordfm.org Web Site: www.wordfm.org. Licensee: Four Rivers Community Broadcasting Corp. Population served: 300,000 Law Firm: Schwartz, Woods & Miller. Format: Adult contemp, Christian, religious. News staff: one; News: 10 hrs wkly. Target aud: General. Spec prog: Country gospel 2 hrs, gospel bluegrass 2 hrs wkly. ♦Charles W. Loughery, pres, gen mgr, engrg mgr & chief of engrg; Nancy K. Loughery, CFO; David Baker, VP & stn mgr; Kristine McClain, progmg mgr.

Shamokin

WBLJ-FM— 1968: 95.3 mhz; 1.25 w. 505 ft TL: N40 45 36 W76 32 19. Stereo. Hrs opn:
Rebroadcasts WBYL(FM) Salladasburg 100%.
Box 3638, Williamsport, 17701-3638. Phone: (570) 327-1400. Fax: (570) 327-8156. E-mail: bill@billcountry.com Web Site: www.billcountry.com. Licensee: Clear Channel Broadcasting Licenses Inc. Group owner: Clear Channel Communications Inc. (acq 10-4-01; $800,000 with co-located AM). Format: Country. ♦Jim Dabney, gen mgr; Joe Daniels, gen sls mgr; Gary Chrisman, prom dir; Ken Sawyer, progmg dir; Kathy Thomas, news dir.

WISL(AM)— January 1948: 1480 khz; 1 kw-U, DA-N. TL: N40 45 53 W76 31 18. Hrs open: 24 550 California Rd., Unit 11, Quakertown, 18951. Phone: (215) 536-6648. Fax: (215) 536-8296. Licensee: Basic Licensing Inc. (acq 11-18-2002). Population served: 175,000 Format: MOR. ♦David C. Gorman, pres, CFO & gen mgr; Kurt Gorman, exec VP.

Sharon

WPIC(AM)— Oct 25, 1938: 790 khz; 1.3 kw-D, 58 w-N. TL: N41 13 10 W80 28 25. Hrs open: 24 2030 Pine Hollow Blvd., Hermitage, 16148. Phone: (330) 783-1000. Fax: (330) 783-0060. Web Site: www.cumulus.com. Licensee: Cumulus Licensing Corp. Group owner: Cumulus Media Inc. (acq 3-15-00; grpsl). Population served: 22,653 Natl. Network: ABC, Jones Radio Networks, Talk Radio Network, Westwood One. Format: News/talk. News staff: one. Target aud: 35 plus. Spec prog: Pol 3 hrs, It 2 hrs, relg 2 hrs, infomercials 12 hrs wkly. ♦Clyde Bass, gen mgr; Bob Greenburg, gen sls mgr & natl sls mgr; Bob Popa, progmg dir; Joe Biro, news dir; Wes Boyd, chief of engrg.

WYFM(FM)—Co-owned with WPIC(AM). Oct 25, 1947: 102.9 mhz; 44 kw. Ant 455 ft TL: N41 13 10 W80 28 25. Stereo. 4040 Simon Rd., Youngstown, 44512. Web Site: www.cumulus.com.445,000 Natl. Network: Westwood One. Format: Classic hits, classic rock. Target aud: 25-54. ♦Joiv Jacubec, sls dir; Scott Kennedy, progmg dir; Lynn Davis, news dir; Dave Messersmith, disc jockey.

Sharpsville

WAKZ(FM)—Licensed to Sharpsville. See Youngstown OH

Shenandoah

***WCIM(FM)**—Not on air, target date: unknown: 91.5 mhz; 5.6 kw. Ant 702 ft TL: N40 50 58 W76 06 55. Hrs open: Box 506, Bath, NY, 14810-0506. Phone: (607) 776-4151. Fax: (607) 776-6929. Licensee: Family Life Ministries Inc. (acq 7-24-2007; $800,000 for CP). ♦Rick Snavely, pres.

Shippensburg

WEEO(AM)— Dec 5, 1961: 1480 khz; 460 w-D, 9 w-N. TL: N40 04 30 W77 32 09. Hrs open: 6 AM-10 PM 37 S. Main St., Chambersburg, 17201. Phone: (717) 709-0801. Fax: (717) 709-0802. Licensee: Shippensburg Broadcasting Inc. (acq 10-30-2005; $65,000). Population served: 35536 Rgnl. Network: Radio Pa. Format: Adult contemp. Target aud: 30 plus. ♦Eric Swidler, pres; Matthew J. Becker, prom dir & prom mgr.

***WSYC-FM**— February 1975: 88.7 mhz; 100 w. 155 ft TL: N40 03 32 W77 31 20. (CP: TL: N40 04 30 W77 31 15). Stereo. Hrs opn: 24 Shippensburg Univ., Cumberland Union Bldg., Shippenburg, 17257. Phone: (717) 532-6006. Fax: (717) 477-4024. Web Site: www.wsyc.org. Licensee: Shippensburg University. Population served: 10,000 Natl. Network: Westwood One. Format: AOR. News staff: 7; News: 3 hrs wkly. Target aud: 16-25; college & area high school students. Spec prog: Black 9 hrs, class 2 hrs, jazz 3 hrs, blues 2 hrs, wkly. ♦Melanie Warfel, gen mgr; Phil Smith, opns dir; Becke Arline, sls dir.

Shiremanstown

WWII(AM)— June 1987: 720 khz; 2 kw-D. TL: N40 11 28 W76 57 09. Stereo. Hrs opn: 6 AM-sunset 8 W. Main St., 17011. Phone: (717) 731-9944. Fax: (717) 761-0665. E-mail: therock@igateway.com Web Site: www.720therock.com. Licensee: Hensley Broadcasting. Format: Christian. News: 2 hrs wkly. Target aud: 25 plus; Christian. Spec prog:

Gospel 2 hrs, polka 7 hrs, Indian one hr, blues 4 hrs wkly. ♦Dean Lebo, gen mgr; Joe Green, sls dir & progmg dir; Tom Sullivan, progmg dir.

Slippery Rock

***WSRU(FM)**— Sept 20, 1991: 88.1 mhz; 100 w. 79 ft TL: N41 03 43 W80 02 35. Stereo. Hrs opn: 14 Box C-211, Univ. Union, 16057. Phone: (724) 738-2655. Phone: (724) 738-2931. Fax: (724) 738-2754. Web Site: organizations.sru.edu/WRSK/index.asp. Licensee: Slippery Rock University. Population served: 10,000 Natl. Network: ABC. Format: Classic rock, progsv, var/div. News staff: one; News: 14 hrs wkly. Target aud: 18-24; on & off campus students. Spec prog: Relg one hr, campus info one hr, sports one hr wkly.Jay Hunsicker, gen mgr & sports cmtr; Carly Dobbins-Bucklad, stn mgr; Erin McConahy, sls dir; Laura Cooper, prom VP; Vinny Karpuska, progmg dir; Bob Kitzinger, mus dir & rsch dir; Anthony Skariot, asst music dir; Kaleena Zohoranacky, news dir & pub affrs dir; Werner Ullrich, chief of engrg; Lynsi Craig, traf mgr; Glenn Arrington, min affrs dir; Kristin Gibbs, spec ev coord; Dan Miller, outdoor ed; Sean Argyle, political ed; Jeff Ankney, spanish dir; Pat Leishman, women's int ed; Stephanie Stewart, disc jockey

Smethport

WQRM(FM)— January 1990: 106.3 mhz; 1.2 kw. 731 ft TL: N41 48 36 W78 23 10. Stereo. Hrs opn: 24 211 W. Main St., 16749. Phone: (814) 887-1977. Phone: (814) 887-2425. Fax: (814) 887-5178. E-mail: wqrm106@comcast.net Licensee: Farm & Home Broadcasting Co. Group owner: Allegheny Mountain Network Stations Rgnl. Network: Allegheny Mtn. Net. Natl. Rep: Dome. Format: Adult contemp, news. News: one hr wkly. Target aud: 18-54; 60% females & 40% males. Spec prog: Relig 2 hrs wkly, Sports 3 hrs wkly. ♦Cary Simpson, pres; Rose Bishop, gen mgr & rgnl sls mgr.

Somerset

WBHV(AM)— June 15, 1981: 1330 khz; 5 kw-D, 35 w-N, DA-1. TL: N39 59 33 W79 05 41. Hrs open: 24 970 Tripoli St., Johnstown, 15902. Phone: (814) 534-8975. Fax: (814) 534-8979. Licensee: 2510 Licenses LLC. (group owner; (acq 2-16-2005; grpsl). Natl. Network: ESPN Radio. Rgnl. Network: Radio Pa. Format: Sports. Target aud: 35-64. ♦Nick Ferrara, opns mgr.

WLKH(FM)— June 15, 1966: 97.7 mhz; 3.5 kw. Ant 430 ft TL: N40 01 31 W79 05 42. Hrs open: 24 970 Tripoli St., Johnstown, 15902. Phone: (814) 534-8975. Fax: (814) 534-8979. Web Site: www.klove.com. Licensee: 2510 Licenses LLC. (acq 5-1-2005; grpsl). Population served: 280,000 Natl. Network: K-Love. Format: Contemp Christian. ♦Nick Ferrara, opns mgr.

WNTW(AM)— Jan 15, 1951: 990 khz; 10 kw-D, 75 w-N, DA-1. TL: N40 01 31 W79 05 42. Hrs open: 24
Rebroadcasts WLYE(AM) Johnstown 100%.
109 Plaza Dr. Suite 2, Johnstown, 15501-1212. Phone: (814) 255-4186. Fax: (814) 255-6145. Licensee: Forever Broadcasting LLC. Group owner: Forever Broadcasting (acq 9-9-97; grpsl). Population served: 325,000 Format: Country. News staff: one; News: 8 hrs wkly. Target aud: 35-55. ♦Carol Logan, pres; Verla Price, gen mgr & gen sls mgr; Mike Stevens, progmg dir; Rick Sheppard, news dir & pub affrs dir; Jim Boxler, chief of engrg; Tegan Hayes, traf mgr.

South Waverly

WPHD(FM)— 2003: 96.1 mhz; 920 w. Ant 612 ft TL: N41 58 04 W76 40 02. Stereo. Hrs opn: 24 734 Chemung St., Horseheads, NY, 14845. Secondary address: 495 Court St., 2nd Fl., Binghamton 13904. Phone: (607) 795-0795. Phone: (607) 772-1005. Fax: (607) 795-1095. Fax: (607) 772-2945. E-mail: geoequinox@aol.com Web Site: cool96oldies.com. Licensee: Fitzgerald and Hawras Partnership (acq 3-22-01). Natl. Rep: Katz Radio. Format: Oldies. Target aud: 35-64; adults. ♦George Harris, gen mgr; Stephen Shimer, opns mgr; Kevin Fitzgerald, chief of engrg.

South Williamsport

WZXR(FM)— June 1, 1968: 99.3 mhz; 6 kw. 1,237 ft TL: N41 12 42 W76 57 16. Stereo. Hrs opn: 1685 Four Mile Dr., Williamsport, 17740. Phone: (570) 323-8200. Fax: (570) 323-5075. Web Site: www.wzxr.com. Licensee: South Williamsport SabreCom Inc. Group owner: Backyard Broadcasting LLC (acq 12-1-02; grpsl). Population served: 400,000 Natl. Network: ABC. Natl. Rep: Christal. Format: AOR, classic rock. News staff: 3; News: 7 hrs wkly. Target aud: 25-54. ♦Barry Drake, pres; Robin Smith, CFO; Dan Farr, gen mgr; Bob Pawlikowski, sls dir; Ted Minier, progmg dir; John Finn, news dir; Tom Atkins, chief of engrg.

Starview

WSJW(FM)— Nov 22, 1971: 92.7 mhz; 700 w. Ant 954 ft TL: N40 04 32 W76 48 03. Stereo. Hrs opn: 24 Box 4368, Lancaster, 17604. Secondary address: 1996 Auction Rd., Manheim 17545. Phone: (717) 653-0800. Phone: (800) 222-1013. Fax: (717) 653-0122. E-mail: bbaldwin@hallradio.com Web Site: smoothjazz927.com. Licensee: Hall Communications Inc. (group owner; acq 1-16-96; $2.3 million). Population served: 1,315,500 Natl. Rep: D & R Radio. Law Firm: Fletcher, Heald & Hildreth. Format: Smooth jazz. Target aud: 25-54; adult. ♦Bonnie H. Rowbotham, chmn; Arthur J. Rowbotham, pres; Bill Baldwin, sr VP & gen mgr; Tom Shannon, opns mgr; Fran Martin, prom dir.

State College

WBHV-FM— Oct 23, 1991: 94.5 mhz; 940 w. Ant 581 ft TL: N40 54 04 W77 50 20. Stereo. Hrs opn: 160 W. Clearview Ave., 16803. Phone: (814) 238-5085. Fax: (814) 238-8993. Web Site: www.b945live.com. Licensee: 2510 Licenses LLC. Group owner: Forever Broadcasting (acq 2-1-2006; $1.2 million). Population served: 100,000 Natl. Rep: Christal. Format: Christian. ♦Nick Ferrara, gen mgr.

WBLF(AM)—See Bellefonte

WGMR(FM)—See Tyrone

***WKPS(FM)**— 1995: 90.7 mhz; 100 w. 85 ft TL: N40 47 58 W77 52 11. Hrs open: 24 James Bldg., 125 Hub-Robeson Ctr., 16802. Phone: (814) 865-7983. Phone: (814) 865-9577. Fax: (814) 865-2751. E-mail: lion-officers@psu.edu Web Site: www.lion-radio.org. Licensee: Board of Trustees of Pennsylvania State University. Population served: 100,000 Format: Var. News: 4 hrs wkly. Target aud: University students. Spec prog: Jazz 11 hrs, Sp 8 hrs wkly. ♦Scott DeBourke, VP; Mike Fecht, stn mgr; Don Hausmann, opns dir.

WMAJ(AM)— 1945: 1450 khz; 1 kw-U. TL: N40 48 32 W77 50 28. Hrs open: 24 2551 Park Center Blvd., 16801. Phone: (814) 237-9800. Fax: (814) 237-2477. Web Site: www.1450espnradio.com. Licensee: Forever Broadcasting LLC. Group owner: Forever Broadcasting (acq 3-10-98; $2.9 million with co-located FM). Population served: 110,000 Natl. Network: ESPN Radio. Natl. Rep: Christal. Format: Sports. Target aud: 30 plus; college educated, upscale. ♦Carol Logan, pres; Andrew Sumereau, gen mgr; Glen Turner, opns mgr; Bob Taylor, progmg dir & chief of engrg; Pat Boland, news dir & sports cmtr.

WQWK(FM)—Co-owned with WMAJ(AM). 1965: 103.1 mhz; 3 kw. Ant -55 ft TL: N40 48 32 W77 50 28. Stereo. 24 E-mail: mail@hot1031.com Web Site: www.hot1031.com. Format: CHR. News staff: one. Target aud: 18-34; college-aged youth, young adults. ♦Joe Trimarchi, sls dir; Mike Martin, prom dir; Glen Turner, progmg dir.

WOWY(FM)—University Park, April 1965: 97.1 mhz; 3 kw. Ant 403 ft TL: N40 48 27 W77 56 29. Stereo. Hrs opn: 160 W. Clearview Ave., 16803. Phone: (814) 238-5085. Fax: (814) 238-7932. Web Site: www.wowyonline.com. Licensee: 2510 Licenses LLC. (acq 2-16-2005; grpsl). Population served: 38,000 Format: Oldies. ♦Nick Ferrara, gen mgr.

***WPSU(FM)**— Dec 6, 1953: 91.5 mhz; 1.7 kw. Ant 1,197 ft TL: N40 48 32 W77 50 28. Stereo. Hrs opn: 24 174 Outreach Bldg, University Park, 16802. Phone: (814) 865-1877. Fax: (814) 865-4043. E-mail: wpsu@psu.edu Web Site: www.wpsu.org. Licensee: Pennsylvania State University. Population served: 100,000 Natl. Network: PRI, NPR. Law Firm: Paul, Hastings, Janofsky & Walker. Wire Svc: AP Format: News, class, public radio. News staff: one; News: 35 hrs wkly. Target aud: General; upscale, educated adults. Spec prog: Folk 10, jazz 3 hrs, blues 2 hrs wkly. ♦Ted Krichels, gen mgr; Greg Petersen, stn mgr; Steve Shipman, opns dir; Ashear Barr, sls dir; Bill Hiergeist, rgnl sls mgr; Sam Komlenic, rgnl sls mgr; Kristine Allen, progmg dir; Carl Fisher, chief of engrg; Leslie Dyer, traf mgr.

WRSC(AM)— May 29, 1961: 1390 khz; 2 kw-D, 1 kw-N, DA-N. TL: N40 48 50 W77 53 30. Hrs open: 2551 Park Center Blvd., 16801. Phone: (814) 237-9800. Fax: (814) 237-2477. E-mail: comments@newsradio1390.com Web Site: www.newsradio1390.com. Licensee: Forever Broadcasting LLC. (group owner; (acq 5-1-2005; $2.65 million with WBUS(FM) Boalsburg). Population served: 112,000 Rgnl. Network: Radio Pa. Format: News/talk. ♦Andrew Sumereau, gen mgr; Glen Turner, opns mgr; Dave Shannon, progmg dir; Pat Boland, news dir.

***WRXV(FM)**— June 18, 2004: 89.1 mhz; 1 w horiz, 4.4 kw vert. Ant 1,099 ft TL: N40 43 56 W78 19 33. Hrs open: 24 925 Houserville Rd., 16801. Phone: (814) 867-1922. Fax: (814) 867-1922. E-mail: info@revfm.net Web Site: www.revfm.net. Licensee: Invisible Allies Ministries. Format: Contemp Christian. ♦Michael Schomer, gen mgr; Erik Lane, stn mgr.

***WTLR(FM)**— Jan 1, 1978: 89.9 mhz; 25 kw. 584 ft TL: N40 53 32 W77 51 49. Stereo. Hrs opn: 24 2020 Cato Ave., 16801. Phone: (814) 237-9857. Licensee: Central Pennsylvania Christian Institute Inc. Population served: 500,000 Wire Svc: AP Format: Christian. News staff: one; News: 9 hrs wkly. Target aud: 30-55; Adults, family oriented. ♦Mark Van Ouse, gen mgr.

***WXFR(FM)**—Not on air, target date: unknown: 88.3 mhz; 1.8 kw vert. Ant 544 ft TL: N40 54 07 W77 50 11. Hrs open: 24 290 Hegenberger Rd., Oakland, CA, 94621. Phone: (916) 641-8191. E-mail: info@familyradio.org Web Site: www.familyradio.com. Licensee: Family Stations Inc. ♦Harold Camping, pres.

WZWW(FM)—Bellefonte, Sept 15, 1986: 95.3 mhz; 790 w. Ant 636 ft TL: N40 53 35 W77 51 48. Stereo. Hrs opn: 24 863 Benner Pike, Suite 200, 16801. Phone: (814) 231-0953. Fax: (814) 231-0950. E-mail: nancy@3wz.com Web Site: www.3wz.com. Licensee: First Media Radio LLC. (group owner; (acq 11-2-2000). Population served: 124,000 Natl. Network: CNN Radio. Rgnl rep: Commercial Media Sales. Format: Adult contemp. News staff: 2; News: 7 hrs wkly. Target aud: 25-54; upscale families. Spec prog: Sports 3 hrs wkly. ♦Alex Kolobielski, pres; Mike McGough, gen mgr; Dave Kurten, gen sls mgr & progmg dir; Steve Jones, mus dir & sports cmtr.

Stroudsburg

***WBYX(FM)**— Oct. 1, 1999: 88.7 mhz; 1 w horiz, 4 kw vert. Ant 794 ft TL: N41 02 40 W75 22 45. Stereo. Hrs opn: 24
Rebroadcasts WBYO(FM) Sellersville 100%.
Box 186, Sellersville, 18960. Phone: (215) 721-2141. Fax: (215) 721-9811. E-mail: wordfm@wordfm.org Web Site: www.wordfm.org. Licensee: Four Rivers Community Broadcasting Corp. Law Firm: Schwartz, Woods & Miller. Format: Adult contemp, Christian, religious. Target aud: 25-45. Spec prog: Bluegrass Gospel;. ♦Nancy Loughery, CFO; David W. Baker, VP; Charles W. Loughery, gen mgr; Kristine McClain, progmg dir.

WSBG(FM)—Listing follows WVPO(AM).

WVPO(AM)— 1947: 840 khz; 250 w-D. TL: N40 58 26 W75 11 43. Hrs open: 22 S. 6th St., 18360. Phone: (570) 421-2100. Fax: (570) 421-2040. Licensee: Nassau Broadcasting II L.L.C. Group owner: Nassau Broadcasting Partners L.P. (acq 2-15-02; grpsl). Population served: 120,000 Format: Adult standards. Target aud: 35 plus. ♦Peter Tonks, CFO; Rick Musselman, gen mgr; Michele Stevens, progmg VP; Rod Bauman, progmg dir; Bob Matthews, news dir; Tony Gervasi, engrg VP; George Guilda, engrg mgr.

WSBG(FM)—Co-owned with WVPO(AM). Oct 1, 1964: 93.5 mhz; 550 w. Ant 764 ft TL: N40 56 56 W57 09 29. Stereo. Web Site: www.lite935.com. Format: Lite rock. Target aud: 20 plus. Spec prog: Modern rock 3 hrs wkly.

Summerdale

***WJAZ(FM)**— Jan 10, 1991: 91.7 mhz; 140 w. 683 ft TL: N40 18 16 W76 55 53. Stereo. Hrs opn: 24
Rebroadcasts WRTI(FM) Philadelphia 100%.
1509 Cecil B. Moore Ave., 2nd Fl., ., Philadelphia, 19122. Phone: (215) 204-8405. Fax: (215) 204-7027. E-mail: comments@wrti.org Web Site: www.wrti.org. Licensee: Temple University of the Commonwealth System of Higher Education (acq 12-88; $5,000; FTR: 12-5-88). Population served: 1,450,000 Rgnl. Network: Radio Pa. Format: Classical, jazz. News staff: one; News: 15 hrs wkly. Target aud: 30-65. ♦Dave Conant, CEO & gen mgr; Vic Scarpato, CFO; Tobias Poole, opns dir; Rick Torpey, gen sls mgr & mktg mgr; Porsche Blakey, prom dir; Jack Moore, progmg dir; Windsor Johnson, news dir; Jeff DePolo, chief of engrg; Lorna Nixon, traf mgr.

Sunbury

WKOK(AM)— 1933: 1070 khz; 10 kw-D, 1 kw-N, DA-N. TL: N40 52 46 W76 49 18. Hrs open: 24 Box 1070, 17801. Secondary address: R.D. 2, County Line Rd., Selinsgrove 17870. Phone: (570) 286-5838. Phone: (570) 743-1841. Fax: (570) 743-7837. Fax: (570) 743-1605. E-mail: wkok@wkok.com Web Site: www.wkok.com. Licensee: Sunbury Broadcasting Corp. (acq 5-33). Population served: 237,000 Natl. Network: CBS, ESPN Radio, AP Radio. Natl. Rep: Roslin. Rgnl rep: Dome. Law Firm: Wilkinson Barker Knauer. Wire Svc: AP Format: News/talk, sports. News staff: 4; News: 168 hrs wkly. Target aud:

35-64. ♦Roger Haddon Jr., CEO, pres & gen mgr; Kevin Herr, opns mgr & news dir; Gayle Fedder, gen sls mgr & mktg mgr; Mark Lawrence, progmg dir.

WQKX(FM)—Co-owned with WKOK(AM). Sept 15, 1948: 94.1 mhz; 16 kw. 879 ft TL: N40 47 07 W76 41 51. Stereo. 24 E-mail: wqkx@wqkx.com Web Site: www.wqkx.com.200,000 Format: CHR. News staff: 4; News: 7 hrs wkly. Target aud: 25-54. ♦Drew Kelly, progmg dir; Rob Senter, mus dir.

Susquehanna

WCDW(FM)— March 1995: 100.5 mhz; 1.35 kw. Ant 692 ft TL: N42 03 10 W75 42 07. Stereo. Hrs opn: 24 495 Court St., 2nd Floor, Binghamton, NY, 13904. Phone: (607) 772-1005. Fax: (607) 772-2945. E-mail: cool100oldies@aol.com Web Site: www.cool100oldies.com. Licensee: Equinox Broadcasting Corp. Natl. Rep: Katz Radio. Format: Oldies. Target aud: 35-64. Spec prog: Polish 5 hrs wkly. ♦George Hawras, pres.

Swarthmore

***WSRN-FM**— Dec 31, 1939: 91.5 mhz; 110 w. 140 ft TL: N39 54 18 W75 21 16. Stereo. Hrs opn: Swarthmore College, 500 College Ave., 19081. Phone: (610) 328-8336. Phone: (610) 328-8335. Web Site: wsrn.swarthmore.edu. Licensee: Swarthmore College. Population served: 500,000 Format: Div. ♦Alex Flurie, gen mgr.

Sweet Valley

***WRGN(FM)**— Oct 15, 1984: 88.1 mhz; 500 w. 239 ft TL: N41 17 54 W76 07 28. (CP: Ant 302 ft.). Hrs opn: 24 2457 State Rt. 118, Hunlock Creek, 18621. Phone: (570) 477-3688. Fax: (570) 477-2310. E-mail: wrgn@epix.net Web Site: www.wrgn.com. Licensee: Gospel Media Institute Inc. Format: Relg. ♦Burl F. Updyke, pres, gen mgr, gen sls mgr & chief of engrg; Shirley J. Updyke, prom dir & progmg dir.

Tafton

***WLKA(FM)**— 2002: 88.3 mhz; 580 w. Ant 968 ft TL: N41 35 36 W75 25 56. Hrs open: 2351 Sunset Blvd., Suite 170-218, Rocklin, CA, 95765. Phone: (916) 251-1600. Fax: (916) 251-1650. Web Site: www.klove.com. Licensee: Educational Media Foundation. (acq 11-17-2006; $675,000). Natl. Network: K-Love. Format: Contemp Christian. ♦Richard Jenkins, pres.

Tamaqua

WMGH-FM— June 14, 1965: 105.5 mhz; 1.4 kw. 485 ft TL: N40 47 14 W76 01 59. Stereo. Hrs opn: 24 Box D, Lansford, 18232. Phone: (570) 668-2992. Phone: (570) 645-2105. Fax: (570) 645-2159. E-mail: wmgh@ptdprolog.net Web Site: www.wmgh.com. Licensee: J-Systems Franchising Corp. (group owner; acq 2-28-87; $300,000; FTR: 12-15-86). Population served: 120,000 Natl. Network: Westwood One, ABC. Format: Adult contemp. Target aud: 25-54; primary women, secondary adults. Spec prog: Oldies 14 hrs, polka 3 hrs wkly. ♦Harold G. Fulmer III, pres; Christopher G. Fulmer, VP & gen sls mgr; Bill Lakatas, gen mgr & progmg dir; Mark Marek, news dir & edit dir; Joe Manjack, chief of engrg & disc jockey; Cheryl Lee, disc jockey; J Z, disc jockey; Nicky Vee, disc jockey.

Tarentum

WZPT(FM)—See New Kensington

Telford

***WBMR(FM)**— June 1967: 91.7 mhz; 115 w. Ant 249 ft TL: N40 18 15 W75 17 39. Hrs open: 300 E. Rock Rd., Allentown, 18103. Phone: (610) 797-4530. Fax: (610) 791-3000. Licensee: United Ministries. (acq 8-30-2002). Format: Relg.

Tioga

WMTT(FM)— May 23, 1991: 94.7 mhz; 820 w. 895 ft TL: N41 54 36 W77 00 40. (CP: 12 kw). Stereo. Hrs opn: 24 734 Chemung St., Horseheads, NY, 14845. Secondary address: 495 Court St., 2nd Fl, Binghamton, NY 13904. Phone: (607) 795-0795. Phone: (607) 772-1005. Fax: (607) 795-1095. Fax: (607) 772-2945. E-mail: themetrocks@aol.com Web Site: www.themetrocks.com. Licensee: Europa Communications Inc. (acq 5-22-92). Population served: 116,966 Natl. Rep: Katz Radio. Format: Classic rock, AOR. News: 2 hrs wkly. Target aud: 25-49. ♦Kevin Fitzgerald, CEO & VP; George Harris, gen mgr & opns dir; Robert Smith, stn mgr; Stephen Shimer, opns mgr.

Titusville

WTIV(AM)— Nov 27, 1955: 1230 khz; 1 kw-U. TL: N41 37 00 W79 41 34. Hrs open: 6 AM-midnight 900 Water St., Downtown Mall, Meadville, 16335. Phone: (814) 432-2188. Fax: (814) 827-1679. Licensee: Forever Broadcasting LLC. Group owner: Forever Broadcasting (acq 7-20-00; grpsl). Population served: 25,000 Rgnl. Network: Radio Pa. Law Firm: Reddy, Begley & McCormick. Format: News/talk info. News: 13 hrs wkly. Target aud: 22-54; mixed. ♦Thomas J. Sauber, gen mgr.

Tobyhanna

WKRF(FM)— Jan 15, 1993: 107.9 mhz; 5.7 kw. 564 ft TL: N41 07 04 W75 22 43. Stereo. Hrs opn: 24
Rebroadcasts WKRZ(FM) Wilkes-Barre 100%.
305 Hwy. 315, Pittston, 18640. Phone: (570) 839-5858. Fax: (570) 883-9851. Web Site: www.wkrf.com. Licensee: Entercom Wilkes-Barre Scranton LLC. Group owner: Entercom Communications Corp. (acq 5-11-00). Population served: 80,000 Natl. Network: Jones Radio Networks. Format: Top-40. News staff: one; News: one hr wkly. Target aud: 25-54. ♦John Burkavage, gen mgr; Jim Rising, stn mgr & opns mgr; Andy Zapotek, gen sls mgr; Bob Demono, natl sls mgr; Michael Ignatz, prom dir; Jerry Padden, progmg dir; Elizabeth Masich, mus dir; Lamar Smith, chief of engrg; Tracy Iannaprone, traf mgr; Joe Thomas, local news ed; Tom Regan, news rptr.

Towanda

WTTC(AM)— 1959: 1550 khz; 500 w-D. TL: N41 45 55 W76 29 10. Hrs open: 204 Desmond St., Sayre, 18840. Phone: (570) 888-7745. Fax: (570) 888-9005. Licensee: WATS Broadcasting Inc. (acq 5-1-96; $175,000 for stock with co-located FM). Population served: 30,000 Natl. Network: Motor Racing Net. Format: Oldies. Target aud: General. ♦Charles C Carver Jr., pres; Charles C. Carver Jr., gen mgr; Meade T. Murtland, stn mgr.

WTTC-FM— November 1959: 95.3 mhz; 3 kw. 125 ft TL: N41 45 55 W76 29 10.6 AM-11 PM ♦Joel Clawson, adv dir.

Trout Run

***WCIT(FM)**— 2001: 90.1 mhz; 350 w. Ant 295 ft TL: N41 27 26 W77 06 55. Hrs open:
Rebroadcasts WCIK(FM) Bath, NY 100%.
Box 506, Bath, NY, 14810. Secondary address: 7634 Campbell Creek Rd., Bath, NY 14810. Phone: (607) 776-4151. Fax: (607) 776-6929. E-mail: mail@fln.org Web Site: www.fln.org. Licensee: Family Life Ministries Inc. Group owner: Family Life Network Natl. Network: Salem Radio Network. Law Firm: Hardy, Carey, Chautin & Balkin, LLP. Wire Svc: Metro Weather Service Inc. Format: Contemp Christian. News staff: 3; News: 14 hrs wkly. Target aud: 30-54. ♦Rick Snavely, pres & gen mgr; Dick Snavely, CFO; Roger Settje, prom mgr; John Owens, progmg dir; Jim Travis, chief of engrg.

Troy

WHGL-FM—See Canton

WTZN(AM)— Mar 3, 1982: 1310 khz; 1 kw-D, 72 w-N. TL: N41 46 51 W76 49 09. Hrs open: Box 100, 170 Redington Ave., 16947. Phone: (570) 297-0100. Fax: (570) 297-3193. E-mail: whgl@ptd.net Web Site: www.wtzn.com. Licensee: Cantroair Communications Inc. (acq 1-7-99; $560,000 for 85% of stock with WHGL-FM Canton). Format: All sports. Target aud: 25-54. ♦Bob Gisler, VP & gen sls mgr; Georgia Pepper, prom mgr; Mike Powers, pres, gen mgr & progmg dir; Michael Dean, mus dir; Kevin Smith, chief of engrg.

Tunkhannock

WEMR(AM)— June 13, 1986: 1460 khz; 5 kw-D, 1.25 kw-N, DA-2. TL: N41 33 46 W75 58 11. Hrs open: 18 Box 230, Dushore, 18614. Phone: (570) 928-7200. Fax: (570) 928-2100. Licensee: GEOS Communications (acq 1-30-2004; $515,000 with co-located FM). Format: News/talk. ♦Ben Smith, gen mgr.

WGMF(FM)—Co-owned with WEMR(AM). Oct 10, 1990: 107.7 mhz; 490 w. Ant 1,138 ft TL: N41 30 45 W76 04 16. Stereo. 24 Format: Adult contemp.

Tyrone

WGMR(FM)—Listing follows WTRN(AM).

WTRN(AM)— Jan 12, 1955: 1340 khz; 1 kw-U. TL: N40 39 48 W78 15 24. Hrs open: 24 Box 247, 16686. Secondary address: Washington Ave. & 1st St. 16686. Phone: (814) 684-3200. Fax: (814) 684-1220. E-mail: amnnet@aol.com Web Site: www.wtrn.net. Licensee: Allegheny Mountain Network. (group owner) Population served: 7,072 Natl. Network: Jones Radio Networks. Rgnl. Network: Allegheny Mtn. Net. Natl. Rep: Dome. Law Firm: Borsari & Paxson. Format: Adult contemp. News: 16 hrs wkly. Target aud: General; total community targeted. Spec prog: Relg 4 hrs wkly. ♦Cary H. Simpson, pres; Peg Baney, gen sls mgr; Rich Saupp, progmg dir; Jean Dixon, news dir; Robert Lynn, chief of engrg.

WGMR(FM)—Co-owned with WTRN(AM). Aug 15, 1961: 101.1 mhz; 8.5 kw. 1,171 ft TL: N40 55 10 W77 58 28. Stereo. 24 2351 Commercial Blvd., State College, 16801. Phone: (814) 238-0717. Fax: (814) 234-3533. Web Site: www.g101fm.com.650,000 Rgnl rep: Dome. Law Firm: Borsari & Paxson. Format: CHR. Target aud: 18-49. ♦John F. Simpson, gen mgr; Ted Swanson, gen sls mgr.

Union City

WCTL(FM)— Apr 23, 1967: 106.3 mhz; 3.4 kw. 430 ft TL: N42 00 04 W79 52 33. Stereo. Hrs opn: 10912 Peach St., Waterford, 16441-9151. Phone: (814) 796-6000. Fax: (814) 796-3200. E-mail: wctl@wctl.org Web Site: www.wctl.org. Licensee: Inspiration Time Inc. (acq 3-72). Population served: 300,000 Natl. Network: USA. Natl. Rep: Salem. Law Firm: Hogan & Hartson. Format: Adult contemp, christian. News staff: one; News: 2.5 hrs wkly. Target aud: 25-54; Christian families. Spec prog: Children one hr wkly. ♦Ed Mattson, pres; Adam Frase, progmg dir & mus dir; Ronald Raymond, gen mgr & progmg dir.

Uniontown

WMBS(AM)— July 15, 1937: 590 khz; 1 kw-U, DA-N. TL: N39 51 35 W79 44 44. Hrs open: 24 44 S. Mt. Vernon Ave., 15401. Phone: (724) 438-3900. Fax: (724) 438-2406. E-mail: bmroziak@aol.com Web Site: www.wmbs590.com. Licensee: Fayette Broadcasting Corp. Population served: 60,000 Natl. Network: CBS Radio, Westwood One. Natl. Rep: Commercial Media Sales. Rgnl rep: West Media Group Law Firm: Pillsbury, Winthrop, Shaw, Pittman, LLP. Wire Svc: AP Wire Svc: Metro Weather Service Inc. Format: Var/div. News staff: one; News: 24 hrs wkly. Target aud: 25 plus; General. Spec prog: Talk shows, polka 3 hrs wkly, Pittsburgh Pirates, Pittsburgh Steelers, Pittsburgh Penguins, Pitt Panthers, Westwood One Sports, CBS News. ♦Bob Pritts, pres; Brian Mroziak, gen mgr & sports cmtr; Doreen Minafee, opns VP & traf mgr; Sandy Tracy, sls VP; Michael Pasqua, gen sls mgr; Jim Morgan, news dir; Timothy Schwer, pub affrs dir; Larry Campbell, chief of engrg; Tim Schwer, women's cmtr & disc jockey.

WPKL(FM)— Dec 20, 1968: 99.3 mhz; 3 kw. Ant 300 ft TL: N39 53 09 W79 46 29. Stereo. Hrs opn: 24 123 Blaine Rd., Brownsville, 15417. Phone: (724) 938-2000. Fax: (724) 938-7842. E-mail: rherring@keymarketradio.com Web Site: oldiesradioonline.com. Licensee: Keymarket Licenses LLC Group owner: Keymarket Communications

LLC (acq 1-17-2001; $475,000 with WYJK(AM) Connellsville). Population served: 150,000 Natl. Rep: Dome. Format: Oldies. Target aud: 25 plus. ♦Gerald Getz, pres; Andrew Powaski, gen mgr, sls dir, gen sls mgr & progmg dir.

University Park

WOWY(FM)—Licensed to University Park. See State College

Upton

WPPT(FM)—See Mercersburg

Villanova

***WXVU(FM)**— August 1991: 89.1 mhz; 710 w. 223 ft TL: N40 03 22 W75 22 30. (CP: 100 w vert, ant 279 ft. TL: N40 01 58 W75 20 15). Hrs opn: Villanova Univeristy, 210 Dougherty Hall, 800 Lancaster Ave., 19085-1699. Phone: (610) 519-7200. Fax: (610) 519-7956. Web Site: wxvu.villanova.edu. Licensee: Villanova University. Population served: 150,000 Format: Hip hop. Target aud: 15-25; youngsters. Spec prog: Black 10 hrs, relg 2 hrs wkly. ♦Brian Golden, gen mgr; Laysan Unger, progmg dir; Kristin Lavin, mus dir; Greg Ebbecke, pub affrs dir.

Warminster

***WRDV(FM)**— Sept 6, 1976: 89.3 mhz; 1 kw horiz, 100 w vert. 118 ft TL: N40 12 19 W75 06 27. Stereo. Hrs opn: 24 Box 2012, 18974. Secondary address: 126 S. York Rd., Hatboro 19040. Phone: (215) 674-8002. Fax: (215) 674-4586. Web Site: wrdv.org. Licensee: Bux-Mont Educational Radio Associates. (acq 3-80). Population served: 50,000 Format: Variety. News: 2 hrs wkly. Target aud: General. Spec prog: C&W 4 hrs, blues 3 hrs, folk 4 hrs, new age 3 hrs, jazz 3 hrs wkly. ♦Charles W. Loughery, pres; Todd H. Allen, gen mgr & progmg dir.

Warren

WNAE(AM)— Dec 31, 1946: 1310 khz; 5 kw-D, 94 w-N. TL: N41 48 50 W79 10 04. Hrs open: 6 AM-midnight Box 824, 16365. Secondary address: 310 2nd Ave. 16365. Phone: (814) 723-1310. Fax: (814) 723-3356. E-mail: info@kibcoradio.com Web Site: www.kibcoradio.com. Licensee: Radio Partners LLC. (acq 9-19-2005; grpsl). Population served: 12,998 Rgnl. Network: Radio Pa. Format: Adult contemp. News staff: one; News: 21 hrs wkly. Target aud: General. ♦David Whipple, VP & gen sls mgr; Denny Haight Jr., VP & engr; W. LeRoy Schneck, gen mgr; Karen White, opns mgr; Dale Bliss, prom dir & sls; Mark Silvis, progmg dir; Robert Seiden, news dir; Dana Simmons, traf mgr.

WRRN(FM)—Co-owned with WNAE(AM). March 1948: 92.3 mhz; 50 kw. 410 ft TL: N41 48 50 W79 10 04. Stereo. 24 Web Site: www.kibcoradio.com. Format: Oldies.

Warwick

***WZZD(FM)**— December 2000: 88.1 mhz; 180 w vert. 587 ft TL: N40 07 45 W75 52 43. Stereo. Hrs opn: 24
Rebroadcasts WBYO(FM) Sellersville 90%.
Box 186, Sellersville, 18960. Phone: (215) 721-2141. Fax: (215) 721-9811. E-mail: wordfm@wordfm.org Web Site: www.wordfm.org. Licensee: Four Rivers Community Broadcasting Corp. Population served: 75,000 Format: Contemp Christian. Spec prog: Bluegrass/gospel 3 hrs wkly. ♦David Baker, gen mgr; Charles W. Loughery, stn mgr.

Washington

WJPA(AM)— Feb 1, 1941: 1450 khz; 1 kw-U. TL: N40 11 23 W80 14 02. Hrs open: 24 98 S. Main St., 15301. Phone: (724) 222-2110. Fax: (724) 228-2299. E-mail: email@wjpa.com Web Site: www.wjpa.com. Licensee: Washington Broadcasting Co. Population served: 220,000 Rgnl. Network: Radio Pa. Format: Oldies. News staff: 2; News: 6 hrs wkly. ♦Michael S. Siegel, pres & gen mgr; Bob Gregg, opns dir, chief of opns & gen sls mgr; Dale Allen, prom mgr; Pete Povich, progmg dir; Margie Konstantinou, mus dir; Jim Jefferson, news dir.

WJPA-FM— Sept 26, 1964: 95.3 mhz; 2.15 kw. 390 ft TL: N40 11 23 W80 14 02. (CP: 4.2 kw). 24 Web Site: www.wjpa.com.

WKZV(AM)— August 1968: 1110 khz; 1 kw-D, DA. TL: N40 13 16 W80 14 34. Hrs open: Daylight 80 E. Chestnut St., 15301. Phone: (724) 228-6678. Fax: (724) 228-6678. Licensee: My-Key Broadcasting Inc. (acq 11-9-92; $100,000; FTR: 11-30-92). Population served: 210,000 Natl. Rep: Dome. Format: Country. Target aud: 35 plus. Spec prog: Pol 3 hrs, polka 2 hrs, Croatian one hr, gospel one hr, racing one hr, relg one hr wkly. ♦Helen C. Supinski, pres; Michael Panjuscek, VP, gen mgr, progmg dir & traf mgr.

***WNJR(FM)**— Nov 26, 1972: 91.7 mhz; 950 w. Ant 112 ft TL: N40 10 13 W80 14 43. Stereo. Hrs opn: 24 60 S. Lincoln St., 15301. Phone: (724) 503-1001 x3025 (gen mgr). Phone: (724) 223-6039 (studio). Fax: (724) 223-5271. E-mail: wnjr@washjeff.edu Web Site: www.wasjeff.edu. Licensee: Washington and Jefferson College. Population served: 1,200,000 Wire Svc: AP News: 4 hrs wkly. Target aud: All ages; college students, staff, community, alumni. ♦Rob Valella, gen mgr; Ryan Ray, sls dir; Christine Briski, prom dir; Megan McCauley, mus dir; Alex Hines, news dir; Cliff Bryson, chief of engrg; Jason Chase, sports cmtr; Mario Sacco, sports cmtr; Steve Capone Jr., stn mgr & sports cmtr.

Waynesboro

WCBG(AM)— Aug 19, 1953: 1380 khz; 1 kw-D. TL: N39 44 20 W77 36 10. Hrs open: Box 788, Greencastle, 17225. Secondary address: 10960 John Wayne Dr., Greencastle 17225. Phone: (717) 597-9200. Fax: (717) 597-9210. Licensee: HJV L.P. Group owner: VerStandig Broadcasting (acq 1-6-97; $1,068,699 with co-located FM). Population served: 10,011 Format: Contemp country. Target aud: 25-54. ♦Marge Martin, gen mgr & gen sls mgr; Don Brake, progmg dir.

WFYN(FM)—Co-owned with WCBG(AM). Feb 3, 1959: 101.5 mhz; 50 kw horiz, 48 kw vert. 230 ft TL: N39 49 44 W77 33 10. Stereo. 250,000 Format: Rock. ♦Marge Martin, mktg mgr; Don Blake, progmg dir.

Waynesburg

WANB-FM— Apr 21, 1978: 103.1 mhz; 970 w. Ant 617 ft TL: N39 52 12 W80 08 01. Stereo. Hrs opn: 24 369 Tower Rd., 15370. Phone: (724) 627-5555. Fax: (724) 627-4021. E-mail: wanb@greenepa.net Web Site: www.wanb103.com. Licensee: Broadcast Communications Inc. (group owner; (acq 4-1-2002; with co-located AM). Rgnl. Network: Radio Pa. Natl. Rep: Dome. Format: Country. Target aud: 20 plus. ♦Judy E. Rastoka, gen mgr; Doug Wilson, progmg dir; Rick Williams, chief of engrg; Marcia Mackey, traf mgr.

WANB(AM)— Sept 27, 1956: 1580 khz; 720 w-D. TL: N39 52 12 W80 08 01. (CP: 1210 khz; 5 kw-D, 710 w-CH). Sunrise-sunset 5,152 ♦Marcia Mackay, traf mgr.

***WCYJ-FM**— July 6, 1979: 88.7 mhz; 18 w. -33 ft TL: N39 53 59 W80 11 07. Hrs open: 24 Waynesburg College, 51 W. College St., 15370. Phone: (724) 627-8191. Fax: (724) 627-4757. E-mail: wcyjgm@hotmail.com Web Site: www.waynesburg.edu. Licensee: Waynesburg College. (acq 7-3-79). Population served: 10,000 Format: Adult contemp. News staff: one. Target aud: 18-25; college & high school students. Spec prog: Oldies 3 hrs, country 3 hrs, Christian 3 hrs, R&B 3 hrs, classic rock 3 hrs wkly. ♦Brad Baker, gen mgr; Mark Perry, stn mgr; Nick Daniels, opns dir.

Wellsboro

WNBT(AM)— May 13, 1955: 1490 khz; 1 kw-U. TL: N41 44 41 W77 17 35. Hrs open: 24 Box 98, 16901. Secondary address: 198-B RR 7 16901. Phone: (570) 724-1490. Phone: (570) 662-7100. Fax: (570) 724-6971. E-mail: wnbt@ynt.net Web Site: www.wnbt.net. Licensee: Farm & Home Broadcasting Co. Group owner: Allegheny Mountain Network Stations Population served: 4,003 Natl. Network: Westwood One. Rgnl. Network: Radio Pa. Natl. Rep: Dome. Law Firm: Borsari & Paxson. Format: Adult standards. News staff: one; News: 10 hrs wkly. Target aud: 45+. ♦Cary Simpson, pres.

WNBT-FM— July 2, 1969: 104.5 mhz; 50 kw. 380 ft TL: N41 44 17 W77 21 50. Stereo. 24 Web Site: www.wnbt.net. Natl. Network: Westwood One. Format: CHR, popular music, adult contemp. News staff: one; News: 2 hrs wkly. Target aud: 18-55.

West Chester

WCHE(AM)— Oct 4, 1963: 1520 khz; 1000 w-D. TL: N39 58 06 W75 37 59. Hrs open: Sunrise-sunset 105 W. Gay St., 19380. Phone: (610) 692-3131. Fax: (610) 692-3133. E-mail: wche@wche1520.com Web Site: www.wche1520.com. Licensee: Chester County Radio Inc. (acq 7-11-97; $230,000). Population served: 500,000 Natl. Network: USA, Westwood One. Law Firm: Pepper & Corazzini. Format: Talk, Alternative Rock. News staff: one; News: 20 hrs wkly. Target aud: 24-64; upscale. Spec prog: Relg 8 hrs, country 2 hrs wkly. ♦David S. Shur, pres, sls dir, progmg dir & news dir; Jay Shur, gen mgr, opns mgr, chief of opns, prom dir & pub affrs dir.

WCOJ(AM)—See Coatesville

***WCUR(FM)**— 1999: 91.7 mhz; 100 w. Ant 108 ft TL: N39 57 02 W75 35 58. Hrs open: West Chester University, Sykes Union Bldg., 19383. Phone: (610) 436-2414. Fax: (610) 436-2477. E-mail: wcur@yahoo.com Web Site: www.wcur.fm. Licensee: Student Services Inc. Format: Diversified.

West Hazleton

WKZN(AM)— 1982: 1300 khz; 5 kw-D, 500 w-N, DA-2. TL: N40 56 26 W76 00 07. Hrs open: 24
Rebroadcasts WILK(AM) Wilks-Barre 100%.
Box 729, 305 Hwy. 315, Pittstown, 18640. Phone: (570) 883-9850. Fax: (570) 883-0832. Web Site: www.wilknewsradio.com. Licensee: Entercom Scranton Wilkes-Barre License LLC. Group owner: Entercom Communications Corp. (acq 12-13-99; grpsl). Population served: 25,000 Rgnl. Network: Radio Pa. Natl. Rep: D & R Radio. Wire Svc: ABC Wire Svc: AP Format: News/talk. News staff: 6; News: 25 hrs wkly. Target aud: General; affluent, educated. Spec prog: Relg one hr wkly. ♦John Burkavage, gen mgr; Jim Rising, opns dir; Andy Zapotek, gen sls mgr; Bob DeMond, natl sls mgr; Casey Consagra, prom mgr; Nancy Kman, progmg dir; Joe Thomas, news dir & sports cmtr; Lamar Smith, chief of engrg; Shannon Ball, traf mgr; Tom Ragan, news rptr; Shadoe Steele, mus critic.

Whitneyville

WLIH(FM)— Mar 15, 1987: 107.1 mhz; 3.3 kw. Ant 298 ft TL: N41 46 13 W77 12 08. Stereo. Hrs opn: 6 AM-Midnight Box 97, 2352 Charleston Rd., Wellsboro, 16901. Phone: (570) 724-4272. Fax: (570) 724-2302. E-mail: wlih107@quik.com Web Site: www.wlih.com. Licensee: Good Christian Radio Broadcasting Inc. Natl. Network: USA. Format: Relg, Christian, news. News: 28 hrs wkly. Target aud: General; serving the Christian community of the county. ♦Robert Makin, pres; Carol Makin, gen mgr.

Wilkes-Barre

WARM(AM)—See Scranton

WBAX(AM)— May 1, 1922: 1240 khz; 1 kw-U. TL: N41 15 13 W75 54 25. Hrs open: 24
Simulcasts with WEJL (AM) Scranton.
149 Penn Ave., Scranton, 18503. Phone: (570) 346-6555. Fax: (570) 346-6038. Web Site: www.wejl-wbax.com. Licensee: The Scranton Times L.P. Population served: 58,856 Format: Sports. News staff: one; News: 5 hrs wkly. Target aud: 18 + men. Spec prog: sports. ♦William R. Lynett, CEO; Jim Loftus, COO & gen mgr; Tim Durkin, sls dir; Jerry Arndt, natl sls mgr; Mark Hoover, mktg dir & prom mgr; Michael Neff, progmg dir; Ruth Miller, news dir; Kevin Fitzgerald, engrg dir; Krista Saar, traf mgr.

WBZU(AM)—See Scranton

***WCLH(FM)**— Feb 6, 1972: 90.7 mhz; 175 w. 1,020 ft TL: N41 10 58 W75 52 21. Stereo. Hrs opn: 24 84 W. South St., Wilkes University, Wilkes Barre, 18766. Phone: (570) 408-5907. Fax: (570) 408-5908. E-mail: wclh@wilkes.edu Web Site: www.wclh.org. Licensee: Wilkes University. Population served: 700,000 Natl. Network: AP Network News. Format: Metal, hip hop, alternative. News: 7 hrs wkly. Target aud: 12-44. Spec prog: Ger 3 hrs, Sp 3 hrs wkly. ♦Renee Loftus, gen mgr; Ariel Cohen, progmg dir.

WEJL(AM)—See Scranton

WGGY(FM)—See Scranton

WICK(AM)—See Scranton

WILK(AM)— Feb 13, 1947: 980 khz; 5 kw-D, 1 kw-N, DA-N. TL: N41 13 42 W75 56 53. Stereo. Hrs opn: 24 305 Hwy. 315, Pittston, 18640. Phone: (570) 883-9800. Fax: (570) 883-9851. E-mail: feedback@thewilknetwork.com Web Site: www.wilknetwork.com. Licensee: Entercom Scranton Wilkes-Barre License LLC. Group owner: Entercom Communications Corp. (acq 12-13-99; grpsl). Population served: 72,500 Format: News/talk. Target aud: 15-54. Spec prog: Relg one hr wkly. ♦Joseph Fields, pres; John Burkavage, gen mgr.

WITK(AM)—See Pittston

WKRZ(FM)— 1947: 98.5 mhz; 8.7 kw. 1,171 ft TL: N41 11 56 W75 49 06. Stereo. Hrs opn:
Rebroadcasts WKRZ(FM) Tobyhanna 100%.
305 Hwy. 315, Box 729, Pittston, 18640. Phone: (570) 883-9850. Fax: (570) 883-9851. Web Site: www.wkrz.com. Licensee: Entercom Scranton Wilkes-Barre License LLC. Group owner: Entercom Communications Corp. (acq 12-13-99; grpsl). Population served: 200,000 Format: CHR. Target aud: 19-54; women. ♦John Burkavage, gen mgr; Jim Rising, opns mgr; Ryan Flynn, gen sls mgr & sls; Tias Schuster, progmg dir; Elizabeth Masich, mus dir; Lamar Smith, chief of engrg.

WMGS(FM)— 1946: 92.9 mhz; 5.3 kw. Ant 1,384 ft TL: N41 10 58 W75 52 26. Stereo. Hrs opn: 24 600 Baltimore Dr., 18702. Phone: (570) 824-9000. Fax: (570) 820-0520. Licensee: Citadel Broadcasting Co. Group owner: Citadel Broadcasting Corp. (acq 7-1-97; grpsl). Format: Soft rock. News staff: one; News: 2 hrs wkly. Target aud: 25-54; adult women. Spec prog: Farm one hr, relg 3 hrs wkly. ♦Taylor Walet, gen mgr.

WNAK(AM)—See Nanticoke

***WRKC(FM)**— Sept 18, 1968: 88.5 mhz; 440 w. -470 ft TL: N41 14 57 W75 52 26. Stereo. Hrs opn: 7 AM-2 AM 133 N. Franklin St., 18711. Phone: (570) 208-5931. Fax: (570) 825-9049. Web Site: www.kings.edu/~wrke/. Licensee: King's College. Population served: 300,000 Format: Jazz, AOR, reading for the blind. Target aud: General; people who need wide-ranging svcs. ♦Pete Phillips, gen mgr.

WWRR(FM)—See Scranton

Wilkinsburg

WPYT(AM)— Aug 25, 1960: 660 khz; 260 w-D. TL: N40 24 47 W79 51 14. Hrs open: Daylight 4736 Penn Ave., Pittsburgh, 15224. Phone: (412) 661-6001. Fax: (412) 661-7195. Licensee: Langer Broadcasting Group L.L.C. (group owner; acq 5-22-98). Rgnl. Network: Radio Pa. Law Firm: Reddy, Begley & McCormick. Format: Talk. ♦Ed Dehart, gen sls mgr; Stephen Zelenko, gen mgr, gen mgr & progmg dir.

Williamsport

WBZD-FM—Muncy, Aug 11, 1983: 93.3 mhz; 1.7 kw. Ant 1,220 ft TL: N41 12 42 W76 57 16. Stereo. Hrs opn: 24 1685 Four Mile Dr., 17701. Phone: (570) 323-8200. Fax: (570) 323-5075. E-mail: bobpawlikowski@byradio.com Web Site: www.wbzd.com. Licensee: South Williamsport SaberCom Inc. Group owner: Backyard Broadcasting LLC (acq 12-1-02; grpsl). Population served: 300,000 Natl. Rep: Christal. Format: Oldies. News staff: one; News: 3 hrs wkly. Target aud: 18-54; adult oriented, mass appeal. ♦Barry Drake, pres; Robin Smith, CFO; Dan Farr, gen mgr; Bob Pawlikowski, gen sls mgr; Ted Minier, progmg dir; Brian Hill, chief of engrg & engr.

***WCRG(FM)**— Feb 20, 2002: 90.7 mhz; 3 kw. Ant -216 ft TL: N41 13 50 W77 08 59. Stereo. Hrs opn: 24
Rebroadcasts WGRC(FM) Lewisburg 100%.
101 Armory Blvd., Lewisburg, 17837. Phone: (570) 523-1190. Fax: (570) 523-1114. E-mail: email@wqrc.com Web Site: www.wgrc.com. Licensee: Salt & Light Media Ministries Inc. Population served: 150,000 Law Firm: Miller & Neely. Wire Svc: AP Format: Contemp Christian. News staff: 3; News: 12 hrs wkly. Target aud: 25-54. ♦Larry Weidman, gen mgr; Lamar Smith, chief of engrg; Linda Dantonio, traf mgr; Jim Diehl, news rptr; John Callahan, news rptr; Chris Miller, engr.

WILQ(FM)— July 31, 1949: 105.1 mhz; 9.2 kw. Ant 1,135 ft TL: N41 11 43 W76 58 18. Stereo. Hrs opn: 24 1685 Four Mile Dr., 17701. Phone: (570) 323-8200. Fax: (570) 323-5075. E-mail: dougd@wilq.com Web Site: www.wilq.com. Licensee: South Williamsport SabreCom Inc. Group owner: Backyard Broadcasting LLC (acq 12-1-02; grpsl). Population served: 350,000 Natl. Rep: Christal. Format: Country. News staff: 4; News: 7 hrs wkly. Target aud: 25 plus; adults in a 10 county area. ♦Barry Drake, pres; Robin Smith, CFO; Dan Farr, gen mgr; Doug Dodge, gen sls mgr; Ted Minier, progmg dir; John Finn, news dir.

WKSB(FM)—Listing follows WRAK(AM).

WLYC(AM)— June 1951: 1050 khz; 1 kw-D, 36 w-N. TL: N41 15 44 W77 01 59. Hrs open: 24 101 Phillips Park Dr., So. Williamsport, 17702-7063. Phone: (570) 327-1300. Fax: (570) 327-1331. E-mail: wlyc1050@yahoo.com Web Site: www.sports1050.com. Licensee: Sentry Communications License LLC (acq 7-19-2005; $75,000). Population served: 350,000 Natl. Network: ESPN Radio, Westwood One. Rgnl rep: . Law Firm: Miller and Neely, PC. Format: Sports. News: Sports news only. Target aud: 25-54; male. ♦James R. McKowne, gen mgr; Jeffrey Andruionis, opns mgr; Christy Andruionis, sls.

***WPTC(FM)**— Sept 3, 1980: 88.1 mhz; 494 w. -101 ft TL: N41 14 11 W77 01 26. Stereo. Hrs opn: 24 One College Ave., 17701. Phone: (570) 326-3761, EXT. 7548. Fax: (570) 320-2423. E-mail: wptc@pct.edu Web Site: www.pct.edu/wptc. Licensee: Pennsylvania College of Technology. (acq 3-14-90). Population served: 99,000 Format: Jazz, modern rock. News: 2 hrs wkly. Target aud: 18-24; college students. Spec prog: Sports 2 hrs, hip hop 5 hrs, metal 5 hrs wkly. ♦Davie Gilmour, pres; Brad Nason, gen mgr.

WRAK(AM)— Apr 10, 1930: 1400 khz; 1 kw-U. TL: N41 14 22 W77 02 27. Hrs open: 24
Rebroadcasts WRKK(AM) Hughesville 100%.
Box 3638, 1559 W. 4th St., 17701. Phone: (570) 327-1400. Fax: (570) 327-8156. E-mail: wrak@wrak.com Web Site: www.wrak.com. Licensee: Clear Channel Radio License Inc. Group owner: Clear Channel Communications Inc. (acq 8-5-98; grpsl). Population served: 37,918 Natl. Network: Westwood One. Format: News/talk, sports. News staff: one; News: 3 hrs wkly. Target aud: 35 plus. ♦James Dabney, gen mgr & gen sls mgr; Tom Scott, mktg dir; Ken Sawyer, progmg dir.

WKSB(FM)—Co-owned with WRAK(AM). Apr 1, 1948: 102.7 mhz; 53 kw. 1,270 ft TL: N41 11 21 W76 58 53. Stereo. 24 E-mail: wksb@wksb.com Web Site: www.wksb.com.71,000 Format: Adult contemp. News staff: one. Target aud: 25-54. ♦Russell Davidson, opns dir; Tom Scott, progmg dir; Tom Turner, asst music dir; Mark Lawrence, pub affrs dir; Dan Milliken, engrg dir; Liz Stroup, traf mgr.

***WRLC(FM)**— Apr 5, 1976: 91.7 mhz; 740 w. -298 ft TL: N41 14 42 W76 59 50. Stereo. Hrs opn: 24 Mass Communication Dept., Lycoming College, 700 College Place, 17701. Phone: (570) 321-4060. Fax: (570) 321-4372. E-mail: koehn@lycoming.edu Web Site: www.lycoming.edu. Licensee: Lycoming College. (acq 1-76). Population served: 35,000 Format: Div, progsv, urban contemp. News: 15 hrs wkly. Target aud: General; Lycoming College & its surrounding communities. Spec prog: Class one hr, gospel 6 hrs, pub affrs 2 hrs, Christian rock 3 hrs, jazz 8 hrs, blues 3 hrs wkly. ♦Steve Koehn, gen mgr.

WRVH(FM)— Aug 16, 1989: 107.9 mhz; 360 w. 1,289 ft TL: N41 12 39 W76 57 17. Stereo. Hrs opn: 24 1685 Four Mile Dr., 17701. Phone: (570) 323-8200. Fax: (570) 327-9138. Licensee: South Williamsport SabreCom Inc. Group owner: Backyard Broadcasting LLC (acq 12-1-02; grpsl). Population served: 311,200 Natl. Network: ABC. Natl. Rep: Christal. Format: Adult contemp. News staff: one; News: one hr wkly. Target aud: 25-54. ♦Dan Farr, gen mgr.

WVRT(FM)—See Mill Hall

***WVYA(FM)**— 2003: 89.7 mhz; 3.3 kw. Ant -16 ft TL: N41 14 54 W77 01 52. Stereo. Hrs opn: 24
Rebroadcasts WVIA-FM Scranton.
100 Wvia Way, Pittston, 18640-6197. Phone: (570) 655-2808. Fax: (570) 655-1180. E-mail: wvianews@epix.net Web Site: www.wvia.org. Licensee: Northeastern Pennsylvania Educational TV Association. Natl. Network: NPR. Law Firm: Dow, Lohnes & Albertson. Wire Svc: AP Format: Class, jazz, news. News staff: one; News: 30 hrs wkly. Target aud: Upscale, mature audience. ♦A. William Kelly, CEO & gen mgr; A William Kelly, pres; Chris Norton, VP; Larry Vojtko, progmg mgr.

WWPA(AM)— May 22, 1949: 1340 khz; 1 kw-U. TL: N41 13 45 W77 00 45. Hrs open: 24 1685 Four Mile Dr., 17701. Phone: (570) 323-8200. Fax: (570) 327-9138. Licensee: South Williamsport SabreCom Inc. Group owner: Backyard Broadcasting LLC (acq 12-1-02; grpsl). Population served: 136,000 Natl. Network: CSN. Natl. Rep: Christal. Format: News/talk. News staff: one; News: 168 hrs wkly. Target aud: 35 plus. Spec prog: Sports. ♦Barry Drake, pres; Robin Smith, CFO; Dan Farr, gen mgr.

WZXR(FM)—See South Williamsport

Wyomissing

***WYTL(FM)**— 2005: 91.7 mhz; 10 w horiz, 320 w vert. Ant 840 ft TL: W40 19 22 W76 11 52. Hrs open: Box 186, Sellersville, 18960. Phone: (215) 721-2141. Fax: (215) 721-9811. Web Site: www.wordfm.org. Licensee: Four Rivers Community Broadcasting Corp. Format: Aduct contemp Christian. ♦David Baker, gen mgr.

York

WARM-FM—Listing follows WSBA(AM).

WOYK(AM)— March 1932: 1350 khz; 5 kw-D, 1 kw-N, DA-N. TL: N39 56 00 W76 49 06. Hrs open: 24 Box 20249, 17402. Phone: (717) 840-0355. Fax: (717) 840-0355. E-mail: teamespn@earthlink.net Web Site: sportsradioespn1350.com. Licensee: WOYK Inc. (acq 12-87). Natl. Network: ESPN Radio. Format: Sports. Target aud: 25-64; men. ♦Douglas George, pres; Vincent Grande, gen mgr; Sam Conrad, opns dir.

WQXA-FM— 1948: 105.7 mhz; 25 kw. Ant 705 ft TL: N39 59 56 W76 41 43. Stereo. Hrs opn: 515 S. 32nd St., Camp Hill, 17011. Phone: (717) 367-7700. Fax: (717) 367-0239. Web Site: www.1057thex.com. Licensee: Citadel Broadcasting Co. (acq 5-29-97; grpsl). Population served: 1,800,000 Format: Active rock. Target aud: 18-49. ♦Cindy Miller, sls dir & gen sls mgr; Claudine DeLorenzo, progmg mgr.

WSBA(AM)— Sept 1, 1942: 910 khz; 5 kw-D, 1 kw-N, DA-2. TL: N39 59 53 W76 44 42. Hrs open: 24 Box 910, 17402-0910. Secondary address: 5989 Susquehanna Plaza Dr. 17406. Phone: (717) 764-1155. Fax: (717) 252-4708. Web Site: www.wsba910.com. Licensee: WSBA Lico Inc. Group owner: Susquehanna Radio Corp. Population served: 1,000,000 Format: News/talk. News staff: 5. Spec prog: Black 3 hrs, farm 4 hrs wkly. ♦Tom Rawker, gen mgr; Bob Popa, prom mgr & chief of engrg; Jim Horn, progmg dir.

WARM-FM—Co-owned with WSBA(AM). Sept 1, 1962: 103.3 mhz; 6.4 kw. 1,305 ft TL: N40 01 38 W76 36 08. Stereo. 24 Web Site: www.warm103.com. Format: Adult contemp. ♦Tom Ranker, VP & stn mgr; Tina Heim, gen sls mgr; Bob Hurbert, prom mgr; Kelly West, progmg dir & disc jockey; Joel Murphy, traf mgr; Gina Koch, spec ev coord; Dennis Wagner, disc jockey; Dennis John Cahill, disc jockey; John London, disc jockey; Larry K. Scott, disc jockey.

WSJW(FM)—See Starview

***WVYC(FM)**— Nov 18, 1976: 99.7 mhz; 370 w. 97 ft TL: N39 56 49 W76 43 47. (CP: 99.7 mhz). Stereo. Hrs opn: 18 York College of Pennsylvania, Country Club Rd., 17405-7199. Phone: (717) 815-1932. Phone: (717) 815-1311. E-mail: tgibson@ycp.edu Web Site: www.ycp.edu/wvyc. Licensee: York College of Pennsylvania. Format: Educ, progsv, rock. News staff: one; News: 3 hrs wkly. Target aud: 14-24; new mus lovers. Spec prog: Class 8 hrs, jazz 8 hrs, Sp one hr wkly. ♦Kevin Peterson, gen mgr.

WYYC(AM)— 1948: 1250 khz; 1 kw-D. TL: N39 59 56 W76 41 43. Stereo. Hrs opn: 24 919 Buckingham Blvd., Elizabethtown, 17022. Phone: (717) 757-9402. Fax: (717) 367-9322. E-mail: wyyc@wilkinsradio.com Web Site: www.wilkinsradio.com. Licensee: Steel City Radio Inc. Group owner: Citadel Broadcasting Corp. (acq 10-11-2005; $250,000). Population served: 1,154,000 Law Firm: Womble, Carlyle, Sandridge & Rice. Format: Chrisitian teaching/talk. News: 10 hrs wkly. Target aud: 35 plus. Spec prog: High school football & basketball, Philadelphia Phillies. ♦Bob Wilkins, pres & stn mgr; LuAnn Wilkins, exec VP; Mitchell Mathis, VP; Bob Moore, gen mgr & stn mgr; Greg Garrett, opns mgr; Roy Landis, engrg VP & engr.

York-Hanover

WYCR(FM)—Licensed to York-Hanover. See Hanover

Youngsville

***WTMV(FM)—** Jan 19, 1999: 88.5 mhz; 100 w. -335 ft TL: N41 51 01 W79 18 41. Stereo. Hrs opn: 24 409 E. Main St., 16371. Phone: (814) 563-4903. Fax: (814) 563-4903. E-mail: wtmv@verizon.net Web Site: www.wtmv.com. Licensee: Living Word of Faith Christian Outreach. Population served: 5,000 Natl. Network: American Family Radio, Moody. Format: Christian. News staff: 5; News: 7-11. Target aud: 21 plus; Christians of all ages. Spec prog: Children 10 hrs, class 2.5 hrs wkly. ♦Rev. Patricia A. Baker, VP & mus dir; Rev. William E. Baker, pres & gen mgr; Kathy Joy, pub affrs dir; Khlare Bracken, mus critic.

Rhode Island

Block Island

WCRI(FM)— June 13, 1994: 95.9 mhz; 6 kw. 174 ft TL: N41 10 21 W71 33 52. Stereo. Hrs opn: 24 19 Railroad Ave, Westerly, 02891. Phone: (401) 596-6795. Fax: (401) 596-6782. E-mail: mail@classical959.com Web Site: www.classical959.com. Licensee: Judson Group Inc. (acq 11-22-2006; $1.6 million with WCNX(AM) Hope Valley). Population served: 50,000 Format: Class. Target aud: General. Spec prog: New age 4 hrs, folk 4 hrs, big band 4 hrs, relg 2 hrs wkly. ♦Christopher S. Jones, pres; Mark Halliday, gen mgr; Michael Abranson, stn mgr.

WJZS(FM)— Oct 3, 1988: 99.3 mhz; 6 kw. Ant 256 ft TL: N41 10 28 W71 34 20. Stereo. Hrs opn: 24 Box 367, Newport, 02480. Phone: (401) 846-1540. Fax: (401) 846-1598. E-mail: blancaster@wjzs.com Web Site: www.wjzs.com. Licensee: Astro Tele-Communications Corp. (acq 8-24-99). Population served: 1,000,000 Law Firm: Shaw Pittman. Format: Adult contemp. Target aud: 30-50; total community. ♦William C. Lancaster, gen mgr; Lynn Abrams, gen sls mgr; Steve Bianchi, progmg dir; Robert Sullivan, news dir; Kate Jennings, pub affrs dir & prom; Maurice B. Polayes, chief of engrg; Lisa Mcguire, traf mgr.

Bristol

***WQRI(FM)—** April 1989: 88.3 mhz; 100 w. 75 ft TL: N41 38 49 W91 15 34. Hrs open: One Old Ferry Rd., Campus Program, 02809. Phone: (401) 254-3283. Phone: (401) 254-3282. Fax: (401) 254-3355. Licensee: Roger Williams University. Format: AOR. ♦Becky Riopel, gen mgr.

Charlestown

WKFD(AM)—Not on air, target date: unknown: 1370 khz; 2.5 kw-D, 5 kw-N, DA-2. TL: N41 22 38 W71 39 50. Hrs open: 24 20 Freeman Pl., Needham, MA, 02192. Phone: (781) 444-4754. Fax: (781) 444-8630. Licensee: Astro Tele-Communications Corp. ♦Maurice B. Polayes, pres.

Coventry

***WCVY(FM)—** Oct 19, 1978: 91.5 mhz; 200 w. 36 ft TL: N41 41 10 W71 35 37. Stereo. Hrs opn: 40 Reservoir Rd., 02816-6404. Phone: (401) 822-9499. Fax: (401) 822-9492. Licensee: Coventry Public Schools. Format: Top-40. Target aud: 12-30. Spec prog: Sports 2 hrs wkly. ♦Jason Murry, stn mgr.

East Greenwich

WARV(AM)—See Warwick

***WRJI(FM)—**Not on air, target date: unknown: 91.5 mhz; 100 w. Ant 157 ft TL: N41 39 35 W71 30 00. Hrs open: 39 Julian St., Providence, 02909. Licensee: Educational Radio for the Public of the New Millennium.

Greenville

WALE(AM)—Licensed to Greenville. See Providence

Hope Valley

WCNX(AM)— Oct 7, 1985: 1180 khz; 1.8 kw-D. TL: N41 31 36 W71 44 35. Stereo. Hrs opn: 19 Railroad Ave., Westerly, 02891. Phone: (401) 596-6795. Fax: (401) 596-6782. Web Site: www.newsradio1180.com. Licensee: Judson Group Inc. (acq 11-22-2006; $1.6 million with WCRI(FM) Block Island). Population served: 750,000 Natl. Network: USA. Law Firm: Smithwick & Belendiuk. Format: Loc news. ♦Christopher S. Jones, pres; Mark Halliday, gen mgr; Mike Abramson, stn mgr & chief of engrg.

Kingston

***WRIU(FM)—** Feb 16, 1964: 90.3 mhz; 3.44 kw. 415 ft TL: N41 29 52 W71 31 42. Stereo. Hrs opn: 24 326 Memorial Union, 02881. Phone: (401) 874-4949. Fax: (401) 874-4349. Web Site: www.wriu.org. Licensee: University of Rhode Island. Population served: 1,000,000 Format: Div, rock. Target aud: Diverse. Spec prog: Folk 15 hrs, gospel 5 hrs, heavy metal 6 hrs, blues 3 hrs, reggae 7 hrs, Sp 3 hrs wkly. ♦James Proctor, gen mgr.

Middletown

WKKB(FM)— Oct 6, 1978: 100.3 mhz; 1.55 kw. Ant 656 ft TL: N41 35 48 W71 11 24. Stereo. Hrs opn: 24 1185 N. Main St., Providence, 02904. Phone: (401) 331-1003. Fax: (401) 521-5077. E-mail: marcklowan@supermaxfm.com Web Site: latina 1003.com. Licensee: Davidson Media Rhode Island Stations LLC. Group owner: Citadel Broadcasting Corp. (acq 1-31-2005; $7.5 million with WAKX(FM) Narragansett Pier). Population served: 40,000 Format: Sp tropical. News staff: one; News: 7 hrs wkly. Target aud: 12+; Latino Americans 1st & 2nd generation. ♦Craig Rapoza, gen mgr; Cesar Salas, gen sls mgr; Enrique Ortaga, progmg VP; Juan Gonzalez, progmg dir.

Narragansett Pier

WRNI-FM— July 15, 1990: 102.7 mhz; 1.95 kw. Ant 226 ft TL: N41 25 27 W71 28 38. Stereo. Hrs opn: 19 One Union Station, Providence, 02903. Phone: (401) 351-2800. Fax: (401) 351-0246. Web Site: www.wrni.org. Licensee: Rhode Island Public Radio Group owner: Citadel Communications Corp. (acq 5-16-2007; $2.56 million). Population served: 1,000,000 Natl. Network: NPR. Format: News/talk. ♦Eugene B. Mihaly, pres; Joe O'Connor, gen mgr.

Newport

WADK(AM)— Nov 6, 1948: 1540 khz; 1 kw-D, 20 w-N. TL: N41 30 13 W71 18 43. Hrs open: 6am-6pm Box 367, 02840. Phone: (401) 846-1540. Fax: (401) 846-1598. E-mail: bangel@wadk.com Web Site: www.wadk.com. Licensee: Astro Tele-Communications Corp. (acq 8-24-99). Population served: 78,000 Natl. Network: ABC, Talk Radio Network, Westwood One. Law Firm: Shaw Pittman. Format: News/talk/sports. News staff: 2; News: 17.5 hrs wkly. Target aud: 35 +. Spec prog: Jazz 7 hrs wkly; gardening 1 hr; Real Estate 1 hr; Law 1 hr; Pet Care 2 hrs; Irish Music 2hrs. ♦Bobb Angel, gen mgr; Lynn Abrams, gen sls mgr; Lisa McGuire, progmg dir & traf mgr; Bob Sullivan, news dir; Art Berluti, pub svc dir.

Pawtucket

WDDZ(AM)— Feb 12, 1950: 550 khz; 1 kw-D, 500 w-N, DA-N. TL: N41 54 20 W71 23 56. Stereo. Hrs opn: 24 hrs 203 Concord St., Suite 453, 02860. Phone: (401) 722-0839. Fax: (401) 722-1459. Licensee: Radio Disney Group LLC. Group owner: ABC Inc. (acq 5-29-2001; $2.05 million). Population served: 30,000 Natl. Network: Radio Disney. Format: Children. Target aud: Children & Teens 3-14, Parents 25-54, esp. moms. ♦Michael Kellogg, gen mgr; Jaccalen Grillo, prom mgr.

Portsmouth

***WJHD(FM)—** Apr 3, 1972: 90.7 mhz; 360 w. 80 ft TL: N41 36 06 W71 16 20. Hrs open: Portsmouth Abbey School, Cory's Ln., 02871. Phone: (401) 683-2000. Fax: (401) 683-5888. Licensee: The Order of St. Benedict. Population served: 12,000 Format: Div. Spec prog: Opera, rock. ♦Edmund Adams, gen mgr.

Providence

WALE(AM)—Greenville, 1948: Stn currently dark. 990 khz; 50 kw-D, 5 kw-N, DA-2. TL: N41 57 18 W71 35 39. Hrs open: 6 AM-midnight 1185 N. Main St., Greenville, 02904. Phone: (401) 521-0990. Fax: (401) 521-5077. Licensee: Cumbre Communications Corp., debtor in possession (acq 8-10-2004). Population served: 2,500,000 Format: Sp. Target aud: . ♦Manolo Pazos, gen mgr.

WBRU(FM)— Feb 21, 1966: 95.5 mhz; 20 kw. 440 ft TL: N41 49 40 W71 22 09. (CP: 50 kw, ant 492 ft. TL: N41 48 28 W71 28 12). Stereo. Hrs opn: 24 88 Benevolent St., 02906. Phone: (401) 272-9550. Fax: (401) 272-9278. E-mail: wbru@wbru.com Web Site: www.wbru.com. Licensee: Brown Broadcasting Service Inc. Population served: 180,000 Format: Alternative Urban contemp. News: 3 hrs wkly. Target aud: 18-34; highly educated professionals. Spec prog: Black 20 hrs, jazz 18 hrs wkly. ♦Anit Jindal, stn mgr; Mark Stackowski, gen sls mgr; Marta La Silva, prom dir; Natalie Rubin, gen mgr & prom dir; Chris Novello, progmg dir; George Mesthos, news dir.

WCTK(FM)—Listing follows WNBH(AM).

***WDOM(FM)—** Mar 15, 1966: 91.3 mhz; 125 w. 130 ft TL: N41 50 39 W71 26 14. Stereo. Hrs opn: 18 Providence College, 02918. Phone: (401) 865-2460. Fax: (401) 865-2822. E-mail: wdom@studentweb.providence.edu Web Site: www.listen.to/wdom. Licensee: Providence College. Population served: 179,213 Format: College alternative. News: one hr wkly. Target aud: General; college students & professionals. Spec prog: Urban contemp 16 hrs, metal 6 hrs, country 2 hrs, classic rock 3 hrs, sports 2 hrs wkly. ♦Scott Seseske, gen mgr, dev dir, pub affrs dir & sports cmtr; Brian Wall, opns dir, mktg dir & progmg dir; Carlin Corrigan, prom dir; Dan Devine, mus dir & mus critic; Jaclyn Schede, asst music dir; Sott Seseske, spec ev coord.

***WELH(FM)—** September 1994: 88.1 mhz; 150 w. 98 ft TL: N41 51 30 W71 19 04. Hrs open: 24 216 Hope St., 02906. Phone: (401) 421-8100. Fax: (401) 751-7674. Web Site: www.wheelerschool.org. Licensee: The Wheeler School. Law Firm: Fletcher, Heald & Hildreth. Format: Div, jazz, Sp. Target aud: General. ♦Dave Schiano, gen mgr.

WHJJ(AM)— Sept 6, 1922: 920 khz; 5 kw-U, DA-N. TL: N41 46 53 W71 19 55. Hrs open: 24 75 Oxford St., 02905. Phone: (401) 781-9979. Fax: (401) 781-9329. Web Site: www.920whjj.com. Licensee: Capstar TX L.P. Group owner: Clear Channel Communications Inc. (acq 8-30-00; grpsl). Population served: 184,000 Natl. Network: CBS. Natl. Rep: Clear Channel. Law Firm: Wilkinson, Barker, Knauer & Quinn. Format: News/talk. Target aud: 35-64. ♦Jim Corwin, gen mgr; Kevin Hickey, sls dir; Bill George, progmg dir.

WHJY(FM)—Co-owned with WHJJ(AM). Mar 14, 1966: 94.1 mhz; 50 kw. 546 ft TL: N41 49 40 W71 22 09. Stereo. Web Site: www.whjy.com.250,000 Format: AOR. Target aud: 18-34; adults. ♦Scott Laudani, progmg dir.

WLKW(AM)—West Warwick, Aug 12, 1986: 1450 khz; 1 kw-U. TL: N41 41 38 W71 31 26. Hrs open: 24 75 Oxford St., 02905. Phone: (401) 467-4366. Fax: (401) 941-2795. E-mail: twall@hallradio.com Licensee: Hall Communications Inc. (group owner; acq 6-4-01; $410,000). Population served: 30,000 Natl. Rep: D & R Radio. Law Firm: Fletcher, Heald and Hildreth. Format: Adult Standards. News staff: one. Target aud: 35-64. Spec prog: Pol 2 hrs wkly. ♦Bonnie Rowbotham, CEO; Arthur Rowbotham, pres; Tom Wall, gen mgr; Steve Giuttari, opns mgr.

WNBH(AM)—New Bedford, MA) May 21, 1921: 1340 khz; 1 kw-U. TL: N41 37 21 W70 55 07. Hrs open: 24
Simulcast with WLKW(AM) West Warwick, RI.
888 Purchase St., New Bedford, MA, 02740. Phone: (508) 979-8003. Phone: (401) 467-4366. Fax: (508) 979-8009. E-mail: twall@hallradio.com Web Site: wnbhradio.com. Licensee: Hall Communications Inc. (group owner; (acq 10-1-66). Population served: 101,777 Natl. Network: ABC. Natl. Rep: D & R Radio. Law Firm: Fletcher, Heald & Hildreth. Format: Btfl mus. News: 3 hrs wkly. Target aud: 35-64. Spec prog: Pol 2 hrs wkly. ♦Bonnie H. Rowbotham, chmn; Arthur J. Rowbotham, pres; Tom Wall, gen mgr & sls dir.

WCTK(FM)—Co-owned with WNBH(AM). Dec 9, 1946: 98.1 mhz; 47.3 kw. 508 ft TL: N41 37 21 W70 55 07. Stereo. 24 75 Oxford St., 02905. Phone: (401) 467-4366. Fax: (401) 941-2795. E-mail: mail@wctk.com Web Site: www.wctk.com.1,594,300 Natl. Rep: D & R Radio. Law Firm: Fletcher, Heald & Hildreth. Format: Country. Target aud: 25-54.

WPMZ(AM)— Apr 15, 1947: 1110 khz; 5 kw-D. TL: N41 49 40 W71 22 09. Hrs open: 1270 Mineral Spring Ave., North Providence, 02904. Phone: (401) 726-8413. Fax: (401) 726-8649. E-mail: wpmz@aol.com Web Site: www.poder1110.com. Licensee: Videomundo Broadcasting Co. L.L.C. (acq 1-27-98; $900,000). Population served: 1,200,000 Format: Sp. Target aud: General. ♦Dilson Mendez, pres; Tony Mendez, gen mgr; Johanna Petrarca, sls dir; Zoilo Garcia, progmg dir.

WPRO(AM)— Oct 16, 1931: 630 khz; 5 kw-U, DA-N. TL: N41 46 28 W71 19 23. Hrs open: 1502 Wampanoag Tr., East Providence, 02915. Phone: (401) 433-4200. Fax: (401) 433-5967. Web Site: www.630wpro.com. Licensee: Citadel Broadcasting Co. Group owner: Citadel Broadcasting Corp. (acq 5-29-97; grpsl). Population served: 179,213 Natl. Rep: McGavren Guild. Format: News/talk, sports. ♦Andrea Scott, gen sls mgr.

WPRO-FM— April 1949: 92.3 mhz; 39 kw. 550 ft TL: N41 48 18 W71 28 24. (CP: 45.4 kw, ant 489 ft.). Stereo. Web Site: www.92wpro.com.346,100 Format: CHR. ♦Steve Maully, rgnl sls mgr; Tony Brisco, progmg dir.

WRNI(AM)— April 1948: 1290 khz; 5 kw-U, DA-2. TL: N41 51 21 W71 26 41. Hrs open: 24 One Union Station, 02903. Phone: (401) 351-2800. Fax: (401) 351-0246. E-mail: info@wrni.org Web Site: www.wrni.org. Licensee: WRNI Foundation (acq 7-1-98). Natl. Network: NPR, PRI. Natl. Rep: Rgnl Reps. Format: News/talk. News staff: 8; News: 80 hrs wkly. Target aud: 25-54; intelligent adults interested in news & politics. ♦Joe O'Connor, gen mgr; Steve Callahan, chief of engrg.

WSKO(AM)— June 2, 1922: 790 khz; 5 kw-U, DA-N. TL: N41 50 03 W71 21 56. Stereo. Hrs opn: 1502 Wampanoag Tr., East Providence, 02915. Phone: (401) 433-4200. Fax: (401) 433-5967. Web Site: www.790thescore.com. Licensee: Citadel Broadcasting Co. Group owner: Citadel Broadcasting Corp. (acq 5-29-97; grpsl). Population served: 1,792,130 Natl. Rep: McGavren Guild. Format: All sports. Target aud: 25-64; upper class, affluent, college educated. ♦Ron St. Pierre, VP; Chris Gardiner, gen mgr; Steve Maully, rgnl sls mgr & prom mgr; Kristen Chudy, prom mgr; Tony Brisco, progmg dir.

WWLI(FM)—Co-owned with WSKO(AM). July 11, 1948: 105.1 mhz; 50 kw. 500 ft TL: N41 48 22 W71 28 12. Stereo. Web Site: www.lite105.com.220,000 Format: Adult contemp. Target aud: 25-54; mid to upper income professionals, general appeal format.

WSNE-FM—See Taunton, MA

WSTL(AM)— June 16, 1946: 1220 khz; 1 kw-D, 166 w-N. TL: N41 49 15 W71 23 07. Hrs open: 24 95 Sagamore Rd., Seekonk, MA, 02771-3428. Phone: (508) 336-4233. Fax: (508) 336-5789. E-mail: patricia.varner@shineradio.us Web Site: shineradio.us. Licensee: New England Christian Media Inc. (acq 10-1-2006; $1.9 million). Population served: 79,000 Format: Christian broadcasting. ♦Patricia Varner, gen mgr; Mike Laliberte, progmg dir.

WWBB(FM)— June 7, 1968: 101.5 mhz; 13.5 kw horiz, 12 kw vert. 951 ft TL: N41 52 13 W71 17 47. Stereo. Hrs opn: 24 75 Oxford St., 3rd Fl., 02905. Phone: (401) 781-9979. Fax: (401) 781-9329. Web Site: www.b101.com. Licensee: Clear Channel Radio Licenses Inc. Group owner: Clear Channel Communications Inc. Population served: 1,278,800 Natl. Network: AP Radio, Premiere Radio Networks. Natl. Rep: Clear Channel. Format: Classic hits. Target aud: 35-54; indispensable & powerful adults. ♦Jim Corwin, gen mgr; Mark Cottey, sls dir & pub affrs dir; Michelle Maker, mktg dir; Steve Lariviere, chief of engrg.

Smithfield

***WJMF(FM)**— Aug 1, 1974: 88.7 mhz; 225 w. 130 ft TL: N41 55 13 W71 32 26. Stereo. Hrs opn: 7 AM-2 AM Box 6, Bryant College, 1150 Douglas Pike, 02917. Phone: (401) 232-6044. Phone: (401) 232-6160. Fax: (401) 232-6748. Web Site: www.wjmf887.com. Licensee: Bryant College of Business Administration. Population served: 10,000 Format: Urban, alternative. News staff: one; News: 12 hrs wkly. Target aud: 16-30; from teenagers to young executives. Spec prog: Folk 4 hrs, gospel 2 hrs, relg 2 hrs wkly. ♦Bryan Adams, gen mgr.

Wakefield-Peacedale

WSKO-FM— June 1995: 99.7 mhz; 2.3 kw. 535 ft TL: N41 25 31 W71 34 59. Hrs open: 24 1502 Wampanoag Trail, East Providence, 02915. Phone: (401) 433-4200. Fax: (401) 437-3297. Web Site: www.790thescore.com. Licensee: Citadel Broadcasting Co. Group owner: Citadel Broadcasting Corp. (acq 8-6-97; $8.5 million with WKKB(FM) Middletown). Population served: 1,100,000 Natl. Network: Westwood One. Law Firm: Wiley, Rein & Fielding. Format: Sports. ♦Barbara Haynes, gen mgr; Duffy Egan, chief of engrg.

Warwick

WARV(AM)— Aug 12, 1959: 1590 khz; 5 kw-U, DA-2. TL: N41 43 40 W71 27 46. Hrs open: 24 19 Luther Ave., 02886. Phone: (401) 737-0700. Fax: (401) 737-1604. E-mail: warv@aol.com Web Site: www.warv.net. Licensee: Blount Communications Inc. Group owner: Blount Communications Group (acq 7-7-78). Population served: 1,500,000 Natl. Network: Salem Radio Network. Format: Relg. Target aud: 25-54; Adults. Spec prog: Black 2 hrs wkly. ♦Deborah C. Blount, exec VP; David O. Young, VP; William A. Blount, pres & gen mgr; Kevin Linegan, opns mgr.

West Warwick

WLKW(AM)—Licensed to West Warwick. See Providence

Westerly

WEEI-FM— Oct 17, 1967: 103.7 mhz; 37 kw. 570 ft TL: N41 34 22 W71 37 55. Stereo. Hrs opn: 24 150 Chestnut St., Providence, 02903. Phone: (401) 751-9334. Fax: (401) 351-8109. Web Site: www.fnxradio.com. Licensee: Entercom Providence License LLC. Group owner: Entercom Communications Corp. (acq 6-15-2004; $14.5 million). Population served: 250,000 Natl. Network: Westwood One. Format: Sports, talk. News staff: one. Target aud: 25-49. ♦David J. Field, CEO; Joseph M. Field, chmn & VP; Joseph Harrington, stn mgr; Rod Morrison, prom dir & prom mgr.

***WKIV(FM)**— Dec 8, 1997: 88.1 mhz; 100 w. Ant 66 ft TL: N41 21 16 W71 46 12. Stereo. Hrs opn: Southern Rhode Island Public Radio, 244 Post Rd., 02891. Phone: (401) 322-9091. Fax: (401) 322-1645. Web Site: www.klove.com. Licensee: Southern Rhode Island Public Radio Broadcasting Inc. Natl. Network: K-Love. Format: Contemp Christian. ♦Chris Di Paola, pres & gen mgr; Vito Di Paola, exec VP; Dade Nunez, opns mgr.

WXNI(AM)— July 1949: 1230 khz; 1 kw-U. TL: N41 21 57 W71 50 11. Hrs open:
Rebroadcasts WRNI(AM) Providence 100%.
One Union Station, Providence, 02903. Phone: (401) 351-2800. Fax: (401) 351-0246. E-mail: info@wrni.org Web Site: www.wrni.org. Licensee: WRNI Foundation (acq 3-26-99). Natl. Network: NPR, PRI. Format: News, talk. News: 80 hrs wkly. Target aud: 25-54; intelligent adults interested in news & politics. ♦Joe O'Connor, gen mgr; Steve Callahan, chief of engrg.

Wickford

WKKB(FM)—See Middletown

Woonsocket

WNRI(AM)— Nov 28, 1954: 1380 khz; 2.5 kw-D, 16 w-N. TL: N42 00 58 W71 29 30. Stereo. Hrs opn: 786 Diamond Hill Rd., 02895. Phone: (401) 769-6925. Fax: (401) 762-0442. E-mail: rogerwnri@prodigy.net Web Site: www.wnri.com. Licensee: Bouchard Broadcasting Inc. (group owner; (acq 10-20-2004; $900,000). Population served: 100,000 Natl. Network: USA. Format: News/talk. Target aud: 35 plus. Spec prog: Fr 4 hrs, Pol 2 hrs, Por 2 hrs wkly. ♦Roger Bouchard, gen mgr.

WOON(AM)— Nov 11, 1946: 1240 khz; 1 kw-U. TL: N41 59 35 W71 30 33. (CP: TL: N41 59 34 W71 30 20). Hrs opn: One Social St., 02895-3136. Phone: (401) 762-1240. Fax: (401) 769-8232. E-mail: email@onworldwide.com Web Site: www.onworldwide.com. Licensee: O-N Radio Inc. (acq 10-19-99). Natl. Network: CBS. Format: Full service. Target aud: 35 plus. Spec prog: Fr one hr, Pol 3 hrs, Black one hr, gospel one hr wkly. ♦Dave Richards, gen mgr.

WWKX(FM)— July 1, 1949: 106.3 mhz; 1.5 kw. 518 ft TL: N41 59 43 W71 26 54. (CP: Ant 520 ft.). Stereo. Hrs opn: 1502 Wampanoaq Tr., East Providence, 02915. Phone: (401) 433-4200. Fax: (401) 433-5967. E-mail: hot1063@hot1063.com Web Site: www.hot1063.com. Licensee: Citadel Broadcasting Co. Group owner: Citadel Communications Corp. (acq 1-24-2005; $16.5 million with WAKX(FM) Narragansett Pier). Population served: 900,000 Natl. Network: Westwood One. Rgnl rep: Christal. Format: Rhythm/dance, CHR. Target aud: 18-49. ♦Barbara Haynes, gen mgr; Duffy Egan, chief of engrg.

South Carolina

Abbeville

WABV(AM)— March 1956: 1590 khz; 1 kw-D, 27 w-N. TL: N34 09 03 W82 23 34. Hrs open: 75 Hwy. 28 South, 29620. Phone: (864) 366-9228. E-mail: info@wabv1590.com Web Site: www.walkerbroadcasting.com/wabv.html. Licensee: Hellinger Broadcasting Inc. (acq 4-14-97). Population served: 100,000 Format: Country. ♦Mark Hellinger, pres; Paul B. Walker Jr., opns mgr & progmg dir.

WZLA-FM— Jan 1, 1990: 92.9 mhz; 6 kw. Ant 243 ft TL: N34 11 13 W82 19 28. Stereo. Hrs opn: 24 Box 548, 29620. Secondary address: 112 N. Main St. 29620. Phone: (864) 366-5785. Fax: (864) 366-9391. E-mail: z93@wctel.net Web Site: z93oldies.com. Licensee: Shelley Reid. Natl. Network: Motor Racing Net, Salem Radio Network. Law Firm: Fletcher, Heald & Hildreth. Format: Oldies. Target aud: 25-65. Spec prog: Gospel 8 hrs wkly. ♦Shelley Reid, pres, gen mgr, opns dir, progmg dir & chief of engrg; Oscar H. Reid Jr., stn mgr & gen sls mgr; Oscar Reid, traf mgr; Wayne Stevenson, sports cmtr.

Aiken

WGUS-FM—See New Ellenton

WKSP(FM)— Sept 17, 1966: 96.3 mhz; 17.5 kw. Ant 846 ft TL: N33 41 06 W81 55 36. Hrs open: 2743 Perimeter Pkwy. Bldg. 100, Suite 300, Augusta, GA, 30909. Phone: (706) 396-6000. Fax: (706) 396-6010. Web Site: www.kiss963.com. Licensee: Capstar TX L.P. Group owner: Clear Channel Communications Inc. (acq 12-19-00; grpsl). Format: Rhythm and blues, oldies. ♦Mark Bass, gen mgr.

WKXC-FM— August 1966: 99.5 mhz; 22.5 kw. 728 ft TL: N33 38 44 W21 55 40. Stereo. Hrs opn: 4051 Jimmie Dyess Pky, Augusta, GA, 30909. Phone: (706) 396-7000. Fax: (706) 396-7092. Web Site: www.kicks99.com. Licensee: WGAC License LLC. Group owner: Beasley Broadcast Group Inc. (acq 4-2-2001; $12 million with WHHD(FM) Clearwater). Population served: 45,000 Format: Country. ♦Kent Dunn, gen mgr; Mark Haddon, gen sls mgr; T. Gentry, progmg dir.

***WLJK(FM)**— 1990: 89.1 mhz; 10 kw. 1,374 ft TL: N33 24 18 W81 50 15. Hrs open: 24
Rebroadcasta WRJA-FM Columbia 100%.
1101 George Rogers Blvd., Columbia, 29201. Phone: (803) 737-3420. Fax: (803) 737-3552. E-mail: gasque@scetv.org Web Site: www.scern.org. Licensee: South Carolina Educational TV Commission. Natl. Network: NPR, PRI. Law Firm: Dow, Lohnes & Albertson. Format: Talk, NPR news. ♦Moss Bresnahan, pres; Paul Zweimiller, VP & stn mgr; Tom Holloway, dev dir.

Allendale

WDOG(AM)— Jan 1, 1966: 1460 khz; 1 kw-D. TL: N33 01 22 W81 19 58. (CP: COL Barnwell. 320 w-D, 45 w-N. TL: N33 13 25 W81 21 35). Hrs opn: 2447 Agusta Hwy., 29810. Phone: (803) 584-3500. Fax: (240) 358-7473. Licensee: Good Radio Broadcasting Inc. Population served: 50,000 Rgnl. Network: S.C. Net. Format: C&W, Black. ♦H. Carl Gooding, pres, gen mgr, gen sls mgr & chief of engrg; Rick Gooding, prom mgr, progmg dir & disc jockey; Lisa Gooding, news dir, traf mgr & women's int ed; Jim Lowe, local news ed & disc jockey; Carl Gooding, farm dir; Ron Lopez, disc jockey.

WDOG-FM— Aug 29, 1983: 93.5 mhz; 3 kw. 300 ft TL: N33 01 22 W81 19 58. Stereo. ♦Lisa Gooding, traf mgr & women's int ed; Jim Lowe, local news ed & disc jockey; Carl Gooding, farm dir; Rick Gooding, disc jockey; Ron Lopez, disc jockey.

Anderson

WAIM(AM)— April 1935: 1230 khz; 1 kw-U. TL: N34 31 52 W82 36 50. Hrs open: 2203 Old Williamston Rd., 29621. Phone: (864) 226-1511. Phone: (864) 225-1230. Fax: (864) 226-1513. E-mail: waimrd@carol.net Licensee: Palmetto Broadcasting Corp FTR: 10-19-92) Population served: 50,000 Format: News/talk. Target aud: 25-64. ♦Rick Driver, gen mgr, gen sls mgr, progmg dir & disc jockey; Craig More, disc jockey; Dan Cooley, disc jockey; Dave Chastain, disc jockey; Dave Shannon, disc jockey; Deb Kent, disc jockey; Terry Mitchell, disc jockey.

WANS(AM)— June 1, 1949: 1280 khz; 5 kw-D, 1 kw-N, DA-N. TL: N34 32 17 W82 41 28. Hrs open: 24 141 Powell Rd., 29625. Phone: (864) 224-9267. Fax: (864) 224-8841. Web Site: www.wans1280.com. Licensee: FM 103 Inc. (acq 10-28-96). Population served: 30,000 Format: Gospel. ♦Ray Morris, gen mgr.

WJMZ-FM— Aug 1, 1963: 107.3 mhz; 100 kw. 1,008 ft TL: N34 42 06 W82 36 20. Stereo. Hrs opn: 24 220 N. Main St., Suite 402, Greenville, 29601. Phone: (864) 235-1073. Fax: (864) 370-3403. Web Site: www.1073jamz.com. Licensee: Cox Radio Inc. Group owner: Cox Communications Inc. (acq 2-1-2001; grpsl). Population served: 1,408,400 Format: Urban contemp.Steve Sinicropi, VP & gen mgr; Bob Grossmall, gen sls mgr; Cathy Tabor, natl sls mgr; Laurie Madden, mktg dir & prom mgr; Doug Davis, mus dir & disc jockey; K.J. Bland, asst music dir; Ed Bailey, news dir; Lemont Bryant, chief of engrg; Kenny Mac Miles, disc jockey; Malcolm Rockhold, disc jockey; Rob Laidbac, disc jockey; Stanlley Toole, disc jockey

WROQ(FM)— 1947: 101.1 mhz; 100 kw. 994 ft TL: N34 38 51 W82 16 13. Stereo. Hrs opn: 25 Garlington Rd., Greenville, 29615. Phone: (864) 271-9200. Fax: (864) 242-1567. Web Site: www.wroq.com. Licensee: Entercom Greenville License LLC. Group owner: Barnstable Broadcasting Inc. (acq 10-7-2005; grpsl). Population served: 675,000 Format: Classic rock. Target aud: 25-54; baby boomers. ♦David J. Field, pres; Sharon Day, gen mgr; Mark Hendrix, progmg dir.

WTBI(AM)—See Pickens

Andrews

WGTN-FM—Licensed to Andrews. See Georgetown

Atlantic Beach

WMIR(AM)— Oct 1, 1997: 1200 khz; 690 w-D. TL: N33 50 10 W78 51 08. Hrs open: Sunrise-sunset 4337 Big Barn Dr., Little River, 29566. Phone: (843) 399-2653. Fax: (843) 399-2659. E-mail: reggiedyson@juno.com Licensee: Atlantic Beach Radio Inc. (acq 1-31-97). Population served: 175,000 Format: Relg. Target aud: 25-65; Urban Black gospel. ♦Dr. Gardner Altman, pres; Reggie Dyson, CEO & gen mgr.

WSEA(FM)— 1998: 100.3 mhz; 12 kw. Ant 476 ft TL: N33 47 03 W78 52 44. Hrs open: 11640 Hwy. 17 Bypass S., Murrells Inlet, 29576-9332. Phone: (843) 651-7869. Fax: (843) 651-3197. Web Site: www.hot100fm.com. Licensee: Cumulus Licensing Corp. Group owner: Cumulus Media Inc. (acq 7-16-98; $1.3 million). Format: CHR. ♦Bill Hazen, gen mgr.

Bamberg

WWBD(FM)—Licensed to Bamberg. See Bamberg-Denmark

Bamberg-Denmark

WVCD(AM)— June 23, 1957: 790 khz; 1 kw-D, 100 w-N. TL: N33 18 50 W81 04 43. Hrs open: Box 678, Denmark, 29042. Phone: (803) 703-7002. Fax: (803) 703-7022. Licensee: Voorhees College (acq 1-29-03; $112,500). Population served: 10,000 Rgnl. Network: S.C. Net. Format: Relg. Spec prog: Farm one hr wkly. ♦Annette Gantt, gen mgr.

WWBD(FM)—Bamberg, May 1967: 95.7 mhz; 6 kw. 308 ft TL: N33 18 50 W81 04 43. Stereo. Hrs opn: 200 Regional Pkwy., Bldg. C, Suite 200, Bamberg, 29118. Phone: (803) 536-1710. Fax: (803) 531-1089. E-mail: advertising@bd957.com Web Site: www.baddog957.com. Licensee: Miller Communications Inc. (group owner; acq 7-31-2003; $850,000). Population served: 20,000 Format: Rock & roll classics. ♦Harold T. Miller Jr., pres.

Batesburg

WBLR(AM)— May 10, 1956: 1430 khz; 5 kw-D, 142 w-N. TL: N33 54 58 W81 31 42. Hrs open: 24 2278 Wortham Lane, Grovetown, GA, 30813. Phone: (706) 309-9610. Fax: (706) 309-9669. E-mail: ctbarinowski@comcast.net Web Site: www.gnnradio.org. Licensee: Barinowski Investment Company Group owner: Good News Network. (acq 8-18-98). Population served: 45,000 Rgnl. Network: S.C. Net. Format: Sp. Target aud: General. ♦Clarence Barinowski, pres & gen mgr.

WZMJ(FM)— Aug 5, 1965: 93.1 mhz; 2.1 kw. Ant 561 ft TL: N33 54 02 W81 24 25. Stereo. Hrs opn:
Simulcast with WOIC(AM) Columbia 100%.
1900 Pineview Rd., Columbia, 29209. Phone: (803) 695-8600. Fax: (803) 695-8605. E-mail: mhanisch@innercity.sc.com Licensee: Urban Radio II L.L.C. Group owner: Inner City Broadcasting (acq 5-30-2003; $11.1 million with WHXT(FM) Orangeburg). Population served: 472,800 Natl. Network: ESPN Radio, Jones Radio Networks. Format: Sports. ♦Steve Patterson, gen mgr; Scott Norton, gen sls mgr; Dave Stewart, progmg dir & chief of engrg.

Beaufort

***WAGP(FM)**— Oct 10, 1987: 88.7 mhz; 6 kw. 302 ft TL: N32 24 05 W80 44 21. Stereo. Hrs opn: 24 Box 119, 29901. Secondary address: 4 Grober Hill Rd., Suite C 29901. Phone: (843) 525-1859. Fax: (843) 522-3691. E-mail: waagp@islc.net Web Site: www.wagp.net. Licensee: The Christian Broadcasting Corp. of Beaufort. Natl. Network: Moody. Format: Relg. News: 20 hrs wkly. Target aud: General; evangelical Christians. ♦Carl J. Broggi, pres; Richard Forschner, gen mgr.

WGZO(FM)—See Parris Island

***WJWJ-FM**— Aug 1, 1980: 89.9 mhz; 47 kw. Ant 1,096 ft TL: N32 42 42 W80 40 54. Hrs open: 24
Rebroadcasts WRJA(FM) Sumter 100%.
1101 George Rogers Blvd., Columbia, 29201. Phone: (803) 737-3420. Fax: (803) 737-3552. E-mail: gasque@scetv.org Web Site: www.etvradio.org. Licensee: South Carolina Educational TV Commission. Natl. Network: NPR. Format: News info. News: 25 hrs wkly. Target aud: General. ♦Moss Brenahan, pres & gen mgr; Tom Fowler, VP; Paul Zweimiller, stn mgr.

WVGB(AM)— 1959: 1490 khz; 1 kw-U. TL: N32 26 08 W80 41 54. Hrs open: 806 Monson St., 29901. Phone: (843) 524-4700. Phone: (843) 524-9742. Fax: (843) 524-1329. E-mail: vgbradio@earthlink.net Licensee: Vivian Broadcasting Inc. (acq 1-21-83). Population served: 95,000 Natl. Network: American Urban. Format: Relg, gospel. Target aud: 18-65; African American. Spec prog: Community progmg, sports. ♦William A. Galloway, pres; Vivian M. Galloway, gen mgr, opns VP & sls VP; Darryl Jamison, stn mgr; Derrick Moon, progmg mgr.

WYKZ(FM)— Aug 8, 1962: 98.7 mhz; 100 kw. 1,001 ft TL: N32 19 50 W80 56 19. Stereo. Hrs opn: 24 245 Alfred St., Savannah, GA, 31408. Phone: (912) 964-7794. Fax: (912) 964-9414. Web Site: www.987theriver.com. Licensee: Capstar TX L.P. Group owner: Clear Channel Communications Inc. (acq 8-30-00; grpsl). Population served: 350,000 Law Firm: Wiley, Rein & Fielding. Format: Light adult contemp. News staff: one; News: 3 hrs wkly. Target aud: 25-54; female. Spec prog: Oldies 5 hrs wkly. ♦Craig Scott, gen mgr.

Belton

***WEPC(FM)**— May 1994: 88.5 mhz; 50 kw. 298 ft TL: N34 23 43 W82 29 49. Hrs open: 24
Rebroadcasts WRAF-FM Toccoa Falls, GA 100%.
Secondary address: 292 Old Clarkesville Hwy., Toccoa Falls, GA 30577. Phone: (706) 282-6030. Phone: (800) 251-8326. Fax: (706) 282-6090. E-mail: tfcrn@tfc.edu Web Site: www.myfavoritestation.net. Licensee: Toccoa Falls College. Format: Christian, MOR, educ. ♦David Cornelius, gen mgr.

WLUA(AM)— October 1956: Stn currently dark. 1390 khz; 1 kw-D, 17 w-N. TL: N34 35 19 W82 32 17. Hrs open: Box 646, 29627. Phone: (864) 338-7742. Fax: (864) 338-7743. Licensee: Robert Earl Bryson (acq 4-2-97; $4,000). Population served: 145,000 Format: Christian. Target aud: General. Spec prog: Black gospel 8 hrs wkly. ♦Robert Earl Bryson, gen mgr.

Belvedere

***WAFJ(FM)**— August 1994: 88.3 mhz; 4.5 kw. 1,387 ft TL: N33 24 29 W81 50 36. Stereo. Hrs opn: 24 102 LeCompte Ave., N. Augusta, 29841. Phone: (803) 819-3125. Fax: (803) 819-3129. E-mail: info@wafj.com Web Site: www.wafj.com. Licensee: Radio Training Network Inc. (acq 1994; $291,000). Format: Contemp Christian. News staff: one. ♦Steve Swanson, gen mgr, progmg dir & mus dir; Cleve Walker, news dir.

Bennettsville

WBSC(AM)— June 1947: 1550 khz; 10 kw-D, 5 kw-N, DA-N. TL: N34 40 52 W79 42 04. Hrs open: Box 1275, 29512-1275. Phone: (843) 479-7121. Fax: (843) 479- 4474. E-mail: wbsc@aol.com Web Site: www.wbsc1550.com. Licensee: D. Mitch Broadcasting Inc. (acq 4-95). Population served: 87,000 Natl. Network: ABC. Natl. Rep: Dora-Clayton. Format: Oldies, gospel. Spec prog: Black 15 hrs wkly. ♦Dwight Johnson, CEO, pres, VP & gen mgr; Richard Gehm, chief of opns.

Bishopville

WAGS(AM)— Feb 24, 1954: 1380 khz; 1 kw-D. TL: N34 12 35 W80 13 34. Hrs open: 6:30 AM-6 PM 142 Wags Dr., 29010. Phone: (803) 484-5415. E-mail: wagsradio@sc.rr.com Licensee: Beaver Communications (acq 11-01-99; $27,500). Population served: 22,000 Natl. Network: USA. Format: Country, bluegrass Live radio, Real people in Real time. News: 6 hrs wkly. Target aud: 25-55 plus. Spec prog: Live remotes-parades, civic events, festivals 2 hrs, relg 7 hrs wkly. ♦James D. Jenkins, gen mgr & chief of opns.

Blackville

WIIZ(FM)— April 1996: 97.9 mhz; 50 kw. 433 ft TL: N33 06 52 W81 23 13. Hrs open: 8968 Marlboro Ave., Barnwell, 29812. Phone: (803) 259-9797. Fax: (803) 541-9700. Licensee: NicWild Communications Inc. (acq 10-9-96; $340,000). Format: Urban contemp. ♦Bobby Nichols, gen mgr & progmg dir.

Bluffton

WGZR(FM)— June 22, 1988: 106.9 mhz; 100 kw. Ant 800 ft TL: N32 13 36 W80 50 53. Stereo. Hrs opn: 24 401 Mall Blvd., Suite 101 D, Savannah, GA, 31406. Phone: (912) 351-9830. Fax: (912) 352-4821. Web Site: luckydogcountry1069.com. Licensee: Monterey Licenses LLC. Group owner: Triad Broadcasting Co. LLC (acq 8-8-2000; grpsl). Population served: 500,000 Natl. Rep: Christal. Format: Country. News staff: one. ♦Robert Leonard, gen mgr.

Blythwood

WBAJ(AM)— 1999: 890 khz; 50 kw-D, 8.5 kw-CH. TL: N34 06 31 W81 04 28. Hrs open: 241-A Riverchase Way, Lexington, 29072. Phone: (803) 794-9673. E-mail: radio@wbaj.net Web Site: www.wbaj.net. Licensee: Family First. (acq 8-29-98; $60,000). Format: Christian. ♦Linda de Romanett, pres & gen mgr.

Bowman

WSPX(FM)— October 1997: 94.5 mhz; 3.5 kw. Ant 434 ft TL: N33 19 13 W80 43 52. Hrs open: 24 Box 1445, Orangeburg, 29116. Secondary address: 1236 Five Chop Rd., Orangeburg 29115. Phone: (803) 539-9450. Fax: (803) 539-9458. E-mail: email@wfmv.com Web Site: wspx@sc.rr.com. Licensee: Glory Communications Inc. (group owner; acq 4-19-01; $400,000). Population served: 175,000 Format: Gospel. ♦Alex Snipes Jr., gen mgr.

Branchville

WGFG(FM)— Dec 13,1993: 105.1 mhz; 2.9 kw. Ant 478 ft TL: N33 26 35 W80 48 16. Hrs open: 24 200 Regional Pkwy., Bldg. C, Suite 200, Orangeburg, 29118. Phone: (803) 536-1710. Fax: (803) 531-1089. E-mail: mail@miller.fm Web Site: www.miller.fm. Licensee: Miller Communications Inc. (group owner; (acq 4-30-2003; $1.25 million with WQKI-FM Orangeburg). Population served: 175,000 Natl. Network: ABC. Format: Oldies. Target aud: 25-64; baby boomers. ♦Harold Miller Jr., pres; Theresa Miller, gen mgr; Russ T. Fender, opns mgr.

Briarcliff Acres

WQSD(FM)— Apr 5, 1975: 107.1 mhz; 50 kw. Ant 492 ft TL: N33 56 14 W78 57 53. Stereo. Hrs opn: 18 4841 Hwy. 17 Bypass S., Myrtle Beach, 29577. Phone: (843) 293-0107. Fax: (843) 293-1717. Web Site: www.thesound1071.com. Licensee: Qantum of Myrtle Beach License Co. LLC. Group owner: Qantum Communications Corp. (acq 7-2-2003; grpsl). Format: Classic Rock. ♦Jimmy Feuger, gen mgr.

Bucksport

WGTR(FM)— June 1, 1993: 107.9 mhz; 20 kw. Ant 784 ft TL: N33 35 45 W79 03 11. Hrs open: 4841 Hwy.17 by-pass South, Myrtle Beach, 29577. Phone: (843) 293-0107. Fax: (843) 293-1717. Web Site: www.gator1079.com. Licensee: Qantum of Myrtle Beach License Co. LLC. Group owner: Qantum Communications Corp. (acq 7-2-2003; grpsl). Population served: 271,400 Format: Country. Spec prog: Motor racing 8 hrs wkly. ♦Jimmy Feuger, gen mgr.

Camden

WCAM(AM)— July 23, 1948: 1590 khz; 1 kw-D, 27 w-N. TL: N34 13 36 W80 40 45. Stereo. Hrs opn: 6 AM-11 PM Box 753, 29020. Secondary address: 5 The Commons Ward Rd., Lugoff 29078. Phone: (803) 438-9002. Fax: (803) 408-2288. Web Site: www.kol1027.com. Licensee: Kershaw Radio Corp. (acq 8-87; $75,000; FTR: 5-5-86). Population served: 42,000 Rgnl. Network: S.C. Net. Format: Nostalgia, adult standards. Target aud: 45 plus. ♦Chris Johnson, gen mgr; Bill Rogers, progmg dir.

WPUB-FM—Co-owned with WCAM(AM). December 1974: 102.7 mhz; 3.3 kw. 299 ft TL: N34 13 31 W80 40 44. Stereo. Web Site: www.kol1027.com.44,000 Format: Oldies. Target aud: 25-55.

WEAF(AM)— Dec 10, 1970: 1130 khz; 1 kw-D, 7 w-N. TL: N34 15 32 W80 34 47. Hrs open: Box 1165, 29021. Phone: (803) 939-9469. Licensee: Jeff Andrulonis Group owner: GHB Radio Group (acq 10-13-2004; $200,000). Format: Gospel. ♦Jeffrey M. Andrulonis, gen mgr.

Cameron

WTQS(AM)—Not on air, target date: unknown: 1490 khz; 1 kw-U. TL: N33 32 54 W80 44 20. Hrs open: Box 2355, West Columbia, 29171. Phone: (803) 939-9530. Fax: (803) 939-9469. Licensee: Glory Communications Inc. (acq 2-10-2006; $50,000 for CP). ♦Alex Snipe, pres.

Cayce

WGCV(AM)— Aug 22, 1958: 620 khz; 2.5 kw-D, 126 w-N. TL: N33 57 34 W81 02 28. Hrs open: 24 2440 Millwood Ave., Colombia, 29205. Phone: (803) 748-9620. Fax: (803) 799-1620. E-mail: wgcvproduction@wgcv.net Web Site: www.wgcv.net. Licensee: Glory Communications Inc. (group owner; acq 10-8-99). Natl. Network: American Urban. Format: Gospel. News staff: one. Target aud: 34-65; Black adults. ♦Alex Snipe, pres & gen mgr; Rev. Isaac Heyward, stn mgr & prom dir; Tezra Haire, gen sls mgr; Tony Green, progmg dir.

WLTY(FM)—Licensed to Cayce. See Columbia

***WYFV(FM)**— Oct 10, 1990: 88.5 mhz; 50 kw. Ant 171 ft TL: N33 54 32 W81 05 57. Stereo. Hrs opn: 24 Bible Broadcasting Network, Charlotte, NC, 28214-7300. Phone: (800) 888-7077. Web Site: bbnradio.org. Licensee: Bible Broadcasting Network Inc. (group owner; acq 6-26-90; FTR: 7-23-90). Format: Relg, bible preaching & teaching. News: 10 hrs wkly. Target aud: General. ♦Lowell Davey, pres.

Charleston

WALC(FM)— Apr 4, 1990: 100.5 mhz; 17.5 kw. Ant 394 ft TL: N32 49 20 W79 58 45. Stereo. Hrs opn: 24 950 Houston Northcutt Blvd., Ste. 201, Mt. Pleasant, 29464. Phone: (843) 884-2534. Fax: (843) 884-1218. Web Site: www.thedrive100.com. Licensee: Citicasters Licenses L.P. Group owner: Clear Channel Communications Inc. (acq 5-4-99; grpsl). Natl. Rep: Katz Radio. Format: Adult contemp. ♦Alene Grevey, gen mgr.

WAVF(FM)—See Hanahan

WEZL(FM)— Oct 3, 1970: 103.5 mhz; 100 kw. Ant 659 ft TL: N32 49 04 W79 50 08. (CP: Ant 987 ft.). Stereo. Hrs opn: 24 950 Houston Northcutt Blvd., Mount Pleasant, 29464. Phone: (843) 884-2534. Fax: (843) 884-1218. Web Site: www.wezlfm.com. Licensee: Citicasters Licenses L.P. Group owner: Clear Channel Communications Inc. (acq 5-4-99; grpsl). Population served: 420,800 Format: C&W. News staff: one; News: 3 to 4 hrs wkly. ♦Paul Smith, gen mgr; Scott Johnson, progmg mgr; Teri Hegel, mus dir & sls; Brian Worboys, traf mgr & outdoor ed; Willie Bennett, news dir & engr.

***WFCH(FM)**— December 1986: 88.5 mhz; 29.6 kw. 305 ft TL: N32 49 04 W79 50 08. Hrs open: Box 1505, Mount Pleasant, 29465. Phone: (843) 881-9450. Phone: (510) 568-6200. Fax: (510) 568-6190. E-mail: famradio@familyradio.com Web Site: www.familyradio.com. Licensee: Family Stations Inc. (group owner) Natl. Network: Family Radio. Format: Relg. ♦Harold Camping, gen mgr; Joe Papp, chief of engrg.

WLTQ(AM)— 1947: 730 khz; 1 kw-D, 100 w-N. TL: N32 46 22 W80 00 58. Hrs open: 24 950 Houston North Cutt Blvd., Suite 201, Mount Pleasant, 29464. Phone: (843) 884-2534. Fax: (843) 884-1218. Licensee: Citicasters Licenses L.P. Group owner: Clear Channel Communications Inc. (acq 5-4-99; grpsl). Population served: 66,945 Format: Standards. ♦Alene Grevey, gen mgr & gen sls mgr; Greg Alan, prom mgr & progmg dir; Willie Bennett, chief of engrg.

WQNT(AM)— 1948: 1450 khz; 1 kw-U. TL: N32 48 15 W79 57 43. Hrs open: 60 Markfield Dr., Suite 4, 29407. Phone: (843) 763-6631. Fax: (843) 766-1239. E-mail: wqnt@kirkmanbroadcasting.com Licensee: Kirkman Broadcasting Inc. (group owner; acq 1995). Population served: 460,000 Natl. Network: CNN Radio. Law Firm: Brian Madden & Assoc. Format: News. News staff: 24 hrs news prog wkly. Target aud: 25-54; Adults. Spec prog: Relg one hr wkly. ♦Gil Kirkman, pres; Stew Williams, opns mgr & progmg dir; Robby Robinson, gen sls mgr; John Dixon, news dir; Wally Momeier, chief of engrg.

WQSC(AM)— 1946: 1340 khz; 1 kw-U. TL: N32 49 07 W79 57 43. Hrs open: 24 60 Markfield Dr., Suite 4, 29407. Phone: (843) 763-6631. Fax: (843) 766-1239. E-mail: wqsc@berkeleyelectric.net Licensee: Kirkman Broadcasting Inc. (group owner; acq 11-1-94). Population served: 66,945 Natl. Network: Westwood One. Rgnl. Network: S.C. Net. Format: Talk. News staff: 1; News: 5 hrs wkly. Target aud: 25-54; Male. Spec prog: Relg 1 hr wkly. ♦Gil Kirkman, pres; Robby Robinson, gen mgr & gen sls mgr; Stew Williams, opns mgr; John Dixon, news dir; Wally Momeier, chief of engrg.

***WSCI(FM)**— 1973: 89.3 mhz; 97 kw. 540 ft TL: N32 47 44 W79 50 27. Stereo. Hrs opn: 24
Rebroadcasts WLTR(FM) Columbia 95%.
1101 George Rogers Blvd., Columbia, 29201. Phone: (803) 737-3420. Fax: (803) 737-3552. E-mail: gasque@scetv.org Web Site: www.etvradio.org. Licensee: South Carolina Educational TV Commission. Population served: 66,945 Natl. Network: NPR, PRI. Format: Class, jazz, news. News staff: 2. Spec prog: Black 5 hrs wkly. ♦Moss Bresnahan, pres; Tom Fowler, sr VP; Paul Zweimiller, stn mgr & engrg mgr; Tom Holloway, sls dir; John Gasque, progmg dir; Hap Griffin, engrg VP; Connie Murray, traf mgr.

WSSX-FM— 1945: 95.1 mhz; 100 kw. 361 ft TL: N32 49 20 W79 58 45. (CP: Ant 1,000 ft.). Stereo. Hrs opn: 24 4230 Faber Place Dr., Ste. 100, No. Charleston, 99405. Phone: (843) 277-1200. Fax: (843) 277-1212. Web Site: www.95fx.com. Licensee: Citadel Broadcasting Co. Group owner: Citadel Broadcasting Corp. (acq 6-9-99; grpsl). Population served: 100,000 Natl. Network: Westwood One. Format: Adult contemp, Top-40. News staff: one. Target aud: 18-34. ♦Paul O'Mailey, gen mgr.

WSUY(FM)— Apr 1, 1948: 96.9 mhz; 100 kw. 1,750 ft TL: N32 55 28 W79 41 58. Stereo. Hrs opn: 4230 Faber Place Drive, Suite 100, North Charleston, 29405. Phone: (843) 277-1200. Fax: (843) 277-1212. Web Site: www.sunny969.com. Licensee: Citadel Broadcasting Co. Group owner: Citadel Broadcasting Corp. Natl. Rep: McGavren Guild. Format: Soft rock. ♦Paul O'Malley, gen mgr; Bocky Gilleath, gen sls mgr; Eric Chaney, progmg dir; J.T. Tucker, chief of engrg.

WTMA(AM)— 1939: 1250 khz; 5 kw-D, 1 kw-N, DA-N. TL: N32 49 20 W79 58 45. Hrs open: 24 4230 Faber Place Dr., Suite 100, N. Charleston, 29405. Phone: (843) 277-1200. Fax: (843) 227-1212. E-mail: mike.edwards@citcomm.com Web Site: www.wtma.com. Licensee: Citadel Broadcasting Co. Group owner: Citadel Broadcasting Corp. (acq 6-9-99; grpsl). Population served: 15,000 Natl. Network: Westwood One, ABC. Wire Svc: AP Format: News/talk. News staff: 2; News: 5 hrs wkly. Target aud: 25-54. Spec prog: USC sports. ♦Paul O'Malley, gen mgr; Mike Edwards, opns dir.

WXLY(FM)—See North Charleston

WXTC(AM)— May 14, 1930: 1390 khz; 5 kw-U, DA-N. TL: N32 49 26 W80 00 06. Stereo. Hrs opn: 24 4230 Faber Place Dr. Suite 100, North Charleston, 29405-6512. Phone: (843) 308-9300. Fax: (843) 566-1222. Web Site: www.wtma.com. Licensee: Citadel Broadcasting Co. Group owner: Citadel Broadcasting Corp. (acq 6-9-99; grpsl). Population served: 66,945 Natl. Rep: McGavren Guild. Law Firm: Haley, Bader & Potts. Format: Black gospel. News staff: one. Target aud: 25-54. ♦Paul O'Malley, gen mgr; Terry Base, stn mgr.

WYBB(FM)—See Folly Beach

WZJY(AM)—See Mt. Pleasant

Cheraw

WCRE(AM)— July 1953: 1420 khz; 1 kw-D, 97 w-N. TL: N34 41 12 W79 53 42. Hrs open: 6 AM-8 PM Box 160, 29520. Phone: (843) 537-7887. Fax: (843) 537-7307. Web Site: www.wcreradio.com. Licensee: Pee Dee Broadcasting LLC (acq 1-28-2004; $50,000). Format: Adult contemp, oldies. News: 12 hrs wkly. Target aud: 25 plus. Spec prog: Black 5 hrs wkly. ♦Jane Pigg, pres.

WJMX-FM— July 17, 1979: 103.3 mhz; 44 kw. 525 ft TL: N34 30 19 W79 54 15. (CP: 50 kw, ant 492 ft.). Stereo. Hrs opn: 181 E. Evans St., Suite 311, Florence, 29501. Phone: (843) 667-9569. Fax: (843) 673-7390. Web Site: www.wjmx.com. Licensee: Qantum of Florence License Co. LLC. Group owner: Qantum Communications Corp. (acq 7-2-2003; grpsl). Population served: 278,000 Rgnl. Network: S.C. Net. Natl. Rep: McGavren Guild. Format: CHR. Target aud: 18-34. ♦Jonathan Brewster, gen mgr & stn mgr; Gary Downes, opns dir; Craig Dallariva, gen sls mgr & engrg dir.

Chester

WBT-FM— Aug 30, 1969: 99.3 mhz; 7.6 kw. 603 ft TL: N34 47 29 W81 16 01. Stereo. Hrs opn: 24
Rebroadcasts WBT(AM) Charlotte 100%.
One Julian Price Pl., Charlotte, NC, 28208. Phone: (704) 374-3500. Fax: (704) 338-3062. Web Site: www.wbt.com. Licensee: Jefferson-Pilot Communications Co. (group owner; acq 1995; $1.5 million). Population served: 1,105,000 Format: News/talk. News staff: 7; News: 6 hrs wkly. Target aud: 25-54; information, sports seekers. Spec prog: Gospel 6 hrs wkly. ♦Rick Jackson, VP & gen mgr; Tom Jackson, opns dir; Terry Mace, dev VP. Co-owned TV: WBTV-TV affil

WGCD(AM)— July 19, 1948: 1490 khz; I kw-U. TL: N34 41 54 W81 12 06. Hrs open: Box 11584, Rock Hill, 29731. Phone: (803) 329-2760. Fax: (803) 329-3317. Web Site: www.rejoynetwork.com. Licensee: Frank Neely. Group owner: Neely Enterprises (acq 5-7-97; $65,000). Format: Gospel. ♦Frank K. Neeley, gen mgr; Frankie Hemphill, stn mgr.

Chesterfield

***WRFE(FM)**—Not on air, target date: unknown: 89.3 mhz; 1.5 kw. Ant 197 ft TL: N34 43 15 W80 05 18. Stereo. Hrs opn: Box 371177, Cayey, PR, 00737. Phone: (787) 529-8917. E-mail: jcmb@coqui.net Licensee: Christian Educational Association. Population served: 203,000 Format: Christian contemp. ♦Aurio Chaparro, pres & gen mgr.

WVSZ(FM)— 1993: 107.3 mhz; 4.5 kw. 328 ft TL: N34 43 12 W80 05 45. Hrs open: Box 307, Rock Hill, 29731. Phone: (843) 286-1071. Phone: (843) 623-3299. Fax: (803) 324-2860. E-mail: almiller@wrhi.com Web Site: www.wrhi.com. Licensee: Our Three Sons Broadcasting L.L.P. (group owner; acq 2-28-97; $142,500). Natl. Network: ABC. Format: Country. ♦Allan M. Miller, gen mgr; Steven Stone, opns mgr; Mike Crowder, news dir.

Clearwater

WHHD(FM)— April 1987: 98.3 mhz; 11.5 kw. Ant 485 ft TL: N33 30 44 W82 04 48. Stereo. Hrs opn: 24 4051 Jimmie Dyess Pky., Augusta, GA, 30909. Phone: (706) 396-7000. Fax: (706) 396-7100. Web Site:

www.hd983.com. Licensee: WGAC License LLC. Group owner: Beasley Broadcast Group Inc. (acq 4-2-2001; $12 million with WKXC-FM Aiken). Format: Soft adult contemp. ♦Kent Dunn, gen mgr; Kent Murphy, gen sls mgr; Chuck Whittaker, progmg dir; Charlie McCoy, chief of engrg.

Clemson

WAHT(AM)— July 27, 1969: 1560 khz; 1 kw-D, 500 w-CH. TL: N34 42 04 W82 49 30. Hrs open: 6 AM-8 PM Box 1560, 202 Lawrence Rd., 29631. Phone: (864) 654-1560. Fax: (864) 654-3300. E-mail: waht@wahtam.com Web Site: www.wahtam.com. Licensee: Golden Corners Broadcasting Inc. (acq 9-28-89; $100,000; FTR: 10-16-89). Population served: 150,000 Wire Svc: CBS Format: Oldies. News staff: one; News: 35 hrs wkly. Target aud: 35-58; older yuppies. ♦George W. Clement, pres & gen mgr; Faye Clement, VP; Jeff Bright, stn mgr, gen sls mgr, progmg dir & chief of engrg.

WCCP-FM—Co-owned with WAHT(AM). Apr 8, 1993: 104.9 mhz; 4.6 kw. Ant 371 ft TL: N34 38 13 W82 42 30. Stereo. 24 Phone: (864) 654-4004. E-mail: info@wccpfm.com Web Site: www.wccpfm.com.900,000 Natl. Network: CBS, Sporting News Radio Network. Format: Sports. Target aud: 25-50. ♦George Clement, CEO; Aly Darby, stn mgr & progmg dir; Barry Clement, opns VP, opns mgr & prom dir; Pam Ponder, gen sls mgr.

***WSBF-FM**— Mar 16, 1961: 88.1 mhz; 3 kw. 200 ft TL: N34 40 42 W82 49 15. Stereo. Hrs opn: Clemson University, 210 Hendrix Student Ctr., 29634. Phone: (864) 656-4010. Fax: (864) 656-4011. E-mail: program@wsbf.net Web Site: www.wsbf.net. Licensee: Clemson University Board of Trustees. Population served: 50,000 Natl. Network: Westwood One. Format: Progsv. Target aud: 16-25.

Clinton

WPCC(AM)— Sept 11, 1957: 1410 khz; 1 kw-D, 100 w-N. TL: N34 26 42 W81 53 24. Hrs open: 24 Box 1455, 29325. Secondary address: 1766 Hwy 72 West, West Clinton 29325. Phone: (864) 833-1410. Fax: (864) 833-2467. E-mail: wpcc@charter.net Web Site: www.sportsradio1410wpcc.com. Licensee: Laurens County Communications Inc. (acq 12-83; $90,000; FTR: 12-5-83). Population served: 250,000 Natl. Network: ESPN Radio. Format: Sports. News staff: one; News: 1 wkly. Target aud: General. Spec prog: Moring Show- The Doghouse, Local and Regional Sports. ♦A. Cruickshanks, pres; Rhonda Cruickshanks, gen mgr, opns mgr & mktg dir; Chris Burgin, progmg dir.

Columbia

WARQ(FM)— Feb 6, 1971: 93.5 mhz; 2.8 kw. 443 ft TL: N34 02 00 W80 58 56. Stereo. Hrs opn: 24 Box 9127, 29290. Secondary address: 1900 Pineview Rd. 29290. Phone: (803) 695-8600. Fax: (803) 695-8605. Web Site: www.warq.com. Licensee: Urban Radio II L.L.C. Group owner: Inner City Broadcasting (acq 8-7-2000; grpsl). Population served: 461,000 Natl. Network: ABC, Westwood One. Format: Rock. News staff: one; News: 5 hrs wkly. Target aud: 18-49. ♦Steve Patterson, gen mgr; Scott Norton, gen sls mgr; Jamie Muldrow, prom dir; Dave Stewart, progmg dir.

WCEO(AM)—Jan 1, 1994: 840 khz; 50 kw-D, DA. TL: N34 12 42 W80 50 05. Stereo. Hrs opn: Sunrise-sunset 108 Columbia N.E. Dr., Suite F, 29223. Phone: (803) 419-7366. Fax: (803) 419-7363. Web Site: latremendaradio.com. Licensee: Norsan Broadcasting WCEO LLC. (group owner; (acq 10-1-2006; $1.6 million). Law Firm: Tom McCoy. Format: Sp. ♦Lino Cruz, stn mgr.

WCOS(AM)— 1939: 1400 khz; 1 kw-U. TL: N34 00 18 W81 00 43. Hrs open: 24 316 Greystone Blvd., 29210-8007. Phone: (803) 343-1100. Fax: (803) 798-5255. Licensee: Capstar TX L.P. Group owner: Clear Channel Communications Inc. (acq 9-1-00; grpsl). Population served: 450,000 Natl. Rep: Clear Channel. Format: Sports. Target aud: Men 25-54. ♦Bobby Martin, gen mgr; Tim McFalls, gen mgr; Gary Barboza, opns mgr & progmg dir; Gary Frakes, prom dir; Gary Robinson, engrg dir; Mary Pais, traf mgr; Christopher Thompson, news rptr.

WCOS-FM— March 1951: 97.5 mhz; 100 kw. 981 ft TL: N34 08 23 W81 03 22. Stereo. 24 Web Site: www.wcosfm.com. Format: Country. News staff: one. Target aud: 25-54. ♦Margret Wallace, sls dir & gen sls mgr; Susan Brown, prom mgr; Ron Brooks, progmg dir & news dir; Glen Garrett, mus dir; Gary Robinson, chief of engrg.

WISW(AM)— June 30, 1954: 1320 khz; 5 kw-D, 2.5 kw-N, DA-N. TL: N34 00 16 W81 04 15. Hrs open: 24 1801 Charleston Hwy., Cayce, 29033. Phone: (803) 796-7600. Fax: (803) 796-5502. Licensee: Citadel Broadcasting Co. Group owner: Citadel Broadcasting Corp. Natl. Rep: Christal. Law Firm: Reddy, Begley & McCormick. Format: Talk Radio. News staff: 5; News: 168 hrs wkly. Target aud: 35-64. Spec prog: Sports. ♦William L. McElveen, pres & gen mgr; Tim Miller, opns mgr; Bill MacAvine, gen sls mgr; Al Conner, progmg dir; Ray Allen, news dir; Ed Noyes, engrg dir.

WOMG(FM)—Co-owned with WISW(AM). Apr 15, 1989: 103.1 mhz; 3 kw. 300 ft TL: N34 03 05 W81 00 07. (CP: 6 kw, ant 308 ft.). Stereo. Format: Oldies. Target aud: 25-54. ♦Ray Allen, prom dir; Al Conner, news dir.

***WLTR(FM)**— July 1, 1976: 91.3 mhz; 96 kw. 761 ft TL: N34 07 07 W80 56 12. Stereo. Hrs opn: 24 1101 George Rogers Blvd., 29201. Phone: (803) 737-3420. Fax: (803) 737-3552. E-mail: gasque@scetv.org Web Site: www.etvradio.org. Licensee: South Carolina Educ. TV Commission. Population served: 450,000 Natl. Network: NPR, PRI. Law Firm: Dow, Lohnes & Albertson. Format: Classical, NPR news. ♦Moss Bresnahan, pres & VP; Paul Zweimiller, pres & stn mgr; Tom Fowler, sr VP; Tom Holloway, dev dir.

WLTY(FM)—Cayce, July 11, 1974: 96.7 mhz; 3.3 kw. 443 ft TL: N34 00 04 W81 02 05. Stereo. Hrs opn: 316 Greystone Blvd., 29250. Phone: (803) 343-1100. Fax: (803) 779-9727. Web Site: www.lite967.com. Licensee: Capstar TX L.P. Group owner: Clear Channel Communications Inc. (acq 8-30-00; grpsl). Natl. Rep: Clear Channel. Format: Light Adult Contemp. News: 6 hrs wkly. Target aud: 25-44; professionals & young adults. ♦Bob Huntley, gen mgr.

WMFX(FM)—Saint Andrews, Jan 23, 1985: 102.3 mhz; 6 kw. 322 ft TL: N34 05 55 W81 04 48. Stereo. Hrs opn: 24 Box 9127, 29290-0127. Secondary address: 1900 Pineview Rd. 29209. Phone: (803) 695-8600. Fax: (803) 695-8605. E-mail: mhanisch@innercity.sc.com Web Site: www.fox102.com. Licensee: Urban Radio II L.L.C. Group owner: Inner City Broadcasting (acq 8-7-2000; grpsl). Population served: 461,000 Format: Classic rock, AOR. News staff: one; News: one hr wkly. Target aud: 18-49. ♦Maggie Hanisch, gen mgr; Scott Norton, gen sls mgr; Jamie Bowman, prom dir; Dave Stewart, progmg dir.

***WMHK(FM)**— Aug 30, 1976: 89.7 mhz; 100 kw. 1,398 ft TL: N34 05 49 W80 45 51. (CP: Ant 1,397 ft.). Stereo. Hrs opn: Box 3122, 29230. Phone: (803) 754-5400. Fax: (803) 714-0849. E-mail: wmhk@wmhk.com Web Site: www.wmhk.com. Licensee: Columbia Bible College Broadcasting Co. Population served: 713,200 Format: Relg. Target aud: 25-44; women. ♦Jerry Grimes, gen mgr.

WNOK(FM)— July 15, 1959: 104.7 mhz; 100 kw. 1,014 ft TL: N34 09 06 W80 54 36. (CP: 96 kw, ant 1,033 ft. TL: N34 09 03 W80 54 36). Stereo. Hrs opn: 24 316 Greystone Blvd, 29210. Phone: (803) 343-1080. Fax: (803) 256-1968. Web Site: www.wnok.com. Licensee: Capstar TX L.P. Group owner: Clear Channel Communications Inc. (acq 8-30-00; grpsl). Population served: 390,000 Format: CHR. Target aud: 18-34; landed gentry. ♦Bob Hentley, gen mgr.

WOIC(AM)— Jan 1, 1947: 1230 khz; 1 kw-U, DA-N. TL: N33 59 34 W81 02 45. Hrs open: Box 9127, 29290. Phone: (803) 776-1013. Fax: (803) 695-8605. Web Site: www.espn1230am.com. Licensee: Urban Radio II L.L.C. Group owner: Inner City Broadcasting (acq 8-7-2000; grpsl). Population served: 461,000 Natl. Network: ABC. Natl. Rep: D & R Radio. Format: Sports. News: 3 hrs wkly. Target aud: 25-54; Male. ♦Steve Patterson, gen mgr.

WQXL(AM)— June 15, 1945: 1470 khz; 5 kw-D, 138 w-N. TL: N34 01 44 W81 02 23. Hrs open: 6 AM-8:30 PM Box 280233, 29230. Phone: (803) 779-7911. Fax: (803) 252-2158. E-mail: wqxl1470@aol.com Licensee: Metro Communications Inc. (acq 7-3-89). Population served: 153,542 Natl. Network: USA. Format: Praise & worship. Target aud: 25-49. ♦John Lastinger, pres; Donna Moore, stn mgr; Olin Jenkins, opns mgr.

WTCB(FM)—See Orangeburg

***WUSC-FM**— Jan 17, 1977: 90.5 mhz; 2.5 kw. 233 ft TL: N34 00 02 W81 01 19. (CP: Ant 253 ft.). Stereo. Hrs opn: 24 Drawer B, Univ. of South Carolina, 1400 Greene St., 29208. Phone: (803) 777-5468. Fax: (803) 777-6482. Web Site: wusc.sc.edu. Licensee: University of South Carolina. Population served: 400,000 Format: Var. News staff: one; News: 3 hrs wkly. Target aud: General; alternative generation. ♦Will Belenger, stn mgr.

WVOC(AM)— July 10, 1930: 560 khz; 5 kw-U, DA-N. TL: N34 02 00 W81 08 32. Stereo. Hrs opn: 24 316 Greystone Blvd., 29210-8007. Phone: (803) 343-1100. Fax: (803) 256-1968. Web Site: www.wvoc.com. Licensee: Capstar TX L.P. Group owner: Clear Channel Communications Inc. (acq 8-30-00; grpsl). Population served: 415,800 Natl. Network: CNN Radio. Rgnl. Network: S.C. Net. Wire Svc: Dow Jones Financial News Services Format: News/talk, sports. News staff: 2; News: 40 hrs wkly. Target aud: 35-64. ♦Tim McFalls, gen mgr.

WXBT(FM)—West Columbia, Aug 5, 1975: 100.1 mhz; 5.9 kw. 331 ft TL: N34 04 08 W81 04 16. Stereo. Hrs opn: 24 316 Greystone Blvd., 29210. Phone: (803) 343-1100. Fax: (803) 252-9267. Licensee: Capstar TX L.P. Group owner: Clear Channel Communications Inc. (acq 8-30-00; grpsl). Population served: 462,000 Natl. Network: CBS. Format: Rhythm and blues, urban contemp. News staff: one; News: 17 hrs wkly. Target aud: 35 plus; mature adults. ♦Bryan Anthony, progmg dir & chief of engrg; Tim McFalls, gen mgr & news dir.

Conway

***WHMC-FM**— Sept 15, 1985: 90.1 mhz; 30 kw. 706 ft TL: N33 57 05 W79 06 31. Hrs open: 24
Rebroadcasts WLTR-FM Columbia 100%.
1101 George Rogers Blvd., Columbia, 29201. Phone: (803) 737-3420. Fax: (803) 737-3552. E-mail: gasque@scetv.org Web Site: www.etvradio.org. Licensee: South Carolina Educational Television Commission. Natl. Network: NPR, PRI. Format: Classical, NPR news. ♦Moss Bresnahan, pres; Paul Zweimiller, VP & stn mgr.

WIQB(AM)— Feb 23, 1977: 1050 khz; 5 kw-D, 473 w-N, DA-2. TL: N33 50 56 W79 05 03. Stereo. Hrs opn: 24 11640 Hwy. 17 Bypass, Murrells Inlet, 29576. Phone: (843) 651-7869. Fax: (843) 397-3197. Licensee: Cumulus Licensing Corp. Group owner: Cumulus Media Inc. (acq 12-29-97; grpsl). Population served: 256,000 Rgnl. Network: S.C. Net. Law Firm: Gardner, Carton & Douglas. Format: Sports. News staff: one. Target aud: 50 plus; affluent retirees. ♦Ron Raybourne, gen mgr; Dave Solomon, progmg dir; Robert Kesler, news dir; Buddy Womack, chief of engrg.

WJXY-FM—Co-owned with WIQB(AM). October 1990: 93.9 mhz; 3.7 kw. 420 ft TL: N33 50 07 W78 52 06. Stereo. 24 Natl. Network: ABC. Format: Sports. News: 2 hrs wkly. Target aud: 18-34. ♦Lou Dickey, pres & opns dir; Dave Solomon, adv dir & pub affrs dir; Lisa Van Horn, pub affrs dir.

WPJS(AM)— August 1945: 1330 khz; 5 kw-D, 500 w-N, DA-N. TL: N33 50 57 W79 04 11. Hrs open: Box 961, 29528. Phone: (843) 248-9040. Fax: (843) 248-6365. Licensee: WPJS Broadcasters Inc. Format: Black gospel. Target aud: 12 plus. ♦P.J. Parrish, gen mgr.

Cross Hill

WYOR(FM)— September 1999: 94.1 mhz; 3.6 kw. Ant 417 ft TL: N34 12 16 W81 54 37. Hrs open: 637 E. Durst Ave., Greenwood, 29649. Phone: (864) 223-8553. Fax: (864) 943-0314. Web Site: www.941thebull.com. Licensee: Peregon Broadcasting LLC (acq 11-6-2006; $800,000 with WCRS(AM) Greenwood). Format: Country. ♦Carl Pundt, gen mgr.

Darlington

WDAR-FM— December 1965: 105.5 mhz; 4.1 kw. 400 ft TL: N34 18 58 W79 53 17. (CP: 17 kw). Stereo. Hrs opn: 24hrs Box 103000, Florence, 29501. Secondary address: 181 E. Evans St., Suite 311, Florence 29506. Phone: (843) 667-4600. Fax: (843) 673-7390. E-mail: production@rootsc.com Licensee: Qantum of Florence License Co. LLC. Group owner: Qantum Communications Corp. (acq 7-2-2003; grpsl). Format: Easy lstng. ♦Jonathan Brewster, gen mgr; Craig Dallariviva, gen sls mgr; Scott Gorman, progmg dir; Thoma Lesieur, news dir; Doug Carter, chief of engrg; Veronica Wingate, traf mgr.

WWRK(AM)—Co-owned with WDAR-FM. 1955: 1400 khz; 1 kw-U. TL: N34 18 58 W79 53 17.200,000 Format: Black gospel. Target aud: 25 plus. ♦Terri Burgess, traf mgr.

Dillon

***WDLL(FM)**—Not on air, target date: unknown: 90.5 mhz; 18 kw vert. Ant 304 ft TL: N34 23 26 W79 35 25. Hrs open: Box 2440, Tupelo, MS, 38803-2440. Phone: (662) 844-8888. Fax: (662) 842-6791. Web Site: www.afr.net. Licensee: American Family Association. ♦Donald E. Wildmon, chmn.

WDSC(AM)— May 22, 1946: 800 khz; 1 kw-D, 382 w-N. TL: N34 22 11 W79 24 08. Hrs open: 24 Box 103000, Florence, 29501. Secondary address: 181 E. Evans St., Florence 29506. Phone: (843) 667-4600. Fax: (843) 673-7390. Licensee: Qantum of Florence License Co. LLC. Group owner: Qantum Communications Corp. (acq 7-2-2003; grpsl). Population served: 100,780 Format: Gospel. Target aud: General. ♦Jonathan Brewster, gen mgr & gen sls mgr.

WEGX(FM)—Co-owned with WDSC(AM). Feb 16, 1954: 92.9 mhz; 100 kw. 1,801 ft TL: N34 21 53 W79 19 49. Stereo. 24 Box 103000, Florence, 29501. 481,400 Format: Gospel. Spec prog: Jazz one hr wkly. ♦Randy Wilcox, progmg dir.

Dorchester Terrace-Brentwood

WTMZ(AM)—Licensed to Dorchester Terrace-Brentwood. See North Charleston

Easley

WELP(AM)— Mar 4, 1951: 1360 khz; 5 kw-D, 36 w-N. TL: N34 50 20 W82 38 24. Hrs open: 24 100 Cross Hill Rd., 29640. Phone: (864) 855-9300. Fax: (864) 855-8444. E-mail: welp@wilkinsradio.com Web Site: www.wilkinsradio.com. Licensee: Upstate Radio Inc. Group owner: Wilkins Communications Network Inc. (acq 1999; $150,000). Population served: 775,000 Law Firm: Womble, Carlyle, Sandridge & Rice. Format: Christian teaching/talk. News staff: 2; News: 22 hrs wkly. Target aud: 35 plus. ♦Bob Wilkins, pres & gen sls mgr; LuAnn Wilkins, exec VP; Mitchell Mathis, VP; Greg Garrett, gen mgr, stn mgr, opns mgr & mus dir; Ted McCall, chief of engrg & engr.

WOLI-FM— 1964: 103.9 mhz; 6 kw. Ant 328 ft TL: N34 50 21 W82 31 37. Stereo. Hrs opn: 25 Garlington Rd., Greenville, 29615. Phone: (864) 271-9200. Fax: (864) 242-1567. Web Site: thewalkonline.com. Licensee: Davidson Media Station WOLI Licensee LLC. Group owner: Entercom Communications Corp. (acq 10-6-2005; grpsl). Population served: 72,900 Format: Christian. ♦Mark Yearout, gen sls mgr; Tom Durney, gen mgr & prom dir; Jerry Massey, engr.

Elloree

WORG(FM)—Licensed to Elloree. See Elloree-Santee

Elloree-Santee

WORG(FM)—Elloree, May 1988: 100.3 mhz; 25 kw. 328 ft TL: N33 21 42 W80 41 05. Stereo. Hrs opn: 24 1675 Chestnut St., Orangeburg, 29115. Phone: (803) 516-8400. Fax: (803) 516-0704. E-mail: worg@worg.com Web Site: www.worg.com. Licensee: Garris Communications Inc. (acq 7-95). Population served: 400,000 Format: Adult contemp. Target aud: 25-54. ♦Marion R. Garris, pres & gen mgr.

Florence

WDSC(AM)—See Dillon

WEGX(FM)—See Dillon

WJMX(AM)— July 13, 1947: 970 khz; 5 kw-D, 3 kw-N, DA-N. TL: N34 13 47 W79 48 07. Stereo. Hrs opn: 24 Box 103000, 29501. Secondary address: Florence Bus. & Tech. Ctr., 181 E. Evans St., Ste. 311 29506. Phone: (843) 667-4600. Phone: (843) 665-0970. Fax: (843) 673-7390. Web Site: www.wjmx.com. Licensee: Qantum of Florence License Co. LLC. Group owner: Qantum Communications Corp. (acq 7-2-03; grpsl). Population served: 300,000 Natl. Network: CBS, AP Radio. Rgnl. Network: S.C. Net. Natl. Rep: McGavren Guild. Format: Talk, news. News staff: one; News: 49 hrs wkly. Target aud: 25-54. Spec prog: Big band 3 hrs wkly. ♦Jonathan Brewster, gen mgr; Craig Dalla Riva, stn mgr.

WJMX-FM—See Cheraw

***WLPG(FM)**— May 15, 1993: 91.7 mhz; 10 kw horiz, 9.2 kw vert. 492 ft TL: N34 07 45 W79 50 06. Hrs open: 24 2278 Wortham Ln., Grovetown, GA, 30802. Phone: (706) 309-9610. Fax: (706) 309-9669. E-mail: ctbarinowski@comcast.net Web Site: www.gnnradio.org. Licensee: Augusta Radio Fellowship Institute Inc. Format: Christian. News: 12 hrs wkly. ♦Clarence Barinowski, gen mgr.

WOLS(AM)— Nov 18, 1937: 1230 khz; 1 kw-U. TL: N34 13 48 W79 44 49. Hrs open: 24 338 E. McIver Rd., 29506. Phone: (843) 665-1230. Fax: (843) 665-8786. E-mail: jjones1990@sc.rr.com Licensee: WOLS Broadcasting Corp. Group owner: GHB Radio Group (acq 9-13-88). Population served: 110,000 Natl. Network: ABC. Rgnl rep: Jim D. Jones Format: Sports/talk. News staff: one; News: 4 hrs wkly. Target aud: 21-101; within a 30 mile radius. Spec prog: Farm one hr, gospel 4 hrs, jazz 4 hrs, talk 16 hrs wkly. ♦Geo. H. Buck Jr., pres; Jeffrey Andrew Lonas, gen mgr.

WYNN(AM)— Nov 5, 1958: 540 khz; 250 w-U. TL: N34 13 05 W79 48 22. Hrs open: 24 2014 N. Irby St., 29501. Phone: (843) 661-5000. Fax: (843) 661-0888. Licensee: Cumulus Licensing Corp. Group owner: Cumulus Media Inc. (acq 12-17-98; with co-located FM). Population served: 150,000 Natl. Network: American Urban. Law Firm: Scott Johnson. Format: Black gospel, blues, Black classics. News staff: one; News: 12 hrs wkly. Target aud: 35 plus; Black. Spec prog: Jazz. ♦Matt Scurry, opns mgr; Rick Howze, gen sls mgr; Ollie Williams, progmg dir; Daniel Tindal, asst music dir.

WYNN-FM— Oct 1, 1964: 106.3 mhz; 1.1 kw. 507 ft TL: N34 14 03 W79 46 52. (CP: 1.7 kw). Stereo. 24 Format: Urban contemp. News staff: one; News: one hr wkly. Target aud: 12-34. ♦Gerald McSwain, progmg dir.

Folly Beach

WYBB(FM)— July 4, 1988: 98.1 mhz; 50 kw. 500 ft TL: N32 39 57 W80 03 11. Stereo. Hrs opn: 24 59 Windermere Blvd., Charleston, 29407. Phone: (843) 769-4799. Fax: (843) 769-4797. Web Site: www.98xonline.com. Licensee: L.M. Communications of South Carolina Inc. Group owner: L.M. Communications Inc. (acq 5-17-88). Population served: 500,000 Natl. Network: ABC. Law Firm: Leventhal, Senter & Lerman. Format: New rock. News: 28 hrs wkly. Target aud: 25-49; men. ♦Lynn Martin, pres; Charlie Cohn, gen mgr; Mike Allen, opns dir.

Forest Acres

WWNQ(FM)— 2005: 94.3 mhz; 2.55 kw. Ant 446 ft TL: N34 00 04 W81 02 05. Hrs open: Double O Radio Corp., 1010 Jervais St., Columbia, 29201. Phone: (212) 486-4446. E-mail: live@countrylegends94.3 Web Site: www.countrylegends943.com. Licensee: Double O South Carolina Corp. (acq 9-10-2004; $4.73 million for CP). Format: Country. ♦Robert Sherman, sr VP; Margaret Wallace, gen mgr.

Fountain Inn

WFIS(AM)— October 1956: 1600 khz; 1 kw-D, 29 w-N. TL: N34 42 28 W82 13 40. Hrs open: 24 Box 156, 29644. Secondary address: 1318 N. Main St. 29644. Phone: (864) 963-5991. Fax: (864) 963-5992. E-mail: wfis16@aol.com Licensee: Golden Strip Broadcasting Inc. (acq 2-19-99; $195,000 for stock). Population served: 45,000 Natl. Network: Westwood One, Jones Radio Networks, ABC. Rgnl. Network: S.C. Net. Natl. Rep: Rgnl Reps. Format: Talk, sports. News staff: one; News: 3 hrs wkly. Target aud: 25-49; working adults. Spec prog: Gospel 4, Black 4 hrs, Christian 3 hrs wkly. ♦Joseph E. LaStringer, gen mgr.

Gaffney

WEAC(AM)—Listing follows WNOW-FM.

WFGN(AM)— 1948: 1180 khz; 2.5 kw-D. TL: N35 02 59 W81 38 42. Hrs open: 6 AM-8 PM 470 Leadmine Rd., 29342. Phone: (864) 489-9430. Fax: (864) 489-9440. Licensee: Hope Broadcasting Inc. (acq 8-7-90; $160,000; FTR: 8-27-90). Population served: 100,000 Format: Relg. ♦Eddie Leroy Bridges Jr., pres; Ed Ridges, gen mgr; Charles Montgomery, opns mgr & disc jockey; Rev. Eula Miller, gen sls mgr; Caroline Allen, disc jockey; Clarence Quarles, disc jockey; Lonnie Dawkins, disc jockey; Marianetta Smith, disc jockey.

WNOW-FM— 1959: 105.3 mhz; 100 kw. Ant 1,190 ft TL: N35 25 05 W81 46 32. Stereo. Hrs opn: 24 Box 1210, 29342. Secondary address: 340 Providence Rd. 29341. Phone: (864) 489-9066. Fax: (864) 489-9069. E-mail: feedback@wagifm.com Web Site: www.wagifm.com. Licensee: Gaffney Broadcasting Inc. (acq 8-2-2005; with co-located AM). Population served: 1,000,000 Natl. Network: CNN Radio. Wire Svc: AP Format: Contemp country, gospel. News staff: one. Target aud: 18-54. Spec prog: Clemson Univ. sports, loc sports, talk 10 hrs wkly. ♦Ronald Owenby, pres & gen mgr; Dennis Fowler, stn mgr & news dir; Ernie Payne, Jr., gen sls mgr; Jonathan Fitch, mus dir; Claudella Moss, traf mgr; Knox Blanton, disc jockey.

WEAC(AM)—Co-owned with WNOW-FM. Sept 28, 1962: 1500 khz; 1 kw-D, 500 w-N. TL: N35 05 18 W81 38 40.13,253 Format: Country. News staff: one; News: 2 hrs wkly. Target aud: 18-54. ♦Knox Blanton, disc jockey; Randy Catoe, disc jockey.

***WYFG(FM)**— Oct 12, 1982: 91.1 mhz; 100 kw. 574 ft TL: N35 06 57 W81 46 42. Stereo. Hrs opn: 24 Bible Broadcasting Network, Charlotte, NC, 28241-7300. Phone: (800) 888-7077. Licensee: Bible Broadcasting Network Inc. (group owner) Population served: 1,000,000 Natl. Network: USA. Format: Relg/Conservative Christian. News staff: one. Target aud: General. ♦Lowell Davey, pres; Stan Schenkel, gen mgr.

Garden City

WWXM(FM)— Sept 25, 1971: 97.7 mhz; 100 kw. Ant 718 ft TL: N33 35 45 W79 03 11. Stereo. Hrs opn: 24 4841 Hwy. 17 By-pass S., Myrtle Beach, 29577. Phone: (843) 293-0107. Fax: (843) 293-1717. Web Site: www.977online.com. Licensee: Qantum of Myrtle Beach License Co. LLC. Group owner: Qantum Communications Corp. (acq 7-2-2003; grpsl). Population served: 200,000 Law Firm: Fletcher, Heald & Hildreth. Format: CHR. News staff: one; News: 2 hrs wkly. Target aud: 18-49. ♦Jimmy Feuger, gen mgr.

Georgetown

WGTN(AM)— July 1, 1949: 1400 khz; 1 kw-U. TL: N33 24 15 W79 19 36. Stereo. Hrs opn: 24 Box 1400, 29442. Phone: (843) 546-1400. Fax: (843) 527-2337. Web Site: www.wgtnradio.com. Licensee: RJ Stalvey. (Dec 2000 Population served: 10,449 Natl. Network: NBC. Format: News/talk. News staff: one; News: 12 hrs wkly. Target aud: 25-54; upscale adult; bus, professional and technical. ♦Rod Stalvey, gen mgr.

WGTN-FM—Andrews, Aug 19, 1985: 100.7 mhz; 3.1 kw. Ant 446 ft TL: N33 24 03 W79 27 30. Stereo. Hrs opn: 24 3926 Wesley St., Suite 301, Myrtle Beach, 29578. Phone: (843) 903-9962. Fax: (843) 903-1797. Licensee: Coastline Communications of Carolina Inc. (acq 10-12-2000; $800,000). Format: Adult Hits. Target aud: 25-54; Adults. ♦Will Isaacs, gen mgr; Jerome Bresson, news dir.

WLMC(AM)— March 1962: 1470 khz; 1 kw-D. TL: N33 22 15 W79 16 39. Stereo. Hrs opn: Box 2865, 29442. Phone: (843) 546-8863. Fax: (843) 546-6281. E-mail: wradio@sc.rr.com Licensee: Cumberland A & A Corp. (acq 2-10-2003; $200,000). Natl. Network: ABC. Format: Gospel, Christian, inspirational. Target aud: 25 plus; African-Americans. Spec prog: Talk 4 hrs wkly. ♦Reggie Dyson, CEO.

WSYN(FM)— May 1, 1973: 106.5 mhz; 50 kw. 530 ft TL: N33 26 20 W79 08 11. Stereo. Hrs opn: 24 11640 Hwy. 17 By-pass S., Murrells Inlet, 29576. Phone: (843) 651-7869. Fax: (843) 651-3197. Web Site: www.sunny1065.net. Licensee: Cumulus Licensing Corp. Group owner: Cumulus Media Inc. (acq 1-27-98). Format: Oldies. ♦Bill Hazen, gen mgr.

WWGS(AM)—Not on air, target date: unknown: 1580 khz; 20 kw-D, 5 kw-N, DA-2. TL: N33 23 25 W79 27 34. Hrs open: Box 1400, 29442. Phone: (843) 546-2337. E-mail: stalvey@aol.com Licensee: R.J. Stalvey. ♦R.J. Stalvey, gen mgr.

WXJY(FM)— Sept 1, 1990: 93.7 mhz; 6 kw. 328 ft TL: N33 16 09 W79 17 49. Stereo. Hrs opn: 24 11640 Highway 17 Bypass, Murrells Inlet, GA, 29576. Phone: (404) 949-0700. Fax: (404) 949-0740. Web Site: www.teammyrtlebeach.com. Licensee: Cumulus Licensing Corp. Group owner: Cumulus Media Inc. (acq 12-29-97; grpsl). Population served: 50,000 Format: Sports. News: 4 hrs wkly. Target aud: 25-49; career-oriented, college-educated adults. ♦Lyne Ryan, gen mgr; Roderick Smith, opns mgr; Kellly Broderick, progmg dir.

Goose Creek

WSCC-FM— May 19, 1983: 94.3 mhz; 25 kw. Ant 328 ft TL: N32 49 04 W79 50 08. Stereo. Hrs opn: 950 Houston Northcutt Blvd., Mt. Pleasant, 29464. Phone: (843) 856-6100. Fax: (843) 884-1218. E-mail: bjkay@clearchannel.com Web Site: www.wscfm.com. Licensee: Clear Channel Broadcasting Licenses Inc. Group owner: Clear Channel Communications Inc. (acq 7-29-2003). Natl. Network: Fox News Radio. Natl. Rep: McGavren Guild. Format: Talk. News staff: 3. Target aud: 25-54. ♦Paul Smith, gen mgr; Willie Bennett, chief of engrg.

Gray Court

WSSL-FM—Licensed to Gray Court. See Greenville

Greenville

WCSZ(AM)—Sans Souci, May 26, 1966: Stn currently dark. 1070 khz; 50 kw-D, 1.5 kw-N, DA-3. TL: N34 55 05 W82 27 21. Stereo. Hrs opn: 24 200 N. Hwy. 25 Bypass, 29617. Phone: (864) 294-1071. Fax: (864) 246-8695. Licensee: WHYZ Radio L.P. (acq 1996; $200,000 for foreclosure). Population served: 700000 Natl. Network: Westwood One, American Urban. Format: Inspirational gospel. Target aud: 25-54; $50,000 plus houshold income, college educated, 60% male, 40% female. ♦Glenn Cherry, CEO; Jerry Young, gen mgr & stn mgr.

***WEPR(FM)**— Sept 3, 1972: 90.1 mhz; 85 kw. 1,184 ft TL: N34 56 26 W82 24 38. Stereo. Hrs opn: 24
Rebroadcasts WLTR(FM) Columbia 100%.
1101 George Rogers Blvd., Columbia, 29201. Phone: (803) 737-3420. Fax: (803) 737-3552. E-mail: gasque@scetv.org Web Site: www.etvradio.org. Licensee: South Carolina Educ. TV Commission. Natl. Network: NPR, PRI. Law Firm: Dow, Lohnes & Albertson. Format: Classical, NPR news. ♦Moss Bresnahan, pres; Tom Fowler, sr VP & prom dir; Paul Zweimiller, stn mgr & engrg mgr; Tom Holloway, dev dir & sls dir; John Gasque, progmg dir; Hap Griffin, engrg VP; Connie Murray, traf mgr.

WESC-FM— March 1948: 92.5 mhz; 100 kw. 2,000 ft TL: N35 08 16 W82 36 31. Stereo. Hrs opn: 24 Box 100, 29602. Secondary address: 7 N. Laurens St., Suite 700 29601. Phone: (864) 242-4660. Fax: (864) 242-8813. Web Site: www.wescfm.com. Licensee: Clear Channel Broadcasting Licenses Inc. Group owner: Clear Channel Communications Inc. (acq 1998; grpsl). Population served: 281,700 Format: Country. ♦Bill McMartin, gen mgr; Bob Hooper, sls dir; Sandra Dill, mktg VP; Vicky Sexton, prom VP; Scott Johnson, progmg dir; John Landrum, mus dir; Roger Davis, news dir; Jim Graham, chief of engrg; Goldia Williams, traf mgr; Charlie Munson, disc jockey; Jessy Howard, disc jockey; John Crenshaw, disc jockey.

WFBC-FM—Listing follows WYRD(AM).

WGVL(AM)— 1950: 1440 khz; 5 kw-U, DA-N. TL: N34 52 06 W82 28 04. Hrs open: 24 Box 100, 29602. Secondary address: Bank of America, 7 N. Laurens St. 29601. Phone: (864) 242-1005. Fax: (864) 271-9775. Fax: (864) 233-7827. Web Site: www.clearchannel.com. Licensee: Capstar TX L.P. Group owner: Clear Channel Communications Inc. (acq 8-30-00; grpsl). Natl. Network: USA, Westwood One, ABC. Natl. Rep: McGavren Guild. Law Firm: Fisher, Wayland, Cooper, Leader & Zaragoza L.L.P. Format: Hispanic. News staff: 2; News: 4 hrs wkly. Target aud: 25-54. ♦Bill McMartin, gen mgr & gen sls mgr; Bruce Logan, progmg dir.

WSSL-FM—Co-owned with WGVL(AM). November 1960: 100.5 mhz; 100 kw. 1,240 ft TL: N34 34 19 W82 06 41.24 Fax: (864) 271-9775.172,200 Format: Country.

WLFJ(AM)— March 1947: 660 khz; 50 kw-D, 10 kw-CH. TL: N34 53 10 W82 28 03. Hrs open: 2420 Wade Hampton Blvd., 29615. Phone: (864) 292-6040. Fax: (864) 292-8428. E-mail: wlfj@wlfj.com Web Site: www.hisradio.com. Licensee: Clear Channel Broadcasting Licenses Inc. Group owner: Clear Channel Communications Inc. (acq 1998; grpsl). Population served: 61,208 Format: Contemp Christian. Target aud: 25-54. ♦Allen Henderson, gen mgr.

***WLFJ-FM**— May 1983: 89.3 mhz; 41 kw. 1,100 ft TL: N34 56 26 W82 24 44. Stereo. Hrs opn: 24 2420 Wade Hampton Blvd., 29615. Phone: (864) 292-6040. Phone: (864) 292-5683. Fax: (864) 292-8428. E-mail: wlfj@wlfj.com Web Site: www.hisradio.com. Licensee: Radio Training Network Inc. (acq 8-31-89). Population served: 850000 Format: Contemp Christian. News: 9 hrs wkly. Target aud: 18-49. ♦Allen Henderson, gen mgr; Rob Dempsey, progmg dir; Ted McCall, chief of engrg.

WMUU(AM)— Sept 15, 1949: 1260 khz; 5 kw-D, 29 w-N. TL: N34 53 16 W82 23 27. Hrs open: 920 Wade Hampton Blvd., 29609. Phone: (864) 242-6240. Fax: (864) 370-3829. E-mail: generalmanaager@wmuu.com Web Site: www.wmuu.com. Licensee: WMUU Inc. (acq 3-27-75). Population served: 61,208 Law Firm: Fletcher, Heald & Hildreth. Format: Relg. Target aud: 35 plus. ♦Paul Wright, gen mgr; Jeff Gainous, prom mgr; Brigette Barrett, progmg dir; Charles Koelsch, mus dir; Joe Norris, chief of engrg; Frank Richards, relg ed.

WMUU-FM— Aug 15, 1960: 94.5 mhz; 100 kw. 1,200 ft TL: N34 56 29 W82 24 41. Stereo. 24 Web Site: www.wmuu.com. Format: Btfl mus. Target aud: 35 plus. Spec prog: Class 14 hrs, relg 20 hrs wkly. ♦Jeff Gainous, prom dir & outdoor ed; Brigette Barrett, progmg mgr; Joe Norris, engrg dir.

WMYI(FM)—Hendersonville, NC) Apr 15, 1958: 102.5 mhz; 20 kw. 1,778 ft TL: N35 13 22 W82 32 57. Stereo. Hrs opn: 24 7 N. Laurens St., Suite 700, 29601-2744. Phone: (864) 235-1025. Fax: (864) 242-2536. Web Site: www.wmyi.com. Licensee: Clear Channel Radio Licenses, Inc. Group owner: Clear Channel Communications Inc. (acq 8-30-00; grpsl). Population served: 150,000 Format: Adult contemp. News staff: one. ♦Bill McMartin, gen mgr.

WPCI(AM)— Feb 8, 1954: 1490 khz; 1 kw-U. TL: N34 51 38 W82 24 31. Hrs open: 840 N. Hwy. 25 Bypass, 29617. Phone: (864) 834-3193, EXT. 35. Phone: (864) 836-3551. Licensee: Hunter Broadcast Group. (acq 12-88; $15,000; FTR: 2-20-89). Format: Rhythm oldies. ♦Randy Mathena, pres & gen mgr.

***WTBI-FM**— June 1991: 91.7 mhz; 3 kw. 328 ft TL: N34 49 43 W82 26 59. Stereo. Hrs opn: 24 3931 White Horse Rd., 29611. Phone: (864) 295-2145. Fax: (864) 295-6313. E-mail: jwatts@tabernacleministries.org Web Site: www.wtbi.org. Licensee: Tabernacle Baptist Bible College. Format: Relg music, educ, gospel. Target aud: General. ♦Charles Garrett, Sr., gen mgr.

WYRD(AM)— May 1933: 1330 khz; 5 kw-U, DA-N. TL: N34 51 18 W82 25 24. Hrs open: 24
Rebroadcasts WORD(AM) Spartanburg.
25 Garlington Rd., 29615. Phone: (864) 271-9200. Fax: (864) 242-1567. Web Site: www.newsradioword.com. Licensee: Entercom Greenville License LLC. Group owner: Entercom Communications Corp. (acq 12-13-99; grpsl). Population served: 85,000 Natl. Network: ABC, Salem Radio Network. Format: News/talk info. Target aud: 30-64. ♦Tom Durney, gen mgr.

WFBC-FM—Co-owned with WYRD(AM). March 1947: 93.7 mhz; 100 kw. 1,850 ft TL: N35 06 40 W82 36 17. Stereo. 24 Web Site: www.b937online.com.650,000 Format: Adult contemp. News staff: 4; News: one hr wkly. Target aud: 35-64. Spec prog: Alternative 2 hrs wkly. ♦Niki Knight, CFO & progmg dir; Heidi Aiken, news dir.

Greenwood

WCRS(AM)— Sept 1, 1941: 1450 khz; 1 kw-U. TL: N34 12 34 W82 09 05. Hrs open: 24 2881 Peachtree Rd. N.E., Apt. 2405, Atlanta, GA, 30305. Phone: (864) 223-1450. Fax: (864) 943-0314. Web Site: 1450wcrsam.com. Licensee: Peregon Broadcasting LLC (acq 11-6-2006; $800,000 with WYOR(FM) Cross Hill). Population served: 100,000 Natl. Network: CBS. Rgnl. Network: S.C. Net. Format: Adult standards, news/talk. News staff: 2; News: 20 hrs wkly. Target aud: 25 plus; middle & upper income adults. ♦Mike Hatfield, opns mgr.

WCZZ(AM)— June 20, 1973: 1090 khz; 5 kw-D, 2.25 kw-CH. TL: N34 09 46 W82 11 41. Stereo. Hrs opn: Sunrise-sunset 210 Montague Ave., 29649. Phone: (864) 223-4300. Fax: (864) 223-4096. E-mail: sunny@sunny103-5.com Licensee: Broomfield Broadcasting LLC (acq 7-28-2005; $1.03 million with co-located FM). Population served: 211,400 Natl. Network: Westwood One. Law Firm: Wiley, Rein & Fielding. Format: Rejoice Black gospel. News: 12 hrs wkly. Target aud: 25-65. ♦John Broomfield, pres; Dave Fezler, exec VP & progmg dir; Rick Prusator, sls VP; Kathleen Prusator, mktg VP & prom VP; Tonya Branyon, traf mgr; Stephanie White, prom.

WZSN(FM)—Co-owned with WCZZ(AM). March 1989: 103.5 mhz; 25 kw. 328 ft TL: N34 09 46 W82 11 41. Stereo. 24 Web Site: www.sunny103-5.com. Natl. Network: Westwood One. Format: Adult contemp. News: 3 hrs wkly. Target aud: 25-54. ♦Stephanie White, prom.

Greer

WCKI(AM)— Mar 3, 1955: 1300 khz; 1 kw-D. TL: N34 55 39 W82 15 42. Hrs open: 6 AM-6 PM Box 170022, Spartanburg, 29301. Phone: (864) 877-8458. Phone: (864) 877-8459. Fax: (864) 877-8500. Licensee: Mediatrix SC Inc. (acq 10-13-2004; $280,000). Population served: 10,642 Format: Christian, talk radio. News: one hr wkly. Target aud: 25-54; working people who spend money. ♦Mike Brannen, pres; Gary Powery, stn mgr.

WOLT(FM)— January 1993: 103.3 mhz; 2.7 kw. Ant 495 ft TL: N34 59 13 W82 09 56. Hrs open: 225 S. Pleasantburg Dr., Suite B3, Greenville, 29607. Phone: (864) 569-0901. Fax: (864) 569-0945. Web Site: www.wolt-fm.com. Licensee: Davidson Media Station WOLT Licensee LLC. Group owner: Entercom Communications Corp. (acq 10-6-2005; grpsl). Format: Oldies. ♦Robert Freese, gen mgr; Bill Prather, opns mgr & progmg dir; Ann Freese, natl sls mgr; Tom Herndon, natl sls mgr.

WPJM(AM)— June 15, 1949: 800 khz; 1 kw-D, 438 w-N. TL: N34 56 59 W82 14 43. Hrs open: 305 N. Tryon St., 29651. Phone: (864) 877-1112. Phone: (864) 877-1821. Fax: (864) 877-0342. Licensee: Full Gospel WPJM 800 AM Radio Inc. (acq 11-28-97; $200,000). Population served: 1,000,000 Rgnl. Network: S.C. Net. Format: Gospel. Target aud: General. ♦Bobby Cohen, pres & gen mgr; J.B. Adams, progmg dir.

Hampton

WBHC-FM— September 1970: 92.1 mhz; 6 kw. Ant 328 ft TL: N32 50 38 W81 07 31. Stereo. Hrs opn: Box 607, 29924. Phone: (803) 943-2831. Fax: (803) 943-5450. Web Site: www.cruise92.com. Licensee: Bocock Communications LLC (acq 6-1-2004; $375,000 with co-located AM). Rgnl. Network: S.C. Net. Format: Adult contemp. Spec prog: Relg 11 hrs wkly. ♦Carrie Michaels, opns mgr, adv dir & traf mgr; John Bocock, pres & gen sls mgr.

WHGS(AM)—Co-owned with WBHC-FM. September 1957: 1270 khz; 700 w-D, 219 w-N. TL: N32 50 38 W81 07 32.153560 Format: Gospel.

Hanahan

WAVF(FM)— Mar 11, 1985: 96.1 mhz; 538 w. 1,443 ft TL: N32 49 04 W79 50 08. Stereo. Hrs opn: 24 2294 Clements Ferry Rd., Charleston, 29492-7729. Phone: (843) 972-1100. Fax: (843) 972-1200. Web Site: www.96wave.com. Licensee: Apex Broadcasting Inc. (group owner; acq 12-5-01; $6 million). . Natl. Rep: Christal. Format: Rock alternative. ♦Dean Pearce, CEO, pres & gen mgr; John Anthony, opns VP.

Hardeeville

WLVH(FM)— Aug 30, 1992: 101.1 mhz; 50 kw. 476 ft TL: N32 05 48 W81 19 17. Stereo. Hrs opn: 24 245 Alfred St., Savannah, GA, 31408. Phone: (912) 964-7794. Fax: (912) 964-9414. E-mail: garyyoung@clearchannel.com Web Site: www.love1011.com. Licensee: Capstar TX L.P. Group owner: Clear Channel Communications Inc. (acq 8-30-00; grpsl). Natl. Network: ABC. Format: Adult urban contemp. News: one hr wkly. Target aud: 25-54; affluent Black adults. ♦Craig Scott, gen mgr.

Hartsville

WBZF(FM)—Listing follows WHSC(AM).

WHSC(AM)— Oct 1, 1946: 1450 khz; 1 kw-U. TL: N34 21 15 W80 04 20. Hrs open: 24 2014 N. Irby St., Florence, 29501. Phone: (843) 661-5000. Fax: (843) 661-0888. Web Site: www.cumulus.com. Licensee: Cumulus Licensing Corp. Group owner: Cumulus Media Inc. (acq 4-20-98; 700,000 with co-located FM). Population served: 95,000 Rgnl. Network: Tobacco, S.C. Net. Law Firm: Reddy, Begley & McCormick. Format: CHR. News: 12 hrs wkly. Target aud: 19-49; those with buying power. Spec prog: Farm 3 hrs, gospel 3 hrs, relg 6 hrs, big band 3 hrs, oldies 6 hrs wkly. ♦Dave McWhorter, gen mgr; Steve Crumbley, opns mgr; Rick Howze, gen sls mgr; Buzz Bowman, progmg dir & disc jockey; Gale Gilbraith, chief of engrg.

WBZF(FM)—Co-owned with WHSC(AM). Nov 19, 1992: 98.5 mhz; 6 kw. 328 ft TL: N34 21 16 W80 04 06. Web Site: www.cumulus.com. Format: Black, relg.

WJDJ(AM)— Dec 4, 1972: 1490 khz; 1 kw-U. TL: N34 21 47 W80 04 28. Hrs open: 6:30 AM-6PM
WAGS—WAGS/WJDJ are simulcast. Program originates at WAGS.
142 Wags Dr., Bishopville, 29010. Phone: (803) 484-5415. E-mail: wagsradio@sc.rr.com Licensee: Beaver Communications (acq 4-26-02). Population served: 24,000 Natl. Network: USA. Format: Country, bluegrass, gospel, Live Radio, Real People in Real Time. News: 6 hrs wkly. Target aud: 28 & up; Adults 28 & up. Spec prog: live remotes 2 hrs wkly, religious 7 hrs wkly. ♦James D. Jenkins, pres & gen mgr.

Hemingway

***WLGI(FM)**— July 1, 1984: 90.9 mhz; 50 kw. 505 ft TL: N33 43 09 W79 19 50. Stereo. Hrs opn: 15 1272 Williams Hill Rd., 29554. Phone: (843) 558-9544. Phone: (843) 558-9100. Fax: (843) 558-5778. Licensee: Louis G. Gregory Baha'i Institute. Population served: 800,000 Law Firm: Reddy, Begley & McCormick. Wire Svc: Weather Wire Format: Gospel, Black, urban contemp. Target aud: General. Spec prog: Jazz.

Hilton Head Island

WFXH(AM)— Feb 14, 1983: 1130 khz; 1 kw-D, 500 w-N, DA-N. TL: N32 12 01 W80 43 27. Stereo. Hrs opn: 24 One Saint Augustine Pl., 29928. Phone: (843) 785-9569. Fax: (843) 842-3369. Web Site: www.adventureradio.com. Licensee: Monterey Licenses LLC. Group owner: Triad Broadcasting Co. LLC (acq 7-18-00; grpsl). Population served: 50,000 Format: ESPN sports & news. News staff: one; News: 5 hrs wkly. Target aud: 35 plus. ♦Robert Leonard, gen mgr.

WFXH-FM— July 14, 1973: 106.1 mhz; 10.5 kw. 794 ft TL: N32 19 50 W80 56 19. (CP: 25 kw, ant 594 ft.). Stereo. 401 Mall Blvd., Suite 101 D, Savannah, GA, 31406. Phone: (912) 351-9830. Fax: (912) 352-4821. E-mail: email@106.1thefox.com Web Site: www.adventureradio.com.100,000 Natl. Rep: Christal. Format: Rock. News staff: 2. Target aud: 18-49; more male than female. ♦Robert Leonard, gen mgr.

WWJN(FM)—Ridgeland, July 15, 1986: 104.9 mhz; 16 kw. Ant 410 ft TL: N32 26 10 W80 55 23. Stereo. Hrs opn: 24 One Saint Augustine Pl., 29928. Phone: (843) 785-9569. Fax: (843) 842-3369. E-mail: wave1049@adventureradio.fm Web Site: www.wave1049.com. Licensee: JB Broadcasting LLC Group owner: Triad Broadcasting Co. LLC (acq 11-2-2006; $800,000). Population served: 500,000 Format: Adult hits. News staff: 2. Target aud: 18-49; general persons. ♦Robert Leonard, gen mgr.

Holly Hill

WJBS(AM)— Dec 1, 1972: 1440 khz; 1 kw-D, 98 w-N. TL: N33 20 23 W80 26 18. Hrs open: Box 1087, 29059. Phone: (803) 496-5352. Fax: (803) 496-2526. Licensee: Eugene Schoebinger. (acq 7-1-85). Population served: 8,000 Format: Gospel. Spec prog: Black 17 hrs, fishing/hunting 2 hrs, farm 2 hrs wkly. ♦Harry Govan, gen mgr & gen sls mgr; Robert Small, prom dir.

Hollywood

WXST(FM)— July 15, 1988: 99.7 mhz; 70 kw. Ant 781 ft TL: N32 49 04 W79 50 08. Stereo. Hrs opn: 24 2294 Clements Ferry Rd., Charleston, 29492-7729. Phone: (843) 972-1100. Fax: (843) 972-1200. Web Site: www.star997.com. Licensee: Apex Broadcasting Inc. (group owner; acq 11-20-01). Population served: 350,000 Natl. Network: Jones Radio Networks. Law Firm: Garvey Schubert Barer. Format: Adult urban contemp. Target aud: 25-54; urban professional. ♦Dean Pearce, CEO, pres & gen mgr; John Anthony, opns mgr; Carl Wine, gen sls mgr & prom dir; Walt Rosen, sls.

Homeland Park

WRIX(AM)— Sept 1, 1986: 1020 khz; 10 kw-D. TL: N34 28 14 W82 38 03. Hrs opn: 102 E. Shockley Ferry Rd., Anderson, 29624. Phone: (864) 224-6733. Fax: (864) 224-0260. Licensee: AM 1020 Inc. (acq 10-28-99). Rgnl. Network: S.C. Net. Format: Relg. Spec prog: Black 7 hrs wkly. ♦Karen Small, pres & gen mgr.

Honea Path

WRIX-FM— June 10, 1977: 103.1 mhz; 6 kw. 392 ft TL: N34 23 43 W82 29 49. Stereo. Hrs opn: 24 102 E. Shockley Ferry Rd., Anderson, 29624. Phone: (864) 224-9749. Fax: (864) 224-0260. Licensee: FM 103 Inc. (acq 10-28-99). Natl. Network: ABC. Rgnl. Network: S.C. Net. Format: News/talk. Spec prog: Talk 20 hrs wkly. ♦Karen Small, gen mgr.

Irmo

WWNU(FM)— May 23, 1987: 92.1 mhz; 15 kw. Ant 427 ft TL: N34 04 55 W81 07 36. Stereo. Hrs opn: 24
Rebroadcasts WKSX(FM) Johnston 100%.
Drawer 1, Johnston, 29832. Secondary address: 121 B N. Main St. 29138. Phone: (803) 275-4444. Fax: (803) 275-3185. Licensee: Double O South Carolina Corp. (acq 11-1-2004; $4.7 million). Rgnl. Network: S.C. Net. Format: Oldies. News: 7 hrs wkly. Target aud: 25-54; upper middle class. ♦Mike Casey, gen mgr; Frank Davis, opns dir & sls VP; Jay West, prom mgr & progmg mgr; Ruth Casey, traf mgr.

Johnsonville

WPDT(FM)— May 1995: 105.1 mhz; 2.95 kw. Ant 472 ft TL: N33 54 36 W79 40 09. Hrs open: 109 N. McAllister St., Lake City, 29560. Phone: (843) 374-5255. Fax: (843) 374-5256. E-mail: wpdt@ftc-i.net Web Site: www.wfmv.com. Licensee: Glory Communications Inc. (group owner; acq 5-20-02). Format: Urban inspiration. ♦Alex Snipes Jr., gen mgr; Tersa Haire, sls dir; Tony Gee, progmg VP.

Johnston

WKSX-FM— Aug 26, 1985: 92.7 mhz; 1.8 kw. Ant 577 ft TL: N33 45 19 W81 50 44. Stereo. Hrs opn: Drawer 1, 29832. Secondary address: 102 Slide Hill Rd. 29832. Phone: (803) 275-4444. Fax: (803) 275-3185. Licensee: Edgefield Saluda Radio Co. Inc. (acq 4-85; $3,586; FTR: 4-8-85). Natl. Rep: Keystone (unwired net). Format: Oldies. ♦Mike Casey, pres & gen mgr; Andy Moore, chief of engrg & traf mgr.

Kershaw

WKSC(AM)— Dec 21, 1961: 1300 khz; 500 w-D. TL: N34 33 30 W80 33 34. Hrs open: 24 Box 516, 203 E. Hilton St., 29067-0516. Phone: (803) 475-8585. Fax: (805) 966-3530. E-mail: wksc@wkscradio.com Web Site: www.wkscradio.com. Licensee: Kershaw Broadcasting Corp. (acq 7-23-03). Population served: 1,818 Natl. Network: ABC. Format: Oldies. Target aud: 35-64. ♦John Griffin, pres; Denise Robinson, stn mgr.

Kiawah Island

WCOO(FM)— Dec 7, 1969: 105.5 mhz; 50 kw. Ant 436 ft TL: N32 39 57 W80 03 11. Stereo. Hrs opn: 24 c/o WYBB(FM), 59 Windermere Blvd., Charleston, 29407. Phone: (843) 769-4799. Fax: (843) 769-4797. Web Site: www.thebridgeat1055.com. Licensee: L.M. Communications II of South Carolina Inc. Group owner: L.M. Communications Inc. (acq 3-30-95; FTR: 6-26-95). Natl. Network: ABC. Law Firm: Leventhal Senter & Lerman. Format: Rhythmic oldies. Target aud: 25-54; general. ♦Lynn Martin, pres; Charlie Cohn, gen mgr; Mike Allen, opns mgr.

Kingstree

WDKD(AM)— July 1949: 1310 khz; 5 kw-D, 67 w-N. TL: N33 42 11 W79 49 08. Hrs open: 24 51 Commerce St., Sumter, 29150. Phone: (803) 775-2321. Fax: (803) 773-4856. Web Site: www.miller.fm. Licensee: Miller Communications Inc. (group owner; (acq 12-18-2001; $1,415,456 assumption of debt with co-located FM). Population served: 39,960 Natl. Network: ABC. Rgnl. Network: S.C. Net. Format: Soft adult contemp. Target aud: 25-54. ♦Harold T. Miller Jr., CEO; Harold T. Miller, Jr., pres; Theresa Miller, VP & gen mgr; Dave Baker, opns VP & progmg VP; John Mcleod, pub affrs dir; Sarah Skinner, traf mgr.

WWKT-FM—Co-owned with WDKD(AM). May 28, 1966: 99.3 mhz; 11 kw. Ant 492 ft TL: N33 54 07 W79 59 52. Stereo. 24 Web Site: www.miller.fm.300,000 Law Firm: Smithwick & Belendiuk. Format: Rhythmic CHR. News staff: one; News: 14 hrs wkly. ♦Johnny Green, progmg dir & disc jockey; Vakenya Brunson, traf mgr; Gary "Thrills" Mills, disc jockey.

WGSS(FM)— 1998: 94.1 mhz; 6 kw. 328 ft TL: N33 43 32 W79 58 19. (CP: 6 kw). Hrs opn: 24 Box 103000, BTC-311, 181 E. Evans St., Florence, 29506. Phone: (843) 667-4600. Phone: (843) 665-0970. Fax: (843) 673-7390. Licensee: Qantum of Florence License Co. LLC. Group owner: Qantum Communications Corp. (acq 7-2-2003; grpsl). Format: Gospel. Target aud: 25-54; urban & caucasian. ♦Jonathan Brewster, gen mgr; Craig Dalla Riva, stn mgr.

Ladson

WJNI(FM)— June 15, 1998: 106.3 mhz; 6 kw. 328 ft TL: N32 55 42 W80 06 13. Stereo. Hrs opn: 5081 Rivers Ave., North Charleston, 29418. Phone: (843) 554-1063. Fax: (843) 554-1088. E-mail: traffic@jabarcommunications.com Web Site: jabarcommunications.com. Licensee: Thomas B. Daniels. Group owner: Jabar Communications. Population served: 800,000 Format: Urban contemp, inspirational. ♦Michael Baynard, gen mgr.

***WKCL(FM)**— Jan 11, 1982: 91.5 mhz; 100 kw. 305 ft TL: N33 00 24 W80 05 17. Stereo. Hrs opn: 24 526 College Park Rd., 29456. Phone: (843) 553-5420. Fax: (843) 553-0636. E-mail: wkcl@msn.com Web Site: www.wkclradio.com. Licensee: Chapel of the Holy Spirit and Holy Spirit Bible College. Population served: 500,000 Format: Contemp MOR, southern gospel. Target aud: General; baby boomers. ♦Carl L. Wiggins Sr., pres & gen mgr.

Lake City

WHYM(AM)— Oct 9, 1953: 1260 khz; 5 kw-D, 55 w-N. TL: N33 51 42 W79 44 15. Hrs opn: 6 AM-6 PM
Simulcasts WOLS(AM) Florence 75%.
Box 1177, ., 29560. Secondary address: 925 E. Main St. 29560. Phone: (843) 665-1230. Licensee: GHB of Lake City Inc. (group owner; (acq 5-28-92; $35,000; FTR: 6-15-92). Population served: 175,000 Natl. Network: ABC. Rgnl. Network: S.C. Net. Rgnl rep: Jim D.Jones Format: Country. News staff: one; News: 5 hrs wkly. Target aud: 35 plus. Spec prog: Loc news. ♦Jeff Andrew Lonis, gen mgr.

WWFN-FM— May 11, 1977: 100.1 mhz; 3.3 kw. 433 ft TL: N33 51 42 W79 44 15. Stereo. Hrs opn: 2014 N. Irby St., Florence, 29501. Phone: (843) 661-5000. Fax: (843) 661-0888. Licensee: Cumulus Licensing Corp. Group owner: Cumulus Media Inc. (acq 3-12-2001; $850,000). Format: Sports. Target aud: 25-54. ♦Rick House, gen mgr.

Lamar

WSIM(FM)— October 1992: 93.7 mhz; 2.8 kw. Ant 485 ft TL: N34 12 12 W79 51 52. Hrs open: 24 51 Commerce St., Sumter, 29151. Phone: (803) 775-2321. Fax: (803) 773-4856. Licensee: Miller Communications Inc. (group owner; (acq 11-14-2000; grpsl). Population served: 18,000 Natl. Network: ABC. Law Firm: Smithwick & Belendiuk. Format: Soft rock. News staff: one. Target aud: 35-64; adults. ♦Harold T. Miller Jr., CEO & pres; Theresa Miller, VP & gen mgr; Dave Baker, opns VP & opns mgr.

Lancaster

WAGL(AM)— Aug 7, 1962: 1560 khz; 50 kw-D, DA. TL: N34 49 53 W80 52 08. Stereo. Hrs opn: 101 S. Woodland Dr., 29720. Phone: (803) 283-8431. Fax: (803) 286-4702. E-mail: waglradio@infoave.net Web Site: www.waglradio.com. Licensee: Palmetto Broadcasting System Inc. Population served: 1,000,000 + Format: Gospel, oldies. ♦B.L. Phillips Jr., pres & gen mgr.

WRHM(FM)— July 27, 1964: 107.1 mhz; 3.3kw. 436 ft TL: N34 48 05 W80 47 51. Stereo. Hrs opn: Box 307, Rock Hill, 29731. Secondary address: 142 N. Confederate Ave., Rock Hill 29730. Phone: (803) 286-1071. Fax: (803) 324-2860. E-mail: almiller@wrhi.com Web Site: fm107.com. Licensee: Our Three Sons Broadcasting L.L.P. (acq 10-1-87). Population served: 800,000 Natl. Network: ABC. Rgnl. Network: S.C. Net. Format: Country, news, sports. Target aud: 25-54. ♦Allan M. Miller, gen mgr; Steven Stone, opns mgr; Mike Crowder, news dir.

Latta

WCMG(FM)— Sept 18, 1970: 94.3 mhz; 10.5 kw. 502 ft TL: N34 11 14 W79 31 23. Stereo. Hrs opn: 24 2014 N. Irby St., Florence, 29501. Phone: (843) 661-5000. Fax: (843) 661-0888. E-mail: mattscurry@cumulus.com Web Site: www.cumulus.com. Licensee: Cumulus Licensing Corp. Group owner: Cumulus Media Inc. (acq 6-99; $525,000). Population served: 96,581 Natl. Network: USA. Law Firm: Smithwick & Belendiuk. Format: Ault urban contemp. News staff: one; News: 8 hrs wkly. Target aud: 21-54; African-American. ♦Matt Scurry, opns mgr; Bill Brooks, gen sls mgr & sls; Ernie Frierson, progmg dir; Gail Gilbreath, chief of engrg; Martha Clark, traf mgr.

Laurens

WLBG(AM)— Mar 1, 1947: 860 khz; 1 kw-D, 12 w-N. TL: N34 30 13 W82 01 06. Hrs open: 24 Box 1289, 315 Hillcrest, 29360. Phone: (864) 984-3544. Fax: (864) 984-3545. E-mail: mail@wlbg.com Web Site: www.wlbg.com. Licensee: Southeastern Broadcast Associates Inc. (acq 8-5-83). Population served: 200,000 Natl. Network: Fox News Radio, Fox Sports. Law Firm: Pepper & Corazzini. Format: Var. News: 4 hrs wkly. Target aud: 30 plus; Black. ♦Emil J. Finley, pres; Michael C. Johnson, dev VP.

Leesville

WBLR(AM)—See Batesburg

Lexington

WLXC(FM)— Aug 31, 1994: 98.5 mhz; 6 kw. 328 ft TL: N33 52 42 W81 12 59. Hrs open: 24 Box 5106, Columbia, 29250. Secondary address: 1801 Charleston Hwy., Suite J, Cayce 29033. Phone: (803) 796-9975. Fax: (803) 796-5502. E-mail: doug.william@citcomm.com Web Site: www.kiss985fm.com. Licensee: Citadel Broadcasting Co. Group owner: Citadel Broadcasting Corp. (acq 5-30-00; grpsl). Population served: 525,000 Law Firm: Fletcher, Heald & Hildreth. Format: Urban contemp. News staff: one; News: one hr wkly. Target aud: 25-54; upward, mobile, higher income. Spec prog: Beach, boogie & blues. ♦William McElveen, gen mgr.

WQVA(AM)— 1983: 1170 khz; 10 kw-D. TL: N33 58 17 W81 16 43. Hrs open: Sunrise-sunset Box 537, Irmo, 29063. Phone: (803) 407-5223. Fax: (803) 407-6160. Web Site: myritmo.com. Licensee: Peregon Communications Inc. Group owner: Levas Communications LLC (acq 4-28-2005; $575,000). Format: Sp. ♦Sergio Perez, gen mgr & stn mgr; Beth Well, gen sls mgr.

Loris

WLSC(AM)— August 1958: 1240 khz; 1 kw-U. TL: N34 02 41 W78 53 39. Hrs open: 6 AM-midnight Box 578, 29569. Phone: (843) 756-1183. Licensee: JARC Broadcasting Inc. (acq 8-15-88). Population served: 1,741 Rgnl. Network: S.C. Net. Natl. Rep: Keystone (unwired net). Format: Full service. Target aud: 21-54. ♦Jerry Jenrette, gen mgr.

WVCO(FM)— Nov 19, 1993: 94.9 mhz; 11 kw. Ant 489 ft TL: N33 59 39 W78 46 16. Hrs open: 24 Box 3689, North Myrtle Beach, 29587. Phone: (843) 445-9491. Fax: (843) 445-9490. Web Site: www.949thesurf.com. Licensee: Carolina Beach Music Broadcasting Corp. (acq 6-18-03; $2.2 million). Population served: 300,000 Format: beach and boogie. News: 2 hrs wkly. Target aud: 25-45. ♦Earl P. Taylor, exec VP, VP & gen mgr; Selene Graham, sr VP.

Manning

WYMB(AM)— July 15, 1957: 920 khz; 2.3 kw-D, 1 kw-N. TL: N33 41 22 W80 16 16. (CP: 920 khz; 2.3 kw-D, 1 kw-N, DA-N. TL: N33 41 22 W80 16 16). Hrs opn: 2014 N. Irby St., Florence, 29501. Phone: (843) 661-5000. Fax: (843) 661-0888. Licensee: Cumulus Licensing Corp. Group owner: Cumulus Media Inc. (acq 3-24-99; with co-located FM). Population served: 4,025 Natl. Network: AP Radio. Natl. Rep: McGavren Guild. Format: Top 40. Target aud: General. ♦Rick House, gen mgr.

Marion

WHLZ(FM)— August 1991: 100.5 mhz; 21.5 kw. 354 ft TL: N34 19 36 W79 32 35. Hrs open: 2014 N. Irby St., Florence, 29501. Phone: (843) 661-5000. Fax: (843) 661-0888. E-mail: ernie.frieson@cumulus.com Web Site: www.cumulus.com. Licensee: Cumulus Licensing Corp. Group owner: Cumulus Media Inc. (acq 3-24-99; $3.8 million with WMXT(FM) Pamplico). Format: Country. ♦Matt Scurry, opns mgr & progmg dir; Gail Gilbreath, chief of engrg & engr; Bill Brooks, sls.

Mauldin

WBZT-FM— Apr 28, 1965: 96.7 mhz; 700 w. Ant 964 ft TL: N34 55 16 W82 24 05. Stereo. Hrs opn: 24 Box 100, Greenville, 29602. Secondary address: 7 N. Laurens St., Suite 700, Greenville 29601. Phone: (864) 242-1005. Fax: (864) 242-8813. E-mail: craigdebolt @clearchannel.com Web Site: www.thebuzzardjustrocks.con. Licensee: Clear Channel Broadcasting Licenses Inc. Group owner: Clear Channel Communications Inc. (acq 12-22-00). Population served: 300,000 Wire Svc: AP Format: Mainstream rock. Target aud: 18-49; males. ♦Marc Chase, exec VP; Bruce Logan, VP; Bill McMartin, gen mgr; Craig Debolt, stn mgr; Scott Johnson, opns mgr.

McClellanville

WAZS-FM— Dec 1, 1994: 98.9 mhz; 50 kw. Ant 492 ft TL: N33 11 20 W79 33 25. Stereo. Hrs opn: 5081 Rivers Ave., North Charleston, 29406. Phone: (843) 554-1063. Fax: (843) 554-1088. E-mail: traffic@jabarcommunications.com Web Site: www.jabarcommunications.com. Licensee: 98.9 Inc. Group owner: Jabar Communications (acq 1-5-2001). Population served: 800,000 Format: Sp top-40. Target aud: 25-54; upscale profesionals, yuppies. ♦Michael Baynard, gen mgr & opns mgr.

Moncks Corner

WIHB(FM)— Apr 16, 1973: 92.5 mhz; 100 kw. Ant 777 ft TL: N32 49 04 W79 50 08. (CP: 56 kw, ant 1,776 ft. TL: N32 55 28 W79 41 58). Stereo. Hrs opn: 24 2294 Clements Ferry Rd., Charleston, 29492-7729. Phone: (843) 972-1100. Fax: (843) 972-1200. Web Site: www.coast925.com. Licensee: Apex Broadcasting Inc. (group owner; (acq 10-17-2001; $3 million). Population served: 1,000,000 Format: Adult contemp. ♦Dean Pearce, CEO, pres, gen mgr & stn mgr; John Anthony, opns VP; Carl Wine, prom dir; Bruce Roberts, chief of engrg; Walt Rosen, sls.

WJKB(AM)— December 1963: 950 khz; 10 kw-D, 6 kw-N, DA-2. TL: N33 12 20 W80 03 54. Hrs open: 24 60 Markfield Dr., Suite 4, Charleston, 29407. Secondary address: 337 E. Main St. 29461. Phone: (843) 763-6631. Fax: (843) 763-1239. Licensee: Kirkman Broadcasting Inc. (group owner; (acq 11-29-2000; $150,000). Population served: 400,000 Rgnl. Network: S.C. Net Law Firm: Brian Madden. Format: Sports. Target aud: 25-54; male. ♦Gil Kirkman, pres; John Dixon, gen mgr & news dir; Stew Williams, opns mgr; Robby Robinson, gen sls mgr; Wally Momeier, chief of engrg.

Mt. Pleasant

WRFQ(FM)— June 1, 1985: 104.5 mhz; 100 kw. 659 ft TL: N32 49 04 W79 50 09. Stereo. Hrs opn: 24 950 Houston Northcutt Blvd., 29464. Phone: (843) 884-2534. Fax: (843) 884-6096. E-mail: kevin@qious.com Web Site: q1045.com. Licensee: Citicasters Licenses L.P. Group owner: Clear Channel Communications Inc. (acq 5-4-99; grpsl). Population served: 500,000 Format: Classic rock. Target aud: 25-54; adults, men. ♦Paul Smith, VP & mktg mgr; Scott Johnson, opns mgr; Tom Bustard, sls dir; Kevin Harbison, progmg dir.

WZJY(AM)— May 21, 1982: 1480 khz; 1 kw-D, 44 w-N. TL: N32 48 59 W79 50 18. Hrs open: 24 5081 Rivers Ave., North Charleston, 29406. Phone: (843) 554-1063. Fax: (843) 554-1088. Licensee: Thomas B. Daniels Group owner: Levas Communications LLC (acq 6-24-2007; $375,000). Format: Rgnl Mexican. ♦Michael Baynard, gen mgr.

Mullins

WJAY(AM)— June 1, 1949: 1280 khz; 5 kw-D, 270 w-N. TL: N34 11 30 W79 18 55. Hrs open: 18 Box 1020, Marion, 29571. Phone: (843) 423-1140. Fax: (843) 423-2829. Web Site: www.wjay.com. Licensee: The Greater Highway Church of Christ. Population served: 50,000 Natl. Network: ABC. Rgnl. Network: S.C. Net, Tobacco. Format: Gospel. News: 8 hrs wkly. Target aud: General. Spec prog: Farm 10 hrs wkly. ♦Curtis Campbell, pres, gen mgr & progmg dir.

Murrell's Inlet

***WMBJ(FM)—** 1997: 88.3 mhz; 1.8 kw vert. Ant 331 ft TL: N33 26 35 W79 08 21. Hrs open: 24
Rebroadcasts WHCB(FM) Bristol, TN 50%.
2420 Wade Hampton Blvd., Greenville, 29615. Phone: (864) 292-6040. Fax: (864) 292-8428. Web Site: www.hisradio.com. Licensee: Radio Training Network Inc. (acq 7-20-99; $5,000 cash). Format: Talk, Christian, educ. ♦Allen Henderson, gen mgr.

WYEZ(FM)— Apr 7, 1991: 94.5 mhz; 25 kw. Ant 328 ft TL: N33 33 13 W79 13 14. Stereo. Hrs opn: 24 Box 2830, Myrtle Beach, 29578. Secondary address: 3926 Wesley St., Suite 301 29579. Phone: (843) 903-9962. Fax: (843) 903-1797. Web Site: www.yes945.com. Licensee: Fidelity Broadcasting Corp. (acq 11-30-2000; $1 million). Natl. Network: CBS, Wall Street. Format: Adult contemp. News staff: one; News: 40 hrs wkly. ♦Matt Sedota, gen mgr; Bob Gauss, engrg dir.

Myrtle Beach

WKZQ-FM— July 3, 1969: 101.7 mhz; 50 kw. 601 ft TL: N33 56 14 W78 57 53. Stereo. Hrs opn: 24 1116 Ocala St., 29577. Phone: (843) 448-1041. Fax: (843) 626-5988. Web Site: www.wkzq.net. Licensee: NM Licensing LLC. Group owner: NextMedia Group L.L.C. (acq 11-26-01; grpsl). Format: Rock. Target aud: 18-34. ♦Steven Dinetz, CEO; Jeff Dinetz, COO; Carl Hirsch, chmn; Skip Weller, pres; Barry Brown, gen mgr; Art Greene, sls dir; Stephanie Nix, prom mgr; Brian Rickman, progmg dir; Paul Matthews, chief of engrg; Trimeshia Jeffery, traf mgr; Kim Johnson, spec ev coord.

WMYB(FM)— Jan 11, 1965: 92.1 mhz; 94 kw. Ant 863 ft TL: N33 35 27 W79 02 55. Stereo. Hrs opn: 24 1116 Ocala St., 29577. Phone: (843) 448-1041. Fax: (843) 626-5988. Web Site: www.wmybstar92.net. Licensee: NM Licensing LLC. Group owner: NextMedia Group L.L.C. (acq 11-26-01; grpsl). Population served: 175000 Law Firm: Cohn & Marks. Format: Adult contemporary. Target aud: 18 plus; adults. ♦Steven Dinetz, CEO; Jeff Dinetz, COO; Carl Hirsch, chmn; Skip Weller, pres; Barry Brown, gen mgr; Art Greene, sls dir; Kim Johnson, sls dir & spec ev coord; Stephanie Nix, prom dir; Bill Catcher, progmg dir; Paul Matthews, chief of engrg; Ginny Batchelder, traf mgr.

WRNN(AM)— Apr 24, 1965: 1450 khz; 5 kw-D, DA. TL: N33 42 20 W78 58 23. Hrs open: 1116 Ocala St., 29577. Phone: (843) 448-1041. Fax: (843) 626-5988. Licensee: NM Licensing LLC. Group owner: NextMedia Group LLC (acq 11-26-2001; grpsl). Population served: 150,000 Format: News/talk. Target aud: 25-60. ♦Steven Dinetz, CEO; Carl Hirsch, chmn; Skip Weller, pres; Barry Brown, gen mgr.

WYAV(FM)— July 1964: 104.1 mhz; 100 kw. Ant 981 ft TL: N33 35 27 W79 02 55. Stereo. Hrs opn: 1116 Ocala St., 29577. Phone: (843) 448-1041. Fax: (843) 626-5988. Web Site: www.wave104.net. Licensee: NM Licensing LLC. Group owner: NextMedia Group L.L.C. (acq 11-26-01; grpsl). Population served: 300,000 Format: Classic rock. Target aud: 18-49. ♦Steven Dinetz, CEO; Jeff Dinetz, COO; Carl Hirsch, chmn; Skip Weller, pres; Barry Brown, gen mgr; Art Greene, sls dir; Kim Johnson, sls dir & spec ev coord; Stephanie Nix, prom dir; Brian Rickman, progmg dir; Paul Matthews, chief of engrg; Trimeshia Jeffery, traf mgr.

New Ellenton

WGUS-FM— December 1989: 102.7 mhz; 4.3 kw. Ant 387 ft TL: N33 30 49 W81 38 03. Hrs open: 24 4051 Jimmie Dyess Pkwy., Augusta, 30909. Phone: (706) 396-7000. Fax: (706) 396-7092. Web Site: www.oldies107.com. Licensee: WGAC License LLC. Group owner: Beasley Broadcast Group (acq 12-22-94; $700,000; FTR: 2-13-95). Format: Oldies. News: 7 hrs wkly. Target aud: 35-64; affluent audience loyal fan base. ♦Kent Dunn, gen mgr; T. Gentry, opns dir; Zach Taylor, progmg dir.

Newberry

WKDK(AM)— October 1946: 1240 khz; 1 kw-U. TL: N34 17 30 W81 37 15. Hrs open: 24 hrs Box 753, 3000 Hazel St., 29108. Phone: (803) 276-2957. Fax: (803) 276-3337. E-mail: jcoggins@wkdk.com Web Site: www.wkdk.com. Licensee: Newberry Broadcasting Co. (acq 1951). Population served: 31,111 Natl. Network: ABC. Rgnl. Network: S.C. Net. Rgnl rep: Dora-Clayton. Format: Adult contemp, oldies. Target aud: General. ♦James P. Coggins, VP & gen mgr; Heather Hawkins, opns mgr.

WKMG(AM)— May 22, 1968: 1520 khz; 1 kw-D. TL: N34 15 12 W81 35 44. Hrs open: Sunrise-sunset 1840 Glenn St. Extention, 29108. Phone: (803) 405-0111. Fax: (803) 276-5677. Licensee: Cornell Blakely (acq 3-20-01; $10,000). Population served: 31,000 Format: Hispanic. Spec prog: Relg 2 hrs, gospel 3 hrs, Sp 10 hrs wkly. ♦Cornell Blakely, gen mgr.

North Augusta

WKZK(AM)— May 9, 1962: 1600 khz; 500 w-D. TL: N34 09 03 W82 23 34. Hrs open: 6 AM-sunset Box 1454, Augusta, GA, 30903.

Secondary address: 2 Milledge Rd., Augusta, GA 30904. Phone: (706) 738-0044. Fax: (706) 481-8442. E-mail: wkzk1600@bellsouth.net Web Site: wkzk.net. Licensee: Gospel Radio Inc. (acq 9-22-83; $190,000; FTR: 10-10-83). Natl. Network: American Urban. Natl. Rep: Dora-Clayton. Format: Black gospel, relg. Target aud: Black adults. ♦Garfield Turner, gen mgr & progmg dir.

WTHB(AM)—See Augusta, GA

WYNF(AM)—Licensed to North Augusta. See Augusta GA

North Charleston

WTMZ(AM)—Dorchester Terrace-Brentwood, Nov 17, 1960: 910 khz; 500 w-U, DA-N. TL: N34 09 03 W82 23 34. Hrs open: 24 60 Markfield Dr., #4, Charleston, 29407. Phone: (843) 763-6631. Fax: (843) 766-1239. Licensee: Kirkman Broadcasting Inc. Group owner: Citadel Broadcasting Corp. (acq 1-5-2005; $500,000). Format: Sports. ♦Gil Kirkman, gen mgr.

WXLY(FM)— July 17, 1962: 102.5 mhz; 100 kw. Ant 659 ft TL: N32 49 04 W79 50 09. Stereo. Hrs opn: 24 950 Houston Northcutt Blvd., 2nd Fl., Mt. Pleasant, 29464. Phone: (843) 856-6100. Fax: (843) 884-1218. E-mail: lisacooper@clearchannel.com Web Site: www.wxly.com. Licensee: Citicasters Licenses L.P. Group owner: Clear Channel Communications Inc. (acq 5-4-99; grpsl). Population served: 405,000 Format: Oldies/classic hits. News staff: 2. Target aud: 25-54. ♦Paul Smith, VP, gen mgr & mktg mgr; Scott Johnson, opns mgr; Tom Bustard, sls dir; Michelle Kelly, gen sls mgr; Chris Rivers, progmg dir; Willie Bennett, chief of engrg.

***WYFH(FM)**— July 7, 1984: 90.7 mhz; 50 kw. 492 ft TL: N32 58 23 W80 13 54. Stereo. Hrs open: 10870 Dorchester Rd., Summerville, 29485. Phone: (843) 875-9095. Web Site: www.bbnradio.org. Licensee: Bible Broadcasting Network Inc. (group owner) Natl. Network: Bible Bcstg Net. Format: Relg, Christian. ♦Dave Phillps, pres.

North Myrtle Beach

WEZV(FM)— Aug 15, 1972: 105.9 mhz; 17 kw. Ant 360 ft TL: N33 49 19 W78 46 18. Stereo. Hrs opn: 24 Box 2830, Myrtle Beach, 29578. Secondary address: 3926 Wesley St., Suite 301, Myrtle Beach 29579. Phone: (843) 903-9962. Fax: (843) 903-1797. Web Site: www.wezv.com. Licensee: Fidelity Broadcasting Corp. (acq 4-1-2000; $2.6 million with WNMB(AM) North Myrtle Beach). Population served: 125000 Format: Easy lstng. Target aud: 35 plus. ♦Matt Sedota, gen mgr & gen sls mgr.

***WKVC(FM)**— Sept 9, 1997: 88.9 mhz; 100 kw vert. 587 ft TL: N34 05 46 W78 28 28. Hrs open: 24 4337 Big Barn Dr., Little River, 29566. Phone: (843) 399-9649. Fax: (843) 399-9031. E-mail: kreeder@klove.com Web Site: www.klove.com. Licensee: Educational Media Foundation. Group owner: EMF Broadcasting (acq 5-11-00; $1.2 million). Natl. Network: K-Love. Format: Relg. Target aud: 25-65; contemp Christian. ♦Richard Jenkins, CEO & pres; Kurt Reeder, gen mgr.

WNMB(AM)— Apr 1, 1983: 900 khz; 500 w-U, DA-2. TL: N33 49 26 W78 45 59. Hrs open: 429 Pine Ave., 29582. Phone: (843) 249-6662. Fax: (843) 249-7823. Licensee: Norman Communications NMB Inc. (acq 6-4-2004; $250,000). Format: Relg. ♦Bill Norman, gen mgr.

Orangeburg

WHXT(FM)— September 1973: 103.9 mhz; 9.2 kw. Ant 531 ft TL: N33 40 13 W80 52 25. Stereo. Hrs opn: 1900 Pineview Rd., Columbia, 29209. Phone: (803) 695-8680. Phone: (803) 376-1039. Fax: (803) 695-8605. E-mail: mhanisch@innercity.sc.com Web Site: www.hot1039fm.com. Licensee: Urban Radio II L.L.C. Group owner: Inner City Broadcasting (acq 5-30-2003; $11.1 million with WZMJ(FM) Batesburg). Population served: 130,100 Format: Urban contemp. ♦Steve Patterson, gen mgr.

WPJK(AM)— Nov 3, 1958: 1580 khz; 1 kw-D. TL: N33 28 43 W80 52 46. Hrs open: Sunrise-sunset 175 Cannon Bridge Rd., 29115. Phone: (803) 534-4848. Fax: (803) 534-0888. Licensee: Radio Orangeburg Partnership. (acq 6-86). Natl. Network: USA. Format: Relg, urban contemp, gospel. ♦Bose Gowdy, pres & gen mgr; Rev. Pinckney Palmer Jr., opns mgr.

WQKI-FM— Oct 10, 1987: 102.9 mhz; 2.7 kw. Ant 492 ft TL: N33 27 55 W80 56 44. Stereo. Hrs opn: 24 200 Regional Pkwy., Bldg. C, Suite 200, 29118. Phone: (803) 536-1710. Fax: (803) 531-1089. E-mail: mail@miller.fm Web Site: miller.fm. Licensee: Miller Communications Inc. (group owner; acq 4-30-03; $1.25 million with WGFG(FM) Branchville). Population served: 175,000 Natl. Network: ABC. Format: Classic rock. News staff: one; News: 10 hrs wkly. Target aud: 25-54. Spec prog: Relg 6 hrs wkly. ♦Harold Miller Jr., pres; Russ T. Fender, opns mgr & progmg dir; Sonny Pagan, sls VP; Theresa Miller, gen mgr & gen sls mgr; Dave Baker, progmg VP; Dave Dalesky, engrg VP & chief of engrg.

***WSSB-FM**— Mar 15, 1985: 90.3 mhz; 90 kw. 225 ft TL: N33 29 55 W80 50 30. Stereo. Hrs opn: 24 Box 7619, Nance B-114, 29117. Phone: (803) 536-8196. Fax: (803) 533-3652. Web Site: www.scsu.edu. Licensee: South Carolina State University. Natl. Network: American Urban, NPR. Format: Urban contemp, gospel, jazz. News staff: one; News: 7 hrs wkly. Target aud: 8-65. Spec prog: Jazz 10 hrs, reggae 4 hrs, blues 2 hrs, rap 4 hrs wkly. ♦Marion White, progmg dir & mus dir; Milton E. McKissick, gen mgr & news dir; Ken Durst, chief of engrg.

WTCB(FM)— July 6, 1967: 106.7 mhz; 100 kw. 787 ft TL: N33 46 52 W80 55 14. Stereo. Hrs opn: Box 5106, Columbia, 29250. Secondary address: 1801 Charleston Hwy., Suite J, Cayce 29033. Phone: (803) 796-7600. Fax: (803) 796-9291. Web Site: www.b106fm.com. Licensee: Citadel Broadcasting Co. Group owner: Citadel Broadcasting Corp. (acq 5-30-00; grpsl). Population served: 771,000 Natl. Rep: Christal. Format: Adult contemp. Target aud: 25-54; affluent, upscale young adults. ♦William L. McElveen, pres & gen mgr; Brent Johns, opns mgr.

Pageland

WRML(FM)— Feb 22, 1975: 102.3 mhz; 6 kw. Ant 213 ft TL: N34 49 04 W80 19 21. Stereo. Hrs opn: Box 2148, Tucker, GA, 30085. Phone: (843) 672-7839. Phone: (704) 442-7222. Fax: (843) 672-1023. Licensee: Robert Broadcasting Inc. (acq 12-18-98). Population served: 100,000 Natl. Network: Salem Radio Network. Format: Sp (Mexician Rgnl). ♦John Griffin, pres.

Pamplico

WMXT(FM)— Nov 1, 1990: 102.1 mhz; 50 kw. 500 ft TL: N34 04 56 W79 37 19. Stereo. Hrs opn: 24 2014 N. Irby St., Florence, 29501-1504. Phone: (843) 661-5000. Fax: (843) 661-0888. E-mail: buzz.bowman@cumulus.com Web Site: www.cumulus.com. Licensee: Cumulus Licensing Corp. Group owner: Cumulus Media Inc. (acq 3-24-99; $3.8 million with WHLZ(FM) Marion). Population served: 178,000 Law Firm: Fletcher, Heald & Hildreth. Format: Classic Rock. News staff: 2; News: 3 hrs wkly. Target aud: 25-54. Spec prog: Beach mus 5 hrs wkly. ♦Matt Scurry, opns mgr; Buzz Bowman, prom dir & progmg dir; Gail Gilbreath, chief of engrg; Bill Brooks, sls.

Parris Island

WGZO(FM)— July 1985: 103.1 mhz; 17.5 kw. 328 ft TL: N32 26 10 W80 55 23. Stereo. Hrs opn: 24 401 Mall Blvd., Suite 101 D, Savannah, GA, 31406. Phone: (912) 351-9830. Fax: (912) 352-4821. Web Site: www.1031thedrive.com. Licensee: Zip Communications Inc. (acq 7-10-01; $100,000). Population served: 500,000 Natl. Rep: Christal. Format: The Drive. News staff: 2. Target aud: 12-34; young, hip, trendy adults. ♦Robert Leonard, gen mgr.

Pawley's Island

WDAI(FM)— Oct 2, 1993: 98.5 mhz; 6.1 kw. Ant 666 ft TL: N33 35 27 W79 02 55. Stereo. Hrs opn: 24 11640 Highway 17 Bypass, Murrells Inlet, 29576. Phone: (843) 651-7869. Fax: (843) 651-3197. Web Site: www.985kissfm.net. Licensee: Cumulus Licensing Corp. Group owner: Cumulus Media Inc. (acq 1-27-98). Natl. Network: Westwood One. Format: Urban contemporary. News staff: one; News: 6 hrs wkly. Target aud: 25-54. ♦Bill Hazen, gen mgr.

Pickens

WTBI(AM)— Aug 3, 1967: 1540 khz; 10 kw-D. TL: N34 51 37 W82 43 25. Hrs open: Sunrise-sunset
Rebroadcasts WTBI-FM Greenville.
3931 White Horse, Greenville, 29611. Phone: (864) 295-2145. Fax: (864) 295-6313. E-mail: jwatts@tabernacleministries.org Web Site: www.wtbi.org. Licensee: Tabernacle Christian Schools (acq 11-83; $150,000; FTR: 1-30-84). Population served: 94,000 Natl. Network: USA. Format: Christian, educational. Target aud: All ages.Dr. Melvin Aiken, pres; Charles Garrett Sr., stn mgr, opns mgr & gen sls mgr; John Watts, progmg dir & disc jockey; Fay Frazier, traf mgr; Jeremy Chisam, disc jockey; Joey Robertson, disc jockey; Mike Belcher, disc jockey; Mike Stratton, disc jockey; Patricia Hutchins, disc jockey; Raymond Davis, disc jockey; Sam Jones, disc jockey; Sarah Parker, disc jockey; Stephen Davis, disc jockey; Tim Selby, disc jockey

Port Royal

WLOW(FM)— February 1988: 107.9 mhz; 24 kw. Ant 725 ft TL: N32 13 36 W80 50 53. Hrs open: 24 One St. Augustine Pl., Hilton Head Island, 29928. Phone: (843) 785-9569. Fax: (843) 842-3369. E-mail: wlow1079@adventureradio.fm Web Site: www.wlow.com. Licensee: Monterey Licenses LLC. Group owner: Triad Broadcasting Co. LLC (acq 7-18-2000; grpsl). Population served: 500,000 Natl. Rep: Christal. Format: The coast. Target aud: 45 plus; active, affluent, older. ♦Robert Leonard, gen mgr.

Ravenel

WMGL(FM)— February 1986: 101.7 mhz; 3 kw. 482 ft TL: N32 46 44 W80 10 37. (CP: 6.5 kw, ant 689 ft. TL: N32 38 59 W80 19 00). Stereo. Hrs opn: 24 4230 Faber Place Dr., Suite 100, North Charleston, 29405. Phone: (843) 277-1200. Fax: (843) 277-1212. Web Site: www.magic1017.com. Licensee: The Last Bastion Station Trust LLC, as Trustee Group owner: Citadel Broadcasting Corp. (acq 6-12-2007; grpsl). Law Firm: Cole, Raywid & Braverman. Format: New adult contemp. News staff: 2; News: 6 hrs wkly. Target aud: 25-54; upscale adults. ♦Paul O'Malley, gen mgr.

Richburg

***WRBK(FM)**— 1998: 90.3 mhz; 7.5 kw horiz, 7.3 kw vert. Ant 538 ft TL: N34 41 46 W81 01 23. Stereo. Hrs opn: 24 Box 15, Chester, 29706. Phone: (803) 581-9030. Fax: (803) 581-9932. Licensee: Richburg Educational Broadcasters Inc. Format: Beach mus, oldies. Target aud: 30-60; middle aged adults who like beach flavored oldies. ♦Jeff Sigmon, pres & gen mgr.

Ridgeland

WNFO(AM)— 1964: 1430 khz; 1 kw-D, 880 w-N. TL: N32 28 07 W81 00 15. (CP: COL: Sun City-Hilton Head, 213 w-D. TL: N32 21 24 W80 55 23). Hrs opn: Sunrise-sunset Box 6567, Hilton Head Island, 29938. Phone: (843) 785-5769. Fax: (843) 785-8139. Licensee: Walter M. Czura (acq 7-9-91; $22,500; FTR: 7-29-91). Population served: 135,000 Format: Hispanic. Target aud: General; incoming visitors to South Carolina & Spanish speaking people. Spec prog: Spanish. ♦Walter M. Czura, pres & gen mgr.

WWJN(FM)—Licensed to Ridgeland. See Hilton Head Island

Ridgeville

WPAL-FM— September 1968: 100.9 mhz; 25 kw. Ant 328 ft TL: N33 04 16 W80 21 31. Stereo. Hrs opn: 24 Phone: (843) 727-0099. Fax: (843) 974-6002. Licensee: Gresham Communications Inc. (acq 12-12-93; $150,000; FTR: 12-20-93). Population served: 20,000 Rgnl. Network: S.C. Net. Law Firm: Gardner, Carton & Douglas. Format: Urban adult contemp. ♦Judith Aidoo, gen mgr.

Rock Hill

WAGL(AM)—See Lancaster

WAVO(AM)— May 18, 1948: 1150 khz; 1 kw-D, 57 w-N. TL: N34 57 02 W81 00 16. Hrs open: 24
Rebroadcasts WHVN(FM) Charlotte 75%.
Box 1024, 29731. Secondary address: 400 Pineview Rd. 29731. Phone: (803) 327-1150. Phone: (704) 596-4900. Fax: (704) 596-6939. E-mail: bboonewhvn@bellsouth.net Licensee: WHVN Inc. Group owner: GHB Radio Group (acq 2-4-92; $115,000; FTR: 2-24-92). Natl. Network: Moody, USA. Law Firm: Reddy, Begley & McCormick. Format: Relg, talk. News staff: one; News: 15 hrs wkly. Target aud: 25-54; career-oriented. Spec prog: College football & baseball. ♦Tom Gentry, gen mgr, stn mgr & gen sls mgr; Buddy Boone, progmg dir; Brant Hart, mus dir & pub affrs dir; Stu Albert, chief of engrg.

WBT-FM—See Chester

WBZK(AM)—See York

***WNSC-FM**— Jan 3, 1978: 88.9 mhz; 100 kw. 600 ft TL: N34 50 24 W81 01 07. Stereo. Hrs opn: 24 1101 George Rogers Blvd., Columbia, 29201. Phone: (803) 737-3420. Fax: (803) 737-3552. E-mail: gasque@scetv.org Web Site: www.etvradio.org. Licensee: South Carolina Educational Television Commission. Natl. Network: NPR. Format: Jazz. ♦Moss Bresnahan, pres; Paul Zweimiller, stn mgr; Tom Holloway, dev dir.

WRHI(AM)— Dec 14, 1944: 1340 khz; 1 kw-U. TL: N34 54 51 W81 00 42. Hrs open: 24 hours Box 307, 29731. Secondary address: 142 N. Confederate Ave. 29730. Phone: (803) 324-1340. Fax: (803) 324-2860. E-mail: newsroom@wrhi.com Web Site: www.wrhi.com. Licensee: Our Three Sons Broadcasting L.L.P. (group owner; acq 10-1-84). Population served: 225,000 Natl. Network: ABC. Rgnl. Network: S.C. Net. Format: News/talk, sports. News staff: 2; News: 14 hrs wkly. Target aud: 30 plus. ♦Allan M. Miller, gen mgr; Steven Stone, opns mgr; Mike Crowder, news dir.

Saint Andrews

WMFX(FM)—Licensed to Saint Andrews. See Columbia

Saint George

WNKT(FM)— Jan 5, 1971: 107.5 mhz; 100 kw. 984 ft TL: N33 05 11 W80 22 33. Stereo. Hrs opn: 24 4230 Faber Place Dr., Suite 100, N. Charleston, 29405. Phone: (843) 277-1200. Fax: (843) 277-1212. Web Site: www.catcountry1075.com. Licensee: Citadel Broadcasting Co. Group owner: Citadel Broadcasting Corp. (acq 6-9-99; grpsl). Population served: 500,000 Natl. Rep: McGavren Guild. Format: Country. News: one hr wkly. Target aud: 25-54. ♦Paul O'Malley, gen mgr; Bob McNeill, progmg dir; Justin Tucker, chief of engrg; Barbara Young, traf mgr.

WQIZ(AM)— Aug 23, 1962: 810 khz; 5 kw-D. TL: N33 08 51 W80 33 47. Hrs open: 173 Radio Rd., 29477. Phone: (904) 859-0980. Licensee: Radio Properties LLC (acq 6-12-2003; $200,000). Population served: 80,000 Format: Catholic. ♦Paul Danese, gen mgr; Bert Artlip, stn mgr.

Saint Matthews

WIGL(FM)— 1990: 93.9 mhz; 1.75 kw. Ant 607 ft TL: N33 45 46 W80 49 23. Hrs open: 24 200 Regional Pkwy., Bldg. C, Suite 200, Orangeburg, 29118. Phone: (803) 534-2777. Fax: (803) 531-1089. Licensee: Miller Communications Inc. (acq 6-30-2003; $900,000 with co-located AM). Population served: 100,000 Format: Hot adult contemp. ♦Theresa Miller, gen mgr.

WPOG(AM)— Aug 15, 1975: 710 khz; 1 kw-D, DA. TL: N33 37 04 W80 46 50. Hrs open: 6 AM-6 PM 4305 Columbia Rd., Orangeburg, 29118-1268. Phone: (803) 536-4300. Licensee: Grace Baptist Church of Orangeburg (group owner; (acq 10-6-2005; $235,000). Population served: 100,000 Format: Gospel. ♦Gene G. Soult, gen mgr.

Saint Stephen

WTUA(FM)— May 1990: 106.1 mhz; 3 kw. 328 ft TL: N33 29 36 W79 53 21. Hrs open: Box 1240, 29479. Secondary address: 4013 Burns Dr. 29479. Phone: (843) 567-2091. Fax: (843) 567-3088. E-mail: wtuaradio@direcway.com Licensee: Praise Communications Inc. (acq 1-27-2005). Format: Gospel. News staff: one; News: 5 hrs wkly. Target aud: 20-65; African American. ♦Lynette L. Nelson, gen mgr.

Saluda

WJES(AM)— June 12, 1961: 1190 khz; 350 w-D. TL: N33 57 27 W81 47 34. (CP: 1200 khz; 10 kw-D, 4 w-N, 6.1 kw-CH). Hrs opn: 125 N. Main St., 29138. Phone: (864) 445-9537. E-mail: info@1190wjes.com Web Site: www.1190wjes.com. Licensee: Carolina Broadcast Partners LLC (acq 6-26-2006; $100,000). Population served: 3,150 Rgnl. Network: S.C. Net. Format: Country, beach and oldies. ♦Jeffery S. Roper, pres.

Sans Souci

WCSZ(AM)—Licensed to Sans Souci. See Greenville

Scranton

WZTF(FM)— 1991: 102.9 mhz; 2.9 kw. Ant 466 ft TL: N34 00 39 W79 45 24. Hrs open: Box 103000, Florence, 29501. Phone: (843) 667-4600. Phone: (843) 667-0970. Fax: (843) 673-7390. Licensee: Qantum of Florence License Co. LLC. Group owner: Qantum Communications Corp. (acq 7-2-2003; grpsl). Format: Urban contemp. ♦Jonathan Brewster, gen mgr.

Seneca

WHZT(FM)— June 6, 1953: 98.1 mhz; 100 kw. 1,004 ft TL: N34 41 14 W82 59 12. Stereo. Hrs opn: 24 220 N. Main St., Suite 402, Greenville, 26901. Phone: (864) 232-9810. Fax: (864) 370-3403. Web Site: www.hot981.com. Licensee: Cox Radio Inc. Group owner: Cox Communications Inc. (acq 2-1-2001; grpsl). Population served: 800,000 Natl. Network: Westwood One, CBS. Law Firm: Dickstein Shapiro Morin & Oshinsky. Format: CHR. News staff: 2; News: 18 hrs wkly. Target aud: 25-54; affluent adults. ♦Steve Sinicropi, VP & gen mgr; Rob Grossman, gen sls mgr; Cathy Tabor, natl sls mgr; Laurie Madden, mktg VP, prom dir & news dir; Murph Dawg, mus dir & traf mgr; Lemont Bryant, chief of engrg.

WSNW(AM)— June 1, 1949: 1150 khz; 1 kw-D, 58 w-N. TL: N34 41 11 W82 59 27. Hrs open: 24 Box 1251, 29679. Secondary address: 103 Ram Cat Alley 29678. Phone: (864) 882-9769. Fax: (864) 886-0082. E-mail: allgood@gacaradio.com Web Site: www.wsnwradio.com. Licensee: Tugart Properties LLC. Group owner: Georgia-Carolina Radiocasting Companies (acq 9-28-2001). Population served: 100,000 Natl. Network: ABC, CBS Radio. Law Firm: Dan J. Alpert. Wire Svc: AP Format: MOR, local news. News staff: one; News: 12 hrs wkly. Target aud: Adults 35 plus. ♦Art Sutton, pres; Terry Carter, VP; George Allgood, stn mgr.

Simpsonville

WFIS(AM)—See Fountain Inn

WGVC(FM)— July 10, 1989: 106.3 mhz; 25 kw. Ant 328 ft TL: N34 50 33 W82 09 59. Stereo. Hrs opn: 24 25 Garlington Rd., Greenville, 29615-4613. Phone: (864) 271-9200. Fax: (864) 242-1567. Web Site: www.1063charliefm.com. Licensee: Entercom Greenville License LLC. Group owner: Barnstable Broadcasting Inc. (acq 10-7-2005; grpsl). Law Firm: Latham & Watkins. Format: Adult hits. Target aud: 25-54; adults. ♦David J. Field, pres; Sharon Day, gen mgr.

Socastee

WRNN-FM— 1997: 99.5 mhz; 14.5 kw. 430 ft TL: N33 49 30 W78 51 47. Hrs open: 1116 Ocala St., Myrtle Beach, 29577. Phone: (843) 448-1041. Fax: (843) 626-5988. Web Site: www.wrnn.net. Licensee: NM Licensing LLC. Group owner: NextMedia Group L.L.C. (acq 11-26-2001; grpsl). Format: Talk. ♦Steven Dinetz, CEO; Jeff Dinetz, COO; Carl Hirsch, chmn; Skip Weller, pres; Barry Brown, gen mgr; Art Greene, sls dir; Stephanie Nix, prom dir; Dave Priest, progmg dir; Paul Matthews, chief of engrg; Ginny Batchelder, traf mgr; Kim Johnson, spec ev coord.

South Congaree

WFMV(FM)— 1993: 95.3 mhz; 6 kw. 328 ft TL: N33 53 58 W81 13 29. Hrs open: 24 Box 2355, West Columbia, 29171. Secondary address: 2440 Millwood Ave., Columbia 29205. Phone: (803) 939-9530. Fax: (803) 939-9469. E-mail: email@wfmv.com Web Site: www.wfmv.com. Licensee: Glory Communications. Group owner: Glory Communications Inc. Population served: 471,800 Format: Urban inspirational. Target aud: Primary : adult 25-54; secondary: Women 25-54. ♦Alex Snipe Jr., gen mgr; Tezra Haire, gen sls mgr.

Spartanburg

WASC(AM)— Jan 15, 1968: 1530 khz; 1 kw-D, 250 w-CH. TL: N34 56 58 W81 57 33. Hrs open: Box 5686, 29304. Secondary address: 840 Wofford St. 29304. Phone: (864) 585-1530. Fax: (864) 573-7790. Licensee: New South Broadcasting Corp. (acq 2-9-76). Population served: 49,000 Format: Black, Urban Gold. ♦Sam E. Floyd, pres; K. Joseph Sessoms, VP & chief of engrg; K. Joseph Sessmos, gen mgr; Ed Waddell, min affrs dir & farm dir.

WOLI(AM)— Sept 1, 1940: 910 khz; 3.6 kw-D, 960 w-N. TL: N35 01 10 W82 00 36. Hrs open: 24
Rebroadcasts WYRD(AM) Greenville 100%.
6665 Pottery Rd., 29303. Phone: (864) 271-9200. Fax: (864) 241-4225. Web Site: www.newstalkword.com. Licensee: Davidson Media Station WSPA Licensee LLC. Group owner: Entercom Communications Corp. (acq 10-6-2005; grpsl). Natl. Network: CBS, ABC. Format: News/talk, sports. News staff: 4; News: 50 hrs wkly. Target aud: 35-64. Spec prog: Atlanta Falcons football, Univ. of South Carolina football and basketball, relg 5 hrs, sports 12 hrs wkly. ♦Jimmy Vineyard, gen mgr; Jim Kirkland, opns mgr; Tom Durney, gen sls mgr; Kelly Cowen, prom dir; Peter Phiele, progmg dir; Lisa Rollins, news dir & news rptr; Jerry Massey, chief of engrg; Katherine Lambert, traf mgr; Amy Hierder, news rptr; Chris Witmire, news rptr; Shannon Campbell, news rptr; John Boone, sports cmtr; Josh Bartlett, sports cmtr.

WORD(AM)— Feb 17, 1930: 950 khz; 5 kw-U, DA-N. TL: N34 58 53 W81 59 14. Hrs open: 24 25 Garlington Rd., Greenville, 29615. Phone: (864) 271-9200. Fax: (800) 967-9329. Fax: (864) 242-1567. Web Site: www.newsradioword.com. Licensee: Entercom Greenville License L.L.C. Group owner: Entercom Communications Corp. (acq 12-13-99; grpsl). Population served: 600,000 Natl. Network: CBS, Motor Racing Net. Rgnl. Network: S.C. Net. Format: News/talk. News staff: 4; News: 45 hrs wkly. Target aud: 35-64. ♦David J. Field, CEO; Steve Fisher, CFO; Tom Durney, gen mgr; Jim Kirkland, opns mgr.

WSPA-FM— Aug 29, 1946: 98.9 mhz; 100 kw. 1,910 ft TL: N35 10 12 W82 17 27. Stereo. Hrs opn: 24 25 Garlington Rd., Grenville, 29615. Phone: (864) 271-9200. Fax: (864) 370-1473. Web Site: www.magic989online.com. Licensee: Entercom Greenville License LLC. (acq 12-13-99; grpsl). Population served: 1,200,000 Format: Light adult contemp. News: one hr wkly. Spec prog: Relg 3 hrs, jazz 6 hrs, 70s oldies 10 hrs wkly. ♦Jerry Stevens, gen sls mgr; David Patella, natl sls mgr; Michael McKeel, progmg dir; Stephen Hester, traf mgr; Lisa Rollins, news rptr; Jeff Cross, disc jockey; John Gosnell, disc jockey; Lee Alexander, disc jockey; Rick Woodell, disc jockey; Ted Love, disc jockey.

WSPG(AM)— Sept 1, 1952: 1400 khz; 1 kw-U. TL: N34 58 26 W81 55 37. Hrs open: Box 193, 29304. Secondary address: 340 Garner Rd. 29303. Phone: (864) 573-1400. Fax: (864) 573-8699. E-mail: info@spartanburg1400.com Licensee: Fulmer Broadcasting Inc. (acq 12-30-2003; $300,000). Wire Svc: AP Format: News/talk, sports. Target aud: Adult male 25-54. ♦Matthew Y. Fulmer, pres; J. Dwayne Corn, gen mgr; Jan Scruggs, progmg dir & progmg.

Summerton

WLJI(FM)— 1997: 98.3 mhz; 6 kw. 328 ft TL: N33 42 58 W80 20 44. Hrs open: 24
Simulcast of WFMV(FM) South Congaree 100%.
Box 1348, Sumter, 29151. Phone: (803) 774-5512. Fax: (803) 774-5534. Licensee: Glory Communications Inc. (group owner; acq 4-1-97). Format: Urban inspirational. Target aud: Adults 25-54. ♦Alex Snipe Jr., gen mgr; Tony Jamison, opns mgr; Tezra Haire, gen sls mgr.

Summerville

WAZS(AM)— June 7, 1963: 980 khz; 1 kw-D, 131 w-N. TL: N33 01 57 W80 12 00. Hrs open: 5081 Rivers Ave., North Charleston, 29406. Phone: (843) 554-1063. Fax: (843) 554-1088. E-mail: traffic@jaborcommunications.com Web Site: jabarcommunications.com. Licensee: Thomas B. Daniels. Group owner: Jabar Communications (acq 9-1-2000). Population served: 200,000 Format: Jazz. ♦Michael Baynard, gen mgr.

WWWZ(FM)— May 10, 1974: 93.3 mhz; 50 kw. 492 ft TL: N33 06 54 W79 54 25. Hrs open: 4230 Faber Place Drive, Suite 100, North Charleston, 29405. Phone: (843) 277-1200. Fax: (843) 277-1212. Web Site: www.z93jams.com. Licensee: Citadel Broadcasting Co. Group owner: Citadel Broadcasting Corp. (acq 6-9-99; grpsl). Population served: 3,704 Format: Urban contemp. ♦Paul O'Malley, gen mgr; Star Israel, gen sls mgr; Terry Base, progmg dir; Judy Herold, news dir; Justin Tucker, chief of engrg; Stephanie Gaines, women's int ed.

Sumter

WDXY(AM)— May 23, 1960: 1240 khz; 1 kw-U. TL: N33 54 16 W80 19 25. Hrs open: 24 Box 1269, 29151. Secondary address: 51 Commerce St. 29150. Phone: (803) 775-2321. Fax: (803) 773-4856. Web Site: www.newsstalk1240.am. Licensee: Miller Communications Inc. (group owner; acq 1-2-2001; grpsl). Population served: 244,000 Natl. Rep: Rgnl Reps. Law Firm: Smith & Belendiuk. Format: News, talk. News staff: one; News: 6 hrs wkly. Target aud: 35 plus. ♦Harold T. Miller, CEO & pres; Theresa Miller, VP & gen mgr; Dave Baker, opns VP.

WICI(FM)— June 21, 1995: 94.7 mhz; 3 kw. 479 ft TL: N33 51 55 W80 17 09. Hrs open: 24 Box 1269, 29151. Secondary address: 51 Commerce St. 29150. Phone: (803) 773-1859. Fax: (803) 773-4856. E-mail: Tmiller55@aol.com Web Site: www.mix947.fm. Licensee: Miller Communications Inc. (group owner; acq 9-27-01). Population served: 297,000 Natl. Network: American Urban. Law Firm: Smithwick & Belendiuk. Format: Hits of the 80s, 90s & now. News staff: one. Adults 25 to 49 & secondary females 25-54. ♦Harold T. Miller, Jr., CEO & pres; Dave Baker, opns VP & sls dir; Theresa Miller, chmn, VP, gen mgr & gen sls mgr.

WQMC(AM)— Mar 16, 1940: 1290 khz; 1 kw-U, DA-N. TL: N33 55 16 W80 16 59. Hrs open: 16 Richardson-Johnson Learning Resc. Ctr., 100 W. College St., 29150-3599. Phone: (803) 775-1290. Phone: (803) 775-6262. Fax: (803) 773-3687. Fax: (803) 775-2580. Licensee: Morris College. Population served: 25,000 Format: Gospel, talk, christian. News staff: one; News: 4 hrs wkly. Target aud: General; working adults & retirees. Spec prog: Farm 2 hrs, class 15 hrs wkly. ♦Janet Clayton, gen mgr; Fred Brown, progmg dir; Willete Stocker, mus dir.

***WRJA-FM**— Aug 25, 1975: 88.1 mhz; 98 kw. 1,000 ft TL: N33 52 32 W80 16 14. Stereo. Hrs opn: 24 1101 George Rogers Blvd., Columbia, 29201. Phone: (803) 737-3404. Phone: (803) 737-3420. Fax: (803) 737-3552. E-mail: gasque@scetv.org Web Site: www.etvradio.org. Licensee: South Carolina Educational TV Commission. Population served: 24,555 Natl. Network: NPR, PRI. Format: News/talk, jazz. ♦Moss Bresnahan, pres; Tom Fower, sr VP; Paul Zweimiller, stn mgr & chief of engrg; Tom Holloway, sls dir; John Gasque, progmg dir; Connie Murray, traf mgr.

WSSC(AM)— Apr 27, 1953: 1340 khz; 1 kw-U. TL: N33 55 45 W80 19 29. Hrs open: 6 AM-midnight 201 Oswego Rd., 29150. Phone: (803) 469-0288. Fax: (803) 469-0297. Web Site: www.sumterbaptisttemple.org. Licensee: Sumpter Baptist Temple Inc. (acq 2-28-94; $157,500; FTR: 5-2-94). Population served: 26,536 Rgnl. Network: S.C. Net. Format: Christian radio. Target aud: 25-54. ♦Eddie Richardson, pres & gen mgr.

WWDM(FM)— 1961: 101.3 mhz; 100 kw. 1,322 ft TL: N33 52 52 W80 16 14. Stereo. Hrs opn: 24 1900 Pineview Rd., Columbia, 29209. Phone: (803) 695-8600. Fax: (803) 695-8605. E-mail: mhanisch@innercity.sc.com Web Site: www.thebigdm.com. Licensee: Urban Radio II L.L.C. Group owner: Inner City Broadcasting (acq 8-7-2000; grpsl). Population served: 461,000 Natl. Network: ABC, Westwood One. Natl. Rep: D & R Radio. Format: Urban contemp. News staff: one; News: 6 hrs wkly. Target aud: 18-49. ♦Maggie Hanisch, gen mgr; Mike Love, opns dir & progmg dir; Scott Norton, gen sls mgr; Susan Morningstar, VP & prom dir.

Surfside Beach

WYAK-FM— Apr 4, 1977: 103.1 mhz; 12.5 kw. 325 ft TL: N33 34 32 W79 02 29. (CP: 11.5 kw, ant 485 ft. TL: N33 43 22 W79 03 43). Stereo. Hrs opn: 24
Rebroadcasts WVCO(FM) Loris 100%.
11640 Hwy. 17 Bypass, Murrells Inlet, 29576. Phone: (843) 651-7869. Fax: (843) 651-3197. Web Site: www.cumulus.com. Licensee: Cumulus Licensing Corp. Group owner: Cumulus Media Inc. (acq 12-6-00; swap of WYAK-FM for WQSL(FM) & WXQR(FM) Jacksonville, NC). Population served: 250,000 Natl. Network: Westwood One. Law Firm: Fisher, Wayland, Cooper, Leader & Zaragoza. Format: Country. News staff: one; News: 4 hrs wkly. Target aud: 25-54; adults & families of loc towns & tourists. ♦John Sheftic, gen mgr.

Union

WBCU(AM)— Aug 27, 1949: 1460 khz; 1 kw-U, DA-N. TL: N34 43 10 W81 39 44. Hrs open: 24 210 E. Main, 29379. Phone: (864) 427-2411. Phone: (864) 427-2412. Fax: (864) 429-2975. E-mail: chris@wbcuradio.com Web Site: www.wbcuradio.com. Licensee: Union-Carolina Broadcasting Co. Inc. (acq 3-3-2006; $240,000 for all of the stock). Population served: 50,000 Natl. Network: ABC. Rgnl. Network: S.C. Net. Law Firm: Dan J. Alpert. Format: Country standards, loc news, weather, sports. News staff: 2; News: 24 hrs wkly. Target aud: 30 plus; working class adult buyers. Spec prog: Sports 10 hrs, news/talk 10 hrs wkly. ♦James C. Woodson, CEO, pres & gen mgr; Daniel Prince, opns mgr; Linda Comer, gen sls mgr; Kevin Shehan, news dir; Tim Stephens, chief of engrg.

Walhalla

WGOG(FM)— Sept 1, 1991: 96.3 mhz; 6 kw. Ant 302 ft TL: N34 51 33 W83 03 31. Stereo. Hrs opn: 24 Box 10, 29691. Secondary address: 2058 Westminster Hwy. 29691. Phone: (864) 638-3616. Fax: (864) 638-6810. E-mail: wgog@wgog.com Web Site: www.wgog.com. Licensee: Appalachian Broadcasting Co. Inc. (acq 10-16-2001; with co-located AM). Population served: 442,686 Natl. Network: ABC. Rgnl. Network: S.C. Net. Law Firm: Dan J. Alpert. Format: Country. News staff: one; News: 15 hrs wkly. Target aud: 25-54. ♦Douglas M. Sutton Jr., pres; M. Terry Carter, VP; Gary Butts, gen mgr; Wayne Morton, opns mgr; Tim Stephens, progmg dir & chief of engrg; Dick Mangrum, news dir & local news ed.

WWOF(AM)— Apr 15, 1959: 1000 khz; 1 kw-D. TL: N34 44 29 W83 04 18. (CP: COL Lithia Springs, GA. 890 khz; 5 kw-D. TL: N33 48 39 W84 37 02). Hrs opn: Sunrise-sunset Box 10, 29691. Phone: (864) 638-3616. Fax: (864) 638-6810. Licensee: Tugart Properties LLC. Group owner: Georgia-Carolina Radiocasting Companies (acq 10-16-2001; with WGOG(FM) Walhalla, SC). Population served: 168,603 Natl. Network: ABC. Law Firm: Dan J. Alpert. Format: Oldies, talk. News staff: one; News: 6 hrs wkly. Target aud: 25-54; emphasis on women. ♦Douglas M Sutton Jr., pres; Gary Butts, VP & gen mgr; Dick Mangrum, news dir; Tim Stephens, chief of engrg.

Walterboro

WALD(AM)— August 1947: Stn currently dark. 1080 khz; 2.5 kw-D. TL: N32 52 52 W80 41 24. (CP: COL Johnsonville. 16.5 kw-D, 5.1 kw-CH. TL: N33 54 36 W79 40 09). Hrs opn: sunup-sundown Box 2480, 29488. Phone: (843) 538-4780. Fax: (843) 538-5392. E-mail: rswaldradio@lowcountry.com Licensee: Glory Communications, Inc. (acq 5-28-2002; with WBGC(AM) Chipley, FL). Population served: 20,000 Format: Gospel. ♦Jesse Bowers, gen mgr; Annette Gantt, opns mgr; Ronda Simpson, progmg dir.

WALI(FM)— Dec 13, 1991: 93.7 mhz; 6 kw. 345 ft TL: N32 49 54 W80 43 30. Hrs open: 24 724 S. Jefferies Blvd., 29488. Phone: (843) 549-1543. Fax: (843) 549-2711. Licensee: Hess Communications L.L.C. (acq 1996; $285,000). Population served: 37,000 Rgnl. Network: S.C. Net. Rgnl rep: Rgnl Reps. Format: Country, sports. News: 3 hrs wkly. Target aud: General. Spec prog: Gospel 5 hrs wkly.Karl Hess, pres, gen mgr, opns VP, mus dir & chief of engrg; Thomas Heirs, sls dir & adv dir; Belinda Pierpaoli, progmg dir, asst music dir & pub affrs dir; Samantha Hess, mus dir; Susan Linder, spec ev coord; Dennis Hall, disc jockey; Johnathan Spell, disc jockey; Stephanie Bailey, disc jockey; Tina Elliott, disc jockey; Travis Kronman, disc jockey

Wedgefield

WIBZ(FM)— Mar 1, 1985: 95.5 mhz; 4.4 kw. Ant 387 ft TL: N33 56 56 W80 23 34. Stereo. Hrs opn: 24 Box 1269, Sumter, 29151. Secondary address: 51 Commerce St., Sumter 29150. Phone: (803) 773-1859. Fax: (803) 773-4856. Licensee: Miller Communicatins Inc. (group owner; acq 11-14-00; grpsl). Population served: 264,000 Natl. Network: ABC. Law Firm: Smithwick & Belendiuk. Format: Solid gold. News staff: one. Target aud: 18-49. ♦Harold T. Miller, CEO & pres; Dave Baker, opns VP; Theresa Miller, VP & gen sls mgr.

West Columbia

WGCV(AM)—See Cayce

WLTY(FM)—See Columbia

WXBT(FM)—Licensed to West Columbia. See Columbia

Williston

WAAW(FM)— Aug 12, 1994: 94.7 mhz; 2.11 kw. 561 ft TL: N33 28 33 W81 32 57. Stereo. Hrs opn: 24 2166 Park Ave S.E., Aiken, 29801. Phone: (803) 641-6499. Fax: (803) 641-8844. E-mail: frank@rejoiceradio.com Licensee: Frank Neely. Group owner: Neely Enterprises (acq 7-9-02; $700,000). Population served: 68,376 Format: Gospel music. ♦Larry Adamson, stn mgr.

Woodruff

WDRF(AM)— July 7, 1967: 1510 khz; 1 kw-D, 250 w-CH. TL: N34 45 22 W82 03 18. Hrs open: Box 547, 29388. Phone: (864) 476-7184. Fax: (864) 476-0474. Licensee: B&B Media Inc. (acq 8-10-99). Format: Relg. ♦T.C. Lewis, gen mgr.

York

WBZK(AM)— Apr 19, 1956: 980 khz; 3.15 kw-D, 291 w-N, DA-2. TL: N34 59 50 W81 15 09. Hrs open: 24 4201-J Stuart Andrew Blvd., Charlotte, NC, 28217. Phone: (704) 665-8240. Fax: (208) 545-9888. E-mail: ann@wnow.com Web Site: www.wbzk.com. Licensee: Davidson Media Carolinas Stations LLC. Group owner: Davidson Media Group LLC (acq 5-10-2004; grpsl). Population served: 893,630 Natl. Network: ABC. Format: Spanish, Christian. News: 6 hrs wkly. Target aud: 22-54. Spec prog: Chinese 10 hrs, Greek 10 hrs wkly. ♦Peter W. Davidson, pres; Russ Jones, gen mgr; Robert Freeze, opns dir; Humberto Martinez, progmg dir; Winston Hawkins, chief of engrg.

South Dakota

Aberdeen

KBFO(FM)— Feb 20, 1999: 106.7 mhz; 100 kw. 338 ft TL: N45 27 57 W98 20 08. Hrs open: 24 Box 1930, 57401. Secondary address: 13541 386th Ave. 57401. Phone: (605) 225-1560. Fax: (605) 229-4849. Licensee: Armada Media - Aberdeen Inc. Group owner: Clear Channel Communications Inc. (acq 10-31-2006; grpsl). Population served: 70,000 Rgnl rep: Jones Satellite Audio. Format: Hot adult contemp. Target aud: 18-35. ♦DaLime LeGrand, gen mgr; Rob Feller, gen sls mgr; Doug Pitts, progmg mgr.

KGIM(AM)— September 1933: 1420 khz; 1 kw-D, 232 w-N. TL: N45 29 07 W98 29 46. Hrs open: 13541 386th Ave., 57401. Phone: (605) 229-3632. Fax: (605) 229-4849. Licensee: Armada Media - Aberdeen Inc. Group owner: Robert Ingstad Broadcast Properties (acq 10-31-2006; grpsl). Population served: 77,107 Law Firm: Fisher, Wayland, Cooper, Leader & Zaragoza L.L.P. Format: Country, news, sports. News staff: 10 hrs wkly Target aud: 25 plus; general. Spec prog: Weather, farm 12 hrs wkly. ♦Jim Coursolle, pres; Brian Lundquist, gen mgr.

***KKAA(AM)**— Sept 12, 1974: 1560 khz; 10 kw-D, 5 kw-N, DA-2. TL: N45 25 05 W98 28 36. Hrs open: 24 Family Stations Inc., 4135 Northgate Blvd., Suite 1, Sacramento, CA, 95834. Phone: (916) 641-8191. Fax: (916) 641-8238. Licensee: Family Stations Inc. Group owner: Clear Channel Communications Inc. (acq 11-30-2004; $75,000 with KQKD(AM) Redfield). Population served: 100,000 Format: Relg. ♦Harold Camping, pres.

KLRJ(FM)— September 1979: 94.9 mhz; 100 kw. Ant 446 ft TL: N45 27 57 W98 20 08. Stereo. Hrs opn: 5 AM-1 AM
Rebroadcasts KLVR(FM) Santa Rosa, CA.
2351 Sunset Blvd., Suite 170-218, Rocklin, CA, 95765. Phone: (916) 251-1600. Fax: (916) 251-1650. Web Site: www.klove.com. Licensee: Educational Media Foundation. (acq 11-30-2004; $200,000). Population served: 250,000 Natl. Network: K-Love. Format: Christian music. ♦Richard Jenkins, pres; Mike Novak, VP; Keith Whipple, dev dir; David Pierce, progmg mgr; Sam Wallington, engrg dir; Karen Johnson, news rptr; Marya Morgan, news rptr; Richard Hunt, news rptr.

KSDN(AM)— Apr 16, 1947: 930 khz; 5 kw-D, 1 kw-N, DA-2. TL: N45 25 29 W98 31 03. Hrs open: Box 1930, 57402. Secondary address: 13541 386th Ave. 57401. Phone: (605) 225-1560. Fax: (605) 229-4849. Web Site: www.aberdeenradioranch.com. Licensee: Armada Media - Aberdeen Inc. Group owner: Clear Channel Communications Inc. (acq 10-31-2006; grpsl). Population served: 100,000 Natl. Network: ABC. Format: Talk. News staff: one; News: 15 hrs wkly. Target aud: 25-54. Spec prog: Farm 15 hrs wkly. ♦Ron Feller, gen sls mgr; Doug Pitts, progmg dir.

KSDN-FM— Nov 18, 1979: 94.1 mhz; 100 kw. 440 ft TL: N45 25 27 W98 31 00. Stereo. 50,000 Format: Classic rock.

Belle Fourche

KBFS(AM)— July 22, 1959: 1450 khz; 1 kw-U. TL: N44 40 02 W103 51 22. Hrs open: 24
Rebroadcasts KYDT(FM) Sundance, WY 99%.
Box 787, 57717. Phone: (605) 892-2571. Fax: (605) 892-2573. E-mail: kbfs@mato.com Web Site: www.kbfs.com. Licensee: Ultimate Caps Inc. (acq 3-17-94; $95,000; FTR: 6-20-83). Population served: 25,000 Natl. Network: Jones Radio Networks, ESPN Radio, Motor Racing Net, CBS Radio, Westwood One. Rgnl rep: Colorado Avalanche, Colorado Rockies Wire Svc: AP Format: Country, sports, news, talk. News: 20 hrs wkly. Target aud: 25-54; farmers, ranchers, sports fans. Spec prog: Farm 20 hrs, relg 2 hrs wkly. ♦Cynthia A. Grimmelmann, pres; Karl Grimmelmann, exec VP, gen mgr & opns mgr.

KFMH(FM)—Not on air, target date: unknown: 102.1 mhz; 7 kw. Ant -12 ft TL: N44 39 48 W103 51 26. Hrs open: Bad Lands Broadcasting Co. Inc., 288 S. River Rd., Bedford, NH, 03110. Phone: (603) 668-6400. Fax: (603) 668-6470. Licensee: Bad Lands Broadcasting Co. Inc. Group owner: Kona Coast Radio LLC (acq 9-6-2005; $915,000). ♦Steven A. Silberberg, pres.

KZZI(FM)— Sept 22, 1995: 95.9 mhz; 100 kw. Ant 1,788 ft TL: N44 19 35 W103 50 06. Stereo. Hrs opn: 24 2827 E. Colorado Blvd., Spearfish, 57783. Phone: (605) 642-85747. Fax: (605) 642-7849. Web Site: www.kzcountry.com. Licensee: Western South Dakota Broadcasting L.L.C. (acq 1999; $79,006). Population served: 150,000 Natl. Rep: Katz Radio. Format: Country. Target aud: 18-54. ♦Steve Duffy, gen mgr; Les Tuttle, gen sls mgr.

Brandon

KDEZ(FM)—Not on air, target date: unknown: 100.1 mhz; 2.15 kw. Ant 558 ft TL: N43 31 07 W96 32 05. Hrs open: 5100 S. Tennis Ln., Sioux Falls, 57108. Phone: (605) 361-0300. Fax: (605) 361-5410. Licensee: Cumulus Licensing LLC. ♦Don Jacobs, gen mgr.

Brookings

KBRK(AM)— July 28, 1955: 1430 khz; 1 kw-D, 100 w-N. TL: N44 18 13 W96 46 10. (CP: TL: N44 18 12 W96 46 01). Hrs opn: 227 22nd Ave. S., 57006. Phone: (605) 692-1430. Fax: (605) 692-6434. Web Site: www.brookingsradio.com. Licensee: Three Eagles Communications Co. Group owner: Three Eagles Communications Population served: 125,000 Format: Traditional radio today. Spec prog: Farm 9 hrs wkly. ♦Cami Powers, gen mgr.

KBRK-FM— Aug 10, 1968: 93.7 mhz; 36 kw. 571 ft TL: N44 20 22 W96 09 16. Hrs open: 24 227 22nd Ave. S., 57006. Phone: (605) 692-1430. Fax: (605) 692-4441. Web Site: www.b937.com. Licensee: Three Eagles of Huron Inc. Population served: 35,000 Natl. Network: Westwood One. Format: Adult contemp. Target aud: 20-45. ♦Cami Powers, gen mgr.

***KESD(FM)**— July 1967: 88.3 mhz; 50 kw. 623 ft TL: N44 20 10 W97 13 41. Stereo. Hrs opn: Box 2218B Pugsely Ctr., 57007. Phone: (605) 688-4191. Fax: (605) 677-5010. E-mail: sdpr@sdpb.org Web Site: www.sdpb.org. Licensee: South Dakota Board of Directors for Educational Telecommunications. Population served: 278,000 Natl. Network: NPR, PRI. Format: News, class. Target aud: 35-65; upscale, higher educated & arts-oriented. ♦Julie Andersen, pres; Terry Harris, gen mgr & stn mgr; Terry Spencer, dev dir. Co-owned TV: *KESD-TV affil.

KJJQ(AM)—Volga, May 6, 1981: 910 khz; 500 w-U. TL: N44 15 01 W96 57 22. Hrs open: 24 111 Main Ave., 57006. Phone: (605) 692-9125. Fax: (605) 692-6434. E-mail: info@depotradio.com Web Site: www.depotradio.com. Licensee: Three Eagles of Joliet Inc. Group owner: Three Eagles Communications (acq 6-18-2004; grpsl). FTR: . Population served: 43,000 Natl. Network: Westwood One. Rgnl. Network: AgriAmerica, S.D. News Net. Law Firm: Pepper & Corazzini. Format: News. News staff: one; News: 20 hrs wkly. Target aud: 30-60; general. ♦Tom Coughlin, gen mgr; Scott Kwas, opns mgr; Bryan Waltz, progmg dir, pub affrs dir & engrg mgr; Perry Miller, news dir; John Brendall, engrg mgr & chief of engrg.

KKQQ(FM)—Co-owned with KJJQ(AM). Apr 15, 1984: 102.3 mhz; 25 kw. 234 ft TL: N44 15 01 W96 57 22. (CP: Ant 243 ft.). Stereo. Web Site: www.depotradio.com. Format: Country. Target aud: 18-49.

***KSDJ(FM)**— 1993: 90.7 mhz; 1 kw. 148 ft TL: N44 19 01 W96 47 02. Stereo. Hrs opn: 24 Box 2815, Rm. 069-D, 57007-2815. Phone: (605) 688-5559. E-mail: newrock907ksdj@hotmail.com Web Site: www.907ksdj.com. Licensee: South Dakota State University. (group owner) Population served: 25,000 Format: Alternative. News staff: one; News: 5 hrs wkly. Target aud: 17-22; college students. Spec prog: Black 8 hrs, jazz 2 hrs wkly. ♦Peggy Gordon-Miller, pres; Jay Buchholz, gen mgr.

Canton

KYBB(FM)— 1996: 102.7 mhz; 50 kw. 485 ft TL: N43 28 48 W96 41 05. Hrs open: 5100 S. Tennis Ln., Sioux Falls, 57108. Phone: (605) 339-9999. Fax: (605) 339-2735. Web Site: www.81027.com. Licensee: Cumulus Licensing LLC. Group owner: Cumulus Media Inc. (acq 3-29-2004; grpsl). Natl. Rep: Christal. Format: Classic rock. Target aud: 25-49; men. ♦Don Jacobs, gen mgr; Scott Maguire, opns dir; Dan Rahman, progmg dir.

Clear Lake

KDBX(FM)— 1999: 107.1 mhz; 15 kw. 430 ft TL: N44 52 36 W96 52 28. Hrs open: 227 22nd Ave. S., Brookings, 57006. Phone: (605) 692-9125. Fax: (605) 692-6434. Web Site: www.brookingsradio.com. Licensee: Three Eagles of Joliet Inc. Group owner: Waitt Radio Inc. (acq 5-17-2005; $250,000). Format: Classic rock. ♦Cami Powers, gen mgr.

Custer

KAWK(FM)—Listing follows KFCR(AM).

KFCR(AM)— May 1, 1988: 1490 khz; 830 w-U. TL: N43 43 03 W103 35 00. Hrs open: Box 804, 57730. Secondary address: 145 Mount Rushmore Rd. 57730. Phone: (605) 673-5327. Phone: (605) 673-5094. Fax: (605) 673-3079. Licensee: Mount Rushmore Broadcasting Inc. (group owner; (acq 5-6-92; FTR: 5-25-92). Format: Adult contemp.

KAWK(FM)—Co-owned with KFCR(AM). November 1996: 105.1 mhz; 6.5 kw. Ant 1,312 ft TL: N43 44 41 W103 28 52. Box 611, Hot Springs, 57747. Phone: (605) 745-3637. Fax: (605) 745-3517. Licensee: Mt. Rushmore Broadcasting, Inc. Format: Oldies.

Deadwood

KDSJ(AM)— July 2, 1947: 980 khz; 5 kw-D, 1 kw-N, DA-N. TL: N44 22 57 W103 39 44. Hrs open: Box 567, 57732. Phone: (605) 578-1826. Fax: (605) 578-1827. Web Site: www.kdsj980.com. Licensee: Goldrush Broadcasting. (acq 7-1-82). Population served: 100,000 Rgnl. Network: S.D. News Net. Format: Top-40, oldies, news, sports. Target aud: 25-50. ♦Al Decker, pres & gen mgr.

KSQY(FM)— Sept 4, 1982: 95.1 mhz; 100 kw. 1,707 ft TL: N44 19 49 W103 50 10. Stereo. Hrs opn: 24 Box 1680, Rapid City, 57709. Secondary address: 306 E. St. Joe, Rapid City 57709. Phone: (605) 343-0888. Fax: (605) 342-3075. Web Site: www.951ksky.com. Licensee: Haugo Broadcasting Inc. (group owner) Population served: 160,000 Natl. Rep: Midwest Radio. Rgnl rep: Midwest Radio. Format: Triple A. News: 2 hrs wkly. Target aud: 18-49; young, active adults within a 5 state region. ♦Houston Haugo, CEO & pres; Chris Haugo, exec VP & gen mgr.

Dell Rapids

KSQB-FM— Oct 2, 1998: 95.7 mhz; 25 kw. 328 ft TL: N43 45 48 W96 48 27. Stereo. Hrs opn: 24 500 South Phillips Ave., Sioux Falls, 57104. Phone: (605) 331-5350. Fax: (605) 336-0415. Licensee: Backyard Broadcasting South Dakota Licensee LLC. (acq 8-1-2006; grpsl). Population served: 175,000 Natl. Rep: Rgnl Reps. Wire Svc: AP Format: Classic hits. Target aud: 20-40; young, active adults with spending ability. ♦Mark Nelson, opns mgr.

Faith

KPSD(FM)— June 1, 1989: 97.1 mhz; 100 kw. 1,525 ft TL: N45 03 14 W102 15 47. Hrs open:
Rebroadcasts KUSD(FM) Vermillion.
Box 5000, Vermillion, 57069. Secondary address: 555 N. Dakota St. 57069. Phone: (605) 677-5861. Fax: (605) 677-5010. E-mail: sdpr@sdpb.org Web Site: www.sdpb.org. Licensee: South Dakota Board of Directors for Educational Telecommunications. Natl. Network: NPR. Format: Class, jazz, news. ♦Julie Andersen, pres; Terry Harris, gen mgr; Terry Spencer, dev dir; Carol Robertson, prom dir; Matt Weesner, progmg mgr; Stacey Decker, chief of engrg.

Flandreau

KXQL(FM)— October 2000: 107.9 mhz; 21 kw. Ant 761 ft TL: N43 57 56 W96 49 11. Stereo. Hrs opn: 24 500 South Phillips Ave., Sioux Falls, 57104. Phone: ()605 331-5350. Fax: (605) 336-0415. Licensee: Backyard Broadcasting South Dakota Licensee LLC. (acq 8-1-2006; grpsl). Population served: 175,000 Natl. Rep: Rgnl Reps. Wire Svc: AP Format: Country hits. Target aud: 24-54 adults; upbeat country music listeners. ♦Mark Nelson, opns mgr.

Freeman

***KVCF(FM)**— 2002: 90.5 mhz; 9 kw. Ant 807 ft TL: N43 29 22 W97 26 33. Stereo. Hrs opn: 3434 W. Kilbourn Ave., Milwaukee, WI, 53208-3313. Phone: (414) 935-3000. Fax: (414) 935-3015. Web Site: www.vcyamerica.org. Licensee: VCY America Inc. Format: Relg, Christian. ♦Vic Eliason, gen mgr; Jim Schneider, progmg dir; Andy Eliason, chief of engrg.

Gregory

***KVCX(FM)**— May 8, 1982: 101.5 mhz; 100 kw. 640 ft TL: N43 07 41 W99 26 10. Stereo. Hrs opn: 3434 W. Kilbourn Ave., Milwaukee, WI, 53208. Phone: (414) 935-3000. Fax: (414) 935-3015. E-mail: kvcx@vcyamerica.org Web Site: www.vcyamerica.org. Licensee: VCY/America Inc. (group owner; acq 4-87). Natl. Network: USA, Moody. Format: Relg, Christian. ♦Dr. Randall Melchert, pres; Vic Eliason, VP & gen mgr; Jim Schneider, progmg dir & pub affrs dir; Tom Schlueter, mus dir; Gordon Morris, news dir; Andrew Eliason, chief of engrg.

Hot Springs

KZMX(AM)— July 4, 1958: 580 khz; 2.3 kw-D, 310 w-N. TL: N43 27 24 W103 28 34. Hrs open: Box 611, 57747. Secondary address: North Wind Cave Rd. 57747. Phone: (605) 745-3637. Fax: (605) 745-3517. E-mail: themorningshow@email.com Licensee: Mount Rushmore Broadcasting Inc. (group owner; (acq 5-20-93; $45,000 with co-located FM; FTR: 6-14-93). Population served: 5,000 Format: Real country. Spec prog: Farm 6 hrs wkly. ♦Gary Baker, gen mgr, gen sls mgr & progmg dir.

KZMX-FM— Feb 10, 1981: 96.7 mhz; 1.4 kw. Ant 440 ft TL: N43 26 34 W103 27 27. Stereo. Format: Real country.

Huron

KIJV(AM)— July 1, 1947: 1340 khz; 1 kw-U. TL: N44 20 46 W98 12 34. Hrs open: 24 1726 Dakota Ave. S., 57350. Phone: (605) 352-8621. Fax: (605) 352-8622. Licensee: Dakota Communications Ltd. (group owner; acq 3-11-2004; $400,000 with co-located FM). Population served: 14,299 Format: Oldies, talk, sports. News staff: one. Target aud: 35 plus. ♦Duane D. Butt, pres; John Speeney, gen mgr & gen sls mgr; Matt Price, progmg dir; Curt Coleman, news dir; Sheri Barth, sports cmtr; Nick Rottum, disc jockey.

KZNC(FM)—Co-owned with KIJV(AM). Nov 1, 1972: 99.1 mhz; 3 kw. 184 ft TL: N44 20 46 W98 12 34. (CP: 91 kw, ant 804 ft. TL: N44 05 47 W98 37 09). Stereo. Format: Hot country. Target aud: 25-54. ♦Sheri Barth, sports cmtr; Troy Sargent, disc jockey.

KOKK(AM)— Jan 13, 1976: 1210 khz; 5 kw-D, 1 kw-N, DA-2. TL: N44 21 44 W98 09 09. Hrs open: 5:30 AM-midnight Box 931, 57350. Secondary address: 1835 Dakota Ave. 57350. Phone: (605) 352-1933. Fax: (605) 352-0911. E-mail: traffic@kokk.com Web Site: www.kokk.com. Licensee: Dakota Communications Ltd. (group owner). Population served: 14,000 Format: Country, agriculture, info. News staff: one; News: 20 hrs wkly. Target aud: 25 plus. ♦Linda Marcus, gen mgr & gen sls mgr; Jeff Duffy, chief of opns & progmg dir; Sarah Klick, mus dir; Mike Rudd, news dir; Dick Schultz, chief of engrg.

KZKK(FM)—Co-owned with KOKK(AM). 1993: 105.1 mhz; 6 kw. 154 ft TL: N44 21 44 W98 09 09.5:30 AM-midnight Web Site: www.kokk.com.35,000 Format: Adult contemp. News staff: one.

Ipswich

KABD(FM)—Not on air, target date: unknown: 107.7 mhz; 55 kw. Ant 492 ft TL: N45 26 19 W98 40 45. Hrs open: Box 364, Pierre, 57501-0364. Phone: (605) 224-5434. Licensee: Dakota Broadcasting LLC (acq 3-28-2007; $375,000). ♦Duane D. Butt, gen mgr.

Lemmon

KBJM(AM)— Apr 1, 1966: 1400 khz; 1 kw-U. TL: N45 55 05 W102 11 55. Hrs open: 24 Box 540, 57638. Secondary address: 500 First Ave. E. 57638. Phone: (605) 374-5747. Fax: (605) 374-5332. E-mail: kbjm@kbjm.com Web Site: www.kbjm.com. Licensee: Media Associates Inc. (acq 1-17-91; $108,240; FTR: 2-4-91). Population served: 15,000 Natl. Rep: Keystone (unwired net). Format: C&W, oldies, farm. News: 30 hrs wkly. Target aud: General. ♦Mike Schweitzer, pres, gen mgr & gen sls mgr; James Schwab, progmg dir.

Little Eagle

***KLND(FM)**— June 25, 1997: 89.5 mhz; 100 kw. Ant 679 ft TL: N45 44 54 W100 48 30. Stereo. Hrs opn: 6 AM-midnight 11420 SD Hwy. 63, McLaughlin, 57642. Phone: (605) 823-4661. Fax: (605) 823-4660. Web Site: www.klnd.org. Licensee: Seventh Generation Media Services Inc. Natl. Network: PRI. Law Firm: Morrison & Foerster LLP. Format: Var. News staff: one; News: 5 hrs wkly. Target aud: General; tribal people on the Standing Rock & Cheyenne River Nations. Spec prog: Gospel 3 hrs, children 4 hrs, Sp one hr, elders 2 hrs, news/talk 5 hrs wkly. ♦Jana Shields Gipp, chmn; Beau Fontenalla, gen mgr & stn mgr.

Lowry

KMLO(FM)— 1996: 100.7 mhz; 100 kw. 587 ft TL: N45 16 26 W99 58 21. Hrs open:
Rebroadcasts KPLO-FM Reliance 100%.
c/o KMLO-FM, 214 W. Pleasant Dr., Pierre, 57501. Phone: (605) 224-8686. Fax: (605) 224-8984. E-mail: drgprod1@amfmradio.biz Licensee: James River Broadcasting Inc. Group owner: Robert Ingstad Broadcast Properties. Format: Country. ♦Robert Inqstad, pres; Mark A. Swendsen, gen mgr.

***KQSD-FM**— 1994: 91.9 mhz; 100 kw. 725 ft TL: N45 16 34 W99 59 03. Hrs open: Box 5000, Vermillion, 57069. Secondary address: 555 N. Dakota St. 57069. Phone: (605) 677-5861. Fax: (605) 677-5010. Web Site: www.sdpb.org. Licensee: South Dakota Board of Directors for Educational Telecommunications. Format: Class, news, pub affrs, jazz.

Madison

KJAM(AM)— Dec 3, 1959: 1390 khz; 500 w-D, 62 w-N. TL: N44 00 37 W97 10 18. Hrs open: 18 101 S. Egan Ave., 57042. Phone: (605) 256-4515. Fax: (605) 256-6477. Web Site: www.kjamradio.com. Licensee: Three Eagles of Brookings Inc. Group owner: Three Eagles Communications (acq 12-8-99; $1.2 million with co-located FM). Population served: 164,000 Format: C&W, news/talk. News staff: 13; News: 12 hrs wkly. Target aud: 21 plus. Spec prog: National agriculture talk program 11 hrs wkly. ♦Gary Buchanan, pres; Lorin Larsen, gen mgr & dev dir; Jim Hockett, gen sls mgr; Peg Nordling, progmg dir; Sue Bergheim, news dir; Bob Cook, chief of engrg; Dave Borman, farm dir; Joyce Wiesman, women's int ed.

KJAM-FM— Dec 17, 1967: 103.1 mhz; 33 kw. 305 ft TL: N43 59 08 W97 07 41. Stereo. 24 E-mail: manager@kjamradio.com Web Site: www.kjamradio.com.200,000 Format: Country, news. News staff: 2; News: 20 hrs wkly. Target aud: 21 plus. ♦Nicole Nordbye, news rptr; Dan Sudenga, farm dir; James Wyngaard, sports cmtr.

Martin

***KZSD-FM**— July 3, 1991: 102.5 mhz; 100 kw. 754 ft TL: N43 26 06 W101 33 14. Hrs open: Box 5000, Vermillion, 57069. Secondary address: 555 N. Dakota St. 57069. Phone: (605) 677-5861. Fax: (605) 677-5010. E-mail: sdpr@sdpb.org Web Site: www.sdpb.org. Licensee: South Dakota Board of Directors for Educational Telecommunications. Format: Class, jazz, folk, news. ♦Terry Harris, gen mgr.

Milbank

KCGN-FM—Ortonville, MN) Sept 23, 1983: 101.5 mhz; 98 kw. Ant 1,000 ft TL: N45 22 29 W97 02 20. Stereo. Hrs opn: 24 Box 247, Osakis, MN, 56360. Phone: (320) 859-3000. Fax: (320) 859-3010. Web Site: www.praisefm.org. Licensee: Praise Broadcasting Inc. (acq 10-24-2003). Population served: 500,000 Format: Praise & worship, adult contemp Christian. Target aud: 25-44; middle-aged women. ♦David McIver, gen mgr.

KKSD(FM)— Feb 4, 1991: 104.3 mhz; 100 kw. 981 ft TL: N45 10 31 W96 59 15. Stereo. Hrs opn: 24 921 9th Ave., Watertown, 57201-4960. Phone: (605) 882-1480. Fax: (605) 886-2121. Web Site: www.ksdr.com. Licensee: Three Eagles of Joliet Inc. Group owner: Three Eagles Communications (acq 6-18-2004; grpsl). Format: Oldies. Target aud: 25-54. Spec prog: Sports 5 hrs wkly. ♦Nancy Linneman, gen mgr.

KMSD(AM)— Mar 20, 1975: 1510 khz; 5 kw-D. TL: N45 11 42 W96 38 18. Hrs open: PO Box 1005, 57252. Phone: (605) 432-5516. Fax: (605) 432-4231. E-mail: kmsd@tnics.com Licensee: Big Stone Broadcasting Inc. Group owner: Robert Ingstad Broadcast Properties (acq 10-15-99; grpsl). Population served: 126,000 Rgnl. Network: S.D. News Net., AgriAmerica. Wire Svc: UPI Format: News/talk, oldies. Target aud: General. Spec prog: Farm 6 hrs wkly. ♦Jeff Kurtz, gen mgr.

Mitchell

KMIT(FM)— Mar 10, 1975: 105.9 mhz; 100 kw. 549 ft TL: N43 41 25 W98 00 27. Stereo. Hrs opn: 24 Box 520, 57301. Secondary address: 501 S. Ohlman Phone: (605) 996-9667. Fax: (605) 996-0013. E-mail: kmit@kmit.com Web Site: www.kmit.com. Licensee: Saga Communications of South Dakota LLC. Group owner: Saga Communications Inc. (acq 5-1-01; $4.05 million with KUQL(FM) Wessington Springs). Population served: 185,000 Format: Modern country. News staff: 2. Target aud: 18-54. Spec prog: Farm 18 hrs wkly. ♦Nikki Frederickson, gen sls mgr; Lisa Youngstrom, prom mgr; Joel VanDover, progmg dir; John Cyr, chief of engrg; Eric Roozen, disc jockey; Joel Van Dover, disc jockey; Tim Smith, gen mgr & disc jockey.

KORN(AM)— 1947: 1490 khz; 1 kw-U. TL: N43 42 14 W97 59 57. Hrs open: 24 Box 921, 57301. Secondary address: 319 N. Main 57301. Phone: (605) 996-1490. Fax: (605) 996-6680. E-mail: kornnews@waittradio.com Licensee: Sorenson Broadcasting Corp. (group owner; (acq 7-1-97; $1.2 million with co-located FM). Population served: 15,000 Natl. Network: Westwood One, ABC. Wire Svc: AP Format: Talk, news/talk, sports. News staff: one; News: 15 hrs wkly. Target aud: 35 plus; mature adults. Spec prog: Farm 10 hrs wkly. ♦Mary Quass, pres; John Koons, gen mgr, gen sls mgr & news rptr; Clayton Mick, progmg dir; J.P. Skelly, news dir.

KQRN(FM)—Co-owned with KORN(AM). Aug 17, 1980: 107.3 mhz; 100 kw. 450 ft TL: N43 41 46 W98 03 35. Stereo. 24 Web Site: q107radio.com.50,000 Format: Adult Contemp, CHR. News staff: one; News: 4 hrs wkly. Target aud: 10-49; adult female. ♦Steve Morgan, progmg dir.

Mobridge

KOLY(AM)— Aug 10, 1956: 1300 khz; 5 kw-D, 111 w-N. TL: N45 32 07 W100 20 45. Hrs open: 24 Box 400, 118 E. 3rd St., 57601. Phone: (605) 845-3654. Fax: (605) 845-5094. Licensee: James River Broadcasting Co. Group owner: Robert Ingstad Broadcast Properties (acq 7-8-97; $890,742 with co-located FM). Population served: 50,000 Format: Pop standards. News staff: one; News: 21 hrs wkly. Target aud: General. Spec prog: Farm, American Indian. ♦Dawn Konold, gen mgr; Mark Swenden, gen sls mgr; John Schreier, progmg dir & news dir; Rolland Cory, chief of engrg; Pat Morrison, sports cmtr; Sharon Martin, women's int ed; Andy Shumacher, disc jockey.

KOLY-FM— Oct 1, 1973: 99.5 mhz; 56 kw. 560 ft TL: N45 31 50 W100 20 30. (CP: 100 kw, ant 361 ft. TL: N45 32 07 W100 20 45). Stereo. Format: Adult contemp. ♦Cindy Dafnis, opns mgr; John Schreier, news rptr, women's int ed & disc jockey.

Pierpont

***KDSD-FM**— Apr 1, 1984: 90.9 mhz; 70 kw. 1,057 ft TL: N45 29 55 W97 40 35. Stereo. Hrs opn:
Rebroadcasts KUSD(FM) Vermillion.
Box 5000, Vermillion, 57069. Secondary address: 555 N. Dakota St., Vermillion 57069. Phone: (605) 677-5861. Fax: (605) 677-5010. E-mail: sdpr@sdpb.org Web Site: www.sdpb.org. Licensee: South Dakota Board of Directors for Educational Telecommunications. Population served: 152,300 Natl. Network: PRI, NPR. Rgnl. Network: S.D. Pub. Format: News, class, jazz. ♦Terry Harris, gen mgr; Terry Spencer, dev dir.

Pierre

KCCR(AM)— Feb 4, 1959: 1240 khz; 1 kw-U. TL: N44 21 02 W100 19 08. Hrs open: 5:30 AM-midnight Phone: (605) 224-1240. Fax: (605) 224-0095. Licensee: Sorenson Broadcasting Corp. (acq 3-1-72). Population served: 15,000 Natl. Network: CBS. Format: Oldies, News/talk. News staff: 2; News: 24 hrs wkly. Target aud: 35 plus; well-educated, upper income, politically aware business people, retirees, housewives. ♦Dean Sorenson, pres; Steve White, gen mgr; Tanya Martin, gen sls mgr; Dan Myer, progmg dir.

KLXS-FM—Co-owned with KCCR(AM). Apr 15, 1981: 95.3 mhz; 49 kw. Ant 299 ft TL: N44 22 15 W100 24 17. Stereo. Phone: (605) 224-7381. Natl. Network: Westwood One. Format: Adult contemp. Target aud: 18-34; 55% female, 45% male.

KGFX(AM)— 1927: 1060 khz; 10 kw-D, 1 kw-N, DA-2. TL: N44 17 12 W100 20 18. Hrs open: 24 Box 1197, 214 W. Pleasant Dr., 57501. Phone: (605) 224-8686. Fax: (605) 224-8984. Licensee: James River Broadcasting. Group owner: Robert Ingstad Broadcast Properties (acq 11-15-68). Population served: 40,000 Rgnl. Network: S.D. News Net. Law Firm: Shaw Pittman. Format: Country, farm. News staff: one; News: 20 hrs wkly. Target aud: 25-54. ♦Robert E. Ingstad, pres; Mark A. Swendsen, gen mgr & gen sls mgr; Paul Rollie, progmg dir; Pat Callahan, reporter, political ed & disc jockey; Del Fisher, farm dir.

KGFX-FM— Jan 4, 1982: 92.7 mhz; 3 kw. 245 ft TL: N44 22 15 W100 24 17. Stereo. 24 Licensee: Robert E. Ingstad Properties. Format: Adult contemp. Target aud: 25-49. ♦Patrick Callahan, news rptr.

KSQP(AM)—Not on air, target date: unknown: 1450 khz; 1 kw-U. TL: N44 25 54 W100 30 49. Hrs open: 207 E. Capital Ave., Ste. 412, 57501. Phone: (605) 224-4999. Licensee: Patriot Radio of South Dakota Inc. Natl. Network: ABC. Format: Talk. ♦Lee O. Axdahl, pres & gen mgr.

***KVFL(FM)**— 2006: 89.1 mhz; 400 w vert. Ant 371 ft TL: N44 25 33 W100 21 28. Stereo. Hrs opn: 24 3434 W. Kilbourn Ave., Milwaukee, WI, 53208-3313. Phone: (414) 935-3000. Fax: (414) 935-3015. E-mail: kvfl@vcyamerica.org Web Site: www.vcyamerica.org. Licensee: VCY America Inc. Format: Relg, Christian. ♦Vic Eliason, VP & gen mgr; Jim Schneider, progmg dir.

Pine Ridge

KVAR(FM)—Not on air, target date: July 2008: Stn currently dark. 93.7 mhz; 12 kw. Ant 479 ft TL: N42 49 47 W102 39 80. Stereo. Hrs opn: Box 563, Tanner, AL, 35671. Secondary address: 709 Coleman Ave., Athens, AL 35611. Phone: (256) 497-4502. Fax: (443) 342-2478. E-mail: varietyrock@hotmail.com Licensee: Alleycat Communications. Format: Var rock. Target aud: 18-54. ♦Richard W. Dabney, gen mgr.

Porcupine

***KILI(FM)**— 1984: 90.1 mhz; 100 kw. Ant 508 ft TL: N43 10 48 W102 19 25. Hrs open: Box 150, 57772. Secondary address: 901 Lamont Ln. 57772. Phone: (605) 867-5002. Fax: (605) 867-5634. Web Site:

www.kiliradio.org. Licensee: Lakota Communications Inc. Format: Native American, var. ♦Melanie Janis, gen mgr.

Presho

KJBI(FM)—Not on air, target date: unknown: 100.3 mhz; 100 kw. Ant 981 ft TL: N44 03 07 W100 05 03. Hrs open: Box 907, Valley City, ND, 21807. Phone: (701) 845-1490. Fax: (701) 845-1245. Licensee: James River Broadcasting Inc. (acq 7-31-2007; $450,000 for CP). ♦Janice M. Ingstad, gen mgr.

Rapid City

***KASD(FM)**— 2006: 90.3 mhz; 1 kw. Ant 407 ft TL: N44 04 13 W103 15 01. Hrs open:
Rebroadcasts WAFR(FM) Tupelo, MS 100%.
Drawer 2440, Tupelo, MS, 38803. Phone: (662) 844-8888. Fax: (662) 842-6791. Web Site: www.afr.net. Licensee: American Family Association. Natl. Network: American Family Radio. Format: Christian. ♦Marvin Sanders, gen mgr.

KBHB(AM)—See Sturgis

***KBHE-FM**— 1984: 89.3 mhz; 9.8 kw. 410 ft TL: N44 03 09 W103 14 38. Stereo. Hrs opn: Box 5000, Vermillion, 57069. Phone: (605) 677-5861. Fax: (605) 677-5010. Web Site: www.sdpb.org. Licensee: South Dakota Board of Educational Telecommunications. Population served: 122,500 Natl. Network: PRI, NPR. Rgnl. Network: S.D. Pub. Format: Classical Jazz. Spec prog: Sioux one hr wkly. ♦Terry Harris, gen mgr.

KFXS(FM)— Apr 11, 1977: 100.3 mhz; 100 kw. 450 ft TL: N44 04 14 W103 15 01. Stereo. Hrs opn: 24 Box 2480, 57709. Secondary address: 660 Flormann St., Suite 100 57709. Phone: (605) 343-6161. Fax: (605) 343-9012. E-mail: request@foxradio.com Web Site: www.foxradio.com. Licensee: New Rushmore Radio Inc. Group owner: Triad Broadcasting Co. LLC (acq 10-23-2006; grpsl). Natl. Rep: Christal. Law Firm: Shaw Pittman. Format: Classic rock. News staff: one; News: 3 hrs wkly. Target aud: 25-54. ♦Lia Green, gen mgr; Charlie O'Douglas, opns mgr; Jake Michaels, progmg dir; Kay Duda, traf mgr; Gary Peterson, engr.

KIMM(AM)— Mar 16, 1962: 1150 khz; 5 kw-D, 500 w-N, DA-N. TL: N44 04 35 W103 08 49. Hrs open: Box 2480, 57709. Phone: (605) 343-6161. Fax: (605) 343-9012. Licensee: KIMM Radio Inc. (acq 1-7-98; $150,000). Population served: 157,900 Natl. Rep: Christal. Format: Classic country. Target aud: 35-64. Spec prog: Colorado Rockies baseball, farm one hr wkly. ♦Matthew Ward, pres; Ron Hansen, gen mgr; Gary Peterson, chief of opns & chief of engrg; Michael Goodroad, sls dir; Gail Hanson, gen sls mgr; Wayne Janke, progmg dir, news dir & pub affrs dir; Dan Rahman, disc jockey; Scott Bader, disc jockey.

KIQK(FM)—Listing follows KTOQ(AM).

KKLS(AM)— June 7, 1959: 920 khz; 5 kw-D, 100-N, DA-2. TL: N44 03 43 W103 10 29. Stereo. Hrs opn: 24 Box 2480, 57709-2480. Phone: (605) 343-6161. Fax: (605) 343-9012. Web Site: www.kkls.net. Licensee: New Rushmore Radio Inc. Group owner: Triad Broadcasting Co. LLC (acq 10-23-2006; grpsl). Population served: 164,100 Natl. Network: Westwood One. Natl. Rep: Christal. Law Firm: Wilmer Hale. Wire Svc: AP Format: Oldies. Target aud: 35-64. ♦Lia Green, gen mgr; Charlie O'Douglas, opns mgr; Michael Goodroad, sls dir & gen sls mgr.

KKMK(FM)—Co-owned with KKLS(AM). 1971: 93.9 mhz; 100 kw. 650 ft TL: N44 02 48 W103 14 46. Stereo. 24 660 Flormann St., Suite 100, 57701. E-mail: kkmk@rapidnet.comm Web Site: www.kkmk.com. Natl. Rep: Christal. Law Firm: Wilmer Hale. Format: Adult contemp. Target aud: 25-54.

***KLMP(FM)**— Feb 17, 2005: 88.3 mhz; 63 kw. Ant 1,712 ft TL: N44 19 42 W103 50 03. Hrs open: 1853 Fountain Plaza Dr., 57702. Phone: (605) 342-6822. Fax: (605) 342-0854. E-mail: info@klmp.com Web Site: www.klmp.com. Licensee: Bethesda Christian Broadcasting Inc. Natl. Network: USA. Format: Inspirational programs. Target aud: 35 plus; general. ♦Tom Schoenstedt, gen mgr; Joe Meunch, progmg dir; Joe Standish, mus dir; Tracey Krsnak, chief of engrg; Suzanne Happs, traf mgr.

KOTA(AM)— November 1936: 1380 khz; 5 kw-U, DA-N. TL: N44 02 00 W103 11 15. Stereo. Hrs opn: 24 Box 1760, 518 St. Joe, 57709-1760. Phone: (605) 342-2000. Fax: (605) 342-7305. E-mail: leskota@rushmore.com Web Site: www.kotaradio.com. Licensee: Duhamel Broadcasting Enterprises. (group owner; acq 5-54). Population served: 216,000 Natl. Network: CBS. Natl. Rep: Katz Radio. Law Firm: Shaw Pittman. Wire Svc: AP Format: News/talk. News staff: 2; News: 10 hrs wkly. Target aud: 35 plus. ♦William F. Duhamel, pres; Les Tuttle, stn mgr. Co-owned TV: KOTA-TV affil

KOUT(FM)— 1993: 98.7 mhz; 100 kw. 515 ft TL: N44 01 50 W103 15 34. Hrs open: 24 Box 2480, 57709-2480. Secondary address: 660 Flormann St., Suite 100 57701. Phone: (605) 343-6161. Fax: (605) 343-9012. Web Site: www.katcountry.com. Licensee: New Rushmore Radio Inc. Group owner: Triad Broadcasting Co. LLC (acq 10-23-2006; grpsl). Population served: 87,000 Natl. Rep: Christal. Law Firm: Shaw Pittman. Format: Country. Target aud: Adults 25-54. ♦Lia Green, gen mgr; Charlie O'Douglas, opns mgr; Mark Houston, progmg dir; Kay Duda, traf mgr; Gary Peterson, engr.

***KQFR(FM)**— Aug. 5, 2005: 89.9 mhz; 2.3 kw. Ant 1,843 ft TL: N44 19 42 W103 50 03. Hrs open: 24 Family Stations Inc., 4135 Northgate Blvd., Suite 1, Sacramento, CA, 95834. Phone: (916) 641-8191. Fax: (916) 641-8238. E-mail: kebr@jps.net Licensee: Family Stations Inc. (group owner). Format: Relg. ♦Harold Camping, pres; John Rorvik, opns mgr & rgnl sls mgr; Joe Papp, chief of engrg.

KQRQ(FM)— October 2002: 92.3 mhz; 86 kw. Ant 581 ft TL: N44 04 07 W103 15 02. Hrs open: 24 Box 1760, 57709. Phone: (605) 342-2000. Fax: (605) 721-5732. E-mail: leskota@rushmore.com Web Site: www.q923.com. Licensee: New Generation Broadcasting LLC. Natl. Rep: Katz Radio. Wire Svc: AP Format: Classic hits. Target aud: Adults 25-44. ♦Les Tuttle, gen mgr.

KRCS(FM)—See Sturgis

KTOQ(AM)— Sept 26, 1953: 1340 khz; 1 kw-U. TL: N44 04 06 W103 10 11. Hrs open: 24 Box 1680, 306 1/2 E. St. Joseph St., 57709. Phone: (605) 343-0888. Fax: (605) 342-3075. Licensee: Haugo Braodcasting Inc. (group owner; acq 11-20-98; $1.97 million with co-located FM). Population served: 150,000 Natl. Rep: McGavren Guild. Law Firm: Booth, Freret, Imlay & Tepper. Format: Talk. News staff: 2; News: 3 hrs wkly. Target aud: 35 plus; upscale. Spec prog: Farm 2 hrs wkly. ♦Houston Haugo, CEO & pres; Christian Haugo, VP & gen mgr; Georgia McGaa, gen sls mgr; Brad Anderson, news dir & disc jockey; Rose Jeffert, pub affrs dir; Tracy Krsnak, chief of engrg; Phil Amundson, traf mgr.

KIQK(FM)—Co-owned with KTOQ(AM). Jan 7, 1992: 104.1 mhz; 100 kw. 515 ft TL: N44 01 50 W103 15 34. Stereo. 24 Phone: (605) 341-5425. Format: Country. News staff: one; News: 3 hrs wkly. Target aud: 25-54. ♦Cory Ward, prom dir; Phil Amundson, traf mgr; Cliff Miller, disc jockey; Eric Andrews, disc jockey; JD Wright, disc jockey.

KTPT(FM)— Oct 1, 1968: 97.9 mhz; 100 kw horiz. Ant 390 ft TL: N44 02 46 W103 14 41. Stereo. Hrs opn: 24 1853 Fountain Plaza Dr., 57702. Phone: (605) 342-6822. Fax: (605) 342-0854. Web Site: www.979thepoint.com. Licensee: Bethesda Christian Broadcasting Inc. Group owner: Bethesda Christian Broadcasting (acq 6-25-96; $350,000). Population served: 65,000 Format: Contemp Christian. ♦Mark Pluimer, pres & gen mgr; Tom Schoenstedt, stn mgr.

***KWRC(FM)**—Not on air, target date: unknown: 90.9 mhz; 875 w. Ant 1,116 ft TL: N43 44 43 W103 28 52. Hrs open: 4002 N. 3300 E., Twin Falls, ID, 83301. Phone: (208) 734-6633. Fax: (208) 736-1958. Web Site: www.csnradio.com. Licensee: CSN International. ♦Mike Kestler, VP.

KZLK(FM)— 2001: 106.3 mhz; 92 kw. Ant 695 ft TL: N44 04 07 W103 15 02. Stereo. Hrs opn: 24 Box 1760, 57709. Phone: (605) 342-2000. Fax: (605) 721-5732. E-mail: leskota@rushmore.com Web Site: www.1063maxfm.com. Licensee: Steven E. Duffy. Population served: 90,000 Natl. Rep: Katz Radio. Law Firm: Shaw Pittman. Wire Svc: AP Format: Jack. Target aud: Adult 25-54. ♦Les Tuttle, gen mgr.

Redfield

KGIM-FM— Apr 7, 1991: 103.7 mhz; 100 kw. Ant 564 ft TL: N45 12 52 W98 40 54. Stereo. Hrs opn: 5:30 AM-midnight 13541 386th Ave., Aberdeen, 57401. Phone: (605) 229-3632. Fax: (605) 229-4849. Licensee: Armada Media - Aberdeen Inc. (acq 10-31-2006; grpsl). Format: Country. News staff: one. Target aud: 25-49. ♦Jim Coursolle, pres.

KNBZ(FM)— 1999: 97.7 mhz; 62 kw. Ant 190 ft TL: N44 54 30 W98 19 40. Hrs open: Box 1930, Aberdeen, 57402. Secondary address: 13541 386th Ave., Aberdeen 57401. Phone: (605) 229-3632. Fax: (605) 229-4849. Web Site: www.hubcityradio.com. Licensee: Armada Media - Aberdeen Inc. Group owner: Robert Ingstad Broadcast Properties. (acq 10-31-2006; grpsl). Format: Adult contemp. ♦Brian Lundquist, gen mgr; Doc Sebastian, progmg dir.

***KQKD(AM)**— December 1962: 1380 khz; 500 w-D, 140 w-N, DA-2. TL: N44 53 53 W98 30 23. Hrs open: 24 Family Stations Inc., 4135 Northgate Blvd., Suite 1, Sacramento, CA, 95834. Phone: (916) 641-8191. Licensee: Family Stations Inc. Group owner: Robert Ingstad Broadcast Properties (acq 11-30-2004; $75,000 with KKAA(AM) Aberdeen). Population served: 12,000 Format: Relg. ♦Harold Camping, pres.

Reliance

KPLO-FM— January 1986: 94.5 mhz; 95 kw. 1,000 ft TL: N43 57 55 W99 36 11. Stereo. Hrs opn: 24 214 W. Pleasant Dr., Pierre, 57325. Phone: (605) 734-4000. Phone: (605) 224-8686. Fax: (605) 224-8984. E-mail: drgprod1@amfmradio.biz Licensee: James River Broadcasting Co. Group owner: Robert Ingstad Broadcast Properties (acq 8-21-98; $98,000). Format: Country. Target aud: 25-54. Spec prog: Farm 5 hrs wkly. ♦Mark Swendsen, gen mgr.

***KTSD-FM**— 1984: 91.1 mhz; 100 kw. 1,480 ft TL: N43 57 55 W99 35 56. Stereo. Hrs opn:
Rebroadcasts KUSD-FM, Vermillion,SD 89.7%.
Box 5000, Vermillion, 57069. Secondary address: 555 N. Dakota St., Vermillion 57069. Phone: (605) 677-5861. Fax: (605) 677-5010. E-mail: sdpr@sdpb.org Web Site: www.sdpb.org. Licensee: S.D. Board of Educational Telecommunications. Population served: 150,000 Natl. Network: PRI, NPR. Format: Class, jazz , news. Spec prog: Sioux one hr wkly. ♦Terry Harris, gen mgr.

Roscoe

KMOM-FM—Not on air, target date: unknown: 105.5 mhz; 100 kw. Ant 456 ft TL: N45 27 13 W98 48 10. Hrs open: Box 364, Pierre, 57501-0364. Phone: (605) 224-5434. Licensee: Dakota Broadcasting LLC (acq 3-28-2007; $375,000 for CP). ♦Duane D. Butt, gen mgr.

Saint Francis

KINI(FM)—See Crookston, NE

Salem

KIKN-FM— Nov 4, 1993: 100.5 mhz; 100 kw. 981 ft TL: N43 29 18 W97 26 34. Stereo. Hrs opn: 24 5100 S. Tennis Ln., Sioux Falls, 57108. Phone: (605) 361-0300. Fax: (605) 361-5410. Web Site: www.kikn.com. Licensee: Cumulus Licensing LLC. Group owner: Cumulus Media Inc. (acq 4-1-2004; grpsl). Population served: 350,000 Natl. Rep: Christal. Format: New country. News staff: one. Target aud: 18-49. ♦Lew Dickey, pres; Don Jacobs, gen mgr; J.D. Collins, progmg dir.

Sioux Falls

***KAUR(FM)**— Oct 9, 1972: 89.1 mhz; 680 w. 184 ft TL: N43 31 37 W96 44 18. Stereo. Hrs opn: 10 AM-3 AM Box 751, KAUR-FM, Augustana College, 57197. Secondary address: 2001 S. Summit Ave. 57197. Phone: (605) 274-0770. Fax: (605) 336-5465. Web Site: www.kaur.com. Licensee: Augustana College Association. Population served: 150,000 Format: Rock, jazz, alternative. Target aud: General. Spec prog: Folk 2 hrs, world mus 6 hrs, blues 3 hrs wkly. ♦Chuck Carlson, gen mgr.

***KCFS(FM)**— July 1985: 94.5 mhz; 2.35 kw. 190 ft TL: N43 31 57 W96 44 20. Hrs open: Sioux Falls College, 1101 W. 22nd St., 57105. Phone: (605) 331-6691. Fax: (605) 331-6615. E-mail: kcfs@thecoo.edu Web Site: www.usiouxfalls.edu/campus/radio/index.html. Licensee: University of Sioux Falls. Population served: 110,000 Format: Div. Spec prog: Urban 6 hrs wkly. ♦Jesse Logterman, gen mgr; Chris Stafford, stn mgr; Jason Peiser, progmg dir.

***KCSD(FM)**— July 1, 1985: 90.9 mhz; 2.35 kw. 190 ft TL: N43 31 57 W96 44 20. Stereo. Hrs opn: 24 1101 W. 22nd St., 57105. Phone: (605) 331-6690. Fax: (605) 331-6692. E-mail: sdpr@sdpb.org Web Site: www.sdpb.org. Licensee: University of Sioux Falls. Natl. Network: NPR. Rgnl. Network: S.D. Pub. Format: Class. News staff: one; News: 44 hrs wkly. Target aud: 25 plus; educated males & females. Spec prog: Folk 5 hrs, jazz 10 hrs wkly. ♦Janice Davis, stn mgr.

KDLO-FM—See Watertown

KELO-FM— July 11, 1965: 92.5 mhz; 100 kw. Ant 1,820 ft TL: N43 31 07 W96 32 05. Stereo. Hrs opn: 24 500 S. Phillips, 57104. Phone: (605) 331-5350. Fax: (605) 336-0415. Web Site: kelofm.com. Licensee: Backyard Broadcasting South Dakota Licensee LLC. (acq 4-2005;

grpsl). Population served: 368,000 Natl. Rep: Katz Radio. Law Firm: Leventhal, Senter & Lerman. Format: Light adult contemp. ♦Barry Drake, pres; Craig Hodgson, gen mgr.

KELO(AM)— 1937: 1320 khz; 5 kw-U, DA-N. TL: N43 29 17 W96 38 14. Stereo. 24 Web Site: kelo.com. Group owner: Midcontinent Media Inc. 222,000 Natl. Network: Fox News Radio. Natl. Rep: Katz Radio. Format: News/talk, weather. News staff: 6. Target aud: 25-54. ♦Craig R. Hodgson, VP.

KKLS-FM—Listing follows KXRB(AM).

KMXC(FM)—Listing follows KSOO(AM).

***KNWC(AM)**— March 1961: 1270 khz; 5 kw-D, 2.3 kw-N, DA-2. TL: N43 17 07 W96 45 53. Hrs open: 24 6300 S. Tallgrass Ave., 57108-8107. Phone: (605) 339-1270. Fax: (605) 339-1271. E-mail: knwc@knwc.org Web Site: www.knwc.org. Licensee: Northwestern College. Group owner: Northwestern College & Radio (acq 1961). Population served: 150,000 Law Firm: Bryan Cave. Format: Relg, news. News staff: one; News: 24 hrs wkly. Target aud: 35-54. ♦David Martin, opns dir; Jeff Rupp, gen mgr, progmg dir, local news ed & news rptr.

KNWC-FM— Mar 28, 1969: 96.5 mhz; 100 kw. 1,600 ft TL: N43 31 07 W96 32 05. Stereo. 24 Web Site: www.knwc.org.200,000 Law Firm: Bryan Cave. Format: Christian. News staff: one; News: 24 hrs wkly. Target aud: 20-54. ♦Tim Unsinn, prom dir & progmg dir.

KRRO(FM)— May 6, 1969: 103.7 mhz; 38 kw. Ant 394 ft TL: N43 27 28 W96 40 14. Stereo. Hrs open: 24 500 S. Phillips, 57104. Phone: (605) 331-5350. Fax: (605) 336-0415. Web Site: www.krro.com. Licensee: Backyard Broadcasting South Dakota Licensee LLC. (acq 4-2005; grpsl). Population served: 220,000 Natl. Rep: Katz Radio. Law Firm: Shaw Pittman. Wire Svc: AP Format: AOR. Target aud: 25-49; young adults, family-rearing age with disposable income. ♦Barry Drake, pres; Craig Hodgson, gen mgr.

KWSN(AM)—Co-owned with KRRO(FM). May 6, 1948: 1230 khz; 1 kw-U. TL: N43 33 31 W96 46 10.24 Web Site: www.kwsn.com. Group owner: Midcontinent Media Inc. .220,000 Natl. Network: ESPN Radio. Natl. Rep: Katz Radio. Format: All sports. News staff: 2; News: 27 hrs wkly. Target aud: 25-54; adults with disposable income, business leaders.

***KRSD(FM)**— May 11, 1985: 88.1 mhz; 2 kw. 183 ft TL: N43 31 37 W96 44 18. Stereo. Hrs opn: 24 Box 737, Augustana College, 57197. Phone: (605) 335-6666. Phone: (800) 228-7123. Fax: (605) 335-1259. E-mail: mail@mpr.org Web Site: www.mpr.org. Licensee: Minnesota Public Radio. Natl. Network: NPR, PRI. Format: Class, news. News staff: one. Target aud: General. ♦William H. Kling, pres; Michael Olsen, gen mgr; Mike Edgerly, news dir; Vince Fuhs, chief of engrg.

***KSFS(FM)**— 2006: 90.1 mhz; 1 kw. Ant 57 ft TL: N43 32 41 W96 45 45. Hrs open:
Rebroadcasts WAFR(FM) Tupelo, MS 100%.
Box 2440, Tupelo, MS, 38803-2440. Phone: (662) 844-8888. Fax: (662) 842-6791. Web Site: www.afr.net. Licensee: American Family Association. Format: Christian. ♦Marvin Sanders, gen mgr.

KSOO(AM)— 1926: 1140 khz; 10 kw-D, 5 kw-N, DA-N. TL: N43 28 47 W96 41 04. Hrs open: 5100 S. Tennis Ln., Suite 200, 57108. Phone: (605) 339-1140. Fax: (605) 339-2735. Web Site: www.ksoo.com. Licensee: Cumulus Licensing LLC. Group owner: Cumulus Media Inc. (acq 3-29-2004; grpsl). Population served: 150,000 Natl. Rep: Christal. Format: News/talk, sports. Target aud: 35-54. ♦Lew Dickey, pres & gen mgr; Don Jacobs, stn mgr; Brad Peterson, progmg dir; Gene Hetland, news dir; Mike Langford, chief of engrg.

KMXC(FM)—Co-owned with KSOO(AM). Oct 1, 1973: 97.3 mhz; 60 kw. 221 ft TL: N43 35 48 W96 38 20. Stereo. Web Site: www.mix97-3.com. Format: Adult contemp. Target aud: 25-44; females. ♦Scott Maguire, progmg dir.

KSQB(AM)— June 13, 1970: 1520 khz; 500 w-D. TL: N43 33 28 W96 47 46. Hrs open: 500 South Phillips Ave., 57104. Phone: (605) 331-5350. Fax: (605) 336-0415. Licensee: Backyard Broadcasting South Dakota Licensee LLC. (acq 8-1-2006; grpsl). Population served: 150,000 Natl. Rep: Rgnl Reps. Law Firm: Law Offices of Richard J. Hayes. Wire Svc: AP Format: Sports talk. Target aud: 21-49. ♦Mark Nelson, opns dir.

KSQB-FM—See Dell Rapids

KTWB(FM)— May 5, 1990: 101.9 mhz; 34 kw. 580 ft TL: N43 45 11 W96 53 22. Stereo. Hrs opn: 24 500 S. Phillips Ave., 57104. Phone: (605) 331-5350. Fax: (605) 336-0415. E-mail: ktwb@bybradio.com Web Site: www.ktwb.com. Licensee: Backyard Broadcasting South Dakota Licensee LLC. Group owner: Midcontinent Media Inc. (acq 4-2005; grpsl). Population served: 198,530 Natl. Rep: Katz Radio. Law Firm: Richard Hayes. Wire Svc: AP Format: Country. News staff: one; News: 10 hrs wkly. Target aud: 25-54; adult. ♦Barry Drake, pres; Craig Hodgson, gen mgr.

KXQL(FM)—See Flandreau

KXRB(AM)— February 1969: 1000 khz; 10 kw-D, DA. TL: N43 29 13 W96 35 48. Hrs open: 24 5100 S. Tennis Ln, 57108. Phone: (605) 361-0300. Fax: (605) 361-5410. Licensee: Cumulus Licensing LLC. Group owner: Cumulus Media Inc. (acq 3-29-2004; grpsl). Population served: 300,000 Natl. Network: CNN Radio. Rgnl. Network: CNN. Natl. Rep: Christal. Wire Svc: UPI Format: Country, farm. News staff: one; News: 5 hrs wkly. Target aud: 25-54. ♦Lew Dickey, pres; Don Jacobs, gen mgr; Joe Morrison, progmg dir & mus dir; Jerry Dohmen, news dir; Mike Langford, progmg dir & chief of engrg.

KKLS-FM—Co-owned with KXRB(AM). March 1975: 104.7 mhz; 100 kw. 860 ft TL: N43 43 46 W97 05 10. Stereo. Format: Contemp hit. Target aud: 18-49. ♦Andy Erickson, progmg dir & progmg mgr.

Sisseton

KBWS-FM— Dec 28, 1983: 102.9 mhz; 100 kw. 496 ft TL: N45 36 52 W97 24 51. Stereo. Hrs opn: 509 Veterans Ave., 57262. Phone: (605) 698-3471. Fax: (605) 698-3330. E-mail: kbws@tnics.com Web Site: www.phcountry.com. Licensee: Pheasant Country Broadcasting Inc. Group owner: Robert Ingstad Broadcast Properties. Format: Country. Target aud: General. ♦Robert Ingstead, pres; Jeff Kurtz, gen mgr; Randy Peterson, rgnl sls mgr; John Seiber, progmg dir & news dir; Terry Heitman, pub affrs dir; Don Brittnal, chief of engrg; Jamie Rothe, traf mgr.

Spearfish

***KBHU-FM**— Oct 18, 1974: 89.1 mhz; 100 w. 55 ft TL: N44 29 48 W103 52 13. Stereo. Hrs opn: 24 Unit 9003, 1200 University St., 57799. Phone: (605) 642-6011. Phone: (605) 642-6265. Fax: (605) 642-6762. E-mail: kbhufm@hotmail.com Web Site: www.kbhufm.com. Licensee: Black Hills State University. Population served: 12,000 Natl. Network: Westwood One. Wire Svc: AP Format: Alternative. News: 3 hrs wkly. Target aud: 12-35. ♦Dave Diamond, CEO; Kay Schallenkamp, pres; Thomas O. Flickemna, pres; Cody Oliver, gen mgr; Stephanie Bechen, gen mgr; Stephen Webb, stn mgr; Cody Holliwell, dev dir; Erica Morris, sls dir; David Martin, progmg dir.

KDDX(FM)— July 19, 1985: 101.1 mhz; 100 kw. 1,604 ft TL: N44 19 40 W103 50 14. (CP: Ant 1,817 ft. TL: N44 19 36 W103 50 12). Stereo. Hrs opn: 24 2827 E. Colorado Blvd., 57783. Phone: (605) 642-5747. Fax: (605) 642-7849. Web Site: www.xrock.fm. Licensee: Duhamel Broadcasting Enterprises. (group owner; acq 3-16-92; $525,000; FTR: 3-30-92). Population served: 150,000 Format: Active rock. News staff: one; News: 3 hrs wkly. Target aud: 18-49. ♦Les Tuttle, gen mgr; Ted Peiffer, gen sls mgr; Jim Kallas, progmg dir.

***KOAR(FM)**—Not on air, target date: unknown: 90.9 mhz; 800 w. Ant 420 ft TL: N44 29 33 W103 50 05. Hrs open: 5700 West Oaks Blvd., Rocklin, CA, 95765. Phone: (916) 251-1600. Fax: (916) 251-1650. Licensee: Educational Media Foundation. (acq 3-23-2007; grpsl). ♦Richard Jenkins, pres.

KSLT(FM)— Feb 17, 1984: 107.3 mhz; 100 kw. 1,702 ft TL: N44 19 36 W103 50 12. Stereo. Hrs opn: 24 1853 Fountain Plaza Dr., Rapid City, 57702-9315. Phone: (605) 342-6822. Fax: (605) 342-0854. E-mail: kslt@kslt.com Web Site: www.kslt.com. Licensee: Bethesda Christian Broadcasting Inc. Natl. Network: USA. Format: Contemp Christian. News: 9 hrs wkly. Target aud: 25-49; affluent, educated, 60% female, 40% male. Spec prog: Focus on the Family 5 hrs, Insight for living 3 hrs, family news in focus one hr, Revival time one hr wkly. ♦Mark Pluimer, pres; Mark Plumer, gen mgr; John Derrek, rgnl sls mgr; Jon Anderson, progmg dir; Joe Standish, mus dir; Tracey Krsnak, chief of engrg; Suzanne Happs, traf mgr & spec ev coord.

Sturgis

KBHB(AM)— Sept 27, 1962: 810 khz; 21 kw-D. TL: N44 25 23 W103 25 38. Hrs open: Sunrise-sunset Box 99, Hwy. 79 N., 57785. Phone: (605) 347-4455. Fax: (605) 347-5120. Licensee: New Rushmore Radio Inc. Group owner: Triad Broadcasting Co. LLC (acq 10-23-2006; grpsl). Population served: 100,000 Natl. Network: ABC. Rgnl. Network: Agri-Net, AgriAmerica. Format: Farm. News staff: one; News: 17 hrs wkly. Target aud: 35 plus. Spec prog: American Indian one hr, gospel 3 hrs wkly. ♦Dean Kinney, gen mgr & gen sls mgr; Toni Kinney, opns mgr; Gary Matthews, progmg dir; Gary Maki, news dir; Gary Peterson, chief of engrg.

KRCS(FM)—Co-owned with KBHB(AM). Dec 5, 1972: 93.1 mhz; 100 kw. 1,059 ft TL: N44 19 58 W103 32 20. Stereo. Box 2480, Rapid City, 57709. Phone: (605) 343-6161. Fax: (605) 343-9012. Web Site: www.hot931.com. Format: Continuous hit radio. ♦Lia Green, gen mgr; Charlie O'Douglas, opns mgr; Leah Green, sls dir & gen sls mgr; Chad Bower, progmg dir; D. Ray Knight, news dir; Gary Peterson, chief of engrg.

Vermillion

***KAOR(FM)**— September 1986: 91.1 mhz; 120 w. 107 ft TL: N42 47 01 W96 55 26. Stereo. Hrs opn: 18 Contemporary Media & Journalism, 414 E. Clark, 57069-2390. Phone: (605) 677-5477. Fax: (605) 677-4250. E-mail: kaor@usd.edu Web Site: www.usd.edu/kaor. Licensee: The University of South Dakota. Population served: 35,000 Law Firm: Cohn & Marks. Format: AOR, CHR, progsv. News: 2 hrs wkly. Target aud: 16-30; college age students & faculty. Spec prog: American Indian 2 hrs wkly. ♦Ramon Chavez, chmn; Kent Osborne, gen mgr; Don Harris, chief of engrg.

***KUSD(FM)**— Oct 1, 1967: 89.7 mhz; 50 kw horiz, 21.5 kw vert. 518 ft TL: N43 03 00 W96 47 12. (CP: 32 kw, ant 663 ft.). Stereo. Hrs opn: Box 5000, 57069. Secondary address: 555 N. Dakota St. 57069. Phone: (605) 677-5861. Fax: (605) 677-5010. E-mail: sdpr@sdpb.org Web Site: www.sdpb.org. Licensee: South Dakota Board of Directors/Educational Telecommunications. Population served: 150,000 Natl. Network: NPR, PRI. Rgnl. Network: S.D. Pub. Format: Class, news & pub affrs, jazz. Spec prog: Sioux one hr wkly. ♦Julie Anderson, pres; Terry Harris, gen mgr; Terry Spencer, dev dir. Co-owned TV: *KUSD-TV affil.

KVHT(FM)— Nov 16, 1967: 106.3 mhz; 50 kw. 390 ft TL: N42 59 45 W96 49 25. Stereo. Hrs opn: 24 Box 718, Yankton, 57078. Secondary address: 210 W. 3rd St., Yankton 57078. Phone: (605) 665-2600. Fax: (605) 665-8875. E-mail: mix106@kvht.com Web Site: www.kvht.com. Licensee: Culhane Communications Inc. (acq 5-6-93; $340,000 with co-located AM; FTR: 5-24-93) Population served: 350,000 Law Firm: Shaw Pittman. Format: Adult contemp. News staff: one; News: 42 hrs wkly. Target aud: 18 plus. ♦Kevin Culhane, gen mgr; Julie Auch, gen sls mgr & chief of engrg; Randy Eichelburg, opns mgr & progmg dir; Lee Rettig, news dir; Daphne Howley, traf mgr.

Volga

KJJQ(AM)—Licensed to Volga. See Brookings

KKQQ(FM)—Licensed to Volga. See Brookings

Watertown

KDLO-FM— Mar 1, 1968: 96.9 mhz; 100 kw. 1,571 ft TL: N44 57 57 W97 35 22. Stereo. Hrs opn: 24 921 9th Ave. S., 57201. Phone: (605) 886-8444. Fax: (605) 886-9306. E-mail: kwatprod@iw.net Licensee: Three Eagles of Joliet Inc. Group owner: Three Eagles Communications (acq 6-18-2004; grpsl). Population served: 60,000 Natl. Network: USA.

Format: C&W. News staff: one. Target aud: 25-54. ♦Dean Johnson, gen mgr; Bruce Erlandson, opns mgr.

KIXX(FM)—Listing follows KWAT(AM).

***KJBB(FM)**— August 2000: 89.1 mhz; 200 w vert. Ant 20 ft TL: N44 53 57 W97 06 18. Hrs open: 24 501 E. Kemp Ave., 57201. Phone: (605) 884-0156. E-mail: kjbbradio89-1fm@daliypost.com Web Site: www.kjbbfm.com. Licensee: Church Planters of America (acq 7-20-2003). Population served: 20,500 Format: Educational. ♦Ms. Sheila Hawkins, VP; Danny Hawkins, gen mgr.

KSDR(AM)— Apr 16, 1961: 1480 khz; 1 kw-D, 53 w-N. TL: N44 55 58 W97 06 19. Hrs open: 6 AM-midnight 3 E. Kemp, Suite 300, 57201. Phone: (605) 886-5747. Fax: (605) 886-2121. Licensee: Three Eagles of Brookings Inc. Group owner: Three Eagles Communications (acq 6-19-00; $3.25 million with co-located FM). Population served: 50,000 Law Firm: Tierney & Swift. Format: Talk. News staff: 3; News: 15 hrs wkly. Target aud: 25-54. ♦Gary Buchanan, pres; Dean Johnsson, gen mgr & mus dir.

KSDR-FM— Mar 10, 1992: 92.9 mhz; 97 kw. 977 ft TL: N45 10 31 W96 59 15. Stereo. 24 100,000 Format: C&W. News staff: one; News: 12 hrs wkly. Target aud: General; rgnl country stn with wide var of ages. Spec prog: Farm 8 hrs, sports 8 hrs wkly.

KWAT(AM)— Mar 8, 1940: 950 khz; 1 kw-U, DA-N. TL: N44 52 12 W97 06 49. Hrs open: 5 AM-midnight Box 950, 57201. Secondary address: 921 9th Ave S. E. 57201. Phone: (605) 886-8444. Fax: (605) 886-9306. E-mail: radionews@home.com Licensee: Three Eagles of Joliet Inc. Group owner: Three Eagles Communications (acq 6-18-2004; grpsl). Population served: 15,571 Natl. Network: CBS. Format: MOR, news, farm. Target aud: 35 plus. ♦Gary Buchanan, pres; Dean Johnson, VP, gen mgr & gen sls mgr; Bruce Erlandson, opns mgr; Mike Blakenship, mus dir; David Law, news dir; Jim Thoreson, farm dir; Todd Enderson, disc jockey; Wayne Hunter, disc jockey.

KIXX(FM)—Co-owned with KWAT(AM). Sept 29, 1968: 96.1 mhz; 97 kw. 977 ft TL: N45 10 31 W96 59 15. Stereo. 6 AM-1 AM Phone: (605) 886-9696. Format: Adult contemp. Target aud: 25-54. ♦Curt Crawford, progmg dir.

Wessington Springs

KJRV(FM)— 2005: 93.3 mhz; 65 kw. Ant 623 ft TL: N44 11 39 W98 19 05. Hrs open: 1726 Dakota Ave. S., Huron, 57350. Phone: (605) 352-1933. Phone: (605) 352-8623. Fax: (605) 352-1934. E-mail: mlyon@kokk.com Web Site: www.bigjimrocks.com. Licensee: Alpena Broadcasting Co. Format: Classic rock. ♦Linda Marcus, gen mgr; Mike Lyon, gen sls mgr.

KLQT(FM)— 1999: 98.3 mhz; 100 kw. Ant 899 ft TL: N43 45 28 W98 24 39. Hrs open: Box 520, Mitchell, 57301. Secondary address: 501 S. Ohlman, Mitchell 57301. Phone: (605) 996-9667. Phone: (605) 996-1100. Fax: (605) 996-0013. Web Site: www.kool98.com. Licensee: Saga Communications of South Dakota LLC. Group owner: Saga Communications Inc. (acq 5-1-2001; $4.05 million with KMIT(FM) Mitchell). Format: Oldies. ♦Tim Smith, gen mgr; Nikki Frederickson, gen sls mgr; Kory Hartman, progmg dir; John Cyr, chief of engrg.

Winner

KWYR(AM)— Sept 27, 1957: 1260 khz; 5 kw-D, 146 w-N. TL: N43 22 57 W99 54 38. Stereo. Hrs opn: 24 346 Main St., 57580. Secondary address: Box 491 57580. Phone: (605) 842-3333. Fax: (605) 842-3875. E-mail: 937radio@gwtc.net Web Site: www.kwyr.com. Licensee: Midwest Radio Corp. (acq 12-2-2005; $378,000 for stock with co-located FM). Population served: 30,000 Wire Svc: AP Format: Country. News staff: one; News: 14 hrs wkly. Target aud: 25-60. ♦John Driscoll, VP; Scott Schramm, pres & gen mgr.

KWYR-FM— Nov 25, 1971: 93.7 mhz; 100 kw. 560 ft TL: N43 17 46 W99 52 02. Stereo. 24 Phone: (605) 842-3693. Web Site: www.kwyr.com.30,000 Natl. Network: Jones Radio Networks. Format: Adult contemp, CHR. Target aud: 18-45. ♦John Driscoll, gen sls mgr & engrg dir.

Yankton

KKYA(FM)—Listing follows KYNT(AM).

KYNT(AM)— Mar 15, 1955: 1450 khz; 1 kw-U. TL: N42 53 30 W97 25 10. Hrs open: 24 Box 628, 57078. Secondary address: 202 W. 2ndn St. 57078. Phone: (605) 665-7892. Fax: (605) 665-0818. E-mail: kynt1450@yahoo.com Licensee: Sorenson Broadcasting Corp. (group owner; (acq 7-1-73). Population served: 14,500 Natl. Network: Westwood One. Natl. Rep: Katz Radio. Format: Adult contemp. News staff: 2; News: 25 hrs wkly. Target aud: General. Spec prog: Farm 5 hrs, polka one hr, Pol one hr wkly. ♦Dean Sorenson, pres; Bill Holst, gen mgr; Dave Lesher, opns mgr; Dave Bradbury, prom mgr; Barb Steffen, traf mgr; Al Lundy, news rptr; Bob Bakken, reporter; Gene Williams, farm dir; J.J. Glenn, sports cmtr.

KKYA(FM)—Co-owned with KYNT(AM). May 25, 1982: 93.1 mhz; 100 kw. 469 ft TL: N42 43 49 W97 24 13. Stereo. 24 E-mail: kk93fm@hotmail.com 50,000 Rgnl rep: Cindy Weiland Format: Country. News staff: 2; News: 1.5 hrs wkly. ♦Barb Steffen, traf mgr; Al Lundy, news rptr; Bob Bakken, reporter; Gene Williams, farm dir.

WNAX(AM)— November 1922: 570 khz; 5 kw-U, DA-N. TL: N42 54 47 W97 18 58. Hrs open: 24 1609 E. Hwy. 50, 57078. Phone: (605) 665-7442. Fax: (605) 665-8788. E-mail: wnax@wnax.com Web Site: www.wnax.com. Licensee: Saga Communications Inc. (acq 1996). Population served: 3,500,000 Natl. Network: CBS. Rgnl. Network: MNN. Natl. Rep: Katz Radio. Law Firm: Smithwick & Belendiuk. Wire Svc: NOAA Weather Wire Svc: Knight-Ridder/Tribune Information Services Format: News/talk, farm. News staff: 4; News: 23 hrs wkly. Target aud: 35 plus; farmers & agri-businesses. Spec prog: Relg 16 hrs, sports 10 hrs, weather 15 hrs, farm news 35 hrs wkly. wkly. ♦Edward Christian, pres; Bill Holst, gen mgr; Steve Crawford, opns mgr; Jim Reimler, progmg mgr & mus dir; Jerry Oster, news dir.

WNAX-FM— Aug 9, 1973: 104.1 mhz; 97 kw. 981 ft TL: N42 38 24 W97 03 21. Stereo. 24 Web Site: www.wnax.com.400000 Format: Country. News staff: one; News: 3 hrs wkly. Target aud: 25-54.

Tennessee

Alamo

WCTA(AM)— October 1983: 810 khz; 250 w-D, DA. TL: N35 47 59 W89 07 20. Hrs open: 114 S. Johnson St., 38001. Phone: (731) 696-2781. Fax: (731) 696-5006. E-mail: billy@wcta810.com Web Site: www.wcta810.com. Licensee: Billy H. Williams. (acq 5-1-96; $119,933). Population served: 25,000 Format: News, talk. Target aud: 30 plus. Spec prog: Relg 9 hrs wkly. ♦Billy H. Williams, pres & gen mgr; Billy Williams, progmg mgr; Dave Hacker, chief of engrg.

WWGM(FM)— Aug 10, 1989: 93.1 mhz; 25 kw. 443 ft TL: N35 43 31 W89 03 25. Stereo. Hrs opn: 24 25 Stonebrook Pl., Suite 322, Jackson, 38305. Phone: (731) 855-9334. Fax: (731) 855-1600. Web Site: www.gracebroadcasting.com. Licensee: Grace Broadcasting Services Inc. (acq 8-18-97; $800,000). Population served: 150,000 Law Firm: Miller & Miller. Format: Southern gospel, relg. News staff: one; News: 3 hrs wkly. Target aud: 24-54; upscale women. ♦Lacy Ennis, stn mgr & sls VP; Phillip Chambers, prom dir.

Alcoa

WBCR(AM)— Aug 25, 1957: 1470 khz; 1 kw-D. TL: N35 47 47 W83 56 17. Hrs open: Box 130, 37701. Secondary address: 118 Defoe Cir. 37701. Phone: (865) 984-1470. Fax: (865) 983-0890. E-mail: truthradioam1470@yahoo.com Licensee: Blount County Broadcasting Co. (acq 2-20-96). Format: News/talk. News staff: local news/ talk 10 hrs wkly News: one;. Target aud: 35+. ♦Harry Grothjahn, gen mgr.

***WYLV(FM)**— Feb 14, 1993: 89.1 mhz; 4.5 kw. 994 ft TL: N36 00 13 W83 56 35. Stereo. Hrs opn: 24 1621 E. Magnolia Ave., Knoxville, 37917. Phone: (865) 521-8910. Fax: (865) 521-8923. E-mail: info@love89.org Web Site: www.love89.org. Licensee: Foothills Broadcasting Inc. Population served: 600,000 Format: Contemp Christian. ♦David Wells, gen mgr & opns mgr; Marisa Lykins, prom dir; Jonathan Unthank, progmg dir.

Algood

WATX(AM)— Oct 5, 1981: 1590 khz; 1 kw-D, 500 w-N. TL: N36 11 02 W85 25 03. Hrs open: 6 AM-9 PM 259 S. Willow Ave., Cookeville, 38501. Phone: (931) 528-6064. Fax: (931)520-1590. E-mail: jimstapleton @jwcbroadcasting.com Licensee: JWC Broadcasting (group owner; acq 8-3-01). Population served: 150,000 Natl. Network: Salem Radio Network. Natl. Rep: Rgnl Reps. Format: Christian. News staff: 2; News: 34 hrs wkly. Target aud: General. Spec prog: Gospel. ♦Jim Stapleton, gen mgr.

Ashland City

WQSV(AM)— July 14, 1982: 790 khz; 2 kw-D, 35 w-N. TL: N36 17 08 W87 05 00. Hrs open: Box 619, 37015-0619. Secondary address: 208 1/2 N. Main St. 37015-1316. Phone: (615) 792-6789. Fax: (615) 792-7795. E-mail: wqsvradio@bellsouth.net Web Site: www.wqsvam790.com. Licensee: Sycamore Valley Broadcasting Inc. (acq 12-20-91; $55,000; FTR: 1-13-92). Natl. Network: ABC. Rgnl. Network: Tenn. Radio Net. Format: Var. News staff: 4. Target aud: General. ♦Richard Albright, CEO, gen mgr & stn mgr.

Athens

WJSQ(FM)—Listing follows WLAR(AM).

WLAR(AM)— May 15, 1946: 1450 khz; 1 kw-U. TL: N35 26 44 W84 36 43. Hrs open: 24 2110 Oxnard Rd., 37303. Phone: (423) 745-1000. Fax: (423) 745-2000. Web Site: www.1017wlar.com. Licensee: James C. Sliger. (acq 4-18-83; $200,000; FTR: 5-9-83). Population served: 18,600 Format: Contemp country. Spec prog: Farm 2 hrs wkly. ♦James Sigler, gen mgr & gen sls mgr.

WJSQ(FM)—Co-owned with WLAR(AM). Dec 1, 1979: 101.7 mhz; 7.5 kw. 528 ft TL: N35 31 19 W84 27 29. Stereo. 24 Web Site: www.1017wlar.com. Format: Country.

WYXI(AM)— Oct 5, 1966: 1390 khz; 2.5 kw-D, 62 w-N. TL: N35 26 48 W84 34 19. Hrs open: 6 AM-7 PM Box 1390, 112 E. Madison Ave., 37371-1390. Phone: (423) 746-1390. Fax: (423) 744-1390. E-mail: wyxi@bellsouth.net Web Site: wyxi.com. Licensee: Cornerstone Broadcasting Inc. (acq 9-11-86; $75,000; FTR: 7-21-86). Population served: 48,000 Natl. Network: ABC. Rgnl rep: Rgnl Reps Law Firm: Fletcher, Heald & Hildreth, PLC. Format: Talk. News staff: one; News: 10 hrs wkly. Target aud: 25-64; mature middle class. Spec prog: Black one hr, relg 8 hrs wkly. ♦Bob Ketchersid, VP, opns dir & progmg dir; Mark Lefler, pres & stn mgr.

Atwood

WTKB-FM— 1992: 93.7 mhz; 6 kw. Ant 328 ft TL: N35 57 25 W88 41 44. (CP: 15 kw, ant 325 ft). Hrs opn: 24 Box 500, Trenton, 38382. Secondary address: 302 W. Eaton St. 38382. Phone: (731) 855-0098. Fax: (731) 855-1600. E-mail: wtkbwtne@bellsouth.nct Licensee: Grace Broadcasting Services Inc. Group owner: Thunderbolt Broadcasting Co./Gibson County Broadcasting (acq 1-7-2005; grpsl). Population served: 350,000 Format: Christian music. ♦Lacy Ennis, gen mgr.

Bartlett

WMFS(FM)— May 1994: 92.9 mhz; 6 kw. 328 ft TL: N35 10 20 W89 56 40. Hrs open: 24 5904 Ridgeway Center Parkway, Memphis, 38120. Phone: (901) 726-0555. Fax: (901) 725-5101. Web Site: 93xmemphis.com. Licensee: CBS Radio Inc. of Illinois. Group owner: Infinity Broadcasting Corp. (acq 7-25-2001; $7.2 million). Population served: 1,000,000 + Natl. Rep: Interep. Law Firm: Leventhal, Senter & Lerman. Wire Svc: Metro Weather Service Inc. Format: Alternative. News: one hr wkly. Target aud: 18-49; adults. ♦Terry Wood, pres, VP & gen mgr; Kory Myers, gen sls mgr; Scott Speropoulas, natl sls mgr; Rob Cressman, progmg dir.

WMPS(AM)—Licensed to Bartlett. See Memphis

Baxter

WBXE(FM)— October 1995: 93.7 mhz; 25 kw. 328 ft TL: N36 18 53 W85 32 00. Hrs open: 24 259 S. Willow Ave., Cookeville, 38501. Phone: (931) 528-6064. Fax: (931) 520-1590. E-mail: jstapleton @jwcbroadcasting.com Web Site: www.brock937.com. Licensee: JWC Broadcasting (group owner; acq 8-29-01). Natl. Network: Westwood One. Natl. Rep: Rgnl Reps. Format: Rock. News staff: one. Target aud: 18-45; Men 25 plus. ♦Jim Stapleton, gen mgr.

Belle Meade

WNFN(FM)— 1998: 106.7 mhz; 1.1 kw. 774 ft TL: N36 08 27 W86 51 56. Hrs open: 10 Music Circle E., Nashville, 37203. Phone: (615) 321-1067. Fax: (615) 321-5771. E-mail: danielle.haese@cumulus.com Web Site: www.cumulus.com. Licensee: Cumulus Licensing Corp. Group owner: Cumulus Media Inc. (acq 2-12-2002; grpsl). Format: Sports. ♦Michael Dickey, gen mgr; Derick Corbett, progmg dir; Dan Goodman, chief of engrg; Dave Elliott, sls.

Bells

WNWS(AM)—See Brownsville

Benton

WBIN(AM)— May 18, 1977: 1540 khz; 1 kw-D, 2 w-N, 500 w-CH. TL: N35 11 15 W84 38 13. (CP: 1 kw-D, 4 w-N, 500 w-CH. TL: N35 10 50 W84 38 34). Hrs opn: Sunrise-sunset 108 Lifestyle Way, 37307. Phone: (423) 338-2864. E-mail: www.craig@wbinthelight.com Licensee: John A. Sines and L. Jane Sines, JTWROS (acq 12-29-99; $79,000). Population served: 100,000 Format: Relg. Target aud: All ages; Includes baby boomers and seniors. ♦Chris Harding, gen mgr, progmg dir & disc jockey.

WSAA(FM)— November 1996: 93.1 mhz; 5 kw. Ant 358 ft TL: N35 05 40 W84 53 45. Stereo. Hrs opn: 24 Box 9170, Chattanooga, 37412. Phone: (423) 485-8987. Fax: (423) 485-8946. Licensee: LB Radio of Chattanooga, LLC Format: Sp hits. Target aud: 25-54. Spec prog: Pub affrs 2 hrs, talk, Univ. of Tennessee sports, loc sports wkly.

***WTSE(FM)**— 2005: 91.1 mhz; 8.5 kw vert. Ant 466 ft TL: N35 19 25 W84 17 54. Hrs open: Box 5459, Twin Falls, ID, 83303. Phone: (208) 733-3551. Fax: (208) 733-3548. Licensee: Radio Assist Ministry Inc. (acq 9-13-2004; $1 for CP). Law Firm: Sciarrino & Associates. ♦Clark Parrish, pres.

Berry Hill

WVOL(AM)— December 1951: 1470 khz; 5 kw-D, 1 kw-N, DA-2. TL: N36 12 01 W86 46 47. Hrs open: 24 1320 Brickchurch Pike, Nashville, 37207. Phone: (615) 226-9510. Fax: (615) 226-0709. E-mail: wvol1470@aol.com Web Site: www.wvol1470.com. Licensee: Heidelberg Broadcasting LLC. (acq 4-24-00). Population served: 1,636,000 Format: Classic oldies, rhythm and blues. Target aud: 25-54; relg. Spec prog: Gospel 6 hrs wkly. ♦John Heidelberg, chmn, pres, gen mgr, gen sls mgr & min affrs dir; Roderick M. Heidelberg, opns mgr & progmg dir; Watt Harriston, chief of engrg; Betty Fykes, traf mgr.

Blountville

WXSM(AM)— Sept 20, 1967: 640 khz; 10 kw-D, 810 w-N, DA-N. TL: N36 31 19 W81 25 25. Stereo. Hrs opn: 24 Box 8668, Gray, 37615. Phone: (423) 477-1000. Fax: (423) 477-4747. E-mail: SportsMonster@640wxsm.com Web Site: www.640wxsm.com. Licensee: Citadel Broadcasting Co. Group owner: Citadel Broadcasting Corp. (acq 5-30-2000; grpsl). Population served: 500,000 Natl. Network: ESPN Radio. Natl. Rep: Dora-Clayton. Format: Sports. ♦Bill Meade, pres & progmg dir; Don Raines, gen mgr; Debbie Caso, sls dir; Paul Overbay, gen sls mgr; Al LeFevre, chief of engrg.

Bolivar

WBOL(AM)— Oct 19, 1962: 1560 khz; 250 w-D. TL: N35 15 30 W88 58 50. Hrs open: Box 191, 123 W. Market, 38008. Phone: (731) 658-3633. Phone: (731) 658-3690. Fax: (731) 658-3408. E-mail: wojg@hcaol.com Licensee: Shaw's Broadcasting Co. Format: Blues, light jazz, oldies. ♦Johnny W. Shaw, gen mgr; Dewayne Dickerson, gen sls mgr; Opal Shaw, progmg dir.

WOJG(FM)—Co-owned with WBOL(AM). June 1992: 94.7 mhz; 6 kw. 328 ft TL: N35 16 39 W88 55 41. Licensee: Johnny W. Shaw & Opal J. Shaw. Format: Gospel. ♦Dwayne Dickerson, opns dir & sls dir; Dave Hacker, chief of engrg.

WMOD(FM)— Jan 27, 1975: 96.7 mhz; 3 kw. 300 ft TL: N35 15 00 W88 53 28. Stereo. Hrs opn: 24 200 E. Market St. Suite B, 38008. Phone: (731) 658-4320. Fax: (731) 658-7328. Fax: (731) 658-4320. E-mail: wmod@newwarecomm.net Licensee: WMOD Inc. (acq 6-17-97; $320,000). Natl. Network: ABC. Rgnl. Network: Tenn. Radio Net. Rgnl rep: Midsouth. Format: Country. News: 7 hrs wkly. Target aud: 25-55; males & females. ♦D. Richard Teubner, pres & gen mgr; Gail R. Teubner, opns mgr & traf mgr.

Brentwood

WNSR(AM)— Sept 4, 1985: 560 khz; 4.5 kw-D, 75 w-N, DA-2. TL: N35 54 32 W86 46 13. Hrs open: 435 37th Ave. N., Nashville, 37209. Phone: (615) 844-1039. Fax: (615) 777-2284. E-mail: info@wnsr.com Web Site: www.wnsr.com. Licensee: Southern Wabash Communications Middle Tennessee Inc. Group owner: Southern Wabash Communications Corp. (acq 11-25-97; $245,000). Format: Sports. Target aud: 18-54; men. ♦Ted Johnson, gen mgr.

Bristol

***WHCB(FM)**— Aug 10, 1984: 91.5 mhz; 1.5 kw. Ant 2,326 ft TL: N36 26 03 W82 08 03. Stereo. Hrs opn: 24 Box 2061, 37621-2061. Secondary address: 340 Edgemont Ave., Suite 100 37620. Phone: (423) 878-6279. Fax: (423) 878-6520. E-mail: whcb@aecc.org Web Site: www.whcbradio.org.yes Licensee: Appalachian Educational Communication Corp. Population served: 2,000,000 Natl. Network: Moody, Salem Radio Network. Format: Talk, educ, Christian. News staff: one; News: 14 hrs wkly. Target aud: General. Spec prog: Class one hr, Appalachian culture 2 hrs, farm one hr, folk one hr, Sp one hr, Jewish one hr, black 3 hrs, children 10 hrs, gospel 15 hrs wkly. ♦Kenneth C. Hill, pres & gen mgr.

WIGN(AM)— Aug 18, 1962: 1550 khz; 5 kw-D, 6 w-N. TL: N36 33 58 W82 09 30. Hrs open: Sunrise-sunset Box 68, 37621. Phone: (276) 591-5800. Fax: (276) 591-5278. E-mail: wignradio@bvunet.net Web Site: www.wignradio.com. Licensee: Sunshine Broadcasters Inc. (acq 8-4-2005; $245,000 for stock). Population served: 520,000 Format: Southern gospel. News: 3 hrs wkly. Target aud: 25-54 yrs. ♦Rick Mitchell, gen mgr.

WKPT(AM)—See Kingsport

WOPI(AM)— June 15, 1929: 1490 khz; 1 kw-U. TL: N36 35 45 W82 09 42. Hrs open: 24 222 Commerce St., Kingsport, 37660. Secondary address: 288 Delaney St. 37620. Phone: (423) 764-5131. Fax: (423) 246-6261. Fax: (423) 247-9836. E-mail: davidw@wtfm.com Web Site: www.wopi.com. Licensee: Holston Valley Broadcasting Corp. Group owner: Glenwood Communications Corp. (acq 5-16-96; $140,000; FTR: 5-7-90). Population served: 200,000 Natl. Rep: McGavren Guild. Law Firm: Cordon & Kelly. Format: Oldies. News staff: 2. Target aud: 35 plus. ♦George DeVault, pres; Bette Lawson, CFO; David Widener, exec VP & gen mgr; Aaron Teffeteller, progmg dir & pub affrs dir.

WQUT(FM)—See Johnson City

WXBQ-FM— 1945: 96.9 mhz; 67 kw. 2,200 ft TL: N36 25 59 W82 08 11. Stereo. Hrs opn: Box 1389, VA, 24203. Secondary address: 901 E. Valley Dr., VA 24201. Phone: (276) 669-8112. Fax: (276) 669-0541. Web Site: www.wxbq.com. Licensee: Bristol Broadcasting Inc. Group owner: Nininger Stations Population served: 188,600 Natl. Rep: McGavren Guild. Format: Country. Target aud: 25-54. ♦W.L. Nininger, pres; Pete Nininger, gen mgr; Winnie Quaintance, gen sls mgr; Roger Bowldin, prom dir & prom mgr; Bruce Clark, progmg dir; George Dixon, news dir; Chuck Lawson, chief of engrg.

Brownsville

WNWS(AM)— Oct 14, 1963: 1520 khz; 250 w-D. TL: N35 36 30 W89 14 40. Hrs open: Sunrise-sunset Box 198, 42 S. Washington Ave., 38012. Phone: (901) 772-3700. Licensee: The Wireless Group Inc. (group owner; acq 4-80; $320,000 with co-located FM; FTR: 3-31-80). Population served: 20,000 Natl. Network: ABC. Format: Sp. ♦Carlton Veirs, pres & gen mgr; Tanya Garcia, progmg dir.

WTBG(FM)—Co-owned with WNWS(AM). Nov 9, 1965: 95.3 mhz; 5 kw. 150 ft TL: N35 36 30 W89 14 40. (CP: 6 kw, ant 328 ft.). Stereo. 24
Rebroadcasts WNWS-FM Jackson 30%.
Law Firm: Wyatt, Tarrant & Combs. Format: Country, news/talk. News staff: one. Target aud: 25-54. ♦Carlton Veirs, CEO; Kim Bishop, sls VP; Pam McCuan, progmg dir & disc jockey; Elizabeth Timbes, disc jockey; Ivory Ellison, disc jockey; Jerry Wilson, disc jockey; Sandra Awex, disc jockey.

***WRRI(FM)**—Not on air, target date: unknown: 88.3 mhz; 500 w. Ant 125 ft TL: N35 35 33 W89 14 50. Hrs open: 2351 Sunset Blvd., Suite 170-218, Rocklin, CA, 95765. Phone: (916) 251-1600. Fax: (916) 251-1650. Licensee: Educational Media Foundation. (acq 11-1-2006; grpsl). ♦Richard Jenkins, pres.

Bulls Gap

WBGQ(FM)— 2001: 100.7 mhz; 6000 w. Ant 1,260 ft TL: N36 22 48 W83 10 47. Stereo. Hrs opn: 24 Cherokee Broadcasting System, Box 519, Morristown, 37815. Phone: (423) 235-4640. E-mail: www.wbgqfm@planetc.com Web Site: wjdtfm@planetc.com. Licensee: S.J. Trent dba Cherokee Broadcasting System. Population served: 500,000 Natl. Network: CNN Radio. Format: Rhythm mix. Target aud: 18-54; female 65% & male 35%. ♦Clark Quillen, CEO & gen mgr; David Quillen, opns VP.

Byrdstown

WLSQ(FM)—Not on air, target date: unknown: 98.9 mhz; 1.1 kw. Ant 760 ft TL: N36 29 26 W85 01 47. Hrs open: 2310 Armstrong Rd., Linden, 37096. Phone: (931) 589-9583. Licensee: Susan Clinton. ♦Susan Clinton, gen mgr.

Calhoun

WCLE-FM— August 1993: 104.1 mhz; 2.3 kw. Ant 522 ft TL: N35 15 59 W84 50 23. Stereo. Hrs opn: 24 Box 2695, Cleveland, 33730. Secondary address: 1860 Executive Park, Suite E, Cleveland 37312. Phone: (423) 472-6700. Fax: (423) 476-4686. Licensee: Williams Communications Inc. (group owner; acq 8-8-02; $2.4 million with WCLE(AM) Cleveland). Format: Adult contemp. News staff: one. Target aud: 25-54. ♦Paul Fink, gen mgr; Walt Williams III, gen mgr.

Camden

WFWL(AM)— Sept 18, 1956: 1220 khz; 250 w-D, 140 w-N. TL: N36 03 10 W88 05 15. Hrs open: 24 Box 539, 117 Vicksburg Ave., 38320. Phone: (731) 584-7570. Phone: (731) 584-4444. Fax: (731) 584-7553. E-mail: reiddbell@morningcoffeebreak.com Web Site: www.morningcoffeebreak.com. Licensee: Community Broadcasting Services Inc. (acq 4-30-98; $767,000 exercise of option with co-located FM). Population served: 45,000 Rgnl. Network: Tenn. Radio Net. Natl. Rep: Keystone (unwired net). Rgnl rep: Midsouth. Law Firm: Miller & Fields, P.C. Format: C&W. Target aud: 25-49; adult. Spec prog: Gospel 8 hrs wkly. ♦Stan Medlin, pres; Ron Lane, gen mgr, gen sls mgr & prom mgr; Jim Hart, progmg dir, news dir & disc jockey; Larry Nunnery, engrg mgr; Bobby Melton, disc jockey; Reid Bell, disc jockey.

WRJB(FM)—Co-owned with WFWL. June 20, 1976: 98.3 mhz; 3 kw. 300 ft TL: N36 03 25 W88 06 10. Stereo. 24 372,000 Format: Adult contemp. News: 4 hrs wkly. Target aud: 20-50. ♦Stan Medlin, pres & CFO; Charles Ennis, exec VP; Larry Nannery, engrg VP; Vickie Dodson, traf mgr; Jim Hart, disc jockey.

Carthage

WRKM(AM)— June 20, 1959: 1350 khz; 1 kw-D, 91 w-N. TL: N36 14 42 W85 56 44. Hrs open: 12 Box 179, 37030. Secondary address: 104 Z Country Ln. 37030. Phone: (615) 735-1350. Fax: (615) 735-0381. E-mail: am1350@smithcounty.com Web Site: www.wucz-wrkm.com. Licensee: Wood Broadcasting Inc. (acq 11-1-87). Population served: 250,000 Natl. Network: Sporting News Radio Network. Rgnl. Network: Tenn. Radio Net. Format: Sports. News staff: one; News: 2 hrs wkly. Target aud: 35 plus. ♦John Wood, pres, gen mgr & gen sls mgr; Dennis Banka, prom dir & progmg dir; Carl Campbell, chief of engrg; Tracy Banka, traf mgr.

WUCZ-FM—Co-owned with WRKM(AM). July 18, 1975: 104.1 mhz; 6 kw. 300 ft TL: N36 18 43 W85 57 08. Stereo. 24 E-mail: z104@smithcounty.com Web Site: www.wucz-wrkm.com.500,000 Natl. Network: Westwood One. Format: Country. News staff: one; News: 2 hrs wkly. Target aud: 18-35. ♦John Wood, stn mgr & opns mgr; Tracy Banka, traf mgr.

Celina

WVFB(FM)— August 1994: 101.5 mhz; 6 kw. 328 ft TL: N36 33 15 W85 36 39. Stereo. Hrs opn: 6 AM-10PM 341 Radio Station Rd., Tompkinsville, KY, 42167. Phone: (270) 487-6119. Fax: (270) 487-8462. Licensee: Elizabeth Bernice Whittimore (acq 5-5-93; $14,000; FTR: 5-24-93). Format: Country. News staff: news progmg 20 hrs wkly News: 2;. All age group. ♦Bernice Whittimore, gen mgr & gen sls mgr; Elizabeth Whittimore, gen mgr.

Centerville

WNKX(AM)— Nov 16, 1955: 1570 khz; 5 kw-D, 77 w-N. TL: N35 45 29 W87 27 35. Hrs open: 24 Box 280, 37033. Secondary address: 150 Hwy. 50 E. 37033. Phone: (931) 729-5191. Phone: (931) 729-9600. Fax: (931) 729-5467. E-mail: wnkx@countrykix96.com Web Site: www.countrykix96.com. Licensee: Hickman County Broadcasting Co. Inc. (acq 5-7-97; $300,000 with co-located FM). Population served: 20,000 Rgnl. Network: Tenn. Radio Net. Natl. Rep: Dora-Clayton. Rgnl rep: Midsouth. Law Firm: McCampbell & Young. Format: Christian, talk. News staff: one; News: 10 hrs wkly. Target aud: 18-65; general. Spec prog: Farm one hr, gospel 5 hrs wkly. ♦Steve Turner, CEO, chmn, pres, gen mgr & sls dir; John Kimery, CFO; Wanda Turner, exec VP.

WNKX-FM— May 1974: 96.7 mhz; 6 kw. 300 ft TL: N35 49 39 W87 34 02. Stereo. 24 E-mail: kix96fm@bellsouth.net Web Site: www.countrykix96.com.100,000 Natl. Rep: Dora-Clayton. Rgnl rep: Midsouth Law Firm: McCampbell & Young. Format: Country. News staff: 3; News: 30 hrs wkly. Target aud: 6-80. ♦Mickey Bunn, opns mgr, asst music dir, news rptr & farm dir; Steve Turner, stn mgr & progmg mgr; Brent Atkinson, disc jockey; Demetria Smith, disc jockey; Kay Atkinson, disc jockey.

Chattanooga

WAWL-FM—See Red Bank

WDEF-FM— Sept 15, 1964: 92.3 mhz; 100 kw. 1,180 ft TL: N35 08 06 W85 19 25. Stereo. Hrs opn: 24 Box 11008, 37401. Secondary address: 2615 S. Broad St. 37408. Phone: (423) 321-6200. Fax: (423) 321-6264. Licensee: Bahakel Communications. (group owner; acq 1996; grpsl). Natl. Network: CBS. Format: Soft rock. News staff: one; News: 5 hrs wkly. Target aud: 25-54; upscale adults. ♦Gary Downs, gen mgr; Jeff Fontana, gen sls mgr; Danny Howard, progmg dir; James Howard, news dir; Ben Johnston, chief of engrg; Sherry Ball, traf mgr.

WDEF(AM)— Dec 31, 1940: 1370 khz; 5 kw-U, DA-N. TL: N35 02 25 W85 20 22.24 366,100 Format: News/talk, sports. News staff: one; News: 20 hrs wkly. Target aud: 25-64; males.

WDOD(AM)— Apr 13, 1925: 1310 khz; 5 kw-U, DA-N. TL: N35 04 54 W85 20 14. Hrs open: 24 Box 1449, 37401. Secondary address: 2615 Broad St. 37408. Phone: (423) 321-6200. Fax: (423) 321-6270. Licensee: WDOD of Chattanooga Inc. Group owner: Bahakel Communications (acq 6-62). Population served: 250,000 Format: Oldies. Target aud: 35 plus; empty nesters. ♦Gary Downs, gen mgr & gen sls mgr; Danny Howard, progmg VP.

WDOD-FM— February 1960: 96.5 mhz; 100 kw. 1,080 ft TL: N35 09 39 W85 19 11. Stereo. Box 11008, 37401. 375,000 Format: Adult rock. Target aud: 18-54; upscale, contemp adults. ♦Danny Howard, progmg dir.

***WDYN-FM**— June 1, 1968: 89.7 mhz; 100 kw. 205 ft TL: N35 10 17 W85 18 58. Stereo. Hrs opn: 24 1815 Union Ave., 37404. Phone: (423) 493-4382. Phone: (423) 493-4383. Fax: (423) 493-4526. E-mail: wdyn@wdyn.com Web Site: www.wdyn.com. Licensee: Tennessee Temple University. Population served: 1,300,000 Format: Relg. News: 2 hrs wkly. Target aud: General; conservative Christians. ♦Tommy L. Sneed, gen mgr & opns dir.

WFLI(AM)—See Lookout Mountain

WGOW(AM)— 1936: 1150 khz; 5 kw-D, 1 kw-N, DA-N. TL: N35 04 05 W85 20 04. Hrs open: Box 11202, 37401. Secondary address: 821 Pineville Rd. 37405. Phone: (423) 756-6141. Fax: (423) 266-3629. Web Site: www.wgow.com. Licensee: Citadel Broadcasting Co. Group owner: Citadel Broadcasting Corp. (acq 5-30-00; grpsl). Population served: 174,000 Natl. Rep: Christal. Law Firm: Reddy, Begley & McCormick. Format: News/talk. ♦Dan Brown, pres, VP & gen mgr; Bill Lockhart, progmg dir.

WSKZ(FM)—Co-owned with WGOW(AM). November 1960: 106.5 mhz; 100 kw. 1,080 ft TL: N35 09 42 W85 19 06. Stereo. Web Site: www.wskz.com. Format: Adult rock. ♦Kelly McCoy, progmg dir.

WGOW-FM—Soddy-Daisy, July 14, 1977: 102.3 mhz; 6 kw. 287 ft TL: N35 11 45 W85 13 45. Stereo. Hrs opn: Box 11202, 37401. Secondary address: 821 Pineville Rd. 37405. Phone: (423) 756-6141. Fax: (423) 266-3629. Web Site: www.wgow.com. Licensee: Citadel Broadcasting Co. Group owner: Citadel Broadcasting Corp. (acq 5-30-00; grpsl). Format: News/talk. Target aud: 18-54; baby boomers. ♦Dan Brown, gen mgr; Kennard Yamada, sls dir, sls dir & gen sls mgr; Bill Lockhart, progmg dir; Kevin West, news dir; Dave Fisher, chief of engrg.

WJOC(AM)— July 4, 1948: 1490 khz; 1 kw-U. TL: N35 03 07 W85 16 24. Hrs open: 805 Chickamauga Ave., Rossville, GA, 30741. Phone: (706) 861-0800. Phone: (888) 963-9562. Fax: (706) 861-2299. Web Site: www.joy1490.com. Licensee: Sara Margarett Fryar. (acq 8-11-97; $230,000). Population served: 500,000 Natl. Network: USA. Format: Southern gospel, Christian mus. ♦Trey Searcy, gen mgr.

WJTT(FM)—See Red Bank

WLMR(AM)— 1961: 1450 khz; 1 kw-U. TL: N35 02 54 W85 16 26. Hrs open: 24 3809 Ringgold Rd., 37412. Phone: (423) 624-4200. Fax: (423) 624-4722. E-mail: wlmr@wilkinsradio.com Web Site: www.wilkinsradio.com. Licensee: Grace Media Inc. Population served: 475,000 Natl. Network: USA. Law Firm: Womble, Carlyle, Sandridge & Rice. Format: Christian teaching/talk. Target aud: 35 plus. ♦Bob Wilkins, pres; LuAnn J. Wilkins, exec VP & VP; Mitchell Mathis, VP & progmg dir; Mike King, stn mgr; Greg Garrett, opns mgr & prom VP; Phil Patton, chief of engrg & engr.

***WMBW(FM)**— Aug 1, 1969: 88.9 mhz; 100 kw. 1,505 ft TL: N34 57 43 W85 22 40. Stereo. Hrs opn: 24
WMKW 89.3FM; WFCM 91.7FM; WFCM-AM710.
Box 73026, 37407. Secondary address: 1920 E. 24th St. PL. 37404. Phone: (423) 629-8900. Fax: (423) 629-0021. Web Site: www.wmbw.org. E-mail: wmbw@moody.edu Licensee: Moody Bible Institute of Chicago. (group owner; acq 5-18-73). Population served: 500,000 Natl. Network: Moody. Law Firm: Southmayd & Miller. Format: Educ, relg. News: 12 hrs wkly. Target aud: 25-54. Spec prog: Black one hr wkly. ♦Dr. Michael Easley, pres; Leighton LeBoeuf, gen mgr & mktg dir; Andy Napier, prom dir, progmg dir, mus dir, spec ev coord & news rptr; David Morais, chief of engrg; Jim Young, news rptr; Paul Martin, mus dir, news rptr & disc jockey.

WMPZ(FM)—Ringgold, GA) Jan 1, 1995: 93.7 mhz; 4.9 kw. 302 ft TL: N34 53 51 W85 10 25. Hrs opn: 24 1305 Carter St., 37402. Phone: (423) 265-9494. Fax: (423) 266-2335. E-mail: jimii@brewerradio.com Web Site: groove93.com. Licensee: J.L. Brewer Broadcasting L.L.C. (acq 11-96). Natl. Network: ABC. Natl. Rep: D & R Radio. Format: Urban adult contemp. Target aud: 25-54; adults. ♦Jim Brewer II, pres, VP & gen mgr; Keith Landecker, opns mgr.

WNOO(AM)— June 1951: 1260 khz; 5 kw-D. TL: N35 03 08 W85 16 22. Hrs open: 24 Box 25399, 37422-5399. Phone: (423) 698-8617. Fax: (423) 698-8796. Licensee: East Tennessee Radio Group III L.P. Group owner: Willis Broadcasting Corp. (acq 12-23-2004; $265,886). Population served: 169,000 Format: Urban contemp. News staff: 4. Target aud: 25-54; mature Black adults & children. ♦Charles Sanders, stn mgr.

WRXR-FM—See Rossville, GA

WSMC-FM—See Collegedale

***WUTC(FM)**— March 1980: 88.1 mhz; 30 kw. 889 ft TL: N35 12 28 W85 16 46. (CP: 30 kw). Stereo. Hrs opn: 24 615 McCallie Ave., 37403. Phone: (423) 425-4756. Fax: (423) 425-2379. Web Site: www.wutc.org. Licensee: Board of Trustees of University of Tennessee. Population served: 460,000 Natl. Network: NPR, PRI. Format: Jazz, blues , AAA. Target aud: General. ♦John McCormack, gen mgr; Ken Dryden, dev dir & sls dir.

Church Hill

WEYE(FM)—Surgoinsville, November 1990: 104.3 mhz; 4.1 kw. Ant 397 ft TL: N36 32 05 W82 47 52. Stereo. Hrs opn: 24 Box 128, 37642. Secondary address: 439 Richmond St. 37642. Phone: (800) 450-1043. Fax: (423) 357-3635. E-mail: dsandz@yahoo.com Web Site: www.eagle1043fm.com. Licensee: ASRadio LLC (acq 6-21-2005; $1.2 million). Population served: 50,000 Natl. Network: USA. Natl. Rep: Rgnl Reps. Law Firm: Bryan Cave. Format: Country. ♦David W. DeFranzo, stn mgr & progmg dir; Daryl Smith, chief of engrg.

WMCH(AM)— May 8, 1954: 1260 khz; 1 kw-D. TL: N36 31 15 W82 44 54. Hrs opn: Box 128, 37642. Phone: (423) 357-5601. Fax: (423) 357-3635. Web Site: www.weye.us. Licensee: Tri-City Radio L.L.C. (acq 10-11-01). Population served: 150,000 Natl. Network: USA. Format: Gospel. Target aud: 25-54; Adult Audience. Spec prog: Farm one hr wkly. ♦Randall Seaver, gen mgr.

Clarksville

***WAPX-FM**— Oct 1, 1984: 91.9 mhz; 6 kw. Ant 194 ft TL: N36 32 13 W87 21 26. Stereo. Hrs opn: 24 Box 4627, Austin Peay State Univ., 37044. Phone: (931) 221-7378. Fax: (931) 221-7265. Web Site: www.apsu.edu/comm_thea/student_activities/wapxfm.htm. Licensee: Austin Peay State University. Population served: 100,000 Format: Consistently Diverse. News: 8 hrs wkly. Target aud: 18-34; college students & young professionals. Spec prog: Black 6 hrs, jazz 6 hrs wkly. ♦Dr. David Michael von Palko, gen mgr.

***WAYQ(FM)**— Oct 22, 2003: 88.3 mhz; 14 kw. Ant 745 ft TL: N36 17 36 W87 18 21. Hrs opn: 24 1012 McEwen Dr., Franklin, 37067. Phone: (615) 261-9293. Fax: (615) 261-3967. Web Site: www.wayfm.com. Licensee: WAY-FM Media Group Inc. (group owner). Format: Christian hit radio. ♦Teresa White, dev dir; Jeff Brown, progmg dir.

WJZM(AM)— Oct 19, 1941: 1400 khz; 1 kw-U. TL: N36 30 57 W87 20 57. Hrs open: 24 Box 648, 37040. Secondary address: 925 Martin St. 37040. Phone: (931) 645-6414. Fax: (931) 551-8432. E-mail: 14jzm@wjzm.com Web Site: www.wjzm.com. Licensee: Cumberland Radio Partners Inc. Population served: 115,000 Format: News/talk, sports. News: 8 hrs wkly. Target aud: 25-60; blue collar, working women, businessmen. Spec prog: Relg. ♦Hank Bonecutter, gen mgr, gen sls mgr & news dir; John Bastin, opns mgr; Ivan Davis, chief of engrg; Angie Brown, disc jockey; Jeff Lyon, disc jockey; Jimmy Baird, disc jockey; Ken Baxter, disc jockey; Sharon Fewless, disc jockey.

WKFN(AM)— Nov 12, 1954: 540 khz; 1 kw-D, 54.5 w-N. TL: N36 32 28 W87 19 33. Hrs open: 24 1640 Old Russelville Pike, 37043. Phone: (931) 648-7720. Fax: (931) 648-7769. Licensee: Saga Communications of Tuckessee L.L.C. Group owner: Saga Communications Inc. (acq 2-1-2001; grpsl). Population served: 158,200 Format: Contemp Christian. News staff: 2; News: 36 hrs wkly. Target aud: 25-54. ♦Susan Quesenberry, VP, gen mgr & prom mgr.

WQZQ(AM)— Jan 24, 1980: 1550 khz; 2.5 kw-D, 250 w-N, DA-N. TL: N36 32 12 W87 22 24. Hrs open: 24 Box 290099, Nashville, 37217. Phone: (931) 645-1550. Fax: (615) 361-9873. E-mail: andreahutchison @cromwell.radio.com Licensee: Winston Communications Inc. Group owner: The Cromwell Group Inc. (acq 10-17-91). Law Firm: Pepper & Corazzini. Format: Black gospel, talk. Target aud: 24-56; talk radio audience. ♦Bayard Walters, pres; Bob Reich, sr VP & gen mgr; Jim Patrick, opns mgr.

Cleveland

WBAC(AM)— June 18, 1945: 1340 khz; 1 kw-U. TL: N35 09 54 W84 51 13. Hrs open: 24 2640 Commerce Drive N.E., 37311. Phone: (423) 242-7656. Fax: (423) 472-5290. Web Site: www.wbacradio.com. Licensee: J.L. Brewer Broadcasting of Cleveland L.L.C. Group owner: Brewer Broadcasting Corp. (acq 6-2-98; $1.5 million with co-located FM). Population served: 88,850 Natl. Network: ABC. Rgnl. Network: Tenn. Radio Net. Natl. Rep: D & R Radio. Format: News/talk. News staff: one; News: 16 hrs wkly. Target aud: 35-64. ♦Jim Brewer Sr., pres; Jim Brewer II, VP; Charles Sells, gen mgr; John Holland, gen sls mgr; Corky Whitlock, progmg dir; Mike Powers, opns mgr & news dir.

WHJK(FM)—Co-owned with WBAC(AM). Feb 27, 1980: 95.3 mhz; 3.5 kw. Ant 436 ft TL: N35 09 54 W84 51 13. Stereo. 24 Web Site: 953jackfm.com.22,500 Natl. Rep: D & R Radio. ♦Mike Powers, progmg dir; Ed Ramsey, mus dir & pub affrs dir.

WCLE(AM)— May 2, 1957: 1570 khz; 5 kw-D, 84 w-N. TL: N35 10 55 W84 50 55. Hrs open: Box 2695, 37320. Phone: (423) 472-6700. Fax: (423) 476-4686. Licensee: Williams Communications Inc. (group owner; acq 8-8-02; $2.4 million with WCLE-FM Calhoun). Population served: 100,000 Law Firm: Greg Skall. Format: Gospel. Target aud: 25-54. Spec prog: Relg 6 hrs wkly. ♦Walter Williams II, CEO.

WSMC-FM—See Collegedale

WUSY(FM)— Aug 1, 1961: 100.7 mhz; 100 kw. 1,191 ft TL: N35 12 26 W85 17 10. Stereo. Hrs opn: 24 7413 Old Lee Hwy., Chattanooga, 37421. Phone: (423) 892-3333. Fax: (423) 899-7224. Fax: (423) 642-9329. Web Site: www.us101country.com. Licensee: Capstar TX L.P. Group owner: Clear Channel Communications Inc. (acq 8-7-00; grpsl). Format: Contemp country. ♦Sammy George, gen mgr; Rhonda

Rollins, gen sls mgr; Chris Van Dyke, progmg dir; Ed Buice, news dir; Andre Johnson, chief of engrg; Vick Grabitt, traf mgr.

Clifton

WLVS-FM— 2002: 106.5 mhz; 3.8 kw. Ant 416 ft TL: N35 28 41 W88 06 36. Hrs open:
Rebroadcasts WXFL(FM) Florence 100%.
624 Sam Philips St., Florence, AL, 35630. Phone: (256) 764-8121. Fax: (256) 764-8169. Web Site: www.wxfl.com. Licensee: Gold Coast Broadcasting Co. (acq 8-3-00; $75,000 for 51% of CP). Format: Country. ♦Nick Martin, gen mgr; Rocky Reich, sls dir; Fletch Brown, progmg mgr.

Clinton

***WDVX(FM)**— November 1997: 89.9 mhz; 200 w. 1,960 ft TL: N36 11 53 W84 13 51. Hrs open: Box 27568, Knoxville, 37927. Phone: (865) 494-2020. Fax: (865) 494-3299. E-mail: mail@wdvx.com Web Site: www.wdvx.com. Licensee: Cumberland Communities Communications Corp. Format: Americana mus. ♦Tony Lawson, gen mgr.

***WYFC(FM)**— July 4, 1966: 95.3 mhz; 540 w. 669 ft TL: N36 04 21 W84 01 18. Stereo. Hrs opn: 24 11530 Carmel Commons Blvd., Charlotte, NC, 28226-3976. Phone: (865) 938-7843. Fax: (865) 938-7843. Web Site: www.bbnradio.com. Licensee: Bible Broadcasting Network Inc. (group owner; acq 8-18-89; $450,000; FTR: 9-5-89). Format: Traditional Christian. ♦Lowell Davey, pres; Grant Bishop, stn mgr.

WYSH(AM)— November 1960: 1380 khz; 1 kw-D, 500 w-N, DA-N. TL: N36 06 48 W84 08 30. Stereo. Hrs opn: 24 Box 329, 37717. Secondary address: 111 Hillcrest Dr. 37716. Phone: (865) 457-1380. Fax: (865) 457-4440. E-mail: wysham@aol.com Web Site: www.wyshradio.com. Licensee: Clinton Broadcasters Inc. (acq 6-10-91; FTR: 11-19-90). Population served: 518,000 Natl. Network: AP Radio. Rgnl. Network: Tenn. Radio Net. Natl. Rep: Keystone (unwired net). Format: Classic country. News staff: one; News: 15 hrs wkly. Target aud: 25-54; families, blue collar to upper income. Spec prog: Relg 15 hrs wkly. ♦Ronald C. Meredith Jr., pres, gen mgr & opns mgr.

Coalmont

WSGM(FM)— June 21, 1994: 104.7 mhz; 1 kw. 548 ft TL: N35 20 22 W85 46 10. Stereo. Hrs opn: 6 AM-10 PM Box 1269, Fire Tower Rd., Tracy City, 37387. Phone: (931) 592-7777. Fax: (931) 592-7778. E-mail: wsgmfm@hotmail.com Licensee: Cumberland Communication Corp. Population served: 25,100 Law Firm: Donald E. Martin. Format: Div, relg, gospel. News: 30 hrs wkly. Target aud: General; interested in community affrs. ♦Dr. Byron Harbolt, pres, exec VP & stn mgr; Sam Harbolt, sr VP; Geniveve Harbolt, VP; Tom Wiseman, chief of engrg; Gina Brady, disc jockey; Jim McKnight, disc jockey; Rhonda Pickett, disc jockey; Rocky Ruehling, disc jockey.

Collegedale

***WSMC-FM**— November 1961: 90.5 mhz; 100 kw. 554 ft TL: N35 01 20 W85 04 32. (CP: 1,029 ft. TL: N35 15 20 W85 13 34). Stereo. Hrs opn: Box 870, 37315. Phone: (423) 236-2905. Fax: (423) 236-1905. Web Site: www.wsmc.org. Licensee: Southern Adventist University. Population served: 3,000 Natl. Network: PRI, NPR. Format: Class, news. Target aud: 25-54. ♦Gordon Bietz, pres; David Brooks, gen mgr; Myrna Ott, opns dir & opns mgr.

Collierville

WCRV(AM)— Oct 1, 1966: 640 khz; 50 kw-D, 500 w-N, DA-N. TL: N34 59 35 W89 53 58. Hrs open: 24 555 Perkins Rd. Ext., Memphis, 38117. Phone: (901) 763-4640. Fax: (901) 763-4920. Licensee: Bott Broadcasting. (group owner) Natl. Network: USA. Format: Christian info. Target aud: 25-54; family oriented. ♦Richard P. Bott, pres; Richard Bott II, VP; Shirley Gossett, opns mgr; Sunny Caldwell, gen mgr & gen sls mgr; Byron Tyler, progmg dir.

Collinwood

WMSR-FM— July 1991: 94.9 mhz; 7.7 kw. Ant 594 ft TL: N35 01 46 W87 47 07. Hrs open: 24 6615 Pumping Station Rd., Cypress Inn, 38452. Phone: (256) 766-9436. Fax: (256) 760-9454. E-mail: thechief@star94.net Web Site: www.star94.net. Licensee: Malkan Broadcasting L.P. Population served: 250,000 Natl. Network: Fox News Radio. Natl. Rep: Katz Radio. Law Firm: Thompson Hines LLP. Format: CHR. News: 8 hrs wkly. Target aud: 18-49; primary women, secondary adults. ♦Glen Powers, CEO; Ann Southern, CFO; Sherrie Powers, gen mgr; Ryan McWhorter, sls VP.

Colonial Heights

WPWT(AM)— Dec 31, 1984: 870 khz; 10 kw-D. TL: N36 27 40 W82 27 12. Hrs open: Sunrise-sunset Box 2061, Bristol, 37621. Secondary address: 340 Edgemont Ave., Suite 100, Bristol 3720. Phone: (423) 878-6279. Fax: (423) 878-6520. E-mail: wpwt@aecc.org Web Site: www.powertalk870.com. Licensee: Information Communications Corp. (acq 6-29-2001). Population served: 2,000,000 Natl. Network: Fox News Radio, Salem Radio Network, Talk Radio Network, Premiere Radio Networks. Format: talk radio. News staff: one; News: 5 hrs wkly. Target aud: Adults 25-54. Spec prog: Health Education one hr wkly. ♦Kenneth C. Hill, gen mgr; Rusty Cury, sls VP; Jerome Jackson III, prom dir; Mathew Hill, progmg dir; Art Countiss, news dir.

WRZK(FM)— Apr 4, 1997: 95.9 mhz; 7.4 kw. Ant 1,253 ft TL: N36 31 36 W82 35 13. Stereo. Hrs opn: 222 Commerce St., Kingsport, 37660. Phone: (423) 246-9578. Fax: (423) 247-9836. E-mail: dmurray@wrzk.com Web Site: www.wrzk.com. Licensee: Murray Communications. Population served: 396,400 Natl. Network: ABC. Natl. Rep: McGavren Guild. Law Firm: Pepper & Corazzini. Format: Modern rock. News staff: 2. Target aud: 18-44; general. ♦David Widener, exec VP & gen mgr; Scott Onksi, progmg dir.

Columbia

***WAYM(FM)**— 1992: 88.7 mhz; 16.5 kw. 508 ft TL: N35 49 27 W86 49 28. Stereo. Hrs opn: 24 1012 McEwen Dr., Franklin, 37067. Phone: (615) 261-9293. Fax: (615) 261-3967. E-mail: wayfm@wayfm.com Web Site: www.wayfm.com. Licensee: WAY-FM Media Group Inc. (group owner; acq 3-13-91; FTR: 4-1-91). Population served: 1,209,000 Format: Christian hit radio. Target aud: 12-34; females. ♦Teresa White, dev dir; Jeff Brown, progmg.

WKOM(FM)—Listing follows WKRM(AM).

WKRM(AM)— Nov 25, 1946: 1340 khz; 1 kw-U. TL: N35 36 38 W87 03 22. Hrs open: 24 Box 1377, 38402. Secondary address: 315 W. 7th St. 38401. Phone: (931) 388-3636. Fax: (931) 381-1017. Licensee: Robert M. McKay III. (acq 12-26-89). Population served: 60,000 Natl. Network: ABC. Format: Adult contemp. News staff: one; News: 8 hrs wkly. Target aud: 25-54. Spec prog: Relg 4 hrs wkly. ♦Robert M. McKay III, pres & gen mgr.

WKOM(FM)—Co-owned with WKRM(AM). Jan 1, 1967: 101.7 mhz; 4.1 kw. 400 ft TL: N35 37 04 W87 02 34. Stereo. 24 (Acq 9-1-72).100,000 Natl. Network: CBS, Motor Racing Net. Format: Pure Gold. News staff: one; News: 8 hrs wkly. Target aud: 30-50. ♦Robert McKay III, CEO.

WMCP(AM)— Nov 12, 1956: 1280 khz; 5 kw-D, 500 w-N, DA-N. TL: N35 37 08 W86 58 52. Hrs open: 24 Box 711, 38402. Secondary address: 1st Farmer & Merchants Bank Bldg., 816 S. Garden, Suite 306 38401. Phone: (931) 388-3241. Fax: (931) 381-2510. Licensee: Maury County Boosters Corp. (acq 12-79). Population served: 261,998 Rgnl. Network: Tenn. Radio Net. Format: Country. News staff: one; News: 13 hrs wkly. Target aud: 18 plus. Spec prog: Farm 4 hrs weekly. ♦Edna Williford, pres; Mack Shaw, VP & gen mgr.

WMRB(AM)— Aug 14, 1982: 910 khz; 500 w-D, 88 w-N. TL: N35 36 24 W87 01 30. Hrs open: 1014 S. Garden St., 38401. Phone: (931) 381-7100. Fax: (931) 381-0088. Web Site: www.wmrb910am.som. Licensee: Ogilvie Family Ministries Inc. (acq 7-3-97; $50,000). Natl. Rep: Dora-Clayton. Format: Gospel, Black, relg. Target aud: General; Christian families. ♦Trent Ogilvie, gen mgr & opns mgr.

Cookeville

WGIC(FM)—Listing follows WHUB(AM).

WGSQ(FM)—Listing follows WPTN(AM).

***WHRS(FM)**— Oct 1, 1996: 91.7 mhz; 500 w. 384 ft TL: N36 08 34 W85 28 02. Hrs open:
Rebroadcasts WPLN(FM) 90.3, Nashville; 100%.
630 Mainstream Dr., Nashville, 37228-1204. Phone: (615) 760-2903. Fax: (615) 760-2904. E-mail: talkback@wpln.org Web Site: www.wpln.org. Licensee: Nashville Public Radio. Format: Class, news. Spec prog: Bluegrass one hr, song writers one hr wkly. ♦William Ivey, chmn; Robert Gordon, gen mgr.

WHUB(AM)— July 20, 1940: 1400 khz; 1 kw-U. TL: N36 10 25 W85 30 40. Hrs open: 24 698 S. Willow Ave., 38501. Phone: (931) 526-7144. Fax: (931) 528-8400. Licensee: CC Licenses LLC. Group owner: Clear Channel Communications Inc. (acq 11-21-97; grpsl). Population served: 50,000 Natl. Network: CBS. Rgnl. Network: Tenn. Radio Net. Format: Classic country, Southern gospel. News staff: one; News: 18 hrs wkly. Target aud: General. Spec prog: Sports 10 hrs, gospel 11 hrs wkly. ♦Dave Thomas, gen mgr; Marty McFly, opns dir; Jim Stapleton, sls dir & natl sls mgr; Lehra Heidel, mktg dir; Lehar Heidel, prom dir; Jim Herrin, news dir; Mike Dinger, mus dir & chief of engrg; Jennifer Henson, traf mgr.

WGIC(FM)—Co-owned with WHUB(AM). Mar 26, 1964: 98.5 mhz; 50 kw. 492 ft TL: N36 08 34 W85 28 02. Stereo. 20 E-mail: email@magic985.com Web Site: www.magic985.com. Natl. Network: ABC. Natl. Rep: Clear Channel. Format: Hot adult Contemp. News staff: one; News: 8 hrs wkly. ♦Marty McFly, progmg dir; Helen Daniels, traf mgr.

WPTN(AM)— July 10, 1962: 780 khz; 1 kw-D. TL: N36 09 30 W85 31 15. Hrs open: 698 S. Willow, 38501. Phone: (931) 526-7144. Fax: (931) 528-8400. Licensee: CC Licenses LLC. Group owner: Clear Channel Communications Inc. (acq 11-21-97; grpsl). Population served: 80,000 Natl. Rep: Clear Channel. Law Firm: Dow, Lohnes & Albertson. Format: News/talk. Target aud: General. ♦David Roederer, gen mgr; Bruce Welker, sls VP; Lehra Mayfield, prom VP; Marty Selby, progmg VP; Jim Herrin, news dir & disc jockey; Dave Johnson, pub affrs dir & disc jockey; Michelle Freeman, disc jockey.

WGSQ(FM)—Co-owned with WPTN(AM). Mar 8, 1963: 94.7 mhz; 100 kw. 1,319 ft TL: N36 10 26 W85 20 37. Stereo. 24 1,500,000 Format: Country. News staff: 2; News: 28 hrs wkly.

***WTTU(FM)**— May 22, 1972: 88.5 mhz; 2.25 kw. 168 ft TL: N36 10 26 W85 30 12. (CP: 2 kw horiz, ant 164 ft.). Stereo. Hrs opn: 7 AM-1 AM Box 5113, University Ctr., Dixie Ave., 38505. Phone: (931) 372-3688. Fax: (931) 372-6225. E-mail: wttu_88.5@hotmail.com Licensee: Tennessee Technological University. Population served: 50,000 Format: Alternative. News staff: one; News: 2 hrs wkly. Target aud: 14-25. Spec prog: Jazz 3 hrs, metal 3 hrs, folk 3 hrs, hip-hop/rap 3 hrs, blues 3 hrs wkly.

***WWOG(FM)**— 1994: 90.9 mhz; 40 kw. Ant 697 ft TL: N36 11 05 W85 22 30. Hrs open: 24
Rebroadcasts WSGP(FM) Glasgow, KY and WTHL(FM) Somerset, KY 100%.
Box 1423, Somerset, KY, 42502. Secondary address: 93 Rainbow Terr., Somerset 42502. Phone: (606) 679-6300. Fax: (606) 679-1342. Web Site: www.kingofkingsradio.net. Licensee: Somerset Educational Broadcasting Foundation. Format: Conservative, traditional relg, educ. ♦David Carr, gen mgr; Carolyn Jones, progmg dir; Marvin Whittaker, chief of engrg.

Copperhill

WLSB(AM)— Dec 2, 1958: 1400 khz; 1 kw-U. TL: N34 58 04 W84 19 39. Hrs open: 6 AM-10 PM
Simulcast with WYHG(AM) Young Harris, GA.
Box 430, 37317. Phone: (423) 496-3311. Fax: (423) 496-2635. E-mail: wlsb@bellsouth.net Web Site: www.wolfcreekbroadcasting.com. Licensee: Copper Basin Broadcasting Co., Inc. (acq 9-26-2002). Format: Country, bluegrass. ♦Rebecca St. John, stn mgr.

Covington

WKBL(AM)— Aug 16, 1954: 1250 khz; 800 w-D, 106 w-N. TL: N35 35 10 W89 38 35. Hrs open: 24 101 WKBL Dr., 38019. Phone: (901) 476-7129. Fax: (901) 476-7120. E-mail: billy.thomas@us51country.com Web Site: www.us51country.com. Licensee: Covington Broadcasting Inc. (acq 2-22-99; $600,000 with co-located FM). Population served: 100,000 Natl. Network: Jones Radio Networks. Rgnl. Network: Tenn. Radio Net. Format: Classic country. News staff: one; News: 14 hrs wkly. Target aud: 21-55. Spec prog: Black 12 hrs wkly. ♦Bob Lakey, CEO & pres; Jimmy HIcks, gen mgr.

WKBQ(FM)—Co-owned with WKBL(AM). Aug 31, 1965: 93.5 mhz; 6 kw. Ant 328 ft TL: N35 35 12 W89 38 21. Stereo. 24 Web Site: www.us51country.com. Natl. Network: AP Radio, Jones Radio Networks. News staff: one; News: 14 hrs wkly.

Cowan

WZYX(AM)— Mar 10, 1957: 1440 khz; 5 kw-D, 100 w-N. TL: N35 09 39 W86 01 51. Stereo. Hrs opn: 24 540 W. Cumberland St., 37318-0398. Phone: (931) 967-7471. Phone: (931) 967-7472. Fax: (931) 962-1440. Web Site: www.wzyxradio.com. Licensee: Tims Ford Broadcasting Co. Inc. (acq 4-26-2004). Population served: 50,000 Natl. Network: CNN Radio. Wire Svc: NOAA Weather Format: Country, oldies, talk. News staff: one; News: 15 hrs wkly. Target aud: 35-55; middle-of-the-road working people. Spec prog: Talk, gospel 10 hrs, farm 2 hrs, relg 12 hrs wkly. ♦Jeff Pennington, VP; Mary Lou Garner, CEO, pres & stn mgr.

Crossville

WAEW(AM)— 1952: 1330 khz; 1 kw-D. TL: N35 56 59 W85 02 08. (CP: TL: N35 57 01 W85 02 09). Hrs open: 24 961 Miller Ave., 38555. Phone: (931) 707-1102. Fax: (931) 707-1220. Web Site: www.1330waew.com. Licensee: Peg Broadcasting Crossville LLC (group owner; acq 10-1-2003; grpsl). Population served: 35,000 Format: Talk. News staff: 2; News: 13 hrs wkly. Target aud: 35-64; adult. ♦Jeff Shaw, gen mgr.

WCSV(AM)— June 15, 1968: 1490 khz; 1 kw-U. TL: N35 56 46 W85 02 13. (CP: TL: N35 57 01 W85 02 09). Hrs opn: 24 961 Miller Ave., 38555. Phone: (931) 707-1102. Fax: (931) 707-1220. Web Site: www.1490wcsv.com. Licensee: Peg Broadcasting Crossville LLC (group owner; acq 10-1-2003; grpsl). Population served: 35,000 Natl. Network: ABC. Format: Sports. News staff: 2; News: 13 hrs wkly. Target aud: 25-54; men. ♦Jeff Shaw, gen mgr.

***WMKW(FM)**— November 1996: 89.3 mhz; 500 w. 1,395 ft TL: N35 46 38 W84 58 34. Hrs open: 24
Rebroadcasts WMBW(FM) Chattanooga 100%.
1920 E. 24th Street Pl., Chattanooga, 37404. Secondary address: Box 73026, Chattanooga 37407. Phone: 423-629-8900. Fax: (423) 629-0021. Web Site: www.moody.edu. Licensee: The Moody Bible Institute of Chicago. Population served: 50,000 Law Firm: Southmayd & Miller. Format: Educ, relg, news/talk. Target aud: General. ♦Dr. Michael Easley, pres; Leighton LeBoeuf, gen mgr & mktg dir; Andy Napier, prom dir & progmg dir; Paul Martin, mus dir; David Morais, chief of engrg; Jim Young, local news ed.

WOWF(FM)— June 15, 1990: 102.5 mhz; 25 kw. 308 ft TL: N36 01 22 W85 00 07. Stereo. Hrs opn: 24 961 Miller Ave., 38555. Phone: (931) 707-1102. Fax: (931) 707-1220. Web Site: www.1025wowcountry.com. Licensee: Peg Broadcasting Crossville LLC (group owner; acq 1-3-01; $2.5 million). Population served: 138,426 Natl. Network: Jones Radio Networks, Fox News Radio. Rgnl rep: Rgnl Reps. Wire Svc: AP Format: Country. News staff: 2; News: 10 hrs wkly. Target aud: 25 -54; adults. ♦Jeffrey H. Shaw, gen mgr; Steve J. Sweeney, gen sls mgr; Jeffrey Shaw, mktg mgr & pub affrs dir; Gordon Stack, progmg dir & progmg mgr; Christy Lewis, news dir; Kendra Williams, traf mgr.

WPBX(FM)— May 12, 1967: 99.3 mhz; 6 kw. 259 ft TL: N35 57 01 W85 02 09. Stereo. Hrs opn: 24 961 Miller Ave., 38555. Phone: (931) 484-5115. Fax: (931) 707-1220. Web Site: mix993.net. Licensee: Peg Broadcasting Crossville LLC (group owner; acq 10-1-2003: grpsl). Population served: 40,000 Natl. Network: ABC. Natl. Rep: Clear Channel. Format: Adult contemp. News staff: 2; News: 10 hrs wkly. Target aud: 25-54; women. ♦Jeffrey H. Shaw, gen mgr; Steve J. Sweeney, gen sls mgr; Jeffrey Shaw, mktg mgr; Gordon Stack, progmg dir & chief of engrg; Christy Lewis, news dir; Kendra Williams, traf mgr.

Dayton

WALV-FM—Listing follows WDNT(AM).

WDNT(AM)— Dec 6, 1957: 1280 khz; 1 kw-D, 345 w-N. TL: N35 28 12 W85 02 15. Hrs open:
Rebroadcasts WBAC(AM) Cleveland 100%.
2640 Commerce Dr., N.E., Cleveland, 37311. Phone: (423) 242-7656. Fax: (423) 472-5290. Web Site: www.wbacradio.com. Licensee: J.L. Brewer Broadcasting of Cleveland LLC. Group owner: Brewer Broadcasting Corp. (acq 7-1-2002; grpsl). Population served: 28,604 Natl. Network: ABC. Natl. Rep: D & R Radio. Format: MOR. News staff: one; News: 26 hrs wkly. Target aud: 45 plus. ♦James L. Brewer, pres; James L. Brewer II, VP; Charles Sells, gen mgr; Mike Powers, opns mgr; John Holland, gen sls mgr; Corky Whitlock, progmg dir.

WALV-FM—Co-owned with WDNT(AM). July 1, 1976: 104.9 mhz; 1.3 kw. Ant 712 ft TL: N35 29 31 W85 02 59. Stereo. 24 Web Site: alive105.com.165,400 Natl. Rep: D & R Radio. Format: Hot adult contemp. News: one hr wkly. ♦Mike Lee, progmg dir.

Dickson

WDKN(AM)— Jan 1, 1955: 1260 khz; 5 kw-D. TL: N36 06 31 W87 22 14. Hrs open: 6 AM-6:30 PM 106 E. College St., 37055. Phone: (615) 446-4000. Phone: (615) 446-0752. Fax: (615) 446-9681. Web Site: www.wdkn.com. Licensee: Edmission & Eubank Communications Inc. (acq 6-15-87; $220,000; FTR: 5-18-87). Population served: 100,000 Law Firm: McCampbell & Young, P. Format: C&W, news/talk. News: 3 hrs wkly. Target aud: General. Spec prog: Relg 12 hrs wkly. ♦Tommy Edmisson, pres; Oscar Eubank, VP; Leroy Kennell, gen mgr.

***WNRZ(FM)**— Apr 7, 1997: 91.5 mhz; 8 kw. Ant 262 ft TL: N36 00 36 W87 30 47. Stereo. Hrs opn: 24
Rebroadcasts WNAZ-FM Nashville 100%.
333 Murfreesboro Rd., Nashville, 37210. Phone: (615) 248-1689. Fax: (615) 248-7786. Licensee: Trevecca Nazarene University Inc. Population served: 80,000 Format: Christian, progsv contemp. Target aud: 14-28; Christians. ♦Dr. Dan Boone, pres; Mark Myers, CFO; David Deese, gen mgr; Paul Eby, stn mgr; Dave Queen, opns mgr.

Donelson

WCRT(AM)—Licensed to Donelson. See Nashville

Dresden

WCDZ(FM)— Apr 10, 1992: 95.1 mhz; 21.5 kw. Ant 276 ft TL: N36 15 50 W88 40 03. Hrs open: Box 318, Martin, 38232. Phone: (731) 885-0051. Fax: (731) 885-0250. Licensee: Thunderbolt Broadcasting Co. Group owner: Thunderbolt Broadcasting Co./Gibson County Broadcasting (acq 1-28-94; $320,000; FTR: 2-21-94). Population served: 158,033 Law Firm: Womble, Carlyle, Sandridge & Rice. Format: Oldies. News staff: one; News: one hr wkly. Spec prog: Gospel one hs wkly. ♦Paul Tinkle, pres.

Dunlap

WSDQ(AM)— Nov 1, 1980: 1190 khz; 5 kw-D. TL: N35 21 41 W85 22 33. Hrs open: 16 Main St. N., 37327-6129. Phone: (423) 949-4114. Fax: (423) 949-5143. E-mail: wsdq@bledsoe.net Licensee: Rodgson Inc. (acq 7-18-02; $165,000). Rgnl. Network: Tenn. Radio Net. Format: Country, bluegrass. Spec prog: Gospel 7 hrs wkly. ♦Charles Rodgers, pres; Earl Nunley, gen mgr.

Dyer

WLLI-FM— Feb 1, 1995: 94.3 mhz; 6 kw. Ant 328 ft TL: N36 06 12 W89 07 45. Hrs open: 24 122 Radio Rd., Jackson, 38301. Phone: (731) 427-3316. Fax: (731) 427-9338. Licensee: Forever South Licenses LLC. (acq 7-31-2006; grpsl). Population served: 150,000 Format: Country. ♦Verla Price, gen mgr.

Dyersburg

WASL(FM)—Listing follows WTRO(AM).

WTRO(AM)— July 13, 1946: 1450 khz; 1 kw-U. TL: N36 03 02 W89 22 07. Hrs open: 24 Box 100, 38025. Secondary address: One Radio Rd. 38024. Phone: (731) 285-1450. Phone: (731) 285-1339. Fax: (731) 287-0100. Web Site: wtro.wasl.net. Licensee: Dr. Pepper/Pepsi Cola Bottling Co. of Dyersburg Inc. (acq 1991). Population served: 100,000 Rgnl. Network: Yancey Action. Law Firm: Bryan Cave. Format: Oldies. ♦Charles W. Maxey, exec VP, gen mgr & gen sls mgr; Steve James, progmg dir.

WASL(FM)—Co-owned with WTRO(AM). July 1, 1968: 100.1 mhz; 26 kw. Ant 676 ft TL: N36 06 00 W89 29 12. Stereo. 24 Web Site: wtro.wasl.net.300,000 Natl. Network: ABC. Format: Hot adult contemp. News: 5 hrs wkly. Target aud: 18-54.

***WZKV(FM)**— Oct 30, 1992: 90.7 mhz; 100 kw. Ant 373 ft TL: N36 06 00 W89 29 12. Stereo. Hrs opn: 24 5700 West Oaks Blvd., Rocklin, CA, 95765. Phone: (916) 251-1600. Fax: (916) 251-1650. Licensee: Educational Media Foundation. (acq 3-29-2007; $825,000). Population served: 310,000 Law Firm: Davis Wright Tremaine LLP. Format: Christian. ♦Richard Jenkins, pres.

East Ridge

WOGT(FM)— Nov 9, 1990: 107.9 mhz; 25 kw. Ant 328 ft TL: N35 07 33 W85 17 25. Hrs open: 24 Box 11202, Chattanooga, 37401. Phone: (423) 756-6141. Fax: (423) 756-0292. Web Site: www.1079dukefm.com. Licensee: Citadel Broadcasting Co. Group owner: Citadel Broadcasting Corp. (acq 5-30-00; grpsl). Format: Country. ♦Dan Brown, gen mgr.

Elizabethton

WBEJ(AM)— July 1946: 1240 khz; 1 kw-U. TL: N36 20 07 W82 13 03. Hrs open: 24 510 Broad St., 37643-2718. Phone: (423) 542-2184. Fax: (423) 542-3912. E-mail: wbej@planetc.com Web Site: www.wbej.com. Licensee: CB Radio Inc. (acq 9-24-82; $335,000; FTR: 10-18-82). Population served: 500,000 Natl. Network: Westwood One. Format: Country. News staff: one; News: 8 hrs wkly. Target aud: 25-49. ♦Don Crisp, pres; Cleo Reed, VP & gen mgr; David A. Miller, opns dir, dev dir & gen sls mgr.

WTZR(FM)— May 17, 1968: 99.3 mhz; 3.6 kw. Ant 810 ft TL: N36 24 07 W82 12 12. Stereo. Hrs opn: Box 1389, Bristol, VA, 24203. Secondary address: 901 E. Valley Dr., Bristol, VA 24201. Phone: (276) 669-8112. Fax: (276) 669-0541. Web Site: www.mix993.com. Licensee: Bristol Broadcasting Co. Group owner: Bristol Broadcasting Co. Inc. (acq 2-13-97; $3 million). Population served: 390,000 Natl. Rep: Christal. Format: Alternative rock. Target aud: 18-49. ♦W.L. Nininger, pres & gen mgr; Bruce Clark, stn mgr, prom dir & progmg dir; Winnie Quaintance, gen sls mgr; Chuck Lawson, chief of engrg; Anna Honaker, traf mgr.

***WUMC(FM)**— 1999: 90.5 mhz; 500 w. -285 ft TL: N36 17 58 W82 17 28. Hrs open: Box 9, Milligan College, 37682. Phone: (423) 461-8464. Phone: (423) 461-8700. Licensee: Milligan College. Format: Contemp hits, contemp Christian music. ♦Carrie Swanay, gen mgr.

Englewood

WENR(AM)— Apr 21, 1967: 1090 khz; 1 kw-D. TL: N35 25 35 W84 30 57. Hrs open: Box 676, Etowah, 37331-0676. Phone: (423) 263-5555. Fax: (423) 263-2555. Licensee: Paul Wilson dba 1090 Radio, a Tennessee sole proprietorship (acq 10-2-98; $75,000). Rgnl. Network: Tenn. Radio Net. Format: Gospel. ♦Carolyne Wilson, gen mgr.

Erwin

WEMB(AM)— May 17, 1956: 1420 khz; 5 kw-D. TL: N36 06 58 W82 26 49. Hrs open: Box 280, 101 Riverview Rd., 37650. Phone: (423) 743-6123. Phone: (423) 743-6124. Fax: (423) 743-6122. Licensee: WEMB Inc. (acq 4-1-61). Population served: 50,000 Rgnl. Network: Tenn. Radio Net. Format: Country, gospel, sports. Spec prog: Bluegrass 2 hrs, gospel 10 hrs wkly. ♦Jim Crawford, pres & gen mgr; Charles W. Ray, opns mgr, progmg dir & chief of engrg; Kathy Thornberry, news dir; Fred Lance, edit dir.

WXIS(FM)—Co-owned with WEMB(AM). Nov 21, 1968: 103.9 mhz; 2.5 kw. 2,600 ft TL: N36 08 15 W82 23 00. Stereo. Phone: (423) 743-7655.225,000 Natl. Network: ABC. Format: Rhythmic. Target aud: 12-44. ♦Todd Ambrose, progmg dir; Fred Lance, edit dir.

Etowah

WCPH(AM)— 1955: 1220 khz; 1 kw-D, 109 w-N. TL: N35 19 15 W84 30 34. Hrs open: Box 676, 37331. Phone: (423) 263-5555. Fax: (423) 263-2555. Licensee: Starr Mountain Broadcasting Co. (acq 3-18-97; $39,000). Population served: 3,736 Format: News/talk, sports, easy listening mus. ♦Carolyne Wilson, gen mgr.

WLLJ(FM)— 1977: 103.1 mhz; 50 kw. 492 ft TL: N35 27 24 W84 40 43. Stereo. Hrs opn: 24
Rebroadcasts WBDX(FM) Trenton, GA 100%.

Box 9396, Chattanooga, 37412. Secondary address: Box 212, McDonald 37353. Phone: (423) 892-1200. Fax: (423) 892-1633. E-mail: debbie@j103.com Web Site: www.j103.com. Licensee: Friendship Broadcasting LLC. (acq 2-13-98). Population served: 1,000,000 Natl. Network: Salem Radio Network. Format: Adult contemp, Christian. News staff: one; News: 2 hrs wkly. Target aud: 18-49; female. ♦Bob Lubell, CEO & pres; Dave Skinner, CFO; Debbie Lubell, gen mgr & mktg dir; Elizabeth Gearu, chief of engrg & traf mgr; Dawn Manor, local news ed.

Fairview

WPFD(AM)— May 28, 1982: 850 khz; 500 w-D. TL: N36 00 29 W87 08 38. Hrs open: 6 AM-sunset 1074 Hwy. 96 N., 37062. Phone: (615) 799-8585. Fax: (615) 799-2999. Licensee: Robert Lee Martin, trustee. Format: Country. Target aud: 18-54. ♦Sam Warden, pres; Chuck Hussey, gen mgr, mus dir & news dir; John Almon, chief of engrg.

Farragut

WMTY(AM)— Nov 10, 1988: 670 khz; 500 w-D. TL: N35 53 12 W84 14 48. Stereo. Hrs opn: 517 Watts Rd., Knoxville, 37922. Phone: (865) 671-7419. Fax: (865) 675-4859. Licensee: Horne Radio L.L.C. Group owner: Horne Radio Group (acq 1999; $275,000). Natl. Network: USA. Target aud: 25-54; upscale adults. ♦Doug Horne, pres.

Fayetteville

WEKR(AM)— Oct 1, 1948: 1240 khz; 1 kw-U. TL: N35 09 28 W86 35 25. Hrs open: 4:30 AM-10 PM Box 656, 37334. Secondary address: 7 Boonshill Rd. 37334. Phone: (931) 433-3545. Fax: (931) 438-0620. Licensee: Joseph D. Young, Wanda Young & Mary Elizabeth Miller. (acq 2-22-94; $194,000; FTR: 3-21-94). Population served: 30,000 Natl. Network: Motor Racing Net. Rgnl. Network: Tenn. Radio Net. Format: Country, Southern gospel, sports, ESPN radio. News staff: one; News: 5 hrs wkly. Target aud: 25 plus; general. Spec prog: Farm one hr wkly. ♦Joseph D. Young, CEO; Al Lawrence, stn mgr, chief of opns, gen sls mgr, progmg dir, pub affrs dir, spec ev coord & disc jockey; Marie Caldwell, prom dir, adv mgr & spec ev coord; Wayne Thomas, news rptr; Jack Atchley, sports cmtr; Charles Hicks, disc jockey; Jennifer Denise, disc jockey.

WYTM-FM— Mar 27, 1970: 105.5 mhz; 3 kw. 295 ft TL: N35 07 37 W86 34 47. (CP: 2.25 kw, ant 495 ft.). Stereo. Hrs opn: 5 AM-10 PM Box 717, 37334. Phone: (931) 433-1531. Fax: (931) 433-4110. Licensee: Time Broadcasters Inc. Population served: 185,000 Natl. Network: ABC. Format: Country. ♦Joseph D. Young, pres & gen mgr; Debbie Kawiecki, opns mgr.

Franklin

WAKM(AM)— Mar 18, 1953: 950 khz; 5 kw-D, 80 w-N. TL: N35 57 25 W86 50 03. (CP: 2.5 kw-D). Hrs opn: 24 222 Mallory Station Rd., 37065. Phone: (615) 794-1594. Fax: (615) 794-1595. E-mail: wakm950@comcast.net Licensee: Franklin Radio Associates Inc. (acq 10-1-82; $310,600; FTR: 10-18-82). Population served: 128,000 Natl. Network: CNN Radio. Rgnl. Network: Tenn. Radio Net. Format: Country, news/talk. News staff: 2; News: 14 hrs wkly. Target aud: 24 plus; community interested adults. Spec prog: Relg 6 hrs, NASCAR racing 6 hrs wkly. ♦Jim Hayes, pres & chief of engrg; Linda Jackson Carden, sls dir; Tom Lawrence, VP, gen mgr & gen sls mgr; Darrell Williams, progmg dir; Charles Dibrell, news dir.

WHEW(AM)— Feb 1, 1969: 1380 khz; 2.8 kw-D, 500 w-N, DA-N. TL: N35 54 22 W86 54 21. Hrs open: 18 1811 Carters Creek Pike, 37064. Phone: (615) 595-0595. Fax: (615) 591-4007. E-mail: laley1330@laley1380.net Web Site: www.laley1380.net. Licensee: SG Communications Inc. (acq 7-29-99; $208,398). Population served: 300,000 Natl. Network: CNN Radio. Format: Spanish, sports, news/talk. News staff: one. Target aud: General; Hispanics. ♦Victor Guzman, CEO; Salvador Guzman, pres & gen mgr; Claudio Vazquez, sls dir; Enrique Garcia, progmg dir.

WRLT(FM)— Nov 16, 1961: 100.1 mhz; 3 kw. 1,134 ft TL: N36 02 06 W86 50 54. Stereo. Hrs opn: 24 1310 Clinton St., Suite 200, Nashville, 37203. Phone: (615) 242-5600. Fax: (615) 523-2153. E-mail: comments@wrlt.com Web Site: www.wrlt.com. Licensee: Tuned In Broadcasting Inc. (acq 1996). Natl. Network: Westwood One. Natl. Rep: Roslin. Law Firm: Davis, Wright, Tremaine. Format: AAA. Target aud: 25-54. Spec prog: Jazz 2 hrs, blues 2 hrs, loc music 2 hrs, indie rock one hr wkly. ♦Lester Turner Jr., CEO, chmn & pres; Fred Buc, gen mgr & gen sls mgr; Jayson Chalfant, prom dir; David Hall, progmg dir; Tom Hansen, chief of engrg.

Gallatin

WGFX(FM)— Dec 1, 1960: 104.5 mhz; 49 kw. 1,312 ft TL: N36 16 05 W86 47 16. Stereo. Hrs opn: 24 506 2nd Ave. S., Nashville, 37210. Phone: (615) 244-9533. Fax: (615) 259-1271. Web Site: www.104thezone.com. Licensee: Citadel Broadcasting Co. Group owner: Citadel Broadcasting Corp. (acq 4-26-01; grpsl). Population served: 1,000,000 Natl. Rep: Katz Radio. Law Firm: Kaye, Scholer, Fierman, Hays & Handler. Format: Sports/talk. Target aud: 18-49. ♦Ken Bailey, gen mgr.

WHIN(AM)— Aug 2, 1948: 1010 khz; 5 kw-D. TL: N36 26 00 W86 28 00. Stereo. Hrs opn: 24 Box 1685, 37066. Phone: (615) 451-0450. Phone: (615) 451-0451. Fax: (615) 452-9446. E-mail: whinam@comcast.net Licensee: WHIN Inc. (acq 10-84). Population served: 100,000 Format: Country. News staff: one; News: 14 hrs wkly. Target aud: 25-54; upper middle to lower middle income. Spec prog: Black 2 hrs, farm 5 hrs wkly. ♦Jack Williams, pres & gen mgr.

WMRO(AM)— Feb 19, 1994: 1560 khz; 1 kw-D, 3 w-N. TL: N36 24 03 W86 27 03. Hrs open: Daytime Box 1445, 37066. Phone: (615) 451-2131. Fax: (615) 206-4207. E-mail: wmroam@bellsouth.net Licensee: Classic Broadcasting Inc. (acq 10-28-93; $40,000). Population served: 103,000 Natl. Network: CNN Radio. Rgnl rep: . Law Firm: Miller & Neeley. Format: Adult Contemp. News staff: 2; News: 5 hrs wkly. Target aud: 25-65; middle to upper class adults. Spec prog: Religion 20 hrs. wkly. ♦Scott Bailey, gen mgr.

***WVCP(FM)**— Jan 4, 1979: 88.5 mhz; 1 kw. 390 ft TL: N36 28 02 W86 28 35. Stereo. Hrs opn: 24 1480 Nashville Pike, Suite 101, Ramer Bldg., 37066. Phone: (615) 230-3618. Fax: (615) 230-4803. E-mail: holly.nimmo@volstate.edu Web Site: www.volstate.edu. Licensee: Volunteer State Community College. Wire Svc: AP Format: Black, CHR, adult contemp, oldies. News: 5 hrs wkly. Target aud: General. Spec prog: Black 8 hrs, metal 15 hrs, gospel 6 hrs, bluegrass 2 hrs wkly. ♦Dr. Warren R. Nichols, pres; Howard Espravnik, gen mgr; Holly Nimmo, opns dir.

WYXE(AM)— Nov 1, 1966: Stn currently dark. 1130 khz; 2.3 kw-D. TL: N36 24 38 W86 27 16. Hrs open: 1079 E. Trinity Ln., Nashvillle, 37216. Phone: (615) 227-1130. Licensee: Jon Gary Enterprises Inc. (acq 3-31-2003). Format: Sp relg. ♦Richard D. Deck Jr., gen mgr.

Gatlinburg

WSEV-FM— January 1983: 105.5 mhz; 650 w. 964 ft TL: N35 42 13 W83 33 57. Stereo. Hrs opn: 24 415 Middle Creek Rd., Sevierville, 37862. Phone: (865) 525-1060. Fax: (865) 429-2601. E-mail: etrg@seviernct.com Licensee: East Tennessee Radio Group L.P. (acq 3-22-2000; $1.45 million with WSEV(AM) Sevierville). Population served: 65000 Natl. Network: CBS. Rgnl. Network: Tenn. Radio Net. Natl. Rep: Rgnl Reps. Format: Adult Contemporary. News staff: one. Target aud: 25-49; loc adults, tourists, C&W lovers. ♦Bill Burkett, stn mgr; Steve Hartford, progmg dir & news dir.

Germantown

WHBQ-FM— June 1994: 107.5 mhz; 3.9 kw. Ant 407 ft TL: N35 10 30 W89 44 26. Stereo. Hrs opn: 24 6080 Mt. Mariah Rd., Memphis, 38115. Phone: (901) 375-9324. Fax: (901) 375-0041. Web Site: www.q1075.com. Licensee: Flinn Broadcasting Corp. (acq 1997; $4). Format: Top-40. ♦Donald Biggs, gen mgr.

WOWW(AM)—Licensed to Germantown. See Memphis

WPLX(AM)—Licensed to Germantown. See Memphis

WSNA(FM)—Licensed to Germantown. See Memphis

Goodlettsville

WRQQ(FM)— Dec 3, 1999: 97.1 mhz; 43 kw. 518 ft TL: N36 17 50 W86 45 11. Hrs open: 24 10 Music Cir. E., Nashville, 37203. Phone: (615) 321-1067. Fax: (615) 321-5771. E-mail: danielle.haese@cumulus.com Web Site: www.cumulus.com. Licensee: Cumulus Licensing Corp. Group owner: Cumulus Media Inc. (acq 2-12-02; grpsl). Format: Best of the 80s. ♦Michael Dickey, gen mgr; Don Boyd, gen sls mgr; Derick Corbett, progmg dir; Dan Goodman, chief of engrg.

Graysville

WAYB-FM— 1994: 95.7 mhz; 6 kw. Ant 328 ft TL: N35 24 39 W85 07 54. Hrs open: Box 262550, Baton Rouge, LA, 70826. Secondary address: 8919 World Ministry Ave, Baton Rouge, LA 70810. Phone: (225) 768-3688. Phone: (225) 768-8300. Fax: (225) 768-3729. E-mail: kawikfish@yahoo.com Web Site: www.jsm.org. Licensee: Family Worship Center Church Inc. (group owner; acq 5-20-02). Format: Christian. ♦David Whitelaw, COO; Jimmy Swaggart, pres; John Santiago, progmg dir.

Greeneville

WAEZ(FM)— 1956: 94.9 mhz; 100 kw. 1,090 ft TL: N36 04 34 W82 41 28. Stereo. Hrs opn: 24 901 E. Valley Dr., Bristol, VA, 24201. Phone: (276) 669-8112. Fax: (276) 669-0541. Web Site: www.electric949.com. Licensee: Bristol Broadcasting Co. Inc. (group owner; acq 6-22-00). Population served: 175000 Natl. Rep: Rgnl Reps. Format: CHR. News staff: 2. Target aud: 25-49. Spec prog: Univ. of Tennessee football & basketball. ♦Pete Nininger, pres; Bill Hickey, opns mgr.

WGRV(AM)—Listing follows WIKQ(FM).

WIKQ(FM)—Tusculum, February 1996: 103.1 mhz; 6 kw. -223 ft TL: N36 07 40 W82 37 57. Hrs open: 24 Box 278, 37744. Secondary address: 1004 Arnold Rd. 37743. Phone: (423) 639-1831. Fax: (423) 638-1979. Licensee: Radio Greeneville Inc. (group owner; (acq 6-22-2000; $1.8 million with WSMG(AM) Greeneville). Population served: 70,000 Format: Country. News staff: one. Target aud: 25-55; middle to upper middle income. ♦Ron P. Metcalf, gen mgr; Ron P. Metcalfe, pres & opns VP; Brian Stayton, progmg dir; Nathan Humbard, mus dir; Bobby Rader, news dir; Ray C. Elliott, chief of engrg.

WGRV(AM)—Co-owned with WIKQ(FM). 1946: 1340 khz; 1 kw-U. TL: N36 10 10 W82 50 52.24 Phone: (423) 638-4147.55,000 Format: Modern country. News staff: 3. ♦Betty Fletcher, traf mgr; Maxine Humphreys, local news ed & women's int ed; Nancy Ensor, news rptr; Al Wrinn, farm dir; Charlie Hicks, disc jockey; Nathan Humbard, disc jockey; Ray Elliott, disc jockey; Ron Metcalfe, prom mgr, progmg mgr & disc jockey.

WSMG(AM)— Dec 1, 1961: 1450 khz; 1 kw-U. TL: N36 10 30 W82 50 18. Hrs open: 24 Box 278, 37744. Secondary address: 10004 Arnold Rd. 37743. Phone: (423) 638-3188. Fax: (423) 638-1979. Licensee: Radio Greeneville Inc. (group owner; acq 6-22-00; $1.8 million with WIKQ(FM) Tusculum). Population served: 57,000 Format: Oldies. Target aud: General. ♦Ronnie Metcalfe, pres, gen mgr & opns mgr.

Halls Crossroads

WMYL(FM)— Aug 15, 1991: 96.7 mhz; 1.9 kw. Ant 489 ft TL: N36 04 21 W84 01 18. Stereo. Hrs opn: Box 329, Clinton, 37717. Secondary address: 111 Hillcrest Dr., Clinton 37716. Phone: (865) 457-1380. Fax: (865) 457-4440. Licensee: M & M Broadcasting (acq 4-18-2006; $1 million). Population served: 150,000 Natl. Network: ABC, CNN Radio. Rgnl. Network: Ky. Net., Tenn. Radio Net., Va. News Net. Natl. Rep: Rgnl Reps. Format: Country. Target aud: 25-54; community-oriented adults. Spec prog: Farm 2 hrs, gospel 2 hrs, relg 2 hrs wkly. ♦Ronald C. Meredith Jr., gen mgr.

Harriman

WIJV(FM)— Jan 21, 1981: 92.7 mhz; 690 w. Ant 964 ft TL: N35 59 04.8 W84 44 06.7. (CP: 6 kw, ant 328 ft. TL: N35 52 04 W84 25 56).

Stereo. Hrs opn: Box 810, Crossville, 38557. Phone: (931) 484-1057. Licensee: Progressive Media Inc. (group owner; (acq 7-2-2007; $2.4 million). Format: Classic rock. ♦Kirk Tollett, gen mgr, gen sls mgr, progmg dir & chief of engrg; Scott Humphrey, news dir; Jennifer Tollett, traf mgr.

Harrogate

***WLMU(FM)**— Aug 5, 1987: 91.3 mhz; 190 w. 284 ft TL: N36 35 10 W83 39 54. Stereo. Hrs opn: 24 Box 2025, Sigmon Communications Ctr., Hwy. 25 E., 37752. Phone: (423) 869-6331. Web Site: www.913thegap.com. Licensee: Lincoln Memorial University. Population served: 30,000 Format: Country/Bluegrass. Target aud: 25-54. ♦Dr. Nancy Moody, pres; Travis Moody, gen mgr; Dustin McCoy, opns mgr.

WRWB(AM)— Nov 10, 1980: 740 khz; 1 kw-D. TL: N36 34 32 W83 39 37. Hrs open: 24 6965 Cumberland Gap Pkwy., 37752. Phone: (423) 869-6335. Fax: (423) 869-6435. E-mail: wrwb@usa.com Licensee: Pine Hills of Tenn. Inc. (acq 7-86). Population served: 125,000 Wire Svc: AP Format: Talk. News staff: 4; News: 150 hrs wkly. Target aud: 34-65. ♦Tom Amis, gen mgr & opns mgr.

Hartsville

WTNK(AM)— Sept 1, 1966: 1090 khz; 1 kw-D, 2 w-N. TL: N36 23 17 W86 09 55. Hrs open: 24 165 Marlene St., 37074. Phone: (615) 374-2111. Fax: (615) 374-3544. Licensee: G & L Aircasters Inc. (acq 2-14-2003; $160,000). Population served: 500,000 Natl. Rep: Keystone (unwired net). Format: Country. News: 5 hrs wkly. Target aud: 35 plus. Spec prog: Gospel 4 hrs wkly. ♦Gary Frank, CEO, pres, gen mgr, stn mgr & chief of opns; K. K. Wilson, sls dir, gen sls mgr & prom dir; Earl White, progmg dir; Jerry Richmond, news dir; Lisa Frank, COO, CFO, exec VP, pub affrs dir & traf mgr.

Henderson

***WFHU(FM)**— May 22, 1967: 91.5 mhz; 10.5 kw. 300 ft TL: N35 27 50 W88 41 10. Stereo. Hrs opn: 24 158 E. Main St., 38340. Phone: (731) 989-6691. Phone: (731) 989-6749. E-mail: wfhc@fhu.edu Web Site: www.fhu.edu/radio. Licensee: Freed-Hardeman University. Population served: 100,000 Format: Jazz, classic rock, classical. News staff: one; News: 5 hrs wkly. Target aud: General; young adults to senior citizens. Spec prog: Class 10 hrs, gospel 9 hrs, jazz 45 hrs wkly. ♦Milton Sewell, pres; Ron Means, gen mgr.

WFKX(FM)— Feb 1, 1984: 95.7 mhz; 4.4 kw. Ant 383 ft TL: N35 29 52 W88 42 29. Stereo. Hrs opn: 24 111 W. Main St., Jackson, 38301. Phone: (731) 427-9616. Fax: (731) 424-2473. E-mail: cthomas@wwyn.fm Web Site: wfkx.fm. Licensee: Thomas Radio LLC (group owner; (acq 11-9-2001; grpsl). Population served: 230,000 Natl. Network: ABC. Law Firm: Borsari & Paxson. Format: Urban contemp, Black. News: 5 hrs wkly. Target aud: 18-54; the general Black population & contemp women. Spec prog: Gospel 3 hrs wkly. ♦Billy Thomas, pres; Chip Thomas, gen mgr; Jim Smith, chief of engrg.

WHHM-FM— Nov 19, 1990: 107.7 mhz; 50 kw horiz, 49.07 kw vert. Ant 459 ft TL: N35 27 23 W88 37 36. Stereo. Hrs opn: 24 111 W. Main St., Jackson, 38301. Phone: (731) 427-9616. Fax: (731) 424-2473. Web Site: www.star1077.fm. Licensee: Thomas Radio LLC (group owner; (acq 11-9-2001; grpsl). Population served: 370,000 Rgnl. Network: Tenn. Agri. Law Firm: McFadden, Evans & Sill. Wire Svc: AP Format: Var, adult contemp. News staff: one; News: 5 hrs wkly. Target aud: 25-54; adults. Spec prog: Gospel 10 hrs wkly. ♦Chip Thomas, gen mgr; Phil Hickerson, gen sls mgr; Shane Connor, opns mgr & progmg mgr; Jim Smith, chief of engrg.

Hendersonville

WQQK(FM)— Oct 16, 1970: 92.1 mhz; 3 kw. 462 ft TL: N36 17 50 W86 45 11. Stereo. Hrs opn: 24 10 Music Cir. E., Nashville, 37203. Phone: (615) 321-1067. Fax: (615) 321-5771. E-mail: danielle.haese@cumulus.com Web Site: www.cumulus.com. Licensee: Cumulus Licensing LLC. Group owner: Cumulus Media Inc. (acq 3-28-2002; grpsl). Natl. Network: ABC. Law Firm: Pepper & Corazzini. Format: Adult urban contemp. Target aud: 18-49; relg. Spec prog: Gospel 6 hrs wkly. ♦Michael Dickey, gen mgr; Don Boyd, gen sls mgr; Derrick Corbett, progmg dir; Dan Goodman, chief of engrg.

Henry

WMUF-FM— Apr 12, 1999: 104.7 mhz; 2.9 kw. Ant 476 ft TL: N36 08 19 W88 15 52. Stereo. Hrs opn: 24
Rebroadcasts WMUF(AM) Paris 100%.
110 India Rd., Paris, 38242. Phone: (731) 644-9455. Fax: (731) 644-9970. E-mail: wmuf@bellsouth.net Licensee: Benton-Weatherford Broadcasting Inc. of Tennessee (group owner). Population served: 50,000 Natl. Network: ABC. Format: Country. Target aud: 25-54. ♦Gary Benton, pres & gen mgr; Janice Benton, opns VP.

Hohenwald

***WAUO(FM)**— 1998: 90.7 mhz; 950 w. Ant 233 ft TL: N35 33 56 W87 33 27. Hrs open: Box 3206, American Family Radio, Tupelo, MS, 38803. Phone: (662) 844-8888. Phone: (662) 844-8893 (call-in). Fax: (662) 842-6791. E-mail: comments@afr.net Web Site: www.afr.net. Licensee: American Family Association. Group owner: American Family Radio Format: Christian classics. ♦Marvin Sanders, gen mgr.

WMLR(AM)— July 4, 1970: 1230 khz; 1 kw-U. TL: N35 31 22 W87 32 40. Hrs open: 24 184 Switzerland Rd., 38462. Phone: (931) 796-5966. Fax: (931) 796-7353. E-mail: harold@wmlr1230am.com Licensee: Cochran Communication Corp. of Lewis County. (acq 6-4-99; $67,500). Population served: 15,000 Natl. Network: ABC. Format: C&W. Spec prog: Gospel. ♦Harold Cochran, pres & gen mgr; Celeste Cochran, news dir; Benjamin Cochran, weather dir; Josiah Cochran, mus critic & disc jockey.

Humboldt

WIRJ(AM)— Jan 20, 1949: 740 khz; 250 w-D, 50 w-N. TL: N35 48 52 W88 54 51. Hrs open: Box 740, 38343-0740. Secondary address: 2606 East End Dr. 38343. Phone: (731) 784-5000. Fax: (731) 784-2533. Licensee: John F. Warmath. (acq 1996; $45,000). Format: Oldies, talk. ♦John F. Warmath, gen mgr.

WLLI(AM)— July 5, 1972: 1190 khz; 420 w-D. TL: N35 50 41 W88 54 08. Hrs open: 122 Radio Rd., Jackson, 38301. Phone: (731) 427-3316. Fax: (731) 427-9338. Web Site: www.realcountryonline.com. Licensee: Forever South Licenses LLC. (acq 7-31-2006; grpsl). Population served: 200,000 Format: Country. ♦Verla Price, gen mgr.

WLSZ(FM)—Co-owned with WHMT(AM). Jan 19, 1989: 105.3 mhz; 3 kw. Ant 328 ft TL: N35 50 41 W88 54 08. Stereo. 24 Format: CHR. Target aud: 18-49. ♦Dave Hacker, engrg dir.

WZDQ(FM)— Sept 1, 1964: 102.3 mhz; 6 kw. Ant 305 ft TL: N35 45 45 W88 51 42. Stereo. Hrs opn: 24 111 W. Main St., Jackcon, 38301. Phone: (731) 427-9616. Fax: (731) 424-2473. Web Site: www.wzdq.fm. Licensee: Thomas Radio LLC. (group owner; (acq 11-9-2001; grpsl). Population served: 500,000 Natl. Network: CNN Radio. Format: Rock. Target aud: 25-49; middle to upper class. ♦Chip Thomas, gen mgr; Marsha Hulsey, gen sls mgr; Shane Connor, opns mgr & prom dir.

Huntingdon

WDAP(AM)— Oct 21, 1975: 1530 khz; 1 kw-D. TL: N36 00 04 W88 26 02. Hrs open: 7 AM-5 PM 9662 H'way 77, 38344. Phone: (731) 986-9746. Fax: (731) 986-9704. Licensee: Mark C. Johnson (acq 5-23-2002). Population served: 32,000 Format: Country. News staff: 2. Target aud: General. ♦Mark C. Johnson, gen mgr; Sarah Dunning, gen sls mgr; Jay Jackson, news dir.

WVHR(FM)— November 1979: 100.9 mhz; 6 kw. 300 ft TL: N35 57 05 W88 27 47. Stereo. Hrs opn: 24 215 Baker Rd., Huntington, 38344. Phone: (731) 986-0242. Fax: (731) 986-8557. E-mail: wvhr@aeneas.net Licensee: Milan Broadcasting Co. Inc. (acq 5-1-91; $150,000; FTR: 5-27-91). Rgnl. Network: Tenn. Radio Net. Format: Classic hit country. ♦Jerry Vandiver, gen mgr & sls dir; Michael Ray, opns dir, progmg dir & news dir; Dave Hacker, chief of engrg.

Jackson

***WAMP(FM)**— 1995: 88.1 mhz; 750 w. 134 ft TL: N35 39 38 W88 51 30. Hrs open: American Family Radio, Box 3206, Tupelo, MS, 38803. Phone: (662) 844-8888. Phone: (662) 844-8893. Fax: (662) 842-6791. E-mail: comments@afr.net Web Site: www.afr.net. Licensee: American Family Association. Group owner: American Family Radio Format: Christian. ♦Marvin Sanders, gen mgr.

WDXI(AM)— Oct 31, 1948: 1310 khz; 5 kw-D, 1 kw-N, DA-N. TL: N35 39 50 W88 49 20. Hrs open: 24 Box 3845, 38303-3845. Secondary address: 1 Radio Park Dr. 38305-4124. Phone: (731) 427-9611. Phone: (731) 424-1310. Fax: (731) 424-1321. Licensee: Gerald W. Hunt (acq 1-15-93; $480,000 with co-located FM; FTR: 2-8-93). Population served: 55,000 Natl. Rep: D & R Radio. Format: Business news. News staff: one; News: 10 hrs wkly. Target aud: 25 plus. Spec prog: Farm 12 hrs, gospel 16 hrs, sports 16 hrs wkly. ♦Gerald W. Hunt, gen mgr, gen sls mgr & progmg mgr.

WMXX-FM—Co-owned with WDXI(AM). May 9, 1979: 103.1 mhz; 42 kw. Ant 538 ft TL: N35 32 39 W88 47 18. Stereo. Format: Oldies. Target aud: 25-54.

WJAK(AM)— Nov 14, 1954: 1460 khz; 1 kw-D, 128 w-N. TL: N35 38 37 W88 46 24. Hrs open: 24 111 W. Main St., 38301. Phone: (731) 427-9616. Fax: (731) 427-9302. Licensee: Thomas Radio L.L.C. (group owner; (acq 7-21-2004; $318,000). Population served: 100,000 Natl. Network: Moody, USA. Natl. Rep: Rgnl Reps. Format: Urban Gospel. News: 14 hrs wkly. Target aud: 18-54; primarily Black Christian middle-class families with low to moderate income. ♦Chip Thomas, gen mgr.

***WKNP(FM)**— Dec 17, 1990: 90.1 mhz; 17 kw. 528 ft TL: N35 38 46 W88 49 57. Stereo. Hrs opn: 24
Rebroadcasts WKNO-FM Memphis 100%.
Box 241880, Memphis, 38124. Secondary address: 900 Getwell Rd., Memphis 38111. Phone: (901) 325-6544. Fax: (901) 325-6506. Web Site: www.wknofm.org. Licensee: Mid-South Public Communications Foundation. Population served: 120,000 Natl. Network: NPR, PRI. Law Firm: Schwartz, Woods & Miller. Format: Class, news. News staff: one; News: 51 hrs wkly. Target aud: 35 plus. ♦Michael LaBonia, pres; Dan Campbell, gen mgr & stn mgr; Darel Snodgrass, opns mgr; Charles McLarty, dev dir. Co-owned TV: *WKNO-TV affil

WNWS-FM— August 1993: 101.5 mhz; 3 kw. 300 ft TL: N35 38 59 W88 46 11. Stereo. Hrs opn: 24 116 N. Church St., 4th Fl., 38301. Phone: (731) 423-8316. Fax: (731) 423-8304. E-mail: newstalk@wnws.com Web Site: www.wnws.com. Licensee: Radiocorp of Jackson Inc. Group owner: The Wireless Group Inc. (acq 12-6-00; $925,000). Population served: 50,000 Natl. Network: CBS. Format: News/talk. News staff: 2; News: 25 hrs wkly. Target aud: 25 plus; upscale adults. ♦Greg Wood, opns mgr; Larry Wood, gen mgr & progmg dir.

WOGY(FM)—Listing follows WTJS(AM).

WTJS(AM)— 1931: 1390 khz; 5 kw-D, 1 kw-N, DA-N. TL: N35 38 50 W88 50 00. Hrs open: 24 122 Radio Rd., 38301. Phone: (731) 427-3316. Fax: (731) 427-4576. Licensee: Forever South Licenses LLC. Group owner: Clear Channel Communications Inc. (acq 5-12-2006; grpsl). Population served: 290,120 Rgnl. Network: Tenn. Radio Net. Format: News/talk. News staff: 3; News: 30 hrs wkly. Target aud: 35 plus; general. ♦Roger Vestal, gen mgr; Dave Hacker, opns mgr & chief of engrg; Gina Langley, gen sls mgr; Connie Cain, traf mgr; Todd Starnes, news dir & local news ed.

WOGY(FM)—Co-owned with WTJS(AM). 1947: 104.1 mhz; 100 kw. 679 ft TL: N35 38 46 W88 49 57. Stereo. Web Site: eagle104.net.426,852 Format: Country. Target aud: 18-54. ♦Deb Smith, prom dir; Rusty Mac, news dir & news rptr.

WYNU(FM)—See Milan

Jamestown

WCLC(AM)— Oct 28, 1957: 1260 khz; 1 kw-D. TL: N36 26 10 W84 55 42. Hrs open: Box 1509, 38556. Phone: (931) 879-8188. Fax: (931) 879-1733. E-mail: wclc@twlakes.net Licensee: Bible Believers Network Inc. Population served: 40,000 Format: Bible believers network, relg. Spec prog: Farm 2 hrs, bluegrass 3 hrs wkly. ♦Jim Cody, gen mgr, gen sls mgr & progmg dir.

WCLC-FM— 1985: 105.1 mhz; 1.1 kw. 605 ft TL: N36 26 31 W84 55 28. (CP: 2.85 kw, ant 476 ft.). Stereo. 100,000 Format: Relg.

WDEB(AM)— Jan 12, 1968: 1500 khz; 1 kw-D, 500 w-CH. TL: N36 25 31 W84 56 32. Hrs open: Sunrise-sunset Box 69, 38556. Secondary address: 403 Livingston Ave. 38556. Phone: (931) 879-8164. Phone: (931) 879-9332. Fax: (931) 879-7437. E-mail: wdebaudio@twlakes.net Web Site: wderadio.com. Licensee: BAZ Broadcasting Inc. (acq 4-1-72). Population served: 201,225 Rgnl. Network: Tenn. Radio Net. Format: Country, relg. News staff: 7; News: 10 hrs wkly. Target aud: 18-54; household members who spend money in the marketplace. Spec prog: Farm 3 hrs wkly. ♦N.A. Baz, pres, gen mgr, prom dir, adv dir & news dir; Jean Baz, VP; Gary Crocket, progmg dir; Kevin R. Baz, mus dir; Gunther Muhsemann, chief of engrg; Gary Clark, reporter & disc jockey; Gary Crockett, reporter & disc jockey; John B. Mullinix, reporter; Turk Baz, local news ed & reporter.

WDEB-FM— Oct 10, 1972: 103.9 mhz; 1.6 kw. 450 ft TL: N36 25 55 W84 56 33. Stereo. 5 AM-10:15 PM Web Site: wderadio.com E-mail: wdebaudio@twlakes.net. Format: Modern country, gospel. ♦Jean Baz, sls dir; N. A. Baz, rgnl sls mgr, mktg mgr & prom mgr; Cindy Mitchell, women's int ed; John Mullinix, disc jockey; Kevin Baz, disc jockey; Turk Baz, disc jockey.

Jasper

WWAM(AM)— Mar 2, 1987: 820 khz; 5 kw-D. TL: N35 04 23 W85 37 39. Stereo. Hrs opn: Box 279, 37347. Secondary address: 4896 Main St. 37347. Phone: (423) 942-1700. Phone: (931) 592-5588. Fax: (423) 942-1700. Licensee: Shelton Broadcasting System. Natl. Network: USA. Format: Gospel. Target aud: 25-49. Spec prog: Bluegrass one hr, acappella one hr wkly. ♦Rick Shelton, gen mgr.

Jefferson City

WJFC(AM)— Nov 1, 1961: 1480 khz; 500 w-D. TL: N36 06 15 W83 29 10. Hrs open: 6 AM-6 PM Box 430, 37760. Phone: (865) 475-3825. Fax: (865) 475-3800. Licensee: Lakeway Broadcasting LLC (acq 10-12-2006; $100,000). Population served: 25,124 Law Firm: Timothy K. Brady. Format: Country. News staff: one; News: 10 hrs wkly. Target aud: 25 plus; Jefferson, Grainger & Hamblen counties. Spec prog: Farm one hr, relg 4 hrs wkly. ♦M. Edward Stiner Jr., pres; Kenneth C. Hill, gen mgr.

WNRX(FM)— Feb 1, 1976: 99.3 mhz; 3 kw. 654 ft TL: N36 04 28 W83 34 56. Hrs open: 24 415 Middle Creek Rd., Sevierville, 37862. Phone: (865) 453-2844. Fax: (865) 428-2601. Licensee: Citadel Broadcasting Co. Group owner: Citadel Broadcasting Corp. (acq 7-20-2004; $1.65 million). Population served: 600,000 Format: CHR. News staff: one; News: 17 hrs wkly. Spec prog: Gospel 4 hrs, relg 2 hrs wkly.

Jellico

WEKX(FM)— 1993: 102.7 mhz; 630 w. 1,008 ft TL: N36 37 55 W84 08 31. Hrs open: 24 522 Main St., Williamsburg, KY, 40769. Phone: (606) 549-1027. Fax: (606) 549-5565. Licensee: Whitley Broadcasting Co. Inc. (group owner; acq 5-23-02; grpsl). Population served: 500,000 Rgnl rep: Rgnl Reps. Format: Adult contemp. News: 5 hrs wkly. ♦David Estes, gen mgr; Frank Folsom, gen sls mgr & chief of engrg; Rick Campbell, stn mgr & progmg mgr.

WJJT(AM)— Feb 1, 1972: Stn currently dark. 1540 khz; 1 kw-D, 1 w-N, 500 w-CH. TL: N36 34 59 W84 08 10. Hrs open: Box 210, 37762. Phone: (423) 784-1500. Fax: (423) 784-5991. E-mail: wjtam1540@aol.com Licensee: Southeast Broadcast Corp. (acq 7-6-2007; $250,000). Population served: 60,000 Rgnl. Network: Tenn. Radio Net. Format: Gospel. ♦James Kilgore, pres; Betty Douglas, exec VP; Marvin Douglas, gen mgr; Betty Douglas, stn mgr.

Johnson City

WETB(AM)— Oct 1, 1947: 790 khz; 5 kw-D, 72 w-N. TL: N36 19 43 W82 24 39. Hrs open: 6 AM-11 PM Box 4127, 37602. Secondary address: 231 Brandonwood Dr. 37604. Phone: (423) 928-7131. Fax: (423) 928-8392. E-mail: webb@mounet.com Licensee: Mountain Signals Inc. (acq 12-5-90; FTR: 12-31-90). Population served: 50,000 Natl. Network: USA. Format: Gospel. News staff: one; News: 3 hrs wkly. Target aud: General. ♦Paul Gobble Jr., gen mgr & pres; Bob Morrison, stn mgr, opns mgr & progmg dir; Loretta Gouge, gen sls mgr.

***WETS(FM)**— Feb 26, 1974: 89.5 mhz; 66 kw. 2,273 ft TL: N36 26 02 W82 08 08. Stereo. Hrs opn: 24 c/o East Tennessee State University, Box 70630, Ellis Hall, 37614-1709. Phone: (423) 439-6440. Phone: (423) 439-6441. Fax: (423) 439-6449. E-mail: winkler@xtn.net Web Site: www.wets.org. Licensee: East Tennessee State University. Population served: 500,000 Natl. Network: NPR, PRI. Format: Class, folk, news/talk. News: 22 hrs wkly. Target aud: General. Spec prog: Blues 12 hrs, Sp one hr wkly. ♦Paul E. Stanton, pres; Dan Hirschi, progmg dir; Jim Blalock, mus dir; Mitch Sandidge, chief of engrg; Wayne Winkler, gen mgr & min affrs dir; Susan Lachmann, women's int ed.

WJCW(AM)— Dec 13, 1938: 910 khz; 5 kw-D, 1 kw-N, DA-N. TL: N36 24 37 W82 27 13. Hrs open: 24 Box 8668, Gray, 37615. Phone: (423) 477-1000. Fax: (423) 477-4747. E-mail: TalkRadio@wjcw.com Web Site: www.wjcw.com. Licensee: Citadel Broadcasting Co. Group owner: Citadel Broadcasting Corp. (acq 5-30-2000; grpsl). Population served: 57,000 Natl. Network: CBS, ABC. Law Firm: Reddy, Begley & McCormick. Format: Talk. Target aud: 25 plus. Spec prog: Relg 4 hrs wkly. ♦Don Raines, VP & gen mgr; Bob Gordon, opns dir; Debbie Caso, sls dir; Paul Overbay, gen sls mgr; Bob Lawrence, mktg dir & prom dir; Brian Bishop, progmg dir; Richard Lovette, news dir; Al F. LeFevere, chief of engrg.

WQUT(FM)—Co-owned with WJCW(AM). Mar 1, 1948: 101.5 mhz; 100 kw. 1,500 ft TL: N36 16 07 W82 20 21. Stereo. 24 Phone: (423) 477-1015. E-mail: wqut@preferred.com Web Site: www.wqvt.com.375,000 Format: Classic rock. News staff: 2; News: 3 hrs wkly. Target aud: 18-49. ♦Randy Ross, gen sls mgr; Jeri George, prom dir & disc jockey; John Patrick, progmg dir & disc jockey; Susan Rines, traf mgr; "John Boy & Billy", disc jockey; Marc Tragler, disc jockey.

WKTP(AM)—See Jonesborough

WTFM(FM)—See Kingsport

Jonesborough

WKTP(AM)— October 1958: 1590 khz; 5 kw-U, DA-2. TL: N36 19 54 W82 28 27. Hrs open: 24
Rebroadcasts WKPT(AM) Kingsport 90%.
222 Commerce St., Kingsport, 37660. Phone: (423) 246-9578. Phone: (423) 926-9800. Fax: (423) 247-9836. Fax: (423) 246-6261. E-mail: davidw@wtfm.com Web Site: www.wkptam.com. Licensee: Holston Valley Broadcasting Corp. Group owner: Glenwood Communications Corp. (acq 1-25-90; $90,000; FTR: 3-5-90). Population served: 200,000 Natl. Network: ABC. Natl. Rep: McGavren Guild. Law Firm: Cordon & Kelly. Format: Oldies. News staff: 2; News: 24 hrs wkly. Target aud: 35 plus. ♦George Devault, pres; N. David Widener, exec VP, gen mgr & stn mgr; Aaron Teffeteller, progmg dir.

WTZR(FM)—See Elizabethton

Karns

WMYU(FM)— Jan 8, 1989: 93.1 mhz; 1.2 kw. 515 ft TL: N35 58 59 W84 04 37. (CP: 2.4 kw, ant 512 ft. TL: N35 57 46 W84 01 23). Hrs opn: 8419 Kingston Pike, Knoxville, 37919. Phone: (865) 693-1020. Fax: (865) 693-8493. Web Site: www.thepoint931.com. Licensee: Journal Broadcast Corp. Group owner: Journal Communications Inc. (acq 5-19-97). Format: Adult Contemp. ♦Andy Laird, VP, progmg dir & engr; Chris Protzman, gen mgr; Bruce Patrick, progmg dir.

Kingsport

***WCQR-FM**— December 1996: 88.3 mhz; 1.2 kw. 2,132 ft TL: N36 25 53 W82 08 16. Stereo. Hrs opn: 2312 Oak St., Gray, 37615-8039. Phone: (423) 477-5676. Fax: (423) 477-7060. E-mail: office@wcqr.org Web Site: www.wcqr.org. Licensee: Positive Alternative Radio Inc. Group owner: Baker Family Stations/Positive Alternative Radio Inc. Natl. Network: Salem Radio Network. Law Firm: Booth, Freret, Imlay & Tepper. Format: Contemp Christian mus. Target aud: 25-54. ♦Mike Perry, gen mgr.

***WCSK(FM)**— Nov 5, 1984: 90.3 mhz; 195 w. 23 ft TL: N36 31 37 W82 35 12. Stereo. Hrs opn: 1800 Legion Dr., 37664. Phone: (423) 378-2150. Fax: (423) 378-2120. Web Site: www.kptk12.tn.us. Licensee: Kingsport Board of Education. Population served: 70,000 Format: Educ, class, div.

WGOC(AM)— October 1951: 1320 khz; 5 kw-D, 500 w-N, DA-N. TL: N36 33 59 W82 33 22. Hrs open: 24 Box 8668, Gray, 37615. Secondary address: 162 Freehill Rd. 37615. Phone: (423) 477-1000. Fax: (423) 477-4747. E-mail: sportsmonster@640wxsm.com Web Site: www.640wxsm.com. Licensee: Citadel Broadcasting Co. Group owner: Citadel Broadcasting Corp. (acq 5-30-2000; grpsl). Population served: 52,000 Natl. Network: ESPN Radio, CBS Radio. Natl. Rep: Katz Radio. Law Firm: Wiley, Rein & Fielding. Wire Svc: AP Format: sports. ♦Don Raines, gen mgr; Bob Gordon, opns dir; Debbie Caso, sls dir; Paul Overbay, gen sls mgr; Bob Lawrence, mktg dir & prom dir; Al LeFevere, chief of engrg.

WKOS(FM)—Co-owned with WGOC(AM). Feb 21, 1970: 104.9 mhz; 2.75 kw. 492 ft TL: N36 33 14 W82 27 00. Stereo. E-mail: oldies@preferred.com Web Site: www.wkos.com.100,000 Natl. Network: Westwood One. Format: Oldies. Target aud: 25-54. ♦Debbie Caso, gen sls mgr; Greg Price, natl sls mgr; Alan Austin, progmg dir & disc jockey; Bob Lawrence, pub affrs dir; Susan Ritea, traf mgr; Dennis Kelly, disc jockey; Don Gibson, disc jockey.

WHGG(AM)— June 1967: 1090 khz; 10 kw-D. TL: N36 27 40 W82 27 12. Hrs open: 12 Box 2061, Bristol, 37621. Secondary address: 340 Edgemont Ave., Suite 100, Bristol 37620. Phone: (423) 878-6279. Fax: (423) 878-6520. Web Site: www.mighty1090.com. Licensee: Information Communication Corp. (acq 1-1-2006; $250,000 with WABN(AM) Abingdon, VA). Population served: 2,000,000 Format: Oldies. News staff: one; News: 10 hrs wkly. Target aud: 24-55. ♦Kenneth C. Hill, pres & gen mgr; Matthew J. Hill, stn mgr; Rusty Curs, sls dir.

WJCW(AM)—See Johnson City

WKPT(AM)— July 14, 1940: 1400 khz; 1 kw-U. TL: N36 32 37 W82 31 21. Hrs open: 24 222 Commerce St., 37660. Phone: (423) 246-9578. Fax: (423) 247-9836. E-mail: davidw@wtfm.com Web Site: www.wkptam.com. Licensee: Holston Valley Broadcasting Corp. Group owner: Glenwood Communications Corp. (acq 6-1-66). Population served: 650,000 Natl. Rep: McGavren Guild. Law Firm: Cordon & Kelly. Format: Oldies. News staff: 2. Target aud: 35 plus. ♦George Devault, pres; N. David Widener, exec VP & gen mgr; Charles Aesque, gen sls mgr; Aaron Teffeteller, progmg dir & progmg mgr; Duane Nelson, news dir & local news ed; Emily Pridemore, traf mgr; Roger Epperson, news rptr; Janet Johnson, women's int ed.

WTFM(FM)—Co-owned with WKPT(AM). February 1948: 98.5 mhz; 74 kw. Ant 2,241 ft TL: N36 25 54 W82 08 15. Stereo. 24 E-mail: davidw@wtfm.com Web Site: www.wtfm.com.1,000,000 Natl. Network: ABC. Natl. Rep: McGavren Guild. Law Firm: Cordon & Kelly. Format: Adult contemp. News staff: 2. Target aud: 25-54. ♦Tim Loy, sls VP & gen sls mgr; Mark Baker, progmg mgr; Lyle Musser, chief of engrg; Emily Pridimore, traf mgr; Duane Nelson, news rptr; Steve Mann, sports cmtr & disc jockey; Dave Barnett, disc jockey; Elva Marie, disc jockey; Taylro Morgan, disc jockey. Co-owned TV: WKPT-TV affil

WQUT(FM)—Johnson City

Kingston

WBBX(AM)— July 1978: 1410 khz; 500 w-D. TL: N35 52 49 W84 30 56. Hrs open: 8 AM-6 PM Box 389, 37763. Secondary address: 705 Greenwood St. 37763. Phone: (615) 376-6954. Licensee: Pilgrim Pathway Inc. (acq 6-30-92; $35,000; FTR: 7-27-92). Format: Gospel. Target aud: General. ♦Grant Carter, pres.

***WKTS(FM)**— 9/11/2006: 90.1 mhz; 55 w vert. Ant 633 ft TL: N35 45 57 W84 34 33. Hrs open: 24 331 Skyline View Ln., 37763. Phone: (865) 717-3335. E-mail: thebridgefm@yahoo.com Web Site: www.bridgeradiofm.org. Licensee: Foothills Broadcasting, Inc. Format: Christian Contemp. News staff: one; News: 3 hrs wkly. Spec prog: 2 church services, 2hrs. ♦David Wells, gen mgr; Darrin Wilcox, stn mgr.

Kingston Springs

WFFI(FM)— Jan 15, 1993: 93.7 mhz; 1.15 kw. Ant 754 ft TL: N36 08 10 W86 59 04. Hrs open: 24
Simulcasts with WFFH(FM) Smyrna.
402 BNA Dr., Suite 400, Nashville, 37217. Phone: (615) 367-2210. Fax: (615) 367-0758. Web Site: www.94fmthefish.net. Licensee: Caron Broadcasting Inc. Group owner: Salem Communications Corp. (acq 12-18-2002; $5.6 million with WFFH(FM) Smyrna). Natl. Network: Salem Radio Network. Natl. Rep: Salem. Format: Contemp Christian. Target aud: 25-54; adults. ♦Michael S. Miller, gen mgr; Kevin R. Anderson, gen sls mgr; Dick Marsh, prom dir; Kim Bindel, news dir & news dir; Carl Campbell, chief of engrg; Ed Evenson, traf mgr; Vance Dillard, progmg dir & disc jockey.

Knoxville

WETR(AM)— July 5, 1995: 760 khz; 2.5 kw-D. TL: N36 02 34 W84 02 51. (CP: 2.4 kw). Hrs open: Day-time 1621 E. Magnolia Ave., 37917. Phone: (865) 525-1060. Fax: (865) 521-8923. E-mail: info@talkradio760.com Web Site: www.talkradio760.com. Licensee: Thomas H. Moffit Jr. (acq 1995). Population served: 850,000 Natl. Network: Salem Radio

Network, Talk Radio Network. Format: News/talk. Target aud: 25-54; blue collar men & women. ♦David Wells, gen mgr; David Wells, progmg dir.

WIFA(AM)— Jan 21, 1941: 1240 khz; 1 kw-U. TL: N35 57 17 W83 57 04. Hrs open: 24 Box 50840, 37950. Secondary address: 818 N. Cedar Bluff Rd. 37923. Phone: (865) 531-2005. Fax: (865) 531-2006. Web Site: www.1240radio.com. Licensee: Progressive Media Inc. (acq 8-6-2004; $550,000). Format: Adult contemp Christian music and talk. ♦Barry Culberson, pres; Brian Brooks, gen mgr.

WIMZ-FM— October 1949: 103.5 mhz; 100 kw. 1,723 ft TL: N36 08 06 W83 43 29. Stereo. Hrs opn: 1100 Sharps Ridge Rd., 37917. Phone: (865) 525-6000. Fax: (865) 525-2000. E-mail: rcchambers@sccradio.com Web Site: www.wimz.com. Licensee: South Central Communications Corp. (group owner; (acq 2-23-93; $3.5 million with co-located AM; FTR: 3-15-93). Population served: 174,587 Format: Classic rock. ♦Randy Ross, sls dir; Neda Gayle, natl sls mgr; Terry Gillingham, VP & mktg mgr; Randy Chambers, progmg dir; Billy Kidd, mus dir; Nikki Roberts, pub affrs dir; Jeff Cutshaw, traf mgr.

WITA(AM)— Sept 1, 1960: 1490 khz; 1 kw-U. TL: N35 58 11 W83 57 56. Hrs open: 24 hrs 7212 Kingston Pike, 37919. Phone: (865) 588-2974. Phone: (865) 588-2975. E-mail: wita1490@aol.com Web Site: www.wwcr.com. Licensee: RR Broadcast Group Inc. (group owner; (acq 2-18-2005; $425,000). Population served: 500,000 Format: Christian talk. Target aud: General. Spec prog: Black 8 hrs wkly. ♦Rex D. Palmer, pres; Gail Scott, gen mgr; Greg McMahon, opns mgr.

WIVK-FM—Listing follows WNML(AM).

WJXB-FM— Apr 10, 1967: 97.5 mhz; 96 kw. 1,296 ft TL: N36 00 36 W83 55 57. Stereo. Hrs opn: 24 1100 Sharps Ridge Mem Park Dr., 37917. Phone: (865) 525-6000. Fax: (865) 656-3292. E-mail: jjarnigan@sccradio.com Web Site: www.b975.com. Licensee: South Central Communications Corp. (group owner). Population served: 174,589 Wire Svc: UPI Format: Adult contemp. Target aud: 25-54. ♦John D. Engelbrecht, chmn; Craig Jacobus, pres & gen mgr; Randy Ross, sls dir; Neda Gayle, natl sls mgr; Terry Gillingham, opns mgr & mktg mgr; Deborah Cox, prom dir; Jeff Jarnigan, progmg dir; Susan Hollingsworth, traf mgr.

***WKCS(FM)**— December 1952: 91.1 mhz; 250 w. 73 ft TL: N35 59 36 W83 55 24. Hrs open: 8 AM-3:30 PM Fulton High School, 2509 Broadway N.E., 37917. Phone: (865) 594-1259. E-mail: wkcsradio@hotmail.com Licensee: Fulton High School. Population served: 250,000 Format: Oldies. News: 3 hrs wkly. Target aud: 18 plus; University of Tennessee. ♦Russell Mayes, gen mgr.

WKGN(AM)— Sept 28, 1947: 1340 khz; 1 kw-U. TL: N35 57 20 W83 58 14. Stereo. Hrs opn: Box 10005, 37919. Phone: (865) 546-7900. Fax: (865) 546-7965. Licensee: Norsan Consulting and Management Inc. (acq 3-8-2006; $500,000). Natl. Network: Westwood One. Natl. Rep: Roslin. Format: Urban contemp, gospel. Target aud: 18-34; young, mobile adults. Spec prog: Relg 5 hrs, medicine/health one hr wkly. ♦Norberto Sanchez, pres; Robert L. Stewart, gen mgr; Thomas Henderson, progmg dir; Ed Martin, chief of engrg.

WKHT(FM)— November 1991: 104.5 mhz; 6 kw. 394 ft TL: N36 00 36 W83 55 57. (CP: 2.3 kw, ant 528 ft.). Hrs open: 1533 Amhearst Rd., 37909. Phone: (865) 693-1020. Phone: (865) 824-1021. Fax: (865) 824-1880. Web Site: www.1045thebone.com. Licensee: Journal Broadcast Corp. (group owner; (acq 3-4-98; $5.745 million with WQBB(AM) Powell). Natl. Rep: Roslin. Format: Classic rock. Target aud: 35 plus; female. Spec prog: Pub affrs 2 hrs wkly. ♦Chris Protzman, gen mgr; Rich Bailey, opns mgr; Eddy Roy, sls dir, news dir & traf mgr; Dodie Manalac, gen sls mgr; Russ Allen, progmg dir; Mark Lucas, chief of engrg.

WQBB(AM)—Co-owned with WKHT(FM). Aug 15, 1984: 1040 khz; 10 kw-D. TL: N36 02 34 W84 02 51. Stereo. Format: Talk, sports. ♦Dan McKel, sls dir; Bruce Patrick, progmg dir; Eddie Roy, sports cmtr.

WKVL(AM)— Jan 16, 1989: 850 khz; 50 kw-D, DA. TL: N36 04 12 W83 58 19. Stereo. Hrs opn: 517 Watt Rd., 37922. Phone: (865) 675-4105. Fax: (865) 675-4859. Licensee: Horne Radio L.L.C. Group owner: Horne Radio Group (acq 10-15-99; grpsl). Format: Talk. Target aud: 25 plus; educated, informed adults. ♦Brian Tatum, CEO & gen mgr.

WKXV(AM)— February 1953: 900 khz; 1 kw-D, 258 w-N. TL: N35 58 52 W83 59 15. Hrs open: 5106 Middlebrook Pike, 37921. Phone: (865) 558-0900. Fax: (865) 588-5848. Licensee: Ratel Broadcasting Co. Inc. Population served: 174,587 Format: Relg, Southern gospel. ♦Ted H. Lowe Sr., pres; Ted H. Lowe Jr., gen mgr & gen sls mgr; Rick Whisman, news dir; Frank Folsom, chief of engrg; Karyn Treece, mus critic; Lisa Lunsford, women's int ed; L. C. Logan, disc jockey; Mike Hall, disc jockey.

WNFZ(FM)—See Oak Ridge

WNML(AM)— Mar 23, 1953: 990 khz; 10 kw-U, DA-N. TL: N36 02 33 W83 53 59. Stereo. Hrs opn: 24 WIVK711, Box 11167, 37939. Secondary address: 4711 Old Kingston Pike 37919. Phone: (865) 588-6511. Fax: (865) 558-4218. Web Site: www.newstalk99.com. Licensee: Citadel Broadcasting Co. Group owner: Citadel Broadcasting Corp. (acq 4-26-2001; grpsl). Population served: 520,184 Format: News/talk, sports. News staff: 8; News: 28 hrs wkly. Target aud: 25-54. ♦Farid Suleman, CEO; Donna Heffner, CFO; Ed Brantley, gen mgr & gen sls mgr; Mike Hammond, opns mgr; Charles Sells, sls dir; Lisa Rotton, natl sls mgr; Jack Lee Gillette, rgnl sls mgr; Steve Queisser, mktg dir; John Crooks, progmg dir; Tom Graham, news dir; Tim Berry, chief of engrg.

WIVK-FM—Co-owned with WNML(AM). Dec 16, 1965: 107.7 mhz; 91 kw. 2,053 ft TL: N35 48 41 W83 40 10. Stereo. 24 Fax: (423) 588-3725. Web Site: www.wivk.com.200,000 Format: C&W. ♦John Crooks, progmg dir.

WNOX(FM)—Oak Ridge, Apr 20, 1974: 100.3 mhz; 100 kw. Ant 2,001 ft TL: N36 11 53 W84 13 51. Stereo. Hrs opn: Box 11167, 37939. Secondary address: 4711 Old Kingston Pike 37919. Phone: (865) 588-6511. Fax: (865) 588-3725. Web Site: www.wnoxnewstalk.com. Licensee: Oak Ridge FM Inc. Population served: 1,201,600 Natl. Rep: Katz Radio. Wire Svc: AP Format: News/talk. ♦Ed Brantley, VP & gen mgr; Mike Hammond, opns mgr & progmg dir; Jack Lee, gen sls mgr; Laura Hall, mktg dir; Catherine Howell, news dir; Tim Berry, chief of engrg.

WNPZ(AM)— May 21, 1961: Stn currently dark. 1580 khz; 5 kw-D, 1 kw-CH. TL: N35 54 42 W83 53 33. Hrs open: Metropolitan Management Corp. of Tennessee, 4284 Memorial Dr., Suite B, Decatur, GA, 30030. Phone: (404) 525-0100. Licensee: Metropolitan Management Corp. of Tennessee (acq 5-10-02; $280,000). ♦Randal A. Mangham, pres.

WNRX(FM)—See Jefferson City

WRJZ(AM)— Feb 12, 1927: 620 khz; 5 kw-U, DA-N. TL: N35 59 24 W83 50 15. Hrs open: 24 Christian Media Ctr., 1621 E. Magnolia Ave., 37917. Phone: (865) 525-0620. Fax: (865) 521-8910. E-mail: joy62@wrjz.com Web Site: www.wrjz.com. Licensee: Tennessee Media Associates. (acq 10-84). Population served: 650,000 Natl. Network: Salem Radio Network. Natl. Rep: Salem. Format: Christian, talk. News: 5 hrs wkly. Target aud: 25-54; white collar men & woman. ♦Thomas Moffit Jr., pres; David Wells, gen mgr & progmg dir.

***WUOT(FM)**— October 1949: 91.9 mhz; 100 kw. 1,580 ft TL: N36 00 19 W83 56 23. Stereo. Hrs opn: 24 Univ. of Tennessee, 209 Communications Bldg., 37996-0322. Phone: (865) 974-5375. Fax: (865) 974-3941. E-mail: wuot@utk.edu Web Site: www.wuot.org. Licensee: University of Tennessee. Population served: 600,000 Natl. Network: PRI, NPR. Law Firm: Cohn & Marks. Format: Class, jazz, news. News staff: 2; News: 30 hrs wkly. Target aud: 35-54. ♦Regina Dean, gen mgr; Daniel Bevvy, progmg dir; Matt Shafer Powell, news dir; Mike Murvell, chief of engrg.

***WUTK-FM**— Jan 4, 1982: 90.3 mhz; 800 w. 23 ft TL: N35 57 09 W83 55 34. Stereo. Hrs opn: 24 Univ. of Tenn., P-103 Andy Holt Tower, 37996. Phone: (865) 974-2228. Phone: (865) 974-2229. Fax: (865) 974-2814. E-mail: wutk@utk.edu Web Site: www.wutkradio.com. Licensee: University of Tennessee. (acq 8-31-88). Population served: 350,000 Format: New rock. Target aud: 18-45; male/female. ♦Benny Smith, gen mgr, prom dir & progmg dir.

WVLZ(AM)— June 1, 1988: 1180 khz; 10 kw-D, 2.6 kw-CH. TL: N35 58 48 W83 49 09. Hrs open: 802 S. Central Ave., 37902. Phone: (865) 546-4653. Fax: (865) 637-7133. Web Site: www.wvlz.com. Licensee: Kirkland Wireless Broadcasters Inc. (acq 3-25-02; $400,000 with WKCE(AM) Maryville). Population served: 400,000 Format: Sports. Target aud: 30 plus; young, married with small children. ♦John Hodge, gen mgr & stn mgr.

WWST(FM)—See Sevierville

WYFC(FM)—See Clinton

La Follette

WLAF(AM)— May 17, 1953: 1450 khz; 1 kw-U. TL: N36 22 52 W84 07 32. Hrs open: 24 Drawer 1409, 37766. Secondary address: 210 N 5th St 37766. Phone: (423) 562-1450. Phone: (423) 562-3557. Fax: (423) 562-5764. E-mail: wlaf@campbellcounty.com Licensee: Stair Co. Inc. (acq 12-15-88; $125,000; FTR: 1-16-89). Population served: 40,000 Natl. Network: USA. Rgnl. Network: Tenn. Radio Net. Format: Gospel. News staff: one; News: 7 hrs wkly. Target aud: 12+ or 25+. Spec prog: Bluegrass 7 hrs wkly. ♦Jim Stair, pres; Bill Waddell, VP, opns VP & dev VP.

WQLA(AM)— Sept 1, 1983: 960 khz; 1 kw-D. TL: N36 22 02 W84 08 50. Hrs open: Box 1530, 37766. Phone: (423) 566-1000. Fax: (423) 457-5900. Fax: (865) 457-5900. Licensee: La Follette Broadcasters Inc. (acq 8-24-99; with co-located FM). Natl. Rep: Roslin. Format: Southern gospel. Target aud: General. ♦Cliff Jennings, pres & gen mgr; Barbara Nuls, gen sls mgr.

WQLA-FM— Sept 1, 1982: 104.9 mhz; 1.1 kw. 499 ft TL: N36 21 08 W84 05 20. (CP: 900 w). Stereo. 24 E-mail: qq104@ccdogisland.net 35,000 Format: C&W, sports. Target aud: 18 plus. ♦Barbara Nuls, gen sls mgr.

La Vergne

WBUZ(FM)— May 1, 1962: 102.9 mhz; 100 kw. Ant 954 ft TL: N35 48 01 W86 37 17. Stereo. Hrs opn: 24 1824 Murfreesboro Rd., Nashville, 37217. Phone: (615) 399-1029. Fax: (615) 399-1023. E-mail: programming@1029thebuzz.com Web Site: www.1029thebuzz.com. Licensee: WYCQ Inc. Group owner: The Cromwell Group Inc. (acq 11-28-89). Population served: 876,500 Rgnl. Network: Tenn. Agri. Format: New rock. Target aud: 18-34; residents in middle TN. Spec prog: Farm one hr wkly. ♦Bayard Walters, pres; Bob Reich, stn mgr; Shauna Conner, prom dir; Russ Schenck, progmg dir; Jim Patrick, news dir; David Wilson, chief of engrg; Andra Kramer, traf mgr.

Lafayette

WEEN(AM)— Nov 3, 1958: 1460 khz; 1 kw-D, 138 w-N. TL: N36 32 06 W86 00 27. Hrs open: Daytime 231 Chaffin Rd., 37083. Phone: (615) 666-2169. Fax: (615) 666-8056. E-mail: wlct@nctc.com Licensee: Lafayette Broadcasting Co. Inc. (acq 11-1-01). Population served: 40,000 Natl. Network: Salem Radio Network. Rgnl. Network: Tenn. Radio Net. Format: Solid Gospel. Target aud: General; 25-54 year olds. Spec prog: Farm 5 hrs wkly. ♦Ivan Davis, CEO & pres; Randall Swaffer, gen mgr, stn mgr & opns dir.

WLCT(FM)— July 1, 1995: 102.1 mhz; 6 kw. Ant 325 ft TL: N36 32 06 W86 00 27. Stereo. Hrs opn: 5 AM-11 PM 231 Chaffin Rd., 37083. Phone: (615) 666-2169. Fax: (615) 666-8056. E-mail: wlct@nctc.com Licensee: Lafayette Broadcasting Co. Inc. Population served: 50,000 Format: Country. Target aud: 20-60. ♦Melinda White, gen mgr & traf mgr; Randy Swaffer, gen mgr & opns mgr; Jamie Dallas, mktg mgr, sls & mktg; Jamie DAllas, prom.

Lakeland

WMQM(AM)—Licensed to Lakeland. See Memphis

Lawrenceburg

***WAWI(FM)**— 1999: 89.7 mhz; 6 kw. Ant 148 ft TL: N35 16 04 W87 19 25. Hrs open: Box 3206, American Family Radio, Tupelo, MS, 38803. Phone: (662) 844-8888. Fax: (662) 842-6791. E-mail: comments@afr.net Web Site: www.afr.net. Licensee: American Family Association. Group owner: American Family Radio Format: Relg. ♦Marvin Sanders, gen mgr.

WDXE(AM)— July 21, 1951: 1370 khz; 1 kw-D, 44 w-N. TL: N35 15 25 W87 18 24. Hrs open: 6 Public Square, 38464. Phone: (931) 762-4411. Fax: (931) 762-4789. E-mail: wdxe@charter.net Licensee: Lakewood Communications LLC (acq 9-11-02; $450,000 with co-located FM). Population served: 12,000 Rgnl. Network: Tenn. Radio Net. Format: Classic country. ♦Jack Cheatwood, gen mgr, stn mgr & news dir; Ron Fisher, gen sls mgr, progmg dir & disc jockey; Phillip Kemper, chief of engrg; Paula Walker, women's int ed; Ronnie Allen, disc jockey; Sunny Cull, disc jockey.

WDXE-FM— Aug 28, 1964: 106.7 mhz; 6 kw. Ant 292 ft TL: N35 15 25 W87 18 24. Format: Adult contemp. ♦Jack Cheatwood, gen mgr & progmg dir.

WLLX(FM)—Listing follows WWLX(AM).

WWLX(AM)— June 21, 1987: 590 khz; 600 w-D, 133 w-N. TL: N35 12 18 W87 19 39. Stereo. Hrs opn: 24 Box 156, 38464. Secondary address: 1212 N. Locust Ave. 38464. Phone: (931) 762-6200. Fax: (931) 762-6200. E-mail: wwlx@bellsouth.net Licensee: Roger W. Wright dba Prospect Communications. Format: C&W. News staff: one; News: 10 hrs wkly. Target aud: General. Spec prog: Oldies R&R 8 hrs, old country 8 hrs wkly. ♦Janet Wright, gen sls mgr & prom mgr; Dan Hollander, progmg dir & disc jockey; Michele Tankersley, news dir; Roger Wright, pres, gen mgr & chief of engrg; Carolyn Thompson, disc jockey; Dawn Washburn, disc jockey; Eddie Landtroop, disc jockey.

WLLX(FM)—Co-owned with WWLX(AM). May 1991: 97.5 mhz; 2.3 kw. 535 ft TL: N35 12 18 W87 19 39. Stereo. 24 E-mail: wllxradio@lorettotel.net Format: Country. News staff: one. Target aud: 25-54. ♦Janet Wright, mktg mgr & adv mgr; Dan Hollander, mus dir & disc jockey; Roger Wright, pub affrs dir; Carolyn Thompson, disc jockey; Dawn Washburn, disc jockey; Eddie Landtroop, disc jockey.

***WZXX(FM)**— 2005: 88.5 mhz; 300 w. Ant 276 ft TL: N35 15 18 W87 19 30. Hrs open: Box 5459, Twin Falls, ID, 83303-5459. Phone: (208) 733-3551. Fax: (208) 733-3548. Web Site: www.radioassistministry.com. Licensee: Radio Assist Ministry Inc. (group owner). (acq 5-5-2005; $85,000). ♦Matt Austin, gen mgr.

Lebanon

WANT(FM)— Oct 1, 1993: 98.9 mhz; 5 kw. 320 ft TL: N36 12 24 W86 16 02. Stereo. Hrs opn: 24 510 Trousdale Ferry Pike, 37087. Phone: (615) 444-0900. Fax: (615) 443-4235. Web Site: www.wantfm.com. Licensee: Bay-Pointe Broadcasting Co. Inc. Population served: 850,000 Law Firm: Tierney & Swift. Format: Country. News staff: one. Target aud: General. ♦Billy Goodman, opns mgr & news dir; M.J. Lucas, mus dir; Susan H. Bay, pres, gen mgr, progmg dir & pub affrs dir; Albert S. Jarratt Sr., chief of engrg.

WCOR(AM)— Dec 7, 2005: 1490 khz; 1 kw-U. TL: N36 12 26 W86 16 03. Stereo. Hrs opn: 24 510 Trousdale Ferry Pike, 37087. Phone: (615) 444-0900. Fax: (615) 443-4235. E-mail: wamble@bellsouth.net Licensee: Finbar Broadcasting Company Inc. Population served: 150,000 Natl. Network: ABC. Law Firm: Irwin, Campbell & Tannenwald. Format: News/talk, sports. ♦William O. Barry, pres & gen mgr; Harry P. Stephenson, gen sls mgr; Billy Goodman, news dir; Gary M. Brown, chief of engrg.

***WFMQ(FM)**— Dec 15, 1966: 91.5 mhz; 500 w. Ant 82 ft TL: N36 12 13 W86 18 01. (CP: 1 kw, ant 262 ft. TL: 36 12 24 W86 16 02). Stereo. Hrs opn: 24 One Cumberland Sq., 37087-3554. Phone: (615) 444-2562. Fax: (615) 444-2569. E-mail: wfmq@cumberland.edu Web Site: cumberland.edu/campus_life/wfmq. Licensee: Cumberland University. Population served: 185,000 Law Firm: Irwin, Campbell & Tannenwald. Format: Jazz. Spec prog: Class 6 hrs wkly. ♦Dr. Harvill Eaton, pres; Jeremiah McElwain, stn mgr; Albert Jarratt Sr., chief of engrg.

WKDA(AM)— Oct 5, 1949: 900 khz; 5 kw-D, 136 w-N. TL: N36 12 26 W86 16 03. Stereo. Hrs opn: 24 510 Trousdale Ferry Pike, 37087. Phone: (615) 444-0900. Fax: (615) 443-4235. Web Site: www.wantfm.com. Licensee: WCOR Inc. (acq 4-93; FTR: 3-15-93). Population served: 500,000 Law Firm: Irwin, Campbell, & Tannewald. Format: Country, big band. News staff: one. Target aud: General. ♦Susan H. Bay, pres & gen mgr; Billy Goodman, news dir; Coleman Walker, pub affrs dir; Gary Brown, chief of engrg.

WRVW(FM)— Aug 31, 1962: 107.5 mhz; 46 kw. Ant 1,342 ft TL: N36 15 50 W86 47 39. Stereo. Hrs opn: 24 55 Music Sq. W., Nashville, 37203. Phone: (615) 664-2400. Fax: (615) 664-2434. E-mail: programming @1075theriver.com Web Site: www.1075theriver.com. Licensee: Capstar TX L.P. Group owner: Clear Channel Communications Inc. (acq 8-30-00; grpsl). Population served: 1,000,000 Format: CHR. News staff: one; News: 4 hrs wkly. Target aud: 18-49. ♦Gene McKay, gen mgr; Keith Kaufman, opns mgr; Darren Smith, sls dir; Tom Schurr, mktg mgr; Temple Hancock, prom mgr; Rich Davis, progmg dir.

Lenoir City

WBLC(AM)— June 15, 1965: 1360 khz; 1 kw-D, 24 w-N. TL: N35 47 32 W84 17 45. Hrs open: 24 Box 247, 37771. Secondary address: 4787 Browder Hollow Rd. 37771. Phone: (865) 986-5332. Fax: (865) 986-5332. E-mail: wblc3abn@bellsouth.net Licensee: Three Angels Broadcasting Network Inc. (acq 8-13-02; $55,000). Population served: 600,000 Format: Christian, relg. Target aud: 35 plus. ♦Jim Morris, gen mgr & stn mgr.

WKZX-FM—Listing follows WLIL(AM).

WLIL(AM)— May 30, 1950: 730 khz; 1 kw-D, 280 w-N. TL: N35 46 12 W84 16 47. Hrs open: 24 Box 340, 406 E. Broadway, 37771. Phone: (865) 986-7536. E-mail: wlilcountry@aol.com Licensee: B.P. Broadcasters L.L.C. (acq 8-01-2000; $1 million with co-located FM). Population served: 31,189 Natl. Network: CNN Radio. Natl. Rep: Keystone (unwired net). Rgnl rep: Rgnl Reps. Format: Country oldies. News staff: one; News: 20 hrs wkly. Target aud: General; adults. Spec prog: Black one hr, farm one hr, gospel 18 hrs, American Indian one hr wkly. ♦Dale Anthony, gen mgr; Glenn A. McNish Sr., stn mgr; Zollie Cantrell Jr., mktg mgr; Ronald McDonald, prom dir.

WKZX-FM—Co-owned with WLIL(AM). Sept 19, 1967: 93.5 mhz; 6 kw. 165 ft TL: N35 46 12 W84 16 47. Fax: (865) 986-1716. E-mail: wkzx@aol.com Format: Soft adult contemp. News staff: one; News: 20 hrs wkly. Target aud: General; Adults. ♦Mark Herzog, progmg dir; Kevin Potter, chief of engrg; Debra Upton, min affrs dir; Glenn McNish, sports cmtr; Buzz McNish, disc jockey; Russell Mayes, disc jockey.

Lewisburg

WAXO(AM)— Sept 1, 1980: 1220 khz; 1 kw-D. TL: N35 25 42 W86 46 22. Hrs open: 217 W. Commerce St., 37091. Phone: (931) 359-6641. Fax: (931) 270-9290. Web Site: www.waxo.com. Licensee: Marshall County Radio Corp. (acq 9-1-82; $250,000; FTR: 8-23-82). Population served: 25,000 Natl. Network: USA. Format: Country. Spec prog: Gospel 12 hrs wkly. ♦Bob Smartt, pres & gen mgr.

WJJM(AM)— May 15, 1947: 1490 khz; 1 kw-U. TL: N35 27 03 W86 46 57. Hrs open: Box 2025, 37091. Secondary address: 344 E. Church St. 37091. Phone: (931) 359-4511. Fax: (931) 270-9556. E-mail: wjjm@wjjm.com Web Site: www.wjjm.com. Licensee: WJJM Inc. (acq 3-26-2004; $230,000 with co-located FM). Population served: 7,207 Rgnl. Network: Tenn. Radio Net. Natl. Rep: Keystone (unwired net). Law Firm: Fletcher, Heald & Hidreth. Format: Country. News staff: one; News: 1 hr wkly. Target aud: 25-65; manufacturing, business, family programming. ♦Michelle W. Haislip, pres; Lisa Savage, gen mgr; Michelle W. Haislip, stn mgr & sls dir; Jeff Haislip, progmg mgr; Doug Hazelwood, mus dir; Tommy Allen, news dir; Don Roden, chief of engrg.

WJJM-FM— Feb 20, 1969: 94.3 mhz; 5.5 kw. Ant 115 ft TL: N35 27 03 W86 46 57. Stereo. 17 Web Site: www.wjjm.com.20,000 ♦Doug Cheek, adv dir; Chris Bates, disc jockey; Doug Hazelwood, disc jockey; Jennifer St. John, disc jockey; Linda Dugan, disc jockey; Tommy Allen, disc jockey.

Lexington

WDXL(AM)— July 1954: 1490 khz; 1 kw-U. TL: N35 38 05 W88 23 34. Hrs open: 24 Box 279, 38351. Secondary address: 584 Smith Ave. 38351. Phone: (731) 968-3500. Phone: (731) 968-9990. Fax: (731) 968-0380. E-mail: wzlt@netease.net Licensee: Lexington Broadcast Service Inc. (acq 1955). Population served: 21,000 Natl. Network: Jones Radio Networks. Format: Southern gospel. News staff: one; News: 10 hrs wkly. Target aud: 30 plus. Spec prog: Black 4 hrs, gospel 10 hrs wkly. ♦Dan Hughes, gen mgr, gen sls mgr & sports cmtr; Terry Rhodes, progmg dir.

WZLT(FM)—Co-owned with WDXL(AM). September 1964: 99.3 mhz; 5 kw. 150 ft TL: N35 38 05 W88 23 34. Stereo. 24 Phone: (731) 968-9990. E-mail: wzlt@netease.net 30,000 Format: Adult contemp. Target aud: General. ♦Todd Buttrey, progmg dir.

***WIGH(FM)**— Sept 30, 1995: 88.7 mhz; 15 kw. Ant 548 ft TL: N35 42 12 W88 36 10. (CP: 14 kw, ant 538 ft. TL: N35 43 19 W88 36 07). Stereo. Hrs opn: 24 Box 3206, Tupelo, MS, 38803. Secondary address: 107 Park Gate Dr., Tupelo, MS 38801. Phone: (662) 844-8888. Fax: (662) 842-6791. Licensee: American Family Association. Group owner: American Family Radio (acq 5-22-03; $20,000). Population served: 250,000 Natl. Network: American Family Radio. Format: Christian. Target aud: Visually & physically impaired. ♦Marvin Sanders, gen mgr.

Livingston

WLIV(AM)— Nov 26, 1956: 920 khz; 1 kw-D. TL: N36 22 28 W85 18 20. Hrs open: 24 Box 359, 1130 W. Main St., 38570. Phone: (931) 823-1226. Fax: (931) 823-6005. Licensee: Sunny Broadcasting G.P. (acq 1996; $100,000 with co-located FM). Population served: 21,504 Natl. Network: CNN Radio. Rgnl. Network: Tenn. Radio Net. Natl. Rep: Keystone (unwired net). Format: News, Sports. News staff: 2; News: 7 hrs wkly. Target aud: General. Spec prog: Farm 2 hrs, relg 15 hrs, gospel 18 hrs wkly. ♦Millard V. Oakley, pres; Joel Upton, gen mgr; Carolyn Peterman, stn mgr; Craig Cantrell, opns dir; Austin Stinnett, chief of engrg; Shirley Burnette, traf mgr.

WLQK(FM)— December 1966: 95.9 mhz; 20 kw. 784 ft TL: N36 11 36 W85 20 41. Stereo. Hrs opn: 259 S. Willow Ave., Cookeville, 38501. Phone: (931) 526-6064. Fax: (931) 520-1590. E-mail: jimstapleton @jwcbroadcasting.com Web Site: www.literock959.com. Licensee: JWC Broadcasting (group owner; acq 12-18-98). Natl. Rep: Rgnl Reps. Format: Soft rock. Target aud: General. ♦Jim Stapleton, gen mgr & stn mgr.

Lobelville

WFGZ(FM)— October 1974: 94.5 mhz; 22 kw. Ant 715 ft TL: N35 45 56 W87 49 50. Stereo. Hrs opn: 24 25 Stonebrook Place, Suite G 322, Jackson, 38305. Phone: (888) 855-9394. Fax: (731) 855-1600. E-mail: info@gracebroadcasting.com Web Site: www.gracebroadcasting.com. Licensee: Grace Broadcasting Services Inc. (acq 11-28-2003; $487,000). Population served: 250,000 Format: Contemp Christian. News staff: one; News: 14 hrs wkly. Target aud: 18-54; mid to upper income adults with purchasing power. ♦Charles M. Ennis, pres; Lacy Ennis, gen mgr & opns mgr.

Lookout Mountain

WFLI(AM)— Feb 20, 1961: 1070 khz; 50 kw-D, 2.5 kw-N, DA-2. TL: N35 02 42 W85 21 44. Hrs opn: 24 621 O' Grady Dr., Chattanooga, 37419. Phone: (423) 821-3555. Fax: (423) 821-3557. E-mail: flipaul@aol.com Licensee: WFLI Inc. Population served: 400,000 Natl. Network: USA. Format: Relg/Southern Gospel. Target aud: 18-54. Spec prog: College football. ♦Ying Hua Benns, pres & gen mgr; Paul White, stn mgr & opns mgr.

Loretto

WJHX(AM)—See Lexington, AL

Loudon

WFIV-FM— May 20, 1991: 105.3 mhz; 6 kw. 328 ft TL: N35 48 40 W84 16 02. Stereo. Hrs opn: 24 517 Watt Rd., Knoxville, 37922. Phone: (865) 675-4105. Fax: (865) 675-4859. E-mail: homeradio@nxs.net Web Site: www.wkvl.com. Licensee: Horne Radio L.L.C. Group owner: Horne Radio Group (acq 8-29-2001; grpsl). Population served: 350,000 Natl. Network: CBS. Format: AAA. News: 3 hrs wkly. Target aud: 25-50; baby boomers. ♦Douglas A. Horne, pres; Jim Christensen, gen mgr; Shawn Nunally, gen sls mgr; Todd Ethridge, progmg dir; Brian Tatum, chief of engrg; Martha Lee, traf mgr.

WLOD(AM)— Jan 1, 1983: 1140 khz; 1 kw-D. TL: N35 43 35 W84 20 49. Hrs open: Sunrise-sunset 517 Watt Rd., Knoxville, 37922. Phone: (865) 675-4105. Fax: (865) 675-4859. Licensee: Horne Radio LLC. Group owner: Horne Radio Group (acq 8-29-01; grpsl). Natl. Network: ABC. Format: News/Talk. News staff: one. Target aud: 35 plus. ♦Bill Tatum, gen mgr.

WNML-FM— Jan 5, 1989: 99.1 mhz; 6 kw. 328 ft TL: N35 47 10 W84 17 24. Stereo. Hrs opn: 24
Simulcast with WNML (AM) & WNRX (FM) Knoxville.
Box 11167, Knoxville, 37939-1167. Secondary address: 4711 Old Kingston Pike, Knoxville 37919. Phone: (865) 588-6511. Fax: (865) 558-4217. Web Site: www.sportsanimal99.com. Licensee: Citadel Broadcasting Co. Group owner: Citadel Broadcasting Corp. (acq 8-2-00; grpsl). Natl. Network: ABC, Westwood One. Natl. Rep: Katz Radio. Rgnl rep: Rgnl Reps. Wire Svc: AP Format: Sports talk. Target aud: 25-54. ♦Ed Brantley, gen mgr; Mike Hammond, opns mgr; Jack Lee, gen sls mgr; Mickey Dearstone, mktg dir; Tim Berry, chief of engrg.

Madison

WPLN(AM)—Licensed to Madison. See Nashville

WRLT(FM)—See Franklin

Madisonville

WRKQ(AM)— July 12, 1967: 1250 khz; 500 w-D, 86 w-N. TL: N35 30 29 W84 22 45. Hrs open: 6 AM-6 PM Box 489, 37354. Phone: (423) 442-1446. Fax: (423) 440-9636. Web Site: www.wrkq.net. Licensee: Beverly Broadcasting Co. LLC (acq 5-4-2004; $40,000). Population served: 2,858 Natl. Network: CBS Radio. Format: News/talk. News staff: one. Target aud: General. ♦Mike Beverly, pres & gen mgr.

WYGO(FM)— Nov 15, 1992: 99.5 mhz; 2.51 kw. 515 ft TL: N35 30 20 W84 27 21. Stereo. Hrs opn: 24 Box 933, Athens, 37371. Secondary address: 2110 Oxnard Rd., Athens 37303. Phone: (423) 337-0995. Phone: (423) 746-0995. Fax: (423) 745-2000. Licensee: Major Broadcasting Corp. Format: Music of 80s & 90s, hot adult contemp. Target aud: 18-54. ♦Randy Sliger, gen mgr.

Manchester

WFTZ(FM)— Nov 16, 1992: 101.5 mhz; 3 kw. 345 ft TL: N35 23 51 W86 08 39. Stereo. Hrs opn: 24 Box 1015, 37349. Secondary address: 1025 Hillsboro Blvd. 37355. Phone: (931) 723-1015. Phone: (931) 728-3458. Fax: (931) 723-1099. E-mail: kahuna@fantasyradio.com Web Site: www.fantasyradio.com. Licensee: Phase Two Communications Inc. (acq 10-21-91). Population served: 250,000 Natl. Network: ABC. Law Firm: Timothy K. Brady. Format: Adult contemp. News staff: one; News: 4 hrs wkly. Target aud: 25-45; white collar, educated. ♦Roger H. Dotson, CEO, pres, gen mgr, opns mgr & chief of engrg; Marsha T. Dotson, gen sls mgr; Amber Dotson, prom VP & traf mgr; Bill Priestly, progmg dir; Wayne D. Hudgens, news dir & pub affrs dir.

WMSR(AM)— Apr 7, 1957: 1320 khz; 5 kw-D, 79 w-N. TL: N35 28 03 W86 05 42. Hrs open: 6 AM-7 PM 1030 Oakdale St., 37355. Phone: (931) 728-3526. Phone: (931) 728-1320. Fax: (931) 728-3527. Web Site: www.wmsrthegroove.com. Licensee: Coffee County Broadcasting Inc. (acq 8-22-2005; $700,000). Wire Svc: AP Format: Oldies, sports, talk. News staff: one; News: 21 hrs wkly. Target aud: General. Spec prog: High school sports, farm. ♦Scott Vaughn, gen mgr.

WWTN(FM)— June 20, 1962: 99.7 mhz; 100 kw. 2,033 ft TL: N35 28 03 W86 05 42. Stereo. Hrs opn: 10 Music Cir. E., Nashville, 37203. Phone: (615) 321-1067. Fax: (615) 321-5771. Fax: (615) 871-6099. Web Site: www.997wtn.com. Licensee: Cumulus Licensing Corp. Group owner: Cumulus Media Inc. (acq 7-21-2003; $65 million with WSM-FM Nashville). Population served: 1,233,170 Natl. Network: ABC, CBS Radio. Wire Svc: UPI Format: News/talk, sports. Target aud: 25-54; general. ♦Michael Dickey, gen mgr.

Martin

WCMT(AM)— June 8, 1957: 1410 khz; 700 w-D, 58 w-N. TL: N36 21 45 W88 50 56. Hrs open: 24 Box 318, 1410 N. Lindell St., 38237. Phone: (731) 587-9526. Fax: (731) 587-5079. E-mail: ptinkle@crunet.com Web Site: www.wcmt.com. Licensee: Thunderbolt Broadcasting Co. Group owner: Thunderbolt Broadcasting Co./Gibson County Broadcasting (acq 3-1-80; FTR: 2-18-80). Population served: 420,335 Natl. Network: Westwood One, AP Radio. Law Firm: Womble, Carlyle, Sandridge & Rice. Format: News/talk, oldies. News: 10 hrs wkly. Target aud: 25-54; baby boomers. ♦Jimmy Smith, VP; Paul F. Tinkle, CEO, pres & gen mgr.

WCMT-FM— Sept 26, 1967: 101.3 mhz; 22 kw. Ant 308 ft TL: N36 29 00 W88 57 10. Stereo. 24 Box 318, Marten, 38232. Phone: (731) 885-0051.507,170 Law Firm: Womble, Carlyle, Sandridge & Rice. Format: Classic Hits. News: 25 hrs wkly. ♦Paul Tinkle, pres.

***WUTM(FM)**— Sept 1, 1971: 90.3 mhz; 185 w. 250 ft TL: N36 20 28 W88 51 39. Stereo. Hrs opn: 8 AM-midnight (M-F) 220 Gooch Hall, Univ. of Tenn. at Martin, 38238. Phone: (731) 881-7095. Fax: (731) 881-7550. E-mail: wutm@utm.edu Web Site: www.utm.edu%7Ewutm/. Licensee: University of Tennessee. Population served: 10,000 Format: CHR. Target aud: General; Univ. ♦Richard Robinson, gen mgr.

Maryville

WBCR(AM)—See Alcoa

WGAP(AM)— Aug 13, 1947: 1400 khz; 1 kw-U. TL: N35 45 41 W83 58 57. Hrs open: 24 517 Watt Rd., 37922. Phone: (865) 983-4310. Phone: (865) 983-4105. Fax: (865) 983-4314. Fax: (865) 675-4859. Licensee: Horne Radio LLC. Group owner: Horne Radio Group (acq 8-29-01; grpsl). Population served: 412,000 Natl. Network: Motor Racing Net. Natl. Rep: Rgnl Reps. Law Firm: Pepper & Corazzini. Format: Country. News staff: one; News: 18 hrs wkly. Target aud: 25 plus; general. ♦Brian Tatum, gen mgr.

WKCE(AM)— 1989: 1120 khz; 500 w-D. TL: N35 45 08 W83 35 04. Stereo. Hrs opn: 802 S. Central, Knoxville, 37902. Phone: (865) 546-4653. Fax: (865) 637-7133. Licensee: Kirkland Wireless Broadcasters Inc. (acq 3-25-2002; $400,000 with WVLZ(AM) Knoxville). Natl. Network: ESPN Deportes. Format: Sp sports. ♦Rob Robinson, opns mgr.

WQJK(FM)— Feb 2, 1990: 95.7 mhz; 6 kw. Ant 321 ft TL: N35 49 53 W84 01 25. Stereo. Hrs opn: 24 1100 Sharps Ridge Mem Park Dr., Knoxville, 37917. Phone: (865) 525-6000. Fax: (865) 656-4386. E-mail: rchambers@sccradio.com Web Site: www.jackfmknoxville.com. Licensee: South Central Communications Corp. (group owner) Population served: 785,000 Format: JACK-FM. Target aud: 35-59; adults. ♦J.P. Engelbrecht, CEO; Craig Jacobus, pres & gen mgr; Terry Gillingham, opns mgr; Randy Chambers, prom mgr & progmg dir; Judy Dyke, traf mgr.

Maynardville

***WOEZ(FM)**— 2001: 88.3 mhz; 2.85 kw horiz. Ant 1,489 ft TL: N36 00 13 W83 56 34. Hrs open: 1621 E. Magnolia Ave., Knoxville, 37917. Phone: (865) 521-8910. Fax: (865) 521-8923. E-mail: info@ez88.org Web Site: www.ez88.org. Licensee: Foothills Broadcasting Inc. Population served: 650,000 Format: Adult Standards. ♦David Wells, gen mgr; Mike Blakemore, progmg dir; Marisa Lykins, prom.

McKenzie

***WAJJ(FM)**— 2002: 89.3 mhz; 1 kw. Ant 328 ft TL: N36 06 55 W88 30 38. Hrs open: 24 1415 Island Ford Rd., Madisonville, KY, 42431. Phone: (270) 825-3004. Fax: (270) 825-3005. E-mail: comments@wsof.org Web Site: www.wsof.org. Licensee: Madisonville Christian School (acq 7-25-2005; $90,000). Format: Christian educ. ♦Gary Hall, gen mgr.

WHDM(AM)— Jan 29, 1954: 1440 khz; 500 w-D, 91 w-N. TL: N36 07 20 W88 31 31. Hrs open: 110 India Rd., Paris, 38242. Phone: (731) 644-9455. Fax: (731) 644-9970. Licensee: WHDM Broadcasting Inc. (acq 1-4-2002; $69,000). Population served: 5,651 Natl. Network: ABC. Format: Oldies. News staff: one; News: 4 hrs wkly. ♦Gary D. Benton, pres; Janice Benton, gen mgr & opns VP.

WWYN(FM)— Feb 11, 1963: 106.9 mhz; 100 kw. Ant 892 ft TL: N35 54 06 W88 46 55. (CP: ant 886 ft. TL: N35 54 05 W88 46 51). Stereo. Hrs opn: 24 111 W. Main St., Jackson, 38301. Phone: (731) 427-9616. Fax: (731) 424-2773. E-mail: cthomas@wwyn.fm Web Site: www.wwyn.fm. Licensee: Rainbow Media Inc. Group owner: Thomas Radio LLC (acq 11-9-2001). Population served: 386,000 Format: Modern country. News staff: one; News: 4 hrs wkly. Target aud: 25-54; adults. ♦Chip Thomas, gen mgr; Shane Conner, progmg dir; Ellen Bennet, news dir; Jim Smith, chief of engrg.

McKinnon

WTPR-FM— 1992: 101.7 mhz; 1.8 kw. Ant 607 ft TL: N36 24 39 W87 58 06. Stereo. Hrs opn: 24
Rebroadcasts WTPR(AM) Paris 100%.
206 N. Brewer St., Paris, 38242. Phone: (731) 642-7100. Fax: (731) 642-9367.yes Licensee: WENK of Union City Inc. (group owner; (acq 1996; $200,000). Population served: 55,000 Rgnl rep: Rgnl Reps. Law Firm: Shainis & Peltzman. Format: 60s & 70s oldies. News staff: one; News: 12 hrs wkly. Target aud: 35-54. ♦Terry Hailey, gen mgr, progmg dir & engr.

McMinnville

WAKI(AM)— 1947: 1230 khz; 1 kw-U. TL: N35 41 42 W85 46 33. Hrs open: 5 AM-midnight Box 759, 37111. Phone: (931) 473-9253. Fax: (931) 473-4149. E-mail: jeffbarnes@clearchannel.com Licensee: Citicasters Licenses L.P. Group owner: Clear Channel Communications Inc. (acq 10-30-99; grpsl). Population served: 85,000 Law Firm: Timothy K. Brady. Format: News/talk info. News staff: one; News: 24 hrs wkly. Target aud: 25-54; general. Spec prog: Farm 2 hrs wkly. ♦Lowery Mays, pres; David Roederer, gen mgr.

WBMC(AM)— May 1, 1955: 960 khz; 500 w-D. TL: N35 40 00 W85 46 00. Hrs open: 5 AM-8 PM Box 759, 37110. Secondary address: 230 W. Colville St. 37110. Phone: (931) 473-2104. Fax: (931) 473-4149. Licensee: Citicasters Licenses L.P. Group owner: Clear Channel Communications Inc. (acq 10-14-99; grpsl). Population served: 90,000 Natl. Network: ABC. Format: Country, Top-40, gospel. News staff: one; News: 10 hrs wkly. Target aud: General. Spec prog: Farm 5 hrs wkly. ♦Bryan Kell, gen mgr, stn mgr & gen sls mgr; Jeff Barnes, progmg dir; Jay Walker, news dir; Homer Wilson Jr., chief of engrg; Kathy Klasek, traf mgr; Bud Godwin, disc jockey; Jonathan Lee, disc jockey; Kelly Marlowe, disc jockey.

WKZP(FM)—Co-owned with WBMC(AM). Jan 23, 1964: 103.9 mhz; 5.3 kw. 130 ft TL: N35 40 00 W85 46 00. Stereo. 24 Phone: (931) 473-9253. Web Site: hotcountry104.com.225,000 Format: Hot country. Target aud: 18-40. ♦Bryan Kell, sls dir; Jeff Edwards, progmg dir; Kathy Klasek, traf mgr; Jonathan Lee, disc jockey; Taylor Bishop, disc jockey.

***WCPI(FM)**— February 1997: 91.3 mhz; 1.6 kw. 182 ft TL: N35 40 41 W85 45 06. Stereo. Hrs opn: 24 110 S. Court Sq., 37110. Phone: (931) 506-9274. Fax: (931) 507-1005. Licensee: Warren County Education Foundation. Population served: 90,763 Wire Svc: AP Format: Educ. Target aud: 6 plus. ♦Dr. Norman Rone, pres; Gloria Grissom, stn mgr; Mary Cantrell, mktg dir; Richard Myers, chief of engrg.

Memphis

KJMS(FM)— Mar 10, 1965: 101.1 mhz; 100 kw. Ant 449 ft TL: N35 08 00 W90 05 38. (CP: ant 561 ft. TL: N35 13 22 W90 02 36). Stereo. Hrs opn: 24 2650 Thousand Oaks Blvd., Suite 4100, 38118. Phone: (901) 259-1300. Fax: (901) 259-6449. E-mail: jeffreyjones@clearchannel.com Web Site: www.v10ll.com. Licensee: CC Licenses LLC. Group owner: Clear Channel Communications Inc. Format: Urban contemp. Target aud: 18-49. ♦Tim Davies, VP; Ralph Salierno, sls dir; Franklin Gilbert Jr., mktg dir & mus dir; Eileen Collier, progmg dir; Alonzo Pendleton, chief of engrg.

KQPN(AM)—See West Memphis, AR

KWAM(AM)— 1946: 990 khz; 10 kw-D, 450 w-N, DA-2. TL: N35 08 04 W90 05 38. Hrs open: 2650 Thousand Oaks Blvd., Suite 4100, 38118. Phone: (901) 259-1300. Fax: (901) 259-6449. Licensee: Concord Media Group Inc. (acq 11-2-2000; $1 million). Population served: 800,000 Format: Talk/news. Target aud: 25 plus; general. ♦Tim Davies, gen mgr; Jeffrey Jones, gen sls mgr; Leonard Blakely, progmg dir.

WBBP(AM)— Apr 11, 1964: 1480 khz; 5 kw-D, 100 w-N. TL: N35 03 18 W90 05 15. Hrs open: 24 369 GE Patterson Ave., 38126. Phone: (901) 278-7878. Fax: (901) 332-1707. Web Site: www.bbless.org. Licensee: Bountiful Blessings Inc. (acq 10-25-90; $462,000; FTR: 11-19-90). Population served: 942,000 Format: Gospel. Target aud: 25-54; Listeners who enjoy a variety of gospel. ♦Bishop G.E. Patterson, pres & gen mgr; Sterlene Chavers, traf mgr.

WCRV(AM)—See Collierville

WDIA(AM)— June 7, 1947: 1070 khz; 50 kw-D, 5 kw-N, DA-2. TL: N35 16 05 W90 01 03. Hrs open: 24 2650 Thousand Oaks Blvd., Suite 4100, 38118. Phone: (901) 259-1300. Fax: (901) 259-6451. Web Site: www.am1070wdia.com. Licensee: CC Licenses LLC. Group owner: Clear Channel Communications Inc. (acq 1996; grpsl). Population served: 670,000 Natl. Network: ABC. Natl. Rep: Clear Channel. Format: Black urban contemp. Target aud: 25-54; Black adults. Spec prog: Gospel. ♦Tim Davies, gen mgr; Ralph Salierno, sls dir; Franklin Gilbert Jr., prom dir; Bobby O'Jay, progmg dir; Alonzo Pendleton, chief of engrg.

WHRK(FM)—Co-owned with WDIA(AM). Jan 1, 1961: 97.1 mhz; 100 kw. Ant 530 ft TL: N35 13 23 W90 02 33. Stereo. 24 Web Site: www.k97fm.com.751,000 Format: Urban contemp. Target aud: 18-49. ♦Devin Steel, progmg dir. Co-owned TV: WLMT(TV), WPTY-TV affil

WEGR(FM)—Listing follows WREC(AM).

***WEVL(FM)**— May 1, 1976: 89.9 mhz; 9.3 kw. 374 ft TL: N35 08 05 W89 45 38. Stereo. Hrs opn: 20 Box 40952, 38174-0952. Secondary address: 518 S. Main St. 38103. Phone: (901) 528-0560. Phone: (901) 528-0561. E-mail: wevl@wevl.org Web Site: www.wevl.org. Licensee: Southern Communication Volunteers Inc. Population served: 1,000,000 Format: Var, educ, blues. News: 2 hrs wkly. Target aud: General. Spec prog: Jazz 15 hrs, C&W 15 hrs, Fr one hr, Irish 4 hrs, Indian subcontinent one hr wkly. ♦Dan Phillips, pres; Judy Dorsey, stn mgr, opns dir & dev dir.

WGKX(FM)— Jan 10, 1968: 105.9 mhz; 100 kw. Ant 993 ft TL: N35 09 16 W89 49 20. Stereo. Hrs opn: 24 5629 Murray Rd., 38119.

Phone: (901) 682-1106. Fax: (901) 767-9531. Web Site: www.kix106.com. Licensee: Citadel Broadcasting Co. Group owner: Citadel Broadcasting Corp. (acq 3-23-2004; grpsl). Population served: 1,800,000 Natl. Rep: Katz Radio. Format: Country. Target aud: 25-54. ♦Dan Barron, sls dir & chief of engrg; Sheri Sawyer, gen mgr & sls dir; Gennora Reed, gen sls mgr & prom dir; Lance Tidwell, progmg dir; Paula Davis, prom.

WGSF(AM)— February 1984: 1030 khz; 50 kw-D, 1 kw-N, 10 kw-CH. TL: N35 10 59 W89 56 17. Hrs open: 24 3654 Park Ave., 38111. Phone: (901) 454-9948. Fax: (901) 454-1027. Licensee: Arlington Broadcasting Co. Inc. Natl. Network: Westwood One, CBS. Format: Spanish. ♦Daniel Ybarra, gen mgr.

WHBQ(AM)— Mar 18, 1925: 560 khz; 5 kw-D, 1 kw-N, DA-2. TL: N35 15 12 W90 02 51. Hrs open: 24 6080 Mt. Moriah, 38115. Phone: (901) 375-9324. Fax: (901) 375-4117. Web Site: www.sports56whbq.com. Licensee: Flinn Broadcasting Corp. (acq 10-1-88). Population served: 1,735,400 Natl. Network: CBS. Rgnl. Network: Conference Call. Format: Sports. News: 2 hrs wkly. Target aud: 18-54. ♦George S. Flinn, pres; Chris Coates, gen mgr; Eli Savoie, progmg dir & chief of engrg.

***WKNO-FM**— Mar 1, 1972: 91.1 mhz; 100 kw. 580 ft TL: N35 09 17 W89 49 20. Stereo. Hrs opn: 24 Box 241880, 38124. Secondary address: 900 Getwell Rd. 38111. Phone: (901) 325-6544. Fax: (901) 325-6506. Web Site: www.wknofm.org. Licensee: Mid-South Public Communications Foundation. Population served: 623,530 Natl. Network: NPR, PRI. Law Firm: Schwartz, Woods & Miller. Wire Svc: AP Format: Class, news. News staff: one; News: 51 hrs wkly. Target aud: 35 plus. ♦Michael LaBonia, pres; Dan Campbell, gen mgr; Darel Snodgrass, opns mgr; Charles McCarty, dev dir. Co-owned TV: *WKNO-TV affil

WLOK(AM)— Mar 1, 1956: 1340 khz; 1 kw-U. TL: N35 07 01 W90 00 59. Hrs open: 24 363 S. 2nd St., 38103. Phone: (901) 527-9565. Fax: (901) 528-0335. Web Site: www.wlok.com. Licensee: Gilliam Communications Inc. (acq 1-12-77). Population served: 1,000,000 Natl. Rep: McGavren Guild. Format: Gospel, talk. News staff: one. Target aud: 25-54. ♦H. Gilliam Jr., gen mgr; Jerry Bafford, gen sls mgr & news dir; John Rhea, gen sls mgr & natl sls mgr; Kim Harper, progmg dir.

WMC(AM)— Jan 21, 1923: 790 khz; 5 kw-U, DA-N. TL: N35 10 09 W89 53 12. Stereo. Hrs opn: 24 1960 Union Ave., 38104. Phone: (901) 726-0555. Fax: (901) 726-5847. Web Site: www.wmc79.com. Licensee: CBS Radio Stations Inc. Group owner: Infinity Broadcasting Corp. (acq 8-30-2000; $75 million with co-located FM). Population served: 1,008,400 Natl. Rep: CBS Radio. Wire Svc: Metro Weather Service Inc. Format: Country. News staff: one. ♦Terry Wood, sr VP, VP, gen mgr, gen mgr & opns mgr.

WMC-FM— May 22, 1947: 99.7 mhz; 300 kw. 970 ft TL: N35 10 09 W89 53 12. Stereo. 24 Fax: (901) 272-9618. Web Site: www.fm100memphis.com.1,008,400 Format: Adult contemp. Target aud: 25-54; adults.

WMCM(FM)—Co-owned with WRKD(AM). Apr 16, 1968: 103.3 mhz; 20.5 kw. 771 ft TL: N44 07 35 W69 08 18. Stereo. 24 Web Site: www.realcountry1033.com.100,000 Format: Country. News staff: one. Target aud: General. ♦D.J. McCoy, progmg dir; Elaine Knowlton, traf mgr; Don Shields, local news ed, political ed & sports cmtr; Peter K. Orne, rsch dir & edit dir.

WMCR-FM— September 1972: 106.3 mhz; 1.25 kw. Ant 718 ft TL: N43 02 48 W75 39 58.

WMPS(AM)—Bartlett, Aug 19, 1986: 1210 khz; 10 kw-D, 250 w-N, DA-2. TL: N35 18 27 W89 38 21. Hrs open: 24 6080 Mt. Moriah Rd. Ext., 38115. Phone: (901) 375-9324. Fax: (901) 375-0041. Licensee: Arlington Broadcasting Co. Inc. Rgnl. Network: Tenn. Radio Net. Format: Music of Your Life. Target aud: 25 plus. Spec prog: Relg progmg 7 hrs wkly. ♦Fred Flinn, pres; Shea Flinn, gen mgr.

WMQM(AM)—Lakeland, Apr 27, 1955: 1600 khz; 50 kw-D, 35 w-N. TL: N35 10 34 W89 56 10. Hrs open: 24 3704 Whittier, 38108. Secondary address: Sale Office, 1300 WWCR Ave., Nashville 37218. Phone: (901) 327-2500. Fax: (901) 327-2777. Web Site: www.wwcr.com. Licensee: WMQM Inc. Group owner: F.W. Robbert Broadcasting Co. Inc. Population served: 2,000,000 Format: Relg. Target aud: General. ♦Fred P. Werstenberger, pres; George McClintock, gen mgr; David Brown, stn mgr; Adam Lock, opns mgr.

WOWW(AM)—Germantown, October 1955: 1430 khz; 2.5 kw-U, DA-N. TL: N35 04 20 W89 51 40 (day); N35 12 50 W89 47 46 (night). (CP: 2.8 kw-U, DA-2. TL: N34 59 22 W89 51 45 (one-site)). Hrs opn: 6080 Mt. Mariah Rd. Ext., 38115. Phone: (901) 375-9324. Fax: (901) 375-0041. Web Site: www.radiodisney.com. Licensee: Flinn Broadcasting Corp. (acq 9-22-93; $695,000; FTR: 10-11-93). Format: Radio disney. Target aud: 25-54. ♦George S. Flinn, pres; Lonnie Treadaway, gen mgr.

***WPLX(AM)**—Germantown, April 1987: 1170 khz; 1 kw-D. TL: N35 01 28 W89 42 21. Hrs open: Sunrise-sunset 2351 Sunset Blvd., Suite 170-218, Rocklin, CA, 95765. Phone: (916) 251-1600. Fax: (916) 251-1650. E-mail: klove@klove.com Web Site: www.klove.com. Licensee: Educational Media Foundation. Group owner: EMF Broadcasting (acq 10-20-2000; grpsl). Population served: 1,500,000 Natl. Network: K-Love. Law Firm: Shaw Pittman. Format: Contemp christian mus. News staff: 3. Target aud: 25-44; female (Judeo-Christian). ♦Richard Jenkins, pres; Mike Novak, VP; Keith Whipple, dev dir; David Pierce, progmg mgr; Ed Lenane, news dir; Sam Wallington, engrg dir; Karen Johnson, news rptr; Marya Morgan, news rptr; Richard Hunt, news rptr.

***WQOX(FM)**— Apr 8, 1974: 88.5 mhz; 30 kw. 430 ft TL: N35 09 17 W89 49 20. Stereo. Hrs opn: 24 Telecommunications Center, 2485 Union Ave., 38112. Phone: (901) 320-3460. Fax: (901) 454-7673. Web Site: www.wqoxmes/admin/avery/mes.com. Licensee: Board of Education Memphis City Schools. Population served: 1,000,000 Format: Urban adult contemp, educ, pub affrs. News staff: one. Target aud: 12-54; Students, teachers, parents & admin staff. ♦Derek Wagner, gen mgr & opns mgr; Derek A. Wagner, opns mgr; Paul Gubala, progmg mgr & mus dir; Chris Malone, asst music dir & pub affrs dir; Sherman Austin, progmg dir & pub affrs dir; Derick McMillan, chief of engrg.

WREC(AM)— September 1922: 600 khz; 5 kw-U, DA-2. TL: N35 11 51 W90 00 31. Hrs open: 24 2650 Thousand Oaks Blvd., Suite 4100, 38118. Phone: (901) 259-1300. Fax: (901) 259-6451. Web Site: www.wrecradio.com. Licensee: CC Licenses LLC. Group owner: Clear Channel Communications Inc. (acq 1996; grpsl). Population served: 1700000 Natl. Network: Westwood One, ABC. Rgnl. Network: Prog Farm, Tenn. Radio Net. Natl. Rep: Clear Channel. Wire Svc: NWS (National Weather Service) Format: News/talk, sports & info. News staff: 2; News: 5 hrs wkly. Target aud: 35 plus; upscale, professional, males 70%. Spec prog: Farm 3 hrs, relg 4 hrs wkly. ♦Timothy P. Davies, gen mgr; Alonzo Pendleton, opns mgr & chief of engrg; Ralph Salierno, sls dir; Frank Gilbert, prom dir & prom mgr; Steve Versnick, progmg dir.

WEGR(FM)—Co-owned with WREC(AM). March 1967: 102.7 mhz; 100 kw. 970 ft TL: N35 10 52 W89 49 56. Stereo. 24 E-mail: rock103@aol.com Web Site: www.rock103.com.1,700,000 Natl. Rep: Clear Channel. Format: Classic rock. News staff: one; News: 15 hrs wkly. Target aud: 25-54; 25-34 core audience-70% male, 30% female. Spec prog: Rockline, flashback, blues show. ♦Tim Spencer, opns mgr; Felicia Moore, prom mgr. Co-owned TV: WPTY-TV, WLMT(TV) affils

WRVR(FM)—Listing follows WSMB(AM).

WSMB(AM)— March 1925: 680 khz; 10 kw-D, 5 kw-N, DA-N. TL: N35 13 23 W90 02 33. Stereo. Hrs opn: 5904 Ridgeway Ctr. Pkwy., 38120. Phone: (901) 767-0104. Fax: (901) 767-0582. Licensee: Entercom Memphis License LLC. Group owner: Entercom Communications Corp. (acq 12-13-99; grpsl). Population served: 733,300 Format: Progressive talk. News staff: one; News: one hr wkly. Target aud: 35-64; adults. ♦Steve Sandman, gen mgr; Jerry Dean, opns dir, progmg dir & progmg mgr; Steve Mohammed, gen sls mgr; Jim Scott, prom mgr; Debby Hall, news dir; Mike Schwartz, chief of engrg.

WRVR(FM)—Co-owned with WSMB(AM). Sept 15, 1968: 104.5 mhz; 100 kw. 751 ft TL: N35 09 17 W89 49 20. Stereo. E-mail: river104@wrvr.com Web Site: www.wrvr.com.897,700 Format: Adult contemp. News staff: one; News: 2 hrs wkly. ♦Mike Ginsburg, VP; Gary Harkin, gen sls mgr; Jim Scott, prom dir; Bill Bannister, disc jockey; Greg Peters, disc jockey; Kay Manley, disc jockey; Pam Yates, disc jockey; Steve Butler, disc jockey.

WSNA(FM)—Germantown, Apr 15, 1977: 94.1 mhz; 50 kw. Ant 472 ft TL: N34 59 22 W89 51 45. Stereo. Hrs opn: 24 5904 Ridgeway Ctr. Pkwy., 38120. Phone: (901) 767-0104. Fax: (901) 682-2804. Web Site: www.snap941.com. Licensee: Entercom Memphis License LLC. Group owner: Entercom Communications Corp. (acq 12-13-99; grpsl). Population served: 1,200,000 Format: Rhythmic adult contemp. News staff: one; News: 2 hrs wkly. Target aud: 18-49; adults with discretionary income. ♦Steve Sandman, gen mgr; Jerry Dean, opns VP.

***WUMR(FM)**— August 1979: 91.7 mhz; 25 kw. 394 ft TL: N35 09 17 W89 51 28. Stereo. Hrs opn: 6 AM-midnight Univ. of Memphis, 3745 Central Ave., 38152. Phone: (901) 678-3176. Phone: (901) 678-4843. Fax: (901) 678-4331. E-mail: rmcdowll@memphis.edu Licensee: The University of Memphis. (acq 1979). Population served: 1,000,000 Law Firm: Schwartz, Woods & Miller. Format: Jazz, sports. News staff: one; News: 2 hrs wkly. Target aud: 18-49; upscale, college-educated. ♦Robert McDowell, gen mgr.

WXMX(FM)—Millington, Apr 12, 1960: 98.1 mhz; 100 kw. 1,240 ft TL: N35 28 03 W90 11 27. (CP: Ant 768 ft.). Stereo. Hrs opn: 5629 Murray Rd., 38119. Phone: (901) 682-1106. Fax: (901) 767-9531. Web Site: www.981themax.com. Licensee: Citadel Broadcasting Co. Group owner: Citadel Broadcasting Corp. (acq 3-23-2004; grpsl). Format: Classic hits, oldies. Target aud: 25-54; adults, upwardly mobile with above average income. ♦Sherri Sawyer, gen mgr; Dan Baron, sls dir; Gennora Reed, gen sls mgr; Michael Webb, progmg dir; Paula Davis, prom.

***WYPL(FM)**— Apr 17, 1991: 89.3 mhz; 100 kw. Ant 1,253 ft TL: N35 28 03 W90 11 27. Stereo. Hrs opn: 24 Memphis Public Library, 3030 Poplar Ave., 38111. Phone: (901) 415-2752. Fax: (901) 323-7902. Web Site: www.memphislibrary.org. Licensee: Memphis Public Library & Information Center. Law Firm: Reddy, Begley & McCormick. Format: News. News: 110 hrs wkly. Target aud: General. Spec prog: Sp one hr wkly. ♦Tommy Warren, gen mgr.

Middleton

WYDL(FM)— 2001: 100.7 mhz; 25 kw. Ant 328 ft TL: N35 00 13 W88 39 39. Hrs open: 102 N. Cass St., Suite D, Corinth, MS, 38834. Phone: (662) 284-4611. Fax: (662) 284-9609. Web Site: www.wydl.com. Licensee: Flinn Broadcasting Corp. Format: CHR, top-40, adult contemp. ♦Mike Brandt, gen mgr; Wendy Sherrod, gen sls mgr.

Milan

WYNU(FM)— Dec 12, 1964: 92.3 mhz; 100 kw. 991 ft TL: N35 54 06 W88 46 55. Stereo. Hrs opn: 24 122 Radio Rd., Jackson, 38301. Phone: (731) 427-3316. Fax: (731) 427-4576. E-mail: steveburke @clearchannel.com Web Site: www.rock923.net. Licensee: Forever South Licenses LLC. Group owner: Clear Channel Communications Inc. (acq 5-12-2006; grpsl). Population served: 1,000,000 Law Firm: Mullin, Rhyne, Emmons & Topel. Format: Classic rock. News staff: one; News: 5 hrs wkly. Target aud: 18-54; middle/upper income adults with disposable income & buying power. ♦Roger Vestal, gen mgr; Dave Hacker, opns mgr; Gina Langley, gen sls mgr; Steve Burke, progmg dir.

Millington

WLRM(AM)— June 22, 1962: 1380 khz; 2.5 kw-D, 1 kw-N, DA-2. TL: N35 18 56 W89 55 23. Hrs open: 6655 Winchester Rd., Memphis, 38115. Phone: (901) 454-4900. Licensee: CPT & T Radio Station Inc. (acq 12-28-2004; $400,000). Format: Inspirational love. ♦Michelle Price, gen mgr.

WXMX(FM)—Licensed to Millington. See Memphis

Minor Hill

WEUZ(FM)— Sept 2, 1983: 92.1 mhz; 1.2 kw. 460 ft TL: N35 07 18 W87 11 17. Stereo. Hrs opn:
Simulcast with WEUP-FM Moulton, AL; 100%.
2609 Jordan Ln. N.W., Huntsville, AL, 35816. Phone: (256) 837-9387. Fax: (256) 837-9404. E-mail: news@103weup.com Web Site: www.103weup.com. Licensee: Broadcast One Inc. (acq 12-2-93;

$310,000; FTR: 1-3-94). Format: Urban contemp, hip hop, rhythm and blues. ♦Hundley Batts, pres & gen sls mgr; Steve Murry, stn mgr; Tony Jordan, mus dir & news dir; John Hain, chief of engrg; Yvonne Craighead, traf mgr.

Monterey

WKXD(FM)— Mar 3, 1986: 106.9 mhz; 23 kw. 735 ft TL: N36 07 13 W85 14 44. Stereo. Hrs opn: 24 259 S. Willow Ave., Cookeville, 38501. Phone: (931) 528-6064. Fax: (931) 520-1590. E-mail: jstapleton@jwcbroadcasting.com Web Site: www.1069kicksfm.com. Licensee: JWC Broadcasting (group owner; acq 8-3-01). Population served: 700,000 Natl. Network: ABC. Natl. Rep: Rgnl Reps. Format: Rock. News staff: one; News: one hr wkly. Target aud: 18-49; young, adult & affluent audiences. ♦Jim Stapleton, gen mgr.

WLIV-FM— Jan 8, 1997: 104.7 mhz; 1.25 kw. 712 ft TL: N36 15 42 W85 16 35. Stereo. Hrs opn: 24 1130 West Main St., Livingston, 38570. Phone: (931) 823-1226. Fax: (931) 823-6005. Licensee: Sunny Broadcasting G.P. (acq 9-7-99). Population served: 40,000 Natl. Network: Westwood One, Fox News Radio. Wire Svc: AP Format: Country, major sports. News staff: one; News: 4 hrs wkly. Target aud: General. ♦Millard V. Oakley, pres; Joel Upton, gen mgr; Carolyn Peterman, stn mgr; Craig Cantrell, opns dir; Austin Stinnett, chief of engrg; Shirley Burnette, traf mgr; Mark Young, pub svc dir.

Morristown

WCRK(AM)— October 1947: 1150 khz; 5 kw-D, 500 w-N, DA-N. TL: N36 14 11 W83 18 33. Hrs open: 24 Box 220, 37815-0220. Phone: (423) 586-9101. Fax: (423) 581-7756. E-mail: wcrk@lcs.net Web Site: www.wcrk.com. Licensee: Radio Acquisition Corp. (acq 7-13-98; $250,000). Population served: 59,000 Natl. Network: ABC. Natl. Rep: Rgnl Reps. Format: Adult contemp, oldies. News staff: one; News: 45 hrs wkly. Target aud: 25-54; slighty more female, average income $50,000 yearly. ♦S. Herschel Lake, pres; Geraldine Lake, VP; Ed Dodson, gen mgr; Tim Crews, sls dir; Anisa Croxdale, progmg dir; Mike Rypel, news dir; Dan Trombley, engrg mgr & engr; Cynthia Moyers, traf mgr.

WJDT(FM)—See Rogersville

WMTN(AM)— Oct 19, 1957: 1300 khz; 5 kw-D, 100 w-N. TL: N36 12 15 W83 19 57. Hrs open: Box 220, 37815. Phone: (423) 586-7993. Fax: (423) 581-4290. Licensee: Radio Acquisition Corp. Group owner: Horne Radio Group (acq 10-12-2006). Population served: 500,000 Natl. Network: USA. Rgnl. Network: Tenn. Radio Net. Format: News/talk. ♦William D. Buntin, gen mgr; Roddy Woods, stn mgr.

WMXK(FM)— May 31, 1964: 94.1 mhz; 920 w. Ant 771 ft TL: N36 13 40 W83 19 58. Stereo. Hrs opn: Box 220, 37815. Phone: (423) 586-7993. Fax: (423) 581-4290. Licensee: East Tennessee Radio Group L.P. (acq 9-22-2006; $1.1 million with WMTN(AM) Morristown). Population served: 60,000 Format: Rgnl Mexican.

Mount Pleasant

WXRQ(AM)— Dec 15, 1981: 1460 khz; 1 kw-D, 170 w-N. TL: N35 31 21 W87 11 34. Hrs open: 6 AM-8 PM Box 31, 209 Bond St., 38474. Phone: (931) 379-3119. Fax: (931) 379-3129. E-mail: gilliamernest@bellsouth.net Licensee: New Life Broadcasting Inc. (acq 1-17-89; $75,000; FTR: 1-30-89). Population served: 66,000 Natl. Network: USA. Format: Southern gospel. News staff: one; News: 7 hrs wkly. Target aud: General. Spec prog: Black 4 hrs wkly. ♦Donald Paul, pres & gen mgr; Monty Gilliam, gen sls mgr & mus dir; Tim Wright, progmg dir & news dir.

Mountain City

WMCT(AM)— Dec 8, 1967: 1390 khz; 1 kw-D. TL: N36 29 23 W81 47 12. Hrs open: 6 AM-6 PM 1211 N. Church St., 37683. Phone: (423) 727-6701. Fax: (423) 727-9454. E-mail: wmct@tibonline.net Web Site: www.wmct1390.com. Licensee: Johnson County Broadcasting Co. Population served: 20,000 Rgnl. Network: Tenn. Radio Net. Natl. Rep: Keystone (unwired net). Format: C&W, relg. Target aud: 25-50. ♦Fran Atkinson, pres & gen mgr.

Munford

WKIM(FM)— 1948: 98.9 mhz; 40 kw. Ant 1,135 ft TL: N35 28 03 W90 11 27. Stereo. Hrs opn: 5629 Murray Rd., Memphis, 38119. Phone: (901) 680-9898. Fax: (901) 767-9531. Web Site: www.989kimfm.com. Licensee: Citadel Broadcasting Co. Group owner: Citadel Broadcasting Corp. (acq 3-23-2004; grpsl). Population served: 10,100 Natl. Rep: Katz Radio. Format: Adult hits. ♦Sherri Sawyer, gen mgr; Dan Barron, sls dir; Amy Goodman, gen sls mgr; Marvin Nugent, progmg dir; Marvin Emilien, prom.

Murfreesboro

WCJK(FM)— Aug 10, 1963: 96.3 mhz; 52 kw. 1,286 ft TL: N36 15 50 W86 47 38. Stereo. Hrs opn: 24 Box 40506, Nashville, 37204. Secondary address: 504 Rosedale Ave., Nashville 37211. Phone: (615) 259-9696. Fax: (615) 259-4594. Web Site: www.963jackfm.com. Licensee: South Central Communications Corp. (group owner; (acq 2-4-94; $6 million; FTR: 3-28-94). Law Firm: Bryan Cave. Wire Svc: NWS (National Weather Service) Format: Adult contemp hits. News staff: one. Target aud: 25-54. ♦John D. Engelbrecht, CEO; Craig Jacobus, pres; Robert Shirel, CFO; Dennis Gwiazdon, gen mgr; Melissa Fisher, prom dir; Randy Hill, progmg dir.

***WFCM-FM**— September 1997: 91.7 mhz; 1 kw. 902 ft TL: N35 43 52 W86 41 25. Hrs open: 24
Rebroadcasts WMBW(FM) Chattanooga 100%.
1920 E. 24th Street Pl., Chattanooga, 37404. Phone: (423) 629-8900. Fax: (423) 629-0021. E-mail: wfcm@moody.edu Web Site: www.wfcm.org. Licensee: The Moody Bible Institute of Chicago. Natl. Network: Moody. Law Firm: Southmayd & Miller. Format: Educ, relg. Target aud: 25-54. ♦Dr. Michael Easley, pres; Leighton LeBoeuf, gen mgr; Andy Napier, progmg dir; Paul Martin, mus dir; David Morais, chief of engrg.

WGNS(AM)— Dec. 31, 1946: 1450 khz; 1 kw-U. TL: N35 50 26 W86 23 27. Hrs open: 24
100.5 FM.
101.9 FM TV 11 306 S. Church St., 37130-3732. Phone: (615) 893-5373. Fax: (615) 867-6397. E-mail: news@1450wgns.com Web Site: www.wgnsradio.com. Licensee: The Rutherford Group Inc. (acq 1984; FTR: 10-15-84). Population served: 250,000 Natl. Network: ABC. Format: News/talk, sports. News: 80 hrs wkly. Target aud: 25 plus; active adults, "movers & shakers" in economic & educ groupings. Spec prog: Black 7 hrs, farm 3 hrs, relg 6 hrs wkly. ♦Bart Walker, pres & gen mgr; Lee Ann Walker, VP; Scott Walker, sr VP & stn mgr; Melissa McCullough, opns mgr; Jeff Jordan, prom mgr & progmg dir; Gary Brown, chief of engrg. Co-owned TV: WETV-LP

WMGC(AM)— Nov 1, 1953: 810 khz; 5 kw-D, 6 w-N. TL: N35 50 14 W86 25 00. Hrs open:
Rebroadcasts WNSR(AM) Brentwood 55%.
435 37th Ave. N., Nashville, 37209. Phone: (615) 844-1039. Licensee: Radio 810 Nashville Ltd. Group owner: Southern Wabash Communications Corp. (acq 7-10-01). Population served: 550,000 Natl. Network: ABC. Format: Sp. Target aud: 18-54; adults. Spec prog: Relg 4 hrs wkly. ♦Ted Johnson, gen mgr.

***WMOT(FM)**— Apr 9, 1969: 89.5 mhz; 100 kw. Ant 676 ft TL: N36 05 07 W86 26 22. Stereo. Hrs opn: 24 Box 3, Middle Tennessee State Univ., 37132. Phone: (615) 898-2800. Phone: (615) 255-9071. Fax: (615) 898-2774. E-mail: wmot@mtsu.edu Web Site: www.wmot.org. Licensee: Middle Tennessee State University. Population served: 800,000 Natl. Network: NPR, AP Radio. Format: Jazz. News staff: 2; News: 10 hrs wkly. Target aud: 24 plus; general. ♦John L. High, gen mgr; John Egly, opns mgr; Keith Palmer, dev dir.

***WMTS-FM**— 1996: 88.3 mhz; 680 w. Ant 138 ft TL: N35 50 56 W86 21 11. Hrs open: 24 Box 58, Middle Tenn. State Univ., 37132. Phone: (615) 898-5051. Phone: (615) 898-2636. Fax: (615) 898-5682. Web Site: www.wmtsradio.com. Licensee: Middle Tennessee State University. Population served: 50,000 Format: Black, Sp, div. Target aud: 18-26; College age, diverse. Spec prog: Polka 2 hrs, electronic 10 hrs, jazz 4 hrs, funk 2 hrs wkly.

Nashville

WAMB(AM)— Dec 21, 2001: 1200 khz; 50 kw-D, 3.8 kw-CH. TL: N36 12 32 W86 52 21. Stereo. Hrs opn: 24 1617 Lebanon Rd., 37210. Phone: (615) 889-1960. Fax: (615) 902-9108. E-mail: wamb@bellsouth.net Web Site: www.wamb.net. Licensee: Great Southern Broadcasting Co. Inc. (acq 7-1-2006; $2 million). Population served: 750,000 Natl. Network: CNN Radio. Law Firm: Irwin, Campbell & Tannenwald. Format: Adult standards/MOR. ♦Will C. Baird Jr., VP; William O. Barry, pres & gen mgr; Harry P. Stephenson, gen sls mgr; Ronald W. Johnson, progmg dir; Gary M. Brown, chief of engrg; Beth Lane, traf mgr.

WCJK(FM)—See Murfreesboro

WCRT(AM)—Donelson, Apr 12, 1971: 1160 khz; 50 kw-D, 1 kw-N, DA-N. TL: N36 09 49 W86 42 56. Stereo. Hrs opn: 24 Two Lakeview Place, 15 Century Blvd., Suite 101, 37214-3692. Phone: (615) 871-1160. Fax: (615) 871-9355. E-mail: ruselton@bottradionetwork.com Licensee: Bott Communications Inc. (acq 1-11-2006; $5 million). Population served: 1,500,000 Law Firm: Fletcher, Heald & Hildreth. Format: Christian. ♦Richard P. Bott, pres; Richard P. Bott II, VP; Randy Uselton, gen mgr.

WENO(AM)— May 23, 1988: 760 khz; 1 kw-D. TL: N36 08 28 W86 45 23. Hrs open: Sunrise-sunset 333 Murfreesboro Rd., 37210. Phone: (615) 248-1689. Fax: (615) 248-7786. Web Site: www.weno.com. Licensee: WENO Inc. (acq 5-7-90; $300,000; FTR: 5-21-90). Population served: 750,000 Natl. Network: AP Radio. Format: Christian, relg. News: 6 hrs wkly. Target aud: 25-54. ♦Dan Boone, pres; Mark Myers, CFO; David Deese, gen mgr; Dave Queen, stn mgr; Dan Klimkowski, chief of engrg; Jennifer Houchin, traf mgr.

WNAZ-FM—Co-owned with WENO(AM). May 23, 1967: 89.1 mhz; 1.4 kw. Ant 200 ft TL: N36 08 28 W86 45 23. Stereo. 24 Web Site: www.wnaz.com. Licensee: Trevecca Nazarene University Inc.447,877 Format: Progsv Christian. News: 3 hrs wkly. Target aud: 18-30; college & young professionals.

***WFSK-FM**— Apr 14, 1973: 88.1 mhz; 700 w horiz. Ant 6 ft TL: N36 10 00 W86 48 17. Stereo. Hrs opn: 24 Fisk Univ., 1000 17th Ave. N., 37208-3051. Phone: (615) 329-8754. Fax: (615) 329-9305. E-mail: skay@fisk.edu Web Site: www.fisk.edu/wfsk. Licensee: Fisk University. Population served: 750,000 Natl. Network: PRI. Format: Smooth jazz-smooth grooves, talk radio. ♦Sharon Kay, gen mgr; Xuam Lawson, progmg dir; Clinton Hooper, chief of engrg.

WGFX(FM)—See Gallatin

WJXA(FM)— Apr 3, 1976: 92.9 mhz; 100 kw. 1,086 ft TL: N36 07 14 W86 58 07. Stereo. Hrs opn: 24 Box 40506, 37204-0506. Secondary address: 504 Rosedale Ave. 37211. Phone: (615) 259-9393. Phone: (615) 259-0929. Fax: (615) 259-4594. Web Site: mix929.com. Licensee: South Central Communications Corp. (group owner) Law Firm: Bryan Cave. Wire Svc: NWS (National Weather Service) Wire Svc: UPI Format: Adult contemp. Target aud: 25-54.John D. Engelbrecht, CEO; Craig Jacobus, pres; Robert L. Shirel, CFO; Dennis Gwiazdon, gen mgr; Becky Sweeney, sls dir; Neda Gayle, natl sls mgr; Katherine Salmon, prom dir & pub affrs dir; Barbara Bridges, progmg dir & disc jockey; Anna Marie Ritter, news dir; Don Haworth, chief of engrg; Dave Steele, traf mgr; Brian Sargent, disc jockey; Duane Hamilton, disc jockey; Rick Marino, disc jockey

WKDF(FM)— Jan 1, 1967: 103.3 mhz; 100 kw. 1,233 ft TL: N36 02 08 W86 50 56. Stereo. Hrs opn: 506 2nd Ave. S., 37210. Phone: (615) 244-9533. Fax: (615) 259-1271. Web Site: www.1-3WKDF.com. Licensee: Citadel Broadcasting Co. Group owner: Citadel Broadcasting Corp. acq 4-26-01; grpsl). Format: Country. Target aud: 18-34; general. ♦Dave Kelly, gen mgr & progmg dir; Cindy Francis, prom mgr; Bud Ford, progmg dir; Eddy Foxx, mus dir; Cameron Adkins, chief of engrg; Jennifer Boucher, traf mgr.

WLAC(AM)— Nov 24, 1926: 1510 khz; 50 kw-U, DA-N. TL: N36 16 15 W86 45 24. Hrs open: 24 55 Music Sq. W., 37203. Phone: (615) 664-2400. Fax: (615) 664-2457. Web Site: www.1510wlac.com. Licensee: Capstar TX L.P. Group owner: Clear Channel Communications Inc. (acq 8-30-00; grpsl). Population served: 486,000 Natl. Network: Wall Street, ABC. Law Firm: Haley, Bader & Potts. Format: News/talk, relg. News staff: 3; News: 18 hrs wly. Target aud: 35-64; professionals, business owners & managers. Spec prog: Black 20 hrs wkly. ♦Dave Alpert, pres; Keith Kaufman, opns dir & mktg dir; Darren Smith, sls dir; Temple Hancock, prom dir; Bruce Collins, progmg mgr; Mike Gideon, chief of engrg.

WNRQ(FM)—Co-owned with WLAC(AM). 1953: 105.9 mhz; 100 kw. 1,226 ft TL: N30 02 08 W86 50 56. Stereo. 24 Web Site: www.1059.com.1,700,000 Format: Adult contemp. News staff: one; News: 3 hrs wkly. Target aud: 25-54; upwardly mobile adults. Spec prog: Christian 6 hrs wkly. ♦David Alpert, gen mgr & mktg mgr; Keith Kaufman, opns mgr; Temple Hancock, prom mgr.

WNAH(AM)— Dec 24, 1949: 1360 khz; 1 kw-U. TL: N36 11 30 W86 46 26. Hrs open: 24 44 Music Sq. E., 37203. Phone: (615) 254-7611. Fax: (615) 467-8600. E-mail: mail@wnah.com Web Site: www.wnah.com. Licensee: Hermitage Broadcasting Corp. Format: Southern gospel. News: 5 hrs wkly. Target aud: 21-50. ♦Van T. Irwin Jr., pres & gen mgr; Tony Cappuccilli, gen sls mgr; Bill Grist, prom mgr; Hoyt M. Carter Jr., progmg dir & chief of engrg; Bobby Lynn II, mus dir.

WNQM(AM)— July 1, 1948: 1300 khz; 50 kw-D, 5 kw-N, DA-N. TL: N36 12 30 W86 53 38. Hrs open: 24 1300 WWCR Ave., 37218. Phone: (615) 255-1300. Fax: (615) 255-1311. E-mail: wnqm@wnqm1300.com Web Site: www.wnqm1300.com. Licensee: WNQM Inc. Group owner: F.W. Robbert Broadcasting Co. Inc. (acq 1-83; $700,000; FTR: 12-19-83). Population served: 2,000,000 Natl. Network: USA. Format: Relg, Sp. News: 2 hrs wkly. Target aud: General. Spec prog: Sp. ♦Fred P. Werstenberger, pres; Eric Westenberg, gen mgr; Brady Muray, opns mgr.

WNSG(AM)— Aug 15, 1983: 880 khz; 2.5 kw-D. TL: N36 12 43 W86 49 09. Hrs open: 3051 Stokers Ln., 37218. Phone: (615) 255-2876. Phone: (615) 254-8880. Fax: (615) 254-8228. Licensee: Davidson Media Station WMDB Licensee LLC. (acq 9-1-2005; $1.6 million). Format: Gumbo Black. Target aud: 18 plus; general. ♦Peter Davidson, pres; Dr. Morgan Babb, gen mgr; Michael Babb, opns mgr, progmg dir & news dir; Morgan Babb, gen sls mgr.

WNVL(AM)— 1948: 1240 khz; 1 kw-U. TL: N36 09 24 W86 46 15. Hrs open: 24 Cummins Stn, 209 Tenth Ave. S., 37203. Phone: (615) 242-1411. Fax: (615) 242-3823. Licensee: Davidson Media Station WNSG Licensee LLC. (acq 8-1-2005; $2.7 million). Population served: 447,877 Format: Rgnl Mexican. Target aud: 25-54. ♦Peter Davidson, pres & gen mgr; Clarence Kilcrease, gen mgr; Pat Hall-Easley, opns mgr; Brian Hogg, gen sls mgr; Vic Watkins, mus dir; Jay Shoemaker, chief of engrg.

WPLN(AM)—Madison, Sept 14, 1958: 1430 khz; 5 kw-D, 1 kw-N, DA-N. TL: N36 16 19 W86 42 53. 15 kw-D, 1 kw-N, DA-N. Stereo. Hrs opn: 24 630 Mainstream Dr., 37228. Phone: (615) 760-2903. Fax: (615) 760-2904. Web Site: www.wpln.org. Licensee: Nashville Public Radio (acq 2-15-02; $3 million). Population served: 850,000 Law Firm: Tierney & Swift. Format: News/talk. Target aud: General. ♦Rob Gordon, pres & gen mgr.

***WPLN-FM**— Dec 17, 1962: 90.3 mhz; 80 kw. Ant 1,132 ft TL: N36 02 08 W86 50 56. Stereo. Hrs opn: 630 Mainstream Dr., 37228-1204. Phone: (615) 760-2903. Fax: (615) 760-2904. E-mail: talkback@wpln.org Web Site: www.wpln.org. Licensee: Nashville Public Radio. Population served: 1,000,000 Natl. Network: NPR, PRI. Format: Class, cultural, news, bluegrass. Target aud: General. ♦Robert Gordon, gen mgr; Laura Landress, gen sls mgr; Henry Fennell, progmg dir; Will Griffin, mus dir; Anita Bugg, news dir; Tom Knox, chief of engrg & news rptr; Wendy Poston, traf mgr; Nina Cardona, spec ev coord.

WQQK(FM)—See Hendersonville

***WRVU(FM)**— Dec 3, 1971: 91.1 mhz; 14.5 kw. 457 ft TL: N36 08 27 W86 51 56. Stereo. Hrs opn: 24 Box 9100-B, Vanderbilt Univ., 128 Sarratt Student Ctr., 37235. Phone: (615) 322-3691. Phone: (615) 322-7625. Fax: (615) 343-2582. Web Site: www.wrvu.org. Licensee: Vanderbilt Student Communications. Population served: 447,877 Natl. Network: ABC. Format: Progsv rock, jazz, div. Target aud: General; div, adventurous individuals. ♦Jennifer Sexton, gen mgr; David Cash, progmg dir.

WSIX-FM— 1948: 97.9 mhz; 100 kw. 1,140 ft TL: N36 02 49 W86 49 49. Stereo. Hrs opn: 24 55 Music Sq. W., 37203. Phone: (615) 664-2400. Fax: (615) 664-2457. Licensee: Capstar TX L.P. Group owner: Clear Channel Communications Inc. (acq 8-30-00; grpsl). Population served: 1,000,000 Format: Country. News staff: 2; News: one hr wkly. Target aud: 25-54. ♦David Alpert, gen mgr; Keith Kaufman, opns mgr, mktg dir & prom dir; Temple Hancock, prom dir; Mike Moore, progmg dir & pub affrs dir; Al Voecks, news dir; Mike Gideon, chief of engrg.

WSM(AM)— Oct 5, 1925: 650 khz; 50 kw-U. TL: N35 59 50 W86 47 32. Stereo. Hrs opn: 24 2804 Opryland Dr., 37214. Phone: (615) 889-6595. Fax: (615) 458-2445. Web Site: www.wsmonline.com. Licensee: Grand Ole Opry LLC (acq 11-14-2000; grpsl). Population served: 447,877 Natl. Network: ABC. Natl. Rep: Christal. Format: Country. News staff: 12; News: 11 hrs wkly. Target aud: 35 plus; high school graduates, married homeowners, income $25,000 plus. Spec prog: Farm 6 hrs, Grand Ole Opry 12 hrs wkly. ♦Chris Kulick, gen mgr; Bill Hutcherson, gen sls mgr & pub affrs dir.

WSM-FM— Nov 1, 1962: 95.5 mhz; 100 kw. Ant 1,279 ft TL: N36 08 27 W86 51 56. Stereo. Hrs opn: 10 Music Circle E., 37203. Phone: (615) 321-1067. Fax: (615) 321-5808. Web Site: www.955thewolf.com. Licensee: Cumulus Licensing LLC. Group owner: Cumulus Media Inc. (acq 7-21-2003; $65 million with WWTN(FM) Manchester). Population served: 294,200 Format: Country. Target aud: 25-54. ♦Michael Dickey, gen mgr.

WVNS-FM—See Pegram

WVOL(AM)—See Berry Hill

WYFN(AM)— Jan 7, 1927: 980 khz; 5 kw-U, DA-N. TL: N36 12 25 W86 40 25. Hrs open: Box 7300, Charlotte, NC, 28241. Phone: (804) 547-9421. Licensee: Bible Broadcasting Network. (group owner; acq 1-31-91; $600,000; FTR: 2-18-91). Natl. Rep: McGavren Guild. Format: Relg. ♦Aaron Tuttle, gen mgr & stn mgr.

New Johnsonville

***WAYW(FM)**— 2001: 89.9 mhz; 3.1 kw. Ant 466 ft TL: N35 56 17 W87 53 39. Hrs open: 24 1012 McEwen Dr., Franklin, 37067. Phone: (615) 261-9293. Fax: (615) 261-3967. E-mail: waym@wayfm.com Web Site: www.wayfm.com. Licensee: WAY-FM Media Group Inc. (group owner; acq 2-1-01). Population served: 111,815 Format: Christian. ♦Teresa White, dev dir; Jeff Brown, progmg dir.

Newport

WLIK(AM)— Apr 9, 1954: 1270 khz; 5 kw-D, 500 w-N, DA-N. TL: N35 57 49 W83 12 31. Hrs open: 24 640 W. Hwy. 25/70, 37821. Phone: (423) 623-3095. Fax: (423) 623-3096. E-mail: wlik@planetc.com Web Site: wlik.net. Licensee: WLIK Inc. Population served: 300,000 Natl. Network: CNN Radio. Rgnl. Network: Tenn. Radio Net. Format: Oldies. Target aud: General. Spec prog: Relg 18 hrs wkly. ♦Dwight D. Wilkerson, pres, gen mgr & stn mgr; Angie Wilkerson, VP; Johnnie Swann, chief of opns.

WNPC(AM)— September 1978: 1060 khz; 1 kw-D. TL: N35 59 10 W83 10 46. Hrs open: 377 Graham St., 37821. Phone: (423) 623-8743. Phone: (423) 623-8744. Fax: (423) 623-0545. Licensee: Harris Broadcasting Inc. dba WNPC Inc. Natl. Network: ABC. Law Firm: Roberts & Eckard. Format: Country. Target aud: 25-49. ♦Dorothy Ann Harris, pres; Mona Sizemore, gen mgr; Brian Fredette, prom mgr & progmg dir.

WNPC-FM— February 1993: 92.9 mhz; 3.1 kw. 459 ft TL: N35 57 27 W83 05 03. Stereo. 24
Rebroadcasts WNPC(AM) Newport 100%.
♦Jim Phillips, progmg VP.

Norris

WRJK(FM)— April 2001: 106.7 mhz; 1.1 kw. Ant 751 ft TL: N36 07 12 W83 55 30. Stereo. Hrs opn: 24
Rebroadcasts WTXM(FM) Maryville 100%.
Box 27100, Knoxville, 37927-7100. Phone: (865) 525-6000. Fax: (865) 525-2000. E-mail: jjarnigan@sccradio.com Web Site: WWW.JACKFMKNOXVILLE.COM. Licensee: South Central Communications Corp. (group owner; (acq 6-14-2001; $2.5 million). Population served: 607,600 Wire Svc: AP Format: JACK-FM. ♦Craig Jacobus, pres; Judy Berkley, natl sls mgr; Terry Gillingham, VP & mktg mgr; Jeff Jarnigan, progmg dir.

Oak Ridge

WATO(AM)— Feb 1, 1948: 1290 khz; 5 kw-D, 500 w-N, DA-2. TL: N36 03 02 W84 12 38. Hrs open: 24 517 Watt Rd., Knoxville, 37922. Phone: (865) 482-1290. Phone: (865) 675-4105. Fax: (865) 675-4859. Licensee: Horne Radio LLC. Group owner: Horne Radio Group (acq 8-29-01; grpsl). Population served: 518,000 Natl. Network: Westwood One. Format: Oldies. News staff: one. Target aud: 25-54. ♦Alex Carroll, sls dir & chief of engrg; Brian Tatum, stn mgr & progmg dir.

WNFZ(FM)— February 1967: 94.3 mhz; 2.5 kw. 515 ft TL: N35 56 28 W84 09 28. Stereo. Hrs opn: Box 27100, Knoxville, 37927-7100. Secondary address: 1100 Sharps Ridge Rd., Knoxville 37917. Phone: (865) 525-6000. Fax: (865) 525-2000. E-mail: SCOX@SCCRADIO.COM Web Site: WWW.943THEX.COM. Licensee: John A. Pirkle. Population served: 450,000 Format: Alternative Rock. Target aud: General. ♦John W. Pirkle, CEO & chmn; Johnathan W. Pirkle, pres; Terry Gillingham, VP & gen mgr; Jeff Cutshaw, natl sls mgr; Shane Cox, progmg dir.

WNOX(FM)—Licensed to Oak Ridge. See Knoxville

Olive Hill

***WDNX(FM)**— Jan 10, 1975: 89.1 mhz; 100 kw. 249 ft TL: N35 12 30 W88 03 46. Stereo. Hrs opn: 24 HHA Administration Bldg., 3575 Lonesome Pine Rd., Savannah, 38372. Secondary address: WDNX Bldg., 3730 Lonesome Pine Rd., Savannah 38372. Phone: (731) 925-9236. Fax: (731) 925-4238. E-mail: sheriwdnx@yahoo.com Web Site: www.lifetalk.net. Licensee: Rural Life Foundation. Population served: 250,000 Format: Inspirational music, Christian teaching & inspiration. News staff: one; News: 5 hrs wkly. Target aud: General; families. Spec prog: Class 5 hrs, farm 1 hr wkly. ♦Charles Harris, chmn; Steven Dickman, pres, CFO, exec VP, stn mgr & dev mgr; Sheri Durbin, gen sls mgr, mktg mgr & pub affrs dir; Steve Dickman, chief of engrg.

Oliver Springs

WOKI(FM)— Sept 15, 1989: 98.7 mhz; 8 kw. Ant 571 ft TL: N36 06 48 W84 03 44. Hrs open: 24 Box 11167, Knoxville, 37939. Secondary address: 4711 Old Kingston Pike, Knoxville 37919. Phone: (865) 588-6511. Fax: (865) 656-7487. Web Site: www.987earlfm.com. Licensee: Citadel Broadcasting Co. Group owner: Citadel Broadcasting Corp. (acq 4-26-2001; grpsl). Population served: 336,335 Natl. Rep: Katz Radio. Rgnl rep: Rgnl Reps. Wire Svc: AP Format: Oldies. Target aud: 18 plus. ♦Mike Hammond, pres & opns mgr; Ed Brantley, VP & gen mgr; Tammy Browning, gen sls mgr; Shanna Lingerfelt, mktg dir; Joe Stutler, progmg dir.

Oneida

WBNT-FM—Listing follows WOCV(AM).

WOCV(AM)— Aug 1, 1959: 1310 khz; 1 kw-D. TL: N36 30 03 W84 29 24. Hrs open: 6 AM-sunset
Rebroadcasts WBNT-FM Oneida 100%.
Box 4370, 37841. Secondary address: 1126 Buffalo Rd. 37841. Phone: (423) 569-8598. Phone: (423) 569-9268. Fax: (423) 569-5572. E-mail: wbnt@highland.net Web Site: www.hive105.com. Licensee: Oneida Broadcasters Inc. (acq 12-14-2006; $525,000 for 70% of stock with co-located FM). Population served: 19,500 Natl. Network: ABC. Format: Adult contemp, new country. News staff: 4; News: 15 hrs wkly. Target aud: 22-55; male & female. ♦Hillard Mattie, gen mgr, gen sls mgr, prom mgr & progmg mgr; Paul C. Strunk, opns dir, sls dir, progmg dir, news dir & local news ed; Darrel E. Smith, chief of engrg; B.J. Gislason, disc jockey; Jeff Massey, disc jockey; Richard Smith, disc jockey; Robert Harris, disc jockey.

WBNT-FM—Co-owned with WOCV(AM). June 10, 1965: 105.5 mhz; 3 kw. Ant 285 ft TL: N36 30 03 W84 29 24. Stereo. 18 Web Site: www.hive105.com.20,000 Natl. Network: ABC. News staff: 4; News: 15 hrs wkly. Target aud: 16-56; male/female working class-retirees. ♦Hillard Mattie, stn mgr, dev mgr, adv mgr & progmg dir; Paul C. Strunk, opns mgr & mus dir; B.J. Gislason, disc jockey; Jeff Massey, disc jockey; Richard Smith, disc jockey; Robert Harris, disc jockey.

Paris

WAKQ(FM)—Listing follows WTPR(AM).

WLZK(FM)—Listing follows WMUF(AM).

WMUF(AM)— May 9, 1980: 1000 khz; 5 kw-D, DA. TL: N36 18 50 W88 17 33. Hrs open: 110 India Rd., 38242. Phone: (731) 644-9455. Fax: (731) 644-9970. E-mail: wmuf@bellsouth.net Licensee: Benton-Weatherford Broadcasting Inc.of Tennessee (group owner; acq 4-1-85). Population served: 50,000 Natl. Network: ABC. Format: Country. News staff: one; News: 3 hrs wkly. Target aud: 25-54; people with disposable income. Spec prog: Farm 2 hrs wkly. ♦Gary D. Benton, pres; Gary Benton, gen mgr.

WLZK(FM)—Co-owned with WMUF(AM). Nov 1, 1991: 94.1 mhz; 10.5 kw. 328 ft TL: N36 18 50 W88 17 33. Stereo. 24 E-mail: wlzk@bellsouth.net (Acq 3-15-91; FTR: 4-8-91).75,000 Natl. Network: Jones Radio Networks. Format: Adult contemp. News staff: one; News: 7 hrs wkly.

WPRH(FM)—Not on air, target date: unknown: 90.9 mhz; 14.5 kw. Ant 289 ft TL: N36 08 55 W88 08 06. Hrs open: Drawer 2440, Tupelo, MS, 38803. Phone: (662) 844-8888. Fax: (662) 842-6791. Web Site: www.afr.net. Licensee: American Family Association. (acq 1-16-2007). Natl. Network: American Family Radio. ♦Donald E. Wildmon, chmn.

WTPR(AM)— May 7, 1947: 710 khz; 750 w-D. TL: N36 16 47 W88 20 32. Stereo. Hrs opn: Sunrise-sunset
Reboadcasts WTPR-FM Paris 100%.
206 N. Brewer St., 38242. Phone: (731) 642-7100. Fax: (731) 642-9367. E-mail: thailey@wenkwtpr.com Web Site: wenkwtpr.com. Licensee: WENK of Uniion City Inc. (group owner; acq 10-28-89; FTR: 8-14-89). Population served: 125,000 Natl. Network: ABC. Rgnl. Network: Reg reps. Rgnl rep: Rgnl Reps Law Firm: Shaninis & Peltzman. Format: Oldies. News staff: one; News: 12 hrs wkly. Target aud: 35-54. ♦Terry Hailey, pres, gen mgr & progmg dir; Brad Hosford, chief of engrg.

WAKQ(FM)—Co-owned with WTPR(AM). September 1967: 105.5 mhz; 3.7 kw. 419 ft TL: N36 16 45 W88 20 31. Stereo.
Rebroadcasts WWKF(FM) Union City 100%.
Web Site: kf99kg105.com. Format: CHR. News staff: one; News: one hr wkly. Target aud: 12-34. ♦Terry Hailey, mus dir.

Parker's Crossroads

WBFG(FM)— 1999: 96.5 mhz; 6 kw. Ant 328 ft TL: N35 45 33 W88 23 15. Hrs open: Box 279, Lexington, 38351. Secondary address: 584 Smith Ave., Lexington 38351. Phone: (731) 968-9990. Fax: (731) 968-0380. E-mail: wbfg965@yahoo.com Web Site: www.wbfg965.com. Licensee: Crossroads Broadcasting LLC. (acq 4-14-99). Natl. Network: ESPN Radio. Format: Sports. ♦Dan Hughes, gen mgr; Lori Becker, opns mgr.

Parsons

WKJQ(AM)— Oct 3, 1970: 1550 khz; 1 kw-D. TL: N35 39 26 W88 09 07. Hrs open: Box 576, 38363. Secondary address: 109 Iron Hill Rd. 38363. Phone: (731) 847-3011. Fax: (731) 847-4600. Licensee: Clenney Broadcasting Corp. (acq 4-1-89). Population served: 30,000 Law Firm: Robert S. Stone. Format: Relg. Target aud: General. ♦Ralph D. Clenney, pres & gen mgr.

WKJQ-FM— June 4, 1990: 97.3 mhz; 6 kw. 256 ft TL: N35 39 39 W88 07 05. Stereo. 24 27,569 Format: Country. Target aud: 25-54.

Pegram

WVNS-FM— Apr 27, 1964: 102.5 mhz; 100 kw. Ant 974 ft TL: N36 17 36 W87 18 20. Stereo. Hrs opn: 24 Box 150846, Nashville, 37215. Phone: (615) 399-1029. Fax: (615) 436-7737. E-mail: programming @1025theparty.com Web Site: www.v1025.com. Licensee: Montgomery Broadcasting. Group owner: The Cromwell Group Inc. (acq 1990). Population served: 1,000,000 Natl. Rep: McGavren Guild. Law Firm: Womble Carlyle. Format: Hot adult contemp. Target aud: 18-34; S. KY residents. ♦Bayard H. Walters, pres; Beth Murphy, gen mgr; Dean Warfield, opns mgr; David Wilson, chief of engrg.

Pikeville

WUAT(AM)— Dec 19, 1972: 1110 khz; 250 w-D. TL: N35 36 18 W85 11 14. Hrs open: Box 128, 37367. Secondary address: 101 N. Main 37367. Phone: (423) 447-2906. Fax: (423) 447-7309. Web Site: www.wuatradio.com. Licensee: Joyce V. Bownds. (acq 6-28-99; $1,500). Population served: 1,454 Law Firm: Lukas, McGowan, Nace & Gutierrez. Format: Country, gospel, bluegrass. Spec prog: Farm 5 hrs, relg 15 hrs wkly. ♦Joyce Bownds, pres, gen mgr, gen sls mgr & progmg dir.

Portland

WQKR(AM)— July 15, 1980: 1270 khz; 1 kw-D, 59 w-N, DA-2. TL: N36 36 11 W86 32 01. Hrs open: 24 100 Main St., Suite 201, 37148-1218. Phone: (615) 325-3250. Fax: (615) 325-0803. Licensee: Venture Broadcasting LLC (acq 8-8-2005; $50,000). Population served: 250,000 Format: Oldies. Target aud: 25-54. ♦Floyd Howard Johnson, gen mgr.

Powell

WQBB(AM)—Licensed to Powell. See Knoxville

Pulaski

WKSR(AM)— May 6, 1947: 1420 khz; 1 kw-U, DA-N. TL: N35 12 04 W87 03 20. Hrs open: Box 738, 38478. Secondary address: 104 S. Second St. 38478. Phone: (931) 363-2505. Fax: (931) 424-3157. Web Site: www.wksr.com. Licensee: Pulaski Broadcasting Inc. (acq 4-4-80; $481,300; FTR: 4-21-80). Population served: 13,500 Format: Oldies. Target aud: 25-54. ♦Ronnie Rose, gen mgr; Ed Carter, progmg dir.

WKSR-FM— Jan 12, 1970: 98.3 mhz; 3 kw. 453 ft TL: N35 08 47 W87 05 28. Stereo. (Acq 1-1-84; $350,000; FTR: 12-19-83).7,800 Format: Country.

Red Bank

***WAWL-FM**— Sept 12, 1980: 91.5 mhz; 11 kw. 951 ft TL: N35 09 42 W85 19 06. Stereo. Hrs opn: 24 4501 Amnicola Hwy., Chattanooga, 37406. Phone: (423) 697-4470. Phone: (423) 697-4405. Fax: (423) 697-2596. Licensee: Chattanooga State Technical Community College. Population served: 250,000 Format: Alternative, educ. News staff: one. Target aud: 18-34. ♦Dr. James L. Catanzaro, pres; Bob Riley, gen mgr & chief of engrg; Linda Miller, mktg VP; Patty Brown, prom VP; Sandy Smith, adv dir; Don Hixson, progmg dir; Jake Land, pub affrs dir.

WJTT(FM)— November 1972: 94.3 mhz; 4.7 kw. 429 ft TL: N35 07 32 W85 17 23. Stereo. Hrs opn: 24 1305 Carter St., Chattanooga, 37402. Phone: (423) 265-9494. Fax: (423) 266-2335. Web Site: www.power94.com. Licensee: Brewer Broadcasting of Chattanooga Inc. (acq 12-26-93; $1.68 million). Population served: 366,800 Natl. Rep: D & R Radio. Format: Urban contemp. Target aud: 18-49. Spec prog: Relg 4 hrs wkly. ♦Jim L. Brewer Sr., pres; Jim L. Brewer II, VP & gen mgr; Jerry Ware, gen sls mgr; Brad Guagriri, natl sls mgr; Jay Holloway, prom mgr; Keith Landecker, progmg dir; Donna Harrison, news dir; Parks Hall, chief of engrg.

Ripley

***WAUV(FM)**— 2000: 89.7 mhz; 5.3 kw. Ant 394 ft TL: N35 46 31 W89 28 18. Hrs open: Box 3206, American Family Radio, Tupelo, MS, 38803. Phone: (662) 844-8888. Fax: (662) 842-6791. E-mail: comments@afr.net Web Site: www.afr.net. Licensee: American Family Association. Group owner: American Family Radio. Format: Relg. ♦Marvin Sanders, gen mgr.

WKVZ(FM)— Jan 1, 1993: 94.9 mhz; 6 kw. 328 ft TL: N35 48 28 W89 28 27. Hrs open: 2351 Sunset Blvd., Suite 170-218, Rocklin, CA, 95765. Phone: (800) 434-8400. Fax: (916) 251-1650. E-mail: info@klove.com Web Site: www.klove.com. Licensee: Educational Media Foundation. Group owner: EMF Broadcasting (acq 1-25-01; $450,000). Natl. Network: K-Love. Format: Contemp Christian. Target aud: 18-50. ♦Richard Jenkins, pres; Mike Novak, VP; Keith Whipple, dev dir; David Pierce, progmg mgr; Ed Lenane, news dir; Sam Wallington, engrg dir; Karen Johnson, news rptr; Marya Morgan, news rptr; Richard Hunt, news rptr.

WTRB(AM)— Dec 11, 1954: 1570 khz; 1 kw-D, 50 w-N. TL: N35 43 46 W89 32 33. (CP: 28 kw-D, 534 w-N). Hrs opn: 17 Box 410, 372 S. Jefferson St., 38063. Phone: (731) 635-1570. Fax: (731) 635-9722. Licensee: West Tennessee Regional Broadcasting Inc. (group owner). (acq 11-16-2004; $265,000). Population served: 25,000 Rgnl. Network: Tenn. Radio Net. Natl. Rep: Keystone (unwired net). Format: C&W. Spec prog: Gospel 6 hrs wkly. ♦Phillip Ennis, pres; Don Paris, gen mgr, gen sls mgr & disc jockey; April Goodrich, disc jockey; Bob Scarborough, disc jockey; Erin Little, disc jockey; Jerry Wilson, disc jockey; Mickey McLure, disc jockey.

Rockwood

WIHG(FM)—Listing follows WOFE(AM).

WOFE(AM)— May 12, 1957: 580 khz; 1 kw-D, 49 w-N. TL: N35 49 40 W84 39 19. Hrs open: Box 810, Crossville, 38557. Phone: (931) 484-1057. Fax: (931) 707-0580. Licensee: Southern Media Group Inc. (group owner; (acq 6-10-2003; grpsl). Population served: 607,315 Format: Country. ♦Kirk Tollett, gen mgr.

WIHG(FM)—Co-owned with WOFE(AM). July 9, 1991: 105.7 mhz; 930 w. Ant 836 ft TL: N35 51 41 W84 43 11.179,800 Format: Hits of the 70s. Target aud: 18-54.

Rogersville

WJDT(FM)— Dec 1, 1990: 106.5 mhz; 6000 w. Ant 1,378 ft TL: N36 22 51 W83 10 47. Stereo. Hrs opn: 24 Box 519, Morristown, 37815-0519. Secondary address: N. Davy Crockett Pkwy., Morristown 37814. Phone: (423) 235-4640. Phone: (865) 993-3639. E-mail: www.wjdtfm@planetc.com Web Site: www.wjdtfm.com. Licensee: C & S Broadcasting. Population served: 500,000 Natl. Network: CNN Radio. Law Firm: Larry Perry. Format: Country. News: 6 hrs wkly. Target aud: 18-59; female 65%, male 35%. ♦Clark Quillen, pres; David C. Quillen, opns mgr.

WRGS(AM)— Aug 20, 1954: 1370 khz; 1 kw-D, 40 w-N. TL: N36 24 58 W82 59 04. Hrs open: 24 211 Burem Rd., 37857. Phone: (423) 272-3900. Fax: (423) 272-0328. E-mail: stationmanager@wrgsradio.com Web Site: www.wrgsradio.com. Licensee: WRGS Inc. Population served: 50,000 Natl. Network: USA. Rgnl. Network: Tenn. Radio Net. Format: Country. Target aud: General. ♦C. Philip Beal, pres; C. Philip Beale, gen mgr; Jay Phillips, opns mgr, progmg dir & progmg mgr; Craig Stapleton, gen sls mgr; Mavis Livingston, natl sls mgr; Jim Cox, rgnl sls mgr; Megan Collins, mktg dir & pub affrs dir; Chuck Windham, chief of engrg.

Saint Joseph

WMXV(FM)— 1991: 101.5 mhz; 4 kw. Ant 403 ft TL: N35 00 42 W87 30 46. Hrs open: Box 374, 37 Old Jackson Hwy., 38481. Phone: (931) 845-4172. Phone: (256) 757-9455. Fax: (931) 845-4172. Licensee: Urban Radio Licenses LLC. (acq 5-13-2005; grpsl). Format: Classic country, Southern gospel. ♦Rick Brown, gen mgr; Randy Paul, opns dir; Lance Knoll, gen sls mgr; Tony Fowler, prom dir; Jane Hoslan, adv dir & adv mgr; Lonnie Box, progmg dir; Sandi Summers, news dir; Craig Westbrook, chief of engrg.

Savannah

***WAZD(FM)**— 2001: 88.1 mhz; 380 w. Ant 128 ft TL: N35 12 58 W88 14 30. Hrs open: Box 3206, American Family Radio, Tupelo, MS, 38803. Phone: (662) 844-8888. Fax: (662) 842-6791. E-mail: comments@afr.net Web Site: www.afr.net. Licensee: American Family Association. Group owner: American Family Radio Format: Relg. ♦Marvin Sanders, gen mgr.

WKWX(FM)— June 23, 1980: 93.5 mhz; 25 kw. Ant 298 ft TL: N35 17 08 W88 10 03. Stereo. Hrs opn: Box 40, 38372. Secondary address: 695 Wayne Rd. 38372. Phone: (731) 925-9600. Fax: (731) 925-8828. E-mail: wkwx@bellsouth.net Licensee: Melco Inc. (acq 12-19-02). Natl. Network: AP Radio. Format: Country. ♦Steve Carnal, pres; Jane Haggard, gen mgr; Jim Jerrolds, gen sls mgr; Dennis Brown, progmg dir & chief of engrg; Tom Treadway, chief of engrg.

WORM(AM)— June 29, 1956: 1010 khz; 250 w-D, 27 w-N. TL: N35 14 24 W88 14 29. Hrs open: Box 550, 38372. Secondary address: 1207 Bowen Dr. 38372. Phone: (731) 925-4981. Phone: (731) 925-7102. Fax: (731) 925-4981. E-mail: thewormq105@excite.com Licensee: Gerald W. Hunt. Population served: 5,576 Rgnl. Network: Tenn. Radio Net. Format: Pure Gold. ♦Gerald W. Hunt, pres, gen mgr, gen sls mgr & chief of engrg; Dave Morgan, progmg dir; Randy Tucker, mus dir.

WORM-FM— Aug 25, 1966: 101.7 mhz; 3 kw. 175 ft TL: N35 14 24 W88 14 29. Stereo. 5,576 Format: Hot country.

Selmer

WDTM(AM)— Oct 31, 1967: 1150 khz; 1 kw-D. TL: N35 11 27 W88 35 21. Hrs open: 25 Stonebrook Pl., Suite G322, Jackson, 38305. Phone: (731) 663-3931. Fax: (731) 663-9804. Licensee: Grace Broadcasting Services Inc. (acq 7-25-2005; $200,000 with co-located FM). Population served: 113,495 Format: Contemp Christian. ♦Lacy Ennis, pres & gen mgr.

WSIB(FM)—Co-owned with WDTM(AM). January 1990: 93.9 mhz; 6 kw. Ant 328 ft TL: N35 11 27 W88 35 21. Stereo. 24 Web Site: www.gracebroadcasting.com. Format: Inspirational gospel. Target aud: 20-45.

***WXKV(FM)**—Not on air, target date: unknown: 90.5 mhz; 6 kw. Ant 236 ft TL: N35 12 53 W88 32 44. Hrs open:
Rebroadcasts KLVR(FM) Santa Rosa, CA 100%.
2351 Sunset Blvd., Suite 170-218, Rocklin, CA, 95765. Phone: (916) 251-1600. Fax: (916) 251-1650. Web Site: www.klove.com. Licensee: Educational Media Foundation. (acq 11-1-2006; grpsl). Natl. Network: K-Love. Format: Contemp Christian. ♦Richard Jenkins, pres.

WXOQ(FM)— June 15, 1986: 105.5 mhz; 6 kw. Ant 298 ft TL: N35 13 11 W88 40 23. Stereo. Hrs opn: 24 Box 550, Savannah, 38372. Secondary address: 1207 Bowen Dr., Savannah 38372. Phone: (731) 645-9880. Fax: (731) 925-4981. E-mail: thewormQ105@excite.com Licensee: Gerald W. Hunt. (acq 6-28-94; $185,000; FTR: 7-11-94). Natl. Network: Westwood One. Format: Country. News staff: 2. Target aud: General. ♦Gerald W. Hunt, pres, gen mgr & chief of engrg; Dave Morgan, progmg dir; Randy Tucker, mus dir.

Sevierville

WSEV(AM)— Apr 23, 1955: 930 khz; 5 kw-D, 148 w-N. TL: N35 52 42 W83 33 18. Stereo. Hrs opn: 24
Rebroadcasts WSEV-FM Gatlinburg.
415 Middle Creek Rd., 37862. Phone: (865) 453-2844. Fax: (865) 429-2601. Licensee: East Tennessee Radio Group L.P. (acq 3-22-2000; $1.45 million with WSEV-FM Gatlinburg). Population served: 65,000 Natl. Network: CBS. Rgnl. Network: Tenn. Radio Net. Natl. Rep: Rgnl Reps. Format: Adult contemp. News staff: one; News: 6 hrs wkly. Target aud: 25 plus. ♦Paul Sink, gen mgr & stn mgr; Bill Burkett, opns mgr; Steve Hartford, progmg dir & news dir.

WWST(FM)— Feb 3, 1961: 102.1 mhz; 15 kw. 1,979 ft TL: N35 48 41 W83 40 08. Stereo. Hrs opn: 24 Journal Broadcast Group, 1533 Amhearst Rd., Knoxville, 37909-1204. Phone: (865) 693-1020. Phone: (865) 824-1021. Fax: (865) 824-1880. Web Site: www.star1021fm.com. Licensee: Journal Broadcast Corp. Group owner: Journal Communications Inc. (acq 5-19-97). Population served: 350,000 Natl. Network: ABC. Law Firm: Hogan & Hartson. Format: CHR. News: 4 hrs wkly. Target aud: 25-54. ♦Chris Protzman, gen mgr & natl sls mgr; Rich Bailey, opns dir, progmg dir & pub affrs dir; Dan McKee, rgnl sls mgr; Justin Buznedo, prom dir; Scott Bohannon, mus dir; Ashey Adams, disc jockey; Brad Allen, disc jockey; Jerry Agar, disc jockey; Randy Chambers, disc jockey; Rob Carter, disc jockey.

Sewanee

***WUTS(FM)**— May 1972: 91.3 mhz; 200 w. 658 ft TL: N35 12 20 W85 55 07. (CP: 88.5 mhz). Hrs opn: Univ. of the South, 735 University Ave., 37383. Phone: (931) 598-1206. Phone: (931) 598-1112. Fax: (931) 598-1145. E-mail: wuts@sewanee.edu Licensee: University of the South. Population served: 2,341 Format: Div, progsv. Target aud: General; college students. Spec prog: Black 2 hrs, class 4 hrs, country 2 hrs, jazz 4 hrs, blues 3 hrs, reggae 5 hrs, new age 4 hrs, Fr 2 hrs, soul 2 hrs, musicals 2 hrs, bluegrass 2 hrs, punk/hardcore 2 hrs wkly. ♦John Lee, gen mgr; Austin Lacy, mus dir; Greg Banworth, chief of engrg.

Seymour

WJBZ-FM— Mar 31, 1991: 96.3 mhz; 3 kw. 328 ft TL: N35 54 32 W83 40 59. Hrs open: 24 Box 2526, Knoxville, 37901. Phone: (865) 577-4885. Fax: (865) 579-4667. Web Site: www.praise963.com. Licensee: Seymour Communications. Format: Southern gospel. ♦Charlotte Mull, CEO; Doug Hutchison, pres & gen mgr; Mike Clark, opns mgr; Jamie Lewis, gen sls mgr & prom mgr; Tim Guinn, mus dir; Tim Berry, chief of engrg; Staci Beal, traf mgr.

Shelbyville

***WBIA(FM)**— 1999: 88.3 mhz; 250 w. Ant 46 ft TL: N35 28 54 W86 27 28. Hrs open: Drawer 2440, Tupelo, MS, 38803. Phone: (662) 844-8888. Fax: (662) 842-6791. Web Site: www.afr.net. Licensee: American Family Association. Group owner: American Family Radio Format: Christian. ♦Marvin Sanders, gen mgr; John Riley, progmg dir; Joey Moody, chief of engrg.

WLIJ(AM)— Dec 2, 1959: 1580 khz; 5 kw-D, 12 w-N. TL: N35 27 21 W86 27 09. Hrs open: 24 Box 7, 236 Woodland Dr., 37160. Phone: (931) 684-1514. Phone: (931) 684-1515. Fax: (931) 684-3956. Licensee: Hopkins-Hall Broadcasting Inc. (acq 7-19-90; $110,000; FTR: 8-6-90). Population served: 12,262 Law Firm: Garvey, Schubert & Barer. Format: Country, bluegrass, gospel. News staff: one; News: 14 hrs wkly. Target aud: General. Spec prog: Black one hr, farm 3 hrs, relg 11 hrs wkly. ♦Nadine Hopkins, pres; Keith Cook, gen sls mgr & disc jockey; Rusty Reed, gen mgr, progmg dir & news dir; Paul Hopkins, chief of engrg; Hal Ball, disc jockey.

WZNG(AM)— December 1946: 1400 khz; 1 kw-U. TL: N35 28 26 W86 26 45. Hrs open: 24 Box 7, 37162. Secondary address: 236 Woodland Dr. 37160. Phone: (931) 680-1214. Fax: (931) 684-3956. Licensee: Hopkins-Hall Broadcasting Inc. (acq 12-19-96; $250,000). Population served: 36,000 Law Firm: Garvey, Schubert & Barer. Format: Radio America, talk star. Target aud: General; residents of the loc area. Spec prog: Relg 5 hrs wkly. ♦Nadine Hopkins, pres; Paul Hopkins, sr VP; Rusty Reed, gen mgr & stn mgr.

Signal Mountain

WLND(FM)— Aug 29, 1994: 98.1 mhz; 1 kw. Ant 794 ft TL: N35 05 16 W85 21 47. Hrs open: 24 7413 Old Lee Hwy., Chattanooga, 37421. Phone: (423) 892-3333. Fax: (423) 642-0097. E-mail: wkxj@clearchannel.com Web Site: www.thelegendonline.com. Licensee: Capstar TX L.P. Group owner: Clear Channel Communications Inc. (acq 8-15-2000; grpsl). Law Firm: Kaye, Scholer, Fierman, Hays & Handler. Format: Classic country. News staff: one; News: 4 hrs wkly. Target aud: 35-54. ♦Sammy George, gen mgr; Kris Van Dyke, opns mgr & progmg dir; Rhonda Rollins, gen sls mgr.

Smithville

WJLE(AM)— Apr 11, 1964: 1480 khz; 1 kw-D, 34 w-N. TL: N35 55 31 W85 49 14. Hrs open: 16 2606 McMinnville Hwy., 37166-5071. Phone: (615) 597-4265. Fax: (615) 597-6025. E-mail: wjle@dtccom.net Web Site: www.wjle.com. Licensee: Center Hill Broadcasting Corp. Population served: 300,000 Format: Country. News staff: one; News: 14 hrs wkly. Target aud: General. Spec prog: Gospel 15 hrs wkly. ♦W.E. Vanatta, pres & gen mgr; Dwayne Page, sls dir, progmg dir & news dir; Homer Wilson Jr., chief of engrg.

WJLE-FM— 1970: 101.7 mhz; 3 kw. 195 ft TL: N35 55 31 W85 49 14.16 Web Site: www.wjle.com.

Smyrna

***WFCM(AM)**— 1993: 710 khz; 250 w-D. TL: N35 58 31 W86 33 16. Hrs open: Sunrise-sunset
Rebroadcasts WMBW(FM) Chattanooga 100%.
1920 E. 24th Street Pl., Chattanooga, 37404. Secondary address: 615 Potomac Pl. 37167. Phone: (423) 629-8900. Fax: (423) 629-0021. E-mail: wfcm@moody.edu Web Site: www.wfcm.org. Licensee: The Moody Bible Institute of Chicago. (group owner; acq 5-16-97; $162,500). Natl. Network: Moody. Format: Educ, relg, news/talk. ♦Dr. Michael Easley, pres; Wayne Pederson, VP; Leighton LeBoeuf, gen mgr & stn mgr; Andy Napier, progmg dir; Paul Martin, mus dir & news rptr; David Morais, chief of engrg; Andy Narier, news rptr.

WFFH(FM)— Oct 7, 1993: 94.1 mhz; 3.2 kw. Ant 305 ft TL: N36 01 14 W86 38 18. Stereo. Hrs opn: 24
Simulcasts with WFFI(FM) Kingston Springs.
402 BNA Dr., Suite 400, Nashville, 37217. Phone: (615) 367-2210. Fax: (615) 367-0758. Web Site: www.94fmthefish.net. Licensee: Caron Broadcasting Inc. Group owner: Salem Communications Corp. (acq 12-18-02; $5.6 million with WFFI(FM) Kingston Springs). Natl. Network: Salem Radio Network. Natl. Rep: Salem. Format: Contemp Christian. Target aud: 25-54; adults. ♦Michael S. Miller, gen mgr; Dick Marsh, prom dir; Vance Dillard, progmg dir; Kim Bindel, news dir; Carl Campbell, chief of engrg; Ed Evenson, traf mgr.

Soddy-Daisy

WGOW-FM—Licensed to Soddy-Daisy. See Chattanooga

WSDT(AM)— Feb 27, 1970: 1240 khz; 1 kw-U. TL: N35 16 16 W85 10 28. Hrs open: Willis Broadcasting Corp., 645 Church St., Suite 400, Norfolk, VA, 23510. Phone: (757) 622-4600. Fax: (757) 624-6515. Licensee: Willis Broadcasting Corp. (group owner; acq 4-27-99; $65,000). Population served: 35,000 ♦Levi E. Willis II, VP; Katrina Chase, gen mgr.

Somerville

WSTN(AM)— Nov 29, 1982: 1410 khz; 500 w-U, DA-2. TL: N35 14 31 W89 19 03. Hrs open: 24 Box 262550, Baton Rouge, LA, 70826. Secondary address: 8919 World Ministry Ave., Baton Rouge, LA 70826. Phone: (225) 768-3688/8300. Fax: (225) 768-3729. E-mail: kawikfish@yahoo.com Web Site: www-jsm.org. Licensee: Family Worship Center Church Inc. (group owner; (acq 12-4-2002). Format: Relg. ♦David Whitelaw, COO; Jimmy Swaggart, pres; John Santiago, gen mgr & progmg dir.

South Fulton

WCMT-FM—Licensed to South Fulton. See Martin

South Pittsburg

WEPG(AM)— July 9, 1954: 910 khz; 5 kw-D, 95 w-N. TL: N35 00 57 W85 42 00. Hrs open: Box 8, 37380. Secondary address: 105 N. Ash Ave. 37380. Phone: (423) 837-0747. Fax: (423) 837-2974. E-mail: wepgtv6@aol.com Licensee: Stone/Collins Communications Inc. (acq 2-1-02). Population served: 80,000 Format: Country. Target aud: 18-50; females. Spec prog: Gospel 10 hrs wkly. ♦Roger Spears, gen mgr; Glenda Frame, prom dir; Rogers Spears, gen sls mgr & progmg dir.

WUUS-FM— Nov 5, 1990: 97.3 mhz; 16 kw. Ant 856 ft TL: N34 58 21 W85 37 58. Stereo. Hrs opn: 24 Box 8799, Chattanooga, 37414. Secondary address: 7413 Old Lee Hwy., Chattanooga 37414. Phone: (423) 892-3333. Fax: (423) 642-0096. Licensee: 3 Daughters Media Inc. Group owner: Clear Channel Communications Inc. (acq 6-22-2007; grpsl). Format: Hip-Hop. News staff: 2; News: 20 hrs wkly. Target aud: 18-54. ♦Sammy George, gen mgr; Kris Van Dyke, opns mgr; Rhonda Rollins, gen sls mgr; Andre Johnson, engrg dir; Vickie Gravitt, traf mgr.

Sparta

WRKK-FM—Listing follows WSMT(AM).

WSMT(AM)— Apr 26, 1953: 1050 khz; 1 kw-D, 181 w-N. TL: N35 57 00 W85 28 50. Hrs open: 24 520 N. Spring St., 38583-1305. Phone: (931) 836-1055. Phone: (931) 836-2824. Fax: (931) 836-2320. Licensee: CC Licenses LLC. Group owner: Clear Channel Communications Inc. (acq 9-8-99; grpsl). Population served: 200,000 Format: Southern gospel. News staff: one; News: 10 hrs wkly. Target aud: General. Spec prog: Relg 10 hrs wkly. ♦David Roederer, CEO & gen mgr.

WRKK-FM—Co-owned with WSMT(AM). Aug 2, 1964: 105.5 mhz; 1.05 kw. 46 ft TL: N35 51 39 W85 26 40. Stereo. 24 Web Site: rockdog1055.com.400,000 Format: Classic rock. News staff: one. Target aud: 18-34. ♦Don Howard, opns mgr & local news ed.

WTZX(AM)— Nov 26, 1971: 860 khz; 1 kw-D, 9.9 w-N. TL: N35 55 20 W85 26 50. Hrs open: 24
Rebroadcasts WPTN(AM) Cookeville 100%.
520 N. Spring St., 38583-1305. Phone: (931) 836-1055. Fax: (931) 836-2320. Licensee: CC Licenses LLC. Group owner: Clear Channel Communications Inc. (acq 12-27-2001; $85,000). Population served: 200,000 Format: News/talk. ♦David Roedarer, gen mgr; Don Howard, stn mgr, opns mgr, progmg dir, news dir & local news ed; Bruce Walker, sls dir & gen sls mgr; Homer Wilson, chief of engrg.

Spencer

WTRZ(FM)— Aug 1, 1993: 107.3 mhz; 2 kw. Ant 508 ft TL: N35 39 55 W85 31 19. Hrs open: 230 W. Colville St., Mc Minnville, 37110. Phone: (931) 473-9253. Fax: (931) 473-4149. E-mail: bryankell@clearchannel.com Web Site: www.kiss107radio.com. Licensee: Citicasters Licenses L.P. Group owner: Clear Channel Communications Inc. (acq 10-30-99; grpsl). Format: CHR. ♦Bryan Kell, gen mgr & sls; Jeff Edwards, progmg dir; Homer Wilson, chief of engrg.

***WZYZ(FM)**— 2003: 90.1 mhz; 30 w. Ant 590 ft TL: N35 44 03 W85 27 33. Hrs open: 24 120707 Beersheba Hwy., McMinnville, 37110.

Phone: (931) 946-7777. E-mail: questions@wzyz.org Licensee: Church Faith Trinity Assemblies (acq 8-1-02). Format: Relg. ♦Daniel Lawson, gen mgr.

Spring City

WAYA(FM)—Listing follows WXQK(AM).

WXQK(AM)— July 12, 1979: 970 khz; 500 w-D. TL: N35 39 59 W84 52 44. Hrs open: 6 AM-6 PM
Rebroadcasts WBAC(AM) Cleveland 100%.
2640 Commerce Dr. N.E., Cleveland, 37311. Phone: (423) 242-7656. Fax: (423) 472-5290. Web Site: www.wbacradio.com. Licensee: J.L. Brewer Broadcasting of Cleveland LLC. Group owner: Brewer Broadcasting Corp. (acq 7-1-2002; grpsl). Population served: 28,608 Natl. Network: ABC. Natl. Rep: D & R Radio. Format: News/talk. News staff: one; News: 16 hrs wkly. Target aud: 35-64. ♦James L. Brewer, pres; James L. Brewer II, VP; Charles Sells, gen mgr; Mike Powers, opns mgr; John Holland, gen sls mgr; Corky Whitlock, progmg dir.

WAYA(FM)—Co-owned with WXQK(AM). October 1989: 93.9 mhz; 5.5 kw. Ant 574 ft TL: N35 31 50 W84 43 03.24 Web Site: www.catcountry939.com.87,000 Natl. Rep: D & R Radio. Format: Country. News staff: one; News: one hr wkly. Target aud: 25-54. ♦Bobby Byrd, progmg dir.

Springfield

WDBL(AM)— July 24, 1950: 1590 khz; 1 kw-D, 30 w-N. TL: N36 29 43 W86 54 26. (CP: 710 w-D. TL: N36 29 42 W86 54 22). Hrs opn: 19 1640 Old Russellville Pike, Clarksburg, 37043. Phone: (931) 648-7720. Fax: (931) 648-7769. Licensee: Lightning Broadcasting LLC (acq 7-28-2004; $150,000). Population served: 42,000 Rgnl. Network: Tenn. Radio Net. Format: Contemp Christian. News staff: one; News: 16 hrs wkly. Target aud: 18 plus. Spec prog: Farm 10 hrs, gospel 10 hrs wkly. ♦Susan Quesenberry, gen mgr; Lee Logan, opns mgr & progmg dir; J.C. Morrow, mus dir & chief of engrg.

WSGI(AM)— Dec 15, 1982: 1100 khz; 1 kw-D. TL: N36 31 00 W86 53 30. Hrs open: 6 AM-sunset Box 909, 37172. Phone: (615) 384-9744. Fax: (615) 384-9746. E-mail: wsgi1100@yahoo.com Licensee: Lightning Broadcasting LLC (acq 3-2001; $155,000). Format: Variety. Target aud: General. Spec prog: Relg, farm 5 hrs, gospel 16 hrs wkly. ♦Jo Petersen, VP; Neil Petersen, pres & gen mgr; Billy Gray, gen sls mgr & news dir.

Static

WSBI(AM)— Apr 7, 1986: 1210 khz; 1 kw-D. TL: N36 37 22 W85 05 15. Hrs open: Sunrise-sunset Box 160, Byrdstown, 38549. Phone: (606) 387-6625. Fax: (606) 387-8126. E-mail: info@wsbiam.com Web Site: www.wsbiam.com. Licensee: Donnie S. Cox. (acq 11-3-99; $60,000). Natl. Network: USA. Format: Country. News staff: one; News: 15 hrs wkly. Target aud: 25 plus. Spec prog: Gospel 10 hrs, bluegrass one hr wkly. ♦Donnie Cox, gen mgr; Robert Huddleston, chief of engrg.

Surgoinsville

WEYE(FM)—Licensed to Surgoinsville. See Church Hill

Sweetwater

WDEH(AM)— 1955: 800 khz; 1 kw-D, 379 w-N. TL: N35 36 49 W84 27 33. Hrs open: 24 Box 330, 37874. Phone: (423) 337-5025. Fax: (423) 337-5026. E-mail: wlodwdeh@yahoo.com Licensee: Horne Radio L.L.C. Group owner: Horne Radio Group (acq 1999; $425,000 with co-located FM). Population served: 500,000 Format: Gospel.

WLOD-FM—Co-owned with WDEH(AM). September 1967: 98.3 mhz; 6 kw. 135 ft TL: N35 36 49 W84 27 33. Stereo. 24 Phone: (865) 458-4400.250,000 Format: Oldies.

Tazewell

WNTT(AM)— July 1, 1960: 1250 khz; 500 w-D. TL: N36 27 09 W83 34 23. Hrs open: 6 AM-sunset Box 95, 115 Bluetop Rd., 37879-0095. Phone: (423) 626-4203. Fax: (423) 626-3040. E-mail: aileen@wntt1250am.co Licensee: WNTT Inc. (acq 9-1-94; $90,000). Population served: 25,860 Natl. Network: ABC. Format: Country, oldies, news. News: 12 hrs wkly. Target aud: 18-65; general. Spec prog: Gospel, bluegrass. ♦Aileen S. Craft, CEO, gen sls mgr, progmg dir, progmg mgr, news dir & local news ed; Jennifer Duff, opns mgr, chief of opns & disc jockey; Mark England, sls VP & mus dir; Frank Folsom, chief of engrg.

Tiptonville

WTNV(FM)—Not on air, target date: unknown: 97.3 mhz; 6 kw. Ant 328 ft TL: N36 20 59 W89 22 12. Hrs open: 35 Radio Rd., Dyersburg, 38025-0100. Phone: (731) 285-1339. Fax: (731) 287-0100. Licensee: Dr. Pepper Pepsi-Cola Bottling Co. of Dyersburg. ♦W.E. Burks, pres.

Trenton

WTNE-FM— August 1980: 97.7 mhz; 50 kw. Ant 328 ft TL: N36 05 10 W88 54 39. Stereo. Hrs opn: 24 Box 500, 38382. Secondary address: 302 W. Eaton 38382. Phone: (731) 855-0098. Fax: (731) 855-1600. E-mail: wtkbwtne@bellsouth.net Licensee: Grace Broadcasting Services Inc. Group owner: Thunderbolt Broadcasting Co./Gibson County Broadcasting (acq 1-7-2005; grpsl). Population served: 500,000 Format: Country, sports. News staff: one. Target aud: 25-60. Spec prog: High school sports 10 hrs, college football 10 hrs wkly. ♦Sherry Vaughn, gen mgr; Randy Gardner, sls dir; Steve Hilton, progmg dir; Robin Cude, news dir & local news ed; Dave Hacker, chief of engrg; Carrie Allmon, traf mgr.

WTNE(AM)— Dec 9, 1966: 1500 khz; 250 w-D, 6 w-N. TL: N35 58 52 W88 55 32.24 Phone: (731) 855-1500.100,000 Rgnl rep: Midsouth. Format: Adult contemp. Target aud: General; Gibson county, news oriented people. Spec prog: Sports. ♦Dave Hacker, chief of engrg.

Tullahoma

***WAUT-FM**— 1998: 88.5 mhz; 1.9 kw. 177 ft TL: N35 20 30 W86 11 05. Hrs open: Box 3206, American Family Radio, Tupelo, MS, 38803. Phone: (662) 844-8888. Phone: (662) 844-8893. Fax: (662) 842-6791. E-mail: comments@afr.net Web Site: www.afr.net. Licensee: American Family Association. Group owner: American Family Radio Format: Relg. ♦Marvin Sanders, gen mgr.

WHRP(FM)— July 1, 1962: 93.3 mhz; 100 kw. Ant 981 ft TL: N35 02 04 W86 22 52. (CP: COL New Market, AL. 14.5 kw, ant 913 ft. TL: N34 47 36 W86 37 51). Stereo. Hrs opn: 24 1717 Hwy. 72 E., Athens, AL, 35611. Phone: (256) 830-8300. Fax: (256) 232-6842. Web Site: www.whrpfm.net. Licensee: Cumulus Licensing LLC. Group owner: Cumulus Media Inc. (acq 7-21-2003; grpsl). Population served: 600,000 Format: Urban adult contemp. News staff: 3. Target aud: Adult; 25-54. ♦Bill West, gen mgr; Tracy Flesch, gen sls mgr; Wendy Black, gen sls mgr & prom dir; Mark Raymond, progmg dir; Marty Broman, news dir; Josh Bohn, chief of engrg; J.J. Vincent, traf mgr & women's int ed.

WJIG(AM)— Aug 1, 1947: 740 khz; 250 w-D, 67 w-N. TL: N35 20 36 W86 12 00. Hrs open: 24 WJIG AM 740, 607 E. Carroll St., 37388-3951. Phone: (931) 455-7426. Fax: (931) 455-7438. E-mail: wjig@cafes.net Licensee: NRS Enterprises Inc. (acq 2-21-97; $163,000). Population served: 18,000 Natl. Network: Salem Radio Network. Rgnl. Network: Tenn. Radio Net. Format: Southern gospel. ♦Roy Woods, pres; Joyce Woods, gen mgr; Tom Wiseman, chief of engrg; Heath Laws, disc jockey; Steven Arnold, disc jockey.

***WTML(FM)**— 2001: 91.5 mhz; 1.55 kw. Ant 269 ft TL: N35 23 53 W86 08 40. Hrs open:
Rebroadcasts WPLN-FM Nashville 100%.
Nashville Public Radio, 630 Mainstream Dr., Nashville, 37228-1204. Phone: (615) 760-2903. Fax: (615) 760-2904. Web Site: www.wpln.org. Licensee: Nashville Public Radio. Format: Classical, news, bluegrass. ♦Robert Gordon, gen mgr; Scott Smith, opns mgr; Laura Landress, gen sls mgr; Henry Fennell, progmg dir; Will Griffin, mus dir; Anita Bugg, news dir; Tom Knox, chief of engrg; Wendy Poston, traf mgr.

Tusculum

WIKQ(FM)—Licensed to Tusculum. See Greeneville

Union City

WENK(AM)— Oct 26, 1946: 1240 khz; 1 kw-U. TL: N36 25 28 W89 02 17. Stereo. Hrs opn: 24
Rebroadcasts WTPR-FM McKinnon 100%.
1729 Nailling Dr., 38261. Phone: (731) 885-1240. Fax: (731) 885-3405. E-mail: thailey@wenkwtpr.com Licensee: WENK of Union City Inc. Group owner: WENK Broadcast Group Inc. (acq 1-74). Population served: 96,579 Natl. Rep: Rgnl Reps. Law Firm: Shainis & Peltzman. Format: Oldies. News staff: one; News: 15 hrs wkly. Target aud: 35-54. ♦Richard Hall, sls dir; Brad Hosford, chief of engrg; Terry Hailey, pres, gen mgr, progmg dir & edit mgr.

WQAK(FM)— March 1994: 105.7 mhz; 6 kw. Ant 308 ft TL: N36 31 07 W89 05 41. Hrs open: Box 318, Marten, 38232. Phone: (731) 885-0051. Fax: (731) 885-0250. Licensee: Thunderbolt Broadcasting Co. (acq 12-29-2005; $900,000 with WYVY(FM) Union City). Population served: 275,000 Law Firm: Womble, Carlyle, Sandridge & Rice. Format: Classic Hits. ♦Paul F. Tinkle, pres.

***WTAI(FM)**— 2005: 88.9 mhz; 860 w vert. Ant 623 ft TL: N36 24 48 W89 08 59. Hrs open: 5700 West Oaks Blvd., Rocklin, CA, 95765. Phone: (916) 251-1600. Fax: (916) 251-1650. Licensee: Educational Media Foundation. (acq 6-21-2005; $25,000 for CP). Format: Christian. ♦Richard Jenkins, pres.

WYVY(FM)— Sept 20, 1974: 104.9 mhz; 6 kw. Ant 305 ft TL: N36 28 25 W88 56 41. Stereo. Hrs opn: 24 Box 318, Martin, 38237. Secondary address: 223 Westgate Dr. 38261. Phone: (731) 885-0051. Fax: (731) 885-0250. Licensee: Thunderbolt Broadcasting Co. (acq 12-29-2005; $900,000 with WQAK(FM) Union City). Population served: 100,000 Rgnl. Network: Tenn. Radio Net. Law Firm: Womble, Carlyle, Sandridge & Rice. Format: Country. News: 25 hrs wkly. Target aud: 18 plus. ♦Paul F. Tinkle, pres; Don Wilson, stn mgr.

Wartburg

WECO(AM)— Aug 31, 1970: 940 khz; 5 kw-D. TL: N36 05 48 W84 35 31. Hrs open: Box 100, 37887. Secondary address: 305 N. Church St. 37887. Phone: (423) 346-3900. Fax: (423) 346-7686. E-mail: wecoradio@highland.net Licensee: Morgan County Broadcasting Co. Inc. Rgnl. Network: Tenn. Radio Net. Format: Gospel. ♦Sandy Lavender, gen mgr; Gary Stone, progmg dir; Carl Stump, chief of engrg.

WECO-FM— August 1988: 101.3 mhz; 500 w. 770 ft TL: N36 05 25 W78 38 05. (Acq 1-30-89). Format: Country. Target aud: 25-49.

Waverly

WQMV(AM)— Sept 25, 1963: 1060 khz; 1 kw-D, 4 w-N. TL: N36 05 15 W87 51 18. Hrs open: 24 Box 610, 37185. Phone: (931) 296-9768. Fax: (931) 296-9892. E-mail: wqmv@comcast.net Licensee: C & L Broadcasting Corp. (acq 10-16-2003; $60,000). Population served: 6,000 Natl. Network: ABC. Format: Music from the 50s, 60s, 70s, 80s. Target aud: Adults 35-65. ♦Richard Albright, pres.

WVRY(FM)— Sept 26, 1972: 105.1 mhz; 50 kw. Ant 492 ft TL: N36 05 16 W87 51 19. Stereo. Hrs opn: 24 N. Main St., Dickson Phone: (615) 740-9879. Fax: (615) 740-7799. E-mail: info@salemmusicnetwork.com Web Site: www.solidgospel105.com. Licensee: Reach Satellite Network Inc. Group owner: Salem Communications Corp. (acq 3-31-2000; $3.1 million for stock with WBOZ(FM) Woodbury). Population served: 1,500,000 Natl. Network: Salem Radio Network. Rgnl. Network: Tenn. Radio Net. Natl. Rep: Salem. Law Firm: Fletcher, Heald & Hildreth. Format: Gospel. Target aud: 25-54. ♦Jim Cumbee, CEO; Michael S. Miller, gen mgr; Carl Campbell, stn mgr; Wade Schoenemann, opns mgr; Dick Marsh, prom dir; Vance Dillard, progmg dir; Ed Evenson, traf mgr.

Waynesboro

WWON(AM)— Jan 31, 1970: 930 khz; 500 w-D. TL: N35 18 30 W87 44 44. Hrs open: 24 Box 999, 38485. Secondary address: 100 Public Sq. S. 38485. Phone: (931) 722-3631. Fax: (931) 722-3632. E-mail: wwon@netease.net Web Site: am930.net. Licensee: Huntingdon Broadcasting Inc. (acq 4-5-2006; $73,269). Population served: 20000 Law Firm: Mullin, Rhyne, Emmons & Topel. Format: Country. News: 13 hrs wkly. Target aud: 18-54; listeners interested in rgnl & natl issues. ♦Chris Lash, gen mgr.

White Bluff

WQSE(AM)— July 18, 1982: 1030 khz; 1 kw-D, DA-N. TL: N36 08 03 W87 12 58. Hrs open: 201 Hall Ln., 37187. Phone: (615) 797-9785. Fax: (615) 797-9788. E-mail: dvanedjwqse@aol.com Licensee: Canaan Communications Inc. (acq 3-19-03; $85,000). Format: Southern gospel. ♦Duane Jeffrey, pres & gen mgr; Kerry Lampley, sls dir; Mary Jeffrey, mus dir & disc jockey; Shery Swaw, sports cmtr.

Winchester

WCDT(AM)— Mar 8, 1948: 1340 khz; 1 kw-U. TL: N35 10 51 W86 05 34. Hrs open: 24 1201 S. College St., 37398. Phone: (931) 967-2201. Phone: (931) 967-2202. Fax: (931) 967-2201. E-mail: wcdt@bellsouth.net Web Site: www.wcdt1340.com. Licensee: Franklin County Radio & Broadcasting Co. Inc. (acq 8-1-56). Population served: 65,000 Natl. Network: ABC. Rgnl. Network: Tenn. Radio Net. Format: Country. News staff: one; News: 15 hrs wkly. Target aud: General. Spec prog: Farm 15 hrs, Relg 6 hrs wkly. ♦John T. Yarbrough, pres & VP; Tommy Yarbrough, gen mgr; Jeanetta Shields, gen sls mgr, progmg mgr & progmg; Sharon Holder, mktg dir & sls; Karen Shetters, prom mgr & disc jockey; Jan Tavalin, news dir & spec ev coord; Phillip Morris, disc jockey; Karen Shetters, sls.

Woodbury

WBOZ(FM)— Oct 5, 1994: 104.9 mhz; 6 kw. Ant 328 ft TL: N35 49 33 W86 09 28. Stereo. Hrs opn: 24 312 S. Church St., Murfreesboro, 37130. Secondary address: 402 BNA Dr., Suite 400, Nashville 37217. Phone: (615) 890-3233. Fax: (615) 890-2990. E-mail: webmaster@solidgospel.com Web Site: www.solidgospel105.com. Licensee: Reach Satellite Network Inc. Group owner: Salem Communications Corp. (acq 4-1-00; $3.1 million for stock with WVRY(FM) Waverly). Population served: 130,000 Natl. Network: Salem Radio Network. Natl. Rep: Salem. Format: Christian country, southern gospel. News: 14 hrs wkly. Target aud: 35+; adults. Spec prog: Sports 5 hrs wkly. ♦Greg R. Anderson, pres; Michael S. Miller, gen mgr; Kevin R. Anderson, stn mgr; Dick Marsh, prom dir; Vance Dillard, progmg dir; Carl Campbell, chief of engrg.

WBRY(AM)— Oct 24, 1963: 1540 khz; 500 w-D. TL: N35 49 53 W86 06 42. Hrs open: Box 7, 37190. Secondary address: 153 Mile Valley Rd. 37190. Phone: (615) 563-2313. Fax: (615) 563-6229. E-mail: akus@wbry.com Web Site: www.wbry.com. Licensee: Volunteer Broadcasting LLC (acq 3-2-2005; $130,000). Population served: 25,000 Format: Traditional country. News staff: one; News: 7 hrs wkly. Target aud: 25 plus; adults. ♦Doug Combs, pres & gen mgr.

Texas

Abilene

***KACU(FM)**— June 2, 1986: 89.7 mhz; 33 kw. 215 ft TL: N32 28 34 W99 42 22. Stereo. Hrs opn: 24 ACU Box 29106, 79699. Phone: (915) 674-2441. Fax: (915) 674-2417. E-mail: info@kacu.org Web Site: www.kacu.org. Licensee: Abilene Christian University. Natl. Network: NPR. Format: Adult contemp, class, news. News: 42 hrs wkly. Target aud: 35 plus; middle-to-upper income professionals. Spec prog: Jazz 3 hrs wkly. ♦John Best, gen mgr & opns dir; Kim Seidman, dev dir.

***KAGT(FM)**— Nov 5, 2002: 90.5 mhz; 100 kw. Ant 341 ft TL: N32 30 37 W99 44 28. Stereo. Hrs opn: 24 2351 Sunset Blvd., Suite 170-218, Rocklin, CA, 95765. Phone: (916) 251-1600. Fax: (916) 251-1650. Licensee: Educational Media Foundation. (acq 12-31-2006; $450,000). Population served: 200,000 Format: Christian. ♦Richard Jenkins, pres.

***KAQD(FM)**— 1998: 91.3 mhz; 1 kw. Ant 177 ft TL: N32 28 36 W99 44 56. Hrs open: Box 3206, American Family Radio, Tupelo, MS, 38803. Phone: (662) 844-8888. Fax: (662) 844-9176. E-mail: comments@afr.net Web Site: www.afr.net. Licensee: American Family Association. Group owner: American Family Radio Format: Inspirational Christian. ♦Marvin Sanders, gen mgr.

KBCY(FM)—See Tye

KEAN-FM—Listing follows KYYW(AM).

KEYJ-FM— Apr 30, 1961: 107.9 mhz; 100 kw. 670 ft TL: N32 17 06 W99 38 38. Stereo. Hrs opn: 3911 S. First St., 79605. Phone: (325) 677-7225. Phone: (325) 676-7711. Fax: (325) 676-3851. Web Site: www.keyj.com. Licensee: GAP Broadcasting Abilene License LLC. Group owner: Clear Channel Communications Inc. (acq 8-3-2007; grpsl). Format: Alternative rock. Target aud: 18-49; men. ♦Dale Harris, gen mgr; James Cameron, opns mgr; Frank Payne, progmg dir.

KFGL(FM)— September 1974: 100.7 mhz; 100 kw. 1,260 ft TL: N32 24 48 W100 06 25. Stereo. Hrs opn: 3911 S. First St., 79605. Phone: (325) 676-7111. Fax: (325) 676-3851. E-mail: reneegonzalez @clearchannel.com Web Site: www.kissabilene.com. Licensee: GAP Broadcasting Abilene License LLC. Group owner: Clear Channel Communications Inc. (acq 8-3-2007; grpsl). Population served: 200,000 Format: Classic rock. Target aud: 18-34; women. Spec prog: Oldies 2 hrs wkly. ♦Ted Warren, gen mgr; James Cameron, opns mgr; Renee Gonzalez, gen sls mgr.

***KGNZ(FM)**— Mar 7, 1981: 88.1 mhz; 75 kw. 710 ft TL: N32 17 46 W99 43 01. Stereo. Hrs opn: 24 542 Butternut St., 79602. Secondary address: 1001 Cedar Crest St. 79601. Phone: (325) 673-3045. Fax: (325) 672-7938. E-mail: studio@kgnz.com Web Site: www.kgnz.com. Licensee: Christian Broadcasting Co. Natl. Network: USA. Format: Adult contemp, Christian. Spec prog: Black 2 hrs, gospel 2 hrs wkly. ♦Larry Jack Hill, pres & gen mgr; Doug Harris, opns mgr; Randy Martinez, dev dir.

KKHR(FM)— June 1988: 106.3 mhz; 50 kw. Ant 184 ft TL: N32 28 34 W99 42 22. Stereo. Hrs opn: 24 402 Cypress St., Suite 709, 79601. Phone: (325) 672-5442. Fax: (325) 672-6128. E-mail: info@radioabilene.com Web Site: radioabilene.com. Licensee: Canfin Enterprises Inc. (acq 3-25-2005; $684,000). Natl. Rep: Lotus Entravision Reps LLC. Format: Tejano. News: 3 hrs wkly. Target aud: 18-49. ♦Parker Cannon, gen mgr; Ben Gonzalez, progmg dir; James Thompson, chief of engrg.

KNCE(FM)—See Winters

KORQ(FM)—See Baird

KSLI(AM)— June 15, 1957: 1280 khz; 500 w-D, 226 w-N. TL: N32 26 30 W99 43 08. Hrs open: 24 Box 3098, 79604. Secondary address: 3911 S. First St. 79605. Phone: (325) 676-7711. Fax: (325) 676-3851. Licensee: GAP Broadcasting Abilene License LLC. Group owner: Clear Channel Communications Inc. (acq 8-3-2007; grpsl). Population served: 235,000 Natl. Network: Jones Radio Networks. Format: Music of your life. News: 4 hrs wkly. Target aud: 18-49; Hispanic. ♦Dale Harris, gen mgr.

KULL(FM)— Apr 1, 1998: 92.5 mhz; 27.5 kw. Ant 663 ft TL: N32 16 35 W99 35 38. Hrs open: 3911 S. 1st St., 79605. Phone: (325) 677-7225. Fax: (325) 677-3851. Licensee: GAP Broadcasting Abilene License LLC. Group owner: Clear Channel Communications Inc. (acq 8-3-2007; grpsl). Population served: 250,000 Natl. Network: ABC. Format: Oldies. ♦Ted Wrenn, VP; James Cameron, opns mgr.

KWKC(AM)— June 19, 1948: 1340 khz; 1 kw-U. TL: N32 25 14 W99 43 54. Hrs open: 24 Box 3498, 79604-3498. Phone: (325) 672-5442. Fax: (325) 672-6128. E-mail: parker@radioabilene.com Web Site: www.radioabilene.com. Licensee: Canfin Enterprises Inc. Population served: 100,000 Rgnl. Network: Texas State Net. Format: News/talk. Target aud: 25 plus. ♦Parker Cannan, pres, gen mgr & opns VP.

KYYW(AM)— Oct 1, 1936: 1470 khz; 5 kw-D, 1 kw-N, DA-N. TL: N32 29 26 W99 45 02. Hrs open: 3911 S. First St., 79605. Phone: (915) 676-7711. Fax: (915) 676-3851. Web Site: www.keanradio.com. Licensee: GAP Broadcasting Abilene License LLC. Group owner: Clear Channel Communications Inc. (acq 8-3-2007; grpsl). Population served: 250,000 Natl. Network: CBS. Law Firm: Kenkel & Associates. Format: Authentic country. Target aud: 30 plus; upscale, affluent. Spec prog: Farm 3 hrs, relg 5 hrs wkly. ♦Dale Harris, gen mgr; Justin Riggan, gen sls mgr; Renee Gonzalez, gen sls mgr; James Cameron, progmg dir; Gary Smith, engrg dir & chief of engrg.

KEAN-FM—Co-owned with KYYW(AM). July 1, 1969: 105.1 mhz; 100 kw. Ant 810 ft TL: N32 16 35 W99 35 39. Stereo. 24 E-mail: kean@keanradio.com Web Site: www.keanradio.com.250,000 Natl. Network: ABC. Format: Country. ♦Rudy Fernandez, progmg dir.

KZQQ(AM)— Aug 29, 1962: 1560 khz; 500 w-D, 45 w-N. TL: N32 27 21 W99 47 59. Hrs open: Box 3498, 79604-3498. Phone: (325) 672-5442. Fax: (325) 672-6128. Web Site: www.radioabilene.com. Licensee: Canfin Enterprises Inc. Population served: 120000 Format: Sports, talk. Target aud: 25 plus. ♦Jim Christoferson, gen mgr.

Alamo

KJAV(FM)— Aug 17, 1980: 104.9 mhz; 6 kw. Ant 328 ft TL: N26 12 49 W98 05 21. Hrs open: BMP Radio LP, 1201 No. Jackson, Suite 900, McAllen, 78501. Phone: (956) 992-8895. Fax: (956) 992-8897. E-mail: info@bmpradio.com Web Site: www.bmpradio.com. Licensee: BMP RGV License Company L.P. (acq 12-14-2004; $7 million). Format: Sp. ♦Thomas Castro, pres; Jose Luis Munoz, gen mgr; Jeff Koch, opns mgr & progmg dir.

Alamo Heights

KDRY(AM)—Licensed to Alamo Heights. See San Antonio

KLUP(AM)—See Terrell Hills

Albany

KNOS(FM)—Not on air, target date: unknown: 98.9 mhz; 6 kw. Ant 328 ft TL: N32 37 25 W99 26 26. Hrs open: 5842 Westslope Dr., Austin, 78731. Phone: (512) 467-0643. Licensee: Matinee Radio LLC. ♦Robert Walker, gen mgr.

Alice

***KIFR(FM)**—Not on air, target date: unknown: 88.3 mhz; 23.5 kw vert. Ant 292 ft TL: N27 53 55 W98 05 55. Hrs open: 4135 Northgate Blvd., Suite 1, Sacramento, CA, 95834-1226. Phone: (916) 641-8191. Fax: (916) 641-8238. Licensee: Family Stations Inc. ♦Harold Camping, pres; Peggy Renschler, gen mgr.

KNDA(FM)— Jan 1, 1974: 102.9 mhz; 50 kw. 492 ft TL: N27 42 26 W97 46 54. Stereo. Hrs opn: 2001 Saratoga, Suite 100, Corpus Christi, 78417. Phone: (361) 814-1030. Phone: (361) 814-1029. Fax: (361) 814-1036. E-mail: lilrichardabomb@aol.com Licensee: Encarnacion A. Guerra (acq 5-95). Population served: 27,500 Format: Hip-hop, R&B. ♦Pat Rodriguez, gen mgr.

KOPY(AM)— 1947: 1070 khz; 1 kw-U, DA-N. TL: N27 46 39 W98 04 53. Hrs open: 24 Box 731, 78333. Secondary address: 2722 N. Business Hwy. 281 78332. Phone: (361) 664-1884. Fax: (361) 664-1886. Licensee: Alice Broadcast Co. (acq 1-1-96). Population served: 60,000 Format: Country. News staff: 12; News: 2 hrs wkly. Target aud: 18-59. ♦Bobby Pena, gen mgr & stn mgr; Jackie Hinojosa, gen sls mgr, prom mgr, progmg dir & news dir.

KOPY-FM— Jan 20, 1976: 92.1 mhz; 3 kw. 300 ft TL: N27 46 39 W98 04 53. Stereo. 24 Format: Tejano. News: 8 hrs wkly. Target aud: General. ♦Bobby Pena, progmg dir.

KUKA(FM)—See San Diego

Allen

KESN(FM)— Dec 1, 1981: 103.3 mhz; 100 kw. Ant 1,968 ft TL: N33 32 08 W96 49 54. Stereo. Hrs opn: 2221 E. Lamar Blvd., Suite 300, Arlington, 76006. Phone: (817) 695-3523. Fax: (817) 695-3516. Web Site: espn1033.com. Licensee: WBAP-KSCS Operating Ltd. Group owner: ABC Inc. (acq 8-9-00; $18 million). Format: Sports. Target aud: 25-54; males and females. Spec prog: Children 2 hrs wkly. ♦Marce Graves, CEO; Keri Korzeniewski, gen mgr.

Alpine

KALP(FM)— September 1986: 92.7 mhz; 2.37 kw. 328 ft TL: N30 19 09 W103 37 04. Hrs open: 6 AM-10 PM Box 9650, 79831. Secondary address: 500 Hendryx Ave. 79830. Phone: (432) 837-2144. Fax: (915) 837-3984. E-mail: alpineradio@brooksdata.net Licensee: Rio Grande Broadcasting Co. Format: C&W. ♦Gene Ray Hendryx, gen mgr, progmg dir, chief of engrg & pres.

KVLF(AM)— Feb 27, 1947: 1240 khz; 1 kw-U. TL: N30 22 30 W103 39 36. Hrs open: 6 AM-10 PM Drawer 779, 79831. Secondary address: 500 Hendryx Ave. 79831. Phone: (432) 837-2144. Fax: (432) 837-3984. E-mail: alpineradio@brooksdata.net Licensee: Big Bend Broadcasters. Population served: 13,700 Natl. Network: ABC. Format: Div. News: 21 hrs wkly. Spec prog: Sp 10 hrs wkly. ♦Gene Ray Hendryx Jr., pres; Ray Hendryx, gen mgr; Jerry Sotello, gen sls mgr.

Alvin

***KACC(FM)—** Nov 1, 1993: 89.7 mhz; 5.6 kw. 338 ft TL: N29 24 01 W95 12 13. Stereo. Hrs opn: 24 3110 Mustang Rd., 77511. Phone: (281) 756-3765. Fax: (281) 756-3885. E-mail: cforsythe@alvincollege.edu Web Site: www.kaccradio.com. Licensee: Alvin Community College. Law Firm: Garvey, Schubert & Barer. Wire Svc: AP Format: AOR. News staff: one; News: 3 hrs wkly. Target aud: General. ♦A. Rodney Allbright, pres; Cathy Forsythe, gen mgr & prom dir; Mark Moss, chief of opns & progmg dir.

KTEK(AM)— November 1981: 1110 khz; 2.5 kw, DA. TL: N29 22 51 W95 14 15. Hrs open: Sunrise-sunset 6161 Savoy, Suite 1200, Houston, 77036. Phone: (713) 260-3600. Fax: (713) 260-3628. Web Site: www.kkht.com. Licensee: South Texas Broadcasting Inc. Group owner: Salem Communications Corp. (acq 7-2-98; $2.7 million with KYCR(AM) Golden Valley, MN). Natl. Network: USA. Format: Christian talk. News: 6 hrs wkly. Target aud: 25-54; upscale families, 60% women, 40% male. Spec prog: Urdu-Hindi 3 hrs wkly. ♦Chuck Jewell, gen mgr; Paul Baker, opns mgr & mktg mgr; Dan Doster, gen sls mgr; Kendall Cockrell, progmg dir; Marsha Lambeth, progmg dir; Scott VanPelt, pub affrs dir; Sydney Jones, chief of engrg; Lorraine Horton, disc jockey.

Amarillo

***KACV-FM—** Mar 15, 1976: 89.9 mhz; 100 kw. 1,041 ft TL: N35 20 33 W101 49 21. Stereo. Hrs opn: 6 AM-midnight Box 447, 79178. Secondary address: 2408 S. Jackson 79109. Phone: (806) 371-5222. Fax: (806) 345-5576. E-mail: kacvfm90@actx.edu Web Site: www.kacvfm.org. Licensee: Amarillo Junior College District. Population served: 420,000 Format: Alternative/block. Spec prog: Jazz 12 hrs, Texas 6 hrs wkly. ♦Brian Frand, progmg dir. Co-owned TV: *KACV-TV affil.

KARX(FM)—Claude, Apr 12, 1992: 95.7 mhz; 100 kw. 391 ft TL: N35 06 16 W101 39 28. Stereo. Hrs opn: 24 301 S. Polk, Suite 100, 79101. Phone: (806) 342-5200. Fax: (806) 342-5202. E-mail: chris.matchett@cumulus.com Web Site: www.cumulus.com. Licensee: Cumulus Licensing Corp. Group owner: Cumulus Media Inc. (acq 2-2-98; $675,000). Population served: 85,000 Format: Classic rock. News staff: one; News: 10 hrs wkly. Target aud: 25-54; male. ♦Rick Matchett, gen mgr; Stan Ross, sls dir; D'Lisa Pohnert, prom dir; Craig Vaughn, progmg dir; Dale Miller, mus dir; Matt Darby, news dir; J.P. Wolf, chief of engrg.

KATP(FM)— Mar 11, 1976: 101.9 mhz; 100 kw. 935 ft TL: N35 20 33 W101 49 21. Stereo. Hrs opn: 24 6214 W. 34th, 79109. Phone: (806) 355-9777. Fax: (806) 359-0136. E-mail: katp@clearchannel.com Web Site: www.catcountry1019.com. Licensee: GAP Broadcasting Amarillo License LLC. Group owner: Clear Channel Communications Inc. (acq 8-3-2007; grpsl). Population served: 250,000 Format: Classic country. News: 2 hrs wkly. Target aud: 18-49. ♦Mike Ryan, gen mgr; Les Montgomery, dev mgr; Debbie Davis, gen sls mgr.

***KAVW(FM)—** July 1998: 90.7 mhz; 1 kw. 213 ft TL: N35 11 50 W101 49 59. Hrs open: Box 3206, Tupelo, MS, 38803. Phone: (662) 844-8888. Fax: (662) 842-6791. E-mail: comments@afr.net Web Site: www.afr.net. Licensee: American Family Association. Group owner: American Family Radio Format: Inspirational Christian. ♦Marvin Sanders, gen mgr.

KBZD(FM)— March 1994: 99.7 mhz; 21.5 kw. 351 ft TL: N35 06 50 W101 49 16. Hrs open: 3639 Wolflin Ave., 79103. Phone: (806) 355-1044. Fax: (806) 457-0642. Licensee: Tejas Broadcasting Ltd. LLP. Group owner: Amigo Broadcasting L.P. (acq 11-15-2004; grpsl). Format: Tejano/Regional Mexican. ♦Mac Douglas, gen mgr; Brad Gonzalez, gen sls mgr; Israel Salazar, progmg dir; Charlie Singleton, chief of engrg; Emelia Chacon, traf mgr.

KTNZ(AM)—Co-owned with KBZD(FM). 1946: 1010 khz; 5 kw-D, 500 w-N, DA-2. TL: N35 11 03 W101 41 28.200,000 Format: Christian. ♦Israel Salazar, progmg VP & news dir.

KDJW(AM)— Sept 15, 1955: 1360 khz; 500 w-D, 137 w-N. TL: N35 14 49 W101 49 13. Hrs open: 1721 Avondale Ctr., 79106. Phone: (806) 331-2826. Fax: (806) 358-9285. E-mail: ffeedlot@aol.com Web Site: www.kdjw.com. Licensee: Avondale Operating Inc. (acq 12-8-03). Format: Classic country. Target aud: 45 plus; adults with money. ♦Ron Slover, pres & gen mgr; Bill Howe, progmg dir & chief of engrg.

KGNC(AM)— May 19, 1922: 710 khz; 10 kw-U, DA-2. TL: N35 25 12 W101 33 20. Hrs open: 24 Box 710, 79189-0710. Secondary address: 3505 Olsen Blvd., Suite 117 79109. Phone: (806) 355-9801. Fax: (806) 354-8779. Fax: (806) 354-9450. Licensee: Morris Communications Corp. Group owner: Morris Communications Inc. (acq 12-22-97; grpsl). Population served: 400,000 Natl. Network: ABC. Rgnl. Network: Texas State Net. Natl. Rep: Katz Radio. Law Firm: Wiley, Rein & Fielding. Wire Svc: Reuters Format: News/talk, sports. News staff: 3; News: 35 hrs wkly. Target aud: General; upscale adults & agricultural business listeners. Spec prog: Farm 11 hrs, relg 4 hrs wkly.Dan Gorman, gen mgr & edit dir; Tim Butler, opns mgr; Doug Surlens, gen sls mgr; Chris Albracht, progmg dir; Barry King, news dir; Greg Wheeler, pub affrs dir; John Wolfe, chief of engrg; Daron Casler, traf mgr; James Hunt, local news ed; Cammie Clark, news rptr; Cindy Stone, news rptr; Bob Givens, farm dir; Mel Phillips, outdoor ed; Mike Roden, sports cmtr; Mary Lyn Halley, women's int ed

KGNC-FM— Dec 24, 1958: 97.9 mhz; 100 kw. 1,285 ft TL: N35 18 52 W101 50 47. Stereo. 24 300,000 Format: Country. News: one hr wkly. Target aud: 25-54; upscale adults. ♦Dan Gorman, stn mgr & edit dir; Jay Johnson, rgnl sls mgr & mktg mgr; Tim Butler, progmg mgr; Lani Clark, rsch dir; Mary Lyn Halley, spec ev coord.

KIXZ(AM)— June 1947: 940 khz; 5 kw-D, 1 kw-N, DA-2. TL: N35 09 17 W101 45 28. Hrs open: 24 6214 W. 34th., 79109. Phone: (806) 355-9777. Fax: (806) 355-5832. E-mail: kixz@clearchannel.com Web Site: www.newsradio940.com. Licensee: GAP Broadcasting Amarillo License LLC. Group owner: Clear Channel Communications Inc. (acq 8-3-2007; grpsl). Population served: 250,000 Law Firm: Akin, Gump, Strauss, Hauer & Feld. Format: News/talk. News staff: one; News: 8 hrs wkly. Target aud: General. Spec prog: Talk 2 hrs, gospel 6 hrs wkly. ♦Matt Martin, gen mgr; Dusty Cagle, sls dir & gen sls mgr; Lori Crofford, prom dir; David Emmons, progmg dir; Charles Fuller, news dir & chief of engrg; Jennifer Stephenson, traf mgr.

KPRF(FM)—Co-owned with KIXZ(AM). October 1979: 98.7 mhz; 100 kw. Ant 480 ft TL: N35 11 02 W101 58 11. Stereo. 24 E-mail: kprf@clearchannel.com Web Site: www.power987.com.200000 Natl. Network: ABC. Format: Adult hits. News staff: one; News: 2 hrs wkly. Target aud: General. ♦Marshal Blevins, progmg dir; Jennifer Stephenson, traf mgr.

***KJJP(FM)—** Dec 6, 1991: 105.7 mhz; 6 kw. 236 ft TL: N35 12 28 W101 51 18. Hrs open: 24 210 N. 7th St., Garden City, KS, 67846. Phone: (620) 275-7444. Fax: (620) 275-7496. Web Site: www.hppr.org. Licensee: Kanza Society Inc. (acq 8-17-2004; $1.25 million). Format: Christian. Target aud: General. ♦Robert Kirby, progmg dir.

***KJRT(FM)—** Apr 1, 1994: 88.3 mhz; 20 kw. 265 ft TL: N35 11 57 W101 48 43. Hrs open:
Rebroadcasts KPDR(FM) Wheeler 100%.
Box 8088, 5754 Canyon Dr., 79114. Phone: (806) 359-8855. Fax: (806) 354-2039. E-mail: kjrt@kingdomkeys.org Web Site: www.kingdomkeys.org. Licensee: Top o'Texas Educational Broadcasting Foundation. Format: Relg, educ. ♦Ricky Pfeil, gen mgr.

KMML-FM— March 1985: 96.9 mhz; 100 kw. 613 ft TL: N35 17 33 W101 50 48. Stereo. Hrs opn: 24 6214 W. 34th St., 79109. Phone: (806) 355-9777. Fax: (806) 355-5832. E-mail: kmml@kmml.com Web Site: www.969kmml.com. Licensee: GAP Broadcasting Amarillo License LLC. Group owner: Clear Channel Communications Inc. (acq 8-3-2007; grpsl). Population served: 200,000 Natl. Network: ABC. Format: Real country. News staff: one; News: 2 hrs wkly. Target aud: General. ♦Mike Ryan, gen mgr; Les Montgomery, dev mgr & progmg dir; Debbie Davis, gen sls mgr; Lori Crofford, prom dir; Charlie Fuller, chief of engrg; Jennifer Stephenson, traf mgr.

KMXJ-FM— March 1946: 94.1 mhz; 100 kw. 1,082 ft TL: N35 20 33 W101 49 21. Stereo. Hrs opn: 24 6214 W. 34th, 79109. Phone: (806) 355-9777. Fax: (806) 355-5832. E-mail: kmxj@clearchannel.com Web Site: www.mix941kmxj.com. Licensee: GAP Broadcasting Amarillo License LLC. Group owner: Clear Channel Communications Inc. (acq 8-3-2007; grpsl). Population served: 200000 Natl. Network: ABC. Format: Adult contemp. News staff: one; News: 2 hrs wkly. Target aud: General. ♦Mike Ryan, gen mgr; Les Montgomery, dev mgr; Debbie Davies, gen sls mgr; Lori Crofford, prom dir; Johnny McQueen, progmg dir & progmg mgr; Charlie Fuller, chief of engrg; Jennifer Stephenson, traf mgr.

KPUR(AM)— Aug 1, 1949: 1440 khz; 5 kw-D, 1 kw-N, DA-N. TL: N35 07 20 W101 48 09. Stereo. Hrs opn: 24 301 S. Polk, Suite 100, 79101. Phone: (806) 342-5200. Fax: (806) 342-5202. E-mail: rickmatchett@cumulus.com Web Site: www.cumulus.com. Licensee: Cumulus Licensing Corp. Group owner: Cumulus Media Inc. (acq 3-12-98; $820,000 with KPUR-FM Canyon). Population served: 175,000 Format: Talk, sports. Target aud: 25-54. ♦Rick Matchett, gen mgr; Carolyn Reinert, sls dir; O'Lisa Pohnert, prom dir; Matt Darby, progmg dir; J.P. Wolf, chief of engrg.

KQFX(FM)—Borger, March 1975: 104.3 mhz; 100 kw. 590 ft TL: N35 25 54 W101 36 47. Stereo. Hrs opn: 3639B Wolfin Ave., 79103. Phone: (806) 355-1044. Fax: (806) 457-0642. Licensee: Tejas Broadcasting Ltd. LLP. Group owner: Amigo Broadcasting L.P. (acq 11-15-2004; grpsl). Population served: 200,000 Format: Regional. ♦Matt Douglas, gen mgr; Willie Palacios, sls dir & gen sls mgr; Israel Salavar, progmg dir; Charlie Singleton, chief of engrg.

KQIZ-FM— November 1976: 93.1 mhz; 100 kw. 700 ft TL: N35 17 33 W101 50 48. Stereo. Hrs opn: 24 301 S. Polk, Suite 100, 79101. Phone: (806) 342-5200. Fax: (806) 342-5202. E-mail: rick.matchett@cumulus.com Web Site: www.cumulus.com. Licensee: Cumulus Licensing Corp. Group owner: Cumulus Media L.L.C (acq 3-5-98; $3.057 million). Population served: 162,000 Format: CHR. Target aud: 18-44; young families. Spec prog: Relg 2 hrs wkly. ♦Rick Matchett, gen mgr; Carolyn Reinert, sls dir; D'Lisa Pohnert, prom dir; Deanna McGuire, progmg dir; J.P. Wolf, chief of engrg.

KRGN(FM)— Oct 6, 1986: 103.1 mhz; 25 kw. 300 ft TL: N35 16 04 W101 53 06. Hrs open: 24 Box 10050, 79116. Secondary address: 910 S. Lamar 79106. Phone: (806) 376-5746. Fax: (806) 376-4212. E-mail: krgn@flc.org Web Site: www.krgn.org. Licensee: Family Life Broadcasting Inc. Group owner: Family Life Broadcasting System (acq 6-24-98; grpsl). Natl. Network: USA. Format: MOR Christian inspirational, news/talk, educ. News: 4 hrs wkly. Target aud: 28 plus; mature Christian, mainstream evangelical. ♦Steve Wright, stn mgr; Steve Johnson, news dir.

KXGL(FM)— November 1997: 100.9 mhz; 100 kw. Ant 1,305 ft TL: N35 18 53 W101 50 47. Stereo. Hrs opn: 24 1616 S. Kentucky, Suite C-215, 79102. Phone: (806) 351-2345. Fax: (806) 331-3170. E-mail: bobrussell@1009theeagle.com Licensee: JMJ Broadcasting Co. Inc. (acq 1-15-2004). Population served: 184,900 Natl. Rep: Katz Radio. Law Firm: Larry Bernstein. Wire Svc: AP Format: Classic hits. News staff: one; News: 5 hrs wkly. Target aud: 25-54. ♦Herbert W. McCord, pres; Bob Russell, gen mgr; Jamey Karr, opns mgr.

***KXLV(FM)—** August 1989: 89.1 mhz; 3 kw. 328 ft TL: N35 15 39 W101 52 53. Stereo. Hrs open: 24 2351 Sunset Blvd., Suite 170-218, Rocklin, CA, 95765. Phone: (916) 251-1600. Fax: (916) 251-1650. E-mail: klove@klove.com Web Site: www.klove.com. Licensee: Educational Media Foundation. Group owner: EMF Broadcasting (acq 11-4-99; $450,000). Population served: 200000 Natl. Network: K-Love. Law Firm: Shaw Pittman. Format: Contemp Christian music. News staff: 3. Target aud: 25-44; Judeo Christian, female. ♦Richard Jenkins, pres; Mike Novak, VP; Keith Whipple, dev dir; Russ Lloyd, rgnl sls mgr; David Pierce, progmg mgr; Jon Rivers, mus dir; Ed Lenane, news dir; Sam Wallington, engrg dir; Karen Johnson, news rptr; Marya Morgan, news rptr; Richard Hunt, news rptr.

***KXRI(FM)—** November 1993: 91.9 mhz; 2.25 kw. Ant 292 ft TL: N35 14 31 W101 48 43. Hrs open: 24 2351 Sunset Blvd., Suite 170-218, Rocklin, CA, 95765. Phone: (916) 251-1600. Fax: (916) 251-1650. E-mail: info@air1.com Web Site: www.air1.com. Licensee: Educational Media Foundation. Group owner: EMF Broadcasting (acq 5-1-2000; $750,000 with KKLU(FM) Lubbock). Natl. Network: Air 1. Law Firm: Shaw Pittman. Format: Contemp Christian. News staff: 3. Target aud: 18-35; Judeo Christian female. ♦Richard Jenkins, pres; Mike Novak, VP; Keith Whipple, dev dir; Eric Allen, natl sls mgr; David Pierce, progmg mgr; Ed Lenane, news dir; Sam Wallington, engrg dir.

KZIP(AM)— Sept 15, 1955: 1310 khz; 1 kw-D. TL: N35 11 02 W101 58 11. Hrs open: 6 AM-10 PM 3639 B. Wolflin, 79102. Phone: (806) 355-1044. Fax: (806) 352-6525. Licensee: Del Norte Communications Inc. (acq 5-22-2001). Population served: 200,000 Format: Talk radio. News staff: one; News: one hr wkly. Target aud: General. ♦Mac Douglas, gen mgr.

KZRK-FM—Canyon, Sept 30, 1985: 107.9 mhz; 100 kw. 476 ft TL: N35 13 36 W102 00 24. Stereo. Hrs opn: 24 301 S. Polk, Suite 100, 79101. Phone: (806) 342-5200. Fax: (806) 342-5202. E-mail: chris.knight@cumulus.com Web Site: www.kzrk.com. Licensee: Cumulus Licensing Corp. Group owner: Cumulus Media Inc. (acq 3-3-98; $1 million with co-located AM). Population served: 300,000 Natl. Network: Westwood One. Natl. Rep: Roslin. Format: AOR. News staff: one; News: 3 hrs wkly. Target aud: 18-34; general. ♦Rick Matchett, gen mgr; Eric Slayter, opns mgr, prom mgr & progmg dir; Stan Ross, sls dir; D'Lisa Pohnert, mktg dir; Chris Collins, mus dir; J. Curry, asst music dir; J.P. Wolf, chief of engrg; Matt Darby, local news ed.

KZRK(AM)— May 8, 1962: 1550 khz; 1 kw-D, 219 w-N. TL: N34 58 54 W101 57 18.6 AM-6 PM Format: Sports talk.

Andrews

KACT(AM)— Jan 12, 1955: 1360 khz; 1 kw-D. TL: N32 20 50 W102 33 23. Hrs open: Box 524, 79714. Phone: (915) 523-2845. Licensee:

Zia Broadcasting Co. (acq 5-26-76). Population served: 12,500 Format: Country. ♦Lonnie Allsup, pres; Gerald Reid, gen mgr & news dir; Roy Norman, gen sls mgr.

KACT-FM— 1980: 105.5 mhz; 3 kw. 210 ft TL: N32 20 50 W102 33 23. Stereo.

Anson

KTLT(FM)— June 1988: 98.1 mhz; 50 kw. Ant 305 ft TL: N32 39 49 W99 51 18. Stereo. Hrs opn: 24 2525 S. Danville Dr., Abilene, 79605. Phone: (325) 793-9700. Fax: (325) 692-1576. Licensee: Cumulus Licensing Corp. Group owner: Cumulus Media Inc. (acq 1999). Law Firm: Kaye, Scholer, Fierman, Hays & Handler. Format: Alternative rock. ♦Trace Michaels, gen mgr; John Scott, progmg dir; Chris Andrews, chief of engrg.

Aransas Pass

***KKWV(FM)**—Not on air, target date: unknown: 88.1 mhz; 45 kw vert. Ant 469 ft TL: N28 06 26 W97 12 19. Hrs open: 2351 Sunset Blvd., Suite 170-218, Rocklin, CA, 95765. Phone: (916) 251-1600. Fax: (916) 251-1650. Licensee: Educational Media Foundation. (acq 7-23-2007; grpsl). ♦Mike Novak, sr VP.

Arlington

KLTY(FM)—Licensed to Arlington. See Dallas

Athens

***KATG(FM)**— 2006: 88.1 mhz; 80 kw vert. Ant 544 ft TL: N32 02 41 W95 40 37. Hrs open: Box 2440, Tupelo, MS, 38803-2440. Phone: (662) 844-8888. Fax: (662) 842-6791. Web Site: www.afr.net. Licensee: American Family Association. Format: Christian classics. ♦Marvin Sanders, gen mgr.

KCKL(FM)—See Malakoff

KLVQ(AM)— May 17, 1948: 1410 khz; 1 kw-U. TL: N32 10 20 W95 50 36. Hrs open: 24 Box 489, Hwy. 31 E., Malakoff, 75148. Phone: (903) 489-1238. Fax: (903) 489-2671. E-mail: kckl959@yahoo.com Web Site: www.kcklklvq.com. Licensee: Lake Country Radio L.P. Group owner: Routt Radio Companies Inc. (acq 11-5-2005;. $550,000 with KCKL(FM) Malakoff). Population served: 40,000 Natl. Network: Salem Radio Network. Wire Svc: NOAA Weather Wire Svc: UPI Format: Southern gospel. News staff: one; News: 7 hrs wkly. Target aud: 35 plus. Spec prog: Black one hr, relg 8 hrs wkly. ♦John Weeks, gen mgr; Pat Isaacson, opns mgr, sls dir & pub affrs dir; Rich Flowers, progmg dir & news dir; Wayne Blackwelder, chief of engrg.

Atlanta

KNRB(FM)— Dec 22, 1978: 100.1 mhz; 50 kw. Ant 492 ft TL: N33 15 18 W94 05 16. Stereo. Hrs opn: 24 Box 262550, Baton Rouge, LA, 70826. Secondary address: 8919 World Ministry Ave., Baton Rouge, LA 70810. Phone: (225) 768-3688. Phone: (225) 768-8300. Fax: (225) 768-3729. E-mail: kawikfish@yahoo.com Web Site: www.jsm.org. Licensee: Family Worship Center Church Inc. (acq 3-7-2002; grpsl). Population served: 200000 Rgnl rep: Riley. Format: Southern gospel. News: 15 hrs wkly. Target aud: General. ♦David Whitelaw, COO; Jimmy Swaggart, pres; John Santiago, progmg dir.

KPYN(AM)— Oct 18, 1950: 900 khz; 1 kw-D, 33 w-N. TL: N33 04 58 W94 10 58. Hrs open: 24 Box 900, 75531. Secondary address: State Hwy. 43 S. 75551. Phone: (903) 796-2817. Fax: (903) 769-1000. Web Site: www.amen900.com. Licensee: Freed AM Corp. (group owner; (acq 8-3-2005; $100,000). Format: Contemp Christian. Target aud: . ♦Robert A. Delgiorno Jr., pres & gen mgr.

Austin

KAMX(FM)—See Luling

KASE-FM— Mar 30, 1969: 100.7 mhz; 100 kw. 1,100 ft TL: N30 19 10 W97 48 06. Stereo. Hrs opn: 24 Clear Channel Radio KVET-KASE, 3601 South Congress, Bldg. F, 78704. Phone: (512) 684-7300. Fax: (512) 684-7441. Web Site: www.kase101.com. Licensee: Capstar TX L.P. Group owner: Clear Channel Communications Inc. (acq 8-30-00; grpsl). Population served: 507,300 Format: Country. Target aud: 18-44. ♦Dusty Black, gen mgr; Mac Daniels, opns mgr, gen sls mgr & progmg VP; Lise Hudson, sls dir & mktg dir; Rachel Marisay, news dir; Jim Reese, chief of engrg & disc jockey; Suzanne Munoz, traf mgr; Tracy Walker, prom dir & traf mgr.

KVET(AM)—Co-owned with KASE-FM. 1946: 1300 khz; 5 kw-D, 1 kw-N, DA-2. TL: N30 22 31 W97 42 59. Stereo. 24 Web Site: www.sportsradio1300.com. Format: Talk, sports. News staff: 6; News: 25 hrs wkly. Target aud: 25-64. ♦Dusty Black, gen mgr; Trey Poston, progmg dir; Bob Pickett, sports cmtr; Todd Hogan, disc jockey.

***KAZI-FM**— Aug 29, 1982: 88.7 mhz; 1.6 kw. 351 ft TL: N30 16 37 W97 49 34. Hrs open: 24 8906 Wall St., Suite 203, 78754. Phone: (512) 836-9544. Phone: (512) 836-9545. Fax: (512) 836-9563. E-mail: steve@katzfm.org Web Site: www.kazifm.com. Licensee: Austin Community Radio. Population served: 100,000 Law Firm: Haley, Bader & Potts. Format: Gospel, rap, rhythm and blues. News staff: one; News: 12 hrs wkly. Target aud: General; all ages, all ethnic groups. Spec prog: Reggae 6 hrs, blues 6 hrs, gospel 18 hrs, talk 10 hrs wkly. ♦David Bursell, chmn; Sharon Jones, mktg VP, prom mgr & mus dir; Steven Savage, gen mgr & progmg dir; Pepper Thomas, mus dir; James Davis, engrg VP & chief of engrg.

KBPA(FM)—San Marcos, 1971: 103.5 mhz; 95.5 kw. 1,256 ft TL: N30 02 42 W97 52 50. Stereo. Hrs opn: 24 8309 N. Hwy 35, Suite 967, 78753. Phone: (512) 832-4000. Fax: (512) 832-4071. Web Site: www.oldies103austin.com. Licensee: Emmis Austin Radio Broadcasting Co. L.P. Group owner: Emmis Communications Corp. (acq 4-25-2003; grpsl). Population served: 1,000,000 Natl. Rep: Clear Channel. Law Firm: Wiley, Rein & Fielding. Format: Oldies. News staff: one; News: 4 hrs wkly. Target aud: 25-64. ♦Bruce Walden, gen mgr; Jeff Carrol, opns mgr; Brad Copland, gen sls mgr & natl sls mgr; Mike Paterson, prom dir; Bo Chase, progmg dir & progmg mgr; Lisa Melton, news dir & traf mgr; Jim Henkle, chief of engrg.

KFIT(AM)—Lockhart, Feb 1, 1967: 1060 khz; 2 kw-D, DA. TL: N30 19 13 W97 38 59. Hrs open: 6 AM-8 PM Box 160158, 78716. Secondary address: 110 Wild Basin Rd., Suite 375 78746. Phone: (512) 328-8400. Fax: (512) 328-8437. E-mail: kfit1060@texas.net Licensee: KFIT Inc. (acq 6-25-91; $400,000; FTR: 7-15-91). Population served: 500,000 Natl. Network: Westwood One. Law Firm: Dow, Lohnes & Albertson. Format: Gospel. News staff: 2. Target aud: 18-65. ♦Darrell Marshi, CEO; Terre Lewis, gen mgr, stn mgr & opns mgr.

KFON(AM)— 1922: 1490 khz; 1 kw-U. TL: N30 15 13 W97 42 25. Stereo. Hrs opn: 24 912 S. Capital of Texas Hwy., 78746. Phone: (512) 416-1100. Fax: (512) 416-8205. Licensee: BMP Austin License Company L.P. (acq 2-10-2005; grpsl). Population served: 500,000 Format: Rgnl Mexician. Target aud: 18 plus; men. ♦Pedro Gasc, gen mgr.

KGSR(FM)—Bastrop, 1966: 107.1 mhz; 46 kw. 518 ft. TL: N30 07 18 W97 34 45. Stereo. Hrs opn: 24 8309 N. IH 35, 78753. Phone: (512) 832-4000. Fax: (512) 832-4042. Web Site: www.kgsr.com. Licensee: LBJS Broadcasting Co. L.P. Group owner: Emmis Communications Corp. (acq 4-25-03; grpsl). Natl. Rep: McGavren Guild. Format: AAA. News staff: one. Target aud: 25-44; upscale, active, educated adults. Spec prog: Jazz 6 hrs wkly. ♦Beverley Wimer, VP & traf mgr; Scott Gillmore, VP; Bruce Walden, gen mgr; Bob Woche, gen sls mgr; Jyl Hershman-Ross, prom dir; Jody Denberg, progmg dir; Susan Castle, mus dir; Todd Feffries, news dir; Jim Henkel, chief of engrg.

KIXL(AM)—Del Valle, Aug 8, 1959: 970 khz; 1 kw-U, DA-2. TL: N30 19 13 W97 37 25. Hrs open: 24 11615 Angus Rd., Suite 120 B, 78759. Phone: (512) 390-5495. Fax: (512) 241-0510. E-mail: rvillarreal@relevantradio.com Web Site: www.relevantradio.com. Licensee: Starboard Media Foundation Inc. (acq 1-20-2006; $3.58 million). Population served: 1,165,000 Format: Catholic talk. ♦Ruben Villarreal, stn mgr.

KJCE(AM)—Listing follows KKMJ-FM.

KKMJ-FM— Jan 5, 1968: 95.5 mhz; 100 kw horiz, 87 kw vert. Ant 1,000 ft TL: N30 19 23 W97 47 58. Stereo. Hrs opn: 24 4301 Westbank Dr., Escalade B, 3rd Fl., 78746. Phone: (512) 327-9595. Fax: (512) 329-6255. E-mail: jdhiatt@cbs.com Web Site: www.majic.com. Licensee: Texas Infinity Radio L.P. (acq 10-13-98; grpsl). Population served: 1,000,000 Natl. Rep: Katz Radio. Law Firm: Leventhal, Senter & Lerman. Format: Adult contemp. Target aud: 25-54. ♦Clint Culp, sr VP & sls dir; John Hiatt, sr VP & gen mgr.

KJCE(AM)—Co-owned with KKMJ-FM. Aug 12, 1958: 1370 khz; 5 kw-D, 500 w-N. TL: N30 18 16 W97 38 53. Web Site: www.talkradio137am.com. Group owner: Infinity Broadcasting Corp. . Natl. Network: Westwood One, ABC, Salem Radio Network. Format: Talk. News staff: one; News: 10 hrs wkly. Target aud: Males 18-54.

KLBJ(AM)— 1939: 590 khz; 5 kw-D, 1 kw-N, DA-N. TL: N30 14 14 W97 37 44. Hrs open: 24 8309 N. I-35, 78753. Phone: (512) 832-4000. Fax: (512) 832-4081. Web Site: www.590klbj.com. Licensee: LBJS Broadcasting Co. L.P. Group owner: Emmis Communications Corp. (acq 4-25-03; grpsl). Population served: 625,000 Natl. Network: ABC, Wall Street. Natl. Rep: McGavren Guild. Wire Svc: NWS (National Weather Service) Format: News/talk. News staff: 6; News: 14 hrs wkly. ♦Brooke Gallagher, VP; Bruce Walden, gen mgr; Julie Springer, prom dir; Mark Caesar, progmg dir; Hal Kemp, news dir; Jim Henkel, engrg dir.

KLBJ-FM— 1960: 93.7 mhz; 100 kw. 1,050 ft TL: N30 18 36 W97 47 33. Stereo. Web Site: www.klbjfm.com.625,000 Format: Rock. ♦Bob Sinclair, exec VP; Scott Gillmore, opns dir; Jeff Carrol, progmg dir; Loris Lowe, news dir.

***KMFA(FM)**— January 1967: 89.5 mhz; 40 kw. Ant 1,306 ft TL: N30 19 23 W97 47 58. Stereo. Hrs opn: 24 3001 N. Lamar, Suite 100, 78705. Phone: (512) 476-5632. Fax: (512) 474-7463. E-mail: info@kmfa.org Web Site: www.kmfa.org. Licensee: Capitol Broadcasting Association Inc. Population served: 821,600 Law Firm: Garvey, Schubert, Barer. Format: Class. Target aud: General. Spec prog: Educ 2 hrs wkly. ♦Frank Bash, chmn; Jack Allen, gen mgr; Rich Upton, opns mgr.

KPEZ(FM)— Aug 13, 1976: 102.3 mhz; 26 kw. Ant 686 ft TL: N30 13 24 W97 49 39. Stereo. Hrs opn: 24 3601 South Congress, #F, 78704-7213. Phone: (512) 684-7300. Fax: (512) 684-7441. Web Site: www.z1023.com. Licensee: CCB Texas Licenses L.P. Group owner: Clear Channel Communications Inc. (acq 7-24-92). Population served: 900,000 Natl. Rep: Clear Channel. Law Firm: Cohn & Marks. Format: Positive music. News staff: one; News: 3 hrs wkly. Target aud: 25-54; young adults with families, above average income, education. ♦Kim Murray, VP & sls dir; Dusty Black, gen mgr; L. A. Lloyd, progmg dir; Jim Reese, chief of engrg; Andy Hancock, disc jockey & prom.

KTXZ(AM)—See West Lake Hills

***KUT(FM)**— Nov 10, 1958: 90.5 mhz; 100 kw. 680 ft TL: N30 18 51 W97 51 58. Stereo. Hrs opn:
Rebroadcasts KUTX(FM) San Angelo 100%.
1 University Station A0704, Univ.of Texas, 78712-1090. Phone: (512) 471-1631. E-mail: kut@kut.org Web Site: www.kut.org. Licensee: University of Texas at Austin. Population served: 1,000,000 Natl. Network: NPR, PRI. Law Firm: Cohn & Marks. Format: Music & news. News staff: 5; News: 25 hrs wkly. Target aud: 25-54; educated; influential decision makers & arts community. Spec prog: Folk 4 hrs, blues 6 hrs wkly. ♦Stewart Vanderwilt, gen mgr; Sylvia Carson, dev dir; Chris Collins, gen sls mgr & engrg mgr; Hawk Mendenhall, progmg dir; Emily Donahue, news dir.

KVET-FM— 1950: 98.1 mhz; 100 kw. 686 ft TL: N30 13 24 W97 49 39. Stereo. Hrs opn: 24 3601 South Congress, Bldg. F, 78704. Phone: (512) 684-7300. Fax: (512) 684-7441. Web Site: www.kvet.com. Licensee: Capstar TX L.P. Group owner: Clear Channel Communications Inc. (acq 8-30-00; grpsl). Natl. Network: Westwood One. Format: Country. News staff: 4. Target aud: 35-64. ♦John Hogan, pres; Charlie Ranilly, sr VP; Dusty Black, gen mgr; Jason Kane, opns dir & progmg dir; Mel Jones, sls dir; Heather Lonsdale, natl sls mgr; Tracy Walker, prom dir; Janice Williams, mus dir; Chuck Meyer, news dir; Jim Reese, chief of engrg & reporter.

***KVRX(FM)**— November 1994: 91.7 mhz; 3 kw. 85 ft TL: N30 16 00 W97 40 27. Hrs open: Box D, c/o UT Austin, 78713. Phone: (512) 471-5106. Fax: (512) 232-5793. E-mail: kvrx@kvrx.com Web Site: www.kvrx.org. Licensee: University of Texas at Austin. Population served: 900,000 Format: Alternative,indie. Target aud: 18-34; general. Spec prog: Shore frequency with KOOP-FM. ♦Loren Seager, stn mgr; Christopher Ainley, mus dir; Keith Rutledge, mus dir.

***KYLR(FM)**—Hutto, February 1980: 92.1 mhz; 1.65 kw. Ant 449 ft TL: N30 32 04 W97 34 52. Stereo. Hrs opn: 2351 Sunset Blvd., Suite 170-218, Rocklin, CA, 95765. Phone: (916) 251-1600. Fax: (916) 251-1650. Web Site: www.klove.com. Licensee: Educational Media Foundation. (acq 3-31-2006; $6 million with KMLR(FM) Gonzales). Population served: 180,000 Format: Contemp Christian. ♦Richard Jenkins, pres; Mike Novak, VP; Keith Whipple, dev dir; David Pierce, progmg mgr; Ed Lenane, news dir; Sam Wallington, engrg dir; Karen Johnson, news rptr; Marya Morgan, news rptr; Richard Hunt, news rptr.

Azle

KTCY(FM)— June 29, 1967: 101.7 mhz; 92 kw. Ant 2,034 ft TL: N33 26 13 W97 29 05. Stereo. Hrs opn: 24 2425 Olympic Blvd., Suite 600 West, Santa Monica, CA, 90404. Phone: (310) 447-3870. Fax: (310) 447-3899. Licensee: Liberman Broadcasting of Dallas License LLC. Group owner: Entravision Communications Corp. (acq 11-2-2006; grpsl). Population served: 6,153,500 Natl. Rep: Lotus Entravision Reps LLC. Format: Sp contemp. Target aud: 18-34; Hispanics. ♦Scott Savage, gen mgr; Dean James, opns mgr, progmg dir & engrg dir; Ande Woods, gen sls mgr.

Baird

KORQ(FM)— Sept 9, 1999: 95.1 mhz; 100 kw. Ant 872 ft TL: N32 17 06 W99 38 39. Hrs open: 24 1740 N. First, Abilene, 79603. Phone: (325) 437-9596. Fax: (325) 673-1819. E-mail: doudmediagroup@aol.com Web Site: www.95a.fm. Licensee: Doud Media Group LLC Acq 9-2-02 Population served: 500,000 Law Firm: Dennis J. Kelly. Format: CHR. News staff: 2. Target aud: 18-49; women & teens. ♦Richard Doud, gen mgr; Brad Whitaker, gen sls mgr; Mark McGill, progmg dir; James Thompson, engrg dir.

Balch Springs

KSKY(AM)—Licensed to Balch Springs. See Dallas

Ballinger

KKCN(FM)— August 1977: 103.1 mhz; 100 kw. Ant 456 ft TL: N31 39 37 W100 05 23. Hrs open: 24 1301 S. Abe St., San Angelo, 76903. Phone: (325) 655-7161. Fax: (325) 658-7377. Web Site: www.kkcn103.com. Licensee: Double O Texas Corp. Group owner: Encore Broadcasting LLC (acq 3-15-2006; grpsl). Population served: 150,000 Format: Country. ♦John Kerr, gen mgr; Randy Phair, gen sls mgr; Boomer Kingston, progmg dir; Garry Vaughn, engr.

KRUN(AM)— August 1947: 1400 khz; 1 kw-U. TL: N31 43 31 W99 57 42. Hrs open: 24 hrs Box 230, 1920 Hutchings Ave., 76821. Phone: (325) 365-5500. Fax: (325) 365-3407. E-mail: krun1400@hotmail.com Web Site: www.krunam.com. Licensee: Graham Brothers Communications L.L.C. (acq 12-11-98; $395,000 with co-located FM). Population served: 4,203 Natl. Network: ABC. Rgnl. Network: Texas State Net. Format: Country, sports. News staff: one; News: 2 hrs wkly. Target aud: 25-54. Spec prog: Christian 4 hrs wkly. ♦Andy Allen, gen mgr; Glynne Collenbark, opns mgr; Kody Mac, progmg dir; Toby Virden, gen mgr & traf mgr.

Bandera

KEEP(FM)— July 11, 1981: 103.1 mhz; 1.65 kw. 430 ft TL: N29 51 21 W99 05 26. Stereo. Hrs opn: 24
Rebroadcasts KFAN-FM Johnson City 100%.
Box 311, 210 Woodcrest, Fredericksburg, 78624. Phone: (830) 997-2197. Fax: (830) 997-2198. E-mail: txradio@ktc.com Web Site: www.texasrebelradio.com. Licensee: J. & J. Fritz Media Ltd. (group owner; acq 7-99; $108,000). Law Firm: Fletcher, Heald & Hildreth. Format: Americana AAA. News staff: one. Target aud: 25-49. ♦Jayson Fritz, pres, gen mgr & gen sls mgr; Jan Fritz, VP, mktg VP & adv mgr; Gloria Ottmers, opns mgr; Ariana Carruth Fritz, prom mgr; Mac McClennahan, progmg dir; Rick Star, mus dir; Duncan Black, chief of engrg; Robbi Frantzen, local news ed.

Bastrop

KGSR(FM)—Licensed to Bastrop. See Austin

***KHIB(FM)**— 1998: 88.5 mhz; 4 kw. Ant 308 ft TL: N30 12 57 W97 08 31. Hrs open: Houston Christian Broadcasters Inc., 2424 South Blvd., Houston, 77098-5196. Phone: (713) 520-5200. Web Site: www.khcb.org. Licensee: Houston Christian Broadcasters Inc. (acq 1-19-2005; $112,000). Format: Christian. ♦Bruce Munsterman, pres & gen mgr.

Bay City

***KEDR(FM)**—Not on air, target date: unknown: 88.1 mhz; 45 kw. Ant 328 ft TL: N28 42 08 W95 56 41. Hrs open: c/o WBFR(FM), 244 Goodwin Crest Dr., Suite 118, Birmingham, AL, 35209. Phone: (205) 942-3530. Fax: (510) 568-6190. Licensee: Family Stations Inc. ♦Stanley Jackson, gen mgr.

KMKS(FM)— July 27, 1984: 102.5 mhz; 50 kw. 492 ft TL: N28 47 47 W96 09 17. (CP: 100 kw). Stereo. Hrs opn: 24 Box 789, 77404-0789. Secondary address: 2309 5th St. 77414. Phone: (979) 244-4242. Fax: (979) 245-0107. E-mail: kmks@kmks.com Web Site: www.kmks.com. Licensee: Sandlin Broadcasting Co. Inc. Format: Hot C&W. News staff: 4; News: one hr wkly. Target aud: 24-54.Margaret K. Sandlin, pres; Larry Sandlin, gen mgr, opns mgr & chief of engrg; Judith Gardiner, gen sls mgr & mktg mgr; Helen Linley, prom mgr, news dir & disc jockey; C.W. Simon, progmg dir & disc jockey; Teresa Kaufmann, pub affrs dir; Kay Sandlin, local news ed, news rptr, farm dir, political ed, women's int ed & disc jockey; Glen Richards, disc jockey; Rhonda Hart, disc jockey

KXGJ(FM)— Sept 25, 1995: 101.7 mhz; 100 kw. 449 ft TL: N28 43 53 W96 05 26. Hrs open:
Simulcast KQQK Jefferson.
3000 Bering Dr., Houston, 77057. Phone: (713) 315-3400. Fax: (713) 314-3506. Licensee: Liberman Broadcasting of Houston License LLC. Group owner: Liberman Broadcasting Inc. (acq 10-11-2002; $3.15 million with KIOX-FM El Campo). Population served: 670,000 Natl. Network: ABC. Format: Sp. ♦Leonard Liberman, CEO, pres & progmg mgr; Winter Horton, VP & gen mgr; Ezequiel Gonzalez, progmg dir; Meliza Posada, traf mgr; Mike Todd, engr.

***KZBJ(FM)**— 2005: 89.5 mhz; 35 kw. Ant 479 ft TL: N29 08 58 W95 59 14. Hrs open:
Rebroadcasts KSBJ(FM) Humble 100%.
Box 187, Humble, 77347. Phone: (281) 446-5725. Fax: (281) 540-2198. Web Site: www.ksbj.org. Licensee: KSBJ Educational Foundation (acq 5-31-2003). Format: Contemp Christian. ♦Tim McDermott, gen mgr.

Baytown

KWWJ(AM)— October 1947: 1360 khz; 5 kw-D, 1 kw-N, DA-2. TL: N29 46 28 W95 00 55. Stereo. Hrs opn: 24 Box 419, 77522. Secondary address: 4638 Decker Dr. 77522. Phone: (281) 837-8777. Fax: (281) 424-7588. E-mail: d.martin@kwwj.org Web Site: www.kwwj.org. Licensee: Salt of the Earth Broadcasting. (acq 8-88). Population served: 69,000 Natl. Network: American Urban. Format: Relg. News staff: one. Target aud: General. ♦Darrell E. Martin, pres, gen mgr, gen sls mgr, prom mgr & pub affrs dir.

Beaumont

KFNC(FM)— 1948: 97.5 mhz; 100 kw. Ant 1,955 ft TL: N29 41 52 W94 24 09. Stereo. Hrs opn: 24 2700 Post Oak Blvd., Suite 2300, Houston, 77056. Phone: (713) 300-3500. Fax: (713) 300-3585. Licensee: CMP KC Licensing LLC. Group owner: Cumulus Media Inc. Natl. Network: ESPN Radio. Format: Sports. ♦Pat Fant, mktg mgr.

***KGHY(FM)**—Not on air, target date: unknown: 88.5 mhz; 12 kw vert. Ant 367 ft TL: N30 16 23 W93 57 23. Hrs open: Box 34321, Houston, 77234-4321. Phone: (832) 615-7765. Web Site: www.thegospelhiway.org. Licensee: CCS Radio Inc. ♦Otis Dyson, pres.

KIKR(AM)— 1938: 1450 khz; 1 kw-U. TL: N30 03 52 W94 07 12. Hrs open: 755 S. 11th, Suite 102, 77701. Phone: (409) 833-9421. Fax: (409) 833-9296. Web Site: www.cumulus.com. Licensee: Cumulus Licensing Corp. Group owner: Cumulus Media Inc. (acq 3-9-98; grpsl). Population served: 115,919 Natl. Rep: McGavren Guild. Law Firm: Scott Johnson. Format: Sports. Target aud: 25-54. ♦Zanatta Kelley, gen mgr; Jim West, opns dir & mus dir; Mark Guzman, prom mgr; Greg Davis, chief of engrg; Liz Soileau, traf mgr.

***KLBT(FM)**— Aug 17, 2006: 88.1 mhz; 7 kw vert. Ant 476 ft TL: N29 54 52 W94 17 06. Hrs open: Box 5928, 77726. Phone: (409) 833-0045. E-mail: info@thekingsmusician.org Web Site: www.thekingsmusician.org. Licensee: The King's Musician Educational Foundation Inc (acq 1-31-2006; $450,000 for CP). Format: Contemp Christian. ♦Leslie E. Jones, pres.

KLVI(AM)— 1924: 560 khz; 5 kw-U, DA-N. TL: N30 02 42 W93 52 07. Stereo. Hrs opn: 24 2885 Interstate 10 East, 77702. Secondary address: 2885 I-10 E. 77726. Phone: (409) 896-5555. Fax: (409) 896-5599. Web Site: www.klvi.com. Licensee: Clear Channel Group owner: Clear Channel Communications Inc. (acq 8-30-00; grpsl). Population served: 288,600 Law Firm: Fisher, Wayland, Cooper, Leader & Zaragoza L.L.P. Format: News/talk. News staff: 4; News: 5 hrs wkly. Target aud: 25-54; informed professionals.John Hogan, CEO; Lowry Mays, CEO & chmn; Randall Mays, CFO; Charlie Rahilly, sr VP; Mark Kopelman, VP; Vesta Brandt, gen mgr; Trey Poston, opns dir; Jim Love, opns mgr & pub affrs dir; Elizabeth Blackstock, sls dir; Rob Windham, gen sls mgr & natl sls mgr; Shon Hodgkinson, prom dir; Al Caldwell, progmg dir & disc jockey; Neil Harrison, news dir; T. J. Bordelon, chief of engrg; Gaile Darbone, traf mgr; Bob West, disc jockey; Don Briscoe, disc jockey; George Noory, disc jockey; Ira Wilsker, disc jockey; Jack Piper, disc jockey; Michael Reagan, disc jockey; Rush Limbaugh, disc jockey

KQBU-FM—See Houston

KQQK(FM)— July 10, 1967: 107.9 mhz; 100 kw. 1,000 ft TL: N30 02 09 W94 08 31. Stereo. Hrs opn: 24
Simulcasts KXGJ(FM) Matagorda.
3000 Bering Dr., Houston, 77057. Phone: (731) 315-3400. Fax: (713) 314-3506. Web Site: www.xoradio.com. Licensee: Liberman Broadcasting of Houston License LLC. Group owner: Liberman Broadcasting Inc. (acq 10-11-2002; $24 million). Population served: 520,000 Format: Sp, rock. Target aud: 18-49; bilingual Hispanics. ♦Lenard Liberman, CEO & pres; Winter Horton, VP & gen mgr; Brad Branson, gen sls mgr; Ezequiel Gonzalez, progmg dir; Meliza Posada, traf mgr; Mike Todd, engr.

KQXY-FM— September 1966: 94.1 mhz; 100 kw. 1,099 ft TL: N30 06 56 W94 00 00. Stereo. Hrs opn: 24 755 S. 11th St., Suite 102, 77701. Phone: (409) 833-9421. Fax: (409) 833-9296. E-mail: psanders@qt.rr.com Web Site: www.kqxy.com. Licensee: Cumulus Licensing Corp. Group owner: Cumulus Media Inc. (acq 3-9-98; grpsl). Population served: 328,800 Format: Contemp hit. News staff: one; News: 5 hrs wkly. Target aud: 18-49; skewed female. ♦Rick Prusator, gen mgr; Mike Simpson, gen sls mgr; Greg Davis, chief of engrg.

KRCM(AM)— July 1947: 1380 khz; 1 kw-D, 127 w-N. TL: N30 02 09 W94 08 31. (CP: COL Shenandoah. 250 w-D, 69 w-N. TL: N30 11 42 W95 23 25). Hrs opn: 6 AM-6 PM
Rebroadcasts KOLE(AM) Port Arthur 50%.
27 Sawyer St., 77702. Phone: (409) 835-1340. Phone: (409) 835-1340. Fax: (409) 832-5686. E-mail: manager@newsradiofox.com Web Site: www.newsradiofox.com. Licensee: Voice Broadcasting Inc. (acq 1-29-03). Population served: 115,917 Natl. Network: Fox News Radio, Talk Radio Network, USA. Law Firm: Gammon & Grange. Format: News/talk. News staff: 2; News: 40 hrs wkly. ♦Ralph McBride, pres & gen mgr; Brent Bobbitt, gen sls mgr; Jeanette Harvey, progmg dir & traf mgr; Dominick Brascia, progmg mgr; Jeff Roberts, pub affrs dir; Russ Ingram, engr.

KTCX(FM)— 1996: 102.5 mhz; 50 kw. 492 ft TL: N29 59 22 W94 14 44. Hrs open: 755 South 11th St., Suite 102, Box 870, 77704. Phone: (409) 833-9421. Fax: (409) 833-9296. Web Site: www.ktcx.com. Licensee: Cumulus Licensing Corp. Group owner: Cumulus Media Inc. (acq 3-26-98; $3.6 million). Format: Adult urban. ♦Zanetta Kelley, gen mgr; Ed Turner, stn mgr; Jim West, opns mgr; Walter Brickhouse, sls VP; Wes Matejka, sls dir; Marco Camacho, rgnl sls mgr; Mark Guzman, prom dir; Douglas Harris, progmg dir; Adrian Scott, asst music dir; Greg Davis, chief of engrg.

***KTXB(FM)**— Jan 23, 1990: 89.7 mhz; 9 kw. 567 ft TL: N30 09 27 W93 48 06. Hrs open: 24 4135 Northgate Blvd., Suite #1, 77701. Phone: (409) 745-1737. Web Site: www.familyradio.com. Licensee: Family Stations Inc. (group owner) Format: Christian relg. ♦Harold Camping, pres; Martha Tallent, stn mgr & opns mgr.

***KVLU(FM)**— 1974: 91.3 mhz; 40 kw. 450 ft TL: N30 06 40 W94 03 10. Stereo. Hrs opn: 24 Box 10064, 77710. Phone: (409) 880-8164. E-mail: kvlu@hal.lamar.edu Web Site: www.kvlu.org. Licensee: Lamar University. Population served: 350,000 Natl. Network: NPR. Format: Class, jazz, news. Target aud: 35 plus. Spec prog: Sp 5 hrs wkly. ♦Byron Balentine, gen mgr; Melanie Dishman, dev dir.

KYKR(FM)— Feb 1, 1966: 95.1 mhz; 100 kw. 500 ft TL: N30 08 57 W94 07 59. Stereo. Hrs opn: 24 2885 Interstate 10 E., 77702. Secondary address: 2885 Interstate 10 E. 77726. Phone: (409) 896-5555. Fax: (409) 896-5599. Web Site: www.kykr.com. Licensee: Capstar TX L.P. Group owner: Clear Channel Communications Inc. (acq 8-30-2000; grpsl). Population served: 288,600 Natl. Rep: Clear Channel. Law Firm: Shaw Pittman. Wire Svc: AP Format: Country. News staff: 3; News: 2 hrs wkly. Target aud: 18-54. Spec prog: "Sunday In The Country"(gospel, 4 hrs., "Kicker Classics", 4 hrs., Bob Kingsley Country Top 40 Countdown(4 hrs), The Road" (2 hrs.), Jeff Foxworthy Countdown(3 hrs.).John Hogan, CEO; Lowry Mays, chmn; Mark Mays, pres; Randall Mays, CFO; Mark Kopelman, sr VP, VP & opns mgr; Vesta Brandt, gen mgr; Joey Armstrong, opns dir; Elizabeth Blackstock, sls dir; Rick Miles, natl sls mgr; Cutter McIntyre, prom dir; Mickey Ashworth, progmg dir & disc jockey; Harold Mann, news dir; Jim Love, pub affrs dir; Dave Smith, chief of engrg; Gaile Darbone, traf mgr; Vicki Cleveland, spec ev coord; Big Dave Bubba, disc jockey; Blair Garner, disc jockey; Bob Pickett, disc jockey; Chrissie Roberts, disc jockey; Jim King, disc jockey

KZZB(AM)— May 1, 1947: 990 khz; 1 kw-U, DA-1. TL: N30 08 57 W94 07 59. Hrs open: 24 2531 Calder Ave., 77702. Phone: (409) 833-0990. Fax: (409) 833-0995. Web Site: www.kzzbradio.com. Licensee: Martin Broadcasting Inc. (acq 7-28-92; FTR: 8-17-92). Natl. Rep: Christal. Law Firm: Haley, Bader & Potts. Format: Gospel. News staff: one. Target aud: 18-49. ♦Darrell Martin, pres & gen mgr; Willie Mae McIver, progmg dir.

Beeville

KIBL(AM)— Oct 20, 1949: 1490 khz; 1 kw-U. TL: N28 23 08 W97 43 42. Hrs open: 5 AM-10 PM Box 252, McAllen, 78505. Phone: (956) 781-5528. Fax: (956) 686-2999. Licensee: Paulino Bernal. (acq 3-7-97; $50,600). Population served: 560,730 Format: Christian, Sp Christian. ♦Eloy Bernal, gen mgr, progmg dir & disc jockey; John Ross, chief of engrg & disc jockey.

KVFM(FM)—Co-owned with KIBL(AM). 2000: 91.3 mhz; 1 kw vert. Ant 302 ft TL: N28 26 42 W97 45 50. Licensee: Paulino Bernal Evangelism.

KRXB(FM)— Dec 2, 1988: 107.1 mhz; 1.25 kw. Ant 305 ft TL: N28 25 40 W97 45 36. Stereo. Hrs opn: 24 Box 1664, 78014. Phone: (361) 358-4941. Fax: (361) 358-0601. E-mail: krxbfm@nbnet.net Licensee: Shaffer Communications Group Inc. (acq 10-95; $380,000). Natl. Network: Jones Radio Networks. Format: Classic rock. Target aud: 35 plus. ♦Joe Shaffer, pres; Gary Hoffman, gen mgr, opns mgr, progmg dir, news dir & chief of engrg; Jay Stone, gen sls mgr & natl sls mgr; Ryan Contreras, traf mgr.

KTKO(FM)— Dec 12, 1976: 105.7 mhz; 25 kw. 328 ft TL: N28 28 16 W97 48 39. Stereo. Hrs opn: 5 AM-10 PM 2300 S. Washington, 78102. Phone: (361) 358-1490. Fax: (361) 358-7814. E-mail: bebekicker106@lonestarinternet.net Licensee: Texas Gulfwest Broadcasting Inc. (acq 6-01-02; $325,000). Population served: 485,355 Law Firm: Gardner, Carton & Douglas. Wire Svc: NOAA Weather Format: Country. News staff: one; News: 13 hrs wkly. Target aud: 18-64. ♦Bebe Adamez, gen mgr, opns mgr & opns mgr.

Bellaire

KGOW(AM)— June 7, 1961: 1560 khz; 500 w-D. TL: N29 37 15 W95 25 04. Hrs open: 6 AM-sunset 5353 W. Alabama St., Suite 415, Houston, TX, 77056-5922. Phone: (713) 479-5300. Fax: (713) 479-5333. Web Site: 1560thegame.com. Licensee: Gow Communications L.L.C. (acq 4-10-2007; $9 million). Population served: 22,000 Format: Sports. ♦David F. Gow, pres & farm dir; Richard Topper, gen mgr.

Bells

KMKT(FM)— September 1997: 93.1 mhz; 6.8 kw. 626 ft TL: N33 41 31 W96 26 36. Hrs open: 101 E. Main, Suite 255, Denison, 75020. Phone: (903) 465-6200. Fax: (903) 463-9816. E-mail: jason@931kmkt.com Web Site: www.931kmkt.com. Licensee: NM Licensing LLC. Group owner: NextMedia Group L.L.C. (acq 11-26-01; grpsl). Format: Country. ♦Steven Dinetz, CEO; Jeff Dinetz, pres; Sean Stover, CFO; David Smith, gen mgr; Jason Taylor, opns mgr & progmg dir; Anne Oliver, prom dir; Tiffany Reynolds, news dir; Vince Richardson, chief of engrg; David MacMullen, sls.

Bellville

KNUZ(AM)— Aug 8, 1974: 1090 khz; 250 w-D. TL: N29 56 50 W96 15 54. Hrs open:
Rebroadcasts KLTR(FM) Caldwell 100%.
530 W. Main St., Brenham, 77833. Phone: (979) 836-9411. Fax: (979) 836-9435. Licensee: Roy E. Henderson Group owner: Bayport Broadcast Group (acq 4-17-90; $150,000). Format: Adult contemp. ♦Roy Henderson, pres.

Belton

KOOC(FM)— Apr 25, 1970: 106.3 mhz; 11.5 kw. Ant 489 ft TL: N31 03 46 W97 31 54. Stereo. Hrs opn: 24 608 Moody Ln., Temple, 76504. Phone: (254) 773-5252. Fax: (254) 773-0115. Web Site: www.b1063.net. Licensee: Cumulus Licensing LLC. Group owner: Cumulus Media Inc. (acq 2-2-2000; grpsl). Population served: 162,000 Format: Rhythmic. Target aud: 25-54. ♦Bourdon Wooten, gen mgr; Brian Mack, stn mgr & progmg dir; Mikie Cummings, gen sls mgr; Chris Cummings, news dir.

KTON(AM)— Dec 1, 1961: 940 khz; 1 kw-D, DA. TL: N31 02 37 W97 25 46. Hrs open: 24 Box 1387, 76513. Phone: (254) 939-9377. Fax: (254) 939-9458. Web Site: www.countrygold.com. Licensee: JLF Communications LLP (group owner; (acq 2-8-2007; $900,000). Population served: 187,000 Natl. Network: USA. Format: Country gold. Target aud: 25-54. ♦James Harrison, gen mgr, stn mgr & opns VP; Jim Cooper, engrg mgr.

Benbrook

KDXX(FM)— January 1990: 107.1 mhz; 74 kw. Ant 1,050 ft TL: N32 35 10 W97 49 52. Stereo. Hrs opn: 24 7700 Carpenter Fwy., Dallas, 75247. Phone: (214) 525-0400. Fax: (214) 631-1196. Web Site: univision.com. Licensee: KCYT-FM License Corp. Group owner: Univision Radio (acq 9-22-2003; grpsl). Population served: 3,000,000 Law Firm: Gammon & Grange. Format: Sp adult contemp. ♦Frank Carter, gen mgr; Ivonne Flaherty, gen mgr; Andy Lockridge, opns dir; Cipriano Robles, sls dir; Betsy Galleguillos, natl sls mgr; Oscar Espinosa, prom dir; Herminio "Chayan" Ortuno, progmg dir; Patrick Parks, chief of engrg; Myrna Vera, rsch dir; Mirentxu Smith, traf mgr.

Benjamin

KBTY(FM)—Not on air, target date: unknown: 95.3 mhz; 13 kw. Ant 459 ft TL: N33 46 28 W99 48 21. Hrs open: 6117 Lemon Thyme Dr., Alexandria, VA, 22310. Phone: (571) 228-1258. Fax: (703) 299-6626. Licensee: Miriam Media Inc. ♦Darryl K. Delawder, pres & gen mgr.

Big Lake

KPDB(FM)— 2001: 98.3 mhz; 50 kw. Ant 430 ft TL: N31 11 45 W101 25 40. Hrs open: Box 252, McAllen, 78505. Phone: (956) 686-6382. E-mail: radiodesafio@radiodesafio.org Web Site: www.radiodesafio.org. Licensee: Centro Cristiano de Fe Inc. (acq 5-13-02; $300,000). Format: Christian, Sp, relg.

KWTR(FM)— 2004: 104.1 mhz; 500 w. Ant 62 ft TL: N31 11 54 W101 27 45. Hrs open: Box 1041 Phone: (325) 884-3451. Licensee: Woodrow Michael Warren. Group owner: Woodrow Michael Warren Stns. ♦Woodrow Michael Warren, gen mgr.

Big Sandy

***KTAA(FM)**— Nov 6, 1995: 90.7 mhz; 5.8 kw. Ant 515 ft TL: N32 37 50 W94 53 44. Hrs open: 24 10550 Barkley St., Overland Park, KS, 66212. Phone: (913) 642-7770. Fax: (913) 642-1319. E-mail: comments@bottradionetwork.com Web Site: www.bottradionetwork.com. Licensee: Community Broadcasting Inc. (acq 9-6-2006; $450,000). Format: Christian teaching and talk. Target aud: 25-54; adults. ♦Richard P. Bott II, exec VP.

Big Spring

***KBCX(FM)**— 2001: 91.5 mhz; 250 w. Ant 331 ft TL: N32 11 06 W101 27 56. Hrs open: Box 3206, Tupelo, MS, 38803. Phone: (662) 844-8888. Fax: (662) 842-6791. E-mail: comments@afr.net Web Site: www.afr.net. Licensee: American Family Association. Group owner: American Family Radio Format: Inspirational Christian. ♦Marvin Sanders, gen mgr.

KBST(AM)— Dec 23, 1936: 1490 khz; 1 kw-U. TL: N32 15 44 W101 27 37. Stereo. Hrs opn: Box 1632, 79721. Secondary address: 608 Johnson St. 79720. Phone: (432) 267-1490. Fax: (432) 267-1579. E-mail: kbst@crcom.net Licensee: Rhattigan Broadcasting (Texas) LP (group owner; (acq 8-19-2004; grpsl). Population served: 28,735 Natl. Network: Fox Sports. Natl. Rep: Riley. Format: Talk, sports, nostalgic music. Target aud: 25 plus. ♦Mike Rhattigan, CEO & disc jockey; John Weeks, gen mgr; Sam Stephens, gen sls mgr; Mike Henry, news dir; Gary Graham, chief of engrg; Tim Knox, progmg dir & disc jockey.

KBST-FM— 1961: 95.7 mhz; 33 kw. Ant 459 ft TL: N32 13 13 W101 26 25. Stereo. Hrs opn: Box 1632, 79721. Secondary address: 608 Johnson St. 79720. Phone: (915) 267-6391. Fax: (915) 267-1579. E-mail: kbst@kbst.com Web Site: www.kbst.com. Licensee: Rhattigan Broadcasting (Texas) LP. (group owner; (acq 8-19-2004; grpsl). Population served: 33,000 Format: News. ♦John Weeks, gen mgr; Jim East, progmg dir; Steve Jess, news dir.

KBTS(FM)— Aug 14, 1995: 94.3 mhz; 8.3 kw. Ant 561 ft TL: N32 13 13 W101 26 25. Hrs open: Box 1632, 79721. Secondary address: 608 Johnson St. 79720. Phone: (915) 267-6391. Fax: (915) 267-1579. E-mail: kbst@crcom.net Licensee: Rhattigan Broadcasting (Texas) LP (group owner; (acq 6-3-2004; grpsl). Format: Classic rock. ♦John Weeks, gen mgr; Sam Stephens, opns mgr; Tim Knox, progmg dir; Mike Henry, news dir.

KBYG(AM)— 1948: 1400 khz; 1 kw-U. TL: N32 13 22 W101 28 35. Hrs open: 24 2801 Wasson Dr., 79720-7301. Phone: (432) 263-5294. Fax: (432) 263-6351. E-mail: kbyg@apex2000.net Licensee: Ballard Drew. (acq 4-9-90). Population served: 150,000 Rgnl. Network: S.W. Agri-Radio. Format: Oldies, Sp, talk. News staff: one. Target aud: 25-54; Anglo-Hispanic. ♦David M. Pappajohn, gen mgr; Raul Marquez, gen mgr; Jennifer Patton, stn mgr.

***KPBD(FM)**— 2005: 89.3 mhz; 3 kw. Ant 328 ft TL: N32 09 51 W101 25 27. Hrs open: Box 252, McAllen, 78505. Phone: (956) 686-6382. Fax: (956) 686-2999. Licensee: Paulino Bernal Evangelism. ♦Paulino Bernal Jr., pres.

Bishop

KMZZ(FM)— June 15, 1980: 106.9 mhz; 25 kw. Ant 298 ft TL: N27 39 10 W97 54 59. (CP: Ant 246 ft. TL: N27 40 16 W97 44 17). Stereo. Hrs opn: 24 701 Benys Rd., Corpus Christi, 78408. Phone: (361) 289-0999. Fax: (361) 289-0810. Licensee: Gerald Benavides. (acq 11-4-2004; $550,000). Population served: 300,000 Law Firm: Baraff, Koerner & Olender. Format: Relg. News staff: one. Target aud: 18-49; people with buying power. ♦Lionel Davila, gen mgr; Mike Aradillias, sls VP; Jeremy Lopez, progmg dir; George Sanders, chief of engrg.

Bloomington

***KHVT(FM)**— 2006: 91.5 mhz; 46 kw. Ant 482 ft TL: N29 00 04 W97 00 05. Hrs open:
Rebroadcasts KHCB-FM Houston 100%.
KHCB Radio Network, 2424 South Blvd., Houston, 77098-5110. Phone: (713) 520-5200. Web Site: www.khcb.org. Licensee: Houston Christian Broadcasters Inc. (group owner). Format: Gospel. ♦Bruce Munsterman, gen mgr.

KLUB(FM)— December 1992: 106.9 mhz; 25 w. 269 ft TL: N28 42 16 W96 50 08. Stereo. Hrs opn: 24 107 N. Star Dr., Victoria, 77904. Secondary address: Box 3325, Victoria 77904. Phone: (361) 573-0777. Fax: (361) 578-0059. E-mail: klub@clearchannel.com Web Site: www.1069therock.com. Licensee: GAP Broadcasting Victoria License LLC. Group owner: Clear Channel Communications Inc. (acq 8-3-2007; grpsl). Format: Classic rock. News staff: one; News: 4 hrs wkly. Target aud: 25-59; listeners in a growth & acquisition mode. Spec prog: Blues. ♦Jeff Lyon, gen mgr; Natalie Franz, gen sls mgr; Adam West, progmg mgr; James Love, news dir; Charles Smithey, engrg mgr & engr; Becky Snell, traf mgr.

Boerne

KBRN(AM)— May 10, 1982: 1500 khz; 250 w-D. TL: N29 48 44 W98 43 41. Hrs open: 11737 Nelon Dr., Corpus Christi, 78410. Phone: (361) 774-4354. Fax: (361) 241-7945. Licensee: Gerald Benavides (acq 6-25-2004; $200,000). Population served: 250,000 Format: Sp. ♦Gerry Benavides, gen mgr.

Bonham

KFYN(AM)— May 1948: 1420 khz; 250 w-D, 148 w-N. TL: N33 34 40 W96 09 55. Stereo. Hrs opn: 5 AM-1 AM 811 E. Sam Rayburn Dr., 75418-4928. Secondary address: Box 248 75418-4928. Phone: (903) 583-3151. Fax: (903) 583-2728. E-mail: royv@netexas.net Licensee: Vision Media Group Inc. (acq 12-4-02). Population served: 238,000 Rgnl. Network: Texas State Net. Format: Country. Spec prog: Farm 6 hrs, relg 6 hrs, oldies rock 6 hrs wkly. ♦C.L. Carter II, pres & gen mgr; Jeff Davis, opns mgr & progmg dir.

Borger

***KASV(FM)**— 1998: 88.7 mhz; 10 kw horiz, 3 kw vert. 203 ft TL: N35 40 42 W101 23 18. Hrs open: 24
Rebroadcasts KJRT(FM) Amarillo 100%.
Box 8088, Amarillo, 79114. Phone: (806) 359-8855. Fax: (806) 354-2039. Web Site: www.kingdomkeys.org. Licensee: Top O' Texas Ed. Broadcasting. Format: Relg, educ. ♦Ricky Pfeil, gen mgr.

KQFX(FM)—Licensed to Borger. See Amarillo

KQTY(AM)— Jan 10, 1947: 1490 khz; 1 kw-U. TL: N35 41 05 W101 23 20. Stereo. Hrs opn: 24 Box 165, 113 Union, 79007. Phone: (806) 273-7533. Phone: (806) 273-5889. Fax: (806) 273-3727. E-mail: kqtyradio@yahoo.com Web Site: kqtyradio.com. Licensee: Zia Broadcasting. (acq 12-1-79). Population served: 26,800 Format: Country. News: 15 hrs wkly. Target aud: 25-54; blue collar workers with traditional values & beliefs. Spec prog: Relg one hrs, southern gospel 3 hrs, Christian country 5 hrs wkly, Texas country 10 hrs wkly. ♦Lonnie Ausups, CEO; Rick Keefer, gen mgr; George Grover, stn mgr.

KQTY-FM— 1999: 106.7 mhz; 6 kw. 259 ft TL: N35 41 05 W101 23 12. Stereo. Hrs opn: 24 Box 165, 79008-0165. Secondary address: 113 Union 79007. Phone: (806) 273-5889. Fax: (806) 273-3727. E-mail: kqtyradio@yahoo.com Web Site: www.kqtyradio.com. Licensee: Zia Broadcasting Co. Population served: 26,800 Natl. Network: ABC. Format: Country. News: 15 hrs wkly. Target aud: 25-54; Blue collar workers w/traditional values & beliefs. Spec prog: Religious, 1hr; southern gospel, 3hrs; christian country, 5hrs; Texas country, 10hrs. ♦Lonnie Allsups, CEO; George Grover, stn mgr; Rick Keefer, gen mgr & stn mgr.

Bowie

KNTX(AM)— May 29, 1959: 1410 khz; 500 w-D, DA. TL: N33 35 10 W97 48 23. Stereo. Hrs opn: 24 hrs Box 1080, State Hwy 59 & FM 1758, 76230. Phone: (940) 872-2288. Fax: (940) 872-1228. E-mail: chenderson@kntxradio.com Web Site: kntxradio.com. Licensee: Henderson Broadcasting Co. L.P. (acq 3-13-03). Population served: 30,000 Natl. Network: ABC. Rgnl. Network: Texas State Net. Natl. Rep: Riley. Law Firm: Reddy, Begley & McCormick. Format: Oldies. News staff: one; News: 15 hrs wkly. Target aud: 25-54. Spec prog: . Gospel 4 hrs wkly ♦Charley M. Henderson, pres, gen mgr, engrg dir & chief of engrg; Dee Blanton, opns dir & progmg dir; Pamela A. Henderson, VP, sls dir & prom VP; Ken Wood, mus dir; Doris McGuffey, news dir.

Brady

KNEL(AM)— December 1935: 1490 khz; 1 kw-U. TL: N31 07 48 W99 19 21. Hrs open: Box 630, 76825. Secondary address: 117 S. Blackburn 76825. Phone: (915) 597-2119. Fax: (915) 597-1925. E-mail: knel@airmail.net Web Site: www.knelradio.com. Licensee: Farris Broadcasting Inc. (acq 10-12-95; $475,000 with co-located FM). Population served: 15,750 Natl. Network: ABC. Rgnl. Network: Texas State Net. Format: Oldies, rock & roll. Target aud: General. ♦Lynn Farris, pres, gen mgr & gen sls mgr; Tracy Pitcox, progmg dir; Stan Cooper, chief of engrg.

KNEL-FM— Aug 21, 1979: 95.3 mhz; 3 kw. 299 ft TL: N31 07 27 W99 21 34. Stereo. Web Site: www.knelradio.com.15,500 Format: Country. Target aud: General.

Breckenridge

KBWM(FM)—Not on air, target date: unknown: 100.1 mhz; 6 kw. Ant 250 ft TL: N32 47 32 W98 56 24. Hrs open: 3654 W. Jarvis Ave., Skokie, IL, 60076. Phone: (847) 674-0864. Fax: (847) 674-9188. Web Site: www.kmcommunications.com. Licensee: KM Communications Inc. ♦Kevin Joel Bae, VP & gen mgr.

KLXK(FM)—Listing follows KROO(AM).

KROO(AM)— September 1947: 1430 khz; 1 kw-D, 17 w-N. TL: N32 45 11 W98 55 57. Hrs open: 24 Box 711, 76424. Secondary address: 101 E. Walker St., Suite 201 76424. Phone: (254) 559-5766. Fax: (254) 559-6545. E-mail: klxk@brazosnet.com Licensee: Graham Newspapers Inc. (group owner; (acq 4-12-2001; with co-located FM). Population served: 70,000 Format: Oldies. News staff: one; News: 2.5 hrs wkly. Target aud: 35+; adults. ♦Roy Robinson, VP; Don Collett, gen mgr & progmg dir; Jim Jones, news dir & chief of engrg.

KLXK(FM)—Co-owned with KROO(AM). Aug 1, 1982: 93.5 mhz; 50 kw. 446 ft TL: N32 45 31 W98 56 00. Stereo. Natl. Network: ABC. Format: Country. News staff: one; News: 4.5 hrs wkly. Target aud: 25-54; general. Spec prog: Relg 3 hrs wkly. ♦Joe Barker, traf mgr; Charlie Parker, sports cmtr.

Brenham

***KBEX(FM)**—Not on air, target date: unknown: 89.7 mhz; 250 w. Ant 403 ft TL: N30 10 28 W96 27 45. Hrs open: 5700 West Oaks Blvd., Rocklin, CA, 95765. Phone: (916) 251-1600. Fax: (916) 251-1650. Licensee: Educational Media Foundation. (acq 3-23-2007; grpsl). ♦Richard Jenkins, pres.

KLTR(FM)— August 1988: 94.1 mhz; 6 kw. 328 ft TL: N30 08 31 W96 25 00. Stereo. Hrs opn: 24 530 W. Main, 77833-9247. Phone: (979) 836-9411. Fax: (979) 836-9435. E-mail: fbbbrenham@sbcglobal.net Licensee: Fort Bend Broadcasting Co. (group owner; (acq 5-31-2001; $1.5 million). Law Firm: Fletcher, Heald & Hildreth. Format: Adult contemp. News staff: one; News: 18 hrs wkly. Target aud: 18-49. Spec prog: Gospel 10 hrs wkly. ♦Roy Henderson, pres & gen mgr; Ryan Henderson, gen sls mgr; Amber Kyle, progmg dir; Lori Henderson, traf mgr.

KTTX(FM)—Listing follows KWHI(AM).

KWHI(AM)— Apr 15, 1947: 1280 khz; 1 kw-D, 89 w-N. TL: N30 10 05 W96 25 20. Hrs open: Box 1280, 77834. Secondary address: 223 E. Main St. 77833. Phone: (979) 836-3655. Fax: (979) 830-8141. E-mail: mail@kwhi.com Web Site: www.kwhi.com. Licensee: Tom S. Whitehead Inc. (acq 5-1-47). Population served: 100,000 Format: Country, news/talk. News staff: 2; News: 14 hrs wkly. Target aud: 25-54. Spec prog: Polka 2 hrs, relg 3 hrs, farm 3 hrs wkly. ♦Tom S. Whitehead Jr., pres; Tom D. Whitehead, gen mgr, gen sls mgr & mktg dir; Shelly Granke, prom dir; Prentice Mearns, progmg dir; Frank Wagner, news dir & local news ed; Mark Whitehead, chief of engrg; Elizabeth Pomyhal, traf mgr; Ed Pothul, news rptr & sports cmtr; Tom S. Whitehead, edit dir.

KTTX(FM)—Co-owned with KWHI(AM). Sept 15, 1964: 106.1 mhz; 50 kw. 492 ft TL: N30 21 49 W96 34 33. Stereo. Phone: (409) 776-1061. Fax: (409) 774-7545. E-mail: mail@ktex.com Web Site: www.ktex.com.300,000 Format: Contemp country. News staff: one; News: 1.5 hrs wkly. Target aud: 18-49. ♦Tom D. Whitehead, VP, gen sls mgr & natl sls mgr; Carolyn Warmke, rgnl sls mgr; Ken Murray, progmg dir & traf mgr; Shelly Granke, pub affrs dir & disc jockey; Michele Daniels, traf mgr & disc jockey; Ed Pothul, news rptr; Chris Hunter, disc jockey; Cody Roberts, disc jockey; Eric Taylor, disc jockey.

Bridgeport

KBOC(FM)— Aug 2, 1982: Stn currently dark. 98.3 mhz; 6 kw. Ant 226 ft TL: N33 13 28 W97 47 51. Stereo. Hrs opn: 24 4201 Pool Rd., Coleyville, 76034. Phone: (817) 868-2900. Fax: (817) 868-2929. Licensee: Liberman Broadcasting of Dallas License LLC. (acq 11-2-2006; grpsl). Population served: 250,000 ♦Alex Sanchez, gen mgr.

Brookshire

KCHN(AM)— 2001: 1050 khz; 410 w-D, DA. TL: N29 52 45 W96 02 08. Hrs open: 1782 W. Sam Houston Pkwy. N., Houston, 77043. Phone: (713) 490-2538. Fax: (713) 984-1721. Web Site: www.mrbi.net. Licensee: KCHN Licensee LLC. Population served: 700,000 Format: Multi-ethnic. ♦Terry Lowry, gen mgr.

Brownfield

KKUB(AM)— August 1949: 1300 khz; 1 kw-D. TL: N33 10 49 W102 14 51. Hrs open: Box 411, 79316-0411. Secondary address: 1722 Tahoka Rd. 79316. Phone: (806) 637-4531. Fax: (806) 637-4610. Licensee: Dios Llega Al Hombre Ministries (acq 5-10-2001). Population served: 10,387 Rgnl. Network: Texas State Net. Format: Sp. ♦Adolph Hernandez, gen mgr.

KLZK(FM)— Nov 12, 1984: 104.3 mhz; 50 kw. Ant 466 ft TL: N33 25 08 W102 08 58. Stereo. Hrs opn: 24 9800 University Ave., Lubbock, 79423. Phone: (806) 745-3434. Fax: (806) 748-2470. E-mail: idee@ramar.com Web Site: www.stars1043.com. Licensee: Ramar Communications II Ltd. (group owner; (acq 3-26-99; $1.025 million). Format: Soft adult contemp. Target aud: 25-54. ♦Diana Dee, gen sls mgr; Lew Dee, gen mgr & progmg dir.

***KMLU(FM)**—Not on air, target date: unknown: 90.7 mhz; 2 kw. Ant 220 ft TL: N33 08 59 W102 16 46. Hrs open: 2351 Sunset Blvd., Suite 170-218, Rocklin, CA, 95765. Phone: (916) 251-1600. Fax: (916) 251-1650. Licensee: Educational Media Foundation. (acq 11-1-2006; grpsl). ♦Richard Jenkins, pres.

***KPBB(FM)**— 1999: 88.5 mhz; 4.5 kw. Ant 377 ft TL: N33 09 18 W102 16 51. Hrs open: Box 252, McAllen, 78505. Secondary address: 4501 N. McCall Rd., McAllen 78504. Phone: (956) 686-6382. Fax: (956) 686-2999. Licensee: Paulino Bernal Evangelism. Format: Sp, Christian. ♦Paulino Bernal, gen mgr.

Brownsville

***KBNR(FM)**— Apr 10, 1984: 88.3 mhz; 5.5 kw. 289 ft TL: N25 55 10 W97 31 44. Stereo. Hrs opn: 24 Box 5480, 78523-5480. Secondary address: 216 W. Elizabeth 78520. Phone: (956) 542-6933. Fax: (956) 542-0523. E-mail: kbnr@hcjb.org Web Site: www.radiokbnr.org. Licensee: World Radio Network Inc. Population served: 1,100,000 Format: Relg, educ, Sp. News: 3 hrs wkly. Target aud: 20-45; Hispanic, middle & upper income. ♦Ted Haney, pres; Abelardo Limon, VP; Moises Flores, stn mgr.

KKPS(FM)— Jan 17, 1978: 99.5 mhz; 100 kw. 1,034 ft TL: N26 04 53 W97 49 44. Stereo. Hrs opn: 24 801 N. Jackson Rd., McAllen, 78501. Phone: (956) 661-6000. Fax: (956) 661-6082. Licensee: Entravision Holdings L.L.C. Group owner: Entravision Communications Corp. (acq 7-20-00; grpsl). Population served: 700,000 Law Firm: Rosenman & Colin. Format: Sp mus, Tejano. News staff: one. Target aud: 18-49; Hispanic females, young adults. ♦Scott Savage, gen mgr.

KRIO(AM)—See McAllen

KTEX(FM)— January 1975: 100.3 mhz; 100 kw. 1,125 ft TL: N26 03 13 W97 44 39. Stereo. Hrs opn: 24 901 E. Pike Blvd., Weslaco, 78596. Phone: (956) 973-9202. Fax: (956) 973-9335. E-mail: ktexx@aol.com Web Site: www.ktex.net. Licensee: Capstar TX L.P. Group owner: Clear Channel Communications Inc. (acq 8-15-00; grpsl). Population served: 750,000 Format: Country. News staff: one. Target aud: 25-54; male & female. ♦Billy Santiago, VP & opns mgr; Danny Fletcher, gen mgr.

KVNS(AM)— 1999: 1700 khz; 8.8 kw-D, 880 w-N. TL: N25 56 57 W97 33 15. Stereo. Hrs opn: 24 901 E. Pike Blvd., Weslaco, 78596. Phone: (956) 973-9202. Fax: (956) 973-9355. Licensee: Clear Channel Broadcasting Licenses Inc. Group owner: Clear Channel Communications Inc. (acq 12-9-2003; grpsl). Law Firm: Shaw Pittman. Format: News/talk. ♦Hilda Trevino, gen mgr; Gilda Gomez, opns mgr; Edgar C. Trevino, sls dir; John Ross, chief of engrg.

Brownwood

***KBUB(FM)**— Mar 12, 1987: 90.3 mhz; 550 w. 308 ft TL: N31 43 10 W99 00 57. Hrs open: Box 1549, 76804. Phone: (325) 641-2223. Phone: (325) 646-3420. Fax: (325) 643-9772. Licensee: Living Word Church of Brownwood Inc. (acq 12-18-96). Format: Christian praise music, teaching. ♦Angelia Schum, gen mgr.

KBWD(AM)— Aug 17, 1941: 1380 khz; 1 kw-D, 500 w-N. TL: N31 42 36 W98 57 36. Hrs open: 18 Box 280, 76804. Secondary address: 300 Carnegie Blvd. 76801. Phone: (915) 646-3505. Fax: (915) 646-2220. E-mail: upfront@koxe.com Web Site: www.koxe.com. Licensee: Brown County Broadcasting Co. Population served: 25,672 Format: Adult contemp. News staff: 2; News: 8 hrs wkly. Target aud: 25-54. ♦Don Dillard, VP; Barbara McAnally, gen mgr; Bob James, progmg dir, chief of engrg & chief of engrg.

KOXE(FM)—Co-owned with KBWD(AM). May 17, 1975: 101.3 mhz; 100 kw. Ant 577 ft TL: N31 43 45 W99 01 12. Stereo. Web Site: www.koxe.com. Format: C&W.

***KHPU(FM)**— September 1998: 91.7 mhz; 290 w. 571 ft TL: N31 43 32 W99 00 48. Stereo. Hrs opn: 24 1000 Fisk Ave., 76801. Phone: (325) 649-8119. Fax: (325) 649-8901. E-mail: khpu@hputx.edu Web Site: www.hputx.edu. Licensee: Howard Payne University. Population served: 25,000 Law Firm: Cohen & Marks. Format: Christian, jazz, hot adult contemp. Target aud: 18-22; high school through college students. ♦Jim Jones, gen mgr & opns dir.

***KPBE(FM)**— 2000: 89.3 mhz; 6 kw. Ant 328 ft TL: N31 46 37 W98 50 30. Hrs open: Box 252, McAllen, 78505. Secondary address: 4501 N. McCall Rd., McAllen 78504. Phone: (956) 686-6382. Fax: (956) 686-2999. Licensee: Paulino Bernal Evangelism. Format: Sp, Christian. ♦Paulino Bernal, pres & gen mgr.

KPSM(FM)— Apr 11, 1981: 99.3 mhz; 100 kw. Ant 446 ft TL: N31 43 10 W99 00 57. Stereo. Hrs opn: Box 1549, 76804. Secondary address: 901 C.C. Woodson Rd. 76801. Phone: (915) 646-5993. Fax: (915) 643-9772. E-mail: rock@web-access.net Web Site: www.kpsm.net. Licensee: Living Word Church of Brownwood Inc. (acq 1996). Population served: 186,132 Natl. Network: Salem Radio Network. Format: Contemp Christian music. Spec prog: Children 3 hrs, Christian hip hop 5 hrs, Southern Gospel 2 hrs wkly. ♦Jack Ruth, CEO; Angelia Schum, gen mgr; Brigitte Rittenour, stn mgr, opns dir, sls dir & mktg dir; Kevin Koontz, prom VP; Erich Schnitz, progmg dir; Tom Zintgraff, chief of engrg.

KXYL-FM— September 1965: 96.9 mhz; 74 kw. Ant 321 ft TL: N31 42 16 W99 00 05. Stereo. Hrs opn: 24 Box 100, 600 Fisk Ave., 76804. Phone: (325) 646-3535. Fax: (325) 646-5347. Web Site: wattsradio.net. Licensee: Tackett-Boazman Broadcasting LP. (group owner; (acq 4-11-2006; grpsl). Population served: 50,000 Natl. Network: ABC. Law Firm: Cohn & Marks. Format: News/talk. News staff: 3. Target aud: 18+. ♦Cathy Hail, gen mgr & progmg dir; Helen Lehman, opns mgr, prom dir & traf mgr; Ted Wrenn, sls dir; Kyle Dennis, news dir & sports cmtr; Stan Cooper, chief of engrg; Will Prickett, disc jockey.

KXYL(AM)— 1953: 1240 khz; 1 kw-U. TL: N31 42 21 W98 59 45.24 Web Site: wattsradio.net.35,000 Natl. Rep: Roslin. Format: Btfl mus, Sp. News staff: 2; News: 8 hrs wkly. Target aud: 18+; Spanish. Spec prog: Christian Sp 36 hrs wkly. ♦Chema Martinez, progmg dir; Helen Lehman, traf mgr.

Bryan

KAGC(AM)— Dec 27, 1977: 1510 khz; 500 w-D. TL: N30 41 21 W96 21 35. Hrs open: Box 3248, 77805. Secondary address: 2700 Rudder Fwy. S., Suite 5000, College Station 77845. Phone: (979) 695-9595. Fax: (979) 695-1933. E-mail: kagcradio@cox-internet.com Web Site: kagcradio.com. Licensee: Divcon Associates Inc. (acq 3-87). Population served: 133,719 Natl. Network: Salem Radio Network. Law Firm: Verner, Liipfert, Bernhard, McPherson & Hand. Format: Contemp Christian. News: 6 hrs wkly. Target aud: 25-54; upscale, higher income & conservative. Spec prog: Black 2 hrs, Czech music 2 hrs wkly. ♦Bill Hicks, pres; Ben Downs, gen mgr; Keith Kane, opns mgr, prom mgr & progmg dir; Sam Jones, gen sls mgr; Michele McNew, adv dir; Chris Dusterhoff, chief of engrg.

KKYS(FM)— July 28, 1984: 104.7 mhz; 50 kw. 350 ft TL: N30 42 59 W96 22 20. Stereo. Hrs opn: 24 1716 Briarcrest Dr., Ste. 150, 77802. Phone: (979) 846-5597. Fax: (979) 268-9090. Web Site: www.mix1047.com. Licensee: CCB Texas Licenses L.P. Group owner: Clear Channel Communications Inc. (acq 10-10-00; grpsl). Population served: 100,000 Format: Hot adult contemp. News: 15 hrs wkly. Target aud: 18-49; heavy office lstng. ♦Leslie Bass, gen mgr; Greg Zweiacber, sls dir & natl sls mgr; Joseph Sanchez, traf mgr; Steve Sauer, engr.

KNDE(FM)—See College Station

KNFX-FM— Oct 7, 1991: 99.5 mhz; 6 kw. Ant 328 ft TL: N30 39 02 W96 20 58. Hrs open: 24 1716 Briarcrest Dr., Suite 150, 77802. Phone: (979) 846-5597. Fax: (979) 268-9090. Web Site: www.995thefox.com. Licensee: CCB Texas Licenses L.P. Group owner: Clear Channel Communications Inc. (acq 7-20-01; $2.5 million). Population served: 150,000 Law Firm: Leventhal, Senter & Lerman. Format: Class rock. Target aud: General. ♦Leslie Bass, gen mgr & stn mgr; Chuck Knuth, sls dir & disc jockey; Bill Sutton, engrg mgr & engr; Derek Wallace, traf mgr & disc jockey.

KORA-FM—Listing follows KTAM(AM).

KTAM(AM)— Sept 10, 1947: 1240 khz; 1 kw-U. TL: N30 39 02 W96 20 59. Hrs open: Box 3069, 77805. Secondary address: 1240 Villa Maria 77802. Phone: (979) 776-1240. Fax: (979) 776-0123. Licensee: Brazos Valley Communications Ltd. (group owner; (acq 8-31-2006; grpsl). Population served: 50,000 Format: Sp. Target aud: 25 plus. ♦Jim Ray, gen mgr; Glenn Hicks, sls dir; Carolyn Benavides, progmg dir; Gary Graham, engrg dir.

KORA-FM—Co-owned with KTAM(AM). Apr 1, 1966: 98.3 mhz; 900 w. 528 ft TL: N30 39 02 W96 20 57. (CP: 2.3 kw). Stereo. 150,000 Format: C&W. Target aud: 25-54. ♦Amy Mattingly, progmg dir.

KZNE(AM)—See College Station

Buda

KROX-FM— Sept 1, 1984: 101.5 mhz; 12.5 kw. Ant 843 ft TL: N30 19 23 W97 47 58. Stereo. Hrs opn: 24 8309 N. IH 35, Austin, 78753. Phone: (512) 832-4000. Fax: (512) 832-4071. Web Site: www.krox.com. Licensee: LBJS Broadcasting Co. L.P. Group owner: Emmis Communications Corp. (acq 4-25-03; grpsl). Population served: 1,000,000 Natl. Rep: McGavren Guild. Format: Alternative, new rock. Target aud: 18-34; young adults. ♦Bruce Walden, gen mgr; James White, sls dir; Melody Lee, progmg dir; Todd Jeffries, news dir; Jim Henkle, chief of engrg; Lisa Melton, traf mgr.

Burkburnett

KYYI(FM)— June 1, 1989: 104.7 mhz; 100 kw. 1,017 ft TL: N34 05 35 W98 52 44. Stereo. Hrs opn: 24 4302 Callfield Rd., Wichita Falls, 76308. Phone: (940) 691-2311. Fax: (940) 696-2255. E-mail: bear104@bear104.com Web Site: www.bear104.com. Licensee: Cumulus Licensing Corp. Group owner: Cumulus Media Inc. (acq 10-3-97; grpsl). Law Firm: Cohn & Marks. Format: Classic Rock. ♦Lindy Parr, gen mgr; Brent Warner, opns mgr; Johnny Tidwell, gen sls mgr; Keith Vaughn, progmg dir; Jeff Chancey, chief of engrg; Dana Jameson, traf mgr.

Burleson

KHFX(AM)— July 22, 1922: 1460 khz; 5 kw-D, 700 w-N, DA-2. TL: N32 34 43 W97 16 50. Hrs open: 24 PO Box 1629, Cleburne, 76033. Secondary address: 919 No. Main, Cleburne 76033. Phone: (817) 645-6643. Fax: (817) 645-6644. E-mail: info@countrygoldradio.com Web Site: www.countrygoldradio.com. Licensee: M&M Broadcasters Ltd. (group owner; (acq 4-26-99; $450,000). Population served: 105,000 Format: Classic country. ♦Gary Moss, gen mgr; Mike Crow, chief of opns.

Burnet

KBEY(FM)— April 1993: 92.5 mhz; 1.8 kw. Ant 604 ft TL: N30 44 29 W98 19 05. Hrs open: 5226 Hwt. 281 N., Marblefalls, 78654. Secondary address: Hwy. 2147, Ste. 112, Horseshoe Bay Phone: (830) 693-5551. Fax: (830) 693-5107. E-mail: realcountry@kbay.net Web Site: www.radiohillcountry.com. Licensee: Munbilla Broadcasting Properties Ltd. (group owner). Format: Real country. ♦Alan Barrows, gen mgr; Bill Woleben, opns dir.

KHLE(FM)—Listing follows KRHC(AM).

KRHC(AM)— Aug 19, 1963: 1340 khz; 1 kw-U. TL: N30 46 04 W98 13 49. Hrs open: 24 Box 639, Marble Falls, 78654. Phone: (830) 693-5551. Fax: (830) 693-5107. E-mail: comments@khlb.com Web Site: www.khlb.com. Licensee: Munbilla Broadcasting Properties Ltd. (group owner; (acq 12-4-2003; $1 million with co-located FM). Population served: 4,300 Rgnl. Network: Texas State Net. Format: News, info, nostalgia. News staff: 2; News: 3 hrs wkly. Target aud: 35 plus. ♦Margaret Ronquille, gen mgr & gen sls mgr; Paula Reed, progmg dir; Harold Mann, news dir; Gary Graham, chief of engrg & disc jockey.

KHLE(FM)—Co-owned with KRHC(AM). Dec 15, 1978: 106.9 mhz; 4.7 kw horiz, 4.6 kw vert. Ant 358 ft TL: N30 44 12 W98 17 36. Stereo. Licensee: Munbilla Fort Hood Ltd. Natl. Network: ABC. Format: Country. News staff: 2; News: 3 hrs wkly. Target aud: 25-54.

Bushland

***KTXP(FM)**— 2004: 91.5 mhz; 1 kw. Ant 262 ft TL: N35 08 51 W102 05 56. Hrs open: 24
Rebroadcasts KANZ(FM) Garden City 100%.
High Plains Public Radio, 210 N. 7th St., Garden City, KS, 67846. Phone: (806) 659-3730. Fax: (620) 275-7496. E-mail: hppr@hppr.org Web Site: www.hppr.org. Licensee: Kanza Society Inc. Natl. Network: NPR, AP Radio, PRI. Format: News, div. ♦Richard Hicks, gen mgr; Diana Gonzales, dev dir; Debra Stout, prom dir; Bob Kirby, progmg dir; Mary Palmer, mus dir; Chuck Springer, chief of engrg.

Byrne

***KLRW(FM)**— 2004: 88.5 mhz; 500 w vert. Ant 522 ft TL: N31 25 16 W100 32 36. Hrs open: 2351 Sunset Blvd., Suite 170-218, Rocklin, CA, 95765. Phone: (916) 251-1600. Fax: (916) 251-1650. Web Site: www.klove.com. Licensee: Educational Media Foundation. Group owner: EMF Broadcasting (acq 5-8-2003; $75,000 for CP). Natl. Network: K-Love. Format: Christian. ♦Richard Jenkins, pres; Mike Novak, VP; Keith Whipple, dev dir; Eric Allen, natl sls mgr; David Pierce, progmg mgr; Ed Lenane, news dir; Sam Wallington, engrg dir; Karen Johnson, news rptr; Marya Morgan, news rptr; Richard Hunt, news rptr.

Caldwell

***KALD(FM)**—Not on air, target date: unknown: 91.9 mhz; 6 kw. Ant 328 ft TL: N30 33 24 W96 48 29. Hrs open: 2424 South Blvd., Houston, 77098. Phone: (713) 520-5200. E-mail: email@khcb.org Web Site: www.khcb.org. Licensee: Houston Christian Broadcasters Inc. (acq 4-9-2007; $10,000 for CP). ♦Bruce Munsterman, pres & gen mgr.

KHTZ(FM)— 2002: 107.3 mhz; 6 kw. Ant 328 ft TL: N30 33 31 W96 34 50. Hrs open: 24
Simulcast with KMBV(FM) Navasota 100%.
530 W. Main St., Brenham, 77833. Phone: (979) 836-9411. Fax: (979) 836-9435. Web Site: www.lonestarfm.com. Licensee: Roy E. Henderson. Group owner: Bayport Broadcast Group Format: Classic country. Target aud: 25-54. ♦Roy E. Henderson, pres; Ryan Henderson, gen mgr; Matt Matthews, mus dir.

Callisburg

***KPFC(FM)**— April 1998: 91.9 mhz; 300 w. 62 ft TL: N33 40 11 W97 00 50. Hrs open: 24 Box 918, Camp Sweeney, Gainesville, 76241. Phone: (940) 665-2011. Fax: (940) 665-9467. E-mail: kpfc@kpfc.org Web Site: www.kpfc.org. Licensee: Camp Sweeney. Format: Contemporary hits. ♦Dr. Ernie Fernandez, gen mgr; Skip Rigsby, progmg mgr.

Cameron

KJXJ(FM)— Apr 8, 1985: 103.9 mhz; 25 kw. Ant 695 ft TL: N30 44 14 W96 50 14. Stereo. Hrs opn: 24 Box 3069, Bryan, 77805. Secondary address: 1240 E. Villa Maria, Bryan 77802. Phone: (979) 776-1240. Fax: (979) 776-6074. Licensee: Brazos Valley Communications Ltd. Group owner: Equicom Inc. (acq 8-31-2006; grpsl). Natl. Network: ABC. Law Firm: Fletcher, Heald & Hildreth. Format: Alternative rock. Target aud: 12-34; generation x. ♦Dan Ginzel, gen mgr; Amy Mattingly, opns dir; Chuck Knugh, gen sls mgr; Bill Kaufmann, progmg dir & disc jockey; Gary Graham, chief of engrg; Jeanna Vickery, traf mgr; Brian Blades, disc jockey; Don Kelley, disc jockey; Kevin McKay, disc jockey.

KMIL(FM)— 2002: 105.1 mhz; 15 kw. Ant 328 ft TL: N30 51 30 W97 01 47. Hrs open: Box 832, 76520. Phone: (254) 697-6633. Fax: (254) 697-6330. E-mail: kmil@kmil.com Web Site: www.kmil.com. Licensee: Cameron Broadcasting Co. Natl. Network: CBS Radio. Format: Country. ♦Clay Gish, gen mgr.

KTAE(AM)— September 1955: 1330 khz; 500 w-D, 97 w-N. TL: N30 50 48 W96 57 55. Hrs open: Drawer 832, 76520. Secondary address: 901 E. First 76520. Phone: (254) 697-6633. Fax: (254) 697-6330. E-mail: kmil@tlab.net Web Site: www.kmil.com. Licensee: Milam Broadcasting Co. (acq 12-31-97). Population served: 25,000 Natl. Rep: Keystone (unwired net). Format: C&W, Sp. Target aud: General. Spec prog: Gospel 6 hrs, Czech 8 hrs wkly. ♦Joe Smitherman, gen mgr, sls dir & news dir; Sarah Haussecker, traf mgr; Nonito Martinez, spanish dir; A.T. Sheffield, disc jockey; Eric Haussecker, disc jockey.

Camp Wood

KAYG(FM)— 2001: 99.1 mhz; 965 w. Ant 226 ft TL: N29 42 53 W100 00 56. Hrs open: Box 252, McAllen, 78505. Secondary address: 4501 N. McCall Rd., McAllen 78504. Phone: (956) 686-6382. Fax: (956) 686-2999. Licensee: La Radio Cristiana Network Inc. Format: Sp, Christian. ♦Paulino Bernal, pres & gen mgr.

***KHPS(FM)—**Not on air, target date: unknown: 89.7 mhz; 500 w. Ant 361 ft TL: N29 42 53 W100 00 56. Hrs open: 2424 South Blvd., Houston, 77098-5110. Phone: (713) 520-5200. Web Site: www.khcb.org. Licensee: Houston Christian Broadcasters Inc. ♦Bruce Munsterman, pres.

Campbell

KRVA-FM— Aug 1, 1969: 107.1 mhz; 3.6 kw. Ant 423 ft TL: N33 07 31 W95 44 35. Stereo. Hrs opn: LKCM Radio Group L.P., 301 Commerce St., Fort Worth, 76102. Phone: (817) 332-3235. Licensee: LKCM Radio Group L.P. (group owner; (acq 5-21-2004; $1 million with KRVF(FM) Kerens). Population served: 5,000,000 Format: Oldies. News staff: 2. ♦Kevin D. Prigel, gen mgr.

Canton

KRDH(AM)— Sept 12, 1963: 1510 khz; 500 w-D. TL: N32 41 02 W95 29 44. Hrs open: Sunrise-suset Box 868, Forney, 75126. Phone: (903) 567-5566. Fax: (903) 567-5567. Licensee: RDH Land & Cattle Co. Inc. (acq 10-20-2006; $185,000). Population served: 50,000 Format: Contemp Christian. News: 6 hrs wkly. Target aud: General. ♦Eric Jontra, gen mgr & stn mgr; Dee Cox, opns mgr, sls dir & progmg dir.

Canyon

KPUR-FM— Jan 12, 1981: 107.1 mhz; 6 kw. 315 ft TL: N35 05 09 W101 54 48. Stereo. Hrs opn: 24 301 S. Polk, Suite 100, Amarillo, 79101. Phone: (806) 342-5200. Fax: (806) 342-5202. E-mail: rickmatchett@cumulus.com Web Site: www.kpur.com. Licensee: Cumulus Licensing Corp. Group owner: Cumulus Media L.L.C (acq 5-98; $820,000 with KPUR(AM) Amarillo). Population served: 200,000 Format: Oldies. News staff: one; News: 5 hrs wkly. Target aud: 35-55; boomers. ♦Rick Matchett, gen mgr; Eric Stevens, opns mgr, progmg dir & disc jockey; Carolyn Reinert, sls dir; D'Lisa Pohnert, prom dir; J.P. Wolf, chief of engrg; Matt Darby, local news ed; Bart Bailey, disc jockey; J.D. Redman, disc jockey; Johnny Black, disc jockey; Randi Rusk, disc jockey.

***KWTS(FM)—** 1971: 91.1 mhz; 6 kw. 141 ft TL: N34 58 59 W101 55 10. Stereo. Hrs opn: 24 Box 1514, Wt. Stn, 79016. Phone: (806) 651-2797. Phone: (806) 651-2911. Fax: (806) 651-2818. E-mail: kwts@mail.wtamu.edu Web Site: www.wtamu.edu/kwts. Licensee: West Texas A & M University. Population served: 10,922 Format: Rock. News staff: 2; News: 3 hrs wkly. Target aud: 16-25. Spec prog: Class 4 hrs, jazz 3 hrs, Black 3 hrs, techo 5 hrs, acoustic 3 hrs, Britsh rock 3 hrs, Sp 3 hrs wkly. ♦Dr. Leigh Browning, pres; Evan Kolius, gen mgr; Elizabeth Wiseman, sls dir; Andi Law, prom dir; Anthony Smith, progmg dir; Randy Ray, chief of engrg.

KZRK(AM)—Licensed to Canyon. See Amarillo

KZRK-FM—Licensed to Canyon. See Amarillo

Carrizo Springs

KBEN(AM)— Aug 9, 1955: 1450 khz; 1 kw-U. TL: N28 31 15 W99 51 30. Hrs open: 203 S. 4th St., 78834. Licensee: Sylvia Mijares (acq 9-30-97; $41,250). Population served: 15,000 Rgnl. Network: Texas State Net. Law Firm: Borsari & Paxson. Format: Relg, Sp. Target aud: English & Sp listeners. ♦Gordon Baehre, gen mgr.

KCZO(FM)— 1991: 92.1 mhz; 3 kw. 296 ft TL: N28 33 24 W99 53 49. Hrs open: Box 252, McAllen, 78505. Secondary address: 4501 N. Mc Call Rd., McAllen 78504. Phone: (956) 686-6382. Fax: (956) 686-2999. Licensee: Paulino Bernal Evangelism. Format: Sp, Christian. ♦Eloy Bernal, gen mgr.

Carrollton

KJON(AM)— Dec 17, 1970: 850 khz; 5 kw-D, DA. TL: N33 16 42 W96 49 16. Hrs open: 521 E. Bolt St., Fort Worth, 76110. Phone: (817) 923-3424. Fax: (817) 923-3451. Licensee: Chatham Hill Foundation Inc. (acq 12-13-2006; grpsl). Format: Sp Catholic. ♦Bob Prouse, gen mgr.

Carthage

KGAS(AM)— October 1955: 1590 khz; 2.5 kw-D, 130 w-N. TL: N32 09 12 W94 18 52. Hrs open: 215 S. Market St., 75633-2623. Phone: (903) 693-6668. Fax: (903) 693-7188. E-mail: info@kgasradio.com Web Site: www.kgasradio.com. Licensee: Jerry T. Hanszen. (acq 9-1-88). Population served: 20,000 Rgnl. Network: Texas State Net. Wire Svc: NOAA Weather Format: Gospel. News staff: one; News: 20 hrs wkly. Target aud: General. Spec prog: Relg 10 hrs wkly. ♦Jerry T. Hanszen, CEO, gen mgr & stn mgr; Judy McNatt, sls dir; Wanda Hanszen, progmg VP; Alan Mayton, mus dir & news dir; Jack Dillard, farm dir.

KGAS-FM— Aug 1, 1992: 104.3 mhz; 6 kw. 328 ft TL: N32 08 11 W94 23 07. Stereo. Web Site: www.kgasradio.com.80,000 Natl. Network: ABC, Westwood One. ♦Jack Dillard, farm dir.

KTUX(FM)—Licensed to Carthage. See Shreveport LA

Cedar Park

KDHT(FM)— Aug 1, 1961: 93.3 mhz; 100 kw. Ant 1,948 ft TL: N30 43 34 W97 59 23. Stereo. Hrs opn: 8309 N. I-35, Austin, 78753. Phone: (512) 832-4000. Fax: (512) 832-4081. Web Site: www.kxmg.com. Licensee: Emmis Austin Radio Broadcasting Co. L.P. Group owner: Emmis Communications Corp. (acq 4-25-03; grpsl). Population served: 2,000,000 Format: CHR, dance. Target aud: 18-34; women & men who like current music. Spec prog: Pub service 2 hrs, Sp one hr, Latino one hr wkly. ♦Bruce Walden, gen mgr; Jeff Carrol, opns mgr; Brad Copland, gen sls mgr; Bob Lewis, progmg dir; Bradley Grein, mus dir; Todd Jeffries, news dir; Jim Henkel, chief of engrg; Lisa Melton, traf mgr.

Center

KDET(AM)— Feb 22, 1949: 930 khz; 1 kw-D, 36 w-N. TL: N31 50 03 W94 12 53. Hrs open: Box 930, 307 San Augustine St., 75935. Phone: (936) 598-3304. Fax: (936) 598-9537. Licensee: Center Broadcasting Co. Inc. (group owner; acq 3-26-98; grpsl). Population served: 29,000 Rgnl. Network: Texas State Net. Natl. Rep: Riley. Format: Southern gospel, country, Sp. ♦Lori Collins, stn mgr.

KQBB(FM)—Co-owned with KDET(AM). July 5, 1978: 100.5 mhz; 2.05 kw. Ant 567 ft TL: N31 43 34 W94 15 27. Stereo. Format: Country. Shelby County.

Centerville

KKEV(FM)—Not on air, target date: unknown: 103.5 mhz; 6 kw. Ant 283 ft TL: N31 16 46 W95 59 00. Hrs open: 3654 W. Jarvis Ave., Skokie, IL, 60076. Phone: (847) 674-0864. Fax: (847) 674-9188. Web Site: www.kmcommunications.com. Licensee: KM Communications Inc. ♦Kevin Joel Bae, VP & gen mgr.

KUZN(FM)— 2001: 105.9 mhz; 25 kw. Ant 328 ft TL: N31 16 56 W95 53 42. Stereo. Hrs opn: 24 404-A N. May St., Madisonville, 77864. Phone: (936) 348-6344. Fax: (936) 348-6383. E-mail: kuzn@rodzoo.com Licensee: KUZN(FM) Inc. (acq 11-21-2005). Law Firm: Dan Alpert, Esq. Format: Country. Spec prog: Spanish 6 hrs. ♦Gerald R. Proctor, pres & gen mgr; Jeff Michaels, stn mgr; Samantha Elom, opns mgr; Chester Leediker, chief of engrg.

Charlotte

KSAQ(FM)—Not on air, target date: unknown: 102.3 mhz; 6 kw. Ant 328 ft TL: N28 45 29 W98 38 01. Hrs open: 819 S.W. Federal Hwy., Suite 106, Stuart, FL, 34994-2952. Phone: (772) 215-1634. Web Site: www.toweritrust.com. Licensee: Tower Investment Trust Inc. ♦William H. Brothers, pres.

Childress

KCTX(AM)— May 8, 1947: 1510 khz; 250 w-D. TL: N34 25 41 W100 13 47. Hrs open: 6:30 AM-6 PM Box 540, 79201-0540. Secondary address: 1111 16th St. N.W. 79201. Phone: (940) 937-6316. Fax: (940) 937-6551. E-mail: kctx@102online.com Licensee: James G. Boles (acq 1-9-2006 $232,000 with KCTX-FM Childress plus land in Turkey, TX). Population served: 6,870 Rgnl. Network: Voice of the S.W. Format: Oldies. ♦James Boles, gen mgr; Chao Ware, gen sls mgr; J. Scott, progmg dir; Mona Boles, traf mgr.

KCTX-FM— July 1, 1984: 96.1 mhz; 50 kw. 520 ft TL: N34 26 20 W100 13 10. Stereo. Hrs opn: Box 540, 79201. Secondary address: 1111 16th St. N.W. 79201. Phone: (940) 937-6316. Fax: (940) 937-6551. E-mail: kctxradio@gmail.com Licensee: James G. Boles (acq 1-9-2006; $232,000 with KCTX(AM) Childress plus land in Turkey, TX). Natl. Rep: Riley. Format: Country. Target aud: General. Spec prog: Relg 5 hrs wkly. ♦James Boles, gen mgr; Chad Ware, gen sls mgr & news dir; J. Scott, progmg dir.

Clarendon

KEFH(FM)— September 2000: 99.3 mhz; 44 kw. Ant 522 ft TL: N35 04 36 W100 53 33. Hrs open: 24 Box 370, 79226-0370. Phone: (806) 874-2296. Fax: (806) 874-4411. E-mail: kefh@anaonline.com Licensee: RoHo Broadcasting Co. Format: Oldies. ♦Ken Meinhart, gen mgr & progmg dir; Patrick Robertson, rgnl sls mgr; John Wolfe, chief of engrg; Carol Hinton, traf mgr.

Clarksville

KCAR(AM)— Apr 27, 1956: 1350 khz; 500 w-D, 50 w-N. TL: N33 36 41 W95 01 01. Hrs open: 24 228 W. Main St., 75426. Phone: (903) 427-3861. Fax: (903) 427-5524. E-mail: kool985@@neato.net Licensee: FFD Holdings I Inc. Group owner: Petracom Media LLC (acq 1-4-2005; grpsl). Population served: 14,497 Natl. Network: Jones Radio Networks. Rgnl. Network: VSA Radio, Texas State Net. Law Firm: Marjorie Esman. Format: Classic country. News: 10 hrs wkly. Target aud: General; rural, agricultural, middle-aged. Spec prog: Gospel 6 hrs, sports 10 hrs, farm 2 hrs wkly. ♦Mike Monday, progmg dir; Dale Gorsuch, chief of engrg; Tex Phillips, gen mgr, sls dir & disc jockey.

KGAP(FM)—Co-owned with KCAR(AM). Dec 11, 1990: 98.5 mhz; 50 kw. 328 ft TL: N33 35 47 W95 01 03. Stereo. 24 Natl. Network: ABC. Format: Oldies. News: one hr wkly. Target aud: 25-64. ♦Tex Phillips, CEO & sls dir.

Claude

KARX(FM)—Licensed to Claude. See Amarillo

Cleburne

KCLE(AM)— April 1947: 1140 khz; 850 w-D, 710 w-N, DA-2. TL: N32 16 54 W97 24 44. Hrs open: 24 Box 1629, 76033. Secondary address: 305 Milsap Hwy., Mineral Wells 76067. Phone: (817) 645-1140. Fax: (817) 645-3944. Licensee: M&M Broadcasters Ltd. Group owner: First Broadcasting Investment Partners LLC (acq 4-16-2006; grpsl). Population served: 100,000 Rgnl. Network: Texas State Net. Format: Country. ♦Gary L. Moss, pres.

Cleveland

KTHT(FM)— Jan 17, 1993: 97.1 mhz; 100 kw. Ant 1,847 ft TL: N30 32 06 W95 01 04. Stereo. Hrs opn: 1990 Post Oak Blvd., Suite 2300, Houston, 77056. Phone: (713) 622-5533. Fax: (713) 993-9300. Web Site: www.countrylegends971.com. Licensee: Cox Radio Inc. Group owner: Cox Broadcasting (acq 8-15-2000; grpsl). Format: Country classics. ♦Caroline Devine, gen mgr; Judy Lakin, gen sls mgr; John Chaing, progmg dir; Ed Wilson, engr.

Clifton

KWOW(FM)— 1989: 104.1 mhz; 16 kw. Ant 459 ft TL: N31 44 05 W97 19 17. Hrs open: 24 6401 Cobbs Dr. Fl. A, Waco, 76710-2536. Phone: (254) 776-1033. Fax: (254) 776-0642. E-mail: bbehnke@amigobroadcasting.com Licensee: BMP Waco License Company L.P. Group owner: Amigo Broadcasting L.P. (acq 11-9-2004; grpsl). Natl. Rep: Lotus Entravision Reps LLC. Law Firm: Blooston, Mordkofsky, Jackson & Dickens. Format: Sp, Mexican rgnl. Target aud: 18-54; adults. ♦Brad Behnke, VP, gen mgr, sls dir & prom dir; George Lopez, progmg dir; Ed Pryer, chief of engrg; Cynthia Ramirez, traf mgr.

Coahoma

KWDC(FM)— 2006: 105.5 mhz; 5.1 kw. Ant 358 ft TL: N32 21 52 W101 19 35. Hrs open:
Rebroadcasts KSRD(FM) Saint Joseph, MO 100%.
c/o KSRD(FM), 1212 Faraon St., Saint Joseph, MO, 64501. Phone: (816) 233-5773. Fax: (816) 233-5777. Web Site: www.ksrdradio.com. Licensee: Horizon Christian Fellowship. (acq 2-9-2006; grpsl). Format: Christian. ♦Mike MacIntosh, pres; Brian KC Jones, gen mgr.

Cockrell Hill

KRVA(AM)—Licensed to Cockrell Hill. See Dallas

Coleman

KQBZ(FM)—Listing follows KSTA(AM).

KSTA(AM)— Nov 1, 1947: 1000 khz; 250 w-D. TL: N31 51 16 W99 25 36. Hrs open: Box 100, Brownwood, 76804. Phone: (325) 646-3535. Fax: (325) 646-5347. E-mail: ksta@web-access.com Licensee: Tackett-Boazman Broadcasting LP. (group owner; (acq 4-11-2006; grpsl). Population served: 20,000 Natl. Network: Jones Radio Networks. Rgnl. Network: Texas State Net., VSA Radio. Format: Classic country. Target aud: General. Spec prog: Farm 14 hrs, Sp 5 hrs, gospel 7 hrs wkly. ♦Mikey Wayne, gen mgr & progmg dir; Stan Cooper, chief of engrg.

KQBZ(FM)—Co-owned with KSTA(AM). 1974: 102.3 mhz; 12 kw. Ant 689 ft TL: N31 44 54 W99 19 57. Stereo. 9,900 ♦Kyle Dennis, gen mgr & progmg dir.

College Station

***KAMU-FM**— Mar 30, 1977: 90.9 mhz; 32 kw. 340 ft TL: N30 37 48 W96 20 33. (CP: 2.4 kw horiz, 32 kw vert). Stereo. Hrs opn: 6 AM-midnight Moore Communications Ctr., 4244 TAMU, 77843-4244. Phone: (979) 845-5613. Fax: (979) 845-1643. E-mail: kamu@tamu.edu Web Site: kamutamu.edu. Licensee: Texas A&M University. Population served: 120,000 Natl. Network: NPR, PRI. Format: Bluegrass, news, class, jazz. News: 35 hrs wkly. Target aud: General. Spec prog: Folk 3 hrs, new age 5 hrs, international 5 hrs wkly. ♦Rodney L. Zent, gen mgr; Penny Zent, stn mgr; Elaine Hoyak, dev dir; Richard Howard, progmg dir; Ken Nelson, engrg dir; Ed Hadden, chief of engrg; Yildiz McNew, traf mgr.

***KEOS(FM)**— Mar 25, 1995: 89.1 mhz; 100 w vert. 254 ft TL: N30 38 54 W96 23 23. (CP: 1 kw). Stereo. Hrs opn: 24 Box 78, 77841. Secondary address: 207 E. Carson St., Bryan 77801-1404. Phone: (979) 779-5367. Fax: (979) 779-7259. E-mail: keos@keos.org Web Site: www.keos.org. Licensee: Brazos Educational Radio. Population served: 130,000 Natl. Network: PRI. Format: News/talk, educ. News staff: News progmg 25 hrs wkly Target aud: General. Spec prog: Folk 10 hrs, gospel 3 hrs, jazz 3 hrs, Jewish & Israeli 2 hrs wkly. ♦Mark McCann, pres; Linda Gunderson, CFO; Jeff White, gen mgr & progmg dir; Tom Schwerdt, opns dir; Chad Brinkley, dev dir; John Roths, mus dir; George Weber, pub affrs dir; Lance Parr, chief of engrg; Mark Purcell, traf mgr; Leann Weatherby, spec ev coord; Reta Taylor, relg ed; Harvey John Miller, disc jockey.

KKYS(FM)—See Bryan

***KLGS(FM)**— 2007: 89.9 mhz; 8.4 kw vert. Ant 358 ft TL: N30 28 35 W96 25 57. Hrs open:
Rebroadcasts WAFR(FM) Tupelo, MS 100%.
Drawer 2440, Tupelo, MS, 38803. Phone: (662) 844-8888. Fax: (662) 842-6791. Web Site: www.afr.net. Licensee: American Family Association. (acq 1-3-2006; $10 for CP). Format: Christian. ♦Marvin Sanders, gen mgr.

KNDE(FM)—Listing follows KZNE(AM).

KTAM(AM)—See Bryan

KZNE(AM)— Oct 2, 1922: 1150 khz; 1 kw-D, 500 w-N, DA-N. TL: N30 38 05 W96 21 20. Stereo. Hrs opn: 24 2700 Earl Rudder Fwy. S., Suite 5000, 77845. Phone: (979) 846-1150. Fax: (979) 846-1933. Web Site: www.kzne.com. Licensee: Bryan Broadcasting Corp. (group owner; acq 8-7-97; with co-located FM). Population served: 120000 Format: Sports. News staff: 3; News: 58 hrs wkly. Target aud: 25-54. Spec prog: Farm 10 hrs wkly. ♦William R. Hicks, pres; Benjamin D. Downs, gen mgr; Sam Jones, sls dir & gen sls mgr; Louie Belina, progmg dir; Chace Murphy, news dir.

KNDE(FM)—Co-owned with KZNE(AM). Aug 8, 1964: 95.1 mhz; 36 kw. Ant 571 ft TL: N30 41 18 W96 25 35. Stereo. Format: Top 40 hits. ♦Bobby Mason, progmg dir.

WTAW(AM)— May 2000: 1620 khz; 10 kw-D, 1 kw-N. TL: N30 37 54 W96 21 28. Hrs open: Box 3248, Bryan, 77805. Secondary address: 2700 Rudder Fwy., Suite 5000 77845. Phone: (979) 846-1150. Fax: (979) 846-1933. E-mail: radio@wtaw.com Web Site: www.wtaw.com. Licensee: Bryan Broadcasting Corp. (group owner). Format: News/talk. ♦Benjamin D. Downs, gen mgr; Sam Jones, gen sls mgr; Scott Delucia, progmg dir; Chris Dusterhoff, chief of engrg; Alisa Dusterhoff, traf mgr.

Colorado City

KAUM(FM)—Listing follows KVMC(AM).

KVMC(AM)— June 16, 1950: 1320 khz; 1 kw-D. TL: N32 23 15 W100 53 33. Hrs open: Box 990, 79512. Phone: (915) 728-5224. Fax: (915) 728-5224. Web Site: www.realcountryonline.com. Licensee: James G. Baum (acq 2-13-81; $395,000; FTR: 3-9-81). Population served: 5,227 Rgnl. Network: Texas State Net. Format: Country. Spec prog: Farm 8 hrs wkly. ♦James G. Baum, pres & farm dir; Gary Graham, chief of engrg; Linda Baum, progmg dir & farm dir.

KAUM(FM)—Co-owned with KVMC(AM). Mar 29, 1983: 107.1 mhz; 3 kw. Ant 157 ft TL: N32 23 15 W100 53 33. Stereo. Web Site: www.realcountryonline.com.5,227 ♦Jim Baum, farm dir; Bill Baum, disc jockey; Doug Baum, disc jockey.

Columbus

KULM(FM)— Sept 3, 1973: 98.3 mhz; 6 kw. Ant 253 ft TL: N29 42 03 W96 34 24. Stereo. Hrs opn: 24 Box 111, 78934. Secondary address: 325 Radio Ln. 78934. Phone: (979) 732-5766. Fax: (979) 732-6377. E-mail: kulmradio@aol.com Licensee: Roy E. Henderson. Group owner: Fort Bend Broadcasting Co. (acq 2-15-00; grpsl). Population served: 30,000 Natl. Network: ABC. Rgnl. Network: Texas State Net. Format: C&W. News: 12 hrs wkly. Target aud: General. Spec prog: Polka 12 hrs wkly. ♦Roy Henderson, pres; Steve Smith, CFO; Carl Geisler, stn mgr, sls dir & progmg dir; Ray Nelson, chief of engrg; Judy Barrett, traf mgr.

Comanche

KCOM(AM)— Apr 1, 1962: 1550 khz; 250 w-D. TL: N31 53 54 W98 35 14. Hrs open: 24 218 N Austin St., 76442-2429. Phone: (325) 356-2558. Fax: (325) 356-3120. E-mail: kcom@comanchetx.com Web Site: www.kcomam.com. Licensee: CCR-Stephenville III LLC. (acq 8-2-2005; $164,000). Population served: 21,000 Rgnl. Network: Texas State Net. Format: Country Gospel. Target aud: 35-64; Men & Women. Spec prog: Gospel 5 hrs wkly. ♦Joseph Schwartz, pres; Marcus Nettleton, gen mgr & opns mgr; John Barnes, gen sls mgr; Peggy Vineyard, progmg dir, pub affrs dir, traf mgr & disc jockey; Stan Cooper, chief of engrg; Bill Cole, disc jockey.

KYOX(FM)— March 1999: 94.3 mhz; 32 kw. 620 ft TL: N31 54 51 W98 41 48. Stereo. Hrs opn: 218 N. Austin St., 76442. Phone: (915) 356-3090. Fax: (915) 356-3120. E-mail: kyox@comanchetx.com Web Site: www.kyoxfm.com. Licensee: CCR-Stephenville III LLC. Group owner: Cherry Creek Radio LLC (acq 6-10-2004; grpsl). Population served: 94,000 Format: Traditional country. Target aud: 35-64; men & women. ♦Richard Niblett, gen mgr; Jenni Lynn Robinson, gen sls mgr & mus dir; Pam Niblett, progmg dir & news dir; Justin McClure, chief of engrg.

Comfort

KCOR-FM— Feb 26, 1994: 95.1 mhz; 100 kw. Ant 925 ft TL: N29 50 26 W98 49 32. Stereo. Hrs opn: 24 1717 N. E. Loop 410, Suite 400, San Antonio, 78217. Phone: (210) 829-1075. Licensee: Univision Radio License Corp. Group owner: Univision Radio (acq 9-22-2003; grpsl). Population served: 1,367,500 Law Firm: Thompson, Hine & Flory. Format: Sp. News staff: one; News: 4 hrs wkly. Target aud: 25-54; average, middle income with small town & rural lifestyle. ♦Mac Tichenor, pres; Dan Wilson, gen mgr; Rick Thomas, opns dir & opns mgr.

Commerce

***KETR(FM)**— Apr 7, 1975: 88.9 mhz; 100 kw. 400 ft TL: N33 14 17 W95 55 27. Stereo. Hrs opn: 24 Box 4504, Performing Arts Ctr., 2600 S. Neal, 75429. Phone: (903) 886-5848. Fax: (903) 886-5850. E-mail: ketr@ketr.org Web Site: ketr.org. Licensee: Board of Regents Texas A&M University-Commerce. Natl. Network: NPR. Format: Jazz, adult contemp, news. News staff: one; News: 7 hrs wkly. Target aud: 21-66; general. Spec prog: Bluegrass 3 hrs wkly. ♦Beverly Nanos, gen sls mgr; Vicki Holloway, gen mgr, stn mgr & progmg dir; Mark Chapman, mus dir; Kevin Jeffries, news dir; Robert Goodwin, chief of engrg; Deborah Smith, traf mgr; Brad Kellar, news rptr.

***KYJC(FM)**— 2005: 91.3 mhz; 350 w horiz. Ant 174 ft TL: N33 15 37 W95 52 59. Hrs open: CSN International Inc., 4022 N. 3300 E., Twin Falls, ID, 83301. Phone: (208) 734-6633. Fax: (208) 736-1958. Web Site: www.csnradio.com. Licensee: CSN International Inc. Group owner: CSN International (acq 3-6-2003). ♦Mike Stocklin, gen mgr; Don Mills, progmg dir; Kelly Carlson, chief of engrg.

Conroe

***KAFR(FM)**— October 1998: 88.3 mhz; 100 kw vert. Ant 443 ft TL: N30 27 52 W95 30 20. Hrs open: American Family Radio, Box3206, Tupelo, MS, 38803. Phone: (662) 844-8888. Fax: (662) 842-6791. E-mail: comments@afr.net Web Site: www.afr.net. Licensee: American Family Association. Group owner: American Family Radio. Format: Inspirational Christian. ♦Marvin Sanders, gen mgr.

KHPT(FM)— Feb 14, 1965: 106.9 mhz; 91.6 kw. Ant 1,899 ft TL: N30 13 53 W95 07 26. Stereo. Hrs opn: 1990 Post Oak Blvd., Suite 2300, Houston, 77056. Phone: (713) 963-1200. Fax: (713) 622-5457. Web Site: www.1069thepoint.com. Licensee: Cox Radio Inc. Group owner: Cox Broadcasting (acq 8-24-2000; grpsl). Format: '80s music. ♦Mark Krieschen, VP & gen mgr; Beth Lavine, gen sls mgr; Mike Murray, natl sls mgr; Shana Sonnier, mktg dir; Dain Craig, progmg dir.

KJOJ(AM)— Apr 16, 1951: 880 khz; 10 kw-D, 1 kw-N, DA-2. TL: N30 17 38 W95 25 55. Stereo. Hrs opn: Little Saigon Radio, 7080 Southwest Fwy., Houston, 77074. Phone: (713) 271-7888. Fax: (713) 271-9333. Web Site: www.littlesaigonradio.com. Licensee: Liberman Broadcasting of Houston License LLC. Group owner: Liberman Broadcasting Inc. (acq 3-20-2001; grpsl). Format: Ethnic Vietnamese. ♦Winter Horton, gen mgr.

KYOK(AM)— Apr 13, 1981: 1140 khz; 5 kw-D, DA. TL: N30 20 40 W95 27 32. Hrs open: 300 E. Bryant Rd., 77301. Phone: (936) 441-1140. Fax: (936) 788-1140. Web Site: www.kyokradio.com. Licensee: Martin Broadcasting Inc. (acq 2-10-92; $175,000; FTR: 3-2-92). Population served: 35000 Format: Gospel. Target aud: 24-55. ♦Darrell Martin, pres & stn mgr; Nicholas Martin, stn mgr; Roland Booker, sls dir, prom dir & progmg dir; Dave Biondi, chief of engrg.

Copperas Cove

KNCT-FM—See Killeen

KSSM(FM)— Nov 21, 1977: 103.1 mhz; 8.6 kw. 276 ft TL: N31 05 05 W97 57 07. Stereo. Hrs opn: 24 608 Moody Ln., Temple, 76504. Phone: (254) 773-5252. Fax: (254) 773-0115. E-mail: bourdon.wooter@cumulus.com Web Site: www.1031kissfm.com. Licensee: Cumulus Licensing Corp. Group owner: Cumulus Media Inc. (acq 2-2-00). Population served: 250000 Natl. Rep: Interep. Law Firm: Cohn & Marks. Format: Urban adult contemp. News: 3 hrs wkly. Target aud: 25-54. Spec prog: Gospel. ♦Bourdon Wooten, gen mgr; Mikie Cummings, gen sls mgr; Jamie Garrett, prom dir & news dir; Lisa Tanne, prom dir & adv mgr; Mark Raymond, progmg dir; Doug Bernhardt, chief of engrg & engr.

Corpus Christi

***KBNJ(FM)**— January 1985: 91.7 mhz; 5 kw. 500 ft TL: N27 46 43 W97 37 57. Stereo. Hrs opn: 24 Box 270068, 78427. Secondary address: 3766 Saturn Rd. 78413. Phone: (361) 855-0975/76. Fax: (361) 855-0977. E-mail: kbnj@hcjb.org Web Site: www.kbnj.org. Licensee: World Radio Network Inc. (group owner; acq 6-5-84; $36,000; FTR: 5-21-84). Natl. Network: Moody, USA. Format: Relg, educ, English. News: 7 hrs wkly. Target aud: General. ♦Joe Fahl, gen mgr, stn mgr & progmg mgr; Michael Barnes, mus dir & chief of engrg.

KBSO(FM)— 1992: 94.7 mhz; 3 kw. 285 ft TL: N27 49 50 W97 32 34. Hrs open: 701 Benys Rd., 78408. Phone: (361) 289-0999. Fax: (361) 299-0810. E-mail: davilabroadcasti@bizstx.rr.com Web Site: www.texasradio947.com. Licensee: Reina Broadcasting Inc. Format: Texas radio. ♦Manuel Davila Jr., gen mgr.

KCCT(AM)— June 1954: 1150 khz; 1 kw-D, 500 w-N, DA-2. TL: N27 48 01 W97 28 44. Stereo. Hrs opn: 24 701 Benys Rd., 78408. Phone: (361) 289-0999. Fax: (361) 289-0810. E-mail: davilabroadcasti@bizstx.rr.com Licensee: Radio KCCT Inc. (acq 8-15-74). Population served: 240,000 Format: Oldies. News staff: one; News: 14 hrs wkly. Target aud: 25-54; Hispanics. ♦Manuel Davila Jr., pres, VP, gen sls mgr & progmg dir; George Sanders, chief of engrg.

KCTA(AM)— Oct 24, 1959: 1030 khz; 50,000 kw-D. TL: N27 56 01 W97 15 34. Hrs opn: Sunrise-sunset 1602 S. Brownlee Blvd., 78404. Phone: (361) 882-7711. Fax: (361) 882-3038. E-mail: kcta@usawide.net Web Site: www.kctaradio.com. Licensee: Broadcasting Corp. of the Southwest. (acq 1959). Population served: 210,000 Natl. Network: USA. Format: Relg. Target aud: 35 plus. Spec prog: Sp 6 hrs wkly. ♦Bill York, pres & gen mgr; David Freymiller, opns mgr.

KDAE(AM)—Sinton, 1954: 1590 khz; 1 kw-D, 500 w-N, DA-2. TL: N28 01 16 W97 28 14. Stereo. Hrs opn: 24 Box 260715, 78426. Secondary address: 929 N. Padre Island Dr. 78406. Phone: (361) 299-1982. Fax: (361) 299-1049. Web Site: www.radiolibertad.net. Licensee: The Worship Center of Kingsville. (acq 1-11-99). Population served: 353,000 Natl. Network: ABC. Natl. Rep: McGavren Guild. Law Firm: Fisher, Wayland, Cooper, Leader & Zaragoza L.L.P. Format: MOR, Spanish, Christian. News staff: one; News: 2 hrs wkly. Target aud: 35-64. Spec prog: Farm 6 hrs wkly. ♦Rufino Sendejo, gen mgr; A.J. Solis, progmg dir; George Sanders, chief of engrg.

***KEDT-FM**— Mar 2, 1982: 90.3 mhz; 100 kw. 802 ft TL: N27 39 12 W97 33 55. Stereo. Hrs opn: 24 4455 S. Padre Island Dr., Suite 38, 78411-4481. Phone: (361) 855-2213. Fax: (361) 855-3877. E-mail: info@kedt.pbs.org Web Site: www.kedt.org. Licensee: South Texas Public Broadcasting System Inc. Natl. Network: NPR. Law Firm: Schwartz, Woods & Miller. Format: Class, news, jazz. Latin. News staff: one; News: 37 hrs wkly. Target aud: General. Spec prog: Sp 4 hrs wkly. ♦Don Dunlap, pres & gen mgr.

KEYS(AM)—Listing follows KZFM(FM).

KFTX(FM)—See Kingsville

KKBA(FM)—See Kingsville

***KKLM(FM)**— Mar 11, 1991: 88.7 mhz; 8.3 kw. Ant 866 ft TL: N27 44 29 W97 36 09. Stereo. Hrs opn: 24 410 S. Padre Island Dr., Suite 207, 78405. Phone: (361) 289-0887. Fax: (361) 289-0649. E-mail: bcrown@emfbroadcasting.com Web Site: www.klove.com. Licensee: Educational Media Foundation. Group owner: EMF Broadcasting (acq 6-5-02; $500,000). Population served: 350,000 Natl. Network: K-Love. Format: Comtemp Christian. News: 10 hrs wkly. Target aud: 35 plus. ♦Richard Jenkins, pres; Lloyd Parker, gen mgr; Brian Crown, stn mgr & chief of engrg; Mike Novak, progmg dir & news dir; David Pierce, pub affrs dir.

KKTX(AM)—Listing follows KRYS-FM.

KLTG(FM)— Sept 1, 1967: 96.5 mhz; 97 kw. 955 ft TL: N27 44 28 W97 36 08. Stereo. Hrs opn: Box 898, 78403. Phone: (361) 883-1600. Fax: (361) 888-5685. Web Site: www.thebeach965online.com. Licensee: Tejas Broadcasting Ltd. LLP. Group owner: Amigo Broadcasting L.P. (acq 11-15-2004; grpsl). Population served: 500,000 Format: Hot AC. Target aud: 25-54. ♦Gloria Apolinario, gen mgr.

KMIQ(FM)—Robstown, July 23, 1989: 104.9 mhz; 3 kw. Ant 298 ft TL: N27 40 39 W97 38 20. (CP: 104.9 mhz; 50 kw, ant 492 ft. TL: N27 56 08 W97 56 19). Stereo. Hrs opn: Box 270547, 78427. Phone: (361) 289-8877. Fax: (361) 289-7722. Licensee: Cotton Broadcasting. Format: Tejano. Target aud: 18 plus. Spec prog: Relg 6 hrs wkly.

KMJR(FM)—See Portland

KMXR(FM)— January 1970: 93.9 mhz; 100 kw. 840 ft TL: N27 46 50 W97 38 03. Stereo. Hrs opn: 24 501 Tupper Ln., 78417. Phone: (361) 289-0111. Fax: (361) 289-5035. E-mail: oldies939@aol.com Web Site: www.939online.com. Licensee: Capstar TX L.P. Group owner: Clear Channel Communications Inc. (acq 8-30-00; grpsl). Population served: 422,300 Natl. Network: AP Radio. Format: Oldies. News staff: one; News: 5 hrs wkly. Target aud: 25-54. ♦Matt Martin, gen mgr.

KNCN(FM)—Sinton, July 1, 1972: 101.3 mhz; 100 kw. 401 ft TL: N27 55 24 W97 25 26. Stereo. Hrs opn: 24 Radio Plaza, 501 Tupper Ln., 78417. Phone: (361) 289-0111. Fax: (361) 289-5035. E-mail: c101@clearchannel.com Web Site: www.c101.com. Licensee: Capstar TX L.P. Group owner: Clear Channel Communications Inc. (acq 8-30-00; grpsl). Population served: 275,000 Format: Active rock. News: one hr wkly. Target aud: 18-49; active. Spec prog: Coastal Bend Forum one hr, In Concert 2 hrs, In the Studio one hr, Flashback 2 hrs wkly. ♦Matt Martin, gen mgr.

KOUL(FM)—Sinton, May 20, 1968: 103.7 mhz; 100 kw. 941 ft TL: N28 02 05 W97 26 10. Stereo. Hrs opn: Box 898, 78403. Secondary address: 1300 Antelope 78401. Phone: (361) 883-1600. Fax: (361) 883-9303. Licensee: Tejas Broadcasting Ltd. LLP. Group owner: Amigo Broadcasting L.P. (acq 11-15-2004; grpsl). Format: C&W. Target aud: 25-49; general. ♦Chuck Brooks, pres; Bert Clark, opns mgr; KC Sheperd, prom dir; Paul Danitz, stn mgr, gen sls mgr & adv mgr; Glenn Michaels, progmg dir; Lon Gonzalez, news dir; Lisa Del Rey, pub affrs dir; Henry Turner, chief of engrg; Debra Reid, traf mgr.

KRYS-FM— Dec 5, 1982: 99.1 mhz; 100 kw. 1,049 ft TL: N27 45 07 W97 38 18. (CP: 97 kw). Stereo. Hrs opn: 24 Radio Plaza, 501 Tupper Ln., 78417. Phone: (361) 289-0111. Fax: (361) 289-5024. Web Site: www.krysfm.com. Licensee: Capstar TX L.P. Group owner: Clear Channel Communications Inc. (acq 8-30-00; grpsl). Population served: 360,000 Natl. Network: ABC. Format: Country. News staff: one. Target aud: 25-54. ♦Matt Martin, gen mgr; Zee Zepola, sls dir & gen sls mgr; Frank Edwards, prom dir & progmg dir; Lou Ramirez, mus dir; Russell Vaughn, chief of engrg; Pamela Anaya, traf mgr.

KKTX(AM)—Co-owned with KRYS-FM. 2002: 1360 khz; 1 kw-U. TL: N27 48 01 W97 27 41.24 E-mail: scottjohnson@clearchannel.com Web Site: www.1360online.com.300,000 Natl. Network: ABC. Format: News/talk. Target aud: 2-18; children. ♦Matt Martrin, gen mgr; Zee Zepola, sls dir; Scott Johnson, progmg dir; Russell Vaughan, chief of engrg.

KSIX(AM)— September 1947: 1230 khz; 1 kw-U. TL: N27 48 09 W97 27 14. Stereo. Hrs opn: 24 710 Buffalo St., Suite 608, 78416. Phone: (361) 882-5749. Fax: (361) 884-1240. E-mail: info@espn1230ksix.com Web Site: www.espn1230ksix.com. Licensee: Withers Family Texas Holding LP (acq 10-28-02). Population served: 232,000 Natl. Network: ESPN Radio. Format: Sports. ♦Jim Withers, gen mgr; Scott Howe, gen sls mgr; Bill Doerner, progmg dir; Valerie Smith, traf mgr.

KUNO(AM)— May 1950: 1400 khz; 1 kw-U. TL: N27 45 36 W97 26 14. Stereo. Hrs opn: 24 Radio Plaza, 501 Tupper Ln., 78417-9736. Phone: (361) 289-0111. Fax: (361) 289-5035. Licensee: Capstar TX L.P. Group owner: Clear Channel Communications Inc. (acq 8-30-00; grpsl). Population served: 360,000 Format: Sp. News: 17 hrs wkly. Target aud: 25-64. ♦Matt Martin, gen mgr.

KZFM(FM)— Dec 7, 1964: 95.5 mhz; 100 kw. Ant 991 ft TL: N27 39 33 W97 34 12. Stereo. Hrs opn: 24 Box 9757, 78469. Secondary address: 2117 Leopard St. 78408. Phone: (361) 883-3516. Fax: (361) 882-9767. E-mail: thechief@star94.net Web Site: www.hotz95.com. Licensee: Malkan FM Associates L.P. Group owner: Malkan Broadcast Assoc. (acq 1976). Population served: 500,000 Natl. Rep: Katz Radio. Law Firm: Thompson Hine. Format: CHR. Target aud: 18-34; female. ♦Glen Powers, pres & stn mgr; Janice Raleigh, gen sls mgr & prom dir; Gino Flores, prom dir; Ed Ocanas, progmg dir; Arlene Cordell, mus dir; John Gifford, chief of engrg.

KEYS(AM)—Co-owned with KZFM(FM). March 1941: 1440 khz; 1 kw-U, DA-N. TL: N27 47 02 W97 27 29. Phone: (361) 882-7411. Fax: (361) 882-9767. E-mail: johngifford1440@yahoo.com Web Site: www.1440keys.com. Licensee: Malkan AM Associates L.P. (acq 1965).250,000 Natl. Network: ABC. Natl. Rep: Katz Radio. Law Firm: Thompson Hine. Format: News/talk, sports. Target aud: 25 plus; men. ♦Will Diaz, progmg dir.

Corsicana

KAND(AM)— May 17, 1937: 1340 khz; 1 kw-U. TL: N32 06 53 W96 27 47. Hrs open: 24 1504 N. Beaton St., 75110. Secondary address: Box 2998 75151. Phone: (903) 874-7421. Phone: (903) 874-1340. Fax: (903) 874-0789. Web Site: www.kand1340am.com. Licensee: Corsicana Media Inc. (acq 1-16-95; $500,000). Population served: 22,900 Rgnl. Network: Texas State Net. Format: Country, news. News staff: one; News: 25 hrs wkly. Target aud: General. ♦John Whetzell, pres; Mike Taylor, gen mgr; Mary Sikes, gen sls mgr, traf mgr & disc jockey; Bob Belcher, progmg dir & disc jockey; Dick Aldama, news dir; Jim Wiggins, chief of engrg & disc jockey.

Crane

KMMZ(FM)— 1995: 101.3 mhz; 100 kw. Ant 485 ft TL: N31 41 02 W102 19 13. Stereo. Hrs opn: 24 Box 60375, Midland, 79711. Secondary address: 12200 W. I-20 E. 79711. Phone: (432) 563-2266. Fax: (432) 563-2288. E-mail: Sonny@kmmz.net Licensee: Don L. Cook. Population served: 428,000 Law Firm: Thompson, Hine & Flory. Format: Sp Top 40. Target aud: 25-54. ♦Don L. Cook, gen mgr.

KXOI(AM)— December 1959: 810 khz; 1 kw-D, 500 w-N, DA-1. TL: N31 28 39 W102 20 24. Hrs open: 6 AM-midnight Box 2344, Odessa, 79760. Phone: (432) 333-5061. Fax: (432) 333-6067. Licensee: Hispanic Outreach Ministries Inc. Format: Sp. Target aud: General. ♦Rev. Pedro Emiliano, pres; Eli Emiliano, gen mgr & progmg dir; Don Cook, chief of engrg.

Creedmoor

KZNX(AM)— Dec 8, 1962: 1530 khz; 10 kw-D, 12 w-N, 1 kw-CH, DA-3. TL: N30 04 38 W97 38 08. Stereo. Hrs opn: Sunrise-sunset 1050 E. 11th St., Suite 300, Austin, 78702. Phone: (512) 346-8255. Fax: (512) 346-8262. E-mail: controlroom@espnaustin.com Web Site: www.espnaustin.com. Licensee: Simmons-Austin, LS LLC. Group owner: Simmons Media Group (acq 6-2-2004; $2 million). Population served: 1,000,000 Natl. Network: ESPN Radio, Westwood One. Format: Talk, sports. News: 63 hrs wkly. Target aud: 25-54; 51% male, 49% female. ♦Daryl O'Neal, gen mgr; Jon Madani, chief of opns & progmg dir; Lori Hatter, sls dir; Courtney Cleland, prom mgr; J. Cole McClellan, chief of engrg; Flavia Chen, traf mgr.

Crockett

KBHT(FM)— Nov 15, 1982: 93.5 mhz; 50 kw. 479 ft TL: N31 20 03 W95 47 13. Stereo. Hrs opn: 24 Box 430, 206 S. Main St., Grapeland, 75844. Phone: (936) 544-9350. Fax: (936) 544-9695. E-mail: lesia@kbht.com Web Site: www.kbht.com. Licensee: Weston Entertainment L.P. (acq 10-4-2005; $1.43 million). Rgnl. Network: Texas AP. Format: Classic country. News staff: 7. Target aud: 25-54. Spec prog: Gospel 6 hrs wkly. ♦Dennis W Goodman, gen mgr; Jeri Sulewski, sls dir.

***KCKT(FM)**— February 2003: 88.5 mhz; 250 w. Ant 161 ft TL: N31 19 37 W95 28 26. Hrs open: Box 3206, Tupelo, MS, 38803. Phone: (662) 844-8888. Fax: (662) 842-6791. E-mail: comments@afr.net Web Site: www.afr.net. Licensee: American Family Association. Group owner: American Family Radio (acq 1-17-01). Format: Christian. ♦Marvin Sanders, gen mgr.

KIVY(AM)— Nov 11, 1949: 1290 khz; 2.5 kw-D, 175 w-N. TL: N31 18 20 W95 27 06. Hrs open: 24 102 S. Fifth St., 75835. Phone: (936) 544-2171. Phone: (936) 544-KIVY. Fax: (936) 544-4891. Web Site: www.kivy.com. Licensee: Leon Hunt (acq 9-19-02; $1.1 million with co-located FM). Population served: 150,000 Natl. Network: ABC. Rgnl. Network: Texas State Net. Format: Classic oldies. Target aud: General. ♦Leon Hunt, pres, gen mgr, gen sls mgr & progmg dir; Chester Leediker, chief of engrg.

KIVY-FM— June 1, 1970: 92.7 mhz; 50 kw. 497 ft TL: N31 18 18 W95 27 06. Stereo. 24 Web Site: www.kivy.com.250,000 Natl. Network: ABC. Format: Country.

Crystal Beach

KPTI(FM)— November 1989: 105.3 mhz; 6 kw. Ant 180 ft TL: N29 30 07 W94 31 15. Stereo. Hrs opn:
Simulcast with KPTY(FM) Missouri City.
5100 Southwest Fwy., Houston, 77056. Phone: (713) 965-2300. Fax: (713) 965-2401. Web Site: www.party1049.com. Licensee: Tichenor License Corp. Group owner: Univision Radio (acq 9-22-2003; grpsl). Population served: 250,000 Natl. Rep: Katz Radio. Format: Urban hip-hop. Target aud: 18-34; urban /Latin audience. ♦Mark Masepohl, VP & gen mgr.

KSTB(FM)— 1996: 101.5 mhz; 14 kw. 449 ft TL: N29 33 52 W94 23 59. Stereo. Hrs opn: 24 755 S. 11th St., Ste.102, Beaumont, 77701. Phone: (409) 833-9421. Fax: (409) 833-9296. Licensee: Cumulus Licensing Corp. Group owner: Cumulus Media Inc. (acq 5-20-02; $2.5 million). Population served: 600,000 Natl. Rep: Roslin. Law Firm: Fisher, Wayland, Cooper, Leader & Zaragoza. Format: Country. News staff: 2; News: 12 hrs wkly. Target aud: 18-49. ♦Rick Prusater, gen mgr & stn mgr; Jim West, opns VP & progmg dir; Greg Davis, chief of engrg; Liz Ferguson, traf mgr.

Crystal City

KHER(FM)— Sept 5, 1985: 94.3 mhz; 3 kw. 135 ft TL: N28 39 57 W99 48 58. Hrs open: Old Big Wells Hwy., Farm Rd. 65, Box 707, Carri 20 S., 78839. Phone: (830) 374-2203. Phone: (830) 374-5730 (office). Fax: (830) 374-9658. E-mail: kherfm@yahoo.com Licensee: Sylvia Mijares. (acq 9-4-97). Rgnl. Network: Texas State Net. Law Firm: Borsari & Paxson. Format: Sp, news/talk. Target aud: 18-54; 90% Hispanic, 10% non-minority. Spec prog: Relg 2 hrs wkly. ♦Sylvia Mijares, pres & gen mgr; Rudy Gomez, sls VP; Marie Thelma Martinez, news dir & pub affrs dir; Charlie Schmele, chief of engrg; Becky P. Reyes, traf mgr.

Cuero

***KTLZ(FM)**— 2003: Stn currently dark. 89.9 mhz; 5 kw. Ant 243 ft TL: N29 02 23 W97 19 24. Hrs open: 24 Box 5459, Twin Falls, ID, 83303-5459. Phone: (208) 733-3551. Fax: (208) 733-3548. Web Site: www.radioassistministry.com. Licensee: Radio Assist Ministry Inc. (acq 9-21-2004; $50,000). ♦Clark Parrish, pres.

Cypress

KYND(AM)— December 1991: 1520 khz; 3 kw-D, DA. TL: N30 00 37 W95 41 40. Hrs open:
Rebroadcasts KJOJ(AM) Conroe 100%.
Little Saigon Radio, 7080 Southwest Fwy., Houston, 77074. Phone: (713) 271-7888. Fax: (713) 271-9333. E-mail: radio@littlesaigonradio.com Web Site: www.littlesaigonradio.com. Licensee: Matthew Provenzano. Format: Ethnic, Vietnamese. Target aud: General. ♦Matt Provenzano, CEO.

Daingerfield

KNGR(AM)— August 1966: 1560 khz; 1.5 kw-D, 60 w-N. TL: N33 01 35 W94 42 22. Stereo. Hrs opn: 24 Box 474, 75638. Phone: (903) 645-4325. Fax: (903) 645-4357. Web Site: www.kingcountry.org. Licensee: Network Communications Co. (acq 2-21-91; $50,000; FTR: 3-11-91). Population served: 40,000 Format: Gospel. News: 3 hrs wkly. ♦Bob Wilson, chmn; Glory Wilson, gen mgr & progmg dir.

Dalhart

***KTDA(FM)**—Not on air, target date: unknown: 91.7 mhz; 250 w. Ant 128 ft TL: N36 03 20 W102 30 34. Hrs open: Drawer 2440, Tupelo, MS, 38803. Phone: (662) 844-8888. Fax: (662) 842-6791. Licensee: Educational Opportunities Inc.

KXIT(AM)— 1948: 1240 khz; 1 kw-U. TL: N36 05 45 W102 30 38. Stereo. Hrs opn: 24 Box 1359, Hwy. 385 N., 79022. Phone: (806) 249-4747. Licensee: Dalhart Radio Inc. (acq 12-3-01; $325,000 with co-located FM). Population served: 8,500 Rgnl. Network: Texas State Net. Format: Country. Spec prog: Farm 7 hrs wkly. ♦George Chambers, pres & gen mgr; Cheryl Riichard, rgnl sls mgr; Adam Taylor, progmg mgr; Auto Maxin, chief of engrg.

KXIT-FM— 1962: 96.3 mhz; 100 kw. Ant 472 ft TL: N35 53 46 W102 23 03. Stereo. 24 12,500 Format: Classic rock, oldies.

Dallas

KAAM(AM)—Garland, 1973: 770 khz; 10 kw-D, 1 kw-N, DA-2. TL: N33 01 58 W96 34 31. Stereo. Hrs opn: 24 3201 Royalty Row, Irving, 75062. Phone: (972) 445-1700. Fax: (972) 438-6574. E-mail: cbcstand@aol.com Web Site: www.kaamradio.com. Licensee: Dontron Inc. Group owner: Crawford Broadcasting Co. (acq 1979). Population served: 6000000 Format: Adult standards. Target aud: 35 plus; Christian. ♦Don Crawford, pres; Don Crawford Jr., gen mgr.

KBFB(FM)— 1965: 97.9 mhz; 99 kw. 1,611 ft TL: N32 35 15 W96 57 59. Stereo. Hrs opn: 13331 Preston Rd., Suite 1180, 75240. Phone: (972) 331-5400. Fax: (972) 331-5560. Web Site: www.979tlpbeat.com. Licensee: Radio One Licenses LLC. Group owner: Radio One Inc. (acq 2000; grpsl). Population served: 3,000,000 Natl. Rep: CBS Radio. Format: Urban contemp. Target aud: 25-50. ♦Alfred Liggine, pres; George Laughlin, gen mgr; John Candelaria, opns mgr & progmg dir; Shawn Nunn, sls dir; Joe Libios, mktg dir & prom dir; Tony Fields, progmg dir; Don Stevenson, chief of engrg; Rowena Montgomery, traf mgr.

***KCBI(FM)**— May 19, 1976: 90.9 mhz; 100 kw. 1,509 ft TL: N32 35 22 W96 58 10. Stereo. Hrs opn: 24 Box 619000, 75261-9000. Phone: (817) 792-3800. Fax: (817) 277-9929. E-mail: kcbi@kcbi.org Web Site: www.kcbi.org. Licensee: Criswell College. (group owner) Population served: 4,500,000 Natl. Network: AP Radio. Format: Inspirational, Christian. News staff: 4; News: 4 hrs wkly. Target aud: 35-54; Christian families. ♦Ronald L. Harris, CEO; Royce Laycock, chmn; Heidi Graham, pres & sls dir; Todd Chatman, stn mgr; Doug Price, opns VP; James Nance, dev VP; Troy Kriechbaum, prom dir & prom mgr; Marc Anderson, progmg VP & mus dir; L.B. Lyon, news dir; Doug Watson, engrg dir.

KDGE(FM)—See Fort Worth-Dallas

KDMX(FM)— 1965: 102.9 mhz; 100 kw. 1,164 ft TL: N32 34 54 W96 58 32. Hrs open: 24 14001 N. Dallas Pkwy., Suite 300, 75240. Phone: (214) 866-8000. Fax: (214) 866-8201. E-mail: contactus@mix1029.listenernetwork.com Web Site: www.mix1029.com. Licensee: Citicasters Licenses L.P. Group owner: Clear Channel Communications Inc. (acq 5-4-99; grpsl). Format: Hot adult contemp. News: 3 hrs wkly. Target aud: 25-49; upper income females. ♦J.D. Freeman, gen mgr; Pat McMahon, opns dir; Jeff Mitchell, gen sls mgr; Steve Lee, mktg dir; Rick O'Bryan, progmg dir; Louis Sutton, engrg dir.

KEGL(FM)—See Fort Worth

***KERA(FM)**— July 11, 1974: 90.1 mhz; 95 kw. 1,260 ft TL: N32 34 43 W96 57 12. Stereo. Hrs opn: 3000 Harry Hines Blvd., 75201. Phone: (214) 871-1390. Fax: (214) 740-9369. E-mail: kerafm@kera.org Web Site: www.kera.org. Licensee: North Texas Public Broadcasting. Population served: 3,600,000 Natl. Network: NPR, PRI. Law Firm: Arnold & Porter. Wire Svc: AP Format: News/talk, progsv. News staff: 5; News: 80 hrs wkly. Target aud: 35-54; general. ♦Kevin Martin, COO, pres, pres, CFO & exec VP; Barger Tygart, chmn; Jeff Luchsinger, stn mgr; Patricia Lyons, dev VP & engrg dir.

KFJZ(AM)—See Fort Worth

KFLC(AM)—See Fort Worth

KFXR(AM)— 1947: 1190 khz; 50 kw-D, 5 kw-N, DA-2. TL: N32 47 10 W96 57 00. Hrs open: 24 720 N. Saint Paul St., 75201. Phone: (214) 855-0002. Web Site: cowboy1190.com. Licensee: Capstar TX L.P. Group owner: Clear Channel Communications Inc. (acq 3-27-2001; $16 million). Population served: 4,000,000 Format: Classic country. News: 2 hrs wkly. Target aud: 25-54; general. Spec prog: Sports 6 hrs wkly.

KGGR(AM)— June 8, 1947: 1040 khz; 3.3 kw-D, 2.8 kw-CH. TL: N32 46 43 W96 43 51. Hrs open: 5787 S. Hampton Rd., Suite 285, 75232. Phone: (972) 572-5447. Phone: (972) 988-1040. Fax: (214) 330-6133. Web Site: www.kggram.com. Licensee: Mortenson Broadcasting Co. of Texas Inc. Group owner: Mortenson Broadcasting Co. (acq 5-1-96; $1.15 million). Population served: 58,000 Natl. Network: American Urban. Format: Relg, talk, Black. Target aud: 18 plus. ♦Ann Arnold, gen mgr; Christie Wafer, gen sls mgr.

KHKS(FM)—See Denton

KHVN(AM)—See Fort Worth

KJKK(FM)— Dec 25, 1965: 100.3 mhz; 97 kw. Ant 1,883 ft TL: N32 35 05 W96 57 46. Hrs open: 7901 Carpenter Fwy., 75247. Phone: (214) 630-3011. Fax: (214) 905-5052. Web Site: www.wild100.com. Licensee: Texas Infinity Broadcasting L.P. Group owner: Infinity Broadcasting Corp. (acq 11-13-98; grpsl). Population served: 450,000 Natl. Network: ABC. Natl. Rep: CBS Radio. Format: CHR. Target aud: 18-49. ♦Mel Karmazin, CEO; Dave Siebert, gen mgr; David Henry, sls dir; Joel Gough, gen sls mgr; Amy Gomoll, prom dir; Alex Valentine, progmg dir; Bethany Parks, mus dir; Lori Dodd, pub affrs dir; Bob Henke, chief of engrg; Connie Pena, traf mgr.

KKDA(AM)—See Grand Prairie

KKDA-FM— June 8, 1947: 104.5 mhz; 100 kw. 1,585 ft TL: N32 35 22 W96 58 10. Stereo. Hrs opn: 24 Box 530860, Grand Prairie, 75053. Secondary address: 1230 River Bend Dr., Ste. 111 75247. Phone: (972) 263-9911. Fax: (972) 558-0010. E-mail: staff@k104fm.com Web Site: www.k104fm.com. Licensee: Service Broadcasting Group LLC. (acq 5-76). Population served: 500,000 Natl. Rep: Christal. Format: Urban contemp. ♦Hymen Childs, pres; Chuck Smith, gen mgr; Vick Romanick, gen sls mgr; Liz Leos, prom mgr; Skip Cheatham, progmg dir; Sam Putney, news dir; Gary Wachter, chief of engrg; Tommy Page, traf mgr.

KLIF(AM)— June 26, 1922: 570 khz; 5 kw-U, DA-2. TL: N32 56 40 W96 59 25. Hrs open: 24 3500 Maple Ave., Suite 1600, 75219. Phone: (214) 526-2400. Phone: (214) 263-4141. Fax: (214) 520-4343. Web Site: www.klif.com. Licensee: KLIF Lico Inc. Group owner: Susquehanna Radio Corp. (acq 12-15-89). Population served: 4,418,400 Natl. Network: Fox News Radio. Format: News/talk. News staff: 1; News: 15 hrs wkly. Target aud: 25-54; men. ♦John Dickey, COO, exec VP & VP; Lew Dickey, CEO, chmn & pres; Dan Bennett, VP & mktg mgr.

KLLI(FM)— Apr 5, 1968: 105.3 mhz; 97 kw. Ant 1,883 ft TL: N32 35 05 W96 57 46. Stereo. Hrs opn: 24 7901 Carpenter Fwy, 75247. Phone: (214) 630-3011. Fax: (214) 905-5052. Web Site: www.live1053.com. Licensee: Texas Infinity Broadcasting L.P. Group owner: Infinity Broadcasting Corp. (acq 11-13-98; grpsl). Population served: 4,761,200 Natl. Network: Westwood One. Natl. Rep: CBS Radio. Law Firm: Leventhal, Senter & Lerman. Format: Talk. News staff: one; News: 3 hrs wkly. Target aud: 18-49; general. ♦Mel Karmazin, CEO; Brian Purdy, gen mgr; Steve Sullivan, gen sls mgr; Lynn Sornsen, natl sls mgr; Jeff Burkett, prom dir; Gavin Spittle, progmg dir; Bob Henke, chief of engrg; Susan Wade, traf mgr.

KLNO(FM)—See Fort Worth

KLTY(FM)—Arlington, April 1949: 94.9 mhz; 99 kw. Ant 1,666 ft TL: N32 35 19 W96 58 05. Stereo. Hrs opn: 24 6400 Belt Line Rd., Suite 120, Irving, 75063. Phone: (972) 870-9949. Fax: (214) 561-2156. Web Site: www.klty.com. Licensee: Inspiration Media of Texas LLC. Group owner: Salem Communications Corp. Population served: 4,761,200 Natl. Rep: Katz Radio. Format: Adult contemp Christian. News: 2 hrs wkly. Target aud: 25-54; female dominant, family oriented, upscale, conservative. ♦John L. Peroyea, VP & gen mgr.

KLUV(FM)— 1961: 98.7 mhz; 98 kw. 1,584 ft TL: N32 35 22 W96 58 10. Stereo. Hrs open: 24 4131 N. Central Expwy., Suite 700, 75204. Phone: (214) 526-9870. Fax: (214) 443-1570. Fax: (214) 522-5588. Web Site: www.kluv.com. Licensee: Texas CBS Radio Broadcasting L.P. Group owner: CBS Radio (acq 9-17-94; $51 million). Population served: 4,727,800 Format: Oldies. News staff: 2. Target aud: 35-54. ♦Mel Karmazin, pres; David Henry, gen mgr; John Phillips, gen sls mgr; Liz Balon, prom mgr; Jay Cresswell, mus dir; Kathy Jones, news dir & pub affrs dir; Bill Taylor, chief of engrg; Julie Davis, traf mgr.

KMNY(AM)—See Hurst

KMVK(FM)—See Fort Worth

KNIT(AM)— 1952: 1480 khz; 5 kw-D, 1.9 kw-N, DA-2. TL: N32 39 42 W96 39 20. Hrs open: 24 6400 N. Beltline Rd., Suite 110, Irving, 75063. Phone: (214) 561-9673. Fax: (214) 561-9662. Licensee: JCE Licenses LLC. Group owner: Univision Radio (acq 2-3-2006; swap in exchange for WORL(AM) Altamonte Springs, FL). Natl. Network: ESPN Deportes. Format: Sp sports. ♦Pete Thomson, gen mgr.

***KNON(FM)**— Aug 3, 1983: 89.3 mhz; 55 kw. Ant 850 ft TL: N32 35 24 W96 58 21. Stereo. Hrs opn: 24 Box 710909, 75371. Secondary address: 5353 Maples Ave. 75235. Phone: (214) 828-9500. Fax: (214) 823-3051. Web Site: www.knon.org. Licensee: Agape Broadcasting Foundation Inc. (acq 8-83). Population served: 1,000,000 Format: Var. Target aud: General. ♦Dave Chaos, stn mgr; Christian Lee, mus dir; Pamela Parker, sls dir & news dir.

KPLX(FM)—See Fort Worth

KRLD(AM)— October 1926: 1080 khz; 50 kw-U, DA-N. TL: N32 53 25 W96 38 44. Hrs open: 24 1080 Ballpark Way, Arlington, 76011. Phone: (817) 543-5400. Fax: (817) 543-5570. Web Site: www.krld.com. Licensee: Texas Infinity Broadcasting L.P. Group owner: Infinity Broadcasting Population served: 3,352,500 Rgnl. Network: Texas

State Net. Natl. Rep: CBS Radio. Format: News/talk, news, sports. News staff: 35; News: 119 hrs wkly. Target aud: 25-54. Spec prog: Texas Rangers baseball. ♦Jerry Bobo, VP & gen mgr; Tom Bigby, opns dir.

KRVA(AM)—Cockrell Hill, Sept 29, 1947: 1600 khz; 5 kw-D, 1 kw-N, DA-2. TL: N32 44 25 W96 42 38. Hrs open: 24 4965 Preston Park Blvd., Suite 120, Plano, 75093. Licensee: Mortenson Broadcasting Co. of Texas Inc. (group owner; (acq 8-30-2004; $3.5 million). Population served: 5,000,000 Format: Ethnic, Indian, Pakistani. ♦Rehan Siddiqi, gen mgr; Naheed Raheel, progmg dir.

KRVA-FM—See Campbell

KSKY(AM)—Balch Springs, Sept 30, 1941: 660 khz; 10 kw-D, 660 w-N, DA-N. TL: N29 22 51 W95 14 15. Hrs open: 24 6400 N. Beltline, Suite 110, Irving, 75063. Phone: (214) 561-9660. Fax: (214) 561-9662. E-mail: ksky@ksky.com Web Site: www.ksky.com. Licensee: Bison Media Inc. Group owner: Salem Communications Corp. (acq 4-24-2000; $7.5 million plus seller gets KMOM(FM) Fountain, CO). Population served: 1,000,000 Law Firm: Latham & Watkins. Format: News, talk. Target aud: 35-59; middle income white female. Spec prog: High school, college sports. ♦Pete Thomson, CFO, VP & gen mgr; David Darling, opns VP, opns mgr & progmg dir; Bots Johnson, gen sls mgr; Carol White, prom dir; Brian Heise, chief of engrg.

KSOC(FM)—Gainesville, 1958: 94.5 mhz; 100 kw. 1,896 ft TL: N33 33 36 W96 57 35. Stereo. Hrs opn: 24 13331 Preston Rd., Suite 1180, 75234. Phone: (972) 331-5400. Fax: (972) 726-0940. Licensee: Radio One Licenses LLC. Group owner: Radio One Inc. (acq 11-8-01; grpsl). Population served: 4,800,000 Law Firm: Latham & Watkins. Format: Urban / adult contemp. News staff: one; News: one hr wkly. Target aud: 18-44; affluent generation X'ers. ♦George Laughlin, gen mgr; John Candelaria, progmg dir.

KTCK(AM)— 1920: 1310 khz; 5 kw-D, 5 kw-N, DA-2. TL: N32 56 41 W96 56 25. Stereo. Hrs opn: 24 3500 Maple Ave., Suite 1310, 75219. Phone: (214) 526-7400. Fax: (214) 525-2525. Web Site: www.theticket.com. Licensee: KRBE Lico Inc. Group owner: Susquehanna Radio Corp. (acq 1996; $14 million). Population served: 4,500,000 Format: Sports, talk. Target aud: 25-54; men & sport enthusiasts. ♦Dan Bennett, VP, gen mgr & mktg mgr; Jim Quirk, sls dir & mktg mgr; Ken Roberts, gen sls mgr; Jami Williams, natl sls mgr; Jamey Garner, prom dir; Jeff Catlin, progmg dir; Rob Chickering, engrg dir; Kimberly Jolly, traf mgr.

KVCE(AM)—See Highland Park

KVIL(FM)—See Highland Park

***KVTT(FM)**— Jan 26, 1950: 91.7 mhz; 100 kw. 1,099 ft TL: N32 35 24 W96 58 21. Stereo. Hrs opn: 24 11061 Shady Tr., 75229. Phone: (214) 351-6655. Fax: (469) 522-0992. E-mail: kvtt@kvtt.org Web Site: www.kvtt.org. Licensee: Covenant Educational Media Inc. (acq 9-21-2004; $16.5 million). Population served: 4,500,000 Format: Teaching, educ, music. Spec prog: Kaverdad (Hispanic Christian) 6 hrs wkly. ♦Douglas Price, gen mgr & stn mgr; Bryan Reeder, progmg dir & mus dir.

KZPS(FM)— Apr 1, 1948: 92.5 mhz; 100 kw. 1,590 ft TL: N32 35 22 W96 58 10. Stereo. Hrs opn: 24 14001 N. Dallas Pkwy., Suite 300, 75240. Phone: (214) 866-8000. Web Site: www.kzps.com. Licensee: AMFM Texas Licenses L.P. Group owner: Clear Channel Communications Inc. (acq 8-30-00; grpsl). Population served: 325,000 Format: Classic rock. News staff: one. Target aud: 25-44; upscale young adults. ♦J.D. Freeman, gen mgr; Pat McMahon, opns dir; Kelly Kibler, sls dir & progmg VP; Tracy Martin, gen sls mgr & rgnl sls mgr; Steve Lee, mktg dir; Duane Doherty, prom dir & progmg dir; Anna DeHaro, news dir; Louis Sutton, chief of engrg; Sylvia Sanchez, traf mgr.

WBAP(AM)—See Fort Worth

WRR(FM)— 1948: 101.1 mhz; 100 kw. 1,510 ft TL: N32 35 22 W96 58 10. Stereo. Hrs opn: 24 Box 159001, 75315-9001. Secondary address: 1516 First Ave. 75210. Phone: (214) 670-8888. Fax: (214) 670-8394. Web Site: www.wrr101.com. Licensee: City of Dallas. Population served: 300,000 Natl. Network: AP Network News. Natl. Rep: McGavren Guild. Law Firm: Kaye, Scholer, Fierman, Hays & Handler. Format: Classical. News staff: 22. Target aud: 25-54; all ages. Spec prog: Children 2 hrs wkly. ♦Gregory T. Davis, VP & gen mgr.

Decatur

***KDKR(FM)**— 1998: 91.3 mhz; 21 kw vert. Ant 564 ft TL: N33 23 12 W97 33 57. Hrs open: 24 5617 Diamond Oaks Dr. S., Fort Worth, 76117. Phone: (817) 831-9130. E-mail: kdkr@csnradio.com Web Site: www.kdkr.org. Licensee: CSN International (group owner; acq 7-12-2000). Format: Positive easy gospel, relg. ♦Chris Rohloff, gen mgr; Stephanie Rohloff, progmg dir.

KRNB(FM)— Aug 15, 1968: 105.7 mhz; 100 kw. 492 ft TL: N32 11 11 W98 17 26. (CP: Ant 1,673 ft.). Stereo. Hrs opn: 24 c/o KKDA(AM), Grand Prairie, 75053. Phone: (972) 263-9911. Fax: (972) 558-0010. Web Site: www.krnb.com. Licensee: Service Broadcasting Group LLC. (acq 2-28-95; FTR: 5-22-95). Population served: 75,000 Natl. Network: ABC. Format: Adult contemp. ♦Hymen Childs, pres; Chuck Smith, gen mgr; Vick Romanick, gen sls mgr; Liz Leos, prom dir; Sam Weaver, progmg dir; Sam Putney, news dir; Gary Wachter, chief of engrg; Tommy Page, traf mgr.

Del Mar Hills

KVOZ(AM)— Apr 15, 1952: 890 khz; 10 kw-D, 1 kw-N, DA-N. TL: N27 32 57 W99 22 21. Hrs open: Box 252, McAllen, 78505. Phone: (956) 781-5528. Phone: (956) 686-6382. Fax: (956) 686-2999. Web Site: www.laradiochristiana.com. Licensee: Consolidated Radio Inc. (acq 3-27-97). Population served: 3,600,000 Format: Sp gospel. Target aud: 18 plus. ♦Paulino Bernal, gen mgr; Eloy Bernal, stn mgr; Pete Guzman, opns mgr.

Del Rio

***KDLI(FM)**— 2007: 89.9 mhz; 1 kw. Ant 171 ft TL: N29 25 24 W100 54 21. Hrs open:
Rebroadcasts WAFR(FM) Tupelo, MS 100%.
Box 2440, Tupelo, MS, 38803-2440. Phone: (662) 844-8888. Fax: (662) 842-6791. Web Site: www.afr.net. Licensee: American Family Association. Format: Contemp Christian. ♦Marvin Sanders, gen mgr.

KDLK-FM—Listing follows KTJK(AM).

KTDR(FM)— Mar 31, 1986: 96.3 mhz; 100 kw. 490 ft TL: N29 32 25 W101 07 21. Stereo. Hrs opn: 24 307 E. 8th St., 78840. Phone: (830) 775-6291. Phone: (830) 775-6291. Fax: (830) 775-6545. E-mail: production@themix96.com Web Site: www.themix96.com. Licensee: Grande Broadcasting of Del Rio Inc. Population served: 43,000 Law Firm: Borsari & Paxson. Format: Adult contemp. News: 1 hr wkly. Target aud: 25-54; male. Spec prog: Relg 3 hrs wkly. ♦Frank Mendoza, pres & gen mgr; Chris Russell, gen sls mgr & disc jockey; Charlene Duncan, chief of engrg & traf mgr; Rodney Lyman, progmg mgr, news dir & chief of engrg; Margaritta Martinez, rsch dir.

KTJK(AM)— 1947: 1230 khz; 860 w-U. TL: N29 25 45 W100 54 17. Hrs open: 24 Box 1489, 78841-1489. Phone: (830) 775-9583. Fax: (830) 774-4009. Web Site: www.ktjk.com. Licensee: Forum Broadcasting Inc. (acq 12-10-02; with co-located FM). Population served: 50,000 Format: Tejano music format. Target aud: 25-54. ♦Larry Mariner, pres, pres & gen mgr; Rudy Briones, opns mgr; Jay Gonzalez, progmg mgr.

KDLK-FM—Co-owned with KTJK(AM). Aug 15, 1966: 94.1 mhz; 18 kw. Ant 276 ft TL: N29 25 45 W100 54 17. Stereo. 24 Web Site: www.kdlk.com. Natl. Network: Westwood One. Format: Country. Target aud: 18 plus. ♦Helen Mariner, traf mgr.

KWMC(AM)— Aug 20, 1967: 1490 khz; 1 kw-U. TL: N29 22 17 W100 51 55. Hrs open: 24 903 E. Cortinas St., 78840. Phone: (830) 775-3544. Fax: (830) 775-3546. E-mail: kwmc1490@wcsonline.net Licensee: Minerva Garza Valdez. Population served: 37,000 Rgnl. Network: Texas State Net. Wire Svc: NOAA Weather Format: Rock oldies. News staff: 2. Spec prog: Relg 5 hrs wkly. ♦Alfredo Garza, pres, gen mgr & chief of engrg; Minerva Garza-Valdez, VP; Guillermo Garza, stn mgr, prom dir, progmg dir & traf mgr; Javier Martinez, gen sls mgr.

Del Valle

KIXL(AM)—Licensed to Del Valle. See Austin

Denison

KJIM(AM)—See Sherman

***KYFB(FM)**— Jan 19, 2007: 91.5 mhz; 4.5 kw. Ant 220 ft TL: N33 42 10 W96 34 05. Hrs open: 11530 Carmel Commons Blvd., Charlotte, NC, 28226-3976. Phone: (704) 523-5555. Fax: (704) 522-1967. Web Site: www.bbnradio.org. Licensee: Bible Broadcasting Network Inc. Format: Relg. ♦Lowell L. Davey, pres.

Denton

KFZO(FM)— September 1988: 99.1 mhz; 100 kw. Ant 1,168 ft TL: N33 23 22 W97 33 53. Stereo. Hrs opn: 24 7700 Carpenter Fwy., Dallas, 75247. Phone: (214) 525-0400. Phone: (214) 630-8531. Fax: (214) 689-3818. Fax: (214) 631-1196 (sales). Web Site: www.kick991.com. Licensee: KHCK-FM License Corp. Group owner: Univision Radio (acq 9-22-2003; grpsl). Format: Sp, Tejano. News staff: one; News: one hr wkly. Target aud: 25-54; affluent/educated adults. ♦Frank Carter, gen mgr; Andy Lockridge, opns dir; Howard Toole, sls dir; Cipriano Robles, gen sls mgr; Betsy Galleguillos, natl sls mgr; Oscar Espinosa, prom dir; Frank "Pancho" Gonzales, progmg dir; Myrna Vera, mus dir & rsch dir; Patrick Parks, chief of engrg; Mirentxu Smith, traf mgr; Claudia Torrescano, pub svc dir.

KHKS(FM)— 1947: 106.1 mhz; 100 kw. 1,584 ft TL: N32 35 22 W96 58 10. Stereo. Hrs opn: 24 14001 N. Dallas Parkway, Suite 300, Dallas, 75240. Phone: (214) 866-8000. Fax: (214) 866-8588. Web Site: www.1061kissfm.com. Licensee: AMFM Texas Licenses L.P. Group owner: Clear Channel Communications Inc. (acq 8-30-00; grpsl). Population served: 700,000 Law Firm: Reed, Smith, Shaw & McClay. Format: CHR. News staff: one. Target aud: 18-49. ♦J.D. Freeman, gen mgr; Jeff Mitchell, gen sls mgr; Kelly Parker, natl sls mgr; Shawn McCalister, rgnl sls mgr; Steve Lee, mktg dir; Sarah Hannon, prom dir; Patrick Davis, progmg dir; Louis Sutton, engrg dir & chief of engrg; Sylvia Sanchez, traf mgr.

Devine

KRPT(FM)— Nov 17, 1982: 92.5 mhz; 50 kw. Ant 492 ft TL: N28 55 32 W99 02 53. Stereo. Hrs opn: 24 6222 N.W. IH 10, San Antonio, 78201. Phone: (210) 736-9700. Fax: (210) 735-8811. Web Site: www.925theoutlaw.com. Licensee: CCB Texas Licenses L.P. Group owner: Clear Channel Communications Inc. (acq 10-2-98; $1.5 million). Population served: 70,000 Format: Texas music. ♦Tom Glade, gen mgr.

Diboll

KAFX-FM—Licensed to Diboll. See Lufkin

KSML(AM)—Licensed to Diboll. See Lufkin

Dilley

KLMO-FM— 2001: 98.9 mhz; 92 kw. Ant 722 ft TL: N28 56 34 W99 16 47. Hrs open: Dilley Broadcasters, 115 West Ave. D, Robstown, 78320. Phone: (210) 532-9858. Fax: (361) 289-7722. Licensee: Dilley Broadcasters. Format: Sp var.

KVWG-FM— March 1984: 95.3 mhz; 100 w. Ant 121 ft TL: N28 40 23 W99 10 08. Stereo. Hrs opn: Box K, 78061. Licensee: Pearsall Radio Works Ltd. Population served: 37,000

Dimmitt

KDHN(AM)— Dec 22, 1963: 1470 khz; 500 w-D, 149 w-N. TL: N34 35 11 W102 18 35. Hrs open: 704 W. Cleveland St., 79027. Phone: (806) 647-4161. Fax: (806) 647-4715. E-mail: kdhn@highplains.net Licensee: Collins Communications Co. (acq 12-12-84). Population served: 45,000 Format: C&W, relg, Sp. News: 8 hrs wkly. Target aud: General. Spec prog: Sp 17 hrs wkly. ♦Wayne Collins, pres & gen mgr.

KNNK(FM)—Licensed to Dimmitt. See Hereford

Doss

***KGLF(FM)**—Not on air, target date: unknown: Stn currently dark. 88.1 mhz; 6 kw. Ant 328 ft TL: N30 22 22 W99 05 02. Hrs open: 103 Rare Eagle Ct., Austin, 78734. Phone: (512) 608-0486. Licensee: Legacy Austin Broadcasting Foundation Inc. (acq 10-15-2003; $100,000 for CP). ♦Rob Hand, gen mgr.

Dripping Springs

***KLLR(FM)**— 2007: 91.9 mhz; 1.1 kw. Ant 472 ft TL: N30 11 53 W98 00 45. Hrs open:
Rebroadcasts KLVR(FM) Santa Rosa, CA 100%.
5700 West Oaks Blvd., Rocklin, CA, 95765. Phone: (916) 251-1600. Fax: (916) 251-1650. Web Site: www.klove.com. Licensee: Educational Media Foundation. Natl. Network: K-Love. Format: Contemp Christian. ♦Richard Jenkins, pres; Mike Novak, VP & progmg dir; Lloyd Parker,

gen mgr; Ed Lenane, opns dir & news dir; Keith Whipple, dev dir; Eric Allen, natl sls mgr; David Pierce, progmg mgr; Jon Rivers, mus dir; Sam Wallington, engrg dir; Arthur Vassar, traf mgr; Karen Johnson, news rptr; Marya Morgan, news rptr; Richard Hunt, news rptr.

KXBT(FM)— 1984: 104.9 mhz; 2.35 kw. Ant 531 ft TL: N30 11 54 W98 00 46. Stereo. Hrs opn: 24 2211 I H 35, Suite 401, Austin, 78741. Phone: (512) 416-1100. Fax: (512) 416-8205. Licensee: BMP Austin License Company L.P. Group owner: Amigo Broadcasting L.P. (acq 11-9-2004; grpsl). Law Firm: Kenkel & Associates. Format: Sp contemp. Target aud: 25-54; upscale, retired, affluent. ♦Pedro Gasc, gen mgr.

Dublin

KSTV-FM— Aug 15, 1968: 93.1 mhz; 7.9 kw. 580 ft TL: N32 11 12 W98 17 44. Hrs open: 24 Box 289, 3209 W. Washington (Dublin Hwy.), Stephenville, 76401. Phone: (254) 968-2141. Fax: (254) 968-6221. E-mail: kstv@htcomp.net Web Site: www.377net.com. Licensee: CCR-Stephenville III LLC. Group owner: Cherry Creek Radio LLC (acq 6-24-2004; grpsl). Format: Country. ♦Robert Elliot, gen mgr; Robert Haschke, gen sls mgr; Tony Hart, progmg dir; Nyki Wyatt, news dir; Justin McClure, chief of engrg; Troy Stark, traf mgr.

Dumas

KDDD-FM— June 29, 1960: 95.3 mhz; 3 kw. 260 ft TL: N35 51 51 W101 55 45. Stereo. Hrs opn: 24 Box 555, 79029. Secondary address: 408 N. Dumas Ave. 79029. Phone: (806) 935-4141. Fax: (806) 935-3836. Licensee: PBI LLC (acq 6-29-2007; $20,000 for 50% ownership interest with co-located AM). Population served: 20000 Natl. Network: ABC. Format: Oldies. News staff: one; News: 13 hrs wkly. Target aud: 25-65; farmers, community, factory. Spec prog: Farm 5 hrs, gospel 5 hrs wkly. ♦Kandi Bray, gen mgr, gen sls mgr & traf mgr; Steve Bayless, opns mgr & progmg dir; Ali Allison, news dir; Stephen White, chief of engrg.

KDDD(AM)— May 1, 1948: 800 khz; 250 w-D. TL: N35 51 42 W101 55 50.6 AM-sunset E-mail: kddd@amaonline.com 19,771 Format: Country. Target aud: General. ♦Candy Bray, traf mgr.

Eagle Pass

***KEPI(FM)**— May 13, 1995: 88.7 mhz; 1 kw. 180 ft TL: N28 39 26 W100 25 00. Hrs open: 24 Box 873, 2477 El Indio Hwy., 78853. Phone: (830) 757-0887. Fax: (830) 757-8950. E-mail: kepi@hcjb.org Licensee: World Radio Network Inc. Population served: 150,000 Format: Contemp Christian. ♦Amado Rodriguez, stn mgr; Saulo Alonzo Rodriguez, progmg mgr; Gary Lawson, chief of engrg.

KEPS(AM)— August 1957: 1270 khz; 1 kw-D. TL: N28 43 45 W100 29 30. Hrs open: Box 1123, 78852. Secondary address: 127 Kilowatt Dr. 78852. Phone: (830) 773-9247. Fax: (830) 773-9500. Licensee: Rhattigan Broadcasting (Texas) LP (group owner; (acq 8-19-2004; grpsl). Format: Tejano. Target aud: 18-49; middle income, Texas-born Hispanics. ♦Rosa T. De La Garza, gen mgr; Rosa T. De La Garza, gen sls mgr; Jose Perez, progmg dir; Mario Martinez, news dir; Gary Graham, chief of engrg.

KINL(FM)—Co-owned with KEPS(AM). Nov 2, 1971: 92.7 mhz; 20 kw. 184 ft TL: N28 43 57 W100 29 34. Stereo. Format: Oldies 60, 70, 80. ♦Cesar Galindo, progmg dir.

***KEPX(FM)**— Sept 9, 1994: 89.5 mhz; 52 kw. 256 ft TL: N28 39 26 W100 25 00. Stereo. Hrs opn: 24 Box 873, 2477 El Indio Hwy., 78853. Secondary address: Box 3333, McAllen 78853. Phone: (830) 757-0895. Phone: (830) 758-0895. Fax: (830) 757-8950. E-mail: kepx@hcjb.org Web Site: kepx.net. Licensee: World Radio Network Inc. Population served: 500,000 Law Firm: Bryan Cave. Format: Sp, Christian. News: 3 hrs wkly. Target aud: Hispanic; Mexican. ♦Amado Rodriguez, VP & stn mgr.

Eastland

KATX(FM)— Sept 1, 1986: 97.7 mhz; 3 kw. Ant 203 ft TL: N32 23 47 W98 46 26. Stereo. Hrs opn: 24 611 W. Commerce, 76448. Phone: (254) 629-2621. Fax: (254) 629-8520. E-mail: radio@txol.net Licensee: Partnership Broadcasting Inc. (acq 2-1-2000; with co-located AM). Format: Country, talk. News staff: one; News: 8 hrs wkly. Target aud: 25-54; adults. ♦David Bacon, pres; Chuck Statler, exec VP.

KEAS(AM)—Co-owned with KATX(FM). August 1953: 1590 khz; 500 w-D. TL: N32 23 47 W98 46 26.6 am-sunset 3,178 Natl. Network: Westwood One. News staff: one; News: 8 hrs wkly. Target aud: 25-54. Spec prog: Gospel 5 hrs wkly.

Edinburg

KBFM(FM)— February 1972: 104.1 mhz; 100 kw. 990 ft TL: N26 05 59 W97 50 16. Stereo. Hrs opn: 24 901 E. Pike St., Weslaco, 78596. Phone: (956) 973-9202. Fax: (956) 973-9355. E-mail: kbfmm@aol.com Web Site: www.b104.net. Licensee: Capstar TX L.P. Group owner: Clear Channel Communications Inc. (acq 8-15-00; grpsl). Population served: 750,000 Natl. Rep: Christal. Format: CHR. News staff: one. Target aud: 18-34; females. Spec prog: Community affrs. ♦Danny Fletcher, VP & gen mgr; Billy Santiago, opns mgr; Cyndi Torres, rgnl sls mgr; Bobby Macias, mus dir; Gloria Garcia, traf mgr & disc jockey; Ken Meek, chief of engrg & disc jockey.

***KOIR(FM)**— Feb 5, 1983: 88.5 mhz; 3 kw. 285 ft TL: N26 07 49 W98 10 51. Stereo. Hrs opn: 24 4300 S. Business Hwy. 281, 78539-9699. Phone: (956) 380-8100. Phone: (956) 380-3435. Fax: (956) 380-8156. E-mail: correo@radioesparanza.com Licensee: Rio Grande Bible Institute Inc. Population served: 10,000 Law Firm: Bryan Cave. Wire Svc: UPI Format: Relg, educ, Sp. Target aud: General. ♦Gerardo Lorenzo, gen mgr & progmg dir; Jerry Jeske, chief of engrg.

KURV(AM)— October 1947: 710 khz; 1 kw-U, DA-2. TL: N26 19 43 W98 09 35. Hrs opn: 24 1201 N. Jackson Rd., Suite 900, McAllen, 78501. Phone: (956) 383-2777. Fax: (956) 383-2570. E-mail: talk@kurv.com Web Site: www.kurv.com. Licensee: BMP RGV License Co. L.P. Group owner: Border Media Partners LLC (acq 1-9-2004; $7.5 million with KSOX(AM) Raymondville). Population served: 800,000 Natl. Network: CBS. Rgnl. Network: Texas State Net. Law Firm: Gammon & Grange. Format: News/talk, sports. News staff: 2; News: 30 hrs wkly. Target aud: 35-64. Spec prog: Farm 10 hrs wkly. ♦Lance Hawkins, VP & gen mgr; Jim Hearn, opns dir & farm dir; Jane Smith, sls dir; Fred Alfaro, mktg dir & prom dir; Jeff Koch, progmg dir; Tim Sullivan, news dir; Joe Espinosa, chief of engrg.

KVLY(FM)— 1974: 107.9 mhz; 100 kw. 765 ft TL: N26 15 01 W97 55 21. Stereo. Hrs opn: 801 Jackson Rd., McAllen, 78501. Phone: (956) 661-6000. Fax: (956) 661-6082. Licensee: Entravision Holdings L.L.C. Group owner: Entravision Communications Corp. (acq 7-20-00; grpsl). Population served: 750,000 Format: Adult contemp Spanish. Target aud: 25-54. ♦Willie Rosales, gen mgr, gen mgr, stn mgr, opns mgr & gen sls mgr; Alex Duran, progmg dir; Lilly Lopez, mus dir; Shirley Kennedy, news dir; Sonny Cavazos, chief of engrg; Dora Borjas, traf mgr.

Edna

KEZB(FM)— Sept. 20, 1998: 96.1 mhz; 13 kw. Ant 456 ft TL: N29 06 05 W96 27 19. Hrs open: 102 Jason Plaza #2, Victoria, 77901. Phone: (361) 572-0105. Fax: (361) 579-4105. Web Site: www.texasthunderradio.net. Licensee: Fort Bend Broadcasting Co. Inc. Group owner: Fort Bend Broadcasting Co. (acq 1-21-00; grpsl). Format: Country. ♦Ryan Henderson, gen mgr.

KTMR(AM)— July 28, 1980: 1130 khz; 10 kw-D, DA. TL: N29 01 40 W96 40 05. Hrs open: 1302 N. Shepherd Dr., Houston, 77008. Phone: (713) 868-9137. Fax: (713) 868-9631. E-mail: docarango@houston.rr.com Licensee: SIGA Broadcasting Corp. (group owner; acq 5-4-99; $333,750). Format: Tejano. Spec prog: Black 3 hrs, Sp 12 hrs wkly. ♦Gabriel Arango, gen mgr.

El Campo

KIOX-FM— September 1968: 96.9 mhz; 100 kw. 981 ft TL: N29 05 44 W96 27 25. (CP: TL: N28 53 35 W96 21 40). Hrs open: 3000 Bering Dr., Houston, 77057. Web Site: www.laraza.fm. Licensee: Liberman Broadcasting of Houston License LLC. Group owner: Liberman Broadcasting Inc. (acq 10-11-2002; $3.15 million with KXGJ(FM) Bay City). Law Firm: Pepper & Corazzini. Format: Hot country. Target aud: 18-49; college, plant workers, business & agriculture. ♦Cheryl Kirk, gen mgr; Tim Michaels, opns mgr.

KULP(AM)— 1948: 1390 khz; 500 w-D, 180 w-N. TL: N29 12 34 W96 15 50. Hrs open: 6 AM-10 PM Box 390, 77437. Secondary address: 515 E. Jackson St. 77437. Phone: (979) 543-3303. Fax: (979) 543-1546. E-mail: kulp@kulpradio.com Web Site: www.kulpradio.com. Licensee: Wharton County Radio Inc. (acq 5-2-00; $240,000). Population served: 13,500 Format: Classic country, news/talk, sports. News staff: 2; News: 5 hrs wkly. Target aud: 25 plus. Spec prog: Sp 14 hrs, Czech 5 hrs wkly. ♦Bob Buckalew, pres & VP; Mike Wenglar, CFO & engrg dir; Jerry Aulds, gen mgr & sls VP; Stephen Zetsche, opns VP; Clint Robinson, opns mgr, mus dir & chief of engrg; Kate Manrriquez, prom mgr & pub affrs dir; Bob Nason, news dir.

El Paso

KAMA(AM)— July 13, 1972: Stn currently dark. 750 khz; 10 kw-D, 1 kw-N, DA-1. TL: N31 46 21 W106 16 56. Hrs open: 24 South Bldg. 300, 2211 E. Missouri, 79903. Phone: (915) 544-9797. Fax: (915) 544-1247. Web Site: www.netmio.com. Licensee: Tichenor Media System Inc. Group owner: Univision Radio (acq 9-22-2003; grpsl). Population served: 462,000 Format: Sp, oldies. News staff: one; News: 8 hrs wkly. Target aud: 25-54; women. ♦MacHenry Tichenor Jr., pres; Domingo Lopez, gen mgr; Margie LaFluer, rgnl sls mgr; Pedro Skaggs, progmg dir; Karla Hernandez, news dir; Andrew Kiska, chief of engrg; Maria Martinez, traf mgr.

KBNA(AM)— June 1947: 920 khz; 1 kw-D, 360 w-N, DA-N. TL: N31 45 41 W106 26 14. Hrs open: 2211 E. Missouri, Suite 5300, 79903. Phone: (915) 544-9797. Fax: (915) 544-1247. Web Site: www.netmio.com. Licensee: Tichenor License Corp. Group owner: Univision Radio (acq 9-22-2003; grpsl). Population served: 1,000,000 Format: Sp. ♦MacHenry Tichenor Jr., pres; Domingo Lopez, gen mgr; Margie LaFluer, gen sls mgr; Leo Lugo, prom mgr; Mario Castillo, progmg dir; Karla Hernandez, news dir; Andrew Kiska, chief of engrg; Maria Martinez, traf mgr.

KBNA-FM— Aug 15, 1969: 97.5 mhz; 100 kw horiz, 48 kw vert. 1,088 ft TL: N31 47 34 W106 28 47. Stereo. Web Site: www.netmio.com.550,000 ♦Mario Castillo, progmg mgr; Maria Martinez, traf mgr.

KELP(AM)— Apr 10, 1959: 1590 khz; 5 kw-D, 800 w-N. TL: N31 46 12 W106 25 37. (CP: TL: N31 44 38 W106 23 45). Stereo. Hrs opn: 24 6900 Commerce, 79915. Phone: (915) 779-0016. Fax: (915) 779-6641. E-mail: info@kelppradio.com Web Site: www.kelpradio.com. Licensee: McClatchey Broadcasting. (acq 2-14-84; $590,000; FTR: 1-30-84). Population served: 1,500,000 Natl. Network: Salem Radio Network. Format: Christian talk, information. Target aud: 25-54; Christian community of El Paso, Las Cruces, Northern Mexico. Spec prog: Spanish 12 hrs wkly. ♦Arnold McClatchey, pres; Craig Rice, gen mgr & progmg; Joe Olivas, gen sls mgr & sls.

KHEY(AM)— Aug 22, 1929: 1380 khz; 5 kw-D, 500 w-N. TL: N31 45 42 W106 24 36. Hrs open: 24 4045 N. Mesa, 79902. Phone: (915) 351-5400. Fax: (915) 351-3102. Web Site: www.khey1380.com. Licensee: CCB Texas Licenses L.P. Group owner: Clear Channel Communications Inc. (acq 5-29-98; $10.5 million with co-located FM). Population served: 680,000 Rgnl. Network: Texas AP. Natl. Rep: Clear Channel. Format: Sports. Target aud: General. ♦Bill Struck, gen mgr; Karen Daniels-Pearson, gen sls mgr; Chris Lucy, mktg dir; Frank Rodriquez, prom dir; Paul Whittler, progmg dir & pub affrs dir; Enrique Lopez, chief of engrg; Julie Bustillos, traf mgr.

KTSM-FM—Co-owned with KHEY(AM). June 11, 1962: 99.9 mhz; 87 kw. 1,820 ft TL: N31 48 19 W106 28 57. Stereo. Natl. Network: CBS. Format: Adult contemp. ♦Bill Clifton Tole, progmg dir; Sam Cassiano, mus dir; Melissa Kerr, pub affrs dir; Enrique Lopez, engrg dir; Krystal Watkins, traf mgr.

KHEY-FM—Listing follows KTSM(AM).

KHRO(AM)— Feb 11, 2004: 1650 khz; 8.5 kw-D, 850 w-N. TL: N31 45 13 W106 24 58. Hrs open: 5426 N. Mesa St., 79912-5421. Phone: (915) 581-1126. Fax: (915) 532-4970. Licensee: Entravision Holdings LLC. Group owner: Entravision Communications Corp. Format: Talk. ♦David Candelaria, gen mgr; Joe Garcia, opns mgr; Phil Gabbard, sls dir; Leo Lugo, prom dir; Jim Lotspeich, chief of engrg; Susan Fleming, traf mgr.

KINT-FM— July 4, 1975: 93.9 mhz; 96.2 kw. 1,420 ft TL: N31 47 36 W106 28 50. Hrs open: 5426 N. Mesa St., 79912-5421. Phone: (915) 581-1126. Fax: (915) 585-4611. Licensee: Entravision Communications Co. L.L.C. Group owner: Entravision Communications Co. L.L.C. (acq 6-4-97; grpsl). Natl. Rep: Lotus Entravision Reps LLC. Format: Sp, adult contemp. News: 4 hrs wkly. Target aud: 25-54. ♦David Candelaria, gen mgr; Joe Garcia, opns mgr; Phil Gabbard, gen sls mgr & natl sls mgr; Leo Lugo, prom mgr; Abel Rodriguez, pub affrs dir; Ron Haney, chief of engrg.

KSVE(AM)—Co-owned with KINT-FM. June 1958: 1150 khz; 5 kw-D, 380 w-N. TL: N31 45 15 W106 25 11. Stereo. 24 Natl. Network: ESPN Radio. Natl. Rep: Lotus Entravision Reps LLC. Format: Espanol. Target aud: 18-54.

***KKLY(FM)**— May 1, 1985: 89.5 mhz; 175 w. Ant 1,007 ft TL: N31 47 33 W106 28 48. Hrs open: 24 2351 Sunset Blvd., Suite 170-218, Rocklin, CA, 95765. Phone: (916) 251-1600. Fax: (916) 251-1650. E-mail: klove@klove.com Licensee: Educational Media Foundation. Group owner: EMF Broadcasting (acq 11-18-2002; $1 million). Natl. Network: K-Love. Law Firm: Shaw Pittman. Format: Contemp Christian music. News staff: 3. Target aud: 25-44. ♦Richard Jenkins, pres; Mike Novak, VP; Keith Whipple, dev dir; David Pierce, progmg mgr; Ed Lenane, news dir; Sam Wallington, engrg dir; Karen Johnson, news rptr; Marya Morgan, news rptr; Richard Hunt, news rptr.

KLAQ(FM)—Listing follows KROD(AM).

KOFX(FM)— June 6, 1978: 92.3 mhz; 100 kw. 1,860 ft TL: N31 48 55 W106 29 20. Stereo. Hrs opn: 24 5426 N. Mesa, 79912. Phone: (915) 581-1126. Fax: (915) 532-4970. Web Site: www.923thefox.com. Licensee: Entravision Holdings L.L.C. Group owner: Entravision Communications Co. L.L.C. (acq 10-19-99). Population served: 750,000 Format: Oldies. News staff: 2. Target aud: 25-54; upscale. ♦David Candelaria, gen mgr; Joe Garcia, opns mgr; Phil Gabbard, gen sls mgr; Al Jones, progmg dir; Jim Lotspeich, chief of engrg; Susan Graham, traf mgr.

KPAS(FM)—See Fabens

KPRR(FM)— Dec 5, 1969: 102.1 mhz; 100 kw horiz, 66 kw vert. 1,289 ft TL: N31 47 34 W106 28 47. Stereo. Hrs opn: 24 4045 N. Mesa, 79902. Phone: (915) 351-5400. Fax: (915) 351-3102. Web Site: www.kprr.com. Licensee: CCB Texas Licenses L.P. Group owner: Clear Channel Communications Inc. (acq 5-16-96; grpsl). Population served: 550,000 Natl. Rep: Clear Channel. Format: CHR. Target aud: 18-34. ♦L. Lowry Mays, CEO, chmn & pres; Randall T. Mays, CFO; Bill Struck, VP & gen mgr; Michelle Haston, gen sls mgr; Christopher Lucy, mktg dir; Frank Rodriguez, prom dir; Bobby Ramos, progmg dir; Patti Diaz, news dir & local news ed; Andrea Thomas, pub affrs dir; Enriquez Lopez, chief of engrg; Amy Page, traf mgr.

KROD(AM)— June 1, 1940: 600 khz; 5 kw-U, DA-N. TL: N31 54 56 W106 23 33. Hrs open: 24 4150 Pinnacle, 79902. Phone: (915) 544-9550. Fax: (915) 532-6342. Web Site: www.krod.com. Licensee: Regent Broadcasting of El Paso Inc. Group owner: Regent Communications Inc. (acq 12-1-99; grpsl). Population served: 425,000 Rgnl. Network: Texas State Net. Natl. Rep: D & R Radio. Format: News/talk, sports. News staff: one; News: 3 hrs wkly. Target aud: 25-54; adult listeners who grew up on the roots of rock and roll. Spec prog: Dallas Cowboys football, UTEP sports. ♦Brad Dubow, gen mgr; Steve Kaplowitz, progmg dir; Ron Haney, chief of engrg.

KLAQ(FM)—Co-owned with KROD(AM). Oct 1, 1978: 95.5 mhz; 88 kw. 1,390 ft TL: N31 47 47 W106 28 55. Stereo. Web Site: www.klaq.com. Format: Rock/AOR. Target aud: 18-49; adults who grew up on FM rock and roll. ♦Brad Dubow, gen mgr; Mike Ramey, progmg dir; Ron Haney, chief of engrg.

KSII(FM)— Dec 30, 1975: 93.1 mhz; 100 kw. 1,422 ft TL: N31 47 34 W106 28 47. Stereo. Hrs opn: 24 4150 Pinnacle, Suite 120, 79902. Phone: (915) 544-9300. Fax: (915) 544-9536. Web Site: www.ksiiinfo.com. Licensee: Regent Broadcasting of El Paso Inc. Group owner: Regent Communications Inc. (acq 12-1-99; grpsl). Population served: 462,000 Law Firm: Wilmer, Cutler & Pickering. Format: Hot adult contemp. News staff: one; News: 2 hrs wkly. Target aud: 25-54; 60% male, 40% female. ♦Brad Dubow, gen mgr; Kelly Calvillo, gen sls mgr; Chris Elliot, progmg dir; Diana Rivas, pub affrs dir & traf mgr; Robert King, chief of engrg.

***KTEP(FM)**— Sept 14, 1950: 88.5 mhz; 94 kw. 731 ft TL: N31 47 17 W106 28 46. Stereo. Hrs opn: 24 500 W. University Ave., 79968-0556. Phone: (915) 747-5152. Phone: (915) 880-5837. Fax: (915) 747-5641. E-mail: ktep@utep.edu Web Site: www.ktep.org. Licensee: University of Texas at El Paso. Population served: 600,000 Natl. Network: PRI, NPR. Format: Class, jazz, news. News staff: one; News: 39 wkly. Target aud: 35 plus; college educated, upper-income. Spec prog: Gospel 4 hrs, folk 3 hrs wkly. ♦Dennis Woo, opns dir & mus dir; Joe Torres, dev dir & prom dir; Patrick J. Piotrowaski, gen mgr & progmg VP; Louie Saenz, news dir; Norbert Miles, chief of engrg; Norma Martinez, traf mgr.

KTSM(AM)— 1947: 690 khz; 10 kw-U, DA-2. TL: N31 58 11 W106 21 15. Stereo. Hrs opn: 24 4045 N. Mesa, 79902. Phone: (915) 351-5400. Fax: (915) 351-3102. Web Site: www.ktsmradio.com. Licensee: CCB Texas Licenses L.P. Group owner: Clear Channel Communications Inc. (acq 5-16-96; grpsl). Population served: 550,000 Natl. Rep: Clear Channel. Format: News/talk. Target aud: 25-54; men. Spec prog: Relg 3 hrs, radio health journal one hr, El Paso public forum one hr wkly. ♦L. Lowry Mays, CEO, chmn & pres; Randall T. May, CFO; Bill Struck, VP & gen mgr; Karen Daniels-Pearson, gen sls mgr; Christopher Lucy, mktg dir; Frank Rodriquez, prom dir; Tom Connelly, progmg mgr; Melissa Kerr, news dir; Michael Calderon, pub affrs dir; Enrique Lopez, chief of engrg; Krystal Watkins, traf mgr.

KHEY-FM—Co-owned with KTSM(AM). Aug 1, 1974: 96.3 mhz; 100 kw. 1,390 ft TL: N31 47 47 W106 28 55. Stereo. 24 Web Site: www.khey.com. Natl. Network: ABC. Format: Country. Target aud: 25-54. ♦Michelle Haston, sls dir; Steve Gramzay, progmg dir; Bobby Gutierrez, news dir, pub affrs dir & local news ed; Enrique Lopez, engrg dir; Amy Page, traf mgr.

***KVER(FM)**— Jan 1, 1993: 91.1 mhz; 510 w. 1,118 ft TL: N31 47 34 W106 28 47. Stereo. Hrs opn: 24 Box 12008, 79913-0008. Phone: (915) 544-9190. E-mail: kver@hcjb.org Web Site: www.kver.org. Licensee: World Network Radio Inc. (group owner) Format: Sp, relg, educ. Hispanic. ♦Marcos Barraza, stn mgr; Gracel Calleros, progmg dir.

KVIV(AM)— Dec 3, 1949: 1340 khz; 1 kw-U. TL: N31 46 24 W106 24 52. Hrs open: 4900 Montana Ave., 79903. Phone: (915) 565-2999. Fax: (915) 562-3156. E-mail: radiovictoria@mail.com Licensee: El Paso y Juarez Companerismo-Cristiano (acq 10-20-2006). Population served: 450,000 Law Firm: Dow, Lohnes & Albertson. Format: Sp, relg. Target aud: Mexican-American. ♦Alfonso Cabrera, pres & gen mgr; Jesus Cruz, progmg dir.

KXPL(AM)— Sept 16, 1985: 1060 khz; 10 kw-D. TL: N31 48 41 W106 31 53. Hrs open: 2211 E. Missouri Ave., E-237, 79903-3837. Phone: (915) 587-8822. Fax: (915) 587-8602. E-mail: kxpl1060am@yahoo.com Licensee: New Radio System Inc. (acq 7-21-2004). Format: Sp info/news. ♦Maria Elena Lazo, gen mgr; Jose Camacho, progmg dir & traf mgr; Paul Gregg, chief of engrg.

KYSE(FM)— Nov 29, 1958: 94.7 mhz; 97 kw horiz, 65 kw vert. Ant 1,191 ft TL: N31 47 34 W106 28 47. Stereo. Hrs opn: 24 5426 N. Mesa, 79902. Phone: (915) 581-1126. Fax: (915) 532-4970. Web Site: www.superestrella947.com. Licensee: Entravision Holdings L.L.C. Group owner: Entravision Communications Co. L.L.C. (acq 10-19-99). Natl. Rep: Lotus Entravision Reps LLC. Format: Sp, Top-40. Target aud: 18-34. ♦David Candelaria, gen mgr; Joe Garcia, opns dir; Phil Gabbard, gen sls mgr; Susan Graham, traf mgr.

Eldorado

KLDE(FM)—Not on air, target date: unknown: 104.9 mhz; 6 kw. Ant 302 ft TL: N30 51 55 W100 35 36. Hrs open: Box 1766, Gaylord, MI, 49734. Phone: (989) 732-2341. Fax: (989) 732-6202. Licensee: Darby Advertising Inc. ♦Kent D. Smith, pres & gen mgr.

Electra

***KOLI(FM)**— January 1998: 94.9 mhz; 50 kw. 492 ft TL: N34 05 01 W98 59 29. Hrs open: 4302 Callfield Rd., Wichita Falls, 76308. Phone: (940) 691-2311. Fax: (940) 696-2255. Web Site: www.culumus.com. Licensee: Cumulus Licensing Corp. Group owner: Cumulus Media Inc. (acq 8-10-99; $238,400). Format: Classic Country. ♦Lindy Parr, gen mgr; Brent Warner, opns mgr & progmg dir; Andrea Lewis, gen sls mgr; Jim Russell, news dir; Jeff Chan, chief of engrg; Dana Jameson, traf mgr.

Elgin

KXXS(FM)— Aug 14, 1992: 92.5 mhz; 1.6 kw. Ant 449 ft TL: N30 19 00 W97 20 22. Hrs open: 24 7524 N. Lamar Blvd., Suite 200, Austin, 78752. Phone: (512) 453-1491. Fax: (512) 453-6809. Licensee: BMP Austin License Company L.P. (acq 2-10-2005; grpsl). Natl. Network: CNN Radio. Law Firm: Bechtel & Cole. Format: Sp, Tejano. ♦Pedro Gasc, gen mgr.

Fabens

KPAS(FM)— Mar 24, 1979: 103.1 mhz; 3 kw. 300 ft TL: N31 35 42 W106 11 58. Stereo. Hrs opn: 18 Box 371010, El Paso, 79937. Phone: (915) 851-3382. Licensee: Algie A. Felder. (acq 6-27-86; FTR: 5-12-86). Population served: 2,000,000 Natl. Network: USA. Format: Christian. News: 6 hrs wkly. ♦Algie A. Felder, pres & gen mgr.

Fairfield

KNES(FM)— Dec 1, 1983: 99.1 mhz; 940 w. 500 ft TL: N31 41 52 W96 09 44. Stereo. Hrs open: 24 Box 347, 627 W. Commerce, 75840. Phone: (903) 389-5637. Fax: (903) 389-7172. Web Site: www.knesfm.com. Licensee: J & J Communications Inc. (acq 11-19-90; $209,000; FTR: 12-10-90). Natl. Network: Jones Radio Networks. Rgnl. Network: Texas State Net. Natl. Rep: Riley. Format: Country. News staff: one; News: 6 hrs wkly. Target aud: General. Spec prog: Farm 3 hrs, talk 15 hrs, Black 3 hrs, gospel 3 hrs wkly. ♦Joe Reid, gen mgr, rgnl sls mgr & traf mgr; Buzz Russell, progmg dir; Lester Leediker, chief of engrg.

Falfurrias

KDFM(FM)—Not on air, target date: unknown: 103.3 mhz; 3 kw. 328 ft TL: N27 15 28 W98 07 07. Hrs open: Box 252, McAllen, 78505. Phone: (956) 686-6382. Phone: (956) 686-2992. Fax: (956) 686-2999. Licensee: La Radio Cristiana Network Inc. Format: Sp.

KLDS(AM)— Jan 1, 1953: 1260 khz; 500 w-D, 330 w-N. TL: N27 14 11 W98 10 22. Hrs open: 6 AM-midnight Box 401, 78355. Secondary address: 215 W. Adam St. 78355. Phone: (361) 325-1212. Fax: (361) 325-1212. Licensee: The Evangelistic Worship Center (acq 10-27-97; $75,000). Population served: 250,000 Law Firm: Baraff, Koerner & Olender. Format: Christian. Target aud: General. ♦Timothy Trevino, gen mgr & progmg dir; Steve Cantu, chief of engrg.

KPSO-FM— Nov 1, 1983: 106.3 mhz; 6 kw. Ant 184 ft TL: N27 14 11 W98 10 22. Stereo. Hrs opn: 6 AM-10 PM 304 E. Rice St., 78355-3624. Phone: (361) 325-2112. Fax: (361) 325-2112. E-mail: kpso@awesomenet.net Licensee: Brooks Broadcasting Corp. Population served: 100,000 Law Firm: Koerner & Olender. Format: Tejano (Sp), country. News: 15 hrs wkly. Target aud: All groups. ♦Raymond O. Creely, gen mgr & chief of engrg; Steve Cantu, exec VP, gen sls mgr & spanish dir.

Fannett

***KZFT(FM)**— Oct 31, 2003: 90.5 mhz; 35 kw vert. Ant 361 ft TL: N29 53 33 W94 08 06. Hrs open: Drawer 3206, Tupelo, MS, 38803. Phone: (662) 844-8888. Fax: (662) 842-6791. Licensee: American Family Association. Group owner: American Family Radio. Format: Christian. ♦Marvin Sanders, gen mgr.

Farmersville

KFCD(AM)— November 1947: 990 khz; 7 kw-D, 920 w-N, DA-2. TL: N33 07 01 W96 16 47. Stereo. Hrs opn: 24 Box 12345, Dallas, 75225. Phone: (972) 354-1990. Fax: (972) 354-0820. Licensee: Bernard Dallas LLC (acq 12-28-2006; $9 million with KHSE(AM) Wylie). Population served: 385,200 Natl. Network: CNN Radio. Rgnl. Network: Texas State Net. Format: Talk. Target aud: 35 plus; men. ♦Jerry Overton, gen mgr; Dave Marcum, opns mgr, progmg mgr & pub affrs dir; Ed Wodka, rgnl sls mgr & adv mgr; Dave Schum, chief of engrg; Leslie Cooke, traf mgr.

KXEZ(FM)— Sept 1, 1998: 92.1 mhz; 1.95 kw. Ant 584 ft TL: N33 16 31 W96 22 02. Stereo. Hrs opn: Box 940670, Plano, 75094. Phone: (903) 482-6750. Phone: (972) 396-1640. Fax: (972) 396-1643. E-mail: josh@khiy.com Licensee: Metro Broadcasters-Texas Inc. (acq 12-3-98). Natl. Network: Jones Radio Networks. Format: Classic country. ♦Ken Jones, CEO, pres, gen mgr & chief of engrg; Glenda Jones, CFO; Jack Bishop, opns dir & pub affrs dir; Joshua Jones, sls dir, gen sls mgr, mktg VP, mktg dir & prom dir; Hal Mayfield, news dir; Ron Eudaly, engrg dir.

Farwell

KICA-FM— Sept 15, 1984: 98.3 mhz; 50 kw. Ant 223 ft TL: N34 23 22 W103 10 27. (CP: 100 kw, ant 384 ft. TL: N34 29 36 W103 23 46).

Stereo. Hrs opn: 24 1000 Sycamore St., Clovis, NM, 88101. Phone: (505) 762-6200. Fax: (505) 762-8800. Licensee: Tallgrass Broadcasting LLC. (group owner; (acq 4-2-2007; grpsl). Population served: 80,000 Format: World class rock. News staff: one; News: 3 hrs wkly. Target aud: 18-49. ♦Dana Taylor, progmg dir & sports cmtr; Shannon Phillips, mus dir & traf mgr.

KIJN(AM)— Apr 17, 1958: 1060 khz; 10 kw-D, DA. TL: N34 23 14 W103 01 51. Hrs open: Box 458, 79325. Phone: (806) 481-3318. Fax: (806) 481-3835. E-mail: kijn@email.com Licensee: Metropolitan Radio Group Inc. (group owner; acq 9-97; with co-located FM). Population served: 150,000 Format: Christian music, relg, Sp Christian. Target aud: General. ♦Mike Rodriquez, gen mgr; David Pollard, progmg dir.

KIJN-FM— Aug 1, 1985: 92.3 mhz; 100 kw. Ant 354 ft TL: N34 32 26 W102 47 56. Stereo. 24 150,000

KMUL(AM)— July 6, 1956: 830 khz; 1.1 kw-D, 10 w-N. TL: N34 29 42 W103 23 39. Hrs open: 600 W 8th St., Muleshoe, 79347-3330. Phone: (806) 272-4273. Phone: (806) 272-4087. Fax: (806) 272-5067. Licensee: Tallgrass Broadcasting LLC. (group owner; (acq 4-2-2007; grpsl). Population served: 4,525 Format: Sp. Spec prog: Farm 3 hrs wkly. ♦Noe Anzaldua, gen mgr & progmg dir; Martha Alvarado, progmg dir; Rick Keefer, chief of engrg.

Ferris

KDFT(AM)— July 13, 1988: 540 khz; 1 kw-D, 249 w-N, DA-2. TL: N32 30 47 W96 34 28 (D), N32 30 52 W96 34 26 (N). Hrs open: 24 3304 W. Camp Wisdom, Suite 100, Dallas, 75237. Phone: (972) 572-1540. Fax: (972) 572-1263. E-mail: kdft-kmny@mrbi.net Licensee: Way Broadcasting Licensee LLC (acq 4-19-2000; grpsl). Format: Sp Christian. Target aud: 25-59; Sp. ♦Arthur Liu, pres; Francisco Martinez, CFO; John Gabel, VP; Ted Sauceman, gen mgr; Juan Benitez, opns mgr.

Floresville

***KJMA(FM)**— 1993: 89.7 mhz; 9 kw. Ant 138 ft TL: N29 13 55 W98 03 05. Hrs open: 1905 10th St., 78114. Phone: (830) 393-6116. Fax: (830) 393-3817. Web Site: www.grnonline.com. Licensee: La Promesa Foundation. (acq 6-25-2007; $130,000). Population served: 45,000 Natl. Network: EWTN Radio. Format: Catholic radio. ♦Leonard J. Oswald, pres; Cissy Gonzalez, gen mgr.

KTFM(FM)— June 15, 1977: 94.1 mhz; 40 kw. 548 ft TL: N29 11 03 W98 30 49. Stereo. Hrs opn: 24 7800 N.W. I-10, Suite 330, San Antonio, 78230. Phone: (210) 340-1234. Fax: (210) 340-1775. Licensee: BMP San Antonio License Co. L.P. Group owner: Border Media Partners LLC (acq 12-23-2003; $24.4 million with KSAH(AM) Universal City). Population served: 1,400,000 Law Firm: Bechtel & Cole. Format: Mexican rgnl. News staff: one; News: 15 hrs wkly. Target aud: 18-49. ♦Raul Rodriguez, gen mgr; Peggy McCormack, gen sls mgr; Lupe Contreras, prom dir; Manny Herrera, progmg dir; Minnie Ochoa, pub affrs dir; Brett Hudkins, chief of engrg; Rosie Pagan, traf mgr.

Flower Mound

KTYS(FM)— Oct 23, 2006: 96.7 mhz; 92 kw. 2,034 ft TL: N33 26 13 W97 29 05. Stereo. Hrs opn: 24 2221 E. Lomar Blvd., Suite 400, Arlington, 76006. Secondary address: 3405 Loy Lake Rd. 76006. Phone: (817) 695-3500. Fax: (817) 695-3516. Web Site: www.967thetwister.com. Licensee: Radio License Holding IV LLC. Group owner: ABC Inc. (acq 6-12-2007; . grpsl). Population served: 132,000 Natl. Network: ABC. Natl. Rep: McGavren Guild. Format: Country. News: one hr wkly. Target aud: 18-34; adults. ♦Keri Korzeniewski, gen mgr; David Klement, gen sls mgr; Greg Heitzman, natl sls mgr; Robert Shiflet, mktg dir; Jim Bucek, prom dir; Chris Huff, mus dir & disc jockey; Neal Peden, chief of engrg; Arty Watkins, traf mgr.

Floydada

KFLP(AM)— 1951: 900 khz; 250 w-D. TL: N33 58 20 W101 21 00. Hrs open: Box 658, 79235. Phone: (806) 983-5704. Fax: (806) 983-5705. E-mail: kflp@kflp.net Web Site: www.kflp.net. Licensee: Anthony L. Ricketts. (acq 8-4-99; with co-located FM). Population served: 20,000 Natl. Network: USA. Natl. Rep: Interep. Format: Farm. News staff: one; News: 12 hrs wkly. Farmers. ♦Tony St. James, gen mgr & progmg dir.

KFLP-FM— Apr 1, 1985: 106.1 mhz; 25 kw. 233 ft TL: N33 58 07 W101 21 15. Stereo. 24 Web Site: www.kflp.net.20,000 Natl. Rep: Interep. Format: Country. Target aud: 18-49; adults.

Fort Stockton

KFST(AM)— May 8, 1954: 860 khz; 250 w-U. TL: N30 52 37 W102 53 30. Hrs open: 24 954 S US Hwy. 385, 79735. Phone: (432) 336-2228. Phone: (432) 336-5834. Fax: (432) 336-5834. E-mail: kfst@ftstockton.net Licensee: Fort Stockton Radio Co Inc. (acq 1-1-86). Population served: 226,422 Rgnl. Network: Texas State Net. Wire Svc: NOAA Weather Format: Adult contemp, relg. News staff: 2; News: 6 hrs wkly. Target aud: General. ♦Ken Ripley, gen mgr, dev dir, progmg dir, chief of engrg & local news ed.

KFST-FM— November 1974: 94.3 mhz; 3 kw. 236 ft TL: N30 52 37 W102 53 30. Stereo. Format: Country, Sp.

Fort Worth

KBFB(FM)—See Dallas

KDMX(FM)—See Dallas

KDXX(FM)—See Benbrook

KEGL(FM)— April 1959: 97.1 mhz; 99 kw. Ant 1,666 ft TL: N32 35 19 W96 58 05. Stereo. Hrs opn: 24 14001 N. Dallas Pkwy., Suite 300, Dallas, 75240. Phone: (214) 866-8000. Licensee: Citicasters Licenses L.P. Group owner: Clear Channel Communications Inc. (acq 5-4-99; grpsl). Population served: 517,600 Format: Sp. ♦J.D. Freeman, gen mgr; Pat McMahon, opns mgr.

KFJZ(AM)— Feb 15, 1947: 870 khz; 500 w-D. TL: N32 45 42 W97 18 49. Hrs open: 8828 N. Stemmons Fwy., Suite 106, Dallas, 75247. Phone: 214) 634-7780. Fax: (214) 634-7523. E-mail: sarita@radioluz.com Licensee: BMP DFW License Co. L.P. (acq 3-11-2005; $2.5 million). Population served: 3,600,000 Format: Sp. ♦Bob Prouse, gen mgr.

KFLC(AM)— 1922: 1270 khz; 50 kw-D, 5 kw-N, DA-2. TL: N32 43 36 W97 11 30. Stereo. Hrs opn: 24 7700 Carpenter Fwy., Dallas, 75247. Phone: (214) 525-0400. Fax: (214) 631-1196. Web Site: www.univision.com. Licensee: KESS-AM License Corp. Group owner: Univision Radio (acq 9-22-2003; grpsl). Population served: 4,000,000 Format: Sp, news/talk, sports. News staff: 2; News: 11 hrs wkly. Target aud: 25-54. ♦Frank Carter, gen mgr; Andy Lockridge, opns dir; Cipriano Robles, sls dir; Ivonne Flaherty, gen sls mgr; Karen Hocking, natl sls mgr; Oscar Espinosa, prom dir; Herminio (Chayan) Ortuno, progmg dir; Myrna Vera, news dir & rsch dir; Patrick Parks, engrg mgr & chief of engrg; Mirentxu Smith, rsch dir & traf mgr; Claudia Torrescano, pub svc dir.

KFXR(AM)—See Dallas

KHVN(AM)— Dec 6, 1946: 970 khz; 1 kw-D, 270 w-N. TL: N32 47 56 W97 17 43. Hrs open: 24 5787 S. Hampton Rd., Dallas, 75232. Phone: (214) 331-5486. Fax: (214) 331-1908. E-mail: traffic@khvanam.com Web Site: www.khvnam.com. Licensee: Mortenson Broadcasting Co. of Texas Inc. (group owner; (acq 5-31-2002; $4.5 million with KNAX(AM) Fort Worth). Population served: 600,000 Natl. Rep: Interep. Format: Gospel. News staff: one; News: 10 hrs wkly. Target aud: 25-54. ♦Jack Mortenson, CEO & VP; Dion Mortenson, gen mgr.

KJKK(FM)—See Dallas

KKDA-FM—See Dallas

KKGM(AM)— 2002: 1630 khz; 10 kw-D, 1 kw-N. TL: N32 48 35 W97 07 24. Hrs open: 24 hrs 5787 S. Hampton Rd., Suite 108, Dallas, 75232. Phone: (214) 337-5700. Fax: (214) 337-5707. E-mail: traffic@kkgmam.com Web Site: www.kkgmam.com. Licensee: Mortenson Broadcasting Co. of Texas Inc. (group owner; (acq 5-31-2002; with KHVN(AM) Fort Worth). Population served: 6,000,000 Format: Southern gospel, ministry, sports. Target aud: 30-64. ♦Lon Sosh, gen mgr; Jack Davis, progmg dir; Mike Price, chief of engrg.

KLIF(AM)—See Dallas

KLLI(FM)—See Dallas

KLNO(FM)— Dec 24, 1964: 94.1 mhz; 100 kw. Ant 1,585 ft TL: N32 35 22 W96 58 10. Stereo. Hrs opn: 24 7700 Carpenter Hwy., Dallas, 75247. Phone: (214) 525-0400. Fax: (214) 525-0473. Fax: (214) 631-1196 (sales). Licensee: HBC License Corp. Group owner: Univision Radio (acq 9-22-2003; grpsl). Population served: 250000 Natl. Network: ABC. Format: Mexican regional. ♦Frank Carter, gen mgr; Andy Lockridge, opns dir; Cipriane Robles, sls dir; Ivonne Flaherty, gen sls mgr; Karen Hecking, natl sls mgr; Oscar Espinosa, prom dir; Herminio (Chayan) Ortuno, progmg dir; Patrick Parks, chief of engrg; Myrna Vera, rsch dir; Mirentxu Smith, traf mgr; Claudia Torrescano, pub svc dir.

KLTY(FM)—See Dallas

KLUV(FM)—See Dallas

KMVK(FM)— Feb 8, 1965: 107.5 mhz; 16.5 kw. Ant 1,883 ft TL: N32 35 02 W96 57 48. Stereo. Hrs opn: 24 7901 Carpenter Fwy., Dallas, 75247. Phone: (214) 526-9870. Fax: (214) 905-5052. Licensee: Texas CBS Radio Broadcasting L.P. Group owner: CBS Radio (acq 6-26-96; grpsl). Format: Smooth Jazz. Target aud: 25-54. ♦Julie Davis, CEO & traf mgr; Mel Karmazin, pres; David Henry, gen mgr; Rick Frisch, gen sls mgr; Dave Dillon, natl sls mgr; Liz Balon, prom dir; Kurt Johnson, progmg dir; Mark Sanford, mus dir; Bill Taylor, news dir & chief of engrg; Vance Henley, engrg mgr.

KPLX(FM)— Dec 15, 1962: 99.5 mhz; 100 kw. 1,680 ft TL: N32 34 54 W96 58 32. Stereo. Hrs opn: 24 3500 Maple Ave., Suite 1600, Dallas, 75219. Phone: (214) 526-2400. Fax: (214) 520-4343. Web Site: www.995thewolf.com. Licensee: KPLX Lico Inc. Group owner: Susquehanna Radio Corp. (acq 1974). Population served: 4,418,400 Natl. Network: AP Network News. Format: Country. News staff: one; News: 4 hrs wkly. Target aud: 25-54; loyal listeners throughout the day. ♦Dan Bennett, pres & mktg mgr; Jim Quirk, sls dir; Rob Chickering, chief of engrg.

KRLD(AM)—See Dallas

KRVA(AM)—See Dallas

KSCS(FM)—Listing follows WBAP(AM).

KSKY(AM)—See Dallas

***KTCU-FM**— Oct 6, 1964: 88.7 mhz; 3 kw. 320 ft TL: N32 42 40 W97 22 00. Stereo. Hrs opn: 6 AM-1 AM Box 298020, Moudy Bldg., Texas Christian Univ., 76129. Phone: (817) 257-7631. Phone: (817) 257-7634. Fax: (817) 257-7637. E-mail: ktcu@tcu.edu Web Site: www.ktcu.tcu.edu. Licensee: Board of Trustees Texas Christian University. Population served: 1,000,000 Format: Rock, class, jazz. ♦Russell Scott, gen mgr.

KZPS(FM)—See Dallas

WBAP(AM)— May 2, 1922: 820 khz; 50 kw-U. TL: N32 36 38 W97 10 00. Stereo. Hrs opn: 24 2221 E. Lamar, Suite 300, Arlington, 76006. Phone: (817) 695-1820. Fax: (817) 695-0014. Web Site: www.wbap.com. Licensee: Radio License Holding IV LLC. Group owner: ABC Inc. (acq 6-12-2007; grpsl). Population served: 3,500,000 Natl. Rep: ABC Radio Sales. Format: News/talk. News staff: 7; News: 37 hrs wkly. Target aud: 25-54. Spec prog: Dallas Stars, farm 6 hrs wkly. ♦Pete Dits, gen mgr & gen sls mgr; Stephanie Calahan, gen sls mgr; Bob Shomper, progmg dir; Neal Reden, chief of engrg.

KSCS(FM)—Co-owned with WBAP(AM). Mar 8, 1949: 96.3 mhz; 99 kw. 1,610 ft TL: N32 35 15 W96 57 59. Stereo. 24 Web Site:

www.kscs.com.3,000,000 Natl. Rep: ABC Radio Sales. Format: Country. ♦Greg Heitzman, natl sls mgr; Robert Shiflet, mktg dir; Lorrin Palagi, progmg dir.

Fort Worth-Dallas

KDGE(FM)— Apr 10, 1962: 102.1 mhz; 100 kw. Ant 1,591 ft TL: N32 34 54 W96 58 32. Stereo. Hrs opn: 24 14001 N. Dallas Pkwy., Suite 300, Dallas, 75240. Phone: (214) 866-8000. Fax: (214) 866-8101. Web Site: www.kdge.com. Licensee: Capstar TX L.P. Group owner: Clear Channel Communications Inc. (acq 8-30-2000; grpsl). Population served: 3,000,000 Natl. Rep: CBS Radio. Format: Alternative rock. News staff: one; News: 5 hrs wkly. Target aud: 20-44. ♦Brenda Adriance, gen mgr; Tracy Martin Taylor, gen mgr; John Roberts, opns mgr.

Franklin

KZTR(FM)— Nov 7, 1994: 101.9 mhz; 25 kw. Ant 328 ft TL: N30 56 04 W96 26 15. Hrs opn: 24 Box 3069, Bryan, 77802. Phone: (979) 776-1240. Fax: (979) 776-0123. Licensee: Brazos Valley Communications Ltd. (group owner; (acq 7-18-2006; grpsl). Format: Adult contemp. ♦Chuck Knuth, gen mgr & gen sls mgr; Dan Ginzel, opns mgr; Ron Elliott, progmg dir; Gary Graham, engrg dir; Jeanna Vickery, traf mgr.

Frankston

KKBM(AM)— 2004: Stn currently dark. 890 khz; 250 w-D, 2 w-N. TL: N32 01 52 W95 40 07. Hrs open: Box 60991, Palo Alto, CA, 94306. Licensee: JNE Investments Inc.

KOYE(FM)— June 15, 1970: 92.3; 50 kw. Ant 492 ft TL: N32 02 22 W95 24 39. Stereo. Hrs opn: 24 Box 7820, Tyler, 75711. Secondary address: 621 Chase, Tyler 75701. Phone: (903) 581-9966. Fax: (903) 534-5300. E-mail: request@koye.com Web Site: www.koye967.com. Licensee: Access.1 Texas License Company LLC. Group owner: Waller Broadcasting (acq 1-7-2005; grpsl). Population served: 300,000 Natl. Rep: McGavren Guild. Format: Rgnl Mexican. Target aud: 18-49. ♦Rick Guest, gen mgr; Genni Causey, gen sls mgr; Robert Taylor, natl sls mgr; Jessie Duron, progmg dir.

Fredericksburg

***KHCF(FM)**—Not on air, target date: unknown: 91.5 mhz; 3.1 kw. Ant 394 ft TL: N30 11 49 W98 38 19. Hrs open: 2424 South Blvd., Houston, 77098-5196. Phone: (713) 520-5200. Licensee: Houston Christian Broadcasters Inc. ♦Bruce Munsterman, pres.

KNAF(AM)— November 1947: 910 khz; 1 kw-D, 174 w-N. TL: N30 17 12 W98 52 58. Hrs open: Box 311, 78624. Secondary address: 210 Woodcrest 78624. Phone: (830) 997-2197. Fax: (830) 997-2198. E-mail: texasrebelradio@fbgn Licensee: J. & J. Fritz Media Ltd. (group owner; (acq 1-23-91; FTR: 2-11-91). Population served: 200,000 Rgnl. Network: Texas State Net. Format: Country, full service, talk. Spec prog: Farm 5 hrs, Polka 4.5 hrs wkly. ♦Jayson Fritz, pres & gen mgr; Jan Fritz, sr VP; Arziana Carruth, prom dir; Rick Star, progmg dir; Holley Day, mus dir.

KNAF-FM— 2005: 105.7 mhz; 9.1 kw. Ant 538 ft TL: N30 21 49 W98 54 47. Stereo. E-mail: txradio@ktc.com (group owner). Format: Country. Target aud: 18-54. ♦Jayson Fritz, engrg VP.

Freeport

KBRZ(AM)— October 1952: 1460 khz; 500 w-D, 214 w-N. TL: N28 58 59 W95 20 00. Hrs open: 912 Curtis Ave., Pasadena, 77502. Phone: (713) 589-1336. Fax: (713) 589-1335. Licensee: Aleluya Christian Broadcasting Inc. (acq 3-1-01; $700,000). Population served: 85,000 Format: Sp Christian. ♦Ruben Villarreal, gen mgr.

KJOJ-FM— 1987: 103.3 mhz; 100 kw. Ant 994 ft TL: N28 48 57 W95 36 03. Hrs open: 3000 Bering Dr., Houston, 77057. Fax: (713) 315-3565. Web Site: www.laraza.fm. Licensee: Liberman Broadcasting of Houston License LLC. Group owner: Liberman Broadcasting Inc. (acq 3-20-2001; grpsl). Population served: 4,000,000 Format: Rgnl Mexican. ♦Lenard Liberman, CEO & pres; Winter Horton, VP; Gerardo Reyes, gen sls mgr; Cheque Gonzalez, progmg dir; Meliza Posada, traf mgr.

Freer

KBRA(FM)— Jan 18, 1985: 95.9 mhz; 190 w horiz. Ant 466 ft TL: N27 51 17 W98 35 49. Hrs open: 850 Brandon Ave., Jackson, MS, 39209. Phone: (361) 394-6959. Phone: (601) 906-0836. Fax: (601) 420-4114. Licensee: Cobra Broadcasting Co. L.L.C. Population served: 5,000 Target aud: 18 plus.

***KPBN(FM)**— 2004: 90.7 mhz; 700 w. Ant 312 ft TL: N27 48 55 W98 41 45. Hrs open: Box 252, McAllen, 78505. Phone: (956) 686-6382. Fax: (956) 686-2999. Licensee: Paulino Bernal Evangelism. Format: Sp, Christian. ♦Paulino Bernal Jr., pres.

Friona

KGRW(FM)— Nov 1, 1994: 94.7 mhz; 50 kw. Ant 492 ft TL: N34 41 17 W102 56 53. Hrs open:
Rebroadcasts KQFX(FM) Borger 100%.
3639 Wolfin Ave., Amarillo, 79102. Phone: (806) 355-1044. Fax: (806) 457-0642. Licensee: Tejas Broadcasting Ltd. LLP. Group owner: Amigo Broadcasting L.P. (acq 11-15-2004; grpsl). Format: Sp, Tejano. Target aud: 25-54; working class Texas born Hispanic audience. ♦Matt Douglas, gen mgr; Brad Gonzalez, gen sls mgr; Israel Salazar, progmg dir; Charles Singleton, chief of engrg; Emilia Chacon, traf mgr.

Frisco

KATH(AM)— October 1936: 910 khz; 1 kw-D, 500 w-N, DA-2. TL: N33 12 55 W96 53 56. Hrs open: 8828 N. Stemmons Fwy., Suite 106, Dallas, 75247. Phone: (214) 634-7780. Fax: (214) 634-7523. Licensee: Chatham Hill Foundation Inc. Group owner: Amigo Broadcasting L.P. (acq 12-13-2006; grpsl). Format: Sp, sports. Target aud: Ages 18-44; Hispanic sports & music fans. ♦Gus Perez, gen mgr; Fernando Gonzalez, sls dir; Arturo Canizalez, progmg dir; Adriana Balero, traf mgr.

Gainesville

KGAF(AM)— 1947: 1580 khz; 250 w-U, DA-N. TL: N33 37 42 W97 06 25. Hrs open: Box 368, Radio Hill Rd., 76241. Phone: (940) 665-5546. Fax: (940) 665-1580. E-mail: kgafadvertising@ntin.net Licensee: First IV Media Inc. (acq 11-15-74). Population served: 16,210 Format: C&W, news/talk. Spec prog: Farm 2 hrs, sports 3 hrs wkly. ♦Linda Roller, dev dir; Shelley Carson, gen sls mgr; Jody Shotwell, progmg dir; Tom Carson, pres, gen mgr & news dir; Frank Bonner, chief of engrg; Brian Cooper, sports cmtr, disc jockey & disc jockey; Clay Corbin, disc jockey; Darren Allred, disc jockey.

KSOC(FM)—Licensed to Gainesville. See Dallas

Galveston

KGBC(AM)— May 1947: 1540 khz; 1 kw-D, 250 w-N, DA-N. TL: N29 18 51 W94 48 16. Hrs open: 6 AM-10:30 PM 1302 N. Shepherd Dr., Houston, 77008. Phone: (409) 744-1540. Licensee: SIGA Broadcasting Corp. (group owner; acq 5-9-2002; $900,000). Population served: 150,000 Format: Catholic talk. News staff: 2; News: 15 hrs wkly. Target aud: 30 plus; general. ♦Gabriel Arango, pres; Dave Lane, gen mgr, opns dir, progmg dir & news dir; Sylvia Arango, VP & stn mgr.

KHCB(AM)—Licensed to Galveston. See Houston

KOVE-FM— July 2001: 106.5 mhz; 100 kw. Ant 1,322 ft TL: N29 24 40 W94 57 04. Stereo. Hrs opn: 24 5100 Southwest Fwy., Houston, 77056. Phone: (713) 965-2300. Fax: (713) 965-2401. Web Site: www.univision.com. Licensee: HBC License Corp. Group owner: Univision Radio (acq 9-22-2003; grpsl). Population served: 4,000,000 Format: Sp super hits. News staff: one; News: 3 hrs wkly. Target aud: 18-49; assimilated Hispanics. ♦Mark Masepohl, sr VP & VP; Dave Burdette, stn mgr; Arnulfo Ramirez, opns dir; Marie Barden, gen sls mgr; Kim Mercier, natl sls mgr; Frances Jones, prom dir; Nestor Enriquez, prom dir & prom mgr; Angel Basulto, progmg dir; Renzo Heredia, pub affrs dir; Marty Scruggs, engrg dir; Carole Van Matre, rsch dir; Claudetta Wallace, traf mgr.

Ganado

KULF(FM)— Nov 26, 1997: 104.7 mhz; 50 kw. 459 ft TL: N28 55 37 W96 46 54. Stereo. Hrs opn: 24 102 Jason Plaza, Suite 1, Victoria, 77901. Phone: (361) 572-0105. Fax: (361) 579-4105. E-mail: ego@lonestarfm.com Web Site: www.lonestarfm.com. Licensee: Fort Bend Broadcasting Co. (group owner; (acq 5-10-2001; $1.5 million). Format: Country. Target aud: 25-64; skews males. ♦Egon Barthels, gen mgr.

Gardendale

KFZX(FM)— Jan 9, 1984: 102.1 mhz; 100 kw. Ant 984 ft TL: N31 57 55 W102 46 10. Stereo. Hrs opn: 24 1330 E. 8th St., Suite 207, Odessa, 79761. Phone: (915) 563-9102. Fax: (915) 580-9102. Licensee: GAP Broadcasting Midland-Odessa License LLC. Group owner: Clear Channel Communications (acq 8-3-2007; grpsl). Population served: 200,000 Format: Classic rock. Target aud: 25-54. ♦Gloria Apolinario, gen mgr & gen sls mgr; Steve Driscoll, progmg dir; Jesse Grimes, news dir; Amy Parker, pub affrs dir; Rodney Norris, chief of engrg; Shelly Todd, traf mgr.

Garland

KAAM(AM)—Licensed to Garland. See Dallas

Gatesville

***KYAR(FM)**— Apr 6, 1976: 98.3 mhz; 200 w. Ant 239 ft TL: N31 27 07 W97 42 14. Stereo. Hrs opn: 24 2351 Sunset Blvd., Suite 170-218, Rocklin, CA, 95765. Phone: (916) 251-1600. Fax: (916) 251-1650. Licensee: Educational Media Foundation. Group owner: EMF Broadcasting (acq 3-21-2003; $100,000). Population served: 130,000 Natl. Network: Air 1. Rgnl. Network: Texas State Net. Law Firm: Shaw Pittman. Format: Christian. News staff: 3. Target aud: 25-44; Judeo Christian, female. ♦Richard Jenkins, pres; Mike Novak, VP; Keith Whipple, dev dir; David Pierce, progmg mgr; Ed Lenane, news dir; Sam Wallington, engrg dir; Karen Johnson, news rptr; Marya Morgan, news rptr; Richard Hunt, news rptr.

Georgetown

KHFI-FM— Mar 1, 1972: 96.7 mhz; 100 kw. 951 ft TL: N30 19 20 W97 48 03. Stereo. Hrs opn: 24 3601 S. Congress Ave., #F, Austin, 78704-7213. Phone: (512) 684-7300. Fax: (512) 684-7441. Web Site: www.967kissfm.com. Licensee: CCB Texas Licenses L.P. Group owner: Clear Channel Communications Inc. (acq 3-9-93; $3.5 million; FTR: 3-29-93). Population served: 900,000 Natl. Rep: Clear Channel. Format: Hit rock. News staff: one; News: one hr wkly. Target aud: 18-49; adult women. ♦Dusty Black, gen mgr; Laura Cullen, sls dir; Angie Hancock, prom dir; Jay Shannon, progmg dir.

KINV(FM)— Oct 31, 1991: 107.7 mhz; 25 kw. Ant 508 ft TL: N30 42 17 W97 38 32. Hrs open: 24 57 W. South Temple, Suite 107, Salt Lake City, UT, 84101. Phone: (512) 419-1077. Fax: (512) 340-7169. Web Site: www.netmio.com. Licensee: Univision Radio License Corp. Group owner: Univision Radio (acq 9-22-2003; grpsl). Law Firm: Jones, Waldo, Holbrook & McDonough. Format: Sp, adult hits. Target aud: 18-34; male.

Giddings

***KANJ(FM)**— Oct 28, 1999: 91.5 mhz; 8 kw. 335 ft TL: N30 09 56 W96 52 16. Hrs open: 24 hrs
Rebroadcasts KHCB-FM Houston 95%.
2424 South Blvd., Houston, 77098. Phone: (713) 520-5200. Web Site: www.khcb.org. Licensee: Houston Christian Broadcasters Inc. (group owner) Natl. Network: Moody. Format: Christian. ♦Bruce Munsterman, gen mgr.

Gilmer

KFRO-FM— July 24, 1980: 95.3 mhz; 5.9 kw. Ant 666 ft TL: N32 37 50 W94 53 44. Stereo. Hrs opn: 24 Box 5818, Longview, 75608. Secondary address: 481 Loop 281 E., Longview 75608. Phone: (903) 663-3700. Fax: (903) 663-9458. Web Site: www.wallerbroadcasting.com. Licensee: Walller Media LLC. Group owner: Waller Broadcasting (acq 6-15-98; $1.425 million with KFRO(AM) Longview). Population served: 300,000 Natl. Rep: Roslin. Law Firm: Kaye, Scholer, Fierman, Hays & Handler. Format: Oldies. Target aud: 25-49. Spec prog: Jazz 5 hrs wkly. ♦Dudley Waller, CEO, pres & gen mgr; Richard Guest, COO; Debbie Tilley, CFO; Robert Taylor, gen sls mgr; Dru Laborde, progmg dir; Sans Hawkins, engrg dir; Shelley Miller, traf mgr.

Gladewater

KEES(AM)— 1947: 1430 khz; 5 kw-D, 1 kw-N, DA-N. TL: N32 31 46 W94 52 50. Hrs open: 1001 E. Southeast Loop 323, Suite 455, Tyler,

75701-9600. Phone: (903) 593-2519. Fax: (903) 597-8378. Licensee: Gleiser Communications LLC (group owner; acq 11-21-03; grpsl). Population served: 25,000 Natl. Network: Westwood One. Natl. Rep: Riley. Format: Talk. ♦Paul Gleiser, gen mgr & gen sls mgr; Mike LaRoux, chief of engrg; Deborah Harrington, traf mgr.

Glen Rose

KTFW-FM— 1989: 92.1 mhz; 25 kw. 1,417 ft TL: N32 16 31 W98 01 22. Stereo. Hrs opn: Box 1629, 919 N. Main, Cleburne, 76033. Phone: (817) 645-6643. Fax: (817) 645-6644. E-mail: info@countrygoldradio.com Web Site: www.countrygoldradio.com. Licensee: LKCM Radio Group L.P. (group owner; (acq 1-13-2006; $10,142,816). Population served: 1,300,000 Rgnl. Network: Texas State Net. Law Firm: Fletcher, Heald & Hildreth. Format: C&W. Target aud: 40 plus. ♦Gerry Schlegel, pres; George Marti, VP; Gary Moss, gen mgr & chief of engrg; Mike Crow, stn mgr, progmg dir & progmg mgr; Norma Savage, gen sls mgr; Mary Montanez, mus dir & traf mgr.

Goldsmith

KTXO(FM)—Not on air, target date: unknown: 94.7 mhz; 6 kw. Ant 328 ft TL: N31 52 02 W102 39 18. Hrs open: 5842 Westslope Dr., Austin, 78731. Phone: (512) 467-0643. Licensee: Matinee Radio LLC. ♦Robert Walker, pres.

Goliad

KHMC(FM)— 1995: 95.9 mhz; 25 kw. Ant 321 ft TL: N28 40 57 W97 18 50. Hrs open: 24 Box 407, Victoria, 77902. Phone: (361) 575-9533. Fax: (361) 575-9502. E-mail: majictejano@yahoo.com Licensee: Cinco de Mayo Broadcasting. Population served: 180,000 Format: Tejano/Sp. ♦Homer Lopez, gen mgr; Ralph Salezar, gen sls mgr.

Gonzales

KCTI(AM)— Dec 17, 1947: 1450 khz; 1 kw-U. TL: N29 30 35 W97 24 51. Hrs open: 615 St. Paul St., 78629. Phone: (830) 672-3631. Fax: (830) 672-9603. E-mail: kcti@gvec.net Web Site: www.kcti1450.com. Licensee: Gonzales Communications, a Texas L.P. (acq 3-1-95; with co-located FM; FTR: 5-22-95). Population served: 110000 Rgnl. Network: Texas State Net. Format: Country. Target aud: General. Spec prog: Polka 5 hrs, farm 5 hrs, Sp 6 hrs wkly. ♦Marina Mann, CEO & pres; Joe Haynes, gen mgr, opns VP, gen sls mgr, mktg dir, adv dir, progmg dir & disc jockey; L.D. Decker, news dir & disc jockey; Bill Wolebon, chief of engrg; Aaron Allen, disc jockey; John Zavadil, disc jockey.

***KMLR(FM)**— March 1986: 106.3 mhz; 15 kw. Ant 423 ft TL: N29 41 17 W97 40 39. Hrs open: 2351 Sunset Blvd., Suite 170-218, Rocklin, CA, 95765. Phone: (916) 251-1600. Fax: (916) 251-1650. Web Site: www.klove.com. Licensee: Educational Media Foundation. (acq 3-31-2006; $6 million with KYLR(FM) Hutto). Natl. Network: K-Love. Format: Contemp Christian. ♦Richard Jenkins, pres; Mike Novak, VP & progmg dir; Lloyd Parker, gen mgr; Ed Lenane, opns dir & news dir; Keith Whipple, dev dir; Eric Allen, natl sls mgr; David Pierce, progmg mgr; Jon Rivers, mus dir; Sam Wallington, engrg dir; Arthur Vassar, traf mgr; Karen Johnson, news rptr; Marya Morgan, news rptr; Richard Hunt, news rptr.

***KZAR(FM)**—Not on air, target date: unknown: 88.1 mhz; 5 kw. Ant 328 ft TL: N29 30 48 W97 25 44. Hrs open: 2351 Sunset Blvd., Suite 170-218, Rocklin, CA, 95765. Phone: (916) 251-1600. Fax: (916) 251-1650. Licensee: Educational Media Foundation. (acq 2-28-2006; $36,000 for CP). ♦Richard Jenkins, pres; Mike Novak, VP; Keith Whipple, dev dir; David Pierce, progmg mgr; Ed Lenane, news dir; Sam Wallington, engrg dir; Karen Johnson, news rptr; Marya Morgan, news rptr; Richard Hunt, news rptr.

Graham

KSWA(AM)— 1948: 1330 khz; 500 w-D. TL: N33 07 37 W98 35 35. Hrs open: 24 Box 1507, 76450. Secondary address: 620 Oak St. 76450. Phone: (940) 549-1330. Fax: (940) 549-8628. E-mail: jm@kwkq-kswa.com Licensee: Graham Newspapers Inc. (group owner; (acq 1996). Population served: 9,300 Natl. Network: ABC. Law Firm: Garvey, Schubert & Barer. Format: Country legends. News staff: one; News: 10 hrs wkly. Target aud: 35 plus; adults. Spec prog: Bluegrass 2 hrs, gospel 2 hrs, Texas mus 2 hrs wkly. ♦Roy Robinson, VP; Joe Graham, gen mgr; Cindy Lewis, gen sls mgr; Greg Tiller, progmg dir; James M. Jones, news dir & chief of engrg; Joe Barker, traf mgr.

KWKQ(FM)—Co-owned with KSWA(AM). August 1975: 94.7 mhz; 10.5 kw. Ant 485 ft TL: N33 02 30 W98 46 44. Stereo. 24 9,300 Natl. Network: ABC. Format: Classic hits. News staff: one; News: 5 hrs wkly. Target aud: 12-49. Spec prog: Alternative 10 hrs wkly. ♦Greg Tiller, progmg dir, disc jockey & disc jockey.

Granbury

KPIR(AM)— Mar 13, 1980: 1420 khz; 500 w-U, DA-2. TL: N32 27 43 W97 47 19. Hrs open: 24 Box 1558, 76049. Phone: (817) 579-7850. Fax: (817) 579-0192. Web Site: www.kpir.com. Licensee: Pirate Broadcasters Inc. (acq 8-13-02). Population served: 250,000 Format: Real country. News staff: one; News: 21 hrs wkly. Target aud: 25-55; general. Spec prog: Farm one hr wkly. ♦Bob Haschke, gen mgr & sls dir; Shayne Hollinger, progmg mgr; Justin McClure, chief of engrg; Sue Haschke, traf mgr.

Grand Prairie

KKDA(AM)— Aug 1, 1957: 730 khz; 500 w-U. TL: N32 45 52 W96 59 36. Hrs open: Box 530860, 75053. Phone: (972) 263-9911. Fax: (972) 558-0010. Web Site: www.k104fm.com. Licensee: Service Broadcasting Group LLC. (acq 12-22-76). Population served: 500,000 Format: Oldies. ♦Hymen Childs, pres; Chuck Smith, gen mgr; Ken Johnson, gen sls mgr; Willis Johnson, progmg dir; Mike Crittender, engrg mgr; Gary Wachter, chief of engrg; Tony Paiage, traf mgr.

Greenville

KESN(FM)—See Allen

KGVL(AM)— Mar 26, 1946: 1400 khz; 1 kw-U. TL: N33 10 02 W96 05 55. Stereo. Hrs opn: Box 1015, 75403. Secondary address: 1517 Wolfe City Dr. 75401. Phone: (903) 455-1400. Phone: (903) 450-1400. Fax: (903) 455-5485. Licensee: Dynamic Broadcasting LLC (acq 1-11-2005; $500,000). Population served: 70,000 Rgnl. Network: Texas State Net. Law Firm: Dow, Lohnes & Albertson. Format: 70, 80, 90 Country. Target aud: 25 plus. Spec prog: Black one hr, farm 5 hrs, relg 6 hrs wkly. ♦Frank Janda, gen mgr; Jim Patrick, progmg dir; Jason Russell, chief of engrg.

KIKT(FM)— Sept 15, 1978: 93.5 mhz; 1.8 kw. Ant 328 ft TL: N33 11 00 W96 03 19. (CP: COL Cooper. 12.4 kw, ant 407 ft. TL: N33 13 16 W95 41 20). Stereo. Hrs opn: 24 Box 1015, 75403. Secondary address: 1517 Wolfe City Dr. 75401. Phone: (903) 450-0935. Phone: (903) 455-1460. Fax: (903) 455-5485. Web Site: www.kiktradio.com. Licensee: KRBE Lico Inc. (acq 9-2-99; with co-located AM). Natl. Network: ABC. Format: Country. News: 15 hrs wkly. Target aud: General. ♦Frank Janda, gen mgr.

***KTXG(FM)**— 2006: 90.5 mhz; 38 kw. Ant 722 ft TL: N33 19 00 W96 24 27. Hrs open:
Rebroadcasts WAFR(FM) Tupelo, MS 100%.
Drawer 2440, Tupelo, MS, 38801. Phone: (662) 844-8888. Fax: (662) 842-6791. Web Site: www.afr.net. Licensee: American Family Association. Format: Chirstian. ♦Marvin Sanders, gen mgr.

Gregory

KPUS(FM)— 1999: 104.5 mhz; 14 kw. Ant 446 ft TL: N27 52 00 W97 13 09. Stereo. Hrs opn: 24 826 S. Padre Island Dr., Corpus Christi, 78416. Phone: (361) 814-3800. Fax: (361) 855-3770. Web Site: classicrock1045.com. Licensee: Convergent Broadcasting Corpus Christi LP. Group owner: Convergent Broadcasting LLC (acq 1-12-2004; grpsl). Population served: 313,600 Format: Classic rock. Target aud: Adults; 25-54. ♦Mark White, gen mgr; Dallas Garcia, gen sls mgr; Scott Holt, opns mgr & progmg dir; Molly Cox, mus dir; Amanda Moreno, traf mgr.

Groves

KCOL-FM— Sept 17, 1983: 92.5 mhz; 50 kw. 440 ft TL: N30 01 45 W93 52 59. Stereo. Hrs opn: 24 Box 5488, Beaumont, 77726. Phone: (409) 896-5555. Fax: (409) 896-5566. Web Site: www.cool925.com. Licensee: Clear Channel Broadcasting Licenses Inc. Group owner: Clear Channel Communications Inc. (acq 1-29-2004; $4.5 million). Population served: 320,000 Format: Oldies. Target aud: 35 plus. ♦John Hogan, CEO; Randall Mays, CFO; Charlie Rahilly, sr VP; Mark Kopelman, VP; Vesta Brandt, gen mgr; Trey Poston, opns dir.

Groveton

KKUL-FM—Not on air, target date: unknown: 98.1 mhz; 6 kw. Ant 328 ft TL: N31 05 18 W94 58 56. Hrs open: 5842 Westslope Dr., Austin, 78731. Phone: (512) 467-0643. Licensee: Matinee Radio LLC. ♦Robert Walker, pres.

Hallettsville

KHLT(AM)— Sept 5, 1979: 1520 khz; 250 w-D. TL: N29 26 38 W96 57 22. Hrs open: Sunrise-sunset 111 N. Main St., 77964. Phone: (361) 798-4333. Fax: (361) 798-3798. E-mail: texasthunderradio@yahoo.com Licensee: Fort Bend Broadcasting Co. Inc. Group owner: Fort Bend Broadcasting Co. (acq 1-21-00; grpsl). Rgnl. Network: Texas State Net. Wire Svc: NWS (National Weather Service) Format: Country. News staff: one; News: 13 hrs wkly. Target aud: General. Spec prog: Farm 5 hrs, Czech 5 hrs, Ger 5 hrs wkly. ♦Laura Kremling, gen mgr, stn mgr & stn mgr; Travis Kremling, progmg dir & chief of engrg.

KTXM(FM)—Co-owned with KHLT(AM). Oct 29, 1997: 99.9 mhz; 6 kw. 131 ft TL: N29 26 38 W96 57 22.
Rebroadcasts KYKM(FM) Yoakum 100%.
Format: Country.

Haltom City

KDBN(FM)— 1995: 93.3 mhz; 50 kw. 276 ft TL: N32 54 44 W97 11 18. Hrs open: 3500 Maple Ave., 13th Fl., Dallas, 75219. Phone: (214) 526-7400. Fax: (214) 525-2525. Web Site: www.933thebone.com. Licensee: Texas Star Radio Inc. Group owner: Susquehanna Radio Corp. (acq 1-28-97). Format: Classic rock. News: 2 hrs wkly. ♦Lew Dickey, CEO, chmn & pres; John Dickey, exec VP; Dan Bennett, VP & gen mgr; Carmen Lee, gen sls mgr; John Winchestor, prom dir; Jerome Fischer, progmg dir; Hue Beavers, chief of engrg.

Hamilton

KCLW(AM)— May 22, 1948: 900 khz; 250 w-D. TL: N31 43 08 W98 08 39. Hrs open: 24 Box 631, 76531. Secondary address: 115 A N. Rice 76531. Phone: (254) 386-8804 phone/fax. E-mail: info@kclw.com Web Site: www.kclw.com. Licensee: Lasting Value Broadcasting Group Inc. (acq 8-22-00; $380,000). Population served: 5,000 Natl. Network: CBS, Jones Radio Networks. Format: Classic country. News staff: one; News: 6 hrs wkly. Target aud: 18-65. Spec prog: Relg 6 hrs, Sp 12 hrs wkly. ♦Meredith Beal, pres; Sammie Casey, gen mgr; Ronald Beal, opns VP.

Hamlin

KCDD(FM)— Jan 30, 1987: 103.7 mhz; 100 kw. 985 ft TL: N32 43 31 W100 04 19. Stereo. Hrs opn: 2525 S. Danville Dr., Abilene, 79605. Phone: (325) 793-9700. Fax: (915) 692-1576. Web Site: www.power103.com. Licensee: Cumulus Licensing Corp. Group owner: Cumulus Media Inc. (acq 2-13-98; grpsl). Format: CHR. ♦Trace Michaels, gen mgr; Brad Elliott, progmg dir; Chris Andrews, chief of engrg.

Harker Heights

KRMY(AM)—See Killeen

KUSJ(FM)—Licensed to Harker Heights. See Temple

Harlingen

KBTQ(FM)—Listing follows KGBT(AM).

KFRQ(FM)— January 1960: 94.5 mhz; 100 kw. 1,158 ft TL: N26 08 55 W97 49 17. Stereo. Hrs opn: 24 801 N. Jackson Rd., McAllen, 78501. Phone: (956) 661-6000. Fax: (956) 661-6082. Web Site: www.kfrq.com. Licensee: Entravision Holdings LLC. (group owner; acq 1996; $6.1 million with KKPS(FM) Brownsville). Population served: 750,000 Format: Adult rock. News staff: one. Target aud: 25-54. ♦Alex Duran, VP, gen mgr & progmg dir.

KGBT(AM)— 1941: 1530 khz; 50 kw-D, 10 kw-N, 50 kw-CH, DA-2. TL: N26 22 33 W97 53 43. Stereo. Hrs opn: 24 200 S. 10th, Suite 600, McAllen, 78501. Phone: (956) 631-5499. Fax: (956) 631-0090. Web Site: www.netmio.com. Licensee: Tichenor License Corp. ("TLC"). Group owner: Univision Radio (acq 9-22-2003; grpsl). Population served: 525,900 Format: Sp. News staff: 2. Target aud: 18 plus. ♦Joe Morales, gen mgr; Hugo Delacruz, progmg dir; Jorge Garza, chief of engrg.

KBTQ(FM)—Co-owned with KGBT(AM). July 1975: 96.1 mhz; 100 kw. Ant 449 ft TL: N26 10 34 W97 46 59. Stereo. Web Site: www.netmio.com.101,500 Format: Tejano. ♦Alex Quintero, progmg dir.

***KMBH-FM**— Apr 30, 1991: 88.9 mhz; 3 kw. Ant 298 ft TL: N26 10 46 W97 30 06. Stereo. Hrs opn: 24 Box 2147, 78551. Secondary address: 1701 Tennessee 78551. Phone: (956) 421-4111. Fax: (956) 421-4150. E-mail: kmbhkhid@aol.com Web Site: www.kmbh.org. Licensee: RGV Educational Broadcasting Inc. Population served: 950,000 Natl. Network: NPR. Format: News, class, jazz. News: 34 hrs wkly. Target aud: General. Spec prog: Sp 3 hrs wkly. ♦Fr. Pedro Briseno, CEO, pres & gen mgr; Chris Maley, progmg dir. Co-owned TV: *KMBH(TV) affil

Haskell

KVRP-FM— Apr 8, 1981: 97.1 mhz; 100 kw. 531 ft TL: N33 09 40 W99 48 57. Stereo. Hrs opn: 24 Box 1118, 1406 N. First, 79521. Phone: (940) 864-8505. Fax: (940) 864-8001. E-mail: gary@kvrp.com Web Site: www.kvrp.com. Licensee: 1 Chronicles 14 L.P. (acq 8-4-2004; $700,000 with KVRP(AM) Stamford). Population served: 45,000 Rgnl. Network: Texas State Net. Natl. Rep: Katz Radio. Format: Country. News: 5 hrs wkly. Target aud: 25 plus. Spec prog: Farm 5 hrs, relg 6 hrs wkly. ♦Greg Weston, pres & VP; Gary Barrett, gen mgr, gen sls mgr & prom mgr; Josh Roysdon, rgnl sls mgr & news dir; Dave Harrison, progmg dir; Tony St. James, farm dir.

Hearne

KVJM(FM)— May 15, 1985: 103.1 mhz; 5 kw horiz, 4.9 kw vert. Ant 361 ft TL: N30 45 35 W96 28 00. Hrs open: Box 3989, Bryan, 77805. Phone: (979) 779-3337. Fax: (979) 779-3444. E-mail: kvjmv103@aol.com Licensee: Equal Access Media Inc. Format: Sp. Target aud: 18-54. ♦Pluria Marshall Jr., gen mgr & sls VP; Edward Sanchez, stn mgr, mus dir, asst music dir, news dir & pub affrs dir; Plyria Marshall Jr., natl sls mgr; Lester Pace, progmg dir; Ed Loftis, chief of engrg.

Hebronville

***KAZF(FM)**— November 2000: 91.9 mhz; 3 kw. 266 ft TL: N27 21 44 W98 40 09. Hrs open: Box 252, McAllen, 78505. Phone: (956) 686-6382. Fax: (956) 686-2999. E-mail: paulinobernal@laradiocristina.com Web Site: www.laradiocristina.com. Licensee: Paulino Bernal Evangelism. Format: Christian contemp. ♦Gilbert Martinez, stn mgr.

KEKO(FM)— 2003: 101.7 mhz; 6 kw. Ant 328 ft TL: N27 18 46 W98 39 51. Hrs open: Box 1614, Laredo, 78044. Phone: (956) 726-4738. Fax: (928) 569-0456. Web Site: www.lacadenaradioluz.com/keko.htm. Licensee: La Nueva Cadena Radio Luz Inc. Format: Sp Christian. ♦Israel Tellez, pres & gen mgr; Hiram Tellez, progmg mgr.

Helotes

KONO-FM— Feb 18, 1971: 101.1 mhz; 98 kw. 1,368 ft TL: N29 50 26 W98 49 32. Stereo. Hrs opn: 24 8122 Datapoint Dr., Suite 500, San Antonio, 78229. Phone: (210) 615-5400. Fax: (210) 615-5300. Web Site: www.kono101.com. Licensee: Cox Radio Inc. Group owner: Cox Broadcasting (acq 2-12-98; $23 million with KONO(AM) San Antonio). Population served: 1,300,000 Natl. Rep: Katz Radio. Wire Svc: AP Format: Oldies. News staff: one; News: one hr wkly. Target aud: 25-54; total audience appeal. ♦Bob Neil, CEO; Marty Choate, VP & gen mgr; Connie Tyra Kremer, gen sls mgr; Jeff Scott, natl sls mgr; Vera Flores, prom dir; Roger Allen, progmg dir; Chrissie Murnin, news dir; Paul Reynolds, chief of engrg; Connye Rodriguez, traf mgr; Dave Griffith, disc jockey; Dave Rios, disc jockey; Steve Casanova, disc jockey; Steve Sellers, disc jockey.

Hemphill

KPBL(AM)— Feb 16, 1978: Stn currently dark. 1240 khz; 1 kw-U. TL: N31 22 03 W93 50 10. Hrs open: R.R. 5 Box 2095, 75948. Licensee: Phillip Burr Broadcasting Co. Population served: 35,000 ♦Phillip Burr, pres & gen mgr.

KTHP(FM)— November 2000: 103.9 mhz;; 6 kw. Ant 243 ft TL: N31 20 28 W93 50 44. (CP: 4.5 kw, ant 377 ft. TL: N31 25 24 W93 50 30). Hrs opn: 605 San Antonio Ave., Many, LA, 71449. Phone: (318) 256-5177. Fax: (318) 256-0950. Licensee: Baldridge-Dumas Communications Inc. (group owner). Format: Classic country. ♦Rhonda Benson, gen mgr.

Hempstead

KTWL(FM)— 1999: 105.3 mhz; 9.2 kw. Ant 544 ft TL: N30 18 19 W96 01 40. Hrs open:
Simulcasts KLTR(FM) Caldwell 100%.
530 W. Main St., Brenham, 77833. Phone: (979) 836-9411. Fax: (979) 836-9435. Licensee: Farmers Communications. Group owner: Bayport Broadcast Group Law Firm: Robert J. Buenzle. Format: Adult comtemp. ♦Roy E. Henderson, CEO & pres; Steve Britewell, engrg VP.

Henderson

KWRD(AM)— March 1956: 1470 khz; 5 kw-D. TL: N32 10 55 W94 47 49. Hrs open: 1101 Kilgore Dr., 75652. Phone: (903) 657-2324. Fax: (903) 657-6221. E-mail: kwrd@tyler.net Licensee: Jerry Russell dba The Russell Co. (acq 5-7-2001; with KOFY(AM) Gilmer). Population served: 50,000 Rgnl. Network: Texas State Net. Format: C&W, Southern gospel. Target aud: General. Spec prog: Farm 5 hrs wkly. ♦Esther Milton, gen mgr; Henry Dunn, stn mgr & opns mgr.

Hereford

KJNZ(FM)— Dec 12, 2000: 103.5 mhz; 50 kw. Ant 249 ft TL: N34 52 10 W102 34 47. Hrs open: 1220 Broadway, Suite 1035, Lubbock, 79401. Phone: (806) 741-0701. Licensee: Tahoka Radio LLC Group owner: The Formby Stations Law Firm: Shaw Pittman. Format: Rgnl Mexican. ♦Chip Formby, gen mgr.

KNNK(FM)—Dimmitt, June 13, 1998: 100.5 mhz; 43 kw. Ant 489 ft TL: N34 44 49 W102 29 37. Stereo. Hrs opn: 24 Box 1635, 207 S. 25-Mile Ave., 79045-9998. Phone: (806) 363-1005. Fax: (806) 364-0226. E-mail: knnk@wtrt.net Web Site: www.knnk.net. Licensee: James D. Peeler. Population served: 253,002 Natl. Network: Moody. Format: Southern gospel, beautiful music. Target aud: General; mature adults. Spec prog: Soft instrumentals 20 hrs wkly. ♦James "Buddy" D. Peeler, gen mgr.

KPAN(AM)— August 1948: 860 khz; 250 w-D, 231 w-N. TL: N34 47 33 W102 25 45. Hrs open: 24 Box 1757, 218 E. 5th St., 79045. Phone: (806) 364-1860. Fax: (806) 364-5814. E-mail: kpan@kpanradio.com Web Site: www.kpanradio.com. Licensee: KPAN Broadcasters. Group owner: Formby Stations Population served: 350,000 Natl. Network: CBS. Rgnl. Network: Texas State Net. Law Firm: Shaw Pittman. Wire Svc: AP Format: Contemp country. News staff: one; News: 20 hrs wkly. Target aud: General. Spec prog: Tejano 15 hrs, farm 12 hrs wkly. ♦Chip Formby, gen mgr.

KPAN-FM— Sept 1, 1965: 106.3 mhz; 30 kw. 259 ft TL: N34 47 33 W102 25 45. Stereo. 24 Web Site: www.kpanradio.com. Natl. Network: CBS Radio. Law Firm: Shaw Pittman. Target aud: General.

***KRLH(FM)**—Not on air, target date: unknown: 90.9 mhz; 250 w. Ant 147 ft TL: N34 51 18 W102 26 07. Hrs open: 5700 West Oaks Blvd., Rocklin, CA, 95765. Phone: (916) 251-1600. Fax: (916) 251-1650. Licensee: Educational Media Foundation. (acq 3-23-2007; grpsl). ♦Richard Jenkins, pres.

***KWDH(FM)**—Not on air, target date: unknown: 88.7 mhz; 28 kw. Ant 230 ft TL: N34 48 48 W102 23 42. Hrs open: 5331 Mt. Alifan Dr., San Diego, CA, 92177-7480. Phone: (858) 277-4991. Fax: (858) 277-1365. Web Site: www.horizonsd.org. Licensee: Horizon Christian Fellowship. (acq 4-7-2006; $150,000 for CP). ♦Mike MacIntosh, pres.

Hewitt

KDRW(FM)—Not on air, target date: unknown: 106.7 mhz; 21.5 kw. Ant 354 ft TL: N31 24 45 W97 12 40. Hrs open: 1126 West Ave., Richmond, VA, 23220. Phone: (804) 422-3452. Licensee: William W. McCutchen III. ♦William W. McCutchen III, gen mgr.

Highland Park

KVCE(AM)— Mar 1, 1960: 1160 khz; 35 kw-D, 1 kw-N, DA-2. TL: N33 10 37 W97 40 36 (D), N33 02 21 W96 56 34 (N). Hrs open: 24 750 N. St. Paul, 10th Fl., Dallas, 75201. Phone: (469) 341-1141. Fax: (469) 341-0111. Web Site: www.kvceradio.com. Licensee: Dallas Broadcasting LLC Group owner: First Broadcasting Investment Partners LLC (acq 8-23-2006; $9.25 million). Population served: 50,000 Format: Talk. ♦Dan Patrick, gen mgr.

KVIL(FM)—Highland Park-Dallas, Aug 14, 1961: 103.7 mhz; 100 kw. Ant 1,571 ft TL: N32 34 54 W96 58 32. Stereo. Hrs opn: 24 4131 N. Central Expwy., Suite 1200, Dallas, 75204. Phone: (214) 526-9870. E-mail: feedback@kvil.com Web Site: www.kvil.com. Licensee: Texas CBS Radio Broadcasting L.P. Group owner: CBS Radio (acq 7-2-87). Population served: 500,000 Natl. Network: CBS. Format: Light rock. ♦David Henry, gen mgr.

Highland Park-Dallas

KVIL(FM)—Licensed to Highland Park-Dallas. See Highland Park

Highland Village

KWRD-FM— Nov 15, 1988: 100.7 mhz; 100 kw. Ant 1,840 ft TL: N33 33 37 W96 57 34. Stereo. Hrs opn: 18 6400 N. Belt Line Rd., Suite 110, Irving, 75063-6037. Phone: (214) 561-9673. Fax: (214) 561-9662. E-mail: theword@thewordfm.com Web Site: www.thewordfm.com. Licensee: Inspiration Media of Texas LLC. Group owner: Salem Communications Corp. (acq 1-17-2001; grpsl). Population served: 38,000 Format: Christian talk. ♦Pete Thomson, gen mgr; David Darling, opns mgr & progmg; Easy Ezell, gen sls mgr; Carol White, prom dir; Brian Heise, chief of engrg.

Hillsboro

KBRQ(FM)— Oct 20, 1959: 102.5 mhz; 100 kw. Ant 449 ft TL: N31 49 23 W97 09 35. Stereo. Hrs opn: 24 314 W. State Hwy. 6, Waco, 76712. Phone: (254) 776-3900. Fax: (254) 761-6371. E-mail: brenthenslee@clearchannel.com Web Site: www.1025thebear.com. Licensee: Clear Channel Broadcasting Licenses Inc. Group owner: Clear Channel Communications Inc. (acq 11-12-2003; $300,000). Law Firm: Fisher, Wayland, Cooper, Leader & Zaragoza. Format: Classic rock. Target aud: 25-49; men. ♦Evan Armstrong, gen mgr; Zack Owen, opns dir; Vernon Riggs, sls dir & gen sls mgr; Brent Henslee, progmg dir.

KHBR(AM)— May 21, 1948: 1560 khz; 250 w-D. TL: N32 01 00 W97 06 32. Hrs open: 12 Box 569, 76645. Secondary address: 335 Country Club Rd. 76645. Phone: (254) 582-3431. Fax: (254) 582-3800. E-mail: info@khbrhillsboro.com Web Site: www.khbrhillsboro.com. Licensee: KHBR Radio Inc. (acq 1955). Population served: 42,750 Rgnl. Network: Texas State Net. Wire Svc: NOAA Weather Format: Classic country. News: 18 hrs wkly. Target aud: General. Spec prog: Czech 1 1/2 hrs, gospel 6 hrs wkly. ♦Roger Galle, pres; Rick Bailey, gen mgr; Roger Creech, progmg dir.

Hondo

KCWM(AM)— Feb 13, 1970: 1460 khz; 500 w-D, 226 w-N. TL: N29 21 42 W99 07 42. Hrs open: 6 AM-10 PM Box 447, 78861. Secondary address: 1605 Ave. K 78861. Phone: (830) 741-5296. Fax: (830) 426-3368. Licensee: Hondo Communications Inc. (acq 10-11-96). Population served: 35,000 Rgnl. Network: Texas State Net. Natl. Rep: Keystone (unwired net). Format: C&W. News staff: one; News: 20 hrs wkly. Target aud: General. ♦Mike Carr, pres, gen mgr & progmg dir; Tim Copeland, gen sls mgr, mktg dir & prom mgr; Paul McKay, chief of engrg; Tom Fusselmen, traf mgr.

KMFR(FM)— 1993: 105.9 mhz; 6 kw. 328 ft TL: N29 18 48 W99 16 03. Stereo. Hrs opn: 8023 Vantage Dr., Suite 840, San Antonio, 78230. Phone: (888) 522-7437. Fax: (210) 341-1777. Licensee: Hondo RadioWorks Ltd. (acq 12-12-2000; $74,925). Format: Classic rock. ♦John W. Barger, CEO.

Hooks

KPWW(FM)— Dec 22, 1985: 95.9 mhz; 11.5 kw. 449 ft TL: N33 27 25 W94 10 59. (CP: 11.3 kw, ant 485 ft.). Stereo. Hrs opn: 24 2324 Arkansas Blvd., Texarkana, 71854. Phone: (870) 772-3771. Fax: (870) 772-0364. Web Site: www.power959.com. Licensee: GAP Broadcasting Texarkana License LLC. Group owner: Clear Channel Communications Inc. (acq 8-3-2007; grpsl). Population served: 200,000 Natl. Rep: McGavren Guild. Format: Modern Top 40. News staff: one. Target aud: 18-49; contemp adults, upscale middle America. ♦Ron Bird, gen mgr; Phil Robken, natl sls mgr; John Williams, news dir; Wes Spicher, progmg dir & chief of engrg; Cindy Esterling, traf mgr.

Hornsby

***KOOP(FM)**— November 1994: 91.7 mhz; 3.13 kw. 85 ft TL: N30 16 00 W97 40 27. (CP: Ant 197 ft.). Stereo. Hrs opn: 9 AM-7 PM (M-F); 9 AM-10 PM (S, Su) Box 2116, Austin, 78768. Secondary address: 304 E. 5th St., Austin 78768. Phone: (512) 472-1369. Phone: (512) 472-5667. Fax: (512) 472-6149. E-mail: info@koop.org Web Site: www.koop.org. Licensee: Texas Educational Broadcasting Inc. Population served: 750,000 Format: Var of music & info. News: 14 hrs wkly. Spec prog: American Indian one hr, Black 2 hrs, folk 7 hrs, Ger .5 hrs, Pol .5 hrs, Sp 10 hrs wkly. ♦Amy Wright, stn mgr; Lonny Stern, prom dir; Joanna Garfinkel, mus dir.

Houston

KBME(AM)— Oct 16, 1944: 790 khz; 5 kw-U, DA-2. TL: N29 54 54 W95 27 42. Stereo. Hrs opn: 24 200 West Loop S., Suite 300, 77027. Phone: (713) 212-8000. Fax: (713) 212-8790. Web Site: www.790kbme.com. Licensee: AMFM Texas Licenses L.P. Group owner: Clear Channel Communications Inc. (acq 8-30-2000; grpsl). Population served: 425,000 Natl. Network: ESPN Radio. Natl. Rep: Christal. Format: Sports. ♦Mark Copelman, gen mgr; Pam McKay, gen sls mgr; Ken Charles, natl sls mgr & progmg VP; Dan Endom, rgnl sls mgr; Melissa Brezner, mktg dir & prom dir; Tim Collins, progmg dir; Bryan Erickson, news dir; Peggy Tuck, pub affrs dir; David Armstrong, chief of engrg.

KBXX(FM)— January 1958: 97.9 mhz; 100 kw. 1,920 ft TL: N29 34 34 W95 30 36. Stereo. Hrs opn: 24 Greenway Plaza, Suite 1508, 77046. Phone: (713) 623-2108. Fax: (713) 623-0344. E-mail: scorpio@kbxx.com Web Site: www.kbtt.com. Licensee: Radio One Licenses LLC. Group owner: Radio One Inc. (acq 2000). Population served: 3,000,000 Natl. Rep: Clear Channel. Law Firm: Dickstein Shapiro Morin & Oshinsky L.L.P. Format: Hip hop. Target aud: 18-29; females. ♦Ernest Jackson, pres & disc jockey; Carl Hamilton, VP; Mark McMillen, gen mgr; Tom Callococci, opns mgr.

KCOH(AM)— 1952: 1430 khz; 5 kw-D. TL: N29 45 22 W95 16 37. Stereo. Hrs opn: 24 5011 Almeda Rd., 77004. Phone: (713) 522-1001. Fax: (713) 521-0769. E-mail: dsamuel@kcohradio.com Web Site: www.kcohradio.com. Licensee: KCOH Inc. (acq 9-27-76). Population served: 2,250,000 Natl. Network: American Urban, Westwood One. Natl. Rep: Roslin. Law Firm: Cavelll, Mertz & Davis. Format: Black, urban contemp, talk. News staff: 2; News: 15 hrs wkly. Target aud: 25-54; upbeat, knowledgeable, civic & politically minded adults. Spec prog: Sports.Mike Petrizzo, exec VP, gen mgr, gen sls mgr & adv dir; Travis O. Gardner, VP, opns VP, prom mgr, mus dir, min affrs dir & mus critic; Michael Harris, progmg dir, news dir, news dir, local news ed, edit dir, edit mgr, political ed, relg ed & disc jockey; Don Samuel, asst music dir & disc jockey; J.D. Rigmaideu, engrg dir & chief of engrg; Carolyn Bell, traf mgr; Ralph Cooper, sports cmtr; Lisa Berry-Dockery, women's cmtr; Steven R. Talton, disc jockey; Tommy Armstrong, disc jockey; Wash Allen, disc jockey

KEYH(AM)— November 1974: 850 khz; 10 kw-D, 185 w-N, DA-1. TL: N29 39 19 W95 40 19. Hrs opn: 3000 Bering Dr., 77057. Phone: (713) 315-3400. Fax: (713) 315-3506. Licensee: Liberman Broadcasting of Houston License LLC. Group owner: Liberman Broadcasting Inc. (acq 4-22-2003; $5.70 million). Population served: 500,000 Format: Sp. Target aud: 24-65; Hispanic, recent immigrants & primarily Sp speakers. ♦Lenard Liberman, CEO & pres; Winter Horton, gen mgr; Gerardo Reyes, gen sls mgr; Ezequiel Gonzalez, progmg dir; Meliza Posada, news dir & traf mgr; Mike Todd, chief of engrg & engr.

***KHCB(AM)**—Galveston, 1922: 1400 khz; 1 kw-U. TL: N29 17 24 W94 50 12. (CP: COL League City. 1 kw-U, DA-2. TL: N29 25 35 W95 08 00). Hrs opn: 24 2424 South Blvd., 77098-5196. Phone: (713) 520-7900. Web Site: www.radioamistad.net. Licensee: Houston Christian Broadcasters Inc. (group owner; (acq 12-4-90; $150,000). Population served: 250,000 Format: Sp, relg. News staff: one. Target aud: General. Spec prog: Chinese 13 hrs, Vietnamese 4 hrs wkly. ♦Bruce Munsterman, pres, stn mgr & progmg dir; Dolly Martin, progmg mgr & disc jockey; Miguel Jacinto, mus dir, news dir & disc jockey; Dan Wales, chief of engrg; Rebecca Aguilar, disc jockey.

KHCB-FM— Mar 10, 1962: 105.7 mhz; 100 kw. Ant 1,614 ft TL: N29 34 06 W95 29 57. Stereo. 24 Phone: (713) 520-5200. Web Site: www.khcb.org.4,000,000 Format: Christian. Spec prog: Sp 10 hrs, Chinese one hr wkly. ♦Bruce Munsterman, gen mgr & disc jockey; Bonnie C. BeMent, mus dir, news dir & disc jockey; Dan Wales, engrg dir; Jim Lingenfelter, disc jockey; Rex Sanders, disc jockey.

KHJZ-FM— Oct 4, 1959: 95.7 mhz; 100 kw. 1,971 ft TL: N29 34 34 W95 30 36. Stereo. Hrs opn: 24 24 Greenway Plaza, Suite 1900, 77046. Phone: (713) 881-5100. Fax: (713) 881-5250. Web Site: www.khjz.com. Licensee: Texas Infinity Broadcasting L.P. Group owner: Infinity Broadcasting Corp. (acq 11-13-98; grpsl). Population served: 3,348,800 Natl. Network: CBS. Natl. Rep: CBS Radio. Law Firm: Leventhal, Senter & Lerman. Format: Smooth Jazz / NAC. Target aud: 25-54. ♦Laura Morris, VP; Diane Holt, sls dir; Maxine Todd, progmg dir; Dan Woodard, chief of engrg.

KHMX(FM)— 1961: 96.5 mhz; 100 kw. 1,952 ft TL: N29 34 34 W95 30 36. Stereo. Hrs opn: 2000 W. Loop S., Suite 300, 77027. Phone: (713) 212-8000. Fax: (713) 212-8970. Web Site: www.khmx.com. Licensee: Citicasters Licenses L.P. Group owner: Clear Channel Communications Inc. (acq 5-4-99; grpsl). Population served: 4,500,000 Format: Hot adult contemp. Target aud: 25-40. ♦Muriel Funches, VP; Rick Miles, gen sls mgr; Alan Ecklund, natl sls mgr; Marc Sherman, progmg dir; Lori Bradley, mus dir; Emma Villanueva, pub affrs dir; Jamie Turner, traf mgr.

KHPT(FM)—See Conroe

KHTC(FM)—See Lake Jackson

KIKK(AM)—See Pasadena

KILT(AM)— 1948: 610 khz; 5 kw-U, DA-2. TL: N29 55 04 W95 25 33. Hrs opn: 8:30-5:00 24 Greenway Plaza, Suite 1900, 77046. Phone: (713) 881-5100. Phone: (713) 881-5957. Fax: (713) 881-5150. Web Site: www.sportsradio610.com. Licensee: Texas Infinity Broadcasting L.P. Group owner: Infinity Broadcasting Corp. (acq 12-89). Population served: 1,594,086 Natl. Rep: CBS Radio. Format: Sports, talk. ♦Laura Morris, VP; Moose Rosenfeld, gen mgr; Bill Van Rysdam, progmg dir; Dan Woodard, chief of engrg.

KILT-FM— 1961: 100.3 mhz; 100 kw. 1,920 ft TL: N29 34 34 W95 30 36. Stereo. 24 Web Site: www.kilt.com. Format: Country. ♦Nick Peterson, gen sls mgr; Jeff Garrison, progmg dir; Jim Carola, news dir.

KIOL(FM)—La Porte, 1992: 103.7 mhz; 94.86 kw. Ant 1,935 ft TL: N29 56 09 W94 30 38. Hrs opn: 2700 Post Oak Blvd., Suite 2300, 77056. Phone: (713) 300-3500. Fax: (713) 300-3585. Web Site: www.houstonrock1037.com. Licensee: CMP KC Licensing LLC. Group owner: Cumulus Media Inc. (acq 5-13-2004; $32.2 million). Format: Rock. ♦Patrick Fant, gen mgr.

KKBQ-FM—See Pasadena

KKRW(FM)— Jan 1, 1964: 93.7 mhz; 100 kw. 1,779 ft TL: N29 34 27 W95 29 37. Stereo. Hrs opn: 200 West Loop S., Suite 300, 77027. Phone: (713) 212-8000. Fax: (713) 830-8099. Web Site: www.kkrw.com. Licensee: Capstar TX L.P. Group owner: Clear Channel Communications Inc. (acq 8-30-00; grpsl). Population served: 2,905,350 Format: Classic rock. Target aud: 25-54. ♦Vince Richards, progmg dir.

KLAT(AM)— July 31, 1961: 1010 khz; 5 kw-U, DA-2. TL: N29 53 47 W95 17 25. (CP: 3.6 kw-N). Stereo. Hrs opn: 5100 Southwest Fwy., 77056. Phone: (713) 407-1415. Fax: (713) 965-2401. Web Site: www.univision.com. Licensee: Tichenor License Corp. Group owner: Univision Radio (acq 9-22-2003; grpsl). Population served: 1,026,000 Format: Sp, news/talk. Target aud: 25-54; Hispanic. ♦Mark Masepohl, sr VP, gen mgr & sls dir; Dave Burdette, VP & stn mgr; Arnulfo Ramirez, opns mgr; Kim McBride, natl sls mgr & pub affrs dir; Manuel Cardona, rgnl sls mgr; Frances Jones, prom dir; Pilar Torres, prom mgr; Rolando Becerra, progmg dir; Renzo Heredia, news dir & disc jockey; Marty Scruggs, chief of engrg.

KQBU-FM—Co-owned with KLAT(AM). July 4, 1969: 93.3 mhz; 100 kw. Ant 594 ft TL: N30 03 05 W94 31 37. Stereo. 24 Web Site: www.univision.com. Target aud: 25-54; Hispanics. ♦Mark Masepohl, sr VP; Jose Lopez, opns mgr; Arnulfo Ramirez, progmg VP; Carole Van Matre, rsch dir; Claudetta Wallace, traf mgr; Renzo Heredia, prom dir, pub affrs dir & disc jockey.

KLOL(FM)—Listing follows KTRH(AM).

KLTN(FM)— Oct 4, 1960: 102.9 mhz; 100 kw. Ant 1,049 ft TL: N29 45 26 W95 20 18. Stereo. Hrs opn: 24 5100 Southwest Fwy., 77056. Phone: (713) 965-2300. Fax: (713) 965-2401. Web Site: www.univision.com. Licensee: Univision Radio Houston License Corp. Group owner: Univision Radio (acq 9-22-2003; grpsl). Population served: 4,000,000 Format: Rgnl Mexican hits. News staff: one. Target aud: 18-49; Hispanics. ♦Mark Masepohl, sr VP & gen mgr; Dave Burdette, stn mgr & gen sls mgr; Arnulfo Ramirez, opns mgr; Kim Mercier, natl sls mgr; Frances Jones, prom dir; Raul Brindis, progmg dir; Renzo Heredia, news dir & pub affrs dir; Marty Scruggs, chief of engrg.

KLVL(AM)—See Pasadena

KMIC(AM)— 1955: 1590 khz; 5 kw-U, DA-N. TL: N29 50 38 W95 26 51. Hrs open: 24 3050 Post Oak Blvd., Suite 220, 77056. Phone: (713) 552-1590. Fax: (713) 552-1588. Web Site: www.disney.com. Licensee: Radio Disney Group LLC. Group owner: ABC Inc. (acq. 1999). Population served: 3,000,000 Natl. Network: Radio Disney. Format: Family progmg. Target aud: 6-14; 25-49; kids, parents. ♦Chris Martin, gen mgr, gen sls mgr & adv dir; A. D. Rigmaiden, chief of engrg; Johanna Anderson, traf mgr; Laura Pena, prom mgr & spec ev coord.

KMJQ(FM)— Feb 1, 1964: 102.1 mhz; 100 kw. 1,719 ft TL: N29 34 27 W95 29 37. Stereo. Hrs opn: 24 24 Greenway Plaza, Suite 1508, 77046-2467. Secondary address: Box 22900 77227-2900. Phone: (713) 623-2108. Fax: (713) 623-0106. Web Site: www.kmjq.com. Licensee: Radio One Licenses LLC. Group owner: Radio One Inc. (acq 11-8-01; grpsl). Population served: 583,400 Natl. Network: ABC. Natl. Rep: Clear Channel. Law Firm: Wiley, Rein & Fielding. Format: Adult urban contemp. News staff: 2. Target aud: 25-54; African-Americans. Spec prog: Talk 3 hrs wkly.Tom Callococci, pres & opns mgr; Carl Hamilton, VP & disc jockey; Mark McMillen, gen mgr; Jerome Hutchinson, gen sls mgr; Brenda Ford-Jones, natl sls mgr; Cindy Webster, rgnl sls mgr; Bobrie Jefferson, prom mgr; Sam Choice, progmg dir & progmg mgr; Carmen Watkins, news dir; David Ainslie, engrg mgr & chief of engrg; Vickie Duke, traf mgr; J.J. Williams, disc jockey; Jeff Harrison, disc jockey; Kandi Eastman, disc jockey; Larry Jones, disc jockey; Val Wilson, disc jockey

KNTH(AM)— Jan 17, 1968: 1070 khz; 10 kw-D, 5 kw-N, DA-2. TL: N29 59 33 W95 29 33. Hrs open: 6161 Savoy, Suite 1200, 77036. Phone: (713) 260-3600. Fax: (713) 260-3628. Licensee: South Texas Broadcasting Inc. Group owner: Salem Communications Corp. (acq 1-6-95; $2.5 million; FTR: 3-6-95). Natl. Rep: Salem. Format: News/talk. ♦Chuck Jewell, gen mgr; Paul Baker, opns mgr, mktg mgr & progmg dir; Dan Doster, gen sls mgr; Ken Garza, pub affrs dir; Sidney Jones, chief of engrg; Kent McDonald, traf mgr.

KODA(FM)— Nov 9, 1958: 99.1 mhz; 95 kw. 1,920 ft TL: N29 34 34 W95 30 36. Stereo. Hrs opn: 24 2000 West Loop S., Suite 300, 77027. Phone: (713) 212-8000. Fax: (713) 830-8099. Web Site: www.sunny99.com. Licensee: AMFM Texas License L.P. Group owner: Clear Channel Communications Inc. (acq 8-30-00; grpsl). Population served: 3,184,200 Law Firm: Latham & Watkins. Format: Adult contemp. News staff: one; News: 22 hrs wkly. Target aud: 25-54. Spec prog: Jazz 4 hrs wkly. ♦Mark Kopelman, gen mgr; Sandy Capell, gen sls mgr; Vince Richards, progmg dir; Donna McCoy, mus dir.

***KPFT(FM)**— March 1970: 90.1 mhz; 100 kw. 433 ft TL: N29 55 26 W95 32 17. (CP: 28 kw, ant 672 ft.). Stereo. Hrs opn: 24 419 Lovett Blvd., 77006. Phone: (713) 526-4000. Fax: (713) 526-5750. Web Site: www.kpft.org. Licensee: Pacifica Foundation Inc. Group owner: Pacifica Foundation Inc. dba Pacifica Radio Population served: 4,000,000 Natl. Network: PRI. Law Firm: Haley, Bader & Potts. Format: Div, news. Target aud: General. Spec prog: Black 15 hrs. ♦Dwande Bradley, gen mgr; Donna Platt, dev dir; Otis Maclay, progmg dir; Phil Edwards, mus dir; Ernesto Aguilar, news dir; Renee Feltz, news dir; Steve Brightwell, engrg dir.

KPRC(AM)— May 9, 1925: 950 khz; 5 kw-U, DA-N. TL: N29 48 14 W95 16 42. Hrs open: 24 2000 W. Loop S., Suite 300, 77027. Phone: (713) 212-8000. Fax: (713) 212-8970. Web Site: www.950kprc.com. Licensee: CCB Texas Licenses L.P. Group owner: Clear Channel Communications Inc. (acq 3-14-95; FTR: 6-5-95). Population served: 3,580,000 Natl. Network: CBS, Fox News Radio, Westwood One. Rgnl. Network: Texas State Net. Natl. Rep: Clear Channel. Law Firm: Dow, Lohnes & Albertson. Format: News/talk info. News staff: 15; News: 32 hrs wkly. Target aud: 25-54. Spec prog: Gardening 7 hrs, home handyman 6 hrs, automotive 3 hrs wkly. ♦Mark Kopelman, gen mgr; Michael Berry, opns mgr; Paul Lambert, gen sls mgr; Ken Charles, progmg dir; Brian Erickson, news dir; David Armstrong, chief of engrg; Matt Thomas, sports cmtr.

KPTY(FM)—Missouri City, Aug 8, 1968: 104.9 mhz; 2.7 kw. Ant 981 ft TL: N29 45 30 W95 22 03. Stereo. Hrs opn:
Simulcast with KLTO(FM) Crystal Beach.
5100 Southwest Fwy., 77056. Phone: (713) 965-2300. Fax: (713) 965-2401. Web Site: www.party1049.com. Licensee: Tichenor License Corp. Group owner: Univision Radio (acq 9-22-2003; grpsl). Population served: 4,000,000 Format: Urban, hip hop. Target aud: 18-34; urban/Latin audience. ♦Mark Masepohl, sr VP & gen mgr; Dave Burdette, stn mgr; J.D. Gonzales, opns VP; Kim Mercier, natl sls mgr; Mark McMillan, rgnl sls mgr; Frances Jones, prom dir; Nestor Enriquez, prom mgr; Pete Manriquez, progmg dir; Renzo Heredia, pub affrs dir; Marty Scruggs, chief of engrg; Carole Van Matre, rsch dir; Claudetta Wallace, system mgr.

KQUE(AM)— Feb 18, 1948: 1230 khz; 1 kw-U. TL: N29 45 26 W95 20 18. Hrs open: 24
Simacasts KKRW (FM) Houston.
3000 Bering Dr., Suite 1170, 77057. Phone: (713) 315-3400. Fax: (713) 315-3506. Licensee: Liberman Broadcasting of Houston License LLC. Group owner: Liberman Broadcasting Inc. (acq 3-20-2001; grpsl). Population served: 3,941,000 Law Firm: Fisher, Wayland, Cooper, Leader & Zaragoza L.L.P. Format: Mexican rgnl. News: 14 hrs wkly. Target aud: 35 plus; mature, upscale, high-income. ♦Lenard Liberman, CEO & VP; Winter Horton, VP & gen mgr; Cheque Gonzalez, stn mgr & progmg dir; Gerardo Reyes, gen sls mgr & chief of engrg; Ezequiel Gonzalez, progmg dir; Meliza Posada, traf mgr.

KRBE(FM)— Nov 8, 1959: 104.1 mhz; 100 kw. 1,920 ft TL: N29 34 34 W95 30 36. Stereo. Hrs opn: 9801 Westheimer, Suite 700, 77042. Phone: (713) 266-1000. Fax: (713) 954-2344. Web Site: www.104krbe.com. Licensee: KRBE Lico Inc. Group owner: Susquehanna Radio Corp. (acq 11-86; $25 million with co-located AM; FTR: 10-6-86). Population served: 451,000 Format: CHR. Target aud: 18-34; general.Peter Brubaker, chmn; David Kennedy, pres; Nancy Vaeth, sr VP; Mark Shecterle, gen mgr; Amy Dewbre, gen sls mgr; Beth Lavine, natl sls mgr; Mike Paterson, mktg dir; Lesley Brotamante, prom mgr & adv mgr; Tracy Austin, progmg dir & progmg mgr; Leslie Whittle, mus dir; Maria Todd, news dir; Benny Boone, pub affrs dir; Andy Hudack, engrg mgr; Chuck Underwood, chief of engrg

KROI(FM)—See Seabrook

KRTX(AM)—Rosenberg-Richmond, Nov 15, 1948: 980 khz; 1 kw-D, 212 w-N. TL: N29 33 10 W95 47 00. (CP: 5 kw-U, DA-2). Stereo. Hrs opn: 24 5100 Southwest Fwy., 77056. Phone: (713) 965-2300. Fax: (713) 965-2401. Web Site: www.univision.com. Licensee: Tichenor Media System Inc. Group owner: Univision Radio (acq 9-22-2003; grpsl). Population served: 4,000,000 Law Firm: Cohn & Marks. Format: Tejano, Sp, var. News staff: one. Target aud: 25-54; Hispanics. ♦Mark Masepohl, sr VP; Arnulfo Ramirez, opns VP & opns mgr; Marie Barden, gen sls mgr; Kim Mercier, natl sls mgr; Frances Jones, prom dir; Pilar Torres, prom mgr & progmg dir; Ferando Hernandez, progmg dir; Renzo Heredia, mktg dir & pub affrs dir; Marty Scruggs, chief of engrg; Carole Van Matre, rsch dir; Claudetta Wallace, traf mgr.

KTBZ-FM— Nov 1, 1964: 94.5 mhz; 100 kw. 2,000 ft TL: N29 34 34 W95 30 36. Stereo. Hrs opn: 2000 W. Loop S., Suite 300, 77027. Phone: (713) 212-8000. Fax: (713) 212-8970. Web Site: www.thebuzz.com. Licensee: AMFM Texas Licenses L.P. Group owner: Clear Channel Communications Inc. (acq 8-30-00; grpsl). Population served: 350,000 Natl. Rep: D & R Radio. Format: Oldies. Target aud: 25-54; baby boomers. Spec prog: Talk 2 hrs, relg one hr, pub affrs one hr wkly. ♦Ellen Cavanaugh, gen mgr; Jim Trapp, progmg dir.

KTEK(AM)—See Alvin

KTRH(AM)— Mar 29, 1930: 740 khz; 50 kw-U, DA-2. TL: N29 57 57 W94 56 32. Hrs open: 2000 West Loop South, Suite 300, 77027. Phone: (713) 526-5874. Fax: (713) 360-3666. Web Site: www.ktrh.com. Licensee: AMFM Texas Licenses L.P. Group owner: Clear Channel Communications Inc. (acq 8-30-2000; grpsl). Population served: 2,900,000 Natl. Network: ABC. Natl. Rep: Christal. Law Firm: Dow, Lohnes & Albertson. Format: News, sports. Target aud: 25-54. ♦Mark Kopelman, gen mgr; Betty Scott, progmg dir.

KLOL(FM)—Co-owned with KTRH(AM). 1947: 101.1 mhz; 100 kw. Ant 1,920 ft TL: N29 34 34 W95 30 36. Stereo. 2000 West Loop S., Suite 300, 77027. Phone: (713) 212-8000. Fax: (713) 212-8970. Web Site: www.klol.com.425,000 Natl. Rep: Christal. Format: Sp. Target aud: 18-54. ♦Paul Lambert, gen sls mgr; Alan Ecklund, natl sls mgr; Melissa Brezner, mktg VP; Rob Skinner, prom dir; Ken Charles, progmg VP; Vince Richards, progmg dir; Steve Fixx, mus dir; Laurent Fouilloud-Buyat, news dir & pub affrs dir; Bob Stroup, chief of engrg; Dave Andrews, disc jockey.

***KTRU(FM)**— May 20, 1971: 91.7 mhz; 50 kw. 492 ft TL: N30 03 54 W95 16 10. Stereo. Hrs opn: 24 6100 S. Main, 77005. Phone: (713) 348-4098. Fax: (713) 348-4093. E-mail: ktru@ktru.org Web Site: www.noise.ktru.org. Licensee: Rice University. Population served: 2,000,000 Format: Div. Target aud: General. Spec prog: Class 8 hrs, jazz 14 hrs, reggae 6 hrs, folk 3 hrs, 60s mus 3 hrs, new age one hr wkly. ♦Will Robedee, gen mgr; Ben Horne, stn mgr; Allan McHale, news dir.

***KTSU(FM)**— October 1973: 90.9 mhz; 18.5 kw. 285 ft TL: N29 43 25 W95 21 52. Stereo. Hrs opn: 24 3100 Cleburne st., 77004. Phone: (713) 313-7591. Fax: (713) 313-7479. Licensee: Board of Regents Texas Southern University. Population served: 216,600 Format: Educ, div, jazz. News staff: one; News: 10 hrs wkly. Target aud: 25-54. Spec prog: Reggae 8 hrs wkly. ♦Dr. Priscilla Slade, pres; George Thomas, gen mgr; Lindsey Williams, dev mgr, mktg dir, mktg mgr, adv dir & progmg dir; Cheryl Brooks, prom dir; Maurice Hopethompson, news dir; Dave Biondi, chief of engrg.

***KUHF(FM)**— Nov 6, 1950: 88.7 mhz; 100 kw. 1,800 ft TL: N29 34 28 W95 29 37. Stereo. Hrs opn: 24 4343 Elgin, 3rd Fl., 77204-0887. Phone: (713) 743-0887. Fax: (713) 743-0868. E-mail: kuhf@kuhf.org Web Site: www.kuhf.org. Licensee: University of Houston. Population served: 3,100,000 Natl. Network: NPR, PRI. Law Firm: Dow, Lohnes & Albertson. Format: Class, news. News staff: 7; News: 25 hrs wkly. Target aud: 25 plus.John Gladney Proffitt, CEO & gen mgr; Debra Fraser, stn mgr, prom dir & progmg mgr; Victor C. Kendell, dev dir; Robert Cahill, prom dir; Kathy Rogers, adv mgr; Dean Dalton, mus dir; Paul Pendergraft, news dir; Chris Hathaway, pub affrs dir; Alex Schneider, chief of engrg; Lance Bean, traf mgr; Rob Cahill, spec ev coord; Capella Tucker, news rptr; Ed Mayberry, news rptr; Jack Williams, news rptr; Jim Bell, news rptr; Laura Johnson, news rptr; Rod Rice, news rptr

KXYZ(AM)— Aug 8, 1930: 1320 khz; 5 kw-U, DA-N. TL: N29 42 37 W95 10 29. Hrs open: 1782 W. Sam Houston Pkwy. N., 77043. Phone: (713) 490-2538. Fax: (713) 984-1721. Licensee: Multicultural Radio Broadcasting Licensee LLC. Group owner: Multicultural Radio Broadcasting Inc. (acq 12-1-2003; grpsl). Population served: 1,594,000 Format: Contemp Sp, talk. ♦Terry Lowry, gen mgr.

Howe

KHYI(FM)— April 1949: 95.3 mhz; 16 kw. Ant 413 ft TL: N33 23 43 W96 35 50. Stereo. Hrs opn: 24 Box 940670, Plano, 75094. Secondary address: 660 N. Central Expwy., Suite 120, Dlano 75074. Phone: (972) 633.0953. Fax: (972) 633-0957. E-mail: ken.jones@kxez.com Web Site: www.khyi.com. Licensee: Metro Broadcasters-Texas Inc. Population served: 4,761,200 Natl. Network: ABC. Law Firm: Fletcher, Heald & Hildreth. Format: Classic country. News staff: one; News: 3 hrs wkly. Target aud: 25-54; affluent, white collar, middle to upper income listeners. ♦Ken Jones, CEO, pres, gen mgr, news dir & chief of engrg; Glenda Jones, CFO; Lou Rogers, opns dir, opns mgr & pub affrs dir; Joshua Jones, sls dir, gen sls mgr, mktg VP, mktg dir & prom dir; Bruce Kidder, progmg dir; Mike Doyal, engrg dir; Ron Eudaly, engrg dir.

Hudson

KLSN(FM)— 2002: 96.3 mhz; 1.35 kw. Ant 695 ft TL: N31 21 55 W94 45 59. Hrs open: Box 111, Livingston, 77351. Phone: (936) 327-8916. Fax: (936) 327-8477. Licensee: Peggy Sue Marsh, administrator (acq 2-27-2007). Format: Country.

Humble

KGOL(AM)— July 18, 1984: 1180 khz; 50 kw-D, 1 kw-N, DA-3. TL: N30 08 21 W95 17 24. Stereo. Hrs opn: 5 AM-midnight 5821 Southwest Fwy., Suite 600, Houston, 77057. Phone: (713) 349-9880. Fax: (713) 349-0647. E-mail: caguilar@entravision.com Licensee: Entravision Holdings LLC. Group owner: Entravision Communications Corp. (acq 7-28-00; grpsl). Law Firm: Luther & Watkins. Format: Foreign/Ethnic. News staff: one. Target aud: General; ethnic & Asian. Spec prog: Hindi 15 hrs wkly. ♦Walter Ulloa, CEO; Jeff Liberman, pres; Carmen Aguilar, gen mgr; David Padgett, opns mgr & progmg mgr; Rick Hunt, engrg VP.

***KSBJ(FM)**— July 6, 1982: 89.3 mhz; 100 kw. Ant 840 ft TL: N30 12 26 W95 05 28. Stereo. Hrs opn: 24 Box 187, 77338. Secondary address: 327 Wilson Rd. 77347. Phone: (281) 446-5725. Fax: (281) 540-2198. Web Site: www.ksbj.org. Licensee: KSBJ Educational Foundation. Population served: 4,000,000 Format: Contemp Christian. News staff: 31; News: 4 hrs wkly. Target aud: 25-49; Christian adults. ♦Tim McDermott, gen mgr; Jason Ray, prom dir; John Hull, progmg dir; Jim Beeler, mus dir & disc jockey; Amanda Carroll, news dir; George Schank, chief of engrg; J.R. Hernandez, spec ev coord; Duane Allen, disc jockey; Jon Hull, disc jockey; Susan O'Donnell, disc jockey; Tom Carter, prom mgr & disc jockey.

Huntington

KSML-FM— March 1, 1994: 101.9 mhz; 24.5 kw. Ant 666 ft TL: N31 22 08 W94 38 45. Stereo. Hrs opn: Yates Broadcasting, 121 Cotton Sq., Lufkin, 75902. Phone: (936) 637-1019. Fax: (409) 632-5722. Web Site: www.kybi.com. Licensee: Yates Broadcasting Corp. Format: Adult contemp. ♦Steven Yates, gen mgr.

Huntsville

***KHCH(AM)**— Oct 4, 1982: 1410 khz; 250 w-D, 87 w-N. TL: N30 42 54 W95 31 42. Hrs open: 24 2424 South Blvd., Houston, 77098. Phone: (713) 520-5200. E-mail: email@khcb.org Web Site: www.khcb.org. Licensee: KHCB Inc. Group owner: Houston Christian Broadcasters Inc. (acq 10-97; $145,000). Format: Sp Christian. ♦Bruce Munsterman, gen mgr.

KHVL(AM)— Nov 3, 1938: 1490 khz; 1 kw-U. TL: N30 41 48 W95 33 08. Hrs open: 24 Box 330, 77342. Secondary address: 622 Interstate 45 S. 77340. Phone: (936) 295-2651. Fax: (936) 295-8201. E-mail: khvlmail@yahoo.com Web Site: www.khvl.com. Licensee: HEH Communications LLC (acq 12-11-00; $1.9 million with co-located FM). Population served: 65,000 Natl. Network: ABC. Format: Oldies. News staff: one. Target aud: 35 plus; general. Spec prog: Black 5 hrs wkly. ♦Steve Everett, gen mgr; Brooke Addams, opns mgr; LeeAn Kelly, news dir; Stacy Selman, traf mgr.

KSAM-FM—Co-owned with KHVL(AM). Aug 1, 1965: 101.7 mhz; 6 kw. 420 ft TL: N30 41 48 W95 33 08. Stereo. 24 E-mail: ksammail@yahoo.com Web Site: www.ksam1017.com.65,000 Format: Country. News staff: one; News: 2 hrs wkly. Target aud: 25-54; general. ♦Stacey Selman, traf mgr; Kooter Roberson, sports cmtr.

***KSHU(FM)**— October 1973: 90.5 mhz; 3 kw. 255 ft TL: N30 42 50 W95 32 58. Stereo. Hrs opn: 6 AM-midnight Box 2207, 1804 Avenue J, 77341. Phone: (936) 294-3939. Phone: (936) 294-1342. Fax: (936) 294-1888. E-mail: rtf_kshu@shsu.edu Web Site: www.shsu.edu/~rtf_kshu. Licensee: Sam Houston State University. Population served: 40,000 Format: Class, CHR, jazz. News: 25 hrs wkly. Target aud: General; rural. Spec prog: Sp 4 hrs wkly. ♦Terry L. Rosati, gen mgr; Matt Orlando, opns dir; Adam Spry, progmg dir; Lowery Woodall, mus dir; Rachel Connner, news dir; Jenna Zibton, pub affrs dir; Steve Sandlin, chief of engrg; Joey Jeffery, traf mgr.

Hurst

KMNY(AM)— April 1947: 1360 khz; 50 kw-D, 890 w-N, DA-2. TL: N32 46 28 W96 57 53. Hrs open: 24 5801 Marvin D. Love Fwy., Suite 409, Dallas, 75237. Phone: (972) 572-1540. Fax: (972) 572-1260. E-mail: kdft-kmny@mrbi.net Licensee: Multicultural Radio Broadcasting Licensee LLC. Group owner: Multicultural Radio Broadcasting Inc. (acq 2-4-2004; grpsl). Natl. Network: CNN Radio. Format: News/talk info. News: 9 hrw wkly. ♦Ted Sauceman, gen mgr.

Hutto

KYLR(FM)—Licensed to Hutto. See Austin

Idalou

KRBL(FM)— Sept 18, 1995: 105.7 mhz; 6 kw. 328 ft TL: N33 40 06 W101 37 52. Hrs open: 916 Main St., Suite 617, Lubbock, 79401. Phone: (806) 749-1057. Fax: (806) 749-1177. Licensee: Triumph Communications Inc. Format: Classic country. Target aud: 24-64.

♦Paul Beane, gen mgr, opns mgr & news dir; Steve Ritchie, gen sls mgr & adv dir; Anthony Garza, progmg dir & chief of engrg; Wanda Byers, traf mgr.

Ingleside

KJKE(FM)— 1996: 107.3 mhz; 14 kw. Ant 446 ft TL: N27 52 00 W97 13 08. Stereo. Hrs opn: 24 826 S. Padre Island Dr., Corpus Christi, 78416. Phone: (361) 814-3800. Fax: (361) 855-3770. Web Site: 1073jakefm.com. Licensee: Convergent Broadcasting Corpus Christi LP. Group owner: Convergent Broadcasting LLC (acq 1-12-2004; grpsl). Population served: 313,600 Format: Classic hits. Target aud: 25-54; adults. ♦Mark White, gen mgr; Dallas Garcia, adv mgr; Scott Holt, progmg dir; William Hooper, engr.

Ingram

KEVE(FM)—Not on air, target date: unknown: 96.5 mhz; 3.4 kw. Ant 327 ft TL: N30 02 27 W99 10 19. Hrs open: 1717 Dixie Hwy., Suite 650, Fort Wright, KY, 41011. Phone: (859) 331-9100. Licensee: Radioactive LLC. ♦Benjamin L. Homel, pres.

***KTXI(FM)**— November 1998: 90.1 mhz; 50 kw. 453 ft TL: N30 06 14 W99 04 36. Stereo. Hrs opn: 24
Rebroadcasts KPAC(FM) San Antonio 75% , KSTX(FM) San Antonio 25%.
Texas Public Radio, 8401 Datapoint Dr., Suite 800, San Antonio, 78229. Phone: (210) 614-8977. Fax: (210) 614-8983. Web Site: www.ktxi.fm. Licensee: Texas Public Radio. Population served: 50000 Natl. Network: NPR, PRI. Law Firm: Garvey, Schubert & Barer. Format: Class, news. Target aud: 25 plus. ♦Dan Skinner, pres & gen mgr; Nathan Cone, opns mgr & progmg dir; Laverne Dittx, dev mgr; Janet Grojean, progmg dir & sls; Randy Anderson, mus dir; Dave Davies, news dir; Wayne Coble, engrg dir.

Iowa Park

KXXN(FM)—Not on air, target date: unknown: 96.3 mhz; 6 kw. Ant 256 ft TL: N33 58 20 W98 45 35. Hrs open: 819 S.W. Federal Hwy., Suite 106, Stuart, FL, 34994. Phone: (772) 215-1634. Web Site: www.toweritrust.com. Licensee: Tower Investment Trust Inc. ♦William H. Brothers, pres.

Jacksboro

KJKB(FM)— Oct 6, 1996: 95.5 mhz; 6 kw. Ant 328 ft TL: N33 19 43 W98 16 46. Hrs open: 24 101 N Main St, Ste 206, 76458. Phone: (940) 567-6600. Fax: (940) 567-6602. E-mail: kjkb@sbcglobal.net Licensee: Hunt Broadcasting Inc. Group owner: On-Air Family LLC (acq 1995; $6,000). Population served: 12,000 Format: Classic Rock. ♦Janice Hunt, CEO & gen mgr; Jim Hunt, exec VP; Debbie Watts, stn mgr; Jerrod Knight, progmg dir.

Jacksonville

***KBJS(FM)**— May 16, 1987: 90.3 mhz; 3 kw. 266 ft TL: N31 58 16 W95 15 51. (CP: 19 kw, ant 1,286 ft. TL: N32 03 40 W95 18 50). Stereo. Hrs opn: Box 193, 75766. Phone: (903) 586-5257. Fax: (903) 586-4986. E-mail: info@kbjs.org Web Site: www.kbjs.org. Licensee: East Texas Media Association Inc. Natl. Network: Moody. Format: Relg, Christian. Spec prog: Black one hr, Sp one hr wkly. ♦Bob Shivery, pres & gen mgr; Randy Featherston, stn mgr; Eddie Baiseri, progmg dir & progmg mgr.

KEBE(AM)— Jan 12, 1947: 1400 khz; 1 kw-U. TL: N31 58 11 W95 15 52. Hrs open: 24 Box 1648, 75766. Secondary address: Radio Ctr., 402 S. Ragsdale 75766. Phone: (903) 586-2527. Fax: (903) 586-1394. Web Site: www.kooi.com. Licensee: Waller Broadcasting Inc. Group owner: Waller Broadcasting (acq 11-58; $75,000). Population served: 12,724 Natl. Rep: McGavren Guild. Format: Classic country. News staff: 2; News: 6 hrs wkly. Target aud: 25-54. Spec prog: Farm 9 hrs wkly. ♦Dudley Waller, CEO, pres & gen mgr; Tina Harper, CFO; Alan Mather, opns dir & opns mgr.

KLJT(FM)— 1993: 102.3 mhz; 50 kw. Ant 492 ft TL: N31 52 18 W95 10 00. Stereo. Hrs opn: 24 Box 1648, 75766. Secondary address: 402 S. Ragsdale 75766. Phone: (903) 586-2527. Fax: (903) 589-0677. Web Site: www.wallerbroadcast.com. Licensee: Waller Media LLC. Group owner: Waller Broadcasting (acq 12-9-02). Natl. Network: ABC. Natl. Rep: McGavren Guild. Law Firm: David Tillotson. Wire Svc: AP Format: Adult favorites. News staff: one; News: 6 hrs wkly. ♦Dudley Waller, CEO & gen mgr.

KOOI-FM— Sept 9, 1967: 106.5 mhz; 100 kw. Ant 1,468 ft TL: N32 03 40 W95 18 50. Stereo. Hrs opn: 24 Box 7820, Tyler, 75711. Secondary address: 210 S. Broadway, Tyler 75702. Phone: (903) 581-9966. Fax: (903) 534-5300. E-mail: kooi@etradiogroup.com Web Site: www.kooi.com. Licensee: Access.1 Texas License Company LLC. (acq 1-7-2005; grpsl). Population served: 848,000 Natl. Network: ABC. Rgnl. Network: Texas State Net. Natl. Rep: McGavren Guild. Format: Adult contemp. Target aud: 25-54. ♦Rick Guest, gen mgr, opns mgr, mktg mgr & mus dir; Genni Causey, gen sls mgr; Robert Taylor, gen sls mgr & natl sls mgr; Paul Orr, progmg dir; Shelley Miller, traf mgr.

Jasper

KCOX(AM)— Aug 6, 1948: 1350 khz; 5 kw-D, 37 w-N. TL: N30 55 11 W93 58 13. Hrs open: 24 Box 2008, 75951. Secondary address: 1408 E. Gibson 75951. Phone: (409) 384-4500. Fax: (409) 384-4525. E-mail: crosstexas@sbcglobal.net Web Site: www.crosstexasmedia.com. Licensee: Lasting Value Radio Inc. (acq 5-11-2000; $902,000 with co-located FM). Population served: 300,000 Natl. Network: Salem Radio Network. Rgnl. Network: Texas State Net. Format: Contemp Christian. Target aud: 18-35. ♦Rick Tallent, gen mgr; Dale Cucancic, gen sls mgr; T.J. Bordelon, opns dir & engrg dir; Barbara Bordelon, traf mgr.

KTXJ-FM—Co-owned with KCOX(AM). November 1964: 102.7 mhz; 26 kw. Ant 440 ft TL: N31 03 36 W93 57 42. (CP: 50 kw, ant 492 ft). Stereo. 24 Web Site: www.crosstexasmedia.com.400,000 Natl. Network: Salem Radio Network. Format: Southern gospel music. Very broad receptive demographic.

KJAS(FM)— 1996: 107.3 mhz; 8 kw. Ant 328 ft TL: N30 58 31 W93 59 24. Stereo. Hrs opn: 24 765 Hemphill St., 75951. Phone: (409) 384-2626. Fax: (409) 383-1979. E-mail: wb5rfk@jas.net Web Site: www.kjas.com. Licensee: DBA Rayburn Broadcasting Co. Population served: 72,000 Law Firm: Booth, Freret, Imlay & Tepper. Format: Adult contemp. News staff: one; News: 4 hrs wkly. Target aud: 24-54; females/buying group. Spec prog: Oldies 4 hrs wkly. ♦Mike Lout, gen mgr; Melaney Dickerson, opns mgr; Debra Foster, sls VP & gen sls mgr; Crystal Mouton, pub affrs dir.

Jefferson

***KHCJ(FM)**— 2003: 91.9 mhz; 3.1 kw. Ant 462 ft TL: N32 49 23 W94 28 32. Hrs open: 24 Houston Christian Broadcasters Inc., 2424 South Blvd., Houston, 77098. Phone: (713) 520-5200. Web Site: www.khcb.org. Licensee: Houston Christian Broadcasters Inc. (group owner). Format: Christian. ♦Bruce E. Munsterman, gen mgr.

KJTX(FM)— October 1990: 104.5 mhz; 2.3 kw. 531 ft TL: N32 49 23 W94 28 32. (CP 4.4 kw, ant 384 ft.). Stereo. Hrs opn: 24 Box 150508, Longview, 75615. Phone: (903) 759-1243. Fax: (903) 759-9725. Web Site: www.kjtx1045fm.com. Licensee: Wisdom Ministries Inc. (acq 4-16-93; $140,000; FTR: 5-3-93). Population served: 500,000 Format: Gospel, Christian. News: 2 hrs wkly. Target aud: 16 plus. ♦Leroy Richardson, pres, pres, gen mgr, progmg dir & chief of engrg; Jocelyn Jordan, traf mgr.

Johnson City

KFAN-FM— 1991: 107.9 mhz; 37.2 kw. 492 ft TL: N30 11 49 W98 38 19. Stereo. Hrs opn: 24 Box 311, 210 Woodcrest, Fredericksburg, 78624. Phone: (830) 997-2197. Fax: (830) 997-2198. E-mail: txradio@ktc.com Web Site: www.texasrebelradio.com. Licensee: J. & J. Fritz Media Ltd. (group owner). Population served: 200,000 Law Firm: Fletcher, Heald + Hildreth. Format: Americana AAA. News staff: one. Target aud: 25-49. Spec prog: Jazz 5 hrs wkly. ♦Jayson Fritz, pres, gen mgr & gen sls mgr; Jan Fritz, sr VP, mktg VP & adv VP; Ariana Carruth, prom VP; Rick Star, progmg dir; Robbie Fish, news dir; Kyle Province, pub affrs dir; Duncan Black, chief of engrg.

Jourdanton

KLEY-FM— 2001: 95.7 mhz; 11 kw. Ant 1,036 ft TL: N28 54 57.4 W98 39 39. Hrs open: 9426 Old Katy Rd., Bldg. 10, Houston, 77055. Phone: (817) 335-5999. Fax: (817) 335-1197. Licensee: BMP San Antonio License Co. L.P. (group owner; (acq 12-23-2004; $7.5 million). Format: Sp. ♦Thomas Castro, gen mgr.

Junction

KMBL(AM)— 1953: 1450 khz; 1 kw-U. TL: N30 29 34 W99 45 41. Hrs open: 24 2125 Sidney Baker, Kerrville, 78028. Secondary address: 214 Pecan St. 76899. Phone: (830) 896-1230. Fax: (830) 792-4142. E-mail: generalmanager@krvl.com Licensee: Kimble County Communications Inc. Group owner: Hill Country Broadcasting Corp. (acq 7-7-98; $165,000 with co-located FM). Population served: 2,654 Natl. Network: Westwood One. Rgnl. Network: Texas State Net. Format: Country. News staff: one; News: 6 hrs wkly. Target aud: General. Spec prog: Farm 6 hrs wkly. ♦Kent Foster, CEO & pres; Monte Spearman, gen mgr, dev dir & mktg dir; Harley Belew, opns dir; Glen Taylor, prom dir; Monte Speaman, adv dir; A.J. Hernandez, progmg dir; Charles Rodaiquez, news dir; Carolyn Anderson, pub affrs dir; Steve Alex, stn mgr, sls dir & engrg VP; Debbie Adams, traf mgr; Laurel Bradford, local news ed.

KOOK(FM)— 1997: 93.5 mhz; 50 kw. 492 ft TL: N30 29 31 W100 02 03. Hrs open: 24 2125 Sidney Baker St, Kerrville, 78028. Phone: (830) 896-1230. Fax: (830) 792-4142. Licensee: Kimble County Communications Inc. Group owner: Hill Country Broadcasting Corp. (acq 8-16-00; grpsl). Natl. Network: ABC. Rgnl. Network: Texas State Net. Format: Country. Spec prog: Gospel 2 hrs wkly. ♦Monte Spearman, gen mgr; Donna Keese, opns mgr.

Karnes City

KTXX(FM)— March 2005: 103.1 mhz; 34 kw. Ant 587 ft TL: N29 00 52 W97 40 02. Hrs open: 2801 Via Fortuna Dr., Suite 675, Austin, 78746. Phone: (512) 329-5843. Fax: (512) 329-5847. Licensee: Palm Broadcasting Co. Format: Talk. ♦Robert Walker, gen mgr.

Keene

***KJCR(FM)**— June 13, 1974: 88.3 mhz; 23 kw. 180 ft TL: N32 24 19 W97 19 55. Stereo. Hrs opn: 6 AM-midnight 304 N. College Dr., 76059. Phone: (817) 556-4788. Fax: (817) 556-4790. Web Site: www.kjcr.org. Licensee: Southwestern Adventist University. Population served: 1,500,000 Law Firm: Donald E. Martin. Format: Relg. News: 8 hrs wkly. Target aud: 18 plus; general. ♦Don Sahly, pres; Randy Yates, gen mgr; Jon Armstrong, opns mgr & progmg dir; Jessica Protasio, sls dir & mus dir; Kristina Pascual, prom dir & news dir; Ron Macomber, chief of engrg.

Kenedy

KTNR(FM)—Licensed to Kenedy. See Kenedy-Karnes City

Kenedy-Karnes City

KAML(AM)— November 1954: 990 khz; 250 w-D, 70 w-N. TL: N28 51 02 W97 52 48. Hrs open: 24 Rt. 1 Box 990, Kenedy, 78119-9719. Secondary address: Box 990, Karnes City 78118. Phone: (830) 583-2990. Fax: (830) 583-0700. Licensee: SIGA Broadcasting Corp. (group owner; (acq 1-17-2002). Population served: 197,000 Natl. Rep: Dome. Wire Svc: Texas News Service Format: Country, news, sports. News staff: 2; News: 8 hrs wkly. Target aud: 24-54; m-f. ♦Gabriel Arango, pres; Clyde Eckols, gen mgr; Clyde S. Eckols, mktg dir & sls; Steve Eckols, progmg dir.

KTNR(FM)—Kenedy, Sept 1, 1982: 92.1 mhz; 3 kw. Ant 220 ft TL: N28 45 35 W97 51 45. (CP: 6 kw, ant 262 ft). Stereo. Hrs opn: Box 1614, Laredo, 78044. Phone: (956) 726-4738. Fax: (956) 722-2184. Web Site: www.lacadenaradioluz.com/ktnr.htm. Licensee: Blue Texas Broadcasting LLC (acq 12-23-2003; $200,000). ♦Israel Tellez Sr., pres & gen mgr; Hiram G. Tellez, opns mgr.

Kerens

KRVF(FM)— May 23, 1979: 106.9 mhz; 21.5 kw. Ant 365 ft TL: N32 06 12 W96 22 33. Stereo. Hrs opn: 1373 S.E. Country Rd. 0070, Coriscana, 75109. Phone: (903) 872-4757. Fax: (903) 885-9107. Licensee: LKCM Radio Group L.P. (group owner; (acq 5-21-2004; $1 million with KRVA-FM Campbell). Population served: 5,000,000 Format: Oldies. ♦Bert Goldman, exec VP; Chris McMurray, gen mgr.

Kermit

KERB(AM)— June 1950: 600 khz; 1 kw-D, DA. TL: N31 50 05 W103 08 10. Hrs open: Box 252, McAllen, 78505. Phone: (956) 781-5528. Fax: (956) 686-2999. Licensee: La Radio Cristiana Network Inc. (acq 5-20-97; $80,000 with co-located FM). Population served: 15,000 Rgnl. Network: Texas State Net. Natl. Rep: Keystone (unwired net). Format: Christian, Sp Christian. ♦Eloy Bernal, gen mgr; Gilbert Martinez, progmg dir.

KERB-FM— 1983: 106.3 mhz; 3 kw. 276 ft TL: N31 50 05 W103 08 10. Web Site: www.laradiocristiana.com.15,000

Kerrville

KCOR-FM—See Comfort

KERV(AM)—Listing follows KRVL(FM).

***KHKV(FM)**— 1998: 91.1 mhz; 300 w. 207 ft TL: N30 02 37 W99 07 17. Hrs open: 24
Rebroadcasts KHCB(AM) Houston-Galveston 95%.
2424 South Blvd., Houston, 77098. Phone: (713) 520-5200. Web Site: www.khcb.org. Licensee: Houston Christian Broadcasters Inc. (group owner). Format: Christian, Sp. ♦Bruce Munsterman, gen mgr.

***KKER(FM)**— Dec 8, 2000: 88.7 mhz; 52 kw. 571 ft TL: N30 03 30 W99 03 50. (CP: 100 kw, ant 380 ft). Hrs opn: 24 Houston Christian Broadcasters Inc., 2424 South Blvd, Houston, 77098. Phone: (713) 520-5200. Web Site: www.khcb.org. Licensee: Houston Christian Broadcasters Inc. (group owner; acq 11-24-00; $3,500 for CP with CP of KHCP(FM) Paris). Natl. Network: Moody. Format: Christian. Spec prog: Sp 6 hrs, Chinese one hr wkly. ♦Bruce Munsterman, gen mgr.

KKVR(FM)—Not on air, target date: unknown: 106.1 mhz; 6 kw. Ant 328 ft TL: N30 02 27 W99 10 19. Hrs open: 24018 Middle Fork, San Antonio, 78258. Phone: (830) 980-7111. Licensee: E-String Wireless Ltd. ♦Bret Huggins, gen mgr.

KRNH(FM)— June 1994: 92.3 mhz; 20 kw. Ant 666 ft TL: N30 03 42 W99 03 43. Hrs open: 24 3505 Fredericksburg Rd., 78028. Phone: (830) 896-4990. Fax: (830) 896-4991. E-mail: phylis@theranchfm92.com Web Site: www.theranchfm92.com. Licensee: Radio Ranch Ltd. (acq 9-6-00; $245,000). Law Firm: Baraff, Koerner & Olender. Format: Country. Target aud: 18-64. Spec prog: Sp 4 hrs, gospel 3 hrs wkly. ♦Mark Grubbs, gen mgr; Leslie Klein, stn mgr.

KRVL(FM)— Sept 12, 1975: 94.3 mhz; 33 kw. Ant 400 ft TL: N30 15 08 W99 08 01. Stereo. Hrs open: 24 2125 Sidney Baker N., 78028. Phone: (830) 896-1230. Fax: (830) 792-4142. Web Site: www.revfmradio.com. Licensee: Barbwire Communications Inc. (acq 8-16-2000; with co-located AM). Population served: 300,000 Natl. Network: ABC. Format: Americana. Target aud: 25-49. Spec prog: Gospel one hr, local church service one hr wkly. ♦Jana Smith, gen mgr; Marti Ashcraft, opns mgr; Glenn Taylor, progmg dir; Diane Philips, traf mgr; Gordon Ames, sls; Monica Smith, sls.

KERV(AM)—Co-owned with KRVL(FM). Nov 5, 1948: 1230 khz; 990 w-U. TL: N30 04 14 W99 11 07. Stereo. 24 Web Site: www.kerv.com. Group owner: Hill Country Broadcasting Corp. 50,000 Natl. Network: ABC. Format: Talk, smooth jazz. News staff: one; News: 3 hrs wkly. Target aud: 45 plus; educated professionals.

Kilgore

KBGE(AM)—Listing follows KKTX-FM.

KKTX-FM— Dec 23, 1976: 96.1 mhz; 50 kw. Ant 492 ft TL: N32 22 14 W94 56 20. Stereo. Hrs open: 3810 Brookside Dr., Tyler, 75701. Phone: (903) 581-0606. Fax: (903) 581-2011. Web Site: www.kktx.com. Licensee: GAP Broadcasting Tyler License LLC. (acq 8-3-2007; grpsl). Population served: 550,000 Format: Classic rock. Target aud: 25-54. ♦Craig Reininger, sls dir; Chris Jones, rgnl sls mgr; Lisa Nix, progmg dir.

KBGE(AM)—Co-owned with KKTX-FM. Dec 26, 1936: 1240 khz; 1 kw-U. TL: N32 25 02 W94 51 15. Web Site: www.kktx.com. Group owner: Clear Channel Communications Inc. 150000 Natl. Rep: Katz Radio, Target Broadcast Sales. Target aud: 25-54.

***KZLO(FM)**— Feb 4, 1991: 88.7 mhz; 63 kw horiz, 79 kw vert. Ant 551 ft TL: N32 20 14 W95 02 41. Stereo. Hrs opn: 19 5700 West Oaks Blvd., Rocklin, CA, 95765. Phone: (916) 251-1600. Fax: (916) 251-1650. Web Site: www.klove.com. Licensee: Educational Media Foundation. (acq 2-15-2007; $2 million). Natl. Network: K-Love. Law Firm: Davis Wright Tremaine LLP. Format: Contemp Christian. ♦Richard Jenkins, pres.

Killeen

KIIZ-FM— Dec 10, 1990: 92.3 mhz; 3 kw. 259 ft TL: N31 06 33 W97 39 00. Stereo. Hrs opn: 24 314 W. State Hwy. 6, Waco, 76712. Phone: (254) 699-5000. Fax: (254) 680-4211. Web Site: www.kiiz.com. Licensee: Capstar TX L.P. Group owner: Clear Channel Communications Inc. (acq 8-30-00; grpsl). Format: Urban contemp. News: 2 hrs wkly. Target aud: 18-49. ♦Tim Thomas, gen mgr & stn mgr; Jim Martin, gen sls mgr; Terry Steele, prom mgr; Julia Conner, news dir & traf mgr; Brett Gilbert, chief of engrg.

***KNCT-FM**— Nov 23, 1970: 91.3 mhz; 50 kw. 1,170 ft TL: N30 59 12 W97 37 47. Stereo. Hrs opn: 24 Box 1800, Central Texas College, 6200 W. Central Texas Expwy., 76542. Phone: (254) 526-1176. Fax: (254) 526-1850. Web Site: www.knct.org. Licensee: Central Texas College. Population served: 240,000 Natl. Network: PRI. Rgnl. Network: Texas AP. Format: Btfl mus, class. Target aud: 45 plus. Spec prog: Jazz 15 hrs wkly, big band 6 hrs wkly. ♦Max Rudolph, gen mgr; Dan Hull, progmg dir; Steve Sulzer, chief of engrg. Co-owned TV: *KNCT(TV) affil

KRMY(AM)— July 4, 1955: 1050 khz; 250 w-D. TL: N31 06 53 W97 42 00. Hrs open: 4638 Decker Dr., Baytown, 77520. Secondary address: 314 N. 2nd St. 76514. Phone: (254) 628-7071. Fax: (254) 634-5263. Licensee: Martin Broadcasting Inc. (group owner; acq 11-89; grpsl; FTR: 11-27-89). Population served: 50,000 Format: Gospel. Target aud: 18-44. ♦Darrell Martin, gen mgr; Horatio Martinez, progmg dir.

KUSJ(FM)—See Temple

Kingsville

KFTX(FM)— May 2, 1970: 97.5 mhz; 100 kw. 1,000 ft TL: N27 30 54 W97 51 58. Stereo. Hrs opn: 24 1520 S. Port Ave., Corpus Christi, 78405. Phone: (361) 883-5987. Fax: (361) 883-3648. Web Site: www.kftx.com. Licensee: Quality Broadcasting Corp. (acq 11-29-2006). Law Firm: Wood, Maines & Brown. Format: Country. News staff: one; News: 2 hrs wkly. Target aud: 25-49; educated, affluent young adults. ♦Bruce Nelson Stratton, gen mgr & stn mgr; Deborah DeSola, sls dir & traf mgr; Kenda West, gen sls mgr & natl sls mgr; Joshua Sandoval, prom VP; Chuck Abel, progmg dir; Austin Daniels, mus dir; Henry Turner, chief of engrg.

KINE(AM)— November 1948: 1330 khz; 1 kw-D, 250 w-N. TL: N27 36 36 W97 47 42. Hrs open: 24 115 W. Avenue D, Rob Towns, 78380. Phone: (361) 855-1330. Fax: (361) 289-7722. Licensee: Cotton Broadcasting. (acq 9-28-90; $50,000; FTR: 10-22-90). Population served: 500,000 Format: Rgnl Sp, relg. Target aud: 25-54. ♦Humberto L. Lopez, CEO; Humberto Lopez, pres; Carlos Lopez, gen mgr; Minerva R. Lopez, dev VP; Ernest Lopez, sls VP; Manuel Lopez, prom VP; Homer Lopez, mus dir; Tommy Greg, chief of engrg.

KKBA(FM)— November 1981: 92.7 mhz; 12.5 kw. 869 ft TL: N27 32 07 W97 53 06. Stereo. Hrs opn: 24 Box 9757, Corpus Christi, 78469. Secondary address: 2117 Leopard St., CorpusChristi 78408. Phone: (361) 883-3516. Fax: (361) 882-9767. E-mail: thechief@star94.net Web Site: www.927kkba.com. Licensee: Malkan Broadcasting L.P. Group owner: Malkan Broadcast Assoc. (acq 9-13-95; FTR: 10-2-95). Population served: 315,000 Natl. Rep: Katz Radio. Law Firm: Thompson Hines, LLP. Format: Adult contemp. News staff: one; News: 20 hrs wkly. Target aud: 25-54. ♦Glen Powers, pres & gen mgr; Janice Raleigh, sls dir; Norma Morales, prom dir; John Gifford, engrg dir & chief of engrg; Bart Allison, progmg.

***KTAI(FM)**— Feb 23, 1970: 91.1 mhz; 100 w. 98 ft TL: N27 31 24 W97 52 42. Stereo. Hrs opn: Noon-12:30 AM (M-F); 4 PM-10 PM (Su) 700 University Blvd., MSC 178, Texas A & M Univ.-Kingsville, 78363. Phone: (361) 593-3489. E-mail: ktaifm@hotmail.com Web Site: www.tamuk.edu/ktai. Licensee: Texas A&M University-Kingsville. Population served: 28,711 Format: Rock. Target aud: 16-25; high school & college ages. Spec prog: Black 8 hrs, gospel 6 hrs, mus from India 3 hrs, mus from Mexico 3 hrs wkly. ♦Rumaldo Juarez, pres & gen mgr.

Knox City

KTSX(FM)—Not on air, target date: unknown: 107.3 mhz; 6 kw. Ant 328 ft TL: N33 25 40 W99 40 00. Hrs open: 2768 Pharmacy Rd., Rio Grande City, 78582. Phone: (956) 487-5621. Licensee: James Falcon. ♦James Falcon, gen mgr.

Krum

KNOR(FM)— Nov 11, 1984: 93.7 mhz; 12 kw. Ant 1,952 ft TL: N33 26 13 W97 29 05. Stereo. Hrs opn: 24 4201 Pool Rd., Colleyville, 76034. Phone: (817) 868-2900. E-mail: Dallasinfo@lbimedia.com Web Site: www.laraza937.com. Licensee: Liberman Broadcasting of Dallas License LLC. Group owner: Liberman Broadcasting Inc. (acq 5-13-2004; $15.5 million). Population served: 100,000 Rgnl. Network: Okla. Radio Net. Natl. Rep: Roslin. Law Firm: Fletcher, Heald & Hildreth. Format: Sp. ♦Alejandro Sanchez, gen mgr.

La Grange

KBUK(FM)—Listing follows KVLG(AM).

KVLG(AM)— June 27, 1959: 1570 khz; 250 w-D, DA-1. TL: N29 52 58 W96 51 57. Hrs open: Box 609, 78945. Secondary address: FM 155 S. 78945. Phone: (979) 968-3173. Phone: (979) 743-4050. Fax: (979) 968-6196. Web Site: www.kvlgkbuk.com. Licensee: Fayette Broadcasting Corp. Population served: 30,000 Rgnl. Network: Texas State Net., VSA Radio. Format: Country. Target aud: General. Spec prog: Ger one hr, Black one hr, farm 4 hrs, Pol/Czech 12 hrs, relg 5 hrs wkly. ♦Roy Cerney, gen mgr & progmg dir.

KBUK(FM)—Co-owned with KVLG(AM). Dec 21, 1970: 104.9 mhz; 3 kw. 203 ft TL: N29 52 57 W96 51 58. (CP: Ant 328 ft.). Stereo. FM Rd. 155, 78945. Phone: (979) 968-3173. Fax: (979) 968-6196. Web Site: www.kvlgkbuk.com.35,000 ♦Roy Cerney, stn mgr.

La Porte

KIOL(FM)—Licensed to La Porte. See Houston

Lake Jackson

KBRZ(AM)—See Freeport

KHTC(FM)— April 1963: 107.5 mhz; 100 kw. Ant 2,000 ft TL: N29 17 16 W95 13 53. Stereo. Hrs open: 24 1990 Post Oak Blvd., Suite 2300, Houston, 77056. Phone: (713) 963-1200. Fax: (713) 622-5457. Web Site: 1075khits.com. Licensee: Cox Radio Inc. Group owner: Cox Television (acq 8-30-2000; grpsl). Population served: 196,100 Format: Oldies. News staff: 2; News: 5 hrs wkly. Target aud: 25-54; college grads from the 60s, 70s & 80s. ♦Mark Krieschen, VP & gen mgr; Rob Hienaman, gen sls mgr; Mike Murray, natl sls mgr; Bill Tatar, mktg dir; Johnny Chiang, progmg dir; Paul Christy, mus dir.

***KYBJ(FM)**— 1995: 91.1 mhz; 5 kw. 459 ft TL: N29 02 37 W95 20 11. Hrs open:
Rebroadcasts KSBJ(FM) Humble 100%.
Box 187, c/o KSBJ(FM), Humble, 77347. Phone: (979) 265-9191. Fax: (800) 966-5925. E-mail: kybj@massnet.net Licensee: Educational Media Foundation of Brazosport. Format: Contemp Christian. Target aud: 25-49. ♦Tim McDermott, gen mgr; Jon Hull, progmg dir; Amanda Carroll, news dir.

Lamesa

***KBKN(FM)**—Not on air, target date: unknown: 91.3 mhz; 250 w. Ant 157 ft TL: N32 45 34 W101 57 09. Hrs open: 1406 E. Garden Ln., Midland, 79702. Phone: (432) 638-1150. Fax: (432) 682-5230. Web Site: www.lapromesa.org. Licensee: La Promesa Foundation. (acq 5-4-2004; $108,000 including six translator stns). Format: Christian. ♦Leonard Oswald, pres & gen mgr.

KPET(AM)— May 21, 1947: 690 khz; 250 w-U. TL: N32 42 27 W101 56 11. Hrs open: 24 Box 1188, One Radio Rd., 79331. Phone: (806) 872-6511. Phone: (806) 872-6537. Fax: (806) 872-6514. E-mail: kpet@pics.net Licensee: KPET Inc. (acq 8-7-91; $150,000). Population served: 16,000 Natl. Network: ABC. Rgnl. Network: Texas State Net. Format: C&W. News staff: one; News: 2 hrs wkly. Target aud: 18-65. ♦Don Sitton, chmn, gen mgr, opns dir & progmg dir; Elaine Githens, gen sls mgr & adv mgr; Grover Clifft, news dir; Anthony Garza, chief of engrg; DeeAnn Martin, traf mgr.

KTXC(FM)— May 1, 1988: 104.7 mhz; 100 kw. 800 ft TL: N32 23 47 W101 57 24. Stereo. Hrs opn: 24 P.O. Box 60403, Midland, 79711. Secondary address: 6 Destra Dr., Suite 6600, Midland 79705. Phone: (915) 570-6670. Fax: (915) 567-9992. Licensee: Graham Brothers Comm. L.L.C. (acq 1999; $270,000). Population served: 500,000 Format: C&W. Target aud: 25-54; college-educated, upper-income families. ♦Terry Graham, pres; Donn Holcomb, gen mgr & gen sls mgr; Tom Rivers, progmg dir.

Lampasas

KCYL(AM)— 1948: 1450 khz; 1 kw-U. TL: N31 04 31 W98 11 02. Hrs open: 5 AM-11 PM 505 N. Key Ave., 76550. Phone: (512) 556-6193. Phone: (512) 556-3671. Fax: (512) 556-2197. E-mail: kcylq102@igg-tx.net Licensee: Ronald K. Witcher. (acq 2-12-85). Population served: 12,000 Rgnl. Network: Texas State Net. Format: C&W. News staff: 3; News: 15 hrs wkly. Target aud: 30 plus; agriculture, farm & ranch. Spec prog: Farm 5 hrs, loc news & community service 14 hrs, sports 8 hrs, relg 10 hrs wkly. ♦Joe Lombardi, prom mgr, progmg dir & disc jockey; Ronnie Witcher, pres, gen mgr, opns mgr, gen sls mgr & chief of engrg; Lela Cooper, traf mgr.

Laredo

***KBNL(FM)**— July 27, 1985: 89.9 mhz; 100 kw. Ant 604 ft TL: N27 39 27 W99 35 10. Stereo. Hrs opn: 24 Box 440029, 78044. Secondary address: 1620 E. Plum St. 78043. Phone: (956) 724-9090/724-9211. Fax: (956) 724-9919. E-mail: kbnlfm@hcjbeat.org Licensee: World Radio Network Inc. (group owner; acq 10-24-85). Format: Relg, Sp. News: 2 hrs wkly. Target aud: General. ♦Arturo Losano, gen mgr.

***KHOY(FM)**— November 1985: 88.1 mhz; 1.8 kw. 348 ft TL: N27 31 14 W99 31 19. Hrs open: 1901 Corpus Christi, 78043. Phone: (956) 722-4167. Fax: (956) 722-4464. Web Site: www.khoy.org. Licensee: Laredo Catholic Communications Inc. Format: Adult contemp, Sp. Target aud: General. ♦Bennett McBride, gen mgr & news dir; Jose Angel Jimenez, progmg dir.

KJBZ(FM)— Dec 29, 1982: 92.7 mhz; 3 kw. 289 ft TL: N27 31 04 W99 31 20. Stereo. Hrs opn: 24 902 E. Calton Rd., 78041. Phone: (956) 726-9393. Fax: (956) 724-9915. Licensee: Encarnacion A. Guerra. (acq 12-14-89; $750,000; FTR: 1-8-90). Format: Tejano. News staff: one; News: 18 hrs wkly. Target aud: General; all ages. ♦Belinda Guerra, VP, mktg dir & prom mgr; Elaine Matthews, opns dir & progmg dir; Roberto Estrada, gen mgr & gen sls mgr; Eric Navarro, mus dir; Joe Martinez, chief of engrg.

KLAR(AM)— 1956: 1300 khz; 1 kw-D, 80 w-N. TL: N27 31 45 W99 31 15. Stereo. Hrs opn: 24 Box 2517, 78044. Secondary address: 3320 Anna Ave. 78040-1070. Phone: (956) 723-1300. Fax: (956) 723-9539. Licensee: Faith and Power Communications Inc. (acq 1996). Population served: 230,000 Format: Sp Christian. News staff: 2; News: 18 hrs wkly. Target aud: 18-54; Hispanic & Anglo middle to upper-middle class. ♦Hector Patino, pres, gen mgr & gen sls mgr.

KLNT(AM)— Apr 20, 1990: 1490 khz; 1 kw-U. TL: N27 29 41 W99 28 16. Hrs open: 24 107 Calle Del Norte, Suite 212, 78041. Phone: (956) 725-1000. Fax: (956) 794-9155. Licensee: BMP 100.5 FM L.P. Group owner: Amigo Broadcasting L.P. (acq 11-9-2004; grpsl). Law Firm: Fisher, Wayland, Cooper, Leader & Zaragoza L.L.P. Format: Rgnl Mexican. ♦Thomas H. Castro, pres; Raul Rodriguez, gen mgr; Ruben Villareal, opns dir; Joe Flores, sls.

KNEX(FM)— 1992: 106.1 mhz; 3 kw. 213 ft TL: N27 33 12 W99 24 17. Hrs open: 505 Houston St., 78040. Phone: (956) 725-1491. Phone: (956) 725-1492. Fax: (956) 725-3424. Licensee: BMP 100.5 FM L.P. Group owner: Amigo Broadcasting L.P. (acq 11-9-2004; grpsl). Format: CHR, Sp. ♦Miguel A. Villarreal Jr., gen mgr; Jorge Arredondo, gen sls mgr; Ruben Villarreal, opns mgr & progmg dir; Artura Serna, mus dir; Deyla Villarreal, news dir & traf mgr; Arturo Trevino, chief of engrg.

KQUR(FM)— Feb 2, 1972: 94.9 mhz; 100 kw. Ant 1,000 ft TL: N27 31 14 W99 31 19. Stereo. Hrs opn: 505 Houston St., 78040. Phone: (956) 725-1491. Fax: (956) 725-3424. Licensee: Border Broadcasters Inc. Population served: 1000000 Natl. Rep: Roslin. Law Firm: Fisher, Wayland, Cooper, Leader & Zaragoza. Format: Classic hits. Target aud: 25-54; general. ♦Miguel Villarreal, gen mgr; Ruben Villarreal, opns dir & opns mgr; Jorge Redando, gen sls mgr; Deyla Villarreal, progmg dir & traf mgr; Arturo Trevino, chief of engrg; Al Guevara, progmg.

KRRG(FM)— October 1982: 98.1 mhz; 100 kw. Ant 737 ft TL: N27 31 14 W99 31 19. Stereo. Hrs opn: 902 E. Calton Rd., 78041. Phone: (956) 724-9800. Fax: (956) 724-9915. Web Site: www.krrg.com. Licensee: Guerra Enterprises (acq 11-20-92; $1.2 million; FTR: 12-21-92). Natl. Rep: D & R Radio. Format: Adult contemp, CHR. ♦Belinda Guerra, pres & VP; Elaine Matthews, opns dir & progmg dir; Roberto Estrada, gen mgr, stn mgr & gen sls mgr; Joe Martinez, chief of engrg; Erica Navarro, traf mgr.

Leakey

KBDK(FM)—Not on air, target date: unknown: 93.1 mhz; 50 kw. Ant 397 ft TL: N29 46 41 W99 45 06. Hrs open: 3654 W. Jarvis Ave., Skokie, IL, 60076. Phone: (847) 674-0864. Licensee: KM Communications Inc. ♦Kevin Joel Bae, VP.

KBLT(FM)— June 10, 1997: 104.3 mhz; 1 kw. 594 ft TL: N29 41 34 W99 48 56. Hrs open: Box 56, 78873. Secondary address: 935 East Main, Uvalde 78801. Phone: (830) 278-3693. Fax: (830) 278-2329. Licensee: Radio Cactus Ltd. (acq 10-23-00; $60,916 for 51% of stock with KBNU(FM) Uvalde). Population served: 10,000 Format: Contemp Christian. Target aud: General. ♦John Furr, pres & chief of engrg; Regenia Tumbarello, gen mgr.

Leander

KHHL(FM)— May 16, 1976: 98.9 mhz; 25 kw. Ant 538 ft TL: N30 23 26 W97 50 13. Stereo. Hrs opn: 24 2211 S. I H 35th Suite 401, 7524 N. Lamar Blvd., Suite 200, Austin, 78741. Phone: (512) 416-1100. Fax: (512) 416-8205. Licensee: BMP Austin License Company L.P. Group owner: Amigo Broadcasting L.P. (acq 11-9-2004; grpsl). Population served: 926,300 Natl. Network: ABC, Westwood One. Natl. Rep: D & R Radio. Law Firm: Wilkinson Barker Knauer. Format: Rgnl Mexican. Target aud: 25-54; adults. ♦Paul Danitz, gen mgr.

Levelland

KJDL-FM— Nov 8, 1983: 105.3 mhz; 25 kw. Ant 298 ft TL: N33 34 54 W102 23 48. Stereo. Hrs opn: 1603 13th St., Suite 210, Lubbock, 79401. Phone: (806) 744-6864. Fax: (806) 744-8018. Web Site: 1053jack.com. Licensee: Walker FM Holdings LLC (acq 7-25-2007; $900,000). Format: Adult hits. ♦Dave Walker, gen mgr.

KLVT(AM)— August 1949: 1230 khz; 1 kw-U. TL: N33 35 49 W102 23 09. Hrs open: Box 967, 79336. Secondary address: 611 N. West Ave. 79336. Phone: (806) 894-3134. Fax: (806) 894-3135. E-mail: klvtvl@aol.com Licensee: Profit Programming of Northern Texas (acq 1-7-2007; $200,000). Rgnl. Network: Texas State Net. Format: Classic country. ♦Jody Rose, gen mgr, opns mgr & progmg dir; Anthony Garza, chief of engrg.

Lewisville

KESS-FM— Apr 10, 1999: 107.9 mhz; 100 kw. Ant 981 ft TL: N33 19 42 W97 03 56. Hrs open: 24 7700 John Carpenter Fwy., Dallas, 75247. Phone: (214) 525-0400. Fax: (214) 631-1154. Web Site: www.univisionradio.com. Licensee: KECS-FM License Corp. Group owner: Univision Radio (acq 9-22-2003; grpsl). Format: Mexican rgnl. ♦Frank Carter, gen mgr; Andy Lockridge, opns dir; Cipriano Robles, sls dir; Karen Hocking, natl sls mgr; Oscar Espinosa, prom dir; Herminio (Chayan) Ortuno, progmg dir.

Liberty

KSHN-FM— November 1977: 99.9 mhz; 26 kw. 679 ft TL: N30 03 05 W94 31 37. Stereo. Hrs opn: 24 2099 Sam Houston St., 77575-4817. Phone: (936) 336-5793. Fax: (936) 336-5250. E-mail: kshn@kshn.com Web Site: www.kshn.com. Licensee: Trinity River Valley Broadcasting Co. (acq 11-77). Population served: 700,000 Rgnl. Network: Texas State Net. Format: Adult contemp, oldies, country. News staff: one; News: 36 hrs wkly. Target aud: 34 plus; Adults. Spec prog: Black 3 hrs, bluegrass 2 hrs, relg 5 hrs wkly.Bill Buchanan, CEO, pres, gen mgr, gen sls mgr, adv mgr & sports cmtr; Eric Latz, progmg dir, news rptr & disc jockey; Tiffany York, dev dir & news dir; Barbara Moss, pub affrs dir & traf mgr; James Stephenson, chief of engrg; Kevin Ladd, reporter; Bill Buchannan, edit dir; Larry Wilburn, outdoor ed; Larry Wazeck, sports cmtr; Allen Wayne, disc jockey; Cassie Litlon, disc jockey; Keleen Carlson, disc jockey

Littlefield

KZZN(AM)— 1947: 1490 khz; 1 kw-U. TL: N33 56 17 W102 20 38. Hrs open: Box 967, Levelland, 79336. Phone: (806) 385-4474. Phone: (806) 385-1490. Fax: (806) 385-6229. Licensee: Profit Programming of Northern Texas (acq 1-7-2007). Population served: 20,000 Natl. Network: USA. Rgnl. Network: Texas State Net., Texas Agribus. Format: Country, gospel. Target aud: General; try to reach all ages. Spec prog: Farm 7 hrs, relg 5 hrs wkly. ♦Paul Beane, gen mgr; Mike Rader, opns mgr & progmg dir; Emil Macha, sls dir; Anthony Garza, chief of engrg.

Livingston

KETX-FM— Sept 1, 1970: 92.3 mhz; 32 kw. Ant 607 ft TL: N30 44 18 W94 55 26. Stereo. Hrs opn: 115 Radio Rd., 77351. Phone: (936) 327-8916. Fax: (936) 327-8477. Licensee: Peggy Sue Marsh, administrator (acq 2-27-2007; with co-located AM). Population served: 7,000 Law Firm: Eugene T. Smith. Format: Country. Target aud: General.

KETX(AM)— June 28, 1957: 1440 khz; 5 kw-D. TL: N30 44 23 W94 55 30.6 AM-midnight 32,000

Llano

KAJZ(FM)— 2000: Stn currently dark. 96.3 mhz; 2.9 kw. Ant 459 ft TL: N30 41 12 W98 34 16. Hrs open: 3102 Oak Lawn Ave., Dallas, 75219. Phone: (512) 383-1112. Licensee: Rawhide Radio LLC. Group owner: Univision Radio (acq 9-22-2003; grpsl). ♦B. Shane Fox, gen mgr.

KITY(FM)— 2004: 102.9 mhz; 2 kw. Ant 495 ft TL: N30 40 37 W98 33 59. Hrs open: 1809 Lightsey Rd., Austin, 78704. Phone: (512) 444-9268. Licensee: Bryan A. King (acq 3-9-2004; grpsl). Natl. Network: CNN Radio, Westwood One. Format: Oldies. ♦Bryan King, gen mgr.

Lockhart

KFIT(AM)—Licensed to Lockhart. See Austin

Lometa

KACQ(FM)— 1996: 101.9 mhz; 6 kw. 328 ft TL: N31 14 33 W98 19 19. Hrs open: 505 N. Key Ave., Lampasas, 76550. Phone: (512) 556-6193. Fax: (512) 556-2197. Licensee: Debra L. Witcher. Format: Country. ♦Norma Spinner, sls dir; Joe Lombardi, progmg dir; Lela Cooper, news dir & traf mgr; Ronnie Witcher, gen mgr & chief of engrg.

Longview

KFRO(AM)— Feb 6, 1935: 1370 khz; 1 kw-U, DA-N. TL: N32 30 07 W94 42 12. Stereo. Hrs opn: 24 4408 N US Highway 259, 75605-7703. Phone: (903) 663-9800. Fax: (903) 663-9458. Web Site: www.kykx.com. Licensee: Access.1 Texas License Company LLC. Group owner: Waller Broadcasting (acq 1-7-2005; grpsl). Population served: 150,000 Natl. Network: ABC, Westwood One. Rgnl. Network: Texas State Net. Law Firm: Kaye, Scholer, Fierman, Hays & Handler L.L.P. Format: News/talk, sports. News staff: one; News: 30 hrs wkly. Target aud: 25-54; general. Spec prog: Black 3 hrs wkly. ♦Sydney L. Small, CEO; Chesley Maddox-Dorsey, pres; Debbie Tilley, CFO; Richard Guest, gen mgr; Robert Taylor, sls dir; Dru Laborde, progmg dir; Sans Hawkins, engrg dir; Shelley Miller, traf mgr.

KYKX(FM)— July 1, 1974: 105.7 mhz; 100 kw. 1,156 ft TL: N32 36 04 W94 52 15. Stereo. Hrs opn: 24 Box 5818, 481 E. Loop 281, 75608-5818. Phone: (903) 663-9800. Phone: (903) 663-3700. Fax:

(903) 663-9458. Web Site: www.kykx.com. Licensee: Access.1 Texas License Company LLC. Group owner: Waller Broadcasting (acq 1-7-2005; grpsl). Population served: 250,000 Natl. Rep: McGavren Guild. Format: Modern country. News staff: one; News: 6 hrs wkly. Target aud: General. ♦Richard Guest, gen mgr; Ginger Nimmons, gen sls mgr; Dru LaBorde, progmg dir; Tom Metzger, gen mgr & news dir; Sans Hawkins, chief of engrg; Shirley Bread, traf mgr.

Lorenzo

KKCL(FM)— 1989: 98.1 mhz; 50 kw. 431 ft TL: N33 36 32 W101 43 45. Stereo. Hrs opn: 24 4413 82nd St., Suite 300, Lubbock, 79424. Phone: (806) 798-9880. Web Site: 98kool.com. Licensee: GAP Broadcasting Lubbock License LLC. Group owner: Clear Channel Communications Inc. (acq 8-3-2007; grpsl). Population served: 350,000 Natl. Network: ABC. Format: Oldies. News staff: 8; News: 14 hrs wkly. Target aud: 25-54; upscale 55% male, 45% female. Spec prog: Talk 17 hrs wkly. ♦Scott Parsons, gen mgr.

Los Ybanez

KYMI(FM)— December 1990: 98.5 mhz; 50 kw. 459 ft TL: N32 43 22 W102 01 50. Stereo. Hrs opn: 24 Box 15, 1919 County Rd. M., 79331-7939. Phone: (806) 872-6554. Fax: (806) 872-6244. Licensee: Israel Ybanez. (acq 3-8-90; FTR: 4-2-90). Population served: 20,000 Natl. Network: Westwood One. Format: Sp, relg, Christian. News staff: one; News: 10 hrs wkly. Target aud: 25-49. Spec prog: Gospel 16 hrs, news/talk 10 hrs wkly. ♦Israel Ybanez, pres & opns VP; Mary Ybanez, opns mgr; Genaro Guerrera, progmg dir; Patrick Park, chief of engrg.

Lovelady

KHMR(FM)—Not on air, target date: unknown: 104.3 mhz; 10.5 kw. Ant 508 ft TL: N31 11 30 W95 29 32. Hrs open: 3654 W. Jarvis Ave., Skokie, IL, 60076. Phone: (847) 674-0864. Fax: (847) 674-9188. Web Site: www.kmcommunications.com. Licensee: KM Communications Inc. ♦Kevin Joel Bae, VP & gen mgr.

Lubbock

***KAMY(FM)**— Oct 1, 1990: 90.1 mhz; 200 w. 492 ft TL: N33 30 08 W101 52 20. Stereo. Hrs opn: 24 5124-C 69th St., 79424. Phone: (806) 794-1766. Fax: (806) 798-3251. E-mail: kamy@flc.org Web Site: kamyfm.org. Licensee: Family Life Broadcasting Inc. Group owner: Family Life Broadcasting System (acq 6-24-98; grpsl). Population served: 207,000 Format: Christian. Target aud: 28 plus; 35-54 female; Christian community of Lubbock. ♦Dave Borowsky, prom dir; Don Webster, gen mgr, gen sls mgr & prom mgr.

KBZO(AM)— April 1953: 1460 khz; 1 kw-D, 250 w-N. TL: N33 32 53 W101 49 24. Hrs open: 24 1220 Broadway, Ste. 600, 79401. Phone: (806) 763-6051. Fax: (806) 744-8363. E-mail: jsauceda@cntravision.com Web Site: www.kbzo-am.entravision.com. Licensee: Entravision Holdings LLC. Group owner: Entravision Communications Corp. (acq 10-7-99). Population served: 250,000 Natl. Rep: Lotus Entravision Reps LLC. Format: Mexican rgnl. Target aud: Hispanic. ♦Jose Sauceda, gen mgr.

KDAV(AM)— May 14, 1947: 1590 khz; 1 kw-U, DA-2. TL: N33 31 16 W101 46 28. Hrs open: 1714 Buddy Holly Ave., 79401. Phone: (806) 744-5859. Phone: (806) 770-5328. Fax: (806) 744-5888. E-mail: radio@door.net Web Site: www.kdav.com. Licensee: Renaissance Broadcasting Inc. (acq 7-29-98; $150,000). Population served: 250,000 Wire Svc: ESSA Weather Service Format: Oldies/Rock & Roll. News staff: one; News: 5 hrs. wkly. Target aud: 50+. ♦Bill Clement, pres; Bud Andrews, gen mgr.

KEJS(FM)— 1993: 106.5 mhz; 34 kw. 587 ft TL: N33 30 08 W101 52 20. Hrs open: 1607 13th St., 79401. Phone: (806) 747-5951. Fax: (806) 747-3524. E-mail: ebarton@kejsfm.com Licensee: Barton Broadcasting Co. Format: Tejano. ♦Ernest Barton, gen mgr; Debra Alcorte, mktg dir & traf mgr; Gilbert Esparza, progmg dir.

KFMX-FM—Listing follows KKAM(AM).

KFYO(AM)— Sept 6, 1927: 790 khz; 5 kw-D, 1 kw-N, DA-3. TL: N33 27 50 W101 55 30. Stereo. Hrs opn: 24 Box 53120, 79464-4670. Secondary address: 4413 82nd St., Suite 300 79424. Phone: (806) 794-7979. Fax: (806) 794-1660. Licensee: GAP Broadcasting Lubbock License LLC. Group owner: Clear Channel Communications Inc. (acq 8-3-2007; grpsl). Population served: 186,000 Natl. Network: CBS. Rgnl. Network: Texas State Net. Format: News/talk. News staff: 2; News: 12 hrs wkly. Target aud: General. Spec prog: Farm 15 hrs, relg 6 hrs wkly. ♦Scott Parsons, gen mgr; Wes Nessman, opns dir; Matt Martin, sls dir; Robert Snyder, progmg dir & traf mgr; Roger Taylor, chief of engrg; Jim Stewart, farm dir.

KZII-FM—Co-owned with KFYO(AM). Mar 10, 1982: 102.5 mhz; 100 kw. 850 ft TL: N33 31 05 W101 51 25. Stereo. 24 Format: CHR. News staff: one; News: 5 hrs wkly. Target aud: 18-49. ♦Kidd Carson, progmg dir.

KJAK(FM)—Slaton, Feb 12, 1978: 92.7 mhz; 100 kw. 584 ft TL: N33 32 32 W101 50 14. Stereo. Hrs opn: 24 Box 6490, 79493. Phone: (806) 745-6677. Fax: (806) 745-8140. E-mail: kjak@door.net Web Site: www.kjak.com. Licensee: G.O. Williams Oil Co. Inc. dba Williams Broadcasting Group (acq 6-19-81; FTR: 7-13-81). Population served: 750,000 Format: Christian. Target aud: General; Christians & those looking for answers to everyday problems. Spec prog: Sports 5 hrs wkly. ♦Woody Van Dyke, gen mgr, gen sls mgr, prom mgr & mus dir; Bob Howell, news dir; Roger Taylor, chief of engrg.

KJDL(AM)— Nov 15, 1966: 1420 khz; 500 w-U, DA-N. TL: N33 36 49 W101 52 30. Hrs open: 18 1603 13th St., Suite 210, 79401. Phone: (806) 765-8114. Fax: (806) 763-0428. E-mail: dwalker@newsradio1420.com Web Site: www.newsradio1420.com. Licensee: Walker Broadcasting & Communications Ltd. (acq 9-14-2005; $350,000). Population served: 179,000 Rgnl. Network: Texas State Net. Format: News/talk. ♦David Walker, pres & gen mgr; Bill Enloe, chief of engrg; Helen Castro, prom mgr, progmg dir & traf mgr.

KJTV(AM)—Listing follows KXTQ-FM.

KKAM(AM)— Jan 1, 1955: 1340 khz; 1 kw-U. TL: N33 33 24 W101 51 46. Hrs open: 4413 82nd St., Suite 300, 79424-3366. Phone: (806) 798-7078. Fax: (806) 798-7052. Web Site: www.kfmx.com. Licensee: GAP Broadcasting Lubbock License LLC. Group owner: Clear Channel Communications Inc. (acq 8-3-2007; grpsl). Natl. Network: ABC, CBS. Format: Sports. ♦Scott Parsons, gen mgr; Wes Nessman, opns dir; Matt Martin, sls dir; Mark Finkner, progmg dir.

KFMX-FM—Co-owned with KKAM(AM). Aug 1, 1966: 94.5 mhz; 100 kw. 817 ft TL: N33 31 05 W101 51 25. Stereo. 376,900 Format: AOR. ♦Wes Nessman, progmg dir.

***KKLU(FM)**— Oct 24, 1993: 90.9 mhz; 13.5 kw. Ant 236 ft TL: N33 32 30 W101 49 16. Stereo. Hrs opn: 24 5700 W. Oaks Blvd., Rocklin, 95765. Phone: (916) 251-1600. Fax: (916) 251-1650. Web Site: www.klove.com. Licensee: Educational Media Foundation. Group owner: EMF Broadcasting (acq 5-1-2000; $750,000 with KXRI(FM) Amarillo). Natl. Network: K-Love. Format: Contemp Christian mus. Target aud: All ages. ♦Richard Jenkins, pres & gen mgr; Mike Novak, progmg VP.

KLLL-FM— Mar 1, 1958: 96.3 mhz; 100 kw. 817 ft TL: N33 31 05 W101 51 25. Stereo. Hrs opn: 24 33 Briercroft Office Park., 70412. Phone: (806) 762-3000. Fax: (806) 770-5363. Web Site: www.klll.com. Licensee: Wilks License Co. -Lubbock LLC. Group owner: NextMedia Group L.L.C. (acq 8-19-2005; grpsl). Population served: 400,000 Natl. Network: ABC. Format: Country. Target aud: 25-54. ♦Scott Harris, gen mgr; Jeff Scott, opns mgr & progmg dir; Jay Richards, gen sls mgr; Rick Gilbert, prom dir & prom mgr; Kelly Greene, mus dir; Stacey James, news dir; Randy Hayes, chief of engrg; Julia Aguilar, traf mgr & women's int ed.

***KOHM(FM)**— January 1973: 89.1 mhz; 70 kw. Ant 567 ft TL: N33 34 55 W101 53 25. Stereo. Hrs open: 24 1901 University Ave., Suite 603-B, Texas Tech Univ., 79410. Secondary address: Box 45891 79409. Phone: (806) 742-3100. Fax: (806) 742-3716. E-mail: kohm@ttv.edu Web Site: www.kohm.org. Licensee: Texas Tech University. (acq 11-87). Population served: 250,000 Natl. Network: NPR, PRI. Format: Classical. ♦Derrick Ginter, gen mgr; Sherril Skibell, dev dir; Clinton Barrick, progmg dir. Co-owned TV: *KTXT-TV affil.

KONE(FM)— 1975: 101.1 mhz; 100 kw. 882 ft TL: N33 30 08 W101 52 20. Stereo. Hrs opn: 24 33 Briercroft Office Park, 79412. Phone: (806) 762-3000. Fax: (806) 762-8419. Web Site: www.cr101.com. Licensee: Wilks License Co.-Lubbock LLC. Group owner: NextMedia Group L.L.C. (acq 8-19-2005; grpsl). Population served: 400,000 Natl. Network: ABC. Format: Soft adult contemp, classic rock. Target aud: 25-54. ♦Scott Harris, gen mgr; Jeff Scott, opns mgr, progmg VP & progmg dir; Jay Richards, gen sls mgr; Rick Gilbert, prom dir; Kelly Greene, mus dir; Stacey James, news dir; Randy Hayes, chief of engrg; Julia Aguilar, traf mgr.

KQBR(FM)— July 15, 1964: 99.5 mhz; 100 kw. 817 ft TL: N33 31 05 W101 51 25. Stereo. Hrs opn: Box 53120, 4413 82nd St., Suite 300, 79424. Phone: (806) 798-7078. Fax: (806) 798-7052. E-mail: jacquineal@clearchannel.com Web Site: www.kqbr.com. Licensee: GAP Broadcasting Lubbock License LLC. Group owner: Clear Channel Communications Inc. (acq 8-3-2007; grpsl). Population served: 125000 Format: Country. ♦Scott Parsons, gen mgr; Wes Nessmann, opns dir; Leslie Tucker, gen sls mgr; Tina Hill, gen sls mgr; Jackie Neal, progmg dir; Landon King, news dir; Roger Taylor, chief of engrg; Kris Torres, traf mgr.

KRFE(AM)— Sept 19, 1953: 580 khz; 500 w-D, 290 w-N, DA-2. TL: N33 32 00 W101 49 14. Hrs open: 24 6602 Martin Luther King Blvd., 79404. Phone: (806) 745-1197. Fax: (806) 745-1088. Web Site: www.krfeam580.com. Licensee: KRFE Radio Inc. (acq 2-94). Population served: 300,000 Natl. Network: ABC. Format: Easy lstng, news/talk. News staff: one; News: 5 hrs wkly. Target aud: 40 plus. Spec prog: News/talk 15 hrs wkly. ♦Wade Wilkes, gen mgr & prom VP.

***KTXT-FM**— Apr 1, 1961: 88.1 mhz; 35 kw. 423 ft TL: N33 34 55 W101 53 25. Stereo. Hrs opn: 24 Box 43081, Texas Tech Univ., 79409-3081. Phone: (806) 742-3388. Fax: (806) 742-2434. Web Site: www.ktxt.net. Licensee: Texas Tech University. Population served: 250,000 Format: Div. ♦Nick Carissimi, stn mgr & chief of engrg; Ali Rana, progmg dir; Sheri Lewis, adv.

KXTQ-FM— November 1963: 93.7 mhz; 100 kw. 740 ft TL: N33 30 57 W101 50 54. Stereo. Hrs opn: 24 Box 3757, 79452. Secondary address: 9800 University Ave. 79423. Phone: (806) 745-3434. Fax: (806) 748-2470. E-mail: cheinz@ramarcom.com Web Site: www.magic937fm.com. Licensee: Ramar Communications II Ltd. (group owner; (acq 9-93; $362,500). Population served: 375,000 Natl. Rep: Univision Radio National Sales. Law Firm: Leventhal, Senter & Lerman. Format: Tejano, Sp. Target aud: 18-49. ♦Brad Moran, pres; Chuck Heinz, gen mgr; Connie Hayes, sls dir & gen sls mgr; Eddie Moreno, progmg dir & progmg mgr; Susie Gonzales, traf mgr; Tee Thomas, chief of engrg & traf mgr. Co-owned TV: KJTV-TV affil.

KJTV(AM)—Co-owned with KXTQ-FM. Nov 1, 1946: 950 khz; 5 kw-D, 500 w-N, DA-2. TL: N33 34 53 W101 49 38.24 Web Site: www.magic937fm.com. Format: News/talk. News staff: one; News: 147 hrs wkly. Target aud: 25-54. Spec prog: Sports. ♦Chuck Heinz, sls dir.

Lufkin

KAFX-FM—Diboll, June 29, 1960: 95.5 mhz; 100 kw. Ant 567 ft TL: N31 24 28 W94 45 53. Stereo. Hrs opn: 24 Box 2209, 75902-2209. Secondary address: 1216 S. 1st St. 75901-4716. Phone: (936) 639-4455. Fax: (936) 639-5540. E-mail: johnnylathrop@clearchannel.com Web Site: kfox95.com. Licensee: GAP Broadcasting Lufkin License LLC. Group owner: Clear Channel Communications Inc. (acq 8-3-2007; grpsl). Population served: 250,000 Format: Hot adult contemp. News staff: one; News: 2 hrs wkly. Target aud: 25-44; female. ♦Johnny Lathrop, gen mgr; Tami Koonce, sls dir & natl sls mgr.

***KAVX(FM)**— Dec 25, 1998: 91.9 mhz; 20 kw. 787 ft TL: N31 22 08 W94 38 43. Stereo. Hrs opn: 24 Box 151340, 75915-1340. Secondary address: 151 Holmes Rd., Lurkin 75904. Phone: (936) 639-5673. Fax: (936) 639-5677. E-mail: alross@kavx.org Web Site: www.kavx.org. Licensee: Lufkin Educational Broadcasting Foundation. Natl. Network: USA. Law Firm: Gammon & Grange. Format: Teaching & talk. Target aud: 35 plus. ♦Dwyan Calvert, chmn & gen mgr; Al Ross, stn mgr, opns mgr & prom mgr.

***KLDN(FM)**— May 2, 1991: 88.9 mhz; 50 kw. Ant 649 ft TL: N31 24 28 W94 45 53. Stereo. Hrs opn: 24
Rebroadcasts KDAQ(FM) Shreveport, LA 100%.
Box 5250, Shreveport, LA, 71135. Phone: (318) 797-5150. Phone: (800) 552-8502. Fax: (318) 797-5265. E-mail: listenermail@redriverradio.org Web Site: www.redriverradio.org. Licensee: Board of Supervisors of Louisiana State University. Natl. Network: NPR, PRI. Format: Classical, news, jazz. Target aud: 25+. ♦Kermit Poling, gen mgr; Rick Shelton, opns mgr.

KRBA(AM)— May 3, 1938: 1340 khz; 1 kw-U. TL: N31 21 51 W94 43 09. Hrs open: 24 Box 1345, 75901. Secondary address: 121 Cotton Sq. 75901. Phone: (936) 634-6661. Fax: (936) 632-5722. E-mail: kybi@lcc.net Web Site: www.krba.net. Licensee: Stephen W. Yates. Population served: 30,000 Format: Var/div. News staff: one; News: 7 hrs wkly. Target aud: General. ♦Stephen Yates, gen mgr; Kevin Sims, progmg dir & disc jockey; Jeremy Chance, traf mgr & disc jockey.

KYBI(FM)—Co-owned with KRBA(AM). May 1, 1978: 100.1 mhz; 25 kw. Ant 699 ft TL: N31 24 28 W94 45 53. Stereo. 24 Phone: (936) 634-5100. Web Site: www.kybi.com.150,000 Format: Adult contemp. Target aud: 25-54. ♦Jeremy Chance, traf mgr; J.P. Heath, local news ed & news rptr.

KSML(AM)—Diboll, June 2, 1957: 1260 khz; 4.5 kw-D, 72 w-N. TL: N31 21 53 W94 43 08. Hrs open: 24 121 Cotton Sq., 75901. Phone: (936) 632-8444. Fax: (936) 632-8451. Licensee: Stephen W. & Karla Yates. (acq 5-95; FTR: 5-22-95). Format: Sp. News staff: one; News: 14 hrs wkly. ♦Stephen W. Yates, pres & gen mgr; Oscar Chavez, progmg dir; Steve Comer, chief of engrg.

***KSWP(FM)**— Aug 31, 1985: 90.9 mhz; 380 w. 174 ft TL: N31 23 17 W94 46 43. (CP: 90.9 mhz, 30 kw, ant 787 ft.). Stereo. Hrs opn: 24 151 Holmes Rd., 75904. Phone: (936) 639-5673. Fax: (936) 639-5677. Web Site: www.kswp.org. Licensee: Lufkin Educational Broadcasting Foundation. Natl. Network: USA. Format: Christian mus. Spec prog: Pub affrs talk show 2 hrs wkly. ♦Dwyan Calvert, pres & gen mgr; Al Ross, opns mgr & progmg dir.

KYKS(FM)— July 9, 1976: 105.1 mhz; 100 kw. 1,066 ft TL: N31 22 08 W94 38 45. Stereo. Hrs opn: 24 Box 2209, 75901. Secondary address: 1216 S. First St. 75901. Phone: (936) 639-4455. Fax: (936) 632-5957. Fax: (936) 639-5540. E-mail: dannymerrell@kicks105.com Web Site: www.kicks105.com. Licensee: GAP Broadcasting Lufkin License LLC. Group owner: Clear Channel Communications Inc. (acq 8-3-2007; grpsl). Population served: 250,000 Law Firm: Fletcher, Heald & Hildreth. Format: Country. News staff: one. Target aud: 25-54. ♦Larry Gunter, gen mgr; Johnny Lathrop, sls dir; Danny Merrell, prom dir, progmg dir & pub affrs dir; Sean Ericson, mus dir; Brandy Abney, traf mgr.

Luling

KAMX(FM)— Mar 22, 1987: 94.7 mhz; 99 kw. 1,305 ft TL: N30 19 23 W97 47 58. Stereo. Hrs opn: 24 4301 Westbank Dr., Escalade B—3rd Fl., Bldg. B. Suite 350, Austin, 78746. Phone: (512) 327-9595. Fax: (512) 329-6255. E-mail: jdhiatt@cbs.com Web Site: mix947.com. Licensee: Texas Infinity Radio L.P. Group owner: Infinity Broadcasting Corp. (acq 11-13-98; grpsl). Population served: 700,000 Natl. Rep: Katz Radio. Law Firm: Levanthal, Senter & Lerman. Format: Modern adult contemp. Target aud: 18-49; upscale adults. Spec prog: Pub affrs 2 hrs wkly. ♦Clint Culp, sr VP & sls dir; John Hiatt, sr VP, mktg dir & mktg mgr.

Lytle

***KZLV(FM)**— Jan 20, 1990: 91.3 mhz; 2.95 kw. 302 ft TL: N29 14 39 W98 44 27. Stereo. Hrs opn: 24 1566 N.E. Loop 410, San Antonio, 78209. Phone: (210) 824-9100. Fax: (210) 824-8870. Web Site: www.klove.com. Licensee: Educational Media Foundation. Group owner: EMF Broadcasting (acq 4-28-99). Natl. Network: K-Love. Format: Adult contemp, Christian. Target aud: 25-49; professional adult & parents. ♦Dick Jenkins, pres; Lloyd Parker, gen mgr; Ed Lenane, opns dir.

Mabank

KTXV(AM)—Not on air, target date: unknown: 890 khz; 20 kw-D, 250 w-N, DA-2. TL: N32 17 13 W95 58 39. Hrs open: Box 60991, Palo Alto, CA, 94306. Licensee: JNE Investments Inc.

Madisonville

KAGG(FM)— Dec 5, 1989: 96.1 mhz; 50 kw. 500 ft TL: N30 48 02 W96 07 00. Stereo. Hrs opn: 24 1730 Briarcrest Dr., Suite 150, Bryan, 77802. Phone: (979) 268-9696. Fax: (979) 268-9090. E-mail: info@aggie96.com Web Site: www.aggie96.com. Licensee: CCB Texas Licenses L.P. Group owner: Clear Channel Communications Inc. (acq 10-10-00; grpsl). Law Firm: Fletcher, Heald & Hildreth. Format: Country. ♦Jan Stott, gen mgr; Kathy Vaughan, gen sls mgr; Nathan Peacock, gen sls mgr; Jennifer Allen, progmg dir; Ed Loftus, chief of engrg; Stacy Colvin, traf mgr.

***KHML(FM)**— 2006: 91.5 mhz; 95 kw. Ant 341 ft TL: N31 06 39.6 W95 57 08.6. Hrs open:
Rebroadcasts KHCB-FM Houston 100%.
2424 South Blvd., Houston, 77098-5110. Phone: (713) 520-5200. Web Site: www.khcb.org. Licensee: Houston Christian Broadcasters Inc. Format: Christian. ♦Bruce Munsterman, pres & gen mgr.

KKLB(FM)—Not on air, target date: unknown: 101.3 mhz; 3.1 kw. Ant 462 ft TL: N31 00 00 W95 59 58. Hrs open: 3500 Maple Ave., Suite 1320, Dallas, 75219-1622. Phone: (214) 363-6030. Licensee: Katherine Pyeatt. ♦Katherine Pyett, gen mgr.

KMVL(AM)— October 1989: 1220 khz; 500 w-D, 12 w-N. TL: N30 57 56 W95 53 52. Hrs open: 24 102 W. Main, 77864. Phone: (409) 348-9200. Fax: (409) 348-9201. E-mail: kmvlradio@ev1.net Web Site: www.kmvl.net. Licensee: Hunt Broadcasting. (acq 7-17-91; FTR: 8-5-91). Rgnl. Network: Texas State Net. Format: Adult standards. News staff: one; News: 15 hrs wkly. Target aud: General. ♦Leon Hunt, gen mgr; Chester Leediker, chief of engrg.

KMVL-FM— April 1997: 100.5 mhz; 13 kw. 449 ft TL: N31 00 42 W96 02 27.24 Web Site: www.kmvl.net. Natl. Network: ABC. Format: Country. Target aud: 25-49.

Malakoff

KCKL(FM)— Aug 8, 1983: 95.9 mhz; 6 kw. Ant 295 ft TL: N32 08 48 W95 58 25. Stereo. Hrs opn: 24 Box 489, Hwy. 31 E., 75148. Phone: (903) 489-1238. Fax: (903) 489-2671. E-mail: kcklklvq@tvec.net Web Site: www.kcklklvq.com. Licensee: Lake Country Radio L.P. Group owner: Routt Radio Companies Inc. (acq 11-5-2005; $550,000 with KLVQ(AM) Athens). Population served: 100,000 Natl. Network: ABC. Format: Real country. News staff: one; News: 10 hrs wkly. Target aud: 25-54; country/city folk, weekenders & visitors to Cedar Creek Lake. Spec prog: Relg 7 hrs wkly. ♦Adabeth Routt, gen mgr; Pat Isaacson, opns mgr & gen sls mgr; Mike Lallande, progmg dir; Rich Flowers, news dir; Wayne Blackwelder, chief of engrg.

Manor

KELG(AM)— Apr 22, 1981: 1440 khz; 800 w-D, 500 w-N, DA-2. TL: N30 19 36 W97 32 35. Hrs open: 24 2111 S. IH 35, Suite 401, Austin, 78741. Phone: (512) 416-1100. Fax: (512) 445-5817. E-mail: kelg@austintejas.com Web Site: www.kelg.com. Licensee: BMP Austin License Company L.P. (acq 2-10-2005; grpsl). Law Firm: Bechtel & Cole. Format: Rgnl Mexican. ♦Thomas Castro, pres; Pedro Gasc, gen mgr; Mike Lozano, progmg dir; Steve Freeman, chief of engrg; Dan Saldana, traf mgr.

Marble Falls

***KBMD(FM)**— 2002: 88.5 mhz; 6 kw. Ant 89 ft TL: N30 33 12 W98 15 30. Hrs open: 24
EWTN.
1903 S. Lemesa Rd., Midland, 78701. Phone: (432) 638-1150. Fax: (432) 682-5230. E-mail: robertd@grnonline.com Web Site: www.grnonline.com. Licensee: La Promesa Foundation. (acq 2-24-2005; $130,000). Format: Catholic relg. News staff: 2. ♦Leonard Oswald, pres; Dick Bigelow, gen mgr.

Marion

KBIB(AM)—Licensed to Marion. See San Antonio

Markham

KZRC(FM)— August 2000: 92.5 mhz; 6 kw. Ant 328 ft TL: N28 52 26 W96 08 22. Stereo. Hrs opn: 24 PO Box 547, Bay City, 77404. Phone: (979) 323-7771. Fax: (775) 719-2182. E-mail: kzrc@kzrc.com Web Site: www.kzrc.com. Licensee: Edwards Broadcasting Co. (acq 7-24-2007; $400,000). Format: AOR, class rock, adult contemp. Target aud: 25-54; White equally mixed gender. Spec prog: Black gospel 4 hrs wkly. ♦Jerry Ball, pres; Ernest Cunnar, gen mgr; Deora Ramsey, progmg dir.

Marlin

KBBW(AM)—See Waco

KLRK(FM)— Apr 2, 1977: 92.9 mhz; 3 kw. 500 ft TL: N31 19 31 W96 54 36. (CP: 50 kw, ant 492 ft. TL: 31 24 45 W97 12 40). Stereo. Hrs opn: 24 220 S. 2nd St., Apt. 282, Waco, 76701. Phone: (254) 772-0930. Fax: (254) 753-0499. Web Site: star929fm.com. Licensee: Simmons Austin, LS LLC. Group owner: Simmons Media Group (acq 6-4-2004; grpsl). Population served: 109000 Natl. Rep: Roslin. Law Firm: Leventhal, Senter & Lerman. Format: Bright adult contemp. News staff: one. Target aud: 25-49. ♦Daryl O'Neal, gen mgr; Rob Reed, opns mgr; Bill LeGrande, sls dir; Dustin Drew, progmg dir; Cole McClellan, chief of engrg; Flavia Chen, traf mgr.

Marshall

***KBWC(FM)**— March 1977: 91.1 mhz; 100 w. 110 ft TL: N32 32 12 W94 22 29. Stereo. Hrs opn: 24 711 Wiley Ave., 75670. Phone: (903) 927-3266. Phone: (903) 927-3307. Fax: (903) 935-0153. Licensee: Wiley College. Population served: 40,000 Natl. Network: American Urban. Format: Urban/mix. Target aud: 18-34. ♦Shanon Levingston, gen mgr.

KCUL(AM)— Oct 7, 1957: 1410 khz; 500 w-D, 90 w-N, DA-2. TL: N32 29 30 W94 21 52. Hrs open: 5:30 AM-11 PM Box 7820, Tyler, 75711. Secondary address: 621 Chase Dr., Tyler 75701. Phone: (903) 581-9966. Fax: (903) 534-5300. Licensee: Access. 1 Texas License Co. LLC. Group owner: Access.1 Communications Corp. (acq 5-9-00; grpsl). Population served: 86,000 Natl. Network: Fox News Radio. Rgnl. Network: Texas State Net. Format: News, classic country. News staff: one; News: 15 hrs wkly. Target aud: General. Spec prog: Farm 3 hrs wkly. ♦Rick Quest, gen mgr.

KCUL-FM— Jan 1, 1992: 92.3 mhz; 5.8 kw. Ant 328 ft TL: N32 32 26 W94 24 03. Stereo. 24 175,000 Format: Mexican rgnl. Target aud: 25 plus.

KMHT(AM)— 1947: 1450 khz; 1 kw-U. TL: N32 33 50 W94 21 04. Hrs open: 2323 Jefferson Ave., 75670. Phone: (903) 923-8000. Phone: (903) 935-6018. Fax: (903) 935-2481. E-mail: kmht@marshalltx.com Licensee: Hanszen Broadcast Group Inc. (acq 9-17-2002; $400,000 with co-located FM). Population served: 27,345 Format: Country. ♦Jerry Hanszen, pres & sr VP; Chris Paddie, gen mgr; Alaina Pool, progmg dir; Rodney Andrews, chief of engrg.

KMHT-FM— Sept 4, 1977: 103.9 mhz; 3 kw. 300 ft TL: N32 33 50 W94 21 04.50,000

Mart

***KSUR(FM)**— 2007: 88.9 mhz; 100 kw vert. Ant 623 ft TL: N31 23 02 W97 16 38. Hrs open:
Rebroadcasts WAFR(FM) Tupelo, MS 100%.
6304 Gardendale Dr., Waco, 76710. Phone: (254) 772-1900. Web Site: www.kbderadio.net. Licensee: American Family Association. Format: Christian. ♦Marvin Sanders, gen mgr.

Mason

KHLB(FM)—Not on air, target date: unknown: 102.5 mhz; 50 kw. Ant 456 ft TL: N30 42 03 W99 13 59. Hrs open: 5526 N. Hwy. 281, Marble, 78654. Phone: (830) 693-5551. Fax: (830) 593-5107. Licensee: Munbilla Broadcasting Properties Ltd. (group owner). Format: Country. ♦Sabrina Preiss, gen mgr; Ben Shields, progmg dir; Bill Woleban, chief of engrg.

KOTY(FM)— 2004: 95.7 mhz; 50 kw. Ant 436 ft TL: N30 33 53 W99 27 13. Hrs open: 1809 Lightsey Rd., Austin, 78704. Phone: (512) 444-9268. Licensee: Bryan A. King (acq 2-13-2004; grpsl). ♦Bryan King, gen mgr.

McAllen

KGBT-FM— 1964: 98.5 mhz; 100 kw. 997 ft TL: N26 07 14 W97 49 18. (CP: Ant 997 ft.). Stereo. Hrs opn: 200 S. 10th, Suite 600, 78501. Phone: (956) 631-5499. Fax: (956) 631-0090. Web Site: www.netmio /com/radio/kgbt-fm. Licensee: Tichenor License Corp. Group owner: Univision Radio (acq 9-22-2003; grpsl). Population served: 500,000 Format: Mexican rgnl. ♦Mac Tichenor Jr., pres; Joe Morales, gen mgr; Angela Navarrete, gen sls mgr; Hugo de la Cruze, progmg dir; Jorge Garza, chief of engrg; Odie Francisco Chavez, news dir & traf mgr.

***KHID(FM)**— July 16, 1992: 88.1 mhz; 2.1 kw. 253 ft TL: N26 21 44 W98 19 26. Stereo. Hrs opn: 24
Rebroadcasts KMBH-FM Harlingen.
Box 2147, Harlingen, 78551. Secondary address: 1701 E. Tennessee

Ave., Harlingen 78550. Phone: (956) 421-4111. Fax: (956) 421-4150. E-mail: kmbhkhid@aol.com Web Site: www.kmbh.org. Licensee: RGV Educational Broadcasting Inc. Population served: 950,000 Format: News, class, jazz. News: 34 hrs wkly. Target aud: General. ♦Pedro Briseno, gen mgr.

KIRT(AM)—See Mission

KJAV(FM)—See Alamo

KRIO(AM)— 1947: 910 khz; 5 kw-U, DA-2. TL: N26 18 02 W98 12 38. Hrs open: 24 4300 S. Business 281, Edinburg, 78539. Phone: (956) 380-3435. Fax: (956) 380-8156. E-mail: correo@radioesperanza.com Web Site: www.radioesperanza.com. Licensee: Rio Grande Bible Institute Inc. (acq 5-30-86). Population served: 750,000 Law Firm: Bryan Cave. Wire Svc: UPI Format: Relg, educ, Sp. News: 5 hrs wkly. Target aud: General. ♦Larry Windle, pres; Gerardo Lorenzo, gen mgr & progmg dir; Jerry Joske, chief of engrg.

KVLY(FM)—See Edinburg

KVMV(FM)— March 1972: 96.9 mhz; 100 kw. 1,160 ft TL: N26 04 53 W97 49 44. Stereo. Hrs opn: 24 Box 3333, 78502. Secondary address: 715 E. Thomas Dr., Pharr 78502. Phone: (956) 787-9700. Fax: (956) 787-9783. Web Site: www.kvmv.org. Licensee: World Radio Network Inc. (group owner; acq 8-27-84). Population served: 600,000 Natl. Network: Moody. Format: Contemp Christian. News: 4 hrs wkly. Target aud: 30-65; general. ♦James Gamblin, gen mgr & progmg dir; Bob Malone, mus dir.

McCamey

KPBM(FM)—Not on air, target date: unknown: 95.3 mhz; 3 kw. 758 ft TL: N31 12 42 W102 16 29. Hrs open: Box 252, McAllen, 78502. Licensee: Paulino Bernal. ♦Paulino Bernal, gen mgr.

McCook

***KCAS(FM)**— Jan 1, 2001: 91.5 mhz; 2.5 kw. Ant 358 ft TL: N26 28 51 W98 23 45. Stereo. Hrs opn: 24 Faith Baptist Church Inc., 4301 N. Shary Rd., Mission, 78574. Secondary address: P. O. Box 8106, Mission 78572. Phone: (956) 424-9098. Fax: (956) 581-7786. E-mail: mail@kcasradio.org Web Site: www.kcasradio.org. Licensee: Faith Baptist Church Inc. Population served: 650,000 Natl. Network: USA. Format: Relg. News: 12 hrs wkly. Target aud: 30-85; male & female. ♦Joel Mangin, gen mgr, opns mgr & progmg dir; Jerry Jeske, chief of engrg.

McKinney

***KNTU(FM)**— November 1969: 88.1 mhz; 100 kw. 443 ft TL: N33 17 24 W97 08 10. Stereo. Hrs opn: 6 AM-midnight Box 310881, Denton, 76203. Secondary address: 1179 Union Cir., Suite 262, Denton 76201. Phone: (940) 565-3688. Phone: (940) 565-3459. Fax: (940) 565-2518. E-mail: kntu@unt.edu Web Site: kntu.fm. Licensee: University of North Texas. Population served: 5,000,000 Natl. Network: AP Radio. Rgnl. Network: Texas State Net. Wire Svc: AP Format: Jazz. News: 8 hrs wkly. Target aud: 18 plus. Spec prog: Class 6 hrs, Sp 6 hrs, new mus 3 hrs, pub affrs 2 hrs wkly. ♦Russ Campbell, gen mgr; Mark Lambert, opns dir; Aaron Brodie, news dir & chief of engrg.

McQueeney

KLTO-FM— July 1989: Stn currently dark. 97.7 mhz; 100 kw. Ant 981 ft TL: N29 20 45 W97 38 44. Hrs open: 1777 N.E. Loop 410, San Antonio, 78217. Phone: (210) 829-1075. Fax: (210) 824-9971. Licensee: Rawhide Radio LLC. Group owner: Univision Radio (acq 9-22-2003; grpsl). Format: Sp. ♦Dan Wilson, pres & gen mgr.

Memphis

KLSR-FM— 1982: 105.3 mhz; 100 kw. 485 ft TL: N34 41 13 W100 30 23. Stereo. Hrs opn: 24 Box 400, 114 N. 7th, 79245. Phone: (806) 259-3511. Licensee: Davis Broadcast Company Inc. (acq 7-15-86; $78,348 with co-located AM; FTR: 7-28-86) Population served: 50,000 Format: C&W, div, contemp. Spec prog: Sp 6 hrs, good time oldies 60s & 70s 10 hrs, relg 5 hrs wkly. ♦Donna Davis, pres; Joe Davis, gen mgr.

Mercedes

KHKZ(FM)— Sept 10, 1982: 106.3 mhz; 1.65 kw. 649 ft TL: N26 13 50 W98 20 18. Stereo. Hrs opn: 24 901 E. Pike Blvd., Weslaco, 78596. Phone: (866) 973-1041. Fax: (956) 544-0311. Licensee: Clear Channel Broadcasting Licenses Inc. Group owner: Clear Channel Communications Inc. (acq 12-9-2003; grpsl). Natl. Network: USA. Law Firm: Fisher, Wayland, Cooper, Leader & Zaragoza. Format: Hot adult contemp. News staff: 3. Target aud: General. Spec prog: Black 3 hrs, southern gospel 2 hrs, Christian rock 3 hrs wkly. ♦Danny Fletcher, gen mgr; Billy Santiago, opns mgr; Cyndia Torres, gen sls mgr; J. Contu, progmg dir; Ken Meek, chief of engrg; Gloria Garcia, traf mgr.

Meridian

KSCG(FM)—Not on air, target date: unknown: 95.3 mhz; 25 kw. Ant 328 ft TL: N32 02 06 W97 50 02. Hrs open: 115 W. 3rd St., Fort Worth, 76102. Phone: (817) 332-0959. Fax: (817) 332-4630. Licensee: LKCM Radio Group LP. ♦Kevin Prigel, VP.

Merkel

KHXS(FM)— Nov 4, 1983: 102.7 mhz; 100 kw. 1,486 ft TL: N32 22 00 W99 58 42. Stereo. Hrs opn: 24 2525 S. Danville, Abilene, 79608. Phone: (325) 793-9700. Fax: (325) 692-1576. Web Site: www.102thebear.com. Licensee: Cumulus Licensing Corp. Group owner: Cumulus Media Inc. (acq 6-15-98; $1.6 million). Format: Classic rock. ♦Trace Michaels, gen mgr; Kelly Jay, opns mgr; Kim Creshaw, gen sls mgr; Justin Case, prom dir; John Miller, progmg dir; Cindy Stephens, news dir; Chris Andrews, chief of engrg; Lori Barrett, traf mgr.

KMXO(AM)— June 1, 1963: 1500 khz; 250 w-D. TL: N32 28 17 W100 00 19. Hrs open: 604 N. 2nd St., 79536. Phone: (325) 928-3060. Fax: (325) 928-4683. Licensee: Ray R. Silva. Format: Chirstian. ♦Zacarias Serrato, gen mgr.

Mertzon

***KMEO(FM)**— 2006: 91.9 mhz; 6.5 kw vert. Ant 522 ft TL: N31 25 16 W100 32 36. Hrs open: Box 2440, Tupelo, MS, 38803. Phone: (662) 844-8888. Fax: (662) 842-6791. Web Site: www.afr.net. Licensee: American Family Association. (acq 8-9-2005). Format: Christian classics. ♦Marvin Sanders, gen mgr.

Mesquite

***KEOM(FM)**— Sept 4, 1984: 88.5 mhz; 61 kw. 514 ft TL: N32 45 46 W96 38 04. Stereo. Hrs opn: 24 2600 Motley Dr., Suite 300, 75150. Phone: (972) 288-6411. E-mail: jgriffin@mosquiteisd.org Web Site: www.keom.fm. Licensee: Mesquite Independent School District. Rgnl. Network: Texas State Net. Format: Div. Target aud: General; citizens of Mesquite & surrounding area. ♦James Griffin, stn mgr.

Mexia

KRQX(AM)— May 21, 1956: 1590 khz; 500 w-D, 128 w-N. TL: N31 41 10 W96 27 18. Hrs open: 24 Box 1590, 76667. Secondary address: 1006-B Milam St. 76667. Phone: (254) 562-5328. Fax: (254) 562-6729. E-mail: radio@kycxfm.com Licensee: Simmons Austin, LS LLC. (acq 8-19-2005; $390,000 with co-located FM). Population served: 30,000 Natl. Network: ABC. Rgnl. Network: Texas State Net. Format: Country. Target aud: General; 20-59. Spec prog: Farm 12 hrs, Gospel 3 hrs wkly. ♦Susan Cholopisa, gen mgr; Bill Ferris, opns dir & gen sls mgr; Jan Phillips, news dir; Brandi Garza, traf mgr; Dave Campbell, sports cmtr.

KWGW(FM)—Co-owned with KRQX(AM). Aug 29, 1983: 104.9 mhz; 2.85 kw. Ant 482 ft TL: N31 38 39 W96 36 51. Stereo. 30,000 Natl. Network: CBS. Law Firm: Roy F. Perkins. ♦Bill Ferris, chief of opns; Susan Cholopisa, stn mgr & progmg dir; Brandi Garza, traf mgr; Dave Campbell, sports cmtr.

Midland

***KAQQ(FM)**— Sept 1, 2006: 90.9 mhz; 1.5 kw. Ant 430 ft TL: N31 54 32 W102 04 01. Stereo. Hrs opn:
Rebroadcasts WAZQ(FM) Key West, FL 100%.
6910 N.W. 2nd Terr., Boca Raton, FL, 33487. Phone: (561) 912-9002. E-mail: bill@qfmonline.com Web Site: qfmonline.com. Licensee: Educational Public Radio Inc. Format: Hot adult contemp. Target aud: 18-54; adults. ♦Bill Lacy, pres.

KCHX(FM)— Aug 15, 1988: 106.7 mhz; 100 kw. 613 ft TL: N31 54 53 W101 57 49. Stereo. Hrs opn: 24 1330 E. 8th St., Suite 207, Odessa, 79761. Phone: (432) 563-9102. Fax: (432) 580-9102. Web Site: www.mymix1067.com. Licensee: GAP Broadcasting Midland-Odessa License LLC. Group owner: Clear Channel Communications Inc. (acq 8-3-2007; grpsl). Population served: 350,000 Format: Adult contemp. News: 2 hrs wkly. Target aud: 25-54; general. ♦Gloria Apolinario, gen mgr; Laura Florez, gen sls mgr; Rob Norris, chief of engrg & engr.

KCRS(AM)— Dec 20, 1935: 550 khz; 5 kw-D, 1 kw-N, DA-2. TL: N32 04 10 W102 01 46. Hrs open: 24 1330 E. 8th St., Suite 207, Odessa, 79761. Phone: (432) 563-9102. Fax: (432) 580-9102. Web Site: www.newstalkkcrs.com. Licensee: GAP Broadcasting Midland-Odessa License LLC. Group owner: Clear Channel Communications Inc. (acq 8-3-2007; grpsl). Population served: 200,000 Rgnl. Network: Texas State Net. Law Firm: Dow, Lohnes & Albertson. Format: News/talk. News staff: 2; News: 30 hrs wkly. Target aud: 25-54. ♦Gloria Apolinario, gen mgr; Robert Hallmark, opns mgr, prom dir, progmg dir, pub affrs dir & spec ev coord; Steve Driscoll, opns mgr; Jesse Grimes, news dir, local news ed & news rptr; Rod Norris, engrg mgr; Shelly Todd, traf mgr.

KCRS-FM— May 25, 1976: 103.3 mhz; 100 kw. 920 ft TL: N32 05 11 W102 17 11.24 Web Site: www.1033kissfm.net.200,000 Format: Adult contemp. ♦Ric Elliott, progmg dir; Jesse Grimes, pub affrs dir, local news ed, news rptr & edit dir; Shelly Todd, traf mgr; Robert Hallmark, spec ev coord.

KJBC(AM)— Aug 6, 1950: 1150 khz; 1 kw-D. TL: N31 58 55 W102 03 30. Hrs open: 24
EWTN.
1903 S. Lamesa Rd., 79701. Phone: (432) 638-1150. Fax: (432) 682-5230. E-mail: robertd@gmonline.com Web Site: www.gmonline.com. Licensee: La Promesa Foundation. (acq 2-11-2002; $85,000). Population served: 250,000 Format: Catholic progmg. News staff: 3. ♦Robert Dominguez, gen mgr; Toya Hall, progmg dir.

KMCM(FM)—See Odessa

KMND(AM)— Nov 27, 1963: 1510 khz; 500 w-D. TL: N31 37 48 W102 04 53. (CP: 2.4 kw-D). Hrs opn: Bldg. #2, 11300 Hwy. 191, 79707. Phone: (432) 563-5636. Fax: (432) 563-3823. E-mail: jmesher@aol.com Web Site: www.kmnd.com. Licensee: Cumulus Licensing Corp. Group owner: Cumulus Media Inc. (acq 12-17-98; grpsl). Natl. Network: ESPN Radio. Format: Sports. Spec prog: Jazz one hr wkly. ♦Kent Cooper, gen mgr; Mike Baer, sls dir & gen sls mgr; Robi Burns, progmg dir; Garry Vaughn, chief of engrg.

KNFM(FM)—Co-owned with KMND(AM). Nov 2, 1959: 92.3 mhz; 100 kw. Ant 985 ft TL: N32 05 51 W102 17 21. Stereo. Web Site: www.lonestar92.com.300,000 Format: Country. ♦George DeMarco, gen mgr; John Moesch, opns mgr & progmg dir; Spencer Bennett, progmg dir; Aleese Fielder, sls; Tonya Calloway, prom; Robbie Green, engr.

***KPBJ(FM)**— 2005: 90.1 mhz; 1.85 kw. Ant 417 ft TL: N31 57 39 W101 54 25. Hrs open:
Rebroadcasts KCZO(FM) Carrizo Springs 100%.
Box 252, McAllen, 78505. Phone: (956) 686-6382. Fax: (956) 686-2999. Licensee: Paulino Bernal Evangelism. Format: Sp. ♦Paulino Bernal Jr., pres.

KQRX(FM)—Licensed to Midland. See Odessa

KWEL(AM)— April 1957: 1070 khz; 2.5 kw-D. TL: N31 57 44 W102 04 07. Hrs open: 6 AM-9 PM 1611 W. College Ave., 79701. Secondary address: 310 W. Wall, Ste 104 79701. Phone: (432) 620-9393. Fax: (432) 620-9591. E-mail: craiganderson@kwel.com Web Site: www.kwel.com. Licensee: Faustino Quiroz. (acq 5-1-93; $140,000; FTR: 4-5-93). Population served: 220,000 Natl. Network: ABC. Format: Talk, news. News: 60 hrs wkly. Target aud: General; Adults 35+. ♦Craig Anderson, CEO, gen mgr & progmg dir; Doris Anderson, traf mgr, traf mgr & opns; Garry Vaughn, engr; Jason Moore, engr.

KZBT(FM)— 1974: 93.3 mhz; 100 kw. 500 ft TL: N31 57 30 W102 03 59. Stereo. Hrs opn: 24 11300 Hwy 191, Bldg. 2, 79707. Phone: (432) 563-5499. Fax: (915) 563-5530. Web Site: www.b93.net. Licensee: Cumulus Licensing Corp. Group owner: Cumulus Media Inc. (acq 12-17-98; grpsl). Population served: 220,000 Format: Contemporary Hit/Top-40. News staff: one; News: 2 hrs wkly. Target aud: 18-44. ♦George DeMarco, gen mgr; John Moesch, opns mgr; Aleese Fielder, gen sls mgr; Rebecca Cruz, prom dir; Leo Caro, progmg dir; Robbie Green, chief of engrg.

Mineola

KMOO-FM— Sept 1, 1977: 99.9 mhz; 6 kw. Ant 295 ft TL: N32 45 04 W95 33 18. Stereo. Hrs opn: 24 Box 628, 75773. Secondary address:

Hwy. 69 N. 75773. Phone: (903) 569-3823. Fax: (903) 569-6641. E-mail: jason@kmoo.com Web Site: www.kmoo.com. Licensee: Hightower Radio Inc. (acq 5-26-98; $600,000 for stock). Rgnl. Network: Texas State Net. Law Firm: Wiley Rein LLP. Format: Country. News staff: one; News: 3 hrs wkly. Target aud: 25-64. ♦Jason Hightower, pres, gen mgr, gen sls mgr & progmg dir; Amy Castleberry, opns dir; Kenny Smith, prom dir; Pat Thurman, pub affrs dir.

Mineral Wells

KFWR(FM)— Mar 1, 1970: 95.9 mhz; 80 kw. Ant 1,079 ft TL: N32 39 50 W98 09 47. Stereo. Hrs opn: 24 115 W. 3rd St., Fort Worth, 76102. Phone: (817) 332-0959. Fax: (817) 348-8373. Web Site: 959theranch.com. Licensee: LKCM Radio Group L.P. (group owner; acq 9-30-02; $6 million). Population served: 1,000,000 Format: Tex country. Target aud: 25-54; local, Texas country. ♦Gerry Schlegel, pres; Joel Gough, sls dir; Rick Lovett, progmg dir.

KJSA(AM)— Dec 1, 1946: 1120 khz; 250 w-D. TL: N32 47 12 W98 05 53. Hrs open: 6 AM-sunset 305 Millsap Hwy., 76067. Phone: (940) 325-1140. Fax: (940) 325-1164. Licensee: M&M Broadcasters Ltd. (group owner; (acq 2-22-2006; grpsl). Population served: 75,000 Format: Classic country. ♦Chuck McKay, gen mgr.

Mirando City

KBDR(FM)— Apr 1, 1993: 100.5 mhz; 42 kw. Ant 551 ft TL: N27 21 17 W99 13 52. Stereo. Hrs opn: 24 107 Calle Del Norte, Suite 102, Laredo, 78041. Phone: (956) 725-1000. Fax: (956) 718-1000. E-mail: ss@bmpradio.com Licensee: BMP 100.5 FM LP. Group owner: Border Media Partners LLC (acq 5-30-2003; $8 million with KBUC(FM) Raymondville). Population served: 1000000 Format: Regional Sp. Target aud: 18-45; upper-income bracket. ♦Tom Castro, CEO; Hugo Del Pozzo, CFO; Steve Stephenson, VP & gen mgr; Nestor Cobos, stn mgr; Issac Carrillo, opns mgr & sls dir; Robert Garcia, prom dir; Joe Flores, adv dir; Rogelio Botello Rios, progmg dir; Joe Espinoza, chief of engrg.

Mission

KGBT-FM—See McAllen

KIRT(AM)— Feb 23, 1958: 1580 khz; 1 kw-D, 302 w-N. TL: N26 17 36 W89 19 50. Hrs open: Box 1058, McAllen, 78505-1058. Phone: (956) 686-2111. Fax: (956) 668-0370. E-mail: kirtradio@aol.com Licensee: Bravo Broadcasting Co. Inc. (acq 10-25-01). Population served: 20,000 Rgnl. Network: Texas State Net. Format: Sp. ♦Humberto Pedraza, gen mgr; Armando Pedraza, progmg dir; John Pankratz, chief of engrg; Rosie Pedraza, gen sls mgr & traf mgr.

KQXX-FM— 1989: 105.5 mhz; 3 kw. 300 ft TL: N26 13 50 W98 20 18. Stereo. Hrs opn: 24
Rebroadcasts KTJN(FM) Brownsville 100%.
901 E. Pike Blvd., Weslaco, 78596. Phone: (956) 973-9202. Fax: (956) 544-0311. E-mail: billysantiago@clearchannel.com Web Site: www.oldies1055.net. Licensee: Clear Channel Broadcasting Licenses Inc. Group owner: Clear Channel Communications Inc. (acq 12-9-2003; grpsl). Population served: 300,000 Format: Oldies. News staff: 3. ♦Danny Fletcher, gen mgr; Billy Santiago, opns mgr; Cyndia Torres, gen sls mgr; Ken Meek, chief of engrg; Gloria Garcia, traf mgr.

Missouri City

KPTY(FM)—Licensed to Missouri City. See Houston

Monahans

KBAT(FM)—Licensed to Monahans. See Odessa

KCKM(AM)— Mar 12, 1947: 1330 khz; 5 kw-D, 1 kw-N, DA-N. TL: N31 38 45 W103 00 04. Hrs open: 24 Box 3069, Odessa, 79760-3069. Phone: (432) 943-2588. Fax: (432) 943-7314. Licensee: Sandhills Communication Inc. (acq 2001; $175,000). Population served: 8,333 Natl. Network: CBS. Rgnl. Network: Texas State Net. Natl. Rep: Riley. Law Firm: Roberts & Eckard. Format: Oldies. News staff: one. Target aud: 25-54; general. Spec prog: Gospel 3 hrs wkly. ♦Rick Anderson, gen mgr; David McCaffity, progmg dir; Allen Martin, news dir; Dexter Nichols, sports cmtr.

Mount Pleasant

KIMP(AM)— Oct 8, 1948: 960 khz; 1 kw-D, 75 w-N. TL: N33 09 54 W95 00 27. Hrs open: Box 990, 75456. Secondary address: 1798 U.S. Hw. 67 West 75455. Phone: (903) 572-8726. Fax: (903) 572-7232. Web Site: easttexasradio.com. Licensee: East Texas Broadcasting Inc. (group owner; acq 11-21-91; $850,000 with co-located FM; FTR: 12-16-91). Population served: 50,000 Natl. Network: ABC. Rgnl. Network: Texas State Net. Format: Classic country, Sp. News staff: 10; News: 10 hrs wkly. Target aud: General. Spec prog: Sp 25 hrs. ♦John Mitchell, chmn; Bud Kitchens, pres, VP & gen mgr; Darrin Tripp, opns dir & progmg dir; Bryan Frimesth, gen sls mgr; Clint Cooper, news dir, pub affrs dir & farm dir; Bill Hughes, chief of engrg; Justice Thornburg, traf mgr; Jesse Carillo, spanish dir; Tammy Ray, disc jockey.

Muenster

KZZA(FM)— Dec 23, 1991: 106.7 mhz; 75 kw. Ant 2,034 ft TL: N33 26 13 W97 29 05. Stereo. Hrs opn: 24 4201 Pool Rd., Coleyville, 76034. Fax: (817) 868-2900. Fax: (817) 868-2929. Web Site: casa1067.com. Licensee: Liberman Broadcasting of Dallas License LLC. Group owner: Entravision Communications Corp. (acq 11-2-2006; grpsl). Population served: 6,153,500 Natl. Rep: Eastman Radio. Law Firm: Thompson Hine. Format: Latino urban. Target aud: 18-34. ♦Alex Sanchez, gen mgr.

Muleshoe

KMUL-FM— Feb 6, 1966: 103.1 mhz; 6 kw. Ant 75 ft TL: N34 13 39 W102 44 10. Stereo. Hrs opn: 1000 Sycamore St., Clovis, NM, 88101. Phone: (505) 762-6200. Fax: (505) 762-8800. Licensee: Tallgrass Broadcasting LLC. (acq 4-2-2007; grpsl). Population served: 4,525 Format: Country.

Nacogdoches

KJCS(FM)— May 1967: 103.3 mhz; 100 kw. 476 ft TL: N31 34 51 W94 40 16. Stereo. Hrs opn: 910 North St., 75961. Phone: (936) 559-8800. Fax: (936) 559-8801. Licensee: Radio Licensing Inc. Population served: 105,000 Natl. Network: ABC. Format: Country. Spec prog: Gospel 3 hrs wkly. ♦Bill Vance Jr., gen mgr; Carolyn Gage, stn mgr & natl sls mgr; Della Huse, gen sls mgr; Lou Bennett, progmg dir & pub affrs dir; Gwen Jordan, traf mgr.

***KSAU(FM)**— July 5, 1975: 90.1 mhz; 3.5 kw. 450 ft TL: N31 37 45 W94 40 44. Stereo. Hrs opn: 10 AM-2 AM Box 13048, 75962. Phone: (936) 468-4000. Fax: (936) 468-1331. E-mail: ksau@sfasu.edu Web Site: www.sfasu.edu/ksau. Licensee: Stephen F. Austin State University. Population served: 120,000 Natl. Network: ABC. Wire Svc: AP Format: Jazz, progsv, new age. News: 3 hrs wkly. Target aud: 18-54. Spec prog: Blues 2 hrs, reggae 2 hrs, urban contemp 4 hrs, classic rock 14 hrs, contemp Christian 2 hrs wkly. ♦Sherry Williford, gen mgr.

KSFA(AM)— June 2, 1947: 860 khz; 1 kw-D, 175 w-N. TL: N31 31 36 W94 39 29. Stereo. Hrs opn: 1216 South First, Lufkin, 75901. Phone: (936) 639-4455. Fax: (936) 639-4440. Web Site: www.ksfa860.com. Licensee: GAP Broadcasting Lufkin License LLC. Group owner: Clear Channel Communications Inc. (acq 8-3-2007; grpsl). Population served: 140,250 Format: News/talk. Target aud: 25 plus; upscale, upper income level men. Spec prog: Houston Astros baseball, farm 7 hrs wkly. ♦Larry Gunter, gen mgr; Danny Merrell, progmg dir.

KTBQ(FM)—Co-owned with KSFA(AM). July 15, 1967: 107.7 mhz; 50 kw. 492 ft TL: N31 42 30 W94 41 18. Stereo. 24 Web Site: www.q1077.com.100,000 Format: Classic rock. News staff: one. Target aud: 18-49; upscale women.

KYKS(FM)—See Lufkin

Navasota

KMBV(FM)— Mar 1, 1989: 92.5 mhz; 6 kw. Ant 263 ft TL: N30 24 58 W96 04 43. Stereo. Hrs opn: 24
Simulcast with KHTZ(FM) Caldwell 100%.
530 W. Main St., Brenham, 77833. Phone: (979) 836-9411. Fax: (979) 836-9435. Web Site: www.lonestarfm.com. Licensee: Fort Bend Broadcasting Co. (group owner; (acq 5-31-2001; $900,000). Population served: 200,000 Natl. Network: Jones Radio Networks. Format: Classic country. News: one hr wkly. Target aud: 25-54. Spec prog: Gospel 4 hrs wkly. ♦Ryan Henderson, gen mgr.

KWBC(AM)— Sept 21, 1960: Stn currently dark. 1550 khz; 250 w-D, 26 w-N. TL: N30 22 48 W96 06 01. (CP: COL College Station. 1.4 kw-D, 24 w-N, DA-D. TL: N30 37 54 W96 21 28). Hrs opn: 303 E. Washington, Suite A, 77868. Phone: (936) 825-9007. Fax: (936) 825-1019. E-mail: news@navasotanews.com Web Site: navasotanews.com. Licensee: Bryan Broadcasting Corp. (acq 7-31-2007; $275,000). Population served: 92,000 Format: Local news/talk. ♦Ben Downs, gen mgr; Dave Hill, stn mgr; Tom Turner, news dir; Chris Dusterhoff, chief of engrg.

Nederland

KBED(AM)— Jan 11, 1969: 1510 khz; 5 kw-D, DA-D. TL: N30 03 35 W93 58 49. Hrs open: Sunrise-sunset 755 S. 11th St., Suite 102, Beaumont, 77704. Phone: (409) 833-9421. Fax: (409) 833-9296. Web Site: www.cumulus.com. Licensee: Cumulus Licensing Corp. Group owner: Cumulus Media Inc. (acq 3-9-98; grpsl). Population served: 15,000 Natl. Network: ESPN Radio. Format: Sports. News staff: one; News: 5 hrs wkly. Target aud: 18-49; males. Spec prog: Relg 4 hrs wkly. ♦Zanetta Kelley, gen mgr; Wes Matejka, sls dir; Mark Guzman, prom dir & spec ev coord; Jim West, progmg dir; Richard Core, news dir; Greg Davis, chief of engrg; Liz Ferguson, traf mgr.

New Boston

KEWL-FM— July 1995: 95.1 mhz; 25 kw. Ant 325 ft TL: N33 26 15 W94 25 11. Hrs opn: 24 1323 College Dr., Texarkana, 75503. Phone: (903) 793-1109. Fax: (903) 794-4717. Web Site: www.kool951.com. Licensee: FFD Holdings I Inc. Group owner: Petracom Media LLC (acq 1-4-2005; grpsl). Format: Oldies. Target aud: 35-64. ♦Mike Basso, gen mgr.

KNBO(AM)— Nov 16, 1969: 1530 khz; 2.5 kw-D. TL: N33 28 56 W94 25 25. Hrs open: 6 AM-6 PM 1198 Daniels Chaper Rd., 75570. Phone: (903) 628-2561. Licensee: Bowie County Broadcasting Co. Inc. (acq 11-7-2005). Population served: 100,000 Format: MOR, Christian. Target aud: General. ♦Martha P. Knox, gen mgr; Jef Rogers, chief of opns; Carmen Johnson, progmg dir.

KTTY(FM)—Not on air, target date: unknown: 105.1 mhz; 4.3 kw. Ant 387 ft TL: N33 28 00 W94 27 48. Hrs open: 819 S.W. Federal Hwy., Suite 106, Stuart, FL, 34994-2952. Phone: (772) 215-1634. Web Site: www.toweritrust.com. Licensee: Tower Investment Trust Inc. ♦William H. Brothers, pres.

KZRB(FM)— Nov 16, 1997: 103.5 mhz; 50 kw. 492 ft TL: N33 24 54 W94 38 10. Stereo. Hrs opn: 24 710 W. Ave. A, Hooks, 75561. Phone: (903) 547-3223. Fax: (903) 547-3095. Web Site: kzrb.com. Licensee: B & H Broadcasting System Inc. (acq 4-16-93; $1.5 million; FTR: 5-3-93). Population served: 450,000 Natl. Network: ABC, CNN Radio. Rgnl. Network: Texas State Net. Format: Adult contemp, oldies, urban contemp. News staff: one; News: 9 hrs wkly. Target aud: 25-54; all age buyers. ♦Ray C. Bursey Jr., CEO, pres, gen mgr, opns mgr & sls VP; Ray C. Bursey, prom VP; Larry Stuart, progmg VP & mus dir; Wayne Duncan, engrg VP.

New Braunfels

KGNB(AM)— Apr 1, 1950: 1420 khz; 1 kw-D, 196 w-N. TL: N29 39 45 W98 10 29. Hrs open: 24 1540 Loop 337 N., 78130. Phone: (830) 625-7311. Fax: (830) 625-7336. Web Site: www.kgnb.com. Licensee: New Braunfels Communications Inc. Population served: 175,000 Natl. Network: CNN Radio, Sporting News Radio Network. Law Firm:

Southmayd & Miller, P.C. Format: Loc news, sports talk. News staff: 3; News: 26 hrs wkly. Target aud: 25-54; men. ♦Bill Rainer, CEO; Mattson Rainer, VP & gen mgr.

KNBT(FM)—Co-owned with KGNB(AM). Nov 22, 1968: 92.1 mhz; 6 kw. 300 ft TL: N29 43 50 W98 07 15. Stereo. 24 Web Site: www.knbtfm.com. Format: Americana. Target aud: 25-54. Spec prog: Relg 3 hrs wkly.

New Ulm

KNRG(FM)— 1999: 92.3 mhz; 6 kw. Ant 328 ft TL: N29 53 50 W96 32 35. Stereo. Hrs opn: Box 111, Columbus, 78934. Phone: (979) 732-5766. Fax: (979) 732-6377. Licensee: New Ulm Broadcasting Co. Group owner: Bayport Broadcast Group Population served: 30,000 Format: Classic hits. ♦Carl Geisler, gen mgr.

Nolanville

KLFX(FM)— 1994: 107.3 mhz; 980 w. Ant 581 ft TL: N31 05 23 W97 35 55. Hrs open: 100 W. Central Texas Expwy., Suite 306, Harker Heights, 76548. Phone: (254) 699-5000. Fax: (254) 680-4212. E-mail: klfx@klfx.com Web Site: 1073rocks.com. Licensee: Clear Channel Broadcasting Licenses Inc. Group owner: Clear Channel Communications Inc. (acq 1-15-2004; $2.6 million). Format: Rock. ♦Evan Armstrong, gen mgr.

Odem

KLHB(FM)— Feb 18, 1985: 98.3 mhz; 50 kw. Ant 433 ft TL: N27 47 26 W97 27 02. Stereo. Hrs opn: 24 1300 Antelope, Corpus Christi, 78401. Phone: (361) 883-1600. Fax: (361) 883-9303. E-mail: johnnyo@johnnyoradio.com Web Site: club983.com. Licensee: Tejas Broadcasting Ltd. LLP. Group owner: Amigo Broadcasting L.P. (acq 11-15-2004; grpsl). Population served: 400,000 Format: Sp. Target aud: 18-49; progsv, affluent, middle class Hispanics. ♦Paul Danitz, gen mgr; Bert Clark, opns mgr.

Odessa

KBAT(FM)—Monahans, Nov 1, 1983: 99.9 mhz; 100 kw. Ant 574 ft TL: N31 45 40 W102 31 28. Stereo. Hrs opn: 24 11300 Hwy. 191, Bldg. 2, Midland, 79707. Phone: (432) 563-5499. Fax: (432) 563-5530. Licensee: Cumulus Licensing Corp. Group owner: Cumulus Media Inc. (acq 12-17-98; grpsl). Law Firm: Jones, Waldo, Holbrook & McDonough. Format: Contemp Christian. Target aud: 25-54; general. ♦George DeMarco, gen mgr; John Moesch, opns mgr; Aleese Fielder, gen sls mgr; Brian Hill, prom dir & mus dir; Kevin Chase, progmg dir; Robbie Green, chief of engrg.

***KBMM(FM)**— 2004: 89.5 mhz; 25 kw. Ant 535 ft TL: N31 40 35 W102 21 32. Hrs open: Drawer 3206, Tupelo, MS, 38803. Phone: (662) 844-8888. Web Site: afr.net. Licensee: American Family Association. Group owner: American Family Radio Format: Christian. ♦Marvin Sanders, gen mgr.

***KFLB(AM)**— Jan 29, 1947: 920 khz; 1 kw-D, 500 w-N, DA-1. TL: N31 49 14 W102 25 42. Hrs open: 24 808 Tower Dr., Suite 6, 79761. Phone: (432) 580-5352. Fax: (432) 332-1044. E-mail: kflb@flc.org Web Site: www.flr.org. Licensee: Family Life Broadcasting System. (group owner; (acq 6-24-98; grpsl). Population served: 260,000 Format: Christian, relg. Target aud: 34-59; females. ♦Dave Borowsky, gen mgr; Merri Jo Leonard, opns dir & opns mgr.

KFLB-FM— Sept 1, 1989: 90.5 mhz; 28 kw. 453 ft TL: N31 53 50 W102 33 57.24 Web Site: www.flc.org. Licensee: Family Life Broadcasting Inc.260,000

KFZX(FM)—See Gardendale

KHKX(FM)— July 1, 1977: 99.1 mhz; 100 kw. Ant 407 ft TL: N32 03 10 W102 17 38. Stereo. Hrs opn: 24 Box 9400, Midland, 79708. Phone: (432) 520-9912. Fax: (432) 520-0112. Licensee: Double O Texas Corp. (group owner; (acq 2006; grpsl). Population served: 850,000 Natl. Network: USA. Format: Country. News: one hr wkly. Target aud: 25-54; general. ♦Tommy Vascocu, gen mgr.

KMCM(FM)— January 1961: 96.9 mhz; 100 kw. 500 ft TL: N32 05 13 W102 17 12. Stereo. Hrs opn: 3303 N. Midkiff Rd., Suite 115, Midland, 79705. Phone: (432) 520-9912. Fax: (432) 520-0112. Web Site: www.97gold.com. Licensee: Double O Texas Corp. (group owner; (acq 2006; grpsl). Population served: 220,000 Natl. Rep: Katz Radio. Law Firm: Wiley, Rein & Fielding. Format: Golden oldies. News: 2 hrs wkly. Target aud: 25-54; men. ♦Terry Bond, pres; Tommy R. Vascocu, gen mgr; Cathy Fowler, opns mgr.

KMRK-FM— Aug 23, 1991: 96.1 mhz; 27.5 kw. Ant 948 ft TL: N32 05 11 W102 17 10. Stereo. Hrs opn: 24 1330 E. 8th St., Suite 207, 79761. Phone: (432) 563-9102. Fax: (432) 580-9102. Fax: (915) 580-4800. E-mail: gloriaapolinario@clearchannel.com Web Site: www.wild961.com. Licensee: GAP Broadcasting Midland-Odessa License LLC. Group owner: Clear Channel Communications Inc. (acq 8-3-2007; grpsl). Population served: 200,000 Natl. Network: American Urban. Wire Svc: AP Format: CHR, urban hip hop. News staff: one; News: 2 hrs wkly. Target aud: 18-34. ♦Gloria Apolinario, gen mgr; Steve Driscoll, opns mgr.

KNFM(FM)—See Midland

***KOCV(FM)**— Jan 6, 1964: 91.3 mhz; 5 kw. 300 ft TL: N31 51 30 W102 23 00. (CP: Ant 289 ft.). Stereo. Hrs opn: 6 AM-midnight Odessa College, 201 W. University Blvd., 79764. Phone: (432) 580-9130. Fax: (915) 337-0529. E-mail: kocv@oc.odessa.edu Web Site: www.odessa.edu. Licensee: Odessa College. Population served: 78,380 Natl. Network: NPR. Format: News, eclectic, class. News staff: one; News: 33 hrs wkly. Target aud: 35 plus; educated, affluent adults. Spec prog: Jazz 4 hrs, opera 4 hrs, folk 4 hrs, blues 4 hrs, bluegrass 2 hrs, Celtic 4 hrs wkly. ♦Chad Hauris, gen mgr, opns mgr & mus dir; Doug Cole, gen mgr.

KODM(FM)— 1965: 97.9 mhz; 100 kw. 361 ft TL: N31 47 40 W102 10 44. Stereo. Hrs opn: 24 11300 Hwy. 191, Bldg. 2, Midland, 79707. Phone: 432) 563-5499. Fax: (432) 563-5330. Web Site: www.kodm.com. Licensee: Cumulus Licensing Corp. Group owner: Cumulus Media Inc. (acq 12-17-98; grpsl). Population served: 220,000 Natl. Network: ABC. Format: Lite rock, adult contemp. Target aud: 25-54; women. ♦George DeMarco, gen mgr; John Moesch, opns mgr; Aleese Fielder, gen sls mgr; Tonya Calloway, progmg dir; Robbie Green, chief of engrg.

KOZA(AM)— Jan 20, 1947: Stn currently dark. 1230 khz; 1 kw-U. TL: N31 49 52 W102 22 09. Hrs open: 24 1319 S. Crane Ave., 79763. Phone: (432) 332-1230. Fax: (432) 335-0064. Licensee: Stellar Media Inc. (acq 4-20-89). Population served: 225,000 Format: Sp. Target aud: 18-54. ♦Benjamin Velasquez, gen mgr; Daniel Melendez, progmg dir.

KQLM(FM)— Mar 11, 1996: 107.9 mhz; 100 kw. 846 ft TL: N32 05 51 W102 17 21. Hrs open: 24 1319 S. Crane, 79763. Phone: (432) 333-1227. Fax: 432) 335-0064. Licensee: Stellar Media Inc. (acq 12-30-02). Population served: 250,000 Law Firm: Haley, Bader & Potts. Format: Sp, contemp. News: 10 hrs wkly. Target aud: General; Hispanics/Latinos. ♦Benjamin Velasquez, CEO & gen mgr.

KQRX(FM)—Midland, Oct 20, 1995: 95.1 mhz; 10.35 kw. Ant 505 ft TL: N32 03 10 W102 17 38. Hrs open: 24 3303 N. Midkiff, Suite 115, Midland, 79705. Phone: (432) 520-9510. Fax: (432) 520-9505. Web Site: www.boblivesintaxes.com. Licensee: Double O Texas Corp. (group owner; (acq 2006; grpsl). Population served: 250,000 Natl. Rep: Katz Radio. Law Firm: Fletcher, Heald & Hildreth. Format: Adult hits. ♦Tommy Vascocu, gen mgr; Michael Todd, progmg dir.

KRIL(AM)— June 1946: 1410 khz; 1 kw-U, DA-N. TL: N31 49 00 W102 21 00. Hrs open:
Rebroadcasts KMND(AM) Midland 100%.
11300 Hwy 191, Bldg. 2, Midland, 79707. Phone: (432) 563-5636. Fax: (432) 563-3823. Web Site: www.kmnd.com. Licensee: Cumulus Licensing Corp. Group owner: Cumulus Media Inc. (acq 8-10-99; $110,000). Population served: 200,000 Natl. Network: ESPN Radio. Format: Sports. Target aud: 25 plus; higher educ level, higher income level. ♦Kent Cooper, gen mgr; Robie Burns, progmg dir; Gary Vaugn, chief of engrg.

KXOI(AM)—See Crane

Olney

KAHA(FM)—Not on air, target date: unknown: 104.3 mhz; 50 kw. Ant 472 ft TL: N33 15 20 W98 49 53. Hrs open: 2768 Pharmacy Rd., Rio Grande Ctiy, 78582. Phone: (956) 487-5621. Licensee: James Falcon. ♦James Falcon, gen mgr.

Orange

KIOC(FM)— Feb 28, 1977: 106.1 mhz; 100 kw. 1,225 ft TL: N30 09 31 W93 59 11. Stereo. Hrs opn: 24 2885 Interstate 10 E., Beaumont, 77702. Phone: (409) 896-5555. Fax: (409) 896-5599. E-mail: bigdog106@bigdog106.com Web Site: www.bigdog106.com. Licensee: Capstar TX L.P. Group owner: Clear Channel Communications Inc. (acq 8-30-00; grpsl). Population served: 375,000 Format: Classic rock. Target aud: 18-49; adults who have discretionary income. ♦John Hogan, CEO; Lowry Mays, chmn; Mark Mays, pres; Randall Mays, CFO; Charlie Rahilly, sr VP; Mark Kopelman, VP; Vesta Brandt, gen mgr; Gaile Darbone, opns dir & traf mgr; Elizabeth Blackstock, sls dir; Mike Davis, progmg dir & disc jockey; Jim Love, pub affrs dir; Shon Hodgkinson, chief of engrg & spec ev coord.

KKMY(FM)— 1972: 104.5 mhz; 100 kw. 440 ft TL: N30 08 07 W93 50 39. (CP: 98 kw, ant 984 ft. TL: N30 08 04 W93 56 59). Stereo. Hrs opn: 2885 Interstate 10 E., Beaumont, 77702. Phone: (409) 896-5555. Fax: (409) 896-5500. Web Site: www.mix1045.com. Licensee: Capstar TX L.P. Group owner: Clear Channel Communications Inc. (acq 8-30-00; grpsl). Population served: 500,000 Law Firm: Fisher, Wayland, Cooper, Leader & Zaragoza. Format: Adult contemp. Target aud: 25-54; at-work lstng audience.John Hogan, CEO; Lowry Mays, chmn; Mark Mays, pres; Randall Mays, CFO; Charlie Rahilly, sr VP; Mark Kopelman, VP; Vesta Brandt, gen mgr; Trey Poston, opns dir & progmg dir; Gaile Darbone, opns mgr & traf mgr; Elizabeth Blackstock, sls dir; Rod Windham, natl sls mgr; Shon Hodgkinson, prom dir, progmg dir & spec ev coord; Neil Harrison, news dir; Jim Love, pub affrs dir; T. J. Bordelon, chief of engrg; D.L. Powell, disc jockey; Margie Maybe, disc jockey; Nunee Oakes, disc jockey; Sam Jodi, disc jockey

KOGT(AM)— January 1948: 1600 khz; 1 kw-U, DA-N. TL: N30 08 25 W93 45 11. Hrs open: Box 1667, 77631-1667. Secondary address: 5304 Meeks Dr. 77632. Phone: (409) 883-4381. Fax: (409) 883-7996. E-mail: news@kogt.com Web Site: www.kogt.com. Licensee: G-CAP Communications Inc. (acq 8-7-92; FTR: 8-24-92). Population served: 90,000 Rgnl. Network: Texas State Net. Format: C&W, news, sports. Target aud: 25 plus. ♦Gary Stelly, pres, gen mgr & progmg dir; Richard Corder, gen sls mgr & disc jockey; Glenn Earle, news dir; Russ Ingram, engrg dir; Iva Key Odom, traf mgr; Clay Williams, disc jockey; Reg Russell, disc jockey; Terry Lyons, disc jockey.

Ore City

KAZE(FM)— May 1991: 106.9 mhz; 8.2 kw. 502 ft TL: N32 41 54 W94 37 04. Hrs open: 212 Grande Blvd., Suite B-100, Tyler, 75703. Phone: (903) 581-5259. Fax: (903) 939-3473. E-mail: chelle@theblaze.cc Web Site: www.theblaze.cc. Licensee: Reynolds Radio Inc. (group owner; acq 1-9-97). Format: Rhythmic Contemporary Hit Radio. ♦Rusty Reynolds, pres; Rick Reynolds, gen mgr; Robin George, gen sls mgr; Charlie O'Douglas, progmg dir; Marcus Love, mus dir; James McWain, engrg dir; Chelle Wright-Peterson, traf mgr.

Overton

KPXI(FM)— Oct 8, 1961: 100.7 mhz; 8.1 kw. Ant 571 ft TL: N32 09 07 W95 03 27. Stereo. Hrs opn: 24 6400 N. Belt Line Rd., Suite 110, Irving, 75063-6037. Phone: (214) 561-9673. Fax: (214) 561-9662. Web Site: www.thewordfm.com. Licensee: Inspiration Media of Texas LLC. Group owner: Sunburst Media L.P. (acq 11-6-2000; with KWRD-FM Highland Village). Population served: 1,500,000 Format: Christian talk. ♦Pete Thomson, gen mgr; Easy Ezell, gen sls mgr.

Ozona

KYXX(FM)— Nov 25, 1976: 94.3 mhz; 13 kw. 456 ft TL: N30 42 43 W101 07 30. Stereo. Hrs opn: HC 65 Box 50, Sonora, 76950. Phone: (325) 387-3553. Fax: (325) 387-3554. E-mail: khoskyxx@sonoratx.net Licensee: Ozona Broadcasting Inc. Group owner: Hill Country Broadcasting Corp. (acq 8-16-00; grpsl). Population served: 63884 Format: Real country. News: 5 hrs wkly. Target aud: 12-50 plus; ranchers, oil field, general, travelers. ♦Kent Foster, pres; Monte Spearman, gen mgr; Donna Keese, stn mgr; Eddy Smith, engr.

Palacios

KROY(FM)— November 1996: 99.7 mhz; 50 kw. 331 ft TL: N28 43 53 W96 05 26. Hrs open: 102 Jason Plaza, Suite 2, Victoria, 77901. Phone: (361) 572-0105. Fax: (361) 798-3798. Licensee: Fort Bend Broadcasting Co. (group owner; acq 1-22-99). Format: Texas country. ♦Egon Barthels, gen mgr; Ryan Henderson, opns mgr; Lori Beusin, gen sls mgr; Robi Austynn, progmg dir; Ray Nelson, chief of engrg; Kim Brazil, traf mgr.

Palestine

KNET(AM)— Jan 2, 1936: 1450 khz; 1 kw-U. TL: N31 46 26 W95 37 00. Hrs open: 24 Box 3649, 75802. Secondary address: 800 W. Palestine Ave. 75801. Phone: (903) 729-6077. Fax: (903) 729-4742. E-mail: traffic@kyyk.com Web Site: www.kyyk.com. Licensee: Tomlinson-Leis Communications L.P. (acq 9-30-2005; $1.2 million with co-located FM). Population served: 45,300 Law Firm: Woble, Carlyle, Sandridge

& Rice. Format: Classic country. News staff: one; News: 6 hrs wkly. Target aud: 35 plus. Spec prog: Relg 7 hrs, farm 6 hrs wkly. ♦Edward B. Tomlinson II, pres; Jason Hightower, gen mgr; Tamie Armstrong, opns dir & sls dir; Dave Peterson, progmg dir.

KYYK(FM)—Co-owned with KNET(AM). Aug 20, 1976: 98.3 mhz; 5 kw. Ant 728 ft TL: N31 55 33 W95 38 48. Stereo. 24 Web Site: www.kyyk.com.115,000 Format: Country. News staff: one; News: 2 hrs wkly. Target aud: 18-54.

***KYFP(FM)**— May 15, 2000: 89.1 mhz; 100 kw. Ant 485 ft TL: N32 00 12 W95 43 06. Stereo. Hrs opn: 24 Bible Broadcasting Network, 8030 Arrowridge Blvd., Charlotte, NC, 28241. Phone: (704) 523-5555. Fax: (704) 522-1967. Web Site: www.bbnradio.org. Licensee: Bible Broadcasting Network Inc. Group owner: Bible Broadcasting Network Law Firm: Smithwick & Belendiuk PC. Format: Relg. ♦Lowell Davey, pres; Richard Johnson, opns mgr; Hank larrior, progmg dir; Ron Muffley, chief of engrg.

Pampa

***KAVO(FM)**— July 1998: 90.9 mhz; 17 kw. Ant 364 ft TL: N35 33 08 W101 02 42. Hrs open:
Rebroadcasts WAFR(FM) Tupelo, MS 100%.
Box 3206, Tupelo, MS, 38803. Phone: (662) 844-8888. Fax: (662) 842-6791. E-mail: comments@afr.net Web Site: www.afr.net. Licensee: American Family Association. Group owner: American Family Radio Natl. Network: American Family Radio. Format: Inspirational Christian. ♦Marvin Sanders, gen mgr.

KGRO(AM)— 1947: 1230 khz; 1 kw-U. TL: N35 34 39 W100 57 08. Hrs open: Box 1779, 79066-1779. Phone: (806) 669-6809. Fax: (806) 669-0662. E-mail: kgrokomx@pampa.com Licensee: Pampa Broadcasters Inc. (acq 8-1-67). Population served: 30,000 Natl. Network: Jones Radio Networks. Format: Adult contemp. Target aud: 18-45. ♦James Hughes, pres; Darrell Sehorn, gen mgr, gen sls mgr & progmg dir; Donny Hooper, news dir & sports cmtr; Greg Campbell, chief of engrg; Linda Sehorn, traf mgr; Jimmy Story, disc jockey.

KOMX(FM)—Co-owned with KGRO(AM). May 18, 1981: 100.3 mhz; 32 kw. 300 ft TL: N35 34 39 W100 57 08. Stereo. 125,000 Natl. Network: ABC. Format: Country. Target aud: 20 plus. ♦Linda Sehorn, traf mgr; Donny Hooper, sports cmtr; Jimmy Story, disc jockey.

Paris

KBUS(FM)— June 3, 1985: 101.9 mhz; 50 kw. 500 ft TL: N33 37 15 W95 32 50. Stereo. Hrs opn: 24 5409 90th St., Lubbock, 79424-4305. Phone: (903) 785-1068. Fax: (903) 785-7176. E-mail: jyoung@easttexasradio.com Web Site: www.easttexasradio.com. Licensee: East Texas Broadcasting Inc. (group owner; acq 5-11-01; grpsl). Population served: 50,000 Law Firm: Pepper & Corazzini. Format: Classic rock, news. News staff: one; News: 20 hrs wkly. Target aud: 25-54. Spec prog: Farm 6 hrs wkly. ♦Bud Kitchens, pres; Jimmy Young, gen mgr; Trey Elliott, opns mgr; Jay James, progmg dir & pub affrs dir; Dave Johnson, news dir; Deanna Thorpe, traf mgr.

***KHCP(FM)**— Jan 10, 2001: 89.3 mhz; 21 kw. Ant 354 ft TL: N33 49 36 W95 27 49. Hrs open: 24 Houston Christian Broadcasters Inc, 2424 South Blvd, Houston, 77098. Phone: (713) 520-5200. Web Site: www.khcb.org. Licensee: Houston Christian Broadcasters Inc. (group owner; acq 11-24-00; $3,500 for CP with CP of KKER(FM) Kerrville). Format: Christian. ♦Bruce Munsterman, gen mgr.

KOYN(FM)— Oct 6, 1988: 93.9 mhz; 50 kw. 492 ft TL: N33 49 36 W95 27 49. Stereo. Hrs opn: 24 Box 1038, 75461. Secondary address: 2810 Pine Mill Rd. 75461. Phone: (903) 785-1068. Fax: (903) 785-7176. E-mail: jyoung@easttexasradio.com Web Site: www.easttexasradio.com. Licensee: East Texas Broadcasting Inc. (group owner; acq 5-11-01; grpsl). Population served: 150,000 Natl. Network: USA. Law Firm: Pepper & Corazzini. Format: Country. News staff: 2; News: 3 hrs wkly. Target aud: 12 plus. ♦Bud Kitchens, exec VP; Jimmy Young, gen mgr & stn mgr; Trey Elliott, opns mgr; Jay James, progmg dir; Dave Johnson, news dir; Deanna Thorpe, traf mgr.

KPLT(AM)— Nov 19, 1936: 1490 khz; 1 kw-U. TL: N33 38 07 W95 33 14. Hrs open: 24 Box 9, 75461. Secondary address: 2305 S.E. 3rd St. 75461. Phone: (903) 785-1068. Fax: (903) 785-7176. Web Site: www.kpltfm.com. Licensee: East Texas Broadcasting Inc. (group owner; acq 5-11-01; grpsl). Population served: 55,000 Natl. Network: ABC. Rgnl. Network: Texas State Net. Law Firm: Pepper & Corazzini. Format: Classic country. Target aud: General. Spec prog: Gospel 15 hrs wkly. ♦John Mitchell, pres; Bob Gipson, exec VP, gen mgr, sls dir & gen sls mgr; Kim Good, prom dir & prom mgr; Trey Elliott, opns dir, opns mgr, progmg VP & progmg dir; Dave Johnson, news dir; Christy Storey, pub affrs dir; Bill Hughes, engrg dir & chief of engrg.

KPLT-FM— Aug 14, 1966: 107.7 mhz; 50 kw. Ant 492 ft TL: N33 44 55 W95 24 53. Web Site: www.kpltfm.com.150,000 Natl. Network: ABC. Format: Hot adult contemp. Target aud: 18-35; heavy female/listen at work.

KZHN(AM)— September 1950: 1250 khz; 500 w-D, 95 w-N. TL: N33 43 21 W95 32 50. Stereo. Hrs opn: 24 4140 North Main St, 75460. Secondary address: 402 Munson Ave, Rockwall 75087. Phone: (903) 784-1234. Fax: (903) 784-2344. E-mail: kzhn@koyote.com /kzhn@hughes.net Web Site: www.kzhnthemusiczone.com. Licensee: Eiffel Tower Broadcasting (acq 10-19-2005). Natl. Network: USA. Rgnl rep: Eiffel Tower Broadcasting Law Firm: Womble-Carlyle, Sandridge & Rice. Wire Svc: AP Format: Country gold classics. News staff: 4. Target aud: 24-54+. Spec prog: Bluegrass, Gospel, Sunday preachers. ♦Larry Ryan, CEO, pres & gen mgr; B.J. Clayton, gen mgr; MaryAnn Ryan, gen sls mgr; Crystal Jewel, mus dir; Dawn Mitchell, news dir; Jesse Gilbert, chief of engrg.

Pasadena

***KFTG(FM)**— February 1981: 88.1 mhz; 440 w. Ant 110 ft TL: N29 40 02 W95 09 17. Hrs open: 24 912 Curtis Ave., 77502. Phone: (713) 589-1336. Fax: (713) 589-1335. Licensee: Aleluya Christian Broadcasting Inc. (acq 3-21-03; $482,500). Population served: 250,000 Format: Southern gospel. ♦Roberto R. Villarreal, gen mgr.

KHJZ-FM—See Houston

KIKK(AM)— October 1957: 650 khz; 250 w-D. TL: N29 41 18 W95 10 29. Hrs open: Sunrise-sunset Suite 1900, 24 Greenway Plaza, Houston, 77046. Phone: (713) 881-5100. Fax: (713) 881-5250. Web Site: www.businessradio650.com. Licensee: Texas Infinity Broadcasting L.P. Group owner: Infinity Broadcasting Corp. (acq 10-20-93; FTR: 11-8-93). Population served: 123,280 Natl. Network: CBS. Natl. Rep: CBS Radio. Format: Business radio. Target aud: 25-44. ♦Laura Morris, VP & gen mgr; Josh Mednick, sls dir; Dan Blanchard, gen sls mgr & natl sls mgr; Richard Topper, natl sls mgr; Pam Kehoe, mktg dir & prom mgr; Brent Clanton, progmg dir; Dan Woodard, chief of engrg; Judy Hart, traf mgr.

KKBQ-FM— August 1962: 92.9 mhz; 100 kw. 1,919 ft TL: N29 34 34 W95 30 36. Stereo. Hrs opn: 24 1990 Post Oak Blvd. #2300, Houston, 77056. Phone: (713) 961-0093. Fax: (713) 993-9300. Web Site: www.KKBQ.com. Licensee: Cox Radio Inc. Group owner: Cox Broadcasting (acq 8-7-2000; grpsl). Population served: 3,458,300 Law Firm: Reed, Smith, Shaw & McClay. Format: Country. Target aud: 25-54.Caroline Devine, gen mgr; Doug Abernethy, sls dir; Judy Lakin, gen sls mgr; Mike Murray, natl sls mgr; Bill Tatar, mktg dir; Remo Mazzini, prom dir; Christi Brooks, mus dir; Mike Mollett, pub affrs dir; Jed Wilkinson, chief of engrg; Emily Gerald, traf mgr; "Cactus Jack" Talley, disc jockey; Beau Bodine, disc jockey; Dave E. Crockett, disc jockey; Elizabeth Rose, disc jockey

KLVL(AM)— May 5, 1950: 1480 khz; 1 kw-D, 500 w-N, DA-N. TL: N29 41 02 W95 11 09. Hrs open: 1302 N. Shepherd Dr., Houston, 77008. Phone: (713) 665-8994. Phone: (713) 868-6166. Fax: (713) 868-9631. E-mail: diddierugalde@hotmail.com Web Site: www.klvl.com. Licensee: SIGA Broadcasting Corp. (group owner; acq 5-16-97; $1.25 million). Population served: 500,000 Format: Sp var. Spec prog: Black 4 hrs wkly. ♦Dr. Gabriel Arango, pres; Hector Guevara, gen mgr.

Pearsall

KRIO-FM— Aug 4, 2002: 104.1 mhz; 100 kw. Ant 981 ft TL: N28 44 53 W98 50 14. Hrs open: 24 7800 IH 10 W., San Antonio, 78230. Phone: (210) 340-1234. Fax: (210) 340-1234. Licensee: BMP San Antonio License Co. L.P. (acq 7-23-2004; $10.25 million). Law Firm: Fletcher, Heald & Hildreth. Format: Sp contemp. ♦Paula Furr, CFO; John Barger, gen mgr.

KSAG(FM)—Not on air, target date: unknown: 103.3 mhz; 6 kw. Ant 328 ft TL: N29 01 04 W99 09 25. Hrs open: 819 S.W. Federal Hwy., Suite 106, Stuart, FL, 34994. Phone: (772) 215-1634. Web Site: www.toweritrust.com. Licensee: Tower Investment Trust Inc. ♦William H. Brothers, pres.

KVWG(AM)— Nov 3, 1962: 1280 khz; 500 w-D. TL: N28 53 13 W99 06 40. Hrs open: Box K, 78061. Secondary address: 205 S. Walnut St. 78061. Phone: (830) 334-8900. Fax: (830) 334-3448. Licensee: Pearsall Radio Works Ltd. (acq 10-20-98; $200,000 with co-located FM). Population served: 47,000 Rgnl. Network: Texas State Net. Format: Classic country, farm. ♦J.R. Gully, gen mgr, gen sls mgr, prom mgr, progmg dir, news dir & chief of engrg.

Pecan Grove

KREH(AM)— 1952: 900 khz; 2.5 kw-D, 10 w-N. TL: N29 38 38 W96 05 46. Hrs open: Sunrise-sunset 5821 Southwest Fwy., Suite 600, Houston, 77057. Phone: (713) 917-0050. Fax: (713) 917-0213. E-mail: info@radio.viet.com Web Site: www.radiosaigonhouston.com. Licensee: Bustos Media Holdings LLC. Group owner: Bustos Media Holdings (acq 6-11-02). Format: Oldies, var, country. ♦Thuy Vu, gen mgr.

Pecos

KGEE(FM)— 1999: Stn currently dark. 97.3 mhz; 100 kw. Ant 413 ft TL: N31 30 54 W103 11 25. Hrs open: 11300 Hwy. 191, Bldg. 2, Midland, 79707. Phone: (432) 563-5636. Fax: (432) 563-3823. Licensee: Cumulus Licensing LLC. (acq 6-11-2002; $1 million). ♦Kent Cooper, gen mgr.

KIUN(AM)— Oct 23, 1935: 1400 khz; 1 kw-U. TL: N31 26 09 W103 30 14. Hrs open: 24 Box 469, 79772. Phone: (432) 445-2497. Fax: (432) 445-4092. E-mail: kiun@valornet.com Web Site: www.98xfm.com. Licensee: Pecos Radio Co. (acq 3-16-2006; with co-located FM). Population served: 13,000 Rgnl. Network: Texas State Net. Law Firm: Sanchez Law Firm. Format: Country. Target aud: General. ♦Bill Cole, gen mgr, progmg dir & chief of engrg.

KPTX(FM)—Co-owned with KIUN(AM). Aug 3, 1981: 98.3 mhz; 9.5 kw. Ant 423 ft TL: N31 29 56 W103 19 50. Stereo. 6 AM-10 PM Web Site: www.98xfm.com. Licensee: Parday Inc. Natl. Network: ABC. Law Firm: Sanchez Law Firm. Format: Adult contemp. Target aud: 25 plus; adult.

Perryton

KEYE(AM)— Nov 19, 1948: 1400 khz; 1 kw-U. TL: N36 23 20 W100 49 37. Hrs open: Box 630, 79070. Phone: (806) 435-5458. Fax: (806) 435-5393. E-mail: keye@arn.net Web Site: www.keye.net. Licensee: Perryton Radio Inc. (acq 8-69). Population served: 40,000 Rgnl. Network: Texas State Net. Format: Country. Spec prog: Farm 2 hrs, relg 3 hrs wkly. ♦Chris Samples, gen mgr, gen sls mgr & progmg dir; Lynlee Mullins, traf mgr.

KEYE-FM— January 1978: 96.1 mhz; 8.5 kw. 400 ft TL: N36 21 54 W100 46 48. Stereo. Web Site: www.keye.net.50,000 Format: Oldies.

Pflugerville

KOKE(AM)— 2001: 1600 khz; 5 kw-D, 700 w-N, DA-2. TL: N30 20 44 W97 32 46. Hrs open: 2211 S. IH 35, Suite 401, Austin, 78741. Phone: (512) 416-1100. Fax: (512) 416-8205. Licensee: BMP Austin License Company L.P. Group owner: Amigo Broadcasting L.P. (acq 11-9-2004; . grpsl). Format: Sp, news/talk info. ♦Paul Danitz, gen mgr; Tim Harper, rgnl sls mgr; Javier Salgado, progmg dir.

Pharr

KVJY(AM)— February 1985: 840 khz; 5 kw-D, 1 kw-N, DA-2. TL: N26 19 00 W98 06 16. Stereo. Hrs opn: 1201 No. Jackson, Suite 900, McAllen, 78501. Phone: (212) 966-1059. Phone: (956) 992-8895. Fax: (956) 992-8897. Web Site: www.radiounica.com. Licensee: BMP RGV License Co. L.P. Group owner: Multicultural Radio Broadcasting Inc.

(acq 3-31-2005; grpsl). Format: Country. ♦Thomas Castro, pres; Jose Luis Munoz, gen mgr; Jeff Koch, opns dir & progmg dir.

Pilot Point

KZMP-FM— Oct 17, 1983: 104.9 mhz; 15.7 kw. Ant 1,755 ft TL: N33 33 37 W96 57 34. Stereo. Hrs opn: 4201 Pool Rd., Coleyville, 76034. Phone: (817) 868-2900. Fax: (817) 868-2929. Licensee: Liberman Broadcasting of Dallas License LLC. Group owner: Entravision Communications Corp. (acq 11-2-2006; grpsl). Population served: 6,153,500 Format: Rgnl Mexican. ♦Alex Sanchez, gen mgr.

Pittsburg

KDVE(FM)— Dec 15, 1986: 103.1 mhz; 10 kw. Ant 672 ft TL: N32 52 50 W94 58 13. Stereo. Hrs opn: 24 Box 1648, Jacksonville, 75766. Secondary address: 402 S. Ragsdale, Jacksonville 75766. Phone: (903) 586-2527. Fax: (903) 589-0677. Licensee: Waller Media LLC. (group owner; (acq 8-24-2005; $975,000 with KXAL-FM Tatum). Population served: 280,000 Format: Sp adult contemp. News staff: one; News: 6 hrs wkly. ♦Dudley Waller, gen mgr.

***KGWP(FM)**— 2003: 91.1 mhz; 800 w vert. Ant 128 ft TL: N33 01 41 W95 02 57. Stereo. Hrs opn: 24 1511 Jefferson Ave., Mount Pleasant, 75455. Phone: (951) 675-8661. Fax: (903) 575-1984. Licensee: Andres Serranos Ministries Inc. (acq 4-5-2006; $83,332). Format: Christian Sp contemp. News staff: one; News: 10 hrs wkly. Target aud: 30-55+. ♦Rafael Garcia, pres.

KSCN(FM)— Mar 1, 1999: 96.9 mhz; 14 kw. 390 ft TL: N33 00 31 W95 04 14. Stereo. Hrs opn: 24 Box 990, Mount Pleasant, 75456. Secondary address: 1798 US Hwy. 67 W., Mount Pleasant 75455. Phone: (903) 572-8726. Fax: (903) 572-7232. E-mail: bud@easttexasradio.com Web Site: www.easttexasradio.com. Licensee: East Texas Broadcasting Inc. (group owner). Population served: 68,000 Format: Country. News staff: 2; News: 3 hrs wkly. Target aud: 25-54; general. ♦John Mitchell, chmn; Bud Kitchens, pres & gen mgr; Craig Morgan, opns dir; Bryan Friesth, gen sls mgr; Darrin Tripp, progmg dir; Clint Cooper, news dir & pub affrs dir; Bill Hughes, chief of engrg; Justice Thornburg, traf mgr.

Plains

***KPHS(FM)**— Nov 14, 1977: 90.3 mhz; 220 w. 135 ft TL: N33 11 16 W102 49 20. Hrs open: 8:30 AM-3:15 PM Box 479, 79355. Phone: (806) 456-7401. Fax: (806) 456-4325. Licensee: Plains Independent School District. Format: Educ. ♦Rennetta O'Quinn, gen mgr.

Plainview

***KBAH(FM)**— Mar 18, 2004: 90.5 mhz; 75 kw. Ant 426 ft TL: N34 03 58 W101 42 16. Hrs open: Box 3206, Tupelo, MS, 38803. Phone: (662) 844-8888 ext. 204. Web Site: afr.net. Licensee: American Family Association. Group owner: American Family Radio Population served: 285,000 Format: Christian classics. ♦Marvin Sanders, gen mgr.

KKYN-FM—Listing follows KVOP(AM).

***KPMB(FM)**—Not on air, target date: unknown: 88.5 mhz; 3 kw. Ant 282 ft TL: N34 13 14 W101 42 52. Hrs open: Box 252, McAllen, 78505. Phone: (956) 686-6382. Fax: (956) 686-2999. Licensee: Paulino Bernal Evangelism. Format: Sp relg.

KREW(AM)— Aug 14, 1944: 1400 khz; 1 kw-U. TL: N34 12 20 W101 42 59. Hrs open: 24 3218 N. Quincy, 79072. Secondary address: Box 1420 79072. Phone: (806) 293-2661. Fax: (806) 293-5732. Web Site: kkyn.net. Licensee: Rhattigan Broadcasting (Texas) LP (group owner; (acq 8-19-2004; grpsl). Population served: 60,000 Rgnl. Network: Texas State Net. Wire Svc: AP Format: Sp, oldies. News: 7 hrs wkly. Target aud: Adults 35+; baby boomers. ♦John Weeks, gen mgr; Cherie Griffith, sls dir; Jerry Larsen, gen sls mgr; Brandy Haines, progmg dir; Tom Hall, stn mgr & news dir; Gary Graham, chief of engrg.

KRIA(FM)—Co-owned with KREW(AM). 1999: 106.9 mhz; 50 kw. 469 ft TL: N34 15 47 W101 40 30. Web Site: kkyn.net. Natl. Network: CSN. Law Firm: Fisher, Wayland, Cooper, Leader & Zaragoza. Format: Adult contemp. News: 10 hrs wkly. Target aud: Hispanic; 18-49. ♦Jim Ray, CEO; Dimas Garcia, progmg dir.

KSTQ-FM— September 1961: 97.3 mhz; 100 kw. Ant 440 ft TL: N34 15 45 W101 40 05. Stereo. Hrs opn: 24 Box 3757, Lubbock, 79452. Secondary address: 9800 University, Lubbock 79423. Phone: (806) 745-3434. Fax: (806) 748-2470. Licensee: Ramar Communications II Ltd. (group owner; (acq 7-12-2002; $750,000). Population served: 350,000 Natl. Rep: Univision Radio National Sales. ♦Chuck Heinz, gen mgr; Connie Hayes, gen sls mgr; Eddie Moreno, progmg dir; Gilbert Saldana, mus dir; Lee Thomas, chief of engrg; Susie Gonsales, traf mgr.

KVOP(AM)— Oct 1, 1974: 1090 khz; 5 kw-D, 500 w-N, DA-2. TL: N34 05 32 W101 38 26. Hrs open: 24 3218 N. Quincy, 79072. Secondary address: Box 147 79072. Phone: (806) 296-2771. Fax: (806) 293-5732. Licensee: Rhattigan Broadcasting (Texas) LP (group owner; (acq 8-19-2004; grpsl). Natl. Rep: Katz Radio. Format: Talk. News: 10 hrs wkly. Target aud: Adults; 25-54. Spec prog: Farm 12 hrs wkly. ♦John Weeks, gen mgr; Chekie Griffith, sls dir; Brandy Haines, progmg dir; Tom Hall, stn mgr & news dir; Gary Graham, chief of engrg.

KKYN-FM—Co-owned with KVOP(AM). 1987: 106.9 mhz; 50 kw. Ant 469 ft TL: N34 15 47 W101 40 30. Stereo. 24 Web Site: www.kkyn.net. Format: Country. News staff: one; News: 5 hrs wkly. Target aud: 35 plus. ♦Jerry Larsen, stn mgr; Tom Hall, mktg dir, prom dir & pub affrs dir.

***KWLD(FM)**— 1952: 91.5 mhz; 370 w. 105 ft TL: N34 11 14 W101 43 32. Stereo. Hrs opn: 24 1900 W. 7th St., #230, ., PLainview, 79072. Phone: (806) 291-1091. Fax: (806) 291-1963. E-mail: kwld@wbu.edu Web Site: www.wbu.edu. Licensee: Wayland Baptist University. Population served: 40,000 Natl. Network: USA. Format: CHR, Christian music, jazz. News: 14 hrs wkly. Target aud: 15-30; high school through college, young adult, afternoon & evening. ♦Paul Armes, pres; Jim Smith, CFO; Bill Hardage, exec VP; Betty Donaldson, VP; Claude Lusk, VP; Steve Long, gen mgr; Paul Sutton, stn mgr & progmg dir; David Carr, chief of engrg.

Plano

KMKI(AM)— July 15, 1999: 620 khz; 5 kw-D, 4.5 kw-N, DA-2. TL: N33 14 34 W96 32 29. Stereo. Hrs opn: 24 2221 E. Lamar Blvd., Suite 300, Arlington, 76006. Phone: (817) 695-1333. Fax: (817) 695-3556. Web Site: www.radiodisney.com. Licensee: Radio Disney Dallas LLC. Group owner: ABC Inc. (acq 9-4-98; $12.1 million). Population served: 2,000,000 Natl. Network: Radio Disney. Natl. Rep: Interep. Format: Top-40. Target aud: 18-34; 25-49; women adults. ♦Keri Littlefield, pres & gen mgr; Robin Jones, opns VP; Jamie Ramsey, gen sls mgr; Greg Heitzman, natl sls mgr; Molly Borsh, prom dir.

Pleasant Valley

KZAM(FM)—Not on air, target date: unknown: 98.7 mhz; 6 kw. Ant 279 ft TL: N33 58 24.5 W98 39 22.7. Hrs open: 2768 Pharmacy Rd., Rio Grande City, 78582. Phone: (956) 487-5621. Licensee: James Falcon. ♦James Falcon, gen mgr.

Pleasanton

KWMF(AM)— Feb 8, 1951: 1380 khz; 4 kw-D, 160 w-N, DA-D. TL: N29 00 00 W98 31 50. Hrs open: 5 AM-midnight 1903 S. Lamesa Rd., Midland, 79701-1706. Phone: (888) 784-3476. Fax: (432) 684-5588. Web Site: www.grnonline.com. Licensee: La Promesa Foundation. Group owner: Border Media Partners LLC (acq 12-13-2006; grpsl). Population served: 30,000 Rgnl. Network: Texas State Net. Format: Christian, Sp. ♦Robert Dominguez, gen mgr.

Point Comfort

KJAZ(FM)— Dec 10, 1998: 94.1 mhz; 25 kw. 194 ft TL: N28 46 08 W96 42 39. Stereo. Hrs opn: 24 102 Jason Plaza, Suite 2, Victoria, 77901. Phone: (361) 572-0105. Fax: (361) 798-3798. Licensee: Fort Bend Broadcasting Co. Inc. Group owner: Fort Bend Broadcasting Co. (acq 4-13-2001; $400,000). Format: Classic rock. Target aud: Baby Boomers; Active, Affluent Adults. ♦Ryan Henderson, gen mgr & progmg dir.

Port Arthur

***KDEI(AM)**— August 1934: 1250 khz; 5 kw-D, 1 kw-N, DA-N. TL: N29 57 04 W93 52 46. Hrs open: 24 601 Washington St., Alexandria, LA, 71301. Phone: (318) 561-6145. Fax: (318) 449-9954. E-mail: info.usa@radiomaria.org Web Site: www.radiomaria.us. Licensee: Radio Maria Inc. (group owner; acq 9-20-99). Population served: 57,371 Natl. Network: American Urban. Format: Christian, Relg, talk. News: 10.5 hrs wkly. Target aud: General; isolated and under-represented groups in society, sick, elderly etc. ♦Dale dePerrodil, gen sls mgr; Duane Stenzel, gen mgr & progmg dir; Danny Brou, chief of engrg & disc jockey.

KOLE(AM)— 1947: 1340 khz; 1 kw-U. TL: N29 54 15 W93 56 10. Hrs open: 24
Rebroadcasts KRCM(AM) Beaumont 50%.
27 Sawyer St., Beaumont, 77702. Phone: (409) 835-2222. Phone: (866) 835-1340. Fax: (409) 832-5686. E-mail: mamager@newsradiofox.com Web Site: www.newsradiofox.com. Licensee: CityGate Media Inc. (acq 10-95; $80,000). Population served: 57,371 Natl. Network: USA, Fox News Radio, Talk Radio Network. Law Firm: Gammon & Grange. Format: News/talk. News staff: 2; News: 40 hrs wkly. Target aud: 25 plus. ♦Ralph McBride, pres & gen mgr; Brent Bobbitt, gen sls mgr; Dominick Brascia, progmg dir & progmg mgr; John St.John, news dir; Jeff Roberts, pub affrs dir; Jeanette Harvey, traf mgr; Russ Ingram, chief of engrg & engr.

KQBU-FM—Licensed to Port Arthur. See Houston

KTJM(FM)— Apr 15, 1963: 98.5 mhz; 100 kw. Ant 1,952 ft TL: N30 03 05 W94 31 37. Stereo. Hrs opn: 3000 Bering Dr., Houston, 77057. Phone: (281) 493-2900. Fax: (281) 596-9608. Web Site: www.laraza.fm. Licensee: Liberman Broadcasting of Houston License LLC. Group owner: Liberman Broadcasting Inc. (acq 3-20-2001; grpsl). Format: Rgnl Mexican. ♦Cheque Gonzalez, gen mgr.

Port Isabel

KZPL(FM)— 1992: 101.1 mhz; 3 kw. Ant 360 ft TL: N25 57 52 W97 14 38. Stereo. Hrs opn: 24 801 Jackson Rd., McAllen, 78501. Phone: (956) 661-6000. Fax: (956) 661-6081. Licensee: Entravision Holdings L.L.C. Group owner: Entravision Communications Co. L.L.C. (acq 7-20-2000; grpsl). Population served: 350,000 Natl. Network: Westwood One. Law Firm: Fletcher, Heald & Hildreth. Format: Sp contemp. News staff: one; News: one hr wkly. Target aud: 25-55. ♦Willie Rosales, gen mgr & gen sls mgr; Mando Sanroman, progmg dir; Sonny Cabazos, chief of engrg; Dora Borjas, traf mgr.

Port Lavaca

KITE(FM)—Licensed to Port Lavaca. See Victoria

Port Neches

KBPO(AM)— June 13, 1959: 1150 khz; 500 w-D, 63 w-N, DA-2. TL: N30 04 45 W93 57 05. Hrs open: 419 Stadium Rd., Port Arthur, 77642. Phone: (409) 460-0029. Fax: (409) 983-5858. Licensee: Vision Latina Broadcasting Inc. (acq 9-8-93; $75,000; FTR: 9-27-93). Population served: 27,000 Natl. Network: Fox Sports. Format: Sports. ♦Eloy Castro, pres & gen mgr; Marco Mata, gen sls mgr; Patricia Montenegro, progmg dir; Richard Ryele, chief of engrg; Don Hebert, disc jockey; Jeremy Ryan, disc jockey; Lauri Grantham, disc jockey.

Port O'Connor

***KHPO(FM)**— 2007: 91.9 mhz; 18 kw. Ant 308 ft TL: N28 25 44 W96 26 54. Hrs open:
Rebroadcasts KHCB-FM Houston 100%.
2424 South Blvd., Houston, 77098-5110. Phone: (713) 520-5200. Web Site: www.khcb.org. Licensee: Houston Christian Broadcasters Inc. Format: Christian. ♦Bruce Munsterman, pres.

Portland

KMJR(FM)— Dec 15, 1979: 105.5 mhz; 1.9 kw. 354 ft TL: N27 47 48 W97 23 51. Stereo. Hrs opn: 24 1300 Antelope, Corpus Christi, 78401. Phone: (361) 883-1600. Fax: (361) 888-5685. Licensee: Tejas Broadcasting Ltd. LLP. Group owner: Amigo Broadcasting L.P. (acq 11-15-2004; grpsl). Format: Rgnl Mexican. Target aud: 18-49; general. ♦Eddie Alonzo, gen mgr; Julie Garza, progmg dir; Lon Gonzalez, news dir; Henry Turner, chief of engrg; Debbie Reid, traf mgr.

KOUL(FM)—See Corpus Christi

***KSGR(FM)**— October 2000: 91.1 mhz; 3 kw. Ant 298 ft TL: N28 00 06 W97 15 01. Hrs open: 3001 Rodd Field Rd., Corpus Christi, 78414. Phone: (361) 814-7775. Fax: (361) 814-7779. Web Site: www.ksgr.org. Licensee: CSN International (group owner; acq 6-8-99). Format: Contemp Christian. ♦Jim Sheperd, gen mgr.

Post

KPOS(FM)— May 1, 1991: 107.3 mhz; 22 kw. Ant 748 ft TL: N33 13 23 W101 26 26. Hrs open: 24
Rebroadcasts KLRD(FM) Yucaipa, CA 100%.

2351 Sunset Blvd., Suite 170-218, Rocklin, CA, 95765. Phone: (916) 251-1600. Fax: (916) 251-1650. Licensee: Educational Media Foundation. Group owner: EMF Broadcasting (acq 5-21-2004; $550,000). Population served: 200,000 Natl. Network: Air 1. Format: Christian contemp. ♦Richard Jenkins, pres; Mike Novak, VP & progmg dir; Lloyd Parker, gen mgr; Ed Lenane, opns dir & news dir; Keith Whipple, dev dir; Eric Allen, natl sls mgr; David Pierce, progmg mgr; Jon Rivers, mus dir; Sam Wallington, engrg dir; Arthur Vassar, traf mgr; Karen Johnson, news rptr; Marya Morgan, news rptr; Richard Hunt, news rptr.

Prairie View

***KPVU(FM)**— Nov 26, 1981: 91.3 mhz; 98.3 kw. 410 ft TL: N30 05 21 W95 59 46. Stereo. Hrs opn: 24 Box 156, Hilliard Hall, 77446. Phone: (936) 857-4511. Phone: (936) 857-4515. Fax: (936) 857-2729. Web Site: www.pvamu.edu. Licensee: Prairie View A&M University. Population served: 250,000 Format: Smooth jazz, adult contemp, gospel. News staff: News progmg 14 hrs wkly Target aud: 18 plus. Spec prog: Black 6 hrs wkly. ♦Larry Coleman, gen mgr; Charles Porter, opns mgr, progmg dir, news dir & sports cmtr; Gwen Johnson, dev dir, sls dir, mktg mgr, prom dir & adv mgr; Dave Cassels, chief of engrg; Wayne Turner, disc jockey.

Premont

KMFM(FM)— 1989: 104.9 mhz; 3 kw. 299 ft TL: N27 22 19 W98 11 21. (CP: 100.7 mhz, 25 kw, ant 285 ft. TL: N27 28 30 W98 03 23). Hrs opn: Box 252, McAllen, 78502. Phone: (956) 686-6382. Fax: (956) 686-2999. Licensee: Radio Cristiana Network. Format: Sp, relg. ♦Eloy Bernal, gen mgr; Paulino Bernal Jr., stn mgr & progmg dir; John Ross, chief of engrg.

Quanah

KREL(AM)— May 11, 1951: 1150 khz; 500 w-D, DA. TL: N34 18 58 W99 44 49. Hrs open: 6 AM-sunset 750 N. Saint Paul St., Dallas, 75201. Phone: (940) 663-5711. Licensee: First Broadcasters Investment Partners LLC. (group owner; (acq 6-13-2005; grpsl). Population served: 85,000 Format: Classic country. Target aud: General.

KWFB(FM)— Sept 1, 1982: 100.9 mhz; 50 kw. Ant 492 ft TL: N34 15 21 W99 30 05. Stereo. Hrs opn: 24 Box 29, 221 N. Main St., 79252. Phone: (940) 663-6363. Fax: (940) 663-6364. E-mail: kixc@broadcast.net Web Site: www.kixc.com. Licensee: KIXC-FM L.L.C. Format: Adult hits. News: 10 hrs wkly. Target aud: General. Spec prog: Farm 3 hrs, relg 2 hrs wkly. ♦Glen Ingram, pres & gen sls mgr; Michael Reeves, gen mgr; John White, opns dir & progmg dir.

Ralls

KCLR(AM)— May 31, 1963: 1530 khz; 5 kw-D, 1 kw-CH. TL: N33 40 00 W101 22 44. Hrs open: Box 252, McAllen, 78505. Phone: (956) 686-6382. Fax: (956) 686-2999. E-mail: paulinobernal@hotmail.com Web Site: www.laradiocristiana.com. Licensee: Paulino Bernal (acq 10-19-2001). Population served: 1,962 Format: Sp, Christian. ♦Paulino Bernal, gen mgr; Eloy Bernal, stn mgr; Pete Guzman, opns mgr.

Ranger

KCUB-FM— July 1, 1990: 98.5 mhz; 5.8 kw. Ant 335 ft TL: N32 20 48 W98 42 50. Stereo. Hrs opn: 24 471 N. Harbin Dr., Suite 102, Stephenville, 76401. Phone: (254) 968-7459. Fax: (254) 968-6258. E-mail: john@mandatoryfm.com Web Site: www.mandatoryfm.com. Licensee: Mandatory Broadcasting Inc. (acq 5-17-2007; $600,000). Natl. Network: Jones Radio Networks. Rgnl. Network: Texas State Net. Format: Texas country. News staff: 0. Target aud: 20-65; all-important age group of today's buying public. ♦John Hollinger, gen mgr; Jonathon Boev, gen sls mgr & progmg dir; Jim Rhodes, chief of engrg.

Raymondville

KBIC(FM)— October 1996: 105.7 mhz; 1.8 kw. 426 ft TL: N26 26 37 W97 42 08. Hrs opn: 24 Box 1290, Weslaco, 78599. Phone: (956) 968-7777. Fax: (956) 968-5143. Web Site: www.radiovida.com. Licensee: Christian Ministries of the Valley Inc. (acq 2-4-93; FTR: 3-1-93). Format: Sp, relg. ♦Enrique Garza, gen mgr.

KBUC(FM)— 1979: 102.1 mhz; 17.9 kw. Ant 758 ft TL: N26 38 09 W97 50 10. Stereo. Hrs opn: 6 AM-midnight 1201 N. Jackson Rd., Suite 900, McAllen, 78501. Phone: (956) 992-8895. Fax: (956) 992-8897. Licensee: BMP RGV License Company L.P. Group owner: Border Media Partners LLC (acq 5-30-2003; $8 million with KBDR(FM) Mirando City). Format: Regional Mexican. ♦Jose Luis Munoz, gen mgr; Rogelio Botelleo Rios, opns mgr; Maria Alvarez, gen sls mgr; Joe Espinoza, chief of engrg; Angela Pina, traf mgr.

KSOX(AM)— June 1, 1957: 1240 khz; 1 kw-U. TL: N26 27 28 W97 46 55. Hrs open: 6 AM-midnight 2921 North Closner, Edinburg, 78541. Phone: (956) 992-8895. Fax: (956) 992-8897. Licensee: BMP RGV License Co. L.P. Group owner: Border Media Partners LLC (acq 1-9-2004; $7.5 million with KURV(AM) Edinburg). Population served: 7,987 Format: Sports. Target aud: 25-55. ♦Angela Pina, traf mgr.

Refugio

KTKY(FM)— Oct 5, 1979: Stn currently dark. 106.1 mhz; 25 kw. Ant 328 ft TL: N28 08 15 W97 12 45. (CP: COL Taft. 50 kw, ant 446 ft. TL: N27 52 00 W97 13 08). Stereo. Hrs opn: 710 Buffalo St., Suite 608, Corpus Christi, 78401. Phone: (314) 345-1030. Phone: (361) 882-5749. Fax: (361) 884-1240. E-mail: jim@koplar.com Licensee: Pacific Broadcasting of Missouri L.L.C. (acq 4-24-98; $725,000). Population served: 400,000 ♦James G. Withers, gen mgr.

Reno

KLOW(FM)—Not on air, target date: unknown: 98.9 mhz; 5.9 kw. Ant 331 ft TL: N33 38 54 W95 36 12. Hrs open: 819 S.W. Federal Hwy., Suite 106, Stuart, FL, 34994-2952. Phone: (772) 215-1634. Web Site: www.toweritrust.com. Licensee: Tower Investment Trust Inc. ♦William H. Brothers, pres.

Richardson

KKLF(AM)— 1999: 1700 khz; 10 kw-D, 1 kw-N. TL: N33 25 23 W96 39 45. Hrs open:
Rebroadcasts KLIF(AM) Dallas 100%.
3500 Maple Ave., Suite 1310, Dallas, 75219. Phone: (214) 526-7400. Fax: (214) 525-2525. Web Site: www.klif.com. Licensee: KRBE Lico Inc. Group owner: Susquehanna Radio Corp. (acq 4-30-98). Format: Talk.

Richmond

KPTY(FM)—See Houston

KRTX(AM)—See Houston

Rio Grande City

KQBO(FM)— April 1985: 107.5 mhz; 1.41 kw. 420 ft TL: N26 25 47 W98 49 25. Stereo. Hrs opn: 5 AM-midnight 102 KCTM-FM 103 Rd., 78582-9805. Phone: (956) 487-8224. Fax: (815) 361-6185. Licensee: Gustavo Valadez Jr. (acq 6-13-03). Format: Latin pop. News: 5 hrs wkly. Target aud: 18-45. ♦Gustavo "Gus" Valadez Jr., pres.

Robinson

KHCK-FM— Nov 1, 1972: 107.9 mhz; 6 kw. Ant 328 ft TL: N31 30 33 W97 10 03. Stereo. Hrs opn: 24 10801 N. Mopac Expwy. 2-250, Austin, 78759-5457. Phone: (214) 525-0400. Phone: (512) 419-1077. Web Site: www.netmio.com. Licensee: KICI-FM License Corp. Group owner: Univision Radio (acq 9-22-2003; grpsl). Population served: 35,000 Format: Mexican, rgnl. Target aud: 18-54. ♦Mac Tichenor, pres; Tim McCoy, gen mgr; Chris Munoz, gen sls mgr; Oscar Rios, progmg dir; Samantha Martinez, traf mgr.

Robstown

***KLUX(FM)**— Mar 17, 1985: 89.5 mhz; 60 kw. Ant 954 ft TL: N27 46 50 W97 38 03. Stereo. Hrs opn: 24 1200 Lantana, Corpus Christi, 78407. Phone: (361) 289-6437/289-2487. Fax: (361) 289-1420. E-mail: klux@goccn.org Web Site: www.klux.org. Licensee: Diocesan Telecommunications Corp. Population served: 500,000 Natl. Network: USA. Law Firm: Ross & Hardies. Format: Easy lstng. News staff: one; News: 13 hrs wkly. Target aud: 35 plus; total persons. Spec prog: Sp 3 hrs wkly. ♦Rev. Msgr. Michael Howell, chmn; Marty Wind, exec VP & gen mgr; Russ Martin, opns dir, opns mgr & dev dir.

KMIQ(FM)—Licensed to Robstown. See Corpus Christi

KROB(AM)— Feb 22, 1963: 1510 khz; 500 w-D. TL: N27 46 39 W97 37 55. Hrs open: 400 SPID, Suite 107, Corpus Christi, 78405. Phone: (361) 774-4354. Fax: (361) 299-6002. E-mail: krobam1510@sbcglobal.net Licensee: B Communications Joint Venture (acq 1-4-02). Format: Oldies. ♦Jerry Benavides, pres; Jerry Benevides, gen mgr; Ben Benavides, sls dir; Bob Pena, progmg dir; Gary Graham, chief of engrg; Peter Hemphill, traf mgr.

KSAB(FM)— Oct 13, 1966: 99.9 mhz; 96 kw. 955 ft TL: N27 44 28 W97 36 08. Stereo. Hrs opn: 501 Tupper Ln., Radio Plaza, Corpus Christi, 78417. Phone: (361) 289-0111. Fax: (361) 289-5035. E-mail: ksabfm@aol.com Web Site: www.ksabfm.com. Licensee: Capstar TX L.P. Group owner: Clear Channel Communications Inc. (acq 8-30-00; grpsl). Format: Tejano, Sp. ♦Matt Martin, gen mgr; Dan Pena, prom dir & progmg dir.

Rockdale

KRXT(FM)— Feb 27, 1989: 98.5 mhz; 6 kw. 328 ft TL: N30 38 32 W97 02 13. Stereo. Hrs opn: 24 hrs 1095 W. Highway 79, 76567. Phone: (512) 446-6985. Fax: (512) 446-6987. E-mail: krxtl@tlab.net Web Site: www.krxt.com. Licensee: KRXT Inc. Rgnl. Network: Texas State Net. Format: Country. News staff: one; News: 20 hrs newa progmg wkly. Target aud: General. Spec prog: Spanish, Czech. ♦Charles W. McGregor, pres, gen mgr & stn mgr.

Rockport

KKPN(FM)— October 1986: 102.3 mhz; 50 kw. 371 ft TL: N28 00 03 W97 04 34. Stereo. Hrs opn: 24 826 S. Padre Island Dr., Corpus Christi, 78416. Phone: (361) 814-3800. Fax: (361) 855-3770. Web Site: planet1023.com. Licensee: Convergent Broadcasting Corpus Christi LP. Group owner: Convergent Broadcasting LLC (acq 1-12-2004; grpsl). Population served: 313,600 Format: CHR/Top40. Target aud: Adult; 18-49. ♦Mark White, gen mgr; Dallas Garcia, gen sls mgr & adv mgr; Scott Holt, progmg dir; William Hooper, chief of engrg.

KTKY(FM)—See Refugio

Rocksprings

KDRX(FM)—Not on air, target date: unknown: 106.9 mhz; 50 kw. Ant 492 ft TL: N29 58 30 W100 25 15. Hrs opn: 819 S.W. Federal Hwy., Suite 106, Stuart, FL, 34994. Phone: (772) 286-5586. Licensee: William H. Brothers. ♦William H. Brothers, gen mgr.

KHES(FM)—Not on air, target date: unknown: 92.5 mhz; 25 kw. Ant 328 ft TL: N29 58 34 W100 25 00. Hrs open: 11700 S.W. Tangerine Ct., Palm City, FL, 34990. Phone: (772) 215-1634. Licensee: Gary S. Hess. ♦Gary S. Hess, gen mgr.

Rollingwood

KJCE(AM)—Licensed to Rollingwood. See Austin

Roma

KBMI(FM)— Apr 30, 1983: 97.7 mhz; 3 kw. 298 ft TL: N26 24 22 W99 00 37. Hrs open: 18 1201 N. Jackson Rd., Suite 900, McAllen, 78501. Phone: (956) 992-8895. Fax: (956) 992-8897. Licensee: Horizon

Broadcasting Inc. (acq 10-26-98; $119,742). Population served: 50,000 Natl. Network: CNN Radio. Format: Country. Target aud: General; Sp speaking audience. ♦Arturo Gonzalez, gen mgr, progmg dir & chief of engrg.

Rosenberg-Richmond

KRTX(AM)—Licensed to Rosenberg-Richmond. See Houston

Round Rock

KFMK(FM)— October 1998: 105.9 mhz; 4.5 kw. 1,302 ft TL: N30 19 23 W97 47 58. Hrs open: 3601 South Congress, Bldg. F, Austin, 78704. Phone: (512) 684-7300. Fax: (512) 684-7441. Web Site: www.jammin1059.com. Licensee: Capstar TX L.P. Group owner: Clear Channel Communications Inc. (acq 8-30-00; grpsl). Format: Rhythmic adult contemp. ♦Mack Daniels, chief of opns; Debbie Harris, gen sls mgr; George Bradshaw, engr.

***KNLE-FM**— Aug 17, 1981: 88.1 mhz; 3 kw. 233 ft TL: N30 26 58 W98 48 48. Stereo. Hrs opn: 24 Box 907, 78680. Secondary address: 12703 Research Dr., Suite 222, Austin 78680. Phone: (512) 257-8881. Fax: (512) 257-8880. E-mail: webmaster@candle88.com Web Site: www.candle88.com. Licensee: Ixoye Productions Inc. (acq 6-23-03). Population served: 1,000,000 Format: Adult contemp, CHR. News: 6 hrs wkly. Target aud: 18-49; primarily female. Spec prog: Children 4 hrs wkly. ♦Sherland Priest, gen mgr, progmg dir, news dir & chief of engrg.

KZNX(AM)—See Creedmoor

Rudolph

***KTER(FM)**—Not on air, target date: unknown: 90.7 mhz; 2.4 kw. 282 ft TL: N26 41 13 W97 45 52. Hrs open: 24 Faith Pleases God Church Corp., 4501 West Expwy. 83, Harlingen, 78552. Phone: (956) 412-5600. Fax: (956) 428-7556. Licensee: Faith Pleases God Church Corp. Format: Educ, Christian, Sp. Target aud: General. Spec prog: Children 4 hrs wkly. ♦Aracelis Ortiz, CEO; Clark Ortiz, pres; Ricardo Mejia, gen mgr; Tonya Porter, opns VP.

Rusk

KTLU(AM)— 1955: 1580 khz; 840 w-D, 165 w-N. TL: N31 49 12 W95 10 19. Hrs open: 24 Box 475, 75785. Secondary address: 618 N. Main St. 75785. Phone: (903) 586-7771. Phone: (903) 683-2257. Fax: (903) 683-5104. E-mail: kwrw@mediactr.com Licensee: E.H. Whitehead. Population served: 20,000 Natl. Network: ABC. Rgnl. Network: Texas State Net. Format: Oldies. News staff: one; News: 3 hrs wkly. Target aud: 35-65. Spec prog: Sp 10 hrs wkly. ♦Marie Whitehead, pres; Robert Gonzalez, gen mgr.

KWRW(FM)—Co-owned with KTLU(AM). July 1, 1981: 97.7 mhz; 14.5 kw. 407 ft TL: N31 49 12 W95 10 19. Stereo. 24 400,000 Target aud: 25-54. Spec prog: Sp 10 hrs wkly.

San Angelo

KCLL(FM)— Aug 17, 1995: 100.1 mhz; 50 kw. Ant 385 ft TL: N31 31 49 W100 29 05. Hrs open: 24 2824 Sherwood Way, 76901. Phone: (325) 658-2995. Fax: (325) 659-2239. E-mail: kyzzfm@cs.com Licensee: Foster Communications Co. Inc. (group owner; (acq 5-19-2004; $450,000). Format: Tejano. Target aud: 18-49; Hispanics, demographics. ♦Fred M. Key, pres; Audrey Carver Luna, gen mgr, natl sls mgr, prom mgr & pub affrs dir; Wilburn Luna, CFO & opns mgr; Doug Smith, gen sls mgr; Freddy Maskill, prom mgr; Juan Vela, progmg dir & progmg mgr; Jeff Rottman, news dir; Richard Whitworth, chief of engrg; Freddy Maskill, traf mgr.

KCRN-FM— Feb 1, 1965: 93.9 mhz; 100 kw. Ant 649 ft TL: N31 42 11 W100 19 20. Stereo. Hrs opn: 24 Box 32, 76902-0032. Secondary address: 17 S. Chadbourne, Suite 500 76903. Phone: (325) 655-6917. Fax: (325) 655-7806. E-mail: mmohr@kcrn.org Web Site: www.kcrn.org. Licensee: Criswell College. (acq 6-18-91; $350,000 with co-located AM; FTR: 7-8-91). Population served: 200,000 Format: Inspirational Christian, relg. News: 2 hrs wkly. Target aud: 25 plus. ♦Mark Mohr, stn mgr; Keith Mayo, chief of engrg.

KCRN(AM)— 1947: 1340 khz; 1 kw-U. TL: N31 28 43 W100 27 50.24 Web Site: www.kcrn.org. (group owner; 100,000 News: 2 hrs wkly. Target aud: 35-54; adults with children in the home.

KDCD(FM)— June 1, 1980: 92.9 mhz; 100 kw. 729 ft TL: N31 26 08 W100 34 08. Stereo. Hrs opn: 3434 Sherwood Way, 76901. Phone: (915) 947-0899. Phone: (325) 947-0899. Fax: (915) 947-0996. E-mail: kdcd@wcc.net Web Site: www.cdcountry.fm. Licensee: Regency Broadcasting Inc. (acq 8-10-92; $186,000; FTR: 9-21-92). Population served: 180,000 Format: Young country. Target aud: 18-49. Spec prog: Relg 2 hrs wkly. ♦Beth Auldridge, CEO, sr VP & gen mgr; Jack Auldridge, chmn & pres; Jack Auldridge Jr., CFO & VP; J. Pat McKaye, opns mgr; JoAnna Alexander, sls VP, gen sls mgr & mktg dir; Lynn Ashley, prom dir, progmg dir & pub affrs dir; Len Martinez, chief of engrg; Jill Martinez, traf mgr.

KELI(FM)— November 1986: 98.7 mhz; 100 kw. 1,290 ft TL: N31 22 01 W100 02 48. Stereo. Hrs opn: 24 1301 S. Abe St., 76903. Phone: (915) 655-7161. Fax: (915) 658-7377. Web Site: www.k-lite987.com. Licensee: Double O Texas Corp. Group owner: Encore Broadcasting LLC (acq 3-15-2006; grpsl). Format: Hot adult contemp. News staff: one; News: 6 hrs wkly. Target aud: 25-54. Spec prog: Relg 6 hrs wkly. ♦John Kerr, exec VP, gen mgr & adv dir; Randy Phair, gen sls mgr; Garry Vaushr, prom mgr & engr; Boomer Kingston, progmg dir.

KGKL(AM)— Dec 4, 1928: 960 khz; 5 kw-D, 1 kw-N, DA-N. TL: N31 29 39 W100 24 55. Hrs open: Box 1878, 76902. Secondary address: 1301 S. Abe 76903. Phone: (325) 655-7161. Fax: (325) 658-7377. Web Site: kgkl960.com. Licensee: Double O Texas Corp Group owner: Encore Broadcasting LLC (acq 3-15-2006; grpsl). Population served: 100,000 Natl. Rep: Katz Radio. Format: News, talk, sports. News: 10 hrs wkly. Target aud: 35 plus. Spec prog: Farm 6 hrs wkly. ♦John Kerr, gen mgr; Boomer Kingsten, opns mgr.

KGKL-FM— Dec 24, 1965: 97.5 mhz; 100 kw. 500 ft TL: N31 29 46 W100 24 50. Stereo. 24 Web Site: www.kgkl975.com.100,000 Natl. Network: ABC. Law Firm: Kenkel & Associates. Format: Country. News: 3 hrs wkly. Target aud: 25-54.

KIXY-FM—Listing follows KKSA(AM).

KKSA(AM)— Nov 28, 1954: 1260 khz; 540 w-D, 71 w-N. TL: N31 29 14 W100 26 57. Stereo. Hrs opn: 24 Box 2191, 76902. Secondary address: KIXY Complex, 2824 Sherwood Way 76902. Phone: (325) 949-2112. Fax: (325) 944-0851. Web Site: www.kksa-am.com. Licensee: Foster Communications Company Inc. Group owner: Foster Communications Co. (acq 4-9-84). Population served: 125,000 Natl. Network: Westwood One, CBS. Rgnl. Network: Texas State Net. Natl. Rep: McGavren Guild. Law Firm: Leventhal, Senter & Lerman. Wire Svc: UPI Format: News/talk, sports. News staff: one; News: 20 hrs wkly. Target aud: 25-54. ♦Fred M. Key, CEO & pres; Jay Michaels, opns mgr; Doug Smith, gen sls mgr.

KIXY-FM—Co-owned with KKSA(AM). October 1966: 94.7 mhz; 100 kw. 446 ft TL: N31 29 14 W100 26 57. Stereo. 24 2824 Sherwood Way, 76901. Phone: (325) 949-3333. E-mail: kixy@kixyfm.com Web Site: www.kixyfm.com. Natl. Network: CNN Radio. Natl. Rep: McGavren Guild. Format: Top-40, adult contemp. News staff: one. Target aud: 18-49. ♦David Carr, progmg dir; Shannon J. Roach, CFO & traf mgr.

***KLTP(FM)**—Not on air, target date: unknown: 90.9 mhz; 15 kw vert. Ant 144 ft TL: N31 24 45 W100 25 54. Hrs open: 2351 Sunset Blvd., Suite 170-218, Rocklin, CA, 95765. Phone: (916) 251-1600. Fax: (916) 251-1650. Licensee: Educational Media Foundation. (acq 7-23-2007; grpsl). ♦Mike Novak, sr VP.

KMDX(FM)— Dec 5, 1998: 106.1 mhz; 50 kw. Ant 456 ft TL: N31 26 08 W100 34 08. Hrs open: 3434 Sherwood Way, 76901. Phone: (325) 947-0899. Fax: (325) 947-0996. E-mail: mixguy@wcc.net Web Site: www.mix106.fm. Licensee: Regency Broadcasting Inc. Format: CHR, rock/AOR. ♦Jack Auldridge, CEO & pres; Beth Auldridge, chmn, sr VP & gen mgr; Jack Auldridge Jr., CFO & VP; J. Pat McKaye, opns mgr, prom dir & mus dir; JoAnna Alexander, gen sls mgr & adv VP; J.Pat McKaye, progmg VP; Lynn Ashley, news dir; Len Martinez, engrg VP; Jill Marinez, traf mgr.

***KNAR(FM)**— 2006: 89.3 mhz; 1 kw. Ant 800 ft TL: N31 41 59 W100 26 30. Hrs open:
Rebroadcasts KLRD(FM) Yucaipa, CA 100%.
2351 Sunset Blvd., Suite 170-218, Rocklin, CA, 95765. Phone: (916) 251-1600. Fax: (916) 251-1650. Web Site: www.air1.com. Licensee: Educational Media Foundation. (acq 9-22-2005; $40,000 for CP). Natl. Network: Air 1. Format: Christian. ♦Richard Jenkins, pres; Mike Novak, VP; Keith Whipple, dev dir; David Pierce, progmg mgr; Ed Lenane, news dir; Sam Wallington, engrg dir; Karen Johnson, news rptr; Marya Morgan, news rptr; Richard Hunt, news rptr.

KSJT-FM— Oct 7, 1985: 107.5 mhz; 100 kw. 656 ft TL: N31 26 19 W100 34 18. Stereo. Hrs opn: 24 209 W. Beauregard Ave., 76903. Phone: (325) 655-1717. Fax: (325) 6557-0601. Licensee: La Unica Broadcasting Co. Format: Sp. Target aud: 18-55. ♦Louis Perez, pres; Armando Martinez, stn mgr; Cody Austin, gen sls mgr; Jesus Zapata, progmg dir; Arturo Madrid, news dir; Dania Salas, traf mgr.

***KUTX(FM)**— Apr 1, 1996: 90.1 mhz; 5 kw. 909 ft TL: N31 35 21 W100 31 00. Hrs open:
Rebroadcasts KUT(FM) Austin 100%.
1 University Station A 0704, Univ. of Texas, Austin, 78712-1090. Phone: (512) 471-1631. Licensee: University of Texas at Austin. Population served: 100,000 Natl. Network: NPR, PRI. Law Firm: Cohn & Marks. Format: Music & news. Target aud: 25-54; educated opinions, leaders and arts community. Spec prog: Folk 4 hrs, blues 6 hrs wkly. ♦Stewart Vanderwilt, gen mgr.

KWFR(FM)— November 1995: 101.9 mhz; 100 kw. 807 ft TL: N31 29 29 W100 26 03. Hrs open: 24 Box 2191, 76902. Secondary address: KIXY Complex, 2824 Sherwood Way 76901. Phone: (325) 949-2112. Fax: (325) 944-0851. Web Site: www.kwfrfm.com. Licensee: Foster Communications Co. Inc. (group owner; acq 12-1-94; $219,000 with KFXJ(FM) Abilene; FTR: 2-13-95). Natl. Rep: McGavren Guild. Law Firm: Leventhal, Senter & Lerman. Format: Classic rock. News staff: one. ♦Fred M. Key, pres; Jay Michaels, opns mgr; Doug Smith, gen sls mgr; Chase O'Reily, progmg dir; Jeff Rottman, news dir; Adolph Ganza, chief of engrg.

San Antonio

KAHL(AM)— 1948: 1310 khz; 5 kw-D, 280 w-N, DA-2. TL: N29 24 53 W98 20 36. Hrs open: 24 3740 Colony Dr., Suite 200, 78230. Phone: (210) 341-1310. Fax: (210) 694-5456. Licensee: Tichenor License Corp. Group owner: Univision Radio (acq 9-22-2003; grpsl). Population served: 1,367,500 Law Firm: Cohn & Marks. Format: Adult standards. ♦John Barger, gen mgr.

KXTN-FM—Co-owned with KAHL(AM). Dec 31, 1967: 107.5 mhz; 100 kw. Ant 1,514 ft TL: N29 16 29 W98 15 52. Stereo. 1777 N.E. Loop 410, Suite 400, 78217. Phone: (210) 829-1075. Fax: (210) 822-2372. Web Site: www.kxtn.com.1,367,500 Format: Tejano, Sp. Target aud: 25-49; contemp Sp, affluent, upscale. ♦Rosemary Scott, progmg dir & rsch dir; Norma Perez, traf mgr.

KAJA(FM)—Listing follows WOAI(AM).

***KBIB(AM)**—Marion, Sept 21, 1989: 1000 khz; 250 w-D, DA. TL: N29 34 09 W98 09 47. Hrs open: 290 N. Santa Clara Rd., Marion, 78124. Phone: (830) 914-2083. E-mail: kbibam@juno.com Web Site: www.kbib.org. Licensee: Hispanic Community College. Population served: 2,000,000 Law Firm: Wiley, Rein & Fielding. Format: Relg, Sp. Target aud: General. ♦Pastor Ken Hutchinson, gen mgr.

KCHL(AM)— June 1960: 1480 khz; 2.5 kw-D, 90 w-N, DA-2. TL: N29 24 45 W98 24 52. Hrs open: 15 1211 W. Hein Rd., 78220. Phone: (210) 337-1480. Fax: (210) 333-0081. E-mail: 1480@netscape.net Licensee: Martin Broadcasting Inc. (group owner; acq 6-4-92; FTR: 6-22-92). Population served: 1,000,000 Natl. Rep: McGavren Guild. Law Firm: Latham & Watkins. Format: Gospel. News staff: one. Target aud: 25-54. ♦Darrel Martin, gen mgr; "Skud R. Jones, gen sls mgr; "Skud R." Jones, progmg dir; Brett Huggins, chief of engrg.

KCOR(AM)— Feb 1, 1946: 1350 khz; 5 kw-U, DA-N. TL: N29 31 27 W98 37 05. Hrs open: 1777 N.E. Loop 410, Suite 400, 78217. Phone: (210) 821-6548. Fax: (210) 804-7820. Web Site: www.netmio.com. Licensee: Tichenor License Corp. Group owner: Univision Radio (acq 9-22-2003; grpsl). Population served: 1,367,500 Law Firm: Cohn & Marks. Format: Sp, news/talk. Target aud: 25-54; adults. ♦McHenry Tichenor, CEO & pres; Gary Stone, COO; Jeff Hinson, CFO; Mark Masepohz, sr VP; Dan Wilson, gen mgr; JD Gonzalez, opns mgr; Ernie Quinones, gen sls mgr; Robert De La Garza, prom dir; Rogelio Leal, progmg dir; Frank Cortez, pub affrs dir; Bret Huggins, chief of engrg; Rosemary Scott, rsch dir; Norma Perez, traf mgr.

KROM(FM)—Co-owned with KCOR(AM). June 1947: 92.9 mhz; 100 kw. 1,016 ft TL: N29 11 03 W98 30 49. Stereo. Fax: (210) 804-7825. Web Site: www.netmio.com.1,367,500 Law Firm: Cohn & Marks. Format: Regional Mexican, Sp. Target aud: 19-49; male. ♦Jd Gonzalez, opns dir; Rosemary Scott, rsch dir; Norma Perez, traf mgr.

KCYY(FM)—Listing follows KKYX(AM).

KDRY(AM)—Alamo Heights, Nov 8, 1963: 1100 khz; 11 kw-D, 1 kw-N, DA-N. TL: N29 33 26 W98 22 35. Hrs open: 24 16414 San Pedro Ave., Suite 575, 78232-2246. Phone: (210) 545-1100. Fax: (210) 545-1139. Web Site: www.am1100.com. Licensee: KDRY Radio Inc. Population served: 888,199 Format: Relg teaching. Target aud: General. Spec prog: Southern Gospel. ♦Diane Rainey, gen mgr.

KEDA(AM)— Mar 17, 1966: 1540 khz; 5 kw-D, 1 kw-N, DA-N. TL: N29 21 30 W98 21 05. Stereo. Hrs opn: 510 S. Flores St., 78204. Phone: (210) 226-5254. Phone: (210) 226-5810. Fax: (210) 227-7937. E-mail: kedakid@aol.com Licensee: D & E Broadcasting Co. (acq 3-7-66). Population served: 250,000 Rgnl. Network: Texas State Net. Format: Tex Mex, Cajun. Target aud: 25-54. Spec prog: Salsa 4 hrs wkly. ♦Madeline Davila, pres & disc jockey; Alberto P. Davila, VP, gen mgr, natl sls mgr & mktg VP; Ricardo P. Davila, progmg dir; Bret Huggins, chief of engrg; Danny Casanova, disc jockey; Eloy Espinoza, disc jockey; Nolda Saenz, disc jockey; Richard Davila, disc jockey; Susan Quijano, traf mgr & disc jockey.

KFIT EXP STN— 1989: 1060 khz; 1 kw-D, DA. TL: N29 17 32 W98 31 57. Hrs open: 6 AM-8 PM
Rebroadcasts KFIT(AM) Lockhart.
Box 160158, Austin, 78716. Secondary address: 110 Wild Basin Rd., Suite 375, Austin 78716. Phone: (512) 328-8400. Fax: (512) 328-8437. Licensee: KFIT Inc. Population served: 500,000 Format: Gospel. ♦Rev. Darrell Martin, gen mgr; Terri Lewis, progmg dir.

KISS-FM— December 1946: 99.5 mhz; 97.7 kw. Ant 1,486 ft TL: N29 16 29 W98 15 52. Stereo. Hrs opn: 8122 Datapoint Dr., Suite 600, 78229. Phone: (210) 646-0105. Fax: (210) 646-9711. E-mail: virgil.thompson@cox.com Web Site: www.kissrocks.com. Licensee: Cox Radio Inc. Group owner: Cox Broadcasting (acq 8-4-97; grpsl). Population served: 1,400,000 Natl. Rep: Christal. Format: AOR. Target aud: 18-44; men. ♦Virgil Thompson, gen mgr; Janis Maxymof, gen sls mgr; Jennifer Schultz, prom dir; Kevin Vargas, progmg dir; C.J. Cruz, mus dir & asst music dir; Steve Hahn, news dir & pub affrs dir; Richard Schuh, chief of engrg.

KJXK(FM)—Listing follows KTSA(AM).

KKYX(AM)— 1926: 680 khz; 50 kw-D, 10 kw-N, DA-N. TL: N29 30 03 W98 49 54. Stereo. Hrs opn: 24 8122 Datapoint, # 500, 78229. Phone: (210) 615-5400. Fax: (210) 615-5300. Web Site: www.kkyx.com. Licensee: Cox Radio Inc. Group owner: Cox Broadcasting (acq 3-28-97; grpsl). Population served: 1,300,000 Format: Country. News staff: one; News: 3 hrs wkly. Target aud: 35-64. Spec prog: Pub affrs 2 hrs wkly. ♦Bob Neil, CEO; Ben Reed, VP & gen mgr; Marty Choate, gen sls mgr; Jim Bratt, natl sls mgr; Julie Busse, mktg dir; Jim Kinney, prom dir; George King, progmg dir & progmg mgr; Chrissie Murnin, news dir & pub affrs dir; Paul Reynolds, chief of engrg; Connye Rodriguez, traf mgr.

KCYY(FM)—Co-owned with KKYX(AM). June 25, 1966: 100.3 mhz; 100 kw. 984 ft TL: N29 31 25 W98 43 25. Stereo. 24 Web Site: www.y100fm.com.1,300,000 News staff: one. Target aud: 25-54. ♦Alyce Ian, pub affrs dir; Connye Rodriguez, traf mgr.

KONO(AM)— January 1927: 860 khz; 5 kw-D, 1 kw-N, DA-N. TL: N29 26 14 W98 25 19. Stereo. Hrs opn: 24 8122 Datapoint Dr., Suite 500, 78229. Phone: (210) 615-5400. Fax: (210) 615-5339. Web Site: www.kono101.com. Licensee: Cox Radio Inc. Group owner: Cox Broadcasting (acq 2-12-98; $23 million with KONO-FM Helotes). Population served: 1,300,000 Format: Oldies. News staff: one; News: one hr wkly. Target aud: 25-64; total audience appeal. ♦Marty Choate, VP & gen mgr; Connie Tyra-Kremer, gen sls mgr; Roger Allen, progmg dir; Paul Reynolds, chief of engrg.

KONO-FM—See Helotes

***KPAC(FM)**— Nov 7, 1982: 88.3 mhz; 100 kw. 656 ft TL: N29 31 25 W98 43 25. Stereo. Hrs opn: 24 8401 Datapoint Dr., Suite 800, 78229. Phone: (210) 614-8977. Fax: (210) 614-8983. Web Site: www.tpr.org. Licensee: Texas Public Radio. Population served: 1,500,000 Natl. Network: PRI. Law Firm: Garvey, Schubert & Barer. Format: Class. News: 3 hrs wkly. Target aud: 25 plus; educated, upscale financially, mature, influential opinion leaders. ♦Dan Skinner, pres & gen mgr; Laverne Ditts, dev dir; Nathan Cone, progmg dir; Randy Anderson, mus dir; Wayne Coble, engrg dir; Janet Grojean, sls.

KPWT(FM)—See Terrell Hills

KQXT(FM)— Nov 19, 1967: 101.9 mhz; 100 kw. 700 ft TL: N29 25 08 W98 29 00. Stereo. Hrs opn: 24 6222 N.W. IH-10, 78201. Phone: (210) 736-9700. Fax: (210) 736-9776. Fax: (210) 735-8811. Licensee: CCB Texas Licenses L.P. Group owner: Clear Channel Communications Inc. (acq 1-27-93; $8 million; FTR: 3-8-93). Population served: 985,000 Rgnl rep: Clear Channel. Format: Soft adult contemp. News staff: one; News: 2 hrs wkly. Target aud: 25-54; core target is women 30-44. Spec prog: Contemp jazz 4 hrs, relg 2 hrs, pub affrs one hr wkly.L. Lowry Mays, CEO; Mark Mays, COO & pres; Randall Mays, CFO; Linda Hardy, gen mgr & gen sls mgr; Tom Glade, gen mgr; Mike McDonald, sls dir; Marian Holdsworth, natl sls mgr; Sue Nicholas, natl sls mgr; Tim Kiesling, mktg dir; Bill Rohde, prom dir; Ed Scarborough, progmg dir; Stan Kelly, news dir; Dan Walthers, chief of engrg; Diane Travis, disc jockey; Tom Graye, disc jockey

KRDY(AM)— Nov 13, 1961: 1160 khz; 10 kw-D, 1 kw-N, DA-2. TL: N29 32 11 W98 41 08. Stereo. Hrs opn: 24 84 N.E. Loop 410, Suite 143, 78216. Phone: (210) 530-5360. Fax: (210) 530-5304. Web Site: psc.disney.go.com/radiodisney/mystation/sanantonio. Licensee: Radio Disney Group LLC. Group owner: ABC Inc. (acq 5-30-2003; $3.2 million). Natl. Network: Radio Disney. Format: Children. ♦Fred Stockwell, gen mgr.

***KRTU(FM)**— Jan 22, 1976: 91.7 mhz; 8.9 kw. 120 ft TL: N29 27 51 W98 28 56. Stereo. Hrs opn: 24 Trinity University, One Trinity Place, 78212-7200. Phone: (210) 999-8917. Fax: (210) 999-8355. E-mail: krtu@trinity.edu Web Site: www.krtu.org. Licensee: Trinity University. Population served: 1,200,000 Law Firm: Cohn & Marks. Format: Jazz. News: 5 hrs wkly. Target aud: 35-64; people from all walks of life who love jazz music. Spec prog: Christian rock 2 hrs; Blues 2 hrs wkly. ♦Dr. William G. Christ, gen mgr; Dr. Rob Huesea, stn mgr; Ryan Weber, opns mgr; Chris Helfrich, dev dir; Aaron Prado, progmg dir; Brett Huggins, chief of engrg.

KSAH(AM)—See Universal City

KSLR(AM)— Dec 26, 1926: 630 khz; 5 kw-U, DA-2. TL: N29 23 24 W98 21 00. Hrs open: 24 9601 McAllister Fwy., Suite 1200, 78216-4686. Phone: (210) 344-8481. Fax: (210) 340-1213. E-mail: kslr@kslr.com Web Site: www.kslr.com. Licensee: Salem Media of Texas Inc. Group owner: Salem Communications Corp. (acq 8-6-94). Population served: 1,421,729 Natl. Network: Salem Radio Network. Format: Christian, educ teaching, talk. News staff: one. Target aud: 18-54; women & families. Spec prog: Sp 18 hrs wkly. ♦David Ziebell, gen mgr; Baron Wiley, opns mgr; James Herring, gen mgr & gen sls mgr.

***KSTX(FM)**— October 1988: 89.1 mhz; 100 kw. 656 ft TL: N29 31 33 W98 43 21. Stereo. Hrs opn: 24 8401 Datapoint Dr., Suite 800, 78229. Phone: (210) 614-8977. Fax: (210) 614-8983. Web Site: www.tpr.org. Licensee: Texas Public Radio. Population served: 1,500,000 Natl. Network: NPR, PRI. Law Firm: Garvey, Schubert & Barer. Wire Svc: AP Format: News & info. News staff: 5; News: 67 hrs wkly. Target aud: 25 plus; educated, upscale financially, influential opinion leaders. Spec prog: Jazz 6 hrs, var talk 6 hrs, folk 5 hrs, blues 6 hrs wkly. ♦Dan Skinner, pres & gen mgr; Nathan Cone, opns mgr & progmg dir; Laverne Pitts, dev dir & progmg dir; Dave Davies, news dir; Wayne Coble, news dir & engrg dir; Janet Grojean, sls.

***KSYM-FM**— Sept 15, 1966: 90.1 mhz; 5.7 kw. 128 ft TL: N29 26 50 W98 29 55. Stereo. Hrs opn: 24 1300 San Pedro Ave., 78212-4299. Phone: (210) 733-2787. Fax: (210) 733-2801. E-mail: ksym@accd.edu Web Site: www.ksym.org. Licensee: San Antonio College. Population served: 1,000,000 Format: AAA, Texas mus., new alternative. Target aud: 12-54; depending on block format. ♦John Onderdonk, gen mgr; Marlene Romo, sls dir; Michael Botsford, progmg dir; Shalom Topps, mus dir; Victor Pfau, chief of engrg.

KTKR(AM)— May 10, 1984: 760 khz; 50 kw-D, 1 kw-N, DA-2. TL: N29 26 58 W98 18 33. Stereo. Hrs opn: 24 6222 N.W. IH-10, 78201. Phone: (210) 736-9700. Fax: (210) 735-8811. Web Site: www.ticketssports.com. Licensee: CCB Texas Licenses L.P. Group owner: Clear Channel Communications Inc. (acq 6-16-93; $800,000; FTR: 7-5-93). Natl. Network: Westwood One. Natl. Rep: Clear Channel. Format: Sports. Target aud: 25-49; male. ♦L.L. Mays, CEO, chmn & opns dir; Randall Mays, CFO; Tom Glade, gen mgr; Tom Gebhart, gen sls mgr; Nate Lundy, progmg dir.

KTSA(AM)— May 9, 1922: 550 khz; 5 kw-U, DA-N. TL: N29 29 41 W98 24 52 (D), N29 29 46 W98 24 54 (N). Hrs open: 24 4050 Eisenhauer Rd., 78218. Phone: (210) 528-5500. Fax: (210) 599-5588. Web Site: www.ktsa.com. Licensee: BMP San Antonio License Co. L.P. Group owner: Infinity Broadcasting Corp. (acq 10-19-2006; $45 million with co-located FM). Population served: 654,153 Law Firm: Cohn & Marks. Format: News/talk. News staff: 11; News: 35 hrs wkly. Target aud: 25-54. ♦Reid Reker, gen mgr; Ann Edwards, opns mgr; Tim Germadnik, pres & progmg dir.

KJXK(FM)—Co-owned with KTSA(AM). 1969: 102.7 mhz; 100 kw horiz, 70 kw vert. 670 ft TL: N29 25 09 W98 29 06. Stereo. Web Site: www.1027krock.com. Format: CHR, top-40. Target aud: 12 plus. ♦John Cook, progmg dir.

KXXM(FM)— May 5, 1964: 96.1 mhz; 100 kw. 479 ft TL: N29 38 00 W98 37 50. (CP: 99 kw, ant 328 ft.). Stereo. Hrs opn: 24 6222 N.W. I-10, 78201. Phone: (210) 736-9700. Fax: (210) 736-8811. E-mail: mix961@mic961.com Web Site: www.mix961.com. Licensee: CCB Texas Licenses L.P. Group owner: Clear Channel Communications Inc. (acq 6-19-98; $15 million). Population served: 654,153 Format: CHR. News staff: one; News: 3 hrs wkly. Target aud: 18-34; females. ♦Tom Glade, VP & gen mgr; Mike Hall, gen sls mgr; Tim Kiesling, mktg dir; Cesar Campa, prom dir.

***KYFS(FM)**— Nov 7, 1982: 90.9 mhz; 100 kw. 476 ft TL: N29 40 20 W98 14 43. Stereo. Hrs opn: 24 9330 Corporate Dr., Suite 808, Selma, 78154. Phone: (210) 651-9093. Fax: (210) 651-9093. E-mail: kyfs@bbnradio.org Web Site: www.bbnradio.org. Licensee: Bible Broadcasting Network Inc. (group owner; acq 11-20-91; $75,000; FTR: 12-9-91). Population served: 1,200,000 Law Firm: Smithwick & Belendiuk. Format: Christian. Target aud: 2 plus. ♦John D. Woolery, gen mgr & opns mgr.

KZDC(AM)— Jan 1, 1953: 1250 khz; 1 kw-U, DA-N. TL: N29 24 29 W98 26 39. Stereo. Hrs opn: 24 7800 W. IH 10, Suite 330, 10017. Phone: (210) 280-4000. Fax: (210) 822-9668. Web Site: www.radiounica.com. Licensee: BMP San Antonio License Co. L.P. Group owner: Multicultural Radio Broadcasting Inc. (acq 2005; grpsl). Format: Mexican rgnl. ♦Scott Keebler, gen mgr; Theo Alvarado, progmg dir; Roy Pressman, chief of engrg.

KZEP-FM— Oct 1, 1966: 104.5 mhz; 100 kw. 735 ft TL: N29 25 09 W98 29 06. Stereo. Hrs opn: 24 427 E. 9th St., 78215. Phone: (210) 226-6444. Fax: (210) 225-5736. E-mail: kzepp@kzep.com Web Site: www.kzep.com. Licensee: Texas Lotus Corp. Group owner: Lotus Communications Corp. Population served: 140,000 Natl. Rep: D & R Radio. Format: Classic rock. News staff: one; News: 3 hrs wkly. Target aud: 25-54; males. ♦Jay A. Levine, pres, VP & gen mgr; Trish Levine, prom dir; Craig Chambers, progmg dir; Tom Scheppke, mus dir; Dave Delgado, news dir; Eddie Miles, chief of engrg; Becky Talamandes, traf mgr.

WOAI(AM)— Sept 29, 1922: 1200 khz; 50 kw-U. TL: N29 30 05 W98 07 09. Hrs open: 24 6222 N.W. IH-10, 78201. Phone: (210) 736-9700. Fax: (210) 735-8811. Web Site: www.woai.com. Licensee: CCB Texas Licenses L.P. Group owner: Clear Channel Communications Inc. (acq 1975). Population served: 1,203,100 Natl. Network: Fox News Radio. Natl. Rep: Clear Channel. Wire Svc: AP Format: News/talk. News staff: 13; News: 20 hrs wkly. Target aud: 35-64; general. ♦L. Lowry Mays, pres & CEO; Matt Martin, gen mgr; George King, opns dir; Mike McDonald, sls dir; Tom Gebhart, gen sls mgr; Marian Holdsworth, natl sls mgr; Sue Nicholas, natl sls mgr; Callie Hoch, mktg dir; Tomm Rivers, prom mgr; Nate Lundy, progmg dir; Jim Forsyth, news dir; Dan Walthers, chief of engrg; Jessica Broadbent, traf mgr.

KAJA(FM)—Co-owned with WOAI(AM). 1951: 97.3 mhz; 100 kw. 984 ft TL: N29 25 20 W98 29 22. Stereo. 24 6222 N.W. IH-10, 78201. Web Site: www.kj97.com.175,000 Natl. Rep: Clear Channel. Format: Country. News: 2 hrs wkly. Target aud: 18-54. ♦Dean Phillips, gen sls mgr; Clayton Allen, prom mgr, progmg dir & disc jockey; Irene Gaitan, traf mgr; Jamie Martin, disc jockey; Lou Ramirez, disc jockey; Randy Carroll, disc jockey.

San Augustine

KQSI(FM)— Dec 29, 1993: 92.5 mhz; 1.4 kw. Ant 220 ft TL: N31 31 44 W94 05 59. Hrs open: 24
Rebroadcasts KDET(AM) Center 100%.
Box 930, Center, 75935. Phone: (936) 275-3242. Fax: (936) 598-9537. Licensee: Center Broadcasting Co. Inc. (group owner; acq 3-26-98; grpsl). Population served: 25,000 Natl. Network: ABC. Format: C & W, Sp. News staff: one. Target aud: 35-65. ♦Tracy Broadway, gen mgr & gen sls mgr; Rob Rockett, progmg dir; Daniel Christie, news dir; Harlan Riley, chief of engrg; Kim Parker, traf mgr.

San Diego

KUKA(FM)— July 14, 1993: 105.9 mhz; 25 kw. 450 ft TL: N27 45 04 W98 07 28. Stereo. Hrs opn: 24 Box 589, Alice, 78333. Phone: (361) 668-6666. Fax: (361) 668-6661. Licensee: Ideal Media Inc. dba KUKA Tejano FM 106. (acq 12-5-96). Population served: 447,510 Law Firm: Akin, Gump, Strauss, Hauer & Feld. Format: Sp. News staff: one; News: one hr wkly. Target aud: 18-34; middle to upper class.Armando Marroquin Jr., pres, mktg mgr, prom mgr & adv mgr; Teo Pena, gen mgr & disc jockey; Estela Nava, opns mgr, chief of engrg & traf mgr; Zulema Z. Marroquin, dev VP; Javier Villanueva, rgnl sls mgr; Tio Pena, mktg dir; Pedro Vasquez, progmg dir & disc jockey; Peter Vasquez, news dir; Henry Turner, chief of engrg & traf mgr; Ted Pena, reporter & mus critic

San Juan

KUBR(AM)— 1991: 1210 khz; 10 kw-D, 1 kw-N, DA-2. TL: N26 14 41 W98 05 25. Hrs open: 24 Box 252, McAllen, 78505. Phone: (956) 686-6382. Fax: (956) 686-2999. E-mail: paylinobernal@hotmail.com Web Site: www.laradiocristiana.com. Licensee: Radio Christiana Network. Format: Sp, Christian. ♦Paulino Bernal, gen mgr; Eloy Bernal, stn mgr; Pete Guzman, opns mgr.

San Marcos

KBPA(FM)—Licensed to San Marcos. See Austin

***KTSW(FM)**— Apr 15, 1992: 89.9 mhz; 10.5 kw. Ant 213 ft TL: N29 39 20 W98 07 59. Stereo. Hrs opn: 24 Old Main 106, 601 University Dr., 78666-4616. Phone: (512) 245-3485. Fax: (512) 245-3732. E-mail: ktsw@txstate.edu Web Site: www.ktsw.net. Licensee: Texas State University-San Marcos. Law Firm: Dow, Lohnes & Albertson, PLLC. Format: College alternative, news/talk, sports. News: 9 hrs wkly. Target aud: 18-24; college students & young adults. ♦Dan Schumacher, gen mgr & sls; Jayce Beasley, stn mgr; Brian Shelton, prom dir; Evan Hilliard, progmg dir; Kristen Hennessey, mus dir; Tom Bruce, engrg dir; Lotta Bucks, traf mgr.

KUOL(AM)— 1948: 1470 khz; 250 w-U, DA-N. TL: N29 53 53 W97 54 44. Hrs open: 5:30 AM-midnight Box 252, McAllen, 78505. Phone: (956) 686-6382. Fax: (956) 686-2999. E-mail: paulinobernal@hotmail.com Web Site: www.laradiocristiana.com. Licensee: Radio Christiana Network. (acq 3-26-97). Population served: 56,000 Format: Sp, Christian. ♦Paulino Bernal, gen mgr; Eloy Bernal, stn mgr; Pete Guzman, opns mgr.

San Saba

KBAL(AM)— 1954: 1410 khz; 800 w-D, 203 w-N. TL: N31 11 26 W98 42 55. Hrs open: 24 Box 126, 76877. Secondary address: Building D-2, 2402 Broadmoor, Bryon 77802. Phone: (915) 372-5225. Fax: (915) 372-3817. E-mail: kabal@centex.net Web Site: www.kbalradio.com. Licensee: Roy E. Henderson. (acq 2-15-00; grpsl). Population served: 5000 Law Firm: Fletcher, Heald & Hildreth. Format: Adult contemp. News staff: 2; News: 6 hrs wkly. Rural all ages. Spec prog: Farm 2 hrs,gospel 7 hrs wkly. ♦Steve Smith, CFO; Shay Hardy, gen mgr, opns dir, progmg dir & news dir; Sherry Spinks, gen mgr & sls dir.

KBAL-FM— 1996: 106.1 mhz; 3 kw. Ant 20 ft TL: N31 11 26 W98 42 55.24 705 Live Oak, 76877. Web Site: www.kbalradio.com. Format: Country. News staff: one; News: 5 hrs wkly. Target aud: General.

Sanger

KTDK(FM)— December 1989: 104.1 mhz; 6.2 kw. Ant 630 ft TL: N33 28 47 W97 03 22. Stereo. Hrs opn: 24 3500 Maple Ave., Suite 1310, Dallas, 75219. Phone: (214) 526-7400. Fax: (214) 525-2525. Web Site: www.theticket.com. Licensee: KRBE Lico Inc. Group owner: Susquehanna Radio Corp. (acq 4-30-98; $3.683 million). Format: Sports. Target aud: 24-55; men & sports enthusiasts. ♦Dan Bennett, VP & gen mgr; Jim Quirk, sls dir; Kim Roberts, gen sls mgr; Jeff Catlin, progmg dir; Rob Chickering, chief of engrg; Kimberly Jolly, traf mgr.

***KVRK(FM)**— July 8, 1999: 89.7 mhz; 14 kw. 1,699 ft TL: N33 33 36 W96 57 35. Hrs open: Research Educational Foundation Inc., 11061 Shady Tr., Dallas, 75229. Phone: (214) 353-8970. Fax: (214) 351-6809. E-mail: chris@897powerfm.com Web Site: www.897powerfm.com. Licensee: Research Educational Foundation Inc. Format: Christian, rock. ♦Stanley Thomas, gen mgr; Ron Evans, stn mgr; Devin Wickham, opns dir; Krystal Coleman, prom dir; Chris Goodwin, progmg dir; Kent Loney, chief of engrg.

Santa Anna

KBWT(FM)—Not on air, target date: unknown: 105.5 mhz; 19 kw. Ant 266 ft TL: N31 40 20.09 W99 10 57.07. Hrs open: 2801 Via Fortuna Dr., Suite 675, Austin, 78746. Phone: (713) 528-2517. Licensee: Ace Radio Corp. ♦Stephen Hackerman, pres.

Savoy

KQDR(FM)—Not on air, target date: unknown: 107.3 mhz; 2.3 kw. Ant 534 ft TL: N33 37 30 W96 20 03. Hrs open: 1126 West Ave., Richmond, VA, 23220. Phone: (804) 422-3452. Licensee: William W. McCutchen III. ♦William W. McCutchen III, gen mgr.

Schertz

KBBT(FM)— Feb 1, 1976: 98.5 mhz; 97 kw. Ant 991 ft TL: N29 31 25 W98 43 25. Stereo. Hrs opn: 24 1777 N.E. Loop 410, Suite 400, San Antonio, 78217. Phone: (210) 829-1075. Fax: (210) 804-7825. Web Site: www.netmio.com. Licensee: Univision Radio License Corp. Group owner: Univision Radio (acq 9-22-2003; grpsl). Population served: 1,367,500 Format: Hip Hop. ♦Mac Tichenor, CEO & pres; Jeff Hinson, CFO; Dan Wilson, gen mgr; J. D. Gonzalez, opns mgr.

Seabrook

KROI(FM)— Apr 23, 1984: 92.1 mhz; 50 kw. Ant 981 ft TL: N29 16 33 W95 22 45. Stereo. Hrs opn: 24 24 Greenway Plaza, Suite 900, Houston, 77046. Phone: (713) 623-2108. Fax: (713) 623-8166. Licensee: Radio One Licenses LLC. Group owner: Radio One Inc. (acq 7-20-2004; $72.5 million). Format: Gospel. Target aud: General. ♦Alfred C. Liggins III, pres; Scott R. Royster, exec VP; Doug Abernethy, gen mgr.

Seadrift

KMAT(FM)— May 1999: 105.1 mhz; 38.5 kw. Ant 456 ft TL: N28 26 17.1 W96 26 54.6. Stereo. Hrs opn: 24 866 N. Wilcrest, Houston, 77079. Phone: (713) 722-0169. Fax: (713) 468-5773. E-mail: bcordell@houston.rr.com Web Site: www.kmat.cc. Licensee: Cordell Communications Inc. Population served: 89,470 Law Firm: Leventhal, Senter & Lerman. Format: Sp Christian. News: 2 hrs wkly. Target aud: 30-50. ♦Bill Cordell, pres & gen mgr.

Sealy

***KCPC(FM)**—Not on air, target date: unknown: 90.7 mhz; 1.7 kw. Ant 446 ft TL: N29 50 05 W96 16 10. Hrs open: 2424 South Blvd., Houston, 77098-5110. Phone: (713) 520-5200. Web Site: www.khcb.org. Licensee: Houston Christian Broadcasters Inc. ♦Bruce Munsterman, pres.

Seguin

KSMG(FM)— Sept 9, 1970: 105.3 mhz; 100 kw. 1,240 ft TL: N29 16 29 W98 15 52. Stereo. Hrs opn: 24 8122 Datapoint Dr., Suite 600, San Antonio, 78229. Phone: (210) 646-0105. Fax: (210) 646-9711. E-mail: virgil.thompson@cox.com Web Site: www.magic1053.com. Licensee: Cox Radio Inc. Group owner: Cox Broadcasting (acq 8-4-97; grpsl). Population served: 1,400,000 Natl. Rep: Christal. Law Firm: Leventhal, Senter & Lerman. Format: Hot adult contemp. News staff: one; News: 5 hrs wkly. Target aud: 25-49; females. ♦Virgil Thompson, gen mgr; Rory Charitan, gen sls mgr; Robert John, progmg dir; Katrina Curtiss, mus dir; Karen Clauss, news dir & pub affrs dir; Richard Schuh, chief of engrg; Cathy Sheehan, traf mgr.

KWED(AM)— Sept 9, 1948: 1580 khz; 1 kw-D, 253 w-N. TL: N29 34 48 W97 59 05. Hrs open: 24 609 E. Court St., 78155. Phone: (830) 379-2234. Fax: (830) 379-2238. E-mail: contact@kwed1580.com Web Site: www.kwed1580.com. Licensee: Seguin Media Group Ltd. (acq 6-10-02; $940,000). Population served: 175,000 Natl. Network: Westwood One, CNN Radio. Natl. Rep: Rgnl Reps. Law Firm: Southmayd & Miller. Wire Svc: AP Format: Country, news/talk. News staff: 5; News: 30 hrs wkly. Target aud: 35-64. Spec prog: Farm 6 hrs wkly. ♦Hal Widsten, pres, gen mgr & gen sls mgr; Darren Dunn, opns mgr & news dir; Sarah Masterson, pub affrs dir; Richard Schuh, chief of engrg; Rosanna Noriega, traf mgr & sports cmtr; Dave Koehn, sports cmtr; Sennett Rockers, sports cmtr.

Seminole

KIKZ(AM)— Apr 15, 1954: 1250 khz; 1 kw-D, 250 w-N. TL: N32 41 58 W102 38 12. Hrs open: 24 105 N.W. 11th St., 79360. Phone: (915) 758-5878. Fax: (915) 758-5474. E-mail: kikz-kssem@midtech.net Licensee: Gaines County Broadcasting Ltd. (acq 6-9-93; $193,276 with co-located FM; FTR: 7-5-93) Population served: 25,000 Natl. Network: ABC. Rgnl. Network: Texas State Net. Format: Country, Sp. News staff: one; News: 7 hrs wkly. Target aud: General. Spec prog: Farm 5 hrs, Ger one hr, relg 6 hrs wkly. ♦Danny Curtis, gen mgr, prom mgr, progmg dir, engrg mgr, sports cmtr & disc jockey; Mike Elder, gen sls mgr; Audie Cox, news dir, sports cmtr & disc jockey; Theresa Hemphill, disc jockey.

KSEM-FM—Co-owned with KIKZ(AM). Mar 15, 1985: 106.3 mhz; 3 kw. 174 ft TL: N32 41 58 W102 38 12. Stereo. 24 Format: Country. ♦Audie Cox, sports cmtr & disc jockey; Danny Curtis, sports cmtr & disc jockey; Theresa Hemphill, disc jockey.

Seymour

KSEY(AM)— Oct 26, 1950: 1230 khz; 1 kw-U. TL: N33 35 49 W99 16 42. Hrs open: 700 eighth St., Suite 210, Wichita Falls, 76301. Phone: (940) 767-0011. Fax: (940) 767-0164. E-mail: kseytejano1230am@aol.com Licensee: Mark Aulabaugh. (acq 11-95). Population served: 100,000 Wire Svc: NWS (National Weather Service) Format: Sp. ♦Tommy Cobos, gen mgr; Orlando Jariez, progmg dir.

KSEY-FM— June 26, 1981: 94.3 mhz; 3 kw. 112 ft TL: N33 35 49 W99 16 42. (CP: 93.9 mhz, 50 kw, ant 492 ft. TL: N33 42 00 W99 08 12). Stereo. Box 471, 76380. Phone: (940) 889-2637. Fax: (940) 889-2637. E-mail: fmksey@aol.com Web Site: www.radioksey.com. Format: Full service. ♦Mark Aulabaugh, gen mgr; Joe Gaither, progmg dir.

KZNO(FM)—Not on air, target date: unknown: 92.3 mhz; 50 kw. Ant 492 ft TL: N33 34 49 W99 18 00. Hrs open: 2768 Pharmacy Rd., Rio Grande City, 78582. Phone: (956) 487-5621. Licensee: James Falcon. ♦James Falcon, gen mgr.

Shamrock

KBKH(FM)— Sept 15, 1997: 92.9 mhz; 50 kw. Ant 253 ft TL: N35 20 29 W100 14 33. Stereo. Hrs opn: 24 Box 688, 79079-0688. Phone: (806) 256-1221. Fax: (806) 256-1223. E-mail: kbkh@kbkh.com Web Site: www.kbkh.com. Licensee: Terry Keith Hammond (acq 8-26-2002). Population served: 198,000 Rgnl. Network: Texas State Net., ABC. Format: Adult contemp, oldies, news/talk. News staff: one; News: 32 hrs wkly. Target aud: 25-75; adults. Spec prog: Farm 25 hrs, gospel 6 wkly. ♦Bessie Hammond, mus critic; Keith Hammond, pres, gen mgr, mktg dir, engrg dir & disc jockey.

Sherman

KJIM(AM)— Dec 19, 1947: 1500 khz; 1 kw-D, DA. TL: N33 41 30 W96 33 29. Hrs open: 4367 Woodlawn Rd., Denison, 75021-8037. Phone: (903) 893-1197. Fax: (908) 893-1197. Licensee: Bob Mark Allen Productions Inc. Natl. Network: Westwood One, CBS. Format: Original hits of the 50s, 60s, 70s & 80s, news, sports. Target aud: 40-65. ♦Bob Mark Allen, pres.

Silsbee

KAYD-FM— June 21, 1980: 101.7 mhz; 10.5 kw. Ant 502 ft TL: N30 06 54 W93 59 56. Stereo. Hrs open: 755 S. 11th St., Suite 102, Beaumont, 77704. Phone: (409) 833-9421. Fax: (409) 833-9296. Web Site: www.kayd.com. Licensee: Cumulus Licensing LLC. (acq 11-8-2004; $2.1 million). Format: Country. News: 20 hrs wkly. Target aud: 25-54; persons. ♦Zanetta Kelley, gen mgr; Jim West, opns mgr & progmg dir; Wes Matejka, sls dir; Mark Guzman, prom mgr; Liz Ferguson, traf mgr; J.P. White, sports cmtr.

KSET(AM)— Oct 13, 1959: 1300 khz; 500 w-D. TL: N30 21 02 W94 13 39. Hrs open: 24 Box 455, 77656. Phone: (409) 385-2883. Fax: (409) 386-1001. E-mail: kset@kset1300.com Web Site: www.kset1300.com. Licensee: Proctor-Williams Inc. (acq 2-14-01; with co-located FM). Population served: 7271 Rgnl. Network: Texas State Net. Format: All sports. Target aud: General. ♦Dave Collier Sr., CEO & gen mgr.

Sinton

KDAE(AM)—Licensed to Sinton. See Corpus Christi

KNCN(FM)—Licensed to Sinton. See Corpus Christi

KOUL(FM)—Licensed to Sinton. See Corpus Christi

Slaton

KJAK(FM)—Licensed to Slaton. See Lubbock

Snyder

KLYD(FM)— 2003: 98.9 mhz; 5.6 kw. Ant 341 ft TL: N32 45 23 W100 54 09. Hrs open: Box 1008, 79550. Phone: (325) 573-9322. Fax: (325) 573-7445. Licensee: Delbert Foree. Format: Modern rock. ♦Dink Foree, gen mgr.

KSNY(AM)— Dec 22, 1949: 1450 khz; 1 kw-U. TL: N32 43 33 W100 56 30. Hrs open: Box 1008, 79550. Secondary address: 2301 Ave. R 79549. Phone: (325) 573-9322. Fax: (325) 573-7445. E-mail: ksnyfm@snydertex.com Licensee: Snyder Broadcasting Co. (acq 1952). Population served: 11,171 Natl. Network: ABC. Format: Christian. Target aud: General. ♦Bill Jamar, pres; Lydia Foree, VP; Dink Foree, gen mgr, prom mgr, progmg dir & progmg dir.

KSNY-FM— Sept 2, 1980: 101.5 mhz; 35 kw. 500 ft TL: N32 53 29 W101 06 29. Stereo. 24 Format: Country. News staff: one; News: 5 hrs wkly. Target aud: 25-54.

Somerset

KYTY(AM)— Mar 1, 1988: 810 khz; 250 w-U, DA-2. TL: N29 18 48 W98 30 29. Hrs open: 24 Box 701582, San Antonio, 78270. Phone: (210) 545-0810. Fax: (210) 545-6713. Licensee: Maranatha Broadcasting Inc. (Group owner: Clear Channel (acq 3-11-98; $750,000). Population served: 1,200,000 Format: Contemp Christian music. ♦Myron Wade, gen mgr.

Sonora

KHOS-FM— May 1979: 92.1 mhz; 3 kw. Ant 298 ft TL: N30 33 33 W100 37 54. Stereo. Hrs opn: 24 HC65, Box 50, Hwy. 277 S., 76950. Phone: (325) 387-3553. Fax: (325) 387-3554. E-mail: khoskyxx@sonoratx.net Licensee: Sonora Broadcasting Co. Group owner: Hill Country Broadcasting Corp. (acq 8-16-00; grpsl). Population served: 10,000 Natl. Network: ABC. Wire Svc: NOAA Weather Wire Svc: NWS (National Weather Service) Format: Real country. News: 3 hrs wkly. Target aud: 12-50 plus. ♦Kent Foster, pres; Monte Spearman, gen mgr; Donna Keese, stn mgr; Eddy Smith, engr.

South Padre Island

KESO(FM)— Aug 27, 1996: 92.7 mhz; 3 kw. 298 ft TL: N26 04 04 W97 13 16. Stereo. Hrs opn: 24 1004 Padre Blvd., 4th Fl., 78597. Phone: (956) 761-2270. Fax: (956) 761-1656. Web Site: www.alternative927.com. Licensee: BMP RGV License Company L.P. (acq 12-6-2004; $6.6 million with KZSP(FM) South Padre Island). Population served: 500,000 Format: Alternative rock. Target aud: 25-44. ♦Terry Kimball, gen mgr & stn mgr; Jim Wilson, progmg dir.

KZSP(FM)— July 27, 1990: 95.3 mhz; 2.5 kw. 353 ft TL: N26 04 04 W97 13 16. Stereo. Hrs opn: 24 1004 Padre Blvd., 4th Fl., 78597. Phone: (956) 761-2270. Fax: (956) 761-1656. Web Site: www.love953.com. Licensee: BMP RGV License Company L.P. (acq 12-6-2004; $6.6 million with KESO(FM) South Padre Island). Population served: 500,000 Format: Light jazz. News: 2 hrs wkly. Target aud: 25-54; success-oriented adults. ♦Terry Kimball, gen mgr & stn mgr; Jim Wilson, progmg dir & chief of engrg.

Spearman

***KTOT(FM)**— 2003: 89.5 mhz; 100 kw. Ant 1,066 ft TL: N36 03 44 W101 01 56. Hrs open: 24 High Plains Public Radio, 210 N. 7th St., Garden City, KS, 67846. Phone: (620) 275-7444. Fax: (620) 275-7496. E-mail: hppr@hppr.org. Web Site: www.hppr.org. Licensee: Kanza Society Inc. Natl. Network: AP Radio, NPR, PRI. Format: Class music, news. Target aud: 25-80; educated. ♦Richard Hicks, gen mgr; Diana Gonzales, dev dir; Debra Stout, prom dir; Bob Kirby, progmg mgr; Mary Palmer, mus dir; Chuck Springer, chief of engrg.

KXDJ(FM)— Dec 16, 1963: 98.3 mhz; 17.5 kw. Ant 836 ft TL: N36 03 44 W101 01 56. Stereo. Hrs opn: 24 Box 830, Perryton, 79081. Phone: (806) 659-2529. Fax: (806) 648-2652. Licensee: George Chambers (acq 12-15-2003; $280,000). Population served: 77,200 Natl. Network: CBS Radio, AP Radio. Rgnl. Network: Texas State Net. Format: Country. News: 25 hrs wkly. ♦George Chambers, pres & VP; Chris Samples, gen mgr.

Springtown

***KSQX(FM)**— August 1985: 89.1 mhz; 3 kw. Ant 184 ft TL: N32 58 53 W97 42 18. Hrs open: 24 905 Palo Pinto St., Weatherford, 76086. Phone: (817) 341-8950. Fax: (817) 596-9842. E-mail: chb890@swbell.net Web Site: www.kyqx.com. Licensee: CSSI Non-Profit Educational Broadcasting Corp. (acq 5-21-2002). Population served: 2,355.316 Law Firm: Hill & Welch. Format: Oldies, big band, talk shows. News: 5 hrs wkly. ♦Jean Hudgens, gen mgr; John Peterson, progmg dir; Ace Little, news dir.

Stamford

KLGD(FM)— Feb 22, 1999: 106.9 mhz; 40 kw. 548 ft TL: N32 56 16 W99 57 20. Hrs open: 209 S. Danville, Suite B-105, Abilene, 79605. Phone: (915) 691-5400. Fax: (915) 691-5653. E-mail: bruce@texas96.com Web Site: countrylegends.com. Licensee: Texas Gulfwest Communications Corp. (acq 6-26-01). Population served: 150,000 Format: Country. Target aud: 35 plus; adults. ♦Bill Hooten, CEO & gen mgr; Pete Garcia, progmg mgr & mus dir.

KVRP(AM)— July 1947: 1400 khz; 1 kw-U. TL: N32 55 52 W99 47 00. Hrs open: 24 Box 1118, 1406 N. First St., Haskell, 79521. Phone: (940) 864-8505. Fax: (940) 864-8001. E-mail: gary@kvrp.com Web Site: www.kvrp.com. Licensee: 1 Chronicles 14 L.P. (acq 8-4-2004; $700,000 with KVRP-FM Haskell). Population served: 50,000 Rgnl. Network: Texas State Net. Natl. Rep: Katz Radio. Format: Praise & Worship/ Christian. Target aud: 35+. ♦Gregg Weston, pres & gen mgr; Gary Barrett, gen mgr, stn mgr & gen sls mgr; Dave Harrison, progmg dir; Megan Cox, chief of engrg & traf mgr.

Stanton

***KFRI(FM)**— 2005: 88.1 mhz; 100 kw. Ant 457 ft TL: N32 05 44 W101 48 47. Stereo. Hrs opn: 24
Rebroadcasts KLRD(FM) Yucaipa, CA 100%.
2351 Sunset Blvd., Suite 170-218, Rocklin, CA, 95765. Phone: (916) 251-1600. Fax: (916) 251-1650. E-mail: info@air1.com Web Site: www.air1.com. Licensee: Educational Media Foundation. Group owner: EMF Broadcasting. Natl. Network: Air 1. Law Firm: Shaw Pittman. Format: Contemp Christian. News staff: 3. Target aud: 18-35. ♦Richard Jenkins, pres; Mike Novak, VP; Ed Lenane, opns dir & news dir; Keith Whipple, dev dir; David Pierce, progmg mgr; Sam Wallington, engrg dir; Karen Johnson, news rptr; Marya Morgan, news rptr; Richard Hunt, news rptr.

KKJW(FM)— 1998: 105.9 mhz; 37 kw. 400 ft TL: N31 51 19 W101 47 32. Hrs open: 4411 Brookdale Dr., Midland, 79703. Phone: (432) 620-8282. Fax: (432) 620-0498. Licensee: Unique Broadcasting L.L.C. (acq 7-23-97). Format: Classic country. ♦Dick Baze, gen mgr.

Stephenville

***KEQX(FM)**—Not on air, target date: unknown: 89.7 mhz; 6 kw vert. Ant 492 ft TL: N32 07 24 W97 58 48. Stereo. Hrs open: 24 905 Palo Pinto St., Weatherford, 76086. Phone: (817) 341-8950. Fax: (817) 596-9842. Web Site: kyqx.com. Licensee: CSSI Non-Profit Educational Broadcasting Corp. Population served: 203,881 Law Firm: Hill & Welch. Format: Hard country. ♦Jean Hudgens, gen mgr.

***KQXS(FM)**— 2004: 89.1 mhz; 1.2 kw vert. Ant 424 ft TL: N32 16 09 W98 18 51. Hrs open: 905 Palo Pinto St., Weatherford, 76086. Phone: (817) 341-8950. Fax: (817) 596-9842. E-mail: ch6890@swbell.net Web Site: kyqx.com. Licensee: CSSI Non-Profit Educational Broadcasting Corp. (acq 11-25-2002). Population served: 205,570 Law Firm: Hill & Welch. Format: Oldies, big band, news. News staff: 3. ♦Jean Hudgens, gen mgr.

KSTV(AM)— 1947: 1510 khz; 500 w-D. TL: N32 12 08 W98 14 54. Hrs open: Box 289, 3209 W. Washington, 76401. Phone: (254) 968-2141. Fax: (254) 968-6221. Web Site: www.kstvfm.com. Licensee: CCR-Stephenville III LLC. Group owner: Cherry Creek Radio LLC (acq 6-24-2004; grpsl). Population served: 75,000 Format: Mexician hits. Target aud: 54 plus; general. Spec prog: Farm 5 hrs, relg 6 hrs wkly. ♦Robert Elliott, gen mgr; Bob Haschke, gen sls mgr; Jose Perez, prom mgr & progmg dir; Loena Rodriquez, news dir; Justin McClure, chief of engrg; Troy Stark, traf mgr.

Sterling City

KNRX(FM)— Dec 1, 1998: 96.5 mhz; 40 kw. Ant 544 ft TL: N31 35 56 W100 50 42. Hrs open: 24 1301 S. Abe St., San Angelo, 76903. Phone: (325) 655-7161. Fax: (325) 658-7377. Licensee: Double O Texas Corp. Group owner: Encore Broadcasting LLC (acq 3-15-2006; grpsl). Format: Classic rock. Target aud: 25-64. ♦John Kerr, gen mgr; Randy Phair, adv dir; Boomer Kingston, progmg dir; Carry Vaughn, engrg dir.

Sulphur Springs

KSCH(FM)— Aug 30, 1982: 95.9 mhz; 6 kw. 285 ft TL: N33 09 07 W95 36 12. Stereo. Hrs opn: 24
Rebroadcasts KSCN(FM) Pittsburg 90%.
930 S. Gilmer, 75482. Phone: (903) 885-1546. Fax: (903) 572-7232. E-mail: hitmusic@klake.net Web Site: www.easttexasradio.com. Licensee: East Texas Broadcasting Inc. (group owner; acq 9-30-99). Population served: 100,000 Law Firm: Fletcher, Heald & Hildreth. Format: Country. News staff: 3; News: 13 hrs wkly. Target aud: 18-60. ♦J.R. "Bud" Kitchens Jr., pres; Daniel Osuna, gen mgr & gen sls mgr.

KSST(AM)— March 1947: 1230 khz; 1 kw-U. TL: N33 07 00 W95 35 05. Hrs open: 24 Box 284, 75483. Secondary address: 717 Shannon Rd. 75483. Phone: (903) 885-3111. Fax: (903) 885-4160. E-mail: ksst@neto.com Web Site: www.ksstradio.com. Licensee: Hopkins County Broadcasting Co. (acq 1948). Population served: 30,000 Rgnl. Network: Texas State Net. Format: Full service, adult standards. News staff: 2; News: 30 hrs wkly. Target aud: 25-54. ♦Dwayne Grimes, opns dir, gen sls mgr, progmg dir & outdoor ed; Enola Gay, prom dir, mus dir, reporter, women's cmtr & disc jockey; Patsy Bradford, adv dir; Jimmy Rogers, news dir, pub affrs dir, local news ed & sports cmtr; Dolly Kelly, min affrs dir; William Bradford, CEO, pres, gen mgr, chief of engrg & edit mgr; Jacob Wilson, disc jockey.

Sweetwater

KXOX(AM)— November 1939: 1240 khz; 1 kw-U. TL: N32 29 16 W100 23 31. Stereo. Hrs opn: 24 Box 570, 79556. Secondary address: 1801 Hoyt Ln. 79556. Phone: (325) 236-6655. Fax: (325) 235-4391. E-mail: kxox@bigcountry.net Licensee: Stein Broadcasting Inc. (acq 1956). Population served: 14,000 Rgnl. Network: Texas State Net. Format: Country. News: 2 plus hrs wkly. Target aud: 25-54. Spec prog: Farm 5 hrs, Sp 8 hrs, gospel 4 hrs wkly. ♦Jack Stein, pres; Jeff Stein, gen mgr, prom mgr, news dir, local news ed & disc jockey; Gary Graham, chief of engrg; Richard Ferguson, progmg dir & women's int ed.

KXOX-FM— Apr 7, 1976: 96.7 mhz; 2.9 kw. 154 ft TL: N32 29 16 W100 23 31. Stereo. 18 News staff: one. ♦Jeff Stein, local news ed & disc jockey; Lillie Guttierez, spanish dir & disc jockey; Maxine Stein, women's int ed; Mitch Moore, disc jockey; Richard Ferguson, disc jockey.

Tahoka

KAMZ(FM)— 2001: 103.5 mhz; 20 kw. Ant 328 ft TL: N33 19 26 W101 48 15. Hrs open: 24 1220 Broadway, Suite 1035, Lubbock, 79401. Phone: (806) 741-0701. Fax: (806) 741-0705. Licensee: Albert Benavides. Format: Mexican regional. ♦Albert Benavides, gen mgr; Rick Benavides, natl sls mgr; Bob Benavides, mus dir; Bill Enloe, chief of engrg; Connie Ledesima, traf mgr.

KMMX(FM)— Aug 13, 1987: 100.3 mhz; 100 kw. 800 ft TL: N33 26 30 W101 52 42. Stereo. Hrs opn: 24 33 Briercroft Park., Lubbock, 79412. Phone: (806) 762-3000. Fax: (806) 762-8419. Web Site: www.kmmx.com. Licensee: Wilks License Co.-Lubbock LLC. Group owner: NextMedia Group L.L.C. (acq 8-19-2005; grpsl). Population served: 400,000 Natl. Network: ABC. Format: Adult contemp. News staff: one; News: 5 hrs wkly. Target aud: 25-54; females. ♦Scott Harris, gen mgr; Jeff Scott, opns mgr; Jay Richards, gen sls mgr; Damon Scott, progmg dir; Stacey James, news dir; Randy Hayes, chief of engrg; Julie Aguilar, traf mgr.

Tatum

KXAL-FM— Aug 1, 1965: 100.3 mhz; 2.45 kw. Ant 518 ft TL: N32 22 37 W94 34 18. Stereo. Hrs opn: 24 Box 1648, Jacksonville, 75766. Secondary address: 402 Ragsdale, Jacksonville 75766. Phone: (903) 586-2527. Fax: (903) 589-0677. Licensee: Waller Media LLC. Group owner: On-Air Family LLC (acq 8-24-2005; $975,000 with KDVE(FM) Pittsburg). Population served: 250,000 Format: Sp. Target aud: 25-54. ♦Dudley Waller, gen mgr; Chris Ousley, gen sls mgr; Victor Covarrubias, progmg dir.

Taylor

KLQB(FM)— Apr 4, 1975: 104.3 mhz; 48 kw. Ant 492 ft TL: N30 26 04 W97 21 53. Stereo. Hrs opn: 24 4301 Westbank Dr., Escalade B-3rd Fl., Austin, 78746-4400. Phone: (512) 327-9595. Fax: (512) 329-6255. Licensee: CBS Radio Stations Inc. Group owner: Infinity Broadcasting Corp. (acq 11-13-98; grpsl). Population served: 908,000 Natl. Rep: Katz Radio. Law Firm: Leventhal, Senter & Lerman. Format: Rgnl Mexican. ♦Clint Culp, sr VP; John Hiatt, gen mgr & mktg mgr; Rodney Brown, gen sls mgr; Carla Spears, mktg dir; Dusty Hayes, progmg VP; Darell Heckendorf, engrg dir.

KWNX(AM)— Apr 1, 1948: 1260 khz; 1 kw-D. TL: N30 36 19 W97 24 51. Hrs open: 24 1050 E. 11th St., Suite 300, Austin, 78702. Phone: (512) 346-8255. Fax: (512) 346-8262. E-mail: controlroom@espnaustin.com Web Site: www.espnaustin.com. Licensee: Simmons-Austin, LS LLC. Group owner: Simmons Media Group (acq 5-17-2004; $950,000). Population served: 1,000,000 Natl. Network: ESPN Deportes. Format: Sp sports. ♦Daryl O'Neal, gen mgr; Jon Madani, opns dir & progmg dir; Lori Hatter, sls dir; Courtney Cleland, prom mgr; J. Cole McClellan, chief of engrg; Flavia Chen, traf mgr.

Temple

***KBDE(FM)**— 2001: 89.9 mhz; 11.5 kw vert. Ant 489 ft TL: N31 16 05 W97 21 34. Hrs open: 6304 Gardendale Dr., Waco, 76710. Phone: (254) 772-1900. Web Site: www.kbderadio.net. Licensee: American Family Association. Group owner: American Family Radio Format: Christian. ♦Marvin Sanders, gen mgr.

KLTD(FM)— 1995: 101.7 mhz; 16.5 kw. 410 ft TL: N31 16 24 W97 23 31. Hrs open: 108 E. Ave. E., Copperas Cove, 76522. Phone: (254) 773-5252. Fax: (254) 547-2394. Web Site: www.cumulus.com. Licensee: Cumulus Licensing Corp. Group owner: Cumulus Media Inc. (acq 4-20-01; $1.5 million including $50,000 noncompete agreement). Format: Classic Rock. ♦Bourdon Wooten, gen mgr; Mikie Cummings, gen sls mgr; Jamie Garrett, prom dir; Tom Rivers, progmg dir; Chris Cummings, news dir; Doug Bernhardt, chief of engrg; Thalesa Hector-Dixon, traf mgr.

KTEM(AM)— Nov 25, 1936: 1400 khz; 950 w-U. TL: N31 04 01 W97 23 57. Hrs open: 24 608 Moody Ln., 76504. Phone: (254) 773-5252. Fax: (254) 773-0115. E-mail: mailbox@ktem.com Web Site: www.ktem.com. Licensee: Cumulus Licensing Corp. Group owner: Cumulus Media Inc. (acq 3-12-01; $425,000). Population served: 300,000 Natl. Network: CBS. Rgnl. Network: Texas State Net. Format: News/talk, sports. News staff: one; News: 30 hrs wkly. Target aud: 35-64; affluent, educated, politically active. Spec prog: Czech 3 hrs wkly. ♦Bourdon Wooten, gen mgr & natl sls mgr; Mike Cummings, gen sls mgr; Lisa Tanner, prom mgr; Dave Hodges, progmg dir & news dir; Troy Carrell, chief of engrg.

KUSJ(FM)—Harker Heights, June 1987: 105.5 mhz; 930 w. 587 ft TL: N31 05 23 W97 35 55. (CP: 33 kw, ant 600 ft. TL: N30 59 09 W97 37 51). Stereo. Hrs opn: 608 Moody Ln., 76504. Phone: (254) 773-5252. Fax: (254) 773-0015. E-mail: bourdon.wooten@cumulus.com Web Site: us105.com. Licensee: Cumulus Licensing Corp. Group owner: Cumulus Media Inc. (acq 2-2-00; grpsl). Format: Country. Target aud: 25-54. ♦Bourdon Wooten, gen mgr; Mikie Cummings, gen sls mgr; Jamie Garrett, prom dir; Lisa Tanner, prom dir; Doug Bernhardt, chief of engrg & engr.

***KVLT(FM)**— May 1, 2003: 88.5 mhz; 5 kw vert. Ant 617 ft TL: N30 59 08 W97 37 56. Stereo. Hrs opn: 24 American Educational Broadcasting Inc., 3185 S. Highland Dr. #13, Las Vegas, NV, 89109. Secondary address: 3411 Market Loop, Studio,Suite 108 76502. Phone: (254) 791-5251. Fax: (254) 791-0200. E-mail: james@kvltfm.com Licensee: American Educational Broadcasting Inc. Natl. Network: K-Love. Law Firm: Fletcher, Heald & Hildreth. Format: Contemp Christian music. ♦Carl J. Auel, pres; James E. Auel, gen mgr.

Terrell

KPYK(AM)— October 1947: 1570 khz; 250 w-D, 6 w-N. TL: N32 44 35 W96 18 18 (day), N32 45 18 W96 18 58 (night). Hrs open: 24 Box 157, 75160. Secondary address: Town West Plaza, 1412-C W. Moore Ave. 75160. Phone: (972) 524-5795. Fax: (972) 524-5795. E-mail: kpyk@broadcast.net Web Site: www.kpyk .com. Licensee: Mohnkern Electronics Inc. (acq 4-1-92; $25,000 plus assumption of debt; FTR: 3-16-92) Population served: 25,000 Natl. Network: USA. Format: Btfl mus, big band, old radio programs. News staff: one; News: 15 hrs wkly. Target aud: 40 plus; mature adults. Spec prog: , Black 8 hrs, relg 13 hrs, drama 7 hrs wkly. ♦Len Mohnkern, pres, gen mgr, gen sls mgr, news dir & local news ed; Chuck Mohnkern, chief of opns, progmg dir & chief of engrg; Liz Mohnkern, asst music dir & disc jockey; Chris Babler, sports cmtr; Susan Pinson, prom dir, pub affrs dir, traf mgr & women's int ed.

Terrell Hills

KLUP(AM)— Oct 17, 1947: 930 khz; 5 kw-D, 1 kw-N, DA-N. TL: N29 31 06 W98 24 25. Hrs open: 24 9601 McAllister Fwy., Suite 1200, San Antonio, 78216. Phone: (210) 344-8481. Fax: (210) 340-1213. E-mail: myopinion@klup.com Web Site: www.klup.com. Licensee: South Texas Broadcasting Inc. Group owner: Salem Communications Corp. (acq 7-27-00; grpsl). Natl. Network: Salem Radio Network. Format: News/talk. Target aud: 35-64; upper & middle income, empty nesters. ♦Baron Wiley, gen mgr, opns mgr, progmg dir & progmg dir; David Ziebell, gen mgr & gen sls mgr; James Herring, gen sls mgr.

KPWT(FM)— July 18, 1979: 106.7 mhz; 100 kw. Ant 1,017 ft TL: N29 11 03 W98 30 49. Stereo. Hrs opn: 24 8122 Datapoint Dr., # 500, San Antonio, 78229. Phone: (210) 615-5400. Fax: (210) 615-5300. Web Site: power1067fm.com. Licensee: Cox Radio Inc. Group owner: Cox Communications Inc. (acq 3-28-97; grpsl). Population served: 1,300,000 Wire Svc: AP Format: Hip hop. News staff: one. Target aud: 18-34. ♦Bob Neil, CEO; Marty Choate, VP & gen mgr; Mark Bowka, gen sls mgr; Jeff Scott, natl sls mgr; Adam Micheals, prom dir; Doug Bennett, progmg dir; Paul Reynolds, chief of engrg.

Texarkana

KCMC(AM)— Feb 26, 1932: 740 khz; 1 kw-U, DA-1. TL: N33 26 17 W94 08 33. Hrs open: 24 615 Olive St., 75501. Phone: (903) 793-4671. Fax: (903) 792-4261. Licensee: ArkLaTex LLC. (group owner; (acq 12-14-2006; grpsl). Population served: 200,000 Natl. Network: ESPN Radio. Natl. Rep: McGavren Guild. Format: Sports. Target aud: 18-59; sports fans. ♦Mike Simpson, gen mgr; Alex Rain, opns mgr & progmg dir.

KEWL-FM—See New Boston

KHTA(FM)—See Wake Village

KKTK(AM)— 1946: 1400 khz; 1 kw-U. TL: N33 26 28 W94 03 16. Hrs open: 24 1323 College Dr., 75503. Phone: (903) 793-1100. Fax: (903) 794-4717. Licensee: FFD Holdings I Inc. Group owner: Petracom Media LLC (acq 1-4-2005; grpsl). Natl. Rep: Katz Radio. Format: Oldies. Target aud: 35 plus. ♦Mike Basso, gen mgr.

KKYR-FM— July 15, 1965: 102.5 mhz; 100 kw. 445 ft TL: N33 22 24 W94 01 00. Stereo. Hrs opn: 2324 Arkansas Blvd., AR, 71854. Phone: (870) 772-3771. Fax: (870) 772-0364. E-mail: wesspicher @gapbroadcasting.com Web Site: www.kkyr.com. Licensee: GAP Broadcasting Texarkana License LLC. Group owner: Clear Channel Communications Inc. (acq 8-3-2007; grpsl). Population served: 150,000 Format: Country. Target aud: General. ♦Ron Bird, gen mgr; Mitzi Dowd, gen sls mgr; Mario Garcia, progmg dir; John Williams, news dir; Wes Spicher, chief of engrg.

KOSY(AM)—See Texarkana, AR

KRMD(AM)—See Shreveport, LA

KRMD-FM—See Shreveport, LA

KTAL-FM— 1945: 98.1 mhz; 100 kw horiz, 61 kw vert. 1,360 ft TL: N32 54 11 W94 00 22. Hrs open: 24 208 N. Thomas Dr., Shreveport, LA, 71137. Phone: (318) 222-3122. Fax: (318) 459-1493. Web Site: www.98rocks.fm. Licensee: Access. 1 Louisiana Holding Co. LLC. Group owner: Access.1 Communications Corp. (acq 12-20-02; grpsl). Format: Classic rock. News staff: 1. Target aud: 25-54. ♦Cary Camp, gen mgr; Don Zimmerman, gen sls mgr; Greg Hanson, progmg dir; Eddie Thurmand, chief of engrg.

KTFS(AM)— Oct 23, 1961: 940 khz; 2.5 kw-D, 11 w-N. TL: N33 24 28 W94 02 45. Hrs open: 6 AM-midnight 615 Olive St., 75501. Phone: (903) 793-4671. Fax: (903) 792-4261. Licensee: ArkLaTex LLC. (group owner; (acq 12-14-2006; grpsl). Population served: 200,000 Natl. Network: CBS Radio, CNN Radio. Natl. Rep: McGavren Guild. Format: News/talk. News staff: one; News: 2 hrs wkly. Target aud: 35 plus. Spec prog: Rush Limbaugh, Black gospel 2 hrs wkly. ♦Mike Simpson, gen mgr; Jay Calhoun, opns dir & news dir.

***KTXK(FM)**— Feb 1, 1984: 91.5 mhz; 5.2 kw. 335 ft TL: N33 23 33 W94 14 44. Stereo. Hrs opn: 24 2500 N. Robinson, 75599. Phone: (903) 838-4541. Fax: (903) 832-5030. E-mail: ktxktc@yahoo.com Licensee: Texarkana College. Population served: 270,000 Natl. Network: PRI, NPR. Format: Btfl mus, class. News staff: one; News: 35 hrs wkly. Target aud: 35 plus. Spec prog: Jazz 15 hrs wkly. ♦Steve Mitchell, pres, gen mgr & chief of opns.

Texas City

KYST(AM)— November 1947: 920 khz; 5 kw-D, 1 kw-N, DA-2. TL: N29 25 03 W94 56 12. Hrs open: 7322 S.W. Fwy., Suite 500, Houston, 77074. Phone: (713) 779-9292. Fax: (713) 779-1651. Web Site: www.radiodeporte.com. Licensee: Hispanic Broadcasting Inc. (acq 10-1-93; $548,000; FTR: 10-18-93). Population served: 76,400 Format: Sp. ♦Cruz Velasquez, gen mgr.

Thorndale

KLGO(FM)— Sept 30, 2005: 99.3 mhz; 6 kw. Ant 328 ft TL: N30 29 23 W97 17 56. Stereo. Hrs opn: 24 117 E. 3rd St., Taylor, 76574. Phone: (512) 367-9300. Fax: (512) 352-5425. E-mail: gbender@klgo.net Web Site: www.theword993.com. Licensee: Jackson Lake Broadcasting Co. Natl. Network: Moody, Salem Radio Network. Law Firm: Hardy, Carey & Chautin. Format: Christian, talk. News: 10 hrs wkly. Spec prog: Black 6 hrs wkly. ♦Dean Clark, opns dir; Gene Bender, gen mgr & gen sls mgr.

Three Rivers

KEMA(FM)— 2003: 94.5 mhz; 48 kw. Ant 492 ft TL: N28 43 10 W98 02 34. Hrs open: 1610 Woodstead Ct., Suite 350, The Woodlands, 77380. Phone: (512) 383-1112. Licensee: Roy E. Henderson (acq 3-24-2000; $25,000 for CP). ♦Roy E. Henderson, gen mgr.

Tom Bean

KLAK(FM)— Jan 6, 1984: 97.5 mhz; 32 kw. Ant 617 ft TL: N33 28 30 W96 26 45. Stereo. Hrs opn: 24 1700 Redbud Blvd., McKinney, 75069. Phone: (972) 542-9755. Phone: (866) 416-2995. Fax: (972) 838-1330. E-mail: webrequest@975klak.com Web Site: www.975klak.com. Licensee: NM Licensing LLC. Group owner: NextMedia Group L.L.C. (acq 11-26-2001; grpsl). Population served: 1,100,000 Natl. Network: ABC. Format: Adult contemp. News staff: one; News: 7 hrs wkly. Target aud: 25-54; women. ♦Steven Dinetz, CEO; Jeff Dinetz, pres; Sean Stover, CFO; David Smith, gen mgr; Jennifer Isbell, prom dir.

Tomball

KSEV(AM)— Dec 1, 1986: 700 khz; 25 kw-D, 1 kw-N, DA-2. TL: N30 11 34 W95 35 40. Hrs open:
Simulcasts KPRC(AM) Houston.
11451 Katy Fwy., Suite 215, Houston, 77079. Phone: (281) 588-4800. Fax: (832) 358-9556. E-mail: thevoice@ksevradio.com Web Site: www.ksevradio.com. Licensee: Liberman Broadcasting of Houston License LLC. Group owner: Liberman Broadcasting Inc. (acq 3-20-2001; grpsl). Population served: 1,700,000 Format: News/talk. ♦Dan Patrick, VP & gen mgr; Louis Wright, opns mgr; Bonnie English, sls dir; Pam McKay, gen sls mgr; Doug Roach, progmg dir; Chuck McLeod, chief of engrg; Carl Grijolla, traf mgr.

Tulia

KBTE(FM)— Apr 1, 1991: 104.9 mhz; 96.6 kw. Ant 977 ft TL: N33 57 35 W101 35 22. Hrs open: 24 33 Briercroft Office Park, Lubbock, 79412. Phone: (806) 762-3000. Fax: (806) 770-5363. Licensee: Wilks License Co.-Lubbock LLC. (acq 8-29-2005; $1,265,000). Format: Rhythm and blues, hip hop. Target aud: General. Spec prog: Loc

sports. ♦Jay Richard, gen mgr; Jeff Scott, opns mgr; Robbie Cruise, gen mgr & progmg dir; Randy Hayes, chief of engrg; Laura Joyner, traf mgr.

KTUE(AM)— November 1954: 1260 khz; 1 kw-D, 53 w-N. TL: N34 31 34 W101 46 56. Hrs open: 24 Box 252, McAllen, 78505-0252. Phone: (956) 686-6382. Fax: (956) 686-2999. Licensee: Paulino Bernal (acq 9-24-2004). Population served: 600,000 Rgnl. Network: Texas State Net. Format: Sp relg. Target aud: General. ♦Paulino Bernal, gen mgr.

Tye

KBCY(FM)— October 1983: 99.7 mhz; 100 kw. 744 ft TL: N32 24 39 W100 06 26. Stereo. Hrs opn: 24 Box 3157, 2525 S. Danvile, Abilene, 79608. Phone: (325) 793-9700. Fax: (325) 692-1576. Web Site: www.kbcy.com. Licensee: Cumulus Licensing Corp. Group owner: Cumulus Media Inc. (acq 2-13-98; grpsl). Population served: 125,000 Format: Country. News: 21 hrs wkly. Target aud: 25-49; upscale adults. Spec prog: Relg 5 hrs wkly. ♦Kelly Jay, opns mgr; Kim Crenshaw, gen sls mgr; Trace Michaels, gen mgr & mktg mgr; Justin Case, prom dir; Doc Alexander, progmg dir; Chris Andrews, chief of engrg; Lori Barrett, traf mgr.

KWFA(AM)—Not on air, target date: unknown: 1030 khz; 5 kw-D, 370 w-N, DA-2. TL: N32 27 37 W99 50 03. Hrs open: 6720 Lakeview Dr., Carmichael, CA, 95608. Licensee: Marlene V. Borman. ♦Marlene V. Borman, gen mgr.

Tyler

KDOK(FM)— November 1975: 92.1 mhz; 9.6 kw. Ant 443 ft TL: N32 22 28 W95 16 24. Stereo. Hrs opn: 24 Box 92, 75710-0092. Phone: (903) 593-2519. Phone: (903) 592-5200. Fax: (903) 597-4141. Web Site: www.kdok.com. Licensee: Gleiser Communications LLC acq 11-21-2003; grpsl). Format: Oldies. News staff: one; News: 2 hrs wkly. Target aud: 35 plus. Spec prog: Tyler Junior Collete football, Saturday Night Big Band Dance Party. ♦Paul Berry, progmg dir & news dir; Mark Lavoux, chief of engrg; Deborah Harrington, traf mgr & political ed; Paul Gleiser, gen mgr, opns mgr, gen sls mgr & traf mgr; Barry Davis, news rptr; Bill Davis, disc jockey.

KGLD(AM)— May 11, 1956: 1330 khz; 1 kw-D, 77 w-N. TL: N32 22 35 W95 15 55. Stereo. Hrs opn: 24 Box 1330, 75710-1330. Secondary address: 1001 E. S.E. Loop 323 75701. Phone: (903) 593-2519. Fax: (903) 597-8378. Web Site: www.kdok.com. Licensee: Salt of the Earth Broadcasting Inc. (acq 4-13-2004; $160,000). Population served: 127,000 Natl. Network: ABC. Law Firm: Gardner, Carton & Douglas. Format: Oldies. Spec prog: S. ♦Paul Gleiser, gen mgr; Roger Gray, progmg dir; Mike LaRoux, chief of engrg; Deborah Harrington, traf mgr.

***KGLY(FM)**— June 1988: 91.3 mhz; 12 kw. 462 ft TL: N32 21 06 W95 16 00. Stereo. Hrs open: 24 Box 8525, 75711. Secondary address: 2721 E. Erwin St. 75708. Phone: (903) 593-5863. Fax: (903) 593-2663. E-mail: kkgly@kgly.com Web Site: www.encouraagementfm.com. Licensee: Educational Radio Foundation of East Texas Inc. Natl. Network: Moody, USA. Format: Relg. News: 3 hrs wkly. Target aud: 35 plus. ♦Dan Bolin, gen mgr; John Paul Little, stn mgr; Leah Coombs, mus dir; Sans Hawkins, chief of engrg.

KISX(FM)—See Whitehouse

KKUS(FM)— 1990: 104.1 mhz; 50 kw. 492 ft TL: N32 29 40 W95 28 55. Stereo. Hrs opn: 24 Box 7820, 75711. Secondary address: 210 S Broadway 75702. Phone: (903) 581-9966. Fax: (903) 534-5300. E-mail: kkus@etradiogroup.com Web Site: www.theranch.fm. Licensee: Access.1 Texas License Company LLC. Group owner: Waller Broadcasting. (acq 1-7-2005; grpsl). Population served: 300,000 Natl. Network: Fox News Radio. Natl. Rep: McGavren Guild. Format: Classic Country. Target aud: 35+. ♦Tom Perryman, gen mgr; Genni Causey, gen sls mgr; Robert Taylor, natl sls mgr; Chuck McKinley, progmg dir.

KNUE(FM)— Dec 31, 1964: 101.5 mhz; 100 kw. 1,074 ft TL: N32 15 35 W94 57 02. Stereo. Hrs open: 24 3810 Brookside, 75701. Phone: (903) 581-0606. Fax: (903) 581-2011. Web Site: www.knue.com. Licensee: GAP Broadcasting Tyler License LLC. Group owner: Clear Channel Communications Inc. (acq 8-3-2007; grpsl). Population served: 750,000 Format: Country. Target aud: General. ♦Steve Joos, gen mgr; Craig Reininger, sls dir; Chris Jones, prom dir & adv dir; Michael Gibson, progmg dir & mus dir; Dave Goldman, news dir; Laura Conway, traf mgr.

KTBB(AM)— Aug 28, 1947: 600 khz; 5 kw-D, 2.5 kw-N, DA-2. TL: N32 16 18 W95 12 23. Stereo. Hrs opn: 24 Box 6, 75710-0006. Phone: (903) 593-2519. Fax: (903) 593-4918. Web Site: www.ktbb.com. Licensee: Gleiser Communications LLC (group owner; acq 11-21-2003; grpsl). Population served: 533,000 Rgnl. Network: Texas State Net. Law Firm: Gardner, Carton & Douglas. Format: Full service, news/talk, sports & info. News staff: 7; News: 70 hrs wkly. Target aud: 35 plus. Spec prog: Gospel 5 hrs wkly. ♦Paul L. Gleiser, CEO, pres, gen mgr & sls dir; Roger Gray, progmg dir; Mike LaRoux, chief of engrg; Deborah Harrington, traf mgr.

KTYL-FM— February 1966: 93.1 mhz; 82 kw. Ant 938 ft TL: N32 15 35 W94 57 02. Stereo. Hrs opn: 24 3810 Brookside Dr., 75701-9420. Phone: (903) 581-0606. Fax: (903) 581-2011. Web Site: www.mix931.com. Licensee: GAP Broadcasting Tyler License LLC. Group owner: Clear Channel Communications Inc. (acq 8-3-2007; grpsl). Population served: 825,000 Format: Hot adult contemp. Target aud: 18-54. ♦Steve Joos, gen mgr; Craig Reininger, sls dir; Chris Jones, gen sls mgr; Jeff Evans, progmg dir.

***KVNE(FM)**— Oct 15, 1983: 89.5 mhz; 100 kw. 899 ft TL: N32 32 21 W95 13 16. (CP: 96 kw). Stereo. Hrs opn: 24 Box 8525, 75711. Phone: (903) 593-5863. Web Site: www.kvne.com. Licensee: Educational Radio Foundation of East Texas Inc. Format: Relg. News: 6 hrs wkly. Target aud: 20-45; families. Spec prog: Gospel 4 hrs, children 4 hrs, Sp 2 hrs wkly. ♦Mike Harper, stn mgr.

KYZS(AM)— 1930: 1490 khz; 1 kw-U. TL: N32 22 30 W95 16 05. Stereo. Hrs open: 24 Box 6, 75710. Phone: (903) 593-2519. Fax: (903) 597-4141. Web Site: www.kdok.com. Licensee: Gleiser Communications LLC (group owner; acq 11-21-2003; grpsl). Format: ESPN radio-sports. ♦Paul Gleiser, gen mgr & opns mgr; Robert Gray, progmg dir; Robert LaRoux, chief of engrg; Deborah Harrington, traf mgr.

KZEY(AM)— 1958: 690 khz; 1 kw-D, 92 w-N, DA-2. TL: N32 22 52 W95 20 52. Hrs open: 24 Box 4248, Lake Park Dr., 75712. Phone: (903) 593-1744. Fax: (903) 593-2666. Licensee: Community Broadcast Group Inc. Population served: 150,000 Format: Urban contemp. News staff: one. Target aud: General. Spec prog: Gospel 16 hrs wkly. ♦Esther Milton, gen mgr, gen sls mgr & mus dir; Darrell Bowdre, news dir.

Universal City

KSAH(AM)— Nov 1, 1986: 720 khz; 10 kw-D, 1 kw-N, DA-2. TL: N29 31 51 W98 10 39. Stereo. Hrs opn: 24 7800 1-10 West, Suite 330, San Antonio, 78230. Phone: (210) 340-1234. Fax: (210) 340-1775. Licensee: BMP San Antonio License Co. L.P. Group owner: Border Media Partners LLC (acq 12-23-2003; $24.4 million with KTFM(FM) Floresville). Format: Sp. News staff: one; News: 2 hrs wkly. Target aud: 18-49. ♦Peggy McCormick, gen mgr & natl sls mgr; Raoul Rodriquez, gen mgr; Lupe Contreras, prom dir; Manny Herrera, progmg dir; Brett Huggins, chief of engrg; Rose Pagan, traf mgr.

University Park

KTNO(AM)— 1938: 1440 khz; 15 kw-D, 350 w-N, DA-2. TL: N32 45 02 W96 43 22. Hrs open: 24 5787 S. Hampton Rd., Suite 340, Dallas, 75232. Phone: (214) 330-5866. Fax: (214) 330-9885. Web Site: www.ktnoam.com. Licensee: Mortenson Broadcasting Co. of Texas Inc. (group owner; (acq 8-8-97; $650,000). Population served: 39,874 Natl. Network: ABC. Format: Sp, talk info, Christian. News staff: 2; News: 40 hrs wkly. Target aud: 25 plus. ♦Jose Alfredo Castillo, gen mgr, gen sls mgr & progmg dir; Mike Benhauser, chief of engrg; Erica Garcia, traf mgr.

KZMP(AM)— 1999: 1540 khz; 32 kw-D, 750 kw-N, DA-2. TL: N32 48 45 W97 00 30. Hrs open: 24 4201 Pool Rd., Coleyville, 76034. Phone: (817) 868-2900. Fax: (817) 868-2929. Licensee: Liberman Broadcasting of Dallas License LLC. Group owner: Entravision Communications Corp. (acq 11-2-2006; grpsl). Population served: 6,153,500 Natl. Rep: Lotus Entravision Reps LLC. Law Firm: Wiley, Rein & Fielding. Format: Ranchero. ♦Alex Sanchez, gen mgr.

Uvalde

KBNU(FM)— 1996: 93.9 mhz; 25 kw horiz, 14.3 kw vert. Ant 292 ft TL: N29 16 34 W99 41 44. Stereo. Hrs opn: 24 Rebroadcast KBLT(FM) Leakey 100%. 1010 Garner Field Rd., 78801. Phone: (830) 278-3693. Fax: (830) 278-2329. E-mail: kbradioranch@hotmail.com Web Site: wwwkbnu.fm. Licensee: Radio Cactus Ltd. (acq 10-23-00; $60,916 for 51% of stock with KBLT(FM) Leakey). Law Firm: John Mc Veigh. Format: Classic country. News staff: 0. Target aud: General; 18-49. ♦John Furr, CEO, progmg dir & chief of engrg; Paula Furr, CFO; Regenia Tumbarello, gen mgr, gen sls mgr & prom dir.

KUVA(FM)— Aug 20, 1984: 102.3 mhz; 3 kw. 280 ft TL: N29 11 46 W99 46 48. Stereo. Hrs opn: 18 Box 758, 1400 Batesville Rd., 78801. Phone: (830) 278-2555. Fax: (830) 278-9461. E-mail: production@uvalderadio.com Web Site: www.uvalderadio.com. Licensee: Rhattigan Broadcasting (Texas) LP (group owner; acq 6-3-2004; grpsl). Population served: 100,000 Law Firm: Baraff, Koerner & Olender. Format: Sp, Tejano. News staff: one; News: 6 hrs wkly. Target aud: 16-60; Hispanic. ♦Justin Rue, gen mgr.

KVOU(AM)— Apr 4, 1947: 1400 khz; 1 kw-U. TL: N29 11 16 W99 46 36. Hrs open: 6 AM-midnight Box 758, 78802-0758. Secondary address: 1400 Batesville Rd. 78801. Phone: (830) 278-2555/2557. Fax: (830) 278-9461. E-mail: production@uvalde.com Web Site: www.uvalderadio.com. Licensee: Rhattigan Broadcasting (Texas) LP (group owner; (acq 8-19-2004; grpsl). Population served: 85,800 Rgnl. Network: VSA Radio. Law Firm: Baraff, Koerner & Olender. Format: Oldies. News staff: one; News: 15 hrs wkly. Target aud: 25-54; general. Spec prog: Farm 12 hrs wkly. ♦Kevin L. Bonner, gen mgr.

KVOU-FM— Sept 9, 1976: 104.9 mhz; 25 kw. 263 ft TL: N29 11 16 W99 46 36. Stereo. Hrs opn: Box 758, 78802-0758. Secondary address: 1400 Batesville Rd. 78801. Phone: (830) 278-2555. Phone: (830) 278-2557. Fax: (830) 278-9461. E-mail: production@uvalderadio.com Web Site: www.uvalderadio.com. Licensee: Rhattigan Broadcasting (Texas) LP (group owner; (acq 8-19-2004; grpsl). Natl. Network: ABC. Law Firm: Baraff, Koerner & Olender. Format: Country. News: 15 hrs wkly. Target aud: 18-49. Spec prog: High school play-by-play sports, Texas A&M football. ♦Justin Rue, gen mgr.

Vernon

KVWC(AM)— July 1939: 1490 khz; 1 kw-U. TL: N34 09 12 W99 16 09. Hrs open: 6 AM-10 PM Box 1419, 76385. Secondary address: 302 E. Wilbarger 76384. Phone: (940) 552-6221. Fax: (940) 553-4222. E-mail: kvwc@kvwc.com Web Site: www.kvwc.com. Licensee: KVWC Inc. (acq 6-1-61). Population served: 25,000 Natl. Rep: Riley. Format: Oldies, farm, country. News staff: one; News: 10 hrs wkly. Target aud: General; Wilbarger & surrounding counties. Spec prog: Gospel 16 hrs wkly. ♦Mike Klappenbach, pres, gen mgr, progmg dir & chief of engrg.

KVWC-FM— Apr 10, 1972: 103.1 mhz; 6 kw. Ant 141 ft TL: N34 09 12 W99 16 09. Stereo. 6 AM-10 PM Web Site: www.kvwc.com.100,000

Victoria

***KAYK(FM)**— October 2003: 88.5 mhz; 50 kw vert. Ant 282 ft TL: N28 46 43 W97 02 51. Hrs open: Drawer 2440, Tupelo, MS, 38803. Phone: (662) 844-8888. Web Site: www.afr.net. Licensee: American Family Association. Group owner: American Family Radio Population served: 122,000 Format: Christian. ♦Marvin Sanders, gen mgr.

KBAR-FM— Feb 2, 1989: 100.9 mhz; 6 kw. Ant 272 ft TL: N28 47 20 W97 03 00. Hrs open: Box 3487, 77903. Secondary address: 3613 N. Main St. 77901. Phone: (361) 576-6111. Fax: (361) 572-0014. Licensee: Victoria RadioWorks Ltd. (group owner; (acq 1999; $27,500). Population served: 1,000,000 Format: CHR. ♦Cindy Cox, gen mgr.

KITE(FM)—Port Lavaca, Aug 1, 1976: 93.3 mhz; 100 kw. Ant 318 ft TL: N28 42 22 W96 48 03. Stereo. Hrs opn: Box 3487, 77903. Secondary address: 3613 N. Main St. 77903. Phone: (361) 576-6111. Fax: (361) 572-0014. Licensee: Victoria RadioWorks Ltd. (group owner; (acq 10-29-98; $500,000). Population served: 500,000 Natl. Rep: McGavren Guild. Format: Oldies. Spec prog: Farm one hr wkly. ♦Cindy Cox, gen mgr.

KIXS(FM)— Dec 4, 1980: 107.9 mhz; 100 kw. 362 ft TL: N28 46 03 W96 59 11. Stereo. Hrs opn: Box 3325, 77903. Secondary address: 107 North Star Dr. 77904. Phone: (361) 573-0777. Fax: (361) 578-0059. E-mail: kixs@clearchannel.com Web Site: www.kixs.com. Licensee: GAP Broadcasting Victoria License LLC. Group owner: Clear Channel Communications Inc. (acq 8-3-2007; grpsl). Population served: 187,000 Format: C&W. Target aud: 25-49; listeners in a growth & acquisition mode. ♦Jeff Lyon, gen mgr; Natalie Franz, gen sls mgr & natl sls mgr; James Love, prom dir, news dir & pub affrs dir; Eric Sharp, progmg dir; Joe Bob Burris, mus dir & disc jockey; Charles Smithey, chief of engrg; Joe Friday, disc jockey; Mark Kingery, disc jockey; Michelle Lee, disc jockey; Tammie Austin, disc jockey.

KNAL(AM)— Apr 16, 1948: 1410 khz; 500 w-U, DA-N. TL: N28 46 48 W97 00 08. Hrs open: 24 Box 3487, 77903. Secondary address: 3613 N. Main St. 77901. Phone: (361) 576-6111. Fax: (361) 572-0014. Licensee: Victoria RadioWorks Ltd. (group owner; (acq 2-1-2002; $100,000). Population served: 100,000 Law Firm: Fletcher, Heald & Hildreth. Format: Adult standards. ♦Cindy Cox, gen mgr.

KQVT(FM)— Dec 1, 1990: 92.3 mhz; 6 kw. Ant 298 ft TL: N28 46 04 W96 59 12. Stereo. Hrs opn: 24 107 North Star Dr., 77904. Phone: (361) 573-0777. Fax: (361) 578-0059. E-mail: kqvt@clearchannel.com Web Site: www.kqvt.com. Licensee: GAP Broadcasting Victoria License LLC. Group owner: Clear Channel Communications Inc. (acq 8-3-2007; grpsl). Population served: 100,000 Format: Adult contemp. ♦Jeff Lyon, gen mgr; Natalie Franz, gen sls mgr & natl sls mgr; James Love, prom dir, prom dir, news dir & pub affrs dir; J.P. Stone, progmg mgr; Charles Smithey, chief of engrg.

KTXN-FM— Dec 1, 1994: 98.7 mhz; 100 kw. 253 ft TL: N28 48 46 W97 03 45. Stereo. Hrs opn: 24 302 Sam Houston Dr., 77901. Phone: (361) 573-2121. Fax: (361) 573-5872. Web Site: www.texasmix.com. Licensee: Cosmopolitan Enterprises of Victoria Inc. Population served: 58,035 Natl. Network: Westwood One. Law Firm: Dennis J. Kelly. Format: Texas blues, classic rock, zydeco. News staff: 2; News: 2 hrs wkly. Target aud: 18-54; general. ♦Steve Coffman, gen mgr, opns mgr, sls dir & progmg dir; Steve Pingel, stn mgr; Bob Nancy, news dir; Jim Koenig, engrg dir.

KVIC(FM)—Listing follows KVNN(AM).

KVNN(AM)— January 1940: 1340 khz; 1 kw-U. TL: N28 49 49 W97 00 33. Hrs open: 24 Box 3487, 77903. Secondary address: 3613 N. Main St. 77901. Phone: (361) 576-6111. Fax: (361) 572-0014. Licensee: Victoria RadioWorks Ltd. (group owner; (acq 10-26-98; $2.1 million with co-located FM). Population served: 300,000 Rgnl. Network: Texas State Net. Natl. Rep: McGavren Guild. Format: Traditional country. Target aud: 25-54; adults. ♦Cindy Cox, gen mgr.

KVIC(FM)—Co-owned with KVNN(AM). Apr 8, 1976: 95.1 mhz; 13 kw. Ant 459 ft TL: N28 46 55 W96 56 29. Stereo. 24 Format: Adult contemp, CHR. Target aud: 18-49.

***KVRT(FM)**— 1995: 90.7 mhz; 30 kw. 328 ft TL: N28 46 55 W96 56 30. Hrs open:
Rebroadcasts KEDT-FM Corpus Christi 100%.
4455 S. Padre Island Dr., Suite 38, Corpus Christi, 78411-1690. Phone: (361) 855-2213. Fax: (361) 855-3877. Licensee: South Texas Public Broadcasting System Inc. Format: Class, news, jazz. ♦Don Dunlap, chmn, pres & gen mgr; Myra Lombardo, exec VP; Anita Herbert, sls dir; Davita Underbrink, prom dir; Jeff Felts, progmg dir; Bill Clough, news dir; Cody Blount, chief of engrg; Michelle Salazar, traf mgr.

***KXBJ(FM)**— Sept 1, 1994: 89.3 mhz; 18.5 kw. 336 ft TL: N28 49 20 W96 58 20. Stereo. Hrs opn:
Rebroadcasts KSBJ(FM) Humble 100%.
Box 187, Humble, 77347. Secondary address: 2207 Wildwood St. 77901. Phone: (361) 574-8936. Fax: (361) 575-0175. E-mail: kxbj@vipx.org Web Site: www.kxbj.org. Licensee: Educational Media Foundation of Victoria. Population served: 72,596 Natl. Network: USA. Law Firm: Hardy & Carey. Format: Contemp Christian mus, relg. Target aud: 25-54. ♦Dr. Billy Powell, pres; Tim McDermott, gen mgr; Bard Letsinger, stn mgr; Jon Hull, progmg dir.

Waco

KBBW(AM)— April 1953: 1010 khz; 10 kw-D, 2.5 kw-N, DA-2. TL: N31 34 09 W97 00 00. Hrs open: 1019 Washington St., 76701. Phone: (254) 757-1010. Fax: (254) 752-5339. E-mail: info@1010kbw.com Licensee: Steve Williams dba American Broadcasting of Texas. (acq 6-16-86; FTR: 5-12-86). Population served: 4,500,000 Format: Christian. Target aud: 25-54. ♦Elizabeth Layne, stn mgr & opns dir; Ryan Williams, opns dir; Allen Newton, gen sls mgr; Jeremy Beutel, prom dir & news dir; Steve Williams, pres, gen mgr & progmg dir; Dave Fricker, chief of engrg.

KBCT-FM— Aug 1, 1996: 94.5 mhz; 3.2 kw. Ant 453 ft TL: N31 30 31 W97 10 03. Hrs open: 24 4701 W. Waco Dr., 76710. Phone: (254) 388-5945. E-mail: mail@lonestar.com Web Site: www.lonestar94.com. Licensee: Kennelwood Broadcasting Co. Inc. Population served: 225,405 Format: News/talk. ♦Jerry Lenamon, pres & gen mgr.

KBGO(FM)— Sept 6, 1959: 95.7 mhz; 24 kw. 505 ft TL: N31 30 51 W97 11 43. Stereo. Hrs opn: 24 314 W. State Hwy. 6, 76712. Phone: (254) 776-3900. Fax: (254) 761-6371. E-mail: brenthenslee @clearchannel.com Web Site: www.oldies95online.com. Licensee: Capstar TX L.P. Group owner: Clear Channel Communications Inc. (acq 8-30-00; grpsl). Population served: 450,000 Format: Oldies. Target aud: 35-64. ♦Evan Armstrong, gen mgr; Zack Owen, opns dir & opns mgr; Vernon Riggs, sls dir; Brett Henslee, progmg dir & chief of engrg.

KBRQ(FM)—See Hillsboro

KQRL(AM)— Oct 8, 1962: 1580 khz; 1 kw-D, 500 w-N, DA-2. TL: N31 31 04 W97 05 16. Hrs open: 24 2200 S. 2nd St., Apt. 2B2, 76701-2250. Phone: (254) 772-0930. Fax: (254) 772-1580. E-mail: productioni@hot.rr.net Licensee: Simmons-Austin, LS LLC. Group owner: Simmons Media Group (acq 6-4-2004; grpsl). Population served: 102,000 Natl. Rep: Roslin. Law Firm: Leventhal, Senter & Lerman. Format: Country. News staff: one; News: 2 hrs wkly. Target aud: 25-54. Spec prog: Black 4 hrs, Sp 14 hrs wkly. ♦Daryl O'Neal, gen mgr; Bill Le Grand, sls dir; Tom Barfield, opns mgr & progmg dir.

KRZI(AM)— 2001: 1660 khz; 10 kw-D, 1 kw-N. TL: N31 24 46 W97 12 18. Hrs open: 220 S. 2nd St., Apt. 2B2, 76701-2250. Phone: (254) 772-0930. Fax: (254) 753-0499. E-mail: production@hot.rr.com Web Site: www.1660espn.com. Licensee: Simmons Austin, LS LLC. Group owner: Simmons Media Group (acq 6-4-2004; grpsl). Natl. Network: ESPN Radio. Format: Sports. ♦Daryl O'Neal, gen mgr; Bill Le Grand, sls dir; Tom Barfield, opns mgr & progmg dir.

***KVLW(FM)**— 2005: 88.1 mhz; 16.5 kw vert. Ant 1,096 ft TL: N31 18 53 W97 19 36. Hrs open: American Educational Broadcasting Inc., 3185 S. Highland Dr., Suite 13, Las Vegas, NV, 89109. Secondary address: 3411 Market Loop, Studio, Suite 108, Temple 76502. Phone: (254) 791-5251. Fax: (254) 791-0200. Licensee: American Educational Broadcasting Inc. Natl. Network: K-Love. Law Firm: Fletcher, Heald & Hildreth. Format: Contemp Christian music. ♦Carl Auel, pres; James E. Auel, gen mgr.

***KWBU-FM**— Mar 15, 1966: 107.1 mhz; 3 kw. 190 ft TL: N31 31 51 W97 09 10. Hrs open: 7-1 am Box 97368, KWBU, Baylor Univ., 76798. Phone: (254) 710-4470. Phone: (254) 710-6909. Fax: (254) 710-1563. E-mail: kwbu@baylor.edu Web Site: www.baylor.edu/kwbu. Licensee: Baylor University. Population served: 146600 Format: Class, div. News staff: one; News: 10 hrs wkly. Target aud: under 45; college age. Spec prog: Jazz 10 hrs wkly. ♦Frank Fallon, pres & gen mgr; Brian Potter, sls dir & adv dir; Aubrey Abbott, prom dir; Kateigh Axness, mus dir; Aron Watman, asst music dir; Lauren Lewis, news dir; Ron Stephens, engrg dir; Jeff Knox, sports cmtr.

KWTX(AM)— May 1, 1946: 1230 khz; 5 kw-D, 250 w-N, DA-2. TL: N31 31 42 W97 07 14. Hrs open: 24 314 W. State Hwy. 6, 76712. Phone: (254) 776-3900. Fax: (254) 761-6371. Web Site: newstalk1230.com. Licensee: Capstar TX L.P. Group owner: Clear Channel Communications Inc. (acq 8-30-2000; grpsl). Natl. Network: ABC. Natl. Rep: Clear Channel. Format: News/talk. Target aud: 35-64; adults. ♦Michael Oppenheimer, gen mgr; Zack Owen, opns mgr; Evan Armstrong, gen sls mgr; Gloria Norris, natl sls mgr; Max Watson, progmg dir; Steve Keating, chief of engrg.

KWTX-FM— Dec 1, 1970: 97.5 mhz; 97 kw. 1,568 ft TL: N31 19 19 W97 18 58. Stereo. 24 Web Site: www.975online.com.143,000 Format: CHR. Target aud: 25-49; women. ♦Jay Charles, progmg dir.

WACO-FM— June 1960: 99.9 mhz; 90 kw. 1,660 ft TL: N31 20 15 W97 18 37. Stereo. Hrs opn: 314 W. State Hwy. 6, 76712. Phone: (254) 776-3900. Fax: (254) 761-6371. E-mail: info@waco100.com Web Site: www.waco100.com. Licensee: Capstar TX L.P. Group owner: Clear Channel Communications Inc. (acq 8-30-00; grpsl). Population served: 500,000 Format: Country. ♦Evan Armstrong, gen mgr & gen sls mgr; Zack Owen, opns dir & progmg dir; Brett Gilbert, chief of engrg; Darla Walson, traf mgr.

Wake Village

***KHTA(FM)**— Sept 22, 2000: 92.5 mhz; 25 kw. Ant 328 ft TL: N33 24 53 W93 58 12. Hrs open: 24 Houston Christian Broadcasters Inc., 2424 South Blvd., Houston, 77098. Phone: (713) 520-5200. E-mail: khcb@nol.net Web Site: www.khcb.org. Licensee: Houston Christian Broadcasters Inc. (group owner) Format: Bible teaching, inspirational mus. Target aud: General; all ages, families. ♦Bruce Munsterman, pres & stn mgr.

Waskom

KQHN(FM)— 1968: 97.3 mhz; 42 kw. Ant 533 ft TL: N32 29 36 W93 45 55. Stereo. Hrs opn: 24 Box 5459, Bossier City, LA, 71171. Secondary address: 270 Plaza Loop, Bossier City 71111. Phone: (318) 549-8500. Fax: (318) 549-8505. Web Site: mixfm973.com. Licensee: Cumulus Licensing LLC. Group owner: Cumulus Media Inc. (acq 11-1-2002; $1.75 million). Rgnl. Network: Ark. Radio Net. Format: Hot adult contemp. ♦Phil Robkin, gen mgr, sls dir & gen sls mgr; Danielle Kaiser, rgnl sls mgr; Gary Robinson, prom dir & progmg dir; Dani Coate, news dir; Jasen Bragg, engrg dir.

Waxahachie

KBEC(AM)— June 1955: 1390 khz; 480 w-D, 260 w-N, DA-2. TL: N32 26 45 W96 48 15. Hrs open: 6 AM-11 PM 711 Ferris Ave., 75165. Phone: (972) 923-1390. Phone: (972) 938-1390. Fax: (972) 935-0871. E-mail: info@kbec.com Web Site: www.kbec.com. Licensee: Faye and Richard Tuck Inc. Population served: 260,000 Rgnl. Network: Texas State Net. Format: Classic country. News: 15 hrs wkly. Target aud: 25-54. Spec prog: Farm 3 hrs, Pol 2 hrs wkly. ♦Ken Roberts, gen mgr, stn mgr, progmg dir, spec ev coord, sports cmtr & disc jockey; Barry Wolverton, sls dir, gen sls mgr & disc jockey; Mark Miller, news rptr, outdoor ed & disc jockey; Mike O'Daniel, sports cmtr & disc jockey.

Weatherford

***KMQX(FM)**—Not on air, target date: unknown: 88.5 mhz; 5.5 kw. Ant 492 ft TL: N32 49 17 W98 09 36. Hrs open: 24 905 Palo Pinto St., Weatherton, 76086. Phone: (817) 341-2337. Fax: (817) 598-1661. E-mail: chb890@swbell.net Web Site: www.kyqx.com. Licensee: CSSI Non-Profit Educational Broadcast Inc. Population served: 1,361,541 Law Firm: Hill & Welch. ♦Charles H. Beard, pres; Jean Hudgens, gen mgr.

***KYQX(FM)**— Jan 5, 1986: 89.5 mhz; 4.5 kw. Ant 518 ft TL: N32 51 05 W98 06 31. Hrs open: 24 905 Palo Pinto St., 76086. Phone: (817) 341-8950. Phone: (817) 341-2337. Fax: (817) 596-9842. E-mail: qxfmnews@yahoo.com Web Site: www.kyqx.com. Licensee: CSSI Non Profit Educational Broadcasting Corp. (acq 8-14-98; $55,000). Population served: 1,943,645 Law Firm: Tim Welch. Format: Lite rock. News staff: 3; News: 7 hrs wkly. ♦Charles Beard, pres; Jean Hudgen, gen mgr; John Peterson, opns mgr & progmg dir.

KZEE(AM)— Aug 12, 1956: 1220 khz; 500 w-D, 8 w-N. TL: N32 47 09 W97 47 55. Hrs open: Box 54803, Hurst, 76054. Phone: (817) 594-1220. Phone: (817) 849-1971. Fax: (817) 427-3931. Web Site: www.radio1220am.com. Licensee: Tarrant Radio Broadcasting Inc. (acq 9-5-01; $800,000). Population served: 25,000 Format: Christian Gospel. ♦Parvez Malik, pres; Cima Hernandez, gen mgr.

Wellington

KXME(FM)—Not on air, target date: unknown: 98.5 mhz; 25 kw. Ant 328 ft TL: N34 48 37 W100 19 46. Hrs open: 2768 Pharmacy Rd., Rio Grande City, 78582. Phone: (956) 487-5621. Licensee: James Falcon. ♦James Falcon, gen mgr.

Wells

KVLL-FM— 1993: 94.7 mhz; 50 kw. Ant 384 ft TL: N31 06 47 W94 48 32. Hrs open: 24 Box 2209, Lufkin, 75902-2209. Secondary address: 1216 S. 1st St., Lufkin 75901-4716. Phone: (936) 639-4455. Fax: (936) 639-5540. Licensee: GAP Broadcasting Lufkin License LLC. (acq 2007; $750,000). Format: Oldies. ♦Johnny Lathrop, gen mgr.

Weslaco

KHKZ(FM)—See Mercedes

KRGE(AM)— 1926: 1290 khz; 5 kw-U, DA-N. TL: N26 12 36 W97 54 33. Hrs open: 24 Box 1290, Mile 6 3/4 W. Business 83, 78599. Phone: (956) 968-7777. Fax: (956) 968-5143. E-mail: egarza@radiovida.com Web Site: www.radiovida.com. Licensee: Christian Ministries of the Valley. (acq 1-31-91; FTR: 2-18-91). Population served: 20,007 Format: Christian, Sp. News: 3 hrs wkly. Target aud: 18-34. ♦Enrique Garza, gen mgr.

West Lake Hills

KTXZ(AM)— June 9, 1982: 1560 khz; 2.5 kw-U, DA-2. TL: N30 21 38 W97 39 11. Hrs open: 24 2211 S. IH 35, Suite 401, Austin, 78741. Phone: (512) 416-1100. Fax: (512) 453-6809. Web Site: www.ktxz.com. Licensee: BMP Austin License Company L.P. (acq 2-10-2005; grpsl). Natl. Network: CNN Radio, Westwood One. Format: Sp contemp. News: 14 hrs wkly. Target aud: 18-54; bilingual, Hispanic, male & female. Spec prog: Christian mus 6 hrs wkly. ♦Paul Danitz, gen mgr.

West Odessa

***KLVW(FM)**— 2001: 88.7 mhz; 100 kw. Ant 426 ft TL: N31 50 53 W102 27 04. Hrs open: 24 2351 Sunset Blvd., Suite 170-218, Rocklin, CA, 95765. Phone: (916) 251-1600. Fax: (916) 251-1650. E-mail: klove@klove.com Web Site: www.klove.com. Licensee: Educational Media Foundation. Group owner: EMF Broadcasting. Natl. Network: K-Love. Law Firm: Shaw Pittman. Format: Contemp Christian. News staff: 3. Target aud: 25-44; Judeo Christian, female. ♦Richard Jenkins, pres; Mike Novak, VP; Keith Whipple, dev dir; David Pierce, progmg mgr; Ed Lenane, news dir; Sam Wallington, engrg dir; Karen Johnson, news rptr; Marya Morgan, news rptr; Richard Hunt, news rptr.

Wharton

KANI(AM)— June 17, 1962: 1500 khz; 500 w-U, DA-N. TL: N29 19 22 W96 03 32. Hrs open: Box 350, 77488. Secondary address: 215 E. Milam St. 77488. Phone: (979) 532-3800. Fax: (979) 532-8510. Licensee: Martin Broadcasting Inc. Population served: 43,000 Rgnl. Network: Texas State Net. Format: Religious. Spec prog: Sp 3 hrs, Pol 6 hrs wkly. ♦Sandra Stuart, gen mgr, gen sls mgr, progmg dir & news dir.

Wheeler

***KPDR(FM)**— Aug 31, 1986: 90.5 mhz; 10 kw. 482 ft TL: N35 25 57 W100 16 31. Stereo. Hrs opn: 24 Box 8088, 5754 Canyon Dr., Amarillo, 79114. Phone: (806) 359-8855. Fax: (806) 354-2039. E-mail: kjrt@kingdomkeys.org Web Site: www.kingdomkeys.org. Licensee: Top O' Texas Educational Broadcasting Foundation. Natl. Network: USA. Law Firm: Dow, Lohnes & Albertson. Format: Relg, educ. Target aud: General. Spec prog: Sp 5 hrs wkly. ♦Ricky Pfeil, pres & gen mgr.

White Oak

KAJK(FM)— May 17, 2002: 99.3 mhz; 34 kw. Ant 541 ft TL: N32 35 17 W94 58 53. Hrs open: 24 212 Grande Blvd., B 100, Tyler, 75703. Phone: (903) 581-5259. Fax: (903) 939-3473. E-mail: chelle@theblaze.cc Web Site: www.993jackfm.com. Licensee: Reynolds Radio Inc. (group owner) Format: Adult hits. ♦Robin George, gen sls mgr; Charlie O'Douglas, progmg dir; Chelle Wright-Peterson, traf mgr.

Whitehouse

KISX(FM)— Aug 15, 1982: 107.3 mhz; 50 kw. Ant 500 ft TL: N32 17 19 W95 11 56. Stereo. Hrs opn: 24 3810 Brookside Dr., Tyler, 75701. Phone: (903) 581-0606. Fax: (903) 581-2011. E-mail: stevejoos@clearchannel.com Web Site: kiss107i.com. Licensee: GAP Broadcasting Tyler License LLC. Group owner: Clear Channel Communications Inc. (acq 8-3-2007; grpsl). Population served: 750,000 Format: CHR. Target aud: 18-44. ♦Steve Joos, gen mgr; Craig Reininger, sls dir; Larry Thompson, progmg dir.

Whitesboro

KMAD-FM— June 1, 1985: 102.5 mhz; 18 kw. Ant 672 ft TL: N33 41 31 W96 26 36. Stereo. Hrs opn: 24 101 E. Main St., Suite 255, Denison, 75021. Phone: (903) 463-6800. Fax: (903) 463-9816. Web Site: www.theclassicrockexperience.com. Licensee: NM Licensing LLC. Group owner: NextMedia Group L.L.C. (acq 11-26-2001; grpsl). Format: Classic rock. ♦David Smith, gen mgr; Jennifer Isbell, prom dir; Jason Taylor, progmg dir.

Wichita Falls

KBZS(FM)— Nov 15, 1984: 106.3 mhz; 2.4 kw. 423 ft TL: N33 53 18 W98 34 08. (CP: 15.5 kw, ant 899 ft. TL: N33 53 23 W98 33 31). Stereo. Hrs opn: 24 2525 Kell Blvd., Suite 200, 76308. Phone: (940) 763-1111. Fax: (940) 322-3166. Web Site: www.1063thebuzz.com. Licensee: GAP Broadcasting Wichita Falls License LLC. Group owner: Clear Channel Communications Inc. (acq 8-3-2007; grpsl). Natl. Network: Westwood One. Natl. Rep: McGavren Guild. Format: AOR, adult contemp. Target aud: 18-49. ♦George Laughlin, pres; Kim Dodds, gen mgr; Chris Walters, opns mgr & news dir; Melissa Detrick, sls dir; Kara Tucker, prom dir; Liz Ryan, progmg dir; Scott Maingi, chief of engrg.

KLUR(FM)— Apr 14, 1963: 99.9 mhz; 100 kw. Ant 808 ft TL: N33 54 04 W98 32 21. Stereo. Hrs opn: 4302 Callfield Rd., 76308. Phone: (940) 691-2311. Fax: (940) 696-2255. Web Site: www.klur.com. Licensee: Cumulus Licensing Corp. Group owner: Cumulus Media Inc. (acq 10-3-97; grpsl). Population served: 97,564 Format: New country. ♦Jim Marks, gen mgr & natl sls mgr; Lindy Parr, sls dir; Andrea Lewis, rgnl sls mgr; Zach Morton, progmg dir; Jeff Chancey, chief of engrg.

***KMCU(FM)**—Not on air, target date: unknown: 88.7 mhz; 5 w horiz, 3 kw vert. Ant 253 ft TL: N33 56 30 W98 34 06. Stereo. Hrs opn: 24 KCCU,KLW,KOCU,KYCU Lawton,Clinton,Ardmore,Altus, Oaklahoma, 90%.
c/o KCCU(FM), Cameron University, 2800 W. Gore Blvd., Lawton, OK, 73505. Phone: (580) 581-2425. Fax: (580) 581-5571. E-mail: kccu@cameron.edu Web Site: www.cameron.edu/kccu/. Licensee: Cameron University. Natl. Network: NPR, PRI. Format: NPR news-classical & jazz music. News staff: 2; News: 36 hrs wkly. Target aud: 35 plus. ♦Mark Norman, gen mgr; Michael Leal, stn mgr, progmg dir & spec ev coord; Kristin Gordon, opns dir, pub affrs dir & mus critic; Nadia Sikes, adv mgr; Terry Anderson, adv mgr; Debbie Taylor, news dir; Charlie Thurston, chief of engrg.

***KMOC(FM)**— July 9, 1987: 89.5 mhz; 3 kw. 672 ft TL: N33 54 04 W98 32 21. (CP: 10 kw). Hrs opn: Box 41, 76307. Secondary address: 1040 W. Wenonah St. 76307. Phone: (940) 767-3303. Fax: (940) 723-5807. E-mail: kmocfm@wf.net Web Site: www.kmocfm.com. Licensee: Christian Service Foundation Inc. Population served: 250,000 Natl. Network: Moody. Format: Christian. Target aud: 25-54. ♦Daniel Boyd, progmg dir; Keith Sanderson, stn mgr & progmg dir; Della Pool, mus dir; Delvin Kinser, news dir.

KNIN-FM— May 12, 1975: 92.9 mhz; 100 kw. 930 ft TL: N33 54 04 W98 32 21. Stereo. Hrs opn: 24 2525 Kell Blvd., Suite 200, 76308. Phone: (940) 763-1111. Fax: (940) 322-3166. E-mail: info@929nin.com Web Site: www.929nin.com. Licensee: GAP Broadcasting Wichita Falls License LLC. Group owner: Clear Channel Communications Inc. (acq 8-3-2007; grpsl). Population served: 356,600 Natl. Rep: McGavren Guild. Format: CHR. Target aud: 18-49. ♦Kim Dodds, gen mgr; Chris Walters, opns mgr; Melissa Detrick, sls dir; Kara Tucker, prom dir; Liz Ryan, progmg dir; Vicki Vox, asst music dir; Scott Maingi, chief of engrg; Pamela Tracy, traf mgr.

KQXC-FM— Jan 7, 1994: 103.9 mhz; 4.5 kw. Ant 315 ft TL: N33 56 30 W98 34 07. Stereo. Hrs opn: 24 4302 Callfield Rd., 76308. Phone: (940) 691-2311. Fax: (940) 696-2255. Web Site: www.cumulus.com. Licensee: Cumulus Licensing Corp. Group owner: Cumulus Media Inc. (acq 10-3-97; grpsl). Population served: 105,000 Natl. Network: ABC. Format: Rhythmic CHR. Target aud: 18-45; males. ♦Jim Marks, gen mgr, opns mgr & natl sls mgr; Belda Holt, rgnl sls mgr; Susan Adkins, mktg dir & prom dir; Zach Morton, progmg dir; Jeff Chancey, chief of engrg.

KWFS(AM)— 1948: 1290 khz; 5 kw-D, 250 w-N. TL: N33 57 38 W98 33 42. Hrs open: 24 2525 Kell Blvd., Suite 200, 76308. Phone: (940) 763-1111. Fax: (940) 322-3166. Web Site: www.newstalk1290.com. Licensee: GAP Broadcasting Wichita Falls License LLC. Group owner: Clear Channel Communications Inc. (acq 8-3-2007; grpsl). Population served: 100,000 Rgnl. Network: Texas State Net. Natl. Rep: Clear Channel. Format: News/talk. News staff: one:; News: 12 hrs wkly. Target aud: 25 plus. Spec prog: Sp 4 hrs wkly. ♦Chris Walters, pres, opns mgr & progmg dir; Kim Dodds, gen mgr; Melissa Detrick, gen sls mgr; Kara Tucker, prom dir; Scott Maingi, chief of engrg; Joe Tom White, local news ed & farm dir.

KWFS-FM— 1961: 102.3 mhz; 100 kw. Ant 449 ft TL: N33 53 51 W98 32 32. Stereo. 24 E-mail: info@lonestar1023.com Web Site: www.lonestar1023.com.323,000 Format: Country. News: one hr wkly. Target aud: 18-49. ♦Joe Tom White, local news ed & farm dir.

***KZKL(FM)**— Sept 1, 1993: 90.5 mhz; 7 kw. Ant 430 ft TL: N33 53 50 W98 32 33. Hrs open: 24 2351 Sunset Blvd., Suite 170-218, Rocklin, CA, 95765. Phone: (916) 251-1600. Fax: (916) 251-1650. Web Site: www.klove.com. Licensee: Educational Media Foundation. (acq 10-20-2005; $600,000). Population served: 200,000 Natl. Network: K-Love. Format: Contemp Christian. ♦Richard Jenkins, pres; Mike Novak, VP; Keith Whipple, dev dir; David Pierce, progmg mgr; Ed Lenane, news dir; Sam Wallington, engrg dir; Karen Johnson, news rptr; Marya Morgan, news rptr; Richard Hunt, news rptr.

Willis

KVST(FM)— 1998: 99.7 mhz; 2.55 kw. Ant 504 ft TL: N30 26 55 W95 31 48. Hrs open: 1212 S. Frazier, Conroe, 77301. Phone: (936) 788-1035. Fax: (936) 788-2525. Web Site: www.kvst.com. Licensee: New Wavo Communication Group Inc. (acq 7-16-98; $158,218). Format: Country. ♦Ben Amato, pres & gen mgr; William Boggs, gen sls mgr; Larry Galla, progmg mgr; Mike Shilo, news dir; Dade Moore, engrg dir; John Erle, traf mgr; Linda Lott, traf mgr.

Winfield

KALK(FM)— Sept 27, 1987: 97.7 mhz; 22.5 kw. 328 ft TL: N33 11 01 W95 12 32. Stereo. Hrs opn: 24 Box 990, Mount Pleasant, 75456. Secondary address: 1798 US Hwy. 67 W., Mount Pleasant 75455. Phone: (903) 577-9770. Phone: (903) 572-8726. Fax: (903) 572-7232. E-mail: hitmusic@klake.net Web Site: www.easttexasradio.com. Licensee: East Texas Broadcasting Inc. (group owner; acq 1999; $600,000). Population served: 68,000 Format: Hot adult contemp. News staff: one; News: one hr wkly. Target aud: 18-54; younger, upwardly mobile, white collar. Spec prog: Gospel one hr wkly. ♦Bud Kitchens, pres, VP & gen mgr; John Mitchell, chmn & pres; Craig Morgan, opns dir.

Winnie

KKHT-FM— Dec 1, 1987: 100.7 mhz; 100 kw. Ant 1,952 ft TL: N30 03 05 W94 31 37. Stereo. Hrs opn: 24 6161 Savoy, Suite 1200, Houston, 77036. Phone: (713) 260-3600. Fax: (713) 260-3628. E-mail: comments@kkht.com Web Site: www.kkht.com. Licensee: Salem Media of Illinois LLC. Group owner: Univision Radio (acq 1-7-2005; with WIND(AM) Chicago, IL and KNIT(AM) Dallas in exchange for WPPN(FM) Des Plaines, IL). Natl. Rep: Salem. Format: Christian talk. ♦Chuck Jewell, gen mgr; Paul Baker, opns mgr, mktg mgr & progmg dir; Dan Doster, gen sls mgr; Marsha Lambeth, mus dir; Ken Garza, pub affrs dir; Sidney Jones, chief of engrg; Kent McDonald, traf mgr.

Winnsboro

KWNS(FM)— Sept 1, 1983: 104.7 mhz; 2.75 kw. Ant 492 ft TL: N32 56 32 W95 18 53. Stereo. Hrs opn: 24 Box 54, 215 Market St., 75494. Phone: (903) 342-3501. E-mail: kwns-fm@cox-internet.com Licensee: Lottie L. Foster, executor of estate of Richard E. Foster. Format: Southern gospel. News staff: one. Target aud: 35-70. ♦Lottie Foster, pres & gen mgr.

Winona

KBLZ(FM)—Not on air, target date: unknown: 102.7 mhz; 9.3 kw. 531 ft TL: N32 23 09 W95 06 43. Hrs open: Box 11196, College Station, 77842. Phone: (903) 581-5259. Phone: (903) 759-1061. Fax: (903) 939-3473. Web Site: www.theblaze.cc. Licensee: S.O. 2,000 LLC. Group owner: Reynolds Radio Inc. (acq 8-26-99). Format: Urban Contemporary. ♦Rick Reynolds, gen mgr.

Winters

KNCE(FM)— Nov 1, 1981: 96.1 mhz; 50 kw. Ant 492 ft TL: N32 12 52 W99 53 22. Stereo. Hrs opn: 24 1740 N. 1st St., Abilene, 79603. Phone: (325) 437-9596. Fax: (325) 673-1819. Licensee: Doud Media Group LLC Population served: 425,000 Law Firm: Dennis Kelly.

Format: Country. Target aud: 25-54; women. ♦Richard Doud, gen mgr; Mark McGill, opns dir; Brad Whitaker, gen sls mgr; James Thompson, engrg dir.

Wolfforth

KAIQ(FM)— 2000: 95.5 mhz; 100 kw. Ant 676 ft TL: N33 31 03 W101 51 24. Hrs open: 24 1220 Broadway, Suite 500, Lubbock, 79401. Phone: (806) 763-6051. Fax: (806) 744-8363. Licensee: Entravision Holdings LLC. (acq 2-10-2005; $1.5 million). Natl. Rep: Lotus Entravision Reps LLC. Format: Spanish CHR. Target aud: Hispanic. ♦Jose Sauceda, gen mgr.

Woodville

KWUD(AM)— Jan 4, 1968: 1490 khz; 1 kw-U. TL: N30 44 52 W94 25 56. Hrs open: 24 Box 129, 75979. Secondary address: 105 E Wheat 75979. Phone: (409) 283-2777. Fax: (409) 283-2283. E-mail: kwud@cmaaccess.com Licensee: Carroll Texas Broadcasting Ltd. Group owner: Jimmy Ray Carroll Stns (acq 10-5-01). Population served: 22,000 Natl. Network: ABC, Jones Radio Networks. Rgnl. Network: Texas State Net. Format: Oldies, country, other. News staff: one; News: 12 hrs wkly. Target aud: 18-55. Spec prog: Farm 2 hrs, gospel 6 hrs, relg 3 hrs wkly. ♦Carol Carroll, gen sls mgr; Traci Authement, gen mgr, gen sls mgr & progmg dir; Chester Leediker, chief of engrg.

Wylie

KHSE(AM)— 2004: 700 khz; 250 w-U, DA-2. TL: N33 01 58 W96 17 56. Hrs open: 24 Box 12345, Dallas, 75225. Phone: (214) 369-2990. Web Site: www.humtumlive.com. Licensee: Bernard Dallas LLC (acq 12-28-2006; $9 million with KFCD(AM) Farmersville). Population served: 4,700,695 Format: Ethnic, Indian, Pakistani. ♦Josiah M. Daniel III, gen mgr.

Yoakum

KYKM(FM)— January 1982: 92.5 mhz; 3 kw. 300 ft TL: N29 21 03 W97 11 32. Stereo. Hrs opn: 24 111 N. Main St., Halletsville, 77964. Phone: (361) 798-4333. Fax: (361) 798-3798. E-mail: texasthunderradio@yahoo.com Licensee: Fort Bend Broadcasting Co. Inc. Group owner: Fort Bend Broadcasting Co. (acq 1-21-00; grpsl). Population served: 75,000 Format: Country. News staff: one; News: 12 hrs wkly. Target aud: General. Spec prog: Polka 9 hrs wkly. ♦Laura Kremling, gen mgr, opns mgr, dev mgr, progmg dir & traf mgr; Ray Nelson, chief of engrg; Bobby Pauliska, local news ed, sports cmtr & disc jockey; Dutch Schorre, disc jockey.

Yorktown

KGGB(FM)—Not on air, target date: unknown: 96.3 mhz; 6 kw. Ant 328 ft TL: N29 02 43 W97 24 23. Hrs open: 11737 Nelon Dr., Corpus Christi, 78410. Phone: (361) 241-7944. Fax: (361) 241-7945. Licensee: Gerald Benavides. ♦Gerald Benavides, gen mgr.

Zapata

KBAW(FM)—Not on air, target date: unknown: 93.5 mhz; 25 kw. 328 ft TL: N26 54 43 W99 17 09. Hrs open: 2702 Pine St., Laredo, 78043. Phone: (956) 726-4738. E-mail: radiooluz@border.net Web Site: www.lacadenaradioluz.com/kbaw-93.htm. Licensee: La Nueva Cadena Radio Luz Inc. Format: Sp, Christian. ♦Isreal Tellez, stn mgr.

Utah

Blanding

KBDX(FM)—Not on air, target date: unknown: 92.7 mhz; 594 w horiz, 255 w vert. Ant 3,405 ft TL: N37 50 24 W109 27 41. Hrs open: 8am-5pm 74-5605 Luhia St. B7, Kailua-Kona, HI, 96740. Phone: (808) 329-8090. Fax: (808) 443-0888. E-mail: info@lava105.com Web Site: www.lava105.com. Licensee: Skynet, Hawaii LLC (acq 12-10-03; $300,000). Format: Oldies. ♦Joe Williams, gen mgr; Chip Begay, opns mgr.

Bountiful

KJMY(FM)— Mar 15, 1988: 99.5 mhz; 39 kw. Ant 2,952 ft TL: N40 36 29 W112 09 33. Hrs open: 24 2801 S. Decker Lake Dr., Salt Lake City, 84119. Phone: (801) 908-1300. Fax: (801) 908-1449. Web Site: www.my995fm.com. Licensee: Citicasters Licenses L.P. Group owner: Clear Channel Communications Inc. (acq 1999; grpsl). Population served: 1,000,400 Natl. Rep: Clear Channel, Katz Radio. Law Firm: Hogan & Hartson. Format: Modern alternative, retro classics. Target aud: 18-49; adults. ♦Stu Stanek, gen mgr; Bill Betts, opns mgr; Bill Matthews, sls dir & prom dir; Emily Hunt, gen sls mgr; Mark Christiansen, progmg dir.

Brian Head

KREC(FM)— Nov 14, 1988: 98.1 mhz; 56 kw. 2,526 ft TL: N37 32 32 W113 04 05. Stereo. Hrs opn: 24 750 W. Ridgeview Dr., Suite 204, Saint George, 84770. Phone: (435) 673-3579. Fax: (435) 673-8900. E-mail: star98fm@bonnevillesg.com Licensee: CCR-St. George IV LLC. Group owner: Bonneville International Corp. (acq 8-10-2006; grpsl). Population served: 112,000 Format: Soft adult contemp. Target aud: 25-54. ♦Don Shelline, gen mgr; Kevin Fry, gen sls mgr; Gary Smith, chief of engrg.

Brigham City

KEGH(FM)— Oct 20, 1972: 106.9 mhz; 81 kw. Ant 2,165 ft TL: N41 47 03 W112 13 55. Stereo. Hrs opn: 2801 S. Decker Lake Dr., Salt Lake City, 84119. Phone: (801) 908-4100. Fax: (801) 908-4122. Licensee: Simmons-SLC, LS LLC. Group owner: Simmons Media Group (acq 4-19-2004; $3.95 million). Format: Hip hop, rhythm and blues.

KXOL(AM)— July 1, 1998: 1660 khz; 10 kw-D, 1 kw-N. TL: N41 18 54 W112 04 43. Hrs open: 24 515 S. 700 E., Salt Lake City, 84102. Phone: (801) 325-3126. Fax: (801) 731-9666. Licensee: Simmons-SLC, LS, LLC. Group owner: Simmons Media Group (acq 4-1-2003; $925,000 with KSOS(AM) Brigham City). Format: Oldies. ♦Bev Snyder, opns mgr.

Cedar City

KCIN(FM)— May 10, 1974: 94.9 mhz; 55 kw. Ant -121 ft TL: N37 45 51 W113 06 15. Stereo. Hrs opn: 750 W. Ridgeview Dr., Suite 204, Saint George, 84770. Phone: (435) 673-3579. Fax: (435) 673-8900. Licensee: CCR-St. George IV LLC. (acq 5-3-2006; grpsl). Format: CHR. ♦Steve Hess, gen mgr.

KNNZ(AM)— 1971: Stn currently dark. 940 khz; 10 kw-D. TL: N37 45 51 W113 06 15. Hrs open: 4600 E. Oxford Pl., Englewood, CO, 80113. Phone: (303) 781-5101. Web Site: www.publicradiocapital.org. Licensee: PRC St. George-I LLC (group owner; (acq 6-30-2006; $125,000). Population served: 9,595 ♦Marc Hand, pres, gen mgr & gen sls mgr.

KSUB(AM)— July 4, 1937: 590 khz; 5 kw-D, 1 kw-N, DA-N. TL: N37 41 55 W113 10 44. Hrs open: 24 251 W. Hilton Dr., Saint George, 84770. Phone: (435) 586-5900. Fax: (435) 673-8228. Web Site: www.590ksub.com. Licensee: CCR-St. George IV LLC. (group owner; (acq 5-3-2006; grpsl). Population served: 20,000 Natl. Network: CBS. Rgnl rep: Target Radio. Format: News/talk, info. News staff: one; News: 15 hrs wkly. Target aud: 35-65; adults. Spec prog: Relg, loc talk, farm 6 hrs wkly. ♦Brent Miner, gen mgr; Steve Miner, progmg dir; Dan Hobson, chief of engrg.

KXBN(FM)—Co-owned with KSUB(AM). Oct 15, 1976: 92.5 mhz; 41.6 kw. Ant 1,690 ft TL: N37 38 41 W113 22 28. Stereo. 24 50,000 Natl. Network: Jones Radio Networks. Format: Oldies. News: 5 hrs wkly.

***KSUU(FM)**— October 1966: 91.1 mhz; 10 kw. -462 ft TL: N37 38 55 W113 05 32. Stereo. Hrs opn: 6 AM-midnight (winter); 10 AM-10 PM (summer) 351 W. Ctr., 84720. Phone: (435) 865-8224. Fax: (435) 865-8352. E-mail: ksuu@suu.edu Web Site: www.suu.edu/ksuu. Licensee: Southern Utah University. Population served: 18,000 Format: CHR. News: 2 hrs wkly. Target aud: 12-34; children, university students. Spec prog: News, class 3 hrs, rhythm and blues 4 hrs, rock 4 hrs wkly. ♦Cal Rollins, stn mgr; Alex May, progmg dir; Alisia Brooks, mus dir; Camie Stables, news dir; Lance Jackson, chief of engrg.

Centerville

KXRV(FM)— Dec 24, 1979: 105.7 mhz; 25 kw. Ant 3,739 ft TL: N40 39 34 W112 12 05. Stereo. Hrs opn: 24 2801 S. Decker Lake Dr., Salt Lake City, 84119. Phone: (801) 908-1300. Fax: (801) 908-1569. Web Site: www.river1057.com. Licensee: Citicasters Licenses L.P. Group owner: Clear Channel Communications Inc. (acq 2-27-2004; $22 million with KOSY-FM Spanish Fork). Population served: 175,885 Natl. Rep: Clear Channel, Katz Radio. Format: Adult rock. Target aud: 25-54; adults. ♦Stu Stanek, gen mgr; Bill Betts, opns mgr; Bill Matthews, sls dir & chief of engrg; Kimberly Dickerson, gen sls mgr; Frank Bell, progmg dir.

KXTA(AM)— Dec 1, 1957: 1600 khz; 5 kw-D, 1 kw-N, DA-N. TL: N40 54 08 W111 55 40. Stereo. Hrs opn: 24 2722 S. Redwood Rd., Suite 1, Salt Lake City, 84119. Phone: (801) 908-8777. Fax: (801) 908-8782. Licensee: Bustos Media of Utah License LLC. Group owner: Bustos Media Holdings (acq 9-1-2004; $1.5 million). Format: Sp. ♦Jose Tovar, gen mgr.

Coalville

KJQN(FM)— 2004: 103.1 mhz; 89 kw horiz. Ant 2,122 ft TL: N40 52 16 W110 59 43. Hrs open: Simmons Media Group, 515 South 700 East, Salt Lake City, 84102. Phone: (801) 524-2600. Fax: (801) 524-6002. Web Site: www.simmonsmedia.com. Licensee: Simmons-SLC, LS LLC. Group owner: Simmons Media Group (acq 5-20-2004; $4.4 million for CP). Format: Alternative. ♦G. Craig Hanson, gen mgr.

KOAY(FM)— Sept 6, 2005: 97.5 mhz; 89 kw horiz. Ant 2,122 ft TL: N40 52 16 W110 59 43. Hrs open: 515 S. 700 E., Suite 1C, Salt Lake City, 84102. Phone: (801) 524-2600. Fax: (801) 521-9234. Web Site: www.975oasis.com. Licensee: 3 Point Media - Franklin LLC. (acq 11-28-2001; $1.5 million for CP). Format: Contemp Christian. ♦Steve Johnson, gen mgr.

Delta

KMGR(FM)— Sept 5, 1989: 95.9 mhz; 100 kw horiz. Ant 961 ft TL: N39 43 58 W111 56 34. Stereo. Hrs opn: 24 3 Point Media - Delta LLC, 980 N. Michigan Ave., Suite 1880, Chicago, IL, 60611. Phone: (312) 204-9900. Licensee: 3 Point Media - Delta LLC. (acq 8-1-2003; $1.25 million). ♦Bruce Buzil, gen mgr.

KNAK(AM)— Feb 25, 1974: 540 khz; 1 kw-U. TL: N39 20 12 W112 33 21. Hrs open: 24 Box 636, 84624-0626. Secondary address: 1259 N. 100 W., American Fork 84003. Phone: (435) 864-5111. Fax: (801) 406-0067. Web Site: www.radioforthefamily.com. Licensee: Accent Radio Inc. (acq 3-16-2006; $185,000). Population served: 50,000 Format: Relg. News: 15 hrs wkly. Target aud: 20-50. Spec prog: Farm 5 hrs wkly. ♦Jedidiah Harrison, pres; Curt Crosby, gen sls mgr, news dir & farm dir; Sam Bushman, gen mgr, stn mgr & progmg dir; Julie Bushman, traf mgr.

Elsinore

KCYQ(FM)— 1978: 97.7 mhz; 43 kw. Ant 2,883 ft TL: N38 32 30 W112 03 31. Stereo. Hrs opn: Box 40, Manti, 84642. Phone: (435) 896-4456. Fax: (435) 896-9333. Web Site: www.kcyq.com. Licensee: Mid-Utah Radio Inc. (acq 3-1-2006; swap for KLGL(FM) Richfield). Format: Country hits. ♦Marianne Barton, pres; Douglas Barton, gen mgr; Dave Gunderson, sls VP; J.D. Fox, progmg VP & mus dir; Kirk Williams, engrg VP.

Ephraim

***KAGJ(FM)**—Not on air, target date: unknown: 89.5 mhz; 100 w. -321 ft TL: N39 21 37 W111 34 54. Hrs open: Snow College, 150 E. College Ave., 84627. Phone: (435) 283-7425/7000. E-mail: kagj_fm@hotmail.com Web Site: www.snow.edu/~kage. Licensee: Snow College. Format: Classic rock with a kick. ♦Gary Chidester, gen mgr.

Garland

KYLZ(FM)—See Tremonton

Hurricane

KURR(FM)—Not on air, target date: unknown: 103.1 mhz; 93 kw. Ant 2,034 ft TL: N36 50 49 W113 29 28. Hrs open: 515 South 700 East, #1C, Salt Lake City, 84102. Phone: (801) 524-2600. Fax: (801) 524-6002. Licensee: Western Broadcasting, LS LLC. ♦Bret Leifson, gen mgr.

Kanab

KPLD(FM)— 1986: 101.1 mhz; 99 kw. 786 ft TL: N36 43 18 W112 12 57. Stereo. Hrs opn: 24 204 Playa Della Rosita, Washington, 84780. Phone: (435) 628-3643. Fax: (435) 673-1210. E-mail: kony@infowest.com Licensee: Marathon Media Group L.L.C. (acq 1999; $1.75 million with KUNF(AM) Washington). Format: Hot adult contemp. ♦Carl Lamar, VP & gen mgr.

Levan

KQMB(FM)— 2001: 96.7 mhz; 67 kw horiz. Ant 1,919 ft TL: N39 20 12 W111 27 06. Hrs open: 1454 W. Business Park Dr., Orem, 84058. Phone: (801) 224-1400. Fax: (801) 224-1524. Licensee: Zeta Holdings LLC. Format: Hot adult contemp. ♦Robert H. Morey, gen mgr.

Logan

KBLQ-FM—Listing follows KLGN(AM).

KGNT(FM)—See Smithfield

KLGN(AM)— March 1968: 1390 khz; 5 kw-D, 500 w-N, DA-N. TL: N41 44 04 W111 51 13. Hrs open: Box 3369, 84323. Secondary address: 810 W. 200 N. 84321. Phone: (435) 752-1390. Fax: (435) 752-1392. Web Site: www.1390.com. Licensee: Sun Valley Radio Inc. (group owner; acq 12-27-91; $572,279 with co-located FM). Population served: 140,000 Natl. Network: Westwood One, CBS. Format: Adult standards, memories MOR. Target aud: 45 plus. Spec prog: Talk. ♦Kent Frandsen, pres; Jay Eubanks, gen mgr, gen sls mgr, mktg mgr & prom mgr; Michael Carver, opns mgr & progmg dir; Dan Baker, chief of engrg.

KBLQ-FM—Co-owned with KLGN(AM). August 1977: 92.9 mhz; 50 kw. 154 ft TL: N41 52 18 W111 48 31. Stereo. Web Site: www.q92.fm. Format: Adult contemp. Target aud: 25-54; general. Spec prog: Gospel 8 hrs, jazz 4 hrs wkly. ♦Laurie Gill, traf mgr; Bill Walter, disc jockey; Michael Steel, disc jockey; Mindy Carey, disc jockey.

***KUSR(FM)**— March 1999: 89.5 mhz; 800 w. -617 ft TL: N41 44 44 W111 48 16. Hrs open: 24 hours Utah Public Radio, 8505 Old Main Hill, 84322-8505. Phone: (435) 797-3138. Fax: (435)797-3150. E-mail: upr@upr.usu.edu Web Site: www.upr.org. Licensee: Utah State University of Agricultural and Applied Science. Natl. Network: NPR. Format: Classical, News/Talk. ♦Richard Meng, gen mgr.

***KUSU-FM**— April 1953: 91.5 mhz; 90 kw. 1,140 ft TL: N41 53 11 W112 04 17. Stereo. Hrs opn: 24 Utah Public Radio, 8505 Old Main Hill, 84322-8505. Phone: (435) 797-3138. Phone: (800) 826-1495. Fax: (435) 797-3150. E-mail: upr@upr.usu.edu Web Site: www.upr.org. Licensee: Utah State University. Population served: 250,000 Natl. Network: NPR, PRI. Law Firm: Dow, Lohnes & Albertson. Format: Class, news/talk. News staff: 2. Target aud: General. ♦Richard Meng, gen mgr; Lee Austin, progmg dir; Nora Zambreno, pub affrs dir; Clifford J. Smith, chief of engrg; Craig Hislop, reporter.

KVFX(FM)—Listing follows KVNU(AM).

KVNU(AM)— Nov 20, 1938: 610 khz; 5 kw-D, 1 kw-N, DA-N. TL: N41 40 30 W111 56 06. Hrs open: 24 Box 267, 84323-0267. Secondary address: 810 W. 200 N. 84321. Phone: (435) 752-5141. Fax: (435) 753-5555. E-mail: kvnu@cvradio.com Web Site: 610kvnu.com. Licensee: Sun Valley Radio Inc. (group owner; acq 1996; $900,000 with co-located FM). Population served: 100,000 Natl. Network: ABC. Format: News/talk. News staff: 2; News: 15 hrs wkly. Target aud: General. Spec prog: Farm 2 hrs, relg 2 hrs wkly. ♦Al Lewis, gen mgr, progmg dir, outdoor ed & sports cmtr; James Murdock, gen sls mgr; Bill Walter, chief of engrg; Jennie Christensen, news dir & local news ed; Eric Frandsen, news rptr; Heather Bailey, reporter.

KVFX(FM)—Co-owned with KVNU(AM). Nov 11, 1974: 94.5 mhz; 70 kw. 1,148 ft TL: N41 53 50 W111 57 39. Stereo. 24 85,000 Format: CHR. News: 2 hrs wkly. Target aud: 18-35. ♦Blair Carter, progmg dir & progmg mgr; Kenton Frat Boy, disc jockey.

***KZCL(FM)**—Not on air, target date: unknown: 90.5 mhz; 300 w. Ant 1,246 ft TL: N41 53 43 W112 04 43. Hrs open: 1971 West North Temple, Salt Lake City, 84116. Phone: (801) 363-1818. Fax: (801) 533-9136. Licensee: Listeners Community Radio of Utah Inc. ♦Donna Land Maldonado, gen mgr.

Manti

KAUU(FM)— December 1978: 105.1 mhz; 48 kw horiz. Ant 2,244 ft TL: N39 45 37 W111 34 38. Stereo. Hrs opn: 24 hours 515 S. 700 E., Suite 1C, Salt Lake City, 84102. Phone: (801) 524-2600. Fax: (801) 521-9234. Licensee: Millcreek Broadcasting L.L.C. (group owner; (acq 4-17-2001). Format: Alternative. ♦Douglas Barton, gen mgr; Sam Penrod, mus dir.

KMTI(AM)— June 7, 1976: 650 khz; 10 kw-D, 1 kw-N, DA-2. TL: N39 17 39 W111 38 13. Hrs open: 24 Box 40, 1600 W. 500 N., 84642. Phone: (435) 835-7301. Fax: (435) 835-2250. Web Site: www.kmtiradio.com. Licensee: Sanpete County Broadcasting Co. Population served: 100,000 Law Firm: Rosenman & Colin. Format: Country, news, full service. News staff: one; News: 20 hrs wkly. Target aud: 25-50. Spec prog: Farm 5 hrs wkly. ♦Douglas Barton, pres & gen mgr; Dave Gunderson, gen sls mgr; Larry Masco, progmg dir; Mike Traina, news dir; Kirk Williams, chief of engrg.

Midvale

KSL-FM— 1995: 102.7 mhz; 25 kw. Ant 3,739 ft TL: N40 39 34 W112 12 05. Stereo. Hrs opn: 24
Rebroadcasts KSL(AM) Salt Lake City 100%.
55 North 300 West, Salt Lake City, 84180. Phone: (801) 575-5555. Fax: (801) 526-1070. Web Site: www.ksl.com. Licensee: Bonneville Holding Co. Group owner: Bonneville International Corp. (acq 12-5-2003; grpsl). Format: News/talk, sports. ♦Chris Redgrave, gen mgr.

Moab

KCYN(FM)— Sept 20, 1998: 97.1 mhz; 29 kw. Ant 1,292 ft TL: N38 31 37 W109 18 21. Stereo. Hrs opn: 24 Box 1119, 84532. Secondary address: 1030 S. Bowling Alley Ln. #3 84532. Phone: (435) 259-1035. Fax: (435) 259-1037. E-mail: kcyn@precisiom.net Web Site: www.kcynfm.com. Licensee: Moab Communications LLC. (acq 8-15-97). Wire Svc: Metro Weather Service Inc. Format: Country. News staff: one; News: 12 hrs wkly. Target aud: 18-54. ♦Phillip Mueller, gen mgr & gen sls mgr; Kenneth Meyer, chief of engrg; Christina Backes, traf mgr.

***KZMU(FM)**— April 1992: 90.1 mhz; 400 w. Ant 1,279 ft TL: N38 31 37 W109 18 21. Stereo. Hrs opn: 24 Box 1076, 84532. Secondary address: 1734 Rocky Rd. 84532. Phone: (435) 259-5968. Phone: (435) 259-8824. Fax: (435) 259-8763. E-mail: kzmu@citylink.net Web Site: www.kzmu.org. Licensee: Moab Public Radio. Population served: 10,000 Format: Var/div, public radio. News staff: one; News: 8 hrs wkly. Target aud: General. Spec prog: Asian one hr, American Indian 5 hrs, Black 3 hrs, Sp one hr, folk 6 hrs, blues 19 hrs wkly. ♦Jeff Flanders, gen mgr; Christy Williams, progmg dir; Glen Peart, mus dir & asst music dir; Bob Owen, engrg dir.

Monroe

KMXD(FM)— 8/01/07: 100.5 mhz; 33 kw. Ant 3,257 ft TL: N38 23 08 W112 19 57. Hrs open: 24 Box 40, Manti, 84642. Phone: (435) 835-7301. Fax: (435) 835-2250. Licensee: Sanpete County Broadcasting Co. Format: Soft Adult Contemp. ♦Douglas L. Barton, pres & gen mgr.

Monticello

KRZX(FM)—Not on air, target date: unknown: 106.1 mhz; 100 kw. Ant 1,378 ft TL: N38 31 36 W109 18 26. Hrs open: Box 36148, Tucson, AZ, 85740. Phone: (520) 797-4434. Licensee: Skywest Media L.L.C. ♦Ted Tucker, gen mgr.

Murray

KJQS(AM)— Nov 8, 1948: 1230 khz; 1 kw-U. TL: N40 39 57 W111 54 26. Stereo. Hrs opn: 24 434 Bearcat Dr., Salt Lake City, 84115. Phone: (801) 485-6700. Fax: (801) 487-5369. Licensee: Citadel Broadcasting Co. Group owner: Citadel Broadcasting Corp. (acq 2-29-00; $104,202). Format: Sports. ♦Eric Hauenstein, gen mgr; Terry Mathis, gen sls mgr; Kelly Hammer, progmg dir; Richard Bauer, chief of engrg; Liz Mills, traf mgr.

Naples

KCUA(FM)— 1993: 92.5 mhz; 840 w. Ant 1,660 ft TL: N40 32 16 W109 41 57. Hrs open: Box 1372, Park City, 84060. Phone: (801) 412-6080. Fax: (435) 645-0963. Licensee: 3 Point Media - Coalville LLC. (acq 5-28-2004; $1.7 million). Format: Classic Rock. ♦Joe Evans, stn mgr.

Nephi

KUDE(FM)— May 9, 1990: 103.9 mhz; 74 kw horiz. Ant 2,244 ft TL: N39 45 37 W111 34 38. Stereo. Hrs opn: 24
Rebroadcasts KUDD(FM) Roy 100%.
2835 East 3300 South, Salt Lake City, 84109. Phone: (801) 412-6040. Fax: (801) 412-6041. Licensee: Millcreek Broadcasting L.L.C. (group owner; (acq 4-17-2001). Population served: 244,000 Format: Adult contemp. Target aud: 18-45. ♦Randy Rodgers, gen mgr; Brian Michel, opns mgr; Lutisha Merrill, gen sls mgr; Scott St. John, mktg mgr; Kevin Terry, engrg VP.

North Ogden

***KNKL(FM)**— Jan 29, 2004: 88.7 mhz; 7.3 kw vert. Ant 984 ft TL: N41 35 30 W112 14 57. Hrs open: 24 2351 Sunset Blvd., Suite 170-218, Rocklin, CA, 95765. Phone: (916) 251-1600. Fax: (916) 251-1650. E-mail: klove@klove.com Web Site: www.klove.com. Licensee: Educational Media Foundation. Group owner: EMF Broadcasting. Population served: 449,000 Natl. Network: K-Love. Law Firm: Shaw Pittman. Format: Contemp Christian. News staff: 3. Target aud: 25-44; female-Judeo Christian.Richard Jenkins, pres; Mike Novak, VP & progmg dir; Lloyd Parker, gen mgr; Ed Lenane, opns dir & news dir; Keith Whipple, dev dir; Eric Allen, natl sls mgr; Dan Beck, rgnl sls mgr; Chris Joyce, prom dir; David Pierce, progmg mgr; Sam Wallington, engrg dir; Arthur Vassar, traf mgr; Karen Johnson, news rptr; Marya Morgan, news rptr; Richard Hunt, news rptr

North Salt Lake City

KALL(AM)— Sept 22, 1981: 700 khz; 50 kw-D, 1 kw-N, DA-2. TL: N40 53 29 W111 56 28 (D), N40 53 32 W111 56 28 (N). Stereo. Hrs opn: 24 2801 S. Decker Lake Dr., Salt Lake City, 84119-2330. Phone: (801) 908-1300. Fax: (801) 908-1459. Web Site: www.hotticket700.com. Licensee: Utah Radio Acquisition LLC Group owner: Clear Channel Communications Inc. (acq 3-17-2006; $4.1 million). Population served: 2,000,000 Natl. Rep: Clear Channel. Format: Sports/talk. News staff: 1; News: 20 hrs wkly. Target aud: 18-49; men. ♦Stu Stanek, gen mgr; Bill Betts, opns mgr; Bill Mathews, sls dir; Steve Pearson, natl sls mgr; Jason Wilmot, progmg dir.

Oakley

KEGA(FM)— 2003: 101.5 mhz; 89 kw horiz, 38 kw vert. Ant 2,122 ft TL: N40 52 16 W110 59 43. Hrs open: Simmons Media Group, 515 South 700 E. #1C, Salt Lake City, 84102. Phone: (801) 524-2600. Fax: (801) 521-8100. Web Site: www.1015theeagle.com. Licensee: Simmons-SLC, LS LLC. Group owner: Simmons Media Group (acq 4-4-2001; grpsl). Population served: 10,000 Format: Country. ♦Craig Hanson, pres; Stephen Johnson, gen mgr.

Ogden

KBER(FM)— July 13, 1976: 101.1 mhz; 25 kw. 3,740 ft TL: N40 39 35 W112 12 05. Stereo. Hrs opn: 24 434 Bearcat Dr., Salt Lake City, 84115. Phone: (801) 485-6700. Fax: (801) 487-5369. Web Site: www.kber.com. Licensee: Citadel Broadcasting Co. Group owner:

Citadel Broadcasting Corp. (acq 1996; $7.7 million). Natl. Rep: Katz Radio. Format: AOR. News staff: one. Target aud: 18-49; men. ♦Eric Hauenstein, gen mgr; Zandi Wilcox, gen sls mgr; Diane Curtis, natl sls mgr; Joel Smith, prom dir & mus dir; Kelly Hamer, progmg dir; Richie Bauer, engrg dir.

KBZN(FM)— 1978: 97.9 mhz; 26 kw. 3,770 ft TL: N40 39 35 W112 12 05. Hrs open: 257 E. 200 S., Suite 400, Salt Lake City, 84111. Phone: (801) 364-9836. Fax: (801) 364-8068. E-mail: breeze@kbzn.com Web Site: www.kbzn.com. Licensee: Capitol Broadcasting Inc. (acq 4-5-91; FTR: 4-29-91). Format: Smooth jazz, new age. ♦John Webb, gen mgr; Cris Winn, gen sls mgr; John Sichler, disc jockey; Rob Riesen, opns dir & disc jockey.

KENZ(FM)— Aug 1, 1964: 101.9 mhz; 25 kw. Ant 3,739 ft TL: N40 39 34 W112 12 05. Stereo. Hrs opn: 24 2835 E. 3300 S., Suite 800, Salt Lake City, 84107. Phone: (801) 412-6040. Fax: (801) 412-6041. Web Site: www.1019popfm.com. Licensee: Citadel Broadcasting Co. Group owner: Citadel Broadcasting Corp. (acq 7-30-2004; $16 million). Population served: 1,000,400 Natl. Network: ABC. Format: Modern country. News staff: one; News: 6 hrs wkly. Target aud: 25-54. ♦Randy Rodgers, gen mgr.

KLO(AM)— 1924: 1430 khz; 5 kw-U, DA-N. TL: N41 10 44 W112 04 09. (CP: TL: N41 02 48 W112 01 38). Hrs opn: 4155 Harrison Blvd., Suite 206, 84003-2463. Phone: (801) 627-1430. Fax: (801) 627-0317. Web Site: www.kloradio.com. Licensee: KLO Broadcasting Co. Population served: 1,159,700 Format: Talk. ♦John Webb, pres & gen mgr; Dan Jessop, opns mgr & progmg dir; Jan Bagley, gen sls mgr; Patrick Gleason, chief of engrg; Arlene Harris, traf mgr.

KOGN(AM)— April 1948: 1490 khz; 1 kw-U. TL: N41 14 23 W111 58 58. Hrs open: 24 1506 Gibson Ave., 84404. Phone: (801) 395-5600. Fax: (801) 395-1490. Licensee: AM Radio 1490 Inc. (acq 4-10-2006; $520,000). Population served: 69,478 Natl. Network: CNN Radio, Westwood One. Law Firm: Dan J. Alpert. Format: Adult standard, CNN radio news. ♦E. Morgan Skinner Jr., CEO & pres; Dick Carter, stn mgr.

KSVN(AM)— Jan 1, 1946: 730 khz; 1 kw-D, 66 w-N. TL: N41 11 17 W112 04 52. Hrs open: 24 4215 W. 4000 S., West Haven, 84401. Phone: (801) 292-1799. Fax: (801) 731-4445. Licensee: Azteca Broadcasting Corp. (group owner; acq 2-1-86). Population served: 1,200,000 Format: Rgnl Mexican. ♦Alex Collantes, pres, gen mgr & progmg dir; Maria Coria, gen sls mgr & traf mgr.

***KWCR-FM**— May 21, 1966: 88.1 mhz; 3 kw. -470 ft TL: N41 11 30 W111 56 37. (CP: Ant 315 ft.). Stereo. Hrs opn: 24 2188 University Cir., 84408-2188. Phone: (801) 626-8800. Fax: (801) 626-6935. E-mail: kwcrradio@mail.weber.edu Web Site: www.weber.edu/kwcr. Licensee: Weber State University Board of Trustees. Population served: 200,000 Wire Svc: UPI Format: Contemp hits, rock. News: 4 hrs wkly. Target aud: 18-26; college students, male & female. Spec prog: Relg 3 hrs, gospel 3 hrs, Sp 16 hrs wkly. ♦Mark Howard, gen mgr & sls dir.

***KYFO-FM**— June 1983: 95.5 mhz; 100 kw. Ant 718 ft TL: N41 14 59 W112 14 11. Stereo. Hrs opn: 24 11530 Carmel Commons Blvd., Charlotte, NC, 28226. Phone: (801) 394-8833. Phone: (801) 773-5858. Web Site: www.bbnradio.org. Licensee: Bible Broadcasting Network. (acq 1994). Population served: 1,000,000 Format: Christian. ♦Lowell Davey, pres; Tom Gearhart, gen mgr.

Orem

KKAT-FM— Nov 15, 1978: 107.5 mhz; 45 kw. Ant 2,850 ft TL: N40 16 48 W111 56 05. Stereo. Hrs opn: 24 434 Bearcat Dr., Salt Lake City, 84115-2520. Phone: (801) 485-6700. Fax: (801) 487-5369. Web Site: www.1075.com. Licensee: Citadel Broadcasting Co. Group owner: Citadel Broadcasting Corp. (acq 12-18-96). Format: Country. Target aud: 25-54. ♦Eric Hauenstein, gen mgr; Diane Curtis, adv mgr; Bruce Jones, progmg dir; Kurt Johnson, prom.

***KOHS(FM)**— October 1994: 91.7 mhz; 1.75 kw. -831 ft TL: N40 17 48 W111 41 04. (CP: Ant -869 ft. TL: N40 17 32 W111 40 56). Stereo. Hrs opn: 175 S. 400 E., 84058. Phone: (801) 224-9236. Fax: (801) 538-5690. Licensee: Orem HI. Sch. Population served: 350,000 Format: Alternative.

KSRR(AM)—See Provo

Park City

***KPCW(FM)**— July 2, 1980: 91.9 mhz; 105 w. Ant -23 ft TL: N40 40 59 W111 31 22. Stereo. Hrs opn: Box 1372, 84060. Secondary address: KPCW City Hall Bldg., 445 Marsac 84060. Phone: (435) 649-9004. Fax: (435) 645-9063. E-mail: letters@kpcw.org Web Site: www.kpcw.org. Licensee: Community Wireless of Park City. Population served: 13,000 Format: AAA, news. Spec prog: Class 17 hrs, C&W 18 hrs, jazz 12 hrs wkly. ♦Blair Feulner, gen mgr; Karen Thomas, progmg dir; Leslie Thatcher, news dir; Dennis Silver, chief of engrg.

Parowan

KENT(AM)— September 2003: 1400 khz; 1 kw-U. TL: N37 48 22 W112 56 40. Hrs open: Box 1450, St. George, 84771-1450. Secondary address: 210 North 1000 East, St. George 84770. Phone: (435) 477-2000. Fax: (435) 477-1400. Licensee: AM Radio 1400 Inc. Group owner: Diamond Broadcasting Corp. (acq 11-16-2004). Natl. Network: CNN Radio, Westwood One. Law Firm: Dan J. Alpert. Format: Adult standards.

Payson

KTCE(FM)— November 1993: 92.1 mhz; 125 w. Ant 2,155 ft TL: N40 05 21 W111 49 15. Hrs open: 2835 E. 3300 S., Salt Lake City, 84603. Phone: (801) 412-6040. Fax: (801) 412-6041. Licensee: Moenkopi Communications Inc. (acq 9-28-2005). Format: Hot adult contemp.

Pleasant Grove

***KPGR(FM)**— May 1976: 88.1 mhz; 115 w. -1,128 ft TL: N40 21 48 W111 43 30. Stereo. Hrs opn: 6:30 AM-10 PM 700 E. 200 S., 84062. Phone: (801) 785-5747. Phone: (801) 785-8700. Fax: (801) 785-8744. Web Site: www.kpgr.tripod.com. Licensee: Alpine School District. Population served: 25,000 Format: Var. Target aud: 12-18; students. Spec prog: All Pleasant Grove High football, basketball, baseball games 4 hrs wkly.

Price

KARB(FM)— July 1977: 98.3 mhz; 7 kw. Ant -105 ft TL: N39 36 33 W110 48 50. Stereo. Hrs opn: Box 875, 84501. Secondary address: 1899 North Carbonville Rd. 84501. Phone: (435) 637-1167. Fax: (435) 637-1177. E-mail: koal@emerytelcom.net Web Site: www.koal.net. Licensee: Eastern Utah Broadcasting Co. Format: Country. ♦Tom Anderson, gen mgr.

KOAL(AM)— October 1936: 750 khz; 10 kw-U, 6.8 kw-N, DA-N. TL: N39 34 02 W110 47 53. Hrs open: Box 875, 84501. Phone: (435) 637-1167. Fax: (435) 637-1177. E-mail: koal@castlenet.com Web Site: www.koal.net. Licensee: Eastern Utah Broadcasting Co. Population served: 34,900 Rgnl. Network: Intermountain Farm/Ranch Network. Format: News/talk, sports. Spec prog: Farm 5 hrs wkly. ♦Keith Mason, progmg dir; Thomas Anderson, pres, gen mgr & chief of engrg.

KSLL(AM)— Sept 6, 1980: 1080 khz; 10 kw-D. TL: N39 33 43 W110 46 36. Stereo. Hrs opn: Box 1080, 84501. Secondary address: 163 E. 100 N. 84501. Phone: (435) 637-1080. Fax: (435) 637-8191. E-mail: kusa@emerytelcom.net Web Site: www.kusaonline.com. Licensee: Against the Wind Broadcasting Inc. (acq 6-21-02; $250,000 with co-located FM). Population served: 100,000 Format: Country. Target aud: General. ♦Randy J. Timothy, pres; David B. Smith, gen mgr & progmg dir; Dennis Silver, chief of engrg.

KWSA(FM)—Co-owned with KSLL(AM). December 1985: 100.9 mhz; 3 kw. 111 ft TL: N39 32 42 W110 48 56. Stereo. 24 Web Site: www.kusaonline.com.50,000 Format: Adult contemp.

Provo

***KBYU-FM**— November 1960: 89.1 mhz; 32 kw. 2,913 ft TL: N40 36 28 W112 09 33. Stereo. Hrs opn: 24 C302 Harris Fine Arts Ctr., 84602. Phone: (801) 422-3552. Fax: (801) 422-0922. E-mail: kbyu@byu.edu Web Site: www.kbyu.org. Licensee: Brigham Young University. Population served: 1,099,000 Natl. Network: PRI. Format: Class, news/talk. News staff: 2; News: 5 hrs wkly. Target aud: 35 plus. ♦Derek Marquis, CEO; Walter B. Rudolph, gen mgr; James Bell, mktg dir & prom dir; Eric Glissmeyer, progmg dir & mus dir; Wes Sims, news dir; Lynn Edwards, engrg dir; Christine Nokleby, prom. Co-owned TV: *KBYU-TV affil.

***KEYY(AM)**— December 1949: 1450 khz; 1 kw-U. TL: N40 13 49 W111 41 12. Hrs open: 24 307 S. 1600 W., 84601-3932. Phone: (801) 374-5210. Fax: (801) 374-2910. E-mail: mail@keyy.com Web Site: www.keyy.com. Licensee: Biblical Ministries Worldwide. (acq 5-10-88). Population served: 400,000 Natl. Network: Moody, Salem Radio Network. Law Firm: Garvey, Schubert & Barer. Format: Christian. News staff: 0; News: 13 hrs wkly. Target aud: General. Spec prog: Sp 5 hrs wkly. ♦Steven A. Barsuhn, gen mgr.

KHTB(FM)— November 1979: 94.9 mhz; 47 kw. Ant 2,798 ft TL: N40 16 58 W111 56 11. Stereo. Hrs opn: 24 2835 E. 3300 S., Suite 800, Salt Lake City, 84107. Phone: (801) 412-6040. Fax: (801) 412-6041. Web Site: theblazeonline.com. Licensee: 3 Point Media-Salt Lake City LLC. (acq 2-23-2004; $26 million with KPQP(FM) Ogden). Natl. Rep: Interep. Format: Rock. Target aud: 18-34; adults. ♦Randy Rodgers, gen mgr; Jody Adam, gen sls mgr; Kayvon Motie, prom dir; John Sichler, engr.

KOVO(AM)— Sept 12, 1939: 960 khz; 5 kw-D, 1 kw-N, DA-N. TL: N40 12 44 W111 40 13. (CP: COL Bluffdale. 50 kw-D, 940 w-N, DA-2. TL: N40 35 06 W112 04 20). Hrs opn: 24 515 S. 700 E., Suite 1-C, Salt Lake, 84102. Phone: (801) 542-2600. Fax: (801) 521-9234. Web Site: www.1280thezone.com. Licensee: Simmons-SLC, LS LLC. Group owner: Simmons Media Group (acq 4-19-2004; $1 million). Population served: 500,000 Natl. Rep: D & R Radio. Format: All sports. Target aud: 35 plus; upper income affluent males & females 35-65. ♦Craig Hanson, pres; Randy Rogers, gen mgr; Ryan Hatch, progmg dir.

KSRR(AM)— Nov 24, 1947: 1400 khz; 1 kw-U. TL: N40 15 29 W111 42 24. Hrs open: 24 Box 828, Orem, 84058. Secondary address: 1454 W. Business Park Dr., Orem 84058. Phone: (801) 224-1400. Licensee: Zeta Holdings LLC (acq 8-27-97). Population served: 550,000 Law Firm: Womble, Carlyle, Sandridge & Rice. Format: Adult contemp. News: one hr wkly. Target aud: 18-54. ♦Robert H. Morey, gen mgr.

KXRK(FM)— Feb 14, 1968: 96.3 mhz; 38 kw. Ant 2,952 ft TL: N40 36 28 W112 09 26. Stereo. Hrs opn: 24 515 South 700 East, Suite 1C, Salt Lake City, 84102. Phone: (801) 524-2600. Fax: (801) 521-9234. E-mail: xmail@x96.com Web Site: www.x96.com. Licensee: Simmons-SLC, LS LLC. Group owner: Simmons Media Group (acq 4-4-2001; grpsl). Population served: 1,200,000 Law Firm: Fletcher, Heald & Hildreth. Format: Alternative. Target aud: 18-34; young, affluent executives.Craig Hanson, pres; Bruce Thomas, CFO; Stephen C. Johnson, gen mgr; Alan Hague, opns dir; Mike Lund, gen sls mgr; Kris Burton, natl sls mgr; Natalie Divino, mktg dir & prom dir; Scott Matthews, chief of engrg; Rachel Wilson, traf mgr; Bill Allred, disc jockey; Chet Tapp, disc jockey; Gina Barberi, disc jockey; Kerry Jackson, disc jockey; Todd Nukem, progmg dir & disc jockey

Randolph

KDUT(FM)— 2001: 102.3 mhz; 89 kw horiz. Ant 2,122 ft TL: N40 52 16 W110 59 43. Hrs open: 24 2722 S. Redwood Rd., Suite 1, Salt Lake City, 84119. Phone: (801) 908-8777. Fax: (801) 908-8782. Web Site: www.bustosmedia.com. Licensee: Bustos Media of Utah License LLC. Group owner: Bustos Media Holdings (acq 7-1-2004; $9 million). Format: Sp CHR. ♦Edward Distel, gen mgr.

Richfield

KLGL(FM)— 2000: 93.7 mhz; 66 kw. Ant 2,355 ft TL: N39 19 17 W111 46 11. Stereo. Hrs opn: 24 Box 40, Manti, 84642. Phone: (435) 835-7301. Fax: (435) 835-2250. Web Site: www.klgl.com. Licensee: Sanpete County Broadcasting Co. (acq 3-1-2006; swap for KCYQ(FM) Elsinore). Population served: 40,000 Natl. Network: Westwood One. Format: Super hits. News staff: one; News: 11 hrs wkly. Target aud: General. ♦Michael Ray, mus dir.

KSVC(AM)— September 1947: 980 khz; 5 kw-D, 1 kw-N, DA-N. TL: N38 45 40 W112 04 35. Hrs open: 24 390 E. Annabella Rd., 84701. Phone: (435) 896-4456. Fax: (435) 896-9333. E-mail: ksvcnews@ksvcradio.com Web Site: www.ksvcradio.com. Licensee: Mid-Utah Radio Inc. (acq 9-15-94; $275,000 with co-located FM; FTR: 10-24-94) Population served: 62,000 Law Firm: Borsari & Paxson. Format: News/talk, sports. News: 18 hrs wkly. Target aud: 18-54. Spec prog: Farm one hr wkly. ♦Kevin Kitchen, gen mgr, sls dir, mus dir & news dir; Kirk Williams, chief of engrg.

Roosevelt

KIFX(FM)— Dec 14, 1987: 98.5 mhz; 2.65 kw. 1,853 ft TL: N40 31 15 W109 42 17. (CP: 3.19 kw, ant 1,689 ft. TL: N40 32 16 W109 41 57). Stereo. Hrs opn: 24 The Fox 98.5, Rt. 2, Box 2384, 84066. Secondary address: 2242 E. 1000 S. 84066. Phone: (435) 722-5011. Phone: (435) 789-5101. Fax: (435) 722-5012. Web Site: www.hitsandfavorites.com. Licensee: Evans Broadcasting Inc. (acq 5-31-91; $283,750; FTR: 6-24-91). Rgnl rep: Art Moore. Format: Adult contemp. News staff: one; News: 5 hrs wkly. Target aud: 21-45. ♦Joseph L. Evans, pres & gen mgr; Teddie Evans, VP; Vickie Reary, opns dir; Teena Christopherson, gen sls mgr; Earl Hawkins, progmg dir; Jean Liddell, news dir; Steve Sprouce, chief of engrg.

KNEU(AM)— Jan 6, 1978: 1250 khz; 5 kw-D, 129 w-N. TL: N40 17 13 W109 57 32. Hrs open: 5 AM-11 PM Rt. 2, Box 2384, Ballard, 84066. Secondary address: 1800 E. 800S, Ballard 84066. Phone: (435)

722-5011. Phone: (435) 789-5101. Fax: (435) 722-5012. E-mail: radio@ubtanet.com Licensee: Country Gold Broadcasting. (acq 2-84; $419,419; FTR: 2-20-84). Population served: 32,000 Format: C&W. News staff: one; News: 10 hrs wkly. Target aud: 25-54. ♦Joseph L. Evans, pres & gen mgr; Teddie Evans, VP & gen sls mgr; Bob Fox, gen mgr; Tenna Christopherson, gen sls mgr; Earl Hawkins, progmg dir & news dir; Jim Leonard, chief of engrg.

KXRQ(FM)— Dec 18, 1998: 94.3 mhz; 17.5 kw. Ant 1,863 ft TL: N40 31 15 W109 42 25. Stereo. Hrs opn: 1420 E. 2850 S., Suite 200, Vernal, 84078. Phone: (435) 722-0940. Phone: (435) 781-1100. Fax: (435) 781-1500. E-mail: cruise@channelx94.com Web Site: www.kxrq.com. Licensee: Uinta Broadcasting L.C. (acq 2-15-01; $450,000). Population served: 60,000 Wire Svc: Metro Weather Service Inc. Format: CHR. hot adult contemp. Target aud: 25-54. Spec prog: Relg 8 hrs wkly. ♦Charles D. Hahl, gen mgr; Charles Hall, opns dir, engrg dir & chief of engrg; Ray Wanty, gen sls mgr & prom dir; Mark Christiansen, progmg dir; Diane Hall, traf mgr; Karine Nelson, local news ed.

Roy

***KANN(AM)**— September 1961: 1120 khz; 10 kw-D, 1 kw-N, DA-2. TL: N41 03 31 W112 04 10. Hrs open: 24 Box 3880, Ogden, 84409. Secondary address: 2500 W. 3700 S., Syracuse 84075. Phone: (801) 776-0249. E-mail: bobalzu6arat@aol.com Web Site: www.sosradio.net. Licensee: Faith Communications Corp. Population served: 1,100,000 Format: Christian. News: 6 hrs wkly. Target aud: 25-44; young families. ♦Jack French, pres; Bob Alzugarat, gen mgr; Brad Staley, opns mgr.

KUDD(FM)— September 1986: 107.9 mhz; 67 kw. Ant 2,383 ft TL: N41 15 27 W112 26 24. Hrs open: 24
Rebroadcasts KUDD(FM) Nephi.
2835 E. 3300 S., Suite 800, Salt Lake City, 84107. Phone: (801) 412-6040. Fax: (801) 412-6041. Licensee: Millcreek Broadcasting L.L.C. (group owner; (acq 4-17-2001; grpsl). Natl. Rep: Interep. Law Firm: Robert Olender. Format: Modern adult contemp. Target aud: General. ♦Randy Rodgers, gen mgr; Lutisha Merrill, gen sls mgr; Scott St. John, prom VP; Brian Michel, progmg dir; Kevin Terry, chief of engrg.

Saint George

***KAER(FM)**— 2006: 89.5 mhz; 7 kw. Ant 1,820 ft TL: N36 50 49 W113 29 28. Hrs open: 24
Rebroadcasts KLRD(FM) Yucaipa, CA 100%.
2351 Sunset Blvd., SUlte 170-218, Rocklin, CA, 95765. Phone: (916) 251-1600. Fax: (916) 251-1650. E-mail: info@air1.com Web Site: www.air1.com. Licensee: Educational Media Foundation. Group owner: EMF Broadcasting. Natl. Network: Air 1. Law Firm: Shaw Pittman. Format: Contemp Christian. News staff: 3. Target aud: 19-35; Judeo Christian female. ♦Richard Jenkins, pres; Lloyd Parker, gen mgr; Ed Lenane, opns dir & news dir; Keith Whipple, dev dir; Eric Allen, natl sls mgr; Mike Novak, progmg dir; David Pierce, progmg mgr; Sam Wallington, engrg dir.

KDXU(AM)— July 3, 1957: 890 khz; 10 kw-U, DA-N. TL: N37 04 04 W113 31 04. Hrs open: 24 750 W. Ridgeview Dr., Suite 204, 84770. Phone: (435) 673-3579. Fax: (435) 673-8900. Licensee: CCR-St. George IV LLC. Group owner: Bonneville International Corp. (acq 8-10-2006; grpsl). Population served: 70,000 Format: News/talk. Target aud: 25-54. Spec prog: Relg 4 hrs wkly. ♦Joseph D. Schwartz, CEO; Don Shelline, gen mgr; Kevin Fry, gen sls mgr; Bryan Hyde, progmg dir; Gary Smith, chief of engrg; Michelle Mathews, traf mgr.

KSNN(FM)—Co-owned with KDXU(AM). June 15, 1973: 93.5 mhz; 3 kw. -125 ft TL: N37 06 54 W113 34 23. Stereo. Format: Adult contemp. Target aud: 12-49. ♦Bryan Benware, progmg dir; Michelle Mathews, traf mgr.

KONY(FM)— Nov 12, 1994: 99.9 mhz; 89 kw. Ant 2,053 ft TL: N36 50 49 W113 29 28. Stereo. Hrs opn: 24 Box 910850, 84791. Phone: (435) 628-3643. Fax: (435) 673-1210. E-mail: kony@infowest.com Licensee: Canyon Media Corp. Group owner: Legacy Communications Corp. Population served: 80,000 Format: Country. Target aud: 25-64; male & female. ♦M. Kent Frandsen, pres; Carl Lamar, gen mgr.

***KRDC-FM**— 1975: 91.7 mhz; 105 w. -312 ft TL: N37 06 16 W113 33 55. Stereo. Hrs opn: 225 S. 700 E., 84770. Phone: (435) 652-7891. Phone: (435) 652-7500. Web Site: www.dixie.edu. Licensee: Dixie College. Population served: 15,000 Format: Alternative, urban contemp, specialty shows. Target aud: 14-25; college & high school students. Spec prog: Class 15 hrs, jazz 10 hrs wkly. ♦Lex De Azevedo, gen mgr; Paul Graves, stn mgr.

***KSGU(FM)**— 2005: 90.3 mhz; 2 kw. Ant 1,820 ft TL: N36 50 49 W113 29 28. Stereo. Hrs opn:
Rebroadcasts KNPR(FM) Las Vegas, NV 100%.
1289 S. Torrey Pines Dr., Las Vegas, NV, 89146. Phone: (702) 258-9895. Fax: (702) 258-5646. Web Site: www.ksgu.org. Licensee: Nevada Public Radio (acq 3-10-2005; $250,000 for CP). Population served: 120,000 Format: All news and info. ♦Lamar Marchese, gen mgr.

KUNF(AM)—Washington, June 6, 1982: 1210 khz; 10 kw-D, 250 w-N. TL: N37 08 38 W113 30 03. Hrs open: 24 750 W. Ridge View Dr., Suite 204, St. George, 84770. Phone: (435) 673-3579. Phone: (435) 673-9398. Fax: (435) 673-8900. Licensee: CCR-St. George IV LLC. Group owner: Bonneville International Corp. (acq 8-10-2006; grpsl). Natl. Network: ESPN Radio. Format: Sports. Target aud: 18-64. ♦Don Shelline, gen mgr; Bob Paterson, stn mgr & progmg dir; Kevin Fry, gen sls mgr; Jessica Paterson, pub affrs dir; Gary Smith, chief of engrg; Michelle Matthews, traf mgr.

KZHK(FM)— January 1997: 95.9 mhz; 96.6 kw. 1,965 ft TL: N36 50 50 W113 29 28. Hrs open: 204 Playa Della Rosita, Washington, 84780. Phone: (435) 628-3643. Fax: (435) 673-1210. E-mail: kony@infowest.com Licensee: Marvin Kent Frandsen. Group owner: Sun Valley Radio Inc. Format: Classic rock. ♦M.K. Frandsen, pres; Carl Lamar, gen mgr & news dir; John Van Wagoner, gen sls mgr; Aaronee Allen, mktg dir, prom dir & pub affrs dir; Marty Lane, stn mgr & progmg dir; Kelton Lloyd, chief of engrg.

KZNU(AM)— Oct 9, 1957: 1450 khz; 1 kw-U. TL: N37 05 02 W113 33 26. Stereo. Hrs opn: 24 Box 910850, 84791. Phone: (435) 628-3643. Phone: (435) 673-1210. Fax: (435) 628-6636. Licensee: Canyon Media Corp. Group owner: Legacy Communications Corp. (acq 7-31-2004). Natl. Network: Fox News Radio. Natl. Rep: Katz Radio. Rgnl rep: Kathy Bingham Format: News/talk. ♦Carl Lamar, gen mgr.

Salt Lake City

KALL(AM)—See North Salt Lake City

KBEE(FM)— 1947: 98.7 mhz; 40 kw. 2,932 ft TL: N40 36 30 W112 09 34. Stereo. Hrs opn: 434 Bearcat Dr., 84115. Phone: (801) 485-6700. Fax: (801) 487-5369. Web Site: www.b987.com. Licensee: Citadel Broadcasting Co. Group owner: Citadel Broadcasting Corp. (acq 7-18-97; $2,873,027 with co-located AM). Format: Adult contemp. Target aud: General. ♦Eric Hauenstein, gen mgr; Ed Hill, opns mgr; Jim Bratt, gen sls mgr; Jaelyn Carillo, prom dir; Rusty Keys, progmg dir; Richie Bauer, chief of engrg; Susan Wasescha, traf mgr.

KFNZ(AM)—Co-owned with KBEE(FM). 1923: 1320 khz; 50 kw-D, 200 w-N. TL: N40 38 36 W111 55 24. Stereo. 175,885 Format: Sports. ♦Zandi Wilcox, gen sls mgr; Joel Smith, prom dir; Jeff Austin, progmg dir; Julie Allen, traf mgr; Dave Coons, sports cmtr; Steve Brown, sports cmtr.

KBJA(AM)—Sandy, June 2001: 1640 khz; 10 kw-D, 1 kw-N. TL: N40 42 47 W111 55 53. Hrs open: 24 10348 S. Redwood Rd., South Jordan, 84095. Phone: (801) 254-7699. Phone: (801) 254-7688. Licensee: United Broadcasting Co. Inc. Format: Sp var. News staff: 5; News: 41 1/2 hrs wkly. Target aud: Hispanic adults; adult Hispanic market.David C. Kifuri, gen mgr, opns dir, adv dir & progmg dir; Jose L. Rivera, stn mgr, sls dir & news dir; Magdalena Garcia, dev dir; Patricia Rivera, mktg dir; Enrique Corona, prom dir, mus dir & sports cmtr; Jessica M. Kifuri, asst music dir; Elizabeth Amores, news dir & pub affrs dir; Jose A. Sanchez, pub affrs dir; Dennis Silver, chief of engrg; Nelson Moran, traf mgr; Alfredo Barbosa, spec ev coord; Claudia Elena Redd, local news ed; David Kifuri, political ed

***KCPW-FM**— 1992: 88.3 mhz; 750 w. Ant -587 ft TL: N40 45 33 W111 49 48. Hrs open: 24 Box 510730, 84151-0730. Phone: (801) 359-5279. Fax: (801) 746-2708. E-mail: news@kcpw.org Web Site: www.kcpw.org. Licensee: Community Wireless of Park City Inc. Natl. Network: NPR, PRI. Format: News/talk. Target aud: 25 plus. ♦Vicki Mann, gen mgr; Bryan Schott, stn mgr & mus dir.

KDYL(AM)—South Salt Lake, Sept 2, 1967: 1060 khz; 10 kw-D, 149 w-N. TL: N40 32 18 W112 04 38. Hrs open: 24 Box 57760, 84157. Secondary address: 3606 South 500 West 84115. Phone: (801) 262-5624. Fax: (801) 266-1510. E-mail: kdylam@aros.net Web Site: www.kdylam.com. Licensee: Holiday Broadcasting Co. Group owner: Carlson Communications International Population served: 1,500,000 Natl. Network: Westwood One. Wire Svc: CNN Format: Adult standards. News: 13 hrs wkly. Target aud: 35-64; general. ♦R. Steve Carlson, sr VP; Brent J. Carlson, VP; Ralph J. Carlson, CEO, pres & gen mgr; Ralph J Carlson, stn mgr; R. Steve Carlson, opns VP.

KENZ(FM)—See Ogden

KJQS(AM)—See Murray

KKAT(AM)— Nov 15, 1955: 860 khz; 10 kw-D, 195.8 w-N, 3 kw-CH. TL: N40 42 47 W111 55 53. Stereo. Hrs opn: 434 Bearcat Dr., 84115. Phone: (801) 485-6700. Fax: (801) 487-5369. Licensee: Citadel Broadcasting Co. Group owner: Citadel Broadcasting Corp. Natl. Rep: Christal. Format: Country. Target aud: 25-54. ♦Larry Wilson, CEO & chmn; Bob Proffitt, sr VP; Eric Hauenstein, gen mgr; Susie Harris Carlson, gen sls mgr; Rusty Keys, prom mgr & progmg dir; Richie Bauer, chief of engrg.

KUBL-FM—Co-owned with KKAT(AM). July 31, 1965: 93.3 mhz; 25 kw. Ant 3,739 ft TL: N40 39 34 W112 12 05. Stereo. Web Site: www.kbull93.com. Format: Country 90s. ♦Ed Hill, opns mgr & progmg dir; Terry Mathis, sls dir & gen sls mgr; Randi P' Poll, prom VP & prom dir; Richie Bauer, chief of engrg; Julie Johnson, traf mgr.

KMRI(AM)—West Valley City, Nov 16, 1956: 1550 khz; 10 kw-D, 500 w-N. TL: N40 43 29 W112 00 43. Hrs open: 24 314 S. Redwood Rd., 84104. Secondary address: Box 352 84110. Phone: (801) 886-1550. Fax: (801) 973-7145. E-mail: kmri1550@aol.com Licensee: KMRI Radio L.L.C. (acq 12-11-97; $500,000). Population served: 1,300,000 Law Firm: Wood, Maines & Borwn. Format: Christian, rgnl Mexican music. News staff: 2; News: 40 hrs wkly. Target aud: 16-50. ♦Pat Openshaw, pres; Dennis Ermel, gen mgr; Micah Coleman, opns dir; Jessica Lockwood, sls dir & chief of engrg; Isaac Velasquez, progmg dir.

KNRS(AM)— Aug 1, 1938: 570 khz; 5 kw-U, DA-2. TL: N40 49 09 W111 55 56. Hrs open: 24 2801 S. Decker Lake Dr., 84119. Phone: (801) 908-1300. Fax: (801) 908-1459. Web Site: www.knrs.com. Licensee: Citicasters Licenses L.P. Group owner: Clear Channel Communications Inc. (acq 1999; grpsl). Population served: 1,400,400 Natl. Rep: Clear Channel. Format: News/talk. News staff: one; News: 2 hrs wkly. Target aud: 25-54; adults. ♦Stu Stanek, gen mgr; Bill Betts, opns mgr; Bill Mathews, sls dir; Jason Wilmot, progmg dir; Jim Vandiver, sls. Co-owned TV: KTVX(TV)

KODJ(FM)— Dec 1, 1968: 94.1 mhz; 40 kw. Ant 3,060 ft TL: N40 36 22 W112 09 49. Stereo. Hrs opn: 24 2801 S. Decker Lake Dr., 84119. Phone: (801) 908-1300. Fax: (801) 908-1429. Web Site: www.kodj.com. Licensee: Citicasters Licenses L.P. Group owner: Clear Channel Communications Inc. (acq 5-4-99; grpsl). Population served: 1,361,800 Natl. Rep: Clear Channel, Katz Radio. Format: Hits of the 60s & 70s. News: 2 hrs wkly. Target aud: 25-54; adults. ♦Stu Stanek, gen mgr; Bill Betts, opns mgr; Kimberly Dickerson, gen sls mgr; Rob Boshard, progmg dir. Co-owned TV: KUTV(TV) affil

***KRCL(FM)**— Dec 3, 1979: 90.9 mhz; 16.5 kw. 3,770 ft TL: N40 39 35 W112 12 05. Stereo. Hrs opn: 24
Rebroadcasts KZMU(FM) Moab 50%.
1971 W. North Temple, 84116-3046. Phone: (801) 363-1818. Fax: (801) 533-9136. E-mail: mailman@krcl.org Web Site: www.krcl.org. Licensee: Listeners Community Radio of Utah Inc. Population served: 45,000 Format: Div, educ, folk. News staff: one; News: 3 hrs wkly. Target aud: General. Spec prog: Black 20 hrs, Sp 9 hrs, American Indian 4 hrs, Asian 4 hrs, Polynesian one hr, wkly. ♦Donna Land Maldonado, pres & gen mgr; Kami St. John, dev VP, dev dir & mktg dir; Troy Mumm, opns dir & progmg dir; Doug Young, mus dir; Gena Edualson, pub affrs dir; Felix Gonzalez, engrg mgr; Lewis Downey, chief of engrg.

KRSP-FM— Aug 21, 1968: 103.5 mhz; 25 kw. Ant 3,739 ft TL: N40 39 34 W112 12 05. Stereo. Hrs opn: 55 North 300 West, 84180. Phone: (801) 575-5555. Fax: (801) 526-1070. Web Site: www.arrow1035.com. Licensee: Bonneville Holding Co. Group owner: Bonneville International

Corp. (acq 12-5-2003; grpsl). Population served: 241,100 Format: Classic rock. Target aud: 18-34. ♦Chris Redgrave, gen mgr.

KSFI(FM)— Dec 26, 1946: 100.3 mhz; 25 kw. Ant 3,740 ft TL: N40 39 34 W112 12 05. Stereo. Hrs opn: 55 North 300 West, 84180. Phone: (801) 575-5555. Fax: (801) 526-1070. Web Site: www.fm100.com. Licensee: Bonneville Holding Co. Group owner: Bonneville International Corp. (acq 12-5-2003; grpsl). Format: Adult contemp. Target aud: 25-54; general. ♦Chris Redgrave, gen mgr; Paulette Cary, sls dir; Dain Craig, progmg dir; Christa Lee Durrant, mus dir; Peggy Ijams, news dir; Trina Bodily, traf mgr.

KSL(AM)— May 6, 1922: 1160 khz; 50 kw-U. TL: N40 46 46 W112 05 56. Stereo. Hrs opn: 24 Box 1160, 84110-1160. Secondary address: Broadcast House, 55 N. 300 W., Salt lake City 84180. Phone: (801) 575-7600. Fax: (801) 575-7625. Web Site: www.ksl.com. Licensee: Bonneville International Corp. (group owner) Population served: 1,180,000 Natl. Network: CBS. Wire Svc: Reuters Wire Svc: UPI Format: News/talk, sports. News staff: 12. Target aud: 25-54.Bruce Reese, CEO; Richard Mecham, pres & gen mgr; Robert Johnson, CFO; Chris Redgrave, gen mgr & gen sls mgr; Lora Woodbury, natl sls mgr & rsch dir; Paulette Cary, rgnl sls mgr; Rochelle Beatty, prom mgr; Rod Arquette, progmg dir; Janine Baker, pub affrs dir & traf mgr; John Dehnel, chief of engrg; Greg Wrubell, sports cmtr; Doug Wright, disc jockey . Co-owned TV: KSL-TV affil.

KSOP(AM)—South Salt Lake, Feb 1, 1955: 1370 khz; 5 kw-D, 500 w-N, DA-N. TL: N40 43 12 W111 55 42. Hrs open: 24 Box 25548, 84119. Secondary address: 1285 W. 2320 S. 84119. Phone: (801) 972-1043. Fax: (801) 974-0868. Web Site: www.goldcountryam1370.com. Licensee: KSOP Inc. Population served: 175,885 Natl. Rep: D & R Radio. Format: Classic country. News staff: one. Target aud: 25-54. ♦Greg Hilton, pres, gen mgr & gen sls mgr; Don Hilton, progmg dir; Debbie Turpin, mus dir; Dick Jacobson, news dir; Bill Traue, chief of engrg; John Greenwell, farm dir & disc jockey; Bill Buckly, disc jockey; Kim Hall, disc jockey; Larry Hunter, disc jockey; Phil Pond, disc jockey.

KSOP-FM— Dec 10, 1964: 104.3 mhz; 25 kw. 3,650 ft TL: N40 39 35 W112 12 05. Stereo. 24 Web Site: www.ksopcountry.com. Natl. Rep: D & R Radio. Target aud: 18 plus. ♦Bill Buckley, disc jockey; John Greenwell, disc jockey; Ken Carlin, disc jockey; Phil Pond, disc jockey.

KTKK(AM)—Sandy, May 13, 1960: 630 khz; 1 kw-D, 500 w-N, DA-2. TL: N40 41 30 W111 55 30. Hrs open: 24 10348 S. Redwood Rd., South Jordan, 84095. Phone: (801) 253-4883. Fax: (801) 253-9085. E-mail: webmaster@k-talk.com Web Site: www.k-talk.com. Licensee: United Broadcasting Co. (acq 12-1-63). Population served: 175,885 Format: Talk. Target aud: 35 plus. ♦Richard Perry, pres & gen mgr; Janet Kelly, sls dir, prom dir & sls; Tom Draschil, progmg dir & news dir; Dennis Silver, engrg dir.

***KUER(FM)**— June 4, 1960: 90.1 mhz; 38 kw. 2,900 ft TL: N40 36 30 W112 09 34. Stereo. Hrs opn: Univ. of Utah, 101 Wasatch Dr., 84112. Phone: (801) 581-6625. Fax: (801) 581-5426. E-mail: radiowest@kuer.org Web Site: www.kuer.org. Licensee: University of Utah. Population served: 60,000 Natl. Network: NPR, PRI. Format: News, class, jazz. Spec prog: Gospel 3 hrs wkly. ♦John Greene, gen mgr; Jenny Brundin, news dir.

***KUFR(FM)**— Dec 14, 1989: 91.7 mhz; 220 w. Ant -318 ft TL: N40 46 09 W111 53 12. Hrs open: 24 136 E. S. Temple, Suite 1630, 84111. Phone: (801) 359-3147. Fax: (801) 359-8112. Web Site: www.familyradio.com. Licensee: Family Stations Inc. (group owner) Format: Christian relg. ♦Harold Camping, pres & gen mgr; Roger Crawford, stn mgr & chief of engrg; James Abrahamson, opns mgr; Thad McKinney, rgnl sls mgr.

KWDZ(AM)— 1945: 910 khz; 5 kw-D, 1 kw-N, DA-2. TL: N40 30 48 W112 00 23. Stereo. Hrs opn: 24 2801 S. Decker Lake Dr., Suite 100, 84119. Phone: (801) 908-5152. Fax: (801) 908-7844. Web Site: www.radiodisney.com. Licensee: Radio Disney Group LLC. Group owner: ABC Inc. (acq 4-30-03; $3.7 million). Population served: 1,361,800 Natl. Network: Radio Disney. Format: Children. ♦Celia Willette, gen mgr; Meradyth Moore, mktg dir & prom dir; Barry McClellen, chief of engrg; Shari Levy, traf mgr.

KXRV(FM)—See Centerville

KXTA(AM)—See Centerville

KZHT(FM)— Feb 1, 1961: 97.1 mhz; 25 kw. Ant 3,739 ft TL: N40 39 34 W112 12 05. Stereo. Hrs opn: 24 2801 S. Decker Lake Dr., 84119. Phone: (801) 908-1300. Fax: (801) 908-1389. Web Site: www.971zht.com. Licensee: CC Licenses LLC. Group owner: Clear Channel Communications Inc. (acq 7-10-2000). Population served: 534,700 Natl. Rep: Katz Radio. Format: CHR. Target aud: 18-49. ♦Bill Betts, gen mgr & opns mgr; Bill Mathews, sls dir; Emily Hunt, gen sls mgr; Stacy Sappenfield, prom dir; Jeff McCartney, progmg dir.

KZNS(AM)— February 1945: 1280 khz; 5 kw-D, 500 w-N, DA-N. TL: N40 44 47 W111 54 42. Hrs open:
Rebroadcasts KOVO(AM) Provo 100%.
515 S. 700 East, Suite 1C, 84102. Phone: (801) 524-2600. Fax: (801) 521-9234. Web Site: www.1280kzn.com. Licensee: Simmons-SLC, LS LLC. Group owner: Simmons Media Group (acq 4-4-2001; grpsl). Natl. Network: Westwood One, CNN Radio. Rgnl. Network: CNN. Natl. Rep: CBS Radio. Format: Sports talk. Target aud: 35 plus; retired, affluent, responsible & loyal. ♦David Simmons, chmn; G. Craig Hanson, pres; Stephen C. Johnson, gen mgr; Amanda Traeger, sls dir & gen sls mgr; Eric Ray, news dir; Scott Matthews, chief of engrg.

Sandy

KBJA(AM)—Licensed to Sandy. See Salt Lake City

KTKK(AM)—Licensed to Sandy. See Salt Lake City

Smithfield

KGNT(FM)— February 1983: 103.9 mhz; 3 kw. Ant -131 ft TL: N41 48 44 W111 47 31. Hrs open: 810 W. 200 N., Logan, 84321. Phone: (435) 752-1390. Fax: (435) 752-1392. Web Site: www.thegiant.com. Licensee: Frandsen Media Co. LLC. Group owner: Sun Valley Radio Inc. (acq 2-4-02; $775,000). Population served: 150,000 Natl. Network: CBS, Westwood One. Law Firm: Dan J. Alpert. Format: Oldies. Target aud: 18-49; 55% female, 45% male middle class. ♦Jay Eubanks, stn mgr; Lori Gill, gen sls mgr; David Denton, progmg dir & news dir; Paul Anderson, chief of engrg.

South Jordan

KUUU(FM)— Sept 1, 1979: 92.5 mhz; 500 w. Ant 3,929 ft TL: N40 39 35.2 W112 12 04.7. Stereo. Hrs opn: 24 2835 E. 3300 S., Salt Lake City, 84109. Phone: (801) 412-6040. Fax: (801) 412-6041. Licensee: Millcreek Broadcasting L.L.C. (group owner; (acq 4-17-2001; grpsl). Population served: 28,000 Natl. Rep: Interep. Format: Hip hop, rhythm. News staff: one; News: 15 hrs wkly. Target aud: General. ♦Randy Rodgers, gen mgr; Brian Michel, opns mgr; Lutisha Merrill, gen sls mgr; Scott St. John, prom mgr; Kevin Cruise, mus dir; Kevin Terry, engrg VP.

South Salt Lake

KDYL(AM)—Licensed to South Salt Lake. See Salt Lake City

KSOP(AM)—Licensed to South Salt Lake. See Salt Lake City

Spanish Fork

KHQN(AM)— July 24, 1960: 1480 khz; 1 kw-D. TL: N40 04 30 W111 39 42. Hrs open: 8628 S. State St., 84660. Phone: (801) 798-3559. Licensee: Robyn Howell (acq 8-9-2006). Population served: 9,560 Format: New age, progsv, relg. Spec prog: Farm 4 hrs wkly. ♦Christine Warden, gen mgr.

KOSY-FM— Nov 1, 1967: 106.5 mhz; 25 kw. Ant 3,739 ft TL: N40 39 34 W112 12 05. Hrs open: 24 2801 S. Decker Lake Dr., Salt Lake City, 84119. Phone: (801) 908-1300. Fax: (801) 908-1459. Web Site: www.kosy.com. Licensee: Citicasters Licenses L.P. Group owner: Clear Channel Communications Inc. (acq 2-27-2004; $22 million with KXRV(FM) Centerville). Natl. Rep: Clear Channel, Katz Radio. Format: Soft adult contemp. Target aud: 25-44; women. ♦Stu Stanek, gen mgr; Bill Betts, opns mgr; Bill Matthews, sls dir; Jim Vandiver, gen sls mgr; Steve Clem, progmg dir.

Spanish Valley

KCPX(AM)—Not on air, target date: 06/07: 1490 khz; 1 kw-U. TL: N38 28 04 W109 26 18. Hrs open: 24 Box 1119, Moab, 84532. Secondary address: 1030 S. Bowling Alley Ln #3, Moab 84532. Phone: (435) 259-1035. Fax: (435) 259-1037. Licensee: Moab Communications LLC. Format: Talk. ♦Ralph J. Carlson, pres; Phillip Mueller, gen mgr.

Taylorsville

KUTR(AM)— May 9, 2005: 820 khz; 50 kw-D, 2.5 kw-N, 50 kw-CH, DA-2. TL: N40 19 48 W112 04 09. Hrs open: 55 North 300 West, Salt Lake City, 84180. Phone: (801) 575-5555. Fax: (801) 526-1070. Web Site: www.utaham820.com. Licensee: Bonneville Holding Co. Group owner: Bonneville International Corp. (acq 12-5-2003; grpsl). Format: Contemp Christian. ♦Chris Redgrave, VP & stn mgr; Paulette Cary, gen sls mgr; Rod Arquette, opns VP & progmg VP; John Dehnel, chief of engrg.

Tooele

KCPW(AM)— July 3, 1956: 1010 khz; 50 kw-D, 13 w-N. TL: N40 32 36 W112 18 33. Hrs open: 24 Box 510730, Salt Lake City, 84151-0730. Phone: (801) 359-5279. Fax: (801) 746-2708. Web Site: www.kiqn1010.com. Licensee: Community Wireless of Park City Inc. (acq 10-1-2003). Population served: 12,539 Natl. Network: NPR, PRI. Rgnl. Network: Metronews Radio Net. Natl. Rep: Katz Radio. Format: News/talk info. ♦Blair Feulner, gen mgr.

Tremonton

KNFL(AM)— Jan 27, 2006: 1470 khz; 1 kw-D, 880 w-N, DA-N. TL: N41 34 42 W112 06 03. Hrs open: 24 1506 Gibson Ave., Odgen, 84404. Phone: (435) 734-2600. Fax: (435) 734-1470. E-mail: kognradio@comcast.net Licensee: AM Radio 1470 Inc. Group owner: Diamond Broadcasting Corp. (acq 12-6-2004; grpsl). Natl. Network: CNN Radio, Westwood One. Law Firm: Dan J. Alpert. Format: Adult standards/CNN radio news. Target aud: 35 plus. ♦E. Morgan Skinner Jr., CEO; Richard Carter, stn mgr.

KYLZ(FM)— July 1, 1983: 104.9 mhz; 99 kw. 1,059 ft TL: N41 43 34 W112 12 33. Stereo. Hrs opn: 24 Box 3369, Logan, 84321. Phone: (435) 752-1390. Fax: (435) 752-1392. Licensee: 3 Point Media - Utah LLC. (acq 11-28-2001; $1.73 million). Population served: 300,000 Natl. Network: CBS, Jones Radio Networks. Law Firm: Dan J. Alpert. Format: Classic country. News staff: 3. Target aud: 25-54; 60% female, 40% male middle to upper class. ♦Kent Frandsen, pres; Jay Eubanks, gen mgr & stn mgr; Lori Gill, gen sls mgr & traf mgr; David Denton, progmg dir & news dir; Paul Anderson, chief of engrg.

Vernal

KLCY-FM—Listing follows KVEL(AM).

KVEL(AM)— Jan 19, 1947: 920 khz; 4.5 kw-D, 1 kw-N, DA-N. TL: N40 29 30 W109 31 45. Hrs open: 24 Box 307, 2425 N. Vernal Ave., 84078. Phone: (435) 789-0920. Phone: (435) 789-1059. Fax: (435) 789-6977. E-mail: kvel@ubtanet.com Licensee: Ashley Communications Inc. (acq 8-11-98; $10,000 for stock with co-located FM). Population served: 30,000 Law Firm: Reddy, Begley & McCormick. Format: Sports, news/talk. News staff: one; News: 20 hrs wkly. Target aud: 35-64; affluent, upscale. Spec prog: Farm 2 hrs, relg, Sp, pub affrs one hr wkly. ♦Steve Evans, gen mgr & gen sls mgr; Clay Johnson, progmg dir; Steve Sprouse, chief of engrg.

KLCY-FM—Co-owned with KVEL(AM). May 1, 1975: 105.9 mhz; 2.9 kw. Ant 413 ft TL: N40 24 50 W109 35 34. (CP: 105.5 mhz; 3.3 kw, ant 1,699 ft. TL: N40 32 16 W109 41 57). 24 Natl. Network: ABC, Jones Radio Networks. Format: Eagle country. News: 2 hrs wkly. Target aud: 18-54; active.

Washington

KUNF(AM)—Licensed to Washington. See Saint George

Wellington

KRPX(FM)— 2006: 95.3 mhz; 6 kw. Ant -138 ft TL: N39 36 33 W110 48 50. Hrs open: Box 875, Price, 84501. Phone: (435) 637-1167. Fax: (435) 637-1177. Licensee: College Creek Media LLC. Format: Light rock hits. ♦Neal J. Robinson, pres; Tom Anderson, gen mgr.

West Jordan

KLLB(AM)— 1982: 1510 khz; 10 kw-D. TL: N40 33 06 W111 58 17. Hrs open: 868 E. 5900 South, Salt Lake City, 84107. Phone: (801) 487-0247. Fax: (801) 262-6200. Licensee: United Security Financial Inc. (acq 6-18-91; $180,001; FTR: 7-8-91). Format: Gospel. ♦Lois Johnson, gen mgr; Joel Cosby, sls dir; D.J. Stone, progmg dir; Darrell Cosby, chief of engrg; Shon Thomas, traf mgr.

West Valley City

KMRI(AM)—Licensed to West Valley City. See Salt Lake City

Woodruff

KYMV(FM)— June 2002: 100.7 mhz; 88 kw horiz. Ant 2,122 ft TL: N40 52 16 W110 59 43. Hrs open: 515 S. 700 E., Suite 1C, Salt Lake City, UT, 84102. Phone: (801) 524-2600. Fax: (801) 521-9234. Web Site: movin1007.com. Licensee: Simmons-SLC, LS LLC. Group owner: Simmons Media Group (acq 4-4-2001; grpsl). Format: Rhythmic adult contemp. ♦Stephen Johnson, gen mgr; Jacquie Louie, gen sls mgr; Naziol Nazarina, prom dir; Alan Hague, progmg VP.

Vermont

Addison

WUSX(FM)— 1999: 93.7 mhz; 21 kw. Ant 354 ft TL: N44 13 15 W73 24 37. Hrs open: 372 Dorset St., South Burlington, 05403. Phone: (802) 863-1010. Fax: (802) 861-7256. Licensee: Addison Broadcasting Co. Inc. Group owner: Northeast Broadcasting Company Inc. (acq 12-19-2000; $434,000). Natl. Network: Jones Radio Networks. Format: Classic country. ♦Rich Delancy, gen mgr; Rich Delancey, gen sls mgr; J.J. Prieve, progmg dir; Chris Fells, news dir; Mike Raymond, chief of engrg.

Barre

***WCMD-FM**— Aug 1, 1998: 89.9 mhz; 940 w. 590 ft TL: N44 07 32 W72 28 36. Stereo. Hrs opn: 24
Rebroadcasts WCMK(FM) Bolton 99%.
140 Main St., Essex Junction, 05452. Secondary address: Box 8310, Essex 05451-8310. Phone: (802) 878-8885. Fax: (802) 879-6835. E-mail: cmi.radio@verizon.net Web Site: thelightradio.net. Licensee: Christian Ministries Inc. Natl. Network: Moody. Law Firm: Joseph E. Dunne III. Format: Inspirational, Christian. News: 15 hrs wkly. Target aud: General; Christian, middle income. ♦Mark Kinsley, pres; Richard McClary, gen mgr; Karlo Salminen, opns dir & progmg dir; Darlene Lamos, gen sls mgr; Peter Morton, chief of engrg.

WORK(FM)—Listing follows WSNO(AM).

WSKI(AM)—See Montpelier

WSNO(AM)— Oct 13, 1959: 1450 khz; 1 kw-U. TL: N44 11 40 W72 30 52. Hrs open: 24 41 Jacques St., 05641. Phone: (802) 476-4168. Fax: (802) 479-5893. Web Site: www.wsno1450.net. Licensee: Nassau Broadcasting III L.L.C. Group owner: Nassau Broadcasting Partners L.P. (acq 8-2-2004; grpsl). Population served: 40,000 Natl. Network: CBS. Format: News/talk, sports. ♦Ken Barlow, gen mgr; Jim Severance, progmg dir.

WORK(FM)—Co-owned with WSNO(AM). Aug 5, 1974: 107.1 mhz; 1.5 kw. Ant 410 ft TL: N44 09 30 W72 28 46. Web Site: www.1071workfm.com.180,000 Format: Classic hits. ♦T.J. Michaels, progmg dir.

Barton

WJPK(FM)—Not on air, target date: unknown: 100.3 mhz; 100 w. Ant 525 ft TL: N44 45 57 W72 09 10. Hrs open: Box 97, Lyndonville, 05851. Phone: (802) 626-9800. Fax: (802) 626-8500. Licensee: Vermont Broadcast Associates Inc. ♦Bruce James, gen mgr.

Bellows Falls

WZLF(FM)— November 1981: 107.1 mhz; 1 kw. 530 ft TL: N43 12 33 W72 19 58. Stereo. Hrs opn: 24
Rebroadcasts WSSH(FM) Marlboro 100%.
Box 1230, Claremont, NH, 03743. Phone: (603) 542-7735. Fax: (603) 542-8721. Web Site: www.bobcountrysm.com. Licensee: Nassau Broadcasting III L.L.C. Group owner: Nassau Broadcasting Partners L.P. (acq 8-2-2004; grpsl). Natl. Rep: Roslin. Format: Country. News staff: one; News: 5 hrs wkly. Target aud: 25-54; general. Spec prog: Farm one hr wkly. ♦Courtney Galluzzo, gen mgr, gen sls mgr & rgnl sls mgr; Doug Daniels, opns mgr; Heath Cole, progmg dir & news dir.

Bennington

WBTN(AM)— Sept 23, 1953: 1370 khz; 1 kw-D. TL: N42 54 19 W73 12 32. Hrs open: WBTN Svc., 982 Mansion Dr., 05201. Secondary address: 407 Harwood Hill 05201. Phone: (802) 442-6321. Fax: (802) 442-3112. E-mail: wbtn@svcedu Licensee: Southern Vermont College (acq 8-20-03). Population served: 47,950 Natl. Network: Westwood One. Wire Svc: AP Format: News/talk, music; student progmg. Target aud: 24-54. ♦Ben Runnels, pres & chief of opns; Rich Ryder, gen mgr; Megan Williams, traf mgr.

WBTN-FM— Nov 4, 1978: 94.3 mhz; 3 kw. 110 ft TL: N42 56 52 W73 10 36. Stereo. Hrs opn: 5:30 AM-midnight 365 Troy Ave., Colchester, 05446. Phone: (802) 655-9451. Fax: (802) 655-2799. E-mail: contact@vpr.net Web Site: www.vpr.net. Licensee: Vermont Public Radio. (acq 11-24-99; $901,000 with co-located AM). Format: Classical. News staff: one; News: 4 hrs wkly. Target aud: 18-45. ♦Mark Vogelzang, pres, gen mgr, stn mgr & disc jockey.

Berlin

WWFY(FM)— Apr 2, 1975: 100.9 mhz; 5.2 kw. Ant 718 ft TL: N44 07 38 W72 28 48. Stereo. Hrs opn: 24 41 Jacques St., Barre, 05641. Phone: (802) 476-4168. Fax: (802) 479-5893. Web Site: www.froggy1009.com. Licensee: Nassau Broadcasting III L.L.C. Group owner: Nassau Broadcasting Partners L.P. (acq 8-2-2004; grpsl). Population served: 100,000 Law Firm: Gardner, Carton & Douglas. Format: Country. Target aud: 18-49; young professionals. ♦John Gales, gen mgr; Jim Severance, sls VP & progmg dir.

Bolton

***WGLY-FM**— 1996: 91.5 mhz; 2 kw. Ant 935 ft TL: N44 21 53 W72 55 52. Stereo. Hrs opn: 24
Rebroadcasts WCMD(FM) Barre 100%.
140 Main St., Essex Junction, 05452. Secondary address: Box 8310, Essex 05451-8310. Phone: (802) 878-8885. Fax: (802) 879-6835. E-mail: cmi.radio@verizon.net Web Site: thelightradio.net. Licensee: Christian Ministries Inc. Natl. Network: Moody. Law Firm: Joseph E. Dunne III. Format: Inspirational. ♦Mark Kinsley, pres; Richard McClary, gen mgr; Karlo Salminen, opns dir, dev mgr & progmg dir; Darlene Lamos, gen sls mgr; Peter Morton, chief of engrg.

Brandon

WEXP(FM)— May 2000: 101.5 mhz; 350 w. Ant 1,305 ft TL: N43 39 31 W73 06 26. Stereo. Hrs opn: 24 1 Scale Ave., Suite 84, Rutland, 05761-4459. Phone: (802) 773-9264. Fax: (802) 747-0553. Web Site: www.101thefox.com. Licensee: Nassau Broadcasting III L.L.C. Group owner: Vox Radio Group L.P. (acq 1-21-2005; $2.5 million with WVAY(FM) Wilmington). Natl. Network: Westwood One. Format: Rock/AOR, classic rock. Target aud: 25-54; male. ♦John Gales, gen mgr; Glenn Novak, gen sls mgr; Kemy Chambers, prom dir; Kelly Kowalski, progmg dir.

Brattleboro

WINQ(FM)—See Winchester, NH

WKVT(AM)— Nov 29, 1959: 1490 khz; 1 kw-U. TL: N42 50 51 W72 34 56. Hrs open: 24 458 Williams St., 05301. Phone: (802) 254-2343. Fax: (802) 254-6683. Web Site: www.wkvt.com. Licensee: Saga Communications of New England LLC. Group owner: Saga Communications Inc. (acq 5-1-02; grpsl). Population served: 12,239 Natl. Network: CBS. Wire Svc: AP Format: News/talk. News staff: one; News: 30 hrs wkly. Target aud: 35-64; news & info oriented adults. ♦Mike Trombly, gen mgr; Peter Case, progmg dir & news dir.

WKVT-FM— 1980: 92.7 mhz; 6 kw. 610 ft TL: N42 53 45 W72 39 49. Stereo. 24 100,000 Natl. Network: AP Radio. Format: Classic rock. News staff: one; News: 8 hrs wkly. Target aud: 18-44.

WTSA(AM)— Apr 19, 1950: 1450 khz; 1 kw-U. TL: N42 52 13 W72 33 35. Hrs open: 24 Box 819, 05302. Secondary address: 827 Western Ave. 05301. Phone: (802) 254-4577. Fax: (802) 257-4644. E-mail: info@wtsa.net Web Site: www.wtsa.net. Licensee: Tri-State Broadcasters Inc. (acq 7-1-86; grpsl; FTR: 5-26-86). Population served: 12,239 Natl. Rep: D & R Radio. Law Firm: Cohn & Marks. Format: Sports. News staff: one. Target aud: General. ♦John Kilduff, pres & opns dir; Tim Johnson, news dir & chief of engrg.

WTSA-FM— Dec 15, 1975: 96.7 mhz; 5.2 kw. 167 ft TL: N42 53 21 W72 36 47. Stereo. Web Site: www.wtsa.net. Licensee: Tri-State Broadcasters Inc.50,000 Format: Adult contemp. Target aud: 12 plus. Spec prog: Oldies 16 hrs wkly.

WYRY(FM)—See Keene, NH

Brighton

WVTI(FM)—Not on air, target date: unknown: 106.9 mhz; 1.42 kw. Ant 679 ft TL: N44 47 02 W71 53 14. Hrs open: Vermont Public Radio, 365 Troy Ave., Colchester, 05446. Phone: (802) 655-9451. Fax: (802) 655-2799. Web Site: www.vpr.net. Licensee: Vermont Public Radio. ♦Mark Vogelzang, pres & gen mgr.

Bristol

WTNN(FM)— 2007: 97.5 mhz; 8.7 kw. Ant 518 ft TL: N44 24 23.1 W73 08 12.8. Hrs open: 4049 Williston Rd., South Burlington, 05403. Phone: (802) 864-9750. Fax: (802) 864-9777. E-mail: comments@eaglecountry975.com Web Site: www.eaglecountry975.com. Licensee: Impact Radio Inc. Format: Country. ♦Arthur V. Belendiuk, pres; John Fuller, gen mgr.

Burlington

WCAT(AM)— Apr 19, 1954: 1390 khz; 5 kw-U, DA-N. TL: N44 29 47 W73 12 49. Stereo. Hrs opn: 24 372 Dorset St., South Burlington, 05403. Phone: (802) 655-6753. Fax: (802) 860-4721. Web Site: www.wcat1390.com. Licensee: Radio Broadcasting Services Inc. Group owner: Radio Vermont Group Inc. (acq 8-3-2006; $400,000). Population served: 300,000 Natl. Network: ESPN Radio. Natl. Rep: McGavren Guild. Wire Svc: AP Format: Sports. ♦Steven A. Silberberg, pres; Richard C. DeLancey Sr., gen mgr; J.J. Prieve, progmg dir; Chris Fells, sports cmtr.

WEZF(FM)— July 19, 1968: 92.9 mhz; 46 kw. 2,703 ft TL: N44 31 40 W72 48 58. Stereo. Hrs opn: 24 265 Hegeman Ave., Colchester, 05446. Secondary address: Box 1093 05402-1093. Phone: (802) 655-0093. Fax: (802) 655-0478. Web Site: www.star929.com. Licensee: Capstar TX L.P. Group owner: Clear Channel Communications Inc. (acq 8-30-00; grpsl). Population served: 100,000 Natl. Rep: Clear Channel. Law Firm: Wiley, Rein & Fielding. Format: Adult contemp. News staff: one; News: 7 hrs wkly. Target aud: 25-54. ♦Karen Marshall, gen mgr; Gale Parmalee, opns mgr.

WIZN(FM)—Vergennes, Nov 15, 1983: 106.7 mhz; 50 kw. 373 ft TL: N44 18 40 W73 14 34. Stereo. Hrs opn: 24 Box 4489, 05406. Phone: (802) 860-2440. Fax: (802) 860-1818. E-mail: wizn@wizn.com Web Site: www.wizn.com. Licensee: Hall Communications Inc. Group owner: Deer River Broadcasting Group (acq 10-31-2005; $17 million). Natl. Rep: Katz Radio. Format: Rock/AOR, live. Target aud: 18-49. Spec prog: Oldies 3 hrs, reggae one hr, progsv one hr, blues 3 hrs wkly. ♦Jennifer McCann, gen mgr; Tracy Ovitt, gen sls mgr; Matt Grasso, progmg dir & mus dir.

WJOY(AM)— Sept 14, 1946: 1230 khz; 1 kw-U. TL: N44 27 03 W73 11 51. Hrs open: 24 Box 4489, 05406-4489. Secondary address: 70 Joy Dr., South Burlington 05403. Phone: (802) 658-1230. Fax: (802) 862-0786. E-mail: wjoy@hallradio.com Web Site: www.wjoy.com. Licensee: Hall Communications Inc. (group owner; (acq 12-1-83; FTR: 12-5-83). Natl. Network: Westwood One. Natl. Rep: D & R Radio. Law Firm: Fletcher, Heald & Hildreth. Format: News, easy lstng. News staff: one; News: 4 hrs wkly. Target aud: General; affluent, empty nesters, well educated. ♦Bonnie Rowbotham, chmn; Arthur J. Rowbotham, pres; Richard P. Reed, exec VP; Bill Baldwin, sr VP; Dan Dubonnet, gen mgr; Steve Pelkey, opns dir, progmg dir & disc jockey; Lee Bodette, gen sls mgr; Wendy Naylor, prom dir; Ginny McGehee, news dir & disc jockey; Dennis Snyder, chief of engrg.

WOKO(FM)—Co-owned with WJOY(AM). June 26, 1962: 98.9 mhz; 100 kw. 307 ft TL: N44 27 03 W73 11 51. Stereo. 24 Web Site:

www.woko.com. Format: Country. ♦Dan Dubonnet, VP; Bill Sargent, mus dir & disc jockey; Laura Lacasse, traf mgr; Ginny McGehee, local news ed; C.K. Coin, disc jockey; Cal Daniels, disc jockey; Nick Morgan, disc jockey; Thom Richards, disc jockey.

***WRUV(FM)**— Oct 3, 1965: 90.1 mhz; 460 w. 145 ft TL: N44 28 37 W73 11 59. (CP: Ant 131 ft.). Stereo. Hrs opn: 24 Univ. of Vermont, Billings Student Ctr., 05405. Phone: (802) 656-4399. Phone: (802) 656-0796. Fax: (802) 656-2281. E-mail: wruv@zoo.uvm.edu Web Site: www.wruv.org. Licensee: University of Vermont & State Agricultural College. Population served: 250,000 Format: Div, educ, jazz. Spec prog: Non-commercial free format, progmg varies. ♦Jake Davignon, stn mgr.

WTWK(AM)—See Plattsburgh, NY

WVMT(AM)— May 20, 1922: 620 khz; 5 kw-U, DA-N. TL: N44 29 47 W73 12 49. Hrs open: 24 Box 620, 118 Malletts Bay Ave, Colchester, 05446. Phone: (802) 655-1620. Fax: (802) 655-1329. E-mail: paulg@95triplex.com Web Site: www.am620wvmt.com. Licensee: Sison Broadcasting Inc. (acq 3-3-97; $2,939,014 with WXXX(FM) South Burlington). Population served: 250,000 Natl. Network: CBS. Natl. Rep: McGavren Guild. Format: News/talk & sports. News staff: 2; News: 14 hrs wkly. Target aud: 35-65. ♦Paul S. Goldman, pres & gen mgr; Mark Esbjerg, opns mgr; Christine Miller, gen sls mgr & prom mgr.

***WVPS(FM)**— Oct 15, 1980: 107.9 mhz; 48.8 kw. Ant 2,716 ft TL: N44 31 32 W72 48 58. Stereo. Hrs opn: 24 365 Troy Ave., Colchester, 05446. Phone: (802) 655-9451. Fax: (802) 655-2799. Fax: (802) 655-9117. E-mail: contract@vpr.net Web Site: www.vpr.net. Licensee: Vermont Public Radio. Natl. Network: NPR, PRI. Law Firm: Haley, Bader & Potts. Format: Class, jazz, news. News staff: 9; News: 43 hrs wkly. Target aud: General. Spec prog: Switchboard call-in progmg 3 hrs, opera 5 hrs, folk 4 hrs, children 5 hrs wkly. ♦Mark Vogelzang, CEO, pres & gen mgr; Brian Donahue, CFO; Victoria St. John, opns dir; Robin Turnau, dev dir; Jody Evans, progmg dir; Walter Parker, mus dir; John Van Hoesen, news dir; Richard Parker, chief of engrg; Betty Smith, spec ev coord.

Castleton

***WIUV(FM)**— Oct 1, 1976: 91.3 mhz; 227 w. -235 ft TL: N43 36 29 W73 10 54. Hrs open: Castleton State College, 86 Seminary St., 05735. Phone: (802) 468-5611. Phone: (802) 468-1264. Fax: (802) 468-5237. Web Site: www.castleton.edu. Licensee: Board of Trustees. Population served: 55,000 Format: Progsv, variety. Target aud: General; smart people. Spec prog: Jazz 10 hrs, reggae 6 hrs, rap-urban 5 hrs, folk 4 hrs, Sp one hr wkly. ♦Robert Gershon, gen mgr.

Colchester

***WWPV-FM**— Aug 10, 1973: 88.7 mhz; 100 w. 82 ft TL: N44 29 38 W73 09 51. Hrs open: 8 AM-2 PM St. Michaels College, Box 274, Winooski Park, 05439. Phone: (802) 654-2334. Fax: (802) 654-2336. E-mail: wwpv@smcvt.edu Web Site: personalweb.smcvt.edu/wwpv. Licensee: Board of Trustees, St. Michaels College. Format: Free-form. Target aud: 14-65; varies by time of day & progmg. ♦Mike McCarthy, stn mgr; Jon Van Luling, progmg dir.

Danville

WDOT(FM)— 1996: 95.7 mhz; 3.8 kw. Ant 246 ft TL: N44 24 58 W72 03 32. Hrs open:
Rebroadcasts WNCS(FM) Montpelier 90%.
Box 374, St. Johnsbury, 05819. Phone: (802) 748-4055. Phone: (877) 367-6468. Fax: (802) 748-6939. E-mail: klm@pointfm.com Web Site: www.pointfm.com. Licensee: Montpelier Broadcasting Inc. Group owner: Northeast Broadcasting Company Inc. (acq 1996; $152,500 for CP). Format: AAA. ♦Kim Buckminster, gen mgr & gen sls mgr; Terry Lieberman, gen mgr; Jamie Canfield, progmg dir; John Hosford, chief of engrg.

Derby Center

WMOO(FM)— Apr 1, 1991: 92.1 mhz; 2.25 kw. 619 ft TL: N44 58 23 W72 04 30. Stereo. Hrs opn: 24 Box 92, Derby/Newport Rd., 05829. Phone: (802) 766-9236. Fax: (802) 766-8067. E-mail: dprudhomme @nassaubroadcasting.com Web Site: www.northstarhits.com. Licensee: Nassau Broadcasting III L.L.C. (acq 12-22-2004; $2.35 million with WIKE(AM) Newport). Population served: 65,000 Law Firm: Shaw Pittman. Wire Svc: AP Format: Hot adult contemp. News staff: one; News: 16 hrs wkly. Target aud: General. Spec prog: Community events 8 hrs wkly. ♦Dawn Prudhomme, opns mgr; Doug Weldon, progmg dir.

Essex Junction

WVVT(AM)—Not on air, target date: unknown: 670 khz; 50 kw-D, 300 w-N, 20 kw-CH, DA-3. TL: N44 29 40 W73 08 37. Hrs open: 16 Doe Run, Pittstown, NJ, 08867. Phone: (908) 730-7959. Fax: (908) 730-7408. Licensee: Charles A. Hecht and Alfredo Alonso. ♦Charles A. Hecht, gen mgr.

Hartford

WWOD(FM)— Mar 15, 1992: 104.3 mhz; 5.6 kw. Ant 495 ft TL: N43 39 15 W72 21 32. Stereo. Hrs opn: 24 106 N. Main St., West Lebanon, NH, 03784. Phone: (603) 298-2953. Phone: (603) 542-7735. Fax: (603) 298-7554. E-mail: info@bestoldies104.com Web Site: www.bestoldies104.com. Licensee: Family Broadcasting Inc. Group owner: Nassau Broadcasting Partners L.P. (acq 8-2-2004; grpsl). Natl. Network: USA. Law Firm: May & Dunne. Format: Oldies. News staff: one. Target aud: 25 plus. Spec prog: Children 3 hrs, country gospel 2 hrs wkly. ♦Camille Losapio, gen mgr & natl sls mgr; Heath Cole, progmg dir.

Johnson

***WJSC-FM**— July 16, 1972: 90.7 mhz; 200 w. Ant -489 ft TL: N44 38 29 W72 40 20. Stereo. Hrs opn: Box 75, c/o Johnson State College., 05656. Phone: (802) 635-1355. Phone: (802) 635-1434. Fax: (802) 635-1202. E-mail: wjsc907@hotmail.com Web Site: www.wjsc.findhere.org. Licensee: Board of Trustees, Vermont State College. Population served: 15,000 Format: Alternative, div. Spec prog: Class 3 hrs, C&W 3 hrs wkly. ♦Andrew Frappier, gen mgr.

Killington

WEBK(FM)— Aug 4, 1993: 105.3 mhz; 50 kw. 2,240 ft TL: N43 38 22 W72 50 12. Stereo. Hrs opn: 24 Box 30, Rutland, 05702. Phone: (802) 775-7500. Fax: (802) 775-7555. E-mail: webk@catamountradio.com Licensee: 6 Johnson Road Licenses Inc. Group owner: Pamal Broadcasting Ltd. (acq 10-19-2001; grpsl). Population served: 180,000 Law Firm: Verner, Liipfert, Bernhard, McPherson & Hand. Format: AAA. News: one hr wkly. Target aud: 24-48; baby boomers. ♦Harry Weinhagen, gen mgr; Paul Hatin, sls dir & gen sls mgr; Sheila Bigelow, mktg dir & mktg mgr; Spider Glen, progmg VP & progmg dir; Peter Morton, chief of engrg.

Lyndon

WGMT(FM)— May 19, 1990: 97.7 mhz; 600 w. 1,883 ft TL: N44 34 15 W71 53 40. Stereo. Hrs opn: 24 Box 97, 10 Church St., Lyndonville, 05851. Phone: (802) 626-9800. Phone: (802) 626-0977. Fax: (802) 626-8500. E-mail: wgmt@kingcon.com Web Site: www.kingcon.com. Licensee: Vermont Broadcast Associates Inc. Population served: 75,000 Natl. Network: CNN Radio. Natl. Rep: Roslin. Law Firm: Bryan Cave. Format: Adult contemp. News staff: 2; News: 5 hrs wkly. Target aud: 22-54; families, more female, disposable income, mobile. ♦Bruce James, pres & gen mgr; Steve Nichols, gen sls mgr; Mike Barrett, progmg dir; Don Smith, chief of engrg.

Lyndonville

***WWLR(FM)**— Feb 4, 1977: 91.5 mhz; 3 kw. -75 ft TL: N44 32 04 W72 01 36. Stereo. Hrs opn: 24 hrs Box F, Lyndon State College, 05851. Phone: (802) 626-6214. Fax: (802) 626-4806. E-mail: impulse915@hotmail.com Web Site: www.lsc.vsc.edu. Licensee: Board of Trustees, Vermont State Colleges. Population served: 100,000 Format: Rock/AOR. Target aud: Everyone. Spec prog: Class 2 hrs, jazz 3 hrs wkly. ♦P.J. Cioffi, gen mgr; Jim Champine, opns dir.

Manchester

WEQX(FM)— November 1984: 102.7 mhz; 1.25 kw. 2,490 ft TL: N43 09 58 W73 06 59. Stereo. Hrs opn: 24 Box 1027, 05254. Secondary address: 161 Elm St., Manchester Center 05255. Phone: (802) 362-4800. Phone: (802) 362-4875. Fax: (802) 362-5555. Fax: (802) 362-4885. E-mail: eqx@weqx.com Web Site: www.weqx.com. Licensee: Northshire Communications Inc. Natl. Network: AP Radio. Format: Alternative. News: 5 hrs wkly. Target aud: 25-44. Spec prog: Jazz 4 hrs, AAA 4 hrs, locl 2 hrs, new music 3 hrs wkly. ♦A. Brooks Brown, pres & gen mgr; Melinda Brown, VP & opns mgr; Tim Bronson, progmg dir.

Marlboro

WRSY(FM)— July 1996: 101.5 mhz; 120 w. Ant 745 ft TL: N42 50 46 W72 41 16. Hrs open:
Rebroadcasts WRSI(FM) Turners Falls, MA 100%.
Box 268, Northhampton, MA, 01061. Secondary address: 15 Hampton Ave., Northampton, MA 01061. Phone: (413) 586-7400. Fax: (413) 585-0927. E-mail: dj@wrsi.com Web Site: www.wrsi.com. Licensee: Saga Communications of New England LLC. Group owner: Saga Communications Inc. (acq 2-13-2004; grpsl). Format: AAA. Target aud: 35 plus; women. ♦Diane O'Connell, opns dir & gen sls mgr; Scott Howard, prom dir; Sean O'Mealy, gen mgr & news dir; Howard Frost, chief of engrg.

Middlebury

WFAD(AM)— Dec 24, 1965: 1490 khz; 1 kw-U. TL: N43 59 57 W73 09 35. Hrs open: 24 372 Dorset St., South Burlington, 05403. Phone: (802) 388-9000. Fax: (802) 388-3000. Web Site: www.wtwk1070.com. Licensee: Addison Broadcasting Co. Inc. Group owner: Northeast Broadcasting Company Inc. (acq 6-22-2001). Population served: 45,000 Natl. Network: ESPN Radio. Law Firm: Mullin, Rhyne, Emmons & Topel. Format: Sports. ♦Bob Rowe, VP; Richard C. DeLancey Sr., gen mgr; J.J. Prieve, progmg dir; Mike Raymond, chief of engrg.

***WRMC-FM**— May 1949: 91.1 mhz; 2.9 kw. Ant -30 ft TL: N44 00 25 W73 10 40. Stereo. Hrs opn: 24 Middlebury College, 05753. Phone: (802) 443-2471. Phone: (802) 443-6324. Fax: (802) 443-5108. E-mail: wrmc@wrmc.middlebury.edu Web Site: www.wrmc.middlebury.edu. Licensee: President and Fellows of Middlebury College. Population served: 100,000 Natl. Network: AP Radio. Format: Div, progsv. News: 8 hrs wkly. Target aud: General. Spec prog: Urban contemp 12 hrs, class 15 hrs, folk 15 hrs, relg one hr, blues 10 hrs, jazz 10 hrs, Sp one hr wkly. ♦Ryan Abernnathey, gen mgr.

Montpelier

WNCS(FM)— June 13, 1977: 104.7 mhz; 1.9 kw. 2,093 ft TL: N44 18 14 W72 37 18. Stereo. Hrs opn: 24 169 River St., 05602-3724. Phone: (802) 223-2396. Fax: (802) 223-1520. Web Site: www.pointfm.com. Licensee: Montpelier Broadcasting Co. Inc. Group owner: Northeast Broadcasting Company Inc. (acq 2-12-87). Population served: 350,000 Format: AAA. Target aud: 25-40; above-average income & educated, baby boomers. Spec prog: Folk 4 hrs, jazz 5 hrs wkly. ♦Steven Silberberg, pres; Terry Lieberman, gen mgr; Caroline Scribner, gen sls mgr; Jamie Canfield, progmg dir; John Hosford, chief of engrg.

WORK(FM)—See Barre

WSKI(AM)— Dec 7, 1947: 1240 khz; 1 kw-U. TL: N44 14 40 W72 32 47. Stereo. Hrs opn: 169 River St., 05602. Phone: (802) 223-5275. Fax: (802) 223-1520. E-mail: terry@pointfm.com Web Site: www.pointfm.com. Licensee: Galloway Communications Inc. Group owner: Northeast Broadcasting Company Inc. (acq 5-2-00; grpsl). Population served: 85,000 Format: Oldies. Target aud: 35-64; 60% female, 40% male. ♦Terry Lieberman, gen mgr; Caroline Scribner, gen sls mgr; Jamie Canfield, progmg dir & progmg mgr; John Hosford, chief of engrg.

WSNO(AM)—See Barre

Morrisville

WLVB(FM)— August 1993: 93.9 mhz; 5.4 kw. 121 ft TL: N44 34 24 W72 38 11. Hrs open: Box 94, 05661. Phone: (802) 888-4294. Fax: (802) 888-8523. E-mail: wlvb@radiovermont.com Web Site: www.wlvbradio.com. Licensee: Radio Vermont Inc. Group owner: Radio Vermont Group Inc. Format: Country. ♦Ken Squier, pres; Eric Michaels, gen mgr; Craig Ladd, opns mgr.

Newport

WIKE(AM)— Oct 12, 1952: 1490 khz; 1 kw-U. TL: N44 56 28 W72 13 35. Stereo. Hrs opn: 24 Box 1490, 05855. Secondary address: Derby Newport Rd., Derby 05829. Phone: (802) 766-9236. Fax: (802) 766-8067. Web Site: www.northstarhits.com. Licensee: Nassau Broadcasting III L.L.C. (acq 12-22-2004; $2.35 million with WMOO(FM) Derby Center). Population served: 55,000 Law Firm: Covington & Burling. Format: Country. News staff: one; News: 4 hrs wkly. Target aud: 18 plus. Spec prog: Loc info/entertainment 5 hrs wkly. ♦William J. Macek, gen mgr; Dawn Prudhomme, opns mgr.

Northfield

***WNUB-FM**— Dec 8, 1967: 88.3 mhz; 285 w. -387 ft TL: N44 08 32 W72 39 31. Stereo. Hrs opn: 24 158 Harmon Dr., Comm. Ctr., Norwich Univ., 05663. Phone: (802) 485-2483. Fax: (802) 485-2565. E-mail: wnub@norwich.edu Web Site: www.norwich.edu. Licensee: The Trustees of Norwich University. Population served: 7,500 Format: Rock/AOR, Triple A, Modern. Target aud: 15-40. ♦Doug Smith, gen mgr.

Norwich

***WNCH(FM)**— 2004: 88.1 mhz; 100 w horiz, 1.6 kw vert. Ant 2,257 ft TL: N43 26 15 W72 27 08. Hrs open: 24 Vermont Public Radio, 365 Troy Ave., Colchester, 05446. Phone: (802) 655-9451. Fax: (802) 655-2799. Web Site: www.vpr.net. Licensee: Vermont Public Radio. Natl. Network: NPR. Format: Cultural music svc. ♦Mark Vogelzang, pres & gen mgr.

Plainfield

***WGDR(FM)**— May 11, 1973: 91.1 mhz; 800 w. -350 ft TL: N44 17 04 W72 26 28. Stereo. Hrs opn: Box 336, Goddard College, 05667. Phone: (802) 454-7762. Phone: (802) 454-9962. Fax: (802) 454-1451. E-mail: wgdr@goddard.edu Web Site: www.wgdr.org. Licensee: Goddard College Corp. Population served: 50,000 Format: Div. Target aud: Multiple. ♦Christine Farren, gen mgr & spec ev coord; Greg Hooker, gen mgr; Bert Klunder, opns mgr; Jen Isaacs, mus dir.

Poultney

WVNR(AM)— Aug 1, 1981: 1340 khz; 1 kw-U. TL: N43 30 16 W73 12 11. Hrs open: 5:30 AM-midnight Box 568, East Poultney, 05741. Secondary address: 1214 Rt. 30 S. 05764. Phone: (802) 287-9030. E-mail: wvnrwnyv@yahoo.com Licensee: Pine Tree Broadcasting Co. (acq 4-86). Format: Adult contemp, country, oldies. News staff: one; News: 3 hrs wkly. Target aud: 25-54; active, community oriented, working and professional, and families. Spec prog: Big band 3 hrs, loc sports 6 hrs, swap shop one hr, Polish one hr, gospel one hr wkly. ♦Michael J. Leech, pres; Judith E. Leech, exec VP & gen mgr.

Putney

***WCMK(FM)**— 2003: 91.9 mhz; 150 w. Ant 758 ft TL: N42 58 28 W72 36 12. Hrs open: Box 8310, Essex, 05451-8310. Phone: (802) 878-8885. Fax: (802) 879-6835. E-mail: cmi.radio@verizon.net Web Site: thelightradio.net. Licensee: Christian Ministries Inc. Format: Christian. ♦Ric McClary, gen mgr.

Randolph

WCVR-FM—Listing follows WTSJ(AM).

WTSJ(AM)— Nov 26, 1968: 1320 khz; 1 kw-D, 66 w-N. TL: N43 56 21 W72 38 13. Hrs open: 62 Radio Dr., 05060. Phone: (802) 728-4411. Fax: (802) 728-4013. E-mail: randolphradio@clearchannel.com Licensee: Capstar TX L.P. Group owner: Clear Channel Communications Inc. (acq 12-22-2000; grpsl). Population served: 25,000 Law Firm: Fisher, Wayland, Cooper, Leader & Zaragoza L.L.P. Format: News/talk. Target aud: 18-49. ♦Tim Plante, gen mgr; Karen Warner, stn mgr.

WCVR-FM—Co-owned with WTSJ(AM). Oct 25, 1982: 102.1 mhz; 11 kw. 436 ft TL: N43 57 20 W72 36 10. Stereo. 24 Web Site: www.champrocks.com.315,000 Format: Classic rock. News: 6 hrs wkly. Target aud: 25-54; 50/50 men and women.

Randolph Center

***WVTC(FM)**— Aug 29, 1983: 90.7 mhz; 300 w horiz. 203 ft TL: N43 56 07 W72 36 10. Stereo. Hrs opn: 24 Vermont Technical College, Box 500, 05061. Phone: (802) 728-1550. Fax: (802) 728-1550. Web Site: www.wvtc.net. Licensee: Vermont State Colleges Vermont Technical College. Population served: 2,115 Format: Punk, rap, rock, alternative. Target aud: General. ♦Marcus Jacobus-Petter, stn mgr.

Royalton

WRJT(FM)— 1996: 103.1 mhz; 1.35 kw. 682 ft TL: N43 46 28 W72 23 55. Hrs open:
Rebroadcasts WNCS(FM) Montpelier 90%.
c/o WNCS(FM), 169 River St., Montpelier, 05602. Phone: (802) 223-2396. Fax: (802) 223-1520. Web Site: www.pointfm.com. Licensee: Lisbon Communications Inc. Group owner: Northeast Broadcasting Company Inc. (acq 11-30-01). Format: AAA. ♦Ed Flanagan, gen mgr; Tanya Stepasiuk, prom dir; Mark Miller, progmg dir; Jon Hosford, chief of engrg.

Rupert

WMNV(FM)— Apr 10, 1990: 104.1 mhz; 4.3 kw horiz. Ant 200 ft TL: N43 16 01 W73 15 21. Stereo. Hrs opn: 24
Rebroadcasts WHAZ(AM) Troy, NY 100%.
30 Park Ave., Cohoes, NY, 12047-3330. Phone: (518) 237-1330. Fax: (518) 235-4468. E-mail: info@whaz.com Web Site: www.whaz.com. Licensee: Capital Media Corp. (group owner; acq 4-15-97). Population served: 500,000 Format: Adult Christian. Target aud: 25-75. Spec prog: Gospel, relg. ♦Paul F. Lotters, pres & gen mgr; Steven L. Klob, opns dir & dev dir; Rex P. Gregory, progmg dir & mus dir.

Rutland

***WFTF(FM)**— Jan 10, 1987: 90.5 mhz; 720 w. -560 ft TL: N43 37 09 W72 59 04. Hrs open: 24 2 Meadow Ln., 05701. Phone: (802) 775-0358. Phone: (802) 773-2863. Web Site: www.cbcvt.org. Licensee: Calvary Bible Church. Population served: 25,000 Natl. Network: Moody. Format: Christian. ♦Ronald Systo, pres & gen mgr.

WJEN(FM)— October 1988: 94.5 mhz; 6 kw. 389 ft TL: N43 36 49 W73 01 33. Stereo. Hrs opn: 24
Rebroadcasts WJAN(FM) Manchester 90%.
Box 30, 05702. Phone: (802) 775-7500. Fax: (802) 775-7555. E-mail: catcountrymornings@hotmail.com Web Site: www.catcountry.net. Licensee: 6 Johnson Road Licenses Inc. Group owner: Pamal Broadcasting Ltd. (acq 10-19-2001; grpsl). Natl. Network: ABC, Westwood One, Motor Racing Net. Format: New country. News: one hr wkly. Target aud: 18-49; adult, income $35,000 plus, homeowners. Spec prog: Sports 4 hrs wkly. ♦Harry Weinhagen, gen mgr; Paul Hatin, stn mgr; Nicole Daigle, gen sls mgr; Jason Davis, prom mgr; Skip Carlton, progmg dir.

WJJR(FM)— Mar 25, 1971: 98.1 mhz; 1.15 kw. 2,591 ft TL: N43 36 17 W72 49 14. Stereo. Hrs opn: 24 Box 30, 05702. Secondary address: 67 Merchants Row, Ruthland 05702. Phone: (802) 775-7500. Fax: (802) 775-7555. E-mail: wjjr@catamountradio.com Web Site: www.wjjr.net. Licensee: 6 Johnson Road Licenses Inc. Group owner: Pamal Broadcasting Ltd. (acq 10-19-2001; grpsl). Population served: 847,178 Law Firm: Hogan & Hartson. Format: Adult contemp. News staff: one; News: one hr wkly. Target aud: 25-54; in-office managerial, professional. Spec prog: News, pub affrs one hr wkly. ♦Harry Weinhagen, gen mgr; Debbie Grembowiez, gen sls mgr; Ed Kelly, prom dir & news dir; Terry Jayl, progmg dir.

***WRVT(FM)**— Jan 10, 1989: 88.7 mhz; 2.77 kw. 1,328 ft TL: N43 39 32 W73 06 25. Hrs open:
Rebroadcasts WVPS(FM) Burlington 100%.
365 Troy Ave., Colchester, 05446. Phone: (802) 655-9451. Fax: (802) 655-2799. Fax: (802) 655-1801. E-mail: contact@vpr.net Web Site: www.vpr.net. Licensee: Vermont Public Radio. Natl. Network: NPR, PRI. Format: Class, jazz, news. Target aud: General. Spec prog: Switchboard call-in 3 hrs, folk 4 hrs wkly. ♦Mark Vogelzang, pres & gen mgr; Victoria St. John, opns dir; Robin Turnau, dev VP & dev dir; Jody Evans, progmg dir; Walter Parker, mus dir; John Van Hoesen, news dir; Rich Parker, chief of engrg.

WSYB(AM)— Dec 10, 1930: 1380 khz; 5 kw-D, 1 kw-N, DA-D. TL: N43 35 35 W72 59 25. Hrs open: Box 940, 05702-0940. Secondary address: 250 Dorr Dr. 05701. Phone: (802) 775-5597. Fax: (802) 775-6637. Licensee: 6 Johnson Road Licenses Inc. Group owner: Clear Channel Communications Inc. (acq 4-1-2007; grpsl). Population served: 70,000 Natl. Rep: McGavren Guild. Format: News/talk. Target aud: 35-64. ♦Dave Ryeron, progmg dir; Glen Dudley, chief of engrg.

WZRT(FM)—Co-owned with WSYB(AM). 1974: 97.1 mhz; 1.15 kw. 2,591 ft TL: N43 36 17 W72 49 14. Stereo. 24 58,000 Format: Adult contemp. News staff: 2; News: 5 hrs wkly. Target aud: 18-49.

Saint Albans

WLFE-FM—Listing follows WRSA(AM).

WRSA(AM)— 1930: 1420 khz; 1 kw-D, 110 w-N. TL: N44 50 12 W73 04 57. Hrs open: 24 Box 712, 2 Main St., 05478. Phone: (802) 524-2133. Fax: (802) 527-1450. E-mail: wlfe.fm@tverizonr.net Licensee: Champlain Communications Corp. Group owner: Northeast Broadcasting Company Inc. (acq 9-18-98; $500,000 with co-located FM). Population served: 60,000 Format: Talk radio. News staff: one; News: 20 hrs wkly. Target aud: General. ♦Mike Kmack, gen mgr, opns mgr & progmg dir.

WLFE-FM—Co-owned with WRSA(AM). April 1970: 102.3 mhz; 440 w. 800 ft TL: N44 46 56 W73 03 54. Stereo. 24 270000 Format: Country. News staff: one; News: 5 hrs wkly. Target aud: 18-54.

Saint Johnsbury

***WCKJ(FM)**— Aug 1, 1998: 90.5 mhz; 1 kw. 738 ft TL: N44 24 40 W71 58 13. Stereo. Hrs opn:
Rebroadcasts WGLY(FM) Bolton 100%.
Box 8310, Essex, 05451-8310. Secondary address: 140 Main St., Essec Junction 05452. Phone: (802) 878-8885. Fax: (802) 879-6835. E-mail: cmi.radio@verizon.net Web Site: thelightradio.net. Licensee: Christian Ministries Inc. Natl. Network: Moody. Law Firm: Joseph E. Dunne III. Format: Inspirational. ♦Ric McClary, gen mgr.

WKXH(FM)—Listing follows WSTJ(AM).

WSTJ(AM)— July 10, 1949: 1340 khz; 1 kw-U. TL: N44 25 06 W71 59 45. Hrs open: 24 Box 249, 1303 Concord Ave., 05819. Phone: (802) 748-1340. Fax: (802) 748-2361. E-mail: kix105@kix1055.com Licensee: Vermont Broadcast Associates Inc. (acq 4-3-98; $630,000 with co-located FM). Population served: 37,500 Natl. Network: ABC, Jones Radio Networks. Law Firm: Bryan Cave. Format: Adult standards. News staff: one; News: 20 hrs wkly. Target aud: 35-75; adults and work places. ♦Bruce James, pres & gen mgr; Candis Leopold, opns mgr; Dave Labounty, progmg dir; Don Smith, chief of engrg.

WKXH(FM)—Co-owned with WSTJ(AM). Aug 1, 1985: 105.5 mhz; 400 w. 712 ft TL: N44 24 38 W71 58 13. Stereo. 24 Phone: (802) 748-2345.160,000 Natl. Network: ABC, Westwood One. Format: Hot country. News staff: 2. Target aud: 24-55; families.

***WVPA(FM)**— 1999: 88.5 mhz; 290 w vert. Ant 1,863 ft TL: N44 34 15 W71 53 38. Hrs open: Vermont Public Radio, 365 Troy Ave., Colchester, 05446. Phone: (802) 655-9451. Fax: (802) 655-2799. E-mail: mvogelzang@vpr.net Web Site: www.vpr.net. Licensee: Vermont Public Radio. Natl. Network: NPR. Format: Class, jazz, news. ♦Mark Vogelzang, gen mgr; Victoria St. John, opns dir; Robin Turnau, dev VP; Jody Evans, progmg dir; Walter Parker, mus dir; John Van Hoesen, news dir; Richard Parker, chief of engrg.

South Burlington

WXXX(FM)— Nov 16, 1984: 95.5 mhz; 25 kw. 236 ft TL: N44 30 35 W73 11 05. Stereo. Hrs opn: 24 Box 620, Colchester, 05446. Secondary address: Malletts Bay Ave., Colchester 05446. Phone: (802) 655-9550. Fax: (802) 655-1329. E-mail: chris@95triplex.com Web Site: www.95triplex.com. Licensee: Sison Broadcasting Inc. (acq 3-3-97; $2,939,014 with WVMT(AM) Burlington). Population served: 250,000 Natl. Rep: McGavren Guild. Format: CHR. News staff: one; News: one hr wkly. Target aud: 18-49. ♦Mark Esbjerg, opns mgr; Paul Goldman, gen mgr & gen sls mgr; Ben Hamilton, progmg dir; Chantal Paulino, news dir.

Springfield

WCFR(AM)— May 26, 1954: 1480 khz; 5 kw-D. TL: N43 16 54 W72 29 21. Hrs open: 24 19 Main St., 05156-2914. Phone: (603) 885-1480.

Phone: (603) 448-0500. Fax: (603) 448-6601. Fax: (603) 526-9372. E-mail: realoldies1480@vermontel.net Web Site: www.wcfr1480.com. Licensee: KOOR Communications Inc. (group owner; (acq 12-19-2001; $75,000). Population served: 65,000 Format: Oldies. News staff: one; News: 7 hrs wkly. Target aud: 45 plus; 35-54. ♦Bob Vinikoor, pres; Ray Lemire, gen mgr, opns dir & progmg dir.

WTSM(FM)— Jan 1, 1972: 93.5 mhz; 3 kw. Ant 300 ft TL: N43 16 54 W72 29 21. Stereo. Hrs opn:
Rebroadcasts WXXK(FM) Lebanon, NH 100%.
31 Hanover St., Suite 4, Lebanon, NH, 03766. Phone: (603) 448-1400. Fax: (603) 448-1755. E-mail: michaelbarrett@clearchannel.com Web Site: www.kixx.com. Licensee: Clear Channel Broadcasting Licenses Inc. Group owner: Clear Channel Communications Inc. (acq 2-16-2001; $2 million with WMXR(FM) Woodstock). Population served: 120,000 Format: Country. Target aud: 25-54. ♦Tim Plante, gen mgr; Michael Barrett, opns dir & progmg dir; Chris Olsen, gen sls mgr.

Stowe

WCVT(FM)— Feb 28, 1977: 101.7 mhz; 130 w. 2,066 ft TL: N44 25 14 W72 49 42. Stereo. Hrs opn: 24 Box 3536, Mountain Rd., 05672. Secondary address: 9 Stowe St., Waterbury 05676. Phone: (802) 244-1764. Fax: (802) 244-1771. E-mail: wcvt@classicvermont.com Web Site: www.wcvtradio.com. Licensee: Radio Vermont Classics L.L.C. Group owner: Radio Vermont Group Inc. (acq 6-19-97; $450,000). Population served: 185,000 Natl. Rep: McGavren Guild. Format: Classical. Target aud: 25-54; educated, upscale adults & families with active lifestyles. Spec prog: Children one hr wkly. ♦Eric Michaels, gen mgr; Thomas B. Beardsley, stn mgr; Frankie Allen, opns dir.

Sunderland

***WVTQ(FM)**— May 1, 1991: 95.1 mhz; 96 w. Ant 2,398 ft TL: N43 09 58 W73 07 02. Stereo. Hrs opn: 24
Rebroadcasts WNCH(FM) Norwich 100%.
Vermont Public Radio, 365 Troy Ave., Colchester, 05446. Phone: (802) 655-9451. Fax: (802) 655-2799. Web Site: www.vpr.net. Licensee: Vermont Public Radio Group owner: Pamal Broadcasting Ltd. (acq 11-15-2006; $625,000). Natl. Network: NPR. Format: Classical. Target aud: 18-49. ♦Mark Vogelzang, pres & gen mgr.

Vergennes

WIZN(FM)—Licensed to Vergennes. See Burlington

Warren

WDEV-FM— Aug 11, 1989: 96.1 mhz; 3 kw. 4,000 ft TL: N44 07 37 W72 55 43. Stereo. Hrs opn: 24
Rebroadcasts WDEV(AM) Waterbury 100%.
Box 550, Waterbury, 05676. Secondary address: 9 Stowe St., Waterbury 05676. Phone: (802) 244-7321. Fax: (802) 244-1771. E-mail: wdev@radiovermont.com Web Site: www.wdevradio.com. Licensee: Radio Vermont Inc. Group owner: Radio Vermont Group Inc. (acq 10-15-92; $643,000 with WKDR(AM) Burlington; FTR: 11-23-92) Population served: 300,000 Natl. Network: ABC. Law Firm: Wiley, Rein & Fielding. Format: News/talk, sports, music. News staff: one; News: 2 hrs wkly. Target aud: 25-54; affluent, upscale baby boomer generation. ♦Ken D. Squier, pres; Eric Michaels, gen mgr; Fred Hill, gen sls mgr; Jack Donovan, progmg dir; Rick Haskell, news dir; Tom Laffin, chief of engrg.

Waterbury

WDEV(AM)— July 16, 1931: 550 khz; 5 kw-D, 1 kw-N, DA-2. TL: N44 21 17 W72 45 07. Hrs open: Box 550, 9 Stowe St., 05676. Phone: (802) 244-7321. Fax: (802) 244-1771. E-mail: wdev@radiovermont.com Web Site: www.wdevradio.com. Licensee: Radio Vermont Inc. Group owner: Radio Vermont Group Inc. (acq 1969). Population served: 250,000 Format: News, sports, div. Spec prog: Class one hr wkly. ♦Eric Michaels, gen mgr; Fred Hill, gen sls mgr; Jack Donovan, progmg dir; Rich Haskell, news dir; Tom Laffin, chief of engrg.

WWMP(FM)— Feb 14, 1985: 103.3 mhz; 3 kw. Ant 912 ft TL: N44 21 52 W72 55 53. Stereo. Hrs opn: 24 372 Dorset St., South Burlington, 05403. Phone: (802) 863-1010. Fax: (802) 860-4721. Licensee: Radio Broadcasting Services Inc. Group owner: Northeast Broadcasting Company Inc. (acq 5-4-2000). Natl. Network: USA. Law Firm: Joseph E. Dunne III. Format: Variety hits. News staff: one; News: 16 hrs wkly. Target aud: 25-54. Spec prog: Children 3 hrs wkly. ♦Rich Delancey, gen mgr & stn mgr; J.J. Prieve, progmg dir; Mike Raymond, chief of engrg.

Wells River

WTWN(AM)— Oct 3, 1976: 1100 khz; 5 kw-D. TL: N44 08 55 W72 04 02. Hrs open: Sunrise-sunset Box 675, 1047 Rt. 302, 05081. Phone: (802) 757-3311. Fax: (802) 757-2774. E-mail: wtwngch@kingcon.com Web Site: www.wtwnradio.com. Licensee: Puffer Broadcasting Inc. (acq 10-3-73). Population served: 50,000 Natl. Network: Moody. Natl. Rep: Roslin. Law Firm: Shaw Pittman. Format: Relg. News: 9 hrs wkly. Target aud: 25 plus. Spec prog: Children, gospel. ♦Stephen J. Puffer, pres & gen mgr; Glenn Hatch, stn mgr & mktg dir; Teresa Puffer, opns mgr.

White River Junction

WNHV(AM)— Feb 28, 1963: 910 khz; 1 kw-D, 84 w-N. TL: N43 37 19 W72 21 04. Hrs open: 24 106 N. Main St., West Lebanon, NH, 03784. Phone: (603) 298-0332. Fax: (603)727-0134. E-mail: espnthescore@aol.com Licensee: Nassau Broadcasting III L.L.C. Group owner: Nassau Broadcasting Partners L.P. (acq 8-2-2004; grpsl). Population served: 8,000 Natl. Rep: Roslin. Format: All sports. Target aud: 18-55. ♦Shirley Clark, gen mgr.

WXLF(FM)—Co-owned with WNHV(AM). Feb 1, 1969: 95.3 mhz; 3 kw. 225 ft TL: N43 39 14 W72 17 43. Stereo. Web Site: www.bobcountryfm.com.70,000 Format: Hot country. News: one hr wkly. Target aud: 35-54. Spec prog: Jazz.

WTSL(AM)—See Hanover, NH

WXXK(FM)—See Lebanon NH

Wilmington

WTHK(FM)— June 1, 1989: 100.7 mhz; 135 w. Ant 1,460 ft TL: N42 57 33 W72 55 22. Stereo. Hrs opn: 24
Simulcast with WEXP(FM) Brandon 100%.
1 Scale Ave., Skuite 84, Rutland, 05761-4459. Secondary address: Box 850, West Dover 05356. Phone: (802) 464-1350. Fax: (802) 464-1112. Web Site: www.101thefox.com. Licensee: Nassau Broadcasting III L.L.C. (group owner; (acq 1-21-2005; $2.5 million with WEXP(FM) Brandon). Population served: 100,000 Natl. Rep: McGavren Guild. Law Firm: Dan Alpert. Format: Classic rock. Target aud: 25-54; residents, tourists, upscale Mt. ♦John Gales, gen mgr & stn mgr; Kelly Kowalski, progmg dir.

Windsor

***WVPR(FM)**— Aug 13, 1977: 89.5 mhz; 1.78 kw. 2,160 ft TL: N43 26 17 W72 27 08. Stereo. Hrs opn:
Rebroadcasts WVPS(FM) Burlington 100%.
365 Troy Ave., Colchester, 05446. Phone: (802) 655-9451. Fax: (802) 655-2799. Fax: (802) 655-9117. Web Site: www.vpr.net. Licensee: Vermont Public Radio. Population served: 90,000 Natl. Network: NPR, PRI. Law Firm: Haley, Bader & Potts. Format: Jazz, class, news. Target aud: General. Spec prog: Switchboard call-in 3 hrs, folk 6 hrs wkly. ♦Mark Vogelzang, CEO, pres & gen mgr; Brian Donahue, CFO; Michael Crane, opns dir; Victoria St. John, opns mgr; Robin Turnau, dev dir; Walter Parker, mus dir; Steve Young, news dir; Richard Parker, chief of engrg; Wayne Perry, chief of engrg; Jody Evans, progmg.

Woodstock

***WGLV(FM)**— 2003: 91.7 mhz; 100 w. Ant 2,276 ft TL: N43 38 22 W72 50 12. Hrs open:
Rebroadcasts WGLY-FM Bolton 100%.
Box 8310, Essex, 05451-8310. Secondary address: 140 Main St., Essex Junction 05452. Phone: (802) 878-8885. Fax: (802) 879-6835. E-mail: cmi.radio@verizon.net Web Site: thelightradio.net. Licensee: Christian Ministries Inc. Natl. Network: Salem Radio Network, Moody. Format: Inspirational. ♦Ric McClary, gen mgr.

WMXR(FM)— Apr 18, 1989: 93.9 mhz; 670 w. Ant 682 ft TL: N43 36 17 W72 28 03. Stereo. Hrs opn: 24
Rebroadcasts WVRR(FM) Newport, NH 100%.
31 Hanover St., Suite 4, Lebanon, NH, 03766. Phone: (603) 448-1400. Fax: (603) 448-1755. E-mail: pd@rock1017.com Web Site: www.wvrrfm.com. Licensee: Clear Channel Broadcasting Licenses Inc. Group owner: Clear Channel Communications Inc. (acq 2-16-2001; $2 million with WTSM(FM) Springfield). Population served: 65,000 Format: Rock. ♦Tim Plante, gen mgr; Chris Olsen, gen sls mgr; Steven Smith, progmg dir.

Virginia

Abingdon

WABN(AM)— Dec 10, 1956: 1230 khz; 1 kw-U. TL: N36 43 07 W81 56 55. Hrs open: 6 AM-midnight Box 1867, 24212. Phone: (276) 676-3806. Fax: (276) 676-3572. Licensee: Information Communication Corp. (acq 1-1-2006; $250,000 with WHGG(AM) Kingsport, TN). Population served: 220,000 Format: Relg. ♦Kenneth C. Hill, pres; Rev. Michael Smith, gen mgr; Lisa Smith, progmg dir; Billy Ray Dotson, chief of engrg; Sue Howington, traf mgr.

WFHG-FM— Dec 10, 1966: 92.7 mhz; 1.8 kw. 371 ft TL: N36 43 07 W81 56 55. Stereo. Hrs opn: 24 Bristol Broadcasting Co. Inc., 901 E. Valley Dr., Bristol, 24201. Phone: (276) 669-8112. Fax: (276) 669-0541. Web Site: www.supertalkwfhg.com. Licensee: Bristol Broadcasting Co. Inc. (group owner; acq 11-30-99; with co-located AM). Population served: 225,000 Format: News talk. News: 2 hrs wkly. Target aud: 12 plus; progsv, contemp & urban. ♦Larissa Sutherland, disc jockey; P.J. Finn, disc jockey; Ryland Sutherland, mus dir & disc jockey.

Accomac

WVES(FM)— Aug 13, 1990: 99.3 mhz; 22 kw. 344 ft TL: N37 47 05 W75 36 16. Stereo. Hrs opn: 24 27214 Muttonhunk Rd., Parksley, 23421. Phone: (757) 665-6500. Fax: (757) 665-6178. E-mail: hotcountry@tassnet.net Web Site: www.wevs.com. Licensee: Chincoteague Broadcasting Corp. (acq 5-18-98; $350,000). Natl. Network: USA, Westwood One. Format: Hot country. News: 3 hrs wkly. Target aud: 25 plus. ♦Stephen Marks, pres; Mark Dodds, gen mgr, stn mgr, opns mgr, gen sls mgr, prom mgr, prom mgr & mus dir; Dave Bralley, news dir; Kelli Spragg, pub affrs dir; Tom Reynolds, chief of engrg.

Alberta

WWDW(FM)— 2001: 103.1 mhz; 2.2 kw. Ant 535 ft TL: N36 52 02 W77 53 31. Hrs open:
Rebroadcasts WSMY(AM) Weldon, NC 100%.
Box 910, Roanoke Rapids, NC, 27870. Phone: (252) 536-0209. Fax: (252) 538-0378. E-mail: info@wsmy1400.com Web Site: www.wsmy1400.com. Licensee: First Media Radio LLC. (group owner; (acq 12-3-2003; grpsl). Format: Gospel. ♦Alan Garrick, gen mgr.

Alexandria

WXTR(AM)—Licensed to Alexandria. See Washington DC

Altavista

WKDE(AM)— Apr 29, 1962: 1000 khz; 1 kw-D. TL: N37 07 20 W79 17 20. Hrs open: Sunrise-sunset Box 390, 200 Frazier Rd., 24517. Phone: (434) 369-5588. Fax: (434) 369-1632. Licensee: DJ Broadcasting Corp. (acq 1-13-92; $300,000 with co-located FM; FTR: 2-10-92) Population served: 250,000 Natl. Network: CNN Radio. Rgnl. Network: Va. News Net. Format: News. News: one hr wkly. Target aud: General. Spec prog: Southern gospel 5 hrs wkly. ♦David Hoehne, pres & gen mgr; Elizabeth Clancy, chief of engrg & traf mgr.

WKDE-FM— June 30, 1969: 105.5 mhz; 6 kw. 328 ft TL: N37 09 37 W79 13 28. Stereo. 24 Phone: (434) 369-1055. E-mail: info@kdcountry.com Web Site: www.kdcountry.com.322,000 Natl. Network: AP Radio. Format: C&W. Target aud: 25-54. Spec prog: Black gospel 5 hrs, bluegrass 10 hrs wkly.

Amherst

WAMV(AM)— Oct 1, 1976: 1420 khz; 2.2 kw-D, 17 w-N. TL: N37 34 29 W79 01 14. Hrs open: Box 1420, 132 School Rd., 24521. Phone: (434) 946-9000. Fax: (434) 946-2201. E-mail: wamvradio@aol.com Licensee: Community First Broadcasters Inc. (acq 4-1-88; $40,000). Population served: 120,000 Natl. Network: USA. Format: Adult standards. Target aud: 50+. ♦Lee Parr, chief of engrg; Bruce Jarvis, disc jockey; Mary Lu Gregg, disc jockey; Robert Langstaff, pres, gen mgr, sls dir, progmg dir, traf mgr & disc jockey.

WYYD(FM)— Jan 27, 1981: 107.9 mhz; 20.5 kw. 1,768 ft TL: N37 28 13 W79 22 30. Stereo. Hrs opn: 24 3305 Old Forest Rd., Lynchburg, 24501. Phone: (434) 385-8298. Fax: (434) 385-7279. Web Site: www.wyyd.cc. Licensee: Capstar TX L.P. Group owner: Clear Channel Communications Inc. (acq 8-30-00; grpsl). Population served: 337,500 Natl. Network: ABC. Natl. Rep: McGavren Guild. Format: Country.

News staff: one. Target aud: 25-54; those with moderate high expendable income. ♦Chris Clendenen, gen mgr; Ron Gaylor, sls dir; Dave Carwile, gen sls mgr; Tom Sweat, natl sls mgr & rgnl sls mgr; Bobbi Crowder, prom dir; Joel Dearing, progmg dir; Jason Osborne, pub affrs dir; Jeff Parker, chief of engrg.

Appalachia

WAXM(FM)—See Big Stone Gap

Appomattox

WOWZ(AM)— June 1, 1974: 1280 khz; 1 kw-D. TL: N37 22 19 W78 50 06. (CP: COL Roanoke. 1290 khz; 10 kw-D, 17 w-N. TL: N37 16 06 W79 54 46). Hrs opn: 6 AM-8 PM Box 552, Forest, 24551. Phone: (434) 534-0400. Fax: (434) 534-0401. E-mail: news1280@msn.com Web Site: news1280.com. Licensee: Perception Media Inc. (acq 1-28-2004; $150,000). Population served: 12,000 Rgnl. Network: Va. News Net. Format: News. ♦Ben Peyton, pres.

WSNZ(FM)— May 17, 1989: 102.7 mhz; 22 kw. Ant 745 ft TL: N37 28 07 W79 00 27. Stereo. Hrs opn: 3305 Old Forest Rd., Lynchburg, 24501. Phone: (434) 385-8298. Fax: (434) 385-8991. E-mail: stevencross@clearchannel.com Web Site: www.magicfm.cc. Licensee: Capstar TX L.P. Group owner: Clear Channel Communications Inc. (acq 8-30-2000; grpsl). Format: Classic hits. Target aud: 35-54. Spec prog: New age 8 hrs wkly. ♦Chris Clendenen, gen mgr; Dave Carwile, gen sls mgr & adv mgr; Tom Sweat, natl sls mgr; Bobbi Crowder, prom dir; Ron Gaylor, adv VP; Sarah Macomber, adv dir; Bill Cahill, progmg VP; Steve Cross, progmg dir; Jeff Parker, chief of engrg.

WTTX-FM— September 1976: 107.1 mhz; 1.7 kw. Ant 426 ft TL: N37 22 19 W78 50 06. Hrs open: Box 637, 24522. Phone: (434) 352-7607. Fax: (434) 352-2451. E-mail: wttx@lynchburg.net Web Site: users.lynchberg.net/jsaves/wttxfm. Licensee: Positive Alternative Radio Inc. (acq 1-11-2006; $1.8 million). Population served: 600,000 Format: Southern gospel. ♦Edward A. Baker, pres; Terry Cook, gen mgr & gen sls mgr; Laura Coflin, mus dir & disc jockey; Glen Reinheimer, chief of engrg; Doug Marshall, disc jockey; Wayne Greene, disc jockey.

Arlington

WAVA(AM)— Nov 7, 1946: 780 khz; 5 kw-D. TL: N38 53 44 W77 08 04. Hrs open: Daytime 1901 N. Moore St., Suite 200, 22209. Phone: (703) 807-2266. Fax: (703) 807-2248. E-mail: comment@wava.com Web Site: www.wava.com. Licensee: Salem Media of Virginia Inc. Group owner: Salem Communications Corp. (acq 1-10-2000). Population served: 6,000,000 Natl. Network: Salem Radio Network. Natl. Rep: Salem. Law Firm: Fletcher Heald & Hildreth. Format: Contemp praise, worship, talk. News: 4 hrs wkly. Target aud: 25-54. ♦Edward Atsinger, CEO & pres; Stu Epperson, chmn; David Evans, CFO; Joe Davis, sr VP; David Ruleman, VP, gen mgr & opns VP; Tom Moyer, stn mgr.

WAVA-FM—Licensed to Arlington. See Washington DC

WZHF(AM)— Apr 7, 1947: 1390 khz; 5 kw-U, DA-2. TL: N38 54 15 W77 09 54. Stereo. Hrs opn: Way Broadcasting Inc., 12216 Parklawn Dr., Suite 203, Rockville, MD, 20852. Phone: (301) 424-9292. Fax: (301) 424-8266. Licensee: Way Broadcasting Licensee LLC. Group owner: Multicultural Radio Broadcasting Inc. (acq 5-30-00; grpsl). Population served: 489,000 Format: Sp. ♦Bill Parris, gen mgr & progmg dir; Raoul Lopez Bastidas, sls dir; David Song, chief of engrg; Bill Goodliff, traf mgr.

Ashland

WHAN(AM)— May 1, 1962: 1430 khz; 1 kw-D, 31 w-N. TL: N37 44 46 W77 29 44. (CP: COL Victoria. 650 khz; 50 kw-D, DA. TL: N37 22 27 W78 00 45). Hrs opn: Sunrise-sunset Box 148, 11337 W. Ashcake Rd., 23005. Phone: (804) 798-1010. Phone: (804) 345-1430. Fax: (804) 798-7933. E-mail: clear@msn.com Web Site: www.whan1430.com. Licensee: Fifth Estate Communications LLC (acq 2-23-98). Population served: 2,934 Natl. Network: USA, Moody. Format: Talk. News: 7 hrs wkly. Target aud: 18-35; general. ♦William Roberts, pres & stn mgr; Roger Reynolds, gen sls mgr & prom dir; Skip Andrews, progmg dir; Brian Edwards, chief of engrg; Rachel Reynolds, traf mgr.

WYFJ(FM)— Dec 7, 1967: 100.1 mhz; 6 kw. Ant 321 ft TL: N37 44 46 W77 29 44. Stereo. Hrs opn: 24 Box 7300, Charlotte, NC, 28241. Phone: (704) 523-5555. Web Site: www.bbnradio.org. Licensee: Bible Broadcasting Network Inc. (group owner; acq 2-1-80). Natl. Network: Bible Bcstg Net. Format: Relg, Christian. News: 9 hrs wkly. ♦Lowell Davey, pres; Randy Adams, gen mgr & stn mgr.

Bassett

WCBX(AM)— Oct 1, 1960: 900 khz; 1.1 kw-D, 180 w-N, DA-2. TL: N36 46 47 W80 00 35. Hrs open: 24 Box 192, Martinsville, 24114. Secondary address: 1675 Grandview Dr., Martinsville 24112. Phone: (276) 638-5235. Fax: (276) 638-6089. Web Site: thesportsaddictnetwork.com. Licensee: Base Communications Inc. (acq 4-17-98). Population served: 750,000 Format: Fox sports radio affiliate. News staff: one; News: 12 hrs wkly. Target aud: 35-55. ♦Edward A. Baker, pres & gen mgr; Brian Sanders, stn mgr; Michael Carter, opns mgr & progmg dir.

Bayside

WBVA(AM)—Licensed to Bayside. See Virginia Beach

Bedford

WBLT(AM)— Feb 9, 1950: 1350 khz; 5 kw-D, 47 w-N. TL: N37 20 51 W79 31 25. Hrs open: Box 348, Forest, 24551. Phone: (540) 586-8245. Licensee: 3 Daughters Media Inc. (acq 11-1-2005; $240,000). Natl. Network: ESPN Radio. Format: Sports. ♦Gary E. Burns, pres.

WLEQ(FM)— Oct 20, 1992: 106.9 mhz; 290 w. Ant 1,276 ft TL: N37 19 14 W79 37 59. Stereo. Hrs opn: 24 Box 11798, Lynchburg, 24506. Secondary address: 19-C Wadsworth St., Lynchburg 24501. Phone: (434) 845-3698. Phone: (866) 431-5253. Fax: (434) 845-2063. Web Site: www.boboldies.com. Licensee: Centennial Broadcasting LLC. Group owner: Cumulus Media Inc. (acq 8-11-2005; $1.9 million). Natl. Network: ABC. Natl. Rep: McGavren Guild. Law Firm: Womble, Carlyle, Sandridge & Rice. Format: Oldies. News staff: 3; News: 14 hrs wkly. Target aud: 36-64. ♦Harry Williams, gen mgr & gen sls mgr; Bob Abbott, opns mgr; Melinda Schamerhorn, traf mgr.

Berryville

WWRE(FM)— May 19, 1980: 105.5 mhz; 3 kw. 300 ft TL: N39 07 03 W77 58 22. Stereo. Hrs opn:
Simulcast with WWRT(FM) Strasburg 100%.
Box 3300, Winchester, 22604. Secondary address: 520 N. Pleasant Valley Rd., Winchester 22601. Phone: (540) 667-2224. Fax: (540) 722-3295. Web Site: www.realclassicrock.fm. Licensee: Mid Atlantic Network Inc. Group owner: Mid Atlantic Network (acq 5-28-97; $850,000 with WWRT(FM) Strasburg). Population served: 280,000 Law Firm: Cole, Raywid & Braverman. Format: Classic Rock. Target aud: 18 plus. ♦John P. Lewis, pres; Chris Lewis, gen mgr & gen sls mgr; Jeff Adams, opns mgr; Don Wilson, progmg dir; Steve Edwards, news dir; Archie McKay, chief of engrg.

Big Stone Gap

WAXM(FM)—Listing follows WLSD(AM).

WLSD(AM)— Aug 20, 1953: 1220 khz; 1 kw-D, 45 w-N. TL: N36 50 26 W82 44 14. Stereo. Hrs opn: Drawer W, 1600 Intermont Heights, 24219. Phone: (276) 523-1700. Phone: (276) 679-1901. Fax: (276) 679-1198. E-mail: 93.5@waxm.com Licensee: Valley Broadcasting Inc. (acq 5-1-80; $359,000; FTR: 5-12-80). Population served: 90,000 Natl. Network: CBS. Law Firm: Jerry Miller. Format: Relg. Target aud: 19-60. ♦Greg Kress, pres; William Stanley, gen mgr, sls VP, gen sls mgr & rgnl sls mgr; Paul Miller, adv dir; Rick Phillips, progmg dir; Jack Starnes, chief of engrg.

WAXM(FM)—Co-owned with WLSD(AM). Apr 8, 1975: 93.5 mhz; 2.45 kw. 1,883 ft TL: N36 54 50 W82 53 40. Stereo. 24 724 Park Ave., Norton, 24273. Phone: (276) 679-1901. Fax: (276) 679-1198.318,000 Natl. Network: CBS. Format: Country. ♦Kim Swecker, prom dir; Tammy Robinson, traf mgr; David Stanley, disc jockey; Steve Lewis, disc jockey.

Blacksburg

WBRW(FM)— December 1964: 105.3 mhz; 3.8 kw. 472 ft TL: N37 11 12 W80 28 54. (CP: 12 kw). Stereo. Hrs opn: 24 7080 Lee Hwy., Rayford, 24141. Phone: (540) 633-5330. Fax: (540) 633-2998. Licensee: Cumulus Licensing LLC. Group owner: Cumulus Media Inc. (acq 3-31-2004; grpsl). Population served: 110,000 Format: Active Rock. News: 10 hrs wkly. Target aud: 18-49. ♦Scott Claytons, opns dir & gen sls mgr; Scott Stevens, opns mgr & sls dir; Courtney Quinn, progmg dir; Marty Gordon, news dir; Dave Dalesky, chief of engrg.

WFNR(AM)— 1973: 710 khz; 10 kw-D, DA. TL: N37 08 01 W80 21 17. Hrs open: 7080 Lee Hwy., Radford, 24141. Phone: (540) 633-5330. Fax: (540) 633-2998. E-mail: wwatson@valleybroadcasting.com Web Site: www.nrvtoday.com. Licensee: Cumulus Licensing LLC. Group owner: Cumulus Media Inc. (acq 3-31-2004; grpsl). Population served: 561,000 Rgnl. Network: Va. News Net. Format: News/talk. Target aud: 25 plus; general. ♦Joe Mule, gen mgr; Scott Stevens, opns mgr.

WKEX(AM)— July 10, 1969: 1430 khz; 1 kw-D, 62 w-N. TL: N37 13 57 W80 26 40. (CP: 5 kw-U, DA-2). Hrs opn: 24 Box 889, 24063. Secondary address: 1501 Lark Ln. 24060. Phone: (540) 951-9791. Fax: (540) 961-2021. E-mail: wkexam@yahoo.com Web Site: www.thesportsadditicnetwork.com. Licensee: Base Communications Inc. Group owner: Baker Family Stations (acq 6-30-98; $60,000). Population served: 100,000 Natl. Network: USA. Format: Sports talk. Target aud: 35 plus; mature adults, all income levels. Spec prog: Gospel 5 hrs, relg 3 hrs wkly, 3 hrs syndicated, sports 4 hrs. ♦Edward Baker, pres; Brian Sanders, gen mgr, dev dir, sls dir & progmg dir; Drake Anderson, opns dir, progmg dir, local news ed, edit dir, political ed, sports cmtr & disc jockey; Andrew Matney, prom dir; Tammy Chase, pub affrs dir; Tim Pauley, engrg dir; Alison Baker, traf mgr; Danny Rogers, spec ev coord & disc jockey; Sidney Bennett, disc jockey.

***WUVT-FM**— Oct 23, 1969: 90.7 mhz; 3 kw. 156 ft TL: N37 13 28 W80 24 30. Stereo. Hrs opn: 24 350 Squires Student Ctr., 24061-0546. Phone: (540) 231-9880. Fax: (208) 692-5239. E-mail: wuvtamfm@vt.edu Web Site: www.wuvt.vt.edu. Licensee: Educational Media Corporation at Virginia Tech. Population served: 29,400 Format: Div. Target aud: general; college students. Spec prog: American Indian 2 hrs, Greek 2 hrs, Chinese 2 hrs, Turkish 2 hrs, African 2 hrs, Latin 2 hrs wkly. ♦Jake Faber, gen mgr & opns mgr.

WWVT(AM)—See Christiansburg

Blackstone

WBBC-FM—Listing follows WKLV(AM).

WKLV(AM)— 1947: 1440 khz; 5 kw-D. TL: N37 03 14 W78 01 15. Stereo. Hrs opn: Sunrise-sunset Box 300, 950 Kenbridge Rd., 23824. Phone: (434) 292-4146. Fax: (434) 292-7669. E-mail: wbbc@meckcom.net Web Site: www.bobcatcountryradio.com. Licensee: Denbar Communications Inc. (acq 7-26-91). Population served: 100,000 Natl. Network: ESPN Radio. Format: Sports. Target aud: 25-54. ♦Dennis Royer, pres, progmg dir & chief of engrg; Dennis Royer Jr., gen sls mgr.

WBBC-FM—Co-owned with WKLV(AM). Nov 17, 1975: 93.5 mhz; 17.5 kw. 394 ft TL: N37 03 14 W78 01 15. (CP: 17.5 kw). Stereo. 24 E-mail: wbbc@bobcatcountryradio.com 1000000 Natl. Network: Westwood One. Format: Hot country. Target aud: 25-54; middle class, with two cars, homeowners.

Bluefield

WBDY(AM)— Nov 17, 1980: 1190 khz; 10 kw-D, DA. TL: N37 16 19 W81 19 05. Hrs open: 900 Bluefield Ave., WV, 24701. Phone: (304) 327-7114. Fax: (304) 325-7850. Licensee: Monterey Licenses LLC. Group owner: Triad Broadcasting Co. LLC (acq 7-18-2000; grpsl). Natl. Rep: Rgnl Reps. Format: ESPN radio. Target aud: 25 plus; working & retired adults with disposable income. ♦John Halford, gen mgr; Dave Crosier, opns dir; Danny Clemons, gen sls mgr; Joseph Echoles, progmg dir; Keith Brown, chief of engrg.

WHKX(FM)—Co-owned with WBDY(AM). December 1970: 106.3 mhz; 3 kw. 1,122 ft TL: N37 15 30 W81 10 36. Stereo. 24 5,286 Natl. Network: Westwood One. Format: Contemp country. News: 17 hrs wkly. ♦Dave Crosier, progmg dir.

Bon Air

WLES(AM)— September 1959: 580 khz; 201 w-D, 27 w-N. TL: N37 30 52 W77 30 28. Hrs open: 24 2202 Jolliff Rd., Chesapeake, 23321. Phone: (757) 488-1010. Web Site: www.830wtru.com. Licensee: Chesapeake-Portsmouth Broadcasting Corp. (acq 7-19-2000). Population served: 18,000 Rgnl. Network: Va. News Net. Format: Talk. ♦Nancy Epperson, pres; Henry Hoot, VP & gen mgr.

Bowling Green

WWUZ(FM)— 1998: 96.9 mhz; 2.8 kw. 472 ft TL: N37 57 56 W77 22 19. Hrs open: 24 616 Armelia St., Fredericksburg, 22401. Phone: (540) 374-5500. Fax: (540) 374-5525. Web Site: www.classicrock969.com. Licensee: The Free Lance-Star Publishing Co. of Fredericksburg, Virginia. Group owner: The Free Lance-Star Publishing Co. (acq 8-23-01; $2.15 million). Population served: 280,300 Natl. Network: AP Radio. Rgnl rep: RMR Wire Svc: AP Format: Classic rock. News staff: one; News: 2.5 hours wkly. Target aud: 35 plus; male. ♦Josiah Rowe III, pres; John Moen, gen mgr; Jim Butler, sls dir; JoAnn Pope, prom dir; Paul Johnson, gen mgr & progmg dir; Frank Hammon, news dir & pub affrs dir; Chris Wilk, chief of engrg.

Bridgewater

WBHB-FM— Mar 3, 1989: 105.1 mhz; 6 kw. Ant 328 ft TL: N38 27 08 W78 54 32. Stereo. Hrs opn: 24 130 Media Lane, Harrisonburg, 22801. Phone: (540) 434-0331. Fax: (540) 434-7087. Web Site: www.valleyradio.com. Licensee: M. Belmont VerStandig Inc. Group owner: VerStandig Broadcasting (acq 1993; $10,000 with WHBG(AM) Harrisonburg; FTR: 9-13-93). Format: Classic rock. News staff: 2. Target aud: 24-54. ♦Susanne Fitzpatrick, gen mgr.

Bristol

WFHG(AM)— January 1947: 980 khz; 5 kw-D, 1 kw-N, DA-N. TL: N36 36 30 W82 09 36. Hrs open: 24 Box 1389, 24203. Secondary address: 901 E. Valley Dr. 24201. Phone: (276) 669-8112. Fax: (276) 669-0541. E-mail: bhagy@wxbq.com Web Site: www.supertalkwfhg.com. Licensee: Bristol Broadcasting Inc. Group owner: Nininger Stations (acq 1972). Natl. Network: Fox Sports. Natl. Rep: McGavren Guild. Law Firm: Shaw Pittman. Format: Sports. ♦W.L. Nininger, pres; Bill Hagy, gen mgr; Jennifer Worley, opns dir; Winnie Quaintance, sls dir; Roger Bouldin, prom dir; Chuck Lawson, chief of engrg; Anna Honaker, traf mgr; Ned Michaels, sports cmtr.

WIGN(AM)—See Bristol, TN

WQUT(FM)—See Johnson City, TN

WXBQ-FM—See Bristol, TN

WZAP(AM)— 1946: 690 khz; 10 kw-D, 14 w-N. TL: N36 37 51 W82 09 53. Hrs open: 6 AM-midnight Box 369, 24203. Secondary address: 11373 Wallace Pike 24202. Phone: (276) 669-6950. Phone: (276) 669-6900. Fax: (276) 669-0794. E-mail: wzapradio@aol.com Web Site: www.wzapradio.com. Licensee: RAM Communications Inc. (acq 1-10-77; $375,000). Population served: 502,000 Natl. Network: USA. Law Firm: Irwin, Campbell, & Tannenwald, P.C. Format: Relg. News: 8 hrs wkly. Target aud: General. ♦R.A. Morris, pres, gen mgr, gen sls mgr & edit dir; Glen Harlow, mus dir; Al Morris, news dir; Joyce Boyd, pub affrs dir & traf mgr; Chuck Lawson, chief of engrg; Tommy Tester, stn mgr, prom mgr, progmg dir & relg ed; Dave Ray, disc jockey; Glenn Harlow, disc jockey; Tim Hickman, disc jockey.

Broadway

WJDV(FM)—Licensed to Broadway. See Broadway-Timberville

Broadway-Timberville

WBTX(AM)— May 18, 1972: 1470 khz; 5 kw-D, DA. TL: N38 37 24 W78 48 52. Hrs open: 6 AM-sunset Box 337, 166 Main St., Broadway, 22815. Phone: (540) 896-8933. Fax: (540) 896-1448. E-mail: info@positive-radio.com Web Site: www.positive-radio.com. Licensee: Massanutten Broadcasting Co. Inc. Population served: 150,000 Natl. Network: USA. Natl. Rep: Salem. Format: Southern gospel. News: 8 hrs wkly. Target aud: 35. Spec prog: Farm one hr wkly. ♦Thomas Watson, gen mgr; Christine Pompeo, gen sls mgr; Jim Snavely, progmg dir, news dir & disc jockey; Bill Fawcett, chief of engrg; Judy Shafer, traf mgr; David Eshleman, disc jockey; Karen Kenney, disc jockey.

WLTK(FM)—Co-owned with WBTX(AM). 1997: 103.3 mhz; 2.1 kw. 544 ft TL: N38 36 31 W78 54 07. Stereo. 24 Phone: (540) 896-9585. Web Site: www.positive-radio.com. (Acq 8-8-2001; exchange for WJDV(FM) Broadway plus $1.25 million).150,000 Natl. Network: USA. Natl. Rep: Salem. Format: Contemp Christian. News: 9 hrs wkly. Target aud: 25-44. ♦Greg Crabtree, progmg dir; Judy Shafer, traf mgr; Dave Wyant, disc jockey; David Eshleman, disc jockey; Karen Kenney, disc jockey.

WJDV(FM)—Broadway, Dec 18, 1989: 96.1 mhz; 2.6 kw. Ant 1,010 ft TL: N38 33 50 W78 57 00. Stereo. Hrs opn: 24 130 Media Lane, Harrisonburg, 22801. Phone: (540) 434-0331. Fax: (540) 434-7084. Web Site: www.valleyradio.com. Licensee: HJV L.P. Group owner: VerStandig Broadcasting (acq 3-12-2001; swap with WLTK(FM) New Market). Population served: 150,000 Format: Lite rock. Target aud: 25-44. ♦Susanne Fitzpatrick, gen mgr.

Brookneal

WODI(AM)— Feb. 1, 1997: 1230 khz; 1 kw-U. TL: N37 02 17 W78 56 30. Hrs open: 24 1230 Radio Rd., 24528-3141. Phone: (434) 376-1230. Fax: (434) 376-9634. E-mail: wodi@wodiradio.com Web Site: www.wodiradio.com. Licensee: D & M Communications Inc. (acq 9-5-96; $47,000). Population served: 80,000 Natl. Network: USA. Law Firm: Ferrot Imley & Booth. Format: Oldies, talk. News staff: one; News: 6 hrs wkly. Target aud: 25-54; general.David L. Marthouse, pres, gen mgr & progmg dir; Anthony R. DeNicola, VP, opns mgr & asst music dir; Anthony W. DeNicola, gen sls mgr & mktg mgr; Dianne D. DeNicola, prom mgr; Brian R. DeNicola, mus dir & mus critic; Bob O'Brien, pub affrs dir; John Gers, engrg dir; Harry Kane, chief of engrg; Clarence Walker, relg ed & disc jockey; Frank Pool, relg ed; Dave Marthouse, disc jockey; Dennis Frank, disc jockey; Steve Tapper, disc jockey; Tony Dee, disc jockey

Buena Vista

WWZW(FM)— 1981: 96.7 mhz; 2 kw. Ant 1,135 ft TL: N37 43 37 W79 18 24. Stereo. Hrs opn: Box 902, Lexington, 24450. Secondary address: 312 S. Main St., Lexington 24450. Phone: (540) 463-2161. Fax: (540) 463-9524. E-mail: info@wrel.com Licensee: First Media Radio LLC. (group owner; (acq 6-21-2004; $1.33 million with WREL(AM) Lexington). Population served: 60,000 Natl. Network: NBC Radio. Format: Adult contemp. Target aud: 25-54. ♦Alex Kolobielski, pres; Dave Peach, gen mgr; Steve Williams, sls dir & progmg dir.

Buffalo Gap

WBOP(FM)—Licensed to Buffalo Gap. See Staunton

Cape Charles

***WAZP(FM)**— 2000: 90.7 mhz; 13 kw. Ant 512 ft TL: N37 10 53 W75 57 47. Hrs open: 24 2351 Sunset Blvd., Suite 170-218, Rocklin, CA, 95765. Phone: (916) 251-1600. Fax: (916) 251-1650. E-mail: klove@klove.com Web Site: www.klove.com. Licensee: Delmarva Educational Association. Natl. Network: K-Love. Law Firm: Shaw Pittman. Format: Contemp Christian. News staff: 3. Target aud: 25-44; Judeo Christian, female. ♦Richard Jenkins, pres; Mike Novak, VP; Keith Whipple, dev dir; David Pierce, progmg mgr; Ed Lenane, news dir; Sam Wallington, engrg dir; Karen Johnson, news rptr; Marya Morgan, news rptr; Richard Hunt, news rptr.

Cedar Bluff

WHQX(FM)— 1989: 107.7 mhz; 550 w. 751 ft TL: N37 09 49 W81 46 06. Hrs open: 24 900 Bluefield Ave., Bluefield, WV, 24701. Phone: (304) 327-7114. Fax: (304) 325-7850. Web Site: www.kickscountry.com. Licensee: Monterey Licenses LLC. Group owner: Triad Broadcasting Co. LLC (acq 7-18-00; grpsl). Format: Country. News staff: one. ♦John Halford, gen mgr & gen sls mgr; Dave Crouiser, progmg dir; Keith Bowman, chief of engrg.

WYRV(AM)— March 1985: 770 khz; 5 kw-D. TL: N37 05 05 W81 46 07. Hrs open: Sunrise-sunset Box 70, 24609. Secondary address: 504 Middlecreek Rd. 24609. Phone: (276) 964-9619. Phone: (276) 964-5167. Fax: (276) 964-9610. E-mail: brad@youradio.net Web Site: www.youradio.net. Licensee: Faith Christian Music Broadcasting Ministries Inc. (acq 4-12-01). Population served: 285,000 Format: Positive Radio (contemp Christian/mainstream). News staff: one; News: 3 hrs wkly. Target aud: 30 plus. ♦Brad Ratliff, gen mgr, adv mgr & progmg dir; Greg Webb, mus dir; Acie T. Rasnake, chief of engrg.

Charles City

***WAUQ(FM)**— 2000: 89.7 mhz; 10 kw. Ant 351 ft TL: N37 25 58 W77 11 38. Hrs open: Box 3206, Tupelo, MS, 38803. Phone: (662) 844-8888. Fax: (662 842-6791. Web Site: www.afr.net. Licensee: American Family Association. Group owner: American Family Radio Format: Christian, inspirational. ♦Marvin Sanders, gen mgr.

Charlottesville

WCHV(AM)— 1930: 1260 khz; 5 kw-D, 2.5 kw-N, DA-2. TL: N38 06 52 W78 27 18. Stereo. Hrs opn: 24 1150 Pepsi Pl., Suite 300, 22901. Phone: (434) 978-4408. Fax: (434) 978-1109. Web Site: www.wchv.com. Licensee: CC Licenses LLC. (group owner; (acq 9-13-2000; $450,000). Population served: 136,000 Format: News/talk. ♦Phil Robken, gen mgr; Regan Keith, opns mgr; Mike Chiumento, gen sls mgr.

WCJZ(FM)— 1995: 107.5 mhz; 210 w. Ant 1,109 ft TL: N37 59 05 W78 28 49. Hrs open: 1150 Pepsi Pl., Suite 300, 22901. Phone: (434) 978-4408. Phone: (434) 964-1075. Fax: (978) 978-1190. Licensee: CC Licenses LLC. (acq 1-12-2005; $5.9 million). Format: Rock, adult contemp, smooth jazz. ♦David Mitchel, gen mgr; Kevin McCabe, gen sls mgr; Kishore Persaud, engrg dir & chief of engrg.

WCNR(FM)—Keswick, Mar 2, 1991: 106.1 mhz; 600 w. Ant 1,023 ft TL: N37 59 06 W78 28 48. Stereo. Hrs opn: 1140 Rose Hill Dr., 22903. Phone: (434) 220-2300. Fax: (434) 220-2304. Web Site: www.1061thecorner.com. Licensee: Saga Communications of Charlottesville LLC. (group owner; (acq 11-2-2006; $2.9 million). Law Firm: Smithwick & Belendiuk, PC. Format: AAA. ♦Brad Savage, gen mgr & progmg dir; John Kappes, stn mgr; Rick Dainels, opns mgr.

WHTE-FM—Ruckersville, Mar 29, 1990: 101.9 mhz; 6 kw. 228 ft TL: N38 13 06 W78 22 03. Hrs open: 24 1150 Pepsi Pl., Suite 300, 22901. Phone: (434) 978-4408. Fax: (434) 978-0723. E-mail: hot1019@clearchannel.com Web Site: www.1019hot.com. Licensee: CC Licenses LLC. Group owner: Clear Channel Communications Inc. (acq 8-6-99; grpsl). Population served: 300,000 Natl. Network: AP Network News. Format: Pop contemp hit radio. News: 5 hrs wkly. Target aud: 18-24. ♦Phil Robken, gen mgr; Regan Keith, opns mgr; Mike Chiumento, gen sls mgr.

WINA(AM)— September 1949: 1070 khz; 5 kw-U, DA-N. TL: N38 05 22 W78 30 14. Hrs open: 24 1140 Rose Hill Dr., 22903. Phone: (434) 220-2300. Fax: (434) 220-2304. Web Site: www.wina.com. Licensee: Saga Communications of Charlottesville LLC. (group owner; (acq 1-6-2005; grpsl). Population served: 150,000 Natl. Network: CBS. Natl. Rep: Katz Radio. Law Firm: Smithwick & Belendiuk. Format: News/talk, sports. News staff: 4; News: 20 hrs wkly. Target aud: General. ♦Dennis Mockler, VP, gen mgr & progmg dir; Rick Daniels, opns mgr; John Kappes, gen sls mgr; Jay James, progmg dir.

WQMZ(FM)—Co-owned with WINA(AM). Oct 1954: 95.1 mhz; 6 kw. 144 ft TL: N38 02 54 W78 28 12. Stereo. Web Site: www.literockz951.com.150,000 Format: Adult contemp. ♦Les Sinclair, progmg dir.

WKAV(AM)— October 1957: 1400 khz; 1 kw-U. TL: N38 01 49 W78 29 22. Hrs open: 24 1150 Pepsi Pl., Suite 300, 22901. Phone: (434) 978-4408. Fax: (434) 978-1109. Web Site: www.wkav.com. Licensee: CC Licenses LLC. Group owner: Clear Channel Communications Inc. (acq 6-13-2000; $450,000). Population served: 150,000 Format: Sports. Target aud: 35 plus. ♦Phil Robken, gen mgr; Regan Keith, opns mgr; Mike Chiumento, gen sls mgr.

***WNRN(FM)**— September 1996: 91.9 mhz; 320 w. 1,066 ft TL: N37 58 55 W78 29 03. Hrs open: 24
Rebroadcasts WNRS-FM Sweetbriar 85%.
2250 Old Ivy Rd., Suite 2, 22903. Phone: (434) 971-4096. Fax: (434) 971-6562. E-mail: wnrn@wnrn.org Web Site: www.wnrn.org. Licensee: Stu-Comm Inc. Population served: 500,000 Law Firm: Davis Wright Tremaine. Format: Alternative, AAA, urban contemp. Target aud: 18-49; educated, upscale young professionals and students. Spec prog: Folk 19 hrs, techno 6 hrs, industrial 2 hrs, punk 2 hrs wkly, urban 16 hrs wkly. ♦Mike Friend, gen mgr, dev dir, progmg dir & chief of engrg; Anne Williams, mktg dir; Mike Momson, mus dir; Steve Mendenhall, news dir.

WSUH(FM)—See Crozet

***WTJU(FM)**— May 10, 1957: 91.1 mhz; 600 w. 1,066 ft TL: N37 58 55 W78 29 03. Stereo. Hrs opn: 24 Box 400811, 22904-4811. Secondary

address: 2nd Floor 22904. Phone: (434) 924-0885. Fax: (434) 924-8996. E-mail: wtju@virginia.edu Web Site: wtju.radio.virginia.edu. Licensee: University of Virginia. Population served: 330,000 Natl. Network: PRI. Format: Var/div. News: 5 hrs wkly. Target aud: General; from rural population to college educated. Spec prog: Black 8 hrs, children 2 hrs, folk 20 hrs, Gospel 1 hr Sp 2 hrs wkly. ♦Chuck Taylor, gen mgr & stn mgr.

WUVA(FM)— June 22, 1979: 92.7 mhz; 6 kw. 3,000 ft TL: N37 59 06 W78 28 51. Stereo. Hrs opn: 24 1928 Arlington Blvd., Suite 312, 22903. Phone: (434) 817-6880. Fax: (434) 817-6884. E-mail: info@92.7kissfm.com Web Site: www.92.7kissfm.com. Licensee: WUVA Inc. Population served: 118,500 Natl. Rep: Katz Radio. Format: Adult urban contemp. News staff: 20; News: 20 hrs wkly. Target aud: 18-54. ♦Sharon Sant, gen mgr, sls, mktg & adv; Tanisha Thompson, opns mgr, progmg dir & progmg.

WVAX(AM)— April 2006: 1450 khz; 1 kw-U. TL: N38 02 54 W78 28 12. Hrs open: 24 1140 Rose Hill Dr., 22903. Phone: (434) 220-2300. Fax: (434) 220-2304. Web Site: www.wvax.com. Licensee: Saga Communications of Charlottesville LLC. (acq 11-22-2005; $150,000 for CP). Natl. Network: CNN Radio. Natl. Rep: Katz Radio. Law Firm: Smithwick & Belendiuk, P.C. Format: Progsv talk. News staff: 4. Target aud: 25-54; adult. ♦John Kappes, gen sls mgr; Jay James, progmg dir; Rob Graham, news dir.

***WVTU(FM)**— Jan 8, 1991: 89.3 mhz; 195 w horiz, 160 w vert. 1,696 ft TL: N38 03 58 W78 47 54. (CP: 3.2 kw). Stereo. Hrs opn:
Rebroadcasts WVTF(FM) Roanoke 100%.
3520 Kingsburg Ln., Roanoke, 24014. Phone: (540) 989-8900. Fax: (540) 776-2727. Web Site: www.wvtf.org. Licensee: Virginia Tech. Foundation Inc. Population served: 150,000 Natl. Network: NPR, PRI. Rgnl. Network: Va. News Net. Law Firm: Dow, Lohnes & Albertson. Format: Class, jazz, npr (simulcast @ wvtf). ♦Glenn Gleixner, gen mgr & dev dir; Seth Williamson, mus dir; Rick Mattoni, news dir; Paxton Durham, chief of engrg.

***WVTW(FM)**— 1997: 88.5 mhz; 120 w. 1,089 ft TL: N37 58 49 W78 29 21. Hrs open:
Rebroadcasts WVTF(FM) Roanoke 100%.
3520 Kingsburg Ln., Roanoke, 24014. Phone: (540) 989-8900. Fax: (540) 776-2727. Web Site: www.wvtf.org. Licensee: Virginia Tech Foundation Inc. Natl. Network: NPR. Format: Class, jazz, NPR. ♦Glenn Gleixner, gen mgr & dev dir; Seth Williamson, mus dir; Rick Mattoni, news dir; Paxton Durham, chief of engrg.

WWWV(FM)— 1959: 97.5 mhz; 8.9 kw. Ant 1,132 ft TL: N37 59 05 W78 28 49. Stereo. Hrs opn: 1140 Rose Hill Dr., 22903. Phone: (434) 220-2300. Fax: (434) 220-2304. Web Site: www.3wv.com. Licensee: Saga Communications of Charlottesville LLC. (group owner; (acq 1-6-2005; grpsl). Population served: 136,000 Natl. Rep: Katz Radio. Law Firm: Smithwick & Belendiuk. Format: Rock. Target aud: 18-49. ♦John Kappes, gen sls mgr; Rick Daniels, opns mgr & progmg dir.

Chase City

WJYK(AM)— Jan 18, 1959: 980 khz; 500 w-D. TL: N36 48 22 W78 26 22. Hrs open: Box 8, 23924. Phone: (434) 372-0803. Web Site: battagliacommunications.echurchnetwork.net/JoyAM980. Licensee: Stephen C. Battaglia Sr. & Janis G. Battaglia (acq 1-5-2006; $51,000). Format: Contemp Christian Classic. ♦Stephen C. Battaglia Sr., gen mgr.

***WMVE(FM)**—Not on air, target date: July 2007: 90.1 mhz; 8 kw. Ant 371 ft TL: N36 46 29 W78 20 41. Stereo. Hrs opn: 24
Rebroadcasts WCVE-FM Richmond.
23 Sesame St., Richmond, 23235. Phone: (804) 320-1301. Fax: (804) 320-8729. Web Site: www.ideastations.org/radio. Licensee: Commonwealth Public Broadcasting Corp. Population served: 38,000 Natl. Network: NPR, PRI. Wire Svc: AP ♦Bill Miller, gen mgr.

Chatham

WKBY(AM)— June 8, 1966: 1080 khz; 1 kw-D. TL: N36 46 54 W79 23 29. Hrs open: 12932 U.S. Hwy. 29, 24531. Phone: (434) 432-8108. Fax: (434) 432-1523. E-mail: wkby1080@gamewood.net Licensee: William L. Bonner. (acq 11-15-90; $250,000; FTR: 12-3-90). Population served: 225,000 Format: Black, gospel.William L. Bonner, pres; Van Jay, gen mgr; Lois Stephens, gen sls mgr, local news ed, reporter, relg ed, women's int ed & disc jockey; Vickie Pritchett, progmg dir, traf mgr & disc jockey; Tim Walker, chief of engrg & rsch dir; Dr. H.G. McGhee, min affrs dir; Everett C. Peace, political ed; Nick Pierce, sports cmtr & disc jockey; Mike Carter, disc jockey

Cheriton

***WWIP(FM)**— 2005: 89.1 mhz; 20 kw. Ant 449 ft TL: N37 10 53 W75 57 47. Hrs open: 2202 Jolliff Rd., Chesapeake, 23321. Phone: (757) 465-1603. Fax: (757) 488-7761. Licensee: Delmarva Educational Association. Format: Christian music. ♦Nancy A. Epperson, pres; Henry Hoot, gen mgr.

Chesapeake

WCDG(FM)—See Moyock, NC

WCPK(AM)— 1967: 1600 khz; 5 kw-D, 27 kw-N. TL: N36 48 10 W76 16 58. Hrs open: 24 645 Church St., Suite 400, Norfolk, 23501. Phone: (757) 622-4600. Fax: (757) 624-6515. E-mail: martinpepian@aol.com Licensee: Christian Broadcasting of Chesapeake Inc. (acq 10-17-97; $200,000). Population served: 1,200,000 Law Firm: Winston & Strawn. Format: Gospel. News staff: one. Target aud: 25-54; white collar, upscale professionals. ♦L.E. Willis, pres; Katrina Chase, gen mgr; Jonathan Willis, opns dir, opns mgr & progmg dir; Alvin Rooks, sls dir & gen sls mgr; Martin Culpeper, mktg dir & prom dir; James Phillips, news dir; Terry Love, chief of engrg.

***WFOS(FM)**— Sept 14, 1973: 88.7 mhz; 15 kw. 172 ft TL: N76 18 03 W36 43 18. Stereo. Hrs opn: 1617 Cedar Rd., 23320-7111. Phone: (757) 547-1036. Phone: (757) 547-0134. Fax: (757) 547-0160. Licensee: Chesapeake School Board. Population served: 30,000 Format: Big band, blues, oldies, class, educ. Target aud: High school. ♦W. Randolph Nichols, pres; Richie Babb, gen mgr.

WPYA(FM)— Nov 30, 1973: 93.7 mhz; 100 kw. 997 ft TL: N36 32 57 W76 11 21. Stereo. Hrs opn: 24 500 Dominion Tower, 999 Waterside Dr., Norfolk, 23510. Phone: (757) 640-8500. Fax: (757) 640-8552. Web Site: bob-fm.com. Licensee: Commonwealth Radio L.L.C. Group owner: Sinclair Communications Inc. (acq 1996; $8.1 million with WTAR(AM) Norfolk). Population served: 1,400,000 Natl. Rep: McGavren Guild, Interep. Format: Adult hits. News staff: one; News: 2 hrs wkly. Target aud: 25-54. ♦Bob Sinclair, pres; Lisa Sinclair, gen mgr; Dave Morgan, stn mgr & opns mgr; Jeff Kautz, gen sls mgr; Ginger Power, natl sls mgr; Donna Agresto, prom dir; Jay West, progmg dir.

Chester

WDYL(FM)— December 1968: 101.1 mhz; 4 kw. Ant 367 ft TL: N37 26 21 W77 25 57. Stereo. Hrs opn: 812 Moorefield Park Dr., Richmond, 23236. Phone: (804) 330-5700. Fax: (804) 330-4079. Web Site: y101rocks.com. Licensee: Cox Radio Inc. Group owner: Cox Communications Inc. (acq 2-1-2001; grpsl). Wire Svc: UPI Format: New rock/alternative rock. ♦Mike Murphy, progmg dir; Jon Bennett, chief of engrg.

WGGM(AM)— September 1964: 820 khz; 10 kw-D, 1 kw-N, DA-2. TL: N37 22 58 W77 25 21. Hrs open: 24 4301 W. 100 Rd., 23831. Phone: (804) 717-2000. Fax: (804) 717-2009. E-mail: pscott4u@yahoo.com Web Site: www.am820.net. Licensee: Hoffman Communications Inc. (acq 10-76). Population served: 900,000 Natl. Network: USA. Law Firm: Steve Yelverton. Wire Svc: Metro Weather Service Inc. Format: Gospel. Target aud: 25-49. ♦Hubert Hoffman, pres; Paul Scott Bulifant, VP & gen mgr.

Chincoteague

WCTG(FM)— 2004: 96.5 mhz; 5.3 kw. Ant 344 ft TL: N37 55 14 W75 23 07. Hrs open: 24 Sebago Broadcasting L.L.C., 6139 Franklin Park Rd., McLean, 22101. Phone: (703) 761-5013. Fax: (703) 761-5023. Licensee: Sebago Broadcasting Co. L.L.C. Law Firm: Gammon & Grange. Format: Classic rock, oldies. ♦A. Wray Fitch III, pres.

Christiansburg

WBRW(FM)—See Blacksburg

WFNR-FM— 1990: 100.7 mhz; 3 kw. 328 ft TL: N37 08 01 W80 21 17. (CP: Ant 453 ft.). Hrs opn: 17 7080 Lee Hwy., Radford, 24141. Phone: (540) 633-5330. Fax: (540) 633-2998. Web Site: www.nrvtoday.com. Licensee: Cumulus Licensing LLC. Group owner: Cumulus Media Inc. (acq 3-31-2004; grpsl). Population served: 561000 Rgnl. Network: Va. News Net. Format: News talk. News: 7 hrs wkly. Target aud: 25 plus. ♦Ronald Walton, gen mgr; Scott Stevens, opns mgr; Wes Watson, progmg dir; Dave Dalesky, chief of engrg.

WWVT(AM)— October 1954: 1260 khz; 2.8 kw-D, 28 w-N. TL: N37 09 11 W80 24 57. Hrs open: 6am-sunset 3520 Kingsbury Ln., Roanoke, 24014. Phone: (540) 989-8900. Fax: (540) 776-2727. E-mail: mail@wvtf.org Web Site: www.wvtf.org. Licensee: Virginia Tech Foundation Inc. (acq 5-22-98). Population served: 110,000 Natl. Network: NPR, PRI. Format: Talk. News: 60 hrs wkly. Target aud: 30-54; business professionals. ♦Glenn Gleixner, gen mgr; Rick Mattioni, progmg dir; Paxton Durham, chief of engrg.

Churchville

WNLR(AM)— Mar 9, 1962: 1150 khz; 2.5 kw-D, 30 w-N. TL: N38 12 39 W79 07 53. Hrs open: Box 400, 24421. Secondary address: Rt. 250 W. Phone: (540) 885-8600. Phone: (540) 885-1150. Fax: (540) 886-8624. E-mail: wnlr@nlministries.org Web Site: www.positive-radio.com. Licensee: New Life Ministries Inc. (acq 12-8-93; $200,000; FTR: 1-3-94). Population served: 100,000 Natl. Network: Moody, USA. Format: Relg, adult contemp. Target aud: 22-55; females with family size above average. ♦Bill Garvey, pres; Tom Watson, gen mgr; Russ Whitesell, opns mgr.

Claremont

WPMH(AM)— Aug 19, 1997: 670 khz; 20 kw-D, 3 w-N, DA-2. TL: N37 10 29 W76 53 49. Hrs open: Sunrise-sunset 2202 Jolliff Rd., Chesapeake, 23321. Phone: (757) 465-6700. Fax: (757) 488-7761. Web Site: wpmhradio.com. Licensee: Chesapeake-Portsmouth Broadcasting Corp. (acq 3-2-2001; $950,000). Format: Christian talk. ♦Henry Hoot, gen mgr & opns mgr.

Clarksville

WLUS-FM— Jan 1, 1984: 98.3 mhz; 17.5 kw. Ant 394 ft TL: N36 44 24 W78 44 49. Stereo. Hrs opn: 24 Box 1603, Oxford, NC, 27565. Secondary address: 615 B Lewis St., Oxford, NC 27565. Phone: (919) 693-7900. Fax: (919) 693-9585. Web Site: www.bestcountryaround.com. Licensee: Lakes Media Holding Company LLC. (group owner; (acq 2-1-2005; grpsl). Population served: 250,000 Natl. Network: ABC. Format: Country. News staff: 3; News: 2 hrs wkly. Target aud: 25-54; male & female. ♦Thomas C. Birch, pres; Jerry E. Brown, VP & gen mgr; Mike Elliott, opns mgr & gen sls mgr; Melissa P. Wilkerson, gen sls mgr & news dir; John Hart, chief of engrg.

Clifton Forge

WXCF(AM)— Oct 19, 1950: 1230 khz; 1 kw-U, DA-1. TL: N37 49 18 W79 48 46. Hrs open: 24 Box 710, 1047 Ingalls St., 24422. Phone: (540) 862-5751. Phone: (540) 962-1133. Fax: (540) 862-2120. Web Site: www.bigcountry101.com. Licensee: Quorum Radio Partners of Virginia Inc. (acq 5-03; with co-located FM). Population served: 29,000 Format: Adult contemp. News: one hr wkly. Target aud: 40 & up. ♦Marcia Smith, gen sls mgr & prom dir; Michael Stone, pres, gen mgr, sls dir & progmg dir; Lawrence Mason, chief of engrg.

WXCF-FM— Nov 20, 1982: 103.9 mhz; 150 w. Ant 1,909 ft TL: N37 54 12 W79 52 15. Stereo. 29,000 Format: Hits of the 70s, 80s & 90s. News staff: one. Target aud: 18-54; Targeting mostly women 25-54. ♦Dennis Royer Jr., disc jockey.

Clinchco

WDIC(AM)— May 1961: 1430 khz; 5 kw-D. TL: N37 08 42 W82 23 22. Stereo. Hrs opn: 24 2298 Rose Ridge, Clintwood, 24228-7738. Phone: (276) 835-8626. Fax: (276) 835-8627. E-mail: wdic@wdicradio.com Licensee: Dickenson County Broadcasting Corp. Group owner: Richard W. Edwards (acq 1-84; $366,850; FTR: 1-30-84). Population served: 17,000 Natl. Network: ABC. Rgnl rep: Rgnl Reps. Law Firm: Smithwick & Belendiuk. Format: Country. News staff: 3; News: 8 hrs wkly. Target aud: General. Spec prog: Farm one hr, relg 12 hrs wkly. ♦Richard W. Edwards, pres; Rufus E. Nickles, gen mgr, adv mgr & adv; Betty N. Fleming, opns mgr & traf mgr; Tammy Hill, progmg mgr & progmg.

WDIC-FM— July 2, 1989: 92.1 mhz; 2.5 kw. Ant 505 ft TL: N37 08 42 W82 23 22. Stereo. 24 Web Site: www.wdicradio.com.35,000 Natl. Network: ABC. Natl. Rep: Rgnl Reps. Law Firm: Smithwick & Belendiuk. Format: Oldies, news, loc sports. News: 3 hrs wkly. Target aud: 25-55; baby boomers.

Coeburn

WGCK(FM)— Apr 15, 1991: 99.7 mhz; 1.95 kw. Ant 1,168 ft TL: N37 03 15 W82 38 34. Stereo. Hrs opn: Box 729, Whitesburg, KY, 41858. Secondary address: 32 Cowan St., Whitesburg 41858. Phone: (606) 633-9430. Fax: (606) 633-3314. E-mail: wvsg@msn.com Licensee: Letcher County Broadcasting Inc. (acq 11-22-2005; $250,000). Law Firm: Miller & Neely. Format: Adult contemp. ♦Ernestine Kincer, pres; G.C. Kincer, gen mgr.

Collinsville

WFIC(AM)— Mar 1, 1970: 1530 khz; 1 kw-D, 250 w-CH. TL: N36 42 56 W79 55 15. Hrs open: Box 192, Martinsville, 24114. Secondary address: 1675 Grandview Rd., Martinsville 24112. Phone: (276) 638-5235. Fax: (276) 638-6089. E-mail: eddie@spiritfm.com Licensee: BASE Communications Inc. Group owner: Baker Family Stations (acq 9-3-97; $60,000). Population served: 250,000 Natl. Network: USA. Format: Southern Gospel. Target aud: 35 plus; general. ♦Brian Sanders, gen mgr; Michael Carter, opns mgr & gen sls mgr; Wendall Minter, progmg dir.

Colonial Beach

WGRQ(FM)— May 3, 1986: 95.9 mhz; 2.4 kw. Ant 525 ft TL: N38 13 45 W77 07 10. Stereo. Hrs opn: 24 4414 Lafayette Blvd. #100, Fredericksburg, 22408. Phone: (540) 891-9696. Fax: (540) 891-1656. E-mail: tcooper@oldies959.com Web Site: www.oldies959.com. Licensee: Telemedia Broadcasting Inc. (acq 1-20-88). Natl. Network: ABC. Natl. Rep: Roslin. Format: Oldies. News staff: one; News: 3 hrs wkly. Target aud: 25 plus; baby boomers. ♦Carl W. Hurlebaus, pres; Thomas P. Cooper, gen mgr; Bill Keeler, sls dir; Branch Harper, prom dir; Jim Herring, progmg dir; Cathy Sato, pub affrs dir & traf mgr; Paul Hayden, spec ev coord; Keefe Coble, news rptr.

Colonial Heights

WDZY(AM)—Licensed to Colonial Heights. See Richmond

WKHK(FM)—Licensed to Colonial Heights. See Richmond

Covington

WIQO-FM—Listing follows WKEY(AM).

WKEY(AM)— May 23, 1941: 1340 khz; 1 kw-U. TL: N37 46 09 W79 58 59. Hrs open: 24 Box 710, 508 W. Oak St., 24426. Phone: (540) 962-1133. Web Site: www.big country101.com. Licensee: Quorum Radio Partners of Virginia Inc., debtor-in-possession Group owner: Quorum Radio Partners of Virginia Inc. (acq 4-20-2005; with co-located FM). Population served: 75,000 Law Firm: Fletcher, Heald & Hildreth. Format: Oldies. News staff: one; News: 12 hrs wkly. Target aud: 25 plus; younger country listeners. Spec prog: Black one hr, gospel 2 hrs wkly. ♦Marcia Smith, gen sls mgr; Michael Stone, pres, gen mgr & progmg dir; Dwight Rohr, news dir; Lawrence Mason, chief of engrg.

WIQO-FM—Co-owned with WKEY(AM). October 1964: 100.9 mhz; 560 w. 1,059 ft TL: N37 47 36 W79 55 57.24 Web Site: www.bigcountry101.com.85,000 Format: Country. News staff: one; News: 10 hrs wkly. Target aud: 25-54; general. ♦Pat Pleasant, disc jockey.

Crewe

WPZZ(FM)— June 9, 1949: 104.7 mhz; 100 kw horiz, 84 kw vert. Ant 981 ft TL: N37 10 15 W77 57 16. Stereo. Hrs opn: 2809 Emerywood Pkwy., Suite 300, Richmond, 23294. Phone: (804) 672-9299. Fax: (804) 672-9314. Licensee: Radio One Licenses LLC. Group owner: Radio One Inc. (acq 11-8-2001; grpsl). Population served: 1,725 Natl. Rep: Eastman Radio. Law Firm: Hogan & Hartson. Format: Gospel ; Contemporary. Target aud: 25-54; adults. ♦Alfred Liggins, CFO; Linda Forem, VP & gen mgr; Al Payne, opns mgr; Yvonne Hagen, gen sls mgr; Bobby Walden, natl sls mgr; Dottie Brooks, mktg dir; June Grant, prom dir; Kevin Gardner, progmg dir; Clovia Lawrence, pub affrs dir; Chris Lawless, chief of engrg.

WSVS(AM)— Apr 7, 1947: 800 khz; 5 kw-D, 275 w-N. TL: N37 11 43 W78 10 01. Hrs open: 24 Box 47, 800 Melody Ln., 23930. Phone: (434) 645-7734. Phone: (434) 645-7735. Fax: (434) 645-1701. E-mail: wsvs@meckcom.net Licensee: Colonial Broadcasting of Crewe Inc. Population served: 2,200 Natl. Network: Family Radio. Format: C&W. News staff: one; News: 7 hrs wkly. Target aud: 35 plus. ♦John Wilson, CEO & pres; Francis Wood, gen mgr; Elliott Irving, stn mgr & disc jockey; Eddie Higgins, opns dir; Steve Winn, progmg dir.

Crozet

***WMRY(FM)**— May 1995: 103.5 mhz; 270 w. 1,515 ft TL: N37 57 00 W78 43 38. Stereo. Hrs opn: 24
Rebroadcasts WMRA(FM) Harrisonburg 100%.
983 Reservoir St., harrisonburg, 22801. Phone: (540) 568-6221. Fax: (540) 568-3814. Web Site: www.wmra.org. Licensee: James Madison University Board of Visitors. Population served: 235,000 Natl. Network: NPR, PRI. Format: Class, news. News staff: one; News: 32 hrs wkly. Target aud: 35-64; well educated adults. Spec prog: Folk 8 hrs, blues 4 hrs wkly. ♦Tom DuVal, gen mgr; Dan Easley, opns mgr; Diane Halke, dev dir; William Fawcett, chief of engrg.

WSUH(FM)— September 1980: 102.3 mhz; 4.9 kw. 360 ft TL: N38 04 47 W78 44 22. Stereo. Hrs opn: 1150 Pepsi Pl., Ste 300, Charlottesville, 22901. Phone: (434) 978-4408. Fax: (434) 978-0723. Licensee: CC Licenses LLC. Group owner: Clear Channel Communications Inc. (acq 8-6-99). Format: Classic hits of the 60s & 70s. Target aud: 35-54; general. ♦Mike Chiumento, stn mgr & gen sls mgr; Phil Robken, gen mgr & progmg dir.

Culpeper

***WARN(FM)**— 1997: 91.5 mhz; 930 w. 121 ft TL: N38 27 15 W77 59 10. Hrs open: Box 3206, American Family Radio, Tupelo, MS, 38803. Phone: (662) 844-8888. Fax: (662) 842-6791. Web Site: www.afr.net. Licensee: American Family Association. Group owner: American Family Radio Format: Christian, inspirational. ♦Marvin Sanders, gen mgr.

WCVA(AM)— February 1949: 1490 khz; 1 kw-U. TL: N38 29 04 W77 59 22. Hrs open: 24 Box 271, Orange, 22960. Phone: (540) 672-1000. Fax: (540) 672-0282. Licensee: Piedmont Communications Inc. (group owner; acq 11-21-2003; grpsl). Population served: 28,000 Rgnl. Network: Va. News Net. Format: Nostalgia. News staff: one; News: 10 hrs wkly. Target aud: General. ♦John Schick, gen mgr.

WJMA-FM—Co-owned with WCVA(AM). Dec 4, 1971: 103.1 mhz; 3 kw. 300 ft TL: N38 29 04 W77 59 22. (CP: 3.3 kw). Stereo. Web Site: www.wjmafm.com.118,000 Format: Modern country.

***WPER(FM)**— 1999: 89.9 mhz; 41 kw. Ant 417 ft TL: N38 40 42 W77 47 18. Hrs open: Box 113, Warrenton, 20188. Phone: (540) 347-4825. Fax: (540) 347-3562. E-mail: info@wper.org Web Site: www.wper.org. Licensee: Positive Alternative Radio Inc. Group owner: Baker Family Stations (Positive Radio Group) Law Firm: Booth, Freret, Imlay & Tepper. Format: Contemp Christian. ♦Frankie Morea, gen mgr.

Danville

WAKG(FM)—Listing follows WBTM(AM).

WBTM(AM)— May 24, 1930: 1330 khz; 5 kw-D, 1 kw-N, DA-N. TL: N36 36 36 W79 25 47. Stereo. Hrs opn: 24 Box 1629, 24543. Secondary address: 710 Grove St. 24541. Phone: (434) 793-4411. Fax: (434) 797-3918. Web Site: www.wbtm1330.com. Licensee: Piedmont Broadcasting Corp. Population served: 150,000 Format: Oldies. News staff: 2; News: 5 hrs wkly. Target aud: 24 plus; young working adults. ♦Bob Ashby, CEO, chmn, pres & gen mgr; Mike Wimmer, natl sls mgr; Carol Metz, prom mgr; Alex Vardavas, progmg dir; Chuck Vipperman, news dir; Johnny Cole, chief of engrg.

WAKG(FM)—Co-owned with WBTM(AM). June 3, 1968: 103.3 mhz; 100 kw. 630 ft TL: N36 44 28 W79 23 05. Stereo. 24 Phone: (434) 797-4290. Web Site: www.wakg.com.250,000 Format: Modern country. Target aud: 18-54. ♦Alan Rowe, disc jockey; Carol Metz, disc jockey; Jimmy Allen, disc jockey; Sherri Crowder, progmg mgr & disc jockey.

WDVA(AM)— June 29, 1947: 1250 khz; 5 kw-U, DA-N. TL: N36 34 53 W79 26 33. Stereo. Hrs opn: 24 One Radio Ln., 24541. Phone: (434) 797-1250. Phone: (434) 797-1266. Fax: (434) 797-1255. Licensee: Mitchell Communications Inc. (acq 6-28-93; FTR: 7-19-93). Population served: 1,900,000 Natl. Network: CBS. Law Firm: Latham & Watkins. Format: Gospel. News: 12 hrs wkly. Target aud: 18 plus. ♦C.G. Hairston, pres & gen mgr.

WILA(AM)— Aug 25, 1957: 1580 khz; 1 kw-D. TL: N36 34 03 W79 22 50. Hrs open: Box 3444, 24543. Secondary address: 865 Industrial Ave. 24541. Phone: (434) 792-2133. Fax: (434) 792-2134. E-mail: wilaradio@gamewood.net Licensee: Tol-Tol Communications Inc. (acq 12-29-92; $250,000; FTR: 1-25-93). Population served: 100,000 Natl. Network: American Urban. Format: Black, gospel, oldies. Target aud: General; ethnic (Black) and citizens who enjoy div progmg. ♦Lawrence A. Toller, pres, gen mgr, sls dir, progmg dir & pub affrs dir.

***WOKD-FM**— 1998: 91.1 mhz; 18 kw. 466 ft TL: N56 44 30 W79 23 07. Hrs open:
Rebroadcasts WPAR(FM) Salem 100%.
Box 889, Blacksburg, 24063. Phone: (540) 961-2377. Fax: (540) 951-5282. E-mail: mail@spiritfm.com Web Site: www.spiritfm.com. Licensee: Positive Alternative Radio Inc. Group owner: Baker Family Stations Format: Christian adult contemp. ♦Barry Armstrong, gen mgr.

Deltaville

WTYD(FM)— Jan 5, 1999: 92.3 mhz; 2.4 kw. Ant 525 ft TL: N37 29 37 W76 26 30. Hrs open: 24 5000 New Point Rd., Suite 2102, Williamsburg, 23188. Phone: (757) 565-1079. Fax: (757) 565-7094. Web Site: www.tideradio.com. Licensee: Bullseye Broadcasting LLC (acq 3-3-2005). Wire Svc: AP Format: Triple A. News staff: one. ♦Tom Davis, pres & gen mgr; Derek Mason, gen sls mgr; Betsy Balkcom, prom mgr; Amy Miller, mus dir; Lori Starks, traf mgr.

Dillwyn

WBNN-FM— July 2000: 105.3 mhz; 6 kw. Ant 328 ft TL: N37 34 50 W78 37 18. Stereo. Hrs opn: 24 Box 846, 23936. Secondary address: 18498 N. Madison Hwy. 23936. Phone: (434) 983-6621. Fax: (434) 983-6772. E-mail: mail@bigcountry1053.com Web Site: www.bigcountry1053.com. Licensee: WKGM Inc. Group owner: Baker Family Stations Natl. Network: CNN Radio, Premiere Radio Networks. Law Firm: Booth, Freret, Imlay & Tepper P. Format: Country. Target aud: 18-49 and 25-54. Spec prog: Relg. ♦Vernon H. Baker, CEO; Brian Sanders, VP; Greg Breeden, gen mgr & progmg dir; Nancy McCaig, sls.

Dublin

WPIN(AM)— 1995: 810 khz; 4.2 kw-D. TL: N37 07 55 W80 37 07. Hrs open: Box 889, Blacksburg, 24063. Phone: (540) 961-2377. Fax: (540) 961-2021. Licensee: Dublin Radio. Law Firm: Booth, Freret, Imlay & Tepper. Format: Fox sports. ♦Edward A. Baker, pres & gen mgr; Brian Sanders, stn mgr.

***WPIN-FM**— 1994: 91.5 mhz; 100 w horiz, 90 w vert. 1,204 ft TL: N37 01 27 W80 44 47. Hrs open:
Rebroadcasts WPAR(FM) Salem 100%.
Box 889, Blacksburg, 24060. Phone: (540) 552-8073. Fax: (540) 951-5282. Web Site: www.spiritfm.com. Licensee: Positive Alternative Radio Inc. Group owner: Baker Family Stations (Positive Radio Group) Population served: 100,000 Law Firm: Booth, Freret, Imlay & Tepper. Format: Christian adult contemp. ♦Barry Armstrong, CEO & gen mgr; Vernon H. Baker, chmn; Edward A. Baker, pres.

Duffield

WDUF(AM)— Aug 12, 1986: Stn currently dark. 1120 khz; 1 kw-D. TL: N36 42 30 W82 47 30. Hrs open: 1230 Wincrest Dr., 24244. Phone: (276) 431-4357. Licensee: Deanna D. Smith (acq 6-19-2007). ♦James Smith, gen mgr & stn mgr.

Dumfries-Triangle

WPWC(AM)— Dec 22, 1961: 1480 khz; 1 kw-D, 500 w-N, DA-2. TL: N38 34 06 W77 20 20. Hrs open: 14416 Jefferson Davis Hwy., Suite

20, Woodbridge, 22191. Phone: (703) 494-0100. Phone: (703) 490-1579. Fax: (703) 490-1579. E-mail: radiofiesta1480@yahoo.com Licensee: JMK Communications Inc. (acq 1-19-00; $900,000). Population served: 550,000 Format: Sp. ♦Grant Chang, pres; Carlos Aragon, gen mgr; Clasa Marshall, gen sls mgr; Jose Luis Liriano, progmg dir & traf mgr.

Earlysville

WKTR(AM)— Feb 17, 1991: 840 khz; 8.2 kw-D, DA. TL: N38 15 57 W78 24 53. Hrs open: Daytime Box 309, Spotswood Business Park Cir., Quinque, 22965. Secondary address: Spotswood Business Park Circle, Quinque 22965. Phone: (434) 985-8585. Fax: (757) 365-0412. E-mail: larrycobb@cox.net Licensee: Rural Radio Service. Group owner: Baker Family Stations Law Firm: Booth, Freret, Imlay & Tepper. Format: Relg. ♦Edward A. Baker, pres; Larry W. Cobb, VP & gen mgr.

Edinburg

***WOTC(FM)**— Apr 1, 1994: 88.3 mhz; 1 kw. 403 ft TL: N38 48 12 W78 41 23. Hrs open: 408 Stony Creek Rd., 22824. Phone: (540) 984-8665 ext 206. Fax: (540) 984-9877. Web Site: www.valleybaptistchurch.net. Licensee: Valley Baptist Church-Christian School. Natl. Network: USA. Format: Educ, relg, news. Target aud: General; relg, children. ♦Ed Dorrin, gen mgr.

Elkton

WACL(FM)— Mar 6, 1989: 98.5 mhz; 900 w. Ant 1,607 ft TL: N38 23 36 W78 46 14. Stereo. Hrs opn: 24 207 University Blvd., Harrisonburg, 22801. Phone: (540) 434-1777. Fax: (540) 432-9968. E-mail: ksdavis@clearchannel.com Web Site: www.98rockme.com. Licensee: Capstar TX L.P. Group owner: Clear Channel Communications Inc. (acq 3-12-01; grpsl). Population served: 76,000 Format: Mainstream rock. News staff: one; News: one hr wkly. Target aud: 35-49. ♦Steve Davis, gen mgr.

Emory

***WEHC(FM)**— Nov 15, 1994: 90.7 mhz; 100 w. 95 ft TL: N36 46 20 W81 49 56. Hrs open: 8 AM-midnight Box 947, Keller, Garnand Dr., 24327-0947. Phone: (276) 944-6822. Phone: (276) 944-4161. Fax: (276) 944-6934. E-mail: tdkeller@ehc.edu Web Site: www.ehcweb.ehc.edu/masscomm/wehc. Licensee: Emory and Henry College. Format: Div, progsv. News: 3 hrs wkly. Target aud: College community. ♦Dr. Teresa Keller, gen mgr.

Emporia

WEVA(AM)— Nov 4, 1952: 860 khz; 1 kw-D. TL: N36 41 56 W77 32 55. Hrs open: 24 Box 1056, 705 Washington St., 23847. Secondary address: 705 Washington Street 23847. Phone: (434) 634-2133. Fax: (434) 634-5050. E-mail: info@wevaradio.com Web Site: www.wevaradio.com. Licensee: Colonial Media Corp. Dec. 2001 Population served: 30,000 Natl. Network: CBS, Westwood One. Law Firm: Wilkinson, Barker, Knauer & Quinn. Format: Adult contemp, news/talk. News: 14 hrs wkly. Target aud: 25 plus; Adults 25-64. Spec prog: Gospel 6 hrs, garden 3 hrs, news 14 hrs. ♦James Vavtrout, CEO; George A. Sperry, gen mgr, sls VP & prom mgr; Andy Lucy, progmg mgr, mus dir & disc jockey; Joseph Wetherbee, chief of engrg; Willis L. Stone, local news ed & edit dir; Donna Foxx, disc jockey; Jim Wood, disc jockey; Mallory Avent, disc jockey.

***WJYA(FM)**— January 1999: 89.3 mhz; 2 kw vert. Ant 443 ft TL: N36 46 04 W77 43 39. Stereo. Hrs opn: 24 20276 A Timberlake Rd., Lynchburg, 24502. Phone: (434) 237-9798. Fax: (434) 237-1025. E-mail: office@spiritfm.com Web Site: www.spiritfm.com. Licensee: Positive Alternative Radio Inc. (group owner; (acq 12-30-2005; grpsl). Law Firm: Booth, Freret, Imlay & Tepper. Format: Contemp Christian. ♦Barry Armstrong, gen mgr.

WYTT(FM)— 2003: 99.5 mhz; 1.27 kw. Ant 501 ft TL: N36 39 20 W77 34 22. Hrs open: Box 910, Roanoke Rapids, NC, 27890. Phone: (252) 536-0597. Fax: (252)538-0378. Web Site: www.firstmediarr.com. Licensee: First Media Radio LLC. Group owner: The MainQuad Group. (acq 12-3-2003; grpsl). Format: Hits & Oldies. ♦Alan Garrick, gen mgr & progmg dir; Frank White, chief of engrg.

Ettrick

WLFV(FM)— 2001: 93.1 mhz; 5.2 kw. Ant 348 ft TL: N37 16 21 W77 33 59. Hrs open: 300 Arboretum Pl., Suite 590, Richmond, 23236. Phone: (804) 327-9902. Fax: (804) 327-9911. Licensee: MLB-Richmond IV LLC. (acq 12-13-2005; grpsl). Format: Country. ♦Clyde Bass, gen mgr; Laura Lee Bathje, opns mgr; Bill Keeler, gen sls mgr; Mike Murphy, progmg dir.

Exmore

WROX-FM—Licensed to Exmore. See Virginia Beach

Fairfax

WDCT(AM)— Sept 25, 1955: 1310 khz; 5 kw-D, 500 w-N, DA-2. TL: N38 51 08 W77 18 57. Hrs open: 6 AM-midnight 3251 Old Lee Hwy., Suite 506, 22030. Phone: (703) 273-4000. Fax: (703) 273-1015. E-mail: 1310@radiowashingtonnews.com Web Site: www.radiowashingtonnews.com. Licensee: Family Radio Ltd. (acq 1995). Population served: 3,500,000 Natl. Network: Moody. Format: Korean. Target aud: 25-54; 60% female, 40% male. ♦Kenneth Shin, gen mgr & chief of engrg; Young Jang, sls dir; Ronnie Shin, progmg dir.

Fairlawn

WKNV(AM)— 1998: 890 khz; 10 kw-D, DA. TL: N37 08 26 W80 36 49. Hrs open: Box 889, Blacksburg, 24063. Phone: (540) 961-2377. Fax: (540) 951-5282. Licensee: Base Communications Inc. Group owner: Baker Family Stations Format: Relg. ♦Edward A. Baker, pres & gen mgr; Brian Sanders, stn mgr & progmg dir; Winston Hawkins, chief of engrg.

Falls Church

WFAX(AM)—Licensed to Falls Church. See Washington DC

Falmouth

WGRX(FM)— May 17, 2001: 104.5 mhz; 2.7 kw. Ant 492 ft TL: N38 16 31 W77 32 34. Stereo. Hrs opn: 4414 Lafayette Blvd. #100, Fredericksburg, 22408. Phone: (540) 891-9696. Fax: (540) 891-1656. E-mail: wgrx@thunder1045.com Web Site: www.thunder1045.com. Licensee: Telemedia Broadcasting Inc. (acq 4-17-01; $1.8 million for two-thirds). Population served: 220,000 Natl. Rep: Roslin. Format: Country. News: 3 hrs wkly. Target aud: 18-49; male audience, working class. ♦Carl W. Hurlebaus, pres; Thomas P. Cooper, gen mgr; Bill Keeler, sls dir; Tim Stone, progmg dir; Cathy Sato, pub affrs dir & traf mgr; Paul Hayden, spec ev coord; Keefe Coble, news rptr.

Farmville

WFLO(AM)— August 1947: 870 khz; 1 kw-D. TL: N37 19 35 W78 23 09. Hrs open: Sunrise-sunset Box 367, 1582 Cumberland Rd., 23901. Phone: (434) 392-4195. Fax: (434) 392-1823. E-mail: wflo@moonstar.com Web Site: www.wflo.net. Licensee: Colonial Broadcasting Co. Inc. (acq 4-71). Population served: 100,000 Rgnl. Network: Va. News Net. Natl. Rep: Salem. Wire Svc: AP Format: C&W, news/talk. News staff: one; News: 4 hrs wkly. Target aud: 25 plus. Spec prog: Relg 10 hrs wkly. ♦John D. Wilson, pres; Henry Fulcher, sr VP; Francis E. Wood Jr., gen mgr; Chris Wood, opns dir & progmg dir; Chris Brochon, prom dir & traf mgr; Elliott Irving, news dir, local news ed & news rptr; Polly Davis, sls.

WFLO-FM— May 1961: 95.7 mhz; 50 kw. 492 ft TL: N37 19 35 W78 23 09. Stereo. 24 Web Site: www.wflo.net.200,000 Natl. Network: Jones Radio Networks, AP Radio. Format: Adult contemp. News staff: one; News: 4 hrs wkly. Target aud: 25 plus. ♦John Wilson, chmn; Elliott Irving, news rptr & disc jockey; Bill McKay, disc jockey; Chris Wood, disc jockey; Francis Wood, disc jockey; Henry Fulcher, disc jockey.

***WMLU(FM)**— 1988: 91.3 mhz; 1 w horiz, 150 w vert. Ant 72 ft TL: N37 17 50 W78 23 42. Hrs open: 24 Longwood University, 201 High St., Virginia, 23909. Phone: (434) 395-2475. Phone: (434) 395-2792. Fax: (434) 395-2378. E-mail: wlcx@longwood.edu Web Site: http://lancer.longwood.edu/org/wlcx. Licensee: Longwood University. Population served: 5,000 Format: Div, progsv. Target aud: 18-23; div college students. ♦Matt Taylor, gen mgr; Bryan Lee, progmg VP & progmg dir; John Gross, chief of engrg; Mike Gravitt, local news ed & news rptr; Kevin Donovan, mus critic.

WPAK(AM)— June 15, 1978: 1490 khz; 1 kw-U. TL: N37 18 47 W78 23 41. Hrs open: 446 Plank Rd., 23901. Phone: (434) 392-8114. Fax: (434) 392-8115. Licensee: Great Virginia Venture Inc. (acq 5-11-98; $201,000). Population served: 100,000 Format: Relg. Spec prog: Farm one hr, gospel 12 hrs wkly. ♦George H. Granger, pres; Mark Neimand, gen mgr.

WVHL(FM)— Sept 1, 1997: 92.9 mhz; 6 kw. 328 ft TL: N37 17 06 W78 29 39. Hrs open: Drawer T, 116 North St., 23901. Secondary address: 116 North St. 23901. Phone: (434) 392-9393. Fax: (434) 392-6091. E-mail: v93@wvhl.net Web Site: www.wvhl.net. Licensee: The Farmville Herald Inc. Natl. Network: ABC. Format: Country. Target aud: General. ♦Steve Wall, gen mgr; Sherry Massaro, opns mgr.

Ferrum

***WFFC(FM)**— January 1989: 89.9 mhz; 1.1 kw. Ant 679 ft TL: N36 54 50 W79 57 07. Stereo. Hrs opn: 16 WFFC Radio, Ferrum College, 24088. Phone: (540) 365-4482. Fax: (540) 365-5589. E-mail: wffc@ferrum.edu Web Site: www.radioif.org. Licensee: Virginia Tech Foundation Inc. (acq 1-23-2004; $10). Format: Class, progsv. ♦John wojtowicz, stn mgr.

Fieldale

WODY(AM)— July 1, 1993: 1160 khz; 5 kw-D, 250 w-N, DA-2. TL: N36 42 36 W79 57 58. Hrs open: 24 Box 192, Martinsville, 24114. Secondary address: 1675 Grandview R., Martinsville 24112. Phone: (276) 638-5235. Fax: (276) 638-6089. E-mail: mcarter@thesportsaddictnetwork.com Web Site: thesportsaddictnetwork.com. Licensee: Base Communications Inc. Group owner: Baker Family Stations (acq 4-17-98). Population served: 250,000 Natl. Network: USA. Format: Sports. Target aud: 25-54. ♦Vernon H. Baker, pres & CFO; Edward A. Baker, exec VP & gen mgr; Brian Sanders, stn mgr & sls dir; Michael Carter, opns mgr; Winston Hawkins, chief of engrg.

Floyd

WGFC(AM)— Apr 20, 1985: 1030 khz; 1 kw-D. TL: N36 55 53 W80 16 34. Hrs open: Sunrise-sunset Box 495, 24091. Secondary address: 401 Shooting Creek Rd. S.E. 24091. Phone: (540) 745-9811. Fax: (540) 745-9812. E-mail: wgfc@wgfcradio.com Web Site: www.wgfcradio.com. Licensee: New Life Christian Communications Inc. (acq 7-1-02). Population served: 110,000 Format: Bluegrass, southern gospel. News staff: 2; News: 10 hrs wkly. Target aud: General; loc community interest. ♦Jackie Goad, VP & progmg dir; R. Leon Goad, CEO, pres & gen mgr; Leon Goad, chief of engrg.

Fort Lee

WKLR(FM)—Licensed to Fort Lee. See Richmond

Franklin

WLQM(AM)— Oct 13, 1956: 1250 khz; 1 kw-D. TL: N36 40 57 W76 55 43. Hrs open: Box 735, 23851. Secondary address: 320 Franklin St. 23851. Phone: (757) 562-3135. Fax: (757) 562-2345. E-mail: wlqm@wlqmradio.com Web Site: www.wlqmradio.com. Licensee: Franklin Broadcasting Corp. (acq 8-10-59). Format: Urban Gospel. Target aud: 25-49. Spec prog: Gospel 12 hrs wkly. ♦Peter E. Clark, pres; Michael E. Clark, gen mgr; Tim Parsons, opns VP; Johnny Hart, gen sls mgr

& prom mgr; Michael Clark, natl sls mgr; Walter Hale, rgnl sls mgr & progmg dir; Peter Clark, news dir & local news ed; Mickel Pruden, chief of engrg.

WLQM-FM— January 1988: 101.7 mhz; 6 kw. 469 ft TL: N36 41 17 W77 00 58. Web Site: www.wlqmradio.com. Natl. Network: ABC. Law Firm: Pepper & Corazzini. Format: Contemp country. ♦Tim Parsons, progmg dir; Peter Clark, local news ed.

Fredericksburg

WBQB(FM)—Listing follows WFVA.

WFLS-FM—Listing follows WYSK(AM).

WFVA(AM)— Sept 8, 1939: 1230 khz; 1 kw-U. TL: N38 16 50 W77 26 11. Hrs open: 24 Box 269, 1914 Mimosa St., 22405. Phone: (540) 373-7721. Fax: (540) 899-3879. E-mail: wfva@am1230wfva.com Web Site: www.am1230wfva.com. Licensee: Mid-Atlantic Network Inc. (group owner) Population served: 125,000 Rgnl. Network: Va. News Net. Law Firm: Cole, Raywid & Braverman. Format: MOR, talk. News staff: 3; News: 15 hrs wkly. Target aud: 35 plus. Spec prog: Washington Redskins football. ♦John P. Lewis, pres; Shawn Sloan, gen mgr; Brian Demay, opns dir; Kat Kammer, gen sls mgr; Jessica Beadel, prom dir; Rod Spencer, progmg mgr; Veronica Robinson, news dir; John Diamantis, chief of engrg.

WGRQ(FM)—See Colonial Beach

***WJYJ(FM)—** May 6, 1983: 90.5 mhz; 38 kw. Ant 500 ft TL: N38 11 48 W77 33 45. Stereo. Hrs opn: 24 Box 113, Warrenton, 20188. Phone: (540) 347-4825. Fax: (540) 347-3562. E-mail: info@wper.org Web Site: www.wper.org. Licensee: Positive Alternative Radio Inc. (group owner; (acq 12-30-2005; grpsl). Format: Contemp Christian. ♦Frankie Morea, gen mgr & stn mgr.

WYSK(AM)— July 15, 1960: 1350 khz; 1 kw-D, 37 w-N. TL: N38 18 46 W77 26 20. Hrs open: 24 616 Amelia St., 22401. Phone: (540) 373-1500. Fax: (540) 374-5525. Web Site: www.wysk.com. Licensee: The Free Lance-Star Publishing Co. (group owner). Population served: 85,000 Format: Latino. News staff: 3; News: 12 hrs wkly. Target aud: General. ♦John Moen, gen mgr; Jim Butler, sls dir; Jo Anne Pope, prom dir; Chris Wilk, chief of engrg; Sandy Ridgeway, traf mgr.

WFLS-FM—Co-owned with WYSK(AM). June 12, 1962: 93.3 mhz; 50 kw. 492 ft TL: N38 18 46 W77 26 20. Stereo. 24 Web Site: www.wfls.com.225,000 Format: Country. News staff: 4. Target aud: 25-54; adults. ♦Paul Johnson, opns dir & progmg mgr; Frank Hammon, news dir & pub affrs dir.

Front Royal

WFQX(FM)— Jan 17, 1973: 99.3 mhz; 3 kw. 295 ft TL: N39 00 11 W78 20 28. Stereo. Hrs opn: 24 510 Pegasus Ct., Winchester, 22602. Phone: (540) 662-5101. Fax: (540) 662-8610. Web Site: www.993thefox.com. Licensee: Capstar TX L.P. Group owner: Clear Channel Communications Inc. (acq 8-30-00; grpsl). Population served: 127,000 Format: Rock/AOR. News staff: 3; News: one hr wkly. Target aud: 25-49; men. ♦Jim Shea, CEO; Chuck Peterson, gen mgr; David Miller, opns mgr & progmg mgr; Marcella Vance, sls dir; Justin Maglione, prom dir, prom dir & mus dir; Ben Gates, pub affrs dir; Mark Kesner, chief of engrg; Krissy Groves, traf mgr; Elwood King, disc jockey; Max James, disc jockey.

WFTR(AM)— Sept 19, 1948: 1450 khz; 1 kw-U. TL: N38 54 31 W78 10 37. Hrs open: 24 Box 192, 22630. Secondary address: 1106 Elm St. 22630. Phone: (540) 635-4121. Fax: (540) 635-9387. E-mail: sales@royalbroadcasting.net Web Site: www.realcountryonline.com. Licensee: Royal Broadcasting Inc. (acq 8-15-00; $950,000 with co-located FM). Population served: 40,000 Natl. Network: ABC. Rgnl rep: Commercial Media Sales. Format: Country. News staff: one; News: 20 hrs wkly. Target aud: 25-54. ♦Andrew Shearer, CEO; Lonnie Hill, opns mgr; Mike O'Dell, COO, gen mgr, gen sls mgr & progmg dir; Kathy Willis, traf mgr.

WZRV(FM)—Co-owned with WFTR(AM). 1981: 95.3 mhz; 6 kw. 300 ft TL: N38 58 29 W78 12 09. Stereo. 24 Phone: (540) 665-9595. Web Site: www.oldiesradioonline.com.100,000 Natl. Network: ABC. Format: Oldies. News staff: one. Target aud: 25-54. ♦Mike O'Dell, sls dir; Mario Retrosi, news dir; Kathy Willis, traf mgr; Randy Woodward, progmg mgr & disc jockey.

Galax

WBRF(FM)— Dec 15, 1961: 98.1 mhz; 100 kw. 1,756 ft TL: N36 34 50 W80 58 23. Stereo. Hrs opn: 24 Box 838, 24333. Secondary address: 325 Poplar Knob Rd. 24333. Phone: (276) 236-9273. Fax: (276) 236-7198. E-mail: brc98prod@adelphia.net Web Site: www.blueridgecountry98.com. Licensee: Blue Ridge Radio Inc. (acq 4-19-85; FTR: 12-31-84). Natl. Network: CBS. Format: Country. Target aud: General. ♦Debby Sizer Stringer, gen mgr & gen sls mgr; Betty Liddle, progmg dir & traf mgr; Jason Blevins, mus dir; John Mullins, chief of engrg; Ray Bass, opns mgr & pub svc dir.

***WOKG(FM)—** 2005: 90.3 mhz; 2.7 kw. Ant 538 ft TL: N36 39 27 W80 54 22. Hrs open: Box 889, Blacksburg, 24063. Licensee: Positive Alternative Radio Inc. Group owner: Baker Family Stations.

WWWJ(AM)— Feb 1, 1947: 1360 khz; 5 kw-D. TL: N36 39 48 W80 54 52. Hrs open: 6 AM-10 PM Box 270, 325 Poplar Knob Rd., 24333. Phone: (276) 236-2921. Fax: (276) 236-2922. Licensee: Twin County Broadcasting Corp. (acq 4-19-85; $200,000; FTR: 4-8-85). Population served: 50,373 Natl. Network: CBS. Rgnl. Network: Va. News Net. Format: Gospel. News: one hr wkly. Target aud: 18 plus. Spec prog: Sp 20 hrs, farm one hr wkly. ♦Deborah E. Stringer, pres & gen mgr; J. Brice Parks, gen sls mgr; Carole Bonn, progmg dir & traf mgr; John Mullins, chief of engrg; Joe Hicks, disc jockey; Joel Bonn, disc jockey; Samantha Farmer, mus dir & disc jockey.

Gate City

WGAT(AM)— July 24, 1959: 1050 khz; 1 kw-D, 266 w-N. TL: N36 37 59 W82 34 56. Hrs open: 24 117 E. Jackson, Suite 2, 24251. Phone: (276) 386-7025. Fax: (276) 386-7025. E-mail: wgatradio@earthlink.com Licensee: Tri-Cities Broadcasting Corp. (acq 11-28-90; $70,000; FTR: 12-17-90). Population served: 700,000 Natl. Network: Salem Radio Network. Law Firm: Dow, Lohnes & Albertson. Format: Sports, Southern Gospel Music. News staff: one; News: 10 hrs wkly. Target aud: 25 plus. ♦Alan Giles, pres; Mike Long, gen mgr.

Glade Spring

WFYE(FM)—Not on air, target date: unknown: 102.7 mhz; 6 kw. Ant -14 ft TL: N36 48 45.38 W81 33 20.38. Hrs open: 5835 Lawrence Dr., Indianapolis, IN, 46226. Phone: (317) 541-0417. Fax: (317) 541-0417. E-mail: asradio@aol.com Licensee: ASRadio LLC. ♦Alan Sneed, gen mgr.

Glen Allen

WTOX(AM)— 2004: 1480 khz; 6.3 kw-D, 1.5 kw-N, DA-2. TL: N37 40 56 W77 33 49. Hrs open: 24 308 W. Broad St., Richmond, 23220. Phone: (804) 643-0990. Fax: (804) 474-5070. E-mail: df@radiorichmond.com Web Site: www.radiorichmond.com. Licensee: Davidson Media Station WTOX Licensee LLC. Group owner: 4M Communications Inc. (acq 5-13-2005; grpsl). Population served: 650,000 Format: Rgnl Mexican. Target aud: 25 plus; Hispanic. ♦Peter Davidson, pres; Michael Mazurksy, gen mgr; Tim Hurley, opns dir; Demetrio Flores, progmg dir.

Gloucester

WXGM(AM)— Jan 20, 1957: 1420 khz; 740 w-D. TL: N37 24 36 W76 32 52. Stereo. Hrs opn: 24 Box 634, 6267 Professional Dr., 23061. Phone: (804) 693-2105. Phone: (804) 693-9946. Fax: (804) 693-2182. E-mail: office@xtra99.com Web Site: www.xtra99.com. Licensee: WXGM Inc. (acq 7-91; FTR: 6-22-81). Population served: 200,000 Natl. Network: ABC. Rgnl. Network: Agri-Net, Va. News Net. Law Firm: Verner, Liipfert, Bernhard, McPherson & Hand. Format: Adult contemp. News staff: one; News: 7 hrs wkly. Target aud: 25-54. Spec prog: Farm 2 hrs, relg 2 hrs wkly. ♦Thomas W. Robinson, pres & gen mgr; Harvey King, opns mgr & progmg dir; Iris Lassister, gen sls mgr; Herman King, news dir; Bill Swartz, chief of engrg.

WXGM-FM— July 29, 1991: 99.1 mhz; 6 kw. 328 ft TL: N37 24 36 W76 32 52. Stereo. 24 Web Site: www.xtra99.com. Format: Adult contemp.

Goochland

WZEZ(FM)— 2001: 100.5 mhz; 4.8 kw. Ant 262 ft TL: N37 38 16 W77 54 21. Hrs open: 24 4301 W. 100 Rd., Chester, 23831. Phone: (804) 717-2000. Fax: (804) 717-2009. E-mail: wzez@cavtel.net Web Site: www.wzezradio.com. Licensee: Hubert N. Hoffman III, executor (acq 9-12-02). Natl. Network: USA. Wire Svc: Metro Weather Service Inc. Format: Nostalgia. ♦Paul Scott, gen mgr.

Gretna

WMNA(AM)— Aug 11, 1956: 730 khz; 1 kw-D, 28 w-N, DA. TL: N36 55 31 W79 19 50. Hrs open: 6 AM-10 PM
Rebroadcasts WLNI(FM) Lynchburg 80%.
Box 730, 677 Zion Rd., 24557. Phone: (434) 656-1234. Fax: (434) 847-5709. Web Site: www.wlni.com. Licensee: 3 Daughters Media Inc. Group owner: Burns Media Strategies Inc. (acq 11-14-2002; $300,000 with co-located FM). Population served: 111,000 Rgnl. Network: Agri-Net. Law Firm: Womble, Carlyle, Sandridge & Rice. Format: Country. News staff: one; News: 19 hrs wkly. Target aud: 18-55; family groups. Spec prog: Farm 8 hrs, Black 2 hrs, bluegrass 20 hrs, sports 10 hrs, gospel 15 hrs wkly. ♦Gary E. Burns, CEO, pres & gen mgr; Mike Slenski, gen mgr; Charlotte Wells, progmg dir & disc jockey; Dale Cook, chief of engrg; Melissa Eckhert, traf mgr; Bob Haynes, disc jockey; Devin Taylor, disc jockey; Ron Franklin, disc jockey.

WMNA-FM— Feb 28, 1959: 106.3 mhz; 6 kw. Ant 260 ft TL: N36 55 31 W79 19 50. Stereo. 5:30 AM-10 PM
Rebroadcasts WLNI(FM) Lynchburg 85%.
Format: Talk. News: 10 hrs wkly. Target aud: 24-54; active & mature. ♦Melissa Eckhert, traf mgr; Brian Wegan, disc jockey; Charlotte Wells, disc jockey; Dave Lewis, disc jockey; Larry Richmond, disc jockey; Mari White, disc jockey; Rich Roth, progmg dir, women's int ed & disc jockey.

Grundy

WMJD(FM)—Listing follows WNRG(AM).

WNRG(AM)— Nov 16, 1955: 940 khz; 5 kw-D, 14 w-N. TL: N37 18 08 W82 07 04. Hrs open: Box 2045, 24614. Secondary address: Rt. 460 W. 24614. Phone: (276) 935-7227. Fax: (276) 935-2587. E-mail: wnrg940@hotmail.com Web Site: www.wnrg-wnjd-tv7.com. Licensee: Peggy Sue Broadcasting Corp. (group owner; acq 3-29-2004; $200,000 with co-located FM). Population served: 40,000 Natl. Network: ABC, Salem Radio Network. Format: Southern gospel. News staff: 2. Spec prog: Farm 5 hrs wkly. ♦Dirk Hall, gen mgr, opns mgr, gen sls mgr & progmg dir.

WMJD(FM)—Co-owned with WNRG(AM). June 21, 1965: 100.7 mhz; 2.3 kw. Ant 535 ft TL: N37 18 08 W82 07 04.24 Phone: (276) 935-7227. E-mail: wmjd97@hotmail.com Web Site: www.wnrg-wmjd-tv7.com.50,000 Natl. Network: ABC. Format: Classic country. News staff: 5. Target aud: 24-59. ♦Ron Cole, stn mgr; Elliott Stewart, mus dir & news dir; Marie Fair, traf mgr & disc jockey; Bink Rush, disc jockey; Jesse Wagner, disc jockey.

Hampden-Sydney

***WWHS-FM—** Oct 11, 1972: 92.1 mhz; 10 w. 140 ft TL: N37 14 19 W78 27 48. (CP: Ant 216 ft.). Stereo. Hrs opn: 6 AM-2 AM Box 606, Hampden-Sydney College, 23943. Phone: (434) 223-6009. E-mail: wwhs@hsc.edu Web Site: people.hsc.edu. Licensee: President & Board of Trustees of Hampden-Sydney College. Population served: 3,700 Format: CHR, div, progsv. News: 5 hrs wkly. Target aud: 18-25; college community. Spec prog: Blues 2 hrs, class 2 hrs, jazz 6 hrs, reggae 4 hrs, funk 4 hrs wkly. ♦Elijah Wallace, gen mgr.

Hampton

***WHOV(FM)—** Mar 5, 1964: 88.1 mhz; 2 kw horiz, 8 kw vert. 200 ft TL: N37 01 03 W76 20 13. Stereo. Hrs opn: 17 Dept. of Mass Media Arts, Hampton Univ., 23668. Phone: (757) 727-5407. Phone: (757) 727-5711. Fax: (757) 727-5084. E-mail: info@whovfm.com Web Site: www.whovfm.com. Licensee: Hampton University. Population served: 200,000 Format: Var/div. News: 2 hrs wkly. Target aud: 18-54. Spec prog: Sp 12 hrs, blues 3 hrs, reggae 4 hrs wkly. ♦Leon Scott, pres; Jay Wright, gen mgr, gen sls mgr, adv mgr & progmg dir; Rebecca Milbourne, opns mgr; Robert Grau, chief of engrg.

WLRT(AM)— July 1, 1948: 1490 khz; 1 kw-U. TL: N37 01 46 W76 22 35. Hrs open: 24 2845 N. Armistead Ave., Suite C, 23666. Phone: (757) 766-9262. Fax: (757) 766-7439. Web Site: www.1021thegame.com.racetalklive.com Licensee: Hampton Radio II Inc. (acq 12-9-86; $485,000; FTR: 10-20-86). Population served: 1,000,000 Natl. Network: Fox Sports. Law Firm: Hazes Associates. Format: Sports. ♦Joseph Russo, pres; George Greenlaw, gen sls mgr; Keith Bennett, progmg dir; Ralph Stevens, chief of engrg.

WWDE-FM— June 1, 1962: 101.3 mhz; 50 kw. 499 ft TL: N36 49 41 W76 15 05. Stereo. Hrs opn: 236 Clearfield Ave., Suite 206, Virginia Beach, 23462. Phone: (757) 497-2000. Fax: (757) 456-5458. Web Site: www.2wd.com. Licensee: Entercom Norfolk License LLC. Group owner: Entercom Communications Corp. (acq 12-13-99; grpsl). Population served: 200,000 Natl. Rep: D & R Radio. Law Firm: Bryan Cave.

Format: Adult contemp. News: one hr wkly. ♦David J. Field, CEO, pres & opns dir; Steve Fisher, CFO & exec VP; John C. Donlevie, exec VP; Steve Godofsky, VP; Skip Schmidt, gen mgr; Don London, opns mgr.

Harrisonburg

***WEMC(FM)**— 1955: 91.7 mhz; 1.85 kw. 190 ft TL: N38 28 20 W78 52 57. Stereo. Hrs opn: 24 983 Reservior St., 22801. Phone: (540) 568-3812. E-mail: wemc@emu.edu Web Site: www.emu.edu/wemc. Licensee: Eastern Mennonite University. Population served: 250,000 Natl. Network: NPR. Format: News & info. News: 123 hrs wkly. Target aud: General. ♦Thomas DuVal, gen mgr.

WHBG(AM)— August 1956: 1360 khz; 5 kw-D, 9 w-N. TL: N38 27 04 W78 54 29. Stereo. Hrs opn: 24 130 Media Ln., 22801. Phone: (540) 434-0331. Fax: (540) 434-7087. Web Site: www.valleyradio.com. Licensee: M. Belmont VerStandig Inc. Group owner: VerStandig Broadcasting Population served: 120,779 Format: Sports. Target aud: 24-50. ♦Susanne FitzPatrick, gen mgr.

WKCY(AM)— May 11, 1967: 1300 khz; 5 kw-D. TL: N38 27 52 W78 50 53. Hrs opn: Box 1107, 22801. Secondary address: 207 University Blvd. 22801. Phone: (540) 434-1777. Fax: (540) 432-9968. Web Site: www.goodradio.com. Licensee: Capstar TX L.P. Group owner: Clear Channel Communications Inc. (acq 3-12-01; grpsl). Population served: 54,200 Natl. Network: USA. Format: Talk. Target aud: 55 plus. Spec prog: Relg 2 hrs wkly. ♦Steve Davis, gen mgr; Susie Smith, gen sls mgr; Steve Knupp, progmg dir; David Burman, news dir; Jeff Caudell, chief of engrg.

WKCY-FM— November 1980: 104.3 mhz; 50 kw. 409 ft TL: N38 23 40 W79 08 26. Stereo. 24 Web Site: www.goodradio.com. Format: Country. Target aud: 25-54. ♦Dennis Hughes, progmg dir.

***WMRA(FM)**— June 18, 1975: 90.7 mhz; 10.5 kw. 1,046 ft TL: N38 33 40 W78 56 56. Stereo. Hrs opn: 24 983 Reservoir St., 22801. Phone: (540) 568-6221. Fax: (540) 568-3814. Web Site: www.wmra.org. Licensee: James Madison University Board of Visitors. Population served: 250,000 Natl. Network: PRI, NPR. Format: Class, news. News staff: one; News: 32 hrs wkly. Target aud: 35-64; well-educated. Spec prog: Folk 8 hrs, blues 4 hrs wkly. ♦Tom DuVal, gen mgr; Dan Easley, opns mgr; Diane Halke, dev dir; William Fawcett, chief of engrg.

WQPO(FM)—Listing follows WSVA(AM).

WSVA(AM)— June 9, 1935: 550 khz; 5 kw-D, 1 kw-N, DA-N. TL: N38 27 04 W78 54 29. Hrs opn: 24 Box 752, 22803. Secondary address: 130 Media Ln. 22801. Phone: (540) 434-0331. Web Site: www.valleyradio.com. Licensee: M. Belmont VerStandig Inc. Group owner: VerStandig Broadcasting (acq 4-17-87). Population served: 100,000 Rgnl. Network: Va. News Net. Format: News/talk. News staff: 5. Target aud: 35 plus. Spec prog: Farm 8 hrs wkly. ♦John D. VerStandig, pres; Susanne Mowbray, gen mgr; Dennis Burchill, sls dir; Frank Wilt, progmg dir; Ellsworth Neff, chief of engrg.

WQPO(FM)—Co-owned with WSVA(AM). Dec 3, 1946: 100.7 mhz; 50 kw. Ant 492 ft TL: N38 27 08 W78 54 32. Stereo. 24 Web Site: www.valleyradio.com.36,500 Format: CHR. News staff: 3. ♦Dennis Burchill, sls dir; Jeremy Lee, progmg dir.

***WXJM(FM)**— September 1990: 88.7 mhz; 390 w. 62 ft TL: N38 26 22 W78 52 21. Stereo. Hrs opn: 24 Anthony Seeger Hall, James Madison Univ., 22807. Phone: (540) 568-6878. Phone: (540) 568-3425. Fax: (540) 568-7907. E-mail: wxjm@jmv.edu Web Site: www.jmu.edu/wxjm. Licensee: Board of Trustees of James Madison University. (acq 9-1-89). Format: Progsv. News: 7 hrs wkly. Target aud: General. Spec prog: Jazz 14 hrs, relg 2 hrs, Sp 3 hrs wkly. ♦Carissa Page, gen mgr; Nathan Marsh, opns mgr; Jess Wodward, progmg dir.

Heathsville

***WCNV(FM)**— 2007: 89.1 mhz; 3.8 kw. Ant 318 ft TL: N37 54 22 W76 29 09. Stereo. Hrs opn: Community Idea Stations, 23 Sesame St., Richmond, 23235. Phone: (804) 320-1301. Fax: (804) 320-8729. Web Site: www.ideastations.org/radio. Licensee: Commonwealth Public Broadcasting Corp. Population served: 25,000 Natl. Network: NPR, PRI. Wire Svc: AP Format: News/talk. ♦Bill Miller, gen mgr.

Highland Springs

WCLM(AM)— May 18, 1959: 1450 khz; 960 w-U. TL: N37 32 39 W77 20 47. Hrs open: 24 3165 Hull St., Richmond, 23224. Phone: (804) 231-2186. Fax: (804) 231-2186. Web Site: www.wclmradio.com. Licensee: World Media Broadcast Co. (acq 10-25-94; FTR: 11-14-94). Population served: 200,000 Format: Var/div. Target aud: 25-65. Spec prog: Gospel, blues 5 hrs wkly, country, Top 40.George Lacey, dev dir; Kimberly Osacio, gen sls mgr; Jim Grainger, chief of engrg; Preston T. Brown, CEO, pres, VP, gen mgr, opns VP, progmg dir & rsch dir; Jean Trimble, traf mgr & disc jockey; Curtis Bowman, disc jockey; Don Payten, disc jockey; Jay Love, disc jockey; John Trimble, disc jockey; Kimberly Ocasio, disc jockey; Les Greene, disc jockey

***WHCE(FM)**— Sept 29, 1980: 91.1 mhz; 3 kw horiz. Ant 105 ft TL: N37 32 18 W77 19 27. Hrs open: Henrico County Schools, 100 Tech Dr., 23075. Phone: (804) 328-4075. Phone: (804) 328-4078. Fax: (804) 328-4074. Licensee: Henrico County Schools. Format: CHR. Target aud: 12-20; teenagers, young adults. ♦Bob Kaufman, gen mgr.

Hillsville

WHHV(AM)— Sept 16, 1961: 1400 khz; 1 kw-U. TL: N36 45 00 W80 43 20. Hrs open: 6 AM-midnight Box 648, 24333. Secondary address: 343 Virginia St. 24343. Phone: (276) 728-9114. Fax: (276) 728-9968. E-mail: whhv@whhvradio.com Web Site: www.whhvradio.com. Licensee: New Life Christian Communications Inc. (acq 4-2-01). Population served: 78,000 Format: Gospel. News: 20 hrs wkly. Target aud: General. Spec prog: Farm one hr wkly. ♦Leon Goad, pres & chief of engrg; Jackie Goad, VP, progmg dir & news dir; R. Leon Goad, gen mgr & adv dir.

Hopewell

WHAP(AM)— Jan 16, 1949: 1340 khz; 1 kw-U. TL: N37 17 46 W77 18 50. Hrs open: 24 150 S. Mesa Dr., 23860. Phone: (804) 452-4999. Web Site: whapradio.com. Licensee: P.T. Brown Broadcast Company Inc. (acq 5-17-2007; $150,000). Population served: 150,000 Rgnl. Network: Va. News Net. Format: Gospel, oldies, var. Target aud: 25-65. ♦Preston Brown, CEO; Judy Brown, gen mgr.

Hot Springs

***WCHG(FM)**— September 1995: 107.1 mhz; 160 w. 1,407 ft TL: N38 01 53 W79 46 52. Stereo. Hrs opn: 6 AM-10 PM
Rebroadcasts WVMR(AM) Frost, WV 50%.
Drawer G, 24445. Phone: (540) 839-5400. Fax: (540) 839-5403. E-mail: wchg@tds.net Licensee: Pocahontas Communications Cooperative Corp. (acq 11-23-93; $2,000; FTR: 12-13-93). Population served: 5,000 Rgnl. Network: Va. News Net. Format: Country plus. News: 15 hrs wkly. Target aud: General. ♦Cheryl Kinderman, gen mgr; Heather Dooley, news dir; Chuck Niday, chief of engrg.

Jonesville

WJNV(FM)— 2000: 99.1 mhz; 4 kw. Ant 403 ft TL: N36 42 05 W83 10 14. Hrs open: Box 996, 24263. Phone: (276) 346-2000. Fax: (276) 346-2049. E-mail: wjnvfm@naxs.net Licensee: Regina Kay Moore. Format: Country.

Keswick

WCNR(FM)—Licensed to Keswick. See Charlottesville

Kilmarnock

WKWI(FM)— Sept 1, 1975: 101.7 mhz; 3 kw. Ant 328 ft TL: N37 43 26 W76 23 27. Stereo. Hrs opn: 24 Box 819, 101 Radio Rd., 22482. Phone: (804) 435-1414. Phone: (804) 435-1313. Fax: (804) 435-0484. E-mail: wkwi@rivnet.net Licensee: Two Rivers Communications Inc. (acq 4-7-2004; $900,000). Population served: 40,000 Natl. Network: AP Network News. Rgnl. Network: Va. News Net. Format: Adult contemp. Target aud: 25-64. Spec prog: Farm 2 hrs, Black 6 hrs, Gospel 5 hrs wkly. ♦William C. Sherard, pres; Charlie Lassitor, gen mgr; Carl Christiansen, sls dir; Joanne Chewning, mus dir; Joe Patton, chief of engrg.

Lawrenceville

WHFD(FM)— Sept 1, 1991: 105.5 mhz; 6 kw. 154 ft TL: N36 45 10 W77 51 49. Hrs open: Box 4, 23868. Secondary address: 2162 Plank Rd. 23868. Phone: (434) 848-9433. Fax: (434) 848-9434. E-mail: whfd1055@aol.com Web Site: www.whfdradio.com. Licensee: Willis Broadcasting Corp. (group owner; acq 4-27-99; $350,000 with co-located AM). Format: Gospel. ♦Katrina Chase, gen mgr.

Lebanon

WLRV(AM)— Oct 28, 1974: 1380 khz; 1 kw-D, 63 w-N. TL: N36 55 18 W82 06 16. Hrs open: 24 Box 939, 303 W. Main St., 24266-0939. Phone: (276) 889-1380. Fax: (276) 889-1388. E-mail: wrlv@youmax.com Web Site: www.wlrv.com. Licensee: Gary W. Ward Broadcasting Corp. (acq 12-13-99; $161,250). Population served: 32,000 Natl. Network: USA. Format: Bluegrass, gospel, soft 60's rock & roll. Target aud: 25 plus. ♦Gary W. Ward, pres & gen mgr; Mike Lowe, opns dir, prom dir, progmg dir & news dir; Barbara Jessee, gen sls mgr, adv dir & adv mgr; Rick Lang, chief of engrg.

WXLZ-FM— Feb 1, 1993: 107.3 mhz; 530 w. 774 ft TL: N36 50 38 W82 11 04. Stereo. Hrs opn: 24 Box 1299, 24266. Secondary address: Russell County Industrial Park, WXLZ Dr. 24266. Phone: (276) 889-1073. Fax: (276) 889-3677. E-mail: wxlz1073@mounet.com Web Site: www.wxlz.net. Licensee: Yeary Broadcasting Inc. Population served: 231,000 Natl. Network: CBS. Format: Modern country. News staff: 2; News: 3 hrs wkly. Target aud: 18 plus; students, farmers, miners, executives & rural residents. ♦Lannis Yeary, pres & gen mgr; Gary Scott, progmg dir & disc jockey; Richard Quillen, mus dir; Ron Keane, chief of engrg & local news ed; Pat Jenkins, spec ev coord & disc jockey; Anthony Steven, sports cmtr; Ryland Sutherland, disc jockey.

Leesburg

WAGE(AM)— Mar 6, 1958: 1200 khz; 5 kw-D, 1 kw-N, DA-N. TL: N39 06 36 W77 35 03. Hrs open: 24 711 Wage Dr. S.W., 20175. Phone: (703) 777-1200. Fax: (703) 777-7431. E-mail: wage@wage.com Web Site: www.wage.com. Licensee: Radio WAGE Inc. (acq 3-80). Population served: 100,000 Rgnl. Network: Va. News Net. Law Firm: Leventhal, Senter & Lerman. Format: News, full service, talk. News staff: one; News: 114 hrs wkly. Target aud: 25 plus; above average income, families, homeowners. ♦Grenville Emmet III, pres; Dene Hill, stn mgr; Karen E. Snoots, natl sls mgr & mktg dir; Chris King, progmg VP; Tim Jon, news dir; Fran Little, chief of engrg; Jeremy Huber, traf mgr, sports cmtr & disc jockey; Ron Kitemiller, sports cmtr.

Lexington

***WLUR(FM)**— Feb 27, 1967: 91.5 mhz; 175 w. Ant -75 ft TL: N37 47 17 W79 26 36. Stereo. Hrs opn: 6:30 AM-2 AM WLUR, Washington and Lee Univ., 24450-0303. Phone: (540) 458-4017. Fax: (540) 458-4079. E-mail: wlur@wlu.edu Web Site: wlur.wlu.edu. Licensee: Washington & Lee University. Population served: 12,000 Format: Div. News: 2 hrs wkly. Target aud: General; college students, city and county residents. ♦Tom Burish, pres; Jeremy Franklin, gen mgr; Angela Ernst, opns dir & prom dir; Benjamin Losi, opns dir; Amy McCamphill, mus dir; Derrick Barksdale, mus dir.

***WMRL(FM)**— June 1992: 89.9 mhz; 100 w. 196 ft TL: N37 42 22 W79 26 11. Stereo. Hrs opn: 24
Rebroadcasts WMRA(FM) Harrisonburg 100%.
983 Reseroir St., Harrisonburg, 22801. Phone: (540) 568-6221. Fax: (540) 568-3814. Web Site: www.wmra.org. Licensee: James Madison University Board of Visitors. Population served: 15,000 Natl. Network: NPR, PRI. Format: Class, news. News staff: one; News: 32 hrs wkly. Target aud: 35-64; well-educated. Spec prog: Folk 8 hrs, blues 4 hrs wkly. ♦Tom DuVal, gen mgr; Diane Halke, dev dir.

WREL(AM)— Nov 14, 1948: 1450 khz; 1 kw-U. TL: N37 46 00 W79 25 56. Hrs open: 24 312 Main St., 24450. Phone: (540) 463-2161. Fax: (540) 463-9524. E-mail: info@wrel.com Web Site: www.wrel.com. Licensee: First Media Radio LLC. (group owner; (acq 6-21-2004; $1.33 million with WWZW-FM Buena Vista). Population served: 30,280 Natl. Rep: Keystone (unwired net). Format: News/talk, sports. News staff: one; News: 6 hrs wkly. Target aud: 35 plus. ♦James Putgrese, gen mgr; Scott Lancey, sls dir & gen sls mgr; Russ Brown, progmg dir & chief of engrg; Jim Bresnahan, news dir; Debra Reed, traf mgr.

Louisa

WOJL(FM)— July 10, 1980: 105.5 mhz; 3.3 kw. 325 ft TL: N38 01 37 W78 01 05. Stereo. Hrs opn: 24 Box 277, 23093. Secondary address: 21128 Louisa Rd. 23093. Phone: (540) 967-1142. Fax: (540) 967-1150. Web Site: www.louisa.net. Licensee: Piedmont Communications Inc. (group owner; acq 12-1-2003; $550,000). Population served: 215,000 Format: Country. News: 2 hrs wkly. Target aud: 25-54. Spec prog: Farm 2 hrs, bluegrass 14 hrs, relg 3 hrs, gospel 3 hrs wkly. ♦John Schick, pres; J. David Watt, gen mgr & progmg dir; James Granger, chief of engrg; Tommy Nelson, sports cmtr.

Luray

WMXH-FM—Listing follows WRAA(AM).

WRAA(AM)— October 1962: 1330 khz; 1 kw-D, 40 w-N. TL: N38 39 34 W78 29 28. Hrs open: 130 University Blvd., Suite B, Harrisonburg, 22801. Phone: (540) 801-1057. Fax: (540) 564-2873. E-mail: production @easyradioinc.com Licensee: EZ Radio Inc. (acq 6-2-88; $585,000 with co-located FM). Population served: 50,000 Rgnl. Network: Va. News Net., Agri-Net. Natl. Rep: Keystone (unwired net). Law Firm: Tharrington, Smith & Hargrove. Format: Country. Target aud: 18-49; middle to upper income, mobile. Spec prog: Relg 7 hrs wkly. ♦Jason Cave, pres, gen sls mgr, progmg dir & chief of engrg; Joshua Cave, VP.

WMXH-FM—Co-owned with WRAA(AM). Oct 16, 1979: 105.7 mhz; 440 w. 1,079 ft TL: N38 30 41 W78 29 15. Stereo. Web Site: www.stardust1057.com. Format: MOR, music of your life. Target aud: 18-54; working people with disposable income interested in mus, news & sports. Spec prog: Relg 5 hrs wkly.

WYFT(FM)— October 1986: 103.9 mhz; 6 kw. Ant 302 ft TL: N38 38 17 W78 24 06. Stereo. Hrs opn: Bible Broadcasting Network, 11530 Carmel Commons Blvd., Charlotte, NC, 28226. Phone: (704) 523-5555. Fax: (704) 522-1967. E-mail: wyft@bbnradio.org Web Site: www.bbnradio.org. Licensee: Bible Broadcasting Network Inc. (group owner; (acq 12-22-86). Format: Relg. News: 13 hrs wkly. Target aud: General. ♦Lowell Davey, pres.

Lynchburg

WBRG(AM)— Sept 6, 1956: 1050 khz; 4000 w-D, 100 w-N. TL: N37 25 15 W79 06 55. Hrs open: 24 Box 1079, 24505. Secondary address: 239 Ragland Rd., Madison Heights 24572. Phone: (434) 845-5916. E-mail: wbrg@rev.net Licensee: Tri-County Broadcasting Inc. (acq 7-1-67). Population served: 670,000 Natl. Network: ABC, Westwood One, Motor Racing Net. Rgnl. Network: Va. News Net Format: News/talk, sports. News staff: 2; News: 12 hrs. wkly. Target aud: 25-54; College educated, professional/management, high household income, married with children. ♦Brent Epperson, gen mgr.

WJJX(FM)— Aug 1, 1964: 101.7 mhz; 3.4 kw. Ant 300 ft TL: N37 25 37 W79 07 26. Stereo. Hrs opn:
Rebroadcasts WJJS(FM) Vinton, 100%.
3807 Brandon Ave. S.W., Suite 2350, Roanoke, 24018. Phone: (540) 725-1220. Fax: (540) 725-1245. E-mail: cisqo@wjjs.com Web Site: www.wjjs.com. Licensee: Capstar TX L.P. (acq 8-30-2000; grpsl). Population served: 150,000 Format: Top-40. Target aud: 18-54; general. ♦Chris Clendenen, gen mgr; David Lee Michaels, progmg dir.

WKPA(AM)— July 7, 1988: 1390 khz; 4.7 kw-D, 34 w-N. TL: N37 27 52 W79 07 21. Hrs open: 942 Kyle Ave., Roanoke, 24012. Secondary address: 2043 10th St., Roanoke 24012. Phone: (540) 343-5597. Fax: (540) 345-4064. Web Site: www.radiowkpa.com. Licensee: Seven Hills Media Inc. Law Firm: Booth, Freret, Imlay & Tepper P. Format: Relg. Target aud: General. ♦Dorothy Durrett, gen mgr, opns dir, progmg dir, chief of engrg & traf mgr; Zeke Leonard, sls dir; Sharon M. Moran, prom mgr; Buddy Durrett, mus dir.

WLLL(AM)— Nov 1, 1963: 930 khz; 9 kw-D, 42 w-N. TL: N37 24 25 W79 13 57. Hrs open: Box 11375, 24506. Secondary address: 105 Whitehall Rd. 24501. Phone: (434) 385-9555. Fax: (434) 385-6073. E-mail: wlllam930@aol.com Web Site: www.wlllradio.com. Licensee: Hubbards Advertising Agency Inc. (acq 1-28-02). Population served: 68,000 Format: Gospel. ♦Fletcher Hubbard, pres, gen mgr, sls dir & progmg dir; Susanna Hubbard, traf mgr.

WLNI(FM)— Feb 2, 1994: 105.9 mhz; 6 kw. Ant 266 ft TL: N37 25 37 W79 07 26. Hrs open: 24 Box 11798, 24506. Secondary address: 19-C Wadsworth St. 24501. Phone: (434) 845-5463. Fax: (434) 845-2063. E-mail: wlni@wlni.com Web Site: www.wlni.com. Licensee: Centennial Broadcasting LLC. Group owner: Burns Media Strategies Inc. (acq 1-7-2005; grpsl). Natl. Network: ABC, Westwood One, Fox News Radio, Talk Radio Network. Natl. Rep: McGavren Guild. Law Firm: Womble, Carlyle, Sandridge & Rice. Format: News/talk, sports. News staff: 3; News: 20 hrs wkly. Target aud: 25-54. ♦Harry Williams, gen mgr; Bob Abbott, opns mgr; Sandi Conner, prom dir; Mari White, news dir & local news ed; Melinda Schamerhorn, traf mgr.

WLVA(AM)— Apr 21, 1930: 590 khz; 5 kw-D, 1 kw-N, DA-2. TL: N37 25 39 W79 13 23. Stereo. Hrs opn: 24 Box 552, Forest, 24551. Phone: (434) 534-0400. Fax: (434) 534-0401. E-mail: news1280@msn.com Licensee: Truth Broadcasting Corp. (group owner; (acq 11-10-2005; $275,000). Population served: 250,000 Format: All news. Target aud: 25-54; upscale, decision makers & professionals. ♦Vic Bosiger, gen mgr & progmg dir.

***WRVL(FM)**— June 19, 1981: 88.3 mhz; 50 kw. 1,082 ft TL: N37 11 50 W79 21 07. Stereo. Hrs opn: 24 1971 University Blvd., 24502-2269. Phone: (434) 582-3688. Fax: (434) 582-2994. E-mail: wrvl@liberty.edu Web Site: www.wrvlfm.com. Licensee: Liberty University. Law Firm: Fletcher, Heald & Hildreth. Format: Educ, relg. News staff: one; News: 8 hrs wkly. Target aud: 25 plus. Spec prog: Liberty Univ. football & basketball. ♦David Young, sr VP; Jerry Edwards, gen mgr, stn mgr & mus dir; Mark Edwards, mus dir; Erick Petersen, pub affrs dir; Vangie Alban, pub affrs dir; Rob Branch, engrg dir; Chris Wygal, chief of engrg.

WVBE-FM— 1948: 100.1 mhz; 20 kw. 646 ft TL: N37 20 56 W79 10 06. (CP: Ant 328 ft. TL: N37 28 06 W79 05 50). Stereo. Hrs opn: 24
Rebroadcasts WVBE(AM) Roanoke 90%.
Box 92, Roanoke, 24022. Secondary address: 3934 Electric Rd. S.W., Roanoke 24018. Phone: (540) 774-9200. Fax: (540) 774-5667. E-mail: info@vibe100.com Web Site: www.vibe100.com. Licensee: Mel Wheeler Inc. (group owner; acq 3-12-97; $7.5 million with WXLK(FM) Roanoke). Population served: 200,000 Format: Urban contemp. News: one hr wkly. Target aud: 25-54; skew women, skew black. ♦Leonard Wheeler, CEO, pres & gen mgr; Kathy Rilee, rsch dir.

WVGM(AM)— Feb 22, 1962: 1320 khz; 1 kw-U. TL: N37 25 37 W79 07 26. Hrs open: 24
Rebroadcast WGMN(AM) Roanoke.
3807 Brandon Ave. S.W., Suite 2350, Roanoke, 24018. Phone: (540) 725-1220. Fax: (540) 725-1245. E-mail: stevencross@clearchannel.com Web Site: www.espnnow.net. Licensee: 3 Daughters Media Inc. Group owner: Clear Channel Communications Inc. (acq 6-22-2007; grpsl). Population served: 150,000 Natl. Network: ESPN Radio. Natl. Rep: Katz Radio. Format: Sports talk. News: 10 hrs wkly. ♦Chris Clendenen, gen mgr; Steve Curtis, progmg dir; Jeff Parker, chief of engrg.

***WWMC(FM)**— February 1993: 90.9 mhz; 100 w. 604 ft TL: N37 20 56 W79 10 05. Stereo. Hrs opn: 24 1971 Univ. Blvd., 24502. Phone: (434) 582-3691. Fax: (434) 582-7461. E-mail: wwmcfm@liberty.edu Web Site: www.thelightonline.com. Licensee: Liberty University Inc. Format: Christian, sports, var/div. News: 5 hrs wkly. Target aud: 16-30; high school, college and young adult; general: 12-45. ♦Jamie Hall, stn mgr.

WYYD(FM)—See Amherst

WZZU(FM)— Sept 1, 1970: 97.9 mhz; 570 w. Ant 1,925 ft TL: N37 33 46 W79 11 38. Stereo. Hrs opn: 24 19-C Wadsworth St., 24501. Secondary address: Box 11798 24506. Phone: (434) 845-3698. Fax: (434) 845-2063. E-mail: babbott@centenialbroadcasting.com Licensee: Centennial Broadcasting LLC. (group owner; (acq 11-23-2004; $4.15 million with WZZI(FM) Vinton). Natl. Network: Fox News Radio. Natl. Rep: McGavren Guild. Format: Rock. News staff: 3. Spec prog: Relg one hr wkly. ♦Harry Williams, gen mgr, stn mgr & gen sls mgr; Bob Abbott, opns mgr & progmg dir; Melinda Schamerhorn, traf mgr.

Manassas

WJFK-FM—Licensed to Manassas. See Washington DC

WKDV(AM)— Oct 1, 1957: 1460 khz; 5 kw-U, DA-2. TL: N38 45 00 W77 30 49. Hrs open: 24 9540 Godwin Dr., 20110. Phone: (703) 330-8022. Web Site: www.metroradioinc.com. Licensee: Metro Radio Inc. Group owner: Multicultural Radio Broadcasting Inc. (acq 8-1-2005; exchange for WFBR(AM) Glen Burnie, MD). Population served: 3,500,000 Format: Sp talk var. ♦Kelly Koonce, gen mgr.

WTWP-FM—See Warrenton

Marion

WHGB(AM)— Apr 25, 1962: Stn currently dark. 1330 khz; 5 kw-D, 31 w-N. TL: N36 49 11 W81 28 12. Hrs open: Box 2061, Bristol, TN, 37621-2061. Phone: (423) 878-6391. Fax: (423) 878-6520. Licensee: Appalachian Educational Communication Corp. (acq 11-3-2003; $35,000). Population served: 78,158 ♦Kenneth C. Hill, pres & gen mgr.

WMEV(AM)— Dec 12, 1948: 1010 khz; 1 kw-D, 30 w-N. TL: N36 51 23 W81 30 21. Hrs open: 24 1041 Radio Hill Rd., 24354. Phone: (276) 783-3151. Phone: (276) 783-9400 (STUDIO). Fax: (276) 783-3152. E-mail: supercountry@fm94.com Web Site: www.fm94.com. Licensee: Holston Valley Broadcasting Corp. Group owner: Glenwood Communications Corp. (acq 7-1-98; $1.65 million with co-located FM). Population served: 500,000 Natl. Network: Motor Racing Net, Salem Radio Network. Rgnl rep: Rgnl Reps Format: Southern Gospel. News: 3 hrs wkly. Target aud: 18-54. Spec prog: Gospel 2 hrs, relg 8 hrs wkly. ♦George E. DeVault Jr., pres; Jim Mabe, opns dir & disc jockey; Anita Dixon, sls dir; N. David Widener, gen mgr & natl sls mgr; Lynn Rutledge, progmg dir & disc jockey; Duane Nelson, news dir; Evelyn Payne, traf mgr; Henry Thomas, disc jockey.

WMEV-FM— June 21, 1961: 93.9 mhz; 100 kw. 1,480 ft TL: N36 54 08 W81 32 33. Stereo. Format: Hot country. ♦N. David Widener, exec VP; Anita Dixon, sls VP; Lyle Musser, chief of engrg; Everly Payne, traf mgr; Henry Thomas, disc jockey; Jim Mabe, disc jockey; Lynn Rutledge, disc jockey.

WOLD-FM— Mar 14, 1968: 102.5 mhz; 440 w. Ant 1,204 ft TL: N36 54 10 W81 32 27. Stereo. Hrs opn: 24 Box 1047, 24354-1047. Phone: (276) 783-7100. Licensee: Emerald Sound Inc. Population served: 185,000 Natl. Network: CNN Radio. Format: Adult contemp. ♦Robert S. Dix, pres & gen mgr; Patricia Ann Dix, stn mgr & opns mgr.

***WVTR(FM)**— Nov 22, 1991: 91.9 mhz; 4.5 kw. 1,489 ft TL: N36 44 52 W81 18 15. Hrs open:
Rebroadcasts WVTF(FM) Roanoke 100%.
c/o WVTF, 3520 Kingsbury Ln., Roanoke, 24014. Phone: (540) 989-8900. Fax: (540) 776-2727. Web Site: www.wvtf.org. Licensee: Virginia Tech Foundation Inc. Natl. Network: NPR. Format: Class, jazz, NPR. ♦Glenn Gleixner, stn mgr & dev dir.

WZVA(FM)— Sept 2, 1996: 103.5 mhz; 1.35 kw. -36 ft TL: N36 52 07 W81 26 07. Hrs open: 24 Box 85, 24354. Phone: (276) 783-4042. Fax: (276) 783-2120. E-mail: staff@z-103.com Web Site: www.z-103.com. Licensee: T.E.C.O. Broadcasting Inc. (acq 11-7-97; $125,000). Population served: 55,000 Format: CHR, adult contemp. Target aud: 18-49; women & men. ♦Tom Copenhaver, CEO; Darla Gross, stn mgr.

Martinsville

WHEE(AM)— Aug 4, 1954: 1370 khz; 5 kw-D, 500 w-N. TL: N36 41 09 W79 54 14. Hrs open: Drawer 3551, 24115. Secondary address: 40 Franklin St. 24112. Phone: (276) 632-9811. Phone: (276) 632-5433. Fax: (276) 632-9813. E-mail: news@whee.net Web Site: www.wheeradio.com. Licensee: Martinsville Media Inc. (acq 1-5-98; $200,000 for stock). Population served: 125,000 Natl. Network: CBS. Law Firm: Wilkinson, Barker, Knauer & Quinn. Format: Talk. Target aud: 17-60; agriculture & mfg area audience. ♦Bill Wyatt, pres, gen mgr, gen sls mgr, progmg dir, traf mgr & disc jockey; T.L. Walker, chief of engrg; Lewis Compton, farm dir & disc jockey; Danny Wyatt, disc jockey; Teddy Thomas, prom mgr & disc jockey.

WMVA(AM)— December 1941: 1450 khz; 1 kw-U. TL: N36 42 00 W79 51 07. Hrs open: Box 3831, 24112-0545. Phone: (276) 632-2152. Fax: (276) 632-4500. Licensee: Martinsville Media Inc. Format: Adult contemp. ♦Bill Wilson, pres & gen mgr.

***WPIM(FM)**— 1997: 90.5 mhz; 4 kw. 326 ft TL: N36 42 16 W79 50 06. Hrs open:
Rebroadcasts WPIR(FM) Salem 100%.
Box 929, Blacksburg, 24063. Phone: (540) 552-8073. Fax: (540) 951-5282. E-mail: mail@spiritfm.com Web Site: www.spiritfm.com. Licensee: Positive Alternative Radio Inc. Group owner: Baker Family Stations (Positive Radio Group) Law Firm: Booth, Freret, Imlay & Tepper. Format: Christian adult contemp. ♦Vernon H. Baker, CEO & chmn; Edward A. Baker, pres & VP; Barry Armstrong, gen mgr & stn mgr.

WROV-FM— January 1950: 96.3 mhz; 13.8 kw. 2,076 ft TL: N36 43 00 W79 51 07. Stereo. Hrs opn: 3807 Brandon Ave., Suite 2350,

Roanoke, 24018. Phone: (540) 725-1220. Fax: (540) 725-1245. Web Site: www.96-3rov.com. Licensee: Capstar TX L.P. Group owner: Clear Channel Communications Inc. (acq 8-30-00; grpsl). Population served: 1,000,000 Natl. Rep: D & R Radio. Format: AOR. Target aud: 18-49; general. ♦Chris Clendenen, gen mgr; Ron Gaylor, gen sls mgr; Aaron Roberts, progmg dir; Ed Kilbane, news dir; Jeff Parker, chief of engrg.

Mechanicsville

WCDX(FM)— Oct 7, 1985: 92.1 mhz; 4.5 kw. Ant 770 ft TL: N37 42 50 W77 30 23. Stereo. Hrs opn: 2809 Emerywood Pkwy., Suite 300, Richmond, 23294. Phone: (804) 672-9299. Fax: (804) 672-9314. Fax: (804) 672-9316. Licensee: Radio One Licenses LLC. Group owner: Radio One Inc. (acq 11-8-2001; grpsl). Natl. Rep: Eastman Radio. Law Firm: Fletcher, Heald & Hildreth. Format: Urban contemp. Target aud: Ages 18-44. ♦Linda Forem, VP & gen mgr; Al Payne, opns mgr; Brian Robertson, gen sls mgr; Bobby Walden, natl sls mgr; Dottie Brooks, mktg dir; Dawna Covington, prom dir & prom mgr; Reggie Baker, progmg dir & mus dir; Clovia Lawrence, pub affrs dir; Chris Lawless, chief of engrg.

Midlothian

WWLB(FM)— Nov 22, 1971: 98.9 mhz; 4.8 kw. Ant 746 ft TL: N37 36 52 W77 30 56. Stereo. Hrs opn: 5 AM-midnight Box 271, Orange, 22960-0157. Phone: (540) 672-1000. Fax: (540) 672-0282. Licensee: MLB-Richmond IV LLC. Group owner: The MainQuad Group (acq 12-13-2005; grpsl). Population served: 185,000 Format: Var. ♦John Schick, gen mgr.

Moneta

WCQV(AM)— November 1991: 880 khz; 900 w-D. TL: N37 10 00 W70 37 50. Hrs open: 12 1848 Clay St. S.E., Roanoke, 24013. Phone: (540) 343-7109. Fax: (540) 343-2306. Web Site: www.wcqv.com. Licensee: Perception Media Group Inc. (group owner; acq 1999; $75,000). Population served: 211,000 Format: Adult standards. Target aud: 35 plus; mostly middle aged, affluent, cosmopolitan. Spec prog: Relg 5 hrs wkly. ♦Ben Peyton, pres, stn mgr & sls dir; Barbara Evans, progmg dir; Dale Cook, chief of engrg; Blair Peyton, traf mgr.

Monterey

***WVLS(FM)**— September 1995: 89.7 mhz; 200 w. 1,460 ft TL: N38 20 39 W79 35 47. Stereo. Hrs open: 6 AM-10 PM
Rebroadcasts WVMR(AM) Frost, WV 60%.
Rt. 28, Dunmore, WV, 24934. Phone: (304) 799-6004. Fax: (304) 799-7444. E-mail: wvls@htcnet.net Web Site: wvls.cfw.com. Licensee: Pocahontas Communications Cooperative Corp. Population served: 2,500 Format: Country. Target aud: General. ♦Cheryl Kinderman, gen mgr; Carson Raleton, stn mgr; Shaun Harvey, progmg dir & mus dir; Heather Dooley, news dir; Chuck Niday, chief of engrg.

Mount Jackson

WSIG(FM)— October 1988: 96.9 mhz; 7 kw. Ant 558 ft TL: N38 36 31 W78 54 07. Hrs open: 24 1866 E. Market St., Suite 325, Harrisonburg, 22801. Phone: (540) 432-1063. Fax: (540) 433-9267. E-mail: wsig@shentel.net Licensee: Vox Communications Group LLC. (group owner; (acq 8-31-2005; $2 million). Population served: 70,000 Natl. Network: AP Radio. Law Firm: Southmayd & Miller. Format: Classic country. Spec prog: Bluegrass 6 hrs, gospel 4 hrs wkly. ♦Tom Manley, gen mgr.

WSVG(AM)— Apr 23, 1954: 790 khz; 1 kw-D, 40 w-N. TL: N38 46 15 W78 37 17. Hrs open: Sunrise-sunset Box 425, 22842. Phone: (540) 477-4443. Fax: (540) 477-4407. Licensee: Hometown Broadcasting of Mt. Jackson LLC (acq 2-25-2005). Population served: 35,000 Rgnl. Network: Agri-Net. Format: Big band. ♦Alan Arehart, gen mgr; Patty Shaffer, gen mgr.

Narrows

WZFM(FM)— 1992: 101.3 mhz; 210 w. Ant 1,200 ft TL: N37 17 54 W80 48 36. Hrs open: Box 889, Blacksburg, 24063. Phone: (540) 726-2765. Fax: (540) 961-2021. Licensee: WZFM LLC. (acq 5-19-2006; $600,000). Format: Oldies. ♦Brian Sanders, gen mgr.

Narrows-Pearisburg

WNRV(AM)— August 1953: 990 khz; 5 kw-D, 10 w-N. TL: N37 20 39 W80 46 36. Hrs open: 24
Rebroadcast WWWR(AM) Roanoke 90%.
1848 Clay St. S.E., Roanoke, 24013. Phone: (540) 343-7109. Fax: (540) 343-2306. Web Site: www.radio3wr.com. Licensee: Perception Media Group Inc. (group owner; acq 6-2-99). Population served: 41,000 Format: Gospel. ♦Ben Peyton, pres; Barbara Evans, stn mgr, sls dir & progmg dir; Blair Peyton, mus dir; Dale Cook, chief of engrg; Blair Peyton, traf mgr.

Nassawadox

***WJCN(FM)**— 2005: 90.1 mhz; 450 w vert. Ant 199 ft TL: N37 33 27 W75 49 44. Hrs open: 20276 A Timberlake Rd., Lynchburg, 24502. Phone: (434) 237-9798. Fax: (434) 237-1025. E-mail: office@spiritfm.com Web Site: www.spiritfm.com. Licensee: Positive Alternative Radio Inc. (group owner). (acq 12-30-2005; grpsl). Law Firm: Booth, Freret, Imlay & Tepper. Format: Contemp Christian. ♦Barry Armstrong, gen mgr.

New Market

WLTK(FM)—Licensed to New Market. See Broadway-Timberville

Newport News

WCMS(AM)—Listing follows WGH-FM.

WGH-FM— November 1948: 97.3 mhz; 74 kw. 415 ft TL: N36 57 47 W76 24 42. Stereo. Hrs opn: 5589 Greenwich Rd., Virginia Beach, 23462. Phone: (757) 671-1000. Fax: (757) 671-1010. Web Site: www.eagle97.com. Licensee: MHR License LLC. Group owner: Barnstable Broadcasting Inc. (acq 3-24-2005; grpsl). Format: Contemp country. Target aud: 25-34. ♦Andy Graham, gen mgr; Frankie Roman, mktg dir & prom mgr; John Shomby, progmg dir; Keith O'Malley, chief of engrg.

WCMS(AM)—Co-owned with WGH-FM. October 1928: 1310 khz; 5 kw-U, DA-N. TL: N36 57 47 W76 24 42.24 Web Site: www.espnradio.com. Format: Sports talk. Target aud: General. ♦Anthony Mercurio, mktg dir & progmg dir.

WNVZ(FM)—See Norfolk

WTJZ(AM)— November 1947: 1270 khz; 1.5 kw-D, 900 w-N, DA-N. TL: N37 01 52 W76 22 00. Hrs open: 24 Box 610, Hampton, 23669-0610. Secondary address: 553 Michigan, Hampton 23669. Phone: (757) 723-1270. E-mail: wtjz1270@aol.com Licensee: Chesapeake-Portsmouth Broadcasting Corp. (acq 1999; $380,000). Population served: 8,000 Format: Gospel, inspirational. ♦Jerome Barber, gen mgr & opns mgr; Felecia Benet, progmg dir.

Norfolk

WCMS(AM)—See Newport News

WGH-FM—See Newport News

WGPL(AM)—See Portsmouth

***WHRO-FM**— 1990: 90.3 mhz; 23 kw. 630 ft TL: N36 48 32 W76 30 13. Stereo. Hrs opn: 20 5200 Hampton Blvd., 23508. Phone: (757) 889-9400. Fax: (757) 489-0007. E-mail: info@whro.org Web Site: www.whro.org. Licensee: Hampton Roads Educational Telecommunications Association Inc. Natl. Network: NPR, PRI. Format: Class, fine arts. Target aud: 35 plus; well-educated, executives, leaders. ♦Joseph Widoff, CEO & pres; Regina Brayboy, COO; Carol Vollbrecht, CFO; John Heimerl, VP & stn mgr; Heather Fleming Mazzoni, opns mgr & chief of opns; Virginia Thumm, dev dir. Co-owned TV: *WHRO-TV affil.

***WHRV(FM)**— 1974: 89.5 mhz; 8.8 kw. Ant 1,148 ft TL: N36 48 31 W76 30 13. Stereo. Hrs opn: 24 5200 Hampton Blvd., 23508. Phone: (757) 889-9400. Fax: (757) 489-0007. Web Site: www.whro.org. Licensee: Hampton Roads Educational Telecommunications Association, Inc. (acq 2-86). Population served: 1,600,000 Natl. Network: NPR, PRI. Law Firm: Dow Lohnes. Wire Svc: AP Format: News/talk, jazz, alternative. News staff: one; News: 105 hrs wkly. Target aud: 35 plus. Spec prog: Progsv 14 hrs, folk 7 hrs wkly. ♦Regina Brayboy, COO; Carol Volbrecht, CFO; John Heimerl, gen mgr; Heather Mazzon, chief of opns & progmg mgr; Virginia Thumm, dev VP & dev dir. Co-owned TV: *WHRO-TV affil.

WJOI(AM)— 1949: 1230 khz; 1 kw-U. TL: N36 50 03 W76 16 12. Hrs open: 870 Greenbriar Cir., Suite 399, Chesapeake, 23320. Phone: (757) 366-9900. Fax: (757) 366-0022. Licensee: Tidewater Communications LLC. Group owner: Saga Communications Inc. (acq 9-15-86). Population served: 1,400,000 Natl. Rep: McGavren Guild. Format: Adult Standards. ♦Dave Paulus, pres, VP & gen mgr; Don Crowder, chief of engrg.

WNOR(FM)—Co-owned with WJOI(AM). 1961: 98.7 mhz; 46 kw. Ant 518 ft TL: N36 50 04 W76 16 11. Stereo. Web Site: www.fm99.com. Law Firm: Smithwick & Belendiuk. Format: Rock/AOR. ♦Harvey Najen, progmg dir.

WKUS(FM)—Licensed to Norfolk. See Portsmouth

WLRT(AM)—See Hampton

WNIS(AM)— Sept 21, 1923: 790 khz; 5 kw-U, DA-1. TL: N37 04 23 W76 17 28. Stereo. Hrs opn: 24 500 Dominion Tower, 999 Waterside Dr., 23510. Phone: (757) 640-8500. Fax: (757) 640-8552. Fax: (757) 622-6397. E-mail: wnis@wnis.com Web Site: www.WNIS.com. Licensee: Sinclair Communications Inc. (group owner). Population served: 1,400,000 Natl. Network: ABC, Westwood One. Rgnl. Network: Va. News Net. Natl. Rep: McGavren Guild, Interep. Wire Svc: AP Format: News/talk. Target aud: 25-54. ♦Bob Sinclair, CEO & gen mgr; Lisa Sinclair, stn mgr; Dave Morgan, opns mgr; Juli Zobel, gen sls mgr; Ginger Power, natl sls mgr; Donna Agresto, prom dir; Jay West, progmg dir.

***WNSB(FM)**— Mar 22, 1980: 91.1 mhz; 8.1 kw. 422 ft TL: N36 46 32 W76 23 11. (CP: 18 kw, ant 299 ft. TL: N36 45 23 W76 23 06). Stereo. Hrs opn: 700 Park Ave., Suite 129, 23504-8015. Phone: (757) 823-9672. Fax: (757) 823-2385. E-mail: wnsb@nsu.edu Web Site: www.hot91.com. Licensee: Norfolk State University Board of Visitors. Population served: 38,000 Natl. Network: NPR. Format: Urban. Target aud: 18-24. ♦Dr. Emmanuel Onyedike, gen mgr; Edward Turner, stn mgr.

WNVZ(FM)— July 1967: 104.5 mhz; 50 kw. 480 ft TL: N37 02 20 W76 18 30. Stereo. Hrs opn: 236 Clearfield Ave., Suite 206, Virginia Beach, 23462. Phone: (757) 497-2000. Fax: (757) 497-7158. Web Site: www.z104.com. Licensee: Entercom Norfolk License LLC. Group owner: Entercom Communications Corp. (acq 12-13-99; grpsl). Population served: 983,800 Natl. Rep: D & R Radio. Format: CHR. News staff: News progmg one hr wkly Target aud: 18-34; females.David J. Field, CEO & pres; Steve Fisher, CFO & exec VP; Steve Godofsky, VP; Skip Schmidt, gen mgr; Don London, opns mgr & progmg dir; Mark Warlaumont, sls dir; Bernardo Nogueira, natl sls mgr; Nathan James, prom dir; Suzanne McGovern, adv dir; Ernie Warinner, mus dir & chief of engrg; Jay West, mus dir & disc jockey; John C. Donlevie, engrg VP; Joanna Sheppardson, traf mgr; Brian Thomas, disc jockey; Mike Klien, disc jockey; Nick Taylor, disc jockey; Tricia Harris, pub affrs dir & disc jockey

WOWI(FM)— June 1948: 102.9 mhz; 50 kw. 500 ft TL: N36 45 23 W76 23 06. Stereo. Hrs opn: 1003 Norfolk Sq., 23502. Phone: (757) 466-0009. Fax: (757) 466-7043. Web Site: www.103jamz.cc. Licensee: CC Licenses LLC. Group owner: Clear Channel Communications Inc. (acq 1996; grpsl). Population served: 307,951 Natl. Rep: McGavren Guild. Format: Urban contemp. Target aud: 18-34. ♦Lowry Mays, CEO, chmn & pres; Reggie Jordan, VP & stn mgr; Eric Mychaels, opns mgr, progmg dir & progmg mgr; Terry Ratliff, sls dir; Bob Rischitelli, gen sls mgr; Toni Bailey Jones, prom mgr; D.J. Law, mus dir; Cheryl Wilkerson, news dir; Doc Christian, pub affrs dir; Greg Gabriel, chief of engrg.

WPCE(AM)—See Portsmouth

WRJR(AM)—See Portsmouth

WTAR(AM)— September 1952: 850 khz; 50 kw-D, 25 kw-N, DA-2. TL: N36 51 39 W76 21 13. Hrs open: 24 500 Dominion Tower, 999 Waterside Dr., 23510. Phone: (757) 640-8500. Fax: (757) 640-8552. Web Site: www.wtar.com. Licensee: Sinclair Communications Inc. (group owner; (acq 9-87; $725,000; FTR: 9-21-87). Population served: 1,400,000 Natl. Rep: McGavren Guild, Interep. Format: Sports. Target aud: 25-54; 35 plus. ♦Bob Sinclair, CEO & gen mgr; Lisa Sinclair, gen mgr & stn mgr; Juli Zobel, gen sls mgr; Ginger Power, natl sls mgr; Donna Agresto, prom dir; Jay West, progmg dir.

WTJZ(AM)—See Newport News

WVAB(AM)—Virginia Beach, Dec 10, 1954: 1550 khz; 5 kw-D, 249 w-N. TL: N36 49 20 W76 05 30. Hrs open: Box 3640, Northfolk, 23510. Phone: (757) 456-9822. Fax: (757) 460-2728. E-mail: wvab@email.com Licensee: Ronald W. Cowan Jr. Natl. Network: CNN Radio. Format: News. ♦Ronald W. Cowan Jr., gen mgr, CEO & pres.

WVBW(FM)—See Suffolk

WVKL(FM)— Sept 21, 1961: 95.7 mhz; 40 kw. Ant 881 ft TL: N36 48 56 W76 28 00. Stereo. Hrs opn: 236 Clearfield Ave., Suite 206, Virginia Beach, 23462. Phone: (757) 497-2000. Fax: (757) 456-5458. Web Site: www.957rnb.com. Licensee: Entercom Norfolk License LLC. Group owner: Entercom Communications Corp. (acq 12-13-99; grpsl). Population served: 1,200,000 Format: Rhythm and blues. ♦David J. Field, CEO & pres; Steve Fisher, CFO & exec VP; John C. Donleive, exec VP; Steve Godofsky, VP; Skip Schmidt, gen mgr; Don London, opns mgr; Karen Parker-Chesson, news dir.

WVXX(AM)— July 1, 1954: 1050 khz; 5 kw-D, 358 w-N, DA-2. TL: N36 49 44 W76 12 26. Hrs open: Radisson Hotel, 700 Monticello Ave., Suite 301, 23510. Phone: (757) 627-9899. Fax: (757) 627-0123. Licensee: Davidson Media Station WVXX Licensee LLC. Group owner: Barnstable Broadcasting Inc. (acq 2-10-2005; $975,000). Population served: 1,210,900 Format: Sp contemp. ♦Andy Hindlin, pres & gen mgr.

WWDE-FM—See Hampton

WXMM(FM)— Oct 1, 1962: 100.5 mhz; 50 kw. Ant 500 ft TL: N36 49 44 W76 12 26. Stereo. Hrs opn: 24 5589 Greenwich Rd., Suite 200, Virginia Beach, 23462. Phone: (757) 671-1000. Fax: (757) 671-1010. Web Site: www.maxfm.fm. Licensee: MHR License LLC. (acq 3-24-2005; grpsl). Natl. Rep: Christal. Format: Rock. ♦Eric Martel, pres; Vonneva Carter, gen mgr.

WYFI(FM)— Oct 2, 1971: 99.7 mhz; 50 kw. 456 ft TL: N36 49 41 W76 15 05. Stereo. Hrs opn: 11530 Carmel Commons Blvd., Charlotte, NC, 28226. Phone: (757) 420-9505. Fax: (757) 420-9505. E-mail: wyfi@bbnradio.org Web Site: www.bbnradio.org. Licensee: Bible Broadcasting Network Inc. (group owner; acq 12-24-70). Format: Relg. ♦Lowell Davey, pres; Dennis Gast, gen mgr.

WYRM(AM)— Apr 6, 1976: Stn currently dark. 1110 khz; 50 kw-D, DA. TL: N36 56 34 W76 31 56. Hrs open: day 700 Monticello Ave., Suite 305, 23510. Phone: (757) 622-9256. Fax: (757) 622-9253. E-mail: wyrm1110@hotmail.com Web Site: www.wyrmradio.com. Licensee: Word Broadcasting Network Inc. (group owner; acq 7-29-2003; $1.25 million with WYMM(AM) Jacksonville, FL). Population served: 1,379,700 Format: Relig teaching. News: 2-5 hrs wkly. ♦Larry Cobb, gen mgr.

Norton

WNVA(AM)— March 1946: 1350 khz; 5 kw-D. TL: N36 57 58 W82 35 17. Hrs open: Sunrise-sunset Box 500, 24273. Phone: (276) 328-2244. Fax: (276) 328-0024. E-mail: wnva@mounet.com Licensee: Radio-Wise Inc. Population served: 60,000 Rgnl. Network: Rgnl reps. Rgnl rep: Regnl Reps Format: Adult contemp. News: 2 hrs wkly. Target aud: 25-65; adults & young adults. Spec prog: Gospel 15 hrs, relg 2 hrs wkly. ♦William G. Stallard, VP; William G. Stallard, gen mgr; Deborah Baker, opns dir & sls dir; Gerald Hibbitts, chief of engrg.

WNVA-FM— July 25, 1969: 106.3 mhz; 1.65 kw. 613 ft TL: N36 57 58 W82 35 17. Stereo. 24 45,000 Natl. Network: Jones Radio Networks. Rgnl rep: Regnl Reps Format: CHR country. News: 10 hrs wkly. Target aud: 18-45; young adults. ♦Debbie Baker, sls dir.

Onley-Onancock

WESR(AM)— Jan 23, 1958: 1330 khz; 5 kw-D, 51 w-N. TL: N37 43 02 W75 41 01. Hrs open: Box 100, Tasley, 23441. Secondary address: 22479 Front St., Accomac 23301. Phone: (757) 787-3852. Fax: (757) 787-3819. Web Site: www.west.net. Licensee: Eastern Shore Radio Inc. (acq 1-23-97; $148,300 for stock with co-located FM). Population served: 6,000 Natl. Network: ABC. Natl. Rep: Dome. Format: Country classics, talk. ♦Charles Russell, gen mgr; Bill LeCato, progmg dir.

WESR-FM— 1968: 103.3 mhz; 50 kw. 320 ft TL: N37 43 02 W75 41 01. Stereo. Web Site: www.wesr.net.46,000 Natl. Network: ABC. Format: Adult contemp.

Orange

WVCV(AM)— Sept 10, 1949: 1340 khz; 1 kw-U. TL: N38 15 14 W78 07 15. Hrs open: Box 271, 22960-0157. Secondary address: 207 Spicers Mill Rd. 22960. Phone: (540) 672-1000. Fax: (540) 672-0282. E-mail: advertising@wjmafm.com Licensee: Piedmont Communications Inc. (group owner; acq 2-18-93; $30,000 with co-located FM; FTR: 3-8-93). Population served: 162,400 Natl. Network: CBS. Format: Oldies, talk. Target aud: 18-54. Spec prog: Gospel 2 hrs, relg one hr, news 18 hrs wkly. ♦John Schick, gen mgr & min affrs dir; Gary Harrison, opns mgr, prom dir, progmg dir, pub affrs dir & chief of engrg; Ann Velez, gen sls mgr; Phil Goodwin, news dir; Joe Boucher, traf mgr; Red Shipley, sports cmtr.

Pearisburg

WNRV(AM)—See Narrows-Pearisburg

Pennington Gap

WSWV(AM)— June 1, 1959: 1570 khz; 2.3 kw-D, 191 w-N. TL: N36 44 02 W83 02 34. Hrs open: 24 Box 630, 203 W. Morgan Ave., 24277. Phone: (276) 546-2520. Fax: (276) 546-1356. E-mail: wswv@optidynamic.com Web Site: www.wswv.net. Licensee: B C Broadcasting Co. Inc. (acq 6-6-2005; $105,000 with co-located FM). Population served: 26,000 Natl. Network: AP Network News. Natl. Rep: Rgnl Reps. Format: Adult contemp, relg. Target aud: 24-54; young, working adults. ♦Kathy Laufer, gen mgr & adv VP; Nicole Clontz, stn mgr, opns dir, gen sls mgr, progmg VP, mus dir, traf mgr & disc jockey; Mary Lou Clontz, asst music dir, relg ed, disc jockey & mktg dir; Mike Cook, chief of engrg; Wayne Nickodam, sports cmtr; Daryl Combs, women's int ed & disc jockey; Johnny Woliver, disc jockey; Rob McLaughlin, disc jockey.

WSWV-FM— 1973: 105.5 mhz; 3.5 kw. 276 ft TL: N36 44 02 W83 02 34. Stereo. 24 Phone: (276) 546-2521. Web Site: www.wswv.net.26,000 Format: Adult contemp. ♦Nicole Clontz, opns mgr, prom mgr, progmg dir, traf mgr & disc jockey; Kathy Laufer, adv mgr; Rob McLaughlin, mktg dir & mus dir; Johnny Woliver, asst music dir & disc jockey; Mary Lou Clontz, relg ed & disc jockey; Daryl Combs, disc jockey; Rob McCloughlin, disc jockey.

Petersburg

WARV-FM— December 1992: 100.3 mhz; 4.7 kw. 328 ft TL: N37 08 57 W77 24 54. Hrs open:
Simulcast with WBBT-FM Powhatan.
300 Arboretum Pl., Suite 590, Richmond, 23236. Phone: (804) 327-9902. Fax: (804) 327-9911. Licensee: MLB-Richmond IV LLC. (acq 12-13-2006; grpsl). Format: Oldies. ♦William McCutchen, exec VP; Kevin Lein, gen mgr & gen sls mgr; Joey Butler, opns dir & progmg dir; Mike McClain, prom mgr; Kelly Fever, pub affrs dir; Frank White, chief of engrg.

WKJM(FM)—Listing follows WROU(AM).

WROU(AM)— May 7, 1945: 1240 khz; 1 kw-U. TL: N37 14 01 W77 22 36. Hrs open: 24 4301 W. Hundred Rd., Chester, 23831. Phone: (804) 717-2000. Fax: (804) 717-2009. E-mail: wgcv@cavte.net Licensee: Radio One Licenses LLC. Group owner: Radio One Inc. (acq 7-26-99; grpsl). Population served: 36,103 Natl. Rep: McGavren Guild. Format: News/talk. Target aud: 25-54. ♦Paul Scott, gen mgr.

WKJM(FM)—Co-owned with WROU(AM). Oct 1, 1966: 99.3 mhz; 3 kw. Ant 328 ft TL: N37 14 01 W77 22 36. Stereo. 2809 Emory Wood Pkwy., Suite 300, Richmond, 23294. Phone: (804) 672-9299. Fax: (804) 672-9314.700,000 Format: Urban adult contemp. Target aud: 25-54; Black adults.

***WVST-FM**— July 12, 1987: 91.3 mhz; 2.2 kw vert. Ant 167 ft TL: N37 14 15 W77 24 55. Stereo. Hrs opn: 19 Box 9067, 130 Harris Hall, Virginia State Univ., 23806. Phone: (804) 524-5000. Fax: (804) 524-5826. Web Site: www.vsu.edu/wvst. Licensee: Virginia State University. Population served: 60,000,000 Format: Diversified, jazz. Target aud: 25 plus; general. Spec prog: Gospel 13 hrs wkly. ♦Dr. Moadab, gen mgr; Yolie Thomas, stn mgr & opns mgr; Hugh Mannah, chief of opns.

Poquoson

WNRJ(FM)— April 2001: 106.1 mhz; 2.6 kw. Ant 502 ft TL: N37 04 24 W76 17 33. Stereo. Hrs opn: 24 500 Dominion Tower, 999 Waterside Dr., Norfolk, 23510. Phone: (757) 640-8500. Fax: (757) 640-8552. Web Site: www.zone1061.com. Licensee: Commonwealth Broadcasting L.L.C. (acq 8-24-2001; $1.883 million for CP). Population served: 1,400,000 Natl. Rep: McGavren Guild, Interep. Format: CHR, top-40. Target aud: 18-44. ♦Lisa Sinclair, gen mgr; Jeanett Xenakis, gen sls mgr; Ginger Power, natl sls mgr; Donna Agresto, prom dir; Jay Michaels, progmg dir.

Portsmouth

WGPL(AM)— January 1942: 1350 khz; 5 kw-U, DA-2. TL: N36 53 00 W76 22 22. Hrs open: 24 645 Church St., Suite 400, Norfolk, 23501. Phone: (757) 622-4600. Fax: (757) 624-6515. E-mail: martinpepion@aol.com Licensee: Christian Broadcasting of Norfolk Inc. Group owner: Willis Broadcasting Corp. Population served: 1,309,500 Format: Gospel. News staff: 2; News: 6 hrs wkly. Target aud: 25-54. ♦L. E. Willis, pres; Katrina Chase, gen mgr & opns mgr; Alvin Rooks, sls dir; Martha Culpeper, mktg dir & prom dir; Jonathan Willis, progmg dir; Terry Love, chief of engrg.

WHKT(AM)— 1999: 1650 khz; 10 kw-D, 1 kw-N. TL: N36 48 10 W76 16 58. Hrs open: 5041 Corporate Woods Dr., Ste 165, Virginia Beach, 23462. Phone: (757) 519-9171. Fax: (757) 519-9147. Web Site: www.radiodisney.com. Licensee: Radio Disney Group LLC. Group owner: ABC Inc. (acq 6-5-2002; $1.08 million with WRJR(AM) Portsmouth). Format: Radio Disney. ♦Monica Rae Clanin, mktg mgr, prom mgr & pub affrs dir; Richard Bowen, stn mgr, opns mgr, gen sls mgr & chief of engrg.

WKUS(FM)—Norfolk, Aug 3, 1962: 105.3 mhz; 50 kw. 499 ft TL: N36 48 43 W76 27 49. Stereo. Hrs opn: Clear Channel Communications, Inc., 1003 Norfolk Sq., Norfolk, 23502. Phone: (757) 466-0009. Fax: (757) 466-4043. E-mail: vibe@clearchannel.com Web Site: www.vibezone.com. Licensee: CC Licenses LLC. Group owner: Clear Channel Communications Inc. (acq 1996; grpsl). Natl. Rep: Roslin. Format: Adult urban. Target aud: 25-54. ♦Janet Armstead, gen mgr; Michelle Smith, prom mgr.

WPCE(AM)— Jan 11, 1964: 1400 khz; 1 kw-U. TL: N36 49 45 W76 19 23. Hrs open: 24 645 Church St., Suite 400, Norfolk, 23501. Phone: (757) 622-4600. Fax: (757) 624-6515. E-mail: martinpepion@aol.com Licensee: Christian Broadcasting of Portsmouth Inc. (group owner; (acq 3-4-92; grpsl; FTR: 3-23-92). Population served: 307,951 Format: Inspirational. ♦L.E. Willis, pres; Katrina Chase, gen mgr; Jonathan Willis, opns mgr & progmg dir; Alvin Rooks, sls dir; Cindy Perkins, gen sls mgr; Martin Culpeper, mktg dir & prom dir; James Phillips, mus dir & news dir; Terry Love, chief of engrg.

WRJR(AM)— Jan 9, 1972: 1010 khz; 5 kw-D, 449 w-N, DA-2. TL: N36 49 20 W76 26 38. Hrs open: 2202 Jolliff Rd., Chesapeake, 23321. Phone: (757) 488-1010. Fax: (757) 488-7761. Licensee: Radio Disney Group LLC. Group owner: ABC Inc. (acq 6-5-2002; $1.08 million with WHKT(AM) Portsmouth). Population served: 75,000 Format: Contemp Christian. ♦Henry W. Hoot, gen mgr.

Pound

WDXC(FM)— 1990: 102.3 mhz; 280 w. 992 ft TL: N37 09 07 W82 37 57. (CP: 35 kw, ant 1,315 ft.). Stereo. Hrs opn: 24 12552 Orby Cantrell Hwy., 24279. Phone: (276) 796-5411. Fax: (276) 796-5412. E-mail: wdxxc@tgtel.com Web Site: www.wdxcfm.com. Licensee: WDXC Radio Inc. (acq 6-90; FTR: 6-4-90). Population served: 300,000 Format: Country. Target aud: General. ♦Howard Cornett, pres, gen mgr & sls VP; Jackie Cornett, exec VP.

Powhatan

WBBT-FM— 1999: 107.3 mhz; 1.4 kw. Ant 679 ft TL: N37 30 15 W77 42 14. Hrs open: 300 Arboretum Pl., Suite 590, Richmond, 23236. Phone: (804) 327-9902. Fax: (804) 327-9911. Web Site: www.oldies1073.net. Licensee: MLB-Richmond IV LLC. Group owner: The MainQuad Group (acq 12-13-2005; grpsl). Format: Oldies. ♦Clyde Bass, gen mgr; Laura Lee Bathje, opns mgr; Bill Keeler, gen sls mgr; Mike Murphy, progmg dir.

Pulaski

WPSK-FM— Dec 1, 1967: 107.1 mhz; 25 kw. 1,207 ft TL: N37 01 28 W80 44 47. Stereo. Hrs opn: 24 7080 Lee Hwy., Radford, 24141. Phone: (540) 633-5330. Fax: (540) 633-2998. Web Site: nrvtoday.com. Licensee: Cumulus Licensing LLC. Group owner: Cumulus Media Inc.

(acq 3-31-2004; grpsl). Population served: 163,000 Format: Country. News staff: one; News: 10 hrs wkly. Target aud: 25-54. ♦Ronald Walton, gen mgr; Scott Stevens, opns mgr; Sean Summer, progmg dir; Dave Dalesky, chief of engrg.

Quantico

WPWC(AM)—See Dumfries-Triangle

Radford

WRAD(AM)— 1950: 1460 khz; 5 kw-D, 500 w-N, DA-N. TL: N37 08 35 W80 34 38. Stereo. Hrs opn: 5 AM-11 PM Box 3788, 24143. Secondary address: 7080 Lee Hwy. 24141. Phone: (540) 633-5330. Fax: (540) 633-6300. Licensee: Cumulus Licensing LLC. Group owner: Cumulus Media Inc. (acq 3-31-2004; grpsl). Population served: 141,000 Format: Classic hits of the 60s, 70s & 80s, sports. Target aud: 25 plus. ♦Ron Walton, gen mgr; Scott Stevens, gen sls mgr; David Dalesky, chief of engrg.

WWBU(FM)—Co-owned with WRAD(AM). 1965: 101.7 mhz; 3 kw. 66 ft TL: N37 08 33 W80 34 39. Stereo. 50,000 Format: Classic country. ♦Randy Thompson, gen sls mgr.

***WVRU(FM)**— Oct 9, 1978: 89.9 mhz; 500 w. 15 ft TL: N37 08 26 W80 33 11. Stereo. Hrs opn: 24 Box 6973, 24142. Secondary address: 236 Porterfield 24142. Phone: (540) 831-5171. Phone: (540) 831-6059. Fax: (540) 831-5893. E-mail: wvru@radford.edu Web Site: www.wvru.org. Licensee: Radford University. Population served: 48,000 Format: Jazz, triple A, BBC. News: 6 hrs wkly. Target aud: General. Spec prog: Jazz 19 hrs, class 15 hrs, Black 6 hrs, folk one hr, blues 7 hrs, oldies 5 hrs, public affrs 6 hrs, new age 3 hrs wkly. ♦Ashlee B. Claud, gen mgr, dev VP & progmg VP; Jonathan Benfield, opns dir; Randy McCallister, engrg VP.

Richlands

WGTH-FM— Jan 3, 1977: 105.5 mhz; 450 w. 800 ft TL: N37 09 20 W81 46 11. Hrs opn: 24 Box 370, 24641. Phone: (276) 964-2502. Fax: (276) 964-4500. E-mail: wgth@wgth.net Web Site: www.wgth.net. Licensee: High Knob Broadcasters Inc. Natl. Network: Salem Radio Network. Format: Relg, southern gospel. News: 8 hrs wkly. Target aud: General. ♦Marcus Boyd, opns dir; Charlene Pinkerton, gen sls mgr; Ron Brown, pres, gen mgr & mus dir; Mike Luttrel, chief of engrg.

WGTH(AM)— Oct 5, 1951: 540 khz; 1 kw-D, 97 w-N. TL: N37 05 01 W81 46 58.24 (Acq 2-28-95; FTR: 5-22-95).50,000 News: 10 hrs wkly. Target aud: General.

WRIC-FM— November 1989: 97.7 mhz; 1.35 kw. Ant 702 ft TL: N37 09 04 W81 53 56. Stereo. Hrs opn: 24 Box 838, 1600 Front St., Suite 202, 24641. Phone: (276) 964-4066. Phone: (276) 963-4400. Fax: (276) 963-4927. E-mail: wric@netscope.net Licensee: Peggy Sue Broadcasting Corp. (group owner; acq 12-1-98; $190,000). Population served: 50,000 Natl. Network: USA. Rgnl. Network: Va. News Net. Law Firm: Wilkinson, Barker, Knauer & Quinn. Format: Hot adult contemp. News staff: 3; News: 2 hrs wkly. Target aud: 24-54; young adult professionals. ♦Henry Beam, pres; Dirk Hall, sr VP, gen mgr, gen sls mgr, mktg VP & prom VP; Jennie Casey, opns mgr & traf mgr; Dave Mann, chief of opns, progmg mgr, news dir & spec ev coord; Wayne Boone, chief of engrg.

Richmond

WBTJ(FM)— May 1957: 106.5 mhz; 7.6 kw. Ant 1,233 ft TL: N37 30 14 W77 41 53. Stereo. Hrs opn: 24 3245 Basic Rd, 23228. Phone: (804) 474-0000. Fax: (804) 474-0090. E-mail: sheilahbelle@clearchannel.com Web Site: www.1065thebeat.com. Licensee: Capstar TX L.P. Group owner: Clear Channel Communications Inc. (acq 8-30-00; grpsl). Population served: 249,621 Natl. Rep: Christal. Law Firm: Pepper & Corazzini. Format: Oldies, urban contemp. News staff: one; News: 5 hrs wkly. Target aud: 18-44. ♦Mark Mays, pres; Tracy Onskell, sls dir; Ruth Jones, gen sls mgr; Ronda Steers, natl sls mgr; I.L. Scout, prom dir; Aaron Maxwell, progmg dir; Mike Street, mus dir; Sheilah Belle, news dir, pub affrs dir, min affrs dir & news rptr; Mike Fleming, engrg VP; Jon Bennett, chief of engrg; Scott Steven, traf mgr.

WBTK(AM)— September 1926: 1380 khz; 5 kw-U, DA-2. TL: N37 37 13 W77 26 57. Hrs open: 9401 Courthouse Rd., Suite 307, Chesterfield, 23832. Phone: (804) 717-5600. Fax: (804) 717-5602. Licensee: Mount Rich Media LLC. Group owner: Salem Communications Corp. (acq 5-30-2006; $1.5 million). Population served: 556,200 Format: Sp Christian. Target aud: 35 plus; older, upscale. ♦David Ruleman, VP; David Jackson, gen mgr & gen sls mgr; Dave Terry, opns mgr & prom dir; Glen Motto, progmg dir & pub affrs dir.

WCDX(FM)—See Mechanicsville

WCLM(AM)—See Highland Springs

***WCVE(FM)**— May 6, 1988: 88.9 mhz; 17.5 kw. Ant 840 ft TL: N37 34 00 W77 28 36. Stereo. Hrs opn: 23 Sesame St., 23235. Phone: (804) 320-1301. Fax: (804) 320-8729. Web Site: www.ideastations.org/radio. Licensee: Commonwealth Public Broadcasting Corp. Population served: 1,000,000 Natl. Network: NPR, PRI. Wire Svc: AP Format: News/talk/classical. News staff: one; News: 35 hrs wkly. Target aud: 35 plus. Spec prog: Folk 6 hrs, blues 3 hrs, jazz 18 hrs wkly. ♦Bill Miller, VP & stn mgr; Peter Solomon, opns mgr; Lisa Tait, dev VP & dev dir. Co-owned TV: *WCVE-TV affil.

***WDCE(FM)**— Sept 7, 1977: 90.1 mhz; 100 w. 118 ft TL: N37 34 48 W77 32 35. Stereo. Hrs opn: 24 Univ. of Richmond, Box 85, 23173. Phone: (804) 289-8698. Fax: (804) 289-8996. Licensee: University of Richmond. Population served: 300,000 Format: Progsv, div, new mus. News: 3 hrs wkly. Target aud: 15-30. Spec prog: Class 3 hrs, jazz 9 hrs, relg 3 hrs wkly. ♦Dan Inglis, gen mgr.

WDZY(AM)—Colonial Heights, 1955: 1290 khz; 5 kw-D, 41 w-N. TL: N37 15 30 W77 23 40. (CP: 25 kw-D). Hrs opn: 24 413 Stuart Cir., Suite 110, 23220. Phone: (804) 353-7200. Fax: (804) 353-2633. Web Site: www.radiodisney.com/wdzy/290. Licensee: Radio Disney Group LLC. Group owner: ABC Inc. (acq 8-22-00; grpsl). Format: Family progmg. News: 7 hrs wkly. Target aud: 2-12, 25-54; children & women. ♦Laura Haemker, stn mgr & gen sls mgr; Amy Garelick, prom mgr.

WFTH(AM)— June 16, 1964: 1590 khz; 5 kw-D, 19 w-N. TL: N37 30 02 W77 27 28. Hrs open: 227 E. Belt Blvd., 23224. Phone: (804) 233-0765. Fax: (804) 233-3725. Web Site: www.faith1590.com. Licensee: Tri-City Christian Radio Inc. (acq 3-22-90; $450,000; FTR: 4-16-90). Format: Gospel. ♦Jack Johnson, pres; Mary Johnson, VP; Bryant Johnson, gen mgr, stn mgr & opns mgr.

WGGM(AM)—See Chester

WKHK(FM)—Colonial Heights, Nov 17, 1972: 95.3 mhz; 17.5 kw. 393 ft TL: N37 26 21 W77 25 57. Stereo. Hrs opn: 24 Gateway Crossing, 351 Tilghmen Rd., Salisbury, 21804. Phone: (410) 742-1923. Fax: (410) 742-2329. Web Site: k9sscountry.com. Licensee: Cox Radio Inc. Group owner: Cox Broadcasting (acq 8-31-00; grpsl). Law Firm: Hogan & Hartson. Format: Country. News staff: one; News: 8 hrs wkly. Target aud: 25-54. ♦Doug Hillard, gen mgr; Frank Hamilton, sls dir; Dixie Penner, prom dir; Kenny Love, progmg dir; Walt Barcus, mus dir, mus dir & pub affrs dir; Jon Bennett, chief of engrg; Marie Merrill, traf mgr.

WKJM(FM)—See Petersburg

WKJS(FM)— 1996: 99.3 mhz; 2.3 kw. Ant 531 ft TL: N37 30 52 W77 30 28. Hrs open:
100% simulcast 99.3& 105.7(WKJM).
2809 Emerywood Pkwy., Suite 300, 23294. Phone: (804) 672-9299. Fax: (804) 672-9314. Web Site: www.wkjs-fm.firstmediaworks.com. Licensee: Radio One Licenses LLC. Group owner: Radio One Inc. (acq 11-8-2001; grpsl). Natl. Network: ABC. Natl. Rep: Eastman Radio. Law Firm: Fletcher, Heald & Hildreth. Format: Urban adult contemp. Target aud: 25-54; general. ♦Linda Forem, VP & gen mgr; Al Payne, opns mgr; Dennis Gettis, gen sls mgr; Bobby Walden, natl sls mgr; Dottie Brooks, mktg dir; Dawna Covington, prom dir; Clovia Lawrence, pub affrs dir; Chris Lawless, chief of engrg.

WKLR(FM)—Fort Lee, July 29, 1963: 96.5 mhz; 50 kw. 453 ft TL: N37 20 22 W77 24 31. Stereo. Hrs opn: 24 812 Moorefield Park Dr., Suite 300, 23236. Phone: (804) 330-5700. Fax: (320) 330-4079. Web Site: www.965theplanet.com. Licensee: Cox Radio Inc. Group owner: Cox Broadcasting (acq 8-31-00; grpsl). Population served: 800,000 Format: Classic rock. News staff: one; News: 5 hrs wkly. Target aud: 25-49. ♦James Kennedy, chmn; Bob Willoughby, sls dir; Kevin Meek, gen sls mgr; Ronda Steers, natl sls mgr; Micki Long, mktg VP; Scott Weimer, progmg dir; Leslie Taylor, pub affrs dir; Scott Swingle, engrg dir; Jon Bennett, engrg mgr & chief of engrg; Dick Hungate, disc jockey; Sam Giles, disc jockey.

WLEE(AM)— May 4, 1951: 990 khz; 1 kw-D, 13 w-N. TL: N37 31 40 W77 22 48. (CP: COL East Highland Park. 4 kw-D, 2 kw-N, DA-2. TL: N37 37 08 W77 25 27). Hrs opn: 24 308 W. Broad St., 23220. Phone: (804) 643-0990. Fax: (804) 474-5070. E-mail: info@radiorichmond.com Web Site: www.radiorichmond.com. Licensee: Davidson Media Station WLEE Licensee LLC. Group owner: 4M Communications Inc. (acq 5-13-2005; grpsl). Population served: 800200 Natl. Network: CNN Radio. Format: All news. Target aud: 25-64; white collar, upscale professionals. ♦Peter Davidson, pres; Bryan Hill, exec VP & gen sls mgr; Michael Mazursky, gen mgr; Dee Daniels, progmg dir; Tim Hurley, opns mgr & progmg mgr.

WMXB(FM)— Dec 23, 1961: 103.7 mhz; 18.5 kw. 750 ft TL: N37 30 31 W77 34 37. Stereo. Hrs opn: 24 812 Moorefield Park Dr., Suite 300, 23236. Phone: (804) 330-5700. Fax: (804) 330-4079. Fax: (804) 323-1524 sls. E-mail: tim.baldwin@cox.com Web Site: www.mix1037.com. Licensee: Cox Radio Inc. Group owner: Cox Broadcasting (acq 8-00; grpsl). Population served: 766,100 Natl. Network: ABC. Natl. Rep: McGavren Guild. Format: Adult contemp. News staff: one; News: 8 hrs wkly. Target aud: 25-54; predominately female. ♦James Kennedy, CEO & COO; Steve McCall, gen mgr; Amy DeVries, gen sls mgr; Tim Baldwin, progmg dir.

WREJ(AM)— May 8, 1964: 1540 khz; 10 kw-D, DA-D. TL: N37 37 08 W77 25 27. Hrs open: 24 306 W. Broad St., 23220. Phone: (804) 643-0990. Fax: (804) 474-5070. E-mail: blwestbrook@radiorichmond.com Web Site: www.radiorichmond.com. Licensee: Davidson Media Station WREJ Licensee LLC. (acq 5-13-2005; grpsl). Population served: 750,000 Natl. Network: ABC. Rgnl. Network: Va. News Net. Format: Urban inspirational. Target aud: 35-64; African-American concerned about financial, civic & economic issues. ♦Peter Davidson, pres; Michael Mazurksy, gen mgr; Tim Hurley, opns dir; Bryan Hill, gen sls mgr; B.L. Westbrook, progmg dir.

***WRIH(FM)**— 2007: 88.1 mhz; 2 kw vert. Ant 410 ft TL: N37 43 00 W77 38 02. Hrs open:
Rebroadcasts WAFR(FM) Tupelo, MS 100%.
Drawer 2440, Tupelo, MS, 38803. Phone: (662) 844-8888. Fax: (662) 842-6791. Web Site: www.afr.net. Licensee: American Family Association. Natl. Network: American Family Radio. Format: Contemp Christian. ♦Marvin Sanders, gen mgr.

WRNL(AM)— Nov 15, 1937: 910 khz; 5 kw-U, DA-N. TL: N37 36 52 W77 30 49. Hrs open: 24 3245 Basie Rd., 23228. Phone: (804) 345-1140. Fax: (804) 474-0168. Web Site: www.sportsradio910.com. Licensee: CC Licenses LLC. Group owner: Clear Channel Communications Inc. (acq 8-10-93; $9.75 million with co-located FM; FTR: 8-30-93) Population served: 249,621 Rgnl. Network: Va. News Net. Format: Sports. Target aud: 25-54; men. ♦Ruth Jones, VP & gen mgr; James Levy, gen sls mgr; June Snead, prom dir; Tom Parker, progmg dir.

WROU(AM)—See Petersburg

WRVA(AM)— Nov 2, 1925: 1140 khz; 50 kw-U, DA-1. TL: N37 24 13 W77 18 59. Hrs open: 24 3245 Basie Rd., 23228. Phone: (804) 345-1140. Phone: (804) 474-0000. Fax: (804) 474-0168. E-mail: tomparker@clearchannel.com Web Site: www.wrva.com. Licensee: CC Licenses LLC. (group owner: Clear Channel Communications Inc. (acq 6-26-92; grpsl; FTR: 7-20-92). Population served: 249,621 Natl. Network: ABC. Natl. Rep: Clear Channel. Format: News/talk. News staff: 10; News: 24 hrs wkly. Target aud: 35-54. Spec prog: Relg 10 hrs, computers 2 hrs, gardening 3 hrs, home care 2 hrs, legal one hr wkly. ♦Ruth Jones, sr VP & gen mgr; Dan O'Shea, gen sls mgr; Tom Parker, opns dir, natl sls mgr, progmg dir & chief of engrg.

WRVQ(FM)— Aug 4, 1948: 94.5 mhz; 200 kw. Ant 455 ft TL: N37 24 13 W77 18 59. Stereo. Hrs opn: 3245 Basie Rd., 23228. Phone: (804) 474-0000. Fax: (804) 474-0090. E-mail: billcahill@clearchannel.com Web Site: www.q94radio.com. Licensee: CC Licenses LLC. Group owner: Clear Channel Communications Inc. Natl. Network: ABC. Format: Mainstream. Target aud: 18-44; women. ♦Ruth Jones, gen mgr; Bill Cahill, opns mgr & traf mgr; Tracy Driskol, gen sls mgr; Wayne Coy, progmg dir; Mike Fleming, chief of engrg.

WRXL(FM)— Mar 4, 1949: 102.1 mhz; 20 kw. Ant 790 ft TL: N37 36 52 W77 30 56. Stereo. Hrs opn: 24 3245 Basie Rd., 23228. Phone: (804) 474-0000. Fax: (804) 474-0092. Web Site: www.1021thex.com. Licensee: CC Licenses LLC. Format: AOR, oldies new rock. Target aud: 25-44. ♦Ruth Jones, gen mgr; Bill Cahill, opns mgr.

WTVR-FM— February 1946: 98.1 mhz; 50 kw. 1,004 ft TL: N37 34 00 W77 28 36. Stereo. Hrs opn: 3245 Basie Rd., 23228. Phone: (804) 474-0000. Fax: (804) 474-0090. E-mail: billcahill@clearchannel.com Web Site: www.lite98.com. Licensee: CC Licenses LLC. Group owner: Clear Channel Communications Inc. (acq 1996; $18 million with co-located AM). Natl. Rep: Clear Channel. Format: Adult contemp. ♦Ruth Jones, gen mgr; Bill Cahill, opns mgr & progmg dir; Rhonda Reeser, gen sls mgr; Adam Stubbs, prom dir; Mike Fleming, chief of engrg.

WVNZ(AM)— September 1955: 1320 khz; 5 kw-D, DA. TL: N37 28 00 W77 27 08. Hrs open: 24 306 W. Broad St., 23220. Phone: (804) 643-0990. Fax: (804) 474-5070. E-mail: df@radiorichmond.com Web Site: www.radiorichmond.com. Licensee: Davidson Media Station WVNZ Licensee LLC. Group owner: 4M Communications Inc. (acq 5-13-2005; grpsl). Population served: 650000 Natl. Network: ABC. Format: Sp. News staff: 3; News: 15 hrs wkly. Target aud: 35 plus; Hispanic population. ♦Peter Davidson, pres; Michael Mazursky, gen mgr; Tim Hurley, opns mgr; Carolyn Resendiz, gen sls mgr & progmg dir; Demetrio Flores, progmg dir.

WXGI(AM)— Oct 1, 1947: 950 khz; 5 kw-D, 64 w-N. TL: N37 30 52 W77 30 28. Hrs open: 5:30 AM-midnight 701 German School Rd., 23225. Phone: (804) 233-7666. Fax: (804) 233-7681. E-mail: info@espn950am.com Web Site: espn950am.com. Licensee: Red Zebra Broadcasting Licensee (Richmond) LLC. (acq 9-27-2006; $1.4 million). Population served: 249,621 Natl. Network: ESPN Radio. Format: All sports. Target aud: 30 plus. ♦Bruce Gilbert, CEO; Sharon Eichenlaub, CFO; Howard H. Keller, opns dir.

WYFJ(FM)—See Ashland

Roanoke

WFIR(AM)— June 20, 1924: 960 khz; 5 kw-U, DA-N. TL: N37 15 20 W79 57 20. Hrs open: 24 3934 Electric Rd., 24018. Secondary address: Box 92 24022. Phone: (540) 345-1511. Fax: (504) 342-2270. Web Site: www.960wfir.com. Licensee: Mel Wheeler Inc. (group owner; acq 3-31-00; with co-located FM). Population served: 250,000 Natl. Network: ABC, Fox News Radio. Natl. Rep: Katz Radio. Format: News/talk. Target aud: 25-54; adult, skewed male 35-54. ♦Anne Booze, gen sls mgr; Jim Murphy, progmg dir; Becky Bruce, news dir.

WSLC-FM—Co-owned with WFIR(AM). November 1948: 94.9 mhz; 100 kw. Ant 1,979 ft TL: N37 11 41 W80 09 22. Stereo. 24 Phone: (540) 774-0201. Fax: (540) 774-5667. Web Site: www.949starcountry.com.1,054,204 Natl. Rep: Katz Radio. Format: Country. Target aud: 25-54. ♦Stan Reynolds, gen sls mgr; Rachel Rodes, prom dir; Brett Sharp, progmg dir.

WGMN(AM)— 1946: 1240 khz; 1 kw-U. TL: N37 16 12 W79 58 14. Hrs open: 3807 Brandon Ave., Suite 2350, 24018. Phone: (540) 725-1220. Fax: (540) 725-1245. Web Site: www.espnradio.com. Licensee: 3 Daughters Media Inc. Group owner: Clear Channel Communications Inc. (acq 6-22-2007; grpsl). Population served: 92,115 Rgnl. Network: Va. News Net. Natl. Rep: D & R Radio. Format: Sports. Target aud: 25-54. ♦Tex Meyer, gen mgr; Aaron Roberts, gen sls mgr & progmg dir; Tammy Cazad, gen sls mgr & progmg dir; Jeff Parker, chief of engrg.

WKBA(AM)—Vinton, Oct 9, 1961: 1550 khz; 10 kw-D, DA. TL: N37 17 24 W79 55 22. Hrs open: 2043 10th St. N.E., 24012. Phone: (540) 343-5597. Fax: (540) 345-4064. E-mail: ddurrett@radiowkba.com Web Site: www.radiowkba.com. Licensee: Tinker Creek Broadcasters Inc. (acq 2-1-83). Population served: 250,000 Law Firm: Booth, Freret, Imlay & Tepper P. Format: Relg. Target aud: General. Spec prog: Black 10 hrs wkly. ♦Dorothy Durrett, stn mgr; Dale Cook, chief of engrg.

WPAR(FM)—See Salem

WRIS(AM)— Feb 28, 1953: 1410 khz; 5 kw-D, 72 w-N. TL: N37 16 47 W79 59 29. Hrs open: 24 Box 6099, 219 Luckett St. N.W., 24017. Phone: (540) 342-1410. Phone: (540) 342-7811. Fax: (540) 342-5952. E-mail: wrisam@aol.com Licensee: WRIS L.L.C. (acq 1-15-98). Population served: 250,000 Rgnl. Network: Va. News Net. Law Firm: Blair, Joyce & Silva. Format: Relg, inspirational. News staff: one. Target aud: 35-75. ♦Lloyd Gochenour, pres & gen mgr; Russ Brown, opns mgr.

WROV-FM—See Martinsville

***WRXT(FM)**— July 31, 1994: 90.3 mhz; 5.5 kw. Ant 1,112 ft TL: N37 23 09 W79 40 10. Hrs open: 20276A Timberlake Road, Lynchburg, 24502. Phone: (434) 237-9798. Fax: (434) 237-1025. Web Site: www.spiritfm.com. Licensee: Positive Alternative Radio Inc. Group owner: Baker Family Stations (acq 4-15-2002). Format: Contemp Christian. ♦Barry Armstrong, gen mgr; Brian Sumner, mus dir.

WSLQ(FM)— Nov 1, 1947: 99.1 mhz; 200 kw. 1,985 ft TL: N37 11 42 W80 09 22. Stereo. Hrs opn: 24 3934 Electric Rd., 24018. Secondary address: Box 92 24018. Phone: (540) 387-0234. Web Site: www.q99fm.com. Licensee: Mel Wheeler Inc. Population served: 1,062,000 Natl. Rep: Katz Radio. Format: Adult contemp. News: 1 hr wkly. Target aud: 25-54; skew female. ♦Leonard Wheeler, pres; Jim Murphy, progmg dir.

WSNV(FM)—See Salem

WVBE(AM)— Oct 1, 1940: 610 khz; 5 kw-D, 1 kw-N, DA-2. TL: N37 18 11 W80 02 33. Hrs open: 24
Rebroadcasts WVBE Lynchburg 100%.
3934 Electric Rd., 24018. Secondary address: Box 92 24022. Phone: (540) 774-9200. Fax: (540) 774-5667. E-mail: info@vibe100.com Web Site: www.vibe100.com. Licensee: Mel Wheeler Inc. (group owner; (acq 10-1-76). Population served: 420,000 Natl. Rep: Katz Radio. Format: Urban adult contemp. Target aud: Adult 25-54; skew female, skew black. ♦Leonard Wheeler, pres & gen mgr; Stan Reynolds, gen sls mgr; Walt Ford, progmg dir.

***WVTF(FM)**— Aug 1, 1973: 89.1 mhz; 100 kw. 1,970 ft TL: N37 11 56 W80 09 02. Stereo. Hrs opn: 24 3520 Kingsbury Ln., 24014-1348. Phone: (540) 989-8900. Fax: (540) 776-2727. E-mail: wvtf@vt.edu Web Site: www.wvtf.org. Licensee: Virginia Tech Foundation Inc. (acq 1-1-82). Population served: 850,000 Natl. Network: PRI, NPR. Rgnl. Network: Va. News Net. Law Firm: Dow, Lohnes & Albertson. Format: Class, jazz, news. News staff: 3; News: 39 hrs wkly. Target aud: General. ♦Glenn Gleixner, gen mgr & dev dir.

WWWR(AM)— April 1957: 910 khz; 1 kw-D, 84 w-N. TL: N37 16 06 W79 54 46. Hrs open: 24 1848 Clay St. S.E., 24013. Phone: (540) 343-7109. Fax: (540) 343-2306. E-mail: tonybroom@3wradio.com Web Site: www.3wradio.com. Licensee: Perception Media Group Inc. (group owner; acq 4-25-91; $150,000; FTR: 5-13-91). Natl. Network: USA. Format: Gospel. Target aud: 35-64. ♦Ben Peyton, pres; Barbara Evans, stn mgr; Tony Broom, opns mgr; Dale Cook, chief of engrg.

WXLK(FM)— Dec 17, 1960: 92.3 mhz; 93 kw. Ant 2,050 ft TL: N37 11 56 W80 09 01. Stereo. Hrs opn: Box 92, 24022. Secondary address: 3934 Electric Rd. S.W. 24018. Phone: (540) 774-9200. Fax: (540) 774-5667. Web Site: www.k92radio.com. Licensee: Mel Wheeler Inc. (group owner; acq 3-12-97; $7.5 million with WVBE-FM Lynchburg). Population served: 1,000,000 Format: CHR. Target aud: 18-44; women. ♦Leonard Wheeler, CEO, pres & gen mgr; Kathy Rilee, mktg dir.

WZBL(FM)— November 1993: 104.9 mhz; 14.5 kw. Ant 925 ft TL: N37 22 23 W79 55 40. Hrs open: 24 3305 Old Forest Rd., Lynchburg, 24501. Phone: (434) 385-8298. Fax: (434) 385-8991. E-mail: joeldearing @clearchannel.com Web Site: www.magicfm.cc. Licensee: Capstar TX L.P. Group owner: Clear Channel Communications Inc. (acq 8-30-2000; grpsl). Format: Adult contemp. News staff: 2. ♦Chris Clendenen, pres & gen mgr; Dave Carwile, sls dir & natl sls mgr; Joel Dearing, progmg dir.

Rocky Mount

WYTI(AM)— Mar 31, 1957: 1570 khz; 2.5 kw-D, 220 w-N. TL: N36 58 37 W79 53 45. Hrs open: 6 AM-9 PM Box430, 275 Glenwood Dr., 24151-0430. Phone: (540) 483-9955. Phone: (540) 483-2166. Fax: (540) 483-7802. E-mail: wyti@cablenet-va.com Web Site: www.wytiradio.com. Licensee: WYTI Inc. (acq 1-71). Population served: 37,500 Natl. Network: ABC. Format: Traditional country, bluegrass, gospel. Target aud: 30 plus; general. Spec prog: NASCAR races, relg 10 hrs wkly. ♦Susan Mullins, exec VP, gen mgr & opns mgr; William E. Jefferson, pres & stn mgr.

Ruckersville

WHTE-FM—Licensed to Ruckersville. See Charlottesville

Rural Retreat

WLOY(AM)— May 15, 1985: 660 khz; 550 w-D. TL: N36 55 17 W81 14 34. Hrs open: Box 660, 24368. Phone: (276) 228-3185. Fax: (276) 228-9261. E-mail: wyve.wxby@wiredog.com Licensee: Three Rivers Media Corp. (acq 10-16-2006; $125,000). Rgnl. Network: Va. News Net. Natl. Rep: Rgnl Reps. Law Firm: Brooks, Pierce, McClendon, Humphrey & Leonard. Format: Contemp Christian. News staff: one. Target aud: 25-54. ♦Gary W. Hagerich, pres.

WXBX(FM)— June 11, 1992: 95.3 mhz; 6 kw. Ant 400 ft TL: N36 55 17 W81 14 34. Stereo. Hrs opn: 24 Box 1247, Three Rivers Media Corp., 110 W. Spiller St., Wytheville, 24382. Phone: (276) 228-3185. Fax: (276) 228-9261. E-mail: wyve/wxbx@wiredog.com Web Site: www.wyve.com. Licensee: Three Rivers Media Corp. (acq 10-01-98; $200,000). Population served: 50,000 Natl. Network: AP Network News, Jones Radio Networks. Rgnl rep: Rgnl Reps Law Firm: Brooks, Pierce, McLendon, Humphrey & Leonard. Format: Oldies. News staff: one; News: 12 hrs wkly. Target aud: 25+. ♦Gary W. Hagerich, CEO, pres & gen mgr; Tammy Chase, opns mgr & news dir; Sam Parks, chief of engrg; Kristy Wrobel, traf mgr.

Rustburg

***WWEM(FM)**—Not on air, target date: unknown: 91.7 mhz; 1.15 kw. Ant 748 ft TL: N37 17 07 W79 05 26. Hrs open: Box 905, Spotsylvania, 22553. Phone: (540) 582-9700. Web Site: www.wwedfm.org. Licensee: Educational Media Corp. ♦Peter D. Stover, pres & gen mgr.

Saint Paul

WXLZ(AM)— Nov 3, 1981: 1140 khz; 2.5 kw-D. TL: N36 52 15 W82 18 21. Hrs open: Sunrise-sunset
Rebroadcasts WXLZ-FM Lebanon 80%.
Box 250, Mew Rd., Castlewood, 24224. Phone: (276) 762-5595. Fax: (276) 889-3677. E-mail: wxlz1073@mounet.com Web Site: www.wxlz.net. Licensee: Yeary Broadcasting Inc. Population served: 108,000 Natl. Network: CBS. Format: Relg. News staff: 3; News: 3 hrs wkly. Target aud: 25 plus; students, farmers, miners & rural area residents. Spec prog: Gospel 15 hrs wkly. ♦Lannis Yeary, CEO; Donna Yeary, exec VP; Gary Scott, stn mgr, progmg dir & local news ed.

Salem

***WPAR(FM)**— April 1994: 91.3 mhz; 3.3 kw horiz, 3 kw vert. 902 ft TL: N37 22 23 W79 55 40. Stereo. Hrs opn: 24 Box 889, Blacksburg, 24063. Phone: (540) 961-2377. Fax: (540) 951-5282. E-mail: mail@spiritfm.com Web Site: www.spiritfm.com. Licensee: Positive Alternative Radio Inc. Group owner: Baker Family Stations (acq 5-90; FTR: 5-21-90). Natl. Network: USA. Law Firm: Booth, Freret, Imlay & Tepper. Format: Christian adult contemp. Target aud: 25-54; adults with families. ♦Barry Armstrong, gen mgr.

WSNV(FM)— Mar 7, 1969: 93.5 mhz; 5.8 kw. 98 ft TL: N37 16 47 W79 59 29. Stereo. Hrs opn: 24 3807 Brandon Ave. S.W., Suite 2350, Roanoke, 24018. Phone: (540) 725-1220. Fax: (540) 725-1245. Web Site: www.j93.com. Licensee: Capstar TX L.P. Group owner: Clear Channel Communications Inc. (acq 8-30-00; grpsl). Population served: 200,000 Natl. Network: Westwood One. Format: Soft adult contemp. News: one hr wkly. Target aud: 24 plus. ♦Chris Clendenen, gen mgr & gen sls mgr; Steve Cross, progmg dir; Ed Kilbane, mus dir & news dir.

WTOY(AM)— Sept 7, 1956: 1480 khz; 5 kw-D. TL: N37 16 21 W80 04 52. Hrs open: 504 23rd St., Roanoke, 24017. Phone: (540) 344-9869. Fax: (540) 344-0976. Licensee: Ward Broadcasting Corp. (acq 3-2-92). Population served: 220,000 Format: Adult urban contemp. ♦Irving L. Ward Sr., gen mgr & pres.

Saltville

WXMY(AM)— Nov 5, 1981: 1600 khz; 5 kw-D. TL: N36 51 43 W81 43 29. (CP: 10 kw-D, 10 w-N, DA-2). Hrs opn: 6 AM-sunset Box 5555, Chilhowie, 24319. Phone: (276) 496-0016. Fax: (276) 496-0005. E-mail: wxmy@aol.com Web Site: www.1600wxmy.com. Licensee: Continental Media Group LLC (acq 4-11-2001; $62,000). Format: Classic country, bluegrass, gospel. Target aud: 25-54. ♦Jeff Raynor, gen mgr, stn mgr & gen sls mgr; J.C. Heath, mus dir; Wendy Raynor, traf mgr.

Smithfield

WKGM(AM)— Dec 18, 1974: 940 khz; 10 kw-D, 3 kw-N, DA-N. TL: N36 57 16 W76 37 48. Hrs open: 24 Box 339, 23431. Secondary address: 13379 Great Spring Rd. 23430. Phone: (757) 357-9546. Phone: (757) 622-9546. Fax: (757) 365-0412. E-mail: wkgm@hotmail.com Licensee: WKGM Inc. Group owner: Baker Family Stations Population served: 1,700,000 Format: Relg. News: one hr wkly. Target aud: 25 plus. Spec prog: Farm 2 hrs, Ger one hr, Sp one hr, gospel 5 hrs wkly. ♦Vernon H. Baker, pres; Larry W. Cobb, VP & gen mgr.

South Boston

WAJL(AM)—Not on air, target date: unknown: 1400 khz; 1 kw-U. TL: N36 42 35 W78 52 28. Hrs open: Box 127, Semora, NC, 27343-0127. Phone: (336) 234-8416. Fax: (434) 572-9245. Licensee: Linda Waller-Barton. ♦Linda Waller-Barton, gen mgr.

WHLF(FM)— Sept 1, 1992: 95.3 mhz; 6 kw. Ant 246 ft TL: N36 42 24 W78 55 28. Stereo. Hrs opn: 24 Box 526, 1210 Porter Ln., 24592. Phone: (434) 572-2988. Fax: (434) 572-1662. E-mail: whlf@whlf.com Web Site: www.whlf.com. Licensee: JLC Properties Inc. (acq 4-1-93; FTR: 4-19-93). Population served: 30,000 Natl. Network: ABC. Wire Svc: AP Format: Adult contemp hits, CHR. News staff: one; News: 8 hrs wkly. Target aud: 25-54; those that have spendable income. ♦John L. Cole III, pres; Nick Long, gen mgr; Kelly Redd, progmg dir.

WQOK(FM)—Licensed to South Boston. See Raleigh NC

WSBV(AM)— 1980: 1560 khz; 2.5 kw-D, DA-1. TL: N36 42 24 W78 52 28. Hrs open: Box 778, 1180 Plywood Tr., 24592. Phone: (434) 572-4418. Fax: (434) 572-9245. Licensee: Linda Waller-Barton Format: Contemp Christian, Multi-Cultural. Target aud: 40 plus. Spec prog: Farm one hr wkly. ♦Linda Waller-Barton, gen mgr; April Warf, stn mgr; James W. Barton, opns dir.

South Hill

WKSK-FM—Listing follows WSHV(AM).

WSHV(AM)— Nov 1, 1953: 1370 khz; 5 kw-D. TL: N36 44 39 W78 09 42. Hrs open: 6 AM-6 PM Box 216, 23970. Phone: (434) 447-8997. Phone: (434) 447-4007. Fax: (434) 447-4789. E-mail: wshv@hotmail.com Licensee: Lakes Media Holding Company LLC. Group owner: Joyner Radio Inc. (acq 2-1-2005; grpsl). Population served: 30,650 Natl. Network: ABC. Rgnl. Network: Tobacco. Format: Black. News staff: one. Target aud: General. ♦Jerry E. Brown, sr VP; Greg Thirft, gen mgr; Robert Wilson, progmg mgr; Robby McMulian, news dir; John Hart, chief of engrg; Heather Skuggen, disc jockey; Jay Phillippi, disc jockey; Paul Hoefler, disc jockey.

WKSK-FM—Co-owned with WSHV(AM). Dec 23, 1966: 101.9 mhz; 6 kw. Ant 315 ft TL: N36 44 39 W78 09 42. Stereo. E-mail: wjws@yahoo.com Natl. Network: ABC. Format: Country. News staff: one; News: 2 hrs wkly. Target aud: General; adults 25-54. ♦Frank Malone, farm dir, women's int ed & disc jockey; Greg Thrift, disc jockey; Robert Wilson, disc jockey; Ron Major, disc jockey; Steve Howell, progmg dir & disc jockey.

Spotsylvania

***WWED(FM)**— 2005: 89.5 mhz; 380 w vert. Ant 433 ft TL: N38 11 48 W77 33 45. Hrs open: Box 905, 22553. Phone: (540) 582-9700. Fax: (530) 325-4033. Licensee: Educational Media Corp. Natl. Network: Moody. Format: Christian music. ♦Peter Stover, pres & gen mgr.

WYSK-FM— Mar 31, 1988: 99.3 mhz; 3 kw. 295 ft TL: N38 08 31 W77 41 38. Hrs open: 24 616 Amelia St., Fredericksburg, 22401. Phone: (540) 582-2405. Phone: (540) 374-5500. Fax: (540) 374-5525. Web Site: www.wysk.com. Licensee: The Free Lance-Star Publishing Co. (group owner; acq 4-19-93; $200,000; FTR: 5-3-93). Population served: 90,000 Format: Rock alternative. News staff: 3. Target aud: 25-49. ♦Josiah P Rowe III, pres; John Moen, gen mgr; James T. Butler, sls dir & gen sls mgr; Joann Pope, prom dir; Paul Johnson, opns dir & progmg dir; Chris Wilk, chief of engrg.

Stanleytown

WZBB(FM)— March 1989: 99.9 mhz; 3.6 kw. 722 ft TL: N36 54 50 W79 57 07. Stereo. Hrs opn: 24 10899 Virginia Ave., Bassett, 24055. Phone: (540) 489-9999. Fax: (276) 629-8399. E-mail: kristib@wzbbfm.com Web Site: www.wzbbfm.com. Licensee: WNLB Radio Inc. Rgnl. Network: Va. News Net. Format: Country. Target aud: 21 plus. ♦Donny Brook, pres; Glenn Lynch, VP; Kristi Banks, gen mgr; Amy Coleman, stn mgr; Craig Richards, progmg dir; Lisa Layne, news dir.

Staunton

WBOP(FM)—Buffalo Gap, 1988: 95.5 mhz; 6 kw. Ant 308 ft TL: N38 10 55 W79 13 34. Stereo. Hrs opn: 24 Box 2460, Harrisonburg, 22801. Secondary address: 639 N. Main St., Mt. Crawford 22841. Phone: (540) 432-1063. Fax: (540) 433-9267. E-mail: business@magic1063fm.com Web Site: www.magic955fm.com. Licensee: Vox Communications Group LLC. (acq 8-31-2005; $900,000). Population served: 150,000 Format: Oldies. ♦Randy Thompson, gen mgr.

WCYK-FM— September 1984: 99.7 mhz; 3.3 kw. 1,692 ft TL: N38 03 52 W78 48 18. Stereo. Hrs opn: 24 1150 Pepsi Pl., Suite 300, Charlottesville, 22901. Phone: (434) 978-4408. Fax: (434) 978-0723. Web Site: www.country997.com. Licensee: CC Licenses LLC. Group owner: Clear Channel Communications Inc. (acq 8-6-99; grpsl). Population served: 750,000 Natl. Network: Motor Racing Net, Westwood One. Natl. Rep: Clear Channel, Katz Radio. Format: Country. Target aud: 25-54. ♦Phil Robken, gen mgr & natl sls mgr; Mike Chiumento, gen sls mgr & chief of engrg; Regan Keith, opns mgr & progmg dir.

WKCI(AM)—See Waynesboro

WKDW(AM)— April 1954: 900 khz; 2.5 kw-D, 128 w-N. TL: N38 10 27 W79 04 12. Stereo. Hrs opn: 24 Box 2189, 24401. Phone: (540) 886-2376. Fax: (540) 885-8662. E-mail: wkdw@ntelos.net Web Site: www.goodradio.com. Licensee: CC Licenses LLC. Group owner: Clear Channel Communications Inc. (acq 11-15-2000; grpsl). Population served: 22,200 Law Firm: Holland & Knight. Format: Country. News staff: one; News: 15 hrs wkly. Target aud: 25-54. Spec prog: Farm one hr, blue grass one hr wkly. ♦Steve Davis, gen mgr; Kris Losh, progmg dir; Jeff Caudell, chief of engrg.

WSVO(FM)—Co-owned with WKDW(AM). May 29, 1959: 93.1 mhz; 2.8 kw. 338 ft TL: N38 10 27 W79 04 12. (CP: TL: N38 10 32 W79 04 12). Stereo. Web Site: www.goodradio.com. Format: Oldies. News staff: one; News: 15 hrs wkly. Target aud: 35-54.

WNLR(AM)—See Churchville

WTON(AM)— Mar 9, 1946: 1240 khz; 1 kw-U. TL: N38 08 30 W79 02 33. Hrs open: 24 Box 1085, 24402-1085. Secondary address: 304 W. Beverly St. 24401. Phone: (540) 885-5188. Fax: (540) 885-1240. E-mail: star94@ntelos.net Licensee: High Impact Communications Inc. (acq 3-1-96; $1 million). Population served: 120,200 Natl. Network: CBS, ESPN Radio. Law Firm: Reddy & Begley. Format: All sports. News: one hr wkly. Target aud: 18-49. ♦Brenda Ratcliff, gen mgr; J. Gary Ratcliff, pres & gen mgr; Barry Bland, sls VP; Cass Johnson, progmg dir.

WTON-FM— November 1990: 94.3 mhz; 330 w. 2,263 ft TL: N38 09 55 W79 18 51. Stereo. 24 360,000 Natl. Network: CBS. Rgnl rep: Rgnl Reps Law Firm: Reddy & Begley. Format: Classic hits. News: 3 hrs wkly. Target aud: 18-49; 60% women, 40% men.

Stephens City

WKSI-FM—Listing follows WMRE(AM).

Strasburg

WWRT(FM)— Jan 1, 1987: 104.9 mhz; 3 kw. 219 ft TL: N39 01 22 W78 25 35. Stereo. Hrs opn: 24
Simulcast with WWRE(FM) Berryville 100%.
Box 3300, Winchester, 22604. Secondary address: 520 N. Pleasant Valley Rd., Winchester 22601. Phone: (540) 667-2224. Fax: (540) 722-3295. Web Site: www.realclassicrock.fm. Licensee: Mid Atlantic Network Inc. Group owner: Mid Atlantic Network (acq 7-8-97; $850,000 with WWRE(FM) Berryville). Law Firm: Cole, Raywid & Braverman. Format: Classic Rock. Target aud: 18 plus. ♦John P. Lewis, pres; Chris Lewis, gen mgr; Jeff Adams, opns mgr; Don Wilson, progmg dir; Steve Edwards, news dir.

Stuart

WHEO(AM)— Oct 12, 1959: 1270 khz; 5 kw-D. TL: N36 37 25 W80 15 50. Hrs open: 6 AM-sunset 3824 Wayside Rd., 24171. Phone: (276) 694-3114. Fax: (276) 694-2241. E-mail: wheo@sitesstar.net Web Site: www.wheo.net. Licensee: Mountain View Communications Inc. (acq 6-86). Population served: 75,000 Natl. Network: CNN Radio. Rgnl. Network: Va. News Net. Format: Country. News staff: one; News: 25 hrs wkly. Target aud: General. Spec prog: Farm 4 hrs, relg 10 hrs, loc news 10 hrs wkly. ♦Dean Goad, pres; Jamie Clark, VP, opns mgr, news dir & traf mgr; La Vergne Collins, gen sls mgr; Richard Rogers, progmg dir.

Suffolk

WAFX(FM)— Dec 12, 1983: 106.9 mhz; 100 kw. 984 ft TL: N36 48 16 W76 45 17. Stereo. Hrs opn: 870 Greenbriar Cir., Suite 399, Chesapeake, 23320. Phone: (757) 366-9900. Fax: (757) 366-0022. E-mail: mbeck@tciradio.com Web Site: www.1069thefox.com. Licensee: Tidewater Communications LLC. Group owner: Saga Communications Inc. (acq 3-15-94; $4 million; FTR: 5-9-94). Natl. Rep: McGavren Guild. Format: Classic rock. Target aud: 18-49. ♦Dave Paulus, gen mgr; Barry Haugh, gen sls mgr; Mike Beck, progmg dir; Leila Rice, news dir; Don Crowder, chief of engrg.

WVBW(FM)— December 1965: 92.9 mhz; 50 kw. 480 ft TL: N36 52 35 W76 23 28. Stereo. Hrs opn: 24 5589 Greenwich Rd., Suite 200, Virginia Beach, 23462. Phone: (757) 671-1000. Fax: (757) 671-1010. Web Site: www.929thewave.com. Licensee: MHR License LLC. Group owner: Barnstable Broadcasting Inc. (acq 3-24-2005; grpsl). Population served: 1,200,000 Natl. Rep: Christal. Format: Adult contemp. Target aud: 25-54; women. Spec prog: Relg 2 hrs, Sunday Morning Magazine one hr wkly. ♦Vonneva Carter, gen mgr & sls dir; Jim Long, prom dir & news dir; Mike Allen, progmg dir; Keith O'Malley, chief of engrg.

Sweet Briar

***WNRS-FM**— 1980: 89.9 mhz; 30 w. Ant 1,942 ft TL: N37 33 50 W79 11 34. Stereo. Hrs opn: 24
Rebroadcasts WNRN(FM) Charlottesville 100%.
Box 143, 24595. Phone: (434) 381-6187. Fax: (434) 381-6173. E-mail: wnrs@sbc.edu Web Site: wnrs-fm.sbc.edu. Licensee: Sweet Briar College. Population served: 75,000 Format: Modern rock. Target aud: General. ♦Tess Drahman, gen mgr; Sarah Liston, stn mgr; Carolina Muglia, mus dir.

Tappahannock

***WRAR(AM)**— Nov 1, 1970: 1000 khz; 300 w-D. TL: N37 52 27 W76 43 37. Hrs open: Box 1393, 22560. Phone: (804) 443-6572. Licensee: A.C.T.I.O.N. Inc. (acq 7-19-2006). Population served: 15,000 Format: Life talk. ♦Geoffrey Coleman, pres.

WRAR-FM— July 26, 1971: 105.5 mhz; 6 kw. Ant 328 ft TL: N37 52 27 W76 43 37. Stereo. Hrs opn: 24 Box 1023, 22560. Secondary address: 156 Prince St. 22560. Phone: (804) 443-4321. Fax: (804) 443-1055. E-mail: rich@wrarfm.com Web Site: www.wrarfm.com. Licensee: Real Media Inc. (acq 7-1-2006; $1.9 million). Natl. Network: ABC. Natl. Rep: Rgnl Reps. Wire Svc: AP Format: Adult contemp. ♦Billy Flynn, progmg dir; Tom Davis, news dir; Terry Brooks, traf mgr.

Tasley

WESR-FM—See Onley-Onancock

Tazewell

WKQY(FM)—Listing follows WTZE(AM).

WTZE(AM)— Apr 22, 1966: 1470 khz; 5 kw-D. TL: N37 07 57 W81 33 21. Hrs open: 6 AM-sunset 900 Bluefield Ave., Bluefield, WV, 24701. Phone: (304) 327-7114. Fax: (304) 325-7850. Licensee: Monterey Licenses LLC. Group owner: Triad Broadcasting Co. LLC (acq 7-18-00; grpsl). Population served: 85,000 Natl. Network: Motor Racing Net. Format: News/talk. News: 5 hrs wkly. Target aud: 21-54. Spec prog: Gospel 5 hrs wkly. ♦John Halford, gen mgr; Dave Crosier, opns dir; Joseph Echoles, progmg dir; Keith Bowman, chief of engrg.

WKQY(FM)—Co-owned with WTZE(AM). Sept 1, 1968: 100.1 mhz; 4.2 kw. 395 ft TL: N37 08 00 W81 35 43. Stereo. 24 Format: Oldies. News staff: one. Target aud: 25 plus.

Vinton

WJJS-FM— 1994: 106.1 mhz; 6 kw. Ant 95 ft TL: N37 17 03 W79 59 14. Hrs open: 3807 Brandon Ave. S.W., Suite 2350, Roanoke, 24018. Phone: (540) 725-1220. Fax: (540) 725-1245. Web Site: www.wjjs.cc. Licensee: Capstar TX L.P. Group owner: Clear Channel Communications Inc. (acq 8-30-2000; grpsl). Format: Top-40. ♦Chris Clendenen, gen mgr; Ron Gaylor, gen sls mgr & rgnl sls mgr; David Lee Michaels, progmg dir; Ed Kilbane, news dir; Jeff Parker, chief of engrg.

WKBA(AM)—Licensed to Vinton. See Roanoke

WZZI(FM)— Dec 12, 1995: 101.5 mhz; 630 w. Ant 705 ft TL: N37 21 57 W79 52 01. Stereo. Hrs opn: 24
Rebroadcasts WZZU(FM) Lynchburg 100%.
210 1st St., Suite 240, Roanoke, 24011. Phone: (540) 344-2800. Fax: (540) 344-4001. Web Site: www.rocktheplanet.fm. Licensee: Centennial Broadcasting LLC. (group owner; (acq 11-23-2004; $4.15 million with WZZU(FM) Lynchburg). Natl. Network: Fox News Radio. Natl. Rep: McGavren Guild. Law Firm: Womble, Carlyle, Sandridge & Rice. Format: Rock. News staff: one; News: 2 hrs wkly. ♦Allen B. Shaw, pres; Bob Abbott, gen sls mgr; Dale Cook, chief of engrg; Melinda Schamerhorn, traf mgr.

Virginia Beach

WBVA(AM)—Bayside, May 1999: 1450 khz; 1 kw-U. TL: N36 51 29 W76 09 28. Hrs open:
WPMH.
2202 Jolliff Rd., Chesapeake, 23321. Phone: (757) 465-1603. Fax: (757) 488-7761. Licensee: Ronald W. Cowan Jr. (acq 5-21-2001). Format: News/Talk. ♦Ronald W. Cowan Jr., CEO & pres; Henry Hoot, gen mgr.

***WJLZ(FM)**— Feb 12, 1989: 88.5 mhz; 1.2 kw. Ant 118 ft TL: N36 50 30.7 W76 05 37. Stereo. Hrs opn: 24 3500 Virginia Beach Blvd., Suite 201, 23452. Phone: (757) 498-9632. Fax: (757) 498-8609. E-mail: info@currentfm.com Web Site: www.currentfm.com. Licensee: Virginia Beach Educational Broadcasting Foundation Inc. Population served: 500,000 Format: Christian CHR. Target aud: General. ♦William M. Verebely Jr., pres; Anne Verebely, gen mgr; J.P. Morgan, prom dir & progmg dir.

WPTE(FM)— May 5, 1984: 94.9 mhz; 50 kw. 499 ft TL: N36 48 38 W76 16 57. Stereo. Hrs opn: 24 236 Clearfield Ave., Suite 206, 23462. Phone: (757) 497-2000. Fax: (757) 456-5458. Web Site: www.pointradio.com. Licensee: Entercom Norfolk License LLC. Group owner: Entercom Communications Corp. (acq 12-13-99; grpsl). Population served: 1,200,000 Natl. Rep: D & R Radio. Format: Modern adult comtemp. News: one hr wkly. Target aud: 18-49. ♦David J. Field, CEO; David J. Field, pres; Skip Schmidt, gen mgr.

WPYA(FM)—See Chesapeake

WROX-FM—Exmore, 1986: 96.1 mhz; 23 kw. Ant 722 ft TL: N37 15 45 W76 00 45. Stereo. Hrs opn: 24 500 Dominion Tower, 999 Waterside Dr., Norfolk, 23510. Phone: (757) 640-8500. Fax: (757) 640-8552. Web Site: www.96x.fm. Licensee: Sinclair Telecable Inc. Group owner: Sinclair Communications Inc. (acq 9-28-93; $1.3 million; FTR: 10-25-93). Population served: 1,400,000 Natl. Rep: McGavren Guild, Interep. Format: Modern rock. News staff: one. Target aud: 18-34; men. ♦Bob Sinclair, pres; Lisa Sinclair, gen mgr; Dave Morgan, opns mgr; Jeanette Xenakis, gen sls mgr; Ginger Power, natl sls mgr; Donna Agresto, prom dir; Michele Diamond, progmg dir.

WVAB(AM)—Licensed to Virginia Beach. See Norfolk

WXMM(FM)—See Norfolk

WXTG(FM)— 2002: 102.1 mhz; 6 kw. Ant 328 ft TL: N36 45 07 W76 08 57. Hrs open: 545 South Birdneck Rd., Suite 100, 23450. Phone: (757) 422-6421. Fax: (757) 422-6437. Web Site: www.1021thegame.com. Licensee: Red Zebra Broadcasting Licensee (Norfolk) LLC. Group owner: On Top Communications Inc. (acq 12-4-2006; $4.25 million). Natl. Network: Fox Sports. Format: Sports. ♦Martin Snead, gen mgr; Bart Horton, gen sls mgr; Parish Brown, progmg dir; Terry Love, chief of engrg.

Warrenton

WKCW(AM)— Dec 7, 1957: 1420 khz; 10 kw-D, 17 w-N. TL: N38 45 05 W77 44 38. Hrs open: 6 AM-sunset 320 B. Maple Ave. E., Vienna, 22180. Phone: (540) 347-1421. Phone: (703) 319-3400. Fax: (703) 319-3390. Web Site: www.wkcw1420am.com. Licensee: Metro Radio Inc. (group owner; (acq 1-2-2004; $400,000). Population served: 48,471 Format: Sp tropical. News staff: one; News: 2 hrs wkly. Target aud: 25-54; mature adults with discretionary income of $30,000 plus. Spec prog: Bluegrass, farm one hr, rel 6 hrs wkly. ♦Bruce A. Houston, pres; Bill Parris, gen mgr; Tom Casey, opns mgr.

WPRZ(AM)— Nov 21, 1957: 1250 khz; 5 kw-D, 32 w-N, DA-2. TL: N38 43 52 W77 46 42. Hrs open: 24 7351 Hunton St., 20187-2222. Phone: (540) 349-1250. Fax: (540) 349-2726. E-mail: info@wprz.org Web Site: www.wprz.org. Licensee: Praise Communications Inc. (acq 12-83; $400,000). Population served: 250,000 + Natl. Network: USA, Moody, Salem Radio Network. Law Firm: Bentley Law Office. Format: Christian mus & programs. News: 5 hrs wkly. Target aud: 25-54; Christian. Spec prog: Children 7 hrs wkly. ♦Sally L. Buchanan, gen mgr & gen sls mgr; Steve W. Buchanan, pres, opns mgr & sls dir; Kandy Flippo, traf mgr.

WTWP-FM— Mar 28, 1966: 107.7 mhz; 29 kw. Ant 646 ft TL: N38 44 31 W77 50 07. Stereo. Hrs opn: 24
Rebroadcasts WTWP(AM) Washington, DC 100%.
3400 Idaho Ave. N.W., Washington, DC, 20016. Phone: (202) 895-5000. Fax: (202) 895-5016. Web Site: www.washingtonpostradio.com. Licensee: Bonneville Holding Co. Group owner: Bonneville International Corp. (acq 4-27-98). Population served: 3,300,000 Natl. Network: CBS. Rgnl. Network: Va. News Net. Natl. Rep: Katz Radio. Format: Newstalk. Target aud: 25-54. ♦Bruce Reese, CEO; Bob Johnson, COO; Joel Oxley, sr VP, gen mgr & mktg mgr; Matt Mills, sls dir; Greg Tantum, progmg dir.

WWXX(FM)— Nov 2, 1978: 94.3 mhz; 3 kw. Ant 397 ft TL: N38 40 42 W77 47 18. Stereo. Hrs opn: 24 8121 Georgia Ave., Suite 1050, Silver Springs, 20910. Phone: (301) 562-5800. Fax: (301) 589-9772. Web Site: www.triplexespnradio.com. Licensee: Red Zebra Broadcasting Licensee LLC. Group owner: Mega Communications Inc. (acq 5-9-2006; grpsl). Population served: 58,000 Natl. Network: ESPN Radio. Format: Sports. ♦Bruce Gilbert, CEO.

Warsaw

WNNT-FM— Mar 1, 1967: 100.9 mhz; 3 kw. 305 ft TL: N37 56 39 W76 45 05. Stereo. Hrs opn: Box 877, 22572. Secondary address: 194 Islington Rd. 22572. Phone: (804) 333-4900. Phone: (804) 333-3711. Fax: (804) 333-4531. E-mail: rivercountry@rivercountry1009.com Web Site: www.rivercountry1009.com. Licensee: Northern Neck & Tidewater Communications Inc. (acq 9-16-93; $400,000; FTR: 10-11-93). Population served: 75,000 Format: Country. ♦Lin Wadsworth, gen mgr & gen sls mgr; Mark Bryant, progmg dir; Frank Miner, chief of engrg.

Waynesboro

WKCI(AM)— Mar 10, 1965: 970 khz; 5 kw-D, 1 kw-N, DA-2. TL: N38 05 12 W78 54 42. Hrs open: 24
Rebroadcasts WKCY(AM) Harrisonburg.
207 University Blvd., Harrisonburg, 22801. Phone: (540) 434-1777. Fax: (540) 432-9968. Web Site: www.shenandoahradio.com. Licensee: CC Licenses LLC. Group owner: Clear Channel Communications Inc. (acq 11-16-2000; grpsl). Population served: 250,000 Rgnl. Network: Capitol Radio Net., Va. RFD. Format: News/talk. News staff: one; News: 2 hrs wkly. Target aud: 25-54. Spec prog: Black 3 hrs, farm 2 hrs wkly. ♦Steve Davis, gen mgr; Steve Knupp, opns mgr & progmg VP; Susie Smith, gen sls mgr; Mark Ness, chief of engrg.

***WPVA(FM)**— 1999: 90.1 mhz; 2.5 kw. Ant 961 ft TL: N38 01 16 W78 52 38. Hrs open: 20276 A Timberlake Rd., Lynchburg, 24502. Phone: (434) 237-9798. Fax: (434) 237-1025. E-mail: office@spiritfm.com Web Site: www.spiritfm.com. Licensee: Positive Alternative Radio Inc. (group owner; (acq 12-30-2005; grpsl). Law Firm: Booth, Freret, Imlay & Tepper. Format: Contemp Christian. ♦Barry Armstrong, gen mgr.

West Point

WBQK(FM)— July 1991: 107.9 mhz; 4 kw. Ant 328 ft TL: N37 27 00 W76 48 46. Stereo. Hrs opn: 24 5000 New Point Rd., Suite 2102, Williamsburg, 23188. Phone: (757) 565-1079. Fax: (757) 565-7094. Web Site: www.tideradio.com. Licensee: Davis Media LLC (acq 6-17-2005; $1.13 million). Population served: 250,000 Format: Classical. ♦Thomas G. Davis, pres & gen mgr; Derek Mason, gen sls mgr; Betsy Balkcom, prom mgr; Amy Miller, mus dir; Lori Starks, traf mgr.

White Stone

WIGO-FM— Sept 1, 1995: 104.9 mhz; 6 kw. Ant 282 ft TL: N37 43 26 W76 23 27. Hrs open: Box 896, Urbanna, 22578. Phone: (804) 758-9635. Fax: (804) 758-5835. E-mail: windyradio@windy105fm.com Licensee: Two Rivers Communications Inc. (acq 3-31-2006; $700,000). Natl. Network: Westwood One, CNN Radio. Format: Country. Target aud: 25-54. ♦William C. Sherard, pres; Richard Swift, exec VP; Mitt Younts, gen mgr; Carter Mills, stn mgr; Sharon Lambertti, opns dir; Laurel Taylor, gen sls mgr & engrg dir.

Williamsburg

***WCWM(FM)**— Sept 28, 1959: 90.9 mhz; 13.5 kw. Ant 269 ft TL: N37 21 16 W76 59 58. Stereo. Hrs opn: 24 Campus Ctr., College of Wiliam & Mary, Box 8793, 23186. Phone: (757) 221-3287. Fax: (757) 221-2118. E-mail: wcwm@wm.edu Web Site: www.wm.edu/so/wcwm. Licensee: College of William & Mary. Population served: 15,000 Natl. Network: Moody. Format: Div, alternative new mus, progsv. News: 3 hrs wkly. Target aud: General. Spec prog: Jazz 13 hrs, class 11 hrs, reggae 6 hrs, blues 3 hrs wkly. ♦Anne Gessler, gen mgr & stn mgr.

WMBG(AM)— Jan 1, 1958: 740 khz; 500 w-D, 8 w-N. TL: N37 16 37 W76 45 07. Hrs open: 24 1005 Richmond Rd., 23185. Phone: (757) 229-7400. E-mail: info@wmbgradio.com Licensee: Williamsburg's Radio Station Inc. Population served: 12,000 Natl. Network: CNN Radio. Format: Adult standards. News staff: one; News: one hr wkly. Target aud: 45 plus; wealthy & mature in Williamsburg. Spec prog: Gospel 5 hrs wkly. ♦Bob Sheeran, opns dir; Greg Granger, pres, gen mgr, stn mgr & opns mgr.

Winchester

WINC(AM)— June 15, 1941: 1400 khz; 1 kw-U. TL: N39 11 12 W78 09 06. Hrs open: 24 Box 3300, 22604. Secondary address: 520 N. Pleasant Valley Rd. 22601. Phone: (540) 667-2224. Fax: (540) 722-3295. Web Site: www.winc.fm. Licensee: Mid-Atlantic Network Inc. (group owner) Population served: 75,000 Natl. Network: Westwood One, CBS. Law Firm: Cole, Raywid & Braverman. Format: News/talk, sports. News staff: 3; News: 128 hrs wkly. Target aud: 25 plus; mid-to-upscale active adults. ♦John Lewis, pres; Chris Lewis, gen mgr & gen sls mgr; Jeff Adams, opns mgr & progmg dir; Steve Edwards, news dir; Archie McKay, chief of engrg; Pam Christian, traf mgr.

WINC-FM— October 1946: 92.5 mhz; 22 kw. 1,424 ft TL: N38 57 21 W78 01 28. Stereo. 24 Web Site: www.winc.fm. Natl. Network: Westwood One. Format: Hot adult contemp. News staff: 4; News: 6 hrs wkly. Target aud: 18-49; active, upscale listeners. ♦Pam Christian, mktg dir.

WTFX(AM)— Jan 27, 1961: 610 khz; 500 w-U, DA-2. TL: N39 11 53 W78 13 13. Hrs open: 24 510 Pegasus Ct., 22602. Phone: (540) 662-5101. Fax: (540) 662-8610. Web Site: sportstalk610.com. Licensee: Capstar TX L.P. Group owner: Clear Channel Communications Inc. (acq 8-30-00; grpsl). Population served: 127,500 Natl. Network: ABC, USA. Format: Christian. News staff: one; News: 20 hrs wkly. Target aud: General; educated, affluent. Spec prog: Sports. ♦Jim Shea, pres; Chuck Peterson, gen mgr; David Miller, opns mgr & progmg dir; Marcellla T. Vance, sls dir; Justin Maglione, prom dir; Harve Allen, progmg VP; Ben Gates, pub affrs dir; Mark Kesner, chief of engrg; Krissy Groves, traf mgr.

WUSQ-FM—Co-owned with WTFX(AM). Dec 10, 1965: 102.5 mhz; 31 kw. 630 ft TL: N39 10 38 W78 15 53. Stereo. 24 Web Site: www.wusq.com.127,000 Format: Modern country. News: one hr wkly. Target aud: 25-54; males. ♦Krissy Groves, traf mgr; Chris Mitchell, disc jockey; Chuck Carroll, disc jockey; Mike Montgomery, disc jockey.

***WTRM(FM)**— July 1986: 91.3 mhz; 5.6 kw. Ant 1,401 ft TL: N39 11 02 W78 23 15. Stereo. Hrs opn: 24 Box 3438, 22604. Phone: (540) 869-4997. Fax: (540) 869-7173. E-mail: wtrm@wtrm.org Web Site: www.wtrm.org. Licensee: Timber Ridge Ministries Inc. Population served: 1,200,000 Natl. Network: USA. Law Firm: Lauren A. Colby. Format: Southern gospel. Target aud: General. ♦Leona Choy, pres; Richard Choy, CEO, VP & gen mgr.

Windsor

WJCD(FM)— May 1990: 107.7 mhz; 1.7 kw. Ant 620 ft TL: N36 48 32 W76 30 13. Stereo. Hrs opn: 24 1003 Norfolk Sq., Norfolk, 23502. Phone: (757) 466-0009. Fax: (757) 466-9523. E-mail: contact@litefmnorfolk.com Web Site: www.smoothjazz1077.com. Licensee: CC Licenses LLC. Group owner: Clear Channel Communications Inc.

(acq 9-10-96; grpsl). Natl. Network: CNN Radio. Format: Smooth jazz. ♦Reggie Jordan, gen mgr; Terry Ratliff, sls dir; Travis Dylan, progmg dir; Greg Gabriel, chief of engrg.

Wise

***WISE-FM**— Aug 1, 1999: 90.5 mhz; 220 w. 669 ft TL: N36 57 39 W82 30 56. Stereo. Hrs opn: 24 Office of College Relations, One College Ave., 24293. Phone: (540) 328-0300. Fax: (540) 328-0255. E-mail: wisefm@virginia.edu Web Site: www.wisefm.org. Licensee: Clinch Valley College of the University of Virginia. Population served: 25,000 Natl. Network: NPR, PRI. Rgnl. Network: Va. News Net. Format: Class, educ, news. News: 45 hrs wkly. Target aud: General. Spec prog: Jazz 9 hrs, Celtic 2 hrs wkly. ♦Jay Lemons (chancellor), CEO; Scott Pippin, stn mgr, dev dir, mktg dir, progmg dir & pub affrs dir; Don Mussell, engrg mgr.

WNVA-FM—See Norton

Woodbridge

WJZW(FM)—Licensed to Woodbridge. See Washington DC

Woodstock

WAMM(AM)— Oct 9, 1981: Stn currently dark. 1230 khz; 1 kw-U. TL: N38 51 11 W78 31 30. Hrs open: 24 Box 542, 22664. Licensee: Jason M. Rodriguez (acq 1-5-2006; $300,000). Format: Sp. ♦Bernard Boston, gen mgr; Alan Arehart, stn mgr; Craig Orndorff, progmg dir; Michael Reed, sls dir & chief of engrg.

WAZR(FM)— Oct 18, 1985: 93.7 mhz; 8.5 kw. Ant 420 ft TL: N38 37 04 W78 42 39. Stereo. Hrs opn: 24 207 University Blvd., Harrisonburg, 22801. Phone: (540) 434-1777. Fax: (540) 432-9968. Web Site: www.937kissfm.com. Licensee: CC Licenses LLC. Group owner: Clear Channel Communications Inc. (acq 6-3-2002; $1.35 million including five-year noncompete agreement). Population served: 150,000 Natl. Network: Jones Radio Networks. Format: CHR. ♦Steve Davis, gen mgr; Susie Smith, gen sls mgr; Steve Knupp, progmg dir; Mark Ness, chief of engrg.

Wytheville

WYVE(AM)— Sept 21, 1949: 1280 khz; 2.5 kw-D, 164 w-N. TL: N36 57 54 W81 04 55. Stereo. Hrs opn: 24 Box 1247, 110 W. Spiller St., Suite 1, 24382. Phone: (276) 228-3185. Phone: (276) 228-6308. Fax: (276) 228-9261. E-mail: trmedia@msn.com Web Site: www.wyve. Licensee: Three Rivers Media Corp. (acq 10-1-98; $250,000). Population served: 37,700 Natl. Network: AP Network News, Jones Radio Networks. Rgnl. Network: Va. News Net. Rgnl rep: Rgnl Reps Law Firm: Brooks, Pierce, McLendon, Humphrey & Leonard. Format: C&W, loc news, sports, info. News staff: one; News: 10 hrs wkly. Target aud: 25 plus; general. Spec prog: Gospel 4 hrs wkly. ♦Gary W. Hagerich, CEO, pres & stn mgr; Danny Gordon, opns dir; Sam Parks, chief of engrg; Kristy Wrobel, traf mgr.

Yorktown

WXEZ(FM)— July 4, 1975: 94.1 mhz; 50 kw. 500 ft TL: N37 29 37 W76 26 30. (CP: 40 kw, ant 531 ft.). Stereo. Hrs open: 24 5589 Greenwich Rd., Suite 200, Virginia Beach, 23462. Phone: (757) 671-1000. Fax: (757) 518-9364. E-mail: dmurray@wxez941.com Web Site: www.wxez941.com. Licensee: MHR License LLC. Group owner: Barnstable Broadcasting Inc. (acq 3-24-2005; grpsl). Natl. Rep: Christal. Format: Urban adult contemp. News staff: 8. Target aud: 35-64. ♦Eric Martel, pres; Vonneva Carter, gen mgr; Cynthia Weatherspoon, gen sls mgr; Mary Stott, mktg dir; Dale Murray, progmg dir.

***WYCS(FM)**— February 1966: 91.5 mhz; 1.3 kw horiz, 20 kw vert. Ant 371 ft TL: N37 12 17 W76 30 07. Stereo. Hrs opn: 24 Box 1924, Tulsa, OK, 74101. Phone: (757) 886-7490. Phone: (918) 455-5693. Fax: (757) 886-7491. E-mail: mail@oasisnetwork.org Web Site: www.oasisnetwork.org. Licensee: Creative Educational Media Corp. Inc. Format: Relg. Target aud: General. ♦David Ingles, pres; Greg Roth, gen mgr.

Washington

Aberdeen

KBKW(AM)— Aug 1, 1949: 1450 khz; 1 kw-U. TL: N46 56 59 W123 49 13. Hrs open: 24 Box 1198, 98520. Secondary address: 1520 Simpson Ave. 98520. Phone: (360) 533-3000. Fax: (360) 532-1456. E-mail: bossbill@jodesha.com Web Site: www.jodesha.com. Licensee: Jodesha Broadcasting Inc. (group owner; acq 2-28-03; $750,000 with KSWW(FM) Montesano). Population served: 52,200 Natl. Network: AP Network News. Wire Svc: AP Format: news/talk 24/7. News staff: one; News: 13. Target aud: 25-54. ♦Wm J. Wolfenbarger, pres; Bill Wolfenbarger, gen mgr; Gabrielle Jordan, opns dir.

KDDS-FM—Elma, 1981: 99.3 mhz; 41 kw. Ant 2,034 ft TL: N47 19 12 W123 20 41. Stereo. Hrs opn: 1400 W. Main St., Auburn, 98001. Phone: (253) 735-9700. Fax: (253) 735-7424. Web Site: www.radiolagrand.com. Licensee: Bustos Media of Seattle License LLC. (acq 9-28-2005; $20 million). Format: Rgnl Mexican. ♦Amador Bustos, pres; Jose Diaz, gen mgr; Cesar Valdiosera, progmg dir.

KDUX-FM—Listing follows KXRO(AM).

KWOK(AM)—See Hoquiam

KXRO(AM)— May 28, 1928: 1320 khz; 5 kw-D, 1 kw-N, DA-N. TL: N46 57 28 W123 48 26. Hrs open: 24 1308 Coolidge Rd., 98520. Phone: (360) 533-1320. Fax: (360) 532-0935. Web Site: www.kxro.com. Licensee: Morris Communications Corp. Group owner: Morris Communications Inc. (acq 10-15-98; grpsl). Population served: 60,000 Natl. Network: CBS. Natl. Rep: McGavren Guild. Law Firm: Covington & Burling. Format: News/talk. News staff: 2; News: 6 hrs wkly. Target aud: 35-64. ♦Donna Rosi, gen mgr & gen sls mgr; Pat Anderson, opns dir & progmg dir; Liz Miller, news dir, pub affrs dir & news rptr; Jay White, chief of engrg; Lorrie Larson, traf mgr; Ian Cope, sports cmtr; James Michael Powers, disc jockey; Rick Dee, disc jockey.

KDUX-FM—Co-owned with KXRO(AM). Oct 4, 1964: 104.7 mhz; 48 kw. 360 ft TL: N46 56 00 W123 43 49. Stereo. 24 Web Site: www.kdux.com. Natl. Rep: Interep, McGavren Guild. Format: Classic rock. News staff: one. Target aud: 25-54. ♦Pat Anderson, opns mgr, progmg dir & disc jockey; Lorrie Larson, traf mgr; Liz Miller, news rptr; Ian Cope, sports cmtr; James Michael Powers, disc jockey; Rick Dee, disc jockey.

KXXK(FM)—See Olympia

Airway Heights

KXLX(AM)— October 1986: 700 khz; 10 kw-D, 600 w-N, DA-N. TL: N47 36 31 W117 22 25. Hrs open: 500 W. Boone Ave., Spokane, 99201. Phone: (509) 324-4000. Fax: (509) 324-8992. Licensee: QueenB Radio Inc. (acq 9-1-2005; $236,000). Natl. Network: ESPN Radio. Natl. Rep: Katz Radio. Format: Sports. ♦Stephen Herling, exec VP; Chris Garras, gen mgr; Teddi Gibbon, stn mgr; Roger Nelson, opns mgr & progmg mgr; Dick Brantley, rgnl sls mgr; Bud Nameck, progmg dir.

Anacortes

KWLE(AM)— Dec 18, 1957: 1340 khz; 1 kw-U. TL: N48 29 44 W122 36 15. Hrs open: 24 Box 96, 25th & Commercial Ave., WA, 98221. Phone: (360) 293-3141. Fax: (360) 293-9463. E-mail: klki@ Web Site: www.klki.com. Licensee: San Juan Communications Inc. (acq 6-29-2007; $760,000). Population served: 120,000 Natl. Network: Westwood One, CBS, ABC. Format: Sports, news/talk, adult contemp, big band. News staff: 3; News: 30 hrs wkly. Target aud: 25-60. Spec prog: Sp 6 hrs, relg one hr wkly. ♦Jennifer Uteda, pres; William T. Berry, gen mgr; Lynn Mc Mullen, opns VP; Dedrick Allen, opns mgr; Glen Harris, progmg dir.

Asotin

KCLK(AM)—Licensed to Asotin. See Clarkston

KCLK-FM—See Clarkston

Auburn

***KGRG-FM**— December 1974: 89.9 mhz; 250 w. Ant 367 ft TL: N47 15 23 W122 13 07. Stereo. Hrs opn: 24 12401 S.E. 320th St., 98092-3699. Phone: (253) 833-9111. Fax: (253) 288-3439. E-mail: tkrause@greenriver.edu Web Site: www.kgrg.com. Licensee: Green River Community College. Population served: 500,000 Format: Modern rock. News: 2 hrs wkly. Target aud: 16-34. Spec prog: Loc mus 3 hrs, rap 3 hrs, metal 2 hrs, reggae/ska 4 hrs wkly. ♦Tom Evans Krause, gen mgr; Katie Phelan, sls dir & gen sls mgr; Ian Reas, prom dir & progmg dir; Jacob Nuss, progmg dir; Chris Leir, mus dir; Alex Farnham, news dir; Jon Kasprick, chief of engrg; Kristin Nelson, traf mgr.

Auburn-Federal Way

KWMG(AM)— Aug 6, 1958: 1210 khz; 27.5 kw-D, 10 kw-N, DA-2. TL: N47 18 20 W122 14 53. Stereo. Hrs opn: 24 1400 W. Main St., Auburn, 98001. Phone: (253) 735-9700. Fax: (253) 735-7424. Licensee: Bustos Media of Washington License LLC. Group owner: Entercom Communications Corp. (acq 1-21-2005; $6 million). Population served: 2,500,000 Format: Rgnl Mexican. ♦Amador Bustos, pres; Jose Diaz, gen mgr; Cesar Valdiosera, progmg dir.

Bellevue

***KASB(FM)**— Mar 22, 1971: 89.3 mhz; 10 w. 289 ft TL: N47 36 17 W122 11 47. Hrs open: 10416 E. Wolverine Way, 98004-6698. Phone: (425) 456-7119. Fax: (425) 456-7110. E-mail: kasb89@hotmail.com Web Site: www.bsd405.org. Licensee: Bellevue School District No. 405. Population served: 50,000 Format: Alternative, news. Spec prog: News magazine 3 hrs, sports 6 hrs wkly. ♦Wes Zujko, gen mgr.

***KBCS(FM)**— Feb 3, 1973: 91.3 mhz; 7.9 kw. 216 ft TL: N47 35 07 W122 08 39. Stereo. Hrs opn: 24 3000 Landerholm Cir. S.E., 98007. Phone: (425) 564-2427. Fax: (425) 564-5697. E-mail: kbcs@ctc.edu Web Site: kbcs.fm. Licensee: Bellevue Community College. Population served: 1,000,000 Format: Jazz, folk, world mus. News: 4 hrs wkly. Target aud: General. ♦Steve Ramsey, gen mgr; Bruce Wirth, opns dir, mus dir & pub affrs dir; Sabrina Roach, dev dir; Robert Jefferson, progmg dir; Sam Roffe, chief of engrg.

KQMV(FM)— November 1964: 92.5 mhz; 58 kw. Ant 2,342 ft TL: N47 30 14 W121 58 29. Stereo. Hrs opn: 3650 131st Ave. S.E., Suite 550, 98006. Phone: (425) 653-9462. Fax: (425) 653-9464. E-mail: info@movin925.fm Web Site: www.movin925.fm. Licensee: Bellevue Radio Inc. Group owner: Sandusky Radio Population served: 250,000 Natl. Rep: Christal. Format: Rhythmic adult contemp. News: one hr wkly. Target aud: 25-49; working women & families. ♦Norman Rau, pres; Marc S. Kaye, VP & gen mgr; Lois Mares, sls dir & gen sls mgr; Annie O'Dell, rgnl sls mgr & prom mgr; Maynard Cohen, progmg dir.

KXPA(AM)— March 1958: 1540 mhz; 5 kw-U, DA-N. TL: N47 35 29 W122 10 56. Hrs open: 114 Lakeside Ave., Seattle, 98122-6542. Phone: (206) 292-7800. Fax: (206) 292-2140. Licensee: Multicultural Radio Broadcasting Licensee LLC. Group owner: Multicultural Radio Broadcasting Inc. (acq 2-13-98; grpsl). Format: Sp, ethnic, div. ♦Arthur Liu, pres; Lisa Shepherd, gen mgr; Dennis Hartley, opns mgr & progmg dir.

Bellingham

KAFE(FM)—Listing follows KPUG(AM).

KARI(AM)—See Blaine

KBAI(AM)— Apr 4, 1958: 930 khz; 1 kw-D, 500 w-N, DA-N. TL: N48 47 53 W122 28 01. Hrs open: 24 2219 Yew Street Rd., 98229. Phone: (360) 734-9790. Fax: (360) 733-4551. Licensee: Saga Broadcasting

LLC. Group owner: Saga Communications Inc. (acq 3-8-99; $1 million). Population served: 140,000 Natl. Rep: Tacher. Law Firm: Garvey, Schubert & Barer. Format: Adult contemp. News staff: one; News: 5 hrs wkly. Target aud: 25-35. ♦Ed Christian, pres & local news ed; Rick Staeb, gen mgr.

KGMI(AM)— 1927: 790 khz; 5 kw-D, 1 kw-N, DA-N. TL: N48 41 09 W122 26 43. Hrs open: 2219 Yew Street Rd., 98229. Phone: (360) 734-9790. Web Site: www.kgmi.com. Licensee: Saga Broadcasting LLC. Group owner: Saga Communications Inc. (acq 9-24-98; $8 million with co-located FM). Population served: 147,000 Natl. Rep: McGavren Guild. Format: News/talk. Target aud: 35-64. ♦Ed Christian, pres; Rick Staeb, gen mgr & natl sls mgr; Krista Kay, prom dir & news dir; Brett Bonner, progmg dir; Will Vos, chief of engrg; Doug Lange, sports cmtr; Steve Ricci, women's int ed.

KISM(FM)—Co-owned with KGMI(AM). March 1960: 92.9 mhz; 50 kw. 2,440 ft TL: N48 40 48 W122 50 24. Stereo. Web Site: www.kism.com.246,000 Format: Classic rock. Target aud: 25-44. ♦Carol Dooley, progmg dir & mus dir.

KPUG(AM)— Feb 29, 1948: 1170 khz; 10 kw-D, 5 kw-N, DA-N. TL: N40 46 34 W122 26 21. Hrs open: 24 2219 Yew Street Rd., 98226-8855. Phone: (360) 734-9790. Phone: (360) 734-5233. Web Site: www.1170kpug.com. Licensee: Saga Broadcasting LLC. Group owner: Saga Communications Inc. (acq 10-30-98; $5,825,000 with co-located FM). Population served: 125,000 Natl. Rep: McGavren Guild. Format: Sports/talk. News staff: 3; News: 24 hrs wkly. Target aud: 25-54. ♦Ed Chrtistian, pres & opns dir; Rick Staeb, gen mgr; Doug Lange, progmg dir; Will Vos, engrg dir & chief of engrg.

KAFE(FM)—Co-owned with KPUG(AM). July 2, 1965: 104.3 mhz; 60 kw. 2,310 ft TL: N48 40 48 W122 50 24. Stereo. 24 Web Site: www.kafe.com.2,000,000 Format: Adult contemp. ♦Scotty Kvipers, prom dir; Don Hurley, progmg dir & chief of engrg; Bill Baker, news rptr; Mikelanne Burk, news rptr; Steve Sandmeyer, news rptr; Jeff Nelson, disc jockey; Krista Kay, disc jockey; Like Caus, disc jockey; Lynn Roberts, disc jockey.

***KUGS(FM)**— Jan 29, 1974: 89.3 mhz; 100 w. Ant 384 ft TL: N48 44 11 W122 28 47. (CP: 700 w, ant 482 ft). Stereo. Hrs opn: 7 AM-2 AM Western Wash. Univ., 700 Viking Union Bldg., 98225. Phone: (360) 650-4771. Phone: (360) 650-5847. Fax: (360) 650-6507. Web Site: www.kugs.org. Licensee: Western Washington University. Population served: 150,000 Format: Progsv, news/talk. News: 17 hrs wkly. Target aud: 18-34; college students & adults. Spec prog: Black 10 hrs, Hawaiian 2 hrs wkly. ♦Jamie Hoover, gen mgr; Taylor Napolsky, dev dir; Jenae Norman, prom dir; Oliver Anderson, progmg dir; Alena Feeney, mus dir; Jen Hartman, news dir & pub affrs dir.

***KZAZ(FM)**— Sept 1, 1991: 91.7 mhz; 120 w. 334 ft TL: N48 48 04 W122 27 40. Stereo. Hrs opn: 24
Rebroadcasts KRFA-FM Pullman 95%.
1609 Broadway, Suite E, 98225. Phone: (360) 738-9170. Fax: (360) 738-4605. E-mail: nwpr@wsu.edu Web Site: www.nwpr.org. Licensee: Washington State University. (acq 7-29-97). Population served: 60,000 Natl. Network: NPR, PRI. Law Firm: Dow, Lohnes & Albertson. Format: Class, jazz, NPR news. News: 40 hrs wkly. Target aud: 25-54; general. Spec prog: Folk 8 hrs, world music 7 hrs wkly. ♦Karen Olstad, gen mgr; Roger Johnson, stn mgr; Mary Hawkins, progmg dir; Robin Rilette, mus dir; Scott Weatherly, opns mgr, chief of engrg & traf mgr.

Benton City

KMMG(FM)— Aug 1, 1974: 96.7 mhz; 1.4 kw. Ant 692 ft TL: N46 15 33 W119 21 55. Stereo. Hrs opn: 24 2730 West Lewis #B, Pasco, 99301. Phone: (509) 543-3334. Fax: (509) 452-0541. Licensee: Bustos Media of Eastern Washington License LLC. (acq 11-18-2004; grpsl). Population served: 45,000 Format: Rgnl Mexican. News staff: one; News: 22 hrs wkly. Target aud: 25 plus. ♦Bob Berry, gen mgr; Bob Berrt, gen sls mgr; Martin Ortiz, progmg dir; Keith Teske, chief of engrg.

Blaine

KAFE(FM)—See Bellingham

KARI(AM)— Feb 12, 1960: 550 khz; 5 kw-D, 2.5 kw-N, DA-2. TL: N48 57 15 W122 44 36. Hrs open: 24 4840 Lincoln Rd., 98230. Phone: (360) 371-5500. Phone: (604) 536-7733. Fax: (360) 371-7617. E-mail: kari@kari55.com Web Site: www.kari55.com. Licensee: Way Broadcasting Licensee LLC (acq 7-20-00; $3 million with KVRI(AM) Blaine). Population served: 3,750,000 Format: Relg, news/talk. Target aud: 35 plus. Spec prog: Ger 2 hrs, Ukrainian one hr, Arabic one hr wkly. ♦Arthur Liu, pres; Yvonne Liu, VP; Gary Nawman, gen mgr & opns mgr.

KVRI(AM)— Jan 1, 2001: 1600 khz; 50 kw-D, 10 kw-N. TL: N48 57 15 W122 44 36. Hrs open: 4840 Lincoln Rd., 98230. Phone: (360) 371-5500. Fax: (360) 371-7617. E-mail: gary@kari55.com Web Site: www.kari55.com. Licensee: Way Broadcasting Licensee LLC (acq 5-23-00; with KARI(AM) Blaine). Format: Punjabi. ♦Arthur Liu, pres; Yvonne Liu, VP; Gary Nawman, gen mgr & opns mgr.

Bremerton

KBRO(AM)— May 1947: 1490 khz; 1 kw-U. TL: N47 33 52 W122 39 26. Hrs open: 6 AM-midnight Box 4024, Seattle, 98104. Phone: (206) 583-0811. Fax: (206) 467-9425. E-mail: kbro@worldnet.att.net Licensee: Seattle Streaming Radio LLC. (acq 7-22-2005; $900,000 with KNTB(AM) Lakewood). Population served: 3,000,000 Format: Hits of the 80s. News staff: one; News: 20 hrs wkly. Target aud: 25-54; Kitsap county residents. Spec prog: Samoan 4 hrs, Filipino 4 hrs wkly. ♦Chris Hanley, gen mgr.

KRWM(FM)— Aug 22, 1964: 106.9 mhz; 49 kw. Ant 1,299 ft TL: N47 32 39 W122 06 29. Stereo. Hrs opn: 24 3650 131 Ave. S.E., Suite 550, Bellevue, 98006. Phone: (425) 373-5545. Fax: (425) 653-1188. Web Site: www.warm1069.com. Licensee: Seascape Radio Inc. Group owner: Sandusky Radio (acq 9-12-96; $29.25 million). Natl. Rep: Christal. Format: Soft adult contemp. Target aud: 35-54; educated, upscale professionals, family oriented, white/blue collar. ♦Marc Kaye, VP, gen mgr & chief of engrg; Susan huffman, gen sls mgr; Heather Gardner, prom dir; Laura Dane, progmg dir.

Burbank

KUJ-FM— 1997: 99.1 mhz; 52 kw. Ant 1,263 ft TL: N46 05 58 W119 07 40. Hrs open: 16 830 N. Columbia Center Blvd., Suite B-2, Kennewick, 99336. Phone: (509) 783-0783. Fax: (509) 735-8627. Web Site: www.power991fm.com. Licensee: New Northwest Broadcasters LLC (group owner; (acq 6-22-2004; $1.68 million). Format: Rhythmic CHR. Target aud: 18-34. ♦Don Morin, gen mgr; Curt Cartier, opns mgr.

KVAN(AM)— 2007: 1560 khz; 10 kw-D, 700 w-N, DA-2. TL: N46 10 11 W119 01 32. Hrs open: 3544 W. Court, Pasco, 99301. Phone: (509) 840-4797. Licensee: Compadres LC (acq 8-1-2007). ♦Angel Castaneda, gen mgr.

Burien-Seattle

KGNW(AM)— Oct 10, 1970: 820 khz; 50 kw-D, 5 kw-N, DA-2. TL: N47 26 00 W121 28 02. Hrs open: 24 2201 6th Ave., Ste 1500, Seattle, 98121-1840. Phone: (206) 443-8200. Fax: (206) 777-1133. Web Site: www.kgnw.com. Licensee: Inspiration Media Inc. Group owner: Salem Communications Corp. (acq 1984). Population served: 60,000 Natl. Rep: Salem. Format: Relg, Christian talk. News: 3 hrs wkly. Target aud: 35 plus. Spec prog: Talk, women, loc affrs, health. ♦Stuart Epperson, chmn; Edward G. Atsinger III, pres; Joe Gonzalez, gen mgr; Chuck Olmstead, opns mgr & progmg dir; Doug Rice, gen sls mgr; Monty Passmore, chief of engrg.

Camas

KNRK(FM)—Licensed to Camas. See Portland OR

Cashmere

KYSN(FM)—See Wenatchee

KZPH(FM)— 1993: 106.7 mhz; 6 kw. 513 ft TL: N47 30 35 W120 31 24. Stereo. Hrs opn: 231 N. Wenatchee Ave., Wenatchee, 98801. Phone: (509) 665-6565. Fax: (509) 663-1150. Web Site: www.therock1067.com. Licensee: CCR-Wenatchee IV LLC. Group owner: Fisher Broadcasting Company (acq 10-31-2006; grpsl). Natl. Rep: McGavren Guild. Law Firm: Fisher, Wayland, Cooper, Leader & Zaragoza. Format: Classic rock. Target aud: 25-54. ♦Jim Senst, gen mgr; Leona Frank, gen sls mgr; Dave Keefer, progmg dir.

Castle Rock

KRQT(FM)— January 1994: 107.1 mhz; 740 w. 1,732 ft TL: N46 20 35 W123 05 54. Hrs open: 1130 14th Ave., Longview, 98632. Phone: (360) 425-1500. Fax: (360) 425-1500. Web Site: www.theclassicrockexperience.com. Licensee: Bicoastal Longview LLC. Group owner: Entercom Communications Corp. (acq 1-27-2005; grpsl). Natl. Network: ABC. Format: Classic rock & roll. News: one. ♦Julie Laird, stn mgr; Kevin Taylor, opns mgr; Sam Lee, gen sls mgr; Phil Blair, news dir; Dawn Crowe, traf mgr.

Centralia

KCED(FM)—Licensed to Centralia. See Centralia-Chehalis

KNBQ(FM)—Licensed to Centralia. See Centralia-Chehalis

Centralia-Chehalis

***KCED(FM)**—Centralia, Feb 17, 1975: 91.3 mhz; 1 kw. -72 ft TL: N46 42 54 W122 57 39. (CP: 1.2 kw, ant 131 ft.). Stereo. Hrs opn: 600 W. Locust St., Centralia, 98531-4099. Phone: (360) 736-9391 x343. Fax: (360) 330-7509. Fax: www.centralia.ctc.edu. Licensee: Board of Trustees, Centralia College. Population served: 50,000 Format: Variety/diversified, Sp. Spec prog: Sports 3 hrs wkly. ♦Wade Fisher, gen mgr & progmg dir.

KELA(AM)— Nov 1, 1937: 1470 khz; 5 kw-D, 1 kw-N. TL: N46 41 47 W122 57 23. Hrs open: 24 1635 S. Gold St., Centralia, 98531. Phone: (360) 736-3321. Phone: (360) 748-3321. Fax: (360) 736-0150. E-mail: johndimeoe@clearchannel.com Web Site: www.kelaam.com. Licensee: Citicasters Licenses L.P. Group owner: Clear Channel Communications Inc. (acq 5-4-99; grpsl). Population served: 62,500 Natl. Rep: Clear Channel. Format: News/talk, sports. News staff: 2; News: 28 hrs wkly. Target aud: 35 plus. Spec prog: Big band 3 hrs, class 2 hrs wkly. ♦John DiMeo Jr., gen mgr; Larry Miner, gen sls mgr; Steve Richert, progmg dir; Doug Adamson, news dir; Dan Smith, chief of engrg; Bill Carter, disc jockey; Steve Williams, disc jockey.

KNBQ(FM)—Co-owned with KELA(AM). Aug 24, 1965: 102.9 mhz; 70 kw. Ant 2,191 ft TL: N46 58 31 W123 08 16. Stereo. 24 Web Site: www.kmnt.com.500,000 Format: Country. News staff: 2; News: 7 hrs wkly. Target aud: 18 plus. ♦Deborah Lemmons, progmg dir; Angie Forester, disc jockey; Bill Carter, disc jockey; Debbie Dunlap, disc jockey; Jonny Day, disc jockey.

KITI(AM)—Chehalis-Centralia, October 1954: 1420 khz; 5 kw-U, DA-2. TL: N46 42 08 W122 55 58. Hrs open: 24 1133 Kresky, Centralia, 98531. Phone: (360) 736-1355. Fax: (360) 736-4761. E-mail: mshannon@live95.com Licensee: Premier Broadcasters Inc. (acq 10-77). Population served: 60,000 Law Firm: Leventhal, Senter & Lerman. Format: Oldies. News staff: one; News: 15 hrs wkly. Target aud: 25-54. ♦Rod Etherton, pres & gen mgr.

Chehalis

***KACS(FM)**— Aug 18, 1993: 90.5 mhz; 6 kw. Ant 187 ft TL: N46 43 52 W123 01 28. (CP: 6 kw). Stereo. Hrs opn: 24 2451 N.E. Kresky, Unit A, 98532. Phone: (360) 740-9436. Fax: (360) 740-9415. E-mail: manager@kacs.org Web Site: www.kacs.org. Licensee: Chehalis Valley Educational Foundation. Population served: 350,000 Law Firm: Donald E. Martin. Format: Relg. News: 6 hrs wkly. Target aud: 45-54. ♦Kerry O'Connor, chmn; Cameron Beirle, gen mgr, stn mgr & progmg dir.

KITI(AM)—See Centralia-Chehalis

KMNT(FM)— 2005: 104.3 mhz; 2.35 kw. Ant 1,056 ft TL: N46 33 18 W123 03 27. Hrs open: 1635 S. Gold St., Centralia, 98531. Phone: (360) 330-0777. Fax: (360) 736-0150. Web Site: www.kmnt.com. Licensee: Citicasters Licenses L.P. Format: Country. ♦John DiMeo Jr., gen mgr.

***KSWS(FM)**—Not on air, target date: unknown: 88.9 mhz; 50 w horiz, 1 kw vert. Ant 1,004 ft TL: N46 33 16 W123 03 26. Hrs open: Educational Telecommunications and Technology, Box 642530, Pullman, 99164-2530. Phone: (509) 335-6511. Web Site: www.nwpr.org. Licensee: Washington State University. ♦Dennis Haarsager, gen mgr.

Chehalis-Centralia

KITI(AM)—Licensed to Chehalis-Centralia. See Centralia-Chehalis

Chelan

KOZI(AM)— Mar 1, 1957: 1230 khz; 1 kw-U. TL: N47 51 00 W120 00 20. Hrs open: 24 Box 819, 98816. Secondary address: 123 E. Johnson 98816. Phone: (509) 682-4033. Fax: (509) 682-4035. Web Site: www.kozi.com. Licensee: Icicle Broadcasting Co. (group owner; (acq 8-26-99; grpsl). Population served: 80,000 Natl. Rep: Target Broadcast Sales. Law Firm: Haley, Bader & Potts. Format: Adult contemp, news/talk. News staff: 3; News: 32 hrs wkly. Target aud: General. Spec prog: Farm 3 hrs, Sp 5 hrs wkly. ♦Harriet Bullitt, pres;

Gary Mathews, gen mgr; Joe Fiala, stn mgr; Steve Byquist, opns mgr & progmg dir; Vicky Chandler, gen sls mgr; Michael Dickes, mus dir & farm dir; Clint Strand, news dir.

KOZI-FM— Aug 26, 1981: 93.5 mhz; 590 w. Ant 1,040 ft TL: N47 51 07 W119 52 18. Stereo. Web Site: www.kozi.com. Format: Adult contemp news/talk.

Cheney

***KEWU-FM**— Apr 3, 1964: 89.5 mhz; 10 kw. 1,407 ft TL: N47 34 43 W117 17 50. Stereo. Hrs opn: 6 AM-1 AM (M-F); 9 AM-1 AM (S, Su) Electronic Media & Film, 104 RTV Bldg., 99004-2495. Phone: (509) 359-2440. Fax: (509) 359-2850. E-mail: efarriss@mail.ewu.edu Licensee: Eastern Washington University Board of Trustees. Format: Jazz. ♦Marvin Smith, gen mgr; Elizabeth Farriss, progmg dir.

KEYF-FM— May 4, 1986: 101.1 mhz; 100 kw. 1,607 ft TL: N47 35 35 W117 17 46. Stereo. Hrs opn: 24 1601 E. 57th, Spokane, 99223. Phone: (509) 448-1000. Phone: (509) 232-1011. Fax: (509) 448-7015. Web Site: www.oldies1011.com. Licensee: Citadel Broadcasting Co. Group owner: Citadel Broadcasting Corp. (acq 4-22-99; grpsl). Population served: 360,000 Format: Music of the 70s & 80s. News staff: one; News: 15 hrs wkly. Target aud: 25-54. ♦Don Morin, gen mgr; Tim Cotter, progmg dir; Larry Weir, news dir; Dave Ratener, chief of engrg; Brenda Anderson, traf mgr.

Clarkston

KCLK-FM— 1974: 94.1 mhz; 100 kw. 1,233 ft TL: N46 27 27 W117 06 03. Stereo. Hrs opn: 24 403 C St., Lewiston, ID, 83501. Phone: (208) 743-6564. Fax: (208) 798-0110. E-mail: kclkfm@aol.com Licensee: Pacific Empire Radio Corp. Group owner: Pacific Empire Communications Corp. (acq 9-26-2000; grpsl). Population served: 120,000 Natl. Network: Westwood One. Format: Mainstream country. News staff: one. Target aud: 25 plus. ♦Mark Bolland, pres & sls dir; Jay Mlazgar, gen sls mgr; Mark Bone, pres & progmg dir; Dave Forsman, chief of engrg.

KCLK(AM)— Mar 2, 1971: 1430 khz; 5 kw-D, 1 kw-N, DA-2. TL: N46 18 59 W117 02 24.24 1859 Fifth Ave., Asotin, 99403. Phone: (509) 758-3362. E-mail: kclkam@aol.com 120,000 Format: Sports, talk. News staff: one. Target aud: 15 plus; sports interested. ♦Dan Craig, progmg dir; Leslie Gatherer, chief of engrg & traf mgr.

***KJCF(FM)**—Not on air, target date: unknown: 89.3 mhz; 5 w. Ant 840 ft TL: N46 26 24 W117 15 54. Hrs open: Box 391, Twin Falls, ID, 83303. Fax: (208) 734-6633. Web Site: www.csnradio.com. Licensee: CSN International. ♦Mike Stocklin, gen mgr; Don Mills, progmg dir; Kelly Carlson, chief of engrg.

***KNWV(FM)**— July 11, 1995: 90.5 mhz; 250 w. 1,063 ft TL: N46 27 26 W117 06 00. Hrs open: 24
Rebroadcasts KRFA-FM Moscow, ID 100%.
Box 642530, 382 Murrow Ctr., Pullman, 99164-2530. Phone: (509) 335-6500. Fax: (509) 335-3772. E-mail: nwpr@wsu.edu Web Site: www.nwpr.org. Licensee: Washington State University. Law Firm: Dow, Lohnes & Albertson. Format: Class, news. News staff: one; News: 37 hrs wkly. ♦Karen Olstad, COO & gen mgr; Dennis Haarsager, gen mgr; Roger Johnson, stn mgr & sls dir; Scott Weatherly, opns mgr; Sarah McDaniel, dev dir; Mary Hawkins, progmg dir; Robin Rilette, mus dir; Ralph Hogan, engrg dir; Rachael McDonald, news rptr.

KQQQ(AM)—See Pullman

KRLC(AM)—See Lewiston, ID

KVAB(FM)— July 15, 1997: 102.9 mhz; 440 w. 1,171 ft TL: N46 27 27 W117 06 03. Hrs open: 24 403 C St., Lewiston, ID, 83501. Phone: (509) 758-3362. Fax: (509) 758-4986. E-mail: kvabfm@aol.com Web Site: kvabfm.com. Licensee: Pacific Empire Radio Corp. Group owner: Pacific Empire Communications Corp. (acq 9-26-2000; grpsl). Population served: 50,000 Natl. Network: Westwood One. Natl. Rep: Tacher. Format: Classic rock. Target aud: 25-55. ♦Mark Bolland, CEO & pres; Jay Mlazgar, gen mgr; Mark Bone, progmg dir.

Cle Elum

KXAA(FM)— Nov., 2002: 93.7 mhz; 6 kw. Ant 95 ft TL: N47 09 06 W120 47 23. Stereo. Hrs opn: 24 115 N Harris Ave., 98922. Phone: (509) 674-0937. Fax: (509) 674-4042. E-mail: kxaprod@aol.com Licensee: Wheeler Broadcasting Inc. (group owner; acq 5-28-2004; exercise of option). Population served: 22,000 Natl. Network: Jones Radio Networks. Format: Classic Hits. ♦Jeri Trantham, opns dir; Mark Wheeler, gen mgr & gen sls mgr.

Colfax

KCLX(AM)—Listing follows KRAO-FM.

KMAX(AM)— 1998: 840 khz; 10 kw-D, 280 w-N. TL: N46 54 50 W117 19 28. Hrs open: 24 Box 8849, Moscow, ID, 83843. Phone: (208) 882-2551. Fax: (208) 883-3571. E-mail: hauser@inlandradio.com Licensee: Inland Northwest Broadcasting LLC. (group owner). (acq 6-28-2005; grpsl). Natl. Network: Westwood One. Format: Talk radio. News staff: 2; News: 15 hrs wkly. Target aud: 25-60; established professional aged people - young marrieds couples. ♦Gary Cummings, gen mgr; Ben Bonfield, gen sls mgr; Darin Sievert, progmg dir; Glen Vaagen, news dir; Steve Franko, chief of engrg.

KRAO-FM— Oct 10, 1994: 102.5 mhz; 2.2 kw. 1,073 ft TL: N46 51 44 W117 10 20. Stereo. Hrs opn: 24 Box 710, 840 Fairview Rd. W., 99111. Phone: (509) 397-6371. Fax: (509) 397-4752. E-mail: kzzlkrow@stjohncable.com Web Site: www.palousecountry.com. Licensee: Inland Northwest Broadcasting LLC. (group owner; (acq 6-28-2005; grpsl). Population served: 65,000 Format: Classic rock. Target aud: 18-45; college, young budding professionals to professional business people. ♦Ben Bonfield, gen mgr & gen sls mgr; Gary Cummings, gen mgr; Johnny Mann, progmg dir & spec ev coord; Steve Franco, chief of engrg & farm dir; Terry Linderman, traf mgr; Robert Hauser, sls.

KCLX(AM)—Co-owned with KRAO-FM. 1950: 1450 khz; 1 kw-U. TL: N46 52 17 W117 22 37. Phone: (509) 397-3441. Web Site: www.palousecountry.com. Natl. Rep: Farmakis. Format: Classic country. Target aud: 25 plus; agricultural-urban. Spec prog: Farm 5 hrs, sports 7 hrs wkly. ♦Gary Cummings, gen mgr & stn mgr; Robert Hauser, opns mgr & rgnl sls mgr; Steve Grubbs, disc jockey.

College Place

***KGTS(FM)**— Oct 5, 1963: 91.3 mhz; 7 kw. 1,250 ft TL: N45 59 20 W118 10 29. Stereo. Hrs opn: 24 204 S. College Ave., 99324. Phone: (509) 527-2991. Fax: (509) 527-2611. Web Site: www.plr.org. Licensee: Walla Walla College. Population served: 200,000 Format: Christian contemp. Target aud: 35-64. ♦Jon Dybdahl, pres; Kevin Krueger, gen mgr; Don Godman, opns dir; Elizabeth Nelson, progmg dir; Walter Cox, chief of engrg.

Colville

KCRK-FM—Listing follows KCVL(AM).

KCVL(AM)— Nov 15, 1955: 1240 khz; 1 kw-U. TL: N48 31 15 W117 54 28. Hrs open: 24 Box 111, 187 Mantz & Ricky Rd., 99114. Phone: (509) 684-5031. Fax: (509) 684-5034. Web Site: www.kcvl.com. Licensee: North Country Broadcasting. (acq 1996). Population served: 16,000 Format: Country. ♦Eric Carpenter, pres, pres & gen mgr; Mike Eakins, sls dir.

KCRK-FM—Co-owned with KCVL(AM). Oct 13, 1981: 92.1 mhz; 3 kw. Ant -790 ft TL: N48 31 15 W117 54 28. Stereo. 24 Phone: (509) 684-5032. Natl. Network: Westwood One. Format: Adult contemp.

Davenport

***KKRS(FM)**— 1998: 97.3 mhz; 5.1 kw. Ant 722 ft TL: N47 35 14 W117 53 26. Hrs open: 24 12720 W. Sunset Hwy., Suite C, Airway Heights, 99001. Phone: (509) 244-5577. Fax: (509) 244-2232. E-mail: kkrs@csnradio.com Licensee: CSN International. (group owner; acq 1999; $111,425). Format: Christian talk, educ, music. ♦Barney Dasovich, gen mgr.

Dayton

KZHR(FM)— August 1992: 92.5 mhz; 54 kw. Ant 1,243 ft TL: N46 19 14 W117 58 46. Hrs open: 24 Box 2623, Tri Cities, 99302. Secondary address: 2823 W. Lewis St., Pasco 99301. Phone: (509) 546-0313. Fax: (509) 546-2678. E-mail: gonzalo@kzhr.com Web Site: www.kzhr.com. Licensee: CCR-Tri Cities IV LLC. Group owner: Cherry Creek Radio LLC (acq 12-19-2003; grpsl). Population served: 300,000 Natl. Rep: Interep, McGavren Guild. Format: Mexican rgnl. News staff: one; News: 7 hrs wkly. Target aud: 25-54; adult. ♦Scott Smith, gen mgr; Gonzalo Cortez, stn mgr & progmg dir; Art Blum, chief of engrg; Daena Medina, traf mgr.

Deer Park

KAZZ(FM)— September 1983: 107.1 mhz; 25 kw. Ant 328 ft TL: N48 01 45 W117 35 57. Stereo. Hrs opn: 24 505 W Riverside, Ste 101, Spokane, 99201. Phone: (509) 252-8440. Fax: (509) 252-8453. E-mail: christa@pro-activecomm.com Licensee: Proactive Communications Inc. (group owner; (acq 3-16-2006; $1.75 million). Format: Classic hits. Target aud: 35-64. ♦Christa McDonald, gen mgr; Steve Kicklighter, opns mgr; Toby Howell, progmg dir.

Dishman

KEYF(AM)— Oct 3, 1984: 1050 khz; 5 kw-D, 260 w-N. TL: N47 36 27 W117 21 40. Hrs open: 24 1601 E. 57th Ave, Spokane, 99223. Phone: (509) 448-1000. Fax: (509) 448-7015. Licensee: Citadel Broadcasting Co. Group owner: Citadel Broadcasting Corp. (acq 4-22-99; grpsl). Population served: 360,000 Format: Adult standards. News staff: one; News: 15 hrs wkly. Target aud: 25-54. ♦Don Morin, gen mgr; Bob Castle, progmg dir; Larry Weir, news dir; Dave Ratener, chief of engrg; Brenda Anderson, traf mgr.

KSPO(FM)— 1996: 106.5 mhz; 6 kw. 328 ft TL: N47 41 39 W117 20 3. Hrs open: 24 Box 31000, Spokane, 99223. Phone: (509) 443-1000. E-mail: kspo@kspo.com Web Site: www.kspo.com. Licensee: Thomas W. Read dba Classical Broadcasting. (acq 1996; $100,000). Population served: 1,000,000 Natl. Network: USA, Salem Radio Network. Law Firm: Cohen & Marks. Format: Relg, talk. Target aud: 35 plus. ♦Melinda Read, sr VP; Thomas W. Read, pres & gen mgr.

East Wenatchee

***KFIO(FM)**—Not on air, target date: unknown: 88.1 mhz; 600 w. Ant -131 ft TL: N47 22 52 W120 17 16. Hrs open: Box 31000, Spokane, 99223. Phone: (509) 443-1000. Licensee: Douglas County Educational Radio Association. ♦Thomas W. Read, pres.

KYSN(FM)—Licensed to East Wenatchee. See Wenatchee

Eatonville

KFNK(FM)— 1995: 104.9 mhz; 17 kw. Ant 407 ft TL: N46 50 24 W122 15 27. Hrs open: 24 351 Elliott Ave. W., 3rd Fl., Seattle, 98119. Phone: (206) 494-2000. Fax: (206) 286-2376. E-mail: bobcase@clearchannel.com Web Site: www.funkymonkey1049.fm. Licensee: Ackerley Broadcasting Operations LLC. Group owner: Clear Channel Communications Inc. (acq 2-12-2003; $4.5 million). Population served: 250000 Format: Alternative rock. ♦Michele Grosenick, gen mgr; Bob Case, opns mgr; Allison Hesse, gen sls mgr; Jay Kelly, progmg dir; Steven Kilbreath, news dir; Doug Irwin, chief of engrg.

Edmonds

KCIS(AM)—Listing follows KCMS(FM).

KCMS(FM)— Mar 11, 1960: 105.3 mhz; 54 kw. Ant 1,263 ft TL: N47 32 40 W122 06 26. Stereo. Hrs opn: 24 19303 Fremont Ave. N., Seattle, 98133. Phone: (206) 546-7350. Fax: (206) 546-7372. E-mail: comments@spirit1053.com Web Site: www.spirit1053.com. Licensee: CRISTA Ministries. Population served: 300,000 Law Firm: Fletcher, Heald & Hildretch. Wire Svc: AP Wire Svc: U.S. Newswire Format: Adult contemp, Christian music. News: one hr wkly. Target aud: 25-44.

♦Melene Thompson, VP & gen mgr; Cindy Bowers, sls dir; Michael Tedesco, prom dir; Scott Valentine, progmg dir.

KCIS(AM)—Co-owned with KCMS(FM). 1954: 630 khz; 5 kw-D, 2.5 kw-N, DA-N. TL: N47 46 06 W122 21 07.24 E-mail: comments@kcisradio.com Web Site: www.kcisradio.com. Group owner: Crista Broadcasting 50,000 Natl. Network: USA, Moody, AP Radio. Law Firm: Fletcher, Heald & Hildreth. Format: Christian, inspirational. News staff: one; News: 40 hrs wkly. Target aud: 45-64. ♦Mark Holland, progmg dir.

Ellensburg

***KCSH(FM)**— 1998: 88.9 mhz; 380 w. 548 ft TL: N47 10 02 W120 45 50. Hrs open: 111 W. 6th Ave., Studio B, 98926. Phone: (509) 964-2061. Fax: (509) 964-2825. Web Site: www.lifetalk.net. Licensee: Lifetalk Broadcasting Association. Format: Relg, inspirational. ♦Kermit Netteburg, pres; Philip Follet, CEO & pres; Don Zacharias, gen mgr.

***KCWU(FM)**— Apr 30, 1999: 88.1 mhz; 500 w. Ant -194 ft TL: N47 00 21 W120 30 55. Stereo. Hrs opn: 24 Central Washington Univ., 400 E. University Way, 98926-7594. Phone: (509) 963-2283, 2282. Fax: (509) 963-1688. E-mail: kcwu@cwu.edu Web Site: www.881theburg.com. Licensee: Trustees of Central Washington University. Population served: 18,000 Law Firm: Morrison & Foerster. Format: Modern rock, alternative, div. News: 5 hrs wkly. Target aud: 15-49; 18-34 core. ♦Chris Hull, gen mgr & dev dir.

***KNWR(FM)**— June 1992: 90.7 mhz; 5 kw. 2,552 ft TL: N47 15 48 W120 23 31. Hrs open: 24
Rebroadcasts KRFA-FM Moscow, ID 100%.
Box 642530, 382 Murrow Ctr., Pullman, 99164-2530. Phone: (509) 335-6500. Fax: (509) 335-3772. E-mail: nwpr@wsu.edu Web Site: www.nwpr.org. Licensee: Washington State University. Natl. Network: NPR, PRI. Format: Class, news. News staff: one; News: 37 hrs wkly. Target aud: General. Spec prog: Jazz, folk. ♦Karen Olstad, COO & gen mgr; Dennis Haarsager, gen mgr; Roger Johnson, stn mgr & sls dir; Scott Weatherly, opns mgr; Sarah McDaniel, dev dir; Mary Hawkins, progmg dir; Robin Rilette, mus dir; Ralph Hogan, engrg dir; Rachael McDonald, news rptr.

KQBE(FM)— Nov 24, 1983: 103.1 mhz; 2 kw. Ant 1,273 ft TL: N47 00 21 W120 30 55. Stereo. Hrs opn: Box 1032, 98926. Secondary address: 109 E. Third Ave., Suite 5 98926. Phone: (509) 962-2823. Fax: (509) 962-5105. E-mail: kqbe@elltel.net Licensee: Peak Communications Inc. (acq 5-1-89; $265,000; FTR: 5-1-89). Format: 60s, 70s, 80s, & today. Target aud: 18-49. ♦Tina Bond, opns mgr & gen sls mgr; Angela Hill, progmg dir & news dir.

KXLE(AM)— 1946: 1240 khz; 1 kw-U. TL: N47 00 00 W120 31 40. Hrs open: 24 1311 Vantage Hwy., 98926. Phone: (509) 925-1488. Fax: (509) 962-7882. E-mail: kxle@elltel.net Licensee: KXLE Inc. (acq 6-82). Population served: 400,000 Natl. Network: CBS. Natl. Rep: Tacher. Law Firm: Bosari & Paxson. Wire Svc: AP Format: News/talk, sports. News staff: 4; News: 40 hrs wkly. Target aud: 25-54; adults. ♦Sol M. Tacher, pres; Brad Tacher, VP & gen mgr; Patti Burke, prom mgr; Dennis Leach, news dir; Kevin Whitaker, engrg dir; Frances Moen, traf mgr.

KXLE-FM— 1972: 95.3 mhz; 51 kw. 423 ft TL: N47 09 49 W120 47 34. Stereo. 24 235,000 Rgnl rep: Tacher Law Firm: Bosari & Paxson. Format: Country. News staff: one; News: 6 hrs wkly. Target aud: 18-59; male & female. ♦Sol Tacher, CEO; Brad Tacher, gen mgr; Patti Burke, prom dir; Steve Scellick, progmg dir; Robert Lowery, news dir; Frances Moen, traf mgr.

Elma

KDDS-FM—Licensed to Elma. See Aberdeen

Enumclaw

KGRG(AM)— Mar 1, 1992: 1330 khz; 500 w-D, 26 w-N. TL: N47 12 53 W121 58 19. Hrs open: 24 12401 S.E. 320 St., Auburn, 98092-3699. Phone: (360) 802-1330. Phone: (253) 288-3388. Fax: (253) 288-3460. E-mail: kgrg1@hotmail.com Licensee: Green River Foundation (acq 9-17-96; $40,000). Population served: 84,000 Format: Classic alternative. ♦Tom Evans Krause, gen mgr.

Ephrata

KTAC(FM)—Listing follows KTBI(AM).

KTBI(AM)— Aug 17, 1950: 810 khz; 50 kw-D. TL: N47 21 22 W119 28 56. Hrs open: Sunrise-sunset
Simulcast with KTAC(FM) Ephrata.
Box 31000, Spokane, 99223. Phone: (509) 754-2000. Fax: (509) 448-3811. E-mail: ktbi@ktbi.com Web Site: www.ktbi.com. Licensee: Tacoma Broadcasters Inc. Population served: 530,831 Law Firm: Pepper & Corazzini. Format: Relg, talk. Target aud: 35 plus. Spec prog: Farm 15 hrs wkly. ♦Melinda Read, VP; John Tillman, opns mgr; Thomas W. Read, pres, gen mgr & progmg dir; George Frese, engrg dir.

KTAC(FM)—Co-owned with KTBI(AM). 1998: 93.9 mhz; 18 kw. 384 ft TL: N47 19 13 W119 34 22.24
Simulcast with KTBI(AM) Ephrata.
Box 31000, Spokane, 99223. E-mail: ktac@ktac.com Web Site: www.ktac.com. Licensee: TRMR Inc.

KULE(AM)— 1952: 730 khz; 1 kw-D, 29 w-N. TL: N47 19 01 W119 33 46. Hrs open: Box 2888, Yakima, 98907. Secondary address: 910 Basin S.W. 98823. Phone: (509) 457-1000. Fax: (509) 452-0541. E-mail: zorro@radiozorro.com Web Site: www.radiozorro.com. Licensee: Bustos Media of Eastern Washington License LLC. (group owner; (acq 11-18-2004; grpsl). Natl. Network: Westwood One, CBS. Law Firm: Timothy K. Brady. Format: News/talk, sports. News staff: one; News: 20 hrs wkly. Target aud: 25-54. Spec prog: Farm 1 hr wkly. ♦Amador S. Bustos, pres; Bob Berry, gen mgr; Keith Teske, opns mgr; Judith McInnis, prom dir; Martin Ortiz, progmg dir.

KULE-FM— Dec 25, 1982: 92.3 mhz; 26 kw. 460 ft TL: N47 19 14 W119 34 21. Stereo. Fax: (509) 754-4110. E-mail: kule@kule.com Web Site: www.radiozorro.com.200,000 Natl. Network: CBS. Format: Country. ♦Bob Berry, gen mgr; Keith Teske, opns VP; Tom Vinup, news dir.

Everett

KRKO(AM)— Aug 1, 1920: 1380 khz; 5 kw-U, DA-N. TL: N47 55 32 W122 11 19. Stereo. Hrs opn: 24 2707 Colby Ave. #1380, 98201. Phone: (425) 304-1381. Fax: (425) 304-1382. E-mail: andrew.skotdal@krko.com Web Site: www.krko.com. Licensee: S & R Broadcasting. (acq 1987). Natl. Network: Westwood One, ESPN Radio. Law Firm: Cohn & Marks. Format: Talk, sports. News staff: one; News: 24 hrs wkly. Target aud: men 25-54; residents of Northern Puget Sound. ♦Andrew P. Skotdal, pres & gen mgr; Andy Skotdal, opns mgr & progmg dir; Tony Stevens, gen sls mgr.

***KSER(FM)**— Feb 9, 1991: 90.7 mhz; 5.8 kw. 302 ft TL: N48 01 28 W122 06 41. Stereo. Hrs opn: 24 2623 Wetmore Ave, 98201. Phone: (425) 303-9070. Fax: (425) 303-9075. E-mail: info@kser.com Web Site: www.kser.org. Licensee: KSER Foundation. (acq 1996). Population served: 140,000 Natl. Network: PRI. Law Firm: Bechtel & Cole. Format: Div. News staff: one; News: 17 hrs wkly. Target aud: General; public radio. Spec prog: American Indian 2 hrs, folk 2 hrs, jazz 2 hrs, Reggae 2 hrs, soca 2 hrs, Blues 6 hrs wkly. ♦Bruce Wirth, gen mgr.

KWYZ(AM)— July 21, 1957: 1230 khz; 1 kw-U. TL: N47 58 06 W122 10 24. Hrs open:
Rebroadcasts KSUH(AM) Puyallup.
807 S. 336 St., Federal Way, 98003. Phone: (253) 815-1212. Fax: (253) 815-1913. Web Site: www.radiohankook.com. Licensee: Jean J. Suh dba Radio Hankook. (acq 1999; $480,000). Population served: 84,000 Natl. Rep: Roslin. Format: Korean. ♦Doris Haan, gen mgr.

Ferndale

KRPI(AM)— May 1963: 1550 khz; 50 kw-D, 10 kw-N, DA-2. TL: N48 50 35 W122 36 05. Hrs open: 24 Box 3213, 98248. Secondary address: 5538 Imhoff Rd. 98248. Phone: (360) 384-5117. Fax: (360) 380-4202. E-mail: krpi@krpiradio.com Web Site: www.krpiradio.net. Licensee: BBC Broadcasting Inc. (acq 4-19-02; $600,000). Population served: 2,000,000 Format: East Indian. News: 13.75 hrs wkly. Target aud: 35+; East Indian adults. ♦Suki Badh, gen mgr; Andy Struiksma, stn mgr.

Forks

KBDB-FM—Listing follows KBIS(AM).

KBIS(AM)— October 1967: 1490 khz; 1 kw-U. TL: N47 57 16 W124 23 20. Hrs open: 24 Box 450, 98331. Secondary address: 260 Cedar 9833331. Phone: (360) 374-6233. Fax: (360) 374-6852. E-mail: kllm@centurytel.net Licensee: First Broadcasting Investment Partners LLC. (group owner; (acq 12-18-2003; $300,000 with co-located FM). Rgnl rep: Tacher. Format: Rock, CHR. News staff: 2; News: 2 hrs wkly. Target aud: 25-54; general. Spec prog: NFL, Sonics, college & high school football, Mariners 20 hrs, gospel 6.5 hrs wkly. ♦Gary Lawrence, pres; Al Monroe, gen mgr, progmg dir & chief of engrg; Marcia Nearhoff, gen sls mgr; Arthur W. George, min affrs dir.

KBDB-FM—Co-owned with KBIS(AM). 1985: 103.9 mhz; 3 kw. -75 ft TL: N47 57 16 W124 23 20. Format: CHR.

Gig Harbor

***KGHP(FM)**— Aug 30, 1988: 89.9 mhz; 1.5 kw. 190 ft TL: N47 14 29 W122 46 14. Stereo. Hrs opn: 24 14105 62nd Ave. N.W., 98329. Phone: (253) 857-3513. Phone: (253) 857-3589. Fax: (253) 853-5841. E-mail: info@kghp.org Web Site: www.kghp.wednet.edu. Licensee: Peninsula School District No. 401. Population served: 85,000 Format: AAA, eclectic, community info. News: 10 hrs wkly. Target aud: General; diverse community audience. ♦Leland Smith, gen mgr, news dir & pub affrs dir; Theresa Evans, progmg dir & asst music dir; Keith Stiles, chief of engrg.

Goldendale

KLCK(AM)— Sept 4, 1984: 1400 khz; 1 kw-U. TL: N45 49 14 W120 50 15. Hrs open: Box 305, 514 S. Columbus, 98620. Phone: (541) 296-9102. Fax: (541) 298-7775. E-mail: klck@gorge.net Web Site: www.klck1400.com. Licensee: Klickitat Valley Broadcasting Services Inc. (acq 12-18-01; $400,000 with KYYT(FM) Goldendale). Format: Oldies. ♦Danny V. Manciu, pres & gen mgr; Jeanne Malcolm, gen sls mgr; Kevin Malcolm, progmg dir; Julian Notestine, news dir; Cole Malcolm, chief of engrg.

KYYT(FM)— Jan 6, 1992: 102.3 mhz; 1.8 kw. 574 ft TL: N45 48 02 W120 47 35. Hrs open: Box 1023, The Dalles, 97058. Secondary address: 620 E. 3rd St., The Dalles 97058. Phone: (541) 296-9102. Fax: (541) 298-7775. E-mail: kyyt@gorge.net Web Site: www.y102country.com. Licensee: Haystack Broadcasting Inc. (acq 12-18-01; $400,000 with KLCK(AM) Goldendale). Format: Country. ♦Danny V. Manciu, pres & gen mgr; Betsy Hadden, gen sls mgr; Kevin B. Malcolm, progmg dir.

Grand Coulee

KEYG(AM)— 1979: 1490 khz; 1 kw-U. TL: N47 52 58 W118 58 20. Hrs open: 24 Drawer K, 99133. Phone: (509) 633-2020. Phone: (509) 633-1490. Fax: (509) 633-1014. E-mail: keygfm@bigdam.net Web Site: www.keygfm.com. Licensee: Wheeler Broadcasting Inc. (group owner; (acq 12-6-85). Population served: 75,000 Format: Country. News staff: one; News: 4 hrs wkly. Target aud: 25 plus. Spec prog: Class 4 hrs, mus to remember 4 hrs, big band 4 hrs wkly. ♦Verl D. Wheeler, CEO; Mark Wheeler, gen mgr, gen sls mgr & progmg dir; Mike Helgerson, chief of engrg.

KEYG-FM— Feb 10, 1984: 98.5 mhz; 100 kw horiz, 85 kw vert. 994 ft TL: N47 49 18 W118 55 59. Stereo. 24 350,000 Format: Classic hits.

Grandview

KARY-FM— Aug 21, 1989: 100.9 mhz; 7.8 kw. Ant 1,210 ft TL: N46 29 12 W120 00 05. Stereo. Hrs opn: 24 1200 Chesterly Dr., Suite 160, Yakima, 98902-7345. Phone: (509) 248-2900. Fax: (509) 452-9661. Web Site: cherryfm.com. Licensee: New Northwest Broadcasters LLC (group owner; acq 10-20-98; grpsl). Population served: 140,000 Format: Oldies. Target aud: 25-54. Spec prog: Relg 8 hrs wkly. ♦Pete Benedetti, CEO; Joe Benedetti, pres & gen mgr; Dewey Boynton, sls dir & progmg dir; Kevin Miskimins, gen sls mgr & progmg dir.

Hoquiam

KWOK(AM)— Nov 16, 1961: 1490 khz; 1 kw-U. TL: N46 58 22 W123 51 10. Stereo. Hrs opn: 24 1308 Coolidge Rd., Aberdeen, 98520. Phone: (360) 533-1320. Fax: (360) 532-0935. Licensee: Morris Communications Corp. Group owner: Morris Communications Inc. (acq 2-18-00; $650,000 with KXXK(FM) Hoquiam-Aberdeen). Population served: 68000 Format: Sports/talk-ESPN. Target aud: 26-65. ♦Donna Rosi, gen mgr; Donna Rosa, gen sls mgr; Pat Anderson, opns mgr & progmg dir; Harvey Brooks, chief of engrg.

Hoquiam-Aberdeen

KXXK(FM)—Licensed to Hoquiam-Aberdeen. See Olympia

Ilwaco

KVAS(FM)— 2000: 103.9 mhz; 10 kw. Ant 171 ft TL: N46 18 51 W124 03 07. Hrs open: 1006 W. Marine Dr., Astoria, OR, 97103. Phone: (503) 325-2911. Fax: (503) 325-5570. E-mail: kvas@newnw.com Web Site: www.kvasfm.com. Licensee: New Northwest Broadcasters LLC (group owner; acq 8-25-99; $250,000 for CP). Format: Country. ♦Paul Mitchell, gen mgr; Brian Riffe, gen sls mgr; Tom Freel, opns mgr, progmg dir & news dir.

Kelso

KLOG(AM)— Oct 8, 1949: 1490 khz; 1 kw-U. TL: N46 07 00 W122 53 07. Stereo. Hrs opn: 24 Box 90, 98626. Secondary address: 506 Cowlitz Way W. 98626. Phone: (360) 636-0110. Fax: (360) 577-6949. Web Site: www.klog.com. Licensee: Washington Interstate Broadcasting Co. Inc. (acq 6-5-02; with co-located FM). Population served: 40,296 Law Firm: Richard Hayes. Format: Adult contemp. News staff: 2. Target aud: 25-54. ♦Joel Hanson, gen mgr; Bill Dodd, progmg dir.

KUKN(FM)—Co-owned with KLOG(AM). July 7, 1962: 105.5 mhz; 700 w. Ant 859 ft TL: N46 09 52 W122 51 13. Stereo. 24 Web Site: www.kukn.com. (Acq 4-12-02; in exchange for KLYK(FM) Kelso.).90,000 Format: Country. News staff: 2; News: 2 hrs wkly. ♦Joel Hanson, natl sls mgr & rgnl sls mgr; Jadd Curtis, prom dir & disc jockey; Beth Jensen, pub affrs dir; Ray Byers, news rptr; Kirc Rolland, sports cmtr; Bill Dodd, disc jockey; John Mitchel, disc jockey; Ray Bartley, disc jockey.

KLYK(FM)—Licensed to Kelso. See Longview

***KTJC(FM)**— 2004: 91.1 mhz; 17 w horiz, 8 kw vert. Ant 620 ft TL: N46 19 46 W122 57 50. Hrs open: 803 Vandercook Way, Longview, 98632-4039. Phone: (360) 501-5852. Fax: (208) 736-1958. E-mail: ktjc@csnradio.com Web Site: csnradio.com. Licensee: CSN International (group owner). Format: Christian. ♦Lee Flory, stn mgr & progmg dir.

Kennewick

***KBLD(FM)**— 1998: 91.7 mhz; 800 w. 836 ft TL: N46 14 10 W119 19 15. Hrs open: 412 S. Vancouver St., 99336. Phone: (509) 585-8902. Fax: (509) 586-0521. E-mail: kbld@csnradio.com Web Site: www.kbld.com. Licensee: CSN International (group owner; acq 6-12-97; $14,120). Format: Christian. ♦Marty Atkins, gen mgr & progmg dir.

KONA(AM)—Licensed to Kennewick. See Richland-Pasco-Kennewick

KONA-FM—Licensed to Kennewick. See Richland-Pasco-Kennewick

KTCR(AM)—Licensed to Kennewick. See Richland-Pasco-Kennewick

KTCV(FM)—Licensed to Kennewick. See Richland-Pasco-Kennewick

Kirkland

***KARR(AM)**— 1964: 1460 khz; 5 kw-D, 2.5 kw-N, DA-2. TL: N47 40 24 W122 10 07. Hrs open: 24 Box 883, 98083-0883. Phone: (510) 568-6200. E-mail: farradio@familyradio.com Web Site: www.familyradio.com. Licensee: Family Stations Inc. (group owner; (acq 10-22-86; $50,000; FTR: 8-11-86). Format: Relg. News: 5 hrs wkly. Target aud: General; conservative Christians & evangelicals. ♦Harold Camping, pres; Bob Walther, gen mgr & opns mgr; William Thornton, VP & dev VP.

KJAQ(FM)—See Seattle

Lacey

KBRD(AM)— June 1, 1986: 680 khz; 250 w-D. TL: N47 00 41 W122 49 53. Hrs open: 6 AM-6 PM Box 7034, Olympia, 98507. Phone: (360) 491-6800. Licensee: BJ & Skip's for the Music (acq 11-1-2005; with KLDY(AM) Lacey). Population served: 300,000 Natl. Network: AP Radio. Format: Hits of the 20s, 30s, 40s & 50s. Target aud: General. Spec prog: Jazz one hr wkly. ♦Adrian DeBee, gen mgr.

KLDY(AM)— Sept 22, 1983: 1280 khz; 1 kw-D, 500 w-N. TL: N47 03 44 W122 49 49. Stereo. Hrs opn: 24 Box 7034, Olympia, 98507. Secondary address: 125 N. Turner, Olympia 98507. Phone: (360) 491-6800. Web Site: www.kldyradio.com. Licensee: BJ & Skip's for the Music (acq 11-1-2005; with KBRD(AM) Lacey). Format: The musical arts & classical. Target aud: General. ♦Adrian DeBee, gen mgr.

Lakewood

KLAY(AM)—Licensed to Lakewood. See Tacoma

KNTB(AM)—Licensed to Lakewood. See Tacoma

Leavenworth

KOHO-FM— 1998: 101.1 mhz; 6 kw. -872 ft TL: N47 35 32 W120 38 35. Hrs open: 7475 KOHO Pl., 98826. Phone: (509) 548-1011. Fax: (509) 548-3222. E-mail: gmathews@kohoradio.com Web Site: www.kohoradio.com. Licensee: Icicle Broadcasting Inc. (group owner; acq 8-26-99; grpsl). Format: Bluegrass, jazz, classical. ♦Gary Mathews, gen mgr; Heather Winters, gen sls mgr; Michael Dickes, progmg dir; Ian Dunn, news dir; Traci Ellingson, traf mgr.

Long Beach

KAST-FM— May 1987: 99.7 mhz; 6 kw. Ant 233 ft TL: N46 18 51 W124 03 07. Stereo. Hrs opn: 24 1006 W. Marine Dr., Astoria, OR, 97103. Phone: (503) 325-2911. Fax: (503) 325-5570. E-mail: kaqx@newnw.com Web Site: www.kaqx943.com. Licensee: New Northwest Broadcasters LLC. (group owner; (acq 8-24-99; grpsl). Natl. Rep: Tacher. Format: Retro. News staff: one; News: 6 hrs wkly. Target aud: 25-54; 35 plus female, some college students. ♦Paul Mitchell, gen mgr; Tom Freel, opns mgr; Bob Castle, progmg dir.

Longview

KBAM(AM)— Aug 15, 1955: 1270 khz; 5 kw-D, 83 w-N. TL: N46 10 58 W122 57 28. Hrs open: 1130 14th Ave., 98632. Phone: (360) 423-1210. Fax: (360) 423-1554. E-mail: dcrowe@entercom.com Web Site: www.realcountryonline.com. Licensee: Bicoastal Longview LLC. Group owner: Entercom Communications Corp. (acq 1-27-2005; grpsl). Population served: 80,000 Natl. Network: CBS. Natl. Rep: McGavren Guild. Format: Real country. ♦Julie Laird, gen mgr, gen sls mgr & progmg dir; Phil Blair, news dir; Doug Fisher, chief of engrg; Dawn Crowe, traf mgr.

KEDO(AM)— May 1938: 1400 khz; 1 kw-U. TL: N46 08 57 W122 58 29. Hrs open: 24 Broadcast Ctr., 1130 14th Ave., 98632. Phone: (360) 425-1500. Fax: (360) 423-1554. Web Site: www.oldiesradioonline.com. Licensee: Bicoastal Longview LLC. Group owner: Entercom Communications Corp. (acq 1-27-2005; grpsl). Population served: 86,000 Rgnl rep: Art Moore. Law Firm: Leventhal, Senter & Lerman. Format: News, oldies. News staff: one; News: 15 hrs wkly. Target aud: 25-54. Spec prog: Pub affrs 2 hrs wkly. ♦Gayle Kessinger, gen mgr, stn mgr, progmg mgr & news dir; Doug Fisher, chief of engrg.

KLYK(FM)—Co-owned with KEDO(AM). Aug 7, 1991: 94.5 mhz; 3 kw. Ant 476 ft TL: N46 16 49 W122 52 34. Stereo. 24 Web Site: www.todaysbesthits.com. Natl. Network: Westwood One. Format: Hot adult contemp. News: 2 hrs wkly. Target aud: 18-49. ♦Gayle Kessinger, progmg dir.

***KJVH(FM)**— 1988: 89.5 mhz; 100 w. 780 ft TL: N46 09 52 W122 51 13. Hrs open: 136 E. S. Temple, Suite 1630, Salt Lake City, UT, 84111. Phone: (801) 359-3147. Fax: (801) 359-8112. Web Site: www.familyradio.com. Licensee: Family Stations Inc. (group owner) Format: Christian relg. ♦Harold Camping, pres; Roger Crawford, gen mgr & stn mgr.

KLOG(AM)—See Kelso

KUKN(FM)—Licensed to Longview. See Kelso

***KWYQ(FM)**— Oct 22, 1987: 90.3 mhz; 500 kw. 735 ft TL: N46 10 12 W122 56 43. Stereo. Hrs opn: 24 Box 1000, Kelso, 98626. Phone: (360) 578-1929. Fax: (360) 636-1357. Web Site: wayfm.com. Licensee: WAY-FM Media Group Inc. (group owner; (acq 8-1-2003 with KWYA(FM) Astoria, OR). Format: Contemp Christian. Target aud: 19-34. ♦Robert D. Augsburg, pres; Danny Houle, gen mgr.

Lynden

KWPZ(FM)— Nov 8, 1960: 106.5 mhz; 68 kw. Ant 2,332 ft TL: N48 40 45 W122 50 31. Stereo. Hrs opn: 24 1843 Front St., Suite A, 98264. Phone: (360) 354-5596. Fax: (360) 354-7517. E-mail: comments@praise1065.com Web Site: www.praise1065.com. Licensee: CRISTA Ministries Inc. Group owner: CRISTA Broadcasting (acq 12-80). Population served: 3,000,000 Format: Relg, Christian. Target aud: 25-54; female. ♦James Gwinn, pres; Melene Thompson, VP & gen mgr; Marvin Mickley, stn mgr & opns mgr; Jim Bouma, opns mgr; Roger Burke, sls dir & gen sls mgr; Lynette Schulz, prom mgr.

Mabton

KMNA(FM)— 1997: Stn currently dark. 98.7 mhz; 11.5 kw. Ant 874 ft TL: N46 28 33 W120 08 37. Hrs open: 152101 W. Country Rd. 12, Prosser, 99350. Phone: (509) 789-1209. Fax: (509) 786-1181. Licensee: MBProsser Licensee LLC. Group owner: Moon Broadcasting (acq 2-6-2004; $1.9 million). Format: Sp contemp. ♦Frank Allec, gen mgr.

Manson

KZAL(FM)— 2007: 94.7 mhz; 10.3 kw. Ant 518 ft TL: N47 51 16 W120 09 59. Hrs open: 7475 KOHO Pl., Leavenworth, 98826. Phone: (509) 548-1011. Fax: (509) 548-3222. E-mail: koho@kohoradio.com Licensee: Icicle Broadcasting Inc. Format: Smooth jazz. ♦Harriet Bullitt, pres; Gary Mathews, gen mgr.

McCleary

KGY-FM— October 1992: 96.9 mhz; 2.33 kw. 1,056 ft TL: N47 05 08 W123 11 19. Hrs open: 24 Box 1249, Olympia, 98507. Phone: (360) 943-1240. Fax: (360) 352-1222. E-mail: kgysales@kgyradio.com Web Site: www.realcountryonline.com. Licensee: KGY Inc. Population served: 500,000 Law Firm: Haley, Bader & Potts. Format: Country. News staff: 2. Target aud: 35 plus. ♦Dick Pust, gen mgr; Dennis Brown, gen sls mgr; Tom Trotzer, chief of engrg; Jeanna Spain, traf mgr.

Medical Lake

KTSL(FM)— Mar 7, 1989: 101.9 mhz; 12 kw. 495 ft TL: N47 41 30 W117 46 00. Stereo. Hrs opn: 24 1212 N. Washington, Suite 124, Spokane, 99201. Phone: (509) 326-9500. Fax: (509) 326-1560. E-mail: spirit101.9@spirit1019.com Web Site: www.spirit1019.com. Licensee: Pamplin Broadcasting-Washington Inc. Group owner: Pamplin Broadcasting (acq 7-24-98; $1.3 million). Population served: 375,000 Law Firm: Fisher, Wayland, Cooper, Leader & Zaragoza. Format: Contemp Christian. News staff: one. Target aud: 18-49; young families. ♦Karen Dineen, gen mgr; Jim Prophet, prom dir; Bryan O'Neil, progmg dir & chief of engrg.

Mercer Island

***KMIH(FM)**— February 1970: 104.5 mhz; 30 w. Ant 226 ft TL: N47 34 21 W122 13 01. Stereo. Hrs opn: 24 9100 S.E. 42nd, 98040-4107. Phone: (206) 236-3296. Fax: (206) 236-3342. Web Site: www.x104.fm. Licensee: Mercer Island School District No. 400. Population served: 540,000 Law Firm: Womble, Carlyle, Sandridge & Rice. Format: CHR. Target aud: 18-34; Female. ♦Nick De Vogel, gen mgr; Chelsea Doran, gen sls mgr; Rich Brown, progmg dir & chief of engrg.

Mercer Island-Seattle

KIXI(AM)—Licensed to Mercer Island-Seattle. See Seattle

Montesano

KSWW(FM)— 1998: 102.1 mhz; 50 kw. Ant 440 ft TL: N46 56 30 W123 47 07. Stereo. Hrs opn: 24 Box 1198, Aberdeen, 98520. Secondary address: 1520 Simpson Ave., Aberdeen 98520. Phone: (360) 533-3000. Fax: (360) 532-1456. Web Site: www.jodesha.com. Licensee: Jodesha Broadcasting Inc. (group owner; acq 2-28-03; $750,000 with KBKW(AM) Aberdeen). Population served: 85,000 Natl. Network: ABC. Rgnl rep: Tacher Law Firm: David Tillotson. Wire Svc: AP Format: Adult contemp. News staff: 1; News: 7 hrs wkly. Target aud: 25-54. ♦William J. Wolfenbarger, pres & gen mgr; Gabrielle Jordan, opns mgr; Heidi Persson, gen sls mgr; Kim Kreiss, prom dir.

Moses Lake

KBSN(AM)— November 1947: 1470 khz; 5 kw-D, 1 kw-N, DA-2. TL: N47 06 16 W119 17 32. Hrs open: Drawer B, 98837. Secondary address: 2241 W. Main 98837. Phone: (509) 765-3441. Fax: (509) 766-0273. E-mail: kddrmsales@atnet.net Licensee: KSEM Inc. Population served: 15,000 Format: News/talk, sports, farm. Target aud: General. Spec prog: Sp 9 hrs wkly. ♦Jim Davis, gen mgr, opns mgr & prom mgr; Stacey Lehman, gen sls mgr; Gary Roberts, progmg dir; Andy Patrick, news dir; Will Vos, chief of engrg; Dennis Clay, outdoor ed; Emilio Vela, spanish dir; Dave Heaverlo, sports cmtr.

KDRM(FM)—Co-owned with KBSN(AM). Oct 1, 1980: 99.3 mhz; 3 kw. 275 ft TL: N47 05 54 W119 17 47. Stereo. Format: Hot adult contemp. Target aud: 18-34. ♦Dennis Clay, prom mgr & outdoor ed; Dave Heaverlo, progmg dir & sports cmtr; Emilio Vela, gen mgr & spanish dir.

***KLWS(FM)**— Apr 10, 1997: 91.5 mhz; 7.2 kw. Ant 686 ft TL: N47 18 50 W119 34 55. Hrs open: 24
Rebroadcasts KWSU(AM) Pullman 100%.
Box 642530, Murrow Communications Ctr., Washington State Univ., Rm. 382, Pullman, 99164-2530. Phone: (509) 335-6500. Fax: (509) 335-6577. E-mail: nwpr@wsu.edu Web Site: www.nwpr.org. Licensee: Washington State University. Law Firm: Dow, Lohnes & Albertson. Format: News & views. ♦Dennis Haarsager, gen mgr; Roger Johnson, stn mgr; Scott Weatherly, opns mgr; Mary Hawkins, progmg dir; Robin Rilette, mus dir.

***KMLW(FM)**— May 4, 1997: 88.3 mhz; 4 kw. 817 ft TL: N46 56 31 W119 25 41. Hrs open:
Rebroadcasts KMBI-FM Spokane 100%.
5408 S. Freya, Spokane, 99223. Secondary address: 820 N. LaSalle Blvd., Chicago, IL 60610. Phone: (509) 448-2555. Phone: (312) 329-4301. Fax: (509 448-6855. Fax: (312) 329-4468. E-mail: kmbi@moody.edu Web Site: www.kmbi.fm. Licensee: Moody Bible Institute of Chicago. (group owner) Format: Relg. Target aud: 35-54; Christians. Spec prog: Class one hr wkly. ♦Rich Monteith, gen mgr & stn mgr; Pete Fretwell, progmg dir; Gordon Canaday, chief of engrg.

KWIQ(AM)—Moses Lake North, Feb 20, 1956: 1020 khz; 2 kw-D, 440 w-N, DA-D. TL: N47 09 48 W119 21 39. Hrs open: 24 Box 79, Wenatchee, 98807. Secondary address: 11768 Kittleson Rd. 98837. Phone: (509) 765-1761. Phone: (509) 663-5186. Fax: (509) 765-8901. Licensee: Morris Communications Corp. Group owner: Morris Communications Inc. (acq 10-15-98; grpsl). Population served: 85000 Natl. Network: ESPN Radio. Natl. Rep: Katz Radio. Format: Sports. News staff: one; News: 6 hrs wkly. Target aud: 18-49; men. ♦Gary Patrick, gen mgr; Jeff Dahlstrom, sls dir & gen sls mgr; John Windus, progmg dir; Jay White, chief of engrg.

KWIQ-FM— May 22, 1968: 100.3 mhz; 100 kw. Ant 167 ft TL: N47 06 09 W119 14 26. Stereo. 24 Web Site: www.kwiq.com.105,000 Natl. Rep: Katz Radio. Rgnl rep: Allied Radio Partners. Format: Country. News staff: one. ♦Jeff Dahlstrom, natl sls mgr.

Moses Lake North

KWIQ(AM)—Licensed to Moses Lake North. See Moses Lake

Mount Vernon

KAPS(AM)— Mar 17, 1963: 660 khz; 10 kw-D, 1 kw-N. TL: N48 26 22 W122 20 45. Hrs open: 24 2029 Freeway Dr., 98273. Phone: (360) 424-0660. Fax: (360) 424-1660. E-mail: country@kapsradio.com Web Site: www.kapsradio.com. Licensee: Valley Broadcasters Inc. (acq 8-13-93; FTR: 8-30-93). Population served: 100,000 Rgnl rep: McGaven Guild Format: Country. News staff: one; News: hourly. Target aud: General. ♦Jim Keane, pres; Jerry Keane, gen sls mgr; Mike Yeoman, progmg dir.

KBRC(AM)— Dec 11, 1946: 1430 khz; 5 kw-D, 1 kw-N, DA-N. TL: N48 25 22 W122 21 10. Stereo. Hrs opn: 24 Box 250, 98273. Secondary address: 2029 Freeway Dr. 98273. Phone: (360) 424-4278. Phone: (360) 424-1430. Fax: (360) 424-1660. E-mail: oldies@kbrcradio.com Web Site: kbrcradio.com. Licensee: Valley Broadcasting. (acq 1996; Population served: 154,000 Natl. Network: ABC. Rgnl rep: Tacher. Law Firm: Covington & Burling. Format: Oldies. News staff: 2. Target aud: 25-54. Spec prog: Sp 3 hrs, farm 5 hrs wkly. ♦James Keane, pres & gen mgr; Jerry Keane, gen sls mgr; Mike Yoeman, progmg dir; Kirk Tollifson, news dir; Mike Gilbert, chief of engrg; Julia Rasmussen, traf mgr.

***KMWS(FM)**— May 4, 1973: 90.1 mhz; 100 w hoirz. Ant -161 ft TL: N48 26 13 W122 18 36. Hrs open: 24
Rebroadcasts KWSU(AM) Pullman 100%.
Northwest Public Radio, Box 642530, Pullman, 99164-2530. Phone: (509) 335-6536. E-mail: dahmen@wsu.edu Web Site: www.nwpr.org. Licensee: Washington State University (acq 3-31-2003). Population served: 70,000 Format: News/talk. ♦Dennis Haahnsager, gen mgr.

***KSVR(FM)**— 2002: 91.7 mhz; 170 w. Ant 669 ft TL: N48 23 49 W122 18 26. Hrs open: Skagit Valley College, 2405 E. College Way, 98273. Phone: (360) 416-7711. Fax: (360) 416-7822. E-mail: mail@ksvr.org Web Site: www.ksvr.org. Licensee: Board of Trustees of Skagit Valley College (acq 4-19-00). Format: Progsv, Sp, news/talk. ♦Rip Robbins, gen mgr, opns dir & progmg dir; Bill McCuskey, chief of engrg.

Naches

KQSN(FM)— November 2000: 99.3 mhz; 790 w. Ant 899 ft TL: N46 36 02 W120 52 06. Hrs open: 4010 Summitview Ave., Yakima, 98908. Phone: (509) 972-3461. Fax: (509) 972-3540. Licensee: Capstar TX L.P. Group owner: Clear Channel Communications Inc. (acq 3-12-2001; $1.3 million). Format: Sp. ♦Larry Miner, gen mgr; Connie Johnston, gen sls mgr; Ron Harris, progmg dir; Lance Tormey, news dir; Bill Glenn, chief of engrg.

KZTA(FM)—Licensed to Naches. See Yakima

Newport

KPWL(AM)—Not on air, target date: unknown: 1370 khz; 50 kw-D, 350 w-N, DA-N. TL: N48 10 30 W117 01 27. Hrs open: 110 Green Meadows, Abilene, TX, 79605. Phone: (325) 829-6850. Licensee: Scott Powell. ♦Scott Powell, gen mgr.

KQQB-FM— December 1989: 104.5 mhz; 87 kw hoirz. Ant 1,046 ft TL: N48 23 09 W117 14 15. Stereo. Hrs opn: 24 505 W. Riverside, Suite 101, Spokane, 99201. Phone: (509) 252-8440. Fax: (509) 252-8453. E-mail: christa@pro-activecomm.com Web Site: www.live1045.com. Licensee: ProActive Communications Inc. (acq 9-28-2005; $2 million). Format: Top-40. Target aud: 18-34. ♦Gerald D. Clifton, CEO; Christa McDonald, gen mgr; Steve Kicklighter, opns mgr & progmg dir.

***KUBS(FM)**— Sept 10, 1973: 91.5 mhz; 150 w. -538 ft TL: N48 10 23 W117 03 15. Hrs open: Box 70, Newport High School, 99156-0070. Phone: (509) 447-4931. Fax: (509) 447-4354. Licensee: Newport Consolidated School District #56415. Population served: 2,400 Format: Var radio.

Nile

***KSBC(FM)**— 2003: Stn currently dark. 88.1 mhz; 200 w. Ant -1,145 ft TL: N46 50 02 W120 56 13. Stereo. Hrs opn: 24 2351 Sunset Blvd., Suite 170-218, Rocklin, CA, 95765. Phone: (916) 251-1600. Fax: (916) 251-1650. E-mail: klove@klove.com Web Site: www.klove.com. Licensee: Educational Media Foundation. Group owner: EMF Broadcasting (acq 10-2-2003; grpsl). Natl. Network: K-Love. Law Firm: Shaw Pittman. Format: Contemp Christian. News staff: 3. Target aud: 25-44; Judeo Christian female. ♦Richard Jenkins, pres; Mike Novak, VP; Lloyd Parker, gen mgr; Ed Lenane, opns dir & news dir; Keith Whipple, dev dir; David Pierce, progmg mgr; Sam Wallington, engrg dir; Karen Johnson, news rptr; Marya Morgan, news rptr; Richard Hunt, news rptr.

Oak Harbor

KWDB(AM)— Dec 14, 1984: 1110 khz; 500 w-D. TL: N48 17 27 W122 42 28. (CP: 1520 khz, 1 kw-D. TL: N48 16 55 W122 42 26). Hrs opn: 6 AM-6 PM Box 1455, 3170 D N. Heller Rd., 98277. Phone: (360) 675-7320. Phone: (360) 240-1520. Fax: (360) 675-0166. E-mail: kwdb@kwdb.com Web Site: www.kwdb.com. Licensee: West Beach Broadcasting Corp. (acq 3-20-00; $55,000). Population served: 68,500 Natl. Network: Fox News Radio. Format: Adult standards, big band, oldies. News staff: 2; News: 12 hrs wkly. Target aud: 20-60. Spec prog: Gospel 6 hrs wkly. ♦Richard Bell, gen mgr; Rich Ulrich, gen sls mgr.

Ocean Park

***KLOP(FM)**— 2006: 88.1 mhz; 550 w. Ant 1,044 ft TL: 46 41 46 W123 46 17. Hrs open: 24
Rebroadcasts KLVR(FM) Santa Rosa, CA 100%.
2351 Sunset Blvd., Suite 170-218, Rocklin, CA, 95765. Phone: (916) 251-1600. Fax: (916) 251-1650. E-mail: klove@klove.com Web Site: www.klove.com. Licensee: Educational Media Foundation. Group owner: EMF Broadcasting. Natl. Network: K-Love. Law Firm: Shaw Pittman. Format: Contemp Christian. News staff: 3. Target aud: 25-44; Judeo Christian, female. ♦Richard Jenkins, pres; Mike Novak, VP & progmg dir; Lloyd Parker, gen mgr; Ed Lenane, opns dir & news dir; Keith Whipple, dev dir; Eric Allen, natl sls mgr; David Pierce, progmg mgr; Jon Rivers, mus dir; Sam Wallington, engrg dir; Arthur Vassar, traf mgr; Karen Johnson, news rptr; Marya Morgan, news rptr; Richard Hunt, news rptr.

Ocean Shores

KANY(FM)—Not on air, target date: unknown: 93.5 mhz; 6 kw. Ant 226 ft TL: N46 53 04 W124 00 44. Hrs open: Box 1198, Aberdeen, 98520. Phone: (360) 533-3000. Licensee: Jodesha Broadcasting Inc. (acq 3-16-2007; $600,000 for CP). ♦William J. Wolfenbarger, pres.

Olympia

***KAOS(FM)**— Jan 1, 1973: 89.3 mhz; 1.5 kw. -19 ft TL: N47 04 22 W122 58 51. Stereo. Hrs opn: 24 CAB 301, 98505. Phone: (360) 867-6895. Fax: (360) 866-6797. E-mail: kaos@evergreen.edu Web Site: www.kaosradio.org. Licensee: Evergreen State College. Population served: 100,000 Natl. Network: PRI. Format: Div. News: 5 hrs wkly. Target aud: 18-35. Spec prog: Folks 16 hrs, Asian 3 hrs, American Indian 3 hrs, Sp 6 hrs wkly. ♦Donna DiBianco, opns mgr; Jerry Drummond, gen mgr, dev dir & dev dir; Bryan Johnson, mus dir & pub affrs dir.

KGTK(AM)—Licensed to Olympia. See Tacoma

KGY(AM)— Apr 15, 1922: 1240 khz; 1 kw-U. TL: N47 03 31 W122 54 09. Hrs open: 24 Box 1249, 98507. Secondary address: 1700 Marine Dr. N.E. 98501. Phone: (360) 943-1240. Fax: (360) 352-1222. E-mail: kgysales@kgyradio.com Web Site: www.kgyradio.com. Licensee: KGY Inc. (acq 1-15-39). Population served: 130,000 Wire Svc: AP Format: Full service, adult contemp. News staff: 2. Target aud: General. ♦Dick Pust, gen mgr; Dennis Brown, gen sls mgr; Larry Baile, progmg dir & progmg mgr; Jeanna Spain, mus dir & traf mgr; Ian Fox, news dir; Tom Trotzer, chief of engrg.

KLDY(AM)—See Lacey

***KPLI(FM)**— September 2006: 90.1 mhz; 100 w. Ant -59 ft TL: N47 02 20 W122 54 00. Hrs open:
Rebroadcasts KPLU-FM Tacoma 100%.
121st and Park Ave. S., Tacoma, 98447-0003. Phone: (253) 536-5009. Fax: (253) 535-8332. Licensee: Pacific Lutheran University Inc. (acq 9-27-2005; $400,000). Population served: 60,000 Natl. Network: NPR, PRI. Law Firm: Dow, Lohnes & Albertson. Format: News, jazz, blues. News staff: 7; News: 54 hrs wkly. ♦Paul Stankavich, gen mgr; Jeff Bauman, opns dir; Nancy Knudson, dev dir; Joey Cohn, progmg dir; Nick Francis, mus dir; Erin Hennessey, news dir; Lowell Kiesow, chief of engrg.

KXXK(FM)—Hoquiam-Aberdeen, Sept 3, 1965: 95.3 mhz; 3 kw. 449 ft TL: N46 55 53 W123 44 02. (CP: 5 kw, ant 436 ft. TL: N46 56 30 W123 47 07). Stereo. Hrs opn: 24 1308 Coolidge Rd., Aberdeen, 98520. Phone: (360) 533-1320. Fax: (360) 532-0935. Licensee: Morris Communications Corp. Group owner: Morris Communications Inc. (acq 2-18-00; $650,000 with KWOK(AM) Hoquiam). Population served: 69000 Format: Hot country. Target aud: 26-54. ♦Donna Rosi, gen mgr & gen sls mgr; Patrick Anderson, opns mgr & progmg dir; Ian Cope, news dir; Harvey Brooks, chief of engrg.

KXXO(FM)— Jan 16, 1990: 96.1 mhz; 72 kw. Ant 2,099 ft TL: N46 38 07 W122 28 01. Stereo. Hrs opn: 24 Box 7937, 98507. Secondary address: Rockway/Leland Bldg., 119 N. Washington Ave. 98507. Phone: (360) 943-9937. Phone: (206) 624-3712. Fax: (360) 352-3643. E-mail: admin@mixx96.com. Web Site: www.mixx96.com. Licensee: 3 Cities Inc. Population served: 1,325,000 Format: Adult contemp. News staff: one; News: one hr wkly. Target aud: 25-54; general. ♦David Rauh, pres & gen mgr; Toni Holm, sr VP & stn mgr; Brian Butler, gen sls mgr; Sanrica Marquez, prom dir; John Foster, progmg dir; Tim Vik, chief of engrg; Liz Rendon, traf mgr.

Omak

KNCW(FM)—Listing follows KOMW(AM).

KOMW(AM)— Sept 30, 1947: 680 khz; 5 kw-D. TL: N48 23 40 W119 32 00. Hrs open: Box 151, 98841. Secondary address: 320 Emery St. 98841. Phone: (509) 826-0100. Fax: (509) 826-3929. Web Site: www.komw.net. Licensee: North Cascades Broadcasting Inc. (group owner; acq 7-90). Population served: 35,000 Natl. Network: ABC. Format: Adult standards, talk. Target aud: 18-45. Spec prog: Farm 2 hrs, Sp 2 hrs wkly. ♦John P. Andrist, CEO, pres, gen mgr, prom dir & prom mgr; Rebecca L. Andrist, CFO & sr VP; Rick Duck, gen sls mgr; Chris Schmidt, progmg dir; Steve Hardy, news dir; Randy Gates, pub affrs dir; Jerry Robinson, chief of engrg.

KNCW(FM)—Co-owned with KOMW(AM). Apr 10, 1978: 92.7 mhz; 4.1 kw. 941 ft TL: N48 19 12 W119 32 18. Stereo. 24 Web Site: www.komw.net.25,000 Format: Country. Target aud: General.

***KQWS(FM)**— Jan 6, 1999: 90.1 mhz; 3 kw. Ant 2,457 ft TL: N48 44 37 W119 37 16. Hrs open: 24
Rebroadcasts KWSU(AM) Pullman 100%.
Box 642530, Murrow Communications Ctr., Washington State Univ., Pullman, 99164-2530. Phone: (509) 335-6500. Fax: (509)335-3772. E-mail: nwpr@wsu.edu Web Site: www.nwpr.org. Licensee: Washington State University. Law Firm: Dow, Lohnes & Albertson. Format: News, class. News staff: one; News: 60 hrs wkly. ♦Dennis Haarsager, gen mgr; Roger Johnson, stn mgr; Scott Weatherly, opns mgr; Sarah McDaniel, dev dir; Mary Hawkins, progmg dir; Ralph Hogan, engrg dir & chief of engrg.

KZBE(FM)— June 22, 1998: 104.3 mhz; 3.5 kw. 981 ft TL: N48 19 12 W119 32 18. Hrs open: 24 Box 151, 98841. Phone: (509) 826-0100. Fax: (509) 826-3929. E-mail: news@komw.net Web Site: www.komw.net. Licensee: North Cascades Broadcasting Inc. (group owner; acq 10-14-97; $47,606). Natl. Network: ABC. Format: Contemp hit/Top-40. News staff: 2; News: 3 hrs wkly. Target aud: 18 plus. ♦John P. Andrist, CEO, pres & gen mgr; Rebecca L. Andrist, CFO, sr VP & news dir; Rick Duck, gen sls mgr; Chris Schmidt, progmg dir; Jerry Robinson, chief of engrg.

Opportunity

KIXZ-FM— Apr 1, 1961: 96.1 mhz; 56 kw. 2,378 ft TL: N47 34 11 W117 05 00. Stereo. Hrs opn: 808 East Sprague, Spokane, 99202. Phone: (509) 242-2400. Fax: (509) 242-1160. Web Site: www.kix961.com. Licensee: Capstar TX L.P. Group owner: Clear Channel Communications Inc. (acq 8-30-00; grpsl). Population served: 260,000 Law Firm: Shaw Pittman. Format: Country. Target aud: 25-54.

KTRW(AM)— November 1955: 630 khz; 530 w-D, 53 w-N. TL: N47 36 31 W117 22 25. Hrs open: 24 Box 31000, Spokane, 99223. Phone: (509) 443-1000. E-mail: ktw@fabulous.com Web Site: www.ktrw.com. Licensee: Mutual Broadcasting System LLC Group owner: Morgan Murphy Stations (acq 9-1-2005; $375,000). Population served: 1,100,000 Natl. Network: Fox News Radio, Salem Radio Network, USA. Law Firm: Cohn & Marks. Format: Nostalgia, big band, talk. ♦Thomas W. Read, gen mgr.

Othello

KOLW(FM)— February 1992: 97.5 mhz; 4.6 kw. Ant 656 ft TL: N46 45 55 W119 16 49. (CP: COL Basin City. 50 kw, ant 620 ft. TL: N46 17 23 W119 25 28). Stereo. Hrs opn: 24 45 Campbell Rd., Walla Walla, 99362. Phone: (509) 527-1000. Fax: (509) 529-5534. E-mail: kzln@teleu98.com Licensee: Capstar TX L.P. (acq 11-9-2004; exchange for KHTO(FM) Milton-Freewater, OR). Population served: 225,000 Natl. Network: Jones Radio Networks. Rgnl rep: Wheeler Broadcasting. Format: CHR. Target aud: 25-49; 60% male, 40% female. ♦Mark Wheeler, gen mgr.

KRSC(AM)— Sept 1, 1957: 1400 khz; 1 kw-U. TL: N46 49 29 W119 11 26. Hrs open: 24 128 S. 1st Ave., 99344. Phone: (509) 488-0606. Fax: (509) 488-0909. Licensee: Alexandra Communications Inc. (acq 8-1-2007). Population served: 10,000 Format: Sp, Mexican, sports, talk. ♦Betsy Gomez, gen mgr.

Pasco

KEYW(FM)— June 30, 1986: 98.3 mhz; 3 kw. 197 ft TL: N46 08 48 W119 05 59. Stereo. Hrs opn: 25 2621 W. A St., 99301. Phone: (509) 547-9791. Fax: (509) 547-8509. E-mail: thekey@keyw.com Web Site: www.keyw.com. Licensee: Capstar TX L.P. Group owner: Clear Channel Communications Inc. (acq 2-15-02; grpsl). Format: Adult contemp. Target aud: 18-49. ♦Eric Van Winkle, gen mgr; Grant Linnen, gen sls mgr; Faith Martin, prom dir & news dir; Paul Drake, progmg dir; Chuck Ince, chief of engrg.

KFLD(AM)—Licensed to Pasco. See Richland-Pasco-Kennewick

KGDN(FM)—Licensed to Pasco. See Richland-Pasco-Kennewick

KGSG(FM)— Apr 1, 1997: 93.7 mhz; 600 w. 958 ft TL: N46 04 59 W119 09 38. Hrs open: Box 2852, 99302. Phone: (509) 547-5196. Fax: (509) 547-5203. Web Site: www.kgsg.com. Licensee: Gospel Music Broadcasting Corp. (acq 3-26-98). Format: Southern gospel. ♦Martin L. Gibbs, pres, gen mgr & chief of engrg; Sharon Harmon, progmg dir.

***KOLU(FM)**— Sept 1, 1971: 90.1 mhz; 3.99 kw. 93 ft TL: N46 14 59 W119 09 10. (CP: 7.5 kw, ant 1,000 ft.). Stereo. Hrs opn: 4921 W. Wernett, 99301. Phone: (509) 547-2062. Fax: (509) 544-0340. Web Site: www.riverviewbaptist.org. Licensee: Riverview Baptist Christian Schools. Format: Relg. ♦John Paisley, gen mgr.

Port Angeles

***KNWP(FM)**— Mar 23, 1998: 90.1 mhz; 1.6 kw. Ant 197 ft TL: N48 09 03 W123 40 09. Hrs open: 24
Rebroadcasts KRFA-FM Moscow, ID 100%.
Box 642530, 382 Murrow Ctr., Pullman, 99164-2530. Phone: (509) 335-6500. Fax: (509) 335-3772. E-mail: nwpr@wsu.edu Web Site: www.nwpr.org. Licensee: Washington State University. Law Firm: Dow, Lohnes & Albertson. Format: Classical, news. News staff: one; News: 37 hrs wkly. ♦Dennis Haarsager, gen mgr; Roger Johnson, stn mgr; Scott Weatherly, opns mgr; Sarah McDaniel, dev dir; Mary Hawkins, progmg dir; Ralph Hogan, engrg dir & chief of engrg.

KONP(AM)— 1945: 1450 khz; 1 kw-U. TL: N48 07 19 W123 26 13. Hrs open: Box 1450, 313 W. First, 98362. Phone: (360) 457-1450. Fax: (360) 457-9114. E-mail: office@konp.com Web Site: www.konp.com. Licensee: Radio Pacific Inc. (acq 2-12-02; $850,000). Population served: 38,736 Format: News/talk. Target aud: 28-54. ♦Brown Maloney, chmn; Stan Comeau, gen sls mgr; Todd Ortloff, gen mgr, opns mgr, progmg dir & chief of engrg.

***KVIX(FM)**— Mar 22, 2005: 89.3 mhz; 600 w. Ant 489 ft TL: N48 09 03 W123 40 09. Hrs open:
Rebroadcasts KPLU-FM Tacoma 100%.
c/o KPLU-FM, 121st and Park Ave., Tacoma, 98447. Phone: (253) 535-7758. Fax: (253) 535-8332. E-mail: info@kplu.org Web Site: www.kplu.org. Licensee: Pacific Lutheran University Inc. Natl. Network: NPR, PRI. Law Firm: Dow, Lohnes & Albertson. Format: News, jazz, blues. ♦Paul Stankavich, gen mgr; Jeff Bauman, opns dir; Nancy Knudson, dev dir; Joey Cohn, progmg dir; Nick Francis, mus dir; Erin Hennessey, news dir; Lowell Kiesow, chief of engrg.

Prosser

KLES(FM)— Sept 6, 1962: 101.7 mhz; 3.5 kw. Ant 865 ft TL: N46 11 12 W119 45 13. Stereo. Hrs opn: 152101 W. County Rt. 12, 99350. Phone: (509) 786-4532. Fax: (509) 786-1181. Licensee: MBProsser Licensee LLC. Group owner: Moon Broadcasting (acq 2-15-2000; $750,000). Population served: 8,000 Natl. Rep: Target Broadcast Sales. Format: Hispanic. Target aud: 18-49. ♦Frank Allec, gen mgr, opns mgr & gen sls mgr; Rubin Muniz, gen sls mgr; Juan Tejeda, prom mgr; Piomar Marin, progmg dir, news dir & disc jockey; Karen Zackula, traf mgr; Jamil Esqueviel, disc jockey; Victor Gonzalez, disc jockey.

KZXR(AM)—Co-owned with KLES(FM). Dec 14, 1956: 1310 khz; 5 kw-D, 66 w-N. TL: N46 14 03 W119 48 49. Phone: (509) 786-1310. (Acq 2-24-2000; $500,000).5,000 Format: News/talk, sports. Target aud: 25-54. ♦Frank Allec, gen mgr, opns mgr, gen sls mgr & news dir; Todd Summers, progmg dir & chief of engrg; Karen Zackula, traf mgr.

Pullman

KHTR(FM)—Listing follows KQQQ(AM).

KQQQ(AM)— 1938: 1150 khz; 11 kw-D, 27 w-N. TL: N46 43 36 W117 12 23. Hrs open: 24 801 Old Wawawai Rd., 99163. Secondary address: Box 1 99163. Phone: (509) 332-6551. Fax: (509) 332-5151. E-mail: khtr@aol.com Web Site: www.hot104.net. Licensee: Radio Palouse Inc. (group owner; acq 12-74). Population served: 49,000 Rgnl rep: Allied Radio Partners. Law Firm: David Tillotson. Format: News/talk. News staff: one; News: 28 hrs wkly. Target aud: General. Spec prog: Farm 3 hrs, loc news 10 hrs wkly. ♦Bill Weed, gen mgr; Larry Weir, opns mgr; Rod Schwartz, gen sls mgr; Evan Ellis, news dir; Steve Franko, chief of engrg.

KHTR(FM)—Co-owned with KQQQ(AM). 1967: 104.3 mhz; 24 kw. 1,669 ft TL: N46 48 40 W116 54 55. Stereo. Box 1, 99163. Secondary address: 801 Old Wawawai Rd. 99163. Web Site: www.hot104.net.125,000 Format: CHR. News staff: one; News: 24 hrs wkly. Target aud: General. ♦Jeremy West, mus dir.

KRFA-FM—See Moscow, ID

***KRLF(FM)**— July 1, 1991: 88.5 mhz; 420 w vert. 794 ft TL: N46 38 01 W117 05 13. Stereo. Hrs opn: 24 S.W. 345 Kimball, 99163. Phone: (509) 332-3545. Fax: (509) 332-5433. E-mail: krlf@lffmtc.org Web Site: Krlf.org. Licensee: Living Faith Fellowship Educational Ministries. Population served: 100,000 Law Firm: Gammon & Grange. Format: Christian hot adult contemp. News: 18 hrs wkly. Target aud: 13-52. Spec prog: Alternative one hr, children 5 hrs, CHR one hr, rock one hr wkly. ♦Phillip J. Vance, pres; Frank Younce, stn mgr, engrg mgr & chief of engrg; Ruth Younce, progmg dir.

KUUX(AM)—Not on air, target date: unknown: 650 khz; 3 kw-D, 250 w-N, DA-N. TL: N46 46 03 W117 11 03. Hrs open: Box 1, 99163. Phone: (509) 332-6551. Fax: (509) 332-5151. Licensee: Radio Palouse Inc. (group owner). ♦Bill Weed, gen mgr & progmg dir.

***KWSU(AM)**— June 1922: 1250 khz; 5 kw-U. TL: N46 41 47 W117 14 44. Hrs open: 24 Box 642530, Murrow Communications Ctr., Washington State Univ., Rm. 382, 99164-2530. Phone: (509) 335-6500. Fax: (509) 335-6577. E-mail: nwpr@wsu.edu Web Site: www.nwpr.org. Licensee: Washington State Univ. Population served: 12,000 Natl. Network: NPR. Law Firm: Dow, Lohnes & Albertson. Format: News, class. News staff: one; News: 60 hrs wkly. Target aud: General. Spec prog: Jazz 14 hrs wkly. ♦Karen Olstad, COO; Dennis Haarsager, gen mgr; Roger Johnson, stn mgr; Scott Weatherly, opns mgr; Sarah McDaniel, dev dir & dev mgr; Mary Hawkins, progmg dir & progmg mgr. Co-owned TV: *KWSU-TV affil.

***KZUU(FM)**— Sept 21, 1979: 90.7 mhz; 800 w. 105 ft TL: N46 43 51 W117 09 08. Stereo. Hrs opn: 24 CUB Rm 311, Washington State Univ., 99164-7204. Phone: (509) 335-2208. Fax: (509) 335-3772. Licensee: Washington State University Board of Regents. Format: Rock, jazz, div. News staff: one; News: 5 hrs wkly. Target aud: 18-49; college students. Spec prog: Jazz 12 hrs, Black 12 hrs, folk 4 hrs, Sp 3 hrs, new mus 10 hrs, environmental protection 2 hrs wkly. ♦Mike Guay, gen mgr; Lori Stewart, prom dir; Jackie Kaiser, progmg dir.

KZZL-FM— Nov 1991: 99.5 mhz; 81.4 kw. 1,059 ft TL: N46 40 52 W116 58 19. Stereo. Hrs opn: 24 1114 N. Almon St., Moscow, 83843. Phone: (509) 397-3441. Fax: (509) 397-4752. E-mail: kzzlkrow@stjohncable.com Web Site: www.palousecountry.com. Licensee: Inland Northwest Broadcasting LLC. (group owner). (acq 6-28-2005; grpsl). Population served: 500,000 Law Firm: Dow, Lohnes & Albertson. Format: Country. News staff: one; News: 2 hrs wkly. Target aud: 25 plus. ♦Gary Cummings, gen mgr; Ben Bonfield, gen sls mgr; Ryan Chambers, progmg dir; Steve Franco, news dir & chief of engrg.

Puyallup

KSUH(AM)— Dec 1, 1951: 1450 khz; 1 kw-U. TL: N47 10 41 W122 16 24. (CP: 1440 khz; 5 kw-D, 2 kw-N, DA-2). Hrs opn: 24 807 S. 336 St., Federal Way, 98003. Phone: (253) 815-1212. Fax: (253) 815-1913. Web Site: www.radiohankook.com. Licensee: Jean J. Suh. (acq 4-4-97; $350,000). Population served: 84,000 Format: Korean, var/div. News staff: 2. Target aud: 25-65; working folks. ♦Doris Haan, gen mgr; Nancy Haan, stn mgr.

Quincy

KWNC(AM)— Sept 10, 1957: 1370 khz; 1 kw-D, 40 w-N. TL: N47 16 15 W119 51 13. Hrs open: Box159, Wenatchee, 98807. Phone: (509) 787-4461. Fax: (509) 664-6799. E-mail: news@kwnc.com Licensee: Wescoast Broadcasting Co. Inc. (acq 3-8-99). Population served: 11,370 Format: News, farm. News staff: 3; News: 168 hrs wkly. Target aud: 35 plus; farm-oriented. Spec prog: Farm 5 hrs wkly. ♦Jim Wallace Jr., pres & gen mgr; Debbie Capestrini, opns dir & progmg dir; Steve Hair, news dir.

KWWW-FM— Aug 29, 1985: 96.7 mhz; 440 w. 1,079 ft TL: N47 19 13 W199 48 00. Stereo. Hrs opn: 24 231 N. Wenatchee Ave., Wenatchee, 98801. Phone: (509) 665-6565. Fax: (509) 663-1150. Web Site: www.kw3.com. Licensee: CCR-Wenatchee IV LLC. Group owner: Fisher Broadcasting Company. (acq 10-31-2006; grpsl). Population served: 100,000 Natl. Rep: McGavren Guild. Law Firm: Shaw Pittman. Format: CHR, 80s & 90s. Target aud: 18-49. ♦Jim Senst, gen mgr & adv mgr; Leona Frank, sls dir; Dave Herald, gen sls mgr & natl sls mgr; Dale Roth, prom mgr & progmg dir; Dave Bernstein, news dir; Lisa Rodriguez, pub affrs dir; Manuel Garcia, chief of engrg; Jennifer Busboug, rsch dir; Jose Luis High, spanish dir.

KZML(FM)— October 1998: 95.9 mhz; 2.51 kw. Ant 1,046 ft TL: N47 19 13 W119 47 59. Hrs open: 24
Rebroadcasts KZTA(FM) Naches 100%.
Box 2888, Yakima, 98907. Secondary address: 706 Butterfield Rd., Yakima 98901. Phone: (509) 457-1000. Fax: (509) 452-0541. E-mail: zorro@radiozorro.com Web Site: www.bustosmedia.com. Licensee: Bustos Media of Eastern Washington License LLC. (group owner; (acq 11-18-2004; grpsl). Rgnl rep: Tacher Format: Mexican regional. News staff: one; News: 3 hrs wkly. Target aud: 18-35; Hispanic. ♦Bob Berry, gen mgr & gen sls mgr; Keith Teske, opns mgr & chief of engrg; Martin Ortiz, progmg dir; Judith McInnis, traf mgr.

Raymond

KFMY(FM)— Oct 26, 1984: 97.7 mhz; 44 kw. Ant 1,322 ft TL: N46 54 05 W123 25 07. Stereo. Hrs opn: 24 Box 7489, Olympia, 98507. Secondary address: 1803 State Ave. N.E., Olympia 98506. Phone: (360) 918-9000. Fax: (360) 704-3146. Web Site: www.977theeagle.com. Licensee: South Sound Broadcasting LLC (acq 3-1-2003; $2.28 million). Population served: 85,000 Natl. Network: ABC. Natl. Rep: Tacher. Format: Classic hits. News: 10 hrs wkly. Target aud: 25-54; adults. ♦Bill Bradley, gen mgr; Ed Bruno, sls VP; Craig Sullivan, progmg dir; Jeff Turnbow, opns mgr & news dir.

KJET(FM)— July 1999: 105.7 mhz; 58 kw. Ant 518 ft TL: N46 56 30 W123 47 07. Stereo. Hrs opn: 24 Box 1198, Aberdeen, 98520. Secondary address: 1520 Simpson Ave., Aberdeen 98520. Phone: (360) 538-3000. Fax: (360) 532-1456. E-mail: info@jodesha.com Web Site: www.jodesha.com. Licensee: Jodesha Broadcasting Inc. (group owner) Population served: 85,000 Natl. Network: ABC. Natl. Rep: Tacher. Law Firm: David Tillotson. Wire Svc: AP Format: Adult top-40. News staff: one; News: 7 hrs wkly. Target aud: 18-49. ♦William J. Wolfenbarger, pres & gen mgr; Gabrielle Jordan, opns mgr.

Renton

KRIZ(AM)— Feb 2, 1982: 1420 khz; 1 kw-D, 500 w-N, DA-2. TL: N47 26 25 W122 12 09. Hrs open: 24 2600 S. Jackson St., Seattle, 98144. Secondary address: Box 22462, Seattle 98122-0462. Phone: (206) 323-3070. Fax: (206) 322-6518. E-mail: ztwins@aol.com Web Site: www.ztwins.com. Licensee: KRIZ Broadcasting Inc. (acq 2-84; $400,000; FTR: 3-5-84). Population served: 3,000,000 Natl. Network: American Urban. Format: Black oldies, blues. Target aud: 18 plus. Spec prog: Relg 18 hrs wkly. ♦Christopher H. Bennett, pres & stn mgr; Gloria V. Bennett, VP; Frank P. Barrow, chief of opns; Frank Barrow, progmg dir.

KYIZ(AM)— 1998: 1620 khz; 10 kw-D, 1 kw-N. TL: N47 26 25 W122 12 09. Hrs open: 2600 S. Jackson St., Seattle, 98144. Secondary address: Box 22462, Seattle 98144. Phone: (206) 323-3070. Fax: (206) 322-6518. E-mail: ztwins@aol.com Web Site: www.ztwins.com. Licensee: KRIZ Broadcasting Inc. Format: Urban contemp, rhythm & blues. ♦Christopher H. Bennett, pres & gen mgr; Gloria V. Bennett, VP, gen mgr & stn mgr; Frank P. Barrow, chief of opns; Frank Barrow, progmg dir; Priscilla Hailey, news dir.

Richland

KALE(AM)—Licensed to Richland. See Richland-Pasco-Kennewick

KEGX(FM)—Licensed to Richland. See Richland-Pasco-Kennewick

KFAE-FM—Licensed to Richland. See Richland-Pasco-Kennewick

KIOK(FM)—Licensed to Richland. See Richland-Pasco-Kennewick

KORD-FM—Licensed to Richland. See Richland-Pasco-Kennewick

Richland-Pasco-Kennewick

KALE(AM)—Richland, Apr 1, 1950: 960 khz; 5 kw-D, 1 kw-N, DA-N. TL: N46 14 34 W119 10 48. Hrs open: 24 830 N. Columbia Center Blvd., Suite B-2, Kennewick, 99336. Phone: (509) 783-0783. Fax: (509) 735-8627. Web Site: www.am960.com. Licensee: New Northwest Broadcasters LLC (group owner; acq 12-10-99; grpsl). Population served: 157,000 Natl. Rep: D & R Radio. Law Firm: Haley, Bader & Potts. Format: ESPN sports. News staff: one. Target aud: 25 plus; general. ♦Don Morin, gen mgr.

KIOK(FM)—Co-owned with KALE(AM). Oct 3, 1978: 94.9 mhz; 100 kw. 1,250 ft TL: N46 05 47 W119 11 36. Stereo. Web Site: www.thundercountry949.com.250,000 Format: Country.

KEGX(FM)—Listing follows KTCR(AM).

***KFAE-FM**—Richland, July 1982: 89.1 mhz; 100 kw. 1,148 ft TL: N46 05 43 W119 11 41. Stereo. Hrs opn: 24
Rebroadcasts KRFA-FM Moscow, ID 100%.
Box 642530, Murrow Communications Ctr., Washington State Univ., Rm 382, Pullman, 99164-2530. Secondary address: Washington State Univ. at Tri-Cities, 100 Sprout Rd., Richland 99164-2530. Phone: (509) 335-6500. Fax: (509) 335-3772. E-mail: nwpr@wsu.edu Web Site: www.nwpr.org. Licensee: Washington State University. Population served: 243,000 Natl. Network: PRI, NPR. Law Firm: Dow, Lohnes & Albertson. Format: Class, news. News staff: one; News: 37 hrs wkly. Target aud: General. Spec prog: Folk, jazz 15 hrs wkly. ♦Dennis Haarsager, gen mgr; Roger Johnson, stn mgr; Scott Weatherly, opns dir; Sarah McDaniel, dev dir; Mary Hawkins, progmg dir. Co-owned TV: *KTNW(TV) affil.

KFLD(AM)—Pasco, July 28, 1956: 870 khz; 10 kw-U. TL: N46 13 41 W119 07 32. Hrs open: 24 Box 2485, Pasco, 99301. Secondary address: 2621 W.A. St., Pasco 99301. Phone: (509) 547-9791. Fax: (509) 547-8509. Web Site: www.sportsradio870.com. Licensee: Capstar TX L.P. Group owner: Clear Channel Communications Inc. (acq 2-15-01; grpsl). Population served: 285,400 Rgnl rep: Art Moore. Format: Sports. News staff: one; News: 3 hrs wkly. Target aud: 18-64. ♦Eric Van Winkle, gen mgr; Grant Linnen, gen sls mgr; Curt Cartier, progmg dir; Chuck Ince, chief of engrg.

KORD-FM—Co-owned with KFLD(AM). Oct 15, 1965: 102.7 mhz; 100 kw. 1,100 ft TL: N46 05 47 W119 11 36. Stereo. Web Site: www.1027kord.com. Format: Country. Target aud: 25-54. ♦Paul Drake, progmg dir.

KGDN(FM)—Pasco, February 1992: 101.3 mhz; 2.75 kw. 1,000 ft TL: N46 05 47 W119 11 36. Hrs open: 24 Box 3258, Tri Cities, 99302. Secondary address: 830 N. Columbia Center Blvd., Suite B3, Pasco 99336. Phone: (509) 783-8600. Fax: (509) 448-3811. E-mail: kgdn@kgdn.com Web Site: www.kgdn.com. Licensee: West Pasco Fine Arts Radio. Law Firm: Pepper & Corazzini. Format: Christian. Target aud: 35 plus. ♦Thomas W. Read, gen mgr; Bill Glenn, stn mgr, opns dir & engrg dir; Melinda Read, sls dir; Joseph Spinelli, progmg dir.

KONA(AM)—Kennewick, January 1948: 610 khz; 5 kw-U, DA-2. TL: N46 13 41 W119 04 07. Stereo. Hrs opn: 24 2823 W. Lewis, Pasco, 99301. Secondary address: Box 2623, Tri Cities 99302. Phone: (509) 547-1618. Fax: (509) 546-2678. E-mail: kona@konaradio.com Web Site: www.konaradio.com. Licensee: CCR-Tri Cities IV LLC. (group owner; (acq 12-19-2003; grpsl). Population served: 200,000 Law Firm: Pepper & Corazzini. Format: News/talk. News staff: 2; News: 25 hrs wkly. Target aud: 25-54. Spec prog: Farm 3 hrs, sports 8 hrs wkly. ♦Dennis W. Goodman, gen mgr; Scott Smith, gen sls mgr; Todd Nevard, prom dir, prom dir & progmg dir; Dennis Shannon, news dir; Art Blum, chief of engrg; Bob Martin, disc jockey; Michael McDonnal, disc jockey; Rusty Faust, disc jockey.

KONA-FM— Aug 1, 1969: 105.3 mhz; 100 kw. 1,180 ft TL: N46 05 48 W119 11 36. Stereo. 24 Web Site: www.konaradio.com. Rgnl rep: Allied Radio Partners. Format: Light adult contemp. News staff: 2; News: 3 hrs wkly. ♦Dennis W. Goodman, COO, exec VP & stn mgr; Scott Smith, sls VP & sls dir; Todd Nevard, progmg VP & disc jockey; Linda Howard, traf mgr; Dennis Shannon, local news ed & news rptr; Willy Contretas, spanish dir; Mike McDonnal, sports cmtr; Bob Martin, disc jockey; Michael McDonnal, disc jockey; Rusty Faust, disc jockey.

KTCR(AM)—Kennewick, August 1945: 1340 khz; 1 kw-U. TL: N46 13 16 W119 11 20. Hrs open: 830 N. Columbia Ctr. Blvd., Suite B-2, Kennewick, 99336. Phone: (509) 783-0783. Fax: (509) 735-8627. Web Site: www.ktcr.com. Licensee: New Northwest Broadcasters LLC (group owner; acq 12-10-99; grpsl). Population served: 125,000 Natl. Rep: Christal. Format: News/talk. Target aud: 25-64. Spec prog: Sports. ♦Don Morin, gen mgr.

KEGX(FM)—Co-owned with KTCR(AM). June 10, 1992: 106.5 mhz; 100 kw. Ant 1,392 ft TL: N46 05 58 W119 07 40. Stereo. Web Site: www.kegx.com.250,000 Format: Classic rock. Target aud: 25-54.

***KTCV(FM)**—Kennewick, Dec 10, 1984: 88.1 mhz; 320 w. 92 ft TL: N46 13 05 W119 12 17. Stereo. Hrs opn: 12 5929 W. Metaline, Kennewick, 99336. Phone: (509) 734-3621. Phone: (509) 734-3622. Fax: (509) 734-3609. E-mail: dailed@ksd.org Web Site: www.ktcv.net. Licensee: Kennewick School District No. 17. Format: Alternative rock. ♦Ed Dailey, gen mgr.

Rock Island

KAAP(FM)— Sept 19, 1990: 99.5 mhz; 5 kw. 167 ft TL: N47 22 52 W120 17 15. (CP: 5.3 kw, ant -82 ft.). Stereo. Hrs opn: 24 231 N. Wenatchee Ave., Wenatchee, 98801. Phone: (509) 665-6565. Fax: (509) 663-1150. Web Site: www.applefm.com. Licensee: CCR-Wenatchee IV LLC. Group owner: Fisher Broadcasting Company (acq 10-31-2006; grpsl). Population served: 100,000 Natl. Rep: McGavren Guild. Law Firm: Shaw Pittman. Format: Adult contemp. News staff: one. Target aud: 25-54. ♦Jim Senst, gen mgr & mktg dir; Leona Frank, gen sls mgr; Todd Johnson, prom mgr; Joe Bowers, engrg dir; Manuel Garcia, chief of engrg; Jennifer Busboug, rsch dir; Lisa Rodriguez, traf mgr; Jose Luis High, spanish dir.

Roy

***KWFJ(FM)**— September 1995: 89.7 mhz; 1 kw. 98 ft TL: N46 57 59 W122 32 56. Hrs open: Box 401, 98580. Secondary address: 9006 320 St. S. 98580. Phone: (206) 843-1692. E-mail: cbcroy@cbcroy.org Licensee: Calvary Baptist Church. Natl. Network: Bible Bcstg Net. Format: Christian. ♦Bernie Brill, gen mgr.

Royal City

KRCW(FM)— 1995: 96.3 mhz; 19.5 kw. 790 ft TL: N46 45 55 W119 16 51. Stereo. Hrs opn: 24 508 W. Lewis St., Pasco, CA, 99301. Phone: (509) 545-0700. Fax: (509) 543-4100. Web Site: www.campesina.com. Licensee: Farmworker Educational Radio Network. Population served: 250,000 Law Firm: Borsari & Paxson. Format: Rgnl Mexican. Target aud: 25-54; Hispanic market. ♦Anthony Chavez, pres & gen mgr; Paul Chavez, VP; Armando Ameta, stn mgr; Pepe Escavilla, opns dir & progmg dir; Cesar Chavez Jr., news dir; David Whitehead, chief of engrg.

KWDR(FM)—Not on air, target date: unknown: 93.5 mhz; 210 w. Ant 1,666 ft TL: N46 48 25 W119 33 20. Hrs open: 5331 Mt. Alifan Dr., San Diego, CA, 92111. Phone: (858) 277-4991. Fax: (858) 277-1365. Web Site: www.horizonsd.org/radio.asp. Licensee: Horizon Christian Fellowship. ♦Mike MacIntosh, pres.

Seattle

***KBLE(AM)**— 1948: 1050 khz; 5 kw-D, 440 w-N. TL: N47 33 41 W122 21 34. Hrs open: 6 AM-midnight Box 2482, Kirkland, 98083. Phone: (425) 867-2340. E-mail: info@sacredheartradio.org Web Site: www.kble.com. Licensee: Sacred Heart Radio Inc. (acq 1-11-01). Population served: 2,068,900 Law Firm: Pepper & Corazzini. Format: Relg. ♦Ron Belter, gen mgr & opns mgr.

KBSG-FM—See Tacoma

KCMS(FM)—See Edmonds

KDOW(AM)— Mar 31, 2003: 1680 khz; 10 kw-D, 1 kw-N. TL: N47 39 20 W122 31 05. Hrs open: 24 2815 2nd Ave., Suite 550, 98121. Phone: (206) 443-8200. Fax: (206) 777-1133. Licensee: Inspiration Media Inc. Group owner: Salem Communications Corp. Format: Sp contemp. ♦Joe Gonzalez, gen mgr; Charles Olmstead, opns dir & progmg dir.

***KEXP-FM**— 1972: 90.3 mhz; 3.3 kw. Ant 692 ft TL: N47 36 58 W122 18 28. Stereo. Hrs opn: 24 113 Dexter Ave. N., 98109. Phone: (206) 520-KEXP. Fax: (206) 520-5899. Web Site: www.kexp.org. Licensee: Regents of University of Washington. Population served: 2,000,000 Natl. Network: NPR. Law Firm: Dow,Lohnes & Albertson. Format: Progsv, div, alternative. Target aud: 18-44; educated, culturally interested, active outdoors, prof/mngr/tech positions. ♦Tom Mara, gen mgr; Jack Walters, opns mgr; Courtney Miller, mktg dir; Kevin Cole, progmg dir; Mike McCormick, pub affrs dir; Jamie Alls, chief of engrg.

KGNW(AM)—See Burien-Seattle

KHHO(AM)—Tacoma, August 1942: 850 khz; 10 kw-D, 1 kw-N, DA-2. TL: N47 13 56 W122 23 22. Stereo. Hrs opn: 24 351 Elliott Ave. W., Suite 300, 98119. Phone: (206) 494-2000. Fax: (206) 286-2376. Web Site: khho-am.clearchannel.com. Licensee: Ackerley Broadcasting Operations LLC. Group owner: Clear Channel Communications Inc. (acq 6-14-2002; grpsl). Population served: 800,000 Format: Sports. Target aud: 25-55. ♦Michele Grosenick, pres & gen mgr; Sean Shannon, sls dir & gen sls mgr; Rich Moore, progmg dir; Doug Irwin, chief of engrg; Amy Spino, traf mgr.

KING-FM— 1947: 98.1 mhz; 58 kw. 2,342 ft TL: N47 30 55 W122 58 29. Stereo. Hrs opn: 24 10 Harrison St., Suite 100, 98109. Phone: (206) 691-2981. Fax: (206) 691-2982. E-mail: web@king.org Web Site: www.king.org. Licensee: Classic Radio Inc. (acq 2-92; $9.75 million with co-located AM). Natl. Rep: Katz Radio. Format: Class music. News: 2 hrs wkly. ♦Jennifer Ridewood, gen mgr; Shawna Keen, prom mgr; Bob Goldfarb, progmg dir; Buzz Anderson, chief of engrg.

KIRO(AM)— 1927: 710 khz; 50 kw-U, DA-N. TL: N47 23 55 W122 26 01. Hrs open: 24 1820 Eastlake Ave. E., 98102-3711. Phone: (206) 726-7000. Fax: (206) 726-5446. Web Site: www.710kiro.com. Licensee: Entercom Seattle License LLC. Group owner: Entercom Communications Corp. (acq 3-6-97; grpsl). Natl. Network: CBS. Format: News/talk, sports. News staff: 20. Target aud: 25-54. ♦David Pridemore, gen mgr; Dennis McCormick, gen sls mgr; Tom Lendening, progmg dir.

KISW(FM)— 1950: 99.9 mhz; 100 kw. 1,150 ft TL: N47 32 41 W122 06 28. Stereo. Hrs opn: 24 1100 Olive Way, Suite 1650, 98101. Phone: (206) 285-7625. Fax: (206) 215-9355. E-mail: rcastle@entercom.com Web Site: www.kisw.com. Licensee: Entercom Seattle License LLC. Group owner: Entercom Communications Corp. (acq 1996). Population served: 2,844,400 Natl. Rep: D & R Radio. Format: Rock/AOR. Target aud: 18-49; men. ♦David Field, pres; Amy Griesheimer, VP & gen mgr; Ron Steinman, gen sls mgr; Dave Richards, progmg dir; Dwight Small, engrg dir & chief of engrg; Joyce Jinka, traf mgr.

KIXI(AM)—Mercer Island-Seattle, 1947: 880 khz; 50 kw-D, 10 kw-N, DA-2. TL: N47 34 59 W122 10 52. Hrs open: 3650 131st Ave. S.E., Suite 550, Bellevue, 98006. Phone: (425) 653-9462. Fax: (425) 653-1088. E-mail: bobb@kixi.com Web Site: www.kixi.com. Licensee: Bellevue Radio Inc. Group owner: Sandusky Radio (acq 11-15-91; $3.5 million; FTR: 12-3-91). Population served: 530,831 Natl. Rep: Christal. Format: Adult standards. Target aud: 35 plus; mature active adults. ♦Marc S. Kaye, VP & gen mgr; Lois Mares, gen sls mgr & rgnl sls mgr; Julie Judge, natl sls mgr; Bob Brooks, progmg dir.

KJAQ(FM)— 1959: 96.5 mhz; 100 kw. 1,223 ft TL: N47 32 39 W122 06 32. Stereo. Hrs opn: 24 1000 Dexter Ave. N., Suite 100, 98109. Phone: (206) 805-1100. Fax: (206) 805-0920. Web Site: www.965thepoint.com. Licensee: Infinity Radio Holdings Inc. Group owner: Infinity Broadcasting Corp. (acq 11-13-98; grpsl). Target aud: 25-54. ♦Lisa McDonald, gen mgr; Nils Olsen, gen sls mgr; Jim Trapp, progmg dir; Tom McGinley, chief of engrg.

KJR(AM)— 1921: 950 khz; 5 kw-U, DA-N. TL: N47 34 57 W122 21 46. Hrs open: 351 Elliot Ave. W., Suite 300, 98119. Phone: (206) 285-2295. Fax: (206) 286-2376. Web Site: www.kjram.com. Licensee: Ackerley Broadcasting Operations LLC. Group owner: Clear Channel Communications Inc. (acq 6-14-2002; grpsl). Population served: 2,800,000 Natl. Rep: D & R Radio. Format: Sports. Target aud: 25-54. ♦Michelle Grosnick, VP & gen mgr; Sean Shannon, sls dir & adv mgr; Gus Swanson, mktg dir; Gina Gray, prom dir; Rich Moore, progmg dir; Tom Benton, pub affrs dir; Doug Irwin, chief of engrg; Amy Spino, traf mgr.

KJR-FM— May 25, 1960: 95.7 mhz; 100 kw. 1,150 ft TL: N47 32 41 W122 06 28. Stereo. Phone: (206) 494-2000. Web Site: www.957kjrfm.com. Format: Classic hits. Target aud: 30-44. ♦Rick Carter, gen sls mgr; Valerie Koch, prom dir; Bob Case, progmg dir & disc jockey; Stephen Kilbreath, mus dir & news dir; Amy Spino, traf mgr; Heidi May, disc jockey; Pat Cashman, disc jockey; Ric Hansen, disc jockey.

KKDZ(AM)— May 15 1993: 1250 khz; 5 kw-U, DA-N. TL: N47 33 41 W122 21 34. Hrs opn: 24 200 First Ave. W., Suite 104, 98119. Phone: (206) 281-5300. Fax: (206) 281-8881. Web Site: www.radiodisney.com. Licensee: WMAL Inc. Group owner: ABC Inc. (acq 1-21-98; $1.2 million). Population served: 530,831 Format: Children's. News staff: 5. Target aud: Kids 6-14; Moms 25-49. ♦Bob Nordberg, gen mgr; Laura Dunham, prom mgr.

KKNW(AM)—Listing follows KWJZ(FM).

KKOL(AM)— 1922: 1300 khz; 50 kw-D, 47 kw-N, DA-2. TL: N47 14 56 W122 24 18. Hrs open: 24 2815 2nd Ave., Suite 550, 98121. Phone: (206) 443-8200. Fax: (206) 777-1133. Licensee: Inspiration Media Inc. Group owner: Salem Communications Corp. (acq 4-17-97; $2 million). Population served: 530,831 Natl. Rep: Salem. Format: News/talk. Target aud: 35 plus; men & women. ♦Joe Gonzalez, gen mgr; Charles Olmstead, opns dir, progmg dir & progmg mgr; Doug Rice, gen sls mgr; Monty Passmore, chief of engrg.

KKWF(FM)— 1946: 100.7 mhz; 57 kw horiz, 52 kw vert. Ant 2,342 ft TL: N47 30 14 W121 58 29. Stereo. Hrs opn: 24 1100 Olive Way, Suite 1650, 98101. Phone: (206) 285-7625. Fax: (206) 381-0997. Licensee: Entercom Seattle License LLC. Group owner: Entercom Communications Corp. Format: Country. News: 15 hrs wkly. ♦Steve Oshin, VP & gen mgr; Melissa Forrest, opns mgr; Ron Steinman, gen sls mgr; Dave Richards, progmg dir; Dwight Small, chief of engrg.

KLFE(AM)— Sept 10, 1956: 1590 khz; 5 kw-U, DA-N. TL: N47 39 19 W122 31 06. Hrs open: 24 2815 2nd Ave., Suite 550, 98121. Phone: (206) 443-8200. Fax: (206) 777-1133. Licensee: Inspiration Media Inc. Group owner: Salem Communications Corp. (acq 1994; $500,000). Population served: 530,831 Natl. Rep: Salem. Format: Christian talk. Target aud: 25-54. Spec prog: Ethiopian 2 hrs, Russian 12 hrs, Saimoan 4 hrs wkly. ♦Joe Gonzalez, gen mgr; Charles A. Olmstead, opns mgr.

KMPS-FM— July 8, 1961: 94.1 mhz; 57 kw. 2,342 ft TL: N47 30 14 W121 58 29. Stereo. Hrs opn: Box 24888, 98124. Secondary address: 1000 Dexter Ave., N., Suite 100 98109. Phone: (206) 805-0941. Fax: (206) 805-0911. E-mail: email@kmps.com Web Site: www.kmps.com. Licensee: Infinity Radio Holdings Inc. Group owner: Infinity Broadcasting Corp. (acq 11-13-98; grpsl). Format: Country. News: one. Target aud: General. ♦Dave McDonald, gen mgr; Becky Brenner, opns mgr & progmg dir; Rod Krebs, gen sls mgr; Don Riggs, news dir; Tom McGinley, chief of engrg.

KMTT(FM)—Tacoma, June 2, 1958: 103.7 mhz; 58 kw. 2,343 ft TL: N47 30 14 W121 58 29. Stereo. Hrs opn: 24 1100 Olive Way, Suite 1650, 98101-1827. Phone: (206) 233-1037. Fax: (206) 233-8979. E-mail: studio@kmtt.com Web Site: www.kmtt.com. Licensee: Entercom Seattle License L.L.C. Group owner: Entercom Communications Corp. (acq 6-73; with co-located AM). Natl. Rep: D & R Radio. Format: AAA. News staff: one; News: 3 hrs wkly. Target aud: 25-49. ♦David Field, pres; Steve Oshin, gen mgr; Traci Gregory, gen sls mgr; Jennifer Orr, prom dir; Shaun Stewart, progmg dir; Mike West, news dir; Dwight Smalls, chief of engrg; Joyce Jinka, traf mgr.

KNDD(FM)— Mar 9, 1985: 107.7 mhz; 100 kw. 1,194 ft TL: N47 32 35 W122 06 25. (CP: 57.3 kw, ant 2,342 ft.). Stereo. Hrs opn: 24 1100 Olive Way, Suite 1650, 98101. Phone: (206) 622-3251. Fax: (206) 682-8349. Web Site: www.1077theend.com. Licensee: Entercom Seattle License L.L.C. Group owner: Entercom Communications Corp. (acq 1996). Natl. Rep: D & R Radio. Format: Alternative. News staff: one. Target aud: 18-34; well educated active adults. ♦Amy Griesheimer, VP & gen mgr; Jennifer Wisbey, gen sls mgr; Phil Manning, progmg dir; Dwight Smalls, chief of engrg.

***KNHC(FM)**— Jan 25, 1971: 89.5 mhz; 8.5 kw. Ant 1,220 ft TL: N47 32 35 W122 06 25. Stereo. Hrs opn: 24 10750 30th Ave. N.E., Suite 219, 98125. Phone: (206) 252-3800. Phone: (206) 421-8989. Fax: (206) 252-3805. E-mail: info@c895worldwide.com Web Site: www.c895worldwide.com. Licensee: Seattle Public Schools. Population served: 3,500,000 Law Firm: Wilmer, Cutler & Pickering. Wire Svc: AP Format: CHR, educ. News: 9 hrs wkly. Target aud: 18-34; male & female. Spec prog: Black, gospel 6 hrs, gothic/industrial 6 hrs wkly. ♦Gregg Neilson, gen mgr; Richard Dalton, opns mgr & gen sls mgr; Jon McDaniel, progmg dir.

KOMO(AM)— 1926: 1000 khz; 50 kw-U, DA-N. TL: N47 27 54 W122 26 27. Hrs open: 24 140 4th Ave. N., 98109. Phone: (206) 404-4000. Fax: (206) 404-3646. E-mail: comments@KOMO1000news.com Web Site: www.KOMO1000news.com. Licensee: Fisher Broadcasting - Seattle Radio L.L.C. Group owner: Fisher Broadcasting Company Population served: 3,204,000 Natl. Network: ABC. Law Firm: Pillsbury, Winthrop, Shaw & Pittman. Wire Svc: AP Format: News. News staff: 50; News: 168 hrs wkly. Target aud: 25-54. ♦Colleen Brown, CEO; Larry Roberts, gen mgr; Joe Heslet, gen sls mgr; Gary Greenberg, natl sls mgr; Charles Gouge, rgnl sls mgr; Jen Pivak, prom dir; Dennis Kelly, progmg dir & news dir; Brian Calvert, news dir; John Barrett, chief of engrg; Julie Ross, traf mgr. Co-owned TV: KOMO-TV affil.

KPLZ(FM)— Sept 1, 1959: 101.5 mhz; 99 kw. 1,263 ft TL: N47 32 40 W122 06 26. Stereo. Hrs opn: 24 Fisher Plaza, 140 Fourth Ave. N., Suite 340, 98109. Phone: (206) 404-4000. Fax: (206) 404-3644. E-mail: lroberts@fisherradio.com Web Site: www.star1015.com. Licensee: Fisher Broadcasting - Seattle Radio L.L.C. Group owner: Fisher Broadcasting Company (acq 5-5-94; with co-located AM). Population served: 3,700,000 Natl. Rep: Eastman Radio. Law Firm: Pillsbury, Winthrop. Wire Svc: AP Format: Adult contemp. News staff: 2; News: one hr wkly. Target aud: 25-54; women. ♦Larry Roberts, gen mgr; Bryce Phillippy, gen sls mgr & natl sls mgr; Gary Greenberg, natl sls mgr; Jennifer Pirak, prom dir; Kent Phillips, progmg dir & disc jockey; John Barrett, chief of engrg; Lindsey Fields, traf mgr & disc jockey.

KVI(AM)—Co-owned with KPLZ(FM). 1926: 570 khz; 5 kw-U. TL: N47 25 19 W122 25 44.24 Phone: (206) 404-3050. Fax: (206) 404-3650. E-mail: dkelly@fisherradio.com Web Site: www.570kvi.com.3,700,000 Natl. Network: Fox News Radio. Natl. Rep: Eastman Radio. Law Firm: Pillsbury, Winthrop & Shaw Pittman. Format: Talk. News staff: 2; News: 20 hrs wkly. Target aud: 25-54. ♦Joe Heslet, gen sls mgr; Gary Greenberg, natl sls mgr; Jen Pirak, prom mgr; Dennis Kelley, progmg dir; Anna Johnson, traf mgr. Co-owned TV: KOMO-TV affil

KPTK(AM)— 1927: 1090 khz; 50 kw-U, DA-2. TL: N47 23 38 W122 25 25. Hrs open: 24 1000 Dexter Ave. N., Suite 100, 98109. Phone: (206) 805-1090. Fax: (206) 805-0911. Web Site: www.am1090seattle.com. Licensee: Infinity Radio Holdings Inc. Group owner: Infinity Broadcasting Corp. (acq 11-13-98; grpsl). Population served: 530,831 Format: Progressive talk. ♦Dave McDonald, gen mgr; Jim Trapp, progmg dir; Tom McGinley, chief of engrg; Missy Wise, traf mgr.

KQMV(FM)—See Bellevue

KTTH(AM)— 1925: 770 khz; 50 kw-D, 5 kw-N, DA-2. TL: N47 23 38 W122 25 25. Hrs open: 24 1820 Eastlake Ave. E., 98102-3711. Phone: (206) 726-7000. Fax: (206) 726-5446. Licensee: Entercom Seattle License LLC. Group owner: Entercom Communications Corp. (acq 3-6-97; grpsl). Population served: 400,000 Law Firm: Fletcher, Heald & Hildreth. Format: News. News staff: 6. Target aud: 25-54; adults. ♦David Pridemore, gen mgr; Ken Berry, stn mgr.

KUBE(FM)— May 6, 1964: 93.3 mhz; 100 kw. 1,291 ft TL: N47 32 39 W122 06 29. Stereo. Hrs opn: 24 351 Elliott Ave. W. #300, 98119. Phone: (206) 285-2295. Fax: (206) 286-2376. Web Site: www.kube93.com. Licensee: Ackerley Broadcasting Operations LLC. Group owner: Clear Channel Communications Inc. (acq 6-14-2002; grpsl). Population served: 1,750,000 Format: Rythmic dance, CHR. ♦Michele Grosenick, pres & gen mgr; Shellie Hart, opns mgr; Sean Shannon, gen sls mgr; Eric Powers, progmg dir; Doug Irwin, chief of engrg; Amy Spino, traf mgr.

***KUOW-FM**— Jan 16, 1952: 94.9 mhz; 100 kw. Ant 730 ft TL: N47 36 58 W122 18 28. Stereo. Hrs opn: 24 4518 University Way N.E., Suite 310, 98105. Phone: (206) 543-2710. Fax: (206) 616-9125. E-mail: letters@kuow.org Web Site: www.kuow.org. Licensee: University of Washington. Population served: 2,500,000 Natl. Network: NPR, PRI. Law Firm: Ernest Sanchez. Format: News, info. News staff: 15; News: 60 hrs wkly. Target aud: 25-54; highly educated, influential,decision makers. Spec prog: Sp 2 hrs, jazz 5 hrs wkly. ♦Wayne Roth, gen mgr; Dane Johnson, opns dir; Marcia Scholl, dev dir; Jeff Hansen, progmg dir; Guy Nelson, news dir; Terry Denbrook, chief of engrg.

KWJZ(FM)— Nov 1, 1954: 98.9 mhz; 100 kw. 1,110 ft TL: N47 32 41 W122 06 28. (CP: 58 kw, ant 2,342 ft. TL: N47 30 14 W121 58 29). Stereo. Hrs opn: 24 3650 131st Ave. S.E., Suite 550, Bellevue, 98006. Phone: (425) 373-5536. Fax: (425) 653-1133. Web Site: www.kwjz.com. Licensee: Orca Radio Inc. Group owner: Sandusky Radio (acq 1996; $26 million with co-located AM). Population served: 3,084,700 Natl. Network: Westwood One. Natl. Rep: Christal. Law Firm: Wiley, Rein & Fielding. Format: Smooth jazz, new adult contemp. News staff: 2; News: 9 hrs wkly. Target aud: 25-54; younger active, mid to upper income adults. ♦Marc Kaye, gen mgr; Susan Hoffman, sls dir; Ann Marie Mulholland, gen sls mgr; Cindy Gilsdorf, mktg dir & prom dir; Carol Handley, progmg dir; Dianna Rose, mus dir; George Bisso, chief of engrg; Alan Hines, traf mgr.

KKNW(AM)—Co-owned with KWJZ(FM). 1926: 1150 khz; 10 kw-U, DA-N. TL: N47 35 11 W122 11 11.24 Fax: (425) 373-5507. Web Site: www.newschannel1150.com.530,831 Format: News/talk. News staff: one; News: 10 hrs wkly. Target aud: 35-64; active, well educated adults with middle to upper income. Spec prog: Loc sports 15 hrs, Russian 5 hrs wkly. ♦Eric Burris, opns mgr & progmg dir; Erik Krema, stn mgr & gen sls mgr; Alan Hines, traf mgr.

KZOK-FM— December 1964: 102.5 mhz; 100 kw. 1,170 ft TL: N47 32 35 W122 06 25. (CP: 58 kw, ant 2,342 ft.). Stereo. Hrs opn: 24 1000 Dexter Ave. N., Suite 100, 98109. Phone: (206) 805-1100. Fax: (206) 441-1411. Web Site: www.kzok.com. Licensee: Infinity Radio Holdings Inc. Group owner: Infinity Broadcasting Corp. (acq 11-13-98; grpsl). Population served: 2,000,000 Format: Classic rock. Target aud: 25-49; adults with a primary, men. ♦Carey Curelop, gen mgr & opns mgr; Dave McDonald, pres & gen mgr.

Selah

KBBO(AM)—Licensed to Selah. See Yakima

Shelton

KMAS(AM)— Sept 21, 1962: 1030 khz; 10 kw-D, 1 kw-N. TL: N47 13 17 W123 04 46. Hrs open: 24 Box 760, 210 W. Cota St., 98584. Phone: (360) 426-1030. Fax: (360) 427-5268. E-mail: kmas@kmas.com Web Site: www.kmas.com. Licensee: Olympic Broadcasting Inc. (acq 5-1-2006; $725,000). Population served: 250,000 Natl. Network: ABC. Rgnl rep: Tacher. Wire Svc: AP Format: Adult contemp. News staff: 3; News: 20 hrs wkly. Target aud: 25-64. Spec prog: Sp 2.5 hrs, Christian 4 hrs wkly. ♦Dale Hubbard, pres & gen mgr.

KRXY(FM)— October 1998: 94.5 mhz; 710 w. 954 ft TL: N47 08 18 W123 08 28. Hrs open: 24 2124 Pacific Ave. S.E., Olympia, 98506-4753. Phone: (360) 236-1010. Fax: (360) 236-1133. E-mail: krxy@krxy.com Web Site: www.krxy.com. Licensee: Premier Broadcasters Inc. Group owner: Premier Group Law Firm: Leventhal, Senter, Lerman. Format: Top-40, Hits of the 80s & 90s. ♦Derek Shannon, gen mgr & progmg dir; Paul Walker, news dir.

Silverdale

KITZ(AM)— Oct 26, 1948: 1400 khz; 1 kw-D, 890 w-N. TL: N47 37 45 W122 39 52. Hrs open: 1700 Mile Hill Dr., Suite 201A, Port Orchard, 98366. Phone: (360) 876-1400. Fax: (360) 876-7920. E-mail: info@kittz1400.com Web Site: kitz1400.com. Licensee: KITZ Radio Inc. (acq 12-8-2000; $500,000 for 60%). Population served: 200,000 Natl. Network: Westwood One. Law Firm: Pepper & Corazzini. Format: Megatalk. Target aud: 35-64; adults 25+. ♦Alan Gottlieb, chmn & pres; Paul Lyle, gen mgr; Kevin Corcoran, opns VP.

South Bend

KLSY(FM)—Not on air, target date: unknown: 107.9 mhz; 790 w. Ant 905 ft TL: N46 41 44 W123 46 17. Hrs open: Box 53248, Bellevue, 98015-3248. Phone: (425) 861-0136. Licensee: South Sound Broadcasting LLC. ♦Gregory J. Smith, gen mgr.

Spokane

***KAGU(FM)**— Mar 16, 1988: 88.7 mhz; 100 w. -141 ft TL: N47 40 06 W117 24 05. Stereo. Hrs opn: 502 E. Boone Ave., 99258. Phone: (509) 328-4220. Fax: (509) 324-5718. Web Site: www.gonzaga.edu/kagu/. Licensee: Gonzaga University Telecommunications Association. (acq 12-4-91). Format: Adult contemp. Spec prog: Class 2 hrs, jazz 2 hrs, folk 2 hrs, drama 2 hrs wkly. ♦Fr. Robert Lyons, gen mgr; Matt Caputo, prom dir.

KBBD(FM)— 1988: 103.9 mhz; 5.5 kw. Ant 1,417 ft TL: N47 36 04 W117 17 53. Stereo. Hrs opn: 24 1601 E. 57th, 99223. Phone: (509) 448-1000. Fax: (509) 448-7015. Web Site: www.1039bobfm.com. Licensee: Citadel Broadcasting Co. Group owner: Citadel Broadcasting Corp. (acq 1999; $4.15 million). Natl. Network: ABC. Law Firm: Pepper & Corazzini. Format: Hits of the 80s & 90s. Target aud: 18-34. ♦Don Morin, gen mgr; Frank Jackson, gen mgr & progmg dir; Cary Rolfe, opns mgr; Larry Weir, news dir; Dave Ratener, chief of engrg; Brenda Anderson, traf mgr.

KCDA(FM)—Post Falls, ID) June 29, 1979: 103.1 mhz; 9.4 kw. 2,450 ft TL: N47 34 14 W117 04 55. Stereo. Hrs opn: 24 808 E. Sprague, 99202. Phone: (509) 242-2400. Fax: (509) 448-4043. Web Site: www.mix1031.com. Licensee: Capstar TX L.P. Group owner: Clear Channel Communications Inc. (acq 10-18-00; $4.7 million). Population served: 480,000 Natl. Rep: Roslin. Law Firm: Pepper & Corazzini. Format: Alternative music. News staff: 2; News: one hr wkly. Target aud: 25-54; active & affluent. ♦Kosta Panidis, gen mgr.

KDRK-FM—Listing follows KGA(AM).

***KEEH(FM)**— July 1, 1991: 104.9 mhz; 10.5 kw. Ant 1,548 ft TL: N47 34 45 W117 17 51. Stereo. Hrs opn: 24
Rebroadcasts KGTS(FM) College Place 100%.
Box 19039, 99219. Secondary address: 3715 S. Grove Rd. 99219. Phone: (509) 456-4870. Fax: (509) 838-4882. E-mail: keeh@plr.org Web Site: www.plr.org. Licensee: Upper Columbia Media Association (acq 9-8-93; $148,000; FTR: 10-4-93). Population served: 400,000 Natl. Network: USA. Format: Christian. Target aud: General. ♦John Dolrymple, gen mgr.

KEYF-FM—See Cheney

KEZE(FM)— Dec 25, 1992: 96.9 mhz; 6 kw. 535 ft TL: N47 41 39 W117 20 03. Hrs open: 24 500 W. Boone Ave., 99201. Phone: (509) 324-4000. Fax: (509) 324-8992. Web Site: www.wired969.com. Licensee: QueenB Radio Inc. Group owner: Morgan Murphy Stations. Natl. Network: USA. Natl. Rep: Katz Radio. Format: Rhythmic CHR. Target aud: 25 plus; general. ♦Elizabeth M. Burns, pres; Steve Herling, exec VP; Teddie Gibbon, stn mgr; Ken Hopkins, opns mgr; Maynard Cohen, progmg dir. Co-owned TV: KXLY-TV affil

KGA(AM)— 1926: 1510 khz; 50 kw-U, DA-N. TL: N47 35 44 W117 22 15. Hrs open: 24 1601 E. 57th St., 99223-3000. Phone: (509) 448-1000. Fax: (509) 448-7015. Web Site: www.1510kga.com. Licensee: Citadel Broadcasting Co. Group owner: Citadel Broadcasting Corp. (acq 5-18-92; grpsl, including co-located FM; FTR: 6-8-92). Population served: 170,516 Format: News/talk. News staff: 3; News: 183 hrs wkly. Target aud: 25-54; general. Spec prog: Farm one hr wkly. ♦Don Morin, gen mgr; Cary Rolfe, opns mgr; Regina Winkler, gen sls mgr & natl sls mgr; Bob Castle, progmg mgr; Dave Ratener, chief of engrg.

KDRK-FM—Co-owned with KGA(AM). 1965: 93.7 mhz; 60 kw. Ant 2,424 ft TL: N47 34 14 W117 04 55. Stereo. 24 Web Site: www.catcountry94.com. Format: Country. ♦Cary Rolfe, progmg dir. Co-owned TV: KHQ-TV affil

KISC(FM)—Listing follows KQNT(AM).

KJRB(AM)— 1947: 790 khz; 5 kw-U, DA-N. TL: N47 36 16 W117 23 11. Hrs open: 24 E. 1601 57th, 99223. Phone: (509) 448-1000. Fax: (509) 448-7015. Web Site: www.790kfan.com. Licensee: Citadel Broadcasting Co. Group owner: Citadel Broadcasting Corp. (acq 9-20-93; $125,000; FTR: 10-11-93). Population served: 296,400 Format: Sports, talk. News staff: one; News: 168 hrs wkly. Target aud: 18-34. ♦Don Morin, gen mgr; Cary Rolfe, opns mgr; Joe Via, sls dir & prom dir; Bob Castle, progmg dir; Dave Ratener, news dir & chief of engrg.

KZBD(FM)—Co-owned with KJRB(AM). Nov 8, 1965: 105.7 mhz; 100 kw. 1,910 ft TL: N47 34 44 W117 17 46. Stereo. Web Site: www.1057thebuzzard.com. (Acq 3-22-93; $2.75 million with co-located AM; FTR: 4-12-93). 706,000 Natl. Network: ABC. Format: Classic rock. Target aud: 35-64; adults.

KKZX(FM)—Listing follows KPTQ(AM).

***KMBI-FM**— July 1, 1974: 107.9 mhz; 64 kw. 2,380 ft TL: N47 34 15 W117 05 00. Stereo. Hrs opn: 24 5408 S. Freya St., 99223. Phone: (509) 448-2555. Fax: (509) 448-6855. E-mail: kmbi@moody.edu Web Site: www.kmbi.org. Licensee: Moody Bible Institute. Group owner: The Moody Bible Institute of Chicago Population served: 300,000 Law Firm: Southmayd & Miller. Format: Relg. News: 9 hrs wkly. Target aud: 35-54; Christian men & women. ♦Rich Monteith, gen mgr, stn mgr & opns mgr; Gordon Canaday, chief of engrg; Bret Bremberg, disc jockey; Derek Cutlip, disc jockey; Shelly Hogeweide, disc jockey; Steve Stewart, disc jockey.

KMBI(AM)— July 12, 1959: 1330 khz; 5 kw-D. TL: N47 36 17 W117 21 27.6 AM-sunset (Acq 6-74).184,000 News: 6 hrs wkly. Target aud: 35-54; Christian men & women. ♦D. Gary Leonard, stn mgr; Steve Stewart, pub affrs dir.

***KPBX-FM**— 1970: 91.1 mhz; 56 kw. 2,380 ft TL: N47 34 13 W117 05 00. Stereo. Hrs opn: 24 2319 N. Monroe St., 99205. Phone: (509) 328-5729. Fax: (509) 328-5764. E-mail: rkunkel@kpbx.org Web Site: www.kpbx.org. Licensee: Spokane Public Radio Inc. Population served: 700,000 Natl. Network: NPR, PRI. Format: Class, news, jazz. News staff: 3; News: 50 hrs wkly. Target aud: General; educated. Spec prog: Jazz, folk, world mus, new age/space, new mus. ♦Richard Kunkel, CEO, pres & gen mgr; Brian Flick, opns dir & progmg dir; Kathy Sackett, dev dir & sls dir; John Vlahovich, news dir; Jerry Olson, chief of engrg.

KPTQ(AM)— 1965: 1280 khz; 5 kw-D, DA. TL: N47 36 27 W117 21 40. Hrs open: 6 AM-9 PM 808 E. Sprague Ave., 99202. Phone: (509) 242-2400. Fax: (509) 242-2581. Web Site: www.kaqq1280.com. Licensee: Capstar TX L.P. Group owner: Clear Channel Communications Inc. (acq 8-30-2000; grpsl). Population served: 510,000 Natl. Network: USA. Natl. Rep: D & R Radio. Rgnl rep: Tacher. Law Firm: Arent, Fox, Kintner, Plotkin & Kahn. Format: Progressive talk. News: one hr wkly. Target aud: General. ♦Garth Trimble, gen sls mgr; Kosta Panidis, gen mgr & gen sls mgr; Barry Watkins, progmg dir.

KKZX(FM)—Co-owned with KPTQ(AM). Oct 10, 1975: 98.9 mhz; 100 kw. 1,614 ft TL: N47 35 35 W117 17 46. Stereo. 24 Web Site: www.kkzx.com.350,000 Format: Classic rock. News: one hr wkly. Target aud: 25-54. Spec prog: Blues 2 hrs wkly. ♦Jon McGann, progmg dir.

KQNT(AM)— 1922: 590 khz; 5 kw-U. TL: N47 36 59 W117 22 12. Hrs open: 808 E. Spague Ave., 99202. Phone: (509) 242-2400. Fax: (509) 242-2581. Web Site: www.newstalk590.com. Licensee: Capstar TX L.P. Group owner: Clear Channel Communications Inc. (acq 8-30-00; grpsl). Population served: 170,516 Law Firm: Fisher, Wayland, Cooper, Leader & Zaragoza L.L.P. Wire Svc: Reuters Format: News/talk. Target aud: 35-64. Spec prog: Farm 6 hrs wkly. ♦Kosta Panidis, gen mgr & gen sls mgr; Garth Trimble, gen sls mgr; Jerry Jensen, natl sls mgr; Joe Chabala, prom dir; Dean Allen, progmg dir; Harv Clark, news dir; Kent Abendroth, chief of engrg; Wey Simpson, farm dir.

KISC(FM)—Co-owned with KQNT(AM). May 1, 1966: 98.1 mhz; 94 kw. 2,030 ft TL: N47 34 53 W117 17 47. Stereo. Web Site: www.literockkiss.com. Format: Adult contemp. ♦Rob Harder, progmg dir; Dawn Marcel, disc jockey; Ian Richards, disc jockey; John Chrisopher Kowsky, disc jockey; Mark Holman, disc jockey; Stormy Morgan, disc jockey.

KSBN(AM)— September 1921: 1230 khz; 1 kw-U. TL: N47 39 30 W117 25 08. Hrs open: 24 506 W. 1st Ave., 99201. Phone: (509) 838-4000. Fax: (509) 838-4800. E-mail: ksbn@ksbn.net Web Site: www.ksbn.net. Licensee: KSBN Radio Inc. (acq 6-95). Population served: 400,000 Wire Svc: Bloomberg Financial Format: Business, financial, talk, news. News staff: one; News: 24 hrs wkly. Target aud: 30-65; upscale; business owners. ♦Alan Gottlieb, chmn; Angela Watkins, gen mgr, progmg dir & news dir; Patrick Carey, gen sls mgr; Conrad Agate, chief of engrg.

***KSFC(FM)**— March 1973: 91.9 mhz; 100 w. 92 ft TL: N47 40 37 W117 27 31. Hrs open: 2319 N. Monroe St., 99205. Phone: (509) 328-5729. Fax: (509) 328-5764. E-mail: rkunkel@kpbx.org Web Site: www.ksfc.org. Licensee: Spokane Public Radio Inc. Population served: 100,000 Format: News. Target aud: Curious. Spec prog: American Indian 5 hrs, black 2 hrs wkly. ♦Richard Kunkel, pres & gen mgr; Kathy Sackett, dev dir; Brian Flick, progmg dir; John Vlahovich, news dir; Jerry Olson, engrg dir & chief of engrg.

KTTO(AM)— 1947: 970 khz; 5 kw-D, 1 kw-N, DA-N. TL: N47 36 59 W117 21 55. Hrs open: 24 4419 N. Hawthorne St., 99205-1399. Phone: (509) 327-3695. Fax: (509) 327-5171. Web Site: www.spokanecatholicradio.com. Licensee: Sacred Heart Radio Inc. (acq 9-29-2005;. $850,000). Population served: 700,900 Format: Catholic. ♦Sr. Patricia Proctor, gen mgr.

***KWRS(FM)**— Sept 16, 1991: 90.3 mhz; 10 w. -89 ft TL: N47 45 30 W117 25 00. Stereo. Hrs opn: 17 Stn 40, Whitworth College, 99251. Phone: (509) 777-4575. Fax: (509) 777-3710. E-mail: kwrsuc@mail.witworth.edu Web Site: www.whitworth.edu/kwrs. Licensee: Whitworth College. Population served: 100,000 Format: CHR, progsv, rock/AOR. News staff: one; News: 2 hrs wkly. Target aud: 15-35; mostly college & high school students. Spec prog: Black 12 hrs, folk 8 hrs, relg 4 hrs, Sp 2 hrs wkly. ♦Katie Thompson, gen mgr; Sara Edlin-Marlowe, stn mgr.

KXLY(AM)— October 1922: 920 khz; 5 kw-U. TL: N47 36 30 W117 22 25. Hrs open: 24 W. 500 Boone Ave., 99201. Phone: (509) 324-4000. Fax: (509) 324-8992. E-mail: kxly@kyly920.com Web Site: www.kxly.com. Licensee: Spokane Radio Inc. Group owner: Morgan Murphy Stations (acq 3-21-62). Population served: 700,000 Natl. Network: CBS, Wall Street. Natl. Rep: Katz Radio. Format: News/talk. News staff: 25; News: 30 hrs wkly. Target aud: Adults 35 plus; upper end education & income levels. Spec prog: Sports talk 5 hrs, sports play-by-play 20 hrs, local talk 15 hrs wkly. ♦Stephen R. Herling, VP; Chris Garras, gen mgr; Teddie Gibbon, stn mgr; Roger Nelson, opns mgr & sls dir; Dick Brantley, rgnl sls mgr & mktg dir; Gina Mauro, prom dir.

KXLY-FM— September 1959: 99.9 mhz; 37 kw. Ant 2,998 ft TL: N47 55 18 W117 06 48. Stereo. 24 Fax: (509) 324-8992. E-mail: classy@classy99.cox Web Site: www.classy99.com. Natl. Network: Westwood One, Jones Radio Networks. Natl. Rep: Katz Radio. Format: Adult contemp. Target aud: 25-64; adults, upper end income & education levels. ♦Tery Garras, sls dir; Joe Via, rgnl sls mgr. Co-owned TV: KXLY-TV affil.

KZZU-FM— September 1955: 92.9 mhz; 81 kw. 2,080 ft TL: N47 35 42 W117 17 53. Stereo. Hrs opn: 500 W. Boone Ave., 99201. Phone: (509) 324-4000. Fax: (509) 324-8992. Web Site: www.kzzu.com. Licensee: QueenB Radio Inc. Group owner: Morgan Murphy Stations (acq 4-1-96; $1.75 million with co-located AM). Population served: 313,700 Natl. Rep: Katz Radio. Format: Hot adult contemp. Target aud: 18-49. ♦Steve Herling, VP; Roger Nelson, gen mgr & mktg mgr; Teddie Gibbon, stn mgr; George Kessler, natl sls mgr & disc jockey; Maynard Cohen, progmg dir & mus dir; Tim Anderson, chief of engrg; Catherine Bruntlett, rsch dir; Jolene Longwill, traf mgr. Co-owned TV: KXLY-TV.

Sumner

KZIZ(AM)— 1990: 1560 khz; 5 kw-D. TL: N47 12 48 W122 13 25. Hrs open: Sunrise-sunset Box 22462, Seattle, 98122-0462. Secondary address: 2600 S. Jackson St., Seattle 98144. Phone: (206) 323-3070. Fax: (206) 322-6518. E-mail: ztwins@aol.com Web Site: www.ztwins.com. Licensee: KRIS Bennett Broadcasting Inc. Population served: 3,000,000 Natl. Network: American Urban. Format: Gospel. Target aud: 12 plus; African-American. Spec prog: Relg 18 hrs wkly. ♦Christopher H. Bennett, gen mgr; Gloria Bennett, stn mgr; Frank P. Barrow, chief of opns; Frank Barrow, progmg dir; Priscilla Hailey, chief of engrg.

Sunnyside

***KAYB(FM)**— 1998: 88.1 mhz; 250 w. -190 ft TL: N46 19 53 W120 00 51. Hrs open: Box 3206, Tupelo, MS, 38803. Phone: (662) 844-8888. Fax: (662) 842-6791. E-mail: comments@afr.net Web Site: www.afr.net. Licensee: American Family Association. Group owner: American Family Radio Format: Christian, inspirational. ♦Marvin Sanders, gen mgr.

KZTS(AM)— September 1950: 1210 khz; 10 kw-D, 1 kw-N. TL: N46 19 49 W120 02 10. Hrs open: 24
Rebroadcasts KYXE(AM) Selah 100%.
Box 2888, Yakima, 98907. Secondary address: 706 Butterfield Rd., Yakima 98901. Phone: (509) 457-1000. Fax: (509) 452-0541. E-mail: zorro@radiozorro.com Web Site: www.radiozorro.com. Licensee: Bustos Media of Eastern Washington License LLC. (group owner; (acq 11-18-2004; grpsl). Population served: 100000 Natl. Rep: Tacher. Format: Sp. News staff: one. Target aud: 25-55; Hispanic. ♦Amador S. Bustos, pres; Bob Berry, gen mgr, gen sls mgr & news dir; Keith Teske, opns dir; Martin Ortiz, progmg dir; Lisa Gonzalez, traf mgr.

Tacoma

KBKS-FM— May 1959: 106.1 mhz; 55 kw. 699 ft TL: N47 18 15 W122 23 44. Stereo. Hrs opn: 1000 Dexter Ave. N., Suite 100, Seattle, 98109. Phone: (206) 805-1061. Fax: (206) 805-0920. Web Site: www.kiss1061.com. Licensee: Infinity Radio Holdings Inc. Group owner: Infinity Broadcasting Corp. (acq 11-13-98; grpsl). Population served: 400,000 Law Firm: Leventhal, Senter & Lerman. Format: CHR. Target aud: 25-54. ♦Dave McDonald, sr VP, gen mgr & opns mgr; Bill Sigmar, gen sls mgr; Mike Preston, progmg dir; Jackie Cunningham, news dir; Tom McGinley, chief of engrg.

KBSG-FM— Oct 26, 1948: 97.3 mhz; 52 kw. Ant 2,391 ft TL: N47 30 14 W121 58 29. Stereo. Hrs opn: 24 1820 Eastlake Ave. E., Seattle, 98102-3711. Phone: (206) 343-9700. Fax: (206) 623-7677. E-mail: kbsg@kbsg.com Web Site: www.kbsg.com. Licensee: Entercom Seattle License L.L.C. Group owner: Entercom Communications Corp. Population served: 2,500,000 Natl. Rep: D & R Radio. Format: Oldies. News staff: one. Target aud: 25-54. ♦Joseph M. Field, CEO; Steve Fisher, CFO; Kevin McCarthy, VP; Gail Raisio, gen mgr & traf mgr; Jerry McKenna, gen mgr; Jerry Riley, gen sls mgr; Brian Thomas, progmg dir; Tom Pierson, chief of engrg.

KGTK(AM)—Olympia, October 1956: 920 khz; 3 kw-D, 7 w-N. TL: N47 03 44 W122 49 49. Stereo. Hrs opn: 24
Rebroadcasts KITZ(AM) Silverdale 90%.
12500 N.E. Tenth Pl., Bellevue, 98005. Secondary address: 1700 Mile High Dr., Suite 201A, Port Orchard 98366. Phone: (360) 876-1400. Fax: (360) 876-7920. E-mail: info@kitz1400.com Web Site: www.kitz1400.com. Licensee: KITZ Radio Inc. (acq 4-30-2004; $300,000). Population served: 150,000 Natl. Network: USA, Radio America. Rgnl rep: Tacher Format: Talk. News: 7 hrs wkly. Target aud: 18-65; diversified adults. ♦Alan Gottlieb, pres; Julie Versnel, VP; Conn Williamson, gen mgr; Kevin Corcoran, opns mgr; Nichole Engelstad, gen sls mgr.

KHHO(AM)—Licensed to Tacoma. See Seattle

KKMO(AM)— 1922: 1360 khz; 5 kw-U. TL: N47 18 19 W122 26 33. Hrs open: 24 2815 2nd Ave., Suite 550, Seattle, 98121. Phone: (206) 443-8200. Fax: (206) 443-1561. E-mail: reception@inspirationradio.com Web Site: www.kgnw.com. Licensee: Inspiration Media Inc. Group owner: Salem Communications Corp. (acq 8-12-98; $500,000). Population served: 2,137,800 Format: Sp talk and music. Target aud: 35 plus. ♦Joe Gonzalez, gen mgr; Chuck Olmstead, opns dir, opns mgr, progmg dir & progmg mgr; Doug Rice, gen sls mgr; Juanita Jasso, prom dir; Monte Passmore, chief of engrg.

KLAY(AM)—Lakewood, 1991: 1180 khz; 5 kw-D, 1 kw-N, DA-N. TL: N47 09 00 W122 24 38. Hrs open: 24 10025 Lakewood Dr. S.W., Suite B, 98499. Phone: (253) 581-0324. Fax: (253) 581-0326. E-mail: klay11800@qwest.net Web Site: www.klay1180.com. Licensee: Clay Frank Huntington. Population served: 3,000,000 Format: Talk. News staff: 4; News: 11 hrs wkly. Target aud: 30-65. ♦Clay Frank Huntington, pres; Evan Brown, opns dir & pub affrs dir; Bob McCluskey, sls VP; Walker Mattson, progmg VP; Lynn Benson, news dir; Nick Winter, engrg dir.

KMTT(FM)—Licensed to Tacoma. See Seattle

KNTB(AM)—Lakewood, September 1978: 1480 khz; 1 kw-D, 111 w-N, DA-2. TL: N47 09 56 W122 34 32. Hrs open: 6 AM-sunset Box 4024, Seattle, 98104. Phone: (206) 583-0811. Licensee: Seattle Streaming Radio LLC. (acq 7-22-2005; $900,000 with KBRO(AM) Bremerton). Population served: 500,000 Format: Hits of the 60s, 70s & 80s. ♦Chris Hanley, gen mgr.

***KPLU-FM**— November 1966: 88.5 mhz; 58 kw. 2,356 ft TL: N47 28 50 W122 31 58. Stereo. Hrs opn: 24 12180 Park Ave. S., 98447-0885. Secondary address: 2601 4th Ave., Suite 150, Seattle 98121. Phone: (253) 535-7758. Fax: (253) 535-8332. E-mail: kplu@plu.edu Web Site: www.kplu.org. Licensee: Pacific Lutheran University. Population served: 2,844,400 Natl. Network: NPR, PRI. Law Firm: Dow, Lohnes & Albertson. Format: NPR news, jazz, blues. News staff: 7; News: 54 hrs wkly. Target aud: 25-54; upscale, highly educated professionals. ♦Paul Stankavich, gen mgr; Jeff Bauman, opns dir; Nancy Knudsen, dev dir; Brend Goldstein-Young, prom dir & traf mgr; Joey Cohn, progmg dir; Nick Francis, mus dir; Erin Hennessey, news dir; Lowell Kiesow, chief of engrg.

***KUPS(FM)**— Feb 28, 1978: 90.1 mhz; 100 w. 65 ft TL: N47 15 48 W122 28 37. Stereo. Hrs opn: 24 1500 N. Warner, 98416. Phone: (253) 879-3288. E-mail: thesound@ups.edu Web Site: www.kups.net. Licensee: University of Puget Sound. Population served: 75,000 Format: Progsv. Target aud: 18-45. Spec prog: Black 18 hrs, jazz 12 hrs, reggae 6 hrs, world mus 4 hrs, blues 6 hrs, metal 8 hrs wkly. ♦Brenden Goetz, gen mgr; Doug Herstad, chief of opns.

***KVTI(FM)**— Nov 15, 1955: 90.9 mhz; 51 kw. 364 ft TL: N47 09 39 W122 34 35. Stereo. Hrs opn: 24 4500 Steilacoom Blvd. S.W., Lakewood, 98499-4098. Phone: (253) 589-5884. Fax: (253) 589-5797. E-mail: i-91fm@cptc.edu Web Site: www.i91.ctc.edu. Licensee: Clover Park Technical College. Population served: 2,500,000 Law Firm: Garvey, Shubert & Barer. Wire Svc: AP Format: CHR, Top-40. News: 2 hrs wkly. Target aud: 12-34; young adults & teens. Spec prog: Live mus 3 hrs, talk 4 hrs wkly. ♦John L. Mangan, gen mgr & progmg dir; Beth Valiant, mus dir; Al Bednarczyk, chief of engrg.

***KXOT(FM)**— June 1, 1949: 91.7 mhz; 7.9 kw. Ant 553 ft TL: N47 18 15 W122 23 44. Stereo. Hrs opn: 24 113 Dexter Ave. N., Seattle, 98109. Phone: (206) 520-5800. Fax: (206) 520-5899. Web Site: www.kxot.org. Licensee: PRC Tacoma — I LLC (acq 1-31-2005; $5 million). Population served: 154,581 Natl. Network: NPR. Format: News/talk. Target aud: . ♦Tom Mara, gen mgr; Gary Rubin, gen sls mgr; Kevin Cole, progmg dir. Co-owned TV: .

Toppenish

KDBL(FM)— Oct 31, 1977: 92.9 mhz; 17 kw. 843 ft TL: N46 30 15 W120 23 33. Stereo. Hrs opn: 24 4010 Summitview, Yakima, 98908. Phone: (509) 972-3461. Fax: (509) 972-3542. Web Site: www.929thebull.com. Licensee: Citicasters Licenses L.P. Group owner: Clear Channel Communications Inc. (acq 10-26-99; grpsl). Population served: 190,000 Natl. Rep: McGavren Guild. Law Firm: Shaw Pittman. Format: Country. News staff: 2. Target aud: 18-49. ♦Gary Donovan, pres & exec VP; Lzrry Miner, gen mgr; Ron Harris, opns mgr; Rick Michaels, progmg dir.

KYNR(AM)— May 16, 1954: 1490 khz; 1 kw-U. TL: N46 22 33 W120 19 18. Hrs open: 24 Box 151, 98948-0151. Secondary address: 711 King Ln. 98948. Phone: (509) 865-5363. Fax: (509) 865-2129. E-mail: kyn@yakama.com Web Site: www.kynr.com. Licensee: Confederated Tribes and Bands of the Yakama Nation (acq 2-9-01; $300,000). Population served: 198,000 Format: Classic rock, Country, Diversified, Jazz, Oldies, News, Sports, Urban contemp. News: one hr wkly. Target aud: 18-58. Spec prog: American Indian 20 hrs wkly. ♦Lenny Abrams, gen mgr; Reggie George, progmg dir.

Tumwater

KUOW(AM)— August 1987: 1340 khz; 1 kw-U. TL: N47 00 25 W122 55 07. Hrs open: 24
Rebroadcasts KUOW-FM Seattle 100%.
4518 University Way N.E., Suite 310, Seattle, 98105. Phone: (206) 543-2710. Fax: (206) 616-9125. E-mail: letters@kuow.org Web Site: www.kuow.org. Licensee: KUOW/Puget Sound Public Radio (acq 5-18-2006; $500,000). Population served: 102,000 Natl. Network: NPR, PRI. Format: News, info. ♦Wayne Roth, gen mgr & stn mgr.

Twisp

KCSY(FM)— June 1993: 106.3 mhz; 220 w. Ant 1,633 ft TL: N48 19 06 W120 06 46. Hrs open: Box 637, 98856. Phone: (509) 997-5857. Fax: (509) 997-5859. E-mail: sunnyfm@kcsyfm.com Web Site: kcsyfm.com. Licensee: Resort Radio LLC (acq 11-15-2006;. $250,000). Format: Oldies. ♦Dave Bauer, gen mgr; Debbie Griggs, gen sls mgr; Lonnie England, chief of engrg.

Union Gap

KYXE(AM)— Sept 13, 1983: 1020 khz; 4 kw-D, 400 w-N, DA-D. TL: N46 34 17 W120 27 15 (D), N46 34 14 W120 27 15 (N). Hrs open: 24 Box 2888, Yakima, 98907. Secondary address: 706 Butterfield Rd. 98907. Phone: (509) 457-1000. Fax: (509) 452-0541. E-mail: zorro@radiozorro.com Web Site: www.radiozorro.com. Licensee: Bustos Media of Eastern Washington License LLC. (group owner; (acq 11-18-2004; grpsl). Rgnl rep: Tacher. Format: Sp. Target aud: 25-49; Hispanic adults. ♦Bob Berry, gen mgr; Keith Teske, opns mgr; Martin Ortiz, progmg dir & news dir.

Vancouver

KBMS(AM)— 1955: 1480 khz; 1 kw-D, 2.5 kw-N, DA-N. TL: N45 36 06 W122 43 06. Hrs open: 24 Box 251, 98666. Phone: (360) 699-1881. Fax: (360) 699-5370. E-mail: avjkbms@aol.com Licensee: Christopher H. Bennett Broadcasting Co. of WA Inc. Natl. Network: ABC. Format: Urban contemp, talk. Target aud: 25-54; 54-over; male & female. ♦Chris Bennett, gen mgr; Angela Jenkins, stn mgr.

KIJZ(FM)— 2001: 105.9 mhz; 21 kw. Ant 1,542 ft TL: N45 31 21 W122 44 45. Stereo. Hrs opn: Unlimited 4949 S.W. Macadam Ave., Portland, OR, 97201. Phone: (503) 226-0100. Phone: (503) 323-6400. Fax: (503) 802-1640. Web Site: www.kijz.com. Licensee: Citicasters Licenses L.P. Group owner: Clear Channel Communications Inc. (acq 1999; grpsl). Format: Smooth jazz. ♦Robert Dove, gen mgr; Tony Coles, opns mgr.

KKAD(AM)— Aug 10, 1963: 1550 khz; 50 kw-D, 12 kw-N, DA-N. TL: N45 38 47 W122 30 51. Hrs open: 24 6605 S.E. Lake Rd., Portland, OR, 97222. Phone: (503) 223-4321. Fax: (503) 294-0074. E-mail: markail@kpam.com Web Site: www.sunny1550kkad.com. Licensee: Pamplin Broadcasting-Washington Inc. Group owner: Pamplin Broadcasting (acq 11-20-98; $1.65 million). Natl. Network: AP Network News. Natl. Rep: Tacher. Rgnl rep: The Tacher Co., INc. Wire Svc: AP Format: Adult standards/music of your life. News staff: 2; News: 5.6 hrs wkly. Target aud: 35-64. Spec prog: Portland Beaver baseball 18 hrs wkly. ♦Paul Clithero, gen mgr; Mark L. Ail, stn mgr & opns dir; Margaret Evans, gen sls mgr; Jeanne Winters, natl sls mgr; Misty Osko, prom mgr; Paul Duckworth, progmg dir; Bill Gallagher, news dir; Dave Bischoff, chief of engrg; Paul Blanding, traf mgr.

KTRO(AM)— Sept 1, 1946: 910 khz; 5 kw-U, DA-2. TL: N45 33 28 W122 30 09. Hrs open: 24 6400 S.E. Lake Rd., Suite 350, Portland, OR, 97222. Phone: (503) 786-0600. Fax: (503) 786-1551. E-mail: info@931fmktro.com Web Site: www.931fmktro.com. Licensee: Entercom Portland License LLC. Group owner: Entercom (acq 4-23-98; grpsl). Population served: 230,600 Natl. Rep: D & R Radio. Format: Talk. ♦Dennis Hayes, gen mgr; Leslie Pfau, mktg dir & prom dir; Justin Mansfield, progmg dir.

KXMG(AM)—See Portland, OR

KYCH-FM—See Portland, OR

Walla Walla

KGDC(AM)— Dec 6, 1956: 1320 khz; 1 kw-D, 660 w-N. TL: N46 02 13 W118 21 07. Hrs open: 24 38 E. Main St., Suite 11, 99362. Phone: (509) 525-7878. E-mail: comments@kgdcradio.com Licensee: Two Hearts Communications LLC (acq 8-20-01). Population served: 23,619 Format: News/talk. News: 5 hrs wkly. Target aud: General. ♦Rod Fazzari, pres, stn mgr, progmg dir, chief of engrg & traf mgr.

KGTS(FM)—See College Place

KHSS(FM)— Nov 5, 1986: 100.7 mhz; 1.3 kw. 1,374 ft TL: N46 04 04 W118 20 21. (CP: Ant 1,414 ft. TL: N45 59 04 W118 10 08). Stereo. Hrs opn: 24 38 E. Main St., 99362. Phone: (509) 525-7878. Fax: (509) 522-2046. E-mail: comments@khssradio.com Web Site: www.khssradui.com. Licensee: Two Hearts Communications L.L.C. (acq 3-26-98; $160,000). Natl. Rep: Katz Radio. Law Firm: Pepper & Corazzini. Format: Catholic talk. Target aud: 18-34. Spec prog: Relg 3 hrs wkly. ♦Rodney Fazzari, gen mgr & progmg dir; Todd Brandenburg, chief of engrg.

KNLT(FM)— Jan 1, 1980: 95.7 mhz; 100 kw. Ant 1,401 ft TL: N45 59 04 W118 10 08. Stereo. Hrs opn: 830 N. Columbia Center Blvd., Suite B-2, Kennewick, 99336. Phone: (509) 783-0783. Fax: (509) 735-8627. E-mail: curt.cartier@nnbproduction.com Web Site: www.knlt.com. Licensee: New Northwest Broadcasters LLC (group owner; acq 1999). Natl. Rep: Christal. Format: Bob FM 90s, 90s & whatever. Target aud: 25-54. ♦Jim Richmond, VP & gen mgr; Don Morin, gen mgr; Brad Barrett, opns dir & progmg dir; Allison Crawford, news dir & pub affrs dir; Rob Meadows, chief of engrg; Lisa Perez, traf mgr.

***KRKL(FM)**— May 10, 1977: 93.3 mhz; 42 kw. Ant 1,378 ft TL: N45 59 19 W118 10 28. Stereo. Hrs opn: 24 2351 Sunset Blvd., Suite 170-218, Rocklin, CA, 95765. Phone: (916) 251-1600. Fax: (916) 251-1650. Web Site: www.klove.com. Licensee: Educational Media Foundation. Group owner: EMF Broadcasting (acq 4-1-02; $1 million). Population served: 80,000 Natl. Network: K-Love. Law Firm: Shaw Pittman. Format: Contemp Christian. News staff: 3. Target aud: 25-44; Judeo Christian, female. ♦Richard Jenkins, pres; Mike Novak, VP; Keith Whipple, dev dir; David Pierce, progmg mgr; Ed Lenane, news dir; Sam Wallington, engrg dir; Karen Johnson, news rptr; Marya Morgan, news rptr; Richard Hunt, news rptr.

KTEL(AM)— October 1946: 1490 khz; 1 kw-U. TL: N46 20 33 W118 20 20. Hrs open: 24 13 1/2 E. Main St., Suite 202, 99362. Phone: (541) 522-1383. Fax: (509) 522-0211. E-mail: rmckone@uci.net Licensee: WW2 L.L.C. Group owner: Capps Broadcast Group (acq 6-2-03). Population served: 65,000 Natl. Network: Jones Radio Networks, ABC. Rgnl rep: Tacher Format: Oldies. Target aud: 25 plus; general. Spec prog: Farm 5 hrs wkly. ♦Dave Capps, pres; Randy McKone, VP, gen mgr & progmg dir; Colleen Doyle, opns mgr; Liz Halley, sls VP & sls dir; Andrew Holt, news dir; Stacie Cummings, traf mgr.

KUJ(AM)— 1928: 1420 khz; 5 kw-U, DA-N. TL: N46 04 03 W118 24 08. Hrs open: 24 45 Campbell Rd., 99362. Phone: (509) 527-1000. Fax: (509) 529-5534. E-mail: kujam@bmi.net Licensee: Alexandra Communications Inc. (acq 4-13-2001). Population served: 56,000 Natl. Network: Westwood One. Rgnl rep: Tacher. Law Firm: Taylor, Theimann & Aitken. Format: News/talk, sports. News staff: one; News: 15 hrs wkly. Target aud: 25 plus. ♦Cheryl Hodgins, exec VP; Tom Hodgins, CEO & gen mgr.

***KWCW(FM)**— 1971: 90.5 mhz; 160 w. Ant -52 ft TL: N46 04 11 W118 19 51. Stereo. Hrs opn: 24 Whitman College, 99362. Phone: (509) 527-5285. Web Site: www.kwcw.net. Licensee: The Associated Students of Whitman College Radio Committee. Population served: 40,000 Format: Free form. News staff: one; News: 2.5 hrs wkly. ♦Travis Kiefer, gen mgr.

***KWWS(FM)**— Mar 6, 1997: 89.7 mhz; 3.2 kw. 1,345 ft TL: N45 59 04 W118 10 08. Hrs open:
Rebroadcasts KWSU(AM) Pullman 100%.
Box 642530, Murrow Communications Ctr., Washington State Univ., Pullman, 99164-2530. Phone: (509) 335-6500. Fax: (509) 335-6577. E-mail: nwpr@wsu.edu Web Site: www.nwpr.org. Licensee: Washington State University. Law Firm: Dow, Lohnes & Albertson. Format: News/talk. ♦Karen Olstad, COO & gen mgr; Dennis Haarsager, gen mgr; Roger Johnson, stn mgr; Scott Weatherly, opns dir; Sarah McDaniel, dev dir; Mary Hawkins, progmg dir.

KXRX(FM)— August 1977: 97.1 mhz; 50 kw. Ant 1,338 ft TL: N45 59 04 W118 10 08. (CP: 100 kw, ant 1,328 ft. TL: N45 59 04 W118 10 09). Stereo. Hrs opn: 24 2621 W. A St., Pasco, 99301. Phone: (509) 547-9791. Fax: (509) 547-8509. Web Site: www.97rock.fm. Licensee: Capstar TX L.P. Group owner: Clear Channel Communications Inc. (acq 2-15-01; grpsl). Population served: 216,000 Format: Rock. News staff: one; News: one hr wkly. Target aud: 25 plus; middle to upper income level listeners. ♦Eric Van Winkle, gen mgr.

Wapato

***KSOH(FM)**— Mar 6, 1992: 89.5 mhz; 9.5 kw. 974 ft TL: N46 31 42 W120 31 16. Hrs open: Box 5, Yakima, 98907. Secondary address: 1006 S. Fair Ave. 98901. Phone: (509) 248-4673. Fax: (509) 248-1579. E-mail: ksoh@lifetalk.net Web Site: www.lifetalk.net. Licensee: Life Talk Broadcasting Association. Format: Relg. Target aud: 25-50; families, singles needing courage, hope & answers to societal ills. ♦Andrew Fleming, stn mgr.

Wenatchee

KKRT(AM)— Nov 17, 1956: 900 khz; 1 kw-D, 78 w-N. TL: N47 27 45 W120 19 24. Hrs open: 24 Box 79, 2nd Fl., 32 N. Mission St., 98801. Phone: (509) 663-5186. Fax: (509) 663-8779. Web Site: www.kkrt.com. Licensee: Morris Communications Corp. Group owner: Morris Communications Inc. (acq 10-15-98; grpsl). Population served: 61,000 Natl. Network: ESPN Radio. Natl. Rep: Katz Radio. Format: Sports. Target aud: 18-54; men. ♦William Morris III, CEO; Michael Osterhont, exec VP; Gary Patrick, gen mgr; Jeff Dahlstrom, sls dir; John Windus, progmg dir; Jay White, chief of engrg.

KKRV(FM)—Co-owned with KKRT(AM). May 1, 1976: 104.7 mhz; 6.5 kw. Ant 1,322 ft TL: N47 28 44 W120 12 49. Stereo. 24 Web Site: www.kkrv.com.80,000 Format: New country. Target aud: 25-54; women. ♦Shari Alexander, traf mgr.

***KPLW(FM)**— 1996: 89.9 mhz; 6 kw. 1,222 ft TL: N47 19 10 W120 14 17. Hrs open: 24 606 N. Western Ave., 98801. Phone: (509) 665-6641. Fax: (509) 665-3126. E-mail: kplw@plr.org Web Site: www.plr.org. Licensee: Growing Christian Foundation. Population served: 200,000 Format: Relg, Christian. News: 2 hrs wkly. Target aud: 35-54; female. ♦Kevin Krueger, chmn & pres; Sean Ruud, gen mgr.

KPQ(AM)— December 1929: 560 khz; 5 kw-U, DA-N. TL: N47 27 12 W120 19 43. Hrs open: 24 Box 159, 98807-0159. Secondary address: 32 N. Mission 98801. Phone: (509) 663-5121. Fax: (509) 664-6799. E-mail: kpq@crcwnet.com Web Site: www.kpq.com. Licensee: Wescoast Broadcasting Co. Population served: 52,000 Natl. Rep: Tacher. Law Firm: Davis Wright Tremaine. Format: News/talk. News: 165 hrs wkly. Target aud: 35 plus. Spec prog: Farm 3 hrs wkly. ♦Jim Wallace Jr., pres & gen mgr; Debi Campestrini, opns dir & opns mgr; Greg McEwen, gen sls mgr & mktg dir; Steve Hair, news dir; Pete Peterson, chief of engrg; Janette Morris, traf mgr; Tom Cashman, news rptr; Eric Granstrom, sports cmtr.

KPQ-FM— December 1967: 102.1 mhz; 35 kw. 2,655 ft TL: N47 16 28 W120 25 30. Stereo. 24 E-mail: info@thequake1021.com Web Site: www.thequake1021.com.30,000 Format: Classic rock, CHR. News: one hr wkly. Target aud: 25 plus. ♦Kelly Hart, mus dir & local news ed; Janette Morris, traf mgr; Tom Cashman, news rptr; Eric Granstrom, sports cmtr.

KWWX(AM)— 1948: 1340 khz; 1 kw-U. TL: N47 23 50 W120 16 25. Hrs open: 24 231 N. Wenatchee Ave., 98801. Phone: (509) 665-6565. Fax: (509) 663-1150. Web Site: www.lasuperz.com. Licensee: CCR-Wenatchee IV LLC. Group owner: Fisher Broadcasting Company (acq 10-31-2006; grpsl). Population served: 52,000 Natl. Rep: McGavren Guild. Law Firm: Shaw Pittman. Format: Sp. News staff: one; News: 10 hrs wkly. Target aud: General; Hispanic. ♦Steve Miller, gen mgr; Leona Frank, gen sls mgr; Elsa Esparza, progmg dir; Manuel Garcia, chief of engrg.

KYSN(FM)—East Wenatchee, Dec 25, 1980: 97.7 mhz; 7 kw. -150 ft TL: N47 22 52 W120 17 16. Stereo. Hrs opn: 24 231 N. Wenatchee Ave., 98801. Phone: (509) 665-6565. Fax: (509) 663-1150. E-mail: production@nw-tel.net Web Site: www.kysn.com. Licensee: CCR-Wenatchee IV LLC. Group owner: Fisher Broadcasting Company (acq 10-31-2006; grpsl). Format: Country. News staff: one; News: 20 hrs wkly. Target aud: 25-54. Spec prog: Farm one hr, relg one hr wkly. ♦Steven Miller, gen mgr; Leona Frank, gen sls mgr; John Ross, progmg dir; Todd Allen, opns mgr & progmg dir; Dave Bernstein, news dir; Manuel Garcia, chief of engrg; Lisa Rodriguez, traf mgr.

Westport

KABW(FM)—Not on air, target date: unknown: 101.3 mhz; 6 kw. Ant 226 ft TL: N46 53 04 W124 00 44. Hrs open: College Creek Media LLC, 980 N. Michigan Ave., Suite 1880, Chicago, IL, 60611. Phone: (312) 204-9900. Licensee: College Creek Media LLC. ♦Neal J. Robinson, pres.

White Salmon

***KBNO-FM**— 2001: 89.3 mhz; 20 w vert. Ant 1,102 ft TL: N45 43 23 W121 26 42. Hrs open: 2650 Montello Ave., Hood River, OR, 97031. Phone: (541) 386-8810. Web Site: www.hcjb.org/wrn. Licensee: World Radio Network Inc. Format: Relg, Sp. ♦John Estey, gen mgr.

Wilson Creek

KWLN(FM)— November 1994: 103.3 mhz; 25 kw. 243 ft TL: N47 16 40 W119 00 00. Hrs open: 24 Box 79, Wenatchee, 98807. Phone: (509) 663-5186. Fax: (509) 663-8779. E-mail: jlhigh@morris.com Web Site: www.lanuevaradio.com. Licensee: Morris Communications Corp. Group owner: Morris Communications Inc. (acq 10-15-98; grpsl). Natl. Rep: Katz Radio. Law Firm: Wiley, Rein & Fielding. Format: Sp. Target aud: 15 plus. ♦Gary Patrick, gen mgr; Jeff Dahlstrom, gen sls mgr; Jose Luis High, progmg dir & news dir; Sherri Alexander, traf mgr.

Winlock

KITI-FM— Aug 15, 1995: 95.1 mhz; 380 w. Ant 879 ft TL: N46 32 35 W123 01 14. Stereo. Hrs opn: 24 1133 Kresky Rd., Centralia, 98531. Phone: (360) 736-1355. Fax: (360) 736-9108. Fax: (360) 736-4761. E-mail: live95@live95.com Web Site: www.live95.com. Licensee: Premier Broadcasters Inc. Population served: 76,000 Rgnl rep: Allide Law Firm: Leventhal, Senter & Lerman. Format: Hot adult contemp. News staff: one. Target aud: 25-49. ♦Rod Etherton, pres; Rob Etherton, gen mgr; Andy West, opns mgr; Rick Petty, gen sls mgr & natl sls mgr; Matt Shannon, progmg dir; Harvey Brooks, engrg dir & chief of engrg.

Winthrop

KTRT(FM)— 2007: 97.5 mhz; 330 w. Ant 1,650 ft TL: N48 19 06 W120 06 47. Hrs open: 30 Grizzly Mtn Rd., 98862. Phone: (509) 996-3125. Licensee: Tin Can Communications LLC. ♦Richard T. Mills, gen mgr.

Yakima

KATS(FM)—Listing follows KIT(AM).

KBBO(AM)—Selah, 1955: 980 khz; 5 kw-D, 500 w-N, DA-N. TL: N46 36 46 W120 28 24. Hrs open: 24 1200 Chesterley Dr., #160, 98902. Phone: (509) 248-2900. Fax: (509) 452-9661. Licensee: New Northwest Broadcasters LLC (group owner; acq 12-1-98; grpsl). Population served: 193,000 Natl. Network: ABC, USA. Rgnl rep: Allied Radio Partners. Law Firm: Dow, Lohnes & Albertson. Format: Talk, news, sports. News staff: one. Target aud: 35-64. ♦Pete Benedetti, CEO; Trila Bumstead, COO; Brent Phillipy, VP; Lou Barfelli, opns dir, progmg dir & pub affrs dir; Kit Osborne, sls dir; Ron King, natl sls mgr & rgnl sls mgr; Jenifer Wilde, prom dir; Tim Mauch, chief of engrg; Gail Dahl, traf mgr.

KXDD(FM)—Co-owned with KBBO(AM). July 1, 1971: 104.1 mhz; 61 kw. 781 ft TL: N46 30 48 W120 24 05. (CP: 100 kw, ant 1,128 ft. TL: N46 38 27 W120 23 42). Stereo. Format: Country. Target aud: 18-54. ♦Dewey Boynton, opns mgr & mus dir; Stace Whitmire, prom mgr; Gail Dahl, traf mgr.

***KDNA(FM)**— Dec 19, 1979: 91.9 mhz; 18.5 kw. 920 ft TL: N46 31 42 W120 31 03. Stereo. Hrs opn: 6 AM-midnight Box 800, 121 Sunnyside Ave., Granger, 98932. Phone: (509) 854-1900. Phone: (509) 854-1900. Fax: (509) 854-2223. E-mail: info@kdna.org Web Site: www.kdna.org.

Licensee: Northwest Communities Educational Center. Population served: 80,000 Format: Sp, informational. News staff: one; News: 8 hrs wkly. Target aud: General; Sp speaking farm workers. Spec prog: Relg 4 hrs, children 5 hrs, Sp 106 hrs wkly.

KFFM(FM)—Listing follows KUTI(AM).

KHHK(FM)— Dec 1, 1984: 99.7 mhz; 7.6 kw. 584 ft TL: N46 31 53 W120 26 58. Stereo. Hrs opn: 5 AM-midnight 1200 Chesterley Dr., Suite 160, 98902. Phone: (509) 248-2900. Fax: (509) 452-9661. Web Site: newhot997.com. Licensee: New Northwest Broadcasters LLC (group owner; acq 12-1-98; grpsl). Format: Urban contemp. Target aud: 18-34. ♦Pete Benedetti, CEO & pres; Don Morin, gen mgr; Dewey Boynton, opns mgr & progmg dir.

KIT(AM)— Apr 8, 1929: 1280 khz; 5 kw-D, 1 kw-N. TL: N46 34 19 W120 29 41. Stereo. Hrs opn: 24 4010 Summitview Ave., 98908-2966. Phone: (509) 972-3461. Fax: (509) 972-3540. Fax: (509) 972-3542. E-mail: daveetel@clearchannel.com Web Site: www.1280kit.com. Licensee: Citicasters Licenses L.P. Group owner: Clear Channel Communications Inc. (acq 10-26-99; grpsl). Population served: 190,000 Natl. Network: CBS. Natl. Rep: McGavren Guild. Law Firm: Fisher, Wayland, Cooper, Leader & Zaragoza. Format: News/talk. News staff: 2. Target aud: 25-64; professional, mature. Spec prog: Farm 6 hrs wkly. ♦Gary Donovan, exec VP; Cheryl Salomone, gen mgr; Ron Harris, opns mgr; Connie Johnston, sls dir; Dave Ettl, progmg dir; Lance Tormey, news dir; John Wilbanks, chief of engrg.

KATS(FM)—Co-owned with KIT(AM). Dec 15, 1968: 94.5 mhz; 100 kw. 850 ft TL: N46 31 59 W120 30 14. Stereo. 24 E-mail: katsfm@hotmail.com Web Site: www.katsfm.com.165,000 Format: Rock/AOR. News staff: one. Target aud: 20-45. ♦Ron Harris, mus dir & disc jockey; Lance Tormey, local news ed & news rptr; Charlie Bush, disc jockey; Ken Herman, disc jockey; Todd Lyons, disc jockey.

KJOX(AM)— 1947: 1390 khz; 5 kw-D, 500 w-N, DA-2. TL: N46 34 17 W120 27 15. (CP: 400 w-N). Hrs opn: 24 1200 Chesterly Dr., Suite 160, 98902-7345. Phone: (509) 248-2990. Fax: (509) 452-9661. Licensee: New Northwest Broadcasters LLC (group owner; acq 10-20-98; grpsl). Population served: 140,000 Natl. Network: USA. Format: Relg, Christian adult contemp. News staff: 2; News: 12 hrs wkly. Target aud: 35-64; family oriented. Spec prog: Black 2 hrs, gospel 2 hrs wkly. ♦Pete Benedetti, CEO; Trila Bumstead, COO & pres; Trila Houston, CFO; Brent Phillipy, VP; Kit Osborne, sls dir & traf mgr; Ron King, natl sls mgr & rgnl sls mgr; Jenifer Wilde, prom dir; Lou Bartelli, progmg dir & pub affrs dir; Tim Mauch, chief of engrg; Gail Dahl, traf mgr.

KRSE(FM)—Co-owned with KJOX(AM). Aug 18, 1977: 105.7 mhz; 100 kw. Ant 584 ft TL: N46 42 45 W120 37 46. Stereo. 24 Web Site: k105.com. Format: Adult contemp. Target aud: 25-64; upscale listener, primarily women. ♦Gail Dahl, prom dir & traf mgr; Kendall Weaver, opns mgr & progmg dir.

***KNWY(FM)**— Feb 20, 1993: 90.3 mhz; 5 kw. Ant 895 ft TL: N46 31 57 W120 30 37. Hrs open: 24
Rebroadcasts KFAE-FM Richland 100%.
Box 642530, 382 Murrow Ctr., Pullman, 99164-2530. Phone: (509) 335-6500. Fax: (509) 335-3772. E-mail: nwpr@wsu.edu Web Site: www.nwpr.org. Licensee: Washington State University. Law Firm: Dow, Lohnes & Albertson. Format: News, class. News staff: one; News: 37 hrs wkly. ♦Karen Olstad, COO & gen mgr; Dennis Haarsager, gen mgr; Roger Johnson, stn mgr; Scott Weatherly, opns mgr; Sarah McDaniel, dev dir & dev mgr; Mary Hawkins, progmg dir.

KUTI(AM)— Oct 19, 1944: 1460 khz; 5 kw-U, 3.7 kw-N, DA-N. TL: N46 33 29 W121 27 02. Hrs open: 24 4010 Summitview Ave., 98908-2966. Phone: (509) 972-3461. Fax: (509) 972-3540. E-mail: jackbalzer@clearchannel.com Web Site: www.1460kuti.com. Licensee: Citicasters Licenses L.P. Group owner: Clear Channel Communications Inc. (acq 10-26-99; grpsl). Population served: 10,000 Natl. Rep: McGavren Guild. Law Firm: Fisher, Wayland, Cooper, Leader & Zaragoza. Format: Classic Country. News staff: one; News: 2 hrs wkly. Target aud: 35 plus. ♦Gary Donavan, exec VP; Cheryl Salomone, gen mgr; Ron Harris, opns mgr; Jack Balzer, progmg dir; Lance Tormey, news dir; John Wilbanks, chief of engrg; Lueta Bishop, traf mgr.

KFFM(FM)—Co-owned with KUTI(AM). Aug 31, 1970: 107.3 mhz; 100 kw. 1,500 ft TL: N46 38 27 W120 23 42. Stereo. 24 E-mail: steverocha@clearchannel.com Web Site: www.kffm.com.35,000 Natl. Rep: McGavren Guild. Law Firm: Shaw Pittman. Format: CHR. News staff: one; News: one hr wkly. Target aud: 18-34. ♦Steve Rocha, progmg dir; Esther Johnson, traf mgr.

KYAK(AM)— Oct 17, 1962: 930 khz; 10 kw-D, 127 w-N. TL: N46 36 48 W120 28 51. Hrs open: 24 Box 31000, Spokane, 99223. Phone: (509) 452-5925. E-mail: kyak@kyak.com Web Site: kyak.com. Licensee: Thomas W. Read dba Yakima Christian Broadcasting. (acq 6-1-98; $150,000). Natl. Network: Salem Radio Network, USA. Law Firm: Cohen & Marks. Format: Christian. ♦Melinda Read, gen mgr; Bill Glenn, stn mgr.

***KYPL(FM)**— Oct 15, 1997: 91.1 mhz; 26 kw. Ant 797 ft TL: N46 30 48 W120 24 05. Stereo. Hrs opn: 24
Rebroadcasts KGTS(FM) College Place 95%.
606 N. Western Ave., Wenatchee, 98801. Phone: (509) 527-2991. Fax: (509) 527-2611. E-mail: plr@plr.org Web Site: www.plr.org. Licensee: Growing Christian Foundation. Population served: 225,000 Format: Christian contemp. News: 8 hrs wkly. Target aud: 35-64; family-oriented, Christian. ♦Kevin Krueger, chmn & gen mgr; Harry Watts, sls VP & gen sls mgr; Elizabeth Nelson, progmg dir; Walter Cox, engrg dir & chief of engrg.

***KYVT(FM)**— September 1980: 88.5 mhz; 3 kw. -254 ft TL: N46 35 06 W120 31 41. Stereo. Hrs opn: 24 Yakima Valley Technical Ctr., 1116 S. 15th Ave., 98902. Phone: (509) 573-5013. Phone: (509) 573-5000. Fax: (509) 573-5023. Web Site: yvtech.us. Licensee: Yakima School District No. 7. Population served: 210000 Format: Alternative. News staff: one; News: 2 hrs wkly. Target aud: 18-25; student & working people. Spec prog: Urban alternative 3 hrs wkly. ♦John Schieche, pres; Randy Beckstead, gen mgr, chief of opns & dev dir; Andy Ward, prom dir.

KYXE(AM)—See Union Gap

KZTA(FM)—Naches, Oct 25, 1988: 96.9 mhz; 14 kw. Ant 935 ft TL: N46 35 59 W120 52 08. Stereo. Hrs opn: 24 Box 2888, 98901. Secondary address: 706 Butterfield Rd. 98901. Phone: (509) 457-1000. Fax: (509) 452-0541. E-mail: kteske@bustosmedia.com Web Site: www.bustosmedia.com. Licensee: Bustos Media of Eastern Washington License LLC. (group owner; (acq 11-18-2004; grpsl). Natl. Rep: Tacher. Format: Sp. News: 3 hrs wkly. Target aud: 18-35; Hispanic. ♦Amador S. Bustos, pres; Bob Berry, gen mgr; Keith Teske, opns mgr; Jesus Rosales, sls dir & news dir; Judith McInnes, progmg dir.

West Virginia

Arnoldsburg

WSGD(FM)—Not on air, target date: unknown: 100.7 mhz; 6 kw. Ant 328 ft TL: N38 51 52 W81 07 16. Hrs open: 180 Main St., Sutton, 26601. Phone: (304) 765-3676. Licensee: Daniel W. Finch Jr. ♦Daniel W. Finch Jr., gen mgr.

Barrackville

WFGM-FM— July 1993: 93.1 mhz; 2.6 kw. Ant 495 ft TL: N39 31 23 W80 12 00. Hrs open: 24
Rebroadcasts WBTQ-FM Buckhannon 100%.
Box 189, Buckhannon, 26201. Secondary address: WBUC Rd., Buckhannon 26201. Phone: (304) 472-1460. Fax: (304) 472-1528. E-mail: B93@verizon.net Licensee: Descendants Trust, Lauren M. Kelley, trustee Group owner: McGraw/Elliott Group Stations (acq 10-18-2005; $250,000). Population served: 400,000 Natl. Network: ABC. Rgnl rep: Dome. Law Firm: Fisher, Wayland, Cooper, Leader & Zaragoza. Format: Oldies. News staff: one. Target aud: 25-54; active adults. ♦David Collett, gen sls mgr; Ryan Elliott, CFO, gen mgr, gen mgr & progmg dir; Brad Allen, news dir; Dick McGraw, chief of engrg.

Beckley

WCIR-FM—Listing follows WIWS(AM).

WIWS(AM)— Nov 14, 1966: 1070 khz; 10 kw-D. TL: N37 45 18 W81 14 12. Hrs open: 306 S. Karawha St., 25801-5619. Phone: (304) 253-7000. Fax: (304) 255-1044. Licensee: Southern Communications Corp. (acq 1976). Population served: 70,000 Format: Oldies. Target aud: 18-54; traveling motorists. ♦Jay Quesenberry, gen mgr; Rennolt Madrazo, sls dir; Rick Pizer, prom dir & progmg dir; Randy Kerbawy, chief of engrg; Rhonda Pritt, traf mgr.

WCIR-FM—Co-owned with WIWS(AM). June 1971: 103.7 mhz; 5 kw. 1,485 ft TL: N37 56 51 W81 18 32. Stereo. 24 E-mail: 103cir@103cir.com Web Site: www.103cir.com.194,000 Natl. Rep: Katz Radio. Law Firm: Borsari & Paxson. Format: CHR. News staff: 2; News: 2 hrs wkly. Target aud: 25-54. ♦Rhonda Pritt, traf mgr.

***WJJJ(FM)**—Not on air, target date: unknown: 88.1 mhz; 1 kw vert. Ant 1,059 ft TL: N37 35 20 W81 06 52. Hrs open: 420 Montcall Dr., Charleston, 25302. Licensee: Shofar Broadcasting Corp.

WJLS-FM— Nov 6, 1946: 99.5 mhz; 34 kw. 1,050 ft TL: N37 35 23 W81 06 51. Stereo. Hrs opn: 24 Box 5499, 25801. Secondary address: WJLS Bldg., 102 N. Kanawha St. 25801. Phone: (304) 253-7311. Fax: (304) 253-3466. E-mail: dawg@wjls.com Web Site: www.wjls.com. Licensee: First Media Radio LLC. (group owner; (acq 2-1-2002; $3.6 million with co-located AM). Population served: 349,000 Natl. Network: CNN Radio. Rgnl rep: Dome, Rgnl Reps. Wire Svc: AP Format: Country. News staff: 15; News: 10 hrs wkly. Target aud: 25-54. ♦Mark Reid, gen mgr, sls dir & gen sls mgr; Darrell Ramsay, progmg dir; Bob Cannon, news dir; Charles Marlow, chief of engrg; Maria Marvin, traf mgr.

WJLS(AM)— Mar 5, 1939: 560 khz; 4.5 kw-D, 470 w-N, DA-N. TL: N37 45 32 W81 11 12.24 Web Site: www.wjls.com.234,000 Natl. Network: CNN Radio. Natl. Rep: Dome, Rgnl Reps. Format: Relg, southern gospel. News staff: 5; News: 4 hrs wkly. Target aud: 25-54. Spec prog: Sports 3 hrs wkly. ♦Sandi Smith-Milam, progmg dir; Gary Hosey, mus dir.

WOAY(AM)—See Oak Hill

***WVPB(FM)**— May 1, 1974: 91.7 mhz; 10.5 kw. 917 ft TL: N37 53 46 W80 59 21. Stereo. Hrs opn: 24 600 Capitol St., Charleston, 25301. Phone: (304) 556-4900. Fax: (304) 556-4960. E-mail: feedback@wvpubcast.org Web Site: www.wvpubcast.org. Licensee: West Virginia Educational Broadcasting Authority. Population served: 96,000 Natl. Network: NPR, PRI. Format: News, class, jazz. ♦Marilyn DiVita, gen mgr & dev dir; Rita Ray, gen mgr; James Muhammad, progmg dir; Greg Collard, news dir; Jack Wells, chief of engrg; Teresa Willis, traf mgr.

WWNR(AM)— Aug 9, 1946: 620 khz; 5 kw-D, 25 w-N. TL: N37 45 18 W81 14 12. Hrs open: 5 AM-midnight 306 S. Kanawha St., 25801. Phone: (304) 253-7000. Fax: (304) 255-1044. E-mail: wwnr@netphase.net Web Site: www.newstalk620.com. Licensee: Southern Communications Corp. (group owner; acq 1-26-2004). Population served: 265,000 Natl. Network: CBS Radio. Law Firm: Pepper & Corazzini. Format: News/talk. News staff: 2; News: 35 hrs wkly. Target aud: 25-54. ♦R. Shane Southern, pres; Jay Quesenberry, gen mgr, stn mgr & opns mgr; Rennold Madrazo, sls dir; Shane Sothern, gen sls mgr; Rick Rizer, prom dir & progmg dir; Warren Ellison, news dir; Randy Kerbawy, chief of engrg; Rhonda Pritt, traf mgr.

Berkeley Springs

WCST(AM)— Sept 7, 1958: 1010 khz; 250 w-D, 17 w-N. TL: N39 37 00 W78 13 03. Hrs open: 440 Radio Station Ln., 25411. Phone: (304) 258-1010. Fax: (304) 258-1976. E-mail: wdhc@stargate.net Licensee: Capper Broadcasting Co. Population served: 350,000 Format: Country. ♦Michael Fagan, gen mgr, sls dir & progmg dir; Fran Little, chief of engrg; Paige Wagner, traf mgr.

WDHC(FM)—Co-owned with WCST(AM). December 1965: 92.9 mhz; 3.2 kw. Ant 456 ft TL: N39 37 00 W78 13 03. Stereo. 6 AM-midnight ♦Mike Fagan, gen mgr & traf mgr.

Bethany

***WVBC(FM)**— Jan 1, 1967: 88.1 mhz; 1.1 kw. 410 ft TL: N40 12 58 W80 33 31. Stereo. Hrs opn: Bethany House, Bethany College,

26032. Phone: (304) 829-7853. Licensee: Bethany College. Population served: 225,000 Natl. Network: AP Radio. Wire Svc: AP Format: Div, educ, progsv. News: 1 hr wkly. Target aud: 18-34; high school & college students. Spec prog: Christian rock 4 hrs, folk 2 hrs, classic rock 8 hrs, class 2 hrs wkly. ♦Patrick Sutherland, gen mgr.

Bethlehem

WUKL(FM)— Feb 27, 2004: 105.5 mhz; 13.5 kw. Ant 312 ft TL: N40 03 17 W80 42 26. Stereo. Hrs opn: 6 am-midnight Box 448, Bellaire, OH, 43906. Phone: (740) 676-5661. Fax: (740) 676-2742. Web Site: oldiesradioonline.com. Licensee: Keymarket Licenses LLC Group owner: Keymarket Communications LLC (acq 2-4-2004; $1.35 million). Population served: 44,369 Natl. Network: ABC. Format: Oldies. News staff: one. Target aud: 35-54. ♦Gerald A. Getz, pres.

Blennerhassett

***WPJY(FM)**—Not on air, target date: unknown: 88.7 mhz; 9 kw. Ant 374 ft TL: N39 14 06 W81 53 16. Hrs open: Box 889, Blacksburg, VA, 24063. Phone: (540) 552-4252. Fax: (540) 951-5282. Licensee: Positive Alternative Radio Inc. ♦Vernon H. Baker, pres.

Bluefield

WHAJ(FM)—Listing follows WHIS(AM).

WHIS(AM)— June 27, 1929: 1440 khz; 5 kw-D, 500 w-N. TL: N37 16 33 W81 15 06. Stereo. Hrs opn: 900 Bluefield Ave., 24701. Phone: (304) 327-7114. Fax: (304) 325-7850. Licensee: Monterey Licenses LLC. Group owner: Triad Broadcasting Co. LLC (acq 7-18-00; grpsl). Population served: 26,200 Format: News/talk. Target aud: 35 plus; upper income, leaders of the community. ♦John Halford, gen mgr; Danny Clemons, sls dir; Joseph Echles, progmg dir; Keith Bowman, chief of engrg.

WHAJ(FM)—Co-owned with WHIS(AM). Apr 23, 1963: 104.5 mhz; 100 kw. 1,200 ft TL: N37 15 21 W81 10 55. Stereo. Web Site: www.1045.com.151,000 Format: Adult contemp. Target aud: 25-54. ♦Dave Harris, progmg dir; Jackie White, traf mgr.

WKEZ(AM)— May 18, 1948: 1240 khz; 1 kw-U. TL: N37 15 57 W81 11 20. Hrs open: 24 900 Bluefield Ave., 24701. Phone: (304) 327-7114. Fax: (304) 325-7850. Licensee: Monterey Licenses LLC. Group owner: Triad Broadcasting Co. LLC (acq 7-18-2000; grpsl). Population served: 15,921 Format: Sports. Spec prog: Relg 5 hrs wkly. ♦David Benjamin, CEO & pres; John Halford, gen mgr, natl sls mgr & rgnl sls mgr; Danny Clemons, gen sls mgr; Ed Weiland, progmg dir; Keith Bowman, engrg mgr; Patty Davis, traf mgr.

***WPIB(FM)**— September 1995: 90.9 mhz; 740 w horiz, 700 w vert. 1,102 ft TL: N37 15 26 W81 10 43. Hrs open: Box 889, Blacksburg, VA, 24063. Fax: (540) 200-3215. E-mail: office@spiritfm.com Web Site: www.spiritfm.com. Licensee: Positive Alternative Radio Inc. Group owner: Baker Family Stations (Positive Radio Group) (acq 4-22-92). Natl. Network: USA. Format: Adult contemp, contemp Christian. ♦Vernon H. Baker, pres; Edward A. Baker, opns VP.

Bridgeport

WDCI(FM)— June 29, 1991: 104.1 mhz; 3 kw. 328 ft TL: N39 17 59 W80 17 30. (CP: 2.45 kw, ant 518 ft.). Stereo. Hrs opn: 24 Box 360, 26330. Phone: (304) 842-8644. Fax: (304) 842-8653. E-mail: rtgresak@aol.com Licensee: WDCI Radio Inc. (acq 8-18-98; $405,000). Population served: 125,000 Natl. Network: Jones Radio Networks. Rgnl rep: Dome. Law Firm: William D. Silva. Format: Soft adult contemp. Target aud: 25-54. ♦Bruce Wallace, pres & gen mgr; Tom Thompson, dev mgr, gen sls mgr, mktg mgr, prom mgr & adv mgr; Tina Grefak, progmg mgr; Hank Vest, chief of engrg.

Buckhannon

WBRB(FM)—Listing follows WBUC(AM).

WBTQ(FM)— 1984: 93.5 mhz; 16 kw. 417 ft TL: N38 58 11 W80 01 58. Stereo. Hrs opn: 24 189 WBUC Rd., 26201. Phone: (304) 472-1400. Fax: (304) 472-1528. Licensee: Elkins Radio Corp. Group owner: McGraw/Elliott Group Stations (acq 1996; $205,000). Population served: 110,000 Natl. Network: ABC. Rgnl rep: Dome. Law Firm: Reddy, Begley & McCormick, LLP. Format: Adult contemp. News staff: one; News: 3 hrs wkly. Target aud: 18-49. ♦Brian Elliott, gen mgr, pub affrs dir, sls & prom; Richard McGraw, chief of engrg & engr; Wendy Dogas, traf mgr.

WBUC(AM)— Dec 13, 1959: 1460 khz; 5 kw-D, 87 w-N. TL: N39 00 07 W80 15 50. Stereo. Hrs opn: Sunrise-sunset Box 2377, WBUC Rd., Rt. 33, 26201. Phone: (304) 472-1460. Fax: (304) 472-1528. Licensee: Cat Radio Inc. Group owner: McGraw/Elliott Group Stations (acq 9-4-86; $395,000; FTR: 8-4-86). Format: Talk. News staff: one. Target aud: General. Spec prog: Relg 7 hrs wkly. ♦Dick McGraw, CEO & chief of engrg; Karen McGraw, pres; Harry Elliott, CFO; Todd Elliott, gen mgr & opns VP; Brian Elliott, sls VP, gen sls mgr & adv VP; Ron Roth, prom VP & progmg VP; Tamara Cicogna, adv mgr; Nancy Boyce, news dir & pub affrs dir; Jan Harr, traf mgr; Bill Austin, disc jockey.

WBRB(FM)—Co-owned with WBUC(AM). June 16, 1990: 101.3 mhz; 50 kw. 497 ft TL: N38 56 40 W80 10 46. Stereo. 24 400000 Format: Country. News staff: one; News: 4 hrs wkly. Target aud: 25-54; upper middle class, white collar, craftsman. ♦Jan Harr, traf mgr; Steve Shupp, disc jockey; Theresa Brown, disc jockey.

***WVPW(FM)**— September 1968: 88.9 mhz; 14 kw. 840 ft TL: N39 02 04 W80 33 47. Stereo. Hrs opn: 24 600 Capitol St., Charleston, 25301. Phone: (304) 556-4900. Fax: (304) 556-4960. E-mail: feedback@wvpubcast.org Web Site: www.wvpubcast.org. Licensee: West Virginia Education Broadcasting Authority. Natl. Network: NPR, PRI. Format: News, class, jazz. ♦Marilyn DiVita, gen mgr & dev dir; Rita Ray, gen mgr; James Muhammad, progmg dir; Greg Collard, news dir; Jack Wells, engrg dir; Teresa Wills, traf mgr.

***WVWC(FM)**— Sept 15, 1997: 92.1 mhz; 10 w. 85 ft TL: N38 59 24 W80 13 10. Hrs open: Box 167, 59 College Ave., 26201-2999. Phone: (304) 473-8292. Fax: (304) 472-2571. E-mail: c92@wvwc.edu Web Site: www.wvwc.edu/c92. Licensee: West Virginia Wesleyan College. Population served: 6,261 Format: Classic rock, progsv, div. Target aud: 12-25; high school & college audience. Spec prog: Black 4 hrs, class 2 hrs, jazz 8 hrs, relg 4 hrs, blues 2 hrs wkly. ♦Phillips B. Kolsun, gen mgr.

Charles Town

WMRE(AM)— May 28, 1962: 1550 khz; 5 kw-D, DA. TL: N39 16 23 W77 51 56. Hrs open: Sunrise-sunset 510 Pegasus Ct., Winchester, VA, 22602-4596. Phone: (540) 662-5101. Fax: (540) 662-8610. Licensee: AMFM Radio Licenses LLC. Group owner: Clear Channel Communications Inc. (acq 2-16-2001; $1.525 million with co-located FM). Natl. Rep: Roslin. Format: Adult standards. News staff: 2. Target aud: 25-54; adults. Spec prog: Relg. ♦Chuck Peterson, gen mgr; David Miller, opns mgr; Marcella Vance, sls dir; Justin Maglione, prom dir; Maark Kesner, chief of engrg; Krissy Groves, traf mgr.

WKSI-FM—Co-owned with WMRE(AM). Aug 28, 1966: 98.3 mhz; 1.75 kw. Ant 617 ft TL: N39 10 38 W78 15 53. Stereo. 24 Web Site: wxva.com.150,000 Format: Country. Target aud: 25-54. ♦Jay Flanagan, mus dir; Krissy Groves, traf mgr.

Charleston

WBES(AM)—Dunbar, Nov 4, 1946: 1240 khz; 1 kw-U. TL: N38 23 08 W81 42 51. Hrs open: Box 871, 25323. Secondary address: 4250 Washington St. 25313. Phone: (304) 744-7020. Fax: (304) 744-8562. Licensee: Bristol Broadcasting Co. Inc. (group owner; acq 8-90; grpsl). Format: Talk, sports. Target aud: 18-49. ♦Mike Robinson, gen mgr; John Gush, natl sls mgr & rgnl sls mgr; Dan King, rgnl sls mgr.

WVSR-FM—Co-owned with WBES. September 1964: 102.7 mhz; 50 kw. Ant 403 ft TL: N38 21 26 W81 40 05. Stereo. Web Site: www.electric102.com.250,000 Format: CHR.

WCHS(AM)— Sept 15, 1927: 580 khz; 5 kw-U, DA-N. TL: N38 21 49 W81 46 05. Hrs open: 1111 Virginia St. E., 25301. Phone: (304) 342-8131. Fax: (304) 344-4745. E-mail: 58live@wvradio.com Licensee: West Virginia Radio Corp. of Charleston. (acq 6-1-92; $1.74 million with co-located FM; FTR: 6-22-92) Natl. Network: CBS. Natl. Rep: McGavren Guild. Format: News/talk, sports. News staff: 2. Target aud: 25-54. ♦Dale Miller, pres & opns dir; Sean Banks, pres & gen mgr; Noel Richardson, opns VP & chief of engrg; Rick Johnson, opns dir & progmg dir; Sara Shingleyon, traf mgr.

WKWS(FM)—Co-owned with WCHS(AM). Sept 16, 1969: 96.1 mhz; 50 kw. 360 ft TL: N38 21 24 W81 36 19. (CP: Ant 550 ft. TL: N38 21 51 W81 46 05). Stereo. Web Site: www.kick96.com. Format: Hot country hits. ♦Rick Johnson, opns mgr & pub affrs dir; Christian Miller, gen sls mgr; John Anthony, progmg VP; Sara Shingleton, traf mgr.

WKAZ(AM)— 1946: 680 khz; 50 kw-D, 250 w-N, DA-2. TL: N38 19 15 W81 36 31. Hrs open: 24 1111 Virginia St. E., 25301. Phone: (304) 342-8131. Fax: (304) 344-4745. Licensee: West Virginia Radio Corp. of Charleston. (acq 7-14-93; $1.1 million with co-located FM; FTR: 8-9-93) Population served: 225,000 Natl. Network: ABC. Format: Classic country. News staff: one. Target aud: 25-54. ♦Gregg Smith, progmg dir.

WVAF(FM)—Co-owned with WKAZ(AM). Feb 1, 1965: 99.9 mhz; 50 kw. 490 ft TL: N38 19 15 W81 36 31. Stereo. Web Site: www.v100radio.com.300,000 Format: Adult contemp. Target aud: Female skew. ♦Dale Miller, gen sls mgr; Greg Johnson, asst music dir; Nikki Walters, pub affrs dir; Doug Daniels, traf mgr, sports cmtr, sports cmtr & disc jockey; Rich Johnson, progmg dir & traf mgr; Denise Daniels, disc jockey; Noel Richardson, disc jockey.

WKAZ-FM—See Miami

WQBE-FM—Listing follows WVTS(AM).

WSWW(AM)— 1939: 1490 khz; 1 kw-U. TL: N38 21 28 W81 37 00. Hrs open: 24 1111 Virginia St. E., 25301. Phone: (304) 342-8131. Fax: (304) 344-4745. Licensee: West Virginia Radio Corp. (group owner; (acq 6-5-97; $2.15 million with WKAZ-FM Miami). Population served: 250,000 Natl. Rep: D & R Radio. Format: ESPN-All sports/talk. ♦John Raese, chmn; Dale Miller, pres; Sean Banks, gen mgr; Dave Harmon, opns dir & progmg dir; Vince Wardell, gen sls mgr; Noel Richardson, chief of engrg; Sarah Shingleton, traf mgr.

***WVPN(FM)**— May 8, 1979: 88.5 mhz; 50 kw. Ant 299 ft TL: N38 22 32 W81 29 25. Stereo. Hrs open: 24 600 Capitol St., 25301. Phone: (304) 556-4900. Fax: (304) 556-4960. E-mail: feedback@wvpubcast.org Web Site: www.wvpubcast.org. Licensee: West Virginia Educational Broadcasting Authority. Population served: 100,000 Natl. Network: NPR, PRI. Format: News, class, jazz. ♦Rita Ray, gen mgr; Marilyn DiVita, dev mgr; James Muhammad, progmg dir; Laura Harbert-Allen, mus dir; Greg Collard, news dir; Jack Wells, engrg dir; Teresa Wills, traf mgr.

WVTS(AM)— Feb 16, 1957: 950 khz; 5 kw-D, 1 kw-N, DA-N. TL: N38 23 00 W81 42 52. Hrs open: Box 871, 4250 Washington St. W., 25323. Phone: (304) 744-7020. Fax: (304) 744-8562. Licensee: Bristol Broadcasting Co. Inc. (group owner; acq 5-1-64). Population served: 640,500 Natl. Rep: McGavren Guild. Format: Talk. ♦Mike Robinson, gen mgr; John Gush, natl sls mgr & rgnl sls mgr; Dan King, rgnl sls mgr.

WQBE-FM—Co-owned with WVTS(AM). Feb 16, 1957: 97.5 mhz; 50 kw. 500 ft TL: N38 24 22 W81 43 26. Web Site: www.wqbe.com. Format: Educ.

Clarksburg

WGIE(FM)—Listing follows WXKX(AM).

***WKJL(FM)**— October 1992: 88.1 mhz; 32.5 kw. 712 ft TL: N39 18 02 W80 20 37. Stereo. Hrs opn: 19
Rebroadcasts WAIJ(FM) Grantsville, Md. 100%.
Box 540, He's Alive Inc. Corp. Offices, 34 Springs Rd., Grantsville, MD, 21536-0540. Phone: (301) 895-3292. Fax: (301) 895-3293. E-mail: hesalive@hesalive.net Web Site: www.hesalive.net. Licensee: He's Alive Inc. Natl. Network: USA. Format: Gospel, adult contemp, Christian, relg. Target aud: 18-35. ♦Dewayne Johnson, pres; Melissa Flores, gen mgr & stn mgr; Tim Eutin, progmg dir; Stan Falor, chief of engrg; Brandon Hutzell, traf mgr.

WOBG(AM)— Apr 12, 1936: 1400 khz; 1 kw-U. TL: N39 17 46 W80 18 16. Hrs open: 24 1489 Locust Ave., #C, Fairmont, 26554-1337. Secondary address: Old Weatherservice Bldg., Old Rt. 50 E. 26301. Phone: (304) 624-1400. Fax: (304) 624-1402. Licensee: Clarksburg Radio Co. Group owner: Burbach Broadcasting Group (acq 1-11-99; $330,000 with WOBG-FM Salem). Rgnl rep: Commercial Media Sales. Format: Adult standards. Target aud: 35 plus. ♦Nicholas A. Galli, pres; David Bronham, gen mgr & gen sls mgr; Greg Bolyard, progmg dir; Larry Smith, chief of engrg.

WPDX(AM)— Aug 19, 1947: 750 khz; 1 kw-D. TL: N39 14 40 W80 23 05. Hrs open: 7013 Mountain Park Dr., Fairmont, 26554. Phone: (304) 363-3851. Fax: (304) 363-3852. E-mail: wpdx@iolinc.net Licensee: Tschudy Broadcasting Corp. Group owner: Tschudy Broadcast Group (acq 11-5-91; $405,000 with co-located FM; FTR: 12-2-91) Population served: 400,000 Natl. Network: AP Radio. Rgnl rep: Commercial Media Sales Format: Music of your life. Target aud: 35-54; male & female.

WPDX-FM— Aug 19, 1974: 104.9 mhz; 7.4 kw. Ant 597 ft TL: N39 15 22 W80 06 46. Stereo. 24 (Acq 1992).400,000 Natl. Network: AP Radio. Rgnl rep: Commercial Media Sales Format: Classic country. News: 5 hrs wkly. Target aud: 35-54; blue collar.

WWLW(FM)— 1973: 106.5 mhz; 50 kw. 500 ft TL: N39 11 14 W80 32 45. Stereo. Hrs opn: 24 1065 Radio Park Dr., Mt. Clare, 26408. Phone: (304) 623-6546. Fax: (304) 623-6547. Web Site: www.wvmagic.com. Licensee: West Virginia Radio Corp. of Clarksburg. (acq 3-2-93; $1.2 million; FTR: 3-22-93). Population served: 400,000 Format: Adult contemp. News staff: one; News: 2 hrs wkly. ♦Dale B. Miller, pres; Christian Miller, stn mgr; Chad Perry, opns VP, opns dir & progmg dir; Tim Brady, gen sls mgr & news dir; Steve Lough, chief of engrg; Donna Tubolino, traf mgr.

WXKX(AM)— Nov 28, 1946: 1340 khz; 1 kw-U. TL: N39 17 27 W80 18 56. Hrs open: 24 1489 Locust Ave., #C, Fairmont, 26554-1337. Phone: (304) 624-1400. Fax: (304) 624-1402. Licensee: Burbach of DE LLC. Group owner: Burbach Broadcasting Group (acq 11-2-00; $435,000 cash with co-located FM). Population served: 65,000 Natl. Network: ESPN Radio. Natl. Rep: Roslin, Rgnl Reps. Law Firm: Baraff, Koerner & Olender. Format: Sports. News: 14 hrs wkly. Target aud: 35 plus; office workers, retirees, upper income. Spec prog: Pittsburgh Pirates, Alderson-Broadus College basketball, high school football. ♦Nick Galli, pres; Dan Barham, gen mgr; David Branham, gen sls mgr; Greg Bolgard, prom dir; Larry Smith, chief of engrg; Debbie Southern, traf mgr.

WGIE(FM)—Co-owned with WXKX(AM). 1975: 92.7 mhz; 620 w. 600 ft TL: N39 17 27 W80 18 56. Stereo. 24
Rebroadcasts WGYE(FM) Fairmont 100%.
250,000 Format: Country. News: 3 hrs wkly. Target aud: 25-45; young professionals. ♦Dave Sturm, sports cmtr; Debbie Southrn, traf mgr & disc jockey.

Danville

WZAC-FM— Oct 9, 1989: 92.5 mhz; 610 w. 697 ft TL: N38 05 01 W81 48 17. Stereo. Hrs opn: Box 87, 25053. Secondary address: 457 Main St., Madison 25130. Phone: (304) 369-5200. Phone: (304) 369-5201. Fax: (304) 369-5200. Licensee: Price Broadcasting Co. Format: Classic country. Target aud: General. ♦Wayne Price, gen mgr.

Dunbar

WBES(AM)—Licensed to Dunbar. See Charleston

WZJO(FM)— Oct 13, 1988: 94.5 mhz; 9.6 kw. Ant 525 ft TL: N38 25 11 W81 43 24. Stereo. Hrs opn: 24 Box 871, Charleston, 25323. Secondary address: 4250 Washington St., Charleston 25313. Phone: (304) 744-7020. Fax: (304) 744-8562. Web Site: www.mix945online.com. Licensee: Bristol Broadcasting Co. Inc. (group owner; acq 1996; grpsl). Natl. Network: Moody, Westwood One. Natl. Rep: Dome, Rgnl Reps, Katz Radio. Format: Adult contemp hits of the 80s. Spec prog: Class 2 hrs wkly. ♦Mike Robinson, gen mgr; Barrie Hamm, prom mgr; Dave Evans, progmg dir; Randy Justice, chief of engrg.

Elizabeth

WRZZ(FM)— 1986: 106.1 mhz; 3 kw. 469 ft TL: N39 09 48 W81 26 12. Stereo. Hrs opn: 24 5 Rosemar Cir., Parkersburg, 26104. Phone: (304) 485-4565. Fax: (304) 424-6955. Web Site: www.classicrockz106.com. Licensee: Burbach of DE LLC. Group owner: Clear Channel Communications Inc. (acq 8-8-2005; $750,000). Population served: 200,000 Natl. Network: Westwood One. Law Firm: Koerner & Olender. Format: Classic rock. Target aud: 25-54; baby boomer rock listeners. ♦Don Staats, gen mgr.

Elkins

***WBHZ(FM)**— 1999: 91.9 mhz; 280 w. 1,118 ft TL: N38 52 18 W79 55 39. Hrs open: Box 2440, Tupelo, MS, 38803. Phone: (662) 844-8888. Fax: (662) 842-6791. Web Site: www.afr.net. Licensee: American Family Association. Group owner: American Family Radio Format: Christian. ♦Marvin Sanders, gen mgr.

WDNE(AM)— February 1948: 1240 khz; 1 kw-U. TL: N38 55 25 W79 51 33. Hrs open: 21 Box 1337, 26241. Phone: (304) 636-1300. Fax: (304) 636-2200. E-mail: wdne@wvradio.com Licensee: West Virginia Radio Corp. of Elkins. Group owner: West Virginia Radio Corp. (acq 6-17-97; $750,000 with co-located FM). Population served: 8,287 Format: Adult Standards. News staff: one. Target aud: 18 plus. ♦Rick Cooper, gen mgr & gen sls mgr; Howard Swick, prom dir; Roger Taylor, progmg dir; Joe Gaynor, mus dir; Noel Richardson, chief of engrg; Fay Cowgill, traf mgr.

WDNE-FM— June 15, 1985: 98.9 mhz; 3 kw. 328 ft TL: N38 51 53 W79 48 26. (CP: 5.1 kw, ant 725 ft. TL: N38 54 36 W79 47 18). Stereo. 21 Format: Country. ♦Fay Cowgill, traf mgr.

WELK(FM)— Oct 17, 1982: 94.7 mhz; 5 kw. 728 ft TL: N38 54 43 W79 47 19. Stereo. Hrs opn: 24 228 Randolph Ave., 26241. Phone: (304) 636-8800. Fax: (304) 636-8801. E-mail: radiosales@3wlogic.net Licensee: Elkins Radio Corp. Group owner: McGraw/Elliott Group Stations. Population served: 138,000 Natl. Network: CNN Radio. Law Firm: Shaw Pittman. Format: Adult contemp, Top-40. News staff: one; News: 3 hrs wkly. Target aud: 18-49; female. ♦Richard H. McGraw, CEO; Karen McGraw, pres; Harry Elliot, CFO; Todd Elliott, VP, gen mgr & opns dir; Brian Elliott, gen sls mgr; Brad Elliott, progmg dir; Jane Birdsong, news dir & pub affrs dir; Bill Davisson, chief of engrg; Connie Cade, traf mgr.

Fairmont

WKKW(FM)— October 1975: 97.9 mhz; 32 kw. 600 ft TL: N39 25 04 W80 03 44. Stereo. Hrs opn: 1251 Earl L. Core Rd., Morgantown, 26505. Phone: (304) 296-0029. Fax: (304) 296-3876. E-mail: jshaffer@wvradio.com Licensee: Descendants Trust, Lauren M. Kelley, trustee. (acq 9-13-00; $1.5 million). Population served: 280,000 Law Firm: Putbrese, Hunsaker & Trent. Format: Country. Target aud: 25-54; young professionals. ♦Dave Jecklin, gen mgr; Christian Miller, sls dir; John Thomas, progmg dir.

WMMN(AM)— Dec 22, 1928: 920 khz; 5 kw-U, DA-N. TL: N39 28 03 W80 12 20. (CP: 5 kw-D, 200 w-N, DA-N). Hrs opn: Box 1549, 26555. Phone: (304) 366-3700. Fax: (304) 366-3706. Web Site: www.920wmmn.com. Licensee: Fantasia Broadcasting Inc. Population served: 400,000 Rgnl rep: Commercial Media Sales. Format: Sports. Target aud: 45 plus. ♦Nick L. Fantasia, pres, gen mgr & opns mgr.

WRLF(FM)—Listing follows WTCS(AM).

WTCS(AM)— January 1948: 1490 khz; 1 kw-U. TL: N39 28 19 W80 08 27. Hrs open: 24 Box 1549, 26555. Secondary address: 450 Leonard Ave. 26554. Phone: (304) 366-3700. Fax: (304) 366-3706. Licensee: Fairmont Broadcasting Co. (acq 5-1-56). Population served: 26,093 Natl. Network: CNN Radio. Rgnl rep: Commercial Media Sales. Law Firm: Putbrese, Hunsaker & Trent. Format: News/talk. Target aud: 45 plus. Spec prog: It 3 hrs wkly. ♦Nick Fantasia, gen mgr; Bill Dunn, gen sls mgr; Bob Ice, chief of engrg; Jerri Adams, traf mgr.

WRLF(FM)—Co-owned with WTCS(AM). Aug 26, 1989: 94.3 mhz; 3.6 kw. 249 ft TL: N39 28 03 W80 12 20. Stereo. 24 Web Site: www.wrlf.com.110,000 Format: Classic rock. Target aud: 25-55; 60% male, 40% female.

Fisher

WELD(AM)— Aug 1, 1956: 690 khz; 2 kw-D. TL: N39 03 08 W79 00 21. Hrs open: 24 126 Kessel Rd., 26818. Phone: (304) 538-6062. Fax: (304) 538-7032. E-mail: WELD@hardynet.com Web Site: weldamfm.com. Licensee: Thunder Associates LLC (acq 02-01-04 $600,000 with WELD-FM Petersburg). Population served: 130,000 Natl. Network: ABC. Natl. Rep: Dome. Law Firm: Irwin, Campbell, Tannenwald. Format: Oldies. News: 3 hrs wkly. Target aud: 25+; 25+. Spec prog: Farm 3 hrs, relg 10 hrs wkly. ♦Curtis Durst, pres; Sandra Durst, exec VP; Alan Yokum, gen mgr.

WQWV(FM)— July 1998: 103.7 mhz; 310 w. 1,384 ft TL: N39 02 16 W79 05 23. Hrs open: 24 Box 55, Petersburg, 26847. Secondary address: 2 Alt Ave., Petersburg 26847. Phone: (304) 257-4432. Fax: (304) 257-9733. E-mail: wqwv@wqwv.com Web Site: www.wqwv.com. Licensee: McGuire Broadcasting L.L.C. (acq 5-20-99). Population served: 35,000 Natl. Network: CNN Radio. Format: Contemp hit/Top-40, Variety/diverse. ♦Eric McGuire, gen mgr; Kevin Spencer, opns mgr; Angel Blizzard, gen sls mgr.

Fort Gay

***WFGH(FM)**— June 4, 1973: 90.7 mhz; 7.8 kw. 205 ft TL: N38 07 58 W82 35 37. Hrs open: 24 Box 410, 25514. Phone: (304) 648-5752. Fax: (304) 648-5447. E-mail: wfgh907@radio.com Web Site: www.tolisarebels.org/tech/broadcasting.htm. Licensee: Wayne County Board of Education. Population served: 20,000 Wire Svc: AP Format: Oldies, country, gospel. News staff: 2; News: 15 hrs wkly. Target aud: General. Spec prog: Oldies. ♦Hazel B. Damron, progmg dir; Vernon R. Stanfill, gen mgr, opns dir & chief of engrg.

Frost

***WVMR(AM)**— Aug 17, 1981: 1370 khz; 5 kw-D. TL: N38 17 25 W79 55 52. Hrs open: Box 139, Rte. 1, Dunmore, 24934. Phone: (304) 799-6004. Fax: (304) 799-7444. E-mail: amrinet@starband.net Licensee: Pocahontas Communications Cooperative Corp. Population served: 10,000 Format: Country. Target aud: General. Spec prog: Farm 5 hrs, relg 10 hrs, big band 3 hrs, bluegrass 5 hrs wkly. ♦Bill Ellenburg, pres.

Glenville

WVRW(FM)—Not on air, target date: unknown: 107.7 mhz; 1.55 kw. Ant 653 ft TL: N38 54 22.7 W80 49 47.6. Hrs open: 1309 N. River Ave., Weston, 26452. Phone: (304) 269-3020. Licensee: Della Jane Woofter. ♦Della Jane Woofter, gen mgr.

Grafton

***WDKL(FM)**— Sept 10, 1979: 95.9 mhz; 3 kw. 150 ft TL: N39 21 16 W80 01 27. Stereo. Hrs opn: 24 2351 Sunset Blvd., Suite 170-218, Rocklin, CA, 95765. Phone: (916) 251-1600. Fax: (916) 251-1650. E-mail: klove@klove.com Web Site: www.klove.com. Licensee: Educational Media Foundation Group owner: EMF Broadcasting. (acq 5-28-02). Natl. Network: K-Love. Law Firm: Shaw Pittman. Format: Contemp Christian. News staff: 3. Target aud: 25-44; female-Judeo Christian. ♦Richard Jenkins, pres; Mike Novak, VP; Keith Whipple, dev dir; David Pierce, progmg dir; Ed Lenane, news dir; Sam Wallington, engrg dir; Karen Johnson, news rptr; Marya Morgan, news rptr; Richard Hunt, news rptr.

WTBZ(AM)— January 1948: 1260 khz; 500 w-D. TL: N39 21 01 W80 02 40. Hrs open: 24 hrs Box 2, 26354. Phone: (304) 265-2200. Fax: (304) 265-0972. E-mail: wtbz@go.com Licensee: Appalachian Radio LLC (acq 8-12-2002). Law Firm: Reddy, Begley & McCormick LLP. Format: Adult contemp. News: 15 hrs wkly. Target aud: 18-65. ♦Melanie Tocco, gen mgr.

Green Valley

WAMN(AM)— January 1987: 1050 khz; 1.5 kw-D. TL: N37 18 20 W81 07 30. (CP: 1050 khz; 1.43 kw-D, 250 w-N). Hrs opn: Box 6350, Bluefield, 24701. Secondary address: 4415 Blue Prince Rd. 24701. Phone: (304) 327-9266. Phone: (304) 327-9140. Fax: (304) 325-8058. Web Site: www.thesportsaddictnetwork.com. Licensee: WAMN Inc. Group owner: Baker Family Stations (acq 2-8-89). Format: ESPN. ♦Vernon H. Baker, pres; Amy Burnette, gen mgr.

Hinton

WMTD(AM)— Jan 11, 1963: 1380 khz; 1 kw-D. TL: N37 40 49 W80 54 24. Hrs open: 24
Rebroadcasts WAXS(FM) Oak Hill 75%.
306 S. Kanawha St., Beckley, 25801. Phone: (304) 253-7000. Fax: (304) 255-1044. Licensee: Southern Communications Corp. (group owner; (acq 4-19-2000; $107,000 with co-located FM). Population served: 38,000 Natl. Network: CBS. Format: News/talk, oldies. News staff: 3; News: 9 hrs wkly. Target aud: General. ♦R. Shane Southern, pres; Jay Quesenberry, gen mgr; Rennold Madrazo, sls dir; Steve Coleman, rgnl sls mgr; Rick Rizer, progmg VP & chief of engrg; Randy Kerbawy, engrg VP; Rhonda Pritt, traf mgr.

WMTD-FM— Oct 1, 1985: 102.3 mhz; 160 w. 1,008 ft TL: N37 42 56 W80 56 55. (CP: 368 w, ant 1,273 ft.). Stereo. 24 38,000 News staff: one; News: 1 hr wkly. Target aud: Adults 18-49. ♦Rhonda Pritt, traf mgr.

Huntington

WAMX(FM)—Milton, Oct 1, 1980: 106.3 mhz; 560 w. 1,092 ft TL: N38 30 21 W82 12 33. Stereo. Hrs opn: 24 134 4th Ave., 25701. Phone: (304) 525-7788. Fax: (304) 525-3299. E-mail: x1063@x1063.com Web Site: www.x1063.com. Licensee: Capstar TX L.P. Group owner: Clear Channel Communications Inc. (acq 8-30-00; grpsl). Population served: 350,000 Format: Active rock. News staff: one; News: 4 hrs wkly. Target aud: 25-49; males in their late teens to late '40s. ♦Judi Oslund, VP & chief of engrg; Judi Jennings, gen mgr; Kevin Beller, gen sls mgr; Paul Oslund, progmg dir & progmg mgr; Robin Wilds, prom dir & news dir.

WCMI(AM)—See Ashland, KY

WDGG(FM)—See Ashland, KY

WEMM(AM)— 1946: 1470 khz; 5 kw-D, 72 w-N. TL: N38 24 22 W82 29 04. Hrs opn: 24 703 3rd Ave., 25701. Phone: (304) 525-5141. Fax: (304) 525-0748. Web Site: www.wemmam.com. Licensee: Mortenson Broadcasting Co. of West Virginia LLC. Group owner: Mortenson Broadcasting Co. (acq 11-12-03). Format: Southern gospel. Target aud: 25-50; general. ♦Jack Mortenson, pres; Anita Jones, gen mgr.

WEMM-FM— Sept 6, 1971: 107.9 mhz; 50 kw. 500 ft TL: N38 28 33 W82 15 00. Stereo. Hrs opn: 24 703 3rd Ave., 25701. Phone: (304) 525-5141. Phone: (304) 525-9366. Fax: (304) 525-0748. Web Site: www.wemmfm.com. Licensee: Mortenson Broadcasting Co. (group owner). Population served: 2,500,000 Natl. Network: USA. Format: Christian, gospel. News: 4 hrs wkly. Target aud: 35 plus; responsive, loyal, family oriented. ♦Jack M. Mortenson, pres; Anita G. Jones, gen mgr; Alicia Vance, traf mgr.

WKEE-FM—Listing follows WVHU(AM).

WMEJ(FM)—See Proctorville, OH

***WMUL(FM)**— Nov 1, 1961: 88.1 mhz; 1.15 kw. -56 ft TL: N38 25 26 W82 25 39. Stereo. Hrs opn: 6 AM-3 AM Comm. Bldg., One John Marshall Dr., 25755-2635. Phone: (304) 696-6640. Phone: (304) 696-6651. Fax: (304) 696-3232. E-mail: wmul@marshall.edu Web Site: www.marshall.edu/wmul. Licensee: Marshall University Board of Governors (acq 8-7-01). Population served: 83,700 Law Firm: William D. Silva. Wire Svc: AP Format: Div. News: 7 hrs wkly. Target aud: General. ♦Stephen J. Kopp, pres; Dr. Chuck G. Bailey, gen mgr; Alex Reed, stn mgr; Chuck Cook, chief of opns; Melanie Chapman, news dir; Ryan Epling, sports cmtr; Krystle Nichols, prom.

WRVC(AM)— Oct 23, 1923: 930 khz; 5 kw-D, 1 kw-N, DA-N. TL: N38 24 03 W82 29 42. Hrs open: Box 1150, 25713. Secondary address: 401 11th St., Suite 200 25701. Phone: (304) 523-8401. Fax: (304) 523-4848. Web Site: www.wrvc.am. Licensee: Fifth Avenue Broadcasting Co. Inc. Group owner: Kindred Communications Inc. (acq 6-1-70). Population served: 30,700 Natl. Network: ESPN Radio. Law Firm: Arent, Fox, Kintner, Plotkin & Kahn. Format: Sports. News: 40 hrs wkly. Spec prog: Relg 3 hrs wkly. ♦Tom Wolf, pres; Mike Kirtner, gen mgr; Cameron Smith, opns dir; Newman Adkins, gen sls mgr.

WTCR(AM)—See Kenova

WTCR-FM— May 1, 1966: 103.3 mhz; 50 kw. 490 ft TL: N38 25 11 W82 24 06. Hrs open: 9801 Radio Park Rd., Catlettsburg, KY, 41129. Phone: (606) 739-8427. Fax: (606) 739-6009. E-mail: wtcr@clearchannel.com Web Site: wtcr.com. Licensee: Capstar TX L.P. Group owner: Clear Channel Communications Inc. (acq 8-30-00; grpsl). Population served: 300,000 Natl. Network: ABC. Natl. Rep: Rgnl Reps, Katz Radio. Format: Country. ♦Judy Cornett, gen mgr & stn mgr.

WVHU(AM)— July 1947: 800 khz; 5 kw-D, 185 w-N. TL: N38 23 35 W82 28 24. Stereo. Hrs opn: Box 2288, 25724. Secondary address: 134 4th Ave. 25701. Phone: (304) 525-7788. Fax: (304) 525-6281. E-mail: paulswann@clearchanneler Web Site: www.800wvhu.com. Licensee: Capstar TX L.P. Group owner: Clear Channel Communications Inc. (acq 8-30-2000; grpsl). Population served: 257,000 Law Firm: Alan Campbell. Format: News/talk. Target aud: 25-54; older, professional, higher income. ♦Judy Jennings, gen mgr; Matt Tweel, sls dir; Kym Blake, natl sls mgr; Truezy Robinette, prom dir; Paul Swann, progmg dir; Bill Cornwell, news dir; Scott Hensley, chief of engrg.

WKEE-FM—Co-owned with WVHU(AM). November 1947: 100.5 mhz; 53 kw. 561 ft TL: N38 23 35 W82 28 24. Stereo. Web Site: www.wkee.com.257,900 Format: Chr, pop. Target aud: 18-42. ♦Jim Davis, progmg dir; Gary Miller, mus dir; Bryan Atkins, traf mgr.

***WVWV(FM)**— Nov 28, 1977: 89.9 mhz; 8.1 kw. 1,200 ft TL: N38 29 42 W82 12 03. Stereo. Hrs opn: 24 600 Capitol St., Charleston, 25301. Phone: (304) 556-4900. Fax: (304) 556-4981. E-mail: feedback@wvpubcast.org Web Site: www.wvpubcast.org. Licensee: West Virginia Educational Broadcasting Authority. Population served: 100,000 Natl. Network: NPR, PRI. Format: News, class, jazz. ♦Marilyn DiVita, gen mgr & dev dir; Rita Ray, gen mgr; Craig Lanham, progmg dir; Greg Collard, news dir; Jack Wells, chief of engrg; Teresa Willis, traf mgr.

Hurricane

WOKU(AM)— July 2, 1971: 1080 khz; 1 kw-D. TL: N38 26 41 W82 00 54. Hrs open: 10 3006 Mt. Vernon Rd., Suite 1080, 25526. Phone: (304) 757-9661. Fax: (304) 757-9620. E-mail: woku@wv-cis.net Web Site: www.woku.com. Licensee: Big River Radio Inc. Group owner: Baker Family Stations (acq 1996; $20,000). Population served: 800,000 Format: Adult contemporary Christian. ♦Vernon H. Baker, pres; Randy Parsons, gen mgr; Matt Curry, opns mgr, progmg dir & traf mgr; Winston Hawkins, chief of engrg.

***WPJW(FM)**—Not on air, target date: unknown: 91.5 mhz; 3 kw. Ant 302 ft TL: N38 26 41 W82 00 54. Hrs open: Box 889, Blacksburg, VA, 24063. Phone: (540) 552-4252. Fax: (540) 951-6282. Web Site: www.parfm.com. Licensee: Positive Alternative Radio Inc. ♦Vernon H. Baker, pres.

Kenova

WMGA(FM)— 2006: 97.9 mhz; 3.5 kw. Ant 436 ft TL: N38 25 26 W82 32 08. Hrs open: Box 404, Huntington, 25708. Secondary address: 919 Fifth Ave., Suite 210, Huntington 25701. Phone: (304) 399-9603. Fax: (304) 399-9608. Web Site: www.magic979.com. Licensee: Connoisseur Media LLC. Format: Adult contemp. ♦B.J. Nielsen, gen mgr.

WTCR(AM)— August 1954: 1420 khz; 5 kw-D, 500 w-N, DA-N. TL: N38 24 42 W82 36 13. Hrs open: 9801 Radio Park Rd., Catlettsburg, KY, 41129. Phone: (606) 739-8427. Fax: (606) 739-6009. E-mail: wtcr@clearchannel.com Licensee: Capstar TX L.P. Group owner: Clear Channel Communications Inc. (acq 8-30-00; grpsl). Population served: 266,000 Natl. Rep: Rgnl Reps, Katz Radio. Format: Bluegrass, Americana. Target aud: 35 plus. ♦Judy Cornett, VP & stn mgr.

Keyser

WCBC-FM— January 1990: 107.1 mhz; 530 w. 783 ft TL: N39 31 26 W78 51 44. Stereo. Hrs opn: Box 1290, Cumberland, MD, 21501. Phone: (301) 724-5000. Fax: (301) 722-8336. E-mail: wcbc@1270am.com Licensee: Prosperitas Broadcasting System. (acq 9-7-89; $300,000; FTR: 9-25-89). Format: Oldies. ♦David Aydelotte, gen mgr; Mary Clites, gen sls mgr & progmg dir; Bryan Gowans, news dir; Martin White, chief of engrg.

WKLP(AM)— Aug 31, 1965: 1390 khz; 1 kw-D, 74 w-N. TL: N39 26 12 W78 57 21. Hrs open: 24 Box F, Rt. 46 E., 26726. Phone: (304) 788-1662. Fax: (304) 788-1662. E-mail: wqzk@wqzk.com Web Site: www.wqzk.com. Licensee: Starcast Systems Inc. (group owner; (acq 12-29-2006; with co-located FM). Population served: 100,000 Format: MOR. Target aud: 35 plus. Spec prog: Big band. ♦Jack Mullen II, gen mgr; Jack Mullen III, progmg dir; Mark Allen, mus dir; Pat Sullivan, news dir.

WQZK-FM—Co-owned with WKLP(AM). Sept 15, 1973: 94.1 mhz; 15 kw. Ant 801 ft TL: N39 25 08 W78 57 13. Stereo. 24 Web Site: www.wqzk.com.350,000 Format: Classic rock. Target aud: 18-49.

Kingwood

WFSP(AM)— Aug 25, 1967: 1560 khz; 1 kw-D, 250 w-CH. TL: N39 20 01 W79 43 10. Hrs open: Box 567, 26537. Secondary address: Rt. 7, W. 26537. Phone: (304) 329-1780. Fax: (304) 329-1781. E-mail: wfsp@wvdsl.net Web Site: www.prestoncounty.com/wfsp. Licensee: WFSP Inc. (acq 8-24-79). Population served: 50,000 Natl. Network: CBS. Rgnl. Network: Metronews Radio Net. Natl. Rep: Dome. Law Firm: Cohn & Marks LLP. Format: Christion, relg. News staff: one; News: 6 hrs wkly. Target aud: 20 plus. ♦Arthur W. George, pres & min affrs dir; Donna Nestor, opns mgr & women's int ed; Dave Wills, mus dir, mus critic & disc jockey; Kathy Casseday, news dir, pub affrs dir, spec ev coord & local news ed; Chuck Clemence, chief of engrg; Mike Barnett, disc jockey.

WFSP-FM— June 10, 1991: 107.7 mhz; 1.6 kw. 449 ft TL: N39 28 50 W79 43 11. Stereo. 24 Web Site: www.prestoncounty.com/wfsp.70,000 Natl. Network: Westwood One. Natl. Rep: Dome. Law Firm: Cohn & Marks LLP. Format: Oldies. News staff: one; News: 25 hrs wkly. Target aud: 18-45. ♦Arthur W. George, min affrs dir; Kathy Cassedy, spec ev coord & local news ed; Donna Nestor, women's int ed; Dave Wills, disc jockey; Mike Barnett, disc jockey.

WKMM(FM)— Dec 1, 1986: 96.7 mhz; 300 w. 797 ft TL: N39 27 29 W79 35 18. Stereo. Hrs opn: 24 106 E. Main St., 26537. Phone: (304) 329-0967. Fax: (304) 329-2131. E-mail: jcrogan@wkmmfm.com Web Site: www.wkmmfm.com. Licensee: MarPat Corp. (acq 8-1-93; $190,000; FTR: 8-23-93). Population served: 233,000 Natl. Network: Westwood One, CNN Radio. Format: Country. Target aud: 25-55. ♦P.J. Crogan, pres & opns mgr; Dave Price, gen sls mgr; Marty White, chief of engrg; Jane Crogan, traf mgr.

Lewisburg

WKCJ(FM)— October 1981: 103.1 mhz; 25 kw. 781 ft TL: N37 42 43 W80 30 20. Stereo. Hrs opn: 24 Box 610, Rt. 60 W. Harts Run, White Sulphur Springs, 24986. Phone: (304) 536-1310. Phone: (304) 645-1191. Fax: (304) 536-1311. Licensee: Quorum Radio Partners of Virginia Inc., debtor-in-possession. (group owner; (acq 4-20-2005; grpsl). Population served: 50,000 Natl. Rep: Rgnl Reps. Format: Modern & traditional country. News: 12 hrs wkly. Target aud: 25-55. Spec prog: Gospel 6 hrs, relg 4 hrs, farm 3 hrs wkly. ♦Joyce Tucker, gen mgr & sls dir; Chuck Harper, progmg dir & mus dir; Marcia Smith, gen sls mgr & chief of engrg.

WRON-FM—See Ronceverte

Lindside

***WHFI(FM)**— September 1990: 106.7 mhz; 3 kw. 303 ft TL: N37 28 56 W80 39 40. Hrs open: 24 Box 97, Rt. 1, 24951. Phone: (304) 753-9971. Fax: (304) 753-9792. Web Site: www.whfi-fm.com. Licensee: Monroe County Board of Education (acq 5-1-89). Format: MOR. ♦James W. Higginbotham, gen mgr.

Logan

WVOW(AM)— May 1954: 1290 khz; 5 kw-D, 1 kw-N, DA-N. TL: N37 51 28 W81 58 16. Hrs open: 24 Box 1776, 25601. Secondary address: 204 Main St., Suite 201 25601. Phone: (304) 752-5080. Fax: (304) 752-5711. E-mail: amfmwvow@mountain.net Licensee: Logan Broadcasting Corp. Population served: 50,000 Format: Adult contemp. News staff: 2. Target aud: General. ♦Martha Jane Becker, pres, gen sls mgr & prom mgr; Larry Bevins, gen mgr; Rhonda Bryant, progmg dir & traf mgr; Bill France, mus dir; Bob Weisner, news dir; Terry Bucklew, chief of engrg.

WVOW-FM— August 1969: 101.9 mhz; 15 kw horiz, 13.5 kw vert. Ant 830 ft TL: N37 51 24 W81 58 18.24 3,311 Format: Adult contemp. ♦Rhonda Bryant, traf mgr & women's int ed.

Lost Creek

WOTR(FM)— Dec 9, 1991: 96.3 mhz; 6 kw horiz. 302 ft TL: N39 08 43 W80 19 40. Stereo. Hrs opn: 8 AM-10 PM Box 505, 26385. Phone: (304) 745-4243. Licensee: James W. Allman. Population served: 50,000 Law Firm: William D. Silva. Format: Gospel, relg. Target aud: 25-99. ♦Bill Allman, gen mgr; James W. Allman, CEO & chief of opns.

Mannington

WGYE(FM)— December 1992: 102.7 mhz; 3.2 kw. 453 ft TL: N39 32 18 W80 20 16. Hrs open: 24 1489 Locust Ave., Suite C, Fairmont, 26554. Phone: (304) 363-8888. Fax: (304) 367-1885. E-mail: david@froggycountry.net Web Site: www.froggcountry.net. Licensee: Burbach of DE LLC. Group owner: Burbach Broadcasting Group (acq 6-20-2000; grpsl). Population served: 200,000 Format: Country. News staff: one; News: 4 hrs wkly. Target aud: 25-54. ♦Nicholas A. Galli, pres; David Bronham, gen mgr & gen sls mgr; Greg Bolyard, progmg dir & mus dir; Larry Smith, engrg mgr & chief of engrg.

Marmet

***WKVW(FM)**— June 30, 1995: 93.3 mhz; 1.1 kw. Ant 771 ft TL: N38 16 32 W81 31 36. Hrs open: 24 2351 Sunset Blvd., Suite 170-218, Rocklin, CA, 95765-3719. Phone: (916) 251-1600. Fax: (916) 251-1650. Web Site: www.klove.com. Licensee: Educational Media Foundation Group owner: EMF Broadcasting (acq 7-1-2002; $500,000). Natl.

Network: K-Love. Rgnl. Network: Metronews Radio Net. Law Firm: Shaw Pittman. Format: Contemp Christian. News staff: 3. Target aud: 25-44; Judeo Christian, female. ♦Richard Jenkins, pres; MIke Novak, VP; Keith Whipple, dev dir; Eric Allen, natl sls mgr; Mike Novak, progmg dir; David Pierce, progmg mgr; Ed Lenane, news dir; Sam Wallington, engrg dir; Karen Johnson, news rptr; Marya Morgan, news rptr; Richard Hunt, news rptr.

Martinsburg

WEPM(AM)— Oct 13, 1946: 1340 khz; 1 kw-U. TL: N39 27 48 W77 59 11. Hrs open: 1606 W. King St., 25401. Phone: (304) 263-8868. Fax: (304) 263-8906. Licensee: Prettyman Broadcasting Co. (group owner; (acq 1-1-87; $2 million; FTR: 11-10-86). Population served: 15,000 Natl. Network: CBS. Natl. Rep: Katz Radio. Format: News/talk, sports. Target aud: 35 plus. Spec prog: Relg 6 hrs wkly. ♦Yogi Yoder, gen mgr; Chuck Thornton, sls dir; Jay Young, progmg dir; Rodney Rockwell, chief of engrg; Susan Grissinger, traf mgr.

WLTF(FM)—Co-owned with WEPM(AM). 1949: 97.5 mhz; 11.4 kw. Ant 1,036 ft TL: N39 27 33 W78 03 48. Stereo. 24 Web Site: www.lite975.com. Format: Adult contemp. Target aud: 30-49. ♦Stacey Drake, progmg dir.

WRNR(AM)— Apr 16, 1976: 740 khz; 500 w-D, 21 w-N, DA-2. TL: N39 28 25 W77 55 57. Hrs open: 24 Box 709, 1762 Eagle School Rd., 25402. Phone: (304) 263-6586. Fax: (304) 263-3082. Licensee: Shenandoah Communications Inc. Population served: 300,000 Natl. Network: Westwood One, CBS, CNN Radio. Natl. Rep: Rgnl Reps. Law Firm: Cohn & Marks, LLC. Format: News/talk, sports. News staff: 3; News: 28 hrs wkly. Target aud: 35 plus; middle to upper age & income. ♦Richard S. Wachtel, pres, gen mgr & gen sls mgr; Gregg M. Wachtel, exec VP; Matt Miller, opns dir; Tom Tucker, prom dir, progmg dir & news dir; Fran Little, chief of engrg.

***WVEP(FM)**— Feb 11, 1987: 88.9 mhz; 3.6 kw. 1,623 ft TL: N39 08 38 W78 26 09. Stereo. Hrs open: 24 600 Capitol St., Charleston, 25301. Phone: (304) 556-4900. Fax: (304) 556-4981. E-mail: feedback@wvpubcast.org Web Site: www.wvpubcast.org. Licensee: West Virginia Educational Broadcasting Authority. Population served: 100,000 Natl. Network: NPR, PRI. Format: News, jazz, class. ♦Marilyn DiVita, gen mgr & dev dir; Rita Ray, gen mgr; Greg Collard, news dir; Teresa Wills, traf mgr.

Matewan

WHJC(AM)— Dec 2, 1951: 1360 khz; 1 kw-D. TL: N37 37 02 W82 10 04. Hrs open: Box 68, 25678. Secondary address: 156 Radio Hill, McCarr, KY 41544. Phone: (606) 427-7261. Fax: (606) 427-7260. E-mail: pwr1067@bellsouth.net Licensee: Three States Broadcasting Co. Inc. Population served: 40,000 Format: Southern gospel. ♦George D. Warren, pres & news dir; Evelyn Warren, gen mgr & sls dir; Melissa White, gen sls mgr, progmg dir & traf mgr; Russell Laferty, chief of engrg.

WVKM(FM)—Co-owned with WHJC(AM). Aug 30, 1989: 106.7 mhz; 4.3 kw. Ant 751 ft TL: N37 36 49 W82 11 22. Stereo. 24 E-mail: kixx1067@yahoo.com Web Site: kixx1067.com.160,000 Format: Classic rock. ♦Melissa White, traf mgr.

Miami

WKAZ-FM— November 1985: 107.3 mhz; 50 kw. Ant 600 ft TL: N38 16 25 W81 31 27. Stereo. Hrs opn: 24 1111 Virginia St. E., Charleston, 25301. Phone: (304) 342-8131. Fax: (304) 344-4745. Licensee: West Virginia Radio Corp of Charleston. (group owner; (acq 6-5-97; $2.15 million with WCZR(AM) Charleston). Natl. Rep: D & R Radio. Format: Oldies. Target aud: 25-54. ♦Sean Banks, gen mgr; Max Wulf, progmg dir; Noel Richardson, chief of engrg.

Middlebourne

***WRSG(FM)**— 2001: 91.5 mhz; 900 w. Ant 157 ft TL: N39 30 59 W80 54 00. Hrs open: 24 1993 Silver Knight Dr., Sistersville, 26175. Phone: (304) 758-9007 (Studio). Phone: (304) 758-9000 (School Phone). Fax: (304) 758-9006. E-mail: wrsgfm@yahoo.com Web Site: tchs.tyle.k12.wv.us /ths/wrsg/wrsg.htm. Licensee: Tyler County Board of Education. Format: Var.

Milton

WAMX(FM)—Licensed to Milton. See Huntington

WZZW(AM)— June 26, 1973: 1600 khz; 6 kw-D, 26 w-N. TL: N38 25 46 W82 06 21. Hrs open: 134 4th Ave., Huntington, 25701. Phone: (304) 525-7788. Fax: (304) 525-6281. Web Site: www.havejoy.com. Licensee: Capstar TX L.P. Group owner: Clear Channel Communications Inc. (acq 8-30-2000; grpsl). Population served: 15,000 Format: Contemp Christian. Target aud: 25-44; adult, upscale baby boomers. ♦Judy Jennings, VP & gen mgr; Kym York-Blake, gen mgr, gen sls mgr & prom dir; Dixie McDavid, traf mgr.

Montgomery

WMON(AM)— July 14, 1946: 1340 khz; 1 kw-U. TL: N38 10 48 W81 20 06. Hrs open: 100 Kanawha Terr., St. Albans, 25177. Phone: (304) 722-3808. Fax: (304) 727-1300. E-mail: ccrninfo@hotmail.com Web Site: www.ccrnonline.com. Licensee: L.M. Communications of Kentucky LLC. Group owner: L.M. Communications Inc. (acq 4-1-03; grpsl). Population served: 40,000 Natl. Network: Salem Radio Network. Format: Religious talk radio. Target aud: 25 plus. ♦Ron Walton, gen mgr; Chris Colagrosso, progmg dir; Fred Francis, chief of engrg; Emma Allen, traf mgr.

Morgantown

WAJR(AM)— Dec 7, 1940: 1440 khz; 5 kw-D, 500 w-N, DA-2. TL: N39 40 34 W80 00 12. Hrs open: 24 1251 Earl Core Rd., 26505. Phone: (304) 296-0029. Fax: (304) 296-3876. Web Site: www.wajr.com. Licensee: West Virginia Radio Corp. Population served: 400,000 Natl. Network: ABC. Law Firm: Fletcher, Heald & Hildreth. Format: News/talk. News staff: 7; News: 15 hrs wkly. Target aud: 25+. ♦Dale B. Miller, pres & gen mgr; Harvey Kercheval, opns VP; Gary Mertins, sls dir; Tim Loughry, prom dir; Jim Stallings, progmg dir; Shawm Falkenstein, news dir; Kay Murray, pub affrs dir; Noel Richardson, engrg VP; Ralph Messer, chief of engrg; Donna Tubolino, traf mgr.

WVAQ(FM)—Co-owned with WAJR(AM). 1948: 101.9 mhz; 50 kw. 500 ft TL: N39 36 30 W79 59 07. Stereo. 24 Web Site: www.wvaq.com. Format: CHR. News staff: 2; News: one hr wkly. Target aud: 18-49. ♦Hoppy Kercheval, opns dir; Lacy Neff, progmg dir & disc jockey; Eric McGuire, disc jockey; Meghan Dunt, disc jockey.

WCLG(AM)— December 1954: 1300 khz; 2.5 kw-D, 44 w-N. TL: N39 37 40 W79 58 11. Stereo. Hrs opn: Box 885, 26507. Secondary address: 343 High St. 26505. Phone: (304) 292-2222. Fax: (304) 292-2224. Web Site: www.wclg.com. Licensee: Bowers Broadcasting Corp. (acq 12-19-59). Population served: 29,431 Natl. Rep: Dome. Format: Oldies. ♦Garry Bowers, pres & gen mgr; Rebecca Hunn, sls dir; Jeffrey Miller, progmg dir; Ken Tennant, chief of engrg; Lucinda Funk, traf mgr.

WCLG-FM— Sept 28, 1974: 100.1 mhz; 6 kw. 300 ft TL: N39 37 40 W79 58 11. Stereo. Web Site: www.wclg.com. Format: Classic rock. ♦Lucinda Funk, traf mgr.

***WVPM(FM)**— May 27, 1981: 90.9 mhz; 5 kw. 1,440 ft TL: N39 41 45 W79 45 45. Stereo. Hrs opn: 24 600 Capitol St., Charleston, 25301. Phone: (304) 556-4900. Fax: (304) 556-4960. E-mail: feedback@wvpubcast.org Web Site: www.wvpubcast.org. Licensee: West Virginia Educational Broadcasting Authority. Population served: 96,000 Natl. Network: NPR, PRI. Format: News, class, jazz. ♦Marilyn DiVita, gen mgr; James Muhammad, progmg dir.

***WWVU-FM**— Aug 20, 1982: 91.7 mhz; 2.6 kw. 180 ft TL: N39 38 09 W79 56 38. Stereo. Hrs opn: 24 Box 6446, Mountainlair, West Virginia Univ., 26506-6446. Phone: (304) 293-3329. Fax: (304) 293-7363. E-mail: u92@mail.wvu.edu Web Site: u92.wvu.edu. Licensee: West Virginia University Board of Governors (acq 8-10-01). Law Firm: William D. Silva. Format: Educ, div, progsv. News: 2 hrs wkly. Target aud: 18-35; mostly college & high school students. Spec prog: New age 6 hrs, reggae 4 hrs, metal 6 hrs, bluegrass one hr, oldies 8 hrs, big band 2 hrs wkly. ♦Kim Harrison, gen mgr & stn mgr.

WZST(FM)—See Westover

Moundsville

WRKP(FM)— Jan 15, 1990: 96.5 mhz; 1.9 kw. Ant 594 ft TL: N39 50 51 W80 45 23. Stereo. Hrs opn: 24 2002 First St., 26041. Phone: (304) 845-1052. Fax: (304) 845-1054. E-mail: ronking@wrkp.com Web Site: www.wrkp.com. Licensee: RKP International Corp. Natl. Network: USA. Natl. Rep: Salem. Rgnl rep: CMS Format: Contemp Christian mus, talk. News staff: one; News: 5 hrs wkly. Target aud: 12-45; 60% female, 40% male. ♦Ronald W. King, CEO, pres & gen mgr; LuAnn Jerin, opns mgr.

WVLY(AM)— Oct 1, 1950: 1370 khz; 5 kw-D, 20 w-N. TL: N39 54 20 W80 46 42. Hrs open: 24 1143 Main St., Suite 200, Wheeling, 26003. Phone: (304) 233-9859. Fax: (304) 214-9859. E-mail: wvlyradio@aol.com Web Site: www.talkradio1370.com. Licensee: Monroe Communications LLC (acq 1-9-2004; $75,000). Population served: 185,000 Natl. Network: CNN Radio. Format: News/talk. News: 14 hrs wkly. Target aud: 25-54; adults. ♦Howard Monroe, pres, gen mgr, gen sls mgr & progmg dir.

Mount Hope

WTNJ(FM)— June 1, 1980: 105.9 mhz; 4.4 kw. Ant 1,532 ft TL: N37 56 51 W81 18 29. Stereo. Hrs opn: 24 306 S. Karawha St., Beckley, 25801-5619. Phone: (304) 253-7000. Fax: (304) 255-1044. Licensee: Southern Communications Group owner: Southern Communications Corp. (acq 3-12-01; $2.375 million). Population served: 295,300 Format: Country. News staff: one; News: 12 hrs wkly. Target aud: 25-54. Spec prog: NASCAR races, West Virginia Univ. sports. ♦Jay Quesenberry, gen mgr & gen sls mgr; Rick Reiser, opns mgr; Rick Peiser, progmg dir; Warren Ellison, news dir; Randy Kerbawy, chief of engrg; Bill Wise, traf mgr.

Mullens

WPMW(FM)— Sept 30, 1981: 92.7 mhz; 1.65 kw. 443 ft TL: N37 35 39 W81 23 49. Stereo. Hrs opn: 213 Howard Ave., 25882. Phone: (304) 294-4405. Fax: (304) 294-5616. E-mail: ranny@c92.com Web Site: www.c92radio.com. Licensee: West Virginia-Virginia Holding Co. LLC (acq 3-29-2006; $120,000). Natl. Network: ABC. Format: Classic rock. ♦Ranny Parks, gen mgr; Debra Toler, gen sls mgr; Jeff Halsey, progmg dir.

New Martinsville

WETZ(AM)— May 25, 1953: 1330 khz; 1 kw-D, 60 w-N. TL: N39 39 27 W80 51 34. Hrs open: 24 Box 10, 26155. Secondary address: 325 N.Main St. 26155. Phone: (304) 455-1111. Fax: (304) 455-1170. Licensee: Dailey Corp. (group owner; (acq 2-1-2001; grpsl). Population served: 80,000 Natl. Network: ABC. Law Firm: Reddy, Begley & McCormick. Format: Stardust Timeless Classic. News: 10 hrs wkly. Target aud: 25-64. ♦Calvin Dailey Jr., pres; Dennis Gage, gen mgr.

WETZ-FM— December 1977: 103.9 mhz; 2.5 kw. Ant 502 ft TL: N39 39 10 W80 54 47. Stereo. 24 Web Site: www.powercountry104.com. Natl. Network: ABC. News: 10 hrs wkly. Target aud: 25-54; country.

WXCR(FM)— 2002: 92.3 mhz; 3.2 kw. Ant 453 ft TL: N39 40 16 W80 53 04. Stereo. Hrs opn: 24 Box 564, 26155. Secondary address: Box 374, Saints Marys 26170. Phone: (304) 684-3400. Fax: (304) 684-9241. Licensee: Seven Ranges Radio Co. Inc. Population served: 30,000 Law Firm: Reddy, Begley & McCormick. Format: Classic rock. Target aud: 25-54; 70% men, 30% women. ♦Sam Yoho, pres & gen mgr; Lou Petronio, opns mgr.

WYMJ(FM)— Dec 1, 2002: 99.5 mhz; 2.7 kw. Ant 482 ft TL: N39 39 10 W80 54 47. Stereo. Hrs opn: 24 Box 10, 26155. Phone: (304) 455-1111. Fax: (304) 455-1170. Web Site: www.oldiesradioonline.com. Licensee: Dailey Corp. (group owner; acq 2-6-2001; grpsl). Law Firm:

Reddy, Begley & McCormick. Format: Adult contemp. News staff: 3. ♦Dex Gage, gen mgr, opns mgr & gen sls mgr; Ed Wilhelm, chief of engrg.

Oak Hill

WAXS(FM)— 1948: 94.1 mhz; 26 kw. 650 ft TL: N37 57 30 W81 09 03. Stereo. Hrs opn: 24 306 S. Karawha St., Beckley, 25801. Phone: (304) 253-7000. Fax: (304) 255-1044. Licensee: Plateau Broadcasting Inc. Group owner: Southern Communications Corp. (acq 3-12-01; $875,000). Population served: 300,000 Natl. Network: ABC. Format: Oldies. News staff: one; News: 2 hrs wkly. Target aud: General; baby boomers. ♦Jay Quesenberry, gen mgr & gen sls mgr; Rick Reiser, opns mgr & progmg dir; Warren Ellison, adv dir & news dir; Randy Kerbawy, chief of engrg.

WOAY(AM)— Feb 22, 1947: 860 khz; 10 kw-D, 11 w-N, 5 kw-CH. TL: N37 57 30 W81 09 03. Hrs open: 24 Box 140, 25901-0140. Secondary address: 240 Central Ave. 25901-3006. Phone: (304) 465-0534. Fax: (304) 465-1486. E-mail: info@woayradio.com Web Site: www.woayradio.com. Licensee: Mountaineer Media Inc. (acq 12-31-2006; $250,000). Population served: 300,000 Natl. Network: Moody, USA. Format: Relg. News: 10 hrs wkly. Target aud: General. ♦Thomas H. Moffit Jr., pres; Judy Ellison, VP; Eugene Ellison, gen mgr; Stephanie Gibson, chief of engrg & traf mgr.

WTNJ(FM)—See Mount Hope

Parkersburg

WADC(AM)— Apr 9, 1954: 1050 khz; 5 kw-D, 144 w-N. TL: N39 15 29 W81 33 49. Hrs open: 24 Box 4739, 26104-4739. Phone: (304) 485-4565. Fax: (304) 424-6955. Licensee: Burbach of Delaware, LLC. Group owner: Burbach Broadcasting Group (acq 3-19-98; $1.775 million with co-located FM). Population served: 360,000 Natl. Network: CNN Radio. Law Firm: Koerner & Olender. Format: Adult standards. News: 12 hrs wkly. Target aud: 35-64. Spec prog: Relg 3 hrs wkly. ♦Don Staats, gen mgr; Larry Smith, chief of engrg.

WGGE(FM)—Co-owned with WADC(AM). Sept 1, 1965: 99.1 mhz; 11.5 kw. 485 ft TL: N39 15 29 W81 33 49. Stereo. Web Site: www.froggy99.net.500,000 Format: Mainstream country. Target aud: 25-54; loyal modern country listeners. Spec prog: Farm 2 hrs, NASCAR 5 hrs wkly.

WHBR-FM—Listing follows WVNT(AM).

WHNK(AM)— July 12, 1935: 1450 khz; 1 kw-U. TL: N39 17 23 W81 31 36. Hrs open: 24 Box 5559, 6006 Grand Central Ave., Vienna, 26105. Phone: (304) 295-6070. Fax: (304) 295-4389. E-mail: johnchalfant @clearchannel.com Web Site: www.whnk.com. Licensee: CC Licenses LLC. Group owner: Clear Channel Communications Inc. (acq 4-17-2001). Rgnl. Network: Ohio Radio Net. Natl. Rep: Clear Channel. Format: Class country. News staff: one; News: 3 hrs wkly. Target aud: 25-54; middle-aged, middle to upper income. ♦Chuck Poet, gen mgr; John Chalfant, opns mgr; Kirk McCall, gen sls mgr; Rodney Ortiz, progmg dir; Doug Hess, news dir & pub affrs dir; Jerry Kuhn, chief of engrg; Belinda Marcinko, traf mgr.

WVNT(AM)— September 1947: 1230 khz; 1 kw-U. TL: N39 16 56 W81 33 17. Hrs open: 24 Box 4739, 26104-4739. Phone: (304) 485-4565. Fax: (304) 424-6955. Licensee: Burbach of Delaware, LLC. Group owner: Burbach Broadcasting Group (acq 1-1-97; grpsl). Population served: 44,208 Natl. Network: CNN Radio. Natl. Rep: McGavren Guild. Format: News/talk. News staff: one. Target aud: 30-50. Spec prog: Gospel 4 hrs wkly. ♦Don Staats, gen mgr.

WHBR-FM—Co-owned with WVNT(AM). March 1967: 103.1 mhz; 2.1 kw. Ant 561 ft TL: N39 21 00 W81 33 56. Stereo. 24 299,890 Format: Active rock. News staff: one; News: 6 hrs wkly. Target aud: 18-49; active/modern rock listeners.

***WVPG(FM)**— April 4, 1985: 90.3 mhz; 9 kw. 321 ft TL: N39 12 44 W81 35 30. Stereo. Hrs opn: 24 600 Capitol St., Charleston, 25301. Phone: (304) 556-4900. Fax: (304) 556-4960. E-mail: feedback@wvpubcast.org Web Site: www.wvpubcast.org. Licensee: West Virginia Educational Broadcasting Authority. Population served: 96,000 Natl. Network: NPR, PRI. Format: News, class, jazz. News staff: 2. Spec prog: Mountain Stage 2 hrs, children one hr wkly. ♦Rita Ray, CEO; Marilyn DiVita, gen mgr; Peggy Dorsey, dev mgr; Beth Carenbauer, adv mgr; James Muhammad, progmg dir; Laura H. Allen, mus dir; Giles Snyder, news dir & pub affrs dir; David McClanahan, chief of engrg; Glenna Racer, traf mgr; Greg Callard, local news ed.

WXIL(FM)— Sept 1, 1975: 95.1 mhz; 50 kw. 500 ft TL: N39 14 47 W81 28 19. Stereo. Hrs opn: 24 5 Rosemar Cir., 26104. Phone: (304) 485-7425. Phone: (304) 485-4565. Fax: (304) 424-6955. E-mail: productionparkersburg@resultsradiowv.com Web Site: www.95xil.net. Licensee: PBBC Inc. Group owner: Burbach Broadcasting Group (acq 9-1-80; $1 million; FTR: 7-7-80). Population served: 310,000 Natl. Rep: Katz Radio. Format: Hot Adult Contemp. News staff: one; News: 6 hrs wkly. Target aud: 25-54; women. ♦Nicholas A. Galli, chmn & pres; Don Staats, VP & gen mgr; Brian Steel, opns dir & progmg dir; Larry Smith, news dir & chief of engrg; Luke Woytowich, news dir.

Petersburg

***WAUA(FM)**— Dec 1, 1997: 89.5 mhz; 10 kw. Ant 1,056 ft TL: N39 12 07 W79 16 31. Stereo. Hrs opn: 24 600 Capitol St., Charleston, 25301. Phone: (304) 556-4900. Fax: (304) 556-4981. E-mail: feedback@wvpubcast.org Web Site: www.wvpubcast.org. Licensee: West Virginia Educational Broadcasting Authority. Population served: 56,404 Natl. Network: NPR, PRI. Format: Class, news. ♦Rita Ray, gen mgr; Marilyn DeVita, dev dir & dev mgr; James Muhammad, progmg dir; Greg Collard, news dir; Jack Wells, chief of engrg; Teresa Wills, traf mgr.

WELD-FM— Feb 6, 1987: 101.7 mhz; 1.9 kw horiz, 1.85 kw vert. Ant 515 ft TL: N38 58 34 W79 01 13. Stereo. Hrs opn: 24 126 Kessel Rd., Fisher, 26818. Phone: (304) 538-6062. Fax: (304) 538-7032. E-mail: weld@chardynet.com Web Site: www.weldamfm.com. Licensee: Thunder Associates LLC (acq 12-1-03; $600,000 with WELD(AM) Fisher). Population served: 60,000 Natl. Network: ABC. Natl. Rep: Dome, Keystone (unwired net). Law Firm: Irwin, Campbell & Tammewald. Format: Country, relg, farm. News staff: one; News: 3 hrs wkly. Target aud: 25 plus. Spec prog: Gospel 3 hrs wkly. ♦Curtis Durst, pres; Sandra Durst, exec VP; Alan Yokum, gen mgr.

Philippi

***WQAB(FM)**— October 1975: 91.3 mhz; 7.2 kw. 180 ft TL: N39 09 52 W80 02 57. Stereo. Hrs opn: 8 AM-10 PM Box 2097, Withers-Brandon Hall, Alderson-Broaddus College, 26416. Phone: (304) 457-6281. Phone: (304) 457-2916. Fax: (304) 457-6239. Licensee: Alderson-Broaddus College. Population served: 50,000 Natl. Network: AP Radio. Format: Div, CHR, adult contemp. News: 5 hrs wkly. Target aud: 15-40; college students. Spec prog: Jazz 4 hrs, Black 2 hrs, radio drama 2 hrs, children's 2 hrs wkly. ♦Harry Hancock, stn mgr; George Sommer, engrg VP.

Pineville

WWYO(AM)— 1949: 970 khz; 1 kw-D, 26 w-N. TL: N37 35 20 W81 32 25. Stereo. Hrs opn: Box 647, Bluefield, 24701. Secondary address: Rt. 10, One Radio Rd. 24701. Phone: (304) 327-5651. Phone: (304) 732-8552. Fax: (304) 327-5651. E-mail: am970wwyo@citlink.net Web Site: www.am970wwyo.bizland.com. Licensee: MRJ Inc. (acq 4-20-90; $125,000). Population served: 75,000 Format: Southern gospel, country, MOR. Target aud: 25-65; housewives. Spec prog: Folk one hr, sports 18 hrs, educ 2 hrs, community 8 hrs wkly. ♦Rudolph D. Jennings, pres & gen mgr.

Pocatalico

WRVZ(FM)— 1995: 98.7 mhz; 63 w. 617 ft TL: N38 23 53 W81 41 06. Hrs open: 1111 Virginia St. E., Charleston, 25301. Phone: (304) 342-8131. Fax: (304) 344-4745. E-mail: mbuxser@wvradio.com Web Site: wvradioadvertising.com. Licensee: West Virginia Radio Corp. of Charleston. Group owner: West Virginia Radio Corp. (acq 3-12-2001; $800,000). Format: CHR. ♦Dale Miller, pres; Mike Buxser, gen mgr; Courtney Patrick, prom dir; Woody Woods, opns dir & progmg dir; Jeff Jenkins, news dir; Noel Richardson, chief of engrg.

Point Pleasant

WBGS(AM)— 1994: 1030 khz; 10 kw-D, DA. TL: N38 48 42 W82 05 59. Hrs open: 303 8th St., VA, 25550. Phone: (304) 675-2763. Fax: (304) 675-2771. Licensee: Big River Radio Inc. Group owner: Baker Family Stations (Positive Radio Group) Law Firm: Booth, Freret, Imlay & Tepper. Format: Relg, teaching, gospel music. ♦Vernon H. Baker, CEO, chmn & pres; Edward A. Baker, VP; Kevin Nott, gen mgr & progmg dir; Shari Cochron, sls dir; Tom Payne, mus dir; Winston Hawkins, chief of engrg; Kathy Wise, traf mgr.

WBYG(FM)—Co-owned with WBGS(AM). 1994: 99.5 mhz; 4.7 kw. 328 ft TL: N38 47 52 W82 10 07. 303 8th St., 25550. Phone: (304) 675-2763. Fax: (304) 675-2771. Web Site: www.wbyg.com. (Acq 1-28-92). Format: Country. ♦Kathy Wise, gen mgr & traf mgr.

***WPCN(FM)**— Dec 21, 2000: 88.1 mhz; 3 kw. Ant 289 ft TL: N38 50 49 W82 07 50. Hrs open: 24 303 8th Street, 25550. Phone: (304) 675-2727. Fax: (304) 675-2771. E-mail: joyfm881@yahoo.com Web Site: www.joyfm881.com. Licensee: Positive Alternative Radio Inc. Group owner: Baker Family Stations (Positive Radio Group) Law Firm: Booth, Freret, Imley & Tepper. Format: Southern gospel. ♦Randy Parson, gen mgr, opns mgr & progmg dir.

Princeton

WAEY(AM)— December 1947: 1490 khz; 1 kw-U. TL: N37 23 23 W81 05 58. Hrs open: 24 Box 5588, 24740. Secondary address: Lilly Grove Addition, 1 Radio Ln. 24740. Phone: (304) 425-2151. Fax: (304) 487-2016. Licensee: Princeton Broadcasting Inc. Population served: 65,000 Format: Gospel. News staff: one; News: 14 hrs wkly. Target aud: 25 plus; blue collar. ♦Linda Witt, pres; Pat Tolley, VP; Bob Spencer, gen mgr & gen sls mgr; Jason Reed, opns mgr & prom dir; Ron Witt, progmg dir; Wayne Boone, chief of engrg; Amy Mills, traf mgr; Patricia Tolley, min affrs dir.

WSTG(FM)—Co-owned with WAEY(AM). Apr 1, 1973: 95.9 mhz; 6 kw. 285 ft TL: N37 23 23 W81 05 58. (CP: 480 w, ant 1,141 ft. TL: N37 15 30 W81 10 37). Stereo. 24 Web Site: www.star95.com. Licensee: L & P Broadcasting Inc. Format: Adult Top-40s. News staff: one. ♦Amy Mills, traf mgr; Linda Witt, min affrs dir; Charlie Brown, local news ed & disc jockey; Jim Nelson, sports cmtr; Bob Spencer, disc jockey; Jason Reed, disc jockey; Jeff Davis, disc jockey.

WKOY-FM— April 1983: 100.9 mhz; 630 w. 641 ft TL: N37 18 20 W81 07 30. Stereo. Hrs opn: 24 900 Bluefield Ave., Bluefield, 24701. Phone: (304) 327-7114. Fax: (304) 325-7850. Web Site: www.theeaglefm.com. Licensee: Monterey Licenses LLC. Group owner: Triad Broadcasting Co. LLC (acq 7-18-00; grpsl). Natl. Network: ABC. Natl. Rep: Katz Radio, Rgnl Reps. Format: Classic Rock. News staff: one. Target aud: 25 plus. ♦John Halford, gen mgr; Ken Deitz, opns dir & progmg dir; Danny Clemmon, gen sls mgr; P.J. Toler, news dir; Keith Bowman, chief of engrg.

***WPWV(FM)**— September 2003: 90.1 mhz; 2.5 kw vert. Ant 1,040 ft TL: N37 30 35 W81 12 55. Hrs open: Box 3206, Tupelo, MS, 38801. Phone: (662) 844-8888. Fax: (662) 842-6791. Web Site: www.afr.net. Licensee: American Family Association. Group owner: American Family Radio. Format: Christian. ♦Marvin Sanders, gen mgr.

Rainelle

WRLB(FM)— February 1977: 95.3 mhz; 3.1 kw. 460 ft TL: N37 57 28 W80 45 45. (CP: 12.9 kw). Stereo. Hrs opn: 24 Box 1727, Lewisburg, 24901. Phone: (304) 647-3606. Web Site: www.wrlb.com. Licensee: Faith Communications Network Inc. (acq 10-25-01). Population served: 68,000 Format: Inspirational, Christian. Target aud: 25-54.

WRRL(AM)— 1973: 1130 khz; 1 kw-D. TL: N37 57 28 W80 45 45. Hrs open: H.C. 61, Box 383, Danese, 25831. Phone: (304) 438-8537 phone/fax. E-mail: wrrlam@mountain.net Licensee: Faith Mountain Communications Inc. (acq 2-8-01; $60,000). Population served: 45,000 Rgnl. Network: Metronews Radio Net. Format: Gospel, Christian, News/talk. Target aud: 35+. ♦Nancy Whitt, CEO; Allen R. Whitt, pres & gen mgr.

Ravenswood

WMOV(AM)— 1953: 1360 khz; 1 kw-U. TL: N38 57 52 W81 46 09. Hrs open: 6 AM-midnight 527 Gibbs St., 26164. Phone: (304) 273-2544. Fax: (304) 273-3020. E-mail: wmovam@aol.com Licensee: Shay Hill, executor (acq 2-9-2004). Population served: 37,000 Natl. Network: USA. Format: Full service. News staff: one; News: 14 hrs wkly. Target aud: 16 plus; emphasis on 25 plus. Spec prog: Talk 2 hrs, Pol one hr, folk 2 hrs, jazz 2 hrs, bluegrass 10 hrs wkly. ♦Burke Allen, pres; Greg Carter, gen mgr, opns VP & opns dir.

Richwood

WKQV(FM)— 2007: 105.5 mhz; 3.5 kw. Ant 882 ft TL: N38 21 35 W80 38 51. Stereo. Hrs opn: 24 180 Main St., Sutton, 26601. Phone: (304) 765-7373. Fax: (304) 765-7836. E-mail: al@theboss97fm.com Licensee: Summit Media Broadcasting LLC. (acq 3-15-2006; $482,500 for CP). Format: CHR. Target aud: 18-49. ♦Al Sergi, pres & gen mgr.

WVAR(AM)— 1956: 600 khz; 1 kw-D. TL: N38 13 50 W80 32 49. Hrs open: 6 AM-sunset 202 Back Fork St., Webster Springs, 26288. Phone: (304) 847-5141. Fax: (304) 847-5149. Licensee: Summit Media Inc. (group owner; (acq 3-8-2007; $1.24 million with WAFD(FM) Webster Springs). Population served: 3,717 Natl. Rep: Dome. Format: Real country. News staff: one; News: 13 hrs wkly. Target aud: 25 plus;

general. ♦Nunzio A. Sergi, pres; James A. Hardman, gen mgr & progmg dir; Linda Parks, gen sls mgr.

Ridgeley

WDYK(FM)— 2006: 100.5 mhz; 4.6 kw. Ant 374 ft TL: N39 43 02.8 W78 42 42.5. Hrs open: 1050 W. Industrial Blvd., Suite 3B, Cumberland, MD, 21502. Phone: (301) 759-1005. Fax: (301) 759-3124. Web Site: www.cumberlandsmagic.com. Licensee: Radioactive LLC. Format: Adult contemp. ♦Benjamin L. Homel, pres; Christian Miller, gen mgr.

Ripley

WCEF(FM)— Feb 24, 1981: 98.3 mhz; 3 kw. 300 ft TL: N38 46 04 W81 41 09. Stereo. Hrs opn: 24 Box 798, 98 Cedar Lakes Rd., 25271. Phone: (304) 372-9800. Fax: (304) 372-9811. E-mail: shadow@c98.com Web Site: www.c98.com. Licensee: Big River Radio Inc. Group owner: Baker Family Stations (acq 1-31-2003; $762,500). Natl. Network: ABC. Format: Country. Target aud: 25-54. ♦Charmin McCarty, gen mgr & traf mgr; Ric Shannon, gen mgr; Rich Lacey, progmg dir; Larry Koenig, chief of engrg.

***WLKV(FM)**— Mar 26, 1994: 90.7 mhz; 3 kw. Ant 328 ft TL: N38 51 44 W81 41 27. Hrs open: 2351 Sunset Blvd., Suite 170-218, Rocklin, CA, 95765. Phone: (916) 251-1600. Fax: (916) 251-1650. Web Site: www.klove.com. Licensee: Educational Media Foundation. (acq 3-31-2005; $700,000 with WLKP(FM) Belpre, OH). Natl. Network: K-Love. Format: Christian. ♦Richard Jenkins, pres & gen mgr; Mike Novak, VP; Keith Whipple, dev dir; Eric Allen, natl sls mgr; David Pierce, progmg mgr; Ed Lenane, news dir; Sam Wallington, engrg dir; Karen Johnson, news rptr; Marya Morgan, news rptr; Richard Hunt, news rptr.

Romney

WDZN(FM)— Aug 29, 1988: 100.1 mhz; 480 w. 823 ft TL: N39 25 20 W78 47 25. Stereo. Hrs opn: 24 Box 477, Cumberland, MD, 21501-0477. Phone: (301) 724-6000. Fax: (301) 724-0617. Web Site: www.radiodisney.com. Licensee: Charter Equities Inc. Natl. Network: Jones Radio Networks. Law Firm: Baraff, Koerner & Olender. Format: Disney. News staff: one; News: 15 hrs wkly. Target aud: 25-54; adult decision makers. ♦Warren Gregory, pres & gen mgr; Travis Medcalf, sls dir & news dir; Rick Williams, chief of engrg.

***WVSB(FM)**— Mar 30, 1973: 104.1 mhz; 100 w. Ant 781 ft TL: N39 18 56 W78 43 04. Stereo. Hrs opn: 24 301 E. Main St., 26757. Phone: (304) 822-4838. Fax: (304) 822-4896. E-mail: gpark@access.k12.wv.us Web Site: wvsdb.state.k12.wv.us/radio_station.htm. Licensee: West Virginia Schools for the Deaf & Blind. Population served: 22000 Format: Classic country. Target aud: General. Spec prog: West Virginia tourism information. ♦Jane McBride, pres; Connie Newhouse, VP; George S. Park, gen mgr, chief of opns & progmg mgr.

Ronceverte

WRON(AM)— 1947: 1400 khz; 1 kw-U. TL: N37 45 36 W80 27 18. Hrs open: 24 276 Seneca Trail N., 24970. Phone: (304) 645-1400. Phone: (304) 645-1327. Fax: (304) 647-4802. E-mail: radio@wron.net Web Site: www.wron.com. Licensee: Michael J. Kidd dba Greenbrier Radio. (acq 10-1-97; $450,000 with co-located FM). Population served: 35,604 Natl. Network: Premiere Radio Networks, Westwood One, Talk Radio Network. Rgnl. Network: Metronews Radio Net. Natl. Rep: Dome, Rgnl Reps. Wire Svc: AP Format: Talk, news. News: 14 hrs wkly. Target aud: 35 & under. Spec prog: Relg 5 hrs wkly. ♦Michael J. Kidd, stn mgr, opns mgr, news dir & sports cmtr; Michael Kidd, gen sls mgr & prom mgr; Roy Jarrell, progmg dir; Grace Boxwell, pub affrs dir; Larry Carver, chief of engrg, farm dir & disc jockey; Larry Drennen, disc jockey.

WRON-FM— Dec 6, 1983: 97.7 mhz; 1 kw. 800 ft TL: N37 47 54 W80 30 55. Stereo. 24 Web Site: www.wron.com.75,000 Natl. Network: CNN Radio, Jones Radio Networks. Format: Oldies. Target aud: 25-60. ♦Michael J. Kidd, dev mgr, local news ed & sports cmtr; Michael Kidd, mktg mgr; Roy Jarrell, progmg mgr; Larry Carver, engrg mgr & farm dir.

Rupert

WYKM(AM)— Dec 9, 1981: 1250 khz; 5 kw-D, 32 w-N. TL: N37 59 35 W80 41 03. Hrs open: 6 AM-sunset Box 627, 25984. Secondary address: 714 Nicholas St. 25984. Phone: (304) 392-6003. Fax: (304) 392-5352. E-mail: bettydcrookshan@citynet.net Licensee: Mountain State Broadcasting Co. Population served: 34,000 Natl. Network: CBS. Format: Country, gospel. News: 7 hrs wkly. ♦Betty D. Crookshanks, pres, gen mgr & progmg dir; Donald Crookshanks, exec VP.

Saint Albans

WJYP(AM)— Jan 14, 1956: 1300 khz; 1 kw-D, 49 w-N. TL: N38 23 43 W81 51 00. Hrs open: 24 100 Kanawha Terr., 25177. Phone: (304) 722-3308. Fax: (304) 727-1300. E-mail: ccrninfo@hotmail.com Web Site: www.ccrnonline.com. Licensee: WKLC Inc. Group owner: L.M. Communications Inc. (acq 2-23-80). Population served: 250,000 Law Firm: Leventhal, Senter & Lerman. Format: Relg talk radio. Target aud: 18-49. ♦Ron Walton, gen mgr; Chris Colagrosso, progmg dir; Fred Francis, chief of engrg; Emma Allen, traf mgr.

WKLC-FM—Co-owned with WJYP(AM). Jan 1, 1966: 105.1 mhz; 50 kw. 1,663 ft TL: N38 25 15 W81 55 27. Stereo. 24 E-mail: rock105@wklc.com Web Site: www.wklc.com.1,123,200 Natl. Network: ABC. Format: Rock/AOR. ♦Jay Nunley, progmg dir; Dawn Cox, mus dir.

Saint Marys

WJAW(AM)— October 1984: 630 khz; 1 kw-D. TL: N39 23 42 W81 13 49. Hrs open: 24 925 Lancaster St., Marietta, OH, 45750. Phone: (740) 373-1490. Fax: (740) 373-1717. E-mail: jwharff@charter.net Web Site: www.wjawfm.com. Licensee: JAWCO Inc. (acq 2-26-2001; $25,000). Format: Sports. ♦John Wharff III, gen mgr & gen sls mgr; Dan Castelli, opns dir & progmg dir; Ralph Matheny, chief of engrg.

WRRR-FM— Nov 16, 1983: 93.9 mhz; 17 kw. Ant 390 ft TL: N39 22 49 W81 11 36. Stereo. Hrs opn: Box 374, 26170. Phone: (304) 684-3400. Fax: (304) 684-9241. Licensee: Seven Ranges Radio Co. Inc. Population served: 175,000 Law Firm: Irwin, Campbell & Tannenwald, P.C. Format: Adult contemp. News staff: one; News: 9 hrs wkly. Target aud: 25-49. ♦Sam Yoho, pres & gen mgr; Lou Petronio, opns mgr.

Salem

WAJR-FM— 1999: 103.3 mhz; 1.8 kw. Ant 589 ft TL: N39 15 44 W80 28 01. Hrs open: 1065 Radio Park Dr., Mount Clare, 26408-9516. Phone: (304) 623-6546. Fax: (304) 623-6547. Web Site: www.wajrfm.com. Licensee: West Virginia Radio Corp. of Salem. Format: News/talk. ♦John Halford, gen mgr.

WOBG-FM— Nov 1, 1990: 105.7 mhz; 6 kw. 581 ft TL: N39 19 06 W80 26 18. Stereo. Hrs opn: 24 1489 Locust Ave., Fairmont, 26554. Phone: (304) 624-1400. Fax: (304) 624-1402. Licensee: Burbach of DE LLC. Group owner: Burbach Broadcasting Group (acq 5-17-00; grpsl). Population served: 250,000 Rgnl rep: Commercial Media Sales. Format: Classic rock. Target aud: 25-54. ♦Nicholas A. Galli, pres; David Branham, gen mgr; Greg Bolyard, progmg dir; Jon Fox, news dir; Larry Smith, chief of engrg.

Shepherdstown

***WSHC(FM)**— 1974: 89.7 mhz; 950 w. -10 ft TL: N39 25 53 W77 48 18. Stereo. Hrs opn: 24 WSHC-FM, Shepherd College, King St., 25443. Phone: (304) 876-5134. Fax: (304) 876-5405. E-mail: wshc@shepherd.edu Web Site: www.897wshc.org. Licensee: Shepherd College Board of Governors (acq 8-28-01). Population served: 3,500 Natl. Network: ABC. Format: Alternative. News: 7 hrs wkly. Target aud: 18-24; college/young adult. ♦Buck Lam, gen mgr.

South Charleston

WMXE(FM)—Listing follows WSCW(AM).

WSCW(AM)— Dec 13, 1963: 1410 khz; 5 kw-D. TL: N38 22 34 W81 42 13. Hrs open: 100 Kanawha Terr., St. Albans, 25177. Fax: (304) 727-1300. E-mail: ccrninfo@hotmail.com Web Site: www.ccrnonline.com. Licensee: L.M. Communications of Kentucky LLC. Group owner: L.M. Communications Inc. (acq 4-1-03; grpsl). Population served: 252,000 Format: Relg talk radio. Target aud: 25-64. ♦Ron Walton, gen mgr; Chris Colagrasso, progmg dir & disc jockey; Fred Francis, chief of engrg; Emma Allen, traf mgr.

WMXE(FM)—Co-owned with WSCW(AM). July 29, 1985: 100.9 mhz; 3 kw. 285 ft TL: N38 22 34 W81 42 13. Stereo. 24 E-mail: rock105@wklc.com Web Site: www.wmxe.com. Format: Adult contemp, relg. Target aud: 25-54. ♦Mark Atkinson, progmg dir.

***WWLA(FM)**—Not on air, target date: unknown: 89.3 mhz; 300 w. Ant 528 ft TL: N38 26 37 W81 36 08. Hrs open: 188 S. Bellevue, Suite 222, Memphis, TN, 38104. Phone: (901) 726-8970. Fax: (901) 375-0041. Licensee: Broadcasting for the Challenged Inc. ♦George S. Flinn Jr., pres.

Spencer

WVRC(AM)— Sept 12, 1961: 1400 khz; 1 kw-U. TL: N38 48 23 W81 21 40. Hrs open: Box 622, 25276. Phone: (304) 927-3760. Fax: (304) 927-2877. E-mail: mail@wvrcradio.com Web Site: www.wvrcradio.com. Licensee: Star Communications Inc. (acq 9-22-82; $40,000; FTR: 10-11-82). Population served: 15,000 Natl. Rep: Rgnl Reps. Format: Gospel. ♦Larry Koenig, pres & chief of engrg; Bob Edwards, VP, gen mgr, gen sls mgr & progmg dir; Zachary Zdanek, news dir.

WVRC-FM— October 1992: 104.7 mhz; 3 kw. 328 ft TL: N38 47 40 W81 17 36. Web Site: www.wvrcradio.com. (Acq 3-1-91; FTR: 3-25-91). Format: Country.

Summersville

WCWV(FM)— Mar 13, 1983: 92.9 mhz; 11 kw. 900 ft TL: N38 21 37 W80 38 49. Stereo. Hrs opn: 24 713 Main St., 26651. Phone: (304) 872-5202. Fax: (304) 872-6904. E-mail: wcwv@c93net.com Web Site: www.c93net.com. Licensee: R-S Broadcasting Co. Inc. Population served: 1,000,000 Natl. Network: Westwood One. Natl. Rep: Dome. Format: Adult contemp. News: 23 hrs wkly. Target aud: 18-54. Spec prog: Gospel 15 hrs, relg 18 hrs wkly. ♦Michael D. Brown, VP, gen mgr & chief of opns; Wes Brown, gen sls mgr, mktg dir & progmg dir; Fred Francis, engrg dir & chief of engrg; Cassy Holcomb, traf mgr.

WDBS(FM)—See Sutton

***WMLJ(FM)**— 1993: 90.5 mhz; 11 kw. Ant 1,033 ft TL: N38 06 42 W80 35 52. Hrs open: 24
Rebroadcasts WOTJ(FM) Morehead City, NC 90%.
Box 1014, 26651. Phone: (304) 872-4612. Licensee: Grace Missionary Baptist Church. (acq 5-3-93; FTR: 5-24-93). Format: Gospel, children. Target aud: General. Spec prog: Sp one hr wkly. ♦Clyde I. Ebron, pres; Chris Brown, gen mgr; Mike Tyler, chief of engrg.

Sutton

WDBS(FM)—Listing follows WSGB(AM).

WSGB(AM)— Jan 22, 1964: 1490 khz; 1 kw-U. TL: N38 39 11 W80 43 10. Hrs open: 24 180 Main St., 26601. Phone: (304) 765-7373. Fax: (304) 765-7836. E-mail: thebuzz@theboss97fm.com Web Site: www.theboss97fm.com. Licensee: Summit Media Broadcasting L.L.C. (acq 12-30-99; $250,000 with co-located FM). Population served: 26,000 Natl. Network: ABC. Rgnl. Network: Metronews Radio Net. Natl. Rep: Dome. Rgnl rep: Dome, Regnl Reps Format: Adult contemp. Target aud: 18-34; young adults. ♦Al Sergi, pres & gen mgr; Daniel Finch, CFO.

WDBS(FM)—Co-owned with WSGB(AM). Apr 25, 1987: 97.1 mhz; 22 kw. Ant 751 ft TL: N38 27 05 W80 27 14. Stereo. 24 Web Site: theboss97fm.com.200,000 Natl. Network: Jones Radio Networks, AP Radio. Natl. Rep: Rgnl Reps. Format: Country classic, bluegrass. News: 9 hrs wkly. Target aud: 25-54; young adults females/males.

Vienna

WDMX(FM)— May 22, 1989: 100.1 mhz; 1.65 kw. 440 ft TL: N39 20 18 W81 30 01. Stereo. Hrs opn: 24 Box 5559, 6006 Grand Central Ave., 26105. Phone: (304) 295-6070. Phone: (304) 375-6558. Fax: (304) 295-4389. E-mail: oldies@radio1.netassoc.net Web Site: www.wdmx.com. Licensee: CC Licenses LLC. Group owner: Clear Channel Communications Inc. (acq 4-17-2001; grpsl). Natl. Network: ABC. Rgnl. Network: Ohio Radio Net. Natl. Rep: Clear Channel. Format: Oldies. News staff: 2; News: 2 hrs wkly. Target aud: 25-54. ♦Chuck Poet, gen mgr; Jim Grywalsky, opns mgr; Kurt McCall, gen sls mgr; Jim Grywalsky, progmg dir; Doug Hess, news dir & pub affrs dir; Jerry Kuhn, chief of engrg; Belinda Marcinko, traf mgr.

Webster Springs

WAFD(FM)— Feb 1, 1996: 100.3 mhz; 33 kw. Ant 594 ft TL: N38 27 38 W80 25 18. Hrs open: 202 Backfork St., 26288. Phone: (304) 847-5141. Fax: (304) 847-5149. E-mail: pastorhardman@ezla.net Licensee: Summit Media Inc. (group owner). (acq 3-8-2007; $1.24 million with WVAR(AM) Richwood). Format: Southern gospel. ♦James A. Hardman, gen mgr, opns mgr, progmg dir & chief of engrg; Samantha Moats, traf mgr.

Weirton

WEIR(AM)— Sept 15, 1950: 1430 khz; 1 kw-U, DA-2. TL: N40 26 45 W80 37 36. Stereo. Hrs opn: 24 2307 Pennsylvania Ave., 26062. Phone: (304) 723-1444. Fax: (304) 723-1688. E-mail: weir1430@weir.net Web Site: www.unforgettablefavorites.com. Licensee: Priority Communications Ohio L.L.C. Group owner: Priority Communications (acq 12-4-98; $475,000 with WCDK(FM) Cadiz, OH). Population served: 40,000 Natl. Network: Westwood One. Rgnl rep: Dome Law Firm: Pepper & Corazzini. Format: Talk morning, adult standards. News staff: one; News: 25 hrs wkly. Target aud: General. Spec prog: It 3 hrs, Gr 1 hr wkly. ♦Jay Philippone, pres & gen mgr; Judy Vavrek, stn mgr; Tammie Beagle, opns mgr; Hank Siegle, chief of engrg.

Welch

WELA(AM)—Not on air, target date: unknown: 1340 khz; 1 kw-U. TL: N37 25 50 W81 35 33. Hrs open: 115 Farwood Dr., Moreland, OH, 44022. Phone: (216) 381-6037. Licensee: C. Douglas Thomas. ♦C. Douglas Thomas, gen mgr.

WELC(AM)— Aug 19, 1950: 1150 khz; 5 kw-D. TL: N37 25 01 W81 36 58. Hrs open: 6 AM- Sunset Box 949, 24801. Secondary address: U.S. Rt. 52 24801. Phone: (304) 436-2131. Fax: (304) 436-2132. E-mail: mail@welcamfm.com Web Site: www.welcamfm.com. Licensee: Pocahontas Broadcasting Co. Population served: 35,000 Natl. Network: AP Radio. Rgnl. Network: Metronews Radio Net. Rgnl rep: Rgnl Reps. Law Firm: William D. Silva. Wire Svc: AP Format: Adult contemp. News staff: 2. Target aud: 21-54. Spec prog: Relg 15 hrs wkly. ♦Sam Sidote, pres, gen mgr, gen sls mgr & prom mgr; John Sidote, mus dir, news dir & chief of engrg.

WELC-FM— Feb 1, 1990: 102.9 mhz; 1.8 kw. Ant 423 ft TL: N37 25 01 W81 36 58. Stereo. 6 AM-11 PM Web Site: www.welcamfm.com.50,000 News staff: 2. Target aud: 18-49.

West Liberty

***WGLZ(FM)**— Sept 4, 1990: 91.5 mhz; 150 w. 213 ft TL: N40 09 49 W80 36 06. Hrs open: Box 13, West Liberty State College, 26074. Phone: (304) 336-8045. Phone: (304) 336-8037. Fax: (304) 336-8286. Licensee: West Liberty State College. Format: Alternative/mix. ♦Christian H. Lee, stn mgr & chief of engrg.

Weston

WFBY(FM)— Aug 29, 1972: 102.3 mhz; 940 w. 489 ft TL: N39 04 15 W80 31 13. (CP: 10 kw, ant 509 ft.). Hrs opn: 24 1065 Radio Park Dr., Mount Clare, 26408. Phone: (304) 296-0029. Fax: (304) 296-3876. Licensee: West Virginia Radio Corp. (group owner; (acq 1994; $250,000). Population served: 25,000 Law Firm: Putbrese, Hunsaker & Trent. Format: Classic Rock. Target aud: 25-44. ♦Dale Miller, pres & gen mgr; Christian Miller, CFO & stn mgr; Harvey Kercheval, opns VP; Chad Perry, progmg dir; Mark Rogers, mus dir; Steve Lough, chief of engrg.

WHAW(AM)— Feb 14, 1948: 980 khz; 1 kw-D, 50 w-N. TL: N39 02 25 W80 27 16. Hrs open: 24 300 Harrison Ave., 26452. Phone: (304) 269-5555. Fax: (304) 269-4800. E-mail: whaw@aol.com Web Site: www.whawradio.com. Licensee: Stephen R. Peters. (acq 4-16-98). Natl. Network: ABC. Format: ABC True Oldies. News staff: one; News: 2 hrs wkly. Target aud: General. Spec prog: Folk 4 hrs, gospel 18 hrs wkly,n Bluegrass 8 hrs. ♦Stephen R. Peters, gen mgr.

Westover

WZST(FM)— Jan 5, 1983: 100.9 mhz; 3 kw. 198 ft TL: N39 32 44 W79 55 58. Stereo. Hrs opn: 7013 Mountain Park Dr., Fairmont, 26554. Phone: (304) 363-3851. Fax: (304) 363-3852. E-mail: star100radio@aol.com Licensee: Tschudy Communications Corp. Group owner: Tschudy Broadcast Group (acq 5-88). Format: Hot country. Spec prog: Relg mus 2 hrs wkly. ♦Earl Judy, pres; Dick Yoder, gen mgr; Brian Dulaney, opns mgr; Judy King, gen sls mgr; Mike Donota, progmg dir.

Wheeling

WBBD(AM)— May 2, 1941: 1400 khz; 1 kw-U. TL: N40 05 49 W80 42 06. Hrs open: 24 1015 Main St., 26003. Phone: (304) 232-1170. Fax: (304) 234-0041. Licensee: Capstar TX L.P. Group owner: Clear Channel Communications Inc. (acq 8-30-00; grpsl). Population served: 60,000 Format: Big band, adult standards. Target aud: 35 plus. Spec prog: Pol 2 hrs wkly. ♦Scott Deel, gen sls mgr; Scott Miller, VP, gen mgr & mktg mgr; Minda Moticker, prom dir; Chad Tyson, progmg dir; Jack Reese, chief of engrg; Melissa Richie, traf mgr.

WKWK-FM—Co-owned with WBBD(AM). Mar 17, 1948: 97.3 mhz; 50 kw. 470 ft TL: N40 05 49 W80 42 06. Stereo. 24 Web Site: www.wk973.com.100,000 Format: Var/div. Target aud: 25-54. ♦Jim Connor, progmg dir; Steve Novotry, news dir.

WEGW(FM)— October 1966: 107.5 mhz; 10.5 kw. 882 ft TL: N40 03 41 W80 45 08. Stereo. Hrs opn: 24 1015 Main St., 26003. Phone: (304) 232-1170. Fax: (304) 234-0041. Web Site: www.wegw.com. Licensee: Capstar TX L.P. Group owner: Clear Channel Communications Inc. (acq 8-30-00; grpsl). Population served: 50,000 Format: Rock/AOR. Target aud: 25-54. ♦Mark Mays, pres; Scott Miller, VP & gen mgr; Karen Hardy, sls dir & gen sls mgr; Minda Moticker, prom dir; Chad Tyson, progmg dir; Jonathan Nixon, news dir; Jack Rees, chief of engrg.

WKKX(AM)— Apr 7, 1963: 1600 khz; 5 kw-D, 33 w-N. TL: N40 05 26 W80 42 11. Hrs open: 24 Box 231, 26003. Phone: (304) 214-1610. Fax: (304) 232-8488. E-mail: tsanthony@stratuswave.net Web Site: www.espn1600.com. Licensee: RCK 1 Group LLC (acq 7-19-2004; $400,000). Natl. Network: ESPN Radio. Rgnl. Network: Metronews Radio Net. Natl. Rep: Christal. Format: Sports talk. Target aud: 25-54; men. ♦Tom Anthony, gen mgr.

WOVK(FM)—Listing follows WWVA.

***WPHP(FM)**— Apr 4, 1977: 91.9 mhz; 1 kw. 259 ft TL: N40 04 07 W80 39 04. Hrs open: 1976 Parkview Rd., 26003. Phone: (304) 243-0400. Fax: (304) 243-0449. Licensee: Ohio County Board of Education. Population served: 50,000 Format: Top-40. Spec prog: Black 4 hrs, jazz one hr wkly. ♦Carolyn Ihlenfeld, gen mgr.

***WVNP(FM)**— Oct 7, 1981: 89.9 mhz; 25 kw. Ant 499 ft TL: N40 12 58 W80 33 31. Stereo. Hrs opn: 24 600 Capitol St., Charleston, 25301. Phone: (304) 556-4900. Fax: (304) 556-4960. E-mail: feedback@wvpubcast.org Web Site: www.wvpubcast.org. Licensee: West Virginia Educational Broadcasting Authority. Population served: 100,000 Natl. Network: NPR, PRI. Format: News, class, jazz. ♦Marilyn DiVita, gen mgr & dev dir; Rita Ray, gen mgr; James Muhammad, progmg dir; Greg Callard, news dir; Jack Wells, engrg dir; Teresa Wills, traf mgr.

WWVA(AM)— December 1926: 1170 khz; 50 kw-U, DA-2. TL: N40 06 07 W80 52 02. Hrs open: 1015 Main St., 26003. Phone: (304) 232-1170. Fax: (304) 234-0041. Fax: (304) 234-0036. Web Site: www.wwva.com. Licensee: Capstar TX L.P. Group owner: Clear Channel Communications Inc. (acq 8-30-00; grpsl). Population served: 48188 Natl. Rep: McGavren Guild. Format: News/talk. Target aud: 25-54. Spec prog: Farm 2 hrs wkly. ♦Scott Miller, gen mgr; Scott Peel, natl sls mgr; Jim Harrington, progmg dir; Tammie Beagle, news dir; Barb Vaughn, traf mgr.

WOVK(FM)—Co-owned with WWVA. September 1947: 98.7 mhz; 50 kw. 390 ft TL: N40 04 58 W80 46 18. (CP: 15 kw, ant 906 ft., TL: N40 04 48 W80 46 06). Web Site: www.wovk.com. Format: Country. ♦Jim Elliott, progmg dir; Molly Kilgore, traf mgr & disc jockey.

White Sulphur Springs

WSLW(AM)— 1971: 1310 khz; 5 kw-D. TL: N37 48 17 W80 21 03. Hrs open: 6 AM-sunset Box 610, 24986. Secondary address: Rt. 60 W. Harts Run 24986. Phone: (304) 536-1310. Fax: (304) 536-1311. E-mail: radio@wkcjwslw.com Licensee: Quorum Radio Partners of Virginia Inc., debtor-in-possession. (group owner; (acq 4-20-2005; grpsl). Population served: 150,000 Natl. Rep: Rgnl Reps. Format: Adult standards. News: 8 hrs wkly. Target aud: 25-60; secondary 16-25 & 60 plus. ♦Joyce Tucker, gen mgr.

Williamson

WBTH(AM)— Apr 19, 1939: 1400 khz; 1 kw-U. TL: N37 40 09 W82 16 09. Hrs open: Box 2200, Pikeville, KY, 25661. Phone: (606) 235-3600. Phone: (606) 437-4051. Fax: (606) 432-2809. Licensee: East Kentucky Radio Network Inc. (group owner; acq 4-4-00; $630,000 with co-located FM). Population served: 70,000 Rgnl. Network: Metronews Radio Net. Format: Adult contemp, oldies. Target aud: 25-54. ♦Keith Casebolt, gen mgr.

WXCC(FM)—Co-owned with WBTH(AM). Oct 27, 1978: 96.5 mhz; 50 kw. 500 ft TL: N37 40 09 W82 16 09. E-mail: wxcc@mikrotec.com Web Site: www.wxccfm.com.350,000 Format: Contemp country. ♦Johnny Randolph, progmg dir.

Williamstown

WVVV(FM)— 2000: 96.9 mhz; 3.51 kw. 423 ft TL: N39 20 18 W81 30 01. Hrs open: Box 5559, Vienna, 26105. Phone: (304) 295-6070. Fax: (304) 295-4389. Web Site: www.z969radio.net. Licensee: Bennco Inc. (acq 11-2-01; $1.625 million). Format: Var/div. ♦Jack Horton, gen mgr.

Wisconsin

Adams

WDKM(FM)— Oct 8, 1993: 106.1 mhz; 6 kw. 328 ft TL: N43 57 29 W89 49 43. Stereo. Hrs opn: 24 408 Hillwood Ln., Friendship, 53934. Secondary address: 1040 W. Center St. 53910. Phone: (608) 339-3221. Fax: (608) 339-2403. E-mail: heidi@wdkmfm.com Web Site: www.wdkmfm.com. Licensee: Roche-A-Cri Broadcasting. Format: Classic hits. Spec prog: Polka 14 hrs wkly. ♦Drew Smith, stn mgr & progmg dir; Heidi Roekle, gen mgr & gen sls mgr.

***WHAA(FM)**—Not on air, target date: unknown: 89.1 mhz; 28.5 kw. Ant 577 ft TL: N44 01 13 W89 33 29. Hrs open: 821 University Ave., Madison, 53706. Phone: (608) 263-3970. Fax: (608) 263-9763. Web Site: www.wpr.org. Licensee: State of Wisconsin-Educational Communications Board. ♦Phil Corriveau, gen mgr.

Algoma

WBDK(FM)— Nov 12, 1986: 96.7 mhz; 8 kw. 538 ft TL: N44 42 26 W87 24 26. Stereo. Hrs opn: 24 3030 Park Dr., Suite 3, Sturgeon Bay, 54235. Phone: (920) 746-9430. Fax: (920) 746-9433. E-mail: wbdk@itol.com Web Site: www.doorradio.com. Licensee: Nicolet Broadcasting Inc. (group owner; acq 9-3-93; FTR: 9-27-93). Law Firm: Pepper & Corazzini. Format: Oldies of 50's & 60's. News staff: 3. Target aud: 34 plus. ♦Roger Utnehmer, pres; Paul Schmitt, sr VP; Miles Knuteson, gen mgr; Karen Leitzinger, opns mgr & progmg dir; Kathy Robinson, progmg dir & traf mgr; John Focke, news dir.

WRLU(FM)— Aug 1, 1999: 104.1 mhz; 6 kw. 328 ft TL: N44 40 02 W87 23 55. Hrs open: 3030 Park Dr., Suite 3, Sturgeon Bay, 54235. Phone: (920) 746-9430. Fax: (920) 746-9433. Web Site: www.doorradio.com. Licensee: Nicolet Broadcasting Inc. (group owner) Format: Country. News: 3 hrs wkly. ♦Roger Utnehmer, pres; Miles Knuteson, gen mgr, gen mgr & gen sls mgr; Karen Leitzinger, opns mgr & progmg dir; John Focke, news dir; Kathy Robinson, traf mgr.

Allouez

WZNN(FM)— 1996: 106.7 mhz; 25 kw. Ant 328 ft TL: N44 29 03 W87 56 12. Stereo. Hrs opn: 24 810 Victoria St., Green Bay, 54302. Phone: (920) 468-4100. Fax: (920) 468-0250. Licensee: Cumulus Licensing LLC. Group owner: Cumulus Media Inc. (acq 8-7-98; $2.5 million). Population served: 700,000 Format: Alternative. News staff: one;

News: 5 hrs wkly. ♦Greg Jessen, CEO, CEO & gen mgr; Jimmy Clark, CFO & opns mgr; Buck Hein, gen sls mgr; Ted Bare, progmg dir; Chris Gielow, chief of engrg.

Altoona

WDVM(AM)—See Eau Claire

WISM-FM— Nov 15, 1991: 98.1 mhz; 10 kw. Ant 174 ft TL: N44 46 36 W91 28 30. Stereo. Hrs opn: 24 619 Cameron St., Eau Claire, 54703. Phone: (715) 830-4000. Fax: (715) 835-9680. Web Site: www.mix981.com. Licensee: Clear Channel Broadcasting Licenses Inc. Group owner: Clear Channel Communications Inc. (acq 11-1-2002; $2.4 million). Population served: 175,000 Format: Adult contemp. Target aud: 25-54. ♦Rick Hencley, gen mgr; Mike Cushman, opns mgr; Steve Potter, gen sls mgr; Sara Duclos, prom dir; Jim Finn, progmg dir; Keith Edwards, news dir; Theresa Nelson, traf mgr.

Amery

WXCE(AM)— Jan 23, 1978: 1260 khz; 5 kw-U, DA-2. TL: N45 15 25 W92 22 00. Hrs open: 5 AM-noon Box 1260, 54001. Secondary address: 328 S. 100th St. 54001. Phone: (715) 268-7185. Fax: (715) 268-7187. E-mail: wxce@spacestar.net Web Site: www.wxce.com. Licensee: Lake Country Broadcasting Corp. (acq 1-14-99). Population served: 200,000 Natl. Network: ABC. Rgnl. Network: Tribune, Goetz Group, Wisconsin Radio Net. Format: News/talk. News staff: one; News: 20 hrs wkly. Target aud: 35 plus. ♦Darren Van Blaricom, gen mgr, gen sls mgr, rgnl sls mgr & progmg mgr; Greg Marsten, news dir & sports cmtr; Julie Measner, traf mgr.

Antigo

WACD(FM)— 1998: 106.1 mhz; 10 kw. 276 ft TL: N45 06 23 W89 09 09. Hrs open: Box 509, 54409. Phone: (715) 623-4124. Licensee: Results Broadcasting Inc. (group owner; (acq 4-29-2005; $500,000 with WATK(AM) Antigo). Format: Adult standards. ♦Shaughn Novy, gen mgr; K.B. Butler, opns mgr.

WATK(AM)— Mar 15, 1948: 900 khz; 250 w-D, 196 w-N. TL: N45 06 50 W89 08 20. Hrs open: 24 Box 509, N. 2237 Hwy. 45 S., 54409. Phone: (715) 623-4124. Fax: (715) 627-4497. E-mail: wrlo@marathonmedianorth.net Web Site: www.wrlo1053.com. Licensee: Results Broadcasting Inc. (group owner; (acq 4-29-2005; $500,000 with WACD(FM) Antigo). Population served: 25,900 Natl. Network: Jones Radio Networks. Format: Adult standards. News staff: 2; News: 12 hrs wkly. Target aud: 25-59; two-income families. Spec prog: Gospel 2 hrs wkly. ♦Tom Hopfensperger, gen mgr; Duff Damos, opns dir & opns mgr; Shaughn Novy, gen sls mgr; Dave St. Peter, progmg dir & news dir; Cliff Groth, chief of engrg.

WRLO-FM— Nov 11, 1973: 105.3 mhz; 100 kw. Ant 541 ft TL: N45 22 04 W89 08 20. Stereo. Hrs opn: 24 3616 Hwy. 47 N., Rhinelander, 54501-8819. Phone: (715) 362-1975. Fax: (715) 362-1973. Web Site: www.wrlo.com. Licensee: NRG License Sub. LLC. (acq 10-31-2005; grpsl). Population served: 100,000 Format: Classic rock. Target aud: 21-54. ♦Michele Krueger, sls dir; Shaughn Novy, rgnl sls mgr; Duff Damos, progmg dir; John Burton, news dir; Jim Zastrow, chief of engrg.

Appleton

WAPL-FM— Dec 24, 1965: 105.7 mhz; 100 kw. 1,175 ft TL: N44 21 32 W87 59 07. Stereo. Hrs opn: 24 Box 1519, 54912. Secondary address: 2800 E. College Ave. 54915. Phone: (920) 734-9226. Fax: (920) 733-3291. E-mail: wapl@wcinet.com Web Site: www.wapl.com. Licensee: Woodward Communications Inc. (group owner; (acq 3-75). Population served: 305,000 Natl. Rep: McGavren Guild. Law Firm: Hogan & Hartson. Wire Svc: AP Format: Mainstream rock. News: 2 hrs wkly. Target aud: 20-plus; professional and semi-professional adults. ♦Greg Bell, gen mgr, stn mgr & opns mgr; Greg Lawrence, sls dir; Joe Calgaro, progmg dir & disc jockey; Steve Brown, chief of engrg; Kay Taylor, traf mgr; Elwood, disc jockey; Borna Velic, disc jockey; Len Nelson, disc jockey; Rick McNeal, disc jockey; Ross Maxwell, disc jockey; Scott Stevens, disc jockey.

***WEMI(FM)**— 1994: 91.9 mhz; 3.1 kw. Ant 328 ft TL: N44 15 17 W88 26 13. Hrs open: 24 1909 W. 2nd St., 54914. Phone: (920) 749-9456. Fax: (920) 749-0474. Web Site: christianfamilyradio.net. Licensee: Evangel Ministries Inc. Population served: 300,000 Natl. Network: Moody, USA, Salem Radio Network. Law Firm: Leventhal, Senter & Lerman. Format: Relg. News: 10 hrs wkly. Target aud: 35-49; women. Spec prog: Sp one hr wkly. ♦Mary B. Lieb, chmn; Paul Cameron, gen mgr; Andy Kilgas, opns dir & sls dir; Heidi Prahl, dev dir.

***WOVM(FM)**— Mar 10, 1956: 91.1 mhz; 10.5 kw. Ant 120 ft TL: N44 15 42 W88 23 47. Stereo. Hrs opn: 5 AM-midnight Box 13213, Green Bay, 54307-3213. Phone: (888) 233-9616. Fax: (920) 469-3023. Web Site: www.relevantradio.com. Licensee: Starboard Media Foundation Inc. (acq 9-20-2005; $300,000). Population served: 300,000 Rgnl. Network: Wis. Pub. Format: Catholic talk. ♦Mark C. Follett, chmn.

WSCO(AM)— 1952: 1570 khz; 1 kw-D, 331 w-N. TL: N44 13 04 W88 24 33. Hrs open: 24 hrs PO Box 1519, 2800 E. College Ave., 54915. Phone: (920) 733-6639. Fax: (920) 739-0494. Licensee: Woodward Communications Inc. (group owner; acq 12-3-01; $450,000). Population served: 50,000 Natl. Network: Fox Sports, Sporting News Radio Network. Format: Sports. Target aud: 25-54; Male. ♦Greg Bell, gen mgr; John Wanie, stn mgr; Joe Calgaro, progmg dir.

Ashland

WATW(AM)— May 1, 1940: 1400 khz; 1 kw-U. TL: N46 34 23 W90 51 56. (CP: 480 w-U. TL: N46 34 25 W90 51 56). Stereo. Hrs opn: 24 2320 Ellis Ave., 54806. Phone: (715) 682-2727. Phone: (715) 682-2728. Fax: (715) 682-9338. E-mail: productionash@charter.net Web Site: www.watwam.com. Licensee: Heartland Communications License LLC. (group owner; (acq 4-23-2004; grpsl). Population served: 50,000 Natl. Network: ABC. Law Firm: Lauren A. Colby. Wire Svc: Wheeler News Service Format: Hits of the 40s, 50s & 60s, news. News staff: one; News: 17 hrs wkly. Target aud: 40 plus; middle to upper income adults. Spec prog: Relg 5 hrs wkly. ♦Scott Jaeger, gen mgr; Skip Hunter, progmg dir & chief of engrg.

WJJH(FM)—Co-owned with WATW(AM). Aug 1, 1970: 96.7 mhz; 50 kw. 246 ft TL: N46 34 25 W90 51 56. Stereo. 24 Web Site: wjjhfm.com.250,000 Natl. Network: ABC. Format: Classic rock. Target aud: 25-45.

WBSZ(FM)— July 25, 1994: 93.3 mhz; 100 kw. 246 ft Stereo. Hrs opn: 24 2320 Ellis Ave., 54806. Phone: (715) 682-2727. Fax: (715) 682-9338. E-mail: productionash@charter.net Web Site: www.wbszfm.com. Licensee: Heartland Communications License LLC. (group owner; (acq 4-23-2004; grpsl). Population served: 250,000 Natl. Network: ABC, Westwood One. Law Firm: Lauren A. Colby. Format: Hot country. Target aud: 18-49. ♦Scott Jaeger, gen mgr; Rich Canatta, opns VP; Skip Hunter, opns VP & progmg dir.

WEGZ(FM)—See Washburn

Auburndale

***WLBL(AM)**— 1922: 930 khz; 5 kw-D. TL: N44 36 52 W90 02 08. Hrs open: Sunrise-sunset 518 S. 7th Ave., Wausau, 54401-5362. Phone: (715) 261-6298. Fax: (715) 848-28. E-mail: listener@wpr.org Web Site: www.wpr.org. Licensee: State of Wisconsin, Education Communications Board. Natl. Network: NPR, PRI. Format: News/talk, MOR. News staff: 4. ♦Phil Corriveau, gen mgr; Rick Reyer, stn mgr & dev dir.

Balsam Lake

WLMX-FM— Feb 14, 1997: 104.9 mhz; 22 kw. 348 ft TL: N45 25 07 W92 14 34. Stereo. Hrs opn: Box 179, Luck, 54853. Phone: (715) 825-4240. Fax: (715) 825-4244. E-mail: studio@mix105.ws Web Site: mix105.ws. Licensee: Red Rock Radio Corp. (group owner; (acq 9-1-2006; grpsl). Natl. Network: ABC. Rgnl. Network: Wisconsin Radio Net. Format: Adult contemp. News: 5 hrs wkly. Target aud: 18-49; adults. ♦Ro Grignon, pres; Don Welch, VP; Ron Revere, gen mgr.

Baraboo

WOLX-FM—Licensed to Baraboo. See Madison

WRPQ(AM)— June 1967: 740 khz; 250 w-D, 6.4 w-N. TL: N43 27 19 W89 45 13. Stereo. Hrs opn: 24 Box 456, 53913. Secondary address: 407 Oak St. 53913. Phone: (608) 356-3974. Fax: (608) 355-9952. E-mail: jeffsmith@wrpq.com Web Site: www.wrpq.com. Licensee: Baraboo Broadcasting Co. (acq 7-1-91; $125,000; FTR: 7-13-81). Population served: 14,000 Natl. Network: CNN Radio. Rgnl. Network: Wisconsin Radio Net. Law Firm: Koerner & Olender. Format: Adult contemp. News staff: one; News: 8 hrs wkly. Target aud: 25-54. Spec prog: Relg 5 hrs wkly. ♦Jeff Smith, pres & progmg dir; Gregory Buchwald, VP; Mike Knoll, gen mgr; Reggie Zimmerman, sls dir & gen sls mgr; Mary Bahr, engrg VP & traf mgr.

Barron

WAQE-FM— 1999: 97.7 mhz; 15.5 kw. Ant 289 ft TL: N45 32 16 W91 45 50. Hrs open: Box 703, 1859 21st Ave., Rice Lake, 54868. Phone: (715) 234-9059. Fax: (715) 234-6942. E-mail: info@waqe.com Web Site: www.waqe.com. Licensee: TKC Inc. Law Firm: Shaw Pittman. Format: Hits of the 80s, 90s & today. News staff: one; News: 5 hrs wkly. Target aud: 25-54; general. ♦Brian Schultz, gen mgr & stn mgr; Tom Koser, gen mgr; Sondra Maanum, traf mgr.

Beaver Dam

WBEV(AM)— Mar 21, 1951: 1430 khz; 1 kw-U, DA-N. TL: N43 25 43 W88 53 33. Hrs open: 24 Box 902, 533916. Secondary address: 100 Stoddart St. 533916. Phone: (920) 885-4442. Fax: (920) 885-2152. Licensee: Good Karma Broadcasting L.L.C. (group owner; acq 12-2-97; grpsl). Population served: 80,000 Wire Svc: Wheeler News Service Format: Adult contemp, news/talk. News staff: 3; News: 20 hrs wkly. Target aud: 30 plus; general. Spec prog: Farm 8 hrs, sports 18 hrs wkly. ♦Craig Karmazin, pres & gen mgr; Rick Armon, chief of opns & progmg dir; John Moser, gen sls mgr & news dir; Warren Jorgenson, chief of engrg; Deb Iamers, traf mgr.

WXRO(FM)—Co-owned with WBEV(AM). July 15, 1968: 95.3 mhz; 6 kw. 328 ft TL: N43 28 09 W88 49 32. Stereo. 24 70,000 Format: Modern country. News staff: 3; News: 10 hrs wkly. Target aud: 25-54; general. Spec prog: Farm 6 hrs wkly. ♦Craig Karmazin, chmn; John A. Moser, sls dir; John Kraft, farm dir; Rick Armon, prom dir & disc jockey.

Beloit

***WBCR-FM**— Nov 30, 1965: 90.3 mhz; 100 w. 44 ft TL: N42 30 13 W89 01 55. (CP: 130 w). Hrs opn: Beloit College, 700 College St., 53511. Phone: (608) 363-2402. Fax: (608) 363-2718. E-mail: wbcr@stubeloit.edu Web Site: www.beloit.edu/~wbcr/. Licensee: Beloit College. Population served: 100,000 Format: Educ, div. ♦Kyle McKenzie, gen mgr.

WGEZ(AM)— Sept 26, 1948: 1490 khz; 1 kw-U. TL: N42 29 45 W89 01 03. Hrs open: 24 Box 416, 53512. Secondary address: 622 Public Ave. 53511. Phone: (608) 365-8865. Fax: (608) 365-8867. E-mail: wgezam@hotmail.com Web Site: www.wgez1490am.com. Licensee: Alliance Communications Inc. (acq 2-18-2005; $325,000). Population served: 180,000 Natl. Network: CBS, ABC. Wire Svc: AP Format: Oldies. News staff: 10 Target aud: 25-54; baby boomers. ♦Alan Kearns, gen mgr & progmg dir; Jason Ryan, gen sls mgr; Carla Cornell, traf mgr.

WTJK(AM)—South Beloit, IL) May 18, 1948: 1380 khz; 5 kw-U, DA-N. TL: N42 27 34 W89 01 43. Stereo. Hrs opn: 24 1 Parker Place, Suite 485, Janesville, 53545. Phone: (608) 758-9025. Fax: (608) 758-9550. E-mail: keith@gkbradio.com Web Site: www.espn1380.com. Licensee: Good Karma Broadcasting L.L.C. (group owner; acq 9-13-00; $235,000). Population served: 65,000 Natl. Network: ESPN Radio. Natl. Rep: Interep. Format: Sports, info. Target aud: Males 25-54. Spec prog: NASCAR. ♦Keith Williams, gen mgr; Kevin MacDougall, gen sls mgr; Kyle Jacob, progmg dir; Warren Jorgensen, chief of engrg.

Berlin

WBJZ(FM)— July 31, 1972: 104.7 mhz; 5.2 kw. Ant 351 ft TL: N43 53 57 W88 53 37. Hrs opn: 24 112 Watson St., Ripon, 54971. Phone: (920) 748-9205. Fax: (920) 748-5530. Licensee: Caxambas Corp. Population served: 250,000 Format: Smooth jazz. Target aud: 35-54;

upscale, adults. ♦Mike Enfelt, gen mgr & stn mgr; Bill Denkert, gen sls mgr; Jason Mansmith, progmg dir; Lori Petrie, traf mgr.

WISS(AM)— June 28, 1971: 1100 khz; 1 kw-D. TL: N43 56 55 W88 59 09. Hrs open: Sunrise-sunset Box 71, 54923. Secondary address: 112 N. Pearl St. 54923. Phone: (920) 361-3551. Fax: (920) 361-3737. E-mail: production@hometownbroadcasting.com Web Site: www.hometownbroadcasting.com. Licensee: Hometown Broadcasting LLC (acq 12-1-99; $165,000). Population served: 250,000 Format: Classic country, sports, local news/talk. News staff: one; News: 10 hrs wkly. Target aud: 25-54; local community. ♦Tom Boyson, gen mgr; Margaret Corrente, sls dir; Jaime Bellmer, gen sls mgr; Bernie Phillips, progmg dir; Andrew Disterhaft, chief of engrg & traf mgr.

Birnamwood

WYNW(FM)— 2003: 92.9 mhz; 6 kw. Ant 328 ft TL: N44 59 50 W89 22 07. Hrs open: Starboard Network, 2300 Riverside Drive, Green Bay, 54301. Phone: (920) 469-3021. Fax: (920) 469-3023. Web Site: www.relevantradio.com. Licensee: Starboard Media Foundation Inc. Group owner: Relevant Radio (acq 7-9-2002). Format: Catholic radio. ♦Mike Strub, stn mgr.

Black River Falls

WWIS(AM)— Aug 23, 1958: 1260 khz; 580 w-D. TL: N44 19 11 W90 53 31. Hrs open: 6 AM-sunset W11573 Town Creek Rd., 54615. Phone: (715) 284-4391. Fax: (715) 284-9740. E-mail: wwis@wwisradio.com Web Site: www.wwisradio.com. Licensee: WWIS Radio Inc. (acq 5-1-68). Population served: 17,000 Natl. Network: CBS Radio. Rgnl. Network: Brownfield. Law Firm: Miller & Miller, P.C. Wire Svc: Wheeler News Service Format: Oldies. News: 6 hrs wkly. Target aud: General. ♦Nelson Lent, pres & VP; Robert E. Smith, chmn & pres; Robert A. Gabrielson, gen mgr.

WWIS-FM— Jan 21, 1991: 99.7 mhz; 25 kw. Ant 328 ft TL: N44 19 11 W90 53 31. Stereo. Hrs opn: 24 W. 11573 Town Creek Rd., 54615. Phone: (715) 284-4391. Fax: (715) 284-9740. E-mail: wwis@cuttingedge.net Web Site: www.wwisradio.com. Licensee: WWIS Radio Inc. Population served: 295,000 Natl. Network: CBS. Wire Svc: Wheeler News Service Format: Adult contemp. News: 12 hrs wkly. Target aud: 25-55. ♦Robert Smith, pres; Nelson Lent, VP & gen mgr; Robert Gabrielson, gen mgr.

Bloomer

WQRB(FM)— 1993: 95.1 mhz; 8.9 kw. 430 ft TL: N45 01 59 W91 21 09. Stereo. Hrs opn: 24 Box 45, Eau Claire, 54703. Phone: (715) 830-4000. Fax: (715) 835-9680. Web Site: www.b95radio.com. Licensee: Capstar TX L.P. Group owner: Clear Channel Communications Inc. (acq 2000; grpsl). Population served: 180,000 Format: Hot country. Target aud: 25-54. ♦Rick Hencley, gen mgr; Mike Cushman, opns mgr; Steve Potter, gen sls mgr; Mike McKay, progmg dir; Keith Edwards, news dir; Paul Orth, chief of engrg; Trina Butak, traf mgr; Alex Edwards, disc jockey.

Brillion

WDUZ-FM— March 1993: 107.5 mhz; 6 kw. 328 ft TL: N44 15 28 W88 11 43. (CP: 5 kw). Hrs opn: 24 810 Victoria St., Green Bay, 54302. Phone: (920) 468-4100. Fax: (920) 468-0250. Licensee: Cumulus Licensing Corp. Group owner: Cumulus Media Inc. (acq 8-7-98; $2.065 million). Population served: 650,000 Law Firm: Fisher, Wayland, Cooper, Leader & Zaragoza. Format: Sports/talk. Target aud: 18-49; educated, affluent, upper-income. Spec prog: Sp 2 hrs wkly. ♦Greg Jessen, gen mgr; Jimmy Clark, opns mgr, sls dir & progmg dir; Buck Hein, gen sls mgr; Brian Stenzel, prom dir; Chris Gielow, engrg mgr; Guy Dark, disc jockey; Max O'Brien, disc jockey.

Brookfield

WJZX(FM)— Aug 18, 1995: 106.9 mhz; 6 kw. Ant 328 ft TL: N43 09 00 W88 07 25. Hrs open: 5407 W. McKinley Ave., Milwaukee, 53208. Phone: (414) 978-9000. Fax: (414) 978-9001. Web Site: smoothjazz1069.com. Licensee: Saga Communications of Milwaukee LLC. Group owner: Saga Communications Inc. (acq 5-9-97; $5 million with WJMR-FM Menomonee Falls). Natl. Rep: Katz Radio. Law Firm: Smithwick & Belendiuk. Format: Smooth jazz. News staff: one. Target aud: 35-64. ♦Thomas Joerres, pres & gen mgr; Annmarie Topel, gen sls mgr; Anne Zimmer, natl sls mgr; Steve Murphy, progmg dir; Phil Longenecker, chief of engrg; Michelle Golding, traf mgr.

Brule

***WHSA(FM)**— Sept 14, 1952: 89.9 mhz; 38 kw. 550 ft TL: N46 27 59 W91 33 56. Stereo. Hrs opn: 24 P.O. Box 2000, Superior, 54880. Secondary address: 1800 Grand Ave., Superior 54880. Phone: (715) 394-8530. Fax: (715) 394-8404. E-mail: jmunson@uwsuper.edu Web Site: www.wpr.org. Licensee: State of Wisconsin Educational Communications Board. Population served: 200,000 Natl. Network: NPR. Rgnl. Network: Wis. Pub. Format: Class, news/talk. News staff: one; News: 39 hrs wkly. Target aud: 34 plus. Spec prog: Folk 3 hrs, jazz 6 hrs wkly. ♦John A. Munson, gen mgr.

Burlington

***WBSD(FM)**— Apr 7, 1975: 89.1 mhz; 300 w. 107 ft TL: N42 40 14 W88 16 18. Stereo. Hrs opn: 24 400 McCanna prkwy, 53105. Phone: (262) 763-0195. Fax: (262) 763-0207. E-mail: wbsd@wbsdfm.com Web Site: www.wbsdfm.com. Licensee: Burlington Area School District. Population served: 75,000 Format: Alternative, progsv rock. News: One. Target aud: 25-54. Spec prog: Jazz 8 hrs, reggae 3 hrs, ska/punk 2 hrs, metal one hr, blues 4 hrs, folk 5 hrs wkly. ♦Terry Havel, gen mgr; Kevin Fay, opns VP.

Chetek

WATQ(FM)— May 17, 1997: 106.7 mhz; 50 kw. 492 ft TL: N45 14 31 W91 44 43. Stereo. Hrs opn: 24 Box 45, Eau Claire, 54702. Secondary address: 619 Cameron St. 54702. Phone: (715) 830-4000. Fax: (715) 835-9680. Web Site: www.moose106.com. Licensee: Capstar TX L.P. Group owner: Clear Channel Communications Inc. (acq 2000; grpsl). Natl. Network: CBS. Format: Country. News staff: 2. Target aud: 35-64. ♦Rick Hencley, gen mgr; Steve Potter, gen sls mgr; Mike Cushman, progmg dir; Keith Edwards, news dir; Paul Orth, chief of engrg.

Chilton

WMBE(AM)— May 25, 1984: 1530 khz; 250 w-D. TL: N44 01 10 W88 09 32. Hrs open: Sunrise-sunset Box 1450, Fond du Lac, 54936. Secondary address: 354 Winnebago Dr., Fond du Lac 54935. Phone: (920) 921-1071. Fax: (920) 921-0757. E-mail: info@espnradio1530.com Web Site: www.espnradio1530.com. Licensee: Maszka-Pacer Radio Inc. (acq 12-28-90; $4,469; FTR: 1-14-91). Natl. Network: USA. Format: Sports. News staff: 2. Target aud: Males 18-54; Sports fans. ♦R.B. Hopper, gen mgr; Mark Kastein, gen sls mgr & prom mgr; Shawn A. Kiser, progmg dir; Stu Muck, engrg dir; Cindy Konen, traf mgr.

Chippewa Falls

WAXX(FM)—See Eau Claire

WCFW(FM)— Oct 20, 1968: 105.7 mhz; 25 kw. 305 ft TL: N44 52 18 W91 17 11. Stereo. Hrs opn: 24 318 Well St., 54729. Phone: (715) 723-2257. Fax: (715) 723-8276. E-mail: wcfwradio@clearwire.net Licensee: Roland L. Bushland dba Bushland Radio/WCFW. Wire Svc: AP Format: Adult contemp. News: 8 hrs wkly. Target aud: 35 plus; upscale. Spec prog: Relg 2 hrs wkly. ♦Patricia Bushland, gen sls mgr; Roland Bushland, gen mgr & progmg dir.

WEAQ(AM)— Sept 7, 1958: 1150 khz; 5 kw-D. TL: N45 53 05 W91 23 25. Hrs open: Box 1, Eau Claire, 54702. Secondary address: 944 Harlem St., Altoona 54720. Phone: (715) 832-1530. Fax: (715) 832-5329. Web Site: www.espn1150.com. Licensee: Maverick Media of Eau Claire License LLC. Group owner: Maverick Media LLC (acq 6-13-2003; grpsl). Natl. Rep: Katz Radio. Format: Sports/talk. Target aud: 40 plus; general. ♦Gary Rozynek, pres; George Roberts, VP; Bruce Butler, gen mgr; Dave Craig, stn mgr & opns dir; Jim Casey, chief of opns.

Cleveland

WLKN(FM)— Apr 25, 1985: 98.1 mhz; 5.8 kw. Ant 292 ft TL: N43 59 03 W87 45 55. Stereo. Hrs opn: 24 Box 26, 1050 Linden St., 53015. Phone: (920) 693-3103. Fax: (920) 693-3104. E-mail: manager@wlkn.com Web Site: www.wlkn.com. Licensee: Radio K-T Inc. (acq 10-15-99; $980,000). Population served: 218,000 Law Firm: Vinson & Elkins. Wire Svc: AP Format: Adult contemp. News staff: one; News: 7 hrs wkly. Target aud: 25-54; active, upscale. ♦Jack Taddeo, CEO, pres, progmg dir & chief of engrg; David Jetzer, stn mgr, gen sls mgr, gen sls mgr & news dir; Wendy Dekker, traf mgr & farm dir.

Clintonville

WFCL(AM)— Feb 28, 1983: 1380 khz; 5 kw-D, 2.5 kw-N, DA-2. TL: N44 34 00 W88 44 36. Hrs open: 24 1456 E. Green Bay St., Shawano, 54166. Phone: (715) 524-2194. Fax: (715) 524-9980. Licensee: Results Broadcasting Inc. Group owner: Results Broadcasting (acq 1996). Natl. Network: Jones Radio Networks. Law Firm: Miller & Miller, P.C. Format: MOR. News staff: one; News: 12 hrs wkly. Target aud: 25-54. Spec prog: Farm 8 hrs wkly. ♦Eric Voight, pres & gen mgr; Walt Baldwin, news dir & chief of engrg.

WJMQ(FM)—Co-owned with WFCL(AM). Oct 27, 1986: 92.3 mhz; 6 kw. 328 ft TL: N44 34 00 W88 44 36. Stereo. 24 Format: Country. News staff: one; News: 12 hrs wkly. Target aud: 12-plus. ♦Eric Voight, stn mgr.

Columbus

WTLX(FM)— July 16, 1990: 100.5 mhz; 6 kw. 328 ft TL: N43 24 19 W89 06 24. Stereo. Hrs opn: 24 Box 902, Beaver Dam, 53916. Phone: (920) 885-4442. Fax: (920) 885-2152. Web Site: www.100xmadison.com. Licensee: Good Karma Broadcasting L.L.C. (group owner; acq 12-2-97; grpsl). Population served: 200,000 Wire Svc: Wheeler News Service Format: Talk, sports. News staff: one; News: 8 hrs wkly. Target aud: 25-50. ♦Craig Karmazin, pres & gen mgr; Scott Trentadue, stn mgr, stn mgr & pub affrs dir; Rick Armon, opns dir; John Moser, sls dir & gen sls mgr; Warren Jorgensen, chief of engrg; Debbie Lamers, traf mgr.

Cornell

WDRK(FM)— 2001: 99.9 mhz; 25 kw. Ant 328 ft TL: N45 07 22 W91 24 23. Hrs open: 944 Harlem St., Altoona, 54720. Phone: (715) 832-1530. Fax: (715) 832-5329. Web Site: www.thecarprocks.com. Licensee: Maverick Media of Eau Claire License LLC. Group owner: Maverick Media LLC (acq 6-13-2003; grpsl). Population served: 150,000 Format: Rock. News staff: one; News: one hr wkly. ♦George Roberts, VP, gen mgr & mktg mgr; Al Shannon, opns mgr, mus dir & disc jockey; Lynn Bieritz, sls dir; Dan Gainey, natl sls mgr & rgnl sls mgr; Kris Cooper, prom dir & disc jockey; Rick Roberts, stn mgr & progmg mgr; Dan Lea, news dir; Bill Holden, chief of engrg; Morgan McCarthy, disc jockey.

De Forest

WHLK(FM)— 2003: 93.1 mhz; 6 kw horiz, 5.4 kw vert. Ant 321 ft TL: N43 09 34 W89 12 55. Hrs open: Box 44408, Madison, 53744. Secondary address: 730 Rayovc Dr. 53711. Phone: (608) 273-1000. Fax: (608) 271-0400. E-mail: info@thelake.com Web Site: www.thelakemadison.com. Licensee: Mid-West Management Inc. Group owner: The Mid-West Family Broadcast Group (acq 8-13-2002). Natl. Rep: Mass Media Advertising. Format: Classic hits. ♦Tom Walker, gen mgr; Ted Waldbillig, gen sls mgr; Mark Maloney, progmg dir.

De Pere

WKSZ(FM)—Licensed to De Pere. See Green Bay

Delafield

***WHAD(FM)**— May 30, 1948: 90.7 mhz; 72 kw. Ant 682 ft TL: N43 01 42 W88 23 32. Hrs open: 24
Rebroadcasts WHA(AM) Madison 60%.
111 E. Kilbourn Ave., Suite 2375, Milwaukee, 53202. Phone: (414) 227-2040. Fax: (414) 227-2043. E-mail: whad@wpr.org Web Site: www.wpr.org. Licensee: State of Wisconsin Educational Communications Board. Population served: 1,100,000 Natl. Network: NPR. Rgnl. Network: Wis. Pub. Format: Pub radio, news/talk. News staff: one; News: 11 hrs wkly. Target aud: 35-55; general. ♦Bill Estes, stn mgr; Shavonn Brown, gen sls mgr; Chuck Quirmbach, news dir.

Denmark

WPCK(FM)— Sept 1, 1969: 104.9 mhz; 10 kw. Ant 515 ft TL: N44 24 38 W87 34 20. Stereo. Hrs opn: 24 810 Victoria St., Green Bay, 54302. Phone: (920) 468-4100. Fax: (920) 468-0250. E-mail: country@wpkr.com Web Site: www.kicks104.com. Licensee: Cumulus Licensing LLC. Group owner: Cumulus Media Inc. (acq 11-10-2003; $8.1 million with WPKR(FM) Omro). Natl. Rep: Christal. Law Firm: Cohn & Marks. Format: Country. News staff: 2; News: 4 hrs wkly. Target aud: 25-54; adults. Spec prog: Relg 3 hrs wkly. ♦Greg Jessen, gen mgr; Jimmy Clark, VP & opns mgr.

Dickeyville

WVRE(FM)— 2/1/03: 101.1 mhz; 3.7 kw. Ant 423 ft TL: N42 31 43 W90 36 56. Stereo. Hrs opn: 24 Box 659, Dubuque, IA, 52004. Phone: (563) 690-0800. Fax: (563) 588-5688. Licensee: Radio Dubuque Inc. (group owner; acq 8-1-01). Natl. Rep: International Media, Katz Radio. Format: Country. News staff: one. Target aud: Adults; 25-54. ♦Thomas Parsley, gen mgr.

Dodgeville

WDMP(AM)— Nov 1, 1968: 810 khz; 250 w-D, 10 w-N. TL: N42 55 10 W90 08 06. Hrs open: 24 Box 9, 53533. Secondary address: 2163 Hwy. 151 S. 53523. Phone: (608) 935-2302. Fax: (608) 935-3464. Web Site: www.d99point3.com. Licensee: Dodge-Point Broadcasting Co. Population served: 63250 Format: Country. News staff: one; News: 5 hrs wkly. Target aud: General. ♦Louise E. Hamlin, pres; Kurt Reinicke, gen mgr & gen sls mgr; Jennifer Mick, disc jockey; Jon Satterlee, disc jockey; Melanie Rae, disc jockey; Robert Brainerd, news dir & disc jockey.

WDMP-FM— Nov 1, 1968: 99.3 mhz; 1.55 kw. 459 ft TL: N42 55 10 W90 08 06. Stereo. 24 E-mail: mail@d99point3.com Web Site: www.d99point3.com.23,255 News staff: one. ♦Jennifer Mick, disc jockey; Jon Satterlee, disc jockey; Melanie Rae, disc jockey; Robert Brainerd, disc jockey.

Durand

WDMO(FM)— Oct 24, 1973: 95.9 mhz; 1.3 kw. Ant 498 ft TL: N44 34 53 W91 54 44. Stereo. Hrs opn: 313 Main St., Menomonie, 54751. Phone: (715) 231-9500. Fax: (715) 231-9505. Web Site: www.thunder959.com. Licensee: Zoe Communications Inc. (group owner; (acq 7-31-2001; with co-located AM). Population served: 89,673 Format: Country. ♦Bo Landry, opns mgr, progmg dir & news dir; Mike Oberg, chief of engrg; Wendy Oberg, gen mgr, gen sls mgr & traf mgr.

WQOQ(AM)—Co-owned with WDMO(FM). Nov 21, 1968: 1430 khz; 2 kw-D, 152 w-N. TL: N44 38 28 W91 55 22.18 2,103

Eagle River

WERL(AM)— May 23, 1961: 950 khz; 1 kw-D, 51 w-N. TL: N45 58 38 W89 14 52. Stereo. Hrs opn: 24 Box 309, 909 Railroad St., 54521. Phone: (715) 479-4451. Fax: (715) 479-6511. E-mail: wrjo@wrjo.com Web Site: www.wrjo.com. Licensee: Heartland Communications License LLC. (acq 12-7-2004; $2.2 million with co-located FM). Format: Adult standards. News staff: one; News: 7 hrs wkly. Target aud: General. ♦Mary Jo Berner, pres; Jeff Wagner, gen mgr & gen sls mgr; Jeff Litscher, prom dir & progmg dir; Chris Oatman, news dir; Del Dayton, chief of engrg; Lynn Weiland, traf mgr.

WRJO(FM)—Co-owned with WERL(AM). July 31, 1971: 94.5 mhz; 50 kw. 492 ft TL: N45 58 38 W89 14 52. Stereo. 24 Web Site: www.wrjo.com. Format: Oldies, rock and roll.

Eau Claire

WAXX(FM)— February 1965: 104.5 mhz; 100 kw. 1,830 ft TL: N44 39 51 W90 57 41. Stereo. Hrs opn: 24 944 Harlem, Altoona, 54720. Phone: (715) 832-1530. Fax: (715) 832-5329. Web Site: www.todayswaxx1045.com. Licensee: Maverick Media of Eau Claire License LLC. Group owner: Maverick Media LLC (acq 6-13-2003; grpsl). Population served: 500,000 Format: Country. News staff: 4. Target aud: 25-54; metro & rgnl adults. Spec prog: Farm 15 hrs wkly.Gary Rozynek, pres; George Roberts, gen mgr & mktg mgr; George House, stn mgr, opns mgr, progmg dir, mus dir & disc jockey; Lynn Bieritz, sls dir; Dan Gainey, rgnl sls mgr; John Murphy, prom dir; Dan Lea, news dir; Bill Holden, chief of engrg; Lorraine Diener, traf mgr; Bob Bosold, farm dir; Brian O'Neil, disc jockey; Katie Bright, disc jockey; Terry West, pub affrs dir & disc jockey

WAYY(AM)— May 1937: 790 khz; 5 kw-U, DA-N. TL: N44 49 51 W91 26 58. Hrs open: 24 944 Harlem St., Altoona, 54720. Phone: (715) 832-1530. Fax: (715) 832-5329. E-mail: bruce@wayy790.com Web Site: www.wayy790.com. Licensee: Maverick Media of Eau Claire License LLC. Group owner: Maverick Media LLC (acq 6-13-2003; grpsl). Population served: 250,000 Rgnl. Network: Wisconsin Radio Net. Law Firm: Pepper & Corazzini. Format: News/talk. News staff: one. Target aud: 35 plus; general. Spec prog: Farm 5 hrs wkly. ♦George Roberts, gen mgr; Dave Craig, opns mgr; Lynn Bieritz, gen sls mgr; John Murphy, prom dir; Bruce Butler, progmg dir; Dab Lea, news dir; Bill Holden, chief of engrg; Lorraine Diener, traf mgr.

WIAL(FM)—Co-owned with WAYY(AM). 1948: 94.1 mhz; 85 kw. 350 ft TL: N44 49 48 W91 26 48. Stereo. 24 Web Site: www.i94online.com. Format: Adult contemp. News staff: one. Target aud: 18-54. ♦Luc Anthony, prom dir; Rick Roberts, stn mgr, opns mgr & progmg dir; Deb Lindquist, traf mgr.

WBIZ(AM)— Nov 11, 1947: 1400 khz; 1 kw-U. TL: N44 48 48 W91 31 15. Hrs open: 24 Box 45, 54702. Secondary address: 619 Cameron St. 54702. Phone: (715) 830-4000. Fax: (715) 835-9680. Web Site: www.wbiz.com. Licensee: Capstar TX L.P. Group owner: Clear Channel Communications Inc. (acq 2000; grpsl). Population served: 200,000 Natl. Network: CBS. Format: All sports. Target aud: 25-54. ♦Rick Hencley, gen mgr; Mike Cushman, opns mgr; Steve Potter, gen sls mgr; Mike Cush, progmg mgr; Keith Edwards, news dir; Paul Orth, chief of engrg; Theresa Nelson, traf mgr.

WBIZ-FM— December 1967: 100.7 mhz; 100 kw. 740 ft TL: N44 47 58 W91 27 59. Stereo. Web Site: www.z100radio.com.250,000 Format: CHR. ♦Rick Hencley, VP & gen sls mgr; Audrey Phillips, progmg dir; Jare E. Jordan, mus dir & disc jockey; Keith Edwards, news dir.

***WDVM(AM)**— April 1948: 1050 khz; 1 kw-D, 500 w-N. TL: N44 46 36 W91 28 30. Stereo. Hrs opn: Relevant Radio 1050 AM, WDVM, 1752 Bracket Ave., 54701. Phone: (715) 855-1439. Phone: (715) 577-0943. Fax: (715) 855-1471. Web Site: www.relevantradio.com. Licensee: Starboard Media Foundation Inc. Group owner: Relevant Radio (acq 7-6-2001). Population served: 75,000 Format: Talk, religious. News staff: one. Target aud: 35 plus; mature adults. ♦Mark Follett, CEO; Sherry Brownrigg, pres; Raymond P. Jay, stn mgr; Martin Jury, opns dir.

***WHEM(FM)**— Aug 22, 1995: 91.3 mhz; 350 w. Ant 216 ft TL: N44 45 50 W91 31 06. (CP: 300 w, ant 285 ft). Hrs opn: 24 228 E. Lowes Creek Rd., 54701. Phone: (715) 838-9595. E-mail: whem@discover-net.net Web Site: www.whem.com. Licensee: Fourth Dimension Inc. (acq 7-2-93; $2,810; FTR: 8-2-93). Format: Christian. ♦Harlan Reinders, gen mgr & chief of engrg; Phyllis Reinders, progmg dir.

***WUEC(FM)**— Oct 27, 1975: 89.7 mhz; 5.2 kw. 630 ft TL: N44 47 58 W91 27 59. Stereo. Hrs opn: 24 Wisconsin Public Radio, 1221 W. Clairemont Ave., 54701. Phone: (715) 839-3868. Fax: (715) 839-2939. E-mail: kallenbach@wpr.org Web Site: www.wpr.org. Licensee: Board of Regents, University of Wisconsin. Population served: 168,000 Natl. Network: NPR. Rgnl. Network: Wis. Pub. Law Firm: Dow, Lohnes & Albertson. Format: Class, news, jazz. News: 24 hrs wkly. Target aud: General. Spec prog: Folk 3 hrs, blues 3 hrs wkly. ♦Dean Kallenbach, stn mgr; Marvin Spielman, dev dir & mktg dir; Mary Jo Wagner, news dir; Ron Viste, chief of engrg.

***WVCF(FM)**— 1997: 90.5 mhz; 980 w. 279 ft TL: N44 57 29 W91 28 58. Stereo. Hrs opn: 24 VCY/America Inc., 3434 W. Kilbourn Ave., Milwaukee, 53208. Phone: (414) 935-3000. Fax: (414) 935-3015. E-mail: wvcf@vcyamerica.org Web Site: www.vcyamerica.org. Licensee: VCY America Inc. (group owner) Natl. Network: USA, Moody. Format: Relg, Christian. ♦Dr. Randall Melchert, pres; Victor Eliason, VP & gen mgr; Gordon Morris, opns mgr, mus dir & news dir; Jim Schneider, progmg dir.

Elk Mound

WECL(FM)— March 1, 2004: 92.9 mhz; 3.3 kw. 446 ft TL: N44 53 40 W91 35 40. Stereo. Hrs opn: 24 944 Harlem St., Altoona, 54720. Phone: (715) 832-1530. Fax: (715) 832-5329. Web Site: www.929thebigcheese.com. Licensee: Maverick Media of Eau Claire License LLC. Group owner: Maverick Media LLC (acq 6-13-2003; grpsl). Population served: 150,000 Law Firm: Pepper & Corazzini. Target aud: 35-54; Adults. Spec prog: Flashback, 4hrs. ♦George Roberts, gen mgr; Dan Lea, opns dir; Rick Roberts, stn mgr & opns mgr; Lynn Bieritz, sls dir; Dan Gainey, natl sls mgr.

Elm Grove

WGLB(AM)— Dec 6, 1963: 1560 khz; 185 w-D, 250 w-N, DA-2. TL: N43 00 32 W88 02 06. Hrs open: 24 5181 N. 35th St., Milwaukee, 53209. Phone: (414) 527-4365. Fax: (414) 527-4367. E-mail: wglb@wglbam1560.com Web Site: wglbam1560.com. Licensee: Joel J. Kinlow (acq 7-25-95; with co-located FM; FTR: 8-21-95). Population served: 600,000 Format: Gospel. ♦Joel Kinlow, CEO; Joel Kinlow, pres & gen mgr; Willis H. Payne Jr., stn mgr & progmg dir; Bruce Herzog, chief of engrg.

Evansville

WWHG(FM)— Aug 17, 1989: 105.9 mhz; 1.7 kw. Ant 493 ft TL: N42 43 38 W89 15 02. Stereo. Hrs opn: 24 One Parker Pl., Suite 485, Janesville, 53545. Phone: (608) 758-9025. Fax: (608) 758-9550. E-mail: hot1059@hotmail.com Web Site: hot1059.net. Licensee: Good Karma Broadcasting L.L.C. (acq 10-2-97; $1.5 million). Format: Rock. News staff: one; News: 3 hrs wkly. ♦Craig Karmazin, CEO & pres; Keith Williams, gen mgr, stn mgr, gen sls mgr & prom dir; Rick Armon, opns dir & opns mgr; Dan Hunt, progmg dir; Kyle Jacob, news dir; Warren Jorgensen, engrg dir & chief of engrg; Deb Lamers, traf mgr.

Fond du Lac

KFIZ(AM)— July 6, 1922: 1450 khz; 1 kw-U. TL: N43 47 28 W88 28 16. Stereo. Hrs opn: Box 1450, 54936-1450. Secondary address: 254 Winnebago Dr. 54935. Phone: (920) 921-1071. Fax: (920) 921-0757. E-mail: info@kfiz.com Web Site: www.kfiz.com. Licensee: RBH Enterprises Inc. Group owner: Mountain Dog Media (acq 1-23-97; $1 plus assumption of liabilities with co-located FM). Population served: 135,515 Format: News/talk, sports. Target aud: 35-64; Men & Women. Spec prog: Farm 10 hrs wkly. ♦R.B. Hopper, pres & gen mgr.

***WDKV(FM)**— 2005: 91.7 mhz; 20 kw vert. Ant 357 ft TL: N43 39 35 W88 26 26. Hrs opn: 2351 Sunset Blvd., Suite 170-218, Rocklin, CA, 95765. Phone: (916) 251-1600. Fax: (916) 251-1650. Licensee: Educational Media Foundation. (acq 2-21-2006; $350,000). Population served: 192,134 Format: Christian. ♦Richard Jenkins, pres; Keith Whipple, dev dir; David Pierce, progmg dir; Ed Lenane, news dir; Sam Wallington, engrg dir; Karen Johnson, news rptr; Richard Hunt, news rptr.

WFDL-FM—Lomira, April 1993: 97.7 mhz; 17.5 kw. Ant 400 ft TL: N43 39 14 W88 26 25. Hrs open: 24 210 S. Main St., 54935. Phone: (920) 924-9697. Fax: (920) 929-8865. E-mail: info@sunny97-7.com Web Site: www.wfdl.com. Licensee: Radio Plus of Fond du Lac Inc. (acq 4-96). Population served: 150,000 Law Firm: Haley, Bader & Potts. Format: Adult contemp. News staff: one. Target aud: 25-54. ♦Chris Bernier, pres; Terry Davis, VP, gen mgr & gen sls mgr; Mike Enfelt, opns mgr & chief of engrg; Todd Dehring, progmg dir; Greg Stensland, news dir; Kerry Longrie, traf mgr.

WFON(FM)— Oct 5, 1967: 107.1 mhz; 3 kw. Ant 312 ft TL: N43 50 22 W88 22 06. Stereo. Hrs open: 24 254 Winnebago Dr., 54935. Phone: (920) 921-1071. Fax: (920) 921-0757. Web Site: www.k107.com. Licensee: RBH Enterprises Inc. Group owner: Mountain Dog Media (acq 1-23-97). Population served: 500,000 Format: Hot adult contemp. News staff: one; News: 1 hr wkly. Target aud: Women 25-54. ♦Randy Hopper, gen mgr.

WRPN(AM)—Ripon, Sept 15, 1957: 1600 khz; 5 kw-U, DA-2. TL: N43 49 01 W88 50 49. Hrs open: 24 112 Watson St., Ripon, 54971. Phone: (920) 748-5111. Fax: (920) 748-5530. E-mail: wrpn@wrpnam.com Web Site: www.wrpnam.com. Licensee: Radio Broadcasting L.P. Population served: 23247 Natl. Network: CBS, ABC. Natl. Rep: Farmakis. Law Firm: Haley, Bader & Potts. Wire Svc: Wheeler News Service Format: News/talk. News staff: 4; News: 25 hrs wkly. Target aud: 25 plus; Fond du Lac, Green Lake counties. Spec prog: Pol 2 hrs, sports 15 hrs wkly. ♦Tom Biolo, gen mgr; Jason Mansmith, sls dir; Jason Marsmith, progmg dir; Justin Cleveland, news dir; Mike Enfelt, chief of engrg; Jean Hoffmann, traf mgr.

WTCX(FM)—Co-owned with WRPN(AM). Feb 1, 1965: 96.1 mhz; 4 kw. Ant 403 ft TL: N43 49 10 W88 43 20. Stereo. 24 210 South Main St., Fon du Lac, 54935. Phone: (920) 924-9697. Fax: (920) 929-8865. Web Site: www.961themix.com.90,000 Format: Classic hits, new rock. News staff: one; News: 3 hrs wkly. Target aud: 25-54; women. ♦Mike

Enfelt, opns mgr; Terry Davis, VP, gen mgr & gen sls mgr; Gregg Owens, progmg dir; Jean Hoffmann, chief of engrg & traf mgr.

***WVFL(FM)**— 2007: 89.9 mhz; 1 kw vert. Ant 384 ft TL: N43 48 09 W88 20 18. Hrs open:
Rebroadcasts WVCY-FM Milwaukee 100%.
3434 W. Kilbourn Ave., Milwaukee, 53208-3313. Phone: (414) 935-3000. Fax: (414) 935-3015. E-mail: vcy@vcyamerica.org Web Site: www.vcyamerica.org. Licensee: VCY America Inc. Natl. Network: USA. Format: Relg, Christian. ♦Vic Eliason, VP & gen mgr; Jim Schneider, progmg dir; Andy Eliason, chief of engrg.

Forestville

WRKU(FM)— Aug 1, 1999: 102.1 mhz; 6 kw. 328 ft TL: N44 46 58 W87 22 24. Hrs open: 3030 Park Dr., Suite 3, Sturgeon Bay, 54235. Phone: (920) 746-9430. Fax: (920) 746-9433. Web Site: www.doorradio.com. Licensee: Nicolet Broadcasting Inc. (group owner) Format: Oldies. News: 3 hrs wkly. ♦Roger Utnehmer, pres; Miles Knuteson, gen mgr, gen sls mgr & prom dir; Karen Leitzinger, opns mgr & progmg dir; John Focke, news dir; Kathy Robinson, traf mgr.

Fort Atkinson

WFAW(AM)— Jan 24, 1963: 940 khz; 500 w-D, 550 w-N, DA-2. TL: N42 54 24 W88 45 06. Hrs open: Box 94, 53538. Phone: (920) 563-9329. Fax: (920) 563-0315. Licensee: NRG License Sub. LLC. (group owner; (acq 10-31-2005; grpsl). Population served: 9,782 Natl. Rep: McGavren Guild. Format: Sports, news/talk. News staff: one.Mary Quass, CEO & VP; Chuck DuCoty, COO & stn mgr; Tami Gillmore, CFO; Jim Vriezen, gen mgr; Gary Douglas, opns dir & sports cmtr; Shane Sparks, sls VP, gen sls mgr & sports cmtr; Michael Clish, news dir, local news ed & news rptr; Ernie Swanson, chief of engrg; Lynette Furley, traf mgr; Jim Klug, sports cmtr; Lew Wendlandt, sports cmtr; Jim King, disc jockey

WSJY(FM)—Co-owned with WFAW(AM). Sept 4, 1959: 107.3 mhz; 26 kw. 676 ft TL: N42 50 48 W88 51 16. Stereo. 24 Box 2107, Janesville, 54547. Phone: (608) 756-0747. Fax: (608) 755-1252. Natl. Rep: McGavren Guild. Format: Lite adult contemp. News staff: one. Target aud: 25-54. ♦Sonja Untz, traf mgr; Michael Clish, local news ed; Eric Stone, disc jockey; Jennifer Maxwell, disc jockey; Polly Peterson, disc jockey.

Goodman

***WMVM(FM)**— May 3, 1993: Stn currently dark. 91.3 mhz; 422 w. Ant 118 ft TL: N45 37 36 W88 21 28. (CP: 90.7 mhz; 9 kw, ant 156 ft. TL: N45 46 27 W88 24 28). Hrs opn: Box 212, Suring, 54174. Phone: (920) 842-2900. Fax: (920) 842-2704. Web Site: www.wrvm.org. Licensee: WRVM Inc. (acq 9-11-02; $20,000). Format: Christian. ♦Michael A. Cornell, gen mgr; Bryan Hay, gen sls mgr; Dennis Jones, progmg dir; Alan Kilgore, chief of engrg.

Green Bay

WAPL-FM—See Appleton

WDUZ(AM)— June 19, 1947: 1400 khz; 1 kw-U. TL: N44 29 36 W87 59 13. Stereo. Hrs opn: 24 810 Victoria St., 54302. Phone: (920) 468-4100. Fax: (920) 468-0250. Web Site: www.supertalk1400.cumulus.com. Licensee: Cumulus Licensing Corp. Group owner: Cumulus Media Inc. (acq 6-27-2002; $6 million with WQLH(FM) Green Bay). Population served: 174,300 Natl. Network: ESPN Radio. Wire Svc: AP Format: News/talk. News staff: one. Target aud: 18-64; high income, male skew. ♦Greg Jessen, gen mgr; Jimmy Clark, opns mgr; Buck Hein, gen sls mgr & natl sls mgr; Bob Watts, progmg dir; Chris Gielow, chief of engrg.

***WEMY(FM)**— Aug 26, 1974: 91.5 mhz; 710 w. 741 ft TL: N44 21 32 W87 59 07. Stereo. Hrs opn: 24
Rebroadcasts WEMI(FM) Appleton 95%.
1909 W. 2nd St., Appleton, 54914. Phone: (920) 499-9957. Fax: (920) 749-0474. Web Site: www.christianfamilyradio.net. Licensee: Evangel Ministries Inc. (group owner; acq 3-10-98). Population served: 200,000 Natl. Network: Salem Radio Network, USA. Law Firm: Leventhal, Senter & Lerman. Format: Relg. News: 10 hr wkly. Target aud: 35-49; women. Spec prog: Sp one hr wkly. ♦Mary B. Lieb, chmn; Paul Cameron, gen mgr; Andy Kilgas, opns dir & sls dir; Heidi Prahl, dev dir.

***WHID(FM)**— April 1997: 88.1 mhz; 17 kw. 1,023 ft TL: N44 21 32 W87 59 07. Stereo. Hrs opn: 24 821 University Ave., Madison, 53706. Phone: (608) 263-4199. Fax: (608) 263-9763. E-mail: slaatsg@uwgb.edu Web Site: www.wpr.org. Licensee: Board of Regents of University of Wisconsin Systems. Population served: 300,000 Natl. Network: NPR, PRI. Law Firm: Dow, Lohnes & Albertson. Format: Talk. News staff: one; News: 18 hrs wkly. Target aud: 35 plus; socially active, life-long learners. Spec prog: Sp 3 hrs, Asian 2 hrs wkly. ♦Glen Slaats, gen mgr.

WIXX(FM)—Listing follows WTAQ(AM).

WKSZ(FM)—De Pere, Oct 1, 1984: 95.9 mhz; 4.5 kw. 774 ft TL: N44 21 32 W87 59 07. Stereo. Hrs opn: 24 Box 1519, Appleton, 54912. Secondary address: 1263 main St., Suite #225 54301. Phone: (920) 431-0959. Fax: (920) 739-0494. E-mail: wkszcr@wcinet.com Web Site: www.959kissfm.com. Licensee: Woodward Communications Inc. (group owner; acq 1995). Population served: 192,200 Natl. Rep: McGavren Guild. Law Firm: Hogan & Hartson. Format: Top-40. Target aud: 18-34; females. ♦Greg Bell, VP & gen mgr; Kelly Radandt, gen sls mgr; Dayton Kane, progmg dir; Steve Brown, chief of engrg.

WNFL(AM)— Dec 12, 1947: 1440 khz; 5 kw-D, 1 kw-N, DA-2. TL: N44 28 40 W88 00 00. Hrs open: 24 115 S. Jefferson St., 54301. Secondary address: Box 23333 54305. Phone: (920) 435-3771. Fax: (920) 444-1155. Licensee: Midwest Communications Inc. (group owner; acq 12-10-96; grpsl). Population served: 335,000 Natl. Network: Fox Sports, Westwood One. Natl. Rep: Christal. Law Firm: Miller & Neely. Wire Svc: AP Format: Sports. News staff: 4; News: 10 hrs wkly. Target aud: 35-64; people in upper-income level with above average education. Spec prog: Sports 120 hrs wkly. ♦Duke Wright, pres & gen mgr; Gary Tesch, exec VP; Shelley Lukasik, gen sls mgr; Mark Daniels, progmg dir & progmg mgr; Jerry Bader, news dir; Tim Laes, chief of engrg.

***WORQ(FM)**— Feb 1, 1994: 90.1 mhz; 11 kw. Ant 646 ft TL: N44 21 32 W87 59 07. Hrs open: 24 1075 Brookwood Dr., Suite 2C, 54304. Phone: (920) 494-9010. Fax: (920) 494-7602. E-mail: email@q90fm.com Web Site: www.q90fm.com. Licensee: Lakeshore Communications Inc. Population served: 260,000 Natl. Network: USA. Law Firm: Wiley, Rein & Fielding. Format: Christian. News: 7 hrs wkly. Target aud: Under 30; rock and roll generation. ♦Mike LeMay, gen mgr; Jim Raider, gen sls mgr; Jim Kaider, progmg dir; Scott Grathen, chief of engrg; Jeff Rydell, traf mgr.

***WPNE-FM**— Jan 15, 1973: 89.3 mhz; 100 kw. 940 ft TL: N44 24 35 W88 00 05. Hrs open: 24 2420 Nicolet Dr., 54311-7001. Phone: (920) 465-2444. Fax: (920) 465-2576. Web Site: www.wpr.org. Licensee: State of Wisconsin Educational Communications Board. Population served: 300,000 Natl. Network: NPR, PRI. Rgnl. Network: Wis. Pub. Law Firm: Dow, Lohnes & Albertson. Format: Class music, news. News staff: 2; News: 29 hrs wkly. Target aud: 35 plus; socially aware, artistically stimulated. Spec prog: Jazz 10 hrs, folk 8 hrs, American Indian 2 hrs wkly. ♦Glen Slaats, gen mgr.

WQLH(FM)— July 1, 1967: 98.5 mhz; 100 kw. 1,254 ft TL: N44 38 41 W88 08 13. Stereo. Hrs opn: 810 Victoria St., 54302. Phone: (920) 468-4100. Fax: (920) 468-0250. Web Site: star98.cumulus.com. Licensee: Cumulus Licensing Corp. Group owner: Cumulus Media Inc. (acq 6-27-2002; $6 million with WDUZ(AM) Green Bay). Population served: 681,000 Wire Svc: AP Format: Hot AC. Target aud: Adults 25-54. ♦Jimmy Clark, pres, opns mgr & disc jockey; Greg Jessen, gen mgr; Buck Hein, gen sls mgr; Brian Stenzel, prom dir; Jim Clark, progmg dir; Chris Gielow, chief of engrg; Chally Boutatt, disc jockey; Dave DeVille, disc jockey; Doug Kaufman, disc jockey; Steve Davis, mus dir & disc jockey.

WTAQ(AM)— Apr 6, 1925: 1360 khz; 10 kw—D, 5 kw-N, DA-2. TL: N44 25 51 W88 04 51. Stereo. Hrs opn: Box 23333, 54301. Secondary address: 115 South Jefferson St. 54301. Phone: (414) 435-3771. Fax: (414) 455-1155. Licensee: Midwest Communications Inc. (group owner; acq 1975). Population served: 700,000 Natl. Network: CBS. Natl. Rep: Christal. Law Firm: Miller & Miller. Format: News/talk. News staff: 4. Target aud: 30 plus. Spec prog: Farm 5 hrs wkly. ♦D.E. Wright, pres & gen mgr; Gary Tesch, exec VP; Shelley LuKasik, gen sls mgr; Aaron Vorass, prom dir & prom mgr; Jerry Bader, progmg dir; Daniell Binna, news dir; Tim Laes, chief of engrg; Mike Austin, farm dir.

WIXX(FM)—Co-owned with WTAQ(AM). Nov 1, 1960: 101.1 mhz; 100 kw. 1,080 ft TL: N44 24 35 W88 00 05. Stereo. Format: CHR. ♦Mary Kay Wright, gen sls mgr; Jeff McCarthy, progmg VP; Tony Wailckus, progmg dir.

WZNN(FM)—See Allouez

Greenfield

WMCS(AM)—Licensed to Greenfield. See Milwaukee

Hallie

WOGO(AM)—Listing follows WWIB(FM).

WWIB(FM)— Dec 30, 1972: 103.7 mhz; 100 kw. 706 ft TL: N45 06 35 W91 09 43. Stereo. Hrs opn: 24 2396 State Hwy. 53, Suite One, Chippewa Falls, 54729. Phone: (715) 723-1037. Phone: (715) 723-4626. Fax: (715) 723-1348. Web Site: www.wwib.com. Licensee: Stewards of Sound Inc. (acq 10-29-93; with WOGO(AM) Hallie; FTR: 11-15-93). Population served: 980,000 Natl. Network: USA. Format: Adult Christian contemp. News staff: one; News: 16 hrs wkly. Target aud: 25-54. ♦Terri Steward, gen mgr & stn mgr; Steven Slater, gen sls mgr; Greg Steward, progmg mgr; Mark Halvorsen, news dir.

WOGO(AM)—Co-owned with WWIB(FM). June 1985: 680 khz; 2.5 kw-D, 500 w-N, DA-2. TL: N44 53 22 W91 23 03. Stereo. 24 Web Site: www.wwib.com. (Acq 10-29-93; with WWIB(FM) Ladysmith; FTR: 11-15-93).200,000 Natl. Network: USA. Format: News/talk. News staff: one; News: 16 hrs wkly. Target aud: 25-54. ♦Greg Steward, sports cmtr.

Hartford

WTKM-FM— Oct 1, 1973: 104.9 mhz; 5.8 kw. Ant 300 ft TL: N43 16 48 W88 23 02. Stereo. Hrs opn: 24 Box 270216, 53027-0216. Secondary address: 27 N. Main St. 53027. Phone: (262) 673-7800. Fax: (262) 673-5472. E-mail: wtkm@nconnect.net Web Site: www.wtkm.com. Licensee: Kettle Moraine Broadcasting Co. Inc. (acq 3-12-90; grpsl; FTR: 4-2-90). Population served: 1,500,000 Wire Svc: AP Format: Local talk, classic country, polka. News staff: 2; News: 20 hrs wkly. Target aud: 35 plus. ♦Scott Lopas, pres & gen mgr; Tom Shanahan, stn mgr.

WTKM(AM)— 1951: 1540 khz; 500 w-D. TL: N43 16 48 W88 23 02.Sunrise-sunset Phone: (262) 673-3550. Phone: (262) 252-4567. Natl. Rep: Farmakis.

Hayward

WHSM(AM)— Dec 21, 1957: 910 khz; 5 kw-D, 75 w-N. TL: N45 59 07 W91 32 21. Stereo. Hrs opn: 24 16880 W. US Hwy. 63, 54843. Phone: (715) 634-4836. Phone: (800) 845-8984. Fax: (715) 634-8256. E-mail: radio@whsm.com Web Site: www.whsm.com. Licensee: Red Rock Radio Corp. (acq 7-18-2006; grpsl). Population served: 40,000 Format: Music of your life. News staff: one; News: 2 hrs wkly. Target aud: 25-54. ♦Ro Grignon, pres; Don Welch, exec VP, gen mgr & gen sls mgr; Bobi Hopp, opns mgr, progmg dir & traf mgr; Joe Lancello, news dir; Bill Meys, chief of engrg.

WHSM-FM— June 21, 1980: 101.1 mhz; 1.45 kw. 410 ft TL: N45 59 07 W91 32 23. Stereo. Web Site: www.whsm.com. Format: Bright adult contemp, adult hit radio.

WRLS-FM— Apr 16, 1968: 92.3 mhz; 6 kw. 321 ft TL: N46 01 17 W91 30 41. Stereo. Hrs opn: Box 1008, Radio Hill Rd., 54843. Phone: (715) 634-4871. Fax: (715) 634-3025. E-mail: wrls@cheqnet.net Web Site: www.wrlsfm.com. Licensee: Vacationland Broadcasting Inc. (acq 12-9-92; FTR: 1-4-93). Population served: 25,000 Natl. Network: CNN Radio, AP Radio. Wire Svc: AP Format: Adult contemp. News: News progrmg 10 hrs wkly. Target aud: 25 plus; general. ♦Tom Koser, pres; Robert Koser, VP; Steve Kaner, gen mgr; Shad Harper, opns mgr.

Highland

***WHHI(FM)**— Sept 14, 1952: 91.3 mhz; 100 kw. 560 ft TL: N43 02 58 W90 22 00. Hrs open: 24
Rebroadcasts WHA(AM) Madison 100%.
Wisconsin Public Radio, 821 University Ave., Madison, 53706-1496. Phone: (608) 263-3970. Fax: (608) 263-9763. E-mail: schnirring@wpr.org Web Site: www.wpr.org. Licensee: State of Wisconsin Educational Communications Board. Natl. Network: NPR, PRI. Law Firm: Dow, Lohnes & Albertson. Format: Educ, news/talk. News staff: 9. Target aud: 35-54; Skews female: issue oriented talk-variety of perspectives. ♦Phil Corriveau, gen mgr; Mary Kay Sherer, dev dir & prom dir; Steve Johrnston, chief of engrg & traf mgr.

Holmen

WKBH(AM)— July 28, 1984: 1570 khz; 1 kw-D, 500 w-N. TL: N43 55 32 W91 16 02. Hrs open: 24 1407 2nd Ave. N., Onalaska, 54650. Phone: (608) 779-4415. Fax: (608) 779-4419. Web Site: www.relevantradio.com. Licensee: Starboard Media Foundation Inc. Group owner: Relevant Radio (acq 1-17-2003; $210,000). Law Firm: Smithwick & Belendiuk. Format: Catholic talk. News staff: one; News:

35 hrs wkly. Target aud: 40 plus; those interested in Catholic news, talk & opinion. ♦Jack Socha, stn mgr.

Hudson

WDGY(AM)—Licensed to Hudson. See Minneapolis-St. Paul MN

WMIN(AM)— Dec 14, 1983: 740 khz; 1.1 kw-DA. TL: N44 58 05 W92 40 01. Hrs open: 6 AM-7 PM Box 25130, St. Paul, MN, 55125. Phone: (651) 436-4000. Fax: (651) 436-6770. Licensee: Borgen Broadcasting Corp. (acq 12-11-89; $300,000; FTR: 1-1-90). Format: Oldies. Target aud: 25-54. ♦Gregory Borgen, pres & gen mgr; Jeff Borgen, gen sls mgr; Tom Witschen, progmg dir; Paul Orth, chief of engrg.

Hurley

WHRY(AM)— Mar 1, 1985: 1450 khz; 1 kw-U. TL: N46 24 56 W90 09 34. Hrs open: Box 1450, 54534. Secondary address: 209 Harrison St., Ironwood, MI 49938. Phone: (906) 932-5234. Fax: (906) 932-1548. E-mail: wupm@wupm-whry.com Licensee: Big G Little O Inc. Format: Oldies, Hits of the 50s, 60s & 70s. ♦Charles H. Gervasio, pres, gen mgr, sls VP & gen sls mgr; Norma Rigoni, VP; Laura Keller, progmg dir.

Iron River

WNXR(FM)— November 1994: 107.3 mhz; 50 kw. 380 ft TL: N46 31 27 W91 16 18. Stereo. Hrs opn: 24 2320 Ellis Ave., Ashland, 54806. Phone: (715) 372-5400. Fax: (715) 682-9338. E-mail: productionash@charter.net Web Site: www.wnxrfm.com. Licensee: Heartland Communications License LLC. (group owner; (acq 4-23-2004; grpsl). Population served: 500,000 Natl. Network: Westwood One. Format: Hits of the 50s, 60s & 70s. ♦Rich Collins, gen mgr & opns VP; Scott Jaeger, gen mgr; Skip Hunter, progmg dir.

Jackson

WRRD(AM)— May 1, 1964: 540 khz; 400 w-U, DA-2. TL: N43 20 00 W88 09 11. Stereo. Hrs opn: 24 135 S. 84th St., #310, Milwaukee, 53214-1477. Phone: (414) 258-1700. Web Site: www.540theword.com. Licensee: SCA License Corp. Group owner: Salem Communications Corp. (acq 12-18-00; $7 million with WWTC(AM) Minneapolis, MN). Population served: 1,500,000 Law Firm: Lance Riley. Format: Christian talk/ministries. News: 6 hrs wkly. Target aud: 25-54; Christians. Spec prog: Croation 2 hrs, Latino 2 hrs wkly. ♦Bob Emory, gen mgr; Lil Roohara, sls dir; Bob Bradley, disc jockey; David Abrahams, disc jockey; Frank Colbourn, disc jockey; Herb Wittka, disc jockey; Phil Winchester, disc jockey.

Janesville

WCLO(AM)— July 1930: 1230 khz; 1 kw-U. TL: N42 39 35 W89 02 32. Hrs open: Box 5001, 53545. Secondary address: One S. Parker Dr. 53545. Phone: (608) 752-7895. Fax: (608) 752-4438. E-mail: programming@wclo.com Licensee: Southern Wisconsin Broadcasting L.L.C. Group owner: Bliss Communications Inc. Population served: 136,000 Law Firm: Dow, Lohnes & Albertson. Format: News/talk. Target aud: General. ♦Robert Dailey, gen mgr; Ken Scott, progmg dir.

WJVL(FM)—Co-owned with WCLO(AM). October 1947: 99.9 mhz; 11 kw. 502 ft TL: N42 43 47 W89 10 10. Stereo. Web Site: www.wvl.com.50,046 Format: C&W. Target aud: 25-34. ♦Robert S. Dailey, exec VP; Mike O'Brien, gen sls mgr; Tim Bremel, progmg dir; Stan Stricker, news dir; Ja Mielkey, chief of engrg; Jim Thomas, traf mgr.

WSJY(FM)—See Fort Atkinson

***WWJA(FM)**—Not on air, target date: unknown: 91.5 mhz; 2.2 kw. Ant 387 ft TL: N42 43 47 W89 10 10. Hrs open: 4135 Northgate Blvd., Suite 1, Sacramento, CA, 95834-1226. Phone: (916) 641-8191. Fax: (916) 641-8238. Licensee: Family Stations Inc. ♦Peggy L. Renschler, gen mgr.

Kaukauna

WJOK(AM)— Sept 25, 1965: 1050 khz; 1 kw-D, 500 w-N, DA-2. TL: N44 14 51 W88 18 00. Hrs open: 24 1496 Bellevue St., Suite 202, Green Bay, 54311. Phone: (920) 469-3021. Fax: (920) 469-3023. Web Site: www.1050am.org. Licensee: Starboard Media Foundation Inc. Group owner: Relevant Radio (acq 8-28-2001; $500,000). Population served: 867,000 Natl. Network: USA, Moody. Law Firm: Leventhal, Senter & Lerman. Format: Catholic. Target aud: 35-54. Spec prog: Relg 2 hrs wkly. ♦Dave Nier, gen mgr.

WOGB(FM)— 1996: 103.1 mhz; 3.6 kw. 879 ft TL: N44 21 32 W87 59 07. Hrs open: 810 Victoria St., Green Bay, 54302. Phone: (920) 468-4100. Fax: (920) 468-0250. Web Site: www.cumulus.com. Licensee: Cumulus Licensing Corp. Group owner: Cumulus Media L.L.C. (acq 6-30-97; grpsl). Wire Svc: AP Format: Oldies. Target aud: 35-54; affluent baby-boomers, mid 40's. ♦Greg Jessen, gen mgr; Jimmy Clark, opns mgr; Buck Hein, sls dir & gen sls mgr; Dan Markus, progmg dir; Chris Gielow, news dir & chief of engrg.

Kenosha

***WGTD(FM)**— Dec 23, 1975: 91.1 mhz; 5 kw. 134 ft TL: N42 36 28 W87 50 55. Stereo. Hrs opn: 3520 30th Ave., 53144. Phone: (262) 564-3800. Fax: (262) 564-3801. E-mail: coled@gtc.edu Web Site: wgtd.org. Licensee: Gateway Technical College. Population served: 170,000 Natl. Network: NPR. Wire Svc: AP Format: News, classical music. News staff: 2. Target aud: General. Spec prog: Ger one hr. ♦David Cole, gen mgr.

WIIL(FM)—Listing follows WLIP(AM).

WLIP(AM)— May 11, 1947: 1050 khz; 250 w-U. TL: N42 33 10 W87 53 38. Hrs open: 24 8500 Green Bay Rd., Pleasant Prairie, 53158. Phone: (262) 694-7800. Fax: (262) 694-7767. Web Site: www.wlip.com. Licensee: NM Licensing LLC. Group owner: NextMedia Group L.L.C. (acq 11-26-00; grpsl). Population served: 47,000 Rgnl. Network: Wisconsin Radio Net. Format: Talk. News staff: 2. Target aud: 35 plus. ♦Kara Lafond, gen mgr; John Perry, opns mgr & progmg dir; Rory Fraley, sls dir; Stewart Wattles, news dir; Lisa Sladek, traf mgr.

WIIL(FM)—Co-owned with WLIP(AM). 1961: 95.1 mhz; 50 kw. 384 ft TL: N42 33 10 W87 53 38. Stereo. Web Site: www.95wiil.com. Format: Rock. Target aud: 18-54.

WWDV(FM)—See Zion, IL

Kewaunee

WAUN(FM)— 1973: 92.7 mhz; 6 kw. Ant 328 ft TL: N44 29 50 W87 35 12. Hrs open: 5 AM-10 PM 1021 N. Superior Ave., Suite 5, Tomah, 54660. Phone: (920) 388-9286. Fax: (920) 743-9183. Licensee: Magnum Broadcasting Inc. (acq 12-2-98). Population served: 30,000 Wire Svc: AP Format: Smooth jazz. News staff: one; News: 15 hrs wkly. Target aud: 25-54. Spec prog: Czech one hr, farm 10 hrs, relg 3 hrs wkly. ♦Dave Magnum, pres; Frank Devillers, gen mgr & gen sls mgr; Rick Jensen, opns mgr, progmg dir & news dir; Debbie Doyle, traf mgr.

Kiel

***WSTM(FM)**—Not on air, target date: unknown: 91.3 mhz; 100 w hoirz, 1.25 kw vert. Ant 459 ft TL: N43 43 32 W88 03 07. Hrs open: Box 259, Plymouth, 53073. Phone: (920) 893-2661. Fax: (920) 892-2706. E-mail: wjub@excel.net Web Site: wjub.org. Licensee: Jubilation Ministries Inc. Format: Christian. ♦Susan Noordyk, progmg dir.

Kimberly

WHBY(AM)— Dec 1, 1925: 1150 khz; 20 kw-D, 25 kw-N, DA-2. TL: N44 08 20 W88 32 46. Hrs open: 24 Box 1519, Appleton, 54912. Secondary address: 2800 E. College Ave., Appleton 54915. Phone: (920) 733-6639. Fax: (920) 739-0494. E-mail: whby@wcinet.com Web Site: www.whby.com. Licensee: Woodward Communications Inc. (group owner; (acq 3-75). Population served: 305,000 Rgnl. Network: Wisconsin Radio Net. Natl. Rep: McGavren Guild. Wire Svc: AP Format: News/talk. News staff: 3; News: 25 hrs wkly. Target aud: 35 plus; upper middle class, educated. ♦Greg Bell, gen mgr; John Wanie, stn mgr; Dave Edwards, progmg dir; Mary Ann Drewek, progmg dir; Steve Brown, chief of engrg; Pat Schmidt, traf mgr & disc jockey.

La Crosse

KQEG(FM)—La Crescent, MN) Apr 5, 1989: 102.7 mhz; 4.3 kw. 863 ft TL: N43 44 53 W91 17 51. Stereo. Hrs opn: 24 1407 Second Ave. N., Onalaska, 54650. Phone: (608) 782-8335. Fax: (608) 782-8340. Web Site: oldiesradioonline.net. Licensee: White Eagle Broadcasting Inc. Group owner: La Crosse Radio Group (acq 2-4-2000; $2 million). Rgnl rep: O'Malley. Format: Oldies. News: 4 hrs wkly. Target aud: 25-54. ♦Pat Smith, gen mgr.

***WHLA(FM)**— Nov 21, 1950: 90.3 mhz; 100 kw. 1,010 ft TL: N43 48 17 W91 22 06. Hrs open: 24
Rebroadcasts WHA(AM) Madison 95%.
Wisconsin Public Radio, 1725 State St, LaCrosse, 54601. Phone: (608) 785-8380. Fax: (608) 785-5005. E-mail: gaddo@wpr.org Web Site: www.wpr.org. Licensee: State of Wis. Educational Communications Board. Natl. Network: NPR, PRI. Law Firm: Dow, Lohnes & Albertson. Wire Svc: AP Format: Educ, talk. News staff: 3. Target aud: 35-54; Skews Female: issue oriented talk-variety of perspectives. ♦Wendy Wink, CEO; John Gaddo, gen mgr; Marv Spielman, dev dir; Sandra Harris, news dir; Steve Bauder, chief of engrg.

WIZM(AM)— Jan 2, 1923: 1410 khz; 5 kw-U, DA-N. TL: N43 50 48 W91 13 03. Hrs open: 24 Box 99, 201 State St., 54602. Phone: (608) 782-1230. Phone: (608) 796-2505. Fax: (608) 782-1170. E-mail: dickr@mwtbroadcasting.com Web Site: www.familybroadcasting.com. Licensee: Family Radio Inc. Group owner: Mid-West Family Broadcast Group (acq 7-12-71; $500,000). Population served: 300,000 Natl. Network: Westwood One, CBS. Natl. Rep: Christal. Law Firm: Davis Wright Tremaine LLP. Wire Svc: AP Format: News/talk. News staff: 5; News: 18 hrs wkly. Target aud: 35 plus. Spec prog: Asian 2 hrs wkly.Dick Record, pres, gen mgr & edit dir; Howard Gloede, sls dir & gen sls mgr; Bill Black, rgnl sls mgr & reporter; Theresa Timm, rgnl sls mgr; Mike Hayes, prom mgr; Scott Robert Shaw, progmg dir, news dir, news rptr & reporter; Keith Carr, pub affrs dir; Chris O'Hearn, chief of engrg; Peggy Schelbe, traf mgr; Brad Williams, news rptr; Bruce Marcus, reporter; Mitch Reynolds, reporter; Pam Jahnke, farm dir; Bob Schmidt, sports cmtr

WIZM-FM— 1966: 93.3 mhz; 100 kw. 1,000 ft TL: N43 44 23 W91 22 04. Stereo. 24 (Acq 6-15-76).60,000 Format: Top-40. News staff: 5; News: 2 hrs wkly. Target aud: 18-49. ♦Jen O'Brien, progmg dir & disc jockey; Mitch Reynolds, rsch dir & reporter; Peggy Schelbe, traf mgr; Scott Robert Shaw, local news ed; Brad Williams, news rptr; Bruce Marcus, reporter; Brittany Styles, disc jockey; Keith Carr, disc jockey; Samantha Strong, disc jockey; Shaun Lewis, disc jockey.

WKBH-FM—West Salem, Mar 15, 1982: 100.1 mhz; 3.6 kw. 426 ft TL: N43 51 02 W91 12 08. Stereo. Hrs opn: 24 1407 2nd Ave., Onalaska, 54650. Phone: (608) 782-8335. Fax: (608) 782-8340. Licensee: Mississippi Valley Broadcasters LLC. Group owner: La Crosse Radio Group (acq 12-16-2000). Format: Classic rock. News staff: one. Target aud: 25-54. ♦Lee Norman, pres; Todd Wohlert, stn mgr.

WKTY(AM)— May 1948: 580 khz; 5 kw-D, 1 kw-N, DA-2. TL: N43 44 25 W91 12 21. Hrs open: 24 Box 99, 201 State St., 54602. Phone: (608) 782-1230. Fax: (608) 782-1170. E-mail: dickr@mwfbroadcasting.com Web Site: midwestfamilybroadcasting.com. Licensee: Family Radio Inc. Group owner: The Mid-West Family Broadcast Group (acq 1996; $1.3 million). Population served: 500,000 Natl. Rep: Christal. Law Firm: Davis Wright Tremaine LLP. Wire Svc: AP Format: Sports, talk. News staff: 5; News: 18 hrs wkly. Target aud: 25-54. Spec prog: Farm 5 hrs wkly.Dick Record, pres, gen mgr & edit dir; Howard Gloede, sls dir & gen sls mgr; Theresa Timm, gen sls mgr & rgnl sls mgr; Mike Kearns, prom dir & sports cmtr; Scott Robert Shaw, progmg dir & news dir; Keith Carr, pub affrs dir; Chris O'Hearn, chief of engrg; Peggy Schelbe, traf mgr; Brad Williams, news rptr; Bruce Marcus, reporter; Pam Jahnke, farm dir

WRQT(FM)—Co-owned with WKTY(AM). January 1972: 95.7 mhz; 50 kw. 410 ft TL: N43 44 30 W91 18 14. Stereo. 350,000 Format: Active rock. News: 5 hrs wkly. Target aud: 18-49. ♦Jean Taylor, progmg dir; Peggy Schelbe, traf mgr; Brad Williams, news rptr; Bill Black, reporter.

WLFN(AM)— May 1947: 1490 khz; 1 kw-U. TL: N43 49 42 W91 14 27. Stereo. Hrs opn: 24 Box 2017, 54602-2017. Secondary address: 1407 Second Ave. N., Onalaska 54650. Phone: (608) 782-8335. Fax: (608) 782-8340. Licensee: Mississippi Valley Broadcasters L.L.C. Group owner: La Crosse Radio Group. Population served: 96,500 Format: Original Hits. News staff: one; News: 5 hrs wkly. Target aud: 35 plus. ♦Pat Smith, gen mgr; Mike Schmitz, sls dir; Pete Schreier, progmg dir; Lucy Lemar, news dir; Patrick Delaney, chief of engrg; Laurie Lane, traf mgr.

WLXR-FM—Co-owned with WLFN(AM). March 1975: 104.9 mhz; 1.35 kw. 430 ft TL: N43 45 28 W91 17 26. (CP: 3.4 kw). Stereo. Web Site: www.wlxr.com.51,153 Format: Adult contemp. Target aud: 18-49. ♦Debbie Brague, progmg dir.

***WLSU(FM)**— Jan 4, 1971: 88.9 mhz; 8.2 kw. Ant 928 ft TL: N43 48 17 W91 22 06. Stereo. Hrs opn: 24 Wisconson Public Radio, 1725 State St., 54601. Phone: (608) 785-8380. Fax: (608) 785-5005. E-mail: gaddo@wpr.org Web Site: www.wpr.org. Licensee: University of Wisconsin System. Population served: 51,153 Natl. Network: NPR, PRI. Rgnl. Network: Wis. Pub. Law Firm: Dow, Lohnes & Albertson. Wire Svc: AP Format: Class, jazz, news. News staff: 3; News: 40 hrs wkly. Target aud: General. ♦John Gaddo, gen mgr; Marvin Spielman, dev dir; Sandra Harris, news dir; Steve Johnston, pub affrs dir & chief of engrg.

WQCC(FM)— Mar 31, 1994: 106.3 mhz; 12 kw. 476 ft Hrs opn: 24 Box 2017, 54602-2017. Phone: (608) 782-1063. Phone: (608) 782-8335. Fax: (608) 779-5945. E-mail: wlxr/wqcc@aol.com Licensee: Mississippi Valley Broadcasters L.L.C. Group owner: La Crosse Radio Group (acq 12-31-96). Format: Country. ♦Pat Smith, gen mgr; Mike Schmitz, gen sls mgr; John Stevenson, progmg dir; Lucy Lemar, news dir; Patrick Delaney, chief of engrg.

Ladysmith

WJBL(FM)—Listing follows WLDY(AM).

WLDY(AM)— September 1948: 1340 khz; 1 kw-U. TL: N45 27 52 W91 07 26. Hrs open: 24 Box 351, 54848-0351. Secondary address: W8746 Hwy. 8 54848. Phone: (715) 532-5588. Fax: (715) 532-7357. E-mail: wldy@centurytel.net Licensee: Roth Broadcasting Inc. (acq 1-9-2004; $924,722 with co-located FM). Population served: 100,000 Natl. Network: ABC. Wire Svc: UPI Format: Country, news/talk. News staff: one; News: 15 hrs wkly. Target aud: 35 plus; mature audience. Spec prog: Polka 3 hrs wkly. ♦Sandra Roth, pres & gen mgr; David Roth, gen sls mgr & progmg dir; Jocelyn Kilmer, progmg dir; Del Dayton, chief of engrg; Judi Novak, traf mgr; Tom Costello, news dir & local news ed; Robert Krejcarek, farm dir, sports cmtr & disc jockey; Sandy Zajec, women's int ed.

WJBL(FM)—Co-owned with WLDY(AM). October 1984: 93.1 mhz; 4.9 kw. 358 ft TL: N45 27 59 W91 07 23. Stereo. 24 Natl. Network: ABC. Format: Oldies. News staff: one; News: 10 hrs wkly. Target aud: 25-54. ♦Tom Costello, local news ed; Robert Krejcarek, farm dir, sports cmtr & disc jockey; Sandy Zajec, women's int ed.

Lake Geneva

WLKG(FM)— June 6, 1994: 96.1 mhz; 6 kw. 328 ft TL: N42 36 34 W88 26 36. Hrs open: 24 Box 996, 500 Interchange N., 53147. Phone: (262) 249-9600. Fax: (262) 249-9630. E-mail: lake96@wlkg.com Web Site: www.wlkg.com. Licensee: CTJ Communications Ltd. Population served: 200,000 Law Firm: Shaw Pittman. Format: Hot AC. News: 15 hrs wkly. Target aud: 25-54; mainly female. Spec prog: Hits of the 70's 10 hrs, sports 2 hrs wkly. ♦Tom Kwiatkowski, pres; Barb Kwiatkowski, VP; Nancy Douglass, gen mgr.

WZRK(AM)— May 15, 1964: 1550 khz; 1 kw-D, DA. TL: N42 35 40 W88 23 19. Hrs open: Box 965, 53147. Secondary address: 6715 Hwy. 50 53147. Phone: (262) 248-1005. Fax: (262) 248-2002. E-mail: wzrk@relevantradio.com Web Site: www.relevantradio.com. Licensee: Starboard Media Foundation Inc. Group owner: Relevant Radio (acq 5-31-2001). Population served: 10,000 Rgnl. Network: Wis. Pub., Ill. Radio Net. Law Firm: Donald E. Martin. Format: Catholic talk. Target aud: 25 plus. Spec prog: Farm. ♦Ted Ehlen, gen mgr & stn mgr.

Lancaster

WGLR(AM)— Sept 9, 1977: 1280 khz; 500 w-D. TL: N42 50 22 W90 40 19. Hrs open: Box 587, 206 S. Sheridan St., 53813. Phone: (608) 723-7671. Fax: (608) 723-7674. Licensee: QueenB Radio Wisconsin Inc. Group owner: Morgan Murphy Stations (acq 3-18-98; $1.66 million with co-located FM). Format: C&W. Spec prog: Farm 10 hrs wkly. ♦Danny Sullivan, gen mgr; Doug Wagen, opns dir & news dir; Rick Sanson, sls dir & gen sls mgr; Rob Spangler, progmg dir.

WGLR-FM— Sept 9, 1982: 97.7 mhz; 3 kw. 235 ft TL: N42 50 18 W90 40 14. (CP: 25 kw, ant 328 ft.). Stereo. 54,000 Format: Country.

***WJTY(FM)**— Mar 12, 1983: 88.1 mhz; 7 kw horiz, 50 kw vert. Ant 476 ft TL: N42 57 08 W90 25 47. Stereo. Hrs opn: 24 341 S. Washington, 53813. Phone: (608) 723-7888. Fax: (608) 723-4557. E-mail: info@wjty.org Web Site: www.wjty.org. Licensee: Family Life Broadcasting Inc. (acq 3-27-2007; grpsl). Population served: 100,000 Natl. Network: Moody, USA. Format: Relg, contemp, MOR. News staff: one; News: 11 hrs wkly. Target aud: 30-90; families. ♦Tom Bush, gen mgr & progmg dir; Dennis Baldridge, chief of engrg.

Lomira

WFDL-FM—Licensed to Lomira. See Fond du Lac

Madison

***WERN(FM)**— Mar 30, 1947: 88.7 mhz; 20.5 kw. Ant 990 ft TL: N43 03 18 W89 28 42. Stereo. Hrs opn: 24 821 University Ave., 53706-1496. Phone: (608) 263-3970. Phone: (608) 263-4120. Fax: (608) 263-9763. E-mail: schnirring@wpr.org Web Site: www.wpr.org. Licensee: State of Wisconsin Educational Communications Board. Natl. Network: NPR, PRI. Rgnl. Network: Wis. Pub. Format: News, classical. News staff: 9; News: 29 hrs wkly. Target aud: 25-64; persons seeking quality music & intellectual stimulation. ♦Phil Corriveau, gen mgr; Ben Spindler, dev dir; Anders Yokum, progmg dir; Vicki Nonn, mus dir & news dir.

***WHA(AM)**— 1922: 970 khz; 5 kw-D, 51 w-N. TL: N43 02 30 W89 24 31. Hrs open: 24 821 University Ave., 53706. Phone: (608) 263-3970. Fax: (608) 263-9763. E-mail: listener@wpr.org Web Site: www.wpr.org. Licensee: Regents of University of Wisconsin System. Population served: 310,000 Natl. Network: NPR, PRI. Rgnl. Network: Wis. Pub. Wire Svc: NOAA Weather Format: Educ, talk, news. News staff: 9. Target aud: 35-54; male/female, educated, skews female: issue oriented talk-variety of perspectives. ♦Phil Corriveau, gen mgr; Tom Martin-Erickson, opns dir & opns mgr; Ben Spindler, dev dir & dev mgr; Anders Yokum, progmg dir; Vicki Nonn, mus dir. Co-owned TV: *WHA-TV affil

WIBA(AM)— Apr 2, 1925: 1310 khz; 5 kw-U, DA-N. TL: N42 59 53 W89 25 42. Stereo. Hrs open: 2651 S. Fish Hatchery Rd., 53711. Phone: (608) 274-5450. Fax: (608) 274-5521. Web Site: www.wiba.com. Licensee: Capstar TX L.P. Group owner: Clear Channel Communications Inc. (acq 8-30-00; grpsl). Population served: 176,258 Natl. Network: CBS, Wall Street. Law Firm: Dow, Lohnes & Albertson. Format: News/talk. Target aud: 25-64. ♦Jeff Tyler, gen mgr & opns mgr; Kurt Peterson, sls dir; Tim Scott, progmg dir; Josh Wescott, news dir; Tim Wagner, chief of engrg; Marta Keller, traf mgr.

WIBA-FM— Mar 1947: 101.5 mhz; 50 kw. 450 ft TL: N43 03 22 W89 32 07. (CP: Ant 1,013 ft.). Stereo. Web Site: www.wibafm.com. Format: Classic rock. ♦Mike Ferris, progmg dir; Jennie Hibbard, traf mgr.

WLMV(AM)— September 1948: 1480 khz; 5 kw-U, DA-N. TL: N43 01 30 W89 23 48. Stereo. Hrs opn: 24 Box 2058, 53701. Secondary address: 2740 Ski Ln. 53713. Phone: (608) 271-1484. Phone: (608) 273-1000. Fax: (608) 271-0400. Licensee: Mid-West Management Inc. Group owner: The Mid-West Family Broadcast Group. Population served: 50,000 Natl. Rep: McGavren Guild. Law Firm: Shaw Pittman. Format: Sp. Target aud: General; Latino community. ♦Thomas A. Walker, pres & gen mgr; Ted Waldbillig, sls VP & gen sls mgr; Bill Mann, natl sls mgr; Luis Montoto, progmg dir; Robin Colbert, news dir; John Bauer, chief of engrg.

WMGN(FM)—Co-owned with WLMV(AM). September 1948: 98.1 mhz; 38 kw. 581 ft TL: N42 57 46 W89 22 46. Stereo. 24 2740 Ski Ln., 53713. Phone: (608) 271-1000. Fax: (608) 271-8182. E-mail: info@magic98.com Web Site: www.magic98.com.50,000 Natl. Rep: McGavren Guild. Law Firm: Shaw Pittman. Format: Adult contemp. Target aud: 25-54. ♦Bill Mann, sls VP; Pat O'Neill, opns dir & progmg dir; Mark Van Allen, mus dir.

WMAD(FM)—Sauk City, Sept 18, 1964: 96.3 mhz; 5.1 kw. Ant 672 ft TL: N43 12 37 W89 35 57. Stereo. Hrs opn: 24 2651 S. Fish Hatchery Rd., 53711. Phone: (608) 274-5450. Fax: (608) 274-5521. Web Site: www.wmad.com. Licensee: Capstar TX L.P. Group owner: Clear Channel Communications Inc. (acq 8-30-2000; grpsl). Natl. Rep: Christal. Law Firm: Leventhal, Senter & Lerman. Format: Alternative. News staff: one; News: one hr wkly. Target aud: 25-49. ♦Jeff Tyler, gen mgr; Hugh Garret, gen sls mgr; Brad Savage, progmg dir; Joshua Wescott, news dir; Cliff Groth, chief of engrg; Jacqueline Forney, traf mgr.

WNWC(AM)—See Sun Prairie

***WNWC-FM**— Apr 30, 1959: 102.5 mhz; 50 kw. 460 ft TL: N43 02 07 W89 30 25. Stereo. Hrs opn: 5606 Medical Cir., 53719. Phone: (608) 271-1025. Fax: (608) 271-1150. E-mail: wnwc@nwc.edu Web Site: www.wnwc.org. Licensee: Northwestern College. Group owner: Northwestern College & Radio (acq 1-19-73). Population served: 173,258 Natl. Network: AP Radio. Wire Svc: AP Format: Contemp Christian Music. News staff: one; News: 20 hrs wkly. Target aud: 35-45. ♦Greg Walters, gen mgr.

WOLX-FM—Baraboo, Mar 3, 1946: 94.9 mhz; 37 kw. 1,299 ft TL: N43 25 40 W89 39 14. Stereo. Hrs opn: 24 7601 Ganser Way, 53719. Phone: (608) 826-0077. Fax: (608) 826-1244. Web Site: www.wolx.com. Licensee: Entercom Madison Licensee LLP. Group owner: Entercom Communications Corp. (acq 7-10-00; grpsl). Population served: 1,985,856 Natl. Rep: Christal. Law Firm: Rosenman & Colin. Format: Oldies. News staff: 2; News: 2 hrs wkly. Target aud: 25-54. ♦David Field, pres; Ed Schulz, VP; Michael Weber, CFO & opns mgr; Lindsay Wood Davis, mktg mgr.

***WORT(FM)**— Dec 1, 1975: 89.9 mhz; 2 kw. 900 ft TL: N43 03 01 W89 29 15. Stereo. Hrs opn: 24 118 S. Bedford St., 53703. Phone: (608) 256-2695/256-2001. Fax: (608) 256-3704. E-mail: wort@terracom.net Web Site: www.wort-fm.org. Licensee: Back Porch Radio Broadcasting Inc. Population served: 350,000 Format: Div, class. News staff: one; News: 3 hrs wkly. Target aud: 25-34. Spec prog: Black 3 hrs, jazz 15 hrs wkly. ♦Norman Stockwell, opns dir; Rachel Pundsack, dev dir; Sybil Augustine, mus dir; Nathan Moore, news dir.

***WSUM(FM)**— 2003: 91.7 mhz; 5.5 kw. Ant 338 ft TL: N42 54 16 W89 33 20. Hrs open: 24 Box 260020, 53726-0020. Phone: (608) 262-1864. Web Site: www.wsum.org. Licensee: Board of Regents of the University of Wisconsin. Format: Educ. ♦Dave Black, gen mgr.

WTDY(AM)— 1998: 1670 khz; 10 kw-D, 1 kw-N. TL: N43 01 30 W89 23 48. Hrs open: 24 Box 44408, 53744. Secondary address: 730 Rayovac Dr. 53711. Phone: (608) 273-1000. Fax: (608) 271-0400. Fax: (608) 271-8182. E-mail: thebigboss@wtdy.com Web Site: www.wtdy.com. Licensee: Mid-West Management Inc. Group owner: The Mid-West Family Broadcast Group Natl. Rep: McGavren Guild. Law Firm: Shaw Pittman. Format: News/talk. News staff: 5; News: 15 hrs wkly. Target aud: Male 25-54; Young to middle aged males. ♦Tom Walker, pres & gen mgr; John Sylvester, progmg dir; Robin Colbert, news dir; John Bauer, chief of engrg.

WTSO(AM)— January 1948: 1070 khz; 10 kw-D, 5 kw-N, DA-2. TL: N42 59 45 W89 18 50. Hrs open: 24 2651 S. Fishhatchery Rd., 53711. Phone: (608) 274-5450. Fax: (608) 274-5521. Web Site: www.espn1070.com. Licensee: Capstar TX L.P. Group owner: Clear Channel Communications Inc. (acq 8-30-00; grpsl). Population served: 731,900 Rgnl. Network: Wisconsin Radio Net. Law Firm: Dow, Lohnes & Albertson. Format: Sports. News staff: 6. Target aud: 25-54. Spec prog: Farm 20 hrs wkly. ♦Jeff Tyler, gen mgr & opns mgr; Kurt Peterson, sls dir; Tim Scott, progmg mgr; Tim Wagner, chief of engrg; Jennie Hibbard, traf mgr.

WZEE(FM)—Co-owned with WTSO(AM). 1948: 104.1 mhz; 9.4 kw. 1,119 ft TL: N43 03 09 W89 28 42. Stereo. 24 Web Site: www.z104fm.com.285,700 Format: Adult contemp, CHR. News staff: 2; News: one hr wkly. ♦Tommy Bodean, progmg dir.

WTUX(AM)— Aug 14, 1964: 1550 khz; 5 kw-D, DA. TL: N43 00 08 W89 23 08. Hrs open: 24 Box 44408, 53744. Secondary address: 730 Rayovac Dr. 53711. Phone: (608) 273-1000. Fax: (608) 271-0400. Web Site: www.wtux.com. Licensee: Mid-West Management Inc. Group owner: The Mid-West Family Broadcast Group (acq 8-12-97; $6.4 million with WWQM-FM Middleton). Population served: 173,258 Natl. Network: ABC. Natl. Rep: McGavren Guild. Format: Music of your life. Target aud: 25-54. ♦Tom Walker, pres & gen mgr; Ted Waldbillig, sls dir & chief of engrg; Amy Schiefelbein, progmg dir.

WWQM-FM—See Middleton

WXXM(FM)—See Sun Prairie

Manitowoc

WCUB(AM)—Two Rivers, November 1952: 980 khz; 5 kw-U, DA-2. TL: N44 03 50 W87 41 49. Hrs open: 24 Box 1990, 54221-1990. Secondary address: 1915 Mirro Dr. 54220. Phone: (920) 683-6800. Fax: (920)683-6807. Web Site: www.cubradio.com. Licensee: Cub

Radio Inc. (acq 1-1-61). Population served: 33,000 Natl. Rep: Katz Radio. Format: C&W, farm. News staff: 2; News: 16 hrs wkly. Target aud: 35 plus. ♦Lee Davis, pres, gen mgr & gen sls mgr; Kyle Kristofer, progmg dir; Bryan Lundberg, news dir.

WLTU(FM)—Co-owned with WCUB(AM). Sept 1, 1966: 92.1 mhz; 3.7 kw. 420 ft TL: N44 07 31 W87 37 41. Stereo. 24 Web Site: www.cubradio.com.18,000 Natl. Rep: Katz Radio. Format: Oldies. News staff: one. Target aud: 25-64.

WGBW(AM)—See Two Rivers

WOMT(AM)— Nov 8, 1926: 1240 khz; 1 kw-U. TL: N44 07 31 W87 37 41. Hrs open: 24 Box 1385, 3730 Mangin St., 54221-1385. Phone: (920) 682-0351. Fax: (920) 682-1008. E-mail: info@womtradio.com Web Site: www.womtradio.com. Licensee: Seehafer Broadcasting Corp. (acq 1-1-70). Population served: 150,000 Natl. Network: CBS. Law Firm: Miller & Neely. Wire Svc: AP Format: Full service, adult contemp, MOR. News staff: 2; News: 45 hrs wkly. Target aud: 25-64; business executives, males/females. Spec prog: News 18 hrs wkly.Don Seehafer, pres & gen mgr; Ben Jakel, stn mgr; Russ Matar, gen sls mgr; Lindy Lukes, mktg dir; Mark Seehafer, adv VP; Fred Barry, news dir, local news ed & news rptr; Harley Engel, chief of engrg; Theresa Sponholtz, traf mgr; Ron Zimmerman, farm dir, sports cmtr & disc jockey; Damon Ryan, sports cmtr & disc jockey; Dave Dunn, disc jockey; Tim Strews, progmg dir & disc jockey

WQTC-FM—Co-owned with WOMT(AM). Nov 19, 1965: 102.3 mhz; 3 kw. 328 ft TL: N44 07 31 W87 37 41. Stereo. 24 Web Site: www.womtradio.com.150,000 Law Firm: Miller & Neely. Format: Classic hits. News staff: 8; News: 8 hrs wkly. Target aud: 18-49.

Marathon

WKQH(FM)— 1988: 104.9 mhz; 21 kw. Ant 358 ft TL: N44 50 13 W89 45 57. Stereo. Hrs opn: 500 Division St., Stevens Point, 54481. Phone: (715) 341-9800. Fax: (715) 341-0000. E-mail: rmuzzy@1010wspt.com Web Site: www.b1049.com. Licensee: RLM Communications Inc. Group owner: Muzzy Broadcasting L.L.C. (acq 1994; $150,000). Population served: 210,600 Format: Country. News staff: 3. Target aud: 25-54; adult. ♦Richard Muzzy, gen mgr; Rob West, progmg mgr; Scott Krueger, news dir; Jim Zastnow, chief of engrg; Geri Butler, traf mgr.

Marinette

WAGN(AM)—See Menominee, MI

WHYB(FM)—See Menominee, MI

WLST(FM)—Listing follows WMAM(AM).

WMAM(AM)— Oct 8, 1939: 570 khz; 250 w-D, 100 w-N. TL: N45 06 02 W87 37 30. Hrs open: 24 N. 2880 Roosevelt Rd., 54143. Phone: (715) 735-6631. Fax: (715) 732-0125. Licensee: Armada Media - Menominee Inc. (group owner; (acq 12-19-2006; grpsl). Natl. Network: CBS. Rgnl. Network: Wisconsin Radio Net., Goetz Group. Natl. Rep: Michigan Spot Sales. Format: Talk, sports. News staff: one; News: 4 hrs wkly. Target aud: 25 plus; upscale. Spec prog: Milwaukee Brewers, Green Bay Packers, farm 3 hrs wkly. ♦Shawn Katzbeck, gen mgr & gen sls mgr; Jim Medley, progmg dir; Chuck Gennaro, chief of engrg; Lisa Bougie, traf mgr.

WLST(FM)—Co-owned with WMAM(AM). Sept 1, 1976: 95.1 mhz; 100 kw. Ant 436 ft TL: N45 03 48 W87 39 26. Stereo. 24 450,000 Format: Adult contemp. News: 2 hrs wkly. Target aud: 25-49; females. ♦Lisa Bougie, traf mgr.

Marshall

***WJWD(FM)**— 2003: 90.3 mhz; 51 w horiz, 9.9 kw vert. Ant 312 ft TL: N43 20 40 W89 06 10. Hrs open: 152 McCrae Rd., Fall River, 53932. Phone: (920) 484-6220. Fax: (920) 484-3753. E-mail: wjwd@csnradio.com Licensee: CSN International (group owner). Format: Relg. ♦Patrick Lannoye, gen mgr & opns mgr.

Marshfield

WDLB(AM)— Feb 2, 1947: 1450 khz; 1 kw-U. TL: N44 41 49 W90 09 20. Hrs open: 24 Box 630, 1710 N. Central Ave., 54449. Phone: (715) 384-2191. Fax: (715) 387-3588. Web Site: wdlb@mmwi.net. Licensee: Seehafer Broadcasting Corp. (group owner; (acq 6-1-2006; swap with WOSQ(FM) Spencer and WFHR(AM) Wisconsin Rapids for WBCV(FM) Wausau) Population served: 150,000 Rgnl. Network: Goetz Group. Law Firm: Miller & Fields, P.C. Format: News/talk, sports. News staff: 3; News: 13 hrs wkly. Target aud: 25-49; general. Spec prog: Farm 14 hrs wkly. ♦Wayne Ripp, gen mgr; Jay Lepsch, opns mgr, progmg dir & news dir; Arnie Peck, gen sls mgr & adv VP; George Nicholas, engrg dir.

WYTE(FM)— Dec 1, 1965: 106.5 mhz; 100 kw. Ant 800 ft TL: N44 38 41 W89 51 11. Stereo. Hrs opn: 2301 Plover Rd., Plover, 54467. Phone: (715) 341-8838. Fax: (715) 341-9744. E-mail: info@wyte.com Web Site: www.wyte.com. Licensee: NRG License Sub, LLC. (acq 10-31-2005; grpsl). Population served: 300,000 Format: Country. Target aud: 25-54. ♦Benjamin D. Rosenthal, gen mgr; Bob Jung, rgnl sls mgr.

Mauston

WRJC(AM)— Jan 4, 1962: 1270 khz; 500 w-D. TL: N43 49 52 W90 04 51. Hrs open: 24 Box 200, Fairway Ln., 53948. Phone: (608) 847-6565. Fax: (608) 847-6249. E-mail: office@wrsc.com Web Site: wrsc.com. Licensee: WRJC Broadcasting Co. (acq 2-15-86; FTR: 12-23-85). Population served: 110,000 Natl. Network: CBS. Natl. Rep: Rgnl Reps. Law Firm: Womble, Carlyle, Sandridge & Rice. Wire Svc: AP Format: MOR. News staff: one; News: 8 hrs wkly. Target aud: 25 plus; general. ♦Rick Charles, pres, gen mgr & gen sls mgr; Greg Lawrence, prom VP & mus dir; June Gill, news dir; Ken Ebneter, chief of engrg.

WRJC-FM— 1976: 92.1 mhz; 2 kw. Ant 571 ft TL: N43 47 16 W90 11 52. Stereo. 24 E-mail: wrjc@mwj.net Web Site: www.wrjc.com. Licensee: WRJC Inc. Format: Adult contemp. News staff: one; News: 10 hrs wkly. Target aud: 21 plus; adults. ♦Rick Charles, CEO.

Mayville

WMDC(FM)— Oct 31, 1998: 98.7 mhz; 6 kw. 246 ft TL: N43 28 53 W88 28 45. Hrs open: 24 132 N. Main St., 53050. Phone: (920) 387-0000. Fax: (920) 387-2222. E-mail: bigsky@dotnet.com Web Site: www.great98.com. Licensee: Radio Plus Inc. Format: Hits of the 60s & 70s. News staff: one. Target aud: 25-54. ♦Tom Biolo, gen mgr; Norm Grey, gen sls mgr.

Medford

WIGM(AM)— Oct 26, 1941: 1490 khz; 1 kw-U, DA-1. TL: N45 07 55 W90 19 54. Hrs open: 24 Box 59, 54451. Secondary address: 630 S. 8th 54451. Phone: (715) 748-2566. Web Site: www.k99wigm.com. Licensee: WIGM Inc. (acq 6-55). Population served: 175,000 Natl. Network: ESPN Radio. Wire Svc: AP Format: Sports. News staff: one; News: 14 hrs wkly. Target aud: 21 plus. Spec prog: Farm 10 hrs wkly. ♦Brad Dahlvig, pres & gen mgr; Karen Dahlvig, sls dir; Paula Liske, news dir; Del Dayton, chief of engrg.

WKEB(FM)—Co-owned with WIGM(AM). September 1967: 99.3 mhz; 23 kw. 342 ft TL: N45 07 55 W90 19 54. Stereo. 24 300,000 Natl. Network: ABC. Format: Top-40. ♦Brad Dahlvig, CEO; Del Dayton, sls VP & engrg VP; Karen Dahlvig, sr VP, dev dir, sls VP & mktg dir.

Menomonee Falls

WJMR-FM— June 26, 1956: 98.3 mhz; 6 kw. 364 ft TL: N43 02 49 W88 07 25. (CP: Ant 292 ft.). Stereo. Hrs opn: 24 5407 W. McKinley Ave., Milwaukee, 53208-2540. Phone: (414) 978-9000. Fax: (414) 978-9001. Web Site: www.wjmr.com. Licensee: Lakefront Communications LLC. Group owner: Saga Communications Inc. (acq 4-24-97; $5 million with WJZX(FM) Brookfield). Population served: 1,600,000 Natl. Rep: Katz Radio. Law Firm: Smithwick & Belendiuk. Format: Urban adult comtemp. News staff: one; News: 4 hrs wkly. Target aud: 25 plus. ♦Tom Joerres, pres & gen mgr; Roger Williams, gen sls mgr; Anne Zimmer, natl sls mgr; Scott Marshall, prom dir; Lauri Jones, progmg dir; Phil Longenecker, chief of engrg; Michelle Golding, traf mgr.

Menomonie

***WHWC(FM)**— June 28, 1950: 88.3 mhz; 70 kw. Ant 1,050 ft TL: N45 02 47 W91 51 42. Hrs open: 24
Rebroadcasts WHAD(FM) Delafield 90%.
1221 W. Clairemont Ave., Eau-Claire, 54701. Secondary address: 821 University Ave., Madison 53706-1496. Phone: (715) 839-3868. Fax: (715) 839-2939. Web Site: www.wpr.org. Licensee: State of Wisconsin Educational Communications Board. Natl. Network: NPR, PRI. Rgnl. Network: Wis. Pub. Law Firm: Dow, Lohnes & Albertson. Format: Educ, talk. Target aud: 35-54. Spec prog: Folk 7 hrs wkly. ♦Dean Kallenbach, gen mgr & dev dir; Mary Jo Wagner, news dir.

WMEQ(AM)— May 1951: 880 khz; 10 kw-D, 210 w-N. TL: N44 48 48 W91 55 34. Stereo. Hrs opn: 24 Box 45, Eau Claire, 54702. Secondary address: 619 Cameron ST., Eau Claire 54703. Phone: (715) 830-4000. Fax: (715) 835-9680. Web Site: www.wmeq.com. Licensee: Capstar TX L.P. Group owner: Clear Channel Communications Inc. (acq 2000; grpsl). Population served: 15000 Format: News/talk, sports. Target aud: 35 plus. Spec prog: Farm 15 hrs wkly. ♦Mike Cushman, opns mgr; Rick Hencley, gen mgr & gen sls mgr; Jay Moore, progmg dir; Paul Orth, chief of engrg; Dave Dee, disc jockey.

WMEQ-FM— July 19, 1967: 92.1 mhz; 17.5 kw. Ant 718 ft TL: N44 54 59 W91 41 55. Stereo. 24 Web Site: www.rock921.com. Format: Classic rock. Target aud: 25-54. ♦Mike Cushman, opns dir; Rick Hencley, VP & sls dir; Dave Dee, disc jockey.

***WVSS(FM)**— Apr 22, 1969: 90.7 mhz; 590 w. Ant 426 ft TL: N44 54 56 W92 04 34. Stereo. Hrs opn: 24
Rebroadcasts WERN(FM) Madison 90%.
1221 W. Clairemont Ave., Eau-Claire, 54701. Phone: (715) 839-3868. Fax: (715) 839-2939. Licensee: Board of Regents, University of Wisconsin Systems. Population served: 11,275 Rgnl. Network: Wisconsin Radio Net. Format: Class, news/talk. Target aud: 45-64. Spec prog: Folk 6 hrs, jazz 5 hrs wkly. ♦Dean Kallenbach, gen mgr.

Merrill

WJMT(AM)— May 10, 1960: 730 khz; 1 kw-D, 127 w-N. TL: N45 10 45 W89 38 20. Hrs open: 24 120 S. Mill St., 54452-2508. Phone: (715) 536-6262. Fax: (715) 536-6208. Licensee: Quicksilver Broadcasting LLC. Group owner: Badger Communications L.L.C. Population served: 120,000 Natl. Rep: D & R Radio. Format: Adult contemp, MOR, talk. News staff: one. Target aud: 35-59. Spec prog: Relg 3 hrs, farm 7 hrs, Pol 3 hrs wkly. ♦David Winters, pres; Steven Resnick, gen mgr; Christine Vorpagel, gen sls mgr; Nick Summers, progmg dir; Joe Weniger, news dir; Chuck Genarro, chief of engrg.

WMZK(FM)—Co-owned with WJMT(AM). Aug 25, 1968: 104.1 mhz; 24 kw. 617 ft TL: N45 06 14 W89 43 05. Stereo. 24 E-mail: advertising@z104rocks.com Web Site: www.z104rocks.com.350,000 Format: Rock/AOR. News staff: one. Target aud: 18-49.

Middleton

WTUX(AM)—See Madison

WWQM-FM— Oct 20, 1970: 106.3 mhz; 4.5 kw. 374 ft TL: N43 03 03 W89 29 13. Stereo. Hrs opn: 24 Box 2058, Madison, 53701. Phone: (608) 273-1000. Fax: (608) 271-8182. Web Site: www.q106.com. Licensee: Mid-West Management Inc. Group owner: The Mid-West Family Broadcast Group (acq 8-15-97; $6.4 million with WTUX(AM) Madison). Population served: 173,258 Natl. Rep: McGavren Guild. Law Firm: Shaw Pittman. Format: Country. Target aud: 25-54; general. ♦Thomas A. Walker, pres & gen mgr; Ted Waldbillig, gen sls mgr; Mark Grantin, progmg dir & progmg mgr; John Bauer, chief of engrg.

Milladore

***WGNV(FM)**— Feb 13, 1986: 88.5 mhz; 50 kw. 584 ft TL: N44 38 37 W89 50 48. Stereo. Hrs opn: 24

94.1 Antigo.
Box 88, Country Rd. N., 54454. Phone: (715) 457-2988. Fax: (715) 457-2987. E-mail: wgnv@christianfamilyradio.net Web Site: christianfamilyradio.net. Licensee: Evangel Ministries Inc. Natl. Network: Moody, Salem Radio Network. Law Firm: Leventhal, Senter & Leman. Format: Christian adult contemp. News: 10 hrs wkly. Target aud: Women; 35-49. Spec prog: Children 4 hrs wkly. ♦Paul Cameron, gen mgr; Karen Bencke, opns dir; Bill Schumacher, adv mgr & sls; Mark Bystrom, progmg dir; Todd Christopher, mus dir; Vicky Hofkens, engrg dir & traf mgr.

Milwaukee

WHQG(FM)—Listing follows WJYI(AM).

WISN(AM)— 1922: 1130 khz; 50 kw-D, 10 kw-N, DA-2. TL: N42 45 18 W88 04 53. Hrs open: 24 12100 W. Howard Ave., Greenfield, 53228. Phone: (414) 545-8900. Fax: (414) 327-3200. Web Site: www.newstalk1130.com. Licensee: Capstar TX L.P. Group owner: Clear Channel Communications Inc. (acq 8-30-2000; grpsl). Population served: 1,000,000 Natl. Network: Fox News Radio, Premiere Radio Networks. Format: Talk, news. News staff: 4. Target aud: 25-54. ♦Cindy McDowell, gen mgr; Jay Daily, sls dir; Phil Kurth, gen sls mgr; Jerry Bott, progmg dir; Harold Mester, news dir; Al Hajny, chief of engrg; Shannon Lippert, traf mgr.

WQBW(FM)—Co-owned with WISN(AM). January 1961: 97.3 mhz; 15.5 kw. 980 ft TL: N43 06 41 W87 55 38. Stereo. 24 Web Site: www.l973thebrew.com.717,099 Natl. Network: ABC. ♦Randy Wanek, sls dir; Jeff Lynn, progmg dir.

WJYI(AM)— 1955: 1340 khz; 1 kw-U. TL: N43 02 49 W87 58 52. Hrs open: 24 5407 W. McKinley Ave., 53208. Phone: (414) 978-9000. Fax: (414) 978-9001. Web Site: www.joy1340.com. Licensee: Lakefront Communications LLC. Group owner: Saga Communications Inc. (acq 2-23-94; $7 million with co-located FM; FTR: 3-14-94). Population served: 800,000 Natl. Network: CBS. Format: Contemporary Christian. Target aud: 18-54. ♦Tom Joerres, pres & gen mgr; Ryan Salzer, opns mgr; Annie Zimmer, natl sls mgr; Kevin Garrett, rgnl sls mgr; Phil Longenecken, chief of engrg; Michelle Golding, traf mgr.

WHQG(FM)—Co-owned with WJYI(AM). October 1960: 102.9 mhz; 50 kw. Ant 440 ft TL: N43 02 49 W87 58 52. Stereo. 24 Web Site: www.1029thehog.com. Format: Active rock. ♦Ann Marie King, sls dir; Scott Schubert, prom dir; Sean Elliott, progmg dir & disc jockey.

WKKV-FM—See Racine

WKLH(FM)— 1958: 96.5 mhz; 21.24 kw. 810 ft TL: N43 05 48 W87 54 19. Stereo. Hrs opn: 24 5407 W. McKinley Ave., 53208. Phone: (414) 978-9000. Fax: (414) 978-9001. Web Site: www.wklh.com. Licensee: Lakefront Communications LLC. Group owner: Saga Communications Inc. (acq 7-18-90). Population served: 500,000 Natl. Rep: Katz Radio. Law Firm: Smith & Belendiuk. Format: Classic hits. News staff: one. Target aud: 35-54; baby boomers. ♦Tom Joerres, pres & gen mgr; Annmarie Topel, gen sls mgr; Anne Zimmer, natl sls mgr; Scott Schubert, prom dir; Bob Bellini, progmg dir; Carole Caine, news dir; Phil Longenecker, chief of engrg; Michelle Aldridge, traf mgr.

WKTI(FM)—Listing follows WTMJ(AM).

WLDB(FM)— June 1958: 93.3 mhz; 12.6 kw. Ant 992 ft TL: N43 05 15 W87 54 12. Stereo. Hrs opn: 24 N72 W12922 Good Hope Rd., Menomonee Falls, 53051-4441. Phone: (414) 778-1933. Fax: (414) 771-3036. Web Site: www.b933fm.com. Licensee: Milwaukee Radio Alliance L.L.C. (group owner; (acq 9-23-97; grpsl). Format: Soft rock. Target aud: 25-54; general. ♦Willie D. Davis, chmn; William R. Lynett, pres; Bill Hurwitz, gen mgr; Traci Northrup, opns dir & gen sls mgr; Stan Atkinson, progmg dir & news dir; John Church, chief of engrg; Tiffany Dlugi, traf mgr.

WLUM-FM—Listing follows WMCS(AM).

WMCS(AM)—Greenfield, Apr 27, 1947: 1290 khz; 5 kw-U, DA-2. TL: N42 55 11 W87 59 17. Stereo. Hrs opn: 24 4222 W. Capitol Dr., 53216. Phone: (414) 444-1290. Fax: (414) 444-1409. Web Site: www.1290wmcs.com. Licensee: Milwaukee Radio Alliance L.L.C. (group owner; (acq 9-23-97; grpsl). Population served: 75,000 Natl. Network: ESPN Radio. Format: Talk. News staff: one; News: 7 hrs wkly. Target aud: 25-54; upwardly mobile Blacks. Spec prog: Blues 6 hrs, gospel 5 hrs, Sp 5 hrs, church 3.25 hrs wkly. ♦Don Rosette, gen mgr & gen sls mgr; Eric Von, opns mgr; Tyrene Jackson, progmg dir; Keith Murphy, news dir; John Church, chief of engrg; Carolyn Spinks, traf mgr.

WLUM-FM—Co-owned with WMCS(AM). September 1960: 102.1 mhz; 20 kw. Ant 761 ft TL: N43 05 48 W87 54 19. Stereo. N72 W12922 Good Hope Rd., Menomonee Falls, 53051-4441. Phone: (414) 771-1021. Fax: (414) 771-3036. Web Site: www.fm1021milwaukee.com.275,000 Format: Rock. Target aud: 18-49; teens & adults. ♦Bill Hurwitz, gen mgr; Jerry Arndt, sls dir & natl sls mgr; Traci Northrop, gen sls mgr; Tommy Wilde, progmg dir; Stan Atkinson, news dir; John Church, chief of engrg; Tiffany Dlugi, traf mgr.

WMIL(FM)—See Waukesha

***WMSE(FM)**— Mar 14, 1981: 91.7 mhz; 1 kw. 125 ft TL: N43 02 43 W87 54 57. (CP: 3.2 kw, ant 128 ft.). Stereo. Hrs opn: 24 1025 N. Broadway, 53202. Phone: (414) 277-7247. Fax: (414) 277-7149. Web Site: www.wmse.org. Licensee: Milwaukee School of Engineering. Population served: 1,000,000 Format: Mix. Target aud: 18-35; young adults. Spec prog: Black 13 hrs, jazz 15 hrs, It 3 hrs, Sp 3 hrs, class 3 hrs wkly. ♦Tom Crawford, gen mgr.

***WMWK(FM)**— Dec 7, 1990: 88.1 mhz; 170 w. 955 ft TL: N43 05 24 W87 53 47. Stereo. Hrs opn: 24 290 Hegenberger Rd., Oakland, CA, 94621. Secondary address: Box 11552, 1100 E. Capitol Dr., Shorewood 53211. Phone: (414) 964-9794. Phone: (800) 543-1495. Web Site: www.familyradio.com. Licensee: Family Stations Inc. (group owner) Format: Christian. ♦Harold Camping, pres; John Rorvik, gen mgr.

WMYX(FM)—Listing follows WSSP(AM).

WNOV(AM)— Aug 15, 1946: 860 khz; 250 w-D, 5 w-N. TL: N43 02 20 W87 54 17. (CP: TL: N43 04 20 W87 57 07). Hrs opn: Box 06438, 3815 N. Teutonia Ave., 53206. Phone: (414) 449-9668. Fax: (414) 449-9945. E-mail: wnov860@yahoo.com Web Site: www.wnov.com. Licensee: Courier Communications Corp. (acq 1-2-73). Population served: 717,099 Format: Urban contemp. ♦Jerrel W. Jones, CEO & pres; Sandra Robinson, gen mgr & opns mgr; Homer Blow, progmg dir; Amari Brown, news dir.

WOKY(AM)— 1947: 920 khz; 5 kw-D, 1 kw-N, DA-2. TL: N42 58 32 W88 03 56. Stereo. Hrs opn: 24 12100 W. Howard Ave., Greenfield, 53228. Phone: (414) 545-5920. Fax: (414) 546-9654. Web Site: www.mighty92.com. Licensee: Clear Channel Radio Licenses Inc. Group owner: Clear Channel Communications Inc. (acq 3-17-97; $40 million with WMIL(FM) Waukesha). Population served: 230,000 Natl. Network: CBS Radio. Natl. Rep: Clear Channel. Wire Svc: AP Format: Top 40 hits of the 60's. News staff: 3; News: 15 hrs wkly. Target aud: 35-64. ♦Cindy McDowell, VP & stn mgr; Jay Dailey, sls dir; Phil Kurth, sls dir & gen sls mgr; Kristin Wacker, gen sls mgr & prom dir; Jerry Bott, progmg dir; Gregory Jon, mus dir; Terry James, progmg dir & news dir; Al Hajny, chief of engrg; Barb Fagnat, traf mgr.

WRIT-FM— May 10, 1961: 95.7 mhz; 34 kw. 610 ft TL: N43 05 25 W87 54 54. Stereo. Hrs opn: 12100 W. Howard Ave., Greenfield, 53228-0920. Phone: (414) 944-5150. Fax: (414) 329-2587. Licensee: Clear Channel Radio Licenses Inc. Group owner: Clear Channel Communications Inc. (acq 10-97; $14.5 million). Population served: 633845 Natl. Rep: Clear Channel. Format: Adult Hits. Target aud: 25-54; baby boomers. ♦Cindy McDowell, gen mgr; Jeff Lynn, stn mgr & progmg dir; Kerry Wolfe, opns mgr; Keith Bratel, gen sls mgr; Ken Kohl, prom dir; Harold Mester, news dir; Al Hajny, chief of engrg; Marvy Quesnell, traf mgr.

WRRD(AM)—See Jackson

WSSP(AM)— Oct 14, 1935: 1250 khz; 5 kw-U, DA-2. TL: N42 56 44 W88 03 39. Hrs open: 24 11800 W. Grange Ave., Hales Corners, 53130. Phone: (414) 529-1250. Fax: (414) 529-2122. Licensee: Entercom Milwaukee License LLC. Group owner: Entercom Communications Corp. (acq 12-13-99; grpsl). Population served: 110,000 Law Firm: Akin, Gump, Strauss, Hauer & Feld. Format: Christian radio. Target aud: 25-49. Spec prog: Relg 3 hrs, Ger 8 hrs, Sp 3 hrs wkly. ♦Craig Hodgson, gen mgr; Alan Kirshbom, sls dir; Andrea Biebel, natl sls mgr; Jim Morales, prom dir; Glenn Redd, progmg dir & progmg mgr; Michael Clemens, news dir; Chris Tarr, chief of engrg.

WMYX(FM)—Co-owned with WSSP(AM). Nov 1, 1962: 99.1 mhz; 50 kw. 450 ft TL: N42 56 44 W88 03 39. Stereo. Web Site: www.99wmyx.com.1,200,000 Natl. Network: Westwood One. Natl. Rep: D & R Radio. Format: Hot adult contemp. Target aud: 25-49; women. Spec prog: Relg one hr wkly. ♦Tom Gjerdrum, progmg mgr; Jane Matenaer, pub affrs dir.

WTMJ(AM)— July 25, 1927: 620 khz; 50 kw-D, 10 kw-N. TL: N43 01 56 W88 07 54. Stereo. Hrs opn: 720 E. Capitol Dr., 53212. Secondary address: Box 693 53201. Phone: (414) 332-9611. Fax: (414) 967-5378. Web Site: www.620wtmj.com. Licensee: Journal Broadcast Corp. Group owner: Journal Broadcast Group Inc. Population served: 1,820,000 Natl. Network: ABC. Natl. Rep: Christal. Law Firm: Hogan & Hartson. Format: News/talk, sports. Target aud: General. Spec prog: Relg 2 hrs wkly. ♦Doug Kiel, CEO; Jon Schweitzer, gen mgr & chief of engrg; Rick Belcher, opns VP & progmg dir; Jeff Kuether, sls VP; Diana Paul, mktg VP; Dan Shelley, news dir; Randy Price, engrg VP; Susan Petropoullis, traf mgr.

WKTI(FM)—Co-owned with WTMJ(AM). June 1959: 94.5 mhz; 15.5 kw. 911 ft TL: N43 05 29 W87 54 07. Stereo. 24 Web Site: www.wkti.com.305,800 Format: Adult contemp. News staff: one. Target aud: 25-54. ♦Jon Schweitzer, stn mgr; Bob Walker, progmg dir; Lisa Letterman, mktg VP & rsch dir. Co-owned TV: WTMJ-TV affil

***WUWM(FM)**— Sept 24, 1964: 89.7 mhz; 15 kw. 871 ft TL: N43 05 24 W87 53 47. Stereo. Hrs opn: 24 Milwaukee Public Radio, Box 413, 53201. Secondary address: John Plankinton Bldg., 161 W. Wisconsin Ave., Suite LL1000 53202. Phone: (414) 227-3355. Fax: (414) 270-1297. E-mail: wuwm@uwm.edu Web Site: www.wuwm.com. Licensee: Board of Regents of University of Wisconsin. Natl. Network: PRI, NPR. Format: News, AAA. News staff: 12. Target aud: General. ♦Dave Edwards, CEO & gen mgr; Noel Sharproen, dev mgr; Bruce Winter, progmg mgr; Tom May, chief of engrg.

***WVCY-FM**— 1961: 107.7 mhz; 43 kw. Ant 528 ft TL: N42 57 46 W88 04 23. Stereo. Hrs opn: 24 3434 W. Kilbourn Ave., 53208. Phone: (414) 935-3000. Fax: (414) 935-3015. E-mail: wvcyfm@vcyamerica.org Web Site: www.vcyamerica.org. Licensee: VCY/America Inc. (group owner; (acq 1-70). Natl. Network: USA, Moody. Format: Relg, Christian. ♦Dr. Randall Melchert, pres; Victor Eliason, VP & gen mgr; Gordon Morris, opns mgr & news dir; Jim Schneider, progmg dir.

WXSS(FM)—See Wauwatosa

***WYMS(FM)**— Mar 5, 1973: 88.9 mhz; 1.5 kw. Ant 870 ft TL: N43 05 21 W87 53 47. Stereo. Hrs opn: 24 5312 W. Vliet St., 53208. Phone: (414) 475-8890. Fax: (414) 475-8413. E-mail: info@radiomilwaukee.org Web Site: www.radioformilwaukee.org. Licensee: Milwaukee Board of School Directors. Law Firm: Hogan & Hartson. Wire Svc: AP Format: AAA. Target aud: 20-40.

Minocqua

WLKD(AM)— Aug 1, 1978: 1570 khz; 5 kw-D, 500 w-N. TL: N45 49 13 W89 43 27. Stereo. Hrs opn: 24 3616 Hwy. 47 N., Rhinelander, 54501. Phone: (715) 356-9696. Fax: (715) 356-1977. Web Site: www.wmqa.com. Licensee: Raven License Sub. LLC. Group owner: NewRadio Group LLC (acq 10-31-2005; grpsl). Population served: 40,000 Natl. Network: ESPN Radio. Law Firm: Leventhal, Senter & Lerman. Wire Svc: Wheeler News Service Format: All Sports. News staff: one; News: 18 hrs wkly. Target aud: 45 plus. Spec prog: Relg 3 hrs wkly. ♦Casey Kelly, stn mgr; Duff Damos, opns mgr; Michelle Hartzheim, rgnl sls mgr; Mike Ell, prom dir; Mike Wolf, progmg dir & progmg mgr.

WMQA-FM—Co-owned with WLKD(AM). Apr 3, 1975: 95.9 mhz; 25 kw. 289 ft TL: N45 52 14 W89 42 35. Stereo. 24 E-mail: wmqa@n5gnorthwoods.net Web Site: www.wmqa.com. Law Firm: Leventhal, Senter & Lerman. Format: Adult contemp. News staff: one; News: 18 hrs wkly. Target aud: 25-54. ♦Mike Wolf, prom mgr.

Mishicot

WZOR(FM)— Dec 17, 1994: 94.7 mhz; 21.5 kw. Ant 354 ft TL: N44 20 30 W87 47 10. Hrs open: 24 Box 1519, Appleton, 54912. Secondary address: 2727 E. Radio Rd., Appleton 54915. Phone: (920) 734-9226. Fax: (920) 733-2391. E-mail: razor@wcinet.com Web Site: www.razor947.com. Licensee: Woodward Communications Inc. (group owner; acq 1-27-00). Population served: 350000 Natl. Rep: McGavren Guild. Law Firm: Hogan & Hartson. Format: Active rock. Target aud: 18-34; men. ♦Greg Bell, gen mgr; Kelly Radandt, gen sls mgr; Roxanne Steele, progmg dir.

Monroe

WEKZ(AM)— July 27, 1951: 1260 khz; 1 kw-D, 19 w-N. TL: N42 35 41 W89 35 35. Hrs open: 24 W4765 Radio Ln., 53566. Phone: (608) 325-2161. Fax: (608) 325-2164. E-mail: wekz@wekz.com Web Site: www.wekz.com. Licensee: Green County Broadcasting Corp. (acq 4-96; $1,445,000). Population served: 225,000 Natl. Network: ABC. Format: Country classic. News staff: 4; News: 16 hrs wkly. Target aud: 50 plus. Spec prog: Ger 3 hrs, Swiss 3 hrs wkly. ♦Scott Thompson, gen mgr & gen sls mgr; Wyatt Herrmann, progmg dir; Don Jacobson, news dir; Gary Gulalski, pub affrs dir & women's int ed; Todd Hauser, chief of engrg; Chad Colvin, traf mgr; Chuck Polus, news rptr; Don Jacobson Sr., news rptr; Patty Adamson, news rptr; Dan Blum, disc jockey; Glenn Smith, disc jockey.

WEKZ-FM— June 1959: 93.7 mhz; 36 kw. 581 ft TL: N42 34 36 W89 41 34. Stereo. 24 Web Site: www.wekz.com. Licensee: Ronald M. Spielman, Scott A. Thompson dba Green County Broadcasting. (Acq 2-28-96).720,000 Natl. Network: ABC. Format: Adult contemp. News staff: 4; News: 10 hrs wkly. Target aud: 35-55. ♦Scott Thompson, stn mgr & disc jockey; Macey Koehn, traf mgr; Chuck Polus, news rptr; Don Jacobson, news rptr; Patty Adamson, news rptr; Dan Blum, disc jockey; Jim Douglas, disc jockey; Shary Gibson, disc jockey.

Mosinee

WOFM(FM)— Oct 7, 1991: 94.7 mhz; 50 kw. Ant 492 ft TL: N44 59 18 W89 59 42. Stereo. Hrs opn: 24
Simulcast with WIZD(FM) Rudolph 100%.
557 Scott St., Wausau, 54403. Phone: (715) 842-1672. Fax: (715) 848-3158. E-mail: info@947thepeak.com Web Site: www.lovethevalley.com. Licensee: WRIG Inc. Group owner: Midwest Communications Inc. (acq 9-24-97; $35,000 for 70%). Population served: 212,000 Natl. Rep: Christal. Format: Adult hits. News staff: one; News: 2 hrs wkly. Target aud: 25-54; upscale baby boomers. ♦Duke Wright, pres; Gary Tesch, exec VP; Brett Lucht, gen mgr; Jim Schroeder, gen sls mgr; Chad Edwards, progmg dir; Frank Zastrow, engrg VP & chief of engrg; Melanie Comeau, traf mgr.

Mount Horeb

WJQM(FM)— October 2004: 106.7 mhz; 2.9 kw. Ant 479 ft TL: N43 00 19 W89 52 25. Hrs open: Box 44408, Madison, 53744. Secondary address: 730 Rayovac Dr., Madison 53411. Phone: (608) 273-1000. Fax: (608) 273-3588. Web Site: www.madisonjams.com. Licensee: Mid-West Management Inc. (acq 3-20-2003; $2,166,000 for CP). Format: Rhythmic hits. ♦Tom Walker, gen mgr; Ted Waldbillig, sls dir; Mark Maloney, progmg dir.

Mukwonago

WFZH(FM)— 2002: 105.3 mhz; 1.65 kw. Ant 633 ft TL: N42 58 05 W88 11 20. Hrs open: 24 135 S. 84th St., Suite 310, Milwaukee, 53214. Phone: (414) 258-1700. Fax: (414) 266-5353. E-mail: studio@1053thefish.com Web Site: wfzh.salemwebnetwork.com. Licensee: Caron Broadcasting Inc. Group owner: Salem Communications Corp. (acq 10-22-01). Format: Christian. ♦Dave Santrella, gen mgr; Mark Jaycox, gen sls mgr; Danny Clayton, progmg dir; Marie Maltia, traf mgr.

Neenah-Menasha

WNAM(AM)— May 23, 1947: 1280 khz; 5 kw-U. TL: N44 09 36 W88 27 57. Stereo. Hrs opn: 24 491 S. Washburn, Suite 400, Oshkosh, 54904. Phone: (920) 426-3239. Fax: (920) 231-0145. Web Site: www.1280wnam.com. Licensee: Cumulus Broadcasting L.L.C. Group owner: Cumulus Media L.L.C. (acq 6-30-97; grpsl). Population served: 375,000 Rgnl. Network: Goetz Group. Format: Adult Standards. News staff: 3; News: 13 hrs wkly. Target aud: 35 plus. ♦Jeffrey A. Schmidt, gen mgr; Larry Phillip, sls dir.

WNCY-FM— Sept 9, 1977: 100.3 mhz; 45 kw. 489 ft TL: N44 15 27 W88 11 41. Stereo. Hrs opn: 24 Box 23333, Green Bay, 54305. Secondary address: 115 S. Jefferson St., Green Bay 54301. Phone: (920) 435-3771. Fax: (920) 444-1155. Web Site: www.wncy.com. Licensee: Midwest Communications Inc. (group owner; acq 12-10-96; grpsl). Population served: 820,000 Format: Country. ♦D.E. Wright, pres; Jeff McCarthy, VP; Craig Von Able, gen sls mgr; Dan Stone, progmg dir; Jerry Bader, news dir; Tim Laes, engrg dir & chief of engrg.

WROE(FM)— November 1971: 94.3 mhz; 13 kw. 459 ft TL: N44 09 30 W88 17 03. Stereo. Hrs opn: Box 23333, Green Bay, 54305. Phone: (920) 435-3771. Fax: (920) 444-1155. Web Site: www.mci.fm. Licensee: Midwest Communications Inc. (group owner; acq 12-10-96; grpsl). Population served: 27,600 Format: Soft adult contemp. Target aud: 25-54. ♦Duke Wright, CEO & gen mgr; David Fries, gen sls mgr; Jenny Lawrence, progmg dir; Jerry Bader, news dir & traf mgr; Tim Laes, chief of engrg.

WWWX(FM)—See Oshkosh

Neillsville

WCCN(AM)— Sept 22, 1957: 1370 khz; 5 kw-D, 42 w-N. TL: N44 34 18 W90 35 15. Hrs open: 24 Box 387, 1201 E. Division St., 54456. Phone: (715) 743-2222. Phone: (715) 743-3333. Fax: (715) 743-2288. E-mail: 1075therock@tds.net Web Site: memories1370.com. Licensee: Central Wisconsin Broadcasting Inc. (group owner; acq 12-87; FTR: 3-25-91). Population served: 380,400 Natl. Network: ABC. Rgnl. Network: Wisconsin Radio Net. Law Firm: Miller & Neely P.C. Format: Big Band, nostalgia. News staff: one; News: 20 hrs wkly. Target aud: 45 plus. Spec prog: Farm 19 hrs, Polka 2 hrs wkly. ♦J. Kevin Grap, pres; Margaret L. Grap, VP.

WCCN-FM— July 1964: 107.5 mhz; 100 kw. Ant 577 ft TL: N44 35 30 W90 37 09. Stereo. 24 E-mail: 1075therock@tds.net Web Site: www.1075therock.com.1,000,000 Format: Rock. News staff: one. Target aud: 25-40.

WPKG(FM)— February 2004: 92.7 mhz; 3.4 kw. Ant 440 ft TL: N44 35 30 W90 37 09. Hrs open: 24 Box 387, 54456. Phone: (715) 743-3333. Fax: (715) 743-2288. Web Site: 927wpkg.com. Licensee: Central Wisconsin Broadcasting Inc. (group owner). Natl. Network: ABC. Format: Hot adult contemp. Target aud: 18+; Females. ♦J. Kevin Grap, gen mgr.

Nekoosa

WMMA(FM)— 2002: 93.9 mhz; 18 kw. Ant 367 ft TL: N44 13 23 W89 49 46. Hrs open: 24 Box 1103, Wisconsin Rapids, 54495. Secondary address: 321 Market St., Wisconsin Rapids 54494. Phone: (715) 424-5050. Fax: (715) 424-5656. Web Site: www.relevantradio.com. Licensee: Starboard Media Foundation Inc. Group owner: Relevant Radio (acq 12-20-2001; $2.3 million with WIBU(AM) Wisconsin Dells). Format: Relg. ♦Jack O'Keefe, gen mgr.

WUSP(FM)—Not on air, target date: unknown: Stn currently dark. 105.5 mhz; 6 kw. Ant 279 ft TL: N44 21 45 W90 03 58. Hrs open: 2307 Princess Ann St., Greensboro, NC, 27408. Phone: (336) 286-2087. Licensee: Todd P. Robinson Inc. ♦Todd P. Robinson, pres.

New London

WOZZ(FM)— Oct 6, 1967: 93.5 mhz; 50 kw. 528 ft TL: N44 21 35 W88 42 46. (CP: Ant 492 ft.). Stereo. Hrs opn: 24 1500 N. Casaloma Dr., Suite 307, Appleton, 54913-8220. Phone: (920) 733-4990. Fax: (920) 733-5507. Web Site: www.wozz.com. Licensee: Midwest Communications of Iowa Inc. Group owner: Midwest Communications Inc. (acq 6-30-93; $1.85 million with WZBY(FM) Sturgeon Bay; FTR: 7-26-93). Population served: 837,000 Format: Classic rock. News staff: 2; News: 8 hrs wkly. Target aud: 25-44. ♦David Fries, gen mgr, gen sls mgr & gen sls mgr; David Louis, progmg dir.

New Richmond

WIXK(AM)—Licensed to New Richmond. See Minneapolis-St. Paul MN

Oconto

WOCO(AM)— Mar 11, 1966: 1260 khz; 1 kw-D. TL: N45 53 31 W87 57 18. Hrs open: 3829 Hwy.22, 54153. Phone: (920) 834-3540. Fax: (920) 834-3532. E-mail: wocoamfm@bayland.net Licensee: Lamardo Inc. (acq 3-25-99; with co-located FM). Population served: 15,000 Format: C&W, var. Target aud: 29 plus. ♦Larry Kaszynski, gen sls mgr, adv mgr, mus dir, sports cmtr & disc jockey; Terri Kaszynski, rgnl sls mgr, prom mgr, news dir, spec ev coord, local news ed & disc jockey; Dorothy Kaszynski, progmg dir & women's int ed; Walter P. Kaszynski, pres, gen mgr, chief of engrg & farm dir; Walter Kaszynski, disc jockey.

WOCO-FM— Aug 1, 1968: 107.1 mhz; 3 kw. 210 ft TL: N44 53 31 W87 57 18.25,000 Format: Easy lstng. ♦Terri Kaszynski, spec ev coord, local news ed & disc jockey; Walter P. Kaszynski, farm dir; Larry Kaszynski, sports cmtr & disc jockey; Dorothy Kaszynski, women's int ed; Walter Kaszynski, disc jockey.

Omro

WPKR(FM)—Licensed to Omro. See Oshkosh

Oshkosh

WOSH(AM)— Dec 31, 1941: 1490 khz; 1 kw-U. TL: N44 02 46 W88 31 44. Hrs open: 491 S. Washburn St., Suite 400, 54904. Phone: (920) 426-3239. Fax: (920) 231-0145. Web Site: www.cumulus.com. Licensee: Cumulus Broadcasting Inc. Group owner: Cumulus Media LLC. (acq 9-1-97; grpsl). Population served: 53,221 Rgnl. Network: Goetz Group. Natl. Rep: D & R Radio. Format: News/talk, sports. Target aud: 25 plus. ♦Jeffrey Schmidt, gen mgr; Larry Phillip, gen sls mgr; John Stiloski, prom dir; Bob Burnell, progmg dir; Jonathan Krause, news dir; Steve Griesbach, chief of engrg; Alexandra Marohn, traf mgr.

WVBO(FM)—Co-owned with WOSH(AM). Sept 1, 1966: 103.9 mhz; 25 kw. 318 ft TL: N44 02 47 W88 31 44. Stereo.
Rebroadcasts WOGB(FM) Kaukauna 80%.
Web Site: www.1039vbo.com.325,000 Natl. Network: Westwood One. Format: Oldies. Target aud: 35-54. ♦Brian Roberts, progmg dir; Alexandra Marohn, traf mgr.

WPKR(FM)—Omro, July 12, 1990: 99.5 mhz; 50 kw. 420 ft TL: N43 50 51 W88 51 31. Stereo. Hrs opn: 24 491 S. Washburn St., Suite 400, 54904. Phone: (920) 426-3239. Fax: (920) 231-0145. Web Site: wpkr.com. Licensee: Cumulus Licensing LLC. Group owner: Cumulus Media Inc. (acq 11-10-2003; $8.1 million with WPCK(FM) Denmark). Format: Country. News staff: one; News: 2 hrs wkly. Target aud: 25-54. Spec prog: Farm one hr wkly. ♦Jeff Schmidt, gen mgr.

***WRST-FM**— Apr 20, 1966: 90.3 mhz; 960 w. 125 ft TL: N44 01 45 W88 33 08. Stereo. Hrs opn: 800 Algoma Blvd., 54901. Phone: (920) 424-3113. Phone: (920) 424-1234. Fax: (920) 424-1279. Licensee: Board of Regents, University of Wisconsin System. Population served: 60,000 Natl. Network: NPR. Rgnl. Network: Wis. Pub. Wire Svc: Wheeler News Service Format: Div. Target aud: General. ♦Ben Jarman, gen mgr; Kelly Bougneit, stn mgr.

***WVCY(AM)**— July 1, 1969: 690 khz; 250 w-D, 77 w-N, DA-2. TL: N44 04 51 W88 33 53. Hrs open: 24 3434 W. Kilbourn Ave., Milwaukee, 53208. Phone: (414) 935-3000. Fax: (414) 935-3015. E-mail: wvcyam@vcyamerica.org Web Site: www.vcyamerica.org. Licensee: VCY/America Inc. (group owner; acq 1-19-95). Population served: 500,000 Natl. Network: USA. Format: Relg. ♦Dr. Randall Melchert, pres; Vic Eliason, VP & gen mgr.

WWWX(FM)— Jan 30, 1967: 96.9 mhz; 6 kw. 328 ft TL: N44 03 51 W88 31 44. Hrs open: 491 S. Washburn St., Suite 400, 54904. Phone: (920) 426-3239. Fax: (920) 231-0145. Web Site: www.fox969.com. Licensee: Cumulus Broadcasting Inc. Group owner: Cumulus Media Inc. (acq 6-30-97; grpsl). Population served: 325,000 Natl. Network: ABC. Rgnl. Network: Goetz Group. Wire Svc: UPI Format: Rock. Target aud: 25-54. ♦Jeff Schmidt, gen mgr.

Park Falls

WCQM(FM)—Listing follows WNBI(AM).

***WHBM-FM**— Nov 11, 1988: 90.3 mhz; 17.5 kw. 727 ft TL: N45 56 43 W90 16 28. Hrs open: 24
Rebroadcasts WHA (AM) Madison 100%.
518 S. 7th Ave., Wausau, 54401. Phone: (715) 261-6298. Fax: (715) 848-2890. E-mail: reyer@wpr.org Web Site: www.wpr.org. Licensee: State of Wisconsin Educational Communications Board. Rgnl. Network: Wis. Pub. Law Firm: Dow, Lohnes & Albertson. Format: News/talk. News staff: 9. Target aud: 35-54; Skews Female: Issue oriented talk-variety of perspectives. ♦Wendy Wink, CEO; Ted Tobie, CFO; Phil Corriveau, gen mgr; Tom Martin-Erickson, opns dir & opns mgr; Rick Reyer, dev dir; Allen Rieland, engrg dir & chief of engrg.

WNBI(AM)— 1953: 980 khz; 1 kw-D, 105 w-N. TL: N45 55 04 W90 26 58. Hrs open: 24 Box 309, 54552. Secondary address: Hwy. 13 S. 54552. Phone: (715) 762-3221. Fax: (715) 762-2358. E-mail: wnbi@pctcnet.net Licensee: Heartland Comunications License LLC. (group owner; (acq 7-30-2002; $850,000 with co-located FM). Population served: 18,000 Natl. Network: ABC. Format: Sports. News staff: 2; News: 12 hrs wkly. Target aud: 25-54; adults with disposable income. Spec prog: Relg one hr, loc community talk 5 hrs wkly. ♦James Gregori, gen mgr; Joel Karnick, opns dir, progmg dir & news dir; Darla Isham, gen sls mgr; Arthur Dunham, chief of engrg; Kirk Knoll, traf mgr.

WCQM(FM)—Co-owned with WNBI(AM). Apr 13, 1968: 98.3 mhz; 57 kw. 233 ft TL: N45 55 04 W90 26 58.24 Web Site: www.wcqm.com.30,000 Natl. Network: ABC. Format: Country. News staff: one; News: 6 hrs wkly. Target aud: 20-70. ♦Kirk Knoll, traf mgr.

Peshtigo

WSFQ(FM)— Aug 5, 1996: 96.3 mhz; 49 kw. 482 ft TL: N45 07 19 W87 51 07. Stereo. Hrs opn: 24 N. 2880 Roosevelt Rd., Marinette, 54143. Phone: (715) 735-6631. Fax: (715) 732-0125. Web Site: www.badgerbbayarearadio.com. Licensee: Armada Media - Menominee Inc. (group owner; (acq 12-19-2006; grpsl). Natl. Network: ABC. Rgnl. Network: Wisconsin Radio Net. Format: Oldies. News staff: one. Target aud: 49; male. ♦Jim Coursolle, pres; Jim Medley, opns dir; Mike Wolfe, progmg dir.

Platteville

WPVL(AM)— Feb 22, 1955: 1590 khz; 1 kw-D, 500 w-N, DA-N. TL: N42 44 46 W90 28 28. Hrs open: 24 51 Means Dr., 53818. Phone: (608) 349-2000. Fax: (608) 349-2002. Web Site: www.wpvl.com. Licensee: QueenB Radio Wisconsin Inc. Group owner: Morgan Murphy Stations (acq 3-18-98; $825,000 with co-located FM). Population served: 50,000 Natl. Network: ABC, ESPN Radio. Law Firm: Robert Olender. Format: All Sports. News staff: 2; News: 10 hrs wkly. Target aud: 35 plus. Spec prog: Farm 12 hrs, sports 15 hrs wkly. ♦Dan Sullivan, gen mgr.

WPVL-FM— Sept 1, 1966: 107.1 mhz; 4.1 kw. 235 ft TL: N42 44 45 W90 38 27. Stereo. 24 Format: Oldies. News staff: one; News: 10 hrs wkly. Target aud: 30 plus; general adults. ♦Rick Samson, stn mgr.

***WSSW(FM)**— Feb 1, 2007: 89.1 mhz; 60 w. Ant 561 ft TL: N42 45 50.7 W90 24 19.7. Hrs open:
Rebroadcasts WERN(FM) Madison 100%.
Vilas Communications Hall, 821 University Ave., Madison, 53706-1496. Phone: (608) 263-3970. Fax: (608) 263-9763. Web Site: www.wpr.org. Licensee: State of Wisconsin-Educational Communications Board. Natl. Network: NPR. Format: News, classical. ♦Phil Corniveau, gen mgr.

***WSUP(FM)**— Feb 25, 1964: 90.5 mhz; 1 kw. 146 ft TL: N42 43 57 W90 29 09. Stereo. Hrs opn: 20 One Univ. Plaza, 42 Pioneer Tower, 53818. Phone: (608) 342-1165. Phone: (608) 342-1291. Fax: (608) 342-1290. E-mail: wsup@uwplatt.edu Web Site: ums.www.uwplatt.edu/~wsup/. Licensee: Board of Regents, University of Wisconsin System. Law Firm: Dow, Lohnes & Albertson. Format: AOR. News: 10 hrs wkly. Target aud: 18-24; college-age. Spec prog: Class 4 hrs, jazz 3 hrs, alternative 6 hrs, metal 6 hrs, dance 4 hrs wkly. ♦George E. Smith, gen mgr; Laura Lohfink, stn mgr.

Plymouth

WJUB(AM)— April 1954: 1420 khz; 500 w-D, 62 w-N. TL: N43 44 33 W87 56 21. Hrs open: 24 Box 259, N. 5569 State Hwy. 57, 53073-0259. Phone: (920) 893-2661. Phone: (920) 467-4891. Fax: (920) 892-2706. E-mail: 1420amthebreeze@jmiradio.org Web Site: www.1420thebreeze.com. Licensee: Jubilation Ministries Inc. (acq 12-17-90; $185,000; FTR: 1-7-91). Population served: 62,620 Natl. Network: USA. Format: Adult Standards. News: 11 hrs wkly. Target aud: 25-54. Spec prog: Farm 5 hrs wkly. ♦Gerry Krebsbach, pres; William Horsch, gen mgr; David Hendrickson, progmg dir.

WXER(FM)— Oct. 3, 2000: 104.5 mhz; 6 kw. 328 ft TL: N43 43 32 W88 03 07. Stereo. Hrs opn: 24 2100 Washington Ave., Sheboygan Falls, 53081. Phone: (920) 458-2107. Fax: (920) 467-4300. E-mail: wxer@excel.net Web Site: www.1045thepoint.com. Licensee: Midwest Communications Inc. Group owner: Mountain Dog Media (acq 11-1-2005; $2.3 million). Population served: 100,600 Format: Adult contemp. Target aud: 29-54. Spec prog: Ger 3 hrs wkly. ♦Duke E. Wright, pres; Randall B. Hopper, stn mgr; Steve Schouten, opns mgr & gen sls mgr; Patrick Pendergrast, mktg mgr; Dave Riley, progmg dir; Stewart Muck, chief of engrg; Karen Branch, traf mgr.

Port Washington

WPJP(FM)— October 1969: Stn currently dark. 100.1 mhz; 6 kw. Ant 318 ft TL: N43 25 14 W87 59 40. Stereo. Hrs opn: 24 Box 10707, Green Bay, 54307. Phone: (920) 469-3021. Fax: (262) 784-2149. Licensee: Starboard Media Foundation Inc. Group owner: Relevant Radio (acq 5-15-2003; $900,000). Population served: 600,000 ♦Mark Follett, CEO & pres; Neil Robbins, stn mgr.

Portage

WBKY(FM)— 1998: 95.9 mhz; 5.4 kw. Ant 321 ft TL: N43 38 17 W89 34 16. Hrs open: 24 Box 360, 53901. Secondary address: 1420 E. Wisconsin St. 53901. Phone: (608) 635-7341. Fax: (608) 635-7343. Web Site: www.buckycountry959.com. Licensee: Magnum Communications Inc. Natl. Network: AP Radio. Law Firm: Fisher, Wayland, Cooper, Leader & Zaragoza. Format: Country. News staff: 2; News: 7 hrs wkly. ♦Dave Magnum, gen mgr; Rick Jensen, opns mgr; Doug Steele, gen sls mgr; Steve Paterson, progmg dir; Pete Holliday, news dir; Jon Zecherle, chief of engrg; Deb Doyle, traf mgr.

WDDC(FM)—Listing follows WPDR(AM).

WPDR(AM)— July 31, 1952: 1350 khz; 1 kw-D, 41 w-N. TL: N43 31 40 W89 25 52. Hrs open: 24 Box 448, 53901. Secondary address: N6912 Hwy. 51 53901. Phone: (608) 742-1001. Fax: (608) 742-1688. E-mail: wpdr@jvlnet.com Licensee: Zoe Communications Inc. (group owner; acq 2003; $1.1 million with co-located FM). Population served: 40,000 Format: News/talk, adult contemp. News staff: 2; News: 14 hrs wkly. Target aud: 35 plus. Spec prog: Farm 6 hrs wkly. ♦Mike Oberg, pres, gen mgr & chief of engrg; Lyric Klaske, gen mgr & traf mgr; Wendy Oberg, pres, exec VP & gen mgr; Robert Hoffer, sls dir; Susann Gamble, progmg dir & news dir.

WDDC(FM)—Co-owned with WPDR(AM). Nov 8, 1966: 100.1 mhz; 3.3 kw. 300 ft TL: N43 31 40 W89 25 52. (CP: 1.84 kw. TL: N43 31 42 W89 26 01). Stereo. 24 Format: Country. News staff: 2. Target aud: 25-45; office workers, young adults. ♦Kevin Todryk, progmg dir.

Poynette

WHFA(AM)— July 1925: 1240 khz; 1 kw-U. TL: N43 21 38 W89 24 08. Hrs open: 24 1496 Bellevue St., Suite 202, Green Bay, 54311. Phone: (920) 469-3021. Fax: (608) 833-7117. E-mail: whfa@relevantradio.com Web Site: www.relevantradio.com. Licensee: Starboard Media Foundation Inc. Group owner: Relevant Radio (acq 6-28-2001; $1 million). Population served: 275,000 Natl. Network: ABC. Rgnl. Network: Tribune, Wisconsin Radio Net. Format: Catholic. News: 3 hrs wkly. Target aud: 35-64. ♦Martin Jury, opns mgr & dev dir.

Prairie du Chien

WPRE(AM)— Dec 11, 1952: 980 khz; 1 kw-D. TL: N43 03 39 W91 09 26. Hrs open: 24 Box 90, 53821. Secondary address: 640 North Villa Louis Rd. 53821. Phone: (608) 326-2411. Fax: (608) 326-2412. E-mail: wqpcwpre@mwt.net Web Site: www.wpreradio.com. Licensee: Robinson Corp. Group owner: Robinson Corporation (acq 1-7-98; with co-located FM). Population served: 89,164 Natl. Network: Westwood One. Format: Oldies. News staff: one. ♦David Robinson, pres & gen mgr; Jeff Robinson, opns mgr.

WQPC(FM)—Co-owned with WPRE(AM). 1968: 94.3 mhz; 36 kw. 525 ft TL: N43 03 35 W91 06 02. Stereo. 24 Web Site: www.wqpcradio.com.140,718 Format: Country.

Racine

WEZY(FM)—Listing follows WRJN(AM).

WJTI(AM)— June 4, 1950: 1460 khz; 500 w-D, 65 w-N. TL: N42 45 06 W87 49 55. Hrs open: 24 1530 N Cass St., Suite A, Milwaukee, 53202. Phone: (414) 899-9902. Licensee: El Sol Broadcasting LLC (acq 1-30-2007; $467,500). Population served: 1,300,000 Format: Sp. News staff: 3. Target aud: All ages; Spanish. ♦John Torres, gen mgr.

WKKV-FM— August 1948: 100.7 mhz; 50 kw. 500 ft TL: N42 48 18 W88 02 54. Stereo. Hrs opn: 24 12100 W. Howard Ave., Greenfield, 53228. Phone: (414) 321-1007. Fax: (414) 327-3200. Web Site: www.v100.com. Licensee: Clear Channel Radio Licenses Inc. Group owner: Clear Channel Communications Inc. (acq 1996; grpsl). Natl. Network: Premiere Radio Networks, Superadio. Natl. Rep: Clear Channel. Format: Urban rhythm and blues. Target aud: 18-44; general. Spec prog: Gospel 6 hrs wkly. ♦L. Lowry Mays, CEO; Mark Mays, pres; Cindy McDowell, gen mgr; Kerry Wolfe, opns mgr; Randy Wanek, sls dir; Sean John, prom dir; Bailey Coleman, progmg dir & progmg mgr; Terry James, news dir; Al Hajny, chief of engrg; Karen Mason, traf mgr; Doug Banks, disc jockey; Gary Young, disc jockey; Karl Morrow, disc jockey; Reggie Brown, disc jockey.

WRJN(AM)— December 1926: 1400 khz; 1 kw-U. TL: N42 42 39 W87 49 48. Hrs open: 24 4201 Victory Ave., 53405. Phone: (262) 634-3311. Fax: (262) 634-6515. E-mail: wrjn@wi.net Licensee: Racine Broadcasting L.L.C. Group owner: Bliss Communications Inc. (acq 7-11-97; $5 million with co-located FM). Population served: 47,000 Natl. Rep: Christal. Format: News/talk. News staff: 2; News: 40 hrs wkly. Target aud: 35 plus. Spec prog: Class 2 hrs, It 2.5, Serbian 2 hrs wkly. ♦Skip Bliss, pres; Rob Lisser, CFO; Bob Dailey, exec VP; Tim Etes, VP & gen mgr; Ron Richards, opns dir; Leo Edelstein, gen sls mgr & adv mgr; Lew Turner, prom dir; Don Rosen, progmg dir; Tom Karkow, news dir; Bill Lawrence, pub affrs dir; Bob Gorjance, chief of engrg; Curt Vollman, traf mgr.

WEZY(FM)—Co-owned with WRJN(AM). Aug 6, 1962: 92.1 mhz; 6 kw. 275 ft TL: N42 40 55 W87 50 59. (CP: 2.7 kw, ant 495 ft.). Stereo. 24 53,000 Format: Easy Listening. News staff: 2; News: 4 hrs wkly. Target aud: 25-54. ♦Don Rosen, mus dir; Curt Vollman, traf mgr.

Reedsburg

WBDL(FM)— 1997: 102.9 mhz; 3.6 kw. 423 ft TL: N43 35 32 W90 00 42. Hrs open: Box 349, 53959. Phone: (608) 356-3661. Fax: (608) 356-3561. E-mail: wbdl@baraboo.com Licensee: NRG License Sub. LLC. (group owner; (acq 10-31-2005; grpsl). Natl. Network: ABC. Format: Adult contemp. ♦Tommy Bychinski, gen mgr & gen sls mgr; Kevin Kellogg, progmg dir; David Stoeger, news dir.

WNFM(FM)—Listing follows WRDB(AM).

WRDB(AM)— Feb 6, 1953: 1400 khz; 1 kw-U. TL: N43 32 30 W90 02 05. Hrs open: 24 Box 349 E., 53959. Phone: (608) 524-1400. Phone: (608) 524-1049. Fax: (608) 524-2474. E-mail: saukbroad@mwt.net Licensee: NRG License Sub. LLC. (group owner; (acq 10-31-2005; grpsl). Population served: 5,038 Law Firm: Miller & Fields, P.C. Format: Oldies, farm, news. News staff: one; News: 14 hrs wkly. Target aud: 25-54. ♦Tommy Lee Bychinski, gen mgr; Amber Selje, traf mgr.

WNFM(FM)—Co-owned with WRDB(AM). July 16, 1967: 104.9 mhz; 1.6 kw. 449 ft TL: N43 32 30 W90 02 05. (CP: 3.2 kw). Stereo. 24 Format: C&W. News staff: one; News: 10 hrs wkly. Target aud: 25 plus.

Reserve

***WOJB(FM)**— Apr 1, 1982: 88.9 mhz; 100 kw. 604 ft TL: N45 52 16 W91 20 56. Stereo. Hrs opn: 13386 W. Trepania Rd., Hayward, 54843. Phone: (715) 634-2100. Fax: (715) 634-4070. E-mail: generalmanager@wojb.org Web Site: www.wojb.org. Licensee: Lac Courte Oreilles Ojibwe Public Broadcasting Corp. Natl. Network: NPR, PRI. Format: Div. Spec prog: Indian 15 hrs, country 15 hrs, jazz 10 hrs, bluegrass 2 hrs wkly. ♦Carolyn Nayquonabe, gen mgr.

Rhinelander

WHDG(FM)— Sept 1, 1994: 97.5 mhz; 100 kw. 551 ft TL: N45 22 50 W89 11 22. Stereo. Hrs opn: 24 3616 Hwy. 47 N., 54501. Phone: (715) 362-1975. Fax: (715) 362-1973. E-mail: whdg@whdg.com Web Site: www.whdg.com. Licensee: Raven License Sub. LLC. Group owner: NewRadio Group LLC (acq 10-31-2005; grpsl). Law Firm: Leventhal, Senter & Lerman. Wire Svc: Wheeler News Service Format: Country. News staff: one; News: 10 hrs wkly. Target aud: 25-54. ♦Casey Kelly, gen mgr; Duff Damos, gen mgr & opns dir; Bill Mitchell, progmg dir; John Burton, news dir; Al Johnson, chief of engrg; Mark Johnson, traf mgr; Michele Kreuger, gen sls mgr & disc jockey.

WOBT(AM)— Mar 9, 1947: 1240 khz; 950 w-D, 1 kw-N. TL: N45 37 42 W89 23 38. Hrs open: 24 3616 Hwy. 47 N., 54501. Phone: (715) 362-6140. Fax: (715) 362-1973. Licensee: NRG License Sub. LLC. (group owner; (acq 10-31-2005; grpsl). Population served: 25,000 Law Firm: Fisher, Wayland, Cooper, Leader & Zaragoza L.L.P. Format: ESPN sports. News staff: one; News: 10 hrs wkly. Target aud: 25-55. ♦Maxy Quass, pres; Lindshy Wood Dhuis, gen mgr; Michele Krueger, gen sls mgr & adv mgr; Duff Damos, progmg dir; John Burton, news dir; Al Johnson, chief of engrg; Mark Johnson, traf mgr; Mary Spartz, traf mgr.

WRHN(FM)—Co-owned with WOBT(AM). Jan 26, 1966: 100.1 mhz; 25 kw. 385 ft TL: N45 38 08 W89 22 42. (CP: 100 kw, ant 981 ft. TL: N45 24 03 W89 28 54). Stereo. 24 Web Site: www.wrhnfm.com.75,000 Format: Contemp hits. Target aud: 30-55. ♦Mark Johnson, traf mgr; Mary Spatz, traf mgr.

***WXPR(FM)**— Apr 24, 1983: 91.7 mhz; 100 kw. 403 ft TL: N45 46 28 W89 14 54. Stereo. Hrs opn: 5 AM-midnight 303 W. Prospect St.,

54501. Phone: (715) 362-6000. Fax: (715) 362-6007. E-mail: wxpr@wxpr.org Web Site: www.wxpr.org. Licensee: White Pine Community Broadcasting Inc. Population served: 70,000 Natl. Network: NPR. Format: Div, class, folk. Spec prog: Jazz 8 hrs wkly. ♦Mick Fiocchi, pres & gen mgr; Walt Gander, opns mgr & opns mgr; Jessie Dick, dev dir; Ken Krall, news dir; Elmer Goetsch, chief of engrg.

Rice Lake

WAQE(AM)— Aug 6, 1979: 1090 khz; 5 kw-D. TL: N45 32 16 W91 45 50. Stereo. Hrs opn: Box 703, 1859 21st Ave., 54868. Phone: (715) 234-9059. Fax: (715) 234-6942. E-mail: info@wage.com Web Site: www.waqe.com. Licensee: TKC Inc. Group owner: Koser Radio Group (acq 1999). Population served: 187,000 Law Firm: Shaw Pittman. Format: Classic country. News staff: one; News: 5 hrs wkly. Target aud: 24-59; traditional country mus listeners. ♦Brian Schultz, gen mgr, stn mgr & sls dir; Tom Koser, gen mgr; Dane Jensen, sls VP; Mike Bigner, progmg dir; Mike Murrey, chief of engrg; John Roberts, traf mgr.

WKFX(FM)—Co-owned with WAQE(AM). Nov 20, 1980: 99.1 mhz; 44 kw. 522 ft TL: N45 22 23 W91 55 22. (CP: 44 kw, ant 522 ft. TL: N45 22 23 W91 55 22). 24 Box 352, 54868. E-mail: info@fox99.com Web Site: www.fox99.com. Format: Classic hits. News staff: one; News: 3 hrs wkly. Target aud: 18-49; young, upscale adults. ♦Peter Neuser, stn mgr.

WJMC(AM)— 1938: 1240 khz; 1 kw-U. TL: N45 30 27 W91 46 28. Hrs open: 24 Box 703, 1859 21st Ave., 54868. Phone: (715) 234-2131. Fax: (715) 234-6942. E-mail: info@wjmc.com Web Site: www.wjmcradio.com. Licensee: TKC Inc. Group owner: Koser Radio Group (acq 1-1-89). Population served: 35,000 Rgnl. Network: Wisconsin Radio Net. Law Firm: Fisher, Wayland, Cooper, Leader & Zaragoza L.L.P. Format: Farm, news/talk, adult contemp. Target aud: 25-54. ♦Thomas A. Koser, pres & gen mgr; Dane Jensen, stn mgr & gen sls mgr; Mike Bigner, progmg dir; Ken DeNucci, news dir; Mike Murrey, chief of engrg; John Roberts, traf mgr & disc jockey.

WJMC-FM— 1947: 96.1 mhz; 50 kw. Ant 482 ft TL: N45 37 14 W91 44 44. Stereo. 24 Web Site: www.wjmcradio.com.125,000 Format: Hot country. Target aud: 25-54. ♦Dane Jensen, stn mgr; Ken DeNucci, news rptr; Don Tobias, sports cmtr; Mike Bigner, disc jockey.

Richland Center

WRCO(AM)— Oct 18, 1949: 1450 khz; 1 kw-U. TL: N43 18 58 W90 22 31. Hrs open: 24 Box 529, 2111 Bohmann Dr., 53581-0529. Phone: (608) 647-2111. Fax: (608) 647-8025. E-mail: wrco@wrco.com Web Site: www.wrco.com. Licensee: Fruit Broadcasting LLC. (acq 1994). Population served: 100,000 Natl. Network: CBS, Westwood One. Rgnl. Network: Wisconsin Radio Net. Wire Svc: AP Wire Svc: Wheeler News Service Format: Adult Standards. News staff: one; News: 20 hrs wkly. Target aud: 25-54; general. Spec prog: Farm 2.hrs wkly. ♦Ron Fruit, pres, gen mgr & opns mgr; Alice Schulte, gen sls mgr; Phil Nee, progmg dir; Aaron Joyce, news dir; Dennis Baldridge, chief of engrg; Amy Cook, traf mgr.

WRCO-FM— August 1965: 100.9 mhz; 6 kw. 240 ft TL: N43 20 14 W90 22 44. Stereo. 24 Web Site: www.wrco.com.256,000 Natl. Network: CBS Radio. Format: Country, news. News staff: 2; News: 45 hrs wkly. Target aud: General; adult. Spec prog: Farm 18 hrs, Gospel 6 hrs wkly. ♦Alice Schulte, sls dir; Phil Nee, progmg dir; Ron Fruit, pub affrs dir; Dennis Baldridge, chief of engrg; Adam Hess, disc jockey; Ray Schroeder, disc jockey; Tammy Dotson, disc jockey.

Ripon

WRPN(AM)—Licensed to Ripon. See Fond du Lac

***WRPN-FM**— Sept 15, 1957: 90.1 mhz; 231 w. 110 ft TL: N43 50 37 W88 50 31. Stereo. Hrs opn: 24 Box 248, Harwood Memorial Union, 300 Seward St., 54971-0248. Phone: (920) 748-8147. Phone: (920) 748-8115 (college). Fax: (920) 748-7243. E-mail: wrpnfm@yahoo.com Web Site: www.homestead.com. Licensee: Board of Trustees of Ripon College. Natl. Network: CBS, ABC. Natl. Rep: Farmakis. Law Firm: Haley, Bader & Potts. Wire Svc: Wheeler News Service Format: Classic rock, div, progsv. News staff: 4; News: 25 hrs wkly. Spec prog: Pol 2 hrs, sports 15 hrs wkly. ♦Guy McHendry, gen mgr; Joe Laedtke, opns dir.

WTCX(FM)—Licensed to Ripon. See Fond du Lac

River Falls

WEVR(AM)— Sept 14, 1969: 1550 khz; 1 kw-D. TL: N44 53 19 W92 39 04. Hrs open: 6 AM-sunset 178 Radio Rd., 54022. Phone: (715) 425-1111. Phone: (612) 381-1111. Licensee: Hanten Broadcasting Co. Inc. (acq 6-1-74). Population served: 250,000 Natl. Network: USA. Wire Svc: Wheeler News Service Format: Lite adult contemp, sports. News: 5 hrs wkly. Target aud: General. Spec prog: Farm 18 hrs wkly. ♦Carol Hanten, pres & gen mgr.

WEVR-FM— Sept 30, 1970: 106.3 mhz; 6 kw. 328 ft TL: N44 53 19 W92 39 04. Stereo. 6 AM-11 PM Phone: (715) 381-1111.250,000 Format: News. News: 18 hrs wkly. Target aud: General.

***WRFW(FM)**— Nov 2, 1968: 88.7 mhz; 3 kw. 82 ft TL: N44 53 08 W92 39 20. Stereo. Hrs opn: 24 Univ. of Wisconsin River Falls, 306 North Hall, 410 S. 3rd St., MN, 54022. Phone: (715) 425-3886/3887. E-mail: urfw@uwrf.edu Web Site: www.uwrf.edu/wrfw. Licensee: University of Wisconsin System. Population served: 40,000 Natl. Network: NPR. Rgnl rep: Wisconsin Public Radio Format: Var/div. Spec prog: Farm 10 hrs wkly. ♦Rick Burgsteiner, gen mgr; Nick Hassel, prom dir; Adam Lee, progmg dir; Paul Karklus, mus dir; Tara Sowle, news dir.

Rudolph

WIZD(FM)— Sept 30, 1990: 99.9 mhz; 13 kw. Ant 453 ft TL: N44 20 19 W89 38 55. Stereo. Hrs opn: 24
Simulcast with WOFM(FM) Mosinee 100%.
Box 850, 2460 Plover Rd., Plover, 54467. Phone: (715) 344-6050. Phone: (715) 421-4040. Fax: (715) 341-8070. Web Site: www.lovethevalley.com. Licensee: WRIG Inc. Group owner: Midwest Communications Inc. (acq 1999; $1.4 million). Natl. Network: ABC. Natl. Rep: Christal. Wire Svc: AP Format: Adult hits. News staff: 3; News: 3 hrs wkly. Target aud: 35 plus; upper income. Spec prog: Polka 3 hrs wkly. ♦Duke Wright, pres; Brett Lucht, gen mgr; Dave Weir, gen sls mgr; Bob Jung, progmg dir; Frank Zastrow, chief of engrg; John Flesch, traf mgr.

Sauk City

WMAD(FM)—Licensed to Sauk City. See Madison

Schofield

WRIG(AM)— Aug 1, 1958: 1390 khz; 10 kw-U, DA-2. TL: N44 52 42 W89 38 29. Hrs open: 24 Box 2048, Wausau, 54402-2048. Secondary address: 557 Scott St., Wausau 54403. Phone: (715) 842-1672. Fax: (715) 848-3158. E-mail: ken@bigwrig.com Web Site: www.bigwrig.com. Licensee: WRIG Inc. Group owner: Midwest Communications Inc. Population served: 32,806 Format: Oldies. News staff: 2; News: 2 hrs wkly. ♦D.E. Wright, pres; Gary Tesch, exec VP; Brett Lucht, gen mgr; Samantha Milanowski, gen sls mgr; Ken Clark, progmg dir; Frank Zastrow, chief of engrg; John Flesch, traf mgr.

Seymour

WECB(FM)— May 1998: 104.3 mhz; 5.6 kw. Ant 341 ft TL: N44 31 26 W88 19 56. Hrs open: Box 1519, Appleton, 54912. Phone: (920) 734-9226. Fax: (920) 739-0494. Web Site: www.1043thebreeze.com. Licensee: Woodward Communications Inc. (group owner; acq 6-23-2003; $1.75 million). Population served: 152,069 Law Firm: Hogan & Hartson. Format: Soft adult contemp hits of the 70s & 80s. ♦Greg Bell, gen mgr; Kelly Radandt, gen sls mgr; Dayton Kane, progmg dir; Steve Brown, chief of engrg.

Shawano

WOWN(FM)—Listing follows WTCH(AM).

WTCH(AM)— Sept 3, 1948: 960 khz; 1 kw-U, DA-N. TL: N49 46 47 W88 37 53. Hrs open: 24 1456 E. Green Bay St., 54166. Phone: (715) 524-2194. Fax: (715) 524-9980. Web Site: www.wtcham960.com. Licensee: Results Broadcasting Inc. Group owner: Results Broadcasting (acq 12-23-96; $2,704,670 for 50% of stock with co-located FM). Population served: 85,000 Natl. Network: CBS. Format: Classic country. News staff: one; News: 25 hrs wkly. Target aud: 25-54; northeast Wisconsin adults. Spec prog: Farm 21 hrs wkly. ♦Bruce Grassman, chmn, pres & gen mgr; Trisha Peterson, VP.

WOWN(FM)—Co-owned with WTCH(AM). December 1966: 99.3 mhz; 14 kw. Ant 440 ft TL: N44 45 14 W88 20 01. Stereo. 24 185,000 Natl. Network: ABC. Format: Classic hits. News staff: one; News: 13 hrs wkly. Target aud: 20-45. ♦Bruce Grassman, CEO.

Sheboygan

WBFM(FM)—Listing follows WHBL(AM).

WCLB(AM)— January 1956: 950 khz; 500 w-D, DA. TL: N43 44 33 W87 49 00. Hrs open: 18 1102 Fond Du Lac Ave., Sheboygan Falls, 53085. Phone: (920) 467-9950. Fax: (920) 467-4300. E-mail: wbates@mdogmedia.com Web Site: www.sheboyganespn950.com. Licensee: RBH Enterprises Inc. dba Yellow Dog Broadcasting. Group owner: Mountain Dog Media (acq 6-23-00; $700,000 with WXER(FM) Plymouth). Population served: 109,000 Natl. Network: Westwood One. Law Firm: Holland & Knight. Format: Sports. News staff: 2; News: 8 hrs wkly. Target aud: 35-64. ♦Randal B. Mupper, gen mgr; Steve Schouten, gen sls mgr; Wade Bates, progmg dir; Stu Muck, chief of engrg; Karen Branch, traf mgr.

WHBL(AM)— Jan 1, 1926: 1330 khz; 5 kw-D, 1 kw-N, DA-2. TL: N43 43 14 W87 44 04. Hrs open: 24 Box 27, 53082. Secondary address: 2100 Washington Ave. 53081. Phone: (920) 458-2107. Fax: (920) 458-9775. E-mail: studio@whbl.com Licensee: Midwest Communications Inc. (group owner; acq 8-8-00; grpsl). Population served: 100,000 Natl. Network: ABC. Rgnl. Network: Wisconsin Radio Net. Natl. Rep: Rgnl Reps. Wire Svc: AP Format: News/talk. News staff: 2; News: 24 hrs wkly. Target aud: 35-64. Spec prog: Farm 10 hrs wkly. ♦Patrick Pendergast, gen mgr; Mark Smith, gen sls mgr; Nick Reed, progmg dir; Mike Kinzel, news dir; Tim Laes, chief of engrg.

WBFM(FM)—Co-owned with WHBL(AM). Mar 1, 1977: 93.7 mhz; 6 kw. 253 ft TL: N43 43 12 W87 44 04. Stereo. 920 Washinton Ave., 53081. Web Site: wbfmb93radio.com.75,000 Natl. Network: ABC. Rgnl rep: Rgnl Reps Format: Country. News staff: 2; News: 24 hrs wkly. Target aud: 25-54; adults in Sheboygan county/Northern Milwaukee metro area. ♦Eddie Ybarrna, progmg dir.

***WSHS(FM)**— Nov 19, 1971: 91.7 mhz; 180 w. 82 ft TL: N43 46 37 W87 43 08. Stereo. Hrs opn: 24 1042 School Ave., 53083. Phone: (920) 459-3610. Fax: (920) 803-7612. E-mail: wshs@sheboygan.kiz.wi.us Licensee: Sheboygan Area School District. Population served: 48,484 Natl. Network: NPR. Rgnl rep: Glenn Slatts Format: Adult contemp, AOR. Target aud: 18-45; young adults & teens. Spec prog: Hmong 3 hrs, Sp 3 hrs wkly. ♦Ron Rindfleish, pres; Jon Etter, gen mgr.

Sheboygan Falls

WHBZ(FM)— April 1997: 106.5 mhz; 6 kw. 239 ft TL: N43 43 16 W87 44 03. Hrs open: 2100 Washington Ave., Sheboygan, 53081. Phone: (920) 458-2107. Fax: (920) 458-9775. E-mail: thebuzz@whbzfm.com Web Site: www.WHBZ.fm. Licensee: Midwest Communications Inc. (group owner; acq 8-8-00; grpsl). Format: Rock. News staff: 2; News: 10 hrs wkly. Target aud: 18-49; males. ♦Matt Smith, gen mgr; Jennifer Weber, gen sls mgr & progmg dir; Ron Simonet, progmg dir; Mike Kinzel, news dir; Kevin Zimmerman, traf mgr.

Shell Lake

WCSW(AM)— Dec 30, 1967: 940 khz; 1 kw-D. TL: N45 41 36 W91 57 57. Hrs open: Box 190, 54871. Secondary address: 345 Hwy. 63 S. 54871. Phone: (715) 468-9500. Fax: (715) 468-9505. Licensee: Zoe Communications Inc. (group owner; acq 1-1-00; with co-located FM). Population served: 50,000 Natl. Network: ABC. Format: Talk. Target aud: General. Spec prog: Farm 3 hrs wkly. ♦Wendy Oberg, gen mgr; Loren Miller, gen sls mgr; Mike Oberg, progmg dir, news dir & chief of engrg; Stephanie Butenhoff, traf mgr.

WGMO(FM)—Co-owned with WCSW(AM). December 1974: 95.3 mhz; 7.1 kw. 512 ft TL: N45 40 28 W91 58 52. Stereo. 24 Web Site: www.95wgmo.com. Natl. Network: Westwood One. Format: Classic rock. News staff: one. ♦Donna Nelson, traf mgr.

Siren

WXCX(FM)— 2000: 105.7 mhz; 6 kw. Ant 328 ft TL: N45 52 21 W92 27 39. Stereo. Hrs opn: 24 Box 179, Luck, 54853. Secondary address: 2547 Hwy. 35, Suite 3, Luck 54853. Phone: (715) 472-9569. Fax: (715) 472-6939. Web Site: oldies1057.ws. Licensee: Red Rock Radio Corp. (group owner; (acq 9-1-2006; grpsl). Population served: 150,000 Natl. Network: Jones Radio Networks. Natl. Rep: Midwest Radio. Rgnl rep: MidwestRadio Format: Classic hits. News staff: one. Target aud: 35-64. ♦Ro Grignon, pres; Don Welch, VP; Ron Revere, gen mgr.

Sister Bay

***WHDI(FM)**— 2000: 91.9 mhz; 3.4 kw. Ant 476 ft TL: N45 14 19 W87 05 27. Hrs open: 3319 W. Beltline Hwy., Network HQ, Madison, 53713. Phone: (608) 264-9600. Fax: (608) 264-9664. Web Site: www.ecb.org. Licensee: State of Wisconsin-Educational Communications Board. Law Firm: Dow, Lohnes & Albertson. Format: Talk. ♦Wendy Wink, gen mgr; Phil Corrivean, stn mgr.

***WHND(FM)**— Sept. 16, 1999: 89.7 mhz; 3.4 kw. 538 ft TL: N45 14 19 W87 05 27. Hrs open: 24 2420 Nicolet Dr., Green Bay, 54311. Secondary address: 821 University Ave., Madison 53706. Phone: (920) 465-2444. Fax: (920) 465-2576. Web Site: www.wpr.org. Licensee: State of Wisconsin Educational Communications Board. Population served: 50,000 Natl. Network: NPR, PRI. Law Firm: Dow, Lohnes & Albertson. Format: News, class music. News staff: one; News: 29 hrs wkly. Target aud: 35 plus; socially aware, educated & financially secure. Spec prog: American Indian 2 hrs, blues 2 hrs, folk 2 hrs, jazz 10 hrs wkly. ♦Glen Slaats, gen mgr.

WSBW(FM)— 2007: 105.1 mhz; 3.1 kw. Ant 466 ft TL: N45 14 05 W87 05 27. Hrs open: 24 3030 Park Dr., Suite 3, Sturgeon, 54235. Phone: (920) 746-9430. Fax: (920) 746-9433. Licensee: Nicolet Broadcasting Inc. Format: Oldies. News staff: 2; News: news prgmg 10 hrs wkly. ♦Roger Utnehmer, pres.

Sparta

WCOW-FM— Mar 1, 1960: 97.1 mhz; 100 kw. 587 ft TL: N43 58 06 W90 51 35. Stereo. Hrs opn: 24 113 W. Oak St., 54656-1712. Phone: (608) 269-3307. Phone: (608) 269-3100. Fax: (608) 269-5170. Web Site: www.cow97.com. Licensee: Sparta-Tomah Broadcasting Co. Inc. Population served: 139,300 Law Firm: Miller & Miller. Wire Svc: Wheeler News Service Format: Contemp country. News staff: 3; News: 15 hrs wkly. Target aud: 25 plus; rural & city residents. Spec prog: Green Bay Packers. ♦Gary Michaelson, gen sls mgr; Steve Peterson, gen mgr & gen sls mgr; Jake Preston, progmg dir; Clary Harris, news dir; Shelly Holen, traf mgr.

WKLJ(AM)—Co-owned with WCOW-FM. June 1951: 1290 khz; 5 kw-D, 59 w-N. TL: N43 58 06 W90 51 35.6 AM-6 PM Web Site: espn1290.com. (Acq 1-19-89).30,500 Natl. Network: ESPN Radio. Format: Sports. ♦John Papadopoulis, progmg dir.

Spencer

WOSQ(FM)— Sept 20, 1984: 92.3 mhz; 6 kw. Ant 300 ft TL: N44 48 35 W90 21 51. Stereo. Hrs opn: 24 Box 630, Marshfield, 54449. Secondary address: 1710 N. Central Ave. 54449. Phone: (715) 384-2191. Fax: (715) 387-3588. E-mail: wosq@mmwi.net Licensee: Seehafer Broadcasting Corp. (group owner; (acq 6-1-2006; swap with WDLB(AM) Marshfield and WFHR(AM) Wisconsin Rapids for WBCV(FM) Wausau) Rgnl. Network: Goetz Group. Law Firm: Miller & Miller. Wire Svc: Wheeler News Service Wire Svc: UPI Format: Country. News staff: one; News: 7 hrs wkly. Target aud: 25-54. Spec prog: Loc sports play-by-play, farm. ♦Ben Rosenthal, gen mgr; Jon Albrecht, stn mgr & gen sls mgr; Jay Latsch, opns mgr & progmg dir; Mike Warren, news dir.

Spooner

WPLT(FM)—Not on air, target date: unknown: 106.3 mhz; 6 kw. Ant 279 ft TL: N45 47 40 W91 55 03. Hrs open: 24 Box 190, Shell Lake, 54871. Phone: (715) 468-9500. Fax: (715) 468-9505. E-mail: spots@95gmo.com Web Site: www.zoestations.com. Licensee: Zoe Communications Inc. (group owner; acq 12-6-00; $439,000 for CP). Format: Country. ♦Wendy Oberg, gen mgr; Tasha Hagberg, gen sls mgr; Bo Landry, progmg dir; Mike Oberg, chief of engrg.

Stevens Point

WSPT(AM)— 1948: 1010 khz; 1 kw-D. TL: N44 32 17 W89 35 43. Hrs open: 500 Division St., 54481. Phone: (715) 341-9800. Fax: (715) 341-0000. E-mail: rmuzzy@1010wspt.com Web Site: 1010wspt.com. Licensee: Americus Communications L.L.C. Group owner: Muzzy Broadcasting L.L.C. (acq 1996; $1.2 million with co-located FM). Population served: 23,631 Natl. Network: Fox News Radio. Format: News/talk. Target aud: 25-54; upscale, male-orientated. Spec prog: Pol one hr wkly. ♦Richard L. Muzzy, pres & gen mgr.

WSPT-FM— May 1, 1961: 97.9 mhz; 100 kw. Ant 338 ft TL: N44 32 17 W89 35 43. Stereo. Web Site: 979jackfm.com.210,600 Format: Adult contemp. Target aud: 25-54; adult.

***WWSP(FM)**— Sept 28, 1968: 89.9 mhz; 11.5 kw. 325 ft TL: N44 28 55 W89 40 34. Stereo. Hrs opn: 6 AM-2 AM 105 CAL, UWSP, Reserve St., 54481. Phone: (715) 346-3755. Fax: (715) 346-4012. E-mail: wwsp@uwsp.edu Web Site: www.uwsp.edu/stuorg/wwsp. Licensee: Board of Regents, University of Wisconsin System. Law Firm: Dow, Lohnes & Albertson. Wire Svc: Wheeler News Service Wire Svc: UPI Format: Jazz, progsv. Target aud: College students. Spec prog: Hmong one hr, pub affrs 5 hrs, sports 3 hrs, blues 4 hrs wkly. ♦Mark Tolstedt, gen mgr; Courtney Sikorski, stn mgr; Cynthia Atchison, dev dir.

Sturgeon Bay

WDOR(AM)— Sept 8, 1951: 910 khz; 1 kw-D. TL: N44 49 38 W87 21 27. Hrs open: 6 AM-sunset Box 549, 800 S. 15th Ave., 54235. Phone: (920) 487-2822. Phone: (920) 743-4411. Fax: (920) 743-2334. E-mail: email@wdor.com Web Site: wdor.org. Licensee: Door County Broadcasting Inc. Population served: 50,000 Natl. Network: ABC. Law Firm: Pillsbury, Winthrop, Shaw Pittman. Format: Adult contemp. News staff: one; News: 23 hrs wkly. Target aud: 21-50; general. ♦Edward Allen III, sls dir, pres, gen mgr, gen sls mgr & progmg dir; Dan Allen, mus dir; Roger Levendusky, news dir & local news ed; Steve Konopka, engrg dir; Peggy Pfister, traf mgr; David Allen, relg ed; Chad Michaels, sports cmtr.

WDOR-FM— Dec 12, 1966: 93.9 mhz; 77 kw. 640 ft TL: N44 54 23 W87 22 15. Stereo. 5 AM-midnight Web Site: wdor.com.100,000 Natl. Network: ABC. Law Firm: Pillsbury, Winthrop, Shaw Pittman. Format: Sports. News: 25 hrs wkly. Target aud: 20-40; general. Spec prog: Farm 5 hrs wkly.

***WPFF(FM)**— August 1991: 90.5 mhz; 100 kw. 653 ft TL: N44 54 23 W87 22 15. Stereo. Hrs opn: 24 Box 28, 1715 Michigan St., 54235. Phone: (920) 743-7443. E-mail: wpff@wpff.com Web Site: www.wpff.com. Licensee: Family Educational Broadcasting Corp. Population served: 3,200,000 Natl. Network: USA. Law Firm: Shaw Pittman. Format: Christian, CHR. News: 12 hrs wkly. Target aud: 25-49; baby boomers. ♦Mark Schwarzbauer, gen mgr; Andy King, stn mgr.

***WRGX(FM)**— July 1998: 88.5 mhz; 50 kw. Ant 518 ft TL: N44 54 14 W87 22 13. Hrs open: 24 Box 28, 54235. Secondary address: 1723 Michigan St. 54235. Phone: (920) 743-7443. Fax: (920) 743-7543. E-mail: wrgx@wrgx.com Web Site: www.wrgx.com. Licensee: Family Educational Broadcasting Corp. of Door County Wisconsin. Format: Christian; Rock/AOR. Target aud: 13-35; teens & generation x listeners. ♦Dr. Mark Schwarzbaur, CEO & gen mgr.

WSRG(FM)— Apr 18, 1988: 97.7 mhz; 6 kw. 400 ft TL: N44 54 21 W87 22 15. (CP: Ant 554 ft.). Stereo. Hrs opn: 24 1009 Egg Harbor Rd., Suite 113, 54235. Phone: (920) 743-6677. Fax: (920) 743-9183. Web Site: www.wsrgstar97.com. Licensee: Magnum Broadcasting Inc. (acq 11-30-98; $200,000). Population served: 50,000 Natl. Network: Jones Radio Networks. Law Firm: Leventhal, Senter & Lerman. Format: Adult contemp. News staff: 2. Target aud: 25-54. ♦Dave Magnum, gen sls mgr & news dir; Rick Jensen, opns mgr & progmg dir.

WZBY(FM)— Mar 4, 1982: 99.7 mhz; 46 kw. 512 ft TL: N44 38 08 W87 37 37. Stereo. Hrs open: 24 Box 23333, 115 S. Jefferson St., Green Bay, 54305. Secondary address: 115 S. Jefferson St., Green Bay 54301. Phone: (920) 435-3771. Fax: (920) 444-1155. E-mail: studio@wlyd997.com Web Site: www.wild997.com. Licensee: Midwest Communications Inc. (group owner; (acq 6-18-93; $3.5 million with WOZZ(FM) New London; FTR: 7-5-93). Population served: 867,000 Natl. Rep: Christal. Law Firm: Miller & Miller. Format: Adult Rock. News staff: 4; News: 3 hrs wkly. Target aud: 13-34; young, active, hip. ♦Duke Wright, pres & gen mgr; Jennifer Kalies, gen sls mgr; Jenny Lawrence, progmg dir; Jerry Bader, news dir; Tim Laes, chief of engrg.

Sturtevant

WDDW(FM)— June 18, 1993: 104.7 mhz; 4.2 kw. Ant 338 ft TL: N42 51 20 W87 50 41. Hrs open: 24 8500 Green Bay Rd., Pleasant Prairie, 53158. Phone: (262) 694-7800. Fax: (262) 694-7767. Licensee: Bustos Media Operating LLC. Group owner: NextMedia Group L.L.C. (acq 1-6-2006; $10.2 million). Population served: 920,000 Law Firm: Reddy, Begley & McCormick. Format: Rgnl Mexican. Target aud: 18-49. ♦Kira LaFond, gen mgr; John Perry, opns dir, opns mgr & progmg dir; Jerod Bast, gen sls mgr; Lisa Tyler, news dir; Mark Anthony, pub affrs dir & chief of engrg; Lisa Sladek, rsch dir & traf mgr.

Sun Prairie

***WNWC(AM)**— Jan 12, 1982: 1190 khz; 1 kw-D, DA. TL: N43 09 36 W89 12 41. Hrs open: Sunrise to sunset 5606 Medical Cir., Madison, 53719. Phone: (608) 271-1025. Fax: (608) 271-1150. E-mail: wnwc@nwc.edu Web Site: www.life1025.com. Licensee: Northwestern College. Group owner: Northwestern College & Radio (acq 12-19-96; $250,000). Population served: 697,500 Format: Relg, talk. ♦Greg Walters, gen mgr; Brian Christopher, progmg dir.

WXXM(FM)— Apr 12, 1972: 92.1 mhz; 3.7 kw. Ant 410 ft TL: N43 10 10 W89 15 38. Stereo. Hrs opn: 2651 S. Fish Hatchery Rd., Madison, 53711. Phone: (608) 274-5450. Fax: (608) 274-5521. Web Site: www.themic921.com. Licensee: Capstar TX L.P. Group owner: Clear Channel Communications Inc. (acq 8-30-2000; grpsl). Population served: 697,500 Natl. Network: Fox Sports. Format: Progsv talk. ♦Jeff Tyler, gen mgr; Sue Garret, gen sls mgr; Tim Scott, opns mgr & gen sls mgr; Ryan Turany, progmg dir; Josh Westscott, news dir; Cliff Groth, chief of engrg; Jacqueline Forney, traf mgr.

Superior

KKCB(FM)—See Duluth, MN

KRBR(FM)—Listing follows WDSM(AM).

KTCO(FM)—See Duluth, MN

***KUWS(FM)**— Jan 31, 1966: 91.3 mhz; 83 kw. 646 ft TL: N46 47 21 W92 06 51. Stereo. Hrs opn: 24 Box 2000, 54880. Phone: (715) 394-8530. Fax: (715) 394-8404. E-mail: jmunson@facstaffuwsuper.edu Web Site: www.kuws.fm. Licensee: Board of Regents, University of Wisconsin System. Population served: 150,000 Natl. Network: NPR. Rgnl. Network: Wis. Pub. Format: Div, news/talk, educ. News staff: one; News: 20 hrs wkly. Target aud: 12 plus; above average income & education. Spec prog: Alternative 16 hrs, Black 4 hrs, jazz 15 hrs, sports 6 hrs wkly. ♦John Munson, stn mgr; Patrick Olsen, opns mgr.

WDSM(AM)— October 1939: 710 khz; 10 kw-D, 5 kw-N, DA-N. TL: N46 39 14 W92 08 51. Hrs open: 24 715 E. Central Entrance, Duluth, MN, 55811. Phone: (218) 722-4321. Fax: (218) 722-5423. Web Site: www.wdsm.am. Licensee: Midwest Communications Inc. (group owner; acq 8-1-01; grpsl). Population served: 100,578 Natl. Network: Westwood One, CBS. Law Firm: Rosenman & Colin. Format: Talk. Target aud: 25-54. ♦Roxanne Charles, gen mgr; Howie Leathers, gen sls mgr; Mark Fleischer, progmg mgr; John Talcott, chief of engrg.

KRBR(FM)—Co-owned with WDSM(AM). Sept 9, 1979: 102.5 mhz; 100 kw. 600 ft TL: N46 47 21 W92 07 09. Stereo. 24 Web Site: www.krbr.com. Format: Modern rock. News: 3 hrs wkly. Target aud: 18-49. ♦Susan Smith, gen sls mgr & mktg mgr; Mark Fleischer, prom mgr; Dave Walter, news dir.

WGEE(AM)— June 18, 1959: 970 khz; 1 kw-D, 27 w-N. TL: N46 43 28 W92 07 11. Hrs open: 24 715 E. Central Entrance, Midwest Communications, Duluth, MN, 55811. Phone: (218) 722-4321. Fax: (218) 722-5423. E-mail: fleisch@mwcradio.com Licensee: Midwest Communications Inc. (group owner; acq 8-1-01; grpsl). Rgnl rep: Hyett/Ramsland. Law Firm: Rosenman & Colin. Format: Radio Disney. News staff: 6. Target aud: 2-12. ♦Duke Wright, CEO & pres; Gary Tesch, exec VP; Mark Fleischer, opns mgr & progmg dir; Susan Nash, gen sls mgr; Dave Walter, news dir; John Talcott, chief of engrg; Carla McColough, traf mgr.

Suring

***WRVM(FM)**— Sept 17, 1967: 102.7 mhz; 100 kw. 980 ft TL: N44 59 50 W88 23 49. Stereo. Hrs opn: 24 Box 212, Hwy. 32 N., 54174. Phone: (920) 842-2839. Fax: (920) 842-2704. E-mail: wrvmfm@wrvm.org Web Site: www.wrvm.org. Licensee: WRVM Inc. (acq 5-15-68). Population served: 1,500,000 Natl. Network: Moody. Law Firm: Kenkel & Associates. Format: Christian. Target aud: General; family stn with

children's programs. ♦Michael A. Cornell, gen mgr; Brian Hay, gen sls mgr; Dennis Jones, progmg dir; Alan Kilgore, chief of engrg.

Sussex

WKSH(AM)— November 2002: 1640 khz; 10 kw-D, 1 kw-N. TL: N43 04 38 W88 11 32. Hrs open: 24 W.223 N.3251 Shady Ln., Pewaukee, 53072. Phone: (262) 695-9500. Fax: (262) 691-2378. E-mail: debra.l.brate@disney.com Web Site: www.radiodisney.com. Licensee: Radio Disney Group LLC. Group owner: ABC Inc. (acq 9-26—02; $2.6 million). Population served: 1,000,000 Natl. Network: Radio Disney. Natl. Rep: Interep. Format: Family hits. Target aud: Kids 6-14 & Women 25-49. ♦Debra Bratel, gen mgr & stn mgr; Megan DeLaat, prom mgr; Patricia Kraby, prom mgr; Cindy Miresse, sls; Melissa Macco, sls.

Three Lakes

WLSL(FM)— August 1994: 93.7 mhz; 100 kw. Ant 407 ft TL: N45 46 30 W89 14 55. Hrs open: 24 38 W. Davenport St., Rhinelander, 54501. Phone: (715) 369-9575. Fax: (715) 369-9475. E-mail: b93@nnex.net Licensee: Results Broadcasting of Rhinelander Inc. Group owner: Results Broadcasting (acq 3-10-2000; $500,000). Natl. Network: Premiere Radio Networks. Format: Classic hits. News staff: one; News: 4. Target aud: 25-54. Spec prog: NASCAR. ♦Bruce Grassman, CEO; Miles Knuteson, gen mgr & gen sls mgr; Bryan Thomas, progmg dir.

Tomah

WBOG(AM)— Apr 19, 1959: 1460 khz; 1 kw-D, 42 w-N. TL: N43 58 07 W90 30 50. Hrs open: 24 1021 N. Superior Ave., Suite 5, 54660. Phone: (608) 372-9600. Fax: (608) 372-7566. E-mail: info@oldies1460.com Web Site: www.magnumradiogroup.net. Licensee: Magnum Radio Inc. (group owner; acq 1994; $275,000 with co-located FM). Natl. Network: Jones Radio Networks. Rgnl. Network: Wisconsin Radio Net. Format: Oldies. News staff: 2; News: 10 hrs wkly. Target aud: 35 plus. ♦Dave Magnum, pres; Brian Winnekins, opns dir, farm dir & sports cmtr; Diane Pergande, sls dir; Steve Peterson, gen mgr, prom dir & progmg mgr; Clary Harris, news dir; Darrell Sanders, chief of engrg.

WTMB(FM)—Co-owned with WBOG(AM). July 11, 1990: 94.5 mhz; 8.3 kw. Ant 564 ft TL: N43 53 56 W90 29 23. Stereo. 24 Phone: (608) 372-9420. Fax: (608) 372-7566. E-mail: info@buzzcountry.com Web Site: www.magnumradiogroup.net. Format: Classic rock. News staff: 2; News: 6 hrs wkly.

***WVCX(FM)**— Jan 29, 1965: 98.9 mhz; 100 kw. 991 ft TL: N43 51 13 W90 27 28. Stereo. Hrs opn: 24 3434 W. Kilbourn Ave., Milwaukee, 53208. Phone: (414) 935-3000. Fax: (414) 935-3015. E-mail: wvcx@vcyamerica.org Web Site: www.vcyamerica.org. Licensee: VCY/America Inc. (group owner; acq 1984). Natl. Network: USA, Moody. Format: Relg, talk, Christian. ♦Dr. Randall Melchert, pres; Victor Eliason, VP & gen mgr; Jim Schneider, progmg dir; Gordon Morris, news dir; Andy Eliason, chief of engrg.

WXYM(FM)— Mar 11, 1992: 96.1 mhz; 44 kw. 525 ft TL: N44 01 32 W90 48 58. Stereo. Hrs opn: 24 1021 N. Superior Ave., Suite 5, 54660. Phone: (608) 372-9600. Fax: (608) 372-7566. E-mail: info@mixwxym.com Web Site: www.magnumradiogroup.net. Licensee: Magnum Radio Inc. (acq 9-27-91; FTR: 10-14-91). Rgnl. Network: Wisconsin Radio Net. Format: Hot AC. News staff: 2; News: 6 hrs wkly. Target aud: 25-54. ♦Dave Magnum, pres; Steve Peterson, gen mgr; Debbie Doyle, opns mgr.

Tomahawk

WJJQ(AM)— August 1968: 810 khz; 980 w-D. TL: N45 29 27 W89 43 36. Hrs open: 24 Box 10, 81 E. Mohawk Dr., 54487. Phone: (715) 453-4482. Phone: (715) 457-4481. Fax: (715) 453-7169. E-mail: wjjq@wjjq.com Web Site: www.wjjq.com. Licensee: Albert Broadcasting II LLC (acq 6-11-84). Population served: 25,000 Natl. Network: CBS, ESPN Radio, Westwood One. Natl. Rep: Rgnl Reps. Law Firm: Shaw Pittman. Wire Svc: AP Format: Sports, talk. News staff: one; News: 15 hrs wkly. Target aud: 25 plus. ♦Gregg Albert, pres & gen mgr; Margaruite Albert, VP; Tim Albert, prom mgr & progmg dir.

WJJQ-FM— Oct 15, 1984: 92.5 mhz; 25 kw. 259 ft TL: N45 29 27 W89 43 36. Stereo. 24 Secondary address: 81 E. Mohawk Drive 54487. Phone: (715) 453-4482. Fax: (715) 453-7169. E-mail: galbert@wjjq.com Web Site: www.wjjq.com.40,000 Natl. Network: CBS Radio. Rgnl rep: Regional Reps Law Firm: Shaw Pittman. Format: Lite hits, news, sports,oldies. News staff: one; News: 25 hrs wkly. Target aud: 25 plus. ♦Mary Lu Voermans, adv mgr & women's cmtr; Mark Everett, news rptr, sports cmtr & disc jockey; Gregg Albert, disc jockey; Mary Lou Voermans, disc jockey; Michael McGovern, disc jockey; Phil Richard, disc jockey; Tim Albert, mus dir, traf mgr & disc jockey.

Trempealeau

WFBZ(FM)— Nov 24, 1984: 105.5 mhz; 2.1 kw. Ant 531 ft TL: N43 56 33 W91 26 03. Stereo. Hrs opn: 24 113 W. Oak St., Sparta, 54656. Phone: (608) 269-3307. Fax: (608) 269-5170. Web Site: 1055thezoo.com. Licensee: Sparta-Tomah Broadcasting Co. Inc. Group owner: La Crosse Radio Group (acq 10-19-2006; $850,000). Natl. Network: CNN Radio. Natl. Rep: Rgnl Reps. Wire Svc: AP Format: Classic rock, classic hits. ♦Zel S. Rice II, pres; Steve Peterson, gen mgr.

Two Rivers

WCUB(AM)—Licensed to Two Rivers. See Manitowoc

WGBW(AM)— Oct 29, 1951: 1590 khz; 1 kw-D, 33 w-N. TL: N44 10 23 W87 35 37. Hrs open: 24 1414 16th St., 54241-3031. Phone: (920) 794-1800. Fax: (920) 793-1111. E-mail: wgbw@lsol.net Licensee: WTRW Inc. Population served: 70,000 Natl. Network: CNN Radio, Westwood One. Law Firm: Thompson Hine LLC. Format: Oldies. News staff: one; News: 10 hrs wkly. Target aud: 25-54; baby boomers & professionals. ♦Mark Heller, pres.

WLTU(FM)—See Manitowoc

WTRW(FM)—Not on air, target date: unknown: 97.1 mhz; 6 kw. Ant 249 ft TL: N44 10 24 W87 33 38. Hrs open: 1717 Dixie Hwy., Suite 650, Fort Wright, KY, 41011. Phone: (859) 331-9100. Licensee: Radioactive LLC. ♦Benjamin L. Homel, pres.

Verona

WMMM-FM— July 4, 1991: 105.5 mhz; 4.4 kw. 384 ft TL: N42 57 42 W89 29 32. Hrs open: 7601 Ganser Way, Madison, 53719. Phone: (608) 826-0077. Fax: (608) 826-1244. Web Site: www.1055triplem.com. Licensee: Entercom Madison Licensee LLP. Group owner: Entercom Communications Corp. (acq 7-10-00; grpsl). Format: AAA. Target aud: 25-44; upscale, college educated adults. ♦David Field, pres; Steve Fisher, CFO; Lindsay Wood Davis, mktg mgr.

Viroqua

WVRQ(AM)— Feb 25, 1958: 1360 khz; 1 kw-D, 23 w-N. TL: N43 32 04 W90 52 23. Hrs open: 24 E7601A County Rd. SS, 54665. Phone: (608) 637-7200. Fax: (608) 637-7299. E-mail: wvrq@mwt.net Web Site: www.wvrq.com. Licensee: Robinson Corp. Group owner: Robinson Corporation. Population served: 43,762 Natl. Network: ABC. Rgnl. Network: Wisconsin Radio Net. Wire Svc: AP Format: Oldies. News staff: one; News: 14 hrs wkly. Target aud: 25 plus. Spec prog: Farm one hr, relg 6 hrs, polka 6 hrs wkly. ♦David Robinson, pres & gen mgr.

WVRQ-FM— Oct 6, 1967: 102.3 mhz; 3.3 kw. 298 ft TL: N43 31 27 W90 51 51. Stereo. 24 Web Site: www.wvrq.com.38,394 Format: Country. News: 20 hrs wkly. Target aud: 25-54. Spec prog: Bluegrass 2 hrs, farm one hr wkly. ♦Jeff Robinson, opns mgr & progmg dir.

Washburn

***WEGZ(FM)**— Oct 5, 1981: 105.9 mhz; 98 kw. Ant 741 ft TL: N46 41 31 W90 59 27. Stereo. Hrs opn: 24 3434 W. Kilbourn Ave., Milwaukee, 53208. Phone: (414) 935-3000. Fax: (414) 935-3015. E-mail: wegz@vcyamerica.org Web Site: www.vcyamerica.org. Licensee: Keweenaw Bay Broadcasting Inc. Group owner: VCY/America Inc. (acq 4-19-2002). Population served: 225,000 Natl. Network: USA. Format: Religious, Christian music, talk. ♦Dr. Randall Melchert, pres; Victor Eliason, VP & gen mgr; Gordon Morris, opns dir & news dir; Jim Schneider, progmg dir.

Watertown

WJJO(FM)— Aug 1, 1961: 94.1 mhz; 50 kw. 492 ft TL: N43 11 43 W88 45 17. Stereo. Hrs opn: 24 Box 44408, Madison, 53744. Secondary address: 730 Rayovac Dr., Madison 53711. Phone: (608) 273-1000. Fax: (608) 271-8182. Web Site: www.wjjo.com. Licensee: Mid-West Management Inc. Group owner: Mid-West Family Stations (acq 6-18-93; $1.6 million; FTR: 7-5-93). Population served: 450,000 Natl. Rep: McGavren Guild. Format: Rock. Target aud: 21-49; male. ♦Ted Waldbillig, sls dir & prom mgr; Tom Walker, pres, gen mgr & gen sls mgr; Randy Hawke, progmg dir; Blake Patton, mus dir & pub affrs dir; John Bauer, chief of engrg.

WTTN(AM)— Apr 2, 1950: 1580 khz; 1 kw-D, 6 w-N. TL: N43 11 43 W88 45 17. (CP: COL Columbus. 5 kw-D, 4 w-N, 800 w-CH, DA-D. TL: N43 20 05 W89 09 56). Hrs opn: 24 Box 509, 615 E. Main St., 53094. Phone: (920) 261-1580. Fax: (920) 261-0624. Licensee: Good Karma Broadcasting L.L.C. (group owner; acq 8-26-99; $525,000). Population served: 100,000 Natl. Network: CNN Radio. Rgnl. Network: Wisconsin Radio Net. Wire Svc: Wheeler News Service Format: Oldies. News staff: 2; News: 15 hrs wkly. Target aud: 25-64. Spec prog: Relg 4 hrs wkly. ♦Craig . Karmazin, CEO & pres; Scott M. Trentadue, gen mgr; Rick Armon, opns mgr; John Moser, gen sls mgr; Warren Jorgensen, chief of engrg.

Waukesha

WAUK(AM)— Mar 27, 1947: 1510 khz; 23 kw-D, 20 kw-CH, DA-2. TL: N43 01 02 W88 11 43. Stereo. Hrs opn: 6 AM-8:30 PM 770 N. Jefferson St., Milwaukee, 53202. Phone: (414) 273-3776. Fax: (414) 291-3776. E-mail: www.wauksportsradio@msn.com Web Site: www.espn1510.com. Licensee: Good Karma Broadcasting L.L.C. (group owner; (acq 4-30-2004; $2 million). Population served: 1,000,000 Natl. Network: ABC, ESPN Radio. Law Firm: Hough & Cook. Wire Svc: The Sports Network Format: Sports/talk. News staff: 2; News: 6 hrs wkly. Target aud: 25 plus; male. Spec prog: NASCAR racing. ♦Craig Karmazin, pres, stn mgr & gen sls mgr; C.J. Knee, opns dir; Bill Johnson, progmg dir; Warren Jorgenson, traf mgr.

***WCCX(FM)**— Sept 1, 1978: 104.5 mhz; 10 w. 50 ft TL: N43 00 16 W88 13 39. Hrs open: 24 100 N. E. Ave., 53186. Phone: (262) 524-7355. Fax: (262) 650-4950. E-mail: wccx@cc.edu Web Site: wccx.cc.edu. Licensee: Trustees Carroll College. Format: Div. Target aud: General; high school & college students. Spec prog: Sp 8 hrs, metal 3 hrs, rap 3 hrs wkly.

WMIL(FM)— Jan 1, 1982: 106.1 mhz; 12 kw. Ant 997 ft TL: N43 05 46 W87 54 15. Stereo. Hrs opn: 24 12100 W. Howard Ave., Greenfield, 53228. Phone: (414) 545-8900. Phone: (414) 545-5920. Fax: (414) 546-8058. Web Site: www.fm106.com. Licensee: Clear Channel Broadcasting Inc. Group owner: Clear Channel Communications Inc. (acq 3-17-97; $40 million with WOKY(AM) Milwaukee). Population served: 1,400,000 Natl. Rep: Clear Channel. Format: Country. News staff: 4; News: 1.5 hrs wkly. Target aud: 25-54; middle America. ♦L. Lowry Mays, CEO & chmn; Mark Mays, pres; Cindy McDowell, gen mgr; Keith Bratel, sls dir; Colleen Kurth, natl sls mgr & pub affrs dir; Jean Kemp, natl sls mgr & rgnl sls mgr; Enid Parkinson, prom dir; Kerry Wolfe, progmg dir; Mitch Morgan, mus dir.

Waunakee

WCHY(FM)— Apr 20, 1992: 105.1 mhz; 6 kw. 328 ft TL: N43 13 20 W89 18 01. Stereo. Hrs opn: 7601 Ganser Way, Madison, 53719. Phone: (608) 826-0077. Fax: (608) 826-1244. Web Site: www.y105.com. Licensee: Entercom Madison Licensee LLP. Group owner: Entercom Communications Corp. (acq 7-10-2000; grpsl). Format: 70's & 80's. Target aud: 18-49. ♦David Field, pres; Steve Fisher, CFO; Ed Schulz, gen mgr.

Waupaca

WDUX(AM)— Apr 29, 1956: 800 khz; 5 kw-D, 500 w-N, DA-1. TL: N44 21 15 W89 03 29. Hrs open: 5 AM-midnight Box 247, 54981.

Secondary address: 200 Tower Rd. 54981. Phone: (715) 258-5528. Fax: (715) 258-7711. E-mail: mail@wdux.net Web Site: www.wdux.net. Licensee: Laird Broadcasting Co. (acq 12-30-99; grpsl). Natl. Network: ABC, Jones Radio Networks. Rgnl. Network: Tribune, Goetz Group. Natl. Rep: Katz Radio. Format: Classic country. News staff: one; News: 20 hrs wkly. Target aud: 35 plus. Spec prog: Farm 6 hrs wkly. ♦William L. Laird, pres; Tina Grenlie, stn mgr, gen sls mgr & prom mgr; Jack Barry, opns mgr, progmg dir, news dir & sports cmtr; Jan Calvey, traf mgr.

WDUX-FM— Jan 29, 1967: 92.7 mhz; 6 kw. 243 ft TL: N44 21 14 W89 03 44. Stereo. 5 AM-midnight Web Site: www.wdux.net. Natl. Network: ABC, CBS Radio, Westwood One. Format: Adult contemp. Target aud: 25-54. Spec prog: Sports 15 hrs wkly. ♦Rick Winters, progmg mgr; Jan Calvey, traf mgr.

Waupun

WFDL(AM)— May 26, 1966: 1170 khz; 1 kw-D. TL: N43 38 30 W88 43 22. Hrs open: Sunrise to sunset 609 Home Ave., 53963. Phone: (920) 324-4441. Fax: (920) 324-3139. E-mail: wmrh1170@yahoo.com Web Site: www.am1170.com. Licensee: Radio Plus Inc. (acq 7-90; $170,000; FTR: 7-2-90). Population served: 250,000 Natl. Network: CBS Radio. Format: News/talk. Target aud: 35 plus; mature adults. Spec prog: Farm 5 hrs wkly. ♦Chris Bernier, pres; Terry Davis, gen mgr & gen sls mgr; Mike Enfelt, opns mgr, news dir & chief of engrg; Todd Dehring, progmg dir; Greg Stensland, news dir; Kerry Longrie, traf mgr.

Wausau

WBCV(FM)— Feb 1, 1985: 107.9 mhz; 100 kw. Ant 1,019 ft TL: N45 03 33 W89 26 10. Stereo. Hrs opn: 2301 Plover Rd., Plover, 55467. Phone: (715) 341-8838. Fax: (715) 341-9744. Licensee: NRG License Sub, LLC. (acq 5-31-2006; swap for WDLB(AM) Marshfield, WOSQ(FM) Spencer and WFHR(AM) Wisconsin Rapids) Population served: 221,000 Format: Rock hits. ♦Ben Rosenthal, gen mgr.

***WCLQ(FM)**— May 23, 1988: 89.5 mhz; 8.5 kw. 328 ft TL: N44 58 58 W89 36 06. Stereo. Hrs opn: 24 4111 Schofield Ave., Suite 10, Schofield, 54476. Phone: (715) 355-5151. Fax: (715) 359-3128. E-mail: 89q@89q,org Web Site: www.89q.org. Licensee: Christian Life Communications Inc. Natl. Network: USA. Format: Contemp hit/Top-40. News: 9 hrs wkly. Target aud: 18-35; Christian/young family. ♦Coy Sawyer, CFO, gen mgr & chief of opns; Scott Michaels, mus dir; Frank Zastrow, chief of engrg.

WDEZ(FM)— Mar 27, 1964: 101.9 mhz; 100 kw. 489 ft TL: N44 58 58 W89 36 06. (CP: 98 kw, ant 1,076 ft. TL: N44 55 14 W89 41 31). Stereo. Hrs opn: 24 Box 2048, 54402-2048. Phone: (715) 842-1672. Fax: (715) 848-3158. E-mail: wdez@wdez.com Web Site: www.wdez.com. Licensee: WRIG Inc. Group owner: Midwest Communications Inc. Population served: 212,000 Natl. Rep: Christal. Wire Svc: Wheeler News Service Wire Svc: UPI Format: Country. News staff: one; News: 4 hrs wkly. Target aud: 25-54. Spec prog: Farm 4 hrs wkly. ♦D.E. Wright, pres; Gary Tesch, VP; Brett Lucht, gen mgr; Dave Weir, sls dir; Melissa Heise, prom dir; Chad Edwards, progmg dir; Vanessa Ryan, progmg dir & mus dir; Chris Conley, news dir; Frank Zastrow, chief of engrg; Mike Austin, farm dir.

***WHRM(FM)**— June 10, 1949: 90.9 mhz; 77kw. TL: N44 55 14 W89 41 31. Stereo. Hrs opn: 24 518 S. 7th Ave., 54401. Phone: (715) 848-1978. Fax: (715) 848-2890. E-mail: reyer@wpr.org Web Site: www.wpr.org. Licensee: State of Wisconsin Educational Communications Board. Natl. Network: NPR, PRI. Law Firm: Dow, Lohnes & Albertson. Format: Class, educ. News: 29 hrs wkly. Target aud: 25-64; male/female. ♦Wendy Wink, CEO; Ted Tobie, CFO; Phil Corriveau, gen mgr; Rick Reyer, stn mgr & dev dir; Maru Nonn, opns dir; Tom Martin-Erickson, opns dir & chief of opns; Allen Rieland, engrg dir & chief of engrg.

WIFC(FM)—Listing follows WSAU(AM).

***WLBL-FM**— November 1995: 91.9 mhz; 560 w. 823 ft TL: N44 55 14 W89 41 31. Hrs open: 5 AM-midnight (M-F); 6 AM-midnight (S, Su) 518 S. 7th Ave., 54401. Phone: (715) 848-1978. Fax: (715) 848-2890. E-mail: reyer@wpr.org Web Site: www.wpr.org. Licensee: State of Wisconsin-Educational Communications Board. Population served: 70,000 Natl. Network: NPR. Rgnl. Network: Wisconsin Radio Net. Format: Call-in talk. Spec prog: Folk 3 hrs wkly. ♦Wendy Wink, CEO; Phil Corriveau, gen mgr; Rick Reyer, gen mgr, stn mgr, dev dir & dev mgr; Allen Rieland, news dir & chief of engrg.

WRIG(AM)—See Schofield

WSAU(AM)— Jan 30, 1937: 550 khz; 5 kw-U, DA-2. TL: N44 51 26 W89 35 13. Hrs open: 24 Box 2048, 54402-2048. Secondary address: 557 Scott St. 54403. Phone: (715) 842-1672. Fax: (715) 848-3158. E-mail: wsau@wsau.com Web Site: www.wsau.com. Licensee: WRIG Inc. Group owner: Midwest Communications Inc. (acq 1996; $3.5 million with co-located FM). Population served: 113,000 Rgnl. Network: Goetz Group. Natl. Rep: Christal. Format: News/talk. News staff: 2. Target aud: 35-64. Spec prog: Farm 5 hrs, Polish 3 hrs, religious 3 hrs wkly. ♦Brett Lucht, gen mgr; Patrick Snyder, opns mgr & progmg dir; Samantha Milanwwski, stn mgr & sls dir; Carrie Van Deraa, prom dir; Frank Zastrow, chief of engrg.

WIFC(FM)—Co-owned with WSAU(AM). 1947: 95.5 mhz; 98 kw. 1,150 ft TL: N44 55 14 W89 41 31. Stereo. 24 E-mail: wifc@wifc.com Web Site: www.wifc.com.411,000 Format: CHR. News staff: 2. Target aud: 18-49. ♦Brett Lucht, sls dir; Chris Pickett, progmg dir.

WXCO(AM)— Aug 1, 1953: 1230 khz; 1 kw-U. TL: N44 58 10 W89 36 25. Hrs open: 24 Box 778, 54402. Secondary address: 1110 E. Wausau Ave. 54403. Phone: (715) 845-8218. Fax: (715) 845-6582. E-mail: wxco@wxco.com Web Site: www.wxco.com. Licensee: Seehafer Broadcasting Corp. Group owner: Badger Communications L.L.C. (acq 9-26-73). Population served: 100,000 Natl. Network: CBS, Westwood One. Format: News. News staff: one; News: 10 hrs wkly. Target aud: 25 plus; professionals/business. ♦Jeff Cecil, opns dir & progmg dir; Ken Rajek, gen mgr & gen sls mgr; Bob Starr, chief of engrg; Dana Condon, traf mgr.

***WXPW(FM)**— February 1996: 91.9 mhz; 560 w. 823 ft (ST WLBL-FM) TL: N44 55 14 W89 41 31. Hrs open:
Rebroadcasts WXPR(FM) Rhinelander.
303 W. Prospect St., Rhinelander, 54501. Phone: (715) 362-6000. Fax: (715) 362-6007. E-mail: wxpr@wxpr.org Web Site: www.wxpr.org. Licensee: White Pine Community Broadcasting Inc. Format: NPR news, great music. ♦Mick Fiocchi, pres & gen mgr; Jessie Dick, dev dir; Walt Gander, opns mgr & progmg dir; Ken Krall, news dir; Elmer Goetsch, chief of engrg.

Wautoma

WAUH(FM)— 2001: 102.3 mhz; 5.3 kw. Ant 3,490 ft TL: N44 01 54 W89 09 07. Hrs open: Box 492, 103 W. Main St., 54982. Phone: (902) 787-7220. Fax: (902) 787-0128. E-mail: classichits@wauhradio.com Web Site: www.wauhradio.com. Licensee: Hometown Broadcasting LLC (acq 8-9-02). Format: Classic hits. ♦Tom Boyson, gen mgr; Margaret Corrente, gen sls mgr; Greg Mantiss, progmg dir; Andy Disterhaft, chief of engrg; Ryan Cerney, traf mgr.

Wauwatosa

WXSS(FM)— Jan 1, 1961: 103.7 mhz; 19.5 kw. 840 ft TL: N43 05 48 W87 54 19. Stereo. Hrs opn: 24 11800 W. Grange Ave., Hales Corners, 53130. Phone: (414) 529-1250. Fax: (414) 529-2122. Web Site: www.entercom.com. Licensee: Entercom Milwaukee License LLC. Group owner: Entercom Communications Corp. (acq 12-13-99; grpsl). Population served: 717,099 Format: CHR. News staff: one. Target aud: 18-44; women. ♦Alan Kirshbon, sls dir; Andrea Biebel, natl sls mgr; Brian Kelly, progmg dir; Jane Matenaer, news dir; Chris Tarr, chief of engrg.

West Bend

WBKV(AM)— November 1950: 1470 khz; 2.5 kw-U, DA-2. TL: N43 22 14 W88 09 58. Hrs open: 24 Box 933, 53095. Secondary address: 2410 S. Main St., Suite A 53095. Phone: (262) 334-2344. Fax: (262) 334-1512. E-mail: jhodges@westbendradio.com Web Site: wbkvam.com. Licensee: West Bend Broadcasting Inc. Group owner: Bliss Communications Inc. (acq 10-18-70). Population served: 300,000 Natl. Network: CNN Radio. Rgnl. Network: Wisconsin Radio Net. Law Firm: Dow, Lohnes & Albertson. Format: Classic country. News staff: one; News: 20 hrs wkly. Target aud: 35-64; info-hungry adults interested in loc events & classic country music. ♦Skip Bliss, pres; Rob Lisser, CFO; James N. Hodges, VP, gen mgr & natl sls mgr; Paul Clements, gen sls mgr; Bob Bonenfant, progmg dir; Mark Morris, news dir; Jason Mielke, chief of engrg.

WBWI-FM—Co-owned with WBKV(AM). September 1958: 92.5 mhz; 17.5 kw. 565 ft TL: N43 25 45 W88 17 53. Stereo. 24 Web Site: wbwifm.com.1,500,000 News staff: one; News: 2 hrs wkly. Target aud: 25-54; country mus listeners with disposable income. ♦Fuzz Martin, progmg dir.

West Salem

WKBH-FM—Licensed to West Salem. See La Crosse

Westby

WDSW(FM)—Not on air, target date: unknown: 103.9 mhz; 2.75 kw. Ant 477 ft TL: N43 37 36 W90 41 57. Hrs open: 5331 Mount Alifan Dr., San Diego, CA, 92111-2622. Phone: (858) 277-4991. Fax: (858) 277-1365. Licensee: Horizon Christian Fellowship. ♦Michael MacIntosh, pres.

Whitehall

WHTL-FM— Sept 10, 1981: 102.3 mhz; 3 kw. 450 ft TL: N44 24 47 W91 17 03. Stereo. Hrs opn: 24 Box 66, N35609 Hwy. 53, 54773. Phone: (715) 538-4341. Phone: (715) 538-4341. Fax: (715) 538-4360. E-mail: whtl@triwest.net Licensee: The WHTL Group L.L.C. (acq 11-9-2006; $200,000 for stock). Rgnl. Network: Goetz Group. Law Firm: Bosari & Paxson. Wire Svc: AP Format: Oldies. News staff: one; News: 24 hrs wkly. Target aud: 25 plus; adult fans who make household buying decisions. Spec prog: Farm 5 hrs wkly. ♦Tim Harrington, CFO, opns mgr, sls dir & pub affrs dir; Mary Little, progmg mgr & mus dir; Todd A. Harrington, gen mgr & chief of engrg; Marty Little, sports cmtr.

Whitewater

WKCH(FM)— Jan 2, 1998: 106.5 mhz; 6 kw. Ant 200 ft TL: N42 54 24 W88 45 06. Hrs open: Box 94, Fort Atkinson, 53538. Phone: (920) 563-9329. Fax: (920) 563-0315. Licensee: NRG License Sub. LLC. (group owner; (acq 10-31-2005; grpsl). Format: Oldies. ♦Mary Quass, CEO; Tami Billmore, CFO; James Vriezen, gen mgr; Shane Sparks, gen mgr & gen sls mgr; Gary Douglas, opns dir & progmg dir; Michael Clish, news dir; George Nicholas, chief of engrg; Jaimie Flom, traf mgr.

WSLD(FM)— Nov 16, 1992: 104.5 mhz; 6 kw. Ant 328 ft TL: N42 35 47 W88 43 16. Stereo. Hrs opn: 24 Box 709, N. 6534 Hwy. 89, 53190. Phone: (608) 883-6677. Fax: (608) 883-2054. E-mail: wsld@starband.net Web Site: www.1045wsld.com. Licensee: WPW Broadcasting Inc. (group owner; (acq 8-4-99; $700,000). Format: Country. News staff: one; News: 10 hrs wkly. Target aud: General. ♦Don Davis, CEO; Nora Karbash, gen mgr & gen sls mgr; Reggie Michaels, progmg dir; Ryan O'Brien, news dir; Aaron Winski, chief of engrg; Trish Donovan, traf mgr.

***WSUW(FM)**— Jan 10, 1965: 91.7 mhz; 1.3 kw. 185 ft TL: N42 50 10 W88 44 36. Stereo. Hrs opn: 6 AM-2 AM UW Whitewater, 1201 Anderson Library, 53190. Phone: (262) 472-1323. Phone: (262) 472-1314. Fax: (262) 472-5029. E-mail: wsuw@uww.edu Web Site: www.wsuw.org. Licensee: Board of Regents University of Wisconsin System. Population served: 80,000 Format: Hip Hop, Alternative, jazz. Target aud: 18-44. Spec prog: Heavy metal 14 hrs, jazz/world beat 6 hrs, urban contemp 14 hrs wkly. ♦Wilfred Tremblay, gen mgr; Mark Neilsen, mktg dir; Kelly O'Brien, progmg dir.

Whiting

WLJY(FM)— Oct 21, 1985: 96.7 mhz; 50 kw. Ant 492 ft TL: N44 29 24 W89 32 54. Stereo. Hrs opn: 24 2301 Plover Rd., Plover, 54467. Phone: (715) 341-8838. Fax: (715) 341-9744. Web Site: 967wljy.com. Licensee: NRG License Sub. LLC. (group owner; (acq 10-31-2005; grpsl). Population served: 300,000 Law Firm: Latham & Watkins. Format: Soft adult contemp. News staff: one; News: 7 hrs wkly. Target aud: 25-54. ♦Benjamin D. Rosenthal, gen mgr; Jon Albrecht, rgnl sls mgr.

Winneconne

WVBO(FM)—Licensed to Winneconne. See Oshkosh

Wisconsin Dells

WDLS(AM)— May 1969: 900 khz; 1 kw-D, 229 w-N. TL: N43 38 23 W89 43 14. Hrs open: 24 Box 204, 53965. Secondary address: 121 Broadway 53965. Phone: (608) 254-2546. Fax: (608) 745-5771. Web Site: wdlsam.com. Licensee: Magnum Communications Inc. Group owner: Magnum Radio Inc. (acq 2-10-99; $775,000 with co-located FM). Population served: 105,000 Natl. Network: Jones Radio Networks, AP Radio. Format: Country. News staff: one; News: 6 hrs wlky. Spec prog: Medical 5 hrs wkly. ♦Dave Magnum, chmn; JIm Coursole, gen mgr.

WNNO-FM—Co-owned with WDLS(AM). May 1974: 106.9 mhz; 6 kw. Ant 321 ft TL: N43 38 23 W89 43 14. Stereo. 24 Box 360, Portage, 53901. Web Site: mix106wnno.com.105,000 Format: Adult contemp. News staff: one; News: 7 hrs wkly. Target aud: 18-40.

Wisconsin Rapids

WFHR(AM)— Nov 5, 1940: 1320 khz; 5 kw-D, 500 w-N, DA-N. TL: N44 24 56 W89 50 06. Hrs open: Box 8022, 645 25th Ave. N., 54495-8022. Phone: (715) 424-1300. Fax: (715) 424-1347. Web Site: www.wfhrradio.com. Licensee: Seehafer Broadcasting Corp. (group owner; (acq 6-1-2006; swap with WDLB(AM) Marshfield and WOSQ(FM) Spencer for WBCV(FM) Wausau). Population served: 100,000 Natl. Network: CBS. Format: News/talk, info, full service. ♦Donald Seehafer, pres; Wayne Ripp, gen mgr; Joe Steckbauer, gen sls mgr; Greg Gack, progmg dir; Carl Hilke, news dir; Pam Hilke, traf mgr; Car Hilke, news rptr.

WGLX-FM— Aug 1, 1946: 103.3 mhz; 100 kw. Ant 331 ft TL: N44 24 56 W89 50 10. Stereo. Hrs opn: 2301 PLover Road, Plover, 54467. Phone: (715) 341-8838. Fax: (715) 341-9744. Web Site: www.wglx.com. Licensee: NRG License Sub, LLC. (acq 10-31-2005; grpsl). Population served: 275,000 Format: Classic rock. Target aud: 25-49. ♦Benjamin D. Rosenthal, gen mgr; Panama Jack, progmg dir.

Wittenberg

***WVRN(FM)**—Not on air, target date: unknown: 89.9 mhz; 25 kw. Ant 482 ft TL: N44 57 54 W89 00 18. Hrs open: 3434 W. Kilbourn Ave., Milwaukee, 53208-3313. Phone: (414) 935-3000. Fax: (414) 935-3015. Web Site: www.vcyamerica.org. Licensee: VCY America Inc. Format: Relg, Christian. ♦Vic Eliason, VP & gen mgr; Jim Schneider, progmg dir; Andy Eliason, chief of engrg.

Wyoming

Afton

KRSV(AM)— Aug 13, 1985: 1210 khz; 5 kw-D, 250 w-N. TL: N42 43 22 W110 57 39. Hrs open: Box 1210, Wyoming Hwy. 238, 83110. Phone: (307) 885-5778. E-mail: hansenjw@silverstar.com Licensee: Western Wyoming Radio Inc. Natl. Network: ABC. Format: Modern country, loc news, sports. ♦Jerry Hansen, pres & gen mgr; Jennie Hansen, sls dir & progmg dir; Dan Dockstader, news dir.

KRSV-FM— Nov 13, 1985: 98.7 mhz; 3 kw. Ant -289 ft TL: N42 51 02 W110 58 46.

***KUWA(FM)**— July 1, 1998: 91.3 mhz; 400 w. -312 ft TL: N42 51 02 W110 58 46. Hrs open: 24
Rebroadcasts KUWR(FM) Laramie 100%.
Box 3984, Laramie, 82071. Phone: (307) 766-4240. Fax: (307) 766-6184. E-mail: wpr@uwyo.edu Web Site: wyomingpublicradio.net. Licensee: University of Wyoming. Natl. Network: NPR, PRI. Wire Svc: AP Format: Progsv, news/talk, class. News: 4 hrs wkly. Target aud: 25-54 plus; demographic, high income, education. Spec prog: Folk 5 hrs, jazz 5 hrs wkly. ♦Jon Schwartz, gen mgr; Hank Arnold, dev dir; Roger Adams, opns mgr & progmg dir; Bob Beck, news dir; Larry Dean, chief of engrg.

Albin

KKAW(FM)— 2001: 107.3 mhz; 9.3 kw. Ant 531 ft TL: N41 29 31 W104 05 07. Hrs open:
Rebroadcasts KZDR(FM) Cheyenne 100%.
2109 E. 10th St., Cheyenne, 82001. Phone: (307) 638-8921. Fax: (307) 638-8922. Licensee: Chisholm Trail Boadcasting LLC Group owner: Northeast Broadcasting Company Inc. (acq 5-23-2005; $850,000 with KREO(FM) Pine Bluffs). Format: Country.

Basin

KBEN-FM—Not on air, target date: unknown: 103.3 mhz; 64 kw. Ant 2,375 ft TL: N44 48 38 W107 55 18. Hrs open: 288 S. River Rd., Bedford, NH, 03110. Phone: (603) 668-6400. Fax: (603) 668-6470. Licensee: White Park Broadcasting Inc. ♦Steven A. Silberberg, pres.

KZMQ(AM)—See Greybull

Buffalo

KBBS(AM)— Apr 17, 1956: 1450 khz; 1 kw-U. TL: N44 20 33 W106 40 54. Hrs open: 24 1221 Fort St., 82834. Phone: (307) 684-7070. Fax: (307) 684-7676. E-mail: kbbs@vcn.com Licensee: Legend Communications of Wyoming L.L.C. Group owner: Legend Communications LLC (acq 9-1-00; $1.05 million with KLGT(FM) Buffalo). Population served: 25,800 Natl. Rep: McGavren Guild. Format: Oldies, News/talk, Sports. News: 10 hrs wkly. Target aud: 35-64; age group with the most money to spend. ♦Larry Patrick, CEO; Larry Patrick, pres; Ed Cwiklin, gen mgr & progmg dir; Smokey Wildman, gen mgr; Steve Lawrence, stn mgr & opns dir; Charles Dozier, chief of engrg.

***KBUW(FM)**— 2000: 90.5 mhz; 430 w. Ant -197 ft TL: N44 20 50 W106 43 25. Hrs open: Box 3984, Laramie, 82071. Phone: (307) 766-4240. Fax: (307) 766-6184. E-mail: rgriscom@uwyo.edu Web Site: uwadmnweb.uwyo.edu/WPR. Licensee: University of Wyoming. Format: News, AAA. ♦Jon Schwartz, gen mgr; Peg Arnold, dev dir; Bob Beck, news dir; Larry Dean, chief of engrg; Roger Adams, adv dir & traf mgr.

KLGT(FM)— Mar 7, 1983: 92.9 mhz; 100 kw. Ant 358 ft TL: N44 34 32 W106 52 23. Stereo. Hrs opn: 24 1221 Fort St., 82834. Phone: (307) 684-5126. Phone: (307) 684-2584. Fax: (307) 684-7676. E-mail: klgt@vcn.com Licensee: Legend Communications of Wyoming L.L.C. Group owner: Legend Communications L.L.C. (acq 9-1-00). Population served: 64,000 Natl. Network: Jones Radio Networks. Natl. Rep: McGavren Guild. Format: Country. News staff: one; News: 9 hrs wkly. Target aud: 24-54; those with spendable income. Spec prog: Relg one hr wkly. ♦Larry Patrick, CEO & pres; Nicki Williams, CFO; Smokey Wildeman, gen mgr; Ed Cwiklin, opns mgr & progmg dir; Zach Morton, chief of engrg.

Burns

KIGN(FM)— Sept 26, 1990: 101.9 mhz; 50 kw. 492 ft TL: N41 07 01 W104 40 07. Stereo. Hrs opn: 24 101.9 KING FM, 1912 Capitol Ave., Suite 300, Cheyenne, 82001. Phone: (307) 632-4400. Fax: (307) 632-1818. E-mail: kingfm@hotmail.com Web Site: www.kingfm.com. Licensee: Citicasters Licenses L.P. Group owner: Clear Channel Communications Inc. (acq 9-7-99; $1.2 million). Format: Country. News staff: one; News: 3 hrs wkly. Target aud: 25-54. ♦Craig Cochran, gen mgr & gen sls mgr; Tim Davidson, progmg dir; Amy Richardson, news dir & chief of engrg.

Casper

KASS(FM)— Oct 15, 1990: 106.9 mhz; 94 kw. Ant 1,765 ft TL: N42 44 37 W106 18 31. Stereo. Hrs opn: 24 218 N. Wolcott St., 82602. Phone: (307) 265-1984. Fax: (307) 473-7461. E-mail: kass@wyomingradio.com Web Site: www.wyomingradio.com. Licensee: Mount Rushmore Broadcasting Inc. (group owner; (acq 1995; $150,000). Population served: 60,000 Format: Classic rock. ♦Roger Arndt, gen mgr & gen sls mgr; Donny Rood, progmg dir; Steve Fritz, chief of engrg; Jenniey Lyman, traf mgr.

***KCSP-FM**— 1992: 90.3 mhz; 100 kw. 1,922 ft TL: N42 44 24 W106 18 23. Stereo. Hrs opn: 24 6363 Hwy. 50 E., Carson City, 89701. Phone: (775) 883-5647. Fax: (775) 883-5704. Licensee: Western Inspirational Broadcasters Inc. (acq 10-3-90). Population served: 330,000 Natl. Network: AP Radio. Wire Svc: AP Format: Contemp Christian. News: 16 hrs wkly. Target aud: General. ♦Tom Hesse, gen mgr; Tim Wwidemann, opns mgr; Bill Feltner, progmg dir; Janet Santana, mus dir; Paul Lierman, chief of engrg.

KHOC(FM)— 1998: 102.5 mhz; 100 kw. Ant 1,696 ft TL: N42 44 37 W106 18 31. Stereo. Hrs opn: 218 N. Wolcott St., 82601. Phone: (307) 265-1984. Fax: (307) 266-3295. E-mail: info@wyomingradio.com Web Site: www.wyomingradio.com. Licensee: Mount Rushmore Broadcasting Inc. (group owner; (acq 10-29-98; $300,000). Format: Hot adult contemp. ♦Roger Arndt, gen mgr; Donnie Rood, progmg dir; Steve Fritz, engrg dir; Jenniey Lyman, traf mgr.

KKTL(AM)— 1999: 1400 khz; 1 kw-U. TL: N42 51 22 W106 21 41. Hrs open: 150 N. Nichols Ave., 82601. Phone: (307) 266-5252. Fax: (307) 235-9143. E-mail: kktl@clearchannel.com Licensee: Citicasters Licenses L.P. Group owner: Clear Channel Communications Inc. (acq 5-4-99; grpsl). Format: Sports. ♦Robert Price, gen mgr; Donovan Short, opns mgr & progmg dir; Walter Hawn, gen sls mgr; Dave Nutter, chief of engrg; Dave Borino, traf mgr.

***KLWC(FM)**— 2005: 89.1 mhz; 1.5 kw vert. Ant 1,848 ft TL: N42 44 03 W106 20 00. Hrs open:
Rebroadcasts KLVR(FM) Santa Rosa, CA).
2351 Sunset Blvd., Suite 170-218, Rocklin, CA, 95765. Phone: (916) 251-1600. Fax: (916) 251-1650. Web Site: www.klove.com. Licensee: Educational Media Foundation. (acq 1-11-2005; $100,000 for CP with CP for KLRV(FM) Billings, MT). Natl. Network: K-Love. Format: Christian. ♦Richard Jenkins, pres; Mike Novak, VP; Keith Whipple, dev dir; David Pierce, progmg mgr; Ed Lenane, news dir; Sam Wallington, engrg dir; Karen Johnson, news rptr; Marya Morgan, news rptr; Richard Hunt, news rptr.

KMGW(FM)—Listing follows KTWO(AM).

KMLD(FM)— Oct 1, 1967: 94.5 mhz; 63 kw. Ant 1,909 ft TL: N42 44 03 W106 20 00. Stereo. Hrs opn: 218 N. Wolcott St., 82601. Phone: (307) 265-1984. Fax: (307) 473-7461. E-mail: kmld@wyomingradio.com Web Site: www.wyomingradio.com. Licensee: Mt. Rushmore Broadcasting Inc. Group owner: Mount Rushmore Broadcasting Inc. (acq 3-12-2001; grpsl). Format: Oldies. Target aud: 35-64. ♦Roger Arndt, gen mgr; Donny Rood, progmg dir; Steve Fritz, chief of engrg; Jenniey Lyman, traf mgr.

KQLT(FM)— Oct 7, 1983: 103.7 mhz; 97 kw. Ant 1,860 ft TL: N42 44 37 W106 18 31. Stereo. Hrs opn: 24 218 N. Wolcott St., 82601. Phone: (307) 265-1984. Fax: (307) 473-7461. E-mail: kqlt@wyomingradio.com Web Site: www.wyomingradio.com. Licensee: Mount Rushmore Broadcasting Inc. (group owner; (acq 8-17-94; $230,000; FTR: 9-12-94). Natl. Rep: McGavren Guild. Law Firm: Dow, Lohnes & Albertson. Format: Country. Target aud: General. ♦Jan Charles Gray, pres; Roger Arndt, gen mgr; Don Rood, opns mgr; Donny Rood, progmg dir; Steve Fritz, chief of engrg; Jenniey Lyman, traf mgr.

KTRS-FM— January 1997: 104.7 mhz; 18 w. 1,774 ft TL: N42 44 37 W106 18 26. Hrs open: 150 N. Nichols Ave., 82601. Phone: (307) 266-5252. Fax: (307) 235-9143. E-mail: ktrs@clearchannel.com Licensee: Clear Channel Broadcasting Licenses Inc. Group owner: Clear Channel Communications Inc. (acq 3-29-01; grpsl). Format: CHR. Target aud: 12-24. ♦Robert Price, gen mgr; Donovan Short, opns mgr, gen sls mgr & progmg dir; Walter Hawn, gen sls mgr & news dir; Dave Nutter, chief of engrg; Dave Borino, traf mgr.

KTWO(AM)— Jan 2, 1930: 1030 khz; 50 kw-U, DA-N. TL: N42 50 34 W106 13 07. Hrs open: 150 N. Nichols Ave., 82601. Phone: (307) 266-5252. Fax: (307) 235-9143. E-mail: ktwo@clearchanel.com Web Site: www.k2radio.com. Licensee: Citicasters Licenses L.P. Group owner: Clear Channel Communications Inc. (acq 5-4-99; grpsl). Population served: 159,361 Natl. Network: CBS. Law Firm: Cohn & Marks. Format: Talk. Target aud: 25-54. ♦Bob Price, gen mgr, opns mgr & gen sls mgr; Bob Davis, progmg dir; Vicki Daniels, news dir; David Nutter, chief of engrg; Dave Borino, traf mgr.

KMGW(FM)—Co-owned with KTWO(AM). 1998: 96.7 mhz; 2.85 kw. Ant 1,771 ft TL: N42 44 37 W106 18 26. E-mail: kmgw@clearchannel.com Web Site: www.hitsandfavorites.com. Licensee: Clear Channel Broadcasting Licenses Inc. (acq 3-12-01; grpsl). Format: Adult contemp. ♦Donovan Short, progmg dir; Michelle Webster, traf mgr.

***KUWC(FM)**— 2000: 91.3 mhz; 530 w. Ant 1,784 ft TL: N42 44 26 W106 21 34. Hrs open: Box 3984, Laramie, 82071. Phone: (307) 766-4240. Fax: (307) 766-6184. E-mail: jbs@uwyo.edu Web Site: uwadmnweb.uwyo.edu. Licensee: University of Wyoming. Format: News, AAA. ♦Jon Schwartz, gen mgr; Peg Arnold, prom dir; Roger Adams, progmg dir; Larry Dean, mus dir; Bob Beck, news dir.

KVOC(AM)— Sept 29, 1946: 1230 khz; 1 kw-U. TL: N42 50 05 W106 17 44. Hrs open: 218 N. Wolcott St., 82601. Phone: (307) 265-1984. Fax: (307) 473-7461. E-mail: mtrushmore@mrbradio.com Web Site: www.wyomingradio.com. Licensee: Mount Rushmore Broadcasting Inc. (group owner; (acq 6-12-97; $105,000). Population served: 75,000 Natl. Network: ESPN Radio. Format: Sports. ♦Jan Charles Gray, pres; Roger Arndt, gen mgr; Donny Rood, progmg dir; Steve Fritz, chief of engrg; Jenniey Lyman, traf mgr.

KWYX(FM)—Not on air, target date: unknown: 93.5 mhz; 15 kw. Ant 1,712 ft TL: N42 44 30 W106 18 29. Hrs open: Box 11060, Jackson, 83002. Phone: (703) 812-0482. Licensee: Cochise Broadcasting LLC. ♦Ted Tucker, gen mgr.

KWYY(FM)— Nov 30, 1981: 95.5 mhz; 100 kw. 1,920 ft TL: N42 44 37 W106 18 26. Stereo. Hrs opn: 251 W. First St., 82601. Secondary address: 105 North Nichols Phone: (307) 266-5252. Fax: (307) 235-9143. E-mail: kwyy@clearchannel.com Licensee: Clear Channel Broadcasting Licenses Inc. Group owner: Clear Channel Communications Inc. (acq 3-29-01; grpsl). Format: Country. Target aud: 25-54. ♦Robert Price, gen mgr; Donovan Short, opns mgr & progmg dir; Walter Hawn, gen sls mgr; Dave Nutter, chief of engrg; Dave Borino, traf mgr.

Cheyenne

***KAIX(FM)**— 2006: 88.1 mhz; 1 kw. Ant 209 ft TL: N41 09 37 W104 42 13. Hrs open:
Rebroadcasts KLRD(FM) Yucaipa, CA 100%.
5700 West Oaks Blvd., Rocklin, CA, 95765. Phone: (916) 251-1600. Fax: (916) 251-1650. Web Site: www.air1.com. Licensee: Educational Media Foundation. (acq 3-23-2007; grpsl). Natl. Network: Air 1. Format: Alternative rock, div. ♦Richard Jenkins, pres.

KFBC(AM)— 1940: 1240 khz; 1 kw-U. TL: N41 07 17 W104 50 22. Hrs open: 24 1806 Capitol Ave., 82001. Phone: (307) 634-4461. Fax: (307) 632-8586. Licensee: Montgomery Broadcasting L.L.C. (acq 7-1-93; $250,000). Population served: 72,000 Natl. Network: ABC. Format: Full service, news, sports. News staff: 5; News: 25 hrs wkly. Target aud: 35-54. ♦Dave Montgomery, pres & gen mgr; J.D. Harris, stn mgr.

KGAB(AM)—Orchard Valley, 1952: 650 khz; 8.5 kw-D, 500 w-N, DA-N. TL: N41 03 09 W104 49 53. Hrs open: 24 1912 Capitol Ave., Suite 300, 82001. Phone: (307) 632-4400. Fax: (307) 632-1818. Web Site: www.kgab.com. Licensee: Citicasters Licenses L.P. Group owner: Clear Channel Communications Inc. (acq 1999; grpsl). Population served: 1,400,000 Format: News/talk. News staff: 3. Target aud: 25-64. ♦Craig Cochran, gen mgr; Dave Chaffin, opns mgr & progmg dir; Amy Richards, news dir; Quin Morrison, chief of engrg.

KOLZ(FM)—Co-owned with KGAB(AM). August 1961: 100.7 mhz; 100 kw. 490 ft TL: N41 06 01 W105 00 23. Stereo. Web Site: www.kolz.com.135,000 Format: Country. ♦Jeff Brown, progmg dir.

KIGN(FM)—See Burns

KJUA(AM)— 1952: 1380 khz; 1 kw-D, 8 w-N. TL: N41 07 22 W104 48 07. Stereo. Hrs opn: 24 110 East 17th St., Suite 205, 82001. Phone: (307) 635-8787. Fax: (307) 635-8788. E-mail: kjjl@kjjl.com Web Site: www.kjjl.com. Licensee: Christus Broadcasting Inc. Population served: 70,000 Natl. Network: CNN Radio. Law Firm: Booth, Freret, Imlay. Format: Adult contemp. News staff: 2; News: 22 hrs wkly. Target aud: 35 plus; mature adults. Spec prog: Gospel one hr wkly. ♦Paul Montoya, pres.

KKPL(FM)— 1997: 99.9 mhz; 50 kw. Ant 492 ft TL: N40 59 22 W105 03 47. Hrs open: 3201 E. Mulberry St., Unit H, Fort Collins, CO, 80524. Phone: (970) 674-2700. Phone: (970) 492-0999. Fax: (970) 407-0584. Web Site: www.999thepoint.com. Licensee: Regent Broadcasting of Ft. Collins Inc. Group owner: Regent Communications Inc. (acq 1-8-2004; $7.75 million with KARS-FM Laramie). Format: Alternative. ♦Cal Hall, gen mgr; Mark Callaghan, opns mgr; Miles Schallert, sls dir; Mark Callagham, progmg dir; Susan Moore, news dir; Quin Morrison, chief of engrg.

KLEN(FM)— Sept 26, 1983: 106.3 mhz; 6 kw. Ant 325 ft TL: N41 03 09 W104 49 55. Stereo. Hrs opn: 24 1912 Capitol Ave., Suite 300, 82001. Phone: (307) 632-4400. Fax: (307) 632-1818. E-mail: cheyenneaudio @clearchannel.com Web Site: www.1063klen.com. Licensee: Citicasters Licenses L.P. Group owner: Clear Channel Communications Inc. (acq 5-4-99; grpsl). Population served: 60,000 Format: News/talk. News staff: 3. Target aud: 25-54. ♦Craig Cochran, gen mgr & gen sls mgr; Dave Chaffin, opns mgr, gen sls mgr & progmg dir; Amy Richards, prom dir & news dir; Jim Mross, chief of engrg; Lesley Martin, traf mgr.

KQMY(FM)— Sept 1, 1968: 97.9 mhz; 100 kw. 541 ft TL: N41 06 01 W105 00 23. Stereo. Hrs opn: 24 1612 La Porte, Suite 300, Fort Collons, CO, 80521. Phone: (970) 482-5991. Fax: (970) 482-5994. Licensee: Citicasters Licenses L.P. Group owner: Clear Channel Communications Inc. (acq 1999; grpsl). Population served: 350,000 Format: Adult contemp. News staff: 3; News: 2 hrs wkly. Target aud: 25-54. ♦Stu Haskell, gen mgr; Chris Kelly, opns mgr & progmg dir; Kathy Arias, gen sls mgr; Rich Bircumshaw, news dir; Cliff Mikkelson, chief of engrg; Karen Vissers, traf mgr.

KRAE(AM)— Apr 29, 1961: 1480 khz; 1 kw-D, 67 w-N. TL: N41 07 26 W104 49 10. Stereo. Hrs opn: 2109 E. 10th St., 82001. Phone: (307) 638-8921. Fax: (307) 638-8922. E-mail: news@1049krrr.com Licensee: Brahmin Broadcasting Corp. Group owner: Northeast Broadcasting Company Inc. (acq 5-10-2004; grpsl). Population served: 200,000 Natl. Network: ESPN Radio. Format: Sports. Spec prog: Sp 2 hrs wkly. ♦Larry Proietti, gen mgr; Jessica Cooper, sls dir; Larry Proeitti, progmg dir; R.J. Fox, news dir; Rob Thomas, chief of engrg; Sandra Cooper, traf mgr.

KRRR(FM)—Co-owned with KRAE(AM). 1997: 104.9 mhz; 25.5 kw. Ant 115 ft TL: N41 08 04 W104 41 32. Stereo. 72,400 Format: Oldies. Target aud: 25-54. ♦Sandra Cooper, traf mgr.

KZDR(FM)— June 15, 2006: 93.7 mhz; 25 kw. Ant 115 ft TL: N41 08 04 W104 41 32. Hrs open: 2109 E. 10th St., 82001. Phone: (307) 638-8921. Fax: (307) 638-8922. Licensee: White Park Broadcasting Inc. Format: Country. ♦Steven A. Silberberg, pres.

Chugwater

KCUG(FM)—Not on air, target date: unknown: 99.5 mhz; 6 kw. Ant 272 ft TL: N41 46 09 W104 48 57. Hrs open: White Park Broadcasting Inc., 288 S. River Rd., Bedford, NH, 03110. Phone: (603) 668-9999. Fax: (603) 668-6470. Licensee: White Park Broadcasting Inc. ♦Steven A. Silberberg, pres & gen mgr.

***KLWV(FM)**— 2004: 90.9 mhz; 100 kw. Ant 1,183 ft TL: N41 18 39 W105 27 12. Hrs open: 24 2351 Sunset Blvd., Suite 170-218, Rocklin, CA, 95765. Phone: (916) 251-1600. Fax: (916) 251-1650. E-mail: klove@klove.com Web Site: www.klove.com. Licensee: Educational Media Foundation. Group owner: EMF Broadcasting (acq 10-2-03; grpsl). Natl. Network: K-Love. Law Firm: Shaw Pittman. Format: Contemp Christian. News staff: 3. Target aud: 25-44; Judeo Christian, female. ♦Richard Jenkins, pres; Mike Novak, VP; Ed Lenane, opns dir & news dir; Keith Whipple, dev dir; David Pierce, progmg mgr; Sam Wallington, engrg dir; Arthur Vassar, traf mgr; Karen Johnson, news rptr; Marya Morgan, news rptr; Richard Hunt, news rptr.

Clearmont

KLQQ(FM)— 2006: 104.7 mhz; 2.1 kw. Ant 1,122 ft TL: N44 37 20 W107 06 57. Hrs open: Box 5086, Sheridan, 82801. Phone: (307) 672-7421. Fax: (307) 672-2933. Licensee: Lovcom Inc. Format: Top-40. ♦Kim Love, gen mgr.

Cody

KODI(AM)— March 1947: 1400 khz; 1 kw-U. TL: N44 30 30 W109 04 05. Hrs open: 6 AM-midnight Box 1210, 1949 Mountain View Dr., 82414. Phone: (307) 578-5000. Fax: (307) 527-5045. E-mail: ckary@bhrnwy.com Web Site: www.bighornradio.com. Licensee: Legend Communications of Wyoming LLC. Group owner: Legend Communications LLC (acq 6-29-99; $890,000 with co-located FM). Population served: 20,000 Wire Svc: NWS (National Weather Service) Format: News/talk, sports. News staff: one; News: 20 hrs wkly. Target aud: 35 plus. ♦Larry Patrick, pres; Roger Gelder, exec VP; Carol Kary, VP & gen mgr; Cory Ostermiller, gen sls mgr; Tom Morrison, progmg dir & pub affrs dir; Wendy Corr, news dir; Charles Dozier, chief of engrg.

KTAG(FM)—Co-owned with KODI(AM). Nov 30, 1981: 97.9 mhz; 100 kw. 1,901 ft TL: N44 29 44 W109 09 13. Stereo. 6 AM-midnight Web Site: bighornradio.com.40,000 Format: Adult contemp. News: 6 hrs wkly. Target aud: 25-39; upper middle class, young families, suburban w/some college, 60% female. ♦Larry Patrick, CEO; Roger Gelder, COO; Carol Kary, gen mgr; Cory Ostermiller, gen sls mgr; Wendy Corr, news dir.

KWHO(FM)—Not on air, target date: unknown: 96.7 mhz; 2.4 kw. Ant 1,831 ft TL: N44 29 49 W109 09 19. Hrs open: 288 S. River Rd., Bedford, NH, 03110. Phone: (603) 668-6400. Fax: (603) 668-6470. Licensee: White Park Broadcasting Inc. ♦Steven A. Silberberg, pres.

Diamondville

KDWY(FM)— 2000: 105.3 mhz; 16 kw. Ant 886 ft TL: N41 50 18 W110 30 12. Hrs open: c/o Radio Station KMER(AM), Box 432, Kemmerer, 83101. Phone: (307) 877-4422. Fax: (307) 877-5537. E-mail: kmer@onewest.net Licensee: Simmons-SLC, LS LLC. Group owner: Simmons Media Group (acq 4-19-2004; grpsl). Population served: 80,000 Natl. Network: ABC. Format: Country. News staff: one; News: one hr wkly. Target aud: 25-55. ♦Jim Carroll, gen mgr; Jim Thoeny, opns dir.

Douglas

KDAD(FM)—Not on air, target date: unknown: 92.5 mhz; 5.4 kw. Ant 3,188 ft TL: N42 16 05 W105 26 33. Hrs open: 288 S. River Rd., Bedford, NH, 03110. Phone: (603) 668-6470. Licensee: White Park Broadcasting Inc. ♦Steven A. Silberberg, pres & gen mgr.

***KDUW(FM)**— 2000: 91.7 mhz; 450 w. Ant -52 ft TL: N42 44 41 W105 20 09. Hrs open: Box 3984, Laramie, 82071. Phone: (307) 766-4240. Fax: (307) 766-6184. E-mail: hgriscom@uwyo.edu Web Site: www.wyomingpublicradio.org. Licensee: University of Wyoming. Format: News, progsv, class. ♦Jon Schwartz, gen mgr; Peg Arnold, progmg dir; Roger Adams, progmg dir; Larry Dean, chief of engrg.

KKTY(AM)— June 22, 1957: 1470 khz; 1 kw-D, 500 w-N. TL: N42 45 48 W105 23 32. Hrs open: 24 Box 135, 247 Russell Ave., 82633. Phone: (307) 358-3636. Fax: (307) 358-4010. E-mail: kkty@netcommander.com Web Site: www.kktyonline.com. Licensee: Douglas Broadcasting Inc. (acq 2-11-93; $120,000 with co-located FM; FTR: 3-8-93) Population served: 12,000 Natl. Network: Westwood One, CNN Radio. Rgnl rep: Regnl Reps Wire Svc: AP Format: Oldies. News: 28 hrs wkly. Target aud: General. ♦Dennis Switzer, pres & gen mgr.

KKTY-FM— Dec 6, 1982: 99.3 mhz; 813 w. 530 ft TL: N42 43 42 W105 31 46. Stereo. 24 Web Site: www.kktyonline.com. Natl. Network: Jones Radio Networks, CNN Radio. Format: Country. News staff: one; News: 28 hrs wkly. Target aud: General. ♦Becky Heidt, traf mgr.

KTED(FM)—Not on air, target date: unknown: 100.9 mhz; 31 kw. Ant 626 ft TL: N42 45 30 W105 47 49. Hrs open: 288 S. River Rd., Bedford, NH, 03110. Phone: (603) 668-6470. Licensee: White Park Broadcasting Inc. ♦Steven A. Silberberg, pres.

Ethete

***KWRR(FM)**— 2000: 89.5 mhz; 85 kw. Ant 1,820 ft TL: N43 27 30 W108 11 39. Hrs open: Box 327, Kinnear, 82516. Phone: (307) 335-8659. Phone: (307) 335-8658. Fax: (307) 335-8740. E-mail: rproductions@hotmail.com Licensee: Business Council of the Northern Arapaho Tribe. (acq 8-11-98). Format: Variety, Native American. ♦Steven White, gen mgr; Jason Foskey, progmg dir; Lincoln Scott, chief of engrg.

Evanston

KADQ-FM—Not on air, target date: unknown: 98.3 mhz; 37 kw. Ant 567 ft TL: N41 14 14 W110 58 09. Hrs open: College Creek Media LLC, 980 N. Michigan Ave., Suite 1880, Chicago, IL, 60611. Phone: (312) 204-9900. Licensee: College Creek Media LLC. ♦David Stout, gen mgr.

KBMG(FM)— June 1982: 106.1 mhz; 89 kw horiz. Ant 2,122 ft TL: N40 52 16 W110 59 43. Stereo. Hrs opn: 2722 S. Redwood Rd., Suite 1, Salt Lake City, UT, 84119. Phone: (801) 908-8777. Fax: (801) 908-8782. Licensee: Bustos Media of Utah License LLC. Group owner: Bustos Media Holdings (acq 7-1-2004; $3 million). Format: Sp. ♦Edward Distel, gen mgr.

***KCWW(FM)**—Not on air, target date: unknown: 88.1 mhz; 92 w. Ant 1,348 ft TL: N41 21 10 W110 54 29. Hrs open: Box 1372, Park City, UT, 84060-1372. Phone: (435) 649-9004. Fax: (435) 645-9063. Licensee: Community Wireless of Park City Inc. ♦Blair Feulner, gen mgr.

KEVA(AM)— June 27, 1953: 1240 khz; 1 kw-U. TL: N41 15 29 W111 00 51. Hrs open: 24 Box 190, 568 Airport Rd., 82931. Phone: (307) 789-9101. Phone: (307) 789-9102. Fax: (307) 789-8521. E-mail: keva@vcn.com Web Site: 1240keva,com. Licensee: Sagebrush Broadcasting Co. Inc. Group owner: Jimmy Ray Carroll Stns (acq 1-23-2001). Population served: 25,000 Format: Country. Target aud: 25-54. ♦Linda Burris, gen mgr & prom mgr; Linda Burns, gen sls mgr; Bill Smith, prom dir & progmg dir; J.C. Jewett, news dir; Michael Richard, progmg mgr & chief of engrg.

Evansville

KUYO(AM)— Aug 23, 1985: 830 khz; 25 kw-D. TL: N42 52 13 W106 12 12. Hrs open: Sunrise-sunset Box 50607, Casper, 82605-0607. Secondary address: 1423 S. Beverly, Casper 82609. Phone: (307) 577-5896. Web Site: www.kuyo.com. Licensee: Wyoming Christian

Broadcasting Co. (acq 6-1-99; $75,000). Format: Classic Christian, talk. Target aud: 35-65; general. ♦Aaron Remington, VP; Steve Stumbo, pres & gen mgr.

Fort Bridger

KNYN(FM)— 2001: 99.1 mhz; 27.5 kw. Ant 1,604 ft TL: N41 21 10 W110 54 26. Hrs open: Box 190, Evanston, 83931. Phone: (307) 789-9101. Fax: (307) 789-8521. Licensee: M. Kent Frandsen. (acq 6-22-99; $125,000). Format: Adult contemp. ♦Linda Burris, gen mgr & gen sls mgr; Bill Smith, progmg dir; J.C. Jewett, news dir; Michael Richard, chief of engrg.

Fox Farm

KRND(AM)— 1998: 1630 khz; 10 kw-D, 1 kw-N. TL: N41 07 22 W104 48 07. Hrs open: 24 110 E. 17th St., Suite 205, Cheyenne, 82001. Phone: (307) 635-8787. Fax: (307) 635-8788. E-mail: kwy@kwyradio.com Web Site: www.kwyradio.com. Licensee: Christus Broadcasting Inc. Population served: 170,000 Natl. Network: AP Network News. Law Firm: Booth, Freret & Imlay. Format: Classic country. News staff: 1; News: 22 hrs wkly. Target aud: 35 plus; mature adults. ♦Paul Montoya, pres, gen mgr & chief of opns.

Gillette

KAML-FM—Listing follows KIML(AM).

***KAXG(FM)**— Mar 27, 2003: 89.7 mhz; 250 w. Ant 89 ft TL: N44 13 50 W105 27 45. Hrs open: 24
Rebroadcasts KXEI(FM) Havre 100%.
Box 2426, Havre, MT, 59501-2426. Phone: (406) 265-5845. Fax: (406) 265-8860. E-mail: ynop@ynopradio.org Web Site: www.ynopradio.org. Licensee: Hi-Line Radio Fellowship Inc. (acq 3-13-03; $65,000). Natl. Network: Salem Radio Network. Wire Svc: AP Format: Christian Inspiritational. Target aud: General; those looking for inpirational Christian music & progmg. ♦Roger Lonnquist, gen mgr; Brenda Boyum, stn mgr; Brian Jackson, progmg dir.

KGWY(FM)— Jan 5, 1983: 100.7 mhz; 100 kw. 635 ft TL: N44 14 35 W105 32 19. Stereo. Hrs opn: 24 Box 1179, 82717. Secondary address: 2810 Southern Dr. 82718. Phone: (307) 686-2242. Fax: (307) 686-7736. E-mail: thefox@basinsradio.com Web Site: www.basinsradio.com. Licensee: Legend Communications of Wyoming LLC. Group owner: Legend Communications L.L.C. (acq 5-29-01; $1.9 million). Natl. Network: ABC. Format: Country. News staff: one. Target aud: 20-45. ♦Larry Patrick, pres; Don Clonch, gen mgr; John English, opns mgr.

KIML(AM)— Sept 13, 1957: 1270 khz; 5 kw-D, 1 kw-N, DA-N. TL: N44 18 12 W015 59 52. Stereo. Hrs opn: 24 Box 1179, 82717. Phone: (307) 686-2242. Fax: (307) 686-7736. Web Site: www.basinsradio.com. Licensee: Gillette Broadcasting Co. (acq 5-29-01; $1.2 million for stock, including $100,000 consulting agreement, with co-located FM). Population served: 50,000 Natl. Network: AP Radio, ESPN Radio. Wire Svc: AP Format: News, talk, sports. News staff: one; News: 40 hrs wkly. Target aud: 25 plus. ♦Don Clonch, gen mgr; John English, opns mgr & gen sls mgr.

KAML-FM—Co-owned with KIML(AM). May 1976: 96.9 mhz; 100 kw. 456 ft TL: N44 18 10 W105 27 00. Stereo. 24 Web Site: www.basinsradio.com.30,000 Format: Rock. News staff: one; News: one hr wkly. Target aud: 25-54.

***KLOF(FM)**— 2000: 88.9 mhz; 430 w. Ant 440 ft TL: N44 12 34 W105 28 04. Hrs open: 5700 West Oaks Blvd., Rocklin, CA, 95765. Phone: (916) 251-1600. Fax: (916) 251-1650. Licensee: Educational Media Foundation. (acq 3-29-2007; $55,000). Format: Christian. ♦Mike Novak, sr VP.

***KLWD(FM)**— 2001: 91.9 mhz; 1 kw. Ant 226 ft TL: N44 17 00 W105 31 00. Stereo. Hrs opn: 24 Box 1492, 82717. Phone: (307) 682-9553. Fax: (307) 682-8509. E-mail: klwd@csnradio.com Web Site: klwd.vcn.com. Licensee: CSN International (group owner; acq 5-8-00; $10,000 for CP). Population served: 50,000 Format: Christian. ♦Don Wight, pres, gen mgr & progmg dir.

***KUWG(FM)**— 1997: 90.9 mhz; 450 w. 459 ft TL: N44 12 33 W105 28 05. Hrs open:
Rebroadcasts KUWR(FM) Laramie 100%.
Box 3984, Univ. Station, Laramie, 82071. Phone: (307) 766-4240. Fax: (307) 766-6184. E-mail: wpr@uwyo.edu Web Site: uwadmnweb.uwyo.edu. Licensee: University of Wyoming. Format: News, progsv, class. ♦Jon Schwartz, gen mgr; Hank Arnold, dev dir; Roger Adams, progmg dir; Bob Beck, news dir; Larry Dean, chief of engrg.

KXXL(FM)— 2004: 103.9 mhz; 50 kw. Ant 392 ft TL: N44 13 50 W105 27 45. Hrs open: Box 2230, 82717. Phone: (307) 687-1003. Fax: (307) 687-1006. E-mail: koaal1039@collinscom.net Licensee: Keyhole Broadcasting LLC. Format: Classic rock. ♦Deborah Roberts, gen mgr & gen sls mgr; Rob Olsen, progmg dir.

Glendo

KYOD(FM)— July 10, 1999: 100.1 mhz; 100 kw. Ant 456 ft TL: N42 46 13 W105 13 21. Stereo. Hrs opn: 24 1837 Madora Ave., Suite B, Douglas, 82633. Phone: (307) 358-6177. Fax: (307) 358-0978. E-mail: kyod@netcommander.com Web Site: www.kyod.com. Licensee: Canned Ham Communications LLC (acq 5-23-00; $150,000 for CP). Population served: 50,000 Natl. Network: Westwood One. Natl. Rep: Interep. Law Firm: Womble, Carlyle, Sandridge & Rice, PLLC. Format: Talk/Music variety. News staff: one; News: one hr wkly. Target aud: 25-54. Spec prog: Hard rock 2 hrs wkly; Religious, 2hrs wkly. ♦Darrell Woolsey, gen mgr; Mary Woolsey, stn mgr.

Glenrock

KGRK(FM)— 2007: 98.3 mhz; 200 w. Ant -249 ft TL: N42 51 49 W105 52 15. Hrs open: 1063F Big Thompson Canyon Rd., Loveland, CO, 80537. Phone: (970) 669-9200. Licensee: Michael Radio Group LLC. ♦Victor A. Michael Jr., gen mgr.

Green River

KFRZ(FM)—Listing follows KUGR(AM).

KSIT(FM)—See Rock Springs

KUGR(AM)— June 18, 1976: 1490 khz; 1 kw-U. TL: N41 30 56 W109 26 11. Hrs open: 24 Box 970, 82935. Secondary address: 40 Shoshone Ave. 82935. Phone: (307) 875-6666. Fax: (307) 875-5847. E-mail: kugr@sweetwater.net Web Site: www.kugr.net. Licensee: Wagon Wheel Communications Corp. (acq 1-1-79). Population served: 40,000 Natl. Network: CBS, Westwood One. Format: Soft adult contemp. News staff: one; News: 4 hrs wkly. Target aud: 30 plus. Spec prog: Sp 5 hrs wkly. ♦Al Harris, CEO & pres; Steve Core, gen mgr & opns dir; Jeff Driggs, gen sls mgr; Sean Maxwell, progmg dir, chief of engrg & chief of engrg; Adam Dormoth, news dir; Teresa Warren, traf mgr.

KFRZ(FM)—Co-owned with KUGR(AM). Sept 23, 1999: 92.1 mhz; 90 kw. 1,138 ft TL: N41 29 47 W109 20 44.24 E-mail: kugr@sweetwater.net Web Site: www.kfrz.net.60,000 Natl. Network: Westwood One. Format: Country. News staff: one; News: 3 hrs wkly. Target aud: General. ♦Teresa Warren, traf mgr.

KZWB(FM)— 2005: 97.9 mhz; 10.5 kw. Ant 1,073 ft TL: N41 29 47 W109 20 44. Stereo. Hrs opn: 24 40 Shoshone Ave., 82935. Phone: (307) 875-6666. Fax: (307) 875-5847. E-mail: kugr@sweetwater.net Web Site: theradionetwork.net. Licensee: Wagonwheel Communications Corp. Population served: 50,000 Format: Classic hits. ♦Alan W. Harris, pres.

Greybull

KZMQ(AM)— May 20, 1979: 1140 khz; 10 kw-D. TL: N44 27 01 W108 02 56. Hrs open: Box 1210, 1949 Mountain View Dr., Cody, 82414. Phone: (307) 578-5000. Fax: (307) 527-5045. E-mail: rgelder@bhmway.com Web Site: www.bighornradio.com. Licensee: Legend Communications of Wyoming L.L.C. Group owner: Legend Communications LLC (acq 1-27-98; $1.5 million with co-located FM). Population served: 50,000 Format: Real country. Target aud: 25-49. ♦Larry Patrick, pres; Roger Gelder, exec VP & gen mgr; Carol Kary, stn mgr & gen sls mgr; Rita Conners, opns dir & opns mgr; Barbara Greene, sls VP & sls dir; Jerry Dunning, progmg dir; Mack Frost, news dir; Charlie Dozier, chief of engrg.

KZMQ-FM— Feb 21, 1986: 100.3 mhz; 56 kw. 2,443 ft TL: N44 48 41 W107 55 06. Stereo. Format: Country. ♦Carol Kary, gen mgr.

Guernsey

KANT(FM)—Not on air, target date: unknown: 104.1 mhz; 50 kw. Ant 417 ft TL: N42 20 51 W105 01 54. Hrs open: White Park Broadcasting Inc., 288 S. River Rd., Bedford, NH, 03110. Phone: (603) 668-6470. Licensee: White Park Broadcasting Inc. ♦Steven A. Silberberg, pres & gen mgr.

Hanna

KHNA(FM)—Not on air, target date: unknown: 103.3 mhz; 6 kw. Ant 277 ft TL: N41 43 51 W106 27 26. Hrs open: 288 S. River Rd., Bedford, NH, 03110. Phone: (603) 668-6400. Licensee: White Park Broadcasting Inc. ♦Steven A. Silberberg, pres.

KXMP(FM)—Not on air, target date: unknown: 102.1 mhz; 100 kw. Ant 963 ft TL: N41 55 05 W107 06 01. Hrs open: 288 S. River Rd., Bedford, NH, 03110. Phone: (603) 668-6400. Licensee: White Park Broadcasting Inc. ♦Steven A. Silberberg, pres.

Hudson

KTUG(FM)—Not on air, target date: unknown: 105.1 mhz; 6 kw. Ant 13 ft TL: N42 53 49 W108 34 59. Hrs open: 288 S. River Rd., Bedford, NH, 03110. Phone: (603) 668-6400. Fax: (603) 668-6470. Licensee: White Park Broadcasting Inc. ♦Steven A. Silberberg, pres.

Jackson

KJAX(FM)— 2000: 93.3 mhz; 100 kw. Ant 1,069 ft TL: N43 27 40 W110 45 09. Hrs open: c/o KSGT(AM) and KMTN(FM), Box 100, 83001. Phone: (307) 733-2120. Fax: (307) 733-4760. Licensee: Chaparral Broadcasting Inc. Group owner: Chaparral Communications (acq 3-29-2000; $393,787 for stock). Format: Country. ♦Scott Anderson, gen mgr.

***KMLT(FM)**—Not on air, target date: unknown: 88.3 mhz; 2.35 kw. Ant 1,073 ft TL: N43 27 40 W110 45 09. Hrs open: 2351 Sunset Blvd., Suite 170-218, Rocklin, CA, 95765. Phone: (916) 251-1600. Fax: (916) 251-1650. Licensee: Educational Media Foundation. (acq 7-23-2007; grpsl). ♦Mike Novak, sr VP.

KMTN(FM)—Listing follows KSGT(AM).

KSGT(AM)— July 20, 1962: 1340 khz; 1 kw-U. TL: N43 30 22 W110 45 16. Hrs open: Box 100, 83001. Secondary address: 645 S. Cache St. 83001. Phone: (307) 733-2120. Fax: (307) 733-4760. Web Site: www.jacksonholeradio.com. Licensee: Chaparral Broadcasting Inc. (group owner; (acq 11-30-92; $215,000 with KMER(AM) Kemmerer; FTR: 12-21-92) Population served: 8,000 Law Firm: Keck, Mahin & Cate. Format: Country. Target aud: 25 plus. ♦Del Ray, progmg dir; Scott Anderson, gen mgr, opns VP, gen sls mgr, prom VP & chief of engrg; Lynda John, traf mgr.

KMTN(FM)—Co-owned with KSGT(AM). Dec 16, 1974: 96.9 mhz; 48 kw. 940 ft TL: N43 27 42 W110 45 10. Stereo. Web Site: www.jacksonholeradio.com.16,000 Natl. Network: ABC. Format: AOR. Target aud: 18-54. ♦Mark Fishman, progmg dir; Lynda John, traf mgr.

***KURT(FM)**—Not on air, target date: unknown: 89.1 mhz; 2.2 kw. Ant 1,102 ft TL: N43 27 40 W110 45 09. Hrs open: 6080 Mt. Moriah Ext., Memphis, TN, 38115. Phone: (901) 375-9324. Licensee: Broadcasting for the Challenged Inc. ♦George S. Flinn Jr., pres & gen mgr.

***KUWJ(FM)**— November 1992: 90.3 mhz; 3 kw. 1,105 ft TL: N43 27 40 W110 45 09. Stereo. Hrs opn: 5 AM-midnight

Rebroadcasts KUWR(FM) Laramie 100%.
Department 3984, 1000 E. University Ave., Laramie, 82071. Phone: (307) 766-4240. Fax: (307) 766-6184. E-mail: wpr@uwyo.edu Web Site: www.wyomingpublicradio.net. Licensee: University of Wyoming. Population served: 15,000 Natl. Network: NPR, PRI. Format: News, class, progsv. News staff: 3; News: 42 hrs wkly. Target aud: 25-54; educated, upper-income professionals. Spec prog: Folk 10 hrs, jazz 6 hrs, state news 3 hrs wkly. ♦Jon B. Schwartz, gen mgr; Hank Arnold, dev dir; Roger Adams, progmg dir; Bob Beck, news dir; Larry Dean, engrg dir & chief of engrg.

KZJH(FM)— July 13, 1989: 95.3 mhz; 100 kw. 1,056 ft TL: N43 27 40 W110 45 09. Hrs open: 24 Box 2620, 83001. Phone: (307) 733-1770. Fax: (307) 733-4760. E-mail: kz95@blissnet.com Licensee: Chaparral Broadcasting Co. Group owner: Chaparral Communications (acq 8-25-00; $1.1 million). Natl. Network: ABC. Format: Classic rock. News staff: one. Target aud: 18-55. ♦Patricia Karnik, prom dir; Jay Martin, progmg dir; Dee Dee Dudley, news dir; Scott Anderson, gen mgr & chief of engrg.

Kemmerer

KAOX(FM)— Oct. 1, 1999: 107.3 mhz; 13.5 kw. Ant 948 ft TL: N41 50 18 W110 30 11. Stereo. Hrs opn: 24 Box 432, c/o KMER (AM), 83101. Secondary address: 436 Fossil Butte 83101. Phone: (307) 877-4422. Fax: (307) 877-5537. E-mail: kmer@onewest.net Licensee: Simmons-SLC, LS LLC. Group owner: Simmons Media Group (acq 4-19-2004; grpsl). Population served: 80,000 Natl. Network: CBS. Format: Adult standards. News staff: one; News: 2 hrs wkly. Target aud: 35 plus; male / female. ♦Jim Carroll, gen mgr; Jim Thoeny, chief of opns.

KMER(AM)— Dec 7, 1962: 940 khz; 240 w-D, 150 w-N. TL: N41 47 57 W110 32 44. Hrs open: Box 432, 83101. Secondary address: 436 Fossil Butte Dr. 83101. Phone: (307) 877-4422. Fax: (307) 877-5537. E-mail: kmer@onewest.net Licensee: Simmons-SLC, LS LLC. Group owner: Simmons Media Group (acq 5-20-2004; grpsl). Population served: 70,000 Format: Oldies. News staff: one; News: one hr wkly. Target aud: 25-54. Spec prog: News/talk 7 hrs, farm 3 hrs wkly. ♦Jim Carroll, gen mgr; Jim Thoeny, opns dir.

Lander

KDLY(FM)—Listing follows KOVE(AM).

KOVE(AM)— 1947: 1330 khz; 5 kw-D, 1 kw-N, DA-N. TL: N42 50 35 W108 44 38. Hrs open: 1530 Main St., 82520. Phone: (307) 332-5683. Fax: (307) 332-5548. E-mail: radio1@wyoming.com Web Site: www.kovekdly.com. Licensee: Fremont Broadcasting Inc. Population served: 40,000 Natl. Network: CBS. Format: C&W. Target aud: 25 plus. Spec prog: Talk 15 hrs wkly. ♦Joe Kenney, pres, gen mgr, gen sls mgr, progmg dir & traf mgr; Leslie Myers, news dir; Lincoln Scott, chief of engrg; Jay Bo Jackson, disc jockey.

Laramie

***KAIW(FM)**— 2006: 88.9 mhz; 400 w. Ant 645 ft TL: N41 17 46 W105 53 30. Hrs open:
Rebroadcasts KLVR(FM) Santa Rosa, CA 100%.
2351 Sunset Blvd., Suite 170-218, Rocklin, CA, 95765. Phone: (916) 251-1600. Fax: (916) 251-1650. Web Site: www.klove.com. Licensee: Educational Media Foundation. Natl. Network: K-Love. Format: Contemp Christian. ♦Richard Jenkins, pres; Mike Novak, VP; Keith Whipple, dev dir; David Pierce, progmg mgr; Ed Lenane, news dir; Sam Wallington, engrg dir; Karen Johnson, news rptr; Marya Morgan, news rptr; Richard Hunt, news rptr.

KARS-FM— Sept 23, 1974: 102.9 mhz; 100 kw. Ant 1,220 ft TL: N41 18 39 W105 27 12. Stereo. Hrs opn: 3201 E. Mulberry St., Unit H, Fort Collins, CO, 80524. Phone: (970) 674-2700. Phone: (970) 492-0999. Fax: (970) 407-0584. Web Site: www.oldies1029.com. Licensee: Regent Broadcasting of Ft. Collins Inc. Group owner: Regent Communications Inc. (acq 1-8-2004; $7.75 million with KKPL(FM) Cheyenne). Population served: 80,000 Wire Svc: CBS Format: Oldies. Target aud: 18-44. ♦Cal Hall, gen mgr & progmg dir; Miles Schallert, sls dir & sports cmtr; Bill Cody, progmg dir & disc jockey; Susan Moore, news dir; Quin Morrison, chief of engrg.

KCGY(FM)— Nov 7, 1983: 95.1 mhz; 100 kw. 1,070 ft TL: N41 18 34 W105 27 11. Stereo. Hrs opn: 24 Box 1290, 82073. Secondary address: 3525 Soldier Springs Rd. 82070. Phone: (307) 745-4888. Fax: (307) 742-4576. E-mail: andyhoefer@clearchannel.com Licensee: Clear Channel Broadcasting Licenses Inc. Group owner: Clear Channel Communications Inc. (acq 4-15-2002). Population served: 100,000 Natl. Rep: Katz Radio. Wire Svc: AP Format: Mainstream country. Target aud: 25-54. ♦Andrew W. Hoefer, gen mgr; Eric Henderson, gen sls mgr; John Phillips, progmg dir; Matt Schilv, chief of engrg.

KHAT(AM)— Feb 27, 1962: 1210 khz; 10 kw-D, 1 kw-N, DA-N. TL: N41 15 19 W105 33 01. Stereo. Hrs opn: 302 S. 2rd St., Suite 204, 82070. Phone: (307) 745-5208. Fax: (307) 745-8570. Licensee: Appaloosa Broadcasting Co. Group owner: Northeast Broadcasting Company Inc. (acq 3-2-2004; $160,000). Population served: 26,972 Natl. Network: ESPN Radio. Format: Sports. Spec prog: Farm one hr wkly. ♦Mike Schutta, gen mgr.

KIMX(FM)— 2002: 96.7 mhz; 6.5 kw. Ant 932 ft TL: N41 17 07 W105 26 41. Hrs open: 302 S. 2nd St., Suite 204, 82070. Phone: (307) 745-5208. Fax: (307) 745-8570. E-mail: mix967@fiberpipe.net Licensee: Appaloosa Broadcasting Co. Inc. Group owner: Northeast Broadcasting Company Inc. (acq 11-12-2003; $775,000). Format: Adult contemp. ♦Jim O'Reilly, gen mgr.

KOWB(AM)— Feb 20, 1948: 1290 khz; 5 kw-D, 1 kw-N, DA-2. TL: N41 17 02 W105 34 51. Hrs open: 24 Box 1290, 82073. Secondary address: 3525 Soldier Springs Rd. 82070. Phone: (307) 745-4888. Fax: (307) 742-4576. E-mail: andyhoefer@clearchannel.com Licensee: Clear Channel Broadcasting Licenses Inc. Group owner: Clear Channel Communications Inc. Population served: 25,000 Natl. Network: Fox News Radio. Natl. Rep: Katz Radio. Wire Svc: AP Format: News/talk, sports. News staff: one; News: 7 hrs wkly. Target aud: 25-54. ♦Andrew W. Hoefer, gen mgr; Eric Henderson, gen sls mgr; Frank Kelly, progmg dir & progmg mgr; Matt Schilz, chief of engrg.

KRQU(FM)— 2000: 104.5 mhz; 3 kw. Ant 951 ft TL: N41 17 08 W105 26 41. Hrs open: 302 S. 2nd St., Suite 204, 82070. Phone: (307) 745-5208. Fax: (307) 745-8570. E-mail: mix105@fiberpipe.net Licensee: Laramie Mountain Broadcasting LLC. Group owner: Kona Coast Radio LLC (acq 6-20-2002). Population served: 389,300 Format: Classic rock and roll. ♦Jim O'Reilly, gen mgr.

KUSZ(FM)— 2006: 98.7 mhz; 110 w. Ant 1,073 ft TL: N41 18 39 W105 27 12. Hrs open: 302 S. 2nd St., Suite 204, 82070. Phone: (307) 745-5208. Fax: (307) 745-8570. Licensee: Murray Grey Broadcasting Inc. (acq 12-21-2005; $750,000 with KVUW(FM) Wendover, NV). Format: Classic country. ♦Steven A. Silberberg, pres; Jim O'Reilly, gen mgr.

***KUWL(FM)**—Not on air, target date: unknown: 90.1 mhz; 90 w. Ant 1,030 ft TL: N41 18 36 W105 27 17. Hrs open: Box 3984, 82071-3984. Phone: (307) 766-4240. Fax: (307) 766-6184. Licensee: University of Wyoming. ♦Jon Schwartz, gen mgr.

***KUWR(FM)**— Sept 10, 1966: 91.9 mhz; 100 kw. 1,128 ft TL: N41 18 39 W105 27 12. Stereo. Hrs opn: 5 AM-midnight Department 3984, 1000 E. University Ave., 82071. Phone: (307) 766-4240. Fax: (307) 766-6184. E-mail: wpr@uwyo.edu Web Site: www.wyomingpublicradio.org. Licensee: University of Wyoming. Population served: 325,000 Natl. Network: NPR, PRI. Format: News, progsv, class. News staff: 3; News: 42 hrs wkly. Target aud: 25-54; educated, college graduates, professionals, upper income. Spec prog: Folk 10 hrs, jazz 6 hrs, state news 3 hrs wkly. ♦Jon Schwartz, gen mgr; Don Woods, dev VP & mus dir; Peg Arnold, dev dir, sls dir & adv dir; Bob Beck, news dir; Larry Dean, chief of engrg; Roger Adams, prom mgr, progmg dir & traf mgr; Jim Morgan, local news ed & news rptr; Aaron Alpern, news rptr; Renny MacKay, news rptr; Pat Gabriel, disc jockey.

***KUWY(FM)**—Not on air, target date: unknown: 88.5 mhz; 130 w. Ant 1,007 ft TL: N41 18 36 W105 27 17. Hrs open: Box 3984, 82071-3984. Phone: (307) 766-4240. Fax: (307) 766-6184. Licensee: University of Wyoming. ♦Jon Schwartz, gen mgr.

Lost Cabin

KWYW(FM)— 2001: 99.1 mhz; 50 kw. Ant 1,896 ft TL: N43 26 18 W107 59 37. Hrs open: 320 Senior Ave., Thermopolis, 82443. Phone: (307) 864-2119. Fax: (307) 864-3937. E-mail: kthe@directairnet.com Web Site: www.mykwyw.com. Licensee: Jimmy Ray Carroll. Group owner: Jimmy Ray Carroll Stns (acq 6-25-2001; $30,000 for CP). Format: Country. ♦Jimmy Carroll, pres; Cheerie Dorris, gen mgr; Amanda Plant, prom dir; Jeremy James, progmg dir; Mike St. Clair, news dir.

Lovell

KROW(FM)—Not on air, target date: unknown: 107.1 mhz; 64 kw. Ant 2,375 ft TL: N44 48 38 W107 55 18. Hrs open: 288 S. River Rd., Bedford, NH, 03110. Phone: (603) 668-6470. Licensee: White Park Broadcasting Inc. ♦Steven A. Silberberg, pres & gen mgr.

Lusk

KQWY(FM)—Not on air, target date: unknown: 96.3 mhz; 100 kw. Ant 1,958 ft TL: N42 41 05 W104 39 53. Hrs open: 1282 Smallwood Dr., Suite 372, Waldorf, MD, 20603. Phone: (202) 251-7589. Licensee: Alma Corp. ♦Dennis Wallace, pres.

Manville

KOUZ(FM)—Not on air, target date: unknown: 98.9 mhz; 100 kw. Ant 977 ft TL: N42 41 56 W104 35 18. Hrs open: 1282 Smallwood Dr., Suite 372, Waldorf, MD, 20603. Phone: (202) 251-7589. Licensee: Alma Corp. ♦Dennis Wallace, pres.

Marbleton

KFMR(FM)—Not on air, target date: unknown: 95.7 mhz; 7 kw. Ant 1,394 ft TL: N42 19 28 W110 19 12. Hrs open: Box 36148, Tucson, AZ, 85740. Phone: (520) 797-4434. Licensee: Skywest Media L.L.C. ♦Ted Tucker, gen mgr.

Midwest

KRVK(FM)— 2001: 107.9 mhz; 100 kw. Ant 1,948 ft TL: N42 44 37 W106 18 26. Hrs open: 150 N. Nichols, Casper, 82601. Phone: (307) 266-5252. Fax: (307) 235-9143. E-mail: krvk@clearchannel.com Web Site: theriver1079.com. Licensee: Clear Channel Broadcasting Licenses Inc. Group owner: Clear Channel Communications Inc. (acq 3-28-01). Format: Classic rock. ♦Robert Price, gen mgr; Walter Hawn, gen sls mgr; Donovan Short, progmg dir; Dave Nutter, chief of engrg; Dave Borino, traf mgr.

Mills

KHAD(FM)—Not on air, target date: unknown: 105.5 mhz; 3.4 kw. Ant 1,683 ft TL: N42 44 30 W106 18 29. Hrs open: 288 S. River Rd., Bedford, NH, 03110. Phone: (603) 668-6470. Licensee: White Park Broadcasting Inc. ♦Steven A. Silberberg, pres & gen mgr.

Newcastle

KASL(AM)— July 10, 1953: 1240 khz; 1 kw-U. TL: N43 50 47 W104 12 45. Hrs open: 24 933 W. Main St., 82701. Phone: (307) 746-4433. Fax: (307) 746-4435. E-mail: kasl@vcn.com Licensee: Val Rasmuson Cook (acq 4-9-2007; $76,000). Population served: 7,800 Law Firm: Smithwick & Belendiuk. Format: C&W. News staff: one; News: 35 hrs wkly. Target aud: General; town & county residents, children through adults. Spec prog: Farm 5 hrs, relg 2 hrs wkly. ♦Val Cook, CEO, gen mgr, opns mgr & gen sls mgr; Ed Schlup, progmg dir.

KRKI(FM)— 2003: 99.5 mhz; 400 w. Ant 146 ft TL: N43 49 57 W104 13 08. Hrs open: Box 969, 82701. Secondary address: 1807 W. Main St. 82701. Phone: (307) 746-9614. Licensee: Michael Radio Group. ♦Dick Hinker, gen mgr.

***KUWN(FM)**— 1998: 90.5 mhz; 400 w. 203 ft TL: N43 49 57 W104 13 08. Hrs open:
Rebroadcasts KUWR(FM) Laramie 100%.
Department 3984, 1000 E. University Ave., Laramie, 82071. Phone: (307) 766-4240. Fax: (307) 766-6184. E-mail: wpr@uwyo.edu Web Site: www.wyomingpublicradio.net. Licensee: University of Wyoming. Format: News, progsv, class. ♦Jon Schwartz, gen mgr; Peg Arnold, dev dir, sls dir & adv dir; Don Woods, mus dir; Bob Beck, news dir; Larry Dean, chief of engrg; Roger Adams, progmg dir & traf mgr.

Orchard Valley

KGAB(AM)—Licensed to Orchard Valley. See Cheyenne

***KWYC(FM)**— 2005: 90.3 mhz; 20 kw. Ant 425 ft TL: N41 13 01 W104 26 53. Stereo. Hrs opn: 24 CSN International, 4002 N. 3300 E., Twin Falls, ID, 83301. Phone: (208) 734-6633. Fax: (208) 736-1958. Web Site: www.csnradio.com. Licensee: CSN International. (group owner; (acq 1-9-2004; $1 for CP). Format: Contemp Chrisitan talk. ♦Mike Kestler, pres.

Pine Bluffs

KJJL(AM)—Not on air, target date: Dec 2006: 540 khz; 900 w-D, 700 w-N, DA-2. TL: N41 11 13 W104 11 05. Hrs open: 965 S. Irving St.,

Denver, 80219. Phone: (303) 935-1156. Licensee: Timothy C. Cutforth. Format: Family friendly var. ♦Timothy C. Cutforth, gen mgr.

KREO(FM)— December 2000: 105.3 mhz; 400 w. Ant 157 ft TL: N41 09 55 W104 04 31. (CP: 6 kw, ant 249 ft. TL: N41 17 17 W104 00 21). Stereo. Hrs opn: 24
Simulcast with KRRR(FM) Cheyenne.
2109 E. 10th St., Cheyenne, 82001. Phone: (307) 638-8921. Fax: (307) 638-8922. E-mail: news@1049krrr.com Web Site: www.1049krrr.com. Licensee: Chisholm Trail Broadcasting LLC Group owner: Northeast Broadcasting Company Inc. (acq 5-23-2005; $850,000 with KKAW(FM) Albin). Population served: 50,000 Format: Oldies. Target aud: 25-54. ♦Larry Proietti, gen mgr & progmg dir; Dan Conway, gen sls mgr; Ron Krob, chief of engrg.

Pine Haven

KWAP(FM)—Not on air, target date: unknown: 99.1 mhz; 100 kw. Ant 308 ft TL: N44 19 04 W104 46 31. Hrs open: 1453 Lynwood Ave., Fort Myers, FL, 33901. Licensee: Davao LLC. ♦Hursel Lee Adkins Jr., CEO.

Pinedale

KPIN(FM)— December 1997: 101.1 mhz; 211 w. -180 ft TL: N42 51 59 W109 52 08. Hrs open: Box 2000, 82941. Phone: (307) 367-2000. Fax: (307) 367-3300. E-mail: kpin@wyoming.com Licensee: Robert R. Rule dba Rule Communications. Format: Country, oldies. ♦Robert R. Rule, gen mgr.

***KUWX(FM)**— 2000: 90.9 mhz; 450 w. Ant 440 ft TL: N42 50 40 W109 55 24. Hrs open: Department 3984, 1000 E. Wyoming Ave., Laramie, 82071. Phone: (307) 766-4240. Fax: (307) 766-6184. E-mail: wpr@uwyo.edu Web Site: www.wyomingpublicradio.net. Licensee: University of Wyoming. Format: News, progsv, class. ♦Jon Schwartz, gen mgr; Peg Arnold, dev dir, sls dir & adv dir; Don Woods, mus dir; Bob Beck, news dir; Larry Dean, chief of engrg; Roger Adams, progmg dir & traf mgr.

Powell

KCGL(FM)— Nov 26, 2001: 104.1 mhz; 100 kw. Ant 1,942 TL: N44 29 46 W109 09 16. Hrs open: 24 Box 1210, Cody, 82414. Secondary address: 1949 Mountain View Dr., Cody 82414. Phone: (307) 578-5000. Fax: (307) 527-5045. E-mail: rgelder@bhmwy.com Web Site: www.thecornradio.com. Licensee: Legend Communications of Wyoming LLC. Group owner: Legend Communications L.L.C. (acq 4-3-02; $450,000). Format: Classic rock. ♦Larry Patrick, pres; Roger Gelder, exec VP; Carol Kary, VP, gen mgr & gen sls mgr; Tom Morrison, progmg dir; Tom Huge, news dir; Charles Dozier, chief of engrg; Karla McMillen, traf mgr.

KPOW(AM)— Mar 30, 1941: 1260 khz; 5 kw-D, 1 kw-N, DA-N. TL: N44 42 00 W108 46 00. Hrs open: 6 AM-1 AM Box 968, 82435. Secondary address: 912 Ln. 11 1/2 82435. Phone: (307) 754-5183. Phone: (307) 527-5949. Fax: (307) 754-9667. E-mail: kpow-z92@wir.net Licensee: Chaparral Broadcasting Inc. Group owner: Chaparral Communications (acq 11-30-92; $215,000 with co-located FM; FTR: 12-21-92) Population served: 40,000 Law Firm: Keck, Mahin & Cate. Format: News, Talk. News: 38 hrs wkly. Target aud: 29-64. Spec prog: Farm 19 hrs wkly. ♦Scott Anderson, gen mgr, progmg mgr & chief of engrg; Scott Mangold, sls dir, news dir & traf mgr.

***KUWP(FM)**— 2000: 90.1 mhz; 430 w. Ant 1,624 ft TL: N44 35 14 W108 51 08. Hrs open: Department 3984, 1000 E. Wyoming Ave., Laramie, 82071. Phone: (307) 766-4240. Fax: (307) 766-6184(. E-mail: wpr@uwyo.edu Web Site: www.wyomingpublicradio.net. Licensee: University of Wyoming. Format: News, progsv, class. ♦Jon Schwartz, gen mgr; Peg Arnold, dev dir, sls dir & adv dir; Don Woods, mus dir; Bob Beck, news dir; Larry Dean, chief of engrg; Roger Adams, progmg dir & traf mgr.

Rawlins

KIQZ(FM)—Listing follows KRAL(AM).

KRAL(AM)— February 1947: 1240 khz; 1 kw-U. TL: N41 46 55 W107 15 40. Hrs open: 24 2346 W. Spruce, 82301. Phone: (307) 324-3315. Fax: (307) 324-3509. E-mail: jackmorgan@vcn.com Web Site: www.kiqz-kral.com. Licensee: Mount Rushmore Broadcasting Inc. (group owner; (acq 8-6-93; $80,000 with co-located FM; FTR: 8-23-93). Population served: 15,123 Natl. Network: ABC. Rgnl. Network: Jones Satellite Audio. Wire Svc: UPI Format: Adult contemp. Target aud: 14 plus. ♦Jack Morgan, progmg dir & news dir.

KIQZ(FM)—Co-owned with KRAL(AM). Nov 12, 1981: 92.7 mhz; 3 kw. Ant 298 ft TL: N41 46 16 W107 14 15. Stereo. Web Site: www.kiqz-kral.com.10,000

Riverton

***KCWC-FM**— March 1974: 88.1 mhz; 3 kw. 1,449 ft TL: N42 34 59 W108 42 36. Stereo. Hrs opn: 24 2660 Peck Ave., 82501. Phone: (307) 855-2121. Phone: (307) 855-2268. Fax: (307) 856-3893. E-mail: dsmith@cwc.edu Licensee: Central Wyoming College. Format: Jazz, new age, progsv. ♦JoAnne McFarland, pres; Dale Smith, stn mgr.

KTAK(FM)—Listing follows KVOW(AM).

KTRZ(FM)— Dec 4, 1984: 93.1 mhz; 100 kw. 884 ft TL: N42 43 10 W108 08 41. Stereo. Hrs opn: 24 Box 808, 82501. Secondary address: 1002 N. 8th West 82501. Phone: (307) 856-2922. Fax: (307) 856-7552. E-mail: ktrz@tcinc.net Web Site: www.ktrzfm.com. Licensee: Jimmy Ray Carroll. Group owner: Jimmy Ray Carroll Stns (acq 4-1-02). Population served: 55,000 Natl. Network: AP Network News. Law Firm: Pepper & Corazzini. Wire Svc: AP Format: Adult contemp. News staff: one; News: 2 hrs wkly. Target aud: 25 plus; rgnl/loc tourists, agribusiness, core population. ♦Jim Carroll, CEO; Jim Hockett, gen mgr, stn mgr & gen sls mgr.

KVOW(AM)— July 2, 1948: 1450 khz; 1 kw-U. TL: N43 01 35 W108 20 45. Hrs open: 603 E. Pershing Ave., 82501. Phone: (307) 856-2251. Fax: (307) 856-0252. E-mail: kvow@wyoming.com Web Site: www.kvowradio.com. Licensee: Edwards Communications L.C. (group owner; (acq 6-22-99; $875,000 with co-located FM). Population served: 12,000 Format: Info, oldies. Spec prog: Farm 5 hrs wkly. ♦Larry Cross, gen mgr & gen sls mgr; Jeff Kehl, progmg dir & news dir; Lonnie Fairfield, chief of engrg; Tracy Coston, traf mgr; John Gabrielsen, sports cmtr.

KTAK(FM)—Co-owned with KVOW(AM). Dec 15, 1976: 93.9 mhz; 50 kw. 951 ft TL: N42 43 10 W108 08 45. Stereo. E-mail: ktak@wyoming.com Web Site: www.ktakradio.com. Format: Country.

Rock River

KKHI(FM)—Not on air, target date: unknown: 95.9 mhz; 3.4 kw. Ant 440 ft TL: N41 42 24 W105 54 48. Hrs open: 1063 Big Thompson Rd., Apt F, Loveland, CO, 80537-9424. Phone: (970) 669-9200. Fax: (970) 669-0800. Licensee: Kona Coast Radio LLC. ♦Victor A. Michael Jr., gen mgr.

Rock Springs

KQSW(FM)—Listing follows KRKK(AM).

KRKK(AM)— 1938: 1360 khz; 5 kw-D, 1 kw-N, DA-N. TL: N41 37 12 W109 14 20. Hrs open: Box 2128, 82902. Secondary address: 2717 Yellowstone Rd. 82901. Phone: (307) 362-3793. Fax: (307) 362-8727. E-mail: wyoradio@wyoradio.com Web Site: www.wyoradio.com. Licensee: Big Thicket Broadcasting Co. of Wyoming Inc. (acq 12-2-2005; grpsl). Population served: 11,657 Format: Oldies. ♦Bill Luzmoor, pres, pres & progmg dir; Jon Collins, gen mgr & chief of engrg; Tom Ellis, gen sls mgr; Doug Randall, news dir; Tim Walker, traf mgr.

KQSW(FM)—Co-owned with KRKK(AM). January 1977: 96.5 mhz; 100 kw. 1,680 ft TL: N41 25 54 W109 07 01.28,000 Format: Country.

KSIT(FM)— October 1981: 104.5 mhz; 100 kw. 1,630 ft TL: N41 26 00 W109 07 02. Stereo. Hrs opn: 24 Box 2128, 82902. Secondary address: 2717 Yellowstone Rd. 82902. Phone: (307) 362-7034. Phone: (307) 362-3793. Fax: (307) 362-8727. E-mail: wyoradio@wyoradio.com Web Site: www.wyoradio.com. Licensee: Big Thicket Broadcasting Co. of Wyoming Inc. (acq 12-2-2005; grpsl). Population served: 40000 Law Firm: Arent, Fox, Kintner, Plotkin & Kahn. Format: Classic rock. News staff: one; News: 10 hrs wkly. Target aud: 18-45; general. ♦Bill Luzmoor, pres; John Collins, gen mgr, opns mgr, progmg dir & chief of engrg; Tom Ellis, gen sls mgr; Doug Randall, news dir; Kim Walker, traf mgr.

***KUWZ(FM)**— November 1994: 90.5 mhz; 35 kw. Ant 1,679 ft TL: N41 25 39 W109 07 17. Stereo. Hrs opn: Sunrise-sunset
Rebroadcasts KUWR(FM) Laramie 100%.
Department 3984, 1000E. Wyoming Ave., Laramie, 82071. Phone: (307) 766-4240. Fax: (307) 766-6184. E-mail: wpr@uwyo.edu Web Site: www.wyomingpublicradio.net. Licensee: University of Wyoming. (group owner) Population served: 50,000 Natl. Network: NPR, PRI. Format: Class, progsv, news. News staff: 3; News: 52 hrs wkly. Target aud: 25-54; college graduates, professionals, mgrs, upper income. Spec prog: Folk 10 hrs, jazz 6 hrs, state news 3 hrs wkly. ♦Jon Schwartz, gen mgr; Peg Arnold, dev dir, sls dir & adv dir; Don Woods, mus dir; Bob Beck, news dir; Larry Dean, engrg dir & chief of engrg; Roger Adams, progmg dir & traf mgr.

KYCS(FM)— Oct 1, 1986: 95.1 mhz; 100 kw. 1,635 ft TL: N41 29 50 W109 20 36. Stereo. Hrs opn: 24 40 Shoshone Ave., Green River, 82902. Phone: (307) 362-6746. Fax: (307) 875-5847. E-mail: kugr@sweetwater.net Web Site: www.theradionetwork.net. Licensee: Faith Broadcasting Corporation. Population served: 50,000 Format: Top-40. ♦Faith Harris, pres; Steve Core, gen mgr; Jeff Driggs, gen sls mgr; Jasmine Weaver, progmg dir & mus dir; Ron Krob, chief of engrg.

Saratoga

KTGA(FM)—Not on air, target date: unknown: 99.3 mhz; 30 kw. Ant -7 ft TL: N41 28 24 W106 44 56. Hrs open: 3611 Cherry Hill Dr., Greensboro, NC, 27410. Licensee: United States CP LLC.

Sheridan

***KOHR(FM)**— 2004: 88.9 mhz; 500 w. Ant 7 ft TL: N44 47 54 W106 55 51. Hrs open: 24
KXEI(FM) Havre 100%.
Box 2426, Hevre, MT, 59501. Phone: (406) 265-5845. Fax: (406) 265-8860. Licensee: Hi-Line Radio Fellowship Inc. (acq 7-31-2003; $10,000 for CP). Wire Svc: AP Format: Christian Inspirational. ♦Ed Matter, gen mgr; Brenda Boyum, opns mgr; Roger Lonnquist, dev dir; Brian Jackson, progmg dir.

***KPRQ(FM)**— 2006: 88.1 mhz; 450 w. Ant 1,118 ft TL: N44 37 26 W107 07 02. Hrs open: Yellowstone Public Radio, 1500 University Dr., Billings, MT, 59101-0298. Phone: (406) 657-2941. Fax: (406) 657-2977. Web Site: www.yellowstonepublicradio.org. Licensee: Montana State University-Billings. Natl. Network: NPR. Format: News, classical, jazz. ♦Lois Bent, gen mgr.

KROE(AM)— Mar 18, 1961: 930 khz; 5 kw-D, 117 w-N. TL: N44 47 54 W106 55 51. Hrs open: 24 Box 5086, 82801. Secondary address: 1716 KROE Ln. 82801. Phone: (307) 672-7421. Fax: (307) 672-2933. E-mail: kimlove@sheridanmedia.com Web Site: www.sheridanmedia.com. Licensee: Lovcom Inc. (group owner). Population served: 30,000 Natl. Network: CBS. Law Firm: Pepper & Corazzini. Wire Svc: AP Format: News/talk. News staff: 3. Target aud: 25-54; general. ♦Kim Love, pres & gen mgr; Steve Sisson, opns dir, chief of engrg & engr; Jim Schellinger, gen sls mgr; Russ Davidson, progmg dir; Mary Jo Johnson, news dir; Liz Reynolds, traf mgr; Trevor Jackson, sports cmtr.

KZWY(FM)—Co-owned with KROE(AM). December 1977: 94.9 mhz; 75 kw. 1,207 ft TL: N44 37 20 W107 06 57. Stereo. 24 E-mail: info@sheridanmedia.com Web Site: www.sheridanmedia.com. Format: Classic rock. Target aud: 18-49; general. ♦Steve Sisson, chief of opns; Liz Reynolds, traf mgr; Trevor Jackson, sports cmtr.

***KSUW(FM)**— 1998: 91.3 mhz; 450 w. 1,161 ft TL: N44 56 09 W106 55 51. Hrs open:
Rebroadcasts KUWR(FM) Laramie 100%.
Department 3984, 1000 E. Wyoming Ave., Laramie, 82071. Phone: (307) 766-4240. Fax: (307) 766-6184. E-mail: wpr@uwyo.edu Web

Site: www.wyomingpublicradio.net. Licensee: University of Wyoming. Format: News, progsv, class. ♦Jon Schwartz, gen mgr; Peg Arnold, dev dir & adv dir; Don Woods, mus dir; Bob Beck, news dir; Larry Dean, chief of engrg; Roger Adams, progmg dir & traf mgr.

***KVLZ(FM)**—Not on air, target date: unknown: 89.9 mhz; 500 w. Ant -16 ft TL: N44 47 54 W106 55 51. Hrs open: 5700 West Oaks Blvd., Rocklin, CA, 95765. Phone: (916) 251-1600. Fax: (916) 251-1650. Licensee: Educational Media Foundation. (acq 3-23-2007; grpsl). ♦Richard Jenkins, pres.

***KWCF(FM)**— 2005: 88.9 mhz; 1 kw. Ant 961 ft TL: N44 36 10 W106 55 42. Hrs open: Box 1492, Gillette, 82717. Secondary address: CSN International, 3232 W. MacArthur Blvd., Santa Ana, CA 92704-6802. Phone: (307) 682-9553. Fax: (307) 682-8509. E-mail: calvarycomm@vcn.com Licensee: CSN International (group owner). Format: Relg.

KWYO(AM)— July 9, 1934: 1410 khz; 5 kw-D, 500 w-N. TL: N44 46 15 W106 55 37. Hrs open: 24 Box 5086, 82801-1387. Secondary address: 1716 Kroe Ln 82801. Phone: (307) 672-0701. Fax: (307) 672-2933. E-mail: info@sheridanmedia.com Web Site: www.sheridanmedia.com. Licensee: Lovcom Inc. (group owner; acq 9-11-03). Population served: 40,000 Law Firm: Pepper & Corazzini. Format: adult standards. News staff: one; News: 20 hrs wkly. Target aud: 25-54. ♦Kim Love, gen mgr; Steve Sisson, opns mgr; Jim Schellinger, gen sls mgr; Russ Davidson, progmg dir; Ace Young, news dir; Tony Questa, chief of engrg.

KYTI(FM)— September 1978: 93.7 mhz; 75 kw. 1,207 ft TL: N44 37 20 W107 06 57. Stereo. Hrs opn: 24 Box 5086, 1716 KROE Ln., 82801. Phone: (307) 672-7421. Fax: (307) 672-2933. Web Site: www.sheridanmedia.com. Licensee: Lovcom Inc. (group owner; acq 5-15-97). Population served: 30,000 Natl. Network: ABC. Format: Country. News: 4 hrs wkly. Target aud: 25-50. ♦Kim Love, gen mgr; Steve Sisson, opns mgr; Jim Schellinger, gen sls mgr; Russ Davidson, progmg dir; Ace Young, news dir; Tony Questa, chief of engrg.

Story

KZZS(FM)— Nov 13, 2003: 98.3 mhz; 100 kw. Ant 272 ft TL: N44 34 32 W106 52 23. Stereo. Hrs opn: 24 1221 Fort St., Buffalo, 82834. Secondary address: 610 Illinois St., Buffalo 82834. Phone: (307) 684-7070. Fax: (307) 684-7676. E-mail: kbbs@vcn.com Licensee: Legend Communications of Wyoming L.L.C. Group owner: Legend Communications LLC. (acq 5-31-00; $200,000 for CP). Population served: 64,000 Format: Hot adult hits. News staff: one; News: 6 hrs wkly. Target aud: 18-35. ♦Larry Patrick, CEO & pres; Smokey Wildeman, gen mgr; Ed Cwik, opns mgr & progmg dir; Charles Dozier, chief of engrg; Rita Conners, traf mgr.

Sundance

***KUWD(FM)**— 2000: 91.5 mhz; 430 w. Ant 1,591 ft TL: N44 28 35 W104 26 54. Hrs open: Department 3984, 1000 E. Wyoming Ave., Laramie, 82071. Phone: (307) 766-4240. Fax: (307) 766-6184. E-mail: wpr@uwyo.edu Web Site: www.wyomingpublicradio.net. Licensee: University of Wyoming. Format: News, progsv, class. ♦Jon Schwartz, gen mgr; Peg Arnold, dev dir, sls dir & adv dir; Don Woods, mus dir; Bob Beck, news dir; Larry Dean, chief of engrg; Roger Adams, progmg dir & traf mgr.

KYDT(FM)— November 1997: 103.1 mhz; 25.2 kw. 1,650 ft TL: N44 28 35 W104 26 54. Stereo. Hrs opn: 24
Rebroadcasts KBFS(AM) Belle Fourche 99.9%.
Box 787, Belle Fourche, SD, 57717. Phone: (605) 892-2571. Fax: (605) 892-2573. E-mail: wyodak@hotmail.com Web Site: www.kydt.com. Licensee: Ultimate Caps Inc. Population served: 150,000 Natl. Network: ESPN Radio, Westwood One, AP Network News. Rgnl. Network: Jones Satellite Audio. Wire Svc: AP Format: Country, sports, news, talk. News: 25 hrs wkly. Target aud: General; rural, suburban, country mus, sports fans. ♦Cynthia Grimmelmann, pres; Karl Grimmelmann, exec VP, gen mgr & opns VP.

Ten Sleep

KGCL(FM)—Not on air, target date: unknown: 105.1 mhz; 25 kw. Ant -82 ft TL: N44 01 39 W107 21 04. Hrs open: 5074 Dorsey Hall Dr., Suite 205, Ellicott City, MD, 21042. Phone: (410) 740-0250. Fax: (410) 740-7222. Licensee: Legend Communications of Wyoming LLC. ♦W. Lawrence Patrick, gen mgr.

Thayne

KTYN(FM)—Not on air, target date: unknown: 105.9 mhz; 370 w. Ant 2,332 ft TL: N43 06 18 W111 07 17. Hrs open: College Creek Media LLC, 980 N. Michigan Ave., Suite 1880, Chicago, IL, 60611. Phone: (312) 204-9900. Licensee: College Creek Media LLC. ♦Neal J. Robinson, pres & gen mgr.

Thermopolis

KDNO(FM)— Aug 30, 2001: 101.7 mhz; 16.25 kw. Ant 1,901 ft TL: N43 26 18 W107 59 37. Hrs open: Box 591, 82443-0501. Secondary address: 420 Arapahoe 82443. Phone: (307) 864-2119. Fax: (307) 864-3937. E-mail: kthe@directairnet.com Licensee: Carjim LLC. Group owner: Jimmy Ray Carroll Stns (acq 9-13-2001; $20,000 for CP). Format: Classic country. ♦Jim Carroll, pres; Dick Howe, gen mgr, gen sls mgr & progmg dir; Dennis Silver, chief of engrg.

KTHE(AM)— April 1957: 1240 khz; 1 kw-U. TL: N43 38 42 W108 12 15. Hrs open: Box 591, 82443-0591. Secondary address: 420 Arapahoe 82443. Phone: (307) 864-2119. Fax: (307) 864-3937. E-mail: kthe@directairnet.com Licensee: Carjim LLC. Group owner: Jimmy Ray Carroll Stns (acq 5-7-2002). Population served: 3,800 Format: Adult contemp, oldies. Target aud: 25-54; varied. ♦Jim Carroll, pres; Dick Howe, gen mgr, gen sls mgr, gen sls mgr & progmg dir.

KUWT(FM)— 2001: 91.3 mhz; 450 w. Ant 253 ft TL: N43 39 07 W108 15 07. Hrs open: Department 3984, 1000 E. Wyoming Ave., Laramie, 82071. Phone: (307) 766-4240. Fax: 307 766-6184. E-mail: wpr@uwyo.edu Web Site: www.wyomingpublicradio.net. Licensee: University of Wyoming. Format: News, progsv, class. ♦Jon Schwartz, gen mgr; Peg Arnold, dev dir, sls dir & adv dir; Don Woods, mus dir; Bob Beck, news dir; Larry Dean, chief of engrg; Roger Adams, progmg dir & traf mgr.

Torrington

KERM(FM)—Listing follows KGOS(AM).

KGOS(AM)— May 15, 1950: 1490 khz; 1 kw-U. TL: N42 04 20 W104 13 40. Hrs open: 5:30 AM-10:15 PM 760 Radio Rd., 82240. Phone: (307) 532-2158. Fax: (307) 532-2641. Licensee: Mount Rushmore Broadcasting Inc. (group owner) Population served: 30,000 Rgnl rep: Art Moore. Format: Country. Target aud: 20 plus; general. ♦Grant Kath, gen mgr.

KERM(FM)—Co-owned with KGOS(AM). Dec 15, 1976: 98.3 mhz; 3 kw. Ant 300 ft TL: N41 59 41 W104 12 05. Stereo. 30,000

Upton

KRUG(FM)—Not on air, target date: unknown: 104.5 mhz; 100 kw. Ant 682 ft TL: N44 25 17 W104 26 57. Hrs open: 288 S. River Rd., Bedford, NH, 03110. Phone: (603) 668-6400. Fax: (603) 668-6470. Licensee: White Park Broadcasting Inc. ♦Steven A. Silberberg, pres.

KWDU(FM)—Not on air, target date: unknown: 93.5 mhz; 2 kw. Ant 26 ft TL: N44 06 41 W104 37 40. Hrs open: 5331 Mount Alifan Dr., San Diego, CA, 92111-2622. Phone: (858) 277-4991. Fax: (858) 277-1365. Licensee: Horizon Christian Fellowship. ♦Tom Phillips, COO.

Wamsutter

KRAN(FM)—Not on air, target date: unknown: 94.7 mhz; 5.2 kw. Ant 328 ft TL: N41 44 43 W107 49 31. Hrs open: 288 S. River Rd., Bedford, NH, 03110. Phone: (603) 668-6400. Fax: (603) 668-6470. Licensee: White Park Broadcasting Inc. ♦Steven A. Silberberg, pres.

KYPT(FM)—Not on air, target date: unknown: 104.3 mhz; 100 kw. Ant 606 ft TL: N41 33 46 W108 16 56. Hrs open: 288 S. River Rd., Bedford, NH, 03110. Phone: (603) 668-6400. Fax: (603) 668-6470. Licensee: White Park Broadcasting Inc. ♦Steven A. Silberberg, pres.

Warren AFB

KOLT-FM— Aug 4, 1978: 92.9 mhz; 33 kw. Ant 607 ft TL: N41 04 35 W105 12 10. Stereo. Hrs opn: 24
Rebroadcasts KGRE(AM) Greeley, CO 100%.
1020 9th St., Suite 201, Greeley, CO, 80631. Phone: (970) 356-1452. Fax: (970) 356-8522. Web Site: www.tigre1450.com. Licensee: Tracy Corp. Group owner: Tracy Broadcasting Corp. Population served: 125,000 Format: Sp. ♦Michael Tracy, pres; Ricardo Salazar, gen mgr & stn mgr.

West Laramie

***KRWT(FM)**—Not on air, target date: unknown: 89.9 mhz; 100 kw vert. Ant 571 ft TL: N41 54 19 W106 32 37. Hrs open: CSN International, 4002 N. 3300 E., Twin Falls, ID, 83301. Phone: (208) 734-6633. Fax: (208) 736-1958. Web Site: www.csnradio.com. Licensee: CSN International (group owner). Format: Contemp Christian talk. ♦Mike Stocklin, gen mgr; Don Mills, progmg dir; Kelly Carlson, chief of engrg.

Wheatland

KPAD(FM)—Not on air, target date: unknown: 107.5 mhz; 50 kw. Ant 417 ft TL: N42 20 51 W105 01 54. Hrs open: 288 S. River Rd., Bedford, NH, 03110. Phone: (603) 668-6400. Licensee: White Park Broadcasting Inc. ♦Steven A. Silberberg, pres.

KYCN(AM)— Nov 16, 1960: 1340 khz; 250 w-U. TL: N42 02 44 W104 56 47. Hrs open: 24 Box 248, 82201. Secondary address: 450 E. Cole 32201. Phone: (307) 322-5926. Fax: (307) 322-5927. Fax: (307) 322-9300. E-mail: info@kycn-kzew.com Web Site: www.kycn-kzew.com. Licensee: Smith Broadcasting Inc. (acq 12-6-91; with co-located FM). Population served: 12,000 Natl. Network: ABC. Natl. Rep: Target Broadcast Sales. Law Firm: Bryan Cave. Format: Country. News staff: one; News: 14 hrs wkly. Target aud: General. Spec prog: Farm 7 hrs wkly. ♦Catherine Smith, gen sls mgr, mktg mgr, prom mgr & adv mgr; Derek Barton, news dir; Kent G. Smith, pres, gen mgr, mus dir, chief of engrg & traf mgr.

KZEW(FM)—Co-owned with KYCN(AM). February 1985: 101.7 mhz; 3 kw. 156 ft TL: N42 02 44 W104 56 47. Stereo. 24 Web Site: www.kycn-kzew.com. Natl. Network: Jones Radio Networks. Format: Adult contemp. ♦Kent G. Smith, chief of opns, progmg dir, pub affrs dir & disc jockey.

Worland

KKLX(FM)—Listing follows KWOR(AM).

KWOR(AM)— Mar 7, 1946: 1340 khz; 1 kw-U. TL: N44 01 01 W107 58 14. Hrs open: 1340 Radio Dr., 82401. Phone: (307) 347-3231. Fax: (307) 347-4880. E-mail: kwor@rtconnect.net Web Site: www.kworkklx.com. Licensee: KWOR Inc. (acq 3-21-97; $265,000 with co-located FM). Population served: 30,000 Law Firm: Eugene T. Smith. Format: Oldies. Target aud: General. ♦Bill Harrington, gen mgr, progmg dir & progmg dir; Tony Cuesta, chief of engrg; Nancy Harrington, traf mgr.

KKLX(FM)—Co-owned with KWOR(AM). Dec 1, 1980: 96.1 mhz; 50 kw. 400 ft TL: N44 04 06 W107 51 57. Stereo. 45,000 Format: Mus of the 80s & 90s & today. ♦Nancy Harrington, traf mgr.

Wright

KDDV-FM—Not on air, target date: unknown: 101.5 mhz; 100 kw. Ant 1,098 ft TL: N43 59 57 W105 15 15. Hrs open: Box 1179, Gillette, 82717-1179. Phone: (307) 686-2242. Fax: (307) 686-7736. Licensee: Legend Communications of Wyoming LLC. ♦Don Clonch, gen mgr.

KHRW(FM)—Not on air, target date: unknown: 92.7 mhz; 6 kw. Ant 328 ft TL: N43 44 32 W105 28 14. Hrs open: 5331 Mt. Alifan Dr., San Diego, CA, 92111. Phone: (858) 277-4991. Fax: (858) 277-1365. Web Site: www.horizonsd.org/radio.asp. Licensee: Horizon Christian Fellowship. ♦Mike MacIntosh, pres.

Vieques

WVIS(FM)— June 10, 1973: Stn currently dark. 106.1 mhz; 9 kw. Ant 892 ft TL: N17 44 51 W64 50 11. (CP: 32 kw, ant 485 ft. TL: N18 19 19 W65 17 59). Hrs opn: c/o Michael Bahr, Calle Guajataca #153, San Juan, PR, 00926. Phone: (787) 756-5914. Licensee: V.I. Stereo Communications Corporation (PR) (acq 8-14-2007). ♦Michael Gregory Bahr, pres.

American Samoa

Fagaitua

KHZF(FM)—Not on air, target date: unknown: 103.1 mhz; 4.7 kw. Ant 1,463 ft TL: S14 16 12 W170 41 10. Hrs open: 5331 Mt. Alifan Dr., San Diego, CA, 92111. Phone: (858) 277-4991. Fax: (858) 277-1365. Licensee: Horizon Christian Fellowship. (acq 2-9-2006; grpsl). ♦Mike MacIntosh, pres.

Leone

KKHJ(AM)—Not on air, target date: unknown: 900 khz; 5 kw-D, 3 kw-N. TL: S14 20 24 W170 46 22. Hrs open: Box 6758, Pago Pago, 96799. Phone: (684) 633-7793. Fax: (684) 633-4493. Licensee: South Seas Broadcasting Inc. ♦Larry G. Fuss, pres; Joey Cummings, gen mgr.

KNWJ(FM)— 2001: 104.7 mhz; 280 w. Ant 1,499 ft TL: N14 19 21 W170 45 47. Stereo. Hrs opn: 24 Box 997777, Pago Pago, 96799. Phone: (684) 699-8123. Fax: (684) 699-8126. E-mail: info@fm104.org Web Site: www.fm104.org. Licensee: Showers of Blessings Radio (acq 5-26-00; $70,000 for CP). Population served: 30000 Format: Today's Christian hits. ♦Dan Dalle, gen mgr.

WVUV(AM)— Apr 16, 1975: 648 khz; 10 kw-U. TL: S14 21 28 W170 46 36. Hrs open: Box 6758, Pago Pago, 96799. Phone: (684) 633-7793. Fax: (684) 633-4493. Web Site: www.southseasbroadcasting.com. Licensee: South Seas Broadcasting Inc. ♦Joey Cummings, gen mgr.

Pago Pago

KKHJ-FM— May 1, 2000: 93.1 mhz; 420 w. Ant 1,489 ft TL: N14 16 12 W170 41 10. Stereo. Hrs opn: 24 Box 6758, 96799. Phone: (684) 633-7793. Phone: (702) 898-4669. Fax: (684) 633-4493. Web Site: www.93khj.fm. Licensee: South Seas Broadcasting. Group owner: Contemporary Communications. Population served: 55,000 Law Firm: Wood, Maines & Brown, Chartered. Format: CHR. News staff: one. Target aud: General. ♦Larry G. Fuss, pres; Joey Cummings, gen mgr.

KSBS-FM— Apr 14, 1988: 92.1 mhz; 3 kw. -135 ft TL: S14 17 41 W170 39 44. (CP: 15 kw, ant -92 ft.). Stereo. Hrs opn: 6:00am-Midnight Box 793, 96799-0793. Phone: (684) 633-7000. Fax: (684) 633-5727. E-mail: prescott.esther@ksbsfm.com Web Site: www.ksbsfm.com. Licensee: Samoa Technologies Inc. Format: Adult contemp. Target aud: Working adults. ♦Barney Sene, pres; Esther Prescott, gen mgr.

Tafuna

KJAL(AM)— 2005: 585 khz; 5 kw-U. TL: S14 21 28 W170 46 36. Hrs open: Box 218, Pago Pago, 96799. Phone: (684) 699-2253. Licensee: District Council of the Assemblies of God in AS. ♦Vickie Haleck, stn mgr.

Guam

Agat

***KSDA-FM**— Nov 22, 1990: 91.9 mhz; 3.8 kw. 1,000 ft TL: N13 25 53 W144 42 36. Stereo. Hrs opn: 24 290 Chalan Palasyo, Agana Heights, 96910. Phone: (671) 472-5732. Fax: (671) 477-4678. E-mail: mail@joy92.net Web Site: www.joy92.net. Licensee: Good News Broadcasting Corp. (acq 8-9-01; charitable contribution). Population served: 180,000 Format: Inspirational, Christian. News: 7 hrs wkly. Target aud: 25-54. ♦Robert Gibbons, chmn & pres; Matt Dodd, gen mgr & stn mgr.

Barrigada

***KHMG(FM)**— Mar 26, 1996: 88.1 mhz; 8 kw. 472 ft TL: N13 29 17 W144 49 53. Stereo. Hrs opn: 24 Box 23189, 170C Machaute St., 96921. Phone: (671) 477-6341. Fax: (671) 477-7136. E-mail: khmg@harvestministries.net Web Site: www.harvestministries.net. Licensee: Harvest Christian Academy. Population served: 140,000 Law Firm: Garvey, Schubert & Barer. Format: Relg, Christian. News: 5 hrs wkly. Target aud: General; Christian, church and school families. Spec prog: Children 2 hrs wkly. ♦Dr. Marty Herron, pres; John Collier, stn mgr & progmg dir.

Dededo

KGUM-FM— Feb 28, 1999: 105.1 mhz; 12 kw. Ant 502 ft TL: N13 29 17 W144 49 30. Hrs open: 111 W. Santo Papa, Suite 800, Hagatna, 96910. Phone: (671) 477-5700. Fax: (671) 477-3982. Web Site: www.105therock.com. Licensee: Sorensen Pacific Broadcasting Inc. (group owner; acq 6-23-03; grpsl). Format: Rock. ♦Jon Anderson, pres; Rex W. Sorensen, CEO, chmn & gen mgr; Albert Juan, stn mgr.

Hagatna

KGUM(AM)— February 1975: 567 khz; 10 kw-U, DA-1. TL: N13 23 21 E144 45 34. Hrs open: 24 111 W. Chalan Santo Papa, Suite 800, 96910. Phone: (671) 477-5700. Fax: (671) 477-3982. Web Site: www.radiopacific.com. Licensee: Sorensen Pacific Broadcasting Inc. (group owner; (acq 6-23-2003; grpsl). Population served: 165,000 Natl. Network: CBS, Westwood One. Format: News/talk. News staff: 5; News: 20 hrs wkly. Target aud: 35 plus; adults with high income & education. Spec prog: Educ one hr, computer 3 hrs, police one hr, health & fitness one hr, environmental one hr, drug recovery one hr wkly. ♦Rex Sorensen, VP & gen mgr; Ray Gibson, opns mgr.

KZGZ(FM)—Co-owned with KGUM(AM). December 1986: 97.5 mhz; 40 kw. 538 ft TL: N13 29 17 W144 49 30. Stereo. 24 150,000 Format: Hip hop. News: 15 hrs wkly. Target aud: 18-34; young affluent adults.

KISH(FM)— 2003: 102.9 mhz; 25 kw. Ant 535 ft TL: N13 29 17 W144 49 53. Stereo. Hrs opn: 24 Inter-Island Communications Inc., 1868 Halsey Dr., Piti, 96915. Phone: (671) 477-9448. Phone: (671) 477-5474. Fax: (671) 477-6411. Licensee: Inter-Island Communications Inc. (group owner). Population served: 160,000 Format: Chamorro music-language of Guam and Marianas Islands. News staff: one; News: 15 hrs wkly. Target aud: General; indeginous residents of Marianas Islands. ♦Edward H. Poppe Jr., pres & gen mgr; Frances W. Poppe, CFO; Edward H. Poppe III, exec VP; Rosalin Koss, progmg dir.

KOKU(FM)— Apr 28, 1984: 100.3 mhz; 5 kw. 190 ft TL: N13 26 28 W144 42 40. Stereo. Hrs opn: 24 424 W. O'Brien Dr., 107 Julace Ctr., 96910. Fax: (671) 477-5658. E-mail: marketing@hitradio.com Web Site: www.hitradio100.com. Licensee: Moy Communications Inc. (acq 4-29-2004; $350,000). Population served: 149,000 Law Firm: Dow, Lohnes & Albertson. Format: Top-40. Target aud: 18-34; females. ♦Kurt S. Moylan, pres; Rick Nauta, progmg dir & chief of engrg; Vince R. Limuaco, gen mgr, sls & mktg.

***KPRG(FM)**— Jan 27, 1994: 89.3 mhz; 2.8 kw. 485 ft TL: N13 29 17 W144 49 30. Stereo. Hrs opn: 24 KPRG, UoG Stn., Mangilao, 96923. Phone: (671) 734-8930. Fax: (671) 734-2958. E-mail: kprg@kprg.org Web Site: www.kprg.org. Licensee: Guam Educational Radio Foundation. Natl. Network: NPR. Format: News, div, class. ♦Denise Mendiola, gen mgr; Olympia Terral, dev dir; Lydia Taleu, progmg dir & disc jockey.

KSTO(FM)— September 1973: 95.5 mhz; 25 kw. 530 ft TL: N13 29 17 E144 49 53. Stereo. Hrs opn: 24 1868 Halsey Dr., Piti, 96915. Phone: (671) 477-7108. Phone: (671) 477-5786. Fax: (671) 477-6411. E-mail: ksto@ite.net Licensee: Inter-Island Communications Inc. (group owner; acq 11-77). Population served: 160,000 Format: Adult contemp. News staff: one; News: 14 hrs wkly. Target aud: 25-54. Spec prog: Country 12 hrs, gospel 6 hrs wkly. ♦Edward H. Poppe Jr., pres & gen mgr; Frances Poppe, CFO; Edward H. Poppe III, exec VP; Michelle Poppe-Aguon, opns dir; Rosalin Koss, progmg dir.

KTKB-FM— 2003: 101.9 mhz; 50 kw. Ant 492 ft TL: N13 29 16 W144 49 36. Hrs open: KM Broadcasting of Guam L.L.C., 3654 W. Jarvis Ave., Skokie, IL, 60076. Phone: (847) 674-0864. Phone: (671) 647-1019. Fax: (847) 674-9188. Fax: (671) 648-1019. Web Site: www.ktkb.com. Licensee: KM Broadcasting of Guam L.L.C. Group owner: KM Communications Inc. Format: Filipino. ♦Myoung Bae, pres; Kevin Bae, gen mgr; Rolly Manumtag, progmg dir.

KTWG(AM)— August 1975: 801 khz; 10 kw-U. TL: N13 27 07 W144 42 32. Hrs open: 24
Simulcasts KCNM(AM) Saipan, Northern Mariana Islands.
1868 Halsey Dr., Asan, NC, 96910. Phone: (671) 477-5894. Fax: (671) 477-6411. E-mail: ktwg@ktwg.com Web Site: www.ktwg.com. Licensee: Edward H. Poppe Jr. and Frances W. Poppe. Group owner: Inter-Island Communications Inc. (acq 2-20-2002). Population served: 160,000 Format: Relg. News staff: one; News: 5 hrs wkly. Target aud: 25-49; Christian. ♦K. Leilani S. Dahilig, stn mgr.

KUAM(AM)— Mar 14, 1954: 630 khz; 10 kw-U. TL: N13 26 53 E144 45 22. Hrs open: 600 Harmon Loop Rd. #102, Dededo, 96912. Phone: (671) 637-5826. Fax: (671) 637-9865. Web Site: www.kuam.com/i94. Licensee: Pacific Telestations Inc. (acq 9-27-77). Natl. Network: CBS. Format: MOR, adult contemp. Spec prog: Chamorro 5 hrs, Tagalog 5 hrs, Japanese 3 hrs wkly. ♦Joey Calvo, gen mgr.

KUAM-FM— Sept 1, 1966: 93.9 mhz; 2 kw. 950 ft TL: N13 25 53 E144 42 36. (CP: 5.2 kw, ant 948 ft.). Phone: (671) 637-0094. Web Site: www.kuam.com/i94. Format: Urban contemp, rhythm & blues. Co-owned TV: KUAM-TV affil.

KVOG(AM)—Not on air, target date: unknown: 1530 khz; 250 w-U. TL: N13 27 24 W144 40 20. Hrs open: 1100 Alakea, Suite 1800, Honolulu, HI, 96813-2839. Phone: (808) 521-4711. Fax: (808) 538-3269. Licensee: Guam Power II Inc. ♦Wagdy Guirguis, pres.

Tamuning

KTKB(AM)—Not on air, target date: unknown: 675 khz; 10 kw-U. TL: N13 34 22 W144 51 47. Hrs open: KM Broadcasting of Guam L.L.C., 3654 W. Jarvis Ave., Skokie, IL, 60076. Phone: (847) 674-0864. Fax: (671) 648-1019. Licensee: KM Broadcasting of Guam L.L.C. ♦Myoung Bae, pres; Kevin Bae, gen mgr.

Tumon

KIJI(FM)—Not on air, target date: unknown: 104.3 mhz; 850 w. Ant 72 ft TL: N13 30 37.3 E144 48 17.6. Hrs open: La Casa de Colina, 3rd Fl., 200 Chichirica St., Tamuning, 96913-4217. Phone: (671) 647-4703. Licensee: Guam Broadcast Services Inc. ♦Gary W.F. Gumataotao, pres & gen mgr.

Puerto Rico

Adjuntas

WOQI(AM)— 1997: 1020 khz; 1 kw-D, 280 w-N. TL: N18 09 04 W66 42 48. Hrs open: Box 982, 00601. Phone: (787) 829-1453. Fax: (787) 829-1453. E-mail: Coki@coqui.net Licensee: WPAB Inc. (acq. 6-23-01; $450,000). Population served: 20,000 Format: Sp, var/div. News: 15 hrs wkly. Target aud: General; General Public. ♦Alphoso Jimenez-Luchete, gen mgr.

Aguada

WFDT(FM)— 1975: 105.5 mhz; 3 kw. 1,036 ft TL: N18 18 57 W67 10 54. Stereo. Hrs opn: 24 Box 363222, San Juan, 00936. Phone: (787) 758-1300. Fax: (787) 754-1395. Web Site: www.fidelitypr.com. Licensee: Arso Radio Corp. Group owner: Uno Radio Group (acq 4-19-01; $3.2 million). Format: Adult contemp, easy lstng. News staff: 4. Target aud: 25-49; middle class. ♦Luis Soto, pres.

Aguadilla

WABA(AM)— Nov 15, 1951: 850 khz; 5 kw-D, 1 kw-N. TL: N18 24 02 W67 09 27. Hrs open: Box 188, No. 6, Calle Munoz Rivera, 00605. Phone: (787) 891-1230. Fax: (787) 882-2282. E-mail: wabaradio@hotmail.com Web Site: www.waba850.com. Licensee: Aquadilla Radio & TV Corp. Inc. (acq 1973). Format: Adult contemp, Sp. ♦Hector Richard Carbona, pres; Rosipa Pellot, gen mgr; Mereida Nieves, gen sls mgr, news dir & traf mgr; Tito R. Areicaga, adv dir & progmg dir; Juan Rivera, chief of engrg.

WIVA-FM— Apr 16, 1964: 100.3 mhz; 22 kw. Ant 2,014 ft TL: N18 09 07 W66 59 15. Stereo. Hrs opn: 24 Box 3822, Mayaguez, 00681. Phone: (787) 834-2320. Fax: (787) 831-7969. E-mail: ventas@unoradio.com Web Site: www.salsoul.com. Licensee: Arso Radio Corp. Group owner: Uno Radio Group (acq 3-85). Format: Salsa. News staff: one; News: 5 hrs wkly. Target aud: 12 plus. ♦Jesus M. Soto, CEO & pres; Luis A. Soto, pres & chief of engrg; Maida Bedaya, gen mgr; Raymond Totti, gen sls mgr; Anthony Soto, progmg dir.

WTPM(FM)— May 27, 1971: 92.9 mhz; 50 kw. 1,223 ft TL: N18 18 52 W67 10 58. Stereo. Hrs opn: 18 Box 1629, Mayaguez, 00681. Phone: (787) 831-9200. Phone: (787) 834-6340. Fax: (787) 831-9292. Web Site: www.wtpm.com. Licensee: Corp. of the 7th Day Adventists of West Puerto Rico (acq 2-80; $125,000). Population served: 300,000 Format: Sp adult contemp. News: 11 hrs wkly. Target aud: General; traditional Christian groups. Spec prog: English one hr, class 7 hrs wkly. ♦Daniel Ponce, gen mgr & opns dir.

WWNA(AM)— 1956: 1340 khz; 950 w-U. TL: N18 24 00 W67 09 48. Hrs open: Box 7, Moca, 00676. Secondary address: Rd. 111, Aquadilla 00605. Phone: (787) 252-1730. Fax: (787) 868-1340. Licensee: Dominga Barreto Santiago (acq 1-25-2005; $500,000). Format: Sp var. Target aud: 20-55. Spec prog: Jazz 3 hrs wkly. ♦Aureo Matos, gen mgr, gen sls mgr & progmg dir; Ron Cushing, chief of engrg; Felix Gonzalez, disc jockey.

Arecibo

WCMN(AM)— June 24, 1947: 1280 khz; 5 kw-D, 1 kw-N. TL: N18 28 52 W66 41 16. Hrs open: 24 Box 436, 00613. Secondary address: 55 Gonzalo Marin St. 00612. Phone: (787) 878-0070. Phone: (787) 781-6303. Fax: (787) 880-1112. E-mail: wcmn@xsn.net Licensee: Caribbean Broadcasting Corp. Group owner: Uno Radio Group (acq 4-7-2004; $5.75 million for stock with co-located FM). Population served: 450,000 Law Firm: Fisher, Wayland, Cooper, Leader & Zaragoza L.L.P. Format: Sp, news/talk. News staff: 3; News: 50 hrs wkly. Target aud: 30 plus. ♦Byron Mitchell, VP & gen mgr; Maria M. Mitchell, sls dir, prom dir & progmg dir; Juan Rivera, chief of engrg; Jacquelyn Rames, traf mgr.

WCMN-FM— Jan 1, 1967: 107.3 mhz; 50 kw. 3,000 ft TL: N18 14 52 W66 48 43. Stereo. 24 Web Site: www.delta107.com.1,000,000 Format: Sp, Top-40. Target aud: 18-42; young adults. ♦Jacquelyn Rames, traf mgr.

WMIA(AM)— Feb 21, 1957: 1070 khz; 500 w-D, 2.5 kw-N. TL: N18 27 33 W66 45 20. Hrs open: 19 Box 1055, 00613-1055. Secondary address: 1168 Miramar Ave. 00612. Phone: (787) 878-1275. Phone: (787) 878-2727. Fax: 787-878-1275. E-mail: epifanioro@gmail.com Licensee: Abacoa Radio Corp. Population served: 1,250,000 Law Firm: Booth, Freret, Imlay & Tepper. Format: Sp talk, tropical. Target aud: 25 plus; the buying power in the area. ♦Epifanio Rodriguez-Velez, gen mgr & chief of opns.

WNIK(AM)— 1957: 1230 khz; 1 kw-U. TL: N18 27 20 W66 44 24. Hrs open: Box 142041, 00614. Phone: (787) 880-2461. Fax: (787) 879-1011. Licensee: Unik Broadcasting System Corp. (acq 12-10-2004; $335,000). Format: Div, Sp. Target aud: General. ♦Manuel Santiago, gen mgr.

WNIK-FM— July 17, 1965: 106.5 mhz; 25 kw. Ant 20 ft TL: N18 27 20 W66 44 24. Stereo. Hrs opn: Box 556, 00613. Phone: (787) 880-2613. Fax: (787) 879-1011. Licensee: Kelly Broadcasting System Inc. (acq 4-87). Format: Ballads. ♦Raul Santiago, gen mgr.

Barceloneta-Manati

WBQN(AM)— Mar 1, 1975: 1160 khz; 5 kw-D, 2.5 kw-N, DA-D. TL: N18 26 23 W66 33 07. Hrs open: Box 1625, Manati, 00674. Phone: (787) 854-2450. Phone: (787) 854-3738. Fax: (787) 854-3738. E-mail: riveraolmo@hotmail.com Licensee: Radio Borinquen Inc. Format: Top-40 CHR, Sp. Target aud: General. ♦Angel M. Rivera, pres; Luis R. Rivera Jr., gen mgr.

Barranquitas

WOLA(AM)— March 1986: 1380 khz; 1 kw-U. TL: N18 11 01 W66 18 24. Hrs open: Box 669-A, Carr 719 KMo1 Bo Halachal, 00794. Phone: (787) 857-1380. Fax: (787) 857-1381. E-mail: wola@prtc.com Licensee: R and R Broadcasting Format: Sp/tropical. Target aud: General. Spec prog: Jazz 5 hrs wkly. ♦Edgardo Rivera, pres, gen mgr & progmg dir; Edgardo Riviera, gen sls mgr; Jesus R. Gomez, chief of engrg.

Bayamon

WLUZ(AM)— 1966: 1600 khz; 5 kw-U. TL: N18 21 38 W66 09 30. Hrs open: 18 Box 9394, San Juan, 00908-0394. Secondary address: 403 Del Parque, 15 th Fl., Santurce 00912-3709. Phone: (787) 785-1600. Phone: (787) 729-1600. Fax: (787) 785-2094. Fax: (787) 723-8685. E-mail: ttrelles@yahoo.com Licensee: Marketing Promotion Network Inc. (acq 11-1-98; $1.6 million plus $800,000 penthouse). Population served: 2,000,000. Law Firm: Fletcher, Heald & Hildreth, P. L. C. Format: Romantic ballads, nostalgia, Sp, talk. Target aud: 35 plus. Spec prog: Comedy. ♦Tony Trelles, pres; Martha Villanueva, opns dir.

WODA(FM)— Dec 3, 1959: 94.7 mhz; 31 kw. Ant 1,837 ft TL: N18 16 44 W65 51 12. Stereo. Hrs opn: 24 Box 949, Guayanbo, 00970-0949. Secondary address: Amelia Industrial Park, Calle Frances 42, Guaynabo 00968. Phone: (787) 622-9700. Fax: (787) 622-9478. Web Site: www.onda94.com. Licensee: WLDI Inc. Group owner: Spanish Broadcasting System Inc. (acq 11-29-99; grpsl). Law Firm: Booth, Freret, Imlay & Tepper. Format: Sp, Top-40. Target aud: 12-24; males & females, middle/upper socio-economic. ♦Raul Alarcon, pres; Ismael Nieves, gen mgr; Marie Elena Martinez, gen sls mgr; Luis Enrique Rivera, prom VP; Jose Nelson Diaz, progmg dir; Alejandro Luciano, chief of engrg; Demare Ramirez, traf mgr.

WRSJ(AM)— 1947: 1560 khz; 5 kw-D, 750 w-N. TL: N18 24 05 W66 07 14. Hrs open: 1554 Bori St., San Juan, 00927-6113. Phone: (787) 274-1800. Fax: (787) 281-9758. Licensee: International Broadcasting Corp. (group owner; (acq 7-6-2004; $1.45 million with WCHQ(AM) Quebradillas). Format: Sp. ♦Pedro Roman Collazo, pres; Margarita Nazario, gen mgr.

WXYX(FM)— Feb 1, 1979: 100.7 mhz; 50 kw. 1,092 ft TL: N18 16 58 W66 10 47. Stereo. Hrs opn: 24 HC67 Box 15390, 00956-9535. Secondary address: Rd 174, KM 5.0 Bo. Guaraguao 00956-9535. Phone: (787) 785-9390. Phone: (787) 785-9100. Fax: (787) 785-9377. E-mail: info@lax.fm Web Site: www.lax.fm. Licensee: RAAD Broadcasting Corp. Population served: 1,700,000 Format: Top 40 pop. Target aud: 12-34; young teens, adults. ♦Roberto Davila, pres, gen mgr & opns mgr; Eduardo Cora, gen sls mgr; Michelle Torres, prom dir; Herman Davila, progmg dir; Juan Rivera, chief of engrg; Edwardo Carrasguillo, traf mgr.

Cabo Rojo

WMIO(FM)— Jan 10, 1988: 102.3 mhz; 3 kw. Ant 680 ft TL: N17 59 37 W67 10 27. Stereo. Hrs opn: 6 AM-midnight Box 9023916, San Juan, 00902. Phone: (787) 798-7878. Fax: (787) 620-0720. Licensee: Arso Radio Corp. (acq 2-9-2007; $3.25 million). Law Firm: Reddy, Begley & McCormick. Format: Adult contemp. Target aud: 18-45; general. ♦Alan Mejia, gen mgr.

WYAC(AM)— Jan 9, 1970: 930 khz; 2.5 kw-U. TL: N18 06 05 W67 09 17. Hrs open: 21 Box 489, Mayaguez, 00681. Secondary address: Radio Centre, Post & Bosgue Sts., Mayaguez 00684. Phone: (787) 620-9898. Web Site: www.radiopr740.com. Licensee: Bestov Broadcasting Inc. (acq 4-14-99; $3.65 million with co-located FM). Population served: 650,000 Law Firm: Reddy, Begley & McCormick. Format: News/talk, Sp. Target aud: General; Mayaguez county residents. ♦Luis Majia, pres; Francisco Acosta, gen mgr.

Caguas

WBRQ(FM)—See Cidra

WNEL(AM)— July 21, 1947: 1430 khz; 5 kw-U. TL: N18 14 53 W66 01 25. Stereo. Hrs opn: 24 Box 487, 00726-0487. Phone: (787) 744-3131. Fax: (787) 743-0252. E-mail: leon@unoradio.com Licensee: Turabo Radio Corp. Group owner: Uno Radio Group (acq 4-1-73). Law Firm: John P. Bankson Jr. Format: Latin Oldies. News: 15. Target aud: 24 plus. ♦Jesus M. Soto, CEO & chmn; Luis M. Soto, pres; Luis Gonzales, CFO; Elba Esmirria, VP; Luis Leon, gen mgr, gen sls mgr & progmg dir; Tanya Ramos, mktg dir; Jaime Soto, progmg VP; Alberto Pereira, chief of engrg.

WPRM-FM—See San Juan

WVJP(AM)— Nov 24, 1947: 1110 khz; 2.5 kw-D, 500 w-N. TL: N18 13 25 W66 01 11. Hrs open: 24 Box 207, 00726. Secondary address: Tomas de Castro #2 00626. Phone: (787) 743-5790. Fax: (787) 746-6996. Licensee: Borinquen Broadcasting Co. Inc. (acq 7-7-99; $700,000 for 12). Format: Adult contemp, btfl mus, Sp. ♦Jancel Pereira, CEO; Bienvenido Rodriguez, gen mgr; Norma Rodriquez-Trinidad, progmg dir; Jesus R. Gomez, chief of engrg.

WVJP-FM— October 1968: 103.3 mhz; 28 kw. 1,906 ft TL: N18 16 41 W65 51 09. Stereo. Format: Sp, romantic.

Camuy

WDIN(FM)— Aug 15, 1968: 102.9 mhz; 50 kw. 303 ft TL: N18 28 49 W66 51 14. Hrs open: Box 780, 00627. Phone: (787) 743-5790. Fax: (787) 746-6996. Web Site: www.dimension.fm. Licensee: HQ 103 Inc. (acq 5-28-86). Format: Sp variety. ♦Bienvenido Rodriguez, gen mgr & progmg dir; Maggie Lopez, progmg dir.

Canovanas

WGIT(AM)— 2001: 1660 khz; 10 kw-D, 1 kw-N. TL: N18 23 09 W65 55 16. Hrs open: 1554 Calle Bori, San Juan, 00927. Phone: (787) 274-1800. Fax: (787) 281-9758. Licensee: International Broadcasting Corp. (group owner; acq 5-29-03; $1.3 million). Format: Sp, sports, music. ♦Pedro Roman-Collazo, gen mgr; Margarita Nazario, stn mgr.

Carolina

WIDA(AM)— Mar 16, 1964: 1400 khz; 1 kw-U. TL: N18 23 49 W65 56 06. Hrs open: Box 188, 00986. Secondary address: Ignacio Arzuaga 203-7 00987. Phone: (787) 757-1414. Phone: (787) 757-1717. Fax: (787) 769-4103. E-mail: radiovida@cadenaradiovida.com Web Site: www.cadenaradiovida.com. Licensee: Christian Broadcasting Corp. (acq 7-80; $750,000; FTR: 7-28-80). Format: Sp. ♦Dr. Federico Iglesias, gen mgr; Edwin Carrasquillo, progmg dir; Alberto Periera, chief of engrg; Nephtali Marrero, spec ev coord.

WIDA-FM— August 1983: 90.5 mhz; 25 kw. 1,900 ft TL: N18 06 48 W66 03 07. Web Site: www.cadenaradiovida.com. Format: Educ. ♦Nephtali Marrero, spec ev coord.

WVOZ-FM— Mar 3, 1967: 107.7 mhz; 50 kw. 1,636 ft TL: N18 24 10 W66 03 21. (CP: 12 kw, ant 2,758 ft.). Stereo. Hrs opn: 1554 Calle Bori, Rio Piedras, 00927. Phone: (787) 274-1800. Fax: (787) 281-9758. Licensee: International Broadcasting Corp. (group owner). Format: Adult contemp. ♦Pedro Roman-Collazo, pres; Margarita Nazario, gen mgr & progmg dir.

Cayey

WLEY(AM)— Dec 3, 1965: 1080 khz; 250 w-U. TL: N18 06 55 W66 08 28. Hrs open: 19
WSKN 1320 Radio Isla.
100 Gran Bulevar Paseo, Suite 403A, San Juan, 00926. Phone: (787) 292-1700. Fax: (787) 292-1717. Licensee: Media Power Group Inc. (group owner; (acq 9-30-2003; grpsl). Population served: 179,000 Natl. Network: CNN Radio. Format: Talk/News. News staff: 40; News: 24 hrs wkly. Target aud: 35 plus. ♦Eduardo Rivero, pres; Ismael Nieves, VP & gen mgr; Nora Plaza, gen sls mgr; Luis Penchi, progmg VP; Orlando Morales, opns; Fernando Vazquez, prom & adv.

Ceiba

WFAB(AM)— 1993: 890 khz; 250 w-U. TL: N18 12 16 W65 42 40. Hrs open: Box 318, Rio Blanco, 00744. Phone: (787) 874-0890. Fax: (787) 874-0190. Web Site: www.lanaveprdc.net. Licensee: Daniel Rosario Diaz. (acq 12-18-98). Format: Relg. ♦Daniel Rosario Diaz, pres & gen mgr; Jose N. Garcia, stn mgr.

Cidra

WBRQ(FM)— Mar 1, 1972: 97.7 mhz; 5 kw. Ant 876 ft TL: N18 13 30 W66 05 53. Stereo. Hrs opn: 24 Box 6715, Caguas, 00726-9297. Phone: (787) 745-9700. Phone: (787) 745-9770. Fax: (787) 745-9777. E-mail: nuevavida@nuevavidafm.net Web Site: www.nuevavidafm.net. Licensee: New Life Broadcasting Inc. (acq 3-20-01; $3.6 million). Law Firm: Lee J. Peltzman, Shainis & Peltzman. Format: Sp, Contemp Christian. News staff: one; News: one hr wkly. Target aud: 25-54; women. ♦Juan Carlos Matos, pres; Orlando Mercado, gen mgr & progmg dir.

Coamo

WCPR(AM)— 1967: 1450 khz; 1 kw-U. TL: N18 05 29 W66 22 15. Hrs open: 16 Box 1863, 00769. Phone: (787) 825-7061. Fax: (787) 825-1905. Licensee: Coamo Broadcasting Corp. Format: Adult contemp. News: 9 hrs wkly. Target aud: General. ♦Jose David Soler, pres & gen mgr.

Corozal

WORO(FM)— July 1968: 92.5 mhz; 50 kw. 1,197 ft TL: N18 15 09 W66 19 58. Hrs open: 415 Carbonell St., San Juan, 00918. Secondary address: Box 9021967, San Juan 00902. Phone: (787) 751-1380. Fax: (787) 758-9967. Licensee: Catholic Apostolic & Roman Church San Juan Archdiocese. (acq 1981; $1 million; FTR: 3-2-81). Format: Btfl mus. ♦Roberto Octavio Gonzalez, pres; Allan Corales, gen mgr; Elsa Feernandez, sls dir; Carlos Rodriguiz, progmg dir; Jesus Gomez, chief of engrg; Jose Gomez, chief of engrg.

Culebra

***WJVP(FM)**— 1998: 89.3 mhz; 30 kw vert. 577 ft TL: N18 19 37 W65 18 21. Hrs open: Box 40000, Bayamon, 00958. Secondary address: An 167 Calle Granada AM Alahambra, Bayamon 00956. Phone: (787) 288-4336. Phone: (787) 288-4332. Fax: (787) 740-7104. Web Site: www.clamorpr.org. Licensee: Clamor Broadcasting Network Inc. Format: Relg, civic, cultural. ♦Jorde Raschke, gen mgr.

WJZG(FM)— December 1996: 98.7 mhz; 6 kw. Ant 584 ft TL: N18 19 19 W65 17 59. Stereo. Hrs opn: 24 Box 1047, Fajardo, 00738. Phone: (787) 860-1065. Fax: (787) 860-1055. Licensee: Western New Life Inc (acq 5-30-2005; $1.8 million). Population served: 100,000 Law Firm: Scott C. Cinnanmon LLC. Format: Sp, Christian, tropical music. News staff: 4. Target aud: 25-54. ♦Aureo Matos, gen mgr.

Fajardo

WCMA-FM— Feb 15, 1969: 96.5 mhz; 11.5 kw. 2,795 ft TL: N18 18 36 W65 47 41. Hrs open: 24 Box 949, Guaynabo, 00970-0949. Secondary address: Amelia Industrial Park, Calle Frances #42, Guaynabo 00968. Phone: (787) 622-9700. Fax: (787) 622-9478. Web Site: www.spanishbroadcastingsystem.com. Licensee: WCMA Licensing Inc. Group owner: Spanish Broadcasting System Inc. (acq 8-4-98; $8.25 million). Format: Music of the 80s & 90s. ♦Falex Bonnet, gen mgr.

WIOA(FM)—See San Juan

WMDD(AM)— May 31, 1947: 1480 khz; 5 kw-U. TL: N18 21 46 W65 38 24. Hrs open: 24 Box 948, 00738. Phone: (787) 863-0202. Phone: (787) 793-0669. Fax: (787) 863-0166. Licensee: Pan Caribbean Broadcasting de P.R. Inc. (acq 3-19-03; with WZIN(FM) Charlotte Amalie, VI). Format: Tropical. Target aud: 25-49. ♦Rita Friedman, pres & opns mgr.

WYAS(FM)—See Vieques

Guayama

***WCRP(FM)**— 1991: 88.1 mhz; 27 kw. 1,889 ft TL: N18 06 47 W66 03 08. Hrs open: Box 344, 00785-0344. Phone: (787) 864-3658. Fax: (787) 864-6780. Licensee: Ministerio Radial Cristo Viene Pronto Inc. Format: Educ, Christian. ♦Carmita Rodriguez, pres & gen mgr.

WIBS(AM)— Mar 1, 1981: 1540 khz; 1 kw-D. TL: N17 59 44 W66 04 39. Hrs open: Calle Bori 1554, San Juan, 00927. Phone: (787) 274-1800. Fax: (787) 281-9758. Licensee: International Broadcasting Corp. (group owner; acq 12-3-01; $300,000). Format: Sp, tropical. ♦Pedro Roman-Collazo, CEO; Margarita Nazario, gen mgr & progmg dir.

WMEG(FM)— November 1966: 106.9 mhz; 25 kw. 1,994 ft TL: N18 06 48 W66 03 07. Stereo. Hrs opn: Box 949, Guaynabo, 00970-0949. Secondary address: Amelia Industrial Park, Calle Frances 42, Guaynabo 00968. Phone: (787) 622-9700. Fax: (787) 622-9478. Web Site: www.broadcastingsystem.com. Licensee: WMEG Licensing Inc. Group owner: Spanish Broadcasting System Inc. (acq 3-15-99; $16 million with WZET(FM) Hormigueros). Format: Rock, English. ♦Falex Bonnet, gen mgr & gen sls mgr.

WXRF(AM)— July 1948: 1590 khz; 1 kw-U. TL: N17 57 40 W66 08 20. Hrs open: Calle Bori 1554, San Juan, 00927. Phone: (787) 274-1800. Fax: (787) 281-9758. Licensee: International Broadcasting Corp. (acq 10-7-2004; $1,382,961 with WVEO(TV) Aguadilla). Format: Sports, Music. ♦Pedro Roman-Collazo, pres; Margarita Nazario, gen mgr, gen sls mgr & progmg dir.

Guayanilla

WOIZ(AM)— Oct 1, 1986: 1130 khz; 200 w-D, 700 w-N. TL: N18 01 03 W66 46 22. Hrs open: 5 AM-10 PM Box 561130, 00656. Secondary address: 383 Road klmo.4, Bo Magas Arriba 00656. Phone: (787) 835-1130. Phone: (787) 835-3130. Fax: (787) 835-3130. E-mail: radioantillas@yahoo.com Web Site: www.radioantillas.4t.com. Licensee: Radio Antillas of Harriet Broadcasters. Population served: 250,000 Format: Adult contemp, oldies, news/talk. Target aud: 35 plus. ♦Luis A. Rodriguez III, pres, gen mgr, opns mgr & gen sls mgr; Maria de los Angeles Rivera, VP.

Hatillo

WMSW(AM)— 1980: 1120 khz; 5 kw-U. TL: N18 28 15 W66 50 24. Hrs open: Box 140961, Arecibo, 00614. Phone: (787) 879-4094. Fax: (787) 880-0441. Licensee: Aurora Broadcasting Corp. (acq 11-23-2004). Format: News/talk, music. ♦Manuel Santiago Santos, pres & progmg dir; Hector Santiago Santos, VP; Lloyd Santiago Santos, gen sls mgr; Ronald Cushing, chief of engrg.

Hormigueros

WRRH(FM)— 1998: 106.1 mhz; 800 w. Ant 1,932 ft TL: N18 08 33 W66 58 56. Hrs open: Box 1061, 00660. Phone: (787) 849-1061. Fax: (787) 849-6106. E-mail: renacer1061@yahoo.com Web Site: www.renacer1061.com. Licensee: Renacer Broadcasters Corp. Format: Contemp Christian music. ♦Larry W. Ramos, gen mgr & gen sls mgr; Kehmuel Ramos, progmg dir.

WZET(FM)— Oct 12, 1980: 92.1 mhz; 3 kw. 581 ft TL: N18 11 15 W67 07 04. (CP: 2 kw, ant 1,105 ft.). Stereo. Hrs opn: Box 949, Gauynabo, 00970-0949. Secondary address: Amelia Industrial Park, Calle Frances 42, Guaynabo 00968. Phone: (787) 622-9700. Fax: (787) 622-9478. Web Site: www.spanishbroadcastingsystem.com. Licensee: WSMA Licensing Inc. Group owner: Spanish Broadcasting System Inc. (acq 3-15-99; $16 million with WMEG(FM) Guayama). Format: Sp, tropical. ♦Falex Bonnet, gen mgr.

Humacao

WALO(AM)— Feb 11, 1958: 1240 khz; 1 kw-U. TL: N18 08 49 W65 48 49. Hrs open: 19 Box 9230, 00792. Phone: (787) 852-1240. Fax: (787) 852-1280. E-mail: wlo@prtc.net Licensee: Ochoa Broadcasting Corp. (acq 4-70; $400,000). Law Firm: Fletcher, Heald & Hildreth. Wire Svc: CNN Format: Sp, news/talk, music, sports, MOR. News staff: 2; News: 60 hrs wkly. Target aud: Adult 18-54, male 25-54; general. Spec prog: Relg 2 hrs wkly. ♦Efrain Archilla-Roig, CEO, chmn & pres; Beatriz Archilla, gen mgr; Maribel Ortiz-Del Valle, opns dir; Ken Allen, dev dir.

Isabela

WISA(AM)— Oct 19, 1961: 1390 khz; 1 kw-U. TL: N18 30 08 W67 01 38. Hrs open: Box 750, 00662. Phone: (787) 872-0100. Phone: (787) 872-2030. Fax: (787) 872-0802. Licensee: Isabela Broadcasting Inc. (acq 4-87). Format: MOR. ♦David Marda, gen mgr; Edwin Nieves, progmg dir.

WKSA-FM—Co-owned with WISA(AM).Not on air, target date: unknown: 101.5 mhz; 42 kw. Ant -26 ft TL: N18 26 36 W67 08 50. Phone: (787) 798-7878. Format: Ballads.

Juana Diaz

WCGB(AM)— Nov 23, 1967: 1060 khz; 5 kw-D, 500 w-N. TL: N17 59 28 W66 28 32. Hrs open: 5 AM-midnight Box 1414, 00795. Secondary address: Carretera Hwy. 1, KM 112.0 00795. Phone: (787) 837-1060. Fax: (787) 260-1060. E-mail: wcgbam@prtc.net Licensee: Calvary Evangelistic Mission Inc. (acq 12-3-2004; $500,000). Format: Relg, var/div, Sp. News: 10 hrs wkly. Target aud: Adult. ♦Lawrence Trumbower, gen mgr & opns dir.

Juncos

WRRE(AM)— 1971: 1460 khz; 500 w-U, DA-2. TL: N18 12 54 W65 54 33. Hrs open: 24 Box 1460, Las Piedras, 00771. Phone: (787) 561-1460. Phone: (888) 561-1460. Fax: (787) 716-0808. E-mail: sonidosantidad@hotmail.com Web Site: www.sonidosantidad.com. Licensee: Hacienda San Eladio Inc. (acq 4-16-03; $625,000). Format: Sp relg. Target aud: All. ♦Miguel A. Medina, pres & gen mgr.

Lajas

WBSG(AM)— 1986: 1510 khz; 1 kw-U, DA-1. TL: N18 02 11 W67 04 58. (CP: COL San German. 1 kw-U, DA-2). Hrs opn: 16 Box 1689, 00667. Phone: (787) 899-5724. Fax: (787) 899-5475. Licensee: Perry Broadcasting Systems (acq 4-1-2002; $535,500). Format: Sp news/talk. ♦Oscar Vega, gen mgr.

WXLX(FM)— Jan 5, 1994: 103.7 mhz; 50 kw. 456 ft TL: N17 59 37 W67 11 09. Stereo. Hrs opn: 24
Rebroadcasts WXYX(FM) Bayamon 100%.
HC 67, Bayamon, 00956-9535. Phone: (787) 255-2325. Phone: (787) 785-9390. Fax: (787) 785-9377. Licensee: Radio X Broadcasting Corp. (acq 1-20-98; $3 million). Population served: 600,000 Format: CHR. ♦Roberto Davila, pres & gen mgr; Roberto Davila Rios, opns mgr.

Lares

WGDL(AM)— February 1983: 1200 khz; 250 w-D. TL: N18 17 40 W66 53 50. Hrs open: 12 Box 872, 00669. Phone: (787) 897-1200. Phone: (787) 897-3889. Fax: (787) 897-7821. E-mail: wgdl1200@yahoo.com Licensee: Lares Broadcasting Corp. Law Firm: John P. Bankson Jr. Format: Tropical, Sp. News staff: one; News: 20 hrs wkly. Target aud: General. ♦Pedro Hernandez, pres; Julia Bello, gen mgr, gen sls mgr & traf mgr; Angel Perez, progmg dir.

Levittown

***WPLI(FM)**— Oct 1, 1986: 88.5 mhz; 35 w. Ant 121 ft TL: N18 26 55 W66 10 26. Hrs open: Box 371177, Cayey, 00737. Phone: (787) 798-8850. Fax: (787) 798-8851. Web Site: www.plenitudfm.com. Licensee: Family Educational Association Inc. (acq 4-7-2004; $800,000). Format: Relg. ♦Shay Garcya, stn mgr.

Luquillo

WZOL(FM)— 1976: 92.1 mhz; 4.6 kw. Ant 915 ft TL: N18 19 54 W65 41 11. Hrs open: Box 29027, Rio Piedras, 00929. Phone: (787) 767-1005. Fax: (787) 758-1055. Web Site: www.radiosol.org. Licensee: Radio Sol 92, WZOL Inc. Format: Relg. ♦Pedro M. Canales, pres; William H. Irizarry, gen mgr; Raymond Hernandez, chief of engrg; Willie Lopez, progmg dir & traf mgr.

Manati

WBQN(AM)—See Barceloneta-Manati

WMNT(AM)— Dec 59: 1500 khz; 1 kw-D, 250 w-N. TL: N18 26 06 W66 29 54. Hrs open: 16 Box 6, 00674. Secondary address: Delta St. #1305 Caparra Terr., San Juan 00920. Phone: (787) 854-2223. Fax: (787) 781-7647. Fax: (787) 854-2713. E-mail: radio@atenas.com Web Site: www.radioatenas.com. Licensee: Manati Radio Corp. (acq 9-4-97; $200,000 for 100%). Population served: 350,000 Law Firm: Shaw Pittman. Format: News/talk, sports, Sp. News staff: 2; News: 25

hrs wkly. Target aud: 25 plus; men & women. Spec prog: NBA, World Series in Sp. ♦Jose Ariras Dominicci, CEO; Jose A. Ribas-Dominicci, pres; Freddy Ribas, VP, gen mgr & sls dir; Maria Elena Rodriguez, stn mgr.

WNRT(FM)— 1973: 96.9 mhz; 50 kw. Ant 882 ft TL: N18 15 41 W66 32 19. Hrs open: 24 hours Box 13324, San Juan, 00908. Phone: (787) 758-8562. Fax: (787) 758-8833. Web Site: www.radiotriunfo.com. Licensee: La Voz Evangelica de Puerto Rico Inc. Format: Christian. ♦Luis Barajas, pres; Mosses Flores, gen mgr; Moises Flores, opns mgr & dev mgr; Virgen Perez, sls VP; Carlos Vazquez Flecha, progmg dir; Jorge Figueroa, engrg dir.

Maricao

WAEL-FM— July 1970: 96.1 mhz; 24.2 kw. 2,011 ft TL: N18 09 07 W66 49 15. Stereo. Hrs opn: 24 Box 1370, Mayaguez, 00681-1370. Secondary address: 600 Ramirez Pabon St., Guanajibo Homes, Mayaguez 00681. Phone: (787) 832-4560/ 832-0600. Fax: (787) 792-3140. E-mail: waeline@prte.net Web Site: www.waelfm96.com. Licensee: WAEL Inc. Law Firm: Booth, Freret, Imlay & Tepper. Format: Sp, CHR. Target aud: 12-24. ♦Luis Pirallo, opns mgr & progmg dir; Maria del Pilar-Pirallo, pres & gen sls mgr; Ivan Feliu, chief of engrg; Lydia Vargas, traf mgr.

WYEL(AM)—See Mayaguez

Mayaguez

WAEL-FM—See Maricao

WEGM(FM)—See San German

WIOB(FM)— Oct 12, 1947: 97.5 mhz; 25 kw. 990 ft TL: N18 19 33 W67 10 13. (CP: 50 kw). Hrs opn:
Rebroadcasts WIOA(FM) San Juan 80%.
Box 949, Guaynabo, 00970-0949. Secondary address: Amelia Industrial Park, Calle Frances 42, Guaynabo 00968. Phone: (787) 622-9700. Fax: (787) 622-9478. Web Site: www.spanishbroadcastingsystem.com. Licensee: Cadena Estereotempo Inc. Group owner: Spanish Broadcasting System Inc. (acq 11-29-99; grpsl). Population served: 500,000 Law Firm: Hogan & Hartson. Format: Sp, ballads. Target aud: 30-50; women. ♦Falex Bonnet, gen mgr.

WIVA-FM—See Aguadilla

WKJB(AM)— Dec 6, 1946: 710 khz; 10 kw-D, 750 w-N. TL: N18 10 08 W67 09 03. Hrs open: Box 1293, 00681. Phone: (787) 834-6666. Fax: (787) 831-6925. Licensee: WKJB-AM Inc. Format: News/talk. Spec prog: Sp 1 hr wkly. ♦Dennis Bechara, pres; Jose A. Bechara Jr., exec VP & gen mgr; Ada Ramos, gen sls mgr; Eric Graniela, progmg dir; Rafy Aviles, news dir; Pedro Velez Jr., chief of engrg; Johhny Flores, sports cmtr.

WNOD(FM)— 1960: 94.1 mhz; 25 kw. 2,967 ft TL: N18 09 05 W66 59 20. Stereo. Hrs opn: 24
Rebroadcasts WCOM(FM) San Juan 80%.
Box 949, Guaynabo, 00970-0949. Phone: (787) 265-9494. Fax: (787) 622-9700. Web Site: www.lamega.fm. Licensee: WOYE Inc. Group owner: Spanish Broadcasting System Inc. (acq 11-29-99; grpsl). Population served: 1,000,000 Law Firm: Hogan & Hartson. Format: Top-40. News staff: one; News: 10 hrs wkly. Target aud: 18-49; young adults. ♦Raul Alarcon, chmn; Ismael Nieves, gen mgr; Marie Elena Martinez, gen sls mgr & rgnl sls mgr; Luis Enriquez Rivera, mktg dir; Pedro Arroyo, progmg dir; Alejandro Luciano, chief of engrg; Demare Ramirez, chief of engrg & traf mgr.

WORA(AM)— May 12, 1947: 760 khz; 5 kw-U, DA-1. TL: N18 11 30 W67 09 28. Hrs open: Box 363222, San Juan, 00936-3222. Phone: (787) 758-1300. Fax: (787) 751-2319. E-mail: noticias@notiuno.com Web Site: www.notiuno.com. Licensee: Arso Radio Corp. Group owner: Uno Radio Group (acq 5-10-01; grpsl). Format: News. ♦Luis Soto, pres; Elba Esmurria, sls VP; Tanya Ramos, mktg dir; Ray Cruz, progmg dir; Alberto Pereira, chief of engrg.

WPRA(AM)— Oct 16, 1937: 990 khz; 1 kw-U. TL: N18 10 52 W67 10 07. Stereo. Hrs opn: 18 Box 1293, 00681. Phone: (787) 834-6666. Fax: (787) 831-6925. Licensee: WPRA Inc. (acq 1996). Population served: 110,000 Natl. Network: AP Radio. Law Firm: Shainis & Peltzman, Chartead. Format: Top-40, Spanish, talk. Target aud: General. ♦Dennis Bechara, pres; Jose A. Bechara, VP & gen mgr.

***WRUO(FM)**— December 1998: 88.3 mhz; 2 kw. 1,004 ft TL: N18 19 31 W67 10 13. Hrs open: Box 21305 STNC, San Juan, 00931. Phone: (787) 763-4699. Fax: (787) 763-5205. E-mail: lcandelas@wrtu.org Web Site: www.wrtu.org. Licensee: University of Puerto Rico. Format: News, jazz, classical. ♦Laura Candelas, gen mgr.

WTIL(AM)— November 1950: 1300 khz; 1 kw-U. TL: N18 11 00 W67 10 04. Hrs open: 261 Castilla St., Sultana Park, Naywest, 00680. Secondary address: Post & Bosque Sts. 00680. Phone: (787) 652-0633. Fax: (787) 652-1292. Licensee: International Broadcasting Corp. (group owner; acq 5-12-2004; $700,000). Population served: 600,000 Format: Talk, oldies, Sp, adult contemp. Target aud: 35 plus. ♦Jose Ramiez, gen mgr.

WTPM(FM)—See Aguadilla

WUKQ-FM— Jan 15, 1963: 99.1 mhz; 50 kw. 1,963 ft TL: N18 09 05 W66 59 19. Stereo. Hrs opn: Box 364668, San Juan, 00936. Phone: (787) 833-9910. Fax: (787) 833-9911. Licensee: El Mundo Broadcasting Corp. Group owner: Univision Radio (acq 8-1-2003; grpsl). Format: Top 40. Spec prog: Jazz 6 hrs wkly. ♦Huberto E. Biagi, VP & gen mgr; Raul Muxo, gen sls mgr; Miguel Rodrigio, rgnl sls mgr; Carlos Gonzalez, progmg dir & chief of engrg; Antonio Gonzalez Caballero, news dir; Grafton Olivera, engrg dir.

WYEL(AM)— 1949: 600 khz; 5 kw, DA-1. TL: N18 10 46 W67 10 14. (CP: 5 kw). Hrs opn: 4:30 AM-midnight Box 1370, 00681-1370. Secondary address: 600 Ramirez Pabon , Guanajibo Homes 00680. Phone: (787) 832-0600. Phone: (787) 832-4560. Fax: (787) 792-3140. E-mail: mgmt@waelfm96.com Web Site: www.waelfm96.com. Licensee: Univision Radio Puerto Rico Inc. (acq 11-17-2006; $2 million). Law Firm: Booth, Freret, Imlay & Tepper P. Format: Oldies, sports. Target aud: 25 plus. ♦Luis Pirallo, opns mgr, progmg dir & engrg dir; Maria Pirallo, gen sls mgr; Ivan Seliu, engrg dir.

Moca

WZNA(AM)— December 1983: 1040 khz; 5 kw-D, 245 w-N, DA-1. TL: N18 16 38 W67 10 01. Stereo. Hrs opn: Box 6715, Caguas, 00725. Phone: (787) 745-9770. Fax: (787) 745-9777. E-mail: nuevavuda@nuevavidafm.com Web Site: www.nuevavidafm.com. Licensee: Western New Life Inc. (acq 8-9-2004; $950,000). Format: Contemp Christian. ♦Juan Carlos Matos Barreto, pres; Orlando Mercede, gen mgr & opns mgr.

Morovis

WEKO(AM)— December 1981: 1580 khz; 5 kw-D, 2.5 kw-N, DA-D. TL: N18 20 32 W66 25 08. Hrs open: Calle Bori 1554, San Juan, 00927. Phone: (787) 274-1800. Fax: (787) 281-9758. Licensee: International Broadcasting Corp. (group owner; acq 9-29-98; $315,000). Law Firm: Fletcher, Heald & Hildreth. Format: Pop Latin & American mus, news. Target aud: 30 plus. ♦Pedro Roman-Collazo, pres; Margarita Nazario, gen mgr & progmg dir.

Naguabo

WYQE(FM)— December 1994: 92.9 mhz; 3.9 kw. 853 ft TL: N18 16 49 W65 40 12. Stereo. Hrs opn: 24 Box 9300, 00718. Secondary address: Apt. 2-A 00718. Phone: (809) 847-9300. Phone: (809) 874-9300. Fax: (809) 874-9290. E-mail: wyqe@yunque93.com Web Site: www.yunque93.com. Licensee: Fajardo Broadcasting Co. Inc. (acq 12-29-99). Population served: 500,000 Format: Sp tropical. News staff: 2; News: 20 hrs wkly. Target aud: 18 plus; general. ♦Efrain Archilla-Diez, pres, gen mgr & chief of engrg; Raul Rivera, VP, opns mgr, prom dir, progmg mgr & mus dir; Edwin Glass, sls dir; Vanessa Jimenez, natl sls mgr; Victor Cordero, news rptr; Jose Carrion, sports cmtr.

Pastillo

***WJDZ(FM)**— 2006: 90.1 mhz; 200 w. Ant -121 ft TL: N17 59 40 W66 27 33. Hrs open: Box 8072, Ponce, 00732. Licensee: Gamma Community Services Corp.

Patillas

WEXS(AM)— 1991: 610 khz; 250 w-D, 1 kw-N, DA-D. TL: N18 00 36 W66 01 28. Hrs open: 5:30 AM-10 PM Box 640, 00723. Phone: (787) 839-0610. Fax: (787) 839-0960. Licensee: Community Broadcasting Inc. Population served: 100,000 Format: Adult contemp, CHR, news. Target aud: 18-55. Spec prog: Relg 2 hrs, sports 6 hrs wkly. ♦Enrique Garcia, gen mgr.

Penuelas

WENA(AM)—See Yauco

WPPC(AM)— May 25, 1976: 1570 khz; 1 kw-D, 126 w-N. TL: N18 03 47 W66 43 04. Hrs open: 12 Box 9064, Pompanos Stn., Ponce, 00732-9064. Phone: (809) 836-1570. Phone: (809) 848-4670. Fax: (787) 848-4670. E-mail: radiofelicidad@yahoo.com Web Site: www.wppc1570am.org. Licensee: Radio Felicidad Inc. (acq 6-18-81; $125,000; FTR: 7-13-81). Format: MOR, relg, Sp. Target aud: General. ♦Julio Valazquez, pres; Rafael Acosta, chief of engrg.

Ponce

WDEP(AM)— Feb 1, 1973: 1490 khz; 5 kw-D, 1 kw-N. TL: N17 58 52 W66 36 51. Stereo. Hrs opn: 24
WSKN 1320 Radio Isla.
100 Gran Bulevar Paseo, Suite 403A, San Juan, 00926. Phone: (787) 292-1700. Fax: (787) 292-1717. Licensee: Media Power Group Inc. (group owner; (acq 9-30-2003; grpsl). Population served: 421,000 Natl. Network: CNN Radio. Law Firm: Fletcher, Heald & Hildreth, P.L.C. Format: Sp, news/talk, variety. News: 24 hrs wkly. Target aud: P 35-64 P35+. ♦Eduardo Rivero, pres; Ismaez Nieves, VP & gen mgr; Nora Plaza, gen sls mgr; Luis Penchi, progmg VP; Orlado Moraless, opns; Fernando Vazquez, prom.

WIOC(FM)— January 1970: 105.1 mhz; 47 kw. Ant -200 ft TL: N17 59 27 W66 37 45. Hrs open: 24
Rebroadcasts WIOA(FM) San Juan 80%.
Box 949, Guaynabo, 00970-0949. Secondary address: Amelia Industrial Park, Calle Frances 42, Guaynabo 00978. Phone: (787) 622-9700. Fax: (787) 622-9478. Web Site: www.lamega.fm. Licensee: Cadena Estereotempo Inc. Group owner: Spanish Broadcasting System Inc. (acq 11-29-99; grpsl). Population served: 500,000 Law Firm: Hogan & Hartson. Format: Adult contemp. News staff: one; News: one hr wkly. Target aud: 30-50; women. ♦Raul Alarcon, pres; Ismael Nieves, gen mgr; Marie Elena Martinez, gen sls mgr; Luis Enrique Rivera, prom dir; Pedro Arroyo, progmg dir; Alejandro Luciano, chief of engrg; Demare Ramirez, traf mgr.

WISO(AM)— Sept 15, 1953: 1260 khz; 1 kw-U. TL: N17 59 22 W66 37 11. Hrs open: 16 134 Domenech Ave., Hato Rey, 00918-3502. Phone: (787) 763-1066. Fax: (787) 763-4195. E-mail: jblanco25@hotmail.com Web Site: www.waparadio.net. Licensee: Wilfredo G. Blanco Pi. (acq 1996; $500,000). Population served: 500,000 Format: News/talk. News: 26 hrs wkly. Target aud: Adults. ♦Wilfredo G. Blanco, pres & gen mgr; Jorge Blanco, opns mgr, prom mgr, progmg dir, news dir & chief of engrg; Carmen Blanco, gen sls mgr & traf mgr.

WLEO(AM)— Nov 3, 1956: 1170 khz; 250 w-U. TL: N17 58 52 W66 36 51. Hrs open: 24 Box 7213, 00732. Secondary address: WLEO/WZAR, 46 Sector Purto Viejo, Playa De Ponce 00732. Phone: (787) 842-0048. Phone: (787) 841-1011. Fax: (787) 840-0049. Licensee: Uno Radio of Ponce Inc. Group owner: Uno Radio Group (acq 2-18-00; grpsl). Population served: 225,000 Law Firm: Borsari & Paxson. Format: Oldies. News staff: 2; News: 50 hrs wkly. Target aud: 25 plus; mature, blue-collar & professionals. Spec prog: Sports. ♦Jose Juan Santiago, stn mgr; Carlos Conesa, rgnl sls mgr; Ray Cruz, progmg mgr & news dir; Oscar Vega, chief of engrg; Carmen Reyes, traf mgr; Luis Torres, news rptr; Adam Asencio, disc jockey; Jose L. Perez, disc jockey; Nathanael Paradiso, disc jockey.

WZAR(FM)—Co-owned with WLEO(AM). Mar 17, 1966: 101.9 mhz; 14 kw. 2,580 ft TL: N18 01 40 W66 39 14. Stereo. 24 Phone: (787) 842-0048. Fax: (787) 840-0049.1,700,000 Format: Adult contemp, Sp. News: 12 hrs wkly. Target aud: 18-49; blue & white collar, adults, professionals. Spec prog: Talk show 15 hrs wkly. ♦Jose Juan Santiago, opns mgr; Pedro Gonzales, progmg mgr; Rafael Acosta, engrg mgr; Carmen Reyes, traf mgr.

WPAB(AM)— Aug 14, 1940: 550 khz; 5 kw-U. TL: N17 59 27 W66 37 46. Hrs open: 24 Box 7243, 00732-7243. Secondary address: 1643 Ave. Eduardo Ruberte 00716. Phone: (787) 840-5550. Fax: (787) 840-7077. Licensee: WPAB Inc. (acq 8-21-97; $3 million for stock with co-located FM). Population served: 524,675 Natl. Network: CNN Radio. Rgnl rep: Sayda Ortiz Law Firm: Booth, Freret, Imlay & Tepper PC. Wire Svc: AP Format: Sp, news/talk. News staff: 7; News: 15 hrs +. Target aud: 25 plus; concerned adults. ♦Alfonso Gimenez Porrata, CEO & pres; Alfonso Gimenez Lucchetti, VP & gen mgr; Maria Luisa Gimenez-Lucchztti, opns VP; Sayda Ortiz, sls VP & gen sls mgr.

WPRP(AM)— 1936: 910 khz; 5 kw-U. TL: N17 59 49 W66 37 31. Hrs open: Box 7213, 00732. Secondary address: WLEO/WZAR Bldg., Sector Puerto Viejo, Paseo Sauri, Playa de Ponce 00732. Phone: (787) 842-0048. Fax: (787) 840-0049. E-mail: jsantiago@unoradio.com Web Site: www.unoradio.com. Licensee: Arso Radio Corp. Group owner: Uno Radio Group (acq 5-8-01; grpsl). Format: News/talk, Sp.

Target aud: 35 plus. ♦Jose Juan Santiago, pres, gen mgr & opns VP; Carlos Conesa, sls dir & gen sls mgr; Glerys Rivera, prom dir; Ray Cruz, progmg dir & pub affrs dir; Oscar Vega, engrg dir.

***WPUC-FM—** May 17, 1984: 88.9 mhz; 10.8 kw. 2,912 ft TL: N18 10 27 W66 35 32. Stereo. Hrs opn: 4 AM-midnight 2250 Ave. Las Americas, Suite 529, 00717-9997. Phone: (787) 651-2000 ext. 2600. Phone: (787) 844-8809. Fax: (787) 651-2022. E-mail: catolicaradio @catolicaradiopr.com Web Site: www.catolicaradiopr.com. Licensee: Catholic University of Puerto Rico Service Association. Format: Adult contemp. News staff: 2; News: 30 hrs wkly. Target aud: General; professionals, young adults, retirees, mid & upper middle class. Spec prog: Educ. ♦Julio Feliu Ramirez, gen mgr; Jose R Leon, stn mgr & opns mgr; Ediel Montoluo, progmg dir; Rey Moreira, chief of engrg; Josantonio Cornier, disc jockey; Jose A Tizol, disc jockey; Rafael Sambolin, disc jockey; Rolando Mendez, mktg.

WRIO(FM)— 1986: 101.1 mhz; 34 kw. 1,768 ft TL: N18 09 15 W66 33 15. Hrs open: 24 Box 7213, 00732. Phone: (787) 842-0048. Fax: (787) 840-0049. Web Site: www.salsoul.com. Licensee: Arso Radio Corp. Group owner: Uno Radio Group. Format: Sp Salsa. ♦Jose Santiago, gen mgr; Vicente Veldodere, sls dir; Vicente Bergodere, gen sls mgr; Donny Cruz, progmg dir; Alberto Pereira, chief of engrg; Marianna Colon, traf mgr.

WUKQ(AM)— May 1, 1957: 1420 khz; 1 kw-U. TL: N17 59 23 W66 37 21. Hrs open:
Rebroadcasts WKAQ(AM) San Juan 99%.
Box 364668, San Juan, 00936. Phone: (787) 758-5800. Fax: (787) 763-1854. Web Site: www.wkaqradio.com. Licensee: El Mundo Broadcasting Corp. Group owner: Univision Radio (acq 8-1-2003; grpsl). Format: News/talk info. Target aud: 25-55; young professionals, retirees, middle & upper income. Spec prog: Pub affrs interviews. ♦Hubert E. Blagui, gen mgr.

WZMT(FM)— May 1969: 93.3 mhz; 14.5 kw. -225 ft TL: N17 59 26 W66 37 43. Stereo. Hrs opn: 20
Rebroadcasts WZNT(FM) San Juan 100%.
Box 949, Guaynabo, 00970-0949. Secondary address: Amelia Industrial Park, Calle Frances 42, Guaynabo 00968. Phone: (787) 622-9700. Fax: (787) 622-9478. Web Site: www.lamega.fm. Licensee: Potorican American Broadcasting Inc. Group owner: Spanish Broadcasting System Inc. (acq 2000; grpsl). Population served: 250,000 Law Firm: Latham & Watkins. Format: Modern, tropical, Sp. News staff: one; News: 6 hrs wkly. Target aud: 18-49; affluent young adults. ♦Raul Alarcon, pres; Ismael Nieves, gen mgr; Maria Elena Martinez, gen sls mgr & rgnl sls mgr; Luis Enrique Rivera, prom dir; Pedro Arroyo, progmg dir; Alejandro Luciano, chief of engrg; Demare Ramirez, traf mgr.

Quebradillas

WCHQ(AM)— February 1998: 960 khz; 1 kw-D, 1.7 kw-N, DA-2. TL: N18 26 38 W66 57 43. Hrs open:
Rebroadcasts WZNA(AM) Moca 100%.
Box 4039, Carolina, 00984. Phone: (787) 750-4090. Fax: (787) 750-6440. Licensee: International Broadcasting Corp. (group owner; (acq 7-6-2004; $1.45 million with WRSJ(AM) Bayamon). Format: Sp, relg. ♦Luis Rosado, pres; Josue Salgado, progmg dir.

WIDI(FM)— Nov 17, 1974: 98.3 mhz; 3 kw. Ant 1,000 ft TL: N18 23 33 W66 59 46. Stereo. Hrs open: 24 Box 1553, 00678. Phone: (787) 895-2725. Phone: (787) 895-0000. Fax: (787) 895-4198. E-mail: magic973@prtc.net Licensee: Jose J. Arzuaga. (acq 7-10-79). Population served: 754,130 Format: Tropical, oldies, Sp. News: 2 hrs wkly. Target aud: General. ♦Jose J. Arzuaga, pres, exec VP & gen mgr; Idalia Arzuaga, sr VP; Idalia Arrieta, opns VP; Joshua Arzuaga, dev VP; Rosidalia Villafane, dev dir.

Rio Grande

WOYE(FM)— 2003: 97.3 mhz; 800 w. Ant 1,906 ft TL: N18 16 46 W65 51 12. Hrs open: Box 1553, Quebradillas, 00678. Phone: (787) 895-0000. Fax: (787) 895-4198. E-mail: magic973@prtc.net Web Site: www.magic973.com. Licensee: Josantonio Mellado Romero, et al (acq 11-18-2002). Law Firm: Irwin, Campbell & Tannenwald. Format: Oldies. ♦Idalia Arzuagi, gen mgr; Nitza Mercado, opns dir; Eva Cordero, dev dir; Marimel Almodovar, sls dir & adv dir; Joshua Arluaga, progmg dir; Rafael Brito, news dir; Victor Gonsalez, pub affrs dir; Jose Arluaga, engrg dir.

Rio Piedras

WFID(FM)— Nov 17, 1958: 95.7 mhz; 50 kw. 941 ft TL: N18 16 00 W66 05 05. Stereo. Hrs opn: Box 363222, San Juan, 00936-3222. Secondary address: 1581 Ponce DeLeon St. 00926. Phone: (787) 758-1300. Fax: (787) 754-1395. Web Site: www.unoradiogroup.com. Licensee: Madifide Inc. Group owner: Uno Radio Group (acq 3-26-98; $11,537,500). Law Firm: Wiley, Rein & Fielding. Format: Easy lstng, adult contemp. Target aud: 25-49; middle & upper income. ♦Luis Soto, pres; Elba Esmurria, sls VP; Tanya Ramos, mktg VP; Ray Cruz, progmg VP.

WSKN(AM)—See San Juan

WVOZ(AM)—See San Juan

Sabana

WJIT(AM)— Mar. 31, 2000: 1250 khz; 1 kw-N, 250 w-D. TL: N18 25 28 W66 20 17. Hrs open: 16 Box 1148, 00751. Phone: (787) 824-2755. Secondary address: Road #2 km 30.5, Vega Alta 00769. Phone: (787) 449-9304. Fax: (787) 825-1905. Licensee: WJIT Broadcasting Corp. Format: Var. ♦Olga Fernandez, pres; Jose David Soler, progmg dir & chief of engrg; Carlos Ortiz, local news ed.

Sabana Grande

WYKO(AM)— 1990: Stn currently dark. 880 khz; 1 kw-D, 500 w-N. TL: N18 04 21 W66 57 06. Hrs open: 34 Doctor Felix Tio St., 00637. Fax: (787) 873-5795. Licensee: Juan Galiano Rivera (acq 8-3-90; $450,000; FTR: 8-6-90). ♦Juan Galiano Rivera, pres & gen mgr.

Salinas

WHOY(AM)— Apr 6, 1967: 1210 khz; 5 kw-U, DA-2. TL: N17 58 38 W66 18 14. Hrs open: Box 1148, 00751. Phone: (787) 824-2755. Fax: (787) 824-8054. E-mail: whoyam@coquinet.com Licensee: Colon Radio Corp. (acq 1-31-97; $700,000). Law Firm: Fletcher, Heald & Hildreth. Format: Sp. ♦Martin Colon, gen mgr, gen sls mgr, prom mgr, progmg dir & news dir; Rafael Pagan, chief of engrg.

San German

WEGM(FM)— Feb 1, 1969: 95.1 mhz; 25 kw. 1,970 ft TL: N18 08 55 W66 58 54. Hrs open: 24 Box 949, Guaynabo, 00970-0949. Secondary address: Amelia Industrial Park, Calle Frances 42, Guaynabo 00968. Phone: (787) 622-9700. Fax: (787) 622-9478. Web Site: www.lamega.fm. Licensee: WRPC Inc. Group owner: Spanish Broadcasting System Inc. (acq 11-29-99; grpsl). Format: CHR-English. Target aud: 18-49; men. ♦Raul Alarcon, pres; Ismael Nieves, gen mgr; Marie Elena Martinez, gen sls mgr & progmg VP; Luis Enriquez Rivera, prom dir; Pedro Arroyo, progmg dir; Roque Gallart, progmg mgr; Alejandro Luciano, chief of engrg; Demare Ramirez, traf mgr.

***WNNV(FM)—** Nov 14, 1996: 91.7 mhz; 5 kw. 436 ft TL: N18 04 08 W67 02 54. Hrs open: 24 Box 847, Mayaguez, 00681. Phone: (787) 883-7100. Fax: (787) 833-7940. Licensee: Siembra Fertil P.R. Inc. (acq 4-18-2007; $700,000). Population served: 283,016 Format: Contemp Christian. Target aud: 25-49. ♦Miguel Marquez, progmg dir.

WPRA(AM)—See Mayaguez

WSOL(AM)— 1955: 1090 khz; 250 w-D, 730 w-N. TL: N18 04 44 W67 01 18. (CP: TL: N18 08 18 W67 07 43). Stereo. Hrs opn: Box 5000-442, 00683. Phone: (787) 892-2216. Phone: (787) 892-2975. Fax: (787) 264-1090. Licensee: San German Broadcasters Group. Law Firm: Roy F. Perkins. Format: Tropical, Sp, news. Target aud: Adults. Spec prog: Farm 2 hrs wkly. ♦Alfredo Cardona, pres; Luz Maria Rivera, gen mgr; Gloria Silva, stn mgr & opns mgr; Lucy Rivera, opns dir.

San Juan

WAPA(AM)— Jan 15, 1947: 680 khz; 10 kw-U, DA-N. TL: N18 24 17 W65 56 55. Hrs open: 134 Domenech Ave., Hato Rey, 00918-3502. Phone: (787) 763-1066. Fax: (787) 763-4195. E-mail: jblanco25@hotmail.com Web Site: www.waparadio.net. Licensee: Wifredo G. Blanco Pi (acq 2-25-91). Format: Sp, news/talk. ♦Wilfredo G. Blanco, gen mgr; Jorge Blanco, opns mgr.

WBMJ(AM)— July 19, 1968: 1190 khz; 10 kw-D, 5 kw-N, DA-2. TL: N18 21 00 W66 06 50. Hrs open: 24 Box 367000, 00936-7000. Phone: (787) 724-1190. Phone: (787) 724-2727. Fax: (787) 722-5395. Fax: (787) 723-9633. E-mail: radio@vrockradio.org Licensee: Calvary Evangelistic Mission Inc. (acq 11-85). Population served: 2,200,000 Natl. Network: Moody, USA, Salem Radio Network. Format: Relg, talk, MOR (English). News: 7 hrs wkly. Target aud: General; religious community of central Puerto Rico. ♦Janet Luttrell, CEO & VP.

WCAD(FM)— Mar 5, 1968: 105.7 mhz; 50 kw. 1,100 ft TL: N18 16 54 W66 06 46. Stereo. Hrs opn: Box 9024188, 00902-4188. Secondary address: 1667 Fernandez Juncos Ave., San Turce 00910. Phone: (787) 726-6144. Fax: (787) 268-3313. E-mail: alfa@alfarock.com Web Site: www.alfarock.com. Licensee: Broadcasting & Programming Systems of Puerto Rico Inc. Format: Rock (AOR). ♦Ralph Perez, gen mgr, stn mgr & gen sls mgr; Pedro Davila, progmg dir; Ada Cox, pub affrs dir; T. Morales, engrg dir.

WCMA-FM—See Fajardo

WFID(FM)—See Rio Piedras

WIAC(AM)— 1947: 740 khz; 10 kw-U, DA-1. TL: N18 25 25 W66 08 20. Hrs open: 24 Box 9023916, 00902-3916. Phone: (787) 620-9898. Fax: (787) 620-0730. E-mail: tcarrasquillo@radiopr740.com Web Site: www.radiopr740.com. Licensee: Bestov Broadcasting Inc. (acq 1954). Law Firm: John P. Bankson Jr. Format: News,politics. Target aud: General. ♦Luis A. Mejia, pres; Valerie Majia, VP & gen mgr; Luis Penchi, news dir; Rey Moraira, chief of engrg; Johnny Men, traf mgr.

WIAC-FM— Mar 1, 1961: 102.5 mhz; 50 kw. 1,139 ft TL: N18 25 25 W66 08 20. Stereo. Web Site: www.systema102.com. Licensee: Luis A. Mejia. Format: Pop, soft music. ♦Danny Gonzalez, chief of opns; Glenn Valares, sls dir; Valerie Mejia, progmg dir.

WIOA(FM)— Mar 1, 1961: 99.9 mhz; 31 kw. Ant 1,837 ft TL: N18 16 44 W65 51 12. Stereo. Hrs opn: Box 949, Guaynabo, 00970-0949. Secondary address: Amelia Industrial Park, Calle Frances 42, Guaynabo 00968. Phone: (787) 622-9700. Fax: (787) 622-9478. Web Site: www.lamega.fm. Licensee: Cadena Estereotempo Inc. Group owner: Spanish Broadcasting System Inc. (acq 11-29-99; grpsl). Format: Adult contemp, ballads, Sp. Target aud: 18-49; predominantly women. ♦Raul Alarcon, pres; Ismael Nieves, gen mgr; Maria Elena Martinez, gen sls mgr; Luis Enrique Rivera, prom dir; Fernando de Hostas, progmg mgr; Alejandro Luciano, chief of engrg; Demare Ramirez, traf mgr.

***WIPR(AM)—** Jan 26, 1948: 940 khz; 10 kw-U, DA-1. TL: N18 25 36 W66 08 29. Hrs open: 24 Box 190909, 00919. Phone: (787) 766-0505. Fax: (787) 250-7694. Web Site: tutv.puertorico.pr. Licensee: Puerto Rico Corp. for Public Broadcasting. Population served: 3,000,000 Natl. Network: NPR. Format: Newstalk. News staff: 7; News: 7 hrs wkly. ♦Luis Agrait, chmn; Linda Hernandez, pres; Yolanda Zavala, exec VP; Raul Carbonell, gen mgr, progmg dir & progmg mgr; Vilma Reyes, stn mgr; Susan Marte, opns VP; Ileana Rivera, dev dir; Luis Santiago, sls dir & mktg VP; Jorge Gonzalez, engrg dir & chief of engrg; Osvaldo Torres, chief of engrg.

WIPR-FM— June 3, 1960: 91.3 mhz; 125 kw. 2,719 ft TL: N18 06 42 W66 03 05. (CP: 105 kw, ant 2,706 ft.). 24 Web Site: tutv.puertorico.pr. (Acq 8-87). Format: Class.

WKAQ(AM)— Dec 3, 1922: 580 khz; 10 kw-U, DA-1. TL: N18 25 55 W66 08 07. Hrs open: 24 Box 364668, 00936-4668. Secondary address: 383 F.D. Roosevelt Ave., Hato Rey 00918. Phone: (787) 758-5800. Fax: (787) 763-1854. Licensee: El Mundo Broadcasting Corp. Group owner: Univision Radio (acq 8-1-2003; grpsl). Format: News/talk. News staff: 22. Target aud: General. ♦Huberto Biaggi, gen mgr; Marisol Seda, news dir; Grafton Olivera, chief of engrg.

WKAQ-FM— Oct 8, 1958: 104.7 mhz; 50 kw. 1,220 ft TL: N18 16 51 W66 06 38. Stereo. Fax: (787) 756-5220. Web Site: www.kq105fm.com. Format: Top-40. ♦Huberto E. Biaggi, exec VP; Raul Muxo, gen sls mgr; Carlos Gonzalez, progmg dir.

WKVM(AM)— 1951: 810 khz; 50 kw-U, DA-1. TL: N18 21 47 W66 08 13. Hrs open: 415 Carbonell St., Hato Rey, 00918. Secondary address: c/o Arquidiocesis de San Juan, Apartado 1967 00901-1967. Phone: (787) 751-1018. Fax: (787) 758-9967. Licensee: Catholic, Apostolic & Roman Church, San Juan Archdiocese. (acq 3-4-82; $1.01 million; FTR: 1-18-82). Format: Oldies (daytime), relg Catholic (evenings). ♦Roberto Gonzalez, pres; Allan Corales, stn mgr; Mrs. Elsa Fernandez, sls VP; Jose Antonio Cruz, progmg dir; Jose Gomez, chief of engrg; Placido Padilla, traf mgr; Mrs. Judith Felicie Rivera, local news ed; Reverend Efrain Rodriguez, relg ed; Enrigue Liboy, sports cmtr.

WNEL(AM)—See Caguas

WODA(FM)—See Bayamon

WORO(FM)—See Corozal

WOSO(AM)— Nov 21, 1977: 1030 khz; 10 kw-U, DA-1. TL: N18 22 07 W66 15 17. Hrs open: 24 Box 11487, 00910-2587. Phone: (787) 724-4242. Fax: (787) 723-9676. Web Site: www.woso.com. Licensee: Sherman Broadcasting Corp. Natl. Network: Wall Street, CBS. Format: News/talk. News staff: 2; News: 6 hrs wkly. Target aud: 25-49. ♦Sherman Wildman, pres; Sergio Fernandez, gen mgr; Mariano Calderon, opns dir; Sherman Wildmon, progmg dir; Gary Tuominen, news dir; Rodolfo Rivas, chief of engrg; Danette Hudoba, traf mgr.

WPRM-FM— April 1959: 98.5 mhz; 25 kw. Ant 1,904 ft TL: N18 06 47 W66 03 06. Stereo. Hrs opn: 24 Box 487, Caguas, 00726-0487. Phone: (787) 744-3131. Fax: (787) 743-0252. Web Site: www.salsoul.com. Licensee: Arso Radio Corp. Group owner: Uno Radio Group (acq 4-1-73). Population served: 2,500,000 Law Firm: Drinker Biddle & Reath. Format: Salsoul. News staff: one; News: 3 hrs wkly. Target aud: 18-49. ♦Jesus M. Soto, CEO & chmn; Luis A. Soto, pres; Luis A. Gonzalez, CFO; Maida Bedaya, gen mgr; Raymond Totti, gen sls mgr; Anthony Soto, progmg dir.

WQBS(AM)— Nov 1, 1954: 870 khz; 10 kw-U, DA-1. TL: N18 22 17 W66 12 17. Hrs open: 1508 Calle Bori, 00927-6116. Phone: (787) 758-8700. Fax: (787) 765-2965. E-mail: angel@aercobroadcasting.com Licensee: Aerco Broadcasting Corp. (acq 1-11-2005). Format: Div, Sp. ♦Luz Alvarez, gen mgr. Co-owned TV: WSJU-TV

WQII(AM)— 1947: 1140 khz; 10 kw-U, DA-1. TL: N18 21 30 W66 08 05. Hrs open: Cobians Plaza GM01, 1607 Ponce de Leon Ave., Stop 24, Santurce, 00909. Secondary address: Box 906 6590 00906-6590. Phone: (787) 723-4848. Fax: (787) 723-4035. E-mail: postmaster@1140qpr.com Licensee: Communications Council Group Inc. Format: Talk shows (women's). ♦Nieves Gonzalez Avreu, pres; Jorge Rudolfo Marquina, gen mgr; William Padilla, sls dir; Danny Gonzalez, progmg dir; Raymond Hernandez, chief of engrg.

WRSJ(AM)—See Bayamon

***WRTU(FM)**— Feb 8, 1980: 89.7 mhz; 50 kw. 796 ft TL: N18 16 00 W66 05 05. Stereo. Hrs opn: 18 Box 21305, Stn. C, 00931-1305. Secondary address: Mariana Bracetti St., Ponce de Leon Ave., Rio Piedras 00931. Phone: (787) 763-4699. Phone: (787) 464-0000 ext 5728. Fax: (787) 763-5205. E-mail: lcandelas@wrtu.org Web Site: www.wrtu.org. Licensee: University of Puerto Rico. Law Firm: Dow, Lohnes & Albertson. Format: Class, jazz, Sp. News staff: 5; News: 8 hrs wkly. Target aud: General; young, professional & highly educated. Spec prog: News 7 hrs, talk 2 hrs wkly. ♦Laura Candelas, gen mgr.

WSKN(AM)— Oct 15, 1949: 1320 khz; 5 kw-D, 2.3 kw-N. TL: N18 23 00 W66 04 01. Hrs open: 24 100 Gran Bulevar Paseo, Suite 403A, 00926. Phone: (787) 292-1700. Fax: (787) 292-1717. E-mail: noticias@radioisla1320.com Web Site: www.radioisla1320.com. Licensee: Media Power Group Inc. (group owner; (acq 9-30-2003; grpsl). Natl. Network: CNN Radio. Law Firm: Fletcher, Heald & Hildreth P.L.C. Format: Spanish, news, talk. News staff: 40; News: 24 hrs wkly. Target aud: P 35-64 P35+. ♦Eduardo Rivero, pres; Ismael Nieves, VP & gen mgr; Nora Plaza, gen sls mgr; Luis Penchi, progmg VP; Orlando Morales, opns; Fernando Vazquez, prom & adv.

WUNO(AM)— Jan 11, 1960: 630 khz; 5 kw-U, DA-2. TL: N18 26 00 W66 07 29. Hrs open: 24 Box 363222, 00936-3222. Secondary address: 1581 Ponce de Leon St., Rio Piedras 00926. Phone: (787) 758-1300. Fax: (787) 754-1395. E-mail: info@notiuno.com Web Site: www.notiuno.com. Licensee: Madifide Inc. Group owner: Uno Radio Group (acq 5-8-01; grpsl). Format: News/talk. News staff: 22; News: 168 hrs wkly. Target aud: 25 plus. ♦Jesus M. Soto, CEO; Luis Soto, pres; Tanya Ramus, mktg VP; Ray Cruz, progmg VP.

WVOZ(AM)— July 4, 1949: 1520 khz; 25 kw-U, DA-2. TL: N18 21 11 W66 12 09. Hrs open: 16 1554 Calle Bori, 00927. Phone: (787) 764-1077. Phone: (787) 274-1800. Fax: (787) 281-9758. Licensee: Pedro Roman Collazo. Law Firm: Freret & Imlay. Format: Tropical, sports. Target aud: 35 plus; medium and low income individuals. Spec prog: Puerto Rican & Latin hits. ♦Pedro Roman-Collazo, pres; Margarita Nazario, gen mgr.

WZNT(FM)— 1959: 93.7 mhz; 50 kw. 280 ft TL: N18 22 42 W66 07 04. Stereo. Hrs opn: 24 Box 949, Guaynabo, 00970-0949. Secondary address: Amelia Industrial Park, Calle Frances 42, Guaynabo 00968. Phone: (787) 622-9700. Fax: (787) 622-9478. Web Site: www.lamega.fm. Licensee: WZNT Inc. Group owner: Spanish Broadcasting System Inc. (acq 2000; grpsl). Population served: 920,900 Law Firm: Fletcher, Heald & Hildreth. Format: Tropical, salsa & merengue. Target aud: 18-49; male. ♦Raul Alarcon, pres; Ismael Nieves, gen mgr & stn mgr; Marie Elena Martinez, sls VP & gen sls mgr; Luis Enrique Rivera, prom dir; Billie Fourquet, progmg VP; Pedro Arroyo, progmg dir; Nestor Rodriguez, progmg mgr; Alejandro Luciano, chief of engrg; Demare Ramirez, traf mgr.

San Sebastian

WLRP(AM)— Feb 15, 1965: 1460 khz; 500 w-U. TL: N18 20 50 W66 59 56. (CP: 2.5 kw). Hrs opn: 19 Box 1670, 00685. Phone: (787) 896-1460. Fax: (787) 896-8100. E-mail: radioraices@prtc.net Licensee: Las Raices Pepinianas Inc. Format: Adult contemp. ♦Ramon Colon Pratts, pres; Alfredo Perez, gen mgr & gen sls mgr; Ramon E. Pratts, prom mgr; Jose M. Chaparro, mus dir; Juan Felin, chief of engrg.

WNOD(FM)—See Mayaguez

WRSS(AM)— April 1984: 1410 khz; 1 kw-U, DA-1. TL: N18 19 14 W66 58 45. Hrs open: Box 1410, 00685. Secondary address: Segundo Ruez # 52 St. 00685. Phone: (787) 896-2121. Fax: (787) 896-5753. E-mail: tunuevafamilia@hotmail.com Licensee: Angel Vera-Maury (acq 2-9-03; $250,000). Format: Talk, oldies, Sp. News staff: 6; News: over 30 hrs wkly. Target aud: 30 plus. ♦Angel Vera, pres; Cesar Vera, gen mgr; Arturo Soto, gen sls mgr; Nestor Gonzalez, progmg VP & progmg dir.

Utuado

WERR(FM)— Feb 1, 1970: 104.1 mhz; 50 kw. 255 ft TL: N18 17 31 W66 39 28. Stereo. Hrs opn: 24 P.O. Box 29404, San Juan, 00929. Secondary address: San Felipe # 205, Arecibo 00612. Phone: (787) 751-6318. Phone: (787) 751-1310. Fax: (787) 751-6854. E-mail: jrivera@redentor104fm.com Web Site: www.redentor104fm.com. Licensee: Radio Redentor Inc. (acq 6-75). Population served: 2,000,000 Law Firm: Fletcher, Heald & Hildreth. Format: Adult contemp, Sp, Christian. Target aud: General. ♦Rev. Ricardo Aponte, pres; Jesus M. Rivera, gen mgr; Rev. Abimael Reyes, mktg dir & mktg mgr; Rev. Ramon Rivera, engrg dir; Maria Rivera, traf mgr; Omayra Martinez, spec ev coord; Shirley Lopresti, women's cmtr.

WUPR(AM)— Apr 18, 1964: 1530 khz; 1 kw-D, 250 w-N. TL: N18 16 02 W66 42 38. Hrs open: 17 Box 868, 00641. Phone: (787) 894-2460. Fax: (787) 894-4955. Web Site: www.coqui.net. Licensee: Central Broadcasting Corp. Format: Sp, news/talk. News staff: 2; News: 11 hrs wkly. Target aud: 18-49; middle income adults. ♦Jose A. Martinez, pres, gen mgr, gen sls mgr & progmg dir; Manuel E. Andujar, mus dir; Manuel B. Martinez, news dir; Epifanio Rodriguez Velez, chief of engrg.

Vega Baja

WEGA(AM)— October 1971: 1350 khz; 2.5 kw-U, DA-2. TL: N18 28 38 W66 23 43. Hrs open: Box 1488, 00694-1488. Phone: (787) 858-0386. Fax: (787) 855-0916. Licensee: A Radio Company Inc. (acq 9-1-2004; $850,000). Format: Var/div. ♦Gerardo Angulo, pres; Carmelo Santiago, gen mgr; Hector Santiago, sls VP & progmg dir; Lloyd Santiago, mktg VP; Ronald Cushing, chief of engrg.

Vieques

WIVV(AM)— Dec 8, 1956: 1370 khz; 5 kw-D, 1 kw-N. TL: N18 06 19 W65 28 03. Hrs open: 24
Rebroadcasting WBMJ(AM) San Juan 100%.
Box 367000, San Juan, 00936-7000. Phone: (787) 724-1190. Phone: (787) 741-8717. Fax: (787) 722-5395. Fax: (787) 741-8717. E-mail: radio@vrockradio.org Licensee: Calvary Evangelistic Mission Inc. Population served: 1,340,504 Natl. Network: Salem Radio Network. Format: Relg, talk, MOR, Sp. News: 7 hrs wkly. Target aud: General; eastern Puerto Rico & the Leeward Islands. Spec prog: News 7 hrs wkly. ♦Janet L. Luttrell, CEO & VP.

WYAS(FM)— Nov 4, 1978: 98.9 mhz; 50 kw. Ant 751 ft TL: N18 19 39 W65 18 05. Stereo. Hrs opn: Box 1047, Fajardo, 00738. Phone: (787) 860-1065. Fax: (787) 860-1055. Licensee: La Mas Z Radio Inc. (acq 7-6-2004; $1.99 million). Format: Jazz.

Yabucoa

WXEW(AM)— Jan 1, 1978: 840 khz; 5 kw-D, 1 kw-N, DA-N. TL: N18 02 58 W65 52 07. (CP: 5 kw-U). Hrs opn: 19 203 Font Martelo Ave., Humacao, 00971. Secondary address: Box 100 00767. Phone: (787) 893-3065. Phone: (787) 850-0840. Fax: (787) 850-4055. E-mail: victor@victoria840.com Web Site: www.victoria840.com. Licensee: Radio Victoria Inc. (acq 9-19-83). Format: MOR, Sp, talk. ♦Victoria Vargas, pres; Victor M. Calderon, VP & gen mgr; Caly Burmudez, gen sls mgr; Brenda Calderon, mktg dir; Jose Calderon, prom mgr & mus dir; Luis Calderon, progmg dir; Angel Bena, news dir; Alberto Pereira, chief of engrg.

Yauco

WENA(AM)— Nov 11, 1978: 1330 khz; 2 kw-D, 1.4 kw-N, DA-1. TL: N18 02 04 W66 51 48. Stereo. Hrs opn: 24 Box 1338, 25 DeJullo St., Condominio Torres Navel Bldg., 00698. Phone: (787) 267-1330. Phone: (787) 856-1330. Fax: (787) 267-1340. E-mail: wena@cogui.net Web Site: www.yaucoweb.com/wena/. Licensee: Southern Broadcasting Corp. Population served: 55,000 Law Firm: Roy F. Perkins. Format: Adult contemp, CHR, news/talk. News staff: 5; News: 28 hrs wkly. Target aud: 25 plus; young adults & women.Nephtali Rodriguez, pres, gen mgr & traf mgr; Israel Rodriguez, VP & opns VP; Ramon Ramos, dev mgr; Pedro Gregory, sls dir & disc jockey; Juan Diaz, adv mgr; Guillermo Valls, progmg dir & disc jockey; Isaac Pagan, engrg dir; Ronald Cushing, engrg mgr; Israel Muniz, local news ed; Hector Rios, disc jockey; Omar Mercado, disc jockey; Orlando Borrero, disc jockey

WKFE(AM)— Nov 3, 1961: 1550 khz; 250 w-U. TL: N18 01 24 W66 52 02. Hrs open: 24
Rebroadcasts WSKN(AM) San Juan 70%.
100 Gran Bulevar Paseo, Suite 403A, San Juan, 00926. Phone: (787) 292-1700. Fax: (787) 292-1717. Licensee: Media Power Group Inc. (group owner; (acq 9-30-2003; grpsl). Population served: 100,000 Law Firm: Fletcher, Heald & Hildreth. Format: Sp news/talk. News staff: 2; News: 40 hrs wkly. Target aud: P35-64 P35+. ♦Eduardo Rivero, pres; Jose Pagan, gen mgr; Orlando Morales, progmg dir.

Virgin Islands

Charlotte Amalie

WGOD(AM)— 1992: 1090 khz; 250 w-D. TL: N18 18 57 W64 53 02. Hrs open: Box 305012, St. Thomas, 00803. Phone: (340) 774-4498. Fax: (340) 777-9978. Licensee: Three Angels Broadcasting Corp. Inc. (acq 7-5-89). Format: Gospel. ♦Charles Saunders, pres & gen mgr.

WGOD-FM— Sept 1, 1980: 97.9 mhz; 50 kw. 295 ft TL: N18 21 25 W64 58 00. Stereo. (Acq 8-15-85). Format: Relg, educ.

***WIUJ(FM)**— Oct 5, 1979: 102.9 mhz; 1.5 kw. 1,427 ft TL: N18 21 26 W64 56 50. Stereo. Hrs opn: Box 2477, St. Thomas, 00803. Phone: (340) 776-1029. Phone: (340) 777-9485. Fax: (340) 774-0004. Web Site: www.wiuj.com. Licensee: Virgin Islands Youth Development Radio. (acq 8-6-90). Format: Adult contemp, btfl music, big band. News staff: one. Target aud: General. Spec prog: Class 5 hrs, jazz 6 hrs, Fr 4 hrs, Sp 4 hrs wkly. ♦Leo Morone, gen mgr; F. Ottley, opns mgr; Greg Cyntje, progmg dir; Ron Hall, chief of engrg.

WIVI(FM)— Apr 26, 1992: 96.1 mhz; 2.4 kw. 1,500 ft TL: N18 21 33 W64 58 18. Stereo. Hrs opn: 24 Box 304383, St. Thomas, 00803-4383. Phone: (340) 774-1972. Phone: (340) 776-9696. Fax: (340) 776-7060. Web Site: www.pirateradiovi.com. Licensee: Rox Radio Enterprises Inc. (acq 11-20-98; $30,000 for 60% of stock). Population served: 115,000 Format: AAA, classic rock. News: one hr wkly. Target aud: 25-54; general. ♦Lou Lambert, gen mgr; Dorene Carle, gen sls mgr.

WSTA(AM)— Aug 1, 1950: 1340 khz; 1 kw-U. TL: N18 20 10 W64 57 17. Hrs open: 24 Box 1340, # 121 Subbase, St. Thomas, 00804. Secondary address: 121 Sub Base, St. Thomas 00802. Phone: (340) 774-1340. Phone: (340) 777-4500. Fax: (340) 776-1316. E-mail: addie@wsta.com Web Site: www.WSTA.com. Licensee: Ottley Communications Corp. (acq 12-1-84). Population served: 110,000 Natl. Network: ABC, CNN Radio. Law Firm: Miller & Neely, P.C. Format: Div, oldies, adult urban contemp. News staff: 2; News: 20 hrs wkly. Target aud: General. ♦Athneil Ottley, pres; Athniel Ottley, gen mgr; Athneil C. Ottley, stn mgr; Irvin Brown, chief of opns.

WVGN(FM)— 2002: 107.3 mhz; 1.4 kw. Ant 1,565 ft TL: N18 21 31 W64 58 20. Hrs open: 714 Nisky Mail Box, PMB PP-105, St. Thomas, 00802. Phone: (340) 774-2012. Fax: (340) 776-5362. E-mail: npr@wvgn.org Web Site: www.wvgn.org. Licensee: LKK Group Corp. (acq 6-27-02; $290,000). Natl. Network: NPR. Format: News/talk. ♦Patricia Bourne, VP, gen mgr & stn mgr; Victoria Squires, gen sls mgr.

WVJZ(FM)— Mar 15, 1986: 105.3 mhz; 7.7 kw. 1,490 ft TL: N18 21 33 W64 58 18. Stereo. Hrs opn: 24 Box 305678, 13 Crown Bay Fill, St. Thomas, 00803-5678. Secondary address: Box 8209, Bluebeards's Castle, Suite 255, St. Thomas 00801. Phone: (340) 776-5260. Phone: (340) 776-5260. Fax: (340) 776-5357. Fax: (340) 776-5357. E-mail: contact@kasvi.net Web Site: www.wvjz.net. Licensee: Gark LLC. Group owner: Knight Quality Stations (acq 6-4-98). Population served: 100,000 Law Firm: Pepper & Corazzini. Format: Urban contemp. News staff: 2. Target aud: 18-34; young adults, business professionals & college students. ♦Randolph H. Knight, pres; Mark P. Bastin, gen mgr & progmg dir; Jean Greaux Jr., opns dir.

WVWI(AM)— Nov 19, 1962: 1000 khz; 5 kw-D, 1 kw-N. TL: N18 20 11 W64 41 38. Hrs open: 24 Box 305678, 13 Crown Bay Fill, St. Thomas, 00803-5678. Phone: (340) 776-1000. Fax: (340) 776-5357. E-mail: contact@kqsvi.net Web Site: www.wvwi.net. Licensee: Knight Communications of the Virgin Islands Inc. Group owner: Knight Quality Stations (acq 1996; $250,000). Population served: 100,000 Natl. Network: CBS, Westwood One. Law Firm: Pepper & Corazzini. Format: Sports, news/talk. News staff: 2; News: 25 hrs wkly. Target aud: 25-54; middle/upper income professionals. Spec prog: Relg 6 hrs, West Indian one hr, East Indian one hr wkly. ♦Randolph H. Knight, pres; Mark P. Bastin, gen mgr; Jean Greaux Jr., opns dir.

WZIN(FM)— Nov 2, 1976: 104.3 mhz; 44 kw. Ant 1,617 ft TL: N18 21 35 W64 58 19. Stereo. Hrs opn: Nisky Mall Center, PMB 357, St. Thomas, 00802. Phone: (340) 776-1043. Fax: (340) 775-3446. E-mail: info@buzzrocks.com Web Site: www.buzzrocks.com. Licensee: Pan Caribbean Broadcasting de P.R. Inc. (acq 7-11-02; $1 million). Population served: 4,000,000 Law Firm: Arter & Hadden. Format: Alternative rock. Target aud: 18-34. ♦Alan Friedman, VP & gen mgr.

Christiansted

***WIVH(FM)**— July 1993: 90.1 mhz; 1 kw. 731 ft TL: N17 44 10 W64 42 04. Hrs open: 24 2457 Rt. 118, Hunlock Creek, PA, 18621. Secondary address: 5007 Estate Mt. Washington 00820-4565. Phone: (570) 477-3688. Phone: (340) 778-2852. Fax: (340) 719-3076. E-mail: wrgn@epix.net Web Site: www.wrgn.org/wivh.htm. Licensee: Gospel Media Institute Inc. Format: Relg. Target aud: General. ♦Burl F. Updyke, pres & gen mgr; Shirley J. Updyke, progmg dir.

WJKC(FM)— Oct 29, 1983: 95.1 mhz; 50 kw. 886 ft TL: N17 44 07 W64 40 46. (CP: Ant 791 ft.). Stereo. Hrs opn: Box 25680, St.Croix, 00824-1680. Phone: (340) 773-0995. Fax: (340) 773-9093. Web Site: www.viradio.com. Licensee: Radio 95 Inc. Law Firm: Rosenman & Colin. Format: Reggae, urban hip-hop. Target aud: General. ♦Jonathan K. Cohen, pres & gen mgr; Collin Hodge, gen sls mgr; Tom Yarborough, progmg dir; Alvin Gee, news dir.

WMNG(FM)— 1997: 104.9 mhz; 6 kw. 699 ft TL: N17 44 08 W64 40 47. Hrs open: Box 25680, St. Croix, 00824-1680. Phone: (340) 773-0995. Fax: (340) 773-9093. E-mail: jkc95@aol.com Web Site: www.viradio.com. Licensee: Clara Communications Corp. Format: Classic hits. ♦Jonathan K. Cohen, gen mgr; Amanda Cohen, gen sls mgr; Tom Yarbaugh, progmg dir; Herb Schoenbahm, chief of engrg; Celia Jean, traf mgr.

WSTX(AM)— 1952: 970 khz; 5 kw-D, 1 kw-N. TL: N17 45 23 W64 41 38. Hrs open: Box 3279, 00822. Phone: (340) 773-0390. Fax: (340) 773-8515. E-mail: wstx@vitelcom.net Web Site: vipn.vitelcom.net/herbs/. Licensee: Family Broadcasting Inc. Format: Carribean. ♦Kevin Rames, gen mgr.

WSTX-FM— September 1984: 100.3 mhz; 50 kw. Ant 1,030 ft TL: N17 45 20 W64 47 55. Format: Reggae.

WVIQ(FM)— May 17, 1965: 99.5 mhz; 10.5 kw. 1,080 ft TL: N17 45 20 W64 47 55. Stereo. Hrs opn: Box 25680, St. Croix, 00824-1680. Phone: (340) 773-1180. Fax: (340) 773-9093. E-mail: www.jkc95@aol.com Web Site: www.viradio.com. Licensee: JKC Communications of the Virgin Islands Inc. (acq 1999; $590,000). Format: Adult contemp. ♦Jonathan Cohen, gen mgr & gen sls mgr; Tom Yarborough, progmg dir; Alvin Gee, news dir.

WYAC-FM— Feb 26, 1989: 93.5 mhz; 11.5 kw. Ant 735 ft TL: N17 44 08 W64 40 47. Stereo. Hrs opn: 24 Box 25868, 00824. Secondary address: 118 Estate Mt. Welcome 00824. Phone: (340) 773-3693. Fax: (340) 719-1800. E-mail: rogerwmorgan@juno.com Licensee: Philip E. Kuhlman and Ellen N. Kuhlman, joint tenants (acq 7-8-2004; $300,000). Format: Adult contemp. News: one hr wkly. Target aud: General; affluent young adult permanent residents. Spec prog: Relg 3 hrs wkly. ♦Francisce "Francky" Velasquez, stn mgr; Roger Morgan, pres, gen mgr & progmg dir; Arthur Bird, news dir; Herb Schoerbaum, chief of engrg.

Cruz Bay

WWKS(FM)— Feb 3, 1997: 101.3 mhz; 48 kw. 1,302 ft TL: N18 20 17 W64 43 40. Stereo. Hrs opn: 24 Box 305678, St. Thomas, 00803-5678. Secondary address: Box 8209, Bluebeard's Castle, St. Thomas 00801. Phone: (340) 776-4585. Phone: (340) 776-1013. Fax: (340) 776-5357. Fax: (340) 774-4455. Web Site: www.wwks.net. Licensee: Knight V.I. Radio Corp. Group owner: Knight Quality Stations (acq 1996; $225,000). Population served: 100,000 Natl. Network: ABC. Law Firm: Pepper & Corazzini. Format: Urban contemp. News staff: 2; News: 2 hrs wkly. Target aud: 25-54; middle/upper income professionals. Spec prog: West Indian/calypso 25 hrs wkly. ♦Randolph H. Knight, pres; Mark P. Bastin, gen mgr; Jean Greaux Jr., opns dir.

Frederiksted

WAXJ(FM)— 1999: 103.5 mhz; 6 kw. Ant -33 ft TL: N17 43 28 W64 53 03. Hrs open: 24 79-A Castle Coakley, Christiansted, 00820. Phone: (340) 719-1620. Phone: (340) 778-2753. Fax: (340) 778-1686. E-mail: wrra@islands.vi Web Site: www.wrra.vi. Licensee: Reef Broadcasting Inc. (acq 6-27-98). Format: Div music, bilingual. ♦Hugh Pemberton, gen mgr; Beverley Meyers, gen sls mgr; Hugh Pembeton, progmg dir.

WDHP(AM)— May 1999: 1620 khz; 10 kw-D, 1 kw-N. TL: N17 43 28 W64 53 03. Hrs open: 79A Castle Coakley, 00820. Phone: (340) 719-1620. Fax: (340) 778-1686. E-mail: wrra@islands.vi Web Site: www.wrra.vi. Licensee: Reef Broadcasting Inc. Format: Diversified, talk, beautiful music. ♦Beverley Meyers, gen sls mgr; Hugh Pemberton, gen mgr, progmg dir & chief of engrg.

WEVI(FM)— 2003: 101.7 mhz; 900 w. Ant 790 ft TL: N17 43 15 W64 51 26. Hrs open: Box 892, Christiansted, 00821. Phone: (340) 719-1400. Fax: (340) 719-1733. E-mail: info@frontlinemissions.org Web Site: www.frontlinemissions.org. Licensee: Frontline Missions International Inc. (acq 3-12-02). Format: Christian Carribbean music, Bible teachings. ♦Anthony Whitehead, gen mgr & progmg dir.

WMYP(FM)— 2002: 98.3 mhz; 1.9 kw. Ant 915 ft TL: N17 44 51 W64 50 11. Hrs open: 24 Box 8294, Christiansted, 00823. Phone: (340) 772-0098. Fax: (340) 772-9852. E-mail: latino98@viaccess.net Licensee: Amanda Friedman (acq 7-28-2006; $350,000). Format: Sp, tropical/pop. Target aud: 18-49. ♦Jose Martinez, gen mgr.

WRRA(AM)— 1976: 1290 khz; 500 w-D, 250 w-N. TL: N17 43 28 W64 53 03. Hrs open: 24 79A Castle Coakley, Christiansted, 00820. Phone: (340) 778-1620. Fax: (340) 778-1686. E-mail: wrra@islands.vi Web Site: www.wrra.vi. Licensee: Reef Broadcasting Inc. Law Firm: Roy F. Perkins. Format: Gospel. News staff: 2; News: 25 hrs wkly. Target aud: 18-56. Spec prog: Black, jazz 6 hrs, gospel 12 hrs, relg 10 hrs wkly. ♦Beverley Meyers, gen sls mgr; Hugh Pemberton, gen mgr, progmg dir & chief of engrg.

St. Thomas

WYAS(FM)—See Vieques, PR

Mexico

Tijuana

XETRA(AM)—Licensed to Tijuana. See San Diego CA

XETRA-FM—Licensed to Tijuana. See San Diego CA

XHRM-FM—Licensed to Tijuana. See San Diego CA

Federated States of Micronesia

Pohnpei

V6AH(AM)— 1964: 1449 khz; 10 kw-U. Hrs opn: 18 Box 1086, Kolonia Pohnpel, 96941. Phone: (691) 320-2296. Fax: (691) 320-5212. E-mail: v6ah_radio@mail.fm Web Site: www.fm/ppbc. Licensee: Oltrick D. Santos. Format: CHR. News: 20 hrs wkly. Target aud: General. Spec prog: Farm 20 hrs, folk 20 hrs, gospel 2 hrs, relg 2 hrs wkly. ♦Oltrick D. Santos, gen mgr.

Truk

V6AK(AM)— 1962: 1593 khz; 5 kw-U, DA-1. Hrs opn: Box 2222, Weno, Chuuk, 96942. Phone: (691) 330-2596. Licensee: Dept. of Public Affairs/Government. Format: News. Target aud: General. ♦Johnny Esa, gen mgr.

Yap

***V6AI(AM)**— June 9, 1965: 1494 khz; 10 kw-U. Hrs opn: 6 AM-midnight Box 117, Colonia YAP State Western Caroline Islands, 96943. Phone: (691) 350-2174. Fax: (691) 350-4426. E-mail: s-tamagken@yahoo.com Web Site: www.fm/yap/radio.htm. Licensee: Yap State Government. Population served: 10,000 Format: CHR, country, news. News staff: 2; News: 20 hrs wkly. Target aud: General. Spec prog: Yapese 10 hrs, Micronesian 10 hrs, Japanese 5 hrs, Filipino 5 hrs wkly. ♦Sebastian Tamagken, gen mgr; John Gilmatam, progmg dir; Anthony Taveg Jr., mus dir; Jovencio David, chief of engrg; John Hasmai, news rptr; Benjamin Fithingmen, disc jockey; Joseph Yanfag, disc jockey; Mario Mangar, disc jockey; Richard Ritong, disc jockey. Co-owned TV: *WAAB-TV affil

Northern Mariana Islands

Chalan Kanoa-Saipan

***KRNM(FM)**— Feb 28, 1998: 88.1 mhz; 1.8 kw. Ant 125 ft TL: N15 09 05 E145 43 11. Stereo. Hrs opn: 24 Box 501250, Northern Marianas College, Saipan, 96950. Phone: (670) 234-5766. Fax: (670) 235-0915. E-mail: carlp@nmcnet.edu Web Site: www.krnm.org. Licensee: Northern Marianas College. Population served: 75,000 Natl. Network: NPR, PRI. Law Firm: Thomas Crowe. Format: Classical, jazz, news/talk. News: 65 hrs wkly. Spec prog: Chamorro 2 hrs, Korean one hrs, Chinese one hr wkly. ♦Carl Pogue, gen mgr; Joe Servino, engr.

Garapan-Saipan

KCNM(AM)— October 1984: 1080 khz; 5 kw-U. TL: N15 09 00 W145 42 52. Hrs open: 24 Box 500914, Saipan, 96950. Secondary address: Box 20249, Guam Main Facility 96921. Phone: (670) 234-7239. Phone: (670) 234-8644. Fax: (670) 234-0447. E-mail: kzmi-fm@vzpavifica.net

Licensee: Inter-Island Communications Inc. (group owner; (acq 8-6-84). Population served: 60,000 Natl. Network: AP Network News. Format: News/talk. News staff: one; News: 168 hrs wkly. Target aud: General. Spec prog: Chamorro. ♦Harry B. Blalock, gen mgr; Bob Webb, gen sls mgr; Louie Tenorio, progmg dir.

KCNM-FM— 1999: 101.1 mhz; 3.2 kw. Ant 827 ft TL: N15 11 00 W145 44 06. Hrs open: 1868 Halsey Dr., Piti, GU, 96915. Phone: (670) 234-7239. Fax: (671) 477-6411. E-mail: kcnm@ite.net Web Site: www.kcnmkzmi.com. Licensee: Inter-Island Communications Inc. (group owner). Population served: 65,000 Format: Ethnic. News staff: one; News: 16 hrs wkly. ♦Edward H. Poppe Jr., pres; Frances Poppe, CFO; Edward Poppe, gen mgr; Harry Blalock, stn mgr; Lewis Tenorio, progmg dir.

KPXP(FM)— Nov 5, 1992: 99.5 mhz; 6.5 kw. 1,492 ft TL: N15 11 10 W145 44 25. Stereo. Hrs opn: 24 111 W. Chalan Santo Papa St., Hagatna, GU, 96910. Phone: (670) 235-7996. Phone: (670) 235-7997. Fax: (670) 235-7998. E-mail: rex@spbguam.com Web Site: www.power99.com. Licensee: Sorensen Pacific Broadcasting Inc. (group owner; acq 6-23-03; grpsl). Population served: 40,000 Law Firm: Kaye, Scholer, Fierman, Hays & Handler. Format: CHR. News staff: one; News: 14 hrs wkly. Target aud: 14-39; affluent adults. ♦Jon A. Anderson, pres; Rex W. Sorensen, CEO, chmn & CFO; Curtis Dancoe, stn mgr, sls dir & gen sls mgr; Laurence Bejerana, prom mgr; Raymond Gibson, progmg dir; Marvin Palmer, chief of engrg.

KRSI(FM)— July 1992: 97.9 mhz; 4.5 kw. Ant 1,519 ft TL: N15 11 09 E145 44 29. Stereo. Hrs opn: 24 111 W. Chalan Santo Papa St., Hagatna, GU, 96910. Phone: (670) 235-7996. Fax: (670) 235-7998. E-mail: rex@spbguam.com Web Site: www.radiopacific.com. Licensee: Sorensen Pacific Broadcasting Inc. (group owner; acq 6-23-03; grpsl). Law Firm: Cohn & Marks. Format: Classic rock, div, rock. News: one hr wkly. Target aud: 25-49. Spec prog: Blues 6 hrs, reggae 19 hrs, Hawaiian 2 hrs, Chamdru 4 hrs, jazz one hr wkly. ♦Rex Sorensen, CEO & pres; Curtis Dancoe, gen mgr, gen sls mgr & disc jockey; Laurence Bejerana, prom mgr; Raymond Gibson, progmg dir; Marvin Palmer, chief of engrg.

KWAW(FM)— 1999: 100.3 mhz; 1.1 kw. Ant 1,512 ft TL: N15 11 05 W145 44 26. Hrs open: Box 7094, Tamuning, GU, 96931. Phone: (671) 646-7197. Fax: (670) 234-2262. Web Site: www.magic100radio.com. Licensee: Leon Padilla Ganacias. (acq 8-13-98; $25,615). ♦Victoria Ganacias Borja, gen mgr.

KZMI(FM)— 1997: 103.9 mhz; 3.2 kw. 300 ft Stereo. Hrs opn: 24 1868 Halsey Drive, Piti, GU, 96915. Phone: (670) 234-7239. Fax: (670) 234-0447. Web Site: www.itecrmi.com. Licensee: Inter-Island Communications Inc. (group owner). Population served: 65,000 Format: Adult contemp. News staff: one; News: 16 hrs wkly. ♦Edward H. Poppe Jr., pres; Frances Poppe, CFO; Edward H. Poppe III, exec VP; Harry Blalock, stn mgr; Lewis Tenorio, progmg dir.

Directory of Radio Stations in Canada

Alberta

Airdrie

CFIT-FM— Apr 12, 2007: 106.1 mhz; 3.6 kw. TL: N51 17 35 W113 59 30. Hrs open: 159 B East Lake Blvd., T4A 2G2. Phone: (403) 945-3772. Fax: (403) 945-0277. E-mail: contactus@therangeonline.ca Web Site: www.therangeonline.ca. Licensee: Tiessen Media Inc. Format: Eclectic adult contemp. Target aud: 25-54. ♦Jamie Tiessen, gen mgr; Bruce Daniels, opns mgr & progmg dir; Carol Close, gen sls mgr.

Athabasca

CKBA(AM)— Aug 1, 1989: 850 khz; 1 kw-D. Hrs open: 24 2-4907 51st St., Venture Pl., T9S 1E7. Phone: (780) 675-5301. Fax: (780) 675-4938. E-mail: JPeckham@ab.ncc.ca Licensee: 3937844 Canada Inc. Group owner: NewCap Inc. (acq 4-19-02; grpsl). Population served: 10,000 Format: Today's country. News staff: one. Target aud: 25-54. ♦Joanne Peckham, gen mgr.

Blairmore

CJPR-FM— 2004: 94.9 mhz; 760 w. TL: N49 38 02 W114 29 30. Hrs open: Box 840, T0K 0E0. Phone: (403) 562-2806. Fax: (403) 562-8114. Licensee: 3937844 Canada Inc. Format: Full country. ♦Linda Huze, stn mgr; Linda Ransome, stn mgr; Darryl Ferguson, progmg dir.

Bonnyville

CJEG-FM— May 23, 2006: 101.3 mhz; 27 kw. Hrs open: Box 8251, T9N 2J5. Phone: (780) 812-3058. Web Site: www.1013koolfm.com. Licensee: NewCap Inc. Format: CHR. ♦Lise Lacombe, stn mgr & gen sls mgr; R.C. Ryder, progmg mgr & mus dir; Cash Kaye, mus dir; Robb Hunter, news dir; Raymond Green, chief of engrg.

CKLM-FM-1— 2007: 99.7 mhz; 50 kw. Hrs open: Box 21 Atrium Ctr., Lloydminster, T9V 0K2. Phone: (780) 875-5400. Fax: (780) 875-4628. Web Site: www.borderrock.com. Licensee: 912038 Alberta Ltd. Format: Rock. ♦J. Stewart Dent, gen mgr.

Brooks

CIBQ(AM)— Apr 15, 1973: 1340 khz; 1 kw-U, DA-1. Hrs open: 24 Unit 8-403 2nd Ave., West Brooks, T1R 1S3. Phone: (403) 362-3418. Phone: (403) 362-6000 (NEWS). Fax: (403) 362-8168. Web Site: www.eidnet.org/local/Q13. Licensee: 3937844 Canada Inc. Group owner: NewCap Broadcasting Ltd. (acq 4-19-2002; grpsl). Population served: 13,000 Format: Contemp country. News staff: one; News: 12 hrs wkly. Target aud: 25-54. ♦Ron Thompson, gen mgr; John Petrie, stn mgr & gen sls mgr; Brent Young, progmg dir; Sue Stevens, news dir.

CIXF-FM—Co-owned with CIBQ(AM). Oct 11, 2005: 101.1 mhz; 2.2 kw. Format: Adult contemp.

Calgary

CBCX-FM— 2003: 89.7 mhz; 10 kw. Hrs open: 1724 Westmount Blvd. N.W., T2N 3G7. Phone: (403) 521-6000. Web Site: www.cbc.ca. Licensee: CBC. Natl. Network: Espace Musique. Format: Fr. ♦Don Orchard, gen mgr; Henk VanLeeuwen, progmg dir.

***CBR(AM)**— Oct 1, 1964: 1010 khz; 50 kw-U, DA-2. Hrs open: 20 Box 2640, T2P 2M7. Secondary address: 1724 Westmount Blvd. N.W. T2N 3G7. Phone: (403) 521-6000. Fax: (403) 521-6271. Web Site: www.cbc.ca. Licensee: CBC. Format: Info, div, news/talk. News: 24 hrs wkly. ♦Dan Orchard, gen mgr; Randy Winczira, opns mgr; Harry Wagter, mktg mgr; Michelle Everett, mktg mgr; David Perlich, progmg dir; Helen Henderson, progmg mgr; Donna McElligott, news dir.

CBR-FM— Sept 29, 1975: 102.1 mhz; 100 kw. 788 ft Stereo. Natl. Network: CBC Radio Two. Format: Class, blues.

CBRF-FM— 2002: 103.7 mhz; 22 kw. Hrs open:
Rebroadcasts CHFA(AM) Edmonton 100%.
Box 555, Edmonton, T5J 2P4. Secondary address: 1724 Westmount Blvd. N.W. T2N 3G7. Phone: (780) 468-7500. Fax: (780) 468-7849. Web Site: radio-canada.ca/regions/alberta/index.shtml. Licensee: Canadian Broadcasting Corp. Natl. Network: Premiere Chaine. Format: Fr. ♦Francois Pageau, gen mgr.

CFAC(AM)— May 1922: 960 khz; 50 kw-U. Stereo. Hrs open: 24 2723 37th Ave. N.E., T1Y 5R8. Phone: (403) 291-0000. Fax: (403) 291-4368. Licensee: Rogers Broadcasting Ltd. (acq 12-89). Population served: 850,000 Format: All sports. News staff: 3; News: 15 hrs wkly. Target aud: 55 plus. Spec prog: Agriculture, rural 10 hrs wkly. ♦Tony Viner, pres; Gary Miles, exec VP; Kevin McKenna, VP, gen mgr & opns VP; Jim Dunlop, gen sls mgr; Paul Williams, adv dir; Kelly Kirch, progmg dir.

CHFM-FM—Co-owned with CFAC(AM). Aug 29, 1962: 95.9 mhz; 48 kw. Ant 480 ft TL: N51 03 37 W114 10 13. Stereo. 24 Web Site: www.chfm.com. Natl. Rep: Canadian Broadcast Sales. Format: Adult contemp. News staff: one. Target aud: 35-54; females. ♦Tony Viner, CEO; Kevin McKenna, stn mgr; Jennifer Enns, prom dir; Vince Cownden, progmg dir; Darren Robson, mus dir; David Spence, news dir; Tanya Berner, pub affrs dir.

CFEX-FM— Jan 1, 2007: 92.9 mhz; 45 kw. Hrs open: 6940 Fisher Rd. S.E., #200, T2H 0W3. Phone: (403) 670-0210. Fax: (403) 212-1399. Web Site: www.x929.ca. Licensee: Harvard Broadcasting Inc. Format: Alternative rock. ♦Christian Hall, gen mgr.

CFFR(AM)— Jan 10, 1984: 660 khz; 50 kw-U, DA-2. Stereo. Hrs open: 24 2723 37 Ave. N.E., T1Y 5R8. Phone: (403) 291-0000. Fax: (403) 291-5342. Licensee: Rogers (Alberta) Ltd. Group owner: Rogers Broadcasting Ltd. (acq 9-10-99; grpsl). Format: All news. News staff: 5; News: 10 hrs wkly. Target aud: 25-49. Spec prog: Sports 15 hrs wkly. ♦Kevin McKenna, gen mgr; Karen Parsons, prom dir, progmg dir & news dir; Shannon Kotylak, prom dir & progmg dir.

CKIS-FM—Co-owned with CFFR(AM). June 3, 1996: 96.9 mhz; 48 kw. Ant 686 ft TL: N51 02 18 W114 13 28. Format: Classic rock. ♦Gavin Tucker, progmg dir; K. Kirch, mus dir; Jerry Pendree, engrg VP & chief of engrg.

CFGQ-FM— Apr 15, 1982: 107.3 mhz; 100 kw. Ant 638 ft TL: N51 03 54 W114 12 47. Stereo. Hrs open: 630 3rd Ave. S.W., Suite 105, T2P 4L4. Phone: (403) 716-6500. Fax: (403) 716-2111. Web Site: www.q107fm.ca. Licensee: CKIK-FM Ltd. Group owner: Corus Entertainment Inc. (acq 7-6-2000; grpsl). Population served: 760,000 Format: Classic rock. Target aud: 25-44. ♦Garry McKenzie, gen mgr; Doug Young, gen sls mgr; Christian Hall, progmg dir; Natasha Rapchuk, news dir; Wade Wensink, chief of engrg; Judy Rickleton, traf mgr.

CHQR(AM)—Co-owned with CFGQ-FM. November 1964: 770 khz; 50 kw-U, DA-2. Stereo. 24 Web Site: www.qr77.com. (Acq 4-15-70). Natl. Rep: Canadian Broadcast Sales, Dora-Clayton. Format: News/talk, sports. News staff: 11; News: 17 hrs wkly. Target aud: 35 plus. ♦Phil Kallsen, progmg dir; Bill Powers, sports cmtr. Co-owned TV: CICT-TV affil

CFUL-FM— Mar 12, 2007: 90.3 mhz; 100 kw. TL: N51 03 37 W114 10 13. Hrs open: 1110 Centre St. N.E., Suite 100, T2E 2R2. Phone: (403) 271-6366. Fax: (403) 278-6772. E-mail: youshouldplaythis @fuelcalgary.com Web Site: www.fuelcalgary.com. Licensee: Newcap Inc. Format: Adult album alternative. ♦Murray Brookshaw, opns mgr & progmg dir; Stephen Peck, gen mgr & gen sls mgr; Lisa J. Lima, mktg dir & prom dir.

CHKF-FM— Nov 14, 1998: 94.7 mhz; 53 kw. Hrs open: 24 2723-37 Ave. N.E. #109, T1Y 5R8. Phone: (403) 717-1940. Fax: (403) 717-1945. E-mail: general @fm947.com Web Site: www.fm947.com. Licensee: Fairchild Radio (Calgary FM) Ltd. Population served: 150,000 Format: Ethnic. News staff: 2; News: 21 hrs wkly. Target aud: General. ♦Christine Leung, gen mgr & mktg mgr; Perry Chan, progmg dir.

CIBK-FM— Sept 6, 2002: 98.5 mhz; 100 kw. Hrs open: Suite 300, 1110 Cemter St. N., T2E 2R2. Phone: (403) 240-5800. Fax: (403) 240-5801. Web Site: www.vibe985.com. Licensee: Standard Radio Inc. Group owner: Standard Broadcasting Corp. (acq 4-19-2002; grpsl). Format: CHR. ♦Tom Peacock, gen mgr; Stew Meyers, opns mgr & progmg dir; Vinka Dubroja, gen sls mgr; Amanda Nelson, prom dir.

CIQX-FM— Aug 30, 2002: 103.1 mhz; 100 kw. Stereo. Hrs open: 1110 Centre St. N.E., Suite 100, T2E 2R2. Phone: (403) 271-6366. Fax: (403) 278-6772. E-mail: feedback@california103.com Web Site: www.california103.com. Licensee: 3937844 Canada Inc. Group owner: NewCap Broadcasting Ltd. (acq 4-19-02; grpsl). Population served: 900,000 Wire Svc: BN Wire Format: Adult contemp, smooth jazz. News staff: one; News: 35 hrs wkly. Target aud: 35-54. ♦Stephen Peck, gen mgr & gen sls mgr; John Beaudin, progmg dir; Hal Gardiner, news dir; Mike Gratton, chief of engrg.

CJAY-FM—Listing follows CKMX(AM).

CJSI-FM— December 1997: 88.9 mhz; 100 kw. Ant 979 ft TL: N51 03 54 W114 12 47. Hrs open: Suite 100, 4510 Macleod Trail S., T2G 0A4. Phone: (403) 276-1111. Fax: (403) 276-1114. E-mail: shine@homefm.com Web Site: www.cjsi.ca. Licensee: Touch Canada Broadcasting LP. Format: Contemp Christian. ♦Mike Kelly, pres & gen mgr; Mark Imbach, opns mgr.

***CJSW-FM**— Jan 15, 1985: 90.9 mhz; 4 kw. Stereo. Hrs open: 24 Rm. 127- MacEwan Hall, 2500 University Dr. N.W., T2N 1N4. Phone: (403) 220-3904. Fax: (403) 289-8212. Web Site: www.cjsw.com. Licensee: The University of Calgary Student Radio Society. Population served: 1,000,000 Format: Alternative, jazz, community. News: 5 hrs wkly. Target aud: General; young, trendy & well-heeled. Spec prog: Fr one hr, Ger 2 hrs, It one, Sp one hr wkly. ♦Chad Saunders, gen mgr.

CKAV-FM-3—Not on air, target date: unknown: 88.1 mhz; 33 kw. Ant 1,040 ft TL: N51 03 54 W114 12 47. Hrs open: 366 Adelaide St. E., Suite 323, Toronto, ON, M5A 3X9. Phone: (416) 703-1287. Fax: (416) 703-4328. Web Site: www.aboriginalradio.com. Licensee: Aboriginal Voices Radio Inc. ♦Roy Hennessy, opns mgr; Patrice Mousseau, progmg dir.

CKCE-FM— Mar 22, 2007: 101.5 mhz; 48 kw. Hrs open: 535 7th Ave. S.W., T2P 0Y4. Phone: (403) 508-2222. Fax: (403) 508-2224. Web Site: www.calgary1015.com. Licensee: CHUM Ltd. Format: Hot adult contemp. ♦James Stuart, gen mgr; Gavin Mortimer, gen sls mgr; Khazma Tichon, mktg dir & prom dir; Rob Mise, progmg dir.

CKMX(AM)— May 18, 1922: 1060 khz; 50 kw-U, DA-N. TL: N50 54 02 W26 113 52. Stereo. Hrs open: 24 Box 2750, Broadcast House, T2P 4P8. Phone: (403) 240-5800. Fax: (403) 240-5801. Web Site: www.cjay92.com. Licensee: Standard Radio Inc. Group owner: Standard Broadcasting Corp. (acq 6-19-92). Population served: 750,000 Format: Adult contemp. News staff: one. Target aud: 45 plus. Spec prog: Jazz 5 hrs wkly. ♦Tom Peacock, gen mgr & gen sls mgr.

CJAY-FM—Co-owned with CKMX(AM). June 1, 1977: 92.1 mhz; 100 kw. Ant 979 ft TL: N51 03 37 W114 10 13. Stereo. 24 Fax: (403) 242-6956. Web Site: www.cjay92.com. Format: Rock. ♦Ryan Dryden, mktg dir; Bob Harris, progmg dir; Ben Jeffery, mus dir; Ken Pasolli, engrg dir.

CKRY-FM— July 9, 1982: 105.1 mhz; 100 kw. Ant 400 ft Stereo. Hrs open: 24 630 3rd Ave. S.W., Suite 105, T2P 4L4. Phone: (403) 716-2105. Fax: (403) 716-2111. Web Site: www.country105.com. Licensee: Corus Entertainment Inc. (group owner) Population served: 800,000 Natl. Rep: Canadian Broadcast Sales. Format: Country. News staff: 6; News: 7 hrs wkly. Target aud: 25-54. ♦Garry McKenzie, gen mgr & stn mgr.

Camrose

CFCW(AM)— Nov 2, 1954: 790 khz; 50 kw-U, DA-2. Stereo. Hrs open: 24 2394 W. Edmonton Mall, 8882 170th St., Edmonton, T5T 4M2. Phone: (780) 437-4996. Fax: (780) 436-9803. Licensee: Newcap Inc. Group owner: NewCap Inc. Population served: 900,000 Natl. Rep: imsradio. Format: Country. News: 11 hrs wkly. Target aud: 25-54; country mus, sports & hockey listeners in Edmonton region. Spec prog: Sports open line, farm 5 hrs wkly. ♦Randy Lemay, gen mgr; Jackie Rae-Greening, progmg dir.

CFCW-FM— Oct 1, 2005: 98.1 mhz; 50 kw. Hrs open: 5708 48th Ave., T4V 0K1. Phone: (780) 672-9822. Fax: (780) 672-4678. Web Site: www.cfcw.com. Licensee: Newcap Inc. Format: Classic Hits. ♦Bruce Makokis, gen mgr.

Canmore

CHMN-FM— February 1998: 106.5 mhz; 510 w. Hrs open: Peschl's Corner, 749 Railway Ave., T1W 1P2. Phone: (403) 678-2222. Phone: (403) 678-2223. Fax: (403) 678-6844. Licensee: Rogers Broadcasting Ltd. (group owner). Natl. Rep: Canadian Broadcast Sales. Format: Hot adult contemp. ♦Kevin McKenna, gen mgr; Paul Williams, gen sls mgr; Vince Camden, progmg dir; Jeff Hubbard, mus dir.

Drayton Valley

CIBW-FM— 1994: 92.9 mhz; 7.4 kw. Hrs open: 24 Postal Bag 929, T7A 1V3. Phone: (780) 542-9290. Fax: (780) 542-9319. E-mail: bwcprod@telusplanet.net Licensee: Jim Pattison Broadcast Group Ltd. (the general partner) and Jim Pattison Industries Ltd. (the limited partner) carrying on business as Jim Pattison Broadcast Group L.P. Group owner: The Jim Pattison Broadcast Group (acq 9-7-95). Population served: 100,000 Natl. Rep: Canadian Broadcast Sales. Format: Country. News staff: one; News: 16 hrs wkly. Target aud: General; people with money to spend. ♦Paul Mason, gen mgr; Jay Neff, gen sls mgr; Trevor Grinde, progmg dir.

Drumheller

CKDQ(AM)— 1958: 910 khz; 50 kw-U, DA-2. Stereo. Hrs open: 24 Box 1480, T0J 0Y0. Secondary address: 515 Hwy. 10 E. T0J 0Y0. Phone: (403) 823-3384. Fax: (403) 823-7241. E-mail: bbrown@ab.ncc.ca Licensee: 3937844 Canada Inc. Group owner: NewCap Broadcasting Ltd. (acq 4-19-02; grpsl). Population served: 365000 Natl. Rep: CBS Radio. Format: Country. News staff: 2; News: 11 hrs wkly. Target aud: 25-54. Spec prog: Farm 8 hrs, relg 2 hrs wkly. ♦Hugh MacDonald, gen mgr.

Edmonton

CBX(AM)— 1948: 740 khz; 50 kw-U, DA-2. Hrs open: 24 Box 555, T5J 2P4. Secondary address: Edmonton City Ctr., 10062-102 Ave., Suite 123, Alberta T5J 24G. Phone: (780) 468-7500. Fax: (780) 468-7419. E-mail: cbx_edmonton@cbc.ca Web Site: www.edmonton.cbc.ca. Licensee: CBC. Natl. Network: CBC Radio One. Format: Info, news/talk. Target aud: 35 plus; college educated. ♦Mike Linden, VP; Judy Piercey, progmg dir.

CBX-FM—Not on air, target date: unknown: 90.9 mhz; 100 kw. Ant 633 ft Stereo. Web Site: www.edmonton.cdc.ca. Natl. Network: CBC Radio Two. Format: Btfl mus, class, news.

CFBR-FM—Listing follows CFRN(AM).

CFCW(AM)—See Camrose

CFRN(AM)— 1934: 1260 khz; 50 kw-U, DA-N. Hrs open: 18520 Stony Plain Rd., Suite 100, T5S 2E2. Phone: (780) 486-2800. Fax: (780) 489-6927. Web Site: www.cfrn.com. Licensee: Standard Radio Inc. Group owner: Standard Broadcasting Corp. (acq 6-19-92). Format: Sports. Target aud: 45 plus. Spec prog: Adult standards, nostalgia. ♦Marty Forbes, gen mgr; Bryn Griffith, progmg dir.

CFBR-FM—Co-owned with CFRN(AM). Apr 25, 1951: 100.3 mhz; 100 kw. 482 ft Stereo. Web Site: www.thebearrocks.com. Format: Rock, classic rock. ♦Jane Morissey, gen sls mgr; Ryan Zimmerman, progmg dir; Park Warden, mus dir; Bruce Bedford, chief of engrg.

CHBN-FM— Feb 17, 2005: 91.7 mhz; 100 kw. Hrs open: 10212 Jasper Ave. N.W., T5J 5A3. Phone: (780) 424-2222. Fax: (780) 401-1600. Web Site: www.thebounce.ca. Licensee: Edmonton Urban Partnership. Format: Rhythmic CHR. Target aud: 15-39. ♦James Stuart, gen mgr; Giselle Sowa, gen sls mgr; Johnny Staub, progmg dir; Lamya Asiff, news dir; Darcie Harris, chief of engrg.

CHDI-FM— May 9, 2005: 102.9 mhz; 100 kw. Hrs open: 24 5915 Gateway Blvd., T6H 2H3. Phone: (780) 423-2005. Fax: (780) 437-5129. E-mail: al@radiosonic.fm Web Site: www.radiosonic.fm. Licensee: Rogers Broadcasting Ltd. (acq 11-29-2006; grpsl). Format: Modern rock. News staff: one. ♦Diana Parker, gen mgr; Mike Bowman, gen sls mgr; Brent Shelton, prom dir; Al Ford, progmg dir; Kory Read, news dir.

CHED(AM)— Mar 3, 1954: 630 khz; 50 kw-U, DA-N. Stereo. Hrs open: 5204-84 St., T6E 5N8. Phone: (780) 440-6300. Fax: (780) 468-6739. Fax: (780) 469-5937. E-mail: info@630ched.com Web Site: www.630ched.com. Licensee: Corus Premium Television Ltd. Group owner: Corus Entertainment Inc. (acq 7-6-00; grpsl). Format: Sports, talk. ♦Doug Rutherford, gen mgr; Tanya Laughren, prom dir; Syd Smith, progmg dir; Tom Davies, chief of engrg.

CKNG-FM—Co-owned with CHED(AM). Aug 11, 1982: 92.5 mhz; 100 kw. 900 ft Stereo. E-mail: info@power92.com Format: Top-40. ♦Julie James, progmg dir; Greg Cooper, prom.

***CHFA(AM)—** Nov 20, 1949: 680 khz; 10 kw-U, DA-1. Hrs open: 24 Box 555, T5J 2P4. Secondary address: 123 Edmonton City Center T5J 2P4. Phone: (780) 468-7800. Fax: (780) 468-7849. Licensee: CBC. (acq 4-1-74). Format: Div, MOR, news/talk. News staff: 8; News: 8 hrs wkly. Target aud: 20-60; Fr speaking. ♦Francois Pageau, gen mgr; Jack Tyler, chief of engrg.

CHMC-FM— Dec 8, 2005: 99.3 mhz; 100 kw. Hrs open: 5241 Calgary Tr., Suite 700, Centre 104, T6H 5G8. Phone: (780) 433-7877. Fax: (780) 438-8484. E-mail: thofer@rawlco.com Web Site: www.magic99.ca. Licensee: Rawlco Radio Ltd. Format: Jazz, blues and soft rock. ♦Susan Reade, gen mgr & gen sls mgr; Kurt Leavins, progmg dir & mus dir.

CHQT(AM)—Listing follows CISN-FM.

CIRK-FM— 1949: 97.3 mhz; 100 kw. Stereo. Hrs open: 2394 W. Edmonton Mall, 8882 170th St., T5T 4M2. Phone: (780) 437-4996. Fax: (780) 436-9803. Web Site: www.k-rock973.com. Licensee: NewCap Inc. Group owner: NewCap Broadcasting Ltd. (acq 2-17-99; C$10 million). Natl. Rep: imsradio. Format: Classic rock. Target aud: 18-54; mobile adults. ♦Randy Lemey, gen mgr; Ross House, gen sls mgr; Lochlan Croft, progmg dir.

CISN-FM— June 5, 1982: 103.9 mhz; 100 kw. 757 ft Stereo. Hrs open: 5204 84th St., T6E 5N8. Phone: (780) 428-1104. Fax: (780) 469-5937. E-mail: info@cisnfm.com Web Site: www.cisnfm.com. Licensee: Corus Radio Co. Group owner: Corus Entertainment Inc. (acq 7-6-00; grpsl). Format: Contemp country. ♦Doug Rutherford, gen mgr & stn mgr; Neil Cunningham, gen sls mgr; Danielle L'hirrondelle, prom dir; Chris Scheetz, progmg dir; James Stuart, progmg dir; Bob Layton, news dir; Tom Davies, chief of engrg; Danielle Mattiello, traf mgr; Bryan Hall, sports cmtr.

CHQT(AM)—Co-owned with CISN-FM. Aug 19, 1965: 880 khz; 50 kw-U, DA-N. E-mail: info@cool880.com Format: Hits of the 60s & 70s. ♦Tanya Laughren, prom dir; Syd Smith, progmg dir; Danielle Mattiello, traf mgr.

CJCA(AM)— May 22, 1922: 930 khz; 50 kw-U, DA-N. Hrs open: 24 4207 98th St. N.W., Suite 204, T6E 5R7. Phone: (780) 466-4930. Fax: (780) 469-5335. Web Site: www.cjca.ca. Licensee: Touch Canada Broadcasting LP. (acq 4-12-94). Population served: 1,000,000 Format: Contemp Christian. News: 6 hrs wkly. Target aud: 25-54. ♦Jamie Moffat, gen sls mgr & disc jockey; Topher Braithwaite, prom dir; Malcolm Hunt, progmg dir; Gord Craig, news dir; Len Dehek, disc jockey.

CJRY-FM—Co-owned with CJCA(AM). 2004: 105.9 mhz; 100 kw. Ant 633 ft Web Site: www.cjry.ca/cms. Format: Contemp Christian music.

CJSR-FM— 1984: 88.5 mhz; 900 w. Stereo. Hrs open: 24 Room 0-09, Students' Union Bldg., Univ. of Alberta, T6G 2J7. Phone: (780) 492-5244. Fax: (780) 492-3121. E-mail: admin@cjsr.com Web Site: www.cjsr.ualberta.ca. Licensee: The First Alberta Campus Radio Association. Population served: 1,000,000 Format: Alternative progmg. Target aud: General; everyone who looks for something new in the mus industry. Spec prog: American Indian 2 hrs, Black 7 hrs, class 2 hrs, folk 16 hrs, Fr one hr, gospel 2 hrs, Pol 2 hrs, Sp 2 hrs wkly. ♦Desiree Schell, mktg dir; Jay Hannley, progmg dir; Franny Rowlyk, news dir.

CKER-FM— 1996: 101.7 mhz; 100 kw. Stereo. Hrs open: 24 5915 Gateway Blvd., T6H 2H3. Phone: (780) 702-1188. Fax: (780) 437-5129. E-mail: dianaparker@worldfm.ca Web Site: www.worldfm.ca. Licensee: Rogers Broadcasting Ltd. (acq 11-29-2006; grpsl). Population served: 1,000,000 Format: Ethnic, Christian, Chinese. News staff: 2; News: 14 hrs wkly. Target aud: General; ethnic audience (24 languages), and Christian. Spec prog: It 3 hrs, Sp 8 hrs, Por 2 hrs, Ukrainian 10 hrs, Dutch 3 hrs, Pol 6 hrs, E.Indian 7 hrs wkly. ♦Diana Parker, gen mgr; Roger Charest Jr., opns mgr; Cameron Smith, traf mgr.

CKRA-FM— Nov 15, 1979: 96.3 mhz; 100 kw. Ant 757 ft Stereo. Hrs open: 24 2394 W. Edmonton Mall, 8882 170th St., T5T 4M2. Phone: (780) 437-4996. Fax: (780) 436-9803. Web Site: www.bigearl.ca. Licensee: NewCap Inc. Group owner: NewCap Broadcasting Ltd. Population served: 800,000 Natl. Rep: imsradio. Format: Country. News staff: 6; News: 5 hrs wkly. Target aud: 25-54; urban young adults. ♦Randy Lemay, gen mgr; Jackie Rae-Greening, progmg dir.

***CKUA-FM—** June 28, 1948: 94.9 mhz; 100 kw. 400 ft Stereo. Hrs open: 24 hrs 4th Fl., 10526 Jasper Ave., T5J 1Z7. Phone: (780) 428-7595. Fax: (780) 428-7624. E-mail: radio@ckua.org Web Site: www.ckua.com. Licensee: CKUA Radio Foundation. (acq 5-29-95). Population served: 3,000,000 Format: Div, class, jazz. News staff: 4; News: 6 hrs wkly. Target aud: General; Alberta population. ♦Ken Regan, gen mgr & opns mgr; Andrea Louie, gen sls mgr; Brian Dunsmore, progmg dir; Peter North, mus dir; Neil Lutes, chief of engrg; Sharon Cross, traf mgr.

CKUA(AM)— Nov 21, 1927: 580 khz; 10 kw-U, DA-2. TL: N53 20 34 W113 27 27.24 Format: Diversified.

Edson

CFXE-FM— 2007: 94.3 mhz; 11 kw. TL: N53 38 47 W116 32 26. Hrs open: 24 Box 7800, T7E 1V8. Secondary address: 422 50th St. T7E 1T1. Phone: (780) 723-4461. Fax: (780) 723-3765. E-mail: dschuck@fox-radio.ca Web Site: www.thefoxradio.ca. Licensee: 3937844 Canada Inc. Natl. Rep: imsradio. Wire Svc: Canadian Press Format: Adult contemp, CHR. News staff: 3. Target aud: 18-55. Spec prog: Farm 2 hrs wkly. ♦Al Anderson, VP; Dave Schuck, gen mgr & stn mgr; Rob Alexander, progmg dir; Steve Bethge, news dir.

Falher

***CKRP-FM—** Nov 2, 1996: 95.7 mhz; 671 w. TL: N55 44 07 W117 11 34. Stereo. Hrs open:
Rebroadcasts CITE-FM Montreal 65%.
Box 718, Association Canadienne-Francaise de l'Alberta, Regionale de Riviere-la-Paix, T0H 1M0. Phone: (780) 837-2346. Fax: (780) 837-2092. E-mail: ckrpfm@yahoo.ca Licensee: Association canadienne-francaise de l'Alberta-Regionale de Riviere-la-Paix. Format: French, adult contemp, community service. Target aud: French population. ♦Julie Cadieux, pres.

Fort McMurray

CJOK-FM— 2003: 93.3 mhz; 40 kw. TL: N56 41 16 W111 19 55. Hrs open: 24 9912 Franklin Ave., T9H 2K5. Phone: (780) 743-2246. Fax: (780) 791-7250. Web Site: mymcmurray.com. Licensee: Rogers Broadcasting Ltd. (acq 11-29-2006; grpsl). Population served: 60,000 Format: Today's hot country. News staff: 3; News: 4 hrs wkly. Target aud: 25-44. ♦Jim Schneider, gen mgr & gen sls mgr.

CKYX-FM— March 1985: 97.9 mhz; 40 kw. Stereo. Hrs open: 24 9912 Franklin Ave., T9H 2K5. Phone: (780) 743-2246. Fax: (780) 791-7250. Web Site: mymcmurray.com. Licensee: Rogers Broadcasting Ltd. (acq 11-29-2006; grpsl). Population served: 60,000 Format: Classic rock. News staff: 3; News: 5 hrs wkly. Target aud: 18-44. ♦Jim Schneider, gen mgr & gen sls mgr.

Fort Vermilion

CIAM-FM— Jan 27, 2003: 92.7 mhz; 30 w. Hrs open: 24 Box 609, T0H 1N0. Secondary address: 4709 River Rd. T0H 1N0. Phone: (780)-927-2426. Fax: (780) 927-2427. E-mail: ciam@telus.net Web Site: www.ciamradio.com. Licensee: Care Radio Broadcasting Association. Format: Var/div/multingual. ♦Michael Sandstrom, gen mgr; James Neufeld, news dir; Phil Peters, progmg dir & chief of engrg.

Grand Centre (Cold Lake)

CJXK-FM— Sept 3, 2004: 95.3 mhz; 100 kw. Hrs open: B5412 55th St., Cold Lake, T9M 1R5. Phone: (780) 594-2459. Fax: (780) 594-3001. Licensee: 3937844 Canada Inc. Format: Classic rock. ♦Carla Loffler, stn mgr.

Grande Prairie

CFGP-FM— June 20, 1996: 97.7 mhz; 70 kw. TL: N55 27 57 W118 45 32. Stereo. Hrs open: 24 Suite 200, 9835 101st Ave., T8V 5V4. Phone: (780) 539-9700. Fax: (780) 532-1600. Fax: (780) 539-0367 (news). Web Site: www.sunfm.com. Licensee: Rogers Broadcasting Ltd. (acq 11-29-2006; grpsl). Format: Hot CHR-top 40. News staff: 4; News: 4 hrs wkly. Target aud: 25-55. ♦Dave Reid, gen mgr, opns mgr & gen sls mgr; Lisa Kirby, prom dir; Kevin Becker, progmg dir; Daryl Major, news dir; Sam Lowe, chief of engrg.

CFRI-FM— Mar 30, 2007: 104.7 mhz; 100 kw. Hrs open: #1 11002 104th Ave., T8V 7W5. Phone: (780) 357-1047. Fax: (780) 830-7815. Licensee: Vista Radio Ltd. Format: Classic rock. ♦Gordon Gauvin, gen mgr.

CJXX-FM— Nov 1, 2000: 93.1 mhz; 100 kw. TL: N55 03 08 W118 51 59. Hrs open: 24 9817 101st Ave., Suite 202, Grand Prairie, T8V 0X6. Phone: (780) 532-0840. Fax: (780) 538-1266. Fax: (780) 539-6397. E-mail: general@bigcountryxx.com Web Site: www.bigcountryxx.com. Licensee: Jim Pattison Broadcast Group Ltd. (the general partner) and Jim Pattison Industries Ltd. (the limited partner) carrying on business as Jim Pattison Broadcast Group L.P. Group owner: The Jim Pattison Broadcast Group (acq 12-21-2000; grpsl). Population served: 200,000 Wire Svc: BN Wire Format: C&W. News staff: 5; News: 14 hrs wkly. Target aud: 25-49; adults who love country music. ♦Rick Arnish, pres; Ken Norman, gen mgr & progmg dir; Anne Graham, gen sls mgr; Barbara Shannon, prom dir; Candace Boyne, prom dir.

High Level

CKHL-FM— July 1999: 102.1 mhz; 8.765 kw. Hrs open: Box 3759, T0H 1Z0. Phone: (780) 926-4531. Fax: (780) 926-4564. Web Site: www.ylcountry.com. Licensee: 912038 Alberta Ltd. Format: Country. ♦Terry Babiy, gen mgr & sls dir; Chris Black, gen sls mgr; Don Jennings, progmg dir; Karin Koppitz, news dir.

High Prairie

CKVH(AM)— 1990: 1020 khz; 1 kw-D, 400 w-N. Hrs open: Box 2219, T0G 1E0. Phone: (780) 523-5111. Fax: (780) 523-3360. Web Site: www.ncc.ca. Licensee: 3937844 Canada Inc. Group owner: NewCap Broadcasting Ltd. (acq 4-19-02; grpsl). Format: Country. ♦Ron Gendron, gen mgr & gen sls mgr; Rob Alexander, progmg dir.

High River

CFXL-FM—Listing follows CHRB(AM).

CHRB(AM)— Dec 5, 1977: 1140 khz; 50 kw-D, 46 kw-N, DA-2. TL: N50 55 25 W113 49 58. Hrs open: 24 11 5th Ave. S.E., TIV 1G2. Phone: (403) 652-2472. Fax: (403) 652-7861. E-mail: am1140@am1140radio.com Web Site: www.am1140radio.com. Licensee: Golden West Broadcasting Ltd. (group owner). Population served: 50,000 Natl. Rep: Canadian Broadcast Sales. Format: C&W, relg. News staff: 2; News: 10 hrs wkly. Target aud: General. Spec prog: Farm 5 hrs hrs wkly. ♦Elmer Hildebrand, CEO; Lyndon Friesen, sr VP; Keith Leask, stn mgr, gen sls mgr, prom dir & progmg dir; Menno Friesen, sls VP; Don McCracken, news dir; Vern Moores, chief of engrg.

CFXL-FM—Co-owned with CHRB(AM). 2003: 100.9 mhz; 100 kw.

High River-Okotoks

CFXL-FM—Licensed to High River-Okotoks. See High River

Hinton

CFXH-FM— July 2004: 97.5 mhz; 1.2 kw. Hrs open: Phone: (780) 723-4461. Fax: (780) 723-3765. Web Site: www.thefoxradio.ca. Licensee: 3937844 Canada Inc. Format: Classic hits. Target aud: 18-54. ♦Dave Schuck, gen mgr; Rob Alexander, progmg dir; Steve Bethge, news dir.

Lac La Biche

CFWE-FM— 1990: 89.9 mhz; Hrs open: 24 13245 146th St., Edmonton, T5L 4S8. Phone: (780) 447-2393. Fax: (780) 454-2820. E-mail: cfwe@ammsa.com Web Site: www.ammsa.com/cfwe. Licensee: Aboriginal Multi-Media Society of Alberta. Format: Aboriginal, country. Target aud: General; Cree, Blackfoot, Stoney, Dene & English language listeners. ♦Bert Crowfoot, CEO & gen mgr; Al Standerwick, stn mgr.

Lacombe

CJUV-FM— June 28, 2006: 94.1 mhz; 9.5 kw. Hrs open: 4725 49B Ave, T4L 1K1. Phone: (403) 786-0194. Fax: (403) 786-0199. E-mail: ssawyer@telus.net Web Site: www.sunny94.com. Licensee: L.A. Radio Group Inc. Format: Classic hits. Target aud: 35-54. ♦Troy Stevens, pres; Sonia Sawyer, gen sls mgr & opns.

Lethbridge

CFRV-FM— 1979: 107.7 mhz; 100 kw. 600 ft TL: N49 42 23 W112 43 11. Stereo. Hrs open: 24 1015 3rd Ave S., T1J 0J3. Phone: (403) 328-1077. Fax: (403) 380-1539. Web Site: www.1077theriver.ca. Licensee: Rogers Broadcasting Ltd. (group owner) Population served: 200,000 Format: Adult contemp, classic rock, CHR. News staff: 1; News: 2 hrs wkly. Target aud: 18-49; Males. ♦Terry Voth, gen mgr; Tanya Wolford, prom dir; Robin Haggar, progmg dir; Erin Lucas, mus dir.

CHLB-FM— 1997: 95.5 mhz; 100 kw. Stereo. Hrs open: 24 401 Mayor Magrath Dr. S., T1J 3L8. Phone: (403) 329-0955. Fax: (403) 329-0195. E-mail: rbye@country95.fm Web Site: www.country95.fm. Licensee: Jim Pattison Broadcast Group Ltd. (the general partner) and Jim Pattison Industries Ltd. (the limited partner) carrying on business as Jim Pattison Broadcast Group L.P. Group owner: The Jim Pattison Broadcast Group (acq 12-21-2000; grpsl). Format: Country. News staff: 5. Target aud: 25-54.

CJOC-FM— June 2007: 94.1 mhz; 100 kw. Ant 433 ft TL: N49 43 59 W112 57 36. Hrs open: 220 Third Ave. S., Suite 400, T1J 0G9. Phone: (403) 388-2910. Fax: (866) 841-7971. E-mail: info@loungeradio.ca Web Site: www.loungeradio.ca. Licensee: Clear Sky Radio Inc. Format: Adult standards. ♦Paul Larsen, pres & gen mgr.

CJRX-FM— Nov 3, 2000: 106.7 mhz; 100 kw. 600 ft. TL: N49 42 23 W112 43 11. Stereo. Hrs open: 1015 3rd Ave S., T1K 0J3. Phone: (403) 320-1220. Fax: (403) 380-1539. Web Site: www.rock106.ca. Licensee: Rogers Broadcasting Ltd. (group owner) Population served: 200,000 Format: Rock. News staff: 1; News: 2 hrs wkly. Target aud: 25-44; Female. ♦Terry Voth, gen mgr & progmg dir; Scott McGregor, mus dir.

CKVN-FM— 2001: 98.1 mhz; 20 kw. Hrs open: 24 1277 3rd Ave. S., T1J 0K3. Phone: (403) 327-0981. Fax: (403) 328-0095. Web Site: ckmradio.com. Licensee: Golden West Broadcasting Ltd. (acq 8-2-2006). Format: Contemp Christian. News staff: 2; News: 6 hrs wkly. ♦Keith Leask, gen mgr.

CKXU-FM— Apr 8, 2004: 88.3 mhz; 125 w. Stereo. Hrs open: SU 164, 4401 University Dr. W., T1K 3M4. Phone: (403) 329-2180. Fax: (403) 329-2224. E-mail: ckxu@ckxu.com Web Site: www.ckxu.com. Licensee: CKXU Radio Society. Format: Var, Fr. ♦Nicholas Baingo, stn mgr; Jenn Prosser, progmg dir; John Pantherbone, mus dir.

Lloydminster

CKLM-FM— May 18, 2001: 106.1 mhz; 100 kw. Stereo. Hrs open: 24 Box 21 Atrium Ctr., T9V 0K2. Secondary address: 5012 49th St. T9V 0K2. Phone: (780) 875-5400. Fax: (780) 875-4628. Web Site: www.borderrock.com. Licensee: 912038 Alberta Ltd. (acq 8-17-2001). Population served: 125,000 Format: Rock/AOR. News staff: 3. Target aud: 12-54; male. ♦J. Stewart Dent, pres; Anita B. Dent, VP; Doug Zackodnik, gen sls mgr; James Gushnowski, progmg dir.

CKSA-FM— Aug 29, 2003: 95.9 mhz; 100 kw. Hrs open: 5026 50th St., T9V 1P3. Phone: (780) 875-3321. Fax: (780) 875-4704. E-mail: blabrie@newcap.ca Licensee: NewCap Inc. Group owner: Midwest Broadcasting. (acq 12-22-2004; C$6,246,000 with CILR-FM Lloydminster). Format: Country's best mix. ♦Mike Keller, gen mgr; Brian Labrie, stn mgr.

Medicine Hat

CFMY-FM— Feb 1, 1999: 96.1 mhz; 100 kw. Hrs open: Division of the Jim Pattison Broadcast Group, Box 1270, T1A 7H5. Secondary address: Division to the Jim Pattison Broadcast Group, 10 Boundary Rd. S.E. T0J 2P0. Phone: (403) 548-8282. Fax: (403) 548-8270. E-mail: myfm@jpbg.com Web Site: www.my96fm.com. Licensee: Jim Pattison Broadcast Group Ltd. (the general partner) and Jim Pattison Industries Ltd. (the limited partner) carrying on business as Jim Pattison Broadcast Group L.P. Group owner: The Jim Pattison Broadcast Group (acq 12-21-2000; grpsl). Format: Adult contemp. ♦Rick Arnish, pres; Dwaine Dietrich, gen mgr; Ed Lundberg, gen sls mgr; Michael Thibbau, progmg dir; Adrian Bateman, news dir; Carey Downs, chief of engrg.

CHAT-FM— Jan 9, 2006: 94.5 mhz; 100 kw. Hrs open: 24 Division of the Jim Pattison Broadcast Group, Box 1270, T1A 7H5. Secondary address: Division of the Jim Pattison Broadcast Group, 10 Boundary Rd. S.E. T0J 2P0. Phone: (403) 548-8282. Fax: (403) 548-8270. Web Site: www.chat945.com. Licensee: Jim Pattison Broadcast Group Ltd. (the general partner) and Jim Pattison Industries Ltd. (the limited partner) carrying on business as Jim Pattison Broadcast Group Ltd. Natl. Rep: Canadian Broadcast Sales. Format: Country. News staff: 4; News: 14 hrs wkly. Target aud: General. ♦Rick Arnish, pres; Dwaine Dietrich, gen mgr. Co-owned TV: CHAT-TV affil

CJLT-FM— April 2003: 93.7 mhz; 2.3 kw. Hrs open: 901 3rd Ave. S.W., T1A 4Z2. Phone: (403) 529-9599. Fax: (403) 529-2824. E-mail: alive995@telus.net Web Site: www.alivefm.com. Licensee: Lighthouse Broadcasting Ltd. Format: Christian. ♦Scott Raible, pres & opns mgr; Darcee Grange, progmg dir; John Enns, sls.

Okotoks

CFXL-FM—See High River

Olds

CKLJ-FM— Feb 2, 2004: 96.5 mhz; 35 kw. Hrs open: #6, 4526 49th Ave., T4H 1A4. Phone: (403) 556-2628. Fax: (403) 556-2637. E-mail: cklj@telus.net Licensee: CAB-K Broadcasting Ltd. Format: Country. ♦Brian Hepp, gen mgr.

Peace River

CKKX-FM— July 1997: 106.1 mhz; 990 w. Hrs open: Bag Service No. 300, T8S 1T5. Secondary address: 9807 100th Ave T8S 1T5. Phone: (780) 624-2535. Fax: (780) 624-5424. E-mail: reception@ylcountry.com Web Site: www.kix106.net. Licensee: 912038 Alberta Ltd. Format: Hot adult contemp. ♦Cynthia Babiy, VP; Terry Babiy, pres & gen mgr.

CKYL(AM)— Nov 1, 1954: 610 khz; 10 kw-U, DA-2. Hrs open: Bag Service No. 300, T8S 1T5. Phone: (780) 624-2535. Fax: (780) 624-5424. Web Site: www.ylcountry.com. Licensee: Peace River Broadcasting Ltd. (acq 12-15-95). Natl. Rep: Target Broadcast Sales. Format: Country. ♦Terry Babiy, gen mgr.

Red Deer

CFDV-FM— Nov 8, 2004: 106.7 mhz; 100 kw. Hrs open: 2840 Bremner Ave., T4R 1M9. Phone: (403) 343-7105. Fax: (403) 343-2573. E-mail: onair@1067thedrive.fm Web Site: www.1067thedrive.fm. Licensee: Jim Pattison Broadcast Group Ltd. (the general partner) and Jim Pattison Industries Ltd. (the limited partner) carrying on business as Jim Pattison Broadcast Group L.P. Population served: 250,000 Natl. Rep: Target Broadcast Sales. Wire Svc: BN Wire Format: Classic rock. News staff: 3. Target aud: 25-54; adults, primary demo-males. ♦Jim Pattison, CEO; Rick Arnish, pres; Paul Mason, gen mgr; Bryn James, gen sls mgr; Jim Hall, progmg dir; Marlow Weldon, news dir.

CHUB-FM— 1949: 105.5 mhz; 100 kw. Hrs open: 24 2840 Bremner Ave., T4R 1M9. Phone: (403) 343-7105. Fax: (403) 343-2573. Licensee: Jim Pattison Broadcast Group Ltd. (the general partner) and Jim Pattison Industries Ltd. (the limited partner) carrying on business as Jim Pattison Broadcast Group L.P. Group owner: The Jim Pattison Broadcast Group (acq 12-21-2000; grpsl). Population served: 250,000 Natl. Rep: Target Broadcast Sales. Rgnl rep: WTR Media Sales Wire Svc: BN Wire Format: Hot adult contemp. News staff: 3. Target aud: Adults 25-49; primary demo-females. ♦Jim Pattison, CEO; Rick Arnish, pres; Paul Mason, gen mgr; Bryn James, gen sls mgr; Jim Hall, progmg dir.

CIZZ-FM— Nov 1, 1987: 98.9 mhz; 100 kw. 800 ft Stereo. Hrs open: 24 Box 5339, T4N 6W1. Secondary address: 4920 59th St. T4N 2N1. Phone: (403) 343-1303. Fax: (403) 346-1230. E-mail: zedfm@cnewcap.ca Web Site: www.zedfm.com. Licensee: Newcap Inc. Group owner: Corus Entertainment Inc. (acq 8-10-2005; C$8,392,714 with CKGY-FM Red Deer). Population served: 110,000 Wire Svc: BN Wire Format: Adult contemp, CHR. News staff: 6. Target aud: 18-49; male 55%, female 45%. ♦R.C. (Ron) Thompson, gen mgr, stn mgr & gen sls mgr; Sue Stevenson, news dir; Brent Young, progmg.

CKGY-FM— April 2001: 95.5 mhz; 100 kw. 800 ft. Stereo. Hrs open: 24 Bag 5339, T4N 6W1. Secondary address: 4920 59th St. T4N 2N1. Phone: (403) 348-0955. Fax: (403) 346-1230. E-mail: kgcountry@newcap.ca Web Site: www.ckgy.com. Licensee: Newcap Inc. Group owner: Corus Entertainment Inc. (acq 8-10-2005; C$8,392,714 with CIZZ-FM Red Deer). Population served: 120,000 Natl. Rep: Canadian Broadcast Sales. Wire Svc: BN Wire Format: Today's hottest country. News staff: 6. Target aud: 25-54; 50% male, 50% female. ♦R.C. (Ron) Thompson, gen mgr & gen sls mgr.

Rocky Mountain House

CHBW-FM— 1997: 94.5 mhz; 720 w. Hrs open: 4814B 49th St., T4T 1S8. Phone: (403) 844-9450. Fax: (403) 844-4770. E-mail: bigshow@telus.net Licensee: Jim Pattison Broadcast Group Ltd. (the general partner) and Jim Pattison Industries Ltd. (the limited partner) carrying on business as Jim Pattison Broadcast Group L.P. Group owner: The Jim Pattison Broadcast Group. Format: Country. ♦Paul Mason, gen mgr; Barry Simon, stn mgr.

Saint Albert

CFMG-FM— Aug 29, 1994: 104.9 mhz; 100 kw. Hrs open: 24 18520 Stony Plain Rd., Suite 100, Edmonton, T5S 2E2. Phone: (780) 435-1049. Fax: (780) 489-6927. E-mail: cfmg@sri.ca Web Site: www.ezrock1049.com. Licensee: Standard Radio Inc. Group owner: Standard Broadcasting Corp. (acq 4-19-2002; grpsl). Population served: 870,000 Format: Adult contemp. News staff: 2; News: 4 hrs wkly. Target aud: 25-54; middle to upper income families. ♦Marty Forbes, VP, gen mgr & gen sls mgr; Paul Mothersell, gen sls mgr; Karen Paulguaard, prom dir; Steve Moore, progmg dir; Bruce Bedford, engrg dir & chief of engrg.

Saint Paul

CHLW(AM)— 1975: 1310 khz; 10 kw-U, DA-2. Hrs open: 24 201-4341 50th Ave., St. Paul, T0A 3A3. Phone: (780) 645-4425. Fax: (780) 645-2383. E-mail: dwhite@newcap.ca Web Site: www.1310chlw.com. Licensee: 3937844 Canada Inc. Group owner: NewCap Broadcasting Ltd. (acq 4-19-02; grpsl). Population served: 50,000 Format: New country. News staff: one; News: 4 hrs wkly. Target aud: 25-49. Spec prog: Farm 5 hrs, relg 5 hrs wkly. ♦Mike Keller, gen mgr; Danny White, stn mgr; Paul O'Neil, progmg dir.

Siksika

CHDH-FM— 2002: 97.7 mhz; 50 w. Hrs open: Box 1490, T0J 3W0. Phone: (403) 734-5339. Fax: (403) 734-5497. E-mail: siksikamedia @siksikanation.com Licensee: Siksika Communications Society. Format: News/talk, education, native music. ♦Paul Melting Tallow, gen mgr.

Slave Lake

CHSL-FM— Sept 8, 2006: 92.7 mhz; 5.7 kw. Hrs open: 221 3rd Ave. N.W., T0G 2A1. Phone: (780) 849-2569. Fax: (780) 849-4833. E-mail: rbedard@newcap.ca Web Site: www.chsl@newcap.ca. Licensee: 3937844 Canada Inc. Format: Classic hits. ♦Randy Bedard, opns mgr & gen sls mgr.

Stettler

CKSQ(AM)— Dec 15, 1975: 1400 khz; 1 kw-U, DA-2. Hrs open: 24 Box 2050, 4812A - 50 th Street, T0C 2L0. Phone: (403) 742-1400. Fax: (403) 742-0660. E-mail: cksq@newcap.ca. Licensee: 3937844 Canada Inc. Group owner: NewCap Broadcasting Ltd. (acq 4-19-02; grpsl). Population served: 10,000 Format: Country. News staff: one; News: 7 hrs wkly. Target aud: 25-54; 55% female, 45% male. ♦Vicki Leuck, gen mgr, stn mgr & gen sls mgr; Brent Young, progmg dir; Tim Day, news dir; Cliff Wheeler, chief of engrg.

Taber

CJBZ-FM— 2000: 93.3 mhz; 50 kw. Hrs open: 401 Mayor Magrath Dr., Lethbridge, T1J 3L8. Phone: (403) 394-9300. Fax: (403) 329-0195. E-mail: info@b93.fm Web Site: www.b93.fm. Licensee: Jim Pattison Broadcast Group Ltd. (the general partner) and Jim Pattison Industries Ltd. (the limited partner) carrying on business as Jim Pattison Broadcast Group L.P. Group owner: The Jim Pattison Broadcast Group (acq 12-21-2000; grpsl). Format: CHR/adult contemp. News staff: 4. Target aud: 18-44; adults. ♦Rick Arnish, pres; Rod Schween, gen sls mgr; Jarod Neithercut, prom dir; Reid Morgan, progmg dir; Dori Modney, news dir.

Wainwright

CKKY(AM)— February 1984: 830 khz; 10 kw-D, 3.5 kw-N. Hrs open: 24 1037 2nd Ave., 2nd fl, T9W 1K7. Phone: (780) 842-4311. Fax: (780) 842-4636. E-mail: ckky@ab.ncc.ca Licensee: 3937844 Canada Inc. Group owner: NewCap Broadcasting Ltd. (acq 5-20-2002; grpsl). Population served: 100,000 Format: C&W. News staff: 2; News: 14 hrs wkly. Target aud: 20-45; agriculture-related working class. Spec prog: Farm 10 hrs wkly. ♦Ron Prochner, gen mgr & progmg dir.

CKWY-FM—Co-owned with CKKY(AM). 2005: 93.7 mhz; 100 kw. Web Site: wayneradio.com. Format: Adult contemp. ♦Paul O'Neil, progmg dir.

Westlock

CFOK(AM)— Aug 19, 1975: 1370 khz; 10 kw-U, DA-2. Hrs open: 24 10030-106 St., Suite 17, T7P 2K4. Phone: (780) 349-4421. Fax: (780) 349-6259. E-mail: wbetts@newcap.ca Licensee: 3937844 Canada Inc. Group owner: NewCap Broadcasting Ltd. (acq 4-19-02; grpsl). Population served: 35,000 Natl. Rep: Canadian Broadcast Sales. Format: Today's country. News staff: one; News: 15 hrs wkly. Target aud: 25-54. Spec prog: Farm 5 hrs, relg 6 hrs wkly. ♦Dave Schuck, gen mgr; Rob Alexander, stn mgr & progmg dir; Wray Betts, opns VP & gen sls mgr; Steve Bethge, news dir.

Wetaskiwin

CIHS-FM— December 2000: 93.5 mhz; 5.12 kw. Ant 365 ft Hrs open: 24 5206 50th Ave., T9A 0S8. Phone: (780) 361-0245. Fax: (866) 409-2797. E-mail: mail@cihsfm.com Web Site: www.cihsfm.com. Licensee: 902890 Alberta Ltd. Format: Classic country, country gospel & world music. Target aud: 0-100. ♦Dave Dhillon, CEO, chmn, pres & gen mgr; Paula Osha, stn mgr.

CKJR(AM)— 1971: 1440 khz; 10 kw-U, DA-2. Hrs open: 5214 A-50th Ave., T9A-0S8. Phone: (780) 352-0144. Fax: (780) 352-0606. Web Site: www.1440.com. Licensee: 3937844 Canada Inc. Group owner: NewCap Broadcasting Ltd. (acq 4-19-02; grpsl). Format: Hot new country. Spec prog: Greek 2 hrs wkly. ♦David Gilmore, gen sls mgr; Nick Addams, progmg dir.

Whitecourt

CFXW-FM— July 1, 2005: 96.7 mhz; 9 kw. Hrs open: Box 2288, T7S 1A2. Secondary address: 5118 50th St. T7S 1A1. Phone: (780) 778-5101. Fax: (780) 778-5137. Web Site: www.therig.ca. Licensee: 3937844 Canada Inc. Format: Classic rock. ♦Dave Schuck, gen mgr; Randy Turner, gen sls mgr; Rob Alexander, progmg dir; Jeremy Lye, news dir.

CIXM-FM— 2006: 105.3 mhz; 42.3 kw. Hrs open: Box 1050, T7S 1N9. Secondary address: 4912A 50th Ave. T7S 1N9. Phone: (780) 706-1053. Fax: (780) 706-1017. Web Site: www.xm105.com. Licensee: 1097282 Alberta Ltd. (acq 5-11-2006). Natl. Rep: Target Broadcast Sales. Format: Country. ♦Gene Fabro, pres; Neil Shewchuk, stn mgr & gen sls mgr; Ken Singer, opns VP & natl sls mgr; Sarah Ryan, prom dir; Andrew Joseph, progmg dir; Bayne Opseth, chief of engrg.

British Columbia

100 Mile House

CKBX(AM)— July 30, 1971: 840 khz; 1 kw-D, 250 w-N, DA-1. Hrs open: Box 939, V0K 2E0. Phone: (250) 395-3848. Fax: (250) 395-4147. E-mail: spence@ckbx.ca Web Site: www.thewolfpack.ca. Licensee: Vista Radio Ltd. Group owner: Cariboo Central Interior Radio Inc. (acq 1981). Format: Modern country, southern rock. Spec prog: Class one hr wkly. ♦Paul Mann, gen mgr; Tracey Gard, stn mgr.

Abbotsford

CIVL-FM— September 2007: Stn currently dark. 88.5 mhz; 92 w. Stereo. Hrs open: 24 33844 King Rd., V2S 7M8. Phone: (604) 851-6306. E-mail: Bob@civl.ca Web Site: www.civl.ca. Licensee: UCFV Campus and Community Radio Society. Format: Var. Target aud: 18 plus; campus and community radio. ♦Bob Simpson, stn mgr; Swinder Singh, progmg mgr.

CKQC-FM— September 2001: 107.1 mhz; 215 w. Stereo. Hrs open: #520-45715 Hocking Ave., Chilliwack, V2P 6Z6. Phone: (604) 859-5277. Fax: (604) 702-3212. Licensee: Rogers Radio (British Columbia) Ltd. Group owner: Rogers Broadcasting Ltd. Format: Country. ♦Ken Geiger, gen mgr; Janis Correia, gen sls mgr; Murray Olfert, prom dir.

Boston Bar

CKGO-FM-1— July 4, 1980: 106.1 mhz; 91 w. -2,871 ft Hrs open: #520-45715 Hocking Ave., Chilliwack, V2P 6Z6. Phone: (604) 795-5711. Fax: (604) 702-3212. Web Site: www.starfm.com. Licensee: Rogers Broadcasting Ltd. (group owner; acq 9-10-99; grpsl). Format: Hot adult contemp. News staff: 3. ♦Ken Geiger, gen mgr, stn mgr & progmg dir; Janis Correia, gen sls mgr; Murray Olfert, prom mgr.

Burnaby

CFML-FM— 2006: 107.9 mhz; 12 w. Hrs open: Bldg. SE 10, 3700 Willingdon Ave., V5G 3H2. Phone: (604) 432-8934. Phone: (604) 432-8545(sales). Fax: (604) 432-1792. E-mail: allofus@evolution1079.com Web Site: www.evolution1079.com. Licensee: B.C.I.T. Radio Society. Format: AAA, news, feature programs. ♦Brian Antonson, pres & gen mgr.

***CJSF-FM—** Feb 6, 2003: 90.1 mhz; 450 w. Hrs open: 7 AM-2AM CJSF Radio, TC 216, Simon Fraser University, V5A 1S6. Phone: (604) 291-3727. Fax: (604) 291-3695. Web Site: www.cjsf.ca. Licensee: Simon Fraser Campus Radio Society. Population served: 1,500,000 Format: Diversified. Spec prog: Persian 2 hrs, Portugese 2 hrs, Sp 4 hrs wkly. ♦Magnus Thyvold, stn mgr; Charlotte Bourne, progmg dir; Ed Blake, mus dir; Frieda Werden, pub affrs dir.

Burns Lake

CFLD(AM)— November 1965: 760 khz; 1 kw-U. Hrs open: Box 355, Smithers, V0J 2N0. Phone: (250) 692-3414. Fax: (250) 847-9411. E-mail: thepeak@bulkley.net Licensee: Vista Radio Ltd. (group owner) Format: Adult contemp. ♦J.C. Brown, gen mgr.

Campbell River

CFWB(AM)— September 1968: 1490 khz; 1 kw-U. Hrs open: 24 909 Ironwood Rd., V9W 3E5. Phone: (250) 287-7106. Fax: (250) 287-7170. E-mail: cfwb@oberon.ark.com Licensee: CFCP Radio Ltd. Population served: 30,000 Natl. Rep: Target Broadcast Sales. Format: Country. News staff: 3. Target aud: 20-45. ♦Terry Coles, gen mgr.

Castlegar

CKQR-FM— 1998: 99.3 mhz; 333 w. TL: N 49 18 54 W117 37 27. Hrs open: 24 1101 A 4th St., V1N 2A8. Phone: (250) 365-7600. Fax: (250) 365-8480. E-mail: rudy@mountainfm.net Web Site: www.mountainfm.net. Licensee: Vista Radio Ltd. Natl. Rep: Canadian Broadcast Sales. Format: Classic rock. News staff: one; News: 6 hrs wkly. Target aud: 18-49; median target demo, 36 yr old female. ♦Rudy Parachoniak, gen mgr & progmg dir.

Chase

CFCH-FM— Jan 10,2005: 103.5 mhz; 4.7 w. Hrs open: 8am - 8pm Box 1197, V0E 1M0. Phone: (250) 679-8622. Fax: (250) 679-3231. E-mail: cfchradio@cablelan.net Web Site: www.cablelan.net /ronfair/CFCH.html. Licensee: Chase and District Community Radio Society (acq 5-10-2007). Format: Community radio. ♦Ron Fairhurst, gen mgr.

Chetwynd

CHET-FM— 1997: 94.5 mhz; 25 w. Hrs open: 4612 N. Access Rd., V0C 1J0. Phone: (250) 788-9452. Fax: (250) 788-9402. E-mail: info@peacefm.ca Web Site: www.chetchad.com. Licensee: Chetwynd Communications Society. Format: Hits of the 60s thru today, MOR, div. ♦Leo Sabulsky, gen mgr; Mike Sabulsky, chief of engrg; Sadie Hasketh, sls.

Chilliwack

CKCL-FM— Sept 29, 1986: 107.5 mhz; 303 w. Stereo. Hrs open: 24 2440 Ash St., Vancouver, V5Z 4J6. Phone: (604) 877-6357. Fax: (604) 877-4443. Web Site: www.1049clearfm.com. Licensee: Rogers Radio (British Columbia) Ltd. Group owner: Rogers Broadcasting Ltd. (acq 9-10-99; grpsl). Format: MOR. News: 6 hrs wkly. Target aud: 35 plus. ♦Paul Fisher, VP; David Larsen, progmg dir.

CKSR-FM— Aug 31, 2001: 98.3 mhz; 2.34 kw. Stereo. Hrs open: 24 46167 Yale Rd., Unit 309, V2P 2N2. Phone: (604) 795-5711. Web Site: www.starfm.com. Licensee: Rogers Radio (British Columbia) Ltd. Group owner: Rogers Broadcasting Ltd. Natl. Rep: Canadian Broadcast Sales. Format: Light rock. News staff: 3. Target aud: 25-54; general. Spec prog: Farm 2 hrs wkly. ♦Ken Geiger, gen mgr.

Courtenay

CFCP-FM— 1959: 98.9 mhz; 2.685 kw. Stereo. Hrs open: 24 1625-A McPhee Ave., V9N 3A6. Phone: (250) 334-2421. Fax: (250) 334-1977. Web Site: www.jetfm.ca. Licensee: CFCP Radio Ltd. Wire Svc: BN

Wire Format: Classic rock, classic hits. News staff: 4; News: 16 hrs wkly. Target aud: 25-54; women. ♦Raymond Henderson, pres & gen sls mgr.

CKLR-FM— 1998: 97.3 mhz; 4.7 kw. Stereo. Hrs open: 24 801B 29th St., V9N 7Z5. Phone: (250) 703-2200. Fax: (250) 703-9611. E-mail: info@973theeagle.com Web Site: www.islandradio.bc.ca. Licensee: Jim Pattison Broadcast Group Ltd. (the general partner) and Jim Pattison Industries Ltd. (the limited partner), carrying on business as Jim Pattison Broadcast Group L.P. (acq 6-27-2006; grpsl). Population served: 90,000 Natl. Rep: Canadian Broadcast Sales. Format: Classic hits. News staff: 2. Target aud: 25-54; adult. ♦Richard Skinner, opns mgr & rgnl sls mgr; Ryan Mennie, mktg dir; Robyn Nicholson, prom dir & prom mgr; Steve Power, progmg dir; Bill Nation, news dir; Pam Doherty, traf mgr.

Cranbrook

CHBZ-FM— Oct 1, 1995: 104.7 mhz; 1.26 kw. Hrs open: 24 19 9th Ave. S., V1C 2L9. Phone: (250) 426-2224. Fax: (250) 426-5520. E-mail: info@b104.ca Web Site: www.b104.ca. Licensee: Jim Pattison Broadcast Group Ltd. (the general partner) and Jim Pattison Industries Ltd. (the limited partner) carrying on business as Jim Pattison Broadcast Group L.P. Group owner: The Jim Pattison Broadcast Group (acq 2-1-2001; grpsl). Population served: 65,000 Natl. Rep: Target Broadcast Sales. Format: Country. News staff: 3. Target aud: 18-54. ♦Rick Arnish, pres; Rod Schween, gen mgr; Bruce Davis, sls VP; Derek Kortschaga, progmg dir.

CHDR-FM— 2002: 102.9 mhz; 1.6 kw. Hrs open: 24 19 9th Ave. S., V1C 2L9. Phone: (250) 426-2224. Fax: (250) 426-5520. E-mail: info@thedrivefm.ca Web Site: www.thedrivefm.ca. Licensee: Jim Pattison Broadcast Group Ltd. (the general partner) and Jim Pattison Industries Ltd. (the limited partner) carrying on business as Jim Pattison Broadcast Group L.P. Group owner: The Jim Pattison Broadcast Group. Natl. Rep: Target Broadcast Sales. Format: Adult rock. News staff: 3. Target aud: 25-54. ♦Rick Arnish, pres; Rod Schween, gen mgr; Bruce Davis, sls VP; Dave Walker, gen sls mgr; Derek Kortschaga, progmg dir.

Crawford Bay

CBTE-FM— 1988: 89.9 mhz; 135 kw. TL: N49 38 54 W116 50 53. Hrs open:
Rebroadcasts CBTK-FM Kelowna 100%.
c/o CBTK-FM, 243 Lawrence Ave., Kelowna, V1Y 6L2. Phone: (250) 861-3781. Fax: (250) 861-6644. E-mail: kelowna@cbc.ca Web Site: www.vancouver.cbc.ca/daybreaksouth. Licensee: Canadian Broadcasting Corp. Format: News, pub affrs, entertainment. ♦Charlie Cheffins, opns mgr.

Creston

CFKC(AM)— Sept 21, 1968: 1340 khz; 250 w-U, DA-1. Hrs open: 24 1560 2nd Ave., Trail, V1R 1M4. Secondary address: 138-10 Ave. N. V0B 1G0. Phone: (250) 368-5510. Fax: (250) 368-8471. E-mail: kbs@bc.tri.ca Licensee: Standard Radio Inc. Group owner: Standard Broadcasting Corp. (acq 4-19-02; grpsl). Population served: 11,400 Natl. Rep: Target Broadcast Sales. Format: MOR. Spec prog: Ethnic mus 2 hrs wkly. ♦Lee Sterry, gen mgr; Darren Robertson, progmg dir.

CIDO-FM—Not on air, target date: unknown: 97.7 mhz; 20 w. Ant 1,092 ft TL: N49 05 25 W116 22 45. Hrs open: Box 8, V0B 1G0. Phone: (250) 402-6772. E-mail: info@crestonradio.ca Web Site: www.crestonradio.ca. Licensee: Creston Community Radio Society. Format: Var. ♦Bernie LeFrancois, pres.

Dawson Creek

CHAD-FM— 2003: 104.1 mhz; 50 w. Hrs open:
Rebroadcasts CHET-FM Chetwynd 100%.
c/o CHET-FM, Box 214, Chetwynd, V0C 1J0. Phone: (250) 784-1880. Fax: (250) 782-7566. E-mail: info@chetchad.com Web Site: www.chetchad.com. Licensee: Chetwynd Communications Society. Format: MOR, div, oldies. ♦Leo Sabulsky, gen mgr; Mike Sabulsky, chief of engrg.

CJDC(AM)— Dec 15, 1947: 890 khz; 10 kw-U. Hrs open: 24 901-102 Ave., V1G 2B6. Phone: (250) 782-3341. Fax: (250) 782-3154. Licensee: Standard Radio Inc. Group owner: Standard Broadcasting Corp. (acq 4-19-2002; grpsl). Population served: 56,000 Format: Country. News staff: 7. Target aud: General. Spec prog: Rock 4 hrs wkly. ♦Terry Shepard, gen mgr; Kevin Larkin, opns mgr & chief of engrg.

Duncan

CJSU-FM— August 2000: 89.7 mhz; 1.862 kw. Hrs open: 24 130 Trans Canada Highway, V9L 3P7. Phone: (250) 746-0897. Fax: (250) 748-1517. E-mail: cam@897sunfm.com Web Site: www.897sunfm.com. Licensee: Vista Radio Ltd. Natl. Rep: Target Broadcast Sales, Canadian Broadcast Sales. Format: Adult contemp. News staff: 2; News: 12 hrs wkly. Target aud: 35-54. Spec prog: Oldies 12 hrs wkly. ♦Jason Mann, VP; Keith James, gen mgr; Ron Larson, progmg dir; Al Siebring, news dir.

Egmont

CIEG-FM— July 1985: 107.5 mhz; 50 w. 300 ft Hrs open: 24
Rebroadcasts CISQ-FM Squamish 100%.
Box 1068, Squamish, V0N 3G0. Phone: (604) 892-1021. Fax: (604) 892-6383. E-mail: mountainfm@mountainfm.com Web Site: www.mountainfm.com. Licensee: Rogers Broadcasting. Natl. Rep: Canadian Broadcast Sales. Format: Adult contemp. News staff: 3; News: 16 hrs wkly. Target aud: 25-44. ♦Gary Miles, pres; Paul Fisher, VP; Ken Geiger, gen mgr; Janis Correia, gen sls mgr & prom dir.

Fernie

CJDR-FM— Aug 30, 2002: 99.1 mhz; 470 w. Hrs open: 24
Rebroadcasts CHOR-FM Cranbrook 66%.
19 9th Ave. S., Cranbrook, V1C 2L9. Phone: (250) 426-2224. Fax: (250) 426-5520. E-mail: info@thedrivefm.ca Web Site: www.thedrivefm.ca. Licensee: Jim Pattison Broadcast Group Ltd. (the general partner) and Jim Pattison Industries Ltd. (the limited partner) carrying on business as Jim Pattison Broadcast Group L.P. Group owner: The Jim Pattison Broadcast Group. Natl. Rep: Target Broadcast Sales. Format: Adult rock. News staff: one. ♦Rick Arnish, pres; Rod Schween, gen mgr; Bruce Davis, sls VP; Dave Walker, gen sls mgr; Derek Korschaga, progmg dir.

Fort Nelson

CKRX-FM— 1998: 102.3 mhz; 1.8 kw. Hrs open: Box 880, V0C 1R0. Secondary address: 5152 Liard St. V0C 1R0. Phone: (250) 774-2525. Fax: (250) 774-2577. E-mail: kjohnson@sri.ca Licensee: Standard Radio Inc. Group owner: Standard Broadcasting Corp. (acq 4-19-2002; grpsl). Format: CHR. ♦Terry Shepard, gen mgr; Kevin Larkin, progmg dir & progmg mgr.

Fort St. John

CHRX-FM—Not on air, target date: unknown: 98.5 mhz; 50 kw. Hrs open: 10532 Alaska Rd., V1J 1B3. Phone: (250) 785-6634. Fax: (250) 785-4544. Licensee: Standard Radio Inc. Group owner: Standard Broadcasting Corp. (acq 4-19-2002; grpsl). Format: Rock. ♦Terry Shepard, opns mgr.

CKFU-FM— Sept 1, 2003: 100.1 mhz; 50 w. Hrs open: 8 AM-6 PM (M-F) 10423 101st Ave., V1J 2B7. Phone: (250) 787-7100. Fax: (250) 263-9749. E-mail: reception@moosefm.ca Web Site: www.moosefm.ca. Licensee: 663975 B.C. Ltd. Population served: 30,000 Format: Lite rock 80s & 90s, classic hits. News staff: 2; News: 4 hours wkly. Target aud: 24-45; female. ♦Kerry Mann, pres; Russ Beerlling, gen mgr; Adam Reaburn, prom mgr & progmg mgr.

CKNL-FM— 2003: 101.5 mhz; 40 kw. Hrs open: 10532 Alaska Rd., V1J 1B3. Phone: (250) 785-6634. Fax: (250) 785-4544. Licensee: Standard Radio Inc. Group owner: Standard Broadcasting Corp. Format: Classic rock. ♦Angie Cloury, opns mgr.

Gibsons

CISC-FM— October 1984: 107.5 mhz; 820 w. Ant 1,000 ft Hrs open: 24
Rebroadcasts CISQ-FM Squamish.
Box 1068, Squamish, V0N 3G0. Phone: (604) 892-1021. Fax: (604) 892-6383. E-mail: mountainfm@mountainfm.com Web Site: www.mountainfm.com. Licensee: Rogers Broadcasting. Format: Hot adult contemp. News staff: 3; News: 16 hrs wkly. Target aud: 25-44. Spec prog: Magazine show 3 hrs wkly. ♦Gary Miles, pres; Paul Fisher, VP; Ken Geiger, gen mgr & opns mgr; Janis Correia, opns mgr.

Golden

CKGR(AM)— 1973: 1400 khz; 1 kw-D, DA-1. Hrs open: Box 1403, V0A 1H0. Phone: (250) 832-2161. Phone: (250) 344-7177. Fax: (250) 344-7233. Web Site: www.myezrock.com. Licensee: Standard Radio Inc. Group owner: Standard Broadcasting Corp. (acq 4-19-2002; grpsl). Natl. Rep: Target Broadcast Sales. Format: Easy rock. ♦Ron Langridge, gen mgr.

Greenville

CILZ-FM— 1990: Stn currently dark. 96.1 mhz; 7.9 w. Hrs open: Greenville Television Association, 289 Church St., V0J 1X0. Phone: (250) 621-3212. Fax: (250) 621-3320. E-mail: tiffanym@nisgaa.net Licensee: Greenville Television Association. Format: Contemp country.

Greenwood

CKGF-FM-2— 2003: 96.7 mhz; 40 w horiz. Ant 1,886 ft TL: N49 05 29 W118 36 36. Hrs open: 1101 A 4th St., Castlegar, V1N 2A8. Phone: (250) 365-7600. Fax: (250) 365-8480. E-mail: rudy@mountainfm.net Web Site: www.mountainfm.net. Licensee: Boundary Broadcasting Ltd. Format: Classic rock. Target aud: 25-54. ♦Rudy Parachoniak, gen mgr; Rudy Parachoiak, progmg dir.

Hope

CFSR-FM— 2001: 100.5 mhz; 157 w. Hrs open: #520-45715 Hocking Ave., Suite 520, Chilliwack, V2P 6Z6. Phone: (604) 869-9313. Fax: (604) 702-3212. Web Site: www.starfm.com. Licensee: Rogers Radio (British Columbia) Ltd. Group owner: Rogers Broadcasting Ltd. Format: Adult contemp. ♦Ken Geiger, gen mgr; Janis Correia, gen sls mgr; Murray Olfert, prom mgr.

Invermere

CKIR(AM)— December 1989: 870 khz; 1 kw-U. Hrs open:
Rebroadcasts CKGR(AM) Golden.
Box 69 Stn Main, Salmon Arm, V1E 4N2. Phone: (250) 832-2161. Fax: (250) 344-7233. E-mail: ckir@rockies.net Web Site: www.myezrock.com. Licensee: Standard Radio Inc. Group owner: Standard Broadcasting Corp. (acq 4-19-02; grpsl). Format: Easy rock. ♦Ron Langridge, gen mgr.

Kamloops

CFBX-FM— Apr 2, 2001: 92.5 mhz; 420 w. Hrs open: 900 McGill Rd., House 8, V2C 5N3. Phone: (250) 377-3988. Fax: (250) 852-6350. E-mail: radio@tru.ca Web Site: www.thex.ca. Licensee: The Kamloops Campus/Community Radio Society. Population served: 88,000 Format: Div. ♦Brant Zwicker, stn mgr; Steve Marlow, progmg dir.

CHNL(AM)— May 1, 1970: 610 khz; 25 kw-D, 5 kw-N, DA-N. Hrs open: 24 611 Lansdowne St., V2C 1Y6. Phone: (250) 372-2292. Phone: (250) 372-2197. Fax: (250) 372-2293. E-mail: info@radionl.com Licensee: NL Broadcasting Ltd. Population served: 80,000 Format: Adult contemp, oldies. News staff: 5; News: 15.5 hrs wkly. Target aud: 25-54; family oriented, middle-upper class income. ♦Robbie Dunn, pres, gen mgr & adv mgr; Ravinder Dhaliwal, CFO; Peter Angle, gen sls mgr; Jim Reynolds, progmg dir; T. Tyler, mus dir; Jim Harrison, news dir; Dave Coulter, chief of engrg.

CKRV-FM—Co-owned with CHNL(AM). Jan 28, 1984: 97.5 mhz; 5 kw. Stereo. 24 Phone: (250) 372-2197. (Acq 6-10-93; $925,000.).80,000 Format: Top 40. News staff: one; News: 5 hrs wkly. Target aud: 25-54; professionals, office personnel. ♦Tim Thompson, mus dir.

CIFM-FM— 1961: 98.3 mhz; 4.3 kw. Stereo. Hrs open: 460 Pemberton Terr., V2C 1T5. Phone: (250) 372-3322. Fax: (250) 374-0445. E-mail: info@98.3cifm.com Web Site: www.98.3cifm.com. Licensee: Jim Pattison Broadcast Group Ltd. (the general partner) and Jim Pattison Industries Ltd. (the limited partner) carrying on business as Jim Pattison Broadcast Group L.P. Group owner: The Jim Pattison Broadcast Group (acq 1987). Population served: 170,000 Format: Adult rock. News staff: 6; News: 9 hrs wkly. Target aud: 30-40; baby boomers with disposable income. ♦Rick Arnish, pres, gen mgr & stn mgr; Doug Collins, chief of opns; Bruce Uptigrove, gen sls mgr. Co-owned TV: CFJC-TV affil.

CJKC-FM— 2006: 103.1 mhz; 5 kw. Hrs open: 611 Lansdowne St., V2C 1Y6. Phone: (250) 571-1031. Fax: (250) 372-2293. E-mail: info@country103.ca Web Site: www.country103.ca. Licensee: NL Broadcasting Ltd. Format: Country. Target aud: 25-54. ♦Robbie Dunn, pres & gen mgr; Kelly Moore, progmg dir.

CKBZ-FM— 2001: 100.1 mhz; 3.5 kw. Stereo. Hrs open: 460 Pemberton Terr., V2C 1T5. Phone: (250) 372-3322. Fax: (250) 374-0445. E-mail: info@b100.ca Web Site: www.b100.ca. Licensee:

Jim Pattison Broadcast Group Ltd. (the general partner) and Jim Pattison Industries Ltd. (the limited partner) carrying on business as Jim Pattison Broadcast Group L.P. Group owner: The Jim Pattison Broadcast Group. Population served: 170,000 Format: Adult contemp. News staff: 6; News: 10 hrs wkly. Target aud: 25-44; large, loyal audience with disposable income. ♦Rick Arnish, pres & gen mgr; Doug Collins, opns mgr; Bruce Uptigrove, gen sls mgr. Co-owned TV: CFJC-TV affil.

Kelowna

***CBTK-FM—** November 1987: 88.9 mhz; 4.7 kw. 1,676 ft Hrs open: 243 Lawrence Ave., V1Y 6L2. Phone: (250) 861-3781. Fax: (250) 861-6644. E-mail: kelowna@cbc.ca Web Site: www.vancoouver.cbc.ca/daybreaksouth. Licensee: Canadian Broadcasting Corp. Natl. Network: CBC Radio One. Format: News, pub affrs, entertainment. ♦Charlie Cheffins, opns mgr & progmg dir.

CHSU-FM—Listing follows CKFR(AM).

CILK-FM— June 21, 1985: 101.5 mhz; 11 kw. 1,246 ft (CP: 10.3 kw.). Stereo. Hrs open: 24 1598 Pandosy St., V1Y 1P4. Phone: (250) 860-1010. Fax: (250) 860-5754. E-mail: info@silk.fm Web Site: www.silk.fm. Licensee: Standard Radio Inc. (acq 12-20-2006; C$9.25 million). Format: Adult contemp 80s & 90s. News staff: 2; News: 2 hrs wkly. Target aud: 25-54; women. Spec prog: Gospel 3 hrs wkly. ♦Gary Slaight, pres; Rick Dyer, gen mgr & gen sls mgr.

CKFR(AM)— Nov 8, 1971: 1150 khz; 10 kw-U, DA-N. Hrs open: 300-435 Bernard Ave., V1Y 6N8. Phone: (250) 860-8600. Fax: (250) 860-8856. Licensee: Standard Radio Inc. Group owner: Standard Broadcasting Corp. (acq 4-19-2002; grpsl). Population served: 150,000 Format: Oldies. News staff: 3. Target aud: 25-54. Spec prog: Home improvements 2 hrs, WHL hockey play-by-play 10 hrs, American gold 4 hrs wkly. ♦Don Shafer, gen mgr; Dallas Gray, gen sls mgr & mktg mgr; Mark Burley, progmg dir; Howard Alexander, news dir.

CHSU-FM—Co-owned with CKFR(AM). Sept 21, 1995: 99.9 mhz; 13 kw. Web Site: www.thesun.net. Format: Adult comtemp. News staff: 3.

CKLZ-FM—Listing follows CKOV(AM).

CKOV(AM)— Nov 4, 1931: 630 khz; 5 kw-D, 1 kw-N. TL: N49 50 42 W119 29 07. Stereo. Hrs open: 24 3805 Lakeshore Rd., V1W 3K6. Phone: (250) 762-3331. Fax: (250) 762-2141. E-mail: ckov@cnx.net Licensee: Jim Pattison Broadcast Group Ltd. (the general partner) and Jim Pattison Industries Ltd. (the limited partner) carrying on business as Jim Pattison Broadcast Group L.P. Group owner: The Jim Pattison Broadcast Group (acq 6-30-98). Population served: 140,000 Format: News, info, talk. News staff: 5; News: 13 hrs wkly. Target aud: 29-59. ♦Rick Arnish, pres & VP; Bruce Davis, gen mgr & gen sls mgr; Hugh Dixon, prom mgr; Rob Bye, progmg mgr; Matt Cherrille, news dir; Victor Deveall, chief of engrg.

CKLZ-FM—Co-owned with CKOV(AM). 1964: 104.7 mhz; 3.8 kw. 1,611 ft TL: N49 46 06 W119 29 59. Stereo. 24 Phone: (250) 763-1047. E-mail: rob@power104.fm Web Site: www.power104.fam. Format: AOR. News staff: 2; News: 4 hrs wkly. Target aud: 25-49. ♦Rob Bye, mus dir.

Kitimat

CKTK-FM— 2004: 97.7 mhz; 170 w. Hrs open: 4625 Lazelle Ave., Terrace, V8G 1S4. Phone: (250) 635-6316. Fax: (250) 638-6320. Licensee: Standard Radio Inc. Group owner: Standard Broadcasting Corp. Format: CHR. ♦Brian Langston, stn mgr.

Lillooet

CHLS-FM— 2001: 100.5 mhz; 5 w. TL: lillooet. Hrs open: 24 Box 1545, V0K 1V0. Phone: (250) 256-2113. Fax: (250) 256-2113. E-mail: station@radiolillooet.ca Licensee: Radio Lillooet Society. Population served: 5,000 Format: Var/div. News: 5 hrs wkly. Target aud: All ages; coomunity of Lillooet. Spec prog: First nations 16 hrs wkly.

MacKenzie

CHMM-FM— Oct 27, 2003: 103.5 mhz; 900 w. Hrs open: 86 Centennial Ave., Box 547, V0J 2C0. Phone: (250) 997-6277. Fax: (250) 997-6222. E-mail: jd@chmm.ca Licensee: MacKenzie and Area Community Radio Society. Format: Var. ♦J.D. MacKenzie, stn mgr.

Merritt

CJNL(AM)— 1970: 1230 khz; 1 kw-D, 250 w-N. TL: N50 06 29 W120 46 06. Hrs open:
Rebroadcasts CHNL(AM) Kamloops 60%.
Box 1630, V1K 1B8. Secondary address: 2196 Quilchena Ave., Unit 201 V1K 1B8. Phone: (250) 378-4288. Fax: (250) 378-6979. Licensee: Merritt Broadcasting Ltd. (acq 1-30-95; C$214,800). Population served: 10,000 Natl. Rep: Target Broadcast Sales. Format: Adult contemp, oldies, news. News staff: one; News: 2 hrs wkly. Target aud: General; working family all ages. ♦Elizabeth Laird, gen mgr; Brian Wiebe, opns dir & progmg dir; Leslie Carty, mus dir.

Nanaimo

CHLY-FM— Sept 21, 2001: 101.7 mhz; 1.3 kw. Ant 312 ft TL: N49 13 20 W124 00 07. Hrs open: #2-34 Victoria Crescent, V9R 5B8. Phone: (250) 716-3410. Fax: (250) 716-1082. E-mail: stationmanager@chly.ca Web Site: www.chly.ca. Licensee: Radio Malaspina Society. Format: Div. ♦James Booker, progmg dir.

CHWF-FM— Oct 1, 2001: 106.9 mhz; 1.6 kw. TL: Nanaimo. Stereo. Hrs open: 24 hrs 4550 Wellington Rd., V9T 2H3. Phone: (250) 758-1131. Fax: (250) 758-4644. E-mail: info@1069thewolf.com Web Site: www.1069thewolf.com. Licensee: Jim Pattison Broadcast Group Ltd. (the general partner) and Jim Pattison Industries Ltd. (the limited partner), carrying on business as Jim Pattison Broadcast Group L.P. (acq 6-27-2006; grpsl). Population served: 125,000 Natl. Rep: Canadian Broadcast Sales. Format: Rock. News staff: 2. Target aud: 25-54; general. ♦Chris Barron, prom dir; Kent Wilson, mus dir; Heather Mousseau, news dir; Barry Mandziak, chief of engrg; Pam Dogherty, traf mgr; Rob Bye, sls.

CKWV-FM— Jan 2, 1995: 102.3 mhz; 1.3 kw. TL: Nanaimo, sc. Stereo. Hrs open: 24 4550 Wellington Rd., V9T 2H3. Phone: (250) 758-1131. Fax: (250) 758-4644. E-mail: info@1023thewave.com Web Site: www.1023thewave.com. Licensee: Jim Pattison Broadcast Group Ltd. (the general partner) and Jim Pattison Industries Ltd. (the limited partner), carrying on business as Jim Pattison Broadcast Group L.P. (acq 6-27-2006; grpsl). Population served: 125,000 Natl. Rep: Canadian Broadcast Sales. Format: Adult contemp. News staff: 2. Target aud: 25-54; general. ♦Rob Bye, gen mgr; Chris Barron, prom dir; Heather Mousseau, news dir; Barry Mandziak, chief of engrg; Pam Dogherty, traf mgr.

Nelson

CHNV-FM— Apr 18, 2006: 103.5 mhz; 104 w. Hrs open: 312 Hall St., V1L 1Y8. Phone: (250) 352-1902. Fax: (250) 352-0301. Web Site: www.mountainfm.net. Licensee: Vista Radio Ltd. Format: Rock. ♦Paul Mann, gen mgr; Rudy Parachoniak, progmg dir.

***CJLY-FM—** 2002: 93.5 mhz; 70 w. Hrs open: Kootenay Cooperative Radio, Box 767, V1L 5R4. Phone: (250) 352-9600. Fax: (250) 352-9653. E-mail: kcr@kootenaycoopradio.com Web Site: www.kootenaycoopradio.com. Licensee: Kootenay Co-operative Radio. Format: Div. ♦Bill Metcalfe, opns mgr; Terry Brennan, progmg dir.

CKKC-FM— 2006: 106.9 mhz; 920 w. Hrs open: 1560 Second Ave., Trail, V1R 1M4. Phone: (250) 368-5510. Fax: (250) 368-8471. Web Site: www.nelson.kbsradio.ca. Licensee: Standard Radio Inc. Natl. Rep: Target Broadcast Sales. Format: Adult contemp. ♦Kavin Einarson, opns mgr; Kevin Einarson, gen sls mgr; Chris Kuchar, progmg dir.

New Denver

CKZX-FM— October 1981: 93.5 mhz; 100 w. Hrs open:
Rebroadcasts CKKC(AM) Nelson 100%.
c/o Radio Station CKKC(AM), 1560 2nd Ave., Trail, V1R 1M4. Phone: (250) 368-5510. Fax: (250) 368-8471. Web Site: www.kbsradio.ca. Licensee: Standard Radio Inc. Group owner: Standard Broadcasting Corp. (acq 4-19-2002; grpsl). Format: Adult Contemp. ♦Kevin Einarson, gen sls mgr; Larry King, chief of engrg.

New Westminster

CFMI-FM—Licensed to New Westminster. See Vancouver

CKNW(AM)—Licensed to New Westminster. See Vancouver

Osoyoos

CJOR(AM)— December 1966: 1240 khz; 1 kw-U, DA-1. Hrs open: 33 Carmi Ave., Penticton, V2A 3G4. Phone: (250) 492-2800. Fax: (250) 493-0370. Licensee: Standard Radio Inc. Group owner: Standard Broadcasting Corp. (acq 4-19-2002; grpsl). Natl. Rep: Canadian Broadcast Sales. Format: Adult contemp. Target aud: 25-49. Spec prog: Por 3 hrs wkly. ♦Lee Sterry, gen mgr.

Parksville

CHPQ-FM— Feb 11, 2005: 99.9 mhz; 1.1 kw. Stereo. Hrs open: 24 Box 1370, V9P 2H3. Phone: (250) 248-4211. Fax: (250) 248-4210. E-mail: info@thelounge999.com Web Site: www.thelounge999.com. Licensee: Jim Pattison Broadcast Group Ltd. (the general partner) and Jim Pattison Industries Ltd. (the limited partner), carrying on business as Jim Pattison Broadcast Group L.P. (acq 6-27-2006; grpsl). Population served: 45,000 Natl. Rep: Canadian Broadcast Sales. Format: Adult standards. News staff: one. Target aud: 45 plus; adults. ♦Rob Bye, gen mgr; Heather Mousseau, news dir; Pam Doherty, traf mgr.

CIBH-FM— 1999: 88.5 mhz; 960 w. TL: Parksville, BC. Hrs open: 24 Box 1370, V9P 2H3. Phone: (250) 248-4211. Fax: (250) 248-4210. E-mail: info@885thebeach.com info@885thebeach.com Web Site: www.885thebeach.com. Licensee: Jim Pattison Broadcast Group Ltrd. (the general partner) and Jim Pattison Industries Ltd. (the limited partner), carrying on business as Jim Pattison Broadcast Group L.P. (acq 6-27-2006; grpsl). Population served: 45,000 Format: Adult contemp, oldies soft rock. News staff: 2. Target aud: 25-54; Adults. ♦Rob Bye, sls dir; Heather Mouss, news dir; Pam Dogherty, traf mgr.

Pemberton

CFPV-FM—Not on air, target date: unknown: 98.9 mhz; 420 w. Hrs open: McBride Communications & Media Inc., 10760 Fundy Dr., Richmond, V7E 5K7. Phone: (604) 220-8393. Fax: (604) 677-6316. E-mail: info@mcmi.ca Licensee: 0749943 BC Ltd. (Matthew G. McBride). ♦Matthew G. McBride, pres.

CISP-FM— Oct 5, 1982: 104.5 mhz; 400 w. 1,000 ft Hrs open: 24
Rebroadcasts CISQ-FM Squamish 100%.
Box 1068, Squamish, V0N 3G0. Phone: (604) 892-1021. Fax: (604) 892-6383. E-mail: mountainfm@mountainfm.com Web Site: www.mountainfm.com. Licensee: Rogers Broadcasting Ltd. (group owner) Format: Hot adult contemp. News staff: 3; News: 16 hrs wkly. Target aud: 25-45. ♦Gary Miles, pres; Paul Fisher, VP; Ken Geiger, gen mgr; Janis Correia, opns mgr.

Pender Harbour

CIPN-FM— October 1984: 104.7 mhz; 750 w. 1,500 ft Hrs open: 24
Rebroadcasts CISQ-FM Squamish.
Box 1068, Squamish, V0N 3G0. Phone: (604) 892-1021. Phone: (604) 683-8060. Fax: (604) 892-6383. E-mail: mountainfm@mountainfm.com Web Site: www.mountainfm.com. Licensee: Rogers Broadcasting. Format: Hot adult contemp. News staff: 3; News: 16 hrs wkly. ♦Gary Miles, pres; Paul Fisher, VP; Ken Geiger, gen mgr; Janis Correia, opns mgr.

Penticton

CIGV-FM— Oct 18, 1981: 100.7 mhz; 6.3 kw. Ant 2,486 ft TL: N49 42 46 W119 36 26. Stereo. Hrs open: 24 125 Nanaimo Ave. W., V2A 1N2. Phone: (250) 493-6767. Fax: (250) 493-0098. E-mail: info@giantfm.ca Web Site: www.giantfm.ca. Licensee: Great Valleys Radio Ltd. Population served: 242,323 Natl. Rep: Target Broadcast Sales. Wire Svc: Broadcast News Ltd. Format: Country, adult contemp. News staff: 3; News: 16 hrs wkly. Target aud: 18 plus. Spec prog: Class 2 hrs, farm one hr wkly. ♦James Robinson, CEO, pres & gen mgr; Greg Masson, gen sls mgr; Harry Shaw, chief of engrg.

CJMG-FM—Listing follows CKOR(AM).

CKOR(AM)— September 1948: 800 khz; 10 kw-D, 500 w-N. Hrs open: 33 Carmi Ave., V2A 3G4. Phone: (250) 492-2800. Fax: (250) 493-0370. Licensee: Standard Radio Inc. Group owner: Standard Broadcasting Corp. (acq 4-19-2002; grpsl). Format: Gold. ♦Ross Hawse, gen mgr.

CJMG-FM—Co-owned with CKOR(AM). June 1, 1965: 97.1 mhz; 1.8 kw. 755 ft Stereo. Format: AOR.

Port Alberni

CJAV-FM— Sept 2, 2005: 93.3 mhz; 6 kw. Hrs open: 24 3296 Third Ave., V9Y 4E1. Phone: (250) 723-2455. Fax: (250) 723-0797. E-mail: info@933thepeak.com. Web Site: www.933thepeak.com. Licensee: Jim Pattison Broadcast Group Ltd. (the general partner) and Jim Pattison Industries Ltd. (the limited partner), carrying on business as Jim Pattison Broadcast Group L.P. (acq 6-27-2006; grpsl). Population served: 30,000 Format: Classic Hits. Target aud: 25-54; Adults. ♦Chris Talbot, opns mgr & sls; Rob Bye, gen mgr & gen sls mgr; Pam Doherty, traf mgr.

Port Hardy

CFNI(AM)— Sept 1, 1979: 1240 khz; 1 kw-U, DA-1. Hrs open: Box 1000, V0N 2P0. Phone: (250) 949-6500. Phone: (250) 334-2421. Fax: (250) 949-6580. E-mail: cfniradio@cablerocket.com Licensee: CFCP Radio Ltd. Format: Adult contemp, rock mix. News staff: one. Target aud: General. ♦Paul Mann, CEO.

Powell River

CHQB(AM)— Mar 21, 1967: 1280 khz; 1 kw-U, DA-1. TL: N49 48 10 W124 36 10. Hrs open: 24 1034675 Marine Ave, V8A 2L2. Phone: (604) 485-4207. Fax: (604) 485-4210. E-mail: chqb@onelink.ca Web Site: www.coastradio.com. Licensee: CFCP Radio Ltd. (acq 11-18-97). Natl. Network: Bible Bcstg Net. Rgnl rep: Target. Wire Svc: BN Wire Format: Country. News staff: one; News: 14 hrs wkly. Target aud: General. ♦Ida May Mulligan, stn mgr; Joel Lamourieux, progmg dir.

CJMP-FM— 2006: 90.1 mhz; 3.6 w. Hrs open: 4476 A Marine Ave., V8A 2K2. Phone: (604) 485-2688. Fax: (604) 485-2683. E-mail: modelcommunity@prcn.org; modelcommunity@shaw.ca Web Site: www.jumpradiopr.com. Licensee: Powell River Model Community Project for Persons with Disabilities. Format: Var, diverse. ♦Mike Lang, stn mgr.

Prince George

CBYG-FM— 91.5 mhz; 100 kw. Hrs open: 890 Victoria St., Unit 1, V2L 5P1. Phone: (250) 562-6701. Fax: (250) 562-4777. E-mail: daybreaknorth@cbc.ca Web Site: www.vancouver.cbc.ca/daybreaknorth. Licensee: CBC. Natl. Network: CBC Radio One. Format: Current affrs. ♦Faydra Aldridge, gen mgr.

***CFUR-FM**— 2002: 88.7 mhz; 510 w. Hrs open: 3333 University Way, V2N 4Z9. Phone: (250) 960-7664. Fax: (250) 960-5995. E-mail: info@cfur.ca Web Site: www.cfur.ca. Licensee: Education Alternative Radio Society. Population served: 80,000 Format: Div. Target aud: Community of Prince George. ♦Christopher Earl, stn mgr; Joshua Laurin, progmg dir; Bryndis Ogmundson, mus dir; Glen Yakemchuk, engrg dir.

CIRX-FM— Oct 1, 1983: 94.3 mhz; 3.5 kw. Ant 1,145 ft Stereo. Hrs open: 1940 3rd Ave., V2M 1G7. Phone: (250) 564-2524. Fax: (250) 562-6611. Web Site: www.94xfm.com. Licensee: Vista Radio Ltd. (group owner). Format: Rock. Target aud: 18-34. ♦Gary Russell, gen mgr; Sandy Whitwhan, gen sls mgr; Brad Bregani, progmg dir; Bill Fox, news dir; Chris Terpsma, chief of engrg.

CJCI-FM— Aug 5, 2003: 97.3 mhz; 12 kw. Hrs open: 24 The Wolf, 1940 3rd Ave., V2M 1G7. Phone: (250) 564-2524. Fax: (250) 562-6611. E-mail: thewolf@97fm.ca Web Site: www.97fm.ca. Licensee: Vista Radio Ltd. (group owner). Format: Modern country, southern rock. News staff: 3. Target aud: 25-54. ♦Gary Russell, gen mgr; Sandy Whitwhan, gen sls mgr; Bill Fee, news dir; Chris Terpsma, chief of engrg.

CKDV-FM— 2003: 99.3 mhz; 9.3 kw. Hrs open: 24 1810 3rd Ave., 2nd Fl, V2M 1G4. Phone: (250) 564-8861. Fax: (250) 562-8768. E-mail: ckpgmail@ckpg.bc.ca Web Site: www.993thedrive.com. Licensee: Jim Pattison Broadcast Group Ltd. (the general partner) and Jim Pattison Industries Ltd. (the limited partner) carrying on business as Jim Pattison Broadcast Group L.P. Group owner: The Jim Pattison Broadcast Group. Format: Classic rock. ♦Ken Kilcullen, gen mgr; Randy Seabrook, gen sls mgr; Ron Polillo, progmg dir; Mike Woodworth, news dir.

CKKN-FM— Mar 1, 1981: 101.3 mhz; 10 kw. Ant 1,000 ft Stereo. Hrs open: 2nd Fl. 1810 3rd Ave., V2M 1G4. Phone: (250) 564-8861. Fax: (250) 562-8768. Fax: (250) 562-7681. E-mail: ckpgmail@ckpg.bc.ca Web Site: www.1013theriver.com. Licensee: Jim Pattison Broadcast Group Ltd. (the general partner) and Jim Pattison Industries Ltd. (the limited partner) carrying on business as Jim Pattison Broadcast Group L.P. Group owner: The Jim Pattison Broadcast Group (acq 12-21-2000; grpsl). Format: Hot adult contmep. ♦Rick Arnish, pres; Ken Kilcullen, gen mgr; Ron Polillo, progmg mgr; Mike Woodworth, news dir. Co-owned TV: CKPG-TV affil.

Prince Rupert

***CFPR(AM)**— 1936: 860 khz; 10 kw-D, DA-1. Hrs open: 24 222 Third Ave. W., Suite 1, V8J 1L1. Phone: (250) 624-2161. Fax: (250) 627-8594. E-mail: daybreaknorth@vancouver.cbc.ca Web Site: www.cbc.ca/daybreaknorth. Licensee: CBC. (acq 1953). Natl. Network: CBC Radio One. Format: Current affairs, news. Target aud: General. Spec prog: Current affrs. ♦Faydra Aldridge, gen mgr.

CHTK(AM)— 1965: 560 khz; 1 kw-D, 250 w-N. Hrs open: 215 Cowbay Rd., Unit 212, V8J 1A2. Phone: (250) 627-8255. Fax: (250) 624-3100. E-mail: gsimpson@sri.ca Licensee: Standard Radio Inc. Group owner: Standard Broadcasting Corp. (acq 4-19-2002; grpsl). Format: CHR. ♦Mike Lunn, gen mgr.

CIAJ-FM— 2000: 100.7 mhz; 26.5 w. Hrs open: 531 6th Ave. W., V8J 1Z7. Phone: (250) 627-7305. Fax: (250) 624-3476. E-mail: cfirm@citytel.net Licensee: Canadian Christians In Action Ministries. Format: Christian. ♦Prescott Sandhu, gen mgr.

Princeton

CIOR(AM)— June 1972: 1400 khz; 1 kw-U, DA-1. TL: N49 26 50 W120 30 42. Hrs open: 6 AM-6 PM Box 539, 203 8309 Main St., Osoyoos, V0H 1V0. Phone: (250) 492-2800. Fax: (250) 495-7228. Licensee: Standard Radio Inc. (acq 4-19-2002; grpsl). Population served: 5,000 Natl. Rep: Canadian Broadcast Sales. Format: Adult contemp. ♦Lee Sterry, gen mgr & stn mgr.

Qualicum

CIBH-FM—See Parksville

Quesnel

CKCQ-FM— 2004: 100.3 mhz; 1.8 kw. Hrs open: 160 Front St., V2J 2K1. Phone: (250) 992-7046. Fax: (250) 992-2354. Licensee: Vista Radio Ltd. Format: Modern country, southern rock. News staff: 4. ♦Tracey Gard, gen mgr; Ron Nilsen, progmg dir; George Henderson, news rptr.

Revelstoke

CKCR(AM)— Nov 21, 1965: 1340 khz; 1 kw-D, 250 w-N. Hrs open: Box 1420, V0E 2S0. Secondary address: 208 E. 1st. St. V0E 2S0. Phone: (250) 837-2149. Fax: (250) 837-5577. E-mail: ckcr@rctvonline.net Web Site: www.myezrock.com. Licensee: Standard Radio Inc. Group owner: Standard Broadcasting Corp. (acq 4-19-2002; grpsl). Format: Easy rock. ♦Ron Langridge, pres & gen mgr.

Richmond

CHKG-FM— Sept 6, 1997: 96.1 mhz; 46 kw. Hrs open: 24 2090-4151 Hazelbridge Way, V6X 4J7. Phone: (604) 295-1234. Fax: (604) 295-1201. E-mail: general@fm961.com Web Site: www.fm961.com. Licensee: Fairchild Radio (Vancouver FM) Ltd. Population served: 1,300,000 Natl. Rep: Target Broadcast Sales. Format: Ethnic. News staff: 10; News: 10 hrs wky. Target aud: General. ♦Brenda Lo, sr VP & dev VP; George Lee, sr VP & gen mgr; Thomas Fung, chmn & VP; Alan Kwok, opns mgr.

CISL(AM)— May 1, 1980: 650 khz; 10 kw-U, DA-2. Hrs open: No. 20, 11151 Horseshoe Way, V7A 4S5. Phone: (604) 272-6500. Fax: (604) 272-0917. Web Site: www.650cisl.com. Licensee: Standard Radio Inc. Group owner: Standard Broadcasting Corp. (acq 5-8-96; C$18 million with CKZZ-FM Vancouver). Format: Oldies. ♦Gary Slaight, pres; Gary Russell, VP & gen mgr.

CJVB(AM)— June 18, 1972: 1470 khz; 50 kw-U, DA-2. TL: N49 11 36 W123 01 17. Stereo. Hrs open: 24 2090 -4151 Hazelbridge Way, V6X 4J7. Phone: (604) 295-1234. Fax: (604) 295-1201. E-mail: general@am1470.com Web Site: www.am1470.com. Licensee: Fairchild Radio Group Ltd. Population served: 1,300,000 Natl. Rep: Target Broadcast Sales. Rgnl rep: In House Format: Ethnic, Chinese. News staff: 10; News: 23 hrs wkly. Target aud: General. ♦Thomas Fung, chmn & pres; Brenda Lo, sr VP & dev VP; George Lee, sr VP, VP & gen mgr; Alan Kwok, opns mgr; Seme Ho, adv VP; Alfred Lee, progmg dir.

Salmon Arm

CKXR-FM— June 5, 2006: 91.5 mhz; 400 w. Hrs open: 24 Box 69, V1E 4N2. Secondary address: 360 Ross St. V1E 4N2. Phone: (250) 832-2161. Fax: (250) 832-2240. Web Site: www.salmonarm.myezrock.com. Licensee: Standard Radio Inc. Format: Soft rock. Target aud: 25-54. ♦Ron Langridge, gen mgr.

Sechelt

CKAY-FM— May 20, 2006: 91.7 mhz; 600 w. TL: Mount Bensen-Van. Island. Hrs open: 24 1-1877 Field Rd., V0N 3A1. Phone: (604) 741-9170. Fax: (604) 741-9172. E-mail: info@ckay.ca Web Site: www.ckan.ca. Licensee: Westwave Broadcasting Inc. Natl. Rep: imsradio. Rgnl rep: IMS Wire Svc: BN Wire Format: Pop adult. News staff: 3; News: 8 hrs wkly. Target aud: 35+; Adults. ♦Bob Morris, pres & gen mgr; Paul Nattal, sls VP; Sean Ecktoro, news dir.

CKKS-FM— July 1985: 104.7 mhz; 750 w. 2,000 ft Hrs open: 24
Rebroadcasts CISQ-FM Squamish.
Box 1068, Squamish, V0N 3G0. Phone: (604) 892-1021. Fax: (604) 892-6383. E-mail: mountainfm@mountainfm.com Web Site: www.mountainfm.com. Licensee: Rogers Broadcasting Ltd. (group owner). Natl. Rep: Canadian Broadcast Sales. Format: Hits of the 80s & 90s. News staff: 3; News: 16 hrs wkly. Target aud: 25-45. Spec prog: Magazine show 3 hrs wkly. ♦Gary Miles, pres; Ken Geiger, gen mgr, opns mgr & progmg dir; Janis Correia, opns mgr; Paul Fisher, gen sls mgr; janis Correia, sls.

Smithers

CFBV(AM)— Oct 25, 1963: 870 khz; 1 kw-D, 250 w-N. Hrs open: Box 335, V0J 2N0. Secondary address: 1139 Queen St. V0J 2N0. Phone: (250) 847-2521. Fax: (250) 847-9411. Licensee: Vista Radio Ltd. (group owner) Format: The Best of the 70's, 80's, 90's and now. ♦Al Collison, gen mgr.

Squamish

CISQ-FM— Nov 30, 1981: 107.1 mhz; 12.48 kw. Ant 800 ft Hrs open: 24 Box 1068, V0N 3G0. Phone: (604) 892-1021. Fax: (604) 892-6383. Web Site: www.mountainfm.com. Licensee: Rogers Broadcasting. Natl. Rep: Canadian Broadcast Sales. Format: Hot adult contemp. News staff: 3; News: 5 hrs wkly. Target aud: 18-54. Spec prog: Talk 5 hrs wkly. ♦Ken Geiger, gen mgr; Janis Correia, opns mgr.

Summerland

CHOR(AM)— 1972: 1450 khz; 1 kw-U, DA-1. Hrs open: 24
Rebroadcasts CHOR(AM) Summerland 70%.
Box 1170, 13415 Rosedale, V0H 1Z0. Phone: (250) 494-0333. Fax: (250) 493-0370. Web Site: www.thesun.net. Licensee: Standard Radio Inc. (acq 4-19-2002; grpsl). Natl. Rep: Canadian Broadcast Sales. Format: Oldies. Spec prog: Class 3 hrs, jazz 3 hrs wkly. ♦Lee Sterry, gen mgr; Lee Sterry, stn mgr.

Terrace

CFNR-FM— 1995: 92.1 mhz; 43 w. Hrs open: 24 456 B Queensway Dr., V8G 3X6. Phone: (250) 638-8137. Fax: (250) 638-8027. E-mail: cfnrmailbag@monarch.net Web Site: www.mycfnr.com.Starchoice Ch. 851 Licensee: Northern Native Broadcasting (Terrace, B.C.). Population served: 50,000+ Format: Classic rock. News: 10 hrs wkly. Target aud: General; 35+. Spec prog: First nations 9 hrs. ♦Barry Wall, gen mgr & progmg dir; Ron Bartlett, mktg mgr; D.C. Cara, prom.

CFTK(AM)— 1960: 590 khz; 1 kw-U, DA-1. Hrs open: 4625 Lazelle Ave., V8G 1S4. Phone: (250) 635-6316. Fax: (250) 638-6320. Licensee: Standard Radio Inc. Group owner: Standard Broadcasting Corp. (acq 4-19-2002; grpsl). Format: MOR. ♦Brian Langston, gen mgr.

CJFW-FM—Co-owned with CFTK(AM). December 1983: 103.1 mhz; Format: Contemp country. Co-owned TV: CFTK-TV affil.

Tofino

CHMZ-FM— 2005: 90.1 mhz; 170 w. Hrs open: Box 1092, V0R 2Z0. Phone: (250) 725-4411. Fax: (250) 725-4411. E-mail: info@chmzfm.com Web Site: www.chmzfm.com. Licensee: West Island Radio Enterprises General Partnership. Format: Pop-rock and country. ♦Matthew McBride, gen mgr.

Trail

CJAT-FM— 1996: 95.7 mhz; 13.5 kw. Hrs open: 1560 2nd Ave., V1R 1M4. Phone: (250) 368-5510. Fax: (250) 368-8471. E-mail: kbs@sri.ca Web Site: www.kbsradio.ca. Licensee: Standard Radio Inc. Group owner: Standard Broadcasting Corp. (acq 4-19-2002; grpsl). Natl. Rep: Target Broadcast Sales. Format: Adult contemp. Spec prog: Ethnic mus 2 hrs wkly. ♦Kevin Einarson, gen sls mgr; Darren Robertson, progmg dir; Larry King, chief of engrg.

Ucluelet

CIMM-FM— Sept 1, 2006: 99.5 mhz; 180 w. Hrs open: 10760 Fundy Dr., Richmond, V7E 5K7. Phone: (604) 220-8393. Fax: (604) 677-6316. E-mail: info@cimmfm.com Web Site: www.cimmfm.com. Licensee: CIMM-FM Radio Ltd. Format: Var. ♦Matthew McBride, gen mgr.

Vancouver

***CBU(AM)—** 1925: 690 khz; 50 kw-U, DA-1. (Digital radio: 1459.792 mhz). Hrs open: 24 Box 4600, 700 Hamilton St., V6B 4A2. Phone: (604) 662-6000. Fax: (604) 662-6088. Web Site: www.cbc.ca/bc. Licensee: Canadian Broadcasting Corp. Format: News/talk, variety. ♦Joan Anderson, gen mgr; Joan Athey, prom dir; Brett Ballah, news dir; Dave Newbury, engrg dir; Terry Donnelly, news rptr.

CBU-FM— 1947: 105.7 mhz; 100 kw. 1,823 ft (Digital radio: 1459.792 mhz). Stereo. 24 Web Site: www.cbc.ca/bc. Format: Music. ♦Tod Elvidge, mus dir.

CBUF-FM— Dec 1, 1967: 97.7 mhz; 50 kw. Ant 1,823 ft (Digital radio: 1459.792 mhz). Stereo. Hrs open: 700 Hamilton St., V6B 4A2. Secondary address: Box 4600 V6B 4A2. Phone: (604) 662-6169. Fax: (604) 662-6161. Web Site: www.radio-canada.ca/c-b. Licensee: CBC. Natl. Network: Premiere Chaine. Format: Div. ♦Stephane Boisjoly, gen mgr; Mario Deschamps, progmg dir.

CBUX-FM— Sept 22, 2002: 90.9 mhz; 1.28 kw. Hrs open: 700 Hamilton St., V6B 4A2. Phone: (604) 662-6000. Fax: (604) 662-6335. Web Site: www.radio-canada.ca/c-b. Licensee: Canadian Broadcasting Corp. Natl. Network: Espace Musique. Format: Var, Fr. ♦Stephane Boisjoly, gen mgr; Mario Deschamps, progmg dir.

CFBT-FM— February 2002: 94.5 mhz; 46 kw. Hrs open: 24 #A301 - 770 Pacific Blvd., Plaza of Nations, V6B 5E7. Phone: (604) 699-2328. Fax: (604) 484-4912. E-mail: info@thebeat.com Web Site: www.thebeat.com. Licensee: Focus Entertainment Group Inc. (acq 8-3-01). Format: Rythmic CHR, Top 40. News staff: one; News: 2 hrs wkly. ♦Barry Duggan, pres; Jennifer Smith, VP & gen sls mgr; Chris Myers, progmg dir.

CFMI-FM—Listing follows CKNW(AM).

CFOX-FM—Listing follows CHMJ(AM).

***CFRO-FM—** Apr 22, 1975: 102.7 mhz; 5.5 kw. 1,005 ft Hrs open: 110-360 Columbia St., V6A 4J1. Phone: (604) 684-8494. E-mail: program@coopradio.org Web Site: www.coopradio.org. Licensee: Vancouver Co-Op Radio. Format: Community, news/talk, alternative pub affrs. Target aud: General; alternative community. Spec prog: Black 14 hrs, Chinese 2 hrs, Greek one hr, hip hop 17 hrs, jazz 16 hrs, Latin American 7 hrs, Pol 5 hrs wkly. ♦Leela Chinniah, progmg mgr; Rob Gauvin, mus dir; Danjel van Tijn, engrg dir.

CFUN(AM)—Listing follows CHQM-FM.

CHMB(AM)— Dec 10, 1959: 1320 khz; 50 kw-U, DA-2. TL: N49 09 55 W123 02 28. Hrs open: 24 1200 W. 73rd Ave., Suite 100, V6P 6G5. Phone: (604) 263-1320. Fax: (604) 261-0310. E-mail: info@am1320.com Web Site: www.am1320.com. Licensee: Mainstream Broadcasting Corp. (acq 12-14-93; C$1.8 million). Population served: 600,000 Natl. Rep: Canadian Broadcast Sales. Format: Chinese, multicultural. News staff: 8; News: 30 hrs wkly. Target aud: 18-65; multilingual, multicultural, mainly Chinese. Spec prog: American Indian one hr, Korean one hr, Japanese 7 hrs, Vietnamese 2 hrs wkly, Ukranian one hr, Portuguese one hr, Scandinavian one hour, Greek 1/2 hr, Tamil 1/2 hr. ♦James Ho, chmn & pres; George Feng, dev VP & sls VP; Trix Chan, news dir; K.K. Wong, chief of engrg.

CHMJ(AM)— June 1954: 730 khz; 50 kw-U, DA-2. Stereo. Hrs open: 700 W. Georgia St., Suite 2000, V7Y 1K9. Phone: (604) 681-7511. Fax: (604) 331-2722. Licensee: Corus Radio Co. Group owner: Corus Entertainment Inc. (acq 7-6-2000; grpsl). Format: Drive time traffic, talk. ♦J.J. Johnston, gen mgr; Shari Wong, gen sls mgr; Ian Koenigsfest, progmg dir.

CFOX-FM—Co-owned with CHMJ(AM). October 1964: 99.3 mhz; 100 kw. 2,243 ft Stereo. Phone: (604) 684-7221. Format: AOR. ♦Bob Mills, progmg dir.

CHQM-FM— Aug 10, 1960: 103.5 mhz; 53 kw. Ant 2,026 ft (Digital radio: 1463.280 mhz). Stereo. Hrs open: 24 380 W. 2nd Ave., Suite 300, V5Y 1C8. Phone: (604) 871-9000. Fax: (604) 871-2901. E-mail: qmfmmail@qmfm.com Web Site: www.qmfm.com. Licensee: CHUM (Western) Ltd. Group owner: CHUM Ltd. (acq 8-23-69). Format: Adult contemp. News staff: 5; News: 2 hrs wkly. Target aud: 25-54. ♦Barry O'Donnell, gen sls mgr; Carl LeGrice, prom dir; Neil Gallagher, gen mgr & progmg mgr; Clara Carotenuto, mus dir; Dave Youell, chief of engrg.

CFUN(AM)—Co-owned with CHQM-FM. Apr 20, 1922: 1410 khz; 50 kw-U, DA-2. (Digital radio: 1463.280 mhz). 24 E-mail: cfunmail@cfun.cim Web Site: www.cfun.com. (Acq 1-1-73). Format: Talk. Target aud: 25-49; upscale, well-educated females.

***CITR-FM—** Apr 1, 1982: 101.9 mhz; 1.8 kw. 170 ft Stereo. Hrs open: 7:30 AM-4 AM Univ. of British Columbia, 233-6138 Sub Blvd., V6T 1Z1. Phone: (604) 822-3017. Fax: (604) 822-9364. E-mail: citrmgr@ams.ubc.ca Web Site: www.citr.ca. Licensee: Student Radio Society of University of British Columbia. Format: Div. News: 10 hrs wkly. Target aud: General; campus/community. Spec prog: Black 18 hrs, Fr 2 hrs, Sp 2 hrs, East Indian 2 hrs, Greek one hr wkly. ♦Lydia Masemola, stn mgr & chief of opns; Luke Meat, mus dir.

CJJR-FM—Listing follows CKBD(AM).

CJRJ(AM)— November 2006: 1200 khz; 25 kw-U, DA-2. TL: N49 09 55 W123 02 28. Hrs open: 110-3060 Norland Ave., Unit 110, Burnaby, V5B 3A6. Phone: (604) 299-8863. Fax: (604) 299-3088. E-mail: info@rj1200.com Web Site: www.rj1200.com. Licensee: I.T. Productions Ltd. Format: Ethnic, Hindustani, Punjabi. ♦Shushma Datt, CEO; Sudhir Datta, gen mgr.

CKAV-FM-2— 2007: 106.3 mhz; 9 kw. Ant 1,968 ft TL: N49 21 17 W122 57 25. Hrs open: 366 Adelaide St. E., Suite 323, Toronto, ON, M5A 3X9. Phone: (416) 703-1287. Fax: (416) 703-4328. Web Site: www.aboriginalradio.com. Licensee: Aboriginal Voices Radio Inc. Format: Div. ♦Roy Hennessy, opns mgr; Patrice Mousseau, progmg dir.

CKBD(AM)— July 13, 1926: 600 khz; 10 kw-U, DA-N. Stereo. Hrs open: 24 300-1401 W. 8th Ave., V6H 1C9. Phone: (604) 731-6111. Fax: (604) 731-0493. E-mail: 600am@600am.com Web Site: www.600am.com. Licensee: Jim Pattison Broadcast Group Ltd. (the general partner) and Jim Pattison Industries Ltd. (the limited partner) carrying on business as Jim Pattison Broadcast Group L.P. Group owner: The Jim Pattison Broadcast Group (acq 1965). Natl. Rep: Canadian Broadcast Sales. Format: MOR. News staff: 3; News: 6 hrs wkly. Target aud: 45; mature adults. ♦Jim Pattison, CEO & chmn; Gerry Siemens, VP & gen mgr; Mark Rogers, gen sls mgr; Brian Pritchard, rgnl sls mgr; Sheila Dunn, prom dir; Gord Eno, progmg dir; Mark Patric, mus dir; Campbell McCubbin, news dir; Dave Linder, engrg mgr.

CJJR-FM—Co-owned with CKBD(AM). July 1, 1986: 93.7 mhz; 75 kw. 2,250 ft Stereo. 24 1501 W. 8th Ave., V6H 1C9. Phone: (604) 731-7772. Fax: (604) 731-1329. Format: Contemp country. News staff: 2. Target aud: 25-54. ♦Mark Rogers, gen sls mgr; Brian Pritchaad, natl sls mgr; Karen Seaboyer, prom dir; Gordon End, progmg dir; Mark Patric, mus dir; Campbell McCubbin, news dir; Shiral Tobin, pub affrs dir; David Linder, engrg mgr.

CKLG-FM—Listing follows CKWX(AM).

CKNW(AM)—(New Westminster, Sept 1, 1944: 980 khz; 50 kw-U, DA-2. Stereo. Hrs open: 24 700 W. Georgia St., Suite 2000, V7Y 1K9. Phone: (604) 331-2711. Fax: (604) 331-2722. E-mail: info@cknw.com Web Site: www.cknw.com. Licensee: Corus Premium Television Ltd. Group owner: Corus Entertainment Inc. (acq 7-6-2000; grpsl). Natl. Rep: Canadian Broadcast Sales. Format: News/talk, sports. News staff: 20; News: 14 hrs wkly. Target aud: General. ♦J.J. Johnson, gen mgr.

CFMI-FM—Co-owned with CKNW(AM). Mar 22, 1970: 101.1 mhz; 100 kw. 3,500 ft Stereo. Phone: (604) 331-2808. Fax: (604) 331-2727. E-mail: rock101@rock101.com Web Site: www.rock101.com. Format: Classic rock. Target aud: 25-49.

CKST(AM)— Jan 19, 1963: 1040 khz; 50 kw-U, DA-2. Stereo. Hrs open: 24 300-380 W. 2nd Ave., Suite 300, V5Y 1C8. Phone: (604) 871-9000. Fax: (604) 871-2901. E-mail: live@team1040.ca Web Site: www.team1040.ca. Licensee: CHUM Ltd. (group owner; acq 2-10-03). Population served: 2,000,000 Natl. Rep: Canadian Broadcast Sales. Format: All sports. News staff: 4; News: 8 hrs wkly. Target aud: 40 plus; intelligent, socially conscious, older demographic. ♦Neil Gallagher, gen mgr.

CKWX(AM)— Apr 1, 1923: 1130 khz; 50 kw-U, DA-N. TL: N49 09 22 W123 04 00. Hrs open: 2440 Ash St., V5Z 4J6. Phone: (604) 877-4488. Phone: (604) 872-2557. Fax: (604) 877-4494. Web Site: www.news1130.com. Licensee: Rogers Broadcasting Ltd. (group owner; acq 1989). Population served: 2,000,000 Natl. Rep: CBS Radio. Format: All news. News staff: 45; News: 168 hrs wkly. Target aud: 35-54; men. ♦Ted Rogers, CEO & chmn; Tony Viner, pres; Laura Nixon, CFO; Gary Miles, exec VP; Paul Fisher, gen mgr & opns mgr; May Lam, prom mgr; Jacquie Donaldson, progmg dir & news dir; Rick Dal Farra, chief of engrg.

CKLG-FM—Co-owned with CKWX(AM). Mar 1, 1980: 96.9 mhz; 100 kw. 2,500 ft TL: N49 21 29 W122 57 09. Stereo. 24 Web Site: www.969jackfm.com. Format: Adult contemp. ♦May Lam, prom dir; Doreen Copeland, mus dir; Rick dal Farva, engrg mgr.

CKYE-FM— Feb 1, 2006: 93.1 mhz; 4.2 kw. TL: N49 21 17 W122 57 25. Hrs open: 8383A 128th St. #201, Surrey, V3W 4G1. Phone: (604) 598-9311. E-mail: info@redfm.ca Web Site: redfm.ca. Licensee: South Asian Broadcasting Corp. Inc. Format: Ethnic. ♦Bijoy Samuel, gen mgr.

CKZZ-FM— May 1991: 95.3 mhz; 75 kw. Hrs open: #20, 11151 Horseshoe Way, Richmond, V7A 4S5. Phone: (604) 241-0953. Fax: (604) 272-0917. E-mail: input@95crave.com Web Site: www.95crave.com. Licensee: Standard Radio Inc. Group owner: Standard Broadcasting Corp. (acq 5-8-96; C$18 million with CISL(AM) Richmond). Format: Hot adult contemp. ♦Gary Slaight, pres; Gary Russell, VP & gen mgr.

Vanderhoof

CIVH(AM)— November 1973: 1340 khz; 1 kw-D, 500 w-N, DA-1. TL: N54 01 00 W123 59 00. Hrs open: 6-10 AM Box 1370, V0J 3A0. Phone: (250) 567-4914. Fax: (250) 567-4982. E-mail: thewolf@hwy16.com Licensee: Vista Radio Ltd. (group owner) Natl. Rep: Target Broadcast Sales. Format: Modern country & best southern rock. Target aud: General. Spec prog: Relg 5 hrs wkly. ♦Gary Russell, gen mgr; Karen Fridleifson, gen sls mgr; Jacqui Ryks, prom mgr & mus dir; Bill Fee, news dir.

Vernon

CICF-FM— 2001: 105.7 mhz; 46 kw. Hrs open: 2800 31st St., V1T 5H4. Phone: (250) 545-9222. Fax: (250) 542-2083. E-mail: vernonmail@sri.ca Web Site: www.thesun.net/vernon. Licensee: Standard Radio Inc. Group owner: Standard Broadcasting Corp. Format: Hot adult contemp. ♦Gord Leighton, gen mgr; Larry King, chief of engrg.

CKIZ-FM— Nov 8, 2001: 107.5 mhz; 46 kw. Hrs open: 3313 32nd Ave., V1T 2M7. Phone: (250) 545-2141. Fax: (250) 545-9008. E-mail: kissfm@1075kiss.com Web Site: www.1075kiss.com. Licensee: Rogers Broadcasting Ltd. (group owner). Natl. Rep: Canadian Broadcast Sales. Format: News, lite favorites of yesterday & today. ♦Patrick Nicol, VP & gen mgr; Gord Wiens, gen sls mgr; Don Weglo, progmg dir; Duane Schindel, chief of engrg.

Victoria

***CBCV-FM—** Sept 28, 1998: 90.5 mhz; 3 kw. Hrs open: 1025 Pandora Ave., V8V 3P6. Phone: (250) 360-2227. Fax: (250) 360-2600. E-mail: victoria@cbc.ca Web Site: www.vancouver.cbc.ca/ontheisland. Licensee: CBC. Natl. Network: CBC Radio One. Format: Current Affairs. News staff: 11; News: 23 hrs wkly. ♦Peter Hutchinson, stn mgr & progmg mgr.

CFAX(AM)— Sept 4, 1959: 1070 khz; 10 kw-U, DA-1. TL: N48 23 50 W123 18 20. Stereo. Hrs open: 24 Mellor Bldg., 825 Broughton St., V8W 1E5. Phone: (250) 920-4602. Fax: (250) 386-5775. E-mail: tspence@cfax1070.com Web Site: www.cfax1070.com. Licensee: CHUM Ltd. (acq 10-1-2004;. C$7.5 million with co-located FM). Population served: 350,000 Natl. Rep: Target Broadcast Sales. Wire Svc: BN Wire Format: News/talk. News staff: 7; News: 21 hrs wkly. Target aud: 45 plus; 50% males, 50% females. ♦Terry Spence, gen mgr & opns VP; Kevin Bell, sls dir; Shannon Kowalko, adv dir; Al Ferraby, news dir & disc jockey; Greg Bohnert, news dir; Bud Goes, chief of engrg; Frank Stanford, news rptr; Jennifer Cador, news rptr.

CHBE-FM—Co-owned with CFAX(AM). Aug 23, 2002: 107.3 mhz; 20 kw. TL: N48 25 06 W123 30 35. Phone: (250) 382-1073. Fax: (250) 386-5775. Web Site: www.b1073.ca. Format: Hot AC. News: 1.5 hrs. Target aud: 35-49. ♦Brad Edwards, progmg dir.

***CFUV-FM**— Dec 17, 1984: 101.9 mhz; 2.29 kw. 265 ft Stereo. Hrs open: 24 Box 3035, University of Victoria, V8W 3P3. Phone: (250) 721-8607. Phone: (250) 721-8702. Web Site: www.cfuv.uvic.ca. Licensee: University of Victoria Student Radio Society. Format: Div. News: 8.5 hrs wkly. Target aud: General; people tired of commercial radio, on campus & in the community. Spec prog: It 2 hrs, American Indian 1 hrs; Fr 2 hrs, Pol one hr, Sp 3 hrs wkly. ♦Randy Gelling, stn mgr; Jana Grazley, progmg dir; Justin Lanoue, mus dir.

CHTT-FM— September 2000: 103.1 mhz; 50 w. TL: N48 26 52 W123 19 19. Stereo. Hrs open: 9 AM-9 PM 817 Fort St., V8W 1H6. Phone: (250) 382-0900. Fax: (250) 382-4358. Web Site: www.1031jackfm.ca. Licensee: Rogers Broadcasting Ltd. (group owner) Population served: 200,000. Natl. Rep: Canadian Broadcast Sales. Format: Classic hits. News: 5 hrs wkly. ♦Gorde Edlund, gen mgr & progmg dir; Tony Marsh, gen sls mgr.

CILS-FM—Not on air, target date: unknown: 107.9 mhz; 250 w. Hrs open: 200-535 Yates St., V8W 2Z6. Phone: (250) 220-4137. Fax: (250) 388-6280. E-mail: radio@francocentre.com Web Site: www.cilsfm.ca. Licensee: Societe radio communautaire Victoria. Format: French. ♦Jacques P. Vallee, pres.

CIOC-FM— Mar 18, 1965: 98.5 mhz; 100 kw. 567 ft Stereo. Hrs open: 24 817 Fort St., V8W 1H6. Phone: (250) 382-0900. Fax: (250) 382-4358. Web Site: www.ocean985.com. Licensee: Rogers Broadcasting Ltd. (group owner) Population served: 300,000 Format: Lite rock, adult contemp. News staff: one; News: 2 hrs wkly. Target aud: 35-54; female. ♦Kim Hesketh, VP, gen mgr & stn mgr; Tony Marsh, gen sls mgr; Dawn Kaysoe, progmg dir; Dean Fox, chief of engrg.

CJZN-FM— May 2000: 91.3 mhz; 1.766 kw. (Digital radio: 1,472.000 mhz; 2 kw). Stereo. Hrs open: Top Floor, 2750 Quadra St., V8T 4E8. Phone: (250) 475-6611. Fax: (250) 475-3299. E-mail: modernrock@thezone.fm Web Site: www.thezone.fm. Licensee: Jim Pattison Broadcast Group Ltd. (the general partner) and Jim Pattison Industries Ltd. (the limited partner), carrying on business as Jim Pattison Broadcast Group L.P. (acq 11-24-2006; C$15.75 million with CKKQ-FM Victoria). Format: Modern rock. Target aud: 25-64. ♦Dan McAllister, gen mgr & natl sls mgr; John Shields, opns mgr; Brian Blackburn, sls VP.

CKKQ-FM— Dec 12, 1987: 100.3 mhz; 100 kw. Ant 1,620 ft TL: N48 35 41 W123 32 37. (Digital radio: 1,472.000 mhz; 2 kw). Stereo. Hrs open: 24 Top Floor, 2750 Quadra St., V8T 4E8. Phone: (250) 475-0100. Fax: (250) 475-3299. E-mail: thecrew@theQ.fm Web Site: www.theq.fm. Licensee: Jim Pattison Broadcast Group Ltd. (the general partner) and Jim Pattison Industries Ltd. (the limited partner), carrying on business as Jim Pattison Broadcast Group L.P. (acq 11-24-2006; C$15.75 million with CJZN-FM Victoria). Natl. Rep: Canadian Broadcast Sales. Format: Classic Rock. News staff: 2; News: 4 hrs wkly. Target aud: 25-49. ♦Dan McAllister, gen mgr & natl sls mgr; John Shields, opns mgr; Brian Blackburn, sls VP.

***CKMO(AM)**— Sept 4, 2000: 900 khz; 10 kw-U, DA-1. Stereo. Hrs open: 3100 Foul Bay Rd., V8P 5J2. Phone: (250) 370-3658. Fax: (250) 370-3679. E-mail: info@village900.ca Web Site: www.village900.ca. Licensee: CKMO Radio Society. Format: Educ, Roots, Folk, World. News: 5 hrs wkly. Spec prog: Class, folk, country, blues, news/talk, Portugese 3 hrs. ♦Doug Ozeroff, gen mgr.

Whistler

CISW-FM— Feb 25, 1982: 102.1 mhz; 586 w. 2,250 ft Hrs open: Rebroadcasts CISQ-FM Squamish.
Box 1068, Squamish, VON 1B4. Phone: (604) 892-1021. Fax: (604) 892-6383. E-mail: mountainfm@mountainfm.com Web Site: www.mountainfm.com. Licensee: Rogers Broadcasting Ltd. (group owner). Format: Hot adult contemp. ♦Gary Miles, pres; Paul Fisher, VP; Ken Geiger, gen mgr; Janis Correia, opns mgr.

Williams Lake

CFFM-FM—Listing follows CKWL(AM).

CKWL(AM)— Feb 25, 1960: 570 khz; 1 kw-U, DA-2. Hrs open: 24 83 S. First Ave., V2G 1H4. Phone: (250) 392-6551. Fax: (250) 392-4142. Web Site: www.thewolfpack.ca. Licensee: Vista Radio Ltd. Group owner: Cariboo Central Interior Radio Inc. Natl. Rep: Canadian Broadcast Sales, Target Broadcast Sales. Format: Modern country, southern rock. News staff: one. Target aud: 25-54 plus. ♦Terry Coles, pres & gen sls mgr; Paul Mann, exec VP & stn mgr; Tracey Gard, gen sls mgr; Jason Mann, progmg.

CFFM-FM—Co-owned with CKWL(AM). Aug 31, 1987: 97.5 mhz; Stereo. 24 Phone: (250) 398-2336. Web Site: www.cffmthemax.com. (Acq 2006). Natl. Rep: Target Broadcast Sales, Canadian Broadcast Sales. Format: Rock and pop. News staff: 2. Target aud: 18-44.

Manitoba

Altona

CFAM(AM)— Mar 13, 1957: 950 khz; 10 kw-U, DA-2. Hrs open: 24 Box 950, 201-125 Centre Avenue, R0G 0B0. Phone: (204) 324-6464. Fax: (204) 324-8918. E-mail: cfam@gordenwestradio.com Web Site: www.cfamradio.com. Licensee: Golden West Broadcasting Ltd. (group owner) Natl. Rep: Canadian Broadcast Sales. Format: Agriculture, MOR. News staff: 8; News: 12 hrs wkly. Target aud: General. Spec prog: Class 15 hrs wkly. ♦Elmer Hildebrand, CEO & pres; Ang Enns, stn mgr.

Boissevain

CJRB(AM)— 1973: 1220 khz; 10 kw-U. Hrs open: 24 Box 1220, R0K 0E0. Phone: (204) 324-6464. Fax: (204) 324-8918. Licensee: Golden West Broadcasting Ltd. (group owner) Wire Svc: BN Wire Format: Easy lstng, inspirational, agriculture. News staff: one. Target aud: General. ♦E. Hildebrand, pres; Lyndon Friesen, exec VP; Menno Friesen, gen sls mgr; Al Friesen, progmg mgr; Laverne Siemens, engrg dir.

Brandon

CJJJ-FM— May 2003: 106.5 mhz; 930 w. Hrs open: Assiniboine Community College, 1430 Victoria Ave. E., Rm 223, R7A 2A9. Phone: (204) 571-3900. Fax: (204) 726-7014. Licensee: Assiniboine Campus-Community Radio Society Inc. Format: Var. ♦Bob Crighton, stn mgr.

CKLF-FM— June 1, 2000: 94.7 mhz; 100 kw. Stereo. Hrs open: 24 624 14th St. E., R7A 7E1. Phone: (204) 726-8888. Fax: (204) 726-1270. E-mail: tyler@starfmradio.com Web Site: www.starfmradio.com. Licensee: Riding Mountain Broadcasting Ltd. Format: Adult contemp. News staff: 7. Target aud: 25-54; adults. ♦Don Kille, CEO & gen mgr.

CKLQ(AM)— October 1977: 880 khz; 10 kw-U, DA-2. Hrs open: 24 624 14th St. E., R7A 7E1. Phone: (204) 726-8888. Fax: (204) 726-1270. E-mail: qcountry@cklq.mb.ca Web Site: www.ckcq.mb.ca. Licensee: Riding Mountain Broadcasting Ltd. Natl. Rep: Target Broadcast Sales. Format: Country. News staff: 7. Target aud: 35 plus. Spec prog: Farm 18 hrs wkly. ♦Don Kille, gen mgr.

CKXA-FM— February 2000: 101.1 mhz; 100 kw. Hrs open: 2940 Victoria Ave., R7B 3Y3. Phone: (204) 728-1150. Fax: (204) 725-3794. E-mail: hot101@hot101.ca Web Site: www.hot101.ca. Licensee: Standard Radio Inc. Group owner: Standard Broadcasting Corp. (acq 2-1-2002; grpsl). Format: Country. ♦Sharon Taylor, VP & gen mgr; Kevin Grexton, opns mgr.

CKX-FM— Dec 16, 1963: 96.1 mhz; 100 kw. 1,042 ft Stereo. Hrs open: 2940 Victoria Ave., R7B 3Y3. Phone: (204) 728-1150. Phone: (204) 727-1150. Fax: (204) 727-2505. Licensee: Standard Radio Inc. Group owner: Standard Broadcasting Corp. (acq 2-1-2002; grpsl). Format: Classic rock. ♦Lee Sterry, gen mgr; Kevin Grexton, opns mgr & progmg dir; Gyl Tosheck, gen sls mgr; Norine Mitchell, rgnl sls mgr; Donna Smith, prom dir; Angela Greis, asst music dir; Bob Bruce, news dir.

Churchill

CHFC(AM)— Sept 13, 1959: 1230 khz; 250 w-U, DA-1. Hrs open: Rebroadcasts CBW(AM) Winnipeg.
c/o CBC, Winnipeg, R3C 2H1. Phone: (204) 788-3222. Fax: (204) 788-3225. E-mail: communications@winnipeg.cbc.ca Licensee: Canadian Broadcasting Corp. Natl. Network: CBC Radio One. Format: Var/div. ♦John Bertrand, gen mgr.

Cross Lake

CFNC(AM)— 1990: 1490 khz; 50 w. Hrs open: Box 129, R0B 0J0. Phone: (204) 676-2331. Phone: (204) 676-2248. Fax: (204) 676-2911. Licensee: Native Communications Inc. Format: Community. ♦Dina Monias, pres.

Dauphin

CKDM(AM)— Jan 6, 1951: 730 khz; 10 kw-D, 5 kw-N, DA-N. Hrs open: 24 27 3rd Ave. N.E., R7N 0Y5. Phone: (204) 638-3230. Fax: (204) 638-8257/8891. E-mail: 730ckdm@mb-sympatico.ca Web Site: www.730ckdm.com. Licensee: Dauphin Broadcasting Co. Ltd. (acq 11-23-93). News staff: 3. ♦Alan Truman, VP & gen mgr.

Flin Flon

CFAR(AM)— Nov 14, 1937: 590 khz; 10 kw-D, 1 kw-N, DA-2. Hrs open: 24 316 Green St., R8A 0H2. Phone: (204) 687-3469. Phone: (204) 687-8300. Fax: (204) 687-6786. E-mail: cfar@arcticradio.ca Web Site: www.arcticradio.ca. Licensee: Arctic Radio (1982) Ltd. (acq 9-1-82). Format: Adult contemp. News staff: 2; News: 15 hrs wkly. Target aud: General. ♦Tom O'Brien, pres & gen mgr; Maureen Kozar, stn mgr.

Portage la Prairie

CFRY(AM)— Oct 18, 1956: 920 khz; 25 kw-D, 15 kw-N, DA-2. Hrs open: 350 River Rd., R1N 0N6. Phone: (204) 239-5111. Fax: (204) 857-3456. Licensee: Golden West Broadcasting Ltd. Group owner: Golden West Broadcasting Ltd. (acq 7-26-2000; with co-located FM). Format: C&W. Spec prog: Farm 4 hrs wkly. ♦Warren Neufeld, stn mgr.

CFRY-FM— 1996: 93.1 mhz; 27 kw.

CJPG-FM— May 4, 2004: 96.5 mhz; 24 kw. Hrs open: Box 920, R1N 0N6. Secondary address: 350 River Rd. R1N 0N6. Phone: (204) 239-5111. Fax: (204) 857-3456. Web Site: www.mix965fm.com. Licensee: Golden West Broadcasting Ltd. Format: CHR. ♦Warren Neufeld, stn mgr.

Pukatawagan

CFPX-FM— Sept 19, 1971: 98.3 mhz; 34.8 w. Hrs open: Missinnippi River Native, Communications Inc., R0B 1G0. Phone: (204) 553-2155. Fax: (204) 553-2158. Licensee: Missinnippi River Native Communications Inc. Format: Country, rock. ♦John Colomb, gen mgr.

Saint Boniface

***CKSB(AM)**— May 27, 1946: 1050 khz; 10 kw-U. Hrs open: 24 607 Langevin St., R2H 2W2. Phone: (204) 788-3236. Fax: (204) 788-3245. Web Site: www.radiocanada.ca/radio/manitoba. Licensee: Societe Radio Canada. (acq 4-1-73). Format: Div, Fr. Target aud: General. ♦Robert Rabinovitch, CEO; Huguette Le Gall, mktg mgr; Rene Fontaine, progmg dir; Gilles Frechette, progmg mgr; Michel Boucher, news dir.

CKXL-FM— October 1991: 91.1 mhz; 61 kw. Hrs open: 24 340 Provencher Blvd., Winnipeg, R2H 0G7. Phone: (204) 233-4243. Phone: (204) 231-3691. Fax: (204) 233-3646. E-mail: info@envol91.mb.ca Web Site: www.envol91.mb.ca. Licensee: La Radio Communautaire du Manitoba Inc. Population served: 600,000. Rgnl rep: Target Media, George McKringan Format: Soft rock, div, French. Target aud: General; Francophone 20-50. Spec prog: Folk 2 hrs, jazz 4 hrs, Sp 2 hrs, blues 2 hrs, reggae 2 hrs, dance 5 hrs, techno 4 hrs wkly, disco 2 hrs. ♦Eric Zogbi, progmg dir; Claire Hince, sls.

Selkirk

CFQX-FM— Nov 9, 1981: 104.1 mhz; 100 kw. 500 ft (CP: 95.3 mhz.). Stereo. Hrs open: 24 177 Lombard Ave., 3rd Fl., Manitoba, R3B 0W5. Phone: (204) 944-1031. Fax: (204) 943-7687. E-mail: jtrecarten@qx104fm.com Web Site: www.qx104fm.com. Licensee: Standard Radio Inc. Group owner: Standard Broadcasting Corp. (acq 2-1-2002; grpsl). Format: Country. News staff: 2. Target aud: 25-54. ♦Sharon Taylor, VP, gen mgr & gen mgr; Gyl Toshack, gen sls mgr & natl sls mgr; Janet Trecarten, progmg dir.

CICY-FM— 2000: 105.5 mhz; 100 kw. Stereo. Hrs open: 1507 Inkster Blvd., Winnipeg, R2X 1R2. Phone: (204) 772-8255. Fax: (204)

779-5628. E-mail: info@ncifm.com Web Site: www.ncifm.com. Licensee: Native Communication Inc. Format: Div in English, Cree, Saulteaux, Ojibiway languages; CHR, country. News staff: 2. Target aud: 25 and up; aboriginal. ♦Ron Nadeau, CEO; Rita Ducharme, pres; Dave McLeod, gen mgr.

Steinbach

CHSM(AM)— Mar 19, 1964: 1250 khz; 10 kw-U, DA-2. Hrs open: 24 105-32 Brandt St., R5G 2J7. Phone: (204) 326-3737. Fax: (204) 326-2299. Web Site: www.am1250online.com. Licensee: Golden West Broadcasting Ltd. (group owner) Natl. Rep: Canadian Broadcast Sales. Format: MOR. News staff: 3; News: 12 hrs wkly. Target aud: General. Spec prog: Farm. ♦Elmer Hildebrand, pres; Laverne Tappel, gen mgr; Al Friesen, progmg dir; Laverne Siemens, engrg dir & chief of engrg.

CILT-FM—Co-owned with CHSM(AM). Sept 29, 1998: 96.7 mhz; 50 kw. 24 Phone: (204) 346-0000. Web Site: www.lite967online.com. Format: Light adult contemp.

Swan River

CJSB-FM— July 1, 2006: 104.5 mhz; 210 w. TL: N52 06 18 W101 16 10. Hrs open: Box 1268, R0L 1Z0. Phone: (204) 734-6484. Fax: (204) 734-5897. E-mail: onair@cj104radio.com Web Site: www.cj104radio.ca. Licensee: Stillwater Broadcasting Ltd. Format: Var. ♦Bill Gade, gen mgr.

The Pas

CJAR(AM)— 1974: 1240 khz; 1 kw-U, DA-1. Hrs open: Box 2980, R9A 1R7. Phone: (204) 623-5307. Fax: (204) 623-5337. E-mail: cjar@articradio.ca Web Site: www.articradio.ca. Licensee: Arctic Radio Corp. Ltd. Format: Adult contemp, AOR, C&W. Spec prog: Aboriginal 5 hrs wkly. ♦Tom O'Brien, chmn & pres; Jeremy Walchal, gen mgr.

Thompson

***CBWK-FM**— 1980: 100.9 mhz; 9.4 kw. Hrs open: 7 AM-5 PM 7 Selkirk Ave., R8N 0M4. Phone: (204) 677-1680. Fax: (204) 677-9517. E-mail: north@winnwpeg.cbc.ca Web Site: www.winnipeg.cbc.ca. Licensee: CBC. Population served: 50,000 Natl. Network: CBC Radio One. Format: Div., talk. ♦Mark Sislo, gen mgr & progmg dir; Charles Altiman, news dir; Doug MacPherson, chief of engrg.

CHTM(AM)— Mar 29, 1964: 610 khz; 1 kw-U. Hrs open: 24 103 Cree, R8N 0B9. Phone: (204) 778-7361. Fax: (204) 778-5252. E-mail: chtm@arcticradio.ca Web Site: www.articradio.com. Licensee: Arctic Radio (1982) Ltd. Population served: 60,000 Format: Adult contemp, classic rock, country. News staff: 2; News: 15 hrs wkly. Target aud: General. Spec prog: Relg 12 hrs, Cree (American Indian) 10 hrs wkly. ♦Tom O'Brien, pres; Dave Moore, gen mgr, opns mgr & gen sls mgr; Don Barkman, mus dir & local news ed.

CINC-FM— 1994: 96.3 mhz; 86 w. Hrs open: 76 Severn Crescent, R8N 1M6. Phone: (204) 778-8343. Fax: (204) 778-6559. E-mail: info@ncifm.com Web Site: www.ncifm.com. Licensee: Native Communications Inc. Format: Div in English, Cree, Saulteaux, Ojibiway languages. ♦Dave McLeod, gen mgr; Rey St. Jermain, progmg dir.

Winkler

CJEL-FM— Oct 4, 2000: 93.5 mhz; 100 kw. Hrs open: 24 Box 399, R6W 4A6. Secondary address: 201-295 Main St. R6W 4A6. Phone: (204) 331-9300. Fax: (204) 325-2206. E-mail: info@eagle935fm.com Web Site: www.eagle935fm.com. Licensee: Golden West Broadcasting Ltd. (group owner) Format: Adult contemp. ♦Elmer Hildebrand, CEO, chmn & pres; Bill Hildebrand, stn mgr.

Winkler-Morden

CKMW(AM)— Aug 1, 1980: 1570 khz; 10 kw-U, DA-2. Hrs open: 24 Box 399 201-295 Main St., Winkler, R6W 4AG. Phone: (204) 325-9506. E-mail: country1570@goldenwestradio.com Web Site: www.ckmwradio.com. Licensee: Golden West Broadcasting Ltd. (group owner) Natl. Rep: Canadian Broadcast Sales. Format: Country. News staff: 2. Target aud: General. ♦Bill Hildebrand, gen mgr & stn mgr; Elmer Hildebrand, CEO, pres & engrg dir.

Winnipeg

CBW(AM)— Sept 3, 1948: 990 khz; 50 kw-D, 46 kw-N. Hrs open: Box 160, R3C 2H1. Phone: (204) 788-3222. Fax: (204) 788-3227. Web Site: www.cbc.ca. Licensee: Canadian Broadcasting Corp. Natl. Network: CBC Radio One. Format: Div. Spec prog: Farm 6 hrs, class 8 hrs, C&W one hr wkly. ♦John Berfrand, gen mgr.

CBW-FM— Oct 11, 1965: 98.3 mhz; 354 kw. Stereo. Natl. Network: CBC Radio Two. Format: Class, div. Co-owned TV: CBWT(TV) affil

CFEQ-FM— 2003: 107.1 mhz; 920 w. Hrs open: 738 Osborne St., R3L 2C2. Phone: (204) 944-8961. Fax: (204) 772-5854. E-mail: tom@freq107.com Web Site: www.freq107.com. Licensee: Kesitah Inc. Format: Alternative. ♦Tom Hiebert, gen mgr; Jarret Hannah, mktg dir & progmg dir.

CFRW(AM)— Nov 1, 1963: 1290 khz; 10 kw-U, DA-2. Stereo. Hrs open: 1445 Pembina Hwy., R3T 5C2. Phone: (204) 477-5120. Fax: (204) 453-0815. Licensee: CHUM Ltd. (group owner; acq 7-74). Format: Oldies. ♦Brian Stone, gen mgr; Scott Bodnarchuk, gen sls mgr; Howard Kroeger, progmg mgr.

CHIQ-FM—Co-owned with CFRW(AM). Nov 1, 1963: 94.3 mhz; 100 kw. 450 ft Stereo. Format: CHR.

CFWM-FM— June 13, 1996: 99.9 mhz; 100 kw. Hrs open: 24 1445 Pembina Hwy., R3T 5C2. Phone: (204) 477-5120. Fax: (204) 453-8777. E-mail: bryan@chumwinnipeg.com Web Site: www.999bobfm.com. Licensee: CHUM Ltd. (group owner; (acq 12-20-2001; C$7 million swap with CHOM-FM Montreal, PQ). Population served: 672,000 Format: Adult contemp. News staff: one; News: 5 hrs wkly. Target aud: 25-54. ♦Alan Waters, CEO; Jay Switzer, pres; Bryan Stone, VP & gen mgr.

CHNK-FM— Dec 7, 2002: 100.7 mhz; 1.3 kw. Hrs open: 24 3586 Portage Ave., R3K 0Z8. Phone: (204) 889-2586. Fax: (204) 831-1512. Web Site: www.hank.fm. Licensee: CKVN Radiolink System Inc. (acq 12-5-2005; C$1,790,000 for stock). Format: Country. News staff: one. ♦Kevin Klein, gen mgr; Nick Addams, progmg dir & mus dir.

CHVN-FM— Sept 14, 2000: 95.1 mhz; 100 kw. TL: N49 46 15 W97 30 35. Stereo. Hrs open: 24 Box 1812, R3C 3R1. Secondary address: 1111 Chevrier Blvd. R3T 1Y2. Phone: (204) 949-3395. Fax: (204) 949-3349. E-mail: chvn@chvnradio.com Web Site: www.chvnradio.com. Licensee: Golden West Broadcasting Ltd. (acq 2004). Natl. Network: Salem Radio Network. Format: Christian music. Target aud: 18-49; families. Spec prog: Children 1 hr, Gospel 4 hrs, teens 6 hrs wkly. ♦Elmer Hildebrand, CEO; Wade Kehler, gen mgr; Terry Van Veen, progmg dir.

CITI-FM— 1962: 92.1 mhz; 100 kw. Ant 700 ft Stereo. Hrs open: 166 Osborne St., Unit 4, R3L 1Y8. Phone: (204) 788-3400. Fax: (204) 788-3401. E-mail: geoff.poulton@winnipegradio.rogers.com Web Site: www.92citi.ca. Licensee: Rogers Broadcasting Ltd. (group owner; acq 8-20-92). Format: Classic rock. ♦Geoff Poulton, gen mgr; Gayle Zarbatany, progmg dir; Frank Andrews, mus dir.

CJKR-FM—Listing follows CJOB(AM).

CJNU-FM— December 2006: 104.7 mhz; 40 w. Hrs open: Box 2282, Stn Main, R3C 4A6. Phone: (204) 942-2568. Web Site: www.nostalgiawinnipeg.com. Licensee: Nostalgia Broadcasting Cooperative. Format: Nostalgia. ♦Bill Stewart, pres.

CJOB(AM)— 1946: 680 khz; 50 kw-U, DA-N. Hrs open: 24 930 Portage Ave., R3G 0P8. Phone: (204) 786-2471. Fax: (204) 783-4512. Fax: (204) 780-9750. Licensee: Corus Premium Television Ltd. Group owner: Corus Entertainment Inc. (acq 7-6-2000; grpsl). Population served: 600,000 Natl. Rep: Canadian Broadcast Sales. Format: News/talk. News staff: 25; News: 56 hrs wkly. Target aud: 25-54. ♦Garth Buchko, pres, gen mgr, stn mgr & mktg mgr; Sherrie Johnston, opns mgr; Steve Dubois, gen sls mgr; Robin Bonne, prom dir; Paul Graham, mus dir; Vic Grant, progmg dir & news dir; Jack Hoeppner, chief of engrg.

CJKR-FM—Co-owned with CJOB(AM). March 1948: 97.5 mhz; 310 kw. 228 ft Stereo. Fax: (204) 780-9750. Web Site: www.power97.com. Format: Winning best rock. Target aud: 25-44; males. ♦Steve Parsons, progmg dir; Lochlin Cross, mus dir.

CJUM-FM— Sept 4, 1998: 101.5 mhz; 1.2 kw. Stereo. Hrs open: 24 UMFM, 308 University Center, University of Manitoba, R3T 2N2. Phone: (204) 474-6518. Fax: (204) 269-1299. E-mail: station.manager@umfm.com Web Site: www.umfm.com. Licensee: The University of Manitoba Students' Union. Population served: 650,000 Format: Diversified. News: 12 hrs wkly. Target aud: 18-58; All genders, all ages who prefer non-commercial music and culture. ♦Liz Clayton, gen mgr & stn mgr; Jared McKetiak, progmg dir.

CJWV-FM— November 2003: 107.9 mhz; 200 w. Hrs open: 177 McDermot Ave., Suite 200, R3B 0S1. Phone: (204) 942-0629. Fax: (204) 946-0717. E-mail: thewave@shawcable.com Web Site: www.flava1079online.com. Licensee: Harmony Broadcasting Corp. Format: Hip hop, rhythm and blues. ♦Frank Capozzolo, stn mgr.

CJZZ-FM— March 2003: 99.1 mhz; 63.7 kw. Stereo. Hrs open: 30th Fl. CanWest Global Pl., 201 Portage Ave., R3B 3K6. Phone: (204) 253-2665. Fax: (204) 926-1674. Licensee: Corus Premium Television Ltd. Group owner: CanWest Global Communications Corp. (acq 7-6-2007; C$14.5 million with CKBT-FM Kitchener-Waterloo, ON). Format: Smooth jazz. News staff: one. Target aud: 35-45. ♦Brian Wortley, opns mgr & gen sls mgr; Jay Thomas, prom dir; Barry Horne, progmg dir.

CKIC-FM— March 2004: 92.9 mhz; 201 w. Hrs open: W302-160 Princess St., R3B 1K9. Phone: (204) 949-8480. Fax: (204) 949-0057. Licensee: Red River College Radio. Format: Var. ♦Rick Everett, stn mgr.

CKJS(AM)— Mar 25, 1975: 810 khz; 10 kw-U, DA-1. Hrs open: 520 Corydon Ave., R3L 0P1. Phone: (204) 477-1221. Fax: (204) 453-8244. E-mail: info@ckjs.com Web Site: www.ckjs.com. Licensee: Newcap Radio Manitoba Inc. (acq 4-30-2006; C$2.3 million). Rgnl rep: Direct. Format: Ethnic, Christian. Target aud: General. Spec prog: Ger 6 hrs, It 5 hrs, Pol 7 hrs, Sp 3 hrs, Por 8 hrs, Greek one hr, Filipino 20 hrs wkly. ♦K.E. Klein, gen mgr; Michael George, prom dir; Gido Gigliotti, progmg dir.

CKMM-FM— Feb 14, 1980: 103.1 mhz; 70 kw. Ant 676 ft Stereo. Hrs open: 177 Lombard Ave., 3rd. Fl., R3B 0N5. Phone: (204) 944-1031. Fax: (204) 943-7687. E-mail: staylor@hotqx.com Web Site: www.hot103live.com. Licensee: Standard Radio Inc. Group owner: Standard Broadcasting Corp. (acq 2002; grpsl). Format: CHR. ♦Sharon Taylor, gen mgr; Curtis Strange, progmg dir; Jaxon Hawks, mus dir.

***CKUW-FM**— Mar 1, 1999: 95.9 mhz; 450 w. TL: N49 52 51 W97 08 56. Stereo. Hrs open: 24 515 Portage Ave., Rm. 4C M11, R3B 2E9. Phone: (204) 786-9782. Fax: (204) 783-7080. E-mail: ckuw@uwinnipeg.ca Web Site: www.ckuw.ca. Licensee: The Winnipeg Campus/Community Radio Society. Population served: 660,000 Format: Urban contemp, rock/AOR, news/talk. News: 5 hrs wkly. Target aud: General; Our community. Spec prog: Children 2 hrs, class 4 hrs, folk 10 hrs, jazz 10 hrs wkly. ♦Rob Schmidt, stn mgr; Robin Eriksson, progmg dir; Don Baily, mus dir.

CKY-FM— Jan 21, 2004: 102.3 mhz; 70 kw. Hrs open: 166 Osborne St., Unit 4, R3L 1Y8. Phone: (204) 788-3400. Fax: (204) 788-3401. E-mail: geoff.poulton@winnipegradio.rogers.com Web Site: www.102clearfm.com. Licensee: Rogers Broadcasting Ltd. (group owner). Format: Adult contemp. ♦Geoff Poulton, VP & gen mgr; Gayle Zarbatany, progmg dir; Craig Pfeifer, mus dir.

New Brunswick

Balmoral

CIMS-FM— 1994: 103.9 mhz; 7.295 kw. Hrs open: 1991 Ave. CP2561, des Pionniers, E8E 2W7. Phone: (506) 826-1040. Fax: (506) 826-2400. E-mail: cimsfm@nbnet.nb.ca Web Site: www.cimsfm.ca. Licensee: Cooperative Radio Restigouche Ltee. Format: Var. ♦Pierre Morais, gen mgr; Camille Deschenes, rgnl sls mgr.

Bathurst

CKBC-FM— Jan 22, 2004: 104.9 mhz; 20 kw. Stereo. Hrs open: 24 176 Main St., E2A 1A4. Phone: (506) 547-1360. Fax: (506) 547-1367. Licensee: Astral Media Radio Atlantic Inc. Group owner: Astral Media Inc. Population served: 107,000 Natl. Rep: Canadian Broadcast Sales. Format: Adult contemp. News staff: 3; News: 9 hrs wkly. Target aud: 25-49. Spec prog: Fr 9 hrs wkly. ♦Jacques Parisien, pres; John Eddy, exec VP; Jamie Robichaud, gen mgr; Pat Brenan, sls VP.

CKLE-FM— Mar 29, 1990: 92.9 mhz; 100 kw. Stereo. Hrs open: 24 195 Main St., E2A 1A7. Phone: (506) 546-4600. Phone: (506) 546-1122. Fax: (506) 546-6611. E-mail: ckleadmin@mb.aibn.com Licensee: Radio De LaBaie Ltd. Format: CHR. Spec prog: Jazz 2 hrs wkly. ♦Armand Roussy, gen mgr, gen sls mgr & progmg dir.

Blackville

CJFY-FM— 2004: 107.5 mhz; 45 w. Hrs open: 401 Main St., E9B 1T3. Phone: (506) 843-2208. Fax: (506) 843-2603. E-mail: news@life1075.com Web Site: www.life1075.com. Licensee: Miramichi Fellowship Center Inc. Format: Christian music. ♦John D. Stewart, CEO; Shaun Mackenzie, gen mgr.

Campbellton

CKNB(AM)— 1939: 950 khz; 10 kw-D, 1 kw-N, DA-2. Hrs open: 24 Box 340, 74 Water Street, E3N 3G7. Phone: (506) 753-4415. Fax: (506) 789-9505. E-mail: cknb@nb.sympatico.ca Licensee: Maritime Broadcasting System Ltd. Format: Adult contemp, country, CHR. News: 10 hrs wkly. Target aud: General. Spec prog: Fr 18 hrs wkly. ♦Merv Russell, pres; David Montgomery, gen mgr.

Caraquet

CJVA(AM)— Sept 15, 1977: 810 khz; 10 kw-U, DA-2. Hrs open: 18 195 Main St., Bathurst, E2A 1A7. Phone: (506) 727-4605. Fax: (506) 727-6611. Licensee: Radio Acadie Ltd. Natl. Rep: Canadian Broadcast Sales. Format: Div, adult contemp, MOR. Target aud: 25 plus. Spec prog: C&W 15 hrs wkly. ♦Rufino Landry, pres; Armand Roussy, gen mgr.

Edmundston

CFAI-FM— Jan 15, 1991: 101.1 mhz; 4 kw. TL: N47 23 25 W68 18 59. Stereo. Hrs open: 24 165 Blvd. Hebert, E3V 2S8. Phone: (506) 737-5060. Fax: (506) 737-5084. E-mail: cfai@101rock105.com Web Site: www.101rock105.com. Licensee: La Cooperative des Montagnes Ltee. Population served: 45,000 Format: Top-40, soft rock. News: 6 hrs wkly. Target aud: 12-50. ♦Ron Cromier, pres; Serge Parent, gen mgr & dev dir; Jacques Bard, gen sls mgr.

CJEM-FM— July 1998: 92.7 mhz; 40.75 kw. TL: N47 21 47 W68 17 21. Stereo. Hrs open: 24 174 Church St., E3V 1K2. Phone: (506) 735-3351. Fax: (506) 739-5803. E-mail: cjem@nbnet.nb.ca Licensee: Radio Edmundston Inc. Population served: 50,000 Natl. Rep: Canadian Broadcast Sales. Format: Hot adult contemp, CHR. News staff: 2; News: 6 hrs wkly. Target aud: 25-54. ♦Jean Marc Michaud, pres; Murillo Soucy, gen mgr.

Fredericton

CBZF-FM— 2004: 99.5 mhz; 3.2 kw. Hrs open: Box 2200, E3B 5G4. Secondary address: 1160 Regent St. E3B 5G4. Phone: (506) 451-4000. Fax: (506) 451-4170. Web Site: www.cbc.ca. Licensee: Canadian Broadcasting Corp. Natl. Network: CBC Radio One. Format: Div.

***CBZ-FM**— January 1978: 101.5 mhz; 100 kw. Stereo. Hrs open:
Rebroadcasts CBH-FM Halifax, NS 100%.
Box 2200, E3B 5G4. Secondary address: 1160 Regent St. E3B 5G4. Phone: (506) 451-4000. Fax: (506) 451-4170. Web Site: www.cbc.ca. Licensee: Canadian Broadcasting Corp. Natl. Network: CBC Radio Two. Format: Classics & beyond, news/talk. ♦Gary Arsenault, gen mgr.

CFRK-FM— 2005: 92.3 mhz; 93 kw. Hrs open: NewCap Inc., 745 Windmill Rd., Dartmouth, NS, B3B 1C2. Phone: (902) 468-7557. Fax: (902) 468-7558. Web Site: www.ncc.ca. Licensee: Newcap Inc. Format: Classic rock.

CFXY-FM—Listing follows CKHJ(AM).

CHSR-FM— Jan 24, 1961: 97.9 mhz; 250 w. Ant 157 ft Stereo. Hrs open: 7 AM-3 AM Box 4400, Student Union Bldg., Univ. of New Brunswick, E3B 5A3. Phone: (506) 453-4985. Fax: (506) 453-4999. E-mail: chsr@unb.ca Web Site: www.unb.ca/chsr. Licensee: CHSR Broadcasting Inc. Format: Alternative, div. News: 6 hrs wkly. Target aud: General. Spec prog: Fr 2 hrs, ethnic 5 hrs, American Indian one hr, class 6 hrs, jazz 6 hrs, Chinese 3 hrs wkly. ♦Tristis Ward, stn mgr; Linda Pelletier, dev mgr.

CIBX-FM— June 11, 1996: 106.9 mhz; 100 kw. (CP: 78 kw.). Hrs open: 24 206 Rookwood Ave., E3B 2M2. Phone: (506) 455-1069. Fax: (506) 452-2345. Licensee: Astral Media Radio Atlantic Inc. Group owner: Astral Media Inc. (acq 4-19-2002; grpsl). Natl. Rep: Canadian Broadcast Sales. Format: Lite rock. News staff: 4; News: 8 hrs wkly. Target aud: 25-54. ♦John Eddy, pres; Pat Brennan, gen mgr.

CIXN-FM— Apr 8, 2001: 96.5 mhz; 27 w. Hrs open: Joy FM, 60 Bishop Dr., E3C 1B2. Phone: (506) 454-9600. Fax: (506) 443-0991. E-mail: welcome@joyfm.ca Web Site: www.joyfm.ca. Licensee: The Joy FM Network Inc. Format: Christian. ♦Garth McCrea, gen mgr.

***CJPN-FM**— August 1997: 90.5 mhz; 1.56 kw. Stereo. Hrs open: 715 Priestman St., E3B 5W7. Phone: (506) 454-2576. Fax: (506) 453-3958. E-mail: cjpn@nbnet.nb.ca Web Site: www.cjpn.ca. Licensee: Radio Fredericton Inc. Format: Adult contemp. ♦Pierre Dumas, gen mgr.

CJRI-FM— May 18, 2005: 94.7 mhz; 50 w. Hrs open: 151 Main St., E3A 1C6. Phone: (506) 472-0947. Fax: (506) 459-8194. E-mail: gospel@cjri.fm Web Site: www.cjri.fm. Licensee: Faithway Communications Inc. Format: Southern gospel, country gospel, praise music. ♦Ross Ingram, pres.

CKHJ(AM)— Aug 19, 1977: 1260 khz; 10 kw-U, DA-N. Hrs open: 206 Rookwood Ave., E3B 2M2. Phone: (506) 451-9111. Fax: (506) 452-2345. Licensee: Astral Media Radio Atlantic Inc. Group owner: Astral Media Inc. Format: Country. Spec prog: Fr one hr wkly. ♦John Eddy, gen mgr; Pat Brennan, gen sls mgr.

CFXY-FM—Co-owned with CKHJ(AM). July 15, 1983: 105.3 mhz; 78 kw. 800 ft Stereo. Format: C&W, rock.

Fredericton Centre

CKTP-FM— 2002: 95.7 mhz; 50 w. Hrs open: 120 Paul St., Fredericton, E3A 2V8. Phone: (506) 474-1636. Phone: (506) 461-3750. Fax: (506) 454-7187. Web Site: cap.ic.gc.ca/nb/stmarys. Licensee: Maliseet Nation Radio Inc. Format: Talk. ♦Timothy Paul, gen mgr.

Grand Falls

CIKX-FM— 2001: 93.5 mhz; 5.3 kw. Hrs open: 399 Broadway Blvd., E3Z 2K5. Phone: (506) 473-9393. Fax: (506) 473-3893. E-mail: grdprod@radioatl.ca Licensee: Astral Media Radio Atlantic Inc. Group owner: Astral Media Inc. (acq 2003; grpsl). Format: Hot adult contemp. ♦Pat Brennan, gen mgr; Jacques Lafrance, gen sls mgr; Rick McGuire, progmg dir; Ian Scott, news dir.

CKMV-FM— August 2000: 95.1 mhz; 975 w. Stereo. Hrs open: 24
Rebroadcasts CJEM-FM Edmundston 100%.
174 Church St., Edmundston, E3V 1K2. Phone: (506) 735-3351. Fax: (506) 739-5803. E-mail: cjem@nbnet.nb.ca Licensee: Radio Edmundston Inc. Natl. Rep: Canadian Broadcast Sales. Format: Hot adult contemp, CHR. Target aud: 25-54. ♦Jean-Marc Michaud, pres; Murillo Soucy, gen mgr & gen sls mgr; Paul Clavette, progmg dir.

Kedgwick

***CFJU-FM**— 1991: 90.1 mhz; 3 kw. TL: N47 35 05 W67 21 47. Stereo. Hrs open: 24 C.P. 1043, E8B 1Z9. Phone: (506) 235-9000. Fax: (506) 235-9001. E-mail: cfjufm@nbnet.nb.ca Licensee: La Radio Communautaire des Hauts-Plateaux Inc. Population served: 7,000 Format: Div, Fr. News: 8 hrs wkly. Target aud: 25-55. Spec prog: Country 12 hrs wkly. ♦M. Victor St- Pierre, pres; Lucille Theriault, gen mgr.

Miramichi City

CFAN-FM— Jan 10, 2003: 99.3 mhz; 17.8 kw. Hrs open: 24 Box 338, E1V 3M4. Secondary address: 396 Pleasant St. E1V 1X3. Phone: (506) 622-3311. Fax: (506) 627-0335. E-mail: cfan@nb.sympatico.ca Web Site: www.993theriver.com. Licensee: Maritime Broadcasting System Ltd. Group owner: Maritime Broadcasting. Population served: 50,000 Format: Adult contemp. News staff: one; News: 8 hrs wkly. Target aud: General. Spec prog: Relg 4 hrs, folk 2 hrs wkly. ♦Brent Preston, gen mgr.

Moncton

CBA(AM)— 1939: 1070 khz; 50 kw-U. Hrs open: 24 Box 950, 250 University Ave., E1C 8N8. Phone: (506) 853-6666. Fax: (506) 853-6400. Web Site: www.cbc.ca. Licensee: CBC. Natl. Network: CBC Radio One. Format: News, current affrs.

CBA-FM— March 1982: 95.5 mhz; 68 kw.
Rebroadcasts CBH-FM Halifax, NS 100%.
Web Site: www.cbc.ca. Natl. Network: CBC Radio Two. Format: Class.

***CBAF-FM**— 1982: 88.5 mhz; 50 kw. Hrs open: 250 University Ave., E1C 5K3. Phone: (506) 853-6666. Fax: (506) 867-8000. Web Site: www.cbc.radio-canada.ca. Licensee: Radio Canada. Natl. Network: Radio Canada. Format: Var, news/talk. ♦Benoit Quenneville, gen mgr; Cynthia Boudreau, mktg mgr.

***CBAL-FM**— 1983: 98.3 mhz; 67.6 kw. Ant 577 ft TL: N46 08 41 W64 54 14. Stereo. Hrs open: Box 950, E1C 8N8. Phone: (506) 853-6666. Fax: (506) 853-6739. Web Site: www.cbc.radio-canada.ca. Licensee: Societe Radio Canada. Natl. Network: Espace Musique. Format: Var. ♦Benoit Quenneville, gen mgr; Claire Hendy, opns mgr & chief of opns.

CFQM-FM— 1976: 103.9 mhz; 25 kw. Hrs open: 1000 St. George Blvd., E1E 4M7. Phone: (506) 858-1220. Fax: (506) 858-1209. E-mail: magic104@radiomoncton.com Web Site: www.radiomoncton.com. Licensee: Maritime Broadcasting System Ltd. Format: Adult contemp. ♦Dan Barton, gen mgr.

CHOY-FM— Feb 19, 2001: 99.9 mhz; 9.5 kw. Hrs open: 1000 St. George Blvd., E1E 4M7. Phone: (506) 858-1220. Fax: (506) 858-1209. E-mail: choix@radiomoncton.com Web Site: www.radiomoncton.com. Licensee: CHOY-FM Ltee. Group owner: Maritime Broadcasting. (acq 12-19-2005). Format: Fr. ♦Dan Barton, gen mgr.

CITA-FM— January 2001: 105.1 mhz; 880 w. Hrs open: 101 Ilsley Ave., Unit 3, Dartmouth, NS, B3B 158. Phone: (506) 384-1059. Fax: (506) 854-8609. E-mail: harvest@nbnet.nb.ca Web Site: www.citafm.com. Licensee: International Harvesters for Christ Evangelistic Association Inc. Format: Christian music. ♦Jeff Lutes, pres & stn mgr; Paul Girvan, opns dir; Paul Girvan, progmg dir.

CJMO-FM— June 19, 1987: 103.1 mhz; 46.8 kw. Stereo. Hrs open: 24 27 Arsenault Ct., E1E 4J8. Phone: (506) 858-5525. Fax: (506) 858-5539. E-mail: c103@c103.com Web Site: www.c103.com. Licensee: Atlantic Stereo Ltd. Natl. Rep: Canadian Broadcast Sales. Format: Classic rock. News staff: 4; News: 5 hrs wkly. Target aud: 25-54; adults. Spec prog: Jazz 2 hrs wkly. ♦Mark Maheu, pres; David Murray, CFO & opns VP; Hilary Montbourquette, gen mgr.

CJXL-FM— November 2000: 96.9 mhz; 100 kw. Hrs open: 24 27 Arsenault Ct., E1E 4J8. Phone: (506) 858-5525. Fax: (506) 858-5539. E-mail: xl96@xl96.com Web Site: www.xl96.com. Licensee: Atlantic Stereo Ltd. Population served: 150,000 Format: Today's best country. Target aud: 25-54; adults. ♦Mark Maheu, pres; Hilary Montbourquette, gen mgr; Dave Murray, opns VP.

CKCW-FM— January 2001: 94.5 mhz; 19 kw. Hrs open: 1000 St. George Blvd., E1E 4M7. Phone: (506) 858-1220. Fax: (506) 858-1209. E-mail: k94@radiomoncton.com Web Site: www.radiomoncton.com. Licensee: Maritime Broadcasting System Ltd. Format: Today's newest music. Target aud: 18-44; adults. ♦Dan Barton, gen mgr.

CKNI-FM— Oct 11, 2005: 91.9 mhz; 70 kw. Hrs open: 70 Assomption Blvd., E1C 1A1. Phone: (902) 493-7133. Licensee: Rogers Broadcasting Ltd. Format: News/talk. ♦Rael Merson, pres; Jim Hamm, gen mgr.

CKOE-FM— November 2000: 107.3 mhz; 50 w. Ant 82 ft Hrs open: 3030 Mountain Rd., E1G 2W8. Phone: (506) 384-1009. Fax: (506) 383-9699. E-mail: x101fm@radiochristian.com Web Site: www.radiochristian.com. Licensee: Houssen Broadcasting Ltd. Format: Christian hit radio. ♦James Houssen, gen mgr & progmg dir; Don Houssen, gen sls mgr; Steve Raye, news dir.

CKUM-FM— 1982: 93.5 mhz; 250 w. 98 ft TL: N46 06 16 W64 46 57. Stereo. Hrs open: 24 Universite de Moncton, Centre etudiant, 2e etage, E1A 3E9. Phone: (506) 858-3750. Fax: (506) 858-4524. E-mail: routierm@umoncton.ca Web Site: www.radioj935.com. Licensee: Les Medias Acadiens Universitaires Inc. Population served: 100,000 Format: Var/div, Fr. News: one hr wkly. Target aud: 15-30. Spec prog: Jazz 4 hrs wkly. ♦Brian Gallant, pres; Justin Robichaud, VP; Michele Routier, gen mgr; Mylene Dugas, VP & gen sls mgr.

Pokemouche

CKRO-FM— 1988: 97.1 mhz; 50 kw. Hrs open: Radio Peninsule Inc., 142 Rt. 113, E8P 1K7. Phone: (506) 336-9706. Fax: (506) 336-9058. E-mail: info@ckro.ca Licensee: Radio Peninsule Inc. Format: Fr., MOR. Target aud: General. ♦Rachel Savoie, VP; Donald Noel, gen mgr, stn mgr & gen sls mgr; Marilyne McLaughlin, natl sls mgr & prom mgr.

Sackville

***CHMA-FM**— 1985: 106.9 mhz; 50 w. Stereo. Hrs open: Suite 303, 152A Main St., E4L 1B4. Phone: (506) 364-2221. Fax: (506) 536-4230. E-mail: chma@mta.ca Web Site: www.mta.ca/chma. Licensee:

Attic Broadcasting Ltd. (acq 8-24-00). Population served: 15,000 Format: Div. ♦Pierre Malloy, gen mgr & stn mgr.

Saint John

CBD-FM— April 1981: 91.3 mhz; 80 kw. Hrs open: Box 2358, E2L 3V6. Phone: (506) 632-7744. Fax: (506) 632-7761. Web Site: www.nb.cbc.ca. Licensee: Canadian Broadcasting Corp. Natl. Network: CBC Radio One. Format: Talk, info.

CFBC(AM)— Nov 21, 1946: 930 khz; 50 kw-U, DA-2. TL: N45 13 55 W66 06 15. Stereo. Hrs open: 24 226 Union St., E2L 1B1. Phone: (506) 658-5100. Fax: (506) 658-5116. Licensee: Maritime Broadcasting System Ltd. Group owner: Maritime Broadcasting (acq 9-29-98; C$2 million with co-located FM). Population served: 85,000 Format: Oldies. ♦Mark Lee, gen mgr; Dave Clarkson, gen sls mgr; Dan Roman, progmg dir; Brian McLain, news dir; Roger Vautour, chief of engrg.

CJYC-FM—Co-owned with CFBC(AM). Mar 12, 1965: 98.9 mhz; 12 kw. 350 ft TL: N45 18 49 W66 04 43.24 Format: Classic rock. Target aud: 25-34; young, upwardly, mobile, family oriented. ♦Paul Jensen, progmg dir.

CFMH-FM— January 2001: 107.3 mhz; 250 w. Hrs open: c/o Student Services, UNB Saint John, Box 5050, E2L 4L5. Phone: (506) 648-5667. Fax: (506) 648-5541. E-mail: cfmh@unbsj.ca Web Site: www.cfmhradio.com. Licensee: Campus Radio Saint John Inc. Format: Var. ♦Linda Pellties, stn mgr.

CHNI-FM— Oct 11, 2005: 88.9 mhz; 79 kw. Hrs open: 55 Waterloo St., E2L 4V9. Phone: (902) 493-7133. Licensee: Rogers Broadcasting Ltd. Format: News/talk. ♦Rael Merson, pres; Jim Hamm, gen mgr.

CHQC-FM— 2006: 105.7 mhz; 1.85 kw. Hrs open: 24 67 Chemin Ragged Point, E2K 5C3. Phone: (506) 643-6996. Fax: (506) 658-3984. E-mail: info@chqc.ca Licensee: Cooperative radiophonique - La Brise de la Baie Ltee. Format: French variety. ♦Steve Pilotte, pres; Nay Saade, gen mgr, sls & mktg.

CHSJ-FM— Jan 7, 1998: 94.1 mhz; 50.4 kw. Hrs open: 24 Box 2000, 58 King St., E2L 3T4. Phone: (506) 633-3323. Fax: (506) 644-3485. E-mail: chsj@radioabl.com Licensee: Acadia Broadcasting Ltd. (group owner). Natl. Network: CBS Radio. Format: C&W. Target aud: 25-54. ♦Jim MacMullin, gen mgr.

CHWV-FM— 2001: 97.3 mhz; 55 kw. Hrs open: 24 Box 2000, E2L 3T4. Phone: (506) 633-3323. Fax: (506) 644-3485. E-mail: thewave@radioabl.com Licensee: Acadia Broadcasting Ltd. (group owner). Format: Adult contemp. ♦Jim MacMullin, VP.

CINB-FM— Nov 16, 2000: 96.1 mhz; 50 w. Hrs open: NewSong FM, Box 96, E2L 3X1. Phone: (506) 657-9600. Fax: (506) 657-7664. E-mail: programming@newsongfm.com Web Site: www.newsongfm.com. Licensee: New Song Communications Ministries Ltd. (acq 10-18-2005). Format: Contemp Christian. ♦Don Mabee, stn mgr.

CIOK-FM— Aug 10, 1987: 100.5 mhz; 100 kw. 1,650 ft Hrs open: 24 226 Union St., E2L 1B1. Phone: (506) 658-5100. Fax: (506) 658-5116. E-mail: mlee@nb.aibn.com Licensee: Maritime Broadcasting System Ltd. Group owner: Maritime Broadcasting Format: Adult contemp. News staff: 5; News: 7 hrs wkly. Target aud: 25-49; housewives, families, professionals. Spec prog: Real radio 4 hrs wkly. ♦Merv Russell, pres; Mark Lee, gen mgr.

CJEF-FM— Oct 20, 2003: 103.5 mhz; 49.6 w. Ant 200 ft TL: N45 16 31 W65 04 25. (CP: Ant 102 ft). Hrs open: 24 28 King St., Suite 3E, E2L 1G3. Phone: (506) 657-2533. Fax: (506) 642-7408. E-mail: OnAir@thepirate.ca Web Site: thepirate.ca. Licensee: TFG Communications Inc. Format: Comedy, urban, rock. Target aud: 18-34. ♦Geoffrey Rivett, CEO; Gary Stackhouse, gen mgr & progmg dir; John Kierstead, gen sls mgr; Mark Henwood, mus dir; Kathy Stackhouse, traf mgr.

Saint Stephen

CHTD-FM— May 31, 2001: 98.1 mhz; 40 kw. Stereo. Hrs open: 24 112 Milltown Blvd., St. Stephen, E3L 1G6. Phone: (506) 466-1000. Fax: (506) 466-4500. E-mail: mail@thetide.ca Web Site: www.thetide.ca. Licensee: Acadia Broadcasting Ltd. (group owner). Population served: 61,000 Format: Country. News staff: 2. Target aud: 25-54. ♦Jim MacMullin, gen mgr; John Higgins, stn mgr.

Shediac

CJSE-FM— July 26, 1994: 89.5 mhz; 20.445 kw. Hrs open: 96 Rue Providence, E4P2M9. Phone: (506) 532-0080. Fax: (506) 532-0120. E-mail: gilles@cjse.ca Web Site: www.cjse.ca. Licensee: Radio Beausejour Inc. Format: Var. ♦Gilles Arseneault, gen mgr; John Richard, sls VP.

Sussex

CJCW(AM)— June 1975: 590 khz; 1 kw-D, 250 w-N, DA-2. Hrs open: Box 5900, E4E 4N3. Phone: (506) 432-2529. Fax: (506) 433-4900. E-mail: cjcw@nbnet.nb.ca Licensee: Maritime Broadcasting System Ltd. Group owner: Maritime Broadcasting Format: Adult contemp. Target aud: General. Spec prog: Relg 4 hrs wkly. ♦Robert Pace, CEO; Merv Russell, pres; Louis McNamara, gen mgr.

Woodstock

CJCJ-FM— June 1, 2001: 104.1 mhz; 10 kw. Hrs open: Unit One, 131 Queen St., E7M 2M8. Phone: (506) 325-3030. Fax: (506) 325-3031. E-mail: wskprod@radioatl.ca Licensee: Astral Media Radio Atlantic Inc. Group owner: Astral Media Inc. (acq 4-19-2002; grpsl). Format: Adult contemp. ♦Pat Brennan, gen mgr.

Newfoundland

Argentia

CFOZ-FM— 1980: 100.3 mhz; 5 kw. Hrs open: 24
Rebroadcasts CHOZ-FM St. John's.
c/o CHOZ-FM, Box 2020, 446 Logy Bay Rd., St. John's, A1C 5S2. Phone: (709) 726-2922. Fax: (709) 726-3300. Web Site: www.ozfm.com. Licensee: Newfoundland Broadcasting Co. Ltd. Population served: 500,000 Natl. Rep: Canadian Broadcast Sales. Format: CHR, adult contemp, classic rock. News staff: 2. Target aud: 18-49. ♦Geoff Stirling, chmn; Frank Collins, CFO; Doug Neal, gen mgr, chief of opns & chief of engrg; Lorraine Pope, sls dir, gen sls mgr & progmg dir; Jesse Stirling, mktg VP & mktg dir; Scott G. Stirling, CEO, pres & progmg dir; Maurice Fitzgerald, mus dir; Larry Davis, news dir.

Baie Verte

CKIM(AM)— 1979: 1240 khz; 1 kw-D, 500 w-N. Hrs open: Box 620, Grenfell Heights, A2A 2K2. Phone: (709) 489-2192. Fax: (709) 489-8626. Web Site: www.vocm.com. Licensee: NewCap Inc. Group owner: NewCap Broadcasting Ltd. (acq 6-00; grpsl). Format: Rock. ♦Dave Hillier, gen mgr; Denn Dillion, gen sls mgr; Richard King, progmg dir; Roger Barnett, news dir; Harold Steele, chief of engrg.

Bonavista Bay

CBGY(AM)— Aug 25, 1977: 750 khz; 10 kw-U, DA-2. Hrs open:
Rebroadcasts CBG(AM) Gander.
c/o Radio Stn CBG, Box 369, Gander, A1V 1W7. Phone: (709) 256-4311. Fax: (709) 651-2021. Licensee: CBC. Format: News, current affrs. ♦Robert Rabinowitz, CEO; Chris Norman, gen mgr; Larry O'Brian, stn mgr.

CJOZ-FM— 1979: 92.1 mhz; 50 kw. Hrs open: 24
Rebroadcasts CHOZ-FM, St John's.
c/o CHOZ-FM, 446 Logy Bay Rd., Box 2020, St. John's, A1C 5S2. Phone: (709) 726-2922. Fax: (709) 726-3300. Web Site: www.ozfm.com. Licensee: Newfoundland Broadcasting Co. (group owner) Population served: 500,000 Natl. Rep: Canadian Broadcast Sales. Format: CHR, adult contemp, classic rock. News staff: 2. Target aud: 18-49. ♦Geoff Stirling, chmn; Scott Stirling, CEO & pres; Frank Collins, CFO; Doug Neal, gen mgr, chief of opns & chief of engrg; Jesse Stirling, gen sls mgr & mktg dir; Lorraine Pope, gen sls mgr & progmg dir; Maurice Fitzgerald, mus dir; Larry Davis, news dir.

Carbonear

CHVO(AM)— Oct 7, 1980: 560 khz; 5 kw-U, DA-1. Hrs open: 24 One CHVO Dr., A1Y 1A2. Phone: (709) 596-1560. Fax: (709) 596-8626. E-mail: chvo@vocm.com Licensee: NewCap Inc. Group owner: NewCap Broadcasting Ltd. (acq 6-00; grpsl). Format: Country. Spec prog: Irish/Newfoundland 8 hrs wkly. ♦John Steele, pres; John Murphy, gen mgr; Aiden Hibbs, stn mgr; Ron Ryan, gen sls mgr & prom mgr; Ken Ash, progmg dir; Gerry Phalen, news dir; Harold Steele, chief of engrg.

Churchill Falls

CFLC-FM— 1974: 97.9 mhz; 8 w. 50 ft Hrs open: c/o CFCB(AM), Box 570, Corner Brook, A2H 6H5. Phone: (709) 634-3111. Fax: (709) 634-4081. Licensee: NewCap Inc. Group owner: NewCap Broadcasting Ltd. (acq 4-2-01; grpsl). Format: Country. ♦Michael Murphy, gen mgr & stn mgr; Ken Ash, progmg dir & progmg mgr.

Clarenville

CJKK-FM— 1988: 105.3 mhz; 2.07 kw. Hrs open: 24 c/o CHOZ-FM, 446 Logy Bay Rd., St. John's, A1C 5R6. Phone: (709) 726-2922. Fax: (709) 726-3300. Web Site: www.ozfm.com. Licensee: Newfoundland Broadcasting Co. (group owner) Population served: 500,000 Natl. Rep: Canadian Broadcast Sales. Format: CHR, adult contemp, classic rock. News staff: 2. Target aud: 18-49. ♦Geoff Stirling, chmn; Scott Stirling, CEO & pres; Frank Collins, CFO; Doug Neal, gen mgr, opns mgr & chief of opns.

CKVO(AM)— Nov 15, 1974: 710 khz; 10 kw-U. Hrs open:
Rebroadcasts VOCM(AM) St. John's except 9 AM-5 PM (loc progmg).
c/o VOCM(AM), Box 8590, Stn. A, 391 Kenmount Rd., St. John's, A1B 3P5. Phone: (709) 466-2710. Fax: (709) 726-8626/726-4633. E-mail: feedback@vocm.com Web Site: www.vocm.com. Licensee: NewCap Inc. Group owner: NewCap Broadcasting Ltd. (acq 6-00; grpsl). Format: Contemp country. ♦John Murphy, gen mgr; Ken Ash, opns mgr; Dennis Dillon, gen sls mgr; Paul Raynes, progmg dir; Gerry Phelan, news dir; Harold Steele, chief of engrg.

Corner Brook

***CBY(AM)**— Apr 1, 1949: 990 khz; 10 kw-U, DA-1. Hrs open:
Rebroadcasts CBT(AM) Grand Falls-Windsor.
Box 610, A2H 6G1. Phone: (709) 637-1150. Fax: (709) 634-8506. Licensee: CBC. Natl. Network: CBC Radio One. Format: News/talk, var. Target aud: 30 plus; mature. ♦Robert Rabinowitz, CEO; Gordon Lannon, gen mgr.

CFCB(AM)— Oct 3, 1960: 570 khz; 1 kw-D. Hrs open: 24 Box 570, A2H 6H5. Phone: (709) 634-4570. Fax: (709) 634-4081. Licensee: Newcap Inc. Group owner: NewCap Broadcasting Ltd. (acq 4-2-01; grpsl). Population served: 40,000 Natl. Rep: Canadian Broadcast Sales. Format: Country. ♦Harry Steele, pres; J. Steele, VP; Michael Murphy, gen mgr; Darryl Stevens, opns mgr & progmg dir.

CKOZ-FM— 1979: 92.3 mhz; 50 kw. Hrs open: 24
Rebroadcasts CHOZ-FM St. John's.
c/o CHOZ-FM, 446 Logy Bay Rd., Box 2020, St. John's, A1C 5S2. Phone: (709) 726-2922. Fax: (709) 726-3300. Web Site: www.ozfm.com. Licensee: Newfoundland Broadcasting Co. (group owner) Population served: 500,000 Natl. Rep: Canadian Broadcast Sales. Format: CHR, adult contemp, classic rock. News staff: 2. Target aud: 18-49. ♦Geoff Stirling, chmn; Scott Stirling, CEO & pres; Frank Collins, CFO; Doug Neal, gen mgr, chief of opns & chief of engrg; Lorraine Pope, gen sls mgr & progmg dir; Jesse Stirling, mktg dir; Maurice Fitzgerald, mus dir; Larry Davis, news dir.

CKXX-FM— June 20, 1997: 103.9 mhz; 40 kw. Stereo. Hrs open: 24 P.O. Box570, 345 O'Connell Dr., Corner Brook, NL, A2H6H5. Phone: (709) 634-4570. Fax: (709) 634-4081. Web Site: k-rock1039.com. Licensee: NewCap Inc. Group owner: NewCap Broadcasting Ltd. (acq 8-29-90). Population served: 40,000 Natl. Rep: Canadian Broadcast Sales. Format: Rock. News staff: 2. Target aud: 25-54. Spec prog: Oldies 3 hrs wkly. ♦Michael Murphy, gen mgr; Daryl Stevens, opns mgr.

Gander

***CBG(AM)**— 1949: 1400 khz; 4 kw-U. Hrs open: Box 369, A1V 1W7. Phone: (709) 256-4311. Fax: (709) 651-2021. Licensee: CBC. Format: Info. ♦Robert Rabinowitz, CEO; Larry O'Brien, gen mgr.

CKGA(AM)— 1969: 650 khz; 5 kw-U, DA-2. Hrs open: Box 650, A1V 1X2. Phone: (709) 651-3650. Fax: (709) 651-2542. Licensee: NewCap Inc. Group owner: NewCap Broadcasting Ltd. (acq 6-00; grpsl). ♦Dave Hillier, gen mgr; Dennis Dillon, gen sls mgr; Dean Clarke, progmg dir; Robet Tuck, news dir; Harold Steele, chief of engrg.

CKXD-FM— November 2000: 98.7 mhz; 6 kw. Hrs open: Box 650, A1V 1X2. Phone: (709) 651-3650. Fax: (709) 651-2542. E-mail: ckxd.ckga.psa@vocm.com Web Site: www.vocm.com. Licensee: Newcap Inc. Group owner: NewCap Broadcasting Ltd. Format: Classic rock. Spec prog: Newfoundland & Irish 12 hrs wkly. ♦John Murphy, gen mgr; Dennis Dillon, gen sls mgr; Ken Ash, progmg dir; Harold Steele, chief of engrg; Connie Pasher, traf mgr.

Goose Bay

CFGB-FM—See Happy Valley

CFLN(AM)— Aug 1, 1974: 1230 khz; 1 kw-D, 250 w-N, DA-1. Hrs open: Box 160, Station C, Happy Valley Goose Bay, A0P-1C0. Phone: (709) 896-2968. Fax: (709) 896-8708. Licensee: NewCap Inc. Group owner: NewCap Broadcasting Ltd. (acq 4-2-01; grpsl). Format: Adult contemp. ♦Harry Steele, chmn; Robert G. Steele, pres.

Grand Falls

CKCM(AM)— October 1962: 620 khz; 10 kw-U, DA-1. Hrs open: Box 620, Grand Falls-Windsor, A22 2K2. Secondary address: 35 A Grenfell Heights, Grand Falls-Windsor A2A 2K2. Phone: (709) 489-2192. Fax: (709) 489-8626. Web Site: www.wocm.com. Licensee: NewCap Inc. Group owner: NewCap Broadcasting Ltd. (acq 6-00; grpsl). Format: Contemp country. ♦Dave Hillier, gen mgr; Dennis Dillon, gen sls mgr; Richard King, progmg dir; Roger Barnett, news dir; Harold Steele, chief of engrg.

Grand Falls-Windsor

CBT(AM)— July 1, 1949: 540 khz; 10 kw-U. Hrs open: 2 Harris Ave., A2A 2Y4. Phone: (709) 489-2102. Fax: (709) 489-1055. Licensee: CBC. Format: Info. ♦Robert Rabinowitz, CEO; Diane Humber, gen mgr; Chris Norman, stn mgr.

CKXG-FM— 2001: 102.3 mhz; 20 kw. Hrs open: 24 Box 620, A2A 2K2. Phone: (709) 489-2192. Fax: (709) 489-8626. E-mail: vocm.krock.psa@vocm.com Web Site: www.vocm.com. Licensee: NewCap Broadcasting Ltd. Group owner: NewCap Broadcasting Ltd. Format: Classic rock. ♦Dave Hillier, gen mgr; Dennis Dillon, gen sls mgr; Richard King, progmg dir; Harold Steele, chief of engrg; Margot pitcher-Hamlyn, traf mgr.

Happy Valley

***CFGB-FM**— Feb 23, 1959: 89.5 mhz; 1 kw-w. Hrs open: Box 1029, Stn C, 12 Loring Dr., Happy Valley-Goose Bay, A0P 1CO. Phone: (709) 896-2911. Fax: (709) 896-8900. E-mail: labmorning@stjohns.cbc.ca Web Site: www.stjohns.cbc.ca. Licensee: CBC. Natl. Network: CBC Radio One. Format: Info. ♦Diane Humber, gen mgr; Cynthia Wall, progmg mgr; Lorne Burry, engrg mgr; Conrad Lutes, news rptr.

Labrador City

CBDQ-FM— 1997: 96.3 mhz; 255 w. Hrs open:
Rebroadcasts CFGB-FM Happy Valley.
Box 576, A2V 2L3. Phone: (709) 944-3616. Fax: (709) 944-5472. Web Site: www.stjohns.cbc.ca. Licensee: CBC. Format: Talk, info. ♦Diane Humber, gen mgr.

CJRM-FM— Sept 23, 1992: 97.3 mhz; 500 w. 1,998 ft TL: N52 57 01 W66 55 01. Stereo. Hrs open: C.P. 453, 308 Hudson Dr., A2V 2K7. Phone: (709) 944-7600. Phone: (709) 944-2973. Fax: (709) 944-5125. E-mail: cjrm@crrstv.net Licensee: Radio Communautaire du Labrador Inc. (acq 9-4-92). Population served: 17,000 Format: Fr, English, div. Target aud: General; English & Fr speaking audience in Labrador West. ♦Norman Gillespie, pres; Dean Baker, exec VP & mus dir; linda McLean, stn mgr.

Lewisporte

CIFX-FM— 2002: 93.7 mhz; 50 w. Stereo. Hrs open: 24 Box 601, A0G 3A0. Secondary address: 37 George Street AOG 3AO. Phone: (709) 535-6000. Phone: (709) 535-2546. Fax: (709) 535-6600. E-mail: mixfm@nf.sympatico.ca Licensee: Mix FM Inc. Population served: 25,000 Format: Talk, hot adult contemp, CHR. News staff: 2. Target aud: 18-49. ♦Todd Foss, stn mgr, mus dir & progmg; Vicki Fudge, pub affrs dir; Koren Hurley, sls; Angela Brenton, prom; Peter Ginn, engr.

Marystown

CHCM(AM)— 1961: 740 khz; 10 kw-U, DA-N. Hrs open: Box 560, A0E 2M0. Phone: (709) 279-2560. Phone: (709) 279-2426. Fax: (709) 279-3538. Fax: (709) 279-2800. E-mail: chem.frontdesk@vocm.com Licensee: NewCap Inc. Group owner: NewCap Broadcasting Ltd. (acq 5-4-00; grpsl). Format: Country. ♦Russell Murphy, gen mgr, opns mgr & gen sls mgr; Harry Myles, progmg dir; Bob Tower, news dir; Harold Steele, chief of engrg.

CIOZ-FM— 1979: 96.3 mhz; 25 kw. Hrs open: 24
Rebroadcasts CHOZ-FM St. John's.
c/o CHOZ-FM, 446 Logy Bay Rd., Box 2020, St. John's, A1C 5S2. Phone: (709) 726-2922. Fax: (709) 726-3300. Web Site: www.ozfm.com. Licensee: Newfoundland Broadcasting Co. (group owner) Population served: 500,000 Natl. Rep: Canadian Broadcast Sales. Format: CHR, adult contemp, classic rock. News staff: 2. Target aud: 18-49. ♦Geoff Stirling, chmn; Scott Stirling, CEO & pres; Frank Collins, CFO; Doug Neal, gen mgr, chief of opns & chief of engrg; Lorraine Pope, gen sls mgr & progmg dir; Jesse Stirling, mktg VP, mktg dir & prom VP; Maurice Fitzgerald, mus dir & asst music dir; Larry Davis, news dir.

Mount Pearl

***VOAR(AM)**— 1930: 1210 khz; 10 kw, DA-1. TL: N47 32 01 W52 49 01. Hrs open: 24 1041 Topsail Rd., A1N 5E9. Phone: (709) 745-VOAR. Fax: (709)745-1600. E-mail: voar@voar.org Web Site: www.voar.org. Licensee: Seventh-Day Adventist Church in Newfoundland. Population served: 350,000 Format: Relg, gospel. News: 12 hrs wkly. Target aud: 25-44; individuals interested in family life & traditional values. ♦Gary Hodder, pres; Sherry Griffin, stn mgr & opns mgr.

Port au Choix

CFNW(AM)— 1960: 790 khz; 1 kw-U, DA-1. Hrs open: 24 c/o CFCB(AM), 345 O'Connell Dr., Corner Brook, A2H 6H5. Phone: (709) 634-4570. Fax: (709) 634-4081. Licensee: Newcap Inc. Group owner: NewCap Broadcasting Ltd. (acq 4-2-01; grpsl). Population served: 22,000 Format: Country. ♦Harry Steele, pres; Michael Murphy, gen mgr; Darryl Stevens, opns mgr.

Rattling Brook

CHOS-FM— 1979: 95.9 mhz; 50 kw. Hrs open: 24
Rebroadcasts CHOZ-FM St. John's.
c/o CHOZ-FM, 446 Logy Bay Rd., Box 2020, St. John's, A1C 5S2. Phone: (709) 726-2922. Fax: (709) 726-3300. Web Site: www.ozfm.com. Licensee: Newfoundland Broadcasting Co. (group owner) Population served: 500,000 Natl. Rep: Canadian Broadcast Sales. Format: CHR, adult contemp, classic rock. News staff: 2. Target aud: 18-49. ♦Geoff Stirling, chmn; Scott G. Sterling, CEO & pres; Frank Collins, CFO; Doug Neal, gen mgr & chief of opns.

Red Rocks

CKSS-FM— 1994: 96.9 mhz; 520 w. Hrs open: 24
Rebroadcasts CHOZ-FM St. John's 100%.
c/o CHOZ-FM, 466 Logy Bay Rd., Box 2020, St. John's, A1C 5S2. Phone: (709) 726-2922. Fax: (709) 726-3300. Web Site: www.ozfm.com. Licensee: Newfoundland Broadcasting Co. (group owner) Population served: 500,000 Natl. Rep: Canadian Broadcast Sales. Format: CHR, adult contemp, classic rock. News staff: 2. Target aud: 18-49. ♦Geoff Stirling, chmn; Scott Stirling, CEO & pres; Frank Collins, CFO; Doug Neal, gen mgr, chief of opns & chief of engrg; Lorraine Pope, gen sls mgr & progmg dir; Jesse Stirling, mktg dir; Larry Davis, news dir.

Saint Andrews

CFCV-FM— 1974: 97.7 mhz; Hrs open:
Rebroadcasts CFGN(AM) Port-aux-Basques.
60 West St., Stephenville, A2N 1C6. Secondary address: C/o CFGN(AM), Port-aux-Basques A0M 1C0. Phone: (709) 695-2183. Phone: (709) 643-2191. Fax: (709) 695-9614. E-mail: cfsx@vocm.com Web Site: www.vocm.com. Licensee: NewCap Inc. Group owner: NewCap Broadcasting Ltd. Format: Country, oldies. ♦Michael Murphy, gen mgr; Gerry Murphy, gen sls mgr; Larry Bennett, progmg dir; Harold Steele, chief of engrg.

Saint John's

***CBN(AM)**— Apr 1, 1949: 640 khz; 10 kw-U. Hrs open: 19 Box 12010, Stn A, 342-44 Duckworth St., A1B 3T8. Phone: (709) 576-5000. Fax: (709) 576-5205. Web Site: www.cbc.ca. Licensee: CBC. Natl. Network: CBC Radio One. Format: Div. News staff: 6; News: 10 hrs wkly. Target aud: General. Spec prog: Fisheries 3 hrs wkly. ♦Diane Humber, gen mgr & news dir; Larry O'Brien, opns dir; Lori Wheeler, prom mgr; Liz Lacey, mus dir.

CBN-FM— July 1, 1975: 106.9 mhz; 100 kw. 300 ft Stereo. Natl. Network: CBC Radio Two. Format: Class, news/talk. Co-owned TV: *CBNT-TV affil

***CHMR-FM**— January 1986: 93.5 mhz; 50 w. -10 ft Stereo. Hrs open: 24 Memorial University, MUNSU, South Annex, Rm 2009, A1C 5S7. Phone: (709) 737-4777. Phone: (709) 737-4778. Fax: (709) 737-7688. E-mail: chmr@mun.ca Web Site: www.mun.ca/chmr. Licensee: Memorial University of Newfoundland Radio Society Inc. Population served: 200,000 Format: Div, alternative. News: 7 hrs wkly. Target aud: General. Spec prog: Fr 2 hrs, jazz 6 hrs, relg 4 hrs, blues 6 hrs, rap 2 hrs, reggae 2 hrs, Indian one hr wkly.Kathy Rowe, gen mgr, opns mgr, dev mgr, rgnl sls mgr, adv mgr, mus dir, traf mgr, edit dir, edit mgr & women's cmtr; Ernst Rollmann, progmg dir, asst music dir, rsch dir, local news ed, mus critic, political ed & relg ed; Nancy Earle, mktg mgr, prom mgr, progmg dir, asst music dir & news dir; Craig Peterman, chief of engrg; Jay Healey, min affrs dir; Mike Rossiter, news rptr & sports cmtr; Erin McKee, women's int ed

CHOZ-FM— June 15, 1977: 94.7 mhz; 100 kw. 821 ft TL: NN47 31 36 W52 42 50. Stereo. Hrs open: 24 Box 2050, 446 Logy Bay Rd., A1C 5R6. Phone: (709) 722-5015. Fax: (709) 726-3300. E-mail: ntvsales@ntv.ca Web Site: www.ozfm.com. Licensee: Newfoundland Broadcasting Co. Ltd. Group owner: Newfoundland Broadcasting Co. Population served: 500,000 Natl. Rep: Canadian Broadcast Sales. Format: CHR, adult contemp, classic rock. News staff: 2. Target aud: 18-49. ♦Geoff Stirling, chmn; Scott G. Stirling, CEO & pres; Frank Collins, CFO; Jesse Stirling, VP, sls dir, mktg VP, mktg dir & prom mgr; Doug Neal, gen mgr, opns mgr, chief of opns, engrg dir & chief of engrg; Dave Lawrence, stn mgr; Lorraine Pope, natl sls mgr & mus dir; Paul Kinsman, progmg dir; Larry Davis, news dir. Co-owned TV: CJON-TV affil.

CJYQ(AM)—Listing follows CKIX-FM.

CKIX-FM— Oct 15, 1983: 99.1 mhz; 100 kw. Ant 930 ft Stereo. Hrs open: Box 8590, Station A, St. John's, A1B 3P5. Phone: (709) 726-5590. Fax: (709) 726-4633. Web Site: www.991hitsfm.com. Licensee: Newcap Inc. Group owner: NewCap Broadcasting Ltd. (acq 1-17-83). Format: CHR. ♦John Murphey, gen mgr; Randy Snow, progmg dir; Brad Michaels, mus dir.

CJYQ(AM)—Co-owned with CKIX-FM. 1951: 930 khz; 50 kw-U. Format: Country.

CKSJ-FM— 2004: 101.1 mhz; 20 kw. Hrs open: Box 28106, A1B 4J8. Phone: (709) 754-6748. Fax: (709) 754-6749. E-mail: onair@coast1011.com Web Site: www.coast1011.com. Licensee: Coast Broadcasting Ltd. Format: Adult contemp. ♦Andrew Newman, gen mgr.

VOCM(AM)— Oct 19, 1936: 590 khz; 20 kw-U, DA-2. Stereo. Hrs open: 24 391 Kenmount Rd., A1B 3P5. Phone: (709) 726-5590. Fax: (709) 726-4633. E-mail: feedback@vocm.com Web Site: www.vocm.com. Licensee: NewCap Inc. Group owner: NewCap Broadcasting Ltd. (acq 6-00; grpsl). Natl. Rep: Canadian Broadcast Sales. Format: Adult contemp, country, news/talk. News staff: 16. Target aud: 25 plus. ♦John Steele, pres; John Murphy, gen mgr; Ken Ash, opns mgr; Ron Ryan, sls VP; Paul Magee, progmg dir; Gerry Phelan, news dir; Harold Steele, engrg dir; Cathy Ridgely-Ryan, traf mgr.

VOCM-FM— September 1982: 97.5 mhz; 100 kw. Stereo. 24 Web Site: www.k-rock975.com. Format: Classic rock. News: 7 hrs wkly. Target aud: Adults 25-54. ♦Ron Ryan, stn mgr.

***VOWR(AM)**— June 20, 1924: 800 khz; 10 kw-D, 2.5 kw-N, DA-1. TL: N47 34 16 W52 45 13. Hrs open: 24 Box 7430, Patrick St., St. John's, A1E 3Y5. Phone: (709) 579-9233. Fax: (709) 579-9232. E-mail: vowr@vowr.org Web Site: www.vowr.org. Licensee: Wesley United Church Radio Board. Population served: 200,000 Format: Div, oldies, folk. Target aud: 40 plus. Spec prog: Relg 10 hrs, folk 15 hrs wkly. ♦Marvin Barnes, chmn; John Tessier, gen mgr & opns mgr.

Stephenville

CFSX(AM)— Nov 14, 1964: 870 khz; 500 w-U. Hrs open: 24 60 West St., A2N 1C6. Phone: (709) 643-2191. Fax: (709) 643-5025. Web Site: www.vocm.com. Licensee: NewCap Inc. Group owner: NewCap Broadcasting Ltd. (acq 2001; grpsl). Population served: 33,000 Format: Country. News staff: 2. Target aud: General. Spec prog: Relg one hr wkly. ♦Harry Steele, chmn; Robert G. Steele, pres; John Murphy, gen mgr; Gerry Murphy, stn mgr.

CIOS-FM— 1979: 98.5 mhz; 10 kw. Hrs open: 24
Rebroadcasts CHOZ-FM St. John's.
c/o CHOZ-FM, 446 Logy Bay Rd., Box 2020, St. John's, A1C 5S2. Phone: (709) 726-2922. Fax: (709) 726-3300. Web Site: www.ozfm.com. Licensee: Newfoundland Broadcasting Co. (group owner) Population served: 500,000 Format: CHR, adult contemp, classic rock. News staff: 2. Target aud: 18-49. ♦Geoff Stirling, chmn; Scott Stirling, CEO & pres; Frank Collins, CFO; Doug Neal, gen mgr, opns mgr & chief of opns.

Wabush

CFLW(AM)— 1971: 1340 khz; 250 w-U, DA-1. Hrs open:
Rebroadcasts CFCB(AM) Corner Brooks 99%.
Box 6000, 4 Grenfell Dr., A0R 1B0. Phone: (709) 282-3602. Phone: (709) 282-3601. Fax: (709) 282-5543. Licensee: Newcap Inc. Group owner: NewCap Broadcasting Ltd. (acq 4-2-01; grpsl). Format: Country. Target aud: General. ♦ Harry Steele, chmn; Robert G. Steele, pres; Mike Murphy, gen mgr.

Northwest Territories

Hay River

CJCD-FM-1— Sept 15, 1986: 100.1 mhz; 300 w. 175 ft Stereo. Hrs open:
Rebroadcasts CJCD(FM) Yellowknife.
Box 218, Yellowknife, X1A 2N2. Phone: (867) 920-4636. Fax: (867) 920-4033. E-mail: info@cjcd.ca. Web Site: www.cjcd.ca. Licensee: CJCD Radio Ltd. (acq 3-8-00). Natl. Rep: Canadian Broadcast Sales. Format: Hot adult contemp. Target aud: 25-49. ♦ Eileen Dent, pres & gen mgr; Tim Jaworski, gen sls mgr; Joanne Cochrane, progmg dir; Joanne McKenzie, news dir; Jim Pook, chief of engrg; Kirby Marshall, chief of engrg; Mandy Church, traf mgr.

CKHR-FM— January 1979: 107.3 mhz; 32 w. 185 ft Hrs open: Box 4394, X0E 1G3. Phone: (867) 874-2547. Web Site: www.tvradioworld.com. Licensee: Hay River Broadcasting Society. Format: MOR, var. ♦ Al Erickson, pres; Ray Lawson, stn mgr.

Inuvik

***CHAK(AM)**— Nov 26, 1960: 860 khz; 1 kw-D, DA-1. Hrs open: Bag 8, X0E 0T0. Phone: (867) 777-7600. Fax: (867) 777-7640. Licensee: CBC. Format: News/talk. Target aud: General. Spec prog: Inuvialuktun 9 hrs, Gwich'in 9 hrs wkly. ♦ Peter Skinner, gen mgr.

Tuktoyaktuk

CFCT(AM)— 1971: 600 khz; 1 kw-U. Hrs open: c/o Radio Station CHAK, Bag 8, Inuvik, X0E 0T0. Phone: (867) 777-7600. Fax: (867) 777-7640. Licensee: CBC. (acq 1982). Natl. Network: CBC Radio One. Format: Div. Spec prog: Eskimo 5 hrs wkly. ♦ Peter Skinner, gen mgr.

Yellowknife

***CFYK(AM)**— Dec 13, 1958: 1340 khz; 2.5 kw-U. Hrs open: Box 160, X1A 2N2. Phone: (867) 920-5400. Fax: (867) 920-5410 (ADMIN). Fax: (867) 920-5440 PROG. Licensee: CBC. Format: Div, news/talk. Target aud: General. Spec prog: Slavey 8 hrs, Dogrib 4 hrs, Chipewayan 4 hrs wkly. ♦ Peter Skinner, gen mgr; David McNaughton, opns mgr.

CIVR-FM— 2001: 103.5 mhz; 164 w. Hrs open:
Rebroardcasts RFA Ottawa 60%.
Box 1586, X1A 2P2. Phone: (867) 873-3292. E-mail: civr@franco-nord.com Web Site: www.radiotaiga.ca. Licensee: L'Association franco-culturelle de Yellowknife. Format: Fr. News: 2 hrs wkly. Spec prog: Worldbeat 3 hrs, jazz 4 hrs wkly. ♦ Jeff Hipfner, pres; Sylvie Boisclair, gen mgr.

CJCD-FM— 1998: 100.1 mhz; 400 w. TL: N62 27 00 W114 19 00. Hrs open: Box 218, X1A 2N2. Phone: (867) 920-4636. Fax: (867) 920-4033. E-mail: info@cjcd.ca Web Site: www.cjcd.ca. Licensee: CJCD Radio Ltd. (acq 3-8-00). Natl. Rep: Canadian Broadcast Sales. Format: Hot Adult contemp. Target aud: 25-54. ♦ Eileen Dent, CEO, pres & gen mgr.

CKLB-FM— Dec 11, 1985: 101.9 mhz; 130 w. 162 ft Stereo. Hrs open: 7 AM-10 PM (M-F); 11 AM-9 PM (S)
Rebroadcasts CFWE-FM Lac La Biche, Alberta News.
Box 1919, X1A 2P4. Phone: (867) 920-2277. Fax: (867) 920-4205. E-mail: ncs@internorth.com Licensee: Native Communications Society of the Western N.W.T. Population served: 30,000 Format: C&W. News staff: 3. Spec prog: Black one hr wkly. ♦ Elizabeth Biscaye, gen mgr.

Nova Scotia

Amherst

CKDH(AM)— Oct 25, 1957: 900 khz; 1 kw-U, DA-N. Hrs open: Box 670, B4H 4B8. Phone: (902) 667-3875. Fax: (902) 667-4490. E-mail: ckdh@ckdh.net Web Site: www.ckdh.net. Licensee: Maritime Broadcasting System Ltd. (acq 1989). Natl. Rep: Canadian Broadcast Sales. Format: Light rock. Target aud: 25-49 & 18-34; general. Spec prog: C&W 11 hrs, farm 2 hrs wkly. ♦ Dave March, gen mgr & progmg dir; Jeff DeGanz, news dir.

Antigonish

CFXU-FM— 2006: 92.5 mhz; 50 w. Hrs open: Box 948, St. Francis Xavier University, B2G 2X1. Phone: (902) 867-3941. E-mail: thefox@stfx.ca Web Site: www.theu.ca/comm/cfxu. Licensee: Radio CFXU Club. Format: Var. ♦ Caitlin Van Horne, stn mgr; Mike Fredericks, stn mgr; Rose Murphy, progmg mgr; Cameron Brioux, mus dir; John Best, news dir.

CJFX-FM— 2003: 98.9 mhz; 75.39 kw. Stereo. Hrs open: 24 Box 5800, B2G 2R9. Secondary address: 85 Kirk St. Phone: (902) 863-4580. Fax: (902) 863-6300. E-mail: cjfx@cjfx.ca Web Site: www.cjfx.ca. Licensee: Atlantic Broadcasters Ltd. Natl. Rep: Canadian Broadcast Sales. Rgnl rep: Canadian Broadcast Sales Format: CHR, maritime. News staff: 3; News: 24 hrs wkly. Target aud: 18+. ♦ David MacLean, gen mgr.

Barrington

CJLS-FM-1— 1982: 96.3 mhz; 5.5 kw. Hrs open:
Rebroadcasts CJLS-FM Yarmouth.
c/o CJLS(AM), 328 Main St., Suite 201, Yarmouth, B5A 1E4. Phone: (902) 742-7175. Fax: (902) 742-3143. E-mail: cjls@cjls.com Web Site: www.cjls.com. Licensee: Radio CJLS Ltd. (acq 7-1-98). Format: Adult contemp. ♦ Chris Perry, VP & progmg mgr; Ray Zinck, pres & gen mgr; Dave Hall, gen sls mgr; Gary Nickerson, news dir; Jim Harris, engrg dir.

Bedford

CHSB-FM— 2007: 89.1 mhz; 50 w. Hrs open: Box 44073, B4A 3X5. Phone: (902) 835-5966. Licensee: Bedford Baptist Church. Format: Religious services from the Bedford Baptist Church. ♦ Kevin Haggarty, gen mgr.

Bridgewater

CKBW-FM— February 2002: 98.1 mhz; 32 kw. Stereo. Hrs open: 24 215 Dominion St., B4V 2G8. Phone: (902) 543-2401. Fax: (902) 543-1208. E-mail: ckbw@ckbw.com Web Site: www.ckbw.com. Licensee: Acadia Broadcasting Ltd. (group owner). Population served: 80,000 Format: Hot adult contemp. News staff: 2; News: 11 hrs wkly. Target aud: General; rural & small town urban. ♦ John Wiles, gen mgr, opns mgr & progmg dir; Barry Smith, gen sls mgr & natl sls mgr; Brian Tepper, prom dir; Frank Grayney, chief of engrg; Pamela Smith, traf mgr; Greg Lowe, spec ev coord & sports cmtr; Sheldon MacLeod, news dir, pub affrs dir & local news ed; Jonathan Crouse, disc jockey; Leitha Mayson, disc jockey; Mike Richards, disc jockey.

Cheticamp

***CKJM-FM**— 1995: 106.1 mhz; 3 kw. TL: N46 36 55 W61 02 52. Stereo. Hrs open: 24
Rebroadcasts CFIM-FM Iles-De-La Madeleine, PQ 5%.
Box 699, Les Trois Piqnons, Main Rd., B0E 1H0. Phone: (902) 224-1242. Fax: (902) 224-1770. E-mail: info@ckjm.ca Web Site: www.ckjm.ca. Licensee: La Cooperative Radio Cheticamp Ltee. Population served: 5,000 Format: Var, country, Fr. Target aud: General. Spec prog: Gaelic one hr, jazz 3 hrs wkly. ♦ Normand Poirier, pres; Angus Lefort, gen mgr, opns mgr, mktg dir & engrg dir; Pamela Deveau, natl sls mgr & adv dir; Ginette Chiasson, progmg dir.

Comeauville

***CIFA-FM**— Sept 28, 1990: 104.1 mhz; 39.3 w. 475 ft Stereo. Hrs open: 6am to 10pm Box 8, Saulnierville, B0W 2Z0. Phone: (902) 769-2432. Fax: (902) 769-3101. E-mail: info@cifafm.ca Web Site: www.cifa.fm. Licensee: Association Radio Clare. Population served: 60,000 Format: Div, community news. News staff: 1/2. Target aud: General. ♦ Albert Geddry, pres; Dave LeBlau, gen mgr; Paul Lomberd, gen sls mgr; Emile Blinn, progmg dir & engrg mgr.

Dartmouth

CFDR(AM)— Dec 5, 1962: 780 khz; 50 kw-D, 15 kw-N. Hrs open: 2900 Agricola St., Halifax, B3K 6B2. Phone: (902) 453-2524. Fax: (902) 453-3132. Web Site: www.780kixx.ca. Licensee: New Cap Broadcasting Ltd. (group owner) Natl. Rep: Canadian Broadcast Sales. Format: Classic country. ♦ Ted Hyland, gen mgr & gen sls mgr; J. C. Douglas, progmg dir; Rich Horner, news dir; Walter Labucki, chief of engrg.

CFRQ-FM—Co-owned with CFDR(AM). Nov 28, 1983: 104.3 mhz; 100 kw. 400 ft Stereo. Web Site: www.q104.ca. Format: Classic rock & current rock.

Digby

CKDY(AM)— Feb 2, 1970: 1420 khz; 1 kw-U, DA-1. Hrs open:
Rebroadcasts CKEN(AM) Kentville and CKAD(AM) Middleton.
Box 1420, B0V 1A0. Secondary address: 53 Sydney St. B0V 1A0. Phone: (902) 245-2111. Fax: (902) 678-9720. E-mail: programming@avrnetwork.com Web Site: www.avrnetwork.com. Licensee: Maritime Broadcasting System Ltd. (group owner; (acq 8-79). Format: Contemp country. Spec prog: Farm 7 hrs wkly. ♦ Diane Best, gen mgr; Scott Baines, stn mgr; Karen Corey, gen sls mgr; Amanda Misner, progmg dir; Dave Chaulk, news dir; Garth Faulkner, chief of engrg.

Eastern Passage

CFEP-FM— 2002: 94.7 mhz; 50 w. Hrs open: Seaside-FM, Box 196, B3G 1M5. Phone: (902) 465-9900. Fax: (902) 469-0966. E-mail: seasidefm@ns.sympatico.ca Web Site: www.seasidefm.com. Licensee: Seaside Broadcasting Organization. Format: Easy lstng, big band, adult standards. ♦ Wayne Harrett, gen mgr & progmg dir.

Eskasoni Indian Reserve

CICU-FM— 1994: 94.1 mhz; 1 w. Hrs open: Box 7100, Eskasoni, B1W 1A1. Secondary address: 130 Anslum Rd., Eskasoni Indian Reserve, Eskasoni B0A 1J0. Phone: (902) 379-2955. Fax: (902) 379-2966. E-mail: greguj@ns.sympaatico.ca Licensee: Greg Johnson. Format: Micmac-language (32and English-language (75%) progmg. ♦ Greg Johnson, gen mgr & chief of engrg; Linda Johnson, progmg dir.

Glace Bay

CKOA-FM— 2007: 89.7 mhz; 4.7 kw. TL: N46 11 59 W59 58 46. Hrs open: 24 106 Reserve St., B1A 4W5. Phone: (902) 849-4301. Fax: (902) 849-1272. E-mail: info@coastalradio.ca Web Site: www.coastalradio.ca. Licensee: Coastal Community Radio Cooperative Ltd. Format: Pop, rock and dance music.

Halifax

CBAX-FM— September 2003: 91.5 mhz; 77.5 kw. TL: N44 39 03 W63 39 28. Hrs open: c/o CBAL-FM, 250 University Ave., Moncton, NB, E1C 5K3. Phone: (506) 853-6666. Fax: (506) 867-8000. Licensee: Societe Radio-Canada. Format: Fr classical, jazz, world music. ♦ Benoit Quennecille, gen mgr; Andree Girard, progmg dir.

CBHA-FM— 1989: 90.5 mhz; 91 kw. 711 ft Hrs open: Box 3000, B3J 3E9. Phone: (902) 420-8311. Fax: (902) 420-4357. Fax: (902) 420-4429. E-mail: mainstreet@halifax.cbc.ca Web Site: www.cbc.ca. Licensee: CBC. Natl. Network: CBC Radio One. Format: News, jazz, div. ♦ Robert Rabinovitch, CEO, pres & gen mgr; Carole Taylor, chmn; Susan Mitton, opns dir; Nicole Vautour, progmg dir.

***CBH-FM**— June 1, 1976: 102.7 mhz; 81 kw. 711 ft Stereo. Hrs open: 24 Box 3000, B3J 3E9. Phone: (902) 420-8311. Fax: (902) 420-4429. Fax: (902) 420-4089. E-mail: weekender@halifax.cbc.ca Web Site: www.cbc.ca. Licensee: CBC. Natl. Network: CBC Radio One. Format: Div, class. Target aud: General. ♦ Robert Rabinovitch, CEO & chmn; Carole Taylor, chmn; Robert Rabinoitch, pres; Susan Mitton, opns dir; Nicole Vautour, progmg dir.

CHAL-FM—Not on air, target date: unknown: 105.1 mhz; 32 kw. Hrs open: Global Halifax, 14 Akerley Blvd., Dartmouth, B3B 1J3. Phone: (902) 481-7400. Fax: (902) 468-2154. Licensee: CanWest MediaWorks Inc. Format: Easy lstng. ♦ Barry Saunders, gen mgr.

CHFX-FM— Feb 9, 1970: 101.9 mhz; 100 kw. Ant 546 ft Stereo. Hrs open: 24 Box 400, B3J 2R2. Phone: (902) 422-1651. Fax: (902) 422-5330. E-mail: chfx@ns.sympatico.ca Licensee: Maritime Broadcasting System Ltd. (acq 6-94). Population served: 300,000 Wire Svc: Broadcast News Ltd. Format: Country. News staff: 5; News: 6 hrs wkly. Target aud: 25-44. ♦John Gold, mus dir.

CHNS-FM— July 19, 2006: 89.9 mhz; 100 kw. Hrs open: 5121 Sackville St., 3rd Fl., B3J 1K1. Phone: (902) 425-1225. Fax: (902) 422-5330. Web Site: www.899HALFM.com. Licensee: Maritime Broadcasting System Ltd. Format: Classic rock. Target aud: 25-54; adults. ♦Allan Gidyk, gen mgr.

CIOO-FM—Listing follows CJCH(AM).

CJCH(AM)— Nov 4, 1944: 920 khz; 25 kw-U, DA-D. TL: N44 38 10 W63 40 22. Stereo. Hrs open: 24 2900 Agricola St., B3K 6B2. Phone: (902) 453-2524. Fax: (902) 453-3120. Fax: (902) 453-3132. Web Site: www.cjch.net. Licensee: CJCH 920/C100 FM Division of CHUM Ltd. Group owner: CHUM Ltd. Population served: 150,000 Format: Yesterday. News staff: 5. ♦Scott Bodnarchuk, gen mgr; Bill Bodnarchuk, gen sls mgr; Terry Williams, progmg mgr; Earle Mader, mus dir; Rick Howe, news dir & pub affrs dir; Walter Labrucci, engrg dir; Rob Davidson, traf mgr.

CIOO-FM—Co-owned with CJCH(AM). November 1977: 100.1 mhz; 100 kw. 770 ft TL: N44 39 05 W63 39 51. Stereo. 24 Web Site: www.c100.net. Format: Adult contemp. ♦Trent McGrath, prom dir; Rob Davidson, traf mgr.

CJNI-FM— Oct 11, 2005: 95.7 mhz; 65 kw. Hrs open: 6080 Young St., B3K 5L2. Phone: (902) 493-7133. Licensee: Rogers Broadcasting Ltd. Format: News/talk. ♦Rael Merson, pres; Jim Hamm, gen mgr.

***CKDU-FM**— February 1985: 97.5 mhz; 3.2 kw. Ant 300 ft Hrs open: 24 Dalhousie SUB, 6136 University Ave., B3H 4J2. Phone: (902) 494-6479. E-mail: ckdu@ckdu.ca Web Site: www.ckdu.ca. Licensee: CKDU-FM Society Ltd. Population served: 250,000 Format: Div. News staff: one; News: 4 hrs wkly. Target aud: General. ♦Michael Catano, stn mgr.

CKHZ-FM— 9/01/06: 103.5 mhz; 78 kw. Stereo. Hrs open: 24 Evanov Radio Group, 5302 Dundas St. W., Toronto, ON, M9B 1B2. Phone: (416) 213-1035. Fax: (416) 233-8617. Web Site: z103halifax.com. Licensee: HFX Broadcasting Inc. Format: Top 40.

CKRH-FM— May 2007: 98.5 mhz; 2.35 kw. Hrs open: 5527 Cogswell St., B3J 1R2. Phone: (902) 490-2574. Licensee: Cooperative Radio-Halifax-Metro limitee. Format: Fr var.

CKUL-FM— August 1990: 96.5 mhz; 100 kw. Hrs open: 24 2900 Agricola St., B3K 4P5. Phone: (902) 453-4004. Fax: (902) 453-3132. E-mail: fm965@mrg.ca Web Site: www.planetkool.ca. Licensee: Metro Radio Group Inc. Group owner: NewCap Broadcasting Ltd. (acq 12-17-2001). Natl. Rep: Canadian Broadcast Sales. Format: Classic hits 60s & 70s & 80s. News staff: 3; News: 3 hrs wkly. Target aud: 35-54. ♦Mark Maheu, pres; Scott Bodnarchuk, gen mgr; Tom Manton, sls VP; Trent McGrath, prom mgr; Gary Greer, progmg mgr; Rich Horner, news dir & pub affrs dir; Walter Labucki, chief of engrg.

Kentville

CKEN-FM— Mar 14, 1965: 97.7 mhz; 18 kw. Ant 680 ft Stereo. Hrs open: Box 310, B4N 1H5. Secondary address: 29 Oakdene Ave. B4N 1H5. Phone: (902) 678-2111. Fax: (902) 678-9894. E-mail: avr@avrnetwork.com Web Site: www.avrnetwork.com. Licensee: Maritime Broadcasting System Ltd. Group owner: Maritime Broadcasting (acq 1998; grpsl). Natl. Rep: Canadian Broadcast Sales. Format: Country. News staff: 5. Target aud: 18-49. Spec prog: Farm 5 hrs wkly. ♦Dianne Best, gen mgr; Karen Corey, gen sls mgr; Amanda Misner, progmg dir; James Cormier, mus dir; Garth Faulkner, chief of engrg.

CKWM-FM— 2003: 94.9 mhz; 100 kw. Hrs open: Box 310, B4N 1H5. Secondary address: 29 Oakdene Ave. B4N 1H5. Phone: (902) 678-2111. Fax: (902) 678-9894. E-mail: magic949@magic949.ca Web Site: www.magic949.ca. Licensee: Maritime Broadcasting System Ltd. Group owner: Maritime Broadcasting. Format: Adult contemp. ♦Dianne Best, gen mgr; Karen Corey, gen sls mgr; Angela Rose, progmg dir; Garth Faulkner, chief of engrg.

Liverpool

CKBW-FM-1— Sept 15, 1980: 94.5 mhz; 8.7 kw. Ant 250 ft Stereo. Hrs open: 24
Rebroadcasts CKBW-FM Bridgewater 100%.
c/o CKBW, 215 Dominion St., Bridgewater, B4V 2G8. Phone: (902) 543-2401. Fax: (902) 543-1208. E-mail: ckbw@ckbw.com Web Site: www.ckbw.com. Licensee: Acadia Broadcasting Ltd. (group owner; acq 8-31-89). Rgnl rep: Canadian Broadcast Sales. Wire Svc: Broadcast News Ltd. Format: Hot AC. News staff: 2; News: 11 hrs wkly. Target aud: General; rural & small town urban.John Wiles, gen mgr, opns mgr, progmg dir & mus dir; Barry Smith, gen sls mgr & natl sls mgr; Brian Tepper, prom dir; Frank Grayney, chief of engrg; Pamela Smith, traf mgr; Greg Lowe, spec ev coord & sports cmtr; Sheldon MacLeod, news dir, pub affrs dir & local news ed; Lonnie Townsend, news rptr; Jonathan Crouse, disc jockey; Leitha Mayson, disc jockey; Mike Richards, disc jockey

Middleton

CKAD(AM)— 1962: 1350 khz; 1 kw-U, DA-1. Hrs open: Box 550, B0S 1P0. Phone: (902) 825-3429. Fax: (902) 678-9894. E-mail: programming@avrnetwork.com Licensee: Maritime Broadcasting. (group owner; acq 6-26-79). Format: Country. Spec prog: Farm 3 hrs wkly. ♦Diane Best, gen mgr; Scott Baines, stn mgr; Karen Corey, gen sls mgr; Amanda Misner, progmg dir & progmg mgr; Dave Chaulk, news dir; Garth Faulkner, chief of engrg.

New Glasgow

CKEC(AM)— Dec 23, 1953: 1320 khz; 25 kw-U, DA-N. Hrs open: 24 Box 519, CKEC Radio Bldg., 84 Provost St., B2H 5E7. Phone: (902) 752-4200. Phone: (902) 755-1320. Fax: (902) 755-2468. Fax: (902) 928-1320. E-mail: ckec@ckec.com Licensee: D. Freeman. (acq 1964). Natl. Rep: Canadian Broadcast Sales. Format: Adult contemp, classic rock, MOR, country. News staff: 3; News: 15 hrs wkly. Target aud: General. Spec prog: Scottish. ♦D.B. Freeman, CEO; M.D. Freeman, exec VP & gen mgr.

New Tusket

CJLS-FM-2— 1982: 93.5 mhz; 3 kw. Hrs open:
Rebroadcasts CJLS-FM Yarmouth.
c/o Radio Station CJLS(AM), 328 Main St., Suite 201, Yarmouth, B5A 1E4. Phone: (902) 742-7175. Fax: (902) 742-3143. E-mail: cjls@cjls.com Web Site: www.cjls.com. Licensee: Radio CJLS Ltd. Format: Adult contemp. ♦Chris Perry, VP & progmg mgr; Ray Zinck, pres & gen mgr; Dave Hall, gen sls mgr; Gary Nickerson, news dir; Jim Harris, engrg dir.

Port Hawkesbury

CIGO-FM— 2000: 101.5 mhz; 19 kw. TL: N45 39 00 W61 28 00. Hrs open: 609 Church St., Ste 201, B9A 2X4. Phone: (902) 625-1220. Phone: (902) 625-1015. Fax: (902) 625-2664. Fax: (902) 625-6397. E-mail: 1015thehawk@1015thehawk.com Web Site: www.1015thehawk.com. Licensee: MacEachern Broadcasting Ltd. Population served: 50,000 Wire Svc: BN Wire Format: Adult contemp. News staff: 2; News: 4 hrs wkly. Target aud: 18-49; blue collar, high school education, married. Spec prog: Scottish 2 hrs, Irish one hr wkly. ♦Bob MacEachern, pres & stn mgr; Kelly Atchison, progmg mgr; Kevin MacEachern, sls.

Shelburne

CKBW-FM-2— Sept 15, 1980: 93.1 mhz; 8.6 kw. Stereo. Hrs open:
Rebroadcasts CKBW-FM Bridgewater 100%.
c/o CKBW, 215 Dominion St., Bridgewater, B4V 2G8. Phone: (902) 543-2401. Fax: (902) 543-1208. E-mail: ckbw@ckbw.com Web Site: www.ckbw.com. Licensee: Acadia Broadcasting Ltd. (group owner; acq 8-31-89). Rgnl rep: Canadian Broadcast Sales. Wire Svc: Broadcast News Ltd. Format: Hot adult contemp. Target aud: General; rural & small town urban.John Wiles, gen mgr, opns mgr, progmg dir & mus dir; Barry Smith, gen sls mgr & natl sls mgr; Brian Tepper, prom dir; Frank Grayney, chief of engrg; Pamela Smith, traf mgr; Greg Lowe, spec ev coord & sports cmtr; Sheldon MacLeod, news dir, pub affrs dir & local news ed; Jonathan Crouse, disc jockey; Leitha Mayson, disc jockey; Mike Richards, disc jockey

Sydney

CBI(AM)— Nov 1, 1948: 1140 khz; 10 kw-U, DA-2. Hrs open: 285 Alexandra St., B1S 2E8. Phone: (902) 539-5050. Fax: (902) 539-1562. Web Site: www.cbc.ca. Licensee: CBC. Natl. Network: CBC Radio One. Format: Info. ♦Jill Spelliscy, gen mgr.

CBI-FM— July 1977: 105.1 mhz; 20 kw. 400 ft Stereo. Web Site: www.cbc.ca. Natl. Network: CBC Radio Two. Format: Classics, lite classics.

CHER-FM— 2007: 98.3 mhz; 100 kw. Stereo. Hrs open: 318 Charlotte St., B1P 1C5. Phone: (902) 564-5596. Fax: (902) 564-1873. Web Site: www.capebretonradio.com. Licensee: Maritime Broadcasting System Ltd. Natl. Rep: Canadian Broadcast Sales. Format: Classic hits. News staff: 3. ♦Alan Peddle, gen mgr & gen sls mgr; Fred Denney, opns mgr; Phil Thompson, progmg dir; Gary Andrea, news dir; Roy MacIntosh, chief of engrg.

CJCB(AM)— Feb 14, 1929: 1270 khz; 10 kw-U, DA-N. Stereo. Hrs open: 24 Radio Bldg., 318 Charlotte St., B1P 1C8. Phone: (902) 564-5596. Fax: (902) 564-1057. Web Site: www.capebretonradio.com. Licensee: Maritime Broadcasting System Ltd. Population served: 109,000 Format: Today's country. News staff: 3. Target aud: General. ♦Rod Deviller, opns mgr; Alan Peddle, gen sls mgr; Phil Thompson, prom mgr; Roy MacIntosh, engrg mgr & chief of engrg.

CKPE-FM—Co-owned with CJCB(AM). September 1962: 94.9 mhz; 61 kw. 210 ft Stereo. 24 Web Site: www.capebretonradio.com.109,000 Format: Todays best music. News staff: 3. ♦Phil Thompson, prom dir & prom mgr; Joe Purdy, mus dir; Roy MacIntosh, chief of engrg.

CJIJ-FM— June 2, 2003: 99.9 mhz; 50 w. Hrs open: C99 FM Radio, 49 Tupsi Dr., Membertou, B1S 3K6. Phone: (902) 562-0009. Fax: (902) 539-6645. E-mail: c99fm@hotmail.com Web Site: c99fm.homestead.com. Licensee: Membertou Radio Association Inc. Format: Classic hits, classic rock. ♦Peter Christmas Jr., gen mgr; Jay Bedford, gen sls mgr, progmg dir & chief of engrg; Alex Morrison, prom mgr & chief of engrg.

Truro

CINU-FM— 2004: 98.5 mhz; 50 w. Hrs open: Box 25012, B2N 7B8. Secondary address: 883 Prince St. B2N 1H2. Phone: (902) 843-4673. Fax: (902) 662-2879. E-mail: hopefmministries@eastlink.ca Licensee: Hope FM Ministries Ltd. Format: Christian music. ♦Barry Reid, pres & gen mgr.

CKTO-FM— 1965: 100.9 mhz; 50 kw. Ant 189 ft Stereo. Hrs open: 24 187 Industrial Ave., B2N 6V3. Phone: (902) 893-6060. Fax: (902) 893-7771. Licensee: Astral Media Radio Atlantic Inc. Group owner: Astral Media Inc. (acq 4-19-2002; grpsl). Natl. Rep: Canadian Broadcast Sales. Format: Adult contemp, rock. News staff: 3; News: 6 hrs wkly. Target aud: 25-49. ♦John Eddy, exec VP; Mike Worsley, stn mgr; Chris Van Tassel, progmg dir; James Cormier, mus dir; Dave Guy, news dir.

CKTY-FM— 2001: 99.5 mhz; 16.75 kw. Hrs open: 187 Industrial Ave., B2N 6V3. Phone: (902) 893-6060. Fax: (902) 893-7771. Licensee: Astral Media Radio Atlantic Inc. Group owner: Astral Media Inc. (acq 4-19-2002; grpsl). Natl. Rep: Canadian Broadcast Sales. Format: Country. ♦John Eddy, exec VP; Mike Worsley, gen sls mgr; Chris Van Tassel, progmg dir; James Cormier, mus dir; Dave Guy, news dir.

Windsor

CFAB(AM)— 1945: 1450 khz; 1 kw-U. Hrs open: 24 ARV, 29 Oakdene Ave., Box 310, Kensville, B4N 1H5. Phone: (902) 798-2111. Fax: (902) 798-8140. E-mail: avr@avrnetwork.com Web Site: www.avrnetwork.com. Licensee: Maritime Broadcasting System Ltd. Group owner: Maritime Broadcasting. Format: Country. News staff: 5; News: 9 hrs wkly. Target aud: 25-54. ♦Dianne Best, gen mgr; Karen Corey, gen sls mgr; Amanda Misner, progmg dir; Dave Chaulk, news dir; Garth Faulkner, engrg dir & chief of engrg; Scott Baines, stn mgr & news rptr.

Yarmouth

CJLS-FM— 2003: 95.5 mhz; 18 kw. Stereo. Hrs open: 24 328 Main St., Suite 201, B5A 1E4. Phone: (902) 742-7175. Fax: (902) 742-3143. E-mail: CJLS@cjls.com Web Site: www.cjls.com. Licensee: Radio CJLS Ltd. Format: Adult contemp. ♦Ray Zinck, gen mgr; Dave Hall, gen sls mgr; Chris Perry, progmg dir; Jim Harris, chief of engrg.

Nunavut

Baker Lake

CKQN-FM— 1973: 99.3 mhz; 60 w. -50 ft Hrs open: Box 13, X0C 0A0. Phone: (867) 793-2962. Fax: (867-793-2726). Web Site: www.tvradioworld.com. Licensee: Qamani'tuap Naalautaa Society. Natl. Network: CBC Radio One. Format: Eskimo, Inuit.

Iqaluit

CFFB(AM)— Feb 6, 1961: 1230 khz; 1 kw-U, DA-1. Hrs open: Box 490, X0A 0H0. Phone: (867) 979-6100. Fax: (867) 979-6147. E-mail: cbcnorth@cbc/ca Web Site: cbc.ca/north. Licensee: CBC. Format: News/talk, Inuktitut language. ♦Pat Nagle, gen mgr.

CFRT-FM— 1994: 107.3 mhz; 27 w. Hrs open: C.P. 880, X0A 0H0. Phone: (867) 979-4606. Fax: (867) 979-0800. E-mail: cfrt@nunafranc.ca Web Site: www.franconunavut.ca. Licensee: Association des francophones de Nunavut. Format: Fr. ♦Daniel Cuerrier, gen mgr.

CKIQ-FM— May 26, 2003: 99.9 mhz; 537 w. Hrs open: Box 417, X0A 0H0. Secondary address: 1036 Airport Rd. X0A 0H0. Phone: (867) 975-2547. Fax: (867) 975-2598. E-mail: 99.9@ckiq.com Web Site: www.ckiq.ca. Licensee: Nunavut Natautinga Ltd. Natl. Rep: Target Broadcast Sales. Format: Classic rock. ♦Terri Chegwyn, gen mgr.

Rankin Inlet

CBQR-FM— 1988: 105.1 mhz; 87 w. Hrs open: 24 Box 130, X0C 0G0. Phone: (867) 645-2244. Fax: (867) 645-2820. Web Site: www.north.cbc.ca. Licensee: CBC Radio. Natl. Network: CBC Radio One. Format: Adult contemp, talk, div. Spec prog: Inuktitut 10 hrs wkly. ♦Elizabeth Kusugak, gen mgr.

Ontario

Ajax

CJKX-FM— 1994: 95.9 mhz; 19.94 kw. Ant 330 ft Stereo. Hrs open: 24 1200 Airport Blvd., Suite 207, Oshawa, L1J 8P5. Phone: (905) 428-9600. Fax: (905) 571-1150. E-mail: kx96@kx96.fm Web Site: www.kx96.fm. Licensee: Durham Radio Inc. (group owner). Population served: 2,500,000 Format: New country. News staff: 3; News: 1.5 hrs wkly. Target aud: 25-54; . ♦Douglas E. Kirk, pres & gen mgr; Steve Kassay, opns VP & opns mgr; Steve Macaulay, sls VP & gen sls mgr; Stacey Garfield, traf mgr.

Akwesasne

CKON-FM— Oct 1, 1984: 97.3 mhz; 150 w. 150 ft Hrs open: Box 140, Rooseveltown, NY, 13683. Secondary address: Box 1496 K6H 5V5. Phone: (613) 575-2100. Phone: (518) 358-3426 (US). Fax: (613) 575-2566. E-mail: ckon@ckonfm.com Web Site: www.ckonfm.com. Licensee: Mohawk Nation Council. Group owner: Akwesasne Communication Society Format: Div. Target aud: General. Spec prog: Mohawk. ♦Judy Laffin, gen mgr; Larry Edwards, gen mgr & progmg dir.

Alexandria

CHOD-FM—See Cornwall

Aylmer

CHPD-FM— September 2003: 105.9 mhz; 250 w. TL: N42 45 40 W80 56 03. Hrs open: 7 AM- 8 AM; 5 PM- 8 PM 16 Talbot St., N5H 1H4. Phone: (519) 773-8555. Fax: (519) 773-8606. E-mail: radio@mccayl.org Licensee: Aylmer and Area Inter-Mennonite Community Council. Population served: 10,000 Format: Low German. News staff: one; News: one hr wkly. ♦Abe Harms, CEO; Philip Wiebe, chmn & pres; Peter Bergen, VP; Henry Rempel, stn mgr.

Bancroft

CHMS-FM— May 2001: 97.7 mhz; 50 kw. Hrs open: 24 Box 1240, K0L 1C0. Phone: (613) 332-1423. Fax: (613) 332-0841. E-mail: moose977@hbgradio.com Web Site: www.moosefm.com. Licensee: The Haliburton Broadcasting Group Inc. Group owner: Haliburton Broadcasting Group Inc. Population served: 30,000 Format: Hot adult contemp. Target aud: General. Spec prog: Relg one hr, loc talk 5 hrs, sports 2 hrs wkly. ♦Steve Skelly, opns mgr.

Barrie

CFJB-FM— Oct 7, 1988: 95.7 mhz; 41 kw. Ant 500 ft Hrs open: 24 Box 95, 400 Bayfield St., Suite 205, L4M 5A1. Phone: (705) 725-7304. Fax: (705) 721-7842. E-mail: dbingley@rock95.com Web Site: www.rock95.com. Licensee: Rock 95 Broadcasting (Barrie-Orillia) Ltd. (acq 2-7-94). Population served: 280,000 Format: Classic rock, new rock, 80s rock. News staff: 3; News: 4 hrs wkly. Target aud: 18-49; broad-based, well-educated, above-average income. ♦Doug Bingley, CEO, pres & gen mgr; Jim Cowden, VP & gen sls mgr; Tom Harrison, gen sls mgr; Dave Carr, progmg dir.

CHAY-FM— May 21, 1977: 93.1 mhz; 100 kw. Ant 1,000 ft Stereo. Hrs open: 24 Box 937, 1125 Bayfield St. N., L4M 4Y6. Phone: (705) 737-3511. Fax: (705) 737-0603. E-mail: knoel@corusent.com Web Site: www.thenewchay.com. Licensee: Corus Radio Co. Group owner: Corus Entertainment Inc. Population served: 350,000 Format: Adult contemp. News staff: 3; News: 13 hrs wkly. Target aud: 25-64; general. ♦John Hayes, pres; Kim Noel, gen mgr; Frank Allinson, gen sls mgr; Keith Talbert, prom mgr; Darren Stevens, progmg dir.

CIQB-FM— November 1994: 101.1 mhz; 4.3 kw. Hrs open: 24 Box 937, L4M 4S9. Secondary address: 1125 Bayfield St. N. L4M 4S9. Phone: (705) 726-1011. Fax: (705) 726-0022. E-mail: knoel@corusent.com Web Site: www.b101fm.com. Licensee: 591989 B.C. Ltd. Group owner: Corus Entertainment Inc. (acq 3-24-2000; grpsl). Format: Adult contemp. News staff: 3; News: 11 hrs wkly. Target aud: 25-54; women 35-49 & 25-54 & at work people. ♦John Hayes, pres & VP; Kim Noel, gen mgr; Dave Pinder, mktg dir & prom; Darren Stevens, progmg dir.

CJLF-FM— Aug 15, 1999: 100.3 mhz; 18.7 kw. Hrs open: 115 Bell Farm Rd, Unit 111, L4M 5G1. Phone: (705) 735-3370. Fax: (705) 735-3301. Web Site: www.lifeonline.fm. Licensee: Trust Communications Ministries. Format: Christian music. ♦Scott Jackson, stn mgr; Jen Taylor, prom dir; Ben Davy, progmg dir; Christy Burton, news dir.

CKMB-FM— 2001: 107.5 mhz; 20 kw. Hrs open: 400 Bayfield St., Suite 205, L4M 5A1. Phone: (705) 725-7304. Fax: (705) 721-7842. Web Site: www.star1075.com. Licensee: Rock 95 Broadcasting (Barrie-Orillia) Ltd. Format: Top 40/contemp hits. ♦Doug Bingley, CEO, pres, gen mgr & chief of engrg; Jim Cowden, sls VP & mktg VP; Helen Mathers, prom dir; Dale Smith, progmg mgr & chief of engrg.

Belleville

CHCQ-FM— 2001: 100.1 mhz; 21 kw. Stereo. Hrs open: 24 354 Pinnacle St., K8N 3B4. Phone: (613) 966-0955. Fax: (613) 967-2565. Web Site: www.cool100.fm. Licensee: Starboard Communications Ltd. (acq 7-26-02; C$541,351). Population served: 105,000 Natl. Rep: CHUM Radio Sales. Format: Country. Target aud: 25-54; adults. ♦John Sherratt, pres & gen mgr; Darrin Matassa, gen sls mgr; Mark Philbin, progmg dir.

CIGL-FM—Listing follows CJBQ(AM).

CJBQ(AM)— Aug 12, 1946: 800 khz; 10 kw-U, DA-2. Hrs open: Box 488, K8N 5B2. Secondary address: 10 S. Front St. K8N2Y3. Phone: (613) 969-5555. Fax: (613) 969-8122. Web Site: www.cjbq.com. Licensee: Quinte Broadcasting Ltd. (group owner). Format: Country. Spec prog: Farm 3 hrs wkly. ♦Bill Morton, gen mgr & gen sls mgr; Peter Thompson, progmg dir.

CIGL-FM—Co-owned with CJBQ(AM). August 1962: 97.1 mhz; 50 kw. Web Site: www.mix97.com. Format: Hot adult contemp.

CJLX-FM— October 1992: 91.3 mhz; 3.4 kw. Ant 300 ft TL: N44 09 50 W77 23 24. Stereo. Hrs open: 24 Box 4200, Loyalist College, Wallbridge-Loyalist Rd., K8N 5B9. Phone: (613) 966-0923. Fax: (613) 966-1993. E-mail: 91x@loyalistmail.ca Web Site: www.91x.fm. Licensee: Loyalist College Radio Inc. (acq 11-13-90). Population served: 177,000 Natl. Rep: Target Broadcast Sales. Wire Svc: BN Wire Format: Rock, community service. News staff: 15; News: 6 hrs wkly. Target aud: 18-34; primary, ages 50 plus secondary. Spec prog: folk one hr, jazz 3 hrs, Greek one hr, Dutch one hr wkly, classical 2 hr, blues one hr, educational 4 hrs, big band 2 hrs. ♦Greg Schatzmann, CEO & gen mgr; Sandi Ramsey, sls dir & gen sls mgr; Len Arminio, news dir; Tim Rorabeck, chief of engrg.

CJOJ-FM— Dec 1, 1993: 95.5 mhz; 42 kw. Stereo. Hrs open: 24 354 Pinnacle St., K8N 3B4. Phone: (613) 966-0955. Fax: (613) 967-2565. Web Site: www.classichits955.fm. Licensee: Starboard Communications Ltd. (acq 7-26-02; C$1,456,610). Population served: 105,000 Natl. Rep: CHUM Radio Sales. Format: CHR, adult contemp. Target aud: 25-54; adults - skewed females 60%, males 40%. ♦John Sherratt, pres & stn mgr; Darron Matassa, gen sls mgr; Mark Philbin, progmg dir.

CKJJ-FM— Oct 18, 2003: 102.3 mhz; 45 kw. Hrs open: Box 23095, K8P 5J3. Secondary address: 214 Pinnacle St. K8P 3A6. Phone: (613) 966-4822. Fax: (613) 966-3211. E-mail: info@ucbcanada.com Web Site: www.ucbcanada.com. Licensee: United Christian Broadcasters Canada. Format: Christian music. ♦Garry Quinn, gen mgr; Alan Baker, progmg dir.

Bracebridge

CFBG-FM— May 1988: 99.5 mhz; 12 kw. Hrs open: Box 960, 25 Highland St., Haliburton, K0M 1S0. Phone: (705) 645-2218. Fax: (705) 645-6957. E-mail: moose995@hbgradio.com Web Site: www.hbgradio.com. Licensee: The Haliburton Broadcasting Group Inc. Group owner: Haliburton Broadcasting Group Inc. (acq 12-10-97; C$295,000). Format: Hot adult contemp. Target aud: 34-45; older adult contemporary. Spec prog: Jazz 2 hrs, big band one hr, loc magazine one hr wkly. ♦Christopher Grossman, pres & gen mgr; Kimberley Ward, VP & opns mgr; Sean Connon, gen sls mgr.

Brampton

CIAO(AM)— Dec 23, 1953: 530 khz; 1 kw-D, 250 w-N, DA-2. (Digital radio: 1466.768 mhz; 5.084 kw). Hrs open: 24 5302 Dundas S. W., Toronto, M9B-1B2. Phone: (416) 213-1035. Fax: (416) 233-8617. Web Site: www.am530.ca. Licensee: CKMW Radio Ltd. Group owner: Evanov Radio Group (acq 9-26-83). Natl. Rep: Target Broadcast Sales. Format: Ethnic, multilingual. ♦Bill Evanov, pres; Paul Evanov, exec VP.

Brantford

CFWC-FM— 2002: 93.9 mhz; 250 w. Hrs open: 271 Greenwich St., N3S 2X9. Phone: (519) 759-2339. Fax: (519) 753-1157. E-mail: info@power93.ca Web Site: www.power93.ca. Licensee: 1486781 Ontario Ltd. Format: Christian music. ♦Dean Johnston, gen mgr.

CKPC(AM)— December 1923: 1380 khz; 25 kw-U, DA-2. TL: N43 03 05 W80 18 50. Hrs open: 24 571 West St., N3T 5P8. Phone: (519) 759-1000. Fax: (519) 753-1470. E-mail: salesmgr@ckpc.on.ca Web Site: www.ckpc.on.ca. Licensee: Telephone City Broadcast Ltd. Population served: 115,652 Natl. Rep: Target Broadcast Sales. Format: Classic hits. News staff: 7; News: 9 hrs wkly. Target aud: 35-64. ♦Richard D. Buchanan, pres & gen mgr; Peter Jackman, sls VP & gen sls mgr.

CKPC-FM— May 1949: 92.1 mhz; 80 kw. Ant 750 ft Stereo. 24 Web Site: www.ckpc.on.ca.500,000 Natl. Rep: Target Broadcast Sales. Format: Adult contemp. News staff: 7; News: 7 hrs wkly. Target aud: 25-49.

Brockville

CFJR-FM— 2003: 104.9 mhz; 5.6 kw. Hrs open: 601 Stewart Blvd., K6V 5V9. Phone: (613) 345-1666. Fax: (613) 342-2438. E-mail: comments@hometownradio.ca Web Site: www.hometownradio.ca. Licensee: 1708479 Ontario Inc. (group owner). Format: Adult contemp. News staff: 3; News: 5 hrs wkly. Target aud: 35-54; female slant. ♦Jay Smitzer, pres; Greg Hinton, gen mgr & progmg dir; Rick Moran, gen sls mgr; Warren Davies, chief of engrg.

CJPT-FM— July 28, 1988: 103.7 mhz; 100 kw. Ant 495 ft TL: N44 23 58 W75 58 21. Hrs open: 24 601 Stewart Blvd., K6V 5V9. Phone: (613) 345-1666. Fax: (613) 342-2438. E-mail: info@bob.fm Web Site: www.bob.fm. Licensee: 1708479 Ontario Inc. (group owner). Format: Hits of the 80s. News staff: 3. Target aud: 18-44; male. ♦Jay Switzer, pres; Paul Ski, exec VP; Greg Hinton, gen mgr & progmg dir; Rick Moran, rgnl sls mgr; Alison MacLean, prom dir.

Burlington

CIWV-FM—See Hamilton

CJXY-FM— Sept 23, 1976: 107.9 mhz; 26.4 kw. Ant 672 ft TL: N43 23 12 W79 52 34. Stereo. Hrs open: 24 875 Main St. West, Hamilton, L8S 4R1. Phone: (905) 521-9900. Fax: (905) 521-2306. Web Site: www.y108.ca. Licensee: Corus Radio Co. Group owner: Corus Entertainment Inc. Format: Mainstream rock. News staff: one; News: 2 hrs wkly. Target aud: 25-39. ♦Suzanne Carpenter, gen mgr.

Cambridge

CJDV-FM— 1998: 107.5 mhz; 2.5 kw. Stereo. Hrs open: 24 1315 Bishop St. N., Unit 100, N1R 6Z2. Phone: (519) 621-7510. Fax: (519) 621-0165. E-mail: lars@davefm.com Web Site: www.davefm.com. Licensee: 591989 B.C. Ltd. Group owner: Corus Entertainment Inc. (acq 4-2000; grpsl). Population served: 400,000 Format: Hot adult contemp. News staff: 2. Target aud: 18-49. Spec prog: Por 2 hrs wkly. ♦Lars Wunsche, gen mgr; Lucia Zdeb, prom dir; Kneale Mann, progmg dir; Brian Clemens, engrg VP & chief of engrg.

Campbellford

CKOL-FM— 1993: 93.7 mhz; 500 w. Hrs open: Box 551, K0L 1L0. Secondary address: 15 Ragland St. S. K0L 1L0. Phone: (705) 653-1089. Fax: (705) 653-1089. E-mail: ckl-radio@excite.com Licensee: Campbellford Area Radio Association. Format: Div. Spec prog: Gospel 3 hrs, bluegrass 3 hrs wkly. ♦Dave Lockwood, gen mgr.

Cape Croker (Neyaashiinigmiing)

CHFN-FM— 2003: 100.1 mhz; 72 w. Hrs open: 24 RR 5, Wiarton, N0H 2T0. Phone: (519) 534-1003. Fax: (519) 534-4916. E-mail: chfnradio_station@yahoo.ca Web Site: www.nawash.ca/chfn. Licensee: Jessica Nadjiwon, on behalf of a non-profit corporation to be incorporated. Population served: 1,200 Format: Aboriginal news & programs relevant to the Ojibway people. News: 12 hrs wkly. Target aud: 18-65; progmg is div. ♦Jake Linklater, pres; Peter Akiwenzie, VP; Jessica Nadjiwon, gen mgr; Beedahsega Elliott, mktg mgr; Johnathan Pedoniquotte, progmg mgr.

Chatham

CFCO(AM)— 1926: 630 khz; 10 kw-D, 6 kw-N, DA-2. TL: N42 20 03 W82 16 53. Stereo. Hrs open: 24
FM 92.9 CFCO-FM, CFCO-FM, Chatham, Ont, 100%.
Box 100, N7M 5K1. Secondary address: 117 Keil Dr. S. N7M 5K1. Phone: (519) 354-2853. Fax: (519) 354-2880. E-mail: info@630cfco.com Web Site: www.630cfco.com. Licensee: Bea-Ver Communications Inc. (acq 3-20-97). Population served: 130,000 Rgnl rep: Rgnl Reps Format: MOR. News staff: 6; News: 6 hrs wkly. Target aud: 35 plus. Spec prog: Farm 3 hrs, gospel 2 hrs wkly. ♦Carl Veroba, CEO, pres & gen mgr; Doug Kirk, VP.

CKGW-FM— 2007: 89.3 mhz; 16.7 kw. Ant 436 ft TL: N42 26 14 W82 06 23. Hrs open: 40 Centre St., N7M 5W3. Phone: (613) 966-4822. Fax: (613) 966-3211. Web Site: www.ucbchathamkent.com. Licensee: United Christian Broadcasters Canada. Format: Christian music. ♦Garry Quinn, gen mgr.

CKSY-FM— Nov 17, 1999: 94.3 mhz; 50 kw. Ant 495 ft TL: N42 26 14 W82 06 23. Hrs open: 24 Box 100, N7M 5K1. Secondary address: 117 Keil Dr. S. N7M 5K1. Phone: (519) 354-2200. Phone: (519) 354-0311. Fax: (519) 354-2880. E-mail: info@cksyfm.com Web Site: www.cksyfm.com. Licensee: Bea-Ver Communications Inc. Population served: 130,000 Format: Adult contemp. News staff: 5; News: 5 hrs wkly. Target aud: 18-54. Spec prog: Gospel 2 hrs wkly. ♦Carl Veroba, CEO, pres & gen mgr; Doug Kirk, VP; Phil Ceccacci, gen sls mgr; Shannon Snoes, prom dir; Jay Poole, progmg mgr.

CKUE-FM— July 1, 1986: 95.1 mhz; 36.4 kw. Ant 495 ft TL: N42 26 14 W82 06 23. Stereo. Hrs open: 24 Box 100, N7M 5K1. Secondary address: 117 Keil Dr. S. N7M 5K1. Phone: (519) 354-2853. Fax: (519) 354-2880. E-mail: info@therock951.com Web Site: www.therock951.com. Licensee: Bea-Ver Communications Inc. Population served: 130,000 Format: AOR. News staff: 2. Target aud: 18-49. ♦Carl Veroba, CEO, pres & gen mgr; Doug Kirk, VP; Justin Oliphant, prom dir; Walter Ploegman, progmg dir; Ron Wilken, chief of engrg.

Christian Island

CKUN-FM— 2003: 101.3 mhz; 900 w. Ant 156 ft Hrs open: Beausoleil First Nation Band Council no. 30 & 30A, Administration Office, 1 O'Gema St., L0K 1C0. Phone: (705) 247-2456. Phone: (705) 247-2051. Fax: (705) 247-2239. Web Site: www.chimnissing.ca/xtras/radio.html. Licensee: Chimnissing Communications. Format: Div music. ♦Edna King, gen mgr; Kim Anderson, progmg dir & disc jockey; Richard Sutherland, progmg dir, engrg mgr & disc jockey.

Cobourg

CFMZ-FM— 1978: 103.1 mhz; 86.7 kw. Ant 825 ft TL: N01 44 04 W01 78 09. (Digital radio: 1466.768 mhz; 5.084 kw). Stereo. Hrs open: 24 550 Queen St. E., Suite 205, Toronto, M5A 1VZ. Secondary address: Box 1031, One Queen St. K9A 1M8. Phone: (905) 367-5353. Fax: (905) 367-1742. E-mail: info@classical963fmx.com Web Site: www.classical963fm.com. Licensee: MZ Media Inc. (acq 8-31-2006; C$12 million with CFMZ-FM-1 Toronto). Population served: 630,000 Natl. Rep: imsradio. Law Firm: Robson Broadcast Consultants. Format: Class. News staff: 3; News: 4 hrs wkly. Target aud: 35 plus; well-educated, upscale, owners/managers/professionals.Truus Rosenthal, VP; John van Driel, gen mgr & mus dir; Roberta Hunt, opns mgr; Al Kingdon, rgnl sls mgr; Marissa Colalillo, prom dir & spec ev coord; David Franco, news dir; Wassim Saikali, chief of engrg; Ann Pospischil, traf mgr; Paula Citron, mus critic; Arlene Meadows, disc jockey; Bill Anderson, disc jockey; Kerry Stratton, disc jockey; Michael Kramer, disc jockey; Michael Lyons, disc jockey; Mike Duncan, disc jockey

CHUC-FM— August 2006: 107.9 mhz; 6.3 kw. Hrs open: 24 Box 520, K9A 4L3. Secondary address: 7805 Telephone Rd. K9A 4J7. Phone: (905) 372-5401. Fax: (905) 372-6280. E-mail: don.conway@1079thebreeze.com Licensee: Pineridge Broadcasting Inc. Format: Adult contemp. ♦Don Conway, pres.

CKSG-FM— July 18, 2002: 93.3 mhz; 4 kw. Hrs open: 24 Box 520, K9A 4L3. Secondary address: 7805 Telephone Rd. K9A 4J7. Phone: (905) 372-5401. Fax: (905) 372-6280. E-mail: info@star933.com Web Site: www.star933.com. Licensee: Pineridge Broadcasting Inc. Population served: 170,000 Format: Adult contemp. News staff: one; News: one hr wkly. Target aud: 25-54; predominately female. ♦Don Conway, pres & gen mgr.

Cochrane

CHPB-FM— 2004: 98.1 mhz; 50 w. Hrs open: 24 49 Cedar St. South, Timmins, P4N 2Q5. Phone: (705) 267-6070. Fax: (705) 267-6095. E-mail: moose981@hbgradio.com Licensee: The Haliburton Broadcasting Group Inc. Group owner: Haliburton Broadcasting Group Inc. (acq 11-19-2003; with CFIF-FM Iroquois Falls). Format: Adult contemp. News staff: one; News: 1.5 hrs wkly. Target aud: 18-65. ♦Kimberly Ward, VP; Christopher Grossman, gen mgr; Mike Fry, progmg dir; Kent Matheson, mus dir; Wendy Gray, news dir; Penny Proulx, traf mgr; Donna Todd, sls.

Collingwood

CKCB-FM— Mar 29, 1996: 95.1 mhz; 350 w. Hrs open: 24 1400 Hwy. 26 E., L9Y 4W2. Phone: (705) 446-9510. Fax: (705) 444-6776. E-mail: jeaton@thepeakfm.com Web Site: www.thepeakfm.com. Licensee: 591989 B.C. Ltd. Group owner: Corus Entertainment Inc. (acq 3-24-00; grpsl). Format: Adult contemp. News staff: one; News: 11 hrs wkly. Target aud: 25-54. ♦John Eaton, gen mgr; John Nichols, opns mgr, gen sls mgr & progmg dir; Kim Di Girolamo, prom mgr; Dale West, news dir.

Cornwall

CFLG-FM— 1973: 104.5 mhz; 15 kw. 300 ft TL: N45 03 30 W74 44 45. Stereo. Hrs open: 24 237 Water St. E., K6H 1A2. Phone: (613) 932-5180. Fax: (613) 938-0355. E-mail: derrickscott@variety104.com Web Site: www.seawayvalley.com. Licensee: Corus Radio Co. Group owner: Corus Entertainment Inc. (acq 11-19-01; grpsl). Population served: 104,300 Natl. Rep: Canadian Broadcast Sales. Format: Adult contemp. News staff: 4; News: 5 hrs wkly. Target aud: 25-54; predominantly female professionals & housewives. ♦Tim Wieczorek, gen mgr; Rob Sequin, prom dir; Derrick Scott, progmg dir; Candace Livingstone, traf mgr.

CHOD-FM— May 1, 1994: 92.1 mhz; 19.2 kw. Stereo. Hrs open: 24 1111 Montreal Rd., Suite 202, K6H 1E1. Phone: (613) 936-2463. Fax: (613) 936-2568. E-mail: chodfm921@fastmail.fm Licensee: Radio Communautaire Cornwall-Alexandria Inc. Population served: 45,000 Format: Adult pop, French. News staff: one; News: 10 hrs wkly. Target aud: 25-54. Spec prog: Class 4 hrs, jazz 4 hrs wkly. ♦Norman Couture, pres; Marc Charbonneau, gen mgr.

CJSS-FM— 1999: 101.9 mhz; 1.42 kw. TL: N45 03 30 W74 44 45. Hrs open: 24 Box 969, 237 Water St. E., K6H 5V1. Phone: (613) 932-5180. Fax: (613) 938-0355. E-mail: onaircjss@rock1019.com Web Site: www.seawayvalley.com. Licensee: Corus Radio Co. Group owner: Corus Entertainment Inc. (acq 11-19-01; grpsl). Natl. Rep: Canadian Broadcast Sales. Format: Rock. News staff: 4; News: 2 hrs wkly. Target aud: 35 plus; well informed—interested in loc news & info including talk radio. Spec prog: Relg one hr wkly. ♦Tim Wieczorek, gen mgr.

CJUL(AM)—Co-owned with CJSS-FM. Nov 24, 2000: 1220 khz; 1 kw-U. 24 E-mail: onaircjul@seawayvalley.com Format: Oldies. News staff: 4; News: 4 hrs wkly. Target aud: 45 plus.

Dryden

***CJIV-FM**— March 2003: 97.3 mhz; 50 w. Hrs open: 24 Box 112, P8N 2Y7. Phone: (807) 937-9731. Fax: (807) 937-6490. E-mail: cjivradio@yahoo.ca Web Site: www.cjiv973.net. Licensee: Way of Life Broadcasting. Format: Christian. News: 2 hrs wkly. Target aud: All ages; interested in Christian radio bcsts. ♦Gordon Robinson, gen mgr; Jake Letkeman, progmg mgr.

CKDR-FM— Nov 9, 2005: 92.7 mhz; 36.8 kw. Hrs open: Box 580, P8N 2Z3. Phone: (807) 223-2355. Fax: (807) 223-5090. E-mail: mail@ckdr.net Web Site: www.ckdr.net. Licensee: Fawcett Broadcasting Ltd. Format: Adult contemp. Target aud: 25 plus. ♦Bruce Walchuk, gen mgr; Richard McCarthy, opns dir; Randy Pike, news dir.

Elliot Lake

CKNR-FM— Mar 3, 1997: 94.1 mhz; 90 kw. Hrs open: 144 Ontario Ave, P5A-1Y3. Phone: (705) 848-3608. Fax: (705) 848-1378. E-mail: moose941@hbgradio.com Web Site: www.hbgradio.com. Licensee: The Haliburton Broadcasting Group Inc. Group owner: Haliburton Broadcasting Group Inc. (acq 3-12-2004; C$625,000). Natl. Rep: Canadian Broadcast Sales. Format: Light Rock. Target aud: 35-54. ♦Christopher Grossman, pres & gen mgr; Kimberly Ward, VP; Erika MacLellan, opns mgr & sls; Bob Alexander, prom dir & disc jockey; Joe Snider, news rptr; Chris Waschuk, sls.

Englehart

CJBB-FM— January 2000: 103.1 mhz; 1.6 kw. Stereo. Hrs open: 24 Box 665, 50 Third St., P0J 1H0. Phone: (705) 544-1121. Fax: (705) 544-2286. E-mail: cjbb@ntl.sympatico.ca Licensee: 1353151 Ontario Inc. Population served: 39,000 Format: Adult contemp, rock. News staff: 1; News: 5 hrs wkly. ♦Boyd Woods, CEO; Rick Stow, progmg dir; Pat Ferris, traf mgr.

Erin

CHES-FM— 2006: 101.5 mhz; 50 w. Stereo. Hrs open: 24 Box 881, N0B 1T0. Secondary address: 106 Main St. N0B 1T0. Phone: (519) 833-1015. E-mail: info@erinradio.ca Web Site: www.erinradio.ca. Licensee: Erin Community Radio. Population served: 11,000 Format: Var. ♦Jay Mowat, chmn.

Fort Erie

CKEY-FM— May 19, 1991: 101.1 mhz; 19.7 kw. TL: N42 53 52 W78 57 27. Stereo. Hrs open: 24 4668 St. Clair Ave., Niagara Falls, L2E 6X7. Phone: (905) 356-6710. Fax: (905) 356-0696. E-mail: cjrniz@niagara.com Web Site: www.z101.fm. Licensee: CJRN 710 Inc. Population served: 1,385,000 Format: Top-40, CHR. Target aud: 18-44; upper income adults. ♦David J. Dancy, pres & gen sls mgr; Elizabeth Lewis, gen mgr; Heather Vigna, prom dir; Robert White, progmg dir; Mike Ridley, chief of engrg.

Fort Frances

CFOB-FM— June 4, 2002: 93.1 mhz; 21 kw. Hrs open: 24 242 Scott St., P9A 1G7. Phone: (807) 275-5341. Fax: (807) 274-2033. E-mail: alad@b93.ca Licensee: Fawcett Broadcasting Ltd. (group owner). Population served: 30,000 Natl. Rep: TeleRep. Format: Adult contemp. News staff: 2. Target aud: 25-54; International Falls/N. Central MN. ♦Ala Dulas, gen mgr.

Georgina Island

CFGI-FM— 2004: 102.7 mhz; 250 w. Hrs open: 102.7 Nish Radio, Box N-13, Sutton West, L0E 1R0. Phone: (705) 437-3748. Fax: (705) 437-3748. Licensee: Georgina Island First Nations Communications. Format: Var. ♦Sally Charles, gen mgr.

Goderich

CHWC-FM— August 2007: 104.9 mhz; 5.33 kw. Hrs open: Box 280, Owen Sound, N4K 5P5. Phone: (519) 376-2030. Fax: (519) 371-4242. Licensee: Bayshore Broadcasting Corp. Format: Classic adult contemp. Target aud: 35-64. ♦Ross Kentner, gen mgr.

Guelph

CFRU-FM— Jan 28, 1980: 93.3 mhz; 250 w. 1,085 ft TL: N43 32 07 W80 13 25. Stereo. Hrs open: 24 Level 2 Univ. Ctr., Univ. of Guelph, N1G 2W1. Phone: (519) 824-4120, Ext. 5302. Fax: (519) 763-9603. E-mail: info@cfru.ca Web Site: www.cfru.ca. Licensee: University of Guelph Radio-Radio Gryphon. Format: Multicultural, Div. News staff: one; News: 12 hrs wkly. Target aud: General. Spec prog: It one hr, relg one hr, Sp. ♦MacKenzie Jenkins, opns mgr; Lori Guest, progmg dir; Helen Spitzer, mus dir; Jennifer Moore, news dir.

CIMJ-FM—Listing follows CJOY(AM).

CJOY(AM)— June 14, 1948: 1460 khz; 10 kw-U. Stereo. Hrs open: 24 75 Speedvale Ave. E., N1E 6M3. Phone: (519) 824-7000. Fax: (519) 824-4118. E-mail: cjoy@cjoy.com Licensee: 591989 B.C. Ltd. Group owner: Corus Entertainment Inc. (acq 3-24-00; grpsl). Format: Oldies. News staff: 4; News: 8 hrs wkly. Target aud: 25-54. ♦Guus Hazelaar, gen mgr; Larry Mellott, progmg mgr; Mike Stevens, engrg VP.

CIMJ-FM—Co-owned with CJOY(AM). 1969: 106.1 mhz; 50 kw. 249 ft Stereo. 24 E-mail: kkelly@magic106.com Web Site: www.magic106.com.101,000 Format: Adult contemp. News: 6 hrs wkly. Target aud: 18-49. ♦Kevin Kelly, progmg mgr; Curtis Dunat, mus dir.

Haldimand County

CKJN-FM— May 15, 2006: 92.9 mhz; 3.3 kw. Ant 358 ft TL: N42 56 29 W79 50 45. Hrs open: 282 Argyle St. S., Unit 4, Caledonia, N3W 1K7. Phone: (289) 284-1070. Fax: (289) 284-1072. Web Site: www.jaynefm.ca. Licensee: Bel-Roc Communications Inc. (acq 4-18-2007). Natl. Rep: imsradio. Format: Multi-genre.

Haliburton

CFZN-FM— Mar13, 2006: 93.5 mhz; 6 kw. Hrs open: Box 960, K0M 1S0. Secondary address: 152 Highland St., Upper Level K0M 1S0. Phone: (705) 457-3897. Fax: (705) 457-3827. E-mail: moose935@hbgradio.com Web Site: www.moosefm.com. Licensee: The Haliburton Broadcasting Group Inc. Format: Classic rock. ♦Christopher Grossman, pres; Kim Parish, opns mgr.

CKHA-FM— July 2003: 100.9 mhz; 3.4 kw. Hrs open: Box 1125, K0M 1S0. Phone: (705) 457-9603. Fax: (705) 457-9522. E-mail: ccanoefm@bellnet.ca Web Site: www.canoefm.com. Licensee: Haliburton County Community Radio Association. Format: Var. ♦Dave Sovereign, stn mgr.

Hamilton

***CFMU-FM**— Jan 13, 1978: 93.3 mhz; 166 w. 300 ft TL: N47 14 41 W79 54 58. Stereo. Hrs open: 24 McMaster Univ. Student Center, Rm. B119, L8S 4S4. Phone: (905) 525-9140, EXT. 22053. Fax: (905) 529-3208. E-mail: cfrmunews@msu.mcmaster.ca Web Site: cfmu.mcmaster.ca. Licensee: CFMU Radio Inc. (acq 1978). Population served: 400,000 Format: Div, pub affrs, multicultural. News: 15 hrs wkly. Target aud: General; univ students, people with an adventurous outlook towards life. Spec prog: Class 5 hrs, Sp one hr, blues 5 hrs, Canadian Indian one hr, Fr one hr, It one hr wkly. ♦Sandeep Bhandari, stn mgr; James Hayashi-Tennant, progmg dir; Craig Nordemann, mus dir; Stefan Lozinski, prom.

CHAM(AM)— November 1959: 820 khz; 50 kw-U, DA-2. Stereo. Hrs open: 24 883 Upper WentWorth, Suite 401, L9A 4Y6. Phone: (905) 574-1150. Fax: (905) 575-6429. E-mail: info@820cham.com Web Site: www.820cham.com. Licensee: Standard Radio Inc. Group owner: Standard Broadcasting Corp. (acq 4-29-2002; grpsl). Wire Svc: BN Wire Format: Classic country. News staff: 4; News: 26 hrs wkly. Target aud: 25-54. ♦Gary Slaight, CEO & pres; Tom Cooke, VP & gen mgr.

CHML(AM)— May 27, 1927: 900 khz; 50 kw-U, DA-1. Stereo. Hrs open: 24 875 Main St. W., L8S 4R1. Phone: (905) 521-9900. Fax: (905) 521-2306. Licensee: Corus Premium Television Ltd. Group owner: Corus Entertainment Inc. (acq 7-6-2000; grpsl). Natl. Rep: Canadian Broadcast Sales. Format: News/talk, sports. Target aud: 35 plus. ♦Suzanne Carpenter, gen mgr; Greg Hinton, progmg dir; Mike Rose, mus dir.

CING-FM—Co-owned with CHML(AM). Sept 14, 1964: 95.3 mhz; 100 kw. Ant 1,000 ft Stereo. 24 64 Jefferson Ave., Unit 18, Toronto, M6K 3H4. Secondary address: 875 Main St. W., Suite 900 L8S 4R1. Phone: (416) 534-1191. Phone: (905) 521-9900. Fax: (416) 583-4133. Fax: (905) 540-2453. Web Site: www.country953.com. Natl. Rep: Canadian Broadcast Sales. Format: Country. Target aud: 25-54; female. ♦Ginny Townson Sedik, gen sls mgr; Nadia Cerelli-Fiore, prom dir; Steve Parsons, progmg dir; Rick Walters, mus dir; Ted Townsend, engrg dir.

***CIOI-FM**— 1998: 101.5 mhz; 240 w. TL: N43 14 12 W79 53 13. Stereo. Hrs open: 24 Mohawk College, 135 Fennell Ave. W., L8N 3T2. Phone: (905) 575-2175. Fax: (905) 575-2385. Licensee: The Mohawk College Radio Corp. Population served: 400,000 Format: College alternative. News: 6 hrs wkly. Target aud: 17-24; college students & the div communities they represent. Spec prog: Sp 3 hrs, Indian one hr, Assyrian one hr wkly. ♦Les Palango, gen mgr.

CIWV-FM— Sept 1, 2000: 94.7 mhz; 21.4 kw. Ant 446 ft TL: N43 12 21 W79 43 50. Stereo. Hrs open: 24 589 Upper Wellington St., L9A 3P8. Phone: (905) 388-8911. Fax: (905) 388-7947. E-mail: smoothjazz@wave947.fm Web Site: www.wave947.fm. Licensee: Burlingham Communications Inc. Population served: 3,638,000 Natl. Rep: Target Broadcast Sales. Wire Svc: Broadcast News Ltd. Format: Smooth jazz. News staff: 2. Target aud: 35-64. ♦Douglas E. Kirk, chmn, pres & gen mgr; Thomas A. Pippy, CFO; Steve Kassay, opns VP; Simon Constam, gen sls mgr; Cathy Philippo, traf mgr.

CKLH-FM—Listing follows CKOC(AM).

CKOC(AM)— May 20, 1922: 1150 khz; 50 kw-U, DA-2. TL: N43 03 04 W79 48 42. Stereo. Hrs open: 24 883 Upper Wentworth St., Suite 401, L9A 4Y6. Phone: (905) 574-1150. Fax: (905) 575-6429. Web Site: www.oldies1150.com. Licensee: Standard Radio Inc. Group owner: Standard Broadcasting Corp. (acq 4-29-2002; grpsl). Population served: 572,000 Natl. Rep: Canadian Broadcast Sales. Wire Svc: Broadcast News Ltd. Format: Oldies. News staff: 4; News: 3 hrs wkly. Target aud: 25-54. ♦Tom Cooke, gen mgr & gen sls mgr; Christopher Randall, prom dir & prom mgr; Sunni Genesco, asst music dir.

CKLH-FM—Co-owned with CKOC(AM). Oct 7, 1986: 102.9 mhz; 40.3 kw. TL: N43 20 12 W79 52 07. Stereo. E-mail: info@k-litefm.com Web Site: www.l-litefm.com. Format: Adult contemp. Target aud: 25-54; working women, owners, mgrs, professionals.

Hanover

CFBW-FM— Dec 31, 2001: 91.3 mhz; 250 w. Stereo. Hrs open: 24 275 10th St., Suite 4, N4N 1P1. Phone: (519) 364-0200. Fax: (519) 364-5175. E-mail: bluewaterradio@on.aibn.com Web Site: www.bluewaterradio.ca. Licensee: Bluewater Community Radio. Format: Div. Target aud: 12-75; Ontario audience rural agricultural/urban. Spec prog: Blues 2 hrs, gospel 6 hrs, Scottish 2 hrs wkly. ♦Andrew McBride, stn mgr; Carole Plunkett, adv.

Hawkesbury

CHPR-FM— February 1986: 102.1 mhz; 789 w. 70 ft TL: N45 35 01 N45 35 01. Hrs open: 24 115 Principale E., Suite 101, K6A 1A1. Phone: (613) 632-1000. Fax: (450) 562-1902. E-mail: infocouleurfmm@radionord.vom Web Site: www.radionord.com. Licensee: RNC MEDIA Inc. Group owner: Radio Nord Inc. (acq 8-22-89). Format: Adult contemp. News staff: one. Target aud: 25 plus. ♦Pierre R. Brosseau, pres & gen mgr.

Hearst

CHYK-FM-3— 1996: 92.9 mhz; 140 w. Hrs open: 24 Rebroadcasts CHYK-FM Timmins.
49 Cedar St. S., Timmins, P4N 2G5. Phone: (705) 267-6070. Fax: (705) 267-6095. E-mail: chycfm@nbgradio.com Web Site: www.chycfm.com. Licensee: Haliburton Broadcasting Group Inc. (group owner; (acq 8-31-99; grpsl). Natl. Rep: Canadian Broadcast Sales. Format: Hot adult contemp, French, English. News staff: one; News: 1 hrs wkly. Target aud: 18-65. ♦Christopher Grossman, pres & gen mgr; Kimberley Grossman, VP; Sylvain Boucher, progmg dir & mus dir; Gilles Lafortune, news dir; Penny Proulx, traf mgr; Sylvie Beaulieu, sls.

***CINN-FM**— 1988: 91.1 mhz; 5.5 kw. 298 ft TL: N49 38 50 W83 30 50. Hrs open: 6 AM-9 PM Box 2648, 1004, rue Prince, P0L 1N0. Phone: (705) 372-1011. Fax: (705) 362-7411. E-mail: cinnfm@cinnfm.com Web Site: www.cinnfm.com. Licensee: Radio de l'Epinette Noire Inc. Format: Adult contemp. News staff: 2. Target aud: 0-75. ♦Isabelle Lacroix-Breton, pres; Gaitane Morrissette, gen mgr.

Huntsville

CFBK-FM— September 1957: 105.5 mhz; 5 kw. Hrs open: 24 Unit 2, 15 Main St. E., P1H 2C6. Phone: (705) 789-4461. Fax: (705) 789-1269. Licensee: Muskoka-Parry Sound Broadcasting Ltd. (acq 9-13-94). Population served: 30,000 Natl. Network: CHUM Radio Network. Wire Svc: BN Wire Format: Adult contemp. News staff: 3. Target aud: 21yrs +. ♦Ian Byers, CEO, chmn, pres, gen mgr & gen sls mgr; Margaret Byers, opns dir & progmg mgr.

Iroquois Falls

CFIF-FM— Dec 8, 1998: 101.1 mhz; 50 w. Hrs open: 24 49 Cedar St. S., Timmins, P4N 2G5. Phone: (705) 267-6070. Fax: (705) 267-6095. E-mail: moose1011@hbgradio.com Web Site: www.hbgradio.com.no Licensee: The Haliburton Broadcasting Group Inc. Group owner: Haliburton Broadcasting Group Inc. (acq 11-19-2003; with CHPB-FM Cochrane). Population served: 2,500 Format: Classic hits. News staff: one; News: 1.5 hrs wkly. Target aud: 25-54. ♦Christopher Grossman, pres & gen mgr; Mike Fry, progmg dir; Kent Matheson, mus dir; Wendy Gray, news dir; Penny Proulx, traf mgr; Shawn McArthur, sls.

Kaministiquia

CFQK-FM— 2002: 104.5 mhz; 50 w. Hrs open: 584 Red River Rd., Suite 200, Thunder Bay, P7B 1H3. Phone: (807) 768-5048. Fax: (807) 767-7634. E-mail: max104@tbaytel.net Licensee: Northwest Broadcasting Inc. Format: Hit music. ♦Ari Lahdekorpi, gen mgr; Rich Fleming, progmg dir.

Kapuskasing

CKAP-FM— September 2001: 100.9 mhz; 12 kw. Hrs open: Box 960, Haliburton, K0M 1S0. Phone: (705) 335-2379. Fax: (705) 337-6391. Web Site: www.hbgradio.com. Licensee: The Haliburton Broadcasting Group Inc. Group owner: Haliburton Broadcasting Group Inc. Format: CHR. News staff: 2. Target aud: General. ♦Christopher Grossman, pres; Valerie Isaac, opns mgr.

CKGN-FM— October 1993: 89.7 mhz; 3 kw. Hrs open: 24 77 chemin Brunelle Nd., P5N 2M1. Phone: (705) 335-5915. Fax: (705) 335-3508. E-mail: ckgnfm@nt.net Web Site: www.ckgn.ca. Licensee: Radio communautaire KapNord Inc. Population served: 20,000 Format: Fr, var/div, mixed music. News staff: one. Target aud: General. ♦Claude Chabot, gen mgr.

Kenora

CBQX-FM— Mar 28, 1978: 98.7 mhz; 38 kw. Hrs open:
Rebroadcasts CBW(AM) Winnipeg, Man. & CBQT-FM Thunder Bay.
213 Miles St. E., Thunder Bay, P7C 1J5. Phone: (807) 625-5000/(416) 205-3700. Fax: (416) 205-3111. Web Site: www.nwo.cbc.ca. Licensee: CBC. Natl. Network: CBC Radio One. Format: Info. ♦Kelly McInnes, gen mgr & stn mgr.

CJRL-FM— 2004: 89.5 mhz; 40 kw. TL: N49 46 45 W94 27 25. Hrs open: 128 Main St. S., P9N 1S9. Phone: (807) 468-3181. Fax: (807) 468-4188. E-mail: newsroom@895mix.fm Licensee: Fawcett Broadcasting Ltd. Format: Hot adult contemp. Target aud: 25-54. Spec prog: Ukrainian one hr wkly. ♦H.G. Fawcett, pres; Hugh Syrja, gen mgr.

Kettle Point

CKTI-FM— Apr 26, 2004: 107.7 mhz; 420 w. Hrs open: Points' Eagle Radio, R.R. 2, Forest, N0N 1J0. Secondary address: 9111 W. Ipperwash Rd., Unit 6 N0N 1J0. Phone: (519) 786-3883. Fax: (519) 786-2834. E-mail: points_eagle_radio@hotmail.com Web Site: www.angelfire.com/rock3/points_eagle_radio. Licensee: Point Eagle Radio Inc. Format: Country, classic rock. ♦Connie George, opns dir.

Killaloe

CHCR-FM— 1998: 102.9 mhz; 33 w. Hrs open: Box 195, K0J 2A0. Secondary address: 7A Lake St., 2nd Fl. K0J 2A0. Phone: (613) 757-0657. Fax: (613) 757-0818. E-mail: stationmanager@chcr.org Web Site: www.chcr.org. Licensee: Homegrown Community Radio. Format: Div. Spec prog: Canadian fiddle 8 hrs, Fr 8 hrs, Pol 1 hr, traditional bluegrass 6 hrs wkly. ♦Daryl Andermann, gen mgr.

Kincardine

CIYN-FM— March 2006: 95.5 mhz; 5.66 kw. Hrs open: 24 807 Queen St., N2Z 2Y2. Phone: (519) 396-7770. Fax: (519) 396-7771. E-mail: info@thecoastfm.ca Web Site: www.thecoastfm.ca. Licensee: Brian Cooper and Daniel McCarthy, on behalf of a corporation to be incorporated. Population served: 20,000 Format: Adult classic hits. News staff: one. ♦Mike Brough, gen mgr; Lynda Cooper, news dir; Steve Howard, sls.

Kingston

CBBK-FM— May 21, 1979: 92.9 mhz; 1.6 kw. 395 ft TL: N44 17 32 W76 28 50. Stereo. Hrs open: 24 Box 500, Station A, Toronto, M5W 1E6. Phone: (416) 205-3700. Fax: (416) 205-6063. Web Site: www.cbc.ca. Licensee: Canadian Broadcasting Corp. Natl. Network: CBC Radio Two. Format: Public radio. ♦Robert Raeinobitch, CFO & progmg dir.

CFFX(AM)— Aug 31, 1942: 960 khz; 10 kw-D, 5 kw-N, DA-2. Stereo. Hrs open: 170 Queen St., K7K 1B2. Phone: (613) 549-1911. Fax: (613) 544-5508. Web Site: www.oldies960.com. Licensee: 591989 B.C. Ltd. Group owner: Corus Entertainment Inc. (acq 3-24-00; grpsl). Population served: 150,000 Natl. Rep: CBS Radio. Format: Oldies. News staff: one; News: 3 hrs wkly. Target aud: 35-64; female. Spec prog: Toronto Blue Jays, Toronto Maple Leafs, Kingston Hockey. ♦Mike Ferguson, VP & gen mgr; Jim Elyot, opns mgr & prom mgr.

CFMK-FM—Co-owned with CFFX(AM). Aug 31, 1942: 96.3 mhz; 14 kw. Ant 500 ft Stereo. Phone: (613) 544-2340. Web Site: www.963joefm.com. Format: 80s based Gold. News staff: one; News: new progmg 2 hrs wkly. Target aud: 25-54. Co-owned TV: CKWS-TV affil.

CFLY-FM—Listing follows CKLC(AM).

***CFRC-FM**— January 1953: 101.9 mhz; 3 kw. 295 ft TL: N44 17 24 W77 25 55. Stereo. Hrs open: 24 Queens Univ., Lower Carruthers Hall, K7L 3N6. Phone: (613) 533-2121. Fax: (613) 533-6049. E-mail: cfrcops@ams.queensv.ca Web Site: www.cfrc.ca. Licensee: Radio Queen's University. Population served: 130,000 Format: Div. News: 5 hrs wkly. Target aud: General. ♦Eric Beers, stn mgr; Michael Brolley, progmg mgr; Scott Stevens, mus dir.

CIKR-FM— Feb 19, 2001: 105.7 mhz; 24 kw. Stereo. Hrs open: 24 863 Princess, Suite 301, K7L 5N4. Phone: (613) 549-1057. Fax: (613) 549-5302. Web Site: www.krock1057.ca. Licensee: K-Rock 1057 Inc. Population served: 150,000 Format: Rock. News staff: 2; News: 2 hrs wkly. Target aud: 25-54; adults. ♦John P. Wright, pres & gen mgr.

CKLC(AM)— Nov 18, 1953: 1380 khz; 10 kw-U, DA-1. Stereo. Hrs open: 24 993 Princess St., Suite 10, K7L 1H3. Phone: (613) 544-1380. Fax: (613) 546-9751. E-mail: jrobb@kos.net Web Site: www.cklc.com. Licensee: 1708479 Ontario Inc. (acq 5-14-76). Format: All time favorites. News staff: 4; News: 3 hrs wkly. Target aud: 35-55. ♦James Waters, pres; Gary Perrin, gen mgr.

CFLY-FM—Co-owned with CKLC(AM). 1963: 98.3 mhz; 100 kw. 400 ft Stereo. 24 E-mail: flyfm@flyfmkingston.com Web Site: www.flyfmkingston.com. Format: Adult contemp. News staff: 2; News: 5 hrs wkly. Target aud: 25-44.

***CKVI-FM**— 1997: 91.9 mhz; 6.5 w. Hrs open: 235 Frontenac St., K7L 3S7. Phone: (613) 544-7864. Fax: (613) 544-8795. E-mail: ckvi@limestone.on.ca Web Site: www.thecave.ca. Licensee: KCVI Educational Radio Station Inc. Format: Div. ♦Max Lienhard, gen mgr, progmg dir & chief of engrg.

Kirkland Lake

CJKL-FM— 1934: 101.5 mhz; 23 kw. Stereo. Hrs open: 24 Box 430, P2N 3J4. Phone: (705) 567-3366. Fax: (705) 567-6101. E-mail: cjkl@cjklfm.com Web Site: www.cjklfm.com. Licensee: Connelly Communications Corp. Natl. Rep: Canadian Broadcast Sales. Format: Hot Adult contemp. News staff: 2. ♦Ann Connelly, gen sls mgr; Rob Connelly, pres, stn mgr & progmg dir; Elesha Teskey, news dir; Greg Mackle, news dir; Don Elvidge, engrg dir.

Kitchener

CFCA-FM—Licensed to Kitchener. See Waterloo

CHYM-FM—Listing follows CKGL(AM).

CKGL(AM)— 1929: 570 khz; 10 kw-U, DA-1. Hrs open: 305 King St. W., N2G 4E4. Phone: (519) 743-2611. Fax: (519) 743-7510. Licensee: Rogers Broadcasting Ltd. Natl. Network: CBS. Format: News/talk. Target aud: 25-54. ♦Gavin Tucker, gen mgr.

CHYM-FM—Co-owned with CKGL(AM). 1949: 96.7 mhz; 25 kw. 658 ft (CP: 100 kw). Stereo. 24 310,000 Format: Lite rock. ♦Gavin Tucker, progmg dir; Neil Beaumont, mus dir; Mike McCabe, engrg mgr.

CKKW(AM)—Licensed to Kitchener. See Waterloo

CKWR-FM—Licensed to Kitchener. See Waterloo

Kitchener/Paris

CJIQ-FM— Jan 8, 2001: 88.3 mhz; 4 kw. Hrs open: Rm 3B15, Conestoga College, 299 Doon Valley Dr., Kitchener, N2G 4M4. Phone: (519) 748-5220. E-mail: cjiqinfo@cjiq.fm Web Site: www.cjiq.fm. Licensee: Conestoga College Communications Corp. Format: Div. ♦Mark Burley, stn mgr.

Kitchener-Waterloo

CIKZ-FM— Feb 6, 2004: 106.7 mhz; 1.7 kw. Ant 657 ft Hrs open: 490 Dutton Dr., Unit C2, Waterloo, N2L 6H7. Phone: (519) 746-3331. Fax: (519) 746-3364. E-mail: plarche@kicxfm.com Web Site: www.kicx106.com. Licensee: Larche Communications Inc. Format: New country. ♦Paul Larche, pres & gen mgr; Jordan Cooledge, gen sls mgr; Ron Funnell, gen sls mgr; Derm Carnduff, progmg dir.

CJTW-FM— February 2004: 94.3 mhz; 50 w. Stereo. Hrs open: 24 Faith FM, Box 1433, Stn C, Unit 202, Kitchener, N2G 4H6. Secondary address: 659 King St. E., Kitchener N2G 2M4. Phone: (519) 575-9090. Fax: (519) 575-9119. E-mail: info@faithfm.org Web Site: www.faithfm.org. Licensee: Sound of Faith Broadcasting. Format: Christian. ♦Robert Reid, pres; Dave MacDonald, gen mgr; Brad Loveday, progmg dir.

CKBT-FM— January 2004: 91.5 mhz; 3.6 kw. Stereo. Hrs open: 235 King St. E., Kitchener, N2G 4N5. Phone: (519) 741-9915. Fax: (519) 568-6390. Web Site: www.915thebeat.com. Licensee: Corus Premium Television Ltd. Group owner: CanWest Global Communications Corp. (acq 7-6-2007; C$14.5 million with CJZZ-FM Winnipeg, MB). Format: Rhythmic CHR. Target aud: 18-34. ♦David Jones, gen mgr; Sandra Henein, prom mgr; Ed Ringward, sls; Mike Blake, sls.

Leamington

CHYR-FM— Aug 23, 1993: 96.7 mhz; 10.65 kw. Hrs open: 100 Talbot St. E., N8H 1L3. Phone: (519) 326-6171. Fax: (519) 322-1110. E-mail: 96.7@chyr.com Web Site: www.chyr.com. Licensee: Blackburn Radio Inc. Group owner: Blackburn Group Inc. (acq 12-19-94; grpsl). Format: Hot adult contemp. Target aud: 25-54. Spec prog: It 3 hrs, relg 3 hrs wkly. ♦Terry Regier, gen mgr; Cordell Green, progmg dir; Kevin Black, news dir; Tim O'Neil, gen sls mgr, sls & adv.

CJSP-FM—Not on air, target date: unknown: 92.7 mhz; 960 w. Ant 474 ft TL: N42 00 35 W82 33 45. Hrs open: 100 Talbot St. E., N8H 1L3. Phone: (519) 326-6171. Fax: (519) 322-1110. Licensee: Blackburn Radio Inc. Format: Country. Target aud: 25-64. ♦Terry Regier, gen mgr.

Lindsay (city of Kawartha Lakes)

CKLY-FM— May 16, 1998: 91.9 mhz; 5.27 kw. Hrs open: 24 249 Kent St. W., Lindsay, K9V 2Z3. Phone: (705) 324-9103. Fax: (705) 324-4149. E-mail: y92@y92.net Web Site: www.chumlimited.com. Licensee: 1708479 Ontario Inc. (group owner; (acq 12-21-2000; C$800,000). Population served: 70,000 Natl. Rep: Canadian Broadcast Sales. Format: Adult contemp. News staff: 2; News: 14 hrs wkly. Target aud: 30-65. ♦Rick Ringer, opns mgr; Dave Illman, progmg dir & disc jockey; Steve Fawcett, gen mgr, gen sls mgr & engrg mgr.

Little Current

CFRM-FM— 2002: 100.7 mhz; 1.83 kw. Stereo. Hrs open: 24 10 Campbell St. E., P0P 1K0. Phone: (705) 368-1419. Fax: (705) 368-1080. E-mail: radio@manitoulin.net Web Site: www.101rocks.com. Licensee: Manitoulin Radio Communication Inc. Population served: 3,000 Format: Classic rock. News staff: 2. Target aud: General; baby boomer. Spec prog: Blues 2 hrs, country 5 hrs, gospel 2 hrs wkly. ♦Craig Timmermans, CEO & pres; Rick Nelson, stn mgr; Sam Nardi, opns VP; Bob Clark, sls.

London

CBBL-FM— Oct 1, 1978: 100.5 mhz; 22.5 kw. TL: N42 57 20 W81 21 20. Stereo. Hrs open: 24
Rebroadcasts CBL-FM Toronto.
Box 500, Station A, Toronto, M5W 1E6. Phone: (416) 205-3700. Fax: (416) 205-6063. Web Site: www.cbc.ca. Licensee: Canadian Broadcasting Corp. Natl. Network: CBC Radio Two. Format: Public radio. ♦Robert Raeinobitch, CFO & chief of engrg.

CBCL-FM— June 1998: 93.5 mhz; 69.3 kw. Hrs open: 208 Piccadilly St., Unit 4, N6A 1S1. Phone: (519) 667-1990. Fax: (519) 667-1557. Web Site: www.cbc.ca. Licensee: Canadian Broadcasting Corp. Natl. Network: CBC Radio One. Format: Public radio. Spec prog: News 10 hrs wkly. ♦Robert Raeinobitch, CFO & gen mgr.

CFHK-FM—See St. Thomas

CFPL(AM)— September 1922: 980 khz; 10 kw-D, 5 kw-N, DA-2. TL: N42 53 29 W81 12 02. Stereo. Hrs open: 24 380 Wellington St., Rm. 222, N6A 5B5. Phone: (519) 931-6000. Phone: (519) 667-4623. Fax: (519) 438-2415. Web Site: www.am980.net. Licensee: Corus Radio Co. Group owner: Corus Entertainment Inc. Population served: 450,000 Format: Sports, adult contemp, news/talk. News staff: 8; News: 10 hrs wkly. Target aud: 35-54. ♦Dean Sinclair, gen mgr; Bob Fisher, sls dir, gen sls mgr & prom dir; Rick Jackiw, rgnl sls mgr; Gord Harris, progmg dir & news dir; Andy Bingle, engrg dir.

CFPL-FM— 1948: 95.9 mhz; 179 kw. 885 ft TL: N42 57 15 W81 15 58. Stereo. 24 Phone: (519) 433-3696. Web Site: www.fm96.com.788,000 Format: New rock. Target aud: 25-49.

CHJX-FM— 2003: 105.9 mhz; 10 w. Hrs open: 100 Fullarton St., N6A 1K1. Phone: (519) 679-9882. Fax: (519) 679-2459. E-mail: gracefm_administration@skynet.ca Web Site: www.gracefm.ca. Licensee: Sound of Faith Broadcasting. Format: Contemp Christian music. ♦Doug Chaplin, gen mgr.

***CHRW-FM**— Oct 31, 1981: 94.9 mhz; 3.5 kw. Ant 128 ft TL: N43 00 30 W81 16 36. Hrs open: 24 250 Univ. Community Ctr., Univ. of Western Ontario, Room 250, N6A 3K7. Phone: (519) 661-3601. Fax: (519) 661-3372. E-mail: chrwgm@uwo.ca Web Site: www.chrwradiio.com. Licensee: Radio Western Inc. Population served: 350,000 Format: Alternative, multicultural, jazz,blues,metal. News staff: one; News: 5 hrs wkly. ♦Grant Stein, stn mgr & progmg VP; Alicks Girowski, prom dir & mus dir; Zoltan Harastzy, adv mgr & progmg dir.

CHST-FM— Sept 1, 2000: 102.3 mhz; 4.77 kw. Stereo. Hrs open: 24 102.3 Bob FM, 1 Communication Rd., N6A 6E9. Phone: (519) 690-0102. Fax: (519) 686-5942. Web Site: www.1023bob.com. Licensee: 1708479 Ontario Inc. (group owner) Wire Svc: BN Wire Format: Classic hits. News staff: one; News: one hr wkly. ♦Jay Switzer, CEO; Jim Blundell, gen mgr; Ann LaRocque, gen sls mgr; Al Smith, progmg dir. Co-owned TV: CFPL-TV.

CIQM-FM— June 1, 1986: 97.5 mhz; 50 kw. 300 ft Stereo. Hrs open: 24 743 Wellington Rd. S., N6C 4R5. Phone: (519) 686-2525. Fax: (519) 686-3658. Web Site: www.975sri.ca. Licensee: Standard Radio Inc. Group owner: Standard Broadcasting Corp. (acq 4-19-2002; grpsl). Wire Svc: BN Wire Format: Adult contemp. Target aud: 25-54; female. ♦Gary Slaight, CEO; Braden Doerr, exec VP, VP & gen mgr; Barry Smith, opns mgr & progmg mgr; Dan MacGillivray, gen sls mgr.

***CIXX-FM**— Oct 31, 1978: 106.9 mhz; 3 kw. Ant 150 ft Stereo. Hrs open: 24 1460 Oxford St. E., N5V 1W2. Phone: (519) 453-2810. Fax: (519) 453-2250. E-mail: mstoparczyk@1069fm.ca Web Site: www.1069fm.ca. Licensee: Radio Fanshawe Inc. Population served: 350,000 Format: Urban contemp. News staff: 2. Target aud: 12-34; primarily college, univ., high school. Spec prog: Christian 3 hrs, educ 4 hrs hrs wkly. ♦Steve Andruiak, gen mgr; Barry Sutherland, opns mgr, gen sls mgr & prom mgr.

CJBC-FM-4— Sept 3, 1978: 99.3 mhz; 22.5 kw. 91 ft TL: N42 57 20 W81 21 20. Hrs open:
Rebroadcasts CJBC(AM) Toronto.
Box 500, Stn A, Toronto, M5W 1E6. Phone: (416) 205-3311. Fax: (416) 205-7795. Web Site: www.torontocbc.ca. Licensee: Canadian Broadcasting Corp. Natl. Network: Premiere Chaine. Format: Div, Fr. ♦Claire Margetti, gen mgr & progmg dir.

CJBK(AM)— Jan 25, 1967: 1290 khz; 10 kw-U, DA-2. TL: N42 52 08 W81 13 58. Stereo. Hrs open: 24 743 Wellington Rd. S., N6C 4R5. Phone: (519) 686-2525. Fax: (519) 686-3658. Fax: (519) 686-1156. Web Site: www.cjbk.com. Licensee: Standard Radio Inc. Group owner: Standard Broadcasting Corp. (acq 4-19-2000; grpsl). Population served: 330,000 Format: News/talk. News staff: 4. Target aud: 35-54. ♦Gary Slaight, CEO; Braden Doerr, pres & gen mgr.

CJBX-FM—Co-owned with CJBK(AM). Mar 3, 1980: 92.7 mhz; 50 kw. 400 ft Stereo. 24 Web Site: www.bx93.com.330,000 Format: Country. ♦Braden Doerr, VP.

CKSL(AM)— June 1956: 1410 khz; 10 kw-U, DA-2. Stereo. Hrs open: 24 743 Wellington Rd. S., N6C4R5. Phone: (519) 686-2525. Fax: (519) 686-3658. E-mail: comments@oldies1410.com Web Site: www.am1410.ca. Licensee: Standard Radio Inc. Group owner: Standard Broadcasting Corp. (acq 4-19-2002; grpsl). Wire Svc: BN Wire Format:

Adult Standards. News staff: one; News: 1 hr wkly. Target aud: 35-54; adults 35-54. ♦Gary Slaight, CEO; Braden Doerr, gen mgr; Barry Smith, opns mgr; Dan MacGillivray, gen sls mgr.

Marathon

CFNO-FM— July 17, 1982: 93.1 mhz; 50 kw. 879 ft Stereo. Hrs open: Box 1000, P0T 2E0. Secondary address: 93 Evergreen Dr. P0T 2E0. Phone: (807) 229-1010. Fax: (807) 229-1686. E-mail: sales@cfno.fm Licensee: North Superior Broadcasting Ltd. (acq 1982). Format: Adult contemp. Spec prog: C&W 12 hrs wkly.

Midland

CICZ-FM— September 1993: 104.1 mhz; 9.354 kw. Hrs open: Box 609, 355 Cranston Crescent, L4R 4L3. Phone: (705) 526-2268. Fax: (705) 526-3060. E-mail: plarche@kicxfm.com Web Site: www.kicx104fm.com. Licensee: Larche Communications Inc. Format: New country. ♦Mora Austin, gen mgr & gen sls mgr; Paul Larche, pres & gen mgr; Drem Carnduff, progmg dir; Glen Prinz, chief of engrg.

Mississauga

CJMR(AM)— June 17, 1974: 1320 khz; 20 kw-U. (Digital radio: 1466.768 mhz; 5.084 kw). Hrs open: 24 Broadcasting Ctr., 284 Church St., Oakville, L6J 7N2. Phone: (905) 845-2821. Phone: (905) 271-1320. Fax: (905) 842-1250. E-mail: hmcdonald@whiteoaksgroup.ca Licensee: Trafalgar Broadcasting Ltd. Natl. Rep: Target Broadcast Sales. Format: Ethnic. ♦Harry McDonald, sls VP & sls dir; Michael Caine, pres, gen mgr & progmg dir.

Moosonee

***CHMO(AM)**— Feb 29, 1976: 1450 khz; 50 w. TL: N51 16 39 W80 38 40. Hrs open: 6 AM-11 PM Box 400, P0L 1Y0. Secondary address: 28 First St. P0L 1Y0. Phone: (705) 336-2466. Fax: (705) 336-2186. E-mail: jbbtcorp@owlink.net Licensee: James Bay Broadcasting Corp. Population served: 5,000 Format: Div, country, news. News staff: one; News: 10 hrs progm wkly. Target aud: General. Spec prog: Cree Indian. ♦John Kirk, pres; Ernest Hunter, stn mgr & prom mgr; Jack Williams, mus dir; George Witham, chief of engrg.

New Liskeard

CJTT-FM— June 26, 1998: 104.5 mhz; 10 kw. Hrs open: PO Box 1058, P0J 1P0. Secondary address: 55 Whitewood Ave. P0J 1P0. Phone: (705) 647-7334. Fax: (705) 647-8660. E-mail: cjtt@nt.net Licensee: Connelly Communications Corp. (acq 9-79). Format: Mix. News staff: one. ♦Gail Moore, gen mgr.

Newmarket

CKDX-FM— September 1994: 88.5 mhz; 11.3 kw. Hrs open: 5302 Dundas St. W., Etobicoke, M9B 1B2. Phone: (416) 213-1035. Fax: (416) 233-8617. E-mail: gracep@885thejewel.com Web Site: www.885thejewel.com. Licensee: CKDX Radio Ltd. Group owner: Evanov Radio Group (acq 12-21-2000). Format: Adult favorities. ♦Bill Evanov, pres; Bruce Campbell, progmg dir & sls; Gary Gamble, progmg mgr; Grace Pascucci, prom.

Niagara Falls

CFLZ-FM— 1996: 105.1 mhz; 4 kw. Hrs open: Box 710, L2E 6X7. Secondary address: 4668 St. Clair Ave. L2E 6X7. Phone: (905) 356-6710. Fax: (905) 356-0696. E-mail: robwhite@niagara.com Web Site: www.river.fm. Licensee: 788813 Ontario Inc. Natl. Rep: Canadian Broadcast Sales. Format: Modern adult contemp. News staff: 3; News: 5 hrs wkly. ♦David J. Dancy, pres & gen sls mgr; Elizabeth Lewis, gen mgr; Robert White, progmg dir.

North Bay

CFXN-FM— 2006: 106.3 mhz; 10 kw. Hrs open: 118 Main St. E., P1B 1A8. Licensee: The Haliburton Broadcasting Group Inc. Format: Contemp country.

CHUR-FM— 1996: 100.5 mhz; 100 kw. Stereo. Hrs open: Box 3000, P1B 8K8. Phone: (705) 474-2000. Fax: (705) 474-7761. Web Site: www.ezrocknorthbay.com. Licensee: Rogers Broadcasting Ltd. (group owner; acq 4-19-2002; grpsl). Format: Adult contemp, soft rock. Target aud: 25-54. ♦Ted Rogers, CEO & pres; Peter Mckeown, gen mgr & stn mgr.

CKAT(AM)— Mar 3, 1931: 600 khz; 10 kw-D, 5 kw-N, DA-1. Hrs open: 24 Box 3000, P1B 8K8. Phone: (705) 474-2000. Fax: (705) 474-7761. Licensee: Rogers Broadcasting Ltd. (group owner; acq 4-19-02; grpsl). Population served: 80,000 Format: Country. News staff: 5; News: 6 hrs wkly. Target aud: 25-54. ♦Rick Doughty, gen mgr; James Dahlke, gen sls mgr; Peter McKeown, prom dir; Dean Belanger, asst music dir; Clint Thomas, news dir; Csaba Senyi, engrg dir.

CKFX-FM—Co-owned with CKAT(AM). Jan 19, 1967: 101.9 mhz; 100 kw. Ant 350 ft Stereo. 24 Format: Rock. ♦Kevin Ochefski, prom dir; Mike Belanger, progmg dir.

North York

CILQ-FM—Licensed to North York. See Toronto

Oakville

CJYE(AM)— Nov 17, 1956: 1250 khz; 10 kw-D, 5 kw-N, DA-2. Hrs open: Broadcasting Ctr., 284 Church St., L6J 7N2. Phone: (905) 845-2821. Phone: (905) 271-1320. Fax: (905) 842-1250. E-mail: dmillar@joy1250.ca Web Site: www.christianradio.ca/station/cjye. Licensee: Trafalgar Broadcasting Ltd. Format: Contemp Christian music. ♦Harry H. McDonald, sr VP; Michael Caine, pres & gen mgr.

Ohsweken

***CKRZ-FM**— 1991: 100.3 mhz; 250 w. Stereo. Hrs open: 6 AM-11 PM Box 189, N0A 1M0. Secondary address: 1721 Chiefswood Rd., Oashweken Phone: (519) 445-4140. Fax: (519) 445-0177. E-mail: ckrzinfo@ckrz.com Web Site: www.ckrz.com. Licensee: Southern Onkwehon: We Nishinabec Indigenous Communications Society. Format: Div, country, classic contemp rock, blues. News staff: one; News: 4.5 hrs wkly. Target aud: General. ♦Loreen Harris, sls.

Orangeville

CIDC-FM— May 1, 1987: 103.5 mhz; 30.7 kw. Stereo. Hrs open: 24 5302 Dundas St., W., Etobicoke, M9B 1B2. Phone: (416) 213-1035. Fax: (416) 233-8617. E-mail: info@z1035.com Web Site: www.z1035.com. Licensee: Dufferin Communications Inc. Group owner: Evanov Radio Group (acq 9-28-94). Natl. Rep: Canadian Broadcast Sales. Format: Dance, Top-40. News staff: one. Target aud: 18-44. ♦Bill Evanov, pres; Bruce Campbell, gen mgr; Paul Evanov, progmg dir.

Orillia

CICX-FM— Sept 7, 1943: 105.9 mhz; 43 kw. Stereo. Hrs open: 24 7 Progress Dr., Box 550, L3V 6K2. Phone: (705) 326-3511. Fax: (705) 326-1816. E-mail: jack@1059jackfm.com Web Site: www.1059jackfm.com. Licensee: Rogers Broadcasting Ltd. (group owner; acq 4-19-02; grpsl). Format: Hot adult contemp. News staff: 2; News: 2 hrs wkly. Target aud: 25-54; upscale, educated, white collar, female skewed. ♦Gary Miles, CEO; Rael Merson, pres; Rick Doughty, gen mgr; Jack Latimer, progmg dir.

Oshawa

CKDO(AM)— 1946: 1580 khz; 10 kw-U, DA-2. Stereo. Hrs open: 24 1200 Airport Blvd., Suite 207, L1J 8P5. Phone: (905) 571-0949. Fax: (905) 571-1150. Licensee: Durham Radio Inc. (group owner; (acq 4-23-2003; C$3.9 million with co-located FM). Format: Golden oldies. News staff: 4; News: 9 hrs wkly. Target aud: 35-54. Spec prog: Relg one hr wkly. ♦Doug Freeman, gen mgr.

CKGE-FM—Co-owned with CKDO(AM). Sept 12, 1957: 94.9 mhz; 50 kw. 474 ft TL: N43 57 15 W78 48 24. Stereo. 24 Format: Classic rock, news. News staff: 4; News: 5 hrs wkly. Target aud: 45 plus.

Ottawa

***CBOF-FM**— Sept 12, 1974: 90.7 mhz; (Digital radio: 1482.464 mhz). Stereo. Hrs open: Box 3220, Station C, K1Y 1E4. Phone: (613) 724-1200. Phone: (613) 562-8521. Fax: (613) 562-8520. Web Site: www.cbc.radio-canada.ca/regions/ottawa. Licensee: Societe Radio-Canada. Natl. Network: Radio Canada. Format: Var/div.

***CBO-FM**— Jan 7, 1991: 91.5 mhz; 20 kw. (Digital radio: 1482.464 mhz). Hrs open: Box 3220, Station C, K1Y 1E4. Secondary address: Ottawa Broadcast Centre, 181 Queen St. K1P 1K9. Phone: (613) 724-1200. Phone: (613) 562-8422. Fax: (613) 562-8430. Fax: (613) 562-8408. Web Site: www.ottawa.cbc.ca. Licensee: CBC. Natl. Network: CBC Radio One. Format: Var/div. ♦Guylaine Saucier, chmn; Robert Rabinovitch, CEO & pres; Miriam Fry, gen mgr; Gilles R. Tessier, opns mgr.

***CBOQ-FM**— Feb 18, 1947: 103.3 mhz; 70 kw. (Digital radio: 1482.464 mhz). Stereo. Hrs open: Box 3220, Station C, K1Y 1E4. Phone: (613) 724-1200. Phone: (613) 562-8422. Fax: (613) 562-8430. Fax: (613) 562-8408. Web Site: www.ottawa.cbc.ca. Licensee: CBC. Natl. Network: CBC Radio Two. Format: Div, class. ♦Robert Rabinovitch, CEO & pres; Miriam Fry, gen mgr; Gilles R. Tessier, opns dir.

***CBOX-FM**— 1990: 102.5 mhz; 70 kw. Ant 1,077 ft (Digital radio: 1482.464 mhz). Stereo. Hrs open: Box 3220, Station C, K1Y 1E4. Phone: (613) 724-1200. Phone: (613) 562-8521. Fax: (613) 562-8520. Web Site: www.radio-canada.ca/regions/ottawa. Licensee: Societe Radio-Canada. Natl. Network: Radio Canada. Format: Div, class.

CFGO(AM)— June 7, 1964: 1200 khz; 50 kw-U, DA-2. (Digital radio: 1487.696 mhz). Stereo. Hrs open: 24 Team 1200, 87 George St., K1N 9H7. Phone: (613) 789-2486. Fax: (613) 738-2881. Web Site: www.team1200.com. Licensee: CHUM (Ottawa) Ltd. Group owner: CHUM Ltd. (acq 9-10-99; for 87.5%). Population served: 1,000,000 Format: Sports, talk. News staff: 5. Target aud: 18-34; men. ♦Allan Waters, CEO; Chris Gordon, VP, gen mgr & opns mgr; Jack Derouin, gen mgr & gen sls mgr; Don Holtby, sls VP; Brad Boechler, natl sls mgr; Al Macartney, mktg dir; J. R. Ello, prom dir & prom mgr; Dave Mitchell, progmg dir; Steve Winogron, news dir; Harrie Jones, engrg dir & chief of engrg.

CJMJ-FM—Co-owned with CFGO(AM). Aug 13, 1991: 100.3 mhz; 100 kw. (Digital radio: 1487.696 mhz). Stereo. 24 Fax: (613) 750-0100. Web Site: www.majic100.fm.1,000,000 Format: Adult contemp, oldies. Target aud: 25-44; female. ♦Jack Derouin, rgnl sls mgr; Al Macartney, prom mgr; Kent Newson, progmg mgr; Codi Jeffreys, mus dir.

CFRA(AM)— May 3, 1947: 580 khz; 50 kw-D, 10 kw-N, DA-2. (Digital radio: 1487.696 mhz). Hrs open: 24 87 George St., K1N 9H7. Phone: (613) 789-2486. Fax: (613) 523-6423. Fax: (613) 738-5024. Web Site: www.cfra.com. Licensee: CHUM Ltd. (group owner). Natl. Network: ABC. Format: News/talk. News: 24 hrs wkly. Target aud: 35-54. ♦Chris Gordon, VP, gen mgr & opns mgr; Jack Derouin, rgnl sls mgr; Al Macartney, mktg dir; Dave Mitchell, progmg mgr; Steve Winogron, news dir; Linda Ulmer, pub affrs dir & traf mgr; Harrie Jones, engrg mgr & chief of engrg; Daniel Proussalidis, news rptr; Norman Jack, news rptr.

CKKL-FM—Co-owned with CFRA(AM). 1959: 93.9 mhz; 95 kw. Ant 1,077 ft (Digital radio: 1487.696 mhz). Stereo. 24 Fax: (613) 739-4040. Web Site: www.939bobfm.com. Natl. Network: ABC. Format: Hits of the 80s & 90s. News staff: 2; News: one hr wkly. Target aud: 18-34. ♦Chris Gordon, VP, gen mgr, sls VP & progmg dir; Al Macartney, mktg mgr, prom dir, prom mgr & disc jockey; Steve Winogron, asst music dir & news dir; Harrie Jones, engrg mgr & local news ed; Louise Seguin, traf mgr; J.R. Rodenburg, disc jockey; John Mielke, disc jockey; Sandy Sharkey, disc jockey; Steve Gregory, disc jockey.

CHEZ-FM— Mar 25, 1977: 106.1 mhz; 100 kw. Ant 998 ft (Digital radio: 1484.208 mhz). Stereo. Hrs open: 2001 Thurston Dr., K1G 6C9. Phone: (613) 736-2001. Fax: (613) 736-2002. Web Site: www.chez106.com. Licensee: Rogers Broadcasting Ltd. (acq 7-2-99; grpsl). Format: Classic rock. News staff: 3; News: 2 hrs wkly. Target aud: 25-54; Males. ♦Scott Parsons, chmn, pres, VP & gen mgr.

CHLX-FM—See Gatineau, PQ

CHRI-FM— Mar 6, 1997: 99.1 mhz; 25.3 kw. 551 ft TL: N45 13 01 W75 37 51. Stereo. Hrs open: 24 1010 Thomas Spratt Pl., Suite 3, K1G 5L5. Phone: (613) 247-1440. Phone: (613) 247-1886. Fax: (613) 247-7128. E-mail: chri@chri.ca Web Site: www.chri.ca. Licensee: Christian Hit Radio Inc. Population served: 1,300,000 Format: Contemp Christian mus. News staff: one; News: 2 hrs wkly. Target aud: 18-44. Spec prog: Children 2 hrs wkly. ♦Gord Walford, pres; Bill Stevens, gen mgr; Brock Tozer, mus dir.

***CHUO-FM**— May 31, 1991: 89.1 mhz; 18.2 kw. TL: N45 30 11 W75 51 02. Stereo. Hrs open: 24 65 University Rd., Suite 201, K1N 1G7. Phone: (613) 562-5965. Fax: (613) 562-5969. E-mail: info@chuo.fm Web Site: www.chuo.fm. Licensee: Radio Ottawa Inc. Population served: 900,000 Format: English/French. Target aud: General. Spec prog: Ger 2 hrs, jazz 5 hrs, relg 2 hrs, Sp 3 hrs, Chinese 2 hrs, Haitian 2 hrs, African 4 hrs wkly. ♦Marc Gill, gen mgr.

CIHT-FM— February 2003: 89.9 mhz; 27 kw. Hrs open: 1504 Merivale Rd., K2E 6Z5. Phone: (613) 723-8990. Fax: (613) 723-7016. Web Site: www.hot899.com. Licensee: NewCap Inc. Group owner: NewCap Broadcasting Ltd. Format: CHR, Top-40. ♦Scott Broderick, gen mgr; Rob Mise, stn mgr.

CILV-FM— Dec 26, 2005: 88.5 mhz; 2.3 kw. Hrs open: 1504 Merivale Rd., K2E 6Z5. Phone: (613) 688-8888. Fax: (613) 723-7016. E-mail: kmann@newcap.ca Web Site: www.live885.com. Licensee: NewCap Inc. ♦Scott Broderick, gen mgr; Kneale Mann, progmg dir; John Moran, mus dir.

CIMF-FM—See Gatineau, PQ

CISS-FM—Listing follows CIWW(AM).

CIWW(AM)— June 1, 1949: 1310 khz; 50 kw-U, DA-2. (Digital radio: 1484.208 mhz). Hrs open: 2001 Thurston Dr., K1G 6C9. Phone: (613) 736-2001. Fax: (613) 736-2002. Web Site: www.oldies1310.com. Licensee: Rogers Media. Format: Oldies. ♦Scott Anderson, gen mgr.

CISS-FM—Co-owned with CIWW(AM). Oct 29, 1969: 105.3 mhz; 100 kw. Ant 1,077 ft (Digital radio: 1484.208 mhz). Stereo. Web Site: www.1053kissfm.com. Format: Adult contemp, top-40. ♦Al Campagnola, progmg dir.

CJLL-FM— 2003: 97.9 mhz; 6.77 kw. Hrs open: 24 CHIN Radio Ottawa, 30 Murray St., Suite 100, K1N 5M4. Phone: (613) 244-0979. Fax: (613) 244-3858. E-mail: chinottawa@chinradio.com Web Site: www.chinradio.com/ottawa.asp. Licensee: Radio 1540 Ltd. Population served: 1,000,000 Format: Ethnic. Target aud: Ethnic 12 plus; mutlicultural. ♦Edward Ylanen, VP & gen mgr; Ed Ylanen, gen sls mgr; Gary Michaels, progmg dir.

CJRC-FM—(Gatineau, PQ) Apr 16, 2007: 104.7 mhz; 2.9 kw. (Digital radio: 1463.280 mhz). Hrs open: 150, rue d'Edmonton, Gatineau, PQ, J8Y 3S6. Phone: (819) 561-8801. Fax: (819) 561-9439. E-mail: nouvelles@cjrc1150.com Web Site: www.cjrc1150.com. Licensee: 591991 B.C. Ltd. Population served: 200,000 Format: News/talk. News staff: 3; News: 49 hrs wkly. Target aud: 35-64; adult-babyboomers 50% males, 50% females. ♦Kathleen Michaud, sls dir; Sylvie Charette, gen mgr & progmg dir; Louis-Philippe Bruce, news dir.

CJWL-FM— February 2006: 98.5 mhz; 485 w. Hrs open: 127 York St., K1N 5T4. Phone: (613) 241-9850. Fax: (613) 241-9852. Web Site: www.thejewelradio.com. Licensee: Ottawa Media Inc. Format: Adult standards, lite adult contemp. ♦Al Pervin, gen mgr; Al Baldwin, mus dir.

CKAV-FM-9— 2007: 95.7 mhz; 6 kw. Hrs open: 366 Adelaide St. E., Suite 323, Toronto, M5A 3X9. Phone: (416) 703-1287. Fax: (416) 703-4328. Web Site: www.aboriginalradio.com. Licensee: Aboriginal Voices Radio Inc. Format: Div. ♦Roy Hennessy, opns mgr; Patrice Mousseau, progmg dir.

CKCU-FM— Nov 15, 1975: 93.1 mhz; 12 kw. 853 ft TL: N45 30 11 W75 51 02. Stereo. Hrs open: 24 517 Unicentre, 1125 Colonel By Dr., K1S 5B6. Phone: (613) 520-2898. Fax: (613) 520-4060. E-mail: info@ckcufm.com Web Site: www.ckcufm.com. Licensee: Radio Carleton Inc. Population served: 2,000,000 Format: Progsv, div, community, campus. News staff: 2; News: 25 hrs wkly. Target aud: General; alternative rock, spoken word, ethnic audience. Spec prog: Jazz 15 hrs, Black 12 hrs, Fr 2 hrs, Pol one hr, Vietnamese one hr, Canadian Indian 2 hrs, folk 12 hrs, It 1 hr, relg 3 hrs wkly. ♦Matthew Crosier, stn mgr.

***CKDJ-FM**— Oct 3, 1994: 107.9 mhz; 100 w. Hrs open: 1385 Woodroffe Ave., Algonquin College, Rm. N 101, K2G 1V8. Phone: (613) 727-4723, ext. 7740. Fax: (613) 727-7689. Web Site: www.ckdj.net. Licensee: CKDJ-FM Algonquin Radio. Format: Hip Hop / Alternative. Target aud: 17-24; collegel students. ♦Don Crockford, gen mgr; Ryan Lindsay, stn mgr.

CKQB-FM— Sept 1, 1994: 106.9 mhz; 84 kw. (Digital radio: 1487.696 mhz). Stereo. Hrs open: 24 1504 Merivale Rd., K2E 6Z5. Phone: (613) 225-1069. Fax: (613) 226-3381. E-mail: bearinfo@thebear.net Web Site: www.thebear.fm. Licensee: Standard Radio Inc. Format: Mainstream rock. News staff: 3; News: one hr wkly. Target aud: 18-34; professionals. ♦Gary Slaight, pres; Eric Stafford, VP & gen mgr; Scott Broderick, gen sls mgr; Gord Taylor, progmg dir.

Owen Sound

CFOS(AM)— Mar 1, 1940: 560 khz; 7.5 kw-D, 1 kw-N. TL: N44 32 40 W80 54 08. Hrs open: 24 Box 280, N4K 5P5. Phone: (519) 376-2030. Fax: (519) 371-4242. E-mail: bayshore@radioowensound.com Web Site: www.560cfos.ca. Licensee: Bayshore Broadcating Corp. Natl. Rep: Target Broadcast Sales. Wire Svc: BN Wire Format: News/talk, oldies. News staff: 7; News: 12 hrs wkly. Target aud: 35 plus. ♦Ross Kentner, gen mgr; J.D. Moffat, opns mgr; Manny Paiva, news dir.

CIXK-FM—Co-owned with CFOS(AM). Jan 3, 1989: 106.5 mhz; 100 kw. 555 ft TL: N44 44 37 W80 54 16. Stereo. 24 Web Site: www.mix106.ca. Natl. Rep: Target Broadcast Sales. Format: Today's hits, yesterday's classics. News staff: 7. Target aud: 18-40.

CKYC-FM— Sept 4, 2001: 93.7 mhz; 31.6 kw. Hrs open: 24 270 9th St. E., N4K 5P5. Phone: (519) 376-2030. Fax: (519) 371-4242. E-mail: bayshore@radioowensound.com Web Site: www.radioowensound.com. Licensee: Bayshore Broadcasting Corp. Natl. Rep: Target Broadcast Sales. Format: New country. Target aud: 25-54; adult. ♦Deb Shaw, gen mgr; Ross Kentner, gen mgr; Kevin Brown, gen sls mgr.

Parry Sound

CKLP-FM— July 1986: 103.3 mhz; 50 kw. 400 ft (CP: 46.6 kw.). Stereo. Hrs open: 24 60 James St., Suite 301, P2A 1T5. Phone: (705) 746-2163. Fax: (705) 746-4292. E-mail: moose1033@hbgradio.com Web Site: www.moosefm.com/cklp. Licensee: The Haliburton Broadcasting Group Inc. Group owner: Haliburton Broadcasting Group Inc. (acq 11-9-01; C$2,025,000). Population served: 60,000 Natl. Rep: Target Broadcast Sales. Wire Svc: Broadcast News Ltd. Format: Adult contemp. News staff: 2; News: 12 hrs wkly. Target aud: General. Spec prog: Canadian Indian 1 hr, Gospel 1 hr wkly. ♦Christopher Grossman, pres; Dave Keeble, stn mgr; Kimberly Ward-Grossman, opns VP.

Pembroke

CHVR-FM— May 6, 1996: 96.7 mhz; 100 kw. Stereo. Hrs open: 595 Pembroke St. E., K8A 3L7. Phone: (613) 735-9670. Fax: (613) 735-7748. E-mail: music@star96.ca Web Site: www.star96.ca. Licensee: Standard Radio Inc. Group owner: Standard Broadcasting Corp. (acq 4-19-2002; grpsl). Population served: 500,000 Format: Country. Target aud: 25-54. ♦Al Kennedy, gen mgr & gen sls mgr; Rick Johnston, progmg dir.

CIMY-FM— September 2005: 104.9 mhz; 1.62 kw. Hrs open: 215 Pembroke St. E., K8A 3J8. Phone: (613) 735-6936. Fax: (613) 732-4054. Web Site: www.myfmradio.ca/1049/index.htm. Licensee: My Broadcasting Corp. Format: Adult contemp. ♦Andrew Dickson, gen mgr; Jon Pole, progmg dir.

Penetanguishene

***CFRH-FM**— Sept 24, 1999: 88.1 mhz; 8.6 kw. TL: N44 46 10 W79 59 25. Hrs open: 24 C.P. 5099, L9M 2G3. Secondary address: 63 rue Main L9M 2G3. Phone: (705) 549-3116. Fax: (705) 549-6463. E-mail: cfrh@lacle.ca Web Site: www.lacle.ca. Licensee: La Cle d'la Baie en Huronie - Association culturelle francophone. (acq 2-17-99). Population served: 15,000 Format: Fr. Target aud: Francophone; francophone minority in mid-southern Ontario. ♦Michelle Laurin, opns VP & mktg VP; Peter Hominuk, CEO, gen mgr & sls dir.

Perth

CHLK-FM— July 2007: 88.1 mhz; 700 w. TL: N44 54 34 W76 16 51. Hrs open: 43 Wilson St. W., K7H 2N3. Phone: (613) 264-8811. Fax: (613) 264-1119. E-mail: communityradio@perth.igs.net Web Site: www.lake88.ca. Licensee: Perth FM Radio Inc. Format: Adult contemp. ♦Norm Wright, gen mgr.

Peterborough

***CFFF-FM**— 1969: 92.7 mhz; 700 w. Hrs open: 715 George St., N., K9H 3T2. Phone: (705) 741-4011. E-mail: info@trentradio.ca Web Site: www.trentu.ca/trentradio. Licensee: Trent Radio. Population served: 40,000 Format: Div. Target aud: General. ♦John Muir, gen mgr.

CKKK-FM— Nov 24, 2004: 90.5 mhz; 230 w. Hrs open: 24 993 Talwood Dr., K9J 7R8. Phone: (705) 876-0404. Fax: (705) 755-0688. E-mail: info@kaosradio.com Web Site: www.kaosradio.com. Licensee: King's Kids Promotions Outreach Ministries Inc. Population served: 112,000 Format: Non classic religious music. News staff: 5. Target aud: 18-40. ♦Rick Kirschner, gen mgr.

CKPT(AM)— December 1959: 1420 khz; 10 kw-D, 5 kw-N, DA-2. TL: N44 16 13 W78 17 23. Hrs open: 24 Box 177, K9J 6Y8. Phone: (705) 742-8844. Fax: (705) 742-1417. E-mail: radio@ckpt.com Web Site: www.ckpt.com. Licensee: 1708479 Ontario Inc. Group owner: CHUM Ltd. Format: Memories. News staff: 2. Target aud: 25-54; 60% women. ♦Allan Waters, chmn; Jim Waters, pres; Taylor Baiden, CFO; Steve Fawcett, gen mgr; Ray Hebert, prom dir; Angela Rose, progmg dir; George Gall, news dir; Ed Crompton, engrg dir.

CKQM-FM—Co-owned with CKPT(AM). Sept 16, 1977: 105.1 mhz; 50 kw. 301 ft TL: N44 17 36 W78 21 20. Stereo. 24 E-mail: radio@ckqm.com Web Site: www.country105.fm. Format: Country. Target aud: 25-64. ♦Brian Young, mus dir.

CKRU(AM)— Mar 21, 1942: 980 khz; 10 kw-D, 7.5 kw-D, DA-2. Hrs open: 159 King St., K9J 2R8. Phone: (705) 748-6101. Fax: (705) 742-7708. Web Site: www.980kruz.net. Licensee: 591989 B.C. Ltd. Group owner: Corus Entertainment Inc. (acq 3-24-00; grpsl). Format: Oldies. ♦Kathleen McNair, gen mgr; Brian Ellis, progmg dir.

CKWF-FM—Co-owned with CKRU(AM). July 24, 1968: 101.5 mhz; 15.2 kw. Ant 896 ft Stereo. E-mail: info@thewolf.com Web Site: www.thewolf.ca. Format: Classic rock.

Port Elgin

CFPS-FM— 2005: 97.9 mhz; 3.8 kw. Stereo. Hrs open: 24 382 Goderich St., N0H 2C0. Phone: (519) 376-2030. Fax: (519) 371-4242. Web Site: www.98thebeach.ca. Licensee: Bayshore Broadcasting Corp. Format: Adult contemp. Target aud: 18-54. ♦Ross Kentner, gen mgr; Rob Brignell, dev dir & mktg dir; Don Vail, progmg dir; John Divinski, news dir.

Port Hope

CKSG-FM—See Cobourg

Quinte West

CJTN-FM— 2004: 107.1 mhz; 3.64 kw. Hrs open: 24 31 Quinte St., K8V 3S7. Phone: (613) 392-1237. Fax: (613) 394-6430. E-mail: billmorton@mix97.com Web Site: www.cjtn.com. Licensee: Quinte Broadcasting Co. Ltd. Format: Lite rock. Spec prog: Scottish one hr wkly. ♦Bill Morton, pres; Bob Rowbotham, gen mgr & gen sls mgr; Lorne Brooker, prom.

Red Lake

CKDR-5(AM)— Aug 1, 1981: 1340 khz; 250 w-U, DA-1. Hrs open: Rebroadcasts CKDR-FM Dryden 99%.
c/o CKDR-FM, Box 580, Dryden, P8N 2Z3. Phone: (807) 223-2355. Fax: (807) 223-5090. E-mail: mail@ckdr.net Web Site: www.ckdr.net. Licensee: Fawcett Broadcasting Ltd. (group owner) Format: Contemp hits/top 40. ♦Bruce Walchuk, gen mgr.

Renfrew

CHMY-FM— August 2004: 96.1 mhz; 1.66 kw. Hrs open: Box 961, K7V 1R6. Secondary address: 321-B Raglan St. S. K7V 4H4. Phone: (613) 432-6936. Fax: (613) 432-1086. Web Site: www.myfmradio.ca. Licensee: My Broadcasting Corp. Format: Adult contemp. ♦Andrew Dickson, gen mgr.

CJHR-FM— Dec 11, 2006: 98.7 mhz; 14 kw. Hrs open: Box 945, K7V 4H4. Phone: (613) 432-9873. Fax: (613) 432-3686. Web Site: www.valleyheritageradio.ca. Licensee: Valley Heritage Radio. Format: Country and easy lstng. ♦Vic Garbutt, gen mgr.

Saint Catharines

CFBU-FM— 1997: 103.7 mhz; 250 w. Stereo. Hrs open: 24 % Brock University, 500 Glenridge Ave., St. Catharines, L2S 3A1. Phone: (905) 346-2644. E-mail: pd@cfbu.ca Web Site: www.cfbu.ca. Licensee: Brock University Student Radio. Population served: 300,000 Format: Var. Spec prog: American Indian one hr, jazz 4 hrs, Sp 4 hrs, Por 2 hrs, Mandarin 2 hrs, blues 2 hrs, classical one hr wkly. ♦Deborah Cartmer, progmg dir; Jordy Yack, mus dir.

CHRE-FM— Mar 1, 1967: 105.7 mhz; 50 kw. 438 ft Stereo. Hrs open: 24 Box 610, 12 Yates St., L2R 6Z4. Phone: (905) 688-1057. Fax: (905) 684-4800. Web Site: www.1057ezrock.com. Licensee: Standard

Radio Inc. Group owner: Standard Broadcasting Corp. (acq 4-19-2002; grpsl). Format: Soft rock. News staff: 4. Target aud: 25-54. ♦Tom Cooke, gen mgr.

CHSC(AM)— Mar 20, 1967: 1220 khz; 10 kw-U, DA-2. Hrs open: 36 Queenston St., L2R 2Y9. Phone: (905) 682-6692. Fax: (905) 682-9434. E-mail: pssa@1220chsc.ca Web Site: www.1220chsc.ca. Licensee: Pellpropco Inc. (acq 6-19-02; C$725,000). Natl. Rep: Canadian Broadcast Sales. Format: Adult contemp, news. Target aud: 25-54; female. ♦Domick Pellgrino, gen mgr.

CHTZ-FM—Listing follows CKTB(AM).

CKTB(AM)— 1930: 610 khz; 10 kw-D, 5 kw-N, DA-1. Hrs open: Box 977, L2R6Z4. Secondary address: 12 Yates St. L2R6X7. Phone: (905) 984-6610. Fax: (905) 684-4800. E-mail: newsroom@610cktb.com Web Site: www.610cktb.com. Licensee: Standard Radio Inc. Group owner: Standard Broadcasting Corp. (acq 4-19-2002; grpsl). Format: News/talk. Target aud: 35-65. ♦Tom Cook, gen mgr, opns VP & gen sls mgr; Joe Gurney, chief of engrg.

CHTZ-FM—Co-owned with CKTB(AM). February 1949: 97.7 mhz; 50 kw. 414 ft Stereo. Box 977, L2R 6Z4. Phone: (905) 688-0977. Web Site: www.htzfm.com. Format: Rock/AOR.

Sarnia

CBEG-FM— Nov 27, 1977: 90.3 mhz; 50 kw. 375 ft Hrs open: Rebroadcasts CBE(AM) Windsor.
Box 500 Stn A, Toronto, M5W 1E6. Phone: (519) 255-3411. Fax: (519) 255-3443. Web Site: www.windsor.cbc.ca. Licensee: CBC. Natl. Network: CBC Radio One. Format: Info. ♦Janice Stein, stn mgr.

CFGX-FM— Sept 14, 1981: 99.9 mhz; 27 kw. TL: N42 52 12 W82 23 50. Stereo. Hrs open: 24 1415 London Rd., N7S 1P6. Phone: (519) 542-5500. Fax: (519) 542-1520. Web Site: www.foxfm.com. Licensee: Blackburn Radio Inc. Group owner: Blackburn Group Inc. (acq 12-19-94; grpsl). Format: Adult contemp. Target aud: 25-54; females in the workplace. Spec prog: New age 7 hrs wkly. ♦Terry Regier, gen mgr; Ron Dann, opns mgr & mktg mgr; George Hayes, progmg dir; Larry Gordon, news dir.

CHKS-FM— 1999: 106.3 mhz; 35 kw. Hrs open: 1415 London Rd., N7S 1P6. Phone: (519) 542-5500. Fax: (519) 542-1520. E-mail: rock@k106fm.com Web Site: www.k106fm.com. Licensee: Blackburn Radio Inc. Group owner: Blackburn Group Inc. Format: Rock. ♦Terry Regier, gen mgr; Ron Dann, opns mgr & mktg mgr; George Hayes, progmg dir; Larry Gordon, news dir.

CHOK(AM)— July 26, 1946: 1070 khz; 10 kw-U, DA-2. TL: N42 53 30 W82 19 20. Stereo. Hrs open: 24 1415 London Rd., N7S 1P6. Phone: (519) 542-5500. Fax: (519) 542-1520. E-mail: radio@chok.com Web Site: www.chok.com. Licensee: Sarnia Broadcasters (1993) Ltd. Group owner: Blackburn Group Inc. (acq 12-18-98; C$902,600). Population served: 118,000 Natl. Rep: Canadian Broadcast Sales. Format: Baby boomers classics. News staff: 3; News: 11 hrs wkly. Target aud: 25-54. Spec prog: Toronto Blue Jays baseball, Toronto Maple Leaf hockey, Jr. "A" Sting hockey. ♦Terry Regier, gen mgr; Ron Dann, opns mgr, sls dir & mktg mgr; Larry Gordon, prom dir & news dir; George Hays, progmg dir.

Sault Ste. Marie

CHAS-FM— May 15, 1964: 100.5 mhz; 13.9 kw. 103 ft TL: N46 35 40 W84 21 00. Stereo. Hrs open: 642 Great Northern Rd., P6B 4Z9. Phone: (705) 759-9200. Fax: (705) 946-3575. Web Site: www.ezrocksoo.com. Licensee: Rogers Broadcasting Ltd. (group owner; acq 4-19-2002; grpsl). Format: Adult contemp. News staff: 3. Target aud: 25-54; adults. Spec prog: Class 5 hrs, jazz 2 hrs, It 2 hrs wkly. ♦Scott Sexsmith, gen mgr.

CJQM-FM— May 13, 1964: 104.3 mhz; 100 kw. Ant 1,000 ft Stereo. Hrs open: 642 Great Northern Rd., P6B 4Z9. Phone: (705) 759-9200. Fax: (705) 946-3575. Web Site: www.qcountry.ca. Licensee: Rogers Broadcasting Ltd. (group owner; acq 4-19-2002; grpsl). Format: Country. News: 3 hrs wkly. Target aud: 25-54; adults. Spec prog: It 4 hrs wkly. ♦Scott Sexsmith, gen mgr.

Savant Lake

CBQL-FM— January 1977: 104.9 mhz; 78 w. 287 ft Hrs open: Rebroadcasts CBQT-FM Thunder Bay.
213 Miles St. E., Thunder Bay, P7C 1J5. Phone: (807) 625-5000. Phone: (416) 205-3700. Fax: (807) 625-5035. Fax: (416) 205-3311. Licensee: CBC. Natl. Network: CBC Radio One. Format: Info. ♦Tom Grand, stn mgr.

Simcoe

CHCD-FM— 1997: 98.9 mhz; 14.37 kw. Ant 500 Ft Stereo. Hrs open: 24 Box 98, N3Y 4K8. Secondary address: 55 Park Rd. N3Y 4K8. Phone: (519) 426-7700. Fax: (519) 426-8574. Web Site: www.cd989.com. Licensee: CHCD Inc. (acq 2-26-01; C$1.05 million). Population served: 50,000 Natl. Rep: Canadian Broadcast Sales. Format: Adult contemp. News staff: 3. Target aud: Women; 25-54. ♦Jim MacLeod, pres; Blair Daggett, gen mgr; Gerry Hamill, prom mgr; Kate Buick, news dir.

Sioux Lookout

CKDR-2(AM)— May 1977: 1400 khz; 50 w. Hrs open: 24 Rebroadcasts CKDR-FM Dryden 99%.
c/o CKDR-FM, Box 580, Dryden, P8N 2Z3. Phone: (807) 223-2355. Fax: (807) 223-5090. E-mail: mail@ckdr.net Web Site: www.ckdr.net. Licensee: Fawcett Broadcasting. Group owner: Fawcett Broadcasting Ltd. Format: Contemp hits of the 60s, 70s, 80s & 90s. News staff: one. Target aud: 25-54. ♦Bruce Walchuk, gen mgr; Richard McCarthy, opns.

Sioux Narrows

CBQS-FM— May 1977: 95.7 mhz; 1.3 kw. 134 ft Hrs open: Rebroadcasts CBQT-FM Thunder Bay 100%.
c/o CBC Radio, 213 Miles St. E., Thunder Bay, P7C 1J5. Phone: (807) 625-5000, EXT. 5021. Fax: (807) 625-5035. Web Site: www.cbc.ca/ottowa. Licensee: CBC. Format: Public radio. Spec prog: Canadian Indian one hr wkly. ♦Tom Grand, gen mgr & stn mgr.

Smiths Falls

CJET-FM— November 2000: 92.3 mhz; 9.3 kw. Hrs open: Box 630, K7A 2B1. Phone: (613) 283-4630. Fax: (613) 283-7243. E-mail: webmaster@923jackfm.com Web Site: www.923jackfm.com. Licensee: Rogers Broadcasting Ltd., on behalf of CHEZ-FM Inc. Format: Hits from 80s to present. ♦Scott Parsons, gen mgr.

CKBY-FM— Jan 29, 1969: 101.1 mhz; 100 kw. Ant 500 ft Stereo. Hrs open: 24 2001 Thurston Dr, Ottawa, K1G 6C9. Phone: (613) 736-2001. Fax: (613) 736-2002. Web Site: www.y101.fm. Licensee: CHEZ-FM Inc. Group owner: Rogers Broadcasting Ltd. (acq 7-2-99; grpsl). Format: Country. News staff: 2; News: 6 hrs wkly. Target aud: 35-54; female. ♦Scott Parsons, chmn, pres & gen mgr.

St. Thomas

CFHK-FM— July 8, 1994: 103.1 mhz; 50 kw. 492 ft TL: N42 50 57 W81 08 52. Stereo. Hrs open: 24 380 Wellington St., Rm. 222, London, N6A 5B5. Phone: (519) 931-6000. Fax: (519) 679-1967. E-mail: jeff@energy103.ca Web Site: www.energy103.ca. Licensee: Corus Radio Co. Group owner: Corus Entertainment Inc. (acq 8-23-99; grpsl). Population served: 600,000 Format: Top 40. Target aud: 18-39. ♦Dave Farough, gen mgr; Bob Fisher, gen sls mgr & natl sls mgr; Jim McCourtie, progmg dir & progmg mgr.

Stella

CJAI-FM— Apr 1, 2006: 93.7 mhz; 5 w. Hrs open: R.R. 1, K0H 2S0. Phone: (613) 384-8287. Web Site: www.amherstisland.on.ca/AIR/. Licensee: Amherst Island Radio Broadcasting Inc. Format: Var. ♦Terry Culbert, stn mgr.

Stratford

CHGK-FM—Listing follows CJCS(AM).

CJCS(AM)— 1924: 1240 khz; 1 kw-U, DA-1. Hrs open: 376 Romeo St. S., N5A 4T9. Phone: (519) 271-2450. Fax: (519) 271-3102. Web Site: www.cjcsradio.com. Licensee: Raedio Inc. Format: Oldies. Target aud: 25-54. ♦Steve Rae, pres, gen mgr & sls dir; Jim Fewer, prom dir; Eddie Matthews, progmg mgr; Kirk Dickson, news dir; Bill Tofflemire, chief of engrg.

CHGK-FM—Co-owned with CJCS(AM). Sept 2, 2003: 107.7 mhz; 2.805 kw. Format: Adult contemp.

Strathroy

CJMI-FM— Feb 6, 2007: 105.7 mhz; 1.75 kw. Hrs open: 125 Metcalfe St., N7G 1M9. Phone: (519) 246-6936. Fax: (519) 245-6670. Web Site: www.myfmradio.ca/1057/index.php. Licensee: My Broadcasting Corp. Format: Adult contemp/MOR. ♦Jeff Degraw, gen mgr.

Sturgeon Falls

CFSF-FM— Apr 4, 2003: 99.3 mhz; 1.35 kw. Hrs open: 12006 Hwy. 17, Unit 8, P2B 3K8. Phone: (705) 753-6776. Fax: (705) 753-6776. E-mail: joco@bellnet.ca Web Site: www.joco.ca. Licensee: JOCO Communications Inc. Format: Top-40, adult contemp, Fr (20%). News staff: 4. ♦Joseph Cormier, pres.

Sudbury

CBBS-FM— Mar 29, 2001: 90.1 mhz; 50 kw. Hrs open: CBC Radio, 15 MacKenzie St., P3C 4Y1. Phone: (705) 688-3200. Fax: (705) 688-3220. Web Site: www.sudbury.cbc.ca. Licensee: Canadian Broadcasting Corp. Natl. Network: CBC Radio Two. Format: Classical jazz. ♦Kelly McInnes, gen mgr.

CBBX-FM— Mar 29, 2001: 90.9 mhz; 50 kw. Hrs open: c/o CBFX-FM, Box 6000, Montreal, PQ, H3C 3A8. Phone: (514) 597-6000. E-mail: auditoire@radio-canada.ca Web Site: www.cbc.radio-canada.ca. Licensee: Canadian Broadcasting Corp. Natl. Network: Espace Musique. Format: Var. ♦Sylvain LaFrance, VP.

***CBCS-FM**— June 17, 1978: 99.9 mhz; 50 kw. 250 ft Hrs open: 24 15 Mackenzie St., P3C 4Y1. Phone: (705) 688-3200. Fax: (705) 688-3220. Web Site: www.sudbury.cbc.ca. Licensee: Radio-Canada/CBC. Population served: 550,000 Natl. Network: CBC Radio One. Format: Info, news/talk. News staff: 4; News: 3 hrs wkly. Target aud: 30 plus; College/university educated/professional. ♦Kelly McInnes, progmg mgr.

***CBON-FM**— June 19, 1978: 98.1 mhz; 50 kw. 800 ft Stereo. Hrs open: 15 Mackenzie St., P3C 4Y1. Phone: (705) 688-3200. Fax: (705) 688-3220. Web Site: www.radio-canada.ca. Licensee: Radio-Canada/CBC. Natl. Network: CBC Radio One. Format: Fr var. ♦Gui Babineau, gen mgr.

CHNO-FM— February 2000: 103.9 mhz; 11 kw. Stereo. Hrs open: 24 493-B Barrydowne Rd., P3A 3T4. Phone: (705) 560-8323. Fax: (705) 560-7765. E-mail: wwatson@bigdaddy1039.ca Web Site: www.bigdaddy1039.ca. Licensee: NewCap Inc. Group owner: NewCap Broadcasting Ltd. (acq 11-9-01; C$2,843,000). Population served: 160,000 Format: Adult hits. News staff: 2; News: 4 hrs wkly. Target aud: 25-54; middle-income. Spec prog: American Indian 1 hr wkly. ♦Mark Maheu, pres; Wendy Watson, gen mgr; Dave Murray, opns VP; Rick Tompkins, progmg dir.

CHYC-FM— 2000: 98.9 mhz; 1 kw. Hrs open: 493-B Barrydowne Rd., P3A 3T4. Phone: (705) 560-8323. Fax: (705) 560-2492. E-mail: sbncher@nbgradio.com Web Site: www.chycfm.com. Licensee: The Haliburton Broadcasting Group Inc. Group owner: Haliburton Broadcasting Group Inc. Format: CHR, adult contemp. Target aud: General. ♦Christopher Grossman, pres & gen mgr.

CIGM(AM)— Aug 23, 1935: 790 khz; 50 kw-U, DA-2. Hrs open: 24 880 LaSalle Blvd., P3A 1X5. Phone: (705) 566-4480. Fax: (705) 560-7232. E-mail: chris.johnson@sudburyradio.rogers.com Web Site: 790cigm.com. Licensee: Rogers Broadcasting Ltd. (group owner; (acq 4-19-2002; grpsl). Population served: 165,000 Format: New country, sports. News staff: 5. Target aud: Adults 35+. ♦Ted Rogers, pres; Rick Doughty, gen mgr; Gerry Currie, gen sls mgr; Gary Duguay, prom dir & prom mgr; Chris Johnson, progmg dir; Henri Belanger, chief of engrg.

CJRQ-FM—Co-owned with CIGM(AM). September 1965: 92.7 mhz; 100 kw. 889 ft Stereo. 24 E-mail: q92@q92rocks.com Web Site: www.q92rocks.com. Format: Rock. News staff: 5; News: one hr wkly. ♦Bryan Bailey, mus dir.

CJMX-FM— 1980: 105.3 mhz; 100 kw. Ant 780 ft TL: N46 30 02 W81 01 16. Stereo. Hrs open: 24 880 Lasalle Blvd., P3A 1X5. Phone: (705) 566-4480. Fax: (705) 560-7232. Web Site: www.ezrocksudbury.com. Licensee: Rogers Broadcasting Ltd. (group owner; (acq 4-19-2002; grpsl). Population served: 210,000 Natl. Rep: Canadian Broadcast Sales. Format: Adult contemp, soft rock. News staff: 6; News: 1 hr wkly. Target aud: Males 35-44. ♦Stephanie Hunter, pres & mus dir.

CJTK-FM— 1998: 95.5 mhz; 1.4 kw. Stereo. Hrs open: 24 417 Notre Dame Ave., P3C 5K6. Phone: (705) 674-2585. Fax: (705) 688-1081.

E-mail: mail@cjtk.com Web Site: cjtk.com. Licensee: Eternacom Inc. Population served: 240,000 Format: Relg, Christian. News staff: one; News: 2 hrs wkly. Target aud: General. ♦Curtis Belcher, CEO, chmn & pres; Louis Depatie, chief of opns.

***CKLU-FM**— Apr 30, 1997: 96.7 mhz; 1.3 kw. TL: N46 25 29 W81 00 54. Stereo. Hrs open: 7:30 AM-2:30 AM 935 Ramsey Lake Rd., P3E 2C6. Phone: (705) 673-6538. Phone: (705) 675-1151. Fax: (705) 675-4878. E-mail: chef@ckfu.ujyf.ca Licensee: Laurentian Student and Community Radio Corp. Population served: 150,000 Format: News/talk, div, jazz. News: 3 hrs wkly. Target aud: General. Spec prog: It one hr, Polish one hr, Fr 19 hrs, Sp 1 hr, Ger 1 hr wkly. ♦Dan Welch, pres; Lindsey Chrysler, VP; Carl Jorgensen, opns mgr.

CKSO-FM— 2002: 101.1 mhz; 50 w. Hrs open: 24 Box 536, South Porcupine, P0N 1H0. Phone: (705) 235-3072. Fax: (705) 235-3921. E-mail: cksofm@vianet.ca Web Site: www.cksofm.netfirms.com. Licensee: David Jackson, on behalf of a corporation to be incorporated. Format: Christian. ♦David Jackson, gen mgr; Sarah Jackson, mus dir.

Thunder Bay

***CBQ-FM**— July 5, 1984: 101.7 mhz; 23.5 kw. 900 ft Stereo. Hrs open: 24 213 Miles St. E., P7C 1J5. Phone: (807) 625-5000. Fax: (807) 625-5035. Web Site: www.cbc.ca. Licensee: CBC. Natl. Network: CBC Radio One. Format: Talk/news, public radio. ♦Robert Rabinovitch, pres & stn mgr; Tom Grand, gen mgr.

***CBQT-FM**— August 1990: 88.3 mhz; 23.5 kw. Hrs open: 19 213 Miles St. E., P7C 1J5. Phone: (807) 625-5000. Fax: (807) 625-5035. Web Site: www.cbc.ca. Licensee: CBC. Natl. Network: CBC Radio One. Format: News, current affairs. Target aud: General; northwestern Ontario residents. Spec prog: Canadian Indian one hr wkly. ♦Robert Rabinovitch, CEO & progmg mgr; Tom Grand, gen mgr.

CILU-FM— Mar 1, 2005: 102.7 mhz; 100 w. TL: N48 25 14 W89 15 37. Stereo. Hrs open: 24 955 Oliver Rd., Rm. UC 2014A, Lakehead University, P7B 5E1. Phone: (807) 343-8881. Fax: (807) 343-8598. E-mail: info@luradio.ca. Web Site: www.luradio.ca. Licensee: LU Campus Radio Inc. Population served: 113,000 Format: Var. ♦Dave Angell, stn mgr; Jason McKee, mktg mgr; Jason Wellwood, progmg mgr & mus dir; Amy Hadley, news dir.

***CJOA-FM**— Dec 20, 1998: 95.1 mhz; 50 w. Hrs open: 24 63 Carrie St., Rm 42, P7A 4J2. Phone: (807) 344-9525. Fax: (807) 344-9525. E-mail: info@cjoa.org Web Site: www.cjoa.org. Licensee: Thunder Bay Christian Radio. Population served: 130,000 Format: Christian music. All ages. ♦Ray Gauthier, pres; Bonnie Gauthier, gen mgr.

CJSD-FM— October 1948: 94.3 mhz; 93 kw. Ant 1,009 ft Stereo. Hrs open: 24 87 N. Hill St., P7A 5V6. Phone: (807) 346-2600. Fax: (807) 345-9923. E-mail: rock@rock94.fm Web Site: rock94.com. Licensee: C.J.S.D. Inc. (acq 5-25-92). Population served: 113,000 Natl. Rep: Target Broadcast Sales. Format: Adult rock. Target aud: 18-44. ♦H.F. Dougall, pres.

CJUK-FM— August 2001: 99.9 mhz; 37 w. Hrs open: 180 Park Ave., Suite 200, P7B6J4. Phone: (807) 344-2000. Fax: (807) 345-9939. E-mail: magicmail@magic999.fm Web Site: www.magic999.fm. Licensee: Newcap Inc. (acq 5-10-2005; C$2.3 million). Format: Soft rock, adult contemp. ♦Dennis Landriault, pres & gen mgr.

CKPR-FM— June 4, 2007: 91.5 mhz; 100 kw. TL: N48 31 27 W89 06 53. Hrs open: 24 87 N. Hill St., P7A 5V6. Phone: (807) 346-2600. Fax: (807) 345-9923. Web Site: ckpr.com. Licensee: C.J.S.D. Inc. Natl. Rep: Target Broadcast Sales. Format: Adult contemp. Target aud: 25-54; families & office workers. Spec prog: News/talk 10 hrs wkly. ♦H.F. Dougall, pres.

CKTG-FM— March 1996: 105.3 mhz; 100 kw. Hrs open: 24 180 Park Ave., Suite 200, 7V5 6J4. Phone: (807) 346-2006. Fax: (807) 345-9923. E-mail: hits@hot105.fm Web Site: www.thenewhot105.com. Licensee: NewCap Broadcasting Ltd. (group owner) Natl. Rep: Canadian Broadcast Sales. Format: Classic rock. News staff: 5; News: 4 hrs wkly. Target aud: 25 plus. Spec prog: It, relg, Finnish one hr wkly. ♦Bob Templeton, pres; K. Klein, VP, gen mgr & gen mgr.

Tillsonburg

CKOT(AM)— Apr 30, 1955: 1510 khz; 10 kw-D, DA-D. Hrs open: Sunrise-sunset Box 10, 77 Broadway, N4G 4H3. Phone: (519) 842-4281. Fax: (519) 842-4284. Licensee: Tillsonburg Broadcasting Co. Ltd. Population served: 200,000 Natl. Rep: Target Broadcast Sales. Format: Country. News staff: 5; News: 7 hrs wkly. Target aud: 18-50; general. Spec prog: Ger one hr, Hungarian one hr, Belgian one hr, Dutch one hr wkly. ♦John Lamers, pres & gen mgr; Robin Henry, sls VP.

CKOT-FM— Dec 1, 1965: 101.3 mhz; 26 kw. Ant 454 ft Stereo. 24 Format: Easy lstng. Target aud: 30 plus; general. Spec prog: Gospel one hr wkly.

Timmins

***CHIM-FM**— 1996: 102.3 mhz; 84 w. Hrs open: 226 Delnite Rd., P4N 7C2. Phone: (705) 264-2150. E-mail: chimfm@vianet.ca Web Site: www.chimfm.com. Licensee: 1158556 Ontario Ltd. Format: Christian music. ♦Roger de Brabant, chmn; Karen Turner, stn mgr.

CHMT-FM— July 12, 2001: 93.1 mhz; 3.6 kw. Hrs open: 24 49 Cedar St. S, P4N 2G5. Phone: (705) 267-6070. Fax: (705) 267-6095. E-mail: moose931@hbgradio.com Web Site: www.hbgradio.com. Licensee: The Haliburton Broadcasting Group Inc. Group owner: Haliburton Broadcasting Group Inc. Natl. Network: CBS Radio. Format: Classic hits. News staff: 2; News: 1.5 hrs wkly. Target aud: 25-54. ♦Christopher Grossman, pres & gen mgr; Kimberly Ward, VP; Penny Proulx, opns mgr; Mike Fry, progmg dir; Kent Matheson, mus dir; Wendy Gray, news dir; Shawn McArthur, sls.

CHYK-FM— 2000: 104.1 mhz; 3.5 kw. Hrs open: 24 49 Cedar St. S, P4N 2Gs. Phone: (705) 267-6070. Fax: (705) 267-6095. E-mail: chykfm@hbgradio.com Licensee: Haliburton Broadcasting Group Inc. (group owner). Population served: 45,000 Natl. Rep: Canadian Broadcast Sales. Format: Hot adulte contemp, Fr, English. News staff: 1; News: 1 hr wkly. Target aud: 18-65. ♦Christopher Grossman, gen mgr; Kimberly Ward, opns VP; Sean Connon, gen sls mgr; Jim Whealan, natl sls mgr; Sylvain Boucher, progmg dir & mus dir; Gilles Lafortune, news dir; Penny Proulx, traf mgr; Sylvie Bealieu, sls.

CJQQ-FM— Sept 6, 1976: 92.1 mhz; 40 kw. Ant 400 ft Stereo. Hrs open: 24 260 2nd Ave., P4N 8A4. Phone: (705) 264-2351. Fax: (705) 264-2984. Web Site: www.q92timmis.com. Licensee: Rogers Broadcasting Ltd. (group owner; acq 4-19-02;. grpsl). Format: AOR. Target aud: 18-44. ♦Art Pultz, opns mgr; Angelo Lia, gen sls mgr.

CKGB-FM— August 2001: 99.3 mhz; 40 kw. Hrs open: 260 2nd Ave., P4N 8A4. Phone: (705) 264-2351. Fax: (705) 264-2984. Web Site: www.ezrocktimmins.com. Licensee: Rogers Broadcasting Ltd. (group owner; (acq 4-19-2002; grpsl). Format: Easy rock. News staff: 2; News: 4 hrs wkly. Target aud: 35-55. ♦Al Campagnola, gen mgr; Dave Novak, gen sls mgr.

Toronto

CBLA-FM— Apr 19, 1998: 99.1 mhz; 55.1 kw. (Digital radio: Dec 3, 1998: 1461.536 mhz; 5.084 kw). Hrs open: Box 500, Station A, M5W 1E6. Phone: (416) 205-7400. Fax: (416) 205-6336. Web Site: www.cbc.ca. Licensee: CBC. Natl. Network: CBC Radio One. Format: Talk, public radio. ♦Robert Rabinovitch, CEO & pres; Tom Grand, gen mgr.

CBL-FM— 1946: 94.1 mhz; 55.7 kw. 389 ft (Digital radio: Dec 3, 1998: 1461.536 mhz; 5.084 kw). Hrs open: Box 500, Station A., M5W 1E6. Secondary address: Box 3220, Station C., Ottawa K1Y 1E4. Phone: (416) 205-7400. Fax: (416) 205-6336. Web Site: www.cbc.ca. Licensee: CBC. Natl. Network: CBC Radio Two. Format: Class, public radio. ♦Robert Rabinovitch, CEO & pres; Tom Grand, gen mgr.

CFMJ(AM)— July 1, 1957: 640 khz; 50 kw-U, DA-2. (Digital radio: Dec 3, 1998: 1465.024 mhz; 5.084 kw). Stereo. Hrs open: 24 One Dundas St. W., Suite 1600, 5G123. Phone: (416) 221-6400. Fax: (416) 847-3300. Web Site: www.am640toronto.com. Licensee: Corus Premium Corp. Group owner: Corus Entertainment Inc. (acq 7-6-00; grpsl). Natl. Rep: Canadian Broadcast Sales. Format: News/talk. News staff: 12. Target aud: 35-64; upscale, mature, male. Spec prog: Sports/NHL hockey. ♦John Cassidy, CEO; John Hayes, pres; Chris Sisam, gen mgr; Darren Wasylyk, prom dir; Gord Harris, progmg dir; Stephanie Smyth, news dir.

CFMX-FM—Listing follows CFRB(AM).

CFMZ-FM-1— 1988: 96.3 mhz; 24.5 kw. Ant 930 ft TL: N56 43 38 W55 79 22. Stereo. Hrs open: 24
Rebroadcasts CFMX-FM Coburg 100%.
550 Queen St. E., Suite 205, M5A 1V2. Phone: (416) 367-5353. Fax: (416) 367-1742. E-mail: info@classical1963.com Web Site: www.classical963fm.com. Licensee: MZ Media Inc. (acq 8-31-2006; C$12 million with CFMZ-FM Cobourg). Population served: 4,800,000 Natl. Network: BN Audio. Natl. Rep: imsradio. Wire Svc: Standard Broadcast News Format: Class. News staff: 3; News: 4 hrs wkly. Target aud: 35 plus; well educated, upscale, owners/managers/professionals. ♦George Grant, CEO; John Van Driel, gen mgr & progmg VP.

CFNY-FM— Aug 8, 1960: 102.1 mhz; 35 kw. Ant 1,378 ft TL: N43 38 33 W79 23 15. (Digital radio: Dec 3, 1998: 1465.024 mhz; 5.084 kw). Stereo. Hrs open: 24 One Dundas St. W., Suite 1600, M5G 1Z3. Phone: (416) 408-3343. Fax: (416) 847-3300. Web Site: www.edge.ca. Licensee: Corus Radio Co. Group owner: Corus Entertainment Inc. (acq 1995; C$16.75 million). Population served: 5,000,000 Natl. Rep: Canadian Broadcast Sales. Format: Modern rock, progsv. News staff: 2; News: 3 hrs wkly. Target aud: 18-34; self motivated, mus loving, active, young at heart people. ♦John Cassaday, CEO & pres; Heather Shaw, chmn; Tom Peddie, CFO; Chris Sisam, gen mgr; Alan Cross, progmg dir.

CFRB(AM)— Feb 19, 1927: 1010 khz; 50 kw-U, DA-2. (Digital radio: Dec 3, 1998: 1458.048 mhz; 5.084 kw). Stereo. Hrs open: 24 2nd Fl., 2 St. Clair Ave. W., M4V 1L6. Phone: (416) 924-5711. Fax: (416) 872-8683. E-mail: comments@cfrb.com Web Site: www.cfrb.com. Licensee: Standard Radio Inc. Group owner: Standard Broadcasting Corp. (acq 1985). Format: News/talk. News staff: 10; News: 40 hrs wkly. Target aud: 25-64; general. Spec prog: Class 7 hrs, farm 2 hrs wkly. ♦Gary Slaight, CEO; Alan Slaight, chmn; Ian Laurie, CFO; Pat Holiday, gen mgr; Bill Herz, sls dir; G. Scott Johns, rgnl sls mgr; Nancy Ceneviva, mktg dir, prom dir & pub affrs dir; Steve Kowch, opns mgr & progmg dir; Dave Trafford, news dir; Dave Simon, engrg VP; Gail Prentice, rsch dir & traf mgr; Scott Ferguson, sports cmtr.

CFMX-FM—Co-owned with CFRB(AM). July 1, 1961: 99.9 mhz; 40 kw. 1,550 ft TL: N43 38 33 W79 23 15. (Digital radio: Dec 3, 1998: 1458.048 mhz; 5.084 kw). Stereo. Phone: (416) 922-9999. Web Site: www.mix999.com. Format: Hit radio. Target aud: 25-49. ♦Karen Steele, opns mgr, mktg mgr, prom dir & progmg dir; Sarah Commings, mktg mgr & disc jockey; Wayne Webster, mus dir; Anastasia Moshona, asst music dir; Dave Simon, engrg dir; Jane Martindale, rsch dir.

CFTR(AM)— Aug 8, 1962: 680 khz; 50 kw-U. (Digital radio: Dec 3, 1998: 1456.304 mhz; 5.084 kw). Stereo. Hrs open: 24 777 Jarvis St., M4Y 3B7. Phone: (416) 935-8468. Fax: (416) 935-8480. Web Site: www.680news.com. Licensee: Rogers Broadcasting Ltd. Population served: 6,000,000 Natl. Network: ABC. Natl. Rep: Canadian Broadcast Sales. Wire Svc: BN Wire Wire Svc: Bloomberg News Format: News. News staff: 50; News: 168 hrs wkly. Target aud: 25-54; owners, managers, professionals.Anthony P. Viner, CEO & pres; Sandy Sanderson, exec VP; Derek Berghuis, gen mgr, sls VP & gen sls mgr; Vicky Belfiore, mktg dir & prom mgr; John Hinnen, progmg VP; Connie Ricciuti, news dir & rsch dir; Kirk Nesbitt, chief of engrg; Phyllis Antoniandis, traf mgr; Ben Meccer, spec ev coord & local news ed; Carl Hanstke, news rptr; Paul Cook, edit dir; Rudy Blair, mus critic; John Stall, political ed; Bill Cole, sports cmtr

CHFI-FM—Co-owned with CFTR(AM). Feb 8, 1957: 98.1 mhz; 44 kw. 1,815 ft (Digital radio: Dec 3, 1998: 1456.304 mhz; 5.084 kw). Stereo. 24 Phone: (416) 935-8298. Format: Soft adult contemp. Target aud: 25-54. ♦Julie Adam, VP, gen mgr, progmg VP & progmg dir; Victor Dann, gen sls mgr; Vicky Belfiore, prom dir & adv dir; Drew Keith, mus dir; Phyllis Antoniandis, traf mgr; John Hinnen, local news ed; Jim Morris, news rptr; Bill Cole, sports cmtr.

CFXJ-FM— Feb 9, 2001: 93.5 mhz; 1.058 kw. Ant 980 ft (Digital radio: 1454.56 mhz; 5.084 kw). Hrs open: 9 AM-5:30 PM 211 Yonge St., Suite 400, M5B 1M4. Phone: (416) 214-5000. Fax: (416) 214-0660. Web Site: www.flow935.com. Licensee: Milestone Radio Inc. Wire Svc: BN Wire Format: Urban. News staff: one. Target aud: 18-35. ♦Denham Jolly, CEO & pres; Nicole Jolly, opns VP; Vanessa Santos, prom mgr; Wayne Williams, progmg dir; Scott Palmateer, chief of engrg.

CHHA(AM)— Nov 21, 2004: 1610 khz; 10 kw-D, 1 kw-N. TL: N43 42 40 W79 27 11. Hrs open: 30 22 Wenderly Dr., M6B 2N9. Phone: (416) 782-2953, ext 225. Fax: (416) 782-1219. E-mail: sanlorenzo@rogers.com Web Site: www.torontohispano.com. Licensee: San Lorenzo Latin American Community Centre. Format: Sp, ethnic. News: 10 hrs wkly. ♦Herman Astudillo, gen mgr.

CHIN(AM)— 1966: 1540 khz; 50 kw-D, 30 kw-N, DA-2. TL: N43 35 32 W79 39 22. Stereo. Hrs open: 24 622 College St., M6G 1B6. Phone: (416) 531-9991. Fax: (416) 531-5274. E-mail: sales@chinradio.com Web Site: www.chinradio.com.ANIK e-z, KUBAND (digital) Licensee: Radio 1540 Ltd. Format: Ethnic, multilingual (21 languages). News staff: 4; News: 9 hrs wkly. Target aud: 30 plus; immigrants in the Toronto census metropolitan area. ♦Johnny Lombardi, CEO; Lenny Lombardi, pres & exec VP; Theresa Lombardi, sr VP; Joe Mulvihill, gen mgr; Donina Lombardi, pub affrs dir; Michael Evans, engrg dir.

CHIN-FM— 1967: 100.7 mhz; 8.5 kw. 1,700 ft TL: N48 38 33 W79 23 15. (Digital radio: Dec 3, 1998: 1465.024 mhz; 5.084 kw). Stereo. 24 News: 10 hrs wkly. Target aud: 30 plus; multi-ethnic, first & second generation immigrants.

CHKT(AM)— Feb 21, 1951: 1430 khz; 50 kw-U, DA-2. Hrs open: 24 135 East Beaver Creek Rd., Units 7 & 8, Richmond Hill, L4B 1E2. Phone: (905) 763-3360. Fax: (905) 889-9828. Web Site: www.fairchildradio.com. Licensee: Fairchild Radio Group Ltd. (acq 10-3-96; C$1.8 million). Format: Multicultural, Chinese. Chinese and other ethnic groups. ♦Cyril Lai, gen mgr; Maureen Tang, sls dir & mktg dir; River Lee, progmg dir; Louisa Lam, news dir & pub affrs dir.

CHOQ-FM— 2005: 105.1 mhz; 1 kw. Hrs open: 24 425 W. Adelaide St., # 302, M5V 3C1. Phone: (416) 599-2666. Fax: (416) 599-7639. E-mail: choqfm@gmail.com Web Site: www.chocfm.com. Licensee: La Cooperative radiophonique de Toronto inc. Format: Fr variety. ♦Tonia Mori, gen mgr.

CHRY-FM— 1987: 105.5 mhz; 158 w. Hrs open: 24 413 Student Ctr., York University, 4700 Keele St., M3J 1P3. Phone: (416) 736-5293. Fax: (416) 650-8052. E-mail: chry@yorku.ca Web Site: www.yorku.ca/chry. Licensee: CHRY Community Radio Inc. Format: Black, alternative, div. News staff: 4; News: 10 hrs wkly. Target aud: General; campus community. Spec prog: Afghan, African, Black, Chinese, environment, Fr, gospel, jazz, Sp, Hebrew. ♦Susy Glass, gen mgr & stn mgr; Anderson Rouse, opns mgr; Neil Armstrong, progmg dir.

CHTO(AM)—Not on air, target date: unknown: 1690 khz; 1 kw-U. TL: N43 43 26 W79 16 41. Hrs open: 437 Danforth Ave., Suite 307, M4K 1P1. Fax: (450) 963-7229. Licensee: Canadian Hellenic Toronto Radio Inc. Format: Greek, ethnic.

CHUM(AM)— October 1944: 1050 khz; 50 kw-U, DA-2. (Digital radio: Dec 3, 1998: 1456.304 mhz; 5.084 kw). Stereo. Hrs open: 1331 Yonge St., M4T 1Y1. Phone: (416) 925-6666. Fax: (416) 926-4026. Web Site: www.1050chum.com. Licensee: CHUM Ltd. (group owner) Format: Oldies. News staff: 6. ♦Paul Ski, pres; Bill Bodnarchuk, gen mgr; Marc Charlebois, gen sls mgr; Brad Jones, progmg dir; Jeff Howatt, news dir; Larry Keats, chief of engrg.

CHUM-FM— Sept 15, 1963: 104.5 mhz; 40 kw. 1,380 ft (Digital radio: Dec 3, 1998: 1456.304 mhz; 5.084 kw). Stereo. Web Site: www.chumfm.com. Format: Adult contemp. News: 6 hrs wkly. ♦Loretta Tate, prom dir; David Corey, progmg dir.

CHWO(AM)— Jan 8, 2001: 740 khz; 50 kw-U. TL: N43 34 30 W79 49 03. (Digital radio: 1454.56 mhz; 5.084 kw). Hrs open: 24 Box 740, Station A, M5W 4K6. Secondary address: Broadcasting Ctr., 284 Church St., Oakville L6J 7N2. Phone: (905) 845-2821. Phone: (416) 544-0740. Fax: (905) 842-1250. Web Site: www.am740.ca. Licensee: Primetime Radio Inc. Population served: 4,000,000 Format: Adult standard. News: 9 hrs wkly. Target aud: 50 plus. Spec prog: Scottish 2hrs, British 1hr, Irish 1hr. ♦J. E. Caine, CEO; Michael Caine, pres & gen mgr; Jacqui Gerrard, opns mgr.

CIAO(AM)—See Brampton

CILQ-FM—(North York, May 22, 1977: 107.1 mhz; 40 kw. 1,380 ft (Digital radio: Dec 3, 1998: 1465.024 mhz; 5.084 kw). Stereo. Hrs open: 24 1 Dundas St., Suite 1600, M5G 1Z3. Phone: (416) 221-0107. Fax: (416) 847-3300. Web Site: www.q107.com. Licensee: Corus Premium Television Ltd. Group owner: Corus Entertainment Inc. (acq 7-2000; grpsl). Format: Classic rock. News staff: 5; News: 15 hrs wkly. Target aud: 18-44. ♦John Cassaday, CEO; John P Hayes Jr., pres; Chris Sisam, gen mgr; Blair Bartrem, progmg dir.

CIRR-FM— Apr 16, 2007: 103.9 mhz; 50 w. Ant 433 ft TL: N43 42 20 W79 23 44. Hrs open: 65 Wellesley St. E., Suite 201, M4Y 1G7. Phone: (416) 922-1039. Fax: (416) 922-3692. E-mail: info@proudfm.com Web Site: www.proudfm.com. Licensee: Rainbow Media Group Inc. Format: Top-40, talk. ♦Carmela Laurignano, pres; Rob Basile, progmg dir; Sean Moreman, news dir.

CIRV-FM— 1986: 88.9 mhz; 1.88 kw. (Digital radio: 1466.768 mhz; 5.084 kw). Hrs open: 24 1087 Dundas St. W., M6J 1W9. Phone: (416) 537-1088. Phone: (416) 588-2472. Fax: (416) 537-2463. E-mail: info@cirvfm.com Web Site: www.cirvfm.com. Licensee: CIRC Radio Inc. Format: Multicultural/ethnic. News staff: 5. ♦Alberto Elmir, VP; Frank Alvarez, CEO, pres & gen mgr.

***CIUT-FM**— 1986: 89.5 mhz; 15 kw. Stereo. Hrs open: 24 91 St. George St., M5S 2E8. Phone: (416) 978-0909. Fax: (416) 946-7004. E-mail: b.burchell@ciut.fm Web Site: www.ciut.fm. Licensee: University of Toronto Community Radio Inc. Population served: 8,800,000 Wire Svc: Canadian Press Format: Var/div. News: 3 hrs wkly. Target aud: General. Spec prog: Fr 2 hrs, Sp 4 , Punjabi 5 hrs hrs wkly. ♦Brian Burchell, stn mgr; Ken Stowar, progmg dir.

CJAQ-FM— Jan 26, 1993: 92.5 mhz; 9.1 kw. (Digital radio: Dec 3, 1998: 1456.304 mhz; 5.084 kw). Hrs open: 24 777 Jarvis St., M4Y 3B7. Phone: (416) 935-8392. Fax: (416) 935-8410. Web Site: www.925jackfm.com. Licensee: Rogers (Toronto) Ltd. Group owner: Rogers Broadcasting Ltd. (acq 9-10-99; grpsl). Population served: 4,000,000 Format: All hits. News staff: 4; News: 6 hrs wkly. Target aud: 25-54. ♦Gary L. Miles, CEO; Rael Merson, pres; Laura Nixon, CFO; Pat Cardinal, gen mgr.

CJBC(AM)— 1947: 860 khz; 50 kw-U. (Digital radio: Dec 3, 1998: 1461.536 mhz; 5.084 kw). Hrs open: Box 500, Station A, M5W 1E6. Phone: (416) 205-3311. Fax: (416) 205-5622. Web Site: www.cbc.ca. Licensee: CBC. Format: Educ, cultural, var. ♦Alain Dorion, gen mgr.

CJBC-FM— 1993: 90.3 mhz; 5.73 kw. 1,414 ft TL: N43 38 33 W79 23 15. (Digital radio: Dec 3, 1998: 1461.536 mhz; 5.084 kw). Stereo. Phone: (416) 205-2522. Fax: (416) 205-7660. Natl. Network: Radio Canada. Format: Class. ♦Manon Cote, gen mgr.

CJCL(AM)— 1944: 590 khz; 50 kw-U, DA-1. (Digital radio: Dec 3, 1998: 1458.048 mhz; 5.084 kw). Stereo. Hrs open: 24 777 Jarvis St., M4Y 3B7. Phone: (416) 935-0590. Fax: (416) 413-4116. E-mail: contact@fan590.com Web Site: www.fan590.com. Licensee: Rogers Broadcasting Ltd. (group owner; acq 4-19-02; grpsl). Population served: 4,000,000 Format: Sports, talk. Target aud: 25-54; men. ♦Nelson Millmen, stn mgr.

CJEZ-FM— May 24, 1987: 97.3 mhz; 28.9 kw. Ant 1,500 ft (Digital radio: Dec 3, 1998: 1458.048 mhz; 5.084 kw). Hrs open: 2 St. Clair Ave. W., 2nd Fl., M4V 1L6. Phone: (416) 482-0973. Fax: (416) 486-5696. E-mail: info@ezrock.com Web Site: www.ezrock.com. Licensee: Standard Radio Inc. Group owner: Standard Broadcasting Corp. (acq 4-19-2002; grpsl). Population served: 3,000,000 Format: Adult contemp. Target aud: 35-54. ♦Gary Slaight, CEO; Pat Holiday, gen mgr.

CJMR(AM)—See Mississauga

***CJRT-FM**— 1949: 91.1 mhz; 100 kw. 1,300 ft (Digital radio: Dec 3, 1998: 1458.048 mhz; 5.084 kw). Stereo. Hrs open: 24 4 Pardee Ave., Unit 100, M6K 3H5. Phone: (416) 595-0404. Fax: (416) 595-9413. E-mail: info@jazz.fm Web Site: www.jazz.fm. Licensee: CJRT-FM Inc. (acq 1974). Population served: 6,000,000 Wire Svc: Broadcast News Ltd. Format: Jazz. Target aud: 35 plus. ♦Ross Porter, CEO; B. Webber, chmn; Brad Barker, opns dir; Vince De Lilla, sls dir; Stacy MacKenzie, traf mgr; Donnie Tong, engr.

CJSA-FM— 2004: 101.3 mhz; 373 w. Hrs open: Canadian Multicultural Radio, 306 Rexdale Rd., Unit 7, M9W 1R6. Phone: (416) 292-4059. Fax: (416) 292-4574. E-mail: info@cmr24.com Web Site: www.cmr24.com. Licensee: 3885275 Canada Inc. Format: Ethnic. ♦Sivakumaran Sivapaphafundaram, gen mgr.

CKAV-FM— Dec 13, 2002: 106.5 mhz; 1.1 kw. Hrs open: 366 Adelaide St., Suite 323, M5A 3X9. Phone: (416) 703-1287. Fax: (416) 703-4328. E-mail: info@aboriginalradio.com Web Site: www.aboriginalradio.com. Licensee: Aboriginal Voices Radio Inc. Format: Canadian aboriginal and world aboriginal. ♦Mark MacLeod, opns mgr; Roy Hennessy, opns mgr; Patrice Mousseau, progmg dir.

CKHC-FM— 2005: 96.9 mhz; 5 w. Stereo. Hrs open: 24 Radio Humber 96.9fm, 205 Humber College Blvd., M5W 5L7. Phone: (416) 675-6622, ext 4913. Fax: (416) 675-9730. E-mail: radiohumber@humber.ca Web Site: radio.humber.ca. Licensee: Humber College Institute of Technology and Advanced Learning. Format: Var. News: 8 am - 8 pm wkly. ♦Jerry Chomyn, progmg dir.

***CKLN-FM**— July 1983: 88.1 mhz; 250 w. Hrs open: c/o CKLN Radio Inc., 380 Victoria St., M5B 1W7. Phone: (416) 595-1477. Fax: (416) 595-0226. E-mail: stationmanager@ckln.fm Web Site: www.ckln.fm. Licensee: CKLN Radio Inc. Format: Alternative. ♦Tim May, progmg dir.

Vermillion Bay

CKQV-FM— June 2004: 103.3 mhz; 1.6 kw. Hrs open: Box 459, P0V 2V0. Secondary address: 78 Spruce St. P0V 2V0. Phone: (807) 227-9988. Fax: (807) 227-9985. E-mail: info@q104fm.ca Web Site: www.q104fm.ca. Licensee: Norwesto Communications Ltd. Format: Hot adult contemp. ♦Rick Doucet, gen mgr; Ken O'Neil, progmg dir.

Wahta Mohawk Territory near Bala

CFWP-FM— 2003: 98.3 mhz; 1.06 kw. Ant 96 ft Hrs open: 24 Box 711, 2350 Muskoka Rd. 38, Bala, P0C 1A0. Phone: (705) 762-1274. E-mail: hawk98@wahta.com Web Site: www.wahta.com/hawkradio. Licensee: Wahta Communications Society. Format: Var. ♦Cal White, gen mgr.

Wasaga Beach

CHGB-FM— Apr 30, 2007: 97.7 mhz; 200 w. Hrs open: 1383 Mosley St., L9Z 2C5. Phone: (705) 422-0970. Web Site: www.977thebeach.ca. Licensee: Bayshore Broadcasting Corp. Format: Classic adult contemp. ♦Ross Kentner, gen mgr; Rick Ringer, opns mgr.

Waterloo

CFCA-FM—Listing follows CKKW(AM).

CIKZ-FM—See Kitchener-Waterloo

CKKW(AM)—(Kitchener, Aug 1, 1959: 1090 khz; 10 kw-U, DA-2. TL: N43 17 20 W80 24 16. Stereo. Hrs open: 24 255 King St. N., Suite 207, N2J 4V2. Phone: (519) 884-4470. Fax: (519) 884-6482. Web Site: www.oldies1090.com. Licensee: 1708479 Ontario Inc. (group owner; (acq 7-30-93; C$5 million). Format: Oldies. ♦Jay Switzer, pres; Paul Cugliari, gen mgr; John Yost, gen sls mgr; Jay Nijhuis, prom mgr; Pete Travers, progmg dir.

CFCA-FM—Co-owned with CKKW(AM). Apr 3, 1967: 105.3 mhz; 100 kw. 820 ft TL: N43 24 15 W80 38 05. Stereo. 24 Web Site: www.koolfm.com. Format: Classic rock.

***CKMS-FM**— Oct 16, 1977: 100.3 mhz; 250 w. 110 ft Stereo. Hrs open: 6 AM-midnight 200 University Ave. W., N2L 3G1. Phone: (519) 886-2567. Fax: (519) 884-3530. E-mail: ckmsfm@web.ca Web Site: www.ckmsfm.uwaterloo.ca. Licensee: Radio Waterloo Inc. Population served: 300,000 Format: Div, campus. News staff: one. ♦Heather Majaury, stn mgr; Kobe George, progmg.

CKWR-FM—(Kitchener, Mar 23, 1974: 98.5 mhz; 15.2 kw. Ant 576 ft Stereo. Hrs open: 24 375 University Ave. E., N2k 3M7. Phone: (519) 886-9870. Fax: (519) 886-0090. E-mail: general@ckwr.com Web Site: www.ckwr.com. Licensee: Wired World Inc. Population served: 1,000,000 Natl. Rep: CHUM Radio Sales. Wire Svc: BN Wire Format: Adult contemp; speciality & multicultural. News staff: 2; News: 8 hrs wkly. Target aud: 35-64; mature audience. Spec prog: Romanian 2 hrs, Ger 3 hrs, Greek 2 hrs, Serbian 2 hrs, Pol 4 hrs, Por 5 hrs, Sp 4 hrs wkly. ♦Scott Jensen, pres; Clyde Ross, stn mgr & gen sls mgr.

Wawa

CJWA-FM— 1996: 107.1 mhz; 210 w. Hrs open: 55 Broadway Ave., P0S 1K0. Phone: (705) 856-4555. Fax: (705) 856-1520. Licensee: North Superior Broadcasting Ltd. Natl. Rep: Canadian Broadcast Sales. Format: Adult contemp. Target aud: 25-54. ♦Rick Labbe, pres, gen mgr, sls dir & progmg dir; Mark Capeless, progmg dir & news dir; Vern Valois, chief of engrg.

Welland

CIXL-FM— May 20, 1999: 91.7 mhz; 50 kw. TL: N52 42 56 W19 79 16. Stereo. Hrs open: 24 860 Forks Road West, L3B 5R6. Phone: (905) 732-4433. Fax: (905) 732-4780. E-mail: info@giantfm.org Web Site: www.giantfm.com. Licensee: R.B. Communications, LTD. Population served: 500,000 Natl. Rep: Canadian Broadcast Sales. Format: Adult hits. News staff: 2; News: 6 hrs wkly. Target aud: 25-54; adults. ♦Pat St. John, pres; Peter Morena, opns mgr & chief of opns; Brian Salmon, mus dir; Susan Honsberger, traf mgr.

Whitchurch-Stouffville

CIWS-FM—Not on air, target date: unknown: 102.7 mhz; 50 w. Hrs open: Box 59, Stouffville, L4A 7Z4. Secondary address: 6379 Main St., Stouffville L4A 7Z4. Phone: (905) 640-6429. E-mail: jim@whistleradio.com Web Site: www.whistleradio.com. Licensee: WhiStle Community Radio. Format: Var. ♦Jim Priebe, CEO.

Windsor

***CBE(AM)**— July 1, 1950: 1550 khz; 10 kw-U, DA-1. (Digital radio: 1484.208 mhz; 4.369 kw). Hrs open: 825 Riverside Dr. W., N9A 5K9. Phone: (519) 255-3411. Fax: (519) 255-3443. Web Site: www.cbc.ca/windsor. Licensee: CBC. Natl. Network: CBC Radio One. Format: Div, news/talk. Spec prog: Class 4 hrs, jazz 2 hrs wkly.

CBE-FM— Oct 15, 1978: 89.9 mhz; 100 kw. 538 ft (Digital radio: 1484.208 mhz; 4.369 kw). Stereo. Web Site: www.cbc.ca. Natl. Network: CBC Radio Two. Format: Class, div.

***CBEF(AM)**— May 1970: 540 khz; 2.5 kw-D, 5 kw-N, DA-1. Hrs open: 24 825 Riverside Dr. W., Box 1609, N9A 1k7. Phone: (519) 255-3572. Fax: (519) 255-3573. Licensee: CBC. Natl. Network: CBC Radio One. Format: Div. News: 3. ♦Benoit Quenneville, gen mgr.

CIDR-FM—Listing follows CKLW(AM).

CIMX-FM—Listing follows CKWW(AM).

***CJAM-FM**— November 1983: 91.5 mhz; 456 w. Hrs open: 401 Sunset Ave., N9B 3P4. Phone: (519) 971-3606. Fax: (519) 971-3605. E-mail: news@cjam.ca Web Site: www.cjam.ca. Licensee: Student Media, University of Windsor. Population served: 2,500,000 Format: Progsv, info, ethnic. Target aud: General; listeners in Windsor/Detroit area. Spec prog: Black 10 hrs, class 4 hrs, folk 4 hrs, jazz 6 hrs, Pol one hr, Sp one hr wkly. ♦Armondo Correia, pres; Christien Gagnier, stn mgr & progmg dir.

CKLW(AM)— June 1, 1932: 800 khz; 50 kw-U, DA-2. (Digital radio: 1484.208 mhz; 4.369 kw). Hrs open: 24 1640 Ouellette Ave., N8X 1L1. Phone: (519) 258-8888. Fax: (519) 258-0182. Web Site: www.am800cklw.com. Licensee: 1708479 Ontario Inc. (group owner; acq 1-29-93). Natl. Rep: McGavren Guild. Format: News/talk. Target aud: 25-54. ♦Jay Switzer, pres; Eric Proksch, gen mgr; Sandra Neposlan, gen sls mgr; Heidi Baiden, prom dir; Keith Chinnery, progmg dir; Jason Moore, news dir; Jim Valvasori, engrg dir & chief of engrg.

CIDR-FM—Co-owned with CKLW(AM). 1949: 93.9 mhz; 100 kw. Ant 700 ft (Digital radio: 1484.208 mhz; 4.369 kw). Stereo. 24 30100 Telegraph Rd., Suite 460, Bingham Farms, 48025. Phone: (313) 961-9811. Fax: (313) 961-1603. E-mail: feedback@93.9fmradio.com Web Site: www.93.9fmradio.com. Natl. Rep: McGavren Guild. Format: Adult contemp. ♦Christine Copeland, prom dir; Murray Brookshaw, progmg dir.

CKWW(AM)— Mar 29, 1964: 580 khz; 500 w-U, DA-1. (Digital radio: 1484.208 mhz; 4.369 kw). Hrs open: 1640 Ouellette Ave., N8X 1L1. Secondary address: 30100 Telegraph Rd., Suite 460, Bingham Farms, MI 48025. Phone: (519) 258-8888. Phone: (313) 961-9811. Fax: (519) 258-0182. Fax: (313) 961-1603. E-mail: info@am580radio.com Web Site: www.am580radio.com. Licensee: 1708479 Ontario Inc. Group owner: CHUM Ltd. (acq 9-5-85). Natl. Rep: McGavren Guild. Format: Soft adult contemp. News: 2 hrs wkly. Target aud: 45 plus. ♦Eric Proksch, gen mgr; Charlie O'Brien, progmg dir.

CIMX-FM—Co-owned with CKWW(AM). July 10, 1967: 88.7 mhz; 100 kw. 577 ft (Digital radio: 1484.208 mhz; 4.369 kw). Stereo. Format: Modern rock. Target aud: 18-34. ♦Cal Cagro, prom mgr; Murray Brookshaw, progmg dir; Matt Franklin, mus dir.

Wingham

CIBU-FM— Apr 1, 2005: 94.5 mhz; 70.14 kw. Hrs open: 215 Carling Terr., N0G 2W0. Phone: (519) 357-1310. Fax: (519) 357-1897. Web Site: www.945thebull.ca. Licensee: Blackburn Radio Inc. Format: Adult rock. ♦John Weese, gen mgr.

CKNX(AM)— Feb 20, 1926: 920 khz; 10 kw-D, 1 kw-N, DA-2. Hrs open: 24 215 Carling Terr., N0G 2W0. Phone: (519) 357-1310. Fax: (519) 357-1897. E-mail: news@cknxradio.com Licensee: Blackburn Radio Inc. Group owner: Blackburn Group Inc. Format: Country. News staff: 7; News: 10 hrs wkly. Target aud: 35-54. Spec prog: Relg 6 hrs wkly. ♦John Weese, gen mgr.

CKNX-FM— Apr 17, 1977: 101.7 mhz; 65.766 kw. Ant 741 ft Stereo. 24 Format: Adult contemp. Target aud: 25-49. Co-owned TV: CKNX-TV affil.

Woodstock

CIHR-FM— Apr 10, 2006: 104.7 mhz; 1.91 kw. TL: Woodstock. Hrs open: 24 233 Norwich Ave., N4S 3V8. Phone: (519) 537-8400. Fax: (519)537-8600. E-mail: info@ByrnesMedia.com Web Site: 1047.ca. Licensee: Byrnes Communications Inc. Population served: 100,000 Natl. Rep: Target Broadcast Sales. Format: Adult contemp. News staff: 3. Target aud: 25-54; Adults. ♦Chris Byrnes, pres; Michael Jones, gen mgr & gen sls mgr; Dan Henry, progmg dir; Adam Nyp, news dir.

CJFH-FM— 2004: 94.3 mhz; 37 w. Hrs open: 659 King St. E., Suite 207, Box 1433 Stn. "C", Kitchener, N2G 4H6. Phone: (519) 575-9090. Fax: (519) 575-9119. Licensee: Sound of Faith Broadcasting. Format: Christian. ♦Dave MacDonald, gen mgr; Joy Cooper, mus dir.

CKDK-FM— July 1, 1987: 103.9 mhz; 52 kw. 400 ft Stereo. Hrs open: 24 290 Dundas St., N4S 1B2. Phone: (519) 539-1040. Fax: (519) 539-7479. E-mail: dave.farough@corusent.com Web Site: www.thehawk.ca. Licensee: Corus Radio Group owner: Corus Entertainment Inc. (acq 1991). Population served: 100,000 Natl. Rep: Canadian Broadcast Sales. Format: Classic Rock. News staff: 3; News: 9 hrs wkly. Target aud: 25-54. ♦John Cassaday, CEO; John Hayes, pres; Dave Farough, gen mgr; Gord Harris, progmg dir.

Prince Edward Island

Charlottetown

***CBCT-FM**— 1972: 96.1 mhz; 93.5 kw. 540 ft Hrs open: 24 Box 2230, 430 University Ave., C1A 8B9. Phone: (902) 629-6400. Fax: (902) 629-6518. Fax: (902) 629-6520. Web Site: pei.cbc.ca. Licensee: CBC. Population served: 140,000 Natl. Network: CBC Radio One. Format: Talk. News: 11 hrs wkly. ♦Barbara Nymark, gen mgr; John Channing, sls dir; Susan Mitton, progmg dir & news dir; Carolyn Lounsbury, traf mgr; Sara Fraser, news rptr.

CFCY-FM— 2006: 95.1 mhz; 100 kw. Hrs open: Box 1060, C1A 7M4. Secondary address: 5 Prince St. C1A 4P4. Phone: (902) 892-1066. Fax: (902) 566-1338. E-mail: requests@cfcy.pe.ca Web Site: www.951fmcfcy.com. Licensee: Maritime Broadcasting System Ltd. Population served: 140,000 Format: Country. ♦Robert Pace, chmn; Paul Alan, opns mgr.

CHLQ-FM— March 1982: 93.1 mhz; 25 kw. Stereo. Hrs open: Box 1066, C1A 7M4. Secondary address: 5 Prince St. C1A 4P4. Phone: (902) 892-1066. Fax: (902) 566-1338. E-mail: requests@magic93.pe.ca Web Site: www.magic93.fm.ca. Licensee: Maritime Broadcasting System Ltd. Population served: 140,000 Format: Hot adult contemp. ♦Robert Pace, chmn; Paul Alan, opns mgr.

CHTN-FM— 2006: 100.3 mhz; 33 kw. TL: N46 11 22 W63 09 54. Hrs open: 90 University Ave., Suite 320, Atlantic Technology Centre, C1A 4K9. Phone: (902) 569-1003. Fax: (902) 569-8693. Web Site: www.ocean100.com. Licensee: Newcap Inc. Population served: 130,000 Natl. Rep: Canadian Broadcast Sales. Format: Classic hits. News staff: 2; News: 17 hrs wkly. Target aud: 25-54. ♦Jennifer Evans, gen mgr & gen sls mgr; Karen Bell, prom dir; Gerard Murphy, progmg dir; Scott Chapman, news dir.

CKQK-FM— July 25, 2006: 105.5 mhz; 33 kw. TL: N46 12 44 W63 20 32. Hrs open: 90 University Ave., Suite 320, Atlantic Technology Centre, C1A 4K9. Phone: (902) 569-1003. Fax: (902) 569-8693. Web Site: www.krock1055.com. Licensee: Newcap Inc. Format: Classic rock. Target aud: 25-44; male. ♦Jennifer Evans, gen mgr.

Summerside

CJRW-FM— 2000: 102.1 mhz; 11 kw. Hrs open: 5:57 AM-12:15 AM 763 Water St. E., C1N 4J3. Phone: (902) 436-2201. Fax: (902) 436-8573. Licensee: Maritime Broadcasting System Ltd. Group owner: Maritime Broadcasting (acq 8-10-2000; C$650,000 for approximately 92.9% of the common shares). Format: Hit country. News staff: one. Target aud: General. ♦Lois E. Schurman, chmn; Paul M. Schurman, pres; Brent Schurman, VP; Dave Chamberlain, gen mgr; Harry McLellan, gen sls mgr; Gina Cole, prom dir; Trisha Smith, progmg dir; Kevin Warren, mus dir; Ken Kingston, news dir & pub affrs dir; Steve Harvey, engrg dir.

Quebec

Acton Vale

CFID-FM— 2004: 103.7 mhz; 1.65 kw. Hrs open: 24 C.P. 130, J0H 1A0. Phone: (450) 546-1037. Fax: (450) 546-7521. E-mail: info@radio-acton.com Web Site: www.radio-acton.com. Licensee: Radio-Acton inc. Format: Fr, var. ♦Gaetan Chevanelle, gen mgr.

Alma

CFGT(AM)— October 1953: 1270 khz; 10 kw-D, 5 kw-N, DA-2. Hrs open: 460 Sacre- Coeur W., Suite 200, G8B 1L9. Phone: (418) 662-6673. Fax: (418) 662-6070. E-mail: cfgt@antenne6.com Licensee: Groupe Radio Antenne 6 Inc. Group owner: Group Radio Antenne 6 Inc. (acq 8-18-94). Format: Talk AC some country. ♦Marc-Andre Levesque, gen mgr; Lewis Gagnon, sls dir & mus dir; Louis Arcand, progmg dir.

CKYK-FM—Co-owned with CFGT(AM). 1993: 95.7 mhz; 100 kw. TL: N48 24 05 W72 05 23. Phone: (418) 662-6888. Phone: (418) 543-8912. Web Site: kykf.com. Format: Rock of the 80's. ♦Marc-Andre Levesque, pres.

Amqui

CFVM-FM— 2003: 99.9 mhz; 23.8 kw. Hrs open: 111 rue de l'Hopital, G5J 2K1. Phone: (418) 629-2025. Fax: (418) 629-2599. E-mail: cfvm@globetrotter.net Web Site: www.lamatapedia.com/cfvm. Licensee: Astral Media Radio inc. Group owner: Corus Entertainment Inc. (acq 5-30-2005; grpsl). Format: Adult contemp, classic rock, CHR. News staff: 2; News: 6 hrs wkly. Target aud: 18 plus. ♦Adalbert Levesque, gen mgr & sls dir; Jean Lemay, progmg dir; Jean Fournier, engrg mgr; Jennifer Gravel, news rptr; Alain Revard, disc jockey.

Asbestos

CJAN-FM— AM 1972;FM 2001: 99.3 mhz; 11.1 kw. Stereo. Hrs open: 185 du Roi, PE, J1T 1S4. Phone: (819) 879-5439. Phone: (819) 879-5430. Fax: (819) 879-7922. E-mail: info@fm993.ca Web Site: www.fm993.ca. Licensee: Radio Plus B.M.D. inc. Natl. Rep: Target Broadcast Sales. Format: Adult contemp, MOR. News staff: one. Target aud: 35-75; general. ♦Marie-Paule Drouin, pres.

Baie Comeau

CBMI-FM— May 28, 1974: 93.7 mhz; 3 kw. Hrs open: Rebroadcasts CBVE-FM Quebec 100%.
c/o CBVE-FM, 888 Saint-Jean St., Quebec, G1R 5H6. Phone: (418) 691-3613. Fax: (418) 691-3610. Licensee: Canadian Broadcasting Corp. Natl. Network: CBC Radio One. Format: Pub affrs, info. ♦David Kyle, gen mgr.

CHLC-FM— 1996: 97.1 mhz; 4.21 kw. Stereo. Hrs open: 907 Rue de Puyjalon, G5C 1N3. Phone: (418) 589-3771. Fax: (418) 589-9086. E-mail: info@chlc.com Web Site: www.chlc.com. Licensee: 9022-6242 Quebec Inc. (acq 4-29-96). Format: Adult contemp, MOR. Target aud: General. ♦Yvon Savoes, pres; Francois Morache, VP; George Daviault, gen mgr, sls dir & news dir; Mike Mainville, progmg dir; Mark Andre Halle, news dir.

Carleton

CIEU-FM— 1983: 94.9 mhz; 25 kw. 1,466 ft TL: N48 08 27 W66 06 32. Hrs open: 24 1645 Perron Est., G0C 1J0. Phone: (418) 364-7094. Fax: (418) 364-3150. E-mail: cieufm@cieufm.com Web Site: www.cieufm.com. Licensee: Diffusion Communautaire Baie des Chaleurs Inc. Format: CHR, adult contemp. News staff: 2; News: 7 hrs wkly. Target aud: General. Spec prog: Blues 5 hrs, class 3 hrs, folk 3 hrs, jazz 3 hrs wkly. ♦Jacques Veillette, pres; Louis St-Laurent, gen mgr; Carol Boudreau, mus dir; Yues Sigouin, traf mgr; Claude Roy, local news ed.

Chandler

CFMV-FM— June 8, 2005: 96.3 mhz; 5.716 kw. Hrs open: C.P. 99, G0C 1K0. Secondary address: 141 rue Commerciale Ouest G0C 1K0. Phone: (418) 689-4921. Fax: (418) 689-3852. E-mail: cfmv@fm92-1.com Web Site: www.fm92-1.com. Licensee: Radio du Golfe inc. Format: French, rock. ♦Jacques Vallee, gen mgr.

Chapais

CFED(AM)— 1969: 1340 khz; 250 w-U. Hrs open: 24 Rebroadcasts CJMD(AM) Chibougamau 100%.
c/o CHRL, 568 Boul. St. Joseph, Roberval, G8H 2K6. Secondary address: 539 3ieme Rue, Chibougamau G8P 1N8. Phone: (418) 275-1831. Phone: (418) 748-3931. Fax: (418) 275-2475. Fax: (418) 748-3931. Web Site: contacteantenne6.com. Licensee: Groupe Radio Antenne 6 inc. Group owner: Group Radio Antenne 6 Inc. (acq 1-6-93). Format: Pop music. News staff: one; News: 20 hrs wkly. Target aud: General; mainly adults. ♦Marc-Andre Levesque, pres; Louis Arcand, mus dir.

Charlesbourg

CIMI-FM— Aug 10, 2001: 103.7 mhz; 20 w. Hrs open: 4500, Blvd. Henri-Bairassa bur. 106, G1H 3A5. Phone: (418) 841-4445. Phone: (418) 624-0700. Fax: (418) 841-3330. Web Site: www.cimifm.com. Format: Alternative. ♦Stephane Tremblay, pres; Francois Beaule, opns mgr; Stephane Bertrand, sls dir; Dominic Tessier, progmg dir & engrg dir; Annie Bouchard, news dir; Jeff Labrie, news rptr.

Chateauguay

CHAI-FM— 1980: 101.9 mhz; 100 w. Stereo. Hrs open: 25 boul. St. Francis, J6J 1Y6. Phone: (450) 698-3131. Phone: (450) 698-3138. Fax: (450) 698-3330. Licensee: Radio Communautaires de Chateauguay Inc. Population served: 75,000 Format: Adult contemp, CHR. News staff: 2; News: 4 hrs wkly. Target aud: General; all ages. ♦Christian Laberge, pres; Sylvain Poirier, opns dir & progmg dir.

Chibougamau

CJMD(AM)— Nov 21, 1969: 1240 khz; 1 kw-U. TL: N49 54 35 W74 22 08. Hrs open: 24 c/o CHRL, 568 Boul. St. Joseph, Roberval, G8H 2K6. Secondary address: 539 Zieme Rue G8P 1N8. Phone: (418) 275-1831. Phone: (418) 748-3931. Fax: (418) 275-2475. Fax: (418) 748-3931. E-mail: contact@antenne6.com Licensee: Groupe Radio Antenne 6 inc. Group owner: Group Radio Antenne 6 Inc. (acq 1-6-93). Format: Pop music. News staff: one; News: 20 hrs wkly. Target aud: General; mainly adults. ♦Marc-Andre Levesque, pres; Louis Arcand, opns dir.

CKXO-FM—Co-owned with CJMD(AM).Not on air, target date: unknown: 93.5 mhz; 19.8 kw.

Chicoutimi

CBJE-FM— 1976: 102.7 mhz; 30 kw. Ant 294 ft TL: N48 25 29 W71 06 32. Hrs open:
Rebroadcasts CBM(AM) Montreal. Weekdays 6 AM-9 AM rebroadcasts CBVE-FM Quebec City.
CP 6000, c/o CBM(AM), Montreal, H3C 3A8. Phone: (514) 597-4444. Fax: (514) 597-4416. Licensee: CBC. Format: Talk radio. ♦Patricia Pleszczynska, gen mgr; Judith Bleier, opns mgr; Kate Arthur, prom mgr; Sally Caudwell, news dir.

CBJ-FM— 2001: 93.7 mhz; 50 kw. Ant 1,719 ft TL: N48 36 04 W70 49 46. Hrs open: 24
Rebroadcasts CBF-FM Montreal, 70%.
500 rue Des Sagueneens, G7H 6N4. Phone: (418) 696-6600. Fax: (418) 696-6689. Licensee: Canadian Broadcasting Corp. Population served: 295,000 Natl. Network: Premiere Chaine. Format: Var, talk, adlult contemp. News staff: 7. Target aud: 35 and up; adlut, news-oriented. ♦Patrick Boie, gen mgr, prom dir & news dir.

***CBJX-FM—** Sept 20, 1933: 100.9 mhz; 98 kw. Ant 294 ft TL: N48 25 29 W71 06 32. Hrs open: 24
Rebroadcasts CBF-FM Montreal 95%.
500 rue Des Sagueneens, G7H 6N4. Phone: (418) 696-6600. Fax: (418) 696-6689. Licensee: Canadian Broadcasting Corp. Population served: 150,000 Natl. Network: Espace Musique. Format: Classical. ♦Patrick Boie, opns mgr.

CFIX-FM— July 31, 1987: 96.9 mhz; 43.8 kw. Hrs open: 24 267 est, rue Racine, G7H 5K3. Phone: (418) 543-9797. Fax: (418) 543-7968. E-mail: cfix@rock-detente.com Web Site: www.rock-detente.com. Licensee: Astral Media Radio inc. Group owner: Radio Nord (acq 4-19-2002; grpsl). Format: Adult contemp, MOR. ♦Richard Durcotte, gen mgr.

Degelis

CFVD-FM— 1995: 95.5 mhz; 12.47 kw. 300 ft TL: N47 33 02 W68 43 48. Stereo. Hrs open: 24 654 6ieme rue est, Ville Degelis, G5T 1Y1. Phone: (418) 853-2370. Phone: (418) 853-3370. Fax: (418) 853-3321. E-mail: cfvd@fm95.ca Web Site: www.infotemis.com. Licensee: Radio Degelis Inc. Population served: 30,000. Format: CHR, country, adult contemp. News staff: 10. Target aud: General. ♦Gilles Caron, pres & gen mgr; Real Provencher, VP.

Dolbeau-Mistassini

CHVD-FM— 2003: 100.3 mhz; 21.4 kw. Hrs open: 24 1975 Boul Wallberg, Dolbeau, G8L 1J5. Phone: (418) 276-3333. Fax: (418) 276-6755. Licensee: Groupe Radio Antenne 6 Inc. Group owner: Radio Nord (acq 5-2004). Format: MOR. ♦Pierre Broseau, pres; Marc Andre Levesque, opns mgr & sls dir; Louis Arcand, progmg dir & news dir.

CKII-FM— 2004: 101.3 mhz; 250 w. Hrs open: 1709 boul. Wallberg, G8L 1H6. Phone: (418) 239-2544. Fax: (418) 239-0842. Licensee: L'Alliance Laurentienne des metis et indiens sans statut, Local 30 Mistassini inc. Format: Fr. ♦Michel Bouchard, gen mgr.

Donnacona

CKNU-FM— 1997: 100.9 mhz; 1.585 kw. Hrs open: 274 rue Notre-Dame, G0A 1T0. Phone: (418) 285-2568. Fax: (418) 285-5483. Licensee: RNC MEDIA Inc. (acq 12-23-2005). Format: News/talk. ♦Patrice Denerse, opns mgr.

Drummondville

CHRD-FM— 1997: 105.3 mhz; 2.9 kw. Stereo. Hrs open: 24 2070 rue St. Georges, J2C 5G6. Phone: (819) 475-1480. Phone: (819) 478-0099. Fax: (819) 475-5180. E-mail: chrd@hy.cgocable.ca Licensee: Astral Media Radio Inc. Group owner: Astral Media Inc. (acq 8-13-2001). Format: Adult contemp, MOR, news. News staff: 3; News: 15 hrs wkly. Target aud: 18 plus; general. Spec prog: Relg one hr wkly. ♦Joel Rioux, pres, gen mgr, opns dir, sls dir & prom dir; Martin Tremblay, progmg dir; David Rivet, news dir; Michel Cournoyer, chief of engrg; Robert Veilleux, spec ev coord & mus critic; Julie Brisson, women's int ed.

CJDM-FM— Aug 15, 1987: 92.1 mhz; 3 kw. 300 ft Stereo. Hrs open: 24 207 Rue St-Georges, BC, J2C 5G6. Phone: (819) 474-1892. Fax: (819) 474-6610. E-mail: cjdm@cgocable.ca Web Site: www.cjdm.fm. Licensee: Astral Media Radio inc. Group owner: Corus Entertainment Inc. (acq 5-30-2005; grpsl). Population served: 65,000 Natl. Rep: Canadian Broadcast Sales. Format: Adult contemp, Fr. News staff: 2; News: 5 hrs wkly. Target aud: 18-44. ♦Joel Rioux, gen mgr, sls dir & gen sls mgr; Martine Pichette, prom dir; Claude Rene Piette, progmg dir; Alain Rivard, mus dir; Claude Boucher, news dir; Daniel Pelletier, engrg dir.

Fermont

CBMR-FM— 1982: 105.1 mhz; 16 w. Hrs open: 1400 Rene Levesque E., c/o CBM(AM) - A 4, Montreal, H2L 8M2. Phone: (514) 597-4444. Fax: (514) 597-4416. Licensee: Canadian Broadcasting Corp. ♦Patricia Pleszczynska, opns dir; Judith Bleier, opns mgr.

CFMF-FM— 1980: 103.1 mhz; 50 w. 100 ft Stereo. Hrs open: Box 280, G0G 1J0. Phone: (418) 287-5147. Fax: (418) 287-5776. E-mail: diffusion1.ferment@sympatico.ca Web Site: www.diffusionfermont.ca. Licensee: Radio Communautaire de Fermont Inc. Format: Div, Fr, adult contemp. Target aud: 7-55. Spec prog: Jazz one hr, C&W 4 hrs wkly. ♦D. Brouard, pres; Pierre McKinnon, sr VP; Diane Levesque, gen mgr; Johanne Chasse, gen sls mgr; Marjonne Lavoie, progmg mgr; Frederic Harrisson, mus dir; Genevieve Vincent, news dir.

Forestville

CFRP(AM)— 1977: 620 khz; 1 kw-U. Hrs open: 907 Rue de Puyjalon, Baie Cameau, G5C 1N3. Phone: (418) 589-3771. Fax: (418) 589-9086. Licensee: 9022-6242 Quebec Inc. (acq 3-29-96). Format: Adult contemp, MOR. ♦Yvon Savoie, pres; George Baviauet, gen mgr; Lynn Martin, traf mgr.

Fort Coulonge

CHIP-FM— May 2, 1981: 101.7 mhz; 10 kw. 299 ft TL: N45 45 41 W76 35 01. Stereo. Hrs open: 24 Box 820, La Radio du Pontiac, 33 Romain St., Fort Coalonge, J0X 1V0. Phone: (819) 683-3155. Fax: (819) 683-3211. E-mail: chip-fm@qc.aira.com Web Site: www.chipfm.com. Licensee: La Radio du Pontiac Inc. Population served: 105,000 Format: Country in Fr & English. News: 5 hrs wkly. Target aud: 35 yrs & up; rural people. farming communities, small towns. Spec prog: Oldies, rock, gospel 7 hrs, class 2 hrs wkly. ♦Fern Laliberte, pres; Frank Doyle, gen mgr; Gilles Gervais, disc jockey; Martin Rivest, disc jockey; Nathalie Vasiloff, disc jockey.

Gaspe

CJRG-FM— December 1978: 94.5 mhz; 3.8 kw. Ant 1,150 ft Stereo. Hrs open: 162 Jacques Cartier, G4X 1M9. Phone: (418) 368-3511. Fax: (418) 368-1663. Licensee: Radio Gaspesie Inc. Format: MOR. Spec prog: Class 2 hrs, jazz 2 hrs wkly. ♦Jacques Chartier, gen mgr & progmg dir; Paul Minville, gen sls mgr; Richard O'Leary, news dir; Yvan DuPuis, engrg dir.

Gatineau

CFTX-FM— 2006: 96.5 mhz; 1.75 kw. Hrs open: 171-A rue Jean-Proulx, J8Z 1W5. Phone: (819) 770-9650. Web Site: www.tagradio.fm. Licensee: RNC MEDIA Inc. Format: Fr CHR. ♦Robert H. Parent, VP & gen mgr; Eric Brousseau, prom dir; Benoit Vanasse, mus dir.

CHLX-FM— Sept 23, 2002: 97.1 mhz; 12.6 kw. Hrs open: 125 rue Jean-Proulx, J8Z 1T4. Phone: (819) 770-9710. Fax: (819) 770-9740. E-mail: classique@radionord.com Web Site: www.radionord.com/radio-classic/index.html. Licensee: RNC MEDIA Inc. Group owner: Radio Nord Communications Inc. (acq 8-25-2004). Format: Fr, classical, jazz. ♦Jean-Pierre Major, gen mgr; Diane Pelletier, sls dir; Yurs Trottier, progmg dir.

CIMF-FM— Jan 1, 1970: 94.9 mhz; 84 kw horiz. Ant 1,059 ft TL: N45 30 11 W75 51 02. (Digital radio: 1463.280 mhz). Stereo. Hrs open: 15 Taschereau, J8Y 2V6. Phone: (819) 770-2463. Fax: (819) 770-9338. E-mail: cimf@rockdetente Web Site: www.rockdetente.com. Licensee: Astral Media Radio Inc. Group owner: Astral Media Inc. (acq 10-28-2002; grpsl). Format: Soft rock. News staff: 2; News: 3 hrs wkly. ♦Ian Greenberg, pres; Carmen Rodrigue, gen mgr; Claude Raymond, sls dir & gen sls mgr; Eric St-Louis, prom mgr; Patrice Croteau, progmg mgr; Jean-Guy Faucher, mus dir; Mano Aube, news dir; Pierre Sylvestre, chief of engrg.

CJLL-FM—See Ottawa, ON

CJRC-FM—Licensed to Gatineau. See Ottawa ON

CKTF-FM— Mar 11, 1988: 104.1 mhz; 19 kw. Ant 1,077 ft TL: N45 30 11 W75 51 02. (Digital radio: 1463.280 mhz). Stereo. Hrs open: 24 15 rue Taschereau, J8Y 2V6. Phone: (819) 243-5555. Fax: (819) 243-6816. Web Site: www.radioenergie.com. Licensee: Astral Media Radio Inc. Group owner: Astral Media Inc. Format: Dance, top-40, AOR. ♦Carmen Rodrigue, gen mgr; Vincent Pons, sls dir; Melany Gauvin, prom mgr; Astral Musique, mus dir; Pierre Sylvestre, engrg dir.

Harrington Harbour

***CFTH-FM-1—** Oct 30, 1991: 97.7 mhz; 180 w. Stereo. Hrs open: Box 88, Harrington Harbour, Duplessis, G0G 1N0. Phone: (418) 795-3349. Fax: (418) 795-3200. E-mail: cfth@globetrotter.qc.ca Licensee: Radio Communautaire de Harrington Harbour. Format: Adult contemp, Country, Oldies. Target aud: General; five fishing villages. Spec prog: News/talk. ♦Lana Shattler, gen mgr; Quenton Lessard, progmg dir; Rowena Osborne, progmg dir; Betty Strickland, news dir.

Havre-Saint-Pierre

CILE-FM— 1987: 95.1 mhz; 1.496 kw. Ant 201 ft Hrs open: 24 992 Rue du Bouleau, G0G 1P0. Phone: (418) 538-2453. Phone: (418) 538-2451. Fax: (418) 538-3870. E-mail: cilemf@globetrotter.net Web Site: www.cilemf.com. Licensee: Radio & Television Communautaire Havre-St. Pierre. Format: MOR. ♦Berchmens Boudreau, gen mgr & progmg dir; Catherine Ramoisy, news dir; Gerald Gallant, engrg mgr.

Iles-de-la-Madeleine

CFIM-FM— Nov 15, 1981: 92.7 mhz; 6.3 kw. Stereo. Hrs open: 24 C.P. 490, 1172 Chemin Laverniere, Cap-aux-Meules, G0B 1B0. Phone: (418) 986-5233. Fax: (418) 986-5319. E-mail: cfimmf@sympatico.ca Licensee: Diffusion Communautaire des Iles Inc. Diffusion Communautaire des Iles Format: Div, news/talk. Target aud: General. ♦Charles Eugene Cyr, gen mgr; Linda Noel, mktg VP; Suzanne Richard, gen mgr & progmg dir; Helen Fauteux, news dir; Paul Turbide, engrg dir.

Joliette

CJLM-FM— 1996: 103.5 mhz; 3 kw. TL: N45 59 0 W73 25 52. Stereo. Hrs open: 24 540 St. Thomas, J6E 3R4. Phone: (450) 756-1035. Fax: (450) 756-8097. Licensee: Cooperative de Radiodiffusion MF 103.5 de Lanaudiere. Cooperative de Radiodiffusion MF 103.5 de Lanaudiere Format: MOR. Target aud: 25-49. Spec prog: Oldies 6 hrs wkly. ♦Marie Josee, prom VP; Jacques Plante, progmg mgr & news dir; Normand Masse, gen mgr, sls dir & engrg dir.

Jonquiere

CKAJ-FM— Apr 11, 1977: 92.5 mhz; 2.693 kw. Stereo. Hrs open: 6 AM-midnight C.P. 872, G7X 7M8. Secondary address: Pavillon Manicouagan, 3791, De La Fabrique G7X 7W8. Phone: (418) 546-2525. Phone: (418) 546-2526. Fax: (418) 546-2528. E-mail: informations@ckaj.org Web Site: www.ckaj.org. Licensee: Radio Communautaire du Saguenay Inc. Format: Div, classic rock, country. News staff: one; News: 6 hrs wkly. Target aud: 25-54. Spec prog: Fr. ♦Anick Bilodau, pres; Alec Tremblay, opns dir; Pierre Boivin, progmg dir; Henri Girard, chief of engrg.

Kahnawake

***CKRK-FM—** Mar 30, 1981: 103.7 mhz; 250 w. 75 ft Stereo. Hrs open: 18 Box 1050, J0L 1B0. Phone: (450) 638-1313. Phone: (450) 638-1407. Fax: (450) 638-4009. E-mail: programming@k103radio.com Web Site: www.k103radio.com. Licensee: Mohawk Radio Kahnawake Association. (acq 8-4-94). Format: Adult contemp, C&W, contemp hit. News staff: 2; News: 3 hrs wkly. Target aud: General; young adults in Montreal suburban area. Spec prog: Mohawk 10 wkly. ♦Lois Williams, gen sls mgr; Christin Jerome, progmg dir, mus dir & news dir; David Lahache, news rptr; Al Briand, sports cmtr & disc jockey; Marsha Dailleboust, traf mgr & sports cmtr; Blake Stacey, disc jockey.

Kuujjuaq

CKUJ-FM— 1992: 97.3 mhz; 394 w. Hrs open: 10am-12pm; 2-5pm Box 1082, J0M 1C0. Phone: (819) 964-2921. Fax: (819) 964-2229. Licensee: Minister Council of Kuujjuaq. Format: Inuit. ♦Michael Gordon, pres; Mary Gordon, opns dir & news dir.

La Malbaie

***CBV-FM-6—** Sept 20, 1979: 99.3 mhz; 820 w. Ant 108 ft TL: N47 41 02 W70 08 06. Hrs open: 24
Rebroadcasts CBV-FM Quebec 100%.
888 Saint-Jean St., Quebec, G1R 5H6. Phone: (418) 654-1341. Fax: (418) 656-8842. Licensee: CBC. Natl. Network: Premiere Chaine. Format: Current Affairs/News. ♦Susan Campbell, gen mgr; Claude-Saindon, opns mgr; Sally Caldwell, news dir; Gaston LeBlanc, engrg mgr.

La Pocatiere

CHOX-FM— Apr 23, 1992: 97.5 mhz; 25 kw. Stereo. Hrs open: 601 lere rue, Suite 50, G0R1Z0. Phone: (418) 856-1310. Fax: (418) 856-3747. E-mail: choxfm@globetrotter.net Web Site: cibm107.com. Licensee: CHOX-FM Inc. Format: CHR. ♦Guy Simard, pres; Gilles Gosselin, opns mgr & progmg VP; Diane Bouchard, dev VP; Georgette Charent, sls VP; Renee Giard, mktg VP; Michel Cloutier, prom VP; Michel Farvey, mus dir; Jacques Dufour, news dir; Clement Lavoie, engrg VP.

La Tabatiere

CFTH-FM-2— 1991: 98.5 mhz; 70 w. Hrs open: Box 88, Harrington Harbour, G0G 1N0. Phone: (418) 795-3349. Fax: (418) 795-3200. Licensee: Radio communautaire de Harrington Harbour. Format: Var.

La Tuque

CFLM(AM)— Oct 3, 1959: 1240 khz; 1 kw-U, DA-2. Hrs open: 24 C.P. 850, 529 St. Louis, G9X 3P6. Phone: (819) 523-4575. Fax: (819) 676-8000. E-mail: radio.h-m@sympatico.ca Licensee: Radio Haute Mauricle Inc. (acq 1982). Format: Var; adult contemp; CHR. News staff: one. Target aud: General. ♦Rejean LeClerc, pres, gen mgr & opns dir.

Lac Megantic

CJIT-FM— 2002: 106.7 mhz; 4.25 kw. Hrs open: 24 4766 rue Laval, G6B 1C7. Phone: (819) 583-0663. Fax: (819) 583-0665. Web Site: www.cjitfm.com. Licensee: Les Productions du temps perdu inc. (acq 3-12-2007; C$200,000). Natl. Rep: Target Broadcast Sales. Format: Top-40, MOR. ♦Michel Brochu, opns mgr & progmg dir; Rachel Frigon, gen sls mgr; Mathieu Beaumont, news dir; Michel Mathieu, engrg dir.

Lac-Brome

CIDI-FM— Sept 20, 2007: 99.1 mhz; 1.45 kw. Ant 165 ft TL: N45 11 10 W72 35 20. Stereo. Hrs open: Box 3611, 305B Knowlton Rd., Knowlton, J0E 1V0. Phone: (450) 243-6285. Fax: (450) 243-1041. E-mail: deweydurrell@axion.ca Web Site: www.sunnymead.org/cidi. Licensee: Radio Communautaire Missisquoi. Population served: 138,000 Format: 60% music, 40% talk. News staff: 3; News: 21 hrs wkly. Target aud: 18-60.

Lac-Etchemin

***CFIN-FM—** Mar 27, 1992: 100.5 mhz; 6.7 kw. 676 ft TL: N46 24 41 W70 35 44. Stereo. Hrs open: 24 201 Claude-Bilodeau St., G0R 1S0. Phone: (418) 625-3737. Fax: (418) 625-3730. E-mail: cfinfm@sogetel.net Web Site: www.cfinfm.com. Licensee: Radio Bellechasse. Population served: 50,000 Rgnl rep: Target. Format: MOR, Country, Flashback. News staff: 2; News: 30 hrs wkly. Target aud: 35-60. Spec prog: Fr, Sp, class 4 hrs, jazz 6 hrs, relg one hr, country 6 hrs wkly. ♦Marcel Asselin, pres; Raymond Boutin, pres & stn mgr; Isabelle Giasson, progmg dir; Norman Poulin, mus dir, news dir & pub affrs dir.

Lachute

CJLA-FM— Dec 1, 1974: 104.9 mhz; 3 kw. Stereo. Hrs open: 24 11 Argenteuil, J8H 1X8. Phone: (450) 562-3733. E-mail: fusionfm@citenet.net Licensee: RNC MEDIA Inc. (group owner; (acq 8-22-89). Format: Adult contemp. News staff: one; News: 8 hrs wkly. Target aud: 25-59. ♦Pierre Brosseau, pres; Jean-Pierre Major, gen mgr; Marc Dubois, opns mgr & progmg dir; Yves Trottier, gen sls mgr; Olivier Proulx, news dir; Gaston Tousignant, chief of engrg.

Lac-Simon (Louvicourt)

CHUT-FM— 2000: 95.3 mhz; 97.9 w. Hrs open: 1016 rue Wabanonik, Lac-Simon, J0Y 3M0. Phone: (819) 736-4501. Fax: (819) 736-2333. Licensee: Radio communautaire MF Lac Simon inc. Format: Community/aboriginal. ♦Alain Flamand, gen mgr.

Laval

CFAV(AM)— January 2004: 1570 khz; 10 kw-U. Hrs open: Radio Nostalgie, 2040 Autoroute Laval, H7S 2M9. Phone: (450) 680-1570. E-mail: avidtoire@nostalgie1570.com Web Site: www.nostalgie1570.com. Licensee: Gilles Lajoie and Colette Chabot, on behalf of a corporation to be incorporated. Format: Nostalgia. ♦Colette Chabot, gen mgr.

CFGL-FM— September 1968: 105.7 mhz; 41 kw. TL: N45 30 20 W73 35 32. (Digital radio: 1454.56 mhz; 1.594 kw at Laval, 1.4 kw at Montreal). Stereo. Hrs open: 24 2830 Boul. St. Martin E., H7E 5A1. Phone: (450) 664-1500. Phone: (514) 381-5903. Fax: (450) 664-4138. Fax: (450) 664-1651. Web Site: www.rythmefm.com. Licensee: Newco, a wholly owned subsidiary of Cogeco Radio Television Inc. Group owner: Cogeco Inc. Format: Adult contemp. News staff: 2; News: 2 hrs wkly. Target aud: 25-54; those preferring soft & easy lstng hits. ♦Richard LaChance, gen mgr; Sylvain Venne, chief of opns; Daniel Brouillette, prom dir; Andre St-Amand, progmg dir; Lilianne Randall, mus dir; Jean Arcand, engrg dir.

Les Escoumins

CHME-FM— 1994: 94.9 mhz; 5kw. Hrs open: 24 C.P. 730, G0T 1K0. Secondary address: 34 rue de la Reserve G0T 1K0. Phone: (418) 233-2700. Fax: (418) 233-3326. E-mail: chme@b2b2c.ca Licensee: Radio Essipit Haute Cote-Nord inc. Population served: 25,000 Format: Fr. ♦Gilles LaBelle, gen mgr.

Levis

CFOM-FM— 1992: 102.9 mhz; 16.8 kw. Stereo. Hrs open: 2136.ch.Ste-foy, Quebec, G1V 1R8. Phone: (418) 694-1029. Fax: (418) 682-8430. E-mail: radioflashback@cfom1029.com Web Site: www.cfom1029.com. Licensee: 591991 B.C. Ltd. Group owner: Astral Media Inc. (acq 1-21-2005; grpsl). Format: Hits of the 60s, 70s, 80s & 90s. Target aud: 25 plus. ♦Pierre DeMondehare, gen mgr; Jean-Paul Lemire, sls dir; Annie Anglehart, prom dir; Mario Paquin, progmg dir.

Listuguj

CFIC-FM— 2000: 105.1 mhz; 425 w. Hrs open: Box 304, G0C 2R0. Secondary address: 44A Riverside E. G0C 2R0. Phone: (418) 788-5166. Fax: (418) 788-3524. Web Site: www.105hotcountry.com. Licensee: Societe d'Art, de Culture et d'Histoire Micmacs. Format: Country. ♦Gerald Dedam, pres; Chris Dedam, gen mgr & engrg dir; Linda Gilbert, sls dir & news dir.

Longueuil

CHAA-FM— 1987: 103.3 mhz; 64 w. Stereo. Hrs open: 24 91 St. Jean, J4H 2W8. Phone: (450) 646-6800. Fax: (450) 646-7378. E-mail: admin@fm1033.ca Web Site: www.fm1033.ca. Licensee: Radio Communautaire de la Rive-Sud Inc. Population served: 500,000 Natl. Rep: Target Broadcast Sales. Format: Adult contemp. News staff: 2; News: 5 hrs wkly. Target aud: 24-54; general. Spec prog: Fr 18 hrs, retro oldies 9 hrs, Greek 5 hrs, Vietnamese 4 hrs, Sp 3 hrs wkly. ♦Eric Tetreault, chmn & gen mgr; Richard Boileau, sls dir & mktg dir; France Dube, progmg dir.

CHMP-FM— Apr 9, 1977: 98.5 mhz; 40.8 kw. Ant 623 ft Stereo. Hrs open: 211 avenue Gordon, Verdun, H4G 2R2. Phone: (514) 767-2435. Fax: (514) 761-0985. Web Site: www.fm985.ca. Licensee: Diffusion Metromedia CMR Inc. Group owner: Corus Entertainment Inc. (acq 1-26-2001; grpsl). Natl. Rep: Canadian Broadcast Sales. Format: Talk. News: 5 hrs wkly. Target aud: 18-44. ♦Pierre Beland, pres; Pierre Accand, VP; Jacques Papin, gen mgr; David Therrien, sls dir; Michel Lacroix, gen sls mgr; Maurice Tietolman, natl sls mgr; Pierre Tremblay, prom mgr; Denis Fortin, progmg dir; Michel Belleau, mus dir; Real Terrault, chief of engrg.

Louiseville

CHHO-FM— Jan 22, 2007: 103.1 mhz; 1.52 kw. Hrs open: 50-A de la Fabrique, J0K 2W0. Phone: (819) 228-1001. Fax: (819) 228-0330. E-mail: info@ch2ofm.ca Web Site: www.ch2ofm.ca. Licensee: Coop de solidarite radio communautaire de la MRC de Maskinonge. Format: Fr var. ♦Stephane Carbonneau, gen mgr.

Lourdes-de-Blanc-Sablon

CFBS-FM— 1989: 89.9 mhz; 178 w. Hrs open: 7 AM-5 PM C.P. 8, G0G 1W0. Phone: (418) 461-2445. Fax: (418) 461-2425. E-mail: cfbs@globetrotter.qc.ca Licensee: Radio Blanc-Sablon inc. Format: Current affrs. ♦Vicky Driscol, pres; Patrick Bereburbe, progmg dir; Dominique Jones, news dir & engrg mgr.

Magog

CIMO-FM— 1979: 106.1 mhz; 50 kw. Stereo. Hrs open: 6 AM-8 PM 1845 King W., #200, Sherbrooke, J1J 2E4. Phone: (819) 347-1414. Fax: (819) 347-1061. Web Site: www.radioenergie.com. Licensee: Astral Media Radio Inc. Group owner: Astral Media Inc. Format: Top 40. Target aud: 18-34. ♦Nathalie Johnson, gen mgr; Isabelle Gagnon, sls dir; Anne-Marie Bercier, prom dir & progmg dir; Marc Toussaint, news dir; J.P. Maheu, chief of engrg.

Maliotenam

CKAU-FM— 1993: 104.5 mhz; 50 w. Hrs open: C.P. 338, Sept-Iles, G4R 4K6. Phone: (418) 927-2440. Fax: (418) 927-2800. E-mail: dels@globetrotter.net Web Site: www.ckau.com. Licensee: Corporation de Radio Kushapetsheken Apetuamiss Uashat. Format: Var. ♦Yves Rock, gen mgr; Reginald Thomas, adv dir; Mathieu McKenzie, engrg mgr.

Maniwaki

CBOF-1(AM)— Oct 22, 1973: 990 khz; 40 w, DA-1. Hrs open:
Rebroadcasts CBOF-FM Ottawa.
Box 3220, Stn C, Ottawa, ON, K1Y 1E4. Phone: (613) 288-6000. Fax: (613) 288-6560. Web Site: cbc.ca. Licensee: Canadian Broadcasting Corp. Format: Div. ♦Robert Rabinowitz, CEO; Denis Simard, gen mgr.

CFOR-FM— August 1994: 99.3 mhz; 2.4 kw. Hrs open: 24 139 Principale Sud., J9E 1Z8. Phone: (819) 441-0993. Fax: (819) 441-3488. E-mail: cfor993@b2b2c.ca Licensee: 9116-1299 Quebec Inc. (acq 4-22-02). Format: Rock music. News staff: 3. Target aud: 15-45. ♦Laure Voilquin, gen sls mgr; Rock Lepine, pres, gen mgr & progmg dir.

CHGA-FM— Nov 1980: 97.3 mhz; 2.8 kw. Hrs open: 24 163 Laurier Maniwaki, J9E 2K6. Phone: (819) 449-3959. Phone: (819) 449-5590. Fax: (819) 449-7331. E-mail: chga@bellnet.ca Web Site: www.chga.qc.ca. Licensee: Radio Communautaire Type B. Format: Div, adult contemp. News staff: one; News: 15 hrs wkly. Spec prog: Class one hr, jazz 3 hrs, country 8 hrs, folk 5 hrs wkly. ♦Hubert Tremblay, pres; Lise

Morissette, gen mgr; Lise Morisette, opns dir; Gaitam Bussiere, dev dir, gen sls mgr, mktg dir & progmg dir; Linda Lemieux, rgnl sls mgr; Kim Lacaille, mus dir; Georges Vasiloff, engrg dir; Michel Riel, news rptr.

Maniwaki (Kitigan Zibi Anishinabeg Reserve)

CKWE-FM— 1987: 103.9 mhz; 50 w. Hrs open: River Desert Indian Band, Box 309, Maniwaki, J9E 3C9. Phone: (819) 449-5170. Phone: (819) 449-5097. Fax: (819) 449-5673. E-mail: anita.tenasco@kza.qc.ca Licensee: Jean-Guy Whiteduck. Format: Talk, var, community news. ♦Anita Penasco, gen mgr, sls dir & progmg dir; Eleanor Whiteduck, opns mgr.

Maria (Reserve)

CHRG-FM— 1991: 101.7 mhz; 10 w. Hrs open: 24 Box 118, G0C 1Y0. Secondary address: 120 School St. G0C 1Y0. Phone: (418) 759-5424. Fax: (418) 759-5424. E-mail: radio@globetrotter.net Licensee: Douglas Martin. Format: Country, oldies, var/div. ♦Douglas Martin, gen mgr & progmg dir; Veronica Jerome, gen sls mgr & news dir.

Mashteuiatsh (Pointe-Bleue)

CHUK-FM— 1996: 107.3 mhz; 50 w. Hrs open: 24 1491 rue Ouiatchouan, Mashteuiatsh, G0W 2H0. Phone: (418) 275-4684. Fax: (418) 275-7964. Web Site: www.chukfm.ca. Licensee: Corporation Mediatique Teuehikan. Format: Montagnais, Fr. ♦Marc Gill, gen mgr; Jean Denis Gill, progmg dir & progmg dir.

Matagami

CHEF-FM— 2001: 99.9 mhz; 36 w. Hrs open: 110 boulevard Matagami, C.P. 39, J0Y 2A0. Phone: (819) 739-9990. Fax: (819) 739-6003. Licensee: Radio Matagami. Format: Fr, MOR. ♦M. Jean-Claude Constantineau, pres; Marie-Eve C. Gallant, gen mgr; Daniel Cliche, dev mgr; David Chabot, news dir.

Matane

CBGA-FM— 2004: 102.1 mhz; 42.93 kw. Hrs open: 5:30 AM-midnight 155 rue Saint-Sacrement, G4W 1Y9. Phone: (418) 562-0290. Phone: (418) 566-2322. Fax: (418) 562-3555. E-mail: communications_matane@radiocanada.ca Web Site: radio-canada.ca/gaspesie. Licensee: CBC. Natl. Network: Premiere Chaine. Format: CHR, div, news/talk. News staff: 5; News: 3 hrs wkly. Target aud: General. ♦Louis Pelletier, gen mgr; Johanne LaBrie, prom mgr; Richard Morisset, mus dir.

CHOE-FM— May 1991: 95.3 mhz; 30 kw. Stereo. Hrs open: 24 800 Ouest du Phare, G4W 1V7. Phone: (418) 562-8181. Fax: (418) 562-0778. E-mail: choe.routage@globetrotter.net Licensee: Les Communications Matane Inc. Population served: 20,000 Format: Light rock. Target aud: 18-34; young workers. ♦Kenneth Gagne, pres; Kenneth Gagne Jr., gen mgr, chief of opns & progmg dir; Michel Desrosiers, sls dir; Carol St-Pierre, news dir; Jacques Tremblay, chief of engrg.

CHRM-FM— April 2001: 105.3 mhz; 30 kw. Hrs open: 24 800 avenue du Phare Ouest, G4W 1V7. Phone: (418) 562-4141. Fax: (418) 562-0778. Licensee: Les Communications Matane inc. Format: MOR. ♦Kenneth Gagne, pres; Kenneth Gagne Jr., gen mgr, chief of opns & progmg dir; Michel Desrosiers, sls dir; Carol St-Pierre, news dir.

Mont-Laurier

CFLO-FM— 1995: 104.7 mhz; 10.98 kw. Stereo. Hrs open: Rebroadcasts CFLO FM-1 L'Annonciation 100%.
332 de la Madone, J9L 1R9. Phone: (819) 623-5610. Phone: (819) 623-6610. Fax: (819) 623-7406. E-mail: cflofm@cflo.ca Web Site: www.cflo.ca. Licensee: Soneme Inc. Soneme Inc. (acq 1988). Format: Adult contemp, French. News: 3 hrs wkly. Target aud: 24-54. ♦Alain Desjardins, pres & stn mgr.

Montmagny

CFEL-FM— 1987: 102.1 mhz; 25.7 kw. Ant 443 ft TL: N46 56 21 W70 30 29. Stereo. Hrs open: 24 191 Chemen des Poirier, G5V 4L2. Secondary address: 5245 Boulevard De La Rive-Sud Levis, Levis G6V4ZA. Phone: (418) 248-1122. Fax: (418) 248-1951. E-mail: cfel@globetrotter.net Licensee: 5191991 B.C. Ltd. Group owner: Corus Entertainment Inc. (acq 3-24-2000; grpsl). Population served: 30,000 Format: Adult contemp. News staff: one. Target aud: 25-49. ♦Michel Montminy, gen mgr & opns VP; Rene' Nadeau, sls VP & progmg dir.

Montreal

CBF-FM— 1947: 95.1 mhz; 100 kw. 823 ft (Digital radio: 1458.048 mhz; 11.724 kw). Stereo. Hrs open: Box 6000, H3C 3A8. Phone: (514) 597-6000. Fax: auditoire@radio-canada.ca. Web Site: www.cbc.radio-canada.ca. Licensee: CBC. Format: Class. ♦Sylvain LaFrance, VP; Bertrand Emond, gen mgr. Co-owned TV: CBFT(TV) affil.

CBFX-FM— 1998: 100.7 mhz; 100 kw. (Digital radio: 1458.048 mhz; 11.724 kw). Hrs open: Box 6000, H3C 3A8. Phone: (514) 597-6000. Fax: (416) 205-3714. Web Site: www.cbc.radio-canada.ca. Licensee: CBC. Natl. Network: Espace Musique. Format: Var. ♦Sylvain LaFrance, VP; Bertrand Emond, stn mgr & progmg dir; Alain Saulnier, news dir.

CBME-FM— 1998: 88.5 mhz; 16.9 kw. (Digital radio: 1458.048 mhz; 11.724 kw). Hrs open: 24 Box 6000, H3C 3A8. Phone: (514) 597-4444. Fax: (514) 597-4142. Web Site: www.radio-canada.ca. Licensee: Canadian Broadcasting Corp. Natl. Network: CBC Radio One. Format: News, current affrs. ♦Patricia Pleszczynska, stn mgr; Judith Bleier, opns mgr; Sally Caldwell, progmg dir & news dir.

CBM-FM— 1947: 93.5 mhz; 24.6 kw. 823 ft (Digital radio: 1458.048 mhz; 11.724 kw). Stereo. Hrs open: Box 6000, H3C 3A8. Phone: (514) 597-6000. Fax: (514) 597-4416. Licensee: Canadian Broadcasting Corp. Format: Class. ♦Patricia Pleszczynska, stn mgr; Judith Bleier, opns mgr; Patricia Plesczynska, progmg dir. Co-owned TV: CBMT(TV) affil.

CFMB(AM)— Dec 21, 1962: 1280 khz; 50 kw-U, DA-2. TL: N45 19 31 W73 32 55. Hrs open: 24 35 York St., Westmount, H3Z 2Z5. Phone: (514) 483-2362. Fax: (514) 483-1362. E-mail: admin@cfmb.ca Web Site: www.cfmb.ca. Licensee: CFMB Ltee. Population served: 4,000,000 Rgnl rep: Direct Format: Ethnic. News staff: 7; News: 35 hrs. news programing wkly.Andrew Mielewczyk, pres & gen sls mgr; A.M. St. Germain-Stanczykowski, exec VP; Luigi Valente, stn mgr & chief of engrg; Marcello Silveri, rgnl sls mgr; Ivana Bombardieri, prom mgr & women's int ed; Walter Centa, natl sls mgr & progmg dir; Tony Ferrara, mus dir; Nino Di Stefano, news dir & sports cmtr; Teddy Colantonio, pub affrs dir, local news ed & reporter; Silvana Di Flavio, spec ev coord; Elisa Pierna, spanish dir; Denise Agiman, disc jockey; Nick DeVincenzo, disc jockey

CFQR-FM—Listing follows CINW(AM).

CFZZ-FM—(Saint Jean-Iberville, 1992: 104.1 mhz; 1.35 kw. Hrs open: 24 104 rue Richelieu, St. Jean-Sur-Richelieu, J3B 6X3. Phone: (450) 346-0104. Fax: (450) 348-2274. E-mail: lstemarie@boomfm.astral.com Web Site: www.boomfm.com. Licensee: Astral Media Radio inc. Group owner: Corus Entertainment Inc. (acq 5-30-2005; grpsl). Format: Oldies. ♦Leopold Stemarie, gen mgr; Luc Lalonde, mktg dir; Ghislaine Plourde, progmg dir & news dir.

CHOM-FM— July 16, 1963: 97.7 mhz; 47.1 kw. Ant 979 ft TL: N45 30 20 W73 35 32. (Digital radio: 1452.816 mhz; 11.724 kw). Stereo. Hrs open: 24 1411 Du Fort, 3rd Fl., H3H 2R1. Phone: (514) 931-2466. Fax: (514) 846-4747. Web Site: www.chom.com. Licensee: Standard Radio Inc. Group owner: Standard Broadcasting Corp. (acq 12-20-2001; C$15 million in swap for CFWM-FM Winnipeg, MB). Format: Rock. ♦Bob Harris, opns mgr; Jacques Bolduc, sls dir; Matt Cundill, progmg dir; Rob Braide, gen mgr & mus dir; Derek Conton, news dir; Mark Kavanagh, engrg mgr.

CHOU(AM)— 2007: 1450 khz; 1 kw-U. Hrs open: 11876 rue de Meulles, H4J 2E6. Phone: (514) 790-0002. Licensee: 9015-2018 Quebec inc. Format: Ethnic, Arabic. ♦Antoine Karam, pres.

***CIBL-FM**— Apr 26, 1980: 101.5 mhz; 315 w. Stereo. Hrs open: 24 2nd Fl., 1691 Boul. Pie IX, H1V 2C3. Phone: (514) 526-2581. Fax: (514) 526-3583. E-mail: administration@cibl1015.com Web Site: www.cibl1015.com. Licensee: Radio Communautaire Francophone de Montreal Inc. Population served: 2,500,000 Format: Music/talk. News staff: 4; News: 14 hrs wkly. Target aud: General. Spec prog: Black 13 hrs, class 4 hrs, jazz 14 hrs, reggae 4 hrs, world beat 8 hrs wkly. ♦Eric Lefebvre, gen mgr; Genevieve Dore, gen mgr & sls dir.

CINF(AM)—(Verdun, Nov 3, 1946: 690 khz; 50 kw-U, DA-2. Hrs open: 215 Jacques, Bureau 333, H2Y 1M6. Phone: (514) 849-1690. Fax: (514) 849-0733. Web Site: www.info690.com. Licensee: Metromedia CMR Montreal Inc. Group owner: Corus Entertainment Inc. (acq 1-26-01; grpsl). Format: News. ♦Pierre Arcand, pres; Maurice Tietolman, gen mgr; Christian Chalifour, sls dir; Marie Claude Baribault, prom dir; Yvon Vadnais, progmg dir; Kim Bickerdike, chief of engrg.

CKOI-FM—Co-owned with CINF(AM). 1953: 96.9 mhz; 307 kw. 712 ft Stereo. Web Site: www.info690.com. Format: CHR. ♦Andre St. Amard, progmg dir.

CINQ-FM— Jan 27, 1975: 102.3 mhz; 1.29 kw. Ant 180 ft Stereo. Hrs open: 24 5212 Boul. St. Laurent, H2T 1S1. Phone: (514) 495-2597. Fax: (514) 495-2429. Web Site: www.radiocentreville.com. Licensee: Radio Centre-Ville Saint Louis Inc. Format: Multilingual, Fr, world music. News staff: one. Target aud: 25-54; Fr & multilingual. Spec prog: Sp 16, Portugese 13 hrs, Greek 13 hrs, Chinese 5 hrs, Haitian 6 hrs, Creole 12, English 16 hrs wkly. ♦Nadi Mobarak, pres; Evan Kapetanakis, stn mgr; Daniel Moreau, gen sls mgr & adv mgr; Ricardo Costa (English), progmg mgr; Suzanne Charland (Fr), progmg mgr; Robert Laplante, news dir; Marc Provencher, chief of engrg.

CINW(AM)— November 1999: 940 khz; 50 kw-U. Hrs open: 24 215 St. Jacques, Suite 333, H2Y 1M6. Phone: (514) 849-0940. Fax: (514) 849-0733. E-mail: news@940news.com Web Site: www.940news.com. Licensee: Metromedia CMR Broadcasting Inc. Group owner: Corus Entertainment Inc. (acq 1-26-2001; grpsl). Natl. Rep: Canadian Broadcast Sales. Format: News. News staff: 4; News: 168 hrs wkly. Target aud: 35-54. ♦Pierre Beland, pres; Pierre Arcand, exec VP; Maurice Tietolman, gen mgr; George Weiss, sls VP & sls dir; Marie Claude Baribault, prom dir & prom mgr; Yven G. Vadnais, progmg dir & progmg mgr; Kim Bickerdike, chief of engrg.

CFQR-FM—Co-owned with CINW(AM). November 1966: 92.5 mhz; 41.4 kw. 979 ft Stereo. 24 Box 925, H4G 3M1. Secondary address: 211 Gordon Ave, Verdum Quebec H4F-2R2. Phone: (514) 767-9250. Web Site: www.940news.com. Format: Light rock. Target aud: 25-54. ♦Kathie Murphy, prom mgr; Ted Silver, progmg dir.

CIRA-FM— 1994: 91.3 mhz; 36.2 kw. Hrs open: 24 4020 Rue St-Ambroise, #199, H4C 2C7. Phone: (514) 382-3913. Fax: (514) 858-0965. E-mail: cira@radiovm.com Web Site: www.radiovm.com. Licensee: Radio Ville-Marie. Format: Relg, div music. News staff: 15. ♦Jean-Guy Roy, gen mgr; Gaston Pearson, sls dir & progmg dir; Renaude Gregoire, progmg dir; Mario Bard, news dir; Joe Pachecho, engrg mgr; Rogel Landry, chief of engrg.

***CISM-FM**— March 1991: 89.3 mhz; 10 kw. Hrs open: Box 6128, C-1509, 2332 Edouard Montpetit, H3C 3J7. Phone: (514) 343-7511. Fax: (514) 343-2418. E-mail: cism@cam.org Web Site: www.cismfm.qc.ca. Licensee: Communications du Versant Nord. Format: Alternative. ♦Dave Ouellet, gen mgr; Guillaume St-Onge, prom dir; Candide Proulx, progmg dir; Martin Roussy, mus dir; Cecile Boumati, news dir; Luc Guillox, engrg dir.

CITE-FM— May 20, 1977: 107.3 mhz; 42.9 kw. 700 ft (Digital radio: 1452.816 mhz; 11.724 kw). Stereo. Hrs open: 24 1717 Rene Levesque Est, H2L 4T9. Phone: (514) 845-2483. Fax: (514) 288-1073. E-mail: cite@rock-detente.com Web Site: www.rock-detente.com. Licensee: Astral Media Radio Inc. Group owner: Astral Media Inc. (acq 4-19-2002; grpsl). Format: Adult contemp. Target aud: 25-49. ♦Jacques Parisien, pres; Sylvain Langlois, VP; Luc Tremblay, gen mgr.

CJAD(AM)— Dec 8, 1945: 800 khz; 50 kw-D, 10 kw-N, DA-2. (Digital radio: 1454.56 mhz; 1.4 kw in Montreal, 1.594 kw in Laval). Stereo. Hrs open: 24 1411 Rue du Fort, H3H 2R1. Phone: (514) 989-2523. Fax: (514) 989-3868. Web Site: www.cjad.com. Licensee: Standard Radio Inc. Group owner: Standard Broadcasting Corp. Format: News/talk, info. News staff: 15; News: 14 hrs wkly. ♦Rob Braide, VP & gen mgr; Bob Harris, opns mgr; Jacques Bolduc, gen sls mgr; Lisa Faoco, prom mgr; Mike Bendixen, progmg mgr; Derek Conlon, news dir; Mark Kavanagh, engrg dir.

CJFM-FM—Co-owned with CJAD(AM). Oct 1, 1962: 95.9 mhz; 41.2 kw. 979 ft (Digital radio: 1454.56 mhz; 1.4 kw in Montreal, 1.594 kw in Laval). Stereo. Phone: (514) 989-2536. Fax: (514) 989-2554. Web Site: www.themix.com. Format: Adult contemp. ♦Matthew Wood, prom mgr; Bob Harris, progmg mgr; Ray Scott, mus dir; Mark Kavanagh, chief of engrg.

CJLO(AM)— 2007: 1690 khz; 1 kw-U. TL: N45 26 51 W73 37 57. Hrs open: 7141 Sherbrooke St. Ouest, Suite CC-430, H4B 1R6. Fax: (514) 848-7450. Web Site: www.cjlo.com. Licensee: Concordia Student Broadcasting Corp. Format: Var. ♦Chris Quinnell, stn mgr; Amrew Weekes, gen sls mgr; Katie Seline, prom dir & progmg dir; Omar Husain, mus dir; Jessica Hemmerich, news dir; Joshua Mocle, news dir.

CJPX-FM— June 25, 1998: 99.5 mhz; Hrs open: Radio Classique Montreal Inc., Iles Notre Dame, Parc Jean-Drapeau, H3C 1A9. Phone: (514) 871-0995. Fax: (514) 871-0990. Licensee: Radio Classique

Montreal Inc. Format: Class. ♦Jean-Pierre Coallier, CEO; Pierre Barbeau, VP; Francois Pare, gen mgr; Rejean Beaulieu, sls dir; Marie-Claude Goulet, traf mgr.

CJRS(AM)— 2007: 1650 khz; 1 kw-U. Hrs open: 5775 Ave. Victoria, Bureau 103, H3W 2R4. Phone: (514) 738-4100. Web Site: www.radio-shalom.ca/EN/. Licensee: Radio Chalom. Format: Relg. ♦Robert Levy, pres; Greg McLachlan, sls dir.

CJWI(AM)— 2002: 1610 khz; 1 kw-U. Hrs open: 3733 Jarry St. E., H1Z 2G1. Phone: (514) 287-1288. Fax: (514) 287-3299. Licensee: CPAM Radio Union.com inc. Format: Ethnic. ♦Jean Ernest Pierre, gen mgr.

CKAC(AM)— Sept 22, 1922: 730 khz; 50 kw-U, DA-1. (Digital radio: 1452.816 mhz; 11.724 kw). Hrs open: 24 1411 Peel St., BUR. 400, H3A 3L5. Phone: (514) 845-5151. Fax: (514) 845-2229. Web Site: www.ckac.com. Licensee: 591991 B.C. Ltd. Group owner: Astral Media Inc. (acq 1-21-2005; grpsl). Format: Sports. News: 20 hrs wkly. Target aud: 35-54. ♦Sylvain Chamberland, gen mgr; Julie Gagnon, opns dir.

CKDG-FM— January 2004: 105.1 mhz; 141 w. Hrs open: 5:30 AM-1 AM 5899 Park Ave., H2V 4H4. Phone: (514) 273-2481. Fax: (514) 273-3707. Web Site: www.ckdgfm.ca. Licensee: Canadian Hellenic Cable Radio Ltd. Format: Ethnic. ♦John DeParis, pres; Francis Bergeron, gen mgr; Marie Griffths, gen mgr; Marie Grif, progmg dir; Tony Choundalas, news dir; Jean Frechette, engrg mgr.

CKGM(AM)— Dec 7, 1959: 990 khz; 50 kw-U. TL: N45 17 43 W73 43 20. (Digital radio: 1452.816 mhz; 11.724 kw). Hrs open: 1310 Greene Ave., H3Z 2B5. Phone: (514) 931-4487. Fax: (514) 931-4079. Web Site: www.team990.com. Licensee: CHUM Ltd. (group owner; acq 9-5-85). Format: Sports. ♦Lee Hambleton, gen mgr; Wayne Bews, sls dir & natl sls mgr.

CKLX-FM— 2004: 91.9 mhz; 1.9 kw. Ant 633 ft Hrs open: 1 Place Ville Marie, Bureau 1523, H3B 2B5. Phone: (514) 866-8686. Fax: (514) 866-8056. Licensee: 9115-0318 Quebec inc. Group owner: Radio Nord Communications Inc. Format: Jazz and blues. Target aud: 35-64. ♦Jean-Pierre Major, gen mgr.

CKMF-FM— May 11, 1964: 94.3 mhz; 41.4 kw. 979 ft (Digital radio: 1452.816 mhz; 11.724 kw). Stereo. Hrs open: 24 1717 Rene Levesque E., H2L 4T9. Phone: (514) 529-3229. Fax: (514) 529-9308. E-mail: Lsabbatini@radio.astral.com Web Site: www.radioenergie.com. Licensee: Astral Media Radio Inc. Group owner: Astral Media Inc. (acq 1-12-2000; grpsl). Natl. Network: Radiomutuel. Format: CHR. News: one hr wkly. Target aud: 18-34. Spec prog: Disco. ♦Ian Greenburg, CEO; Jacques Parisien, chmn & pres; Luc Sabbatini, exec VP & opns mgr; Charles Benoit, VP; Luc Tremblay, gen mgr; Robert Latreille, stn mgr & engrg dir; Marie Josee Lefelbvre, natl sls mgr; Michel Tartif, rgnl sls mgr; Andre Allara, mktg dir; Johanne Cloutier, mktg mgr; Sylvain Legare, prom dir; Sylvain Simard, progmg dir.

***CKUT-FM**— November 1987: 90.3 mhz; Hrs open: 24 3647 University, H3A 2B3. Phone: (514) 398-6787. Phone: (514) 398-6788. Fax: (514) 398-8261. Web Site: www.ckut.ca. Licensee: Radio McGill Inc. Population served: 3,500,000 Format: Div. News staff: one; News: 6 hrs wkly. Spec prog: Black, Fr, Sp, folk, Gospel. ♦Louise Burns, sls VP & mktg dir; Suhrid Manchanda, prom dir; Steve Guimond, progmg dir, mus dir & pub affrs dir; Alex Moskos, asst music dir; Gretchen King, news dir, local news ed, news rptr & reporter; Valentine Latty, chief of engrg.

Natashquan

CKNA-FM— Jan 30, 1983: 104.1 mhz; 6.56 kw. Hrs open: C.P. 9, G0G 2E0. Secondary address: 29 chemin d'en Haut G0G 2E0. Phone: (418) 726-3284. Phone: (418) 726-3240. Fax: (418) 726-3367. E-mail: ckna@globetrotter.net Web Site: pages.globetrotter.net/ckna/. Licensee: La Radio Communautaire CKNA Inc. Format: MOR. ♦Carmen Rodrigue, gen mgr; Vincent Pons, sls dir; Melanie Gauvin, prom mgr; Charles Benoit, progmg VP; Genevieve Moreau, mus dir.

New Carlisle

CHNC(AM)— Dec 23, 1933: 610 khz; 10 kw-D, 5 kw-N, DA-1. TL: N48 01 19 W65 14 52. Hrs open: 24 Box 610, G0C 1Z0. Secondary address: 153 Rt. 132 G0C 1Z0. Phone: (418) 752-2215. Fax: (418) 752-6939. E-mail: radiochnc@globtrotter.net Licensee: Radio CHNC Ltee. Natl. Network: Radiomedia. Rgnl rep: Lyne Appleby et Gail Morris Format: Adult contemp, Country. Target aud: General; Adult. ♦Francis Remillard, asst music dir; Michel Morin, news dir.

Pikogan

CKAG-FM— 1993: 100.1 mhz; 50 w. Hrs open: 45 rue Migwan, J9T 3A3. Phone: (819) 727-3237. Fax: (819) 732-1569. E-mail: ckagfm@cableamos.com Licensee: Societe de Communication Ikito Pikogan Ltee.

Plessisville

CKYQ-FM— 1996: 95.7 mhz; 1 kw. Hrs open: 24 Box 142, G6L 2Y6. Phone: (819) 362-3737. Fax: (819) 362-3414. E-mail: ckyq-fm@ivic.qc.ca Web Site: www.kyqfm.com. Licensee: Societe CKYQ Radio Media Enr. Format: MOR. ♦Hugh Laroche, news dir; Stephane Dion, pres, stn mgr, mktg dir, progmg dir & chief of engrg.

Pohenegamook

CFVD-FM-2— Sept 10, 1983: 92.1 mhz; 294 w. Stereo. Hrs open: Rebroadcasts CFVD-FM Degelis.
654 6th St. E., Degelis, G5T 1Y1. Phone: (418) 853-3370. Phone: (418) 853-2370. Fax: (418) 853-3321. E-mail: cfvd@fm95.ca Web Site: infotemis.com. Licensee: Radio Degelis Inc. Format: Div, CHR. News staff: 10. Target aud: General. ♦Gilles Caron, pres, gen mgr & dev dir.

Port-Cartier

CIPC-FM— 1995: 99.1 mhz; 13 kw. Stereo. Hrs open: 24 52 Elie Roche Fort, G5B 1N2. Phone: (418) 766-6868. Fax: (418) 766-6870. Web Site: www.laradioactive.com. Licensee: Radio Port-Cartier Inc. Format: Top-40. ♦Yvan Beavlier, gen mgr; Denis Simard, gen sls mgr; Matthew Pineau, progmg dir & progmg mgr; Joanie Hebert Cimon, mus dir; Jean-Hugo Savard, news dir.

Port-Menier

CJBE-FM— 1989: 90.5 mhz; 88 w. Hrs open: C.P. 15, Port-Menier (Ile d'Anticosti), G0G 2Y0. Phone: (418) 535-0292. Fax: (418) 535-0292. E-mail: radioanticosti@hotmail.com Licensee: Radio Anticosti Inc. Format: Var. ♦Sandra Dussault, pres; Denis Tremblay, gen mgr & progmg dir.

Quebec

CBVE-FM— March 1979: 104.7 mhz; 100 kw. 411 ft Hrs open: 5:30-8:30 AM; 4-6 PM 888 Saint-Jean St., G1R 5H6. Phone: (418) 691-3613. Fax: (418) 691-3610. E-mail: quebecam@cbc.ca Web Site: www.cbc.ca. Licensee: Canadian Broadcasting Corp. Natl. Network: CBC Radio One. Format: News/talk, div. ♦Claude Saindon, gen mgr & progmg dir; Judith Bleier, opns dir; Peter Black, news rptr.

***CBV-FM**— 1974: 106.3 mhz; 100 kw. Ant 541 ft TL: N46 51 40 W71 04 46. Hrs open: 24 888 Rue Saint-Jean, GIR 5H6. Phone: (418) 656-8235. Fax: (418) 656-8842. Web Site: www.cbc.ca. Licensee: Societe Radio Canada. Population served: 500,000 Natl. Network: Premiere Chaine. Format: Div, news/talk. News: 15 hrs wkly. Target aud: General. ♦Robert Rabinovitch, pres & gen mgr; Norman LaCombe, news dir; Robert Jacques, engrg mgr. Co-owned TV: *CBVT-TV affil.

***CBVX-FM**— 1998: 95.3 mhz; 100 kw. 541 ft TL: N46 51 40 W71 04 46. Stereo. Hrs open: 888 Saint-Jean St., G1R 5H6. Phone: (418) 654-1341. Fax: (418) 656-8212. Web Site: www.cbc.ca. Licensee: Societe Radio Canada. Natl. Network: Radio Canada. Format: Class. Target aud: General. Spec prog: Jazz 16 hrs, news 7 hrs wkly. ♦Marleine Simard, gen mgr & progmg dir; Real Jean, opns mgr; Clodine Dorval, traf mgr.

CHIK-FM— Aug 1, 1982: 98.9 mhz; 41 kw. 1,355 ft TL: N46 49 22 W71 29 43. Hrs open: 900 d'Youville St., 1st Floor, G1R 3P7. Phone: (418) 687-9900. Fax: (418) 687-3106. Web Site: www.radioenergie.com. Licensee: Astral Media Radio Inc. Group owner: Astral Media Inc. Format: Adult contemp. ♦Daniel Tremblay, gen mgr; Real Marcotte, sls dir; Julie Durand, prom dir; Jean Alexandre, progmg dir; Rejean Bergeron, news dir; Michel Duval, engrg dir.

CHOI-FM— Nov 1, 1949: 98.1 mhz; 40 kw. Ant 250 ft TL: N46 49 17 W71 29 48. Hrs open: 1134 Grande-Allee Ouest, Bureau 300, G1S 1E5. Phone: (418) 687-9810. Fax: (418) 682-8427. Web Site: www.choiradiox.com. Licensee: RNC MEDIA Inc. Format: Fr alternative rock music. Target aud: 18-34; young adults.

CHRC(AM)— Apr 1, 1926: 800 khz; 50 kw-U, DA-1. Hrs open: 24 2136 Chemin Sainte-Foy, Sainte.-Foy, G1V 1R8. Phone: (418) 688-8080. Fax: (418) 670-1234. Web Site: www.chrc.com. Licensee: 591991 B.C. Ltd. Group owner: Astral Media Inc. (acq 1-21-2005; grpsl). Format: Sports. Target aud: 35 plus. Spec prog: French.

***CION-FM**— Sept 19, 1995: 90.9 mhz; 5.69 kw. 1,364 ft TL: N46 49 17 W71 29 48. Stereo. Hrs open: 8 (M-S); 16 (Su) 2511 Chemin Ste-Foy, Suite 200, G1V 1T7. Phone: (418) 659-9090. Phone: (418) 650-1572. Fax: (418) 650-3306. E-mail: cionfm@radiogalilee.qc.ca Licensee: Radio Galilee. Format: Relg, adult contemp, btfl mus. Target aud: General. ♦Alexandre St. Hilaire, pres & rsch dir; Denis Veilleux, gen mgr & progmg dir; Mario Blouin, mus dir; Jacques Fortin, news dir; Daniel Coulombe, engrg mgr.

CITF-FM— July 22, 1982: 107.5 mhz; 37.8 kw. 500 ft Stereo. Hrs open: 24 900 Dyouville St., 1st Fl., G1R 3P7. Phone: (418) 527-3232. Fax: (418) 687-3106. Web Site: www.rockdetente.com. Licensee: Astral Media Radio Inc. Group owner: Astral Media Inc. (acq 4-19-2002; grpsl). Rgnl rep: Radio Plus. Format: Adult contemp. News staff: 2. Target aud: 25-54. ♦Michel Duval, CEO & chief of engrg; Daniel Tremblay, gen mgr; Suzie Baronet, sls dir; Julie Durand, prom dir; Marc Tanguay, progmg dir.

CJEC-FM— August 2003: 91.9 mhz; 14.45 kw. Hrs open: 1305 Chemin Ste-Foy, 4e etage, G1S 4Y5. Phone: (418) 688-0919. Fax: (418) 527-0919. Web Site: www.rythmefm.com. Licensee: Cogeco Diffusion Inc. Group owner: Cogeco Radio-Television Inc. Format: Adult contemp. ♦Louis Audet, pres; Jean-Paul Lemire, gen mgr; Carole Vezina, sls dir; Daniel Plante, mktg dir, prom dir & progmg dir; Lilianne Randall, mus dir; Martin Perkins, engrg dir.

CJMF-FM— Sept 15, 1979: 93.3 mhz; 32.96 kw. Ant 1,275 ft Stereo. Hrs open: 24 1305 Chemin Ste-Foy, 4e etage, G1S 4Y5. Phone: (418) 687-9330. Fax: (418) 687-0211. E-mail: commentaire@cjmf.com Web Site: www.le933.com. Licensee: Cogeco Diffusion Inc. Group owner: Cogeco Radio-Television Inc. (acq 11-87; $8 million). Format: Talk, classic rock. News staff: 3; News: 5 hrs wkly. Target aud: 25-54; mostly males. ♦Louis Audet, pres; Jean-Paul Lemue, gen mgr.

CJSQ-FM—Not on air, target date: unknown: 92.7 mhz; 2.1 kw. Hrs open: Radio-Classique, 6666 Christophe-Colomb, Montreal, H2S 2G8. Fax: (514) 274-3127. Licensee: 9147-2605 Quebec inc. Format: Classical music. ♦Jean-Pierre Coallier, gen mgr.

CKIA-FM— Oct 31, 1984: 88.3 mhz; 350 w. 700 ft TL: N46 48 28 W71 12 57. Hrs open: 24 600 Cote d'Abraham, G1R 1A1. Phone: (418) 529-9026. Fax: (418) 529-4156. E-mail: ckiafm@moso.com Web Site: www.meduse.org/ckiafm. Licensee: Radio Basse-Ville Inc. Format: Classic rock, country, div. News staff: one; News: 2 hrs wkly. Target aud: 18-35; general. Spec prog: Class 3 hrs, jazz 4 hrs, Sp 4 hrs, Haitian 2 hrs, African one hr wkly. ♦Jacynthe Huard, gen mgr; Reynald Poirier, gen sls mgr; Sophie Anne Mailloux, progmg dir & news dir; Denis Roberge, chief of engrg.

***CKRL-FM**— Feb 15, 1973: 89.1 mhz; 1.4 kw. 700 ft TL: N46 48 27 W71 13 02. Stereo. Hrs open: 24 405 3rd Ave., G1L 2W2. Phone: (418) 640-2575. Fax: (418) 640-1588. E-mail: ckrl@ckrl.qc.ca Web Site: www.ckrl.qc.ca. Licensee: CKRL MF 89.1 Inc. Format: Adult contemp, jazz, class, rock. News staff: one. Spec prog: Sp 2 hrs, It 2 hrs, Black 6 hrs, Arab 3 hrs wkly. ♦Dany Fortin, gen mgr & sls dir; Daniel Deslauriers, mktg dir; Bastien Gagnon La France, progmg dir; Daniel Marcoux, mus dir.

Radisson

CIAU-FM— 1996: 103.1 mhz; 17 w. Hrs open: Box 285, J0Y 2X0. Secondary address: 143 rue Jolliet J0Y 2X0. Phone: (819) 638-7033. Phone: (819) 638-1031. Fax: (819) 638-7033. E-mail: ciaufm@lino.com Web Site: www.ciaufm.com. Licensee: Radio communautaire de Radisson (acq 4-23-2004). Population served: 1,000 Format: Var. News staff: one; News: 3 hrs wkly. Spec prog: Fr 8 hrs, Jazz 3 hrs wkly. ♦Eric Hamel, pres, dev dir & sls dir; Martin Beaucage, stn mgr, opns dir, prom dir & adv dir; Martin Beaucaage, progmg dir.

Restigouche

CHRQ-FM— 1991: 106.9 mhz; 31 w. Hrs open: Box 180, G0C 2R0. Phone: (418) 788-2449. Fax: (418) 788-2653. E-mail: chrq1069@globetrotter.net Licensee: Gespegewag Communications Society. Format: English-language & Micmac-language community radio. News staff: 5; News: one hr wkly. Community members all ages. ♦Sandra Bulmer, stn mgr; Chad Gedeon, rgnl sls mgr; Karen Duguay, prom mgr; Steve Clement, progmg mgr.

Rimouski

***CBRX-FM—** Feb 28, 1959: 101.5 mhz; 50 kw. 931 ft Stereo. Hrs open: 273 rue St-Jean Baptiste Ouest, G5L 4J8. Phone: (418) 723-2217. Fax: (418) 723-6126. Licensee: Canadian Broadcasting Corp. Natl. Network: Espace Musique. Format: Class, jazz, talk. ♦Bernard Labarge, pres & news dir; Bernard Lebarge, gen mgr; Bernard Labarbe, progmg dir.

CIKI-FM— Feb 14, 1988: 98.7 mhz; 76 kw. Stereo. Hrs open: 24 875 Boul. St. Germain Ouest, G5L 3T9. Phone: (418) 723-2323. Fax: (418) 722-7508. E-mail: ciki@pqm.net Web Site: www.ciki.fm. Licensee: Astral Media Radio inc. Group owner: Corus Entertainment Inc. (acq 5-30-2005; grpsl). Format: AOR. ♦Bertrand Bellavance, gen mgr; Jean Fournier, opns mgr & chief of engrg; Ghislain Desgardins, sls dir; Francois La Fond, progmg dir; Alain Rivard, mus dir; Martin Bressard, news dir.

CJBR-FM— 2000: 89.1 mhz; 19.4 kw. Hrs open: 273 rue St-Jean Baptiste Ouest, G5L 4J8. Phone: (418) 723-2217. Fax: (418) 723-6126. Licensee: Canadian Broadcasting Corp. Format: MOR, adult contemp, news/talk. Spec prog: Fr. ♦Bernard Labarge, gen mgr & news dir.

CJOI-FM— Oct 22, 2000: 102.9 mhz;; 33.6 kw. Hrs open: 24 875 Boul. St. Germain Ouest, G5L 3T9. Phone: (418) 723-2323. Fax: (418) 722-7508. E-mail: cjoi@pqm.net Web Site: www.cjoi.fm. Licensee: Astral Media Radio inc. Group owner: Corus Entertainment Inc. (acq 5-30-2005; grpsl). Format: MOR. ♦Bertrand Bellavance, gen mgr; Jean Fournier, opns mgr & chief of engrg; Francois La Fond, progmg dir; Martin Bressard, news dir.

Rimouski-Mont Joli

CKMN-FM— June 4, 1990: 96.5 mhz; 6.4 kw. TL: N48 22 32 W68 35 43. Stereo. Hrs open: 24 323 Montee Industrielle, Rimouski, G5M 1A7. Phone: (418) 722-2566. Fax: (418) 724-7815. E-mail: ckmn-fm@cgocable.ca Licensee: La Radio Communautaire du Comte. Wire Svc: CNW Broadcast Format: Adult contemp, CHR, country. News staff: 5. Target aud: 25-55; general. Spec prog: Oldies 3 hrs, classical 3 hrs wkly. ♦Antonini Michaud, exec VP; Jean-Claude Pinel, exec VP; Claude Marmen, mktg dir; Renie Langlois, prom dir & progmg dir; Michel Vallee, chief of engrg.

Riviere au Renard

CJRE-FM— 1979: 97.9 mhz; 56 w. 594 ft Hrs open: 24 162 Jacques Cartier, Gaspe, G4X 1M9. Phone: (418) 368-3511. Fax: (418) 368-1663. Licensee: Radio Gaspesie Inc. Format: MOR. ♦Jacques Chartier, gen mgr; Paul Mainville, sls dir; Richard O'Leary, news dir; Yvan Dupuis, engrg dir.

Riviere du Loup

CIBM-FM— 1966: 107.1 mhz; 100 kw. Ant 244 ft Hrs open: 24 64 Hotel de Ville, G5R 1L5. Phone: (418) 867-1071. Fax: (418) 862-7704. Web Site: www.cibm107.com. Licensee: CIBM Mont-Bleu. Natl. Network: Radiomedia. Format: Pop rock. ♦Guy Simard, gen mgr; Renee Giard, gen sls mgr & prom dir; Daniel St. Pierre, progmg dir; Martin Pelletier, news dir; Clement Lavore, engrg dir.

CIEL-FM— Dec 15, 1994: 103.7 mhz; 60 kw. Ant 1,050 ft TL: N47 34 53 W69 22 20. Hrs open: 64 Hotel-de-Ville, G5R 1L5. Phone: (418) 862-8241. Fax: (418) 862-7704. Web Site: www.103rockdouceur.com/ciel. Licensee: Radio CJFP (1986) Ltee. Format: Adult contemp. ♦Guy Simard, pres & gen mgr; Renee Giard, sls dir, natl sls mgr, rgnl sls mgr & mktg dir; Christian Duchesne, prom dir; Daniel St. Pierre, progmg dir, mus dir, asst music dir & news dir; Clement LaVoie, engrg dir.

Roberval

CHRL-FM— March 1, 2002: 99.5 mhz; 16.6 kw. TL: N48 26 25 W72 06 47. Hrs open: 24 568 Blvd. St. Joseph, G8H 2K6. Phone: (418) 275-1831. Fax: (418) 275-2475. E-mail: chrl@antenne6.com Licensee: Groupe Radio Antenne 6 inc. Group owner: Group Radio Antenne 6 Inc. Format: Div, info, music. News staff: one; News: 20 hrs wkly. Target aud: General; mainly adults. ♦Marc Andre' Levesque, gen mgr; Lewis Gagnon, sls dir; Louis Arcand, progmg dir & news dir.

Rouyn-Noranda

CHIC-FM— 2003: 88.7 mhz; 50 w. Hrs open: C.P. 2185, J9X 5A6. Phone: (819) 797-4242. Fax: (819) 797-3803. Licensee: Communications CHIC (C.H.I.C.). Format: Fr, Christian music. ♦Andre Curadeau, gen mgr; Vic Cimon, stn mgr & opns dir; Jocelyn Cote, sls dir & progmg dir.

CHOA-FM— Sept 21, 1990: 96.5 mhz; 55 kw. 600 ft Stereo. Hrs open: 380 Ave. Murdoch, J9X 1G5. Phone: (819) 762-0741. Fax: (819) 762-2280. Licensee: RNC MEDIA Inc. (group owner). Format: Adult contemp. Target aud: 25-54. ♦Pierre R. Brosseau, CEO & pres; Jean-Pierre Major, gen mgr; Frantz Boivin, sls dir; Jean Gagnon, news dir; Gerald Landry, chief of engrg. Co-owned TV: CFEM-TV, CKRN-TV affils.

CJMM-FM— June 17, 1988: 99.1 mhz; 3.5 kw. Ant 200 ft Stereo. Hrs open: 24 33B Gamble Ouest, J9X 2R3. Phone: (819) 797-2566. Fax: (819) 797-1664. E-mail: mtrottier@radioenergie.astral.com Web Site: www.radioenergie.com. Licensee: Astral Media Radio Inc. Group owner: Astral Media Inc. (acq 1-12-2000; grpsl). Format: Adult contemp. News staff: one; News: 5 hrs wkly. Target aud: 18-44. ♦Marlene Trottier, gen mgr; Louis Kirouac, sls dir; Serge Trudel, prom dir, prom mgr, progmg dir & mus dir; Guy Champoux, news dir; Mathieu Barrette, engrg mgr.

Saguenay

CJAB-FM— May 25, 1979: 94.5 mhz; 44.2 kw. Stereo. Hrs open: 24 267 rue Racine Est., 2ieme etage, Chicoutimi, G7H 1S5. Phone: (418) 545-9450. Fax: (418) 543-7968. Web Site: www.radioenergie.com. Licensee: Astral Media Radio Inc. Group owner: Astral Media Inc. (acq 8-21-92). Natl. Network: Radiomutuel. Format: Pop. ♦Richard Turcotte, gen mgr; Carol Tremblay, sls dir; Katia Boivin, progmg dir; Jean-Francois Cote, news dir; Stephane Villeneuve, engrg mgr.

CKRS(AM)— June 23, 1947: 590 khz; 25 kw-D, 7.5 kw-N. Hrs open: 121 Racine St., Chicoutimi, G7H 5G4. Phone: (418) 545-2577. Fax: (418) 545-9186. Web Site: www.ckrs.com. Licensee: 591991 B.C. Ltd. Group owner: Astral Media Inc. (acq 1-21-2005; grpsl). Population served: 200,000. Natl. Network: Radiomutuel. Format: Oldies, news/talk. Target aud: 25-54. ♦Richard Turcotte, gen mgr; Michel Gaguou, sls dir; Brigette Simard, prom VP; Daniel Coto, prom VP; Eric Arsene;aut, chief of engrg; Michel Thiffault, sports cmtr.

Saint Augustin

CJAS-FM— 1992: 93.5 mhz; 100 w. Hrs open:
Rebroadcasts VOCM(AM) St. John's, NF weekends.
C.P. 100, Comte Duplessis, Riviere Saint-Augustin, G0G 2R0. Phone: (418) 947-2239. Fax: (418) 947-2664. E-mail: sajcr@globetrotter.net Licensee: La Radio Communautaire de Riviere St-Augustin Inc. Format: Adult contemp. ♦Laurette Gallibois, gen mgr, sls dir & mktg dir; Maria Shattler, progmg dir; Lindsey Durepos, mus dir; Rachel Bilodeau, news dir.

Saint Constant

CJMS(AM)— May 1999: 1040 khz; 10 kw-D, 5 kw-N. Hrs open: 24 143 rue St-Pierre, J5A 2G9. Phone: (514) 990-2567. Fax: (450) 632-0528. Web Site: www.cjms.ca. Licensee: 3553230 Canada Inc. (acq 3-29-01). Natl. Network: Radio Unica. Format: Country. News staff: 2. ♦Alex Azoulay, pres; Jean-Francois DuBois, gen mgr.

Saint Gabriel-de-Brandon

CFNJ-FM— Aug 10, 1985: 99.1 mhz; 9.75 kw. Ant 1,300 ft Hrs open: 24 245 Beauvilliers, J0K 2N0. Phone: (450) 835-3437. Phone: (450) 835-3438. Fax: (450) 835-3581. E-mail: cfng99@pandora.gc.ca Web Site: www.cfnj.net. Licensee: Radio Nord-Joli Inc. Format: MOR. Spec prog: Black one hr, class 2 hrs, C&W 4 hrs, jazz 2 hrs. ♦Denis Roch, pres & gen mgr; Denise LaVoie, sls dir; Nicolas Bellemare, progmg dir.

Saint Georges

CHJM-FM— June 22, 1987: 99.7 mhz; 100 kw. 350 ft Stereo. Hrs open: 24 C.P. 100, Saint Georges-de-Beauce, G5Y 5C4. Secondary address: 170 120ieme rue, Saint Georges-de-Beauce G5Y 5C4. Phone: (418) 227-0997. Phone: (418) 228-5535. Fax: (418) 228-0096. E-mail: adminrb@cgocable.ca Web Site: www.mix997.som. Licensee: Radio Beauce Inc. (acq 10-24-02; C$432,000 with CKRB-FM Saint Georges-de-Beauce). Format: Rock. News staff: one; News: 4 hrs wkly. Target aud: 18-35. ♦Guy Simard, pres; Marie Jalbert, gen mgr; Renee Giard, sls dir; Jacques Goulet, natl sls mgr; Louis Poulin, prom mgr; Marcel Rancourt, mus dir; Susanne Bougie, news dir; Gaston Guay, chief of engrg.

Saint Georges-de-Beauce

CKRB-FM— October 1953: 103.5 mhz; 17 kw. Stereo. Hrs open: 24 C.P. 100, G5Y 5C4. Secondary address: 11760 3 E. Ave. G5Y 5C4. Phone: (418) 228-1460. Phone: (418) 228-5535. Fax: (418) 228-0096. E-mail: adminrb@cgocable.ca Web Site: www.ckrb1033.com. Licensee: Radio Beauce Inc. (acq 10-24-02; C$432,000 with CHJM-FM Saint Georges). Format: Adult contemp. News staff: 2; News: 11 hrs wkly. Target aud: 35 plus. ♦Guy Simard, pres; Maurice Marcotte, gen mgr; Renee Giard, sls dir; Jacques Goulet, natl sls mgr; Louis Poulin, prom mgr; Marcel Rancourt, mus dir; Suzanne Bougie, news dir; Gaston Guay, chief of engrg.

Saint Hilarion

CIHO-FM— Oct 10, 1986: 96.3 mhz; Hrs open: 24 315 Cartier Nord, G0A 3V0. Phone: (418) 457-3333. Fax: (418) 457-3518. E-mail: ciho@charlevoix.net Licensee: Radio MF Charlevoix Inc. Format: MOR. Spec prog: Class 2 hrs, jazz 2 hrs wkly. ♦Gervais Desbiens, gen mgr; Rene Belanger, adv mgr; Pierre Beauchesne, progmg dir & engrg mgr; Dave Kid, news dir.

Saint Hyacinthe

CFEI-FM— 1988: 106.5 mhz; 3 kw. Hrs open: 24 855 rue Ste. Marie, J2S 4R9. Phone: (450) 774-6486. Fax: (450) 774-7785. E-mail: cfei@cgocable.ca Web Site: www.boomfm.astral.com. Licensee: Astral Media Radio Inc. Group owner: Astral Media Inc. (acq 8-13-2001). Format: 60s, 70s. ♦Jacques Parisien, pres; Martin Tremblay, gen mgr, opns mgr & sls dir; Jean-Frrancois Herbert, progmg dir; Anddre Lalier, mus dir.

Saint Jean-Iberville

CFZZ-FM—Licensed to Saint Jean-Iberville. See Montreal

Saint Jerome

CIME-FM— Mar 25, 1977: 103.9 mhz; 39.3 kw. Stereo. Hrs open: 24 120 Delagare St., J7Z 2C2. Phone: (450) 431-2463. Fax: (450) 565-9755. E-mail: ventes@cime.fm Licensee: Diffusion Metromedia CMR Inc. Group owner: Corus Entertainment Inc. (acq 1-26-01; grpsl). Format: Adult contemp. News staff: 2. Target aud: 25-54; Adult. ♦John Cassidy, pres; Gilbert Cerat, gen mgr; Etienne Gregoire, mktg dir; Ghislaiu Plourde, progmg dir; Jean Francois Rousseau, news dir.

Saint Pamphile

CJDS-FM— Dec 7, 2001: 94.7 mhz; 24 w. Hrs open: C.P. 550, G0R 3X0. Phone: (418) 356-1303. Fax: (418) 356-2586. E-mail: cjdsradio@globetrotter.net Licensee: 3819914 Canada inc. Format: MOR. ♦Jean-Claude Dignard, pres & gen mgr; Claire Soulieres, opns mgr & sls dir; Ann Dignard, dev dir & progmg dir; J. C. Dignard, pub affrs dir.

Saint Remi

***CHOC-FM—** 1999: 104.9 mhz; 250 w. Hrs open: 93 Ruelachapelle Est, J0L 2L0. Phone: (450) 454-5500. Fax: (450) 454-9435. E-mail: studio@chocfm.com Web Site: www.chocfm.com. Licensee: Radio Communautaire Intergeneration Jardin du Quebec. Format: Community radio. News staff: one; News: 10 hrs wkly. ♦Sylvain Remillard, pres; Richard Vegneault, gen mgr.

Sainte Anne des Monts

CBGN(AM)— 1972: 1340 khz; 1 kw-D, 250 w-N. Hrs open: 155 St. Sacrament St., Matane, G4W 1Y9. Phone: (418) 562-0290. Fax: (418) 566-6068. Licensee: CBC. (acq 9-1-72). Format: Talk, news. ♦Louis Pelletier, gen mgr.

CJMC-FM— March 1996: 100.3 mhz; 2.51 kw. Hrs open: 24 170 Boul. Ste. Anne, G4V 1N1. Phone: (418) 763-5522. Phone: (418) 763-5523. Fax: (418) 763-7211. Web Site: cjmc@quebectel.com. Licensee: Radio du Golfe Inc. Format: MOR. Spec prog: Class 2 hrs, western 2 hrs wkly. ♦Jacques Vallee, pres, opns dir, sls dir & mktg dir; Olivier Vallee, prom dir; Stephane Cyr, progmg dir, mus dir, news dir & chief of engrg.

Sainte Foy

***CHYZ-FM**— Jan 29, 1997: 94.3 mhz; 6 kw. Stereo. Hrs open: 24 Pavillon Pollack, Cite Universitaire, Suite 023, G1K 7P4. Phone: (418) 656-2131 ext. 4595. Fax: (418) 656-3660. E-mail: chyz@public.ulaval.ca Web Site: www.chyz.qc.ca. Licensee: Radio Campus Laval. Population served: 600,000 Format: Fr, div, techno/electronica. News staff: 4; News: 10 hrs wkly. Target aud: 18-30; univ students. Spec prog: Hip-hop/rap 15 hrs wkly. ♦Jean-Philippe Lessard, gen mgr.

Sainte-Marie-de-Beauce

CHEQ-FM— Nov 29, 1998: 101.3 mhz; 4.677 kw. Hrs open: 1068 boul. Vachon N., Suite 101, G6E 1M6. Phone: (418) 387-1013. Fax: (418) 387-3757. E-mail: info@cheqfm.qc.ca Web Site: www.cheqfm.qc.ca. Licensee: 9079-3670 Quebec Inc. (acq 8-16-00). Format: MOR, adult contemp. ♦Jacques Poulin, pres; Sylvie Paulin, gen mgr & opns mgr.

Senneterre

CIBO-FM— 1982: 100.5 mhz; Hrs open: 24 C.P. 1150, J0Y 2M0. Phone: (819) 737-2222. Fax: (819) 737-8599. E-mail: cibofm@yahoo.ca Licensee: Radio communautaire M.F. de Senneterre Inc. Population served: 5,600 Format: Community. ♦Guy Bilodeau, pres & gen mgr.

Sept-Iles

***CBSI-FM**— Nov 1, 1982: 98.1 mhz; 96.7 kw. 350 ft Hrs open: 24 350 rue Smith, bur. 30, G4R 3X2. Phone: (418) 968-0720. Phone: (800) 463-1731. Fax: (418) 968-9219. Fax: (418) 962-1344. E-mail: cbsi@radio-canada.ca Web Site: www.radio-canada.ca/cote-nord. Licensee: CBC. Population served: 8,000 Format: Div, educ, talk. Target aud: Over 30. ♦Pierre Lafreniere, stn mgr.

CKCN-FM— December 1998: 94.1 mhz; 4.88 kw. Hrs open: 437 Arnaud St., G4R 3B3. Phone: (418) 962-3838. Fax: (418) 968-6662. E-mail: ckcn@globetrotter.net Licensee: Radio Sept-Iles Inc. Format: Adult contemp, news/talk. News: 16 hrs wkly. Target aud: 25-54. Spec prog: Country 8 hrs wkly. ♦Pierre Bergeron, pres & gen mgr; Dominique Marquis, stn mgr.

Shawinigan

CFUT-FM— Feb 7, 2005: 91.1 mhz; 5 w. Hrs open: 540 avenue Broadway, G9N 1M3. Phone: (819) 537-0911. E-mail: info@radio911.ca Web Site: www.radio911.ca. Licensee: La radio campus communautaire francophone de Shawinigan inc. Format: Fr var. ♦Pierre-Yves Rousselle, progmg dir.

CKSM(AM)— Apr 30, 1951: 1220 khz; 10 kw-D, 2.5 kw-N, DA-2. Hrs open:
Rebroadcasts CHLN(AM) Trois Rivieres 100%.
Phone: (819) 378-1023. Fax: (819) 378-1360. Licensee: Astral Media Radio Inc. Group owner: Astral Media Inc. Format: Adult contemp, news/talk, sports. Target aud: 40-55 years old. ♦Jean Martin, gen mgr.

Sherbrooke

CFAK-FM— 2003: 88.3 mhz; 490 w. Hrs open: Radio CFAK, 2500 boul. de Universite, local 116, J1K 2R1. Phone: (819) 821-8000 ext 2693. Fax: (819) 821-7930. E-mail: dq@cfak.qc.ca Web Site: www.cfak.qc.ca. Licensee: Comite de la radio etudiante universitaire de Sherbrooke (CREUS). Format: Fr. ♦Steve Bazinet, gen mgr.

CFGE-FM— July 2004: 93.7 mhz; 3.8 kw. TL: Fleurimont. Hrs open: 3720 boulevard Industriel, J1L 1Z9. Phone: (819) 822-0937. Fax: (819) 822-2112. Web Site: www.rythmefm.com/estrie. Licensee: Cogeco Diffusion inc. Format: Adult contemp. News staff: one. Target aud: 25-54. ♦Michel Cloutier, gen mgr; Marc Fabi, gen sls mgr; Dominc D'Anjou, progmg dir.

CFLX-FM— 1984: 95.5 mhz; 1.35 kw. TL: N45 22 50 W71 54 51. Stereo. Hrs open: 24 67 N Wellington St., J1H 5A9. Phone: (819) 566-2787. Fax: (819) 566-7331. E-mail: cflx@cflx.qc.ca Web Site: www.cflx.qc.ca. Licensee: Radio communautaire de l'Estrie. Population served: 100,000 Format: News/talk, div, Fr. News staff: 2; News: 10 hrs wkly. Target aud: 18-35; college degree. Spec prog: Class 7 hrs, jazz 8 hrs, Sp 3 hrs wkly. ♦Daniel Bergeron, pres; Jose Deschenes, gen mgr & opns mgr.

CHLT(AM)— June 1937: 630 khz; 10 kw-D, 5 kw-N. TL: N45 18 16 W71 51 59. Hrs open: 24
Rebroadcasts CKAC(AM) Montreal 75%.
4020 boul. de Portland, J1L 2V6. Phone: (819) 563-6363. Fax: (819) 566-4222. Licensee: 591991 B.C. Ltd. Group owner: Astral Media Inc. (acq 1-21-2005; grpsl). Natl. Network: Radiomedia. Format: News/talk. News staff: 4; News: 20 hrs wkly. Target aud: 18 plus. ♦Marc Fabi, gen mgr & sls dir; Gilles Morin, prom dir; Jocelyn Proulx, progmg dir; Michel Laroche, chief of engrg.

CHLT-FM—Not on air, target date: unknown: 102.1 mhz; 58 kw. Ant 298 ft TL: N45 26 28 W72 00 35. Format: Talk.

CIGR-FM— 2004: 104.5 mhz; 1.3 kw. Hrs open: 4020 Boul. Portland, J1L 2V6. Phone: (819) 829-1045. Fax: (819) 829-1315. E-mail: info@grock.fm Web Site: www.generationrock.fm. Licensee: Groupe Generation Rock. Format: Fr, rock. ♦Jean-Pierre Beaudoin, gen mgr.

CIMO-FM—See Magog

CITE-FM-1— September 1962: 102.7 mhz; 92.8 kw. Ant 1,851 ft Stereo. Hrs open: 1845 King West, Suite 200, J1J 2E4. Phone: (819) 566-6655. Fax: (819) 566-1011. Licensee: Astral Media Radio Inc. (acq 4-19-2002; grpsl). Format: Adult contemp. ♦Natalie Johnson, gen mgr.

CJMQ-FM— 2004: 88.9 mhz; 1.67 kw. Stereo. Hrs open: 6am-12am 2600 College St, J1M 0C8. Phone: (819) 822-9600, EXT. 2689. Phone: (819) 822-9600. Fax: (819) 822-9682. E-mail: cjmqnews@yahoo.ca Web Site: www.cjmq.fm. Licensee: Radio Bishop's Inc. Population served: 170,000 Format: Community. News: 5 hrs wkly. Target aud: General; campus & community. ♦David Teasdale, stn mgr; Joel Heath, prom dir; Maureen Teasdale, progmg dir; Zaheed Bardai, mus dir; David Humble, engrg dir; Wayne Stacey, chief of engrg.

Sorel

CJSO-FM— Sept 27, 1989: 101.7 mhz; 3.5 kw. 327 ft Hrs open: 24 100 boul. Couillard Despres, Sorel-Tracy, J3P 5C1. Phone: (450) 743-2772. Fax: (450) 743-0293. E-mail: cjso@cjso.qc.ca Web Site: www.cjso.qc.ca. Licensee: Radio Diffusion Sorel-Tracy Inc. (acq 4-21-95). Format: Soft rock. News staff: 2; News: 5 hrs wkly. Spec prog: Class 2 hrs wkly. ♦Claude St. Germain, gen mgr; Jean Lemay, progmg dir; Valerie Ferland, mus dir; Jean-Marc Lebeau, news dir; Marie-Theresee Thibeault, traf mgr; Yanick Levesque, sports cmtr; Andre Champagne, disc jockey; Jocelyne Lambert, disc jockey.

Thetford Mines

CFJO-FM— July 15, 1989: 97.3 mhz; 100 kw. Ant 270 ft Hrs open: 24 55 St. Jean Baptiste, Victoriaville, G6P 6T3. Secondary address: 327 Rue Labbe G6G 5S3. Phone: (418) 338-1009. Phone: (819) 752-2785. Fax: (418) 338-0386. Fax: (819) 752-3182. Web Site: www.o973.com. Licensee: Reseau des Appalaches (FM) Ltee. Group owner: Gestion Appalaches inc. Population served: 87,913 Natl. Rep: Target Broadcast Sales. Format: Classic rock, AOR. News staff: 15. Target aud: 18-40. ♦Annie Labbe, pres.

CKLD-FM— 1950: 105.5 mhz; 6 kw. Stereo. Hrs open: 24 Box 69, 327 Rue Labbe, G6G 5S3. Phone: (418) 335-7533. Fax: (418) 335-9009. Web Site: www.passionrock.com. Licensee: Radio Megantic Ltee. Group owner: Gestion Appalaches inc. Population served: 40,384 Natl. Rep: Target Broadcast Sales. Format: Hot adult contemp. News staff: 5. Target aud: 35-64. ♦Annie Labbe, gen mgr & opns dir.

Trois Rivieres

CBF-FM-1— July 21, 1977: 104.3 mhz; 100 kw. 1,000 ft TL: N46 29 27 W72 39 00. Stereo. Hrs open: Box 6000, Montreal, H3C 3A8. Phone: (514) 597-6000. Fax: (514) 597-4510. Web Site: www.cbc.radio-canada.ca. Licensee: CBC French. Natl. Network: Radio Canada. Format: Class, cultural, drama. ♦Sylvain La France, gen mgr; Louise Carriere, progmg dir.

CFOU-FM— Sept 7, 1997: 89.1 mhz; 3 kw. TL: Trois Rivieres, Quebec, Canada. Stereo. Hrs open: 24 Universite du Quebec a Trois-Rivieres, 1002 Pavillon Neree-Beauchemin, 3351 boulevard des Forges, G9A 5H7. Phone: (819) 376-5184. Fax: (819) 376-5239. E-mail: dgcfou@uqtr.uqtr.ca Web Site: www.cfou.ca. Licensee: Radio campus des etudiants de l'Universite du Quebec a Trois-Rivieres. Natl. Network: CBC Radio One. Format: Div. News: 1/2 hour. Target aud: 18-35 yrs. ♦Marc Periard, gen mgr; Francois Marchand, mktg.

CHEY-FM— Aug 22, 1990: 94.7 mhz; 100 kw. Hrs open: RockDetente 94.7, 1500 rue Royale, Bur 260, G9A 6J4. Phone: (819) 376-0947. Fax: (819) 373-5555. Web Site: www.rockdetente.com. Licensee: Astral Media Radio Inc. (acq 4-19-2002; grpsl). Format: Soft rock. Target aud: 30-40; women. ♦Jean Martin, gen mgr; Rene Rivard, sls dir; Damien Miville-Deschenes, prom dir; Eric Lachapelle, progmg dir.

CHLN(AM)— Oct 17, 1937: 550 khz; 10 kw-D, 5 kw-N, DA-2. Hrs open: 1500 Rue Royale, Bur 1200, G9A 4J4. Phone: (819) 374-3556. Fax: (819) 374-3222. Web Site: www.chln550.com. Licensee: 591991 B.C. Ltd. Group owner: Astral Media Inc. (acq 1-21-2005; grpsl). Format: News/talk, sports, MOR, top-40. Spec prog: Pub affrs 8 hrs wkly. ♦Jean Martin, gen mgr; Diane Marchand, sls dir; Claude Bolduc, prom dir & progmg dir; Cluade Bolduc, adv dir.

CIGB-FM— Aug 27, 1979: 102.3 mhz; 5.8 kw. Stereo. Hrs open: 24 1500 Royale, Bureau 260, G9A 6J4. Phone: (819) 378-1023. Fax: (819) 378-1360. Web Site: www.radioenergie.com. Licensee: Astral Media Radio Inc. Group owner: Astral Media Inc. (acq 8-24-90). Population served: 150,000 Format: CHR. News staff: 3; News: 3 hrs wkly. Target aud: 18-30. ♦Jean Martin, gen mgr; Mr. Damien Miville-Deschenes, prom dir; Mr. Danny Champagne, progmg dir.

CJEB-FM— June 8, 2004: 100.1 mhz; 30.61 kw. Ant 1,177 ft Hrs open: 4141 boul. St. Jean, G9B 2M8. Phone: (819) 691-1001. Fax: (819) 691-1002. Web Site: www.rythmefm.com/maur. Licensee: Cogeco Radio-Television inc. Format: Fr adult contemp. ♦Michel Cloutier, gen mgr.

Val d'Or

CHGO-FM— 2000: 104.3 mhz; 100 kw. Hrs open: 1729 3e Ave., J9X 1G5. Phone: (819) 825-0010. Fax: (819) 825-7313. Web Site: www.gofm.net. Licensee: RNC MEDIA Inc. (group owner). Natl. Network: Radiomedia. Format: Classic rock. ♦Pierre R. Brosseau, CEO & pres; Jean-Pierre Major, gen mgr; Ghislain Beaulieu, opns mgr.

CJMV-FM— June 17, 1989: 102.7 mhz; 65 kw. 200 ft Hrs open: 24 173 Perreault St., J9P 2H3. Phone: (819) 825-2568. Fax: (819) 825-2840. Web Site: www.radioenergie.com. Licensee: Astral Media Radio Inc. Group owner: Astral Media Inc. (acq 1-12-2000; grpsl). Population served: 50,000 Format: Top-40. Target aud: 18-49. ♦Ian Greenberg, pres; Marlene Trottier, gen mgr & opns dir.

Valleyfield

CKOD-FM— June 6, 1994: 103.1 mhz; 3 kw. 167 ft TL: N45 16 08 W74 05 50. (CP: 103.1 mhz). Stereo. Hrs open: 24 249 Victoria St., J6T 1A9. Phone: (450) 373-0103. Phone: (450) 452-0103. Fax: (450) 373-4297. E-mail: fm103@ckod.qc.ca Web Site: www.ckod.qc.ca. Licensee: Radio Express Inc. Natl. Rep: Target Broadcast Sales. Format: Adult contemp. News staff: one; News: 7 hrs wkly. Target aud: 18-54; general. ♦Robert Brunet, pres & gen mgr; Martin Leblanc, opns dir & progmg dir; Dean Nevins, sls dir.

Verdun

CINF(AM)—Licensed to Verdun. See Montreal

CKOI-FM—Licensed to Verdun. See Montreal

Victoriaville

CFDA-FM— 1999: 101.9 mhz; 1.35 kw. Hrs open: 24 Box 490, G6P 6T3. Secondary address: 55 St. Jean Baptiste St. G6P 6T3. Phone: (819) 752-5545. Fax: (819) 752-7552. Web Site: www.passionrock.com. Licensee: Radio Victoriaville Ltee. Group owner: Gestion Appalaches inc. Natl. Rep: Target Broadcast Sales. Format: Adult contemp. News staff: 7. Target aud: 35-65. Spec prog: Country 3 hrs, retro/oldies 3 hrs wkly. ♦Annie Labbe, gen mgr.

Ville-Marie

CKVM-FM— 2004: 93.1 mhz; 18.4 kw. TL: N47 19 57 W79 25 38. Stereo. Hrs open: 24 62 Suite Anne, J9V 2B7. Secondary address: 62 Ste. Anne J9V 2B7. Phone: (819) 629-2710. Fax: (819) 622-0716. E-mail: ckvm@ckvm.qc.ca Licensee: Radio Temiscamingue Inc. Natl. Network: Radiomedia. Format: Adult contemp. News staff: one; News: 11 hrs wkly. Target aud: General. ♦Claude Gagnon, pres; Jacquelin Bastien, VP; Serge Lalonde, stn mgr.

Westmount

CKGM(AM)—See Montreal

Windsor

CIAX-FM— 2000: 98.3 mhz; 426 w. Hrs open: 49 Sixth Ave., J1S 1T2. Phone: (819) 845-5900. Phone: (819) 845-2692. Fax: (819) 845-2692. E-mail: unite@qc.aira.ca Licensee: Carrefour Jeunesse Emploi - Comte Johnson. Format: Div. ♦Patrick Levesque, pres & gen mgr; Julie Lupien, progmg dir.

Saskatchewan

Blucher

CFAQ-FM— 2006: 100.3 mhz; 36 w. Hrs open: 2127 St. Andrews Ave., Saskatoon, S7M 0M2. Phone: (306) 290-7222. Web Site: www.saskatoonchristianradio.com. Licensee: Bertor Communications Ltd. Format: Christian music. ♦Robert Orr, gen mgr.

Cumberland House

CJCF-FM— 1990: 89.9 mhz; 30.5 w. Hrs open: Box 100, S0E 0S0. Phone: (306) 888-2176. Phone: (306) 888-4444. Fax: (306) 888-2103. Licensee: Cumberland House Radio & Television Committee Inc. Format: Ethnic. ♦Rachel Fiddler, gen mgr.

Estevan

CHSN-FM—Listing follows CJSL(AM).

CJSL(AM)— August 1959: 1280 khz; 10 kw-U, DA-N. Hrs open: 24 200-1236 Fifth St., S4A 0Z6. Phone: (306) 634-1280. Fax: (306) 634-6364. Licensee: Golden West Broadcasting Ltd. (group owner; (acq 3-95). Natl. Rep: Canadian Broadcast Sales. Wire Svc: BN Wire Format: Contemp country. Target aud: General. Spec prog: Farm 4 hrs, relg 10 hrs wkly. ♦Elmer Hilderbrand, CEO & pres; Laverne Pappel, stn mgr & gen sls mgr.

CHSN-FM—Co-owned with CJSL(AM). 2001: 102.3 mhz; 100 kw. Format: Light rock.

Gravelbourg

***CBKF-1(AM)**— 1952: 540 khz; 5 kw-U, DA-2. Hrs open: 2440 Broad St., Regina, S4P 4A1. Phone: (306) 347-9540. Fax: (306) 347-9635. Web Site: www.cbc.ca. Licensee: CBC French. Format: Div. ♦Rikki Bote, gen mgr; Robert Rabinowitz, CEO & progmg dir.

CFRG-FM— 2003: 93.1 mhz; 48 w. Hrs open: C.P. 176, S0H 1X0. Phone: (306) 648-3103. Fax: (306) 648-3258. E-mail: acfq1@sasktel.net Licensee: Association communautaire fransaskoise de Gravelbourg Inc. Format: Fr.

Hudson Bay

CFMQ-FM— Sept 15, 1994: 98.1 mhz; 38.2 w. TL: N52 54 03 W102 23 31. Stereo. Hrs open: 24
Rebroadcasts CIRK-FM Edmonton, AB and CISN-FM Edmonton, AB 60%.
Box 1272, S0E 0Y0. Phone: (306) 865-3065. Fax: (306) 865-2227. E-mail: cfmq@sk.sympatico.ca Licensee: HB Communications Inc. Population served: 6,000 Format: Easy lstng, div, country. News: one hr wkly. Target aud: General. Spec prog: AOR, talk. ♦Mark Brann, pres; Dan Brann, gen mgr, sls VP, progmg mgr, news dir & engrg mgr.

Kindersley

CFYM(AM)— July 29, 1987: 1210 khz; 1 kw-U. Hrs open:
Rebroadcasts CJYM(AM) Rosetown 95%.
Box 490, Rosetown, S0L 2VO. Phone: (306) 463-4411. Fax: (306) 882-3037. Web Site: www.cjym.com. Licensee: Golden West Broadcasting Ltd. Group owner: Golden West Broadcasting Ltd. (acq 10-21-99). Natl. Rep: Target Broadcast Sales. Format: Hit Gold, adult contemp. Spec prog: Farm news. ♦Elmer Hildebrand, pres.

CKVX-FM—Co-owned with CFYM(AM).Not on air, target date: unknown: 104.9 mhz; 50 w. Format: Contemp country.

La Ronge

***CBKA-FM**— September 1979: 105.9 mhz; 80 w. Hrs open: Box 959, S0J 1L0. Phone: (306) 347-9540. Fax: (306) 425-2270. Web Site: www.sask.cbc.ca. Licensee: CBC. Natl. Network: CBC Radio One. Format: Rgnl current affrs. Spec prog: Cree & Dene 20 hrs wkly. ♦David Kyle, gen mgr.

CJLR-FM— 1990: 89.9 mhz; 216 w. Ant 151 ft Hrs open: 24 Box 1529, S0J 1L0. Phone: (306) 425-4003. Fax: (306) 425-3123. E-mail: mbcradio@mbcradio.com Web Site: www.mbcradio.com. Licensee: Natotawin Broadcasting Inc. (acq 9-2-93). Population served: 50,000 Format: Ethnic, country, div. Target aud: General. Spec prog: Cree & Dene languages. ♦Deborah Charles, CEO & gen mgr; William Dumais, pres; Keith Kratchmer, CFO; Teddy Clark, VP; Dallas Hicks, stn mgr & progmg dir.

Meadow Lake

CFDM-FM— 2001: 105.7 mhz; 46.5 w. Hrs open: Box 8168, Flying Dust First Nation, S9X 1T8. Phone: (306) 236-1445. Fax: (306) 236-2861. E-mail: cfdmradio@hotmail.com Licensee: FDB Broadcasting Inc. Format: Country, top-40. ♦Duwayne Derocher, stn mgr.

CJNS(AM)— Nov 1, 1977: 1240 khz; 1 kw-U, DA-1. Hrs open: Box 1460, North Battleford, S9A 2Z5. Phone: (306) 236-6494. Fax: (306) 236-6141. Licensee: Northwestern Radio Partnership Group owner: Rawlco Radio Ltd. (acq 1-7-2003; grpsl). Format: Contemp country. ♦David Dekker, gen mgr.

CJNS-FM—Not on air, target date: unknown: 102.3 mhz; 45 kw.

Melfort

CJVR-FM—Listing follows CKJH(AM).

CKJH(AM)— Oct 8, 1966: 750 khz; 25 kw-U, DA-N. TL: N52 47 57 W104 35 25. Stereo. Hrs open: 24 Box 750, 611 Main St., S0E 1A0. Phone: (306) 752-2587. Fax: (306) 752-5932. Fax: (306) 752-6339. E-mail: cjvr@cjvr.com Web Site: www.cjvr.com. Licensee: Radio CJVR Ltd. (acq 9-27-90). Natl. Rep: Target Broadcast Sales. Format: Just the hits. News staff: 4; News: 15 hrs wkly. Target aud: General. Spec prog: Relg 9 hrs wkly. ♦Eugene Fabro Sr., chmn; Eugene W. Fabro, pres; Gary Fitz, VP, gen mgr & natl sls mgr; Karen Anderson, prom dir; Bill Wood, progmg dir; Cal Gratton, mus dir; Neil Shewchuk, news dir; Bayne Opseth, chief of engrg.

CJVR-FM—Co-owned with CKJH(AM). March 1, 2002: 105.1 mhz; 100 kw. Format: Country.

Moose Jaw

CHAB(AM)— Apr 22, 1922: 800 khz; 10 kw-U, DA-N. TL: N50 22 38 W105 23 35. Stereo. Hrs open: 24 1704 Main St. N., S6J 1L4. Phone: (306) 694-0800. Fax: (306) 692-8880. E-mail: country800@sk.simpatico.com Licensee: Golden West Broadcasting Ltd. (group owner; (acq 8-20-92). Natl. Rep: Canadian Broadcast Sales. Format: Greatest hits. News staff: 3. Target aud: 25-54. ♦Elmer Hildebrand, CEO & pres; Barry Vice, stn mgr & progmg dir; Abbey White, prom mgr; Rob Carnie, news dir.

CILG-FM—Co-owned with CHAB(AM). 2002: 100.7 mhz; 100 kw. Phone: (306) 692-1007. Format: Country.

Nipawin

CIOT-FM— January 2005: 104.1 mhz; 200 w. Stereo. Hrs open: 24 Box 1240, S0E 1E0. Phone: (306) 862-2468. Fax: (306) 862-2660. E-mail: info@lighthousefm.ca Web Site: www.lighthousefm.ca. Licensee: Wilderness Ministries Inc. Population served: 14,000 Format: Christian. News staff: one. Target aud: 20-55. ♦Rod Petersen, progmg dir; Andrew Hildebrandt, mus dir; Angela Petersen, news rptr; Andrew Clark, progmg.

CJNE-FM— June 2002: 94.7 mhz; 14.8 kw. Hrs open: Box 220, S0E 1E0. Phone: (306) 862-9478. Fax: (306) 862-2334. E-mail: sales@cjnefm.com Web Site: www.cjnefm.com. Licensee: CJNE FM Radio Inc. Format: Classic rock, golden oldies. ♦Norm Rudock, gen mgr & gen sls mgr; Treana Rudock, stn mgr; Les Blair, prom mgr.

North Battleford

CJCQ-FM—Listing follows CJNB(AM).

CJNB(AM)— Jan 28, 1947: 1050 khz; 10 kw-U, DA-1. Hrs open: Box 1460, S9A 2Z5. Phone: (306) 445-2477. Fax: (306) 445-4599. Licensee: Northwestern Radio Partnership. Group owner: Rawlco Radio Ltd. (acq 1-7-2003;. grpsl). Natl. Rep: Canadian Broadcast Sales. Format: Country. Spec prog: Farm 7 hrs, relg 10 hrs wkly. ♦Gord Rawlinson, pres; David Dekker, gen mgr & gen sls mgr; Harry M. Dekker, prom mgr; Doug Harrison, progmg dir; Dave Senft, chief of engrg.

CJCQ-FM—Co-owned with CJNB(AM). September 2001: 97.9 mhz; 100 kw. Format: Pop rock.

Okanese Indian Reserve

CHXL-FM— June 2003: 95.3 mhz; 50 kw. TL: N50 56 34 W103 23 51. Hrs open: Box 940, Balcarres, S0G 0C0. Phone: (306) 334-3331. Fax: (306) 334-2545. Licensee: O.K. Creek Radio Station Inc. Format: Var. ♦William Yuzicapi, stn mgr.

Pinehouse Lake

CFNK-FM— 1996: 89.9 mhz; 7 w. Hrs open: General Delivery, Box 370, S0J 2B0. Phone: (306) 884-2011. Phone: (306) 884-2016. Fax: (306) 884-2365. Web Site: www.cfnk.radiok.sympatico.ca. Licensee: Pinehouse Communications Society Inc. Format: Cree language, adult contemp, oldies. ♦Peter Smith, gen mgr; Vince Natomagan, progmg dir.

Prince Albert

CFMM-FM—Listing follows CKBI(AM).

CHQX-FM— June 18, 2001: 101.5 mhz; 100 kw. Ant 606 ft Stereo. Hrs open: 24 Box 900, S6V 7R4. Phone: (306) 763-7421. Fax: (306) 764-1850. E-mail: mix101@rawlco.com Licensee: Rawlco Radio Ltd. (group owner). Format: Adult rock. ♦Jim Scarrow, gen mgr & opns mgr; Karl Johnson, gen sls mgr.

CKBI(AM)— 1934: 900 khz; 10 kw-U, DA-N. Hrs open: Box 900, S6V 7R4. Phone: (306) 763-7421. Fax: (306) 764-1850. Licensee: Rawlco Radio Ltd. (group owner; acq 1946). Natl. Rep: Canadian Broadcast Sales. Format: Adult contemp, MOR, oldies. Target aud: 34 plus; working women. Spec prog: Farm 2 hrs wkly. ♦Jim Scarrow, gen mgr; Dave Hryhor, rgnl sls mgr; Neil Headrick, progmg mgr; Jeff White, news dir; Dale Zimmerman, engrg mgr.

CFMM-FM—Co-owned with CKBI(AM). Jan 30, 1982: 99.1 mhz; 100 kw. 606 ft Stereo. Format: Classic rock mix 101, contemp hit, news. ♦Garth Kalin, progmg mgr.

Regina

***CBK(AM)**— July 29, 1939: 540 khz; 50 kw-D, DA. Hrs open: 24 Box 540, 2440 Broad St., S4P 4A1. Phone: (306) 347-9540. Fax: (306) 347-9524. E-mail: kyled@cbc.ca Licensee: CBC. Natl. Network: CBC Radio One. Format: Div, news, talk. Spec prog: Farm 5 hrs wkly. ♦Debbie Carpentier, gen mgr & opns dir; David Kyle, stn mgr & news dir; Nigel Simms, stn mgr & news dir.

CBK-FM— May 1, 1977: 96.9 mhz; 100 kw. 501 ft Stereo. Phone: (360) 956-7400. Fax: (306) 956-7417. Web Site: www.sask.cbc.ca. Natl. Network: CBC Radio Two. Format: Var/div, jazz, classical. Co-owned TV: *CBKT(TV) affil

CBKF-FM— Sept 1, 1973: 97.7 mhz; 13.7 kw. 501 ft Hrs open: 24 Box 540, S4P 4A1. Secondary address: 2440 Broad St. S4P-3Z4. Phone: (306) 347-9540. Fax: (306) 347-9493. Web Site: www.cbc.ca. Licensee: Radio Canada. Natl. Network: CBC Radio Two. Format: Div. ♦Rene Fontaine, gen mgr; Anne Brochu, progmg dir & news dir; Steve Tomchuk, engrg mgr. Co-owned TV: CBKF(TV) affil.

CFWF-FM—Listing follows CKRM(AM).

CHMX-FM— Feb 4, 1966: 92.1 mhz; 100 kw. 499 ft Stereo. Hrs open: 24 2060 Halifax St., S4P 1T7. Phone: (306) 546-6200. Web Site: lite92fm.com. Licensee: Harvard Broadcasting Inc. (acq 3-1-81). Format: Adult Contemp. ♦Les Schuster, opns dir & gen sls mgr.

CIZL-FM—Listing follows CJME(AM).

CJME(AM)— July 27, 1926: 980 khz; 10 kw-D, 5 kw-N, DA-2. Hrs open: 24 2401 Saskatchewan Dr., Suite 210, S4P 4H8. Phone: (306) 525-0000. Fax: (306) 347-8557. Licensee: Rawlco Radio Ltd. (group owner; acq 11-30-2001). Format: News/talk. ♦Tom Newton, gen mgr; Keith Black, rgnl sls mgr; Marcie Watson, prom dir; Don Kollins, progmg dir; Murray Wood, news dir.

CIZL-FM—Co-owned with CJME(AM). June 1982: 98.9 mhz; 100 kw. 435 ft Stereo. (Acq 4-67). Format: AOR, classic hits, adult contemp. Target aud: 18-49. ♦Craig Romanyk, rgnl sls mgr; Marci Watsen, prom mgr; Tom Newton, progmg dir.

***CJTR-FM**— Nov 1, 2001: 91.3 mhz; 480 w. TL: N50 27 18 W104 36 30. Stereo. Hrs open: 24 Box 334, Station Main, S4P 3A1. Phone: (306) 525-7274. Fax: (306) 525-9741. E-mail: radius@cjtr.ca Web Site: www.cjtr.ca. Licensee: Radius Communications Inc. Population served: 250,000 Format: Div. News: 10 hrs wkly. Spec prog: American Indian 4 hrs, Black 2 hrs, Chinese one hr, It one hr, Portugese one hr wkly. ♦Rick August, pres; Dave Kuzenko, VP; Keith Colhoun, gen mgr.

CKCK-FM— Aug 9, 2002: 94.5 mhz; 100 kw. Hrs open: 24 2401 Saskatchewan Dr., Suite 210, S4P 4H8. Phone: (306) 525-0000. Fax: (306) 347-8557. Licensee: Rawlco Radio Ltd. (group owner). Format: Classic rock. ♦Gord Rawlinson, pres; Ralph Bird, gen sls mgr; Tom Newton, gen mgr & progmg dir; Michael Zaplitny, news dir; Gord Stankey, engrg mgr; Karen Mains, traf mgr.

CKRM(AM)— July 29, 1922: 620 khz; 10 kw-U, DA-2. Hrs open: 2060 Halifax St., S4P 1T7. Phone: (306) 546-6200. Fax: (306) 781-7338. Web Site: www.620ckrm.com. Licensee: Harvard Broadcasting Inc. (acq 11-30-2001; C$4.2 million with co-located FM). Format: C&W, farm. ♦Mike Olstrom, stn mgr.

CFWF-FM—Co-owned with CKRM(AM). Apr 15, 1982: 104.9 mhz; 100 kw. 400 ft Stereo. Web Site: www.620ckrm.com. (Acq 8-25-95). Natl. Rep: Canadian Broadcast Sales. Format: Classic rock.

Rosetown

CJYM(AM)— Aug 8, 1966: 1330 khz; 10 kw-U, DA-1. Hrs open: 24 Box 490, S0L 2V0. Secondary address: 208 Hwy.4 S0L 2V0. Phone: (306) 882-2686. Fax: (306) 882-3037. Web Site: www.cjym.com. Licensee: Dace Broadcasting Corp. Group owner: Golden West Broadcasting Ltd. (acq 10-21-99). Natl. Rep: Target Broadcast Sales. Format: Classic hits. News staff: 2. Target aud: General. ♦Barb Bell, gen mgr.

Saskatoon

CBKF-2(AM)— Nov 6, 1952: 860 khz; 10 kw-U, DA-2. Hrs open: 144 2nd Ave., S7K 1K5. Phone: (306) 956-7400. Fax: (306) 956-7476. Web Site: www.sask.cbc.ca. Licensee: CBC French. Format: Div. Target aud: General. ♦David Kyle, gen mgr; Robert Rabinowitz, progmg dir.

***CBKS-FM**— July 1, 1978: 105.5 mhz; 98 kw. 586 ft Stereo. Hrs open: 24

Rebroadcasts CBK-FM Regina.

144 2nd Ave. S., S7K 1K5. Phone: (306) 956-7400. Fax: (306) 956-7417. Web Site: www.sask.cbc.ca. Licensee: CBC. Natl. Network: CBC Radio One. Format: Div, jazz, class. ♦David Kyle, stn mgr & progmg dir; Carley Caverly, gen sls mgr; John Calver, news dir; Steve Tomchuk, chief of engrg.

CFCR-FM— 1991: 90.5 mhz; 1.48 kw. Stereo. Hrs open: 6 AM-1 AM Box 7544, 103 3rd Ave. N., S7K 4L4. Phone: (306) 664-6678. E-mail: cfcr@quadrant.net Web Site: www.cfcr.ca. Licensee: Community Radio Society of Saskatoon Inc. Population served: 200,000 Format: Var. News: one hr wkly. Target aud: General. Spec prog: Fr one hr, Ger 2 hrs, It one hr, Pol one hr, Sp 2 hrs wkly. ♦Dianne Deminchuk, pres; Ron Spizziri, gen mgr; Bill Jones, sls dir; Theo Kivol, progmg VP & traf mgr.

CFMC-FM—Listing follows CKOM(AM).

CFQC-FM—Listing follows CJWW(AM).

CJDJ-FM— June 1990: 102.1 mhz; 100 kw. Hrs open: 24 715 Saskatchewan Crescent West, S7M 5V7. Phone: (306) 934-2222. Fax: (306) 477-0002. Licensee: Rawlco Radio Ltd. (group owner; (acq 12-21-2000; C$870,000 for all the issued and outstanding shares). Population served: 200,000 Format: Rock. News staff: 3; News: 4 hrs wkly. Target aud: 25-49; well educated, well paid professionals. Spec prog: Relg 6 hrs wkly. ♦Jamie Wall, pres & gen mgr; Pam Leyland, pres & gen mgr.

CJMK-FM— May 2001: 98.3 mhz; 100 kw. Hrs open: 24 366 3rd Ave. S., S7K 5S5. Phone: (306) 244-1975. Fax: (306) 665-5501. Web Site: www.magic983.fm. Licensee: 629112 Saskatchewan Ltd. (group owner). Format: Adult gold contemp. ♦Elmer Hildebrand, pres; Vic Dubois, gen mgr; Ken McFarlane, gen sls mgr; Steve Chisholm, progmg dir; Matt Bradley, mus dir; Eldon Duchscher, news dir; Kurtis Krowchuk, chief of engrg.

CJWW(AM)— January 1976: 600 khz; 25 kw-D, 8 kw-N, DA-N. Stereo. Hrs open: 24 366 3rd Ave. S., S7K 1M5. Phone: (306) 244-1975. Fax: (306) 665-7730. Fax: (306) 665-5501. Web Site: www.cjwwradio.com. Licensee: 629112 Saskatchewan Ltd. (group owner; acq 12-21-00; C$7,450,000). Population served: 350,000 Natl. Rep: Canadian Broadcast Sales. Format: Country, info. News staff: 7. Target aud: General; 35-64 central; 18+ full coverage. Spec prog: Gospel 3 hrs wkly. ♦Elmer Hildebrand, chmn & pres; Dawn Mann, CFO; V. Dubois, gen mgr; Ken McFarlane, gen sls mgr; Rod Kitter, progmg dir; Jay Richards, mus dir; E. Duchscher, news dir; Steve Shannon, pub affrs dir; Kurtis Krwochuk, chief of engrg.

CFQC-FM—Co-owned with CJWW(AM). Feb 6, 1995: 92.9 mhz; 100 kw. Natl. Rep: Canadian Broadcast Sales. Format: New country. News staff: 2. Target aud: General.

CKOM(AM)— June 8, 1951: 650 khz; 10 kw-U, DA-2. Hrs open: 715 Saskatchewan Crescent W., S7M 5V7. Phone: (306) 934-2222. Fax: (306) 373-7587. Licensee: Rawlco Radio Ltd. (group owner). Format: News/talk. ♦Jamie Wall, gen mgr.

CFMC-FM—Co-owned with CKOM(AM). Dec 12, 1965: 95.1 mhz; 100 kw. Ant 110 ft Stereo. Format: Today's best music.

Shaunavon

CJSN(AM)— Dec 6, 1966: 1490 khz; 1 kw-U. Hrs open: 407 Centre St., S0N 2M0. Phone: (306) 297-2671. Fax: (306) 297-3051. Web Site: www.swiftcurrentonline.com. Licensee: Frontier City Broadcasting. Group owner: Golden West Broadcasting Ltd. (acq 1973). Population served: 1,200,000 Format: C&W, MOR. ♦Deborah Gauger, gen mgr & opns mgr; Darwin Gooding, progmg dir.

Swift Current

CIMG-FM— Oct 20, 1979: 94.1 mhz; 100 kw. 400 ft Stereo. Hrs open: 24 134 Central Ave. N., S9H 0L1. Phone: (306) 773-4605. Fax: (306) 773-6390. E-mail: eaglecontrol@goldenwestradio.com Web Site: www.eagle94.com. Licensee: Golden West Broadcasting Ltd. (group owner; (acq 11-8-95; C$97,500). Population served: 48,500 Natl. Rep: Canadian Broadcast Sales. Format: Classic hits. News staff: 3; News: 4 hrs wkly. Target aud: 18-35. ♦Elmer Hildebrand, CEO; Menno Friesen, VP; Deborah Gauger, stn mgr & gen sls mgr; Ryan Switzer, progmg dir.

CKFI-FM— Nov 5, 2005: 97.1 mhz; 100 kw. Stereo. Hrs open: 134 Central Ave. N., S9H 0L1. Phone: (306) 773-4605. Fax: (306) 773-6390. Licensee: Golden West Broadcasting Ltd. Format: Adult contemp. ♦Deborah Gauger, gen mgr & stn mgr; Ryan Switzer, progmg dir.

CKSW(AM)— June 1, 1956: 570 khz; 10 kw-U, DA-2. Hrs open: 24 134 Central Ave., S9H 0L1. Phone: (306) 773-4605. Fax: (306) 773-6390. Web Site: www.swiftcurrentonline.com. Licensee: Golden West Broadcasting Ltd. Group owner: Golden West Broadcasting Ltd. Natl. Rep: Canadian Broadcast Sales. Format: Country, regl. News staff: 5; News: 9 hrs wkly. Target aud: 25-54. Spec prog: Farm 5 hrs, Ger one hr wkly. ♦Jill Ahrens, gen sls mgr & rgnl sls mgr; Darwin Gooding, progmg dir; Kim Johnston, adv mgr & progmg dir; Dave Funk, chief of engrg.

Weyburn

CFSL(AM)— Aug 16, 1957: 1190 khz; 10 kw-D, 5 kw-N, DA-N. Hrs open: 24 305 Souris Ave., S4H 2K2. Phone: (306) 848-1190. Fax: (306) 842-2720. Licensee: Golden West Broadcasting Ltd. Group owner: Golden West Broadcasting Ltd. (acq 2-16-95). Natl. Rep: Canadian Broadcast Sales. Wire Svc: BN Wire Format: C&W. News staff: 3. Target aud: 25 plus. Spec prog: Farm 3 hrs, relg 9 hrs wkly. ♦Elmer Hildenbrand, CEO; Cameron Birnie, stn mgr & news dir.

CKRC-FM—Co-owned with CFSL(AM).Not on air, target date: unknown: 103.5 mhz; 100 kw. Format: Rock.

White Bear Lake Resort

CIDD-FM— 2002: 97.7 mhz; 46.5 w. Hrs open: The Moose, Box 121, Kenosee Lake, S0C 2S0. Phone: (306) 577-2450. Licensee: White Bear Children's Charity Inc. Format: Div. ♦Lana Littlechief, gen mgr.

Yorkton

CFGW-FM—Listing follows CJGX(AM).

CJGX(AM)— Aug 19, 1927: 940 khz; 50 kw-D, 10 kw-N. Hrs open: 120 Smith St. E., S3N 3V3. Phone: (306) 782-2256. Fax: (306) 783-4994. Licensee: Yorkton Broadcasting Ltd. and Walsh Investments Inc., partners of GX Radio, a gen partnership. (acq 2-15-89). Format: Country. ♦George G. Gallagher, pres; Lyle J. Walsh, gen mgr; Bryan Mireau, engrg dir.

CFGW-FM—Co-owned with CJGX(AM). July 1, 2001: 94.1 mhz; 100 kw. Format: Hot adult contemp.

CJJC-FM— Jan 2, 2006: 100.5 mhz; 44.8 w. Hrs open: 395 Riverview Rd., S3N 3V6. Phone: (306) 786-7625. Fax: (306) 782-4437. E-mail: rocktalk@1005therock.ca Web Site: www.1005therock.ca. Licensee: Dennis M. Dyck, on behalf of a corporation to be incorporated. Format: Christian music. ♦Dennis Dyck, stn mgr; Scott Fitzsimmons, progmg dir.

Zenon Park

CKZP-FM— January 2002: 102.7 mhz; 5.4 w. Hrs open: Box 100, S0E 1W0. Phone: (306) 767-2451. Fax: (306) 767-2548. E-mail: legeru@tsd53.ca Web Site: www.thinkfast.ca/tsd/tsdpromo2/zpradio.htm. Licensee: Radio Zenon Park Inc. Format: Var (English and French progmg). ♦J. Ulysse Leger, gen mgr.

Yukon Territory

Tagish

CFET-FM— June 1, 2003: 106.7 mhz; 50 w. Hrs open: Mile 234, Y0B 1T0. Phone: (867) 667-6397. Fax: (867) 668-2633. E-mail: cfet@tagishtel.ca Web Site: www.tagishtel.ca/radio. Licensee: Robert G. Hopkins. Format: Variety/classic rock. ♦Robert G. Hopkins, gen mgr.

Whitehorse

***CFWH(AM)**— 1958: 570 khz; 5 kw-U, DA-1. Hrs open: 24 3103 Third Ave., Y1A 1E5. Phone: (867) 668-8400. Fax: (867) 668-8408. Licensee: CBC. Natl. Network: CBC Radio One. Format: Talk, info. Spec prog: Fr one hr wkly. ♦Frank Fry, opns dir, opns mgr & progmg dir.

CHON-FM— Feb 1, 1985: 98.1 mhz; 4.261 kw. 250 ft Stereo. Hrs open: 4230 A 4th Ave., Suite 6, Y1A 1K1. Phone: (867) 668-6629. Fax: (867) 668-6612. E-mail: nnby@nnby.net Web Site: www.nnby.net. Licensee: Northern Native Broadcasting. Format: C&W, classic rock. Spec prog: Yukon native language 15 hrs wkly. ♦Shirley Adamson, gen mgr; Christine Genier, gen sls mgr; Les Carpenter, progmg dir; Denis Gerard, engrg dir.

CIAY-FM— 2003: 100.7 mhz; 50 w. Hrs open: 24 91806 Alaska Hwy., Y1A 5B7. Phone: (867) 393-2429. Fax: (867) 393-2439. E-mail: stnmgr@newlifefmyukon.ca Web Site: www.newlifefmyukon.ca. Licensee: Bethany Pentecostal Tabernacle. Population served: 20,000 Format: Christian. ♦Rod Carby, stn mgr; Theresa Aitcheson, gen sls mgr; Ian McDonald, progmg dir.

CKRW(AM)— November 1969: 610 khz; 1 kw-U, DA-1. Hrs open: 203-4103 4th Ave., Suite 203, Y1A 1H6. Phone: (867) 668-6100. Fax: (867) 668-4209. E-mail: marketing@ckrw.com Web Site: www.ckrw.com. Licensee: Klondike Broadcasting Co. Ltd. Natl. Rep: Canadian Broadcast Sales. Format: CHR. ♦Rolf Hougen, CEO; Jennifer Johnstone, gen mgr; Eva Bidrman, gen sls mgr; Ron McFadyen, news dir; Alan Dailey, chief of engrg.

Miscellaneous Radio Services

American Forces Radio & Television Service (AFRTS), Department of Defense, American Forces Info Service, 601 N. Fairfax St., Alexandria, VA, 22314. Phone: (703) 428-0616. Fax: (703) 428-0624. E-mail: afrtdir@hq.afis.osd.mil Web Site: www.afrts.osd.mil.

Melvin W. Russell, dir; Andreas I. Friedrich, deputy dir.

March Air Reserve Base, CA 92518-2017. AFRTS Broadcast Center, 1363 Z St, Bldg. 2730. Phone: (909) 413-2201. Tom Weber, industry liaison; Larry Sichter, mgr affiliate rel.

AFRTS has radio & TV svc 177 countries & on bd U.S. Navy ships. AFRTS stns operate in 15 countries providing rgnl & loc info to large concentrations of U.S. forces. All of the entertainment progmg, U.S sporting events, & natl. & international news is provided to the outlets either directly via international satellites from the AFRTS Broadcast Center at March Air Reserve Base, CA., or through the AFRTS operated stns which insert their rgnl & loc radio drive-time programs & radio & TV news & spot announcements. A rgnl AFRTS svc in Europe delivers the AFRTS fed progmg & rgnl news & info via EUTELSAT to affils located in seven nations as well as directly to cable head-ends, remote transmitters, homes throughout Europe & the Middle East. The worldwide AFRTS-TV progmg consists of: an entertainment svc time shifted for the various parts of the globe & providing the best of U.S. net TV progmg; a news svc providing natl & international news from CNN, Fox News, MSNB, major U.S. networks; a sports ch providing sports news & sporting events from ESPN, ESPN2 & the major U.S. nets & a fourth svc devoted to alternative entertainment progmg primarily oriented on family-type programs from PBS & from U.S.cable TV chs. A fifth & sixth svc were added in 2005 for family entertainmnet & full-time movies. A seventh svc began in 2006 providing additional sporting events. Finally, the Pentagon Channel is also carried by AFRTS. AFRTS-Radio satellite progmg consists of two continuous news info & sporting events svcs, a NPR svc & entertainment svcs with music for practically all tastes & likes of AFRTS' authorized audience which is all active duty military & Department of Defense civilian personnel & their families stationed overseas. Access to the AFRTS worldwide satellite svcs is restricted through the use of the Scientific Atlanta Power-Vu MPEG-2 digital compression encoding system.

Radio Free Asia

Radio Free Asia, 2025 M St. N.W., Suite 300, Washington, DC, 20036. Phone: (202) 530-4900. E-mail: contact@rfa.org Web Site: www.rfa.org.

Richard Richter, pres; Patrick Taylor, CFO; Daniel Southerland, VP progmg/exec editor; Sarah Jackson-Han, dir of communications.

Provides info, news. & commentary about events in the respective countries of Asia & elsewhere. The svc is intended to be a forum for a var of opinions & voices from within Asian nations whose people do not fully enjoy freedom of expression.

Radio Free Europe/Radio Liberty

Radio Free Europe/Radio Liberty, (RFE/RL Inc.). 2025 M St. N.W., Suite 300, Washington, DC, 20036. Phone: (202) 457-6947. Fax: (202) 457-6992. E-mail: contact@rfa.org Web Site: www.rferl.org.

Libby Liu, pres.

110 00 Prague 1, NO Czech Republic, Vinohradska 1. Phone: (420-2) 2112-1111. Anna Rausova, media relations coord.

Bcsts to East Europe in Bulgarian, Czech, Slovak, Romanian, Lithuanian, Estonian, & Latvian; bcsts to the Commonwealth of Independent States in Russian, Ukrainian, Belorussian, Armenian, Azeri, Bosnian, Georgian, Tatar-Bashkir, Kazak, Kirghiz, Tajik, Turkmen, Uzbek. Also bcsts to the former Yugoslavia in Serbian & Croatian, bcsts in Persian to Iran & in Arabic to Iraq.

Broadcasting Board of Governors, 330 Independence Ave. S.W., Rm. 3360, Washington, DC, 20237. Phone: (202) 203-4545. Fax: (202) 203-4585. Web Site: www.bbg.gov/.

The bd makes & supervises grants to Radio Free Europe, Radio Liberty, & Radio Free Asia, the Middle East Bcstg Networks & assures that funds are applied consistently with the broad foreign policy objectives of the U.S. govt. The BBG serves as the governing body for all non-military U.S. bcstg including VOA, OCB, RFE/RL, RFA, & MBN.

U.S. International Radio

Adventist World Radio, 12501 Old Columbia Pike, Silver Spring, MD, 20904-6600. Phone: (301) 680-6304. Fax: (301) 680-6303. E-mail: info@awr.org Web Site: www.awr.org.

Greg Scott, sr VP; Benjamin D. Schoun, pres.

The international radio svc of the General Conference of Seventh-day Adventists, AWR is made up of four rgns: the Americas, Africa, Asia-Pacific, & Europe. AWR programs are produced in more than 50 studios around the world & is currently bcstng, in more than 70 languages, more than 1,200 hrs each week worldwide.

Blue Ridge Communications Inc., Shortwave Radio Station WWRB, c/o Airline Transport Communications, Box 7, Manchester, TN, 37349-0007. Phone: (931) 841-0492. Fax: (931) 728-6087. Web Site: www.wwrb.org.

WWRB Manchester, TN. Worldwide bcstg utilizing 5 shortwave transmitters & 6 major antenna systems (azimuths); more than 10 years of well established global audience.

EWTN Global Catholic Network, 5817 Old Leeds Rd., Irondale, AL, 35210-2164. Phone: (205) 271-2900. Fax: (205) 271-2926. E-mail: radio@ewtn.com Web Site: www.ewtn.com.

Michael Warsaw, pres; Frank Leurck, stn mgr; Thom Price, dir English progmg; Doug Keck, sr VP progmg & production; David Brantley, engrg dir; Scott Hults, VP communications; Terry Boarders, engrg VP; Doug Archer, dir Sp progmg; John Pepe, mktg mgr.

Global Catholic Radio Networks available in English & Sp 24 hours a day, satellite delivered, free of charge.

Family Stations Inc., 10400 N.W. 240th St., Okeechobee, FL, 34972. Phone: (863) 763-0281. Fax: (863) 763-8867. E-mail: fsiyfr@okeechobee.com Web Site: www.familyradio.com.

Harold Camping, pres; Dan Elyea, engrg mgr.

WYFR Okeechobee, Fla. Twelve 100 kw transmitters & two 50 kw transmitters in Florida. Bcstg on various frequencies, in English to Europe, Africa & the Americas (including Caribbean area), in German to Europe, in Russian to East Europe, in Arabic to West Africa, in French to Europe, North Africa & the Americas & in Sp to Southern Europe, Central & South America, in Portuguese to Europe, South America & West Africa, in It to Europe. Format: Relg.

Far East Broadcasting Co. Inc., Box 1, La Mirada, CA, 90637. Phone: (562) 947-4651. Fax: (310) 943-0160. E-mail: febc@febc.org Web Site: www.febc.org.

Dr. Robert S. Fortner, chmn; Gregg J. Harris, pres.

Broadcasts 560 hrs of progmg in 150 languages, to a potential audience of more than 2.5 billion people. FEBC's broadcasts are heard in many countries with limited access to Christian ministry, or where there is tremendous political and cultural opposition to the gospel.

Fundamental Broadcasting Network, c/o Grace Missionary Baptist Church, 520 Roberts Rd., Newport, NC, 28570. Phone: (252) 223-4600. Fax: (252) 223-2201. E-mail: fbn@fbnradio.com Web Site: www.fbnradio.com.

WBOH Newport, N.C. Broadcasts on 5.920 mhz 24 hrs a day. **WTJC Newport, N.C.** Broadcasts on 9.370 mhz 24 hrs a day.

Good News World Outreach, WRNO Worldwide, Box 895, Fort Worth, TX, 76101-0895. Phone: (817) 850-9990. Fax: (817) 850-9994. E-mail: wrnoradio@mailup.net Web Site: www.wrnoworldwide.org.

WRNO New Orleans. 50 kw shortwave transmitter reaching North America, Central America, Europe, & Far East. Format: news, talk (educational, Christian), sports, music.

International Fellowship of Churches Inc., dba IMF World Missions. Radio Station KIMF, 9746 6th St., Rancho Cucamonga, CA, 91730. E-mail: james@plancktech.com

Broadcasts on 5.835 mhz and 11.885 mhz with two 50 kw transmitters.

La Voz de Restauracion Broadcasting Inc., Box 56320, Los Angeles, CA, 90056. Phone: (323) 766-2454. Fax: (323) 766-2458. Web Site: www.restauracion.com.

Rene F. Molina, dir.

KVOH Rancho Simi, CA Format: Sp.

Our Lady's Youth Center, 230 High Valley Rd., Vado, NM, 88072. Phone: (505) 233-2090. Fax: (505) 233-3019.

Operates **KJES Vado, N.M.**

Radio Miami International, 175 Fontainebleau Blvd., Suite 1N4, Miami, FL, 33172. Phone: (305) 559-9764. Fax: (305) 559-8186. E-mail: info@wrmi.net Web Site: www.wrmi.net.

Jeff White, gen mgr.

WRMI Miami. Stn sells block airtime for $1/minute to organizations wanting to reach any part of the Americas in any language. 7,385 & 9,955 & 15,725 khz shortwave, 50 kw power.

Trans World Radio, Box 8700, Cary, NC, 27512-8700. Phone: (919) 460-3700. Fax: (919) 460-3702. E-mail: webmaster@twr.org Web Site: www.twr.org.

David McCreary, dir public rel; David Tucker, pres/CEO.

KTWR Agana, Guam. Guam E-mail: twrguamk@twr.hafa.net.gu Four 100-kw shortwave transmitters to bcst to Australia, Bali, China, the eastern & central part of the Commonwealth of Independent States, Far East, India, Indonesia, Japan, Korea, Myanmar, Southeast Asia. Format: Relg (more than 30 languages).

Trinity Broadcasting Network, Attn Superpower KTBN Radio QSL Mgr., Tustin, CA, 92780. Phone: (801) 250-4111 (office). Web Site: www.tbn.org.

Johnny Mitchell, gen mgr.

TBN, is the worlds largest religious net offering 24 hours commerical-free inspiration progmg that appeals to a wind variety of denominations.

Two If By Sea Broadcasting Corp., 1784 W. Northfield Blvd., Suite 305, Murfreesboro, TN, 37129-1702. E-mail: studio@kaij.us Web Site: kaij.us.

George McClintock, gen mgr; John McClintock, progmg dir; Ted Randall, mktg dir.

KAIJ Dallas, TX

United Nations, Audio-Visual Promotion & Distribution. 405 E. 42nd St., Rm. S-805, HQ-Secretariat Bldg.," INT'L ORG", New York, NY, 10017. Phone: (212) 963-6982. Fax: (212) 963-6869. E-mail: audio-visual@un.org Web Site: www.un.org/av.

Caroline Petit, mgr, prom & distribution & news & media; Susan Farkas, chief of opns.

TV coverage of UN meetings, events, the production, promotion & distribution of UN videos & documentary programs. All major UN events are also recorded on audio for radio distribution. Offices in 55 countries of 192 member-countries.

The Voice of the OAS, 17th & Constitution N.W., Washington, DC, 20006. Phone: (202) 458-3000. Fax: (202) 458-3930. E-mail: informacion-publica@oas.org Web Site: www.oas.org.

Von Martin, producer; Claudio Lessa, producer.

Radio programs with news, interviews, info & music from Latin America. Concentrating on the acitivities of the Organization of American States.

WBCQ Radio, 274 Britton Rd., Monticello, ME, 04760-3110. Phone: (207) 538-9180. Fax: (207) 538-9180 (Call First). E-mail: wbcq@wbcq.com Web Site: www.wbcq.com.

Allan H. Weiner, gen mgr.

WBCQ Monticello, Me. International bcst shortwave stn. Lease & program time available. 5.105 mhz, 7.415 mhz, 9.330 mhz & 17.495 mhz. Serves North, Central, South America & the Carribean.

WJIE International Shortwave, Box 197309, Louisville, KY, 40259. Phone: (502) 968-1220. Fax: (502) 964-3304. E-mail: wjiesw@hotmail.com Web Site: www.wjiesw.com.

Robert W. Rodgers, pres; Greg Holt, VP; Doug Rumsey, dir.

WJIE Millerstown, Ky. On two sw frequencies operating 24 hours daily. Target areas: Europe & Asia. Also operates WJIE-FM on 88.5 mhz with 24.5 kw horiz, 18.5 kw vert in Okolona, Ky.

WMLK Radio, Assemblies of Yahweh, Box C, Bethel, PA, 19507. Phone: (717) 933-4518. Phone: (800) 523-3827. Web Site: www.assembliesofyahweh.com.

Elder Jacob O. Meyer, pres.

Branch offices in Metro-Manila, Phillippines; San Juan, Port of Spain, Trinidad & Tobago; Leeds, England. **WMLK Bethel, PA.** Bcstg to Europe & the Middle East 5 hours, five days a week, Mon-Fri, with relg instruction content.

WNQM Inc., Group owner: F.W. Robbert Broadcasting Inc. 1300 WWCR Ave., Nashville, TN, 37218. Phone: (615) 255-1300. Phone: (800) 238-5576. Fax: (615) 255-1311. E-mail: wwcr@aol.com Web Site: www.wwcr.com.

Brady Murray, ops mgr am; Zach Harper, ops mgr sw.

WWCR Nashville. Frequencies: 3.210 mhz, 5.070 mhz, 5.935 mhz, 7.465 mhz, 9.475 mhz, 12.160 mhz, 15.825 mhz.

World Christian Broadcasting Corp., Operations Center, 605 Bradley Court, Franklin, TN, 37067-8200. Phone: (615) 371-8707. Fax: (615) 371-8791. E-mail: dward@worldchristian.org Web Site: www.knls.org.

Charles H. Caudill, pres/CEO.

Anchor Point, AK 99556. KNLS, Box 473. Phone: (907) 235-8262. (907) 235-8462. Kevin Chambers, chief eng.

KNLS Anchor Point, Alaska (transmission facilities): Relg & secular progmg beamed to Asia, eastern Europe & the Pacific Rim on the international shortwave bands.

World Harvest Radio International, 61300 Ironwood Rd., South Bend, IN, 46614. Phone: (574) 291-8200. Fax: (574) 291-9043. E-mail: whr@lesea.com Web Site: www.whr.org.

Steve Sumrall, pres; Peter Sumrall, VP/gen mgr; Joe Hill, sls & opns mgr; Douglas Garlinger, chief engr.

WHRI Indianapolis. Two 100 kw transmitters serving Europe, Russia, North, Central & South America. **KWHR Naalehu, Hawaii.** Two 100 kw transmitters serving primarily Asia, & also Oceania & Australia/New Zealand. **WHRA Greenbush, Me.** One 250 kw transmitter serving Africa & the Middle East. Shortwave transmitters are available for lease (time sls).

World International Broadcasters Inc., Box 88, Red Lion, PA, 17356. Phone: (717) 246-1681, EXT. 140. Fax: (717) 244-9316.

John H. Norris, pres; Patricia Norris-Slaughter, sec; John C. Norris, dir; Mary Norris-Michel, dir; Fred Wise, dir.

WINB Red Lion, Pa. Shortwave progmg of programs to Western Europe, the Mediterranean, North Africa, Mexico, Philippines, Guam, Formosa & Australia. Format: Relg.

Voice of America

Voice of America, 330 Independence Ave. S.W., Washington, DC, 20237. Phone: (202) 203-4959. Fax: (202) 203-4960. E-mail: voaNews@voanews.com Web Site: www.voanews.com.

Joe O'Connell, dir pub affrs; Danforth W. Austin, dir.

The Voice of America, which first went on the air in 1942, us a multimedia international bcstg svc funded by the U.S govt through Bcstg Bd of Governors. VOA bcsts more than 1,000 hrs of news, info, educ, & cultural progmg every week to an estimated worldwide audience of more than 100 million people. Programs are produced in 44 languages. VOA AM, FM & shortwave transmitters are located at over 30 transmitting sites world-wide.

Satellite Services

SIRIUS Satellite Radio

1221 Avenue of the Americas, 36th Fl, New York, NY 10020. Phone: (212) 584-5100; Fax: (212) 584-5200. Website: www.sirius.com.

Management: Mel Karmazin, CEO; Joseph P. Clayton, chmn of bd; Scott Greenstein, pres, entertainment & sports; James E. Meyer, pres, opns & sls; Patrick L. Donnelly, exec VP and gen counsel; David J. Frear, exec VP and CFO; Andrea Lazar, sr VP, business dev; John H. Schultz, sr VP, human resources.

For $12.95 mthly, annually $142.45 with one month free or $271.95 2 years with 3 months free. SIRIUS bcsts coast to coast, digital-quality channels of satellite radio: over 69 channels of 100% commercial-free music, plus over 60 channels dedicated to sports, news, talk, entertainment, weather & data. SIRIUS also bcsts live play-by-play games of the NFL, NBA & NHL. It is the only radio outlet to provide listeners with every NFL game.

The service can be used in cars, trucks, RVs, homes, offices, stores, & even outdoors. Boaters around the country & up to miles offshore can also tune into SIRIUS.

The receiver product line starts with transportable plug & play radios & continues to high-end receivers with motorized touch-control display screens, as well as radios that are found in new cars & trucks. These units are manufactured by leading consumer electronics brands: Alpine; Altec Lansing; Clarion; SIRIUS/Directed Electronics; Eclipse; Jensen; JVC; Kenwood; Pioneer Electronics; Sony; Tivoli.

DaimlerChrysler, Ford & BMW are SIRIUS' exclusive automotive partners, & their production represents over 40% of the new cars & light trucks sold annually in the United States. SIRIUS radios are currently offered in vehicles from Audi, BMW, Chrysler, Dodge, Ford, Infiniti, Jaguar, Jeep, Land Rover, Lexus, Lincoln, Mazda, Maybach, Mercedes-Benz, Mercury, MINI, Mitsubish Motors, Nissan, Porsche, Subaru, Scion, Toyota, Volkswagen & Volvo.

XM Satellite Radio Headquarters

1500 Eckington Place, NE, Washington, DC 20002. Phone: (202) 380-4500; Fax: (202) 380-4500.

Innovation Center 3161 S.W. 10th St., Deerfield Beach, FL 33442. Phone: (954) 571-4300; Fax: (954) 360-2521. Website: www.xmradio.com. Contact: XM Satellite Radio (866) 962-2557.

OEM Liason Office 39810 Grand River, Suite 180, Novi, MI 48375-2138. Phone: (248) 478-6500; Fax: 248-427-9958.

Japan Office XM Satellite Radio, c/o Eugene Moosa-Mikami, Bellhouse B, 27-18 Honmoku Wada, Nakaku, Yokohama, Japan 231-0827. Phone/Fax: 81-45-621-4519.

XM Studios-New York Economist Building, 111 W. 57th St., New York, NY 10019. Phone: (212) 956-5656.

XM Studios-Nashville Country Music Hall of Fame & Museum, 222 5th Ave. S., Nashville, TN 37203.

Management: Gary Parsons, chmn; Steve Cook, exec VP, sls & mktg; Joseph J. Euteneuer, exec VP & CFO; Dara F. Altman, exec VP, Business & Legal Affairs; Stell Patsiokas, exec VP technology & engrg; Eric Logan, exec VP, progmg, Joseph Titlebaum, gen counsel.

For $12.95 mthly, XM offers coast-to-coast, with over 170 crystal clear digital channels of various entertainment in music, news, sports, talk, comedy, women progmg & entertainment. All music channels are commercial-free. Under XM Family Plan, suscribers get a discount rate of $6.99 mthly for additional radios. XM's equity partners include General Motors; Honda Motors Co., Inc.; Clear Channel, the largest U.S. radio station operator. It has receiver and retailing partnerships with such leading audio manufacturers as Delphi, Pioneer, and with such electronics retailers as Walmart, Best Buy, Circuit City, Sears, Mobile-One and Radio Shack franchise dealers.

XM Satellite Radio Inc. is a wholly owned subsidiary of the publicly traded XM Satellite Radio Holdings Inc. (NASDAQ: XMSR).

U.S. AM Stations by Call Letters

KAAA Kingman AZ
KAAB Batesville AR
KAAM(AM) Garland TX
KAAN Bethany MO
KAAY Little Rock AR
KABC Los Angeles CA
KABI Abilene KS
KABQ Albuquerque NM
*KABR Alamo Community NM
KACE(AM) Bishop CA
KACH Preston ID
KACI The Dalles OR
KACT Andrews TX
KADA Ada OK
KADI(AM) Springfield MO
KADR Elkader IA
KADS Elk City OK
KAFF Flagstaff AZ
KAFY(AM) Bakersfield CA
KAGC Bryan TX
KAGE Winona MN
KAGH Crossett AR
*KAGI Grants Pass OR
KAGO Klamath Falls OR
KAGV(AM) Big Lake AK
KAGY Port Sulphur LA
KAHI Auburn CA
KAHL(AM) San Antonio TX
KAHS(AM) El Dorado KS
KAHZ(AM) Pomona CA
KAIR Atchison KS
KAJO Grants Pass OR
KAKC(AM) Tulsa OK
KAKK(AM) Walker MN
KALE Richland WA
KALI West Covina CA
KALL(AM) North Salt Lake City UT
KALM Thayer MO
KALN Iola KS
KALV Alva OK
KALY Los Ranchos de Albuquerque NM
KAMA El Paso TX
*KAMI(AM) Cozad NE
KAML Kenedy-Karnes City TX
KAMQ Carlsbad NM
KANA(AM) Anaconda MT
KAND Corsicana TX
KANE New Iberia LA
KANI Wharton TX
*KANN Roy UT
KAOI(AM) Kihei HI
KAOK Lake Charles LA
KAOL Carrollton MO
KAPE Cape Girardeau MO
*KAPL(AM) Phoenix OR
KAPR Douglas AZ
KAPS Mount Vernon WA
KAPZ Bald Knob AR
KARI Blaine WA
KARN Little Rock AR
*KARR Kirkland WA
KARS Belen NM
KART Jerome ID
KARV Russellville AR
KASA Phoenix AZ
KASI Ames IA
KASL Newcastle WY
KASM Albany MN
KASO Minden LA
KAST Astoria OR
KATA Arcata CA
KATD Pittsburg CA
KATE Albert Lea MN
KATH(AM) Frisco TX
KATK Carlsbad NM
KATL Miles City MT
KATO Safford AZ
KATQ Plentywood MT
KATZ(AM) Saint Louis MO
KAUS Austin MN
KAVA Pueblo CO
KAVL Lancaster CA
KAVP(AM) Colona CO
KAVT Fresno CA
*KAWC Yuma AZ
KAWL York NE
KAWW(AM) Heber Springs AR
KAXX Eagle River AK
KAYL Storm Lake IA
KAYS Hays KS
KAZA(AM) Gilroy CA
KAZG(AM) Scottsdale AZ
KAZM(AM) Sedona AZ
KAZN Pasadena CA
KBAD Las Vegas NV
KBAI(AM) Bellingham WA
KBAL(AM) San Saba TX
KBAM Longview WA
KBAR Burley ID
*KBBI Homer AK
KBBO(AM) Selah WA
KBBR(AM) North Bend OR
KBBS Buffalo WY
KBBW Waco TX
KBCH Lincoln City OR
KBCK(AM) Deer Lodge MT
KBCL(AM) Bossier City LA
KBCQ(AM) Roswell NM
KBCR Steamboat Springs CO
KBCV(AM) Hollister MO
KBDB(AM) Sparks NV
KBEC Waxahachie TX
KBED(AM) Nederland TX
KBEL Idabel OK
KBEN Carrizo Springs TX
KBET(AM) Winchester NV
KBEW Blue Earth MN
KBFI(AM) Bonners Ferry ID
KBFL(AM) Springfield MO
KBFP(AM) Bakersfield CA
KBFS Belle Fourche SD
KBGE(AM) Kilgore TX
KBGG(AM) Des Moines IA
KBGN Caldwell ID
KBHB Sturgis SD
KBHC Nashville AR
KBHS(AM) Hot Springs AR
*KBIB Marion TX
KBIF Fresno CA
KBIM Roswell NM
KBIS(AM) Forks WA
KBIX Muskogee OK
KBIZ Ottumwa IA
KBJA Sandy UT
KBJD Denver CO
KBJM Lemmon SD
KBJT Fordyce AR
KBKB Fort Madison IA
KBKO Santa Barbara CA
KBKR Baker City OR
KBKW Aberdeen WA
KBLA Santa Monica CA
*KBLE(AM) Seattle WA
KBLF Red Bluff CA
KBLG Billings MT
KBLI(AM) Blackfoot ID
KBLJ(AM) La Junta CO
KBLL Helena MT
KBLU Yuma AZ
KBLY(AM) Idaho Falls ID
KBME Houston TX
KBMO(AM) Benson MN
KBMR Bismarck ND
KBMS Vancouver WA
KBMW Breckenridge MN
KBNA(AM) El Paso TX
KBND Bend OR
KBNN Lebanon MO
KBNO(AM) Denver CO
KBNP Portland OR
KBOA Kennett MO
KBOE Oskaloosa IA
KBOI Boise ID
KBOK(AM) Malvern AR
KBOV(AM) Bishop CA
KBOW Butte MT
KBOZ Bozeman MT
KBPO(AM) Port Neches TX
*KBPS Portland OR
KBRB Ainsworth NE
KBRC(AM) Mount Vernon WA
KBRD Lacey WA
KBRF Fergus Falls MN
KBRH Baton Rouge LA
KBRI Brinkley AR
KBRK Brookings SD
KBRL McCook NE
KBRN Boerne TX
KBRO Bremerton WA
KBRT Avalon CA
KBRV Soda Springs ID
*KBRW(AM) Barrow AK
KBRX O'Neill NE
KBRZ Freeport TX
KBSF Springhill LA
KBSN Moses Lake WA
KBSP(AM) Birch Tree MO
KBSR Laurel MT
KBST Big Spring TX
*KBSU Boise ID
KBSZ Wickenburg AZ
KBTA Batesville AR
KBTC Houston MO
KBTM(AM) Jonesboro AR
KBUF(AM) Holcomb KS
KBUL Billings MT
KBUN Bemidji MN
KBUR Burlington IA
KBUY Ruidoso NM
KBWD Brownwood TX
KBYG Big Spring TX
KBYO Tallulah LA
KBYR(AM) Anchorage AK
KBZO Lubbock TX
KBZY Salem OR
KBZZ(AM) Sparks NV
KCAA(AM) Loma Linda CA
KCAB Dardanelle AR
KCAL Redlands CA
KCAM Glennallen AK
KCAP Helena MT
KCAR Clarksville TX
*KCAT Pine Bluff AR
KCBC Riverbank CA
KCBF Fairbanks AK
KCBL Fresno CA
KCBQ San Diego CA
KCBR Monument CO
KCBS San Francisco CA
KCCB Corning AR
KCCC Carlsbad NM
KCCR Pierre SD
KCCT Corpus Christi TX
KCCV Overland Park KS
KCEE(AM) Cortaro AZ
KCEG(AM) Pueblo CO
KCEO Vista CA
KCFC(AM) Boulder CO
KCFJ(AM) Alturas CA
KCFO Tulsa OK
*KCFR(AM) Denver CO
KCGS Marshall AR
KCHA Charles City IA
KCHE(AM) Cherokee IA
KCHI Chillicothe MO
KCHJ Delano CA
KCHK New Prague MN
KCHL San Antonio TX
KCHN(AM) Brookshire TX
KCHR Charleston MO
KCHS Truth or Consequences NM
*KCHU(AM) Valdez AK
KCID(AM) Caldwell ID
KCII(AM) Washington IA
KCIK(AM) Blue Lake CA
KCIM Carroll IA
KCIS(AM) Edmonds WA
KCJB Minot ND
KCJJ(AM) Iowa City IA
KCKK(AM) Littleton CO
KCKM(AM) Monahans TX
KCKN(AM) Roswell NM
KCKX Stayton OR
KCKY Coolidge AZ
KCLA Pine Bluff AR
KCLE(AM) Cleburne TX
KCLI Clinton OK
KCLK Asotin WA
KCLN Clinton IA
KCLR Ralls TX
KCLV(AM) Clovis NM
KCLW Hamilton TX
KCLX Colfax WA
KCMC Texarkana TX
KCMD(AM) Portland OR
KCMN Colorado Springs CO
KCMO Kansas City MO
KCMX Phoenix OR
KCMY(AM) Carson City NV
KCNI Broken Bow NE
KCNM(AM) Garapan-Saipan NP
KCNN East Grand Forks MN
KCNR(AM) Shasta CA
KCNW(AM) Fairway KS
KCNZ(AM) Cedar Falls IA
KCOB Newton IA
KCOG Centerville IA
KCOH Houston TX
KCOL(AM) Wellington CO
KCOM Comanche TX
KCOR(AM) San Antonio TX
KCOW Alliance NE
KCOX(AM) Jasper TX
KCPS Burlington IA
KCPW(AM) Tooele UT
KCPX(AM) Spanish Valley UT
KCQL Aztec NM
KCRC Enid OK
KCRN San Angelo TX
KCRO Omaha NE
KCRS(AM) Midland TX
KCRT Trinidad CO
KCRV Caruthersville MO
KCRX Roswell NM
KCSJ Pueblo CO
KCSP(AM) Kansas City MO
KCSR Chadron NE
KCST Florence OR
KCTA Corpus Christi TX
KCTC Sacramento CA
KCTE Independence MO
KCTI(AM) Gonzales TX
KCTO(AM) Cleveland MO
KCTX(AM) Childress TX
KCUB Tucson AZ
KCUE Red Wing MN
KCUL Marshall TX
KCUP(AM) Toledo OR
KCUZ Clifton AZ
KCVL Colville WA
KCVR Lodi CA
KCWJ(AM) Blue Springs MO
KCWM Hondo TX
KCXL Liberty MO
KCYL Lampasas TX
KCZZ(AM) Mission KS
KDAC Fort Bragg CA
KDAE Sinton TX
KDAK Carrington ND
KDAL Duluth MN
KDAN(AM) Beatty NV
KDAO Marshalltown IA
KDAP Douglas AZ
KDAV Lubbock TX
KDAZ(AM) Albuquerque NM
KDBM Dillon MT
KDBS Alexandria LA
KDBV(AM) Salinas CA
KDCC Dodge City KS
KDCE Espanola NM
KDDD Dumas TX
KDDR Oakes ND
KDDZ Arvada CO
KDEC Decorah IA
KDEF Albuquerque NM
*KDEI(AM) Port Arthur TX
KDET Center TX
KDEX Dexter MO
KDFN Doniphan MO
KDFT Ferris TX
KDGO Durango CO
KDHL Faribault MN
KDHN Dimmitt TX
KDIA Vallejo CA
KDIF Riverside CA
KDIL(AM) Dillon MT
KDIO Ortonville MN
KDIS(AM) Pasadena CA
KDIX Dickinson ND
KDIZ Golden Valley MN
KDJI Holbrook AZ
KDJQ(AM) Meridian ID
KDJS Willmar MN
KDJW(AM) Amarillo TX
KDKA Pittsburgh PA
KDKD Clinton MO
KDKT(AM) Beulah ND
KDLA De Ridder LA
*KDLG Dillingham AK
KDLM Detroit Lakes MN
KDLR Devils Lake ND
KDLS Perry IA
KDMA Montevideo MN
KDMO Carthage MO
KDMS El Dorado AR
KDNZ(AM) Cedar Falls IA
KDOM Windom MN
KDOW(AM) Seattle WA
KDOX Henderson NV
KDQN De Queen AR
KDRO Sedalia MO
KDRS Paragould AR
KDRY Alamo Heights TX
KDSJ Deadwood SD
KDSN Denison IA
KDTA Delta CO
KDTD(AM) Kansas City KS
KDTH Dubuque IA
KDUN(AM) Reedsport OR
KDUS Tempe AZ
KDUZ Hutchinson MN
KDWA(AM) Hastings MN
KDWN Las Vegas NV
KDXE(AM) North Little Rock AR
KDXU Saint George UT
KDYA Vallejo CA
KDYL(AM) South Salt Lake UT
KDYN Ozark AR
KDZR(AM) Lake Oswego OR
KEAR(AM) San Francisco CA
KEAS Eastland TX
KEBC(AM) Midwest City OK
KEBE Jacksonville TX
*KEBR Rocklin CA
*KECR El Cajon CA
KEDA San Antonio TX
KEDO Longview WA
KEEL(AM) Shreveport LA
KEES Gladewater TX
KEIN Great Falls MT

KEIP(AM) Las Vegas NV
KEJO Corvallis OR
KELA Centralia-Chehalis WA
KELD El Dorado AR
KELE Mountain Grove MO
KELG(AM) Manor TX
KELK Elko NV
KELO(AM) Sioux Falls SD
KELP El Paso TX
KELY Ely NV
KENA(AM) Mena AR
KENI Anchorage AK
KENN Farmington NM
KENO Las Vegas NV
KENT(AM) Parowan UT
KEOR(AM) Catoosa OK
KEPL(AM) Estes Park CO
KEPN(AM) Lakewood CO
KEPS Eagle Pass TX
KERB Kermit TX
KERI(AM) Wasco-Greenacres CA
KERN Bakersfield CA
KERR Polson MT
KERV Kerrville TX
KESM El Dorado Springs MO
KESP(AM) Modesto CA
KESQ Indio CA
KEST San Francisco CA
KETX Livingston TX
KEUN Eunice LA
KEVA Evanston WY
KEVT(AM) Sahuarita AZ
KEWE(AM) Oroville CA
KEWI(AM) Benton AR
KEX Portland OR
KEXO Grand Junction CO
*KEXS(AM) Excelsior Springs MO
KEYE Perryton TX
KEYF(AM) Dishman WA
KEYG Grand Coulee WA
KEYH Houston TX
KEYL Long Prairie MN
*KEYQ Fresno CA
KEYS Corpus Christi TX
*KEYY(AM) Provo UT
KEYZ Williston ND
*KEZJ Twin Falls ID
KEZL(AM) Visalia CA
KEZM Sulphur LA
KEZW Aurora CO
*KEZX(AM) Medford OR
KEZY(AM) San Bernardino CA
KFAB Omaha NE
KFAL Fulton MO
KFAN Minneapolis MN
KFAQ(AM) Tulsa OK
KFAR Fairbanks AK
KFAX San Francisco CA
KFAY Farmington AR
KFBC Cheyenne WY
KFBK Sacramento CA
KFBX(AM) Fairbanks AK
KFCD(AM) Farmersville TX
KFCR Custer SD
KFEL Pueblo CO
KFEQ(AM) Saint Joseph MO
KFFA Helena AR
KFFF(AM) Boone IA
KFFK(AM) Rogers AR
KFFN Tucson AZ
KFGO Fargo ND
KFH(AM) Wichita KS
KFI Los Angeles CA
KFIA Carmichael CA
KFIG Fresno CA
KFIL Preston MN
KFIR Sweet Home OR
KFIT-EX San Antonio TX
KFIT Lockhart TX
KFIV Modesto CA
KFIZ Fond du Lac WI
KFJB Marshalltown IA
KFJZ Fort Worth TX
KFKA(AM) Greeley CO
*KFLB(AM) Odessa TX
KFLC(AM) Fort Worth TX
KFLD Pasco WA
KFLG(AM) Bullhead City AZ
KFLN Baker MT
KFLP(AM) Floydada TX
KFLS Klamath Falls OR
*KFLT Tucson AZ
KFMB San Diego CA
KFMO Park Hills MO
KFMZ(AM) Brookfield MO
KFNN Mesa AZ
KFNS(AM) Wood River IL
*KFNW(AM) West Fargo ND
KFNX Cave Creek AZ
KFNZ Salt Lake City UT
KFON Austin TX
KFOR(AM) Lincoln NE
KFOX(AM) Torrance CA
KFPT(AM) Clovis CA
KFPW Fort Smith AR
KFQD Anchorage AK
KFRA Franklin LA
KFRM(AM) Salina KS
*KFRN Long Beach CA
KFRO Longview TX
KFRU Columbia MO
KFSA Fort Smith AR
KFSD(AM) Escondido CA
KFSG(AM) Roseville CA
KFST Fort Stockton TX
KFTA(AM) Rupert ID
KFTI(AM) Wichita KS
KFTM Fort Morgan CO
KFUN Las Vegas NM
*KFUO(AM) Clayton MO
KFUT(AM) Thousand Palms CA
KFVR Crescent City CA
KFWB Los Angeles CA
KFXD(AM) Boise ID
KFXN Minneapolis MN
KFXR(AM) Dallas TX
KFXX(AM) Portland OR
KFXY(AM) Enid OK
KFXZ(AM) Lafayette LA
KFYI(AM) Phoenix AZ
KFYN Bonham TX
KFYO Lubbock TX
KFYR Bismarck ND
KGA Spokane WA
KGAB Orchard Valley WY
KGAF Gainesville TX
KGAK Gallup NM
KGAL(AM) Lebanon OR
KGAM Palm Springs CA
KGAS Carthage TX
KGBA(AM) Calexico CA
KGBC Galveston TX
KGBT Harlingen TX
KGDC Walla Walla WA
KGDD(AM) Oregon City OR
KGDP(AM) Oildale CA
KGEM(AM) Boise ID
KGEN Tulare CA
KGEO Bakersfield CA
KGEZ Kalispell MT
KGFF Shawnee OK
KGFL Clinton AR
KGFW Kearney NE
KGFX Pierre SD
KGGF Coffeyville KS
KGGN Gladstone MO
KGGR Dallas TX
KGHF Pueblo CO
KGHL Billings MT
KGHS International Falls MN
KGHT Sheridan AR
KGIM Aberdeen SD
KGIR Cape Girardeau MO
KGIW Alamosa CO
KGKL San Angelo TX
KGLA Gretna LA
KGLD Tyler TX
KGLE Glendive MT
KGLN Glenwood Springs CO
KGLO(AM) Mason City IA
KGME(AM) Phoenix AZ
KGMI Bellingham WA
KGMS(AM) Tucson AZ
KGMT Fairbury NE
KGMY Springfield MO
KGNB New Braunfels TX
KGNC Amarillo TX
KGND(AM) Vinita OK
KGNM Saint Joseph MO
KGNO Dodge City KS
*KGNU(AM) Denver CO
KGNW Burien-Seattle WA
KGO San Francisco CA
KGOE Eureka CA
KGOL Humble TX
KGOS Torrington WY
KGOW(AM) Bellaire TX
KGRE Greeley CO
KGRG(AM) Enumclaw WA
KGRN Grinnell IA
KGRO Pampa TX
KGRV Winston OR
KGRZ Missoula MT
KGSO(AM) Wichita KS
KGST Fresno CA
KGTK(AM) Olympia WA
KGTL Homer AK
KGTO Tulsa OK
KGU Honolulu HI
KGUM(AM) Hagatna GU
KGVL Greenville TX
KGVO Missoula MT
KGVW Belgrade MT
KGVY(AM) Green Valley AZ
KGWA Enid OK
KGY Olympia WA
KGYM(AM) Cedar Rapids IA
KGYN Guymon OK
KHAC Tse Bonito NM
KHAR Anchorage AK
KHAS Hastings NE
KHAT(AM) Laramie WY
KHBC(AM) Hilo HI
KHBM Monticello AR
KHBR Hillsboro TX
KHBZ(AM) Honolulu HI
*KHCB Galveston TX
*KHCH(AM) Huntsville TX
KHCM(AM) Honolulu HI
KHDN Hardin MT
KHEY(AM) El Paso TX
KHFX(AM) Burleson TX
KHGG(AM) Van Buren AR
KHHO Tacoma WA
KHIL Willcox AZ
KHIT Reno NV
KHJ(AM) Los Angeles CA
KHLO Hilo HI
KHLT(AM) Hallettsville TX
KHMO Hannibal MO
KHNC Johnstown CO
KHND Harvey ND
KHNR(AM) Honolulu HI
KHOB Hobbs NM
KHOJ(AM) Saint Charles MO
KHOT Madera CA
KHOW Denver CO
KHOZ Harrison AR
KHPI(AM) Moreno Valley CA
KHPP(AM) Waukon IA
KHPY(AM) Moreno Valley CA
KHQN Spanish Fork UT
KHRA(AM) Honolulu HI
KHRO(AM) El Paso TX
KHRT Minot ND
KHSE(AM) Wylie TX
KHSN(AM) Coos Bay OR
KHTK Sacramento CA
KHTS(AM) Canyon Country CA
KHTY(AM) Bakersfield CA
KHUB Fremont NE
KHVH Honolulu HI
KHVL(AM) Huntsville TX
KHVN Fort Worth TX
KHWG(AM) Fallon NV
KIAL(AM) Unalaska AK
KIAM Nenana AK
KIBL Beeville TX
KICA Clovis NM
KICD Spencer IA
KICE(AM) Bend OR
KICS Hastings NE
KICY Nome AK
KID Idaho Falls ID
KIDD Monterey CA
KIDO(AM) Nampa ID
KIDR Phoenix AZ
KIEV(AM) Culver City CA
KIFG Iowa Falls IA
KIFO(AM) Hawthorne NV
KIFW Sitka AK
KIGO(AM) Saint Anthony ID
KIGS Hanford CA
*KIHM(AM) Reno NV
KIHN Hugo OK
KIHR Hood River OR
KIID(AM) Sacramento CA
KIIX(AM) Fort Collins CO
KIJN Farwell TX
KIJV Huron SD
KIKC(AM) Forsyth MT
KIKK Pasadena TX
KIKO Miami AZ
KIKR Beaumont TX
KIKZ Seminole TX
KILJ Mount Pleasant IA
KILR Estherville IA
KILT Houston TX
KIMB Kimball NE
KIML Gillette WY
KIMM Rapid City SD
KIMP(AM) Mount Pleasant TX
KINA Salina KS
KIND Independence KS
KINE Kingsville TX
KINF(AM) Santa Maria CA
KINN Alamogordo NM
KINO Winslow AZ
KINS Eureka CA
KINY Juneau AK
KION(AM) Salinas CA
KIOU Shreveport LA
KIOV Payette ID
KIPA(AM) Hilo HI
KIQI San Francisco CA
KIQQ Barstow CA
KIQS Willows CA
KIRN(AM) Simi Valley CA
KIRO Seattle WA
KIRT Mission TX
KIRV Fresno CA
KIRX Kirksville MO
KIST(AM) Santa Barbara CA
KIT Yakima WA
KITI Chehalis-Centralia WA
KITZ Silverdale WA
KIUL Garden City KS
KIUN Pecos TX
KIUP Durango CO
KIVY Crockett TX
KIWA Sheldon IA
KIXC(AM) Hilo HI
KIXI Mercer Island-Seattle WA
KIXL(AM) Del Valle TX
KIXW Apple Valley CA
KIXZ Amarillo TX
*KIYU Galena AK
KJAA Globe AZ
KJAL(AM) Tafuna AS
KJAM(AM) Madison SD
KJAN Atlantic IA
KJAY Sacramento CA
KJBC(AM) Midland TX
KJBN Little Rock AR
KJCB Lafayette LA
KJCE Rollingwood TX
KJCK(AM) Junction City KS
KJDJ San Luis Obispo CA
KJDL(AM) Lubbock TX
KJDY John Day OR
KJEF(AM) Jennings LA
KJFF(AM) Festus MO
KJFK(AM) Reno NV
KJIM Sherman TX
KJIN(AM) Houma LA
KJJD(AM) Windsor CO
KJJK Fergus Falls MN
KJJL(AM) Pine Bluffs WY
KJJQ Volga SD
KJJR Whitefish MT
KJLL South Tucson AZ
*KJLT North Platte NE
KJME(AM) Fountain CO
KJMJ(AM) Alexandria LA
KJMP(AM) Pierce CO
KJMU(AM) Sand Springs OK
KJNO(AM) Juneau AK
*KJNP North Pole AK
KJOC Davenport IA
KJOJ Conroe TX
KJOK(AM) Yuma AZ
KJOL(AM) Grand Junction CO
KJON(AM) Carrollton TX
KJOP Lemoore CA
KJOX(AM) Yakima WA
KJPG(AM) Frazier Park CA
KJPR(AM) Shasta Lake City CA
KJPW Waynesville MO
KJQS(AM) Murray UT
KJR Seattle WA
KJRB Spokane WA
KJRG Newton KS
KJSA Mineral Wells TX
KJSK Columbus NE
KJSL Saint Louis MO
KJTV(AM) Lubbock TX
KJUA(AM) Cheyenne WY
KJUG Tulare CA
*KKAA(AM) Aberdeen SD
KKAD(AM) Vancouver WA
KKAG(AM) Fargo ND
KKAM Lubbock TX
KKAN Phillipsburg KS
KKAQ Thief River Falls MN
KKAR Omaha NE
KKAT(AM) Salt Lake City UT
KKAY White Castle LA
KKBJ(AM) Bemidji MN
KKBM(AM) Frankston TX
KKCQ Fosston MN
KKDA Grand Prairie TX
KKDD San Bernardino CA
KKDZ Seattle WA
KKEA(AM) Honolulu HI
KKEE(AM) Astoria OR
KKFN Denver CO
KKGM(AM) Fort Worth TX
KKGN(AM) Oakland CA
KKGR East Helena MT
KKHJ(AM) Leone AS
KKHK(AM) Kansas City KS
KKIM Albuquerque NM
KKIN Aitkin MN
KKJL San Luis Obispo CA
KKJY(AM) Albuquerque NM
KKKK(AM) Colorado Springs CO
KKLE(AM) Winfield KS
KKLF(AM) Richardson TX
KKLL(AM) Carthage- MO
KKLO Leavenworth KS
KKLS Rapid City SD
KKMC Gonzales CA
KKML(AM) Colorado Springs CO
KKMO Tacoma WA
KKMS Richfield MN
KKNE(AM) Waipahu HI
KKNO Gretna LA
KKNS(AM) Corrales NM
KKNT(AM) Phoenix AZ
KKNW(AM) Seattle WA
KKNX Eugene OR
KKOB-EX Santa Fe NM
KKOB(AM) Albuquerque NM
KKOH Reno NV
KKOJ Jackson MN
KKOL Seattle WA
KKON(AM) Kealakekua HI
KKOW Pittsburg KS
KKOY Chanute KS
KKOZ Ava MO

KKPC Pueblo CO
KKPZ(AM) Portland OR
KKRT Wenatchee WA
KKRX Lawton OK
KKSA San Angelo TX
*KKSM Oceanside CA
KKSN(AM) Salem OR
KKTK(AM) Texarkana TX
KKTL Casper WY
KKTX(AM) Corpus Christi TX
KKTY Douglas WY
KKUB Brownfield TX
KKUZ Sallisaw OK
KKVV(AM) Las Vegas NV
KKXL Grand Forks ND
KKXX Paradise CA
KKYX San Antonio TX
KKZN(AM) Thornton CO
KKZZ(AM) Ventura CA
KLAA(AM) Orange CA
KLAC Los Angeles CA
KLAD(AM) Klamath Falls OR
KLAM Cordova AK
KLAR Laredo TX
KLAT Houston TX
KLAV(AM) Las Vegas NV
KLAY(AM) Lakewood WA
KLBB(AM) Stillwater MN
KLBJ Austin TX
KLBM La Grande OR
KLBS Los Banos CA
KLCB Libby MT
KLCK Goldendale WA
KLCL Lake Charles LA
KLCN Blytheville AR
KLCY(AM) East Missoula MT
KLDC Brighton CO
KLDS Falfurrias TX
KLDY Lacey WA
KLEA Lovington NM
KLEB Golden Meadow LA
KLEE Ottumwa IA
KLEM(AM) Le Mars IA
KLER Orofino ID
KLEX Lexington MO
KLEY(AM) Wellington KS
KLFD Litchfield MN
KLFE Seattle WA
KLFF(AM) Arroyo Grande CA
KLFJ Springfield MO
KLGA Algona IA
KLGN Logan UT
KLGR Redwood Falls MN
KLHC(AM) Bakersfield CA
KLHT Honolulu HI
KLIB(AM) Roseville CA
KLIC(AM) Monroe LA
KLID Poplar Bluff MO
KLIF(AM) Dallas TX
KLIK(AM) Jefferson City MO
KLIM Limon CO
KLIN Lincoln NE
KLIV San Jose CA
KLIX Twin Falls ID
KLIZ(AM) Brainerd MN
KLKC(AM) Parsons KS
KLLA Leesville LA
KLLB West Jordan UT
KLLK Willits CA
KLLV(AM) Breen CO
KLMR Lamar CO
KLMS(AM) Lincoln NE
KLMX Clayton NM
KLNG(AM) Council Bluffs IA
KLNT Laredo TX
KLO Ogden UT
KLOA Ridgecrest CA
KLOC(AM) Turlock CA
KLOE Goodland KS
KLOG Kelso WA
KLOH Pipestone MN
KLOK San Jose CA
KLOO Corvallis OR
KLPL Lake Providence LA
KLPW(AM) Union MO
KLPZ Parker AZ
KLSD(AM) San Diego CA
KLSQ(AM) Whitney NV
KLTC Dickinson ND
KLTF Little Falls MN
KLTI Macon MO
KLTK South West City MO
KLTT Commerce City CO
KLTX Long Beach CA
KLTZ Glasgow MT
KLUP(AM) Terrell Hills TX
KLVI Beaumont TX
KLVL Pasadena TX
KLVQ(AM) Athens TX
KLVT Levelland TX
KLVZ(AM) Denver CO
KLWJ Umatilla OR
KLWN(AM) Lawrence KS
KLWT Lebanon MO
KLXR(AM) Redding CA
KLXX Bismarck-Mandan ND
KLYC(AM) McMinnville OR
KLYQ Hamilton MT
KLYR Clarksville AR
KLZ Denver CO
KLZS(AM) Eugene OR
KMA Shenandoah IA
KMAD Madill OK
KMAJ Topeka KS
KMAL(AM) Malden MO
KMAM Butler MO
KMAN Manhattan KS
KMAQ Maquoketa IA
KMAS Shelton WA
KMAV Mayville ND
KMAX Colfax WA
KMBD Tillamook OR
*KMBI Spokane WA
KMBL Junction TX
KMBS West Monroe LA
KMBX(AM) Soledad CA
KMBZ Kansas City MO
KMCD Fairfield IA
KMCL(AM) Donnelly ID
KMDO Fort Scott KS
KMED Medford OR
KMER Kemmerer WY
KMET Banning CA
KMFS(AM) Guthrie OK
KMFX Wabasha MN
KMHI Mountain Home ID
KMHL Marshall MN
KMHS Coos Bay OR
KMHT Marshall TX
KMIA(AM) Black Canyon City AZ
KMIC(AM) Houston TX
KMIK Tempe AZ
KMIN Grants NM
KMIS Portageville MO
KMJ Fresno CA
KMJC Mount Shasta CA
KMJM(AM) Cedar Rapids IA
KMKI Plano TX
KMKY Oakland CA
KMLB Monroe LA
KMMJ Grand Island NE
KMMO Marshall MO
KMMS(AM) Bozeman MT
KMND Midland TX
KMNS Sioux City IA
KMNV(AM) Saint Paul MN
KMNY(AM) Hurst TX
KMOG Payson AZ
KMON Great Falls MT
KMOX Saint Louis MO
KMOZ Rolla MO
KMPC(AM) Los Angeles CA
KMPG Hollister CA
KMPH(AM) Modesto CA
KMRB San Gabriel CA
KMRC Morgan City LA
KMRF Marshfield MO
KMRI West Valley City UT
KMRN Cameron MO
KMRS Morris MN
KMRY Cedar Rapids IA
KMSD Milbank SD
KMTA Miles City MT
KMTI Manti UT
KMTL Sherwood AR
KMTX Helena MT
KMUL(AM) Farwell TX
KMUR(AM) Pryor OK
KMUS(AM) Sperry OK
KMUZ Gresham OR
KMVI Wailuku HI
KMVL Madisonville TX
KMVP Phoenix AZ
KMXA Aurora CO
KMXO(AM) Merkel TX
KMYC(AM) Marysville CA
KMZK Billings MT
KMZT(AM) Beverly Hills CA
KNAB Burlington CO
KNAF Fredericksburg TX
KNAK Delta UT
KNAL(AM) Victoria TX
KNAX(AM) McCook NE
KNBO New Boston TX
KNBR San Francisco CA
KNBY Newport AR
KNCB Vivian LA
KNCK Concordia KS
KNCO Grass Valley CA
KNCR Fortuna CA
KNCY Nebraska City NE
KNDC Hettinger ND
KNDI Honolulu HI
KNDK Langdon ND
KNDN Farmington NM
KNDY Marysville KS
KNEA(AM) Jonesboro AR
KNEB Scottsbluff NE
KNED McAlester OK
KNEK Washington LA
KNEL Brady TX
KNEM Nevada MO
KNET Palestine TX
KNEU Roosevelt UT
KNEW Oakland CA
KNFL(AM) Tremonton UT
KNFT Bayard NM
KNFX Austin MN
KNGL McPherson KS
*KNGN McCook NE
KNGR(AM) Daingerfield TX
KNHD Camden AR
KNIA Knoxville IA
KNIM(AM) Maryville MO
KNIR New Iberia LA
KNIT(AM) Dallas TX
KNJY(AM) Boise ID
KNLV Ord NE
KNML(AM) Albuquerque NM
KNMX Las Vegas NM
KNND Cottage Grove OR
KNNS Larned KS
KNNZ(AM) Cedar City UT
KNOC Natchitoches LA
KNOE Monroe LA
*KNOM Nome AK
KNOT(AM) Prescott AZ
KNOX Grand Forks ND
KNPT Newport OR
KNRO(AM) Redding CA
KNRS Salt Lake City UT
KNRV(AM) Englewood CO
KNRY Monterey CA
KNSA Unalakleet AK
KNSI Saint Cloud MN
KNSP(AM) Staples MN
KNSS(AM) Wichita KS
KNST Tucson AZ
KNTB Lakewood WA
KNTH(AM) Houston TX
KNTR(AM) Lake Havasu City AZ
KNTS(AM) Palo Alto CA
KNTX(AM) Bowie TX
KNUI Kahului HI
KNUJ(AM) New Ulm MN
KNUS Denver CO
KNUU(AM) Paradise NV
KNUV(AM) Tolleson AZ
KNUZ Bellville TX
KNWA Bellefonte AR
*KNWC(AM) Sioux Falls SD
KNWH(AM) Twentynine Palms CA
KNWQ(AM) Palm Springs CA
*KNWS Waterloo IA
KNWZ(AM) Coachella CA
KNX Los Angeles CA
KNXN Sierra Vista AZ
KNZR Bakersfield CA
KNZZ(AM) Grand Junction CO
KOA Denver CO
*KOAC Corvallis OR
KOAI(AM) Van Buren AR
KOAK Red Oak IA
KOAL Price UT
KOAQ Terrytown NE
KOBB Bozeman MT
KOBE Las Cruces NM
KOBO Yuba City CA
KOCY(AM) Del City OK
KODI Cody WY
KODL The Dalles OR
KODY North Platte NE
KOEL(AM) Oelwein IA
KOFC Fayetteville AR
KOFE Saint Maries ID
KOFI Kalispell MT
KOFO Ottawa KS
KOGA Ogallala NE
KOGN(AM) Ogden UT
KOGO(AM) San Diego CA
KOGT Orange TX
KOHI(AM) Saint Helens OR
KOHU Hermiston OR
KOIL(AM) Plattsmouth NE
KOIT San Francisco CA
KOJM Havre MT
KOKA Shreveport LA
KOKB Blackwell OK
KOKC(AM) Oklahoma City OK
KOKE Pflugerville TX
KOKK Huron SD
KOKL Okmulgee OK
KOKO Warrensburg MO
KOKP Perry OK
KOKX Keokuk IA
KOLE Port Arthur TX
KOLM(AM) Rochester MN
KOLT Scottsbluff NE
KOLY Mobridge SD
KOMC Branson MO
KOMJ(AM) Omaha NE
KOMO(AM) Seattle WA
KOMW Omak WA
KOMY(AM) La Selva Beach CA
KONA Kennewick WA
KONO San Antonio TX
KONP Port Angeles WA
KOOQ North Platte NE
KOPT(AM) Eugene OR
KOPY Alice TX
KORC Waldport OR
KORE Springfield-Eugene OR
KORL(AM) Honolulu HI
KORN Mitchell SD
KORT Grangeville ID
KOSE Wilson AR
KOSY(AM) Texarkana AR
KOTA Rapid City SD
KOTC Kennett MO
KOTK(AM) Omaha NE
KOTN Pine Bluff AR
KOTS Deming NM
*KOTZ Kotzebue AK
KOUU(AM) Pocatello ID
KOVC Valley City ND
KOVE Lander WY
KOVO Provo UT
KOWB Laramie WY
KOWL(AM) South Lake Tahoe CA
KOWZ(AM) Waseca MN
KOXR Oxnard CA
KOY Phoenix AZ
KOZA Odessa TX
KOZE Lewiston ID
KOZI Chelan WA
KOZN(AM) Bellevue NE
KOZQ Waynesville MO
KOZY Grand Rapids MN
KPAM Troutdale OR
KPAN(AM) Hereford TX
KPAY Chico CA
KPBL Hemphill TX
KPCO Quincy CA
KPDQ Portland OR
KPEL Lafayette LA
KPET Lamesa TX
KPGE Page AZ
KPGM(AM) Pawhuska OK
KPHN Kansas City MO
KPHX Phoenix AZ
KPIG(AM) Piedmont CA
KPIR(AM) Granbury TX
KPJC(AM) Salem OR
KPKE(AM) Gunnison CO
KPLT Paris TX
KPLY(AM) Reno NV
*KPMO(AM) Mendocino CA
KPNS(AM) Duncan OK
KPNW Eugene OR
KPOC(AM) Pocahontas AR
KPOD Crescent City CA
*KPOF Denver CO
KPOJ(AM) Portland OR
KPOK Bowman ND
KPOW Powell WY
KPQ(AM) Wenatchee WA
KPRC Houston TX
KPRK Livingston MT
KPRL Paso Robles CA
KPRM Park Rapids MN
KPRO Riverside CA
KPRT Kansas City MO
KPRV Poteau OK
KPRZ San Marcos-Poway CA
KPSI Palm Springs CA
KPSZ(AM) Des Moines IA
KPTK(AM) Seattle WA
KPTO(AM) Pocatello ID
KPTQ(AM) Spokane WA
KPTR(AM) Cathedral City CA
KPUA Hilo HI
KPUG Bellingham WA
KPUR Amarillo TX
KPWB Piedmont MO
KPWL(AM) Newport WA
KPXQ(AM) Glendale AZ
KPYK Terrell TX
KPYN(AM) Atlanta TX
KPZA(AM) Hot Springs AR
KPZK(AM) Little Rock AR
KQAB Lake Isabella CA
KQAD(AM) Luverne MN
KQAM Wichita KS
KQCV Oklahoma City OK
KQDI Great Falls MT
KQDJ Jamestown ND
KQDS Duluth MN
KQEN(AM) Roseburg OR
KQEQ Fowler CA
KQIK Lakeview OR
KQJZ(AM) Kalispell MT
*KQKD(AM) Redfield SD
KQLO(AM) Sun Valley NV
KQLX Lisbon ND
KQMG Independence IA
KQMS Redding CA
KQNA(AM) Prescott Valley AZ
KQNG Lihue HI
KQNK Norton KS
KQNM(AM) Milan NM
KQNT(AM) Spokane WA
KQPN(AM) West Memphis AR
KQQQ Pullman WA
KQRL(AM) Waco TX
KQSP(AM) Shakopee MN
KQTY Borger TX
KQUE Houston TX
KQV Pittsburgh PA
KQWB(AM) West Fargo ND
KQWC Webster City IA

KQYS(AM) Neosho MO
KQYX(AM) Joplin MO
KRAE Cheyenne WY
KRAI(AM) Craig CO
KRAK(AM) Hesperia CA
KRAL Rawlins WY
KRAM West Klamath OR
KRBA Lufkin TX
KRBI(AM) Saint Peter MN
KRBT Eveleth MN
KRCM Beaumont TX
KRCN(AM) Longmont/Denver CO
KRCO(AM) Prineville OR
KRDD Roswell NM
KRDH(AM) Canton TX
KRDM(AM) Redmond OR
KRDO(AM) Colorado Springs CO
KRDU Dinuba CA
KRDY(AM) San Antonio TX
KRDZ Wray CO
KREA(AM) Honolulu HI
KREB(AM) Bentonville-Bella Vista AR
KREF(AM) Norman OK
KREH Pecan Grove TX
KREI Farmington MO
KREL(AM) Quanah TX
KREW(AM) Plainview TX
KRFE Lubbock TX
KRFO Owatonna MN
KRFS Superior NE
KRFT(AM) De Soto MO
KRGE Weslaco TX
KRGI Grand Island NE
KRGS(AM) Rifle CO
KRHC(AM) Burnet TX
KRHW Sikeston MO
KRIB(AM) Mason City IA
KRIL Odessa TX
KRIO(AM) McAllen TX
KRIZ Renton WA
KRJJ(AM) Brooklyn Park MN
KRJO(AM) Monroe LA
KRKC(AM) King City CA
KRKE(AM) Albuquerque NM
KRKK Rock Springs WY
KRKO(AM) Everett WA
KRKS(AM) Denver CO
KRKY(AM) Granby CO
KRLA(AM) Glendale CA
KRLC Lewiston ID
KRLD Dallas TX
KRLL(AM) California MO
KRLN(AM) Canon City CO
KRLV Las Vegas NV
KRLW Walnut Ridge AR
KRMD Shreveport LA
KRMG Tulsa OK
KRML Carmel CA
KRMO(AM) Cassville MO
KRMP(AM) Oklahoma City OK
KRMS Osage Beach MO
KRMX(AM) Pueblo CO
KRMY Killeen TX
KRND(AM) Fox Farm WY
*KRNI(AM) Mason City IA
KRNR Roseburg OR
KRNT Des Moines IA
KROB(AM) Robstown TX
KROC(AM) Rochester MN
KROD El Paso TX
KROE(AM) Sheridan WY
KROF Abbeville LA
KROO Breckenridge TX
KROP(AM) Brawley CA
KROS Clinton IA
KROX Crookston MN
KRPI(AM) Ferndale WA
KRPL Moscow ID
KRQX Mexia TX
KRRP(AM) Coushatta LA
KRRS Santa Rosa CA
KRRZ Minot ND
KRSA Petersburg AK
KRSC Othello WA
KRSL Russell KS
KRSN(AM) Los Alamos NM
KRSV Afton WY
KRSX(AM) Victorville CA
KRSY(AM) Alamogordo NM
KRTA Medford OR
KRTK Chubbuck ID
KRTN Raton NM
KRTR(AM) Honolulu HI
KRTX Rosenberg-Richmond TX
KRUD(AM) Honolulu HI
KRUI Ruidoso Downs NM
KRUN Ballinger TX
KRUS Ruston LA
KRVA Cockrell Hill TX
*KRVM(AM) Eugene OR
KRVN Lexington NE
KRVT(AM) Claremore OK
KRVZ Springerville-Eagar AZ
KRWB Roseau MN
KRWC Buffalo MN
KRXA(AM) Carmel Valley CA
KRXK(AM) Rexburg ID
KRXR Gooding ID
KRZE Farmington NM
KRZI(AM) Waco TX
KRZY Albuquerque NM
KSAC(AM) Sacramento CA
KSAH(AM) Universal City TX
KSAL(AM) Salina KS
KSAM(AM) Whitefish MT
KSAZ(AM) Marana AZ
KSBN Spokane WA
KSBQ Santa Maria CA
KSCB Liberal KS
KSCJ Sioux City IA
KSCO Santa Cruz CA
KSCR(AM) Eugene OR
KSDN Aberdeen SD
KSDO San Diego CA
*KSDP Sand Point AK
KSDR Watertown SD
KSDT Hemet CA
KSEI Pocatello ID
KSEK Pittsburg KS
KSEL Portales NM
KSEN Shelby MT
KSEO Durant OK
KSET(AM) Silsbee TX
KSEV Tomball TX
KSEY Seymour TX
KSFA Nacogdoches TX
KSFN North Las Vegas NV
KSFO San Francisco CA
KSFT Saint Joseph MO
KSGF(AM) Springfield MO
KSGL Wichita KS
KSGM(AM) Chester IL
KSGT Jackson WY
KSHO(AM) Lebanon OR
KSHP North Las Vegas NV
KSIB Creston IA
KSID Sidney NE
KSIG Crowley LA
KSIM(AM) Sikeston MO
KSIR Brush CO
KSIS(AM) Sedalia MO
KSIV(AM) Clayton MO
KSIW Woodward OK
KSIX Corpus Christi TX
KSJB Jamestown ND
*KSJK Talent OR
KSJX(AM) San Jose CA
KSKE(AM) Buena Vista CO
KSKY(AM) Balch Springs TX
KSL Salt Lake City UT
KSLD(AM) Soldotna AK
KSLG(AM) Saint Louis MO
KSLI(AM) Abilene TX
KSLL(AM) Price UT
KSLO Opelousas LA
KSLR San Antonio TX
KSLV Monte Vista CO
KSMA Santa Maria CA
KSMH(AM) West Sacramento CA
KSML Diboll TX
KSMO Salem MO
KSNM(AM) Las Cruces NM
KSNY Snyder TX
KSOK Arkansas City KS
KSON San Diego CA
KSOO(AM) Sioux Falls SD
KSOP South Salt Lake UT
KSOU Sioux Center IA
KSOX Raymondville TX
KSPA(AM) Ontario CA
KSPD Boise ID
KSPI Stillwater OK
KSPN(AM) Los Angeles CA
KSPT(AM) Sandpoint ID
KSPZ(AM) Ammon ID
KSQB(AM) Sioux Falls SD
KSQP(AM) Pierre SD
KSRA Salmon ID
KSRM Soldotna AK
KSRO Santa Rosa CA
KSRR Provo UT
KSRV Ontario OR
KSSK(AM) Honolulu HI
KSSR Santa Rosa NM
KSST Sulphur Springs TX
KSTA Coleman TX
KSTC Sterling CO
KSTE Rancho Cordova CA
KSTL Saint Louis MO
KSTN Stockton CA
KSTP(AM) Saint Paul MN
KSTV Stephenville TX
KSUB Cedar City UT
KSUE Susanville CA
KSUH Puyallup WA
KSUM Fairmont MN
KSUN Phoenix AZ
KSVA(AM) Albuquerque NM
KSVC Richfield UT
KSVE El Paso TX
KSVN Ogden UT
KSVP Artesia NM
KSWA Graham TX
KSWB Seaside OR
KSWD Seward AK
KSWM Aurora MO
KSWV Santa Fe NM
KSXT(AM) Loveland CO
KSYB(AM) Shreveport LA
*KSYC(AM) Yreka CA
KSYL Alexandria LA
KSZL Barstow CA
KSZN(AM) Milwaukie OR
KTAE(AM) Cameron TX
KTAM Bryan TX
KTAN(AM) Sierra Vista AZ
KTAP Santa Maria CA
KTAR(AM) Phoenix AZ
KTAT Frederick OK
KTBA Tuba City AZ
KTBB Tyler TX
KTBI Ephrata WA
KTBL(AM) Los Ranchos de Albuquerque NM
*KTBR(AM) Roseburg OR
KTBZ(AM) Tulsa OK
KTCH Wayne NE
KTCK Dallas TX
KTCR Kennewick WA
KTCS Fort Smith AR
KTCT San Mateo CA
KTDD(AM) San Bernardino CA
KTEK Alvin TX
KTEL Walla Walla WA
KTEM Temple TX
KTFI Twin Falls ID
KTFJ Dakota City NE
KTFS(AM) Texarkana TX
KTGE Salinas CA
*KTGG Spring Arbor MI
KTGO Tioga ND
KTGR Columbia MO
KTHE Thermopolis WY
KTHH(AM) Albany OR
KTHO South Lake Tahoe CA
KTHS Berryville AR
KTIB Thibodaux LA
KTIC West Point NE
KTIE(AM) San Bernardino CA
KTIK Nampa ID
KTIP Porterville CA
KTIQ(AM) Merced CA
*KTIS Minneapolis MN
KTIX Pendleton OR
KTJK Del Rio TX
KTJS Hobart OK
KTKB(AM) Tamuning GU
KTKK Sandy UT
KTKN Ketchikan AK
KTKR San Antonio TX
KTKT Tucson AZ
KTKZ Sacramento CA
KTLK(AM) Los Angeles CA
KTLO Mountain Home AR
KTLQ Tahlequah OK
KTLR(AM) Oklahoma City OK
KTLU Rusk TX
KTLV(AM) Midwest City OK
KTMC McAlester OK
KTMM(AM) Grand Junction CO
KTMR Edna TX
KTMS Santa Barbara CA
KTMT Ashland OR
KTNC Falls City NE
KTNF(AM) Saint Louis Park MN
KTNM Tucumcari NM
KTNN Window Rock AZ
KTNO(AM) University Park TX
KTNP(AM) Tonopah NV
KTNQ Los Angeles CA
KTNS Oakhurst CA
KTNZ Amarillo TX
KTOB(AM) Petaluma CA
KTOE(AM) Mankato MN
KTOK(AM) Oklahoma City OK
KTON Belton TX
KTOP Topeka KS
KTOQ Rapid City SD
KTOX Needles CA
KTPA Prescott AR
KTPI(AM) Mojave CA
KTRB(AM) San Francisco CA
KTRC(AM) Santa Fe NM
KTRF Thief River Falls MN
KTRH Houston TX
KTRO(AM) Vancouver WA
KTRP(AM) Mount Angel OR
KTRS Saint Louis MO
KTRW(AM) Opportunity WA
KTSA San Antonio TX
KTSM(AM) El Paso TX
KTSN Elko NV
KTTH(AM) Seattle WA
KTTN(AM) Trenton MO
KTTO(AM) Spokane WA
KTTP(AM) Pineville LA
KTTR Rolla MO
KTTT Columbus NE
KTUC Tucson AZ
KTUE Tulia TX
KTUI Sullivan MO
KTUV(AM) Little Rock AR
KTWG(AM) Hagatna GU
KTWO Casper WY
KTXV(AM) Mabank TX
KTXZ West Lake Hills TX
KTYM Inglewood CA
KTZN Anchorage AK
KUAI Eleele HI
KUAM(AM) Hagatna GU
KUAU Haiku HI
*KUAZ(AM) Tucson AZ
KUBA(AM) Yuba City CA
KUBC(AM) Montrose CO
KUBR San Juan TX
KUCU(AM) Farmington NM
KUDO(AM) Anchorage AK
KUGN Eugene OR
KUGR Green River WY
KUGT Jackson MO
KUHL(AM) Lompoc CA
KUIK(AM) Hillsboro OR
KUJ Walla Walla WA
KUKI Ukiah CA
KUKU(AM) Willow Springs MO
KULE Ephrata WA
KULP El Campo TX
KULY Ulysses KS
KUMA Pendleton OR
KUMU Honolulu HI
KUNF(AM) Washington UT
KUNO Corpus Christi TX
KUNX(AM) Santa Paula CA
KUOA(AM) Siloam Springs AR
KUOL San Marcos TX
*KUOM Minneapolis MN
KUOW(AM) Tumwater WA
KUPA(AM) Pearl City HI
KURL Billings MT
KURM Rogers AR
KURS San Diego CA
KURV Edinburg TX
KURY Brookings OR
KUSH(AM) Cushing OK
KUTI(AM) Yakima WA
KUTR(AM) Taylorsville UT
KUTY Palmdale CA
KUUX(AM) Pullman WA
KUVR Holdrege NE
KUYO Evansville WY
KUZZ Bakersfield CA
KVAK(AM) Valdez AK
KVAN(AM) Burbank WA
KVBR Brainerd MN
KVCE(AM) Highland Park TX
KVCK(AM) Wolf Point MT
KVCL Winnfield LA
KVCU Boulder CO
KVDW(AM) England AR
KVEC San Luis Obispo CA
KVEL Vernal UT
KVEN Ventura CA
KVET Austin TX
KVFC Cortez CO
KVFD(AM) Fort Dodge IA
KVGB Great Bend KS
KVI(AM) Seattle WA
KVIN(AM) Ceres CA
*KVIP Redding CA
KVIS Miami OK
KVIV El Paso TX
KVJY Pharr TX
KVKK(AM) Verndale MN
KVLE(AM) Vail CO
KVLF Alpine TX
KVLG La Grange TX
KVLH Pauls Valley OK
KVLV(AM) Fallon NV
KVMA Magnolia AR
KVMC Colorado City TX
KVML(AM) Sonora CA
KVNA(AM) Flagstaff AZ
KVNI Coeur d'Alene ID
KVNN(AM) Victoria TX
KVNR Santa Ana CA
KVNS(AM) Brownsville TX
KVNU Logan UT
KVOC Casper WY
KVOE Emporia KS
KVOG(AM) Hagatna GU
KVOI Tucson AZ
KVOK Kodiak AK
KVOL Lafayette LA
KVOM Morrilton AR
KVON Napa CA
KVOP(AM) Plainview TX
KVOR Colorado Springs CO
KVOT(AM) Taos NM
KVOU Uvalde TX
KVOW Riverton WY
KVOX Moorhead MN
KVOZ Del Mar Hills TX
KVPI Ville Platte LA
KVRC Arkadelphia AR
KVRH Salida CO
KVRI(AM) Blaine WA
KVRP Stamford TX
KVSA McGehee AR
KVSF(AM) Santa Fe NM
KVSH Valentine NE

KVSI Montpelier ID
KVSL Show Low AZ
KVSO(AM) Ardmore OK
KVSV Beloit KS
KVTA Port Hueneme CA
KVTK(AM) Vermillion SD
KVTO Berkeley CA
KVVN Santa Clara CA
KVWC Vernon TX
KVWG(AM) Pearsall TX
KVWM(AM) Show Low AZ
KWAC Bakersfield CA
KWAD Wadena MN
KWAI Honolulu HI
KWAK Stuttgart AR
KWAL Wallace ID
KWAM Memphis TN
KWAT Watertown SD
KWAY Waverly IA
KWBC Navasota TX
KWBE Beatrice NE
KWBG Boone IA
KWBW Hutchinson KS
KWBY Woodburn OR
KWCK Searcy AR
KWDB(AM) Oak Harbor WA
KWDF Ball LA
KWDJ(AM) Ridgecrest CA
KWDZ(AM) Salt Lake City UT
KWEB(AM) Rochester MN
KWED Seguin TX
KWEI(AM) Weiser ID
KWEL Midland TX
KWES(AM) Ruidoso NM
KWEY Weatherford OK
KWFA(AM) Tye TX
KWFM(AM) Tucson AZ
KWFS Wichita Falls TX
*KWG(AM) Stockton CA
KWHI Brenham TX
KWHN(AM) Fort Smith AR
KWHW Altus OK
KWIK Pocatello ID
KWIL Albany OR
KWIP Dallas OR
KWIQ(AM) Moses Lake North WA
KWIX Moberly MO
KWJL(AM) Lancaster CA
KWKA Clovis NM
KWKC Abilene TX
KWKH(AM) Shreveport LA
KWKU(AM) Pomona CA
KWKW Los Angeles CA
KWKY Des Moines IA
KWLA Many LA
*KWLC Decorah IA
KWLE(AM) Anacortes WA
KWLM Willmar MN
KWLO(AM) Waterloo IA
KWLS Pratt KS
KWMC Del Rio TX
KWMF(AM) Pleasanton TX
KWMG(AM) Auburn-Federal Way WA
KWMO Washington MO
KWMT Fort Dodge IA
KWNA Winnemucca NV
KWNC Quincy WA
KWNO Winona MN
KWNX(AM) Taylor TX
KWOA Worthington MN
KWOC Poplar Bluff MO
KWOF Waterloo IA
KWOK(AM) Hoquiam WA
KWON Bartlesville OK
KWOR Worland WY
KWOS(AM) Jefferson City MO
KWPC Muscatine IA
KWPM West Plains MO
KWRD Henderson TX
KWRE(AM) Warrenton MO
KWRF Warren AR
KWRM Corona CA
KWRN Apple Valley CA
KWRO Coquille OR
KWRT(AM) Boonville MO
KWRU(AM) Fresno CA

KWSH Wewoka OK
KWSL Sioux City IA
KWSN(AM) Sioux Falls SD
KWST(AM) El Centro CA
*KWSU Pullman WA
KWSW Eureka CA
KWSX(AM) Stockton CA
*KWTL(AM) Grand Forks ND
KWTO Springfield MO
KWTX Waco TX
KWUD(AM) Woodville TX
KWUF(AM) Pagosa Springs CO
KWVR Enterprise OR
KWWJ Baytown TX
KWWN(AM) Las Vegas NV
KWWX Wenatchee WA
KWXI Glenwood AR
KWXT Dardanelle AR
KWYN Wynne AR
KWYO Sheridan WY
KWYR(AM) Winner SD
KWYS West Yellowstone MT
KWYZ Everett WA
KXAM Mesa AZ
KXAR Hope AR
KXBX Lakeport CA
KXCA(AM) Lawton OK
KXEG(AM) Phoenix AZ
KXEL Waterloo IA
KXEN(AM) Saint Louis MO
KXEO Mexico MO
KXEQ Reno NV
KXEW South Tucson AZ
KXEX Fresno CA
KXGF Great Falls MT
KXGN Glendive MT
KXIC Iowa City IA
KXIT Dalhart TX
KXJK Forrest City AR
KXKS(AM) Albuquerque NM
KXL Portland OR
KXLE Ellensburg WA
KXLJ(AM) Juneau AK
KXLO Lewistown MT
KXLQ Indianola IA
KXLX(AM) Airway Heights WA
KXLY Spokane WA
KXMG(AM) Portland OR
KXMR Bismarck ND
KXMX(AM) Anaheim CA
KXNO(AM) Des Moines IA
KXNT North Las Vegas NV
KXO(AM) El Centro CA
KXOI Crane TX
KXOL Brigham City UT
KXOR(AM) Junction City OR
KXOX Sweetwater TX
KXPA Bellevue WA
*KXPD(AM) Tigard OR
KXPL(AM) El Paso TX
KXPN(AM) Kearney NE
KXPO Grafton ND
KXPS(AM) Thousand Palms CA
KXRA Alexandria MN
KXRB(AM) Sioux Falls SD
KXRE(AM) Manitou Springs CO
KXRO Aberdeen WA
KXSP(AM) Omaha NE
KXSS(AM) Waite Park MN
KXTA(AM) Centerville UT
KXTD Wagoner OK
KXTK(AM) Arroyo Grande CA
KXTL Butte MT
KXTO Reno NV
KXTR(AM) Kansas City KS
KXXA(AM) Conway AR
KXXT(AM) Tolleson AZ
KXXX Colby KS
KXYL(AM) Brownwood TX
KXYZ Houston TX
KXZZ Lake Charles LA
KYAA(AM) Soquel CA
KYAK Yakima WA
KYAL(AM) Sapulpa OK
KYBC(AM) Cottonwood AZ
KYCA(AM) Prescott AZ

KYCN Wheatland WY
KYCR Golden Valley MN
KYCY San Francisco CA
KYDZ(AM) Bellevue NE
KYES(AM) Rockville MN
KYET Williams AZ
*KYFR Shenandoah IA
KYHN(AM) Fort Smith AR
KYIZ Renton WA
KYKK(AM) Hobbs NM
KYKN Keizer OR
KYLS Fredericktown MO
KYLT Missoula MT
KYLW(AM) Lockwood MT
KYMN Northfield MN
KYMO East Prairie MO
KYND Cypress TX
KYNG(AM) Springdale AR
KYNN(AM) Cameron AZ
KYNO Fresno CA
KYNR(AM) Toppenish WA
KYNS(AM) San Luis Obispo CA
KYNT Yankton SD
KYOK(AM) Conroe TX
KYOO Bolivar MO
KYOS Merced CA
KYPA Los Angeles CA
KYRO Potosi MO
KYSM(AM) Mankato MN
KYST Texas City TX
KYTY(AM) Somerset TX
*KYUK Bethel AK
KYUL(AM) Scott City KS
KYUU Liberal KS
KYVA(AM) Gallup NM
KYW Philadelphia PA
KYXE(AM) Union Gap WA
KYYW(AM) Abilene TX
KYZS Tyler TX
KZDC San Antonio TX
KZEE Weatherford TX
KZER(AM) Santa Barbara CA
KZEY Tyler TX
KZGX(AM) Watertown MN
KZHN(AM) Paris TX
KZIM Cape Girardeau MO
KZIP Amarillo TX
KZIZ Sumner WA
KZMP(AM) University Park TX
KZMQ Greybull WY
KZMX Hot Springs SD
KZNE(AM) College Station TX
KZNG Hot Springs AR
KZNS(AM) Salt Lake City UT
KZNT(AM) Colorado Springs CO
KZNU(AM) Saint George UT
KZNX(AM) Creedmoor TX
KZOO Honolulu HI
KZPA Fort Yukon AK
KZQQ(AM) Abilene TX
KZRG(AM) Joplin MO
KZRK(AM) Canyon TX
KZSB(AM) Santa Barbara CA
KZSF San Jose CA
KZSJ(AM) San Martin CA
KZTD(AM) Cabot AR
KZTS(AM) Sunnyside WA
KZUE El Reno OK
KZXR Prosser WA
KZYM(AM) Joplin MO
KZZB Beaumont TX
KZZJ Rugby ND
KZZN Littlefield TX
KZZR(AM) Burns OR
KZZZ(AM) Bullhead City AZ
V6AH Pohnpei FM
*V6AI Yap FM
V6AK Truk FM
WAAM Ann Arbor MI
WAAV Leland NC
WAAX Gadsden AL
WABA Aguadilla PR
WABB Mobile AL
WABC(AM) New York NY
WABF(AM) Fairhope AL
WABG(AM) Greenwood MS

WABH Bath NY
WABI Bangor ME
WABJ Adrian MI
WABL Amite LA
WABN Abingdon VA
WABO Waynesboro MS
WABQ(AM) Painesville OH
WABV(AM) Abbeville SC
WABY(AM) Mechanicville NY
WACA Wheaton MD
WACB Taylorsville NC
WACC Hialeah FL
WACE Chicopee MA
WACK(AM) Newark NY
WACM West Springfield MA
WACQ(AM) Carrville AL
WACT Tuscaloosa AL
WACV Montgomery AL
WADA Shelby NC
WADB Asbury Park NJ
WADC Parkersburg WV
WADE Wadesboro NC
WADK Newport RI
WADM(AM) Decatur IN
WADO New York NY
WADR Remsen NY
WADS Ansonia CT
WADV Lebanon PA
WAEB Allentown PA
WAEC Atlanta GA
WAEW Crossville TN
WAEY Princeton WV
WAFC Clewiston FL
WAFS(AM) Atlanta GA
WAFZ Immokalee FL
WAGE Leesburg VA
WAGF Dothan AL
WAGG(AM) Birmingham AL
WAGL(AM) Lancaster SC
WAGN Menominee MI
WAGR Lumberton NC
WAGS Bishopville SC
WAGY(AM) Forest City NC
WAHT Clemson SC
WAIA(AM) Beaver Dam KY
WAIK Galesburg IL
WAIM Anderson SC
WAIN Columbia KY
WAIS Buchtel OH
WAIT(AM) Willow Springs IL
WAIZ(AM) Hickory NC
WAJD Gainesville FL
WAJL(AM) South Boston VA
WAJQ Alma GA
WAJR Morgantown WV
WAKE Valparaiso IN
WAKI McMinnville TN
WAKK(AM) McComb MS
WAKM Franklin TN
WAKO Lawrenceville IL
WAKR Akron OH
WAKV Otsego MI
WALD Walterboro SC
WALE Greenville RI
WALG Albany GA
WALH Mountain City GA
WALK(AM) East Patchogue NY
WALL Middletown NY
WALO(AM) Humacao PR
WALR Atlanta GA
WALT Meridian MS
WAMA Tampa FL
WAMB(AM) Nashville TN
*WAMC(AM) Albany NY
WAMD Aberdeen MD
WAME(AM) Statesville NC
WAMF(AM) Fulton NY
WAMG(AM) Dedham MA
WAMI Opp AL
WAML Laurel MS
WAMM Woodstock VA
WAMN Green Valley WV
WAMO(AM) Millvale PA
WAMT(AM) Pine Castle-Sky Lake FL
WAMV Amherst VA
WAMW(AM) Washington IN

WAMY Amory MS
WANA(AM) Anniston AL
WANB(AM) Waynesburg PA
WANG Havelock NC
WANI Opelika AL
WANO Pineville KY
WANR Warren OH
WANS Anderson SC
WANY Albany KY
WAOC(AM) Saint Augustine FL
WAOK Atlanta GA
WAOS(AM) Austell GA
WAOV Vincennes IN
WAPA San Juan PR
WAPF(AM) McComb MS
WAPI Birmingham AL
WAPZ Wetumpka AL
WAQE Rice Lake WI
WAQI Miami FL
WARE Ware MA
WARF(AM) Akron OH
WARK Hagerstown MD
WARL(AM) Attleboro MA
WARM Scranton PA
WARR Warrenton NC
WARU Peru IN
WARV Warwick RI
WASB Brockport NY
WASC Spartanburg SC
WASG Atmore AL
WASK Lafayette IN
WASN(AM) Youngstown OH
WASO Covington LA
WASP Brownsville PA
WASR Wolfeboro NH
WATA Boone NC
WATB Decatur GA
WATH Athens OH
WATK Antigo WI
WATN Watertown NY
WATO Oak Ridge TN
WATR(AM) Waterbury CT
WATS Sayre PA
WATT Cadillac MI
WATV(AM) Birmingham AL
WATW Ashland WI
WATX Algood TN
WATZ Alpena MI
WAUB Auburn NY
WAUC Wauchula FL
WAUD Auburn AL
WAUG New Hope NC
WAUK Waukesha WI
WAUR Sandwich IL
WAVA(AM) Arlington VA
WAVG(AM) Jeffersonville IN
WAVL Apollo PA
WAVN Southaven MS
WAVO Rock Hill SC
WAVS Davie FL
WAVU(AM) Albertville AL
WAVZ New Haven CT
WAWK Kendallville IN
WAXO Lewisburg TN
WAXY South Miami FL
WAYE Birmingham AL
WAYN Rockingham NC
*WAYR Orange Park FL
WAYS(AM) Macon GA
WAYY(AM) Eau Claire WI
WAZL Hazleton PA
WAZN(AM) Watertown MA
WAZS(AM) Summerville SC
WAZX Smyrna GA
WAZZ(AM) Fayetteville NC
*WBAA(AM) West Lafayette IN
WBAC(AM) Cleveland TN
WBAE Portland ME
WBAF Barnesville GA
WBAG Burlington-Graham NC
WBAJ Blythwood SC
WBAL Baltimore MD
WBAP Fort Worth TX
WBAT Marion IN
WBAX Wilkes-Barre PA
WBBD Wheeling WV

WBBF(AM) Buffalo NY
WBBK Blakely GA
WBBL Grand Rapids MI
WBBM Chicago IL
WBBP Memphis TN
WBBR New York NY
WBBT Lyons GA
WBBW Youngstown OH
WBBX Kingston TN
WBBZ Ponca City OK
WBCB Levittown-Fairless Hills PA
WBCE Wickliffe KY
WBCF(AM) Florence AL
WBCH Hastings MI
WBCK Battle Creek MI
WBCO(AM) Bucyrus OH
WBCP Urbana IL
WBCR Alcoa TN
WBCU Union SC
WBDY Bluefield VA
WBEC Pittsfield MA
WBEJ(AM) Elizabethton TN
WBEN Buffalo NY
WBES(AM) Dunbar WV
WBEV Beaver Dam WI
WBEX Chillicothe OH
WBEY(AM) Pocomoke City MD
WBFC Stanton KY
WBFD(AM) Bedford PA
WBFJ Winston-Salem NC
WBFN(AM) Battle Creek MI
WBGC Chipley FL
WBGG(AM) Pittsburgh PA
WBGN(AM) Bowling Green KY
WBGR Baltimore MD
WBGS Point Pleasant WV
WBGX(AM) Harvey IL
WBGZ Alton IL
WBHB Fitzgerald GA
WBHF(AM) Cartersville GA
WBHN Bryson City NC
WBHP Huntsville AL
WBHR Sauk Rapids MN
WBHV(AM) Somerset PA
WBHY Mobile AL
WBIB Centreville AL
WBIC Royston GA
WBIG Aurora IL
WBIL Tuskegee AL
WBIN Benton TN
WBIP Booneville MS
WBIS(AM) Annapolis MD
WBIW Bedford IN
WBIX(AM) Natick MA
WBIZ Eau Claire WI
WBKV West Bend WI
WBKZ Jefferson GA
WBLA Elizabethtown NC
WBLC Lenoir City TN
WBLF Bellefonte PA
WBLJ(AM) Dalton GA
WBLL Bellefontaine OH
WBLO(AM) Thomasville NC
WBLR Batesburg SC
WBLT(AM) Bedford VA
WBMC McMinnville TN
WBMD Baltimore MD
WBMJ San Juan PR
WBML Macon GA
WBMQ Savannah GA
WBNC Conway NH
WBNL(AM) Boonville IN
WBNR Beacon NY
WBNS Columbus OH
WBNW Concord MA
WBOB(AM) Jacksonville FL
WBOG(AM) Tomah WI
WBOK New Orleans LA
WBOL Bolivar TN
WBOW(AM) Terre Haute IN
WBOX Bogalusa LA
WBPZ Lock Haven PA
WBQN(AM) Barceloneta-Manati PR
WBRD Palmetto FL
WBRG Lynchburg VA
WBRI(AM) Indianapolis IN
WBRK Pittsfield MA
WBRM Marion NC
WBRN Big Rapids MI
WBRT Bardstown KY
WBRV Boonville NY
WBRY Woodbury TN
WBSA Boaz AL
WBSC Bennettsville SC
WBSG(AM) Lajas PR
WBSL Bay St. Louis MS
WBSM New Bedford MA
WBSR Pensacola FL
WBT Charlotte NC
WBTA Batavia NY
WBTC Uhrichsville OH
WBTE Windsor NC
WBTG Sheffield AL
WBTH Williamson WV
WBTK(AM) Richmond VA
WBTM Danville VA
WBTN Bennington VT
WBTO Linton IN
WBTX Broadway-Timberville VA
WBUC Buckhannon WV
WBUD Trenton NJ
*WBUR(AM) West Yarmouth MA
WBUT Butler PA
WBVA(AM) Bayside VA
WBVP(AM) Beaver Falls PA
WBWL Jacksonville FL
WBXR(AM) Hazel Green AL
WBYE Calera AL
WBYN(AM) Lehighton PA
WBYS Canton IL
WBYU New Orleans LA
WBZ Boston MA
WBZI Xenia OH
WBZK York SC
WBZQ Huntington IN
WBZT(AM) West Palm Beach FL
WBZU(AM) Scranton PA
WCAB(AM) Rutherfordton NC
WCAM Camden SC
WCAO Baltimore MD
WCAP Lowell MA
WCAR Livonia MI
WCAT(AM) Burlington VT
WCAZ Carthage IL
WCBA Corning NY
WCBC Cumberland MD
WCBG(AM) Waynesboro PA
WCBL Benton KY
WCBM Baltimore MD
WCBQ Oxford NC
WCBR Richmond KY
WCBS New York NY
WCBT Roanoke Rapids NC
WCBX(AM) Bassett VA
WCBY Cheboygan MI
WCCC(AM) West Hartford CT
WCCD(AM) Parma OH
WCCF(AM) Punta Gorda FL
WCCM(AM) Salem NH
WCCN Neillsville WI
WCCO Minneapolis MN
WCCS Homer City PA
WCCW(AM) Traverse City MI
WCCY Houghton MI
WCDL(AM) Carbondale PA
WCDO Sidney NY
WCDS(AM) Glasgow KY
WCDT Winchester TN
WCEC(AM) Haverhill MA
WCED DuBois PA
WCEH Hawkinsville GA
WCEM Cambridge MD
WCEO(AM) Columbia SC
WCER Canton OH
WCEV Cicero IL
WCFI(AM) Ocala FL
WCFJ Chicago Heights IL
WCFO(AM) East Point GA
WCFR(AM) Springfield VT
WCGA Woodbine GA
WCGB Juana Diaz PR
WCGC Belmont NC
WCGL Jacksonville FL
WCGO Chicago Heights IL
WCGR Canandaigua NY
WCGW Nicholasville KY
WCHA Chambersburg PA
WCHB Taylor MI
WCHE West Chester PA
WCHI Chillicothe OH
WCHJ Brookhaven MS
WCHK Canton GA
WCHL Chapel Hill NC
WCHM Clarkesville GA
WCHN Norwich NY
WCHO(AM) Washington Court House OH
WCHP Champlain NY
WCHQ(AM) Quebradillas PR
WCHR(AM) Flemington NJ
WCHS Charleston WV
WCHT Escanaba MI
WCHV Charlottesville VA
WCIE Spring Lake NC
WCIL Carbondale IL
WCIN Cincinnati OH
WCIS Morganton NC
WCJU Columbia MS
WCJW(AM) Warsaw NY
WCKA(AM) Jacksonville AL
WCKB Dunn NC
WCKI Greer SC
WCKL Catskill NY
*WCKW(AM) Garyville LA
WCKY(AM) Cincinnati OH
WCLA(AM) Claxton GA
WCLB(AM) Sheboygan WI
WCLC Jamestown TN
WCLD Cleveland MS
WCLE Cleveland TN
WCLG Morgantown WV
WCLM Highland Springs VA
WCLN(AM) Clinton NC
WCLO Janesville WI
WCLT(AM) Newark OH
WCLU(AM) Glasgow KY
WCLW Eden NC
WCLY Raleigh NC
WCMA(AM) Daleville AL
WCMC(AM) Wildwood NJ
WCMD(AM) Cumberland MD
WCMI Ashland KY
WCMN Arecibo PR
WCMP Pine City MN
WCMS(AM) Newport News VA
WCMT(AM) Martin TN
WCMX Leominster MA
WCMY(AM) Ottawa IL
WCNC(AM) Elizabeth City NC
WCND Shelbyville KY
WCNN North Atlanta GA
WCNS Latrobe PA
WCNW Fairfield OH
WCNX(AM) Hope Valley RI
*WCNZ(AM) Marco Island FL
WCOA Pensacola FL
WCOC(AM) Dora AL
WCOG Greensboro NC
WCOH Newnan GA
WCOJ(AM) Coatesville PA
WCOK Sparta NC
WCON Cornelia GA
WCOR(AM) Lebanon TN
WCOS Columbia SC
WCPA Clearfield PA
WCPC Houston MS
WCPH Etowah TN
WCPK Chesapeake VA
WCPM Cumberland KY
WCPR Coamo PR
WCPS(AM) Tarboro NC
WCPT(AM) Crystal Lake IL
WCQV(AM) Moneta VA
WCRA Effingham IL
WCRE Cheraw SC
WCRK(AM) Morristown TN
WCRL Oneonta AL
WCRM Fort Myers FL
WCRN Worcester MA
WCRO Johnstown PA
WCRS Greenwood SC
WCRT(AM) Donelson TN
WCRV Collierville TN
WCSA Ripley MS
WCSI(AM) Columbus IN
WCSJ(AM) Morris IL
WCSL Cherryville NC
WCSM Celina OH
WCSR Hillsdale MI
WCSS Amsterdam NY
WCST Berkeley Springs WV
WCSV(AM) Crossville TN
WCSW Shell Lake WI
WCSZ(AM) Sans Souci SC
WCTA Alamo TN
WCTC(AM) New Brunswick NJ
*WCTF Vernon CT
WCTN Potomac-Cabin John MD
WCTR(AM) Chestertown MD
WCTS Maplewood MN
WCTT Corbin KY
WCUB(AM) Two Rivers WI
*WCUE Cuyahoga Falls OH
WCUG Cuthbert GA
WCUM Bridgeport CT
WCVA Culpeper VA
WCVC Tallahassee FL
WCVG Covington KY
WCVL Crawfordsville IN
WCVP Murphy NC
WCVX(AM) Cincinnati OH
WCWA Toledo OH
WCXH(AM) Monticello ME
WCXI(AM) Fenton MI
WCXJ(AM) Kearsarge PA
WCXN Claremont NC
WCYN Cynthiana KY
WCZZ(AM) Greenwood SC
WDAD Indiana PA
WDAE(AM) Saint Petersburg FL
WDAK Columbus GA
WDAL Dalton GA
WDAN(AM) Danville IL
WDAO Dayton OH
WDAP Huntingdon TN
WDAY Fargo ND
WDBC Escanaba MI
WDBL(AM) Springfield TN
WDBO Orlando FL
WDBQ Dubuque IA
WDBZ(AM) Cincinnati OH
WDCD(AM) Albany NY
WDCF Dade City FL
WDCO(AM) Cochran GA
WDCR(AM) Hanover NH
WDCT Fairfax VA
WDCY Douglasville GA
WDDD Johnston City IL
WDDO Macon GA
WDDV(AM) Venice FL
WDDY(AM) Albany NY
WDDZ(AM) Pawtucket RI
WDEA Ellsworth ME
WDEB Jamestown TN
WDEF Chattanooga TN
WDEH Sweetwater TN
WDEL Wilmington DE
WDEO(AM) Ypsilanti MI
WDEP(AM) Ponce PR
WDER(AM) Derry NH
WDEV Waterbury VT
WDEX Monroe NC
WDFB Junction City KY
WDFN Detroit MI
WDGR Dahlonega GA
WDGY(AM) Hudson WI
WDHP Frederiksted VI
WDIA Memphis TN
WDIC(AM) Clinchco VA
WDIG Steubenville OH
WDIS Norfolk MA
WDIZ Panama City FL
WDJA(AM) Delray Beach FL
WDJL Huntsville AL
WDJO(AM) Florence KY
WDJS Mount Olive NC
WDJZ Bridgeport CT
WDKD Kingstree SC
WDKN Dickson TN
WDLA Walton NY
WDLB Marshfield WI
WDLC Port Jervis NY
WDLK Dadeville AL
*WDLM East Moline IL
WDLR(AM) Delaware OH
WDLS(AM) Wisconsin Dells WI
WDLT Fairhope AL
WDLW Lorain OH
WDLX Washington NC
WDMG Douglas GA
WDMJ Marquette MI
WDMN(AM) Rossford OH
WDMP(AM) Dodgeville WI
WDMV(AM) Walkersville MD
WDNC(AM) Durham NC
WDND(AM) South Bend IN
WDNE Elkins WV
WDNG Anniston AL
WDNT(AM) Dayton TN
WDNY(AM) Dansville NY
WDOC Prestonsburg KY
WDOD Chattanooga TN
WDOE Dunkirk NY
WDOG Allendale SC
WDOR Sturgeon Bay WI
WDOS Oneonta NY
WDOV Dover DE
WDOW Dowagiac MI
WDOX(AM) Raleigh NC
WDPC Dallas GA
WDPN Alliance OH
WDPT(AM) Decatur AL
WDQN Du Quoin IL
WDRC(AM) Hartford CT
WDRD(AM) Newburg KY
WDRF(AM) Woodruff SC
WDRJ(AM) Inkster MI
WDRU(AM) Wake Forest NC
WDSC Dillon SC
WDSK Cleveland MS
WDSL Mocksville NC
WDSM Superior WI
WDSP(AM) De Funiak Springs FL
WDSR Lake City FL
WDSS(AM) Ada MI
WDTK(AM) Detroit MI
WDTM Selmer TN
WDTW(AM) Dearborn MI
WDUF Duffield VA
WDUN Gainesville GA
WDUR Durham NC
WDUX Waupaca WI
WDUZ(AM) Green Bay WI
WDVA Danville VA
WDVH(AM) Gainesville FL
*WDVM(AM) Eau Claire WI
WDWD Atlanta GA
WDWR(AM) Pensacola FL
WDWS(AM) Champaign IL
WDXE Lawrenceburg TN
WDXI Jackson TN
WDXL Lexington TN
WDXR Paducah KY
WDXY Sumter SC
WDYT(AM) Kings Mountain NC
WDYZ(AM) Orlando FL
WDZ(AM) Decatur IL
WDZK Bloomfield CT
WDZY Colonial Heights VA
WEAC Gaffney SC
WEAE Pittsburgh PA
WEAF(AM) Camden SC
WEAL Greensboro NC
WEAM Columbus GA
WEAQ Chippewa Falls WI
WEAV Plattsburgh NY
WEBC Duluth MN
WEBJ Brewton AL
WEBO Owego NY
WEBQ Harrisburg IL

WEBS(AM) Calhoun GA
WEBY(AM) Milton FL
WECK Cheektowaga NY
WECM Milton FL
WECO Wartburg TN
WECR Newland NC
WECU(AM) Winterville NC
WECZ Punxsutawney PA
WEDI(AM) Eaton OH
WEDO McKeesport PA
WEEB(AM) Southern Pines NC
WEED(AM) Rocky Mount NC
WEEF Highland Park IL
WEEI Boston MA
WEEL(AM) Dothan AL
WEEN(AM) Lafayette TN
WEEO(AM) Shippensburg PA
WEEU Reading PA
WEEX Easton PA
WEEZ Laurel MS
WEFL(AM) Tequesta FL
WEGA Vega Baja PR
WEGG Rose Hill NC
WEGO Concord NC
WEGP Presque Isle ME
WEHH Elmira Heights-Horseheads NY
WEIC Charleston IL
WEIM(AM) Fitchburg MA
WEIR Weirton WV
WEIS Centre AL
WEJL Scranton PA
WEKB(AM) Elkhorn City KY
WEKC Williamsburg KY
WEKG Jackson KY
WEKO(AM) Morovis PR
WEKR Fayetteville TN
WEKT Elkton KY
WEKY Richmond KY
WEKZ Monroe WI
WELA(AM) Welch WV
WELB Elba AL
WELC Welch WV
WELD(AM) Fisher WV
WELE Ormond Beach FL
WELG(AM) Ellenville NY
WELI New Haven CT
WELM Elmira NY
WELO Tupelo MS
WELP(AM) Easley SC
WELR Roanoke AL
WELS Kinston NC
WELW Willoughby-Eastlake OH
WELY Ely MN
WELZ Belzoni MS
WEMB Erwin TN
WEMD(AM) Easton MD
WEMG(AM) Camden NJ
WEMJ Laconia NH
WEMM(AM) Huntington WV
WEMR Tunkhannock PA
WENA Yauco PR
WENC Whiteville NC
WENE(AM) Endicott NY
WENG Englewood FL
WENI(AM) Corning NY
WENJ(AM) Atlantic City NJ
WENK Union City TN
WENO Nashville TN
WENR Englewood TN
WENT Gloversville NY
WENU(AM) South Glens Falls NY
WENY Elmira NY
WEOA Evansville IN
WEOK Poughkeepsie NY
WEOL Elyria OH
WEPG South Pittsburg TN
WEPM Martinsburg WV
WEPN(AM) New York NY
WERC Birmingham AL
WERE(AM) Cleveland Heights OH
WERH Hamilton AL
WERL Eagle River WI
WERT Van Wert OH
WESB(AM) Bradford PA
WESO Southbridge MA
WESR Onley-Onancock VA

WEST Easton PA
WESX Salem MA
WESY Leland MS
WETB Johnson City TN
WETC Wendell-Zebulon NC
WETR(AM) Knoxville TN
WETZ New Martinsville WV
WEUP(AM) Huntsville AL
WEUS(AM) Orlovista FL
WEUV(AM) Moulton AL
WEVA(AM) Emporia VA
WEVR River Falls WI
WEW Saint Louis MO
WEWC(AM) Callahan FL
WEWO Laurinburg NC
WEXL Royal Oak MI
WEXS Patillas PR
WEXY Wilton Manors FL
WEZE Boston MA
WEZJ Williamsburg KY
WEZR(AM) Lewiston ME
WEZS Laconia NH
WEZZ(AM) Monroeville AL
WFAB Ceiba PR
WFAD Middlebury VT
WFAI(AM) Salem NJ
WFAM(AM) Augusta GA
WFAN New York NY
WFAS(AM) White Plains NY
WFAU Gardiner ME
WFAW Fort Atkinson WI
WFAX Falls Church VA
WFAY(AM) Fayetteville NC
WFBG Altoona PA
WFBL(AM) Syracuse NY
WFBR(AM) Glen Burnie MD
WFBS(AM) Berwick PA
WFCL Clintonville WI
*WFCM Smyrna TN
WFCV Fort Wayne IN
WFDF(AM) Farmington Hills MI
WFDL(AM) Waupun WI
WFDR Manchester GA
WFEA Manchester NH
WFEB Sylacauga AL
WFED(AM) Silver Spring MD
WFFF Columbia MS
WFFG(AM) Marathon FL
WFFX(AM) East St. Louis IL
WFGI(AM) Charleroi PA
WFGL Fitchburg MA
WFGM(AM) Sandy Springs GA
WFGN Gaffney SC
WFGO(AM) Orono ME
WFGW Black Mountain NC
WFHG(AM) Bristol VA
WFHK Pell City AL
WFHR Wisconsin Rapids WI
WFHT(AM) Avon Park FL
WFIA Louisville KY
WFIC(AM) Collinsville VA
WFIF Milford CT
WFIL(AM) Philadelphia PA
WFIN Findlay OH
WFIR(AM) Roanoke VA
WFIS Fountain Inn SC
WFIW(AM) Fairfield IL
*WFKJ Cashtown PA
WFKN Franklin KY
WFKY Frankfort KY
WFLA(AM) Tampa FL
WFLE Flemingsburg KY
WFLF(AM) Pine Hills FL
WFLI Lookout Mountain TN
WFLL(AM) Fort Lauderdale FL
WFLN(AM) Arcadia FL
WFLO Farmville VA
WFLR Dundee NY
WFLT Flint MI
WFLW Monticello KY
WFMB(AM) Springfield IL
WFMC Goldsboro NC
WFMD(AM) Frederick MD
WFMH(AM) Cullman AL
WFMO Fairmont NC
WFMW Madisonville KY

WFNA(AM) Charlotte NC
WFNC Fayetteville NC
WFNN(AM) Erie PA
WFNO Norco LA
WFNR Blacksburg VA
WFNS(AM) Blackshear GA
WFNT Flint MI
WFNW(AM) Naugatuck CT
WFNY(AM) Gloversville NY
WFNZ Charlotte NC
WFOB Fostoria OH
WFOM Marietta GA
WFOR Hattiesburg MS
WFOY(AM) Saint Augustine FL
WFPA(AM) Fort Payne AL
WFPB Orleans MA
WFPR Hammond LA
WFRA Franklin PA
WFRB Frostburg MD
*WFRF(AM) Tallahassee FL
WFRL Freeport IL
WFRM Coudersport PA
WFRN Elkhart IN
WFRX West Frankfort IL
WFSC Franklin NC
WFSH(AM) Valparaiso-Niceville FL
WFSP Kingwood WV
WFSR Harlan KY
*WFST(AM) Caribou ME
WFTD Marietta GA
WFTG London KY
WFTH Richmond VA
WFTL(AM) West Palm Beach FL
WFTM Maysville KY
WFTN Franklin NH
WFTR Front Royal VA
WFTU(AM) Riverhead NY
WFTW Fort Walton Beach FL
WFUL(AM) Fulton KY
WFUN Ashtabula OH
WFUR Grand Rapids MI
WFVA(AM) Fredericksburg VA
WFWL(AM) Camden TN
WFXH(AM) Hilton Head Island SC
WFXJ(AM) Jacksonville FL
WFXN(AM) Moline IL
WFXY Middlesboro KY
WFYC(AM) Alma MI
WFYL(AM) King of Prussia PA
WGAB Newburgh IN
WGAC Augusta GA
WGAD Gadsden AL
WGAI(AM) Elizabeth City NC
WGAM(AM) Nashua NH
WGAN(AM) Portland ME
WGAP Maryville TN
*WGAS South Gastonia NC
WGAT Gate City VA
WGAU(AM) Athens GA
WGAW Gardner MA
WGBB Freeport NY
WGBF Evansville IN
WGBN New Kensington PA
WGBR Goldsboro NC
WGBW(AM) Two Rivers WI
WGCD Chester SC
WGCH(AM) Greenwich CT
WGCL Bloomington IN
WGCM Gulfport MS
WGCR Brevard NC
WGCV(AM) Cayce SC
WGDL Lares PR
WGDN Gladwin MI
WGEA Geneva AL
WGEE(AM) Superior WI
WGEM Quincy IL
WGEN Geneseo IL
WGES(AM) Saint Petersburg FL
WGET Gettysburg PA
WGEZ Beloit WI
WGFA(AM) Watseka IL
WGFC Floyd VA
WGFP Webster MA
WGFS Covington GA
WGFT(AM) Campbell OH
WGFY Charlotte NC

WGGA Gainesville GA
WGGG Gainesville FL
WGGH Marion IL
WGGM Chester VA
WGGO Salamanca NY
WGHB Farmville NC
WGHC(AM) Clayton GA
WGHN Grand Haven MI
WGHQ Kingston NY
WGHT Pompton Lakes NJ
WGIG(AM) Brunswick GA
WGIL Galesburg IL
WGIN Rochester NH
WGIP Exeter NH
WGIR Manchester NH
WGIT(AM) Canovanas PR
WGIV(AM) Pineville NC
WGJK(AM) Rome GA
WGKA(AM) Atlanta GA
WGL Fort Wayne IN
WGLB(AM) Elm Grove WI
WGLD(AM) Red Lion PA
WGLL Auburn IN
WGLR Lancaster WI
WGMA Spindale NC
WGMI Bremen GA
WGML Hinesville GA
WGMN Roanoke VA
WGN(AM) Chicago IL
WGNC Gastonia NC
*WGNR Anderson IN
WGNS Murfreesboro TN
WGNU(AM) Granite City IL
WGNY Newburgh NY
WGNZ Fairborn OH
WGOC(AM) Kingsport TN
WGOD Charlotte Amalie VI
WGOH Grayson KY
WGOK Mobile AL
WGOL Russellville AL
WGOP(AM) Pocomoke City MD
WGOS High Point NC
WGOV(AM) Valdosta GA
WGOW Chattanooga TN
WGPA(AM) Bethlehem PA
WGPC Albany GA
WGPL Portsmouth VA
WGR Buffalo NY
WGRA Cairo GA
WGRB(AM) Chicago IL
WGRK(AM) Greensburg KY
WGRM Greenwood MS
WGRO Lake City FL
WGRP Greenville PA
WGRV Greeneville TN
WGRY Grayling MI
WGSB Mebane NC
WGSF(AM) Memphis TN
WGSO New Orleans LA
WGSP Charlotte NC
WGST Atlanta GA
WGSV Guntersville AL
WGTA Summerville GA
WGTH Richlands VA
WGTJ(AM) Murrayville GA
WGTK(AM) Louisville KY
WGTM Wilson NC
WGTN Georgetown SC
WGTO Cassopolis MI
WGUL(AM) Dunedin FL
WGUN Atlanta GA
WGUS(AM) Augusta GA
WGVA Geneva NY
WGVL Greenville SC
WGVM Greenville MS
WGVS(AM) Muskegon MI
*WGVU Kentwood MI
WGWM London KY
WGY Schenectady NY
WGYM(AM) Hammonton NJ
WGYV Greenville AL
*WHA Madison WI
WHAG Halfway MD
WHAK Rogers City MI
WHAL(AM) Phenix City AL
WHAM Rochester NY

WHAN Ashland VA
WHAP Hopewell VA
WHAS Louisville KY
WHAT Philadelphia PA
WHAW Weston WV
WHAZ Troy NY
WHB Kansas City MO
WHBB Selma AL
WHBC Canton OH
WHBG Harrisonburg VA
WHBK Marshall NC
WHBL(AM) Sheboygan WI
WHBN(AM) Harrodsburg KY
WHBO(AM) Dunedin FL
WHBQ Memphis TN
WHBS(AM) Moultrie GA
WHBT(AM) Tallahassee FL
WHBU Anderson IN
WHBY(AM) Kimberly WI
WHCG Metter GA
WHCO(AM) Sparta IL
WHCU Ithaca NY
WHDD(AM) Sharon CT
WHDL Olean NY
WHDM McKenzie TN
WHEE Martinsville VA
WHEN Syracuse NY
WHEO Stuart VA
WHEP Foley AL
WHEW(AM) Franklin TN
WHFA(AM) Poynette WI
WHFB(AM) Benton Harbor-St. Joseph MI
WHGB(AM) Marion VA
WHGG(AM) Kingsport TN
WHGH Thomasville GA
WHGS(AM) Hampton SC
WHGT(AM) Chambersburg PA
WHHO Hornell NY
WHHV Hillsville VA
WHIC(AM) Rochester NY
WHIE Griffin GA
WHIM Apopka FL
WHIN Gallatin TN
WHIO Dayton OH
WHIP Mooresville NC
WHIR(AM) Danville KY
WHIS Bluefield WV
WHIT(AM) Hudsonville MI
WHIY(AM) Huntsville AL
WHIZ Zanesville OH
WHJB(AM) Bedford PA
WHJC Matewan WV
WHJJ Providence RI
WHK(AM) Cleveland OH
WHKP(AM) Hendersonville NC
WHKT Portsmouth VA
WHKW(AM) Cleveland OH
WHKY(AM) Hickory NC
WHKZ(AM) Warren OH
WHLD Niagara Falls NY
WHLI Hempstead NY
WHLM(AM) Bloomsburg PA
WHLN Harlan KY
WHLO Akron OH
WHLS Port Huron MI
WHLX(AM) Marine City MI
WHLY(AM) South Bend IN
WHMA Anniston AL
WHMP Northampton MA
WHMQ(AM) Greenfield MA
WHNC Henderson NC
WHNK(AM) Parkersburg WV
WHNP(AM) East Longmeadow MA
WHNR Cypress Gardens FL
WHNY McComb MS
WHNZ(AM) Tampa FL
WHO(AM) Des Moines IA
WHOA(AM) Saraland AL
WHOC Philadelphia MS
WHOG Hobson City AL
WHOL Allentown PA
WHON Centerville IN
WHOO(AM) Kissimmee FL
WHOP(AM) Hopkinsville KY
WHOS Decatur AL

WHOW Clinton IL
WHOY Salinas PR
WHP Harrisburg PA
WHPY Clayton NC
WHRY Hurley WI
WHSC Hartsville SC
WHSM Hayward WI
WHSR Pompano Beach FL
WHSY(AM) Hattiesburg MS
WHTB Fall River MA
WHTC(AM) Holland MI
WHTG Eatontown NJ
WHTH Heath OH
WHTK Rochester NY
WHUB Cookeville TN
WHUC Hudson NY
WHUN Huntingdon PA
WHVN Charlotte NC
WHVO(AM) Hopkinsville KY
WHVR Hanover PA
WHVW Hyde Park NY
WHWH Princeton NJ
WHYL Carlisle PA
WHYM(AM) Lake City SC
WHYN Springfield MA
WIAC San Juan PR
WIAM Williamston NC
WIAN Ishpeming MI
WIBA Madison WI
WIBC(AM) Indianapolis IN
WIBG Ocean City NJ
WIBH Anna IL
WIBM Jackson MI
WIBR Baton Rouge LA
WIBS Guayama PR
WIBW Topeka KS
WIBX Utica NY
WICC Bridgeport CT
WICH Norwich CT
WICK Scranton PA
WICO(AM) Salisbury MD
WICY Malone NY
WIDA Carolina PR
WIDG Saint Ignace MI
WIDS Russell Springs KY
WIDU(AM) Fayetteville NC
WIEL Elizabethtown KY
WIEZ Lewistown PA
WIFA(AM) Knoxville TN
WIFE(AM) Connersville IN
WIFI Florence NJ
WIGG Wiggins MS
WIGM Medford WI
WIGN(AM) Bristol TN
WIGO(AM) Morrow GA
*WIHM Taylorville IL
WIIN Ridgeland MS
WIJD(AM) Prichard AL
WIJR(AM) Highland IL
WIKB Iron River MI
WIKC(AM) Bogalusa LA
WIKE Newport VT
WIL(AM) Saint Louis MO
WILA Danville VA
WILB(AM) Canton OH
WILC Laurel MD
WILD Boston MA
WILE Cambridge OH
WILI(AM) Willimantic CT
WILK Wilkes-Barre PA
*WILL(AM) Urbana IL
WILM Wilmington DE
WILO Frankfort IN
WILS Lansing MI
WILY Centralia IL
WIMA Lima OH
WIMG Ewing NJ
WIMO Winder GA
WIMS Michigan City IN
WINA Charlottesville VA
WINC Winchester VA
WIND Chicago IL
WINE Brookfield CT
WING Dayton OH
WINI(AM) Murphysboro IL
WINK(AM) Fort Myers FL

WINR Binghamton NY
WINS New York NY
WINT(AM) Melbourne FL
WINU(AM) Shelbyville IL
WINV(AM) Beverly Hills FL
WINW Canton OH
WINY Putnam CT
WINZ(AM) Miami FL
WIOD Miami FL
WIOI New Boston OH
WION(AM) Ionia MI
WIOO Carlisle PA
WIOS Tawas City MI
WIOU Kokomo IN
WIOV Reading PA
WIOZ Pinehurst NC
WIP Philadelphia PA
WIPC(AM) Lake Wales FL
*WIPR(AM) San Juan PR
WIPS Ticonderoga NY
WIQB(AM) Conway SC
WIQR Prattville AL
WIRA Fort Pierce FL
WIRB(AM) Level Plains AL
WIRD Lake Placid NY
WIRJ Humboldt TN
WIRL(AM) Peoria IL
WIRO Ironton OH
WIRV Irvine KY
WIRY Plattsburgh NY
WISA Isabela PR
WISE Asheville NC
WISK Americus GA
WISL Shamokin PA
WISN Milwaukee WI
WISO Ponce PR
WISP Doylestown PA
WISR Butler PA
WISS Berlin WI
WIST(AM) New Orleans LA
WISW Columbia SC
WITA Knoxville TN
WITK(AM) Pittston PA
WITS(AM) Sebring FL
WITY Danville IL
WITZ Jasper IN
WIVV Vieques PR
WIWA(AM) Saint Cloud FL
WIWS Beckley WV
WIXC(AM) Titusville FL
WIXE(AM) Monroe NC
WIXI(AM) Jasper AL
WIXK(AM) New Richmond WI
WIXN Dixon IL
WIXT(AM) Little Falls NY
WIYD Palatka FL
WIZE Springfield OH
WIZK(AM) Bay Springs MS
WIZM La Crosse WI
WIZR Johnstown NY
WIZS Henderson NC
WIZZ(AM) Greenfield MA
WJAE Westbrook ME
WJAG Norfolk NE
WJAK Jackson TN
WJAS Pittsburgh PA
WJAT Swainsboro GA
WJAW(AM) Saint Marys WV
WJAX Jacksonville FL
WJAY Mullins SC
WJBB Haleyville AL
WJBC Bloomington IL
WJBD Salem IL
WJBI Batesville MS
WJBM Jerseyville IL
WJBO Baton Rouge LA
WJBS Holly Hill SC
WJBW(AM) Jupiter FL
WJBY(AM) Rainbow City AL
WJCM(AM) Sebring FL
WJCP(AM) North Vernon IN
WJCV Jacksonville NC
WJCW Johnson City TN
WJDA Quincy MA
WJDB Thomasville AL
WJDJ(AM) Hartsville SC

WJDM Elizabeth NJ
WJDX Jackson MS
WJDY(AM) Salisbury MD
WJEH Gallipolis OH
WJEJ Hagerstown MD
WJEM(AM) Valdosta GA
WJEP Ochlocknee GA
WJER Dover-New Philadelphia OH
WJES(AM) Saluda SC
WJET(AM) Erie PA
WJFA(AM) Hilliard FL
WJFC Jefferson City TN
WJFJ Tryon NC
WJFK Baltimore MD
WJGK(AM) Highland NY
WJHX(AM) Lexington AL
WJIB Cambridge MA
WJIG Tullahoma TN
WJIL Jacksonville IL
WJIM(AM) Lansing MI
WJIT Sabana PR
WJJB(AM) Brunswick ME
WJJC Commerce GA
WJJG Elmhurst IL
WJJL Niagara Falls NY
WJJM Lewisburg TN
WJJQ Tomahawk WI
WJJT Jellico TN
WJKB(AM) Moncks Corner SC
*WJKN(AM) Jackson MI
WJKY Jamestown KY
WJLD Fairfield AL
WJLE Smithville TN
WJLG Savannah GA
WJLS Beckley WV
WJMC Rice Lake WI
WJML(AM) Petoskey MI
WJMO(AM) Cleveland OH
WJMP Kent OH
WJMS Ironwood MI
WJMT Merrill WI
WJMX Florence SC
WJNC Jacksonville NC
WJNL(AM) Kingsley MI
WJNO(AM) West Palm Beach FL
WJNT Pearl MS
WJNX(AM) Fort Pierce FL
WJNZ(AM) Kentwood MI
WJOB Hammond IN
WJOC Chattanooga TN
WJOE(AM) Orange-Athol MA
WJOI Norfolk VA
WJOK Kaukauna WI
WJOL Joliet IL
WJON Saint Cloud MN
WJOT Wabash IN
WJOY(AM) Burlington VT
WJPA Washington PA
WJPF Herrin IL
WJPI(AM) Plymouth NC
WJQI(AM) Fort Campbell KY
WJR Detroit MI
WJRD(AM) Tuscaloosa AL
WJRI Lenoir NC
WJRM Troy NC
WJSA Jersey Shore PA
WJSB Crestview FL
WJSM Martinsburg PA
WJSS(AM) Havre de Grace MD
WJST(AM) New Castle PA
WJTB North Ridgeville OH
WJTH Calhoun GA
WJTI(AM) Racine WI
WJTN Jamestown NY
WJTO Bath ME
WJUB(AM) Plymouth WI
WJUN Mexico PA
WJUS Marion AL
WJWB(AM) Gibsonia FL
WJWF(AM) Columbus MS
WJWK Seaford DE
WJWL Georgetown DE
WJXL(AM) Jacksonville Beach FL
WJYI Milwaukee WI
WJYK(AM) Chase City VA
WJYM Bowling Green OH

WJYP(AM) Saint Albans WV
WJYZ Albany GA
WJZM Clarksville TN
WJZN(AM) Augusta ME
WKAC Athens AL
WKAM Goshen IN
WKAN Kankakee IL
WKAQ San Juan PR
*WKAR East Lansing MI
WKAT North Miami FL
WKAV Charlottesville VA
WKAX Russellville AL
WKAZ(AM) Charleston WV
WKBA Vinton VA
WKBC North Wilkesboro NC
WKBF Rock Island IL
WKBH Holmen WI
WKBI Saint Marys PA
WKBK(AM) Keene NH
WKBL(AM) Covington TN
WKBN Youngstown OH
WKBO Harrisburg PA
WKBR Manchester NH
WKBV(AM) Richmond IN
WKBY Chatham VA
WKBZ(AM) Muskegon MI
WKCB Hindman KY
WKCE Maryville TN
WKCI(AM) Waynesboro VA
WKCM Hawesville KY
WKCT(AM) Bowling Green KY
WKCU Corinth MS
WKCW Warrenton VA
WKCY Harrisonburg VA
WKDA(AM) Lebanon TN
WKDE Altavista VA
WKDI Denton MD
WKDK Newberry SC
WKDM(AM) New York NY
WKDO Liberty KY
WKDP Corbin KY
WKDV Manassas VA
WKDW Staunton VA
WKDX Hamlet NC
WKDZ Cadiz KY
WKEI(AM) Kewanee IL
WKEU Griffin GA
WKEW Greensboro NC
WKEX Blacksburg VA
WKEY Covington VA
WKEZ Bluefield WV
WKFB(AM) Jeannette PA
WKFD(AM) Charlestown RI
WKFE Yauco PR
WKFI(AM) Wilmington OH
WKFL Bushnell FL
WKFN(AM) Clarksville TN
*WKGC Panama City Beach FL
WKGM Smithfield VA
WKGN Knoxville TN
WKGX Lenoir NC
WKHB(AM) Irwin PA
WKHM(AM) Jackson MI
WKHZ(AM) Ocean City MD
WKIC Hazard KY
WKII(AM) Solana FL
WKIK La Plata MD
WKIP Poughkeepsie NY
WKIQ Eustis FL
WKIZ Key West FL
WKJB Mayaguez PR
WKJG(AM) Fort Wayne IN
WKJK Louisville KY
WKJQ Parsons TN
WKJR(AM) Rantoul IL
WKKD Aurora IL
WKKP McDonough GA
WKKS Vanceburg KY
WKKX(AM) Wheeling WV
WKLA Ludington MI
WKLB Manchester KY
WKLJ Sparta WI
WKLK Cloquet MN
WKLP Keyser WV
WKLV Blackstone VA
WKLY Hartwell GA

WKMB Stirling NJ
WKMC Roaring Spring PA
WKMG Newberry SC
WKMI Kalamazoo MI
WKMQ(AM) Tupelo MS
WKND(AM) Windsor CT
WKNG Tallapoosa GA
WKNR(AM) Cleveland OH
WKNV Fairlawn VA
WKNW Sault Ste. Marie MI
WKNY Kingston NY
WKOK Sunbury PA
WKOR(AM) Starkville MS
WKOX Framingham MA
WKOZ Kosciusko MS
WKPA Lynchburg VA
WKPR Kalamazoo MI
WKPT Kingsport TN
WKQW Oil City PA
WKRA Holly Springs MS
WKRC Cincinnati OH
WKRD(AM) Louisville KY
WKRK Murphy NC
WKRM Columbia TN
WKRO Cairo IL
WKRS Waukegan IL
WKRT(AM) Cortland NY
WKSC Kershaw SC
WKSH(AM) Sussex WI
WKSK(AM) West Jefferson NC
WKSN Jamestown NY
WKSR Pulaski TN
WKST(AM) New Castle PA
WKTA Evanston IL
WKTE King NC
WKTF(AM) Vienna GA
WKTP Jonesborough TN
WKTQ South Paris ME
WKTR Earlysville VA
WKTX Cortland OH
WKTY(AM) La Crosse WI
WKUN Monroe GA
WKVA Lewistown PA
WKVG Jenkins KY
WKVI Knox IN
WKVL(AM) Knoxville TN
WKVM San Juan PR
WKVQ Eatonton GA
WKVT Brattleboro VT
WKVX Wooster OH
WKWF(AM) Key West FL
WKWL Florala AL
WKWN Trenton GA
WKXG Greenwood MS
WKXI Jackson MS
WKXL(AM) Concord NH
WKXM Winfield AL
WKXO Berea KY
WKXR Asheboro NC
WKXV Knoxville TN
WKY Oklahoma City OK
WKYH(AM) Paintsville KY
WKYK Burnsville NC
WKYO Caro MI
WKYX Paducah KY
WKZI Casey IL
WKZK North Augusta SC
WKZN(AM) West Hazleton PA
WKZO(AM) Kalamazoo MI
WKZV Washington PA
WLAA(AM) Winter Garden FL
WLAC Nashville TN
WLAD Danbury CT
WLAF La Follette TN
WLAG La Grange GA
WLAM(AM) Lewiston ME
WLAN Lancaster PA
WLAP Lexington KY
WLAQ Rome GA
WLAR Athens TN
WLAT(AM) New Britain CT
WLAY Muscle Shoals AL
WLBA Gainesville GA
WLBB(AM) Carrollton GA
WLBE Leesburg FL
WLBG Laurens SC

WLBH Mattoon IL
WLBK De Kalb IL
*WLBL Auburndale WI
WLBN Lebanon KY
WLBQ Morgantown KY
WLBR Lebanon PA
WLBY(AM) Saline MI
WLCC Brandon FL
WLCG Macon GA
WLCK(AM) Scottsville KY
WLCM Charlotte MI
WLCO(AM) Lapeer MI
WLCR(AM) Mt. Washington KY
WLDR(AM) Petoskey MI
WLDS Jacksonville IL
WLDX Fayette AL
WLDY Ladysmith WI
WLEA Hornell NY
WLEC Sandusky OH
WLEE(AM) Richmond VA
WLEM Emporium PA
WLEO(AM) Ponce PR
WLES(AM) Bon Air VA
WLET Toccoa GA
WLEW Bad Axe MI
WLEY Cayey PR
WLFJ(AM) Greenville SC
WLFN La Crosse WI
WLFP(AM) Braddock PA
WLGC Greenup KY
WLGN Logan OH
WLGZ(AM) Rochester NY
WLHN Muncie IN
WLIB New York NY
WLIE(AM) Islip NY
WLIJ Shelbyville TN
WLIK Newport TN
WLIL(AM) Lenoir City TN
WLIM Patchogue NY
WLIP Kenosha WI
WLIQ(AM) Quincy IL
WLIS Old Saybrook CT
WLIV Livingston TN
*WLJN Elmwood Township MI
WLJW(AM) Cadillac MI
WLKD Minocqua WI
WLKF Lakeland FL
WLKM Three Rivers MI
WLKR(AM) Norwalk OH
WLKS West Liberty KY
WLKW(AM) West Warwick RI
WLLH Lowell MA
WLLI(AM) Humboldt TN
WLLL Lynchburg VA
WLLM Lincoln IL
*WLLN Lillington NC
WLLQ(AM) Chapel Hill NC
WLLV Louisville KY
WLLY Wilson NC
WLMC Georgetown SC
WLMR(AM) Chattanooga TN
WLMV(AM) Madison WI
WLNA(AM) Peekskill NY
WLNC Laurinburg NC
WLNL Horseheads NY
WLNO New Orleans LA
WLNR Kinston NC
WLOA(AM) Farrell PA
WLOB Portland ME
WLOC(AM) Munfordville KY
WLOD Loudon TN
WLOE Eden NC
WLOH Lancaster OH
WLOI La Porte IN
WLOK Memphis TN
WLOL(AM) Minneapolis MN
WLON Lincolnton NC
WLOP Jesup GA
WLOR Huntsville AL
WLOU Louisville KY
WLOV Washington GA
WLOY(AM) Rural Retreat VA
WLPA Lancaster PA
WLPO La Salle IL
WLPR Prichard AL
WLQH Chiefland FL
WLQM Franklin VA
WLQR Toledo OH
WLQV Detroit MI
WLQY Hollywood FL
WLRB Macomb IL
WLRC Walnut MS
WLRM(AM) Millington TN
WLRP San Sebastian PR
WLRT(AM) Hampton VA
WLRV Lebanon VA
WLS Chicago IL
WLSB Copperhill TN
WLSC Loris SC
WLSD Big Stone Gap VA
WLSG(AM) Wilmington NC
WLSH Lansford PA
WLSI Pikeville KY
WLSS(AM) Sarasota FL
WLSV Wellsville NY
WLTA Alpharetta GA
WLTG Panama City FL
WLTH Gary IN
WLTN Littleton NH
WLTP(AM) Marietta OH
WLTQ(AM) Charleston SC
WLUA(AM) Belton SC
WLUV Loves Park IL
WLUZ(AM) Bayamon PR
WLVA Lynchburg VA
WLVF Haines City FL
WLVJ(AM) Boynton Beach FL
WLVL Lockport NY
WLVP(AM) Gorham ME
WLVV Mobile AL
WLW Cincinnati OH
WLWI(AM) Montgomery AL
WLWL Rockingham NC
WLXE(AM) Rockville MD
WLXG Lexington KY
WLXN Lexington NC
WLYC Williamsport PA
WLYG(AM) Hanceville AL
WLYJ(AM) Jasper AL
WLYN Lynn MA
WLYV Fort Wayne IN
WMAC Macon GA
WMAF Madison FL
WMAJ State College PA
WMAL Washington DC
WMAM Marinette WI
WMAN Mansfield OH
WMAS Springfield MA
WMAX Bay City MI
WMAY(AM) Springfield IL
WMBA Ambridge PA
WMBD Peoria IL
WMBE Chilton WI
WMBG Williamsburg VA
WMBH(AM) Joplin MO
*WMBI(AM) Chicago IL
WMBM Miami Beach FL
WMBN(AM) Petoskey MI
WMBS Uniontown PA
WMC(AM) Memphis TN
WMCA New York NY
WMCH Church Hill TN
WMCJ(AM) Cullman AL
WMCL McLeansboro IL
WMCP Columbia TN
WMCR Oneida NY
WMCS Greenfield WI
WMCT Mountain City TN
WMCW(AM) Harvard IL
WMDD Fajardo PR
WMDH New Castle IN
*WMDR Augusta ME
WMEL Melbourne FL
WMEN(AM) Royal Palm Beach FL
WMEQ(AM) Menomonie WI
WMER Meridian MS
WMET Gaithersburg MD
WMEV Marion VA
WMFA(AM) Raeford NC
WMFC Monroeville AL
WMFD Wilmington NC
WMFG Hibbing MN
WMFJ Daytona Beach FL
WMFN Zeeland MI
WMFR High Point NC
WMGC Murfreesboro TN
WMGG(AM) Largo FL
WMGJ Gadsden AL
WMGO Canton MS
WMGR Bainbridge GA
WMGW Meadville PA
WMGY Montgomery AL
WMHG(AM) Muskegon MI
WMIA Arecibo PR
WMIC(AM) Sandusky MI
WMID(AM) Atlantic City NJ
WMIK Middlesboro KY
WMIN Hudson WI
WMIQ Iron Mountain MI
WMIR Atlantic Beach SC
WMIS Natchez MS
WMIX Mount Vernon IL
WMIZ Vineland NJ
WMJH Rockford MI
WMJL Marion KY
WMJQ(AM) Ontario NY
WMJR Winchester KY
WMKI(AM) Boston MA
WMKT(AM) Charlevoix MI
WMLB(AM) Avondale Estates GA
WMLC(AM) Monticello MS
WMLM Saint Louis MI
WMLP(AM) Milton PA
WMLR Hohenwald TN
WMLT Dublin GA
WMMB Melbourne FL
WMMG Brandenburg KY
WMMI Shepherd MI
WMML Glens Falls NY
WMMN Fairmont WV
WMMV Cocoa FL
WMMW Meriden CT
WMNA Gretna VA
WMNC Morganton NC
WMNE(AM) Riviera Beach FL
WMNI Columbus OH
WMNT(AM) Manati PR
WMNZ Montezuma GA
WMOA Marietta OH
WMOB Mobile AL
WMOG Brunswick GA
WMOH Hamilton OH
WMOK Metropolis IL
WMON Montgomery WV
WMOP Ocala FL
WMOR Morehead KY
WMOU Berlin NH
WMOV Ravenswood WV
WMOX Meridian MS
*WMPC(AM) Lapeer MI
WMPL Hancock MI
WMPM Smithfield NC
WMPO Middleport-Pomeroy OH
WMPS(AM) Bartlett TN
WMPX Midland MI
WMQM(AM) Lakeland TN
WMRB Columbia TN
WMRC Milford MA
WMRD(AM) Middletown CT
WMRE Charles Town WV
WMRI(AM) Marion IN
WMRK Selma AL
WMRN Marion OH
WMRO(AM) Gallatin TN
WMSA Massena NY
WMSG Oakland MD
WMSH Sturgis MI
WMSK(AM) Morganfield KY
WMSP Montgomery AL
WMSR Manchester TN
WMST(AM) Mt. Sterling KY
WMSW Hatillo PR
WMSX Brockton MA
WMT Cedar Rapids IA
WMTA(AM) Central City KY
WMTC Vancleve KY
WMTD Hinton WV
WMTE Manistee MI
WMTL Leitchfield KY
WMTM Moultrie GA
WMTN Morristown TN
WMTR(AM) Morristown NJ
WMTY(AM) Farragut TN
WMUF Paris TN
WMUU Greenville SC
WMVA(AM) Martinsville VA
WMVB Millville NJ
WMVG Milledgeville GA
WMVO(AM) Mount Vernon OH
WMVP Chicago IL
WMXF(AM) Waynesville NC
WMYF Portsmouth NH
WMYJ(AM) Martinsville IN
WMYM(AM) Miami FL
WMYN Mayodan NC
WMYR Fort Myers FL
WMYT Carolina Beach NC
WNAE Warren PA
WNAH Nashville TN
WNAK Nanticoke PA
WNAM Neenah-Menasha WI
WNAP Norristown PA
WNAT Natchez MS
WNAU New Albany MS
WNAV(AM) Annapolis MD
WNAW North Adams MA
WNAX(AM) Yankton SD
WNBF Binghamton NY
WNBH New Bedford MA
WNBI Park Falls WI
WNBN(AM) Meridian MS
WNBP(AM) Newburyport MA
WNBS Murray KY
WNBT Wellsboro PA
WNBY Newberry MI
WNBZ Saranac Lake NY
WNCA Siler City NC
WNCC(AM) Northern Cambria PA
WNCO Ashland OH
WNCT Greenville NC
WNDB Daytona Beach FL
WNDE Indianapolis IN
WNDI Sullivan IN
WNDZ Portage IN
WNEA Newnan GA
WNEB Worcester MA
*WNED Buffalo NY
WNEG(AM) Toccoa GA
WNEL Caguas PR
WNEM(AM) Bridgeport MI
WNER(AM) Watertown NY
WNES Central City KY
WNEX Macon GA
WNEZ(AM) Manchester CT
WNFL Green Bay WI
WNFO Ridgeland SC
WNFS(AM) White Springs FL
WNGA(AM) Elberton GA
WNGM(AM) Hiawassee GA
WNGO(AM) Mayfield KY
WNHV White River Junction VT
WNIK Arecibo PR
WNIL Niles MI
WNIO(AM) Youngstown OH
WNIS Norfolk VA
WNIV Atlanta GA
WNIX(AM) Greenville MS
WNJC Washington Township NJ
WNKX Centerville TN
WNLA Indianola MS
WNLK(AM) Norwalk CT
WNLR Churchville VA
WNLS Tallahassee FL
WNMA Miami Springs FL
WNMB(AM) North Myrtle Beach SC
WNML(AM) Knoxville TN
WNMT Nashwauk MN
WNNC Newton NC
WNNG(AM) Warner Robins GA
WNNJ Newton NJ
WNNR(AM) Jacksonville FL
WNNW(AM) Lawrence MA
WNNZ Westfield MA
WNOG Naples FL
WNOO Chattanooga TN
WNOP Newport KY
WNOS New Bern NC
WNOV Milwaukee WI
WNOW Mint Hill NC
WNPC Newport TN
WNPL(AM) Golden Gate FL
WNPV Lansdale PA
WNPZ(AM) Knoxville TN
WNQM Nashville TN
WNRG Grundy VA
WNRI Woonsocket RI
WNRP(AM) Gulf Breeze FL
WNRR(AM) Augusta GA
WNRS Herkimer NY
WNRV(AM) Narrows-Pearisburg VA
WNSG(AM) Nashville TN
WNSH Beverly MA
WNSI(AM) Robertsdale AL
WNSR Brentwood TN
WNSS(AM) Syracuse NY
WNST Towson MD
WNSW Newark NJ
WNTA Rockford IL
WNTD Chicago IL
WNTF Bithlo FL
WNTJ(AM) Johnstown PA
WNTK Newport NH
WNTM(AM) Mobile AL
WNTN Newton MA
WNTP(AM) Philadelphia PA
WNTS Beech Grove IN
WNTT Tazewell TN
WNTW(AM) Somerset PA
WNUZ Talladega AL
WNVA Norton VA
WNVL(AM) Nashville TN
WNVR Vernon Hills IL
WNVY Cantonment FL
*WNWC(AM) Sun Prairie WI
WNWF(AM) Destin FL
WNWI Oak Lawn IL
WNWK(AM) Newark DE
WNWN Portage MI
WNWR Philadelphia PA
WNWS Brownsville TN
WNWZ Grand Rapids MI
WNXT Portsmouth OH
*WNYC New York NY
WNYG Babylon NY
WNYH(AM) Huntington NY
WNYY(AM) Ithaca NY
WNZF(AM) Bunnell FL
WNZK Dearborn Heights MI
WNZS(AM) Veazie ME
WNZZ Montgomery AL
WOAD Jackson MS
WOAI San Antonio TX
WOAM(AM) Peoria IL
WOAP Owosso MI
WOAY Oak Hill WV
WOBG Clarksburg WV
WOBL Oberlin OH
WOBM(AM) Lakewood NJ
WOBT Rhinelander WI
WOBX(AM) Wanchese NC
WOC Davenport IA
WOCA Ocala FL
WOCC Corydon IN
WOCN Miami FL
WOCO Oconto WI
WOCV Oneida TN
WODI Brookneal VA
WODJ(AM) Whitehall MI
WODT New Orleans LA
WODY Fieldale VA
WOEG Hazlehurst MS
WOEN(AM) Olean NY
WOF(AM) Andover NJ
WOFC(AM) Murray KY
WOFE(AM) Rockwood TN
WOFX(AM) Troy NY
WOGO Hallie WI
WOGR Charlotte NC
WOHI East Liverpool OH
WOHS Shelby NC

*WOI Ames IA
WOIC(AM) Columbia SC
WOIR Homestead FL
WOIZ Guayanilla PR
WOKA(AM) Douglas GA
WOKB(AM) Ocoee FL
WOKC Okeechobee FL
WOKS Columbus GA
WOKT Cannonsburg KY
WOKU Hurricane WV
WOKV Jacksonville FL
WOKY Milwaukee WI
WOL Washington DC
WOLA Barranquitas PR
WOLB(AM) Baltimore MD
WOLF Syracuse NY
WOLI(AM) Spartanburg SC
WOLS(AM) Florence SC
WOLY Battle Creek MI
WOMI(AM) Owensboro KY
WOMN(AM) Franklinton LA
WOMP Bellaire OH
WOMT(AM) Manitowoc WI
WONA Winona MS
WOND(AM) Pleasantville NJ
WONE Dayton OH
WONG Canton MS
WONN Lakeland FL
WONQ Oviedo FL
WONW Defiance OH
WONX Evanston IL
WOOD Grand Rapids MI
WOOF Dothan AL
WOON(AM) Woonsocket RI
WOPI Bristol TN
WOPP Opp AL
WOQI(AM) Adjuntas PR
WOR New York NY
WORA Mayaguez PR
WORC Worcester MA
WORD(AM) Spartanburg SC
WORL Altamonte Springs FL
WORM Savannah TN
WORV Hattiesburg MS
WOSH Oshkosh WI
WOSO San Juan PR
*WOSU Columbus OH
WOTS Kissimmee FL
*WOUB Athens OH
WOWO Fort Wayne IN
WOWW Germantown TN
WOWZ(AM) Appomattox VA
WOYK York PA
WOYL Oil City PA
WOZK Ozark AL
WPAB Ponce PR
WPAD Paducah KY
WPAK Farmville VA
WPAM Pottsville PA
WPAQ Mount Airy NC
WPAT Paterson NJ
WPAX Thomasville GA
WPAY Portsmouth OH
WPAZ Pottstown PA
WPBC Decatur GA
WPBQ(AM) Flowood MS
WPBR Lantana FL
WPBS(AM) Conyers GA
WPCC Clinton SC
WPCE Portsmouth VA
WPCF(AM) Panama City Beach FL
WPCI Greenville SC
WPCM Burlington NC
WPDC Elizabethtown PA
WPDM Potsdam NY
WPDR Portage WI
WPDX(AM) Clarksburg WV
WPEH Louisville GA
WPEK(AM) Fairview NC
*WPEL Montrose PA
WPEN Philadelphia PA
WPEO Peoria IL
WPEP Taunton MA
WPET Greensboro NC
WPFB Middletown OH
WPFC Port Allen LA
WPFD Fairview TN
WPFJ Franklin NC
WPFR(AM) Terre Haute IN
WPGA Perry GA
WPGC Morningside MD
WPGG(AM) Evergreen AL
*WPGM Danville PA
WPGR(AM) Monroeville PA
WPGS Mims FL
WPGW Portland IN
WPGY(AM) Ellijay GA
WPHB Philipsburg PA
WPHE Phoenixville PA
WPHM Port Huron MI
WPHT Philadelphia PA
WPHX(AM) Sanford ME
WPHY(AM) Trenton NJ
WPIC Sharon PA
WPID Piedmont AL
WPIE Trumansburg NY
WPIN Dublin VA
WPIP Winston-Salem NC
WPIT Pittsburgh PA
WPJK Orangeburg SC
WPJL Raleigh NC
WPJM Greer SC
WPJS Conway SC
WPJX(AM) Zion IL
WPKE Pikeville KY
WPKY Princeton KY
WPLK Palatka FL
WPLM Plymouth MA
WPLN(AM) Madison TN
WPLO Grayson GA
WPLV West Point GA
*WPLX(AM) Germantown TN
WPLY(AM) Mount Pocono PA
WPMB Vandalia IL
WPMH(AM) Claremont VA
WPMP(AM) Pascagoula-Moss Point MS
WPMZ Providence RI
WPNA Oak Park IL
WPNH Plymouth NH
WPNI(AM) Amherst MA
WPNN(AM) Pensacola FL
WPNW(AM) Zeeland MI
WPOG(AM) Saint Matthews SC
WPOL Winston-Salem NC
WPON Walled Lake MI
WPOP Hartford CT
WPPA Pottsville PA
WPPC Penuelas PR
WPPI(AM) Sauk Rapids MN
WPRA Mayaguez PR
WPRD Winter Park FL
WPRE(AM) Prairie du Chien WI
WPRN Butler AL
WPRO Providence RI
WPRP Ponce PR
WPRR(AM) Johnstown PA
WPRS Paris IL
WPRT Prestonsburg KY
WPRX Bristol CT
WPRY Perry FL
WPRZ(AM) Warrenton VA
WPSB(AM) Birmingham AL
WPSE Erie PA
WPSL(AM) Port St. Lucie FL
WPSN(AM) Honesdale PA
WPSO New Port Richey FL
WPSP Royal Palm Beach FL
WPTB(AM) Statesboro GA
WPTF Raleigh NC
WPTK(AM) Pine Island Center FL
WPTL(AM) Canton NC
WPTN Cookeville TN
WPTT McKeesport PA
WPTW Piqua OH
WPTX(AM) Lexington Park MD
WPUL South Daytona FL
WPUT(AM) Brewster NY
WPVL Platteville WI
WPWA Chester PA
WPWC Dumfries-Triangle VA
WPWT(AM) Colonial Heights TN
WPYB Benson NC
WPYR(AM) Baton Rouge LA
WPYT(AM) Wilkinsburg PA
WQAH(AM) Priceville AL
WQAM Miami FL
WQBA Miami FL
WQBB Powell TN
WQBC Vicksburg MS
WQBN Temple Terrace FL
WQBQ Leesburg FL
WQBS San Juan PR
WQCH La Fayette GA
WQCR(AM) Alabaster AL
WQCT Bryan OH
WQEW New York NY
WQFX Gulfport MS
WQHL(AM) Live Oak FL
WQII San Juan PR
WQIZ Saint George SC
WQKR Portland TN
WQLA(AM) La Follette TN
WQLR(AM) Kalamazoo MI
WQLS Ozark AL
WQMC Sumter SC
WQMS(AM) Quitman MS
WQMV(AM) Waverly TN
WQNT(AM) Charleston SC
WQNX Aberdeen NC
WQOP Atlantic Beach FL
WQOQ(AM) Durand WI
WQOR(AM) Olyphant PA
WQPM Princeton MN
WQRX Valley Head AL
WQSC Charleston SC
WQSE White Bluff TN
WQST Forest MS
WQSV Ashland City TN
WQTH(AM) Claremont NH
WQTM(AM) Orlando FL
WQTW Latrobe PA
WQUN(AM) Hamden CT
WQVA(AM) Lexington SC
WQXI Atlanta GA
WQXL Columbia SC
WQXM(AM) Bartow FL
WQXO Munising MI
WQXY Hazard KY
WQYK(AM) Seffner FL
WQZQ(AM) Clarksville TN
WRAA Luray VA
WRAB Arab AL
WRAD Radford VA
WRAG Carrollton AL
WRAK Williamsport PA
WRAM Monmouth IL
*WRAR(AM) Tappahannock VA
WRAW(AM) Reading PA
WRAY Princeton IN
WRBE Lucedale MS
WRBS(AM) Baltimore MD
WRBZ Raleigh NC
WRCA Waltham MA
WRCG Columbus GA
WRCO Richland Center WI
WRCR(AM) Spring Valley NY
WRCS Ahoskie NC
WRCY(AM) Mount Vernon IN
WRDB Reedsburg WI
WRDD Ebensburg PA
WRDT(AM) Monroe MI
WRDW(AM) Augusta GA
WRDZ La Grange IL
WREC(AM) Memphis TN
WREF Ridgefield CT
WREJ Richmond VA
WREL Lexington VA
WREV Reidsville NC
WRFC Athens GA
WRFD(AM) Columbus-Worthington OH
WRFS(AM) Alexander City AL
WRFV(AM) Valdosta GA
WRGA Rome GA
WRGC Sylva NC
WRGM Ontario OH
WRGS Rogersville TN
WRHB(AM) Kendall FL
WRHC Coral Gables FL
WRHI(AM) Rock Hill SC
WRHL Rochelle IL
WRIE Erie PA
WRIG Schofield WI
WRIN Rensselaer IN
WRIS Roanoke VA
WRIV Riverhead NY
WRIX Homeland Park SC
WRJC Mauston WI
WRJD(AM) Durham NC
WRJN Racine WI
WRJR(AM) Portsmouth VA
WRJS(AM) Swainsboro GA
WRJW Picayune MS
WRJX(AM) Jackson AL
WRJZ Knoxville TN
WRKB Kannapolis NC
WRKD Rockland ME
WRKK Hughesville PA
WRKL New City NY
WRKM Carthage TN
WRKO Boston MA
WRKQ Madisonville TN
WRLA(AM) West Point GA
WRLL(AM) Cicero IL
WRLM(AM) Irondale AL
WRLV Salyersville KY
WRLZ Eatonville FL
WRME(AM) Hampden ME
WRMG Red Bay AL
WRMN Elgin IL
WRMQ Orlando FL
WRMS Beardstown IL
WRMT(AM) Rocky Mount NC
WRNA China Grove NC
WRNE(AM) Gulf Breeze FL
WRNI(AM) Providence RI
WRNJ(AM) Hackettstown NJ
WRNL Richmond VA
WRNN(AM) Myrtle Beach SC
WRNR Martinsburg WV
WRNS Kinston NC
WRNY Rome NY
WROA Gulfport MS
WROB(AM) West Point MS
WROC(AM) Rochester NY
WROD Daytona Beach FL
WROK Rockford IL
WROL Boston MA
WROM Rome GA
WRON Ronceverte WV
WROS Jacksonville FL
WROU(AM) Petersburg VA
WROW(AM) Albany NY
WROX Clarksdale MS
WROY Carmi IL
WRPM Poplarville MS
WRPN(AM) Ripon WI
WRPQ(AM) Baraboo WI
WRRA Frederiksted VI
WRRD(AM) Jackson WI
WRRE Juncos PR
WRRL Rainelle WV
WRRZ Clinton NC
WRSA(AM) Saint Albans VT
WRSB Canandaigua NY
WRSC State College PA
WRSJ(AM) Bayamon PR
WRSL Stanford KY
WRSM Sumiton AL
WRSS San Sebastian PR
WRSW Warsaw IN
WRTA Altoona PA
WRTG Garner NC
WRTK(AM) Niles OH
WRTN(AM) Berlin NH
WRTO(AM) Chicago IL
WRUF Gainesville FL
*WRUN(AM) Utica NY
WRUS Russellville KY
WRVA Richmond VA
WRVC Huntington WV
WRVK Mt. Vernon KY
WRVP(AM) Mount Kisco NY
WRWB(AM) Harrogate TN
WRWH Cleveland GA
WRXB Saint Petersburg Beach FL
WRXO(AM) Roxboro NC
WRYM(AM) New Britain CT
*WRYT(AM) Edwardsville IL
WRZN Hernando FL
WSAI(AM) Cincinnati OH
WSAL Logansport IN
WSAM Saginaw MI
WSAN(AM) Allentown PA
WSAO Senatobia MS
WSAR Fall River MA
WSAT Salisbury NC
WSAU Wausau WI
WSB Atlanta GA
WSBA York PA
WSBB New Smyrna Beach FL
WSBC(AM) Chicago IL
WSBI Static TN
WSBM(AM) Florence AL
WSBR Boca Raton FL
WSBS Great Barrington MA
WSBT South Bend IN
WSBV South Boston VA
WSCG(AM) Greenville MI
WSCO(AM) Appleton WI
WSCP Sandy Creek-Pulaski NY
WSCR(AM) Chicago IL
WSCW South Charleston WV
WSDE(AM) Cobleskill NY
WSDO(AM) Sanford FL
WSDQ Dunlap TN
WSDR Sterling IL
WSDS(AM) Salem Township MI
WSDT Soddy-Daisy TN
WSDV(AM) Sarasota FL
WSDX(AM) Brazil IN
WSDZ Belleville IL
WSEG(AM) Savannah GA
WSEL Pontotoc MS
WSEM Donalsonville GA
WSEN(AM) Baldwinsville NY
WSEV Sevierville TN
WSEZ Paoli IN
WSFB Quitman GA
WSFC Somerset KY
WSFE(AM) Burnside KY
WSFN Brunswick GA
WSFW Seneca Falls NY
WSFZ(AM) Jackson MS
WSGB(AM) Sutton WV
WSGF(AM) Augusta GA
WSGH Lewisville NC
WSGI Springfield TN
WSGO Oswego NY
WSGW Saginaw MI
WSHE(AM) Columbus GA
WSHN Fremont MI
WSHO New Orleans LA
*WSHU(AM) Westport CT
WSHV(AM) South Hill VA
WSHY(AM) Lafayette IN
WSIC Statesville NC
WSIP Paintsville KY
WSIR Winter Haven FL
WSIV East Syracuse NY
WSJC Magee MS
WSJM Saint Joseph MI
WSJS Winston-Salem NC
WSKI Montpelier VT
WSKN(AM) San Juan PR
WSKO Providence RI
WSKR Denham Springs LA
WSKW Skowhegan ME
WSKY(AM) Asheville NC
WSLA Slidell LA
WSLB Ogdensburg NY
WSLM Salem IN
WSLW White Sulphur Springs WV
WSM Nashville TN
WSMB(AM) Memphis TN
WSME(AM) Camp Lejeune NC
WSMG Greeneville TN
WSMI Litchfield IL
WSML Graham NC
WSMN Nashua NH

WSMT Sparta TN
WSMX Winston-Salem NC
WSMY Weldon NC
WSNG Torrington CT
WSNJ Bridgeton NJ
WSNL(AM) Flint MI
WSNO Barre VT
WSNR(AM) Jersey City NJ
WSNT(AM) Sandersville GA
WSNW(AM) Seneca SC
WSOK Savannah GA
WSOL San German PR
WSOM Salem OH
WSON Henderson KY
WSOO Sault Ste. Marie MI
WSOS(AM) Saint Augustine Beach FL
WSOY(AM) Decatur IL
WSPC Albemarle NC
WSPD Toledo OH
WSPG(AM) Spartanburg SC
WSPL(AM) Streator IL
WSPQ(AM) Springville NY
WSPR Springfield MA
WSPT Stevens Point WI
WSPY(AM) Geneva IL
WSPZ(AM) Birmingham AL
WSQL Brevard NC
WSQR Sycamore IL
WSRA(AM) Albany GA
WSRC(AM) Fair Bluff NC
WSRF Fort Lauderdale FL
WSRO(AM) Ashland MA
WSRP(AM) Jacksonville NC
WSRQ(AM) Sarasota FL
WSRW Hillsboro OH
WSRY(AM) Elkton MD
WSSC Sumter SC
WSSG(AM) Goldsboro NC
WSSO(AM) Starkville MS
WSSP(AM) Milwaukee WI
WSTA(AM) Charlotte Amalie VI
WSTC(AM) Stamford CT
WSTJ Saint Johnsbury VT
WSTL(AM) Providence RI
WSTN Somerville TN
WSTP(AM) Salisbury NC
WSTT Thomasville GA
WSTU(AM) Stuart FL
WSTV Steubenville OH
WSTX(AM) Christiansted VI
WSUA Miami FL
WSUB Groton CT
*WSUI Iowa City IA
WSVA Harrisonburg VA
WSVG(AM) Mount Jackson VA
WSVM(AM) Valdese NC
WSVS Crewe VA
WSVU(AM) North Palm Beach FL
WSVX(AM) Shelbyville IN
*WSWI Evansville IN
WSWN Belle Glade FL
WSWV Pennington Gap VA
WSWW Charleston WV
WSYB Rutland VT
WSYD Mount Airy NC
WSYL Sylvania GA
WSYR Syracuse NY
WSYW Indianapolis IN
WSYY(AM) Millinocket ME
WTAA(AM) Pleasantville NJ
WTAB Tabor City NC
WTAD Quincy IL
WTAG Worcester MA
WTAL Tallahassee FL
WTAM Cleveland OH
WTAN(AM) Clearwater FL
WTAQ(AM) Green Bay WI
WTAR(AM) Norfolk VA
WTAW(AM) College Station TX
WTAX(AM) Springfield IL
WTAY Robinson IL
WTBC Tuscaloosa AL
WTBF Troy AL
WTBI Pickens SC
WTBN(AM) Pinellas Park FL
WTBO Cumberland MD

WTBQ Warwick NY
WTBZ Grafton WV
WTCA Plymouth IN
WTCH Shawano WI
WTCJ Tell City IN
WTCL Chattahoochee FL
WTCM(AM) Traverse City MI
WTCO Campbellsville KY
WTCR Kenova WV
WTCS Fairmont WV
WTCW Whitesburg KY
WTCY Harrisburg PA
WTDY Madison WI
WTEL(AM) Red Springs NC
WTEM Washington DC
WTFX(AM) Winchester VA
WTGA Thomaston GA
WTGM Salisbury MD
WTHB Augusta GA
WTHE(AM) Mineola NY
WTHQ(AM) Brookport IL
WTHV(AM) Hahira GA
WTIC Hartford CT
WTIF Tifton GA
WTIG Massillon OH
WTIK Durham NC
WTIL Mayaguez PR
WTIQ Manistique MI
WTIR(AM) Cocoa Beach FL
WTIS Tampa FL
WTIV Titusville PA
WTIX(AM) Winston-Salem NC
WTJH East Point GA
WTJK South Beloit IL
WTJS Jackson TN
WTJV(AM) De Land FL
WTJZ Newport News VA
WTKA Ann Arbor MI
WTKE(AM) Fort Walton Beach FL
WTKG Grand Rapids MI
WTKI(AM) Huntsville AL
WTKM(AM) Hartford WI
WTKN(AM) Corinth MS
WTKS(AM) Savannah GA
WTKT(AM) Harrisburg PA
WTKY Tompkinsville KY
WTKZ Allentown PA
WTLA North Syracuse NY
WTLB Utica NY
WTLC Indianapolis IN
WTLK Taylorsville NC
WTLM Opelika AL
WTLN Orlando FL
WTLO Somerset KY
WTLS Tallassee AL
WTMA Charleston SC
WTMC(AM) Wilmington DE
WTME(AM) Rumford ME
WTMJ Milwaukee WI
WTMM(AM) Rensselaer NY
WTMN(AM) Gainesville FL
WTMP Egypt Lake FL
WTMR Camden NJ
WTMY Sarasota FL
WTMZ Dorchester Terrace-Brentwood SC
WTNE Trenton TN
WTNI(AM) Biloxi MS
WTNK(AM) Hartsville TN
WTNL Reidsville GA
WTNS Coshocton OH
WTNT(AM) Bethesda MD
WTNY Watertown NY
WTOB Winston-Salem NC
WTOD Toledo OH
WTOE Spruce Pine NC
WTOF(AM) Bay Minette AL
WTON Staunton VA
WTOR Youngstown NY
WTOT Marianna FL
WTOX(AM) Glen Allen VA
WTOY Salem VA
WTPR(AM) Paris TN
WTPS(AM) Coral Gables FL
WTQS(AM) Cameron SC
WTRB Ripley TN

WTRC Elkhart IN
WTRE Greensburg IN
WTRI Brunswick MD
WTRN Tyrone PA
WTRO Dyersburg TN
WTRP La Grange GA
WTRU(AM) Kernersville NC
WTRX Flint MI
WTSA Brattleboro VT
WTSB(AM) Selma NC
WTSJ(AM) Randolph VT
WTSK Tuscaloosa AL
WTSL Hanover NH
WTSN Dover NH
WTSO Madison WI
WTSV Claremont NH
WTSZ(AM) Eminence KY
WTTB Vero Beach FL
WTTC Towanda PA
WTTF Tiffin OH
WTTI(AM) Dalton GA
WTTL Madisonville KY
WTTM(AM) Lindenwold NJ
WTTN Watertown WI
WTTR(AM) Westminster MD
WTTT(AM) Boston MA
WTUP Tupelo MS
WTUV(AM) Louisville KY
WTUX(AM) Madison WI
WTVB(AM) Coldwater MI
WTVL(AM) Waterville ME
WTVN(AM) Columbus OH
WTWA Thomson GA
WTWB Auburndale FL
WTWD(AM) Plant City FL
WTWG(AM) Columbus MS
WTWK(AM) Plattsburgh NY
WTWN Wells River VT
WTWP(AM) Washington DC
WTWT(AM) Frederick MD
WTWZ Clinton MS
WTXY Whiteville NC
WTYL Tylertown MS
WTYM Kittanning PA
WTYS Marianna FL
WTYX(AM) Watkins Glen NY
WTZE Tazewell VA
WTZN(AM) Troy PA
WTZQ(AM) Hendersonville NC
WTZX Sparta TN
WUAM(AM) Saratoga Springs NY
WUAT Pikeville TN
WUBA(AM) Philadelphia PA
WUBR(AM) Baton Rouge LA
WUCO(AM) Marysville OH
WUFE Baxley GA
WUFF Eastman GA
*WUFL Sterling Heights MI
WUFO Amherst NY
WUKQ(AM) Ponce PR
WULA Eufaula AL
WULM(AM) Springfield OH
WUMP Madison AL
WUNA Ocoee FL
*WUNN(AM) Mason MI
WUNO(AM) San Juan PR
WUNR Brookline MA
WUPE(AM) Pittsfield MA
WUPR Utuado PR
WURD(AM) Philadelphia PA
WURL Moody AL
WUST Washington DC
WUTQ Utica NY
WUUS(AM) Rossville GA
WUVR(AM) Lebanon NH
WVAB Virginia Beach VA
WVAE(AM) Biddeford ME
WVAL Sauk Rapids MN
WVAM Altoona PA
WVAR Richwood WV
WVAX(AM) Charlottesville VA
WVBE(AM) Roanoke VA
WVBF(AM) Middleborough Center MA
WVBG(AM) Vicksburg MS
WVBS Burgaw NC
WVCB Shallotte NC

WVCC(AM) Hogansville GA
WVCD(AM) Bamberg-Denmark SC
WVCH Chester PA
WVCV Orange VA
*WVCY Oshkosh WI
WVEI(AM) Worcester MA
WVEL Pekin IL
WVFN(AM) East Lansing MI
WVGB Beaufort SC
WVGM Lynchburg VA
WVHI Evansville IN
WVHU(AM) Huntington WV
WVIE(AM) Pikesville MD
WVJP Caguas PR
WVJS(AM) Owensboro KY
WVKO Columbus OH
WVKY(AM) Nicholasville KY
WVKZ Schenectady NY
WVLD(AM) Valdosta GA
WVLG(AM) Wildwood FL
WVLK Lexington KY
WVLN Olney IL
WVLY(AM) Moundsville WV
WVLZ(AM) Knoxville TN
WVMC Mount Carmel IL
*WVMR Frost WV
WVMT Burlington VT
WVNA Tuscumbia AL
WVNE(AM) Leicester MA
WVNJ(AM) Oakland NJ
WVNN Athens AL
WVNR Poultney VT
WVNT(AM) Parkersburg WV
WVNZ(AM) Richmond VA
WVOA(AM) Dewitt NY
WVOC Columbia SC
WVOE Chadbourn NC
WVOG New Orleans LA
WVOH Hazlehurst GA
WVOI(AM) Marco Island FL
WVOJ(AM) Fernandina Beach FL
WVOK(AM) Oxford AL
WVOL Berry Hill TN
WVON(AM) Berwyn IL
WVOP Vidalia GA
WVOS Liberty NY
WVOT(AM) Wilson NC
WVOW Logan WV
WVOX New Rochelle NY
WVOZ San Juan PR
WVPO Stroudsburg PA
WVRC Spencer WV
WVRQ(AM) Viroqua WI
WVSA Vernon AL
WVSG(AM) Neon KY
WVSM Rainsville AL
WVTJ(AM) Pensacola FL
WVTL(AM) Amsterdam NY
WVTS(AM) Charleston WV
WVUV(AM) Leone AS
WVVM(AM) Dry Branch GA
WVVT(AM) Essex Junction VT
WVWI Charlotte Amalie VI
WVXX(AM) Norfolk VA
WVZN Columbia PA
WWAB Lakeland FL
WWAM Jasper TN
WWBA Pinellas Park FL
WWBC Cocoa FL
WWBF Bartow FL
WWBG Greensboro NC
WWCA Gary IN
WWCB Corry PA
WWCH Clarion PA
WWCK Flint MI
WWCL Lehigh Acres FL
WWCN North Fort Myers FL
WWCO(AM) Waterbury CT
WWCS Canonsburg PA
WWDB(AM) Philadelphia PA
WWDJ Hackensack NJ
WWDR(AM) Murfreesboro NC
WWFE Miami FL
WWFL Clermont FL
WWGA(AM) Waycross GA
WWGB Indian Head MD

WWGC(AM) Albertville AL
WWGE(AM) Loretto PA
WWGK(AM) Cleveland OH
WWGP(AM) Sanford NC
WWGS(AM) Georgetown SC
WWHN Joliet IL
WWIC Scottsboro AL
WWII Shiremanstown PA
WWIL Wilmington NC
WWIN Baltimore MD
WWIO(AM) Saint Mary's GA
WWIS Black River Falls WI
WWJ Detroit MI
WWJB(AM) Brooksville FL
WWJC Duluth MN
WWJZ Mount Holly NJ
WWKB Buffalo NY
WWKU(AM) Glasgow KY
WWL(AM) New Orleans LA
WWLE Cornwall NY
WWLF(AM) Auburn NY
WWLK Eddyville KY
WWLS Moore OK
WWLV(AM) South Bend IN
WWLX Lawrenceburg TN
WWLZ Horseheads NY
WWMI Saint Petersburg FL
WWMK Cleveland OH
WWNA(AM) Aguadilla PR
WWNB New Bern NC
WWNC Asheville NC
WWNH Madbury NH
WWNL(AM) Pittsburgh PA
WWNN Pompano Beach FL
WWNR Beckley WV
WWNS Statesboro GA
WWNT Dothan AL
WWNZ(AM) Veazie ME
WWOF(AM) Walhalla SC
WWOL Forest City NC
WWON(AM) Waynesboro TN
WWOW Conneaut OH
WWPA Williamsport PA
WWPG Tuscaloosa AL
WWPR Bradenton FL
WWRC(AM) Washington DC
WWRF(AM) Lake Worth FL
WWRK(AM) Darlington SC
WWRL New York NY
WWRU(AM) Jersey City NJ
*WWRV New York NY
WWSC Glens Falls NY
WWSD Quincy FL
WWSJ Saint Johns MI
WWSM Annville-Cleona PA
WWSZ(AM) New Albany IN
WWTC Minneapolis MN
WWTK(AM) Lake Placid FL
WWTM(AM) Decatur AL
WWTR(AM) Bridgewater NJ
WWTX(AM) Wilmington DE
WWVA(AM) Wheeling WV
WWVT Christiansburg VA
WWWC(AM) Wilkesboro NC
WWWE(AM) Hapeville GA
WWWI(AM) Baxter MN
WWWJ Galax VA
WWWL(AM) New Orleans LA
WWWR Roanoke VA
WWWS Buffalo NY
WWXL Manchester KY
WWYO Pineville WV
WWZN(AM) Boston MA
WWZQ Aberdeen MS
WXAG Athens GA
WXAL Demopolis AL
WXAM(AM) Buffalo KY
WXBD Biloxi MS
WXBR(AM) Brockton MA
WXCE Amery WI
WXCF Clifton Forge VA
WXCO Wausau WI
WXCT(AM) Southington CT
WXEM Buford GA
WXEW Yabucoa PR
WXFN Muncie IN

WXGI Richmond VA
WXGM Gloucester VA
WXGO Madison IN
WXIC Waverly OH
WXIT Blowing Rock NC
WXJC(AM) Birmingham AL
WXJO(AM) Gordon GA
WXKL Sanford NC
WXKO Fort Valley GA
WXKS Everett MA
WXKX(AM) Clarksburg WV
WXLA Dimondale MI
WXLI Dublin GA
WXLW Indianapolis IN
WXLZ Saint Paul VA
WXMC Parsippany-Troy Hills NJ
WXMY Saltville VA
WXNC(AM) Monroe NC
WXNH(AM) Jaffrey NH
WXNI(AM) Westerly RI
WXNT(AM) Indianapolis IN
WXOK Baton Rouge LA
WXOZ(AM) Highland IL
WXQK(AM) Spring City TN
WXRA(AM) Georgetown KY
WXRF Guayama PR
WXRL Lancaster NY
WXRQ(AM) Mount Pleasant TN
WXSM(AM) Blountville TN

WXTC Charleston SC
WXTN Lexington MS
WXTR(AM) Alexandria VA
WXVI Montgomery AL
*WXXI(AM) Rochester NY
*WXXY(AM) Dover DE
WXYB Indian Rocks Beach FL
WXYT Detroit MI
WYAC(AM) Cabo Rojo PR
WYAL Scotland Neck NC
WYAM(AM) Hartselle AL
WYBC New Haven CT
WYBG Massena NY
WYBT Blountstown FL
WYCB Washington DC
WYCK Plains PA
WYCV Granite Falls NC
WYDE(AM) Birmingham AL
WYEA Sylacauga AL
WYEL(AM) Mayaguez PR
WYFN Nashville TN
*WYFQ(AM) Charlotte NC
WYFY Rome NY
WYGH Paris KY
WYGL Selinsgrove PA
WYGR Wyoming MI
WYHG(AM) Young Harris GA
WYHL(AM) Meridian MS
WYIS McRae GA

WYJK(AM) Connellsville PA
WYKC Grenada MS
WYKM Rupert WV
WYKO Sabana Grande PR
WYLD New Orleans LA
WYLF(AM) Penn Yan NY
WYLL(AM) Chicago IL
WYLS(AM) York AL
WYMB Manning SC
WYMC Mayfield KY
WYMM(AM) Jacksonville FL
WYMR(AM) Bridgeport AL
WYNC Yanceyville NC
WYND De Land FL
WYNE(AM) North East PA
WYNF(AM) North Augusta SC
WYNN Florence SC
WYNY(AM) Cross City FL
WYOS(AM) Binghamton NY
WYPC Wellston OH
WYRD Greenville SC
WYRE(AM) Essex MD
WYRM(AM) Norfolk VA
WYRN(AM) Louisburg NC
WYRV Cedar Bluff VA
WYSE(AM) Canton NC
WYSH Clinton TN
WYSK Fredericksburg VA
WYSL Avon NY

WYSR(AM) High Point NC
WYTH(AM) Madison GA
WYTI Rocky Mount VA
WYTS(AM) Columbus OH
WYUS Milford DE
WYVE Wytheville VA
WYWY Barbourville KY
WYXC(AM) Cartersville GA
WYXE Gallatin TN
WYXI Athens TN
WYYC(AM) York PA
WYYZ Jasper GA
WYZD Dobson NC
WYZE Atlanta GA
WZAM Ishpeming MI
WZAN Portland ME
WZAP(AM) Bristol VA
WZAZ Jacksonville FL
WZBK(AM) Keene NH
WZBO(AM) Edenton NC
WZCT Scottsboro AL
WZEP(AM) De Funiak Springs FL
WZFN(AM) Dilworth MN
WZGM(AM) Black Mountain NC
WZGX(AM) Bessemer AL
WZHF Arlington VA
WZHR(AM) Zephyrhills FL
WZJY Mt. Pleasant SC
WZKY Albemarle NC

WZMG Pepperell AL
WZNA(AM) Moca PR
WZNG Shelbyville TN
WZNH(AM) Fitzwilliam Depot NH
WZNZ Jacksonville FL
WZOB Fort Payne AL
WZOE Princeton IL
WZON Bangor ME
WZOO(AM) Asheboro NC
WZOQ(AM) Lima OH
WZOT Rockmart GA
WZQK(AM) Brandon MS
WZQZ(AM) Trion GA
WZRC New York NY
WZRH(AM) Dallas NC
WZRK(AM) Lake Geneva WI
WZRX Jackson MS
WZSK(AM) Everett PA
WZTA(AM) Vero Beach FL
WZTQ(AM) Centre AL
WZUM Carnegie PA
WZYX(AM) Cowan TN
WZZA(AM) Tuscumbia AL
WZZB Seymour IN
WZZW Milton WV
WZZX Lineville AL
XETRA Tijuana MEX

U.S. FM Stations by Call Letters

KAAI(FM) Palisade CO
KAAK(FM) Great Falls MT
KAAN-FM Bethany MO
KAAP(FM) Rock Island WA
KAAQ(FM) Alliance NE
KAAR(FM) Butte MT
KAAT(FM) Oakhurst CA
*KAAX(FM) Avenal CA
KABD(FM) Ipswich SD
*KABF(FM) Little Rock AR
KABG(FM) Los Alamos NM
*KABN-FM Kasilof AK
KABQ-FM Santa Fe NM
*KABU(FM) Fort Totten ND
KABW(FM) Westport WA
KABX-FM Merced CA
KABZ(FM) Little Rock AR
*KACC(FM) Alvin TX
KACI-FM The Dalles OR
KACL(FM) Bismarck ND
KACO(FM) Apache OK
KACQ(FM) Lometa TX
*KACS(FM) Chehalis WA
KACT-FM Andrews TX
*KACU(FM) Abilene TX
*KACV-FM Amarillo TX
KACY(FM) Arkansas City KS
KACZ(FM) Riley KS
KADA-FM Ada OK
KADD(FM) Logandale NV
KADI-FM Republic MO
KADL(FM) Imperial NE
KADQ-FM Evanston WY
*KADU(FM) Hibbing MN
*KADV(FM) Modesto CA
KAEH(FM) Beaumont CA
*KAER(FM) Saint George UT
KAFC(FM) Anchorage AK
KAFE(FM) Bellingham WA
KAFF-FM Flagstaff AZ
*KAFH(FM) Great Falls MT
*KAFM(FM) Grand Junction CO
KAFN(FM) Gould AR
*KAFR(FM) Conroe TX
KAFX-FM Diboll TX
KAGB(FM) Waimea HI
KAGE-FM Winona MN
KAGG(FM) Madisonville TX
KAGH-FM Crossett AR
*KAGJ(FM) Ephraim UT
KAGL(FM) El Dorado AR
KAGM(FM) Los Lunas NM
KAGO-FM Klamath Falls OR
*KAGT(FM) Abilene TX
*KAGU(FM) Spokane WA
KAHA(FM) Olney TX
KAHM(FM) Prescott AZ
KAHR(FM) Poplar Bluff MO
*KAIA(FM) Blytheville AR
*KAIB(FM) Shafter CA
*KAIC(FM) Tucson AZ
*KAIG(FM) Dodge City KS
*KAIH(FM) Lake Havasu City AZ
*KAIK(FM) Tillamook OR
KAIM-FM Honolulu HI
*KAIO(FM) Idaho Falls ID
*KAIP(FM) Wapello IA
KAIQ(FM) Wolfforth TX
KAIR-FM Horton KS
*KAIS(FM) Redwood Valley CA
*KAIW(FM) Laramie WY
*KAIX(FM) Cheyenne WY
*KAIZ(FM) Mesquite NV
KAJA(FM) San Antonio TX
*KAJC(FM) Salem OR
KAJK(FM) White Oak TX
KAJL(FM) Adelanto CA
KAJM(FM) Payson AZ
KAJN-FM Crowley LA
KAJR(FM) Indian Wells CA
*KAJT(FM) Ada OK
*KAJX(FM) Aspen CO
KAJZ(FM) Llano TX
*KAKA(FM) Salina KS
KAKJ(FM) Marianna AR
*KAKL(FM) Anchorage AK
KAKN(FM) Naknek AK
*KAKO(FM) Ada OK
KAKQ-FM Fairbanks AK
KAKS(FM) Huntsville AR
KAKT(FM) Phoenix OR
*KAKX(FM) Mendocino CA
*KALA(FM) Davenport IA
KALC(FM) Denver CO
*KALD(FM) Caldwell TX
KALF(FM) Red Bluff CA
KALI-FM Santa Ana CA
KALK(FM) Winfield TX
KALP(FM) Alpine TX
KALQ-FM Alamosa CO
*KALR(FM) Hot Springs AR
KALS(FM) Kalispell MT
KALT-FM Alturas CA
*KALU(FM) Langston OK
*KALW(FM) San Francisco CA
*KALX(FM) Berkeley CA
KALZ(FM) Fowler CA
*KAMB(FM) Merced CA
KAMD-FM Camden AR
KAMJ-FM Gosnell AR
KAML-FM Gillette WY
KAMO-FM Rogers AR
KAMS(FM) Mammoth Spring AR
*KAMU-FM College Station TX
KAMX(FM) Luling TX
*KAMY(FM) Lubbock TX
KAMZ(FM) Tahoka TX
*KANH(FM) Emporia KS
*KANJ(FM) Giddings TX
*KANL(FM) Baker City OR
KANM(FM) Magdalena NM
*KANO(FM) Hilo HI
KANR(FM) Belle Plaine KS
KANS(FM) Emporia KS
KANT(FM) Guernsey WY
*KANU(FM) Lawrence KS
*KANV(FM) Olsburg KS
*KANW(FM) Albuquerque NM
*KANX(FM) Sheridan AR
KANY(FM) Ocean Shores WA
*KANZ(FM) Garden City KS
KAOC(FM) Cavalier ND
KAOD(FM) Babbitt MN
*KAOG(FM) Jonesboro AR
KAOI-FM Wailuku HI
*KAOR(FM) Vermillion SD
*KAOS(FM) Olympia WA
*KAOW(FM) Fort Smith AR
KAOX(FM) Kemmerer WY
KAOY(FM) Kealakekua HI
KAPA(FM) Hilo HI
KAPB-FM Marksville LA
*KAPC(FM) Butte MT
*KAPG(FM) Bentonville AR
*KAPI(FM) Ruston LA
*KAPK(FM) Grants Pass OR
*KAPM(FM) Alexandria LA
KAPW(FM) Cotton Plant AR
*KAQA(FM) Kilauea HI
*KAQD(FM) Abilene TX
*KAQF(FM) Clovis NM
*KAQQ(FM) Midland TX
*KARA(FM) Williams CA
KARB(FM) Price UT
*KARF(FM) Independence KS
*KARG(FM) Poteau OK
*KARH(FM) Forrest City AR
*KARJ(FM) Kuna ID
KARL(FM) Tracy MN
*KARM(FM) Visalia CA
KARN-FM Sheridan AR
KARO(FM) Nyssa OR
KARP-FM Dassel MN
*KARQ(FM) East Sonora CA
KARS-FM Laramie WY
*KARU(FM) Cache OK
KARV-FM Ola AR
KARX(FM) Claude TX
KARY-FM Grandview WA
KARZ(FM) Marshall MN
*KASB(FM) Bellevue WA
*KASD(FM) Rapid City SD
KASE-FM Austin TX
*KASF(FM) Alamosa CO
KASH-FM Anchorage AK
*KASK(FM) Fairfield CA
KASR(FM) Conway AR
KASS(FM) Casper WY
KAST-FM Long Beach WA
*KASU(FM) Jonesboro AR
*KASV(FM) Borger TX
*KATB(FM) Anchorage AK
KATC-FM Colorado Springs CO
KATF(FM) Dubuque IA
*KATG(FM) Athens TX
KATI(FM) California MO
KATJ-FM George CA
KATK-FM Carlsbad NM
KATM(FM) Modesto CA
KATP(FM) Amarillo TX
KATQ-FM Plentywood MT
KATR-FM Otis CO
KATS(FM) Yakima WA
KATT-FM Oklahoma City OK
KATW(FM) Lewiston ID
KATX(FM) Eastland TX
KATY-FM Idyllwild CA
KATZ-FM Alton IL
*KAUF(FM) Kennett MO
*KAUG(FM) Anchorage AK
KAUJ(FM) Grafton ND
KAUL(FM) Ellington MO
KAUM(FM) Colorado City TX
*KAUR(FM) Sioux Falls SD
KAUS-FM Austin MN
KAUU(FM) Manti UT
KAVD(FM) Limon CO
*KAVE(FM) Oakridge OR
KAVH(FM) Eudora AR
KAVJ(FM) Sutherlin OR
*KAVK(FM) Many LA
*KAVO(FM) Pampa TX
KAVV(FM) Benson AZ
*KAVW(FM) Amarillo TX
*KAVX(FM) Lufkin TX
*KAWC-FM Yuma AZ
KAWK(FM) Custer SD
KAWO(FM) Boise ID
*KAWS(FM) Boise ID
KAWV(FM) Lihue HI
KAWW-FM Heber Springs AR
*KAWZ(FM) Twin Falls ID
*KAXE(FM) Grand Rapids MN
*KAXG(FM) Gillette WY
*KAXL(FM) Green Acres CA
*KAXR(FM) Arkansas City KS
*KAXV(FM) Bastrop LA
*KAYA(FM) Hubbard NE
*KAYB(FM) Sunnyside WA
*KAYC(FM) Durant OK
KAYD-FM Silsbee TX
*KAYE-FM Tonkawa OK
KAYG(FM) Camp Wood TX
*KAYH(FM) Fayetteville AR
*KAYK(FM) Victoria TX
KAYL-FM Storm Lake IA
*KAYM(FM) Weatherford OK
KAYN(FM) Gooding ID
*KAYP(FM) Burlington IA
KAYQ(FM) Warsaw MO
*KAYT(FM) Jena LA
KAYW(FM) Meeker CO
KAYX(FM) Richmond MO
*KAZC(FM) Tishomingo OK
KAZE(FM) Ore City TX
*KAZF(FM) Hebronville TX
*KAZI-FM Austin TX
KAZR(FM) Pella IA
*KAZU(FM) Pacific Grove CA
KAZX(FM) Kirtland NM
KAZZ(FM) Deer Park WA
KBAA(FM) Grass Valley CA
KBAC(FM) Las Vegas NM
*KBAH(FM) Plainview TX
KBAJ(FM) Deer River MN
KBAL-FM San Saba TX
*KBAN(FM) De Ridder LA
*KBAQ-FM Phoenix AZ
KBAR-FM Victoria TX
KBAT(FM) Monahans TX
KBAW(FM) Zapata TX
KBAY(FM) Gilroy CA
KBAZ(FM) Hamilton MT
KBBB(FM) Billings MT
KBBD(FM) Spokane WA
KBBE(FM) McPherson KS
*KBBF(FM) Santa Rosa CA
*KBBG(FM) Waterloo IA
KBBK(FM) Lincoln NE
KBBM(FM) Jefferson City MO
KBBN-FM Broken Bow NE
KBBO-FM Houston AK
KBBQ-FM Van Buren AR
KBBT(FM) Schertz TX
KBBU(FM) Modesto CA
KBBX-FM Nebraska City NE
KBBY-FM Ventura CA
KBBZ(FM) Kalispell MT
KBCE(FM) Boyce LA
*KBCM(FM) Blytheville AR
KBCN-FM Marshall AR
KBCO-FM Boulder CO
KBCQ-FM Roswell NM
KBCR-FM Steamboat Springs CO
*KBCS(FM) Bellevue WA
KBCT-FM Waco TX
*KBCU(FM) North Newton KS
*KBCW-FM McAlester OK
*KBCX(FM) Big Spring TX
KBCY(FM) Tye TX
*KBDA(FM) Great Bend KS
KBDB-FM Forks WA
*KBDC(FM) Mason City IA
*KBDD(FM) Winfield KS
*KBDE(FM) Temple TX
*KBDG(FM) Turlock CA
*KBDH(FM) San Ardo CA
KBDK(FM) Leakey TX
KBDN(FM) Bandon OR
*KBDO(FM) Des Arc AR
KBDR(FM) Mirando City TX
KBDS(FM) Taft CA
KBDX(FM) Blanding UT
KBDZ(FM) Perryville MO
KBEA-FM Muscatine IA
KBEB-FM Rayne LA
KBEE(FM) Salt Lake City UT
KBEF(FM) Gibsland LA
KBEK(FM) Mora MN
KBEL-FM Idabel OK
*KBEM-FM Minneapolis MN
KBEN-FM Basin WY
KBEQ-FM Kansas City MO
KBER(FM) Ogden UT
*KBES(FM) Ceres CA
KBEV-FM Dillon MT
KBEW-FM Blue Earth MN
*KBEX(FM) Brenham TX
KBEY(FM) Burnet TX
KBEZ(FM) Tulsa OK
KBFB(FM) Dallas TX
KBFC(FM) Forrest City AR
KBFL-FM Buffalo MO
KBFM(FM) Edinburg TX
KBFO(FM) Aberdeen SD
KBFP-FM Delano CA
*KBFR(FM) Bismarck ND
KBFX(FM) Anchorage AK
*KBGA(FM) Missoula MT
KBGL(FM) Larned KS
*KBGM(FM) Park Hills MO
KBGO(FM) Waco TX
KBGR(FM) Beebe AR
KBGX(FM) Keaau HI
KBGY(FM) Faribault MN
*KBHE-FM Rapid City SD
*KBHG(FM) Alexandria MN
KBHH(FM) Kerman CA
KBHI(FM) Miner MO
KBHL(FM) Osakis MN
*KBHN(FM) Booneville AR
KBHP(FM) Bemidji MN
KBHR(FM) Big Bear City CA
KBHT(FM) Crockett TX
*KBHU-FM Spearfish SD
*KBHW(FM) International Falls MN
*KBHZ(FM) Willmar MN
*KBIA(FM) Columbia MO
KBIC(FM) Raymondville TX
KBIG-FM Los Angeles CA
*KBIL(FM) Park City MT
KBIM-FM Roswell NM
*KBIO(FM) Natchitoches LA
KBIQ(FM) Manitou Springs CO
KBIU(FM) Lake Charles LA
*KBIY(FM) Van Buren MO
*KBJQ(FM) Bronson KS
*KBJS(FM) Jacksonville TX
KBJX(FM) Shelley ID
KBKB-FM Fort Madison IA
*KBKC(FM) Moberly MO
KBKG(FM) Corning AR
KBKH(FM) Shamrock TX
KBKK(FM) Ball LA
KBKL(FM) Grand Junction CO
*KBKN(FM) Lamesa TX
KBKO-FM Bakersfield CA
KBKS-FM Tacoma WA
KBKY(FM) Merced CA
KBKZ(FM) Raton NM
KBLB(FM) Nisswa MN
*KBLD(FM) Kennewick WA
KBLL-FM Helena MT
KBLO(FM) Corcoran CA
KBLP(FM) Lindsay OK
KBLQ-FM Logan UT
KBLR-FM Blair NE
KBLS(FM) North Fort Riley KS
KBLT(FM) Leakey TX
*KBLW(FM) Billings MT
KBLX-FM Berkeley CA
*KBLZ(FM) Winona TX
KBMB(FM) Sacramento CA
*KBMC(FM) Bozeman MT
*KBMD(FM) Marble Falls TX
KBMG(FM) Evanston WY
*KBMH(FM) Holbrook AZ
KBMI(FM) Roma TX
*KBMJ(FM) Heber Springs AR
*KBMK(FM) Bismarck ND
*KBMM(FM) Odessa TX
*KBMP(FM) Enterprise KS
*KBMQ(FM) Monroe LA
KBMV-FM Birch Tree MO
KBMX(FM) Proctor MN
KBNA-FM El Paso TX
KBNG(FM) Ridgway CO
*KBNJ(FM) Corpus Christi TX
*KBNL(FM) Laredo TX
*KBNO-FM White Salmon WA

*KBNR(FM) Brownsville TX
KBNU(FM) Uvalde TX
*KBNV(FM) Fayetteville AR
KBOA-FM Piggott AR
KBOB-FM De Witt IA
KBOC(FM) Bridgeport TX
KBOE-FM Oskaloosa IA
*KBOJ(FM) Worthington MN
KBON(FM) Mamou LA
*KBOO(FM) Portland OR
KBOQ(FM) Carmel CA
KBOS-FM Tulare CA
KBOT(FM) Pelican Rapids MN
KBOX(FM) Lompoc CA
KBOY-FM Medford OR
KBOZ-FM Bozeman MT
KBPA(FM) San Marcos TX
*KBPB(FM) Harrison AR
*KBPG(FM) Montevideo MN
KBPI(FM) Denver CO
*KBPK(FM) Buena Park CA
*KBPN(FM) Brainerd MN
*KBPR(FM) Brainerd MN
*KBPS-FM Portland OR
*KBPU(FM) De Queen AR
*KBPW(FM) Hampton AR
KBQB(FM) Chico CA
*KBQC(FM) Independence KS
KBQF(FM) McFarland CA
KBQI(FM) Albuquerque NM
KBQQ(FM) Pinesdale MT
KBRA(FM) Freer TX
KBRB-FM Ainsworth NE
KBRE(FM) Atwater CA
KBRG(FM) San Jose CA
KBRJ(FM) Anchorage AK
KBRK-FM Brookings SD
KBRQ(FM) Hillsboro TX
*KBRW-FM Barrow AK
KBRX-FM O'Neill NE
*KBSA(FM) El Dorado AR
*KBSB(FM) Bemidji MN
KBSG-FM Tacoma WA
*KBSJ(FM) Jackpot NV
*KBSK(FM) McCall ID
*KBSM(FM) McCall ID
KBSO(FM) Corpus Christi TX
*KBSQ(FM) McCall ID
*KBSS(FM) Sun Valley ID
KBST-FM Big Spring TX
*KBSU-FM Boise ID
*KBSW(FM) Twin Falls ID
*KBSX(FM) Boise ID
*KBSY(FM) Burley ID
KBTA-FM Batesville AR
KBTE(FM) Tulia TX
*KBTL(FM) El Dorado KS
KBTN-FM Neosho MO
KBTO(FM) Bottineau ND
KBTQ(FM) Harlingen TX
KBTS(FM) Big Spring TX
KBTT(FM) Haughton LA
KBTW(FM) Lenwood CA
KBTY(FM) Benjamin TX
KBUA(FM) San Fernando CA
*KBUB(FM) Brownwood TX
KBUC(FM) Raymondville TX
KBUD(FM) Sardis MS
KBUE(FM) Long Beach CA
*KBUG(FM) Malin OR
KBUK(FM) La Grange TX
KBUL-FM Carson City NV
KBUS(FM) Paris TX
*KBUT(FM) Crested Butte CO
*KBUW(FM) Buffalo WY
KBUX(FM) Quartzsite AZ
*KBUZ(FM) Topeka KS
KBVA(FM) Bella Vista AR
KBVB(FM) Barnesville MN
KBVC(FM) Buena Vista CO
*KBVM(FM) Portland OR
*KBVR(FM) Corvallis OR
KBVU-FM Alta IA
*KBWA(FM) Brush CO
*KBWC(FM) Marshall TX
KBWF(FM) San Francisco CA

KBWM(FM) Breckenridge TX
KBWS-FM Sisseton SD
KBWT(FM) Santa Anna TX
KBXB(FM) Sikeston MO
KBXL(FM) Caldwell ID
*KBXO(FM) Coachella CA
KBXR(FM) Columbia MO
KBXX(FM) Houston TX
KBYB(FM) Hope AR
*KBYI(FM) Rexburg ID
KBYN(FM) Arnold CA
KBYO-FM Farmerville LA
*KBYR-FM Rexburg ID
*KBYU-FM Provo UT
KBYZ(FM) Bismarck ND
KBZB(FM) Pioche NV
KBZD(FM) Amarillo TX
KBZE(FM) Berwick LA
KBZI(FM) Deerfield MO
KBZM(FM) Big Sky MT
KBZN(FM) Ogden UT
KBZQ(FM) Lawton OK
KBZS(FM) Wichita Falls TX
KBZT(FM) San Diego CA
KBZU(FM) Albuquerque NM
*KCAC(FM) Camden AR
KCAD(FM) Dickinson ND
*KCAI(FM) Kingman AZ
KCAJ-FM Roseau MN
KCAL-FM Redlands CA
KCAQ(FM) Oxnard CA
KCAR-FM Galena KS
*KCAS(FM) McCook TX
*KCAW(FM) Sitka AK
KCAY(FM) Russell KS
*KCBI(FM) Dallas TX
KCBS-FM Los Angeles CA
*KCBX(FM) San Luis Obispo CA
KCBZ(FM) Cannon Beach OR
*KCCD(FM) Moorhead MN
*KCCK-FM Cedar Rapids IA
KCCL(FM) Placerville CA
*KCCM-FM Moorhead MN
KCCN-FM Honolulu HI
KCCQ(FM) Ames IA
*KCCU(FM) Lawton OK
KCCV-FM Olathe KS
KCCY(FM) Pueblo CO
KCDA(FM) Post Falls ID
KCDD(FM) Hamlin TX
KCDG(FM) Madison MO
KCDL(FM) Cordell OK
KCDQ(FM) Douglas AZ
*KCDS(FM) Deadhorse AK
KCDU(FM) Carmel CA
KCDV(FM) Cordova AK
KCDX(FM) Florence AZ
KCDY(FM) Carlsbad NM
KCDZ(FM) Twentynine Palms CA
*KCEA(FM) Atherton CA
KCEC-FM Wellton AZ
*KCED(FM) Centralia WA
KCEL(FM) California City CA
*KCEP(FM) Las Vegas NV
KCEZ(FM) Los Molinos CA
*KCFA(FM) Arnold CA
*KCFB(FM) Saint Cloud MN
*KCFN(FM) Wichita KS
*KCFP(FM) Pueblo CO
*KCFS(FM) Sioux Falls SD
*KCFV(FM) Ferguson MO
KCFX(FM) Harrisonville MO
*KCFY(FM) Yuma AZ
KCGB-FM Hood River OR
KCGL(FM) Powell WY
KCGM(FM) Scobey MT
KCGN-FM Ortonville MN
KCGQ-FM Gordonville MO
KCGY(FM) Laramie WY
KCHA-FM Charles City IA
KCHC(FM) Willows CA
KCHE-FM Cherokee IA
KCHI-FM Chillicothe MO
*KCHO(FM) Chico CA
KCHQ(FM) Driggs ID
KCHX(FM) Midland TX

KCHZ(FM) Ottawa KS
*KCIC(FM) Grand Junction CO
*KCIE(FM) Dulce NM
*KCIF(FM) Hilo HI
KCII-FM Washington IA
KCIJ(FM) Atlanta LA
KCIL(FM) Houma LA
KCIN(FM) Cedar City UT
*KCIR(FM) Twin Falls ID
KCIV(FM) Mount Bullion CA
KCIX(FM) Garden City ID
KCJC(FM) Dardanelle AR
KCJF(FM) Earle AR
*KCJH(FM) Livingston CA
KCJK(FM) Garden City MO
KCJN(FM) Simmesport LA
*KCJX(FM) Carbondale CO
KCKC(FM) Kansas City MO
KCKL(FM) Malakoff TX
*KCKR(FM) Kaplan LA
KCKS(FM) Concordia KS
*KCKT(FM) Crockett TX
KCLB-FM Coachella CA
*KCLC(FM) Saint Charles MO
KCLD-FM Saint Cloud MN
KCLH(FM) Caledonia MN
KCLK-FM Clarkston WA
KCLL(FM) San Angelo TX
KCLQ(FM) Lebanon MO
KCLR-FM Boonville MO
KCLS(FM) Ely NV
KCLT(FM) West Helena AR
*KCLU(FM) Thousand Oaks CA
KCLV-FM Clovis NM
KCLY(FM) Clay Center KS
KCMB(FM) Baker City OR
*KCME(FM) Manitou Springs CO
*KCMF(FM) Fergus Falls MN
*KCMH(FM) Mountain Home AR
KCMI(FM) Terrytown NE
KCML(FM) Saint Joseph MN
KCMM(FM) Belgrade MT
KCMO-FM Kansas City MO
*KCMP(FM) Northfield MN
KCMQ(FM) Columbia MO
*KCMR(FM) Mason City IA
KCMS(FM) Edmonds WA
KCMT(FM) Oro Valley AZ
KCMX-FM Ashland OR
KCNA(FM) Cave Junction OR
*KCND(FM) Bismarck ND
*KCNE-FM Chadron NE
*KCNJ(FM) Oskaloosa IA
KCNL(FM) Sunnyvale CA
KCNM-FM Garapan-Saipan NP
KCNO(FM) Alturas CA
KCNQ(FM) Kernville CA
*KCNT(FM) Hastings NE
*KCNV(FM) Las Vegas NV
KCNY(FM) Bald Knob AR
KCOB-FM Newton IA
KCOL-FM Groves TX
KCOO(FM) Dunkerton IA
KCOR-FM Comfort TX
*KCOU(FM) Columbia MO
*KCOZ(FM) Point Lookout MO
*KCPB-FM Warrenton OR
*KCPC(FM) Sealy TX
KCPI(FM) Albert Lea MN
*KCPR(FM) San Luis Obispo CA
*KCPW-FM Salt Lake City UT
KCQQ(FM) Davenport IA
*KCRB-FM Bemidji MN
KCRE-FM Crescent City CA
KCRF-FM Lincoln City OR
*KCRH(FM) Hayward CA
*KCRI(FM) Indio CA
KCRK-FM Colville WA
*KCRL(FM) Sunrise Beach MO
KCRN-FM San Angelo TX
KCRR(FM) Grundy Center IA
KCRS-FM Midland TX
KCRT-FM Trinidad CO
*KCRU(FM) Oxnard CA
KCRV-FM Caruthersville MO
*KCRW(FM) Santa Monica CA

KCRX-FM Seaside OR
*KCRY(FM) Mojave CA
KCRZ(FM) Tipton CA
*KCSB-FM Santa Barbara CA
*KCSC(FM) Edmond OK
*KCSD(FM) Sioux Falls SD
*KCSH(FM) Ellensburg WA
KCSI(FM) Red Oak IA
*KCSM(FM) San Mateo CA
*KCSN(FM) Northridge CA
*KCSP-FM Casper WY
*KCSS(FM) Turlock CA
KCST-FM Florence OR
*KCSU-FM Fort Collins CO
KCSY(FM) Twisp WA
KCTN(FM) Garnavillo IA
KCTR-FM Billings MT
KCTT-FM Yellville AR
KCTX-FM Childress TX
KCTY(FM) Wayne NE
KCUA(FM) Naples UT
KCUB-FM Ranger TX
KCUF(FM) El Jebel CO
KCUG(FM) Chugwater WY
*KCUK(FM) Chevak AK
KCUL-FM Marshall TX
*KCUR-FM Kansas City MO
KCUV(FM) Greenwood Village CO
KCVD(FM) New England ND
KCVF(FM) Sarles ND
KCVG(FM) Medina ND
KCVI(FM) Blackfoot ID
*KCVJ(FM) Osceola MO
*KCVK(FM) Otterville MO
KCVM(FM) Hudson IA
*KCVN(FM) Cozad NE
*KCVO-FM Camdenton MO
*KCVQ(FM) Knob Noster MO
KCVR-FM Columbia CA
*KCVS(FM) Salina KS
KCVT(FM) Silver Lake KS
KCVW(FM) Kingman KS
*KCVX(FM) Salem MO
*KCVZ(FM) Dixon MO
*KCWC-FM Riverton WY
KCWD(FM) Harrison AR
KCWN(FM) New Sharon IA
KCWR(FM) Bakersfield CA
*KCWU(FM) Ellensburg WA
*KCWW(FM) Evanston WY
KCXM(FM) Lee's Summit MO
KCXR(FM) Taft OK
KCXX(FM) Lake Arrowhead CA
KCXY(FM) East Camden AR
KCYE(FM) North Las Vegas NV
KCYN(FM) Moab UT
KCYQ(FM) Elsinore UT
KCYS(FM) Seaside OR
KCYY(FM) San Antonio TX
KCZE(FM) New Hampton IA
KCZO(FM) Carrizo Springs TX
KCZQ(FM) Cresco IA
KDAA(FM) Rolla MO
KDAD(FM) Douglas WY
KDAG(FM) Farmington NM
*KDAI(FM) Scottsbluff NE
KDAL-FM Duluth MN
KDAO-FM Eldora IA
KDAP-FM Douglas AZ
*KDAQ(FM) Shreveport LA
KDAR(FM) Oxnard CA
KDAT(FM) Cedar Rapids IA
KDAY(FM) Redondo Beach CA
KDB(FM) Santa Barbara CA
KDBB(FM) Bonne Terre MO
KDBH(FM) Natchitoches LA
KDBI(FM) Emmett ID
KDBL(FM) Toppenish WA
KDBN(FM) Haltom City TX
KDBR(FM) Kalispell MT
KDBX(FM) Clear Lake SD
KDBZ(FM) Anchorage AK
KDCD(FM) San Angelo TX
KDCQ(FM) Coos Bay OR
*KDCR(FM) Sioux Center IA
*KDCV-FM Blair NE

KDDB(FM) Waipahu HI
KDDD-FM Dumas TX
KDDG(FM) Albany MN
KDDK(FM) Franklin LA
KDDQ(FM) Comanche OK
KDDS-FM Elma WA
KDDV-FM Wright WY
KDDX(FM) Spearfish SD
KDEC-FM Decorah IA
KDEL-FM Arkadelphia AR
KDEM(FM) Deming NM
KDEP(FM) Garibaldi OR
KDES-FM Palm Springs CA
KDEW-FM De Witt AR
KDEX-FM Dexter MO
KDEZ(FM) Brandon SD
KDFC-FM San Francisco CA
KDFM(FM) Falfurrias TX
KDFO-FM Delano CA
*KDFR(FM) Des Moines IA
KDGE(FM) Fort Worth-Dallas TX
KDGL(FM) Yucca Valley CA
KDGS(FM) Andover KS
KDHT(FM) Cedar Park TX
*KDHX(FM) Saint Louis MO
*KDIM(FM) Coweta OK
KDIS-FM Little Rock AR
*KDJC(FM) Baker City OR
KDJE(FM) Jacksonville AR
KDJF(FM) Delta Junction AK
KDJK(FM) Mariposa CA
KDJR(FM) De Soto MO
KDJS-FM Willmar MN
KDKB(FM) Mesa AZ
KDKD-FM Clinton MO
KDKK-FM Park Rapids MN
*KDKL(FM) Coalinga CA
*KDKR(FM) Decatur TX
KDKS-FM Blanchard LA
KDLD(FM) Santa Monica CA
KDLE(FM) Newport Beach CA
*KDLI(FM) Del Rio TX
KDLK-FM Del Rio TX
*KDLL(FM) Kenai AK
KDLO-FM Watertown SD
KDLS-FM Perry IA
KDLX(FM) Makawao HI
KDLY(FM) Lander WY
KDMG(FM) Burlington IA
*KDMR(FM) Mitchellville IA
KDMU(FM) Bloomfield IA
KDMX(FM) Dallas TX
*KDNA(FM) Yakima WA
KDND(FM) Sacramento CA
*KDNE(FM) Crete NE
*KDNI(FM) Duluth MN
*KDNK(FM) Glenwood Springs CO
KDNN(FM) Honolulu HI
KDNO(FM) Thermopolis WY
KDNS(FM) Downs KS
*KDNW(FM) Duluth MN
KDOE(FM) Antlers OK
KDOG(FM) North Mankato MN
KDOK(FM) Tyler TX
KDOM-FM Windom MN
KDON-FM Salinas CA
KDOT(FM) Reno NV
*KDOV(FM) Medford OR
KDPM(FM) Cottage Grove OR
*KDPR(FM) Dickinson ND
KDQN-FM De Queen AR
KDRB(FM) Des Moines IA
*KDRE(FM) Sterling CO
KDRF(FM) Albuquerque NM
*KDRH(FM) King City CA
KDRK-FM Spokane WA
KDRM(FM) Moses Lake WA
KDRS-FM Paragould AR
KDRW(FM) Hewitt TX
KDRX(FM) Rocksprings TX
*KDSC(FM) Thousand Oaks CA
*KDSD-FM Pierpont SD
KDSK(FM) Grants NM
KDSN-FM Denison IA
KDSR(FM) Williston ND
KDSS(FM) Ely NV

KDST(FM) Dyersville IA
*KDSU(FM) Fargo ND
KDTR(FM) Florence MT
*KDUB(FM) Dubuque IA
KDUC(FM) Barstow CA
KDUK-FM Florence OR
KDUQ(FM) Ludlow CA
*KDUR(FM) Durango CO
KDUT(FM) Randolph UT
*KDUV(FM) Visalia CA
*KDUW(FM) Douglas WY
KDUX-FM Aberdeen WA
KDVA(FM) Buckeye AZ
KDVE(FM) Pittsburg TX
*KDVI(FM) Devils Lake ND
KDVL(FM) Devils Lake ND
*KDVS(FM) Davis CA
KDVV(FM) Topeka KS
KDWB-FM Richfield MN
*KDWG(FM) Dillon MT
KDWY(FM) Diamondville WY
*KDXL(FM) Saint Louis Park MN
KDXT(FM) Victor MT
KDXX(FM) Benbrook TX
KDXY(FM) Lake City AR
KDYN-FM Ozark AR
KDZA-FM Pueblo CO
KDZN(FM) Glendive MT
KDZY(FM) McCall ID
*KEAF(FM) Fort Smith AR
KEAG(FM) Anchorage AK
KEAL(FM) Taft CA
KEAN-FM Abilene TX
*KEAR-FM Sacramento CA
KEAU(FM) Choteau MT
KEBN(FM) Garden Grove CA
KEBT(FM) Lost Hills CA
*KECG(FM) El Cerrito CA
KECH-FM Sun Valley ID
KECO(FM) Elk City OK
KEDD(FM) Johannesburg CA
KEDG(FM) Alexandria LA
KEDJ(FM) Gilbert AZ
*KEDM(FM) Monroe LA
*KEDP(FM) Las Vegas NM
*KEDR(FM) Bay City TX
*KEDT-FM Corpus Christi TX
*KEEH(FM) Spokane WA
KEEI(FM) Hanapepe HI
KEEP(FM) Bandera TX
KEEY-FM Saint Paul MN
KEEZ-FM Mankato MN
KEFH(FM) Clarendon TX
*KEFR(FM) Le Grand CA
*KEFS(FM) North Powder OR
*KEFX(FM) Twin Falls ID
KEGA(FM) Oakley UT
KEGH(FM) Brigham City UT
KEGI(FM) Jonesboro AR
KEGK(FM) Wahpeton ND
KEGL(FM) Fort Worth TX
*KEGR(FM) Fort Dodge IA
KEGX(FM) Richland WA
KEHK(FM) Brownsville OR
KEJJ(FM) Gunnison CO
KEJL(FM) Eunice NM
KEJS(FM) Lubbock TX
KEJY(FM) Blue Lake CA
KEKA-FM Eureka CA
KEKB(FM) Fruita CO
*KEKL(FM) Mesquite NV
KEKO(FM) Hebronville TX
KEKS(FM) Olpe KS
KELD-FM Hampton AR
KELE-FM Mountain Grove MO
KELI(FM) San Angelo TX
KELN(FM) North Platte NE
KELO-FM Sioux Falls SD
*KELP-FM Mesquite NM
KELR-FM Chariton IA
*KELT(FM) Roswell NM
*KELU(FM) Clovis NM
KEMA(FM) Three Rivers TX
*KEMC(FM) Billings MT
KEMX(FM) Locust Grove OK
KENA-FM Mena AR

KEND(FM) Roswell NM
KENR(FM) Superior MT
*KENW-FM Portales NM
KENZ(FM) Ogden UT
KEOJ(FM) Caney KS
KEOK(FM) Tahlequah OK
*KEOL(FM) La Grande OR
*KEOM(FM) Mesquite TX
*KEOS(FM) College Station TX
*KEPC(FM) Colorado Springs CO
*KEPI(FM) Eagle Pass TX
*KEPX(FM) Eagle Pass TX
*KEQX(FM) Stephenville TX
*KERA(FM) Dallas TX
KERB-FM Kermit TX
KERM(FM) Torrington WY
KERX(FM) Paris AR
*KESD(FM) Brookings SD
KESM-FM El Dorado Springs MO
KESN(FM) Allen TX
KESO(FM) South Padre Island TX
KESR(FM) Shasta Lake City CA
KESS-FM Lewisville TX
KESY(FM) Cuba MO
KESZ(FM) Phoenix AZ
*KETR(FM) Commerce TX
KETT(FM) Mitchell NE
KETX-FM Livingston TX
KEUG(FM) Veneta OR
*KEUL(FM) Girdwood AK
KEUN-FM Eunice LA
KEVE(FM) Ingram TX
KEWB(FM) Anderson CA
KEWL-FM New Boston TX
*KEWU-FM Cheney WA
KEXA(FM) King City CA
KEXL(FM) Norfolk NE
*KEXP-FM Seattle WA
*KEYA(FM) Belcourt ND
KEYB(FM) Altus OK
KEYE-FM Perryton TX
KEYF-FM Cheney WA
KEYG-FM Grand Coulee WA
KEYJ-FM Abilene TX
KEYN-FM Wichita KS
KEYW(FM) Pasco WA
KEZA(FM) Fayetteville AR
KEZB(FM) Edna TX
KEZE(FM) Spokane WA
*KEZF(FM) Eaton CO
KEZJ-FM Twin Falls ID
KEZK-FM Saint Louis MO
KEZN(FM) Palm Desert CA
KEZO-FM Omaha NE
KEZP(FM) Bunkie LA
KEZQ(FM) West Yellowstone MT
KEZR(FM) San Jose CA
KEZS-FM Cape Girardeau MO
KEZZ(FM) Walden CO
*KFAE-FM Richland WA
*KFAI(FM) Minneapolis MN
KFAN-FM Johnson City TX
KFAT(FM) Anchorage AK
KFAV(FM) Warrenton MO
KFBD-FM Waynesville MO
*KFBN(FM) Fargo ND
KFBZ(FM) Haysville KS
*KFCF(FM) Fresno CA
KFCM(FM) Cherokee Village AR
KFDI-FM Wichita KS
*KFDN(FM) Lakewood CO
KFEB(FM) Campbell MO
KFEG(FM) Klamath Falls OR
*KFER(FM) Santa Cruz CA
KFFA-FM Helena AR
KFFB(FM) Fairfield Bay AR
KFFF-FM Boone IA
KFFG(FM) Los Altos CA
KFFM(FM) Yakima WA
*KFFW(FM) Cabool MO
KFFX(FM) Emporia KS
KFGE(FM) Milford NE
KFGI(FM) Crosby MN
KFGL(FM) Abilene TX
KFGY(FM) Healdsburg CA
*KFHC(FM) Ponca NE

KFH-FM Clearwater KS
*KFHL(FM) Wasco CA
KFIL-FM Preston MN
KFIN(FM) Jonesboro AR
*KFIO(FM) East Wenatchee WA
KFIS(FM) Scappoose OR
KFIX(FM) Plainville KS
*KFJC(FM) Los Altos CA
KFJK(FM) Fresno CA
*KFJM(FM) Grand Forks ND
KFKF-FM Kansas City KS
*KFKX(FM) Hastings NE
*KFLB-FM Odessa TX
KFLG-FM Kingman AZ
KFLI(FM) Des Arc AR
*KFLO-FM Blanchard LA
KFLP-FM Floydada TX
*KFLQ(FM) Albuquerque NM
*KFLR-FM Phoenix AZ
KFLS-FM Tulelake CA
*KFLT-FM Tucson AZ
*KFLV(FM) Wilber NE
KFLW(FM) Saint Robert MO
KFLX(FM) Kachina Village AZ
KFLY(FM) Corvallis OR
KFMA(FM) Green Valley AZ
KFMB-FM San Diego CA
KFMC(FM) Fairmont MN
KFMF(FM) Chico CA
KFMH(FM) Belle Fourche SD
KFMI(FM) Eureka CA
KFMJ(FM) Ketchikan AK
KFMK(FM) Round Rock TX
KFML(FM) Little Falls MN
KFMM(FM) Thatcher AZ
KFMN(FM) Lihue HI
KFMQ(FM) Gallup NM
KFMR(FM) Marbleton WY
KFMT-FM Fremont NE
KFMU-FM Oak Creek CO
KFMW(FM) Waterloo IA
KFMX-FM Lubbock TX
KFMY(FM) Raymond WA
KFNC(FM) Beaumont TX
KFNF(FM) Oberlin KS
KFNK(FM) Eatonville WA
*KFNL(FM) Kindred ND
*KFNO(FM) Fresno CA
KFNS-FM Troy MO
KFNV-FM Ferriday LA
*KFNW-FM Fargo ND
KFOG(FM) San Francisco CA
KFPB(FM) Chino Valley AZ
*KFPR(FM) Redding CA
KFPW-FM Barling AR
*KFRB(FM) Bakersfield CA
KFRC-FM San Francisco CA
*KFRD(FM) Butte MT
KFRG(FM) San Bernardino CA
*KFRI(FM) Stanton TX
*KFRJ(FM) China Lake CA
KFRO-FM Gilmer TX
*KFRP(FM) Coalinga CA
KFRQ(FM) Harlingen TX
KFRR(FM) Woodlake CA
*KFRS(FM) Soledad CA
*KFRT(FM) Butte MT
*KFRW(FM) Great Falls MT
KFRX(FM) Lincoln NE
*KFRY(FM) Pueblo CO
KFRZ(FM) Green River WY
KFSE(FM) Kasilof AK
KFSH-FM Anaheim CA
*KFSI(FM) Rochester MN
*KFSK(FM) Petersburg AK
KFSO-FM Visalia CA
*KFSR(FM) Fresno CA
KFST-FM Fort Stockton TX
KFSZ(FM) Munds Park AZ
KFTE(FM) Breaux Bridge LA
*KFTG(FM) Pasadena TX
KFTI-FM Newton KS
KFTK(FM) Florissant MO
KFTT(FM) Bagdad AZ
KFTX(FM) Kingsville TX
KFTZ(FM) Idaho Falls ID

KFUO-FM Clayton MO
KFVR-FM La Junta CO
KFWR(FM) Mineral Wells TX
KFXI(FM) Marlow OK
KFXJ(FM) Augusta KS
KFXR-FM Chinle AZ
KFXS(FM) Rapid City SD
*KFXT(FM) Sulphur OK
KFXV(FM) Kensett AR
KFXX-FM Hugoton KS
KFXZ-FM Opelousas LA
KFYV(FM) Ojai CA
KFYX(FM) Texarkana AR
KFYZ-FM Bennington OK
KFZO(FM) Denton TX
KFZX(FM) Gardendale TX
*KGAC(FM) Saint Peter MN
KGAP(FM) Clarksville TX
KGAS-FM Carthage TX
KGBA-FM Holtville CA
KGB-FM San Diego CA
*KGBI-FM Omaha NE
*KGBM(FM) Randsburg CA
KGBR(FM) Gold Beach OR
KGBT-FM McAllen TX
KGBX-FM Nixa MO
KGBY(FM) Sacramento CA
*KGCB(FM) Prescott AZ
KGCL(FM) Ten Sleep WY
*KGCR(FM) Goodland KS
KGCX(FM) Sidney MT
KGDN(FM) Pasco WA
*KGDP-FM Santa Maria CA
KGDQ(FM) Colorado Springs CO
KGEE(FM) Pecos TX
KGEN-FM Hanford CA
*KGFA(FM) Great Falls MT
*KGFC(FM) Great Falls MT
KGFM(FM) Bakersfield CA
KGFT(FM) Pueblo CO
KGFX-FM Pierre SD
KGFY(FM) Stillwater OK
*KGGA(FM) Gallup NM
KGGB(FM) Yorktown TX
KGGF-FM Fredonia KS
KGGG(FM) Haven KS
KGGI(FM) Riverside CA
KGGL(FM) Missoula MT
KGGM(FM) Delhi LA
KGGO(FM) Des Moines IA
KGHL-FM Billings MT
*KGHP(FM) Gig Harbor WA
*KGHR(FM) Tuba City AZ
*KGHY(FM) Beaumont TX
KGIM-FM Redfield SD
KGKL-FM San Angelo TX
KGKS(FM) Scott City MO
KGLC(FM) Miami OK
*KGLF(FM) Doss TX
KGLI(FM) Sioux City IA
KGLM-FM Anaconda MT
*KGLP(FM) Gallup NM
*KGLT(FM) Bozeman MT
KGLU(FM) Gideon MO
*KGLV(FM) Manhattan KS
KGLX(FM) Gallup NM
*KGLY(FM) Tyler TX
KGMG(FM) Oracle AZ
KGMN(FM) Kingman AZ
KGMO(FM) Cape Girardeau MO
KGMX(FM) Lancaster CA
KGMZ-FM Aiea HI
*KGNA-FM Arnold MO
KGNC-FM Amarillo TX
*KGNN-FM Cuba MO
KGNT(FM) Smithfield UT
*KGNU-FM Boulder CO
*KGNV(FM) Washington MO
*KGNZ(FM) Abilene TX
KGON(FM) Portland OR
KGOR(FM) Omaha NE
KGOT(FM) Anchorage AK
*KGOU(FM) Norman OK
KGOZ(FM) Gallatin MO
KGPQ(FM) Monticello AR
*KGPR(FM) Great Falls MT

KGPZ(FM) Coleraine MN
KGRA(FM) Jefferson IA
KGRC(FM) Hannibal MO
*KGRD(FM) Orchard NE
*KGRG-FM Auburn WA
*KGRI(FM) Lebanon OR
KGRK(FM) Glenrock WY
*KGRM(FM) Grambling LA
KGRP(FM) Cazadero CA
KGRR(FM) Epworth IA
KGRS(FM) Burlington IA
KGRT-FM Las Cruces NM
KGRW(FM) Friona TX
*KGSF(FM) Anderson MO
KGSG(FM) Pasco WA
*KGSP(FM) Parkville MO
KGSR(FM) Bastrop TX
KGTM(FM) Rexburg ID
KGTR(FM) Larned KS
*KGTS(FM) College Place WA
KGTW(FM) Ketchikan AK
*KGUD(FM) Longmont CO
KGUM-FM Dededo GU
*KGVA(FM) Fort Belknap Agency MT
KGVE(FM) Grove OK
*KGWP(FM) Pittsburg TX
KGWY(FM) Gillette WY
KGY-FM McCleary WA
*KGZO(FM) Shafter CA
KHAD(FM) Mills WY
*KHAI(FM) Wahiawa HI
KHAK(FM) Cedar Rapids IA
KHAL(FM) Condon OR
KHAM(FM) Britt IA
*KHAP(FM) Chico CA
KHAY(FM) Ventura CA
KHAZ(FM) Hays KS
KHBM-FM Monticello AR
KHBT(FM) Humboldt IA
KHBZ-FM Oklahoma City OK
KHCA(FM) Wamego KS
*KHCB-FM Houston TX
*KHCC-FM Hutchinson KS
*KHCD(FM) Salina KS
*KHCF(FM) Fredericksburg TX
*KHCJ(FM) Jefferson TX
KHCK-FM Robinson TX
*KHCL(FM) Arcadia LA
*KHCO(FM) Hayden CO
*KHCP(FM) Paris TX
KHCR(FM) Bismarck MO
*KHCS(FM) Palm Desert CA
*KHCT(FM) Great Bend KS
*KHDC(FM) Chualar CA
KHDK(FM) New London IA
KHDR(FM) Lenwood CA
KHDV(FM) Darby MT
*KHDX(FM) Conway AR
*KHED(FM) Arkadelphia AR
KHEI-FM Kihei HI
KHER(FM) Crystal City TX
KHES(FM) Rocksprings TX
KHEY-FM El Paso TX
KHFI-FM Georgetown TX
KHFM(FM) Santa Fe NM
*KHFR(FM) Santa Maria CA
KHGE(FM) Fresno CA
KHGG-FM Waldron AR
*KHGN(FM) Kirksville MO
KHGQ(FM) Quincy CA
KHHK(FM) Yakima WA
KHHL(FM) Leander TX
KHHT(FM) Los Angeles CA
KHHZ(FM) Oroville CA
*KHIB(FM) Bastrop TX
*KHID(FM) McAllen TX
*KHII(FM) Cloudcroft NM
*KHIM(FM) Mangum OK
KHIP(FM) Gonzales CA
KHIX(FM) Carlin NV
*KHJC(FM) Lihue HI
KHJL(FM) Thousand Oaks CA
KHJQ(FM) Susanville CA
KHJZ-FM Houston TX
KHKC-FM Atoka OK
*KHKE(FM) Cedar Falls IA

KHKI(FM) Des Moines IA
KHKK(FM) Modesto CA
*KHKL(FM) Laytonville CA
KHKN(FM) Benton AR
KHKR-FM East Helena MT
KHKS(FM) Denton TX
*KHKV(FM) Kerrville TX
KHKX(FM) Odessa TX
KHKZ(FM) Mercedes TX
KHLA(FM) Jennings LA
KHLB(FM) Mason TX
KHLE(FM) Burnet TX
KHLL(FM) Richwood LA
KHLR(FM) Maumelle AR
KHLS(FM) Blytheville AR
*KHLV(FM) Helena MT
KHMB(FM) Hamburg AR
KHMC(FM) Goliad TX
KHME(FM) Winona MN
*KHMG(FM) Barrigada GU
*KHML(FM) Madisonville TX
KHMR(FM) Lovelady TX
*KHMS(FM) Victorville CA
KHMX(FM) Houston TX
KHMY(FM) Pratt KS
KHNA(FM) Hanna WY
*KHNE-FM Hastings NE
KHNK(FM) Columbia Falls MT
KHNR-FM Honolulu HI
*KHNS(FM) Haines AK
KHOC(FM) Casper WY
KHOD(FM) Des Moines NM
*KHOE(FM) Fairfield IA
KHOK(FM) Hoisington KS
KHOM(FM) Salem AR
KHOP(FM) Oakdale CA
KHOS-FM Sonora TX
KHOT-FM Paradise Valley AZ
KHOV-FM Wickenburg AZ
*KHOY(FM) Laredo TX
KHOZ-FM Harrison AR
KHPA(FM) Hope AR
KHPE(FM) Albany OR
*KHPO(FM) Port O'Connor TX
KHPQ(FM) Clinton AR
*KHPR(FM) Honolulu HI
*KHPS(FM) Camp Wood TX
KHPT(FM) Conroe TX
*KHPU(FM) Brownwood TX
KHQT(FM) Las Cruces NM
KHRD(FM) Weaverville CA
*KHRI(FM) Hollister CA
KHRQ(FM) Baker CA
KHRS(FM) Winthrop MN
KHRT-FM Minot ND
KHRU(FM) Beulah ND
KHRW(FM) Wright WY
KHSL-FM Paradise CA
*KHSR(FM) Crescent City CA
KHSS(FM) Walla Walla WA
KHST(FM) Lamar MO
*KHSU-FM Arcata CA
*KHTA(FM) Wake Village TX
KHTB(FM) Provo UT
KHTC(FM) Lake Jackson TX
KHTE-FM England AR
KHTN(FM) Planada CA
KHTQ(FM) Hayden ID
KHTR(FM) Pullman WA
KHTS-FM El Cajon CA
KHTT(FM) Muskogee OK
KHTZ(FM) Caldwell TX
KHUI(FM) Honolulu HI
KHUM(FM) Garberville CA
KHUS(FM) Bennington NE
KHUT(FM) Hutchinson KS
*KHVT(FM) Bloomington TX
KHWA(FM) Holualoa HI
KHWI(FM) Hilo HI
KHWK(FM) Tonopah NV
KHWY(FM) Essex CA
KHWZ(FM) Ludlow CA
KHXS(FM) Merkel TX
KHYI(FM) Howe TX
KHYL(FM) Auburn CA
*KHYM(FM) Copeland KS
KHYS(FM) Hays KS
KHYT(FM) Tucson AZ
KHYY(FM) Minatare NE
KHYZ(FM) Mountain Pass CA
KHZA(FM) Bunker MO
KHZF(FM) Fagaitua AS
*KHZK(FM) Kotzebue AK
KHZR(FM) Potosi MO
KHZS(FM) Saint Regis MT
KHZY(FM) Overton NE
KHZZ(FM) Sargent NE
*KIAD(FM) Dubuque IA
KIAI(FM) Mason City IA
KIAK-FM Fairbanks AK
KIAQ(FM) Clarion IA
KIBB(FM) Augusta KS
*KIBC(FM) Burney CA
KIBG(FM) Wallace ID
KIBR(FM) Sandpoint ID
KIBS(FM) Bishop CA
KIBT(FM) Fountain CO
*KIBX(FM) Bonners Ferry ID
KIBZ(FM) Crete NE
KICA-FM Farwell TX
*KICB(FM) Fort Dodge IA
KICD-FM Spencer IA
KICK-FM Palmyra MO
KICM(FM) Healdton OK
KICR(FM) Coeur d'Alene ID
KICT-FM Wichita KS
KICX-FM McCook NE
KICY-FM Nome AK
*KIDE(FM) Hoopa CA
KID-FM Idaho Falls ID
*KIDH(FM) Jordan Valley OR
KIDI(FM) Guadalupe CA
KIDN-FM Hayden CO
*KIDS(FM) Grants NM
KIDX(FM) Ruidoso NM
KIFG-FM Iowa Falls IA
KIFM(FM) San Diego CA
*KIFR(FM) Alice TX
KIFS(FM) Ashland OR
KIFX(FM) Roosevelt UT
*KIGC(FM) Oskaloosa IA
KIGL(FM) Seligman MO
KIGN(FM) Burns WY
KIHK(FM) Rock Valley IA
*KIHS(FM) Adel IA
KIHT(FM) Saint Louis MO
KIIK-FM Fairfield IA
KIIM-FM Tucson AZ
KIIS-FM Los Angeles CA
KIIZ-FM Killeen TX
KIJI(FM) Tumon GU
KIJN-FM Farwell TX
KIJZ(FM) Vancouver WA
KIKC-FM Forsyth MT
KIKD(FM) Lake City IA
KIKF(FM) Cascade MT
KIKI-FM Honolulu HI
*KIKL(FM) Lafayette LA
KIKN-FM Salem SD
KIKO-FM Claypool AZ
KIKS-FM Iola KS
KIKT(FM) Greenville TX
KIKV-FM Sauk Centre MN
KIKX(FM) Ketchum ID
*KILI(FM) Porcupine SD
KILJ-FM Mount Pleasant IA
KILO(FM) Colorado Springs CO
KILR-FM Estherville IA
KILS(FM) Minneapolis KS
KILT-FM Houston TX
*KILV(FM) Castana IA
KILX(FM) Hatfield AR
KIMN(FM) Denver CO
KIMX(FM) Laramie WY
KIMY(FM) Watonga OK
KINB(FM) Kingfisher OK
KIND-FM Independence KS
KINE-FM Honolulu HI
KING-FM Seattle WA
*KINI(FM) Crookston NE
KINK(FM) Portland OR
KINL(FM) Eagle Pass TX
KINT-FM El Paso TX
KINV(FM) Georgetown TX
KINX(FM) Great Falls MT
KINZ(FM) Humboldt KS
KIOA(FM) Des Moines IA
KIOC(FM) Orange TX
KIOD(FM) McCook NE
KIOI(FM) San Francisco CA
KIOK(FM) Richland WA
KIOL(FM) La Porte TX
KIOO(FM) Porterville CA
*KIOS-FM Omaha NE
KIOT(FM) Los Lunas NM
KIOW(FM) Forest City IA
KIOX-FM El Campo TX
KIOZ(FM) San Diego CA
*KIPO(FM) Honolulu HI
KIPR(FM) Pine Bluff AR
KIQK(FM) Rapid City SD
KIQO(FM) Atascadero CA
KIQQ-FM Newberry Springs CA
KIQX(FM) Durango CO
KIQZ(FM) Rawlins WY
KIRC(FM) Seminole OK
KIRK(FM) Macon MO
KIRQ(FM) Twin Falls ID
KISC(FM) Spokane WA
KISD(FM) Pipestone MN
KISF(FM) Las Vegas NV
KISH(FM) Hagatna GU
*KISL(FM) Avalon CA
KISM(FM) Bellingham WA
KISN(FM) Belgrade MT
KISQ(FM) San Francisco CA
KISR(FM) Fort Smith AR
KISS-FM San Antonio TX
KIST-FM Santa Barbara CA
*KISU-FM Pocatello ID
KISV(FM) Bakersfield CA
KISW(FM) Seattle WA
KISX(FM) Whitehouse TX
KISZ-FM Cortez CO
*KITA(FM) Bunkie LA
KITE(FM) Port Lavaca TX
KITH(FM) Kapaa HI
KITI-FM Winlock WA
KITN(FM) Worthington MN
KITO-FM Vinita OK
KITS(FM) San Francisco CA
KITT(FM) Soda Springs ID
KITY(FM) Llano TX
KIVA(FM) Santa Rosa NM
KIVY-FM Crockett TX
KIWA-FM Sheldon IA
KIWI(FM) McFarland CA
*KIWR(FM) Council Bluffs IA
KIXA(FM) Lucerne Valley CA
KIXB(FM) El Dorado AR
KIXF(FM) Baker CA
KIXN(FM) Hobbs NM
KIXO(FM) Sulphur OK
KIXQ(FM) Joplin MO
KIXR(FM) Ponca City OK
KIXS(FM) Victoria TX
KIXT(FM) Bay City OR
KIXW-FM Lenwood CA
KIXX(FM) Watertown SD
KIXY-FM San Angelo TX
KIXZ-FM Opportunity WA
KIYS(FM) Jonesboro AR
KIYX(FM) Sageville IA
KIZN(FM) Boise ID
KIZS(FM) Collinsville OK
KIZZ(FM) Minot ND
*KJAB-FM Mexico MO
KJAC(FM) Timnath CO
KJAE(FM) Leesville LA
KJAK(FM) Slaton TX
KJAM-FM Madison SD
KJAQ(FM) Seattle WA
*KJAR(FM) Susanville CA
KJAS(FM) Jasper TX
KJAV(FM) Alamo TX
KJAX(FM) Jackson WY
KJAZ(FM) Point Comfort TX
*KJBB(FM) Watertown SD
KJBI(FM) Presho SD
KJBL(FM) Julesburg CO
KJBR(FM) Marked Tree AR
KJBX(FM) Trumann AR
KJBZ(FM) Laredo TX
*KJCC(FM) Carnegie OK
KJCD(FM) Longmont CO
*KJCF(FM) Clarkston WA
*KJCG(FM) Missoula MT
*KJCH(FM) Coos Bay OR
KJCK-FM Junction City KS
*KJCM(FM) Snyder OK
*KJCQ(FM) Quincy CA
*KJCR(FM) Keene TX
KJCS(FM) Nacogdoches TX
*KJCU(FM) Laytonville CA
*KJCV(FM) Country Club MO
KJCY(FM) Saint Ansgar IA
KJDL-FM Levelland TX
KJDX(FM) Susanville CA
KJDY-FM Canyon City OR
KJEB(FM) New Castle CO
KJEE(FM) Montecito CA
KJEL(FM) Lebanon MO
KJET(FM) Raymond WA
KJEZ(FM) Poplar Bluff MO
KJFA(FM) Santa Fe NM
KJFM(FM) Louisiana MO
*KJFT(FM) Arlee MT
KJFX(FM) Fresno CA
*KJHA(FM) Houston AK
*KJHK(FM) Lawrence KS
*KJIA(FM) Spirit Lake IA
KJIK(FM) Duncan AZ
*KJIL(FM) Copeland KS
*KJIR(FM) Hannibal MO
KJIW-FM Helena AR
KJJJ(FM) Lake Havasu City AZ
KJJK-FM Fergus Falls MN
KJJM(FM) Baker MT
*KJJP(FM) Amarillo TX
KJJY(FM) West Des Moines IA
KJJZ(FM) Indio CA
KJKB(FM) Jacksboro TX
KJKE(FM) Ingleside TX
KJKJ(FM) Grand Forks ND
KJKK(FM) Dallas TX
*KJKL(FM) Selma OR
KJKS(FM) Kahului HI
*KJLF(FM) Butte MT
KJLH-FM Compton CA
KJLL-FM Fountain Valley CA
KJLO-FM Monroe LA
*KJLP(FM) Palmer AK
KJLS(FM) Hays KS
*KJLT-FM North Platte NE
*KJLU(FM) Jefferson City MO
KJLV(FM) Hoxie AR
KJLY(FM) Blue Earth MN
*KJMA(FM) Floresville TX
KJMB(FM) Blythe CA
*KJMC(FM) Des Moines IA
KJMD(FM) Pukalani HI
KJMG(FM) Bastrop LA
KJMH(FM) Lake Arthur LA
KJMK(FM) Webb City MO
KJML(FM) Columbus KS
KJMM(FM) Bixby OK
KJMN(FM) Castle Rock CO
KJMO(FM) Linn MO
KJMS(FM) Memphis TN
KJMT(FM) Calico Rock AR
KJMX(FM) Reedsport OR
KJMY(FM) Bountiful UT
KJMZ(FM) Cache OK
KJNA-FM Jena LA
*KJNP-FM North Pole AK
KJNY(FM) Ferndale CA
KJNZ(FM) Hereford TX
KJOE(FM) Slayton MN
KJOJ-FM Freeport TX
KJOR(FM) Windsor CA
KJOT(FM) Boise ID
*KJOV(FM) Woodward OK
KJOY(FM) Stockton CA
KJPW-FM Waynesville MO
KJQN(FM) Coalville UT
KJQY(FM) La Veta CO
*KJRF(FM) Lawton OK
KJR-FM Seattle WA
*KJRL(FM) Herington KS
*KJRT(FM) Amarillo TX
KJRV(FM) Wessington Springs SD
*KJSM-FM Augusta AR
KJSN(FM) Modesto CA
KJSR(FM) Tulsa OK
*KJTA(FM) Flagstaff AZ
*KJTH(FM) Ponca City OK
*KJTW(FM) Jamestown ND
KJTX(FM) Jefferson TX
*KJTY(FM) Topeka KS
KJUG-FM Tulare CA
KJUL(FM) Moapa Valley NV
KJVC(FM) Mansfield LA
*KJVH(FM) Longview WA
KJWL(FM) Fresno CA
KJXJ(FM) Cameron TX
KJXK(FM) San Antonio TX
KJYE(FM) Grand Junction CO
*KJYL(FM) Eagle Grove IA
KJYO(FM) Oklahoma City OK
*KJZA(FM) Drake AZ
KJZN(FM) San Joaquin CA
KJZS(FM) Sparks NV
KJZY(FM) Sebastopol CA
*KJZZ(FM) Phoenix AZ
KKAC(FM) Vandalia MO
KKAJ-FM Ardmore OK
KKAL(FM) Paso Robles CA
KKAT-FM Orem UT
KKAW(FM) Albin WY
KKBA(FM) Kingsville TX
KKBB(FM) Bakersfield CA
KKBC-FM Baker City OR
KKBD(FM) Sallisaw OK
KKBG(FM) Hilo HI
KKBI(FM) Broken Bow OK
KKBJ-FM Bemidji MN
KKBL(FM) Monett MO
KKBN(FM) Twain Harte CA
KKBQ-FM Pasadena TX
KKBR(FM) Billings MT
KKBS(FM) Guymon OK
KKBZ(FM) Clarinda IA
KKCA(FM) Fulton MO
KKCB(FM) Duluth MN
KKCD(FM) Omaha NE
KKCH(FM) Glenwood Springs CO
KKCI(FM) Goodland KS
*KKCJ(FM) Cannon AFB NM
KKCK(FM) Marshall MN
KKCL(FM) Lorenzo TX
KKCM(FM) Sand Springs OK
KKCN(FM) Ballinger TX
KKCQ-FM Bagley MN
*KKCR(FM) Hanalei HI
KKCS-FM Canon City CO
KKCT(FM) Bismarck ND
KKCV(FM) Rozel KS
KKCW(FM) Beaverton OR
KKCY(FM) Colusa CA
KKDA-FM Dallas TX
KKDC(FM) Dolores CO
KKDM(FM) Des Moines IA
KKDQ(FM) Thief River Falls MN
*KKDU(FM) El Dorado AR
KKDV(FM) Walnut Creek CA
KKDY(FM) West Plains MO
KKED(FM) Fairbanks AK
KKEG(FM) Fayetteville AR
KKEN(FM) Duncan OK
KKEQ(FM) Fosston MN
*KKER(FM) Kerrville TX
KKEV(FM) Centerville TX
KKEX(FM) Preston ID
KKEZ(FM) Fort Dodge IA
KKFC(FM) Coalgate OK
KKFG(FM) Bloomfield NM
*KKFI(FM) Kansas City MO
KKFM(FM) Colorado Springs CO
KKFR(FM) Mayer AZ
KKFS(FM) Lincoln CA

KKFT(FM) Gardnerville-Minden NV
KKGB(FM) Sulphur LA
KKGL(FM) Nampa ID
KKGO(FM) Los Angeles CA
KKHB(FM) Eureka CA
KKHI(FM) Rock River WY
KKHJ-FM Pago Pago AS
KKHQ-FM Oelwein IA
KKHR(FM) Abilene TX
KKHT-FM Winnie TX
KKIA(FM) Ida Grove IA
KKID(FM) Salem MO
KKIK(FM) Horseshoe Bend AR
KKIM-FM Santa Fe NM
KKIN-FM Aitkin MN
KKIQ(FM) Livermore CA
KKIS-FM Soldotna AK
KKIT(FM) Taos NM
KKIX(FM) Fayetteville AR
*KKJA(FM) Redmond OR
KKJG(FM) San Luis Obispo CA
KKJJ(FM) Henderson NV
KKJK(FM) Ravenna NE
KKJM(FM) Saint Joseph MN
KKJO(FM) Saint Joseph MO
KKJQ(FM) Garden City KS
KKJW(FM) Stanton TX
*KKJZ(FM) Long Beach CA
KKKJ(FM) Merrill OR
KKLA-FM Los Angeles CA
KKLB(FM) Madisonville TX
*KKLC(FM) Mount Shasta CA
KKLD(FM) Cottonwood AZ
*KKLG(FM) Newton IA
KKLH(FM) Marshfield MO
KKLI(FM) Widefield CO
*KKLJ(FM) Klamath Falls OR
*KKLM(FM) Corpus Christi TX
KKLN(FM) Atwater MN
*KKLP(FM) La Pine OR
KKLQ(FM) Harwood ND
KKLR(FM) Poplar Bluff MO
KKLS-FM Sioux Falls SD
*KKLT(FM) Texarkana AR
*KKLU(FM) Lubbock TX
KKLV(FM) Turrell AR
*KKLW(FM) Willmar MN
KKLX(FM) Worland WY
*KKLY(FM) El Paso TX
KKLZ(FM) Las Vegas NV
KKMA(FM) Le Mars IA
KKMG(FM) Pueblo CO
KKMI(FM) Burlington IA
KKMJ-FM Austin TX
KKMK(FM) Rapid City SD
KKMR(FM) Arizona City AZ
KKMT(FM) Pablo MT
KKMV(FM) Rupert ID
KKMX(FM) Tri City OR
KKMY(FM) Orange TX
KKND(FM) Port Sulphur LA
KKNG-FM Newcastle OK
KKNN(FM) Delta CO
KKNU(FM) Springfield-Eugene OR
KKOA(FM) Volcano HI
KKOB-FM Albuquerque NM
KKOK-FM Morris MN
KKOR(FM) Gallup NM
KKOT(FM) Columbus NE
KKOW-FM Pittsburg KS
KKOY-FM Chanute KS
KKOZ-FM Ava MO
KKPK(FM) Colorado Springs CO
KKPL(FM) Cheyenne WY
KKPN(FM) Rockport TX
KKPR-FM Kearney NE
KKPS(FM) Brownsville TX
KKPT(FM) Little Rock AR
KKQQ(FM) Volga SD
KKQX(FM) Manhattan MT
KKQY(FM) Hill City KS
KKRB(FM) Klamath Falls OR
KKRC(FM) Granite Falls MN
*KKRD(FM) Enid OK
KKRE(FM) Hollis OK
KKRF(FM) Stuart IA
KKRG(FM) Albuquerque NM
*KKRI(FM) Pocola OK
KKRK(FM) Coffeyville KS
KKRL(FM) Carroll IA
*KKRO(FM) Redding CA
KKRQ(FM) Iowa City IA
*KKRS(FM) Davenport WA
KKRV(FM) Wenatchee WA
KKRW(FM) Houston TX
KKRZ(FM) Portland OR
KKSD(FM) Milbank SD
KKSF(FM) San Francisco CA
KKSI(FM) Eddyville IA
KKSJ(FM) Maurice LA
KKSP(FM) Bryant AR
KKSR(FM) Sartell MN
KKSS(FM) Santa Fe NM
KKST(FM) Oakdale LA
KKSY(FM) Anamosa IA
KKTC(FM) Angel Fire NM
*KKTO(FM) Tahoe City CA
*KKTR(FM) Kirksville MO
KKTX-FM Kilgore TX
KKTY-FM Douglas WY
KKTZ(FM) Lakeview AR
*KKUA(FM) Wailuku HI
KKUL-FM Groveton TX
*KKUP(FM) Cupertino CA
KKUS(FM) Tyler TX
KKUU(FM) Indio CA
*KKVO(FM) Altus OK
KKVR(FM) Kerrville TX
*KKVS(FM) Truth or Consequences NM
KKVU(FM) Stevensville MT
KKWD(FM) Bethany OK
KKWF(FM) Seattle WA
KKWK(FM) Cameron MO
KKWQ(FM) Warroad MN
KKWS(FM) Wadena MN
*KKWV(FM) Aransas Pass TX
KKXK(FM) Montrose CO
KKXL-FM Grand Forks ND
KKXS(FM) Shingletown CA
KKXX-FM Shafter CA
KKYA(FM) Yankton SD
KKYC(FM) Clovis NM
KKYN-FM Plainview TX
KKYR-FM Texarkana TX
KKYS(FM) Bryan TX
KKYY(FM) Whiting IA
KKYZ(FM) Sierra Vista AZ
KKZQ(FM) Tehachapi CA
KKZX(FM) Spokane WA
KKZY(FM) Bemidji MN
KLAA-FM Tioga LA
KLAD-FM Klamath Falls OR
*KLAI(FM) Laytonville CA
KLAK(FM) Tom Bean TX
KLAL(FM) Wrightsville AR
KLAN(FM) Glasgow MT
KLAQ(FM) El Paso TX
KLAW(FM) Lawton OK
KLAX-FM East Los Angeles CA
KLAZ(FM) Hot Springs AR
KLBA-FM Albia IA
KLBC(FM) Durant OK
KLBJ-FM Austin TX
KLBN(FM) Auberry CA
KLBQ(FM) El Dorado AR
*KLBR(FM) Bend OR
*KLBT(FM) Beaumont TX
KLBU(FM) Pecos NM
*KLBV(FM) Steamboat Springs CO
*KLBZ(FM) Bozeman MT
KLCA(FM) Tahoe City CA
*KLCC(FM) Eugene OR
*KLCD(FM) Decorah IA
KLCE(FM) Blackfoot ID
KLCH(FM) Lake City MN
KLCI(FM) Elk River MN
KLCM(FM) Lewistown MT
*KLCO(FM) Newport OR
KLCR(FM) Lakeview OR
*KLCU(FM) Ardmore OK
*KLCV(FM) Lincoln NE
KLCX(FM) Saint Charles MN
KLCY-FM Vernal UT
*KLCZ(FM) Lewiston ID
KLDE(FM) Eldorado TX
KLDG(FM) Liberal KS
KLDJ(FM) Duluth MN
*KLDN(FM) Lufkin TX
KLDR(FM) Harbeck-Fruitdale OR
*KLDV(FM) Morrison CO
KLDZ(FM) Medford OR
KLEA-FM Lovington NM
KLEF(FM) Anchorage AK
KLEN(FM) Cheyenne WY
KLEO(FM) Kahaluu HI
KLER-FM Orofino ID
KLES(FM) Prosser WA
*KLEU(FM) Lewistown MT
KLEY-FM Jourdanton TX
KLEZ(FM) Malvern AR
*KLFC(FM) Branson MO
*KLFF-FM San Luis Obispo CA
*KLFH(FM) Ojai CA
KLFM(FM) Great Falls MT
KLFN(FM) Sunburg MN
*KLFO(FM) Florence OR
*KLFR(FM) Reedsport OR
*KLFS(FM) Van Buren AR
*KLFV(FM) Grand Junction CO
KLFX(FM) Nolanville TX
KLGA-FM Algona IA
KLGD(FM) Stamford TX
KLGL(FM) Richfield UT
KLGO(FM) Thorndale TX
*KLGQ(FM) Grants NM
KLGR-FM Redwood Falls MN
*KLGS(FM) College Station TX
KLGT(FM) Buffalo WY
KLHB(FM) Odem TX
KLHI-FM Kahului HI
*KLHV(FM) Fort Collins CO
KLIL(FM) Moreauville LA
KLIP(FM) Monroe LA
KLIQ(FM) Hastings NE
KLIR(FM) Columbus NE
*KLIT(FM) Hobbs NM
KLIX-FM Twin Falls ID
KLIZ-FM Brainerd MN
*KLJC(FM) Kansas City MO
KLJH(FM) Bayfield CO
KLJR-FM Santa Paula CA
KLJT(FM) Jacksonville TX
*KLJV(FM) Scottsbluff NE
KLJZ(FM) Yuma AZ
*KLKA(FM) Globe AZ
KLKC-FM Parsons KS
KLKK(FM) Clear Lake IA
KLKL(FM) Minden LA
*KLKM(FM) Kalispell MT
KLKO(FM) Elko NV
KLKS(FM) Breezy Point MN
KLKX(FM) Rosamond CA
KLKY(FM) Stanfield OR
KLLC(FM) San Francisco CA
KLLE(FM) North Fork CA
KLLI(FM) Dallas TX
KLLL-FM Lubbock TX
*KLLN(FM) Newark AR
KLLP(FM) Chubbuck ID
*KLLR(FM) Dripping Springs TX
KLLT(FM) Spencer IA
*KLLU(FM) Gallup NM
KLLY(FM) Oildale CA
KLLZ-FM Walker MN
KLMA(FM) Hobbs NM
*KLMF(FM) Klamath Falls OR
KLMG(FM) Jackson CA
KLMJ(FM) Hampton IA
KLMM(FM) Morro Bay CA
KLMO-FM Dilley TX
*KLMP(FM) Rapid City SD
KLMR-FM Lamar CO
*KLMT(FM) Billings MT
KLMY(FM) Lincoln NE
*KLMZ(FM) Fouke AR
*KLNB(FM) Grand Island NE
KLNC(FM) Lincoln NE
*KLND(FM) Little Eagle SD
*KLNE-FM Lexington NE
*KLNI(FM) Decorah IA
KLNN(FM) Questa NM
KLNO(FM) Fort Worth TX
*KLNR(FM) Panaca NV
KLNV(FM) San Diego CA
KLNZ(FM) Glendale AZ
KLOA-FM Ridgecrest CA
KLOB(FM) Thousand Palms CA
*KLOF(FM) Gillette WY
KLOK-FM Greenfield CA
KLOL(FM) Houston TX
*KLON(FM) Rockaway Beach OR
KLOO-FM Corvallis OR
*KLOP(FM) Ocean Park WA
KLOQ-FM Winton CA
KLOR-FM Ponca City OK
KLOS(FM) Los Angeles CA
KLOU(FM) Saint Louis MO
*KLOV(FM) Winchester OR
KLOW(FM) Reno TX
*KLOX(FM) Creston IA
*KLOY(FM) Astoria OR
KLOZ(FM) Eldon MO
*KLPI-FM Ruston LA
KLPL-FM Lake Providence LA
*KLPR(FM) Kearney NE
KLPW-FM Union MO
KLPX(FM) Tucson AZ
KLQB(FM) Taylor TX
KLQL(FM) Luverne MN
KLQP(FM) Madison MN
KLQQ(FM) Clearmont WY
KLQT(FM) Wessington Springs SD
KLQV(FM) San Diego CA
*KLRB(FM) Stuart OK
*KLRC(FM) Siloam Springs AR
*KLRD(FM) Yucaipa CA
*KLRE-FM Little Rock AR
*KLRF(FM) Milton-Freewater OR
*KLRH(FM) Sparks NV
*KLRI(FM) Rigby ID
KLRJ(FM) Aberdeen SD
KLRK(FM) Marlin TX
*KLRM(FM) Melbourne AR
*KLRO(FM) Hot Springs AR
*KLRQ(FM) Clinton MO
KLRR(FM) Redmond OR
*KLRS(FM) Lodi CA
*KLRV(FM) Billings MT
*KLRW(FM) Byrne TX
*KLRX(FM) Jamestown ND
*KLRY(FM) Gypsum CO
KLRZ(FM) Larose LA
*KLSA(FM) Alexandria LA
KLSC(FM) Malden MO
*KLSE-FM Rochester MN
*KLSI(FM) Moss Beach CA
KLSK(FM) Great Falls MT
KLSM(FM) Tallulah LA
KLSN(FM) Hudson TX
*KLSP(FM) Angola LA
KLSR-FM Memphis TX
KLSS-FM Mason City IA
*KLSU(FM) Baton Rouge LA
KLSX(FM) Los Angeles CA
KLSY(FM) South Bend WA
KLSZ-FM Fort Smith AR
KLTA(FM) Breckenridge MN
KLTD(FM) Temple TX
KLTE(FM) Kirksville MO
KLTG(FM) Corpus Christi TX
KLTH(FM) Lake Oswego OR
KLTI-FM Ames IA
KLTN(FM) Houston TX
KLTO-FM McQueeney TX
*KLTP(FM) San Angelo TX
KLTQ(FM) Lincoln NE
KLTR(FM) Brenham TX
*KLTU(FM) Mammoth AZ
KLTW-FM Prineville OR
KLTY(FM) Arlington TX
KLUA(FM) Kailua-Kona HI
KLUB(FM) Bloomington TX
KLUC-FM Las Vegas NV
KLUE(FM) Poplar Bluff MO
*KLUH(FM) Poplar Bluff MO
KLUK(FM) Needles CA
KLUN(FM) Paso Robles CA
KLUR(FM) Wichita Falls TX
KLUV(FM) Dallas TX
*KLUX(FM) Robstown TX
*KLVA(FM) Casa Grande AZ
*KLVB(FM) Red Bluff CA
*KLVC(FM) Magalia CA
KLVE(FM) Los Angeles CA
KLVF(FM) Las Vegas NM
*KLVG(FM) Garberville CA
*KLVH(FM) San Luis Obispo CA
*KLVJ(FM) Julian CA
*KLVK(FM) Fountain Hills AZ
*KLVM(FM) Prunedale CA
*KLVN(FM) Livingston CA
KLVO(FM) Belen NM
*KLVP-FM Cherryville OR
*KLVR(FM) Santa Rosa CA
*KLVS(FM) Grass Valley CA
*KLVU(FM) Sweet Home OR
*KLVV(FM) Ponca City OK
*KLVW(FM) West Odessa TX
*KLVY(FM) Fairmead CA
*KLWC(FM) Casper WY
*KLWD(FM) Gillette WY
*KLWG(FM) Lompoc CA
*KLWS(FM) Moses Lake WA
*KLWV(FM) Chugwater WY
*KLXA-FM Alexandria LA
KLXK(FM) Breckenridge TX
KLXQ(FM) Mountain Pine AR
KLXS-FM Pierre SD
*KLXV(FM) Glenwood Springs CO
KLYD(FM) Snyder TX
KLYK(FM) Kelso WA
KLYR-FM Clarksville AR
*KLYT(FM) Albuquerque NM
KLYV(FM) Dubuque IA
KLYY(FM) Riverside CA
KLZA(FM) Falls City NE
KLZK(FM) Brownfield TX
KLZN(FM) Susanville CA
KLZR(FM) Lawrence KS
*KLZV(FM) Sterling CO
KLZX(FM) Weston ID
KLZY(FM) Honokaa HI
KLZZ(FM) Waite Park MN
KMAC(FM) Gainesville MO
KMAD-FM Whitesboro TX
KMAG(FM) Fort Smith AR
KMAJ-FM Topeka KS
KMAK(FM) Orange Cove CA
KMAQ-FM Maquoketa IA
KMAR-FM Winnsboro LA
KMAT(FM) Seadrift TX
KMAV-FM Mayville ND
KMAX-FM Wellington CO
*KMBH-FM Harlingen TX
*KMBI-FM Spokane WA
*KMBN(FM) Las Cruces NM
KMBQ(FM) Wasilla AK
KMBR(FM) Butte MT
KMBV(FM) Navasota TX
KMBY-FM Seaside CA
KMCH(FM) Manchester IA
KMCJ(FM) Colstrip MT
KMCK(FM) Siloam Springs AR
KMCL-FM McCall ID
KMCM(FM) Odessa TX
KMCN(FM) Clinton IA
KMCO(FM) McAlester OK
KMCQ(FM) The Dalles OR
KMCR(FM) Montgomery City MO
KMCS(FM) Muscatine IA
*KMCU(FM) Wichita Falls TX
*KMCV(FM) High Point MO
KMCX(FM) Ogallala NE
KMDL(FM) Kaplan LA
KMDR(FM) McKinleyville CA
KMDX(FM) San Angelo TX
*KMDY(FM) Keokuk IA
KMDZ(FM) Las Vegas NM
KMEL(FM) San Francisco CA
KMEM-FM Memphis MO

KMEN(FM) Mendota CA
*KMEO(FM) Mertzon TX
KMEZ(FM) Belle Chasse LA
*KMFA(FM) Austin TX
KMFB(FM) Mendocino CA
KMFC(FM) Centralia MO
KMFG(FM) Nashwauk MN
KMFM(FM) Premont TX
KMFR(FM) Hondo TX
KMFX-FM Lake City MN
KMFY(FM) Grand Rapids MN
KMGA(FM) Albuquerque NM
KMGC(FM) Camden AR
KMGE(FM) Eugene OR
KMGI(FM) Pocatello ID
KMGJ(FM) Grand Junction CO
KMGK(FM) Glenwood MN
KMGL(FM) Oklahoma City OK
KMGM(FM) Montevideo MN
KMGN(FM) Flagstaff AZ
KMGO(FM) Centerville IA
KMGQ(FM) Goleta CA
KMGR(FM) Delta UT
KMGV(FM) Fresno CA
KMGW(FM) Casper WY
KMGX(FM) Bend OR
KMGZ(FM) Lawton OK
*KMHA(FM) Four Bears ND
*KMHD(FM) Gresham OR
KMHK(FM) Hardin MT
KMHM(FM) Lutesville MO
KMHT-FM Marshall TX
KMHX(FM) Rohnert Park CA
*KMIH(FM) Mercer Island WA
KMIL(FM) Cameron TX
KMIQ(FM) Robstown TX
KMIT(FM) Mitchell SD
KMIX(FM) Tracy CA
KMJE(FM) Gridley CA
*KMJG(FM) Homer AK
KMJI(FM) Ashdown AR
KMJJ-FM Shreveport LA
KMJK(FM) Lexington MO
KMJM-FM Columbia IL
KMJQ(FM) Houston TX
KMJR(FM) Portland TX
KMJV(FM) Soledad CA
KMJX(FM) Conway AR
KMKF(FM) Manhattan KS
KMKK-FM Kaunakakai HI
*KMKL(FM) North Branch MN
*KMKR(FM) Oakridge OR
KMKS(FM) Bay City TX
KMKT(FM) Bells TX
KMKX(FM) Willits CA
KMLA(FM) El Rio CA
KMLD(FM) Casper WY
KMLE(FM) Chandler AZ
KMLK(FM) El Dorado AR
KMLO(FM) Lowry SD
*KMLR(FM) Gonzales TX
*KMLT(FM) Jackson WY
*KMLU(FM) Brownfield TX
*KMLV(FM) Ralston NE
*KMLW(FM) Moses Lake WA
KMMG(FM) Benton City WA
KMML-FM Amarillo TX
KMMM(FM) Madera CA
KMMO-FM Marshall MO
KMMR(FM) Malta MT
KMMS-FM Bozeman MT
KMMT(FM) Mammoth Lakes CA
KMMX(FM) Tahoka TX
KMMY(FM) Soper OK
KMMZ(FM) Crane TX
KMNA(FM) Mabton WA
*KMNE-FM Bassett NE
*KMNR(FM) Rolla MO
KMNT(FM) Chehalis WA
KMOA(FM) Caliente NV
*KMOC(FM) Wichita Falls TX
KMOD-FM Tulsa OK
KMOE(FM) Butler MO
*KMOJ(FM) Minneapolis MN
KMOK(FM) Lewiston ID
KMOM-FM Roscoe SD

KMON-FM Great Falls MT
KMOO-FM Mineola TX
KMOQ(FM) Baxter Springs KS
KMOR(FM) Bridgeport NE
KMOU(FM) Roswell NM
KMOZ-FM Grand Junction CO
*KMPO(FM) Modesto CA
*KMPQ(FM) Roseburg OR
*KMPR(FM) Minot ND
KMPS-FM Seattle WA
KMQA(FM) East Porterville CA
*KMQX(FM) Weatherford TX
KMRJ(FM) Rancho Mirage CA
KMRK-FM Odessa TX
*KMRL(FM) Buras LA
*KMRO(FM) Camarillo CA
KMRQ(FM) Manteca CA
KMRX(FM) El Dorado AR
*KMSA(FM) Grand Junction CO
*KMSC(FM) Sioux City IA
*KMSE(FM) Rochester MN
*KMSI(FM) Moore OK
*KMSK(FM) Austin MN
*KMSL(FM) Mansfield LA
*KMSM-FM Butte MT
KMSO(FM) Missoula MT
*KMST(FM) Rolla MO
*KMSU(FM) Mankato MN
KMSW(FM) The Dalles OR
KMTB(FM) Murfreesboro AR
*KMTC(FM) Russellville AR
*KMTG(FM) San Jose CA
*KMTH(FM) Maljamar NM
KMTK(FM) Bend OR
KMTN(FM) Jackson WY
KMTS(FM) Glenwood Springs CO
KMTT(FM) Tacoma WA
KMTX-FM Helena MT
KMTY(FM) Holdrege NE
KMTZ(FM) Boulder MT
*KMUD(FM) Garberville CA
*KMUE(FM) Eureka CA
KMUL-FM Muleshoe TX
*KMUN(FM) Astoria OR
*KMUW(FM) Wichita KS
KMVA(FM) Dewey-Humboldt AZ
*KMVC(FM) Marshall MO
KMVE(FM) Mojave CA
KMVK(FM) Fort Worth TX
KMVL-FM Madisonville TX
KMVN(FM) Los Angeles CA
KMVQ-FM San Francisco CA
KMVR(FM) Mesilla Park NM
KMVX(FM) Jerome ID
*KMWR(FM) Brookings OR
*KMWS(FM) Mount Vernon WA
KMXA-FM Minot ND
KMXB(FM) Henderson NV
KMXC(FM) Sioux Falls SD
KMXD(FM) Monroe UT
KMXE-FM Red Lodge MT
KMXF(FM) Lowell AR
KMXG(FM) Clinton IA
KMXH(FM) Alexandria LA
KMXI(FM) Chico CA
KMXJ-FM Amarillo TX
KMXK(FM) Cold Spring MN
KMXL(FM) Carthage MO
KMXN(FM) Osage City KS
KMXP(FM) Phoenix AZ
KMXQ(FM) Socorro NM
KMXR(FM) Corpus Christi TX
KMXS(FM) Anchorage AK
*KMXT(FM) Kodiak AK
KMXV(FM) Kansas City MO
KMXW(FM) Hope ND
KMXX(FM) Imperial CA
KMXY(FM) Grand Junction CO
KMXZ-FM Tucson AZ
KMYI(FM) San Diego CA
KMYK(FM) Osage Beach MO
KMYO-FM Morgan City LA
KMYT(FM) Temecula CA
KMYX-FM Arvin CA
KMYY(FM) Rayville LA
KMYZ-FM Pryor OK

KMZA(FM) Seneca KS
KMZE(FM) Woodward OK
*KMZL(FM) Missoula MT
*KMZO(FM) Hamilton MT
KMZQ(FM) Payson AZ
KMZU(FM) Carrollton MO
KMZZ(FM) Bishop TX
*KNAA(FM) Show Low AZ
KNAB-FM Burlington CO
KNAC(FM) Earlimart CA
*KNAD(FM) Page AZ
KNAF-FM Fredericksburg TX
*KNAG(FM) Grand Canyon AZ
KNAH(FM) Merced CA
*KNAI(FM) Phoenix AZ
KNAN(FM) Nanakuli HI
*KNAQ(FM) Prescott AZ
*KNAR(FM) San Angelo TX
KNAS(FM) Nashville AR
*KNAU(FM) Flagstaff AZ
*KNBA(FM) Anchorage AK
KNBB(FM) Dubach LA
*KNBE(FM) Beatrice NE
*KNBJ(FM) Bemidji MN
KNBQ(FM) Centralia WA
KNBT(FM) New Braunfels TX
*KNBU(FM) Baldwin City KS
KNBZ(FM) Redfield SD
*KNCA(FM) Burney CA
KNCB-FM Vivian LA
*KNCC(FM) Elko NV
KNCE(FM) Winters TX
KNCI(FM) Sacramento CA
*KNCM(FM) Appleton MN
KNCN(FM) Sinton TX
KNCO-FM Grass Valley CA
KNCQ(FM) Redding CA
*KNCT-FM Killeen TX
KNCU(FM) Newport OR
KNCW(FM) Omak WA
KNCY-FM Auburn NE
KNDA(FM) Alice TX
KNDD(FM) Seattle WA
KNDE(FM) College Station TX
KNDK-FM Langdon ND
*KNDL(FM) Angwin CA
KNDR(FM) Mandan ND
KNDY-FM Marysville KS
*KNDZ(FM) McKinleyville CA
KNEB-FM Scottsbluff NE
KNEC(FM) Yuma CO
KNEI-FM Waukon IA
KNEK-FM Washington LA
KNEL-FM Brady TX
KNEN(FM) Norfolk NE
*KNEO(FM) Neosho MO
KNES(FM) Fairfield TX
KNEV(FM) Reno NV
KNEX(FM) Laredo TX
*KNFA(FM) Grand Island NE
KNFM(FM) Midland TX
KNFO(FM) Basalt CO
KNFT-FM Bayard NM
KNFX-FM Bryan TX
*KNGA(FM) Saint Peter MN
*KNGM(FM) Emporia KS
KNGS(FM) Coalinga CA
KNGT(FM) Lake Charles LA
KNGY(FM) Alameda CA
*KNHA(FM) Grand Isle NE
*KNHC(FM) Seattle WA
*KNHM(FM) Bayside CA
*KNHS(FM) Hastings NE
*KNHT(FM) Rio Dell CA
KNID(FM) Alva OK
KNIK-FM Anchorage AK
KNIM-FM Maryville MO
KNIN-FM Wichita Falls TX
*KNIS(FM) Carson City NV
KNIX-FM Phoenix AZ
KNKK(FM) Needles CA
*KNKL(FM) North Ogden UT
KNKN(FM) Pueblo CO
KNKT(FM) Armijo, Albuquerque NM
*KNLB(FM) Lake Havasu City AZ
*KNLE-FM Round Rock TX

KNLF(FM) Quincy CA
*KNLG(FM) New Bloomfield MO
*KNLH(FM) Cedar Hill MO
*KNLK(FM) Santa Rosa NM
*KNLL(FM) Nashville AR
*KNLM(FM) Marshfield MO
*KNLN(FM) Vienna MO
*KNLP(FM) Potosi MO
*KNLQ(FM) Cuba MO
KNLR(FM) Bend OR
KNLT(FM) Walla Walla WA
KNLV-FM Ord NE
*KNMA(FM) Socorro NM
KNMB(FM) Cloudcroft NM
*KNMC(FM) Havre MT
*KNMI(FM) Farmington NM
KNMO(FM) Nevada MO
KNMZ(FM) Alamogordo NM
*KNNB(FM) Whiteriver AZ
KNNG(FM) Sterling CO
KNNK(FM) Dimmitt TX
KNNN(FM) Shasta Lake City CA
KNOB(FM) Healdsburg CA
KNOD(FM) Harlan IA
KNOE-FM Monroe LA
KNOF(FM) Saint Paul MN
*KNOG(FM) Nogales AZ
*KNOM-FM Nome AK
*KNON(FM) Dallas TX
KNOR(FM) Krum TX
KNOS(FM) Albany TX
KNOU(FM) Empire LA
*KNOW-FM Minneapolis-St. Paul MN
KNOX-FM Grand Forks ND
KNPE(FM) Hyannis NE
*KNPR(FM) Las Vegas NV
KNRB(FM) Atlanta TX
KNRG(FM) New Ulm TX
*KNRI(FM) Bismarck ND
KNRJ(FM) Payson AZ
KNRK(FM) Camas WA
KNRQ-FM Eugene OR
KNRX(FM) Sterling City TX
*KNSE(FM) Austin MN
KNSG(FM) Springfield MN
*KNSQ(FM) Mount Shasta CA
*KNSR(FM) Collegeville MN
*KNSU(FM) Thibodaux LA
*KNSW(FM) Worthington-Marshall MN
KNSX(FM) Steelville MO
KNTI(FM) Lakeport CA
KNTK(FM) Weed CA
*KNTN(FM) Thief River Falls MN
KNTO(FM) Chowchilla CA
*KNTU(FM) McKinney TX
KNTY(FM) Shingle Springs CA
KNUE(FM) Tyler TX
KNUJ-FM Sleepy Eye MN
KNUQ(FM) Paauilo HI
KNUW(FM) Santa Clara NM
KNWB(FM) Hilo HI
*KNWC-FM Sioux Falls SD
*KNWD(FM) Natchitoches LA
*KNWF(FM) Fergus Falls MN
*KNWI(FM) Osceola IA
KNWJ(FM) Leone AS
KNWM(FM) Madrid IA
*KNWO(FM) Cottonwood ID
*KNWP(FM) Port Angeles WA
*KNWR(FM) Ellensburg WA
*KNWS-FM Waterloo IA
*KNWV(FM) Clarkston WA
*KNWY(FM) Yakima WA
KNXR(FM) Rochester MN
KNXX(FM) Donaldsonville LA
*KNYD(FM) Broken Arrow OK
KNYE(FM) Pahrump NV
KNYN(FM) Fort Bridger WY
*KNYR(FM) Yreka CA
KNZA(FM) Hiawatha KS
*KOAB-FM Bend OR
*KOAP(FM) Lakeview OR
*KOAR(FM) Spearfish SD
KOAS(FM) Dolan Springs AZ
KOAY(FM) Coalville UT
KOBB-FM Bozeman MT

*KOBC(FM) Joplin MO
*KOBH(FM) Hobbs NM
*KOBK(FM) Baker City OR
KOCD(FM) Wilburton OK
KOCN(FM) Pacific Grove CA
KOCP(FM) Camarillo CA
*KOCU(FM) Altus OK
*KOCV(FM) Odessa TX
KODA(FM) Houston TX
KODJ(FM) Salt Lake City UT
KODM(FM) Odessa TX
KODS(FM) Carnelian Bay CA
*KODV(FM) Barstow CA
KODZ(FM) Eugene OR
KOEA(FM) Doniphan MO
KOEL-FM Cedar Falls IA
KOFH(FM) Nogales AZ
KOFM(FM) Enid OK
KOFX(FM) El Paso TX
KOGA-FM Ogallala NE
*KOGL(FM) Gleneden Beach OR
KOGM(FM) Opelousas LA
*KOGR(FM) Rosedale CA
*KOHL(FM) Fremont CA
*KOHM(FM) Lubbock TX
*KOHN(FM) Sells AZ
KOHO-FM Leavenworth WA
*KOHR(FM) Sheridan WY
*KOHS(FM) Orem UT
KOHT(FM) Marana AZ
*KOIR(FM) Edinburg TX
KOIT-FM San Francisco CA
*KOJI(FM) Okoboji IA
KOJK(FM) Blanchard OK
*KOJO(FM) Lake Charles LA
*KOKF(FM) Edmond OK
KOKO-FM Kerman CA
KOKR(FM) Newport AR
*KOKS(FM) Poplar Bluff MO
KOKU(FM) Hagatna GU
KOKX-FM Keokuk IA
KOKY(FM) Sherwood AR
KOKZ(FM) Waterloo IA
KOLA(FM) San Bernardino CA
KOLB(FM) Firth NE
*KOLI(FM) Electra TX
KOLL-FM Lonoke AR
KOLS(FM) Dodge City KS
KOLT-FM Warren AFB WY
*KOLU(FM) Pasco WA
KOLV(FM) Olivia MN
KOLW(FM) Othello WA
KOLY-FM Mobridge SD
KOLZ(FM) Cheyenne WY
KOMA(FM) Oklahoma City OK
KOMB(FM) Fort Scott KS
KOMC-FM Kimberling City MO
KOMG(FM) Ozark MO
KOMP(FM) Las Vegas NV
KOMR(FM) Sun City AZ
KOMS(FM) Poteau OK
KOMT(FM) Mountain Home AR
KOMX(FM) Pampa TX
KONA-FM Kennewick WA
KOND(FM) Clovis CA
KONE(FM) Lubbock TX
KONI(FM) Lanai City HI
KONO-FM Helotes TX
*KONQ(FM) Dodge City KS
KONV(FM) Overton NV
KONY(FM) Saint George UT
KOOC(FM) Belton TX
KOOI-FM Jacksonville TX
KOOK(FM) Junction TX
KOOL-FM Phoenix AZ
*KOOP(FM) Hornsby TX
KOOS(FM) North Bend OR
KOOU(FM) Hardy AR
*KOOZ(FM) Myrtle Point OR
KOPA(FM) Woodward OK
*KOPB-FM Portland OR
*KOPJ(FM) Sebeka MN
*KOPN(FM) Columbia MO
KOPR(FM) Butte MT
KOPW(FM) Plattsmouth NE
KOPY-FM Alice TX

KOQL(FM) Ashland MO
KOQO-FM Fresno CA
KORA-FM Bryan TX
KORD-FM Richland WA
KORI(FM) Mansfield LA
KORL-FM Waianae HI
*KORM(FM) Astoria OR
KORQ(FM) Baird TX
KORR(FM) American Falls ID
KORT-FM Grangeville ID
KOSB(FM) Perry OK
KOSG(FM) Pawhuska OK
KOSI(FM) Denver CO
*KOSK(FM) Oskaloosa IA
*KOSN(FM) Ketchum OK
KOSO(FM) Patterson CA
KOSP(FM) Willard MO
KOSS(FM) Rosamond CA
KOST(FM) Los Angeles CA
*KOSU(FM) Stillwater OK
KOSY-FM Spanish Fork UT
KOTE(FM) Eureka KS
KOTM-FM Ottumwa IA
*KOTO(FM) Telluride CO
KOTY(FM) Mason TX
KOUL(FM) Sinton TX
KOUT(FM) Rapid City SD
KOUZ(FM) Manville WY
KOVE-FM Galveston TX
*KOWI(FM) Lamoni IA
KOWZ-FM Blooming Prairie MN
KOXE(FM) Brownwood TX
KOYE(FM) Frankston TX
KOYN(FM) Paris TX
KOYT(FM) Elko NV
KOZB(FM) Livingston MT
KOZE-FM Lewiston ID
KOZI-FM Chelan WA
*KOZO(FM) Branson MO
KOZT(FM) Fort Bragg CA
KOZX(FM) Cabool MO
KOZY-FM Gering NE
KOZZ-FM Reno NV
*KPAC(FM) San Antonio TX
KPAD(FM) Wheatland WY
*KPAE(FM) Erwinville LA
KPAK(FM) Alva OK
KPAN-FM Hereford TX
*KPAQ(FM) Plaquemine LA
KPAS(FM) Fabens TX
KPAT(FM) Orcutt CA
KPAW(FM) Fort Collins CO
*KPBB(FM) Brownfield TX
*KPBD(FM) Big Spring TX
*KPBE(FM) Brownwood TX
*KPBJ(FM) Midland TX
KPBM(FM) McCamey TX
*KPBN(FM) Freer TX
KPBQ-FM Pine Bluff AR
KPBR(FM) Joliet MT
*KPBS-FM San Diego CA
*KPBX-FM Spokane WA
*KPCC(FM) Pasadena CA
KPCH(FM) Ruston LA
KPCL(FM) Farmington NM
KPCR(FM) Burlington CO
*KPCS(FM) Princeton MN
*KPCW(FM) Park City UT
KPDB(FM) Big Lake TX
*KPDO(FM) Pescadero CA
KPDQ-FM Portland OR
*KPDR(FM) Wheeler TX
KPEK(FM) Albuquerque NM
KPEL-FM Abbeville LA
KPEN-FM Soldotna AK
KPER(FM) Hobbs NM
KPEZ(FM) Austin TX
*KPFA(FM) Berkeley CA
*KPFB(FM) Berkeley CA
*KPFC(FM) Callisburg TX
*KPFK(FM) Los Angeles CA
KPFM(FM) Mountain Home AR
*KPFR(FM) Pine Grove OR
*KPFT(FM) Houston TX
KPFX(FM) Fargo ND
*KPGB(FM) Pryor MT

KPGG(FM) Ashdown AR
*KPGR(FM) Pleasant Grove UT
*KPGS(FM) Pagosa Springs CO
KPHD(FM) Elko NV
*KPHF(FM) Phoenix AZ
*KPHL(FM) Pahala HI
KPHR(FM) Ortonville MN
*KPHS(FM) Plains TX
KPHT(FM) Rocky Ford CO
KPHW(FM) Kaneohe HI
KPIG-FM Freedom CA
*KPIJ(FM) Junction City OR
KPIN(FM) Pinedale WY
*KPJP(FM) Greenville CA
*KPKJ(FM) Mentmore NM
KPKK(FM) Amargosa Valley NV
KPKR(FM) Parker AZ
KPKX(FM) Phoenix AZ
KPKY(FM) Pocatello ID
KPLA(FM) Columbia MO
KPLD(FM) Kanab UT
*KPLG(FM) Plains MT
*KPLI(FM) Olympia WA
KPLM(FM) Palm Springs CA
KPLN(FM) Lockwood MT
KPLO-FM Reliance SD
KPLT-FM Paris TX
*KPLU-FM Tacoma WA
KPLV(FM) Las Vegas NV
*KPLW(FM) Wenatchee WA
KPLX(FM) Fort Worth TX
KPLZ(FM) Seattle WA
*KPMB(FM) Plainview TX
KPMW(FM) Haliimaile HI
KPMX(FM) Sterling CO
KPNC-FM Ponca City OK
KPND(FM) Sandpoint ID
*KPNE-FM North Platte NE
*KPNO(FM) Norfolk NE
KPNT(FM) Sainte Genevieve MO
KPNY(FM) Alliance NE
KPOA(FM) Lahaina HI
KPOC-FM Pocahontas AR
KPOD-FM Crescent City CA
KPOI-FM Honolulu HI
*KPOO(FM) San Francisco CA
*KPOR(FM) Emporia KS
KPOS(FM) Post TX
KPOW-FM La Monte MO
KPPC(FM) Pocatello ID
KPPK(FM) Rainier OR
KPPL(FM) Poplar Bluff MO
*KPPN(FM) Pollock Pines CA
*KPPR(FM) Williston ND
KPPT-FM Toledo OR
KPPV(FM) Prescott Valley AZ
KPQ-FM Wenatchee WA
KPQX(FM) Havre MT
*KPRA(FM) Ukiah CA
KPRB(FM) Brush CO
KPRC-FM Salinas CA
*KPRD(FM) Hays KS
*KPRE(FM) Vail CO
KPRF(FM) Amarillo TX
*KPRG(FM) Hagatna GU
*KPRH(FM) Montrose CO
KPRI(FM) Encinitas CA
*KPRJ(FM) Jamestown ND
*KPRN(FM) Grand Junction CO
*KPRQ(FM) Sheridan WY
KPRR(FM) El Paso TX
KPRS(FM) Kansas City MO
*KPRU(FM) Delta CO
KPRV-FM Heavener OK
KPRW(FM) Perham MN
*KPRX(FM) Bakersfield CA
KPSA-FM Lordsburg NM
*KPSC(FM) Palm Springs CA
KPSD(FM) Faith SD
*KPSH(FM) Coachella CA
KPSI-FM Palm Springs CA
KPSL-FM Bakersfield CA
KPSM(FM) Brownwood TX
KPSO-FM Falfurrias TX
*KPSU(FM) Goodwell OK
KPTE(FM) Durango CO

KPTI(FM) Crystal Beach TX
KPTL(FM) Ankeny IA
KPTT(FM) Denver CO
KPTX(FM) Pecos TX
KPTY(FM) Missouri City TX
*KPUB(FM) Flagstaff AZ
KPUR-FM Canyon TX
KPUS(FM) Gregory TX
KPVR(FM) Bowling Green MO
KPVS(FM) Hilo HI
*KPVU(FM) Prairie View TX
KPVW(FM) Aspen CO
KPWB-FM Piedmont MO
KPWR(FM) Los Angeles CA
KPWT(FM) Terrell Hills TX
KPWW(FM) Hooks TX
KPXI(FM) Overton TX
KPXP(FM) Garapan-Saipan NP
KPYG(FM) Cambria CA
*KPYR(FM) Craig CO
KPZA-FM Jal NM
KPZE-FM Carlsbad NM
KPZK-FM Cabot AR
*KQAC(FM) Gleneden Beach OR
*KQAI(FM) Roswell NM
KQAK(FM) Bend OR
*KQAL(FM) Winona MN
KQAY-FM Tucumcari NM
KQAZ(FM) Springerville-Eagar AZ
KQBA(FM) Los Alamos NM
KQBB(FM) Center TX
KQBE(FM) Ellensburg WA
KQBK(FM) Booneville AR
KQBL(FM) Billings MT
KQBO(FM) Rio Grande City TX
KQBR(FM) Lubbock TX
KQBT(FM) Rio Rancho NM
KQBU-FM Port Arthur TX
KQBW(FM) Omaha NE
KQBZ(FM) Coleman TX
KQCH(FM) Omaha NE
KQCL(FM) Faribault MN
KQCM(FM) Joshua Tree CA
KQCR-FM Parkersburg IA
KQCS(FM) Bettendorf IA
KQCV-FM Shawnee OK
KQDD(FM) Osceola AR
KQDI-FM Great Falls MT
KQDJ-FM Valley City ND
KQDR(FM) Savoy TX
KQDS-FM Duluth MN
KQDY(FM) Bismarck ND
*KQED-FM San Francisco CA
KQEG(FM) La Crescent MN
*KQEI-FM North Highlands CA
KQEL(FM) Alamogordo NM
KQEO(FM) Idaho Falls ID
KQEW(FM) Fordyce AR
KQFC(FM) Boise ID
*KQFE(FM) Springfield OR
KQFM(FM) Hermiston OR
*KQFR(FM) Rapid City SD
KQFX(FM) Borger TX
KQHC(FM) Burns OR
KQHN(FM) Waskom TX
*KQHR(FM) Hood River OR
KQHT(FM) Crookston MN
KQIB(FM) Idabel OK
KQIC(FM) Willmar MN
KQID(FM) Alexandria LA
KQIK-FM Lakeview OR
KQIS(FM) Basile LA
KQIZ-FM Amarillo TX
KQJK(FM) Roseville CA
KQKI(FM) Bayou Vista LA
KQKK(FM) Walker MN
*KQKL(FM) Selma CA
KQKQ-FM Council Bluffs IA
KQKS(FM) Lakewood CO
KQKY(FM) Kearney NE
KQLA(FM) Ogden KS
KQLB(FM) Los Banos CA
KQLK(FM) De Ridder LA
KQLL-FM Owasso OK
KQLM(FM) Odessa TX
KQLP(FM) Gallup NM

KQLQ(FM) Columbia LA
*KQLR(FM) Whitehall MT
KQLS(FM) Colby KS
KQLT(FM) Casper WY
*KQLU(FM) Belgrade MT
*KQLV(FM) Bosque Farms NM
KQLX-FM Lisbon ND
KQMA-FM Phillipsburg KS
KQMB(FM) Levan UT
*KQMC(FM) Hawthorne NV
KQMG-FM Independence IA
*KQMN(FM) Thief River Falls MN
KQMO(FM) Shell Knob MO
KQMQ-FM Honolulu HI
KQMR(FM) Globe AZ
KQMT(FM) Denver CO
KQMV(FM) Bellevue WA
KQMX(FM) Lost Hills CA
KQMY(FM) Cheyenne WY
*KQNC(FM) Quincy CA
KQNG-FM Lihue HI
KQNK-FM Norton KS
KQNO(FM) Coalinga CA
KQNS-FM Lindsborg KS
KQOB(FM) Enid OK
KQOD(FM) Stockton CA
KQOR(FM) Mena AR
*KQPD(FM) Ardmore OK
KQPI(FM) Aberdeen ID
KQPM(FM) Ukiah CA
KQPR-FM Albert Lea MN
KQPT(FM) Colusa CA
KQQB-FM Newport WA
KQQK(FM) Beaumont TX
KQQL(FM) Anoka MN
KQRA(FM) Brookline MO
KQRB(FM) Windom MN
KQRC-FM Leavenworth KS
*KQRI(FM) Belen NM
KQRK(FM) Ronan MT
KQRN(FM) Mitchell SD
KQRQ(FM) Rapid City SD
KQRS-FM Golden Valley MN
KQRT(FM) Las Vegas NV
KQRV(FM) Deer Lodge MT
KQRX(FM) Midland TX
*KQSC(FM) Santa Barbara CA
*KQSD-FM Lowry SD
KQSI(FM) San Augustine TX
KQSK(FM) Chadron NE
KQSM-FM Bentonville AR
KQSN(FM) Naches WA
KQSR(FM) Yuma AZ
KQSS(FM) Miami AZ
KQST(FM) Sedona AZ
KQSW(FM) Rock Springs WY
KQTA(FM) Homedale ID
KQTH(FM) Tucson AZ
KQTP(FM) Saint Marys KS
KQTY-FM Borger TX
KQTZ(FM) Hobart OK
KQUL(FM) Lake Ozark MO
KQUR(FM) Laredo TX
KQUS-FM Hot Springs AR
*KQVO(FM) Calexico CA
KQVT(FM) Victoria TX
KQWB-FM Moorhead MN
KQWC-FM Webster City IA
*KQWS(FM) Omak WA
KQWY(FM) Lusk WY
KQXC-FM Wichita Falls TX
*KQXE(FM) Eastland TX
KQXL-FM New Roads LA
KQXR(FM) Payette ID
*KQXS(FM) Stephenville TX
KQXT(FM) San Antonio TX
KQXX-FM Mission TX
KQXY-FM Beaumont TX
KQYB(FM) Spring Grove MN
KQYK(FM) Lake Crystal MN
KQZB(FM) Troy ID
KQZR(FM) Craig CO
KQZT(FM) Covelo CA
KQZZ(FM) Devils Lake ND
KRAB(FM) Green Acres CA
KRAI-FM Craig CO

KRAJ(FM) Johannesburg CA
KRAN(FM) Wamsutter WY
KRAO-FM Colfax WA
KRAQ(FM) Jackson MN
KRAT(FM) Altamont OR
KRAV(FM) Tulsa OK
*KRAW(FM) Sterling AK
KRAY-FM Salinas CA
KRAZ(FM) Santa Ynez CA
KRBB(FM) Wichita KS
*KRBD(FM) Ketchikan AK
KRBE(FM) Houston TX
KRBG(FM) Guymon OK
KRBI-FM Saint Peter MN
KRBL(FM) Idalou TX
*KRBM(FM) Pendleton OR
KRBR(FM) Superior WI
KRBV(FM) Los Angeles CA
*KRBW(FM) Ottawa KS
KRBZ(FM) Kansas City MO
*KRCB-FM Santa Rosa CA
*KRCC(FM) Colorado Springs CO
KRCD(FM) Inglewood CA
KRCH(FM) Rochester MN
KRCK-FM Mecca CA
*KRCL(FM) Salt Lake City UT
KRCQ(FM) Detroit Lakes MN
KRCS(FM) Sturgis SD
*KRCU(FM) Cape Girardeau MO
KRCV(FM) West Covina CA
KRCW(FM) Royal City WA
KRCX-FM Marysville CA
KRCY-FM Lake Havasu City AZ
KRDA(FM) Hanford CA
*KRDC-FM Saint George UT
KRDE(FM) Globe AZ
KRDG(FM) Shingletown CA
KRDJ(FM) New Iberia LA
KRDO-FM Security CO
*KRDR(FM) Red River NM
KRDS-FM New Prague MN
KRDX(FM) Vail AZ
KREC(FM) Brian Head UT
KRED-FM Eureka CA
KREJ(FM) Medicine Lodge KS
KREK(FM) Bristow OK
KREO(FM) Pine Bluffs WY
KREP(FM) Belleville KS
KRER(FM) Hamilton City CA
KRES(FM) Moberly MO
KREU(FM) Roland OK
KREZ(FM) Chaffee MO
*KRFA-FM Moscow ID
*KRFC(FM) Fort Collins CO
*KRFH(FM) Marshalltown IA
KRFM(FM) Show Low AZ
KRFO-FM Owatonna MN
KRFS-FM Superior NE
KRFX(FM) Denver CO
KRGI-FM Grand Island NE
KRGN(FM) Amarillo TX
KRGT(FM) Indian Springs NV
KRGY(FM) Aurora NE
*KRHS(FM) Overland MO
KRHV(FM) Big Pine CA
KRIA(FM) Plainview TX
KRIG-FM Nowata OK
KRIO-FM Pearsall TX
KRIT(FM) Parker AZ
KRJB(FM) Ada MN
KRJC(FM) Elko NV
KRJM(FM) Mahnomen MN
KRJT(FM) Elgin OR
KRKA(FM) Erath LA
KRKC-FM King City CA
KRKD(FM) Dermott AR
KRKI(FM) Newcastle WY
*KRKL(FM) Walla Walla WA
KRKN(FM) Eldon IA
KRKQ(FM) Mountain Village CO
KRKR(FM) Lincoln NE
KRKS-FM Boulder CO
KRKT-FM Albany OR
KRKU(FM) McCook NE
KRKV(FM) Las Animas CO
KRKX(FM) Billings MT

KRKY-FM Estes Park CO
KRKZ(FM) Altus OK
*KRLE(FM) Oberlin KS
*KRLF(FM) Pullman WA
*KRLH(FM) Hereford TX
KRLI(FM) Malta Bend MO
*KRLJ(FM) La Junta CO
*KRLP(FM) Windom MN
KRLQ(FM) Hodge LA
*KRLR(FM) Sulphur LA
*KRLS(FM) Knoxville IA
KRLT(FM) South Lake Tahoe CA
*KRLU(FM) Roswell NM
KRLW-FM Walnut Ridge AR
*KRLX(FM) Northfield MN
*KRMB(FM) Bisbee AZ
*KRMC(FM) Douglas AZ
KRMD-FM Shreveport LA
*KRMH(FM) Red Mesa AZ
KRMQ-FM Clovis NM
KRNA(FM) Iowa City IA
KRNB(FM) Decatur TX
*KRNC(FM) Steamboat Springs CO
*KRNE-FM Merriman NE
KRNG(FM) Fallon/ Reno NV
KRNH(FM) Kerrville TX
*KRNL-FM Mount Vernon IA
*KRNM(FM) Chalan Kanoa-Saipan NP
*KRNN(FM) Juneau AK
KRNO(FM) Incline Village NV
KRNQ(FM) Keokuk IA
*KRNU(FM) Lincoln NE
KRNV-FM Reno NV
*KRNW(FM) Chillicothe MO
KRNY(FM) Kearney NE
*KROA(FM) Grand Island NE
KROC-FM Rochester MN
KROG(FM) Grants Pass OR
KROI(FM) Seabrook TX
KROK(FM) South Fort Polk LA
KROM(FM) San Antonio TX
KROQ-FM Pasadena CA
KROR(FM) Hastings NE
*KROU(FM) Spencer OK
KROW(FM) Lovell WY
KROX-FM Buda TX
KROY(FM) Palacios TX
KRPH(FM) Yarnell AZ
KRPM(FM) Billings MT
*KRPR(FM) Rochester MN
*KRPS(FM) Pittsburg KS
KRPT(FM) Devine TX
KRPX(FM) Wellington UT
KRQB(FM) San Jacinto CA
KRQK(FM) Lompoc CA
KRQN(FM) Vinton IA
KRQQ(FM) Tucson AZ
KRQR(FM) Orland CA
KRQS(FM) Alberton MT
KRQT(FM) Castle Rock WA
KRQU(FM) Laramie WY
*KRQZ(FM) Lompoc CA
*KRRC(FM) Portland OR
KRRG(FM) Laredo TX
KRRK(FM) Lake Havasu City AZ
KRRM(FM) Rogue River OR
KRRN(FM) Kingman AZ
KRRO(FM) Sioux Falls SD
KRRQ(FM) Lafayette LA
KRRR(FM) Cheyenne WY
KRRV-FM Alexandria LA
KRRW(FM) Saint James MN
KRRX(FM) Burney CA
KRRY(FM) Canton MO
KRSB-FM Roseburg OR
*KRSC-FM Claremore OK
*KRSD(FM) Sioux Falls SD
KRSE(FM) Yakima WA
KRSH(FM) Healdsburg CA
KRSI(FM) Garapan-Saipan NP
KRSJ(FM) Durango CO
KRSK(FM) Molalla OR
KRSP-FM Salt Lake City UT
KRSQ(FM) Laurel MT
*KRSS(FM) Tarkio MO
KRST(FM) Albuquerque NM
*KRSU(FM) Appleton MN
KRSV-FM Afton WY
*KRSW(FM) Worthington MN
KRSX-FM Yermo CA
KRSY-FM La Luz NM
KRTH(FM) Los Angeles CA
KRTI(FM) Grinnell IA
*KRTM(FM) Temecula CA
KRTN-FM Raton NM
KRTO(FM) Lompoc CA
KRTR-FM Kailua HI
*KRTU(FM) San Antonio TX
KRTY(FM) Los Gatos CA
KRTZ(FM) Cortez CO
*KRUA(FM) Anchorage AK
*KRUC(FM) Las Cruces NM
KRUE(FM) Waseca MN
KRUF(FM) Shreveport LA
KRUG(FM) Upton WY
*KRUI-FM Iowa City IA
KRUP(FM) Dillingham AK
*KRUX(FM) Las Cruces NM
KRUZ(FM) Santa Barbara CA
KRVA-FM Campbell TX
KRVB(FM) Nampa ID
KRVC(FM) Hornbrook CA
KRVE(FM) Brusly LA
KRVF(FM) Kerens TX
KRVG(FM) Glenwood Springs CO
*KRVH(FM) Rio Vista CA
KRVK(FM) Midwest WY
KRVL(FM) Kerrville TX
*KRVM-FM Eugene OR
KRVN-FM Lexington NE
KRVO(FM) Columbia Falls MT
KRVQ(FM) Victor ID
KRVR(FM) Copperopolis CA
*KRVS(FM) Lafayette LA
KRVV(FM) Bastrop LA
KRVX(FM) Wimbledon ND
KRVY-FM Starbuck MN
*KRWG(FM) Las Cruces NM
KRWK(FM) Fargo ND
KRWM(FM) Bremerton WA
KRWN(FM) Farmington NM
KRWP(FM) Stockton MO
KRWQ(FM) Gold Hill OR
*KRWT(FM) West Laramie WY
KRXB(FM) Beeville TX
KRXF(FM) Sunriver OR
KRXL(FM) Kirksville MO
KRXO(FM) Oklahoma City OK
KRXQ(FM) Sacramento CA
KRXT(FM) Rockdale TX
KRXV(FM) Yermo CA
*KRXW(FM) Roseau MN
KRXX(FM) Kodiak AK
KRXY(FM) Shelton WA
KRYD(FM) Norwood CO
KRYE(FM) Rye CO
*KRYI(FM) Rye CO
KRYK(FM) Chinook MT
KRYP(FM) Gladstone OR
KRYS-FM Corpus Christi TX
*KRZA(FM) Alamosa CO
KRZK(FM) Branson MO
KRZN(FM) Billings MT
KRZQ-FM Sparks NV
KRZR(FM) Hanford CA
KRZX(FM) Monticello UT
KRZY-FM Santa Fe NM
KRZZ(FM) San Francisco CA
KSAB(FM) Robstown TX
KSAG(FM) Pearsall TX
KSAJ-FM Abilene KS
*KSAK(FM) Walnut CA
KSAL-FM Salina KS
KSAM-FM Huntsville TX
KSAN(FM) San Mateo CA
KSAQ(FM) Charlotte TX
KSAR(FM) Thayer MO
KSAS-FM Caldwell ID
*KSAU(FM) Nacogdoches TX
KSAY(FM) Fort Bragg CA
*KSBA(FM) Coos Bay OR
*KSBC(FM) Nile WA
KSBH(FM) Coushatta LA
*KSBJ(FM) Humble TX
KSBL(FM) Carpinteria CA
*KSBR(FM) Mission Viejo CA
KSBS-FM Pago Pago AS
KSBV(FM) Salida CO
*KSBX(FM) Santa Barbara CA
KSBZ(FM) Sitka AK
KSCA(FM) Glendale CA
KSCB-FM Liberal KS
KSCF(FM) San Diego CA
KSCG(FM) Meridian TX
KSCH(FM) Sulphur Springs TX
*KSCL(FM) Shreveport LA
KSCN(FM) Pittsburg TX
KSCQ(FM) Silver City NM
KSCR-FM Benson MN
KSCS(FM) Fort Worth TX
*KSCU(FM) Santa Clara CA
*KSCV(FM) Springfield MO
KSCY(FM) Big Sky MT
KSD(FM) Saint Louis MO
*KSDA-FM Agat GU
*KSDB-FM Manhattan KS
*KSDJ(FM) Brookings SD
KSDL(FM) Sedalia MO
KSDM(FM) International Falls MN
KSDN-FM Aberdeen SD
KSDR-FM Watertown SD
*KSDS(FM) San Diego CA
KSDZ(FM) Gordon NE
KSEA(FM) Greenfield CA
KSEC(FM) Bentonville AR
KSED(FM) Sedona AZ
*KSEF(FM) Farmington MO
KSEG(FM) Sacramento CA
KSEH(FM) Brawley CA
KSEK-FM Girard KS
KSEL-FM Portales NM
KSEM-FM Seminole TX
KSEQ(FM) Visalia CA
*KSER(FM) Everett WA
KSES-FM Seaside CA
KSEY-FM Seymour TX
KSEZ(FM) Sioux City IA
*KSFC(FM) Spokane WA
*KSFH(FM) Mountain View CA
KSFI(FM) Salt Lake City UT
KSFM(FM) Woodland CA
KSFQ(FM) White Rock NM
*KSFR(FM) Santa Fe NM
*KSFS(FM) Sioux Falls SD
KSFT-FM South Sioux City NE
KSFX(FM) Roswell NM
KSGC(FM) Tusayan AZ
KSGF-FM Ash Grove MO
*KSGN(FM) Riverside CA
*KSGR(FM) Portland TX
*KSGU(FM) Saint George UT
KSHA(FM) Redding CA
KSHE(FM) Crestwood MO
*KSHI(FM) Zuni NM
KSHK(FM) Kekaha HI
KSHL(FM) Gleneden Beach OR
KSHN-FM Liberty TX
KSHR-FM Coquille OR
*KSHU(FM) Huntsville TX
KSIB-FM Creston IA
KSID-FM Sidney NE
KSII(FM) El Paso TX
KSIL(FM) Hurley NM
KSIQ(FM) Brawley CA
KSIT(FM) Rock Springs WY
*KSIV-FM Saint Louis MO
*KSJD(FM) Cortez CO
*KSJE(FM) Farmington NM
KSJJ(FM) Redmond OR
KSJM(FM) Winfield KS
*KSJN(FM) Minneapolis MN
KSJO(FM) San Jose CA
KSJQ(FM) Savannah MO
*KSJR-FM Collegeville MN
*KSJS(FM) San Jose CA
KSJT-FM San Angelo TX
*KSJV(FM) Fresno CA
*KSJY(FM) Saint Martinville LA
KSJZ(FM) Jamestown ND
*KSKA(FM) Anchorage AK
KSKB(FM) Brooklyn IA
KSKD(FM) Livingston CA
KSKE-FM Vail CO
*KSKF(FM) Klamath Falls OR
KSKG(FM) Salina KS
KSKI-FM Sun Valley ID
KSKK(FM) Staples MN
KSKL(FM) Scott City KS
KSKS(FM) Fresno CA
KSKU(FM) Sterling KS
KSKZ(FM) Copeland KS
*KSLC(FM) McMinnville OR
KSLE(FM) Wewoka OK
KSL-FM Midvale UT
KSLG-FM Hydesville CA
KSLK(FM) Visalia CA
KSLQ-FM Washington MO
KSLS(FM) Liberal KS
KSLT(FM) Spearfish SD
*KSLU(FM) Hammond LA
KSLV-FM Monte Vista CO
KSLX-FM Scottsdale AZ
KSLY-FM San Luis Obispo CA
KSLZ(FM) Saint Louis MO
KSMA-FM Osage IA
KSMB(FM) Lafayette LA
*KSMC(FM) Moraga CA
KSMD(FM) Pangburn AR
KSME(FM) Greeley CO
*KSMF(FM) Ashland OR
KSMG(FM) Seguin TX
KSMJ(FM) Shafter CA
KSML-FM Huntington TX
*KSMR(FM) Winona MN
*KSMS-FM Point Lookout MO
KSMT(FM) Breckenridge CO
*KSMU(FM) Springfield MO
*KSMW(FM) West Plains MO
KSMX(FM) Clovis NM
KSMY(FM) Lompoc CA
KSMZ(FM) Viola AR
KSNA(FM) Rexburg ID
KSND(FM) Monmouth OR
KSNE-FM Las Vegas NV
KSNI-FM Santa Maria CA
KSNN(FM) Saint George UT
KSNO-FM Snowmass Village CO
KSNP(FM) Burlington KS
KSNQ(FM) Twin Falls ID
KSNR(FM) Thief River Falls MN
*KSNS(FM) Medicine Lodge KS
KSNX(FM) Show Low AZ
KSNY-FM Snyder TX
KSOC(FM) Gainesville TX
KSOF(FM) Dinuba CA
*KSOH(FM) Wapato WA
KSOK-FM Winfield KS
KSOL(FM) San Francisco CA
KSOM(FM) Audubon IA
KSON-FM San Diego CA
KSOP-FM Salt Lake City UT
KSOQ-FM Escondido CA
*KSOR(FM) Ashland OR
*KSOS(FM) Las Vegas NV
KSOU-FM Sioux Center IA
*KSPB(FM) Pebble Beach CA
*KSPC(FM) Claremont CA
KSPE-FM Ellwood CA
KSPI-FM Stillwater OK
KSPK(FM) Walsenburg CO
*KSPL(FM) Kalispell MT
KSPN-FM Aspen CO
KSPO(FM) Dishman WA
KSPQ(FM) West Plains MO
KSPW(FM) Sparta MO
KSQB-FM Dell Rapids SD
KSQL(FM) Santa Cruz CA
KSQQ(FM) Morgan Hill CA
*KSQS(FM) Ririe ID
*KSQX(FM) Springtown TX
KSQY(FM) Deadwood SD
KSRA-FM Salmon ID
*KSRD(FM) Saint Joseph MO
KSRF(FM) Poipu HI
*KSRG(FM) Ashland OR
*KSRH(FM) San Rafael CA
*KSRI(FM) Santa Cruz CA
KSRN(FM) Kings Beach CA
*KSRQ(FM) Thief River Falls MN
*KSRS(FM) Roseburg OR
KSRT(FM) Cloverdale CA
KSRV-FM Ontario OR
KSRW(FM) Independence CA
KSRX(FM) Sterling CO
KSRZ(FM) Omaha NE
KSSA(FM) Ingalls KS
KSSB(FM) Calipatria CA
KSSC(FM) Ventura CA
KSSD(FM) Fallbrook CA
KSSE(FM) Arcadia CA
KSSH(FM) Ingalls KS
KSSI(FM) China Lake CA
KSSJ(FM) Fair Oaks CA
KSSK-FM Waipahu HI
KSSM(FM) Copperas Cove TX
KSSN(FM) Little Rock AR
*KSSO(FM) Norman OK
KSSS(FM) Bismarck ND
*KSSU(FM) Durant OK
KSSW(FM) Nashville AR
*KSSX(FM) Chickasha OK
KSSZ(FM) Fayette MO
KSTB(FM) Crystal Beach TX
KSTH(FM) Holyoke CO
KSTJ(FM) Boulder City NV
*KSTK(FM) Wrangell AK
*KSTM(FM) Indianola IA
KSTN-FM Stockton CA
KSTO(FM) Hagatna GU
KSTP-FM Saint Paul MN
KSTQ-FM Plainview TX
KSTR-FM Montrose CO
KSTT-FM Los Osos-Baywood Park CA
KSTV-FM Dublin TX
*KSTX(FM) San Antonio TX
KSTZ(FM) Des Moines IA
*KSUA(FM) Fairbanks AK
*KSUI(FM) Iowa City IA
*KSUL(FM) Port Sulphur LA
KSUP(FM) Juneau AK
*KSUR(FM) Mart TX
*KSUT(FM) Ignacio CO
*KSUU(FM) Cedar City UT
*KSUW(FM) Sheridan WY
KSUX(FM) Winnebago NE
KSVL(FM) Smith NV
*KSVR(FM) Mount Vernon WA
*KSVY(FM) Sonoma CA
*KSWC(FM) Winfield KS
KSWD-FM Seward AK
KSWF(FM) Aurora MO
KSWG(FM) Wickenburg AZ
*KSWH(FM) Arkadelphia AR
KSWI(FM) Atlantic IA
KSWN(FM) McCook NE
*KSWP(FM) Lufkin TX
*KSWS(FM) Chehalis WA
KSWW(FM) Montesano WA
KSXE(FM) Kingsburg CA
KSXY(FM) Middletown CA
KSYC-FM Yreka CA
*KSYD(FM) Reedsport OR
*KSYE(FM) Frederick OK
*KSYM-FM San Antonio TX
KSYN(FM) Joplin MO
KSYR(FM) Benton LA
KSYU(FM) Corrales NM
KSYV(FM) Solvang CA
KSYY-FM Bennett CO
KSYZ-FM Grand Island NE
KSZR(FM) Oro Valley AZ
*KTAA(FM) Big Sandy TX
KTAC(FM) Ephrata WA
*KTAD(FM) Sterling CO
KTAG(FM) Cody WY
*KTAI(FM) Kingsville TX
KTAK(FM) Riverton WY
KTAL-FM Texarkana TX
KTAO(FM) Taos NM
KTAR-FM Glendale AZ

*KTAW(FM) Westcliffe CO
*KTBG(FM) Warrensburg MO
KTBH-FM Kurtistown HI
*KTBJ(FM) Festus MO
KTBQ(FM) Nacogdoches TX
KTBT(FM) Broken Arrow OK
KTBZ-FM Houston TX
*KTCB(FM) Tillamook OR
*KTCC(FM) Colby KS
KTCE(FM) Payson UT
*KTCF(FM) Dolores CO
KTCL(FM) Wheat Ridge CO
KTCM(FM) Kingman KS
KTCN(FM) Eureka Springs AR
KTCO(FM) Duluth MN
KTCS-FM Fort Smith AR
*KTCU-FM Fort Worth TX
*KTCV(FM) Kennewick WA
KTCX(FM) Beaumont TX
KTCY(FM) Azle TX
KTCZ-FM Minneapolis MN
*KTDA(FM) Dalhart TX
*KTDB(FM) Ramah NM
KTDE(FM) Gualala CA
KTDK(FM) Sanger TX
*KTDL(FM) Trinidad CO
KTDR(FM) Del Rio TX
*KTDU(FM) Trimble CO
*KTDV(FM) State Center IA
*KTDX(FM) Frisco CO
KTDY(FM) Lafayette LA
KTDZ(FM) College AK
KTEA(FM) Cambria CA
*KTEC(FM) Klamath Falls OR
KTED(FM) Douglas WY
KTEE(FM) North Bend OR
KTEG(FM) Bosque Farms NM
*KTEI(FM) Placerville CO
*KTEP(FM) El Paso TX
*KTER(FM) Rudolph TX
KTEX(FM) Brownsville TX
KTEZ(FM) Zwolle LA
KTFC(FM) Sioux City IA
KTFG(FM) Sioux Rapids IA
KTFM(FM) Floresville TX
KTFR(FM) Chelsea OK
KTFW-FM Glen Rose TX
KTFX-FM Warner OK
*KTFY(FM) Buhl ID
KTGA(FM) Saratoga WY
KTGL(FM) Beatrice NE
*KTGS(FM) Ada OK
KTGV(FM) Jonesville LA
*KTGW(FM) Fruitland NM
KTHC(FM) Sidney MT
KTHI(FM) Caldwell ID
KTHK(FM) Idaho Falls ID
*KTHM(FM) Red Bluff CA
KTHN(FM) La Junta CO
KTHP(FM) Hemphill TX
KTHQ(FM) Eagar AZ
KTHR(FM) Wichita KS
KTHS-FM Berryville AR
KTHT(FM) Cleveland TX
KTHU(FM) Corning CA
KTHX-FM Dayton NV
KTIC-FM West Point NE
KTIG(FM) Pequot Lakes MN
KTIJ(FM) Elk City OK
KTIL-FM Tillamook OR
*KTIS-FM Minneapolis MN
*KTJC(FM) Kelso WA
KTJJ(FM) Farmington MO
KTJM(FM) Port Arthur TX
*KTJO-FM Ottawa KS
KTJZ(FM) Tallulah LA
KTKB-FM Hagatna GU
KTKC(FM) Springhill LA
KTKE(FM) Truckee CA
*KTKL(FM) Stigler OK
KTKO(FM) Beeville TX
KTKS(FM) Versailles MO
KTKU(FM) Juneau AK
KTKY(FM) Refugio TX
KTKZ-FM Dunnigan CA
KTLB(FM) Twin Lakes IA

*KTLC(FM) Canon City CO
*KTLF(FM) Colorado Springs CO
*KTLI(FM) El Dorado KS
KTLK-FM Minneapolis MN
*KTLN(FM) Thibodaux LA
KTLO-FM Mountain Home AR
KTLS-FM Holdenville OK
KTLT(FM) Anson TX
*KTLW(FM) Lancaster CA
*KTLX(FM) Columbus NE
*KTLZ(FM) Cuero TX
KTMB(FM) Mountain Home ID
KTMC-FM McAlester OK
KTMG(FM) Prescott AZ
*KTMH(FM) Colona CO
*KTMK(FM) Tillamook OR
KTMO(FM) New Madrid MO
KTMQ(FM) Temecula CA
KTMT-FM Medford OR
KTMX(FM) York NE
*KTNA(FM) Talkeetna AK
*KTNE-FM Alliance NE
KTNI-FM Strasburg CO
KTNR(FM) Kenedy TX
KTNT(FM) Eufaula OK
KTNX(FM) Arcadia MO
KTNY(FM) Libby MT
*KTOC-FM Jonesboro LA
KTOH(FM) Kalaheo HI
*KTOL(FM) Leadville CO
KTOM-FM Marina CA
*KTOO(FM) Juneau AK
KTOR(FM) Westwood CA
*KTOT(FM) Spearman TX
KTOY(FM) Texarkana AR
KTOZ-FM Pleasant Hope MO
*KTPF(FM) Salida CO
*KTPH(FM) Tonopah NV
KTPI-FM Tehachapi CA
KTPK(FM) Topeka KS
*KTPL(FM) Pueblo CO
KTPO(FM) Kootenai ID
*KTPR(FM) Fort Dodge IA
*KTPS(FM) Pagosa Springs CO
KTPT(FM) Rapid City SD
KTPZ(FM) Hazelton ID
KTQM-FM Clovis NM
*KTQX(FM) Bakersfield CA
KTRA-FM Farmington NM
KTRI-FM Mansfield MO
KTRJ(FM) Hayden CO
*KTRM(FM) Kirksville MO
KTRN(FM) White Hall AR
KTRQ(FM) Colt AR
KTRR(FM) Loveland CO
KTRS-FM Casper WY
KTRT(FM) Winthrop WA
*KTRU(FM) Houston TX
KTRX(FM) Dickson OK
KTRZ(FM) Riverton WY
*KTSC-FM Pueblo CO
*KTSD-FM Reliance SD
KTSE-FM Patterson CA
*KTSG(FM) Sidney CO
KTSL(FM) Medical Lake WA
KTSM-FM El Paso TX
KTSO(FM) Glenpool OK
KTSR(FM) De Quincy LA
KTST(FM) Oklahoma City OK
*KTSU(FM) Houston TX
*KTSW(FM) San Marcos TX
KTSX(FM) Knox City TX
*KTSY(FM) Caldwell ID
KTTA(FM) Esparto CA
KTTB(FM) Glencoe MN
KTTG(FM) Mena AR
KTTI(FM) Yuma AZ
*KTTK(FM) Lebanon MO
KTTL(FM) Alva OK
KTTN-FM Trenton MO
KTTR-FM Saint James MO
KTTS-FM Springfield MO
KTTX(FM) Brenham TX
KTTY(FM) New Boston TX
KTUF(FM) Kirksville MO
KTUG(FM) Hudson WY

*KTUH(FM) Honolulu HI
KTUI-FM Sullivan MO
KTUM(FM) Tatum NM
KTUN(FM) Eagle CO
KTUX(FM) Carthage TX
KTUZ-FM Okarche OK
*KTVR-FM La Grande OR
KTWA(FM) Ottumwa IA
KTWB(FM) Sioux Falls SD
*KTWD(FM) Wallace ID
KTWL(FM) Hempstead TX
KTWS(FM) Bend OR
KTWV(FM) Los Angeles CA
*KTWX(FM) Walsenburg CO
*KTXB(FM) Beaumont TX
KTXC(FM) Lamesa TX
*KTXG(FM) Greenville TX
*KTXI(FM) Ingram TX
KTXJ-FM Jasper TX
*KTXK(FM) Texarkana TX
KTXM(FM) Hallettsville TX
KTXN-FM Victoria TX
KTXO(FM) Goldsmith TX
*KTXP(FM) Bushland TX
KTXR(FM) Springfield MO
*KTXT-FM Lubbock TX
KTXX(FM) Karnes City TX
KTXY(FM) Jefferson City MO
KTYD(FM) Santa Barbara CA
KTYL-FM Tyler TX
KTYN(FM) Thayne WY
KTYS(FM) Flower Mound TX
KTZA(FM) Artesia NM
KTZR-FM Green Valley AZ
KTZU(FM) Velva ND
KTZZ(FM) Conrad MT
*KUAC(FM) Fairbanks AK
KUAD-FM Windsor CO
*KUAF(FM) Fayetteville AR
KUAL-FM Brainerd MN
KUAM-FM Hagatna GU
*KUAP(FM) Pine Bluff AR
*KUAR(FM) Little Rock AR
*KUAT-FM Tucson AZ
*KUAZ-FM Tucson AZ
KUBB(FM) Mariposa CA
KUBE(FM) Seattle WA
KUBL-FM Salt Lake City UT
*KUBO(FM) Calexico CA
KUBQ(FM) La Grande OR
*KUBS(FM) Newport WA
*KUCA(FM) Conway AR
KUCD(FM) Pearl City HI
*KUCI(FM) Irvine CA
*KUCR(FM) Riverside CA
*KUCV(FM) Lincoln NE
KUDD(FM) Roy UT
KUDE(FM) Nephi UT
KUDL(FM) Kansas City KS
*KUDU(FM) Tok AK
KUEL(FM) Fort Dodge IA
*KUER(FM) Salt Lake City UT
*KUFM(FM) Missoula MT
*KUFN(FM) Hamilton MT
KUFO-FM Portland OR
*KUFR(FM) Salt Lake City UT
KUFX(FM) San Jose CA
*KUGS(FM) Bellingham WA
*KUHB-FM Saint Paul AK
*KUHF(FM) Houston TX
*KUHM(FM) Helena MT
KUIC(FM) Vacaville CA
KUJ-FM Burbank WA
KUJJ(FM) Weston OR
KUJZ(FM) Creswell OR
KUKA(FM) San Diego TX
KUKI-FM Ukiah CA
*KUKL(FM) Kalispell MT
KUKN(FM) Longview WA
KUKU-FM Willow Springs MO
KULE-FM Ephrata WA
KULF(FM) Ganado TX
KULH(FM) Wheeling MO
KULL(FM) Abilene TX
KULM(FM) Columbus TX
KULO(FM) Alexandria MN

*KULV(FM) Ukiah CA
KUMA-FM Pendleton OR
*KUMD-FM Duluth MN
*KUMM(FM) Morris MN
KUMR(FM) Doolittle MO
KUMU-FM Honolulu HI
KUMX(FM) North Fort Polk LA
KUNA-FM La Quinta CA
*KUNC(FM) Greeley CO
*KUND-FM Grand Forks ND
*KUNE(FM) Ottumwa IA
*KUNI(FM) Cedar Falls IA
*KUNJ(FM) Fairfield IA
*KUNM(FM) Albuquerque NM
KUNQ(FM) Houston MO
*KUNR(FM) Reno NV
*KUNV(FM) Las Vegas NV
*KUNY(FM) Mason City IA
*KUNZ(FM) Ottumwa IA
*KUOI-FM Moscow ID
*KUOM-FM Saint Louis Park MN
KUOO(FM) Spirit Lake IA
*KUOP(FM) Stockton CA
*KUOR-FM Redlands CA
*KUOW-FM Seattle WA
KUPD-FM Tempe AZ
KUPH(FM) Mountain View MO
KUPI-FM Idaho Falls ID
KUPL-FM Portland OR
*KUPR(FM) Alamogordo NM
*KUPS(FM) Tacoma WA
KUQQ(FM) Milford IA
KURB(FM) Little Rock AR
*KURE(FM) Ames IA
KURK(FM) Reno NV
KURM-FM South West City MO
KURQ(FM) Grover Beach CA
KURR(FM) Hurricane UT
*KURT(FM) Jackson WY
KURY-FM Brookings OR
KUSB(FM) Hazelton ND
*KUSC(FM) Los Angeles CA
*KUSD(FM) Vermillion SD
*KUSF(FM) San Francisco CA
KUSJ(FM) Harker Heights TX
KUSN(FM) Dearing KS
KUSO(FM) Albion NE
*KUSP(FM) Santa Cruz CA
*KUSR(FM) Logan UT
KUSS(FM) Carlsbad CA
*KUSU-FM Logan UT
*KUSW(FM) Flora Vista NM
KUSZ(FM) Laramie WY
*KUT(FM) Austin TX
*KUTE(FM) Ignacio CO
KUTT(FM) Fairbury NE
*KUTX(FM) San Angelo TX
KUUB(FM) Sun Valley NV
KUUL(FM) East Moline IL
KUUR(FM) Carbondale CO
KUUS(FM) Fairfield MT
*KUUT(FM) Farmington NM
KUUU(FM) South Jordan UT
KUUZ(FM) Lake Village AR
KUVA(FM) Uvalde TX
*KUVO(FM) Denver CO
*KUWA(FM) Afton WY
*KUWC(FM) Casper WY
*KUWD(FM) Sundance WY
*KUWG(FM) Gillette WY
*KUWJ(FM) Jackson WY
*KUWL(FM) Laramie WY
*KUWN(FM) Newcastle WY
*KUWP(FM) Powell WY
*KUWR(FM) Laramie WY
*KUWS(FM) Superior WI
KUWT(FM) Thermopolis WY
*KUWX(FM) Pinedale WY
*KUWY(FM) Laramie WY
*KUWZ(FM) Rock Springs WY
*KUYI(FM) Hotevilla AZ
KUYY(FM) Emmetsburg IA
KUZN(FM) Centerville TX
KUZZ-FM Bakersfield CA
KVAB(FM) Clarkston WA
KVAK-FM Valdez AK

KVAN-FM Pilot Rock OR
KVAR(FM) Pine Ridge SD
KVAS(FM) Ilwaco WA
KVAY(FM) Lamar CO
*KVAZ(FM) Henryetta OK
*KVCF(FM) Freeman SD
KVCK-FM Wolf Point MT
KVCL-FM Winnfield LA
*KVCM(FM) Helena MT
*KVCO(FM) Concordia KS
*KVCR(FM) San Bernardino CA
*KVCX(FM) Gregory SD
*KVCY(FM) Fort Scott KS
*KVDP(FM) Dry Prong LA
KVEG(FM) Mesquite NV
*KVER(FM) El Paso TX
KVET-FM Austin TX
KVFG(FM) Victorville CA
*KVFL(FM) Pierre SD
*KVFM(FM) Beeville TX
KVFX(FM) Logan UT
KVGB-FM Great Bend KS
KVGG(FM) Salome AZ
KVGO(FM) Spring Valley MN
KVGS(FM) Laughlin NV
*KVHS(FM) Concord CA
KVHT(FM) Vermillion SD
KVHU(FM) Judsonia AR
KVIB(FM) Sun City West AZ
KVIC(FM) Victoria TX
KVIL(FM) Highland Park-Dallas TX
*KVIP-FM Redding CA
*KVIR(FM) Bullhead City AZ
*KVIX(FM) Port Angeles WA
*KVJC(FM) Globe AZ
KVJM(FM) Hearne TX
*KVJZ(FM) Vail CO
KVKI-FM Shreveport LA
*KVKL(FM) Las Vegas NV
*KVLB(FM) Bend OR
KVLC(FM) Hatch NM
KVLD(FM) Atkins AR
KVLE-FM Gunnison CO
KVLI-FM Lake Isabella CA
*KVLK(FM) Socorro NM
KVLL-FM Wells TX
KVLO(FM) Humnoke AR
*KVLP(FM) Tucumcari NM
*KVLQ(FM) Lincoln ND
*KVLT(FM) Temple TX
*KVLU(FM) Beaumont TX
KVLV-FM Fallon NV
*KVLW(FM) Waco TX
KVLY(FM) Edinburg TX
*KVLZ(FM) Sheridan WY
KVMA-FM Shreveport LA
KVMI(FM) Arthur ND
*KVMN(FM) Cave City AR
*KVMR(FM) Nevada City CA
*KVMT(FM) Montrose CO
KVMV(FM) McAllen TX
KVMX(FM) Banks OR
KVMZ(FM) Waldo AR
KVNA-FM Flagstaff AZ
*KVNE(FM) Tyler TX
*KVNF(FM) Paonia CO
*KVNO(FM) Omaha NE
*KVOD(FM) Denver CO
KVOE-FM Emporia KS
KVOM-FM Morrilton AR
KVOO-FM Tulsa OK
KVOU-FM Uvalde TX
*KVOV(FM) Carbondale CO
KVOX-FM Moorhead MN
KVPI-FM Ville Platte LA
*KVPR(FM) Fresno CA
*KVRA(FM) Sisters OR
KVRD-FM Cottonwood AZ
KVRE(FM) Hot Springs Village AR
KVRG(FM) Victor ID
KVRH-FM Salida CO
*KVRK(FM) Sanger TX
*KVRN(FM) Marvell AR
KVRO(FM) Stillwater OK
KVRP-FM Haskell TX
*KVRS(FM) Lawton OK

*KVRT(FM) Victoria TX
KVRV(FM) Monte Rio CA
KVRW(FM) Lawton OK
*KVRX(FM) Austin TX
*KVSC(FM) Saint Cloud MN
KVSF-FM Pecos NM
KVSP(FM) Anadarko OK
*KVSS(FM) Omaha NE
KVST(FM) Willis TX
KVSV-FM Beloit KS
*KVTI(FM) Tacoma WA
*KVTT(FM) Dallas TX
KVTY(FM) Lewiston ID
*KVUH(FM) Laytonville CA
KVUU(FM) Pueblo CO
KVUW(FM) Wendover NV
KVVA-FM Apache Junction AZ
KVVF(FM) Santa Clara CA
KVVP(FM) Leesville LA
KVVR(FM) Dutton MT
KVVS(FM) Mojave CA
KVVZ(FM) San Rafael CA
KVWC-FM Vernon TX
KVWG-FM Dilley TX
KVYB(FM) Santa Barbara CA
KVYL(FM) Cal-Nev-Ari NV
KVYN(FM) Saint Helena CA
KWAK-FM Stuttgart AR
KWAP(FM) Pine Haven WY
*KWAR(FM) Waverly IA
KWAV(FM) Monterey CA
KWAW(FM) Garapan-Saipan NP
*KWAX(FM) Eugene OR
KWAY-FM Waverly IA
KWBF-FM North Little Rock AR
*KWBI(FM) Great Bend KS
*KWBU-FM Waco TX
*KWBX(FM) Salem OR
KWBZ(FM) Monroe City MO
KWCA(FM) Weaverville CA
KWCD(FM) Bisbee AZ
*KWCF(FM) Sheridan WY
KWCK-FM Searcy AR
KWCL-FM Oak Grove LA
KWCO-FM Chickasha OK
*KWCR-FM Ogden UT
KWCS(FM) Walsenburg CO
*KWCW(FM) Walla Walla WA
KWCX(FM) Willcox AZ
KWDC(FM) Coahoma TX
KWDE(FM) Eureka MT
*KWDH(FM) Hereford TX
KWDI(FM) Idalia CO
*KWDM(FM) West Des Moines IA
KWDN(FM) Newell IA
KWDP(FM) Prineville OR
KWDQ(FM) Woodward OK
KWDR(FM) Royal City WA
*KWDS(FM) Kettleman City CA
KWDU(FM) Upton WY
KWDV(FM) Valier MT
KWEI-FM Fruitland ID
KWEN(FM) Tulsa OK
*KWER(FM) Waverly IA
KWES-FM Ruidoso NM
KWEY-FM Clinton OK
KWFB(FM) Quanah TX
*KWFC(FM) Springfield MO
*KWFH(FM) Parker AZ
*KWFJ(FM) Roy WA
*KWFL(FM) Roswell NM
KWFR(FM) San Angelo TX
KWFS-FM Wichita Falls TX
KWFX(FM) Woodward OK
KWGB(FM) Colby KS
KWGL(FM) Ouray CO
KWGO(FM) Burlington ND
*KWGS(FM) Tulsa OK
KWGW(FM) Mexia TX
KWHF(FM) Harrisburg AR
KWHK(FM) Hutchinson KS
KWHL(FM) Anchorage AK
KWHO(FM) Cody WY
KWHQ-FM Kenai AK
KWHT(FM) Pendleton OR
KWIC(FM) Topeka KS

KWID(FM) Las Vegas NV
KWIE(FM) Ontario CA
KWIM(FM) Window Rock AZ
KWIN(FM) Lodi CA
KWIQ-FM Moses Lake WA
*KWIT(FM) Sioux City IA
KWIZ(FM) Santa Ana CA
*KWJC(FM) Liberty MO
*KWJG(FM) Kasilof AK
KWJJ-FM Portland OR
KWJK(FM) Boonville MO
*KWJT(FM) Rathdrum ID
KWJZ(FM) Seattle WA
KWKJ(FM) Windsor MO
KWKK(FM) Russellville AR
*KWKL(FM) Grandfield OK
KWKM(FM) Saint Johns AZ
KWKQ(FM) Graham TX
KWKR(FM) Leoti KS
KWKZ(FM) Charleston MO
*KWLD(FM) Plainview TX
KWLF(FM) Fairbanks AK
KWLI(FM) Broomfield CO
KWLN(FM) Wilson Creek WA
KWLR(FM) Maumelle AR
KWLT(FM) North Crossett AR
KWLU(FM) Chester CA
KWLV(FM) Many LA
KWLZ-FM Warm Springs OR
*KWMD(FM) Kasilof AK
KWME(FM) Wellington KS
*KWMR(FM) Point Reyes Station CA
KWMT-FM Tucson AZ
*KWMU(FM) Saint Louis MO
KWMW(FM) Maljamar NM
KWMX(FM) Williams AZ
KWMY(FM) Park City MT
KWNA-FM Winnemucca NV
*KWND(FM) Springfield MO
KWNE(FM) Ukiah CA
KWNG(FM) Red Wing MN
*KWNJ(FM) Bettendorf IA
KWNN(FM) Turlock CA
KWNO-FM Rushford MN
KWNR(FM) Henderson NV
KWNS(FM) Winnsboro TX
KWNZ(FM) Sun Valley NV
KWOA-FM Worthington MN
KWOD(FM) Sacramento CA
*KWOF-FM Hiawatha IA
*KWOI(FM) Carroll IA
KWOL-FM Whitefish MT
KWOW(FM) Clifton TX
KWOX(FM) Woodward OK
KWOZ(FM) Mountain View AR
KWPK-FM Sisters OR
*KWPR(FM) Lund NV
KWPT(FM) Fortuna CA
KWPZ(FM) Lynden WA
KWQW(FM) Boone IA
*KWRB(FM) Bisbee AZ
*KWRC(FM) Rapid City SD
KWRD-FM Highland Village TX
KWRF-FM Warren AR
*KWRI(FM) Bartlesville OK
KWRK(FM) Window Rock AZ
KWRL(FM) La Grande OR
KWRQ(FM) Clifton AZ
*KWRR(FM) Ethete WY
*KWRS(FM) Spokane WA
*KWRV(FM) Sun Valley ID
KWRW(FM) Rusk TX
*KWRX(FM) Redmond OR
KWSA(FM) Price UT
*KWSB-FM Gunnison CO
*KWSC(FM) Wayne NE
*KWSO(FM) Warm Springs OR
*KWTD(FM) Ridgecrest CA
*KWTH(FM) Barstow CA
*KWTM(FM) June Lake CA
KWTO-FM Springfield MO
KWTR(FM) Big Lake TX
*KWTS(FM) Canyon TX
*KWTU(FM) Tulsa OK
*KWTW(FM) Bishop CA
KWTX-FM Waco TX

KWTY(FM) Cartago CA
KWUF-FM Pagosa Springs CO
*KWUR(FM) Clayton MO
KWUZ(FM) Poncha Springs CO
*KWVA(FM) Eugene OR
KWVE(FM) San Clemente CA
*KWVI(FM) Waverly IA
KWVR-FM Enterprise OR
KWVV-FM Homer AK
*KWVZ(FM) Florence OR
*KWWC-FM Columbia MO
KWWK(FM) Rochester MN
KWWR(FM) Mexico MO
*KWWS(FM) Walla Walla WA
KWWV(FM) Santa Margarita CA
KWWW-FM Quincy WA
*KWXC(FM) Grove OK
KWXD(FM) Asbury MO
KWXE(FM) Glenwood AR
KWXX-FM Hilo HI
KWXY-FM Cathedral City CA
*KWYA(FM) Astoria OR
*KWYC(FM) Orchard Valley WY
KWYE(FM) Fresno CA
KWYI(FM) Kawaihae HI
KWYK-FM Aztec NM
KWYL(FM) South Lake Tahoe CA
KWYN-FM Wynne AR
*KWYQ(FM) Longview WA
KWYR-FM Winner SD
KWYS-FM Island Park ID
KWYW(FM) Lost Cabin WY
KWYX(FM) Casper WY
KWYY(FM) Casper WY
KXAA(FM) Cle Elum WA
KXAC(FM) Saint James MN
KXAL-FM Tatum TX
KXAZ(FM) Page AZ
KXBA(FM) Nikiski AK
*KXBC(FM) Garberville CA
*KXBJ(FM) Victoria TX
KXBL(FM) Henryetta OK
KXBN(FM) Cedar City UT
*KXBR(FM) International Falls MN
KXBT(FM) Dripping Springs TX
KXBX-FM Lakeport CA
KXBZ(FM) Manhattan KS
*KXCI(FM) Tucson AZ
KXCM(FM) Twentynine Palms CA
*KXCV(FM) Maryville MO
KXDD(FM) Yakima WA
KXDG(FM) Webb City MO
KXDJ(FM) Spearman TX
KXDL(FM) Browerville MN
KXDR(FM) Hamilton MT
KXDZ(FM) Templeton CA
*KXEI(FM) Havre MT
KXEZ(FM) Farmersville TX
KXFE(FM) Dumas AR
KXFF(FM) Colorado City AZ
KXFG(FM) Sun City CA
KXFM(FM) Santa Maria CA
*KXFR(FM) Socorro NM
KXFT(FM) Manson IA
KXFX(FM) Santa Rosa CA
*KXGA(FM) Glennallen AK
KXGE(FM) Dubuque IA
KXGJ(FM) Bay City TX
KXGL(FM) Amarillo TX
KXGO(FM) Arcata CA
KXGT(FM) Carrington ND
KXHT(FM) Marion AR
KXIA(FM) Marshalltown IA
KXIO(FM) Clarksville AR
KXIT-FM Dalhart TX
KXIX(FM) Bend OR
*KXJH(FM) Linton IN
KXJM(FM) Portland OR
*KXJS(FM) Sutter CA
*KXJZ(FM) Sacramento CA
KXKC(FM) New Iberia LA
KXKK(FM) Park Rapids MN
KXKL-FM Denver CO
*KXKM(FM) McCarthy AK
KXKQ(FM) Safford AZ
KXKS-FM Shreveport LA

KXKT(FM) Glenwood IA
KXKU(FM) Lyons KS
KXKX(FM) Knob Noster MO
KXKZ(FM) Ruston LA
KXLB(FM) Livingston MT
*KXLC(FM) La Crescent MN
KXLE-FM Ellensburg WA
*KXLL(FM) Juneau AK
KXLM(FM) Oxnard CA
KXLP(FM) New Ulm MN
KXLR(FM) Fairbanks AK
KXLS(FM) Lahoma OK
KXLT-FM Eagle ID
*KXLU(FM) Los Angeles CA
*KXLV(FM) Amarillo TX
KXLW(FM) Houston AK
KXLY-FM Spokane WA
KXME(FM) Wellington TX
KXMO-FM Owensville MO
KXMP(FM) Hanna WY
*KXMS(FM) Joplin MO
KXMT(FM) Taos NM
KXNA(FM) Springdale AR
*KXNE-FM Norfolk NE
KXNP(FM) North Platte NE
KXO-FM El Centro CA
KXOJ-FM Sapulpa OK
KXOL-FM Los Angeles CA
KXOO(FM) Elk City OK
KXOQ(FM) Kennett MO
KXOR-FM Thibodaux LA
*KXOT(FM) Tacoma WA
KXOW(FM) Eldorado OK
KXOX-FM Sweetwater TX
KXPC(FM) Lebanon OR
KXPK(FM) Evergreen CO
*KXPR(FM) Sacramento CA
KXPT(FM) Las Vegas NV
KXPZ(FM) Las Cruces NM
KXQL(FM) Flandreau SD
KXRA-FM Alexandria MN
*KXRD(FM) Victorville CA
*KXRI(FM) Amarillo TX
*KXRJ(FM) Russellville AR
KXRK(FM) Provo UT
KXRL(FM) Cherry Valley AR
KXRQ(FM) Roosevelt UT
KXRR(FM) Monroe LA
KXRS(FM) Hemet CA
*KXRT(FM) Idabel OK
KXRV(FM) Centerville UT
KXRX(FM) Walla Walla WA
KXRZ(FM) Alexandria MN
KXSA-FM Dermott AR
KXSB(FM) Big Bear Lake CA
KXSE(FM) Davis CA
KXSM(FM) Hollister CA
*KXSR(FM) Groveland CA
KXTC(FM) Thoreau NM
KXTE(FM) Pahrump NV
*KXTH(FM) Seminole OK
KXTN-FM San Antonio TX
KXTQ-FM Lubbock TX
KXTS(FM) Calistoga CA
KXTT(FM) Maricopa CA
KXTY(FM) Morro Bay CA
KXTZ(FM) Pismo Beach CA
*KXUA(FM) Fayetteville AR
*KXUL(FM) Monroe LA
KXUS(FM) Springfield MO
*KXWA(FM) Loveland CO
*KXWY(FM) Rye CO
KXXI(FM) Gallup NM
KXXK(FM) Hoquiam-Aberdeen WA
KXXL(FM) Gillette WY
KXXM(FM) San Antonio TX
KXXN(FM) Iowa Park TX
KXXO(FM) Olympia WA
*KXXQ(FM) Milan NM
KXXR(FM) Minneapolis MN
KXXS(FM) Elgin TX
KXXY-FM Oklahoma City OK
KXXZ(FM) Barstow CA
KXYL-FM Brownwood TX
KXZM(FM) Felton CA
*KYAF(FM) Firebaugh CA

KYAL-FM Muskogee OK
*KYAR(FM) Gatesville TX
KYBA(FM) Stewartville MN
KYBB(FM) Canton SD
KYBE(FM) Frederick OK
KYBI(FM) Lufkin TX
*KYBJ(FM) Lake Jackson TX
KYBR(FM) Espanola NM
*KYCC(FM) Stockton CA
KYCH-FM Portland OR
*KYCJ(FM) Camino CA
KYCK(FM) Crookston MN
*KYCM(FM) Alamogordo NM
KYCS(FM) Rock Springs WY
*KYCT(FM) Ruidoso NM
*KYCU(FM) Clinton OK
*KYCV(FM) Lovington NM
KYDL(FM) Hot Springs AR
KYDN(FM) Del Norte CA
*KYDS(FM) Sacramento CA
KYDT(FM) Sundance WY
KYEE(FM) Alamogordo NM
KYEL(FM) Danville AR
KYEZ(FM) Salina KS
*KYFB(FM) Denison TX
*KYFL(FM) Monroe LA
KYFM(FM) Bartlesville OK
*KYFO-FM Ogden UT
*KYFP(FM) Palestine TX
*KYFS(FM) San Antonio TX
*KYFW(FM) Wichita KS
KYGL(FM) Texarkana AR
KYGO-FM Denver CO
KYIS(FM) Oklahoma City OK
KYIX(FM) South Oroville CA
*KYJC(FM) Commerce TX
KYJK(FM) Missoula MT
KYKC(FM) Byng OK
KYKD(FM) Bethel AK
*KYKL(FM) Tracy CA
KYKM(FM) Yoakum TX
KYKR(FM) Beaumont TX
KYKS(FM) Lufkin TX
KYKX(FM) Longview TX
KYKY(FM) Saint Louis MO
KYKZ(FM) Lake Charles LA
KYLA(FM) Homer LA
*KYLC(FM) Lake Charles LA
KYLD(FM) San Francisco CA
*KYLR(FM) Hutto TX
KYLS-FM Ironton MO
*KYLU(FM) Tehachapi CA
*KYLV(FM) Oklahoma City OK
KYLZ(FM) Tremonton UT
*KYMC(FM) Ballwin MO
KYMG(FM) Anchorage AK
KYMI(FM) Los Ybanez TX
KYMO-FM East Prairie MO
KYMV(FM) Woodruff UT
KYMX(FM) Sacramento CA
KYNF(FM) Prairie Grove AR
KYNU(FM) Jamestown ND
KYNZ(FM) Lone Grove OK
KYOD(FM) Glendo WY
KYOE(FM) Point Arena CA
KYOO-FM Halfway MO
*KYOR(FM) Newport OR
KYOT-FM Phoenix AZ
KYOX(FM) Comanche TX
KYOY(FM) Kimball NE
*KYPL(FM) Yakima WA
*KYPR(FM) Miles City MT
KYPT(FM) Wamsutter WY
KYQQ(FM) Arkansas City KS
*KYQX(FM) Weatherford TX
KYRK(FM) Houma LA
*KYRM(FM) Yuma AZ
*KYRV(FM) Concordia MO
KYRX(FM) Marble Hill MO
KYSC(FM) Fairbanks AK
KYSE(FM) El Paso TX
KYSF(FM) Bonanza OR
KYSJ(FM) Coos Bay OR
KYSL(FM) Frisco CO
KYSM-FM Mankato MN
KYSN(FM) East Wenatchee WA

KYSR(FM) Los Angeles CA
KYSS-FM Missoula MT
KYTC(FM) Northwood IA
KYTE(FM) Newport OR
KYTI(FM) Sheridan WY
KYTT-FM Coos Bay OR
KYTZ(FM) Walhalla ND
KYUN(FM) Hailey ID
KYUS-FM Miles City MT
KYVA-FM Grants NM
*KYVT(FM) Yakima WA
*KYWA(FM) Wichita KS
*KYWH(FM) Lockwood MT
KYXK(FM) Gurdon AR
KYXX(FM) Ozona TX
KYXY(FM) San Diego CA
KYYA-FM Billings MT
KYYI(FM) Burkburnett TX
KYYK(FM) Palestine TX
KYYS(FM) Kansas City MO
KYYT(FM) Goldendale WA
KYYX(FM) Minot ND
KYYY(FM) Bismarck ND
KYYZ(FM) Williston ND
KYZK(FM) Sun Valley ID
KYZX(FM) Pueblo West CO
KYZZ(FM) Salinas CA
*KZAI(FM) Coolidge AZ
KZAL(FM) Manson WA
KZAM(FM) Pleasant Valley TX
*KZAN(FM) Hays KS
KZAP(FM) Paradise CA
*KZAR(FM) Gonzales TX
KZAT-FM Belle Plaine IA
*KZAZ(FM) Bellingham WA
KZBB(FM) Poteau OK
KZBD(FM) Spokane WA
KZBE(FM) Omak WA
KZBG(FM) Lapwai ID
*KZBJ(FM) Bay City TX
KZBK(FM) Brookfield MO
KZBL(FM) Natchitoches LA
KZBQ(FM) Pocatello ID
KZBR(FM) La Jara CO
KZBT(FM) Midland TX
KZCC(FM) McCloud CA
KZCD(FM) Lawton OK
KZCH(FM) Derby KS
*KZCL(FM) Logan UT
KZCR(FM) Fergus Falls MN
KZDR(FM) Cheyenne WY
KZDX(FM) Burley ID
KZDY(FM) Cawker City KS
KZEL-FM Eugene OR
KZEN(FM) Central City NE
KZEP-FM San Antonio TX
KZEW(FM) Wheatland WY
KZFM(FM) Corpus Christi TX
KZFN(FM) Moscow ID
*KZFR(FM) Chico CA
*KZFT(FM) Fannett TX
KZGL(FM) Flagstaff AZ
KZGZ(FM) Hagatna GU
KZHE(FM) Stamps AR
KZHK(FM) Saint George UT
KZHR(FM) Dayton WA
KZHT(FM) Salt Lake City UT
KZIA(FM) Cedar Rapids IA
KZID(FM) Orofino ID
KZII-FM Lubbock TX
KZIN-FM Shelby MT
KZIO(FM) Two Harbors MN
KZIQ-FM Ridgecrest CA
*KZJB(FM) Pocatello ID
KZJF(FM) Jefferson City MO
KZJH(FM) Jackson WY
KZJK(FM) Saint Louis Park MN
KZKE(FM) Seligman AZ
KZKK(FM) Huron SD
*KZKL(FM) Wichita Falls TX
KZKS(FM) Rifle CO
KZKX(FM) Seward NE
KZKZ-FM Greenwood AR
KZLA(FM) Huron CA
KZLE(FM) Batesville AR
KZLG(FM) Mansura LA
KZLK(FM) Rapid City SD
KZLN(FM) Patterson IA
*KZLO(FM) Kilgore TX
KZLS(FM) Great Bend KS
KZLT-FM East Grand Forks MN
*KZLU(FM) Inyokern CA
*KZLV(FM) Lytle TX
KZLZ(FM) Kearny AZ
KZMA(FM) Naylor MO
KZMC(FM) McCook NE
KZMG(FM) New Plymouth ID
KZMI(FM) Garapan-Saipan NP
KZMK(FM) Sierra Vista AZ
KZML(FM) Quincy WA
KZMN(FM) Kalispell MT
KZMP-FM Pilot Point TX
KZMQ-FM Greybull WY
KZMT(FM) Helena MT
*KZMU(FM) Moab UT
KZMV(FM) Kremmling CO
KZMX-FM Hot Springs SD
KZMY(FM) Bozeman MT
KZMZ(FM) Alexandria LA
*KZNA(FM) Hill City KS
KZNC(FM) Huron SD
KZND-FM Houston AK
*KZNJ(FM) Marion IA
KZNM(FM) Los Alamos NM
KZNN(FM) Rolla MO
KZNO(FM) Seymour TX
KZOK-FM Seattle WA
KZON(FM) Phoenix AZ
KZOQ-FM Missoula MT
KZOR(FM) Hobbs NM
KZOZ(FM) San Luis Obispo CA
KZPE(FM) Ford City CA
KZPH(FM) Cashmere WA
*KZPI(FM) Deming NM
KZPK(FM) Paynesville MN
KZPL(FM) Port Isabel TX
KZPO(FM) Lindsay CA
KZPR(FM) Minot ND
KZPS(FM) Dallas TX
KZQD(FM) Liberal KS
KZRB(FM) New Boston TX
KZRC(FM) Markham TX
KZRD(FM) Dodge City KS
*KZRI(FM) Welches OR
KZRK-FM Canyon TX
KZRM(FM) Chama NM
KZRO(FM) Dunsmuir CA
KZRQ-FM Mount Vernon MO
KZRR(FM) Albuquerque NM
KZRX(FM) Dickinson ND
KZRZ(FM) West Monroe LA
*KZSC(FM) Santa Cruz CA
*KZSD-FM Martin SD
*KZSE(FM) Rochester MN
KZSN(FM) Hutchinson KS
KZSP(FM) South Padre Island TX
KZSQ-FM Sonora CA
KZSR(FM) Onawa IA
KZST(FM) Santa Rosa CA
*KZSU(FM) Stanford CA
KZTA(FM) Naches WA
KZTB(FM) Milton-Freewater OR
*KZTH(FM) Piedmont OK
KZTQ(FM) Carson City NV
KZTR(FM) Franklin TX
KZUA(FM) Holbrook AZ
KZUL-FM Lake Havasu City AZ
*KZUM(FM) Lincoln NE
KZUS(FM) Belt MT
*KZUU(FM) Pullman WA
KZWA(FM) Moss Bluff LA
KZWB(FM) Green River WY
KZWV(FM) Eldon MO
KZWY(FM) Sheridan WY
KZXQ(FM) Reserve NM
KZXY-FM Apple Valley CA
KZYP(FM) Pine Bluff AR
KZYQ(FM) Lake Village AR
KZYR(FM) Avon CO
*KZYX(FM) Philo CA
*KZYZ(FM) Willits CA
KZZA(FM) Muenster TX
KZZE(FM) Eagle Point OR
KZZI(FM) Belle Fourche SD
KZZK(FM) New London MO
KZZL-FM Pullman WA
KZZO(FM) Sacramento CA
KZZP(FM) Mesa AZ
KZZQ(FM) Winterset IA
KZZS(FM) Story WY
KZZT(FM) Moberly MO
KZZU-FM Spokane WA
KZZX(FM) Alamogordo NM
KZZY(FM) Devils Lake ND
WAAC(FM) Valdosta GA
*WAAE(FM) New Bern NC
WAAF(FM) Westborough MA
WAAG(FM) Galesburg IL
WAAI(FM) Hurlock MD
*WAAJ(FM) Benton KY
WAAL(FM) Binghamton NY
WAAO-FM Andalusia AL
*WAAQ(FM) Rogers Heights MI
WAAW(FM) Williston SC
WAAZ-FM Crestview FL
WABB-FM Mobile AL
*WABE(FM) Atlanta GA
WABK-FM Gardiner ME
WABO-FM Waynesboro MS
*WABR-FM Tifton GA
WABX(FM) Evansville IN
WABZ(FM) Sherman IL
WACD(FM) Antigo WI
WACF(FM) Young Harris GA
*WACG-FM Augusta GA
WACL(FM) Elkton VA
WACO-FM Waco TX
WACR-FM Aberdeen MS
WADI(FM) Corinth MS
WAEB-FM Allentown PA
*WAEF(FM) Cordele GA
WAEG(FM) Evans GA
WAEL-FM Maricao PR
*WAER(FM) Syracuse NY
*WAES(FM) Lincolnshire IL
WAEV(FM) Savannah GA
WAEZ(FM) Greeneville TN
WAFC-FM Clewiston FL
WAFD(FM) Webster Springs WV
*WAFG(FM) Fort Lauderdale FL
*WAFJ(FM) Belvedere SC
WAFL(FM) Milford DE
WAFM(FM) Amory MS
WAFN-FM Arab AL
*WAFR(FM) Tupelo MS
WAFT(FM) Valdosta GA
WAFX(FM) Suffolk VA
WAFY(FM) Middletown MD
WAFZ-FM Immokalee FL
WAGF-FM Dothan AL
WAGH(FM) Fort Mitchell AL
*WAGO(FM) Snow Hill NC
*WAGP(FM) Beaufort SC
WAGR-FM Lexington MS
WAGX(FM) Manchester OH
WAHR(FM) Huntsville AL
*WAHS(FM) Auburn Hills MI
WAIB(FM) Tallahassee FL
*WAIC(FM) Springfield MA
WAID(FM) Clarksdale MS
*WAIH(FM) Potsdam NY
*WAII(FM) Hattiesburg MS
*WAIJ(FM) Grantsville MD
WAIL(FM) Key West FL
WAIN-FM Columbia KY
*WAIR(FM) Lake City MI
WAIV(FM) Cape May NJ
*WAJC(FM) Wilson NC
WAJI(FM) Fort Wayne IN
*WAJJ(FM) McKenzie TN
WAJK(FM) La Salle IL
WAJM(FM) Atlantic City NJ
WAJQ-FM Alma GA
WAJR-FM Salem WV
*WAJS(FM) Tupelo MS
WAJV(FM) Brooksville MS
WAJZ(FM) Voorheesville NY
WAKB(FM) Wrens GA
*WAKD(FM) Sheffield AL
WAKG(FM) Danville VA
WAKH(FM) McComb MS
*WAKJ(FM) De Funiak Springs FL
WAKK-FM Centreville MS
*WAKL(FM) Flint MI
WAKO-FM Lawrenceville IL
WAKQ(FM) Paris TN
WAKS(FM) Akron OH
WAKT-FM Callaway FL
WAKU(FM) Crawfordville FL
WAKW(FM) Cincinnati OH
WAKY(FM) Radcliff KY
WAKZ(FM) Sharpsville PA
WALC(FM) Charleston SC
*WALF(FM) Alfred NY
WALI(FM) Walterboro SC
WALK-FM Patchogue NY
*WALN(FM) Carrollton AL
WALR-FM La Grange GA
WALS(FM) Oglesby IL
WALV-FM Dayton TN
WALX(FM) Selma AL
WALY(FM) Bellwood PA
WALZ-FM Machias ME
*WAMC-FM Albany NY
*WAMH(FM) Amherst MA
WAMI-FM Opp AL
WAMJ(FM) Mableton GA
*WAMK(FM) Kingston NY
WAMO-FM Beaver Falls PA
*WAMP(FM) Jackson TN
*WAMQ(FM) Great Barrington MA
WAMR-FM Miami FL
*WAMU(FM) Washington DC
WAMW-FM Washington IN
WAMX(FM) Milton WV
WAMZ(FM) Louisville KY
WANB-FM Waynesburg PA
*WANC(FM) Ticonderoga NY
WANK(FM) Mount Vernon KY
*WANM(FM) Tallahassee FL
WANT(FM) Lebanon TN
WANV(FM) Annville KY
WANY-FM Albany KY
WAOA-FM Melbourne FL
WAOL(FM) Ripley OH
WAOQ(FM) Brantley AL
WAOR(FM) Niles MI
WAOX(FM) Staunton IL
*WAOY(FM) Gulfport MS
*WAPB(FM) Madison FL
*WAPD(FM) Campbellsville KY
WAPE-FM Jacksonville FL
*WAPJ(FM) Torrington CT
WAPL-FM Appleton WI
*WAPN(FM) Holly Hill FL
*WAPO(FM) Mount Vernon IL
*WAPR(FM) Selma AL
*WAPS(FM) Akron OH
*WAPX-FM Clarksville TN
*WAQB(FM) Tupelo MS
WAQE-FM Barron WI
*WAQG(FM) Ozark AL
*WAQL(FM) McComb MS
*WAQQ(FM) Onsted MI
*WAQU(FM) Selma AL
*WAQV(FM) Crystal River FL
WAQX-FM Manlius NY
WAQY(FM) Springfield MA
*WARA(FM) New Washington IN
*WARC(FM) Meadville PA
*WARG(FM) Summit IL
WARH(FM) Granite City IL
WARM-FM York PA
*WARN(FM) Culpeper VA
WARO(FM) Naples FL
WARQ(FM) Columbia SC
WARU-FM Roann IN
WARV-FM Petersburg VA
*WARY(FM) Valhalla NY
WASH(FM) Washington DC
WASJ(FM) Panama City Beach FL
WASK-FM Battle Ground IN
WASL(FM) Dyersburg TN
*WASM(FM) Natchez MS
*WASU-FM Boone NC
*WASW(FM) Waycross GA
WATD-FM Marshfield MA
WATG(FM) Trion GA
*WATI(FM) Vincennes IN
*WATP(FM) Laurel MS
WATQ(FM) Chetek WI
*WATU(FM) Port Gibson MS
*WATY(FM) Folkston GA
WATZ-FM Alpena MI
*WAUA(FM) Petersburg WV
WAUH(FM) Wautoma WI
*WAUI(FM) Shelby OH
*WAUM(FM) Duck Hill MS
WAUN(FM) Kewaunee WI
*WAUO(FM) Hohenwald TN
*WAUQ(FM) Charles City VA
*WAUS(FM) Berrien Springs MI
*WAUT-FM Tullahoma TN
*WAUV(FM) Ripley TN
*WAUZ(FM) Greensburg IN
WAVA-FM Arlington VA
WAVC(FM) Mio MI
WAVF(FM) Hanahan SC
WAVH(FM) Daphne AL
*WAVI(FM) Oxford MS
WAVJ(FM) Princeton KY
WAVK(FM) Marathon FL
*WAVM(FM) Maynard MA
WAVR(FM) Waverly NY
WAVT-FM Pottsville PA
WAVV(FM) Marco FL
WAVW(FM) Stuart FL
*WAVX(FM) Schuyler Falls NY
WAWC(FM) Syracuse IN
*WAWF(FM) Kankakee IL
*WAWH(FM) Dublin GA
*WAWI(FM) Lawrenceburg TN
*WAWJ(FM) Marion IL
*WAWL-FM Red Bank TN
*WAWN(FM) Franklin PA
WAWZ(FM) Zarephath NJ
*WAXG(FM) Mt. Sterling KY
WAXI(FM) Rockville IN
WAXJ(FM) Frederiksted VI
WAXL(FM) Santa Claus IN
WAXM(FM) Big Stone Gap VA
WAXQ(FM) New York NY
*WAXR(FM) Geneseo IL
WAXS(FM) Oak Hill WV
*WAXU(FM) Troy AL
WAXX(FM) Eau Claire WI
WAXZ(FM) Georgetown OH
WAYA(FM) Spring City TN
WAYB-FM Graysville TN
WAYC(FM) Bedford PA
*WAYD(FM) Auburn KY
*WAYF(FM) West Palm Beach FL
*WAYG(FM) Grand Rapids MI
*WAYH(FM) Harvest AL
*WAYJ(FM) Fort Myers FL
*WAYK(FM) Kalamazoo MI
*WAYL(FM) Saint Augustine FL
*WAYM(FM) Columbia TN
*WAYO-FM Benton Harbor MI
*WAYQ(FM) Clarksville TN
*WAYR-FM Brunswick GA
*WAYT(FM) Thomasville GA
WAYV(FM) Atlantic City NJ
*WAYW(FM) New Johnsonville TN
WAYZ(FM) Hagerstown MD
WAZA(FM) Liberty MS
*WAZD(FM) Savannah TN
WAZO(FM) Southport NC
*WAZP(FM) Cape Charles VA
*WAZQ(FM) Key West FL
WAZR(FM) Woodstock VA
WAZS-FM McClellanville SC
*WAZU(FM) Peoria IL
WAZX-FM Cleveland GA
WAZY-FM Lafayette IN
*WBAA-FM West Lafayette IN
WBAB(FM) Babylon NY
WBAD(FM) Leland MS
*WBAI(FM) New York NY
WBAM-FM Montgomery AL

WBAQ(FM) Greenville MS
WBAR-FM Lake Luzerne NY
WBAV-FM Gastonia NC
WBAW-FM Pembroke GA
WBAZ(FM) Bridgehampton NY
WBBA-FM Pittsfield IL
WBBB(FM) Raleigh NC
WBBC-FM Blackstone VA
WBBE(FM) Heyworth IL
WBBG(FM) Niles OH
WBBI(FM) Endwell NY
WBBK-FM Blakely GA
WBBM-FM Chicago IL
WBBN(FM) Taylorsville MS
WBBO(FM) Bass River Township NJ
WBBQ-FM Augusta GA
WBBS(FM) Fulton NY
WBBT-FM Powhatan VA
WBBV(FM) Vicksburg MS
WBCG(FM) Murdock FL
WBCH-FM Hastings MI
WBCI(FM) Bath ME
*WBCJ(FM) Spencerville OH
*WBCL(FM) Fort Wayne IN
WBCM(FM) Boyne City MI
WBCN(FM) Boston MA
*WBCR-FM Beloit WI
WBCT(FM) Grand Rapids MI
WBCV(FM) Wausau WI
*WBCX(FM) Gainesville GA
*WBCY(FM) Archbold OH
WBDB(FM) Ogdensburg NY
WBDC(FM) Huntingburg IN
*WBDG(FM) Indianapolis IN
WBDI(FM) Copenhagen NY
WBDK(FM) Algoma WI
WBDL(FM) Reedsburg WI
WBDR(FM) Cape Vincent NY
WBDX(FM) Trenton GA
WBEA(FM) Southold NY
WBEB(FM) Philadelphia PA
WBEC-FM Pittsfield MA
WBEE-FM Rochester NY
WBEI(FM) Reform AL
*WBEL(FM) Cairo IL
WBEN-FM Philadelphia PA
*WBEQ(FM) Morris IL
*WBER(FM) Rochester NY
*WBEW(FM) Chesterton IN
WBEY-FM Crisfield MD
*WBEZ(FM) Chicago IL
WBFA(FM) Smiths AL
WBFB(FM) Belfast ME
WBFG(FM) Parker's Crossroads TN
*WBFH(FM) Bloomfield Hills MI
*WBFI(FM) McDaniels KY
*WBFJ-FM Winston-Salem NC
WBFM(FM) Sheboygan WI
*WBFO(FM) Buffalo NY
*WBFR(FM) Birmingham AL
WBFX(FM) Grand Rapids MI
*WBFY(FM) Pinehurst NC
WBFZ(FM) Selma AL
WBGA(FM) Saint Simons Island GA
*WBGD(FM) Brick Township NJ
WBGE(FM) Bainbridge GA
WBGF(FM) Belle Glade FL
WBGG-FM Fort Lauderdale FL
WBGK(FM) Newport Village NY
*WBGL(FM) Champaign IL
*WBGM(FM) New Berlin PA
*WBGO(FM) Newark NJ
WBGQ(FM) Bulls Gap TN
*WBGU(FM) Bowling Green OH
WBGV(FM) Marlette MI
*WBGW(FM) Fort Branch IN
*WBGY(FM) Naples FL
WBHB-FM Bridgewater VA
WBHC-FM Hampton SC
WBHD(FM) Olyphant PA
WBHJ(FM) Midfield AL
WBHK(FM) Warrior AL
*WBHM(FM) Birmingham AL
WBHT(FM) Mountain Top PA
WBHV-FM State College PA
*WBHW(FM) Loogootee IN
WBHX(FM) Tuckerton NJ
*WBHY-FM Mobile AL
*WBHZ(FM) Elkins WV
*WBIA(FM) Shelbyville TN
*WBIE(FM) Delphos OH
WBIG-FM Washington DC
WBIK(FM) Pleasant City OH
*WBIM-FM Bridgewater MA
WBIO(FM) Philpot KY
*WBIY(FM) La Belle FL
WBIZ-FM Eau Claire WI
*WBJB-FM Lincroft NJ
*WBJC(FM) Baltimore MD
*WBJD(FM) Atlantic Beach NC
WBJI(FM) Blackduck MN
*WBJV(FM) Steubenville OH
*WBJW(FM) Albion IL
*WBJY(FM) Americus GA
WBJZ(FM) Berlin WI
*WBKE-FM North Manchester IN
*WBKG(FM) Macon GA
*WBKL(FM) Clinton LA
WBKN(FM) Brookhaven MS
WBKR(FM) Owensboro KY
WBKS(FM) Ironton OH
WBKT(FM) Norwich NY
*WBKU(FM) Ahoskie NC
*WBKW(FM) Beekman NY
WBKX(FM) Fredonia NY
WBKY(FM) Portage WI
*WBLD(FM) Orchard Lake MI
WBLE(FM) Batesville MS
WBLI(FM) Patchogue NY
WBLJ-FM Shamokin PA
WBLK(FM) Depew NY
WBLM(FM) Portland ME
WBLS(FM) New York NY
*WBLU-FM Grand Rapids MI
*WBLV(FM) Twin Lake MI
*WBLW(FM) Gaylord MI
WBLX-FM Mobile AL
*WBMF(FM) Crete IL
WBMH(FM) Grove Hill AL
WBMI(FM) West Branch MI
*WBMK(FM) Morehead KY
*WBMR(FM) Telford PA
*WBMT(FM) Boxford MA
*WBMV(FM) Mount Vernon IL
WBMW(FM) Ledyard CT
WBMX(FM) Boston MA
WBMZ(FM) Metter GA
WBNE(FM) Shallotte NC
*WBNH(FM) Pekin IL
*WBNI-FM Roanoke IN
WBNN-FM Dillwyn VA
WBNO-FM Bryan OH
WBNQ(FM) Bloomington IL
WBNS-FM Columbus OH
WBNT-FM Oneida TN
WBNV(FM) Barnesville OH
*WBNY(FM) Buffalo NY
WBNZ(FM) Frankfort MI
WBOB-FM Enfield NC
*WBOI(FM) Fort Wayne IN
WBOJ(FM) Cusseta GA
WBOP(FM) Buffalo Gap VA
WBOQ(FM) Gloucester MA
*WBOR(FM) Brunswick ME
WBOS(FM) Brookline MA
WBOW-FM Terre Haute IN
WBOX-FM Varnado LA
WBOZ(FM) Woodbury TN
WBPC(FM) Ebro FL
WBPM(FM) Saugerties NY
*WBPR(FM) Worcester MA
WBPT(FM) Homewood AL
WBPW(FM) Presque Isle ME
WBQB(FM) Fredericksburg VA
WBQI(FM) Bar Harbor ME
WBQK(FM) West Point VA
WBQQ(FM) Kennebunk ME
WBQW(FM) Scarborough ME
WBQX(FM) Thomaston ME
WBRB(FM) Buckhannon WV
WBRF(FM) Galax VA
*WBRH(FM) Baton Rouge LA
WBRK-FM Pittsfield MA
*WBRO(FM) Marengo IN
WBRQ(FM) Cidra PR
WBRR(FM) Bradford PA
*WBRS(FM) Waltham MA
WBRU(FM) Providence RI
WBRV-FM Boonville NY
WBRW(FM) Blacksburg VA
WBRX(FM) Cresson PA
*WBSB(FM) Anderson IN
*WBSD(FM) Burlington WI
*WBSH(FM) Hagerstown IN
*WBSJ(FM) Portland IN
*WBSL-FM Sheffield MA
*WBSN-FM New Orleans LA
*WBST(FM) Muncie IN
*WBSU(FM) Brockport NY
*WBSW(FM) Marion IN
WBSX(FM) Hazleton PA
WBSZ(FM) Ashland WI
WBTF(FM) Midway KY
WBT-FM Chester SC
WBTG-FM Sheffield AL
WBTI(FM) Lexington MI
WBTJ(FM) Richmond VA
WBTN-FM Bennington VT
WBTO-FM Petersburg IN
WBTP(FM) Clearwater FL
WBTQ(FM) Buckhannon WV
WBTR-FM Carrollton GA
WBTS(FM) Doraville GA
WBTT(FM) Naples Park FL
WBTU(FM) Kendallville IN
WBTY(FM) Homerville GA
WBTZ(FM) Plattsburgh NY
WBUF(FM) Buffalo NY
WBUG-FM Fort Plain NY
WBUK(FM) Ottawa OH
WBUL-FM Lexington KY
*WBUQ(FM) Bloomsburg PA
*WBUR-FM Boston MA
WBUS(FM) Boalsburg PA
WBUV(FM) Moss Point MS
*WBUX(FM) Buxton NC
WBUZ(FM) La Vergne TN
WBVB(FM) Coal Grove OH
*WBVC(FM) Pomfret CT
WBVD(FM) Melbourne FL
WBVE(FM) Bedford PA
WBVI(FM) Fostoria OH
*WBVM(FM) Tampa FL
*WBVN(FM) Carrier Mills IL
WBVR-FM Auburn KY
WBVV(FM) Guntown MS
WBVX(FM) Carlisle KY
*WBWB(FM) Bloomington IN
*WBWC(FM) Berea OH
WBWI-FM West Bend WI
WBWN(FM) Le Roy IL
WBWR(FM) Hilliard OH
WBWZ(FM) New Paltz NY
WBXB(FM) Edenton NC
WBXE(FM) Baxter TN
*WBXL(FM) Baldwinsville NY
WBXQ(FM) Patton PA
WBXX(FM) Battle Creek MI
WBXY(FM) La Crosse FL
WBYA(FM) Islesboro ME
WBYG(FM) Point Pleasant WV
*WBYH(FM) Hawley PA
WBYL(FM) Salladasburg PA
*WBYO(FM) Sellersville PA
WBYP(FM) Belzoni MS
WBYR(FM) Van Wert OH
WBYT(FM) Elkhart IN
*WBYX(FM) Stroudsburg PA
WBYY(FM) Somersworth NH
WBYZ(FM) Baxley GA
WBZA(FM) Rochester NY
WBZB(FM) Westhampton NY
*WBZC(FM) Pemberton NJ
WBZD-FM Muncy PA
WBZE(FM) Tallahassee FL
WBZF(FM) Hartsville SC
WBZG(FM) Peru IL
WBZN(FM) Old Town ME
WBZO(FM) Bay Shore NY
WBZT-FM Mauldin SC
WBZV(FM) Hudson MI
WBZX(FM) Columbus OH
WBZY(FM) Bowdon GA
WBZZ(FM) Malta NY
WCAA(FM) Newark NJ
WCAD(FM) San Juan PR
*WCAI(FM) Woods Hole MA
*WCAL(FM) California PA
*WCAN(FM) Canajoharie NY
WCAT-FM Carlisle PA
WCBC-FM Keyser WV
*WCBE(FM) Columbus OH
WCBH(FM) Casey IL
WCBJ(FM) Campton KY
WCBK-FM Martinsville IN
WCBL-FM Benton KY
*WCBN-FM Ann Arbor MI
WCBS-FM New York NY
*WCBU(FM) Peoria IL
*WCBW-FM East St. Louis IL
WCCC-FM Hartford CT
*WCCE(FM) Buie's Creek NC
WCCG(FM) Hope Mills NC
*WCCH(FM) Holyoke MA
WCCI(FM) Savanna IL
WCCK(FM) Calvert City KY
WCCL(FM) Central City PA
WCCN-FM Neillsville WI
WCCP-FM Clemson SC
WCCQ(FM) Crest Hill IL
WCCR(FM) Clarion PA
*WCCT-FM Harwich MA
*WCCV(FM) Cartersville GA
WCCW-FM Traverse City MI
*WCCX(FM) Waukesha WI
WCDA(FM) Versailles KY
*WCDB(FM) Albany NY
WCDD(FM) Canton IL
WCDG(FM) Moyock NC
WCDK(FM) Cadiz OH
WCDO-FM Sidney NY
WCDQ(FM) Crawfordsville IN
*WCDR-FM Cedarville OH
WCDV(FM) Hammond LA
WCDW(FM) Susquehanna PA
WCDX(FM) Mechanicsville VA
WCDZ(FM) Dresden TN
*WCEB(FM) Corning NY
WCEF(FM) Ripley WV
WCEI-FM Easton MD
*WCEL(FM) Plattsburgh NY
WCEM-FM Cambridge MD
WCEN-FM Hemlock MI
WCEZ(FM) Carthage IL
WCFB(FM) Daytona Beach FL
WCFF(FM) Urbana IL
*WCFG(FM) Springfield MI
*WCFL(FM) Morris IL
*WCFM(FM) Williamstown MA
WCFW(FM) Chippewa Falls WI
WCFX(FM) Clare MI
*WCGN(FM) Calhoun GA
WCGQ(FM) Columbus GA
*WCHC(FM) Worcester MA
*WCHG(FM) Hot Springs VA
WCHO-FM Washington Court House OH
WCHR-FM Manahawkin NJ
*WCHW-FM Bay City MI
WCHX(FM) Lewistown PA
WCHY(FM) Waunakee WI
WCHZ(FM) Harlem GA
WCIB(FM) Falmouth MA
*WCIC(FM) Pekin IL
*WCID(FM) Friendship NY
WCIF(FM) Melbourne FL
*WCIG(FM) Carbondale PA
*WCIH(FM) Elmira NY
*WCII(FM) Spencer NY
*WCIK(FM) Bath NY
WCIL-FM Carbondale IL
*WCIM(FM) Shenandoah PA
WCIR-FM Beckley WV
*WCIT(FM) Trout Run PA
*WCIY(FM) Canandaigua NY
WCIZ-FM Watertown NY
WCJC(FM) Van Buren IN
WCJK(FM) Murfreesboro TN
*WCJL(FM) Morgantown IN
WCJM-FM West Point GA
WCJO(FM) Jackson OH
WCJU-FM Prentiss MS
WCJX(FM) Five Points FL
WCJZ(FM) Charlottesville VA
WCKC(FM) Cadillac MI
WCKF(FM) Ashland AL
WCKG(FM) Elmwood Park IL
*WCKJ(FM) Saint Johnsbury VT
WCKK(FM) Carthage MS
WCKM-FM Lake George NY
WCKQ(FM) Campbellsville KY
WCKR(FM) Hornell NY
WCKS(FM) Fruithurst AL
WCKT(FM) Lehigh Acres FL
WCKX(FM) Columbus OH
WCKY-FM Tiffin OH
*WCKZ(FM) Orland IN
WCLC-FM Jamestown TN
WCLD-FM Cleveland MS
WCLE-FM Calhoun TN
WCLG-FM Morgantown WV
*WCLH(FM) Wilkes-Barre PA
*WCLK(FM) Atlanta GA
WCLN-FM Clinton NC
*WCLQ(FM) Wausau WI
*WCLR(FM) Arlington Heights IL
WCLS(FM) Spencer IN
WCLT-FM Newark OH
WCLU-FM Munfordville KY
WCLV(FM) Lorain OH
WCLX(FM) Westport NY
WCLZ(FM) Brunswick ME
WCMA-FM Fajardo PR
*WCMB-FM Oscoda MI
WCMC-FM Creedmoor NC
*WCMD-FM Barre VT
WCME(FM) Boothbay Harbor ME
WCMF-FM Rochester NY
WCMG(FM) Latta SC
WCMJ(FM) Cambridge OH
*WCMK(FM) Putney VT
*WCML-FM Alpena MI
WCMM(FM) Gulliver MI
WCMN-FM Arecibo PR
*WCMO(FM) Marietta OH
WCMP-FM Pine City MN
WCMQ-FM Hialeah FL
WCMR(FM) Bruce MS
WCMS-FM Hatteras NC
WCMT-FM South Fulton TN
*WCMU-FM Mount Pleasant MI
*WCMW-FM Harbor Springs MI
*WCMZ-FM Sault Ste. Marie MI
WCNA(FM) Potts Camp MS
WCNF(FM) Benton Harbor MI
WCNG(FM) Murphy NC
*WCNI(FM) New London CT
WCNK(FM) Key West FL
*WCNO(FM) Palm City FL
WCNR(FM) Keswick VA
*WCNV(FM) Heathsville VA
*WCNY-FM Syracuse NY
WCOD-FM Hyannis MA
WCOE(FM) La Porte IN
*WCOF(FM) Arcade NY
*WCOG-FM Galeton PA
WCOL-FM Columbus OH
WCON-FM Cornelia GA
WCOO(FM) Kiawah Island SC
WCOS-FM Columbia SC
*WCOT(FM) Jamestown NY
*WCOU(FM) Warsaw NY
*WCOV-FM Clyde NY
WCOW-FM Sparta WI
WCOY(FM) Quincy IL
WCOZ(FM) Laporte PA
*WCPE(FM) Raleigh NC
*WCPI(FM) McMinnville TN
*WCPN(FM) Cleveland OH
WCPR-FM D'Iberville MS

WCPV(FM) Essex NY
WCPZ(FM) Sandusky OH
WCQL(FM) Glens Falls NY
WCQM(FM) Park Falls WI
*WCQR-FM Kingsport TN
*WCQS(FM) Asheville NC
WCRB(FM) Lowell MA
WCRC(FM) Effingham IL
*WCRF(FM) Cleveland OH
*WCRG(FM) Williamsport PA
*WCRH(FM) Williamsport MD
WCRI(FM) Block Island RI
*WCRJ(FM) Jacksonville FL
*WCRP(FM) Guayama PR
WCRQ(FM) Dennysville ME
WCRR(FM) South Bristol Township NY
*WCRT-FM Terre Haute IN
*WCRX(FM) Chicago IL
WCRZ(FM) Flint MI
*WCSB(FM) Cleveland OH
*WCSF(FM) Joliet IL
*WCSG(FM) Grand Rapids MI
WCSJ-FM Morris IL
*WCSK(FM) Kingsport TN
WCSM-FM Celina OH
WCSN-FM Orange Beach AL
*WCSO(FM) Columbus MS
*WCSP-FM Washington DC
WCSR-FM Hillsdale MI
*WCSU-FM Wilberforce OH
WCSX(FM) Birmingham MI
WCSY-FM South Haven MI
WCTB(FM) Fairfield ME
WCTG(FM) Chincoteague VA
WCTH(FM) Plantation Key FL
WCTK(FM) New Bedford MA
WCTL(FM) Union City PA
WCTO(FM) Easton PA
*WCTP(FM) Gagetown MI
WCTQ(FM) Sarasota FL
WCTT-FM Corbin KY
WCTW(FM) Catskill NY
WCTY(FM) Norwich CT
WCTZ(FM) Stamford CT
*WCUC-FM Clarion PA
WCUP(FM) L'Anse MI
*WCUR(FM) West Chester PA
*WCUW(FM) Worcester MA
WCUZ(FM) Bear Lake MI
*WCVE(FM) Richmond VA
*WCVF-FM Fredonia NY
*WCVH(FM) Flemington NJ
*WCVJ(FM) Jefferson OH
*WCVK(FM) Bowling Green KY
*WCVM(FM) Bronson MI
*WCVO(FM) Gahanna OH
WCVP-FM Robbinsville NC
WCVQ(FM) Fort Campbell KY
WCVR-FM Randolph VT
WCVS-FM Virden IL
WCVT(FM) Stowe VT
WCVU(FM) Solana FL
*WCVV(FM) Belpre OH
*WCVY(FM) Coventry RI
*WCVZ(FM) South Zanesville OH
*WCWB(FM) Coldwater MI
*WCWM(FM) Williamsburg VA
*WCWP(FM) Brookville NY
*WCWS(FM) Wooster OH
*WCWT-FM Centerville OH
WCWV(FM) Summersville WV
WCXL(FM) Kill Devil Hills NC
WCXR(FM) Lewisburg PA
WCXU(FM) Caribou ME
WCXV(FM) Van Buren ME
WCXX(FM) Madawaska ME
WCYI(FM) Lewiston ME
*WCYJ-FM Waynesburg PA
WCYK-FM Staunton VA
WCYN-FM Cynthiana KY
WCYO(FM) Irvine KY
*WCYT(FM) Lafayette Township IN
WCYY(FM) Biddeford ME
WCZE(FM) Harbor Beach MI
WCZQ(FM) Monticello IL
WCZR(FM) Vero Beach FL
WCZT(FM) Villas NJ
WCZW(FM) Charlevoix MI
WCZX(FM) Hyde Park NY
WCZY-FM Mount Pleasant MI
WDAC(FM) Lancaster PA
WDAF-FM Liberty MO
WDAI(FM) Pawley's Island SC
WDAQ(FM) Danbury CT
WDAR-FM Darlington SC
WDAS-FM Philadelphia PA
*WDAV(FM) Davidson NC
WDAY-FM Fargo ND
WDBA(FM) DuBois PA
*WDBK(FM) Blackwood NJ
*WDBM(FM) East Lansing MI
WDBN(FM) Wrightsville GA
WDBQ-FM Galena IL
WDBR(FM) Springfield IL
WDBS(FM) Sutton WV
WDBT(FM) Headland AL
*WDBX(FM) Carbondale IL
WDBY(FM) Patterson NY
*WDCB(FM) Glen Ellyn IL
*WDCC(FM) Sanford NC
*WDCE(FM) Richmond VA
WDCG(FM) Durham NC
WDCI(FM) Bridgeport WV
*WDCL-FM Somerset KY
*WDCV-FM Carlisle PA
WDCX(FM) Buffalo NY
WDDC(FM) Portage WI
WDDD-FM Marion IL
WDDH(FM) Saint Marys PA
WDDJ(FM) Paducah KY
WDDK(FM) Greensboro GA
*WDDM(FM) Hazlet NJ
WDDQ(FM) Adel GA
WDDW(FM) Sturtevant WI
WDEB-FM Jamestown TN
WDEC-FM Americus GA
WDEE-FM Reed City MI
WDEF-FM Chattanooga TN
WDEK(FM) De Kalb IL
WDEN-FM Macon GA
WDEO-FM San Carlos Park FL
*WDEQ-FM De Graff OH
*WDET-FM Detroit MI
WDEV-FM Warren VT
WDEZ(FM) Wausau WI
*WDFB-FM Danville KY
*WDFH(FM) Ossining NY
WDFM(FM) Defiance OH
WDFX(FM) Cleveland MS
*WDGC-FM Downers Grove IL
WDGG(FM) Ashland KY
WDGL(FM) Baton Rouge LA
WDGM(FM) Greensboro AL
WDHA-FM Dover NJ
WDHC(FM) Berkeley Springs WV
WDHI(FM) Delhi NY
WDHR(FM) Pikeville KY
WDHT(FM) Springfield OH
WDIC-FM Clinchco VA
WDIF(FM) Marion OH
*WDIH(FM) Salisbury MD
WDIN(FM) Camuy PR
*WDIY(FM) Allentown PA
WDJC-FM Birmingham AL
*WDJM-FM Framingham MA
WDJR(FM) Enterprise AL
*WDJW(FM) Somers CT
WDJX(FM) Louisville KY
WDKB(FM) De Kalb IL
WDKC(FM) Covington PA
WDKF(FM) Englewood OH
*WDKL(FM) Grafton WV
WDKM(FM) Adams WI
WDKR(FM) Maroa IL
WDKS(FM) Newburgh IN
*WDKV(FM) Fond du Lac WI
WDKX(FM) Rochester NY
WDKZ(FM) Salisbury MD
WDLA-FM Walton NY
WDLD(FM) Halfway MD
*WDLG(FM) Thomasville AL
WDLJ(FM) Breese IL
*WDLL(FM) Dillon SC
*WDLM-FM East Moline IL
WDLT-FM Chickasaw AL
WDLZ(FM) Murfreesboro NC
WDME-FM Dover Foxcroft ME
WDMG-FM Ambrose GA
WDMK(FM) Detroit MI
WDML(FM) Woodlawn IL
WDMO(FM) Durand WI
WDMP-FM Dodgeville WI
WDMS(FM) Greenville MS
WDMT(FM) Pittston PA
WDMX(FM) Vienna WV
*WDNA(FM) Miami FL
WDNB(FM) Jeffersonville NY
WDNE-FM Elkins WV
WDNH-FM Honesdale PA
WDNL(FM) Danville IL
*WDNR(FM) Chester PA
WDNS(FM) Bowling Green KY
*WDNX-FM Olive Hill TN
WDNY-FM Dansville NY
WDOD-FM Chattanooga TN
WDOG-FM Allendale SC
WDOH(FM) Delphos OH
WDOK(FM) Cleveland OH
*WDOM(FM) Providence RI
WDOR-FM Sturgeon Bay WI
WDOT(FM) Danville VT
*WDPG(FM) Greenville OH
*WDPR(FM) West Carrollton OH
*WDPS(FM) Dayton OH
*WDPW(FM) Greenville MI
WDQN-FM Du Quoin IL
WDQX(FM) Morton IL
WDQZ(FM) Lexington IL
WDRC-FM Hartford CT
WDRE(FM) Calverton-Roanoke NY
WDRK(FM) Cornell WI
WDRM(FM) Decatur AL
WDRQ(FM) Detroit MI
WDRR(FM) Martinez GA
*WDRS(FM) Dorsey IL
WDRV(FM) Chicago IL
WDSD(FM) Smyrna DE
WDSJ(FM) Greenville OH
WDSN(FM) Reynoldsville PA
*WDSO(FM) Chesterton IN
WDST(FM) Woodstock NY
WDSW(FM) Westby WI
WDSY-FM Pittsburgh PA
WDTL-FM Cleveland MS
*WDTR(FM) Monroe MI
WDTW-FM Detroit MI
*WDUB(FM) Granville OH
WDUK(FM) Havana IL
*WDUQ(FM) Pittsburgh PA
WDUV(FM) New Port Richey FL
WDUX-FM Waupaca WI
WDUZ-FM Brillion WI
WDVD(FM) Detroit MI
WDVE(FM) Pittsburgh PA
WDVH-FM Trenton FL
WDVI(FM) Rochester NY
*WDVL(FM) Danville IN
*WDVR(FM) Delaware Township NJ
*WDVV(FM) Wilmington NC
WDVW(FM) La Place LA
*WDVX(FM) Clinton TN
WDWG(FM) Rocky Mount NC
*WDWN(FM) Auburn NY
WDXB(FM) Jasper AL
WDXC(FM) Pound VA
WDXE-FM Lawrenceburg TN
WDXO(FM) Hazlehurst MS
WDXQ-FM Cochran GA
WDXX(FM) Selma AL
*WDYF(FM) Dothan AL
WDYK(FM) Ridgeley WV
WDYL(FM) Chester VA
*WDYN-FM Chattanooga TN
WDZN(FM) Romney WV
WDZQ(FM) Decatur IL
WDZZ-FM Flint MI
*WEAA(FM) Baltimore MD
WEAI(FM) Lynnville IL
WEAM-FM Buena Vista GA
WEAS-FM Springfield GA
WEAT-FM West Palm Beach FL
*WEAX(FM) Angola IN
*WEAZ(FM) Holly Hill FL
WEBB(FM) Waterville ME
WEBE(FM) Westport CT
*WEBH(FM) Cuthbert GA
WEBK(FM) Killington VT
WEBL(FM) Warner Robins GA
WEBN(FM) Cincinnati OH
WEBQ-FM Eldorado IL
*WEBT(FM) Valley AL
WEBX(FM) Tuscola IL
WEBZ(FM) Mexico Beach FL
WECB(FM) Seymour WI
*WECC-FM Folkston GA
*WECI(FM) Richmond IN
WECL(FM) Elk Mound WI
WECO-FM Wartburg TN
WECR-FM Beech Mountain NC
*WECS(FM) Willimantic CT
*WECW(FM) Elmira NY
WEDG(FM) Buffalo NY
WEDJ(FM) Danville IN
*WEDM(FM) Indianapolis IN
WEDR(FM) Miami FL
*WEDW-FM Stamford CT
*WEEC(FM) Springfield OH
WEEI-FM Westerly RI
*WEEM-FM Pendleton IN
WEEO-FM McConnellsburg PA
WEFG-FM Whitehall MI
*WEFI(FM) Effingham IL
WEFM(FM) Michigan City IN
*WEFR(FM) Erie PA
*WEFT(FM) Champaign IL
WEGC(FM) Sasser GA
WEGE(FM) Lima OH
WEGH(FM) Northumberland PA
WEGI(FM) Oak Grove KY
*WEGL(FM) Auburn AL
WEGM(FM) San German PR
WEGR(FM) Memphis TN
*WEGS(FM) Milton FL
WEGT(FM) Lafayette FL
WEGW(FM) Wheeling WV
WEGX(FM) Dillon SC
*WEGZ(FM) Washburn WI
*WEHC(FM) Emory VA
WEHM(FM) Southampton NY
WEHN(FM) East Hampton NY
WEIB(FM) Northampton MA
*WEIU(FM) Charleston IL
*WEJC(FM) White Star MI
*WEJF(FM) Palm Bay FL
WEJK(FM) Boonville IN
WEJT(FM) Shelbyville IL
WEJZ(FM) Jacksonville FL
*WEKF(FM) Corbin KY
*WEKH(FM) Hazard KY
WEKL(FM) Augusta GA
WEKS(FM) Zebulon GA
*WEKU(FM) Richmond KY
*WEKV(FM) South Webster OH
WEKX(FM) Jellico TN
WEKZ-FM Monroe WI
WELC-FM Welch WV
WELD-FM Petersburg WV
*WELH(FM) Providence RI
*WELJ(FM) Brewton AL
WELK(FM) Elkins WV
*WELL-FM Dadeville AL
WELR-FM Roanoke AL
WELS-FM Kinston NC
WELT(FM) East Dublin GA
WELY-FM Ely MN
*WEMC(FM) Harrisonburg VA
*WEMI(FM) Appleton WI
WEMM-FM Huntington WV
*WEMU(FM) Ypsilanti MI
WEMX(FM) Kentwood LA
*WEMY(FM) Green Bay WI
WEND(FM) Salisbury NC
WENI-FM Big Flats NY
WENN(FM) Hoover AL
WENY-FM Elmira NY
WENZ(FM) Cleveland OH
*WEOS(FM) Geneva NY
WEOW(FM) Key West FL
*WEPC(FM) Belton SC
*WEPR(FM) Greenville SC
*WEPS(FM) Elgin IL
WEQX(FM) Manchester VT
*WERB(FM) Berlin CT
*WERG(FM) Erie PA
WERH-FM Hamilton AL
WERK(FM) Muncie IN
*WERN(FM) Madison WI
WERO(FM) Washington NC
WERQ-FM Baltimore MD
WERR(FM) Utuado PR
*WERS(FM) Boston MA
*WERU-FM Blue Hill ME
WERV-FM Aurora IL
WERX-FM Columbia NC
WERZ(FM) Exeter NH
WESC-FM Greenville SC
WESE(FM) Baldwyn MS
*WESM(FM) Princess Anne MD
*WESN(FM) Bloomington IL
WESP(FM) Dothan AL
WESR-FM Onley-Onancock VA
*WESS(FM) East Stroudsburg PA
*WESU(FM) Middletown CT
*WETA(FM) Washington DC
*WETD(FM) Alfred NY
*WETL(FM) South Bend IN
*WETN(FM) Wheaton IL
*WETS(FM) Johnson City TN
WETZ-FM New Martinsville WV
*WEUL(FM) Kingsford MI
WEUP-FM Moulton AL
WEUZ(FM) Minor Hill TN
*WEVC(FM) Gorham NH
WEVE-FM Eveleth MN
*WEVH(FM) Hanover NH
WEVI(FM) Frederiksted VI
WEVJ(FM) Jackson NH
*WEVL(FM) Memphis TN
*WEVN(FM) Keene NH
*WEVO(FM) Concord NH
WEVR-FM River Falls WI
*WEVS(FM) Nashua NH
WEXC(FM) Greenville PA
WEXP(FM) Brandon VT
WEXT(FM) Amsterdam NY
WEYE(FM) Surgoinsville TN
WEZB(FM) New Orleans LA
WEZF(FM) Burlington VT
WEZG(FM) Morganfield KY
WEZJ-FM Williamsburg KY
WEZL(FM) Charleston SC
WEZN-FM Bridgeport CT
WEZQ(FM) Bangor ME
WEZV(FM) North Myrtle Beach SC
WEZW(FM) Wildwood Crest NJ
WEZX(FM) Scranton PA
WEZY(FM) Racine WI
*WFAE(FM) Charlotte NC
WFAF(FM) Mount Kisco NY
*WFAR(FM) Danbury CT
WFAS-FM Bronxville NY
WFAT(FM) Portage MI
WFAV(FM) Gilman IL
WFBC-FM Greenville SC
WFBE(FM) Flint MI
*WFBF(FM) Buffalo NY
WFBI(FM) Inglis FL
WFBQ(FM) Indianapolis IN
WFBX(FM) Parker FL
WFBY(FM) Weston WV
WFBZ(FM) Trempealeau WI
WFCA(FM) Ackerman MS
WFCC-FM Chatham MA
*WFCF(FM) Saint Augustine FL
WFCG(FM) Tylertown MS
*WFCH(FM) Charleston SC
*WFCI(FM) Franklin IN
WFCJ(FM) Miamisburg OH
*WFCM-FM Murfreesboro TN
*WFCO(FM) Lancaster OH

*WFCR(FM) Amherst MA
*WFCS(FM) New Britain CT
WFCT(FM) Apalachicola FL
WFCX(FM) Leland MI
*WFDD-FM Winston-Salem NC
WFDL-FM Lomira WI
WFDR-FM Woodbury GA
WFDT(FM) Aguada PR
*WFDU(FM) Teaneck NJ
WFDX(FM) Atlanta MI
*WFEN(FM) Rockford IL
WFEX(FM) Peterborough NH
*WFFC(FM) Ferrum VA
WFFF-FM Columbia MS
WFFG-FM Corinth NY
WFFH(FM) Smyrna TN
WFFI(FM) Kingston Springs TN
*WFFL(FM) Panama City FL
*WFFM(FM) Ashburn GA
WFFN(FM) Cordova AL
WFFY(FM) Destin FL
WFGA(FM) Hicksville OH
*WFGB(FM) Kingston NY
WFGE(FM) Murray KY
WFGF(FM) Lima OH
*WFGH(FM) Fort Gay WV
WFGI-FM Johnstown PA
WFGM-FM Barrackville WV
WFGR(FM) Grand Rapids MI
WFGY(FM) Altoona PA
WFGZ(FM) Lobelville TN
*WFHB(FM) Bloomington IN
*WFHE(FM) Hickory NC
WFHG-FM Abingdon VA
*WFHL(FM) New Bedford MA
WFHM-FM Cleveland OH
WFHN(FM) Fairhaven MA
*WFHU(FM) Henderson TN
WFIA-FM New Albany IN
WFID(FM) Rio Piedras PR
*WFIT(FM) Melbourne FL
*WFIU(FM) Bloomington IN
WFIV-FM Loudon TN
WFIW-FM Fairfield IL
*WFIX(FM) Florence AL
WFJA(FM) Sanford NC
WFJO(FM) Folkston GA
WFKB(FM) Boyertown PA
WFKL(FM) Fairport NY
WFKS(FM) Neptune Beach FL
WFKX(FM) Henderson TN
WFKZ(FM) Plantation Key FL
WFLA-FM Midway FL
WFLB(FM) Laurinburg NC
WFLC(FM) Miami FL
WFLE-FM Flemingsburg KY
WFLK(FM) Geneva NY
WFLM(FM) White City FL
WFLO-FM Farmville VA
WFLQ(FM) French Lick IN
WFLR-FM Dundee NY
WFLS-FM Fredericksburg VA
WFLY(FM) Troy NY
WFLZ-FM Tampa FL
WFMB-FM Springfield IL
*WFME(FM) Newark NJ
WFMF(FM) Baton Rouge LA
WFMG(FM) Richmond IN
WFMH-FM Hackleburg AL
WFMI(FM) Southern Shores NC
WFMK(FM) East Lansing MI
WFML(FM) Vincennes IN
WFMM(FM) Sumrall MS
WFMN(FM) Flora MS
WFMP(FM) Coon Rapids MN
*WFMQ(FM) Lebanon TN
WFMS(FM) Indianapolis IN
WFMT(FM) Chicago IL
*WFMU(FM) East Orange NJ
WFMV(FM) South Congaree SC
WFMX(FM) Skowhegan ME
WFMZ(FM) Hertford NC
WFNC-FM Lumberton NC
WFNK(FM) Lewiston ME
*WFNM(FM) Lancaster PA
*WFNP(FM) Rosendale NY
WFNQ(FM) Nashua NH
WFNR-FM Christiansburg VA
WFNX(FM) Lynn MA
*WFOF(FM) Covington IN
WFON(FM) Fond du Lac WI
*WFOS(FM) Chesapeake VA
*WFOT(FM) Lexington OH
WFOX(FM) Norwalk CT
*WFPB-FM Falmouth MA
WFPG(FM) Atlantic City NJ
*WFPK(FM) Louisville KY
*WFPL(FM) Louisville KY
WFPS(FM) Freeport IL
WFQR(FM) Harwich Port MA
*WFQS(FM) Franklin NC
WFQX(FM) Front Royal VA
WFRB-FM Frostburg MD
*WFRC(FM) Columbus GA
WFRD(FM) Hanover NH
WFRE(FM) Frederick MD
*WFRF-FM Monticello FL
WFRG-FM Utica NY
*WFRH(FM) Kingston NY
WFRI(FM) Winamac IN
*WFRJ(FM) Johnstown PA
WFRM-FM Coudersport PA
WFRN-FM Elkhart IN
WFRO-FM Fremont OH
*WFRP(FM) Americus GA
WFRQ(FM) Mashpee MA
WFRR(FM) Walton IN
*WFRS(FM) Smithtown NY
*WFRW(FM) Webster NY
WFRY-FM Watertown NY
*WFSE(FM) Edinboro PA
WFSH-FM Athens GA
*WFSI(FM) Annapolis MD
*WFSK-FM Nashville TN
*WFSL(FM) Thomasville GA
*WFSO(FM) Olivebridge NY
WFSP-FM Kingwood WV
*WFSQ(FM) Tallahassee FL
*WFSS(FM) Fayetteville NC
*WFSU-FM Tallahassee FL
*WFSW(FM) Panama City FL
WFSY(FM) Panama City FL
WFTA(FM) Fulton MS
*WFTF(FM) Rutland VT
*WFTI-FM Saint Petersburg FL
WFTK(FM) Lebanon OH
WFTM-FM Maysville KY
WFTN-FM Franklin NH
WFTZ(FM) Manchester TN
*WFUM-FM Flint MI
WFUN-FM Bethalto IL
WFUR-FM Grand Rapids MI
WFUS(FM) Gulfport FL
*WFUV(FM) New York NY
WFVL(FM) Southern Pines NC
WFWI(FM) Fort Wayne IN
*WFWM(FM) Frostburg MD
*WFWR(FM) Attica IN
WFXA-FM Augusta GA
WFXC(FM) Durham NC
WFXD(FM) Marquette MI
WFXE(FM) Columbus GA
WFXF(FM) Honeoye Falls NY
WFXH-FM Hilton Head Island SC
WFXJ-FM North Kingsville OH
*WFXK(FM) Tarboro NC
WFXM(FM) Gordon GA
WFXN-FM Galion OH
WFXO(FM) Iuka MS
WFXX(FM) Georgiana AL
WFYE(FM) Glade Spring VA
*WFYI-FM Indianapolis IN
WFYN(FM) Waynesboro PA
WFYR(FM) Elmwood IL
WFYV-FM Atlantic Beach FL
WFYY(FM) Bloomsburg PA
WFZH(FM) Mukwonago WI
WFZX(FM) Searsport ME
WFZY(FM) Mount Union PA
WGAC-FM Warrenton GA
*WGAJ(FM) Deerfield MA
*WGAO(FM) Franklin MA
WGAR-FM Cleveland OH
*WGBE(FM) Bryan OH
WGBF-FM Henderson KY
WGBG(FM) Seaford DE
*WGBH(FM) Boston MA
WGBJ(FM) Auburn IN
*WGBK(FM) Glenview IL
WGBT(FM) Eden NC
WGBZ(FM) Cape May Court House NJ
*WGCA-FM Quincy IL
*WGCC-FM Batavia NY
*WGCF(FM) Paducah KY
WGCI-FM Chicago IL
WGCK(FM) Coeburn VA
WGCM-FM Gulfport MS
WGCO(FM) Midway GA
*WGCS(FM) Goshen IN
*WGCU-FM Fort Myers FL
WGCX(FM) Navarre FL
WGCY(FM) Gibson City IL
*WGDE(FM) Defiance OH
WGDN-FM Gladwin MI
WGDQ(FM) Hattiesburg MS
*WGDR(FM) Plainfield VT
WGEL(FM) Greenville IL
WGEM-FM Quincy IL
WGER(FM) Saginaw MI
*WGES-FM Key Largo FL
*WGEV(FM) Beaver Falls PA
WGFA-FM Watseka IL
WGFB(FM) Rockton IL
WGFG(FM) Branchville SC
WGFM(FM) Cheboygan MI
WGFN(FM) Glen Arbor MI
*WGFR(FM) Glens Falls NY
WGFX(FM) Gallatin TN
WGGC(FM) Bowling Green KY
WGGE(FM) Parkersburg WV
WGGI(FM) Benton PA
*WGGL-FM Houghton MI
WGGN(FM) Castalia OH
WGGY(FM) Scranton PA
WGH-FM Newport News VA
WGHN-FM Grand Haven MI
*WGHW(FM) Lockwoods Folly Town NC
*WGIB(FM) Birmingham AL
WGIC(FM) Cookeville TN
WGIE(FM) Clarksburg WV
WGIR-FM Manchester NH
WGIX-FM Gouverneur NY
WGKC(FM) Mahomet IL
WGKL(FM) Gladstone MI
*WGKR(FM) Grand Gorge NY
WGKS(FM) Paris KY
*WGKV(FM) Pulaski NY
WGKX(FM) Memphis TN
WGKY(FM) Wickliffe KY
WGLC-FM Mendota IL
*WGLE(FM) Lima OH
WGLF(FM) Tallahassee FL
WGL-FM Huntington IN
WGLI(FM) Hancock MI
WGLM(FM) West Lafayette IN
WGLO(FM) Pekin IL
WGLQ(FM) Escanaba MI
WGLR-FM Lancaster WI
*WGLS-FM Glassboro NJ
*WGLT(FM) Normal IL
*WGLV(FM) Woodstock VT
WGLX-FM Wisconsin Rapids WI
*WGLY-FM Bolton VT
*WGLZ(FM) West Liberty WV
*WGMC(FM) Greece NY
WGMD(FM) Rehoboth Beach DE
WGMF(FM) Tunkhannock PA
WGMG(FM) Crawford GA
WGMK(FM) Donalsonville GA
WGMM(FM) Corning NY
WGMO(FM) Shell Lake WI
WGMR(FM) Tyrone PA
*WGMS(FM) Hagerstown MD
WGMT(FM) Lyndon VT
WGMX(FM) Marathon FL
*WGMY(FM) Montgomery NY
WGMZ(FM) Glencoe AL
WGNA-FM Albany NY
*WGNB(FM) Zeeland MI
WGNE-FM Palatka FL
WGNG(FM) Tchula MS
WGNI(FM) Wilmington NC
*WGNJ(FM) Saint Joseph IL
WGNL(FM) Greenwood MS
*WGNN(FM) Fisher IL
*WGNR-FM Anderson IN
*WGNV(FM) Milladore WI
WGNX(FM) Colchester IL
WGNY-FM Newburgh NY
WGOD-FM Charlotte Amalie VI
WGOG(FM) Walhalla SC
*WGOJ(FM) Conneaut OH
WGOW-FM Soddy-Daisy TN
WGPB(FM) Rome GA
*WGPH(FM) Vidalia GA
WGPR(FM) Detroit MI
*WGPS(FM) Elizabeth City NC
WGQR(FM) Elizabethtown NC
*WGRC(FM) Lewisburg PA
WGRD-FM Grand Rapids MI
*WGRE(FM) Greencastle IN
WGRF(FM) Buffalo NY
WGRK-FM Greensburg KY
WGRM-FM Greenwood MS
WGRQ(FM) Colonial Beach VA
WGRR(FM) Hamilton OH
*WGRS(FM) Guilford CT
WGRT(FM) Port Huron MI
*WGRW(FM) Anniston AL
WGRX(FM) Falmouth VA
WGRY-FM Grayling MI
*WGSG(FM) Mayo FL
*WGSK(FM) South Kent CT
*WGSL(FM) Loves Park IL
WGSM(FM) Greensburg PA
WGSQ(FM) Cookeville TN
WGSS(FM) Kingstree SC
*WGSU(FM) Geneseo NY
WGSY(FM) Phenix City AL
*WGTD(FM) Kenosha WI
*WGTE-FM Toledo OH
*WGTF(FM) Dothan AL
WGTH-FM Richlands VA
WGTI(FM) Windsor NC
WGTN-FM Andrews SC
WGTR(FM) Bucksport SC
*WGTS(FM) Takoma Park MD
*WGTT(FM) Emeralda FL
WGTX(FM) Truro MA
WGTY(FM) Gettysburg PA
WGTZ(FM) Eaton OH
*WGUC(FM) Cincinnati OH
WGUF(FM) Marco FL
*WGUR(FM) Milledgeville GA
WGUS-FM New Ellenton SC
WGUY(FM) Dexter ME
WGVC(FM) Simpsonville SC
*WGVE(FM) Gary IN
WGVS-FM Whitehall MI
*WGVU-FM Allendale MI
WGVX(FM) Lakeville MN
WGVY(FM) Cambridge MN
WGVZ(FM) Eden Prairie MN
WGWD(FM) Gretna FL
*WGWG(FM) Boiling Springs NC
*WGWR(FM) Liberty NY
WGXL(FM) Hanover NH
WGYE(FM) Mannington WV
WGYI(FM) Oil City PA
WGYL(FM) Vero Beach FL
WGYY(FM) Meadville PA
WGZB-FM Lanesville IN
WGZO(FM) Parris Island SC
WGZR(FM) Bluffton SC
WGZZ(FM) Dadeville AL
*WHAA(FM) Adams WI
*WHAB(FM) Acton MA
*WHAD(FM) Delafield WI
WHAI(FM) Greenfield MA
WHAJ(FM) Bluefield WV
WHAK-FM Rogers City MI
WHAL-FM Horn Lake MS
WHAY(FM) Whitley City KY
WHAZ-FM Hoosick Falls NY
WHBC-FM Canton OH
*WHBM-FM Park Falls WI
WHBQ-FM Germantown TN
WHBR-FM Parkersburg WV
WHBX(FM) Tallahassee FL
WHBZ(FM) Sheboygan Falls WI
*WHCB(FM) Bristol TN
WHCC(FM) Ellettsville IN
*WHCE(FM) Highland Springs VA
*WHCF(FM) Bangor ME
*WHCI(FM) Hartford City IN
*WHCJ(FM) Savannah GA
*WHCL-FM Clinton NY
*WHCM(FM) Palatine IL
WHCN(FM) Hartford CT
*WHCR-FM New York NY
WHCY(FM) Blairstown NJ
WHDG(FM) Rhinelander WI
*WHDI(FM) Sister Bay WI
WHDQ(FM) Claremont NH
WHDR(FM) Miami FL
WHDX(FM) Buxton NC
WHDZ(FM) Buxton NC
WHEB(FM) Portsmouth NH
*WHEM(FM) Eau Claire WI
WHER(FM) Heidelberg MS
WHFB-FM Benton Harbor MI
*WHFC(FM) Bel Air MD
WHFD(FM) Lawrenceville VA
*WHFG(FM) Broussard LA
*WHFH(FM) Flossmoor IL
*WHFI(FM) Lindside WV
WHFM(FM) Southampton NY
*WHFR(FM) Dearborn MI
WHFS(FM) Catonsville MD
WHFX(FM) Darien GA
WHGL-FM Canton PA
*WHGN(FM) Crystal River FL
WHGO(FM) Pascagoula MS
*WHHB(FM) Holliston MA
WHHD(FM) Clearwater SC
WHHH(FM) Indianapolis IN
*WHHI(FM) Highland WI
WHHL(FM) Jerseyville IL
WHHM-FM Henderson TN
WHHR(FM) Vienna GA
*WHHS(FM) Havertown PA
WHHT(FM) Horse Cave KY
WHHY-FM Montgomery AL
WHHZ(FM) Newberry FL
*WHID(FM) Green Bay WI
*WHIF(FM) Palatka FL
*WHIJ(FM) Ocala FL
*WHIL-FM Mobile AL
WHIO-FM Piqua OH
WHIT-FM Hartford MI
WHIZ-FM Zanesville OH
*WHJE(FM) Carmel IN
WHJK(FM) Cleveland TN
*WHJM(FM) Anna OH
WHJT(FM) Clinton MS
WHJX(FM) Baldwin FL
WHJY(FM) Providence RI
WHKB(FM) Houghton MI
*WHKC(FM) Columbus OH
WHKF(FM) Harrisburg PA
WHKL(FM) Crenshaw MS
WHKN(FM) Millen GA
WHKO(FM) Dayton OH
WHKR(FM) Rockledge FL
WHKS(FM) Port Allegany PA
*WHKV(FM) Sylvester GA
WHKX(FM) Bluefield VA
*WHLA(FM) La Crosse WI
WHLC(FM) Highlands NC
WHLF(FM) South Boston VA
WHLG(FM) Port St. Lucie FL
WHLH(FM) Jackson MS
WHLJ(FM) Statenville GA
WHLK(FM) De Forest WI
WHLM-FM Berwick PA
*WHLP(FM) Hanna IN
WHLW(FM) Luverne AL
WHLZ(FM) Marion SC
WHMA-FM Ashland AL

*WHMC-FM Conway SC
WHMD(FM) Hammond LA
WHME(FM) South Bend IN
WHMH-FM Sauk Rapids MN
WHMI-FM Howell MI
*WHMR(FM) Ledbetter KY
WHMS-FM Champaign IL
*WHMX(FM) Lincoln ME
*WHND(FM) Sister Bay WI
WHNN(FM) Bay City MI
WHOD(FM) Jackson AL
WHOF(FM) North Canton OH
WHOG-FM Ormond-by-the-Sea FL
*WHOJ(FM) Terre Haute IN
WHOK-FM Lancaster OH
WHOM(FM) Mt. Washington NH
WHOP-FM Hopkinsville KY
WHOT-FM Youngstown OH
WHOU-FM Houlton ME
*WHOV(FM) Hampton VA
WHOW-FM Clinton IL
WHPA(FM) Gallitzin PA
*WHPC(FM) Garden City NY
WHPD(FM) Dowagiac MI
*WHPE-FM High Point NC
WHPH(FM) Jemison AL
*WHPK-FM Chicago IL
*WHPL(FM) West Lafayette IN
WHPO(FM) Hoopeston IL
*WHPR(FM) Highland Park MI
WHPT(FM) Sarasota FL
WHPZ(FM) Bremen IN
WHQG(FM) Milwaukee WI
WHQQ(FM) Neoga IL
*WHQR(FM) Wilmington NC
WHQT(FM) Coral Gables FL
WHQX(FM) Cedar Bluff VA
WHRB(FM) Cambridge MA
WHRK(FM) Memphis TN
WHRL(FM) Albany NY
*WHRM(FM) Wausau WI
*WHRO-FM Norfolk VA
WHRP(FM) Tullahoma TN
*WHRS(FM) Cookeville TN
*WHRV(FM) Norfolk VA
*WHRW(FM) Binghamton NY
*WHSA(FM) Brule WI
WHSB(FM) Alpena MI
*WHSD(FM) Hinsdale IL
WHSM-FM Hayward WI
*WHSN(FM) Bangor ME
*WHSS(FM) Hamilton OH
WHST(FM) Tawas City MI
WHSX(FM) Edmonton KY
WHTA(FM) Hampton GA
WHTD(FM) Mount Clemens MI
WHTE-FM Ruckersville VA
WHTF(FM) Havana FL
WHTG-FM Eatontown NJ
WHTI(FM) Alexandria IN
WHTL-FM Whitehall WI
WHTO(FM) Iron Mountain MI
WHTQ(FM) Orlando FL
WHTS(FM) Coopersville MI
WHTT-FM Buffalo NY
WHTU(FM) Newton MS
WHTY(FM) Hartford City IN
WHTZ(FM) Newark NJ
WHUD(FM) Peekskill NY
WHUG(FM) Jamestown NY
WHUR-FM Washington DC
*WHUS(FM) Storrs CT
WHVE(FM) Russell Springs KY
*WHVP(FM) Hudson NY
*WHVT(FM) Clyde OH
*WHWC(FM) Menomonie WI
*WHWE(FM) Howe IN
*WHWG(FM) Trout Lake MI
WHWK(FM) Binghamton NY
*WHWL(FM) Marquette MI
*WHWY(FM) Marathon FL
WHXQ(FM) Kennebunkport ME
WHXR(FM) North Windham ME
WHXT(FM) Orangeburg SC
WHYB(FM) Menominee MI
*WHYC(FM) Swanquarter NC
WHYI-FM Fort Lauderdale FL
WHYN-FM Springfield MA
*WHYT(FM) Goodland Township MI
*WHYY-FM Philadelphia PA
*WHYZ(FM) Palm Bay FL
WHZR(FM) Royal Center IN
WHZT(FM) Seneca SC
WHZZ(FM) Lansing MI
*WIAA(FM) Interlochen MI
*WIAB(FM) Mackinaw City MI
WIAC-FM San Juan PR
WIAL(FM) Eau Claire WI
WIAU(FM) Franklin IN
WIBA-FM Madison WI
WIBB-FM Fort Valley GA
*WIBI(FM) Carlinville IL
WIBL(FM) Augusta GA
WIBN(FM) Earl Park IN
WIBT(FM) Shelby NC
WIBV(FM) Mount Vernon IL
WIBW-FM Topeka KS
WIBZ(FM) Wedgefield SC
*WICA(FM) Traverse City MI
*WICB(FM) Ithaca NY
WICI(FM) Sumter SC
WICL(FM) Williamsport MD
*WICN(FM) Worcester MA
WICO-FM Salisbury MD
*WICR(FM) Indianapolis IN
*WICV(FM) East Jordan MI
*WIDA-FM Carolina PR
WIDI(FM) Quebradillas PR
WIDL(FM) Caro MI
*WIDR(FM) Kalamazoo MI
WIFC(FM) Wausau WI
WIFE-FM Rushville IN
*WIFF(FM) Binghamton NY
WIFL(FM) Inglis FL
WIFM-FM Elkin NC
WIFN(FM) Macon GA
WIFO-FM Jesup GA
WIFX-FM Jenkins KY
*WIGH(FM) Lexington TN
WIGL(FM) Saint Matthews SC
WIGO-FM White Stone VA
WIGY(FM) Madison ME
WIHB(FM) Moncks Corner SC
WIHC(FM) Newberry MI
WIHG(FM) Rockwood TN
WIHN(FM) Normal IL
*WIHS(FM) Middletown CT
WIHT(FM) Washington DC
WIII(FM) Cortland NY
WIIL(FM) Kenosha WI
WIIS(FM) Key West FL
*WIIT(FM) Chicago IL
WIIZ(FM) Blackville SC
WIJV(FM) Harriman TN
WIKB-FM Iron River MI
WIKI(FM) Carrollton KY
WIKK(FM) Newton IL
*WIKL(FM) Greencastle IN
WIKQ(FM) Tusculum TN
WIKS(FM) New Bern NC
*WIKV(FM) Plymouth IN
WIKX(FM) Charlotte Harbor FL
WIKY-FM Evansville IN
WIKZ(FM) Chambersburg PA
WILE-FM Byesville OH
WIL-FM Saint Louis MO
WILI-FM Willimantic CT
WILK-FM Avoca PA
*WILL-FM Urbana IL
WILN(FM) Panama City FL
WILQ(FM) Williamsport PA
WILT(FM) Jacksonville NC
WILV(FM) Chicago IL
WILW(FM) Avalon NJ
WILZ(FM) Saginaw MI
WIMC(FM) Crawfordsville IN
WIMI(FM) Ironwood MI
WIMK(FM) Iron Mountain MI
WIMT(FM) Lima OH
WIMX(FM) Gibsonburg OH
WIMZ-FM Knoxville TN
WINC-FM Winchester VA
WINH(FM) Paris IL
WINK-FM Fort Myers FL
WINL(FM) Linden AL
WINN(FM) Columbus IN
WINQ(FM) Winchester NH
WINX-FM Cambridge MD
WIOA(FM) San Juan PR
WIOB(FM) Mayaguez PR
WIOC(FM) Ponce PR
WIOG(FM) Bay City MI
WIOK(FM) Falmouth KY
WIOL(FM) Greenville GA
WIOQ(FM) Philadelphia PA
WIOT(FM) Toledo OH
WIOV-FM Ephrata PA
WIOZ-FM Southern Pines NC
*WIPA(FM) Pittsfield IL
*WIPR-FM San Juan PR
*WIQH(FM) Concord MA
WIQO-FM Covington VA
WIQQ(FM) Leland MS
*WIRE(FM) Lebanon IN
WIRK-FM West Palm Beach FL
*WIRN(FM) Buhl MN
*WIRP(FM) Pennsuco FL
*WIRQ(FM) Rochester NY
*WIRR(FM) Virginia-Hibbing MN
WIRX(FM) Saint Joseph MI
*WISE-FM Wise VA
WISH-FM Galatia IL
WISK-FM Americus GA
WISM-FM Altoona WI
WIST-FM Thomasville NC
*WISU(FM) Terre Haute IN
WISX(FM) Philadelphia PA
*WITC(FM) Cazenovia NY
*WITF-FM Harrisburg PA
WITL-FM Lansing MI
*WITR(FM) Henrietta NY
*WITT(FM) Zionsville IN
*WITX(FM) Beaver Falls PA
WITZ-FM Jasper IN
*WIUJ(FM) Charlotte Amalie VI
*WIUM(FM) Macomb IL
*WIUP-FM Indiana PA
*WIUS(FM) Macomb IL
*WIUV(FM) Castleton VT
*WIUW(FM) Warsaw IL
WIVA-FM Aguadilla PR
WIVG(FM) Tunica MS
*WIVH(FM) Christiansted VI
WIVI(FM) Charlotte Amalie VI
WIVK-FM Knoxville TN
WIVQ(FM) Spring Valley IL
WIVR(FM) Kentland IN
WIVY(FM) Morehead KY
*WIWC(FM) Kokomo IN
WIXO(FM) Peoria IL
*WIXQ(FM) Millersville PA
WIXV(FM) Savannah GA
WIXX(FM) Green Bay WI
WIXY(FM) Champaign IL
WIYN(FM) Deposit NY
WIYY(FM) Baltimore MD
WIZB(FM) Abbeville AL
WIZD(FM) Rudolph WI
WIZF(FM) Erlanger KY
WIZM-FM La Crosse WI
WIZN(FM) Vergennes VT
WJAA(FM) Austin IN
*WJAB(FM) Huntsville AL
WJAD(FM) Leesburg GA
WJAM-FM Orrville AL
WJAQ(FM) Marianna FL
WJAW-FM McConnelsville OH
*WJAZ(FM) Summerdale PA
WJBB-FM Haleyville AL
*WJBC-FM Fernandina Beach FL
WJBD-FM Salem IL
WJBL(FM) Ladysmith WI
WJBQ(FM) Portland ME
WJBR-FM Wilmington DE
WJBT(FM) Green Cove Springs FL
WJBX(FM) Fort Myers Beach FL
WJBZ-FM Seymour TN
*WJCA(FM) Albion NY
*WJCB(FM) Clewiston FL
WJCD(FM) Windsor VA
*WJCE(FM) Elkton MI
*WJCF(FM) Morristown IN
*WJCH(FM) Joliet IL
*WJCJ(FM) Ladoga IN
*WJCK(FM) Piedmont AL
WJCL-FM Savannah GA
*WJCN(FM) Nassawadox VA
*WJCO(FM) Montpelier IN
*WJCQ(FM) Jackson MI
*WJCR-FM Upton KY
*WJCS(FM) Allentown PA
*WJCT-FM Jacksonville FL
*WJCU(FM) University Heights OH
WJCX(FM) Pittsfield ME
*WJCY(FM) Cicero IN
*WJCZ(FM) Milford IL
WJDB-FM Thomasville AL
WJDF(FM) Orange MA
WJDK-FM Seneca IL
WJDQ(FM) Marion MS
WJDR(FM) Prentiss MS
*WJDS(FM) Sparta GA
WJDT(FM) Rogersville TN
WJDV(FM) Broadway VA
*WJDZ(FM) Pastillo PR
WJEC(FM) Vernon AL
*WJED(FM) Dogwood Lakes Estate FL
*WJEF(FM) Lafayette IN
*WJEL(FM) Indianapolis IN
WJEN(FM) Rutland VT
WJEQ(FM) Macomb IL
WJEZ(FM) Dwight IL
WJFD-FM New Bedford MA
*WJFF(FM) Jeffersonville NY
*WJFH(FM) Sebring FL
WJFK-FM Manassas VA
WJFL(FM) Tennille GA
*WJFM(FM) Baton Rouge LA
*WJFP(FM) Fort Pierce FL
*WJFR(FM) Jacksonville FL
WJFX(FM) New Haven IN
WJGA-FM Jackson GA
WJGL(FM) Jacksonville FL
WJGO(FM) Tice FL
*WJHD(FM) Portsmouth RI
WJHM(FM) Daytona Beach FL
*WJHS(FM) Columbia City IN
WJHT(FM) Johnstown PA
*WJIA(FM) Guntersville AL
*WJIC(FM) Zanesville OH
*WJIE-FM Okolona KY
*WJIF(FM) Opp AL
*WJIJ(FM) Norlina NC
*WJIK(FM) Monroeville AL
WJIM-FM Lansing MI
*WJIR(FM) Key West FL
*WJIS(FM) Bradenton FL
WJIV(FM) Cherry Valley NY
WJIW(FM) Greenville MS
WJIZ-FM Albany GA
WJJB-FM Topsham ME
*WJJE(FM) Delaware OH
WJJH(FM) Ashland WI
*WJJJ(FM) Beckley WV
WJJK(FM) Noblesville IN
WJJM-FM Lewisburg TN
WJJN(FM) Columbia AL
WJJO(FM) Watertown WI
WJJQ-FM Tomahawk WI
WJJR(FM) Rutland VT
WJJS-FM Vinton VA
*WJJW(FM) North Adams MA
WJJX(FM) Lynchburg VA
WJJY-FM Brainerd MN
WJJZ(FM) Burlington NJ
*WJKA(FM) Jacksonville NC
WJKC(FM) Christiansted VI
WJKD(FM) Vero Beach FL
WJKI(FM) Bethany Beach DE
WJKK(FM) Vicksburg MS
WJKL(FM) Glendale Heights IL
*WJKN-FM Spring Arbor MI
*WJKQ(FM) Jackson MI
WJKS(FM) Canton NJ
WJKW(FM) Athens OH
WJKX(FM) Ellisville MS
WJLB(FM) Detroit MI
WJLE-FM Smithville TN
*WJLF(FM) Gainesville FL
*WJLH(FM) Flagler Beach FL
WJLK(FM) Asbury Park NJ
WJLQ(FM) Pensacola FL
*WJLR(FM) Seymour IN
WJLS-FM Beckley WV
WJLT(FM) Evansville IN
*WJLU(FM) New Smyrna Beach FL
*WJLY(FM) Ramsey IL
*WJLZ(FM) Virginia Beach VA
WJMA-FM Culpeper VA
WJMC-FM Rice Lake WI
WJMD(FM) Hazard KY
*WJMF(FM) Smithfield RI
WJMG(FM) Hattiesburg MS
WJMH(FM) Reidsville NC
WJMI(FM) Jackson MS
*WJMJ(FM) Hartford CT
WJMK(FM) Chicago IL
WJMM-FM Keene KY
WJMN(FM) Boston MA
WJMQ(FM) Clintonville WI
WJMR-FM Menomonee Falls WI
*WJMU(FM) Decatur IL
WJMX-FM Cheraw SC
WJMZ-FM Anderson SC
*WJNF(FM) Marianna FL
WJNG(FM) Johnsonburg PA
WJNI(FM) Ladson SC
WJNR-FM Iron Mountain MI
WJNS-FM Yazoo City MS
WJNV(FM) Jonesville VA
*WJNY(FM) Watertown NY
WJOD(FM) Asbury IA
*WJOG(FM) Good Hart MI
*WJOH(FM) Raco MI
*WJOJ(FM) Harrisville MI
*WJOM(FM) Eagle MI
WJOT-FM Wabash IN
WJOW(FM) Philipsburg PA
WJOX(FM) Northport AL
WJPA-FM Washington PA
WJPD(FM) Ishpeming MI
*WJPG(FM) Cape May Court House NJ
*WJPH(FM) Woodbine NJ
WJPK(FM) Barton VT
*WJPR(FM) Jasper IN
WJPT(FM) Fort Myers Villas FL
*WJPZ-FM Syracuse NY
WJQB(FM) Spring Hill FL
WJQK(FM) Zeeland MI
WJQM(FM) Mount Horeb WI
WJQZ(FM) Wellsville NY
*WJRC(FM) Lewistown PA
WJRE(FM) Galva IL
*WJRF(FM) Duluth MN
*WJRH(FM) Easton PA
WJRL-FM Ozark AL
WJRR(FM) Cocoa Beach FL
WJRS(FM) Jamestown KY
WJRZ-FM Manahawkin NJ
WJSA-FM Jersey Shore PA
*WJSC-FM Johnson VT
WJSE(FM) Petersburg NJ
WJSG(FM) Hamlet NC
WJSH(FM) Folsom LA
WJSJ(FM) Fernandina Beach FL
*WJSL(FM) Houghton NY
WJSM-FM Martinsburg PA
WJSN-FM Jackson KY
*WJSO(FM) Pikeville KY
*WJSP-FM Warm Springs GA
WJSQ(FM) Athens TN
*WJSR(FM) Birmingham AL
*WJSU(FM) Jackson MS
*WJSV(FM) Morristown NJ
WJSZ(FM) Ashley MI
*WJTA(FM) Kosciusko MS
*WJTF(FM) Panama City FL
*WJTG(FM) Fort Valley GA
WJTK(FM) Columbia City FL

*WJTL(FM) Lancaster PA
WJTT(FM) Red Bank TN
*WJTY(FM) Lancaster WI
WJUC(FM) Swanton OH
*WJUF(FM) Inverness FL
WJUN-FM Mexico PA
WJUX(FM) Monticello NY
*WJVK(FM) Owensboro KY
WJVL(FM) Janesville WI
WJVO(FM) South Jacksonville IL
*WJVP(FM) Culebra PR
*WJVS(FM) Cincinnati OH
*WJWD(FM) Marshall WI
*WJWJ-FM Beaufort SC
*WJWT(FM) Gardner MA
*WJWV(FM) Fort Gaines GA
WJWZ(FM) Wetumpka AL
WJXA(FM) Nashville TN
WJXB-FM Knoxville TN
WJXM(FM) De Kalb MS
WJXN-FM Utica MS
WJXQ(FM) Jackson MI
WJXR(FM) Macclenny FL
WJXY-FM Conway SC
*WJYA(FM) Emporia VA
WJYD(FM) London OH
WJYE(FM) Buffalo NY
WJYF(FM) Nashville GA
*WJYJ(FM) Fredericksburg VA
*WJYO(FM) Fort Myers FL
*WJYW(FM) Union City IN
WJYY(FM) Concord NH
WJZA(FM) Lancaster OH
*WJZB(FM) Starkville MS
WJZD(FM) Long Beach MS
WJZE(FM) Oak Harbor OH
WJZG(FM) Culebra PR
WJZI(FM) Livingston Manor NY
WJZJ(FM) Glen Arbor MI
WJZK(FM) Richwood OH
WJZL(FM) Charlotte MI
WJZQ(FM) Cadillac MI
WJZR(FM) Rochester NY
WJZS(FM) Block Island RI
WJZT(FM) Woodville FL
WJZW(FM) Woodbridge VA
WJZX(FM) Brookfield WI
WJZZ-FM Roswell GA
WKAA(FM) Willacoochee GA
WKAD(FM) Harrietta MI
WKAF(FM) Brockton MA
WKAI(FM) Macomb IL
WKAK(FM) Albany GA
*WKAO(FM) Ashland KY
WKAQ-FM San Juan PR
*WKAR-FM East Lansing MI
WKAY(FM) Knoxville IL
WKAZ-FM Miami WV
WKBB(FM) West Point MS
WKBC-FM North Wilkesboro NC
WKBE(FM) Warrensburg NY
WKBH-FM West Salem WI
WKBI-FM Saint Marys PA
WKBQ(FM) Covington TN
WKBU(FM) New Orleans LA
WKBX(FM) Kingsland GA
WKCA(FM) Owingsville KY
WKCB-FM Hindman KY
*WKCC(FM) Kankakee IL
WKCG(FM) Augusta ME
WKCH(FM) Whitewater WI
WKCI-FM Hamden CT
WKCJ(FM) Lewisburg WV
*WKCL(FM) Ladson SC
WKCN(FM) Lumpkin GA
*WKCO(FM) Gambier OH
WKCQ(FM) Saginaw MI
*WKCR-FM New York NY
*WKCS(FM) Knoxville TN
WKCY-FM Harrisonburg VA
WKDB(FM) Laurel DE
WKDD(FM) Canton OH
WKDE-FM Altavista VA
WKDF(FM) Nashville TN
WKDJ-FM Clarksdale MS
*WKDL(FM) Brockport NY
*WKDN-FM Camden NJ
WKDO-FM Liberty KY
WKDP-FM Corbin KY
WKDQ(FM) Henderson KY
*WKDS(FM) Kalamazoo MI
*WKDU(FM) Philadelphia PA
WKDZ-FM Cadiz KY
WKEA-FM Scottsboro AL
WKEB(FM) Medford WI
WKED-FM Frankfort KY
WKEE-FM Huntington WV
WKEL(FM) Confluence PA
WKEN(FM) Fredonia KY
WKEQ(FM) Somerset KY
*WKES(FM) Lakeland FL
*WKET(FM) Kettering OH
*WKEU-FM The Rock GA
WKEY-FM Key West FL
WKEZ-FM Tavenier FL
*WKFA(FM) Saint Catherine FL
WKFM(FM) Huron OH
WKFR-FM Battle Creek MI
WKFS(FM) Milford OH
WKFX(FM) Rice Lake WI
WKGA(FM) Thomaston AL
WKGB-FM Conklin NY
*WKGC-FM Panama City FL
WKGL-FM Loves Park IL
WKGO(FM) Cumberland MD
WKGR(FM) Fort Pierce FL
WKGS(FM) Irondequoit NY
WKHC(FM) Dahlonega GA
WKHG(FM) Leitchfield KY
WKHI(FM) Fruitland MD
WKHJ(FM) Mountain Lake Park MD
WKHK(FM) Colonial Heights VA
WKHM-FM Brooklyn MI
*WKHN(FM) Hubbard Lake MI
WKHQ-FM Charlevoix MI
*WKHR(FM) Bainbridge OH
*WKHS(FM) Worton MD
WKHT(FM) Knoxville TN
WKHW(FM) Pocomoke City MD
WKHX-FM Marietta GA
WKHY(FM) Lafayette IN
WKIB(FM) Anna IL
WKID(FM) Vevay IN
WKIE(FM) Arlington Heights IL
WKIF(FM) Kankakee IL
WKIK-FM California MD
WKIM(FM) Munford TN
WKIS(FM) Boca Raton FL
WKIT-FM Brewer ME
*WKIV(FM) Westerly RI
WKIX(FM) Kinston NC
WKJC(FM) Tawas City MI
*WKJD(FM) Columbus IN
*WKJL(FM) Clarksburg WV
WKJM(FM) Petersburg VA
WKJQ-FM Parsons TN
WKJS(FM) Richmond VA
WKJT(FM) Teutopolis IL
WKJX(FM) Elizabeth City NC
WKJY(FM) Hempstead NY
WKJZ(FM) Hillman MI
WKKB(FM) Middletown RI
*WKKC(FM) Chicago IL
WKKF(FM) Ballston Spa NY
WKKG(FM) Columbus IN
WKKI(FM) Celina OH
WKKJ(FM) Chillicothe OH
*WKKL(FM) West Barnstable MA
*WKKM(FM) Rust Township MI
WKKO(FM) Toledo OH
WKKQ(FM) Barbourville KY
WKKR(FM) Auburn AL
WKKS-FM Vanceburg KY
WKKT(FM) Statesville NC
WKKV-FM Racine WI
WKKW(FM) Fairmont WV
WKKY(FM) Geneva OH
WKKZ(FM) Dublin GA
WKLA-FM Ludington MI
WKLB-FM Waltham MA
WKLC-FM Saint Albans WV
WKLD(FM) Oneonta AL
WKLG(FM) Rock Harbor FL
WKLH(FM) Milwaukee WI
WKLI-FM Albany NY
WKLK-FM Cloquet MN
WKLL(FM) Frankfort NY
WKLM(FM) Millersburg OH
WKLN(FM) Wilmington OH
WKLO(FM) Hardinsburg IN
WKLQ(FM) Greenville MI
WKLR(FM) Fort Lee VA
WKLS(FM) Atlanta GA
WKLT(FM) Kalkaska MI
WKLU(FM) Brownsburg IN
WKLW-FM Paintsville KY
WKLX(FM) Brownsville KY
WKLZ-FM Petoskey MI
WKMJ-FM Hancock MI
WKMK(FM) Ocean Acres NJ
WKML(FM) Lumberton NC
WKMM(FM) Kingwood WV
WKMO(FM) Hodgenville KY
*WKMS-FM Murray KY
*WKMV(FM) Muncie IN
WKMX(FM) Enterprise AL
*WKMY(FM) Winchendon MA
WKNB(FM) Clarendon PA
*WKNC-FM Raleigh NC
WKNE(FM) Keene NH
*WKNG-FM Heflin AL
*WKNH(FM) Keene NH
*WKNJ-FM Union Township NJ
WKNL(FM) New London CT
WKNN-FM Pascagoula MS
*WKNO-FM Memphis TN
*WKNP(FM) Jackson TN
*WKNS(FM) Kinston NC
WKNU(FM) Brewton AL
WKNZ(FM) Collins MS
WKOA(FM) Lafayette IN
WKOL(FM) Plattsburgh NY
WKOM(FM) Columbia TN
WKOR-FM Columbus MS
WKOS(FM) Kingsport TN
WKOT(FM) Marseilles IL
WKOV-FM Wellston OH
WKOY-FM Princeton WV
*WKPB(FM) Henderson KY
WKPE-FM South Yarmouth MA
*WKPK(FM) Michigamme MI
WKPL(FM) Ellwood City PA
WKPQ(FM) Hornell NY
*WKPS(FM) State College PA
*WKPW(FM) Knightstown IN
*WKPX(FM) Sunrise FL
WKQC(FM) Charlotte NC
WKQH(FM) Marathon WI
WKQI(FM) Detroit MI
WKQL(FM) Brookville PA
WKQQ(FM) Winchester KY
WKQS-FM Negaunee MI
WKQV(FM) Richwood WV
WKQW-FM Oil City PA
WKQX(FM) Chicago IL
WKQY(FM) Tazewell VA
WKQZ(FM) Midland MI
WKRA-FM Holly Springs MS
*WKRB(FM) Brooklyn NY
WKRD-FM Shelbyville KY
WKRF(FM) Tobyhanna PA
WKRH(FM) Minetto NY
WKRI(FM) Cleveland Heights OH
*WKRJ(FM) New Philadelphia OH
WKRK-FM Detroit MI
WKRL-FM North Syracuse NY
WKRO-FM Edgewater FL
WKRQ(FM) Cincinnati OH
WKRR(FM) Asheboro NC
WKRV(FM) Vandalia IL
*WKRW(FM) Wooster OH
WKRX(FM) Roxboro NC
*WKRY(FM) Versailles IN
WKRZ(FM) Wilkes-Barre PA
WKSA-FM Isabela PR
WKSB(FM) Williamsport PA
WKSC-FM Chicago IL
WKSD(FM) Paulding OH
WKSE(FM) Niagara Falls NY
WKSF(FM) Asheville NC
*WKSG(FM) Cedar Creek FL
WKSI-FM Stephens City VA
WKSJ-FM Mobile AL
WKSK-FM South Hill VA
WKSL(FM) Burlington NC
WKSM(FM) Fort Walton Beach FL
WKSO(FM) Natchez MS
WKSP(FM) Aiken SC
WKSQ(FM) Ellsworth ME
WKSR-FM Pulaski TN
WKSS(FM) Hartford CT
WKST-FM Pittsburgh PA
*WKSU-FM Kent OH
*WKSV(FM) Thompson OH
WKSW(FM) Urbana OH
WKSX-FM Johnston SC
WKSZ(FM) De Pere WI
WKTG(FM) Madisonville KY
WKTI(FM) Milwaukee WI
WKTJ-FM Farmington ME
WKTK(FM) Crystal River FL
*WKTL(FM) Struthers OH
WKTM(FM) Soperton GA
WKTN(FM) Kenton OH
*WKTO(FM) Edgewater FL
*WKTS(FM) Kingston TN
WKTU(FM) Lake Success NY
*WKTZ-FM Jacksonville FL
WKUB(FM) Blackshear GA
*WKUE(FM) Elizabethtown KY
WKUL(FM) Cullman AL
WKUS(FM) Norfolk VA
WKUZ(FM) Wabash IN
WKVB(FM) Port Matilda PA
*WKVC(FM) North Myrtle Beach SC
WKVE(FM) Semora NC
*WKVF(FM) Byhalia MS
*WKVH(FM) Monticello FL
WKVI-FM Knox IN
*WKVJ(FM) Dannemora NY
*WKVP(FM) Cherry Hill NJ
*WKVR-FM Huntingdon PA
WKVS(FM) Lenoir NC
WKVT-FM Brattleboro VT
*WKVU(FM) Utica NY
*WKVW(FM) Marmet WV
*WKVY(FM) Somerset KY
WKVZ(FM) Ripley TN
*WKWC(FM) Owensboro KY
WKWI(FM) Kilmarnock VA
WKWK-FM Wheeling WV
*WKWR(FM) Key West FL
WKWS(FM) Charleston WV
*WKWV(FM) Watertown NY
WKWX(FM) Savannah TN
WKWY(FM) Tompkinsville KY
*WKWZ(FM) Syosset NY
WKXA-FM Findlay OH
WKXB(FM) Burgaw NC
WKXC-FM Aiken SC
WKXD(FM) Monterey TN
WKXH(FM) Saint Johnsbury VT
WKXI-FM Magee MS
WKXK(FM) Pine Hill AL
WKXM-FM Winfield AL
WKXN(FM) Greenville AL
WKXP(FM) Kingston NY
WKXQ(FM) Rushville IL
WKXS-FM Leland NC
WKXU(FM) Louisburg NC
WKXW(FM) Trenton NJ
WKXX(FM) Attalla AL
WKXY(FM) Clarksdale MS
WKXZ(FM) Norwich NY
WKYA(FM) Greenville KY
WKYB(FM) Burgin KY
WKYE(FM) Johnstown PA
*WKYJ(FM) Rouses Point NY
WKYL(FM) Lawrenceburg KY
WKYM(FM) Monticello KY
WKYQ(FM) Paducah KY
WKYR-FM Burkesville KY
WKYS(FM) Washington DC
*WKYU-FM Bowling Green KY
WKYW(FM) Frankfort KY
WKYX-FM Golconda IL
WKYZ(FM) Key Colony Beach FL
WKZA(FM) Lakewood NY
WKZB(FM) Stonewall MS
WKZC(FM) Scottville MI
WKZE-FM Salisbury CT
WKZJ(FM) Eufaula AL
WKZL(FM) Winston-Salem NC
*WKZM(FM) Sarasota FL
WKZP(FM) McMinnville TN
WKZQ-FM Myrtle Beach SC
WKZR(FM) Milledgeville GA
WKZS(FM) Covington IN
WKZU(FM) Ripley MS
WKZW(FM) Bay Springs MS
WKZX-FM Lenoir City TN
WKZY(FM) Cross City FL
WKZZ(FM) Tifton GA
*WLAB(FM) Fort Wayne IN
*WLAI(FM) Danville KY
WLAK(FM) Huntingdon PA
WLAN-FM Lancaster PA
WLAV-FM Grand Rapids MI
WLAW(FM) Newaygo MI
WLAY-FM Littleville AL
*WLAZ(FM) Kissimmee FL
WLBC-FM Muncie IN
*WLBF(FM) Montgomery AL
WLBH-FM Mattoon IL
*WLBL-FM Wausau WI
*WLBS(FM) Bristol PA
WLBW(FM) Fenwick Island DE
*WLCA(FM) Godfrey IL
WLCE(FM) Petersburg IL
*WLCH(FM) Lancaster PA
WLCN(FM) Atlanta IL
WLCS(FM) North Muskegon MI
WLCT(FM) Lafayette TN
WLCY(FM) Blairsville PA
WLDA(FM) Fort Rucker AL
WLDB(FM) Milwaukee WI
WLDE(FM) Fort Wayne IN
WLDI(FM) Fort Pierce FL
WLDR-FM Traverse City MI
WLEG(FM) Ligonier IN
WLEL(FM) Ellaville GA
WLEN(FM) Adrian MI
WLEQ(FM) Bedford VA
WLER-FM Butler PA
WLEV(FM) Allentown PA
WLEW-FM Bad Axe MI
WLEY-FM Aurora IL
*WLFA(FM) Asheville NC
*WLFC(FM) Findlay OH
WLFE-FM Saint Albans VT
WLFF(FM) Brookston IN
*WLFJ-FM Greenville SC
*WLFR(FM) Pomona NJ
*WLFS(FM) Port Wentworth GA
WLFV(FM) Ettrick VA
WLFW(FM) Chandler IN
WLFX(FM) Berea KY
WLGC-FM Greenup KY
*WLGH(FM) Leroy Township MI
*WLGI(FM) Hemingway SC
WLGL(FM) Riverside PA
WLGN-FM Logan OH
WLGP(FM) Harkers Island NC
WLGT(FM) Washington NC
WLHC(FM) Robbins NC
WLHK(FM) Shelbyville IN
WLHM(FM) Logansport IN
*WLHS(FM) West Chester OH
WLHT-FM Grand Rapids MI
*WLHW(FM) Casey IL
*WLIC(FM) Frostburg MD
WLIF(FM) Baltimore MD
WLIH(FM) Whitneyville PA
WLIN-FM Durant MS
WLIR-FM Hampton Bays NY
WLIT-FM Chicago IL
*WLIU(FM) Southampton NY
WLIV-FM Monterey TN
WLJA-FM Ellijay GA
WLJC(FM) Beattyville KY

WLJE(FM) Valparaiso IN
*WLJH(FM) Glens Falls NY
WLJI(FM) Summerton SC
*WLJK(FM) Aiken SC
*WLJN-FM Traverse City MI
*WLJP(FM) Monroe NY
*WLJR(FM) Birmingham AL
*WLJS-FM Jacksonville AL
WLJY(FM) Whiting WI
WLJZ(FM) Mackinaw City MI
*WLKA(FM) Tafton PA
*WLKB(FM) Bay City MI
WLKC(FM) Campton NH
WLKE(FM) Bar Harbor ME
WLKG(FM) Lake Geneva WI
WLKH(FM) Somerset PA
WLKI(FM) Angola IN
WLKJ(FM) Portage PA
WLKK(FM) Wethersfield Township NY
*WLKL(FM) Mattoon IL
WLKM-FM Three Rivers MI
WLKN(FM) Cleveland WI
*WLKO(FM) Quitman MS
*WLKP(FM) Belpre OH
WLKQ-FM Buford GA
WLKR-FM Norwalk OH
WLKS-FM West Liberty KY
WLKT(FM) Lexington-Fayette KY
WLKU(FM) Rock Island IL
*WLKV(FM) Ripley WV
WLKX-FM Forest Lake MN
WLKZ(FM) Wolfeboro NH
WLLD(FM) Holmes Beach FL
WLLE(FM) Clinton KY
WLLF(FM) Mercer PA
WLLG(FM) Lowville NY
WLLI-FM Dyer TN
WLLJ(FM) Etowah TN
WLLK-FM Somerset KY
WLLR-FM Davenport IA
WLLT(FM) Polo IL
WLLW(FM) Seneca Falls NY
WLLX(FM) Lawrenceburg TN
WLMD(FM) Bushnell IL
WLME(FM) Cannelton IN
WLMG(FM) New Orleans LA
*WLMH(FM) Morrow OH
WLMI(FM) Kane PA
*WLMS(FM) Lecanto FL
*WLMU(FM) Harrogate TN
*WLMW(FM) Manchester NH
WLMX-FM Balsam Lake WI
WLND(FM) Signal Mountain TN
WLNG(FM) Sag Harbor NY
WLNH-FM Laconia NH
WLNI(FM) Lynchburg VA
WLNK(FM) Charlotte NC
WLNP(FM) Carbondale PA
*WLNX(FM) Lincoln IL
*WLNZ(FM) Lansing MI
WLOB-FM Rumford ME
WLOD-FM Sweetwater TN
WLOF(FM) Attica NY
*WLOG(FM) Markleysburg PA
WLOQ(FM) Winter Park FL
WLOW(FM) Port Royal SC
*WLPE(FM) Augusta GA
WLPF(FM) Ocilla GA
*WLPG(FM) Florence SC
*WLPJ(FM) New Port Richey FL
*WLPS-FM Lumberton NC
*WLPT(FM) Jesup GA
WLPW(FM) Lake Placid NY
WLQB(FM) Ocean Isle Beach NC
WLQI(FM) Rensselaer IN
WLQK(FM) Livingston TN
WLQM-FM Franklin VA
WLQT(FM) Kettering OH
*WLRA(FM) Lockport IL
WLRD(FM) Willard OH
*WLRH(FM) Huntsville AL
*WLRK(FM) Greenville MS
*WLRN-FM Miami FL
WLRQ-FM Cocoa FL
WLRR(FM) Milledgeville GA
WLRS(FM) Shepherdsville KY
WLRW(FM) Champaign IL
WLRX(FM) Charlestown IN
*WLRY(FM) Rushville OH
WLSK(FM) Lebanon KY
WLSL(FM) Three Lakes WI
WLSM-FM Louisville MS
*WLSN(FM) Grand Marais MN
*WLSO(FM) Sault Ste. Marie MI
WLSQ(FM) Byrdstown TN
WLSR(FM) Galesburg IL
WLST(FM) Marinette WI
*WLSU(FM) La Crosse WI
WLSW(FM) Scottdale PA
WLSZ(FM) Humboldt TN
WLTB(FM) Johnson City NY
WLTE(FM) Minneapolis MN
WLTF(FM) Martinsburg WV
WLTI(FM) Syracuse NY
WLTJ(FM) Pittsburgh PA
WLTK(FM) New Market VA
*WLTL(FM) La Grange IL
WLTM(FM) Peachtree City GA
WLTN-FM Lisbon NH
WLTO(FM) Nicholasville KY
WLTQ-FM Venice FL
*WLTR(FM) Columbia SC
WLTS(FM) Centre Hall PA
WLTT(FM) Shallotte NC
WLTU(FM) Manitowoc WI
WLTW(FM) New York NY
WLTY(FM) Cayce SC
WLUE(FM) Louisville KY
*WLUJ(FM) Springfield IL
WLUM-FM Milwaukee WI
WLUP-FM Chicago IL
*WLUR(FM) Lexington VA
WLUS-FM Clarksville VA
*WLUW(FM) Chicago IL
WLVB(FM) Morrisville VT
WLVE(FM) Miami Beach FL
*WLVF-FM Haines City FL
WLVG(FM) Center Moriches NY
WLVH(FM) Hardeeville SC
WLVK(FM) Fort Knox KY
*WLVM(FM) Ironwood MI
WLVQ(FM) Columbus OH
*WLVR(FM) Bethlehem PA
WLVS-FM Clifton TN
*WLVU(FM) Halifax PA
WLVX(FM) Elberton GA
WLVY(FM) Elmira NY
WLWD(FM) Columbus Grove OH
WLWI-FM Montgomery AL
*WLWJ(FM) Petersburg IL
WLXC(FM) Lexington SC
WLXO(FM) Stamping Ground KY
*WLXP(FM) Savannah GA
WLXR-FM La Crosse WI
WLXT(FM) Petoskey MI
WLXV(FM) Cadillac MI
WLXX(FM) Lexington KY
WLYE-FM Glasgow KY
WLYF(FM) Miami FL
WLYT(FM) Hickory NC
WLYU(FM) Lyons GA
WLYX(FM) Valdosta GA
WLZA(FM) Eupora MS
WLZK(FM) Paris TN
WLZL(FM) Annapolis MD
WLZN(FM) Macon GA
WLZQ(FM) South Whitley IN
WLZS(FM) Beaver Springs PA
WLZT(FM) Chillicothe OH
WLZW(FM) Utica NY
WLZX(FM) Northampton MA
WLZZ(FM) Montpelier OH
*WMAB-FM Mississippi State MS
WMAD(FM) Sauk City WI
*WMAE-FM Booneville MS
WMAG(FM) High Point NC
*WMAH-FM Biloxi MS
*WMAO-FM Greenwood MS
WMAS-FM Springfield MA
*WMAU-FM Bude MS
*WMAV-FM Oxford MS
*WMAW-FM Meridian MS
WMAX-FM Holland MI
WMBC(FM) Columbus MS
*WMBI-FM Chicago IL
*WMBJ(FM) Murrell's Inlet SC
*WMBL(FM) Mitchell IN
*WMBR(FM) Cambridge MA
*WMBU(FM) Forest MS
*WMBV(FM) Dixons Mills AL
*WMBW(FM) Chattanooga TN
WMBX(FM) Jensen Beach FL
WMCD(FM) Claxton GA
*WMCE(FM) Erie PA
WMC-FM Memphis TN
WMCG(FM) Milan GA
WMCI(FM) Mattoon IL
WMCM(FM) Rockland ME
*WMCN(FM) Saint Paul MN
*WMCO(FM) New Concord OH
*WMCQ(FM) Muskegon MI
WMCR-FM Oneida NY
*WMCU(FM) Miami FL
*WMCX(FM) West Long Branch NJ
WMDC(FM) Mayville WI
WMDH-FM New Castle IN
WMDJ-FM Allen KY
*WMDR-FM Oakland ME
*WMEA(FM) Portland ME
*WMEB-FM Orono ME
*WMED(FM) Calais ME
WMEE(FM) Fort Wayne IN
*WMEF(FM) Fort Kent ME
WMEG(FM) Guayama PR
*WMEH(FM) Bangor ME
*WMEJ(FM) Proctorville OH
*WMEM(FM) Presque Isle ME
*WMEP(FM) Camden ME
WMEQ-FM Menomonie WI
WMEV-FM Marion VA
*WMEW(FM) Waterville ME
WMEX(FM) Farmington NH
WMEZ(FM) Pensacola FL
WMFC-FM Monroeville AL
*WMFE-FM Orlando FL
WMFG-FM Hibbing MN
*WMFL(FM) Florida City FL
WMFM(FM) Key West FL
*WMFO(FM) Medford MA
WMFQ(FM) Ocala FL
WMFS(FM) Bartlett TN
*WMFT(FM) Tuscaloosa AL
WMFX(FM) Saint Andrews SC
WMGA(FM) Kenova WV
WMGB(FM) Montezuma GA
WMGC-FM Detroit MI
WMGE(FM) Miami Beach FL
WMGF(FM) Mount Dora FL
WMGH-FM Tamaqua PA
WMGI(FM) Terre Haute IN
WMGK(FM) Philadelphia PA
WMGL(FM) Ravenel SC
WMGM(FM) Atlantic City NJ
WMGN(FM) Madison WI
WMGP(FM) Hogansville GA
WMGQ(FM) New Brunswick NJ
WMGS(FM) Wilkes-Barre PA
WMGV(FM) Newport NC
WMGX(FM) Portland ME
WMGZ(FM) Eatonton GA
*WMHB(FM) Waterville ME
*WMHC(FM) South Hadley MA
*WMHD-FM Terre Haute IN
*WMHI(FM) Cape Vincent NY
*WMHK(FM) Columbia SC
*WMHN(FM) Webster NY
*WMHQ(FM) Malone NY
*WMHR(FM) Syracuse NY
*WMHS(FM) Pike Creek DE
*WMHT-FM Schenectady NY
*WMHW-FM Mount Pleasant MI
WMHX(FM) Hershey PA
WMIB(FM) Fort Lauderdale FL
*WMIE(FM) Cocoa FL
WMIK-FM Middlesboro KY
WMIL(FM) Waukesha WI
WMIO(FM) Cabo Rojo PR
WMIS-FM Blackduck MN
WMIT(FM) Black Mountain NC
WMIX-FM Mount Vernon IL
WMJC(FM) Smithtown NY
WMJD(FM) Grundy VA
WMJE(FM) Clarkesville GA
WMJI(FM) Cleveland OH
WMJJ(FM) Birmingham AL
WMJK(FM) Clyde OH
WMJL-FM Marion KY
WMJM(FM) Jeffersontown KY
WMJO(FM) Essexville MI
WMJT(FM) McMillan MI
WMJU(FM) Bude MS
WMJW(FM) Cleveland MS
WMJX(FM) Boston MA
WMJY(FM) Biloxi MS
WMJZ-FM Gaylord MI
WMKB(FM) Earlville IL
WMKC(FM) Saint Ignace MI
WMKD(FM) Pickford MI
WMKJ(FM) Mt. Sterling KY
WMKK(FM) Lawrence MA
*WMKL(FM) Key Largo FL
*WMKM(FM) Sebring FL
*WMKO(FM) Marco FL
WMKR(FM) Pana IL
WMKS(FM) Clemmons NC
*WMKV(FM) Reading OH
*WMKW(FM) Crossville TN
WMKX(FM) Brookville PA
*WMKY(FM) Morehead KY
WMKZ(FM) Monticello KY
WMLF(FM) Watseka IL
*WMLJ(FM) Summersville WV
WMLL(FM) Bedford NH
*WMLN-FM Milton MA
WMLQ(FM) Manistee MI
*WMLS(FM) Grand Marais MN
*WMLU(FM) Farmville VA
WMLV(FM) Butler AL
WMLX(FM) Saint Mary's OH
WMMA(FM) Nekoosa WI
WMMC(FM) Marshall IL
WMME-FM Augusta ME
WMMG-FM Brandenburg KY
WMMJ(FM) Bethesda MD
WMMM-FM Verona WI
WMMO(FM) Orlando FL
WMMQ(FM) East Lansing MI
WMMR(FM) Philadelphia PA
WMMS(FM) Cleveland OH
*WMMT(FM) Whitesburg KY
WMMX(FM) Dayton OH
WMMY(FM) Jefferson NC
WMNA-FM Gretna VA
WMNC-FM Morganton NC
*WMNF(FM) Tampa FL
WMNG(FM) Christiansted VI
*WMNJ(FM) Madison NJ
*WMNR(FM) Monroe CT
WMNV(FM) Rupert VT
WMNX(FM) Wilmington NC
*WMOC(FM) Lumber City GA
WMOD(FM) Bolivar TN
WMOI(FM) Monmouth IL
WMOJ-FM Connersville IN
WMOM(FM) Pentwater MI
WMOO(FM) Derby Center VT
WMOQ(FM) Bostwick GA
WMOR-FM Morehead KY
WMOS(FM) Montauk NY
*WMOT(FM) Murfreesboro TN
WMOZ(FM) Moose Lake MN
*WMPG(FM) Gorham ME
*WMPH(FM) Wilmington DE
WMPI(FM) Scottsburg IN
*WMPN-FM Jackson MS
*WMPR(FM) Jackson MS
WMPZ(FM) Ringgold GA
WMQA-FM Minocqua WI
WMQT(FM) Ishpeming MI
WMQZ(FM) Colchester IL
*WMRA(FM) Harrisonburg VA
WMRF-FM Lewistown PA
*WMRL(FM) Lexington VA
WMRN-FM Marion OH
WMRR(FM) Muskegon Heights MI
WMRS(FM) Monticello IN
*WMRT(FM) Marietta OH
WMRV-FM Endicott NY
WMRX-FM Beaverton MI
*WMRY(FM) Crozet VA
WMRZ(FM) Dawson GA
*WMSB(FM) Senatobia MS
*WMSC(FM) Upper Montclair NJ
*WMSD(FM) Rose Township MI
*WMSE(FM) Milwaukee WI
WMSH-FM Sturgis MI
WMSI(FM) Jackson MS
*WMSJ(FM) Freeport ME
WMSK-FM Sturgis KY
*WMSL(FM) Athens GA
WMSO(FM) Meridian MS
*WMSQ(FM) Marlette MI
WMSR-FM Collinwood TN
*WMSS(FM) Middletown PA
WMSU(FM) Starkville MS
*WMSV(FM) Starkville MS
*WMTB-FM Emmittsburg MD
WMTC-FM Vancleve KY
WMTD-FM Hinton WV
WMTE-FM Manistee MI
WMT-FM Cedar Rapids IA
*WMTH(FM) Park Ridge IL
WMTI(FM) Picayune MS
WMTK(FM) Littleton NH
WMTM-FM Moultrie GA
WMTR-FM Archbold OH
*WMTS-FM Murfreesboro TN
WMTT(FM) Tioga PA
*WMTU-FM Houghton MI
WMTX(FM) Tampa FL
*WMUA(FM) Amherst MA
*WMUB(FM) Oxford OH
*WMUC-FM College Park MD
WMUF-FM Henry TN
*WMUH(FM) Allentown PA
*WMUK(FM) Kalamazoo MI
*WMUL(FM) Huntington WV
*WMUM-FM Cochran GA
WMUS(FM) Muskegon MI
WMUT(FM) Grenada MS
WMUU-FM Greenville SC
WMUV(FM) Brunswick GA
*WMUW(FM) Columbus MS
WMUZ(FM) Detroit MI
*WMVE(FM) Chase City VA
WMVL(FM) Linesville PA
*WMVM(FM) Goodman WI
WMVN(FM) East St. Louis IL
WMVR-FM Sidney OH
*WMVV(FM) Griffin GA
*WMVW(FM) Peachtree City GA
WMVX(FM) Cleveland OH
WMVY(FM) Tisbury MA
*WMWK(FM) Milwaukee WI
*WMWM(FM) Salem MA
WMWV(FM) Conway NH
*WMWX(FM) Miamitown OH
WMXA(FM) Opelika AL
WMXB(FM) Richmond VA
WMXC(FM) Mobile AL
WMXD(FM) Detroit MI
WMXE(FM) South Charleston WV
WMXG(FM) Stephenson MI
WMXH-FM Luray VA
WMXI(FM) Laurel MS
WMXJ(FM) Pompano Beach FL
WMXK(FM) Morristown TN
WMXL(FM) Lexington KY
*WMXM(FM) Lake Forest IL
WMXN-FM Stevenson AL
WMXO(FM) Olean NY
WMXQ(FM) Jacksonville FL
WMXR(FM) Woodstock VT
WMXS(FM) Montgomery AL
WMXT(FM) Pamplico SC
WMXU(FM) Starkville MS
WMXV(FM) Saint Joseph TN
WMXW(FM) Vestal NY
WMXX-FM Jackson TN
WMXY(FM) Youngstown OH

WMXZ(FM) De Funiak Springs FL
WMYB(FM) Myrtle Beach SC
WMYE(FM) Rantoul IL
WMYI(FM) Hendersonville NC
*WMYJ-FM Oolitic IN
WMYK(FM) Peru IN
WMYL(FM) Halls Crossroads TN
WMYP(FM) Frederiksted VI
WMYU(FM) Karns TN
WMYX(FM) Milwaukee WI
WMYY(FM) Schoharie NY
*WMYZ(FM) Clermont FL
WMZK(FM) Merrill WI
WMZQ-FM Washington DC
*WNAA(FM) Greensboro NC
*WNAN(FM) Nantucket MA
*WNAS(FM) New Albany IN
WNAX-FM Yankton SD
*WNAZ-FM Nashville TN
WNBB(FM) Bayboro NC
WNBQ(FM) Mansfield PA
WNBR-FM Windsor NC
WNBT-FM Wellsboro PA
WNBY-FM Newberry MI
WNCB(FM) Gardendale AL
WNCC-FM Franklin NC
WNCD(FM) Youngstown OH
*WNCH(FM) Norwich VT
WNCI(FM) Columbus OH
*WNCK(FM) Nantucket MA
WNCL(FM) Milford DE
WNCO-FM Ashland OH
WNCQ-FM Canton NY
WNCS(FM) Montpelier VT
WNCT-FM Greenville NC
*WNCU(FM) Durham NC
WNCV(FM) Shalimar FL
*WNCW(FM) Spindale NC
WNCX(FM) Cleveland OH
WNCY-FM Neenah-Menasha WI
WNDD(FM) Silver Springs FL
WNDH(FM) Napoleon OH
WNDI-FM Sullivan IN
WNDN(FM) Chiefland FL
WNDT(FM) Alachua FL
WNDV-FM South Bend IN
*WNDY(FM) Crawfordsville IN
*WNEC-FM Henniker NH
*WNED-FM Buffalo NY
*WNEE(FM) Jasper GA
*WNEF(FM) Newburyport MA
*WNEK-FM Springfield MA
WNEV(FM) Friar's Point MS
WNEW(FM) Jupiter FL
*WNFA(FM) Port Huron MI
*WNFB(FM) Lake City FL
WNFK(FM) Perry FL
WNFM(FM) Reedsburg WI
WNFN(FM) Belle Meade TN
*WNFR(FM) Sandusky MI
WNFZ(FM) Oak Ridge TN
WNGC(FM) Toccoa GA
WNGE(FM) Negaunee MI
*WNGN(FM) Argyle NY
*WNGU(FM) Dahlonega GA
WNGZ(FM) Montour Falls NY
WNHT(FM) Churubusco IN
*WNHU(FM) West Haven CT
WNHW(FM) Belmont NH
WNIC(FM) Dearborn MI
*WNIE(FM) Freeport IL
*WNIJ(FM) De Kalb IL
WNIK-FM Arecibo PR
*WNIN-FM Evansville IN
*WNIQ(FM) Sterling IL
WNIR(FM) Kent OH
*WNIU(FM) Rockford IL
*WNIW(FM) La Salle IL
*WNJA(FM) Jamestown NY
*WNJB(FM) Bridgeton NJ
*WNJM(FM) Manahawkin NJ
*WNJN-FM Atlantic City NJ
*WNJP(FM) Sussex NJ
*WNJR(FM) Washington PA
*WNJS-FM Berlin NJ
*WNJT-FM Trenton NJ
*WNJZ(FM) Cape May Court House NJ
WNKI(FM) Corning NY
*WNKJ(FM) Hopkinsville KY
WNKK(FM) Circleville OH
WNKL(FM) Wauseon OH
WNKO(FM) Newark OH
WNKR(FM) Williamstown KY
WNKS(FM) Charlotte NC
WNKT(FM) Saint George SC
*WNKU(FM) Highland Heights KY
*WNKV(FM) Norco LA
WNKX-FM Centerville TN
WNLA-FM Indianola MS
WNLC(FM) East Lyme CT
WNLF(FM) Macomb IL
WNLT(FM) Harrison OH
*WNMC-FM Traverse City MI
*WNMH(FM) Northfield MA
WNML-FM Loudon TN
*WNMU-FM Marquette MI
WNMX-FM Waxhaw NC
WNNH(FM) Henniker NH
WNNJ-FM Newton NJ
WNNK-FM Harrisburg PA
WNNL(FM) Fuquay-Varina NC
WNNO-FM Wisconsin Dells WI
WNNS(FM) Springfield IL
WNNT-FM Warsaw VA
*WNNV(FM) San German PR
WNNX(FM) Atlanta GA
WNOD(FM) Mayaguez PR
WNOE-FM New Orleans LA
WNOI(FM) Flora IL
WNOK(FM) Columbia SC
WNOR(FM) Norfolk VA
WNOU(FM) Indianapolis IN
WNOW-FM Gaffney SC
WNOX(FM) Oak Ridge TN
WNPC-FM Newport TN
WNPQ(FM) New Philadelphia OH
*WNPR(FM) Norwich CT
WNPT-FM Linden AL
WNRJ(FM) Poquoson VA
*WNRK(FM) Norwalk OH
*WNRN(FM) Charlottesville VA
*WNRQ(FM) Nashville TN
*WNRS-FM Sweet Briar VA
WNRT(FM) Manati PR
WNRX(FM) Jefferson City TN
*WNRZ(FM) Dickson TN
*WNSB(FM) Norfolk VA
*WNSC-FM Rock Hill SC
WNSI-FM Atmore AL
WNSL(FM) Laurel MS
WNSN(FM) South Bend IN
WNSP(FM) Bay Minette AL
WNSV(FM) Nashville IL
WNSX(FM) Winter Harbor ME
WNSY(FM) Talking Rock GA
WNTB(FM) Wrightsville Beach NC
WNTC(FM) Drakesboro KY
*WNTE(FM) Mansfield PA
*WNTH(FM) Winnetka IL
*WNTI(FM) Hackettstown NJ
WNTK-FM New London NH
WNTO(FM) Racine OH
WNTQ(FM) Syracuse NY
WNTR(FM) Indianapolis IN
WNUA(FM) Chicago IL
*WNUB-FM Northfield VT
WNUE-FM Titusville FL
WNUQ(FM) Sylvester GA
*WNUR-FM Evanston IL
WNUS(FM) Belpre OH
WNUY(FM) Bluffton IN
WNVA-FM Norton VA
WNVZ(FM) Norfolk VA
*WNWC-FM Madison WI
WNWN-FM Coldwater MI
WNWS-FM Jackson TN
WNWV(FM) Elyria OH
WNXR(FM) Iron River WI
WNXT-FM Portsmouth OH
WNXX(FM) Jackson LA
*WNYC-FM New York NY
*WNYE(FM) New York NY
*WNYK(FM) Nyack NY
WNYN-FM Athol MA
*WNYO(FM) Oswego NY
WNYQ(FM) Hudson Falls NY
WNYR-FM Waterloo NY
*WNYU-FM New York NY
WNYV(FM) Whitehall NY
*WNZN(FM) Lorain OH
*WNZR(FM) Mount Vernon OH
WOAB(FM) Ozark AL
WOAD-FM Pickens MS
WOAH(FM) Glennville GA
*WOAK(FM) La Grange GA
*WOAR(FM) South Vienna OH
*WOAS(FM) Ontonagon MI
WOBB(FM) Tifton GA
*WOBC-FM Oberlin OH
WOBE(FM) Crystal Falls MI
WOBG-FM Salem WV
WOBM-FM Toms River NJ
*WOBN(FM) Westerville OH
*WOBO(FM) Batavia OH
WOBR-FM Wanchese NC
WOBX-FM Manteo NC
WOCE(FM) Ringgold GA
*WOCG(FM) Huntsville AL
WOCL(FM) De Land FL
WOCM(FM) Selbyville DE
WOCN-FM Orleans MA
WOCO-FM Oconto WI
WOCQ(FM) Berlin MD
*WOCR(FM) Olivet MI
*WOCS(FM) Lerose KY
WOCY(FM) Carrabelle FL
WODA(FM) Bayamon PR
WODB(FM) Delaware OH
WODE-FM Easton PA
WODR(FM) Fair Bluff NC
WODS(FM) Boston MA
WODZ-FM Rome NY
*WOEL-FM Elkton MD
*WOES(FM) Ovid-Elsie MI
*WOEZ(FM) Maynardville TN
WOFM(FM) Mosinee WI
*WOFN(FM) Beach City OH
*WOFR(FM) Schoolcraft MI
WOFX-FM Cincinnati OH
WOGB(FM) Kaukauna WI
WOGF(FM) East Liverpool OH
WOGG(FM) Oliver PA
WOGH(FM) Burgettstown PA
WOGI(FM) Charleroi PA
WOGK(FM) Ocala FL
WOGL(FM) Philadelphia PA
*WOGR-FM Salisbury NC
WOGT(FM) East Ridge TN
WOGY(FM) Jackson TN
*WOHC(FM) Chillicothe OH
WOHF(FM) Bellevue OH
*WOHP(FM) Portsmouth OH
WOHT(FM) Grenada MS
*WOI-FM Ames IA
*WOJB(FM) Reserve WI
*WOJC(FM) Crothersville IN
WOJG(FM) Bolivar TN
WOJL(FM) Louisa VA
WOJO(FM) Evanston IL
WOKA-FM Douglas GA
*WOKD-FM Danville VA
WOKE(FM) Garrison KY
*WOKG(FM) Galax VA
WOKI(FM) Oliver Springs TN
WOKK(FM) Meridian MS
*WOKL(FM) Troy OH
WOKN(FM) Southport NY
WOKO(FM) Burlington VT
WOKQ(FM) Dover NH
WOKR(FM) Remsen NY
WOKV-FM Ponte Vedra Beach FL
WOKW(FM) Curwensville PA
WOKZ(FM) Fairfield IL
WOLC(FM) Princess Anne MD
WOLD-FM Marion VA
WOLF-FM Oswego NY
*WOLG(FM) Carlinville IL
WOLI-FM Easley SC
WOLL(FM) Hobe Sound FL
*WOLN(FM) Olean NY
*WOLR(FM) Lake City FL
WOLT(FM) Greer SC
WOLV(FM) Houghton MI
*WOLW(FM) Cadillac MI
WOLX-FM Baraboo WI
WOLZ(FM) Fort Myers FL
WOMC(FM) Detroit MI
WOMG(FM) Columbia SC
*WOMR(FM) Provincetown MA
WOMX-FM Orlando FL
WONA-FM Winona MS
*WONB(FM) Ada OH
*WONC(FM) Naperville IL
WONE-FM Akron OH
*WONU(FM) Kankakee IL
*WONY(FM) Oneonta NY
WOOD-FM Grand Rapids MI
WOOF-FM Dothan AL
WOOZ-FM Harrisburg IL
WORC-FM Webster MA
WORD-FM Pittsburgh PA
WORG(FM) Elloree SC
*WORI(FM) Delhi Hills OH
WORK(FM) Barre VT
WORM-FM Savannah TN
WORO(FM) Corozal PR
*WORQ(FM) Green Bay WI
*WORT(FM) Madison WI
*WORW(FM) Port Huron MI
WORX-FM Madison IN
*WOSB(FM) Marion OH
WOSC(FM) Bethany Beach DE
*WOSE(FM) Coshocton OH
WOSM(FM) Ocean Springs MS
WOSN(FM) Indian River Shores FL
*WOSP(FM) Portsmouth OH
WOSQ(FM) Spencer WI
*WOSR(FM) Middletown NY
*WOSS(FM) Ossining NY
*WOSU-FM Columbus OH
*WOSV(FM) Mansfield OH
*WOTC(FM) Edinburg VA
*WOTJ(FM) Morehead City NC
*WOTL(FM) Toledo OH
WOTR(FM) Lost Creek WV
WOTT(FM) Henderson NY
*WOTW(FM) Monee IL
*WOUB-FM Athens OH
*WOUC-FM Cambridge OH
WOUF(FM) Beulah MI
*WOUH-FM Chillicothe OH
*WOUL-FM Ironton OH
WOUR(FM) Utica NY
*WOUZ(FM) Zanesville OH
*WOVI(FM) Novi MI
WOVK(FM) Wheeling WV
*WOVM(FM) Appleton WI
WOVO(FM) Glasgow KY
WOWE(FM) Vassar MI
WOWF(FM) Crossville TN
WOWI(FM) Norfolk VA
*WOWL(FM) Burnsville MS
WOWN(FM) Shawano WI
WOWQ(FM) DuBois PA
WOWY(FM) University Park PA
WOXD(FM) Oxford MS
WOXL-FM Biltmore Forest NC
WOXO-FM Norway ME
WOXX(FM) Franklin PA
WOXY(FM) Oxford OH
WOYE(FM) Rio Grande PR
WOYS(FM) Apalachicola FL
WOZI(FM) Presque Isle ME
*WOZQ(FM) Northampton MA
WOZW(FM) Goshen IN
WOZZ(FM) New London WI
WPAC(FM) Ogdensburg NY
*WPAE(FM) Centreville MS
WPAL-FM Ridgeville SC
WPAP-FM Panama City FL
*WPAR(FM) Salem VA
*WPAS(FM) Pascagoula MS
WPAT-FM Paterson NJ
WPAW(FM) Winston-Salem NC
WPAY-FM Portsmouth OH
WPBG(FM) Peoria IL
WPBH(FM) Port St. Joe FL
WPBX(FM) Crossville TN
WPBZ(FM) Indiantown FL
*WPCD(FM) Champaign IL
WPCH(FM) Gray GA
*WPCJ(FM) Pittsford MI
WPCK(FM) Denmark WI
WPCL(FM) Northern Cambria PA
*WPCN(FM) Point Pleasant WV
*WPCR-FM Plymouth NH
*WPCS(FM) Pensacola FL
WPCV(FM) Winter Haven FL
WPDA(FM) Jeffersonville NY
WPDH(FM) Poughkeepsie NY
WPDT(FM) Johnsonville SC
WPDX-FM Clarksburg WV
*WPEA(FM) Exeter NH
*WPEB(FM) Philadelphia PA
WPEG(FM) Concord NC
WPEH-FM Louisville GA
*WPEL-FM Montrose PA
*WPER(FM) Culpeper VA
WPEZ(FM) Jeffersonville GA
WPFB-FM Middletown OH
*WPFF(FM) Sturgeon Bay WI
WPFL(FM) Century FL
WPFM-FM Panama City FL
*WPFR-FM Clinton IN
*WPFW(FM) Washington DC
WPFX-FM North Baltimore OH
WPGA-FM Perry GA
WPGB(FM) Pittsburgh PA
WPGC-FM Morningside MD
WPGI(FM) Horseheads NY
*WPGL(FM) Pattersonville NY
*WPGM-FM Danville PA
*WPGT(FM) Roanoke Rapids NC
WPGU(FM) Urbana IL
WPGW-FM Portland IN
WPHD(FM) South Waverly PA
WPHH(FM) Waterbury CT
WPHI-FM Media PA
WPHK(FM) Blountstown FL
*WPHN(FM) Gaylord MI
*WPHP(FM) Wheeling WV
WPHR-FM Auburn NY
*WPHS(FM) Warren MI
WPHX-FM Sanford ME
WPHZ(FM) Mitchell IN
WPIA(FM) Eureka IL
*WPIB(FM) Bluefield WV
WPIG(FM) Olean NY
WPIK(FM) Summerland Key FL
*WPIL(FM) Heflin AL
*WPIM(FM) Martinsville VA
*WPIN-FM Dublin VA
*WPIO(FM) Titusville FL
WPIQ(FM) Manistique MI
*WPIR(FM) Hickory NC
*WPJC(FM) Pontiac IL
WPJP(FM) Port Washington WI
*WPJW(FM) Hurricane WV
*WPJY(FM) Blennerhassett WV
WPKE-FM Coal Run KY
WPKF(FM) Poughkeepsie NY
WPKG(FM) Neillsville WI
WPKL(FM) Uniontown PA
*WPKM(FM) Montauk NY
*WPKN(FM) Bridgeport CT
WPKO-FM Bellefontaine OH
WPKQ(FM) North Conway NH
WPKR(FM) Omro WI
*WPKT(FM) Meriden CT
*WPKV(FM) Nanty Glo PA
WPKX(FM) Enfield CT
WPLA(FM) Jacksonville FL
*WPLH(FM) Tifton GA
*WPLI(FM) Levittown PR
WPLJ(FM) New York NY
WPLM(FM) Plymouth MA
*WPLN-FM Nashville TN
WPLR(FM) New Haven CT
WPLT(FM) Spooner WI

WPLY-FM Walpole NH
WPMA(FM) Buckhead GA
WPMJ(FM) Chillicothe IL
WPMW(FM) Mullens WV
WPMX(FM) Statesboro GA
WPNC-FM Plymouth NC
*WPNE-FM Green Bay WI
WPNG(FM) Pearson GA
WPNH-FM Plymouth NH
*WPNR-FM Utica NY
WPNS(FM) Brodhead KY
*WPOB-FM Plainview NY
WPOC(FM) Baltimore MD
WPOI(FM) Saint Petersburg FL
WPOR(FM) Portland ME
WPOS-FM Holland OH
WPOW(FM) Miami FL
*WPOZ(FM) Union Park FL
WPPL(FM) Blue Ridge GA
WPPN(FM) Des Plaines IL
*WPPR(FM) Demorest GA
WPPT(FM) Mercersburg PA
WPPZ-FM Jenkintown PA
*WPQZ(FM) Muskegon MI
WPRB(FM) Princeton NJ
*WPRC(FM) Princeton IL
WPRF(FM) Reserve LA
*WPRG(FM) Columbia MS
WPRH(FM) Paris TN
*WPRJ(FM) Coleman MI
*WPRK(FM) Winter Park FL
*WPRL(FM) Lorman MS
WPRM-FM San Juan PR
WPRN-FM Lisman AL
WPRO-FM Providence RI
WPRS-FM Waldorf MD
WPRW-FM Martinez GA
*WPSA(FM) Paul Smiths NY
*WPSC-FM Wayne NJ
*WPSF(FM) Clewiston FL
WPSK-FM Pulaski VA
*WPSM(FM) Fort Walton Beach FL
*WPSR(FM) Evansville IN
WPST(FM) Trenton NJ
*WPSU(FM) State College PA
*WPSX(FM) Kane PA
*WPTC(FM) Williamsport PA
WPTE(FM) Virginia Beach VA
*WPTH(FM) Olney IL
WPTI(FM) Louisville KY
*WPTJ(FM) Paris KY
WPTM(FM) Roanoke Rapids NC
WPTQ(FM) Cave City KY
WPTR(FM) Clifton Park NY
*WPTS-FM Pittsburgh PA
WPUB-FM Camden SC
*WPUC-FM Ponce PR
*WPUM(FM) Rensselaer IN
WPUP(FM) Royston GA
WPUR(FM) Atlantic City NJ
*WPVA(FM) Waynesboro VA
WPVL-FM Platteville WI
WPVQ(FM) Greenfield MA
*WPWB(FM) Byron GA
WPWQ(FM) Mount Sterling IL
*WPWV(FM) Princeton WV
WPWX(FM) Hammond IN
WPWZ(FM) Pinetops NC
WPXC(FM) Hyannis MA
WPXN(FM) Paxton IL
WPXY-FM Rochester NY
WPXZ-FM Punxsutawney PA
WPYA(FM) Chesapeake VA
WPYO(FM) Maitland FL
WPYX(FM) Albany NY
WPZE(FM) Fayetteville GA
WPZS(FM) Albemarle NC
WPZX(FM) Pocono Pines PA
WPZZ(FM) Crewe VA
*WQAB(FM) Philippi WV
*WQAC-FM Alma MI
WQAH-FM Addison AL
WQAK(FM) Union City TN
WQAL(FM) Cleveland OH
*WQAQ(FM) Hamden CT
WQAR(FM) Stillwater NY
WQBE-FM Charleston WV
WQBJ(FM) Cobleskill NY
WQBK-FM Rensselaer NY
WQBR(FM) Avis PA
WQBT(FM) Savannah GA
WQBU-FM Garden City NY
WQBW(FM) Milwaukee WI
WQBX(FM) Alma MI
WQBZ(FM) Fort Valley GA
WQCB(FM) Brewer ME
WQCC(FM) La Crosse WI
WQCD(FM) New York NY
WQCM(FM) Greencastle PA
*WQCS(FM) Fort Pierce FL
WQCY(FM) Quincy IL
WQDK(FM) Ahoskie NC
WQDR(FM) Raleigh NC
WQDY-FM Calais ME
*WQED-FM Pittsburgh PA
*WQEJ(FM) Johnstown PA
WQEL(FM) Bucyrus OH
WQEM(FM) Columbiana AL
WQEN(FM) Trussville AL
WQFL(FM) Rockford IL
WQFM(FM) Nanticoke PA
WQFN(FM) Forest City PA
*WQFS(FM) Greensboro NC
WQFX-FM Russell PA
WQGN-FM Groton CT
WQHH(FM) Dewitt MI
WQHK-FM Decatur IN
WQHL-FM Live Oak FL
WQHQ(FM) Ocean City-Salisbury MD
WQHR(FM) Presque Isle ME
WQHT(FM) New York NY
WQHY(FM) Prestonsburg KY
WQHZ(FM) Erie PA
WQIC(FM) Lebanon PA
WQIK-FM Jacksonville FL
WQIL(FM) Chauncey GA
WQIO(FM) Mount Vernon OH
WQJB(FM) State College MS
WQJK(FM) Maryville TN
WQJQ(FM) Kosciusko MS
*WQJU(FM) Mifflintown PA
WQJZ(FM) Ocean Pines MD
WQKC(FM) Seymour IN
*WQKE(FM) Plattsburgh NY
WQKI-FM Orangeburg SC
WQKL(FM) Ann Arbor MI
*WQKO(FM) Howe IN
WQKQ(FM) Carthage IL
WQKS-FM Montgomery AL
WQKT(FM) Wooster OH
*WQKV(FM) Rochester IN
WQKX(FM) Sunbury PA
WQKY(FM) Emporium PA
WQKZ(FM) Ferdinand IN
WQLA-FM La Follette TN
WQLB(FM) Tawas City MI
WQLC(FM) Watertown FL
WQLF(FM) Lena IL
WQLH(FM) Green Bay WI
WQLI(FM) Meigs GA
WQLJ(FM) Oxford MS
WQLK(FM) Richmond IN
*WQLN-FM Erie PA
WQLT-FM Florence AL
WQLV(FM) Millersburg PA
WQLZ(FM) Taylorville IL
WQME(FM) Anderson IN
WQMF(FM) Jeffersonville IN
WQMG-FM Greensboro NC
WQMJ(FM) Forsyth GA
WQMR(FM) Snow Hill MD
WQMT(FM) Chatsworth GA
WQMU(FM) Indiana PA
WQMX(FM) Medina OH
WQMZ(FM) Charlottesville VA
*WQNA(FM) Springfield IL
WQNC(FM) Harrisburg NC
WQNQ(FM) Fletcher NC
WQNR(FM) Tallassee AL
WQNS(FM) Waynesville NC
WQNY(FM) Ithaca NY
WQNZ(FM) Natchez MS
WQOK(FM) South Boston VA
WQOL(FM) Vero Beach FL
WQON(FM) Roscommon MI
*WQOX(FM) Memphis TN
WQPC(FM) Prairie du Chien WI
WQPO(FM) Harrisonburg VA
*WQPR(FM) Muscle Shoals AL
WQPW(FM) Valdosta GA
WQQB(FM) Rantoul IL
WQQK(FM) Hendersonville TN
WQQL(FM) Springfield IL
WQQQ(FM) Sharon CT
WQQR(FM) Mayfield KY
WQRB(FM) Bloomer WI
WQRC(FM) Barnstable MA
*WQRI(FM) Bristol RI
WQRK(FM) Bedford IN
WQRL(FM) Benton IL
WQRM(FM) Smethport PA
*WQRP(FM) Dayton OH
WQRS(FM) Salamanca NY
WQRV(FM) Tuscumbia AL
WQRW(FM) Wellsville NY
WQSA(FM) Unadilla GA
WQSB(FM) Albertville AL
WQSD(FM) Briarcliff Acres SC
*WQSG(FM) Lafayette IN
WQSI(FM) Union Springs AL
WQSL(FM) Jacksonville NC
WQSM(FM) Fayetteville NC
WQSO(FM) Rochester NH
WQSR(FM) Baltimore MD
WQSS(FM) Camden ME
WQST-FM Forest MS
*WQSU(FM) Selinsgrove PA
WQTC-FM Manitowoc WI
WQTE(FM) Adrian MI
*WQTQ(FM) Hartford CT
WQTU(FM) Rome GA
WQTX(FM) Saint Johns MI
WQTY(FM) Linton IN
WQUA(FM) Citronelle AL
*WQUB(FM) Quincy IL
WQUE-FM New Orleans LA
WQUL(FM) West Frankfort IL
WQUS(FM) Lapeer MI
WQUT(FM) Johnson City TN
WQVE(FM) Albany GA
*WQVI(FM) Forest MS
WQWK(FM) State College PA
WQWV(FM) Fisher WV
WQXA-FM York PA
WQXB(FM) Grenada MS
WQXC-FM Otsego MI
WQXE(FM) Elizabethtown KY
WQXK(FM) Salem OH
WQXQ(FM) Central City KY
WQXR-FM New York NY
WQXZ(FM) Pinehurst GA
WQYK-FM Saint Petersburg FL
WQYX(FM) Clearfield PA
WQYZ(FM) Ocean Springs MS
WQZK-FM Keyser WV
WQZL(FM) Belhaven NC
WQZS(FM) Meyersdale PA
WQZX(FM) Greenville AL
WQZY(FM) Dublin GA
WQZZ(FM) Eutaw AL
WRAC(FM) West Union OH
*WRAE(FM) Raeford NC
*WRAF-FM Toccoa Falls GA
WRAK-FM Bainbridge GA
WRAL(FM) Raleigh NC
WRAN(FM) Tower Hill IL
WRAR-FM Tappahannock VA
*WRAS(FM) Atlanta GA
WRAT(FM) Point Pleasant NJ
*WRAU(FM) Ocean City MD
WRAY-FM Princeton IN
WRBA(FM) Springfield FL
*WRBB(FM) Boston MA
*WRBC(FM) Lewiston ME
WRBE-FM Lucedale MS
*WRBH(FM) New Orleans LA
WRBI(FM) Batesville IN
WRBJ-FM Brandon MS
*WRBK(FM) Richburg SC
WRBN(FM) Clayton GA
WRBO(FM) Como MS
WRBP(FM) Hubbard OH
WRBQ-FM Tampa FL
WRBR-FM South Bend IN
WRBS-FM Baltimore MD
WRBT(FM) Harrisburg PA
WRBV(FM) Warner Robins GA
WRBX(FM) Reidsville GA
WRCC(FM) Marshall MI
WRCD(FM) Canton NY
WRCH(FM) New Britain CT
WRCI(FM) Webster NY
*WRCJ-FM Detroit MI
WRCK(FM) Utica NY
WRCL(FM) Frankenmuth MI
*WRCM(FM) Wingate NC
WRCN-FM Riverhead NY
WRCO-FM Richland Center WI
WRCQ(FM) Dunn NC
*WRCT(FM) Pittsburgh PA
*WRCU-FM Hamilton NY
WRCV(FM) Dixon IL
*WRDL(FM) Ashland OH
WRDO(FM) Fitzgerald GA
*WRDR(FM) Freehold Township NJ
WRDU(FM) Wilson NC
*WRDV(FM) Warminster PA
WRDW-FM Philadelphia PA
WRDX(FM) Dover DE
WRDZ-FM Plainfield IN
WREB(FM) Greencastle IN
WRED(FM) Saco ME
*WREH(FM) Cypress Quarters FL
*WREK(FM) Atlanta GA
WREO-FM Ashtabula OH
WREQ(FM) Ridgebury PA
WREZ(FM) Metropolis IL
*WRFE(FM) Chesterfield SC
WRFF(FM) Philadelphia PA
*WRFG(FM) Atlanta GA
*WRFL(FM) Lexington KY
*WRFM(FM) Wadesville IN
WRFQ(FM) Mt. Pleasant SC
*WRFT(FM) Indianapolis IN
*WRFW(FM) River Falls WI
WRFX-FM Kannapolis NC
WRFY-FM Reading PA
*WRGF(FM) Greenfield IN
*WRGN(FM) Sweet Valley PA
WRGO(FM) Cedar Key FL
*WRGP(FM) Homestead FL
WRGR(FM) Tupper Lake NY
*WRGX(FM) Sturgeon Bay WI
WRGZ(FM) Rogers City MI
WRHD(FM) Williamston NC
WRHK(FM) Danville IL
WRHL-FM Rochelle IL
WRHM(FM) Lancaster SC
WRHN(FM) Rhinelander WI
*WRHO(FM) Oneonta NY
WRHQ(FM) Richmond Hill GA
WRHT(FM) Morehead City NC
*WRHU(FM) Hempstead NY
*WRHV(FM) Poughkeepsie NY
WRHY(FM) Centre AL
WRIC-FM Richlands VA
WRIF(FM) Detroit MI
*WRIH(FM) Richmond VA
*WRIJ(FM) Masontown PA
WRIK-FM Metropolis IL
WRIL(FM) Pineville KY
WRIO(FM) Ponce PR
WRIP(FM) Windham NY
WRIT-FM Milwaukee WI
*WRIU(FM) Kingston RI
WRIX-FM Honea Path SC
*WRJA-FM Sumter SC
WRJB(FM) Camden TN
WRJC-FM Mauston WI
*WRJI(FM) East Greenwich RI
WRJK(FM) Norris TN
WRJL-FM Eva AL
WRJM-FM Geneva AL
WRJO(FM) Eagle River WI
WRJT(FM) Royalton VT
WRJY(FM) Brunswick GA
WRKA(FM) Saint Matthews KY
*WRKC(FM) Wilkes-Barre PA
*WRKF(FM) Baton Rouge LA
WRKG(FM) Drew MS
WRKH(FM) Mobile AL
WRKI(FM) Brookfield CT
WRKK-FM Sparta TN
WRKN(FM) Niceville FL
WRKP(FM) Moundsville WV
WRKR(FM) Portage MI
WRKS(FM) New York NY
WRKT(FM) North East PA
WRKU(FM) Forestville WI
WRKW(FM) Ebensburg PA
WRKX(FM) Ottawa IL
WRKY-FM Hollidaysburg PA
WRLB(FM) Rainelle WV
*WRLC(FM) Williamsport PA
WRLD-FM Valley AL
WRLF(FM) Fairmont WV
*WRLI-FM Southampton NY
WRLO-FM Antigo WI
WRLS-FM Hayward WI
WRLT(FM) Franklin TN
WRLU(FM) Algoma WI
WRLV-FM Salyersville KY
WRLX(FM) West Palm Beach FL
WRMA(FM) Fort Lauderdale FL
*WRMB(FM) Boynton Beach FL
*WRMC-FM Middlebury VT
WRMF(FM) Palm Beach FL
WRMJ(FM) Aledo IL
WRML(FM) Pageland SC
WRMM-FM Rochester NY
WRMO(FM) Milbridge ME
WRMS-FM Beardstown IL
*WRMU(FM) Alliance OH
WRNB(FM) Pennsauken NJ
*WRNF(FM) Selma AL
WRNH(FM) Groveton NH
WRNI-FM Narragansett Pier RI
WRNN-FM Socastee SC
WRNO-FM New Orleans LA
WRNQ(FM) Poughkeepsie NY
WRNR-FM Grasonville MD
WRNS-FM Kinston NC
WRNX(FM) Amherst MA
WRNZ(FM) Lancaster KY
WROE(FM) Neenah-Menasha WI
WROG(FM) Cumberland MD
WROI(FM) Rochester IN
WRON-FM Ronceverte WV
WROO(FM) Callahan FL
WROQ(FM) Anderson SC
WROR-FM Framingham MA
WROU-FM West Carrollton OH
WROV-FM Martinsville VA
WROX-FM Exmore VA
WROZ(FM) Lancaster PA
WRPG(FM) Hawkinsville GA
*WRPI(FM) Troy NY
*WRPJ(FM) Port Jervis NY
*WRPN-FM Ripon WI
*WRPR(FM) Mahwah NJ
*WRPS(FM) Rockland MA
WRPW(FM) Colfax IL
WRQK(FM) Canton OH
*WRQM(FM) Rocky Mount NC
WRQN(FM) Bowling Green OH
WRQO(FM) Monticello MS
WRQQ(FM) Goodlettsville TN
WRQR(FM) Wilmington NC
WRQT(FM) La Crosse WI
WRQX(FM) Washington DC
WRR(FM) Dallas TX
WRRB(FM) Arlington NY
*WRRC(FM) Lawrenceville NJ
*WRRG(FM) River Grove IL
WRRH(FM) Hormigueros PR
*WRRI(FM) Brownsville TN
WRRK(FM) Braddock PA
WRRM(FM) Cincinnati OH
WRRN(FM) Warren PA
WRRQ(FM) Windsor NY

WRRR-FM Saint Marys WV
WRRV(FM) Middletown NY
WRRX(FM) Gulf Breeze FL
WRSA-FM Decatur AL
*WRSD(FM) Folsom PA
*WRSE(FM) Elmhurst IL
WRSF(FM) Columbia NC
*WRSG(FM) Middlebourne WV
*WRSH(FM) Rockingham NC
WRSI(FM) Turners Falls MA
WRSR(FM) Owosso MI
*WRST-FM Oshkosh WI
*WRSU-FM New Brunswick NJ
WRSV(FM) Rocky Mount NC
WRSW-FM Warsaw IN
WRSY(FM) Marlboro VT
WRTB(FM) Winnebago IL
*WRTC-FM Hartford CT
*WRTE(FM) Chicago IL
*WRTI(FM) Philadelphia PA
*WRTJ(FM) Coatesville PA
*WRTL(FM) Ephrata PA
WRTM-FM Port Gibson MS
WRTO-FM Goulds FL
*WRTP(FM) Roanoke Rapids NC
*WRTQ(FM) Ocean City NJ
WRTR(FM) Brookwood AL
WRTS(FM) Erie PA
WRTT-FM Huntsville AL
*WRTU(FM) San Juan PR
*WRTX(FM) Dover DE
*WRTY(FM) Jackson Township PA
*WRUC(FM) Schenectady NY
WRUF-FM Gainesville FL
WRUL(FM) Carmi IL
WRUM(FM) Orlando FL
*WRUO(FM) Mayaguez PR
WRUP(FM) Munising MI
*WRUR-FM Rochester NY
*WRUV(FM) Burlington VT
*WRUW-FM Cleveland OH
WRVA-FM Rocky Mount NC
WRVB(FM) Marietta OH
WRVC-FM Catlettsburg KY
*WRVD(FM) Syracuse NY
WRVE(FM) Schenectady NY
WRVF(FM) Toledo OH
*WRVG(FM) Georgetown KY
WRVH(FM) Williamsport PA
WRVI(FM) Valley Station KY
*WRVJ(FM) Watertown NY
*WRVL(FM) Lynchburg VA
*WRVM(FM) Suring WI
*WRVN(FM) Utica NY
*WRVO(FM) Oswego NY
WRVQ(FM) Richmond VA
WRVR(FM) Memphis TN
*WRVS-FM Elizabeth City NC
*WRVT(FM) Rutland VT
*WRVU(FM) Nashville TN
WRVV(FM) Harrisburg PA
WRVW(FM) Lebanon TN
WRVX(FM) Eufaula AL
WRVY-FM Henry IL
WRVZ(FM) Pocatalico WV
*WRWA(FM) Dothan AL
WRWC(FM) Ellenville NY
WRWD-FM Highland NY
*WRWJ(FM) Murrysville PA
WRWK(FM) Delta OH
*WRXC(FM) Shelton CT
WRXK-FM Bonita Springs FL
WRXL(FM) Richmond VA
WRXQ(FM) Coal City IL
WRXR-FM Rossville GA
WRXS(FM) Ocean City MD
*WRXT(FM) Roanoke VA
*WRXV(FM) State College PA
WRXW(FM) Pearl MS
WRXX(FM) Centralia IL
*WRYN(FM) Hickory NC
*WRYP(FM) Wellfleet MA
WRYV(FM) Gallipolis OH
WRZA(FM) Park Forest IL
WRZE(FM) Nantucket MA
WRZI(FM) Vine Grove KY
WRZK(FM) Colonial Heights TN
WRZQ-FM Greensburg IN
WRZR(FM) Loogootee IN
WRZX(FM) Indianapolis IN
WRZZ(FM) Elizabeth WV
WSAA(FM) Benton TN
*WSAE(FM) Spring Arbor MI
WSAG(FM) Linwood MI
*WSAJ-FM Grove City PA
WSAK(FM) Hampton NH
WSAQ(FM) Port Huron MI
*WSBF-FM Clemson SC
WSB-FM Atlanta GA
WSBG(FM) Stroudsburg PA
WSBH(FM) Satellite Beach FL
*WSBU(FM) Saint Bonaventure NY
WSBW(FM) Sister Bay WI
WSBY-FM Salisbury MD
WSBZ(FM) Miramar Beach FL
*WSCB(FM) Springfield MA
WSCC-FM Goose Creek SC
*WSCD-FM Duluth MN
*WSCF-FM Vero Beach FL
WSCG-FM Lakeview MI
WSCH(FM) Aurora IN
*WSCI(FM) Charleston SC
*WSCL(FM) Salisbury MD
*WSCN(FM) Cloquet MN
*WSCS(FM) New London NH
*WSCT(FM) Springfield IL
WSCY(FM) Moultonborough NH
*WSDH(FM) Sandwich MA
*WSDL(FM) Ocean City MD
WSDM-FM Brazil IN
*WSDP(FM) Plymouth MI
WSEA(FM) Atlantic Beach SC
*WSEB(FM) Englewood FL
WSEI(FM) Olney IL
WSEK(FM) Burnside KY
WSEL-FM Pontotoc MS
WSEN-FM Baldwinsville NY
WSEO(FM) Nelsonville OH
WSEV-FM Gatlinburg TN
*WSEW(FM) Sanford ME
WSEY(FM) Oregon IL
WSFL-FM New Bern NC
WSFM(FM) Oak Island NC
WSFQ(FM) Peshtigo WI
WSFR(FM) Corydon IN
*WSFX(FM) Nanticoke PA
WSGA(FM) Hinesville GA
WSGC-FM Elberton GA
WSGD(FM) Arnoldsburg WV
*WSGE(FM) Dallas NC
*WSGG(FM) Norfolk CT
WSGL(FM) Naples FL
WSGM(FM) Coalmont TN
*WSGN(FM) Gadsden AL
*WSGP(FM) Glasgow KY
*WSGR-FM Port Huron MI
WSGS(FM) Hazard KY
WSGW-FM Carrollton MI
WSGY(FM) Pleasant Gap PA
*WSHA(FM) Raleigh NC
*WSHC(FM) Shepherdstown WV
*WSHD(FM) Eastport ME
WSHH(FM) Pittsburgh PA
*WSHJ(FM) Southfield MI
WSHK(FM) Kittery ME
*WSHL-FM Easton MA
WSHP(FM) Attica IN
*WSHR(FM) Lake Ronkonkoma NY
*WSHS(FM) Sheboygan WI
*WSHU(FM) Fairfield CT
WSHW(FM) Frankfort IN
WSHZ(FM) Muskegon MI
*WSIA(FM) Staten Island NY
WSIB(FM) Selmer TN
*WSIE(FM) Edwardsville IL
*WSIF(FM) Wilkesboro NC
WSIG(FM) Mount Jackson VA
WSIM(FM) Lamar SC
WSIP-FM Paintsville KY
*WSIU(FM) Carbondale IL
WSIX-FM Nashville TN
*WSJB-FM Standish ME
WSJD(FM) Princeton IN
WSJF(FM) Saint Augustine Beach FL
*WSJL(FM) Northport AL
WSJO(FM) Egg Harbor City NJ
WSJQ(FM) North Cape May NJ
WSJR(FM) Dallas PA
WSJT(FM) Lakeland FL
WSJW(FM) Starview PA
WSJY(FM) Fort Atkinson WI
WSJZ-FM Sebastian FL
*WSKB(FM) Westfield MA
WSKE(FM) Everett PA
*WSKG-FM Binghamton NY
WSKL(FM) Veedersburg IN
WSKO-FM Wakefield-Peacedale RI
WSKQ-FM New York NY
WSKS(FM) Whitesboro NY
WSKU(FM) Little Falls NY
WSKV(FM) Stanton KY
WSKY-FM Micanopy FL
WSKZ(FM) Chattanooga TN
WSLC-FM Roanoke VA
WSLD(FM) Whitewater WI
*WSLE(FM) Salem IL
*WSLI(FM) Belding MI
*WSLJ(FM) Watertown NY
*WSLL(FM) Saranac Lake NY
WSLM-FM Salem IN
*WSLN(FM) Delaware OH
*WSLO(FM) Malone NY
WSLP(FM) Saranac Lake NY
WSLQ(FM) Roanoke VA
*WSLU(FM) Canton NY
*WSLX(FM) New Canaan CT
WSLY(FM) York AL
*WSMA(FM) Scituate MA
*WSMC-FM Collegedale TN
WSMD-FM Mechanicsville MD
WSM-FM Nashville TN
WSMI-FM Litchfield IL
WSMJ(FM) Baltimore MD
WSMK(FM) Buchanan MI
*WSMR(FM) Sarasota FL
WSMS(FM) Artesia MS
WSMW(FM) Greensboro NC
WSNA(FM) Germantown TN
*WSNC(FM) Winston-Salem NC
*WSND-FM Notre Dame IN
WSNE-FM Taunton MA
WSNI(FM) Winchendon MA
WSNN(FM) Potsdam NY
WSNT-FM Sandersville GA
WSNU(FM) Lock Haven PA
WSNV(FM) Salem VA
WSNX-FM Muskegon MI
WSNY(FM) Columbus OH
WSNZ(FM) Appomattox VA
WSOC-FM Charlotte NC
*WSOE(FM) Elon NC
*WSOF-FM Madisonville KY
*WSOG(FM) Spring Valley IL
WSOL-FM Brunswick GA
*WSOR(FM) Naples FL
WSOS-FM Saint Augustine FL
*WSOU(FM) South Orange NJ
WSOX(FM) Red Lion PA
WSOY-FM Decatur IL
WSPA-FM Spartanburg SC
WSPI(FM) Mount Carmel PA
WSPK(FM) Poughkeepsie NY
*WSPM(FM) Cloverdale IN
*WSPN(FM) Saratoga Springs NY
*WSPS(FM) Concord NH
WSPT-FM Stevens Point WI
WSPX(FM) Bowman SC
WSPY-FM Plano IL
*WSQA(FM) Hornell NY
*WSQC-FM Oneonta NY
*WSQE(FM) Corning NY
*WSQG-FM Ithaca NY
*WSQH(FM) Forest MS
*WSQX-FM Binghamton NY
WSRB(FM) Lansing IL
WSRG(FM) Sturgeon Bay WI
*WSRI(FM) Sugar Grove IL
WSRJ(FM) Honor MI
WSRK(FM) Oneonta NY
WSRM(FM) Coosa GA
*WSRN-FM Swarthmore PA
WSRS(FM) Worcester MA
WSRT(FM) Gaylord MI
*WSRU(FM) Slippery Rock PA
WSRV(FM) Gainesville GA
WSRW-FM Hillsboro OH
*WSRX(FM) Naples FL
WSRZ-FM Coral Cove FL
*WSSB-FM Orangeburg SC
*WSSD(FM) Chicago IL
WSSJ(FM) Rincon GA
*WSSK(FM) Saratoga Springs NY
WSSL-FM Gray Court SC
WSSM(FM) Havelock NC
WSSQ(FM) Sterling IL
WSSR(FM) Joliet IL
*WSSW(FM) Platteville WI
WSSX-FM Charleston SC
*WSTB(FM) Streetsboro OH
*WSTF(FM) Andalusia AL
WSTG(FM) Princeton WV
WSTH-FM Alexander City AL
WSTI-FM Quitman GA
WSTK(FM) Aurora NC
*WSTM(FM) Kiel WI
WSTO(FM) Owensboro KY
WSTQ(FM) Streator IL
WSTR(FM) Smyrna GA
WSTS(FM) Fairmont NC
WSTW(FM) Wilmington DE
WSTX-FM Christiansted VI
WSTZ-FM Vicksburg MS
*WSUC-FM Cortland NY
WSUE(FM) Sault Ste. Marie MI
*WSUF(FM) Noyack NY
WSUH(FM) Crozet VA
WSUL(FM) Monticello NY
*WSUM(FM) Madison WI
WSUN-FM Holiday FL
*WSUP(FM) Platteville WI
WSUS(FM) Franklin NJ
*WSUW(FM) Whitewater WI
WSUY(FM) Charleston SC
*WSVH(FM) Savannah GA
WSVO(FM) Staunton VA
WSWD(FM) Fairfield OH
WSWR(FM) Shelby OH
WSWT(FM) Peoria IL
WSWV-FM Pennington Gap VA
*WSYC-FM Shippensburg PA
WSYE(FM) Houston MS
WSYN(FM) Georgetown SC
WSYR-FM Gifford FL
WSYY-FM Millinocket ME
*WTAC(FM) Burton MI
*WTAI(FM) Union City TN
WTAK-FM Hartselle AL
WTAO(FM) Murphysboro IL
*WTBB(FM) Gadsden AL
WTBF-FM Brundidge AL
WTBG(FM) Brownsville TN
*WTBI-FM Greenville SC
*WTBJ(FM) Oxford AL
WTBK(FM) Manchester KY
WTBM(FM) Mexico ME
WTBX(FM) Hibbing MN
WTCB(FM) Orangeburg SC
*WTCC(FM) Springfield MA
WTCD(FM) Indianola MS
WTCJ-FM Tell City IN
*WTCK(FM) Charlevoix MI
WTCM-FM Traverse City MI
WTCQ(FM) Vidalia GA
WTCR-FM Huntington WV
WTCX(FM) Ripon WI
WTDA(FM) Westerville OH
WTDK(FM) Federalsburg MD
WTDR(FM) Talladega AL
*WTEB(FM) New Bern NC
*WTFH(FM) Helen GA
WTFM(FM) Kingsport TN
WTFX-FM Clarksville IN
WTGA-FM Thomaston GA
WTGB-FM Bethesda MD
WTGE(FM) Baker LA
WTGG(FM) Amite LA
*WTGN(FM) Lima OH
*WTGP(FM) Greenville PA
WTGR(FM) Union City OH
WTGV-FM Sandusky MI
WTGY(FM) Charleston MS
WTGZ(FM) Tuskegee AL
WTHB-FM Waynesboro GA
WTHD(FM) Lagrange IN
WTHG(FM) Hinesville GA
WTHI-FM Terre Haute IN
WTHK(FM) Wilmington VT
*WTHL(FM) Somerset KY
*WTHM(FM) Thomson GA
*WTHN(FM) Sault Ste. Marie MI
WTHO-FM Thomson GA
WTHP(FM) Gibson GA
*WTHS(FM) Holland MI
WTHT(FM) Auburn ME
WTHX(FM) Lebanon Junction KY
WTHZ(FM) Lexington NC
WTIC-FM Hartford CT
WTID(FM) Repton AL
WTIF-FM Omega GA
WTIM-FM Taylorville IL
*WTIP(FM) Grand Marais MN
WTIX-FM Galliano LA
*WTJB(FM) Columbus GA
*WTJT(FM) Baker FL
*WTJU(FM) Charlottesville VA
*WTJY(FM) Asheboro NC
WTKB-FM Atwood TN
*WTKC(FM) Findlay OH
WTKE-FM Holt FL
WTKF(FM) Atlantic NC
WTKK(FM) Boston MA
*WTKL(FM) North Dartmouth MA
WTKM-FM Hartford WI
WTKS-FM Cocoa Beach FL
WTKU-FM Ocean City NJ
WTKV(FM) Oswego NY
WTKW(FM) Bridgeport NY
WTKX-FM Pensacola FL
WTKY-FM Tompkinsville KY
WTLC-FM Greenwood IN
*WTLD(FM) Jesup GA
*WTLG(FM) Starke FL
*WTLI(FM) Bear Creek Township MI
WTLP(FM) Braddock Heights MD
WTLQ-FM Punta Rassa FL
*WTLR(FM) State College PA
WTLT(FM) Naples FL
WTLX(FM) Columbus WI
WTLY(FM) Thomasville GA
WTLZ(FM) Saginaw MI
WTMB(FM) Tomah WI
*WTMD(FM) Towson MD
WTMG(FM) Williston FL
*WTMK(FM) Lowell IN
*WTML(FM) Tullahoma TN
WTMM-FM Mechanicville NY
WTMP-FM Dade City FL
*WTMQ(FM) Lumpkin GA
WTMT(FM) Weaverville NC
*WTMV(FM) Youngsville PA
*WTMW(FM) North Judson IN
WTMX(FM) Skokie IL
WTNE-FM Trenton TN
WTNJ(FM) Mount Hope WV
WTNM(FM) Water Valley MS
WTNN(FM) Bristol VT
*WTNP(FM) Richland MI
WTNR(FM) Holland MI
WTNS-FM Coshocton OH
WTNT-FM Tallahassee FL
WTNV(FM) Tiptonville TN
WTOJ(FM) Carthage NY
WTON-FM Staunton VA
WTOP-FM Washington DC
WTOS-FM Skowhegan ME
WTOT-FM Graceville FL
WTPA(FM) Mechanicsburg PA
WTPL(FM) Hillsboro NH
WTPM(FM) Aguadilla PR
WTPO(FM) New Albany MS

WTPR-FM McKinnon TN
WTPT(FM) Forest City NC
WTQR(FM) Winston-Salem NC
WTRB-FM Sylacauga AL
WTRG(FM) Gaston NC
WTRH(FM) Ramsey IL
*WTRK(FM) Freeland MI
*WTRM(FM) Winchester VA
WTRS(FM) Dunnellon FL
*WTRT(FM) Benton KY
WTRV(FM) Walker MI
WTRW(FM) Two Rivers WI
WTRX-FM Pontiac IL
WTRY-FM Rotterdam NY
WTRZ(FM) Spencer TN
WTSA-FM Brattleboro VT
*WTSC-FM Potsdam NY
*WTSE(FM) Benton TN
*WTSG(FM) Carlinville IL
WTSH-FM Rockmart GA
WTSM(FM) Springfield VT
*WTSR(FM) Trenton NJ
WTSS(FM) Buffalo NY
*WTSU(FM) Montgomery-Troy AL
WTSX(FM) Port Jervis NY
WTTC-FM Towanda PA
WTTH(FM) Margate City NJ
WTTS(FM) Bloomington IN
*WTTU(FM) Cookeville TN
WTTX-FM Appomattox VA
WTUA(FM) Saint Stephen SC
WTUE(FM) Dayton OH
WTUF(FM) Boston GA
WTUG-FM Northport AL
WTUK(FM) Harlan KY
*WTUL(FM) New Orleans LA
*WTUR(FM) Upland IN
WTUV-FM Eminence KY
WTUZ(FM) Uhrichsville OH
WTVR-FM Richmond VA
WTVY-FM Dothan AL
WTWF(FM) Fairview PA
WTWP-FM Warrenton VA
WTWR-FM Luna Pier MI
WTWS(FM) Harrison MI
WTWX-FM Guntersville AL
*WTXR(FM) Toccoa Falls GA
WTXT(FM) Fayette AL
WTYB(FM) Springfield GA
WTYD(FM) Deltaville VA
WTYE(FM) Robinson IL
WTYJ(FM) Fayette MS
WTYL-FM Tylertown MS
WTYS-FM Marianna FL
WTZB(FM) Englewood FL
WTZN-FM Pittsburgh PA
WTZR(FM) Elizabethton TN
*WUAG(FM) Greensboro NC
*WUAL-FM Tuscaloosa AL
*WUAW(FM) Erwin NC
WUBB(FM) York Center ME
WUBE-FM Cincinnati OH
*WUBJ(FM) Jamestown NY
WUBL(FM) Atlanta GA
*WUBS(FM) South Bend IN
WUBT(FM) Russellville KY
WUBU(FM) South Bend IN
*WUCF-FM Orlando FL
WUCL(FM) Meridian MS
*WUCX-FM Bay City MI
WUCZ-FM Carthage TN
*WUDR(FM) Dayton OH
*WUEC(FM) Eau Claire WI
*WUEV(FM) Evansville IN
WUEZ(FM) Carterville IL
WUFF-FM Eastman GA
*WUFM(FM) Columbus OH
*WUFN(FM) Albion MI
*WUFR(FM) Bedford PA
*WUFT-FM Gainesville FL
*WUGA(FM) Athens GA
*WUGN(FM) Midland MI
WUGO(FM) Grayson KY
WUHT(FM) Birmingham AL
WUHU(FM) Smiths Grove KY
WUIN(FM) Carolina Beach NC
*WUIS(FM) Springfield IL
*WUJC(FM) Saint Marks FL
WUJM(FM) Gulfport MS
WUKL(FM) Bethlehem WV
WUKQ-FM Mayaguez PR
WUKS(FM) Saint Pauls NC
*WUKY(FM) Lexington KY
WULF(FM) Hardinsburg KY
WULS(FM) Broxton GA
*WUMB-FM Boston MA
*WUMC(FM) Elizabethton TN
*WUMD(FM) North Dartmouth MA
WUME-FM Paoli IN
*WUMF-FM Farmington ME
*WUML(FM) Lowell MA
*WUMR(FM) Memphis TN
WUMS(FM) University MS
WUMX(FM) Rome NY
*WUNC(FM) Chapel Hill NC
*WUND-FM Manteo NC
*WUNH(FM) Durham NH
*WUNV(FM) Albany GA
*WUNY(FM) Utica NY
*WUOG(FM) Athens GA
*WUOL(FM) Louisville KY
*WUOM(FM) Ann Arbor MI
*WUOT(FM) Knoxville TN
WUPE-FM North Adams MA
WUPF(FM) Gwinn MI
WUPG(FM) Crystal Falls MI
*WUPI(FM) Presque Isle ME
WUPK(FM) Marquette MI
WUPM(FM) Ironwood MI
WUPS(FM) Houghton Lake MI
*WUPX(FM) Marquette MI
WUPY(FM) Ontonagon MI
WUPZ(FM) Republic MI
*WURC(FM) Holly Springs MS
*WURI(FM) Manteo NC
WURK(FM) Elwood IN
*WUSB(FM) Stony Brook NY
*WUSC-FM Columbia SC
*WUSF(FM) Tampa FL
*WUSI(FM) Olney IL
WUSJ(FM) Madison MS
WUSL(FM) Philadelphia PA
*WUSM-FM Hattiesburg MS
WUSN(FM) Chicago IL
*WUSO(FM) Springfield OH
WUSP(FM) Nekoosa WI
WUSQ-FM Winchester VA
*WUSR(FM) Scranton PA
WUSV(FM) Estero FL
WUSW(FM) Hattiesburg MS
WUSX(FM) Addison VT
WUSY(FM) Cleveland TN
WUSZ(FM) Virginia MN
*WUTC(FM) Chattanooga TN
*WUTK-FM Knoxville TN
WUTL(FM) Tallahassee FL
*WUTM(FM) Martin TN
*WUTS(FM) Sewanee TN
WUUF(FM) Sodus NY
WUUS-FM South Pittsburg TN
WUUU(FM) Franklinton LA
WUUZ(FM) Cooperstown PA
WUVA(FM) Charlottesville VA
*WUVT-FM Blacksburg VA
*WUWF(FM) Pensacola FL
*WUWG(FM) Carrollton GA
*WUWM(FM) Milwaukee WI
WUZR(FM) Bicknell IN
WUZZ(FM) Saegertown PA
*WVAC-FM Adrian MI
WVAF(FM) Charleston WV
WVAQ(FM) Morgantown WV
*WVAS(FM) Montgomery AL
WVAZ(FM) Oak Park IL
WVBB(FM) Columbia City IN
*WVBC(FM) Bethany WV
WVBE-FM Lynchburg VA
WVBG-FM Redwood MS
*WVBH(FM) Beach Haven West NJ
WVBO(FM) Winneconne WI
WVBR-FM Ithaca NY
*WVBU-FM Lewisburg PA
*WVBV(FM) Medford Lakes NJ
WVBW(FM) Suffolk VA
WVBZ(FM) High Point NC
*WVCF(FM) Eau Claire WI
*WVCM(FM) Iron Mountain MI
*WVCN(FM) Baraga MI
WVCO(FM) Loris SC
*WVCP(FM) Gallatin TN
*WVCR-FM Loudonville NY
*WVCT(FM) Keavy KY
*WVCX(FM) Tomah WI
*WVCY-FM Milwaukee WI
*WVDA(FM) Valdosta GA
WVEE(FM) Atlanta GA
WVEI-FM Easthampton MA
WVEK-FM Cumberland KY
*WVEP(FM) Martinsburg WV
WVES(FM) Accomac VA
WVEZ(FM) Louisville KY
*WVFA(FM) Lebanon NH
WVFB(FM) Celina TN
WVFJ-FM Manchester GA
*WVFL(FM) Fond du Lac WI
WVFM(FM) Kalamazoo MI
*WVFS(FM) Tallahassee FL
WVGA(FM) Lakeland GA
WVGN(FM) Charlotte Amalie VI
*WVGR(FM) Grand Rapids MI
*WVGS(FM) Statesboro GA
*WVHC(FM) Herkimer NY
WVHL(FM) Farmville VA
*WVHM(FM) Benton KY
WVHR(FM) Huntingdon TN
*WVIA-FM Scranton PA
WVIB(FM) Holton MI
WVIC(FM) Jackson MI
*WVIJ(FM) Port Charlotte FL
*WVIK(FM) Rock Island IL
WVIL(FM) Virginia IL
WVIM-FM Coldwater MS
WVIN-FM Bath NY
WVIP(FM) New Rochelle NY
WVIQ(FM) Christiansted VI
WVIS(FM) Vieques PR
WVIV-FM Highland Park IL
WVIX(FM) Joliet IL
*WVJC(FM) Mount Carmel IL
WVJP-FM Caguas PR
WVJZ(FM) Charlotte Amalie VI
*WVKC(FM) Galesburg IL
WVKF(FM) Shadyside OH
WVKL(FM) Norfolk VA
WVKM(FM) Matewan WV
WVKO-FM Johnstown OH
*WVKR-FM Poughkeepsie NY
WVKS(FM) Toledo OH
WVKX(FM) Irwinton GA
WVLC(FM) Mannsville KY
WVLE(FM) Scottsville KY
WVLF(FM) Norwood NY
WVLI(FM) Kankakee IL
WVLK-FM Richmond KY
*WVLS(FM) Monterey VA
WVLT(FM) Vineland NJ
WVLY-FM Milton PA
*WVMC-FM Mansfield OH
WVMD(FM) Midland MD
*WVME(FM) Meadville PA
WVMG(FM) Normal IL
WVMJ(FM) Conway NH
*WVML(FM) Millersburg OH
*WVMM(FM) Grantham PA
*WVMN(FM) New Castle PA
*WVMS(FM) Sandusky OH
WVMV(FM) Detroit MI
*WVMW-FM Scranton PA
WVMX(FM) Cincinnati OH
WVNA-FM Muscle Shoals AL
*WVNH(FM) Concord NH
WVNI(FM) Nashville IN
*WVNL(FM) Vandalia IL
WVNN-FM Trinity AL
WVNO-FM Mansfield OH
*WVNP(FM) Wheeling WV
WVNS-FM Pegram TN
WVNU(FM) Greenfield OH
WVNV(FM) Malone NY
WVNW(FM) Burnham PA
WVOA-FM Mexico NY
*WVOB(FM) Dothan AL
WVOD(FM) Manteo NC
*WVOF(FM) Fairfield CT
WVOH-FM Hazlehurst GA
WVOK-FM Oxford AL
WVOM(FM) Howland ME
WVOR(FM) Canandaigua NY
WVOS-FM Liberty NY
WVOW-FM Logan WV
WVOZ-FM Carolina PR
*WVPA(FM) Saint Johnsbury VT
*WVPB(FM) Beckley WV
*WVPE(FM) Elkhart IN
*WVPG(FM) Parkersburg WV
*WVPH(FM) Piscataway NJ
*WVPM(FM) Morgantown WV
*WVPN(FM) Charleston WV
*WVPR(FM) Windsor VT
*WVPS(FM) Burlington VT
*WVPW(FM) Buckhannon WV
WVRB(FM) Wilmore KY
WVRC-FM Spencer WV
WVRE(FM) Dickeyville WI
*WVRI(FM) Pavo GA
WVRK(FM) Columbus GA
*WVRN(FM) Wittenberg WI
WVRQ-FM Viroqua WI
WVRR(FM) Newport NH
WVRT(FM) Mill Hall PA
*WVRU(FM) Radford VA
WVRW(FM) Glenville WV
WVRY(FM) Waverly TN
*WVSB(FM) Romney WV
*WVSD(FM) Itta Bena MS
*WVSH(FM) Huntington IN
*WVSI(FM) Mount Vernon IL
WVSR-FM Charleston WV
*WVSS(FM) Menomonie WI
*WVST-FM Petersburg VA
*WVSU-FM Birmingham AL
WVSZ(FM) Chesterfield SC
*WVTC(FM) Randolph Center VT
*WVTF(FM) Roanoke VA
WVTI(FM) Brighton VT
WVTK(FM) Port Henry NY
*WVTQ(FM) Sunderland VT
*WVTR(FM) Marion VA
*WVTU(FM) Charlottesville VA
*WVTW(FM) Charlottesville VA
*WVUA-FM Tuscaloosa AL
*WVUB(FM) Vincennes IN
*WVUD(FM) Newark DE
*WVUM(FM) Coral Gables FL
*WVUR-FM Valparaiso IN
WVVE(FM) Panama City Beach FL
WVVL(FM) Elba AL
WVVR(FM) Hopkinsville KY
*WVVS(FM) Valdosta GA
WVVV(FM) Williamstown WV
*WVWC(FM) Buckhannon WV
*WVWV(FM) Huntington WV
*WVXC(FM) Chillicothe OH
WVXG(FM) Mount Gilead OH
*WVXR(FM) Richmond IN
*WVXU(FM) Cincinnati OH
*WVXW(FM) West Union OH
*WVYA(FM) Williamsport PA
WVYB(FM) Holly Hill FL
*WVYC(FM) York PA
WVZA(FM) Herrin IL
WWAG(FM) McKee KY
WWAV-FM Santa Rosa Beach FL
WWAX(FM) Hermantown MN
WWBB(FM) Providence RI
WWBD(FM) Bamberg SC
WWBE(FM) Mifflinburg PA
WWBL(FM) Washington IN
*WWBM(FM) Yates GA
WWBN(FM) Tuscola MI
WWBR(FM) Big Rapids MI
WWBU(FM) Radford VA
WWBX(FM) Bangor ME
WWCD(FM) Grove City OH
*WWCF(FM) McConnellsburg PA
*WWCJ(FM) Cape May NJ
WWCK-FM Flint MI
WWCM(FM) Standish MI
WWCT(FM) Bartonville IL
*WWCU(FM) Cullowhee NC
WWDC-FM Washington DC
WWDE-FM Hampton VA
WWDG(FM) DeRuyter NY
*WWDL(FM) Lebanon IN
WWDM(FM) Sumter SC
*WWDN(FM) New Whiteland IN
*WWDS(FM) Muncie IN
WWDV(FM) Zion IL
WWDW(FM) Alberta VA
*WWEB(FM) Wallingford CT
*WWEC(FM) Elizabethtown PA
*WWED(FM) Spotsylvania VA
WWEG(FM) Hagerstown MD
WWEL(FM) London KY
*WWEM(FM) Rustburg VA
*WWET(FM) Valdosta GA
*WWEV-FM Cumming GA
WWFG(FM) Ocean City MD
*WWFM(FM) Trenton NJ
WWFN-FM Lake City SC
*WWFP(FM) Brigantine NJ
*WWFR(FM) Stuart FL
WWFS(FM) New York NY
WWFT(FM) Fishers IN
WWFX(FM) Southbridge MA
WWFY(FM) Berlin VT
WWGF(FM) Donalsonville GA
WWGM(FM) Alamo TN
*WWGN(FM) Ottawa IL
WWGO(FM) Charleston IL
WWGR(FM) Fort Myers FL
*WWGV(FM) Grove City OH
WWGY(FM) Grove City PA
WWHA(FM) Oriental NC
WWHC(FM) Oakland MD
WWHG(FM) Evansville WI
*WWHI(FM) Muncie IN
WWHK(FM) Concord NH
WWHP(FM) Farmer City IL
WWHQ(FM) Meredith NH
*WWHR(FM) Bowling Green KY
*WWHS-FM Hampden-Sydney VA
WWHT(FM) Syracuse NY
WWHV(FM) Lynn Haven FL
WWIB(FM) Hallie WI
*WWIL-FM Wilmington NC
WWIN-FM Glen Burnie MD
*WWIO-FM Brunswick GA
*WWIP(FM) Cheriton VA
WWIS-FM Black River Falls WI
WWIZ(FM) Mercer PA
*WWJA(FM) Janesville WI
*WWJD(FM) Pippa Passes KY
WWJK(FM) Jackson MS
WWJM(FM) New Lexington OH
WWJN(FM) Ridgeland SC
WWJO(FM) Saint Cloud MN
WWKA(FM) Orlando FL
WWKC(FM) Caldwell OH
WWKF(FM) Fulton KY
WWKI(FM) Kokomo IN
WWKL(FM) Palmyra PA
*WWKM(FM) Imlay City MI
WWKN(FM) Morgantown KY
WWKR(FM) Hart MI
WWKS(FM) Cruz Bay VI
WWKT-FM Kingstree SC
WWKX(FM) Woonsocket RI
WWKY(FM) Providence KY
WWKZ(FM) Columbus MS
*WWLA(FM) South Charleston WV
WWLB(FM) Midlothian VA
*WWLC(FM) Cross City FL
WWLD(FM) Cairo GA
WWLF-FM Sylvan Beach NY
WWL-FM Kenner LA
WWLI(FM) Providence RI
WWLL(FM) Sebring FL
*WWLO(FM) Lowell IN
*WWLR(FM) Lyndonville VT

WWLS-FM Edmond OK
WWLT(FM) Manchester KY
*WWLU(FM) Lincoln University PA
WWLW(FM) Clarksburg WV
*WWMC(FM) Lynchburg VA
WWMG(FM) Millbrook AL
WWMJ(FM) Ellsworth ME
WWMP(FM) Waterbury VT
WWMR(FM) Saltillo MS
WWMS(FM) Oxford MS
WWMX(FM) Baltimore MD
WWMY(FM) Raleigh NC
WWNF(FM) Goldsboro NC
*WWNJ(FM) Dover Township NJ
WWNK(FM) Farmville NC
*WWNO(FM) New Orleans LA
WWNQ(FM) Forest Acres SC
WWNU(FM) Irmo SC
*WWNW(FM) New Wilmington PA
WWOD(FM) Hartford VT
*WWOG(FM) Cookeville TN
WWOJ(FM) Avon Park FL
WWOT(FM) Altoona PA
*WWOZ(FM) New Orleans LA
*WWPH(FM) Princeton Junction NJ
*WWPJ(FM) Pen Argyl PA
WWPN(FM) Westernport MD
WWPR-FM New York NY
*WWPT(FM) Westport CT
*WWPV-FM Colchester VT
WWQM-FM Middleton WI
WWQQ-FM Wilmington NC
WWRE(FM) Berryville VA
WWRM(FM) Tampa FL
WWRQ-FM Valdosta GA
WWRR(FM) Scranton PA
WWRT(FM) Strasburg VA
WWRX(FM) Pawcatuck CT
WWRZ(FM) Fort Meade FL
WWSE(FM) Jamestown NY
WWSL(FM) Philadelphia MS
WWSN(FM) Waycross GA
*WWSP(FM) Stevens Point WI
WWSR(FM) Wapakoneta OH
WWST(FM) Sevierville TN
*WWSU(FM) Dayton OH
WWSW-FM Pittsburgh PA
WWSY(FM) Seelyville IN
*WWTA(FM) Marion MA
WWTB(FM) Topsail Beach NC
*WWTG(FM) Carpentersville IL
WWTH(FM) Oscoda MI
WWTN(FM) Manchester TN
*WWTS(FM) Logansport IN
WWUF(FM) Waycross GA
*WWUH(FM) West Hartford CT
WWUN-FM Friar's Point MS
WWUS(FM) Big Pine Key FL
WWUZ(FM) Bowling Green VA
WWVA-FM Canton GA
WWVR(FM) West Terre Haute IN
*WWVU-FM Morgantown WV
*WWWA(FM) Winslow ME
WWWD(FM) Bolingbroke GA
WWWI-FM Pillager MN
WWWK(FM) Marathon FL
WWWM-FM Sylvania OH
WWWQ(FM) College Park GA
WWWV(FM) Charlottesville VA
WWWW-FM Ann Arbor MI
WWWX(FM) Oshkosh WI
WWWY(FM) North Vernon IN
WWWZ(FM) Summerville SC
*WWXC(FM) Albany GA
WWXM(FM) Garden City SC
WWXT(FM) Prince Frederick MD
WWXX(FM) Warrenton VA
WWYL(FM) Chenango Bridge NY
WWYN(FM) McKenzie TN
WWYW(FM) Dundee IL
WWYY(FM) Belvidere NJ
WWYZ(FM) Waterbury CT
WWZD-FM New Albany MS
WWZW(FM) Buena Vista VA
WWZY(FM) Long Branch NJ
WXAB(FM) McLain MS
*WXAC(FM) Reading PA
WXAJ(FM) Hillsboro IL
WXAN(FM) Ava IL
*WXBA(FM) Brentwood NY
WXBB(FM) Erie PA
WXBC(FM) Hardinsburg KY
*WXBE(FM) Beaufort NC
WXBM-FM Milton FL
WXBN(FM) Groveton NH
WXBQ-FM Bristol TN
WXBT(FM) West Columbia SC
WXBX(FM) Rural Retreat VA
WXCC(FM) Williamson WV
WXCF-FM Clifton Forge VA
WXCH(FM) Versailles IN
*WXCI(FM) Danbury CT
WXCL(FM) Pekin IL
WXCM(FM) Whitesville KY
WXCR(FM) New Martinsville WV
WXCV(FM) Homosassa Springs FL
WXCX(FM) Siren WI
WXCY(FM) Havre de Grace MD
WXDJ(FM) North Miami Beach FL
*WXDU(FM) Durham NC
WXDX-FM Pittsburgh PA
WXEF(FM) Effingham IL
WXEG(FM) Beavercreek OH
*WXEL(FM) West Palm Beach FL
WXER(FM) Plymouth WI
WXET(FM) Arcola IL
WXEZ(FM) Yorktown VA
WXFL(FM) Florence AL
WXFM(FM) Mount Zion IL
*WXFR(FM) State College PA
WXFX(FM) Prattville AL
WXGL(FM) Saint Petersburg FL
WXGM-FM Gloucester VA
*WXGN(FM) Egg Harbor Township NJ
WXHB(FM) Richton MS
WXHC(FM) Homer NY
*WXHD(FM) Mount Hope NY
*WXHL-FM Christiana DE
WXHT(FM) Madison FL
WXIL(FM) Parkersburg WV
WXIS(FM) Erwin TN
WXIZ(FM) Waverly OH
WXJC-FM Cordova AL
*WXJM(FM) Harrisonburg VA
WXJN(FM) Lewes DE
WXJY(FM) Georgetown SC
WXJZ(FM) Gainesville FL
WXKB(FM) Cape Coral FL
WXKC(FM) Erie PA
WXKE(FM) Fort Wayne IN
WXKQ-FM Whitesburg KY
WXKR(FM) Port Clinton OH
WXKS-FM Medford MA
WXKT(FM) Washington GA
WXKU-FM Austin IN
*WXKV(FM) Selmer TN
WXKW(FM) Millville NJ
WXKY-FM Stanford KY
WXKZ-FM Prestonsburg KY
WXLC(FM) Waukegan IL
WXLF(FM) White River Junction VT
*WXLG(FM) North Creek NY
*WXLH(FM) Blue Mountain Lake NY
WXLK(FM) Roanoke VA
WXLM(FM) Stonington CT
WXLO(FM) Fitchburg MA
WXLP(FM) Moline IL
WXLR(FM) Harold KY
WXLT(FM) Christopher IL
*WXLU(FM) Peru NY
*WXLV(FM) Schnecksville PA
WXLX(FM) Lajas PR
WXLY(FM) North Charleston SC
WXLZ-FM Lebanon VA
WXMA(FM) Louisville KY
WXMD(FM) Pocomoke City MD
WXMG(FM) Upper Arlington OH
WXMK(FM) Dock Junction GA
*WXML(FM) Upper Sandusky OH
WXMM(FM) Norfolk VA
WXMP(FM) Glasford IL
WXMX(FM) Millington TN
WXMZ(FM) Hartford KY
WXNR(FM) Grifton NC
WXNU(FM) Saint Anne IL
WXOF(FM) Yankeetown FL
WXOQ(FM) Selmer TN
*WXOU(FM) Auburn Hills MI
*WXPH(FM) Harrisburg PA
WXPK(FM) Briarcliff Manor NY
*WXPL(FM) Fitchburg MA
*WXPN(FM) Philadelphia PA
*WXPR(FM) Rhinelander WI
WXPT(FM) Powers MI
*WXPW(FM) Wausau WI
*WXPZ(FM) Clyde Township MI
WXQR(FM) Jacksonville NC
WXQW(FM) Gurley AL
*WXRB(FM) Dudley MA
WXRC(FM) Hickory NC
WXRD(FM) Crown Point IN
WXRG(FM) Whitefield NH
*WXRI(FM) Winston-Salem NC
WXRK(FM) New York NY
WXRO(FM) Beaver Dam WI
WXRR(FM) Hattiesburg MS
WXRS-FM Swainsboro GA
WXRT-FM Chicago IL
WXRV(FM) Haverhill MA
WXRX(FM) Belvidere IL
WXRZ(FM) Corinth MS
WXSR(FM) Quincy FL
WXSS(FM) Wauwatosa WI
WXST(FM) Hollywood SC
WXTA(FM) Edinboro PA
WXTB(FM) Clearwater FL
WXTG(FM) Virginia Beach VA
WXTK(FM) West Yarmouth MA
WXTQ(FM) Athens OH
*WXTS-FM Toledo OH
WXTT(FM) Danville IL
WXTU(FM) Philadelphia PA
WXUR(FM) Herkimer NY
*WXUT(FM) Toledo OH
*WXVS(FM) Waycross GA
*WXVU(FM) Villanova PA
*WXVW(FM) Veedersburg IN
WXXB(FM) Delphi IN
WXXC(FM) Marion IN
*WXXE(FM) Fenner NY
WXXF(FM) Loudonville OH
*WXXI-FM Rochester NY
WXXK(FM) Lebanon NH
WXXL(FM) Tavares FL
WXXM(FM) Sun Prairie WI
WXXO(FM) Cambridge Springs PA
WXXQ(FM) Freeport IL
WXXR(FM) Fredericktown OH
WXXS(FM) Lancaster NH
WXXX(FM) South Burlington VT
*WXXY-FM Port Republic NJ
WXXZ(FM) Grand Marais MN
*WXYC(FM) Chapel Hill NC
WXYK(FM) Gulfport MS
WXYM(FM) Tomah WI
WXYX(FM) Bayamon PR
WXZO(FM) Willsboro NY
WXZQ(FM) Piketon OH
WXZZ(FM) Georgetown KY
WYAB(FM) Yazoo City MS
WYAC-FM Christiansted VI
*WYAI(FM) Scotia NY
*WYAJ(FM) Sudbury MA
WYAK-FM Surfside Beach SC
*WYAR(FM) Yarmouth ME
WYAS(FM) Vieques PR
WYAV(FM) Myrtle Beach SC
WYAY(FM) Gainesville GA
*WYAZ(FM) Yazoo City MS
WYBB(FM) Folly Beach SC
WYBC-FM New Haven CT
*WYBF(FM) Radnor Township PA
*WYBH(FM) Fayetteville NC
WYBL(FM) Ashtabula OH
WYBR(FM) Big Rapids MI
*WYBV(FM) Wakarusa IN
WYBZ(FM) Crooksville OH
WYCA(FM) Crete IL
WYCD(FM) Detroit MI
*WYCE(FM) Wyoming MI
WYCL(FM) Pensacola FL
*WYCM(FM) Charlton MA
WYCR(FM) York-Hanover PA
*WYCS(FM) Yorktown VA
WYCT(FM) Pensacola FL
WYCY(FM) Hawley PA
WYDE-FM Cullman AL
WYDL(FM) Middleton TN
*WYDM(FM) Monroe MI
WYDS(FM) Decatur IL
WYEC(FM) Kewanee IL
*WYEP-FM Pittsburgh PA
WYEZ(FM) Murrell's Inlet SC
WYFA(FM) Waynesboro GA
*WYFB(FM) Gainesville FL
*WYFC(FM) Clinton TN
*WYFD(FM) Decatur AL
*WYFE(FM) Tarpon Springs FL
*WYFG(FM) Gaffney SC
*WYFH(FM) North Charleston SC
WYFI(FM) Norfolk VA
WYFJ(FM) Ashland VA
*WYFK(FM) Columbus GA
WYFL(FM) Henderson NC
WYFM(FM) Sharon PA
*WYFO(FM) Lakeland FL
*WYFP(FM) Harpswell ME
*WYFQ-FM Wadesboro NC
*WYFS(FM) Savannah GA
WYFT(FM) Luray VA
*WYFU(FM) Masontown PA
*WYFV(FM) Cayce SC
*WYFW(FM) Winder GA
WYFX(FM) Mount Vernon IN
*WYFZ(FM) Belleview FL
WYGB(FM) Edinburgh IN
WYGC(FM) High Springs FL
WYGE(FM) London KY
*WYGG(FM) Asbury Park NJ
WYGL-FM Elizabethville PA
WYGO(FM) Madisonville TN
WYGS(FM) Columbus IN
WYGY(FM) Fort Thomas KY
WYHT(FM) Mansfield OH
WYJB(FM) Albany NY
*WYJC(FM) Greenville FL
WYJK-FM Bellaire OH
WYJZ(FM) Speedway IN
*WYKL(FM) Crestline OH
WYKR-FM Haverhill NH
WYKS(FM) Gainesville FL
WYKT(FM) Wilmington IL
*WYKV(FM) Ravena NY
WYKX(FM) Escanaba MI
WYKZ(FM) Beaufort SC
WYLD-FM New Orleans LA
WYLK(FM) Lacombe LA
*WYLV(FM) Alcoa TN
WYLZ(FM) Pinconning MI
WYMG(FM) Jacksonville IL
WYMJ(FM) New Martinsville WV
*WYMS(FM) Milwaukee WI
WYMV(FM) Madisonville KY
WYMX(FM) Greenwood MS
WYMY(FM) Goldsboro NC
WYNA(FM) Calabash NC
WYND-FM Hatteras NC
WYNG(FM) Mount Carmel IL
WYNK-FM Baton Rouge LA
WYNN-FM Florence SC
WYNR(FM) Waycross GA
WYNT(FM) Upper Sandusky OH
WYNU(FM) Milan TN
WYNW(FM) Birnamwood WI
WYNZ(FM) Westbrook ME
WYOK(FM) Atmore AL
WYOO(FM) Springfield FL
WYOR(FM) Cross Hill SC
WYOY(FM) Gluckstadt MS
*WYPF(FM) Frederick MD
WYPL(FM) Memphis TN
*WYPR(FM) Baltimore MD
WYPW(FM) Nappanee IN
WYPY(FM) Baton Rouge LA
WYQE(FM) Naguabo PR
*WYQS(FM) Mars Hill NC
WYRB(FM) Genoa IL
WYRK(FM) Buffalo NY
WYRO(FM) McArthur OH
WYRQ(FM) Little Falls MN
*WYRS(FM) Manahawkin NJ
WYRX(FM) Lexington Park MD
WYRY(FM) Hinsdale NH
*WYSA(FM) Wauseon OH
WYSB(FM) Springfield KY
WYSC(FM) McRae GA
WYSF(FM) Birmingham AL
WYSK-FM Spotsylvania VA
*WYSM(FM) Lima OH
*WYSO(FM) Yellow Springs OH
WYSP(FM) Philadelphia PA
WYSS(FM) Sault Ste. Marie MI
WYST(FM) Fairbury IL
*WYSU(FM) Youngstown OH
WYSX(FM) Morristown NY
*WYSZ(FM) Maumee OH
WYTE(FM) Marshfield WI
*WYTF(FM) Indianola MS
*WYTJ(FM) Linton IN
WYTK(FM) Rogersville AL
*WYTL(FM) Wyomissing PA
WYTM-FM Fayetteville TN
*WYTN(FM) Youngstown OH
WYTT(FM) Emporia VA
WYTZ(FM) Bridgman MI
WYUL(FM) Chateaugay NY
WYUM(FM) Mount Vernon GA
WYUU(FM) Safety Harbor FL
WYVK(FM) Middleport OH
WYVN(FM) Saugatuck MI
WYVY(FM) Union City TN
WYXB(FM) Indianapolis IN
WYXL(FM) Ithaca NY
WYYD(FM) Amherst VA
WYYS(FM) Streator IL
WYYU(FM) Dalton GA
WYYX(FM) Bonifay FL
WYYY(FM) Syracuse NY
WYZB(FM) Mary Esther FL
WYZY(FM) Saranac Lake NY
WZAC-FM Danville WV
WZAD(FM) Wurtsboro NY
WZAI(FM) Brewster MA
WZAK(FM) Cleveland OH
WZAQ(FM) Louisa KY
WZAR(FM) Ponce PR
WZAT(FM) Savannah GA
WZAX(FM) Nashville NC
WZBA(FM) Westminster MD
WZBB(FM) Stanleytown VA
*WZBC(FM) Newton MA
WZBD(FM) Berne IN
WZBG(FM) Litchfield CT
WZBH(FM) Georgetown DE
WZBL(FM) Roanoke VA
WZBN(FM) Camilla GA
WZBQ(FM) Carrollton AL
*WZBT(FM) Gettysburg PA
WZBX(FM) Sylvania GA
WZBY(FM) Sturgeon Bay WI
WZBZ(FM) Pleasantville NJ
WZCR(FM) Hudson NY
*WZDG(FM) Scotts Hill NC
WZDM(FM) Vincennes IN
WZDQ(FM) Humboldt TN
WZEB(FM) Ocean View DE
WZEE(FM) Madison WI
WZET(FM) Hormigueros PR
WZEW(FM) Fairhope AL
WZEZ(FM) Goochland VA
WZFJ(FM) Pequot Lakes MN
WZFM(FM) Narrows VA
WZFX(FM) Whiteville NC
WZGA(FM) Helen GA
WZGC(FM) Atlanta GA
*WZGO(FM) Aurora NC
WZHL(FM) New Augusta MS
*WZHN(FM) East Tawas MI
WZHT(FM) Troy AL
WZID(FM) Manchester NH

WZIN(FM) Charlotte Amalie VI
*WZIP(FM) Akron OH
WZIQ(FM) Smithville GA
WZJO(FM) Dunbar WV
WZJS(FM) Banner Elk NC
WZJZ(FM) Port Charlotte FL
WZKB(FM) Wallace NC
WZKF(FM) Salem IN
WZKL(FM) Alliance OH
*WZKM(FM) Waynesboro MS
WZKR(FM) Decatur MS
WZKS(FM) Union MS
*WZKV(FM) Dyersburg TN
WZKX(FM) Bay St. Louis MS
WZKZ(FM) Alfred NY
WZLA-FM Abbeville SC
WZLD(FM) Petal MS
WZLF(FM) Bellows Falls VT
WZLK(FM) Virgie KY
WZLM(FM) Goodwater AL
WZLQ(FM) Tupelo MS
WZLR(FM) Xenia OH
WZLT(FM) Lexington TN
WZLX(FM) Boston MA
*WZLY(FM) Wellesley MA
*WZMB(FM) Greenville NC
WZMJ(FM) Batesburg SC
WZMQ(FM) Key Largo FL
WZMR(FM) Altamont NY
WZMT(FM) Ponce PR
WZMX(FM) Hartford CT
*WZNB(FM) New Bern NC
WZNE(FM) Brighton NY
WZNF(FM) Lumberton MS
WZNJ(FM) Demopolis AL
WZNL(FM) Norway MI
WZNN(FM) Allouez WI
WZNS(FM) Fort Walton Beach FL
WZNT(FM) San Juan PR
WZNX(FM) Sullivan IL
WZNY(FM) Old Forge NY
WZOC(FM) Plymouth IN
WZOE-FM Princeton IL
WZOK(FM) Rockford IL
WZOL(FM) Luquillo PR
WZOM(FM) Defiance OH
WZOO-FM Edgewood OH
WZOR(FM) Mishicot WI
WZOW(FM) New Carlisle IN
WZOZ(FM) Oneonta NY
*WZPE(FM) Bath NC
WZPL(FM) Greenfield IN
WZPN(FM) Farmington IL
WZPR(FM) Nags Head NC
WZPT(FM) New Kensington PA
WZPW(FM) Peoria IL
WZQQ(FM) Hyden KY
*WZRD(FM) Chicago IL
*WZRI(FM) Spring Lake NC
*WZRL(FM) Wade NC
*WZRN(FM) Norlina NC
WZRR(FM) Birmingham AL
*WZRS(FM) Pana IL
WZRT(FM) Rutland VT
*WZRU(FM) Roanoke Rapids NC
WZRV(FM) Front Royal VA
WZRX-FM Fort Shawnee OH
WZSN(FM) Greenwood SC
WZSP(FM) Nocatee FL
WZSR(FM) Woodstock IL
WZST(FM) Westover WV
WZTF(FM) Scranton SC
WZTK(FM) Burlington NC
WZUN(FM) Phoenix NY
WZUP(FM) Rose Hill NC
WZUS(FM) Macon IL
WZUU(FM) Allegan MI
WZVA(FM) Marion VA
WZVN(FM) Lowell IN
WZWW(FM) Bellefonte PA
WZWZ(FM) Kokomo IN
*WZXH(FM) Hagerstown MD
WZXL(FM) Wildwood NJ
*WZXM(FM) Middletown PA
*WZXQ(FM) Chambersburg PA
WZXR(FM) South Williamsport PA
WZXV(FM) Palmyra NY
*WZXX(FM) Lawrenceburg TN
WZYP(FM) Athens AL
WZYQ(FM) Mound Bayou MS
WZYY(FM) Renovo PA
*WZYZ(FM) Spencer TN
*WZZD(FM) Warwick PA
*WZZE(FM) Glen Mills PA
*WZZH(FM) Honesdale PA
WZZI(FM) Vinton VA
WZZK-FM Birmingham AL
WZZL(FM) Reidland KY
WZZN(FM) Chicago IL
WZZO(FM) Bethlehem PA
WZZP(FM) Hopkinsville KY
WZZR(FM) Riviera Beach FL
WZZS(FM) Zolfo Springs FL
WZZT(FM) Morrison IL
WZZU(FM) Lynchburg VA
WZZY(FM) Winchester IN
WZZZ(FM) Portsmouth OH
XETRA-FM Tijuana MEX
XHRM-FM Tijuana MEX

Canadian AM Stations by Call Letters

CBA(AM) Moncton NB
*CBE(AM) Windsor ON
*CBEF(AM) Windsor ON
*CBG(AM) Gander NF
CBGN(AM) Sainte Anne des Monts PQ
CBGY(AM) Bonavista Bay NF
CBI(AM) Sydney NS
*CBK(AM) Regina SK
*CBKF-1(AM) Gravelbourg SK
CBKF-2(AM) Saskatoon SK
*CBN(AM) Saint John's NF
CBOF-1(AM) Maniwaki PQ
*CBR(AM) Calgary AB
CBT(AM) Grand Falls-Windsor NF
*CBU(AM) Vancouver BC
CBW(AM) Winnipeg MB
CBX(AM) Edmonton AB
*CBY(AM) Corner Brook NF
CFAB(AM) Windsor NS
CFAC(AM) Calgary AB
CFAM(AM) Altona MB
CFAR(AM) Flin Flon MB
CFAV(AM) Laval PQ
CFAX(AM) Victoria BC
CFBC(AM) Saint John NB
CFBV(AM) Smithers BC
CFCB(AM) Corner Brook NF
CFCO(AM) Chatham ON
CFCT(AM) Tuktoyaktuk NT
CFCW(AM) Camrose AB
CFDR(AM) Dartmouth NS
CFED(AM) Chapais PQ
CFFB(AM) Iqaluit NU
CFFR(AM) Calgary AB
CFFX(AM) Kingston ON
CFGO(AM) Ottawa ON
CFGT(AM) Alma PQ
CFKC(AM) Creston BC
CFLD(AM) Burns Lake BC
CFLM(AM) La Tuque PQ
CFLN(AM) Goose Bay NF
CFLW(AM) Wabush NF
CFMB(AM) Montreal PQ
CFMJ(AM) Toronto ON
CFNC(AM) Cross Lake MB
CFNI(AM) Port Hardy BC
CFNW(AM) Port au Choix NF
CFOK(AM) Westlock AB
CFOS(AM) Owen Sound ON
CFPL(AM) London ON
*CFPR(AM) Prince Rupert BC
CFRA(AM) Ottawa ON
CFRB(AM) Toronto ON
CFRN(AM) Edmonton AB
CFRP(AM) Forestville PQ
CFRW(AM) Winnipeg MB
CFRY(AM) Portage la Prairie MB
CFSL(AM) Weyburn SK
CFSX(AM) Stephenville NF
CFTK(AM) Terrace BC
CFTR(AM) Toronto ON
CFUN(AM) Vancouver BC
CFWB(AM) Campbell River BC
*CFWH(AM) Whitehorse YT
*CFYK(AM) Yellowknife NT
CFYM(AM) Kindersley SK
CHAB(AM) Moose Jaw SK
*CHAK(AM) Inuvik NT
CHAM(AM) Hamilton ON
CHCM(AM) Marystown NF
CHED(AM) Edmonton AB
*CHFA(AM) Edmonton AB
CHFC(AM) Churchill MB
CHHA(AM) Toronto ON
CHIN(AM) Toronto ON
CHKT(AM) Toronto ON
CHLN(AM) Trois Rivieres PQ
CHLT(AM) Sherbrooke PQ
CHLW(AM) Saint Paul AB
CHMB(AM) Vancouver BC
CHMJ(AM) Vancouver BC
CHML(AM) Hamilton ON
*CHMO(AM) Moosonee ON
CHNC(AM) New Carlisle PQ
CHNL(AM) Kamloops BC
CHOK(AM) Sarnia ON
CHOR(AM) Summerland BC
CHOU(AM) Montreal PQ
CHQB(AM) Powell River BC
CHQR(AM) Calgary AB
CHQT(AM) Edmonton AB
CHRB(AM) High River AB
CHRC(AM) Quebec PQ
CHSC(AM) Saint Catharines ON
CHSM(AM) Steinbach MB
CHTK(AM) Prince Rupert BC
CHTM(AM) Thompson MB
CHTO(AM) Toronto ON
CHUM(AM) Toronto ON
CHVO(AM) Carbonear NF
CHWO(AM) Toronto ON
CIAO(AM) Brampton ON
CIBQ(AM) Brooks AB
CIGM(AM) Sudbury ON
CINF(AM) Verdun PQ
CINW(AM) Montreal PQ
CIOR(AM) Princeton BC
CISL(AM) Richmond BC
CIVH(AM) Vanderhoof BC
CIWW(AM) Ottawa ON
CJAD(AM) Montreal PQ
CJAR(AM) The Pas MB
CJBC(AM) Toronto ON
CJBK(AM) London ON
CJBQ(AM) Belleville ON
CJCA(AM) Edmonton AB
CJCB(AM) Sydney NS
CJCH(AM) Halifax NS
CJCL(AM) Toronto ON
CJCS(AM) Stratford ON
CJCW(AM) Sussex NB
CJDC(AM) Dawson Creek BC
CJGX(AM) Yorkton SK
CJLO(AM) Montreal PQ
CJMD(AM) Chibougamau PQ
CJME(AM) Regina SK
CJMR(AM) Mississauga ON
CJMS(AM) Saint Constant PQ
CJNB(AM) North Battleford SK
CJNL(AM) Merritt BC
CJNS(AM) Meadow Lake SK
CJOB(AM) Winnipeg MB
CJOR(AM) Osoyoos BC
CJOY(AM) Guelph ON
CJRB(AM) Boissevain MB
CJRJ(AM) Vancouver BC
CJRS(AM) Montreal PQ
CJSL(AM) Estevan SK
CJSN(AM) Shaunavon SK
CJUL(AM) Cornwall ON
CJVA(AM) Caraquet NB
CJVB(AM) Richmond BC
CJWI(AM) Montreal PQ
CJWW(AM) Saskatoon SK
CJYE(AM) Oakville ON
CJYM(AM) Rosetown SK
CJYQ(AM) Saint John's NF
CKAC(AM) Montreal PQ
CKAD(AM) Middleton NS
CKAT(AM) North Bay ON
CKBA(AM) Athabasca AB
CKBD(AM) Vancouver BC
CKBI(AM) Prince Albert SK
CKBX(AM) 100 Mile House BC
CKCM(AM) Grand Falls NF
CKCR(AM) Revelstoke BC
CKDH(AM) Amherst NS
CKDM(AM) Dauphin MB
CKDO(AM) Oshawa ON
CKDQ(AM) Drumheller AB
CKDR-2(AM) Sioux Lookout ON
CKDR-5(AM) Red Lake ON
CKDY(AM) Digby NS
CKEC(AM) New Glasgow NS
CKFR(AM) Kelowna BC
CKGA(AM) Gander NF
CKGL(AM) Kitchener ON
CKGM(AM) Montreal PQ
CKGR(AM) Golden BC
CKHJ(AM) Fredericton NB
CKIM(AM) Baie Verte NF
CKIR(AM) Invermere BC
CKJH(AM) Melfort SK
CKJR(AM) Wetaskiwin AB
CKJS(AM) Winnipeg MB
CKKW(AM) Kitchener ON
CKKY(AM) Wainwright AB
CKLC(AM) Kingston ON
CKLQ(AM) Brandon MB
CKLW(AM) Windsor ON
*CKMO(AM) Victoria BC
CKMW(AM) Winkler-Morden MB
CKMX(AM) Calgary AB
CKNB(AM) Campbellton NB
CKNW(AM) New Westminster BC
CKNX(AM) Wingham ON
CKOC(AM) Hamilton ON
CKOM(AM) Saskatoon SK
CKOR(AM) Penticton BC
CKOT(AM) Tillsonburg ON
CKOV(AM) Kelowna BC
CKPC(AM) Brantford ON
CKPT(AM) Peterborough ON
CKRM(AM) Regina SK
CKRS(AM) Saguenay PQ
CKRU(AM) Peterborough ON
CKRW(AM) Whitehorse YT
*CKSB(AM) Saint Boniface MB
CKSL(AM) London ON
CKSM(AM) Shawinigan PQ
CKSQ(AM) Stettler AB
CKST(AM) Vancouver BC
CKSW(AM) Swift Current SK
CKTB(AM) Saint Catharines ON
CKUA(AM) Edmonton AB
CKVH(AM) High Prairie AB
CKVO(AM) Clarenville NF
CKWL(AM) Williams Lake BC
CKWW(AM) Windsor ON
CKWX(AM) Vancouver BC
CKYL(AM) Peace River AB
*VOAR(AM) Mount Pearl NF
VOCM(AM) Saint John's NF
*VOWR(AM) Saint John's NF

Canadian FM Stations by Call Letters

*CBAF-FM Moncton NB
CBA-FM Moncton NB
*CBAL-FM Moncton NB
CBAX-FM Halifax NS
CBBK-FM Kingston ON
CBBL-FM London ON
CBBS-FM Sudbury ON
CBBX-FM Sudbury ON
CBCL-FM London ON
*CBCS-FM Sudbury ON
*CBCT-FM Charlottetown PE
*CBCV-FM Victoria BC
CBCX-FM Calgary AB
CBD-FM Saint John NB
CBDQ-FM Labrador City NF
*CBE-FM Windsor ON
CBEG-FM Sarnia ON
CBF-FM Montreal PQ
CBF-FM-1 Trois Rivieres PQ
CBFX-FM Montreal PQ
CBGA-FM Matane PQ
CBHA-FM Halifax NS
*CBH-FM Halifax NS
CBI-FM Sydney NS
CBJE-FM Chicoutimi PQ
CBJ-FM Chicoutimi PQ
*CBJX-FM Chicoutimi PQ
*CBKA-FM La Ronge SK
CBKF-FM Regina SK
*CBK-FM Regina SK
*CBKS-FM Saskatoon SK
CBLA-FM Toronto ON
CBL-FM Toronto ON
CBME-FM Montreal PQ
CBM-FM Montreal PQ
CBMI-FM Baie Comeau PQ
CBMR-FM Fermont PQ
*CBN-FM Saint John's NF
*CBOF-FM Ottawa ON
*CBO-FM Ottawa ON
*CBON-FM Sudbury ON
*CBOQ-FM Ottawa ON
*CBOX-FM Ottawa ON
*CBQ-FM Thunder Bay ON
CBQL-FM Savant Lake ON
CBQR-FM Rankin Inlet NU
CBQS-FM Sioux Narrows ON
*CBQT-FM Thunder Bay ON
CBQX-FM Kenora ON
CBRF-FM Calgary AB
*CBR-FM Calgary AB
*CBRX-FM Rimouski PQ
*CBSI-FM Sept-Iles PQ
CBTE-FM Crawford Bay BC
*CBTK-FM Kelowna BC
CBUF-FM Vancouver BC
CBU-FM Vancouver BC
CBUX-FM Vancouver BC
CBVE-FM Quebec PQ
*CBV-FM Quebec PQ
*CBV-FM-6 La Malbaie PQ
*CBVX-FM Quebec PQ
CBW-FM Winnipeg MB
*CBWK-FM Thompson MB
CBX-FM Edmonton AB
CBYG-FM Prince George BC
CBZF-FM Fredericton NB
*CBZ-FM Fredericton NB
CFAI-FM Edmundston NB
CFAK-FM Sherbrooke PQ
CFAN-FM Miramichi City NB
CFAQ-FM Blucher SK
CFBG-FM Bracebridge ON
CFBK-FM Huntsville ON
CFBR-FM Edmonton AB
CFBS-FM Lourdes-de-Blanc-Sablon PQ
CFBT-FM Vancouver BC
CFBU-FM Saint Catharines ON
CFBW-FM Hanover ON
CFBX-FM Kamloops BC
CFCA-FM Kitchener ON
CFCH-FM Chase BC
CFCP-FM Courtenay BC
CFCR-FM Saskatoon SK
CFCV-FM Saint Andrews NF
CFCW-FM Camrose AB
CFCY-FM Charlottetown PE
CFDA-FM Victoriaville PQ
CFDM-FM Meadow Lake SK
CFDV-FM Red Deer AB
CFEI-FM Saint Hyacinthe PQ
CFEL-FM Montmagny PQ
CFEP-FM Eastern Passage NS
CFEQ-FM Winnipeg MB
CFET-FM Tagish YT
CFEX-FM Calgary AB
*CFFF-FM Peterborough ON
CFFM-FM Williams Lake BC
*CFGB-FM Happy Valley NF
CFGE-FM Sherbrooke PQ
CFGI-FM Georgina Island ON
CFGL-FM Laval PQ
CFGP-FM Grande Prairie AB
CFGQ-FM Calgary AB
CFGW-FM Yorkton SK
CFGX-FM Sarnia ON
CFHK-FM St. Thomas ON
CFIC-FM Listuguj PQ
CFID-FM Acton Vale PQ
CFIF-FM Iroquois Falls ON
CFIM-FM Iles-de-la-Madeleine PQ
*CFIN-FM Lac-Etchemin PQ
CFIT-FM Airdrie AB
CFIX-FM Chicoutimi PQ
CFJB-FM Barrie ON
CFJO-FM Thetford Mines PQ
CFJR-FM Brockville ON
*CFJU-FM Kedgwick NB
CFLC-FM Churchill Falls NF
CFLG-FM Cornwall ON
CFLO-FM Mont-Laurier PQ
CFLX-FM Sherbrooke PQ
CFLY-FM Kingston ON
CFLZ-FM Niagara Falls ON
CFMC-FM Saskatoon SK
CFMF-FM Fermont PQ
CFMG-FM Saint Albert AB
CFMH-FM Saint John NB
CFMI-FM New Westminster BC
CFMK-FM Kingston ON
CFML-FM Burnaby BC
CFMM-FM Prince Albert SK
CFMQ-FM Hudson Bay SK
*CFMU-FM Hamilton ON
CFMV-FM Chandler PQ
CFMX-FM Toronto ON
CFMY-FM Medicine Hat AB
CFMZ-FM Cobourg ON
CFMZ-FM-1 Toronto ON
CFNJ-FM Saint Gabriel-de-Brandon PQ
CFNK-FM Pinehouse Lake SK
CFNO-FM Marathon ON
CFNR-FM Terrace BC
CFNY-FM Toronto ON
CFOB-FM Fort Frances ON
CFOM-FM Levis PQ
CFOR-FM Maniwaki PQ
CFOU-FM Trois Rivieres PQ
CFOX-FM Vancouver BC
CFOZ-FM Argentia NF
CFPL-FM London ON
CFPS-FM Port Elgin ON
CFPV-FM Pemberton BC
CFPX-FM Pukatawagan MB
CFQC-FM Saskatoon SK
CFQK-FM Kaministiquia ON
CFQM-FM Moncton NB
CFQR-FM Montreal PQ
CFQX-FM Selkirk MB
*CFRC-FM Kingston ON
CFRG-FM Gravelbourg SK
*CFRH-FM Penetanguishene ON
CFRI-FM Grande Prairie AB
CFRK-FM Fredericton NB
CFRM-FM Little Current ON
*CFRO-FM Vancouver BC
CFRQ-FM Dartmouth NS
CFRT-FM Iqaluit NU
CFRU-FM Guelph ON
CFRV-FM Lethbridge AB
CFRY-FM Portage la Prairie MB
CFSF-FM Sturgeon Falls ON
CFSR-FM Hope BC
*CFTH-FM-1 Harrington Harbour PQ
CFTH-FM-2 La Tabatiere PQ
CFTX-FM Gatineau PQ
CFUL-FM Calgary AB
*CFUR-FM Prince George BC
CFUT-FM Shawinigan PQ
*CFUV-FM Victoria BC
CFVD-FM Degelis PQ
CFVD-FM-2 Pohenegamook PQ
CFVM-FM Amqui PQ
CFWC-FM Brantford ON
CFWE-FM Lac La Biche AB
CFWF-FM Regina SK
CFWM-FM Winnipeg MB
CFWP-FM Wahta Mohawk Territory near Bala ON
CFXE-FM Edson AB
CFXH-FM Hinton AB
CFXJ-FM Toronto ON
CFXL-FM High River-Okotoks AB
CFXN-FM North Bay ON
CFXU-FM Antigonish NS
CFXW-FM Whitecourt AB
CFXY-FM Fredericton NB
CFZN-FM Haliburton ON
CFZZ-FM Saint Jean-Iberville PQ
CHAA-FM Longueuil PQ
CHAD-FM Dawson Creek BC
CHAI-FM Chateauguay PQ
CHAL-FM Halifax NS
CHAS-FM Sault Ste. Marie ON
CHAT-FM Medicine Hat AB
CHAY-FM Barrie ON
CHBE-FM Victoria BC
CHBN-FM Edmonton AB
CHBW-FM Rocky Mountain House AB
CHBZ-FM Cranbrook BC
CHCD-FM Simcoe ON
CHCQ-FM Belleville ON
CHCR-FM Killaloe ON
CHDH-FM Siksika AB
CHDI-FM Edmonton AB
CHDR-FM Cranbrook BC
CHEF-FM Matagami PQ
CHEQ-FM Sainte-Marie-de-Beauce PQ
CHER-FM Sydney NS
CHES-FM Erin ON
CHET-FM Chetwynd BC
CHEY-FM Trois Rivieres PQ
CHEZ-FM Ottawa ON
CHFI-FM Toronto ON
CHFM-FM Calgary AB
CHFN-FM Cape Croker (Neyaashiinigmiing) ON
CHFX-FM Halifax NS
CHGA-FM Maniwaki PQ
CHGB-FM Wasaga Beach ON
CHGK-FM Stratford ON
CHGO-FM Val d'Or PQ
CHHO-FM Louiseville PQ
CHIC-FM Rouyn-Noranda PQ
CHIK-FM Quebec PQ
*CHIM-FM Timmins ON
CHIN-FM Toronto ON
CHIP-FM Fort Coulonge PQ
CHIQ-FM Winnipeg MB
CHJM-FM Saint Georges PQ
CHJX-FM London ON
CHKF-FM Calgary AB
CHKG-FM Richmond BC
CHKS-FM Sarnia ON
CHLB-FM Lethbridge AB
CHLC-FM Baie Comeau PQ
CHLK-FM Perth ON
CHLQ-FM Charlottetown PE
CHLS-FM Lillooet BC
CHLT-FM Sherbrooke PQ
CHLX-FM Gatineau PQ
CHLY-FM Nanaimo BC
*CHMA-FM Sackville NB
CHMC-FM Edmonton AB
CHME-FM Les Escoumins PQ
CHMM-FM MacKenzie BC
CHMN-FM Canmore AB
CHMP-FM Longueuil PQ
*CHMR-FM Saint John's NF
CHMS-FM Bancroft ON
CHMT-FM Timmins ON
CHMX-FM Regina SK
CHMY-FM Renfrew ON
CHMZ-FM Tofino BC
CHNI-FM Saint John NB
CHNK-FM Winnipeg MB
CHNO-FM Sudbury ON
CHNS-FM Halifax NS
CHNV-FM Nelson BC
CHOA-FM Rouyn-Noranda PQ
*CHOC-FM Saint Remi PQ
CHOD-FM Cornwall ON
CHOE-FM Matane PQ
CHOI-FM Quebec PQ
CHOM-FM Montreal PQ
CHON-FM Whitehorse YT
CHOQ-FM Toronto ON
CHOS-FM Rattling Brook NF
CHOX-FM La Pocatiere PQ
CHOY-FM Moncton NB
CHOZ-FM Saint John's NF
CHPB-FM Cochrane ON
CHPD-FM Aylmer ON
CHPQ-FM Parksville BC
CHPR-FM Hawkesbury ON
CHQC-FM Saint John NB
CHQM-FM Vancouver BC
CHQX-FM Prince Albert SK
CHRD-FM Drummondville PQ
CHRE-FM Saint Catharines ON
CHRG-FM Maria (Reserve) PQ
CHRI-FM Ottawa ON
CHRL-FM Roberval PQ
CHRM-FM Matane PQ
CHRQ-FM Restigouche PQ
*CHRW-FM London ON
CHRX-FM Fort St. John BC
CHRY-FM Toronto ON
CHSB-FM Bedford NS
CHSJ-FM Saint John NB
CHSL-FM Slave Lake AB
CHSN-FM Estevan SK
CHSR-FM Fredericton NB
CHST-FM London ON
CHSU-FM Kelowna BC
CHTD-FM Saint Stephen NB
CHTN-FM Charlottetown PE
CHTT-FM Victoria BC
CHTZ-FM Saint Catharines ON
CHUB-FM Red Deer AB
CHUC-FM Cobourg ON
CHUK-FM Mashteuiatsh (Pointe-Bleue) PQ
CHUM-FM Toronto ON
*CHUO-FM Ottawa ON
CHUR-FM North Bay ON
CHUT-FM Lac-Simon (Louvicourt) PQ
CHVD-FM Dolbeau-Mistassini PQ
CHVN-FM Winnipeg MB
CHVR-FM Pembroke ON
CHWC-FM Goderich ON
CHWF-FM Nanaimo BC
CHWV-FM Saint John NB
CHXL-FM Okanese Indian Reserve SK
CHYC-FM Sudbury ON
CHYK-FM Timmins ON
CHYK-FM-3 Hearst ON
CHYM-FM Kitchener ON
CHYR-FM Leamington ON
*CHYZ-FM Sainte Foy PQ
CIAJ-FM Prince Rupert BC
CIAM-FM Fort Vermilion AB
CIAU-FM Radisson PQ
CIAX-FM Windsor PQ
CIAY-FM Whitehorse YT
CIBH-FM Parksville BC
CIBK-FM Calgary AB
*CIBL-FM Montreal PQ
CIBM-FM Riviere du Loup PQ
CIBO-FM Senneterre PQ
CIBU-FM Wingham ON
CIBW-FM Drayton Valley AB
CIBX-FM Fredericton NB
CICF-FM Vernon BC
CICU-FM Eskasoni Indian Reserve NS
CICX-FM Orillia ON
CICY-FM Selkirk MB
CICZ-FM Midland ON
CIDC-FM Orangeville ON
CIDD-FM White Bear Lake Resort SK
CIDI-FM Lac-Brome PQ
CIDO-FM Creston BC
CIDR-FM Windsor ON
CIEG-FM Egmont BC
CIEL-FM Riviere du Loup PQ
CIEU-FM Carleton PQ
*CIFA-FM Comeauville NS
CIFM-FM Kamloops BC
CIFX-FM Lewisporte NF
CIGB-FM Trois Rivieres PQ
CIGL-FM Belleville ON
CIGO-FM Port Hawkesbury NS
CIGR-FM Sherbrooke PQ
CIGV-FM Penticton BC
CIHO-FM Saint Hilarion PQ
CIHR-FM Woodstock ON
CIHS-FM Wetaskiwin AB
CIHT-FM Ottawa ON
CIKI-FM Rimouski PQ
CIKR-FM Kingston ON
CIKX-FM Grand Falls NB
CIKZ-FM Kitchener-Waterloo ON
CILE-FM Havre-Saint-Pierre PQ
CILG-FM Moose Jaw SK
CILK-FM Kelowna BC
CILQ-FM North York ON
CILS-FM Victoria BC
CILT-FM Steinbach MB
CILU-FM Thunder Bay ON
CILV-FM Ottawa ON
CILZ-FM Greenville BC
CIME-FM Saint Jerome PQ
CIMF-FM Gatineau PQ
CIMG-FM Swift Current SK
CIMI-FM Charlesbourg PQ
CIMJ-FM Guelph ON
CIMM-FM Ucluelet BC
CIMO-FM Magog PQ
CIMS-FM Balmoral NB
CIMX-FM Windsor ON
CIMY-FM Pembroke ON
CINB-FM Saint John NB
CINC-FM Thompson MB
CING-FM Hamilton ON
*CINN-FM Hearst ON
CINQ-FM Montreal PQ
CINU-FM Truro NS
CIOC-FM Victoria BC
*CIOI-FM Hamilton ON
CIOK-FM Saint John NB
*CION-FM Quebec PQ
CIOO-FM Halifax NS

CIOS-FM Stephenville NF
CIOT-FM Nipawin SK
CIOZ-FM Marystown NF
CIPC-FM Port-Cartier PQ
CIPN-FM Pender Harbour BC
CIQB-FM Barrie ON
CIQM-FM London ON
CIQX-FM Calgary AB
CIRA-FM Montreal PQ
CIRK-FM Edmonton AB
CIRR-FM Toronto ON
CIRV-FM Toronto ON
CIRX-FM Prince George BC
CISC-FM Gibsons BC
*CISM-FM Montreal PQ
CISN-FM Edmonton AB
CISP-FM Pemberton BC
CISQ-FM Squamish BC
CISS-FM Ottawa ON
CISW-FM Whistler BC
CITA-FM Moncton NB
CITE-FM Montreal PQ
CITE-FM-1 Sherbrooke PQ
CITF-FM Quebec PQ
CITI-FM Winnipeg MB
*CITR-FM Vancouver BC
*CIUT-FM Toronto ON
CIVL-FM Abbotsford BC
CIVR-FM Yellowknife NT
CIWS-FM Whitchurch-Stouffville ON
CIWV-FM Hamilton ON
CIXF-FM Brooks AB
CIXK-FM Owen Sound ON
CIXL-FM Welland ON
CIXM-FM Whitecourt AB
CIXN-FM Fredericton NB
*CIXX-FM London ON
CIYN-FM Kincardine ON
CIZL-FM Regina SK
CIZZ-FM Red Deer AB
CJAB-FM Saguenay PQ
CJAI-FM Stella ON
*CJAM-FM Windsor ON
CJAN-FM Asbestos PQ
CJAQ-FM Toronto ON
CJAS-FM Saint Augustin PQ
CJAT-FM Trail BC
CJAV-FM Port Alberni BC
CJAY-FM Calgary AB
CJBB-FM Englehart ON
CJBC-FM Toronto ON
CJBC-FM-4 London ON
CJBE-FM Port-Menier PQ
CJBR-FM Rimouski PQ
CJBX-FM London ON
CJBZ-FM Taber AB
CJCD-FM Yellowknife NT
CJCD-FM-1 Hay River NT
CJCF-FM Cumberland House SK
CJCI-FM Prince George BC
CJCJ-FM Woodstock NB
CJCQ-FM North Battleford SK
CJDJ-FM Saskatoon SK
CJDM-FM Drummondville PQ
CJDR-FM Fernie BC
CJDS-FM Saint Pamphile PQ
CJDV-FM Cambridge ON
CJEB-FM Trois Rivieres PQ
CJEC-FM Quebec PQ
CJEF-FM Saint John NB
CJEG-FM Bonnyville AB
CJEL-FM Winkler MB

CJEM-FM Edmundston NB
CJET-FM Smiths Falls ON
CJEZ-FM Toronto ON
CJFH-FM Woodstock ON
CJFM-FM Montreal PQ
CJFW-FM Terrace BC
CJFX-FM Antigonish NS
CJFY-FM Blackville NB
CJHR-FM Renfrew ON
CJIJ-FM Sydney NS
CJIQ-FM Kitchener/Paris ON
CJIT-FM Lac Megantic PQ
*CJIV-FM Dryden ON
CJJC-FM Yorkton SK
CJJJ-FM Brandon MB
CJJR-FM Vancouver BC
CJKC-FM Kamloops BC
CJKK-FM Clarenville NF
CJKL-FM Kirkland Lake ON
CJKR-FM Winnipeg MB
CJKX-FM Ajax ON
CJLA-FM Lachute PQ
CJLF-FM Barrie ON
CJLL-FM Ottawa ON
CJLM-FM Joliette PQ
CJLR-FM La Ronge SK
CJLS-FM Yarmouth NS
CJLS-FM-1 Barrington NS
CJLS-FM-2 New Tusket NS
CJLT-FM Medicine Hat AB
CJLU-FM Halifax NS
CJLX-FM Belleville ON
*CJLY-FM Nelson BC
CJMC-FM Sainte Anne des Monts PQ
CJMF-FM Quebec PQ
CJMG-FM Penticton BC
CJMI-FM Strathroy ON
CJMJ-FM Ottawa ON
CJMK-FM Saskatoon SK
CJMM-FM Rouyn-Noranda PQ
CJMO-FM Moncton NB
CJMP-FM Powell River BC
CJMQ-FM Sherbrooke PQ
CJMV-FM Val d'Or PQ
CJMX-FM Sudbury ON
CJNE-FM Nipawin SK
CJNI-FM Halifax NS
CJNS-FM Meadow Lake SK
CJNU-FM Winnipeg MB
*CJOA-FM Thunder Bay ON
CJOC-FM Lethbridge AB
CJOI-FM Rimouski PQ
CJOJ-FM Belleville ON
CJOK-FM Fort McMurray AB
CJOZ-FM Bonavista Bay NF
CJPG-FM Portage la Prairie MB
*CJPN-FM Fredericton NB
CJPR-FM Blairmore AB
CJPT-FM Brockville ON
CJPX-FM Montreal PQ
CJQM-FM Sault Ste. Marie ON
CJQQ-FM Timmins ON
CJRC-FM Gatineau PQ
CJRE-FM Riviere au Renard PQ
CJRG-FM Gaspe PQ
CJRI-FM Fredericton NB
CJRL-FM Kenora ON
CJRM-FM Labrador City NF
CJRQ-FM Sudbury ON
*CJRT-FM Toronto ON
CJRW-FM Summerside PE
CJRX-FM Lethbridge AB

CJRY-FM Edmonton AB
CJSA-FM Toronto ON
CJSB-FM Swan River MB
CJSD-FM Thunder Bay ON
CJSE-FM Shediac NB
*CJSF-FM Burnaby BC
CJSI-FM Calgary AB
CJSO-FM Sorel PQ
CJSP-FM Leamington ON
CJSQ-FM Quebec PQ
CJSR-FM Edmonton AB
CJSS-FM Cornwall ON
CJSU-FM Duncan BC
*CJSW-FM Calgary AB
CJTK-FM Sudbury ON
CJTN-FM Quinte West ON
*CJTR-FM Regina SK
CJTT-FM New Liskeard ON
CJTW-FM Kitchener-Waterloo ON
CJUK-FM Thunder Bay ON
CJUM-FM Winnipeg MB
CJUV-FM Lacombe AB
CJVR-FM Melfort SK
CJWA-FM Wawa ON
CJWL-FM Ottawa ON
CJWV-FM Winnipeg MB
CJXK-FM Grand Centre (Cold Lake) AB
CJXL-FM Moncton NB
CJXX-FM Grande Prairie AB
CJXY-FM Burlington ON
CJYC-FM Saint John NB
CJZN-FM Victoria BC
CJZZ-FM Winnipeg MB
CKAG-FM Pikogan PQ
CKAJ-FM Jonquiere PQ
CKAP-FM Kapuskasing ON
CKAU-FM Maliotenam PQ
CKAV-FM Toronto ON
CKAV-FM-2 Vancouver BC
CKAV-FM-3 Calgary AB
CKAV-FM-9 Ottawa ON
CKAY-FM Sechelt BC
CKBC-FM Bathurst NB
CKBT-FM Kitchener-Waterloo ON
CKBW-FM Bridgewater NS
CKBW-FM-1 Liverpool NS
CKBW-FM-2 Shelburne NS
CKBY-FM Smiths Falls ON
CKBZ-FM Kamloops BC
CKCB-FM Collingwood ON
CKCE-FM Calgary AB
CKCK-FM Regina SK
CKCL-FM Chilliwack BC
CKCN-FM Sept-Iles PQ
CKCQ-FM Quesnel BC
CKCU-FM Ottawa ON
CKCW-FM Moncton NB
CKDG-FM Montreal PQ
*CKDJ-FM Ottawa ON
CKDK-FM Woodstock ON
CKDR-FM Dryden ON
*CKDU-FM Halifax NS
CKDV-FM Prince George BC
CKDX-FM Newmarket ON
CKEN-FM Kentville NS
CKER-FM Edmonton AB
CKEY-FM Fort Erie ON
CKFI-FM Swift Current SK
CKFU-FM Fort St. John BC
CKFX-FM North Bay ON
CKGB-FM Timmins ON

CKGE-FM Oshawa ON
CKGF-FM-2 Greenwood BC
CKGN-FM Kapuskasing ON
CKGO-FM-1 Boston Bar BC
CKGW-FM Chatham ON
CKGY-FM Red Deer AB
CKHA-FM Haliburton ON
CKHC-FM Toronto ON
CKHL-FM High Level AB
CKHR-FM Hay River NT
CKHZ-FM Halifax NS
CKIA-FM Quebec PQ
CKIC-FM Winnipeg MB
CKII-FM Dolbeau-Mistassini PQ
CKIQ-FM Iqaluit NU
CKIS-FM Calgary AB
CKIX-FM Saint John's NF
CKIZ-FM Vernon BC
CKJJ-FM Belleville ON
*CKJM-FM Cheticamp NS
CKJN-FM Haldimand County ON
CKKC-FM Nelson BC
CKKK-FM Peterborough ON
CKKL-FM Ottawa ON
CKKN-FM Prince George BC
CKKQ-FM Victoria BC
CKKS-FM Sechelt BC
CKKX-FM Peace River AB
CKLB-FM Yellowknife NT
CKLD-FM Thetford Mines PQ
CKLE-FM Bathurst NB
CKLF-FM Brandon MB
CKLG-FM Vancouver BC
CKLH-FM Hamilton ON
CKLJ-FM Olds AB
CKLM-FM Lloydminster AB
CKLM-FM-1 Bonnyville AB
*CKLN-FM Toronto ON
CKLP-FM Parry Sound ON
CKLR-FM Courtenay BC
*CKLU-FM Sudbury ON
CKLX-FM Montreal PQ
CKLY-FM Lindsay (city of Kawartha Lakes) ON
CKLZ-FM Kelowna BC
CKMB-FM Barrie ON
CKMF-FM Montreal PQ
CKMM-FM Winnipeg MB
CKMN-FM Rimouski-Mont Joli PQ
*CKMS-FM Waterloo ON
CKMV-FM Grand Falls NB
CKNA-FM Natashquan PQ
CKNG-FM Edmonton AB
CKNI-FM Moncton NB
CKNL-FM Fort St. John BC
CKNR-FM Elliot Lake ON
CKNU-FM Donnacona PQ
CKNX-FM Wingham ON
CKOA-FM Glace Bay NS
CKOD-FM Valleyfield PQ
CKOE-FM Moncton NB
CKOI-FM Verdun PQ
CKOL-FM Campbellford ON
CKON-FM Akwesasne ON
CKOT-FM Tillsonburg ON
CKOZ-FM Corner Brook NF
CKPC-FM Brantford ON
CKPE-FM Sydney NS
CKPR-FM Thunder Bay ON
CKQB-FM Ottawa ON
CKQC-FM Abbotsford BC
CKQK-FM Charlottetown PE

CKQM-FM Peterborough ON
CKQN-FM Baker Lake NU
CKQR-FM Castlegar BC
CKQV-FM Vermillion Bay ON
CKRA-FM Edmonton AB
CKRB-FM Saint Georges-de-Beauce PQ
CKRC-FM Weyburn SK
CKRH-FM Halifax NS
*CKRK-FM Kahnawake PQ
*CKRL-FM Quebec PQ
CKRO-FM Pokemouche NB
*CKRP-FM Falher AB
CKRV-FM Kamloops BC
CKRX-FM Fort Nelson BC
CKRY-FM Calgary AB
*CKRZ-FM Ohsweken ON
CKSA-FM Lloydminster AB
CKSG-FM Cobourg ON
CKSJ-FM Saint John's NF
CKSO-FM Sudbury ON
CKSR-FM Chilliwack BC
CKSS-FM Red Rocks NF
CKSY-FM Chatham ON
CKTF-FM Gatineau PQ
CKTG-FM Thunder Bay ON
CKTI-FM Kettle Point ON
CKTK-FM Kitimat BC
CKTO-FM Truro NS
CKTP-FM Fredericton Centre NB
CKTY-FM Truro NS
*CKUA-FM Edmonton AB
CKUE-FM Chatham ON
CKUJ-FM Kuujjuaq PQ
CKUL-FM Halifax NS
CKUM-FM Moncton NB
CKUN-FM Christian Island ON
*CKUT-FM Montreal PQ
*CKUW-FM Winnipeg MB
*CKVI-FM Kingston ON
CKVM-FM Ville-Marie PQ
CKVN-FM Lethbridge AB
CKVX-FM Kindersley SK
CKWE-FM Maniwaki (Kitigan Zibi Anishinabeg Reserve) PQ
CKWF-FM Peterborough ON
CKWM-FM Kentville NS
CKWR-FM Kitchener ON
CKWV-FM Nanaimo BC
CKWY-FM Wainwright AB
CKXA-FM Brandon MB
CKXD-FM Gander NF
CKX-FM Brandon MB
CKXG-FM Grand Falls-Windsor NF
CKXL-FM Saint Boniface MB
CKXO-FM Chibougamau PQ
CKXR-FM Salmon Arm BC
CKXU-FM Lethbridge AB
CKXX-FM Corner Brook NF
CKYC-FM Owen Sound ON
CKYE-FM Vancouver BC
CKY-FM Winnipeg MB
CKYK-FM Alma PQ
CKYM-FM Napanee ON
CKYQ-FM Plessisville PQ
CKYX-FM Fort McMurray AB
CKZP-FM Zenon Park SK
CKZX-FM New Denver BC
CKZZ-FM Vancouver BC
VOCM-FM Saint John's NF

U.S. AM Stations by Frequency

540 khz
KRXA(AM) Carmel Valley CA
*KVIP(AM) Redding CA
WFLF(AM) Pine Hills FL
WDAK(AM) Columbus GA
KWMT(AM) Fort Dodge IA
KNOE(AM) Monroe LA
WGOP(AM) Pocomoke City MD
WETC(AM) Wendell-Zebulon NC
WXNH(AM) Jaffrey NH
KNMX(AM) Las Vegas NM
WLIE(AM) Islip NY
WWCS(AM) Canonsburg PA
WYNN(AM) Florence SC
WKFN(AM) Clarksville TN
KDFT(AM) Ferris TX
KNAK(AM) Delta UT
WGTH(AM) Richlands VA
WRRD(AM) Jackson WI
KJJL(AM) Pine Bluffs WY

550 khz
KTZN(AM) Anchorage AK
WASG(AM) Atmore AL
KFYI(AM) Phoenix AZ
KUZZ(AM) Bakersfield CA
KLLV(AM) Breen CO
KRAI(AM) Craig CO
*WAYR(AM) Orange Park FL
WDUN(AM) Gainesville GA
KMVI(AM) Wailuku HI
KFRM(AM) Salina KS
KTRS(AM) Saint Louis MO
KBOW(AM) Butte MT
WIOZ(AM) Pinehurst NC
WAME(AM) Statesville NC
KFYR(AM) Bismarck ND
WGR(AM) Buffalo NY
WKRC(AM) Cincinnati OH
*KOAC(AM) Corvallis OR
WPAB(AM) Ponce PR
WDDZ(AM) Pawtucket RI
KCRS(AM) Midland TX
KTSA(AM) San Antonio TX
WSVA(AM) Harrisonburg VA
WDEV(AM) Waterbury VT
KARI(AM) Blaine WA
WSAU(AM) Wausau WI

560 khz
KVOK(AM) Kodiak AK
WOOF(AM) Dothan AL
KBLU(AM) Yuma AZ
KSFO(AM) San Francisco CA
KLZ(AM) Denver CO
WQAM(AM) Miami FL
WIND(AM) Chicago IL
WMIK(AM) Middlesboro KY
WHYN(AM) Springfield MA
WFRB(AM) Frostburg MD
WGAN(AM) Portland ME
WRDT(AM) Monroe MI
WEBC(AM) Duluth MN
KWTO(AM) Springfield MO
KMON(AM) Great Falls MT
WGAI(AM) Elizabeth City NC
WCKL(AM) Catskill NY
WFIL(AM) Philadelphia PA
WVOC(AM) Columbia SC
WNSR(AM) Brentwood TN
WHBQ(AM) Memphis TN
KLVI(AM) Beaumont TX
KPQ(AM) Wenatchee WA
WJLS(AM) Beckley WV

567 khz
KGUM(AM) Hagatna GU

570 khz
WAAX(AM) Gadsden AL
KCFJ(AM) Alturas CA
KLAC(AM) Los Angeles CA
WTBN(AM) Pinellas Park FL
KQNG(AM) Lihue HI
WKYX(AM) Paducah KY
WIDS(AM) Russell Springs KY
WTNT(AM) Bethesda MD
WWNC(AM) Asheville NC
WDOX(AM) Raleigh NC
KSNM(AM) Las Cruces NM
WMCA(AM) New York NY
WSYR(AM) Syracuse NY
WKBN(AM) Youngstown OH
WNAX(AM) Yankton SD
KLIF(AM) Dallas TX
KNRS(AM) Salt Lake City UT
KVI(AM) Seattle WA
WMAM(AM) Marinette WI

580 khz
KRSA(AM) Petersburg AK
WBIL(AM) Tuskegee AL
KSAZ(AM) Marana AZ
KMJ(AM) Fresno CA
KUBC(AM) Montrose CO
WDBO(AM) Orlando FL
WGAC(AM) Augusta GA
KIDO(AM) Nampa ID
*WILL(AM) Urbana IL
WIBW(AM) Topeka KS
KJMJ(AM) Alexandria LA
WTAG(AM) Worcester MA
WTCM(AM) Traverse City MI
WELO(AM) Tupelo MS
KANA(AM) Anaconda MT
WKSK(AM) West Jefferson NC
KTMT(AM) Ashland OR
WHP(AM) Harrisburg PA
WKAQ(AM) San Juan PR
KZMX(AM) Hot Springs SD
WOFE(AM) Rockwood TN
KRFE(AM) Lubbock TX
WLES(AM) Bon Air VA
WKTY(AM) La Crosse WI
WCHS(AM) Charleston WV

585 khz
KJAL(AM) Tafuna AS

590 khz
KHAR(AM) Anchorage AK
WRAG(AM) Carrollton AL
KPZA(AM) Hot Springs AR
KTIE(AM) San Bernardino CA
KTHO(AM) South Lake Tahoe CA
KCSJ(AM) Pueblo CO
WAFC(AM) Clewiston FL
WDIZ(AM) Panama City FL
WDWD(AM) Atlanta GA
KSSK(AM) Honolulu HI
KID(AM) Idaho Falls ID
KFNS(AM) Wood River IL
WVLK(AM) Lexington KY
WEZE(AM) Boston MA
WJMS(AM) Ironwood MI
WKZO(AM) Kalamazoo MI
KGLE(AM) Glendive MT
WCAB(AM) Rutherfordton NC
WGTM(AM) Wilson NC
KXSP(AM) Omaha NE
WROW(AM) Albany NY
KUGN(AM) Eugene OR
WARM(AM) Scranton PA
WMBS(AM) Uniontown PA
WWLX(AM) Lawrenceburg TN
KLBJ(AM) Austin TX
KSUB(AM) Cedar City UT
WLVA(AM) Lynchburg VA
KQNT(AM) Spokane WA

600 khz
KVNA(AM) Flagstaff AZ
KOGO(AM) San Diego CA
KCOL(AM) Wellington CO
WICC(AM) Bridgeport CT
WBWL(AM) Jacksonville FL
WMT(AM) Cedar Rapids IA
WKYH(AM) Paintsville KY
WVOG(AM) New Orleans LA
WCAO(AM) Baltimore MD
*WFST(AM) Caribou ME
WCHT(AM) Escanaba MI
WSNL(AM) Flint MI
KGEZ(AM) Kalispell MT
WCVP(AM) Murphy NC
WSJS(AM) Winston-Salem NC
KSJB(AM) Jamestown ND
WSOM(AM) Salem OH
WFRM(AM) Coudersport PA
WYEL(AM) Mayaguez PR
WREC(AM) Memphis TN
KROD(AM) El Paso TX
KERB(AM) Kermit TX
KTBB(AM) Tyler TX
WVAR(AM) Richwood WV

610 khz
WAGG(AM) Birmingham AL
KARV(AM) Russellville AR
KAVL(AM) Lancaster CA
KEAR(AM) San Francisco CA
KVLE(AM) Vail CO
WSNG(AM) Torrington CT
WIOD(AM) Miami FL
WVTJ(AM) Pensacola FL
WPLO(AM) Grayson GA
WCEH(AM) Hawkinsville GA
WRUS(AM) Russellville KY
KDAL(AM) Duluth MN
KCSP(AM) Kansas City MO
KOJM(AM) Havre MT
WFNZ(AM) Charlotte NC
KCSR(AM) Chadron NE
WGIR(AM) Manchester NH
KNML(AM) Albuquerque NM
WTVN(AM) Columbus OH
KRTA(AM) Medford OR
WIP(AM) Philadelphia PA
WEXS(AM) Patillas PR
KILT(AM) Houston TX
KVNU(AM) Logan UT
WVBE(AM) Roanoke VA
WTFX(AM) Winchester VA
KONA(AM) Kennewick WA

620 khz
KGTL(AM) Homer AK
WJHX(AM) Lexington AL
KTAR(AM) Phoenix AZ
KIGS(AM) Hanford CA
KMJC(AM) Mount Shasta CA
KJOL(AM) Grand Junction CO
WDAE(AM) Saint Petersburg FL
WTRP(AM) La Grange GA
KIPA(AM) Hilo HI
KMNS(AM) Sioux City IA
KWAL(AM) Wallace ID
WTUV(AM) Louisville KY
WZON(AM) Bangor ME
WJDX(AM) Jackson MS
WDNC(AM) Durham NC
WSNR(AM) Jersey City NJ
WHEN(AM) Syracuse NY
KPOJ(AM) Portland OR
WKHB(AM) Irwin PA
WGCV(AM) Cayce SC
WRJZ(AM) Knoxville TN
KMKI(AM) Plano TX
WVMT(AM) Burlington VT
WTMJ(AM) Milwaukee WI
WWNR(AM) Beckley WV

630 khz
KJNO(AM) Juneau AK
KIAM(AM) Nenana AK
WAVU(AM) Albertville AL
WJDB(AM) Thomasville AL
KVMA(AM) Magnolia AR
KIDD(AM) Monterey CA
KHOW(AM) Denver CO
WMAL(AM) Washington DC
WBMQ(AM) Savannah GA
WNEG(AM) Toccoa GA
KUAM(AM) Hagatna GU
KFXD(AM) Boise ID
WLAP(AM) Lexington KY
KJSL(AM) Saint Louis MO
WAIZ(AM) Hickory NC
WMFD(AM) Wilmington NC
KLEA(AM) Lovington NM
KPLY(AM) Reno NV
KWRO(AM) Coquille OR
WEJL(AM) Scranton PA
WUNO(AM) San Juan PR
WPRO(AM) Providence RI
KSLR(AM) San Antonio TX
KTKK(AM) Sandy UT
KCIS(AM) Edmonds WA
KTRW(AM) Opportunity WA
WDGY(AM) Hudson WI
WJAW(AM) Saint Marys WV

640 khz
*KYUK(AM) Bethel AK
KFI(AM) Los Angeles CA
WMEN(AM) Royal Palm Beach FL
WVLG(AM) Wildwood FL
WGST(AM) Atlanta GA
*WOI(AM) Ames IA
KTIB(AM) Thibodaux LA
WNNZ(AM) Westfield MA
WMFN(AM) Zeeland MI
KGVW(AM) Belgrade MT
WFNC(AM) Fayetteville NC
WWJZ(AM) Mount Holly NJ
WHLO(AM) Akron OH
WWLS(AM) Moore OK
WXSM(AM) Blountville TN
WCRV(AM) Collierville TN

648 khz
WVUV(AM) Leone AS

650 khz
KENI(AM) Anchorage AK
KSTE(AM) Rancho Cordova CA
KRTR(AM) Honolulu HI
WSRO(AM) Ashland MA
WNMT(AM) Nashwauk MN
WSM(AM) Nashville TN
KIKK(AM) Pasadena TX
KMTI(AM) Manti UT
KUUX(AM) Pullman WA
KGAB(AM) Orchard Valley WY

660 khz
KFAR(AM) Fairbanks AK
WDLT(AM) Fairhope AL
KTNN(AM) Window Rock AZ
KGDP(AM) Oildale CA
WORL(AM) Altamonte Springs FL
WNFS(AM) White Springs FL
WMIC(AM) Sandusky MI
WBHR(AM) Sauk Rapids MN
KEYZ(AM) Williston ND
KCRO(AM) Omaha NE
WFAN(AM) New York NY
WXIC(AM) Waverly OH
KXOR(AM) Junction City OR
WPYT(AM) Wilkinsburg PA
WLFJ(AM) Greenville SC
KSKY(AM) Balch Springs TX
WLOY(AM) Rural Retreat VA
KAPS(AM) Mount Vernon WA

670 khz
*KDLG(AM) Dillingham AK
WYLS(AM) York AL
KWXI(AM) Glenwood AR
KIRN(AM) Simi Valley CA
KLTT(AM) Commerce City CO
WWFE(AM) Miami FL
KPUA(AM) Hilo HI
KBOI(AM) Boise ID
WSCR(AM) Chicago IL
WIEZ(AM) Lewistown PA
WMTY(AM) Farragut TN
WPMH(AM) Claremont VA
WVVT(AM) Essex Junction VT

675 khz
KTKB(AM) Tamuning GU

680 khz
*KBRW(AM) Barrow AK
KNBR(AM) San Francisco CA
WGES(AM) Saint Petersburg FL
WCNN(AM) North Atlanta GA
WCTT(AM) Corbin KY
WDRD(AM) Newburg KY
WRKO(AM) Boston MA
WCBM(AM) Baltimore MD
WDBC(AM) Escanaba MI
KFEQ(AM) Saint Joseph MO
KKGR(AM) East Helena MT
WPTF(AM) Raleigh NC
WRGC(AM) Sylva NC
KWKA(AM) Clovis NM
WINR(AM) Binghamton NY
WISR(AM) Butler PA
WAPA(AM) San Juan PR
WSMB(AM) Memphis TN
KKYX(AM) San Antonio TX
KBRD(AM) Lacey WA
KOMW(AM) Omak WA
WOGO(AM) Hallie WI
WKAZ(AM) Charleston WV

690 khz
WSPZ(AM) Birmingham AL
KEWI(AM) Benton AR
KVOI(AM) Tucson AZ
KRMX(AM) Pueblo CO
KRGS(AM) Rifle CO
WADS(AM) Ansonia CT
WOKV(AM) Jacksonville FL
KHCM(AM) Honolulu HI
KBLI(AM) Blackfoot ID
KGGF(AM) Coffeyville KS
WIST(AM) New Orleans LA
XETRA(AM) Tijuana MEX
WNZK(AM) Dearborn Heights MI
KFXN(AM) Minneapolis MN
KSTL(AM) Saint Louis MO
KOAQ(AM) Terrytown NE
KRCO(AM) Prineville OR
WPHE(AM) Phoenixville PA
KTSM(AM) El Paso TX
KPET(AM) Lamesa TX
KZEY(AM) Tyler TX
WZAP(AM) Bristol VA
*WVCY(AM) Oshkosh WI
WELD(AM) Fisher WV

700 khz
KBYR(AM) Anchorage AK
WEEL(AM) Dothan AL
KMBX(AM) Soledad CA
WJWB(AM) Gibsonia FL
WJOE(AM) Orange-Athol MA
WDMV(AM) Walkersville MD
KNAX(AM) McCook NE
WLW(AM) Cincinnati OH
KGRV(AM) Winston OR
KSEV(AM) Tomball TX
KHSE(AM) Wylie TX
KALL(AM) North Salt Lake City UT
KXLX(AM) Airway Heights WA

710 khz
WNTM(AM) Mobile AL
KAPZ(AM) Bald Knob AR
KMIA(AM) Black Canyon City AZ
KFIA(AM) Carmichael CA
KSPN(AM) Los Angeles CA
KNUS(AM) Denver CO
WAQI(AM) Miami FL
WUFF(AM) Eastman GA
WROM(AM) Rome GA
WEKC(AM) Williamsburg KY
KEEL(AM) Shreveport LA
KCMO(AM) Kansas City MO
WZOO(AM) Asheboro NC
WEGG(AM) Rose Hill NC
KXMR(AM) Bismarck ND
WOR(AM) New York NY
WKJB(AM) Mayaguez PR
WPOG(AM) Saint Matthews SC
WTPR(AM) Paris TN
*WFCM(AM) Smyrna TN
KGNC(AM) Amarillo TX
KURV(AM) Edinburg TX
WFNR(AM) Blacksburg VA
KIRO(AM) Seattle WA
WDSM(AM) Superior WI

720 khz
*KOTZ(AM) Kotzebue AK
WRZN(AM) Hernando FL
WVCC(AM) Hogansville GA
KUAI(AM) Eleele HI
WGN(AM) Chicago IL
WGCR(AM) Brevard NC
WQTH(AM) Claremont NH
KDWN(AM) Las Vegas NV
WVOA(AM) Dewitt NY
KFIR(AM) Sweet Home OR
WWII(AM) Shiremanstown PA
KSAH(AM) Universal City TX

730 khz
WUMP(AM) Madison AL
KQPN(AM) West Memphis AR
WWTK(AM) Lake Placid FL
WSTT(AM) Thomasville GA
*KBSU(AM) Boise ID
KLOE(AM) Goodland KS
WFMW(AM) Madisonville KY
WMTC(AM) Vancleve KY
WASO(AM) Covington LA
WACE(AM) Chicopee MA
WJTO(AM) Bath ME
WVFN(AM) East Lansing MI
KWOA(AM) Worthington MN
KWRE(AM) Warrenton MO
KURL(AM) Billings MT
WFMC(AM) Goldsboro NC
WOHS(AM) Shelby NC
KDAZ(AM) Albuquerque NM
WDOS(AM) Oneonta NY
WJYM(AM) Bowling Green OH
*KEZX(AM) Medford OR
WNAK(AM) Nanticoke PA
WPIT(AM) Pittsburgh PA
WLTQ(AM) Charleston SC
WLIL(AM) Lenoir City TN
KKDA(AM) Grand Prairie TX
KSVN(AM) Ogden UT
WXTR(AM) Alexandria VA
WMNA(AM) Gretna VA
KULE(AM) Ephrata WA
WJMT(AM) Merrill WI

740 khz
WMSP(AM) Montgomery AL
KIDR(AM) Phoenix AZ
KBRT(AM) Avalon CA
KCBS(AM) San Francisco CA
KVOR(AM) Colorado Springs CO
KVFC(AM) Cortez CO
WSBR(AM) Boca Raton FL
WQTM(AM) Orlando FL
KBOE(AM) Oskaloosa IA
WVLN(AM) Olney IL
WNOP(AM) Newport KY
WJIB(AM) Cambridge MA
WPAQ(AM) Mount Airy NC
KKAG(AM) Fargo ND
KATK(AM) Carlsbad NM
WNYH(AM) Huntington NY
KRMG(AM) Tulsa OK
WVCH(AM) Chester PA
WIAC(AM) San Juan PR
WRWB(AM) Harrogate TN
WIRJ(AM) Humboldt TN
WJIG(AM) Tullahoma TN
KTRH(AM) Houston TX
KCMC(AM) Texarkana TX
WMBG(AM) Williamsburg VA
WRPQ(AM) Baraboo WI
WMIN(AM) Hudson WI
WRNR(AM) Martinsburg WV

750 khz
KFQD(AM) Anchorage AK
WSB(AM) Atlanta GA
WTHQ(AM) Brookport IL
WNDZ(AM) Portage IN
KKNO(AM) Gretna LA
WBMD(AM) Baltimore MD
WRME(AM) Hampden ME
WLDR(AM) Petoskey MI
KBNN(AM) Lebanon MO
KERR(AM) Polson MT
WAUG(AM) New Hope NC
KMMJ(AM) Grand Island NE
KHWG(AM) Fallon NV
KSEO(AM) Durant OK
KXL(AM) Portland OR
WQOR(AM) Olyphant PA
KAMA(AM) El Paso TX
KOAL(AM) Price UT
WPDX(AM) Clarksburg WV

760 khz
WURL(AM) Moody AL
KMTL(AM) Sherwood AR
KFMB(AM) San Diego CA
KKZN(AM) Thornton CO
WLCC(AM) Brandon FL
WEFL(AM) Tequesta FL
KGU(AM) Honolulu HI
KCCV(AM) Overland Park KS
WVNE(AM) Leicester MA
WJR(AM) Detroit MI
WCIS(AM) Morganton NC
WCPS(AM) Tarboro NC
KEIP(AM) Las Vegas NV
WCHP(AM) Champlain NY
WORA(AM) Mayaguez PR
WETR(AM) Knoxville TN
WENO(AM) Nashville TN
KTKR(AM) San Antonio TX

770 khz
*KCHU(AM) Valdez AK
WVNN(AM) Athens AL
WHOA(AM) Saraland AL
KCBC(AM) Riverbank CA
WWCN(AM) North Fort Myers FL
WYHG(AM) Young Harris GA
WCGW(AM) Nicholasville KY
KJCB(AM) Lafayette LA
*KUOM(AM) Minneapolis MN
WEW(AM) Saint Louis MO
KATL(AM) Miles City MT
WLWL(AM) Rockingham NC
KKOB(AM) Albuquerque NM
KKOB Exp Stn Santa Fe NM
WABC(AM) New York NY
WTOR(AM) Youngstown NY
WAIS(AM) Buchtel OH
WKFB(AM) Jeannette PA
KAAM(AM) Garland TX
WYRV(AM) Cedar Bluff VA
KTTH(AM) Seattle WA

780 khz
*KNOM(AM) Nome AK
WZZX(AM) Lineville AL
KAZM(AM) Sedona AZ
KCEG(AM) Pueblo CO
WBBM(AM) Chicago IL
WCXH(AM) Monticello ME
WTME(AM) Rumford ME
WIIN(AM) Ridgeland MS
WCKB(AM) Dunn NC
WWOL(AM) Forest City NC
WJAG(AM) Norfolk NE
KKOH(AM) Reno NV
KSPI(AM) Stillwater OK
WPTN(AM) Cookeville TN
WAVA(AM) Arlington VA

790 khz
KCAM(AM) Glennallen AK
WTSK(AM) Tuscaloosa AL
KURM(AM) Rogers AR
KOSY(AM) Texarkana AR
KNST(AM) Tucson AZ
KFPT(AM) Clovis CA
KWSW(AM) Eureka CA
KABC(AM) Los Angeles CA
WLBE(AM) Leesburg FL
WPNN(AM) Pensacola FL
WAXY(AM) South Miami FL
WQXI(AM) Atlanta GA
WSFN(AM) Brunswick GA
WGRA(AM) Cairo GA
KKON(AM) Kealakekua HI
KSPD(AM) Boise ID
KBRV(AM) Soda Springs ID
WRMS(AM) Beardstown IL
KXXX(AM) Colby KS
WKRD(AM) Louisville KY
WSGW(AM) Saginaw MI
KGHL(AM) Billings MT
WBLO(AM) Thomasville NC
KFGO(AM) Fargo ND
KBET(AM) Winchester NV
WTNY(AM) Watertown NY
WLSV(AM) Wellsville NY
WHTH(AM) Heath OH
KWIL(AM) Albany OR
WAEB(AM) Allentown PA
WPIC(AM) Sharon PA
WSKO(AM) Providence RI
WVCD(AM) Bamberg-Denmark SC
WQSV(AM) Ashland City TN
WETB(AM) Johnson City TN
WMC(AM) Memphis TN
KBME(AM) Houston TX
KFYO(AM) Lubbock TX
WSVG(AM) Mount Jackson VA
WNIS(AM) Norfolk VA
KGMI(AM) Bellingham WA
KJRB(AM) Spokane WA
WAYY(AM) Eau Claire WI

800 khz
KINY(AM) Juneau AK
WHOS(AM) Decatur AL
WMGY(AM) Montgomery AL
KAGH(AM) Crossett AR
KVOM(AM) Morrilton AR
KBFP(AM) Bakersfield CA
WLAD(AM) Danbury CT
WPLK(AM) Palatka FL
WJAT(AM) Swainsboro GA
KXIC(AM) Iowa City IA
WKZI(AM) Casey IL
WSHO(AM) New Orleans LA
WNNW(AM) Lawrence MA
KQAD(AM) Luverne MN
WVAL(AM) Sauk Rapids MN
KREI(AM) Farmington MO
WKBC(AM) North Wilkesboro NC
WTMR(AM) Camden NJ
KQCV(AM) Oklahoma City OK
KPDQ(AM) Portland OR
WCHA(AM) Chambersburg PA
WDSC(AM) Dillon SC
WPJM(AM) Greer SC
WDEH(AM) Sweetwater TN
KDDD(AM) Dumas TX
WSVS(AM) Crewe VA
WDUX(AM) Waupaca WI
WVHU(AM) Huntington WV

801 khz
KTWG(AM) Hagatna GU

810 khz
WCKA(AM) Jacksonville AL
KGO(AM) San Francisco CA
KLDC(AM) Brighton CO
WEUS(AM) Orlovista FL
WTHV(AM) Hahira GA
WBIC(AM) Royston GA
WDDD(AM) Johnston City IL
WSYW(AM) Indianapolis IN
WEKG(AM) Jackson KY
WYRE(AM) Essex MD
WMJH(AM) Rockford MI
WHB(AM) Kansas City MO
WSJC(AM) Magee MS
KSWV(AM) Santa Fe NM
WGY(AM) Schenectady NY
WEDO(AM) McKeesport PA
WKVM(AM) San Juan PR
WQIZ(AM) Saint George SC
KBHB(AM) Sturgis SD
WCTA(AM) Alamo TN
WMGC(AM) Murfreesboro TN
KXOI(AM) Crane TX
KYTY(AM) Somerset TX
WPIN(AM) Dublin VA
KTBI(AM) Ephrata WA
WDMP(AM) Dodgeville WI
WJJQ(AM) Tomahawk WI

820 khz
KCBF(AM) Fairbanks AK
WMGG(AM) Largo FL
WAIT(AM) Willow Springs IL
*WSWI(AM) Evansville IN
WTWT(AM) Frederick MD
WWLZ(AM) Horseheads NY
*WNYC(AM) New York NY
*WOSU(AM) Columbus OH
KORC(AM) Waldport OR
WWAM(AM) Jasper TN
WBAP(AM) Fort Worth TX
KUTR(AM) Taylorsville UT
WGGM(AM) Chester VA
KGNW(AM) Burien-Seattle WA

830 khz
*KSDP(AM) Sand Point AK
*KFLT(AM) Tucson AZ
KNCO(AM) Grass Valley CA
KLAA(AM) Orange CA
WACC(AM) Hialeah FL
WJFA(AM) Hilliard FL
WFGM(AM) Sandy Springs GA
KHVH(AM) Honolulu HI
WFNO(AM) Norco LA
WCRN(AM) Worcester MA
WMMI(AM) Shepherd MI
WCCO(AM) Minneapolis MN
KOTC(AM) Kennett MO
WTRU(AM) Kernersville NC
WKTX(AM) Cortland OH
WEEU(AM) Reading PA
KMUL(AM) Farwell TX
KUYO(AM) Evansville WY

840 khz
WBHY(AM) Mobile AL
KMPH(AM) Modesto CA
WRYM(AM) New Britain CT
WPGS(AM) Mims FL
WHGH(AM) Thomasville GA
WHAS(AM) Louisville KY
KWDF(AM) Ball LA
WKDI(AM) Denton MD
KTIC(AM) West Point NE
KXNT(AM) North Las Vegas NV
KKNX(AM) Eugene OR
KSWB(AM) Seaside OR
WVPO(AM) Stroudsburg PA
WXEW(AM) Yabucoa PR
WCEO(AM) Columbia SC
KVJY(AM) Pharr TX
WKTR(AM) Earlysville VA
KMAX(AM) Colfax WA

850 khz
KICY(AM) Nome AK
WXJC(AM) Birmingham AL
KOA(AM) Denver CO
WREF(AM) Ridgefield CT
WRUF(AM) Gainesville FL
WFTL(AM) West Palm Beach FL
WCUG(AM) Cuthbert GA
WPTB(AM) Statesboro GA
KHLO(AM) Hilo HI
KWOF(AM) Waterloo IA
WCPT(AM) Crystal Lake IL
WEEI(AM) Boston MA
WGVS(AM) Muskegon MI
WWJC(AM) Duluth MN
*KFUO(AM) Clayton MO
WQST(AM) Forest MS
WLRC(AM) Walnut MS
WRBZ(AM) Raleigh NC
WYLF(AM) Penn Yan NY
WKNR(AM) Cleveland OH
WNTJ(AM) Johnstown PA
WABA(AM) Aguadilla PR
WPFD(AM) Fairview TN
WKVL(AM) Knoxville TN
KJON(AM) Carrollton TX
KEYH(AM) Houston TX
WTAR(AM) Norfolk VA
KHHO(AM) Tacoma WA

860 khz
WAMI(AM) Opp AL
KWRF(AM) Warren AR
KOSE(AM) Wilson AR
KMVP(AM) Phoenix AZ
KTRB(AM) San Francisco CA
WGUL(AM) Dunedin FL
WAEC(AM) Atlanta GA
WDMG(AM) Douglas GA
KWPC(AM) Muscatine IA
WMRI(AM) Marion IN
KKOW(AM) Pittsburg KS
WSON(AM) Henderson KY
WSBS(AM) Great Barrington MA
WBGR(AM) Baltimore MD
KNUJ(AM) New Ulm MN
WFMO(AM) Fairmont NC
WACB(AM) Taylorsville NC
KARS(AM) Belen NM
KPAM(AM) Troutdale OR
WAMO(AM) Millvale PA
WWDB(AM) Philadelphia PA
WLBG(AM) Laurens SC
WTZX(AM) Sparta TN
KFST(AM) Fort Stockton TX
KPAN(AM) Hereford TX
KSFA(AM) Nacogdoches TX
KONO(AM) San Antonio TX
KKAT(AM) Salt Lake City UT
WEVA(AM) Emporia VA
WNOV(AM) Milwaukee WI
WOAY(AM) Oak Hill WV

870 khz
WQRX(AM) Valley Head AL
KRLA(AM) Glendale CA
KJMP(AM) Pierce CO
KHNR(AM) Honolulu HI
WINU(AM) Shelbyville IL
WMTL(AM) Leitchfield KY
WWL(AM) New Orleans LA
WLVP(AM) Gorham ME
*WKAR(AM) East Lansing MI
KPRM(AM) Park Rapids MN
KAAN(AM) Bethany MO
WZNH(AM) Fitzwilliam Depot NH
KLSQ(AM) Whitney NV
WHCU(AM) Ithaca NY
WQBS(AM) San Juan PR
WPWT(AM) Colonial Heights TN
KFJZ(AM) Fort Worth TX
WFLO(AM) Farmville VA
KFLD(AM) Pasco WA

880 khz
KGHT(AM) Sheridan AR
KKMC(AM) Gonzales CA

WBKZ(AM) Jefferson GA
WIJR(AM) Highland IL
KJJR(AM) Whitefish MT
WRRZ(AM) Clinton NC
WPEK(AM) Fairview NC
WPIP(AM) Winston-Salem NC
KRVN(AM) Lexington NE
KHAC(AM) Tse Bonito NM
WCBS(AM) New York NY
WRFD(AM) Columbus-Worthington OH
KWIP(AM) Dallas OR
KCMX(AM) Phoenix OR
WYKO(AM) Sabana Grande PR
WNSG(AM) Nashville TN
KJOJ(AM) Conroe TX
WCQV(AM) Moneta VA
KIXI(AM) Mercer Island-Seattle WA
WMEQ(AM) Menomonie WI

890 khz
*KBBI(AM) Homer AK
WYAM(AM) Hartselle AL
KLFF(AM) Arroyo Grande CA
KJME(AM) Fountain CO
KDJQ(AM) Meridian ID
WLS(AM) Chicago IL
WAMG(AM) Dedham MA
KGGN(AM) Gladstone MO
WEEZ(AM) Laurel MS
WHNC(AM) Henderson NC
KQLX(AM) Lisbon ND
KTLR(AM) Oklahoma City OK
*WFKJ(AM) Cashtown PA
WFAB(AM) Ceiba PR
WBAJ(AM) Blythwood SC
KVOZ(AM) Del Mar Hills TX
KKBM(AM) Frankston TX
KTXV(AM) Mabank TX
KDXU(AM) Saint George UT
WKNV(AM) Fairlawn VA

900 khz
KZPA(AM) Fort Yukon AK
WATV(AM) Birmingham AL
WGOK(AM) Mobile AL
WOZK(AM) Ozark AL
KHOZ(AM) Harrison AR
KKHJ(AM) Leone AS
KBIF(AM) Fresno CA
KALI(AM) West Covina CA
WJWL(AM) Georgetown DE
WSWN(AM) Belle Glade FL
WMOP(AM) Ocala FL
WJTH(AM) Calhoun GA
WBML(AM) Macon GA
WJLG(AM) Savannah GA
KNUI(AM) Kahului HI
KSGL(AM) Wichita KS
WWLK(AM) Eddyville KY
WFIA(AM) Louisville KY
WLSI(AM) Pikeville KY
WILC(AM) Laurel MD
WJJB(AM) Brunswick ME
*KTIS(AM) Minneapolis MN
KFAL(AM) Fulton MO
WYCV(AM) Granite Falls NC
WAYN(AM) Rockingham NC
WIAM(AM) Williamston NC
KJSK(AM) Columbus NE
WGAM(AM) Nashua NH
WBRV(AM) Boonville NY
WUAM(AM) Saratoga Springs NY
WCER(AM) Canton OH
WCPA(AM) Clearfield PA
WURD(AM) Philadelphia PA
WNMB(AM) North Myrtle Beach SC
WKXV(AM) Knoxville TN
WKDA(AM) Lebanon TN
KPYN(AM) Atlanta TX
KFLP(AM) Floydada TX
KCLW(AM) Hamilton TX
KREH(AM) Pecan Grove TX
WCBX(AM) Bassett VA
WKDW(AM) Staunton VA
KKRT(AM) Wenatchee WA
WATK(AM) Antigo WI
WDLS(AM) Wisconsin Dells WI

910 khz
*KIYU(AM) Galena AK
WZMG(AM) Pepperell AL
KLCN(AM) Blytheville AR
KGME(AM) Phoenix AZ
*KECR(AM) El Cajon CA
KRAK(AM) Hesperia CA
KNEW(AM) Oakland CA
KOXR(AM) Oxnard CA
*KPOF(AM) Denver CO
WLAT(AM) New Britain CT
WTWD(AM) Plant City FL
WRFV(AM) Valdosta GA
*WSUI(AM) Iowa City IA
WAKO(AM) Lawrenceville IL
KINA(AM) Salina KS
WSFE(AM) Burnside KY
WUBR(AM) Baton Rouge LA
WABI(AM) Bangor ME
WGTO(AM) Cassopolis MI
WFDF(AM) Farmington Hills MI
WALT(AM) Meridian MS
KBLG(AM) Billings MT
WSRP(AM) Jacksonville NC
KCJB(AM) Minot ND
KBIM(AM) Roswell NM
WRKL(AM) New City NY
WLTP(AM) Marietta OH
WPFB(AM) Middletown OH
KVIS(AM) Miami OK
KURY(AM) Brookings OR
WAVL(AM) Apollo PA
WBZU(AM) Scranton PA
WSBA(AM) York PA
WPRP(AM) Ponce PR
WTMZ(AM) Dorchester Terrace-Brentwood SC
WOLI(AM) Spartanburg SC
KJJQ(AM) Volga SD
WMRB(AM) Columbia TN
WJCW(AM) Johnson City TN
WEPG(AM) South Pittsburg TN
KNAF(AM) Fredericksburg TX
KATH(AM) Frisco TX
KRIO(AM) McAllen TX
KWDZ(AM) Salt Lake City UT
WRNL(AM) Richmond VA
WWWR(AM) Roanoke VA
WNHV(AM) White River Junction VT
KTRO(AM) Vancouver WA
WHSM(AM) Hayward WI
WDOR(AM) Sturgeon Bay WI

920 khz
KSRM(AM) Soldotna AK
WGOL(AM) Russellville AL
KARN(AM) Little Rock AR
KVIN(AM) Ceres CA
KPSI(AM) Palm Springs CA
KVEC(AM) San Luis Obispo CA
KLMR(AM) Lamar CO
WMEL(AM) Melbourne FL
WGKA(AM) Atlanta GA
WVOH(AM) Hazlehurst GA
*KYFR(AM) Shenandoah IA
WGNU(AM) Granite City IL
WMOK(AM) Metropolis IL
*WBAA(AM) West Lafayette IN
WTCW(AM) Whitesburg KY
WBOX(AM) Bogalusa LA
WMPL(AM) Hancock MI
KDHL(AM) Faribault MN
KWAD(AM) Wadena MN
KWYS(AM) West Yellowstone MT
WPCM(AM) Burlington NC
WPTL(AM) Canton NC
WPHY(AM) Trenton NJ
KSVA(AM) Albuquerque NM
KBAD(AM) Las Vegas NV
*KIHM(AM) Reno NV
WKRT(AM) Cortland NY
WGHQ(AM) Kingston NY
WIRD(AM) Lake Placid NY
WMNI(AM) Columbus OH
KSHO(AM) Lebanon OR
WKVA(AM) Lewistown PA
WHJJ(AM) Providence RI

WYMB(AM) Manning SC
KKLS(AM) Rapid City SD
WLIV(AM) Livingston TN
KBNA(AM) El Paso TX
*KFLB(AM) Odessa TX
KYST(AM) Texas City TX
KVEL(AM) Vernal UT
KGTK(AM) Olympia WA
KXLY(AM) Spokane WA
WOKY(AM) Milwaukee WI
WMMN(AM) Fairmont WV

930 khz
KTKN(AM) Ketchikan AK
KNSA(AM) Unalakleet AK
WEZZ(AM) Monroeville AL
WJBY(AM) Rainbow City AL
KAPR(AM) Douglas AZ
KAFF(AM) Flagstaff AZ
KHJ(AM) Los Angeles CA
KKXX(AM) Paradise CA
KIUP(AM) Durango CO
KRKY(AM) Granby CO
WYUS(AM) Milford DE
WLVF(AM) Haines City FL
WFXJ(AM) Jacksonville FL
WLSS(AM) Sarasota FL
WMGR(AM) Bainbridge GA
KSEI(AM) Pocatello ID
WTAD(AM) Quincy IL
WAUR(AM) Sandwich IL
WHON(AM) Centerville IN
WKCT(AM) Bowling Green KY
WFMD(AM) Frederick MD
WBCK(AM) Battle Creek MI
KKIN(AM) Aitkin MN
KWOC(AM) Poplar Bluff MO
WSFZ(AM) Jackson MS
KLCY(AM) East Missoula MT
*WYFQ(AM) Charlotte NC
WDLX(AM) Washington NC
KOGA(AM) Ogallala NE
WGIN(AM) Rochester NH
WPAT(AM) Paterson NJ
KCCC(AM) Carlsbad NM
WBEN(AM) Buffalo NY
WIZR(AM) Johnstown NY
WEOL(AM) Elyria OH
WKY(AM) Oklahoma City OK
*KAGI(AM) Grants Pass OR
WHLM(AM) Bloomsburg PA
WYAC(AM) Cabo Rojo PR
KSDN(AM) Aberdeen SD
WSEV(AM) Sevierville TN
WWON(AM) Waynesboro TN
KDET(AM) Center TX
KLUP(AM) Terrell Hills TX
WLLL(AM) Lynchburg VA
KBAI(AM) Bellingham WA
KYAK(AM) Yakima WA
*WLBL(AM) Auburndale WI
WRVC(AM) Huntington WV
KROE(AM) Sheridan WY

940 khz
KGMS(AM) Tucson AZ
KWRU(AM) Fresno CA
WINE(AM) Brookfield CT
WLQH(AM) Chiefland FL
WINZ(AM) Miami FL
WMAC(AM) Macon GA
KKNE(AM) Waipahu HI
KPSZ(AM) Des Moines IA
WMIX(AM) Mount Vernon IL
WCND(AM) Shelbyville KY
WYLD(AM) New Orleans LA
WGFP(AM) Webster MA
WHIT(AM) Hudsonville MI
WIDG(AM) Saint Ignace MI
KSWM(AM) Aurora MO
WCPC(AM) Houston MS
KDIL(AM) Dillon MT
WKYK(AM) Burnsville NC
KVSH(AM) Valentine NE
WZOQ(AM) Lima OH
KICE(AM) Bend OR
KWBY(AM) Woodburn OR

WFGI(AM) Charleroi PA
WGRP(AM) Greenville PA
WADV(AM) Lebanon PA
*WIPR(AM) San Juan PR
WECO(AM) Wartburg TN
KIXZ(AM) Amarillo TX
KTON(AM) Belton TX
KTFS(AM) Texarkana TX
KNNZ(AM) Cedar City UT
WNRG(AM) Grundy VA
WKGM(AM) Smithfield VA
WFAW(AM) Fort Atkinson WI
WCSW(AM) Shell Lake WI
KMER(AM) Kemmerer WY

950 khz
KSWD(AM) Seward AK
WNZZ(AM) Montgomery AL
KXJK(AM) Forrest City AR
KFSA(AM) Fort Smith AR
KAHI(AM) Auburn CA
KKFN(AM) Denver CO
WTLN(AM) Orlando FL
WGTA(AM) Summerville GA
WGOV(AM) Valdosta GA
KOEL(AM) Oelwein IA
KNJY(AM) Boise ID
KOZE(AM) Lewiston ID
WNTD(AM) Chicago IL
WXLW(AM) Indianapolis IN
KJRG(AM) Newton KS
WYWY(AM) Barbourville KY
KRRP(AM) Coushatta LA
WROL(AM) Boston MA
WCTN(AM) Potomac-Cabin John MD
WWJ(AM) Detroit MI
KTNF(AM) Saint Louis Park MN
KWOS(AM) Jefferson City MO
WHSY(AM) Hattiesburg MS
KMTX(AM) Helena MT
WPET(AM) Greensboro NC
KNFT(AM) Bayard NM
KDCE(AM) Espanola NM
WHVW(AM) Hyde Park NY
WROC(AM) Rochester NY
WIBX(AM) Utica NY
WDIG(AM) Steubenville OH
*KTBR(AM) Roseburg OR
WNCC(AM) Northern Cambria PA
WPEN(AM) Philadelphia PA
WJKB(AM) Moncks Corner SC
WORD(AM) Spartanburg SC
KWAT(AM) Watertown SD
WAKM(AM) Franklin TN
KPRC(AM) Houston TX
KJTV(AM) Lubbock TX
WXGI(AM) Richmond VA
KJR(AM) Seattle WA
WERL(AM) Eagle River WI
WCLB(AM) Sheboygan WI
WVTS(AM) Charleston WV

960 khz
WERC(AM) Birmingham AL
WLPR(AM) Prichard AL
KCGS(AM) Marshall AR
KKNT(AM) Phoenix AZ
KIXW(AM) Apple Valley CA
KKGN(AM) Oakland CA
WELI(AM) New Haven CT
WGRO(AM) Lake City FL
WSVU(AM) North Palm Beach FL
WJYZ(AM) Albany GA
WRFC(AM) Athens GA
KMA(AM) Shenandoah IA
KSRA(AM) Salmon ID
*WDLM(AM) East Moline IL
WSBT(AM) South Bend IN
WPRT(AM) Prestonsburg KY
KROF(AM) Abbeville LA
WFGL(AM) Fitchburg MA
WTGM(AM) Salisbury MD
WHAK(AM) Rogers City MI
KLTF(AM) Little Falls MN
KZIM(AM) Cape Girardeau MO
WABG(AM) Greenwood MS
KFLN(AM) Baker MT

WZRH(AM) Dallas NC
WRNS(AM) Kinston NC
KNEB(AM) Scottsbluff NE
KNDN(AM) Farmington NM
WEAV(AM) Plattsburgh NY
WKVX(AM) Wooster OH
KGWA(AM) Enid OK
KLAD(AM) Klamath Falls OR
WHYL(AM) Carlisle PA
WPLY(AM) Mount Pocono PA
WATS(AM) Sayre PA
WCHQ(AM) Quebradillas PR
WQLA(AM) La Follette TN
WBMC(AM) McMinnville TN
KIMP(AM) Mount Pleasant TX
KGKL(AM) San Angelo TX
KOVO(AM) Provo UT
WFIR(AM) Roanoke VA
KALE(AM) Richland WA
WTCH(AM) Shawano WI

970 khz
KFBX(AM) Fairbanks AK
WERH(AM) Hamilton AL
WTBF(AM) Troy AL
KNEA(AM) Jonesboro AR
KVWM(AM) Show Low AZ
KHTY(AM) Bakersfield CA
KNWZ(AM) Coachella CA
KESP(AM) Modesto CA
KFEL(AM) Pueblo CO
WNNR(AM) Jacksonville FL
WFLA(AM) Tampa FL
WNIV(AM) Atlanta GA
WVOP(AM) Vidalia GA
KFTA(AM) Rupert ID
WMAY(AM) Springfield IL
WFSR(AM) Harlan KY
WGTK(AM) Louisville KY
KSYL(AM) Alexandria LA
WESO(AM) Southbridge MA
WAMD(AM) Aberdeen MD
WZAN(AM) Portland ME
WZAM(AM) Ishpeming MI
WKHM(AM) Jackson MI
KNFX(AM) Austin MN
WZQK(AM) Brandon MS
KBUL(AM) Billings MT
WRCS(AM) Ahoskie NC
WYSE(AM) Canton NC
WDAY(AM) Fargo ND
*KJLT(AM) North Platte NE
WWDJ(AM) Hackensack NJ
KNUU(AM) Paradise NV
*WNED(AM) Buffalo NY
WCHN(AM) Norwich NY
WFUN(AM) Ashtabula OH
WATH(AM) Athens OH
KCFO(AM) Tulsa OK
KCMD(AM) Portland OR
WBLF(AM) Bellefonte PA
WBGG(AM) Pittsburgh PA
WJMX(AM) Florence SC
WXQK(AM) Spring City TN
KIXL(AM) Del Valle TX
KHVN(AM) Fort Worth TX
WKCI(AM) Waynesboro VA
WSTX(AM) Christiansted VI
KTTO(AM) Spokane WA
*WHA(AM) Madison WI
WGEE(AM) Superior WI
WWYO(AM) Pineville WV

980 khz
KCAB(AM) Dardanelle AR
KNTR(AM) Lake Havasu City AZ
KINS(AM) Eureka CA
*KEYQ(AM) Fresno CA
KFWB(AM) Los Angeles CA
KDBV(AM) Salinas CA
KGLN(AM) Glenwood Springs CO
WSUB(AM) Groton CT
WTEM(AM) Washington DC
WDVH(AM) Gainesville FL
WRNE(AM) Gulf Breeze FL
WTOT(AM) Marianna FL
WHSR(AM) Pompano Beach FL

WKLY(AM) Hartwell GA
WPGA(AM) Perry GA
WUUS(AM) Rossville GA
KSPZ(AM) Ammon ID
KSGM(AM) Chester IL
WITY(AM) Danville IL
WGWM(AM) London KY
KOKA(AM) Shreveport LA
WCAP(AM) Lowell MA
WAKV(AM) Otsego MI
KKMS(AM) Richfield MN
KMBZ(AM) Kansas City MO
WAKK(AM) McComb MS
WKOR(AM) Starkville MS
WAAV(AM) Leland NC
WTIX(AM) Winston-Salem NC
KICA(AM) Clovis NM
KMIN(AM) Grants NM
KVLV(AM) Fallon NV
WOFX(AM) Troy NY
WONE(AM) Dayton OH
WILK(AM) Wilkes-Barre PA
WAZS(AM) Summerville SC
WBZK(AM) York SC
KDSJ(AM) Deadwood SD
WYFN(AM) Nashville TN
KRTX(AM) Rosenberg-Richmond TX
KSVC(AM) Richfield UT
WFHG(AM) Bristol VA
WJYK(AM) Chase City VA
KBBO(AM) Selah WA
WNBI(AM) Park Falls WI
WPRE(AM) Prairie du Chien WI
WCUB(AM) Two Rivers WI
WHAW(AM) Weston WV

990 khz
WEIS(AM) Centre AL
WLDX(AM) Fayette AL
KTKT(AM) Tucson AZ
KATD(AM) Pittsburg CA
KTMS(AM) Santa Barbara CA
KRKS(AM) Denver CO
WXCT(AM) Southington CT
WMYM(AM) Miami FL
WDYZ(AM) Orlando FL
WGML(AM) Hinesville GA
KHBZ(AM) Honolulu HI
KAYL(AM) Storm Lake IA
WCAZ(AM) Carthage IL
WITZ(AM) Jasper IN
WLHN(AM) Muncie IN
KRSL(AM) Russell KS
WGSO(AM) New Orleans LA
WDEO(AM) Ypsilanti MI
KRMO(AM) Cassville MO
WABO(AM) Waynesboro MS
WEEB(AM) Southern Pines NC
WBTE(AM) Windsor NC
KSVP(AM) Artesia NM
WLGZ(AM) Rochester NY
WJEH(AM) Gallipolis OH
WTIG(AM) Massillon OH
KTHH(AM) Albany OR
WNTP(AM) Philadelphia PA
WNTW(AM) Somerset PA
WPRA(AM) Mayaguez PR
WALE(AM) Greenville RI
WNML(AM) Knoxville TN
KWAM(AM) Memphis TN
KZZB(AM) Beaumont TX
KFCD(AM) Farmersville TX
KAML(AM) Kenedy-Karnes City TX
WNRV(AM) Narrows-Pearisburg VA
WLEE(AM) Richmond VA

1000 khz
WDJL(AM) Huntsville AL
WNSI(AM) Robertsdale AL
KFLG(AM) Bullhead City AZ
KCEO(AM) Vista CA
WYBT(AM) Blountstown FL
WJBW(AM) Jupiter FL
WMVP(AM) Chicago IL
WKVG(AM) Jenkins KY
WCMX(AM) Leominster MA
WXTN(AM) Lexington MS
WRTG(AM) Garner NC
WOF(AM) Andover NJ
KKIM(AM) Albuquerque NM
WLNL(AM) Horseheads NY
WCCD(AM) Parma OH
KTOK(AM) Oklahoma City OK
WIOO(AM) Carlisle PA
WWOF(AM) Walhalla SC
KXRB(AM) Sioux Falls SD
WMUF(AM) Paris TN
KSTA(AM) Coleman TX
*KBIB(AM) Marion TX
WKDE(AM) Altavista VA
*WRAR(AM) Tappahannock VA
WVWI(AM) Charlotte Amalie VI
KOMO(AM) Seattle WA

1010 khz
WCOC(AM) Dora AL
KXXT(AM) Tolleson AZ
KCHJ(AM) Delano CA
KIQI(AM) San Francisco CA
KXPS(AM) Thousand Palms CA
KSIR(AM) Brush CO
WJXL(AM) Jacksonville Beach FL
WQYK(AM) Seffner FL
WGUN(AM) Atlanta GA
*KRNI(AM) Mason City IA
WCSI(AM) Columbus IN
KIND(AM) Independence KS
KDLA(AM) De Ridder LA
*WCKW(AM) Garyville LA
WOLB(AM) Baltimore MD
WPPI(AM) Sauk Rapids MN
KCHI(AM) Chillicothe MO
KXEN(AM) Saint Louis MO
WMOX(AM) Meridian MS
WSPC(AM) Albemarle NC
WFGW(AM) Black Mountain NC
WELS(AM) Kinston NC
WNTK(AM) Newport NH
WINS(AM) New York NY
WIOI(AM) New Boston OH
KSZN(AM) Milwaukie OR
WHIN(AM) Gallatin TN
WORM(AM) Savannah TN
KTNZ(AM) Amarillo TX
KLAT(AM) Houston TX
KBBW(AM) Waco TX
KCPW(AM) Tooele UT
WMEV(AM) Marion VA
WRJR(AM) Portsmouth VA
WSPT(AM) Stevens Point WI
WCST(AM) Berkeley Springs WV

1020 khz
KAXX(AM) Eagle River AK
KTNQ(AM) Los Angeles CA
WHDD(AM) Sharon CT
WRHB(AM) Kendall FL
WJEP(AM) Ochlocknee GA
WCIL(AM) Carbondale IL
WPEO(AM) Peoria IL
KJJK(AM) Fergus Falls MN
KOIL(AM) Plattsmouth NE
WIBG(AM) Ocean City NJ
KCKN(AM) Roswell NM
KOKP(AM) Perry OK
KDKA(AM) Pittsburgh PA
WOQI(AM) Adjuntas PR
WRIX(AM) Homeland Park SC
KWIQ(AM) Moses Lake North WA
KYXE(AM) Union Gap WA

1030 khz
KFAY(AM) Farmington AR
KCEE(AM) Cortaro AZ
KJDJ(AM) San Luis Obispo CA
WONQ(AM) Oviedo FL
WEBS(AM) Calhoun GA
WNVR(AM) Vernon Hills IL
KBUF(AM) Holcomb KS
WBZ(AM) Boston MA
WWGB(AM) Indian Head MD
*WUFL(AM) Sterling Heights MI
WCTS(AM) Maplewood MN
KCWJ(AM) Blue Springs MO
WNOW(AM) Mint Hill NC
WDRU(AM) Wake Forest NC
KDUN(AM) Reedsport OR
WOSO(AM) San Juan PR
WGSF(AM) Memphis TN
WQSE(AM) White Bluff TN
KCTA(AM) Corpus Christi TX
KWFA(AM) Tye TX
WGFC(AM) Floyd VA
KMAS(AM) Shelton WA
WBGS(AM) Point Pleasant WV
KTWO(AM) Casper WY

1040 khz
KURS(AM) San Diego CA
KCBR(AM) Monument CO
WLVJ(AM) Boynton Beach FL
WWBA(AM) Pinellas Park FL
WPBS(AM) Conyers GA
KLHT(AM) Honolulu HI
WHO(AM) Des Moines IA
WLCR(AM) Mt. Washington KY
WSGH(AM) Lewisville NC
WCHR(AM) Flemington NJ
WYSL(AM) Avon NY
WJTB(AM) North Ridgeville OH
*KXPD(AM) Tigard OR
WZSK(AM) Everett PA
WZNA(AM) Moca PR
WQBB(AM) Powell TN
KGGR(AM) Dallas TX

1050 khz
WRFS(AM) Alexander City AL
WWIC(AM) Scottsboro AL
KJBN(AM) Little Rock AR
KTBA(AM) Tuba City AZ
KJPG(AM) Frazier Park CA
KCAA(AM) Loma Linda CA
KTCT(AM) San Mateo CA
WJSB(AM) Crestview FL
WROS(AM) Jacksonville FL
WJCM(AM) Sebring FL
WFAM(AM) Augusta GA
WMNZ(AM) Montezuma GA
WDZ(AM) Decatur IL
WTCA(AM) Plymouth IN
WNES(AM) Central City KY
KLPL(AM) Lake Providence LA
KVPI(AM) Ville Platte LA
WMSG(AM) Oakland MD
WFED(AM) Silver Spring MD
WTKA(AM) Ann Arbor MI
KLOH(AM) Pipestone MN
KMIS(AM) Portageville MO
KSIS(AM) Sedalia MO
WTWG(AM) Columbus MS
KMTA(AM) Miles City MT
WFSC(AM) Franklin NC
WLON(AM) Lincolnton NC
WWGP(AM) Sanford NC
WBNC(AM) Conway NH
KTBL(AM) Los Ranchos de Albuquerque NM
WSEN(AM) Baldwinsville NY
WYBG(AM) Massena NY
WEPN(AM) New York NY
WCVX(AM) Cincinnati OH
KKRX(AM) Lawton OK
KGTO(AM) Tulsa OK
KORE(AM) Springfield-Eugene OR
WBUT(AM) Butler PA
WLYC(AM) Williamsport PA
WIQB(AM) Conway SC
WSMT(AM) Sparta TN
KCHN(AM) Brookshire TX
KRMY(AM) Killeen TX
WGAT(AM) Gate City VA
WBRG(AM) Lynchburg VA
WVXX(AM) Norfolk VA
KEYF(AM) Dishman WA
*KBLE(AM) Seattle WA
*WDVM(AM) Eau Claire WI
WJOK(AM) Kaukauna WI
WLIP(AM) Kenosha WI
WAMN(AM) Green Valley WV
WADC(AM) Parkersburg WV

1060 khz
KOAI(AM) Van Buren AR
KDUS(AM) Tempe AZ
KTNS(AM) Oakhurst CA
KRCN(AM) Longmont/Denver CO
WIXC(AM) Titusville FL
WKNG(AM) Tallapoosa GA
KHBC(AM) Hilo HI
KBGN(AM) Caldwell ID
WMCL(AM) McLeansboro IL
WRHL(AM) Rochelle IL
WFLE(AM) Flemingsburg KY
WJKY(AM) Jamestown KY
WLNO(AM) New Orleans LA
WBIX(AM) Natick MA
WHFB(AM) Benton Harbor-St. Joseph MI
KFIL(AM) Preston MN
KBFL(AM) Springfield MO
WKMQ(AM) Tupelo MS
WGSB(AM) Mebane NC
WXNC(AM) Monroe NC
WCOK(AM) Sparta NC
KNLV(AM) Ord NE
KUCU(AM) Farmington NM
KKVV(AM) Las Vegas NV
WILB(AM) Canton OH
KYW(AM) Philadelphia PA
WCGB(AM) Juana Diaz PR
KGFX(AM) Pierre SD
WNPC(AM) Newport TN
WQMV(AM) Waverly TN
KXPL(AM) El Paso TX
KIJN(AM) Farwell TX
KFIT(AM) Lockhart TX
KFIT EXP STN San Antonio TX
KDYL(AM) South Salt Lake UT

1070 khz
WAPI(AM) Birmingham AL
KNX(AM) Los Angeles CA
WKII(AM) Solana FL
*WFRF(AM) Tallahassee FL
KILR(AM) Estherville IA
WIBC(AM) Indianapolis IN
KFTI(AM) Wichita KS
WEKT(AM) Elkton KY
KBCL(AM) Bossier City LA
WBEY(AM) Pocomoke City MD
KVKK(AM) Verndale MN
KHMO(AM) Hannibal MO
KATQ(AM) Plentywood MT
WNCT(AM) Greenville NC
WGOS(AM) High Point NC
WKMB(AM) Stirling NJ
WTWK(AM) Plattsburgh NY
WSCP(AM) Sandy Creek-Pulaski NY
KRAM(AM) West Klamath OR
WKOK(AM) Sunbury PA
WMIA(AM) Arecibo PR
WCSZ(AM) Sans Souci SC
WFLI(AM) Lookout Mountain TN
WDIA(AM) Memphis TN
KOPY(AM) Alice TX
KNTH(AM) Houston TX
KWEL(AM) Midland TX
WINA(AM) Charlottesville VA
WTSO(AM) Madison WI
WIWS(AM) Beckley WV

1080 khz
KUDO(AM) Anchorage AK
WKAC(AM) Athens AL
KGVY(AM) Green Valley AZ
KSCO(AM) Santa Cruz CA
WTIC(AM) Hartford CT
WTPS(AM) Coral Gables FL
WHOO(AM) Kissimmee FL
WFTD(AM) Marietta GA
KWAI(AM) Honolulu HI
KOAK(AM) Red Oak IA
KVNI(AM) Coeur d'Alene ID
*WRYT(AM) Edwardsville IL
WNWI(AM) Oak Lawn IL
WOKT(AM) Cannonsburg KY
WKJK(AM) Louisville KY
WOAP(AM) Owosso MI
KYMN(AM) Northfield MN
KYMO(AM) East Prairie MO
WKGX(AM) Lenoir NC
WWDR(AM) Murfreesboro NC
KNDK(AM) Langdon ND
KCNM(AM) Garapan-Saipan NP
WUFO(AM) Amherst NY
KFXX(AM) Portland OR
WWNL(AM) Pittsburgh PA
WLEY(AM) Cayey PR
WALD(AM) Walterboro SC
KRLD(AM) Dallas TX
KSLL(AM) Price UT
WKBY(AM) Chatham VA
WOKU(AM) Hurricane WV

1090 khz
WWGC(AM) Albertville AL
KAAY(AM) Little Rock AR
KNCR(AM) Fortuna CA
KMXA(AM) Aurora CO
WNVY(AM) Cantonment FL
WBAF(AM) Barnesville GA
KSOU(AM) Sioux Center IA
*KNWS(AM) Waterloo IA
WCRA(AM) Effingham IL
WFCV(AM) Fort Wayne IN
WILD(AM) Boston MA
WBAL(AM) Baltimore MD
WCAR(AM) Livonia MI
WKBZ(AM) Muskegon MI
*KEXS(AM) Excelsior Springs MO
KBOZ(AM) Bozeman MT
WKTE(AM) King NC
WTSB(AM) Selma NC
KTGO(AM) Tioga ND
WKFI(AM) Wilmington OH
KLWJ(AM) Umatilla OR
WSOL(AM) San German PR
WCZZ(AM) Greenwood SC
WENR(AM) Englewood TN
WTNK(AM) Hartsville TN
WHGG(AM) Kingsport TN
KNUZ(AM) Bellville TX
KVOP(AM) Plainview TX
WGOD(AM) Charlotte Amalie VI
KPTK(AM) Seattle WA
WAQE(AM) Rice Lake WI

1100 khz
KFNX(AM) Cave Creek AZ
KAFY(AM) Bakersfield CA
KFAX(AM) San Francisco CA
KNZZ(AM) Grand Junction CO
WWWE(AM) Hapeville GA
WCGA(AM) Woodbine GA
WZFN(AM) Dilworth MN
KKLL(AM) Carthage- MO
KQNM(AM) Milan NM
KWWN(AM) Las Vegas NV
WHLI(AM) Hempstead NY
WTAM(AM) Cleveland OH
WGPA(AM) Bethlehem PA
WSGI(AM) Springfield TN
KDRY(AM) Alamo Heights TX
WTWN(AM) Wells River VT
WISS(AM) Berlin WI

1110 khz
KAGV(AM) Big Lake AK
WTOF(AM) Bay Minette AL
WBIB(AM) Centreville AL
KGFL(AM) Clinton AR
KDIS(AM) Pasadena CA
KLIB(AM) Roseville CA
WTIS(AM) Tampa FL
KAOI(AM) Kihei HI
*WMBI(AM) Chicago IL
WKDZ(AM) Cadiz KY
WCBR(AM) Richmond KY
WOMN(AM) Franklinton LA
KTTP(AM) Pineville LA
WUPE(AM) Pittsfield MA
*WUNN(AM) Mason MI
WJML(AM) Petoskey MI
WKRA(AM) Holly Springs MS
WBT(AM) Charlotte NC

KFAB(AM) Omaha NE
WCCM(AM) Salem NH
KYKK(AM) Hobbs NM
WSFW(AM) Seneca Falls NY
WTBQ(AM) Warwick NY
WGNZ(AM) Fairborn OH
KBND(AM) Bend OR
WJSM(AM) Martinsburg PA
WNAP(AM) Norristown PA
WKZV(AM) Washington PA
WVJP(AM) Caguas PR
WPMZ(AM) Providence RI
WUAT(AM) Pikeville TN
KTEK(AM) Alvin TX
WYRM(AM) Norfolk VA
KWDB(AM) Oak Harbor WA

1120 khz
WHOG(AM) Hobson City AL
KZSJ(AM) San Martin CA
KLIM(AM) Limon CO
WPRX(AM) Bristol CT
WUST(AM) Washington DC
WNWF(AM) Destin FL
WXJO(AM) Gordon GA
WBNW(AM) Concord MA
KMOX(AM) Saint Louis MO
WTWZ(AM) Clinton MS
WSME(AM) Camp Lejeune NC
WBBF(AM) Buffalo NY
KEOR(AM) Catoosa OK
KPNW(AM) Eugene OR
WKQW(AM) Oil City PA
WMSW(AM) Hatillo PR
WKCE(AM) Maryville TN
KJSA(AM) Mineral Wells TX
*KANN(AM) Roy UT
WDUF(AM) Duffield VA

1130 khz
WACQ(AM) Carrville AL
KAAB(AM) Batesville AR
KQNA(AM) Prescott Valley AZ
KRDU(AM) Dinuba CA
KSDO(AM) San Diego CA
WWBF(AM) Bartow FL
WLBA(AM) Gainesville GA
KRUD(AM) Honolulu HI
KILJ(AM) Mount Pleasant IA
WSDX(AM) Brazil IN
KLEY(AM) Wellington KS
WOFC(AM) Murray KY
KWKH(AM) Shreveport LA
WDFN(AM) Detroit MI
KFAN(AM) Minneapolis MN
WQFX(AM) Gulfport MS
WPYB(AM) Benson NC
WCLW(AM) Eden NC
WECR(AM) Newland NC
KBMR(AM) Bismarck ND
WBBR(AM) New York NY
WEDI(AM) Eaton OH
KTRP(AM) Mount Angel OR
WASP(AM) Brownsville PA
WOIZ(AM) Guayanilla PR
WEAF(AM) Camden SC
WFXH(AM) Hilton Head Island SC
WYXE(AM) Gallatin TN
KTMR(AM) Edna TX
WISN(AM) Milwaukee WI
WRRL(AM) Rainelle WV

1140 khz
KSLD(AM) Soldotna AK
WBXR(AM) Hazel Green AL
KQAB(AM) Lake Isabella CA
KNWQ(AM) Palm Springs CA
KHTK(AM) Sacramento CA
KNAB(AM) Burlington CO
WQBA(AM) Miami FL
WRMQ(AM) Orlando FL
KGEM(AM) Boise ID
WVEL(AM) Pekin IL
WAWK(AM) Kendallville IN
WMMG(AM) Brandenburg KY
WRLV(AM) Salyersville KY
WJNZ(AM) Kentwood MI
KCXL(AM) Liberty MO
KPWB(AM) Piedmont MO
KLTK(AM) South West City MO
WAPF(AM) McComb MS
WSAO(AM) Senatobia MS
WRNA(AM) China Grove NC
KSFN(AM) North Las Vegas NV
WCJW(AM) Warsaw NY
KRMP(AM) Oklahoma City OK
WQII(AM) San Juan PR
KSOO(AM) Sioux Falls SD
WLOD(AM) Loudon TN
KCLE(AM) Cleburne TX
KYOK(AM) Conroe TX
WRVA(AM) Richmond VA
WXLZ(AM) Saint Paul VA
KZMQ(AM) Greybull WY

1150 khz
WGEA(AM) Geneva AL
WJRD(AM) Tuscaloosa AL
KCKY(AM) Coolidge AZ
KTLK(AM) Los Angeles CA
KNRV(AM) Englewood CO
WMRD(AM) Middletown CT
WDEL(AM) Wilmington DE
WNDB(AM) Daytona Beach FL
WTMP(AM) Egypt Lake FL
WXKO(AM) Fort Valley GA
WJEM(AM) Valdosta GA
KCPS(AM) Burlington IA
KWKY(AM) Des Moines IA
WGGH(AM) Marion IL
KSAL(AM) Salina KS
WMST(AM) Mt. Sterling KY
WLOC(AM) Munfordville KY
WJBO(AM) Baton Rouge LA
WTTT(AM) Boston MA
KASM(AM) Albany MN
KRMS(AM) Osage Beach MO
WONG(AM) Canton MS
KSEN(AM) Shelby MT
WBAG(AM) Burlington-Graham NC
WGBR(AM) Goldsboro NC
KDEF(AM) Albuquerque NM
*WRUN(AM) Utica NY
*WCUE(AM) Cuyahoga Falls OH
WIMA(AM) Lima OH
KNED(AM) McAlester OK
KAGO(AM) Klamath Falls OR
KXMG(AM) Portland OR
WHUN(AM) Huntingdon PA
WGBN(AM) New Kensington PA
WAVO(AM) Rock Hill SC
WSNW(AM) Seneca SC
KIMM(AM) Rapid City SD
WGOW(AM) Chattanooga TN
WCRK(AM) Morristown TN
WDTM(AM) Selmer TN
KZNE(AM) College Station TX
KCCT(AM) Corpus Christi TX
KSVE(AM) El Paso TX
KJBC(AM) Midland TX
KBPO(AM) Port Neches TX
KREL(AM) Quanah TX
WNLR(AM) Churchville VA
KQQQ(AM) Pullman WA
KKNW(AM) Seattle WA
WEAQ(AM) Chippewa Falls WI
WHBY(AM) Kimberly WI
WELC(AM) Welch WV

1160 khz
WEWC(AM) Callahan FL
WIWA(AM) Saint Cloud FL
WCFO(AM) East Point GA
KHPP(AM) Waukon IA
WYLL(AM) Chicago IL
WDJO(AM) Florence KY
WKCM(AM) Hawesville KY
WMET(AM) Gaithersburg MD
WSKW(AM) Skowhegan ME
WCXI(AM) Fenton MI
KCTO(AM) Cleveland MO
WTEL(AM) Red Springs NC
WJFJ(AM) Tryon NC
WOBM(AM) Lakewood NJ
WVNJ(AM) Oakland NJ
WABY(AM) Mechanicville NY
WPIE(AM) Trumansburg NY
WCCS(AM) Homer City PA
WBYN(AM) Lehighton PA
WBQN(AM) Barceloneta-Manati PR
WCRT(AM) Donelson TN
KVCE(AM) Highland Park TX
KRDY(AM) San Antonio TX
KSL(AM) Salt Lake City UT
WODY(AM) Fieldale VA

1170 khz
*KJNP(AM) North Pole AK
WLYG(AM) Hanceville AL
WACV(AM) Montgomery AL
KCBQ(AM) San Diego CA
KLOK(AM) San Jose CA
KJJD(AM) Windsor CO
*WCTF(AM) Vernon CT
WKFL(AM) Bushnell FL
WAVS(AM) Davie FL
WSOS(AM) Saint Augustine Beach FL
KJOC(AM) Davenport IA
WLBH(AM) Mattoon IL
WDFB(AM) Junction City KY
WDIS(AM) Norfolk MA
WFPB(AM) Orleans MA
KOWZ(AM) Waseca MN
KUGT(AM) Jackson MO
WCXN(AM) Claremont NC
WCLN(AM) Clinton NC
WWTR(AM) Bridgewater NJ
WWLE(AM) Cornwall NY
KFAQ(AM) Tulsa OK
WLEO(AM) Ponce PR
WQVA(AM) Lexington SC
*WPLX(AM) Germantown TN
KPUG(AM) Bellingham WA
WFDL(AM) Waupun WI
WWVA(AM) Wheeling WV

1180 khz
KYET(AM) Williams AZ
KERI(AM) Wasco-Greenacres CA
WZQZ(AM) Trion GA
KORL(AM) Honolulu HI
WLDS(AM) Jacksonville IL
WSQR(AM) Sycamore IL
WGAB(AM) Newburgh IN
WXLA(AM) Dimondale MI
KYES(AM) Rockville MN
WJNT(AM) Pearl MS
KOFI(AM) Kalispell MT
WMYT(AM) Carolina Beach NC
KYDZ(AM) Bellevue NE
WHAM(AM) Rochester NY
WFYL(AM) King of Prussia PA
WCNX(AM) Hope Valley RI
WFGN(AM) Gaffney SC
WVLZ(AM) Knoxville TN
KGOL(AM) Humble TX
KLAY(AM) Lakewood WA

1190 khz
WEUV(AM) Moulton AL
KREB(AM) Bentonville-Bella Vista AR
KNUV(AM) Tolleson AZ
KXMX(AM) Anaheim CA
KDYA(AM) Vallejo CA
KVCU(AM) Boulder CO
WAMT(AM) Pine Castle-Sky Lake FL
WPSP(AM) Royal Palm Beach FL
WAFS(AM) Atlanta GA
WWIO(AM) Saint Mary's GA
KDAO(AM) Marshalltown IA
WOWO(AM) Fort Wayne IN
KVSV(AM) Beloit KS
KNEK(AM) Washington LA
WBIS(AM) Annapolis MD
KKOJ(AM) Jackson MN
KMFX(AM) Wabasha MN
KRFT(AM) De Soto MO
KPHN(AM) Kansas City MO
WBSL(AM) Bay St. Louis MS
WIXE(AM) Monroe NC
KXKS(AM) Albuquerque NM
WSDE(AM) Cobleskill NY
WLIB(AM) New York NY
KEX(AM) Portland OR
WBMJ(AM) San Juan PR
WJES(AM) Saluda SC
WSDQ(AM) Dunlap TN
WLLI(AM) Humboldt TN
KFXR(AM) Dallas TX
WBDY(AM) Bluefield VA
*WNWC(AM) Sun Prairie WI

1200 khz
KYAA(AM) Soquel CA
WPTK(AM) Pine Island Center FL
WRTO(AM) Chicago IL
WBCE(AM) Wickliffe KY
WKOX(AM) Framingham MA
WCHB(AM) Taylor MI
KYOO(AM) Bolivar MO
WXIT(AM) Blowing Rock NC
WSML(AM) Graham NC
*KFNW(AM) West Fargo ND
WJGK(AM) Highland NY
WTLA(AM) North Syracuse NY
WRKK(AM) Hughesville PA
WKST(AM) New Castle PA
WGDL(AM) Lares PR
WMIR(AM) Atlantic Beach SC
WAMB(AM) Nashville TN
WOAI(AM) San Antonio TX
WAGE(AM) Leesburg VA

1210 khz
WQLS(AM) Ozark AL
KEVT(AM) Sahuarita AZ
KQEQ(AM) Fowler CA
*KEBR(AM) Rocklin CA
KPRZ(AM) San Marcos-Poway CA
WNMA(AM) Miami Springs FL
WDGR(AM) Dahlonega GA
KZOO(AM) Honolulu HI
WILY(AM) Centralia IL
WSKR(AM) Denham Springs LA
WJNL(AM) Kingsley MI
WDAO(AM) Dayton OH
KGYN(AM) Guymon OK
WPHT(AM) Philadelphia PA
WHOY(AM) Salinas PR
KOKK(AM) Huron SD
WMPS(AM) Bartlett TN
WSBI(AM) Static TN
KUBR(AM) San Juan TX
KUNF(AM) Washington UT
KWMG(AM) Auburn-Federal Way WA
KZTS(AM) Sunnyside WA
KRSV(AM) Afton WY
KHAT(AM) Laramie WY

1220 khz
WAYE(AM) Birmingham AL
WABF(AM) Fairhope AL
KVSA(AM) McGehee AR
KHTS(AM) Canyon Country CA
KNTS(AM) Palo Alto CA
KWKU(AM) Pomona CA
KLVZ(AM) Denver CO
WQUN(AM) Hamden CT
WJAX(AM) Jacksonville FL
WOTS(AM) Kissimmee FL
WSRQ(AM) Sarasota FL
WZOT(AM) Rockmart GA
KJAN(AM) Atlantic IA
KQMG(AM) Independence IA
WLPO(AM) La Salle IL
WKRS(AM) Waukegan IL
WSLM(AM) Salem IN
KOFO(AM) Ottawa KS
WFKN(AM) Franklin KY
WPHX(AM) Sanford ME
WBCH(AM) Hastings MI
KLBB(AM) Stillwater MN
KOMC(AM) Branson MO
KGIR(AM) Cape Girardeau MO
KLPW(AM) Union MO
WOEG(AM) Hazlehurst MS
WDYT(AM) Kings Mountain NC
WREV(AM) Reidsville NC
WENC(AM) Whiteville NC
KDDR(AM) Oakes ND
WZBK(AM) Keene NH
WGNY(AM) Newburgh NY
WHKW(AM) Cleveland OH
WERT(AM) Van Wert OH
KTLV(AM) Midwest City OK
KPJC(AM) Salem OR
WJUN(AM) Mexico PA
WSTL(AM) Providence RI
WFWL(AM) Camden TN
WCPH(AM) Etowah TN
WAXO(AM) Lewisburg TN
KMVL(AM) Madisonville TX
KZEE(AM) Weatherford TX
WLSD(AM) Big Stone Gap VA
WFAX(AM) Falls Church VA

1230 khz
KIFW(AM) Sitka AK
KVAK(AM) Valdez AK
WAUD(AM) Auburn AL
WKWL(AM) Florala AL
WJBB(AM) Haleyville AL
WBHP(AM) Huntsville AL
WRJX(AM) Jackson AL
WNUZ(AM) Talladega AL
WTBC(AM) Tuscaloosa AL
KFPW(AM) Fort Smith AR
KBTM(AM) Jonesboro AR
KAAA(AM) Kingman AZ
KOY(AM) Phoenix AZ
KATO(AM) Safford AZ
KINO(AM) Winslow AZ
KGEO(AM) Bakersfield CA
KSZL(AM) Barstow CA
KBOV(AM) Bishop CA
KXO(AM) El Centro CA
KDAC(AM) Fort Bragg CA
KYPA(AM) Los Angeles CA
KPRL(AM) Paso Robles CA
KLXR(AM) Redding CA
*KWG(AM) Stockton CA
KEXO(AM) Grand Junction CO
KKPC(AM) Pueblo CO
KBCR(AM) Steamboat Springs CO
KSTC(AM) Sterling CO
WNEZ(AM) Manchester CT
WGGG(AM) Gainesville FL
WONN(AM) Lakeland FL
WMAF(AM) Madison FL
WSBB(AM) New Smyrna Beach FL
WDWR(AM) Pensacola FL
WWSD(AM) Quincy FL
WBZT(AM) West Palm Beach FL
WNRR(AM) Augusta GA
WBLJ(AM) Dalton GA
WXLI(AM) Dublin GA
WNGM(AM) Hiawassee GA
WFOM(AM) Marietta GA
WSOK(AM) Savannah GA
WWGA(AM) Waycross GA
KFJB(AM) Marshalltown IA
KBAR(AM) Burley ID
KORT(AM) Grangeville ID
KRXK(AM) Rexburg ID
WJBC(AM) Bloomington IL
WFXN(AM) Moline IL
WHCO(AM) Sparta IL
WJOB(AM) Hammond IN
WSAL(AM) Logansport IN
WTCJ(AM) Tell City IN
WHIR(AM) Danville KY
WWKU(AM) Glasgow KY
WHOP(AM) Hopkinsville KY
WANO(AM) Pineville KY
KLIC(AM) Monroe LA
WBOK(AM) New Orleans LA
KSLO(AM) Opelousas LA
WNAW(AM) North Adams MA
WESX(AM) Salem MA
WNEB(AM) Worcester MA
WRBS(AM) Baltimore MD
WCMD(AM) Cumberland MD
WTKG(AM) Grand Rapids MI
WGRY(AM) Grayling MI

WIKB(AM) Iron River MI
*WMPC(AM) Lapeer MI
WSOO(AM) Sault Ste. Marie MI
WMSH(AM) Sturgis MI
WKLK(AM) Cloquet MN
KGHS(AM) International Falls MN
KYSM(AM) Mankato MN
KMRS(AM) Morris MN
KTRF(AM) Thief River Falls MN
KWNO(AM) Winona MN
KZYM(AM) Joplin MO
KLWT(AM) Lebanon MO
KWIX(AM) Moberly MO
WTKN(AM) Corinth MS
WSSO(AM) Starkville MS
KOBB(AM) Bozeman MT
KHDN(AM) Hardin MT
KXLO(AM) Lewistown MT
KLCB(AM) Libby MT
WSKY(AM) Asheville NC
WFAY(AM) Fayetteville NC
WMFR(AM) High Point NC
WLNR(AM) Kinston NC
WNNC(AM) Newton NC
WCBT(AM) Roanoke Rapids NC
KDIX(AM) Dickinson ND
KTNC(AM) Falls City NE
KHAS(AM) Hastings NE
WMOU(AM) Berlin NH
WTSV(AM) Claremont NH
WCMC(AM) Wildwood NJ
KRSY(AM) Alamogordo NM
KOTS(AM) Deming NM
KYVA(AM) Gallup NM
KFUN(AM) Las Vegas NM
KBCQ(AM) Roswell NM
KELY(AM) Ely NV
KLAV(AM) Las Vegas NV
KJFK(AM) Reno NV
WECK(AM) Cheektowaga NY
WENY(AM) Elmira NY
WMML(AM) Glens Falls NY
WHUC(AM) Hudson NY
WIXT(AM) Little Falls NY
WFAS(AM) White Plains NY
WDBZ(AM) Cincinnati OH
WYTS(AM) Columbus OH
WIRO(AM) Ironton OH
WCWA(AM) Toledo OH
KADA(AM) Ada OK
WBBZ(AM) Ponca City OK
KKEE(AM) Astoria OR
KZZR(AM) Burns OR
KHSN(AM) Coos Bay OR
KMUZ(AM) Gresham OR
KQIK(AM) Lakeview OR
*KSJK(AM) Talent OR
KCUP(AM) Toledo OR
WBVP(AM) Beaver Falls PA
WEEX(AM) Easton PA
WKBO(AM) Harrisburg PA
WCRO(AM) Johnstown PA
WBPZ(AM) Lock Haven PA
WTIV(AM) Titusville PA
WNIK(AM) Arecibo PR
WXNI(AM) Westerly RI
WAIM(AM) Anderson SC
WOIC(AM) Columbia SC
WOLS(AM) Florence SC
KWSN(AM) Sioux Falls SD
WMLR(AM) Hohenwald TN
WAKI(AM) McMinnville TN
KSIX(AM) Corpus Christi TX
KTJK(AM) Del Rio TX
KQUE(AM) Houston TX
KERV(AM) Kerrville TX
KLVT(AM) Levelland TX
KOZA(AM) Odessa TX
KGRO(AM) Pampa TX
KSEY(AM) Seymour TX
KSST(AM) Sulphur Springs TX
KWTX(AM) Waco TX
KJQS(AM) Murray UT
WABN(AM) Abingdon VA
WODI(AM) Brookneal VA
WXCF(AM) Clifton Forge VA

WFVA(AM) Fredericksburg VA
WJOI(AM) Norfolk VA
WAMM(AM) Woodstock VA
WJOY(AM) Burlington VT
KOZI(AM) Chelan WA
KWYZ(AM) Everett WA
KSBN(AM) Spokane WA
WCLO(AM) Janesville WI
WXCO(AM) Wausau WI
WVNT(AM) Parkersburg WV
KVOC(AM) Casper WY

1240 khz

WEBJ(AM) Brewton AL
WULA(AM) Eufaula AL
WBCF(AM) Florence AL
WMGJ(AM) Gadsden AL
WLYJ(AM) Jasper AL
KVRC(AM) Arkadelphia AR
KTLO(AM) Mountain Home AR
KWAK(AM) Stuttgart AR
KJAA(AM) Globe AZ
KPOD(AM) Crescent City CA
KJOP(AM) Lemoore CA
KNRY(AM) Monterey CA
KLOA(AM) Ridgecrest CA
KSAC(AM) Sacramento CA
KEZY(AM) San Bernardino CA
KSON(AM) San Diego CA
KSMA(AM) Santa Maria CA
KSUE(AM) Susanville CA
KRDO(AM) Colorado Springs CO
KDGO(AM) Durango CO
KSLV(AM) Monte Vista CO
KCRT(AM) Trinidad CO
WWCO(AM) Waterbury CT
WBGC(AM) Chipley FL
WYNY(AM) Cross City FL
WKIQ(AM) Eustis FL
WINK(AM) Fort Myers FL
WMMB(AM) Melbourne FL
WFOY(AM) Saint Augustine FL
WBHB(AM) Fitzgerald GA
WGGA(AM) Gainesville GA
WLAG(AM) La Grange GA
WDDO(AM) Macon GA
WWNS(AM) Statesboro GA
WPAX(AM) Thomasville GA
WTWA(AM) Thomson GA
KDEC(AM) Decorah IA
*KWLC(AM) Decorah IA
KBIZ(AM) Ottumwa IA
KICD(AM) Spencer IA
KMCL(AM) Donnelly ID
KMHI(AM) Mountain Home ID
KWIK(AM) Pocatello ID
KOFE(AM) Saint Maries ID
WSBC(AM) Chicago IL
WEBQ(AM) Harrisburg IL
WTAX(AM) Springfield IL
WSDR(AM) Sterling IL
WHBU(AM) Anderson IN
KIUL(AM) Garden City KS
KFH(AM) Wichita KS
WLLV(AM) Louisville KY
WFTM(AM) Maysville KY
WPKE(AM) Pikeville KY
WSFC(AM) Somerset KY
KASO(AM) Minden LA
KANE(AM) New Iberia LA
WHMQ(AM) Greenfield MA
*WBUR(AM) West Yarmouth MA
WCEM(AM) Cambridge MD
WJEJ(AM) Hagerstown MD
WEZR(AM) Lewiston ME
WSYY(AM) Millinocket ME
WATT(AM) Cadillac MI
WCBY(AM) Cheboygan MI
WIAN(AM) Ishpeming MI
WJIM(AM) Lansing MI
WMFG(AM) Hibbing MN
WJON(AM) Saint Cloud MN
KLIK(AM) Jefferson City MO
KNEM(AM) Nevada MO
KFMO(AM) Park Hills MO
WWZQ(AM) Aberdeen MS

WPBQ(AM) Flowood MS
WGRM(AM) Greenwood MS
WGCM(AM) Gulfport MS
WMIS(AM) Natchez MS
WAVN(AM) Southaven MS
KMZK(AM) Billings MT
KLTZ(AM) Glasgow MT
KLYQ(AM) Hamilton MT
KBLL(AM) Helena MT
KSAM(AM) Whitefish MT
WSQL(AM) Brevard NC
WHVN(AM) Charlotte NC
WCNC(AM) Elizabeth City NC
WJNC(AM) Jacksonville NC
WPJL(AM) Raleigh NC
WWWC(AM) Wilkesboro NC
KDLR(AM) Devils Lake ND
KFOR(AM) Lincoln NE
KODY(AM) North Platte NE
WFTN(AM) Franklin NH
WSNJ(AM) Bridgeton NJ
KAMQ(AM) Carlsbad NM
KCLV(AM) Clovis NM
KALY(AM) Los Ranchos de Albuquerque NM
KDAN(AM) Beatty NV
KELK(AM) Elko NV
WGBB(AM) Freeport NY
WGVA(AM) Geneva NY
WJTN(AM) Jamestown NY
WVOS(AM) Liberty NY
WNBZ(AM) Saranac Lake NY
WVKZ(AM) Schenectady NY
WATN(AM) Watertown NY
WBBW(AM) Youngstown OH
WHIZ(AM) Zanesville OH
KVSO(AM) Ardmore OK
KADS(AM) Elk City OK
KBEL(AM) Idabel OK
KOKL(AM) Okmulgee OK
KEJO(AM) Corvallis OR
KTIX(AM) Pendleton OR
KRDM(AM) Redmond OR
KQEN(AM) Roseburg OR
WRTA(AM) Altoona PA
WIOV(AM) Reading PA
WYGL(AM) Selinsgrove PA
WBAX(AM) Wilkes-Barre PA
WALO(AM) Humacao PR
WOON(AM) Woonsocket RI
WLSC(AM) Loris SC
WKDK(AM) Newberry SC
WDXY(AM) Sumter SC
KCCR(AM) Pierre SD
WBEJ(AM) Elizabethton TN
WEKR(AM) Fayetteville TN
WIFA(AM) Knoxville TN
WNVL(AM) Nashville TN
WSDT(AM) Soddy-Daisy TN
WENK(AM) Union City TN
KVLF(AM) Alpine TX
KXYL(AM) Brownwood TX
KTAM(AM) Bryan TX
KXIT(AM) Dalhart TX
KPBL(AM) Hemphill TX
KBGE(AM) Kilgore TX
KSOX(AM) Raymondville TX
KXOX(AM) Sweetwater TX
WROU(AM) Petersburg VA
WGMN(AM) Roanoke VA
WTON(AM) Staunton VA
WSKI(AM) Montpelier VT
KCVL(AM) Colville WA
KXLE(AM) Ellensburg WA
KGY(AM) Olympia WA
WOMT(AM) Manitowoc WI
WHFA(AM) Poynette WI
WOBT(AM) Rhinelander WI
WJMC(AM) Rice Lake WI
WKEZ(AM) Bluefield WV
WBES(AM) Dunbar WV
WDNE(AM) Elkins WV
KFBC(AM) Cheyenne WY
KEVA(AM) Evanston WY
KASL(AM) Newcastle WY
KRAL(AM) Rawlins WY

KTHE(AM) Thermopolis WY

1250 khz

WZOB(AM) Fort Payne AL
WAPZ(AM) Wetumpka AL
KOFC(AM) Fayetteville AR
KPZK(AM) Little Rock AR
KBSZ(AM) Wickenburg AZ
KHIL(AM) Willcox AZ
KHOT(AM) Madera CA
KZER(AM) Santa Barbara CA
KNWH(AM) Twentynine Palms CA
KLLK(AM) Willits CA
WQHL(AM) Live Oak FL
WHNZ(AM) Tampa FL
WSRA(AM) Albany GA
WYTH(AM) Madison GA
KDNZ(AM) Cedar Falls IA
WSPL(AM) Streator IL
WGL(AM) Fort Wayne IN
WRAY(AM) Princeton IN
KKHK(AM) Kansas City KS
WVKY(AM) Nicholasville KY
WLCK(AM) Scottsville KY
WARE(AM) Ware MA
WNEM(AM) Bridgeport MI
KBRF(AM) Fergus Falls MN
KCUE(AM) Red Wing MN
KBTC(AM) Houston MO
WHNY(AM) McComb MS
KIKC(AM) Forsyth MT
WGHB(AM) Farmville NC
WKDX(AM) Hamlet NC
WBRM(AM) Marion NC
KTFJ(AM) Dakota City NE
WKBR(AM) Manchester NH
WMTR(AM) Morristown NJ
WIPS(AM) Ticonderoga NY
WCHO(AM) Washington Court House OH
KCST(AM) Florence OR
WLEM(AM) Emporium PA
*WPEL(AM) Montrose PA
WEAE(AM) Pittsburgh PA
WYYC(AM) York PA
WJIT(AM) Sabana PR
WTMA(AM) Charleston SC
WKBL(AM) Covington TN
WRKQ(AM) Madisonville TN
WNTT(AM) Tazewell TN
KZHN(AM) Paris TX
*KDEI(AM) Port Arthur TX
KZDC(AM) San Antonio TX
KIKZ(AM) Seminole TX
KNEU(AM) Roosevelt UT
WDVA(AM) Danville VA
WLQM(AM) Franklin VA
WPRZ(AM) Warrenton VA
*KWSU(AM) Pullman WA
KKDZ(AM) Seattle WA
WSSP(AM) Milwaukee WI
WYKM(AM) Rupert WV

1260 khz

WYDE(AM) Birmingham AL
KCCB(AM) Corning AR
KBHC(AM) Nashville AR
KMZT(AM) Beverly Hills CA
KOIT(AM) San Francisco CA
*WSHU(AM) Westport CT
WWRC(AM) Washington DC
WNWK(AM) Newark DE
WFTW(AM) Fort Walton Beach FL
WSUA(AM) Miami FL
WIYD(AM) Palatka FL
WUFE(AM) Baxley GA
WBBK(AM) Blakely GA
WTJH(AM) East Point GA
KFFF(AM) Boone IA
KBLY(AM) Idaho Falls ID
KWEI(AM) Weiser ID
WSDZ(AM) Belleville IL
WNDE(AM) Indianapolis IN
KBRH(AM) Baton Rouge LA
WMKI(AM) Boston MA
WPNW(AM) Zeeland MI
KROX(AM) Crookston MN

KDUZ(AM) Hutchinson MN
KSGF(AM) Springfield MO
WGVM(AM) Greenville MS
WCSA(AM) Ripley MS
WKXR(AM) Asheboro NC
WZBO(AM) Edenton NC
KIMB(AM) Kimball NE
WBUD(AM) Trenton NJ
KTRC(AM) Santa Fe NM
WBNR(AM) Beacon NY
WNSS(AM) Syracuse NY
WWMK(AM) Cleveland OH
WNXT(AM) Portsmouth OH
KWSH(AM) Wewoka OK
KLYC(AM) McMinnville OR
WRIE(AM) Erie PA
WPHB(AM) Philipsburg PA
WISO(AM) Ponce PR
WMUU(AM) Greenville SC
WHYM(AM) Lake City SC
KWYR(AM) Winner SD
WNOO(AM) Chattanooga TN
WMCH(AM) Church Hill TN
WDKN(AM) Dickson TN
WCLC(AM) Jamestown TN
KSML(AM) Diboll TX
KLDS(AM) Falfurrias TX
KKSA(AM) San Angelo TX
KWNX(AM) Taylor TX
KTUE(AM) Tulia TX
WCHV(AM) Charlottesville VA
WWVT(AM) Christiansburg VA
WXCE(AM) Amery WI
WWIS(AM) Black River Falls WI
WEKZ(AM) Monroe WI
WOCO(AM) Oconto WI
WTBZ(AM) Grafton WV
KPOW(AM) Powell WY

1270 khz

WGSV(AM) Guntersville AL
WIJD(AM) Prichard AL
KDJI(AM) Holbrook AZ
KXBX(AM) Lakeport CA
KFUT(AM) Thousand Palms CA
KJUG(AM) Tulare CA
WRLZ(AM) Eatonville FL
WNOG(AM) Naples FL
WNLS(AM) Tallahassee FL
WYXC(AM) Cartersville GA
WSHE(AM) Columbus GA
WJJC(AM) Commerce GA
KNDI(AM) Honolulu HI
KTFI(AM) Twin Falls ID
WEIC(AM) Charleston IL
WKBF(AM) Rock Island IL
WFRN(AM) Elkhart IN
WWCA(AM) Gary IN
WXGO(AM) Madison IN
KSCB(AM) Liberal KS
WAIN(AM) Columbia KY
WFUL(AM) Fulton KY
KVCL(AM) Winnfield LA
WSPR(AM) Springfield MA
WCBC(AM) Cumberland MD
WMKT(AM) Charlevoix MI
WXYT(AM) Detroit MI
WWWI(AM) Baxter MN
KWEB(AM) Rochester MN
KGNM(AM) Saint Joseph MO
KOZQ(AM) Waynesville MO
WMLC(AM) Monticello MS
WCGC(AM) Belmont NC
WMPM(AM) Smithfield NC
KLXX(AM) Bismarck-Mandan ND
WTSN(AM) Dover NH
WMIZ(AM) Vineland NJ
KINN(AM) Alamogordo NM
KBZZ(AM) Sparks NV
WHLD(AM) Niagara Falls NY
WDLA(AM) Walton NY
WILE(AM) Cambridge OH
WUCO(AM) Marysville OH
KRVT(AM) Claremore OK
KAJO(AM) Grants Pass OR
WLBR(AM) Lebanon PA

WHGS(AM) Hampton SC
*KNWC(AM) Sioux Falls SD
WLIK(AM) Newport TN
WQKR(AM) Portland TN
KEPS(AM) Eagle Pass TX
KFLC(AM) Fort Worth TX
WTJZ(AM) Newport News VA
WHEO(AM) Stuart VA
KBAM(AM) Longview WA
WRJC(AM) Mauston WI
KIML(AM) Gillette WY

1280 khz
WPID(AM) Piedmont AL
WWPG(AM) Tuscaloosa AL
KNBY(AM) Newport AR
KXEG(AM) Phoenix AZ
KXTK(AM) Arroyo Grande CA
*KFRN(AM) Long Beach CA
KWSX(AM) Stockton CA
KBNO(AM) Denver CO
WJWK(AM) Seaford DE
WDSP(AM) De Funiak Springs FL
WIPC(AM) Lake Wales FL
WTMY(AM) Sarasota FL
WLCG(AM) Macon GA
KCOB(AM) Newton IA
WBIG(AM) Aurora IL
WGBF(AM) Evansville IN
KSOK(AM) Arkansas City KS
WCPM(AM) Cumberland KY
WODT(AM) New Orleans LA
WEIM(AM) Fitchburg MA
WFAU(AM) Gardiner ME
WFYC(AM) Alma MI
WWTC(AM) Minneapolis MN
KVOX(AM) Moorhead MN
KDKD(AM) Clinton MO
KYRO(AM) Potosi MO
WSAT(AM) Salisbury NC
WYAL(AM) Scotland Neck NC
KCNI(AM) Broken Bow NE
KRZE(AM) Farmington NM
KDOX(AM) Henderson NV
WADO(AM) New York NY
WHTK(AM) Rochester NY
WONW(AM) Defiance OH
KPRV(AM) Poteau OK
*KRVM(AM) Eugene OR
WFBS(AM) Berwick PA
WHVR(AM) Hanover PA
WJST(AM) New Castle PA
WCMN(AM) Arecibo PR
WANS(AM) Anderson SC
WJAY(AM) Mullins SC
WMCP(AM) Columbia TN
WDNT(AM) Dayton TN
KSLI(AM) Abilene TX
KWHI(AM) Brenham TX
KVWG(AM) Pearsall TX
KZNS(AM) Salt Lake City UT
WOWZ(AM) Appomattox VA
WYVE(AM) Wytheville VA
KLDY(AM) Lacey WA
KPTQ(AM) Spokane WA
KIT(AM) Yakima WA
WGLR(AM) Lancaster WI
WNAM(AM) Neenah-Menasha WI

1290 khz
WOPP(AM) Opp AL
WBTG(AM) Sheffield AL
WYEA(AM) Sylacauga AL
KDMS(AM) El Dorado AR
KUOA(AM) Siloam Springs AR
KCUB(AM) Tucson AZ
KPAY(AM) Chico CA
KAZA(AM) Gilroy CA
KKDD(AM) San Bernardino CA
KZSB(AM) Santa Barbara CA
WCCC(AM) West Hartford CT
WWTX(AM) Wilmington DE
WCFI(AM) Ocala FL
WPCF(AM) Panama City Beach FL
WJNO(AM) West Palm Beach FL
WCHK(AM) Canton GA
WTKS(AM) Savannah GA
KOUU(AM) Pocatello ID
WIRL(AM) Peoria IL
KWLS(AM) Pratt KS
WCBL(AM) Benton KY
WKLB(AM) Manchester KY
KJEF(AM) Jennings LA
WNIL(AM) Niles MI
WLBY(AM) Saline MI
KBMO(AM) Benson MN
KALM(AM) Thayer MO
WJBI(AM) Batesville MS
WNBN(AM) Meridian MS
WTYL(AM) Tylertown MS
KGVO(AM) Missoula MT
WHKY(AM) Hickory NC
WJCV(AM) Jacksonville NC
WXKL(AM) Sanford NC
KKAR(AM) Omaha NE
WKBK(AM) Keene NH
WNBF(AM) Binghamton NY
WOMP(AM) Bellaire OH
WHIO(AM) Dayton OH
KUMA(AM) Pendleton OR
WFBG(AM) Altoona PA
WRNI(AM) Providence RI
WQMC(AM) Sumter SC
WATO(AM) Oak Ridge TN
KIVY(AM) Crockett TX
KRGE(AM) Weslaco TX
KWFS(AM) Wichita Falls TX
WDZY(AM) Colonial Heights VA
WRRA(AM) Frederiksted VI
WMCS(AM) Greenfield WI
WKLJ(AM) Sparta WI
WVOW(AM) Logan WV
KOWB(AM) Laramie WY

1300 khz
WBSA(AM) Boaz AL
WTLS(AM) Tallassee AL
WKXM(AM) Winfield AL
KWCK(AM) Searcy AR
KROP(AM) Brawley CA
KYNO(AM) Fresno CA
*KPMO(AM) Mendocino CA
KAZN(AM) Pasadena CA
KKML(AM) Colorado Springs CO
WAVZ(AM) New Haven CT
WTIR(AM) Cocoa Beach FL
WFFG(AM) Marathon FL
WQBN(AM) Temple Terrace FL
WMTM(AM) Moultrie GA
WNEA(AM) Newnan GA
WIMO(AM) Winder GA
KGLO(AM) Mason City IA
KLER(AM) Orofino ID
WRDZ(AM) La Grange IL
WFRX(AM) West Frankfort IL
WBZQ(AM) Huntington IN
WBOW(AM) Terre Haute IN
WLXG(AM) Lexington KY
WIBR(AM) Baton Rouge LA
KSYB(AM) Shreveport LA
WJDA(AM) Quincy MA
WJFK(AM) Baltimore MD
WOOD(AM) Grand Rapids MI
WQPM(AM) Princeton MN
KMMO(AM) Marshall MO
WOAD(AM) Jackson MS
WSSG(AM) Goldsboro NC
WLNC(AM) Laurinburg NC
WSYD(AM) Mount Airy NC
KBRL(AM) McCook NE
WPNH(AM) Plymouth NH
WIMG(AM) Ewing NJ
KCMY(AM) Carson City NV
WAMF(AM) Fulton NY
WXRL(AM) Lancaster NY
WTMM(AM) Rensselaer NY
WRCR(AM) Spring Valley NY
WJMO(AM) Cleveland OH
WMVO(AM) Mount Vernon OH
KAKC(AM) Tulsa OK
*KAPL(AM) Phoenix OR
KACI(AM) The Dalles OR
WWCH(AM) Clarion PA
WKZN(AM) West Hazleton PA
WTIL(AM) Mayaguez PR
WCKI(AM) Greer SC
WKSC(AM) Kershaw SC
KOLY(AM) Mobridge SD
WMTN(AM) Morristown TN
WNQM(AM) Nashville TN
KVET(AM) Austin TX
KKUB(AM) Brownfield TX
KLAR(AM) Laredo TX
KSET(AM) Silsbee TX
WKCY(AM) Harrisonburg VA
KKOL(AM) Seattle WA
WCLG(AM) Morgantown WV
WJYP(AM) Saint Albans WV

1310 khz
WHEP(AM) Foley AL
WJUS(AM) Marion AL
WQAH(AM) Priceville AL
KBOK(AM) Malvern AR
KXAM(AM) Mesa AZ
KIQQ(AM) Barstow CA
KFVR(AM) Crescent City CA
KMKY(AM) Oakland CA
KFKA(AM) Greeley CO
WICH(AM) Norwich CT
WYND(AM) De Land FL
WAUC(AM) Wauchula FL
WPBC(AM) Decatur GA
WOKA(AM) Douglas GA
WPLV(AM) West Point GA
KOKX(AM) Keokuk IA
KDLS(AM) Perry IA
KLIX(AM) Twin Falls ID
WTLC(AM) Indianapolis IN
KYUL(AM) Scott City KS
WTTL(AM) Madisonville KY
WDOC(AM) Prestonsburg KY
KEZM(AM) Sulphur LA
KMBS(AM) West Monroe LA
WORC(AM) Worcester MA
WLOB(AM) Portland ME
WDTW(AM) Dearborn MI
WCCW(AM) Traverse City MI
KRBI(AM) Saint Peter MN
KBSP(AM) Birch Tree MO
KZRG(AM) Joplin MO
KEIN(AM) Great Falls MT
WISE(AM) Asheville NC
WGSP(AM) Charlotte NC
WTIK(AM) Durham NC
KNOX(AM) Grand Forks ND
KGMT(AM) Fairbury NE
WADB(AM) Asbury Park NJ
WEMG(AM) Camden NJ
WXMC(AM) Parsippany-Troy Hills NJ
KKNS(AM) Corrales NM
WRSB(AM) Canandaigua NY
WRVP(AM) Mount Kisco NY
WTLB(AM) Utica NY
WDPN(AM) Alliance OH
KNPT(AM) Newport OR
WBFD(AM) Bedford PA
WTZN(AM) Troy PA
WNAE(AM) Warren PA
WDKD(AM) Kingstree SC
WDOD(AM) Chattanooga TN
WDXI(AM) Jackson TN
WOCV(AM) Oneida TN
KZIP(AM) Amarillo TX
KTCK(AM) Dallas TX
KAHL(AM) San Antonio TX
WDCT(AM) Fairfax VA
WCMS(AM) Newport News VA
KZXR(AM) Prosser WA
WIBA(AM) Madison WI
WSLW(AM) White Sulphur Springs WV

1320 khz
WPSB(AM) Birmingham AL
WAGF(AM) Dothan AL
KYHN(AM) Fort Smith AR
KRLW(AM) Walnut Ridge AR
*KAWC(AM) Yuma AZ
KSDT(AM) Hemet CA
*KKSM(AM) Oceanside CA
KCTC(AM) Sacramento CA
WATR(AM) Waterbury CT
WLQY(AM) Hollywood FL
WBOB(AM) Jacksonville FL
WDDV(AM) Venice FL
WHIE(AM) Griffin GA
KNIA(AM) Knoxville IA
KMAQ(AM) Maquoketa IA
WKAN(AM) Kankakee IL
KLWN(AM) Lawrence KS
WBRT(AM) Bardstown KY
WCVG(AM) Covington KY
WNGO(AM) Mayfield KY
KNCB(AM) Vivian LA
WARL(AM) Attleboro MA
WICO(AM) Salisbury MD
WILS(AM) Lansing MI
WDMJ(AM) Marquette MI
KOZY(AM) Grand Rapids MN
KSIV(AM) Clayton MO
WRJW(AM) Picayune MS
WAGY(AM) Forest City NC
WCOG(AM) Greensboro NC
WKRK(AM) Murphy NC
KHRT(AM) Minot ND
KOLT(AM) Scottsbluff NE
WDER(AM) Derry NH
KRDD(AM) Roswell NM
WHHO(AM) Hornell NY
WLOH(AM) Lancaster OH
WOBL(AM) Oberlin OH
KCLI(AM) Clinton OK
KSCR(AM) Eugene OR
WTKZ(AM) Allentown PA
WGET(AM) Gettysburg PA
WJAS(AM) Pittsburgh PA
WSKN(AM) San Juan PR
WISW(AM) Columbia SC
KELO(AM) Sioux Falls SD
WGOC(AM) Kingsport TN
WMSR(AM) Manchester TN
KVMC(AM) Colorado City TX
KXYZ(AM) Houston TX
KFNZ(AM) Salt Lake City UT
WVGM(AM) Lynchburg VA
WVNZ(AM) Richmond VA
WTSJ(AM) Randolph VT
KXRO(AM) Aberdeen WA
KGDC(AM) Walla Walla WA
WFHR(AM) Wisconsin Rapids WI

1330 khz
KXLJ(AM) Juneau AK
WPRN(AM) Butler AL
WZCT(AM) Scottsboro AL
KXXA(AM) Conway AR
KJLL(AM) South Tucson AZ
KWKW(AM) Los Angeles CA
KLBS(AM) Los Banos CA
KJPR(AM) Shasta Lake City CA
WJNX(AM) Fort Pierce FL
WWAB(AM) Lakeland FL
WEBY(AM) Milton FL
WCVC(AM) Tallahassee FL
WLBB(AM) Carrollton GA
WMLT(AM) Dublin GA
WGTJ(AM) Murrayville GA
KWLO(AM) Waterloo IA
WKTA(AM) Evanston IL
WRAM(AM) Monmouth IL
WNTA(AM) Rockford IL
WVHI(AM) Evansville IN
WTRE(AM) Greensburg IN
KNSS(AM) Wichita KS
WKDP(AM) Corbin KY
WMOR(AM) Morehead KY
KVOL(AM) Lafayette LA
WRCA(AM) Waltham MA
WJSS(AM) Havre de Grace MD
WTRX(AM) Flint MI
WLOL(AM) Minneapolis MN
KUKU(AM) Willow Springs MO
WNIX(AM) Greenville MS
WANG(AM) Havelock NC
KGAK(AM) Gallup NM
*WWRV(AM) New York NY
WMJQ(AM) Ontario NY
WEBO(AM) Owego NY
WSPQ(AM) Springville NY
WHAZ(AM) Troy NY
WGFT(AM) Campbell OH
WFIN(AM) Findlay OH
WYPC(AM) Wellston OH
WELW(AM) Willoughby-Eastlake OH
KKPZ(AM) Portland OR
WFNN(AM) Erie PA
WBHV(AM) Somerset PA
WENA(AM) Yauco PR
WPJS(AM) Conway SC
WYRD(AM) Greenville SC
WAEW(AM) Crossville TN
KTAE(AM) Cameron TX
KSWA(AM) Graham TX
KINE(AM) Kingsville TX
KCKM(AM) Monahans TX
KGLD(AM) Tyler TX
WBTM(AM) Danville VA
WRAA(AM) Luray VA
WHGB(AM) Marion VA
WESR(AM) Onley-Onancock VA
KGRG(AM) Enumclaw WA
*KMBI(AM) Spokane WA
WHBL(AM) Sheboygan WI
WETZ(AM) New Martinsville WV
KOVE(AM) Lander WY

1340 khz
WFMH(AM) Cullman AL
WSBM(AM) Florence AL
WMRK(AM) Selma AL
WFEB(AM) Sylacauga AL
KBTA(AM) Batesville AR
KZNG(AM) Hot Springs AR
*KCAT(AM) Pine Bluff AR
KIKO(AM) Miami AZ
KPGE(AM) Page AZ
KATA(AM) Arcata CA
KACE(AM) Bishop CA
KPTR(AM) Cathedral City CA
KCBL(AM) Fresno CA
KOMY(AM) La Selva Beach CA
KTPI(AM) Mojave CA
KTOX(AM) Needles CA
KEWE(AM) Oroville CA
KYNS(AM) San Luis Obispo CA
KIST(AM) Santa Barbara CA
*KCFR(AM) Denver CO
KTMM(AM) Grand Junction CO
KVRH(AM) Salida CO
WYBC(AM) New Haven CT
WYCB(AM) Washington DC
WTAN(AM) Clearwater FL
WWFL(AM) Clermont FL
WROD(AM) Daytona Beach FL
WDSR(AM) Lake City FL
WPBR(AM) Lantana FL
WTYS(AM) Marianna FL
WITS(AM) Sebring FL
WFSH(AM) Valparaiso-Niceville FL
WGAU(AM) Athens GA
WALR(AM) Atlanta GA
WSGF(AM) Augusta GA
WOKS(AM) Columbus GA
WBBT(AM) Lyons GA
WALH(AM) Mountain City GA
WTIF(AM) Tifton GA
KROS(AM) Clinton IA
KACH(AM) Preston ID
WSOY(AM) Decatur IL
WJPF(AM) Herrin IL
WJOL(AM) Joliet IL
WBIW(AM) Bedford IN
WTRC(AM) Elkhart IN
WXFN(AM) Muncie IN
KDTD(AM) Kansas City KS
KSEK(AM) Pittsburg KS
WCMI(AM) Ashland KY
WBGN(AM) Bowling Green KY
WKCB(AM) Hindman KY
WNBS(AM) Murray KY
WEKY(AM) Richmond KY
KRMD(AM) Shreveport LA

WGAW(AM) Gardner MA
WNBH(AM) New Bedford MA
WBRK(AM) Pittsfield MA
*WMDR(AM) Augusta ME
WNZS(AM) Veazie ME
WLEW(AM) Bad Axe MI
WBBL(AM) Grand Rapids MI
WCSR(AM) Hillsdale MI
WMTE(AM) Manistee MI
WAGN(AM) Menominee MI
WMBN(AM) Petoskey MI
WEXL(AM) Royal Oak MI
KVBR(AM) Brainerd MN
KDLM(AM) Detroit Lakes MN
KRBT(AM) Eveleth MN
KROC(AM) Rochester MN
KWLM(AM) Willmar MN
KXEO(AM) Mexico MO
KLID(AM) Poplar Bluff MO
KSMO(AM) Salem MO
KADI(AM) Springfield MO
WKOZ(AM) Kosciusko MS
WAML(AM) Laurel MS
KCAP(AM) Helena MT
KQJZ(AM) Kalispell MT
KPRK(AM) Livingston MT
KYLT(AM) Missoula MT
WJRI(AM) Lenoir NC
WAGR(AM) Lumberton NC
WCBQ(AM) Oxford NC
WADE(AM) Wadesboro NC
WLSG(AM) Wilmington NC
WPOL(AM) Winston-Salem NC
KPOK(AM) Bowman ND
KXPO(AM) Grafton ND
KHUB(AM) Fremont NE
KGFW(AM) Kearney NE
KSID(AM) Sidney NE
WDCR(AM) Hanover NH
WWNH(AM) Madbury NH
WMID(AM) Atlantic City NJ
KCQL(AM) Aztec NM
KSSR(AM) Santa Rosa NM
KVOT(AM) Taos NM
KTSN(AM) Elko NV
KRLV(AM) Las Vegas NV
KXEQ(AM) Reno NV
WWLF(AM) Auburn NY
WENT(AM) Gloversville NY
WKSN(AM) Jamestown NY
WLVL(AM) Lockport NY
WMSA(AM) Massena NY
WALL(AM) Middletown NY
WIRY(AM) Plattsburgh NY
WNCO(AM) Ashland OH
*WOUB(AM) Athens OH
WIZE(AM) Springfield OH
WSTV(AM) Steubenville OH
KIHN(AM) Hugo OK
KEBC(AM) Midwest City OK
KJMU(AM) Sand Springs OK
KLOO(AM) Corvallis OR
KWVR(AM) Enterprise OR
KIHR(AM) Hood River OR
KBBR(AM) North Bend OR
WYJK(AM) Connellsville PA
WOYL(AM) Oil City PA
WHAT(AM) Philadelphia PA
WYCK(AM) Plains PA
WRAW(AM) Reading PA
WTRN(AM) Tyrone PA
WWPA(AM) Williamsport PA
WWNA(AM) Aguadilla PR
WQSC(AM) Charleston SC
WRHI(AM) Rock Hill SC
WSSC(AM) Sumter SC
KIJV(AM) Huron SD
KTOQ(AM) Rapid City SD
WBAC(AM) Cleveland TN
WKRM(AM) Columbia TN
WGRV(AM) Greeneville TN
WKGN(AM) Knoxville TN
WLOK(AM) Memphis TN
WCDT(AM) Winchester TN
KWKC(AM) Abilene TX
KRHC(AM) Burnet TX
KAND(AM) Corsicana TX
KVIV(AM) El Paso TX
KKAM(AM) Lubbock TX
KRBA(AM) Lufkin TX
KOLE(AM) Port Arthur TX
KCRN(AM) San Angelo TX
KVNN(AM) Victoria TX
WKEY(AM) Covington VA
WHAP(AM) Hopewell VA
WVCV(AM) Orange VA
WSTA(AM) Charlotte Amalie VI
WVNR(AM) Poultney VT
WSTJ(AM) Saint Johnsbury VT
KWLE(AM) Anacortes WA
KTCR(AM) Kennewick WA
KUOW(AM) Tumwater WA
KWWX(AM) Wenatchee WA
WLDY(AM) Ladysmith WI
WJYI(AM) Milwaukee WI
WXKX(AM) Clarksburg WV
WEPM(AM) Martinsburg WV
WMON(AM) Montgomery WV
WELA(AM) Welch WV
KSGT(AM) Jackson WY
KYCN(AM) Wheatland WY
KWOR(AM) Worland WY

1350 khz

WELB(AM) Elba AL
WGAD(AM) Gadsden AL
KZTD(AM) Cabot AR
KLHC(AM) Bakersfield CA
KTDD(AM) San Bernardino CA
KSRO(AM) Santa Rosa CA
KGHF(AM) Pueblo CO
WNLK(AM) Norwalk CT
WINY(AM) Putnam CT
WMMV(AM) Cocoa FL
WDCF(AM) Dade City FL
WCRM(AM) Fort Myers FL
WFNS(AM) Blackshear GA
WRWH(AM) Cleveland GA
WNNG(AM) Warner Robins GA
KRNT(AM) Des Moines IA
KRLC(AM) Lewiston ID
KTIK(AM) Nampa ID
WOAM(AM) Peoria IL
WJBD(AM) Salem IL
WIOU(AM) Kokomo IN
KMAN(AM) Manhattan KS
WLOU(AM) Louisville KY
WWWL(AM) New Orleans LA
WGDN(AM) Gladwin MI
KCHK(AM) New Prague MN
KDIO(AM) Ortonville MN
WCMP(AM) Pine City MN
KCHR(AM) Charleston MO
KWMO(AM) Washington MO
WKCU(AM) Corinth MS
WQNX(AM) Aberdeen NC
WZGM(AM) Black Mountain NC
WHIP(AM) Mooresville NC
WLLY(AM) Wilson NC
KBRX(AM) O'Neill NE
WEZS(AM) Laconia NH
WHWH(AM) Princeton NJ
KABQ(AM) Albuquerque NM
WCBA(AM) Corning NY
WRNY(AM) Rome NY
WARF(AM) Akron OH
WCSM(AM) Celina OH
WCHI(AM) Chillicothe OH
KPNS(AM) Duncan OK
KTLQ(AM) Tahlequah OK
WOYK(AM) York PA
WEGA(AM) Vega Baja PR
WRKM(AM) Carthage TN
KCAR(AM) Clarksville TX
KCOX(AM) Jasper TX
KCOR(AM) San Antonio TX
WBLT(AM) Bedford VA
WYSK(AM) Fredericksburg VA
WNVA(AM) Norton VA
WGPL(AM) Portsmouth VA
WPDR(AM) Portage WI

1360 khz

WIXI(AM) Jasper AL
WMOB(AM) Mobile AL
WMFC(AM) Monroeville AL
WELR(AM) Roanoke AL
KLYR(AM) Clarksville AR
KFFA(AM) Helena AR
KPXQ(AM) Glendale AZ
KFIV(AM) Modesto CA
KWDJ(AM) Ridgecrest CA
KLSD(AM) San Diego CA
KHNC(AM) Johnstown CO
WDRC(AM) Hartford CT
WHNR(AM) Cypress Gardens FL
WCGL(AM) Jacksonville FL
WKAT(AM) North Miami FL
WHCG(AM) Metter GA
WGJK(AM) Rome GA
KMJM(AM) Cedar Rapids IA
KBKB(AM) Fort Madison IA
KSCJ(AM) Sioux City IA
WLBK(AM) De Kalb IL
WVMC(AM) Mount Carmel IL
WGFA(AM) Watseka IL
KAHS(AM) El Dorado KS
WFLW(AM) Monticello KY
KNIR(AM) New Iberia LA
KBYO(AM) Tallulah LA
WLYN(AM) Lynn MA
WKYO(AM) Caro MI
WKMI(AM) Kalamazoo MI
KKBJ(AM) Bemidji MN
KRWC(AM) Buffalo MN
KMRN(AM) Cameron MO
KELE(AM) Mountain Grove MO
WFFF(AM) Columbia MS
WCHL(AM) Chapel Hill NC
*KNGN(AM) McCook NE
WNNJ(AM) Newton NJ
WNJC(AM) Washington Township NJ
KBUY(AM) Ruidoso NM
WYOS(AM) Binghamton NY
WOEN(AM) Olean NY
WSAI(AM) Cincinnati OH
WWOW(AM) Conneaut OH
KOHU(AM) Hermiston OR
KUIK(AM) Hillsboro OR
WPTT(AM) McKeesport PA
WPPA(AM) Pottsville PA
WELP(AM) Easley SC
WBLC(AM) Lenoir City TN
WNAH(AM) Nashville TN
KDJW(AM) Amarillo TX
KACT(AM) Andrews TX
KWWJ(AM) Baytown TX
KKTX(AM) Corpus Christi TX
KMNY(AM) Hurst TX
WWWJ(AM) Galax VA
WHBG(AM) Harrisonburg VA
KKMO(AM) Tacoma WA
WTAQ(AM) Green Bay WI
WVRQ(AM) Viroqua WI
WHJC(AM) Matewan WV
WMOV(AM) Ravenswood WV
KRKK(AM) Rock Springs WY

1370 khz

WBYE(AM) Calera AL
KAWW(AM) Heber Springs AR
KTPA(AM) Prescott AR
KWRM(AM) Corona CA
KPCO(AM) Quincy CA
KZSF(AM) San Jose CA
KGEN(AM) Tulare CA
WOCA(AM) Ocala FL
WCOA(AM) Pensacola FL
WZTA(AM) Vero Beach FL
WGHC(AM) Clayton GA
WLOP(AM) Jesup GA
WFDR(AM) Manchester GA
WLOV(AM) Washington GA
KUPA(AM) Pearl City HI
KDTH(AM) Dubuque IA
WLLM(AM) Lincoln IL
WGCL(AM) Bloomington IN
WLTH(AM) Gary IN
KGNO(AM) Dodge City KS
KALN(AM) Iola KS
WJQI(AM) Fort Campbell KY
WGOH(AM) Grayson KY
WTKY(AM) Tompkinsville KY
WVIE(AM) Pikesville MD
WDEA(AM) Ellsworth ME
WLJW(AM) Cadillac MI
WGHN(AM) Grand Haven MI
KSUM(AM) Fairmont MN
KWRT(AM) Boonville MO
KCRV(AM) Caruthersville MO
WMGO(AM) Canton MS
KXTL(AM) Butte MT
*WLLN(AM) Lillington NC
WGIV(AM) Pineville NC
WTAB(AM) Tabor City NC
*KWTL(AM) Grand Forks ND
KAWL(AM) York NE
WFEA(AM) Manchester NH
WALK(AM) East Patchogue NY
WELG(AM) Ellenville NY
*WXXI(AM) Rochester NY
WSPD(AM) Toledo OH
KAST(AM) Astoria OR
WWCB(AM) Corry PA
WPAZ(AM) Pottstown PA
WKMC(AM) Roaring Spring PA
WIVV(AM) Vieques PR
WKFD(AM) Charlestown RI
WDEF(AM) Chattanooga TN
WDXE(AM) Lawrenceburg TN
WRGS(AM) Rogersville TN
KFRO(AM) Longview TX
KJCE(AM) Rollingwood TX
KSOP(AM) South Salt Lake UT
WHEE(AM) Martinsville VA
WSHV(AM) South Hill VA
WBTN(AM) Bennington VT
KPWL(AM) Newport WA
KWNC(AM) Quincy WA
WCCN(AM) Neillsville WI
*WVMR(AM) Frost WV
WVLY(AM) Moundsville WV

1380 khz

WRAB(AM) Arab AL
WGYV(AM) Greenville AL
WVSA(AM) Vernon AL
KDXE(AM) North Little Rock AR
KLPZ(AM) Parker AZ
KWJL(AM) Lancaster CA
KTKZ(AM) Sacramento CA
WFNW(AM) Naugatuck CT
WTMC(AM) Wilmington DE
WWRF(AM) Lake Worth FL
WELE(AM) Ormond Beach FL
WWMI(AM) Saint Petersburg FL
WAOK(AM) Atlanta GA
KCIM(AM) Carroll IA
KCII(AM) Washington IA
WTJK(AM) South Beloit IL
WKJG(AM) Fort Wayne IN
KCNW(AM) Fairway KS
WMTA(AM) Central City KY
WMJR(AM) Winchester KY
WPYR(AM) Baton Rouge LA
WSCG(AM) Greenville MI
WPHM(AM) Port Huron MI
KLIZ(AM) Brainerd MN
KAGE(AM) Winona MN
KSLG(AM) Saint Louis MO
WNLA(AM) Indianola MS
WTOB(AM) Winston-Salem NC
KUVR(AM) Holdrege NE
WMYF(AM) Portsmouth NH
WABH(AM) Bath NY
WKDM(AM) New York NY
WDLW(AM) Lorain OH
KXCA(AM) Lawton OK
KMUS(AM) Sperry OK
KSRV(AM) Ontario OR
WTYM(AM) Kittanning PA
WMLP(AM) Milton PA
WCBG(AM) Waynesboro PA
WOLA(AM) Barranquitas PR
WNRI(AM) Woonsocket RI
WAGS(AM) Bishopville SC
WYNF(AM) North Augusta SC
KOTA(AM) Rapid City SD
*KQKD(AM) Redfield SD
WYSH(AM) Clinton TN
WHEW(AM) Franklin TN
WLRM(AM) Millington TN
KRCM(AM) Beaumont TX
KBWD(AM) Brownwood TX
KHEY(AM) El Paso TX
KWMF(AM) Pleasanton TX
WLRV(AM) Lebanon VA
WBTK(AM) Richmond VA
WSYB(AM) Rutland VT
KRKO(AM) Everett WA
WFCL(AM) Clintonville WI
WMTD(AM) Hinton WV
KJUA(AM) Cheyenne WY

1390 khz

WHMA(AM) Anniston AL
KDQN(AM) De Queen AR
KFFK(AM) Rogers AR
KLTX(AM) Long Beach CA
KLOC(AM) Turlock CA
*KGNU(AM) Denver CO
WFHT(AM) Avon Park FL
WAJD(AM) Gainesville FL
WISK(AM) Americus GA
WTNL(AM) Reidsville GA
KCLN(AM) Clinton IA
WGRB(AM) Chicago IL
WFIW(AM) Fairfield IL
WZZB(AM) Seymour IN
KNCK(AM) Concordia KS
WANY(AM) Albany KY
WKIC(AM) Hazard KY
KFRA(AM) Franklin LA
WPLM(AM) Plymouth MA
WEGP(AM) Presque Isle ME
WLCM(AM) Charlotte MI
KRFO(AM) Owatonna MN
KXSS(AM) Waite Park MN
KJPW(AM) Waynesville MO
WROA(AM) Gulfport MS
WMER(AM) Meridian MS
WEED(AM) Rocky Mount NC
WADA(AM) Shelby NC
WJRM(AM) Troy NC
KRRZ(AM) Minot ND
KENN(AM) Farmington NM
KHOB(AM) Hobbs NM
WEOK(AM) Poughkeepsie NY
WRIV(AM) Riverhead NY
WFBL(AM) Syracuse NY
WBLL(AM) Bellefontaine OH
WMPO(AM) Middleport-Pomeroy OH
WNIO(AM) Youngstown OH
KCRC(AM) Enid OK
KKSN(AM) Salem OR
WLAN(AM) Lancaster PA
WRSC(AM) State College PA
WISA(AM) Isabela PR
WLUA(AM) Belton SC
WXTC(AM) Charleston SC
KJAM(AM) Madison SD
WYXI(AM) Athens TN
WTJS(AM) Jackson TN
WMCT(AM) Mountain City TN
KULP(AM) El Campo TX
KBEC(AM) Waxahachie TX
KLGN(AM) Logan UT
WZHF(AM) Arlington VA
WKPA(AM) Lynchburg VA
WCAT(AM) Burlington VT
KJOX(AM) Yakima WA
WRIG(AM) Schofield WI
WKLP(AM) Keyser WV

1400 khz

WWTM(AM) Decatur AL
WXAL(AM) Demopolis AL
WJLD(AM) Fairfield AL
WFPA(AM) Fort Payne AL
WANI(AM) Opelika AL
KELD(AM) El Dorado AR
KCLA(AM) Pine Bluff AR

KWYN(AM) Wynne AR
KSUN(AM) Phoenix AZ
KRVZ(AM) Springerville-Eagar AZ
KTUC(AM) Tucson AZ
KJOK(AM) Yuma AZ
KVTO(AM) Berkeley CA
KESQ(AM) Indio CA
KQMS(AM) Redding CA
KKJL(AM) San Luis Obispo CA
KUNX(AM) Santa Paula CA
KUKI(AM) Ukiah CA
KEZL(AM) Visalia CA
KRLN(AM) Canon City CO
KDTA(AM) Delta CO
KFTM(AM) Fort Morgan CO
KBLJ(AM) La Junta CO
KWUF(AM) Pagosa Springs CO
WSTC(AM) Stamford CT
WILI(AM) Willimantic CT
WFLL(AM) Fort Lauderdale FL
WIRA(AM) Fort Pierce FL
WTKE(AM) Fort Walton Beach FL
WZAZ(AM) Jacksonville FL
WPRY(AM) Perry FL
WSDO(AM) Sanford FL
WZHR(AM) Zephyrhills FL
WAJQ(AM) Alma GA
WLTA(AM) Alpharetta GA
WNGA(AM) Elberton GA
WNEX(AM) Macon GA
WHBS(AM) Moultrie GA
WCOH(AM) Newnan GA
WSEG(AM) Savannah GA
KCOG(AM) Centerville IA
KADR(AM) Elkader IA
KVFD(AM) Fort Dodge IA
KART(AM) Jerome ID
KRPL(AM) Moscow ID
KSPT(AM) Sandpoint ID
WDWS(AM) Champaign IL
WGIL(AM) Galesburg IL
WEOA(AM) Evansville IN
WBAT(AM) Marion IN
KVOE(AM) Emporia KS
KAYS(AM) Hays KS
WCYN(AM) Cynthiana KY
WIEL(AM) Elizabethtown KY
WFTG(AM) London KY
WFPR(AM) Hammond LA
KAOK(AM) Lake Charles LA
KWLA(AM) Many LA
WHTB(AM) Fall River MA
WLLH(AM) Lowell MA
WHMP(AM) Northampton MA
WWIN(AM) Baltimore MD
WJZN(AM) Augusta ME
WVAE(AM) Biddeford ME
WWNZ(AM) Veazie ME
WBFN(AM) Battle Creek MI
WDTK(AM) Detroit MI
*WLJN(AM) Elmwood Township MI
WCCY(AM) Houghton MI
WQXO(AM) Munising MI
WSAM(AM) Saginaw MI
WSJM(AM) Saint Joseph MI
WKNW(AM) Sault Ste. Marie MI
KEYL(AM) Long Prairie MN
KMHL(AM) Marshall MN
KMNV(AM) Saint Paul MN
KFRU(AM) Columbia MO
KJFF(AM) Festus MO
KSIM(AM) Sikeston MO
KGMY(AM) Springfield MO
WBIP(AM) Booneville MS
WJWF(AM) Columbus MS
WYKC(AM) Grenada MS
WFOR(AM) Hattiesburg MS
WKXI(AM) Jackson MS
KBCK(AM) Deer Lodge MT
KXGN(AM) Glendive MT
KXGF(AM) Great Falls MT
WKEW(AM) Greensboro NC
WMFA(AM) Raeford NC
WSIC(AM) Statesville NC
WMXF(AM) Waynesville NC
WSMY(AM) Weldon NC
KQDJ(AM) Jamestown ND
KBRB(AM) Ainsworth NE
KCOW(AM) Alliance NE
KLIN(AM) Lincoln NE
WTSL(AM) Hanover NH
WLTN(AM) Littleton NH
WOND(AM) Pleasantville NJ
KVSF(AM) Santa Fe NM
KCHS(AM) Truth or Consequences NM
KTNM(AM) Tucumcari NM
KSHP(AM) North Las Vegas NV
KBDB(AM) Sparks NV
KTNP(AM) Tonopah NV
KWNA(AM) Winnemucca NV
*WAMC(AM) Albany NY
WWWS(AM) Buffalo NY
WDNY(AM) Dansville NY
WSLB(AM) Ogdensburg NY
WMAN(AM) Mansfield OH
WPAY(AM) Portsmouth OH
KWON(AM) Bartlesville OK
KTMC(AM) McAlester OK
KREF(AM) Norman OK
KNND(AM) Cottage Grove OR
KJDY(AM) John Day OR
KBCH(AM) Lincoln City OR
WEST(AM) Easton PA
WJET(AM) Erie PA
WTCY(AM) Harrisburg PA
WWGE(AM) Loretto PA
WKBI(AM) Saint Marys PA
WICK(AM) Scranton PA
WRAK(AM) Williamsport PA
WIDA(AM) Carolina PR
WCOS(AM) Columbia SC
WWRK(AM) Darlington SC
WGTN(AM) Georgetown SC
WSPG(AM) Spartanburg SC
KBJM(AM) Lemmon SD
WJZM(AM) Clarksville TN
WHUB(AM) Cookeville TN
WLSB(AM) Copperhill TN
WKPT(AM) Kingsport TN
WGAP(AM) Maryville TN
WZNG(AM) Shelbyville TN
KRUN(AM) Ballinger TX
KBYG(AM) Big Spring TX
KUNO(AM) Corpus Christi TX
*KHCB(AM) Galveston TX
KGVL(AM) Greenville TX
KEBE(AM) Jacksonville TX
KIUN(AM) Pecos TX
KEYE(AM) Perryton TX
KREW(AM) Plainview TX
KVRP(AM) Stamford TX
KTEM(AM) Temple TX
KKTK(AM) Texarkana TX
KVOU(AM) Uvalde TX
KENT(AM) Parowan UT
KSRR(AM) Provo UT
WKAV(AM) Charlottesville VA
WHHV(AM) Hillsville VA
WPCE(AM) Portsmouth VA
WAJL(AM) South Boston VA
WINC(AM) Winchester VA
KLCK(AM) Goldendale WA
KEDO(AM) Longview WA
KRSC(AM) Othello WA
KITZ(AM) Silverdale WA
WATW(AM) Ashland WI
WBIZ(AM) Eau Claire WI
WDUZ(AM) Green Bay WI
WRJN(AM) Racine WI
WRDB(AM) Reedsburg WI
WOBG(AM) Clarksburg WV
WRON(AM) Ronceverte WV
WVRC(AM) Spencer WV
WBBD(AM) Wheeling WV
WBTH(AM) Williamson WV
KKTL(AM) Casper WY
KODI(AM) Cody WY

1410 khz
WLVV(AM) Mobile AL
WIQR(AM) Prattville AL
WZZA(AM) Tuscumbia AL
KTCS(AM) Fort Smith AR
KERN(AM) Bakersfield CA
KRML(AM) Carmel CA
KUHL(AM) Lompoc CA
KMYC(AM) Marysville CA
KCAL(AM) Redlands CA
KIIX(AM) Fort Collins CO
WPOP(AM) Hartford CT
WDOV(AM) Dover DE
WMYR(AM) Fort Myers FL
WQBQ(AM) Leesburg FL
WHBT(AM) Tallahassee FL
WKKP(AM) McDonough GA
WYIS(AM) McRae GA
WLAQ(AM) Rome GA
KGRN(AM) Grinnell IA
KLEM(AM) Le Mars IA
WRMN(AM) Elgin IL
*WIHM(AM) Taylorville IL
WSHY(AM) Lafayette IN
KKLO(AM) Leavenworth KS
KGSO(AM) Wichita KS
WHLN(AM) Harlan KY
KDBS(AM) Alexandria LA
WMSX(AM) Brockton MA
WHAG(AM) Halfway MD
WNWZ(AM) Grand Rapids MI
KLFD(AM) Litchfield MN
KRWB(AM) Roseau MN
WDSK(AM) Cleveland MS
WEGO(AM) Concord NC
WRJD(AM) Durham NC
WVCB(AM) Shallotte NC
KDKT(AM) Beulah ND
KOOQ(AM) North Platte NE
WHTG(AM) Eatontown NJ
WDOE(AM) Dunkirk NY
WELM(AM) Elmira NY
WENU(AM) South Glens Falls NY
WNER(AM) Watertown NY
WING(AM) Dayton OH
KBNP(AM) Portland OR
WLSH(AM) Lansford PA
KQV(AM) Pittsburgh PA
WRSS(AM) San Sebastian PR
WPCC(AM) Clinton SC
WBBX(AM) Kingston TN
WCMT(AM) Martin TN
WSTN(AM) Somerville TN
KLVQ(AM) Athens TX
KNTX(AM) Bowie TX
*KHCH(AM) Huntsville TX
KCUL(AM) Marshall TX
KRIL(AM) Odessa TX
KBAL(AM) San Saba TX
KNAL(AM) Victoria TX
WRIS(AM) Roanoke VA
WIZM(AM) La Crosse WI
WSCW(AM) South Charleston WV
KWYO(AM) Sheridan WY

1420 khz
WACT(AM) Tuscaloosa AL
KBHS(AM) Hot Springs AR
KPOC(AM) Pocahontas AR
KMOG(AM) Payson AZ
KTAN(AM) Sierra Vista AZ
KSTN(AM) Stockton CA
WLIS(AM) Old Saybrook CT
WDJA(AM) Delray Beach FL
WBRD(AM) Palmetto FL
WAOC(AM) Saint Augustine FL
WRCG(AM) Columbus GA
WATB(AM) Decatur GA
WPEH(AM) Louisville GA
WLET(AM) Toccoa GA
WKWN(AM) Trenton GA
KKEA(AM) Honolulu HI
WOC(AM) Davenport IA
KIGO(AM) Saint Anthony ID
WINI(AM) Murphysboro IL
WIMS(AM) Michigan City IN
KJCK(AM) Junction City KS
KULY(AM) Ulysses KS
WHBN(AM) Harrodsburg KY
WVJS(AM) Owensboro KY
KPEL(AM) Lafayette LA
WBSM(AM) New Bedford MA
WBEC(AM) Pittsfield MA
WFLT(AM) Flint MI
WKPR(AM) Kalamazoo MI
KTOE(AM) Mankato MN
KRLL(AM) California MO
KQYS(AM) Neosho MO
WQBC(AM) Vicksburg MS
WIGG(AM) Wiggins MS
WMYN(AM) Mayodan NC
*WGAS(AM) South Gastonia NC
WVOT(AM) Wilson NC
KOTK(AM) Omaha NE
WASR(AM) Wolfeboro NH
WNRS(AM) Herkimer NY
WACK(AM) Newark NY
WLNA(AM) Peekskill NY
WHK(AM) Cleveland OH
KTJS(AM) Hobart OK
KMHS(AM) Coos Bay OR
WCOJ(AM) Coatesville PA
WCED(AM) DuBois PA
WUKQ(AM) Ponce PR
WCRE(AM) Cheraw SC
KGIM(AM) Aberdeen SD
WEMB(AM) Erwin TN
WKSR(AM) Pulaski TN
KFYN(AM) Bonham TX
KPIR(AM) Granbury TX
KJDL(AM) Lubbock TX
KGNB(AM) New Braunfels TX
WAMV(AM) Amherst VA
WXGM(AM) Gloucester VA
WKCW(AM) Warrenton VA
WRSA(AM) Saint Albans VT
KITI(AM) Chehalis-Centralia WA
KRIZ(AM) Renton WA
KUJ(AM) Walla Walla WA
WJUB(AM) Plymouth WI
WTCR(AM) Kenova WV

1430 khz
WFHK(AM) Pell City AL
WRMG(AM) Red Bay AL
KHBM(AM) Monticello AR
KWST(AM) El Centro CA
KFIG(AM) Fresno CA
KJAY(AM) Sacramento CA
KMRB(AM) San Gabriel CA
KVVN(AM) Santa Clara CA
KEZW(AM) Aurora CO
WTMN(AM) Gainesville FL
WOIR(AM) Homestead FL
WLKF(AM) Lakeland FL
WLTG(AM) Panama City FL
WGFS(AM) Covington GA
WDAL(AM) Dalton GA
KASI(AM) Ames IA
WEEF(AM) Highland Park IL
WCMY(AM) Ottawa IL
WXNT(AM) Indianapolis IN
WXAM(AM) Buffalo KY
WYMC(AM) Mayfield KY
KMRC(AM) Morgan City LA
WPNI(AM) Amherst MA
WXKS(AM) Everett MA
WNAV(AM) Annapolis MD
WION(AM) Ionia MI
KNSP(AM) Staples MN
KKOZ(AM) Ava MO
KAOL(AM) Carrollton MO
WIL(AM) Saint Louis MO
WDEX(AM) Monroe NC
WMNC(AM) Morganton NC
WDJS(AM) Mount Olive NC
WRXO(AM) Roxboro NC
KRGI(AM) Grand Island NE
WNSW(AM) Newark NJ
KCRX(AM) Roswell NM
WENE(AM) Endicott NY
WFOB(AM) Fostoria OH
WCLT(AM) Newark OH
KALV(AM) Alva OK
KTBZ(AM) Tulsa OK
KYKN(AM) Keizer OR
WVAM(AM) Altoona PA
WNEL(AM) Caguas PR
WBLR(AM) Batesburg SC
WNFO(AM) Ridgeland SC
KBRK(AM) Brookings SD
WOWW(AM) Germantown TN
WPLN(AM) Madison TN
KROO(AM) Breckenridge TX
KEES(AM) Gladewater TX
KCOH(AM) Houston TX
KLO(AM) Ogden UT
WHAN(AM) Ashland VA
WKEX(AM) Blacksburg VA
WDIC(AM) Clinchco VA
KCLK(AM) Asotin WA
KBRC(AM) Mount Vernon WA
WBEV(AM) Beaver Dam WI
WQOQ(AM) Durand WI
WEIR(AM) Weirton WV

1440 khz
WLWI(AM) Montgomery AL
KTUV(AM) Little Rock AR
KAZG(AM) Scottsdale AZ
KVON(AM) Napa CA
KDIF(AM) Riverside CA
KINF(AM) Santa Maria CA
KRDZ(AM) Wray CO
WWCL(AM) Lehigh Acres FL
WPRD(AM) Winter Park FL
WGMI(AM) Bremen GA
WGIG(AM) Brunswick GA
WDCO(AM) Cochran GA
KCHE(AM) Cherokee IA
KPTO(AM) Pocatello ID
WIBH(AM) Anna IL
WPRS(AM) Paris IL
WGEM(AM) Quincy IL
WROK(AM) Rockford IL
WPGW(AM) Portland IN
KMAJ(AM) Topeka KS
WCDS(AM) Glasgow KY
WYGH(AM) Paris KY
WEZJ(AM) Williamsburg KY
KMLB(AM) Monroe LA
WVEI(AM) Worcester MA
WJAE(AM) Westbrook ME
WMAX(AM) Bay City MI
WDOW(AM) Dowagiac MI
WDRJ(AM) Inkster MI
KDIZ(AM) Golden Valley MN
WRBE(AM) Lucedale MS
WSEL(AM) Pontotoc MS
WBLA(AM) Elizabethtown NC
WLXN(AM) Lexington NC
KKXL(AM) Grand Forks ND
WMVB(AM) Millville NJ
WNYG(AM) Babylon NY
WFNY(AM) Gloversville NY
WJJL(AM) Niagara Falls NY
WSGO(AM) Oswego NY
WRGM(AM) Ontario OH
WHKZ(AM) Warren OH
KMED(AM) Medford OR
KODL(AM) The Dalles OR
WCDL(AM) Carbondale PA
WNPV(AM) Lansdale PA
WGLD(AM) Red Lion PA
WGVL(AM) Greenville SC
WJBS(AM) Holly Hill SC
WZYX(AM) Cowan TN
WHDM(AM) McKenzie TN
KPUR(AM) Amarillo TX
KEYS(AM) Corpus Christi TX
KETX(AM) Livingston TX
KELG(AM) Manor TX
KTNO(AM) University Park TX
WKLV(AM) Blackstone VA
WNFL(AM) Green Bay WI
WHIS(AM) Bluefield WV
WAJR(AM) Morgantown WV

1449 khz
V6AH(AM) Pohnpei FM

1450 khz
KLAM(AM) Cordova AK
KIAL(AM) Unalaska AK
WDNG(AM) Anniston AL
WZGX(AM) Bessemer AL
WDLK(AM) Dadeville AL
WWNT(AM) Dothan AL
WTKI(AM) Huntsville AL
WLAY(AM) Muscle Shoals AL
KNHD(AM) Camden AR
KENA(AM) Mena AR
KYNN(AM) Cameron AZ
KDAP(AM) Douglas AZ
KNOT(AM) Prescott AZ
KVSL(AM) Show Low AZ
KWFM(AM) Tucson AZ
KCIK(AM) Blue Lake CA
KFSD(AM) Escondido CA
KGAM(AM) Palm Springs CA
KTIP(AM) Porterville CA
KEST(AM) San Francisco CA
KVML(AM) Sonora CA
KVEN(AM) Ventura CA
KOBO(AM) Yuba City CA
KGIW(AM) Alamosa CO
KSKE(AM) Buena Vista CO
KAVP(AM) Colona CO
KGRE(AM) Greeley CO
WCUM(AM) Bridgeport CT
WOL(AM) Washington DC
WILM(AM) Wilmington DE
WWJB(AM) Brooksville FL
WMFJ(AM) Daytona Beach FL
WOCN(AM) Miami FL
WBSR(AM) Pensacola FL
WSDV(AM) Sarasota FL
WSTU(AM) Stuart FL
WTAL(AM) Tallahassee FL
WGPC(AM) Albany GA
WBHF(AM) Cartersville GA
WCON(AM) Cornelia GA
WKEU(AM) Griffin GA
WMVG(AM) Milledgeville GA
WVLD(AM) Valdosta GA
KMRY(AM) Cedar Rapids IA
KBFI(AM) Bonners Ferry ID
KVSI(AM) Montpelier ID
KIOV(AM) Payette ID
*KEZJ(AM) Twin Falls ID
WCEV(AM) Cicero IL
WRLL(AM) Cicero IL
WKEI(AM) Kewanee IL
WFMB(AM) Springfield IL
WLYV(AM) Fort Wayne IN
WAVG(AM) Jeffersonville IN
WASK(AM) Lafayette IN
WAOV(AM) Vincennes IN
KWBW(AM) Hutchinson KS
WTCO(AM) Campbellsville KY
WWXL(AM) Manchester KY
WDXR(AM) Paducah KY
WLKS(AM) West Liberty KY
KSIG(AM) Crowley LA
KNOC(AM) Natchitoches LA
WBYU(AM) New Orleans LA
WNBP(AM) Newburyport MA
WMAS(AM) Springfield MA
WTBO(AM) Cumberland MD
WRKD(AM) Rockland ME
WKTQ(AM) South Paris ME
WATZ(AM) Alpena MI
WHTC(AM) Holland MI
WMIQ(AM) Iron Mountain MI
WIBM(AM) Jackson MI
WKLA(AM) Ludington MI
WNBY(AM) Newberry MI
WHLS(AM) Port Huron MI
KATE(AM) Albert Lea MN
KBUN(AM) Bemidji MN
KBMW(AM) Breckenridge MN
WELY(AM) Ely MN
KNSI(AM) Saint Cloud MN
KYLS(AM) Fredericktown MO
KQYX(AM) Joplin MO
KIRX(AM) Kirksville MO
KOKO(AM) Warrensburg MO
KWPM(AM) West Plains MO
WROX(AM) Clarksdale MS
WCJU(AM) Columbia MS
WYHL(AM) Meridian MS
WNAT(AM) Natchez MS
WROB(AM) West Point MS
KMMS(AM) Bozeman MT
KQDI(AM) Great Falls MT
KYLW(AM) Lockwood MT
KGRZ(AM) Missoula MT
KVCK(AM) Wolf Point MT
WATA(AM) Boone NC
WGNC(AM) Gastonia NC
WIZS(AM) Henderson NC
WHKP(AM) Hendersonville NC
WNOS(AM) New Bern NC
WCIE(AM) Spring Lake NC
KZZJ(AM) Rugby ND
KWBE(AM) Beatrice NE
WKXL(AM) Concord NH
WENJ(AM) Atlantic City NJ
WCTC(AM) New Brunswick NJ
KRZY(AM) Albuquerque NM
KLMX(AM) Clayton NM
KOBE(AM) Las Cruces NM
KSEL(AM) Portales NM
KWES(AM) Ruidoso NM
KIFO(AM) Hawthorne NV
KHIT(AM) Reno NV
WENI(AM) Corning NY
WWSC(AM) Glens Falls NY
WHDL(AM) Olean NY
WKIP(AM) Poughkeepsie NY
WYFY(AM) Rome NY
WJER(AM) Dover-New Philadelphia OH
WMOH(AM) Hamilton OH
WLEC(AM) Sandusky OH
KWHW(AM) Altus OK
KGFF(AM) Shawnee OK
KSIW(AM) Woodward OK
KLZS(AM) Eugene OR
KFLS(AM) Klamath Falls OR
KLBM(AM) La Grande OR
*KBPS(AM) Portland OR
WPSE(AM) Erie PA
WFRA(AM) Franklin PA
WDAD(AM) Indiana PA
WPAM(AM) Pottsville PA
WMAJ(AM) State College PA
WJPA(AM) Washington PA
WCPR(AM) Coamo PR
WLKW(AM) West Warwick RI
WQNT(AM) Charleston SC
WCRS(AM) Greenwood SC
WHSC(AM) Hartsville SC
WRNN(AM) Myrtle Beach SC
KBFS(AM) Belle Fourche SD
KSQP(AM) Pierre SD
KYNT(AM) Yankton SD
WLAR(AM) Athens TN
WLMR(AM) Chattanooga TN
WTRO(AM) Dyersburg TN
WSMG(AM) Greeneville TN
WLAF(AM) La Follette TN
WGNS(AM) Murfreesboro TN
KIKR(AM) Beaumont TX
KBEN(AM) Carrizo Springs TX
KCTI(AM) Gonzales TX
KMBL(AM) Junction TX
KCYL(AM) Lampasas TX
KMHT(AM) Marshall TX
KNET(AM) Palestine TX
KSNY(AM) Snyder TX
*KEYY(AM) Provo UT
KZNU(AM) Saint George UT
WBVA(AM) Bayside VA
WVAX(AM) Charlottesville VA
WFTR(AM) Front Royal VA
WCLM(AM) Highland Springs VA
WREL(AM) Lexington VA
WMVA(AM) Martinsville VA
WSNO(AM) Barre VT
WTSA(AM) Brattleboro VT
KBKW(AM) Aberdeen WA
KCLX(AM) Colfax WA
KONP(AM) Port Angeles WA
KSUH(AM) Puyallup WA
KFIZ(AM) Fond du Lac WI
WHRY(AM) Hurley WI
WDLB(AM) Marshfield WI
WRCO(AM) Richland Center WI
WHNK(AM) Parkersburg WV
KBBS(AM) Buffalo WY
KVOW(AM) Riverton WY

1460 khz
WMCJ(AM) Cullman AL
WHAL(AM) Phenix City AL
KTYM(AM) Inglewood CA
KION(AM) Salinas CA
KRRS(AM) Santa Rosa CA
KCNR(AM) Shasta CA
KZNT(AM) Colorado Springs CO
WQXM(AM) Bartow FL
WZEP(AM) De Funiak Springs FL
WNPL(AM) Golden Gate FL
WZNZ(AM) Jacksonville FL
WXEM(AM) Buford GA
KHRA(AM) Honolulu HI
KXNO(AM) Des Moines IA
WROY(AM) Carmi IL
WIXN(AM) Dixon IL
WKJR(AM) Rantoul IL
WKAM(AM) Goshen IN
WJCP(AM) North Vernon IN
KKOY(AM) Chanute KS
WEKB(AM) Elkhorn City KY
WRVK(AM) Mt. Vernon KY
WXOK(AM) Baton Rouge LA
KBSF(AM) Springhill LA
WXBR(AM) Brockton MA
WEMD(AM) Easton MD
WBRN(AM) Big Rapids MI
WPON(AM) Walled Lake MI
KDWA(AM) Hastings MN
KDMA(AM) Montevideo MN
KKAQ(AM) Thief River Falls MN
KHOJ(AM) Saint Charles MO
WELZ(AM) Belzoni MS
WRKB(AM) Kannapolis NC
WEWO(AM) Laurinburg NC
WHBK(AM) Marshall NC
KLTC(AM) Dickinson ND
KXPN(AM) Kearney NE
WIFI(AM) Florence NJ
KENO(AM) Las Vegas NV
WDDY(AM) Albany NY
WVOX(AM) New Rochelle NY
WHIC(AM) Rochester NY
WBNS(AM) Columbus OH
WABQ(AM) Painesville OH
KZUE(AM) El Reno OK
KCKX(AM) Stayton OR
WMBA(AM) Ambridge PA
WTKT(AM) Harrisburg PA
WEMR(AM) Tunkhannock PA
WRRE(AM) Juncos PR
WLRP(AM) San Sebastian PR
WDOG(AM) Allendale SC
WBCU(AM) Union SC
WJAK(AM) Jackson TN
WEEN(AM) Lafayette TN
WXRQ(AM) Mount Pleasant TN
KHFX(AM) Burleson TX
KBRZ(AM) Freeport TX
KCWM(AM) Hondo TX
KBZO(AM) Lubbock TX
WKDV(AM) Manassas VA
WRAD(AM) Radford VA
*KARR(AM) Kirkland WA
KUTI(AM) Yakima WA
WJTI(AM) Racine WI
WBOG(AM) Tomah WI
WBUC(AM) Buckhannon WV

1470 khz
WPGG(AM) Evergreen AL
KNXN(AM) Sierra Vista AZ
KUTY(AM) Palmdale CA
KIID(AM) Sacramento CA
KEPL(AM) Estes Park CO
WMMW(AM) Meriden CT
WHBO(AM) Dunedin FL
WWNN(AM) Pompano Beach FL
WXAG(AM) Athens GA
WCLA(AM) Claxton GA
WRGA(AM) Rome GA
KWSL(AM) Sioux City IA
KWAY(AM) Waverly IA
WCFJ(AM) Chicago Heights IL
WMBD(AM) Peoria IL
*WGNR(AM) Anderson IN
KAIR(AM) Atchison KS
KYUU(AM) Liberal KS
WBFC(AM) Stanton KY
KLCL(AM) Lake Charles LA
WAZN(AM) Watertown MA
WJDY(AM) Salisbury MD
WTTR(AM) Westminster MD
WLAM(AM) Lewiston ME
WFNT(AM) Flint MI
KRJJ(AM) Brooklyn Park MN
KFMZ(AM) Brookfield MO
KMAL(AM) Malden MO
WCHJ(AM) Brookhaven MS
WNAU(AM) New Albany MS
WVBS(AM) Burgaw NC
WWBG(AM) Greensboro NC
WJPI(AM) Plymouth NC
WTOE(AM) Spruce Pine NC
KHND(AM) Harvey ND
WNYY(AM) Ithaca NY
WPDM(AM) Potsdam NY
WLQR(AM) Toledo OH
KVLH(AM) Pauls Valley OK
KGND(AM) Vinita OK
WSAN(AM) Allentown PA
WLOA(AM) Farrell PA
WQXL(AM) Columbia SC
WLMC(AM) Georgetown SC
WBCR(AM) Alcoa TN
WVOL(AM) Berry Hill TN
KYYW(AM) Abilene TX
KDHN(AM) Dimmitt TX
KWRD(AM) Henderson TX
KUOL(AM) San Marcos TX
KNFL(AM) Tremonton UT
WBTX(AM) Broadway-Timberville VA
WTZE(AM) Tazewell VA
KELA(AM) Centralia-Chehalis WA
KBSN(AM) Moses Lake WA
WBKV(AM) West Bend WI
WEMM(AM) Huntington WV
KKTY(AM) Douglas WY

1480 khz
WYMR(AM) Bridgeport AL
WRLM(AM) Irondale AL
WABB(AM) Mobile AL
KTHS(AM) Berryville AR
KPHX(AM) Phoenix AZ
KGOE(AM) Eureka CA
KYOS(AM) Merced CA
KVNR(AM) Santa Ana CA
KSBQ(AM) Santa Maria CA
KAVA(AM) Pueblo CO
WKND(AM) Windsor CT
WFLN(AM) Arcadia FL
WVOI(AM) Marco Island FL
WUNA(AM) Ocoee FL
*WKGC(AM) Panama City Beach FL
WYZE(AM) Atlanta GA
WGUS(AM) Augusta GA
KLEE(AM) Ottumwa IA
KRXR(AM) Gooding ID
WSPY(AM) Geneva IL
WJBM(AM) Jerseyville IL
WPFR(AM) Terre Haute IN
WRSW(AM) Warsaw IN
KCZZ(AM) Mission KS
KQAM(AM) Wichita KS
WHVO(AM) Hopkinsville KY
WVSG(AM) Neon KY
WTLO(AM) Somerset KY
KIOU(AM) Shreveport LA
WSAR(AM) Fall River MA
*WGVU(AM) Kentwood MI
WSDS(AM) Salem Township MI
WIOS(AM) Tawas City MI
KAUS(AM) Austin MN
KKCQ(AM) Fosston MN
WGFY(AM) Charlotte NC
WSRC(AM) Fair Bluff NC
WPFJ(AM) Franklin NC
WYRN(AM) Louisburg NC
KLMS(AM) Lincoln NE
WLEA(AM) Hornell NY
WZRC(AM) New York NY
WADR(AM) Remsen NY
WHBC(AM) Canton OH
WCIN(AM) Cincinnati OH
WCNS(AM) Latrobe PA
WUBA(AM) Philadelphia PA
WISL(AM) Shamokin PA
WEEO(AM) Shippensburg PA
WMDD(AM) Fajardo PR
WZJY(AM) Mt. Pleasant SC
KSDR(AM) Watertown SD
WJFC(AM) Jefferson City TN
WBBP(AM) Memphis TN
WJLE(AM) Smithville TN
KNIT(AM) Dallas TX
KLVL(AM) Pasadena TX
KCHL(AM) San Antonio TX
KHQN(AM) Spanish Fork UT
WPWC(AM) Dumfries-Triangle VA
WTOX(AM) Glen Allen VA
WTOY(AM) Salem VA
WCFR(AM) Springfield VT
KNTB(AM) Lakewood WA
KBMS(AM) Vancouver WA
WLMV(AM) Madison WI
KRAE(AM) Cheyenne WY

1490 khz
WANA(AM) Anniston AL
WDPT(AM) Decatur AL
WIRB(AM) Level Plains AL
WHBB(AM) Selma AL
KWXT(AM) Dardanelle AR
KXAR(AM) Hope AR
KDRS(AM) Paragould AR
KOTN(AM) Pine Bluff AR
KZZZ(AM) Bullhead City AZ
KCUZ(AM) Clifton AZ
KYCA(AM) Prescott AZ
KFFN(AM) Tucson AZ
KWAC(AM) Bakersfield CA
KMET(AM) Banning CA
KGBA(AM) Calexico CA
KRKC(AM) King City CA
KTOB(AM) Petaluma CA
KBLF(AM) Red Bluff CA
KBKO(AM) Santa Barbara CA
KOWL(AM) South Lake Tahoe CA
*KSYC(AM) Yreka CA
KCFC(AM) Boulder CO
KPKE(AM) Gunnison CO
KXRE(AM) Manitou Springs CO
WGCH(AM) Greenwich CT
WWPR(AM) Bradenton FL
WTJV(AM) De Land FL
WAFZ(AM) Immokalee FL
WMBM(AM) Miami Beach FL
WECM(AM) Milton FL
WTTB(AM) Vero Beach FL
WSIR(AM) Winter Haven FL
WMOG(AM) Brunswick GA
WCHM(AM) Clarkesville GA
WYYZ(AM) Jasper GA
WKUN(AM) Monroe GA
WSFB(AM) Quitman GA
WSNT(AM) Sandersville GA
WSYL(AM) Sylvania GA
WRLA(AM) West Point GA
KBUR(AM) Burlington IA
WDBQ(AM) Dubuque IA
KXLQ(AM) Indianola IA
KRIB(AM) Mason City IA
KCID(AM) Caldwell ID
KRTK(AM) Chubbuck ID
WKRO(AM) Cairo IL
WDAN(AM) Danville IL
WFFX(AM) East St. Louis IL

WPNA(AM) Oak Park IL
WZOE(AM) Princeton IL
WKBV(AM) Richmond IN
WDND(AM) South Bend IN
KKAN(AM) Phillipsburg KS
KTOP(AM) Topeka KS
WFKY(AM) Frankfort KY
WCLU(AM) Glasgow KY
WFXY(AM) Middlesboro KY
WOMI(AM) Owensboro KY
WSIP(AM) Paintsville KY
WIKC(AM) Bogalusa LA
KEUN(AM) Eunice LA
KJIN(AM) Houma LA
KRUS(AM) Ruston LA
WCEC(AM) Haverhill MA
WMRC(AM) Milford MA
WACM(AM) West Springfield MA
WARK(AM) Hagerstown MD
WBAE(AM) Portland ME
WTVL(AM) Waterville ME
WABJ(AM) Adrian MI
WTIQ(AM) Manistique MI
WMPX(AM) Midland MI
WODJ(AM) Whitehall MI
KXRA(AM) Alexandria MN
KQDS(AM) Duluth MN
KLGR(AM) Redwood Falls MN
KDMO(AM) Carthage MO
KTTR(AM) Rolla MO
KDRO(AM) Sedalia MO
WXBD(AM) Biloxi MS
WCLD(AM) Cleveland MS
WHOC(AM) Philadelphia MS
WTUP(AM) Tupelo MS
WVBG(AM) Vicksburg MS
KDBM(AM) Dillon MT
KBSR(AM) Laurel MT
WDUR(AM) Durham NC
WLOE(AM) Eden NC
WAZZ(AM) Fayetteville NC
WWNB(AM) New Bern NC
WRMT(AM) Rocky Mount NC
WSTP(AM) Salisbury NC
WSVM(AM) Valdese NC
WWIL(AM) Wilmington NC
KNDC(AM) Hettinger ND
KOVC(AM) Valley City ND
KOMJ(AM) Omaha NE
WRTN(AM) Berlin NH
WEMJ(AM) Laconia NH
WUVR(AM) Lebanon NH
WTAA(AM) Pleasantville NJ
KRSN(AM) Los Alamos NM
KRTN(AM) Raton NM
KRUI(AM) Ruidoso Downs NM
WCSS(AM) Amsterdam NY
WBTA(AM) Batavia NY
WKNY(AM) Kingston NY
WICY(AM) Malone NY
WDLC(AM) Port Jervis NY
WCDO(AM) Sidney NY
WOLF(AM) Syracuse NY
WTYX(AM) Watkins Glen NY
WBEX(AM) Chillicothe OH
WERE(AM) Cleveland Heights OH
WOHI(AM) East Liverpool OH
WMOA(AM) Marietta OH
WMRN(AM) Marion OH
KMFS(AM) Guthrie OK
KBIX(AM) Muskogee OK
KBKR(AM) Baker City OR
KRNR(AM) Roseburg OR
KBZY(AM) Salem OR
WESB(AM) Bradford PA
WAZL(AM) Hazleton PA
WPRR(AM) Johnstown PA
WLPA(AM) Lancaster PA
WBCB(AM) Levittown-Fairless Hills PA
WMGW(AM) Meadville PA
WNBT(AM) Wellsboro PA
WDEP(AM) Ponce PR
WVGB(AM) Beaufort SC
WTQS(AM) Cameron SC
WGCD(AM) Chester SC
WPCI(AM) Greenville SC
WJDJ(AM) Hartsville SC
KFCR(AM) Custer SD
KORN(AM) Mitchell SD
WOPI(AM) Bristol TN
WJOC(AM) Chattanooga TN
WCSV(AM) Crossville TN
WITA(AM) Knoxville TN
WCOR(AM) Lebanon TN
WJJM(AM) Lewisburg TN
WDXL(AM) Lexington TN
KFON(AM) Austin TX
KIBL(AM) Beeville TX
KBST(AM) Big Spring TX
KQTY(AM) Borger TX
KNEL(AM) Brady TX
KWMC(AM) Del Rio TX
KHVL(AM) Huntsville TX
KLNT(AM) Laredo TX
KZZN(AM) Littlefield TX
KPLT(AM) Paris TX
KYZS(AM) Tyler TX
KVWC(AM) Vernon TX
KWUD(AM) Woodville TX
KOGN(AM) Ogden UT
KCPX(AM) Spanish Valley UT
WCVA(AM) Culpeper VA
WPAK(AM) Farmville VA
WLRT(AM) Hampton VA
WKVT(AM) Brattleboro VT
WFAD(AM) Middlebury VT
WIKE(AM) Newport VT
KBRO(AM) Bremerton WA
KBIS(AM) Forks WA
KEYG(AM) Grand Coulee WA
KWOK(AM) Hoquiam WA
KLOG(AM) Kelso WA
KYNR(AM) Toppenish WA
KTEL(AM) Walla Walla WA
WGEZ(AM) Beloit WI
WLFN(AM) La Crosse WI
WIGM(AM) Medford WI
WOSH(AM) Oshkosh WI
WSWW(AM) Charleston WV
WTCS(AM) Fairmont WV
WAEY(AM) Princeton WV
WSGB(AM) Sutton WV
KUGR(AM) Green River WY
KGOS(AM) Torrington WY

1494 khz
*V6AI(AM) Yap FM

1500 khz
WQCR(AM) Alabaster AL
WVSM(AM) Rainsville AL
WKAX(AM) Russellville AL
KIEV(AM) Culver City CA
KSJX(AM) San Jose CA
WFIF(AM) Milford CT
WTWP(AM) Washington DC
WKIZ(AM) Key West FL
WPSO(AM) New Port Richey FL
WDPC(AM) Dallas GA
WSEM(AM) Donalsonville GA
WAYS(AM) Macon GA
KUMU(AM) Honolulu HI
WGEN(AM) Geneseo IL
WPMB(AM) Vandalia IL
WPJX(AM) Zion IL
WBRI(AM) Indianapolis IN
WAKE(AM) Valparaiso IN
WKXO(AM) Berea KY
WMJL(AM) Marion KY
WOLY(AM) Battle Creek MI
WLQV(AM) Detroit MI
KSTP(AM) Saint Paul MN
KDFN(AM) Doniphan MO
WQMS(AM) Quitman MS
WSMX(AM) Winston-Salem NC
WGHT(AM) Pompton Lakes NJ
*KABR(AM) Alamo Community NM
WBZI(AM) Xenia OH
WASN(AM) Youngstown OH
KPGM(AM) Pawhuska OK
WMNT(AM) Manati PR
WEAC(AM) Gaffney SC
WDEB(AM) Jamestown TN
WTNE(AM) Trenton TN
KBRN(AM) Boerne TX
KMXO(AM) Merkel TX
KJIM(AM) Sherman TX
KANI(AM) Wharton TX

1510 khz
KFNN(AM) Mesa AZ
KIRV(AM) Fresno CA
KSPA(AM) Ontario CA
KPIG(AM) Piedmont CA
KCKK(AM) Littleton CO
WWBC(AM) Cocoa FL
KIFG(AM) Iowa Falls IA
WXOZ(AM) Highland IL
WWHN(AM) Joliet IL
WLRB(AM) Macomb IL
WJOT(AM) Wabash IN
KNNS(AM) Larned KS
KAGY(AM) Port Sulphur LA
WWZN(AM) Boston MA
*WJKN(AM) Jackson MI
KCTE(AM) Independence MO
KMRF(AM) Marshfield MO
WEAL(AM) Greensboro NC
KTTT(AM) Columbus NE
WRNJ(AM) Hackettstown NJ
WFAI(AM) Salem NJ
WPUT(AM) Brewster NY
WLGN(AM) Logan OH
WLKR(AM) Norwalk OH
WWSM(AM) Annville-Cleona PA
WPGR(AM) Monroeville PA
WBSG(AM) Lajas PR
WDRF(AM) Woodruff SC
KMSD(AM) Milbank SD
WLAC(AM) Nashville TN
KAGC(AM) Bryan TX
KRDH(AM) Canton TX
KCTX(AM) Childress TX
KMND(AM) Midland TX
KBED(AM) Nederland TX
KROB(AM) Robstown TX
KSTV(AM) Stephenville TX
KLLB(AM) West Jordan UT
KGA(AM) Spokane WA
WAUK(AM) Waukesha WI

1520 khz
WTLM(AM) Opelika AL
KMPG(AM) Hollister CA
KVTA(AM) Port Hueneme CA
WHIM(AM) Apopka FL
WXYB(AM) Indian Rocks Beach FL
WEXY(AM) Wilton Manors FL
WDCY(AM) Douglasville GA
WKVQ(AM) Eatonton GA
KSIB(AM) Creston IA
WHOW(AM) Clinton IL
WLUV(AM) Loves Park IL
WKVI(AM) Knox IN
WSVX(AM) Shelbyville IN
WLGC(AM) Greenup KY
WRSL(AM) Stanford KY
KFXZ(AM) Lafayette LA
WIZZ(AM) Greenfield MA
WTRI(AM) Brunswick MD
WMLM(AM) Saint Louis MI
WLKM(AM) Three Rivers MI
KOLM(AM) Rochester MN
KRHW(AM) Sikeston MO
WDSL(AM) Mocksville NC
WGMA(AM) Spindale NC
WARR(AM) Warrenton NC
KMAV(AM) Mayville ND
WWKB(AM) Buffalo NY
WTHE(AM) Mineola NY
WQCT(AM) Bryan OH
WINW(AM) Canton OH
WJMP(AM) Kent OH
WDMN(AM) Rossford OH
KOKC(AM) Oklahoma City OK
KGDD(AM) Oregon City OR
WCHE(AM) West Chester PA
WVOZ(AM) San Juan PR
WKMG(AM) Newberry SC
KSQB(AM) Sioux Falls SD
WNWS(AM) Brownsville TN
KYND(AM) Cypress TX
KHLT(AM) Hallettsville TX

1530 khz
KVDW(AM) England AR
KHPI(AM) Moreno Valley CA
KFBK(AM) Sacramento CA
KCMN(AM) Colorado Springs CO
WDJZ(AM) Bridgeport CT
WENG(AM) Englewood FL
WYMM(AM) Jacksonville FL
WTTI(AM) Dalton GA
KVOG(AM) Hagatna GU
KDSN(AM) Denison IA
WJJG(AM) Elmhurst IL
WLIQ(AM) Quincy IL
KQNK(AM) Norton KS
WVBF(AM) Middleborough Center MA
WCTR(AM) Chestertown MD
WFGO(AM) Orono ME
WLCO(AM) Lapeer MI
WYGR(AM) Wyoming MI
KQSP(AM) Shakopee MN
KMAM(AM) Butler MO
WRPM(AM) Poplarville MS
WLLQ(AM) Chapel Hill NC
WOBX(AM) Wanchese NC
WJDM(AM) Elizabeth NJ
WCKY(AM) Cincinnati OH
KXTD(AM) Wagoner OK
WYNE(AM) North East PA
WUPR(AM) Utuado PR
WASC(AM) Spartanburg SC
WDAP(AM) Huntingdon TN
KZNX(AM) Creedmoor TX
KGBT(AM) Harlingen TX
KNBO(AM) New Boston TX
KCLR(AM) Ralls TX
WFIC(AM) Collinsville VA
WMBE(AM) Chilton WI

1540 khz
WRSM(AM) Sumiton AL
KDYN(AM) Ozark AR
KASA(AM) Phoenix AZ
KMPC(AM) Los Angeles CA
KREA(AM) Honolulu HI
KXEL(AM) Waterloo IA
WSMI(AM) Litchfield IL
WBNL(AM) Boonville IN
WADM(AM) Decatur IN
WLOI(AM) La Porte IN
WMYJ(AM) Martinsville IN
KNGL(AM) McPherson KS
KLKC(AM) Parsons KS
WGRK(AM) Greensburg KY
KGLA(AM) Gretna LA
WACA(AM) Wheaton MD
*KTGG(AM) Spring Arbor MI
KBOA(AM) Kennett MO
WKXG(AM) Greenwood MS
WOGR(AM) Charlotte NC
WTXY(AM) Whiteville NC
WYNC(AM) Yanceyville NC
WGIP(AM) Exeter NH
WDCD(AM) Albany NY
WSIV(AM) East Syracuse NY
WBCO(AM) Bucyrus OH
WWGK(AM) Cleveland OH
WRTK(AM) Niles OH
WBTC(AM) Uhrichsville OH
WNWR(AM) Philadelphia PA
WECZ(AM) Punxsutawney PA
WIBS(AM) Guayama PR
WADK(AM) Newport RI
WTBI(AM) Pickens SC
WBIN(AM) Benton TN
WJJT(AM) Jellico TN
WBRY(AM) Woodbury TN
KGBC(AM) Galveston TX
KEDA(AM) San Antonio TX
KZMP(AM) University Park TX
WREJ(AM) Richmond VA
KXPA(AM) Bellevue WA
WTKM(AM) Hartford WI

1550 khz
WLOR(AM) Huntsville AL
*KUAZ(AM) Tucson AZ
KWRN(AM) Apple Valley CA
KXEX(AM) Fresno CA
KYCY(AM) San Francisco CA
WDZK(AM) Bloomfield CT
WNZF(AM) Bunnell FL
WRHC(AM) Coral Gables FL
WAMA(AM) Tampa FL
WTHB(AM) Augusta GA
WAZX(AM) Smyrna GA
WKTF(AM) Vienna GA
KIWA(AM) Sheldon IA
WJIL(AM) Jacksonville IL
WCSJ(AM) Morris IL
WOCC(AM) Corydon IN
WCVL(AM) Crawfordsville IN
WMDH(AM) New Castle IN
WNDI(AM) Sullivan IN
KDCC(AM) Dodge City KS
KKLE(AM) Winfield KS
WIRV(AM) Irvine KY
WMSK(AM) Morganfield KY
WPFC(AM) Port Allen LA
WNTN(AM) Newton MA
WSRY(AM) Elkton MD
WSHN(AM) Fremont MI
KAPE(AM) Cape Girardeau MO
KSFT(AM) Saint Joseph MO
KLFJ(AM) Springfield MO
WCLY(AM) Raleigh NC
WBFJ(AM) Winston-Salem NC
KICS(AM) Hastings NE
KKJY(AM) Albuquerque NM
KXTO(AM) Reno NV
WCGR(AM) Canandaigua NY
WUTQ(AM) Utica NY
WDLR(AM) Delaware OH
KMAD(AM) Madill OK
KYAL(AM) Sapulpa OK
WLFP(AM) Braddock PA
WITK(AM) Pittston PA
WTTC(AM) Towanda PA
WKFE(AM) Yauco PR
WBSC(AM) Bennettsville SC
WIGN(AM) Bristol TN
WQZQ(AM) Clarksville TN
WKJQ(AM) Parsons TN
KZRK(AM) Canyon TX
KCOM(AM) Comanche TX
KWBC(AM) Navasota TX
KMRI(AM) West Valley City UT
WKBA(AM) Vinton VA
WVAB(AM) Virginia Beach VA
KRPI(AM) Ferndale WA
KKAD(AM) Vancouver WA
WZRK(AM) Lake Geneva WI
WTUX(AM) Madison WI
WEVR(AM) River Falls WI
WMRE(AM) Charles Town WV

1560 khz
WZTQ(AM) Centre AL
WCMA(AM) Daleville AL
KNZR(AM) Bakersfield CA
KIQS(AM) Willows CA
WINV(AM) Beverly Hills FL
WINT(AM) Melbourne FL
WPGY(AM) Ellijay GA
KLNG(AM) Council Bluffs IA
WBYS(AM) Canton IL
WSEZ(AM) Paoli IN
WRIN(AM) Rensselaer IN
KABI(AM) Abilene KS
WQXY(AM) Hazard KY
WKDO(AM) Liberty KY
WPAD(AM) Paducah KY
WSLA(AM) Slidell LA
WKIK(AM) La Plata MD
WNWN(AM) Portage MI
KBEW(AM) Blue Earth MN
WMBH(AM) Joplin MO
KLTI(AM) Macon MO
KTUI(AM) Sullivan MO
WYZD(AM) Dobson NC
WQEW(AM) New York NY

WTNS(AM) Coshocton OH
WCNW(AM) Fairfield OH
WTOD(AM) Toledo OH
KOCY(AM) Del City OK
KKUZ(AM) Sallisaw OK
WRSJ(AM) Bayamon PR
WAHT(AM) Clemson SC
WAGL(AM) Lancaster SC
*KKAA(AM) Aberdeen SD
WBOL(AM) Bolivar TN
WMRO(AM) Gallatin TN
KZQQ(AM) Abilene TX
KGOW(AM) Bellaire TX
KNGR(AM) Daingerfield TX
KHBR(AM) Hillsboro TX
KTXZ(AM) West Lake Hills TX
WSBV(AM) South Boston VA
KVAN(AM) Burbank WA
KZIZ(AM) Sumner WA
WGLB(AM) Elm Grove WI
WFSP(AM) Kingwood WV

1570 khz
WCRL(AM) Oneonta AL
KBRI(AM) Brinkley AR
KCVR(AM) Lodi CA
KPRO(AM) Riverside CA
KTGE(AM) Salinas CA
KSXT(AM) Loveland CO
WTWB(AM) Auburndale FL
WVOJ(AM) Fernandina Beach FL
WOKC(AM) Okeechobee FL
WIGO(AM) Morrow GA
KUAU(AM) Haiku HI
KMCD(AM) Fairfield IA
KQWC(AM) Webster City IA
WBGZ(AM) Alton IL
WFRL(AM) Freeport IL
WBGX(AM) Harvey IL
WTAY(AM) Robinson IL
WGLL(AM) Auburn IN
WILO(AM) Frankfort IN
WWSZ(AM) New Albany IN
KNDY(AM) Marysville KS
WLBQ(AM) Morgantown KY
WKKS(AM) Vanceburg KY
WABL(AM) Amite LA
KLLA(AM) Leesville LA
WNSH(AM) Beverly MA
WPEP(AM) Taunton MA
WNST(AM) Towson MD
WWCK(AM) Flint MI
WFUR(AM) Grand Rapids MI
KYCR(AM) Golden Valley MN
KAKK(AM) Walker MN
KBCV(AM) Hollister MO
KLEX(AM) Lexington MO
WIZK(AM) Bay Springs MS
WONA(AM) Winona MS
WNCA(AM) Siler City NC
WTLK(AM) Taylorsville NC
WECU(AM) Winterville NC
WVTL(AM) Amsterdam NY
WFLR(AM) Dundee NY
WFTU(AM) Riverhead NY
WPTW(AM) Piqua OH
WANR(AM) Warren OH
KTAT(AM) Frederick OK
KMUR(AM) Pryor OK
*WPGM(AM) Danville PA
WISP(AM) Doylestown PA
WQTW(AM) Latrobe PA
WPPC(AM) Penuelas PR
KVTK(AM) Vermillion SD
WNKX(AM) Centerville TN
WCLE(AM) Cleveland TN
WTRB(AM) Ripley TN
KVLG(AM) La Grange TX
KPYK(AM) Terrell TX
WSWV(AM) Pennington Gap VA
WYTI(AM) Rocky Mount VA
WSCO(AM) Appleton WI
WKBH(AM) Holmen WI
WLKD(AM) Minocqua WI

1580 khz
WVOK(AM) Oxford AL
KHGG(AM) Van Buren AR
KMIK(AM) Tempe AZ
KBLA(AM) Santa Monica CA
KKKK(AM) Colorado Springs CO
WNTF(AM) Bithlo FL
WTCL(AM) Chattahoochee FL
WSRF(AM) Fort Lauderdale FL
WCCF(AM) Punta Gorda FL
WEAM(AM) Columbus GA
KCHA(AM) Charles City IA
WKKD(AM) Aurora IL
WDQN(AM) Du Quoin IL
WBCP(AM) Urbana IL
WIFE(AM) Connersville IN
WHLY(AM) South Bend IN
WAMW(AM) Washington IN
WXRA(AM) Georgetown KY
WPKY(AM) Princeton KY
KXZZ(AM) Lake Charles LA
WPGC(AM) Morningside MD
WWSJ(AM) Saint Johns MI
KDOM(AM) Windom MN
KTGR(AM) Columbia MO
KESM(AM) El Dorado Springs MO
KNIM(AM) Maryville MO
WAMY(AM) Amory MS
WORV(AM) Hattiesburg MS
WESY(AM) Leland MS
WPMP(AM) Pascagoula-Moss Point MS
WZKY(AM) Albemarle NC
*KAMI(AM) Cozad NE
WGYM(AM) Hammonton NJ
WLIM(AM) Patchogue NY
WVKO(AM) Columbus OH
KOKB(AM) Blackwell OK
KGAL(AM) Lebanon OR
WVZN(AM) Columbia PA
WRDD(AM) Ebensburg PA
WANB(AM) Waynesburg PA
WEKO(AM) Morovis PR
WWGS(AM) Georgetown SC
WPJK(AM) Orangeburg SC
WNPZ(AM) Knoxville TN
WLIJ(AM) Shelbyville TN
KGAF(AM) Gainesville TX
KIRT(AM) Mission TX
KTLU(AM) Rusk TX
KWED(AM) Seguin TX
KQRL(AM) Waco TX
WILA(AM) Danville VA
WTTN(AM) Watertown WI

1590 khz
WVNA(AM) Tuscumbia AL
KBJT(AM) Fordyce AR
KYNG(AM) Springdale AR
KLIV(AM) San Jose CA
KKZZ(AM) Ventura CA
KRSX(AM) Victorville CA
WPSL(AM) Port St. Lucie FL
WRXB(AM) Saint Petersburg Beach FL
WPUL(AM) South Daytona FL
WALG(AM) Albany GA
WQCH(AM) La Fayette GA
WRJS(AM) Swainsboro GA
WTGA(AM) Thomaston GA
KIXC(AM) Hilo HI
KWBG(AM) Boone IA
WONX(AM) Evanston IL
WAIK(AM) Galesburg IL
WNTS(AM) Beech Grove IN
WRCY(AM) Mount Vernon IN
KVGB(AM) Great Bend KS
WLBN(AM) Lebanon KY
KKAY(AM) White Castle LA
WFBR(AM) Glen Burnie MD
WKHZ(AM) Ocean City MD
WTVB(AM) Coldwater MI
WHLX(AM) Marine City MI
KCNN(AM) East Grand Forks MN
KDJS(AM) Willmar MN
KDEX(AM) Dexter MO
KPRT(AM) Kansas City MO
KMOZ(AM) Rolla MO
WZRX(AM) Jackson MS
WBHN(AM) Bryson City NC
WVOE(AM) Chadbourn NC
WCSL(AM) Cherryville NC
WHPY(AM) Clayton NC
WYSR(AM) High Point NC
KTCH(AM) Wayne NE
WSMN(AM) Nashua NH
KQLO(AM) Sun Valley NV
WAUB(AM) Auburn NY
WASB(AM) Brockport NY
WGGO(AM) Salamanca NY
WAKR(AM) Akron OH
WSRW(AM) Hillsboro OH
KWEY(AM) Weatherford OK
KMBD(AM) Tillamook OR
WZUM(AM) Carnegie PA
WHGT(AM) Chambersburg PA
WPWA(AM) Chester PA
WPSN(AM) Honesdale PA
WCXJ(AM) Kearsarge PA
WXRF(AM) Guayama PR
WARV(AM) Warwick RI
WABV(AM) Abbeville SC
WCAM(AM) Camden SC
WATX(AM) Algood TN
WKTP(AM) Jonesborough TN
WDBL(AM) Springfield TN
KGAS(AM) Carthage TX
KEAS(AM) Eastland TX
KELP(AM) El Paso TX
KMIC(AM) Houston TX
KDAV(AM) Lubbock TX
KRQX(AM) Mexia TX
KDAE(AM) Sinton TX
WFTH(AM) Richmond VA
KLFE(AM) Seattle WA
WIXK(AM) New Richmond WI
WPVL(AM) Platteville WI
WGBW(AM) Two Rivers WI

1593 khz
V6AK(AM) Truk FM

1600 khz
WHIY(AM) Huntsville AL
WXVI(AM) Montgomery AL
KNWA(AM) Bellefonte AR
KYBC(AM) Cottonwood AZ
KXEW(AM) South Tucson AZ
KGST(AM) Fresno CA
KAHZ(AM) Pomona CA
KTAP(AM) Santa Maria CA
KUBA(AM) Yuba City CA
KEPN(AM) Lakewood CO
*WXXY(AM) Dover DE
WQOP(AM) Atlantic Beach FL
WKWF(AM) Key West FL
WOKB(AM) Ocoee FL
WMNE(AM) Riviera Beach FL
WAOS(AM) Austell GA
KLGA(AM) Algona IA
KGYM(AM) Cedar Rapids IA
WCGO(AM) Chicago Heights IL
WMCW(AM) Harvard IL
WBTO(AM) Linton IN
WARU(AM) Peru IN
KMDO(AM) Fort Scott KS
WAIA(AM) Beaver Dam KY
WTSZ(AM) Eminence KY
KLEB(AM) Golden Meadow LA
WUNR(AM) Brookline MA
WHNP(AM) East Longmeadow MA
WLXE(AM) Rockville MD
WAAM(AM) Ann Arbor MI
WMHG(AM) Muskegon MI
KZGX(AM) Watertown MN
KATZ(AM) Saint Louis MO
KTTN(AM) Trenton MO
WIDU(AM) Fayetteville NC
WTZQ(AM) Hendersonville NC
KDAK(AM) Carrington ND
KNCY(AM) Nebraska City NE
KRFS(AM) Superior NE
KRKE(AM) Albuquerque NM
WEHH(AM) Elmira Heights-Horseheads NY
WWRL(AM) New York NY
WMCR(AM) Oneida NY
WULM(AM) Springfield OH
WTTF(AM) Tiffin OH
KUSH(AM) Cushing OK
KOPT(AM) Eugene OR
KOHI(AM) Saint Helens OR
WHOL(AM) Allentown PA
WHJB(AM) Bedford PA
WPDC(AM) Elizabethtown PA
WJSA(AM) Jersey Shore PA
WLUZ(AM) Bayamon PR
WFIS(AM) Fountain Inn SC
WKZK(AM) North Augusta SC
WMQM(AM) Lakeland TN
KRVA(AM) Cockrell Hill TX
KOGT(AM) Orange TX
KOKE(AM) Pflugerville TX
KXTA(AM) Centerville UT
WCPK(AM) Chesapeake VA
WXMY(AM) Saltville VA
KVRI(AM) Blaine WA
WRPN(AM) Ripon WI
WZZW(AM) Milton WV
WKKX(AM) Wheeling WV

1620 khz
KSMH(AM) West Sacramento CA
WNRP(AM) Gulf Breeze FL
WWLV(AM) South Bend IN
KOZN(AM) Bellevue NE
WTAW(AM) College Station TX
WDHP(AM) Frederiksted VI
KYIZ(AM) Renton WA

1630 khz
WRDW(AM) Augusta GA
KCJJ(AM) Iowa City IA
KKGM(AM) Fort Worth TX
KRND(AM) Fox Farm WY

1640 khz
KDIA(AM) Vallejo CA
WTNI(AM) Biloxi MS
KFXY(AM) Enid OK
KDZR(AM) Lake Oswego OR
KBJA(AM) Sandy UT
WKSH(AM) Sussex WI

1650 khz
KWHN(AM) Fort Smith AR
KFOX(AM) Torrance CA
KBJD(AM) Denver CO
KCNZ(AM) Cedar Falls IA
KHRO(AM) El Paso TX
WHKT(AM) Portsmouth VA

1660 khz
KTIQ(AM) Merced CA
*WCNZ(AM) Marco Island FL
KXTR(AM) Kansas City KS
WQLR(AM) Kalamazoo MI
WFNA(AM) Charlotte NC
KQWB(AM) West Fargo ND
WWRU(AM) Jersey City NJ
WGIT(AM) Canovanas PR
KRZI(AM) Waco TX
KXOL(AM) Brigham City UT

1670 khz
KHPY(AM) Moreno Valley CA
KNRO(AM) Redding CA
WVVM(AM) Dry Branch GA
WTDY(AM) Madison WI

1680 khz
KAVT(AM) Fresno CA
WLAA(AM) Winter Garden FL
KRJO(AM) Monroe LA
WDSS(AM) Ada MI
WTTM(AM) Lindenwold NJ
KDOW(AM) Seattle WA

1690 khz
KFSG(AM) Roseville CA
KDDZ(AM) Arvada CO
WMLB(AM) Avondale Estates GA
WVON(AM) Berwyn IL
WPTX(AM) Lexington Park MD

1700 khz
WEUP(AM) Huntsville AL
KBGG(AM) Des Moines IA
KVNS(AM) Brownsville TX
KKLF(AM) Richardson TX

U.S. FM Stations by Frequency

87.9 mhz
*KSFH(FM) Mountain View CA

88.1 mhz
*KRUA(FM) Anchorage AK
*KCUK(FM) Chevak AK
*KCDS(FM) Deadhorse AK
*WAYH(FM) Harvest AL
*WSJL(FM) Northport AL
*KAPG(FM) Bentonville AR
*KARH(FM) Forrest City AR
*KBPW(FM) Hampton AR
*KUYI(FM) Hotevilla AZ
*KLTU(FM) Mammoth AZ
*KNNB(FM) Whiteriver AZ
*KCFY(FM) Yuma AZ
*KECG(FM) El Cerrito CA
*KFCF(FM) Fresno CA
*KLWG(FM) Lompoc CA
*KKJZ(FM) Long Beach CA
*KCRY(FM) Mojave CA
*KNSQ(FM) Mount Shasta CA
*KQNC(FM) Quincy CA
*KEAR-FM Sacramento CA
*KSRH(FM) San Rafael CA
*KZSC(FM) Santa Cruz CA
*KJAR(FM) Susanville CA
*KDNK(FM) Glenwood Springs CO
*KAFM(FM) Grand Junction CO
*KFDN(FM) Lakewood CO
*KPGS(FM) Pagosa Springs CO
*WESU(FM) Middletown CT
*WMNR(FM) Monroe CT
*WMHS(FM) Pike Creek DE
*WJIS(FM) Bradenton FL
*WEAZ(FM) Holly Hill FL
*WRGP(FM) Homestead FL
*WCRJ(FM) Jacksonville FL
*WBGY(FM) Naples FL
*WHIJ(FM) Ocala FL
*WUWF(FM) Pensacola FL
*WAYF(FM) West Palm Beach FL
*WLXP(FM) Savannah GA
*WAYT(FM) Thomasville GA
*WJSP-FM Warm Springs GA
*KHMG(FM) Barrigada GU
*KHPR(FM) Honolulu HI
*KUNJ(FM) Fairfield IA
*KICB(FM) Fort Dodge IA
*KBBG(FM) Waterloo IA
*KTFY(FM) Buhl ID
*WESN(FM) Bloomington IL
*WWTG(FM) Carpentersville IL
*WCRX(FM) Chicago IL
*WSSD(FM) Chicago IL
*WBMF(FM) Crete IL
*WAXR(FM) Geneseo IL
*WLTL(FM) La Grange IL
*WAES(FM) Lincolnshire IL
*WLRA(FM) Lockport IL
*WPTH(FM) Olney IL
*WLWJ(FM) Petersburg IL
*WSOG(FM) Spring Valley IL
*WETN(FM) Wheaton IL
*WNTH(FM) Winnetka IL
*WDVL(FM) Danville IN
*WVPE(FM) Elkhart IN
*WHCI(FM) Hartford City IN
*WMBL(FM) Mitchell IN
*WJCF(FM) Morristown IN
*WNAS(FM) New Albany IN
*WKRY(FM) Versailles IN
*KBTL(FM) El Dorado KS
*KBCU(FM) North Newton KS
*KJTY(FM) Topeka KS
*WAYD(FM) Auburn KY
*WTRT(FM) Benton KY
*WDFB-FM Danville KY
*WRFL(FM) Lexington KY
*WAXG(FM) Mt. Sterling KY
*WKVY(FM) Somerset KY
*KAYT(FM) Jena LA
*KPAQ(FM) Plaquemine LA
*WMBR(FM) Cambridge MA
*WFHL(FM) New Bedford MA
*WCHC(FM) Worcester MA
*WYPR(FM) Baltimore MD
*WMUC-FM College Park MD
*WYPF(FM) Frederick MD
*WBFH(FM) Bloomfield Hills MI
*WBLW(FM) Gaylord MI
*WHYT(FM) Goodland Township MI
*WHPR(FM) Highland Park MI
*WKHN(FM) Hubbard Lake MI
*WLGH(FM) Leroy Township MI
*WKPK(FM) Michigamme MI
*WDTR(FM) Monroe MI
*WPQZ(FM) Muskegon MI
*WSDP(FM) Plymouth MI
*WAAQ(FM) Rogers Heights MI
*WYCE(FM) Wyoming MI
*KRLX(FM) Northfield MN
*KVSC(FM) Saint Cloud MN
*KRLP(FM) Windom MN
*KBOJ(FM) Worthington MN
*KLFC(FM) Branson MO
*KCOU(FM) Columbia MO
*KYRV(FM) Concordia MO
*KDHX(FM) Saint Louis MO
*WURC(FM) Holly Springs MS
*WMAW-FM Meridian MS
*KFRT(FM) Butte MT
*KGVA(FM) Fort Belknap Agency MT
*WCQS(FM) Asheville NC
*WPIR(FM) Hickory NC
*WGHW(FM) Lockwoods Folly Town NC
*WKNC-FM Raleigh NC
*KCNT(FM) Hastings NE
*KFHC(FM) Ponca NE
*KMLV(FM) Ralston NE
*WYGG(FM) Asbury Park NJ
*WNJS-FM Berlin NJ
*WJPG(FM) Cape May Court House NJ
*WNJT-FM Trenton NJ
*KUSW(FM) Flora Vista NM
*KGGA(FM) Gallup NM
*KIDS(FM) Grants NM
*KNMA(FM) Socorro NM
*KRNM(FM) Chalan Kanoa-Saipan NP
*KCEP(FM) Las Vegas NV
*WXBA(FM) Brentwood NY
*WCWP(FM) Brookville NY
*WUBJ(FM) Jamestown NY
*WGWR(FM) Liberty NY
*WGMY(FM) Montgomery NY
*WARY(FM) Valhalla NY
*WFRW(FM) Webster NY
*WZIP(FM) Akron OH
*WBGU(FM) Bowling Green OH
*WWGV(FM) Grove City OH
*WBCJ(FM) Spencerville OH
*WDPR(FM) West Carrollton OH
*KDIM(FM) Coweta OK
*KMSI(FM) Moore OK
*KKRI(FM) Pocola OK
*KDJC(FM) Baker City OR
*KLBR(FM) Bend OR
*KWVA(FM) Eugene OR
*KLFO(FM) Florence OR
*KGRI(FM) Lebanon OR
*KMPQ(FM) Roseburg OR
*WDIY(FM) Allentown PA
*WEFR(FM) Erie PA
*WTGP(FM) Greenville PA
*WXPH(FM) Harrisburg PA
*WRWJ(FM) Murrysville PA
*WBGM(FM) New Berlin PA
*WPEB(FM) Philadelphia PA
*WSRU(FM) Slippery Rock PA
*WRGN(FM) Sweet Valley PA
*WZZD(FM) Warwick PA
*WPTC(FM) Williamsport PA
*WCRP(FM) Guayama PR
*WELH(FM) Providence RI
*WKIV(FM) Westerly RI
*WSBF-FM Clemson SC
*WRJA-FM Sumter SC
*KRSD(FM) Sioux Falls SD
*WUTC(FM) Chattanooga TN
*WAMP(FM) Jackson TN
*WFSK-FM Nashville TN
*WAZD(FM) Savannah TN
*KGNZ(FM) Abilene TX
*KKWV(FM) Aransas Pass TX
*KATG(FM) Athens TX
*KEDR(FM) Bay City TX
*KLBT(FM) Beaumont TX
*KGLF(FM) Doss TX
*KZAR(FM) Gonzales TX
*KHOY(FM) Laredo TX
*KTXT-FM Lubbock TX
*KHID(FM) McAllen TX
*KNTU(FM) McKinney TX
*KFTG(FM) Pasadena TX
*KNLE-FM Round Rock TX
*KFRI(FM) Stanton TX
*KVLW(FM) Waco TX
*KWCR-FM Ogden UT
*KPGR(FM) Pleasant Grove UT
*WHOV(FM) Hampton VA
*WRIH(FM) Richmond VA
*WNCH(FM) Norwich VT
*KFIO(FM) East Wenatchee WA
*KCWU(FM) Ellensburg WA
*KTCV(FM) Kennewick WA
*KSBC(FM) Nile WA
*KLOP(FM) Ocean Park WA
*KAYB(FM) Sunnyside WA
*WHID(FM) Green Bay WI
*WJTY(FM) Lancaster WI
*WMWK(FM) Milwaukee WI
*WJJJ(FM) Beckley WV
*WVBC(FM) Bethany WV
*WKJL(FM) Clarksburg WV
*WMUL(FM) Huntington WV
*WPCN(FM) Point Pleasant WV
*KAIX(FM) Cheyenne WY
*KCWW(FM) Evanston WY
*KCWC-FM Riverton WY
*KPRQ(FM) Sheridan WY

88.3 mhz
*WJCK(FM) Piedmont AL
*WAPR(FM) Selma AL
*KBCM(FM) Blytheville AR
*KXUA(FM) Fayetteville AR
*KABF(FM) Little Rock AR
*KNAI(FM) Phoenix AZ
*KPHF(FM) Phoenix AZ
*KYCJ(FM) Camino CA
*KDKL(FM) Coalinga CA
*KMUE(FM) Eureka CA
*KAXL(FM) Green Acres CA
*KLVN(FM) Livingston CA
*KLVC(FM) Magalia CA
*KUCR(FM) Riverside CA
*KSDS(FM) San Diego CA
*KCLU(FM) Thousand Oaks CA
*KPYR(FM) Craig CO
*KLHV(FM) Fort Collins CO
*KPRH(FM) Montrose CO
*KTPL(FM) Pueblo CO
*WAZQ(FM) Key West FL
*WBIY(FM) La Belle FL
*WLMS(FM) Lecanto FL
*WJNF(FM) Marianna FL
*WIRP(FM) Pennsuco FL
*WMKM(FM) Sebring FL
*WTLG(FM) Starke FL
*WPOZ(FM) Union Park FL
*WPPR(FM) Demorest GA
*WAWH(FM) Dublin GA
*WNEE(FM) Jasper GA
*WLPT(FM) Jesup GA
*KCCK-FM Cedar Rapids IA
*KKLG(FM) Newton IA
*KUNE(FM) Ottumwa IA
*KMSC(FM) Sioux City IA
*KARJ(FM) Kuna ID
*WCLR(FM) Arlington Heights IL
*WZRD(FM) Chicago IL
*WDGC-FM Downers Grove IL
*WAWF(FM) Kankakee IL
*WIUS(FM) Macomb IL
*WHCM(FM) Palatine IL
*WPJC(FM) Pontiac IL
*WPRC(FM) Princeton IL
*WJLY(FM) Ramsey IL
*WFEN(FM) Rockford IL
*WQNA(FM) Springfield IL
*WEAX(FM) Angola IN
*WDSO(FM) Chesterton IN
*WNIN-FM Evansville IN
*WLAB(FM) Fort Wayne IN
*WKMV(FM) Muncie IN
*WARA(FM) New Washington IN
*WWDN(FM) New Whiteland IN
*KBJQ(FM) Bronson KS
*KVCO(FM) Concordia KS
*KYFW(FM) Wichita KS
*WSGP(FM) Glasgow KY
*WOCS(FM) Lerose KY
*WRBH(FM) New Orleans LA
*KAPI(FM) Ruston LA
*WBMT(FM) Boxford MA
*WIQH(FM) Concord MA
*WGAO(FM) Franklin MA
*WRPS(FM) Rockland MA
*WRAU(FM) Ocean City MD
*WYAR(FM) Yarmouth ME
*WCBN-FM Ann Arbor MI
*WXOU(FM) Auburn Hills MI
*WLVM(FM) Ironwood MI
*WAYK(FM) Kalamazoo MI
*WAQQ(FM) Onsted MI
*WNFA(FM) Port Huron MI
*WSHJ(FM) Southfield MI
*WEJC(FM) White Star MI
*KBPN(FM) Brainerd MN
*KJAB-FM Mexico MO
*KWND(FM) Springfield MO
*WAFR(FM) Tupelo MS
*KJCG(FM) Missoula MT
*KPGB(FM) Pryor MT
*WGWG(FM) Boiling Springs NC
*WGPS(FM) Elizabeth City NC
*WUAW(FM) Erwin NC
*KBMK(FM) Bismarck ND
*KLNB(FM) Grand Island NE
*KLJV(FM) Scottsbluff NE
*WEVS(FM) Nashua NH
*WVBH(FM) Beach Haven West NJ
*WBGO(FM) Newark NJ
*KLYT(FM) Albuquerque NM
*KLRH(FM) Sparks NV
*WBKW(FM) Beekman NY
*WVCR-FM Loudonville NY
*WFSO(FM) Olivebridge NY
*WXLU(FM) Peru NY
*WSBU(FM) Saint Bonaventure NY
*WLIU(FM) Southampton NY
*WAER(FM) Syracuse NY
*WCOU(FM) Warsaw NY
*WBWC(FM) Berea OH
*WJVS(FM) Cincinnati OH
*WLFC(FM) Findlay OH
*WMRT(FM) Marietta OH
*WOHP(FM) Portsmouth OH
*WAUI(FM) Shelby OH
*WOAR(FM) South Vienna OH
*WXTS-FM Toledo OH
*WXUT(FM) Toledo OH
*KAZC(FM) Tishomingo OK
*KSRG(FM) Ashland OR
*KBVM(FM) Portland OR
*WGEV(FM) Beaver Falls PA
*WDCV-FM Carlisle PA
*WZXQ(FM) Chambersburg PA
*WWEC(FM) Elizabethtown PA
*WRCT(FM) Pittsburgh PA
*WXFR(FM) State College PA
*WLKA(FM) Tafton PA
*WRUO(FM) Mayaguez PR
*WQRI(FM) Bristol RI
*WAFJ(FM) Belvedere SC
*WMBJ(FM) Murrell's Inlet SC
*KESD(FM) Brookings SD
*KLMP(FM) Rapid City SD
*WRRI(FM) Brownsville TN
*WAYQ(FM) Clarksville TN
*WCQR-FM Kingsport TN
*WOEZ(FM) Maynardville TN
*WMTS-FM Murfreesboro TN
*WBIA(FM) Shelbyville TN
*KIFR(FM) Alice TX
*KJRT(FM) Amarillo TX
*KBNR(FM) Brownsville TX
*KAFR(FM) Conroe TX
*KJCR(FM) Keene TX
*KPAC(FM) San Antonio TX
*KCPW-FM Salt Lake City UT
*WOTC(FM) Edinburg VA
*WRVL(FM) Lynchburg VA
*WNUB-FM Northfield VT
*KMLW(FM) Moses Lake WA
*WHWC(FM) Menomonie WI
*KMLT(FM) Jackson WY

88.5 mhz
*KAKL(FM) Anchorage AK
*KTNA(FM) Talkeetna AK
*WLJR(FM) Birmingham AL
*WJIA(FM) Guntersville AL
*WBHY-FM Mobile AL
*KLKA(FM) Globe AZ
*KFLT-FM Tucson AZ
*KWTW(FM) Bishop CA
*KVUH(FM) Laytonville CA
*KSBR(FM) Mission Viejo CA
*KCSN(FM) Northridge CA
*KPSC(FM) Palm Springs CA
*KJCQ(FM) Quincy CA
*KQED-FM San Francisco CA
*KLVH(FM) San Luis Obispo CA
*KQKL(FM) Selma CA
*KHMS(FM) Victorville CA
*KGNU-FM Boulder CO
*KCIC(FM) Grand Junction CO
*KRNC(FM) Steamboat Springs CO
*KTDU(FM) Trimble CO
*KVJZ(FM) Vail CO
*WVOF(FM) Fairfield CT
*WEDW-FM Stamford CT
*WAMU(FM) Washington DC
*WJCB(FM) Clewiston FL
*WWLC(FM) Cross City FL
*WMFL(FM) Florida City FL
*WHYZ(FM) Palm Bay FL
*WFCF(FM) Saint Augustine FL
*WKPX(FM) Sunrise FL
*WMNF(FM) Tampa FL
*WRAS(FM) Atlanta GA
*WTMQ(FM) Lumpkin GA
*WVDA(FM) Valdosta GA
*KURE(FM) Ames IA
*KALA(FM) Davenport IA
*KIAD(FM) Dubuque IA
*KBDC(FM) Mason City IA
*KDCR(FM) Sioux Center IA
*KBSY(FM) Burley ID
*WBEL(FM) Cairo IL
*WHPK-FM Chicago IL

*WHFH(FM) Flossmoor IL
*WGBK(FM) Glenview IL
*WHSD(FM) Hinsdale IL
*WBNH(FM) Pekin IL
*WGCA-FM Quincy IL
*WTMK(FM) Lowell IN
*WQKV(FM) Rochester IN
*WCRT-FM Terre Haute IN
*WXVW(FM) Veedersburg IN
*KBQC(FM) Independence KS
*KAKA(FM) Salina KS
*WEKF(FM) Corbin KY
*WBMK(FM) Morehead KY
*WJIE-FM Okolona KY
*WJFM(FM) Baton Rouge LA
*WFCR(FM) Amherst MA
*WWTA(FM) Marion MA
*WHCF(FM) Bangor ME
*WSEW(FM) Sanford ME
*WGVU-FM Allendale MI
*WJOM(FM) Eagle MI
*WCTP(FM) Gagetown MI
*WJKQ(FM) Jackson MI
*WIAB(FM) Mackinaw City MI
*WOAS(FM) Ontonagon MI
*WKKM(FM) Rust Township MI
*KNCM(FM) Appleton MN
*KCRB-FM Bemidji MN
*KBEM-FM Minneapolis MN
*KGSF(FM) Anderson MO
*KLJC(FM) Kansas City MO
*KMST(FM) Rolla MO
*WMUW(FM) Columbus MS
*WUSM-FM Hattiesburg MS
*WJSU(FM) Jackson MS
*WXBE(FM) Beaufort NC
*WZNB(FM) New Bern NC
*WRTP(FM) Roanoke Rapids NC
*WZDG(FM) Scotts Hill NC
*WHYC(FM) Swanquarter NC
*WFDD-FM Winston-Salem NC
*KEYA(FM) Belcourt ND
*KLCV(FM) Lincoln NE
*WNJP(FM) Sussex NJ
*KPKJ(FM) Mentmore NM
*KEKL(FM) Mesquite NV
*WPOB-FM Plainview NY
*WRUR-FM Rochester NY
*WCII(FM) Spencer NY
*WKWZ(FM) Syosset NY
*WMUB(FM) Oxford OH
*WYSA(FM) Wauseon OH
*WYSU(FM) Youngstown OH
*KZTH(FM) Piedmont OK
*KTKL(FM) Stigler OK
*KSBA(FM) Coos Bay OR
*KQAC(FM) Gleneden Beach OR
*KPIJ(FM) Junction City OR
*KLMF(FM) Klamath Falls OR
*KLRF(FM) Milton-Freewater OR
*KAVE(FM) Oakridge OR
*KWRX(FM) Redmond OR
*KAIK(FM) Tillamook OR
*WMCE(FM) Erie PA
*WLVU(FM) Halifax PA
*WYFU(FM) Masontown PA
*WXPN(FM) Philadelphia PA
*WRKC(FM) Wilkes-Barre PA
*WTMV(FM) Youngsville PA
*WPLI(FM) Levittown PR
*WEPC(FM) Belton SC
*WYFV(FM) Cayce SC
*WFCH(FM) Charleston SC
*WTTU(FM) Cookeville TN
*WVCP(FM) Gallatin TN
*WZXX(FM) Lawrenceburg TN
*WQOX(FM) Memphis TN
*WAUT-FM Tullahoma TN
*KHIB(FM) Bastrop TX
*KGHY(FM) Beaumont TX
*KPBB(FM) Brownfield TX
*KLRW(FM) Byrne TX
*KCKT(FM) Crockett TX
*KOIR(FM) Edinburg TX
*KTEP(FM) El Paso TX
*KBMD(FM) Marble Falls TX
*KEOM(FM) Mesquite TX
*KPMB(FM) Plainview TX
*KVLT(FM) Temple TX
*KAYK(FM) Victoria TX
*KMQX(FM) Weatherford TX
*WVTW(FM) Charlottesville VA
*WJLZ(FM) Virginia Beach VA
*WVPA(FM) Saint Johnsbury VT
*KRLF(FM) Pullman WA
*KPLU-FM Tacoma WA
*KYVT(FM) Yakima WA
*WGNV(FM) Milladore WI
*WRGX(FM) Sturgeon Bay WI
*WVPN(FM) Charleston WV
*KUWY(FM) Laramie WY

88.7 mhz
*KJHA(FM) Houston AK
*WELL-FM Dadeville AL
*WRWA(FM) Dothan AL
*WQPR(FM) Muscle Shoals AL
*KBPU(FM) De Queen AR
*KNAU(FM) Flagstaff AZ
*KISL(FM) Avalon CA
*KUBO(FM) Calexico CA
*KSPC(FM) Claremont CA
*KZLU(FM) Inyokern CA
*KMPO(FM) Modesto CA
*KAIS(FM) Redwood Valley CA
*KQSC(FM) Santa Barbara CA
*KXJS(FM) Sutter CA
*KYLU(FM) Tehachapi CA
*KRZA(FM) Alamosa CO
*KCME(FM) Manitou Springs CO
*WNHU(FM) West Haven CT
*WMYZ(FM) Clermont FL
*WKTO(FM) Edgewater FL
*WAYJ(FM) Fort Myers FL
*WJFR(FM) Jacksonville FL
*WFRP(FM) Americus GA
*WMOC(FM) Lumber City GA
*WJDS(FM) Sparta GA
*KLNI(FM) Decorah IA
*KRFH(FM) Marshalltown IA
*KIGC(FM) Oskaloosa IA
*KWDM(FM) West Des Moines IA
*WPCD(FM) Champaign IL
*WLUW(FM) Chicago IL
*WSIE(FM) Edwardsville IL
*WRSE(FM) Elmhurst IL
*WCSF(FM) Joliet IL
*WSRI(FM) Sugar Grove IL
*WGVE(FM) Gary IN
*WICR(FM) Indianapolis IN
*WBHW(FM) Loogootee IN
*KGLV(FM) Manhattan KS
*WMMT(FM) Whitesburg KY
*KRVS(FM) Lafayette LA
*KBMQ(FM) Monroe LA
*WIAA(FM) Interlochen MI
*WMLS(FM) Grand Marais MN
*KMSE(FM) Rochester MN
*KXMS(FM) Joplin MO
*KTRM(FM) Kirksville MO
*WYTF(FM) Indianola MS
*WJZB(FM) Starkville MS
*KLKM(FM) Kalispell MT
*WXDU(FM) Durham NC
*WRAE(FM) Raeford NC
*WAGO(FM) Snow Hill NC
*WNCW(FM) Spindale NC
*KFBN(FM) Fargo ND
*KLNE-FM Lexington NE
*WRSU-FM New Brunswick NJ
*WXXY-FM Port Republic NJ
*WPSC-FM Wayne NJ
*KWPR(FM) Lund NV
*KUNR(FM) Reno NV
*WBFO(FM) Buffalo NY
*WHCL-FM Clinton NY
*WRHU(FM) Hempstead NY
*WSQA(FM) Hornell NY
*WPKM(FM) Montauk NY
*WNYK(FM) Nyack NY
*WRHV(FM) Poughkeepsie NY
*WFNP(FM) Rosendale NY
*WKYJ(FM) Rouses Point NY
*WHJM(FM) Anna OH
*WOBO(FM) Batavia OH
*WOFN(FM) Beach City OH
*WUFM(FM) Columbus OH
*WJCU(FM) University Heights OH
*KAJT(FM) Ada OK
*KLVV(FM) Ponca City OK
*KWTU(FM) Tulsa OK
*KLOY(FM) Astoria OR
*KLVP-FM Cherryville OR
*KBVR(FM) Corvallis OR
*KOAP(FM) Lakeview OR
*KJKL(FM) Selma OR
*WWLU(FM) Lincoln University PA
*WWCF(FM) McConnellsburg PA
*WZXM(FM) Middletown PA
*WSYC-FM Shippensburg PA
*WBYX(FM) Stroudsburg PA
*WCYJ-FM Waynesburg PA
*WJMF(FM) Smithfield RI
*WAGP(FM) Beaufort SC
*WAYM(FM) Columbia TN
*WIGH(FM) Lexington TN
*KAZI-FM Austin TX
*KASV(FM) Borger TX
*KKLM(FM) Corpus Christi TX
*KEPI(FM) Eagle Pass TX
*KTCU-FM Fort Worth TX
*KWDH(FM) Hereford TX
*KUHF(FM) Houston TX
*KKER(FM) Kerrville TX
*KZLO(FM) Kilgore TX
*KLVW(FM) West Odessa TX
*KMCU(FM) Wichita Falls TX
*KNKL(FM) North Ogden UT
*WFOS(FM) Chesapeake VA
*WXJM(FM) Harrisonburg VA
*WWPV-FM Colchester VT
*WRVT(FM) Rutland VT
*KAGU(FM) Spokane WA
*WERN(FM) Madison WI
*WRFW(FM) River Falls WI
*WPJY(FM) Blennerhassett WV

88.9 mhz
*KEUL(FM) Girdwood AK
*KMJG(FM) Homer AK
*KJLP(FM) Palmer AK
*WMFT(FM) Tuscaloosa AL
*KKDU(FM) El Dorado AR
*KAOW(FM) Fort Smith AR
*KAIC(FM) Tucson AZ
*KAWC-FM Yuma AZ
*KUCI(FM) Irvine CA
*KTLW(FM) Lancaster CA
*KXLU(FM) Los Angeles CA
*KFPR(FM) Redding CA
*KOGR(FM) Rosedale CA
*KXPR(FM) Sacramento CA
*KUSP(FM) Santa Cruz CA
*KRTM(FM) Temecula CA
*KDUV(FM) Visalia CA
*KCJX(FM) Carbondale CO
*KEZF(FM) Eaton CO
*KRFC(FM) Fort Collins CO
*WJMJ(FM) Hartford CT
*WQCS(FM) Fort Pierce FL
*WDNA(FM) Miami FL
*WFSU-FM Tallahassee FL
*WYFE(FM) Tarpon Springs FL
*WMSL(FM) Athens GA
*WWIO-FM Brunswick GA
*WBKG(FM) Macon GA
*WGUR(FM) Milledgeville GA
*WKEU-FM The Rock GA
*KHJC(FM) Lihue HI
*KIHS(FM) Adel IA
*KSTM(FM) Indianola IA
*KDMR(FM) Mitchellville IA
*KJIA(FM) Spirit Lake IA
*KAIP(FM) Wapello IA
*KWVI(FM) Waverly IA
*KLCZ(FM) Lewiston ID
*KEFX(FM) Twin Falls ID
*WEIU(FM) Charleston IL
*WIIT(FM) Chicago IL
*WEPS(FM) Elgin IL
*WMXM(FM) Lake Forest IL
*WLNX(FM) Lincoln IL
*WOTW(FM) Monee IL
*WVSI(FM) Mount Vernon IL
*WWGN(FM) Ottawa IL
*WRRG(FM) River Grove IL
*WARG(FM) Summit IL
*WJCJ(FM) Ladoga IN
*WSND-FM Notre Dame IN
*WMYJ-FM Oolitic IN
*WJYW(FM) Union City IN
*KPRD(FM) Hays KS
*KTJO-FM Ottawa KS
*WKYU-FM Bowling Green KY
*WEKU(FM) Richmond KY
*WERS(FM) Boston MA
*WEAA(FM) Baltimore MD
*WMDR-FM Oakland ME
*WDBM(FM) East Lansing MI
*WJCE(FM) Elkton MI
*WAKL(FM) Flint MI
*WBLU-FM Grand Rapids MI
*KNSR(FM) Collegeville MN
*KRNW(FM) Chillicothe MO
*KSEF(FM) Farmington MO
*KJLU(FM) Jefferson City MO
*WMAU-FM Bude MS
*WMSB(FM) Senatobia MS
*KFRD(FM) Butte MT
*KGFC(FM) Great Falls MT
*KYWH(FM) Lockwood MT
*WUND-FM Manteo NC
*WSHA(FM) Raleigh NC
*KLRX(FM) Jamestown ND
*KMPR(FM) Minot ND
*KNBE(FM) Beatrice NE
*KVSS(FM) Omaha NE
WAJM(FM) Atlantic City NJ
*WMNJ(FM) Madison NJ
*WBZC(FM) Pemberton NJ
*WMCX(FM) West Long Branch NJ
*KHII(FM) Cloudcroft NM
*KNMI(FM) Farmington NM
*KLLU(FM) Gallup NM
*KRUC(FM) Las Cruces NM
*KNPR(FM) Las Vegas NV
*WCIY(FM) Canandaigua NY
*WITC(FM) Cazenovia NY
*WCVF-FM Fredonia NY
*WNYO(FM) Oswego NY
*WRPJ(FM) Port Jervis NY
*WFRS(FM) Smithtown NY
*WSIA(FM) Staten Island NY
*WSLJ(FM) Watertown NY
*WRDL(FM) Ashland OH
*WMWX(FM) Miamitown OH
*WLRY(FM) Rushville OH
*WBJV(FM) Steubenville OH
*WSTB(FM) Streetsboro OH
*WCSU-FM Wilberforce OH
*KARU(FM) Cache OK
*KWXC(FM) Grove OK
*KYLV(FM) Oklahoma City OK
*KOBK(FM) Baker City OR
*KKLJ(FM) Klamath Falls OR
*KYOR(FM) Newport OR
*KQFE(FM) Springfield OR
*WFSE(FM) Edinboro PA
*WFRJ(FM) Johnstown PA
*WWNW(FM) New Wilmington PA
*WQSU(FM) Selinsgrove PA
*WBYO(FM) Sellersville PA
*WPUC-FM Ponce PR
*WKVC(FM) North Myrtle Beach SC
*WNSC-FM Rock Hill SC
*WMBW(FM) Chattanooga TN
*WTAI(FM) Union City TN
*KETR(FM) Commerce TX
*KMBH-FM Harlingen TX
*KLDN(FM) Lufkin TX
*KSUR(FM) Mart TX
*WCVE(FM) Richmond VA
*KSWS(FM) Chehalis WA
*KCSH(FM) Ellensburg WA
*WLSU(FM) La Crosse WI
*WYMS(FM) Milwaukee WI
*WOJB(FM) Reserve WI
*WVPW(FM) Buckhannon WV
*WVEP(FM) Martinsburg WV
*KLOF(FM) Gillette WY
*KAIW(FM) Laramie WY
*KOHR(FM) Sheridan WY
*KWCF(FM) Sheridan WY

89.1 mhz
*WKNG-FM Heflin AL
*WLBF(FM) Montgomery AL
*KUAR(FM) Little Rock AR
*KLVK(FM) Fountain Hills AZ
*KUAZ-FM Tucson AZ
*KCEA(FM) Atherton CA
*KPRX(FM) Bakersfield CA
*KODV(FM) Barstow CA
*KHAP(FM) Chico CA
*KXBC(FM) Garberville CA
*KCJH(FM) Livingston CA
*KCRU(FM) Oxnard CA
*KUOR-FM Redlands CA
*KBBF(FM) Santa Rosa CA
*KBWA(FM) Brush CO
*KTLC(FM) Canon City CO
*KRLJ(FM) La Junta CO
*KVMT(FM) Montrose CO
*WNPR(FM) Norwich CT
*WXHL-FM Christiana DE
*WUFT-FM Gainesville FL
*WLAZ(FM) Kissimmee FL
*WFSW(FM) Panama City FL
*WSMR(FM) Sarasota FL
*WBCX(FM) Gainesville GA
*KWOF-FM Hiawatha IA
*KWAR(FM) Waverly IA
*KAWS(FM) Boise ID
*WNIE(FM) Freeport IL
*WVJC(FM) Mount Carmel IL
*WONC(FM) Naperville IL
*WGLT(FM) Normal IL
*WSPM(FM) Cloverdale IN
*WBOI(FM) Fort Wayne IN
*WAUZ(FM) Greensburg IN
*WWLO(FM) Lowell IN
*KMUW(FM) Wichita KS
*KFLO-FM Blanchard LA
*KVDP(FM) Dry Prong LA
*WBSN-FM New Orleans LA
*KLPI-FM Ruston LA
*KRLR(FM) Sulphur LA
*WHAB(FM) Acton MA
*WGMS(FM) Hagerstown MD
*WLKB(FM) Bay City MI
*WWKM(FM) Imlay City MI
*WIDR(FM) Kalamazoo MI
*WPHS(FM) Warren MI
*WEMU(FM) Ypsilanti MI
*KCLC(FM) Saint Charles MO
*KWFC(FM) Springfield MO
*WMBU(FM) Forest MS
*WPAS(FM) Pascagoula MS
*KUFM(FM) Missoula MT
*WRYN(FM) Hickory NC
*KVLQ(FM) Lincoln ND
*KHNE-FM Hastings NE
*KDAI(FM) Scottsbluff NE
*WEVO(FM) Concord NH
*WWCJ(FM) Cape May NJ
*WFDU(FM) Teaneck NJ
*WWFM(FM) Trenton NJ
*KANW(FM) Albuquerque NM
*KQAI(FM) Roswell NM
*WDWN(FM) Auburn NY
*WBSU(FM) Brockport NY
*WCID(FM) Friendship NY
*WNYU-FM New York NY
*WMHT-FM Schenectady NY
*WJPZ-FM Syracuse NY
*WOUC-FM Cambridge OH
*WJJE(FM) Delaware OH
*WOUL-FM Ironton OH
*WNZN(FM) Lorain OH
*WLMH(FM) Morrow OH

*WUSO(FM) Springfield OH
*WKSV(FM) Thompson OH
*KWRI(FM) Bartlesville OK
*KYCU(FM) Clinton OK
*KXTH(FM) Seminole OK
*KSMF(FM) Ashland OR
*KMHD(FM) Gresham OR
*KLFR(FM) Reedsport OR
*WBYH(FM) Hawley PA
*WFNM(FM) Lancaster PA
*WLOG(FM) Markleysburg PA
*WSFX(FM) Nanticoke PA
*WYBF(FM) Radnor Township PA
*WRXV(FM) State College PA
*WXVU(FM) Villanova PA
*WLJK(FM) Aiken SC
*KVFL(FM) Pierre SD
*KAUR(FM) Sioux Falls SD
*KBHU-FM Spearfish SD
*KJBB(FM) Watertown SD
*WYLV(FM) Alcoa TN
*WNAZ-FM Nashville TN
*WDNX(FM) Olive Hill TN
*KXLV(FM) Amarillo TX
*KEOS(FM) College Station TX
*KOHM(FM) Lubbock TX
*KYFP(FM) Palestine TX
*KSTX(FM) San Antonio TX
*KSQX(FM) Springtown TX
*KQXS(FM) Stephenville TX
*KBYU-FM Provo UT
*WWIP(FM) Cheriton VA
*WCNV(FM) Heathsville VA
*WVTF(FM) Roanoke VA
*KFAE-FM Richland WA
*WHAA(FM) Adams WI
*WBSD(FM) Burlington WI
*WSSW(FM) Platteville WI
*KLWC(FM) Casper WY
*KURT(FM) Jackson WY

89.3 mhz
*KATB(FM) Anchorage AK
*WALN(FM) Carrollton AL
*WLRH(FM) Huntsville AL
*WJIK(FM) Monroeville AL
*KAYH(FM) Fayetteville AR
*KKLT(FM) Texarkana AR
*KAIH(FM) Lake Havasu City AZ
*KNAQ(FM) Prescott AZ
*KPFB(FM) Berkeley CA
*KOHL(FM) Fremont CA
*KVPR(FM) Fresno CA
*KPJP(FM) Greenville CA
*KCRI(FM) Indio CA
*KNDZ(FM) McKinleyville CA
*KAKX(FM) Mendocino CA
*KLSI(FM) Moss Beach CA
*KQEI-FM North Highlands CA
*KPCC(FM) Pasadena CA
*KPDO(FM) Pescadero CA
*KMTG(FM) San Jose CA
*KLFF-FM San Luis Obispo CA
*KUVO(FM) Denver CO
*KLBV(FM) Steamboat Springs CO
*KTAW(FM) Westcliffe CO
*WRTC-FM Hartford CT
*WSGG(FM) Norfolk CT
*WPFW(FM) Washington DC
*WRMB(FM) Boynton Beach FL
*WKFA(FM) Saint Catherine FL
*WPIO(FM) Titusville FL
*WBJY(FM) Americus GA
*WRFG(FM) Atlanta GA
*WECC-FM Folkston GA
*KPRG(FM) Hagatna GU
*KIPO(FM) Honolulu HI
*KJMC(FM) Des Moines IA
*KUOI-FM Moscow ID
*WKKC(FM) Chicago IL
*WDLM-FM East Moline IL
*WNUR-FM Evanston IL
*WZRS(FM) Pana IL
*WIPA(FM) Pittsfield IL
*WGNJ(FM) Saint Joseph IL
*WJEL(FM) Indianapolis IN
*WYTJ(FM) Linton IN
*WIKV(FM) Plymouth IN
*WVXR(FM) Richmond IN
*WNKJ(FM) Hopkinsville KY
*WFPL(FM) Louisville KY
*WGCF(FM) Paducah KY
*WRKF(FM) Baton Rouge LA
*WAMH(FM) Amherst MA
*WUMD(FM) North Dartmouth MA
*WHSN(FM) Bangor ME
*WMSJ(FM) Freeport ME
*WTLI(FM) Bear Creek Township MI
*WHFR(FM) Dearborn MI
*WMSQ(FM) Marlette MI
*WBLD(FM) Orchard Lake MI
*WJKN-FM Spring Arbor MI
*WGNB(FM) Zeeland MI
*KCMP(FM) Northfield MN
*KOPJ(FM) Sebeka MN
*KRSW(FM) Worthington MN
*KTBJ(FM) Festus MO
*KCUR-FM Kansas City MO
*WAII(FM) Hattiesburg MS
*WATU(FM) Port Gibson MS
*KLMT(FM) Billings MT
*KLBZ(FM) Bozeman MT
*WXYC(FM) Chapel Hill NC
*WSOE(FM) Elon NC
*WTEB(FM) New Bern NC
*WZRI(FM) Spring Lake NC
*WBFJ-FM Winston-Salem NC
*KUND-FM Grand Forks ND
*KZUM(FM) Lincoln NE
*KXNE-FM Norfolk NE
*WNJB(FM) Bridgeton NJ
*WDDM(FM) Hazlet NJ
*KELP-FM Mesquite NM
*WSKG-FM Binghamton NY
*WGSU(FM) Geneseo NY
*WLJP(FM) Monroe NY
*WMHN(FM) Webster NY
*WVXC(FM) Chillicothe OH
*WCSB(FM) Cleveland OH
*WYSM(FM) Lima OH
*WYSZ(FM) Maumee OH
*WMKV(FM) Reading OH
*WKRW(FM) Wooster OH
*KALU(FM) Langston OK
*KCCU(FM) Lawton OK
*KSSO(FM) Norman OK
*KLRB(FM) Stuart OK
*KOGL(FM) Gleneden Beach OR
*KVRA(FM) Sisters OR
*KLOV(FM) Winchester OR
*WJCS(FM) Allentown PA
*WRTJ(FM) Coatesville PA
*WQED-FM Pittsburgh PA
*WRDV(FM) Warminster PA
*WJVP(FM) Culebra PR
*WSCI(FM) Charleston SC
*WRFE(FM) Chesterfield SC
*WLFJ-FM Greenville SC
*KBHE-FM Rapid City SD
*WMKW(FM) Crossville TN
*WAJJ(FM) McKenzie TN
*WYPL(FM) Memphis TN
*KPBD(FM) Big Spring TX
*KPBE(FM) Brownwood TX
*KNON(FM) Dallas TX
*KSBJ(FM) Humble TX
*KHCP(FM) Paris TX
*KNAR(FM) San Angelo TX
*KXBJ(FM) Victoria TX
*WVTU(FM) Charlottesville VA
*WJYA(FM) Emporia VA
*KASB(FM) Bellevue WA
*KUGS(FM) Bellingham WA
*KJCF(FM) Clarkston WA
*KAOS(FM) Olympia WA
*KVIX(FM) Port Angeles WA
*KBNO-FM White Salmon WA
*WPNE-FM Green Bay WI
*WWLA(FM) South Charleston WV

89.5 mhz
*KABN-FM Kasilof AK
*WBFR(FM) Birmingham AL
*WGTF(FM) Dothan AL
*WRNF(FM) Selma AL
*KCAC(FM) Camden AR
*KBMJ(FM) Heber Springs AR
*KJZA(FM) Drake AZ
*KBAQ-FM Phoenix AZ
*KBES(FM) Ceres CA
*KARQ(FM) East Sonora CA
*KSMC(FM) Moraga CA
*KVMR(FM) Nevada City CA
*KLFH(FM) Ojai CA
*KPBS-FM San Diego CA
*KPOO(FM) San Francisco CA
*KSBX(FM) Santa Barbara CA
*KAIB(FM) Shafter CA
*KPRA(FM) Ukiah CA
*KXRD(FM) Victorville CA
*KTCF(FM) Dolores CO
*KPRN(FM) Grand Junction CO
*KTSC-FM Pueblo CO
*WPKN(FM) Bridgeport CT
*WKSG(FM) Cedar Creek FL
*WGSG(FM) Mayo FL
*WFIT(FM) Melbourne FL
*WSRX(FM) Naples FL
*WPCS(FM) Pensacola FL
*WYFK(FM) Columbus GA
*WNGU(FM) Dahlonega GA
*WYFS(FM) Savannah GA
*WYFW(FM) Winder GA
*KHKE(FM) Cedar Falls IA
*KLCD(FM) Decorah IA
*KEGR(FM) Fort Dodge IA
*KCNJ(FM) Oskaloosa IA
*KTSY(FM) Caldwell ID
*KLRI(FM) Rigby ID
*WNIJ(FM) De Kalb IL
*WJMU(FM) Decatur IL
*WDRS(FM) Dorsey IL
*WEFI(FM) Effingham IL
*WIUW(FM) Warsaw IL
*WBSB(FM) Anderson IN
*WBEW(FM) Chesterton IN
*WFCI(FM) Franklin IN
*WWTS(FM) Logansport IN
*WBKE-FM North Manchester IN
*KHCD(FM) Salina KS
*WKPB(FM) Henderson KY
*KITA(FM) Bunkie LA
*KYFL(FM) Monroe LA
*WNCK(FM) Nantucket MA
*WSKB(FM) Westfield MA
*WSCL(FM) Salisbury MD
*WAHS(FM) Auburn Hills MI
*WCMU-FM Mount Pleasant MI
*WOVI(FM) Novi MI
*WOFR(FM) Schoolcraft MI
*KBHG(FM) Alexandria MN
*WJRF(FM) Duluth MN
*KBPG(FM) Montevideo MN
*KQAL(FM) Winona MN
*KNLH(FM) Cedar Hill MO
*KOPN(FM) Columbia MO
*KCFV(FM) Ferguson MO
*KOKS(FM) Poplar Bluff MO
*WMAE-FM Booneville MS
*WPRG(FM) Columbia MS
*WYAZ(FM) Yazoo City MS
*WTJY(FM) Asheboro NC
*WLPS-FM Lumberton NC
*KPPR(FM) Williston ND
*WKVP(FM) Cherry Hill NJ
*WSOU(FM) South Orange NJ
*KENW-FM Portales NM
*KVLK(FM) Socorro NM
*KCNV(FM) Las Vegas NV
*WCOF(FM) Arcade NY
*WSLU(FM) Canton NY
*WSLL(FM) Saranac Lake NY
*WUNY(FM) Utica NY
*WBCY(FM) Archbold OH
*WCVV(FM) Belpre OH
*WDPS(FM) Dayton OH
*WQRP(FM) Dayton OH
*WHSS(FM) Hamilton OH
*WFOT(FM) Lexington OH
*WVMS(FM) Sandusky OH
*WVXW(FM) West Union OH
*KTGS(FM) Ada OK
*KJCC(FM) Carnegie OK
*KWGS(FM) Tulsa OK
*KTEC(FM) Klamath Falls OR
*KEFS(FM) North Powder OR
*KPFR(FM) Pine Grove OR
*KTCB(FM) Tillamook OR
*WDNR(FM) Chester PA
*WAWN(FM) Franklin PA
*WITF-FM Harrisburg PA
*WNTE(FM) Mansfield PA
*WWPJ(FM) Pen Argyl PA
*KLND(FM) Little Eagle SD
*WETS(FM) Johnson City TN
*WMOT(FM) Murfreesboro TN
*KMFA(FM) Austin TX
*KZBJ(FM) Bay City TX
*KEPX(FM) Eagle Pass TX
*KKLY(FM) El Paso TX
*KBMM(FM) Odessa TX
*KLUX(FM) Robstown TX
*KTOT(FM) Spearman TX
*KVNE(FM) Tyler TX
*KYQX(FM) Weatherford TX
*KMOC(FM) Wichita Falls TX
*KAGJ(FM) Ephraim UT
*KUSR(FM) Logan UT
*KAER(FM) Saint George UT
*WHRV(FM) Norfolk VA
*WWED(FM) Spotsylvania VA
*WVPR(FM) Windsor VT
*KEWU-FM Cheney WA
*KJVH(FM) Longview WA
*KNHC(FM) Seattle WA
*KSOH(FM) Wapato WA
*WCLQ(FM) Wausau WI
*WAUA(FM) Petersburg WV
*KWRR(FM) Ethete WY

89.7 mhz
*KXKM(FM) McCarthy AK
*KBHN(FM) Booneville AR
*KUAP(FM) Pine Bluff AR
*KRMH(FM) Red Mesa AZ
*KNCA(FM) Burney CA
*KLRS(FM) Lodi CA
*KFJC(FM) Los Altos CA
*KLVM(FM) Prunedale CA
*KGBM(FM) Randsburg CA
*KSGN(FM) Riverside CA
*KHFR(FM) Santa Maria CA
*KARM(FM) Visalia CA
*KEPC(FM) Colorado Springs CO
*KXWA(FM) Loveland CO
*KTPS(FM) Pagosa Springs CO
*KRYI(FM) Rye CO
*WDJW(FM) Somers CT
*WMCU(FM) Miami FL
*WJLU(FM) New Smyrna Beach FL
*WVFS(FM) Tallahassee FL
*WUSF(FM) Tampa FL
*WMUM-FM Cochran GA
*WTXR(FM) Toccoa Falls GA
*WWBM(FM) Yates GA
*KIWR(FM) Council Bluffs IA
*KDUB(FM) Dubuque IA
*KRUI-FM Iowa City IA
*KRNL-FM Mount Vernon IA
*WCBW-FM East St. Louis IL
*WONU(FM) Kankakee IL
*WBMV(FM) Mount Vernon IL
*WLUJ(FM) Springfield IL
*WRGF(FM) Greenfield IN
*WHWE(FM) Howe IN
*WUBS(FM) South Bend IN
*WISU(FM) Terre Haute IN
*WTUR(FM) Upland IN
*KNBU(FM) Baldwin City KS
*KANH(FM) Emporia KS
*KBDA(FM) Great Bend KS
KHYS(FM) Hays KS
*WAAJ(FM) Benton KY
*WNKU(FM) Highland Heights KY
*WDCL-FM Somerset KY
*KAVK(FM) Many LA
*KBIO(FM) Natchitoches LA
*WGBH(FM) Boston MA
*WTMD(FM) Towson MD
*WMED(FM) Calais ME
*WMHB(FM) Waterville ME
*WTAC(FM) Burton MI
*WJOJ(FM) Harrisville MI
*WJCQ(FM) Jackson MI
*WLNZ(FM) Lansing MI
*WOCR(FM) Olivet MI
*KBSB(FM) Bemidji MN
*KCMF(FM) Fergus Falls MN
*WLSN(FM) Grand Marais MN
*KMSU(FM) Mankato MN
*KUMM(FM) Morris MN
*KPCS(FM) Princeton MN
*KYMC(FM) Ballwin MO
*KOZO(FM) Branson MO
*KJCV(FM) Country Club MO
*KKTR(FM) Kirksville MO
*KCVQ(FM) Knob Noster MO
*KNLP(FM) Potosi MO
*KMNR(FM) Rolla MO
*WPAE(FM) Centreville MS
*WZKM(FM) Waynesboro MS
*KBIL(FM) Park City MT
*KQLR(FM) Whitehall MT
*WCPE(FM) Raleigh NC
*WDVV(FM) Wilmington NC
*KNRI(FM) Bismarck ND
*WNJN-FM Atlantic City NJ
*WDVR(FM) Delaware Township NJ
*WRDR(FM) Freehold Township NJ
*WGLS-FM Glassboro NJ
*KUUT(FM) Farmington NM
*KMBN(FM) Las Cruces NM
*KTDB(FM) Ramah NM
*WALF(FM) Alfred NY
*WKVJ(FM) Dannemora NY
*WEOS(FM) Geneva NY
*WITR(FM) Henrietta NY
*WNJA(FM) Jamestown NY
*WFGB(FM) Kingston NY
*WRHO(FM) Oneonta NY
*WSSK(FM) Saratoga Springs NY
*WRUC(FM) Schenectady NY
*WOSU-FM Columbus OH
*WTKC(FM) Findlay OH
*WKSU-FM Kent OH
*KJTH(FM) Ponca City OK
*KWYA(FM) Astoria OR
*KLCC(FM) Eugene OR
*WQEJ(FM) Johnstown PA
*WVYA(FM) Williamsport PA
*WRTU(FM) San Juan PR
*WMHK(FM) Columbia SC
*KUSD(FM) Vermillion SD
*WDYN-FM Chattanooga TN
*WAWI(FM) Lawrenceburg TN
*WAUV(FM) Ripley TN
*KACU(FM) Abilene TX
*KACC(FM) Alvin TX
*KTXB(FM) Beaumont TX
*KBEX(FM) Brenham TX
*KHPS(FM) Camp Wood TX
*KJMA(FM) Floresville TX
*KVRK(FM) Sanger TX
*KEQX(FM) Stephenville TX
*WAUQ(FM) Charles City VA
*WVLS(FM) Monterey VA
*KWFJ(FM) Roy WA
*KWWS(FM) Walla Walla WA
*WUEC(FM) Eau Claire WI
*WUWM(FM) Milwaukee WI
*WHND(FM) Sister Bay WI
*WSHC(FM) Shepherdstown WV
*KAXG(FM) Gillette WY

89.9 mhz
*KAUG(FM) Anchorage AK
*KUAC(FM) Fairbanks AK
*WTBB(FM) Gadsden AL

*WTSU(FM) Montgomery-Troy AL
*WAKD(FM) Sheffield AL
*KVMN(FM) Cave City AR
*KVIR(FM) Bullhead City AZ
*KZAI(FM) Coolidge AZ
*KJTA(FM) Flagstaff AZ
*KNDL(FM) Angwin CA
*KCRH(FM) Hayward CA
*KWDS(FM) Kettleman City CA
*KJCU(FM) Laytonville CA
*KEFR(FM) Le Grand CA
*KPPN(FM) Pollock Pines CA
*KFER(FM) Santa Cruz CA
*KCRW(FM) Santa Monica CA
*KFRS(FM) Soledad CA
*KTMH(FM) Colona CO
*KFRY(FM) Pueblo CO
*KTAD(FM) Sterling CO
*KPRE(FM) Vail CO
*WQTQ(FM) Hartford CT
*WAPJ(FM) Torrington CT
*WWEB(FM) Wallingford CT
*WJCT-FM Jacksonville FL
*WUCF-FM Orlando FL
*WCNO(FM) Palm City FL
*WJTF(FM) Panama City FL
*WTFH(FM) Helen GA
*WTHM(FM) Thomson GA
*KAYP(FM) Burlington IA
*KZNJ(FM) Marion IA
*KWER(FM) Waverly IA
*KBSK(FM) McCall ID
*KWJT(FM) Rathdrum ID
*KAWZ(FM) Twin Falls ID
*WLCA(FM) Godfrey IL
*WLKL(FM) Mattoon IL
*WCBU(FM) Peoria IL
*WOJC(FM) Crothersville IN
*WHLP(FM) Hanna IN
*WBRO(FM) Marengo IN
*WATI(FM) Vincennes IN
*WYBV(FM) Wakarusa IN
*WHPL(FM) West Lafayette IN
*KAIG(FM) Dodge City KS
*KRPS(FM) Pittsburg KS
*WRVG(FM) Georgetown KY
*WSOF-FM Madisonville KY
*KLXA-FM Alexandria LA
*WWNO(FM) New Orleans LA
*KSJY(FM) Saint Martinville LA
*KDAQ(FM) Shreveport LA
*WSCB(FM) Springfield MA
*WOEL-FM Elkton MD
*WMTB-FM Emmittsburg MD
*WERU-FM Blue Hill ME
*WAYO-FM Benton Harbor MI
*WAYG(FM) Grand Rapids MI
*WTHS(FM) Holland MI
*WKDS(FM) Kalamazoo MI
*WLJN-FM Traverse City MI
*WHWG(FM) Trout Lake MI
*KMOJ(FM) Minneapolis MN
*KRPR(FM) Rochester MN
KQRB(FM) Windom MN
*KGNA-FM Arnold MO
*KFFW(FM) Cabool MO
*KMCV(FM) High Point MO
*KAUF(FM) Kennett MO
*KGNV(FM) Washington MO
*WMAB-FM Mississippi State MS
*KGPR(FM) Great Falls MT
*KUKL(FM) Kalispell MT
*KBGA(FM) Missoula MT
*WDAV(FM) Davidson NC
*WRVS-FM Elizabeth City NC
*KDVI(FM) Devils Lake ND
*KDPR(FM) Dickinson ND
*KJTW(FM) Jamestown ND
*KFLV(FM) Wilber NE
*WNJM(FM) Manahawkin NJ
*WJPH(FM) Woodbine NJ
*KYCM(FM) Alamogordo NM
*KUNM(FM) Albuquerque NM
*WFBF(FM) Buffalo NY
*WKCR-FM New York NY
*WXLG(FM) North Creek NY
*WSUF(FM) Noyack NY
*WRVO(FM) Oswego NY
*WDPG(FM) Greenville OH
*WLHS(FM) West Chester OH
*KWKL(FM) Grandfield OK
*KBPS-FM Portland OR
*KKJA(FM) Redmond OR
*WVIA-FM Scranton PA
*WTLR(FM) State College PA
*WJWJ-FM Beaufort SC
*KQFR(FM) Rapid City SD
*WDVX(FM) Clinton TN
*WEVL(FM) Memphis TN
*WAYW(FM) New Johnsonville TN
*KACV-FM Amarillo TX
*KLGS(FM) College Station TX
*KTLZ(FM) Cuero TX
*KDLI(FM) Del Rio TX
*KBNL(FM) Laredo TX
*KTSW(FM) San Marcos TX
*KBDE(FM) Temple TX
*WPER(FM) Culpeper VA
*WFFC(FM) Ferrum VA
*WMRL(FM) Lexington VA
*WVRU(FM) Radford VA
*WNRS-FM Sweet Briar VA
*WCMD-FM Barre VT
*KGRG-FM Auburn WA
*KGHP(FM) Gig Harbor WA
*KPLW(FM) Wenatchee WA
*WHSA(FM) Brule WI
*WVFL(FM) Fond du Lac WI
*WORT(FM) Madison WI
*WWSP(FM) Stevens Point WI
*WVRN(FM) Wittenberg WI
*WVWV(FM) Huntington WV
*WVNP(FM) Wheeling WV
*KVLZ(FM) Sheridan WY
*KRWT(FM) West Laramie WY

90.1 mhz

*KRAW(FM) Sterling AK
*WOCG(FM) Huntsville AL
*WDLG(FM) Thomasville AL
KXRL(FM) Cherry Valley AR
*KBNV(FM) Fayetteville AR
*KLRO(FM) Hot Springs AR
*KRMB(FM) Bisbee AZ
*KWFH(FM) Parker AZ
*KTQX(FM) Bakersfield CA
*KBPK(FM) Buena Park CA
*KZFR(FM) Chico CA
*KCBX(FM) San Luis Obispo CA
*KZSU(FM) Stanford CA
*KYCC(FM) Stockton CA
*KSAK(FM) Walnut CA
*KLRD(FM) Yucaipa CA
*KVOD(FM) Denver CO
*KHCO(FM) Hayden CO
*KUTE(FM) Ignacio CO
*WRXC(FM) Shelton CT
*WGSK(FM) South Kent CT
*WECS(FM) Willimantic CT
*WCSP-FM Washington DC
*WTJT(FM) Baker FL
*WGCU-FM Fort Myers FL
*WJUF(FM) Inverness FL
*WKWR(FM) Key West FL
*WABE(FM) Atlanta GA
*WXVS(FM) Waycross GA
*WOI-FM Ames IA
*KNWO(FM) Cottonwood ID
*WTSG(FM) Carlinville IL
*WEFT(FM) Champaign IL
*WMBI-FM Chicago IL
*WAWJ(FM) Marion IL
*WFYI-FM Indianapolis IN
*KXJH(FM) Linton IN
*WRFM(FM) Wadesville IN
*KHCC-FM Hutchinson KS
*WHMR(FM) Ledbetter KY
*WJSO(FM) Pikeville KY
*WJCR-FM Upton KY
*WYCM(FM) Charlton MA
*WRYP(FM) Wellfleet MA
*WCAI(FM) Woods Hole MA
*WMEA(FM) Portland ME
*WUCX-FM Bay City MI
*WXPZ(FM) Clyde Township MI
*WCWB(FM) Coldwater MI
*WNMU-FM Marquette MI
*WLSO(FM) Sault Ste. Marie MI
*KNSE(FM) Austin MN
*KSJR-FM Collegeville MN
*KADU(FM) Hibbing MN
*KSRQ(FM) Thief River Falls MN
*KKFI(FM) Kansas City MO
*KBKC(FM) Moberly MO
*KRHS(FM) Overland MO
*KSCV(FM) Springfield MO
*WMPR(FM) Jackson MS
*KBLW(FM) Billings MT
*KNMC(FM) Havre MT
*KHLV(FM) Helena MT
*WZPE(FM) Bath NC
*WCCE(FM) Buie's Creek NC
*WNAA(FM) Greensboro NC
*WJKA(FM) Jacksonville NC
*WZRU(FM) Roanoke Rapids NC
*KFKX(FM) Hastings NE
*KRDR(FM) Red River NM
*KRLU(FM) Roswell NM
*KQMC(FM) Hawthorne NV
*WIFF(FM) Binghamton NY
*WGMC(FM) Greece NY
*WRCU-FM Hamilton NY
*WMHQ(FM) Malone NY
*WXHD(FM) Mount Hope NY
*WUSB(FM) Stony Brook NY
*WKWV(FM) Watertown NY
*WOHC(FM) Chillicothe OH
*WORI(FM) Delhi Hills OH
*WXML(FM) Upper Sandusky OH
*WOUZ(FM) Zanesville OH
*KOCU(FM) Altus OK
*KCSC(FM) Edmond OK
*KSOR(FM) Ashland OR
*KQHR(FM) Hood River OR
*KKLP(FM) La Pine OR
*KAJC(FM) Salem OR
*WIUP-FM Indiana PA
*WPSX(FM) Kane PA
*WVMN(FM) New Castle PA
*WRTI(FM) Philadelphia PA
*WCIT(FM) Trout Run PA
*WJDZ(FM) Pastillo PR
*WHMC-FM Conway SC
*WEPR(FM) Greenville SC
*KILI(FM) Porcupine SD
*KSFS(FM) Sioux Falls SD
*WKNP(FM) Jackson TN
*WKTS(FM) Kingston TN
*WZYZ(FM) Spencer TN
*KERA(FM) Dallas TX
*KPFT(FM) Houston TX
*KTXI(FM) Ingram TX
*KAMY(FM) Lubbock TX
*KPBJ(FM) Midland TX
*KSAU(FM) Nacogdoches TX
*KUTX(FM) San Angelo TX
*KSYM-FM San Antonio TX
*KZMU(FM) Moab UT
*KUER(FM) Salt Lake City UT
*WMVE(FM) Chase City VA
*WJCN(FM) Nassawadox VA
*WDCE(FM) Richmond VA
*WPVA(FM) Waynesboro VA
*WIVH(FM) Christiansted VI
*WRUV(FM) Burlington VT
*KMWS(FM) Mount Vernon WA
*KPLI(FM) Olympia WA
*KQWS(FM) Omak WA
*KOLU(FM) Pasco WA
*KNWP(FM) Port Angeles WA
*KUPS(FM) Tacoma WA
*WORQ(FM) Green Bay WI
*WRPN-FM Ripon WI
*WPWV(FM) Princeton WV
*KUWL(FM) Laramie WY
*KUWP(FM) Powell WY

90.3 mhz

*KNBA(FM) Anchorage AK
*WBHM(FM) Birmingham AL
*WDYF(FM) Dothan AL
*KLRM(FM) Melbourne AR
*KLFS(FM) Van Buren AR
*KNAG(FM) Grand Canyon AZ
*KBMH(FM) Holbrook AZ
*KFLR-FM Phoenix AZ
*KMRO(FM) Camarillo CA
*KBXO(FM) Coachella CA
*KDVS(FM) Davis CA
*KFNO(FM) Fresno CA
*KLAI(FM) Laytonville CA
*KAZU(FM) Pacific Grove CA
*KUSF(FM) San Francisco CA
*KBUT(FM) Crested Butte CO
*KTDX(FM) Frisco CO
*KLFV(FM) Grand Junction CO
*WWPT(FM) Westport CT
*WJLH(FM) Flagler Beach FL
*WAFG(FM) Fort Lauderdale FL
*WYJC(FM) Greenville FL
*WLVF-FM Haines City FL
*WEJF(FM) Palm Bay FL
*WAEF(FM) Cordele GA
*WHCJ(FM) Savannah GA
*KCIF(FM) Hilo HI
*KTUH(FM) Honolulu HI
*KWIT(FM) Sioux City IA
*KBSU-FM Boise ID
*KZJB(FM) Pocatello ID
*WUSI(FM) Olney IL
*WQUB(FM) Quincy IL
*WVIK(FM) Rock Island IL
*WKJD(FM) Columbus IN
*WFOF(FM) Covington IN
*WBCL(FM) Fort Wayne IN
*KBUZ(FM) Topeka KS
*WMKY(FM) Morehead KY
*WKWC(FM) Owensboro KY
*WBRH(FM) Baton Rouge LA
*KYLC(FM) Lake Charles LA
*KEDM(FM) Monroe LA
*WCCT-FM Harwich MA
*WZBC(FM) Newton MA
*WAIJ(FM) Grantsville MD
*WDIH(FM) Salisbury MD
*WBLV(FM) Twin Lake MI
*KFAI(FM) Minneapolis MN
*KCCD(FM) Moorhead MN
*KMKL(FM) North Branch MN
*KWUR(FM) Clayton MO
*KGNN-FM Cuba MO
*KNLG(FM) New Bloomfield MO
*KGSP(FM) Parkville MO
*KLUH(FM) Poplar Bluff MO
*KCRL(FM) Sunrise Beach MO
*WMAH-FM Biloxi MS
*WMAV-FM Oxford MS
*KJFT(FM) Arlee MT
*KMZO(FM) Hamilton MT
*WFHE(FM) Hickory NC
*WKNS(FM) Kinston NC
*WBFY(FM) Pinehurst NC
*KMNE-FM Bassett NE
*KRNU(FM) Lincoln NE
*WNJZ(FM) Cape May Court House NJ
*WRPR(FM) Mahwah NJ
*WVPH(FM) Piscataway NJ
*WKNJ-FM Union Township NJ
*WMSC(FM) Upper Montclair NJ
*KELU(FM) Clovis NM
*KLGQ(FM) Grants NM
*WAMC-FM Albany NY
*WKRB(FM) Brooklyn NY
*WCIH(FM) Elmira NY
*WHPC(FM) Garden City NY
*WJSL(FM) Houghton NY
*WHCR-FM New York NY
*WDFH(FM) Ossining NY
*WAIH(FM) Potsdam NY
*WRVD(FM) Syracuse NY
*WCDR-FM Cedarville OH
*WCPN(FM) Cleveland OH
*WOTL(FM) Toledo OH
*KLCU(FM) Ardmore OK
*KVRS(FM) Lawton OK
*KTVR-FM La Grande OR
*KSLC(FM) McMinnville OR
*KLON(FM) Rockaway Beach OR
*KWBX(FM) Salem OR
*KZRI(FM) Welches OR
*WESS(FM) East Stroudsburg PA
*WJTL(FM) Lancaster PA
*WARC(FM) Meadville PA
*WXLV(FM) Schnecksville PA
*WRIU(FM) Kingston RI
*WSSB-FM Orangeburg SC
*WRBK(FM) Richburg SC
*KASD(FM) Rapid City SD
*WCSK(FM) Kingsport TN
*WUTK-FM Knoxville TN
*WUTM(FM) Martin TN
*WPLN-FM Nashville TN
*KBUB(FM) Brownwood TX
*KEDT-FM Corpus Christi TX
*KBJS(FM) Jacksonville TX
*KPHS(FM) Plains TX
*KSGU(FM) Saint George UT
*WOKG(FM) Galax VA
*WHRO-FM Norfolk VA
*WRXT(FM) Roanoke VA
*KWYQ(FM) Longview WA
*KEXP-FM Seattle WA
*KWRS(FM) Spokane WA
*KNWY(FM) Yakima WA
*WBCR-FM Beloit WI
*WHLA(FM) La Crosse WI
*WJWD(FM) Marshall WI
*WRST-FM Oshkosh WI
*WHBM-FM Park Falls WI
*WVPG(FM) Parkersburg WV
*KCSP-FM Casper WY
*KUWJ(FM) Jackson WY
*KWYC(FM) Orchard Valley WY

90.5 mhz

*KXGA(FM) Glennallen AK
*KWMD(FM) Kasilof AK
*KAOG(FM) Jonesboro AR
*KLRE-FM Little Rock AR
*KNLL(FM) Nashville AR
*KUAT-FM Tucson AZ
*KHSU-FM Arcata CA
*KIBC(FM) Burney CA
*KVHS(FM) Concord CA
*KADV(FM) Modesto CA
*KWMR(FM) Point Reyes Station CA
*KSJS(FM) San Jose CA
*KGDP-FM Santa Maria CA
*KKTO(FM) Tahoe City CA
*KVOV(FM) Carbondale CO
*KTLF(FM) Colorado Springs CO
*KCSU-FM Fort Collins CO
*WPKT(FM) Meriden CT
*WVUM(FM) Coral Gables FL
*WREH(FM) Cypress Quarters FL
*WYFB(FM) Gainesville FL
*WANM(FM) Tallahassee FL
*WBVM(FM) Tampa FL
*WUOG(FM) Athens GA
*WPWB(FM) Byron GA
*WFRC(FM) Columbus GA
*WTLD(FM) Jesup GA
*WVRI(FM) Pavo GA
*KPHL(FM) Pahala HI
*KHOE(FM) Fairfield IA
*KOSK(FM) Oskaloosa IA
*KAIO(FM) Idaho Falls ID
*WRTE(FM) Chicago IL
*WAPO(FM) Mount Vernon IL
*WMTH(FM) Park Ridge IL
*WNIU(FM) Rockford IL
*WSCT(FM) Springfield IL
*WIKL(FM) Greencastle IN
*WWDS(FM) Muncie IN
*WPUM(FM) Rensselaer IN
*KBMP(FM) Enterprise KS
*KZNA(FM) Hill City KS
*KRBW(FM) Ottawa KS

*WVHM(FM) Benton KY
*WUOL(FM) Louisville KY
*WTHL(FM) Somerset KY
*KTLN(FM) Thibodaux LA
*WSMA(FM) Scituate MA
*WICN(FM) Worcester MA
*WCRH(FM) Williamsport MD
*WKHS(FM) Worton MD
*WMEP(FM) Camden ME
*WKAR-FM East Lansing MI
*WPHN(FM) Gaylord MI
*KDNI(FM) Duluth MN
*KGAC(FM) Saint Peter MN
*KWWC-FM Columbia MO
*KXCV(FM) Maryville MO
*KSMS-FM Point Lookout MO
*WCSO(FM) Columbus MS
*WQVI(FM) Forest MS
*WAQL(FM) McComb MS
*KJLF(FM) Butte MT
*WASU-FM Boone NC
*WBUX(FM) Buxton NC
*WWCU(FM) Cullowhee NC
*WYQS(FM) Mars Hill NC
*WZRN(FM) Norlina NC
*WDCC(FM) Sanford NC
*WWIL-FM Wilmington NC
*WAJC(FM) Wilson NC
*WSNC(FM) Winston-Salem NC
*KCND(FM) Bismarck ND
*WSPS(FM) Concord NH
*WPEA(FM) Exeter NH
*WVFA(FM) Lebanon NH
*WWFP(FM) Brigantine NJ
*WXGN(FM) Egg Harbor Township NJ
*WCVH(FM) Flemington NJ
*WBJB-FM Lincroft NJ
*WVBV(FM) Medford Lakes NJ
*WJSV(FM) Morristown NJ
*KCIE(FM) Dulce NM
*KSOS(FM) Las Vegas NV
*WBXL(FM) Baldwinsville NY
*WHRW(FM) Binghamton NY
*WSUC-FM Cortland NY
*WXXE(FM) Fenner NY
*WJFF(FM) Jeffersonville NY
*WBER(FM) Rochester NY
*WHVT(FM) Clyde OH
*WCBE(FM) Columbus OH
*WVML(FM) Millersburg OH
*KNYD(FM) Broken Arrow OK
*KSSX(FM) Chickasha OK
*KAYM(FM) Weatherford OK
*KORM(FM) Astoria OR
*KVLB(FM) Bend OR
*KLCO(FM) Newport OR
*WERG(FM) Erie PA
*WVBU-FM Lewisburg PA
*WDUQ(FM) Pittsburgh PA
*WIDA-FM Carolina PR
*WUSC-FM Columbia SC
*WDLL(FM) Dillon SC
*KVCF(FM) Freeman SD
*WSMC-FM Collegedale TN
*WUMC(FM) Elizabethton TN
*WXKV(FM) Selmer TN
*KAGT(FM) Abilene TX
*KUT(FM) Austin TX
*KZFT(FM) Fannett TX
*KTXG(FM) Greenville TX
*KSHU(FM) Huntsville TX
*KFLB-FM Odessa TX
*KBAH(FM) Plainview TX
*KPDR(FM) Wheeler TX
*KZKL(FM) Wichita Falls TX
*KZCL(FM) Logan UT
*WJYJ(FM) Fredericksburg VA
*WPIM(FM) Martinsville VA
*WISE-FM Wise VA
*WFTF(FM) Rutland VT
*WCKJ(FM) Saint Johnsbury VT
*KACS(FM) Chehalis WA
*KNWV(FM) Clarkston WA
*KWCW(FM) Walla Walla WA
*WVCF(FM) Eau Claire WI
*WSUP(FM) Platteville WI
*WPFF(FM) Sturgeon Bay WI
*WMLJ(FM) Summersville WV
*KBUW(FM) Buffalo WY
*KUWN(FM) Newcastle WY
*KUWZ(FM) Rock Springs WY

90.7 mhz

*WGRW(FM) Anniston AL
*WVAS(FM) Montgomery AL
*WVUA-FM Tuscaloosa AL
*KEAF(FM) Fort Smith AR
*KVRN(FM) Marvell AR
*KNAA(FM) Show Low AZ
*KALX(FM) Berkeley CA
*KFRP(FM) Coalinga CA
*KFSR(FM) Fresno CA
*KHRI(FM) Hollister CA
*KPFK(FM) Los Angeles CA
*KZYX(FM) Philo CA
*KTHM(FM) Red Bluff CA
*KSRI(FM) Santa Cruz CA
*KYKL(FM) Tracy CA
*KGUD(FM) Longmont CO
*KTEI(FM) Placerville CO
*KDRE(FM) Sterling CO
*KTDL(FM) Trinidad CO
*WMFE-FM Orlando FL
*WKGC-FM Panama City FL
*WXEL(FM) West Palm Beach FL
*WWXC(FM) Albany GA
*WACG-FM Augusta GA
*WAYR-FM Brunswick GA
*WUWG(FM) Carrollton GA
*WMVV(FM) Griffin GA
*WFSL(FM) Thomasville GA
*KKUA(FM) Wailuku HI
*KWOI(FM) Carroll IA
*KOJI(FM) Okoboji IA
*KBSQ(FM) McCall ID
*KCIR(FM) Twin Falls ID
*WVKC(FM) Galesburg IL
*WBEQ(FM) Morris IL
*WAZU(FM) Peoria IL
*WPSR(FM) Evansville IN
*WKPW(FM) Knightstown IN
*WQSG(FM) Lafayette IN
*WMHD-FM Terre Haute IN
*KPOR(FM) Emporia KS
*KJHK(FM) Lawrence KS
*KYWA(FM) Wichita KS
*WCVK(FM) Bowling Green KY
*WPTJ(FM) Paris KY
*KLSA(FM) Alexandria LA
*WWOZ(FM) New Orleans LA
*WTCC(FM) Springfield MA
*WKKL(FM) West Barnstable MA
*WSDL(FM) Ocean City MD
*WAUS(FM) Berrien Springs MI
*WNFR(FM) Sandusky MI
*WNMC-FM Traverse City MI
*KBPR(FM) Brainerd MN
*WTIP(FM) Grand Marais MN
*KZSE(FM) Rochester MN
*KOBC(FM) Joplin MO
*KHGN(FM) Kirksville MO
*KTTK(FM) Lebanon MO
*KWMU(FM) Saint Louis MO
*WATP(FM) Laurel MS
*KGFA(FM) Great Falls MT
*KYPR(FM) Miles City MT
*WFAE(FM) Charlotte NC
*WNCU(FM) Durham NC
*WOTJ(FM) Morehead City NC
*WZRL(FM) Wade NC
*KABU(FM) Fort Totten ND
*KFJM(FM) Grand Forks ND
*KNFA(FM) Grand Island NE
*KVNO(FM) Omaha NE
*WEVN(FM) Keene NH
*WLMW(FM) Manchester NH
*WYRS(FM) Manahawkin NJ
*KQRI(FM) Belen NM
*KKCJ(FM) Cannon AFB NM
*KRWG(FM) Las Cruces NM
*KSFR(FM) Santa Fe NM
*WETD(FM) Alfred NY
*WGCC-FM Batavia NY
*WFUV(FM) New York NY
*WPGL(FM) Pattersonville NY
*WPNR-FM Utica NY
*WGLE(FM) Lima OH
*WVMC-FM Mansfield OH
*WMCO(FM) New Concord OH
*WNRK(FM) Norwalk OH
*WKTL(FM) Struthers OH
*KFXT(FM) Sulphur OK
*KAYE-FM Tonkawa OK
*KJOV(FM) Woodward OK
*KANL(FM) Baker City OR
*KMWR(FM) Brookings OR
*KBOO(FM) Portland OR
*WRTL(FM) Ephrata PA
*WVMM(FM) Grantham PA
*WPKV(FM) Nanty Glo PA
*WKPS(FM) State College PA
*WCLH(FM) Wilkes-Barre PA
*WCRG(FM) Williamsport PA
*WJHD(FM) Portsmouth RI
*WYFH(FM) North Charleston SC
*KSDJ(FM) Brookings SD
*WZKV(FM) Dyersburg TN
*WAUO(FM) Hohenwald TN
*KAVW(FM) Amarillo TX
*KTAA(FM) Big Sandy TX
*KMLU(FM) Brownfield TX
*KPBN(FM) Freer TX
*KTER(FM) Rudolph TX
*KCPC(FM) Sealy TX
*KVRT(FM) Victoria TX
*WUVT-FM Blacksburg VA
*WAZP(FM) Cape Charles VA
*WEHC(FM) Emory VA
*WMRA(FM) Harrisonburg VA
*WJSC-FM Johnson VT
*WVTC(FM) Randolph Center VT
*KNWR(FM) Ellensburg WA
*KSER(FM) Everett WA
*KZUU(FM) Pullman WA
*WHAD(FM) Delafield WI
*WVSS(FM) Menomonie WI
*WFGH(FM) Fort Gay WV
*WLKV(FM) Ripley WV

90.9 mhz

*WELJ(FM) Brewton AL
*WJAB(FM) Huntsville AL
*KBSA(FM) El Dorado AR
*KLLN(FM) Newark AR
*KWRB(FM) Bisbee AZ
*KGCB(FM) Prescott AZ
*KHDC(FM) Chualar CA
*KPSH(FM) Coachella CA
*KWTM(FM) June Lake CA
*KXJZ(FM) Sacramento CA
*KGZO(FM) Shafter CA
*KBDG(FM) Turlock CA
*KASF(FM) Alamosa CO
*KTOL(FM) Leadville CO
*KVNF(FM) Paonia CO
*KXWY(FM) Rye CO
*WCNI(FM) New London CT
*WETA(FM) Washington DC
*WAQV(FM) Crystal River FL
*WKTZ-FM Jacksonville FL
*WGES-FM Key Largo FL
*WJIR(FM) Key West FL
*WSOR(FM) Naples FL
*WJWV(FM) Fort Gaines GA
*WOAK(FM) La Grange GA
*WRAF-FM Toccoa Falls GA
*WVVS(FM) Valdosta GA
*KKCR(FM) Hanalei HI
*KUNI(FM) Cedar Falls IA
*KLOX(FM) Creston IA
*KMDY(FM) Keokuk IA
*WDCB(FM) Glen Ellyn IL
*WILL-FM Urbana IL
*WBDG(FM) Indianapolis IN
*WBSW(FM) Marion IN
*WCJL(FM) Morgantown IN
*KHCT(FM) Great Bend KS
*WKUE(FM) Elizabethtown KY
*WEKH(FM) Hazard KY
*KSLU(FM) Hammond LA
*KIKL(FM) Lafayette LA
*WBUR-FM Boston MA
*WMEH(FM) Bangor ME
*WMPG(FM) Gorham ME
*WQAC-FM Alma MI
*WSLI(FM) Belding MI
*WTCK(FM) Charlevoix MI
*WRCJ-FM Detroit MI
*WTRK(FM) Freeland MI
*WMSD(FM) Rose Township MI
*WCFG(FM) Springfield MI
*WIRR(FM) Virginia-Hibbing MN
*KKLW(FM) Willmar MN
*KRCU(FM) Cape Girardeau MO
*KNLN(FM) Vienna MO
*KTBG(FM) Warrensburg MO
*KSMW(FM) West Plains MO
*WMAO-FM Greenwood MS
*WAQB(FM) Tupelo MS
*KQLU(FM) Belgrade MT
*KLRV(FM) Billings MT
*KDWG(FM) Dillon MT
*KSPL(FM) Kalispell MT
*WQFS(FM) Greensboro NC
*WURI(FM) Manteo NC
*WRQM(FM) Rocky Mount NC
*WSIF(FM) Wilkesboro NC
*KNHA(FM) Grand Isle NE
*KPNO(FM) Norfolk NE
*WSCS(FM) New London NH
*KSJE(FM) Farmington NM
*KLIT(FM) Hobbs NM
*KSHI(FM) Zuni NM
*WCDB(FM) Albany NY
*WLJH(FM) Glens Falls NY
*WSQG-FM Ithaca NY
*WCOT(FM) Jamestown NY
*WAMK(FM) Kingston NY
*WSLO(FM) Malone NY
*WONY(FM) Oneonta NY
*WAVX(FM) Schuyler Falls NY
*WJNY(FM) Watertown NY
*WGBE(FM) Bryan OH
*WGUC(FM) Cincinnati OH
*WCVJ(FM) Jefferson OH
*WFCO(FM) Lancaster OH
*WNZR(FM) Mount Vernon OH
*WCWS(FM) Wooster OH
*KKVO(FM) Altus OK
*KOKF(FM) Edmond OK
*KXRT(FM) Idabel OK
*KJCH(FM) Coos Bay OR
*KIDH(FM) Jordan Valley OR
*KSKF(FM) Klamath Falls OR
*KRBM(FM) Pendleton OR
*KCPB-FM Warrenton OR
*WITX(FM) Beaver Falls PA
*WZZH(FM) Honesdale PA
*WJRC(FM) Lewistown PA
*WHYY-FM Philadelphia PA
*WLGI(FM) Hemingway SC
*KDSD-FM Pierpont SD
*KWRC(FM) Rapid City SD
*KCSD(FM) Sioux Falls SD
*KOAR(FM) Spearfish SD
*WWOG(FM) Cookeville TN
WPRH(FM) Paris TN
*KAMU-FM College Station TX
*KCBI(FM) Dallas TX
*KRLH(FM) Hereford TX
*KTSU(FM) Houston TX
*KKLU(FM) Lubbock TX
*KSWP(FM) Lufkin TX
*KAQQ(FM) Midland TX
*KAVO(FM) Pampa TX
*KLTP(FM) San Angelo TX
*KYFS(FM) San Antonio TX
*KRCL(FM) Salt Lake City UT
*WWMC(FM) Lynchburg VA
*WCWM(FM) Williamsburg VA
*KVTI(FM) Tacoma WA
*WHRM(FM) Wausau WI
*WPIB(FM) Bluefield WV
*WVPM(FM) Morgantown WV
*KLWV(FM) Chugwater WY
*KUWG(FM) Gillette WY
*KUWX(FM) Pinedale WY

91.1 mhz

*KSKA(FM) Anchorage AK
*WEGL(FM) Auburn AL
*WJSR(FM) Birmingham AL
*WVSU-FM Birmingham AL
*WAQU(FM) Selma AL
*WAXU(FM) Troy AL
*KMTC(FM) Russellville AR
*KANX(FM) Sheridan AR
*KNLB(FM) Lake Havasu City AZ
*KNOG(FM) Nogales AZ
*KFRJ(FM) China Lake CA
*KLVY(FM) Fairmead CA
*KMUD(FM) Garberville CA
*KCSM(FM) San Mateo CA
*KRCB-FM Santa Rosa CA
*KDSC(FM) Thousand Oaks CA
*KWSB-FM Gunnison CO
*KLDV(FM) Morrison CO
*WSHU(FM) Fairfield CT
*WBVC(FM) Pomfret CT
*WJED(FM) Dogwood Lakes Estate FL
*WJFP(FM) Fort Pierce FL
*WPSM(FM) Fort Walton Beach FL
*WKES(FM) Lakeland FL
*WUJC(FM) Saint Marks FL
*WREK(FM) Atlanta GA
*WSVH(FM) Savannah GA
*WABR-FM Tifton GA
*KANO(FM) Hilo HI
*KWNJ(FM) Bettendorf IA
*KTPR(FM) Fort Dodge IA
*KUNZ(FM) Ottumwa IA
*KISU-FM Pocatello ID
*KBSS(FM) Sun Valley ID
*WDBX(FM) Carbondale IL
*WIBI(FM) Carlinville IL
*WKCC(FM) Kankakee IL
*WGSL(FM) Loves Park IL
WYGS(FM) Columbus IN
*WGCS(FM) Goshen IN
*WBSH(FM) Hagerstown IN
*WEDM(FM) Indianapolis IN
*WCYT(FM) Lafayette Township IN
*WIRE(FM) Lebanon IN
*WVUB(FM) Vincennes IN
*KANZ(FM) Garden City KS
*KCFN(FM) Wichita KS
*WKAO(FM) Ashland KY
*KLSU(FM) Baton Rouge LA
*KOJO(FM) Lake Charles LA
*KXUL(FM) Monroe LA
*WNKV(FM) Norco LA
*WMUA(FM) Amherst MA
*WNAN(FM) Nantucket MA
*WJJW(FM) North Adams MA
*WTKL(FM) North Dartmouth MA
*WKMY(FM) Winchendon MA
*WHFC(FM) Bel Air MD
*WBOR(FM) Brunswick ME
XETRA-FM Tijuana MEX
*WOLW(FM) Cadillac MI
*WFUM-FM Flint MI
*WGGL-FM Houghton MI
*WPCJ(FM) Pittsford MI
*KXLC(FM) La Crescent MN
*KNOW-FM Minneapolis-St. Paul MN
*KCCM-FM Moorhead MN
*KBGM(FM) Park Hills MO
*KSMU(FM) Springfield MO
*WASM(FM) Natchez MS
*WMSV(FM) Starkville MS
*KLEU(FM) Lewistown MT
*KMZL(FM) Missoula MT
*WZGO(FM) Aurora NC
*WYBH(FM) Fayetteville NC
*WPGT(FM) Roanoke Rapids NC
*WRSH(FM) Rockingham NC
*KTNE-FM Alliance NE
*KDCV-FM Blair NE
*KUCV(FM) Lincoln NE
*WVNH(FM) Concord NH

*WWNJ(FM) Dover Township NJ
*WFMU(FM) East Orange NJ
*KAQF(FM) Clovis NM
*KEDP(FM) Las Vegas NM
*KVKL(FM) Las Vegas NV
*KAIZ(FM) Mesquite NV
*WSQE(FM) Corning NY
*WHVP(FM) Hudson NY
*WOSS(FM) Ossining NY
*WTSC-FM Potsdam NY
*WSPN(FM) Saratoga Springs NY
*WRMU(FM) Alliance OH
*WRUW-FM Cleveland OH
*WOSE(FM) Coshocton OH
*WDUB(FM) Granville OH
*WOSB(FM) Marion OH
*KQPD(FM) Ardmore OK
*KAYC(FM) Durant OK
*KKRD(FM) Enid OK
*KJRF(FM) Lawton OK
*KWAX(FM) Eugene OR
*KAPK(FM) Grants Pass OR
*KTMK(FM) Tillamook OR
*WUFR(FM) Bedford PA
*WBUQ(FM) Bloomsburg PA
*WZBT(FM) Gettysburg PA
*WSAJ-FM Grove City PA
*WRTY(FM) Jackson Township PA
*WMSS(FM) Middletown PA
*WYFG(FM) Gaffney SC
*KTSD-FM Reliance SD
*KAOR(FM) Vermillion SD
*WTSE(FM) Benton TN
*WKCS(FM) Knoxville TN
*WKNO-FM Memphis TN
*WRVU(FM) Nashville TN
*KWTS(FM) Canyon TX
*KQXE(FM) Eastland TX
*KVER(FM) El Paso TX
*KHKV(FM) Kerrville TX
*KTAI(FM) Kingsville TX
*KYBJ(FM) Lake Jackson TX
*KBWC(FM) Marshall TX
*KGWP(FM) Pittsburg TX
*KSGR(FM) Portland TX
*KSUU(FM) Cedar City UT
*WTJU(FM) Charlottesville VA
*WOKD-FM Danville VA
*WHCE(FM) Highland Springs VA
*WNSB(FM) Norfolk VA
*WRMC-FM Middlebury VT
*WGDR(FM) Plainfield VT
*KTJC(FM) Kelso WA
*KPBX-FM Spokane WA
*KYPL(FM) Yakima WA
*WOVM(FM) Appleton WI
*WGTD(FM) Kenosha WI

91.3 mhz

*WVOB(FM) Dothan AL
*WFIX(FM) Florence AL
*WHIL-FM Mobile AL
*WTBJ(FM) Oxford AL
*KUCA(FM) Conway AR
*KUAF(FM) Fayetteville AR
*KGHR(FM) Tuba City AZ
*KXCI(FM) Tucson AZ
*KFRB(FM) Bakersfield CA
*KWTH(FM) Barstow CA
*KIDE(FM) Hoopa CA
*KDRH(FM) King City CA
*KCPR(FM) San Luis Obispo CA
*KSVY(FM) Sonoma CA
*KUOP(FM) Stockton CA
*KNYR(FM) Yreka CA
*KMSA(FM) Grand Junction CO
*KLRY(FM) Gypsum CO
*KSUT(FM) Ignacio CO
*KTPF(FM) Salida CO
*KLZV(FM) Sterling CO
*KTWX(FM) Walsenburg CO
*WWUH(FM) West Hartford CT
*WVUD(FM) Newark DE
*WYFZ(FM) Belleview FL
*WAKJ(FM) De Funiak Springs FL
*WSEB(FM) Englewood FL
*WOLR(FM) Lake City FL
*WLRN-FM Miami FL
*WHIF(FM) Palatka FL
*WCGN(FM) Calhoun GA
*WATY(FM) Folkston GA
*WJTG(FM) Fort Valley GA
*KDFR(FM) Des Moines IA
*WNIW(FM) La Salle IL
*WIUM(FM) Macomb IL
*WJCZ(FM) Milford IL
*WSLE(FM) Salem IL
*WFHB(FM) Bloomington IN
*WHJE(FM) Carmel IN
*WNDY(FM) Crawfordsville IN
*WJCO(FM) Montpelier IN
*WWHI(FM) Muncie IN
*WTMW(FM) North Judson IN
*WCKZ(FM) Orland IN
*KAXR(FM) Arkansas City KS
*KRLE(FM) Oberlin KS
*KANV(FM) Olsburg KS
*WUKY(FM) Lexington KY
*WKMS-FM Murray KY
*WHFG(FM) Broussard LA
*KSCL(FM) Shreveport LA
*WSHL-FM Easton MA
*WXPL(FM) Fitchburg MA
*WDJM-FM Framingham MA
*WCUW(FM) Worcester MA
*WESM(FM) Princess Anne MD
*WMEW(FM) Waterville ME
*WCHW-FM Bay City MI
*WZHN(FM) East Tawas MI
*WJOG(FM) Good Hart MI
*WCSG(FM) Grand Rapids MI
*WOES(FM) Ovid-Elsie MI
*WSGR-FM Port Huron MI
*KRSU(FM) Appleton MN
*KMSK(FM) Austin MN
*KNBJ(FM) Bemidji MN
*KBIA(FM) Columbia MO
*KBIY(FM) Van Buren MO
*WMPN-FM Jackson MS
*KAPC(FM) Butte MT
*WLFA(FM) Asheville NC
*WFQS(FM) Franklin NC
*WZMB(FM) Greenville NC
*WHQR(FM) Wilmington NC
*WXRI(FM) Winston-Salem NC
*KMHA(FM) Four Bears ND
*KAYA(FM) Hubbard NE
*KLPR(FM) Kearney NE
*WUNH(FM) Durham NH
*WEVH(FM) Hanover NH
*WKNH(FM) Keene NH
*WRTQ(FM) Ocean City NJ
*WTSR(FM) Trenton NJ
*KYCV(FM) Lovington NM
*KYCT(FM) Ruidoso NM
*KNIS(FM) Carson City NV
*KBSJ(FM) Jackpot NV
*WXLH(FM) Blue Mountain Lake NY
*WBNY(FM) Buffalo NY
*WOLN(FM) Olean NY
*WVKR-FM Poughkeepsie NY
*WRLI-FM Southampton NY
*WCNY-FM Syracuse NY
*WAPS(FM) Akron OH
*WOUB-FM Athens OH
*WGTE-FM Toledo OH
*WYSO(FM) Yellow Springs OH
*KAKO(FM) Ada OK
*KRSC-FM Claremore OK
*KOAB-FM Bend OR
*WLVR(FM) Bethlehem PA
*WCIG(FM) Carbondale PA
*WQLN-FM Erie PA
*WLCH(FM) Lancaster PA
*WGRC(FM) Lewisburg PA
*WYEP-FM Pittsburgh PA
*WXAC(FM) Reading PA
*WIPR-FM San Juan PR
*WDOM(FM) Providence RI
*WLTR(FM) Columbia SC
*WLMU(FM) Harrogate TN
*WCPI(FM) McMinnville TN
*WUTS(FM) Sewanee TN
*KAQD(FM) Abilene TX
*KVLU(FM) Beaumont TX
*KVFM(FM) Beeville TX
*KYJC(FM) Commerce TX
*KDKR(FM) Decatur TX
*KNCT-FM Killeen TX
*KBKN(FM) Lamesa TX
*KZLV(FM) Lytle TX
*KOCV(FM) Odessa TX
*KPVU(FM) Prairie View TX
*KGLY(FM) Tyler TX
*WMLU(FM) Farmville VA
*WVST-FM Petersburg VA
*WPAR(FM) Salem VA
*WTRM(FM) Winchester VA
*WIUV(FM) Castleton VT
*KBCS(FM) Bellevue WA
*KCED(FM) Centralia WA
*KGTS(FM) College Place WA
*WHEM(FM) Eau Claire WI
*WMVM(FM) Goodman WI
*WHHI(FM) Highland WI
*WSTM(FM) Kiel WI
*KUWS(FM) Superior WI
*WQAB(FM) Philippi WV
*KUWA(FM) Afton WY
*KUWC(FM) Casper WY
*KSUW(FM) Sheridan WY
KUWT(FM) Thermopolis WY

91.5 mhz

*KSUA(FM) Fairbanks AK
*KWJG(FM) Kasilof AK
*WSTF(FM) Andalusia AL
*WSGN(FM) Gadsden AL
*WUAL-FM Tuscaloosa AL
*WEBT(FM) Valley AL
*KAIA(FM) Blytheville AR
*KALR(FM) Hot Springs AR
*KCMH(FM) Mountain Home AR
*KJZZ(FM) Phoenix AZ
*KNHM(FM) Bayside CA
*KKUP(FM) Cupertino CA
*KASK(FM) Fairfield CA
*KSJV(FM) Fresno CA
*KRQZ(FM) Lompoc CA
*KUSC(FM) Los Angeles CA
*KKRO(FM) Redding CA
*KYDS(FM) Sacramento CA
*KZYZ(FM) Willits CA
*KAJX(FM) Aspen CO
*KRCC(FM) Colorado Springs CO
*KSJD(FM) Cortez CO
*KUNC(FM) Greeley CO
*WGRS(FM) Guilford CT
*WPSF(FM) Clewiston FL
*WMIE(FM) Cocoa FL
*WGTT(FM) Emeralda FL
*WJYO(FM) Fort Myers FL
*WAPN(FM) Holly Hill FL
*WHWY(FM) Marathon FL
*WLPJ(FM) New Port Richey FL
*WJFH(FM) Sebring FL
*WFSQ(FM) Tallahassee FL
*WPRK(FM) Winter Park FL
*WWEV-FM Cumming GA
*WGPH(FM) Vidalia GA
*KUNY(FM) Mason City IA
*KBSX(FM) Boise ID
*KBYR-FM Rexburg ID
*WLHW(FM) Casey IL
*WBEZ(FM) Chicago IL
*WCIC(FM) Pekin IL
*WNIQ(FM) Sterling IL
*WFWR(FM) Attica IN
*WJCY(FM) Cicero IN
*WJHS(FM) Columbia City IN
*WUEV(FM) Evansville IN
*WGRE(FM) Greencastle IN
*WRFT(FM) Indianapolis IN
*WWDL(FM) Lebanon IN
*WECI(FM) Richmond IN
*WJLR(FM) Seymour IN
*KANU(FM) Lawrence KS
*KSNS(FM) Medicine Lodge KS
*WVCT(FM) Keavy KY
*WBFI(FM) McDaniels KY
*KBAN(FM) De Ridder LA
*KPAE(FM) Erwinville LA
*KGRM(FM) Grambling LA
*WTUL(FM) New Orleans LA
*KSUL(FM) Port Sulphur LA
*KNSU(FM) Thibodaux LA
*WBIM-FM Bridgewater MA
*WUML(FM) Lowell MA
*WMFO(FM) Medford MA
*WMLN-FM Milton MA
*WNMH(FM) Northfield MA
*WSDH(FM) Sandwich MA
*WMHC(FM) South Hadley MA
*WZLY(FM) Wellesley MA
*WBJC(FM) Baltimore MD
*WRBC(FM) Lewiston ME
*WSJB-FM Standish ME
*WVCM(FM) Iron Mountain MI
*WUPX(FM) Marquette MI
*WMHW-FM Mount Pleasant MI
*WJOH(FM) Raco MI
*WICA(FM) Traverse City MI
*KNWF(FM) Fergus Falls MN
*KCFB(FM) Saint Cloud MN
*KNGA(FM) Saint Peter MN
*KQMN(FM) Thief River Falls MN
*KSIV-FM Saint Louis MO
*WLRK(FM) Greenville MS
*WAVI(FM) Oxford MS
*KAFH(FM) Great Falls MT
*KPLG(FM) Plains MT
*WBJD(FM) Atlantic Beach NC
*WUNC(FM) Chapel Hill NC
*KPRJ(FM) Jamestown ND
*KRNE-FM Merriman NE
*KIOS-FM Omaha NE
*WDBK(FM) Blackwood NJ
*KFLQ(FM) Albuquerque NM
*KRUX(FM) Las Cruces NM
*KNCC(FM) Elko NV
*KUNV(FM) Las Vegas NV
*WSQX-FM Binghamton NY
*WVHC(FM) Herkimer NY
*WNYE(FM) New York NY
*WXXI-FM Rochester NY
*WRPI(FM) Troy NY
*WKHR(FM) Bainbridge OH
*WHKC(FM) Columbus OH
*WBIE(FM) Delphos OH
*WKRJ(FM) New Philadelphia OH
*WOBC-FM Oberlin OH
*WOSP(FM) Portsmouth OH
*KSYE(FM) Frederick OK
*KVAZ(FM) Henryetta OK
*KWVZ(FM) Florence OR
*KOPB-FM Portland OR
*KSRS(FM) Roseburg OR
*WCIM(FM) Shenandoah PA
*WPSU(FM) State College PA
*WSRN-FM Swarthmore PA
*WCVY(FM) Coventry RI
*WRJI(FM) East Greenwich RI
*WKCL(FM) Ladson SC
*WHCB(FM) Bristol TN
*WNRZ(FM) Dickson TN
*WFHU(FM) Henderson TN
*WFMQ(FM) Lebanon TN
*WAWL-FM Red Bank TN
*WTML(FM) Tullahoma TN
*KBCX(FM) Big Spring TX
*KHVT(FM) Bloomington TX
*KTXP(FM) Bushland TX
*KYFB(FM) Denison TX
*KHCF(FM) Fredericksburg TX
*KANJ(FM) Giddings TX
*KHML(FM) Madisonville TX
*KCAS(FM) McCook TX
*KWLD(FM) Plainview TX
*KTXK(FM) Texarkana TX
*KUSU-FM Logan UT
*WARN(FM) Culpeper VA
*WPIN-FM Dublin VA
*WLUR(FM) Lexington VA
*WYCS(FM) Yorktown VA
*WGLY-FM Bolton VT
*WWLR(FM) Lyndonville VT
*KLWS(FM) Moses Lake WA
*KUBS(FM) Newport WA
*WEMY(FM) Green Bay WI
*WWJA(FM) Janesville WI
*WPJW(FM) Hurricane WV
*WRSG(FM) Middlebourne WV
*WGLZ(FM) West Liberty WV
*KUWD(FM) Sundance WY

91.7 mhz

*WYFD(FM) Decatur AL
*WPIL(FM) Heflin AL
*WAQG(FM) Ozark AL
*KBDO(FM) Des Arc AR
*KRMC(FM) Douglas AZ
*KPUB(FM) Flagstaff AZ
*KNAD(FM) Page AZ
*KCHO(FM) Chico CA
*KXSR(FM) Groveland CA
*KHCS(FM) Palm Desert CA
*KBDH(FM) San Ardo CA
*KALW(FM) San Francisco CA
*KFHL(FM) Wasco CA
*KTSG(FM) Sidney CO
*KOTO(FM) Telluride CO
*WXCI(FM) Danbury CT
*WHUS(FM) Storrs CT
*WRTX(FM) Dover DE
*WMPH(FM) Wilmington DE
*WJBC-FM Fernandina Beach FL
*WJLF(FM) Gainesville FL
*WAPB(FM) Madison FL
*WMKO(FM) Marco FL
*WEGS(FM) Milton FL
*WFFL(FM) Panama City FL
*WVIJ(FM) Port Charlotte FL
*WFTI-FM Saint Petersburg FL
*WWFR(FM) Stuart FL
*WUNV(FM) Albany GA
*WUGA(FM) Athens GA
*WLPE(FM) Augusta GA
*WCCV(FM) Cartersville GA
*WTJB(FM) Columbus GA
*WMVW(FM) Peachtree City GA
*WWET(FM) Valdosta GA
*KSUI(FM) Iowa City IA
*KBSM(FM) McCall ID
*KRFA-FM Moscow ID
*KSQS(FM) Ririe ID
*KBSW(FM) Twin Falls ID
*WBJW(FM) Albion IL
*WBGL(FM) Champaign IL
*WVNL(FM) Vandalia IL
*WJPR(FM) Jasper IN
*WIWC(FM) Kokomo IN
*WEEM-FM Pendleton IN
*WBSJ(FM) Portland IN
*WETL(FM) South Bend IN
*KZAN(FM) Hays KS
*KCVS(FM) Salina KS
*WWHR(FM) Bowling Green KY
*WAPD(FM) Campbellsville KY
*WJVK(FM) Owensboro KY
*WWJD(FM) Pippa Passes KY
*KAPM(FM) Alexandria LA
*KLSP(FM) Angola LA
*KMSL(FM) Mansfield LA
*KNWD(FM) Natchitoches LA
*WGAJ(FM) Deerfield MA
*WJWT(FM) Gardner MA
*WAVM(FM) Maynard MA
*WNEF(FM) Newburyport MA
*WMWM(FM) Salem MA
*WBSL-FM Sheffield MA
*WZXH(FM) Hagerstown MD
*WSHD(FM) Eastport ME
*WCML-FM Alpena MI
*WUOM(FM) Ann Arbor MI
*WMCQ(FM) Muskegon MI
*KAXE(FM) Grand Rapids MN
*KLSE-FM Rochester MN
*WMCN(FM) Saint Paul MN
*KNSW(FM) Worthington-Marshall MN
*KCVO-FM Camdenton MO

*KJIR(FM) Hannibal MO
*KMVC(FM) Marshall MO
*KNEO(FM) Neosho MO
*KCOZ(FM) Point Lookout MO
*KCVX(FM) Salem MO
*WSQH(FM) Forest MS
*WAOY(FM) Gulfport MS
*WVSD(FM) Itta Bena MS
*WJTA(FM) Kosciusko MS
*WPRL(FM) Lorman MS
*WAJS(FM) Tupelo MS
*KEMC(FM) Billings MT
*KUHM(FM) Helena MT
*WBKU(FM) Ahoskie NC
*WSGE(FM) Dallas NC
*KBFR(FM) Bismarck ND
*KNHS(FM) Hastings NE
*KPNE-FM North Platte NE
*WNEC-FM Henniker NH
*WPCR-FM Plymouth NH
*WLFR(FM) Pomona NJ
*KUPR(FM) Alamogordo NM
*KZPI(FM) Deming NM
*KTGW(FM) Fruitland NM
*KGLP(FM) Gallup NM
*KOBH(FM) Hobbs NM
*KELT(FM) Roswell NM
*KVLP(FM) Tucumcari NM
*KLNR(FM) Panaca NV
*KTPH(FM) Tonopah NV
*WICB(FM) Ithaca NY
*WFRH(FM) Kingston NY
*WOSR(FM) Middletown NY
*WSQC-FM Oneonta NY
*WRVJ(FM) Watertown NY
*WVXU(FM) Cincinnati OH
*WOSV(FM) Mansfield OH
*WYTN(FM) Youngstown OH
*WJIC(FM) Zanesville OH
*KPSU(FM) Goodwell OK
*KARG(FM) Poteau OK
*KOSU(FM) Stillwater OK
*KEOL(FM) La Grande OR
*KDOV(FM) Medford OR
*WMUH(FM) Allentown PA
*WLBS(FM) Bristol PA
*WCUC-FM Clarion PA
*WIXQ(FM) Millersville PA
*WKDU(FM) Philadelphia PA
*WVMW-FM Scranton PA
*WJAZ(FM) Summerdale PA
*WBMR(FM) Telford PA
*WNJR(FM) Washington PA
*WCUR(FM) West Chester PA
*WRLC(FM) Williamsport PA
*WYTL(FM) Wyomissing PA
*WNNV(FM) San German PR
*WLPG(FM) Florence SC
*WTBI-FM Greenville SC
*WHRS(FM) Cookeville TN
*WUMR(FM) Memphis TN
*WFCM-FM Murfreesboro TN
*KVRX(FM) Austin TX
*KHPU(FM) Brownwood TX
*KBNJ(FM) Corpus Christi TX
*KTDA(FM) Dalhart TX
*KVTT(FM) Dallas TX
*KOOP(FM) Hornsby TX
*KTRU(FM) Houston TX
*KRTU(FM) San Antonio TX
*KOHS(FM) Orem UT
*KRDC-FM Saint George UT
*KUFR(FM) Salt Lake City UT
*WEMC(FM) Harrisonburg VA
*WWEM(FM) Rustburg VA
*WGLV(FM) Woodstock VT
*KZAZ(FM) Bellingham WA
*KBLD(FM) Kennewick WA
*KSVR(FM) Mount Vernon WA
*KXOT(FM) Tacoma WA
*WDKV(FM) Fond du Lac WI
*WSUM(FM) Madison WI
*WMSE(FM) Milwaukee WI
*WXPR(FM) Rhinelander WI
*WSHS(FM) Sheboygan WI
*WSUW(FM) Whitewater WI
*WVPB(FM) Beckley WV
*WWVU-FM Morgantown WV
*KDUW(FM) Douglas WY

91.9 mhz

*KBRW-FM Barrow AK
*KDLL(FM) Kenai AK
*KUHB-FM Saint Paul AK
*KUDU(FM) Tok AK
*WGIB(FM) Birmingham AL
*WMBV(FM) Dixons Mills AL
*WLJS-FM Jacksonville AL
*WJIF(FM) Opp AL
*KHED(FM) Arkadelphia AR
*KBPB(FM) Harrison AR
*KASU(FM) Jonesboro AR
*KXRJ(FM) Russellville AR
*KVJC(FM) Globe AZ
*KCAI(FM) Kingman AZ
*KOHN(FM) Sells AZ
*KYRM(FM) Yuma AZ
*KHSR(FM) Crescent City CA
*KHKL(FM) Laytonville CA
*KSPB(FM) Pebble Beach CA
*KWTD(FM) Ridgecrest CA
*KVCR(FM) San Bernardino CA
*KCSB-FM Santa Barbara CA
*KLVR(FM) Santa Rosa CA
*KCSS(FM) Turlock CA
*KDUR(FM) Durango CO
*KLXV(FM) Glenwood Springs CO
*KCFP(FM) Pueblo CO
*WSLX(FM) New Canaan CT
*WHGN(FM) Crystal River FL
*WMKL(FM) Key Largo FL
*WYFO(FM) Lakeland FL
*WKVH(FM) Monticello FL
*WAYL(FM) Saint Augustine FL
*WSCF-FM Vero Beach FL
*WCLK(FM) Atlanta GA
*WEBH(FM) Cuthbert GA
*WLFS(FM) Port Wentworth GA
*WVGS(FM) Statesboro GA
*WASW(FM) Waycross GA
*KSDA-FM Agat GU
*KAQA(FM) Kilauea HI
*KTDV(FM) State Center IA
*KWRV(FM) Sun Valley ID
*WSIU(FM) Carbondale IL
*WJCH(FM) Joliet IL
*WUIS(FM) Springfield IL
*WQKO(FM) Howe IN
*WVSH(FM) Huntington IN
*WJEF(FM) Lafayette IN
*WHOJ(FM) Terre Haute IN
*WITT(FM) Zionsville IN
*KTCC(FM) Colby KS
*KONQ(FM) Dodge City KS
*KNGM(FM) Emporia KS
*KWBI(FM) Great Bend KS
*KARF(FM) Independence KS
*KSDB-FM Manhattan KS
*KBDD(FM) Winfield KS
*WFPK(FM) Louisville KY
*KAXV(FM) Bastrop LA
*KMRL(FM) Buras LA
*KCKR(FM) Kaplan LA
*WUMB-FM Boston MA
*WFPB-FM Falmouth MA
*WOZQ(FM) Northampton MA
*WAIC(FM) Springfield MA
*WCFM(FM) Williamstown MA
*WBPR(FM) Worcester MA
*WFWM(FM) Frostburg MD
*WGTS(FM) Takoma Park MD
*WYFP(FM) Harpswell ME
*WMEB-FM Orono ME
*WDPW(FM) Greenville MI
*WMTU-FM Houghton MI
*WORW(FM) Port Huron MI
*WTNP(FM) Richland MI
*KXBR(FM) International Falls MN
*KBHZ(FM) Willmar MN
*KNLQ(FM) Cuba MO
*KWJC(FM) Liberty MO
*KNLM(FM) Marshfield MO
*KSRD(FM) Saint Joseph MO
*WOWL(FM) Burnsville MS
*WAUM(FM) Duck Hill MS
*KGLT(FM) Bozeman MT
*KFRW(FM) Great Falls MT
*KUFN(FM) Hamilton MT
*WFSS(FM) Fayetteville NC
*WAAE(FM) New Bern NC
*WRCM(FM) Wingate NC
*KDSU(FM) Fargo ND
*KCNE-FM Chadron NE
*KTLX(FM) Columbus NE
*KDNE(FM) Crete NE
*KWSC(FM) Wayne NE
*WBGD(FM) Brick Township NJ
*WNTI(FM) Hackettstown NJ
*KNLK(FM) Santa Rosa NM
*KXFR(FM) Socorro NM
*WNGN(FM) Argyle NY
*WCEB(FM) Corning NY
*WSHR(FM) Lake Ronkonkoma NY
*WCEL(FM) Plattsburgh NY
*WRVN(FM) Utica NY
*WLKP(FM) Belpre OH
*WOUH-FM Chillicothe OH
*WGDE(FM) Defiance OH
*WKCO(FM) Gambier OH
*WMEJ(FM) Proctorville OH
*KSSU(FM) Durant OK
*KBCW-FM McAlester OK
*KMUN(FM) Astoria OR
*KRVM-FM Eugene OR
*KWSO(FM) Warm Springs OR
*WCAL(FM) California PA
*WVME(FM) Meadville PA
*KQSD-FM Lowry SD
*WAPX-FM Clarksville TN
*WUOT(FM) Knoxville TN
*KXRI(FM) Amarillo TX
*KALD(FM) Caldwell TX
*KPFC(FM) Callisburg TX
*KLLR(FM) Dripping Springs TX
*KAZF(FM) Hebronville TX
*KHCJ(FM) Jefferson TX
*KAVX(FM) Lufkin TX
*KMEO(FM) Mertzon TX
*KHPO(FM) Port O'Connor TX
*KPCW(FM) Park City UT
*WNRN(FM) Charlottesville VA
*WVTR(FM) Marion VA
*WCMK(FM) Putney VT
*KSFC(FM) Spokane WA
*KDNA(FM) Yakima WA
*WEMI(FM) Appleton WI
*WHDI(FM) Sister Bay WI
*WLBL-FM Wausau WI
*WXPW(FM) Wausau WI
*WBHZ(FM) Elkins WV
*WPHP(FM) Wheeling WV
*KLWD(FM) Gillette WY
*KUWR(FM) Laramie WY

92.1 mhz

KBBO-FM Houston AK
WJJN(FM) Columbia AL
WKUL(FM) Cullman AL
WZEW(FM) Fairhope AL
WERH-FM Hamilton AL
KHPQ(FM) Clinton AR
KDQN-FM De Queen AR
KKEG(FM) Fayetteville AR
KSBS-FM Pago Pago AS
KFMA(FM) Green Valley AZ
KZUA(FM) Holbrook AZ
KSGC(FM) Tusayan AZ
KPSL-FM Bakersfield CA
KOND(FM) Clovis CA
KSOQ-FM Escondido CA
KQCM(FM) Joshua Tree CA
KCCL(FM) Placerville CA
KKDV(FM) Walnut Creek CA
KJMN(FM) Castle Rock CO
KTHN(FM) La Junta CO
WLBW(FM) Fenwick Island DE
WFFY(FM) Destin FL
WAFZ-FM Immokalee FL
WJXR(FM) Macclenny FL
WNFK(FM) Perry FL
WLTQ-FM Venice FL
WRLX(FM) West Palm Beach FL
WDDQ(FM) Adel GA
WBTR-FM Carrollton GA
WSGC-FM Elberton GA
WJGA-FM Jackson GA
WPEH-FM Louisville GA
WHHR(FM) Vienna GA
KHWA(FM) Holualoa HI
KCHE-FM Cherokee IA
KUEL(FM) Fort Dodge IA
KRLS(FM) Knoxville IA
*KIBX(FM) Bonners Ferry ID
KPPC(FM) Pocatello ID
WQKQ(FM) Carthage IL
WWGO(FM) Charleston IL
WFPS(FM) Freeport IL
*WBST(FM) Muncie IN
WROI(FM) Rochester IN
WZDM(FM) Vincennes IN
KREP(FM) Belleville KS
KMZA(FM) Seneca KS
WBVX(FM) Carlisle KY
WKEN(FM) Fredonia KY
WTKY-FM Tompkinsville KY
KSYR(FM) Benton LA
KTSR(FM) De Quincy LA
KLIL(FM) Moreauville LA
KVCL-FM Winnfield LA
*WOMR(FM) Provincetown MA
*WUPI(FM) Presque Isle ME
WPHX-FM Sanford ME
WOUF(FM) Beulah MI
WIDL(FM) Caro MI
WHPD(FM) Dowagiac MI
WGHN-FM Grand Haven MI
WTWS(FM) Harrison MI
WCSR-FM Hillsdale MI
WQTX(FM) Saint Johns MI
WMIS-FM Blackduck MN
WWAX(FM) Hermantown MN
WYRQ(FM) Little Falls MN
KLQP(FM) Madison MN
KRUE(FM) Waseca MN
KKOZ-FM Ava MO
KMOE(FM) Butler MO
KMFC(FM) Centralia MO
*KCVZ(FM) Dixon MO
KSDL(FM) Sedalia MO
WBKN(FM) Brookhaven MS
WKXY(FM) Clarksdale MS
WJMG(FM) Hattiesburg MS
WMSU(FM) Starkville MS
WUMS(FM) University MS
WJNS-FM Yazoo City MS
WMNC-FM Morganton NC
WCDG(FM) Moyock NC
WRSV(FM) Rocky Mount NC
KZRX(FM) Dickinson ND
KHZZ(FM) Sargent NE
WFEX(FM) Peterborough NH
WVLT(FM) Vineland NJ
KATK-FM Carlsbad NM
KJZS(FM) Sparks NV
WSEN-FM Baldwinsville NY
WCKR(FM) Hornell NY
WVTK(FM) Port Henry NY
WRNQ(FM) Poughkeepsie NY
WLNG(FM) Sag Harbor NY
WDLA-FM Walton NY
WOHF(FM) Bellevue OH
WYVK(FM) Middleport OH
WBIK(FM) Pleasant City OH
WWSR(FM) Wapakoneta OH
WROU-FM West Carrollton OH
KTBT(FM) Broken Arrow OK
KFXI(FM) Marlow OK
KMZE(FM) Woodward OK
KWVR-FM Enterprise OR
*KMKR(FM) Oakridge OR
KVAN-FM Pilot Rock OR
*KSYD(FM) Reedsport OR
WKPL(FM) Ellwood City PA
WJHT(FM) Johnstown PA
WSNU(FM) Lock Haven PA
WPPT(FM) Mercersburg PA
WQFM(FM) Nanticoke PA
WWKL(FM) Palmyra PA
*WPTS-FM Pittsburgh PA
WZET(FM) Hormigueros PR
WZOL(FM) Luquillo PR
WBHC-FM Hampton SC
WWNU(FM) Irmo SC
WMYB(FM) Myrtle Beach SC
WQQK(FM) Hendersonville TN
WEUZ(FM) Minor Hill TN
KOPY-FM Alice TX
KCZO(FM) Carrizo Springs TX
KXEZ(FM) Farmersville TX
KTFW-FM Glen Rose TX
*KYLR(FM) Hutto TX
KTNR(FM) Kenedy TX
KNBT(FM) New Braunfels TX
KROI(FM) Seabrook TX
KHOS-FM Sonora TX
KDOK(FM) Tyler TX
KTCE(FM) Payson UT
WDIC-FM Clinchco VA
*WWHS-FM Hampden-Sydney VA
WCDX(FM) Mechanicsville VA
WMOO(FM) Derby Center VT
KCRK-FM Colville WA
WLTU(FM) Manitowoc WI
WRJC-FM Mauston WI
WMEQ-FM Menomonie WI
WEZY(FM) Racine WI
WXXM(FM) Sun Prairie WI
*WVWC(FM) Buckhannon WV
KFRZ(FM) Green River WY

92.3 mhz

WLWI-FM Montgomery AL
KIPR(FM) Pine Bluff AR
KWCD(FM) Bisbee AZ
KTAR-FM Glendale AZ
KRED-FM Eureka CA
KHHT(FM) Los Angeles CA
KSJO(FM) San Jose CA
KHJQ(FM) Susanville CA
KJYE(FM) Grand Junction CO
KSTH(FM) Holyoke CO
KVRH-FM Salida CO
WCMQ-FM Hialeah FL
WWKA(FM) Orlando FL
WMOQ(FM) Bostwick GA
WAEG(FM) Evans GA
WSGA(FM) Hinesville GA
WLZN(FM) Macon GA
WQLI(FM) Meigs GA
KSSK-FM Waipahu HI
KKHQ-FM Oelwein IA
KIZN(FM) Boise ID
KRVQ(FM) Victor ID
WZPW(FM) Peoria IL
WTTS(FM) Bloomington IN
WFWI(FM) Fort Wayne IN
WPWX(FM) Hammond IN
KFTI-FM Newton KS
KCCV-FM Olathe KS
WYGE(FM) London KY
WZAQ(FM) Louisa KY
WDVW(FM) La Place LA
KMYY(FM) Rayville LA
WERQ-FM Baltimore MD
WWHC(FM) Oakland MD
WMME-FM Augusta ME
WZUU(FM) Allegan MI
WMXD(FM) Detroit MI
WJPD(FM) Ishpeming MI
KXRA-FM Alexandria MN
WIL-FM Saint Louis MO
KSAR(FM) Thayer MO
KTTN-FM Trenton MO
WOHT(FM) Grenada MS
KYUS-FM Miles City MT
KQRK(FM) Ronan MT
WKRR(FM) Asheboro NC
WQSL(FM) Jacksonville NC
WZPR(FM) Nags Head NC
KCVG(FM) Medina ND

KEZO-FM Omaha NE
WGXL(FM) Hanover NH
KRST(FM) Albuquerque NM
KOMP(FM) Las Vegas NV
KSVL(FM) Smith NV
WXRK(FM) New York NY
WFLY(FM) Troy NY
WKRI(FM) Cleveland Heights OH
WCOL-FM Columbus OH
KREU(FM) Roland OK
KGON(FM) Portland OR
*WKVR-FM Huntingdon PA
WNBQ(FM) Mansfield PA
WLGL(FM) Riverside PA
WRRN(FM) Warren PA
WPRO-FM Providence RI
KQRQ(FM) Rapid City SD
WDEF-FM Chattanooga TN
WYNU(FM) Milan TN
KOFX(FM) El Paso TX
KIJN-FM Farwell TX
KOYE(FM) Frankston TX
KRNH(FM) Kerrville TX
KIIZ-FM Killeen TX
KETX-FM Livingston TX
KCUL-FM Marshall TX
KNFM(FM) Midland TX
KNRG(FM) New Ulm TX
KZNO(FM) Seymour TX
KQVT(FM) Victoria TX
WTYD(FM) Deltaville VA
WXLK(FM) Roanoke VA
KULE-FM Ephrata WA
WJMQ(FM) Clintonville WI
WRLS-FM Hayward WI
WOSQ(FM) Spencer WI
WXCR(FM) New Martinsville WV

92.5 mhz
WXJC-FM Cordova AL
WVNN-FM Trinity AL
KWYN-FM Wynne AR
KTHQ(FM) Eagar AZ
KMYX-FM Arvin CA
KBRE(FM) Atwater CA
KSRW(FM) Independence CA
KKAL(FM) Paso Robles CA
KGBY(FM) Sacramento CA
KWLI(FM) Broomfield CO
KCRT-FM Trinidad CO
WWYZ(FM) Waterbury CT
WNDT(FM) Alachua FL
WUSV(FM) Estero FL
WPAP-FM Panama City FL
WYUU(FM) Safety Harbor FL
WFJO(FM) Folkston GA
WKZZ(FM) Tifton GA
WEKS(FM) Zebulon GA
KLHI-FM Kahului HI
KJJY(FM) West Des Moines IA
WDEK(FM) De Kalb IL
WKXQ(FM) Rushville IL
WCFF(FM) Urbana IL
WZWZ(FM) Kokomo IN
KQMA-FM Phillipsburg KS
KCVT(FM) Silver Lake KS
WBKR(FM) Owensboro KY
*KHCL(FM) Arcadia LA
KVPI-FM Ville Platte LA
WXRV(FM) Haverhill MA
WXMD(FM) Pocomoke City MD
XHRM-FM Tijuana MEX
WJSZ(FM) Ashley MI
WFDX(FM) Atlanta MI
WBGV(FM) Marlette MI
WLAW(FM) Newaygo MI
*WIRN(FM) Buhl MN
KQRS-FM Golden Valley MN
KXKK(FM) Park Rapids MN
KKWQ(FM) Warroad MN
*KSMR(FM) Winona MN
KSYN(FM) Joplin MO
KELE-FM Mountain Grove MO
KPPL(FM) Poplar Bluff MO
KAYX(FM) Richmond MO
WESE(FM) Baldwyn MS
WQST-FM Forest MS
WQYZ(FM) Ocean Springs MS
KAAR(FM) Butte MT
KPQX(FM) Havre MT
KWMY(FM) Park City MT
WYFL(FM) Henderson NC
WKGB-FM Conklin NY
WBEE-FM Rochester NY
WZKL(FM) Alliance OH
WOFX-FM Cincinnati OH
WVKS(FM) Toledo OH
KPRV-FM Heavener OK
KKRE(FM) Hollis OK
KOMA(FM) Oklahoma City OK
KLAD-FM Klamath Falls OR
WQMU(FM) Indiana PA
WJUN-FM Mexico PA
WXTU(FM) Philadelphia PA
WORO(FM) Corozal PR
WESC-FM Greenville SC
WIHB(FM) Moncks Corner SC
KELO-FM Sioux Falls SD
KULL(FM) Abilene TX
KBEY(FM) Burnet TX
KZPS(FM) Dallas TX
KRPT(FM) Devine TX
KXXS(FM) Elgin TX
KCOL-FM Groves TX
KZRC(FM) Markham TX
KMBV(FM) Navasota TX
KHES(FM) Rocksprings TX
KQSI(FM) San Augustine TX
*KHTA(FM) Wake Village TX
KYKM(FM) Yoakum TX
KXBN(FM) Cedar City UT
KCUA(FM) Naples UT
KUUU(FM) South Jordan UT
WINC-FM Winchester VA
KQMV(FM) Bellevue WA
KZHR(FM) Dayton WA
WJJQ-FM Tomahawk WI
WBWI-FM West Bend WI
WZAC-FM Danville WV
KDAD(FM) Douglas WY

92.7 mhz
WAFN-FM Arab AL
WKZJ(FM) Eufaula AL
WJBB-FM Haleyville AL
WTDR(FM) Talladega AL
KLYR-FM Clarksville AR
KASR(FM) Conway AR
KRRN(FM) Kingman AZ
KAJL(FM) Adelanto CA
KNGY(FM) Alameda CA
KBQB(FM) Chico CA
KJLL-FM Fountain Valley CA
KKUU(FM) Indio CA
KTOM-FM Marina CA
KMFB(FM) Mendocino CA
KZIQ-FM Ridgecrest CA
KZSQ-FM Sonora CA
KHJL(FM) Thousand Oaks CA
KKCH(FM) Glenwood Springs CO
WGMD(FM) Rehoboth Beach DE
WJBT(FM) Green Cove Springs FL
WEOW(FM) Key West FL
WAVW(FM) Stuart FL
WKKZ(FM) Dublin GA
WBGA(FM) Saint Simons Island GA
KHWI(FM) Hilo HI
KLGA-FM Algona IA
KTWA(FM) Ottumwa IA
KORT-FM Grangeville ID
KSRA-FM Salmon ID
WKIE(FM) Arlington Heights IL
WLSR(FM) Galesburg IL
WVZA(FM) Herrin IL
WKIF(FM) Kankakee IL
WQLZ(FM) Taylorville IL
WXKU-FM Austin IN
WZBD(FM) Berne IN
WSDM-FM Brazil IN
KANR(FM) Belle Plaine KS
KILS(FM) Minneapolis KS
WRVC-FM Catlettsburg KY
WMIK-FM Middlesboro KY
WHVE(FM) Russell Springs KY
*WBKL(FM) Clinton LA
KBYO-FM Farmerville LA
KLPL-FM Lake Providence LA
KJVC(FM) Mansfield LA
WMVY(FM) Tisbury MA
WWXT(FM) Prince Frederick MD
WQDY-FM Calais ME
WOXO-FM Norway ME
WJZL(FM) Charlotte MI
WDZZ-FM Flint MI
WYVN(FM) Saugatuck MI
KLOZ(FM) Eldon MO
KSJQ(FM) Savannah MO
WKRA-FM Holly Springs MS
KVCK-FM Wolf Point MT
WQNC(FM) Harrisburg NC
*KFNL(FM) Kindred ND
KBRB-FM Ainsworth NE
KUSO(FM) Albion NE
WOBM-FM Toms River NJ
KDSK(FM) Grants NM
KRSY-FM La Luz NM
KQAY-FM Tucumcari NM
KDSS(FM) Ely NV
KHWK(FM) Tonopah NV
KWNA-FM Winnemucca NV
WENY-FM Elmira NY
WQBU-FM Garden City NY
*WGFR(FM) Glens Falls NY
WXUR(FM) Herkimer NY
WRRV(FM) Middletown NY
WBDB(FM) Ogdensburg NY
WQEL(FM) Bucyrus OH
*WCVZ(FM) South Zanesville OH
KTRX(FM) Dickson OK
KKBS(FM) Guymon OK
KQHC(FM) Burns OR
KGBR(FM) Gold Beach OR
KNCU(FM) Newport OR
KRXF(FM) Sunriver OR
KMSW(FM) The Dalles OR
WCCR(FM) Clarion PA
WJSM-FM Martinsburg PA
WSJW(FM) Starview PA
WKSX-FM Johnston SC
KGFX-FM Pierre SD
WIJV(FM) Harriman TN
KALP(FM) Alpine TX
KIVY-FM Crockett TX
KINL(FM) Eagle Pass TX
KKBA(FM) Kingsville TX
KJBZ(FM) Laredo TX
KJAK(FM) Slaton TX
KESO(FM) South Padre Island TX
KBDX(FM) Blanding UT
WFHG-FM Abingdon VA
WUVA(FM) Charlottesville VA
WKVT-FM Brattleboro VT
KNCW(FM) Omak WA
WAUN(FM) Kewaunee WI
WPKG(FM) Neillsville WI
WDUX-FM Waupaca WI
WGIE(FM) Clarksburg WV
WPMW(FM) Mullens WV
KIQZ(FM) Rawlins WY
KHRW(FM) Wright WY

92.9 mhz
KFAT(FM) Anchorage AK
WBLX-FM Mobile AL
WTUG-FM Northport AL
KVRE(FM) Hot Springs Village AR
KAFF-FM Flagstaff AZ
KWMT-FM Tucson AZ
KFGY(FM) Healdsburg CA
KJEE(FM) Montecito CA
KXFG(FM) Sun City CA
KFSO-FM Visalia CA
KKPK(FM) Colorado Springs CO
WDSD(FM) Smyrna DE
WIKX(FM) Charlotte Harbor FL
WMFQ(FM) Ocala FL
WZGC(FM) Atlanta GA
WAAC(FM) Valdosta GA
KATF(FM) Dubuque IA
KKIA(FM) Ida Grove IA
WRPW(FM) Colfax IL
WSEI(FM) Olney IL
WNDV-FM South Bend IN
WSKL(FM) Veedersburg IN
KMXN(FM) Osage City KS
WLXX(FM) Lexington KY
KHLA(FM) Jennings LA
KTKC(FM) Springhill LA
WBOX-FM Varnado LA
WBOS(FM) Brookline MA
WEZQ(FM) Bangor ME
WJZQ(FM) Cadillac MI
*WSCD-FM Duluth MN
*KFSI(FM) Rochester MN
KKJM(FM) Saint Joseph MN
KGRC(FM) Hannibal MO
KLSC(FM) Malden MO
KOMG(FM) Ozark MO
KKID(FM) Salem MO
WDTL-FM Cleveland MS
WDXO(FM) Hazlehurst MS
KLFM(FM) Great Falls MT
KEZQ(FM) West Yellowstone MT
KYYY(FM) Bismarck ND
KKXL-FM Grand Forks ND
KTGL(FM) Beatrice NE
KTZA(FM) Artesia NM
KYBR(FM) Espanola NM
KRWN(FM) Farmington NM
KSCQ(FM) Silver City NM
KMXQ(FM) Socorro NM
KURK(FM) Reno NV
WBUF(FM) Buffalo NY
WBPM(FM) Saugerties NY
WEHM(FM) Southampton NY
WGTZ(FM) Eaton OH
KBEZ(FM) Tulsa OK
KDCQ(FM) Coos Bay OR
WLTJ(FM) Pittsburgh PA
WMGS(FM) Wilkes-Barre PA
WTPM(FM) Aguadilla PR
WYQE(FM) Naguabo PR
WZLA-FM Abbeville SC
WEGX(FM) Dillon SC
KSDR-FM Watertown SD
WMFS(FM) Bartlett TN
WJXA(FM) Nashville TN
WNPC-FM Newport TN
KLRK(FM) Marlin TX
KKBQ-FM Pasadena TX
KDCD(FM) San Angelo TX
KROM(FM) San Antonio TX
KBKH(FM) Shamrock TX
KNIN-FM Wichita Falls TX
KBLQ-FM Logan UT
WVHL(FM) Farmville VA
WVBW(FM) Suffolk VA
WEZF(FM) Burlington VT
KISM(FM) Bellingham WA
KZZU-FM Spokane WA
KDBL(FM) Toppenish WA
WYNW(FM) Birnamwood WI
WECL(FM) Elk Mound WI
WDHC(FM) Berkeley Springs WV
WCWV(FM) Summersville WV
KLGT(FM) Buffalo WY
KOLT-FM Warren AFB WY

93.1 mhz
WGMZ(FM) Glencoe AL
KZLE(FM) Batesville AR
*KHDX(FM) Conway AR
KKHJ-FM Pago Pago AS
KLJZ(FM) Yuma AZ
KXGO(FM) Arcata CA
KCBS-FM Los Angeles CA
KOSO(FM) Patterson CA
KKXX-FM Shafter CA
KMGJ(FM) Grand Junction CO
WKRO-FM Edgewater FL
WHDR(FM) Miami FL
WBBK-FM Blakely GA
WEAS-FM Springfield GA
WGAC-FM Warrenton GA
KQMQ-FM Honolulu HI
KMCS(FM) Muscatine IA
KZMG(FM) New Plymouth ID
WXRT-FM Chicago IL
WYDS(FM) Decatur IL
WTFX-FM Clarksville IN
WNOU(FM) Indianapolis IN
KHMY(FM) Pratt KS
WMKZ(FM) Monticello KY
WDHR(FM) Pikeville KY
KQID(FM) Alexandria LA
WHYN-FM Springfield MA
WPOC(FM) Baltimore MD
WMGX(FM) Portland ME
WDRQ(FM) Detroit MI
WIMK(FM) Iron Mountain MI
KXLP(FM) New Ulm MN
KWJK(FM) Boonville MO
KBDZ(FM) Perryville MO
WGDQ(FM) Hattiesburg MS
WYAB(FM) Yazoo City MS
KGCX(FM) Sidney MT
WPAW(FM) Winston-Salem NC
KRVN-FM Lexington NE
WPAT-FM Paterson NJ
WEZW(FM) Wildwood Crest NJ
KPLV(FM) Las Vegas NV
WNTQ(FM) Syracuse NY
WZAK(FM) Cleveland OH
WFGF(FM) Lima OH
WNTO(FM) Racine OH
KRYP(FM) Gladstone OR
WQYX(FM) Clearfield PA
WZMJ(FM) Batesburg SC
KRCS(FM) Sturgis SD
KKYA(FM) Yankton SD
WWGM(FM) Alamo TN
WSAA(FM) Benton TN
WMYU(FM) Karns TN
KQIZ-FM Amarillo TX
KMKT(FM) Bells TX
KSTV-FM Dublin TX
KSII(FM) El Paso TX
KBDK(FM) Leakey TX
KTYL-FM Tyler TX
WLFV(FM) Ettrick VA
WSVO(FM) Staunton VA
WHLK(FM) De Forest WI
WJBL(FM) Ladysmith WI
WFGM-FM Barrackville WV
KTRZ(FM) Riverton WY

93.3 mhz
KXBA(FM) Nikiski AK
KVAK-FM Valdez AK
KMJI(FM) Ashdown AR
KKSP(FM) Bryant AR
KAGL(FM) El Dorado AR
KDKB(FM) Mesa AZ
KXAZ(FM) Page AZ
KBHR(FM) Big Bear City CA
KRHV(FM) Big Pine CA
KNTO(FM) Chowchilla CA
KHTS-FM El Cajon CA
KRZZ(FM) San Francisco CA
KZOZ(FM) San Luis Obispo CA
KJDX(FM) Susanville CA
KKDC(FM) Dolores CO
KLMR-FM Lamar CO
KTCL(FM) Wheat Ridge CO
*WFAR(FM) Danbury CT
WROO(FM) Callahan FL
WGWD(FM) Gretna FL
WNCV(FM) Shalimar FL
WFLZ-FM Tampa FL
WVFJ-FM Manchester GA
KIOA(FM) Des Moines IA
WPBG(FM) Peoria IL
WTRH(FM) Ramsey IL
WBTU(FM) Kendallville IN
WQTY(FM) Linton IN
WDNS(FM) Bowling Green KY
WKYQ(FM) Paducah KY
WQUE-FM New Orleans LA
WSNE-FM Taunton MA
WKQZ(FM) Midland MI

KBLB(FM) Nisswa MN
KMXV(FM) Kansas City MO
KIGL(FM) Seligman MO
KNSX(FM) Steelville MO
WSYE(FM) Houston MS
KYYA-FM Billings MT
KGGL(FM) Missoula MT
WTPT(FM) Forest City NC
*WOGR-FM Salisbury NC
WERO(FM) Washington NC
KSJZ(FM) Jamestown ND
KHUS(FM) Bennington NE
WNHW(FM) Belmont NH
KKOB-FM Albuquerque NM
*WCAN(FM) Canajoharie NY
WFKL(FM) Fairport NY
WWSE(FM) Jamestown NY
WBWZ(FM) New Paltz NY
WSLP(FM) Saranac Lake NY
WCIZ-FM Watertown NY
WLZT(FM) Chillicothe OH
WAKW(FM) Cincinnati OH
WNCD(FM) Youngstown OH
KKNG-FM Newcastle OK
KKNU(FM) Springfield-Eugene OR
WQZS(FM) Meyersdale PA
WBZD-FM Muncy PA
WMMR(FM) Philadelphia PA
WZMT(FM) Ponce PR
WWWZ(FM) Summerville SC
KJRV(FM) Wessington Springs SD
WHRP(FM) Tullahoma TN
KDHT(FM) Cedar Park TX
KDBN(FM) Haltom City TX
KZBT(FM) Midland TX
KQBU-FM Port Arthur TX
KITE(FM) Port Lavaca TX
KUBL-FM Salt Lake City UT
WFLS-FM Fredericksburg VA
KUBE(FM) Seattle WA
*KRKL(FM) Walla Walla WA
WBSZ(FM) Ashland WI
WIZM-FM La Crosse WI
WLDB(FM) Milwaukee WI
*WKVW(FM) Marmet WV
KJAX(FM) Jackson WY

93.5 mhz

KDJF(FM) Delta Junction AK
WMLV(FM) Butler AL
KBKG(FM) Corning AR
KBFC(FM) Forrest City AR
KKTZ(FM) Lakeview AR
KSNX(FM) Show Low AZ
KNAC(FM) Earlimart CA
KXSM(FM) Hollister CA
KWIE(FM) Ontario CA
KDAY(FM) Redondo Beach CA
KLKX(FM) Rosamond CA
KKBN(FM) Twain Harte CA
KMKX(FM) Willits CA
KALQ-FM Alamosa CO
WZBH(FM) Georgetown DE
WBGF(FM) Belle Glade FL
WKEY-FM Key West FL
WPBH(FM) Port St. Joe FL
WLJA-FM Ellijay GA
WVOH-FM Hazlehurst GA
KPOA(FM) Lahaina HI
KQNG-FM Lihue HI
KQCS(FM) Bettendorf IA
KKMI(FM) Burlington IA
WVIX(FM) Joliet IL
WEBX(FM) Tuscola IL
WLFW(FM) Chandler IN
WHTY(FM) Hartford City IN
WKHY(FM) Lafayette IN
KOTE(FM) Eureka KS
KLKC-FM Parsons KS
KWME(FM) Wellington KS
WMMG-FM Brandenburg KY
WAIN-FM Columbia KY
KGGM(FM) Delhi LA
KJAE(FM) Leesville LA
WFQR(FM) Harwich Port MA
WCTB(FM) Fairfield ME

WBCM(FM) Boyne City MI
WKMJ-FM Hancock MI
WHMI-FM Howell MI
KSCR-FM Benson MN
KITN(FM) Worthington MN
KMYK(FM) Osage Beach MO
*KRSS(FM) Tarkio MO
WHJT(FM) Clinton MS
KWDE(FM) Eureka MT
KLAN(FM) Glasgow MT
WLQB(FM) Ocean Isle Beach NC
*WYFQ-FM Wadesboro NC
KKOT(FM) Columbus NE
WMWV(FM) Conway NH
KWES-FM Ruidoso NM
KADD(FM) Logandale NV
WZCR(FM) Hudson NY
WVBR-FM Ithaca NY
WVIP(FM) New Rochelle NY
WOKR(FM) Remsen NY
WQRW(FM) Wellsville NY
WBNV(FM) Barnesville OH
WRQN(FM) Bowling Green OH
KRKZ(FM) Altus OK
KWFX(FM) Woodward OK
KHAL(FM) Condon OR
KQIK-FM Lakeview OR
WHPA(FM) Gallitzin PA
WTPA(FM) Mechanicsburg PA
WSBG(FM) Stroudsburg PA
WDOG-FM Allendale SC
WARQ(FM) Columbia SC
WKBQ(FM) Covington TN
WKZX-FM Lenoir City TN
WKWX(FM) Savannah TN
KLXK(FM) Breckenridge TX
KBHT(FM) Crockett TX
KIKT(FM) Greenville TX
KOOK(FM) Junction TX
KBAW(FM) Zapata TX
KSNN(FM) Saint George UT
WAXM(FM) Big Stone Gap VA
WBBC-FM Blackstone VA
WSNV(FM) Salem VA
WYAC-FM Christiansted VI
WTSM(FM) Springfield VT
KOZI-FM Chelan WA
KANY(FM) Ocean Shores WA
KWDR(FM) Royal City WA
WOZZ(FM) New London WI
WBTQ(FM) Buckhannon WV
KWYX(FM) Casper WY
KWDU(FM) Upton WY

93.7 mhz

KAFC(FM) Anchorage AK
WDJC-FM Birmingham AL
WRJM-FM Geneva AL
KISR(FM) Fort Smith AR
KJBR(FM) Marked Tree AR
KHBM-FM Monticello AR
KRQQ(FM) Tucson AZ
KCLB-FM Coachella CA
KXZM(FM) Felton CA
KSKS(FM) Fresno CA
KQJK(FM) Roseville CA
KDB(FM) Santa Barbara CA
KJZY(FM) Sebastopol CA
KRAI-FM Craig CO
KSBV(FM) Salida CO
WZMX(FM) Hartford CT
WSTW(FM) Wilmington DE
WTLT(FM) Naples FL
WOGK(FM) Ocala FL
WGYL(FM) Vero Beach FL
WPEZ(FM) Jeffersonville GA
WMPZ(FM) Ringgold GA
KKRL(FM) Carroll IA
KZBQ(FM) Pocatello ID
WTRX-FM Pontiac IL
WQKC(FM) Seymour IN
WFRR(FM) Walton IN
KAIR-FM Horton KS
KYEZ(FM) Salina KS
WDGG(FM) Ashland KY
KRDJ(FM) New Iberia LA

KXKS-FM Shreveport LA
WMKK(FM) Lawrence MA
WRMO(FM) Milbridge ME
WRCL(FM) Frankenmuth MI
WBCT(FM) Grand Rapids MI
WKAD(FM) Harrietta MI
KXXR(FM) Minneapolis MN
KTUF(FM) Kirksville MO
KSD(FM) Saint Louis MO
WMJY(FM) Biloxi MS
WQLJ(FM) Oxford MS
KOBB-FM Bozeman MT
KTZZ(FM) Conrad MT
WNTB(FM) Wrightsville Beach NC
WDAY-FM Fargo ND
KIZZ(FM) Minot ND
KOLB(FM) Firth NE
WXBN(FM) Groveton NH
KXXI(FM) Gallup NM
KLKO(FM) Elko NV
KWNZ(FM) Sun Valley NV
*WCOV-FM Clyde NY
WBLK(FM) Depew NY
*WYAI(FM) Scotia NY
WFCJ(FM) Miamisburg OH
WQIO(FM) Mount Vernon OH
KSPI-FM Stillwater OK
KTMT-FM Medford OR
WBUS(FM) Boalsburg PA
WSJR(FM) Dallas PA
WTZN-FM Pittsburgh PA
WZNT(FM) San Juan PR
WXJY(FM) Georgetown SC
WFBC-FM Greenville SC
WSIM(FM) Lamar SC
WALI(FM) Walterboro SC
KBRK-FM Brookings SD
KVAR(FM) Pine Ridge SD
KWYR-FM Winner SD
WTKB-FM Atwood TN
WBXE(FM) Baxter TN
WFFI(FM) Kingston Springs TN
KLBJ-FM Austin TX
KKRW(FM) Houston TX
KNOR(FM) Krum TX
KXTQ-FM Lubbock TX
KLGL(FM) Richfield UT
WPYA(FM) Chesapeake VA
WAZR(FM) Woodstock VA
WUSX(FM) Addison VT
KXAA(FM) Cle Elum WA
KGSG(FM) Pasco WA
KDRK-FM Spokane WA
WEKZ-FM Monroe WI
WBFM(FM) Sheboygan WI
WLSL(FM) Three Lakes WI
KZDR(FM) Cheyenne WY
KYTI(FM) Sheridan WY

93.9 mhz

WYTK(FM) Rogersville AL
WQSI(FM) Union Springs AL
KAMJ-FM Gosnell AR
KMGN(FM) Flagstaff AZ
KRIT(FM) Parker AZ
KFMF(FM) Chico CA
KEXA(FM) King City CA
KMVN(FM) Los Angeles CA
KBBU(FM) Modesto CA
KRLT(FM) South Lake Tahoe CA
KYSL(FM) Frisco CO
WKYS(FM) Washington DC
WLVE(FM) Miami Beach FL
WDRR(FM) Martinez GA
WMTM-FM Moultrie GA
KUAM-FM Hagatna GU
KIKI-FM Honolulu HI
KLUA(FM) Kailua-Kona HI
KIAI(FM) Mason City IA
KSOU-FM Sioux Center IA
WCEZ(FM) Carthage IL
WLIT-FM Chicago IL
WYEC(FM) Kewanee IL
WABZ(FM) Sherman IL
*WPFR-FM Clinton IN
WWFT(FM) Fishers IN

KDGS(FM) Andover KS
KZRD(FM) Dodge City KS
WSEK(FM) Burnside KY
WKTG(FM) Madisonville KY
KMXH(FM) Alexandria LA
WRSI(FM) Turners Falls MA
WCYI(FM) Lewiston ME
WAVC(FM) Mio MI
WNBY-FM Newberry MI
KKRC(FM) Granite Falls MN
WTBX(FM) Hibbing MN
KSSZ(FM) Fayette MO
KGKS(FM) Scott City MO
KJMK(FM) Webb City MO
KSPQ(FM) West Plains MO
WGRM-FM Greenwood MS
WRXW(FM) Pearl MS
WKSL(FM) Burlington NC
KSWN(FM) McCook NE
KRTN-FM Raton NM
WDNY-FM Dansville NY
*WNYC-FM New York NY
WKXZ(FM) Norwich NY
*WQKE(FM) Plattsburgh NY
WLWD(FM) Columbus Grove OH
KIMY(FM) Watonga OK
KPDQ-FM Portland OR
WTWF(FM) Fairview PA
WKBI-FM Saint Marys PA
WJXY-FM Conway SC
WIGL(FM) Saint Matthews SC
KKMK(FM) Rapid City SD
WSIB(FM) Selmer TN
WAYA(FM) Spring City TN
KMXR(FM) Corpus Christi TX
KINT-FM El Paso TX
KOYN(FM) Paris TX
KCRN-FM San Angelo TX
KBNU(FM) Uvalde TX
WMEV-FM Marion VA
WLVB(FM) Morrisville VT
WMXR(FM) Woodstock VT
KTAC(FM) Ephrata WA
WMMA(FM) Nekoosa WI
WDOR-FM Sturgeon Bay WI
WRRR-FM Saint Marys WV
KTAK(FM) Riverton WY

94.1 mhz

WZBQ(FM) Carrollton AL
WXQW(FM) Gurley AL
KKPT(FM) Little Rock AR
KRDE(FM) Globe AZ
KXKQ(FM) Safford AZ
KISV(FM) Bakersfield CA
*KPFA(FM) Berkeley CA
KNCO-FM Grass Valley CA
KSLG-FM Hydesville CA
KBKY(FM) Merced CA
KLMM(FM) Morro Bay CA
KMYI(FM) San Diego CA
KWDI(FM) Idalia CO
KKXK(FM) Montrose CO
KEZZ(FM) Walden CO
WAKU(FM) Crawfordville FL
WSJT(FM) Lakeland FL
WTYS-FM Marianna FL
WMEZ(FM) Pensacola FL
WSOS-FM Saint Augustine FL
WQBT(FM) Savannah GA
WSTR(FM) Smyrna GA
KRNA(FM) Iowa City IA
KBXL(FM) Caldwell ID
WMIX-FM Mount Vernon IL
WGFA-FM Watseka IL
*WBNI-FM Roanoke IN
KDNS(FM) Downs KS
KFKF-FM Kansas City KS
WLYE-FM Glasgow KY
KRLQ(FM) Hodge LA
WEMX(FM) Kentwood LA
WWKR(FM) Hart MI
WVIC(FM) Jackson MI
WUPK(FM) Marquette MI
KKLN(FM) Atwater MN
KFML(FM) Little Falls MN

KPVR(FM) Bowling Green MO
KRKX(FM) Billings MT
KOPR(FM) Butte MT
WTHZ(FM) Lexington NC
WWHA(FM) Oriental NC
KQCH(FM) Omaha NE
KNEB-FM Scottsbluff NE
WFTN-FM Franklin NH
KZRR(FM) Albuquerque NM
KZOR(FM) Hobbs NM
KMXB(FM) Henderson NV
WZNE(FM) Brighton NY
WZNY(FM) Old Forge NY
WNYV(FM) Whitehall NY
WHBC-FM Canton OH
WVMX(FM) Cincinnati OH
KTSO(FM) Glenpool OK
KZCD(FM) Lawton OK
KXIX(FM) Bend OR
*KOOZ(FM) Myrtle Point OR
WYSP(FM) Philadelphia PA
WQKX(FM) Sunbury PA
WNOD(FM) Mayaguez PR
WHJY(FM) Providence RI
WYOR(FM) Cross Hill SC
WGSS(FM) Kingstree SC
KSDN-FM Aberdeen SD
WSNA(FM) Germantown TN
WMXK(FM) Morristown TN
WLZK(FM) Paris TN
WFFH(FM) Smyrna TN
KMXJ-FM Amarillo TX
KQXY-FM Beaumont TX
KLTR(FM) Brenham TX
KDLK-FM Del Rio TX
KTFM(FM) Floresville TX
KLNO(FM) Fort Worth TX
KJAZ(FM) Point Comfort TX
KODJ(FM) Salt Lake City UT
WXEZ(FM) Yorktown VA
KCLK-FM Clarkston WA
KMPS-FM Seattle WA
WIAL(FM) Eau Claire WI
WJJO(FM) Watertown WI
WQZK-FM Keyser WV
WAXS(FM) Oak Hill WV

94.3 mhz

WIZB(FM) Abbeville AL
WQZX(FM) Greenville AL
KAMO-FM Rogers AR
KSMZ(FM) Viola AR
KFPB(FM) Chino Valley AZ
KBUX(FM) Quartzsite AZ
KDUC(FM) Barstow CA
KCRE-FM Crescent City CA
KEBN(FM) Garden Grove CA
KLMG(FM) Jackson CA
KOKO-FM Kerman CA
KBUA(FM) San Fernando CA
KILO(FM) Colorado Springs CO
KMAX-FM Wellington CO
WYBC-FM New Haven CT
WNFB(FM) Lake City FL
WGMX(FM) Marathon FL
WZZR(FM) Riviera Beach FL
WLEL(FM) Ellaville GA
WTHP(FM) Gibson GA
KEEI(FM) Hanapepe HI
KDLX(FM) Makawao HI
KTPZ(FM) Hazelton ID
KSNA(FM) Rexburg ID
WRMS-FM Beardstown IL
WPMJ(FM) Chillicothe IL
WJKL(FM) Glendale Heights IL
WKYX-FM Golconda IL
WMKR(FM) Pana IL
WSSQ(FM) Sterling IL
WREB(FM) Greencastle IN
WZOC(FM) Plymouth IN
WIFE-FM Rushville IN
KCVW(FM) Kingman KS
WULF(FM) Hardinsburg KY
WIFX-FM Jenkins KY
WEGI(FM) Oak Grove KY
WTIX-FM Galliano LA

WZAI(FM) Brewster MA
WINX-FM Cambridge MD
WCYY(FM) Biddeford ME
WFCX(FM) Leland MI
WZNL(FM) Norway MI
KKIN-FM Aitkin MN
KULO(FM) Alexandria MN
KDOM-FM Windom MN
KATI(FM) California MO
WKZW(FM) Bay Springs MS
WXRZ(FM) Corinth MS
WBAD(FM) Leland MS
WWNK(FM) Farmville NC
*WJIJ(FM) Norlina NC
WZKB(FM) Wallace NC
WJLK(FM) Asbury Park NJ
WILW(FM) Avalon NJ
KYEE(FM) Alamogordo NM
KDEM(FM) Deming NM
WLVY(FM) Elmira NY
WKXP(FM) Kingston NY
WMJC(FM) Smithtown NY
WKKI(FM) Celina OH
WKKJ(FM) Chillicothe OH
WDIF(FM) Marion OH
KXOO(FM) Elk City OK
KTIL-FM Tillamook OR
WLNP(FM) Carbondale PA
WQCM(FM) Greencastle PA
WBXQ(FM) Patton PA
WUZZ(FM) Saegertown PA
WWNQ(FM) Forest Acres SC
WSCC-FM Goose Creek SC
WCMG(FM) Latta SC
WLLI-FM Dyer TN
WJJM-FM Lewisburg TN
WNFZ(FM) Oak Ridge TN
WJTT(FM) Red Bank TN
KBTS(FM) Big Spring TX
KYOX(FM) Comanche TX
KHER(FM) Crystal City TX
KFST-FM Fort Stockton TX
KRVL(FM) Kerrville TX
KYXX(FM) Ozona TX
KSEY-FM Seymour TX
KXRQ(FM) Roosevelt UT
WTON-FM Staunton VA
WWXX(FM) Warrenton VA
WBTN-FM Bennington VT
WROE(FM) Neenah-Menasha WI
WQPC(FM) Prairie du Chien WI
WRLF(FM) Fairmont WV

94.5 mhz

WYSF(FM) Birmingham AL
WHOD(FM) Jackson AL
KFPW-FM Barling AR
KJIW-FM Helena AR
KOOL-FM Phoenix AZ
KCNO(FM) Alturas CA
KSEH(FM) Brawley CA
KSPE-FM Ellwood CA
KBAY(FM) Gilroy CA
KGEN-FM Hanford CA
KMYT(FM) Temecula CA
KWNE(FM) Ukiah CA
KJEB(FM) New Castle CO
*WERB(FM) Berlin CT
WCFB(FM) Daytona Beach FL
WARO(FM) Naples FL
WFBX(FM) Parker FL
WBYZ(FM) Baxley GA
WFDR-FM Woodbury GA
KKEZ(FM) Fort Dodge IA
KHTQ(FM) Hayden ID
WLRW(FM) Champaign IL
WRZR(FM) Loogootee IN
KSKL(FM) Scott City KS
WIBW-FM Topeka KS
WMXL(FM) Lexington KY
KSMB(FM) Lafayette LA
KRUF(FM) Shreveport LA
WJMN(FM) Boston MA
WKSQ(FM) Ellsworth ME
WCEN-FM Hemlock MI
WTNR(FM) Holland MI

WLJZ(FM) Mackinaw City MI
WELY-FM Ely MN
KSTP-FM Saint Paul MN
KRXL(FM) Kirksville MO
KKLR(FM) Poplar Bluff MO
WCMR(FM) Bruce MS
WJZD(FM) Long Beach MS
KMON-FM Great Falls MT
WGBT(FM) Eden NC
WCMS-FM Hatteras NC
WKXS-FM Leland NC
KQDY(FM) Bismarck ND
KLIQ(FM) Hastings NE
WPST(FM) Trenton NJ
KKOR(FM) Gallup NM
KMOA(FM) Caliente NV
KOYT(FM) Elko NV
KUUB(FM) Sun Valley NV
*WNED-FM Buffalo NY
*WYKV(FM) Ravena NY
WYYY(FM) Syracuse NY
WDKF(FM) Englewood OH
WXKR(FM) Port Clinton OH
KEMX(FM) Locust Grove OK
KJDY-FM Canyon City OR
KMGE(FM) Eugene OR
WDAC(FM) Lancaster PA
WWSW-FM Pittsburgh PA
WBHV-FM State College PA
WSPX(FM) Bowman SC
WMUU-FM Greenville SC
WYEZ(FM) Murrell's Inlet SC
KPLO-FM Reliance SD
*KCFS(FM) Sioux Falls SD
WFGZ(FM) Lobelville TN
KSOC(FM) Gainesville TX
KFRQ(FM) Harlingen TX
KTBZ-FM Houston TX
KFMX-FM Lubbock TX
KEMA(FM) Three Rivers TX
KBCT-FM Waco TX
KVFX(FM) Logan UT
WRVQ(FM) Richmond VA
WJEN(FM) Rutland VT
KLYK(FM) Kelso WA
KRXY(FM) Shelton WA
KATS(FM) Yakima WA
WRJO(FM) Eagle River WI
WKTI(FM) Milwaukee WI
WTMB(FM) Tomah WI
WZJO(FM) Dunbar WV
KMLD(FM) Casper WY

94.7 mhz

KZND-FM Houston AK
WTBF-FM Brundidge AL
KKLV(FM) Turrell AR
KFLG-FM Kingman AZ
KEWB(FM) Anderson CA
KSSJ(FM) Fair Oaks CA
*KYAF(FM) Firebaugh CA
KTWV(FM) Los Angeles CA
KLOB(FM) Thousand Palms CA
KRKS-FM Boulder CO
WRDX(FM) Dover DE
WSYR-FM Gifford FL
WDEC-FM Americus GA
KWXX-FM Hilo HI
KUMU-FM Honolulu HI
KMCN(FM) Clinton IA
KMCH(FM) Manchester IA
WZZN(FM) Chicago IL
WFBQ(FM) Indianapolis IN
WFIA-FM New Albany IN
KSKU(FM) Sterling KS
WQQR(FM) Mayfield KY
WKLW-FM Paintsville KY
WBIO(FM) Philpot KY
WYLK(FM) Lacombe LA
WMAS-FM Springfield MA
WTGB-FM Bethesda MD
WCSX(FM) Birmingham MI
*WCVM(FM) Bronson MI
KCLH(FM) Caledonia MN
KNSG(FM) Springfield MN
KSKK(FM) Staples MN

KSHE(FM) Crestwood MO
KTTS-FM Springfield MO
WWJK(FM) Jackson MS
WQDR(FM) Raleigh NC
KNOX-FM Grand Forks ND
KNEN(FM) Norfolk NE
*WFME(FM) Newark NJ
KKIM-FM Santa Fe NM
*WMHI(FM) Cape Vincent NY
WYUL(FM) Chateaugay NY
WIYN(FM) Deposit NY
WBAR-FM Lake Luzerne NY
WSNY(FM) Columbus OH
KHBZ-FM Oklahoma City OK
KRRM(FM) Rogue River OR
WBRX(FM) Cresson PA
WXBB(FM) Erie PA
WMTT(FM) Tioga PA
WODA(FM) Bayamon PR
WICI(FM) Sumter SC
WAAW(FM) Williston SC
WOJG(FM) Bolivar TN
WGSQ(FM) Cookeville TN
KBSO(FM) Corpus Christi TX
KYSE(FM) El Paso TX
KGRW(FM) Friona TX
KTXO(FM) Goldsmith TX
KWKQ(FM) Graham TX
KAMX(FM) Luling TX
KIXY-FM San Angelo TX
KVLL-FM Wells TX
KNRK(FM) Camas WA
KZAL(FM) Manson WA
WZOR(FM) Mishicot WI
WOFM(FM) Mosinee WI
WELK(FM) Elkins WV
KRAN(FM) Wamsutter WY

94.9 mhz

WKSJ-FM Mobile AL
KHLR(FM) Maumelle AR
KYNF(FM) Prairie Grove AR
KMXZ-FM Tucson AZ
KHRQ(FM) Baker CA
KPYG(FM) Cambria CA
KXTT(FM) Maricopa CA
KBZT(FM) San Diego CA
KYLD(FM) San Francisco CA
KBOS-FM Tulare CA
WMGE(FM) Miami Beach FL
WTNT-FM Tallahassee FL
WWRM(FM) Tampa FL
WUBL(FM) Atlanta GA
WHKN(FM) Millen GA
KGGO(FM) Des Moines IA
KRVB(FM) Nampa ID
KPKY(FM) Pocatello ID
WRHK(FM) Danville IL
WDKB(FM) De Kalb IL
WAAG(FM) Galesburg IL
WYNG(FM) Mount Carmel IL
KCKS(FM) Concordia KS
KSBH(FM) Coushatta LA
WPRF(FM) Reserve LA
WSYY-FM Millinocket ME
WCNF(FM) Benton Harbor MI
WUPG(FM) Crystal Falls MI
WMMQ(FM) East Lansing MI
WKJZ(FM) Hillman MI
WKZC(FM) Scottville MI
KCPI(FM) Albert Lea MN
KMXK(FM) Cold Spring MN
KQDS-FM Duluth MN
KLCH(FM) Lake City MN
KCMO-FM Kansas City MO
*WKVF(FM) Byhalia MS
WKOR-FM Columbus MS
KYSS-FM Missoula MT
KTZU(FM) Velva ND
*KJLT-FM North Platte NE
WHOM(FM) Mt. Washington NH
KWYK-FM Aztec NM
KBIM-FM Roswell NM
WKLL(FM) Frankfort NY
*WONB(FM) Ada OH
WSWD(FM) Fairfield OH

WQMX(FM) Medina OH
*WEKV(FM) South Webster OH
KCBZ(FM) Cannon Beach OR
KTEE(FM) North Bend OR
*WRSD(FM) Folsom PA
WRBT(FM) Harrisburg PA
WOGG(FM) Oliver PA
WHKS(FM) Port Allegany PA
WVCO(FM) Loris SC
KLRJ(FM) Aberdeen SD
WMSR-FM Collinwood TN
WAEZ(FM) Greeneville TN
WKVZ(FM) Ripley TN
KLTY(FM) Arlington TX
*KOLI(FM) Electra TX
KQUR(FM) Laredo TX
KCIN(FM) Cedar City UT
KHTB(FM) Provo UT
WSLC-FM Roanoke VA
WPTE(FM) Virginia Beach VA
KIOK(FM) Richland WA
*KUOW-FM Seattle WA
WOLX-FM Baraboo WI
KZWY(FM) Sheridan WY

95.1 mhz

WRTT-FM Huntsville AL
WXFX(FM) Prattville AL
KAMS(FM) Mammoth Spring AR
KVIB(FM) Sun City West AZ
KTTI(FM) Yuma AZ
*KAAX(FM) Avenal CA
KMXI(FM) Chico CA
KMDR(FM) McKinleyville CA
KHOP(FM) Oakdale CA
KFRG(FM) San Bernardino CA
KBBY-FM Ventura CA
KATC-FM Colorado Springs CO
KKNN(FM) Delta CO
WRKI(FM) Brookfield CT
WBPC(FM) Ebro FL
WAPE-FM Jacksonville FL
WBVD(FM) Melbourne FL
WCHZ(FM) Harlem GA
WMGB(FM) Montezuma GA
WACF(FM) Young Harris GA
KAOI-FM Wailuku HI
KMAQ-FM Maquoketa IA
KCZE(FM) New Hampton IA
KLER-FM Orofino ID
WUEZ(FM) Carterville IL
WDZQ(FM) Decatur IL
WVLI(FM) Kankakee IL
WAJI(FM) Fort Wayne IN
WVNI(FM) Nashville IN
*WVUR-FM Valparaiso IN
KICT-FM Wichita KS
WGGC(FM) Bowling Green KY
WFLE-FM Flemingsburg KY
*WXRB(FM) Dudley MA
WXTK(FM) West Yarmouth MA
WRBS-FM Baltimore MD
WFBE(FM) Flint MI
KBVB(FM) Barnesville MN
KWOA-FM Worthington MN
KMXL(FM) Carthage MO
KTKS(FM) Versailles MO
WJDQ(FM) Marion MS
WQNZ(FM) Natchez MS
WONA-FM Winona MS
KMMS-FM Bozeman MT
*KXEI(FM) Havre MT
KTHC(FM) Sidney MT
WNKS(FM) Charlotte NC
WRNS-FM Kinston NC
KRKR(FM) Lincoln NE
WAYV(FM) Atlantic City NJ
KSYU(FM) Corrales NM
KNUW(FM) Santa Clara NM
KNYE(FM) Pahrump NV
WFXF(FM) Honeoye Falls NY
WVXG(FM) Mount Gilead OH
KQCV-FM Shawnee OK
KSND(FM) Monmouth OR
KLTW-FM Prineville OR
WZZO(FM) Bethlehem PA

WIKZ(FM) Chambersburg PA
WWGY(FM) Grove City PA
WEGM(FM) San German PR
WSSX-FM Charleston SC
KSQY(FM) Deadwood SD
WCDZ(FM) Dresden TN
KORQ(FM) Baird TX
KYKR(FM) Beaumont TX
KNDE(FM) College Station TX
KCOR-FM Comfort TX
KQRX(FM) Midland TX
KEWL-FM New Boston TX
KVIC(FM) Victoria TX
WQMZ(FM) Charlottesville VA
WJKC(FM) Christiansted VI
*WVTQ(FM) Sunderland VT
KITI-FM Winlock WA
WQRB(FM) Bloomer WI
WIIL(FM) Kenosha WI
WLST(FM) Marinette WI
WXIL(FM) Parkersburg WV
KCGY(FM) Laramie WY
KYCS(FM) Rock Springs WY

95.3 mhz

WFFN(FM) Cordova AL
WRLD-FM Valley AL
KCXY(FM) East Camden AR
KVHU(FM) Judsonia AR
KERX(FM) Paris AR
KCDQ(FM) Douglas AZ
KOZT(FM) Fort Bragg CA
KBHH(FM) Kerman CA
KRTY(FM) Los Gatos CA
KLLY(FM) Oildale CA
KXTZ(FM) Pismo Beach CA
KUIC(FM) Vacaville CA
KSLV-FM Monte Vista CO
WKDB(FM) Laurel DE
WOLZ(FM) Fort Myers FL
WXCV(FM) Homosassa Springs FL
WPYO(FM) Maitland FL
WSRM(FM) Coosa GA
WJYF(FM) Nashville GA
KQMG-FM Independence IA
KIFG-FM Iowa Falls IA
KOKX-FM Keokuk IA
KCSI(FM) Red Oak IA
KCII-FM Washington IA
KPND(FM) Sandpoint ID
KECH-FM Sun Valley ID
WRXX(FM) Centralia IL
WRKX(FM) Ottawa IL
WMYE(FM) Rantoul IL
WRTB(FM) Winnebago IL
WLFF(FM) Brookston IN
WUME-FM Paoli IN
WNDI-FM Sullivan IN
KINZ(FM) Humboldt KS
KHCA(FM) Wamego KS
WIKI(FM) Carrollton KY
WEZG(FM) Morganfield KY
WVRB(FM) Wilmore KY
KQKI(FM) Bayou Vista LA
WHRB(FM) Cambridge MA
WPVQ(FM) Greenfield MA
WALZ-FM Machias ME
*WWWA(FM) Winslow ME
WUBB(FM) York Center ME
WQTE(FM) Adrian MI
WBXX(FM) Battle Creek MI
WCFX(FM) Clare MI
WAOR(FM) Niles MI
WGVS-FM Whitehall MI
WXXZ(FM) Grand Marais MN
KNOF(FM) Saint Paul MN
KDJS-FM Willmar MN
KAGE-FM Winona MN
KDKD-FM Clinton MO
KXMO-FM Owensville MO
WAFM(FM) Amory MS
WVIM-FM Coldwater MS
WADI(FM) Corinth MS
WRKG(FM) Drew MS
WZNF(FM) Lumberton MS
WOBR-FM Wanchese NC

KSEL-FM Portales NM
KRJC(FM) Elko NV
WGIX-FM Gouverneur NY
WBKT(FM) Norwich NY
WHFM(FM) Southampton NY
WKTN(FM) Kenton OH
WKLM(FM) Millersburg OH
WLKR-FM Norwalk OH
WZLR(FM) Xenia OH
KMGZ(FM) Lawton OK
KKBC-FM Baker City OR
KURY-FM Brookings OR
KUJZ(FM) Creswell OR
KLCR(FM) Lakeview OR
WZWW(FM) Bellefonte PA
WDNH-FM Honesdale PA
WBLJ-FM Shamokin PA
WTTC-FM Towanda PA
WJPA-FM Washington PA
WFMV(FM) South Congaree SC
KLXS-FM Pierre SD
WTBG(FM) Brownsville TN
WHJK(FM) Cleveland TN
*WYFC(FM) Clinton TN
KBTY(FM) Benjamin TX
KNEL-FM Brady TX
KVWG-FM Dilley TX
KDDD-FM Dumas TX
KFRO-FM Gilmer TX
KHYI(FM) Howe TX
KPBM(FM) McCamey TX
KSCG(FM) Meridian TX
KZSP(FM) South Padre Island TX
KRPX(FM) Wellington UT
WKHK(FM) Colonial Heights VA
WZRV(FM) Front Royal VA
WXBX(FM) Rural Retreat VA
WHLF(FM) South Boston VA
WXLF(FM) White River Junction VT
KXLE-FM Ellensburg WA
KXXK(FM) Hoquiam-Aberdeen WA
WXRO(FM) Beaver Dam WI
WGMO(FM) Shell Lake WI
WRLB(FM) Rainelle WV
KZJH(FM) Jackson WY

95.5 mhz

WHMA-FM Ashland AL
WTVY-FM Dothan AL
WFMH-FM Hackleburg AL
WJDB-FM Thomasville AL
KYOT-FM Phoenix AZ
KBOQ(FM) Carmel CA
KLOS(FM) Los Angeles CA
KZCC(FM) McCloud CA
KRVG(FM) Glenwood Springs CO
KRKQ(FM) Mountain Village CO
KPHT(FM) Rocky Ford CO
WLDI(FM) Fort Pierce FL
WNDD(FM) Silver Springs FL
WBTS(FM) Doraville GA
WIXV(FM) Savannah GA
KSTO(FM) Hagatna GU
KAIM-FM Honolulu HI
KZAT-FM Belle Plaine IA
KJCY(FM) Saint Ansgar IA
KGLI(FM) Sioux City IA
WFUN-FM Bethalto IL
WNUA(FM) Chicago IL
WGLO(FM) Pekin IL
WFMS(FM) Indianapolis IN
KOLS(FM) Dodge City KS
KQNS-FM Lindsborg KS
KNDY-FM Marysville KS
WQHY(FM) Prestonsburg KY
KRRQ(FM) Lafayette LA
WPGC-FM Morningside MD
WJJB-FM Topsham ME
WKQI(FM) Detroit MI
WJZJ(FM) Glen Arbor MI
KKZY(FM) Bemidji MN
KBEK(FM) Mora MN
KRDS-FM New Prague MN
KAAN-FM Bethany MO
KTOZ-FM Pleasant Hope MO
KJEZ(FM) Poplar Bluff MO

WHLH(FM) Jackson MS
WOXD(FM) Oxford MS
KMBR(FM) Butte MT
KMHK(FM) Hardin MT
*WHPE-FM High Point NC
WPWZ(FM) Pinetops NC
KYNU(FM) Jamestown ND
KSDZ(FM) Gordon NE
KHFM(FM) Santa Fe NM
KWNR(FM) Henderson NV
KNEV(FM) Reno NV
WYJB(FM) Albany NY
WPLJ(FM) New York NY
WFHM-FM Cleveland OH
WHOK-FM Lancaster OH
KWEY-FM Clinton OK
KWEN(FM) Tulsa OK
KXJM(FM) Portland OR
WFGI-FM Johnstown PA
WBYL(FM) Salladasburg PA
WBRU(FM) Providence RI
WIBZ(FM) Wedgefield SC
WSM-FM Nashville TN
KKMJ-FM Austin TX
KZFM(FM) Corpus Christi TX
KAFX-FM Diboll TX
KLAQ(FM) El Paso TX
KJKB(FM) Jacksboro TX
KAIQ(FM) Wolfforth TX
*KYFO-FM Ogden UT
WBOP(FM) Buffalo Gap VA
WXXX(FM) South Burlington VT
WIFC(FM) Wausau WI
KWYY(FM) Casper WY

95.7 mhz

WBHJ(FM) Midfield AL
KSEC(FM) Bentonville AR
KSSN(FM) Little Rock AR
KWKM(FM) Saint Johns AZ
KUSS(FM) Carlsbad CA
KJFX(FM) Fresno CA
KPAT(FM) Orcutt CA
KALF(FM) Red Bluff CA
KBWF(FM) San Francisco CA
KPTT(FM) Denver CO
WKSS(FM) Hartford CT
WBTP(FM) Clearwater FL
WGCX(FM) Navarre FL
WXDJ(FM) North Miami Beach FL
WHOG-FM Ormond-by-the-Sea FL
WIOL(FM) Greenville GA
WATG(FM) Trion GA
WQPW(FM) Valdosta GA
KKSY(FM) Anamosa IA
KSWI(FM) Atlantic IA
KQWC-FM Webster City IA
KEZJ-FM Twin Falls ID
WCRC(FM) Effingham IL
WSEY(FM) Oregon IL
WJDK-FM Seneca IL
WSHP(FM) Attica IN
WQMF(FM) Jeffersonville IN
WYPW(FM) Nappanee IN
KCHZ(FM) Ottawa KS
WCCK(FM) Calvert City KY
KLKL(FM) Minden LA
WKBU(FM) New Orleans LA
KROK(FM) South Fort Polk LA
WWMJ(FM) Ellsworth ME
WLHT-FM Grand Rapids MI
*WHWL(FM) Marquette MI
*WCMB-FM Oscoda MI
KDAL-FM Duluth MN
KQYK(FM) Lake Crystal MN
KKOK-FM Morris MN
KWWR(FM) Mexico MO
WTGY(FM) Charleston MS
WHAL-FM Horn Lake MS
KCGM(FM) Scobey MT
WXRC(FM) Hickory NC
WKML(FM) Lumberton NC
KNDK-FM Langdon ND
KCVD(FM) New England ND
*KROA(FM) Grand Island NE
WZID(FM) Manchester NH

KPCL(FM) Farmington NM
KPER(FM) Hobbs NM
WAQX-FM Manlius NY
WPIG(FM) Olean NY
WIMX(FM) Gibsonburg OH
WHIO-FM Piqua OH
WVKF(FM) Shadyside OH
KKAJ-FM Ardmore OK
KXLS(FM) Lahoma OK
KBOY-FM Medford OR
WMRF-FM Lewistown PA
WBHD(FM) Olyphant PA
WBEN-FM Philadelphia PA
WFID(FM) Rio Piedras PR
WWBD(FM) Bamberg SC
KSQB-FM Dell Rapids SD
WAYB-FM Graysville TN
WFKX(FM) Henderson TN
WQJK(FM) Maryville TN
KBST-FM Big Spring TX
KARX(FM) Claude TX
KHJZ-FM Houston TX
KLEY-FM Jourdanton TX
KOTY(FM) Mason TX
KBGO(FM) Waco TX
WFLO-FM Farmville VA
WVKL(FM) Norfolk VA
WDOT(FM) Danville VT
KJR-FM Seattle WA
KNLT(FM) Walla Walla WA
WRQT(FM) La Crosse WI
WRIT-FM Milwaukee WI
KFMR(FM) Marbleton WY

95.9 mhz

KXLR(FM) Fairbanks AK
WKXN(FM) Greenville AL
WTWX-FM Guntersville AL
WTGZ(FM) Tuskegee AL
KWHF(FM) Harrisburg AR
KUUZ(FM) Lake Village AR
KKLD(FM) Cottonwood AZ
KFSH-FM Anaheim CA
KBYN(FM) Arnold CA
KXXZ(FM) Barstow CA
KOCP(FM) Camarillo CA
KRSH(FM) Healdsburg CA
KAJR(FM) Indian Wells CA
KSKD(FM) Livingston CA
KNLF(FM) Quincy CA
KIDN-FM Hayden CO
WFOX(FM) Norwalk CT
WOSC(FM) Bethany Beach DE
WSJZ-FM Sebastian FL
WRBA(FM) Springfield FL
WQZY(FM) Dublin GA
KPVS(FM) Hilo HI
KSRF(FM) Poipu HI
KCHA-FM Charles City IA
KILR-FM Estherville IA
KIIK-FM Fairfield IA
KCOB-FM Newton IA
KLZX(FM) Weston ID
WERV-FM Aurora IL
*WOLG(FM) Carlinville IL
WHOW-FM Clinton IL
WDQN-FM Du Quoin IL
WNLF(FM) Macomb IL
WMLF(FM) Watseka IL
WIAU(FM) Franklin IN
WEFM(FM) Michigan City IN
WWSY(FM) Seelyville IN
WKID(FM) Vevay IN
WKUZ(FM) Wabash IN
KWHK(FM) Hutchinson KS
KCAY(FM) Russell KS
KSOK-FM Winfield KS
WFTM-FM Maysville KY
WGKY(FM) Wickliffe KY
KZLG(FM) Mansura LA
KMAR-FM Winnsboro LA
WATD-FM Marshfield MA
WBEC-FM Pittsfield MA
WWIN-FM Glen Burnie MD
WICL(FM) Williamsport MD
WRED(FM) Saco ME

WLKM-FM Three Rivers MI
KQCL(FM) Faribault MN
WLKX-FM Forest Lake MN
WWWI-FM Pillager MN
KYLS-FM Ironton MO
KTRI-FM Mansfield MO
KKBL(FM) Monett MO
WCNA(FM) Potts Camp MS
WBBN(FM) Taylorsville MS
KHNK(FM) Columbia Falls MT
KLCM(FM) Lewistown MT
WPNC-FM Plymouth NC
WCVP-FM Robbinsville NC
WRAT(FM) Point Pleasant NJ
KZRM(FM) Chama NM
KANM(FM) Magdalena NM
KIVA(FM) Santa Rosa NM
KKIT(FM) Taos NM
WFLR-FM Dundee NY
WCQL(FM) Glens Falls NY
WVOS-FM Liberty NY
WJKW(FM) Athens OH
WNPQ(FM) New Philadelphia OH
WYNT(FM) Upper Sandusky OH
KYBE(FM) Frederick OK
KKBD(FM) Sallisaw OK
KOPA(FM) Woodward OK
KIXT(FM) Bay City OR
WGGI(FM) Benton PA
WAKZ(FM) Sharpsville PA
WCRI(FM) Block Island RI
KZZI(FM) Belle Fourche SD
WRZK(FM) Colonial Heights TN
WLQK(FM) Livingston TN
KBRA(FM) Freer TX
KHMC(FM) Goliad TX
KPWW(FM) Hooks TX
KCKL(FM) Malakoff TX
KFWR(FM) Mineral Wells TX
KSCH(FM) Sulphur Springs TX
KMGR(FM) Delta UT
KZHK(FM) Saint George UT
WGRQ(FM) Colonial Beach VA
KZML(FM) Quincy WA
WKSZ(FM) De Pere WI
WDMO(FM) Durand WI
WMQA-FM Minocqua WI
WBKY(FM) Portage WI
*WDKL(FM) Grafton WV
WSTG(FM) Princeton WV
KKHI(FM) Rock River WY

96.1 mhz

*KNOM-FM Nome AK
WXFL(FM) Florence AL
WRKH(FM) Mobile AL
WQKS-FM Montgomery AL
KMRX(FM) El Dorado AR
KCWD(FM) Harrison AR
KLPX(FM) Tucson AZ
KWRK(FM) Window Rock AZ
KSIQ(FM) Brawley CA
KMVE(FM) Mojave CA
KSQQ(FM) Morgan Hill CA
KYMX(FM) Sacramento CA
KRQB(FM) San Jacinto CA
KSLY-FM San Luis Obispo CA
KKXS(FM) Shingletown CA
KSLK(FM) Visalia CA
KIBT(FM) Fountain CO
KSME(FM) Greeley CO
KSTR-FM Montrose CO
WRXK-FM Bonita Springs FL
WTMP-FM Dade City FL
WEJZ(FM) Jacksonville FL
WHBX(FM) Tallahassee FL
WKLS(FM) Atlanta GA
KMXG(FM) Clinton IA
KCVM(FM) Hudson IA
KNWM(FM) Madrid IA
KID-FM Idaho Falls ID
WQQB(FM) Rantoul IL
WQLK(FM) Richmond IN
KANS(FM) Emporia KS
WKKQ(FM) Barbourville KY
WSTO(FM) Owensboro KY

WLXO(FM) Stamping Ground KY
KRVE(FM) Brusly LA
KYKZ(FM) Lake Charles LA
WSRS(FM) Worcester MA
WQHR(FM) Presque Isle ME
WHNN(FM) Bay City MI
WMAX-FM Holland MI
KQPR-FM Albert Lea MN
KGPZ(FM) Coleraine MN
KQHT(FM) Crookston MN
*KLRQ(FM) Clinton MO
WLZA(FM) Eupora MS
WIVG(FM) Tunica MS
WBBB(FM) Raleigh NC
WIBT(FM) Shelby NC
KYYZ(FM) Williston ND
*KINI(FM) Crookston NE
KICX-FM McCook NE
KQBW(FM) Omaha NE
WTTH(FM) Margate City NJ
WJYE(FM) Buffalo NY
WLVG(FM) Center Moriches NY
WVLF(FM) Norwood NY
WPKF(FM) Poughkeepsie NY
WODZ-FM Rome NY
WMTR-FM Archbold OH
WKFM(FM) Huron OH
KXXY-FM Oklahoma City OK
KITO-FM Vinita OK
KZEL-FM Eugene OR
KSRV-FM Ontario OR
KLKY(FM) Stanfield OR
WCTO(FM) Easton PA
WKST-FM Pittsburgh PA
WSOX(FM) Red Lion PA
WPHD(FM) South Waverly PA
WAEL-FM Maricao PR
WAVF(FM) Hanahan SC
KIXX(FM) Watertown SD
KCTX-FM Childress TX
KEZB(FM) Edna TX
KBTQ(FM) Harlingen TX
KKTX-FM Kilgore TX
KAGG(FM) Madisonville TX
KMRK-FM Odessa TX
KEYE-FM Perryton TX
KXXM(FM) San Antonio TX
KNCE(FM) Winters TX
WJDV(FM) Broadway VA
WROX-FM Exmore VA
WIVI(FM) Charlotte Amalie VI
WDEV-FM Warren VT
KXXO(FM) Olympia WA
KIXZ-FM Opportunity WA
WLKG(FM) Lake Geneva WI
WJMC-FM Rice Lake WI
WTCX(FM) Ripon WI
WXYM(FM) Tomah WI
WKWS(FM) Charleston WV
KKLX(FM) Worland WY

96.3 mhz

KXLW(FM) Houston AK
KHLS(FM) Blytheville AR
KTTG(FM) Mena AR
KSWG(FM) Wickenburg AZ
KFMI(FM) Eureka CA
KXOL-FM Los Angeles CA
KUBB(FM) Mariposa CA
KLZN(FM) Susanville CA
KXCM(FM) Twentynine Palms CA
WHUR-FM Washington DC
WXOF(FM) Yankeetown FL
WJIZ-FM Albany GA
KRTR-FM Kailua HI
KRNQ(FM) Keokuk IA
WLCN(FM) Atlanta IL
WBBM-FM Chicago IL
WJAA(FM) Austin IN
WNHT(FM) Churubusco IN
WHHH(FM) Indianapolis IN
KZDY(FM) Cawker City KS
KZCH(FM) Derby KS
KSSH(FM) Ingalls KS
KACZ(FM) Riley KS
WIVY(FM) Morehead KY

WXKY-FM Stanford KY
WRZE(FM) Nantucket MA
WLOB-FM Rumford ME
WDVD(FM) Detroit MI
WLXT(FM) Petoskey MI
KTTB(FM) Glencoe MN
KIHT(FM) Saint Louis MO
WUSJ(FM) Madison MS
KRZN(FM) Billings MT
KBAZ(FM) Hamilton MT
WRHT(FM) Morehead City NC
WPLY-FM Walpole NH
KBZU(FM) Albuquerque NM
KKLZ(FM) Las Vegas NV
WQXR-FM New York NY
WAJZ(FM) Voorheesville NY
WLVQ(FM) Columbus OH
WJSA-FM Jersey Shore PA
WKQW-FM Oil City PA
WKSP(FM) Aiken SC
WGOG(FM) Walhalla SC
WCJK(FM) Murfreesboro TN
WJBZ-FM Seymour TN
KXIT-FM Dalhart TX
KTDR(FM) Del Rio TX
KHEY-FM El Paso TX
KSCS(FM) Fort Worth TX
KLSN(FM) Hudson TX
KXXN(FM) Iowa Park TX
KAJZ(FM) Llano TX
KLLL-FM Lubbock TX
KGGB(FM) Yorktown TX
KXRK(FM) Provo UT
WROV-FM Martinsville VA
KRCW(FM) Royal City WA
WSFQ(FM) Peshtigo WI
WMAD(FM) Sauk City WI
WOTR(FM) Lost Creek WV
KQWY(FM) Lusk WY

96.5 mhz

KKIS-FM Soldotna AK
WMJJ(FM) Birmingham AL
KHTE-FM England AR
KDAP-FM Douglas AZ
KRFM(FM) Show Low AZ
KBKO-FM Bakersfield CA
KYDN(FM) Del Norte CA
KXXY(FM) San Diego CA
KOIT-FM San Francisco CA
KLCA(FM) Tahoe City CA
KFLS-FM Tulelake CA
KXPK(FM) Evergreen CO
KJBL(FM) Julesburg CO
WTIC-FM Hartford CT
WJTK(FM) Columbia City FL
WZNS(FM) Fort Walton Beach FL
WPOW(FM) Miami FL
WHTQ(FM) Orlando FL
WPCH(FM) Gray GA
WJCL-FM Savannah GA
KSOM(FM) Audubon IA
WMT-FM Cedar Rapids IA
KOZE-FM Lewiston ID
KLIX-FM Twin Falls ID
WKIB(FM) Anna IL
WZPN(FM) Farmington IL
WKOT(FM) Marseilles IL
WAZY-FM Lafayette IN
WGZB-FM Lanesville IN
WTGG(FM) Amite LA
KFTE(FM) Breaux Bridge LA
KVKI-FM Shreveport LA
WQHH(FM) Dewitt MI
WFAT(FM) Portage MI
WKLK-FM Cloquet MN
KJJK-FM Fergus Falls MN
KWWK(FM) Rochester MN
KRBZ(FM) Kansas City MO
KSPW(FM) Sparta MO
WKDJ-FM Clarksdale MS
WXHB(FM) Richton MS
KDZN(FM) Glendive MT
WOXL-FM Biltmore Forest NC
WFLB(FM) Laurinburg NC
KBYZ(FM) Bismarck ND

KRGI-FM Grand Island NE
WMLL(FM) Bedford NH
KLMA(FM) Hobbs NM
KBKZ(FM) Raton NM
WBKX(FM) Fredonia NY
WVNV(FM) Malone NY
WCMF-FM Rochester NY
WAKS(FM) Akron OH
WFTK(FM) Lebanon OH
KECO(FM) Elk City OK
KMMY(FM) Soper OK
KRAV(FM) Tulsa OK
KBDN(FM) Bandon OR
KWLZ-FM Warm Springs OR
WKYE(FM) Johnstown PA
*WPEL-FM Montrose PA
WRDW-FM Philadelphia PA
WCMA-FM Fajardo PR
*KNWC-FM Sioux Falls SD
WDOD-FM Chattanooga TN
WBFG(FM) Parker's Crossroads TN
KLTG(FM) Corpus Christi TX
KHMX(FM) Houston TX
KEVE(FM) Ingram TX
KNRX(FM) Sterling City TX
WCTG(FM) Chincoteague VA
WKLR(FM) Fort Lee VA
KJAQ(FM) Seattle WA
WKLH(FM) Milwaukee WI
WRKP(FM) Moundsville WV
WXCC(FM) Williamson WV
KQSW(FM) Rock Springs WY

96.7 mhz

WMXA(FM) Opelika AL
WKXK(FM) Pine Hill AL
KYDL(FM) Hot Springs AR
KOKR(FM) Newport AR
KDYN-FM Ozark AR
KRCY-FM Lake Havasu City AZ
KWMX(FM) Williams AZ
KALZ(FM) Fowler CA
KNOB(FM) Healdsburg CA
KUNA-FM La Quinta CA
KMRQ(FM) Manteca CA
KZAP(FM) Paradise CA
KCAL-FM Redlands CA
KWIZ(FM) Santa Ana CA
KLJR-FM Santa Paula CA
KSYV(FM) Solvang CA
KUUR(FM) Carbondale CO
WCTZ(FM) Stamford CT
WDXQ-FM Cochran GA
WLTM(FM) Peachtree City GA
WLYX(FM) Valdosta GA
KLBA-FM Albia IA
KKEX(FM) Preston ID
WGNX(FM) Colchester IL
WSSR(FM) Joliet IL
WKGL-FM Loves Park IL
WIHN(FM) Normal IL
WCVS-FM Virden IL
WHTI(FM) Alexandria IN
WBWB(FM) Bloomington IN
WCOE(FM) La Porte IN
WORX-FM Madison IN
WFML(FM) Vincennes IN
KGTR(FM) Larned KS
KBBE(FM) McPherson KS
WANV(FM) Annville KY
WBVR-FM Auburn KY
KMYO-FM Morgan City LA
KWCL-FM Oak Grove LA
WCEI-FM Easton MD
WDLD(FM) Halfway MD
WCME(FM) Boothbay Harbor ME
*WUFN(FM) Albion MI
WLXV(FM) Cadillac MI
WMJT(FM) McMillan MI
WUPZ(FM) Republic MI
WRGZ(FM) Rogers City MI
KKCQ-FM Bagley MN
KDOG(FM) North Mankato MN
KKSR(FM) Sartell MN
KCMQ(FM) Columbia MO
KAHR(FM) Poplar Bluff MO

WFFF-FM Columbia MS
WUJM(FM) Gulfport MS
WSEL-FM Pontotoc MS
KISN(FM) Belgrade MT
KZIN-FM Shelby MT
WKJX(FM) Elizabeth City NC
WNCC-FM Franklin NC
WKRX(FM) Roxboro NC
KQZZ(FM) Devils Lake ND
WLTN-FM Lisbon NH
WQSO(FM) Rochester NH
KNMB(FM) Cloudcroft NM
KMDZ(FM) Las Vegas NM
KHIX(FM) Carlin NV
WPTR(FM) Clifton Park NY
WYSX(FM) Morristown NY
WOLF-FM Oswego NY
WTSX(FM) Port Jervis NY
WXZO(FM) Willsboro NY
WCMJ(FM) Cambridge OH
WCSM-FM Celina OH
WBVI(FM) Fostoria OH
WKOV-FM Wellston OH
KBEL-FM Idabel OK
KCRF-FM Lincoln City OR
WVNW(FM) Burnham PA
WFRM-FM Coudersport PA
*WPGM-FM Danville PA
WLLF(FM) Mercer PA
WLTY(FM) Cayce SC
WBZT-FM Mauldin SC
KZMX-FM Hot Springs SD
WMOD(FM) Bolivar TN
WNKX-FM Centerville TN
WMYL(FM) Halls Crossroads TN
KTYS(FM) Flower Mound TX
KHFI-FM Georgetown TX
KXOX-FM Sweetwater TX
KQMB(FM) Levan UT
WWZW(FM) Buena Vista VA
WTSA-FM Brattleboro VT
KMMG(FM) Benton City WA
KWWW-FM Quincy WA
WBDK(FM) Algoma WI
WJJH(FM) Ashland WI
WLJY(FM) Whiting WI
WKMM(FM) Kingwood WV
KMGW(FM) Casper WY
KWHO(FM) Cody WY
KIMX(FM) Laramie WY

96.9 mhz

KYSC(FM) Fairbanks AK
WRSA-FM Decatur AL
WDJR(FM) Enterprise AL
KWLR(FM) Maumelle AR
KSSW(FM) Nashville AR
KMXP(FM) Phoenix AZ
KQZT(FM) Covelo CA
KHDR(FM) Lenwood CA
KEBT(FM) Lost Hills CA
KWAV(FM) Monterey CA
KSEG(FM) Sacramento CA
KCCY(FM) Pueblo CO
KBCR-FM Steamboat Springs CO
WINK-FM Fort Myers FL
WJGL(FM) Jacksonville FL
WKEZ-FM Tavenier FL
WRDO(FM) Fitzgerald GA
WAKB(FM) Wrens GA
KFMN(FM) Lihue HI
KIAQ(FM) Clarion IA
KKGL(FM) Nampa ID
WLBH-FM Mattoon IL
WXLP(FM) Moline IL
WWDV(FM) Zion IL
WHPZ(FM) Bremen IN
WKLO(FM) Hardinsburg IN
KKOW-FM Pittsburg KS
KFIX(FM) Plainville KS
WDDJ(FM) Paducah KY
WGKS(FM) Paris KY
KZMZ(FM) Alexandria LA
WTKK(FM) Boston MA
WBPW(FM) Presque Isle ME
WLAV-FM Grand Rapids MI

WBTI(FM) Lexington MI
WWCM(FM) Standish MI
KMFY(FM) Grand Rapids MN
KZBK(FM) Brookfield MO
KUPH(FM) Mountain View MO
WTCD(FM) Indianola MS
WXAB(FM) McLain MS
KQRV(FM) Deer Lodge MT
WYMY(FM) Goldsboro NC
WKKT(FM) Statesville NC
KZKX(FM) Seward NE
KCMI(FM) Terrytown NE
WFPG(FM) Atlantic City NJ
KDAG(FM) Farmington NM
WRRB(FM) Arlington NY
WGRF(FM) Buffalo NY
WEHN(FM) East Hampton NY
WOUR(FM) Utica NY
*WOKL(FM) Troy OH
WNKL(FM) Wauseon OH
WLRD(FM) Willard OH
KXOW(FM) Eldorado OK
KQOB(FM) Enid OK
KROG(FM) Grants Pass OR
WRRK(FM) Braddock PA
WLAN-FM Lancaster PA
WREQ(FM) Ridgebury PA
WNRT(FM) Manati PR
WSUY(FM) Charleston SC
KDLO-FM Watertown SD
WXBQ-FM Bristol TN
KMML-FM Amarillo TX
KXYL-FM Brownwood TX
KIOX-FM El Campo TX
KVMV(FM) McAllen TX
KMCM(FM) Odessa TX
KSCN(FM) Pittsburg TX
WWUZ(FM) Bowling Green VA
WSIG(FM) Mount Jackson VA
KGY-FM McCleary WA
KZTA(FM) Naches WA
KEZE(FM) Spokane WA
WWWX(FM) Oshkosh WI
WVVV(FM) Williamstown WV
KAML-FM Gillette WY
KMTN(FM) Jackson WY

97.1 mhz

WWMG(FM) Millbrook AL
KJMT(FM) Calico Rock AR
KAMD-FM Camden AR
KTZR-FM Green Valley AZ
KLSX(FM) Los Angeles CA
KTSE-FM Patterson CA
*KULV(FM) Ukiah CA
KSEQ(FM) Visalia CA
KZBR(FM) La Jara CO
WASH(FM) Washington DC
WSUN-FM Holiday FL
WOSN(FM) Indian River Shores FL
WSRV(FM) Gainesville GA
KNWB(FM) Hilo HI
WDRV(FM) Chicago IL
WLHK(FM) Shelbyville IN
KGGG(FM) Haven KS
WKEQ(FM) Somerset KY
WXCM(FM) Whitesville KY
WEZB(FM) New Orleans LA
*WLIC(FM) Frostburg MD
WQJZ(FM) Ocean Pines MD
WWBX(FM) Bangor ME
WKRK-FM Detroit MI
WGLQ(FM) Escanaba MI
KYCK(FM) Crookston MN
KTCZ-FM Minneapolis MN
KFTK(FM) Florissant MO
KNIM-FM Maryville MO
KAYQ(FM) Warsaw MO
WOKK(FM) Meridian MS
KKBR(FM) Billings MT
KALS(FM) Kalispell MT
WQMG-FM Greensboro NC
WYND-FM Hatteras NC
KYYX(FM) Minot ND
KELN(FM) North Platte NE
KBCQ-FM Roswell NM

KXPT(FM) Las Vegas NV
WQHT(FM) New York NY
WREO-FM Ashtabula OH
WBVB(FM) Coal Grove OH
WBNS-FM Columbus OH
KYAL-FM Muskogee OK
KYCH-FM Portland OR
WBHT(FM) Mountain Top PA
WOWY(FM) University Park PA
KPSD(FM) Faith SD
WRQQ(FM) Goodlettsville TN
WHRK(FM) Memphis TN
KTHT(FM) Cleveland TX
KEGL(FM) Fort Worth TX
KVRP-FM Haskell TX
KCYN(FM) Moab UT
KZHT(FM) Salt Lake City UT
WZRT(FM) Rutland VT
KXRX(FM) Walla Walla WA
WCOW-FM Sparta WI
WTRW(FM) Two Rivers WI
WDBS(FM) Sutton WV

97.3 mhz

KEAG(FM) Anchorage AK
WNCB(FM) Gardendale AL
KDEW-FM De Witt AR
KPKR(FM) Parker AZ
KQNO(FM) Coalinga CA
KNCQ(FM) Redding CA
KSON-FM San Diego CA
KLLC(FM) San Francisco CA
KBCO-FM Boulder CO
WZBG(FM) Litchfield CT
WFLC(FM) Miami FL
WSKY-FM Micanopy FL
WRAK-FM Bainbridge GA
WAEV(FM) Savannah GA
KHKI(FM) Des Moines IA
KGRR(FM) Epworth IA
KHDK(FM) New London IA
KLCE(FM) Blackfoot ID
WRUL(FM) Carmi IL
WFYR(FM) Elmwood IL
WTIM-FM Taylorville IL
WMEE(FM) Fort Wayne IN
KKJQ(FM) Garden City KS
WYGY(FM) Fort Thomas KY
WJSN-FM Jackson KY
KJMG(FM) Bastrop LA
KMDL(FM) Kaplan LA
KDBH(FM) Natchitoches LA
WJFD-FM New Bedford MA
WJDF(FM) Orange MA
WMJO(FM) Essexville MI
WDEE-FM Reed City MI
*KDNW(FM) Duluth MN
KRVY-FM Starbuck MN
KCXM(FM) Lee's Summit MO
KCDG(FM) Madison MO
KYRX(FM) Marble Hill MO
KXUS(FM) Springfield MO
WFMN(FM) Flora MS
WKSO(FM) Natchez MS
WFMM(FM) Sumrall MS
WKBC-FM North Wilkesboro NC
WMNX(FM) Wilmington NC
KRGY(FM) Aurora NE
KBLR-FM Blair NE
WXKW(FM) Millville NJ
KKSS(FM) Santa Fe NM
KZTQ(FM) Carson City NV
WYXL(FM) Ithaca NY
WMYY(FM) Schoharie NY
WZAD(FM) Wurtsboro NY
WJZE(FM) Oak Harbor OH
KOJK(FM) Blanchard OK
KSHR-FM Coquille OR
*WZZE(FM) Glen Mills PA
WRVV(FM) Harrisburg PA
WPCL(FM) Northern Cambria PA
WOYE(FM) Rio Grande PR
KMXC(FM) Sioux Falls SD
WKJQ-FM Parsons TN
WUUS-FM South Pittsburg TN
WTNV(FM) Tiptonville TN

KGEE(FM) Pecos TX
KSTQ-FM Plainview TX
KAJA(FM) San Antonio TX
KQHN(FM) Waskom TX
WGH-FM Newport News VA
*KKRS(FM) Davenport WA
KBSG-FM Tacoma WA
WQBW(FM) Milwaukee WI
WKWK-FM Wheeling WV

97.5 mhz
WZLM(FM) Goodwater AL
WABB-FM Mobile AL
KQUS-FM Hot Springs AR
KMVA(FM) Dewey-Humboldt AZ
KSZR(FM) Oro Valley AZ
KABX-FM Merced CA
KLYY(FM) Riverside CA
KRUZ(FM) Santa Barbara CA
KWUZ(FM) Poncha Springs CO
KSRX(FM) Sterling CO
WPCV(FM) Winter Haven FL
WUFF-FM Eastman GA
WPZE(FM) Fayetteville GA
WHLJ(FM) Statenville GA
KZGZ(FM) Hagatna GU
KHNR-FM Honolulu HI
KBVU-FM Alta IA
*KTWD(FM) Wallace ID
WDLJ(FM) Breese IL
WHMS-FM Champaign IL
WBBA-FM Pittsfield IL
WZOK(FM) Rockford IL
KJCK-FM Junction City KS
WZZP(FM) Hopkinsville KY
WAMZ(FM) Louisville KY
KTJZ(FM) Tallulah LA
WICO-FM Salisbury MD
WIGY(FM) Madison ME
WYTZ(FM) Bridgman MI
WKLT(FM) Kalkaska MI
WJIM-FM Lansing MI
*WYDM(FM) Monroe MI
WEFG-FM Whitehall MI
KDKK-FM Park Rapids MN
KNXR(FM) Rochester MN
KOEA(FM) Doniphan MO
KJMO(FM) Linn MO
KNMO(FM) Nevada MO
WWMS(FM) Oxford MS
KOZB(FM) Livingston MT
KKCT(FM) Bismarck ND
KQSK(FM) Chadron NE
WOKQ(FM) Dover NH
WJJZ(FM) Burlington NJ
KPHD(FM) Elko NV
KVEG(FM) Mesquite NV
WHAZ-FM Hoosick Falls NY
WALK-FM Patchogue NY
WFRY-FM Watertown NY
WONE-FM Akron OH
WVNU(FM) Greenfield OH
WTGR(FM) Union City OH
KPAK(FM) Alva OK
KMOD-FM Tulsa OK
KNLR(FM) Bend OR
KSHL(FM) Gleneden Beach OR
WDDH(FM) Saint Marys PA
WIOB(FM) Mayaguez PR
WCOS-FM Columbia SC
WJXB-FM Knoxville TN
WLLX(FM) Lawrenceburg TN
KFNC(FM) Beaumont TX
KBNA-FM El Paso TX
KFTX(FM) Kingsville TX
KGKL-FM San Angelo TX
KLAK(FM) Tom Bean TX
KWTX-FM Waco TX
KOAY(FM) Coalville UT
WWWV(FM) Charlottesville VA
WQOK(FM) South Boston VA
WTNN(FM) Bristol VT
KOLW(FM) Othello WA
KTRT(FM) Winthrop WA
WHDG(FM) Rhinelander WI
WQBE-FM Charleston WV
WLTF(FM) Martinsburg WV
KDLY(FM) Lander WY

97.7 mhz
WKKR(FM) Auburn AL
WHPH(FM) Jemison AL
WKLD(FM) Oneonta AL
WKGA(FM) Thomaston AL
WKXM-FM Winfield AL
*KJSM-FM Augusta AR
KAVV(FM) Benson AZ
*KQVO(FM) Calexico CA
KWIN(FM) Lodi CA
KFFG(FM) Los Altos CA
KRCK-FM Mecca CA
KVVS(FM) Mojave CA
KVRV(FM) Monte Rio CA
KHHZ(FM) Oroville CA
KSMJ(FM) Shafter CA
KZYR(FM) Avon CO
WCTY(FM) Norwich CT
WAFL(FM) Milford DE
WYYX(FM) Bonifay FL
WAVK(FM) Marathon FL
WTLQ-FM Punta Rassa FL
WMGZ(FM) Eatonton GA
WGPB(FM) Rome GA
WTCQ(FM) Vidalia GA
WWUF(FM) Waycross GA
KCRR(FM) Grundy Center IA
KHBT(FM) Humboldt IA
KOTM-FM Ottumwa IA
KZBG(FM) Lapwai ID
WMOI(FM) Monmouth IL
WLCE(FM) Petersburg IL
WSTQ(FM) Streator IL
WQUL(FM) West Frankfort IL
WOZW(FM) Goshen IN
WLQI(FM) Rensselaer IN
WCLS(FM) Spencer IN
KSNP(FM) Burlington KS
WWKY(FM) Providence KY
KNBB(FM) Dubach LA
KAPB-FM Marksville LA
WKAF(FM) Brockton MA
*WYAJ(FM) Sudbury MA
WSNI(FM) Winchendon MA
WYRX(FM) Lexington Park MD
WCXU(FM) Caribou ME
WNSX(FM) Winter Harbor ME
WMRX-FM Beaverton MI
WOLV(FM) Houghton MI
WMLQ(FM) Manistee MI
WTGV-FM Sandusky MI
KLGR-FM Redwood Falls MN
KPOW-FM La Monte MO
KHZR(FM) Potosi MO
KQMO(FM) Shell Knob MO
WRBJ-FM Brandon MS
WTYJ(FM) Fayette MS
WTYL-FM Tylertown MS
KGLM-FM Anaconda MT
WKIX(FM) Kinston NC
WGTI(FM) Windsor NC
KMTY(FM) Holdrege NE
KBBX-FM Nebraska City NE
KLVO(FM) Belen NM
KPSA-FM Lordsburg NM
WEXT(FM) Amsterdam NY
WENI-FM Big Flats NY
WCZX(FM) Hyde Park NY
WILE-FM Byesville OH
WGGN(FM) Castalia OH
WAXZ(FM) Georgetown OH
WCJO(FM) Jackson OH
*WTGN(FM) Lima OH
WOXY(FM) Oxford OH
KICM(FM) Healdton OK
*KHIM(FM) Mangum OK
KRAT(FM) Altamont OR
KACI-FM The Dalles OR
WLER-FM Butler PA
WVRT(FM) Mill Hall PA
WLKH(FM) Somerset PA
WBRQ(FM) Cidra PR
WWXM(FM) Garden City SC
KNBZ(FM) Redfield SD
WTNE-FM Trenton TN
KATX(FM) Eastland TX
KLTO-FM McQueeney TX
KBMI(FM) Roma TX
KWRW(FM) Rusk TX
KALK(FM) Winfield TX
KCYQ(FM) Elsinore UT
WRIC-FM Richlands VA
WGMT(FM) Lyndon VT
KYSN(FM) East Wenatchee WA
KFMY(FM) Raymond WA
WAQE-FM Barron WI
WGLR-FM Lancaster WI
WFDL-FM Lomira WI
WSRG(FM) Sturgeon Bay WI
WRON-FM Ronceverte WV

97.9 mhz
WRVX(FM) Eufaula AL
WVOK-FM Oxford AL
WJWZ(FM) Wetumpka AL
KTLO-FM Mountain Home AR
KUPD-FM Tempe AZ
KPOD-FM Crescent City CA
KLAX-FM East Los Angeles CA
KTTA(FM) Esparto CA
KMGV(FM) Fresno CA
KLUK(FM) Needles CA
KYZZ(FM) Salinas CA
KISZ-FM Cortez CO
WPKX(FM) Enfield CT
WXTB(FM) Clearwater FL
WFKS(FM) Neptune Beach FL
WRMF(FM) Palm Beach FL
WJZT(FM) Woodville FL
WDMG-FM Ambrose GA
WIBB-FM Fort Valley GA
KKBG(FM) Hilo HI
*KOWI(FM) Lamoni IA
*KCMR(FM) Mason City IA
KSEZ(FM) Sioux City IA
KQFC(FM) Boise ID
WLUP-FM Chicago IL
WXEF(FM) Effingham IL
WBBE(FM) Heyworth IL
*WGNR-FM Anderson IN
WSLM-FM Salem IN
KWGB(FM) Colby KS
KRBB(FM) Wichita KS
WZQQ(FM) Hyden KY
KQLK(FM) De Ridder LA
WIYY(FM) Baltimore MD
WBEY-FM Crisfield MD
WJBQ(FM) Portland ME
WJLB(FM) Detroit MI
WGRD-FM Grand Rapids MI
WIHC(FM) Newberry MI
WEVE-FM Eveleth MN
KICK-FM Palmyra MO
KBXB(FM) Sikeston MO
KFBD-FM Waynesville MO
KXDG(FM) Webb City MO
WCPR-FM D'Iberville MS
WBAQ(FM) Greenville MS
WHTU(FM) Newton MS
KVVR(FM) Dutton MT
KDXT(FM) Victor MT
WNBB(FM) Bayboro NC
WPEG(FM) Concord NC
WTRG(FM) Gaston NC
KHRU(FM) Beulah ND
*KFNW-FM Fargo ND
KNPE(FM) Hyannis NE
KRSI(FM) Garapan-Saipan NP
WSKQ-FM New York NY
WPXY-FM Rochester NY
WSKS(FM) Whitesboro NY
WRIP(FM) Windham NY
WNCI(FM) Columbus OH
KJMZ(FM) Cache OK
KWLS-FM Edmond OK
KZBB(FM) Poteau OK
KNRQ-FM Eugene OR
KZTB(FM) Milton-Freewater OR
*KRRC(FM) Portland OR
WXTA(FM) Edinboro PA
WBSX(FM) Hazleton PA
WIIZ(FM) Blackville SC
KTPT(FM) Rapid City SD
WSIX-FM Nashville TN
KGNC-FM Amarillo TX
KBFB(FM) Dallas TX
KBXX(FM) Houston TX
KODM(FM) Odessa TX
KBZN(FM) Ogden UT
WZZU(FM) Lynchburg VA
WGOD-FM Charlotte Amalie VI
WSPT-FM Stevens Point WI
WKKW(FM) Fairmont WV
WMGA(FM) Kenova WV
KQMY(FM) Cheyenne WY
KTAG(FM) Cody WY
KZWB(FM) Green River WY

98.1 mhz
KLEF(FM) Anchorage AK
KWLF(FM) Fairbanks AK
WTXT(FM) Fayette AL
*KVIP-FM Redding CA
KIFM(FM) San Diego CA
KISQ(FM) San Francisco CA
KKJG(FM) San Luis Obispo CA
KRXV(FM) Yermo CA
KKFM(FM) Colorado Springs CO
KAYW(FM) Meeker CO
*WQAQ(FM) Hamden CT
WKZE-FM Salisbury CT
WOCM(FM) Selbyville DE
WTKE-FM Holt FL
WQHL-FM Live Oak FL
WNUE-FM Titusville FL
WMRZ(FM) Dawson GA
WELT(FM) East Dublin GA
WMGP(FM) Hogansville GA
KAWV(FM) Lihue HI
KHAK(FM) Cedar Rapids IA
KGTM(FM) Rexburg ID
WZOE-FM Princeton IL
WIBN(FM) Earl Park IN
WRAY-FM Princeton IN
KSKZ(FM) Copeland KS
KUSN(FM) Dearing KS
KUDL(FM) Kansas City KS
WBUL-FM Lexington KY
WDGL(FM) Baton Rouge LA
WCTK(FM) New Bedford MA
WCXV(FM) Van Buren ME
WGFN(FM) Glen Arbor MI
*WEUL(FM) Kingsford MI
WKCQ(FM) Saginaw MI
KBEW-FM Blue Earth MN
WWJO(FM) Saint Cloud MN
KOZX(FM) Cabool MO
KYKY(FM) Saint Louis MO
WMXI(FM) Laurel MS
WQSM(FM) Fayetteville NC
WOBX-FM Manteo NC
KFGE(FM) Milford NE
KBAC(FM) Las Vegas NM
KBUL-FM Carson City NV
WHWK(FM) Binghamton NY
WKDD(FM) Canton OH
*WUDR(FM) Dayton OH
WDFM(FM) Defiance OH
KFYZ-FM Bennington OK
KVRO(FM) Stillwater OK
KCYS(FM) Seaside OR
WFGY(FM) Altoona PA
WOGL(FM) Philadelphia PA
WYBB(FM) Folly Beach SC
WHZT(FM) Seneca SC
WXMX(FM) Millington TN
WLND(FM) Signal Mountain TN
KTLT(FM) Anson TX
KVET-FM Austin TX
KKUL-FM Groveton TX
KRRG(FM) Laredo TX
KKCL(FM) Lorenzo TX
KTAL-FM Texarkana TX
KREC(FM) Brian Head UT
WBRF(FM) Galax VA
WTVR-FM Richmond VA
WJJR(FM) Rutland VT
KING-FM Seattle WA
KISC(FM) Spokane WA
WISM-FM Altoona WI
WLKN(FM) Cleveland WI
WMGN(FM) Madison WI

98.3 mhz
WDLT-FM Chickasaw AL
WAGH(FM) Fort Mitchell AL
WKEA-FM Scottsboro AL
WTRB-FM Sylacauga AL
KQSM-FM Bentonville AR
KFCM(FM) Cherokee Village AR
KOHT(FM) Marana AZ
KKFR(FM) Mayer AZ
KQSS(FM) Miami AZ
KZLA(FM) Huron CA
KXBX-FM Lakeport CA
KDAR(FM) Oxnard CA
KWNN(FM) Turlock CA
KRCV(FM) West Covina CA
KEJJ(FM) Gunnison CO
KATR-FM Otis CO
WDAQ(FM) Danbury CT
WILI-FM Willimantic CT
WWRZ(FM) Fort Meade FL
WRTO-FM Goulds FL
WGCO(FM) Midway GA
WQXZ(FM) Pinehurst GA
KJMD(FM) Pukalani HI
KWQW(FM) Boone IA
KDZY(FM) McCall ID
KSNQ(FM) Twin Falls ID
WCCQ(FM) Crest Hill IL
WWHP(FM) Farmer City IL
WRIK-FM Metropolis IL
WRAN(FM) Tower Hill IL
WRDZ-FM Plainfield IN
WZZY(FM) Winchester IN
WQXE(FM) Elizabethtown KY
WOKE(FM) Garrison KY
WHAY(FM) Whitley City KY
KZRZ(FM) West Monroe LA
WHAI(FM) Greenfield MA
WSMD-FM Mechanicsville MD
WTWR-FM Luna Pier MI
WRUP(FM) Munising MI
WLCS(FM) North Muskegon MI
*WCMZ-FM Sault Ste. Marie MI
WCSY-FM South Haven MI
WBJI(FM) Blackduck MN
KQYB(FM) Spring Grove MN
WCKK(FM) Carthage MS
WDFX(FM) Cleveland MS
WJDR(FM) Prentiss MS
KBEV-FM Dillon MT
WDLZ(FM) Murfreesboro NC
WSFM(FM) Oak Island NC
WIST-FM Thomasville NC
WLGT(FM) Washington NC
KXGT(FM) Carrington ND
KBBN-FM Broken Bow NE
WLNH-FM Laconia NH
WMGQ(FM) New Brunswick NJ
WTKU-FM Ocean City NJ
WVIN-FM Bath NY
WKJY(FM) Hempstead NY
WSUL(FM) Monticello NY
*WPSA(FM) Paul Smiths NY
WTRY-FM Rotterdam NY
WQRS(FM) Salamanca NY
WYBL(FM) Ashtabula OH
WPKO-FM Bellefontaine OH
WXXR(FM) Fredericktown OH
*WKET(FM) Kettering OH
WLGN-FM Logan OH
KTWS(FM) Bend OR
KLDR(FM) Harbeck-Fruitdale OR
KPPK(FM) Rainier OR
WOGI(FM) Charleroi PA
WWBE(FM) Mifflinburg PA
WIDI(FM) Quebradillas PR
WHHD(FM) Clearwater SC
WLJI(FM) Summerton SC

KLQT(FM) Wessington Springs SD
WRJB(FM) Camden TN
WKSR-FM Pulaski TN
WLOD-FM Sweetwater TN
KPDB(FM) Big Lake TX
KBOC(FM) Bridgeport TX
KORA-FM Bryan TX
KULM(FM) Columbus TX
KICA-FM Farwell TX
*KYAR(FM) Gatesville TX
KLHB(FM) Odem TX
KYYK(FM) Palestine TX
KPTX(FM) Pecos TX
KXDJ(FM) Spearman TX
KARB(FM) Price UT
WLUS-FM Clarksville VA
WKSI-FM Stephens City VA
WMYP(FM) Frederiksted VI
KEYW(FM) Pasco WA
WJMR-FM Menomonee Falls WI
WCQM(FM) Park Falls WI
WCEF(FM) Ripley WV
KADQ-FM Evanston WY
KGRK(FM) Glenrock WY
KZZS(FM) Story WY
KERM(FM) Torrington WY

98.5 mhz

WINL(FM) Linden AL
KURB(FM) Little Rock AR
KRDX(FM) Vail AZ
KWXY-FM Cathedral City CA
KDFO-FM Delano CA
KSAY(FM) Fort Bragg CA
KRXQ(FM) Sacramento CA
KUFX(FM) San Jose CA
KYGO-FM Denver CO
KAAI(FM) Palisade CO
WGBG(FM) Seaford DE
WKTK(FM) Crystal River FL
WFSY(FM) Panama City FL
WDEO-FM San Carlos Park FL
WSBH(FM) Satellite Beach FL
WSB-FM Atlanta GA
WLPF(FM) Ocilla GA
KDNN(FM) Honolulu HI
KOEL-FM Cedar Falls IA
KQKQ-FM Council Bluffs IA
KLLP(FM) Chubbuck ID
KZID(FM) Orofino ID
WPIA(FM) Eureka IL
WXXQ(FM) Freeport IL
WINH(FM) Paris IL
WQKZ(FM) Ferdinand IN
WMYK(FM) Peru IN
KSAJ-FM Abilene KS
WYLD-FM New Orleans LA
WBMX(FM) Boston MA
WEBB(FM) Waterville ME
WNWN-FM Coldwater MI
WUPS(FM) Houghton Lake MI
*KTIS-FM Minneapolis MN
KTJJ(FM) Farmington MO
KWKJ(FM) Windsor MO
WZLQ(FM) Tupelo MS
KGHL-FM Billings MT
KBBZ(FM) Kalispell MT
WDWG(FM) Rocky Mount NC
KRKU(FM) McCook NE
WKMK(FM) Ocean Acres NJ
KABG(FM) Los Alamos NM
KLUC-FM Las Vegas NV
WCTW(FM) Catskill NY
WCKM-FM Lake George NY
WKSE(FM) Niagara Falls NY
WNYR-FM Waterloo NY
WBZB(FM) Westhampton NY
WRRM(FM) Cincinnati OH
WNCX(FM) Cleveland OH
*WCMO(FM) Marietta OH
KACO(FM) Apache OK
KVOO-FM Tulsa OK
WKEL(FM) Confluence PA
WGYI(FM) Oil City PA
WKRZ(FM) Wilkes-Barre PA
WYCR(FM) York-Hanover PA

WPRM-FM San Juan PR
WBZF(FM) Hartsville SC
WLXC(FM) Lexington SC
WDAI(FM) Pawley's Island SC
WGIC(FM) Cookeville TN
WTFM(FM) Kingsport TN
KGAP(FM) Clarksville TX
KYMI(FM) Los Ybanez TX
KGBT-FM McAllen TX
KTJM(FM) Port Arthur TX
KCUB-FM Ranger TX
KRXT(FM) Rockdale TX
KBBT(FM) Schertz TX
KXME(FM) Wellington TX
KIFX(FM) Roosevelt UT
WACL(FM) Elkton VA
KEYG-FM Grand Coulee WA
WQLH(FM) Green Bay WI

98.7 mhz

WBHK(FM) Warrior AL
KLBQ(FM) El Dorado AR
KPKX(FM) Phoenix AZ
KYSR(FM) Los Angeles CA
KSXY(FM) Middletown CA
KLOQ-FM Winton CA
KRTZ(FM) Cortez CO
WNLC(FM) East Lyme CT
WMZQ-FM Washington DC
WKGR(FM) Fort Pierce FL
WLLD(FM) Holmes Beach FL
WCNK(FM) Key West FL
WYCT(FM) Pensacola FL
WISK-FM Americus GA
WBTY(FM) Homerville GA
KMGO(FM) Centerville IA
KSMA-FM Osage IA
WFMT(FM) Chicago IL
WNNS(FM) Springfield IL
WQME(FM) Anderson IN
WASK-FM Battle Ground IN
KFH-FM Clearwater KS
WHOP-FM Hopkinsville KY
WKDO-FM Liberty KY
KKST(FM) Oakdale LA
WVMV(FM) Detroit MI
WFGR(FM) Grand Rapids MI
WGLI(FM) Hancock MI
KQWB-FM Moorhead MN
KISD(FM) Pipestone MN
KWTO-FM Springfield MO
WNEV(FM) Friar's Point MS
WJKK(FM) Vicksburg MS
KXDR(FM) Hamilton MT
WSMW(FM) Greensboro NC
WILT(FM) Jacksonville NC
KACL(FM) Bismarck ND
KSID-FM Sidney NE
WBYY(FM) Somersworth NH
WINQ(FM) Winchester NH
WCZT(FM) Villas NJ
*KMTH(FM) Maljamar NM
KKVS(FM) Truth or Consequences NM
WGMM(FM) Corning NY
WRKS(FM) New York NY
WPAC(FM) Ogdensburg NY
WLZW(FM) Utica NY
*WYKL(FM) Crestline OH
*WSLN(FM) Delaware OH
WYRO(FM) McArthur OH
KYTT-FM Coos Bay OR
KUBQ(FM) La Grande OR
KARO(FM) Nyssa OR
KUPL-FM Portland OR
WSGY(FM) Pleasant Gap PA
WJZG(FM) Culebra PR
WYKZ(FM) Beaufort SC
KOUT(FM) Rapid City SD
WOKI(FM) Oliver Springs TN
KPRF(FM) Amarillo TX
KLUV(FM) Dallas TX
KZAM(FM) Pleasant Valley TX
KELI(FM) San Angelo TX
KTXN-FM Victoria TX
KBEE(FM) Salt Lake City UT
WNOR(FM) Norfolk VA

KMNA(FM) Mabton WA
WMDC(FM) Mayville WI
WRVZ(FM) Pocatalico WV
WOVK(FM) Wheeling WV
KRSV-FM Afton WY
KUSZ(FM) Laramie WY

98.9 mhz

KYMG(FM) Anchorage AK
WBAM-FM Montgomery AL
KWLU(FM) Chester CA
KCVR-FM Columbia CA
KSOF(FM) Dinuba CA
KHWY(FM) Essex CA
KRVC(FM) Hornbrook CA
KSOL(FM) San Francisco CA
KKMG(FM) Pueblo CO
WGUF(FM) Marco FL
WBCG(FM) Murdock FL
WMMO(FM) Orlando FL
WBZE(FM) Tallahassee FL
WQMT(FM) Chatsworth GA
KITH(FM) Kapaa HI
KGRA(FM) Jefferson IA
KQCR-FM Parkersburg IA
WJEZ(FM) Dwight IL
WISH-FM Galatia IL
WHQQ(FM) Neoga IL
WLKU(FM) Rock Island IL
WZKF(FM) Salem IN
KKRK(FM) Coffeyville KS
KQRC-FM Leavenworth KS
WSIP-FM Paintsville KY
WUUU(FM) Franklinton LA
WORC-FM Webster MA
WSBY-FM Salisbury MD
WCLZ(FM) Brunswick ME
WKLZ-FM Petoskey MI
WOWE(FM) Vassar MI
KTCO(FM) Duluth MN
KZPK(FM) Paynesville MN
KFLW(FM) Saint Robert MO
WAJV(FM) Brooksville MS
*WLKO(FM) Quitman MS
KAAK(FM) Great Falls MT
WNBR-FM Windsor NC
KKPR-FM Kearney NE
KBZB(FM) Pioche NV
WBZA(FM) Rochester NY
WXMG(FM) Upper Arlington OH
WBYR(FM) Van Wert OH
WMXY(FM) Youngstown OH
KYIS(FM) Oklahoma City OK
KWDP(FM) Prineville OR
WQKY(FM) Emporium PA
WQLV(FM) Millersburg PA
WUSL(FM) Philadelphia PA
WYAS(FM) Vieques PR
WAZS-FM McClellanville SC
WSPA-FM Spartanburg SC
WLSQ(FM) Byrdstown TN
WANT(FM) Lebanon TN
WKIM(FM) Munford TN
KNOS(FM) Albany TX
KTUX(FM) Carthage TX
KLMO-FM Dilley TX
KHHL(FM) Leander TX
KLOW(FM) Reno TX
KLYD(FM) Snyder TX
WWLB(FM) Midlothian VA
WOKO(FM) Burlington VT
KWJZ(FM) Seattle WA
KKZX(FM) Spokane WA
*WVCX(FM) Tomah WI
WDNE-FM Elkins WV
KOUZ(FM) Manville WY

99.1 mhz

KRUP(FM) Dillingham AK
WDGM(FM) Greensboro AL
WAHR(FM) Huntsville AL
KMAG(FM) Fort Smith AR
KSMD(FM) Pangburn AR
KVMZ(FM) Waldo AR
KOFH(FM) Nogales AZ
KTMG(FM) Prescott AZ
KFMM(FM) Thatcher AZ

KJNY(FM) Ferndale CA
KGGI(FM) Riverside CA
KSQL(FM) Santa Cruz CA
KXFM(FM) Santa Maria CA
*KARA(FM) Williams CA
KMTS(FM) Glenwood Springs CO
KUAD-FM Windsor CO
WPLR(FM) New Haven CT
WWOJ(FM) Avon Park FL
WQIK-FM Jacksonville FL
WEDR(FM) Miami FL
WDEN-FM Macon GA
KAGB(FM) Waimea HI
KSKB(FM) Brooklyn IA
KUPI-FM Idaho Falls ID
KTMB(FM) Mountain Home ID
WXTT(FM) Danville IL
*KJIL(FM) Copeland KS
*KTLI(FM) El Dorado KS
KSEK-FM Girard KS
WCBL-FM Benton KY
WHSX(FM) Edmonton KY
WJMM-FM Keene KY
WWKN(FM) Morgantown KY
KXKC(FM) New Iberia LA
WPLM-FM Plymouth MA
WLZL(FM) Annapolis MD
WLKE(FM) Bar Harbor ME
WSMK(FM) Buchanan MI
WFMK(FM) East Lansing MI
WIKB-FM Iron River MI
KEEZ-FM Mankato MN
KLLZ-FM Walker MN
KFUO-FM Clayton MO
KYOO-FM Halfway MO
WYMX(FM) Greenwood MS
WKNN-FM Pascagoula MS
KCMM(FM) Belgrade MT
KHZS(FM) Saint Regis MT
WVOD(FM) Manteo NC
WZFX(FM) Whiteville NC
KCAD(FM) Dickinson ND
WNNH(FM) Henniker NH
WXRG(FM) Whitefield NH
WAWZ(FM) Zarephath NJ
KCLV-FM Clovis NM
KGLX(FM) Gallup NM
KXMT(FM) Taos NM
KKFT(FM) Gardnerville-Minden NV
WAAL(FM) Binghamton NY
WHKO(FM) Dayton OH
WFRO-FM Fremont OH
KODZ(FM) Eugene OR
WRKW(FM) Ebensburg PA
WUKQ-FM Mayaguez PR
KZNC(FM) Huron SD
WNML-FM Loudon TN
KAYG(FM) Camp Wood TX
KRYS-FM Corpus Christi TX
KFZO(FM) Denton TX
KNES(FM) Fairfield TX
KODA(FM) Houston TX
KHKX(FM) Odessa TX
WXGM-FM Gloucester VA
WJNV(FM) Jonesville VA
WSLQ(FM) Roanoke VA
KUJ-FM Burbank WA
WMYX(FM) Milwaukee WI
WKFX(FM) Rice Lake WI
WGGE(FM) Parkersburg WV
KNYN(FM) Fort Bridger WY
KWYW(FM) Lost Cabin WY
KWAP(FM) Pine Haven WY

99.3 mhz

WMFC-FM Monroeville AL
KVLD(FM) Atkins AR
KAPW(FM) Cotton Plant AR
KZYP(FM) Pine Bluff AR
KMZQ(FM) Payson AZ
KKBB(FM) Bakersfield CA
KJWL(FM) Fresno CA
*KLVS(FM) Grass Valley CA
KMXX(FM) Imperial CA
KVYN(FM) Saint Helena CA
KNNN(FM) Shasta Lake City CA

KJOY(FM) Stockton CA
KPCR(FM) Burlington CO
WLRQ-FM Cocoa FL
WJBX(FM) Fort Myers Beach FL
WFBI(FM) Inglis FL
WEBZ(FM) Mexico Beach FL
WCON-FM Cornelia GA
WKCN(FM) Lumpkin GA
WBAW-FM Pembroke GA
KFFF-FM Boone IA
KKBZ(FM) Clarinda IA
KDST(FM) Dyersville IA
KWAY-FM Waverly IA
WDUK(FM) Havana IL
WAJK(FM) La Salle IL
WXFM(FM) Mount Zion IL
WSCH(FM) Aurora IN
WKVI-FM Knox IN
WCJC(FM) Van Buren IN
KWIC(FM) Topeka KS
WWKF(FM) Fulton KY
WVLE(FM) Scottsville KY
KPCH(FM) Ruston LA
WLZX(FM) Northampton MA
WKTJ-FM Farmington ME
WBQQ(FM) Kennebunk ME
WATZ-FM Alpena MI
WBNZ(FM) Frankfort MI
WMSH-FM Sturgis MI
WJQK(FM) Zeeland MI
KXRZ(FM) Alexandria MN
KWNO-FM Rushford MN
KKDQ(FM) Thief River Falls MN
KCLR-FM Boonville MO
KCGQ-FM Gordonville MO
KUNQ(FM) Houston MO
WBVV(FM) Guntown MS
WHER(FM) Heidelberg MS
KMXE-FM Red Lodge MT
WQDK(FM) Ahoskie NC
WZAX(FM) Nashville NC
KETT(FM) Mitchell NE
KHZY(FM) Overton NE
WFRD(FM) Hanover NH
WZBZ(FM) Pleasantville NJ
KVLV-FM Fallon NV
KRGT(FM) Indian Springs NV
WRWC(FM) Ellenville NY
WLLG(FM) Lowville NY
WSNN(FM) Potsdam NY
WLLW(FM) Seneca Falls NY
WTNS-FM Coshocton OH
WNXT-FM Portsmouth OH
KADA-FM Ada OK
KCDL(FM) Cordell OK
KGVE(FM) Grove OK
KLOR-FM Ponca City OK
WOXX(FM) Franklin PA
WHKF(FM) Harrisburg PA
WZXR(FM) South Williamsport PA
WPKL(FM) Uniontown PA
WJZS(FM) Block Island RI
WBT-FM Chester SC
WWKT-FM Kingstree SC
WPBX(FM) Crossville TN
WTZR(FM) Elizabethton TN
WNRX(FM) Jefferson City TN
WZLT(FM) Lexington TN
KPSM(FM) Brownwood TX
KEFH(FM) Clarendon TX
KLGO(FM) Thorndale TX
KAJK(FM) White Oak TX
WVES(FM) Accomac VA
WFQX(FM) Front Royal VA
WKJM(FM) Petersburg VA
WKJS(FM) Richmond VA
WYSK-FM Spotsylvania VA
KDDS-FM Elma WA
KDRM(FM) Moses Lake WA
KQSN(FM) Naches WA
WDMP-FM Dodgeville WI
WKEB(FM) Medford WI
WOWN(FM) Shawano WI
KKTY-FM Douglas WY
KTGA(FM) Saratoga WY

99.5 mhz
WZRR(FM) Birmingham AL
KBTA-FM Batesville AR
KHMB(FM) Hamburg AR
KAKS(FM) Huntsville AR
KDIS-FM Little Rock AR
KMTB(FM) Murfreesboro AR
KIIM-FM Tucson AZ
KRPH(FM) Yarnell AZ
KLOK-FM Greenfield CA
KNTI(FM) Lakeport CA
KKLA-FM Los Angeles CA
KHYZ(FM) Mountain Pass CA
KMRJ(FM) Rancho Mirage CA
KQMT(FM) Denver CO
WIHT(FM) Washington DC
WJBR-FM Wilmington DE
WAFC-FM Clewiston FL
WKSM(FM) Fort Walton Beach FL
WAIL(FM) Key West FL
WBXY(FM) La Crosse FL
WQYK-FM Saint Petersburg FL
WKAA(FM) Willacoochee GA
KHUI(FM) Honolulu HI
KHAM(FM) Britt IA
KDAO-FM Eldora IA
KKMA(FM) Le Mars IA
KZZQ(FM) Winterset IA
KQPI(FM) Aberdeen ID
KWEI-FM Fruitland ID
WUSN(FM) Chicago IL
WDQZ(FM) Lexington IL
WCOY(FM) Quincy IL
WZPL(FM) Greenfield IN
KHAZ(FM) Hays KS
WKDP-FM Corbin KY
WKDQ(FM) Henderson KY
KNGT(FM) Lake Charles LA
WRNO-FM New Orleans LA
WCRB(FM) Lowell MA
WVMD(FM) Midland MD
WJCX(FM) Pittsfield ME
WYCD(FM) Detroit MI
WNGE(FM) Negaunee MI
WYSS(FM) Sault Ste. Marie MI
*KBHW(FM) International Falls MN
*KSJN(FM) Minneapolis MN
KPRW(FM) Perham MN
KHCR(FM) Bismarck MO
KADI-FM Republic MO
KMCJ(FM) Colstrip MT
KBLL-FM Helena MT
WXNR(FM) Grifton NC
WMAG(FM) High Point NC
KUTT(FM) Fairbury NE
WEVJ(FM) Jackson NH
KMGA(FM) Albuquerque NM
KXPZ(FM) Las Cruces NM
*KWFL(FM) Roswell NM
KPXP(FM) Garapan-Saipan NP
WTKW(FM) Bridgeport NY
WDCX(FM) Buffalo NY
*WBAI(FM) New York NY
WRVE(FM) Schenectady NY
WOKN(FM) Southport NY
WGAR-FM Cleveland OH
WAOL(FM) Ripley OH
KRBG(FM) Guymon OK
KXBL(FM) Henryetta OK
KBZQ(FM) Lawton OK
KAGO-FM Klamath Falls OR
KWJJ-FM Portland OR
KJMX(FM) Reedsport OR
WLTS(FM) Centre Hall PA
*WUSR(FM) Scranton PA
WKXC-FM Aiken SC
WRNN-FM Socastee SC
KOLY-FM Mobridge SD
WYGO(FM) Madisonville TN
KKPS(FM) Brownsville TX
KNFX-FM Bryan TX
KPLX(FM) Fort Worth TX
KQBR(FM) Lubbock TX
KISS-FM San Antonio TX
KJMY(FM) Bountiful UT
WYTT(FM) Emporia VA
WVIQ(FM) Christiansted VI
KZZL-FM Pullman WA
KAAP(FM) Rock Island WA
WPKR(FM) Omro WI
WJLS-FM Beckley WV
WYMJ(FM) New Martinsville WV
WBYG(FM) Point Pleasant WV
KCUG(FM) Chugwater WY
KRKI(FM) Newcastle WY

99.7 mhz
KMBQ(FM) Wasilla AK
WOOF-FM Dothan AL
KXTY(FM) Morro Bay CA
KIOO(FM) Porterville CA
KMVQ-FM San Francisco CA
KTOR(FM) Westwood CA
KPTE(FM) Durango CO
WJKD(FM) Vero Beach FL
WNNX(FM) Atlanta GA
KXFT(FM) Manson IA
KBEA-FM Muscatine IA
WXAJ(FM) Hillsboro IL
WSHW(FM) Frankfort IN
WDJX(FM) Louisville KY
KMJJ-FM Shreveport LA
WIMI(FM) Ironwood MI
*WUGN(FM) Midland MI
KXDL(FM) Browerville MN
KKCK(FM) Marshall MN
KMAC(FM) Gainesville MO
KYYS(FM) Kansas City MO
KBTN-FM Neosho MO
KTTR-FM Saint James MO
WJMI(FM) Jackson MS
KKMT(FM) Pablo MT
WRFX-FM Kannapolis NC
KOGA-FM Ogallala NE
WNTK-FM New London NH
WBHX(FM) Tuckerton NJ
WJUX(FM) Monticello NY
WBGK(FM) Newport Village NY
WZXV(FM) Palmyra NY
WBZX(FM) Columbus OH
WKSD(FM) Paulding OH
KNID(FM) Alva OK
KMTK(FM) Bend OR
WSPI(FM) Mount Carmel PA
WSHH(FM) Pittsburgh PA
*WVYC(FM) York PA
WSKO-FM Wakefield-Peacedale RI
WXST(FM) Hollywood SC
WWTN(FM) Manchester TN
WMC-FM Memphis TN
KBZD(FM) Amarillo TX
KROY(FM) Palacios TX
KBCY(FM) Tye TX
KVST(FM) Willis TX
WGCK(FM) Coeburn VA
WYFI(FM) Norfolk VA
WCYK-FM Staunton VA
KAST-FM Long Beach WA
KHHK(FM) Yakima WA
WWIS-FM Black River Falls WI
WZBY(FM) Sturgeon Bay WI

99.9 mhz
KFMJ(FM) Ketchikan AK
WRJL-FM Eva AL
WMXC(FM) Mobile AL
WQNR(FM) Tallassee AL
*KSWH(FM) Arkadelphia AR
KTCS-FM Fort Smith AR
KGPQ(FM) Monticello AR
KWCK-FM Searcy AR
KESZ(FM) Phoenix AZ
KRCX-FM Marysville CA
KCIV(FM) Mount Bullion CA
KOLA(FM) San Bernardino CA
KTYD(FM) Santa Barbara CA
KEKB(FM) Fruita CO
KVUU(FM) Pueblo CO
WEZN-FM Bridgeport CT
WKIS(FM) Boca Raton FL
WEGT(FM) Lafayette FL
WGNE-FM Palatka FL
WSNT-FM Sandersville GA
WQSA(FM) Unadilla GA
KJKS(FM) Kahului HI
KTOH(FM) Kalaheo HI
KCWN(FM) New Sharon IA
KZDX(FM) Burley ID
WWCT(FM) Bartonville IL
WOOZ-FM Harrisburg IL
WRZA(FM) Park Forest IL
WTHI-FM Terre Haute IN
KWKR(FM) Leoti KS
KSKG(FM) Salina KS
WVLC(FM) Mannsville KY
WMTC-FM Vancleve KY
KTDY(FM) Lafayette LA
KTEZ(FM) Zwolle LA
WNYN-FM Athol MA
WQRC(FM) Barnstable MA
*WHHB(FM) Holliston MA
WFRE(FM) Frederick MD
WWFG(FM) Ocean City MD
WTHT(FM) Auburn ME
WHFB-FM Benton Harbor MI
WPIQ(FM) Manistique MI
WHAK-FM Rogers City MI
KAUS-FM Austin MN
KVOX-FM Moorhead MN
KCML(FM) Saint Joseph MN
WUSZ(FM) Virginia MN
KBFL-FM Buffalo MO
KIRK(FM) Macon MO
KZMA(FM) Naylor MO
KFAV(FM) Warrenton MO
WSMS(FM) Artesia MS
KBOZ-FM Bozeman MT
WKSF(FM) Asheville NC
WKXB(FM) Burgaw NC
WHDX(FM) Buxton NC
WCMC-FM Creedmoor NC
KMXA-FM Minot ND
KGOR(FM) Omaha NE
KKTC(FM) Angel Fire NM
KTQM-FM Clovis NM
KXTC(FM) Thoreau NM
WIII(FM) Cortland NY
WBTZ(FM) Plattsburgh NY
WLQT(FM) Kettering OH
WKKO(FM) Toledo OH
WTUZ(FM) Uhrichsville OH
KRKT-FM Albany OR
KWRL(FM) La Grande OR
WQBR(FM) Avis PA
WODE-FM Easton PA
WXKC(FM) Erie PA
WIOA(FM) San Juan PR
KTSM-FM El Paso TX
KTXM(FM) Hallettsville TX
KSHN-FM Liberty TX
KMOO-FM Mineola TX
KBAT(FM) Monahans TX
KSAB(FM) Robstown TX
WACO-FM Waco TX
KLUR(FM) Wichita Falls TX
KONY(FM) Saint George UT
WZBB(FM) Stanleytown VA
KISW(FM) Seattle WA
KXLY-FM Spokane WA
WDRK(FM) Cornell WI
WJVL(FM) Janesville WI
WIZD(FM) Rudolph WI
WVAF(FM) Charleston WV
KKPL(FM) Cheyenne WY

100.1 mhz
KYKD(FM) Bethel AK
KWHQ-FM Kenai AK
*KMXT(FM) Kodiak AK
WGSY(FM) Phenix City AL
WDXX(FM) Selma AL
KVNA-FM Flagstaff AZ
KGMN(FM) Kingman AZ
KNGS(FM) Coalinga CA
KZRO(FM) Dunsmuir CA
KGBA-FM Holtville CA
*KLVJ(FM) Julian CA
KHWZ(FM) Ludlow CA
KZST(FM) Santa Rosa CA
KQOD(FM) Stockton CA
KKZQ(FM) Tehachapi CA
WVVE(FM) Panama City Beach FL
WZJZ(FM) Port Charlotte FL
WQMJ(FM) Forsyth GA
WSSJ(FM) Rincon GA
WNSY(FM) Talking Rock GA
WXKT(FM) Washington GA
KUYY(FM) Emmetsburg IA
KCTN(FM) Garnavillo IA
KITT(FM) Soda Springs ID
WKAI(FM) Macomb IL
WGLC-FM Mendota IL
WJBD-FM Salem IL
WNUY(FM) Bluffton IN
WFLQ(FM) French Lick IN
WFRI(FM) Winamac IN
WMDJ-FM Allen KY
WKQQ(FM) Winchester KY
KRVV(FM) Bastrop LA
WUPE-FM North Adams MA
WWFX(FM) Southbridge MA
*WBRS(FM) Waltham MA
*WUMF-FM Farmington ME
WHOU-FM Houlton ME
WCUZ(FM) Bear Lake MI
WBCH-FM Hastings MI
WVIB(FM) Holton MI
KOLV(FM) Olivia MN
WZFJ(FM) Pequot Lakes MN
KKWK(FM) Cameron MO
KDJR(FM) De Soto MO
KBBM(FM) Jefferson City MO
KOMC-FM Kimberling City MO
WQXB(FM) Grenada MS
KMMR(FM) Malta MT
KZOQ-FM Missoula MT
KATQ-FM Plentywood MT
WBXB(FM) Edenton NC
KYOY(FM) Kimball NE
WPNH-FM Plymouth NH
WJRZ-FM Manahawkin NJ
KTHX-FM Dayton NV
WDST(FM) Woodstock NY
WNIR(FM) Kent OH
WXZQ(FM) Piketon OH
WSWR(FM) Shelby OH
KYFM(FM) Bartlesville OK
KYKC(FM) Byng OK
WWOT(FM) Altoona PA
WBRR(FM) Bradford PA
WQFN(FM) Forest City PA
WQIC(FM) Lebanon PA
WWFN-FM Lake City SC
WXBT(FM) West Columbia SC
KDEZ(FM) Brandon SD
WASL(FM) Dyersburg TN
WRLT(FM) Franklin TN
KNRB(FM) Atlanta TX
KBWM(FM) Breckenridge TX
KYBI(FM) Lufkin TX
KCLL(FM) San Angelo TX
WYFJ(FM) Ashland VA
WVBE-FM Lynchburg VA
WKQY(FM) Tazewell VA
WPJP(FM) Port Washington WI
WDDC(FM) Portage WI
WRHN(FM) Rhinelander WI
WKBH-FM West Salem WI
WCLG-FM Morgantown WV
WDZN(FM) Romney WV
WDMX(FM) Vienna WV
KYOD(FM) Glendo WY

100.3 mhz
KICY-FM Nome AK
*KJNP-FM North Pole AK
WAOQ(FM) Brantley AL
WGZZ(FM) Dadeville AL
WQRV(FM) Tuscumbia AL
KDJE(FM) Jacksonville AR
KQMR(FM) Globe AZ
KJMB(FM) Blythe CA
KWPT(FM) Fortuna CA
KRQK(FM) Lompoc CA
KRBV(FM) Los Angeles CA
KMAK(FM) Orange Cove CA
KHGQ(FM) Quincy CA
KBRG(FM) San Jose CA
KIMN(FM) Denver CO
WBIG-FM Washington DC
WRKN(FM) Niceville FL
WRUM(FM) Orlando FL
WCTH(FM) Plantation Key FL
WOBB(FM) Tifton GA
KOKU(FM) Hagatna GU
KAPA(FM) Hilo HI
KCCN-FM Honolulu HI
KDRB(FM) Des Moines IA
KQXR(FM) Payette ID
KATZ-FM Alton IL
WIXY(FM) Champaign IL
WILV(FM) Chicago IL
WCCI(FM) Savanna IL
WLKI(FM) Angola IN
WMOJ-FM Connersville IN
WYGB(FM) Edinburgh IN
KQLS(FM) Colby KS
KTCM(FM) Kingman KS
KDVV(FM) Topeka KS
*KSWC(FM) Winfield KS
WVVR(FM) Hopkinsville KY
KRRV-FM Alexandria LA
KLRZ(FM) Larose LA
WKIT-FM Brewer ME
WNIC(FM) Dearborn MI
WGRY-FM Grayling MI
WUPF(FM) Gwinn MI
KTLK-FM Minneapolis MN
KSNR(FM) Thief River Falls MN
*KCVJ(FM) Osceola MO
KURM-FM South West City MO
KUKU-FM Willow Springs MO
WNSL(FM) Laurel MS
KLSK(FM) Great Falls MT
WLGP(FM) Harkers Island NC
WVBZ(FM) High Point NC
KZEN(FM) Central City NE
WHEB(FM) Portsmouth NH
WHTZ(FM) Newark NJ
KPEK(FM) Albuquerque NM
KWAW(FM) Garapan-Saipan NP
WDHI(FM) Delhi NY
WWLF-FM Sylvan Beach NY
WKBE(FM) Warrensburg NY
WCLT-FM Newark OH
*KJCM(FM) Snyder OK
KCXR(FM) Taft OK
KRWQ(FM) Gold Hill OR
KKRZ(FM) Portland OR
WHGL-FM Canton PA
WGYY(FM) Meadville PA
WPHI-FM Media PA
WIVA-FM Aguadilla PR
WKKB(FM) Middletown RI
WSEA(FM) Atlantic Beach SC
WORG(FM) Elloree SC
KJBI(FM) Presho SD
KFXS(FM) Rapid City SD
WNOX(FM) Oak Ridge TN
KTEX(FM) Brownsville TX
KJKK(FM) Dallas TX
KILT-FM Houston TX
KOMX(FM) Pampa TX
KCYY(FM) San Antonio TX
KMMX(FM) Tahoka TX
KXAL-FM Tatum TX
KSFI(FM) Salt Lake City UT
WARV-FM Petersburg VA
WSTX-FM Christiansted VI
WJPK(FM) Barton VT
KWIQ-FM Moses Lake WA
WNCY-FM Neenah-Menasha WI
WAFD(FM) Webster Springs WV
KZMQ-FM Greybull WY

100.5 mhz
KBFX(FM) Anchorage AK
WLDA(FM) Fort Rucker AL
WJOX(FM) Northport AL
KEGI(FM) Jonesboro AR
KZHE(FM) Stamps AR

KMQA(FM) East Porterville CA
KTDE(FM) Gualala CA
KMEN(FM) Mendota CA
KPSI-FM Palm Springs CA
KZZO(FM) Sacramento CA
KXDZ(FM) Templeton CA
KRSJ(FM) Durango CO
KCUF(FM) El Jebel CO
WRCH(FM) New Britain CT
WOYS(FM) Apalachicola FL
WHHZ(FM) Newberry FL
WWWQ(FM) College Park GA
WXRS-FM Swainsboro GA
KDEC-FM Decorah IA
*KBYI(FM) Rexburg ID
KQZB(FM) Troy ID
WRVY-FM Henry IL
WYMG(FM) Jacksonville IL
WWKI(FM) Kokomo IN
WSJD(FM) Princeton IN
KIBB(FM) Augusta KS
WLUE(FM) Louisville KY
WSGW-FM Carrollton MI
WTRV(FM) Walker MI
*WSCN(FM) Cloquet MN
KXAC(FM) Saint James MN
KSWF(FM) Aurora MO
KKCA(FM) Fulton MO
KMEM-FM Memphis MO
WBLE(FM) Batesville MS
WRTM-FM Port Gibson MS
KJJM(FM) Baker MT
WXXK(FM) Lebanon NH
KSFX(FM) Roswell NM
KKJJ(FM) Henderson NV
WDVI(FM) Rochester NY
WYJK-FM Bellaire OH
WKXA-FM Findlay OH
KATT-FM Oklahoma City OK
KDPM(FM) Cottage Grove OR
KQFM(FM) Hermiston OR
WYGL-FM Elizabethville PA
WJNG(FM) Johnsonburg PA
WCDW(FM) Susquehanna PA
WALC(FM) Charleston SC
WSSL-FM Gray Court SC
WHLZ(FM) Marion SC
KIKN-FM Salem SD
KQBB(FM) Center TX
KNNK(FM) Dimmitt TX
KMVL-FM Madisonville TX
KBDR(FM) Mirando City TX
KMXD(FM) Monroe UT
WZEZ(FM) Goochland VA
WXMM(FM) Norfolk VA
WTLX(FM) Columbus WI
WKEE-FM Huntington WV
WDYK(FM) Ridgeley WV

100.7 mhz

*KXLL(FM) Juneau AK
WCKF(FM) Ashland AL
KLSZ-FM Fort Smith AR
KAWW-FM Heber Springs AR
KJIK(FM) Duncan AZ
KSLX-FM Scottsdale AZ
KIBS(FM) Bishop CA
KTHU(FM) Corning CA
KATJ-FM George CA
KPRC-FM Salinas CA
KFMB-FM San Diego CA
KVVZ(FM) San Rafael CA
KHAY(FM) Ventura CA
KMOZ-FM Grand Junction CO
KGFT(FM) Pueblo CO
WHYI-FM Fort Lauderdale FL
WFLA-FM Midway FL
WJLQ(FM) Pensacola FL
WMTX(FM) Tampa FL
WMUV(FM) Brunswick GA
WEAM-FM Buena Vista GA
WLRR(FM) Milledgeville GA
*KJYL(FM) Eagle Grove IA
KKRQ(FM) Iowa City IA
KAYN(FM) Gooding ID
KIBG(FM) Wallace ID
WRXQ(FM) Coal City IL
WVMG(FM) Normal IL
WBYT(FM) Elkhart IN
WMGI(FM) Terre Haute IN
KHOK(FM) Hoisington KS
WKLX(FM) Brownsville KY
WCYO(FM) Irvine KY
WYPY(FM) Baton Rouge LA
KZBL(FM) Natchitoches LA
WZLX(FM) Boston MA
WZBA(FM) Westminster MD
WTBM(FM) Mexico ME
WOBE(FM) Crystal Falls MI
WSRJ(FM) Honor MI
WITL-FM Lansing MI
WWTH(FM) Oscoda MI
KIKV-FM Sauk Centre MN
KGMO(FM) Cape Girardeau MO
KMZU(FM) Carrollton MO
KBZI(FM) Deerfield MO
KFNS-FM Troy MO
WDMS(FM) Greenville MS
KXLB(FM) Livingston MT
WZJS(FM) Banner Elk NC
WRVA-FM Rocky Mount NC
KKLQ(FM) Harwood ND
*KGBI-FM Omaha NE
WZXL(FM) Wildwood NJ
KLVF(FM) Las Vegas NM
*KXXQ(FM) Milan NM
WOTT(FM) Henderson NY
WHUD(FM) Peekskill NY
*WKVU(FM) Utica NY
WMMS(FM) Cleveland OH
*WEEC(FM) Springfield OH
KTFR(FM) Chelsea OK
KMGX(FM) Bend OR
KPPT-FM Toledo OR
WLEV(FM) Allentown PA
*WCOG-FM Galeton PA
WZPT(FM) New Kensington PA
WXYX(FM) Bayamon PR
WGTN-FM Andrews SC
KMLO(FM) Lowry SD
WBGQ(FM) Bulls Gap TN
WUSY(FM) Cleveland TN
WYDL(FM) Middleton TN
KFGL(FM) Abilene TX
KASE-FM Austin TX
KWRD-FM Highland Village TX
KPXI(FM) Overton TX
KKHT-FM Winnie TX
KYMV(FM) Woodruff UT
WFNR-FM Christiansburg VA
WMJD(FM) Grundy VA
WQPO(FM) Harrisonburg VA
WTHK(FM) Wilmington VT
KKWF(FM) Seattle WA
KHSS(FM) Walla Walla WA
WBIZ-FM Eau Claire WI
WKKV-FM Racine WI
WSGD(FM) Arnoldsburg WV
KOLZ(FM) Cheyenne WY
KGWY(FM) Gillette WY

100.9 mhz

KCDV(FM) Cordova AK
KAKN(FM) Naknek AK
*KFSK(FM) Petersburg AK
WALX(FM) Selma AL
KDEL-FM Arkadelphia AR
KTCN(FM) Eureka Springs AR
KWKK(FM) Russellville AR
KHOM(FM) Salem AR
KZMK(FM) Sierra Vista AZ
KQSR(FM) Yuma AZ
KAEH(FM) Beaumont CA
KSSB(FM) Calipatria CA
KXTS(FM) Calistoga CA
KRAJ(FM) Johannesburg CA
KMIX(FM) Tracy CA
KNEC(FM) Yuma CO
WKNL(FM) New London CT
WXJZ(FM) Gainesville FL
WJAQ(FM) Marianna FL
WLYU(FM) Lyons GA
WPGA-FM Perry GA
WTHB-FM Waynesboro GA
WCJM-FM West Point GA
KWDN(FM) Newell IA
WHPO(FM) Hoopeston IL
WZUS(FM) Macon IL
WBZG(FM) Peru IL
WQFL(FM) Rockford IL
WBDC(FM) Huntingburg IN
WPGW-FM Portland IN
WYJZ(FM) Speedway IN
KCLY(FM) Clay Center KS
WLSK(FM) Lebanon KY
KHLL(FM) Richwood LA
WRNX(FM) Amherst MA
WAAI(FM) Hurlock MD
WYNZ(FM) Westbrook ME
WWBR(FM) Big Rapids MI
*WICV(FM) East Jordan MI
WQXC-FM Otsego MI
WYLZ(FM) Pinconning MI
KOWZ-FM Blooming Prairie MN
WCMP-FM Pine City MN
KRRY(FM) Canton MO
KTUI-FM Sullivan MO
WJXN-FM Utica MS
WKBB(FM) West Point MS
WPZS(FM) Albemarle NC
WIFM-FM Elkin NC
WSTS(FM) Fairmont NC
WFMI(FM) Southern Shores NC
KAUJ(FM) Grafton ND
KEJL(FM) Eunice NM
KRZQ-FM Sparks NV
WKLI-FM Albany NY
WPGI(FM) Horseheads NY
WKRL-FM North Syracuse NY
WCDO-FM Sidney NY
WBNO-FM Bryan OH
WMJK(FM) Clyde OH
WJAW-FM McConnelsville OH
WXIZ(FM) Waverly OH
KGLC(FM) Miami OK
KPNC-FM Ponca City OK
KXOJ-FM Sapulpa OK
*KBUG(FM) Malin OR
WAYC(FM) Bedford PA
WVLY-FM Milton PA
WRKT(FM) North East PA
WPAL-FM Ridgeville SC
WVHR(FM) Huntingdon TN
KXGL(FM) Amarillo TX
KWFB(FM) Quanah TX
KBAR-FM Victoria TX
KWSA(FM) Price UT
WIQO-FM Covington VA
WNNT-FM Warsaw VA
WWFY(FM) Berlin VT
KARY-FM Grandview WA
WRCO-FM Richland Center WI
WKOY-FM Princeton WV
WMXE(FM) South Charleston WV
WZST(FM) Westover WV
KTED(FM) Douglas WY

101.1 mhz

KAKQ-FM Fairbanks AK
KRXX(FM) Kodiak AK
WYDE-FM Cullman AL
WVVL(FM) Elba AL
WTID(FM) Repton AL
KWBF-FM North Little Rock AR
*KLRC(FM) Siloam Springs AR
KRRK(FM) Lake Havasu City AZ
KNRJ(FM) Payson AZ
KHYL(FM) Auburn CA
KWYE(FM) Fresno CA
KRTH(FM) Los Angeles CA
KWCA(FM) Weaverville CA
KOSI(FM) Denver CO
WWDC-FM Washington DC
WJRR(FM) Cocoa Beach FL
WAVV(FM) Marco FL
WYOO(FM) Springfield FL
WTGA-FM Thomaston GA
WAFT(FM) Valdosta GA
KORL-FM Waianae HI
KXIA(FM) Marshalltown IA
KMCL-FM McCall ID
WKQX(FM) Chicago IL
WMVN(FM) East St. Louis IL
WXMP(FM) Glasford IL
WLZQ(FM) South Whitley IN
KEOJ(FM) Caney KS
KFNF(FM) Oberlin KS
WIZF(FM) Erlanger KY
WSGS(FM) Hazard KY
WUBT(FM) Russellville KY
KBON(FM) Mamou LA
WNOE-FM New Orleans LA
KRMD-FM Shreveport LA
WFRQ(FM) Mashpee MA
WQMR(FM) Snow Hill MD
WWPN(FM) Westernport MD
WRIF(FM) Detroit MI
WUPY(FM) Ontonagon MI
WQON(FM) Roscommon MI
KBHP(FM) Bemidji MN
KLQL(FM) Luverne MN
KHME(FM) Winona MN
KCFX(FM) Harrisonville MO
WLIN-FM Durant MS
KZMT(FM) Helena MT
WQZL(FM) Belhaven NC
WZTK(FM) Burlington NC
KQDJ-FM Valley City ND
KDSR(FM) Williston ND
KLIR(FM) Columbus NE
WGIR-FM Manchester NH
KVLC(FM) Hatch NM
KSFQ(FM) White Rock NM
KCNM-FM Garapan-Saipan NP
KPKK(FM) Amargosa Valley NV
WBUG-FM Fort Plain NY
WCBS-FM New York NY
WWCD(FM) Grove City OH
WHOT-FM Youngstown OH
KWOX(FM) Woodward OK
KUFO-FM Portland OR
KAVJ(FM) Sutherlin OR
WBEB(FM) Philadelphia PA
WGMR(FM) Tyrone PA
WRIO(FM) Ponce PR
WROQ(FM) Anderson SC
WLVH(FM) Hardeeville SC
KDDX(FM) Spearfish SD
KJMS(FM) Memphis TN
WRR(FM) Dallas TX
KONO-FM Helotes TX
KLOL(FM) Houston TX
KONE(FM) Lubbock TX
KZPL(FM) Port Isabel TX
KPLD(FM) Kanab UT
KBER(FM) Ogden UT
WDYL(FM) Chester VA
KEYF-FM Cheney WA
KOHO-FM Leavenworth WA
WVRE(FM) Dickeyville WI
WIXX(FM) Green Bay WI
WHSM-FM Hayward WI
KPIN(FM) Pinedale WY

101.3 mhz

KGOT(FM) Anchorage AK
WAGF-FM Dothan AL
WBFA(FM) Smiths AL
KARV-FM Ola AR
KPBQ-FM Pine Bluff AR
KATY-FM Idyllwild CA
KSTT-FM Los Osos-Baywood Park CA
KIOI(FM) San Francisco CA
KIQX(FM) Durango CO
KWCS(FM) Walsenburg CO
WKCI-FM Hamden CT
WNCL(FM) Milford DE
WKYZ(FM) Key Colony Beach FL
WHLG(FM) Port St. Lucie FL
WTMG(FM) Williston FL
WQIL(FM) Chauncey GA
KSIB-FM Creston IA
KKYY(FM) Whiting IA
KUUL(FM) East Moline IL
WMCI(FM) Mattoon IL
WVIL(FM) Virginia IL
WFMG(FM) Richmond IN
*WBAA-FM West Lafayette IN
KFDI-FM Wichita KS
WMJM(FM) Jeffersontown KY
WMSK-FM Sturgis KY
KKGB(FM) Sulphur LA
WKCG(FM) Augusta ME
WBFX(FM) Grand Rapids MI
WSUE(FM) Sault Ste. Marie MI
KDWB-FM Richfield MN
KTXR(FM) Springfield MO
WMUT(FM) Grenada MS
WMSO(FM) Meridian MS
WBBV(FM) Vicksburg MS
KRYK(FM) Chinook MT
KIKC-FM Forsyth MT
WWQQ-FM Wilmington NC
KMOR(FM) Bridgeport NE
KLZA(FM) Falls City NE
WYKR-FM Haverhill NH
KKRG(FM) Albuquerque NM
KRNG(FM) Fallon/ Reno NV
WBRV-FM Boonville NY
WCPV(FM) Essex NY
WRMM-FM Rochester NY
WQAR(FM) Stillwater NY
WNCO-FM Ashland OH
WAGX(FM) Manchester OH
KLAW(FM) Lawton OK
KMCO(FM) McAlester OK
WROZ(FM) Lancaster PA
WGGY(FM) Scranton PA
WWDM(FM) Sumter SC
WCMT-FM South Fulton TN
WECO-FM Wartburg TN
KOXE(FM) Brownwood TX
KMMZ(FM) Crane TX
KKLB(FM) Madisonville TX
KNCN(FM) Sinton TX
WWDE-FM Hampton VA
WZFM(FM) Narrows VA
WWKS(FM) Cruz Bay VI
KGDN(FM) Pasco WA
KABW(FM) Westport WA
WBRB(FM) Buckhannon WV

101.5 mhz

WQEM(FM) Columbiana AL
KBGR(FM) Beebe AR
KMLK(FM) El Dorado AR
KAVH(FM) Eudora AR
KLEZ(FM) Malvern AR
KZON(FM) Phoenix AZ
KIXF(FM) Baker CA
KGFM(FM) Bakersfield CA
KEKA-FM Eureka CA
KMJE(FM) Gridley CA
*KAMB(FM) Merced CA
*KRVH(FM) Rio Vista CA
KGB-FM San Diego CA
KTKE(FM) Truckee CA
KTNI-FM Strasburg CO
WLYF(FM) Miami FL
WTKX-FM Pensacola FL
WXSR(FM) Quincy FL
WPOI(FM) Saint Petersburg FL
WSOL-FM Brunswick GA
WKHX-FM Marietta GA
KAOY(FM) Kealakekua HI
KKSI(FM) Eddyville IA
KCVI(FM) Blackfoot ID
KATW(FM) Lewiston ID
WBNQ(FM) Bloomington IL
WCIL-FM Carbondale IL
WKKG(FM) Columbus IN
*WBGW(FM) Fort Branch IN
WNSN(FM) South Bend IN
KIKS-FM Iola KS
KSLS(FM) Liberal KS
KMKF(FM) Manhattan KS
WVLK-FM Richmond KY
WRZI(FM) Vine Grove KY
WYNK-FM Baton Rouge LA
WMJZ-FM Gaylord MI

WJNR-FM Iron Mountain MI
WMTE-FM Manistee MI
WWBN(FM) Tuscola MI
KFGI(FM) Crosby MN
KRJM(FM) Mahnomen MN
KCGN-FM Ortonville MN
KRRW(FM) Saint James MN
KPLA(FM) Columbia MO
WWUN-FM Friar's Point MS
WTPO(FM) New Albany MS
WHDZ(FM) Buxton NC
WRAL(FM) Raleigh NC
KSSS(FM) Bismarck ND
KROR(FM) Hastings NE
WRNH(FM) Groveton NH
WWHQ(FM) Meredith NH
WKXW(FM) Trenton NJ
KRMQ-FM Clovis NM
KQLP(FM) Gallup NM
KVSF-FM Pecos NM
KIDX(FM) Ruidoso NM
WRCD(FM) Canton NY
WXHC(FM) Homer NY
WMXO(FM) Olean NY
WPDH(FM) Poughkeepsie NY
*WCWT-FM Centerville OH
WRYV(FM) Gallipolis OH
WRVF(FM) Toledo OH
*WOBN(FM) Westerville OH
KIZS(FM) Collinsville OK
KFLY(FM) Corvallis OR
WDKC(FM) Covington PA
WORD-FM Pittsburgh PA
WFYN(FM) Waynesboro PA
WKSA-FM Isabela PR
WWBB(FM) Providence RI
*KVCX(FM) Gregory SD
WVFB(FM) Celina TN
WNWS-FM Jackson TN
WQUT(FM) Johnson City TN
WFTZ(FM) Manchester TN
WMXV(FM) Saint Joseph TN
KROX-FM Buda TX
KSTB(FM) Crystal Beach TX
KSNY-FM Snyder TX
KNUE(FM) Tyler TX
KEGA(FM) Oakley UT
WBQB(FM) Fredericksburg VA
WZZI(FM) Vinton VA
WEXP(FM) Brandon VT
WRSY(FM) Marlboro VT
KPLZ(FM) Seattle WA
WIBA-FM Madison WI
KDDV-FM Wright WY

101.7 mhz
KPEN-FM Soldotna AK
*KSTK(FM) Wrangell AK
WBEI(FM) Reform AL
WMXN-FM Stevenson AL
KBYB(FM) Hope AR
KVLO(FM) Humnoke AR
KVOM-FM Morrilton AR
KCTT-FM Yellville AR
KKYZ(FM) Sierra Vista AZ
KQAZ(FM) Springerville-Eagar AZ
KXSB(FM) Big Bear Lake CA
KCDU(FM) Carmel CA
KSBL(FM) Carpinteria CA
KRER(FM) Hamilton City CA
KKIQ(FM) Livermore CA
KXFX(FM) Santa Rosa CA
KTUN(FM) Eagle CO
WZEB(FM) Ocean View DE
WTOT-FM Graceville FL
WDVH-FM Trenton FL
WCZR(FM) Vero Beach FL
WQVE(FM) Albany GA
WYUM(FM) Mount Vernon GA
WTHO-FM Thomson GA
WRBV(FM) Warner Robins GA
KBKB-FM Fort Madison IA
KAYL-FM Storm Lake IA
WRCV(FM) Dixon IL
WGEL(FM) Greenville IL
WTYE(FM) Robinson IL
WURK(FM) Elwood IN
WLDE(FM) Fort Wayne IN
WIVR(FM) Kentland IN
KVOE-FM Emporia KS
KREJ(FM) Medicine Lodge KS
WKYM(FM) Monticello KY
WKRD-FM Shelbyville KY
WFNX(FM) Lynn MA
WBRK-FM Pittsfield MA
WFZX(FM) Searsport ME
*WPRJ(FM) Coleman MI
WHZZ(FM) Lansing MI
WMRR(FM) Muskegon Heights MI
KLDJ(FM) Duluth MN
KRCH(FM) Rochester MN
WHMH-FM Sauk Rapids MN
KGOZ(FM) Gallatin MO
KHST(FM) Lamar MO
KLPW-FM Union MO
WYOY(FM) Gluckstadt MS
WZHL(FM) New Augusta MS
KZUS(FM) Belt MT
KTNY(FM) Libby MT
WVRR(FM) Newport NH
WJKS(FM) Canton NJ
KLEA-FM Lovington NM
KQBT(FM) Rio Rancho NM
KCLS(FM) Ely NV
WLOF(FM) Attica NY
WFLK(FM) Geneva NY
WNYQ(FM) Hudson Falls NY
WLTB(FM) Johnson City NY
*WGKV(FM) Pulaski NY
WBEA(FM) Southold NY
WNKO(FM) Newark OH
WHOF(FM) North Canton OH
WKSW(FM) Urbana OH
KEOK(FM) Tahlequah OK
KTFX-FM Warner OK
KLRR(FM) Redmond OR
WCCL(FM) Central City PA
WMVL(FM) Linesville PA
WKZQ-FM Myrtle Beach SC
WMGL(FM) Ravenel SC
WJSQ(FM) Athens TN
WKOM(FM) Columbia TN
WTPR-FM McKinnon TN
WORM-FM Savannah TN
WJLE-FM Smithville TN
KTCY(FM) Azle TX
KXGJ(FM) Bay City TX
KEKO(FM) Hebronville TX
KSAM-FM Huntsville TX
KAYD-FM Silsbee TX
KLTD(FM) Temple TX
WLQM-FM Franklin VA
WKWI(FM) Kilmarnock VA
WJJX(FM) Lynchburg VA
WWBU(FM) Radford VA
WEVI(FM) Frederiksted VI
WCVT(FM) Stowe VT
KLES(FM) Prosser WA
WELD-FM Petersburg WV
KDNO(FM) Thermopolis WY
KZEW(FM) Wheatland WY

101.9 mhz
WHHY-FM Montgomery AL
KIYS(FM) Jonesboro AR
KMXF(FM) Lowell AR
KLXQ(FM) Mountain Pine AR
KVGG(FM) Salome AZ
KOQO-FM Fresno CA
KSCA(FM) Glendale CA
KNTY(FM) Shingle Springs CA
KGDQ(FM) Colorado Springs CO
WJHM(FM) Daytona Beach FL
WWGR(FM) Fort Myers FL
WBGE(FM) Bainbridge GA
WAZX-FM Cleveland GA
WPNG(FM) Pearson GA
WOCE(FM) Ringgold GA
WJFL(FM) Tennille GA
KTKB-FM Hagatna GU
KUCD(FM) Pearl City HI
*KNWS-FM Waterloo IA
KDBI(FM) Emmett ID
WTMX(FM) Skokie IL
WQQL(FM) Springfield IL
WKLU(FM) Brownsburg IN
WARU-FM Roann IN
KKQY(FM) Hill City KS
WPNS(FM) Brodhead KY
WQXQ(FM) Central City KY
KNOE-FM Monroe LA
WLMG(FM) New Orleans LA
WCIB(FM) Falmouth MA
WLIF(FM) Baltimore MD
WPOR(FM) Portland ME
WOZI(FM) Presque Isle ME
*WDET-FM Detroit MI
WKQS-FM Negaunee MI
WLDR-FM Traverse City MI
KQKK(FM) Walker MN
KZWV(FM) Eldon MO
WFTA(FM) Fulton MS
WZYQ(FM) Mound Bayou MS
KRSQ(FM) Laurel MT
WBAV-FM Gastonia NC
WIKS(FM) New Bern NC
KBTO(FM) Bottineau ND
KRWK(FM) Fargo ND
KLTQ(FM) Lincoln NE
KTAO(FM) Taos NM
KWID(FM) Las Vegas NV
WZKZ(FM) Alfred NY
WJIV(FM) Cherry Valley NY
WHUG(FM) Jamestown NY
WQCD(FM) New York NY
WKRQ(FM) Cincinnati OH
WRBP(FM) Hubbard OH
KTST(FM) Oklahoma City OK
KCMX-FM Ashland OR
KINK(FM) Portland OR
KUJJ(FM) Weston OR
WAVT-FM Pottsville PA
WZAR(FM) Ponce PR
KTWB(FM) Sioux Falls SD
KATP(FM) Amarillo TX
KZTR(FM) Franklin TX
KSML-FM Huntington TX
KACQ(FM) Lometa TX
KBUS(FM) Paris TX
KWFR(FM) San Angelo TX
KQXT(FM) San Antonio TX
KENZ(FM) Ogden UT
WHTE-FM Ruckersville VA
WKSK-FM South Hill VA
KTSL(FM) Medical Lake WA
WDEZ(FM) Wausau WI
WVOW-FM Logan WV
WVAQ(FM) Morgantown WV
KIGN(FM) Burns WY

102.1 mhz
KDBZ(FM) Anchorage AK
WQUA(FM) Citronelle AL
WDRM(FM) Decatur AL
KENA-FM Mena AR
KOKY(FM) Sherwood AR
KCMT(FM) Oro Valley AZ
KAHM(FM) Prescott AZ
KPRI(FM) Encinitas CA
KZPE(FM) Ford City CA
KRKC-FM King City CA
KCEZ(FM) Los Molinos CA
KDFC-FM San Francisco CA
KRKY-FM Estes Park CO
WKLG(FM) Rock Harbor FL
WWAV-FM Santa Rosa Beach FL
WQLC(FM) Watertown FL
WWWD(FM) Bolingbroke GA
WGMG(FM) Crawford GA
WZAT(FM) Savannah GA
WNUQ(FM) Sylvester GA
KTBH-FM Kurtistown HI
KUQQ(FM) Milford IA
KCHQ(FM) Driggs ID
KIRQ(FM) Twin Falls ID
WDNL(FM) Danville IL
WQLF(FM) Lena IL
WIBV(FM) Mount Vernon IL
WALS(FM) Oglesby IL
KZSN(FM) Hutchinson KS
WLJC(FM) Beattyville KY
WLLE(FM) Clinton KY
WKYL(FM) Lawrenceburg KY
KQIS(FM) Basile LA
KDKS-FM Blanchard LA
WAQY(FM) Springfield MA
WGUY(FM) Dexter ME
WLEW-FM Bad Axe MI
*WMUK(FM) Kalamazoo MI
KCAJ-FM Roseau MN
KEEY-FM Saint Paul MN
KQRA(FM) Brookline MO
KCKC(FM) Kansas City MO
KJFM(FM) Louisiana MO
WUCL(FM) Meridian MS
WRQO(FM) Monticello MS
KBUD(FM) Sardis MS
*KBMC(FM) Bozeman MT
WJMH(FM) Reidsville NC
KPNY(FM) Alliance NE
KZMC(FM) McCook NE
WSAK(FM) Hampton NH
KTRA-FM Farmington NM
KRNV-FM Reno NV
*WJCA(FM) Albion NY
WDNB(FM) Jeffersonville NY
WZUN(FM) Phoenix NY
WAVR(FM) Waverly NY
WDOK(FM) Cleveland OH
WIMT(FM) Lima OH
WRVB(FM) Marietta OH
KHKC-FM Atoka OK
WOWQ(FM) DuBois PA
WIOQ(FM) Philadelphia PA
WMXT(FM) Pamplico SC
KFMH(FM) Belle Fourche SD
WLCT(FM) Lafayette TN
WWST(FM) Sevierville TN
KPRR(FM) El Paso TX
KDGE(FM) Fort Worth-Dallas TX
KFZX(FM) Gardendale TX
KMJQ(FM) Houston TX
KBUC(FM) Raymondville TX
WRXL(FM) Richmond VA
WXTG(FM) Virginia Beach VA
WCVR-FM Randolph VT
KSWW(FM) Montesano WA
KPQ-FM Wenatchee WA
WRKU(FM) Forestville WI
WLUM-FM Milwaukee WI
KXMP(FM) Hanna WY

102.3 mhz
*KHNS(FM) Haines AK
WAMI-FM Opp AL
WELR-FM Roanoke AL
KTRQ(FM) Colt AR
KCJC(FM) Dardanelle AR
KQEW(FM) Fordyce AR
KWRQ(FM) Clifton AZ
KJJJ(FM) Lake Havasu City AZ
KZXY-FM Apple Valley CA
KJLH-FM Compton CA
KBLO(FM) Corcoran CA
KJJZ(FM) Indio CA
KJSN(FM) Modesto CA
KYOE(FM) Point Arena CA
KNTK(FM) Weed CA
KSMT(FM) Breckenridge CO
KCUV(FM) Greenwood Village CO
KVLE-FM Gunnison CO
KSPK(FM) Walsenburg CO
WXLM(FM) Stonington CT
WTRS(FM) Dunnellon FL
WMBX(FM) Jensen Beach FL
WIBL(FM) Augusta GA
WLKQ-FM Buford GA
WWLD(FM) Cairo GA
WKZR(FM) Milledgeville GA
WQTU(FM) Rome GA
KMKK-FM Kaunakakai HI
KCZQ(FM) Cresco IA
KXGE(FM) Dubuque IA
KZSR(FM) Onawa IA
KICR(FM) Coeur d'Alene ID
WRMJ(FM) Aledo IL
WYCA(FM) Crete IL
WEBQ-FM Eldorado IL
WDQX(FM) Morton IL
WRHL-FM Rochelle IL
WKJT(FM) Teutopolis IL
WXLC(FM) Waukegan IL
WGBJ(FM) Auburn IN
WLHM(FM) Logansport IN
WCBK-FM Martinsville IN
WZOW(FM) New Carlisle IN
WBTO-FM Petersburg IN
WCYN-FM Cynthiana KY
WUGO(FM) Grayson KY
WXMA(FM) Louisville KY
WCLU-FM Munfordville KY
WLLK-FM Somerset KY
KBCE(FM) Boyce LA
WGTX(FM) Truro MA
WMMJ(FM) Bethesda MD
WCXX(FM) Madawaska ME
WYBR(FM) Big Rapids MI
WHKB(FM) Houghton MI
WGRT(FM) Port Huron MI
*WTHN(FM) Sault Ste. Marie MI
KRCQ(FM) Detroit Lakes MN
KBXR(FM) Columbia MO
KDEX-FM Dexter MO
KJPW-FM Waynesville MO
WGCM-FM Gulfport MS
WIQQ(FM) Leland MS
WWSL(FM) Philadelphia MS
WKZU(FM) Ripley MS
KEAU(FM) Choteau MT
WECR-FM Beech Mountain NC
WWNF(FM) Goldsboro NC
WFNC-FM Lumberton NC
WPTM(FM) Roanoke Rapids NC
KRNY(FM) Kearney NE
WWHK(FM) Concord NH
WXXS(FM) Lancaster NH
WAIV(FM) Cape May NJ
WSUS(FM) Franklin NJ
KKYC(FM) Clovis NM
KVUW(FM) Wendover NV
WBAB(FM) Babylon NY
WKKF(FM) Ballston Spa NY
WVOR(FM) Canandaigua NY
WRGR(FM) Tupper Lake NY
WFXN-FM Galion OH
WPOS-FM Holland OH
WKLN(FM) Wilmington OH
KDOE(FM) Antlers OK
KKEN(FM) Duncan OK
KKCM(FM) Sand Springs OK
KWDQ(FM) Woodward OK
KEHK(FM) Brownsville OR
KCRX-FM Seaside OR
WCAT-FM Carlisle PA
WQHZ(FM) Erie PA
WDMT(FM) Pittston PA
WMIO(FM) Cabo Rojo PR
WRML(FM) Pageland SC
WMFX(FM) Saint Andrews SC
KKQQ(FM) Volga SD
WZDQ(FM) Humboldt TN
WGOW-FM Soddy-Daisy TN
KPEZ(FM) Austin TX
KSAQ(FM) Charlotte TX
KQBZ(FM) Coleman TX
KLJT(FM) Jacksonville TX
KKPN(FM) Rockport TX
KUVA(FM) Uvalde TX
KWFS-FM Wichita Falls TX
KDUT(FM) Randolph UT
WSUH(FM) Crozet VA
WDXC(FM) Pound VA
WLFE-FM Saint Albans VT
KYYT(FM) Goldendale WA
WQTC-FM Manitowoc WI
WVRQ-FM Viroqua WI
WAUH(FM) Wautoma WI
WHTL-FM Whitehall WI
WMTD-FM Hinton WV
WFBY(FM) Weston WV

102.5 mhz
KIAK-FM Fairbanks AK
WESP(FM) Dothan AL
WDXB(FM) Jasper AL
KPZK-FM Cabot AR
KAFN(FM) Gould AR
KNIX-FM Phoenix AZ
KCNQ(FM) Kernville CA
KDUQ(FM) Ludlow CA
KDON-FM Salinas CA
KSNI-FM Santa Maria CA
KSFM(FM) Woodland CA
KQZR(FM) Craig CO
KTRR(FM) Loveland CO
WHPT(FM) Sarasota FL
WPIK(FM) Summerland Key FL
WAMJ(FM) Mableton GA
WEBL(FM) Warner Robins GA
WYNR(FM) Waycross GA
KSTZ(FM) Des Moines IA
KMGI(FM) Pocatello ID
KIBR(FM) Sandpoint ID
*WGNN(FM) Fisher IL
WJRE(FM) Galva IL
WPHZ(FM) Mitchell IN
WMDH-FM New Castle IN
KACY(FM) Arkansas City KS
KKCI(FM) Goodland KS
KBLS(FM) North Fort Riley KS
KKCV(FM) Rozel KS
WLTO(FM) Nicholasville KY
WFMF(FM) Baton Rouge LA
WKLB-FM Waltham MA
WOLC(FM) Princess Anne MD
WQSS(FM) Camden ME
WIOG(FM) Bay City MI
WCMM(FM) Gulliver MI
WBZV(FM) Hudson MI
KMFX-FM Lake City MN
KQIC(FM) Willmar MN
KIXQ(FM) Joplin MO
KEZK-FM Saint Louis MO
KKDY(FM) West Plains MO
WJKX(FM) Ellisville MS
WAGR-FM Lexington MS
KMSO(FM) Missoula MT
WERX-FM Columbia NC
WMYI(FM) Hendersonville NC
WKXU(FM) Louisburg NC
WIOZ-FM Southern Pines NC
KDVL(FM) Devils Lake ND
KIOT(FM) Los Lunas NM
WBAZ(FM) Bridgehampton NY
WTSS(FM) Buffalo NY
WUMX(FM) Rome NY
WZOO-FM Edgewood OH
WHIZ-FM Zanesville OH
KTNT(FM) Eufaula OK
WDVE(FM) Pittsburgh PA
WRFY-FM Reading PA
WIAC-FM San Juan PR
WXLY(FM) North Charleston SC
*KZSD-FM Martin SD
WOWF(FM) Crossville TN
WVNS-FM Pegram TN
KMKS(FM) Bay City TX
KTCX(FM) Beaumont TX
KBRQ(FM) Hillsboro TX
KZII-FM Lubbock TX
KHLB(FM) Mason TX
KKYR-FM Texarkana TX
KMAD-FM Whitesboro TX
WOLD-FM Marion VA
WUSQ-FM Winchester VA
KRAO-FM Colfax WA
KZOK-FM Seattle WA
*WNWC-FM Madison WI
KRBR(FM) Superior WI
KHOC(FM) Casper WY

102.7 mhz
*KRNN(FM) Juneau AK
WCKS(FM) Fruithurst AL
KWLT(FM) North Crossett AR
KBBQ-FM Van Buren AR
KSSI(FM) China Lake CA
KHGE(FM) Fresno CA
KIIS-FM Los Angeles CA
*KLVB(FM) Red Bluff CA
KBIQ(FM) Manitou Springs CO
WPHK(FM) Blountstown FL
WRGO(FM) Cedar Key FL
WXHT(FM) Madison FL
WXBM-FM Milton FL
WMXJ(FM) Pompano Beach FL
WHKR(FM) Rockledge FL
WPMA(FM) Buckhead GA
WYSC(FM) McRae GA
WBDX(FM) Trenton GA
KDDB(FM) Waipahu HI
KYTC(FM) Northwood IA
WJEQ(FM) Macomb IL
WZZT(FM) Morrison IL
WVAZ(FM) Oak Park IL
WLEG(FM) Ligonier IN
WBOW-FM Terre Haute IN
KLDG(FM) Liberal KS
WVEK-FM Cumberland KY
WMJL-FM Marion KY
WYSB(FM) Springfield KY
WKWY(FM) Tompkinsville KY
KJNA-FM Jena LA
WQSR(FM) Baltimore MD
WHTD(FM) Mount Clemens MI
WMOM(FM) Pentwater MI
KQEG(FM) La Crescent MN
KTIG(FM) Pequot Lakes MN
*KNTN(FM) Thief River Falls MN
KQUL(FM) Lake Ozark MO
WCNG(FM) Murphy NC
WGNI(FM) Wilmington NC
KFRX(FM) Lincoln NE
WJSE(FM) Petersburg NJ
KSTJ(FM) Boulder City NV
WBDR(FM) Cape Vincent NY
WWFS(FM) New York NY
WRCI(FM) Webster NY
WEBN(FM) Cincinnati OH
WCPZ(FM) Sandusky OH
KJYO(FM) Oklahoma City OK
KCNA(FM) Cave Junction OR
KYTE(FM) Newport OR
WKSB(FM) Williamsport PA
WRNI-FM Narragansett Pier RI
WPUB-FM Camden SC
WGUS-FM New Ellenton SC
KYBB(FM) Canton SD
WEKX(FM) Jellico TN
WEGR(FM) Memphis TN
KTXJ-FM Jasper TX
KHXS(FM) Merkel TX
KJXK(FM) San Antonio TX
KBLZ(FM) Winona TX
KSL-FM Midvale UT
WSNZ(FM) Appomattox VA
WFYE(FM) Glade Spring VA
WEQX(FM) Manchester VT
KORD-FM Richland WA
*WRVM(FM) Suring WI
WVSR-FM Charleston WV
WGYE(FM) Mannington WV

102.9 mhz
WKXX(FM) Attalla AL
WNPT-FM Linden AL
KHOZ-FM Harrison AR
KARN-FM Sheridan AR
KQST(FM) Sedona AZ
KBLX-FM Berkeley CA
KWTY(FM) Cartago CA
KIWI(FM) McFarland CA
KXLM(FM) Oxnard CA
KLQV(FM) San Diego CA
KWYL(FM) South Lake Tahoe CA
WDRC-FM Hartford CT
WMXQ(FM) Jacksonville FL
WJGO(FM) Tice FL
WMJE(FM) Clarkesville GA
WVRK(FM) Columbus GA
WPMX(FM) Statesboro GA
KISH(FM) Hagatna GU
KLZY(FM) Honokaa HI
KZIA(FM) Cedar Rapids IA
KTFG(FM) Sioux Rapids IA
KWYS-FM Island Park ID
KMVX(FM) Jerome ID
WSOY-FM Decatur IL
WMKB(FM) Earlville IL
WLME(FM) Cannelton IN
WXXB(FM) Delphi IN
WGL-FM Huntington IN
KHUT(FM) Hutchinson KS
KIND-FM Independence KS
KQTP(FM) Saint Marys KS
WANK(FM) Mount Vernon KY
WLKS-FM West Liberty KY
KMEZ(FM) Belle Chasse LA
KAJN-FM Crowley LA
KVMA-FM Shreveport LA
WPXC(FM) Hyannis MA
WKIK-FM California MD
WROG(FM) Cumberland MD
WCRQ(FM) Dennysville ME
WBLM(FM) Portland ME
WWWW-FM Ann Arbor MI
WFUR-FM Grand Rapids MI
WMKC(FM) Saint Ignace MI
WLTE(FM) Minneapolis MN
KMFG(FM) Nashwauk MN
KEZS-FM Cape Girardeau MO
KMMO-FM Marshall MO
WMSI(FM) Jackson MS
WWMR(FM) Saltillo MS
KCTR-FM Billings MT
WLYT(FM) Hickory NC
WELS-FM Kinston NC
WWMY(FM) Raleigh NC
KWGO(FM) Burlington ND
KADL(FM) Imperial NE
KBRX-FM O'Neill NE
KNFT-FM Bayard NM
KIXN(FM) Hobbs NM
KAZX(FM) Kirtland NM
KLBU(FM) Pecos NM
WNCQ-FM Canton NY
*WMHR(FM) Syracuse NY
WCLX(FM) Westport NY
WDHT(FM) Springfield OH
KQIB(FM) Idabel OK
KYSF(FM) Bonanza OR
KSJJ(FM) Redmond OR
WOKW(FM) Curwensville PA
WMGK(FM) Philadelphia PA
WYFM(FM) Sharon PA
WDIN(FM) Camuy PR
WQKI-FM Orangeburg SC
WZTF(FM) Scranton SC
KBWS-FM Sisseton SD
WBUZ(FM) La Vergne TN
KNDA(FM) Alice TX
KDMX(FM) Dallas TX
KLTN(FM) Houston TX
KITY(FM) Llano TX
WOWI(FM) Norfolk VA
*WIUJ(FM) Charlotte Amalie VI
KNBQ(FM) Centralia WA
KVAB(FM) Clarkston WA
WHQG(FM) Milwaukee WI
WBDL(FM) Reedsburg WI
WELC-FM Welch WV
KARS-FM Laramie WY

103.1 mhz
KMXS(FM) Anchorage AK
KSBZ(FM) Sitka AK
WEUP-FM Moulton AL
KXSA-FM Dermott AR
KFFA-FM Helena AR
KHGG-FM Waldron AR
KHZF(FM) Fagaitua AS
KFTT(FM) Bagdad AZ
KCDX(FM) Florence AZ
KKCY(FM) Colusa CA
KDLE(FM) Newport Beach CA
KAAT(FM) Oakhurst CA
KEZN(FM) Palm Desert CA
KLUN(FM) Paso Robles CA
KDLD(FM) Santa Monica CA
KTPI-FM Tehachapi CA
KVFG(FM) Victorville CA
KHRD(FM) Weaverville CA
KSPN-FM Aspen CO
KAVD(FM) Limon CO
WMXZ(FM) De Funiak Springs FL
WPBZ(FM) Indiantown FL
WFKZ(FM) Plantation Key FL
WAIB(FM) Tallahassee FL
WLOQ(FM) Winter Park FL
WFXA-FM Augusta GA
*WPLH(FM) Tifton GA
KDMG(FM) Burlington IA
KCDA(FM) Post Falls ID
WVIV-FM Highland Park IL
WAKO-FM Lawrenceville IL
WCSJ-FM Morris IL
WGFB(FM) Rockton IL
WKZS(FM) Covington IN
WHME(FM) South Bend IN
WXCH(FM) Versailles IN
KEKS(FM) Olpe KS
WPKE-FM Coal Run KY
WGRK-FM Greensburg KY
WGBF-FM Henderson KY
WWLT(FM) Manchester KY
WRKA(FM) Saint Matthews KY
KQLQ(FM) Columbia LA
WRNR-FM Grasonville MD
WAFY(FM) Middletown MD
WDME-FM Dover Foxcroft ME
WGDN-FM Gladwin MI
WQUS(FM) Lapeer MI
KFIL-FM Preston MN
KDAA(FM) Rolla MO
WMBC(FM) Columbus MS
WOSM(FM) Ocean Springs MS
KRVO(FM) Columbia Falls MT
*KVCM(FM) Helena MT
*WUAG(FM) Greensboro NC
WLHC(FM) Robbins NC
KRVX(FM) Wimbledon ND
KNCY-FM Auburn NE
KKJK(FM) Ravenna NE
KHQT(FM) Las Cruces NM
WHRL(FM) Albany NY
*WCIK(FM) Bath NY
WBZO(FM) Bay Shore NY
WTOJ(FM) Carthage NY
WGNY-FM Newburgh NY
WZOZ(FM) Oneonta NY
WVKO-FM Johnstown OH
WNDH(FM) Napoleon OH
WRAC(FM) West Union OH
KOFM(FM) Enid OK
KRSB-FM Roseburg OR
WILK-FM Avoca PA
WQFX-FM Russell PA
WQWK(FM) State College PA
WANB-FM Waynesburg PA
WOMG(FM) Columbia SC
WRIX-FM Honea Path SC
WGZO(FM) Parris Island SC
WYAK-FM Surfside Beach SC
KJAM-FM Madison SD
WLLJ(FM) Etowah TN
WMXX-FM Jackson TN
WIKQ(FM) Tusculum TN
KRGN(FM) Amarillo TX
KKCN(FM) Ballinger TX
KEEP(FM) Bandera TX
KSSM(FM) Copperas Cove TX
KPAS(FM) Fabens TX
KVJM(FM) Hearne TX
KTXX(FM) Karnes City TX
KMUL-FM Muleshoe TX
KDVE(FM) Pittsburg TX
KVWC-FM Vernon TX
KJQN(FM) Coalville UT
KURR(FM) Hurricane UT
WWDW(FM) Alberta VA
WJMA-FM Culpeper VA
WRJT(FM) Royalton VT
KQBE(FM) Ellensburg WA
WOGB(FM) Kaukauna WI
WKCJ(FM) Lewisburg WV
WHBR-FM Parkersburg WV
KYDT(FM) Sundance WY

103.3 mhz
WMXS(FM) Montgomery AL
KIXB(FM) El Dorado AR
KWOZ(FM) Mountain View AR
KZKE(FM) Seligman AZ
KBAA(FM) Grass Valley CA
KZPO(FM) Lindsay CA
KATM(FM) Modesto CA
KVYB(FM) Santa Barbara CA
*KSCU(FM) Santa Clara CA
KTMQ(FM) Temecula CA
KUKI-FM Ukiah CA
*KPRU(FM) Delta CO
KJQY(FM) La Veta CO
WQQQ(FM) Sharon CT
WVYB(FM) Holly Hill FL
WVEE(FM) Atlanta GA
WWSN(FM) Waycross GA
KSHK(FM) Kekaha HI
WJOD(FM) Asbury IA
KAZR(FM) Pella IA
KTFC(FM) Sioux City IA
KSAS-FM Caldwell ID
KFTZ(FM) Idaho Falls ID
WIVQ(FM) Spring Valley IL
WRZX(FM) Indianapolis IN
WAXL(FM) Santa Claus IN
KJLS(FM) Hays KS
WXZZ(FM) Georgetown KY
WCDV(FM) Hammond LA
KBIU(FM) Lake Charles LA
WODS(FM) Boston MA
WMCM(FM) Rockland ME
WKFR-FM Battle Creek MI
WFXD(FM) Marquette MI
WQLB(FM) Tawas City MI
*KUMD-FM Duluth MN
KZCR(FM) Fergus Falls MN
KPRS(FM) Kansas City MO
KLOU(FM) Saint Louis MO
WZKR(FM) Decatur MS
KDTR(FM) Florence MT
WKVS(FM) Lenoir NC
WMGV(FM) Newport NC
KUSB(FM) Hazelton ND
WPRB(FM) Princeton NJ
KDRF(FM) Albuquerque NM
WEDG(FM) Buffalo NY
WMXW(FM) Vestal NY
*WCRF(FM) Cleveland OH
*WDEQ-FM De Graff OH
WMLX(FM) Saint Mary's OH
KJSR(FM) Tulsa OK
KKCW(FM) Beaverton OR
WKQL(FM) Brookville PA
WARM-FM York PA
WVJP-FM Caguas PR
WJMX-FM Cheraw SC
WOLT(FM) Greer SC
WKDF(FM) Nashville TN
KESN(FM) Allen TX
KDFM(FM) Falfurrias TX
KJOJ-FM Freeport TX
KCRS-FM Midland TX
KJCS(FM) Nacogdoches TX
KSAG(FM) Pearsall TX
WAKG(FM) Danville VA
WLTK(FM) New Market VA
WESR-FM Onley-Onancock VA
WWMP(FM) Waterbury VT
KWLN(FM) Wilson Creek WA
WGLX-FM Wisconsin Rapids WI
WTCR-FM Huntington WV
WAJR-FM Salem WV
KBEN-FM Basin WY
KHNA(FM) Hanna WY

103.5 mhz
KWVV-FM Homer AK
WLAY-FM Littleville AL
KZYQ(FM) Lake Village AR
KLNZ(FM) Glendale AZ
KTEA(FM) Cambria CA
KOST(FM) Los Angeles CA
KHSL-FM Paradise CA

KBMB(FM) Sacramento CA
KRAY-FM Salinas CA
KRFX(FM) Denver CO
WTOP-FM Washington DC
WJKI(FM) Bethany Beach DE
WAKT-FM Callaway FL
WMIB(FM) Fort Lauderdale FL
WFUS(FM) Gulfport FL
WJAD(FM) Leesburg GA
*KHAI(FM) Wahiawa HI
KNEI-FM Waukon IA
WKSC-FM Chicago IL
WXLT(FM) Christopher IL
WIKK(FM) Newton IL
WAWC(FM) Syracuse IN
KQLA(FM) Ogden KS
WAKY(FM) Radcliff KY
KLAA-FM Tioga LA
*WCCH(FM) Holyoke MA
WMUZ(FM) Detroit MI
WTCM-FM Traverse City MI
KUAL-FM Brainerd MN
KYSM-FM Mankato MN
*KRXW(FM) Roseau MN
KWXD(FM) Asbury MO
KLUE(FM) Poplar Bluff MO
WRBO(FM) Como MS
KZMY(FM) Bozeman MT
WRCQ(FM) Dunn NC
KZZY(FM) Devils Lake ND
KXNP(FM) North Platte NE
KISF(FM) Las Vegas NV
WQBJ(FM) Cobleskill NY
WKTU(FM) Lake Success NY
WUUF(FM) Sodus NY
WJQZ(FM) Wellsville NY
WGRR(FM) Hamilton OH
WJZA(FM) Lancaster OH
KVSP(FM) Anadarko OK
KLDZ(FM) Medford OR
KWHT(FM) Pendleton OR
WHLM-FM Berwick PA
WOGH(FM) Burgettstown PA
WLAK(FM) Huntingdon PA
WEZL(FM) Charleston SC
WZSN(FM) Greenwood SC
WIMZ-FM Knoxville TN
KKEV(FM) Centerville TX
KJNZ(FM) Hereford TX
KZRB(FM) New Boston TX
KBPA(FM) San Marcos TX
KAMZ(FM) Tahoka TX
KRSP-FM Salt Lake City UT
*WMRY(FM) Crozet VA
WZVA(FM) Marion VA
WAXJ(FM) Frederiksted VI

103.7 mhz
WAAO-FM Andalusia AL
WQEN(FM) Trussville AL
KABZ(FM) Little Rock AR
KZGL(FM) Flagstaff AZ
KODS(FM) Carnelian Bay CA
KMLA(FM) El Rio CA
*KLVG(FM) Garberville CA
KRZR(FM) Hanford CA
KIQQ-FM Newberry Springs CA
KSCF(FM) San Diego CA
KKSF(FM) San Francisco CA
KBNG(FM) Ridgway CO
WRUF-FM Gainesville FL
WQOL(FM) Vero Beach FL
WULS(FM) Broxton GA
WBOJ(FM) Cusseta GA
WVKX(FM) Irwinton GA
WBMZ(FM) Metter GA
WPUP(FM) Royston GA
KNUQ(FM) Paauilo HI
KLKK(FM) Clear Lake IA
WLLR-FM Davenport IA
KXKT(FM) Glenwood IA
KSKI-FM Sun Valley ID
KVRG(FM) Victor ID
WFAV(FM) Gilman IL
WDBR(FM) Springfield IL
*WFIU(FM) Bloomington IN
WHZR(FM) Royal Center IN
KEYN-FM Wichita KS
WCBJ(FM) Campton KY
WPTQ(FM) Cave City KY
WKED-FM Frankfort KY
WFGE(FM) Murray KY
KBTT(FM) Haughton LA
WXCY(FM) Havre de Grace MD
WCZE(FM) Harbor Beach MI
WHIT-FM Hartford MI
WHYB(FM) Menominee MI
KKBJ-FM Bemidji MN
KLZZ(FM) Waite Park MN
KJEL(FM) Lebanon MO
WUSW(FM) Hattiesburg MS
KBBB(FM) Billings MT
KUUS(FM) Fairfield MT
WSOC-FM Charlotte NC
WBNE(FM) Shallotte NC
WRHD(FM) Williamston NC
WKNE(FM) Keene NH
WPKQ(FM) North Conway NH
WMGM(FM) Atlantic City NJ
WNNJ-FM Newton NJ
KNMZ(FM) Alamogordo NM
KYVA-FM Grants NM
KPZA-FM Jal NM
KLNN(FM) Questa NM
WQNY(FM) Ithaca NY
WCKY-FM Tiffin OH
KOCD(FM) Wilburton OK
KXPC(FM) Lebanon OR
WRTS(FM) Erie PA
WCXR(FM) Lewisburg PA
WEEO-FM McConnellsburg PA
WXLX(FM) Lajas PR
WEEI-FM Westerly RI
KGIM-FM Redfield SD
KRRO(FM) Sioux Falls SD
KCDD(FM) Hamlin TX
KVIL(FM) Highland Park-Dallas TX
KIOL(FM) La Porte TX
KOUL(FM) Sinton TX
WMXB(FM) Richmond VA
KMTT(FM) Tacoma WA
WWIB(FM) Hallie WI
WXSS(FM) Wauwatosa WI
WCIR-FM Beckley WV
WQWV(FM) Fisher WV
KQLT(FM) Casper WY

103.9 mhz
KTDZ(FM) College AK
*KHZK(FM) Kotzebue AK
WJRL-FM Ozark AL
KPGG(FM) Ashdown AR
KCJF(FM) Earle AR
KKIX(FM) Fayetteville AR
KEDJ(FM) Gilbert AZ
KRCD(FM) Inglewood CA
KEDD(FM) Johannesburg CA
KCXX(FM) Lake Arrowhead CA
KKFS(FM) Lincoln CA
KDJK(FM) Mariposa CA
KMBY-FM Seaside CA
KBDS(FM) Taft CA
KSYC-FM Yreka CA
KYZX(FM) Pueblo West CO
KSNO-FM Snowmass Village CO
WXKB(FM) Cape Coral FL
WPPL(FM) Blue Ridge GA
WDDK(FM) Greensboro GA
WRPG(FM) Hawkinsville GA
WTYB(FM) Springfield GA
KCOO(FM) Dunkerton IA
KUOO(FM) Spirit Lake IA
WXAN(FM) Ava IL
WWYW(FM) Dundee IL
WNOI(FM) Flora IL
WQCY(FM) Quincy IL
WRBI(FM) Batesville IN
WIMC(FM) Crawfordsville IN
WXRD(FM) Crown Point IN
WXKE(FM) Fort Wayne IN
WRBR-FM South Bend IN
*KHYM(FM) Copeland KS
KOMB(FM) Fort Scott KS
KNZA(FM) Hiawatha KS
WNTC(FM) Drakesboro KY
WWEL(FM) London KY
WPTI(FM) Louisville KY
WXKQ-FM Whitesburg KY
WKPE-FM South Yarmouth MA
WOCQ(FM) Berlin MD
WTLP(FM) Braddock Heights MD
WVOM(FM) Howland ME
WLEN(FM) Adrian MI
*WCMW-FM Harbor Springs MI
WRSR(FM) Owosso MI
KBHL(FM) Osakis MN
KTNX(FM) Arcadia MO
KCHI-FM Chillicothe MO
KGLU(FM) Gideon MO
KRLI(FM) Malta Bend MO
KMCR(FM) Montgomery City MO
WCLD-FM Cleveland MS
WWKZ(FM) Columbus MS
KZMN(FM) Kalispell MT
WNNL(FM) Fuquay-Varina NC
WWTB(FM) Topsail Beach NC
KVMI(FM) Arthur ND
KOZY-FM Gering NE
KNLV-FM Ord NE
KRFS-FM Superior NE
KGRT-FM Las Cruces NM
KZMI(FM) Garapan-Saipan NP
WFAS-FM Bronxville NY
WVOA-FM Mexico NY
WSRK(FM) Oneonta NY
WQBK-FM Rensselaer NY
WRCN-FM Riverhead NY
WDKX(FM) Rochester NY
*WANC(FM) Ticonderoga NY
WXEG(FM) Beavercreek OH
WTDA(FM) Westerville OH
KOSG(FM) Pawhuska OK
WALY(FM) Bellwood PA
WPPZ-FM Jenkintown PA
WLMI(FM) Kane PA
WCOZ(FM) Laporte PA
WWIZ(FM) Mercer PA
WLSW(FM) Scottdale PA
WOLI-FM Easley SC
WHXT(FM) Orangeburg SC
WXIS(FM) Erwin TN
WDEB-FM Jamestown TN
WKZP(FM) McMinnville TN
KJXJ(FM) Cameron TX
KTHP(FM) Hemphill TX
KMHT-FM Marshall TX
KQXC-FM Wichita Falls TX
KUDE(FM) Nephi UT
KGNT(FM) Smithfield UT
WXCF-FM Clifton Forge VA
WYFT(FM) Luray VA
KBDB-FM Forks WA
KVAS(FM) Ilwaco WA
KBBD(FM) Spokane WA
WDSW(FM) Westby WI
WVBO(FM) Winneconne WI
WETZ-FM New Martinsville WV
KXXL(FM) Gillette WY

104.1 mhz
KBRJ(FM) Anchorage AK
WYOK(FM) Atmore AL
KILX(FM) Hatfield AR
KPOC-FM Pocahontas AR
KQTH(FM) Tucson AZ
KBOX(FM) Lompoc CA
KHKK(FM) Modesto CA
KJOR(FM) Windsor CA
KFRR(FM) Woodlake CA
KBVC(FM) Buena Vista CO
KNAB-FM Burlington CO
KFMU-FM Oak Creek CO
WPHH(FM) Waterbury CT
WWUS(FM) Big Pine Key FL
WTKS-FM Cocoa Beach FL
WGLF(FM) Tallahassee FL
WRJY(FM) Brunswick GA
WRBN(FM) Clayton GA
WALR-FM La Grange GA
WRBX(FM) Reidsville GA
KLTI-FM Ames IA
KORR(FM) American Falls ID
WMQZ(FM) Colchester IL
WHHL(FM) Jerseyville IL
WBWN(FM) Le Roy IL
WIKY-FM Evansville IN
WLBC-FM Muncie IN
KGGF-FM Fredonia KS
WCKQ(FM) Campbellsville KY
KYRK(FM) Houma LA
KJLO-FM Monroe LA
WBCN(FM) Boston MA
WPRS-FM Waldorf MD
*WVGR(FM) Grand Rapids MI
WSAG(FM) Linwood MI
KSDM(FM) International Falls MN
KBOT(FM) Pelican Rapids MN
KZJK(FM) Saint Louis Park MN
KSGF-FM Ash Grove MO
KZJF(FM) Jefferson City MO
KMHM(FM) Lutesville MO
WZKS(FM) Union MS
KHKR-FM East Helena MT
WCXL(FM) Kill Devil Hills NC
WTQR(FM) Winston-Salem NC
KIBZ(FM) Crete NE
KCDY(FM) Carlsbad NM
KABQ-FM Santa Fe NM
WHTT-FM Buffalo NY
WWYL(FM) Chenango Bridge NY
WQAL(FM) Cleveland OH
WPAY-FM Portsmouth OH
KMGL(FM) Oklahoma City OK
KFIS(FM) Scappoose OR
KWPK-FM Sisters OR
WAEB-FM Allentown PA
WNNK-FM Harrisburg PA
WPXZ-FM Punxsutawney PA
WERR(FM) Utuado PR
WYAV(FM) Myrtle Beach SC
KIQK(FM) Rapid City SD
WNAX-FM Yankton SD
WCLE-FM Calhoun TN
WUCZ-FM Carthage TN
WOGY(FM) Jackson TN
KWTR(FM) Big Lake TX
KWOW(FM) Clifton TX
KBFM(FM) Edinburg TX
KRBE(FM) Houston TX
KRIO-FM Pearsall TX
KTDK(FM) Sanger TX
KKUS(FM) Tyler TX
WMNV(FM) Rupert VT
KXDD(FM) Yakima WA
WRLU(FM) Algoma WI
WZEE(FM) Madison WI
WMZK(FM) Merrill WI
WDCI(FM) Bridgeport WV
*WVSB(FM) Romney WV
KANT(FM) Guernsey WY
KCGL(FM) Powell WY

104.3 mhz
*KTOO(FM) Juneau AK
WZYP(FM) Athens AL
WQZZ(FM) Eutaw AL
WHLW(FM) Luverne AL
*KLMZ(FM) Fouke AR
KBCN-FM Marshall AR
KAJM(FM) Payson AZ
KXSE(FM) Davis CA
KHIP(FM) Gonzales CA
KBIG-FM Los Angeles CA
KBQF(FM) McFarland CA
KSHA(FM) Redding CA
KMXY(FM) Grand Junction CO
KJCD(FM) Longmont CO
WIFL(FM) Inglis FL
WWHV(FM) Lynn Haven FL
*WKZM(FM) Sarasota FL
WEAT-FM West Palm Beach FL
WAJQ-FM Alma GA
WBBQ-FM Augusta GA
WKHC(FM) Dahlonega GA
KIJI(FM) Tumon GU
KPHW(FM) Kaneohe HI
KRKN(FM) Eldon IA
KAWO(FM) Boise ID
WCBH(FM) Casey IL
WJMK(FM) Chicago IL
WLRX(FM) Charlestown IN
KCAR-FM Galena KS
KVGB-FM Great Bend KS
WXBC(FM) Hardinsburg KY
WEZJ-FM Williamsburg KY
KEZP(FM) Bunkie LA
WSMJ(FM) Baltimore MD
WABK-FM Gardiner ME
*WVCN(FM) Baraga MI
WOMC(FM) Detroit MI
WCZY-FM Mount Pleasant MI
KLKS(FM) Breezy Point MN
KZLT-FM East Grand Forks MN
KVGO(FM) Spring Valley MN
KZIO(FM) Two Harbors MN
KDBB(FM) Bonne Terre MO
KBEQ-FM Kansas City MO
KXOQ(FM) Kennett MO
KKAC(FM) Vandalia MO
WMJU(FM) Bude MS
WGNL(FM) Greenwood MS
WQNQ(FM) Fletcher NC
WJSG(FM) Hamlet NC
WFXK(FM) Tarboro NC
KCYE(FM) North Las Vegas NV
WAXQ(FM) New York NY
WFRG-FM Utica NY
WOGF(FM) East Liverpool OH
WNLT(FM) Harrison OH
WJZK(FM) Richwood OH
KKMX(FM) Tri City OR
WKNB(FM) Clarendon PA
WSKE(FM) Everett PA
KKSD(FM) Milbank SD
WEYE(FM) Surgoinsville TN
KQFX(FM) Borger TX
KLZK(FM) Brownfield TX
KGAS-FM Carthage TX
KBLT(FM) Leakey TX
KHMR(FM) Lovelady TX
KAHA(FM) Olney TX
KLQB(FM) Taylor TX
KSOP-FM Salt Lake City UT
WKCY-FM Harrisonburg VA
WZIN(FM) Charlotte Amalie VI
WWOD(FM) Hartford VT
KAFE(FM) Bellingham WA
KMNT(FM) Chehalis WA
KZBE(FM) Omak WA
KHTR(FM) Pullman WA
WECB(FM) Seymour WI
KYPT(FM) Wamsutter WY

104.5 mhz
KMGC(FM) Camden AR
KWXE(FM) Glenwood AR
KTRN(FM) White Hall AR
KZUL-FM Lake Havasu City AZ
KCEC-FM Wellton AZ
KIQO(FM) Atascadero CA
KVLI-FM Lake Isabella CA
KBTW(FM) Lenwood CA
KFOG(FM) San Francisco CA
KKCS-FM Canon City CO
WFYV-FM Atlantic Beach FL
WKAK(FM) Albany GA
WYYU(FM) Dalton GA
KDAT(FM) Cedar Rapids IA
*WBVN(FM) Carrier Mills IL
WFMB-FM Springfield IL
WJJK(FM) Noblesville IN
KFXJ(FM) Augusta KS
WLKT(FM) Lexington-Fayette KY
KNOU(FM) Empire LA
KBEF(FM) Gibsland LA
WNXX(FM) Jackson LA
KLSM(FM) Tallulah LA
WXLO(FM) Fitchburg MA
WKHJ(FM) Mountain Lake Park MD
WSNX-FM Muskegon MI

WILZ(FM) Saginaw MI
KJLY(FM) Blue Earth MN
KUMR(FM) Doolittle MO
KSLQ-FM Washington MO
WXRR(FM) Hattiesburg MS
WQJB(FM) State College MS
KKVU(FM) Stevensville MT
WSTK(FM) Aurora NC
WHLC(FM) Highlands NC
WCCG(FM) Hope Mills NC
WRQR(FM) Wilmington NC
*KCVN(FM) Cozad NE
KSRZ(FM) Omaha NE
WVMJ(FM) Conway NH
KKFG(FM) Bloomfield NM
KZXQ(FM) Reserve NM
KDOT(FM) Reno NV
WTMM-FM Mechanicville NY
WLZZ(FM) Montpelier OH
WQKT(FM) Wooster OH
KMYZ-FM Pryor OK
KMCQ(FM) The Dalles OR
WXXO(FM) Cambridge Springs PA
WRFF(FM) Philadelphia PA
WNBT-FM Wellsboro PA
WRFQ(FM) Mt. Pleasant SC
WGFX(FM) Gallatin TN
WKHT(FM) Knoxville TN
WRVR(FM) Memphis TN
KKDA-FM Dallas TX
KPUS(FM) Gregory TX
KJTX(FM) Jefferson TX
KKMY(FM) Orange TX
KZEP-FM San Antonio TX
WGRX(FM) Falmouth VA
WNVZ(FM) Norfolk VA
*KMIH(FM) Mercer Island WA
KQQB-FM Newport WA
WAXX(FM) Eau Claire WI
WXER(FM) Plymouth WI
*WCCX(FM) Waukesha WI
WSLD(FM) Whitewater WI
WHAJ(FM) Bluefield WV
KRQU(FM) Laramie WY
KSIT(FM) Rock Springs WY
KRUG(FM) Upton WY

104.7 mhz
KKED(FM) Fairbanks AK
*KCAW(FM) Sitka AK
WZZK-FM Birmingham AL
KQBK(FM) Booneville AR
KFLI(FM) Des Arc AR
KOOU(FM) Hardy AR
KTOY(FM) Texarkana AR
KNWJ(FM) Leone AS
KZZP(FM) Mesa AZ
KHUM(FM) Garberville CA
KCAQ(FM) Oxnard CA
KDES-FM Palm Springs CA
KHTN(FM) Planada CA
KNNG(FM) Sterling CO
KSKE-FM Vail CO
WAAZ-FM Crestview FL
WSGL(FM) Naples FL
WRBQ-FM Tampa FL
WFLM(FM) White City FL
WFSH-FM Athens GA
WTHG(FM) Hinesville GA
KONI(FM) Lanai City HI
KIKX(FM) Ketchum ID
WLMD(FM) Bushnell IL
*WCFL(FM) Morris IL
WNSV(FM) Nashville IL
WFRN-FM Elkhart IN
WITZ-FM Jasper IN
*KVCY(FM) Fort Scott KS
KXBZ(FM) Manhattan KS
WJMD(FM) Hazard KY
WJSH(FM) Folsom LA
KORI(FM) Mansfield LA
KNEK-FM Washington LA
WOCN-FM Orleans MA
WAYZ(FM) Hagerstown MD
WQHQ(FM) Ocean City-Salisbury MD
WBFB(FM) Belfast ME
WHXQ(FM) Kennebunkport ME
WYKX(FM) Escanaba MI
WKJC(FM) Tawas City MI
KCLD-FM Saint Cloud MN
KREZ(FM) Chaffee MO
KKLH(FM) Marshfield MO
KRES(FM) Moberly MO
WJIW(FM) Greenville MS
KBZM(FM) Big Sky MT
WKQC(FM) Charlotte NC
WZUP(FM) Rose Hill NC
KMXW(FM) Hope ND
KNDR(FM) Mandan ND
KTEG(FM) Bosque Farms NM
KMOU(FM) Roswell NM
KJUL(FM) Moapa Valley NV
WBBS(FM) Fulton NY
WMOS(FM) Montauk NY
WSPK(FM) Poughkeepsie NY
*WIRQ(FM) Rochester NY
WTUE(FM) Dayton OH
WKKY(FM) Geneva OH
WIOT(FM) Toledo OH
KIXR(FM) Ponca City OK
KSLE(FM) Wewoka OK
KCMB(FM) Baker City OR
KDUK-FM Florence OR
KFEG(FM) Klamath Falls OR
WPGB(FM) Pittsburgh PA
WKAQ-FM San Juan PR
WNOK(FM) Columbia SC
KKLS-FM Sioux Falls SD
WSGM(FM) Coalmont TN
WMUF-FM Henry TN
WLIV-FM Monterey TN
KKYS(FM) Bryan TX
KYYI(FM) Burkburnett TX
KULF(FM) Ganado TX
KTXC(FM) Lamesa TX
KWNS(FM) Winnsboro TX
WPZZ(FM) Crewe VA
WNCS(FM) Montpelier VT
KDUX-FM Aberdeen WA
KKRV(FM) Wenatchee WA
WBJZ(FM) Berlin WI
WDDW(FM) Sturtevant WI
WVRC-FM Spencer WV
KTRS-FM Casper WY
KLQQ(FM) Clearmont WY

104.9 mhz
WOAB(FM) Ozark AL
WSLY(FM) York AL
KAGH-FM Crossett AR
KHPA(FM) Hope AR
KDXY(FM) Lake City AR
KXNA(FM) Springdale AR
KCLT(FM) West Helena AR
KWCX(FM) Willcox AZ
KWIM(FM) Window Rock AZ
KLOA-FM Ridgecrest CA
KMHX(FM) Rohnert Park CA
KYIX(FM) South Oroville CA
KCNL(FM) Sunnyvale CA
KCRZ(FM) Tipton CA
KRYD(FM) Norwood CO
KRYE(FM) Rye CO
*WIHS(FM) Middletown CT
WHTF(FM) Havana FL
WYGC(FM) High Springs FL
WCVU(FM) Solana FL
WFXE(FM) Columbus GA
WMCG(FM) Milan GA
KBOB-FM De Witt IA
KLMJ(FM) Hampton IA
KBOE-FM Oskaloosa IA
KLLT(FM) Spencer IA
WXRX(FM) Belvidere IL
KMJM-FM Columbia IL
WFIW-FM Fairfield IL
WPXN(FM) Paxton IL
WXCL(FM) Pekin IL
WINN(FM) Columbus IN
WERK(FM) Muncie IN
WAXI(FM) Rockville IN
KFFX(FM) Emporia KS
KSAL-FM Salina KS
WKYW(FM) Frankfort KY
WXLR(FM) Harold KY
WJRS(FM) Jamestown KY
WKHG(FM) Leitchfield KY
WAVJ(FM) Princeton KY
WSKV(FM) Stanton KY
WKKS-FM Vanceburg KY
KNXX(FM) Donaldsonville LA
*KTOC-FM Jonesboro LA
KZWA(FM) Moss Bluff LA
*WRBB(FM) Boston MA
WBOQ(FM) Gloucester MA
WQBX(FM) Alma MI
*WAIR(FM) Lake City MI
WRCC(FM) Marshall MI
KRFO-FM Owatonna MN
KPWB-FM Piedmont MO
WAKK-FM Centreville MS
WFXO(FM) Iuka MS
WBUV(FM) Moss Point MS
WCJU-FM Prentiss MS
KIKF(FM) Cascade MT
WYNA(FM) Calabash NC
WFMZ(FM) Hertford NC
WQNS(FM) Waynesville NC
KCTY(FM) Wayne NE
KTMX(FM) York NE
WYRY(FM) Hinsdale NH
WLKZ(FM) Wolfeboro NH
WSJO(FM) Egg Harbor City NJ
KMVR(FM) Mesilla Park NM
KVYL(FM) Cal-Nev-Ari NV
WZMR(FM) Altamont NY
*WKDL(FM) Brockport NY
WNGZ(FM) Montour Falls NY
WWKC(FM) Caldwell OH
*WCVO(FM) Gahanna OH
WEGE(FM) Lima OH
WCLV(FM) Lorain OH
KKWD(FM) Bethany OK
KREK(FM) Bristow OK
KRIG-FM Nowata OK
*WJRH(FM) Easton PA
WRKY-FM Hollidaysburg PA
WWRR(FM) Scranton PA
WCCP-FM Clemson SC
WWJN(FM) Ridgeland SC
WALV-FM Dayton TN
WKOS(FM) Kingsport TN
WQLA-FM La Follette TN
WYVY(FM) Union City TN
WBOZ(FM) Woodbury TN
KJAV(FM) Alamo TX
KXBT(FM) Dripping Springs TX
KLDE(FM) Eldorado TX
KBUK(FM) La Grange TX
KWGW(FM) Mexia TX
KPTY(FM) Missouri City TX
KZMP-FM Pilot Point TX
KMFM(FM) Premont TX
KMIQ(FM) Robstown TX
KBTE(FM) Tulia TX
KVOU-FM Uvalde TX
KYLZ(FM) Tremonton UT
WZBL(FM) Roanoke VA
WWRT(FM) Strasburg VA
WIGO-FM White Stone VA
WMNG(FM) Christiansted VI
KFNK(FM) Eatonville WA
*KEEH(FM) Spokane WA
WLMX-FM Balsam Lake WI
WPCK(FM) Denmark WI
WTKM-FM Hartford WI
WLXR-FM La Crosse WI
WKQH(FM) Marathon WI
WNFM(FM) Reedsburg WI
WPDX-FM Clarksburg WV
KRRR(FM) Cheyenne WY

105.1 mhz
KTKU(FM) Juneau AK
WQSB(FM) Albertville AL
KMJX(FM) Conway AR
KFLX(FM) Kachina Village AZ
KLBN(FM) Auberry CA
KRTO(FM) Lompoc CA
KKGO(FM) Los Angeles CA
KOCN(FM) Pacific Grove CA
KNCI(FM) Sacramento CA
KXKL-FM Denver CO
WPFL(FM) Century FL
WHQT(FM) Coral Gables FL
WOMX-FM Orlando FL
WASJ(FM) Panama City Beach FL
WKUB(FM) Blackshear GA
WLVX(FM) Elberton GA
WZGA(FM) Helen GA
KGUM-FM Dededo GU
KINE-FM Honolulu HI
KCCQ(FM) Ames IA
KJOT(FM) Boise ID
KVTY(FM) Lewiston ID
WOJO(FM) Evanston IL
WTAO(FM) Murphysboro IL
WGEM-FM Quincy IL
WEJT(FM) Shelbyville IL
WQHK-FM Decatur IN
WHCC(FM) Ellettsville IN
KZQD(FM) Liberal KS
WTUK(FM) Harlan KY
WRNZ(FM) Lancaster KY
WLRS(FM) Shepherdsville KY
KPEL-FM Abbeville LA
KTGV(FM) Jonesville LA
*WAMQ(FM) Great Barrington MA
*WNEK-FM Springfield MA
WTOS-FM Skowhegan ME
WGFM(FM) Cheboygan MI
WMGC-FM Detroit MI
KLTA(FM) Breckenridge MN
KKCB(FM) Duluth MN
WGVX(FM) Lakeville MN
KARL(FM) Tracy MN
KCRV-FM Caruthersville MO
KCJK(FM) Garden City MO
KOSP(FM) Willard MO
WQJQ(FM) Kosciusko MS
KQBL(FM) Billings MT
KWOL-FM Whitefish MT
WDCG(FM) Durham NC
WSSM(FM) Havelock NC
KAOC(FM) Cavalier ND
KWMW(FM) Maljamar NM
KJFA(FM) Santa Fe NM
KQRT(FM) Las Vegas NV
WWDG(FM) DeRuyter NY
WWPR-FM New York NY
WKOL(FM) Plattsburgh NY
WUBE-FM Cincinnati OH
WQXK(FM) Salem OH
KBLP(FM) Lindsay OK
KTMC-FM McAlester OK
KOSB(FM) Perry OK
KRSK(FM) Molalla OR
KAKT(FM) Phoenix OR
WIOV-FM Ephrata PA
WILQ(FM) Williamsport PA
WIOC(FM) Ponce PR
WWLI(FM) Providence RI
WGFG(FM) Branchville SC
WPDT(FM) Johnsonville SC
KAWK(FM) Custer SD
KZKK(FM) Huron SD
WCLC-FM Jamestown TN
WVRY(FM) Waverly TN
KEAN-FM Abilene TX
KMIL(FM) Cameron TX
KYKS(FM) Lufkin TX
KTTY(FM) New Boston TX
KMAT(FM) Seadrift TX
KAUU(FM) Manti UT
WAVA-FM Arlington VA
WBHB-FM Bridgewater VA
WSBW(FM) Sister Bay WI
WCHY(FM) Waunakee WI
WKLC-FM Saint Albans WV
KTUG(FM) Hudson WY
KGCL(FM) Ten Sleep WY

105.3 mhz
*KRBD(FM) Ketchikan AK
WDBT(FM) Headland AL
WBFZ(FM) Selma AL
KJLV(FM) Hoxie AR
KAKJ(FM) Marianna AR
KQOR(FM) Mena AR
KZLZ(FM) Kearny AZ
KHOV-FM Wickenburg AZ
KBFP-FM Delano CA
KIOZ(FM) San Diego CA
KITS(FM) San Francisco CA
KRDG(FM) Shingletown CA
KRSX-FM Yermo CA
KZKS(FM) Rifle CO
WJSJ(FM) Fernandina Beach FL
WYKS(FM) Gainesville FL
WZSP(FM) Nocatee FL
WBZY(FM) Bowdon GA
WSTI-FM Quitman GA
WRHQ(FM) Richmond Hill GA
KBGX(FM) Keaau HI
KLYV(FM) Dubuque IA
KNOD(FM) Harlan IA
KIWA-FM Sheldon IA
WKAY(FM) Knoxville IL
WAOX(FM) Staunton IL
WJLT(FM) Evansville IN
WKOA(FM) Lafayette IN
WMPI(FM) Scottsburg IN
KJML(FM) Columbus KS
KFBZ(FM) Haysville KS
WOVO(FM) Glasgow KY
WXKZ-FM Prestonsburg KY
WWL-FM Kenner LA
KLIP(FM) Monroe LA
KCJN(FM) Simmesport LA
KNCB-FM Vivian LA
WFRB-FM Frostburg MD
WSHK(FM) Kittery ME
WKHM-FM Brooklyn MI
WHTS(FM) Coopersville MI
WGVY(FM) Cambridge MN
KYBA(FM) Stewartville MN
KYMO-FM East Prairie MO
KZNN(FM) Rolla MO
WACR-FM Aberdeen MS
KMTX-FM Helena MT
WODR(FM) Fair Bluff NC
KZPR(FM) Minot ND
KLNC(FM) Lincoln NE
KIOD(FM) McCook NE
*KGRD(FM) Orchard NE
KHOD(FM) Des Moines NM
WDRE(FM) Calverton-Roanoke NY
*WGKR(FM) Grand Gorge NY
WKPQ(FM) Hornell NY
WYHT(FM) Mansfield OH
KJMM(FM) Bixby OK
KDDQ(FM) Comanche OK
KINB(FM) Kingfisher OK
WYCY(FM) Hawley PA
WDAS-FM Philadelphia PA
WNOW-FM Gaffney SC
WLSZ(FM) Humboldt TN
WFIV-FM Loudon TN
KPTI(FM) Crystal Beach TX
KLLI(FM) Dallas TX
KTWL(FM) Hempstead TX
KJDL-FM Levelland TX
KLSR-FM Memphis TX
KSMG(FM) Seguin TX
WBRW(FM) Blacksburg VA
WBNN-FM Dillwyn VA
WKUS(FM) Norfolk VA
WVJZ(FM) Charlotte Amalie VI
WEBK(FM) Killington VT
KCMS(FM) Edmonds WA
KONA-FM Kennewick WA
WRLO-FM Antigo WI
WFZH(FM) Mukwonago WI
KDWY(FM) Diamondville WY
KREO(FM) Pine Bluffs WY

105.5 mhz
WNSP(FM) Bay Minette AL
WENN(FM) Hoover AL

WVNA-FM Muscle Shoals AL
KYEL(FM) Danville AR
KPFM(FM) Mountain Home AR
KNAS(FM) Nashville AR
KBOA-FM Piggott AR
KWAK-FM Stuttgart AR
KWRF-FM Warren AR
*KLVA(FM) Casa Grande AZ
KRVR(FM) Copperopolis CA
KTKZ-FM Dunnigan CA
KKHB(FM) Eureka CA
KIDI(FM) Guadalupe CA
KBUE(FM) Long Beach CA
KFYV(FM) Ojai CA
KOSS(FM) Rosamond CA
KJZN(FM) San Joaquin CA
KRDO-FM Security CO
KJAC(FM) Timnath CO
WQGN-FM Groton CT
WFCT(FM) Apalachicola FL
WOLL(FM) Hobe Sound FL
WWWK(FM) Marathon FL
WYZB(FM) Mary Esther FL
WBTT(FM) Naples Park FL
WDUV(FM) New Port Richey FL
WSJF(FM) Saint Augustine Beach FL
WZBN(FM) Camilla GA
WIFO-FM Jesup GA
WIFN(FM) Macon GA
WRXR-FM Rossville GA
KPMW(FM) Haliimaile HI
KELR-FM Chariton IA
KILJ-FM Mount Pleasant IA
KDLS-FM Perry IA
KTHK(FM) Idaho Falls ID
WREZ(FM) Metropolis IL
WCZQ(FM) Monticello IL
WJVO(FM) South Jacksonville IL
WYKT(FM) Wilmington IL
WZSR(FM) Woodstock IL
WQRK(FM) Bedford IN
WTHD(FM) Lagrange IN
WLJE(FM) Valparaiso IN
WWVR(FM) West Terre Haute IN
KVSV-FM Beloit KS
KKOY-FM Chanute KS
WLVK(FM) Fort Knox KY
WKYA(FM) Greenville KY
WMKJ(FM) Mt. Sterling KY
KBKK(FM) Ball LA
KEUN-FM Eunice LA
KDDK(FM) Franklin LA
WVEI-FM Easthampton MA
WDKZ(FM) Salisbury MD
WBYA(FM) Islesboro ME
WWCK-FM Flint MI
WGKL(FM) Gladstone MI
WMKD(FM) Pickford MI
WBMI(FM) West Branch MI
KDDG(FM) Albany MN
KBAJ(FM) Deer River MN
KMGM(FM) Montevideo MN
KRBI-FM Saint Peter MN
KESM-FM El Dorado Springs MO
KZZT(FM) Moberly MO
KKJO(FM) Saint Joseph MO
WNLA-FM Indianola MS
WVBG-FM Redwood MS
WTNM(FM) Water Valley MS
WABO-FM Waynesboro MS
KRQS(FM) Alberton MT
WXQR(FM) Jacksonville NC
WFJA(FM) Sanford NC
KMAV-FM Mayville ND
KFMT-FM Fremont NE
WJYY(FM) Concord NH
WGBZ(FM) Cape May Court House NJ
WDHA-FM Dover NJ
KZZX(FM) Alamogordo NM
*KQLV(FM) Bosque Farms NM
KSIL(FM) Hurley NM
WLPW(FM) Lake Placid NY
WSKU(FM) Little Falls NY
WTKV(FM) Oswego NY
WDBY(FM) Patterson NY
WXTQ(FM) Athens OH
*WGOJ(FM) Conneaut OH
WMVR-FM Sidney OH
WWWM-FM Sylvania OH
WCHO-FM Washington Court House OH
KWCO-FM Chickasha OK
KKFC(FM) Coalgate OK
KGFY(FM) Stillwater OK
KDEP(FM) Garibaldi OR
KCGB-FM Hood River OR
KEUG(FM) Veneta OR
WMKX(FM) Brookville PA
WCHX(FM) Lewistown PA
WMGH-FM Tamaqua PA
WFDT(FM) Aguada PR
WDAR-FM Darlington SC
WCOO(FM) Kiawah Island SC
KMOM-FM Roscoe SD
WYTM-FM Fayetteville TN
WSEV-FM Gatlinburg TN
WBNT-FM Oneida TN
WAKQ(FM) Paris TN
WXOQ(FM) Selmer TN
WRKK-FM Sparta TN
KACT-FM Andrews TX
KWDC(FM) Coahoma TX
KUSJ(FM) Harker Heights TX
KQXX-FM Mission TX
KMJR(FM) Portland TX
KBWT(FM) Santa Anna TX
WKDE-FM Altavista VA
WWRE(FM) Berryville VA
WHFD(FM) Lawrenceville VA
WOJL(FM) Louisa VA
WSWV-FM Pennington Gap VA
WGTH-FM Richlands VA
WRAR-FM Tappahannock VA
WKXH(FM) Saint Johnsbury VT
KUKN(FM) Longview WA
WUSP(FM) Nekoosa WI
WFBZ(FM) Trempealeau WI
WMMM-FM Verona WI
WUKL(FM) Bethlehem WV
WKQV(FM) Richwood WV
KHAD(FM) Mills WY

105.7 mhz

KNIK-FM Anchorage AK
WQAH-FM Addison AL
WCSN-FM Orange Beach AL
WZHT(FM) Troy AL
KRKD(FM) Dermott AR
KFXV(FM) Kensett AR
KMCK(FM) Siloam Springs AR
KVRD-FM Cottonwood AZ
KOAS(FM) Dolan Springs AZ
KXRS(FM) Hemet CA
KQMX(FM) Lost Hills CA
KVVF(FM) Santa Clara CA
KVAY(FM) Lamar CO
KWGL(FM) Ouray CO
KPMX(FM) Sterling CO
WHJX(FM) Baldwin FL
*WFRF-FM Monticello FL
WWLL(FM) Sebring FL
*WFFM(FM) Ashburn GA
WEKL(FM) Augusta GA
WWVA-FM Canton GA
KOKZ(FM) Waterloo IA
WIXO(FM) Peoria IL
WUZR(FM) Bicknell IN
WYXB(FM) Indianapolis IN
WTCJ-FM Tell City IN
*KJRL(FM) Herington KS
WTUV-FM Eminence KY
WLGC-FM Greenup KY
WTBK(FM) Manchester KY
KVVP(FM) Leesville LA
WROR-FM Framingham MA
WHFS(FM) Catonsville MD
*WHMX(FM) Lincoln ME
WOOD-FM Grand Rapids MI
WCUP(FM) L'Anse MI
WGVZ(FM) Eden Prairie MN
KRAQ(FM) Jackson MN
KXKX(FM) Knob Noster MO
KPNT(FM) Sainte Genevieve MO
WJXM(FM) De Kalb MS
WAKH(FM) McComb MS
KKQX(FM) Manhattan MT
KWDV(FM) Valier MT
WMKS(FM) Clemmons NC
WRSF(FM) Columbia NC
WGQR(FM) Elizabethtown NC
KSUX(FM) Winnebago NE
WLKC(FM) Campton NH
WCHR-FM Manahawkin NJ
KOZZ-FM Reno NV
WMRV-FM Endicott NY
WBZZ(FM) Malta NY
WMJI(FM) Cleveland OH
WZOM(FM) Defiance OH
WBWR(FM) Hilliard OH
KTTL(FM) Alva OK
*KROU(FM) Spencer OK
KQAK(FM) Bend OR
KKKJ(FM) Merrill OR
WLKJ(FM) Portage PA
WQXA-FM York PA
WCAD(FM) San Juan PR
WIHG(FM) Rockwood TN
WQAK(FM) Union City TN
*KJJP(FM) Amarillo TX
KTKO(FM) Beeville TX
KRNB(FM) Decatur TX
KNAF-FM Fredericksburg TX
*KHCB-FM Houston TX
KRBL(FM) Idalou TX
KYKX(FM) Longview TX
KBIC(FM) Raymondville TX
KXRV(FM) Centerville UT
WMXH-FM Luray VA
KJET(FM) Raymond WA
KZBD(FM) Spokane WA
KRSE(FM) Yakima WA
WAPL-FM Appleton WI
WCFW(FM) Chippewa Falls WI
WXCX(FM) Siren WI
WOBG-FM Salem WV

105.9 mhz

KSWD-FM Seward AK
WNSI-FM Atmore AL
WRTR(FM) Brookwood AL
WRHY(FM) Centre AL
KLAZ(FM) Hot Springs AR
KHOT-FM Paradise Valley AZ
KFJK(FM) Fresno CA
KPWR(FM) Los Angeles CA
KRAZ(FM) Santa Ynez CA
KQPM(FM) Ukiah CA
KALC(FM) Denver CO
WHCN(FM) Hartford CT
WXJN(FM) Lewes DE
WOCL(FM) De Land FL
WTZB(FM) Englewood FL
WBGG-FM Fort Lauderdale FL
WILN(FM) Panama City FL
WXMK(FM) Dock Junction GA
WVGA(FM) Lakeland GA
KPOI-FM Honolulu HI
KZLN(FM) Patterson IA
KTLB(FM) Twin Lakes IA
KCIX(FM) Garden City ID
WCKG(FM) Elmwood Park IL
WOKZ(FM) Fairfield IL
WGKC(FM) Mahomet IL
WMMC(FM) Marshall IL
WJOT-FM Wabash IN
KSSA(FM) Ingalls KS
KLZR(FM) Lawrence KS
WKYB(FM) Burgin KY
WRVI(FM) Valley Station KY
KBZE(FM) Berwick LA
KFXZ-FM Opelousas LA
WBCI(FM) Bath ME
WKHQ-FM Charlevoix MI
WDMK(FM) Detroit MI
KWNG(FM) Red Wing MN
KKWS(FM) Wadena MN
KHRS(FM) Winthrop MN
KZZK(FM) New London MO
KGBX-FM Nixa MO
KULH(FM) Wheeling MO
WHGO(FM) Pascagoula MS
WOAD-FM Pickens MS
KPBR(FM) Joliet MT
KYJK(FM) Missoula MT
WTMT(FM) Weaverville NC
KCVF(FM) Sarles ND
KAAQ(FM) Alliance NE
KQKY(FM) Kearney NE
KKCD(FM) Omaha NE
WCAA(FM) Newark NJ
KRZY-FM Santa Fe NM
WJZR(FM) Rochester NY
WLTI(FM) Syracuse NY
WPFB-FM Middletown OH
WWJM(FM) New Lexington OH
KQTZ(FM) Hobart OK
KIRC(FM) Seminole OK
KRJT(FM) Elgin OR
WJOW(FM) Philipsburg PA
WXDX-FM Pittsburgh PA
WPZX(FM) Pocono Pines PA
WEZV(FM) North Myrtle Beach SC
KMIT(FM) Mitchell SD
WGKX(FM) Memphis TN
WNRQ(FM) Nashville TN
KUZN(FM) Centerville TX
KMFR(FM) Hondo TX
KFMK(FM) Round Rock TX
KUKA(FM) San Diego TX
KKJW(FM) Stanton TX
KLCY-FM Vernal UT
WLNI(FM) Lynchburg VA
WJZW(FM) Woodbridge VA
KIJZ(FM) Vancouver WA
WWHG(FM) Evansville WI
*WEGZ(FM) Washburn WI
WTNJ(FM) Mount Hope WV
KTYN(FM) Thayne WY

106.1 mhz

WSTH-FM Alexander City AL
WBMH(FM) Grove Hill AL
WTAK-FM Hartselle AL
KFFB(FM) Fairfield Bay AR
KIKO-FM Claypool AZ
KFSZ(FM) Munds Park AZ
*KCFA(FM) Arnold CA
KRRX(FM) Burney CA
KRAB(FM) Green Acres CA
KPLM(FM) Palm Springs CA
KMEL(FM) San Francisco CA
KWWV(FM) Santa Margarita CA
KNFO(FM) Basalt CO
WRRX(FM) Gulf Breeze FL
WUTL(FM) Tallahassee FL
WKTM(FM) Soperton GA
*WHKV(FM) Sylvester GA
WNGC(FM) Toccoa GA
KLEO(FM) Kahaluu HI
KLSS-FM Mason City IA
KIYX(FM) Sageville IA
KZFN(FM) Moscow ID
KKMV(FM) Rupert ID
WSMI-FM Litchfield IL
WYYS(FM) Streator IL
WDKS(FM) Newburgh IN
WWWY(FM) North Vernon IN
KXKU(FM) Lyons KS
WMOR-FM Morehead KY
KXRR(FM) Monroe LA
WCOD-FM Hyannis MA
WKGO(FM) Cumberland MD
*WMEM(FM) Presque Isle ME
WJXQ(FM) Jackson MI
WHST(FM) Tawas City MI
KLCI(FM) Elk River MN
KJOE(FM) Slayton MN
KOQL(FM) Ashland MO
KWKZ(FM) Charleston MO
WMTI(FM) Picayune MS
WMXU(FM) Starkville MS
KQDI-FM Great Falls MT
WMMY(FM) Jefferson NC
WNMX-FM Waxhaw NC
WRDU(FM) Wilson NC
KQLX-FM Lisbon ND
WHDQ(FM) Claremont NH
KPZE-FM Carlsbad NM
KFMQ(FM) Gallup NM
WNKI(FM) Corning NY
WPDA(FM) Jeffersonville NY
WBLI(FM) Patchogue NY
WVNO-FM Mansfield OH
WBBG(FM) Niles OH
KKBI(FM) Broken Bow OK
KQLL-FM Owasso OK
KIXO(FM) Sulphur OK
KLOO-FM Corvallis OR
WLZS(FM) Beaver Springs PA
WISX(FM) Philadelphia PA
WRRH(FM) Hormigueros PR
WVIS(FM) Vieques PR
WFXH-FM Hilton Head Island SC
WTUA(FM) Saint Stephen SC
KTTX(FM) Brenham TX
KHKS(FM) Denton TX
KFLP-FM Floydada TX
KKVR(FM) Kerrville TX
KNEX(FM) Laredo TX
KIOC(FM) Orange TX
KTKY(FM) Refugio TX
KMDX(FM) San Angelo TX
KBAL-FM San Saba TX
KRZX(FM) Monticello UT
WCNR(FM) Keswick VA
WNRJ(FM) Poquoson VA
WJJS-FM Vinton VA
KBKS-FM Tacoma WA
WDKM(FM) Adams WI
WACD(FM) Antigo WI
WMIL(FM) Waukesha WI
WRZZ(FM) Elizabeth WV
KBMG(FM) Evanston WY

106.3 mhz

KSUP(FM) Juneau AK
WKNU(FM) Brewton AL
WBTG-FM Sheffield AL
KZKZ-FM Greenwood AR
KOLL-FM Lonoke AR
KYGL(FM) Texarkana AR
KRLW-FM Walnut Ridge AR
KMGX(FM) Oracle AZ
KOMR(FM) Sun City AZ
KEJY(FM) Blue Lake CA
KGRP(FM) Cazadero CA
KMGQ(FM) Goleta CA
KSXE(FM) Kingsburg CA
KGMX(FM) Lancaster CA
KNAH(FM) Merced CA
KALI-FM Santa Ana CA
KMJV(FM) Soledad CA
KCHC(FM) Willows CA
KPRB(FM) Brush CO
KZMV(FM) Kremmling CO
KWUF-FM Pagosa Springs CO
KKLI(FM) Widefield CO
WJPT(FM) Fort Myers Villas FL
WNEW(FM) Jupiter FL
WZMQ(FM) Key Largo FL
WCIF(FM) Melbourne FL
WSBZ(FM) Miramar Beach FL
WJQB(FM) Spring Hill FL
WTUF(FM) Boston GA
WGMK(FM) Donalsonville GA
WQBZ(FM) Fort Valley GA
WOAH(FM) Glennville GA
WKBX(FM) Kingsland GA
KPTL(FM) Ankeny IA
KQTA(FM) Homedale ID
KBJX(FM) Shelley ID
WQRL(FM) Benton IL
WYRB(FM) Genoa IL
WGCY(FM) Gibson City IL
WSRB(FM) Lansing IL
WVBB(FM) Columbia City IN
WCDQ(FM) Crawfordsville IN
WUBU(FM) South Bend IN
WANY-FM Albany KY
WXMZ(FM) Hartford KY

WKMO(FM) Hodgenville KY
WRIL(FM) Pineville KY
WCDA(FM) Versailles KY
KKSJ(FM) Maurice LA
KXOR-FM Thibodaux LA
WEIB(FM) Northampton MA
WCEM-FM Cambridge MD
WBQW(FM) Scarborough ME
WSCG-FM Lakeview MI
WKLA-FM Ludington MI
WGER(FM) Saginaw MI
WMXG(FM) Stephenson MI
KRJB(FM) Ada MN
WMFG-FM Hibbing MN
KPHR(FM) Ortonville MN
KRZK(FM) Branson MO
KHZA(FM) Bunker MO
WZLD(FM) Petal MS
WGNG(FM) Tchula MS
KDBR(FM) Kalispell MT
WLTT(FM) Shallotte NC
KLMY(FM) Lincoln NE
WMTK(FM) Littleton NH
WFNQ(FM) Nashua NH
WHCY(FM) Blairstown NJ
WHTG-FM Eatontown NJ
KAGM(FM) Los Lunas NM
WFAF(FM) Mount Kisco NY
WMCR-FM Oneida NY
WYZY(FM) Saranac Lake NY
WCDK(FM) Cadiz OH
WJYD(FM) London OH
WBUK(FM) Ottawa OH
*KGOU(FM) Norman OK
KZZE(FM) Eagle Point OR
WLCY(FM) Blairsville PA
WFZY(FM) Mount Union PA
WQRM(FM) Smethport PA
WCTL(FM) Union City PA
WWKX(FM) Woonsocket RI
WYNN-FM Florence SC
WJNI(FM) Ladson SC
WGVC(FM) Simpsonville SC
KZLK(FM) Rapid City SD
KVHT(FM) Vermillion SD
KKHR(FM) Abilene TX
KOOC(FM) Belton TX
KPSO-FM Falfurrias TX
*KMLR(FM) Gonzales TX
KPAN-FM Hereford TX
KERB-FM Kermit TX
KHKZ(FM) Mercedes TX
KSEM-FM Seminole TX
KBZS(FM) Wichita Falls TX
WHKX(FM) Bluefield VA
WMNA-FM Gretna VA
WNVA-FM Norton VA
KCSY(FM) Twisp WA
WQCC(FM) La Crosse WI
WWQM-FM Middleton WI
WEVR-FM River Falls WI
WPLT(FM) Spooner WI
WAMX(FM) Milton WV
KLEN(FM) Cheyenne WY

106.5 mhz

KWHL(FM) Anchorage AK
WAVH(FM) Daphne AL
WZNJ(FM) Demopolis AL
WJEC(FM) Vernon AL
KBVA(FM) Bella Vista AR
KELD-FM Hampton AR
KKIK(FM) Horseshoe Bend AR
KKMR(FM) Arizona City AZ
KALT-FM Alturas CA
KIXA(FM) Lucerne Valley CA
KMMT(FM) Mammoth Lakes CA
KWOD(FM) Sacramento CA
KLNV(FM) San Diego CA
KEZR(FM) San Jose CA
KEAL(FM) Taft CA
KFVR-FM La Junta CO
WBMW(FM) Ledyard CT
WOCY(FM) Carrabelle FL
WCJX(FM) Five Points FL
WOKV-FM Ponte Vedra Beach FL
WCTQ(FM) Sarasota FL
WZIQ(FM) Smithville GA
WZBX(FM) Sylvania GA
KCQQ(FM) Davenport IA
WARH(FM) Granite City IL
WXNU(FM) Saint Anne IL
WWBL(FM) Washington IN
KYQQ(FM) Arkansas City KS
WKDZ-FM Cadiz KY
WRLV-FM Salyersville KY
WNKR(FM) Williamstown KY
KCIJ(FM) Atlanta LA
KQXL-FM New Roads LA
WWMX(FM) Baltimore MD
WKHW(FM) Pocomoke City MD
WQCB(FM) Brewer ME
*WMEF(FM) Fort Kent ME
WVFM(FM) Kalamazoo MI
KFMC(FM) Fairmont MN
*KDXL(FM) Saint Louis Park MN
*KUOM-FM Saint Louis Park MN
KLFN(FM) Sunburg MN
WDAF-FM Liberty MO
KTMO(FM) New Madrid MO
WAID(FM) Clarksdale MS
WSFL-FM New Bern NC
WEND(FM) Salisbury NC
KMCX(FM) Ogallala NE
WMEX(FM) Farmington NH
WBBO(FM) Bass River Township NJ
KEND(FM) Roswell NM
KSNE-FM Las Vegas NV
WPYX(FM) Albany NY
WYRK(FM) Buffalo NY
WKRH(FM) Minetto NY
WMVX(FM) Cleveland OH
WRWK(FM) Delta OH
WDSJ(FM) Greenville OH
KTLS-FM Holdenville OK
KYSJ(FM) Coos Bay OR
WFYY(FM) Bloomsburg PA
WDSN(FM) Reynoldsville PA
WNIK-FM Arecibo PR
WSYN(FM) Georgetown SC
WSKZ(FM) Chattanooga TN
WLVS-FM Clifton TN
WJDT(FM) Rogersville TN
KOVE-FM Galveston TX
KOOI-FM Jacksonville TX
KEJS(FM) Lubbock TX
KOSY-FM Spanish Fork UT
WBTJ(FM) Richmond VA
KSPO(FM) Dishman WA
KWPZ(FM) Lynden WA
KEGX(FM) Richland WA
WYTE(FM) Marshfield WI
WHBZ(FM) Sheboygan Falls WI
WKCH(FM) Whitewater WI
WWLW(FM) Clarksburg WV

106.7 mhz

KGTW(FM) Ketchikan AK
WKMX(FM) Enterprise AL
KHKN(FM) Benton AR
KJBX(FM) Trumann AR
KPPV(FM) Prescott Valley AZ
KSMY(FM) Lompoc CA
KRQR(FM) Orland CA
KROQ-FM Pasadena CA
KJUG-FM Tulare CA
KBPI(FM) Denver CO
WRMA(FM) Fort Lauderdale FL
WXXL(FM) Tavares FL
WOKA-FM Douglas GA
WYAY(FM) Gainesville GA
KNAN(FM) Nanakuli HI
KRTI(FM) Grinnell IA
KIKD(FM) Lake City IA
KYUN(FM) Hailey ID
KTPO(FM) Kootenai ID
WPPN(FM) Des Plaines IL
WPWQ(FM) Mount Sterling IL
WZNX(FM) Sullivan IL
WTLC-FM Greenwood IN
WYFX(FM) Mount Vernon IN
WGLM(FM) West Lafayette IN
KFXX-FM Hugoton KS
KQNK-FM Norton KS
WLFX(FM) Berea KY
WHHT(FM) Horse Cave KY
WZZL(FM) Reidland KY
KYLA(FM) Homer LA
KUMX(FM) North Fort Polk LA
KKND(FM) Port Sulphur LA
KBEB-FM Rayne LA
WMJX(FM) Boston MA
WHXR(FM) North Windham ME
WDTW-FM Detroit MI
WSRT(FM) Gaylord MI
WHTO(FM) Iron Mountain MI
KAOD(FM) Babbitt MN
WJJY-FM Brainerd MN
KAUL(FM) Ellington MO
KZRQ-FM Mount Vernon MO
WWZD-FM New Albany MS
WSTZ-FM Vicksburg MS
KPLN(FM) Lockwood MT
KBQQ(FM) Pinesdale MT
WUIN(FM) Carolina Beach NC
WKVE(FM) Semora NC
KYTZ(FM) Walhalla ND
KEXL(FM) Norfolk NE
WSJQ(FM) North Cape May NJ
KZNM(FM) Los Alamos NM
WBDI(FM) Copenhagen NY
WKGS(FM) Irondequoit NY
WLTW(FM) New York NY
WRRQ(FM) Windsor NY
WFGA(FM) Hicksville OH
WSRW-FM Hillsboro OH
KTUZ-FM Okarche OK
KLTH(FM) Lake Oswego OR
WAMO-FM Beaver Falls PA
WMHX(FM) Hershey PA
WTCB(FM) Orangeburg SC
KBFO(FM) Aberdeen SD
WNFN(FM) Belle Meade TN
WDXE-FM Lawrenceburg TN
WRJK(FM) Norris TN
KQTY-FM Borger TX
KDRW(FM) Hewitt TX
KCHX(FM) Midland TX
KZZA(FM) Muenster TX
KPWT(FM) Terrell Hills TX
WJFK-FM Manassas VA
WIZN(FM) Vergennes VT
KZPH(FM) Cashmere WA
WZNN(FM) Allouez WI
WATQ(FM) Chetek WI
WJQM(FM) Mount Horeb WI
*WHFI(FM) Lindside WV
WVKM(FM) Matewan WV

106.9 mhz

KFSE(FM) Kasilof AK
WBPT(FM) Homewood AL
KXIO(FM) Clarksville AR
KXFE(FM) Dumas AR
KYXK(FM) Gurdon AR
KDVA(FM) Buckeye AZ
KCEL(FM) California City CA
KQLB(FM) Los Banos CA
KFRC-FM San Francisco CA
KDGL(FM) Yucca Valley CA
KNKN(FM) Pueblo CO
WCCC-FM Hartford CT
WKZY(FM) Cross City FL
WZZS(FM) Zolfo Springs FL
KWYI(FM) Kawaihae HI
KDMU(FM) Bloomfield IA
KIHK(FM) Rock Valley IA
KMOK(FM) Lewiston ID
WSWT(FM) Peoria IL
WDML(FM) Woodlawn IL
WXXC(FM) Marion IN
KBGL(FM) Larned KS
KTPK(FM) Topeka KS
WVEZ(FM) Louisville KY
WYMV(FM) Madisonville KY
KEDG(FM) Alexandria LA
WWEG(FM) Hagerstown MD
WRXS(FM) Ocean City MD
WBQX(FM) Thomaston ME
WUPM(FM) Ironwood MI
WMUS(FM) Muskegon MI
*WSAE(FM) Spring Arbor MI
KARP-FM Dassel MN
WMOZ(FM) Moose Lake MN
KROC-FM Rochester MN
KTXY(FM) Jefferson City MO
WHKL(FM) Crenshaw MS
WRBE-FM Lucedale MS
WKZB(FM) Stonewall MS
KSCY(FM) Big Sky MT
*KMSM-FM Butte MT
WMIT(FM) Black Mountain NC
WFVL(FM) Southern Pines NC
KHRT-FM Minot ND
KEGK(FM) Wahpeton ND
KHYY(FM) Minatare NE
KOPW(FM) Plattsmouth NE
WSCY(FM) Moultonborough NH
*WKDN-FM Camden NJ
KRNO(FM) Incline Village NV
KONV(FM) Overton NV
WPHR-FM Auburn NY
WKZA(FM) Lakewood NY
WRQK(FM) Canton OH
*WWSU(FM) Dayton OH
WMRN-FM Marion OH
KTIJ(FM) Elk City OK
KHTT(FM) Muskogee OK
KCST-FM Florence OR
KKRB(FM) Klamath Falls OR
*WRIJ(FM) Masontown PA
WZYY(FM) Renovo PA
WEZX(FM) Scranton PA
WMEG(FM) Guayama PR
WGZR(FM) Bluffton SC
WWYN(FM) McKenzie TN
WKXD(FM) Monterey TN
KMZZ(FM) Bishop TX
KLUB(FM) Bloomington TX
KHLE(FM) Burnet TX
KHPT(FM) Conroe TX
KRVF(FM) Kerens TX
KAZE(FM) Ore City TX
KKYN-FM Plainview TX
KRIA(FM) Plainview TX
KDRX(FM) Rocksprings TX
KLGD(FM) Stamford TX
KEGH(FM) Brigham City UT
WLEQ(FM) Bedford VA
WAFX(FM) Suffolk VA
WVTI(FM) Brighton VT
KRWM(FM) Bremerton WA
WJZX(FM) Brookfield WI
WNNO-FM Wisconsin Dells WI
KASS(FM) Casper WY

107.1 mhz

KCNY(FM) Bald Knob AR
KTHS-FM Berryville AR
KXHT(FM) Marion AR
KDRS-FM Paragould AR
KFYX(FM) Texarkana AR
KVVA-FM Apache Junction AZ
KSSE(FM) Arcadia CA
KCWR(FM) Bakersfield CA
KSRT(FM) Cloverdale CA
KSSD(FM) Fallbrook CA
KMMM(FM) Madera CA
KNKK(FM) Needles CA
KSES-FM Seaside CA
KESR(FM) Shasta Lake City CA
KSSC(FM) Ventura CA
KPVW(FM) Aspen CO
KLJH(FM) Bayfield CO
KSYY-FM Bennett CO
WIIS(FM) Key West FL
WCKT(FM) Lehigh Acres FL
WAOA-FM Melbourne FL
WFXM(FM) Gordon GA
WTSH-FM Rockmart GA
WTLY(FM) Thomasville GA
WYFA(FM) Waynesboro GA
KDSN-FM Denison IA
*KNWI(FM) Osceola IA
KRQN(FM) Vinton IA
KTHI(FM) Caldwell ID
KQEO(FM) Idaho Falls ID
WEAI(FM) Lynnville IL
WSPY-FM Plano IL
WPGU(FM) Urbana IL
WKRV(FM) Vandalia IL
WEJK(FM) Boonville IN
WEDJ(FM) Danville IN
WZVN(FM) Lowell IN
KMOQ(FM) Baxter Springs KS
*WLAI(FM) Danville KY
WKCB-FM Hindman KY
WUHU(FM) Smiths Grove KY
KFNV-FM Ferriday LA
WHMD(FM) Hammond LA
KWLV(FM) Many LA
KOGM(FM) Opelousas LA
WFHN(FM) Fairhaven MA
WTDK(FM) Federalsburg MD
WQKL(FM) Ann Arbor MI
WCKC(FM) Cadillac MI
WSAQ(FM) Port Huron MI
WTLZ(FM) Saginaw MI
WIRX(FM) Saint Joseph MI
WFMP(FM) Coon Rapids MN
KKEQ(FM) Fosston MN
KMGK(FM) Glenwood MN
KBMV-FM Birch Tree MO
KBHI(FM) Miner MO
WBYP(FM) Belzoni MS
WKNZ(FM) Collins MS
WXYK(FM) Gulfport MS
WLSM-FM Louisville MS
WFXC(FM) Durham NC
KSFT-FM South Sioux City NE
WERZ(FM) Exeter NH
*WEVC(FM) Gorham NH
WWYY(FM) Belvidere NJ
WWZY(FM) Long Branch NJ
KNKT(FM) Armijo, Albuquerque NM
KTUM(FM) Tatum NM
WXPK(FM) Briarcliff Manor NY
WFFG-FM Corinth NY
WLIR-FM Hampton Bays NY
WJZI(FM) Livingston Manor NY
WNUS(FM) Belpre OH
WNKK(FM) Circleville OH
WDOH(FM) Delphos OH
WBKS(FM) Ironton OH
WKFS(FM) Milford OH
KLBC(FM) Durant OK
KYNZ(FM) Lone Grove OK
*KLVU(FM) Sweet Home OR
WGSM(FM) Greensburg PA
WEXC(FM) Greenville PA
*WQJU(FM) Mifflintown PA
WLIH(FM) Whitneyville PA
WQSD(FM) Briarcliff Acres SC
WRHM(FM) Lancaster SC
KDBX(FM) Clear Lake SD
KGSR(FM) Bastrop TX
KRXB(FM) Beeville TX
KDXX(FM) Benbrook TX
KRVA-FM Campbell TX
KPUR-FM Canyon TX
KAUM(FM) Colorado City TX
*KWBU-FM Waco TX
WTTX-FM Appomattox VA
*WCHG(FM) Hot Springs VA
WPSK-FM Pulaski VA
WORK(FM) Barre VT
WZLF(FM) Bellows Falls VT
KRQT(FM) Castle Rock WA
KAZZ(FM) Deer Park WA
WFON(FM) Fond du Lac WI
WOCO-FM Oconto WI
WPVL-FM Platteville WI
WCBC-FM Keyser WV
KROW(FM) Lovell WY

107.3 mhz

WQLT-FM Florence AL
KQDD(FM) Osceola AR
KFXR-FM Chinle AZ
KXFF(FM) Colorado City AZ

KURQ(FM) Grover Beach CA
KIXW-FM Lenwood CA
*KNHT(FM) Rio Dell CA
KSTN-FM Stockton CA
KTRJ(FM) Hayden CO
KRKV(FM) Las Animas CO
WRQX(FM) Washington DC
WPLA(FM) Jacksonville FL
WYCL(FM) Pensacola FL
WXGL(FM) Saint Petersburg FL
WMCD(FM) Claxton GA
WCGQ(FM) Columbus GA
KGRS(FM) Burlington IA
KIOW(FM) Forest City IA
WDDD-FM Marion IL
WDKR(FM) Maroa IL
WRZQ-FM Greensburg IN
WRSW-FM Warsaw IN
KTHR(FM) Wichita KS
WCTT-FM Corbin KY
WTHX(FM) Lebanon Junction KY
WTGE(FM) Baker LA
WAAF(FM) Westborough MA
WBZN(FM) Old Town ME
WKLQ(FM) Greenville MI
WXPT(FM) Powers MI
KNUJ-FM Sleepy Eye MN
KESY(FM) Cuba MO
KMJK(FM) Lexington MO
WFCG(FM) Tylertown MS
KINX(FM) Great Falls MT
WTKF(FM) Atlantic NC
WCLN-FM Clinton NC
WBOB-FM Enfield NC
KBBK(FM) Lincoln NE
WPUR(FM) Atlantic City NJ
WRWD-FM Highland NY
WCRR(FM) South Bristol Township NY
WRCK(FM) Utica NY
WYBZ(FM) Crooksville OH
WNWV(FM) Elyria OH
WJUC(FM) Swanton OH
KVRW(FM) Lawton OK
KOMS(FM) Poteau OK
KOOS(FM) North Bend OR
WDBA(FM) DuBois PA
WEGH(FM) Northumberland PA
WCMN-FM Arecibo PR
WJMZ-FM Anderson SC
WVSZ(FM) Chesterfield SC
KQRN(FM) Mitchell SD
KSLT(FM) Spearfish SD
WTRZ(FM) Spencer TN
KHTZ(FM) Caldwell TX
KJKE(FM) Ingleside TX
KJAS(FM) Jasper TX
KTSX(FM) Knox City TX
KLFX(FM) Nolanville TX
KPOS(FM) Post TX
KQDR(FM) Savoy TX
KISX(FM) Whitehouse TX
WXLZ-FM Lebanon VA
WBBT-FM Powhatan VA
WVGN(FM) Charlotte Amalie VI
KFFM(FM) Yakima WA
WSJY(FM) Fort Atkinson WI
WNXR(FM) Iron River WI
WKAZ-FM Miami WV
KKAW(FM) Albin WY
KAOX(FM) Kemmerer WY

107.5 mhz

KASH-FM Anchorage AK
KOMT(FM) Mountain Home AR
KSED(FM) Sedona AZ
KHYT(FM) Tucson AZ
KQPT(FM) Colusa CA
KXO-FM El Centro CA
KPIG-FM Freedom CA
KRDA(FM) Hanford CA
KLVE(FM) Los Angeles CA
KQKS(FM) Lakewood CO
WAMR-FM Miami FL
WWGF(FM) Donalsonville GA
WTIF-FM Omega GA
WJZZ-FM Roswell GA
WDBN(FM) Wrightsville GA
KHEI-FM Kihei HI
*KILV(FM) Castana IA
KKDM(FM) Des Moines IA
KYZK(FM) Sun Valley ID
WGCI-FM Chicago IL
WDBQ-FM Galena IL
WABX(FM) Evansville IN
KSCB-FM Liberal KS
WIOK(FM) Falmouth KY
WZLK(FM) Virgie KY
KCIL(FM) Houma LA
KJMH(FM) Lake Arthur LA
KXKZ(FM) Ruston LA
WFCC-FM Chatham MA
WKHI(FM) Fruitland MD
WFNK(FM) Lewiston ME
WGPR(FM) Detroit MI
WCCW-FM Traverse City MI
KLIZ-FM Brainerd MN
KBGY(FM) Faribault MN
KARZ(FM) Marshall MN
KFEB(FM) Campbell MO
KWBZ(FM) Monroe City MO
WMJW(FM) Cleveland MS
WKXI-FM Magee MS
KRPM(FM) Billings MT
KENR(FM) Superior MT
WAZO(FM) Southport NC
WKZL(FM) Winston-Salem NC
KJKJ(FM) Grand Forks ND
KSMX(FM) Clovis NM
KQBA(FM) Los Alamos NM
KXTE(FM) Pahrump NV
WBBI(FM) Endwell NY
WBLS(FM) New York NY
WCKX(FM) Columbus OH
WZRX-FM Fort Shawnee OH
WFXJ-FM North Kingsville OH
WZZZ(FM) Portsmouth OH
*KOSN(FM) Ketchum OK
KIFS(FM) Ashland OR
KVMX(FM) Banks OR
WBVE(FM) Bedford PA
WFKB(FM) Boyertown PA
WNKT(FM) Saint George SC
WHBQ-FM Germantown TN
WRVW(FM) Lebanon TN
KMVK(FM) Fort Worth TX
KHTC(FM) Lake Jackson TX
KQBO(FM) Rio Grande City TX
KSJT-FM San Angelo TX
KXTN-FM San Antonio TX
KKAT-FM Orem UT
WCJZ(FM) Charlottesville VA
WDUZ-FM Brillion WI
WCCN-FM Neillsville WI
WEGW(FM) Wheeling WV
KPAD(FM) Wheatland WY

107.7 mhz

WUHT(FM) Birmingham AL
WFXX(FM) Georgiana AL
WPRN-FM Lisman AL
KLAL(FM) Wrightsville AR
KSRN(FM) Kings Beach CA
KSAN(FM) San Mateo CA
KIST-FM Santa Barbara CA
KCDZ(FM) Twentynine Palms CA
*WFCS(FM) New Britain CT
WWRX(FM) Pawcatuck CT
WMGF(FM) Mount Dora FL
WHFX(FM) Darien GA
WPRW-FM Martinez GA
WEGC(FM) Sasser GA
KKOA(FM) Volcano HI
KICD-FM Spencer IA
WYST(FM) Fairbury IL
WLLT(FM) Polo IL
WSFR(FM) Corydon IN
WMRS(FM) Monticello IN
*KGCR(FM) Goodland KS
KMAJ-FM Topeka KS
WKCA(FM) Owingsville KY
WBQI(FM) Bar Harbor ME
WHSB(FM) Alpena MI
WMQT(FM) Ishpeming MI
WRKR(FM) Portage MI
KBMX(FM) Proctor MN
KLCX(FM) Saint Charles MN
*KCVK(FM) Otterville MO
KSLZ(FM) Saint Louis MO
KRWP(FM) Stockton MO
WAZA(FM) Liberty MS
KMTZ(FM) Boulder MT
WUKS(FM) Saint Pauls NC
KSYZ-FM Grand Island NE
WTPL(FM) Hillsboro NH
*WRRC(FM) Lawrenceville NJ
WGNA-FM Albany NY
*WECW(FM) Elmira NY
WLKK(FM) Wethersfield Township NY
WMMX(FM) Dayton OH
WXXF(FM) Loudonville OH
WSEO(FM) Nelsonville OH
WPFX-FM North Baltimore OH
KRXO(FM) Oklahoma City OK
KUMA-FM Pendleton OR
WUUZ(FM) Cooperstown PA
WGTY(FM) Gettysburg PA
WGMF(FM) Tunkhannock PA
WVOZ-FM Carolina PR
KABD(FM) Ipswich SD
WHHM-FM Henderson TN
WIVK-FM Knoxville TN
KINV(FM) Georgetown TX
KTBQ(FM) Nacogdoches TX
KPLT-FM Paris TX
WHQX(FM) Cedar Bluff VA
WTWP-FM Warrenton VA
WJCD(FM) Windsor VA
KNDD(FM) Seattle WA
*WVCY-FM Milwaukee WI
WVRW(FM) Glenville WV
WFSP-FM Kingwood WV

107.9 mhz

WJAM-FM Orrville AL
KEZA(FM) Fayetteville AR
KFIN(FM) Jonesboro AR
KMLE(FM) Chandler AZ
KUZZ-FM Bakersfield CA
KSEA(FM) Greenfield CA
*KKLC(FM) Mount Shasta CA
KLLE(FM) North Fork CA
KDND(FM) Sacramento CA
KWVE(FM) San Clemente CA
KPAW(FM) Fort Collins CO
KBKL(FM) Grand Junction CO
KDZA-FM Pueblo CO
WEBE(FM) Westport CT
WNDN(FM) Chiefland FL
WSRZ-FM Coral Cove FL
WMFM(FM) Key West FL
WPFM-FM Panama City FL
WIRK-FM West Palm Beach FL
WHTA(FM) Hampton GA
WWRQ-FM Valdosta GA
KGMZ-FM Aiea HI
KKRF(FM) Stuart IA
KFMW(FM) Waterloo IA
KXLT-FM Eagle ID
WXET(FM) Arcola IL
WLEY-FM Aurora IL
WCDD(FM) Canton IL
WNTR(FM) Indianapolis IN
WJFX(FM) New Haven IN
WAMW-FM Washington IN
KZLS(FM) Great Bend KS
KSJM(FM) Winfield KS
WKYR-FM Burkesville KY
WCVQ(FM) Fort Campbell KY
WWAG(FM) McKee KY
WBTF(FM) Midway KY
KRKA(FM) Erath LA
WXKS-FM Medford MA
*WFSI(FM) Annapolis MD
WFMX(FM) Skowhegan ME
*WVAC-FM Adrian MI
WCZW(FM) Charlevoix MI
WCRZ(FM) Flint MI
WSHZ(FM) Muskegon MI
KQQL(FM) Anoka MN
KLTE(FM) Kirksville MO
KCLQ(FM) Lebanon MO
WFCA(FM) Ackerman MS
WZKX(FM) Bay St. Louis MS
KHDV(FM) Darby MT
WLNK(FM) Charlotte NC
WNCT-FM Greenville NC
KPFX(FM) Fargo ND
KTIC-FM West Point NE
WRNB(FM) Pennsauken NJ
*WWPH(FM) Princeton Junction NJ
KQEL(FM) Alamogordo NM
KBQI(FM) Albuquerque NM
KVGS(FM) Laughlin NV
WWHT(FM) Syracuse NY
WENZ(FM) Cleveland OH
WODB(FM) Delaware OH
KEYB(FM) Altus OK
KHPE(FM) Albany OR
*WHHS(FM) Havertown PA
WDSY-FM Pittsburgh PA
WKVB(FM) Port Matilda PA
WKRF(FM) Tobyhanna PA
WRVH(FM) Williamsport PA
WGTR(FM) Bucksport SC
WLOW(FM) Port Royal SC
KXQL(FM) Flandreau SD
WOGT(FM) East Ridge TN
KEYJ-FM Abilene TX
KQQK(FM) Beaumont TX
KZRK-FM Canyon TX
KVLY(FM) Edinburg TX
KFAN-FM Johnson City TX
KESS-FM Lewisville TX
KQLM(FM) Odessa TX
KHCK-FM Robinson TX
KIXS(FM) Victoria TX
KUDD(FM) Roy UT
WYYD(FM) Amherst VA
WBQK(FM) West Point VA
*WVPS(FM) Burlington VT
KLSY(FM) South Bend WA
*KMBI-FM Spokane WA
WBCV(FM) Wausau WI
WEMM-FM Huntington WV
KRVK(FM) Midwest WY

Canadian AM Stations by Frequency

530 khz
CIAO(AM) Brampton ON

540 khz
CBT(AM) Grand Falls-Windsor NF
*CBEF(AM) Windsor ON
*CBKF-1(AM) Gravelbourg SK
*CBK(AM) Regina SK

550 khz
CHLN(AM) Trois Rivieres PQ

560 khz
CHTK(AM) Prince Rupert BC
CHVO(AM) Carbonear NF
CFOS(AM) Owen Sound ON

570 khz
CKWL(AM) Williams Lake BC
CFCB(AM) Corner Brook NF
CKGL(AM) Kitchener ON
CKSW(AM) Swift Current SK
*CFWH(AM) Whitehorse YT

580 khz
CKUA(AM) Edmonton AB
CFRA(AM) Ottawa ON
CKWW(AM) Windsor ON

590 khz
CFTK(AM) Terrace BC
CFAR(AM) Flin Flon MB
CJCW(AM) Sussex NB
VOCM(AM) Saint John's NF
CJCL(AM) Toronto ON
CKRS(AM) Saguenay PQ

600 khz
CKBD(AM) Vancouver BC
CFCT(AM) Tuktoyaktuk NT
CKAT(AM) North Bay ON
CJWW(AM) Saskatoon SK

610 khz
CKYL(AM) Peace River AB
CHNL(AM) Kamloops BC
CHTM(AM) Thompson MB
CKTB(AM) Saint Catharines ON
CHNC(AM) New Carlisle PQ
CKRW(AM) Whitehorse YT

620 khz
CKCM(AM) Grand Falls NF
CFRP(AM) Forestville PQ
CKRM(AM) Regina SK

630 khz
CHED(AM) Edmonton AB
CKOV(AM) Kelowna BC
CFCO(AM) Chatham ON
CHLT(AM) Sherbrooke PQ

640 khz
*CBN(AM) Saint John's NF
CFMJ(AM) Toronto ON

650 khz
CISL(AM) Richmond BC
CKGA(AM) Gander NF
CKOM(AM) Saskatoon SK

660 khz
CFFR(AM) Calgary AB

680 khz
*CHFA(AM) Edmonton AB
CJOB(AM) Winnipeg MB
CFTR(AM) Toronto ON

690 khz
*CBU(AM) Vancouver BC
CINF(AM) Verdun PQ

710 khz
CKVO(AM) Clarenville NF

730 khz
CHMJ(AM) Vancouver BC
CKDM(AM) Dauphin MB
CKAC(AM) Montreal PQ

740 khz
CBX(AM) Edmonton AB
CHCM(AM) Marystown NF
CHWO(AM) Toronto ON

750 khz
CBGY(AM) Bonavista Bay NF
CKJH(AM) Melfort SK

760 khz
CFLD(AM) Burns Lake BC

770 khz
CHQR(AM) Calgary AB

780 khz
CFDR(AM) Dartmouth NS

790 khz
CFCW(AM) Camrose AB
CFNW(AM) Port au Choix NF
CIGM(AM) Sudbury ON

800 khz
CKOR(AM) Penticton BC
*VOWR(AM) Saint John's NF
CJBQ(AM) Belleville ON
CKLW(AM) Windsor ON
CJAD(AM) Montreal PQ
CHRC(AM) Quebec PQ
CHAB(AM) Moose Jaw SK

810 khz
CKJS(AM) Winnipeg MB
CJVA(AM) Caraquet NB

820 khz
CHAM(AM) Hamilton ON

830 khz
CKKY(AM) Wainwright AB

840 khz
CKBX(AM) 100 Mile House BC

850 khz
CKBA(AM) Athabasca AB

860 khz
*CFPR(AM) Prince Rupert BC
*CHAK(AM) Inuvik NT
CJBC(AM) Toronto ON
CBKF-2(AM) Saskatoon SK

870 khz
CKIR(AM) Invermere BC
CFBV(AM) Smithers BC
CFSX(AM) Stephenville NF

880 khz
CHQT(AM) Edmonton AB
CKLQ(AM) Brandon MB

890 khz
CJDC(AM) Dawson Creek BC

900 khz
*CKMO(AM) Victoria BC
CKDH(AM) Amherst NS
CHML(AM) Hamilton ON
CKBI(AM) Prince Albert SK

910 khz
CKDQ(AM) Drumheller AB

920 khz
CFRY(AM) Portage la Prairie MB
CJCH(AM) Halifax NS
CKNX(AM) Wingham ON

930 khz
CJCA(AM) Edmonton AB
CFBC(AM) Saint John NB
CJYQ(AM) Saint John's NF

940 khz
CINW(AM) Montreal PQ
CJGX(AM) Yorkton SK

950 khz
CFAM(AM) Altona MB
CKNB(AM) Campbellton NB

960 khz
CFAC(AM) Calgary AB
CFFX(AM) Kingston ON

980 khz
CKNW(AM) New Westminster BC
CFPL(AM) London ON
CKRU(AM) Peterborough ON
CJME(AM) Regina SK

990 khz
CBW(AM) Winnipeg MB
*CBY(AM) Corner Brook NF
CBOF-1(AM) Maniwaki PQ
CKGM(AM) Montreal PQ

1010 khz
*CBR(AM) Calgary AB
CFRB(AM) Toronto ON

1020 khz
CKVH(AM) High Prairie AB

1040 khz
CKST(AM) Vancouver BC
CJMS(AM) Saint Constant PQ

1050 khz
*CKSB(AM) Saint Boniface MB
CHUM(AM) Toronto ON
CJNB(AM) North Battleford SK

1060 khz
CKMX(AM) Calgary AB

1070 khz
CFAX(AM) Victoria BC
CBA(AM) Moncton NB
CHOK(AM) Sarnia ON

1090 khz
CKKW(AM) Kitchener ON

1130 khz
CKWX(AM) Vancouver BC

1140 khz
CHRB(AM) High River AB
CBI(AM) Sydney NS

1150 khz
CKFR(AM) Kelowna BC
CKOC(AM) Hamilton ON

1190 khz
CFSL(AM) Weyburn SK

1200 khz
CJRJ(AM) Vancouver BC
CFGO(AM) Ottawa ON

1210 khz
*VOAR(AM) Mount Pearl NF
CFYM(AM) Kindersley SK

1220 khz
CJRB(AM) Boissevain MB
CJUL(AM) Cornwall ON
CHSC(AM) Saint Catharines ON
CKSM(AM) Shawinigan PQ

1230 khz
CJNL(AM) Merritt BC
CHFC(AM) Churchill MB
CFLN(AM) Goose Bay NF
CFFB(AM) Iqaluit NU

1240 khz
CJOR(AM) Osoyoos BC
CFNI(AM) Port Hardy BC
CJAR(AM) The Pas MB
CKIM(AM) Baie Verte NF
CJCS(AM) Stratford ON
CJMD(AM) Chibougamau PQ
CFLM(AM) La Tuque PQ
CJNS(AM) Meadow Lake SK

1250 khz
CHSM(AM) Steinbach MB
CJYE(AM) Oakville ON

1260 khz
CFRN(AM) Edmonton AB
CKHJ(AM) Fredericton NB

1270 khz
CJCB(AM) Sydney NS
CFGT(AM) Alma PQ

1280 khz
CHQB(AM) Powell River BC
CFMB(AM) Montreal PQ
CJSL(AM) Estevan SK

1290 khz
CFRW(AM) Winnipeg MB
CJBK(AM) London ON

1310 khz
CHLW(AM) Saint Paul AB
CIWW(AM) Ottawa ON

1320 khz
CHMB(AM) Vancouver BC
CKEC(AM) New Glasgow NS
CJMR(AM) Mississauga ON

1330 khz
CJYM(AM) Rosetown SK

1340 khz
CIBQ(AM) Brooks AB
CFKC(AM) Creston BC
CKCR(AM) Revelstoke BC
CIVH(AM) Vanderhoof BC
CFLW(AM) Wabush NF
*CFYK(AM) Yellowknife NT
CKDR-5(AM) Red Lake ON
CFED(AM) Chapais PQ
CBGN(AM) Sainte Anne des Monts PQ

1350 khz
CKAD(AM) Middleton NS

1370 khz
CFOK(AM) Westlock AB

1380 khz
CKPC(AM) Brantford ON
CKLC(AM) Kingston ON

1400 khz
CKSQ(AM) Stettler AB
CKGR(AM) Golden BC
CIOR(AM) Princeton BC
*CBG(AM) Gander NF
CKDR-2(AM) Sioux Lookout ON

1410 khz
CFUN(AM) Vancouver BC
CKSL(AM) London ON

1420 khz
CKDY(AM) Digby NS
CKPT(AM) Peterborough ON

1430 khz
CHKT(AM) Toronto ON

1440 khz
CKJR(AM) Wetaskiwin AB

1450 khz
CHOR(AM) Summerland BC
CFAB(AM) Windsor NS
*CHMO(AM) Moosonee ON
CHOU(AM) Montreal PQ

1460 khz
CJOY(AM) Guelph ON

1470 khz
CJVB(AM) Richmond BC

1490 khz
CFWB(AM) Campbell River BC
CFNC(AM) Cross Lake MB
CJSN(AM) Shaunavon SK

1510 khz
CKOT(AM) Tillsonburg ON

1540 khz
CHIN(AM) Toronto ON

1550 khz
*CBE(AM) Windsor ON

1570 khz
CKMW(AM) Winkler-Morden MB
CFAV(AM) Laval PQ

1580 khz
CKDO(AM) Oshawa ON

1610 khz
CHHA(AM) Toronto ON
CJWI(AM) Montreal PQ

1650 khz
CJRS(AM) Montreal PQ

1690 khz
CHTO(AM) Toronto ON
CJLO(AM) Montreal PQ

Canadian FM Stations by Frequency

88.1 mhz
CKAV-FM-3 Calgary AB
*CFRH-FM Penetanguishene ON
CHLK-FM Perth ON
*CKLN-FM Toronto ON

88.3 mhz
CKXU-FM Lethbridge AB
CJIQ-FM Kitchener/Paris ON
*CBQT-FM Thunder Bay ON
CKIA-FM Quebec PQ
CFAK-FM Sherbrooke PQ

88.5 mhz
CJSR-FM Edmonton AB
CIVL-FM Abbotsford BC
CIBH-FM Parksville BC
*CBAF-FM Moncton NB
CKDX-FM Newmarket ON
CILV-FM Ottawa ON
CBME-FM Montreal PQ

88.7 mhz
*CFUR-FM Prince George BC
CKYM-FM Napanee ON
CIMX-FM Windsor ON
CHIC-FM Rouyn-Noranda PQ

88.9 mhz
CJSI-FM Calgary AB
*CBTK-FM Kelowna BC
CHNI-FM Saint John NB
CIRV-FM Toronto ON
CJMQ-FM Sherbrooke PQ

89.1 mhz
CHSB-FM Bedford NS
*CHUO-FM Ottawa ON
*CKRL-FM Quebec PQ
CJBR-FM Rimouski PQ
CFOU-FM Trois Rivieres PQ

89.3 mhz
CKGW-FM Chatham ON
*CISM-FM Montreal PQ

89.5 mhz
CJSE-FM Shediac NB
*CFGB-FM Happy Valley NF
CJRL-FM Kenora ON
*CIUT-FM Toronto ON

89.7 mhz
CBCX-FM Calgary AB
CJSU-FM Duncan BC
CKOA-FM Glace Bay NS
CKGN-FM Kapuskasing ON

89.9 mhz
CFWE-FM Lac La Biche AB
CBTE-FM Crawford Bay BC
CHNS-FM Halifax NS
CIHT-FM Ottawa ON
*CBE-FM Windsor ON
CFBS-FM Lourdes-de-Blanc-Sablon PQ
CJCF-FM Cumberland House SK
CJLR-FM La Ronge SK
CFNK-FM Pinehouse Lake SK

90.1 mhz
*CJSF-FM Burnaby BC
CJMP-FM Powell River BC
CHMZ-FM Tofino BC
*CFJU-FM Kedgwick NB
CBBS-FM Sudbury ON

90.3 mhz
CFUL-FM Calgary AB
CBEG-FM Sarnia ON
CJBC-FM Toronto ON
*CKUT-FM Montreal PQ

90.5 mhz
*CBCV-FM Victoria BC
*CJPN-FM Fredericton NB
CBHA-FM Halifax NS
CKKK-FM Peterborough ON
CJBE-FM Port-Menier PQ
CFCR-FM Saskatoon SK

90.7 mhz
*CBOF-FM Ottawa ON

90.9 mhz
*CJSW-FM Calgary AB
CBX-FM Edmonton AB
CBUX-FM Vancouver BC
CBBX-FM Sudbury ON
*CION-FM Quebec PQ

91.1 mhz
CKXL-FM Saint Boniface MB
*CINN-FM Hearst ON
*CJRT-FM Toronto ON
CFUT-FM Shawinigan PQ

91.3 mhz
CJZN-FM Victoria BC
CBD-FM Saint John NB
CJLX-FM Belleville ON
CFBW-FM Hanover ON
CIRA-FM Montreal PQ
*CJTR-FM Regina SK

91.5 mhz
CBYG-FM Prince George BC
CKXR-FM Salmon Arm BC
CBAX-FM Halifax NS
CKBT-FM Kitchener-Waterloo ON
*CBO-FM Ottawa ON
CKPR-FM Thunder Bay ON
*CJAM-FM Windsor ON

91.7 mhz
CHBN-FM Edmonton AB
CKAY-FM Sechelt BC
CIXL-FM Welland ON

91.9 mhz
CKNI-FM Moncton NB
*CKVI-FM Kingston ON
CKLY-FM Lindsay (city of Kawartha Lakes) ON
CKLX-FM Montreal PQ
CJEC-FM Quebec PQ

92.1 mhz
CJAY-FM Calgary AB
CFNR-FM Terrace BC
CITI-FM Winnipeg MB
CJOZ-FM Bonavista Bay NF
CKPC-FM Brantford ON
CHOD-FM Cornwall ON
CJQQ-FM Timmins ON
CJDM-FM Drummondville PQ
CFVD-FM-2 Pohenegamook PQ
CHMX-FM Regina SK

92.3 mhz
CFRK-FM Fredericton NB
CKOZ-FM Corner Brook NF
CJET-FM Smiths Falls ON

92.5 mhz
CKNG-FM Edmonton AB
CFBX-FM Kamloops BC
CFXU-FM Antigonish NS
CJAQ-FM Toronto ON
CKAJ-FM Jonquiere PQ
CFQR-FM Montreal PQ

92.7 mhz
CIAM-FM Fort Vermilion AB
CHSL-FM Slave Lake AB
CJEM-FM Edmundston NB
CKDR-FM Dryden ON
CJSP-FM Leamington ON
CJBX-FM London ON
*CFFF-FM Peterborough ON
CJRQ-FM Sudbury ON
CFIM-FM Iles-de-la-Madeleine PQ
CJSQ-FM Quebec PQ

92.9 mhz
CFEX-FM Calgary AB
CIBW-FM Drayton Valley AB
CKIC-FM Winnipeg MB
CKLE-FM Bathurst NB
CKJN-FM Haldimand County ON
CHYK-FM-3 Hearst ON
CBBK-FM Kingston ON
CFQC-FM Saskatoon SK

93.1 mhz
CJXX-FM Grande Prairie AB
CKYE-FM Vancouver BC
CFRY-FM Portage la Prairie MB
CKBW-FM-2 Shelburne NS
CHAY-FM Barrie ON
CFOB-FM Fort Frances ON
CFNO-FM Marathon ON
CKCU-FM Ottawa ON
CHMT-FM Timmins ON
CHLQ-FM Charlottetown PE
CKVM-FM Ville-Marie PQ
CFRG-FM Gravelbourg SK

93.3 mhz
CJOK-FM Fort McMurray AB
CJBZ-FM Taber AB
CJAV-FM Port Alberni BC
CKSG-FM Cobourg ON
CFRU-FM Guelph ON
*CFMU-FM Hamilton ON
CJMF-FM Quebec PQ

93.5 mhz
CIHS-FM Wetaskiwin AB
*CJLY-FM Nelson BC
CKZX-FM New Denver BC
CJEL-FM Winkler MB
CIKX-FM Grand Falls NB
CKUM-FM Moncton NB
*CHMR-FM Saint John's NF
CJLS-FM-2 New Tusket NS
CFZN-FM Haliburton ON
CBCL-FM London ON
CFXJ-FM Toronto ON
CKXO-FM Chibougamau PQ
CBM-FM Montreal PQ
CJAS-FM Saint Augustin PQ

93.7 mhz
CJLT-FM Medicine Hat AB
CKWY-FM Wainwright AB
CJJR-FM Vancouver BC
CIFX-FM Lewisporte NF
CKOL-FM Campbellford ON
CKYC-FM Owen Sound ON
CJAI-FM Stella ON
CBMI-FM Baie Comeau PQ
CBJ-FM Chicoutimi PQ
CFGE-FM Sherbrooke PQ

93.9 mhz
CJLU-FM Halifax NS
CFWC-FM Brantford ON
CKKL-FM Ottawa ON
CIDR-FM Windsor ON

94.1 mhz
CJUV-FM Lacombe AB
CJOC-FM Lethbridge AB
CHSJ-FM Saint John NB
CICU-FM Eskasoni Indian Reserve NS
CKNR-FM Elliot Lake ON
CBL-FM Toronto ON
CKCN-FM Sept-Iles PQ
CIMG-FM Swift Current SK
CFGW-FM Yorkton SK

94.3 mhz
CFXE-FM Edson AB
CIRX-FM Prince George BC
CHIQ-FM Winnipeg MB
CKSY-FM Chatham ON
CJTW-FM Kitchener-Waterloo ON
CJSD-FM Thunder Bay ON
CJFH-FM Woodstock ON
CKMF-FM Montreal PQ
*CHYZ-FM Sainte Foy PQ

94.5 mhz
CHAT-FM Medicine Hat AB
CHBW-FM Rocky Mountain House AB
CHET-FM Chetwynd BC
CFBT-FM Vancouver BC
CKCW-FM Moncton NB
CKBW-FM-1 Liverpool NS
CIBU-FM Wingham ON
CJRG-FM Gaspe PQ
CJAB-FM Saguenay PQ
CKCK-FM Regina SK

94.7 mhz
CHKF-FM Calgary AB
CKLF-FM Brandon MB
CJRI-FM Fredericton NB
CHOZ-FM Saint John's NF
CFEP-FM Eastern Passage NS
CIWV-FM Hamilton ON
CJDS-FM Saint Pamphile PQ
CHEY-FM Trois Rivieres PQ
CJNE-FM Nipawin SK

94.9 mhz
CJPR-FM Blairmore AB
*CKUA-FM Edmonton AB
CKWM-FM Kentville NS
CKPE-FM Sydney NS
*CHRW-FM London ON
CKGE-FM Oshawa ON
CIEU-FM Carleton PQ
CIMF-FM Gatineau PQ
CHME-FM Les Escoumins PQ

95.1 mhz
CHVN-FM Winnipeg MB
CKMV-FM Grand Falls NB
CKUE-FM Chatham ON
CKCB-FM Collingwood ON
*CJOA-FM Thunder Bay ON
CFCY-FM Charlottetown PE
CILE-FM Havre-Saint-Pierre PQ
CBF-FM Montreal PQ
CFMC-FM Saskatoon SK

95.3 mhz
CJXK-FM Grand Centre (Cold Lake) AB
CKZZ-FM Vancouver BC
CING-FM Hamilton ON
CHUT-FM Lac-Simon (Louvicourt) PQ
CHOE-FM Matane PQ
*CBVX-FM Quebec PQ
CHXL-FM Okanese Indian Reserve SK

95.5 mhz
CHLB-FM Lethbridge AB
CKGY-FM Red Deer AB
CBA-FM Moncton NB
CJLS-FM Yarmouth NS
CJOJ-FM Belleville ON
CIYN-FM Kincardine ON
CJTK-FM Sudbury ON
CFVD-FM Degelis PQ
CFLX-FM Sherbrooke PQ

95.7 mhz
*CKRP-FM Falher AB
CJAT-FM Trail BC
CKTP-FM Fredericton Centre NB
CJNI-FM Halifax NS
CFJB-FM Barrie ON
CKAV-FM-9 Ottawa ON
CBQS-FM Sioux Narrows ON
CKYK-FM Alma PQ
CKYQ-FM Plessisville PQ

95.9 mhz
CHFM-FM Calgary AB
CKSA-FM Lloydminster AB
*CKUW-FM Winnipeg MB
CHOS-FM Rattling Brook NF
CJKX-FM Ajax ON
CFPL-FM London ON
CJFM-FM Montreal PQ

96.1 mhz
CFMY-FM Medicine Hat AB
CILZ-FM Greenville BC
CHKG-FM Richmond BC
CKX-FM Brandon MB
CINB-FM Saint John NB
CHMY-FM Renfrew ON
*CBCT-FM Charlottetown PE

96.3 mhz
CKRA-FM Edmonton AB
CINC-FM Thompson MB
CBDQ-FM Labrador City NF
CIOZ-FM Marystown NF
CJLS-FM-1 Barrington NS
CFMK-FM Kingston ON
CFMZ-FM-1 Toronto ON
CFMV-FM Chandler PQ
CIHO-FM Saint Hilarion PQ

96.5 mhz
CKLJ-FM Olds AB
CJPG-FM Portage la Prairie MB
CIXN-FM Fredericton NB
CKUL-FM Halifax NS
CFTX-FM Gatineau PQ
CKMN-FM Rimouski-Mont Joli PQ
CHOA-FM Rouyn-Noranda PQ

96.7 mhz
CFXW-FM Whitecourt AB
CKGF-FM-2 Greenwood BC
CILT-FM Steinbach MB
CHYM-FM Kitchener ON
CHYR-FM Leamington ON
CHVR-FM Pembroke ON
*CKLU-FM Sudbury ON

96.9 mhz
CKIS-FM Calgary AB
CKLG-FM Vancouver BC
CJXL-FM Moncton NB
CKSS-FM Red Rocks NF
CKHC-FM Toronto ON
CFIX-FM Chicoutimi PQ
CKOI-FM Verdun PQ
*CBK-FM Regina SK

97.1 mhz
CJMG-FM Penticton BC
CKRO-FM Pokemouche NB
CIGL-FM Belleville ON
CHLC-FM Baie Comeau PQ
CHLX-FM Gatineau PQ
CKFI-FM Swift Current SK

97.3 mhz
CIRK-FM Edmonton AB
CKLR-FM Courtenay BC
CJCI-FM Prince George BC
CHWV-FM Saint John NB
CJRM-FM Labrador City NF
CKON-FM Akwesasne ON
*CJIV-FM Dryden ON
CJEZ-FM Toronto ON
CKUJ-FM Kuujjuaq PQ
CHGA-FM Maniwaki PQ
CFJO-FM Thetford Mines PQ

97.5 mhz
CFXH-FM Hinton AB
CKRV-FM Kamloops BC
CFFM-FM Williams Lake BC
CJKR-FM Winnipeg MB
VOCM-FM Saint John's NF
*CKDU-FM Halifax NS
CIQM-FM London ON
CHOX-FM La Pocatiere PQ

97.7 mhz
CFGP-FM Grande Prairie AB
CHDH-FM Siksika AB
CIDO-FM Creston BC
CKTK-FM Kitimat BC
CBUF-FM Vancouver BC
CFCV-FM Saint Andrews NF
CKEN-FM Kentville NS
CHMS-FM Bancroft ON
CHTZ-FM Saint Catharines ON
CHGB-FM Wasaga Beach ON
*CFTH-FM-1 Harrington Harbour PQ
CHOM-FM Montreal PQ
CBKF-FM Regina SK
CIDD-FM White Bear Lake Resort SK

97.9 mhz
CKYX-FM Fort McMurray AB
CHSR-FM Fredericton NB
CFLC-FM Churchill Falls NF
CJLL-FM Ottawa ON
CFPS-FM Port Elgin ON
CJRE-FM Riviere au Renard PQ
CJCQ-FM North Battleford SK

98.1 mhz
CFCW-FM Camrose AB
CKVN-FM Lethbridge AB
CHTD-FM Saint Stephen NB
CKBW-FM Bridgewater NS
CHPB-FM Cochrane ON
*CBON-FM Sudbury ON
CHFI-FM Toronto ON
CHOI-FM Quebec PQ
*CBSI-FM Sept-Iles PQ
CFMQ-FM Hudson Bay SK
CHON-FM Whitehorse YT

98.3 mhz
CKSR-FM Chilliwack BC
CIFM-FM Kamloops BC
CFPX-FM Pukatawagan MB
CBW-FM Winnipeg MB
*CBAL-FM Moncton NB
CHER-FM Sydney NS
CFLY-FM Kingston ON
CFWP-FM Wahta Mohawk Territory near Bala ON
CIAX-FM Windsor PQ
CJMK-FM Saskatoon SK

98.5 mhz
CIBK-FM Calgary AB
CHRX-FM Fort St. John BC
CIOC-FM Victoria BC
CIOS-FM Stephenville NF
CKRH-FM Halifax NS
CINU-FM Truro NS
CKWR-FM Kitchener ON
CJWL-FM Ottawa ON
CFTH-FM-2 La Tabatiere PQ
CHMP-FM Longueuil PQ

98.7 mhz
CKXD-FM Gander NF
CBQX-FM Kenora ON
CJHR-FM Renfrew ON
CIKI-FM Rimouski PQ

98.9 mhz
CIZZ-FM Red Deer AB
CFCP-FM Courtenay BC
CFPV-FM Pemberton BC
CJYC-FM Saint John NB
CJFX-FM Antigonish NS
CHCD-FM Simcoe ON
CHYC-FM Sudbury ON
CHIK-FM Quebec PQ
CIZL-FM Regina SK

99.1 mhz
CJDR-FM Fernie BC
CJZZ-FM Winnipeg MB
CKIX-FM Saint John's NF
CHRI-FM Ottawa ON
CBLA-FM Toronto ON
CIDI-FM Lac-Brome PQ
CIPC-FM Port-Cartier PQ
CJMM-FM Rouyn-Noranda PQ
CFNJ-FM Saint Gabriel-de-Brandon PQ
CFMM-FM Prince Albert SK

99.3 mhz
CHMC-FM Edmonton AB
CKQR-FM Castlegar BC
CKDV-FM Prince George BC
CFOX-FM Vancouver BC
CFAN-FM Miramichi City NB
CKQN-FM Baker Lake NU
CJBC-FM-4 London ON
CFSF-FM Sturgeon Falls ON
CKGB-FM Timmins ON
CJAN-FM Asbestos PQ
*CBV-FM-6 La Malbaie PQ
CFOR-FM Maniwaki PQ

99.5 mhz
CIMM-FM Ucluelet BC
CBZF-FM Fredericton NB
CKTY-FM Truro NS
CFBG-FM Bracebridge ON
CJPX-FM Montreal PQ
CHRL-FM Roberval PQ

99.7 mhz
CKLM-FM-1 Bonnyville AB
CHJM-FM Saint Georges PQ

99.9 mhz
CHSU-FM Kelowna BC
CHPQ-FM Parksville BC
CFWM-FM Winnipeg MB
CHOY-FM Moncton NB
CJIJ-FM Sydney NS
CKIQ-FM Iqaluit NU
CFGX-FM Sarnia ON
*CBCS-FM Sudbury ON
CJUK-FM Thunder Bay ON
CFMX-FM Toronto ON
CFVM-FM Amqui PQ
CHEF-FM Matagami PQ

100.1 mhz
CKFU-FM Fort St. John BC
CKBZ-FM Kamloops BC
CIOO-FM Halifax NS
CJCD-FM-1 Hay River NT
CJCD-FM Yellowknife NT
CHCQ-FM Belleville ON
CHFN-FM Cape Croker (Neyaashiinigmiing) ON
CKAG-FM Pikogan PQ
CJEB-FM Trois Rivieres PQ

100.3 mhz
CFBR-FM Edmonton AB
CKCQ-FM Quesnel BC
CKKQ-FM Victoria BC
CFOZ-FM Argentia NF
CJLF-FM Barrie ON
*CKRZ-FM Ohsweken ON
CJMJ-FM Ottawa ON
*CKMS-FM Waterloo ON
CHTN-FM Charlottetown PE
CHVD-FM Dolbeau-Mistassini PQ
CJMC-FM Sainte Anne des Monts PQ
CFAQ-FM Blucher SK

100.5 mhz
CFSR-FM Hope BC
CHLS-FM Lillooet BC
CIOK-FM Saint John NB
CBBL-FM London ON
CHUR-FM North Bay ON
CHAS-FM Sault Ste. Marie ON
*CFIN-FM Lac-Etchemin PQ
CIBO-FM Senneterre PQ
CJJC-FM Yorkton SK

100.7 mhz
CIGV-FM Penticton BC
CIAJ-FM Prince Rupert BC
CHNK-FM Winnipeg MB
CFRM-FM Little Current ON
CHIN-FM Toronto ON
CBFX-FM Montreal PQ
CILG-FM Moose Jaw SK
CIAY-FM Whitehorse YT

100.9 mhz
CFXL-FM High River-Okotoks AB
*CBWK-FM Thompson MB
CKTO-FM Truro NS
CKHA-FM Haliburton ON
CKAP-FM Kapuskasing ON
*CBJX-FM Chicoutimi PQ
CKNU-FM Donnacona PQ

101.1 mhz
CIXF-FM Brooks AB
CFMI-FM New Westminster BC
CKXA-FM Brandon MB
CFAI-FM Edmundston NB
CKSJ-FM Saint John's NF
CIQB-FM Barrie ON
CKEY-FM Fort Erie ON
CFIF-FM Iroquois Falls ON
CKBY-FM Smiths Falls ON
CKSO-FM Sudbury ON

101.3 mhz
CJEG-FM Bonnyville AB
CKKN-FM Prince George BC
CKUN-FM Christian Island ON
CKOT-FM Tillsonburg ON
CJSA-FM Toronto ON
CKII-FM Dolbeau-Mistassini PQ
CHEQ-FM Sainte-Marie-de-Beauce PQ

101.5 mhz
CKCE-FM Calgary AB
CKNL-FM Fort St. John BC
CILK-FM Kelowna BC
CJUM-FM Winnipeg MB
*CBZ-FM Fredericton NB
CIGO-FM Port Hawkesbury NS
CHES-FM Erin ON
*CIOI-FM Hamilton ON
CJKL-FM Kirkland Lake ON
CKWF-FM Peterborough ON
*CIBL-FM Montreal PQ
*CBRX-FM Rimouski PQ
CHQX-FM Prince Albert SK

101.7 mhz
CKER-FM Edmonton AB
CHLY-FM Nanaimo BC
*CBQ-FM Thunder Bay ON
CKNX-FM Wingham ON
CHIP-FM Fort Coulonge PQ
CHRG-FM Maria (Reserve) PQ
CJSO-FM Sorel PQ

101.9 mhz
*CITR-FM Vancouver BC
*CFUV-FM Victoria BC
CHFX-FM Halifax NS
CKLB-FM Yellowknife NT
CJSS-FM Cornwall ON
*CFRC-FM Kingston ON
CKFX-FM North Bay ON
CHAI-FM Chateauguay PQ
CFDA-FM Victoriaville PQ

102.1 mhz
*CBR-FM Calgary AB
CKHL-FM High Level AB
CISW-FM Whistler BC
CHPR-FM Hawkesbury ON
CFNY-FM Toronto ON
CJRW-FM Summerside PE
CBGA-FM Matane PQ
CFEL-FM Montmagny PQ
CHLT-FM Sherbrooke PQ
CJDJ-FM Saskatoon SK

102.3 mhz
CKRX-FM Fort Nelson BC
CKWV-FM Nanaimo BC
CKY-FM Winnipeg MB
CKXG-FM Grand Falls-Windsor NF
CKJJ-FM Belleville ON
CHST-FM London ON
*CHIM-FM Timmins ON
CINQ-FM Montreal PQ
CIGB-FM Trois Rivieres PQ
CHSN-FM Estevan SK
CJNS-FM Meadow Lake SK

102.5 mhz
*CBOX-FM Ottawa ON

102.7 mhz
*CFRO-FM Vancouver BC
*CBH-FM Halifax NS
CFGI-FM Georgina Island ON
CILU-FM Thunder Bay ON
CIWS-FM Whitchurch-Stouffville ON
CBJE-FM Chicoutimi PQ
CITE-FM-1 Sherbrooke PQ
CJMV-FM Val d'Or PQ
CKZP-FM Zenon Park SK

102.9 mhz
CHDI-FM Edmonton AB
CHDR-FM Cranbrook BC
CKLH-FM Hamilton ON
CHCR-FM Killaloe ON
CFOM-FM Levis PQ
CJOI-FM Rimouski PQ

103.1 mhz
CIQX-FM Calgary AB
CJKC-FM Kamloops BC
CJFW-FM Terrace BC
CHTT-FM Victoria BC
CKMM-FM Winnipeg MB
CJMO-FM Moncton NB
CFMZ-FM Cobourg ON
CJBB-FM Englehart ON
CFHK-FM St. Thomas ON
CFMF-FM Fermont PQ
CHHO-FM Louiseville PQ
CIAU-FM Radisson PQ
CKOD-FM Valleyfield PQ

103.3 mhz
*CBOQ-FM Ottawa ON
CKLP-FM Parry Sound ON
CKQV-FM Vermillion Bay ON
CHAA-FM Longueuil PQ

103.5 mhz
CFCH-FM Chase BC
CHMM-FM MacKenzie BC
CHNV-FM Nelson BC
CHQM-FM Vancouver BC
CJEF-FM Saint John NB
CKHZ-FM Halifax NS
CIVR-FM Yellowknife NT
CIDC-FM Orangeville ON
CJLM-FM Joliette PQ
CKRB-FM Saint Georges-de-Beauce PQ
CKRC-FM Weyburn SK

103.7 mhz
CBRF-FM Calgary AB
CJPT-FM Brockville ON
CFBU-FM Saint Catharines ON
CFID-FM Acton Vale PQ
CIMI-FM Charlesbourg PQ
*CKRK-FM Kahnawake PQ
CIEL-FM Riviere du Loup PQ

103.9 mhz
CISN-FM Edmonton AB
CIMS-FM Balmoral NB
CFQM-FM Moncton NB
CKXX-FM Corner Brook NF
CHNO-FM Sudbury ON
CIRR-FM Toronto ON
CKDK-FM Woodstock ON
CKWE-FM Maniwaki (Kitigan Zibi Anishinabeg Reserve) PQ

CIME-FM Saint Jerome PQ

104.1 mhz
CHAD-FM Dawson Creek BC
CFQX-FM Selkirk MB
CJCJ-FM Woodstock NB
*CIFA-FM Comeauville NS
CICZ-FM Midland ON
CHYK-FM Timmins ON
CKTF-FM Gatineau PQ
CKNA-FM Natashquan PQ
CFZZ-FM Saint Jean-Iberville PQ
CIOT-FM Nipawin SK

104.3 mhz
CFRQ-FM Dartmouth NS
CJQM-FM Sault Ste. Marie ON
CBF-FM-1 Trois Rivieres PQ
CHGO-FM Val d'Or PQ

104.5 mhz
CISP-FM Pemberton BC
CJSB-FM Swan River MB
CFLG-FM Cornwall ON
CFQK-FM Kaministiquia ON
CJTT-FM New Liskeard ON
CHUM-FM Toronto ON
CKAU-FM Maliotenam PQ
CIGR-FM Sherbrooke PQ

104.7 mhz
CFRI-FM Grande Prairie AB
CHBZ-FM Cranbrook BC
CKLZ-FM Kelowna BC
CIPN-FM Pender Harbour BC
CKKS-FM Sechelt BC
CJNU-FM Winnipeg MB
CIHR-FM Woodstock ON
CJRC-FM Gatineau PQ
CFLO-FM Mont-Laurier PQ
CBVE-FM Quebec PQ

104.9 mhz
CFMG-FM Saint Albert AB
CKBC-FM Bathurst NB
CFJR-FM Brockville ON
CHWC-FM Goderich ON
CIMY-FM Pembroke ON
CBQL-FM Savant Lake ON
CJLA-FM Lachute PQ
*CHOC-FM Saint Remi PQ
CKVX-FM Kindersley SK
CFWF-FM Regina SK

105.1 mhz
CKRY-FM Calgary AB
CITA-FM Moncton NB
CHAL-FM Halifax NS
CBI-FM Sydney NS
CBQR-FM Rankin Inlet NU
CFLZ-FM Niagara Falls ON
CKQM-FM Peterborough ON
CHOQ-FM Toronto ON
CBMR-FM Fermont PQ
CFIC-FM Listuguj PQ
CKDG-FM Montreal PQ
CJVR-FM Melfort SK

105.3 mhz
CIXM-FM Whitecourt AB
CFXY-FM Fredericton NB
CJKK-FM Clarenville NF
CFCA-FM Kitchener ON
CISS-FM Ottawa ON
CJMX-FM Sudbury ON
CKTG-FM Thunder Bay ON
CHRD-FM Drummondville PQ
CHRM-FM Matane PQ

105.5 mhz
CHUB-FM Red Deer AB
CICY-FM Selkirk MB
CFBK-FM Huntsville ON
CHRY-FM Toronto ON
CKQK-FM Charlottetown PE
CKLD-FM Thetford Mines PQ
*CBKS-FM Saskatoon SK

105.7 mhz
CBU-FM Vancouver BC
CICF-FM Vernon BC
CHQC-FM Saint John NB
CIKR-FM Kingston ON
CHRE-FM Saint Catharines ON
CJMI-FM Strathroy ON
CFGL-FM Laval PQ
CFDM-FM Meadow Lake SK

105.9 mhz
CJRY-FM Edmonton AB
CHPD-FM Aylmer ON
CHJX-FM London ON
CICX-FM Orillia ON
*CBKA-FM La Ronge SK

106.1 mhz
CFIT-FM Airdrie AB
CKLM-FM Lloydminster AB
CKKX-FM Peace River AB
CKGO-FM-1 Boston Bar BC
*CKJM-FM Cheticamp NS
CIMJ-FM Guelph ON
CHEZ-FM Ottawa ON
CIMO-FM Magog PQ

106.3 mhz
CKAV-FM-2 Vancouver BC
CFXN-FM North Bay ON
CHKS-FM Sarnia ON
*CBV-FM Quebec PQ

106.5 mhz
CHMN-FM Canmore AB
CJJJ-FM Brandon MB
CIXK-FM Owen Sound ON
CKAV-FM Toronto ON
CFEI-FM Saint Hyacinthe PQ

106.7 mhz
CJRX-FM Lethbridge AB
CFDV-FM Red Deer AB
CIKZ-FM Kitchener-Waterloo ON
CJIT-FM Lac Megantic PQ
CFET-FM Tagish YT

106.9 mhz
CHWF-FM Nanaimo BC
CKKC-FM Nelson BC
CIBX-FM Fredericton NB
*CHMA-FM Sackville NB
*CBN-FM Saint John's NF
*CIXX-FM London ON
CKQB-FM Ottawa ON
CHRQ-FM Restigouche PQ

107.1 mhz
CKQC-FM Abbotsford BC
CISQ-FM Squamish BC
CFEQ-FM Winnipeg MB
CILQ-FM North York ON
CJTN-FM Quinte West ON
CJWA-FM Wawa ON
CIBM-FM Riviere du Loup PQ

107.3 mhz
CFGQ-FM Calgary AB
CHBE-FM Victoria BC
CKOE-FM Moncton NB
CFMH-FM Saint John NB
CKHR-FM Hay River NT
CFRT-FM Iqaluit NU
CHUK-FM Mashteuiatsh (Pointe-Bleue) PQ
CITE-FM Montreal PQ

107.5 mhz
CKCL-FM Chilliwack BC
CIEG-FM Egmont BC
CISC-FM Gibsons BC
CKIZ-FM Vernon BC
CJFY-FM Blackville NB
CKMB-FM Barrie ON
CJDV-FM Cambridge ON
CITF-FM Quebec PQ

107.7 mhz
CFRV-FM Lethbridge AB
CKTI-FM Kettle Point ON
CHGK-FM Stratford ON

107.9 mhz
CFML-FM Burnaby BC
CILS-FM Victoria BC
CJWV-FM Winnipeg MB
CJXY-FM Burlington ON
CHUC-FM Cobourg ON
*CKDJ-FM Ottawa ON

Radio Formats Defined

AAA (or Triple A)—Adult Album Alternative. Eclectic choice of music ranging from hard rock to folk music.

Adult Contemporary—Recent popular songs, with a few oldies. The songs tend to be upbeat and soft. News and talk segments are prominent during rush hour "drive times." Also known as **Light Rock**.

Agriculture & Farm—News, weather and features of interest to farmers and others involved in agriculture.

Albanian.

Album-Oriented Rock—Popular rock music from past and present rock albums. Also see **Rock/AOR**.

Alternative—Rock music first popularized in the late 80s and early 90s. Also known as **Progressive**.

American Indian—Programming for North American Indians; includes native language (i.e. Navajo) broadcasts.

Arabic.

Armenian.

Beautiful Music—Uninterrupted, instrumental soft music. There is usually very little talk and few commercials. Also known as **Easy Listening**.

Big Band—Popular music from the 30s and 40s. Primarily instrumental works by bands such as Glen Miller's Orchestra and Tommy Dorsey. Also see **Nostalgia**.

Black—Music, talk and news targeted at Black listeners. Music at these stations is similar to **Urban Contemporary** stations, but this format caters more directly to the interests and tastes of Black audiences.

Bluegrass—Related formats are **Country** and **Folk**.

Blues—Some **Jazz** and **Progressive** stations also program blues music.

Children—Programming for children, usually for educational purposes. Includes music, informational programming, and news presented for young people.

Chinese.

Christian.

Classic Rock—Popular rock music of the 60s, 70s and 80s. Also see **Rock/AOR.**

Classical—Classical music, often long pieces played without interruption. Announcers provide extended commentary and criticism on the pieces. Special features, such as live concerts, are common. Primarily a noncommercial FM format.

Comedy—Recorded stand-up comics and/or old radio comedy series. A rare format.

Contemporary Hit/Top-40—Current hot selling records. Usually a playlist of 20 to 40 songs continuously played throughout the day. DJs are often upbeat "personalities." News and information are given light coverage.

Country—Country music, ranging from older traditional country and western to today's "Hit Country" sounds. The amount of news and talk on country stations varies widely from station to station.

Croatian.

Czech.

Disco—High-energy dance music first popular in the 70s. Also see **Black** and **Urban Contemporary**.

Discussion.

Diversified—See **Variety/Diverse**.

Drama/Literature—Dramatic readings, poetry, and broadcasts of live dramatic performances. A rare format in the U.S. and Canada.

Easy Listening—Similar to **Beautiful Music**, but may include some soft rock.

Educational—Informative and instructional programming, such as over-the-air college courses. Primarily a noncommercial format.

Eskimo.

Ethnic—Programming for ethnic minorities, mostly in foreign languages.

Farsi.

Filipino.

Finnish.

Folk—Played full-time on very few stations, American folk music is also heard on noncommercial **Variety** stations. Also see **Bluegrass**.

Foreign Language/Ethnic—In addition to the specific language categories (i.e. French, German), this format denotes multilingual stations and others catering to ethnic minorities.

French.

Full Service—Mixture of music, news and talk with a general target audience.

German.

Golden Oldies—Hit songs of the 50s. Also see **Oldies**.

Gospel—Especially popular in the South, evangelical music is programmed on many **Religious** format stations.

Greek.

Hardcore.

Hebrew.

Hindi.

Hungarian.

Inspirational.

Irish.

Italian.

Japanese.

Jazz—Primarily a noncommercial FM format. Some Classical stations program jazz music features.

Jewish.

Korean.

Light Rock—See **Adult Contemporary**.

Lithuanian.

MOR (Middle-of-the-Road)—Traditional AM format featuring a balanced mix of music, news and talk. Songs are usually popular standards. Announcers are often personalities who try to keep the listener interested and informed. News, both local and national, plays an important role at most MOR stations; coverage of sporting events and other features of interest to the community is common.

Native American.

New Age—Soft "fusion" (a form mixing elements of jazz and rock), often played as background entertainment. As the name implies, this format is a recent development.

New Wave—A type of rock music which gained popularity in the early 80s, often performed by United Kingdom musicians.

News—Continous coverage of local, national and international news, including sports, weather forecasts and features.

News/Talk—Combination of news and talk formats. One of these elements may receive more emphasis. Also see **News** and **Talk**.

Nostalgia—Popular tunes from the 30s, 40s and 50s. Nostalgia stations often feature on-air personalities, and usually have heavy news and information coverage.

Oldies—Hit songs from the 50s, 60s and 70s. Usually played by upbeat DJs, with news, talk and special features (chart countdowns, trivia contests, etc.) playing an important role.

Other—Programming which falls outside the categories listed here.

Polish.

Polka—Music for the traditional dance. Most polka format stations are located in Wisconsin.

Portuguese.

Progressive—Progressive stations play many types of music, often including avant-garde music not played on conventional stations. Primarily a noncommercial format, common among college radio stations. Also known as **Alternative**.

Public Affairs—Community interest programming (ie: broadcasts of city council meetings.) Many noncommercial, **News**, and **Talk** stations cover local issues on news features or talk shows.

Reggae—Jamaican music. Often played on **Progressive** stations.

Religious—Inspirational/spiritual talk and music. Most religious stations air Christian sermons or songs. Also see **Gospel**.

Rock/AOR—Rock music from the 60s to the present. Album-oriented rock features music "sweeps" or uninterrrupted sets. News plays a secondary role. Also see **Classic Rock**.

Russian.

Sacred.

Scottish.

Serbian.

Slovak.

Slovenian.

Spanish.

Sports— Play-by-play and taped coverage, sports news, interviews, discussion.

Talk—Topical programs on various subjects. Includes health, finance, and community issues. Listener call-in and interview shows are common, and the host's personality tends to be an important element. Many talk stations air national satellite-delivered talk programs. News, sports and weather are usually emphasized during "drive times." Also see **News** and **News/Talk.**

Tejano—Bicultural programming including Spanish programming, popular in Texas, particularly near the Mexican border. Interest surged in this type of Spanish music during the early 90s.

Ukranian.

Top-40—See **Contemporary Hit/Top-40**.

Underground—The opposite of mainstream, this music is produced and appreciated by those outside the establishment.

Urban Contemporary—Dance music, often from a variety of genres (i.e. rhythm & blues, rap). Most Urban Contemporary stations emphasize music by Black artists. Also see **Black** and **Disco**.

Variety/Diverse—A station listing four or more formats. Typical of noncommerical stations.

Vietnamese.

Women—Programming for women. Emphasis on news and information, pertaining to women's issues.

U.S. and Canada Radio Programming Formats

	United States					Canada				
Format	Total	AM	FM	Com	Non	Total	AM	FM	Com	Non
Adult Contemp	1763	257	1506	1649	114	237	27	210	229	8
Agriculture	56	45	11	56	0	3	3	0	3	0
Albanian	0	0	0	0	0	0	0	0	0	0
Album-Oriented Rock	87	0	87	75	12	6	1	5	6	0
Alternative	260	3	257	110	150	13	0	13	6	7
American Indian	5	2	3	2	3	3	0	3	3	0
Arabic	1	1	0	1	0	1	1	0	1	0
Armenian	0	0	0	0	0	0	0	0	0	0
Beautiful Music	58	25	33	42	16	2	0	2	1	1
Big Band	59	41	18	48	11	1	0	1	1	0
Black	74	48	26	61	13	1	0	1	1	0
Bluegrass	33	22	11	24	9	0	0	0	0	0
Blues	88	25	63	62	26	5	0	5	2	3
Children	44	41	3	41	3	0	0	0	0	0
Chinese	5	5	0	5	0	4	3	1	4	0
Christian	1339	298	1041	488	851	34	3	31	31	3
Classic Rock	629	18	611	590	39	73	2	71	72	1
Classical	489	10	479	43	446	30	2	28	15	15
Comedy	1	0	1	1	0	1	0	1	1	0
Contemporary Hit/Top-40	595	25	570	512	83	75	12	63	73	2
Country	2108	611	1497	2078	30	142	62	80	135	7
Croation	0	0	0	0	0	0	0	0	0	0
Czech	0	0	0	0	0	0	0	0	0	0
Disco	2	0	2	2	0	0	0	0	0	0
Discussion	0	0	0	0	0	0	0	0	0	0
Diversified	275	21	254	27	248	88	16	72	51	37
Drama/Literature	0	0	0	0	0	1	0	1	1	0
Easy Listening	47	17	30	39	8	6	1	5	6	0
Educational	236	18	218	19	217	8	3	5	3	5
Eskimo	0	0	0	0	0	1	0	1	1	0
Ethnic	79	47	32	68	11	23	10	13	21	2
Farsi	1	1	0	1	0	0	0	0	0	0
Filipino	2	1	1	2	0	1	0	1	1	0
Finnish	0	0	0	0	0	0	0	0	0	0
Folk	15	1	14	1	14	1	1	0	0	1
Foreign/Ethnic	51	22	29	46	5	9	4	5	9	0
French	1	1	0	1	0	51	1	50	44	7
Full Service	61	48	13	57	4	0	0	0	0	0
German	0	0	0	0	0	1	0	1	1	0
Golden Oldies	17	9	8	16	1	3	2	1	3	0
Gospel	632	448	184	537	95	3	1	2	2	1
Greek	3	3	0	3	0	1	1	0	1	0
Hardcore	0	0	0	0	0	0	0	0	0	0
Hebrew	0	0	0	0	0	0	0	0	0	0
Hindi	0	0	0	0	0	1	0	1	1	0
Hungarian	0	0	0	0	0	0	0	0	0	0
Inspirational	100	15	85	24	76	2	1	1	2	0
Irish	0	0	0	0	0	0	0	0	0	0
Italian	1	1	0	1	0	0	0	0	0	0

Format	United States Total	AM	FM	Com	Non	Canada Total	AM	FM	Com	Non
Japanese	1	1	0	1	0	0	0	0	0	0
Jazz	348	13	335	50	298	17	1	16	7	10
Jewish	0	0	0	0	0	1	1	0	1	0
Korean	11	11	0	11	0	0	0	0	0	0
Light Rock	39	3	36	36	3	13	1	12	13	0
Lithuanian	0	0	0	0	0	0	0	0	0	0
MOR	189	141	48	169	20	43	14	29	41	2
Native American	9	2	7	2	7	3	0	3	3	0
New Age	21	2	19	6	15	0	0	0	0	0
New Wave	0	0	0	0	0	0	0	0	0	0
News	743	264	479	321	422	31	11	20	24	7
News/talk	1262	1000	262	1069	193	57	35	22	40	17
Nostalgia	89	63	26	81	8	5	3	2	4	1
Oldies	1064	450	614	1033	31	47	29	18	45	2
Other	220	88	132	174	46	20	3	17	16	4
Polish	5	5	0	5	0	0	0	0	0	0
Polka	5	3	2	4	1	0	0	0	0	0
Portugese	3	1	2	1	2	0	0	0	0	0
Progressive	142	6	136	14	128	3	0	3	2	1
Public Affairs	68	18	50	16	52	21	4	17	13	8
Reggae	2	0	2	2	0	0	0	0	0	0
Religious	809	364	445	411	398	9	4	5	7	2
Rock/AOR	518	11	507	384	134	69	8	61	67	2
Russian	2	2	0	2	0	0	0	0	0	0
Sacred	0	0	0	0	0	0	0	0	0	0
Scottish	0	0	0	0	0	0	0	0	0	0
Serbian	0	0	0	0	0	0	0	0	0	0
Slovak	0	0	0	0	0	0	0	0	0	0
Slovenian	0	0	0	0	0	0	0	0	0	0
Smooth Jazz	54	2	52	46	8	4	0	4	4	0
Soul	11	3	8	11	0	0	0	0	0	0
Spanish	753	410	343	690	63	1	1	0	1	0
Sports	993	874	119	982	11	16	16	0	16	0
Talk	761	602	159	686	75	26	8	18	18	8
Tejano	24	5	19	24	0	0	0	0	0	0
Top-40	78	9	69	75	3	15	0	15	15	0
Triple A	107	2	105	54	53	2	0	2	2	0
Ukranian	0	0	0	0	0	0	0	0	0	0
Underground	1	0	1	0	1	0	0	0	0	0
Urban Contemporary	352	56	296	294	58	6	0	6	4	2
Variety/Diverse	330	74	256	109	221	70	6	64	56	14
Vietnamese	5	5	0	5	0	0	0	0	0	0
Women	2	2	0	2	0	0	0	0	0	0

Programming on Radio Stations in the U.S.

Adult Contemp

KDBZ(FM) Anchorage AK
KMXS(FM) Anchorage AK
KYMG(FM) Anchorage AK
*KBRW-FM Barrow AK
KTDZ(FM) College AK
KCDV(FM) Cordova AK
*KDLG(AM) Dillingham AK
KYSC(FM) Fairbanks AK
*KEUL(FM) Girdwood AK
KWVV-FM Homer AK
KINY(AM) Juneau AK
KTKN(AM) Ketchikan AK
KRXX(FM) Kodiak AK
KAKN(FM) Naknek AK
KSWD-FM Seward AK
KIFW(AM) Sitka AK
KKIS-FM Soldotna AK
KVAK-FM Valdez AK
KMBQ(FM) Wasilla AK
WKXX(FM) Attalla AL
WMJJ(FM) Birmingham AL
WYDE(AM) Birmingham AL
WYSF(FM) Birmingham AL
WMLV(FM) Butler AL
WDLT-FM Chickasaw AL
WYDE-FM Cullman AL
WRSA-FM Decatur AL
WAGF-FM Dothan AL
WOOF-FM Dothan AL
WKZJ(FM) Eufaula AL
WQLT-FM Florence AL
WFPA(AM) Fort Payne AL
WLDA(FM) Fort Rucker AL
WCKS(FM) Fruithurst AL
WFXX(FM) Georgiana AL
WZLM(FM) Goodwater AL
WAHR(FM) Huntsville AL
WHOD(FM) Jackson AL
WWMG(FM) Millbrook AL
*WBHY-FM Mobile AL
WMXC(FM) Mobile AL
WMXS(FM) Montgomery AL
WTUG-FM Northport AL
WMXA(FM) Opelika AL
WCSN-FM Orange Beach AL
WVOK-FM Oxford AL
WOAB(FM) Ozark AL
WGSY(FM) Phenix City AL
WPID(AM) Piedmont AL
WBEI(FM) Reform AL
WALX(FM) Selma AL
WTRB-FM Sylacauga AL
KDEL-FM Arkadelphia AR
KMJI(FM) Ashdown AR
KCNY(FM) Bald Knob AR
KBTA-FM Batesville AR
KZLE(FM) Batesville AR
KAMD-FM Camden AR
*KUCA(FM) Conway AR
KBKG(FM) Corning AR
KDMS(AM) El Dorado AR
KLBQ(FM) El Dorado AR
KMRX(FM) El Dorado AR
KEZA(FM) Fayetteville AR
KHMB(FM) Hamburg AR
KHOZ(AM) Harrison AR
KILX(FM) Hatfield AR
KAWW-FM Heber Springs AR
KFFA-FM Helena AR
KBHS(AM) Hot Springs AR
KLAZ(FM) Hot Springs AR
KYDL(FM) Hot Springs AR
KVHU(FM) Judsonia AR
KFXV(FM) Kensett AR
KKTZ(FM) Lakeview AR
KURB(FM) Little Rock AR
KHLR(FM) Maumelle AR
KGPQ(FM) Monticello AR
KOMT(FM) Mountain Home AR
KSSW(FM) Nashville AR
KDRS-FM Paragould AR
KBOA-FM Piggott AR
KOTN(AM) Pine Bluff AR
KPOC(AM) Pocahontas AR
KPOC-FM Pocahontas AR
KYNF(FM) Prairie Grove AR
KWKK(FM) Russellville AR
KOKY(FM) Sherwood AR
KJBX(FM) Trumann AR
KSMZ(FM) Viola AR
KLAL(FM) Wrightsville AR
KSBS-FM Pago Pago AS
KVVA-FM Apache Junction AZ
KDVA(FM) Buckeye AZ
KIKO-FM Claypool AZ
KWRQ(FM) Clifton AZ
KMVA(FM) Dewey-Humboldt AZ
KJIK(FM) Duncan AZ
KVNA-FM Flagstaff AZ
KFLX(FM) Kachina Village AZ
KZUL-FM Lake Havasu City AZ
KXAZ(FM) Page AZ
KESZ(FM) Phoenix AZ
KMXP(FM) Phoenix AZ
KPKX(FM) Phoenix AZ
KSUN(AM) Phoenix AZ
*KGCB(FM) Prescott AZ
KNOT(AM) Prescott AZ
KPPV(FM) Prescott Valley AZ
KWKM(FM) Saint Johns AZ
KQST(FM) Sedona AZ
KRFM(FM) Show Low AZ
KZMK(FM) Sierra Vista AZ
KOMR(FM) Sun City AZ
KTBA(AM) Tuba City AZ
KMXZ-FM Tucson AZ
KSGC(FM) Tusayan AZ
KWIM(FM) Window Rock AZ
KLJZ(FM) Yuma AZ
KQSR(FM) Yuma AZ
KAJL(FM) Adelanto CA
KZXY-FM Apple Valley CA
KHYL(FM) Auburn CA
KBLX-FM Berkeley CA
KJMB(FM) Blythe CA
*KBPK(FM) Buena Park CA
*KYCJ(FM) Camino CA
KCDU(FM) Carmel CA
KSBL(FM) Carpinteria CA
KBQB(FM) Chico CA
KMXI(FM) Chico CA
KQPT(FM) Colusa CA
KCRE-FM Crescent City CA
KPOD(AM) Crescent City CA
KXSE(FM) Davis CA
KXO-FM El Centro CA
KFSD(AM) Escondido CA
KHWY(FM) Essex CA
KFMI(FM) Eureka CA
KSAY(FM) Fort Bragg CA
KJLL-FM Fountain Valley CA
KALZ(FM) Fowler CA
KFJK(FM) Fresno CA
KHGE(FM) Fresno CA
KJWL(FM) Fresno CA
KMGQ(FM) Goleta CA
KNCO-FM Grass Valley CA
KMJE(FM) Gridley CA
KTDE(FM) Gualala CA
KATY-FM Idyllwild CA
KSRW(FM) Independence CA
KRCD(FM) Inglewood CA
KRKC-FM King City CA
KXBX-FM Lakeport CA
KGMX(FM) Lancaster CA
KKIQ(FM) Livermore CA
*KCJH(FM) Livingston CA
KSKD(FM) Livingston CA
KUHL(AM) Lompoc CA
KBIG-FM Los Angeles CA
KCBS-FM Los Angeles CA
KLVE(FM) Los Angeles CA
KMVN(FM) Los Angeles CA
KOST(FM) Los Angeles CA
KYSR(FM) Los Angeles CA
KSTT-FM Los Osos-Baywood Park CA
KEBT(FM) Lost Hills CA
KMMT(FM) Mammoth Lakes CA
KJSN(FM) Modesto CA
KMPH(AM) Modesto CA
KWAV(FM) Monterey CA
KLMM(FM) Morro Bay CA
*KLSI(FM) Moss Beach CA
KHYZ(FM) Mountain Pass CA
KTNS(AM) Oakhurst CA
KLLY(FM) Oildale CA
KFYV(FM) Ojai CA
KSPA(AM) Ontario CA
KXLM(FM) Oxnard CA
KEZN(FM) Palm Desert CA
KKAL(FM) Paso Robles CA
KLUN(FM) Paso Robles CA
KOSO(FM) Patterson CA
*KLVM(FM) Prunedale CA
KLXR(AM) Redding CA
KSHA(FM) Redding CA
KZIQ-FM Ridgecrest CA
KMHX(FM) Rohnert Park CA
KOSS(FM) Rosamond CA
KQJK(FM) Roseville CA
KYMX(FM) Sacramento CA
KZZO(FM) Sacramento CA
KVYN(FM) Saint Helena CA
KFMB-FM San Diego CA
KMYI(FM) San Diego CA
KSCF(FM) San Diego CA
KIOI(FM) San Francisco CA
KKSF(FM) San Francisco CA
KLLC(FM) San Francisco CA
KMVQ-FM San Francisco CA
KOIT(AM) San Francisco CA
KOIT-FM San Francisco CA
KEZR(FM) San Jose CA
KINF(AM) Santa Maria CA
KSBQ(AM) Santa Maria CA
KLJR-FM Santa Paula CA
KZST(FM) Santa Rosa CA
KESR(FM) Shasta Lake City CA
KSYV(FM) Solvang CA
KZSQ-FM Sonora CA
KRLT(FM) South Lake Tahoe CA
KTHO(AM) South Lake Tahoe CA
KJOY(FM) Stockton CA
*KYCC(FM) Stockton CA
KHJQ(FM) Susanville CA
KLCA(FM) Tahoe City CA
KHJL(FM) Thousand Oaks CA
KLOB(FM) Thousand Palms CA
KCRZ(FM) Tipton CA
KCDZ(FM) Twentynine Palms CA
KWNE(FM) Ukiah CA
KUIC(FM) Vacaville CA
KBBY-FM Ventura CA
KKZZ(AM) Ventura CA
*KHMS(FM) Victorville CA
KKDV(FM) Walnut Creek CA
KJOR(FM) Windsor CA
KRXV(FM) Yermo CA
KGIW(AM) Alamosa CO
KPRB(FM) Brush CO
KKPK(FM) Colorado Springs CO
KRTZ(FM) Cortez CO
KRAI-FM Craig CO
KALC(FM) Denver CO
KIMN(FM) Denver CO
KOSI(FM) Denver CO
KPTT(FM) Denver CO
KIQX(FM) Durango CO
KPTE(FM) Durango CO
KEPL(AM) Estes Park CO
KFTM(AM) Fort Morgan CO
KJYE(FM) Grand Junction CO
KMXY(FM) Grand Junction CO
KSTH(FM) Holyoke CO
KTRR(FM) Loveland CO
KBIQ(FM) Manitou Springs CO
KSLV-FM Monte Vista CO
KWUF-FM Pagosa Springs CO
KJMP(AM) Pierce CO
KVUU(FM) Pueblo CO
KBNG(FM) Ridgway CO
KZKS(FM) Rifle CO
KPHT(FM) Rocky Ford CO
KVRH-FM Salida CO
KPMX(FM) Sterling CO
KKLI(FM) Widefield CO
KNEC(FM) Yuma CO
WEZN-FM Bridgeport CT
WDAQ(FM) Danbury CT
WNLC(FM) East Lyme CT
WTIC-FM Hartford CT
WBMW(FM) Ledyard CT
WZBG(FM) Litchfield CT
WRCH(FM) New Britain CT
WINY(AM) Putnam CT
WQQQ(FM) Sharon CT
WCTZ(FM) Stamford CT
WEBE(FM) Westport CT
WILI(AM) Willimantic CT
WASH(FM) Washington DC
WRQX(FM) Washington DC
*WXHL-FM Christiana DE
WRDX(FM) Dover DE
WKDB(FM) Laurel DE
WAFL(FM) Milford DE
WZEB(FM) Ocean View DE
WJBR-FM Wilmington DE
WSTW(FM) Wilmington DE
WOYS(FM) Apalachicola FL
*WJIS(FM) Bradenton FL
*WKSG(FM) Cedar Creek FL
WBTP(FM) Clearwater FL
WHQT(FM) Coral Gables FL
WKZY(FM) Cross City FL
*WAQV(FM) Crystal River FL
WKTK(FM) Crystal River FL
WCFB(FM) Daytona Beach FL
WROD(AM) Daytona Beach FL
WMXZ(FM) De Funiak Springs FL
WBPC(FM) Ebro FL
WWRZ(FM) Fort Meade FL
WINK-FM Fort Myers FL
WJPT(FM) Fort Myers Villas FL
WSYR-FM Gifford FL
WCMQ-FM Hialeah FL
WVYB(FM) Holly Hill FL
WIFL(FM) Inglis FL
WEJZ(FM) Jacksonville FL
WMBX(FM) Jensen Beach FL
*WAZQ(FM) Key West FL
WKEY-FM Key West FL
WNFB(FM) Lake City FL
WAVK(FM) Marathon FL
WGMX(FM) Marathon FL
WAVV(FM) Marco FL
*WJNF(FM) Marianna FL
WTOT(AM) Marianna FL
WMMB(AM) Melbourne FL
WAMR-FM Miami FL
WLYF(FM) Miami FL
WSBZ(FM) Miramar Beach FL
WMGF(FM) Mount Dora FL
WBCG(FM) Murdock FL
WSGL(FM) Naples FL
WTLT(FM) Naples FL
WDUV(FM) New Port Richey FL
*WLPJ(FM) New Port Richey FL
*WHIJ(FM) Ocala FL
WMMO(FM) Orlando FL
WOMX-FM Orlando FL
*WHIF(FM) Palatka FL
*WEJF(FM) Palm Bay FL
WRMF(FM) Palm Beach FL
*WCNO(FM) Palm City FL
WFSY(FM) Panama City FL
WVVE(FM) Panama City Beach FL
WBSR(AM) Pensacola FL
WJLQ(FM) Pensacola FL
WHLG(FM) Port St. Lucie FL
WKLG(FM) Rock Harbor FL
WSOS-FM Saint Augustine FL
WRXB(AM) Saint Petersburg Beach FL
WSDV(AM) Sarasota FL
WWLL(FM) Sebring FL
WNCV(FM) Shalimar FL
WCVU(FM) Solana FL
WBZE(FM) Tallahassee FL
WMTX(FM) Tampa FL
WWRM(FM) Tampa FL
WJGO(FM) Tice FL
WDDV(AM) Venice FL
WGYL(FM) Vero Beach FL
WJKD(FM) Vero Beach FL
WEAT-FM West Palm Beach FL
WRLX(FM) West Palm Beach FL
WDEC-FM Americus GA
WSB-FM Atlanta GA
WBBQ-FM Augusta GA
WBGE(FM) Bainbridge GA
WRAK-FM Bainbridge GA
WMUV(FM) Brunswick GA
WSOL-FM Brunswick GA
WMCD(FM) Claxton GA
WRBN(FM) Clayton GA
WGMG(FM) Crawford GA
WYYU(FM) Dalton GA
WXMK(FM) Dock Junction GA
WGMK(FM) Donalsonville GA
WMGZ(FM) Eatonton GA
WRDO(FM) Fitzgerald GA
WZGA(FM) Helen GA
WSGA(FM) Hinesville GA
WJGA-FM Jackson GA
WALR-FM La Grange GA
*WBKG(FM) Macon GA
WQLI(FM) Meigs GA
WJYF(FM) Nashville GA
WLTM(FM) Peachtree City GA
WPNG(FM) Pearson GA
WPGA-FM Perry GA
WSTI-FM Quitman GA
WRHQ(FM) Richmond Hill GA
WQTU(FM) Rome GA
WEGC(FM) Sasser GA
WPMX(FM) Statesboro GA
WTLY(FM) Thomasville GA
WTWA(AM) Thomson GA
WKZZ(FM) Tifton GA
WBDX(FM) Trenton GA
WQSA(FM) Unadilla GA
WQPW(FM) Valdosta GA
WTCQ(FM) Vidalia GA
WWSN(FM) Waycross GA
KSTO(FM) Hagatna GU
KUAM(AM) Hagatna GU
KUAI(AM) Eleele HI
KHBC(AM) Hilo HI
KKBG(FM) Hilo HI
KPVS(FM) Hilo HI
KWXX-FM Hilo HI
KINE-FM Honolulu HI
KSSK(AM) Honolulu HI
KUMU-FM Honolulu HI
KLEO(FM) Kahaluu HI
KJKS(FM) Kahului HI
KRTR-FM Kailua HI
KLUA(FM) Kailua-Kona HI
KWYI(FM) Kawaihae HI
KAOY(FM) Kealakekua HI

KFMN(FM) Lihue HI
KQNG-FM Lihue HI
KAOI-FM Wailuku HI
KSSK-FM Waipahu HI
KLGA(AM) Algona IA
KLGA-FM Algona IA
KLTI-FM Ames IA
KJAN(AM) Atlantic IA
KSKB(FM) Brooklyn IA
KBUR(AM) Burlington IA
KGRS(FM) Burlington IA
KKMI(FM) Burlington IA
KKRL(FM) Carroll IA
KMRY(AM) Cedar Rapids IA
WMT-FM Cedar Rapids IA
KCOG(AM) Centerville IA
KELR-FM Chariton IA
KCHA-FM Charles City IA
KCHE-FM Cherokee IA
KKBZ(FM) Clarinda IA
KMCN(FM) Clinton IA
KMXG(FM) Clinton IA
KQKQ-FM Council Bluffs IA
KCZQ(FM) Cresco IA
*KLOX(FM) Creston IA
KDEC-FM Decorah IA
KDSN(AM) Denison IA
KDSN-FM Denison IA
KDRB(FM) Des Moines IA
KSTZ(FM) Des Moines IA
KATF(FM) Dubuque IA
KDAO-FM Eldora IA
KADR(AM) Elkader IA
KUYY(FM) Emmetsburg IA
KIIK-FM Fairfield IA
KIOW(FM) Forest City IA
KKEZ(FM) Fort Dodge IA
KUEL(FM) Fort Dodge IA
KGRN(AM) Grinnell IA
KLMJ(FM) Hampton IA
KCVM(FM) Hudson IA
KHBT(FM) Humboldt IA
KQMG(AM) Independence IA
KQMG-FM Independence IA
KIFG(AM) Iowa Falls IA
KOKX(AM) Keokuk IA
KRLS(FM) Knoxville IA
KLEM(AM) Le Mars IA
KMCH(FM) Manchester IA
*KBDC(FM) Mason City IA
KLSS-FM Mason City IA
KHDK(FM) New London IA
KCWN(FM) New Sharon IA
KSMA-FM Osage IA
KTWA(FM) Ottumwa IA
KQCR-FM Parkersburg IA
KSOU-FM Sioux Center IA
KGLI(FM) Sioux City IA
KUOO(FM) Spirit Lake IA
KAYL-FM Storm Lake IA
KCII(AM) Washington IA
*KNWS-FM Waterloo IA
KWAY-FM Waverly IA
KQWC-FM Webster City IA
KORR(FM) American Falls ID
KLCE(FM) Blackfoot ID
KTHI(FM) Caldwell ID
KLLP(FM) Chubbuck ID
KMCL(AM) Donnelly ID
KXLT-FM Eagle ID
KCIX(FM) Garden City ID
KMVX(FM) Jerome ID
KATW(FM) Lewiston ID
KMCL-FM McCall ID
KLER-FM Orofino ID
*KSQS(FM) Ririe ID
KSRA(AM) Salmon ID
KSRA-FM Salmon ID
KBJX(FM) Shelley ID
WXET(FM) Arcola IL
WUEZ(FM) Carterville IL
WHMS-FM Champaign IL
WLRW(FM) Champaign IL
WILV(FM) Chicago IL
WLIT-FM Chicago IL
WNUA(FM) Chicago IL

KMJM-FM Columbia IL
WDNL(FM) Danville IL
WDEK(FM) De Kalb IL
WDKB(FM) De Kalb IL
WLBK(AM) De Kalb IL
WDQN(AM) Du Quoin IL
WJEZ(FM) Dwight IL
WMVN(FM) East St. Louis IL
WXEF(FM) Effingham IL
WEBQ-FM Eldorado IL
WYST(FM) Fairbury IL
WFIW-FM Fairfield IL
WNOI(FM) Flora IL
WISH-FM Galatia IL
WSPY(AM) Geneva IL
WYRB(FM) Genoa IL
WXMP(FM) Glasford IL
WBBE(FM) Heyworth IL
WLDS(AM) Jacksonville IL
WSSR(FM) Joliet IL
WVIX(FM) Joliet IL
WYEC(FM) Kewanee IL
WKAY(FM) Knoxville IL
WAJK(FM) La Salle IL
WSRB(FM) Lansing IL
WAKO(AM) Lawrenceville IL
WAKO-FM Lawrenceville IL
WKAI(FM) Macomb IL
WMMC(FM) Marshall IL
WLBH-FM Mattoon IL
WREZ(FM) Metropolis IL
WRIK-FM Metropolis IL
WMOI(FM) Monmouth IL
*WCFL(FM) Morris IL
WYNG(FM) Mount Carmel IL
*WBMV(FM) Mount Vernon IL
WXFM(FM) Mount Zion IL
WNSV(FM) Nashville IL
WVMG(FM) Normal IL
WVAZ(FM) Oak Park IL
WCMY(AM) Ottawa IL
WRKX(FM) Ottawa IL
WRZA(FM) Park Forest IL
WPXN(FM) Paxton IL
*WCIC(FM) Pekin IL
WSWT(FM) Peoria IL
WLLT(FM) Polo IL
*WGCA-FM Quincy IL
WTAY(AM) Robinson IL
WTYE(FM) Robinson IL
WRHL-FM Rochelle IL
WGFB(FM) Rockton IL
WJBD-FM Salem IL
WJDK-FM Seneca IL
WABZ(FM) Sherman IL
WTMX(FM) Skokie IL
WIVQ(FM) Spring Valley IL
WNNS(FM) Springfield IL
WAOX(FM) Staunton IL
WSSQ(FM) Sterling IL
WRAN(FM) Tower Hill IL
WKRV(FM) Vandalia IL
WPMB(AM) Vandalia IL
WGFA-FM Watseka IL
WXLC(FM) Waukegan IL
WRTB(FM) Winnebago IL
WZSR(FM) Woodstock IL
WQME(FM) Anderson IN
*WEAX(FM) Angola IN
WLKI(FM) Angola IN
WZBD(FM) Berne IN
WNUY(FM) Bluffton IN
WBNL(AM) Boonville IN
WCDQ(FM) Crawfordsville IN
WIKY-FM Evansville IN
WVHI(AM) Evansville IN
WAJI(FM) Fort Wayne IN
WGL(AM) Fort Wayne IN
WMEE(FM) Fort Wayne IN
WSHW(FM) Frankfort IN
WKAM(AM) Goshen IN
WZPL(FM) Greenfield IN
WRZQ-FM Greensburg IN
WKLO(FM) Hardinsburg IN
WNTR(FM) Indianapolis IN
WITZ(AM) Jasper IN

WITZ-FM Jasper IN
WKVI(AM) Knox IN
WKVI-FM Knox IN
WZWZ(FM) Kokomo IN
*WIRE(FM) Lebanon IN
WLEG(FM) Ligonier IN
WSAL(AM) Logansport IN
WZVN(FM) Lowell IN
WORX-FM Madison IN
WEFM(FM) Michigan City IN
WPHZ(FM) Mitchell IN
WMRS(FM) Monticello IN
*WWDS(FM) Muncie IN
WYPW(FM) Nappanee IN
WMDH(AM) New Castle IN
WPGW(AM) Portland IN
WFMG(FM) Richmond IN
WAXL(FM) Santa Claus IN
WZZB(AM) Seymour IN
WNSN(FM) South Bend IN
WLZQ(FM) South Whitley IN
WYJZ(FM) Speedway IN
WAWC(FM) Syracuse IN
WBOW-FM Terre Haute IN
WAKE(AM) Valparaiso IN
*WVUB(FM) Vincennes IN
WZDM(FM) Vincennes IN
*WRFM(FM) Wadesville IN
WAMW(AM) Washington IN
WAMW-FM Washington IN
WGLM(FM) West Lafayette IN
WZZY(FM) Winchester IN
KZDY(FM) Cawker City KS
KKOY-FM Chanute KS
KCLY(FM) Clay Center KS
KQLS(FM) Colby KS
KCKS(FM) Concordia KS
KSKZ(FM) Copeland KS
KOLS(FM) Dodge City KS
*KTLI(FM) El Dorado KS
KANS(FM) Emporia KS
KFFX(FM) Emporia KS
KVOE(AM) Emporia KS
KGGF-FM Fredonia KS
KKJQ(FM) Garden City KS
KKCI(FM) Goodland KS
KZLS(FM) Great Bend KS
KGGG(FM) Haven KS
KJLS(FM) Hays KS
KFBZ(FM) Haysville KS
KIND-FM Independence KS
KIKS-FM Iola KS
KUDL(FM) Kansas City KS
KSCB-FM Liberal KS
KSLS(FM) Liberal KS
KQNS-FM Lindsborg KS
KBBE(FM) McPherson KS
KBLS(FM) North Fort Riley KS
KQNK(AM) Norton KS
KQNK-FM Norton KS
KQLA(FM) Ogden KS
KLKC-FM Parsons KS
KHMY(FM) Pratt KS
KCAY(FM) Russell KS
KRSL(AM) Russell KS
KMAJ-FM Topeka KS
KHCA(FM) Wamego KS
KRBB(FM) Wichita KS
WKKQ(FM) Barbourville KY
*WTRT(FM) Benton KY
WKLX(FM) Brownsville KY
WCKQ(FM) Campbellsville KY
WQXQ(FM) Central City KY
WCTT-FM Corbin KY
WQXE(FM) Elizabethtown KY
WCVQ(FM) Fort Campbell KY
WJQI(AM) Fort Campbell KY
WKED-FM Frankfort KY
WUGO(FM) Grayson KY
WHLN(AM) Harlan KY
WHOP-FM Hopkinsville KY
WHHT(FM) Horse Cave KY
WZQQ(FM) Hyden KY
WRNZ(FM) Lancaster KY
WKHG(FM) Leitchfield KY
WMXL(FM) Lexington KY

WLUE(FM) Louisville KY
WVEZ(FM) Louisville KY
WXMA(FM) Louisville KY
WYMV(FM) Madisonville KY
WFXY(AM) Middlesboro KY
WMOR-FM Morehead KY
WCLU-FM Munfordville KY
WOFC(AM) Murray KY
*WGCF(FM) Paducah KY
WKLW-FM Paintsville KY
*WWJD(FM) Pippa Passes KY
WQHY(FM) Prestonsburg KY
WHVE(FM) Russell Springs KY
WUHU(FM) Smiths Grove KY
WLLK-FM Somerset KY
WYSB(FM) Springfield KY
WCDA(FM) Versailles KY
KRVE(FM) Brusly LA
KQLK(FM) De Ridder LA
KBYO-FM Farmerville LA
KDDK(FM) Franklin LA
WDVW(FM) La Place LA
KTDY(FM) Lafayette LA
KBIU(FM) Lake Charles LA
KZLG(FM) Mansura LA
WLMG(FM) New Orleans LA
KOGM(FM) Opelousas LA
KVKI-FM Shreveport LA
KLSM(FM) Tallulah LA
KNEK-FM Washington LA
KMBS(AM) West Monroe LA
KZRZ(FM) West Monroe LA
KTEZ(FM) Zwolle LA
WQRC(FM) Barnstable MA
WBMX(FM) Boston MA
WMJX(FM) Boston MA
WJIB(AM) Cambridge MA
WEIM(AM) Fitchburg MA
WXLO(FM) Fitchburg MA
WSBS(AM) Great Barrington MA
WHAI(FM) Greenfield MA
WFQR(FM) Harwich Port MA
WXRV(FM) Haverhill MA
WCOD-FM Hyannis MA
WATD-FM Marshfield MA
WFRQ(FM) Mashpee MA
WMRC(AM) Milford MA
*WMLN-FM Milton MA
WNAW(AM) North Adams MA
WJDF(FM) Orange MA
WOCN-FM Orleans MA
WBEC-FM Pittsfield MA
WBRK-FM Pittsfield MA
WPLM(AM) Plymouth MA
WPLM-FM Plymouth MA
*WRPS(FM) Rockland MA
WESX(AM) Salem MA
WHYN-FM Springfield MA
WMAS(AM) Springfield MA
WMAS-FM Springfield MA
WSNE-FM Taunton MA
WSNI(FM) Winchendon MA
WSRS(FM) Worcester MA
WNAV(AM) Annapolis MD
WLIF(FM) Baltimore MD
WWMX(FM) Baltimore MD
WMMJ(FM) Bethesda MD
WTRI(AM) Brunswick MD
WCEM-FM Cambridge MD
WCTR(AM) Chestertown MD
WKGO(FM) Cumberland MD
WCEI-FM Easton MD
WKHI(FM) Fruitland MD
WWEG(FM) Hagerstown MD
WILC(AM) Laurel MD
WAFY(FM) Middletown MD
WKHJ(FM) Mountain Lake Park MD
WMSG(AM) Oakland MD
WQHQ(FM) Ocean City-Salisbury MD
WXMD(FM) Pocomoke City MD
WKCG(FM) Augusta ME
WQDY-FM Calais ME
WCXU(FM) Caribou ME
WCRQ(FM) Dennysville ME
WDME-FM Dover Foxcroft ME
WKSQ(FM) Ellsworth ME

WKTJ-FM Farmington ME
WFAU(AM) Gardiner ME
WHOU-FM Houlton ME
WALZ-FM Machias ME
WCXX(FM) Madawaska ME
WCXH(AM) Monticello ME
WMGX(FM) Portland ME
WQHR(FM) Presque Isle ME
XHRM-FM Tijuana MEX
WLEN(FM) Adrian MI
WQBX(FM) Alma MI
WHSB(FM) Alpena MI
WJSZ(FM) Ashley MI
WLEW-FM Bad Axe MI
WBXX(FM) Battle Creek MI
WIOG(FM) Bay City MI
WCNF(FM) Benton Harbor MI
WYBR(FM) Big Rapids MI
WKHM-FM Brooklyn MI
WLXV(FM) Cadillac MI
WIDL(FM) Caro MI
*WPRJ(FM) Coleman MI
WHTS(FM) Coopersville MI
WNIC(FM) Dearborn MI
WDVD(FM) Detroit MI
WMGC-FM Detroit MI
WFMK(FM) East Lansing MI
WGLQ(FM) Escanaba MI
WCRZ(FM) Flint MI
WMJZ-FM Gaylord MI
WSRT(FM) Gaylord MI
WGHN(AM) Grand Haven MI
WGHN-FM Grand Haven MI
WLHT-FM Grand Rapids MI
WOOD-FM Grand Rapids MI
WCSR(AM) Hillsdale MI
WCSR-FM Hillsdale MI
WSRJ(FM) Honor MI
WBZV(FM) Hudson MI
WIMI(FM) Ironwood MI
WUPM(FM) Ironwood MI
WMQT(FM) Ishpeming MI
WVFM(FM) Kalamazoo MI
WBTI(FM) Lexington MI
WKLA-FM Ludington MI
WMJT(FM) McMillan MI
WHTD(FM) Mount Clemens MI
WCZY-FM Mount Pleasant MI
WSHZ(FM) Muskegon MI
WKQS-FM Negaunee MI
WZNL(FM) Norway MI
WWTH(FM) Oscoda MI
WMOM(FM) Pentwater MI
WGRT(FM) Port Huron MI
WQON(FM) Roscommon MI
WTGV-FM Sandusky MI
WSOO(AM) Sault Ste. Marie MI
WMXG(FM) Stephenson MI
WLDR-FM Traverse City MI
WTRV(FM) Walker MI
KDDG(FM) Albany MN
KCPI(FM) Albert Lea MN
KAUS(AM) Austin MN
KKBJ-FM Bemidji MN
KKZY(FM) Bemidji MN
KOWZ-FM Blooming Prairie MN
WJJY-FM Brainerd MN
KLTA(FM) Breckenridge MN
KLKS(FM) Breezy Point MN
KXDL(FM) Browerville MN
KRWC(AM) Buffalo MN
WKLK-FM Cloquet MN
KMXK(FM) Cold Spring MN
KROX(AM) Crookston MN
KDLM(AM) Detroit Lakes MN
KDAL-FM Duluth MN
KZLT-FM East Grand Forks MN
WEVE-FM Eveleth MN
WLKX-FM Forest Lake MN
KMGK(FM) Glenwood MN
*WTIP(FM) Grand Marais MN
KMFY(FM) Grand Rapids MN
WWAX(FM) Hermantown MN
KLCH(FM) Lake City MN
KFML(FM) Little Falls MN
KQAD(AM) Luverne MN

KEEZ-FM Mankato MN
KKCK(FM) Marshall MN
KTCZ-FM Minneapolis MN
KTLK-FM Minneapolis MN
WLTE(FM) Minneapolis MN
KMGM(FM) Montevideo MN
KDOG(FM) North Mankato MN
KYMN(AM) Northfield MN
KTIG(FM) Pequot Lakes MN
KPRW(FM) Perham MN
KBMX(FM) Proctor MN
*KCFB(FM) Saint Cloud MN
KCML(FM) Saint Joseph MN
KSTP-FM Saint Paul MN
KKSR(FM) Sartell MN
KNUJ-FM Sleepy Eye MN
KNSG(FM) Springfield MN
KSKK(FM) Staples MN
KRVY-FM Starbuck MN
KYBA(FM) Stewartville MN
KQKK(FM) Walker MN
KQIC(FM) Willmar MN
KAGE-FM Winona MN
KITN(FM) Worthington MN
KBMV-FM Birch Tree MO
KWJK(FM) Boonville MO
KFMZ(AM) Brookfield MO
KZBK(FM) Brookfield MO
KKWK(FM) Cameron MO
KRRY(FM) Canton MO
KMXL(FM) Carthage MO
*KNLH(FM) Cedar Hill MO
KREZ(FM) Chaffee MO
KPLA(FM) Columbia MO
KBZI(FM) Deerfield MO
KLOZ(FM) Eldon MO
KZWV(FM) Eldon MO
KAUL(FM) Ellington MO
*KCFV(FM) Ferguson MO
KMAC(FM) Gainesville MO
KCJK(FM) Garden City MO
KYOO-FM Halfway MO
KGRC(FM) Hannibal MO
KTXY(FM) Jefferson City MO
KCKC(FM) Kansas City MO
KIRK(FM) Macon MO
KLSC(FM) Malden MO
KXEO(AM) Mexico MO
KMCR(FM) Montgomery City MO
KUPH(FM) Mountain View MO
KZMA(FM) Naylor MO
KGBX-FM Nixa MO
KTOZ-FM Pleasant Hope MO
KAHR(FM) Poplar Bluff MO
*KNLP(FM) Potosi MO
KADI-FM Republic MO
KDAA(FM) Rolla MO
KGNM(AM) Saint Joseph MO
KKJO(FM) Saint Joseph MO
KEZK-FM Saint Louis MO
KYKY(FM) Saint Louis MO
KGKS(FM) Scott City MO
KSDL(FM) Sedalia MO
*KWND(FM) Springfield MO
KTTN(AM) Trenton MO
*KBIY(FM) Van Buren MO
KSLQ-FM Washington MO
KJPW-FM Waynesville MO
KJMK(FM) Webb City MO
KULH(FM) Wheeling MO
WKZW(FM) Bay Springs MS
WMJY(FM) Biloxi MS
WMJU(FM) Bude MS
*WOWL(FM) Burnsville MS
*WKVF(FM) Byhalia MS
WMGO(AM) Canton MS
WAID(FM) Clarksdale MS
WFFF-FM Columbia MS
WMBC(FM) Columbus MS
WZKR(FM) Decatur MS
WLIN-FM Durant MS
WLZA(FM) Eupora MS
WFTA(FM) Fulton MS
WGNL(FM) Greenwood MS
WYMX(FM) Greenwood MS
WUJM(FM) Gulfport MS
WSYE(FM) Houston MS
WNLA-FM Indianola MS
WIQQ(FM) Leland MS
WLSM-FM Louisville MS
WMSO(FM) Meridian MS
WKSO(FM) Natchez MS
WOXD(FM) Oxford MS
WQLJ(FM) Oxford MS
WRXW(FM) Pearl MS
WWSL(FM) Philadelphia MS
WRTM-FM Port Gibson MS
WKZB(FM) Stonewall MS
WZLQ(FM) Tupelo MS
WUMS(FM) University MS
WJKK(FM) Vicksburg MS
KGLM-FM Anaconda MT
KBBB(FM) Billings MT
KRPM(FM) Billings MT
KYYA-FM Billings MT
KOBB(AM) Bozeman MT
KZMY(FM) Bozeman MT
KOPR(FM) Butte MT
KRYK(FM) Chinook MT
KVVR(FM) Dutton MT
KLAN(FM) Glasgow MT
KXGN(AM) Glendive MT
KAAK(FM) Great Falls MT
KXDR(FM) Hamilton MT
KOJM(AM) Havre MT
KMTX-FM Helena MT
KALS(FM) Kalispell MT
KBSR(AM) Laurel MT
KTNY(FM) Libby MT
KATL(AM) Miles City MT
KMSO(FM) Missoula MT
KYJK(FM) Missoula MT
KQRK(FM) Ronan MT
KTHC(FM) Sidney MT
KKVU(FM) Stevensville MT
KENR(FM) Superior MT
KEZQ(FM) West Yellowstone MT
WECR-FM Beech Mountain NC
WSQL(AM) Brevard NC
WKSL(FM) Burlington NC
WKQC(FM) Charlotte NC
WLNK(FM) Charlotte NC
*WWCU(FM) Cullowhee NC
WFXC(FM) Durham NC
WKJX(FM) Elizabeth City NC
WIFM-FM Elkin NC
WBAV-FM Gastonia NC
WSMW(FM) Greensboro NC
WANG(AM) Havelock NC
WSSM(FM) Havelock NC
WMYI(FM) Hendersonville NC
WTZQ(AM) Hendersonville NC
WLYT(FM) Hickory NC
WMAG(FM) High Point NC
WCXL(FM) Kill Devil Hills NC
WLNC(AM) Laurinburg NC
WTHZ(FM) Lexington NC
WDEX(AM) Monroe NC
WDLZ(FM) Murfreesboro NC
WZAX(FM) Nashville NC
*WAAE(FM) New Bern NC
WNNC(AM) Newton NC
WPNC-FM Plymouth NC
WRAL(FM) Raleigh NC
*WZRU(FM) Roanoke Rapids NC
WAYN(AM) Rockingham NC
WEND(FM) Salisbury NC
WIOZ-FM Southern Pines NC
WRGC(AM) Sylva NC
WFXK(FM) Tarboro NC
WADE(AM) Wadesboro NC
WERO(FM) Washington NC
WLGT(FM) Washington NC
WGNI(FM) Wilmington NC
*WWIL-FM Wilmington NC
*WRCM(FM) Wingate NC
*WSNC(FM) Winston-Salem NC
KYYY(FM) Bismarck ND
KWGO(FM) Burlington ND
KDIX(AM) Dickinson ND
KKXL(AM) Grand Forks ND
KHND(AM) Harvey ND
KMXW(FM) Hope ND
KNDK-FM Langdon ND
KNDR(FM) Mandan ND
KIZZ(FM) Minot ND
KMXA-FM Minot ND
KQDJ-FM Valley City ND
KYTZ(FM) Walhalla ND
KBRB-FM Ainsworth NE
KRGY(FM) Aurora NE
KWBE(AM) Beatrice NE
KLIR(FM) Columbus NE
*KINI(FM) Crookston NE
KRGI(AM) Grand Island NE
KSYZ-FM Grand Island NE
KHAS(AM) Hastings NE
KLIQ(FM) Hastings NE
KMTY(FM) Holdrege NE
KBBK(FM) Lincoln NE
KLTQ(FM) Lincoln NE
KICX-FM McCook NE
KSWN(FM) McCook NE
KEXL(FM) Norfolk NE
KNEN(FM) Norfolk NE
KELN(FM) North Platte NE
*KJLT-FM North Platte NE
KOMJ(AM) Omaha NE
KQCH(FM) Omaha NE
KSRZ(FM) Omaha NE
KSID-FM Sidney NE
KTMX(FM) York NE
WBNC(AM) Conway NH
WVMJ(FM) Conway NH
WFTN-FM Franklin NH
WGXL(FM) Hanover NH
*WNEC-FM Henniker NH
WKNE(FM) Keene NH
WLNH-FM Laconia NH
WLTN-FM Lisbon NH
WZID(FM) Manchester NH
WHOM(FM) Mt. Washington NH
WFNQ(FM) Nashua NH
WBYY(FM) Somersworth NH
WASR(AM) Wolfeboro NH
WJLK(FM) Asbury Park NJ
WAYV(FM) Atlantic City NJ
WFPG(FM) Atlantic City NJ
WBBO(FM) Bass River Township NJ
WHCY(FM) Blairstown NJ
WAIV(FM) Cape May NJ
WHTG-FM Eatontown NJ
WSJO(FM) Egg Harbor City NJ
WIMG(AM) Ewing NJ
WSUS(FM) Franklin NJ
WWZY(FM) Long Branch NJ
WMGQ(FM) New Brunswick NJ
WKMK(FM) Ocean Acres NJ
WIBG(AM) Ocean City NJ
WPAT-FM Paterson NJ
WOBM-FM Toms River NJ
WBHX(FM) Tuckerton NJ
WCZT(FM) Villas NJ
WMIZ(AM) Vineland NJ
WVLT(FM) Vineland NJ
WAWZ(FM) Zarephath NJ
*KYCM(FM) Alamogordo NM
KKJY(AM) Albuquerque NM
KKOB-FM Albuquerque NM
KMGA(FM) Albuquerque NM
KPEK(FM) Albuquerque NM
KWYK-FM Aztec NM
KAMQ(AM) Carlsbad NM
KCDY(FM) Carlsbad NM
KSMX(FM) Clovis NM
KTQM-FM Clovis NM
KSYU(FM) Corrales NM
KDEM(FM) Deming NM
KKOR(FM) Gallup NM
KZOR(FM) Hobbs NM
KSNM(AM) Las Cruces NM
KLVF(FM) Las Vegas NM
KMVR(FM) Mesilla Park NM
KLBU(FM) Pecos NM
KLNN(FM) Questa NM
KRTN(AM) Raton NM
KBIM-FM Roswell NM
KSSR(AM) Santa Rosa NM
KSCQ(FM) Silver City NM
KKIT(FM) Taos NM
KHAC(AM) Tse Bonito NM
KQAY-FM Tucumcari NM
KZMI(FM) Garapan-Saipan NP
KHIX(FM) Carlin NV
KELK(AM) Elko NV
KLKO(FM) Elko NV
KCLS(FM) Ely NV
KVLV-FM Fallon NV
KKJJ(FM) Henderson NV
KMXB(FM) Henderson NV
KRNO(FM) Incline Village NV
KKVV(AM) Las Vegas NV
KPLV(FM) Las Vegas NV
KSNE-FM Las Vegas NV
*KSOS(FM) Las Vegas NV
KADD(FM) Logandale NV
KJUL(FM) Moapa Valley NV
KNEV(FM) Reno NV
KJZS(FM) Sparks NV
WYJB(FM) Albany NY
*WBXL(FM) Baldwinsville NY
WVIN-FM Bath NY
WBAZ(FM) Bridgehampton NY
WFAS-FM Bronxville NY
WJYE(FM) Buffalo NY
WTSS(FM) Buffalo NY
WVOR(FM) Canandaigua NY
WTOJ(FM) Carthage NY
WCTW(FM) Catskill NY
WLVG(FM) Center Moriches NY
WGMM(FM) Corning NY
WNKI(FM) Corning NY
WDNY-FM Dansville NY
WWDG(FM) DeRuyter NY
WFLR-FM Dundee NY
WENY-FM Elmira NY
WLVY(FM) Elmira NY
WBKX(FM) Fredonia NY
*WHPC(FM) Garden City NY
WENT(AM) Gloversville NY
WHLI(AM) Hempstead NY
WKJY(FM) Hempstead NY
WKPQ(FM) Hornell NY
WNYH(AM) Huntington NY
WCZX(FM) Hyde Park NY
WYXL(FM) Ithaca NY
WWSE(FM) Jamestown NY
WLTB(FM) Johnson City NY
WIZR(AM) Johnstown NY
WKNY(AM) Kingston NY
WSKU(FM) Little Falls NY
WBZZ(FM) Malta NY
WMSA(AM) Massena NY
WSUL(FM) Monticello NY
WBWZ(FM) New Paltz NY
WLTW(FM) New York NY
WPLJ(FM) New York NY
WGNY-FM Newburgh NY
WKXZ(FM) Norwich NY
WVLF(FM) Norwood NY
WMXO(FM) Olean NY
WMCR(AM) Oneida NY
WMCR-FM Oneida NY
WSRK(FM) Oneonta NY
WOLF-FM Oswego NY
WALK-FM Patchogue NY
WDBY(FM) Patterson NY
WHUD(FM) Peekskill NY
WYLF(AM) Penn Yan NY
WZUN(FM) Phoenix NY
WIRY(AM) Plattsburgh NY
WPDM(AM) Potsdam NY
WRNQ(FM) Poughkeepsie NY
WOKR(FM) Remsen NY
WDVI(FM) Rochester NY
WLGZ(AM) Rochester NY
WRMM-FM Rochester NY
WNBZ(AM) Saranac Lake NY
WSLP(FM) Saranac Lake NY
WYZY(FM) Saranac Lake NY
WRVE(FM) Schenectady NY
WCDO(AM) Sidney NY
WCDO-FM Sidney NY
WMJC(FM) Smithtown NY
WHFM(FM) Southampton NY
WRCR(AM) Spring Valley NY
WSPQ(AM) Springville NY
WQAR(FM) Stillwater NY
WWLF-FM Sylvan Beach NY
WLTI(FM) Syracuse NY
WYYY(FM) Syracuse NY
WRGR(FM) Tupper Lake NY
WLZW(FM) Utica NY
WMXW(FM) Vestal NY
WNYR-FM Waterloo NY
WAVR(FM) Waverly NY
WQRW(FM) Wellsville NY
WNYV(FM) Whitehall NY
WRIP(FM) Windham NY
WZKL(FM) Alliance OH
WREO-FM Ashtabula OH
WJKW(FM) Athens OH
*WOUB-FM Athens OH
WXTQ(FM) Athens OH
WBNV(FM) Barnesville OH
WPKO-FM Bellefontaine OH
WCMJ(FM) Cambridge OH
WHBC-FM Canton OH
WKDD(FM) Canton OH
WCSM-FM Celina OH
WKKI(FM) Celina OH
WLZT(FM) Chillicothe OH
*WJVS(FM) Cincinnati OH
WKRQ(FM) Cincinnati OH
WRRM(FM) Cincinnati OH
WVMX(FM) Cincinnati OH
WFHM-FM Cleveland OH
WMVX(FM) Cleveland OH
WQAL(FM) Cleveland OH
WBNS-FM Columbus OH
WNCI(FM) Columbus OH
WSNY(FM) Columbus OH
WTNS-FM Coshocton OH
WMMX(FM) Dayton OH
WDFM(FM) Defiance OH
WZOO-FM Edgewood OH
WKXA-FM Findlay OH
WBVI(FM) Fostoria OH
WFOB(AM) Fostoria OH
WFRO-FM Fremont OH
*WCVO(FM) Gahanna OH
WVNU(FM) Greenfield OH
WNLT(FM) Harrison OH
WRBP(FM) Hubbard OH
WKTN(FM) Kenton OH
WLQT(FM) Kettering OH
WAGX(FM) Manchester OH
WVNO-FM Mansfield OH
WYHT(FM) Mansfield OH
WMOA(AM) Marietta OH
WKLM(FM) Millersburg OH
WQIO(FM) Mount Vernon OH
WNDH(FM) Napoleon OH
WWJM(FM) New Lexington OH
WHOF(FM) North Canton OH
WLKR-FM Norwalk OH
WKSD(FM) Paulding OH
WNXT-FM Portsmouth OH
WNTO(FM) Racine OH
WMLX(FM) Saint Mary's OH
WCPZ(FM) Sandusky OH
WMVR-FM Sidney OH
*WCVZ(FM) South Zanesville OH
*WKTL(FM) Struthers OH
WWWM-FM Sylvania OH
WTTF(AM) Tiffin OH
WRVF(FM) Toledo OH
WYNT(FM) Upper Sandusky OH
WERT(AM) Van Wert OH
WKOV-FM Wellston OH
*WYSO(FM) Yellow Springs OH
WMXY(FM) Youngstown OH
WHIZ-FM Zanesville OH
KTTL(FM) Alva OK
KYFM(FM) Bartlesville OK
KOJK(FM) Blanchard OK
KSEO(AM) Durant OK
KQTZ(FM) Hobart OK
KTLS-FM Holdenville OK
KQIB(FM) Idabel OK

KXLS(FM) Lahoma OK
KBZQ(FM) Lawton OK
KMGZ(FM) Lawton OK
KYNZ(FM) Lone Grove OK
KMGL(FM) Oklahoma City OK
KYIS(FM) Oklahoma City OK
*KXTH(FM) Seminole OK
KGFF(AM) Shawnee OK
*KJCM(FM) Snyder OK
KSPI-FM Stillwater OK
KBEZ(FM) Tulsa OK
KRAV(FM) Tulsa OK
KOCD(FM) Wilburton OK
KMZE(FM) Woodward OK
KCMX-FM Ashland OR
KVMX(FM) Banks OR
KKCW(FM) Beaverton OR
KMGX(FM) Bend OR
KYSF(FM) Bonanza OR
KURY-FM Brookings OR
KEHK(FM) Brownsville OR
KCBZ(FM) Cannon Beach OR
KFLY(FM) Corvallis OR
KDPM(FM) Cottage Grove OR
*KLCC(FM) Eugene OR
KMGE(FM) Eugene OR
KCST-FM Florence OR
KGBR(FM) Gold Beach OR
KLDR(FM) Harbeck-Fruitdale OR
KQFM(FM) Hermiston OR
KCGB-FM Hood River OR
KKRB(FM) Klamath Falls OR
KWRL(FM) La Grande OR
KQIK-FM Lakeview OR
KLYC(AM) McMinnville OR
KTMT-FM Medford OR
KRSK(FM) Molalla OR
KSND(FM) Monmouth OR
KYTE(FM) Newport OR
KOOS(FM) North Bend OR
*KAVE(FM) Oakridge OR
KSRV-FM Ontario OR
KUMA-FM Pendleton OR
KYCH-FM Portland OR
KLTW-FM Prineville OR
KPPK(FM) Rainier OR
KLRR(FM) Redmond OR
*KLFR(FM) Reedsport OR
*KMPQ(FM) Roseburg OR
KWPK-FM Sisters OR
KMCQ(FM) The Dalles OR
KODL(AM) The Dalles OR
KKMX(FM) Tri City OR
KEUG(FM) Veneta OR
*KWSO(FM) Warm Springs OR
WLEV(FM) Allentown PA
WAYC(FM) Bedford PA
WZWW(FM) Bellefonte PA
WFYY(FM) Bloomsburg PA
WRRK(FM) Braddock PA
WESB(AM) Bradford PA
WXXO(FM) Cambridge Springs PA
WIKZ(FM) Chambersburg PA
*WZXQ(FM) Chambersburg PA
WCCR(FM) Clarion PA
WQYX(FM) Clearfield PA
WWCB(AM) Corry PA
WFRM-FM Coudersport PA
WOKW(FM) Curwensville PA
WQKY(FM) Emporium PA
WXBB(FM) Erie PA
WXKC(FM) Erie PA
*WRSD(FM) Folsom PA
WQFN(FM) Forest City PA
WOXX(FM) Franklin PA
WGET(AM) Gettysburg PA
WRVV(FM) Harrisburg PA
WMHX(FM) Hershey PA
WRKY-FM Hollidaysburg PA
WCCS(AM) Homer City PA
WDNH-FM Honesdale PA
WLAK(FM) Huntingdon PA
WFGI-FM Johnstown PA
WKYE(FM) Johnstown PA
WLAN-FM Lancaster PA
WROZ(FM) Lancaster PA

WCOZ(FM) Laporte PA
WQTW(AM) Latrobe PA
WQIC(FM) Lebanon PA
*WGRC(FM) Lewisburg PA
WMRF-FM Lewistown PA
WSNU(FM) Lock Haven PA
WNBQ(FM) Mansfield PA
*WRIJ(FM) Masontown PA
*WMSS(FM) Middletown PA
WVRT(FM) Mill Hall PA
WQLV(FM) Millersburg PA
WVLY-FM Milton PA
WSPI(FM) Mount Carmel PA
*WRWJ(FM) Murrysville PA
WQFM(FM) Nanticoke PA
WZPT(FM) New Kensington PA
*WWNW(FM) New Wilmington PA
WNCC(AM) Northern Cambria PA
WKQW-FM Oil City PA
WBEB(FM) Philadelphia PA
WBEN-FM Philadelphia PA
WDAS-FM Philadelphia PA
WISX(FM) Philadelphia PA
*WPTS-FM Pittsburgh PA
WSHH(FM) Pittsburgh PA
WHKS(FM) Port Allegany PA
WPPA(AM) Pottsville PA
WPXZ-FM Punxsutawney PA
WZYY(FM) Renovo PA
WDSN(FM) Reynoldsville PA
WKMC(AM) Roaring Spring PA
WKBI(AM) Saint Marys PA
WKBI-FM Saint Marys PA
WATS(AM) Sayre PA
WLSW(FM) Scottdale PA
WWRR(FM) Scranton PA
*WBYO(FM) Sellersville PA
WEEO(AM) Shippensburg PA
WQRM(FM) Smethport PA
*WBYX(FM) Stroudsburg PA
WMGH-FM Tamaqua PA
WGMF(FM) Tunkhannock PA
WTRN(AM) Tyrone PA
WCTL(FM) Union City PA
WNAE(AM) Warren PA
*WCYJ-FM Waynesburg PA
WNBT-FM Wellsboro PA
WKSB(FM) Williamsport PA
WRVH(FM) Williamsport PA
WARM-FM York PA
WFDT(FM) Aguada PR
WABA(AM) Aguadilla PR
WTPM(FM) Aguadilla PR
WLUZ(AM) Bayamon PR
WMIO(FM) Cabo Rojo PR
WNEL(AM) Caguas PR
WVJP(AM) Caguas PR
WVOZ-FM Carolina PR
WCPR(AM) Coamo PR
WXRF(AM) Guayama PR
WOIZ(AM) Guayanilla PR
WIOB(FM) Mayaguez PR
WTIL(AM) Mayaguez PR
WEXS(AM) Patillas PR
WIOC(FM) Ponce PR
*WPUC-FM Ponce PR
WZAR(FM) Ponce PR
WFID(FM) Rio Piedras PR
WIOA(FM) San Juan PR
WLRP(AM) San Sebastian PR
WRSS(AM) San Sebastian PR
WERR(FM) Utuado PR
WENA(AM) Yauco PR
WJZS(FM) Block Island RI
WWLI(FM) Providence RI
WYKZ(FM) Beaufort SC
WCAM(AM) Camden SC
WLTY(FM) Cayce SC
WALC(FM) Charleston SC
WSSX-FM Charleston SC
WCRE(AM) Cheraw SC
WHHD(FM) Clearwater SC
WORG(FM) Elloree SC
WZSN(FM) Greenwood SC
WBHC-FM Hampton SC
WDKD(AM) Kingstree SC

WSIM(FM) Lamar SC
WIHB(FM) Moncks Corner SC
WYEZ(FM) Murrell's Inlet SC
WMYB(FM) Myrtle Beach SC
WKDK(AM) Newberry SC
WTCB(FM) Orangeburg SC
WMGL(FM) Ravenel SC
WWJN(FM) Ridgeland SC
WIGL(FM) Saint Matthews SC
WGVC(FM) Simpsonville SC
WSPA-FM Spartanburg SC
KBFO(FM) Aberdeen SD
KBRK-FM Brookings SD
KFCR(AM) Custer SD
KZKK(FM) Huron SD
KQRN(FM) Mitchell SD
KOLY-FM Mobridge SD
KGFX-FM Pierre SD
KLXS-FM Pierre SD
KKMK(FM) Rapid City SD
KZLK(FM) Rapid City SD
KNBZ(FM) Redfield SD
KELO-FM Sioux Falls SD
KMXC(FM) Sioux Falls SD
KVHT(FM) Vermillion SD
KIXX(FM) Watertown SD
KWYR-FM Winner SD
KYNT(AM) Yankton SD
WBGQ(FM) Bulls Gap TN
WCLE-FM Calhoun TN
WRJB(FM) Camden TN
WKRM(AM) Columbia TN
WGIC(FM) Cookeville TN
WPBX(FM) Crossville TN
WALV-FM Dayton TN
WASL(FM) Dyersburg TN
WLLJ(FM) Etowah TN
*WVCP(FM) Gallatin TN
WSEV-FM Gatlinburg TN
WSNA(FM) Germantown TN
WHHM-FM Henderson TN
WEKX(FM) Jellico TN
WMYU(FM) Karns TN
WTFM(FM) Kingsport TN
WJXB-FM Knoxville TN
WDXE-FM Lawrenceburg TN
WKZX-FM Lenoir City TN
WZLT(FM) Lexington TN
WYGO(FM) Madisonville TN
WFTZ(FM) Manchester TN
WMC-FM Memphis TN
WRVR(FM) Memphis TN
WYDL(FM) Middleton TN
WCRK(AM) Morristown TN
WKIM(FM) Munford TN
WCJK(FM) Murfreesboro TN
WJXA(FM) Nashville TN
WNRQ(FM) Nashville TN
WBNT-FM Oneida TN
WOCV(AM) Oneida TN
WLZK(FM) Paris TN
WVNS-FM Pegram TN
WSEV(AM) Sevierville TN
WTNE(AM) Trenton TN
*KACU(FM) Abilene TX
*KGNZ(FM) Abilene TX
KNDA(FM) Alice TX
KMXJ-FM Amarillo TX
KKMJ-FM Austin TX
KNUZ(AM) Bellville TX
KLTR(FM) Brenham TX
KLZK(FM) Brownfield TX
KBWD(AM) Brownwood TX
*KHPU(FM) Brownwood TX
KKYS(FM) Bryan TX
KRHC(AM) Burnet TX
*KETR(FM) Commerce TX
KLTG(FM) Corpus Christi TX
KDMX(FM) Dallas TX
KRNB(FM) Decatur TX
KTDR(FM) Del Rio TX
KAFX-FM Diboll TX
KVLY(FM) Edinburg TX
KINT-FM El Paso TX
KSII(FM) El Paso TX
KTSM-FM El Paso TX

KFST(AM) Fort Stockton TX
KZTR(FM) Franklin TX
KSOC(FM) Gainesville TX
KTWL(FM) Hempstead TX
KHMX(FM) Houston TX
KODA(FM) Houston TX
KSML-FM Huntington TX
KLJT(FM) Jacksonville TX
KOOI-FM Jacksonville TX
KJAS(FM) Jasper TX
KKBA(FM) Kingsville TX
*KHOY(FM) Laredo TX
KRRG(FM) Laredo TX
KJDL-FM Levelland TX
KSHN-FM Liberty TX
KONE(FM) Lubbock TX
KYBI(FM) Lufkin TX
KAMX(FM) Luling TX
*KZLV(FM) Lytle TX
KZRC(FM) Markham TX
KLRK(FM) Marlin TX
KLSR-FM Memphis TX
KHKZ(FM) Mercedes TX
*KAQQ(FM) Midland TX
KCHX(FM) Midland TX
KCRS-FM Midland TX
KQRX(FM) Midland TX
KZRB(FM) New Boston TX
KODM(FM) Odessa TX
KKMY(FM) Orange TX
KAZE(FM) Ore City TX
KGRO(AM) Pampa TX
KPLT-FM Paris TX
KPTX(FM) Pecos TX
KDVE(FM) Pittsburg TX
KRIA(FM) Plainview TX
*KPVU(FM) Prairie View TX
KWFB(FM) Quanah TX
KFMK(FM) Round Rock TX
*KNLE-FM Round Rock TX
KELI(FM) San Angelo TX
KIXY-FM San Angelo TX
KQXT(FM) San Antonio TX
KBAL(AM) San Saba TX
KSMG(FM) Seguin TX
KBKH(FM) Shamrock TX
KJIM(AM) Sherman TX
KMMX(FM) Tahoka TX
KLAK(FM) Tom Bean TX
KTYL-FM Tyler TX
KQVT(FM) Victoria TX
KVIC(FM) Victoria TX
KQHN(FM) Waskom TX
KAJK(FM) White Oak TX
KBZS(FM) Wichita Falls TX
KALK(FM) Winfield TX
KREC(FM) Brian Head UT
KPLD(FM) Kanab UT
KQMB(FM) Levan UT
KBLQ-FM Logan UT
KMXD(FM) Monroe UT
KUDE(FM) Nephi UT
*KPCW(FM) Park City UT
KTCE(FM) Payson UT
KWSA(FM) Price UT
KSRR(AM) Provo UT
KIFX(FM) Roosevelt UT
KXRQ(FM) Roosevelt UT
KUDD(FM) Roy UT
KSNN(FM) Saint George UT
KBEE(FM) Salt Lake City UT
KBEE(FM) Salt Lake City UT
KSFI(FM) Salt Lake City UT
KOSY-FM Spanish Fork UT
KYMV(FM) Woodruff UT
WAVA-FM Arlington VA
WWZW(FM) Buena Vista VA
WCJZ(FM) Charlottesville VA
WQMZ(FM) Charlottesville VA
WNLR(AM) Churchville VA
WXCF(AM) Clifton Forge VA
WXCF-FM Clifton Forge VA
WGCK(FM) Coeburn VA
*WPIN-FM Dublin VA
WEVA(AM) Emporia VA
WFLO-FM Farmville VA

WBQB(FM) Fredericksburg VA
WXGM(AM) Gloucester VA
WXGM-FM Gloucester VA
WWDE-FM Hampton VA
WKWI(FM) Kilmarnock VA
WOLD-FM Marion VA
WZVA(FM) Marion VA
WMVA(AM) Martinsville VA
WNVA(AM) Norton VA
WESR-FM Onley-Onancock VA
WSWV(AM) Pennington Gap VA
WSWV-FM Pennington Gap VA
WRIC-FM Richlands VA
WMXB(FM) Richmond VA
WTVR-FM Richmond VA
WSLQ(FM) Roanoke VA
WZBL(FM) Roanoke VA
*WPAR(FM) Salem VA
WSNV(FM) Salem VA
WHLF(FM) South Boston VA
WVBW(FM) Suffolk VA
WRAR-FM Tappahannock VA
WPTE(FM) Virginia Beach VA
WINC-FM Winchester VA
*WIUJ(FM) Charlotte Amalie VI
WVIQ(FM) Christiansted VI
WYAC-FM Christiansted VI
WTSA-FM Brattleboro VT
WEZF(FM) Burlington VT
WMOO(FM) Derby Center VT
WGMT(FM) Lyndon VT
WVNR(AM) Poultney VT
WJJR(FM) Rutland VT
WZRT(FM) Rutland VT
KWLE(AM) Anacortes WA
KQMV(FM) Bellevue WA
KAFE(FM) Bellingham WA
KBAI(AM) Bellingham WA
KRWM(FM) Bremerton WA
KOZI(AM) Chelan WA
KOZI-FM Chelan WA
KCRK-FM Colville WA
KCMS(FM) Edmonds WA
KLOG(AM) Kelso WA
KLYK(FM) Kelso WA
KONA(AM) Kennewick WA
KONA-FM Kennewick WA
KTSL(FM) Medical Lake WA
KIXI(AM) Mercer Island-Seattle WA
KSWW(FM) Montesano WA
KDRM(FM) Moses Lake WA
KWDB(AM) Oak Harbor WA
KGY(AM) Olympia WA
KXXO(FM) Olympia WA
KEYW(FM) Pasco WA
*KRLF(FM) Pullman WA
KAAP(FM) Rock Island WA
KPLZ(FM) Seattle WA
KUBE(FM) Seattle WA
KWJZ(FM) Seattle WA
KMAS(AM) Shelton WA
*KAGU(FM) Spokane WA
KEZE(FM) Spokane WA
KISC(FM) Spokane WA
KXLY-FM Spokane WA
KZZU-FM Spokane WA
KITI-FM Winlock WA
KJOX(AM) Yakima WA
KRSE(FM) Yakima WA
WISM-FM Altoona WI
WACD(FM) Antigo WI
WATK(AM) Antigo WI
WLMX-FM Balsam Lake WI
WRPQ(AM) Baraboo WI
WBEV(AM) Beaver Dam WI
WWIS-FM Black River Falls WI
WCFW(FM) Chippewa Falls WI
WLKN(FM) Cleveland WI
WIAL(FM) Eau Claire WI
WFON(FM) Fond du Lac WI
WSJY(FM) Fort Atkinson WI
WQLH(FM) Green Bay WI
WHSM-FM Hayward WI
WRLS-FM Hayward WI
WLFN(AM) La Crosse WI
WLXR-FM La Crosse WI

WLKG(FM) Lake Geneva WI
*WJTY(FM) Lancaster WI
WFDL-FM Lomira WI
WMGN(FM) Madison WI
WZEE(FM) Madison WI
WOMT(AM) Manitowoc WI
WLST(FM) Marinette WI
WRJC-FM Mauston WI
WJMT(AM) Merrill WI
*WGNV(FM) Milladore WI
WKTI(FM) Milwaukee WI
WMYX(FM) Milwaukee WI
WMQA-FM Minocqua WI
WEKZ-FM Monroe WI
WOFM(FM) Mosinee WI
WNAM(AM) Neenah-Menasha WI
WROE(FM) Neenah-Menasha WI
WPKG(FM) Neillsville WI
WXER(FM) Plymouth WI
WPDR(AM) Portage WI
WBDL(FM) Reedsburg WI
WJMC(AM) Rice Lake WI
WEVR(AM) River Falls WI
WEVR-FM River Falls WI
WIZD(FM) Rudolph WI
WECB(FM) Seymour WI
WOWN(FM) Shawano WI
*WSHS(FM) Sheboygan WI
WSPT-FM Stevens Point WI
WDOR(AM) Sturgeon Bay WI
WSRG(FM) Sturgeon Bay WI
WXYM(FM) Tomah WI
WDUX-FM Waupaca WI
WNNO-FM Wisconsin Dells WI
WHAJ(FM) Bluefield WV
*WPIB(FM) Bluefield WV
WDCI(FM) Bridgeport WV
WBTQ(FM) Buckhannon WV
WVAF(FM) Charleston WV
*WKJL(FM) Clarksburg WV
WOBG(AM) Clarksburg WV
WWLW(FM) Clarksburg WV
WZJO(FM) Dunbar WV
WDNE(AM) Elkins WV
WELK(FM) Elkins WV
WTBZ(AM) Grafton WV
WMGA(FM) Kenova WV
WVOW(AM) Logan WV
WVOW-FM Logan WV
WLTF(FM) Martinsburg WV
WYMJ(FM) New Martinsville WV
WXIL(FM) Parkersburg WV
*WQAB(FM) Philippi WV
WDYK(FM) Ridgeley WV
WRRR-FM Saint Marys WV
WMXE(FM) South Charleston WV
WCWV(FM) Summersville WV
WSGB(AM) Sutton WV
WELC(AM) Welch WV
WELC-FM Welch WV
WBTH(AM) Williamson WV
KHOC(FM) Casper WY
KMGW(FM) Casper WY
KJUA(AM) Cheyenne WY
KQMY(FM) Cheyenne WY
KTAG(FM) Cody WY
KNYN(FM) Fort Bridger WY
KAOX(FM) Kemmerer WY
KIMX(FM) Laramie WY
KIQZ(FM) Rawlins WY
KRAL(AM) Rawlins WY
KTRZ(FM) Riverton WY
KTHE(AM) Thermopolis WY
KZEW(FM) Wheatland WY
KKLX(FM) Worland WY

Agriculture

KSIR(AM) Brush CO
KNAB(AM) Burlington CO
KNAB-FM Burlington CO
KSPK(FM) Walsenburg CO
KDSN(AM) Denison IA
WSMI(AM) Litchfield IL
WLBH(AM) Mattoon IL
WMCL(AM) McLeansboro IL

WSLM(AM) Salem IN
WSLM-FM Salem IN
KXXX(AM) Colby KS
KLOE(AM) Goodland KS
KNDY(AM) Marysville KS
KFRM(AM) Salina KS
KSUM(AM) Fairmont MN
KDHL(AM) Faribault MN
WYRQ(FM) Little Falls MN
KMHL(AM) Marshall MN
KOLV(FM) Olivia MN
KLOH(AM) Pipestone MN
KCUE(AM) Red Wing MN
KKOZ(AM) Ava MO
KKOZ-FM Ava MO
KAOL(AM) Carrollton MO
KMZU(FM) Carrollton MO
KGLE(AM) Glendive MT
KMON(AM) Great Falls MT
WDAY(AM) Fargo ND
KNOX(AM) Grand Forks ND
KZZJ(AM) Rugby ND
KCSR(AM) Chadron NE
KJSK(AM) Columbus NE
KRVN(AM) Lexington NE
KNEB-FM Scottsbluff NE
KTIC(AM) West Point NE
WRFD(AM) Columbus-Worthington OH
KWHW(AM) Altus OK
KOKK(AM) Huron SD
KBJM(AM) Lemmon SD
KGFX(AM) Pierre SD
KXRB(AM) Sioux Falls SD
KBHB(AM) Sturgis SD
KWAT(AM) Watertown SD
WNAX(AM) Yankton SD
KRUN(AM) Ballinger TX
KVWG-FM Dilley TX
KFLP(AM) Floydada TX
KVWG(AM) Pearsall TX
KVWC(AM) Vernon TX
KVWC-FM Vernon TX
KBSN(AM) Moses Lake WA
KWNC(AM) Quincy WA
WRDB(AM) Reedsburg WI
WJMC(FM) Rice Lake WI
WCUB(AM) Two Rivers WI
WELD-FM Petersburg WV

Album-Oriented Rock

KDJE(FM) Jacksonville AR
KWBF-FM North Little Rock AR
KERX(FM) Paris AR
KRAB(FM) Green Acres CA
KHDR(FM) Lenwood CA
KHWZ(FM) Ludlow CA
KTYD(FM) Santa Barbara CA
KMBY-FM Seaside CA
KBPI(FM) Denver CO
WYYX(FM) Bonifay FL
WFBX(FM) Parker FL
WTKX-FM Pensacola FL
WNDD(FM) Silver Springs FL
WVRK(FM) Columbus GA
WPEZ(FM) Jeffersonville GA
KGUM-FM Dededo GU
KFMW(FM) Waterloo IA
KRVB(FM) Nampa ID
WXRX(FM) Belvidere IL
*WARG(FM) Summit IL
WLRX(FM) Charlestown IN
*WMHD-FM Terre Haute IN
KACY(FM) Arkansas City KS
KQRC-FM Leavenworth KS
KDVV(FM) Topeka KS
WZZP(FM) Hopkinsville KY
WLRS(FM) Shepherdsville KY
WMVY(FM) Tisbury MA
WAAF(FM) Westborough MA
WIYY(FM) Baltimore MD
WDLD(FM) Halfway MD
WBLM(FM) Portland ME
WIMK(FM) Iron Mountain MI

WJXQ(FM) Jackson MI
WUPK(FM) Marquette MI
WRKR(FM) Portage MI
KQDS-FM Duluth MN
KQRA(FM) Brookline MO
KCMQ(FM) Columbia MO
KSHE(FM) Crestwood MO
KXOQ(FM) Kennett MO
KZRQ-FM Mount Vernon MO
WSMS(FM) Artesia MS
WFXO(FM) Iuka MS
WXQR(FM) Jacksonville NC
WSFL-FM New Bern NC
KJKJ(FM) Grand Forks ND
WFRD(FM) Hanover NH
WHOM(FM) Mt. Washington NH
*WMNJ(FM) Madison NJ
KZRR(FM) Albuquerque NM
KEND(FM) Roswell NM
*WDWN(FM) Auburn NY
*WGCC-FM Batavia NY
*WBSU(FM) Brockport NY
*WCWP(FM) Brookville NY
WEDG(FM) Buffalo NY
WFXF(FM) Honeoye Falls NY
*WNYU-FM New York NY
WQBK-FM Rensselaer NY
WCMF-FM Rochester NY
*WARY(FM) Valhalla NY
WRQK(FM) Canton OH
WEBN(FM) Cincinnati OH
WDOK(FM) Cleveland OH
WTUE(FM) Dayton OH
WBYR(FM) Van Wert OH
KATT-FM Oklahoma City OK
KHBZ-FM Oklahoma City OK
*KRVM-FM Eugene OR
KLRR(FM) Redmond OR
WLLF(FM) Mercer PA
WMMR(FM) Philadelphia PA
*WQSU(FM) Selinsgrove PA
WQSD(FM) Briarcliff Acres SC
WFXH-FM Hilton Head Island SC
WMFX(FM) Saint Andrews SC
KRRO(FM) Sioux Falls SD
WLQK(FM) Livingston TN
KQQK(FM) Beaumont TX
KFMX-FM Lubbock TX
KISS-FM San Antonio TX
KBZS(FM) Wichita Falls TX
*KWCR-FM Ogden UT
KHTB(FM) Provo UT
WWWV(FM) Charlottesville VA
WXMM(FM) Norfolk VA

Alternative

*KRUA(FM) Anchorage AK
*KXLL(FM) Juneau AK
*WEGL(FM) Auburn AL
WZEW(FM) Fairhope AL
*WVUA-FM Tuscaloosa AL
*KCAC(FM) Camden AR
*KXUA(FM) Fayetteville AR
KXNA(FM) Springdale AR
*KZAI(FM) Coolidge AZ
KEDJ(FM) Gilbert AZ
KFMA(FM) Green Valley AZ
KOHT(FM) Marana AZ
KQAZ(FM) Springerville-Eagar AZ
*KAIC(FM) Tucson AZ
KRDX(FM) Vail AZ
*KSPC(FM) Claremont CA
*KKUP(FM) Cupertino CA
*KFSR(FM) Fresno CA
KSLG-FM Hydesville CA
KDLE(FM) Newport Beach CA
KROQ-FM Pasadena CA
KMRJ(FM) Rancho Mirage CA
*KUCR(FM) Riverside CA
KWOD(FM) Sacramento CA
KBZT(FM) San Diego CA
*KUSF(FM) San Francisco CA
*KSCU(FM) Santa Clara CA
KDLD(FM) Santa Monica CA
*KAIB(FM) Shafter CA

KCNL(FM) Sunnyvale CA
KKZQ(FM) Tehachapi CA
KFRR(FM) Woodlake CA
*KLRD(FM) Yucaipa CA
*KMSA(FM) Grand Junction CO
KTCL(FM) Wheat Ridge CO
*WXCI(FM) Danbury CT
*WQAQ(FM) Hamden CT
WROO(FM) Callahan FL
WJRR(FM) Cocoa Beach FL
*WVUM(FM) Coral Gables FL
WOCL(FM) De Land FL
WSUN-FM Holiday FL
WPBZ(FM) Indiantown FL
WIIS(FM) Key West FL
WXSR(FM) Quincy FL
*WKPX(FM) Sunrise FL
*WUOG(FM) Athens GA
WNNX(FM) Atlanta GA
*WRFG(FM) Atlanta GA
WAEG(FM) Evans GA
WPCH(FM) Gray GA
*WGUR(FM) Milledgeville GA
*WVGS(FM) Statesboro GA
*WPLH(FM) Tifton GA
*WVDA(FM) Valdosta GA
*WVVS(FM) Valdosta GA
KUCD(FM) Pearl City HI
*KHAI(FM) Wahiawa HI
KORL-FM Waianae HI
KBVU-FM Alta IA
*KICB(FM) Fort Dodge IA
*KSTM(FM) Indianola IA
*KIGC(FM) Oskaloosa IA
*KMSC(FM) Sioux City IA
*KWDM(FM) West Des Moines IA
KQXR(FM) Payette ID
KPPC(FM) Pocatello ID
KCDA(FM) Post Falls ID
KSKI-FM Sun Valley ID
WWCT(FM) Bartonville IL
*WPCD(FM) Champaign IL
WKQX(FM) Chicago IL
WXRT-FM Chicago IL
*WRSE(FM) Elmhurst IL
*WGBK(FM) Glenview IL
*WLCA(FM) Godfrey IL
*WIUS(FM) Macomb IL
*WRRG(FM) River Grove IL
*WARG(FM) Summit IL
*WHJE(FM) Carmel IN
*WJHS(FM) Columbia City IN
*WGRE(FM) Greencastle IN
*WCYT(FM) Lafayette Township IN
*WLAI(FM) Danville KY
*WFPK(FM) Louisville KY
KNXX(FM) Donaldsonville LA
*KSLU(FM) Hammond LA
WNXX(FM) Jackson LA
*KXUL(FM) Monroe LA
*KLPI-FM Ruston LA
*KSCL(FM) Shreveport LA
*KNSU(FM) Thibodaux LA
WBCN(FM) Boston MA
WBOS(FM) Brookline MA
*WDJM-FM Framingham MA
WFNX(FM) Lynn MA
*WUMD(FM) North Dartmouth MA
*WKKL(FM) West Barnstable MA
*WSKB(FM) Westfield MA
*WMTB-FM Emmittsburg MD
*WFWM(FM) Frostburg MD
WRXS(FM) Ocean City MD
*WTMD(FM) Towson MD
*WHSN(FM) Bangor ME
WCLZ(FM) Brunswick ME
WPHX-FM Sanford ME
XETRA-FM Tijuana MEX
*WQAC-FM Alma MI
WQKL(FM) Ann Arbor MI
WOUF(FM) Beulah MI
*WHFR(FM) Dearborn MI
*WDET-FM Detroit MI
*WDBM(FM) East Lansing MI
WGRD-FM Grand Rapids MI
*WTHS(FM) Holland MI

*WUPX(FM) Marquette MI
*WOVI(FM) Novi MI
*WPHS(FM) Warren MI
*WYCE(FM) Wyoming MI
*KUMM(FM) Morris MN
*KUOM-FM Saint Louis Park MN
*KSRQ(FM) Thief River Falls MN
*KTRM(FM) Kirksville MO
*KMVC(FM) Marshall MO
KZZK(FM) New London MO
*KGSP(FM) Parkville MO
KNSX(FM) Steelville MO
WHOC(AM) Philadelphia MS
*WMSV(FM) Starkville MS
WUMS(FM) University MS
*KGLT(FM) Bozeman MT
*KMSM-FM Butte MT
KBAZ(FM) Hamilton MT
*KBGA(FM) Missoula MT
*WASU-FM Boone NC
*WZMB(FM) Greenville NC
WSFM(FM) Oak Island NC
*WKNC-FM Raleigh NC
WEND(FM) Salisbury NC
*WDCC(FM) Sanford NC
*WZRL(FM) Wade NC
*KFJM(FM) Grand Forks ND
*KRNU(FM) Lincoln NE
*KWSC(FM) Wayne NE
WFEX(FM) Peterborough NH
WPNH-FM Plymouth NH
WBBO(FM) Bass River Township NJ
*WDBK(FM) Blackwood NJ
WHTG-FM Eatontown NJ
*WMNJ(FM) Madison NJ
WKMK(FM) Ocean Acres NJ
*WLFR(FM) Pomona NJ
*WTSR(FM) Trenton NJ
*WMSC(FM) Upper Montclair NJ
*WPSC-FM Wayne NJ
KTEG(FM) Bosque Farms NM
*KQAI(FM) Roswell NM
KXTE(FM) Pahrump NV
KRZQ-FM Sparks NV
WHRL(FM) Albany NY
WXPK(FM) Briarcliff Manor NY
WZNE(FM) Brighton NY
*WBSU(FM) Brockport NY
*WBNY(FM) Buffalo NY
WEDG(FM) Buffalo NY
*WITC(FM) Cazenovia NY
*WGSU(FM) Geneseo NY
WLIR-FM Hampton Bays NY
*WNYK(FM) Nyack NY
*WDFH(FM) Ossining NY
WBTZ(FM) Plattsburgh NY
*WQKE(FM) Plattsburgh NY
*WTSC-FM Potsdam NY
*WBER(FM) Rochester NY
*WIRQ(FM) Rochester NY
*WSIA(FM) Staten Island NY
WXEG(FM) Beavercreek OH
*WCSB(FM) Cleveland OH
WKRI(FM) Cleveland Heights OH
WRWK(FM) Delta OH
*WHSS(FM) Hamilton OH
WOXY(FM) Oxford OH
*WXUT(FM) Toledo OH
*WOBN(FM) Westerville OH
*KRSC-FM Claremore OK
KMYZ-FM Pryor OK
KOPA(FM) Woodward OK
*KSBA(FM) Coos Bay OR
*KBVR(FM) Corvallis OR
KNRQ-FM Eugene OR
KROG(FM) Grants Pass OR
*KSLC(FM) McMinnville OR
*KWBX(FM) Salem OR
*KVRA(FM) Sisters OR
*WBUQ(FM) Bloomsburg PA
*WESS(FM) East Stroudsburg PA
*WFSE(FM) Edinboro PA
*WWEC(FM) Elizabethtown PA
*WXPH(FM) Harrisburg PA
*WARC(FM) Meadville PA
*WXPN(FM) Philadelphia PA

WJOW(FM) Philipsburg PA
WXDX-FM Pittsburgh PA
WDMT(FM) Pittston PA
*WUSR(FM) Scranton PA
*WVMW-FM Scranton PA
*WSRU(FM) Slippery Rock PA
WCHE(AM) West Chester PA
*WCLH(FM) Wilkes-Barre PA
WBRU(FM) Providence RI
*WJMF(FM) Smithfield RI
WAVF(FM) Hanahan SC
*KAUR(FM) Sioux Falls SD
*KBHU-FM Spearfish SD
WMFS(FM) Bartlett TN
*WTTU(FM) Cookeville TN
WTZR(FM) Elizabethton TN
WNFZ(FM) Oak Ridge TN
*WAWL-FM Red Bank TN
KEYJ-FM Abilene TX
*KACV-FM Amarillo TX
KTLT(FM) Anson TX
*KVRX(FM) Austin TX
KGSR(FM) Bastrop TX
KROX-FM Buda TX
KJXJ(FM) Cameron TX
KDGE(FM) Fort Worth-Dallas TX
*KTSW(FM) San Marcos TX
KESO(FM) South Padre Island TX
KJMY(FM) Bountiful UT
KJQN(FM) Coalville UT
KAUU(FM) Manti UT
*KOHS(FM) Orem UT
KXRK(FM) Provo UT
*KRDC-FM Saint George UT
*WNRN(FM) Charlottesville VA
*WHRV(FM) Norfolk VA
WYSK-FM Spotsylvania VA
*WCWM(FM) Williamsburg VA
WZIN(FM) Charlotte Amalie VI
*WJSC-FM Johnson VT
WEQX(FM) Manchester VT
*WVTC(FM) Randolph Center VT
WRJT(FM) Royalton VT
*KASB(FM) Bellevue WA
KFNK(FM) Eatonville WA
*KCWU(FM) Ellensburg WA
KGRG(AM) Enumclaw WA
*KTCV(FM) Kennewick WA
*KEXP-FM Seattle WA
KNDD(FM) Seattle WA
*KYVT(FM) Yakima WA
WZNN(FM) Allouez WI
*WBSD(FM) Burlington WI
WMAD(FM) Sauk City WI
*WSUW(FM) Whitewater WI
WQZK-FM Keyser WV
*WSHC(FM) Shepherdstown WV
*WGLZ(FM) West Liberty WV
KKPL(FM) Cheyenne WY

American Indian

KNDN(AM) Farmington NM
KHAC(AM) Tse Bonito NM
*KWSO(FM) Warm Springs OR
*KILI(FM) Porcupine SD
*KWRR(FM) Ethete WY

Arabic

KXMX(AM) Anaheim CA

Beautiful Music

*KJNP-FM North Pole AK
WJRD(AM) Tuscaloosa AL
KFLG(AM) Bullhead City AZ
KBUX(FM) Quartzsite AZ
KWXY-FM Cathedral City CA
KLXR(AM) Redding CA
*WJMJ(FM) Hartford CT
WVOI(AM) Marco Island FL
*WNEE(FM) Jasper GA
WNNG(AM) Warner Robins GA
KWLO(AM) Waterloo IA
KQWC(AM) Webster City IA
KXLT-FM Eagle ID
WJIL(AM) Jacksonville IL
KVSV-FM Beloit KS
WMST(AM) Mt. Sterling KY
WNBH(AM) New Bedford MA
*WYAR(FM) Yarmouth ME
WMUZ(FM) Detroit MI
WMPX(AM) Midland MI
KLKS(FM) Breezy Point MN
WBAQ(FM) Greenville MS
WROA(AM) Gulfport MS
WELO(AM) Tupelo MS
WNMX-FM Waxhaw NC
*KMTH(FM) Maljamar NM
*KENW-FM Portales NM
*WHPC(FM) Garden City NY
WTLB(AM) Utica NY
WSRW(AM) Hillsboro OH
WSRW(AM) Hillsboro OH
*WFCO(FM) Lancaster OH
*KSSX(FM) Chickasha OK
KCST(AM) Florence OR
WLNP(FM) Carbondale PA
*WPGM(AM) Danville PA
*WPGM-FM Danville PA
*WPEL-FM Montrose PA
WNAK(AM) Nanticoke PA
WNIK-FM Arecibo PR
WVJP(AM) Caguas PR
WVJP-FM Caguas PR
WORO(FM) Corozal PR
WKSA-FM Isabela PR
WIAC-FM San Juan PR
WIOA(FM) San Juan PR
WMUU-FM Greenville SC
KXYL(AM) Brownwood TX
KNNK(FM) Dimmitt TX
*KNCT-FM Killeen TX
KBBT(FM) Schertz TX
KPYK(AM) Terrell TX
*KTXK(FM) Texarkana TX
*WIUJ(FM) Charlotte Amalie VI
WDHP(AM) Frederiksted VI
WBTN(AM) Bennington VT
WSTJ(AM) Saint Johnsbury VT
*KBCS(FM) Bellevue WA
WHSM(AM) Hayward WI

Big Band

KGOT(FM) Anchorage AK
*KBRW-FM Barrow AK
WAUD(AM) Auburn AL
KFPW(AM) Fort Smith AR
KFLG(AM) Bullhead City AZ
*KCEA(FM) Atherton CA
KTEA(FM) Cambria CA
KIDD(AM) Monterey CA
KEZW(AM) Aurora CO
KNAB(AM) Burlington CO
WMMB(AM) Melbourne FL
WKII(AM) Solana FL
KCLN(AM) Clinton IA
KWLO(AM) Waterloo IA
KQWC(AM) Webster City IA
WAIK(AM) Galesburg IL
WPMB(AM) Vandalia IL
WGFA(AM) Watseka IL
WFRX(AM) West Frankfort IL
KSGL(AM) Wichita KS
*WICN(FM) Worcester MA
*WYAR(FM) Yarmouth ME
WMRX-FM Beaverton MI
WCBY(AM) Cheboygan MI
WMPX(AM) Midland MI
KLKS(FM) Breezy Point MN
KOMC-FM Kimberling City MO
KRLI(FM) Malta Bend MO
KBFL(AM) Springfield MO
WELO(AM) Tupelo MS
WPAQ(AM) Mount Airy NC
WNOS(AM) New Bern NC
WPNH(AM) Plymouth NH
WOBM(AM) Lakewood NJ
*KNCC(FM) Elko NV
KHIT(AM) Reno NV
WPTR(FM) Clifton Park NY
WALK(AM) East Patchogue NY
WHUC(AM) Hudson NY
WABY(AM) Mechanicville NY
WSGO(AM) Oswego NY
WKIP(AM) Poughkeepsie NY
WUAM(AM) Saratoga Springs NY
*WKHR(FM) Bainbridge OH
WJAS(AM) Pittsburgh PA
WKDA(AM) Lebanon TN
*WOEZ(FM) Maynardville TN
*KSQX(FM) Springtown TX
*KQXS(FM) Stephenville TX
KPYK(AM) Terrell TX
*WFOS(FM) Chesapeake VA
WSVG(AM) Mount Jackson VA
*WIUJ(FM) Charlotte Amalie VI
WSTJ(AM) Saint Johnsbury VT
KWLE(AM) Anacortes WA
KWDB(AM) Oak Harbor WA
KTRW(AM) Opportunity WA
WCCN(AM) Neillsville WI
WBBD(AM) Wheeling WV

Black

WATV(AM) Birmingham AL
WXAL(AM) Demopolis AL
WMGJ(AM) Gadsden AL
WLOR(AM) Huntsville AL
WMFC(AM) Monroeville AL
WAPZ(AM) Wetumpka AL
*KABF(FM) Little Rock AR
KAKJ(FM) Marianna AR
KTYM(AM) Inglewood CA
*KSRH(FM) San Rafael CA
WZAZ(AM) Jacksonville FL
WWAB(AM) Lakeland FL
WVTJ(AM) Pensacola FL
WYZE(AM) Atlanta GA
WFXA-FM Augusta GA
WWLD(FM) Cairo GA
WOKS(AM) Columbus GA
WBHB(AM) Fitzgerald GA
WJGA-FM Jackson GA
WBBT(AM) Lyons GA
WLCG(AM) Macon GA
*KIGC(FM) Oskaloosa IA
*WIIT(FM) Chicago IL
WVAZ(FM) Oak Park IL
KRUS(AM) Ruston LA
KTKC(FM) Springhill LA
KBYO(AM) Tallulah LA
WBGR(AM) Baltimore MD
WFBR(AM) Glen Burnie MD
WPRS-FM Waldorf MD
WEFG-FM Whitehall MI
WCHJ(AM) Brookhaven MS
WCLD(AM) Cleveland MS
WCPC(AM) Houston MS
WXTN(AM) Lexington MS
WMIS(AM) Natchez MS
WFMO(AM) Fairmont NC
WIDU(AM) Fayetteville NC
WFMC(AM) Goldsboro NC
*WNAA(FM) Greensboro NC
WHNC(AM) Henderson NC
WEGG(AM) Rose Hill NC
WWIL(AM) Wilmington NC
*WNEC-FM Henniker NH
*KCEP(FM) Las Vegas NV
WSIV(AM) East Syracuse NY
WTHE(AM) Mineola NY
WBLS(FM) New York NY
*WHCR-FM New York NY
WLIB(AM) New York NY
*WBGU(FM) Bowling Green OH
WGFT(AM) Campbell OH
WCIN(AM) Cincinnati OH
WASN(AM) Youngstown OH
*WIXQ(FM) Millersville PA
WNAP(AM) Norristown PA
WDAS-FM Philadelphia PA
WDOG(AM) Allendale SC
WDOG-FM Allendale SC
WPJS(AM) Conway SC
WWRK(AM) Darlington SC
WBZF(FM) Hartsville SC
*WLGI(FM) Hemingway SC
WASC(AM) Spartanburg SC
WMRB(AM) Columbia TN
*WVCP(FM) Gallatin TN
WFKX(FM) Henderson TN
WDIA(AM) Memphis TN
*WMTS-FM Murfreesboro TN
KGGR(AM) Dallas TX
KCOH(AM) Houston TX
WKBY(AM) Chatham VA
WILA(AM) Danville VA
WSHV(AM) South Hill VA

Bluegrass

*WPIL(FM) Heflin AL
WRMG(AM) Red Bay AL
KCGS(AM) Marshall AR
*WAMU(FM) Washington DC
WALH(AM) Mountain City GA
WYHG(AM) Young Harris GA
*WAAJ(FM) Benton KY
*WMMT(FM) Whitesburg KY
WTWZ(AM) Clinton MS
WPYB(AM) Benson NC
WKTE(AM) King NC
WKGX(AM) Lenoir NC
WDSL(AM) Mocksville NC
WPAQ(AM) Mount Airy NC
WMPM(AM) Smithfield NC
WADV(AM) Lebanon PA
WJDJ(AM) Hartsville SC
*WDVX(FM) Clinton TN
WLSB(AM) Copperhill TN
WSDQ(AM) Dunlap TN
*WLMU(FM) Harrogate TN
*WPLN-FM Nashville TN
WUAT(AM) Pikeville TN
WLIJ(AM) Shelbyville TN
*WTML(FM) Tullahoma TN
*KAMU-FM College Station TX
WGFC(AM) Floyd VA
WLRV(AM) Lebanon VA
WYTI(AM) Rocky Mount VA
WXMY(AM) Saltville VA
KOHO-FM Leavenworth WA
WTCR(AM) Kenova WV
WDBS(FM) Sutton WV

Blues

WQZZ(FM) Eutaw AL
WJLD(AM) Fairfield AL
WZEW(FM) Fairhope AL
WKXN(FM) Greenville AL
WDBT(FM) Headland AL
*WJAB(FM) Huntsville AL
WKXK(FM) Pine Hill AL
WAPZ(AM) Wetumpka AL
KOHT(FM) Marana AZ
KAJM(FM) Payson AZ
KRML(AM) Carmel CA
KJLH-FM Compton CA
*KKUP(FM) Cupertino CA
*KSDS(FM) San Diego CA
KQKS(FM) Lakewood CO
WTMP-FM Dade City FL
WKIQ(AM) Eustis FL
WFLM(FM) White City FL
WVKX(FM) Irwinton GA
WLZN(FM) Macon GA
WNUQ(FM) Sylvester GA
KUAM-FM Hagatna GU
KATZ-FM Alton IL
*WSSD(FM) Chicago IL
WFFX(AM) East St. Louis IL
*WDCB(FM) Glen Ellyn IL
*WGLT(FM) Normal IL
WTLC-FM Greenwood IN
WHHH(FM) Indianapolis IN
WTLO(AM) Somerset KY
KBRH(AM) Baton Rouge LA
KMEZ(FM) Belle Chasse LA
KJMH(FM) Lake Arthur LA
KRUS(AM) Ruston LA
WKAF(FM) Brockton MA
WATD-FM Marshfield MA
*WESM(FM) Princess Anne MD
WRED(FM) Saco ME
*WUCX-FM Bay City MI
WDMK(FM) Detroit MI
WJNZ(AM) Kentwood MI
*WLNZ(FM) Lansing MI
*WNMC-FM Traverse City MI
*WEMU(FM) Ypsilanti MI
*KMVC(FM) Marshall MO
*KCOZ(FM) Point Lookout MO
WBSL(AM) Bay St. Louis MS
WONG(AM) Canton MS
WROX(AM) Clarksdale MS
WTYJ(FM) Fayette MS
*WVSD(FM) Itta Bena MS
WKXI(AM) Jackson MS
*WMPR(FM) Jackson MS
WEEZ(AM) Laurel MS
WESY(AM) Leland MS
WNBN(AM) Meridian MS
WZYQ(FM) Mound Bayou MS
WMIS(AM) Natchez MS
WABO(AM) Waynesboro MS
WQMG-FM Greensboro NC
WQNC(FM) Harrisburg NC
WCPS(AM) Tarboro NC
WENC(AM) Whiteville NC
*KCEP(FM) Las Vegas NV
WRKS(FM) New York NY
WJZR(FM) Rochester NY
WDAO(AM) Dayton OH
WXMG(FM) Upper Arlington OH
*KRVM-FM Eugene OR
*KMHD(FM) Gresham OR
WKSP(FM) Aiken SC
WYNN(AM) Florence SC
WVOL(AM) Berry Hill TN
WBOL(AM) Bolivar TN
*WUTC(FM) Chattanooga TN
*WEVL(FM) Memphis TN
WEUZ(FM) Minor Hill TN
*KAZI-FM Austin TX
KBFB(FM) Dallas TX
KTXN-FM Victoria TX
*WFOS(FM) Chesapeake VA
*WMRY(FM) Crozet VA
WVKL(FM) Norfolk VA
*KPLI(FM) Olympia WA
*KVIX(FM) Port Angeles WA
KRIZ(AM) Renton WA
*KPLU-FM Tacoma WA
WKKV-FM Racine WI

Children

KDIS-FM Little Rock AR
KMIK(AM) Tempe AZ
KAVT(AM) Fresno CA
KMKY(AM) Oakland CA
KDIS(AM) Pasadena CA
KIID(AM) Sacramento CA
KKDD(AM) San Bernardino CA
KDDZ(AM) Arvada CO
WDZK(AM) Bloomfield CT
WAJD(AM) Gainesville FL
WBWL(AM) Jacksonville FL
WMYM(AM) Miami FL
WDYZ(AM) Orlando FL
WDWD(AM) Atlanta GA
WNEX(AM) Macon GA
WPGA(AM) Perry GA
WRDZ(AM) La Grange IL
KQAM(AM) Wichita KS
WDRD(AM) Newburg KY
WBYU(AM) New Orleans LA
*WMDR(AM) Augusta ME
WDSS(AM) Ada MI
KDIZ(AM) Golden Valley MN
KPHN(AM) Kansas City MO
WGFY(AM) Charlotte NC
WCOG(AM) Greensboro NC

KOIL(AM) Plattsmouth NE
KALY(AM) Los Ranchos de Albuquerque NM
WWLF(AM) Auburn NY
WQEW(AM) New York NY
WOLF(AM) Syracuse NY
WWMK(AM) Cleveland OH
KOCY(AM) Del City OK
KMUS(AM) Sperry OK
WWCS(AM) Canonsburg PA
*WWCF(FM) McConnellsburg PA
WDDZ(AM) Pawtucket RI
KMIC(AM) Houston TX
KRDY(AM) San Antonio TX
KWDZ(AM) Salt Lake City UT
WDZY(AM) Colonial Heights VA
KKDZ(AM) Seattle WA
WGEE(AM) Superior WI
*WMLJ(FM) Summersville WV

Chinese

KAZN(AM) Pasadena CA
KAHZ(AM) Pomona CA
KSON(AM) San Diego CA
WKDM(AM) New York NY
WZRC(AM) New York NY

Christian

KAFC(FM) Anchorage AK
*KAKL(FM) Anchorage AK
KYKD(FM) Bethel AK
KAGV(AM) Big Lake AK
*KHZK(FM) Kotzebue AK
KAKN(FM) Naknek AK
KIAM(AM) Nenana AK
KICY-FM Nome AK
*KJLP(FM) Palmer AK
WIZB(FM) Abbeville AL
WAVU(AM) Albertville AL
*WGRW(FM) Anniston AL
WASG(AM) Atmore AL
WDJC-FM Birmingham AL
*WGIB(FM) Birmingham AL
*WELJ(FM) Brewton AL
*WALN(FM) Carrollton AL
WQUA(FM) Citronelle AL
*WELL-FM Dadeville AL
WKWL(AM) Florala AL
*WTBB(FM) Gadsden AL
*WJIA(FM) Guntersville AL
*WAYH(FM) Harvest AL
WBXR(AM) Hazel Green AL
*WOCG(FM) Huntsville AL
WIXI(AM) Jasper AL
WLYJ(AM) Jasper AL
WBHY(AM) Mobile AL
*WBHY-FM Mobile AL
WXVI(AM) Montgomery AL
*WAQG(FM) Ozark AL
WIJD(AM) Prichard AL
WJBY(AM) Rainbow City AL
*WAQU(FM) Selma AL
*WAKD(FM) Sheffield AL
WBTG(AM) Sheffield AL
*WAXU(FM) Troy AL
*WMFT(FM) Tuscaloosa AL
*KJSM-FM Augusta AR
KBGR(FM) Beebe AR
*KAPG(FM) Bentonville AR
*KBCM(FM) Blytheville AR
*KBHN(FM) Booneville AR
KKSP(FM) Bryant AR
KWXT(AM) Dardanelle AR
*KBNV(FM) Fayetteville AR
KOFC(AM) Fayetteville AR
*KLMZ(FM) Fouke AR
KZKZ-FM Greenwood AR
*KBPW(FM) Hampton AR
*KBPB(FM) Harrison AR
*KBMJ(FM) Heber Springs AR
*KALR(FM) Hot Springs AR
*KLRO(FM) Hot Springs AR
KJLV(FM) Hoxie AR
KJBR(FM) Marked Tree AR
*KVRN(FM) Marvell AR
*KMTC(FM) Russellville AR
*KLRC(FM) Siloam Springs AR
*KKLT(FM) Texarkana AR
KKLV(FM) Turrell AR
*KLFS(FM) Van Buren AR
KNWJ(FM) Leone AS
*KWRB(FM) Bisbee AZ
*KLVA(FM) Casa Grande AZ
KCKY(AM) Coolidge AZ
*KZAI(FM) Coolidge AZ
*KRMC(FM) Douglas AZ
*KJTA(FM) Flagstaff AZ
*KLVK(FM) Fountain Hills AZ
*KVJC(FM) Globe AZ
*KNLB(FM) Lake Havasu City AZ
*KNOG(FM) Nogales AZ
*KFLR-FM Phoenix AZ
KXEG(AM) Phoenix AZ
*KGCB(FM) Prescott AZ
KNXN(AM) Sierra Vista AZ
KTBA(AM) Tuba City AZ
*KFLT(AM) Tucson AZ
KGMS(AM) Tucson AZ
KWIM(FM) Window Rock AZ
*KCFY(FM) Yuma AZ
*KYRM(FM) Yuma AZ
KFSH-FM Anaheim CA
*KWTH(FM) Barstow CA
*KWTW(FM) Bishop CA
KWLU(FM) Chester CA
*KPSH(FM) Coachella CA
*KDKL(FM) Coalinga CA
*KARQ(FM) East Sonora CA
*KLVY(FM) Fairmead CA
KIRV(AM) Fresno CA
KWRU(AM) Fresno CA
*KLVG(FM) Garberville CA
KKMC(AM) Gonzales CA
*KLVS(FM) Grass Valley CA
KSDT(AM) Hemet CA
*KHRI(FM) Hollister CA
*KLVJ(FM) Julian CA
*KWTM(FM) June Lake CA
*KWDS(FM) Kettleman City CA
*KDRH(FM) King City CA
*KTLW(FM) Lancaster CA
*KHKL(FM) Laytonville CA
KKFS(FM) Lincoln CA
*KLVN(FM) Livingston CA
*KRQZ(FM) Lompoc CA
*KFRN(AM) Long Beach CA
KKLA-FM Los Angeles CA
*KLVC(FM) Magalia CA
*KAMB(FM) Merced CA
KCIV(FM) Mount Bullion CA
*KKLC(FM) Mount Shasta CA
KGDP(AM) Oildale CA
*KLFH(FM) Ojai CA
KDAR(FM) Oxnard CA
*KHCS(FM) Palm Desert CA
*KLVM(FM) Prunedale CA
KNLF(FM) Quincy CA
*KGBM(FM) Randsburg CA
*KLVB(FM) Red Bluff CA
*KKRO(FM) Redding CA
*KVIP(AM) Redding CA
*KVIP-FM Redding CA
*KWTD(FM) Ridgecrest CA
*KSGN(FM) Riverside CA
KDBV(AM) Salinas CA
KWVE(FM) San Clemente CA
*KLFF-FM San Luis Obispo CA
*KLVH(FM) San Luis Obispo CA
KPRZ(AM) San Marcos-Poway CA
*KSRI(FM) Santa Cruz CA
*KGDP-FM Santa Maria CA
KSBQ(AM) Santa Maria CA
*KLVR(FM) Santa Rosa CA
*KQKL(FM) Selma CA
KYIX(FM) South Oroville CA
*KYLU(FM) Tehachapi CA
*KRTM(FM) Temecula CA
*KYKL(FM) Tracy CA
*KULV(FM) Ukiah CA
*KHMS(FM) Victorville CA
*KXRD(FM) Victorville CA
*KARM(FM) Visalia CA
*KSAK(FM) Walnut CA
*KFHL(FM) Wasco CA
KALI(AM) West Covina CA
*KARA(FM) Williams CA
KIQS(AM) Willows CA
*KLRD(FM) Yucaipa CA
KLJH(FM) Bayfield CO
*KBWA(FM) Brush CO
*KTLC(FM) Canon City CO
*KTMH(FM) Colona CO
*KTLF(FM) Colorado Springs CO
KLTT(AM) Commerce City CO
KDTA(AM) Delta CO
KLZ(AM) Denver CO
*KPOF(AM) Denver CO
*KTCF(FM) Dolores CO
*KLHV(FM) Fort Collins CO
*KLXV(FM) Glenwood Springs CO
KJOL(AM) Grand Junction CO
*KLFV(FM) Grand Junction CO
*KLRY(FM) Gypsum CO
*KFDN(FM) Lakewood CO
*KTOL(FM) Leadville CO
*KXWA(FM) Loveland CO
KBIQ(FM) Manitou Springs CO
KCBR(AM) Monument CO
*KLDV(FM) Morrison CO
*KTPS(FM) Pagosa Springs CO
*KTEI(FM) Placerville CO
*KFRY(FM) Pueblo CO
KGFT(FM) Pueblo CO
*KTPL(FM) Pueblo CO
*KRYI(FM) Rye CO
*KTPF(FM) Salida CO
*KTSG(FM) Sidney CO
*KLBV(FM) Steamboat Springs CO
*KDRE(FM) Sterling CO
*KLZV(FM) Sterling CO
*KTAD(FM) Sterling CO
*KTDU(FM) Trimble CO
*WIHS(FM) Middletown CT
*WSGG(FM) Norfolk CT
WXCT(AM) Southington CT
*WXHL-FM Christiana DE
*WKSG(FM) Cedar Creek FL
*WMYZ(FM) Clermont FL
WAKU(FM) Crawfordville FL
*WWLC(FM) Cross City FL
*WAQV(FM) Crystal River FL
*WHGN(FM) Crystal River FL
*WREH(FM) Cypress Quarters FL
*WAKJ(FM) De Funiak Springs FL
*WSEB(FM) Englewood FL
*WJLH(FM) Flagler Beach FL
*WMFL(FM) Florida City FL
*WAFG(FM) Fort Lauderdale FL
*WAYJ(FM) Fort Myers FL
WCRM(AM) Fort Myers FL
*WJYO(FM) Fort Myers FL
*WPSM(FM) Fort Walton Beach FL
*WJLF(FM) Gainesville FL
*WYJC(FM) Greenville FL
*WAPN(FM) Holly Hill FL
*WEAZ(FM) Holly Hill FL
*WCRJ(FM) Jacksonville FL
WROS(AM) Jacksonville FL
*WGES-FM Key Largo FL
*WMKL(FM) Key Largo FL
*WJIR(FM) Key West FL
*WKWR(FM) Key West FL
*WLAZ(FM) Kissimmee FL
*WBIY(FM) La Belle FL
*WOLR(FM) Lake City FL
*WLMS(FM) Lecanto FL
*WJNF(FM) Marianna FL
*WMCU(FM) Miami FL
*WEGS(FM) Milton FL
*WKVH(FM) Monticello FL
*WSOR(FM) Naples FL
*WSRX(FM) Naples FL
WGCX(FM) Navarre FL
*WLPJ(FM) New Port Richey FL
*WHIJ(FM) Ocala FL
*WAYR(AM) Orange Park FL
WTLN(AM) Orlando FL
*WHIF(FM) Palatka FL
*WEJF(FM) Palm Bay FL
*WHYZ(FM) Palm Bay FL
*WCNO(FM) Palm City FL
*WFFL(FM) Panama City FL
*WIRP(FM) Pennsuco FL
WVTJ(AM) Pensacola FL
WTBN(AM) Pinellas Park FL
WTWD(AM) Plant City FL
WOKV-FM Ponte Vedra Beach FL
*WAYL(FM) Saint Augustine FL
*WSMR(FM) Sarasota FL
*WFRF(AM) Tallahassee FL
WTAL(AM) Tallahassee FL
*WBVM(FM) Tampa FL
*WPOZ(FM) Union Park FL
*WSCF-FM Vero Beach FL
*WAYF(FM) West Palm Beach FL
*WBJY(FM) Americus GA
WFSH-FM Athens GA
*WMSL(FM) Athens GA
WAEC(AM) Atlanta GA
WNIV(AM) Atlanta GA
*WLPE(FM) Augusta GA
WGMI(AM) Bremen GA
*WAYR-FM Brunswick GA
WPMA(FM) Buckhead GA
*WPWB(FM) Byron GA
WQIL(FM) Chauncey GA
WCHM(AM) Clarkesville GA
WPBS(AM) Conyers GA
*WAEF(FM) Cordele GA
WWGF(FM) Donalsonville GA
*WAWH(FM) Dublin GA
*WECC-FM Folkston GA
WTHP(FM) Gibson GA
*WMVV(FM) Griffin GA
*WTFH(FM) Helen GA
*WNEE(FM) Jasper GA
*WLPT(FM) Jesup GA
WVFJ-FM Manchester GA
WIGO(AM) Morrow GA
WGTJ(AM) Murrayville GA
WNEA(AM) Newnan GA
WJEP(AM) Ochlocknee GA
WLPF(FM) Ocilla GA
WTIF-FM Omega GA
*WVRI(FM) Pavo GA
*WLFS(FM) Port Wentworth GA
*WLXP(FM) Savannah GA
*WYFS(FM) Savannah GA
WZIQ(FM) Smithville GA
*WHKV(FM) Sylvester GA
*WAYT(FM) Thomasville GA
*WTXR(FM) Toccoa Falls GA
WBDX(FM) Trenton GA
*WGPH(FM) Vidalia GA
*WASW(FM) Waycross GA
*KSDA-FM Agat GU
*KHMG(FM) Barrigada GU
*KCIF(FM) Hilo HI
KAIM-FM Honolulu HI
KGU(AM) Honolulu HI
*KIHS(FM) Adel IA
KSKB(FM) Brooklyn IA
*KAYP(FM) Burlington IA
*KILV(FM) Castana IA
KLNG(AM) Council Bluffs IA
*KLOX(FM) Creston IA
KPSZ(AM) Des Moines IA
*KIAD(FM) Dubuque IA
*KJYL(FM) Eagle Grove IA
*KWOF-FM Hiawatha IA
*KMDY(FM) Keokuk IA
KNWM(FM) Madrid IA
KCWN(FM) New Sharon IA
*KKLG(FM) Newton IA
*KNWI(FM) Osceola IA
KJCY(FM) Saint Ansgar IA
*KYFR(AM) Shenandoah IA
KSOU(AM) Sioux Center IA
KTFC(FM) Sioux City IA
KTFG(FM) Sioux Rapids IA
*KJIA(FM) Spirit Lake IA
*KAIP(FM) Wapello IA
*KNWS(AM) Waterloo IA
KWOF(AM) Waterloo IA
*KWVI(FM) Waverly IA
KZZQ(FM) Winterset IA
KSPD(AM) Boise ID
*KTFY(FM) Buhl ID
KBGN(AM) Caldwell ID
KBXL(FM) Caldwell ID
*KTSY(FM) Caldwell ID
KRTK(AM) Chubbuck ID
*KAIO(FM) Idaho Falls ID
*KARJ(FM) Kuna ID
*KZJB(FM) Pocatello ID
*KLRI(FM) Rigby ID
*KSQS(FM) Ririe ID
*KAWZ(FM) Twin Falls ID
*KCIR(FM) Twin Falls ID
*KEFX(FM) Twin Falls ID
*WCLR(FM) Arlington Heights IL
WRMS(AM) Beardstown IL
*WBEL(FM) Cairo IL
*WIBI(FM) Carlinville IL
*WBVN(FM) Carrier Mills IL
WCBH(FM) Casey IL
WKZI(AM) Casey IL
*WMBI-FM Chicago IL
*WBMF(FM) Crete IL
WYCA(FM) Crete IL
*WDRS(FM) Dorsey IL
WDQN-FM Du Quoin IL
*WRYT(AM) Edwardsville IL
*WEFI(FM) Effingham IL
WPIA(FM) Eureka IL
*WAXR(FM) Geneseo IL
WJKL(FM) Glendale Heights IL
WIJR(AM) Highland IL
*WAWF(FM) Kankakee IL
*WONU(FM) Kankakee IL
WLLM(AM) Lincoln IL
*WAWJ(FM) Marion IL
*WJCZ(FM) Milford IL
*WOTW(FM) Monee IL
*WCFL(FM) Morris IL
*WAPO(FM) Mount Vernon IL
*WBMV(FM) Mount Vernon IL
*WPTH(FM) Olney IL
*WWGN(FM) Ottawa IL
*WZRS(FM) Pana IL
*WCIC(FM) Pekin IL
*WLWJ(FM) Petersburg IL
*WPJC(FM) Pontiac IL
*WPRC(FM) Princeton IL
*WGCA-FM Quincy IL
*WJLY(FM) Ramsey IL
WKBF(AM) Rock Island IL
WLKU(FM) Rock Island IL
*WFEN(FM) Rockford IL
WQFL(FM) Rockford IL
*WGNJ(FM) Saint Joseph IL
*WSLE(FM) Salem IL
*WLUJ(FM) Springfield IL
*WSCT(FM) Springfield IL
*WSRI(FM) Sugar Grove IL
*WIHM(AM) Taylorville IL
*WETN(FM) Wheaton IL
*WGNR(AM) Anderson IN
*WGNR-FM Anderson IN
WQME(FM) Anderson IN
WHPZ(FM) Bremen IN
*WPFR-FM Clinton IN
WFRN(AM) Elkhart IN
WFRN-FM Elkhart IN
*WBCL(FM) Fort Wayne IN
WFCV(AM) Fort Wayne IN
*WLAB(FM) Fort Wayne IN
WLYV(AM) Fort Wayne IN
*WIKL(FM) Greencastle IN
*WAUZ(FM) Greensburg IN
*WQKO(FM) Howe IN
WBRI(AM) Indianapolis IN
*WQSG(FM) Lafayette IN
*KXJH(FM) Linton IN
*WWLO(FM) Lowell IN
WVNI(FM) Nashville IN
WFIA-FM New Albany IN
*WARA(FM) New Washington IN

WGAB(AM) Newburgh IN
*WIKV(FM) Plymouth IN
*WVXR(FM) Richmond IN
*WQKV(FM) Rochester IN
*WJLR(FM) Seymour IN
WHME(FM) South Bend IN
*WHOJ(FM) Terre Haute IN
WPFR(AM) Terre Haute IN
*WJYW(FM) Union City IN
*WTUR(FM) Upland IN
*WATI(FM) Vincennes IN
*WRFM(FM) Wadesville IN
WFRR(FM) Walton IN
*KAXR(FM) Arkansas City KS
KEOJ(FM) Caney KS
*KHYM(FM) Copeland KS
*KJIL(FM) Copeland KS
*KTLI(FM) El Dorado KS
*KNGM(FM) Emporia KS
*KBMP(FM) Enterprise KS
KCNW(AM) Fairway KS
*KVCY(FM) Fort Scott KS
*KBDA(FM) Great Bend KS
*KWBI(FM) Great Bend KS
*KJRL(FM) Herington KS
*KARF(FM) Independence KS
*KBQC(FM) Independence KS
KCVW(FM) Kingman KS
KKLO(AM) Leavenworth KS
KZQD(FM) Liberal KS
*KGLV(FM) Manhattan KS
*KSNS(FM) Medicine Lodge KS
KCCV-FM Olathe KS
*KRBW(FM) Ottawa KS
*KTJO-FM Ottawa KS
KCCV(AM) Overland Park KS
KKCV(FM) Rozel KS
*KCVS(FM) Salina KS
KCVT(FM) Silver Lake KS
KHCA(FM) Wamego KS
*KCFN(FM) Wichita KS
*KYFW(FM) Wichita KS
*KBDD(FM) Winfield KS
*WAYD(FM) Auburn KY
*WTRT(FM) Benton KY
*WCVK(FM) Bowling Green KY
*WAPD(FM) Campbellsville KY
WOKT(AM) Cannonsburg KY
WMTA(AM) Central City KY
*WDFB-FM Danville KY
*WLAI(FM) Danville KY
*WRVG(FM) Georgetown KY
*WSGP(FM) Glasgow KY
WKCB(AM) Hindman KY
*WNKJ(FM) Hopkinsville KY
*WHMR(FM) Ledbetter KY
WFIA(AM) Louisville KY
WWLT(FM) Manchester KY
WMIK-FM Middlesboro KY
*WBMK(FM) Morehead KY
*WJIE-FM Okolona KY
*WJVK(FM) Owensboro KY
*WGCF(FM) Paducah KY
*WWJD(FM) Pippa Passes KY
*WKVY(FM) Somerset KY
WXKY-FM Stanford KY
WRVI(FM) Valley Station KY
WMTC(AM) Vancleve KY
WMTC-FM Vancleve KY
WVRB(FM) Wilmore KY
WMJR(AM) Winchester KY
*KAPM(FM) Alexandria LA
KJMJ(AM) Alexandria LA
*KLXA-FM Alexandria LA
*KHCL(FM) Arcadia LA
*KAXV(FM) Bastrop LA
*WJFM(FM) Baton Rouge LA
KBCL(AM) Bossier City LA
*WBKL(FM) Clinton LA
*KBAN(FM) De Ridder LA
*KVDP(FM) Dry Prong LA
KBEF(FM) Gibsland LA
KKNO(AM) Gretna LA
KYLA(FM) Homer LA
*KTOC-FM Jonesboro LA
*KIKL(FM) Lafayette LA
*KOJO(FM) Lake Charles LA
*KYLC(FM) Lake Charles LA
*KMSL(FM) Mansfield LA
*KAVK(FM) Many LA
*KBMQ(FM) Monroe LA
KLIC(AM) Monroe LA
*KYFL(FM) Monroe LA
*KBIO(FM) Natchitoches LA
KNIR(AM) New Iberia LA
*WBSN-FM New Orleans LA
WLNO(AM) New Orleans LA
WVOG(AM) New Orleans LA
*WNKV(FM) Norco LA
KUMX(FM) North Fort Polk LA
*KSUL(FM) Port Sulphur LA
KHLL(FM) Richwood LA
*KAPI(FM) Ruston LA
*KSJY(FM) Saint Martinville LA
KSYB(AM) Shreveport LA
*KRLR(FM) Sulphur LA
WROL(AM) Boston MA
*WYCM(FM) Charlton MA
WFGL(AM) Fitchburg MA
WCMX(AM) Leominster MA
*WTKL(FM) North Dartmouth MA
*WRYP(FM) Wellfleet MA
WRBS(AM) Baltimore MD
WRBS-FM Baltimore MD
WKDI(AM) Denton MD
*WYPF(FM) Frederick MD
*WLIC(FM) Frostburg MD
*WAIJ(FM) Grantsville MD
WWGB(AM) Indian Head MD
*WDIH(FM) Salisbury MD
WWPN(FM) Westernport MD
*WHCF(FM) Bangor ME
WBCI(FM) Bath ME
*WFST(AM) Caribou ME
*WMSJ(FM) Freeport ME
*WHMX(FM) Lincoln ME
*WMDR-FM Oakland ME
WJCX(FM) Pittsfield ME
*WWWA(FM) Winslow ME
*WUFN(FM) Albion MI
*WVCN(FM) Baraga MI
WOLY(AM) Battle Creek MI
*WLKB(FM) Bay City MI
*WTLI(FM) Bear Creek Township MI
*WCVM(FM) Bronson MI
*WTAC(FM) Burton MI
WLJW(AM) Cadillac MI
WLCM(AM) Charlotte MI
*WPRJ(FM) Coleman MI
WMUZ(FM) Detroit MI
WHPD(FM) Dowagiac MI
*WLJN(AM) Elmwood Township MI
*WAKL(FM) Flint MI
WSNL(AM) Flint MI
*WTRK(FM) Freeland MI
*WCTP(FM) Gagetown MI
*WBLW(FM) Gaylord MI
*WAYG(FM) Grand Rapids MI
*WCSG(FM) Grand Rapids MI
WCZE(FM) Harbor Beach MI
*WJOJ(FM) Harrisville MI
*WWKM(FM) Imlay City MI
*WVCM(FM) Iron Mountain MI
*WAYK(FM) Kalamazoo MI
*WLGH(FM) Leroy Township MI
WCAR(AM) Livonia MI
*WUGN(FM) Midland MI
*WDTR(FM) Monroe MI
*WMCQ(FM) Muskegon MI
*WPCJ(FM) Pittsford MI
*WKKM(FM) Rust Township MI
*WJKN-FM Spring Arbor MI
*WSAE(FM) Spring Arbor MI
*WUFL(AM) Sterling Heights MI
*WLJN-FM Traverse City MI
*WEJC(FM) White Star MI
WDEO(AM) Ypsilanti MI
WJQK(FM) Zeeland MI
*KBHG(FM) Alexandria MN
*KDNW(FM) Duluth MN
*WJRF(FM) Duluth MN
KBGY(FM) Faribault MN
WLKX-FM Forest Lake MN
KKEQ(FM) Fosston MN
*KADU(FM) Hibbing MN
*KBHW(FM) International Falls MN
*KXBR(FM) International Falls MN
WCTS(AM) Maplewood MN
*KTIS(AM) Minneapolis MN
WLOL(AM) Minneapolis MN
*KMKL(FM) North Branch MN
KCGN-FM Ortonville MN
KBHL(FM) Osakis MN
KTIG(FM) Pequot Lakes MN
WZFJ(FM) Pequot Lakes MN
KKMS(AM) Richfield MN
*KFSI(FM) Rochester MN
*KCFB(FM) Saint Cloud MN
KKJM(FM) Saint Joseph MN
*KKLW(FM) Willmar MN
KQRB(FM) Windom MN
*KBOJ(FM) Worthington MN
KPVR(FM) Bowling Green MO
*KLFC(FM) Branson MO
KOMC(AM) Branson MO
*KFFW(FM) Cabool MO
*KCVO-FM Camdenton MO
KKLL(AM) Carthage- MO
KMFC(FM) Centralia MO
KSIV(AM) Clayton MO
*KLRQ(FM) Clinton MO
*KCVZ(FM) Dixon MO
*KTBJ(FM) Festus MO
*KOBC(FM) Joplin MO
KZYM(AM) Joplin MO
*KLJC(FM) Kansas City MO
*KAUF(FM) Kennett MO
KLTE(FM) Kirksville MO
*KCVQ(FM) Knob Noster MO
*KTTK(FM) Lebanon MO
*KNLM(FM) Marshfield MO
*KBKC(FM) Moberly MO
*KCVJ(FM) Osceola MO
*KCVK(FM) Otterville MO
*KBGM(FM) Park Hills MO
*KOKS(FM) Poplar Bluff MO
KHZR(FM) Potosi MO
KADI-FM Republic MO
KAYX(FM) Richmond MO
KGNM(AM) Saint Joseph MO
*KSRD(FM) Saint Joseph MO
KJSL(AM) Saint Louis MO
*KSIV-FM Saint Louis MO
KXEN(AM) Saint Louis MO
*KCVX(FM) Salem MO
*KSCV(FM) Springfield MO
*KWFC(FM) Springfield MO
*KWND(FM) Springfield MO
*KCRL(FM) Sunrise Beach MO
*KRSS(FM) Tarkio MO
KULH(FM) Wheeling MO
WCMR(FM) Bruce MS
*WKVF(FM) Byhalia MS
WDFX(FM) Cleveland MS
WHJT(FM) Clinton MS
WKNZ(FM) Collins MS
*WPRG(FM) Columbia MS
*WCSO(FM) Columbus MS
*WMBU(FM) Forest MS
WQST-FM Forest MS
*WQVI(FM) Forest MS
*WSQH(FM) Forest MS
WWUN-FM Friar's Point MS
*WLRK(FM) Greenville MS
*WAOY(FM) Gulfport MS
WCPC(AM) Houston MS
*WYTF(FM) Indianola MS
WSJC(AM) Magee MS
WMER(AM) Meridian MS
*WAVI(FM) Oxford MS
*WPAS(FM) Pascagoula MS
*WLKO(FM) Quitman MS
*WMSB(FM) Senatobia MS
WSAO(AM) Senatobia MS
*WAFR(FM) Tupelo MS
*WAJS(FM) Tupelo MS
WLRC(AM) Walnut MS
*KJFT(FM) Arlee MT
*KQLU(FM) Belgrade MT
*KLMT(FM) Billings MT
*KLRV(FM) Billings MT
KMZK(AM) Billings MT
*KFRD(FM) Butte MT
*KFRT(FM) Butte MT
*KAFH(FM) Great Falls MT
*KFRW(FM) Great Falls MT
*KGFA(FM) Great Falls MT
*KMZO(FM) Hamilton MT
*KHLV(FM) Helena MT
KALS(FM) Kalispell MT
*KLKM(FM) Kalispell MT
*KLEU(FM) Lewistown MT
*KYWH(FM) Lockwood MT
*KMZL(FM) Missoula MT
*KBIL(FM) Park City MT
*KQLR(FM) Whitehall MT
*WBKU(FM) Ahoskie NC
*WLFA(FM) Asheville NC
WSKY(AM) Asheville NC
*WZGO(FM) Aurora NC
*WXBE(FM) Beaufort NC
WFGW(AM) Black Mountain NC
WMIT(FM) Black Mountain NC
WVBS(AM) Burgaw NC
WOGR(AM) Charlotte NC
*WYFQ(AM) Charlotte NC
WCSL(AM) Cherryville NC
WHPY(AM) Clayton NC
WCLN-FM Clinton NC
WPFJ(AM) Franklin NC
WSSG(AM) Goldsboro NC
WJSG(FM) Hamlet NC
WKDX(AM) Hamlet NC
WLGP(FM) Harkers Island NC
WJCV(AM) Jacksonville NC
WTRU(AM) Kernersville NC
*WGHW(FM) Lockwoods Folly Town NC
WDJS(AM) Mount Olive NC
WECR(AM) Newland NC
*WJIJ(FM) Norlina NC
*WBFY(FM) Pinehurst NC
WGIV(AM) Pineville NC
*WRAE(FM) Raeford NC
WPJL(AM) Raleigh NC
*WPGT(FM) Roanoke Rapids NC
*WRTP(FM) Roanoke Rapids NC
WKVE(FM) Semora NC
WNCA(AM) Siler City NC
*WAGO(FM) Snow Hill NC
*WGAS(AM) South Gastonia NC
*WZRI(FM) Spring Lake NC
WJRM(AM) Troy NC
WJFJ(AM) Tryon NC
WADE(AM) Wadesboro NC
WDRU(AM) Wake Forest NC
*WDVV(FM) Wilmington NC
*WWIL-FM Wilmington NC
*WAJC(FM) Wilson NC
WVOT(AM) Wilson NC
*WRCM(FM) Wingate NC
WBFJ(AM) Winston-Salem NC
*WBFJ-FM Winston-Salem NC
WPIP(AM) Winston-Salem NC
*KBMK(FM) Bismarck ND
*KNRI(FM) Bismarck ND
*KDVI(FM) Devils Lake ND
*KFBN(FM) Fargo ND
*KFNW-FM Fargo ND
*KWTL(AM) Grand Forks ND
KKLQ(FM) Harwood ND
*KFNL(FM) Kindred ND
*KVLQ(FM) Lincoln ND
KNDR(FM) Mandan ND
KHRT-FM Minot ND
*KNBE(FM) Beatrice NE
*KAMI(AM) Cozad NE
*KCVN(FM) Cozad NE
*KLNB(FM) Grand Island NE
KMMJ(AM) Grand Island NE
*KNFA(FM) Grand Island NE
*KROA(FM) Grand Island NE
*KNHA(FM) Grand Isle NE
*KAYA(FM) Hubbard NE
*KJLT(AM) North Platte NE
KCRO(AM) Omaha NE
*KGBI-FM Omaha NE
*KVSS(FM) Omaha NE
*KGRD(FM) Orchard NE
KHZY(FM) Overton NE
*KMLV(FM) Ralston NE
*KLJV(FM) Scottsbluff NE
*KFLV(FM) Wilber NE
*WVNH(FM) Concord NH
*WVFA(FM) Lebanon NH
*WLMW(FM) Manchester NH
*WVBH(FM) Beach Haven West NJ
*WJPG(FM) Cape May Court House NJ
*WKVP(FM) Cherry Hill NJ
*WXGN(FM) Egg Harbor Township NJ
*WRDR(FM) Freehold Township NJ
WWDJ(AM) Hackensack NJ
*WVBV(FM) Medford Lakes NJ
*WFME(FM) Newark NJ
WNSW(AM) Newark NJ
WIBG(AM) Ocean City NJ
WXMC(AM) Parsippany-Troy Hills NJ
WKMB(AM) Stirling NJ
WAWZ(FM) Zarephath NJ
KKIM(AM) Albuquerque NM
*KLYT(FM) Albuquerque NM
KSVA(AM) Albuquerque NM
KXKS(AM) Albuquerque NM
*KQRI(FM) Belen NM
*KQLV(FM) Bosque Farms NM
*KKCJ(FM) Cannon AFB NM
KAMQ(AM) Carlsbad NM
*KAQF(FM) Clovis NM
*KELU(FM) Clovis NM
*KZPI(FM) Deming NM
*KNMI(FM) Farmington NM
KPCL(FM) Farmington NM
*KTGW(FM) Fruitland NM
*KLGQ(FM) Grants NM
*KMBN(FM) Las Cruces NM
*KELP-FM Mesquite NM
*KRLU(FM) Roswell NM
KKIM-FM Santa Fe NM
*KVLK(FM) Socorro NM
*KXFR(FM) Socorro NM
KHAC(AM) Tse Bonito NM
KSFQ(FM) White Rock NM
*KNIS(FM) Carson City NV
KRNG(FM) Fallon/ Reno NV
KKVV(AM) Las Vegas NV
*KSOS(FM) Las Vegas NV
*KAIZ(FM) Mesquite NV
*KIHM(AM) Reno NV
*KLRH(FM) Sparks NV
WDCD(AM) Albany NY
*WJCA(FM) Albion NY
*WCOF(FM) Arcade NY
*WNGN(FM) Argyle NY
WNYG(AM) Babylon NY
*WCIK(FM) Bath NY
*WIFF(FM) Binghamton NY
WASB(AM) Brockport NY
*WKDL(FM) Brockport NY
WDCX(FM) Buffalo NY
*WFBF(FM) Buffalo NY
*WCIY(FM) Canandaigua NY
WRSB(AM) Canandaigua NY
*WMHI(FM) Cape Vincent NY
*WCOV-FM Clyde NY
*WKVJ(FM) Dannemora NY
WSIV(AM) East Syracuse NY
*WCIH(FM) Elmira NY
*WCID(FM) Friendship NY
*WLJH(FM) Glens Falls NY
*WGKR(FM) Grand Gorge NY
WHAZ-FM Hoosick Falls NY
*WHVP(FM) Hudson NY
*WCOT(FM) Jamestown NY
*WFGB(FM) Kingston NY
*WGWR(FM) Liberty NY
*WMHQ(FM) Malone NY
*WLJP(FM) Monroe NY
WRVP(AM) Mount Kisco NY
WMCA(AM) New York NY

*WWRV(AM) New York NY
*WNYK(FM) Nyack NY
WZXV(FM) Palmyra NY
*WPGL(FM) Pattersonville NY
*WRPJ(FM) Port Jervis NY
*WGKV(FM) Pulaski NY
*WYKV(FM) Ravena NY
*WSSK(FM) Saratoga Springs NY
WMYY(FM) Schoharie NY
*WAVX(FM) Schuyler Falls NY
*WYAI(FM) Scotia NY
*WFRS(FM) Smithtown NY
*WCII(FM) Spencer NY
*WMHR(FM) Syracuse NY
WHAZ(AM) Troy NY
*WKVU(FM) Utica NY
*WCOU(FM) Warsaw NY
*WKWV(FM) Watertown NY
*WMHN(FM) Webster NY
WRCI(FM) Webster NY
*WBCY(FM) Archbold OH
WJKW(FM) Athens OH
*WCVV(FM) Belpre OH
*WLKP(FM) Belpre OH
WCER(AM) Canton OH
*WVXC(FM) Chillicothe OH
WAKW(FM) Cincinnati OH
*WCRF(FM) Cleveland OH
WFHM-FM Cleveland OH
*WGOJ(FM) Conneaut OH
*WYKL(FM) Crestline OH
*WQRP(FM) Dayton OH
*WORI(FM) Delhi Hills OH
*WTKC(FM) Findlay OH
*WCVO(FM) Gahanna OH
WPOS-FM Holland OH
*WCVJ(FM) Jefferson OH
*WFCO(FM) Lancaster OH
*WTGN(FM) Lima OH
*WYSM(FM) Lima OH
*WVMC-FM Mansfield OH
WUCO(AM) Marysville OH
*WYSZ(FM) Maumee OH
WFCJ(FM) Miamisburg OH
*WVML(FM) Millersburg OH
*WNZR(FM) Mount Vernon OH
WNPQ(FM) New Philadelphia OH
*WMEJ(FM) Proctorville OH
WDMN(AM) Rossford OH
*WLRY(FM) Rushville OH
*WVMS(FM) Sandusky OH
*WAUI(FM) Shelby OH
*WOAR(FM) South Vienna OH
*WEKV(FM) South Webster OH
*WBCJ(FM) Spencerville OH
*WEEC(FM) Springfield OH
*WBJV(FM) Steubenville OH
*WOKL(FM) Troy OH
WHKZ(AM) Warren OH
*WYSA(FM) Wauseon OH
*WVXW(FM) West Union OH
WKLN(FM) Wilmington OH
*WJIC(FM) Zanesville OH
*KAKO(FM) Ada OK
*KQPD(FM) Ardmore OK
*KWRI(FM) Bartlesville OK
*KARU(FM) Cache OK
*KJCC(FM) Carnegie OK
KTFR(FM) Chelsea OK
*KAYC(FM) Durant OK
*KOKF(FM) Edmond OK
KXOO(FM) Elk City OK
*KWKL(FM) Grandfield OK
*KXRT(FM) Idabel OK
*KJRF(FM) Lawton OK
*KVRS(FM) Lawton OK
KEMX(FM) Locust Grove OK
KGLC(FM) Miami OK
KTLV(AM) Midwest City OK
*KYLV(FM) Oklahoma City OK
KPGM(AM) Pawhuska OK
*KKRI(FM) Pocola OK
*KJTH(FM) Ponca City OK
*KLVV(FM) Ponca City OK
*KARG(FM) Poteau OK
KKCM(FM) Sand Springs OK
*KXTH(FM) Seminole OK
KQCV-FM Shawnee OK
*KTKL(FM) Stigler OK
*KAYM(FM) Weatherford OK
*KJOV(FM) Woodward OK
KHPE(FM) Albany OR
KWIL(AM) Albany OR
*KLOY(FM) Astoria OR
*KWYA(FM) Astoria OR
*KANL(FM) Baker City OR
KNLR(FM) Bend OR
*KVLB(FM) Bend OR
*KMWR(FM) Brookings OR
*KLVP-FM Cherryville OR
KYSJ(FM) Coos Bay OR
KYTT-FM Coos Bay OR
*KAPK(FM) Grants Pass OR
*KIDH(FM) Jordan Valley OR
*KKLJ(FM) Klamath Falls OR
*KKLP(FM) La Pine OR
*KGRI(FM) Lebanon OR
*KBUG(FM) Malin OR
*KLRF(FM) Milton-Freewater OR
*KYOR(FM) Newport OR
*KEFS(FM) North Powder OR
KARO(FM) Nyssa OR
*KAPL(AM) Phoenix OR
*KPFR(FM) Pine Grove OR
KKPZ(AM) Portland OR
KPDQ(AM) Portland OR
KPDQ-FM Portland OR
*KLON(FM) Rockaway Beach OR
KPJC(AM) Salem OR
*KWBX(FM) Salem OR
KFIS(FM) Scappoose OR
*KJKL(FM) Selma OR
KORE(AM) Springfield-Eugene OR
*KLVU(FM) Sweet Home OR
*KAIK(FM) Tillamook OR
KLWJ(AM) Umatilla OR
*KZRI(FM) Welches OR
*KLOV(FM) Winchester OR
KGRV(AM) Winston OR
WAVL(AM) Apollo PA
*WGEV(FM) Beaver Falls PA
*WCIG(FM) Carbondale PA
WHGT(AM) Chambersburg PA
*WZXQ(FM) Chambersburg PA
WPWA(AM) Chester PA
WVCH(AM) Chester PA
WDBA(FM) DuBois PA
*WAWN(FM) Franklin PA
*WCOG-FM Galeton PA
*WVMM(FM) Grantham PA
WEXC(FM) Greenville PA
WKBO(AM) Harrisburg PA
*WBYH(FM) Hawley PA
*WFRJ(FM) Johnstown PA
WDAC(FM) Lancaster PA
*WJTL(FM) Lancaster PA
WBYN(AM) Lehighton PA
*WGRC(FM) Lewisburg PA
*WJRC(FM) Lewistown PA
*WRIJ(FM) Masontown PA
*WYFU(FM) Masontown PA
*WZXM(FM) Middletown PA
*WQJU(FM) Mifflintown PA
*WRWJ(FM) Murrysville PA
*WPKV(FM) Nanty Glo PA
*WBGM(FM) New Berlin PA
*WVMN(FM) New Castle PA
WPCL(FM) Northern Cambria PA
WWKL(FM) Palmyra PA
WORD-FM Pittsburgh PA
WPIT(AM) Pittsburgh PA
WWNL(AM) Pittsburgh PA
WKVB(FM) Port Matilda PA
WLKJ(FM) Portage PA
WRAW(AM) Reading PA
WREQ(FM) Ridgebury PA
*WBYO(FM) Sellersville PA
WWII(AM) Shiremanstown PA
WLKH(FM) Somerset PA
WBHV-FM State College PA
*WRXV(FM) State College PA
*WTLR(FM) State College PA
*WBYX(FM) Stroudsburg PA
*WLKA(FM) Tafton PA
*WCIT(FM) Trout Run PA
WCTL(FM) Union City PA
*WZZD(FM) Warwick PA
WLIH(FM) Whitneyville PA
*WCRG(FM) Williamsport PA
*WYTL(FM) Wyomissing PA
WYYC(AM) York PA
*WTMV(FM) Youngsville PA
WBRQ(FM) Cidra PR
WJZG(FM) Culebra PR
*WCRP(FM) Guayama PR
WRRH(FM) Hormigueros PR
WNRT(FM) Manati PR
WZNA(AM) Moca PR
WPPC(AM) Penuelas PR
*WNNV(FM) San German PR
WERR(FM) Utuado PR
WSTL(AM) Providence RI
*WKIV(FM) Westerly RI
WLUA(AM) Belton SC
*WAFJ(FM) Belvedere SC
WBAJ(AM) Blythwood SC
*WRFE(FM) Chesterfield SC
WELP(AM) Easley SC
WOLI-FM Easley SC
*WLPG(FM) Florence SC
*WYFG(FM) Gaffney SC
WLMC(AM) Georgetown SC
WLFJ(AM) Greenville SC
*WLFJ-FM Greenville SC
WCKI(AM) Greer SC
*WMBJ(FM) Murrell's Inlet SC
WTBI(AM) Pickens SC
WQMC(AM) Sumter SC
WSSC(AM) Sumter SC
WBZK(AM) York SC
KLRJ(FM) Aberdeen SD
*KVCF(FM) Freeman SD
*KVCX(FM) Gregory SD
*KVFL(FM) Pierre SD
*KASD(FM) Rapid City SD
KTPT(FM) Rapid City SD
*KNWC-FM Sioux Falls SD
*KSFS(FM) Sioux Falls SD
KSLT(FM) Spearfish SD
*KJBB(FM) Watertown SD
*WYLV(FM) Alcoa TN
WATX(AM) Algood TN
WTKB-FM Atwood TN
WBIN(AM) Benton TN
*WHCB(FM) Bristol TN
WNKX(AM) Centerville TN
WJOC(AM) Chattanooga TN
WLMR(AM) Chattanooga TN
*WAYQ(FM) Clarksville TN
WKFN(AM) Clarksville TN
*WNRZ(FM) Dickson TN
WCRT(AM) Donelson TN
*WZKV(FM) Dyersburg TN
*WUMC(FM) Elizabethton TN
WLLJ(FM) Etowah TN
*WPLX(AM) Germantown TN
WAYB-FM Graysville TN
*WAUO(FM) Hohenwald TN
*WAMP(FM) Jackson TN
*WCQR-FM Kingsport TN
*WKTS(FM) Kingston TN
WFFI(FM) Kingston Springs TN
WIFA(AM) Knoxville TN
WITA(AM) Knoxville TN
WRJZ(AM) Knoxville TN
WBLC(AM) Lenoir City TN
*WIGH(FM) Lexington TN
WFGZ(FM) Lobelville TN
*WAJJ(FM) McKenzie TN
WENO(AM) Nashville TN
*WNAZ-FM Nashville TN
*WAYW(FM) New Johnsonville TN
WKVZ(FM) Ripley TN
WDTM(AM) Selmer TN
*WXKV(FM) Selmer TN
*WBIA(FM) Shelbyville TN
WFFH(FM) Smyrna TN
WDBL(AM) Springfield TN
*WTAI(FM) Union City TN
WBOZ(FM) Woodbury TN
*KAGT(FM) Abilene TX
*KAQD(FM) Abilene TX
*KGNZ(FM) Abilene TX
KTEK(AM) Alvin TX
*KAVW(FM) Amarillo TX
*KJJP(FM) Amarillo TX
KTNZ(AM) Amarillo TX
*KXLV(FM) Amarillo TX
*KXRI(FM) Amarillo TX
KLTY(FM) Arlington TX
*KATG(FM) Athens TX
KPYN(AM) Atlanta TX
*KHIB(FM) Bastrop TX
*KZBJ(FM) Bay City TX
*KLBT(FM) Beaumont TX
*KTXB(FM) Beaumont TX
KIBL(AM) Beeville TX
KPDB(FM) Big Lake TX
*KTAA(FM) Big Sandy TX
*KBCX(FM) Big Spring TX
*KPBB(FM) Brownfield TX
*KBUB(FM) Brownwood TX
*KHPU(FM) Brownwood TX
*KPBE(FM) Brownwood TX
KPSM(FM) Brownwood TX
KAGC(AM) Bryan TX
*KLRW(FM) Byrne TX
KAYG(FM) Camp Wood TX
KRDH(AM) Canton TX
KCZO(FM) Carrizo Springs TX
KWDC(FM) Coahoma TX
*KLGS(FM) College Station TX
*KAFR(FM) Conroe TX
*KKLM(FM) Corpus Christi TX
*KCKT(FM) Crockett TX
*KCBI(FM) Dallas TX
*KDLI(FM) Del Rio TX
KIXL(AM) Del Valle TX
*KLLR(FM) Dripping Springs TX
*KEPI(FM) Eagle Pass TX
*KEPX(FM) Eagle Pass TX
KELP(AM) El Paso TX
*KKLY(FM) El Paso TX
KPAS(FM) Fabens TX
KLDS(AM) Falfurrias TX
*KZFT(FM) Fannett TX
KIJN(AM) Farwell TX
KIJN-FM Farwell TX
KDFT(AM) Ferris TX
KBRZ(AM) Freeport TX
*KPBN(FM) Freer TX
*KYAR(FM) Gatesville TX
*KANJ(FM) Giddings TX
*KMLR(FM) Gonzales TX
*KTXG(FM) Greenville TX
*KAZF(FM) Hebronville TX
KEKO(FM) Hebronville TX
KWRD-FM Highland Village TX
*KHCB-FM Houston TX
*KSBJ(FM) Humble TX
*KHCH(AM) Huntsville TX
*KYLR(FM) Hutto TX
*KBJS(FM) Jacksonville TX
KCOX(AM) Jasper TX
*KHCJ(FM) Jefferson TX
KERB(AM) Kermit TX
KERB-FM Kermit TX
*KHKV(FM) Kerrville TX
*KKER(FM) Kerrville TX
*KZLO(FM) Kilgore TX
*KYBJ(FM) Lake Jackson TX
*KBKN(FM) Lamesa TX
KLAR(AM) Laredo TX
KBLT(FM) Leakey TX
KYMI(FM) Los Ybanez TX
*KAMY(FM) Lubbock TX
*KKLU(FM) Lubbock TX
*KSWP(FM) Lufkin TX
*KZLV(FM) Lytle TX
*KHML(FM) Madisonville TX
*KBMD(FM) Marble Falls TX
*KSUR(FM) Mart TX
KVMV(FM) McAllen TX
KMXO(AM) Merkel TX
*KMEO(FM) Mertzon TX
KBAT(FM) Monahans TX
KNBO(AM) New Boston TX
*KBMM(FM) Odessa TX
*KFLB(AM) Odessa TX
*KFLB-FM Odessa TX
KPXI(FM) Overton TX
*KAVO(FM) Pampa TX
*KHCP(FM) Paris TX
*KGWP(FM) Pittsburg TX
*KBAH(FM) Plainview TX
*KWLD(FM) Plainview TX
KWMF(AM) Pleasanton TX
*KHPO(FM) Port O'Connor TX
*KSGR(FM) Portland TX
KPOS(FM) Post TX
KCLR(AM) Ralls TX
*KTER(FM) Rudolph TX
KCRN(AM) San Angelo TX
KCRN-FM San Angelo TX
*KNAR(FM) San Angelo TX
KSLR(AM) San Antonio TX
*KYFS(FM) San Antonio TX
KUBR(AM) San Juan TX
KUOL(AM) San Marcos TX
*KVRK(FM) Sanger TX
KMAT(FM) Seadrift TX
KDAE(AM) Sinton TX
KJAK(FM) Slaton TX
KSNY(AM) Snyder TX
KYTY(AM) Somerset TX
KVRP(AM) Stamford TX
*KFRI(FM) Stanton TX
*KBDE(FM) Temple TX
*KVLT(FM) Temple TX
KLGO(FM) Thorndale TX
KTNO(AM) University Park TX
*KAYK(FM) Victoria TX
*KXBJ(FM) Victoria TX
KBBW(AM) Waco TX
*KVLW(FM) Waco TX
*KHTA(FM) Wake Village TX
KRGE(AM) Weslaco TX
*KLVW(FM) West Odessa TX
*KMOC(FM) Wichita Falls TX
*KZKL(FM) Wichita Falls TX
KKHT-FM Winnie TX
KBAW(FM) Zapata TX
KOAY(FM) Coalville UT
*KNKL(FM) North Ogden UT
*KYFO-FM Ogden UT
*KEYY(FM) Provo UT
*KANN(AM) Roy UT
*KAER(FM) Saint George UT
*KUFR(FM) Salt Lake City UT
KUTR(AM) Taylorsville UT
KMRI(AM) West Valley City UT
WAVA(AM) Arlington VA
WAVA-FM Arlington VA
WYFJ(FM) Ashland VA
*WAZP(FM) Cape Charles VA
WYRV(AM) Cedar Bluff VA
*WAUQ(FM) Charles City VA
WJYK(AM) Chase City VA
*WWIP(FM) Cheriton VA
WPMH(AM) Claremont VA
*WARN(FM) Culpeper VA
*WPER(FM) Culpeper VA
*WOKD-FM Danville VA
*WPIN-FM Dublin VA
*WJYA(FM) Emporia VA
*WJYJ(FM) Fredericksburg VA
*WWMC(FM) Lynchburg VA
*WPIM(FM) Martinsville VA
WNRV(AM) Narrows-Pearisburg VA
*WJCN(FM) Nassawadox VA
WLTK(FM) New Market VA
WRJR(AM) Portsmouth VA
WBTK(AM) Richmond VA
*WRIH(FM) Richmond VA
*WRXT(FM) Roanoke VA
WLOY(AM) Rural Retreat VA
*WPAR(FM) Salem VA
WSBV(AM) South Boston VA
*WWED(FM) Spotsylvania VA
*WJLZ(FM) Virginia Beach VA

WPRZ(AM) Warrenton VA
*WPVA(FM) Waynesboro VA
WTFX(AM) Winchester VA
WEVI(FM) Frederiksted VI
*WCMD-FM Barre VT
*WCMK(FM) Putney VT
WMNV(FM) Rupert VT
*WFTF(FM) Rutland VT
KGNW(AM) Burien-Seattle WA
*KGTS(FM) College Place WA
*KKRS(FM) Davenport WA
KCIS(AM) Edmonds WA
KCMS(FM) Edmonds WA
*KTJC(FM) Kelso WA
*KBLD(FM) Kennewick WA
*KJVH(FM) Longview WA
*KWYQ(FM) Longview WA
KWPZ(FM) Lynden WA
KTSL(FM) Medical Lake WA
*KSBC(FM) Nile WA
*KLOP(FM) Ocean Park WA
KGDN(FM) Pasco WA
*KRLF(FM) Pullman WA
*KWFJ(FM) Roy WA
KLFE(AM) Seattle WA
*KEEH(FM) Spokane WA
*KAYB(FM) Sunnyside WA
*KRKL(FM) Walla Walla WA
*KPLW(FM) Wenatchee WA
KJOX(AM) Yakima WA
KYAK(AM) Yakima WA
*KYPL(FM) Yakima WA
*WHEM(FM) Eau Claire WI
*WVCF(FM) Eau Claire WI
*WDKV(FM) Fond du Lac WI
*WVFL(FM) Fond du Lac WI
*WMVM(FM) Goodman WI
*WORQ(FM) Green Bay WI
WWIB(FM) Hallie WI
WRRD(AM) Jackson WI
WJOK(AM) Kaukauna WI
*WSTM(FM) Kiel WI
WZRK(AM) Lake Geneva WI
*WNWC-FM Madison WI
*WGNV(FM) Milladore WI
WJYI(AM) Milwaukee WI
*WMWK(FM) Milwaukee WI
WSSP(AM) Milwaukee WI
*WVCY-FM Milwaukee WI
WFZH(FM) Mukwonago WI
*WPFF(FM) Sturgeon Bay WI
*WRGX(FM) Sturgeon Bay WI
*WRVM(FM) Suring WI
*WVCX(FM) Tomah WI
*WEGZ(FM) Washburn WI
*WVRN(FM) Wittenberg WI
*WPIB(FM) Bluefield WV
*WKJL(FM) Clarksburg WV
*WBHZ(FM) Elkins WV
*WDKL(FM) Grafton WV
WEMM(AM) Huntington WV
WEMM-FM Huntington WV
WOKU(AM) Hurricane WV
WFSP(AM) Kingwood WV
*WKVW(FM) Marmet WV
WZZW(AM) Milton WV
WRKP(FM) Moundsville WV
*WPWV(FM) Princeton WV
WRLB(FM) Rainelle WV
WRRL(AM) Rainelle WV
*WLKV(FM) Ripley WV
*KCSP-FM Casper WY
*KLWC(FM) Casper WY
*KLWV(FM) Chugwater WY
KUYO(AM) Evansville WY
*KAXG(FM) Gillette WY
*KLOF(FM) Gillette WY
*KLWD(FM) Gillette WY
*KAIW(FM) Laramie WY
*KWYC(FM) Orchard Valley WY
*KRWT(FM) West Laramie WY

Classic Rock

KBFX(FM) Anchorage AK
KLAM(AM) Cordova AK
KXLR(FM) Fairbanks AK
KSUP(FM) Juneau AK
KRXX(FM) Kodiak AK
KSLD(AM) Soldotna AK
*WJSR(FM) Birmingham AL
WZRR(FM) Birmingham AL
WERH-FM Hamilton AL
WTAK-FM Hartselle AL
WRKH(FM) Mobile AL
WQKS-FM Montgomery AL
WVNA-FM Muscle Shoals AL
WJRL-FM Ozark AL
WXFX(FM) Prattville AL
WMXN-FM Stevenson AL
*KSWH(FM) Arkadelphia AR
KMJX(FM) Conway AR
KCJF(FM) Earle AR
KAGL(FM) El Dorado AR
KKEG(FM) Fayetteville AR
KXJK(AM) Forrest City AR
KCWD(FM) Harrison AR
KEGI(FM) Jonesboro AR
KHBM-FM Monticello AR
KLXQ(FM) Mountain Pine AR
KWLT(FM) North Crossett AR
KYGL(FM) Texarkana AR
KBBQ-FM Van Buren AR
KCUZ(AM) Clifton AZ
KMGN(FM) Flagstaff AZ
KCDX(FM) Florence AZ
KRRK(FM) Lake Havasu City AZ
KZUL-FM Lake Havasu City AZ
KWKM(FM) Saint Johns AZ
KSLX-FM Scottsdale AZ
KFMM(FM) Thatcher AZ
KHYT(FM) Tucson AZ
KLPX(FM) Tucson AZ
KALT-FM Alturas CA
KXGO(FM) Arcata CA
KHRQ(FM) Baker CA
KRHV(FM) Big Pine CA
KOCP(FM) Camarillo CA
KWTY(FM) Cartago CA
KTHU(FM) Corning CA
KDFO-FM Delano CA
KZRO(FM) Dunsmuir CA
KJFX(FM) Fresno CA
KBAY(FM) Gilroy CA
KHIP(FM) Gonzales CA
KVLI-FM Lake Isabella CA
KHDR(FM) Lenwood CA
KCBS-FM Los Angeles CA
KDJK(FM) Mariposa CA
KHKK(FM) Modesto CA
KVRV(FM) Monte Rio CA
KLUK(FM) Needles CA
KIOO(FM) Porterville CA
KLKX(FM) Rosamond CA
KSEG(FM) Sacramento CA
KGB-FM San Diego CA
KFRC-FM San Francisco CA
KUFX(FM) San Jose CA
KZOZ(FM) San Luis Obispo CA
KSAN(FM) San Mateo CA
KXFM(FM) Santa Maria CA
KTMQ(FM) Temecula CA
KHRD(FM) Weaverville CA
KTOR(FM) Westwood CA
KDGL(FM) Yucca Valley CA
KKFM(FM) Colorado Springs CO
KQZR(FM) Craig CO
KKNN(FM) Delta CO
KQMT(FM) Denver CO
KRFX(FM) Denver CO
KTUN(FM) Eagle CO
KPAW(FM) Fort Collins CO
KRVG(FM) Glenwood Springs CO
KVLE-FM Gunnison CO
KLMR-FM Lamar CO
KSTR-FM Montrose CO
KYZX(FM) Pueblo West CO
KSBV(FM) Salida CO
KCRT-FM Trinidad CO
KMAX-FM Wellington CO
WRKI(FM) Brookfield CT
WFOX(FM) Norwalk CT
WJKI(FM) Bethany Beach DE
WGBG(FM) Seaford DE
WNDT(FM) Alachua FL
WFYV-FM Atlantic Beach FL
WWUS(FM) Big Pine Key FL
WRXK-FM Bonita Springs FL
WCJX(FM) Five Points FL
WBGG-FM Fort Lauderdale FL
WKGR(FM) Fort Pierce FL
WXCV(FM) Homosassa Springs FL
*WJUF(FM) Inverness FL
WKYZ(FM) Key Colony Beach FL
WAIL(FM) Key West FL
WAIL(FM) Key West FL
WARO(FM) Naples FL
WRKN(FM) Niceville FL
WHTQ(FM) Orlando FL
WHOG-FM Ormond-by-the-Sea FL
WXGL(FM) Saint Petersburg FL
WHPT(FM) Sarasota FL
WNDD(FM) Silver Springs FL
WRBA(FM) Springfield FL
WGLF(FM) Tallahassee FL
WXOF(FM) Yankeetown FL
WDMG-FM Ambrose GA
WKLS(FM) Atlanta GA
WZGC(FM) Atlanta GA
WEKL(FM) Augusta GA
WVRK(FM) Columbus GA
WQBZ(FM) Fort Valley GA
WSRV(FM) Gainesville GA
WIOL(FM) Greenville GA
WTHG(FM) Hinesville GA
WMGP(FM) Hogansville GA
WBBT(AM) Lyons GA
WBMZ(FM) Metter GA
WOCE(FM) Ringgold GA
WIXV(FM) Savannah GA
*WKEU-FM The Rock GA
WWRQ-FM Valdosta GA
WDBN(FM) Wrightsville GA
KPOI-FM Honolulu HI
KLKK(FM) Clear Lake IA
KCQQ(FM) Davenport IA
KGGO(FM) Des Moines IA
KXGE(FM) Dubuque IA
KKSI(FM) Eddyville IA
KGRR(FM) Epworth IA
KCRR(FM) Grundy Center IA
KKRQ(FM) Iowa City IA
KRNA(FM) Iowa City IA
KGRA(FM) Jefferson IA
KRNQ(FM) Keokuk IA
KKMA(FM) Le Mars IA
KUQQ(FM) Milford IA
KSEZ(FM) Sioux City IA
KJOT(FM) Boise ID
KQEO(FM) Idaho Falls ID
KWYS-FM Island Park ID
KIKX(FM) Ketchum ID
KTPO(FM) Kootenai ID
KKGL(FM) Nampa ID
KMGI(FM) Pocatello ID
KPKY(FM) Pocatello ID
KOFE(AM) Saint Maries ID
KECH-FM Sun Valley ID
KSNQ(FM) Twin Falls ID
KRVQ(FM) Victor ID
KLZX(FM) Weston ID
WDLJ(FM) Breese IL
WDRV(FM) Chicago IL
WRXQ(FM) Coal City IL
WRHK(FM) Danville IL
WXTT(FM) Danville IL
WMKB(FM) Earlville IL
KUUL(FM) East Moline IL
WYMG(FM) Jacksonville IL
*WMXM(FM) Lake Forest IL
WQLF(FM) Lena IL
WKGL-FM Loves Park IL
WJEQ(FM) Macomb IL
WGKC(FM) Mahomet IL
WXLP(FM) Moline IL
WZZT(FM) Morrison IL
WIKK(FM) Newton IL
WGLO(FM) Pekin IL
WPBG(FM) Peoria IL
WBZG(FM) Peru IL
WZNX(FM) Sullivan IL
WCVS-FM Virden IL
WQUL(FM) West Frankfort IL
WWDV(FM) Zion IL
WHTI(FM) Alexandria IN
WSHP(FM) Attica IN
WJAA(FM) Austin IN
*WHJE(FM) Carmel IN
WSFR(FM) Corydon IN
WXRD(FM) Crown Point IN
WABX(FM) Evansville IN
WFWI(FM) Fort Wayne IN
WXKE(FM) Fort Wayne IN
WOZW(FM) Goshen IN
WHTY(FM) Hartford City IN
WFBQ(FM) Indianapolis IN
WQMF(FM) Jeffersonville IN
WKHY(FM) Lafayette IN
WRZR(FM) Loogootee IN
WZOW(FM) New Carlisle IN
WMYK(FM) Peru IN
WRSW-FM Warsaw IN
WWVR(FM) West Terre Haute IN
KFXJ(FM) Augusta KS
KKRK(FM) Coffeyville KS
KZRD(FM) Dodge City KS
KOTE(FM) Eureka KS
KCAR-FM Galena KS
KSEK-FM Girard KS
KVGB-FM Great Bend KS
KINZ(FM) Humboldt KS
KWKR(FM) Leoti KS
KILS(FM) Minneapolis KS
KWIC(FM) Topeka KS
KTHR(FM) Wichita KS
WLFX(FM) Berea KY
WDNS(FM) Bowling Green KY
WCBJ(FM) Campton KY
WPTQ(FM) Cave City KY
WVEK-FM Cumberland KY
WTBK(FM) Manchester KY
WQQR(FM) Mayfield KY
WKYM(FM) Monticello KY
WKQQ(FM) Winchester KY
KZMZ(FM) Alexandria LA
KCIJ(FM) Atlanta LA
WDGL(FM) Baton Rouge LA
KLIP(FM) Monroe LA
WKBU(FM) New Orleans LA
KKGB(FM) Sulphur LA
WNYN-FM Athol MA
WZLX(FM) Boston MA
WROR-FM Framingham MA
*WGAO(FM) Franklin MA
*WSDH(FM) Sandwich MA
WKPE-FM South Yarmouth MA
WAQY(FM) Springfield MA
*WYAJ(FM) Sudbury MA
WTGB-FM Bethesda MD
*WMTB-FM Emmittsburg MD
WDLD(FM) Halfway MD
WSMD-FM Mechanicsville MD
WKHW(FM) Pocomoke City MD
WZBA(FM) Westminster MD
WBQI(FM) Bar Harbor ME
WKIT-FM Brewer ME
WQDY-FM Calais ME
WQSS(FM) Camden ME
*WSHD(FM) Eastport ME
WWMJ(FM) Ellsworth ME
WCTB(FM) Fairfield ME
WHXQ(FM) Kennebunkport ME
WSHK(FM) Kittery ME
WFNK(FM) Lewiston ME
WRMO(FM) Milbridge ME
WOZI(FM) Presque Isle ME
WFZX(FM) Searsport ME
WNSX(FM) Winter Harbor ME
WZUU(FM) Allegan MI
WFDX(FM) Atlanta MI
WLEW-FM Bad Axe MI
WCSX(FM) Birmingham MI
WCKC(FM) Cadillac MI
WGFM(FM) Cheboygan MI
WMMQ(FM) East Lansing MI
WBFX(FM) Grand Rapids MI
WLAV-FM Grand Rapids MI
WWKR(FM) Hart MI
WKJZ(FM) Hillman MI
WOLV(FM) Houghton MI
WIMK(FM) Iron Mountain MI
WFCX(FM) Leland MI
WUPK(FM) Marquette MI
WRCC(FM) Marshall MI
WKQZ(FM) Midland MI
WRUP(FM) Munising MI
WMRR(FM) Muskegon Heights MI
WIHC(FM) Newberry MI
WAOR(FM) Niles MI
*WOVI(FM) Novi MI
WRSR(FM) Owosso MI
WRKR(FM) Portage MI
WILZ(FM) Saginaw MI
WYVN(FM) Saugatuck MI
WSUE(FM) Sault Ste. Marie MI
WQLB(FM) Tawas City MI
WLKM-FM Three Rivers MI
KQPR-FM Albert Lea MN
KXRA-FM Alexandria MN
KKLN(FM) Atwater MN
KLIZ-FM Brainerd MN
KFGI(FM) Crosby MN
KQDS-FM Duluth MN
KFMC(FM) Fairmont MN
KQCL(FM) Faribault MN
KQRS-FM Golden Valley MN
WXXZ(FM) Grand Marais MN
KRAQ(FM) Jackson MN
KARZ(FM) Marshall MN
KKCK(FM) Marshall MN
KMFG(FM) Nashwauk MN
KXLP(FM) New Ulm MN
KWNG(FM) Red Wing MN
KRCH(FM) Rochester MN
*KRPR(FM) Rochester MN
KRWB(AM) Roseau MN
KLFN(FM) Sunburg MN
KLZZ(FM) Waite Park MN
KLLZ-FM Walker MN
KWOA-FM Worthington MN
KTNX(FM) Arcadia MO
KGMO(FM) Cape Girardeau MO
KSHE(FM) Crestwood MO
KDFN(AM) Doniphan MO
KMAC(FM) Gainesville MO
KCFX(FM) Harrisonville MO
KRXL(FM) Kirksville MO
KPOW-FM La Monte MO
KKLH(FM) Marshfield MO
KWBZ(FM) Monroe City MO
KZZK(FM) New London MO
KMYK(FM) Osage Beach MO
KJEZ(FM) Poplar Bluff MO
KIHT(FM) Saint Louis MO
KIGL(FM) Seligman MO
KXUS(FM) Springfield MO
KFBD-FM Waynesville MO
KXDG(FM) Webb City MO
KSPQ(FM) West Plains MO
WRKG(FM) Drew MS
WMUT(FM) Grenada MS
WXRR(FM) Hattiesburg MS
WWJK(FM) Jackson MS
WZNF(FM) Lumberton MS
WCNA(FM) Potts Camp MS
WSTZ-FM Vicksburg MS
KJJM(FM) Baker MT
KBZM(FM) Big Sky MT
KRKX(FM) Billings MT
KMBR(FM) Butte MT
KTZZ(FM) Conrad MT
KQDI-FM Great Falls MT
KZMT(FM) Helena MT
KBBZ(FM) Kalispell MT
KZMN(FM) Kalispell MT
KLCM(FM) Lewistown MT
KKQX(FM) Manhattan MT
KMTA(AM) Miles City MT
KZOQ-FM Missoula MT
KWMY(FM) Park City MT

KMXE-FM Red Lodge MT
KGCX(FM) Sidney MT
WKRR(FM) Asheboro NC
*WZPE(FM) Bath NC
WCLN(AM) Clinton NC
WRFX-FM Kannapolis NC
WSFL-FM New Bern NC
WRVA-FM Rocky Mount NC
WBNE(FM) Shallotte NC
WOBR-FM Wanchese NC
WQNS(FM) Waynesville NC
*WSIF(FM) Wilkesboro NC
WRQR(FM) Wilmington NC
WNTB(FM) Wrightsville Beach NC
KBYZ(FM) Bismarck ND
KSSS(FM) Bismarck ND
KPFX(FM) Fargo ND
KRWK(FM) Fargo ND
KTZU(FM) Velva ND
KTGL(FM) Beatrice NE
KMOR(FM) Bridgeport NE
KBBN-FM Broken Bow NE
KKOT(FM) Columbus NE
KFMT-FM Fremont NE
KROR(FM) Hastings NE
KRKR(FM) Lincoln NE
KRKU(FM) McCook NE
KQBW(FM) Omaha NE
KBRX(AM) O'Neill NE
WMLL(FM) Bedford NH
WHDQ(FM) Claremont NH
WWHK(FM) Concord NH
WSAK(FM) Hampton NH
WMTK(FM) Littleton NH
WWHQ(FM) Meredith NH
WMGM(FM) Atlantic City NJ
WCHR-FM Manahawkin NJ
WNNJ-FM Newton NJ
WZXL(FM) Wildwood NJ
KNMZ(FM) Alamogordo NM
KZRM(FM) Chama NM
KEJL(FM) Eunice NM
KDAG(FM) Farmington NM
KXXI(FM) Gallup NM
KMDZ(FM) Las Vegas NM
KIOT(FM) Los Lunas NM
KIDX(FM) Ruidoso NM
KTUM(FM) Tatum NM
KRSI(FM) Garapan-Saipan NP
KKLZ(FM) Las Vegas NV
KOZZ-FM Reno NV
KURK(FM) Reno NV
WPYX(FM) Albany NY
*WGCC-FM Batavia NY
WAAL(FM) Binghamton NY
WTKW(FM) Bridgeport NY
WGRF(FM) Buffalo NY
WKGB-FM Conklin NY
WIII(FM) Cortland NY
*WECW(FM) Elmira NY
WBBI(FM) Endwell NY
WCPV(FM) Essex NY
WLPW(FM) Lake Placid NY
WMOS(FM) Montauk NY
WNGZ(FM) Montour Falls NY
WFAF(FM) Mount Kisco NY
WAXQ(FM) New York NY
WJJL(AM) Niagara Falls NY
*WRHO(FM) Oneonta NY
WTKV(FM) Oswego NY
WPDH(FM) Poughkeepsie NY
WRCN-FM Riverhead NY
WQRS(FM) Salamanca NY
WBPM(FM) Saugerties NY
WLLW(FM) Seneca Falls NY
WRGR(FM) Tupper Lake NY
WRCK(FM) Utica NY
WLKK(FM) Wethersfield Township NY
WQEL(FM) Bucyrus OH
*WCWT-FM Centerville OH
WOFX-FM Cincinnati OH
WNCX(FM) Cleveland OH
WLVQ(FM) Columbus OH
WFXN-FM Galion OH
WRYV(FM) Gallipolis OH
*WDUB(FM) Granville OH
WBWR(FM) Hilliard OH
WRBP(FM) Hubbard OH
*WKET(FM) Kettering OH
WEGE(FM) Lima OH
WXXF(FM) Loudonville OH
WAGX(FM) Manchester OH
WYRO(FM) McArthur OH
*WMWX(FM) Miamitown OH
*WLMH(FM) Morrow OH
WNKO(FM) Newark OH
WPFX-FM North Baltimore OH
WFXJ-FM North Kingsville OH
WBIK(FM) Pleasant City OH
WXKR(FM) Port Clinton OH
WZZZ(FM) Portsmouth OH
*WKTL(FM) Struthers OH
WZLR(FM) Xenia OH
KRKZ(FM) Altus OK
KWCO-FM Chickasha OK
KDDQ(FM) Comanche OK
KCDL(FM) Cordell OK
KTRX(FM) Dickson OK
*KHIM(FM) Mangum OK
KTMC-FM McAlester OK
KRXO(FM) Oklahoma City OK
KLOR-FM Ponca City OK
WBBZ(AM) Ponca City OK
KKBD(FM) Sallisaw OK
KJSR(FM) Tulsa OK
KWDQ(FM) Woodward OK
KBDN(FM) Bandon OR
KTWS(FM) Bend OR
KLOO-FM Corvallis OR
KZEL-FM Eugene OR
KAGO-FM Klamath Falls OR
KFEG(FM) Klamath Falls OR
KUBQ(FM) La Grande OR
KLCR(FM) Lakeview OR
KCRF-FM Lincoln City OR
KBOY-FM Medford OR
KGON(FM) Portland OR
KCRX-FM Seaside OR
KMSW(FM) The Dalles OR
KPPT-FM Toledo OR
KWLZ-FM Warm Springs OR
WBVE(FM) Bedford PA
*WBUQ(FM) Bloomsburg PA
WBUS(FM) Boalsburg PA
WMKX(FM) Brookville PA
*WCAL(FM) California PA
WUUZ(FM) Cooperstown PA
WWCB(AM) Corry PA
WBRX(FM) Cresson PA
WODE-FM Easton PA
WQHZ(FM) Erie PA
WQCM(FM) Greencastle PA
*WKVR-FM Huntingdon PA
WJNG(FM) Johnsonburg PA
WCXR(FM) Lewisburg PA
WCHX(FM) Lewistown PA
WRKT(FM) North East PA
WBXQ(FM) Patton PA
WMGK(FM) Philadelphia PA
WPZX(FM) Pocono Pines PA
WQFX-FM Russell PA
WEZX(FM) Scranton PA
WYFM(FM) Sharon PA
*WSRU(FM) Slippery Rock PA
WZXR(FM) South Williamsport PA
WMTT(FM) Tioga PA
*WNJR(FM) Washington PA
*WRLC(FM) Williamsport PA
WQXA-FM York PA
*WRIU(FM) Kingston RI
WROQ(FM) Anderson SC
WWBD(FM) Bamberg SC
WBZT-FM Mauldin SC
WRFQ(FM) Mt. Pleasant SC
WYAV(FM) Myrtle Beach SC
WQKI-FM Orangeburg SC
WMXT(FM) Pamplico SC
WMFX(FM) Saint Andrews SC
KSDN-FM Aberdeen SD
KYBB(FM) Canton SD
KDBX(FM) Clear Lake SD
KFXS(FM) Rapid City SD
KJRV(FM) Wessington Springs SD
*WAPX-FM Clarksville TN
WIJV(FM) Harriman TN
*WFHU(FM) Henderson TN
WQUT(FM) Johnson City TN
WIMZ-FM Knoxville TN
WKHT(FM) Knoxville TN
WEGR(FM) Memphis TN
WYNU(FM) Milan TN
WRKK-FM Sparta TN
KFGL(FM) Abilene TX
KRXB(FM) Beeville TX
KBTS(FM) Big Spring TX
KLUB(FM) Bloomington TX
KNFX-FM Bryan TX
KYYI(FM) Burkburnett TX
KARX(FM) Claude TX
KXIT-FM Dalhart TX
KZPS(FM) Dallas TX
KWMC(AM) Del Rio TX
KICA-FM Farwell TX
KFZX(FM) Gardendale TX
KPUS(FM) Gregory TX
KDBN(FM) Haltom City TX
KBRQ(FM) Hillsboro TX
KMFR(FM) Hondo TX
KKRW(FM) Houston TX
*KSHU(FM) Huntsville TX
KJKE(FM) Ingleside TX
KJKB(FM) Jacksboro TX
KBGE(AM) Kilgore TX
KKTX-FM Kilgore TX
KONE(FM) Lubbock TX
KZRC(FM) Markham TX
KHXS(FM) Merkel TX
KTBQ(FM) Nacogdoches TX
KNRG(FM) New Ulm TX
KIOC(FM) Orange TX
KBUS(FM) Paris TX
KJAZ(FM) Point Comfort TX
KWFR(FM) San Angelo TX
KZEP-FM San Antonio TX
KNRX(FM) Sterling City TX
KLTD(FM) Temple TX
KTAL-FM Texarkana TX
KNAL(AM) Victoria TX
KTXN-FM Victoria TX
KMAD-FM Whitesboro TX
*KAGJ(FM) Ephraim UT
KCUA(FM) Naples UT
KZHK(FM) Saint George UT
KRSP-FM Salt Lake City UT
WWRE(FM) Berryville VA
WWUZ(FM) Bowling Green VA
WBHB-FM Bridgewater VA
WCTG(FM) Chincoteague VA
WKLR(FM) Fort Lee VA
WWRT(FM) Strasburg VA
WAFX(FM) Suffolk VA
WIVI(FM) Charlotte Amalie VI
WKVT-FM Brattleboro VT
WCVR-FM Randolph VT
*WVTC(FM) Randolph Center VT
WTHK(FM) Wilmington VT
KDUX-FM Aberdeen WA
KISM(FM) Bellingham WA
KZPH(FM) Cashmere WA
KRQT(FM) Castle Rock WA
KVAB(FM) Clarkston WA
KRAO-FM Colfax WA
*KGHP(FM) Gig Harbor WA
KFMY(FM) Raymond WA
KJR-FM Seattle WA
KZOK-FM Seattle WA
KKZX(FM) Spokane WA
KZBD(FM) Spokane WA
KYNR(AM) Toppenish WA
KPQ-FM Wenatchee WA
WRLO-FM Antigo WI
WJJH(FM) Ashland WI
WRJO(FM) Eagle River WI
WECL(FM) Elk Mound WI
WMEQ-FM Menomonie WI
WKLH(FM) Milwaukee WI
WQBW(FM) Milwaukee WI
WOZZ(FM) New London WI
WTCX(FM) Ripon WI
WGMO(FM) Shell Lake WI
WTMB(FM) Tomah WI
WFBZ(FM) Trempealeau WI
WAUH(FM) Wautoma WI
WKBH-FM West Salem WI
WGLX-FM Wisconsin Rapids WI
*WVWC(FM) Buckhannon WV
WRZZ(FM) Elizabeth WV
WRLF(FM) Fairmont WV
WQZK-FM Keyser WV
WVKM(FM) Matewan WV
WCLG-FM Morgantown WV
WPMW(FM) Mullens WV
WXCR(FM) New Martinsville WV
WKOY-FM Princeton WV
WOBG-FM Salem WV
WFBY(FM) Weston WV
KASS(FM) Casper WY
KXXL(FM) Gillette WY
KZJH(FM) Jackson WY
KRQU(FM) Laramie WY
KRVK(FM) Midwest WY
KCGL(FM) Powell WY
KSIT(FM) Rock Springs WY
KZWY(FM) Sheridan WY

Classical

KLEF(FM) Anchorage AK
*KBRW-FM Barrow AK
*KUAC(FM) Fairbanks AK
*WBHM(FM) Birmingham AL
*WRWA(FM) Dothan AL
*WSGN(FM) Gadsden AL
*WLRH(FM) Huntsville AL
*WLJS-FM Jacksonville AL
*WHIL-FM Mobile AL
*WTSU(FM) Montgomery-Troy AL
*WQPR(FM) Muscle Shoals AL
*WAPR(FM) Selma AL
*WUAL-FM Tuscaloosa AL
*KBSA(FM) El Dorado AR
*KUAF(FM) Fayetteville AR
*KASU(FM) Jonesboro AR
*KLRE-FM Little Rock AR
*KNAU(FM) Flagstaff AZ
*KNAG(FM) Grand Canyon AZ
*KBAQ-FM Phoenix AZ
*KNAA(FM) Show Low AZ
*KUAT-FM Tucson AZ
*KAWC-FM Yuma AZ
*KPRX(FM) Bakersfield CA
KMZT(AM) Beverly Hills CA
*KQVO(FM) Calexico CA
KBOQ(FM) Carmel CA
*KCHO(FM) Chico CA
*KVPR(FM) Fresno CA
*KXSR(FM) Groveland CA
*KUSC(FM) Los Angeles CA
*KCSN(FM) Northridge CA
*KPSC(FM) Palm Springs CA
*KNHT(FM) Rio Dell CA
*KXPR(FM) Sacramento CA
*KPBS-FM San Diego CA
KDFC-FM San Francisco CA
*KCBX(FM) San Luis Obispo CA
KDB(FM) Santa Barbara CA
*KQSC(FM) Santa Barbara CA
*KSBX(FM) Santa Barbara CA
*KRCB-FM Santa Rosa CA
*KUOP(FM) Stockton CA
*KKTO(FM) Tahoe City CA
*KDSC(FM) Thousand Oaks CA
KVEN(AM) Ventura CA
*KNYR(FM) Yreka CA
*KAJX(FM) Aspen CO
*KCJX(FM) Carbondale CO
*KVOV(FM) Carbondale CO
*KPRU(FM) Delta CO
*KVOD(FM) Denver CO
*KCME(FM) Manitou Springs CO
*KCFP(FM) Pueblo CO
*KPRE(FM) Vail CO
*WSHU(FM) Fairfield CT
*WGRS(FM) Guilford CT
*WJMJ(FM) Hartford CT
*WMNR(FM) Monroe CT
*WSLX(FM) New Canaan CT
*WRXC(FM) Shelton CT
*WGSK(FM) South Kent CT
WCCC(AM) West Hartford CT
*WRTX(FM) Dover DE
WTMP-FM Dade City FL
*WGCU-FM Fort Myers FL
*WQCS(FM) Fort Pierce FL
*WUFT-FM Gainesville FL
*WJCT-FM Jacksonville FL
*WMKO(FM) Marco FL
WKAT(AM) North Miami FL
*WMFE-FM Orlando FL
*WUWF(FM) Pensacola FL
*WFSQ(FM) Tallahassee FL
*WUSF(FM) Tampa FL
*WXEL(FM) West Palm Beach FL
*WPRK(FM) Winter Park FL
*WUNV(FM) Albany GA
*WUGA(FM) Athens GA
*WABE(FM) Atlanta GA
*WACG-FM Augusta GA
*WWIO-FM Brunswick GA
*WUWG(FM) Carrollton GA
*WMUM-FM Cochran GA
*WTJB(FM) Columbus GA
*WNGU(FM) Dahlonega GA
*WPPR(FM) Demorest GA
*WJWV(FM) Fort Gaines GA
WGPB(FM) Rome GA
WWIO(AM) Saint Mary's GA
*WSVH(FM) Savannah GA
*WFSL(FM) Thomasville GA
*WABR-FM Tifton GA
*WWET(FM) Valdosta GA
*WJSP-FM Warm Springs GA
*WXVS(FM) Waycross GA
*KPRG(FM) Hagatna GU
*KANO(FM) Hilo HI
*KHPR(FM) Honolulu HI
*KKUA(FM) Wailuku HI
*WOI-FM Ames IA
*KWOI(FM) Carroll IA
*KHKE(FM) Cedar Falls IA
*KLCD(FM) Decorah IA
*KHOE(FM) Fairfield IA
*KTPR(FM) Fort Dodge IA
*KSUI(FM) Iowa City IA
*KOWI(FM) Lamoni IA
*KOJI(FM) Okoboji IA
*KWIT(FM) Sioux City IA
*KBSU-FM Boise ID
*KIBX(FM) Bonners Ferry ID
*KNWO(FM) Cottonwood ID
*KBSM(FM) McCall ID
*KRFA-FM Moscow ID
*KBYI(FM) Rexburg ID
*KWRV(FM) Sun Valley ID
*KBSW(FM) Twin Falls ID
*WSIU(FM) Carbondale IL
WFMT(FM) Chicago IL
*WNIE(FM) Freeport IL
*WNIW(FM) La Salle IL
*WIUM(FM) Macomb IL
*WVSI(FM) Mount Vernon IL
*WUSI(FM) Olney IL
*WCBU(FM) Peoria IL
*WIPA(FM) Pittsfield IL
*WQUB(FM) Quincy IL
*WVIK(FM) Rock Island IL
*WNIU(FM) Rockford IL
*WUIS(FM) Springfield IL
*WNIQ(FM) Sterling IL
*WILL-FM Urbana IL
*WIUW(FM) Warsaw IL
*WETN(FM) Wheaton IL
*WBSB(FM) Anderson IN
*WFIU(FM) Bloomington IN
*WNIN-FM Evansville IN
*WBSH(FM) Hagerstown IN
*WFYI-FM Indianapolis IN
*WICR(FM) Indianapolis IN
*WBSW(FM) Marion IN
*WBST(FM) Muncie IN

*WSND-FM Notre Dame IN
*WCKZ(FM) Orland IN
*WBSJ(FM) Portland IN
*WECI(FM) Richmond IN
*WBNI-FM Roanoke IN
*WBAA-FM West Lafayette IN
*KANH(FM) Emporia KS
*KANZ(FM) Garden City KS
*KHCT(FM) Great Bend KS
*KZAN(FM) Hays KS
*KZNA(FM) Hill City KS
*KHCC-FM Hutchinson KS
KXTR(AM) Kansas City KS
*KANU(FM) Lawrence KS
*KANV(FM) Olsburg KS
*KRPS(FM) Pittsburg KS
*KHCD(FM) Salina KS
*WKYU-FM Bowling Green KY
*WEKF(FM) Corbin KY
*WKUE(FM) Elizabethtown KY
*WEKH(FM) Hazard KY
*WKPB(FM) Henderson KY
*WUOL(FM) Louisville KY
*WEKU(FM) Richmond KY
*WDCL-FM Somerset KY
WSKV(FM) Stanton KY
*KLSA(FM) Alexandria LA
*WRKF(FM) Baton Rouge LA
*WWNO(FM) New Orleans LA
*KDAQ(FM) Shreveport LA
*KTLN(FM) Thibodaux LA
*WFCR(FM) Amherst MA
*WGBH(FM) Boston MA
WHRB(FM) Cambridge MA
WFCC-FM Chatham MA
WCRB(FM) Lowell MA
*WBJC(FM) Baltimore MD
*WFWM(FM) Frostburg MD
*WSCL(FM) Salisbury MD
*WMEH(FM) Bangor ME
*WMED(FM) Calais ME
*WMEP(FM) Camden ME
*WMEF(FM) Fort Kent ME
WBQQ(FM) Kennebunk ME
*WMEA(FM) Portland ME
*WMEM(FM) Presque Isle ME
WBQW(FM) Scarborough ME
WBQX(FM) Thomaston ME
*WMEW(FM) Waterville ME
*WYAR(FM) Yarmouth ME
*WCML-FM Alpena MI
*WAUS(FM) Berrien Springs MI
*WICV(FM) East Jordan MI
*WKAR-FM East Lansing MI
*WBLU-FM Grand Rapids MI
*WGGL-FM Houghton MI
*WIAA(FM) Interlochen MI
*WMUK(FM) Kalamazoo MI
*WIAB(FM) Mackinaw City MI
*WNMU-FM Marquette MI
*WCMU-FM Mount Pleasant MI
*WCMB-FM Oscoda MI
WKLZ-FM Petoskey MI
*WCMZ-FM Sault Ste. Marie MI
WWCM(FM) Standish MI
*WBLV(FM) Twin Lake MI
*KRSU(FM) Appleton MN
*KCRB-FM Bemidji MN
*KBPR(FM) Brainerd MN
*KSJR-FM Collegeville MN
*KSJR-FM Collegeville MN
*WSCD-FM Duluth MN
*KCMF(FM) Fergus Falls MN
*WMLS(FM) Grand Marais MN
*KSJN(FM) Minneapolis MN
*KCCM-FM Moorhead MN
*KLSE-FM Rochester MN
*KGAC(FM) Saint Peter MN
*KQMN(FM) Thief River Falls MN
*WIRR(FM) Virginia-Hibbing MN
*KRSW(FM) Worthington MN
*KRCU(FM) Cape Girardeau MO
*KRNW(FM) Chillicothe MO
KFUO-FM Clayton MO
*KBIA(FM) Columbia MO
*KSEF(FM) Farmington MO
*KXMS(FM) Joplin MO
*KKTR(FM) Kirksville MO
*KWJC(FM) Liberty MO
*KXCV(FM) Maryville MO
*KSMS-FM Point Lookout MO
*KMST(FM) Rolla MO
*KSMU(FM) Springfield MO
*KSMW(FM) West Plains MO
*WMAH-FM Biloxi MS
*WMAE-FM Booneville MS
*WMAU-FM Bude MS
*WMAO-FM Greenwood MS
*WUSM-FM Hattiesburg MS
*WMPN-FM Jackson MS
*WMAW-FM Meridian MS
*WMAB-FM Mississippi State MS
*WMAV-FM Oxford MS
*KEMC(FM) Billings MT
*KBMC(FM) Bozeman MT
*KAPC(FM) Butte MT
*KUFN(FM) Hamilton MT
*KNMC(FM) Havre MT
*KUHM(FM) Helena MT
*KUKL(FM) Kalispell MT
KPRK(AM) Livingston MT
*KUFM(FM) Missoula MT
*WCQS(FM) Asheville NC
*WDAV(FM) Davidson NC
*WFQS(FM) Franklin NC
*WTEB(FM) New Bern NC
*WZRN(FM) Norlina NC
*WCPE(FM) Raleigh NC
*WHQR(FM) Wilmington NC
*WFDD-FM Winston-Salem NC
*KCND(FM) Bismarck ND
*KDPR(FM) Dickinson ND
*KUND-FM Grand Forks ND
*KPRJ(FM) Jamestown ND
*KMPR(FM) Minot ND
*KPPR(FM) Williston ND
*KTNE-FM Alliance NE
*KMNE-FM Bassett NE
*KCNE-FM Chadron NE
*KHNE-FM Hastings NE
*KLNE-FM Lexington NE
*KUCV(FM) Lincoln NE
*KRNE-FM Merriman NE
*KXNE-FM Norfolk NE
*KPNE-FM North Platte NE
*KIOS-FM Omaha NE
*KVNO(FM) Omaha NE
*WWCJ(FM) Cape May NJ
*WWNJ(FM) Dover Township NJ
*WRTQ(FM) Ocean City NJ
WPRB(FM) Princeton NJ
*WWFM(FM) Trenton NJ
*KSJE(FM) Farmington NM
*KRWG(FM) Las Cruces NM
*KMTH(FM) Maljamar NM
*KENW-FM Portales NM
KHFM(FM) Santa Fe NM
*KRNM(FM) Chalan Kanoa-Saipan NP
*KNCC(FM) Elko NV
*KCNV(FM) Las Vegas NV
KXPT(FM) Las Vegas NV
*KUNR(FM) Reno NV
KSVL(FM) Smith NV
*KTPH(FM) Tonopah NV
WEXT(FM) Amsterdam NY
*WSKG-FM Binghamton NY
*WNED-FM Buffalo NY
*WSQE(FM) Corning NY
*WJSL(FM) Houghton NY
*WSQG-FM Ithaca NY
*WNJA(FM) Jamestown NY
*WKCR-FM New York NY
*WNYC-FM New York NY
WQXR-FM New York NY
*WSQC-FM Oneonta NY
*WRHV(FM) Poughkeepsie NY
*WXXI-FM Rochester NY
*WMHT-FM Schenectady NY
*WCNY-FM Syracuse NY
*WUNY(FM) Utica NY
*WJNY(FM) Watertown NY
*WGBE(FM) Bryan OH
*WGUC(FM) Cincinnati OH
WMJK(FM) Clyde OH
*WOSU-FM Columbus OH
*WOSE(FM) Coshocton OH
*WGDE(FM) Defiance OH
*WDPG(FM) Greenville OH
*WKSU-FM Kent OH
*WGLE(FM) Lima OH
WCLV(FM) Lorain OH
*WOSV(FM) Mansfield OH
*WMRT(FM) Marietta OH
*WOSB(FM) Marion OH
*WKRJ(FM) New Philadelphia OH
*WNRK(FM) Norwalk OH
*WOSP(FM) Portsmouth OH
*WKSV(FM) Thompson OH
*WGTE-FM Toledo OH
*WDPR(FM) West Carrollton OH
*WKRW(FM) Wooster OH
*WYSU(FM) Youngstown OH
*KOCU(FM) Altus OK
*KLCU(FM) Ardmore OK
*KYCU(FM) Clinton OK
*KCSC(FM) Edmond OK
*KCCU(FM) Lawton OK
*KBCW-FM McAlester OK
*KOSU(FM) Stillwater OK
*KWTU(FM) Tulsa OK
*KSOR(FM) Ashland OR
*KSRG(FM) Ashland OR
*KWAX(FM) Eugene OR
*KWVZ(FM) Florence OR
*KQHR(FM) Hood River OR
*KLMF(FM) Klamath Falls OR
*KOOZ(FM) Myrtle Point OR
*KBPS-FM Portland OR
KCMD(AM) Portland OR
*KWRX(FM) Redmond OR
*KSRS(FM) Roseburg OR
*WDIY(FM) Allentown PA
*WMCE(FM) Erie PA
*WQLN-FM Erie PA
*WSAJ-FM Grove City PA
*WITF-FM Harrisburg PA
*WRTY(FM) Jackson Township PA
*WQEJ(FM) Johnstown PA
*WPSX(FM) Kane PA
*WWPJ(FM) Pen Argyl PA
*WRTI(FM) Philadelphia PA
*WQED-FM Pittsburgh PA
*WVIA-FM Scranton PA
*WPSU(FM) State College PA
*WJAZ(FM) Summerdale PA
*WVYA(FM) Williamsport PA
*WRUO(FM) Mayaguez PR
*WIPR-FM San Juan PR
*WRTU(FM) San Juan PR
WCRI(FM) Block Island RI
*WSCI(FM) Charleston SC
*WLTR(FM) Columbia SC
*WHMC-FM Conway SC
*WEPR(FM) Greenville SC
*KESD(FM) Brookings SD
KPSD(FM) Faith SD
*KQSD-FM Lowry SD
*KDSD-FM Pierpont SD
*KBHE-FM Rapid City SD
*KTSD-FM Reliance SD
*KCSD(FM) Sioux Falls SD
*KRSD(FM) Sioux Falls SD
*KUSD(FM) Vermillion SD
*WSMC-FM Collegedale TN
*WHRS(FM) Cookeville TN
*WFHU(FM) Henderson TN
*WKNP(FM) Jackson TN
*WETS(FM) Johnson City TN
*WETS(FM) Johnson City TN
*WCSK(FM) Kingsport TN
*WUOT(FM) Knoxville TN
*WKNO-FM Memphis TN
*WPLN-FM Nashville TN
*WTML(FM) Tullahoma TN
*KACU(FM) Abilene TX
*KMFA(FM) Austin TX
*KVLU(FM) Beaumont TX
*KAMU-FM College Station TX
*KEDT-FM Corpus Christi TX
WRR(FM) Dallas TX
*KTEP(FM) El Paso TX
*KTCU-FM Fort Worth TX
*KMBH-FM Harlingen TX
*KUHF(FM) Houston TX
*KSHU(FM) Huntsville TX
*KTXI(FM) Ingram TX
*KNCT-FM Killeen TX
*KOHM(FM) Lubbock TX
*KLDN(FM) Lufkin TX
*KHID(FM) McAllen TX
*KOCV(FM) Odessa TX
*KPAC(FM) San Antonio TX
*KTOT(FM) Spearman TX
*KTXK(FM) Texarkana TX
*KVRT(FM) Victoria TX
*KWBU-FM Waco TX
*KMCU(FM) Wichita Falls TX
*KUSR(FM) Logan UT
*KUSU-FM Logan UT
*KPCW(FM) Park City UT
*KBYU-FM Provo UT
*KUER(FM) Salt Lake City UT
*WVTU(FM) Charlottesville VA
*WVTW(FM) Charlottesville VA
*WFOS(FM) Chesapeake VA
*WFFC(FM) Ferrum VA
*WMRA(FM) Harrisonburg VA
*WMRL(FM) Lexington VA
*WVTR(FM) Marion VA
*WHRO-FM Norfolk VA
*WVTF(FM) Roanoke VA
WBQK(FM) West Point VA
*WISE-FM Wise VA
WBTN-FM Bennington VT
*WVPS(FM) Burlington VT
*WNCH(FM) Norwich VT
*WRVT(FM) Rutland VT
*WVPA(FM) Saint Johnsbury VT
WCVT(FM) Stowe VT
*WVTQ(FM) Sunderland VT
*WVPR(FM) Windsor VT
*KZAZ(FM) Bellingham WA
*KNWV(FM) Clarkston WA
*KNWR(FM) Ellensburg WA
KLDY(AM) Lacey WA
KOHO-FM Leavenworth WA
*KQWS(FM) Omak WA
*KNWP(FM) Port Angeles WA
*KWSU(AM) Pullman WA
*KFAE-FM Richland WA
KING-FM Seattle WA
*KPBX-FM Spokane WA
*KNWY(FM) Yakima WA
*WHSA(FM) Brule WI
*WUEC(FM) Eau Claire WI
*WPNE-FM Green Bay WI
*WGTD(FM) Kenosha WI
*WHLA(FM) La Crosse WI
*WLSU(FM) La Crosse WI
*WERN(FM) Madison WI
*WORT(FM) Madison WI
*WVSS(FM) Menomonie WI
*WSSW(FM) Platteville WI
*WXPR(FM) Rhinelander WI
*WHND(FM) Sister Bay WI
*WHRM(FM) Wausau WI
*WVPB(FM) Beckley WV
*WVPW(FM) Buckhannon WV
*WVPN(FM) Charleston WV
*WVWV(FM) Huntington WV
*WVEP(FM) Martinsburg WV
*WVPM(FM) Morgantown WV
*WVPG(FM) Parkersburg WV
*WAUA(FM) Petersburg WV
*WVNP(FM) Wheeling WV
*KUWA(FM) Afton WY
*KDUW(FM) Douglas WY
*KUWG(FM) Gillette WY
*KUWJ(FM) Jackson WY
*KUWR(FM) Laramie WY
*KUWN(FM) Newcastle WY
*KUWX(FM) Pinedale WY
*KUWP(FM) Powell WY
*KUWZ(FM) Rock Springs WY
*KPRQ(FM) Sheridan WY
*KSUW(FM) Sheridan WY
*KUWD(FM) Sundance WY
KUWT(FM) Thermopolis WY

Comedy

KXDJ(FM) Spearman TX

Contemporary Hit/Top-40

KFAT(FM) Anchorage AK
KGOT(FM) Anchorage AK
KAKQ-FM Fairbanks AK
KWLF(FM) Fairbanks AK
KSLD(AM) Soldotna AK
WZYP(FM) Athens AL
WZBQ(FM) Carrollton AL
WAGF-FM Dothan AL
WKMX(FM) Enterprise AL
WKZJ(FM) Eufaula AL
WABB-FM Mobile AL
WHHY-FM Montgomery AL
WQEN(FM) Trussville AL
WSLY(FM) York AL
*KSWH(FM) Arkadelphia AR
KMRX(FM) El Dorado AR
KHTE-FM England AR
KISR(FM) Fort Smith AR
KIYS(FM) Jonesboro AR
KKPT(FM) Little Rock AR
KMXF(FM) Lowell AR
*KVRN(FM) Marvell AR
KMCK(FM) Siloam Springs AR
KKHJ-FM Pago Pago AS
KCDQ(FM) Douglas AZ
KOHT(FM) Marana AZ
KZZP(FM) Mesa AZ
KIKO(AM) Miami AZ
KIDR(AM) Phoenix AZ
KVIB(FM) Sun City West AZ
KRQQ(FM) Tucson AZ
KLJZ(FM) Yuma AZ
KEWB(FM) Anderson CA
KBKO-FM Bakersfield CA
KISV(FM) Bakersfield CA
KDUC(FM) Barstow CA
KXSB(FM) Big Bear Lake CA
KSIQ(FM) Brawley CA
KBFP-FM Delano CA
KHTS-FM El Cajon CA
KFMI(FM) Eureka CA
KSSD(FM) Fallbrook CA
KWPT(FM) Fortuna CA
*KOHL(FM) Fremont CA
KWYE(FM) Fresno CA
KQCM(FM) Joshua Tree CA
KOKO-FM Kerman CA
KEXA(FM) King City CA
KSXE(FM) Kingsburg CA
KWIN(FM) Lodi CA
KIIS-FM Los Angeles CA
KPWR(FM) Los Angeles CA
KDUQ(FM) Ludlow CA
KSXY(FM) Middletown CA
KVVS(FM) Mojave CA
*KSMC(FM) Moraga CA
*KSFH(FM) Mountain View CA
*KLFH(FM) Ojai CA
KCAQ(FM) Oxnard CA
KPSI-FM Palm Springs CA
KHTN(FM) Planada CA
KNLF(FM) Quincy CA
*KRVH(FM) Rio Vista CA
KGGI(FM) Riverside CA
KDND(FM) Sacramento CA
KRAY-FM Salinas CA
KYZZ(FM) Salinas CA
KMEL(FM) San Francisco CA
KYLD(FM) San Francisco CA
KWWV(FM) Santa Margarita CA
KDLD(FM) Santa Monica CA
KLJR-FM Santa Paula CA
KNNN(FM) Shasta Lake City CA

KBDS(FM) Taft CA
KBOS-FM Tulare CA
KWNN(FM) Turlock CA
KCDZ(FM) Twentynine Palms CA
KWNE(FM) Ukiah CA
*KDUV(FM) Visalia CA
KSEQ(FM) Visalia CA
*KASF(FM) Alamosa CO
KUUR(FM) Carbondale CO
KMGJ(FM) Grand Junction CO
WQGN-FM Groton CT
WKCI-FM Hamden CT
WKSS(FM) Hartford CT
WWRX(FM) Pawcatuck CT
WILI-FM Willimantic CT
WIHT(FM) Washington DC
*WMPH(FM) Wilmington DE
WFCT(FM) Apalachicola FL
WXKB(FM) Cape Coral FL
WJHM(FM) Daytona Beach FL
WFFY(FM) Destin FL
WRLZ(AM) Eatonville FL
WHYI-FM Fort Lauderdale FL
WLDI(FM) Fort Pierce FL
WZNS(FM) Fort Walton Beach FL
WHTF(FM) Havana FL
WVYB(FM) Holly Hill FL
WIFL(FM) Inglis FL
WAPE-FM Jacksonville FL
WEOW(FM) Key West FL
WXHT(FM) Madison FL
WAOA-FM Melbourne FL
WEDR(FM) Miami FL
WMYM(AM) Miami FL
WPOW(FM) Miami FL
WBTT(FM) Naples Park FL
WILN(FM) Panama City FL
WPFM-FM Panama City FL
WPRY(AM) Perry FL
WWMI(AM) Saint Petersburg FL
WFLZ-FM Tampa FL
WXXL(FM) Tavares FL
V6AH(AM) Pohnpei FM
*V6AI(AM) Yap FM
WQVE(FM) Albany GA
WCGQ(FM) Columbus GA
WXMK(FM) Dock Junction GA
WKKZ(FM) Dublin GA
WELT(FM) East Dublin GA
WOAH(FM) Glennville GA
WVFJ-FM Manchester GA
WDRR(FM) Martinez GA
WMGB(FM) Montezuma GA
WAEV(FM) Savannah GA
WZAT(FM) Savannah GA
WSTR(FM) Smyrna GA
WJFL(FM) Tennille GA
KOKU(FM) Hagatna GU
KPMW(FM) Haliimaile HI
KNWB(FM) Hilo HI
KIKI-FM Honolulu HI
KPHW(FM) Kaneohe HI
KNUQ(FM) Paauilo HI
KSRF(FM) Poipu HI
KJMD(FM) Pukalani HI
KDDB(FM) Waipahu HI
KCCQ(FM) Ames IA
KSWI(FM) Atlantic IA
KZAT-FM Belle Plaine IA
KZIA(FM) Cedar Rapids IA
KKDM(FM) Des Moines IA
KLYV(FM) Dubuque IA
KRTI(FM) Grinnell IA
KCJJ(AM) Iowa City IA
KBEA-FM Muscatine IA
KSMA-FM Osage IA
KOTM-FM Ottumwa IA
KIWA-FM Sheldon IA
KTPZ(FM) Hazelton ID
KFTZ(FM) Idaho Falls ID
KVTY(FM) Lewiston ID
KZFN(FM) Moscow ID
KZMG(FM) New Plymouth ID
WKIB(FM) Anna IL
WERV-FM Aurora IL
WBNQ(FM) Bloomington IL
WCDD(FM) Canton IL
WCIL-FM Carbondale IL
WQKQ(FM) Carthage IL
WLRW(FM) Champaign IL
WBBM-FM Chicago IL
WKSC-FM Chicago IL
WRPW(FM) Colfax IL
WSOY-FM Decatur IL
WYDS(FM) Decatur IL
WOJO(FM) Evanston IL
WISH-FM Galatia IL
WVZA(FM) Herrin IL
WXAJ(FM) Hillsboro IL
WVLI(FM) Kankakee IL
WEAI(FM) Lynnville IL
*WLKL(FM) Mattoon IL
WZPW(FM) Peoria IL
WLCE(FM) Petersburg IL
*WGCA-FM Quincy IL
WQQB(FM) Rantoul IL
WZOK(FM) Rockford IL
WIVQ(FM) Spring Valley IL
WDBR(FM) Springfield IL
WSTQ(FM) Streator IL
WKRV(FM) Vandalia IL
WBWB(FM) Bloomington IN
WNHT(FM) Churubusco IN
WINN(FM) Columbus IN
WIMC(FM) Crawfordsville IN
WXXB(FM) Delphi IN
*WFCI(FM) Franklin IN
*WHWE(FM) Howe IN
*WVSH(FM) Huntington IN
*WBDG(FM) Indianapolis IN
*WEDM(FM) Indianapolis IN
WHHH(FM) Indianapolis IN
WNOU(FM) Indianapolis IN
WAZY-FM Lafayette IN
*WCYT(FM) Lafayette Township IN
WLHM(FM) Logansport IN
WXXC(FM) Marion IN
*WNAS(FM) New Albany IN
WJFX(FM) New Haven IN
WDKS(FM) Newburgh IN
WUME-FM Paoli IN
WRDZ-FM Plainfield IN
WZKF(FM) Salem IN
WNDV-FM South Bend IN
WMGI(FM) Terre Haute IN
*WVUB(FM) Vincennes IN
KDGS(FM) Andover KS
KIBB(FM) Augusta KS
KMOQ(FM) Baxter Springs KS
*KTCC(FM) Colby KS
KZCH(FM) Derby KS
KLZR(FM) Lawrence KS
KMXN(FM) Osage City KS
*KTJO-FM Ottawa KS
KACZ(FM) Riley KS
KSAL-FM Salina KS
KSKU(FM) Sterling KS
*KYWA(FM) Wichita KS
WWKF(FM) Fulton KY
WZQQ(FM) Hyden KY
WTHX(FM) Lebanon Junction KY
WLKT(FM) Lexington-Fayette KY
WDJX(FM) Louisville KY
WFTM-FM Maysville KY
WBTF(FM) Midway KY
WEGI(FM) Oak Grove KY
WSTO(FM) Owensboro KY
WDDJ(FM) Paducah KY
*WGCF(FM) Paducah KY
WZLK(FM) Virgie KY
KQID(FM) Alexandria LA
KQIS(FM) Basile LA
WFMF(FM) Baton Rouge LA
KTSR(FM) De Quincy LA
KRKA(FM) Erath LA
WYLK(FM) Lacombe LA
KSMB(FM) Lafayette LA
KNOE-FM Monroe LA
WEZB(FM) New Orleans LA
KRUF(FM) Shreveport LA
WFHN(FM) Fairhaven MA
*WGAO(FM) Franklin MA
WBOQ(FM) Gloucester MA
WXKS-FM Medford MA
*WSDH(FM) Sandwich MA
*WYAJ(FM) Sudbury MA
WTRI(AM) Brunswick MD
WPGC-FM Morningside MD
WMME-FM Augusta ME
WWBX(FM) Bangor ME
WBZN(FM) Old Town ME
WJBQ(FM) Portland ME
WRED(FM) Saco ME
*WSJB-FM Standish ME
*WAHS(FM) Auburn Hills MI
WKFR-FM Battle Creek MI
WYBR(FM) Big Rapids MI
*WBFH(FM) Bloomfield Hills MI
WSMK(FM) Buchanan MI
WKHQ-FM Charlevoix MI
WCFX(FM) Clare MI
*WPRJ(FM) Coleman MI
WWCK(AM) Flint MI
WWCK-FM Flint MI
WRCL(FM) Frankenmuth MI
WKMJ-FM Hancock MI
*WHPR(FM) Highland Park MI
WMAX-FM Holland MI
WUPS(FM) Houghton Lake MI
WHMI-FM Howell MI
WUPM(FM) Ironwood MI
WVIC(FM) Jackson MI
WHZZ(FM) Lansing MI
WJIM-FM Lansing MI
WTWR-FM Luna Pier MI
WSNX-FM Muskegon MI
WKQS-FM Negaunee MI
*WORW(FM) Port Huron MI
WYSS(FM) Sault Ste. Marie MI
KXRZ(FM) Alexandria MN
*KBSB(FM) Bemidji MN
KCLH(FM) Caledonia MN
KTTB(FM) Glencoe MN
WTBX(FM) Hibbing MN
KDWB-FM Richfield MN
KROC-FM Rochester MN
KCLD-FM Saint Cloud MN
KSYN(FM) Joplin MO
KMXV(FM) Kansas City MO
*KWJC(FM) Liberty MO
KKBL(FM) Monett MO
*KCLC(FM) Saint Charles MO
KSLZ(FM) Saint Louis MO
KSPW(FM) Sparta MO
KWKJ(FM) Windsor MO
WYOY(FM) Gluckstadt MS
WXYK(FM) Gulfport MS
WNSL(FM) Laurel MS
WQYZ(FM) Ocean Springs MS
KISN(FM) Belgrade MT
KBEV-FM Dillon MT
KIKC(AM) Forsyth MT
KOFI(AM) Kalispell MT
KRSQ(FM) Laurel MT
KPLN(FM) Lockwood MT
WQZL(FM) Belhaven NC
WNKS(FM) Charlotte NC
WDCG(FM) Durham NC
WQSM(FM) Fayetteville NC
WQSL(FM) Jacksonville NC
WRHT(FM) Morehead City NC
WKBC-FM North Wilkesboro NC
WRHD(FM) Williamston NC
WKZL(FM) Winston-Salem NC
KKCT(FM) Bismarck ND
KKXL-FM Grand Forks ND
*KCNT(FM) Hastings NE
KFRX(FM) Lincoln NE
WJYY(FM) Concord NH
WKNE(FM) Keene NH
WXXS(FM) Lancaster NH
WGBZ(FM) Cape May Court House NJ
WSJQ(FM) North Cape May NJ
*WBZC(FM) Pemberton NJ
WZBZ(FM) Pleasantville NJ
WPST(FM) Trenton NJ
KYEE(FM) Alamogordo NM
KDRF(FM) Albuquerque NM
*KNMI(FM) Farmington NM
KAZX(FM) Kirtland NM
KHQT(FM) Las Cruces NM
KBCQ-FM Roswell NM
KKSS(FM) Santa Fe NM
KPXP(FM) Garapan-Saipan NP
KZTQ(FM) Carson City NV
KLUC-FM Las Vegas NV
KQRT(FM) Las Vegas NV
KVEG(FM) Mesquite NV
KWNZ(FM) Sun Valley NV
*WBXL(FM) Baldwinsville NY
WKKF(FM) Ballston Spa NY
*WXBA(FM) Brentwood NY
*WBSU(FM) Brockport NY
*WKRB(FM) Brooklyn NY
WTSS(FM) Buffalo NY
*WCIY(FM) Canandaigua NY
WBDR(FM) Cape Vincent NY
WYUL(FM) Chateaugay NY
WWYL(FM) Chenango Bridge NY
WBDI(FM) Copenhagen NY
WNKI(FM) Corning NY
WDHI(FM) Delhi NY
WLVY(FM) Elmira NY
WMRV-FM Endicott NY
*WCID(FM) Friendship NY
WFNY(AM) Gloversville NY
WKGS(FM) Irondequoit NY
WKTU(FM) Lake Success NY
WKZA(FM) Lakewood NY
WSKU(FM) Little Falls NY
WYSX(FM) Morristown NY
WQHT(FM) New York NY
WKSE(FM) Niagara Falls NY
WBDB(FM) Ogdensburg NY
*WOSS(FM) Ossining NY
WBLI(FM) Patchogue NY
WPKF(FM) Poughkeepsie NY
WSPK(FM) Poughkeepsie NY
WPXY-FM Rochester NY
WENU(AM) South Glens Falls NY
*WJPZ-FM Syracuse NY
WNTQ(FM) Syracuse NY
WWHT(FM) Syracuse NY
WFLY(FM) Troy NY
WAJZ(FM) Voorheesville NY
WKBE(FM) Warrensburg NY
WCIZ-FM Watertown NY
WSKS(FM) Whitesboro NY
*WONB(FM) Ada OH
*WZIP(FM) Akron OH
WYJK-FM Bellaire OH
WWMK(AM) Cleveland OH
*WUFM(FM) Columbus OH
WLWD(FM) Columbus Grove OH
WGTZ(FM) Eaton OH
WXXR(FM) Fredericktown OH
WBKS(FM) Ironton OH
WRVB(FM) Marietta OH
WDIF(FM) Marion OH
*WYSZ(FM) Maumee OH
WYVK(FM) Middleport OH
WKFS(FM) Milford OH
WXZQ(FM) Piketon OH
WVKF(FM) Shadyside OH
WVKS(FM) Toledo OH
WWSR(FM) Wapakoneta OH
WNKL(FM) Wauseon OH
*WYSA(FM) Wauseon OH
*WCWS(FM) Wooster OH
WHOT-FM Youngstown OH
KKWD(FM) Bethany OK
KTBT(FM) Broken Arrow OK
*KSSU(FM) Durant OK
KTIJ(FM) Elk City OK
KHTT(FM) Muskogee OK
KJYO(FM) Oklahoma City OK
KZBB(FM) Poteau OK
KIFS(FM) Ashland OR
KXIX(FM) Bend OR
KQHC(FM) Burns OR
KDUK-FM Florence OR
KLDR(FM) Harbeck-Fruitdale OR
KKRB(FM) Klamath Falls OR
*KEOL(FM) La Grande OR
KTEE(FM) North Bend OR
*KMKR(FM) Oakridge OR
KKRZ(FM) Portland OR
KXJM(FM) Portland OR
KEUG(FM) Veneta OR
WAEB-FM Allentown PA
WHOL(AM) Allentown PA
WWOT(FM) Altoona PA
*WCUC-FM Clarion PA
WRTS(FM) Erie PA
*WZZE(FM) Glen Mills PA
WNNK-FM Harrisburg PA
WJHT(FM) Johnstown PA
WLAN-FM Lancaster PA
*WNTE(FM) Mansfield PA
WBHT(FM) Mountain Top PA
WBHD(FM) Olyphant PA
WIOQ(FM) Philadelphia PA
WRDW-FM Philadelphia PA
WKST-FM Pittsburgh PA
WAVT-FM Pottsville PA
*WYBF(FM) Radnor Township PA
WRFY-FM Reading PA
WAKZ(FM) Sharpsville PA
WQWK(FM) State College PA
WQKX(FM) Sunbury PA
WKRF(FM) Tobyhanna PA
WGMR(FM) Tyrone PA
*WXVU(FM) Villanova PA
WNBT-FM Wellsboro PA
WKRZ(FM) Wilkes-Barre PA
WKSB(FM) Williamsport PA
WQXA-FM York PA
WYCR(FM) York-Hanover PA
WCMN-FM Arecibo PR
WBQN(AM) Barceloneta-Manati PR
WODA(FM) Bayamon PR
WXLX(FM) Lajas PR
WAEL-FM Maricao PR
WPRA(AM) Mayaguez PR
WUKQ-FM Mayaguez PR
WEKO(AM) Morovis PR
WEXS(AM) Patillas PR
WEGM(FM) San German PR
WENA(AM) Yauco PR
*WCVY(FM) Coventry RI
WPRO-FM Providence RI
WWKX(FM) Woonsocket RI
WSEA(FM) Atlantic Beach SC
WSSX-FM Charleston SC
WJMX-FM Cheraw SC
WNOK(FM) Columbia SC
WWXM(FM) Garden City SC
WFBC-FM Greenville SC
WHSC(AM) Hartsville SC
WWKT-FM Kingstree SC
WHZT(FM) Seneca SC
KQRN(FM) Mitchell SD
KQRQ(FM) Rapid City SD
KKLS-FM Sioux Falls SD
KRCS(FM) Sturgis SD
*KAOR(FM) Vermillion SD
KWYR-FM Winner SD
WMSR-FM Collinwood TN
*WUMC(FM) Elizabethton TN
*WVCP(FM) Gallatin TN
WAEZ(FM) Greeneville TN
WLSZ(FM) Humboldt TN
WNRX(FM) Jefferson City TN
WRVW(FM) Lebanon TN
*WUTM(FM) Martin TN
WYDL(FM) Middleton TN
WAKQ(FM) Paris TN
WWST(FM) Sevierville TN
WTRZ(FM) Spencer TN
KPRF(FM) Amarillo TX
KQIZ-FM Amarillo TX
KXGL(FM) Amarillo TX
KORQ(FM) Baird TX
KQXY-FM Beaumont TX
*KPFC(FM) Callisburg TX
KDHT(FM) Cedar Park TX
KZFM(FM) Corpus Christi TX
KMMZ(FM) Crane TX
KJKK(FM) Dallas TX
KHKS(FM) Denton TX

KBFM(FM) Edinburg TX
KPRR(FM) El Paso TX
KWKQ(FM) Graham TX
KCDD(FM) Hamlin TX
KPWW(FM) Hooks TX
KRBE(FM) Houston TX
KNEX(FM) Laredo TX
KQUR(FM) Laredo TX
KRRG(FM) Laredo TX
KZII-FM Lubbock TX
KZBT(FM) Midland TX
KMRK-FM Odessa TX
KAZE(FM) Ore City TX
*KWLD(FM) Plainview TX
KKPN(FM) Rockport TX
*KNLE-FM Round Rock TX
KIXY-FM San Angelo TX
KMDX(FM) San Angelo TX
KJXK(FM) San Antonio TX
KXXM(FM) San Antonio TX
KSCH(FM) Sulphur Springs TX
KBAR-FM Victoria TX
KVIC(FM) Victoria TX
KWTX-FM Waco TX
KISX(FM) Whitehouse TX
KNIN-FM Wichita Falls TX
KQXC-FM Wichita Falls TX
KAIQ(FM) Wolfforth TX
KEGH(FM) Brigham City UT
KCIN(FM) Cedar City UT
*KSUU(FM) Cedar City UT
KVFX(FM) Logan UT
*KWCR-FM Ogden UT
KXRQ(FM) Roosevelt UT
KZHT(FM) Salt Lake City UT
KUUU(FM) South Jordan UT
*WWHS-FM Hampden-Sydney VA
WQPO(FM) Harrisonburg VA
*WHCE(FM) Highland Springs VA
WJJX(FM) Lynchburg VA
WZVA(FM) Marion VA
WNVZ(FM) Norfolk VA
WNVA-FM Norton VA
WNRJ(FM) Poquoson VA
WXLK(FM) Roanoke VA
WHTE-FM Ruckersville VA
WHLF(FM) South Boston VA
WJJS-FM Vinton VA
WAZR(FM) Woodstock VA
WORK(FM) Barre VT
*WWLR(FM) Lyndonville VT
*WVTC(FM) Randolph Center VT
WXXX(FM) South Burlington VT
KUJ-FM Burbank WA
KBDB-FM Forks WA
KBIS(AM) Forks WA
*KMIH(FM) Mercer Island WA
KZBE(FM) Omak WA
KOLW(FM) Othello WA
KHTR(FM) Pullman WA
KWWW-FM Quincy WA
*KNHC(FM) Seattle WA
KUBE(FM) Seattle WA
*KWRS(FM) Spokane WA
KBKS-FM Tacoma WA
*KVTI(FM) Tacoma WA
KNLT(FM) Walla Walla WA
KPQ-FM Wenatchee WA
KFFM(FM) Yakima WA
WBIZ-FM Eau Claire WI
WIXX(FM) Green Bay WI
WZEE(FM) Madison WI
WQTC-FM Manitowoc WI
WKEB(FM) Medford WI
WRHN(FM) Rhinelander WI
*WPFF(FM) Sturgeon Bay WI
*WCLQ(FM) Wausau WI
WIFC(FM) Wausau WI
WXSS(FM) Wauwatosa WI
WCIR-FM Beckley WV
WVSR-FM Charleston WV
WQWV(FM) Fisher WV
WKEE-FM Huntington WV
WVAQ(FM) Morgantown WV
WRVZ(FM) Pocatalico WV
WSTG(FM) Princeton WV
WKQV(FM) Richwood WV
*WPHP(FM) Wheeling WV
KTRS-FM Casper WY
KDLY(FM) Lander WY
KYCS(FM) Rock Springs WY
KZZS(FM) Story WY

Country

KASH-FM Anchorage AK
KBRJ(FM) Anchorage AK
*KCUK(FM) Chevak AK
KLAM(AM) Cordova AK
*KDLG(AM) Dillingham AK
KIAK-FM Fairbanks AK
KTKU(FM) Juneau AK
KWHQ-FM Kenai AK
KGTW(FM) Ketchikan AK
KVOK(AM) Kodiak AK
*KJNP(AM) North Pole AK
*KJNP-FM North Pole AK
KRSA(AM) Petersburg AK
KSBZ(FM) Sitka AK
KPEN-FM Soldotna AK
KVAK(AM) Valdez AK
WQAH-FM Addison AL
WQSB(FM) Albertville AL
WSTH-FM Alexander City AL
WAAO-FM Andalusia AL
WRAB(AM) Arab AL
WHMA-FM Ashland AL
WYOK(FM) Atmore AL
WKKR(FM) Auburn AL
WZZK-FM Birmingham AL
WAOQ(FM) Brantley AL
WKNU(FM) Brewton AL
WPRN(AM) Butler AL
WEIS(AM) Centre AL
WRHY(FM) Centre AL
WBIB(AM) Centreville AL
WKUL(FM) Cullman AL
WGZZ(FM) Dadeville AL
WDRM(FM) Decatur AL
WTVY-FM Dothan AL
WELB(AM) Elba AL
WVVL(FM) Elba AL
WDJR(FM) Enterprise AL
WPGG(AM) Evergreen AL
WLDX(AM) Fayette AL
WTXT(FM) Fayette AL
WKWL(AM) Florala AL
WXFL(FM) Florence AL
WZOB(AM) Fort Payne AL
WNCB(FM) Gardendale AL
WGEA(AM) Geneva AL
WQZX(FM) Greenville AL
WBMH(FM) Grove Hill AL
WTWX-FM Guntersville AL
WFMH-FM Hackleburg AL
WJBB-FM Haleyville AL
WERH(AM) Hamilton AL
*WPIL(FM) Heflin AL
WCKA(AM) Jacksonville AL
WDXB(FM) Jasper AL
WINL(FM) Linden AL
WNPT-FM Linden AL
WZZX(AM) Lineville AL
WPRN-FM Lisman AL
WKSJ-FM Mobile AL
WBAM-FM Montgomery AL
WLWI-FM Montgomery AL
WKLD(FM) Oneonta AL
WAMI(AM) Opp AL
WAMI-FM Opp AL
WOPP(AM) Opp AL
WOAB(FM) Ozark AL
WFHK(AM) Pell City AL
WRMG(AM) Red Bay AL
WELR-FM Roanoke AL
WGOL(AM) Russellville AL
WKEA-FM Scottsboro AL
WWIC(AM) Scottsboro AL
WDXX(FM) Selma AL
WRSM(AM) Sumiton AL
WTDR(FM) Talladega AL
WQRV(FM) Tuscumbia AL
WQSI(FM) Union Springs AL
KPGG(FM) Ashdown AR
KEWI(AM) Benton AR
KHKN(FM) Benton AR
KQSM-FM Bentonville AR
KTHS(AM) Berryville AR
KTHS-FM Berryville AR
KHLS(FM) Blytheville AR
KLYR(AM) Clarksville AR
KLYR-FM Clarksville AR
KXIO(FM) Clarksville AR
KHPQ(FM) Clinton AR
KAGH(AM) Crossett AR
KAGH-FM Crossett AR
KYEL(FM) Danville AR
KCJC(FM) Dardanelle AR
KWXT(AM) Dardanelle AR
KDQN-FM De Queen AR
KDEW-FM De Witt AR
KXSA-FM Dermott AR
KXFE(FM) Dumas AR
KCXY(FM) East Camden AR
KIXB(FM) El Dorado AR
KKIX(FM) Fayetteville AR
KQEW(FM) Fordyce AR
KBFC(FM) Forrest City AR
KMAG(FM) Fort Smith AR
KTCS-FM Fort Smith AR
KWXE(FM) Glenwood AR
KYXK(FM) Gurdon AR
KWHF(FM) Harrisburg AR
KHOZ-FM Harrison AR
KFFA(AM) Helena AR
KHPA(FM) Hope AR
KKIK(FM) Horseshoe Bend AR
KQUS-FM Hot Springs AR
KFIN(FM) Jonesboro AR
KDXY(FM) Lake City AR
KSSN(FM) Little Rock AR
KVMA(AM) Magnolia AR
KBOK(AM) Malvern AR
KAMS(FM) Mammoth Spring AR
KBCN-FM Marshall AR
KENA-FM Mena AR
KVOM-FM Morrilton AR
KPFM(FM) Mountain Home AR
KTLO(AM) Mountain Home AR
KWOZ(FM) Mountain View AR
KMTB(FM) Murfreesboro AR
KOKR(FM) Newport AR
KDYN(AM) Ozark AR
KDYN-FM Ozark AR
KPBQ-FM Pine Bluff AR
KHOM(FM) Salem AR
KWCK-FM Searcy AR
KZHE(FM) Stamps AR
KFYX(FM) Texarkana AR
KOSY(AM) Texarkana AR
KVMZ(FM) Waldo AR
KRLW-FM Walnut Ridge AR
KWRF(AM) Warren AR
KWRF-FM Warren AR
KWYN(AM) Wynne AR
KWYN-FM Wynne AR
KAVV(FM) Benson AZ
KWCD(FM) Bisbee AZ
KMLE(FM) Chandler AZ
KFXR-FM Chinle AZ
KFPB(FM) Chino Valley AZ
KVRD-FM Cottonwood AZ
KDAP-FM Douglas AZ
KTHQ(FM) Eagar AZ
KAFF(AM) Flagstaff AZ
KAFF-FM Flagstaff AZ
KRDE(FM) Globe AZ
KZUA(FM) Holbrook AZ
KFLG-FM Kingman AZ
KGMN(FM) Kingman AZ
KJJJ(FM) Lake Havasu City AZ
KQSS(FM) Miami AZ
KPGE(AM) Page AZ
KLPZ(AM) Parker AZ
KMOG(AM) Payson AZ
KNIX-FM Phoenix AZ
KTMG(FM) Prescott AZ
KBUX(FM) Quartzsite AZ
KXKQ(FM) Safford AZ
KSED(FM) Sedona AZ
KIIM-FM Tucson AZ
KSWG(FM) Wickenburg AZ
KHIL(AM) Willcox AZ
KTNN(AM) Window Rock AZ
KINO(AM) Winslow AZ
*KAWC(AM) Yuma AZ
KTTI(FM) Yuma AZ
KCNO(FM) Alturas CA
KBYN(FM) Arnold CA
KIXF(FM) Baker CA
KCWR(FM) Bakersfield CA
KUZZ(AM) Bakersfield CA
KUZZ-FM Bakersfield CA
KIBS(FM) Bishop CA
KROP(AM) Brawley CA
KUSS(FM) Carlsbad CA
KKCY(FM) Colusa CA
KPOD-FM Crescent City CA
KWST(AM) El Centro CA
KSOQ-FM Escondido CA
KEKA-FM Eureka CA
KRED-FM Eureka CA
KSKS(FM) Fresno CA
KATJ-FM George CA
KFGY(FM) Healdsburg CA
*KIDE(FM) Hoopa CA
KRVC(FM) Hornbrook CA
KCNQ(FM) Kernville CA
KRKC(AM) King City CA
KIXW-FM Lenwood CA
KCAA(AM) Loma Linda CA
KKGO(FM) Los Angeles CA
KRTY(FM) Los Gatos CA
KTOM-FM Marina CA
KUBB(FM) Mariposa CA
KATM(FM) Modesto CA
*KSMC(FM) Moraga CA
KPLM(FM) Palm Springs CA
KHSL-FM Paradise CA
KYOE(FM) Point Arena CA
KALF(FM) Red Bluff CA
KNCQ(FM) Redding CA
KLOA-FM Ridgecrest CA
KWDJ(AM) Ridgecrest CA
KNCI(FM) Sacramento CA
KFRG(FM) San Bernardino CA
KTDD(AM) San Bernardino CA
KSON-FM San Diego CA
KBWF(FM) San Francisco CA
KKJG(FM) San Luis Obispo CA
KSLY-FM San Luis Obispo CA
KSNI-FM Santa Maria CA
KRAZ(FM) Santa Ynez CA
KNTY(FM) Shingle Springs CA
KXFG(FM) Sun City CA
KJDX(FM) Susanville CA
KTPI-FM Tehachapi CA
KJUG(AM) Tulare CA
KJUG-FM Tulare CA
KFLS-FM Tulelake CA
KKBN(FM) Twain Harte CA
KXCM(FM) Twentynine Palms CA
KQPM(FM) Ukiah CA
KUKI-FM Ukiah CA
KHAY(FM) Ventura CA
KVFG(FM) Victorville CA
KSYC-FM Yreka CA
KALQ-FM Alamosa CO
KWLI(FM) Broomfield CO
KBVC(FM) Buena Vista CO
KNAB-FM Burlington CO
KKCS-FM Canon City CO
KAVP(AM) Colona CO
KATC-FM Colorado Springs CO
KISZ-FM Cortez CO
KRAI(AM) Craig CO
KYGO-FM Denver CO
KRSJ(FM) Durango CO
KEKB(FM) Fruita CO
KMTS(FM) Glenwood Springs CO
KRKY(AM) Granby CO
KMOZ-FM Grand Junction CO
KPKE(AM) Gunnison CO
KTRJ(FM) Hayden CO
KJBL(FM) Julesburg CO
KTHN(FM) La Junta CO
KLMR(AM) Lamar CO
KVAY(FM) Lamar CO
KCKK(AM) Littleton CO
KSLV(AM) Monte Vista CO
KKXK(FM) Montrose CO
KUBC(AM) Montrose CO
KRYD(FM) Norwood CO
KATR-FM Otis CO
KWGL(FM) Ouray CO
KWUF(AM) Pagosa Springs CO
KCCY(FM) Pueblo CO
KBCR-FM Steamboat Springs CO
KNNG(FM) Sterling CO
KCRT(AM) Trinidad CO
KSKE-FM Vail CO
KSPK(FM) Walsenburg CO
KUAD-FM Windsor CO
WPKX(FM) Enfield CT
WCTY(FM) Norwich CT
WWYZ(FM) Waterbury CT
WMZQ-FM Washington DC
WXJN(FM) Lewes DE
WDSD(FM) Smyrna DE
WWOJ(FM) Avon Park FL
WQXM(AM) Bartow FL
WPHK(FM) Blountstown FL
WKIS(FM) Boca Raton FL
WAKT-FM Callaway FL
WOCY(FM) Carrabelle FL
WIKX(FM) Charlotte Harbor FL
WAFC-FM Clewiston FL
WAAZ-FM Crestview FL
WJSB(AM) Crestview FL
WYNY(AM) Cross City FL
WDSP(AM) De Funiak Springs FL
WZEP(AM) De Funiak Springs FL
WTRS(FM) Dunnellon FL
WKRO-FM Edgewater FL
WUSV(FM) Estero FL
WWGR(FM) Fort Myers FL
WDVH(AM) Gainesville FL
WGWD(FM) Gretna FL
WNRP(AM) Gulf Breeze FL
WFUS(FM) Gulfport FL
WYGC(FM) High Springs FL
WPLA(FM) Jacksonville FL
WQIK-FM Jacksonville FL
WCNK(FM) Key West FL
WCKT(FM) Lehigh Acres FL
WQHL-FM Live Oak FL
WMAF(AM) Madison FL
WJAQ(FM) Marianna FL
WTYS(AM) Marianna FL
WYZB(FM) Mary Esther FL
WXBM-FM Milton FL
*WBGY(FM) Naples FL
WSVU(AM) North Palm Beach FL
WCFI(AM) Ocala FL
WOGK(FM) Ocala FL
WOKC(AM) Okeechobee FL
WWKA(FM) Orlando FL
WGNE-FM Palatka FL
WIYD(AM) Palatka FL
WPAP-FM Panama City FL
WYCT(FM) Pensacola FL
WNFK(FM) Perry FL
WCTH(FM) Plantation Key FL
WHKR(FM) Rockledge FL
WQYK-FM Saint Petersburg FL
WCTQ(FM) Sarasota FL
WAVW(FM) Stuart FL
WAIB(FM) Tallahassee FL
WTNT-FM Tallahassee FL
WDVH-FM Trenton FL
WQLC(FM) Watertown FL
WIRK-FM West Palm Beach FL
WPCV(FM) Winter Haven FL
WZZS(FM) Zolfo Springs FL
*V6AI(AM) Yap FM
WKAK(FM) Albany GA
WAJQ-FM Alma GA
WISK-FM Americus GA
WUBL(FM) Atlanta GA
WIBL(FM) Augusta GA

WBAF(AM) Barnesville GA
WBYZ(FM) Baxley GA
WKUB(FM) Blackshear GA
WPPL(FM) Blue Ridge GA
WTUF(FM) Boston GA
WMOQ(FM) Bostwick GA
WRJY(FM) Brunswick GA
WJTH(AM) Calhoun GA
WBTR-FM Carrollton GA
WQMT(FM) Chatsworth GA
WRWH(AM) Cleveland GA
WDCO(AM) Cochran GA
WDXQ-FM Cochran GA
WCON(AM) Cornelia GA
WCON-FM Cornelia GA
WCUG(AM) Cuthbert GA
WKHC(FM) Dahlonega GA
WSEM(AM) Donalsonville GA
WOKA-FM Douglas GA
WQZY(FM) Dublin GA
WXLI(AM) Dublin GA
WUFF(AM) Eastman GA
WSGC-FM Elberton GA
WPGY(AM) Ellijay GA
*WATY(FM) Folkston GA
WYAY(FM) Gainesville GA
WHIE(AM) Griffin GA
WKLY(AM) Hartwell GA
WCEH(AM) Hawkinsville GA
WVOH-FM Hazlehurst GA
WYYZ(AM) Jasper GA
WIFO-FM Jesup GA
WKBX(FM) Kingsland GA
WQCH(AM) La Fayette GA
WPEH(AM) Louisville GA
WPEH-FM Louisville GA
WKCN(FM) Lumpkin GA
WLYU(FM) Lyons GA
WDEN-FM Macon GA
WKHX-FM Marietta GA
WMCG(FM) Milan GA
WKZR(FM) Milledgeville GA
WHKN(FM) Millen GA
WYUM(FM) Mount Vernon GA
WALH(AM) Mountain City GA
WCOH(AM) Newnan GA
WTSH-FM Rockmart GA
WSNT-FM Sandersville GA
WJCL-FM Savannah GA
WXRS-FM Swainsboro GA
WSYL(AM) Sylvania GA
WZBX(FM) Sylvania GA
WKNG(AM) Tallapoosa GA
WTHO-FM Thomson GA
WOBB(FM) Tifton GA
WTIF(AM) Tifton GA
WNGC(FM) Toccoa GA
WAAC(FM) Valdosta GA
WEBL(FM) Warner Robins GA
WXKT(FM) Washington GA
WYNR(FM) Waycross GA
WCJM-FM West Point GA
WKAA(FM) Willacoochee GA
WYHG(AM) Young Harris GA
WEKS(FM) Zebulon GA
KUAI(AM) Eleele HI
KHCM(AM) Honolulu HI
KDLX(FM) Makawao HI
KKOA(FM) Volcano HI
KKNE(AM) Waipahu HI
KLBA-FM Albia IA
WJOD(FM) Asbury IA
KSOM(FM) Audubon IA
KDMG(FM) Burlington IA
KOEL-FM Cedar Falls IA
KHAK(FM) Cedar Rapids IA
KMGO(FM) Centerville IA
KIAQ(FM) Clarion IA
KSIB(AM) Creston IA
KSIB-FM Creston IA
WLLR-FM Davenport IA
KDSN(AM) Denison IA
KHKI(FM) Des Moines IA
KDST(FM) Dyersville IA
KRKN(FM) Eldon IA
KILR-FM Estherville IA
KIOW(FM) Forest City IA
KWMT(AM) Fort Dodge IA
KBKB-FM Fort Madison IA
KCTN(FM) Garnavillo IA
KXKT(FM) Glenwood IA
KLMJ(FM) Hampton IA
KKIA(FM) Ida Grove IA
KNIA(AM) Knoxville IA
KIKD(FM) Lake City IA
KMCH(FM) Manchester IA
KMAQ(AM) Maquoketa IA
KXIA(FM) Marshalltown IA
KIAI(FM) Mason City IA
KILJ(AM) Mount Pleasant IA
KMCS(FM) Muscatine IA
KCZE(FM) New Hampton IA
KCOB(AM) Newton IA
KCOB-FM Newton IA
KYTC(FM) Northwood IA
KKHQ-FM Oelwein IA
KBOE(AM) Oskaloosa IA
KLEE(AM) Ottumwa IA
KCSI(FM) Red Oak IA
KOAK(AM) Red Oak IA
KIHK(FM) Rock Valley IA
KICD-FM Spencer IA
KKRF(FM) Stuart IA
KNEI-FM Waukon IA
KWAY(AM) Waverly IA
KJJY(FM) West Des Moines IA
KKYY(FM) Whiting IA
KIZN(FM) Boise ID
KQFC(FM) Boise ID
KICR(FM) Coeur d'Alene ID
KCHQ(FM) Driggs ID
KORT(AM) Grangeville ID
KORT-FM Grangeville ID
KYUN(FM) Hailey ID
KID-FM Idaho Falls ID
KTHK(FM) Idaho Falls ID
KUPI-FM Idaho Falls ID
KART(AM) Jerome ID
KMOK(FM) Lewiston ID
KRLC(AM) Lewiston ID
KDZY(FM) McCall ID
KVSI(AM) Montpelier ID
KMHI(AM) Mountain Home ID
KLER(AM) Orofino ID
KOUU(AM) Pocatello ID
KZBQ(FM) Pocatello ID
KKEX(FM) Preston ID
KKMV(FM) Rupert ID
KSRA(AM) Salmon ID
KSRA-FM Salmon ID
KIBR(FM) Sandpoint ID
KBRV(AM) Soda Springs ID
KITT(FM) Soda Springs ID
KEZJ-FM Twin Falls ID
KVRG(FM) Victor ID
KWAL(AM) Wallace ID
WRMJ(FM) Aledo IL
WIBH(AM) Anna IL
WLCN(FM) Atlanta IL
WRMS-FM Beardstown IL
WLMD(FM) Bushnell IL
WRUL(FM) Carmi IL
WIXY(FM) Champaign IL
KSGM(AM) Chester IL
WUSN(FM) Chicago IL
WCCQ(FM) Crest Hill IL
WDZQ(FM) Decatur IL
WRCV(FM) Dixon IL
WDQN(AM) Du Quoin IL
WCRC(FM) Effingham IL
WFYR(FM) Elmwood IL
WOKZ(FM) Fairfield IL
WFPS(FM) Freeport IL
WXXQ(FM) Freeport IL
WAAG(FM) Galesburg IL
WJRE(FM) Galva IL
WGEL(FM) Greenville IL
WEBQ(AM) Harrisburg IL
WOOZ-FM Harrisburg IL
WDUK(FM) Havana IL
WRVY-FM Henry IL
WHPO(FM) Hoopeston IL
WAKO(AM) Lawrenceville IL
WAKO-FM Lawrenceville IL
WBWN(FM) Le Roy IL
WSMI(AM) Litchfield IL
WSMI-FM Litchfield IL
WLUV(AM) Loves Park IL
WZUS(FM) Macon IL
WDDD-FM Marion IL
WMCI(FM) Mattoon IL
WMCL(AM) McLeansboro IL
WGLC-FM Mendota IL
WMOK(FM) Metropolis IL
WFXN(AM) Moline IL
WRAM(AM) Monmouth IL
WIBV(FM) Mount Vernon IL
WMIX-FM Mount Vernon IL
WALS(FM) Oglesby IL
WMKR(FM) Pana IL
WINH(FM) Paris IL
WXCL(FM) Pekin IL
WIRL(AM) Peoria IL
WBBA-FM Pittsfield IL
WCOY(FM) Quincy IL
WXNU(FM) Saint Anne IL
WJBD(AM) Salem IL
WCCI(FM) Savanna IL
WJVO(FM) South Jacksonville IL
WFMB-FM Springfield IL
WKJT(FM) Teutopolis IL
WSCH(FM) Aurora IN
WXKU-FM Austin IN
WRBI(FM) Batesville IN
WSDM-FM Brazil IN
WLFF(FM) Brookston IN
WLFW(FM) Chandler IN
WKKG(FM) Columbus IN
WKZS(FM) Covington IN
WADM(AM) Decatur IN
WQHK-FM Decatur IN
WYGB(FM) Edinburgh IN
WBYT(FM) Elkhart IN
WHCC(FM) Ellettsville IN
WQKZ(FM) Ferdinand IN
WFLQ(FM) French Lick IN
WREB(FM) Greencastle IN
WTRE(AM) Greensburg IN
WBDC(FM) Huntingburg IN
WFMS(FM) Indianapolis IN
WAVG(AM) Jeffersonville IN
WBTU(FM) Kendallville IN
WIVR(FM) Kentland IN
*WKPW(FM) Knightstown IN
WWKI(FM) Kokomo IN
WCOE(FM) La Porte IN
WKOA(FM) Lafayette IN
WSHY(AM) Lafayette IN
WTHD(FM) Lagrange IN
WBTO(AM) Linton IN
WCBK-FM Martinsville IN
WRCY(AM) Mount Vernon IN
WMDH-FM New Castle IN
WARU(AM) Peru IN
WPGW-FM Portland IN
WRAY-FM Princeton IN
*WECI(FM) Richmond IN
WQLK(FM) Richmond IN
WARU-FM Roann IN
WHZR(FM) Royal Center IN
WIFE-FM Rushville IN
WSLM(AM) Salem IN
WMPI(FM) Scottsburg IN
WQKC(FM) Seymour IN
WLHK(FM) Shelbyville IN
WCLS(FM) Spencer IN
WNDI(AM) Sullivan IN
WNDI-FM Sullivan IN
WTHI-FM Terre Haute IN
WLJE(FM) Valparaiso IN
WCJC(FM) Van Buren IN
WXCH(FM) Versailles IN
WKID(FM) Vevay IN
WFML(FM) Vincennes IN
WKUZ(FM) Wabash IN
WWBL(FM) Washington IN
KSOK(AM) Arkansas City KS
KREP(FM) Belleville KS
KSNP(FM) Burlington KS
KCLY(FM) Clay Center KS
KWGB(FM) Colby KS
KXXX(AM) Colby KS
KNCK(AM) Concordia KS
KUSN(FM) Dearing KS
KDNS(FM) Downs KS
KVOE-FM Emporia KS
KOTE(FM) Eureka KS
KKJQ(FM) Garden City KS
KLOE(AM) Goodland KS
KHAZ(FM) Hays KS
KNZA(FM) Hiawatha KS
KKQY(FM) Hill City KS
KHOK(FM) Hoisington KS
KBUF(AM) Holcomb KS
KAIR-FM Horton KS
KHUT(FM) Hutchinson KS
KZSN(FM) Hutchinson KS
KFKF-FM Kansas City KS
KLDG(FM) Liberal KS
KXKU(FM) Lyons KS
KXBZ(FM) Manhattan KS
KNDY(AM) Marysville KS
KNDY-FM Marysville KS
KFTI-FM Newton KS
KFNF(FM) Oberlin KS
KOFO(AM) Ottawa KS
KKOW(AM) Pittsburg KS
KKOW-FM Pittsburg KS
KQTP(FM) Saint Marys KS
KSKG(FM) Salina KS
KYEZ(FM) Salina KS
KMZA(FM) Seneca KS
KTPK(FM) Topeka KS
WIBW-FM Topeka KS
KULY(AM) Ulysses KS
KFDI-FM Wichita KS
KFTI(AM) Wichita KS
KSOK-FM Winfield KS
WANY(AM) Albany KY
WANY-FM Albany KY
WMDJ-FM Allen KY
WDGG(FM) Ashland KY
WBVR-FM Auburn KY
WBRT(AM) Bardstown KY
WGGC(FM) Bowling Green KY
WMMG(AM) Brandenburg KY
WMMG-FM Brandenburg KY
WKYR-FM Burkesville KY
WSEK(FM) Burnside KY
WKDZ-FM Cadiz KY
WCCK(FM) Calvert City KY
WIKI(FM) Carrollton KY
WLLE(FM) Clinton KY
WAIN-FM Columbia KY
WKDP-FM Corbin KY
WCPM(AM) Cumberland KY
WCYN-FM Cynthiana KY
WHSX(FM) Edmonton KY
WFLE(AM) Flemingsburg KY
WFLE-FM Flemingsburg KY
WLVK(FM) Fort Knox KY
WYGY(FM) Fort Thomas KY
WKYW(FM) Frankfort KY
WFKN(AM) Franklin KY
WFUL(AM) Fulton KY
WLYE-FM Glasgow KY
WGOH(AM) Grayson KY
WGRK(AM) Greensburg KY
WGRK-FM Greensburg KY
WLGC-FM Greenup KY
WULF(FM) Hardinsburg KY
WXBC(FM) Hardinsburg KY
WTUK(FM) Harlan KY
WXLR(FM) Harold KY
WHBN(AM) Harrodsburg KY
WKCM(AM) Hawesville KY
WSGS(FM) Hazard KY
WKDQ(FM) Henderson KY
WKMO(FM) Hodgenville KY
WVVR(FM) Hopkinsville KY
WCYO(FM) Irvine KY
WJSN-FM Jackson KY
WJKY(AM) Jamestown KY
WJRS(FM) Jamestown KY
WMTL(AM) Leitchfield KY
WBUL-FM Lexington KY
WLXX(FM) Lexington KY
WKDO(AM) Liberty KY
WKDO-FM Liberty KY
WWEL(FM) London KY
WZAQ(FM) Louisa KY
WAMZ(FM) Louisville KY
WPTI(FM) Louisville KY
WFMW(AM) Madisonville KY
WKLB(AM) Manchester KY
WVLC(FM) Mannsville KY
WWAG(FM) McKee KY
WMKZ(FM) Monticello KY
WMOR(AM) Morehead KY
WEZG(FM) Morganfield KY
WMSK(AM) Morganfield KY
WLBQ(AM) Morgantown KY
WRVK(AM) Mt. Vernon KY
WLOC(AM) Munfordville KY
WFGE(FM) Murray KY
WBKR(FM) Owensboro KY
WKCA(FM) Owingsville KY
WKYQ(FM) Paducah KY
WSIP-FM Paintsville KY
WBIO(FM) Philpot KY
WDHR(FM) Pikeville KY
WRIL(FM) Pineville KY
WRUS(AM) Russellville KY
WRLV(AM) Salyersville KY
WRLV-FM Salyersville KY
WVLE(FM) Scottsville KY
WRSL(AM) Stanford KY
WSKV(FM) Stanton KY
WMSK-FM Sturgis KY
WKWY(FM) Tompkinsville KY
WTKY(AM) Tompkinsville KY
WTKY-FM Tompkinsville KY
WKKS(AM) Vanceburg KY
WKKS-FM Vanceburg KY
WLKS-FM West Liberty KY
*WMMT(FM) Whitesburg KY
WTCW(AM) Whitesburg KY
WHAY(FM) Whitley City KY
WEZJ-FM Williamsburg KY
WNKR(FM) Williamstown KY
KRRV-FM Alexandria LA
WABL(AM) Amite LA
WTGE(FM) Baker LA
KBKK(FM) Ball LA
WYNK-FM Baton Rouge LA
WYPY(FM) Baton Rouge LA
KQKI(FM) Bayou Vista LA
WBOX(AM) Bogalusa LA
KSBH(FM) Coushatta LA
KEUN-FM Eunice LA
WOMN(AM) Franklinton LA
WUUU(FM) Franklinton LA
KLEB(AM) Golden Meadow LA
WFPR(AM) Hammond LA
WHMD(FM) Hammond LA
KCIL(FM) Houma LA
KJNA-FM Jena LA
KMDL(FM) Kaplan LA
KNGT(FM) Lake Charles LA
KYKZ(FM) Lake Charles LA
KJAE(FM) Leesville LA
KVVP(FM) Leesville LA
KBON(FM) Mamou LA
KJVC(FM) Mansfield LA
KORI(FM) Mansfield LA
KWLV(FM) Many LA
KAPB-FM Marksville LA
KJLO-FM Monroe LA
KDBH(FM) Natchitoches LA
KXKC(FM) New Iberia LA
WNOE-FM New Orleans LA
KFXZ-FM Opelousas LA
KSLO(AM) Opelousas LA
KMYY(FM) Rayville LA
KXKZ(FM) Ruston LA
KRMD-FM Shreveport LA
KWKH(AM) Shreveport LA
KXKS-FM Shreveport LA
KLAA-FM Tioga LA
WBOX-FM Varnado LA

KVPI(AM) Ville Platte LA
KNCB(AM) Vivian LA
KNCB-FM Vivian LA
KVCL-FM Winnfield LA
KMAR-FM Winnsboro LA
WPVQ(FM) Greenfield MA
WCTK(FM) New Bedford MA
WESO(AM) Southbridge MA
WKLB-FM Waltham MA
WGFP(AM) Webster MA
WPOC(FM) Baltimore MD
WKIK-FM California MD
WINX-FM Cambridge MD
WBEY-FM Crisfield MD
WROG(FM) Cumberland MD
WFRE(FM) Frederick MD
WFRB-FM Frostburg MD
WAYZ(FM) Hagerstown MD
WXCY(FM) Havre de Grace MD
WAAI(FM) Hurlock MD
WKIK(AM) La Plata MD
WWHC(FM) Oakland MD
WWFG(FM) Ocean City MD
WICO-FM Salisbury MD
WTHT(FM) Auburn ME
WLKE(FM) Bar Harbor ME
WBFB(FM) Belfast ME
WQCB(FM) Brewer ME
WTBM(FM) Mexico ME
WSYY-FM Millinocket ME
WOXO-FM Norway ME
*WMDR-FM Oakland ME
WPOR(FM) Portland ME
WBPW(FM) Presque Isle ME
WMCM(FM) Rockland ME
WEBB(FM) Waterville ME
WTVL(AM) Waterville ME
WUBB(FM) York Center ME
WQTE(FM) Adrian MI
WATZ-FM Alpena MI
WWWW-FM Ann Arbor MI
WLEW(AM) Bad Axe MI
WCUZ(FM) Bear Lake MI
WHFB-FM Benton Harbor MI
WWBR(FM) Big Rapids MI
WBCM(FM) Boyne City MI
WYTZ(FM) Bridgman MI
WKYO(AM) Caro MI
WNWN-FM Coldwater MI
WDTW-FM Detroit MI
WYCD(FM) Detroit MI
WYKX(FM) Escanaba MI
WCXI(AM) Fenton MI
WFBE(FM) Flint MI
WGDN-FM Gladwin MI
WBCT(FM) Grand Rapids MI
WGRY-FM Grayling MI
WCMM(FM) Gulliver MI
WTWS(FM) Harrison MI
WBCH(AM) Hastings MI
WBCH-FM Hastings MI
WCEN-FM Hemlock MI
WTNR(FM) Holland MI
WHKB(FM) Houghton MI
WJNR-FM Iron Mountain MI
WJMS(AM) Ironwood MI
WJPD(FM) Ishpeming MI
WSCG-FM Lakeview MI
WCUP(FM) L'Anse MI
WITL-FM Lansing MI
WLCO(AM) Lapeer MI
WBGV(FM) Marlette MI
WFXD(FM) Marquette MI
WHYB(FM) Menominee MI
WAVC(FM) Mio MI
WMUS(FM) Muskegon MI
WLAW(FM) Newaygo MI
WNBY(AM) Newberry MI
WUPY(FM) Ontonagon MI
WLDR(FM) Petoskey MI
WYLZ(FM) Pinconning MI
WSAQ(FM) Port Huron MI
WRGZ(FM) Rogers City MI
WKCQ(FM) Saginaw MI
WMKC(FM) Saint Ignace MI
WMLM(AM) Saint Louis MI

WSDS(AM) Salem Township MI
WMIC(AM) Sandusky MI
WKZC(FM) Scottville MI
WKJC(FM) Tawas City MI
WLKM(AM) Three Rivers MI
WTCM-FM Traverse City MI
WBMI(FM) West Branch MI
KRJB(FM) Ada MN
KKIN-FM Aitkin MN
KASM(AM) Albany MN
KAUS-FM Austin MN
KKCQ-FM Bagley MN
KBVB(FM) Barnesville MN
KBHP(FM) Bemidji MN
WBJI(FM) Blackduck MN
KBEW-FM Blue Earth MN
KBMW(AM) Breckenridge MN
KRWC(AM) Buffalo MN
KGPZ(FM) Coleraine MN
KROX(AM) Crookston MN
KARP-FM Dassel MN
KRCQ(FM) Detroit Lakes MN
KKCB(FM) Duluth MN
KTCO(FM) Duluth MN
KLCI(FM) Elk River MN
KBRF(AM) Fergus Falls MN
KJJK-FM Fergus Falls MN
KSDM(FM) International Falls MN
KKOJ(AM) Jackson MN
KMFX-FM Lake City MN
WYRQ(FM) Little Falls MN
KEYL(AM) Long Prairie MN
KLQL(FM) Luverne MN
KLQP(FM) Madison MN
KYSM-FM Mankato MN
KDMA(AM) Montevideo MN
KVOX-FM Moorhead MN
KKOK-FM Morris MN
KNUJ(AM) New Ulm MN
KBLB(FM) Nisswa MN
KOLV(FM) Olivia MN
KDIO(AM) Ortonville MN
KRFO-FM Owatonna MN
KPRM(AM) Park Rapids MN
KXKK(FM) Park Rapids MN
KZPK(FM) Paynesville MN
KBOT(FM) Pelican Rapids MN
WWWI-FM Pillager MN
WCMP-FM Pine City MN
KLOH(AM) Pipestone MN
KFIL(AM) Preston MN
KFIL-FM Preston MN
WQPM(AM) Princeton MN
KLGR(AM) Redwood Falls MN
KWWK(FM) Rochester MN
KWNO-FM Rushford MN
WWJO(FM) Saint Cloud MN
KRRW(FM) Saint James MN
KEEY-FM Saint Paul MN
KRBI(AM) Saint Peter MN
KIKV-FM Sauk Centre MN
WVAL(AM) Sauk Rapids MN
KJOE(FM) Slayton MN
KQYB(FM) Spring Grove MN
KNSP(AM) Staples MN
KKAQ(AM) Thief River Falls MN
KKDQ(FM) Thief River Falls MN
KSNR(FM) Thief River Falls MN
KARL(FM) Tracy MN
KVKK(AM) Verndale MN
WUSZ(FM) Virginia MN
KMFX(AM) Wabasha MN
KKWS(FM) Wadena MN
KWAD(AM) Wadena MN
KKWQ(FM) Warroad MN
KDJS-FM Willmar MN
KDOM(AM) Windom MN
KDOM-FM Windom MN
KAGE(AM) Winona MN
KWXD(FM) Asbury MO
KSWF(FM) Aurora MO
KKOZ(AM) Ava MO
KAAN(AM) Bethany MO
KAAN-FM Bethany MO
KYOO(AM) Bolivar MO
KCLR-FM Boonville MO

KWRT(AM) Boonville MO
KRZK(FM) Branson MO
KMAM(AM) Butler MO
KMOE(FM) Butler MO
KATI(FM) California MO
KRLL(AM) California MO
KEZS-FM Cape Girardeau MO
KAOL(AM) Carrollton MO
KMZU(FM) Carrollton MO
KCRV(AM) Caruthersville MO
KRMO(AM) Cassville MO
KCHR(AM) Charleston MO
KWKZ(FM) Charleston MO
KDKD-FM Clinton MO
KESY(FM) Cuba MO
KDEX(AM) Dexter MO
KDEX-FM Dexter MO
KOEA(FM) Doniphan MO
KESM(AM) El Dorado Springs MO
KESM-FM El Dorado Springs MO
KTJJ(FM) Farmington MO
KFAL(AM) Fulton MO
KGOZ(FM) Gallatin MO
KBTC(AM) Houston MO
KUNQ(FM) Houston MO
KYLS-FM Ironton MO
KZJF(FM) Jefferson City MO
KIXQ(FM) Joplin MO
KBEQ-FM Kansas City MO
KOTC(AM) Kennett MO
KTUF(FM) Kirksville MO
KXKX(FM) Knob Noster MO
KCLQ(FM) Lebanon MO
KJEL(AM) Lebanon MO
KLWT(AM) Lebanon MO
KJFM(FM) Louisiana MO
KLTI(AM) Macon MO
KMMO(AM) Marshall MO
KMMO-FM Marshall MO
KMEM-FM Memphis MO
KWWR(FM) Mexico MO
KRES(FM) Moberly MO
KELE(AM) Mountain Grove MO
KELE-FM Mountain Grove MO
KBTN-FM Neosho MO
KNEM(AM) Nevada MO
KNMO(FM) Nevada MO
KTMO(FM) New Madrid MO
KOMG(FM) Ozark MO
KICK-FM Palmyra MO
KBDZ(FM) Perryville MO
KPWB-FM Piedmont MO
KKLR(FM) Poplar Bluff MO
KPPL(FM) Poplar Bluff MO
KYRO(AM) Potosi MO
KMOZ(AM) Rolla MO
KZNN(FM) Rolla MO
KSD(FM) Saint Louis MO
WIL(AM) Saint Louis MO
WIL-FM Saint Louis MO
KKID(FM) Salem MO
KSMO(AM) Salem MO
KSJQ(FM) Savannah MO
KDRO(AM) Sedalia MO
KBXB(FM) Sikeston MO
KRHW(AM) Sikeston MO
KTTS-FM Springfield MO
KRWP(FM) Stockton MO
KTUI-FM Sullivan MO
KSAR(FM) Thayer MO
KTTN-FM Trenton MO
KLPW-FM Union MO
KKAC(FM) Vandalia MO
KTKS(FM) Versailles MO
KFAV(FM) Warrenton MO
KWRE(AM) Warrenton MO
KAYQ(FM) Warsaw MO
KKDY(FM) West Plains MO
WBLE(FM) Batesville MS
WIZK(AM) Bay Springs MS
WZKX(FM) Bay St. Louis MS
WBYP(FM) Belzoni MS
WBIP(AM) Booneville MS
WZQK(AM) Brandon MS
WBKN(FM) Brookhaven MS
WCKK(FM) Carthage MS

WAKK-FM Centreville MS
WKDJ-FM Clarksdale MS
WDTL-FM Cleveland MS
WMJW(FM) Cleveland MS
WFFF(AM) Columbia MS
WKOR-FM Columbus MS
WADI(FM) Corinth MS
WQST(AM) Forest MS
WDMS(FM) Greenville MS
WABG(AM) Greenwood MS
WQXB(FM) Grenada MS
WYKC(AM) Grenada MS
WGCM(AM) Gulfport MS
WHER(FM) Heidelberg MS
WCPC(AM) Houston MS
WFXO(FM) Iuka MS
WMSI(FM) Jackson MS
WAGR-FM Lexington MS
WRBE(AM) Lucedale MS
WUSJ(FM) Madison MS
WJDQ(FM) Marion MS
WAKH(FM) McComb MS
WOKK(FM) Meridian MS
WUCL(FM) Meridian MS
WQNZ(FM) Natchez MS
WWZD-FM New Albany MS
WWMS(FM) Oxford MS
WKNN-FM Pascagoula MS
WRJW(AM) Picayune MS
WJDR(FM) Prentiss MS
WKZU(FM) Ripley MS
WQJB(FM) State College MS
WBBN(FM) Taylorsville MS
WTYL(AM) Tylertown MS
WTYL-FM Tylertown MS
WBBV(FM) Vicksburg MS
WABO(AM) Waynesboro MS
WABO-FM Waynesboro MS
WIGG(AM) Wiggins MS
WONA(AM) Winona MS
WONA-FM Winona MS
KFLN(AM) Baker MT
KCTR-FM Billings MT
KGHL(AM) Billings MT
KGHL-FM Billings MT
KBOZ-FM Bozeman MT
KAAR(FM) Butte MT
*KMSM-FM Butte MT
KIKF(FM) Cascade MT
KHNK(FM) Columbia Falls MT
KBCK(AM) Deer Lodge MT
KQRV(FM) Deer Lodge MT
KDBM(AM) Dillon MT
KHKR-FM East Helena MT
KIKC-FM Forsyth MT
KLTZ(AM) Glasgow MT
KDZN(FM) Glendive MT
KMON(AM) Great Falls MT
KMON-FM Great Falls MT
KPQX(FM) Havre MT
KBLL-FM Helena MT
KPBR(FM) Joliet MT
KDBR(FM) Kalispell MT
KXLO(AM) Lewistown MT
KLCB(AM) Libby MT
KMMR(FM) Malta MT
KYUS-FM Miles City MT
KGGL(FM) Missoula MT
KYSS-FM Missoula MT
KATQ(AM) Plentywood MT
KATQ-FM Plentywood MT
KERR(AM) Polson MT
KCGM(FM) Scobey MT
KZIN-FM Shelby MT
KVCK-FM Wolf Point MT
WQDK(FM) Ahoskie NC
WKXR(AM) Asheboro NC
WKSF(FM) Asheville NC
WNBB(FM) Bayboro NC
WPYB(AM) Benson NC
WKYK(AM) Burnsville NC
WSME(AM) Camp Lejeune NC
WPTL(AM) Canton NC
WSOC-FM Charlotte NC
WRSF(FM) Columbia NC
WCMC-FM Creedmoor NC

WWNK(FM) Farmville NC
WAGY(AM) Forest City NC
WNCC-FM Franklin NC
WJSG(FM) Hamlet NC
WCMS-FM Hatteras NC
WYND-FM Hatteras NC
WIZS(AM) Henderson NC
WMMY(FM) Jefferson NC
WKTE(AM) King NC
WKIX(FM) Kinston NC
WRNS(AM) Kinston NC
WRNS-FM Kinston NC
WKGX(AM) Lenoir NC
WKVS(FM) Lenoir NC
WKXU(FM) Louisburg NC
WKML(FM) Lumberton NC
WBRM(AM) Marion NC
WDSL(AM) Mocksville NC
WIXE(AM) Monroe NC
WMNC(AM) Morganton NC
WMNC-FM Morganton NC
WKRK(AM) Murphy NC
WZPR(FM) Nags Head NC
WECR(AM) Newland NC
WKBC(AM) North Wilkesboro NC
WLQB(FM) Ocean Isle Beach NC
WWHA(FM) Oriental NC
WQDR(FM) Raleigh NC
WPTM(FM) Roanoke Rapids NC
WCVP-FM Robbinsville NC
WDWG(FM) Rocky Mount NC
WKRX(FM) Roxboro NC
WRXO(AM) Roxboro NC
WCAB(AM) Rutherfordton NC
WWGP(AM) Sanford NC
WADA(AM) Shelby NC
WMPM(AM) Smithfield NC
WCOK(AM) Sparta NC
WKKT(FM) Statesville NC
WTAB(AM) Tabor City NC
WACB(AM) Taylorsville NC
WIST-FM Thomasville NC
WKSK(AM) West Jefferson NC
WWQQ-FM Wilmington NC
WRDU(FM) Wilson NC
WPAW(FM) Winston-Salem NC
WTIX(AM) Winston-Salem NC
WTQR(FM) Winston-Salem NC
KVMI(FM) Arthur ND
*KEYA(FM) Belcourt ND
KBMR(AM) Bismarck ND
KQDY(FM) Bismarck ND
KBTO(FM) Bottineau ND
KPOK(AM) Bowman ND
KDAK(AM) Carrington ND
KAOC(FM) Cavalier ND
KDLR(AM) Devils Lake ND
KZZY(FM) Devils Lake ND
KCAD(FM) Dickinson ND
KLTC(AM) Dickinson ND
KXPO(AM) Grafton ND
KNOX-FM Grand Forks ND
KNDC(AM) Hettinger ND
KSJB(AM) Jamestown ND
KYNU(FM) Jamestown ND
KNDK(AM) Langdon ND
KQLX-FM Lisbon ND
KMAV(AM) Mayville ND
KMAV-FM Mayville ND
KCJB(AM) Minot ND
KYYX(FM) Minot ND
KDDR(AM) Oakes ND
KZZJ(AM) Rugby ND
KTGO(AM) Tioga ND
KOVC(AM) Valley City ND
KEYZ(AM) Williston ND
KYYZ(FM) Williston ND
KBRB(AM) Ainsworth NE
KUSO(FM) Albion NE
KAAQ(FM) Alliance NE
KNCY-FM Auburn NE
KBLR-FM Blair NE
KCNI(AM) Broken Bow NE
KZEN(FM) Central City NE
KCSR(AM) Chadron NE
KQSK(FM) Chadron NE

KUTT(FM) Fairbury NE
KSDZ(FM) Gordon NE
KRGI-FM Grand Island NE
KRNY(FM) Kearney NE
KRVN(AM) Lexington NE
KRVN-FM Lexington NE
KIOD(FM) McCook NE
KFGE(FM) Milford NE
KXNP(FM) North Platte NE
KMCX(FM) Ogallala NE
KBRX-FM O'Neill NE
KNEB(AM) Scottsbluff NE
KNEB-FM Scottsbluff NE
KZKX(FM) Seward NE
KSID(AM) Sidney NE
KRFS-FM Superior NE
KVSH(AM) Valentine NE
KTCH(AM) Wayne NE
KTIC(AM) West Point NE
KTIC-FM West Point NE
KSUX(FM) Winnebago NE
WNHW(FM) Belmont NH
WOKQ(FM) Dover NH
WYKR-FM Haverhill NH
WYRY(FM) Hinsdale NH
WXXK(FM) Lebanon NH
WSCY(FM) Moultonborough NH
WNTK(AM) Newport NH
WPKQ(FM) North Conway NH
WINQ(FM) Winchester NH
WOF(AM) Andover NJ
WPUR(FM) Atlantic City NJ
KZZX(FM) Alamogordo NM
KBQI(FM) Albuquerque NM
KRST(FM) Albuquerque NM
KKTC(FM) Angel Fire NM
KTZA(FM) Artesia NM
KNFT-FM Bayard NM
KARS(AM) Belen NM
KATK-FM Carlsbad NM
KLMX(AM) Clayton NM
KNMB(FM) Cloudcroft NM
KCLV-FM Clovis NM
KKYC(FM) Clovis NM
KOTS(AM) Deming NM
KTRA-FM Farmington NM
KGLX(FM) Gallup NM
KYVA(AM) Gallup NM
KMIN(AM) Grants NM
KIXN(FM) Hobbs NM
KPER(FM) Hobbs NM
KRSY-FM La Luz NM
KGRT-FM Las Cruces NM
KFUN(AM) Las Vegas NM
KQBA(FM) Los Alamos NM
KAGM(FM) Los Lunas NM
KWMW(FM) Maljamar NM
KSEL-FM Portales NM
*KTDB(FM) Ramah NM
KBKZ(FM) Raton NM
KCKN(AM) Roswell NM
KMOU(FM) Roswell NM
KWES-FM Ruidoso NM
KSSR(AM) Santa Rosa NM
KMXQ(FM) Socorro NM
KCHS(AM) Truth or Consequences NM
KTNM(AM) Tucumcari NM
KBUL-FM Carson City NV
KCMY(AM) Carson City NV
KRJC(FM) Elko NV
KDSS(FM) Ely NV
KHWG(AM) Fallon NV
KVLV(AM) Fallon NV
KWNR(FM) Henderson NV
KJUL(FM) Moapa Valley NV
KCYE(FM) North Las Vegas NV
KUUB(FM) Sun Valley NV
KBET(AM) Winchester NV
KWNA-FM Winnemucca NV
WGNA-FM Albany NY
WZKZ(FM) Alfred NY
WHWK(FM) Binghamton NY
WBRV(AM) Boonville NY
WBRV-FM Boonville NY
WYRK(FM) Buffalo NY
WNCQ-FM Canton NY

WECK(AM) Cheektowaga NY
WFFG-FM Corinth NY
WFLR(AM) Dundee NY
WRWC(FM) Ellenville NY
WBUG-FM Fort Plain NY
WAMF(AM) Fulton NY
WBBS(FM) Fulton NY
WFLK(FM) Geneva NY
WRWD-FM Highland NY
WCKR(FM) Hornell NY
WPGI(FM) Horseheads NY
WQNY(FM) Ithaca NY
WHUG(FM) Jamestown NY
WDNB(FM) Jeffersonville NY
WKXP(FM) Kingston NY
WXRL(AM) Lancaster NY
WVOS(AM) Liberty NY
WVOS-FM Liberty NY
WLLG(FM) Lowville NY
WVNV(FM) Malone NY
WBGK(FM) Newport Village NY
WBKT(FM) Norwich NY
WPIG(FM) Olean NY
WDOS(AM) Oneonta NY
WTSX(FM) Port Jervis NY
WSNN(FM) Potsdam NY
WBEE-FM Rochester NY
WUMX(FM) Rome NY
WSCP(AM) Sandy Creek-Pulaski NY
WUUF(FM) Sodus NY
WCRR(FM) South Bristol Township NY
WOKN(FM) Southport NY
WSPQ(AM) Springville NY
WFRG-FM Utica NY
WDLA-FM Walton NY
WCJW(AM) Warsaw NY
WFRY-FM Watertown NY
WTYX(AM) Watkins Glen NY
WLSV(AM) Wellsville NY
WNYV(FM) Whitehall NY
WNCO-FM Ashland OH
WNUS(FM) Belpre OH
WAIS(AM) Buchtel OH
WWKC(FM) Caldwell OH
WKKJ(FM) Chillicothe OH
WUBE-FM Cincinnati OH
WNKK(FM) Circleville OH
WGAR-FM Cleveland OH
WCOL-FM Columbus OH
WTNS(AM) Coshocton OH
WHKO(FM) Dayton OH
WZOM(FM) Defiance OH
WOGF(FM) East Liverpool OH
WEDI(AM) Eaton OH
WKKY(FM) Geneva OH
WAXZ(FM) Georgetown OH
WSRW-FM Hillsboro OH
WKFM(FM) Huron OH
WCJO(FM) Jackson OH
WHOK-FM Lancaster OH
WFGF(FM) Lima OH
WIMT(FM) Lima OH
WLGN(AM) Logan OH
WLGN-FM Logan OH
WMRN-FM Marion OH
WQMX(FM) Medina OH
WPFB(AM) Middletown OH
WPFB-FM Middletown OH
WLZZ(FM) Montpelier OH
WSEO(FM) Nelsonville OH
WCLT-FM Newark OH
WOBL(AM) Oberlin OH
WPAY-FM Portsmouth OH
WAOL(FM) Ripley OH
WQXK(FM) Salem OH
WCKY-FM Tiffin OH
WKKO(FM) Toledo OH
WTOD(AM) Toledo OH
WTUZ(FM) Uhrichsville OH
WTGR(FM) Union City OH
WKSW(FM) Urbana OH
WCHO-FM Washington Court House OH
WXIZ(FM) Waverly OH
WRAC(FM) West Union OH
WKFI(AM) Wilmington OH

WQKT(FM) Wooster OH
WBZI(AM) Xenia OH
KADA-FM Ada OK
KEYB(FM) Altus OK
KWHW(AM) Altus OK
KNID(FM) Alva OK
KKAJ-FM Ardmore OK
KHKC-FM Atoka OK
KREK(FM) Bristow OK
KKBI(FM) Broken Bow OK
KYKC(FM) Byng OK
KWEY-FM Clinton OK
KKFC(FM) Coalgate OK
KKEN(FM) Duncan OK
KLBC(FM) Durant OK
KECO(FM) Elk City OK
KOFM(FM) Enid OK
KTNT(FM) Eufaula OK
KYBE(FM) Frederick OK
KGVE(FM) Grove OK
KGYN(AM) Guymon OK
KICM(FM) Healdton OK
KPRV-FM Heavener OK
KXBL(FM) Henryetta OK
KTJS(AM) Hobart OK
KBEL-FM Idabel OK
KLAW(FM) Lawton OK
KBLP(FM) Lindsay OK
KMAD(AM) Madill OK
KFXI(FM) Marlow OK
KMCO(FM) McAlester OK
KNED(AM) McAlester OK
KKNG-FM Newcastle OK
KRIG-FM Nowata OK
KTST(FM) Oklahoma City OK
KXXY-FM Oklahoma City OK
KPNC-FM Ponca City OK
KOMS(FM) Poteau OK
KPRV(AM) Poteau OK
KIRC(FM) Seminole OK
KGFY(FM) Stillwater OK
*KLRB(FM) Stuart OK
KIXO(FM) Sulphur OK
KEOK(FM) Tahlequah OK
KVOO-FM Tulsa OK
KWEN(FM) Tulsa OK
KGND(AM) Vinita OK
KITO-FM Vinita OK
KTFX-FM Warner OK
KWEY(AM) Weatherford OK
KWSH(AM) Wewoka OK
KWFX(FM) Woodward OK
KWOX(FM) Woodward OK
KRKT-FM Albany OR
KTHH(AM) Albany OR
KCMB(FM) Baker City OR
KMTK(FM) Bend OR
KURY(AM) Brookings OR
KZZR(AM) Burns OR
KJDY-FM Canyon City OR
KSHR-FM Coquille OR
KNND(AM) Cottage Grove OR
KUJZ(FM) Creswell OR
KWVR-FM Enterprise OR
KCST-FM Florence OR
KSHL(FM) Gleneden Beach OR
KGBR(FM) Gold Beach OR
KRWQ(FM) Gold Hill OR
KOHU(AM) Hermiston OR
KIHR(AM) Hood River OR
KJDY(AM) John Day OR
KLAD-FM Klamath Falls OR
KQIK(AM) Lakeview OR
KXPC(FM) Lebanon OR
*KBUG(FM) Malin OR
KNCU(FM) Newport OR
KSRV(AM) Ontario OR
KWHT(FM) Pendleton OR
KAKT(FM) Phoenix OR
KCMD(AM) Portland OR
KUPL-FM Portland OR
KWJJ-FM Portland OR
KRCO(AM) Prineville OR
KSJJ(FM) Redmond OR
KJMX(FM) Reedsport OR
KRRM(FM) Rogue River OR

KRNR(AM) Roseburg OR
KRSB-FM Roseburg OR
KCYS(FM) Seaside OR
KKNU(FM) Springfield-Eugene OR
KCKX(AM) Stayton OR
KFIR(AM) Sweet Home OR
WFGY(FM) Altoona PA
WWSM(AM) Annville-Cleona PA
WQBR(FM) Avis PA
WGGI(FM) Benton PA
WLCY(FM) Blairsville PA
WOGH(FM) Burgettstown PA
WVNW(FM) Burnham PA
WBUT(AM) Butler PA
WHGL-FM Canton PA
WCAT-FM Carlisle PA
WIOO(AM) Carlisle PA
WFGI(AM) Charleroi PA
WOGI(FM) Charleroi PA
WKNB(FM) Clarendon PA
WWCH(AM) Clarion PA
WFRM(AM) Coudersport PA
WDKC(FM) Covington PA
WSJR(FM) Dallas PA
WOWQ(FM) DuBois PA
WCTO(FM) Easton PA
WXTA(FM) Edinboro PA
WYGL-FM Elizabethville PA
WLEM(AM) Emporium PA
WIOV-FM Ephrata PA
WSKE(FM) Everett PA
WTWF(FM) Fairview PA
WGTY(FM) Gettysburg PA
WWGY(FM) Grove City PA
WHVR(AM) Hanover PA
WRBT(FM) Harrisburg PA
WHUN(AM) Huntingdon PA
WNTJ(AM) Johnstown PA
WLMI(FM) Kane PA
WADV(AM) Lebanon PA
WGYY(FM) Meadville PA
WPPT(FM) Mercersburg PA
WJUN-FM Mexico PA
WWBE(FM) Mifflinburg PA
WFZY(FM) Mount Union PA
WGYI(FM) Oil City PA
WKQW(AM) Oil City PA
WOGG(FM) Oliver PA
WXTU(FM) Philadelphia PA
WPHB(AM) Philipsburg PA
WDSY-FM Pittsburgh PA
WSGY(FM) Pleasant Gap PA
WLGL(FM) Riverside PA
WDDH(FM) Saint Marys PA
WBYL(FM) Salladasburg PA
WGGY(FM) Scranton PA
WYGL(AM) Selinsgrove PA
WBLJ-FM Shamokin PA
*WSRU(FM) Slippery Rock PA
WNTW(AM) Somerset PA
WKZV(AM) Washington PA
WCBG(AM) Waynesboro PA
WANB(AM) Waynesburg PA
WANB-FM Waynesburg PA
WILQ(FM) Williamsport PA
WABV(AM) Abbeville SC
WKXC-FM Aiken SC
WDOG(AM) Allendale SC
WDOG-FM Allendale SC
WAGS(AM) Bishopville SC
WGZR(FM) Bluffton SC
WGTR(FM) Bucksport SC
WEZL(FM) Charleston SC
WVSZ(FM) Chesterfield SC
WCOS-FM Columbia SC
WYOR(FM) Cross Hill SC
WWNQ(FM) Forest Acres SC
WEAC(AM) Gaffney SC
WNOW-FM Gaffney SC
WSSL-FM Gray Court SC
WESC-FM Greenville SC
WJDJ(AM) Hartsville SC
WHYM(AM) Lake City SC
WRHM(FM) Lancaster SC
WHLZ(FM) Marion SC
WYNF(AM) North Augusta SC

WNKT(FM) Saint George SC
WJES(AM) Saluda SC
WYAK-FM Surfside Beach SC
WBCU(AM) Union SC
WGOG(FM) Walhalla SC
WALI(FM) Walterboro SC
KGIM(AM) Aberdeen SD
KBFS(AM) Belle Fourche SD
KZZI(FM) Belle Fourche SD
KXQL(FM) Flandreau SD
KZMX(AM) Hot Springs SD
KZMX-FM Hot Springs SD
KOKK(AM) Huron SD
KZNC(FM) Huron SD
KBJM(AM) Lemmon SD
KMLO(FM) Lowry SD
KJAM(AM) Madison SD
KJAM-FM Madison SD
KMIT(FM) Mitchell SD
KGFX(AM) Pierre SD
KIMM(AM) Rapid City SD
KIQK(FM) Rapid City SD
KOUT(FM) Rapid City SD
KGIM-FM Redfield SD
KPLO-FM Reliance SD
KIKN-FM Salem SD
KTWB(FM) Sioux Falls SD
KXRB(AM) Sioux Falls SD
KBWS-FM Sisseton SD
KKQQ(FM) Volga SD
KDLO-FM Watertown SD
KSDR-FM Watertown SD
KWYR(AM) Winner SD
KKYA(FM) Yankton SD
WNAX(AM) Yankton SD
WNAX-FM Yankton SD
WJSQ(FM) Athens TN
WLAR(AM) Athens TN
WMOD(FM) Bolivar TN
WXBQ-FM Bristol TN
WTBG(FM) Brownsville TN
WFWL(AM) Camden TN
WUCZ-FM Carthage TN
WVFB(FM) Celina TN
WNKX-FM Centerville TN
WUSY(FM) Cleveland TN
WLVS-FM Clifton TN
*WDVX(FM) Clinton TN
WYSH(AM) Clinton TN
WMCP(AM) Columbia TN
WGSQ(FM) Cookeville TN
WHUB(AM) Cookeville TN
WLSB(AM) Copperhill TN
WKBL(AM) Covington TN
WKBQ(FM) Covington TN
WZYX(AM) Cowan TN
WOWF(FM) Crossville TN
WDKN(AM) Dickson TN
WSDQ(AM) Dunlap TN
WLLI-FM Dyer TN
WOGT(FM) East Ridge TN
WBEJ(AM) Elizabethton TN
WEMB(AM) Erwin TN
WPFD(AM) Fairview TN
WEKR(AM) Fayetteville TN
WYTM-FM Fayetteville TN
WAKM(AM) Franklin TN
WHIN(AM) Gallatin TN
WGRV(AM) Greeneville TN
WMYL(FM) Halls Crossroads TN
*WLMU(FM) Harrogate TN
WTNK(AM) Hartsville TN
WMUF-FM Henry TN
WMLR(AM) Hohenwald TN
WLLI(AM) Humboldt TN
WDAP(AM) Huntingdon TN
WVHR(FM) Huntingdon TN
WOGY(FM) Jackson TN
WDEB(AM) Jamestown TN
WDEB-FM Jamestown TN
WJFC(AM) Jefferson City TN
WIVK-FM Knoxville TN
WQLA-FM La Follette TN
WEEN(AM) Lafayette TN
WLCT(FM) Lafayette TN
WDXE(AM) Lawrenceburg TN

WLLX(FM) Lawrenceburg TN
WWLX(AM) Lawrenceburg TN
WANT(FM) Lebanon TN
WKDA(AM) Lebanon TN
WLIL(AM) Lenoir City TN
WAXO(AM) Lewisburg TN
WJJM(AM) Lewisburg TN
WJJM-FM Lewisburg TN
WGAP(AM) Maryville TN
WWYN(FM) McKenzie TN
WBMC(AM) McMinnville TN
WKZP(FM) McMinnville TN
WGKX(FM) Memphis TN
WMC(AM) Memphis TN
WLIV-FM Monterey TN
WMCT(AM) Mountain City TN
WKDF(FM) Nashville TN
WSIX-FM Nashville TN
WSM(AM) Nashville TN
WSM-FM Nashville TN
WNPC(AM) Newport TN
WNPC-FM Newport TN
WBNT-FM Oneida TN
WOCV(AM) Oneida TN
WMUF(AM) Paris TN
WKJQ-FM Parsons TN
WUAT(AM) Pikeville TN
WKSR-FM Pulaski TN
WTRB(AM) Ripley TN
WOFE(AM) Rockwood TN
WJDT(FM) Rogersville TN
WRGS(AM) Rogersville TN
WMXV(FM) Saint Joseph TN
WKWX(FM) Savannah TN
WORM-FM Savannah TN
WXOQ(FM) Selmer TN
WLIJ(AM) Shelbyville TN
WLND(FM) Signal Mountain TN
WJLE(AM) Smithville TN
WJLE-FM Smithville TN
WEPG(AM) South Pittsburg TN
WAYA(FM) Spring City TN
WSBI(AM) Static TN
WEYE(FM) Surgoinsville TN
WNTT(AM) Tazewell TN
WTNE-FM Trenton TN
WIKQ(FM) Tusculum TN
WYVY(FM) Union City TN
WECO-FM Wartburg TN
WWON(AM) Waynesboro TN
WCDT(AM) Winchester TN
WBRY(AM) Woodbury TN
KEAN-FM Abilene TX
KYYW(AM) Abilene TX
KOPY(AM) Alice TX
KALP(FM) Alpine TX
KATP(FM) Amarillo TX
KDJW(AM) Amarillo TX
KGNC-FM Amarillo TX
KMML-FM Amarillo TX
KACT(AM) Andrews TX
KACT-FM Andrews TX
KASE-FM Austin TX
KVET-FM Austin TX
KKCN(FM) Ballinger TX
KRUN(AM) Ballinger TX
KMKS(FM) Bay City TX
KYKR(FM) Beaumont TX
KTKO(FM) Beeville TX
KMKT(FM) Bells TX
KTON(AM) Belton TX
KFYN(AM) Bonham TX
KQTY(AM) Borger TX
KQTY-FM Borger TX
KNEL-FM Brady TX
KLXK(FM) Breckenridge TX
KTTX(FM) Brenham TX
KWHI(AM) Brenham TX
KTEX(FM) Brownsville TX
KOXE(FM) Brownwood TX
KORA-FM Bryan TX
KHFX(AM) Burleson TX
KBEY(FM) Burnet TX
KHLE(FM) Burnet TX
KHTZ(FM) Caldwell TX
KMIL(FM) Cameron TX
KTAE(AM) Cameron TX
KGAS-FM Carthage TX
KDET(AM) Center TX
KQBB(FM) Center TX
KUZN(FM) Centerville TX
KCTX-FM Childress TX
KCAR(AM) Clarksville TX
KCLE(AM) Cleburne TX
KTHT(FM) Cleveland TX
KQBZ(FM) Coleman TX
KSTA(AM) Coleman TX
KAUM(FM) Colorado City TX
KVMC(AM) Colorado City TX
KULM(FM) Columbus TX
KCOM(AM) Comanche TX
KYOX(FM) Comanche TX
KRYS-FM Corpus Christi TX
KAND(AM) Corsicana TX
KBHT(FM) Crockett TX
KIVY-FM Crockett TX
KSTB(FM) Crystal Beach TX
KXIT(AM) Dalhart TX
KFXR(AM) Dallas TX
KDLK-FM Del Rio TX
KRPT(FM) Devine TX
KVWG-FM Dilley TX
KDHN(AM) Dimmitt TX
KSTV-FM Dublin TX
KDDD(AM) Dumas TX
KATX(FM) Eastland TX
KEAS(AM) Eastland TX
KEZB(FM) Edna TX
KIOX-FM El Campo TX
KULP(AM) El Campo TX
KHEY-FM El Paso TX
*KOLI(FM) Electra TX
KNES(FM) Fairfield TX
KPSO-FM Falfurrias TX
KXEZ(FM) Farmersville TX
KTYS(FM) Flower Mound TX
KFLP-FM Floydada TX
KFST-FM Fort Stockton TX
KPLX(FM) Fort Worth TX
KSCS(FM) Fort Worth TX
KNAF(AM) Fredericksburg TX
KNAF-FM Fredericksburg TX
KGAF(AM) Gainesville TX
KULF(FM) Ganado TX
KTFW-FM Glen Rose TX
KCTI(AM) Gonzales TX
KSWA(AM) Graham TX
KPIR(AM) Granbury TX
KGVL(AM) Greenville TX
KIKT(FM) Greenville TX
KHLT(AM) Hallettsville TX
KTXM(FM) Hallettsville TX
KCLW(AM) Hamilton TX
KUSJ(FM) Harker Heights TX
KVRP-FM Haskell TX
KTHP(FM) Hemphill TX
KWRD(AM) Henderson TX
KPAN(AM) Hereford TX
KHBR(AM) Hillsboro TX
KCWM(AM) Hondo TX
KILT-FM Houston TX
KHYI(FM) Howe TX
KLSN(FM) Hudson TX
KSAM-FM Huntsville TX
KRBL(FM) Idalou TX
KEBE(AM) Jacksonville TX
KMBL(AM) Junction TX
KOOK(FM) Junction TX
KAML(AM) Kenedy-Karnes City TX
KRNH(FM) Kerrville TX
KRVL(FM) Kerrville TX
KFTX(FM) Kingsville TX
KBUK(FM) La Grange TX
KVLG(AM) La Grange TX
KPET(AM) Lamesa TX
KTXC(FM) Lamesa TX
KCYL(AM) Lampasas TX
KLVT(AM) Levelland TX
KSHN-FM Liberty TX
KZZN(AM) Littlefield TX
KETX(AM) Livingston TX
KETX-FM Livingston TX
KACQ(FM) Lometa TX
KYKX(FM) Longview TX
KLLL-FM Lubbock TX
KQBR(FM) Lubbock TX
KYKS(FM) Lufkin TX
KAGG(FM) Madisonville TX
KMVL-FM Madisonville TX
KCKL(FM) Malakoff TX
KCUL(AM) Marshall TX
KMHT(AM) Marshall TX
KMHT-FM Marshall TX
KHLB(FM) Mason TX
KLSR-FM Memphis TX
KRQX(AM) Mexia TX
KWGW(FM) Mexia TX
KNFM(FM) Midland TX
KMOO-FM Mineola TX
KFWR(FM) Mineral Wells TX
KJSA(AM) Mineral Wells TX
KIMP(AM) Mount Pleasant TX
KMUL-FM Muleshoe TX
KJCS(FM) Nacogdoches TX
KMBV(FM) Navasota TX
KHKX(FM) Odessa TX
KOGT(AM) Orange TX
KYXX(FM) Ozona TX
KROY(FM) Palacios TX
KNET(AM) Palestine TX
KYYK(FM) Palestine TX
KOMX(FM) Pampa TX
KOYN(FM) Paris TX
KPLT(AM) Paris TX
KZHN(AM) Paris TX
KKBQ-FM Pasadena TX
KVWG(AM) Pearsall TX
KREH(AM) Pecan Grove TX
KIUN(AM) Pecos TX
KEYE(AM) Perryton TX
KVJY(AM) Pharr TX
KSCN(FM) Pittsburg TX
KKYN-FM Plainview TX
KREL(AM) Quanah TX
KCUB-FM Ranger TX
KRXT(FM) Rockdale TX
KBMI(FM) Roma TX
KDCD(FM) San Angelo TX
KGKL-FM San Angelo TX
KAJA(FM) San Antonio TX
KCYY(FM) San Antonio TX
KKYX(AM) San Antonio TX
KQSI(FM) San Augustine TX
KBAL-FM San Saba TX
KWED(AM) Seguin TX
KIKZ(AM) Seminole TX
KSEM-FM Seminole TX
KAYD-FM Silsbee TX
KOUL(FM) Sinton TX
KSNY-FM Snyder TX
KHOS-FM Sonora TX
KLGD(FM) Stamford TX
KKJW(FM) Stanton TX
*KEQX(FM) Stephenville TX
KSCH(FM) Sulphur Springs TX
KXOX(AM) Sweetwater TX
KXOX-FM Sweetwater TX
KKYR-FM Texarkana TX
KBCY(FM) Tye TX
KKUS(FM) Tyler TX
KNUE(FM) Tyler TX
KBNU(FM) Uvalde TX
KVOU-FM Uvalde TX
KVWC(AM) Vernon TX
KIXS(FM) Victoria TX
KVNN(AM) Victoria TX
KQRL(AM) Waco TX
WACO-FM Waco TX
KBEC(AM) Waxahachie TX
KLUR(FM) Wichita Falls TX
KWFS-FM Wichita Falls TX
KVST(FM) Willis TX
KNCE(FM) Winters TX
KWUD(AM) Woodville TX
KYKM(FM) Yoakum TX
KCYQ(FM) Elsinore UT
KMTI(AM) Manti UT
KCYN(FM) Moab UT
KEGA(FM) Oakley UT
KENZ(FM) Ogden UT
KKAT-FM Orem UT
KARB(FM) Price UT
KSLL(AM) Price UT
KNEU(AM) Roosevelt UT
KONY(FM) Saint George UT
KKAT(AM) Salt Lake City UT
KSOP-FM Salt Lake City UT
KUBL-FM Salt Lake City UT
KSOP(AM) South Salt Lake UT
KYLZ(FM) Tremonton UT
KLCY-FM Vernal UT
WVES(FM) Accomac VA
WKDE-FM Altavista VA
WYYD(FM) Amherst VA
WAXM(FM) Big Stone Gap VA
WBBC-FM Blackstone VA
WHKX(FM) Bluefield VA
WHQX(FM) Cedar Bluff VA
WLUS-FM Clarksville VA
WDIC(AM) Clinchco VA
WKHK(FM) Colonial Heights VA
WIQO-FM Covington VA
WSVS(AM) Crewe VA
WJMA-FM Culpeper VA
WAKG(FM) Danville VA
WBNN-FM Dillwyn VA
WLFV(FM) Ettrick VA
WGRX(FM) Falmouth VA
WFLO(AM) Farmville VA
WVHL(FM) Farmville VA
WLQM-FM Franklin VA
WFLS-FM Fredericksburg VA
WFTR(AM) Front Royal VA
WBRF(FM) Galax VA
WMNA(AM) Gretna VA
WMJD(FM) Grundy VA
WKCY-FM Harrisonburg VA
*WCHG(FM) Hot Springs VA
WJNV(FM) Jonesville VA
WXLZ-FM Lebanon VA
WOJL(FM) Louisa VA
WRAA(AM) Luray VA
WMEV-FM Marion VA
*WVLS(FM) Monterey VA
WSIG(FM) Mount Jackson VA
WGH-FM Newport News VA
WNVA-FM Norton VA
WESR(AM) Onley-Onancock VA
WDXC(FM) Pound VA
WPSK-FM Pulaski VA
WWBU(FM) Radford VA
WSLC-FM Roanoke VA
WYTI(AM) Rocky Mount VA
WXMY(AM) Saltville VA
WKSK-FM South Hill VA
WZBB(FM) Stanleytown VA
WCYK-FM Staunton VA
WKDW(AM) Staunton VA
WKSI-FM Stephens City VA
WHEO(AM) Stuart VA
WNNT-FM Warsaw VA
WIGO-FM White Stone VA
WUSQ-FM Winchester VA
WYVE(AM) Wytheville VA
WUSX(FM) Addison VT
WZLF(FM) Bellows Falls VT
WWFY(FM) Berlin VT
WTNN(FM) Bristol VT
WOKO(FM) Burlington VT
WLVB(FM) Morrisville VT
WIKE(AM) Newport VT
WVNR(AM) Poultney VT
WJEN(FM) Rutland VT
WLFE-FM Saint Albans VT
WKXH(FM) Saint Johnsbury VT
WTSM(FM) Springfield VT
WXLF(FM) White River Junction VT
KNBQ(FM) Centralia WA
KMNT(FM) Chehalis WA
KCLK-FM Clarkston WA
KCLX(AM) Colfax WA
KCVL(AM) Colville WA
KYSN(FM) East Wenatchee WA
KXLE-FM Ellensburg WA
KXLE-FM Ellensburg WA
KULE-FM Ephrata WA
KYYT(FM) Goldendale WA
KEYG(AM) Grand Coulee WA
KXXK(FM) Hoquiam-Aberdeen WA
KVAS(FM) Ilwaco WA
KBAM(AM) Longview WA
KUKN(FM) Longview WA
KGY-FM McCleary WA
KWIQ-FM Moses Lake WA
KAPS(AM) Mount Vernon WA
KNCW(FM) Omak WA
KIXZ-FM Opportunity WA
KZZL-FM Pullman WA
KIOK(FM) Richland WA
KORD-FM Richland WA
KKWF(FM) Seattle WA
KMPS-FM Seattle WA
KDRK-FM Spokane WA
KDBL(FM) Toppenish WA
KYNR(AM) Toppenish WA
KKRV(FM) Wenatchee WA
KUTI(AM) Yakima WA
KXDD(FM) Yakima WA
WRLU(FM) Algoma WI
WBSZ(FM) Ashland WI
WXRO(FM) Beaver Dam WI
WISS(AM) Berlin WI
WQRB(FM) Bloomer WI
WATQ(FM) Chetek WI
WJMQ(FM) Clintonville WI
WPCK(FM) Denmark WI
WVRE(FM) Dickeyville WI
WDMP(AM) Dodgeville WI
WDMP-FM Dodgeville WI
WDMO(FM) Durand WI
WQOQ(AM) Durand WI
WAXX(FM) Eau Claire WI
WTKM(AM) Hartford WI
WTKM-FM Hartford WI
WJVL(FM) Janesville WI
WQCC(FM) La Crosse WI
WLDY(AM) Ladysmith WI
WGLR(AM) Lancaster WI
WGLR-FM Lancaster WI
WKQH(FM) Marathon WI
WYTE(FM) Marshfield WI
WWQM-FM Middleton WI
WEKZ(AM) Monroe WI
WNCY-FM Neenah-Menasha WI
WIXK(AM) New Richmond WI
WOCO(AM) Oconto WI
WPKR(FM) Omro WI
WCQM(FM) Park Falls WI
WBKY(FM) Portage WI
WDDC(FM) Portage WI
WQPC(FM) Prairie du Chien WI
WNFM(FM) Reedsburg WI
WHDG(FM) Rhinelander WI
WAQE(AM) Rice Lake WI
WJMC-FM Rice Lake WI
WRCO-FM Richland Center WI
WTCH(AM) Shawano WI
WBFM(FM) Sheboygan WI
WCOW-FM Sparta WI
WOSQ(FM) Spencer WI
WPLT(FM) Spooner WI
WCUB(AM) Two Rivers WI
WVRQ-FM Viroqua WI
WMIL(FM) Waukesha WI
WDUX(AM) Waupaca WI
WDEZ(FM) Wausau WI
WBKV(AM) West Bend WI
WBWI-FM West Bend WI
WSLD(FM) Whitewater WI
WDLS(AM) Wisconsin Dells WI
WJLS-FM Beckley WV
WCST(AM) Berkeley Springs WV
WDHC(FM) Berkeley Springs WV
WBRB(FM) Buckhannon WV
WKAZ(AM) Charleston WV
WKWS(FM) Charleston WV
WGIE(FM) Clarksburg WV
WPDX-FM Clarksburg WV
WZAC-FM Danville WV
WDNE-FM Elkins WV

WKKW(FM) Fairmont WV
*WFGH(FM) Fort Gay WV
*WVMR(AM) Frost WV
WTCR-FM Huntington WV
WKMM(FM) Kingwood WV
WKCJ(FM) Lewisburg WV
WGYE(FM) Mannington WV
WTNJ(FM) Mount Hope WV
WGGE(FM) Parkersburg WV
WHNK(AM) Parkersburg WV
WELD-FM Petersburg WV
WWYO(AM) Pineville WV
WBYG(FM) Point Pleasant WV
WVAR(AM) Richwood WV
WCEF(FM) Ripley WV
*WVSB(FM) Romney WV
WYKM(AM) Rupert WV
WVRC-FM Spencer WV
WDBS(FM) Sutton WV
WZST(FM) Westover WV
WOVK(FM) Wheeling WV
WXCC(FM) Williamson WV
KRSV(AM) Afton WY
KRSV-FM Afton WY
KKAW(FM) Albin WY
KLGT(FM) Buffalo WY
KIGN(FM) Burns WY
KQLT(FM) Casper WY
KWYY(FM) Casper WY
KOLZ(FM) Cheyenne WY
KZDR(FM) Cheyenne WY
KDWY(FM) Diamondville WY
KKTY-FM Douglas WY
KEVA(AM) Evanston WY
KRND(AM) Fox Farm WY
KGWY(FM) Gillette WY
KFRZ(FM) Green River WY
KZMQ(AM) Greybull WY
KZMQ-FM Greybull WY
KJAX(FM) Jackson WY
KSGT(AM) Jackson WY
KOVE(AM) Lander WY
KCGY(FM) Laramie WY
KUSZ(FM) Laramie WY
KWYW(FM) Lost Cabin WY
KASL(AM) Newcastle WY
KPIN(FM) Pinedale WY
KTAK(FM) Riverton WY
KQSW(FM) Rock Springs WY
KYTI(FM) Sheridan WY
KYDT(FM) Sundance WY
KDNO(FM) Thermopolis WY
KERM(FM) Torrington WY
KGOS(AM) Torrington WY
KYCN(AM) Wheatland WY

Disco

KNRJ(FM) Payson AZ
WVBB(FM) Columbia City IN

Diversified

*KUAC(FM) Fairbanks AK
KCAM(AM) Glennallen AK
*KXGA(FM) Glennallen AK
*KRNN(FM) Juneau AK
*KTOO(FM) Juneau AK
*KRBD(FM) Ketchikan AK
*KMXT(FM) Kodiak AK
*KXKM(FM) McCarthy AK
*KNOM(AM) Nome AK
*KNOM-FM Nome AK
*KSDP(AM) Sand Point AK
*KCAW(FM) Sitka AK
*KTNA(FM) Talkeetna AK
*KCHU(AM) Valdez AK
*KSTK(FM) Wrangell AK
*WLJR(FM) Birmingham AL
*KVMN(FM) Cave City AR
*KKDU(FM) El Dorado AR
*KABF(FM) Little Rock AR
KVSA(AM) McGehee AR
KNRJ(FM) Payson AZ
KBUX(FM) Quartzsite AZ
*KAIC(FM) Tucson AZ
*KNNB(FM) Whiteriver AZ
*KALX(FM) Berkeley CA
*KPFA(FM) Berkeley CA
*KPFB(FM) Berkeley CA
*KZFR(FM) Chico CA
*KSPC(FM) Claremont CA
*KECG(FM) El Cerrito CA
*KMUE(FM) Eureka CA
*KFSR(FM) Fresno CA
*KMUD(FM) Garberville CA
KNOB(FM) Healdsburg CA
*KCRI(FM) Indio CA
*KUCI(FM) Irvine CA
*KLAI(FM) Laytonville CA
*KPFK(FM) Los Angeles CA
*KKSM(AM) Oceanside CA
*KUCR(FM) Riverside CA
*KBDH(FM) San Ardo CA
*KALW(FM) San Francisco CA
*KPOO(FM) San Francisco CA
*KUSF(FM) San Francisco CA
*KCPR(FM) San Luis Obispo CA
*KUSP(FM) Santa Cruz CA
*KZSC(FM) Santa Cruz CA
*KAIB(FM) Shafter CA
*KCSS(FM) Turlock CA
*KZYZ(FM) Willits CA
*KLRD(FM) Yucaipa CA
KVCU(AM) Boulder CO
*KRCC(FM) Colorado Springs CO
*KBUT(FM) Crested Butte CO
*KGNU(AM) Denver CO
*KDUR(FM) Durango CO
*KUNC(FM) Greeley CO
*KRLJ(FM) La Junta CO
*KRNC(FM) Steamboat Springs CO
*WPKN(FM) Bridgeport CT
*WRTC-FM Hartford CT
*WSLX(FM) New Canaan CT
*WHUS(FM) Storrs CT
*WNHU(FM) West Haven CT
*WVUD(FM) Newark DE
*WFCF(FM) Saint Augustine FL
*WVFS(FM) Tallahassee FL
*WMNF(FM) Tampa FL
*WREK(FM) Atlanta GA
*WVDA(FM) Valdosta GA
*KPRG(FM) Hagatna GU
*KKCR(FM) Hanalei HI
*KAQA(FM) Kilauea HI
*KHAI(FM) Wahiawa HI
*KRUI-FM Iowa City IA
*KWAR(FM) Waverly IA
*KBSQ(FM) McCall ID
*WESN(FM) Bloomington IL
*WDBX(FM) Carbondale IL
*WHPK-FM Chicago IL
*WDGC-FM Downers Grove IL
*WEPS(FM) Elgin IL
*WMXM(FM) Lake Forest IL
*WLRA(FM) Lockport IL
*WHCM(FM) Palatine IL
*WQNA(FM) Springfield IL
*WILL(AM) Urbana IL
*WNTH(FM) Winnetka IL
*WFHB(FM) Bloomington IN
*WPSR(FM) Evansville IN
*WUEV(FM) Evansville IN
*WHWE(FM) Howe IN
*WRFT(FM) Indianapolis IN
WMRS(FM) Monticello IN
*WBKE-FM North Manchester IN
*WVUR-FM Valparaiso IN
*KBTL(FM) El Dorado KS
*KANZ(FM) Garden City KS
*KTJO-FM Ottawa KS
KKAN(AM) Phillipsburg KS
*WOCS(FM) Lerose KY
*KLSP(FM) Angola LA
*WHAB(FM) Acton MA
*WSHL-FM Easton MA
*WUML(FM) Lowell MA
*WMLN-FM Milton MA
*WOZQ(FM) Northampton MA
*WNMH(FM) Northfield MA
*WMWM(FM) Salem MA
*WBSL-FM Sheffield MA
*WAIC(FM) Springfield MA
*WNEK-FM Springfield MA
*WTCC(FM) Springfield MA
*WSKB(FM) Westfield MA
*WCHC(FM) Worcester MA
*WCUW(FM) Worcester MA
*WKHS(FM) Worton MD
*WERU-FM Blue Hill ME
*WBOR(FM) Brunswick ME
*WMPG(FM) Gorham ME
*WRBC(FM) Lewiston ME
*WMEB-FM Orono ME
*WUPI(FM) Presque Isle ME
*WMHB(FM) Waterville ME
*WYAR(FM) Yarmouth ME
*WVAC-FM Adrian MI
*WCBN-FM Ann Arbor MI
*WBFH(FM) Bloomfield Hills MI
*WKDS(FM) Kalamazoo MI
*WOCR(FM) Olivet MI
*WNMC-FM Traverse City MI
KDAL(AM) Duluth MN
*KFAI(FM) Minneapolis MN
KOLV(FM) Olivia MN
*KCVO-FM Camdenton MO
*KWUR(FM) Clayton MO
*KOPN(FM) Columbia MO
*KCVZ(FM) Dixon MO
*KCVQ(FM) Knob Noster MO
*KCVJ(FM) Osceola MO
*KCVK(FM) Otterville MO
KLUE(FM) Poplar Bluff MO
*KMNR(FM) Rolla MO
*KMST(FM) Rolla MO
*KDHX(FM) Saint Louis MO
*KCVX(FM) Salem MO
*WMUW(FM) Columbus MS
*KGLT(FM) Bozeman MT
*KYPR(FM) Miles City MT
*WXYC(FM) Chapel Hill NC
*WXDU(FM) Durham NC
*WSOE(FM) Elon NC
WLHC(FM) Robbins NC
*WZRL(FM) Wade NC
*WSIF(FM) Wilkesboro NC
*KMHA(FM) Four Bears ND
*KFKX(FM) Hastings NE
*KZUM(FM) Lincoln NE
*WSPS(FM) Concord NH
WAJM(FM) Atlantic City NJ
*WDVR(FM) Delaware Township NJ
*WFMU(FM) East Orange NJ
*WCVH(FM) Flemington NJ
*WGLS-FM Glassboro NJ
*WNTI(FM) Hackettstown NJ
*KUNM(FM) Albuquerque NM
*KCIE(FM) Dulce NM
KSIL(FM) Hurley NM
*KQAI(FM) Roswell NM
*KWPR(FM) Lund NV
KBZB(FM) Pioche NV
*WCDB(FM) Albany NY
*WALF(FM) Alfred NY
*WKRB(FM) Brooklyn NY
*WCVF-FM Fredonia NY
*WRCU-FM Hamilton NY
*WPKM(FM) Montauk NY
*WXHD(FM) Mount Hope NY
*WFUV(FM) New York NY
*WONY(FM) Oneonta NY
*WNYO(FM) Oswego NY
*WVKR-FM Poughkeepsie NY
WFTU(AM) Riverhead NY
*WRUR-FM Rochester NY
*WUSB(FM) Stony Brook NY
*WKWZ(FM) Syosset NY
*WPNR-FM Utica NY
*WMCO(FM) New Concord OH
*WOBC-FM Oberlin OH
*WJCU(FM) University Heights OH
*WCWS(FM) Wooster OH
*KPSU(FM) Goodwell OK
*KMUN(FM) Astoria OR
*KWVA(FM) Eugene OR
*KTEC(FM) Klamath Falls OR
*KLCO(FM) Newport OR
*KBOO(FM) Portland OR
*KRRC(FM) Portland OR
*KVRA(FM) Sisters OR
*WMUH(FM) Allentown PA
*WDCV-FM Carlisle PA
*WESS(FM) East Stroudsburg PA
*WMCE(FM) Erie PA
*WRSD(FM) Folsom PA
*WTGP(FM) Greenville PA
*WHHS(FM) Havertown PA
*WPSX(FM) Kane PA
*WLCH(FM) Lancaster PA
*WIXQ(FM) Millersville PA
*WSFX(FM) Nanticoke PA
*WPEB(FM) Philadelphia PA
WPHE(AM) Phoenixville PA
*WRCT(FM) Pittsburgh PA
*WYBF(FM) Radnor Township PA
*WXLV(FM) Schnecksville PA
*WSRN-FM Swarthmore PA
*WCUR(FM) West Chester PA
*WRLC(FM) Williamsport PA
WNIK(AM) Arecibo PR
WQBS(AM) San Juan PR
*WRIU(FM) Kingston RI
*WJHD(FM) Portsmouth RI
*WELH(FM) Providence RI
*WJWJ-FM Beaufort SC
*KCFS(FM) Sioux Falls SD
*WAPX-FM Clarksville TN
*WAPX-FM Clarksville TN
WSGM(FM) Coalmont TN
*WMTS-FM Murfreesboro TN
*WRVU(FM) Nashville TN
*KTXP(FM) Bushland TX
*KPFT(FM) Houston TX
*KTRU(FM) Houston TX
*KTSU(FM) Houston TX
*KOHM(FM) Lubbock TX
*KTXT-FM Lubbock TX
KLSR-FM Memphis TX
*KEOM(FM) Mesquite TX
*KWBU-FM Waco TX
*KRCL(FM) Salt Lake City UT
*WUVT-FM Blacksburg VA
*WEHC(FM) Emory VA
*WMLU(FM) Farmville VA
*WWHS-FM Hampden-Sydney VA
*WLUR(FM) Lexington VA
*WVST-FM Petersburg VA
*WDCE(FM) Richmond VA
*WCWM(FM) Williamsburg VA
WSTA(AM) Charlotte Amalie VI
WAXJ(FM) Frederiksted VI
WDHP(AM) Frederiksted VI
*WRUV(FM) Burlington VT
*WJSC-FM Johnson VT
*WWLR(FM) Lyndonville VT
*WRMC-FM Middlebury VT
*WGDR(FM) Plainfield VT
WDEV(AM) Waterbury VT
KXPA(AM) Bellevue WA
*KCED(FM) Centralia WA
*KCWU(FM) Ellensburg WA
*KSER(FM) Everett WA
*KAOS(FM) Olympia WA
*KZUU(FM) Pullman WA
*KEXP-FM Seattle WA
KYNR(AM) Toppenish WA
*KWCW(FM) Walla Walla WA
*WRST-FM Oshkosh WI
*WOJB(FM) Reserve WI
*KUWS(FM) Superior WI
*WVBC(FM) Bethany WV
*WVWC(FM) Buckhannon WV
*WVMR(AM) Frost WV
*WMUL(FM) Huntington WV
*WWVU-FM Morgantown WV
*WQAB(FM) Philippi WV
*KAIX(FM) Cheyenne WY

Easy Listening

*WVAS(FM) Montgomery AL
KLEZ(FM) Malvern AR
KTLO-FM Mountain Home AR
*KNAI(FM) Phoenix AZ
KAHM(FM) Prescott AZ
KGFM(FM) Bakersfield CA
*KGUD(FM) Longmont CO
WAVV(FM) Marco FL
*WKGC(AM) Panama City Beach FL
WKEZ-FM Tavenier FL
WGPC(AM) Albany GA
*KCMR(FM) Mason City IA
KILJ-FM Mount Pleasant IA
WGCY(FM) Gibson City IL
WLLM(AM) Lincoln IL
WGFA(AM) Watseka IL
KVSV-FM Beloit KS
WJEJ(AM) Hagerstown MD
WEZQ(FM) Bangor ME
WEZR(AM) Lewiston ME
WCCY(AM) Houghton MI
WMLQ(FM) Manistee MI
WCZY-FM Mount Pleasant MI
WMBN(AM) Petoskey MI
WIOS(AM) Tawas City MI
KYMO(AM) East Prairie MO
KTXR(FM) Springfield MO
WBAQ(FM) Greenville MS
WGCM-FM Gulfport MS
KHDN(AM) Hardin MT
WHLC(FM) Highlands NC
WEZS(AM) Laconia NH
KLEA(AM) Lovington NM
*WMEJ(FM) Proctorville OH
KORC(AM) Waldport OR
WLTS(FM) Centre Hall PA
WFDT(FM) Aguada PR
WFID(FM) Rio Piedras PR
WDAR-FM Darlington SC
WEZV(FM) North Myrtle Beach SC
WCPH(AM) Etowah TN
*WDNX(FM) Olive Hill TN
KRFE(AM) Lubbock TX
*KLUX(FM) Robstown TX
WJOY(AM) Burlington VT
WOCO-FM Oconto WI
WEZY(FM) Racine WI

Educational

*KXKM(FM) McCarthy AK
*WSTF(FM) Andalusia AL
*WLJR(FM) Birmingham AL
WQEM(FM) Columbiana AL
*WYFD(FM) Decatur AL
*WDYF(FM) Dothan AL
*WVOB(FM) Dothan AL
*WLBF(FM) Montgomery AL
*WTBJ(FM) Oxford AL
*KVMN(FM) Cave City AR
KOFC(AM) Fayetteville AR
KJBN(AM) Little Rock AR
*KCMH(FM) Mountain Home AR
*KWRB(FM) Bisbee AZ
*KRMC(FM) Douglas AZ
*KPUB(FM) Flagstaff AZ
*KNOG(FM) Nogales AZ
*KNNB(FM) Whiteriver AZ
*KALX(FM) Berkeley CA
*KIBC(FM) Burney CA
*KHAP(FM) Chico CA
*KECR(AM) El Cajon CA
*KECG(FM) El Cerrito CA
*KMUE(FM) Eureka CA
*KMUD(FM) Garberville CA
*KEFR(FM) Le Grand CA
*KFRN(AM) Long Beach CA
*KAKX(FM) Mendocino CA
*KADV(FM) Modesto CA
*KSMC(FM) Moraga CA
*KSGN(FM) Riverside CA
*KUSF(FM) San Francisco CA
*KBBF(FM) Santa Rosa CA
*KZSU(FM) Stanford CA
*KCLU(FM) Thousand Oaks CA
*KBUT(FM) Crested Butte CO
*KCIC(FM) Grand Junction CO

WADS(AM) Ansonia CT
*WERB(FM) Berlin CT
*WFAR(FM) Danbury CT
*WQTQ(FM) Hartford CT
*WFCS(FM) New Britain CT
*WVUD(FM) Newark DE
*WTJT(FM) Baker FL
*WJED(FM) Dogwood Lakes Estate FL
*WJFP(FM) Fort Pierce FL
WXYB(AM) Indian Rocks Beach FL
*WJIR(FM) Key West FL
*WKES(FM) Lakeland FL
*WYFO(FM) Lakeland FL
*WMCU(FM) Miami FL
*WHIJ(FM) Ocala FL
*WPCS(FM) Pensacola FL
*WVIJ(FM) Port Charlotte FL
*WKZM(FM) Sarasota FL
*WOAK(FM) La Grange GA
*WYFS(FM) Savannah GA
*WVGS(FM) Statesboro GA
*KKCR(FM) Hanalei HI
*KCIF(FM) Hilo HI
*KAQA(FM) Kilauea HI
*KHJC(FM) Lihue HI
*KHOE(FM) Fairfield IA
*KRUI-FM Iowa City IA
*KBBG(FM) Waterloo IA
*KWAR(FM) Waverly IA
*KCIR(FM) Twin Falls ID
*KEFX(FM) Twin Falls ID
*WBGL(FM) Champaign IL
*WHPK-FM Chicago IL
*WKKC(FM) Chicago IL
*WMBI-FM Chicago IL
*WZRD(FM) Chicago IL
*WEPS(FM) Elgin IL
*WGNN(FM) Fisher IL
*WGBK(FM) Glenview IL
*WKCC(FM) Kankakee IL
*WLRA(FM) Lockport IL
*WVJC(FM) Mount Carmel IL
*WWGN(FM) Ottawa IL
*WHCM(FM) Palatine IL
*WPSR(FM) Evansville IN
*WGVE(FM) Gary IN
*WHWE(FM) Howe IN
*WQKO(FM) Howe IN
*WRFT(FM) Indianapolis IN
*WWHI(FM) Muncie IN
*WNAS(FM) New Albany IN
*WJLR(FM) Seymour IN
*WETL(FM) South Bend IN
*WMHD-FM Terre Haute IN
KDCC(AM) Dodge City KS
*KONQ(FM) Dodge City KS
*KANZ(FM) Garden City KS
*KZAN(FM) Hays KS
*KZNA(FM) Hill City KS
*WDFB-FM Danville KY
*WSOF-FM Madisonville KY
*WBFI(FM) McDaniels KY
*WTHL(FM) Somerset KY
*KVDP(FM) Dry Prong LA
*KPAE(FM) Erwinville LA
*WRBH(FM) New Orleans LA
WBNW(AM) Concord MA
*WXRB(FM) Dudley MA
*WOZQ(FM) Northampton MA
*WRPS(FM) Rockland MA
*WSDH(FM) Sandwich MA
*WFSI(FM) Annapolis MD
*WOEL-FM Elkton MD
*WGTS(FM) Takoma Park MD
*WERU-FM Blue Hill ME
*WYAR(FM) Yarmouth ME
*WBFH(FM) Bloomfield Hills MI
*WRCJ-FM Detroit MI
*WIDR(FM) Kalamazoo MI
*WKDS(FM) Kalamazoo MI
*WPCJ(FM) Pittsford MI
*KMSK(FM) Austin MN
*KMSU(FM) Mankato MN
KTIG(FM) Pequot Lakes MN
*KCFB(FM) Saint Cloud MN
*KVSC(FM) Saint Cloud MN
*KRHS(FM) Overland MO
*KMNR(FM) Rolla MO
*WPAE(FM) Centreville MS
*KGLT(FM) Bozeman MT
*KMSM-FM Butte MT
*KDWG(FM) Dillon MT
*KGPR(FM) Great Falls MT
*KBGA(FM) Missoula MT
*WPIR(FM) Hickory NC
*WRSH(FM) Rockingham NC
WBFJ(AM) Winston-Salem NC
*KABU(FM) Fort Totten ND
*KTLX(FM) Columbus NE
*KCNT(FM) Hastings NE
*KIOS-FM Omaha NE
*WVFA(FM) Lebanon NH
*WSCS(FM) New London NH
*WGLS-FM Glassboro NJ
*WFME(FM) Newark NJ
*WVPH(FM) Piscataway NJ
*KTDB(FM) Ramah NM
*KSHI(FM) Zuni NM
*KNIS(FM) Carson City NV
*KNCC(FM) Elko NV
*WXBA(FM) Brentwood NY
*WFBF(FM) Buffalo NY
*WHPC(FM) Garden City NY
*WBAI(FM) New York NY
*WNYE(FM) New York NY
*WONY(FM) Oneonta NY
*WOSS(FM) Ossining NY
*WPSA(FM) Paul Smiths NY
*WPOB-FM Plainview NY
*WFRS(FM) Smithtown NY
*WCII(FM) Spencer NY
*WFRW(FM) Webster NY
*WRDL(FM) Ashland OH
*WHVT(FM) Clyde OH
*WDEQ-FM De Graff OH
*WKET(FM) Kettering OH
*WLMH(FM) Morrow OH
*WMCO(FM) New Concord OH
*WOBC-FM Oberlin OH
*WLHS(FM) West Chester OH
*KKVO(FM) Altus OK
*KOSN(FM) Ketchum OK
KQCV(AM) Oklahoma City OK
*KOSU(FM) Stillwater OK
KBNP(AM) Portland OR
*KBPS(AM) Portland OR
*WLCH(FM) Lancaster PA
*WMSS(FM) Middletown PA
*WRCT(FM) Pittsburgh PA
WREQ(FM) Ridgebury PA
*WVYC(FM) York PA
*WIDA-FM Carolina PR
*WJVP(FM) Culebra PR
*WCRP(FM) Guayama PR
*WEPC(FM) Belton SC
*WYFV(FM) Cayce SC
*WTBI-FM Greenville SC
*WMBJ(FM) Murrell's Inlet SC
WTBI(AM) Pickens SC
*KJBB(FM) Watertown SD
*WHCB(FM) Bristol TN
*WMBW(FM) Chattanooga TN
*WWOG(FM) Cookeville TN
*WMKW(FM) Crossville TN
*WCSK(FM) Kingsport TN
*WCPI(FM) McMinnville TN
*WEVL(FM) Memphis TN
*WQOX(FM) Memphis TN
*WFCM-FM Murfreesboro TN
*WDNX(FM) Olive Hill TN
*WAWL-FM Red Bank TN
*WFCM(AM) Smyrna TN
*KJRT(FM) Amarillo TX
KRGN(FM) Amarillo TX
*KASV(FM) Borger TX
*KBNR(FM) Brownsville TX
*KEOS(FM) College Station TX
*KBNJ(FM) Corpus Christi TX
*KVTT(FM) Dallas TX
*KOIR(FM) Edinburg TX
*KVER(FM) El Paso TX
*KTSU(FM) Houston TX
*KAVX(FM) Lufkin TX
KRIO(AM) McAllen TX
*KPHS(FM) Plains TX
*KTER(FM) Rudolph TX
KSLR(AM) San Antonio TX
*KPDR(FM) Wheeler TX
*KRCL(FM) Salt Lake City UT
*WFOS(FM) Chesapeake VA
*WOTC(FM) Edinburg VA
*WRVL(FM) Lynchburg VA
*WISE-FM Wise VA
WGOD-FM Charlotte Amalie VI
*WRUV(FM) Burlington VT
*KKRS(FM) Davenport WA
*KNHC(FM) Seattle WA
*WBCR-FM Beloit WI
*WHHI(FM) Highland WI
*WHLA(FM) La Crosse WI
*WHA(AM) Madison WI
*WSUM(FM) Madison WI
*WHWC(FM) Menomonie WI
*KUWS(FM) Superior WI
*WHRM(FM) Wausau WI
*WVBC(FM) Bethany WV
WQBE-FM Charleston WV
*WWVU-FM Morgantown WV
KVOW(AM) Riverton WY

Ethnic

KAAB(AM) Batesville AR
*KTQX(FM) Bakersfield CA
KIQQ(AM) Barstow CA
*KUBO(FM) Calexico CA
*KHDC(FM) Chualar CA
KQEQ(AM) Fowler CA
*KSJV(FM) Fresno CA
KSRN(FM) Kings Beach CA
*KVUH(FM) Laytonville CA
KBTW(FM) Lenwood CA
KRQK(FM) Lompoc CA
*KMPO(FM) Modesto CA
KLIB(AM) Roseville CA
KTGE(AM) Salinas CA
KRZZ(FM) San Francisco CA
KSQL(FM) Santa Cruz CA
KOBO(AM) Yuba City CA
WLQY(AM) Hollywood FL
WXYB(AM) Indian Rocks Beach FL
WRHB(AM) Kendall FL
WMGE(FM) Miami Beach FL
WRUM(FM) Orlando FL
WTIS(AM) Tampa FL
*WREK(FM) Atlanta GA
WATB(AM) Decatur GA
WVVM(AM) Dry Branch GA
WAZX(AM) Smyrna GA
*WWBM(FM) Yates GA
*KKCR(FM) Hanalei HI
KAPA(FM) Hilo HI
KDNN(FM) Honolulu HI
KHUI(FM) Honolulu HI
KINE-FM Honolulu HI
KNDI(AM) Honolulu HI
KORL(AM) Honolulu HI
KRUD(AM) Honolulu HI
KITH(FM) Kapaa HI
KMKK-FM Kaunakakai HI
*KAQA(FM) Kilauea HI
KAGB(FM) Waimea HI
WLEY-FM Aurora IL
WCEV(AM) Cicero IL
WONX(AM) Evanston IL
WEEF(AM) Highland Park IL
WPJX(AM) Zion IL
WNDZ(AM) Portage IN
WTUV(AM) Louisville KY
KLCL(AM) Lake Charles LA
KANE(AM) New Iberia LA
WSHO(AM) New Orleans LA
WHTB(AM) Fall River MA
WRCA(AM) Waltham MA
WAZN(AM) Watertown MA
WNZK(AM) Dearborn Heights MI
WEW(AM) Saint Louis MO
WKRA(AM) Holly Springs MS
WKRA-FM Holly Springs MS
WIVG(FM) Tunica MS
WYMY(FM) Goldsboro NC
WWTR(AM) Bridgewater NJ
WSNR(AM) Jersey City NJ
WCAA(FM) Newark NJ
WPAT(AM) Paterson NJ
*KABR(AM) Alamo Community NM
KCNM-FM Garapan-Saipan NP
KWIP(AM) Dallas OR
WNWR(AM) Philadelphia PA
KCHN(AM) Brookshire TX
KRVA(AM) Cockrell Hill TX
KJOJ(AM) Conroe TX
KXTQ-FM Lubbock TX
KZMP-FM Pilot Point TX
KEDA(AM) San Antonio TX
KHSE(AM) Wylie TX
WYSK(AM) Fredericksburg VA
WSBV(AM) South Boston VA
KXPA(AM) Bellevue WA
KVRI(AM) Blaine WA
KRPI(AM) Ferndale WA

Farsi

KIRN(AM) Simi Valley CA

Filipino

KTKB-FM Hagatna GU
KNDI(AM) Honolulu HI

Folk

*WDJW(FM) Somers CT
*WGCS(FM) Goshen IN
*WUMB-FM Boston MA
*WFPB-FM Falmouth MA
*WNEF(FM) Newburyport MA
WFPB(AM) Orleans MA
*WBPR(FM) Worcester MA
*WICN(FM) Worcester MA
*KGPR(FM) Great Falls MT
*KZSD-FM Martin SD
*WETS(FM) Johnson City TN
*KRCL(FM) Salt Lake City UT
*WMRY(FM) Crozet VA
*KBCS(FM) Bellevue WA
*WXPR(FM) Rhinelander WI

Foreign/Ethnic

KVTO(AM) Berkeley CA
*KBES(FM) Ceres CA
KSRT(FM) Cloverdale CA
KBIF(AM) Fresno CA
KOQO-FM Fresno CA
KLOK-FM Greenfield CA
KIGS(AM) Hanford CA
KSQQ(FM) Morgan Hill CA
KLYY(FM) Riverside CA
KEST(AM) San Francisco CA
KEST(AM) San Francisco CA
KSOL(FM) San Francisco CA
KMRB(AM) San Gabriel CA
KSJX(AM) San Jose CA
KALI-FM Santa Ana CA
WDJZ(AM) Bridgeport CT
*WPFW(FM) Washington DC
WUST(AM) Washington DC
WAVS(AM) Davie FL
WSRF(AM) Fort Lauderdale FL
WRTO-FM Goulds FL
WWRF(AM) Lake Worth FL
WAMR-FM Miami FL
WHSR(AM) Pompano Beach FL
WAZX-FM Cleveland GA
KISH(FM) Hagatna GU
KCCN-FM Honolulu HI
*KIPO(FM) Honolulu HI
KZOO(AM) Honolulu HI
KPOA(FM) Lahaina HI
WOJO(FM) Evanston IL
WNWI(AM) Oak Lawn IL
WPNA(AM) Oak Park IL
KTCM(FM) Kingman KS
WUNR(AM) Brookline MA
WLYN(AM) Lynn MA
WJFD-FM New Bedford MA
*WTCC(FM) Springfield MA
WKTX(AM) Cortland OH
WVJP-FM Caguas PR
WMDD(AM) Fajardo PR
WZET(FM) Hormigueros PR
WSOL(AM) San German PR
WPRM-FM San Juan PR
WVOZ(AM) San Juan PR
KHER(FM) Crystal City TX
KFZO(FM) Denton TX
*KEPX(FM) Eagle Pass TX
KGRW(FM) Friona TX
KGOL(AM) Humble TX
KQLM(FM) Odessa TX
KMFM(FM) Premont TX

French

KLEB(AM) Golden Meadow LA

Full Service

WTLS(AM) Tallassee AL
*KHDX(FM) Conway AR
KHTS(AM) Canyon Country CA
WLAD(AM) Danbury CT
WTIC(AM) Hartford CT
WICH(AM) Norwich CT
WILI(AM) Willimantic CT
WDEL(AM) Wilmington DE
WMT(AM) Cedar Rapids IA
KROS(AM) Clinton IA
KDTH(AM) Dubuque IA
WJBC(AM) Bloomington IL
WSPY-FM Plano IL
WSQR(AM) Sycamore IL
WZZB(AM) Seymour IN
WZZY(FM) Winchester IN
WCLU(AM) Glasgow KY
KRLQ(FM) Hodge LA
WNAW(AM) North Adams MA
WOCN-FM Orleans MA
WBRK(AM) Pittsfield MA
WNAV(AM) Annapolis MD
WTVB(AM) Coldwater MI
WDBC(AM) Escanaba MI
WHTC(AM) Holland MI
WION(AM) Ionia MI
WSAM(AM) Saginaw MI
WSJM(AM) Saint Joseph MI
WJJY-FM Brainerd MN
KLFD(AM) Litchfield MN
WJON(AM) Saint Cloud MN
KQRV(FM) Deer Lodge MT
KUSO(FM) Albion NE
KBRL(AM) McCook NE
KELK(AM) Elko NV
WENT(AM) Gloversville NY
WSYR(AM) Syracuse NY
WRIP(FM) Windham NY
WHBC(AM) Canton OH
*WCDR-FM Cedarville OH
WBEX(AM) Chillicothe OH
*WOHC(FM) Chillicothe OH
WHIO(AM) Dayton OH
*WOHP(FM) Portsmouth OH
KBCH(AM) Lincoln City OR
KEX(AM) Portland OR
WBVP(AM) Beaver Falls PA
WCNS(AM) Latrobe PA
WKVA(AM) Lewistown PA
WEEU(AM) Reading PA
WOON(AM) Woonsocket RI
WLSC(AM) Loris SC
KNAF(AM) Fredericksburg TX
KSEY-FM Seymour TX
KSST(AM) Sulphur Springs TX
KTBB(AM) Tyler TX
WAGE(AM) Leesburg VA
WOMT(AM) Manitowoc WI

WFHR(AM) Wisconsin Rapids WI
WMOV(AM) Ravenswood WV
KFBC(AM) Cheyenne WY

Golden Oldies

KLSZ-FM Fort Smith AR
WYBT(AM) Blountstown FL
WRZN(AM) Hernando FL
WOKA(AM) Douglas GA
WDQX(FM) Morton IL
WQCY(FM) Quincy IL
*WXRB(FM) Dudley MA
WGOP(AM) Pocomoke City MD
WGTO(AM) Cassopolis MI
KOZY(AM) Grand Rapids MN
KBEK(FM) Mora MN
WIBT(FM) Shelby NC
WCHN(AM) Norwich NY
KSEO(AM) Durant OK
WIBZ(FM) Wedgefield SC
KMCM(FM) Odessa TX
WEIR(AM) Weirton WV

Gospel

*KJHA(FM) Houston AK
KAKN(FM) Naknek AK
KICY(AM) Nome AK
WAVU(AM) Albertville AL
WRFS(AM) Alexander City AL
WHMA(AM) Anniston AL
WRAB(AM) Arab AL
WAGG(AM) Birmingham AL
WAYE(AM) Birmingham AL
*WELJ(FM) Brewton AL
WBYE(AM) Calera AL
WEIS(AM) Centre AL
WZTQ(AM) Centre AL
WBIB(AM) Centreville AL
WQEM(FM) Columbiana AL
WXJC-FM Cordova AL
WMCJ(AM) Cullman AL
WDLK(AM) Dadeville AL
WXAL(AM) Demopolis AL
WAGF(AM) Dothan AL
*WVOB(FM) Dothan AL
WRJL-FM Eva AL
WJLD(AM) Fairfield AL
WKWL(AM) Florala AL
WGEA(AM) Geneva AL
WJBB(AM) Haleyville AL
WERH(AM) Hamilton AL
WLYG(AM) Hanceville AL
*WKNG-FM Heflin AL
*WPIL(FM) Heflin AL
WDJL(AM) Huntsville AL
WEUP(AM) Huntsville AL
*WJAB(FM) Huntsville AL
WLOR(AM) Huntsville AL
*WOCG(FM) Huntsville AL
WRJX(AM) Jackson AL
WIXI(AM) Jasper AL
WHLW(FM) Luverne AL
WGOK(AM) Mobile AL
WLVV(AM) Mobile AL
WEZZ(AM) Monroeville AL
WMFC(AM) Monroeville AL
WMGY(AM) Montgomery AL
WURL(AM) Moody AL
WEUV(AM) Moulton AL
*WJIF(FM) Opp AL
WQLS(AM) Ozark AL
WKXK(FM) Pine Hill AL
WQAH(AM) Priceville AL
WLPR(AM) Prichard AL
WJBY(AM) Rainbow City AL
WVSM(AM) Rainsville AL
WRMG(AM) Red Bay AL
WKAX(AM) Russellville AL
WZCT(AM) Scottsboro AL
WBTG-FM Sheffield AL
WNUZ(AM) Talladega AL
WTSK(AM) Tuscaloosa AL
WWPG(AM) Tuscaloosa AL

WZZA(AM) Tuscumbia AL
WBIL(AM) Tuskegee AL
*WEBT(FM) Valley AL
WJEC(FM) Vernon AL
WYLS(AM) York AL
KNWA(AM) Bellefonte AR
KBRI(AM) Brinkley AR
KPZK-FM Cabot AR
KNHD(AM) Camden AR
KWXT(AM) Dardanelle AR
KVDW(AM) England AR
*KAYH(FM) Fayetteville AR
KFSA(AM) Fort Smith AR
KTCS(AM) Fort Smith AR
KWXI(AM) Glenwood AR
*KBPB(FM) Harrison AR
KJIW-FM Helena AR
KVLO(FM) Humnoke AR
KNEA(AM) Jonesboro AR
KAAY(AM) Little Rock AR
*KABF(FM) Little Rock AR
KJBN(AM) Little Rock AR
KPZK(AM) Little Rock AR
KCGS(AM) Marshall AR
KENA(AM) Mena AR
*KLLN(FM) Newark AR
*KCAT(AM) Pine Bluff AR
KTPA(AM) Prescott AR
KGHT(AM) Sheridan AR
KOSE(AM) Wilson AR
*KYCJ(FM) Camino CA
*KCJH(FM) Livingston CA
*KYCC(FM) Stockton CA
KDYA(AM) Vallejo CA
KLDC(AM) Brighton CO
WDJZ(AM) Bridgeport CT
WYCB(AM) Washington DC
*WXXY(AM) Dover DE
WHIM(AM) Apopka FL
WFHT(AM) Avon Park FL
WSWN(AM) Belle Glade FL
WNVY(AM) Cantonment FL
*WJCB(FM) Clewiston FL
*WJED(FM) Dogwood Lakes Estate FL
WKIQ(AM) Eustis FL
*WJBC-FM Fernandina Beach FL
WIRA(AM) Fort Pierce FL
WTMN(AM) Gainesville FL
WRNE(AM) Gulf Breeze FL
WLVF(AM) Haines City FL
*WLVF-FM Haines City FL
WZAZ(AM) Jacksonville FL
WDSR(AM) Lake City FL
WGRO(AM) Lake City FL
WWAB(AM) Lakeland FL
WTYS-FM Marianna FL
WMBM(AM) Miami Beach FL
WOKB(AM) Ocoee FL
WRMQ(AM) Orlando FL
*WHYZ(FM) Palm Bay FL
WBRD(AM) Palmetto FL
WVTJ(AM) Pensacola FL
WPUL(AM) South Daytona FL
*WTLG(FM) Starke FL
WEXY(AM) Wilton Manors FL
WSIR(AM) Winter Haven FL
WZHR(AM) Zephyrhills FL
WJYZ(AM) Albany GA
WAJQ(AM) Alma GA
WXAG(AM) Athens GA
WAFS(AM) Atlanta GA
WYZE(AM) Atlanta GA
WGUS(AM) Augusta GA
WTHB(AM) Augusta GA
WGMI(AM) Bremen GA
WULS(FM) Broxton GA
WEAM-FM Buena Vista GA
WRWH(AM) Cleveland GA
WOKS(AM) Columbus GA
WSHE(AM) Columbus GA
WCON(AM) Cornelia GA
WCUG(AM) Cuthbert GA
WDPC(AM) Dallas GA
WTTI(AM) Dalton GA
WSEM(AM) Donalsonville GA
WMLT(AM) Dublin GA

WTJH(AM) East Point GA
WUFF-FM Eastman GA
WLJA-FM Ellijay GA
WPZE(FM) Fayetteville GA
WBHB(AM) Fitzgerald GA
WQMJ(FM) Forsyth GA
*WJTG(FM) Fort Valley GA
WXKO(AM) Fort Valley GA
WTHV(AM) Hahira GA
WKLY(AM) Hartwell GA
WVOH(AM) Hazlehurst GA
WGML(AM) Hinesville GA
WVKX(FM) Irwinton GA
WYYZ(AM) Jasper GA
WBKZ(AM) Jefferson GA
*WTLD(FM) Jesup GA
*WMOC(FM) Lumber City GA
WDDO(AM) Macon GA
WLCG(AM) Macon GA
WKKP(AM) McDonough GA
WHCG(AM) Metter GA
WKUN(AM) Monroe GA
WMNZ(AM) Montezuma GA
WIGO(AM) Morrow GA
WHBS(AM) Moultrie GA
WMTM(AM) Moultrie GA
WALH(AM) Mountain City GA
WRBX(FM) Reidsville GA
WTNL(AM) Reidsville GA
WZOT(AM) Rockmart GA
WROM(AM) Rome GA
WBIC(AM) Royston GA
WSOK(AM) Savannah GA
WRJS(AM) Swainsboro GA
WHGH(AM) Thomasville GA
WSTT(AM) Thomasville GA
WLET(AM) Toccoa GA
WJEM(AM) Valdosta GA
WKTF(AM) Vienna GA
WTHB-FM Waynesboro GA
WIMO(AM) Winder GA
WYHG(AM) Young Harris GA
KTFC(FM) Sioux City IA
KTFG(FM) Sioux Rapids IA
*KBBG(FM) Waterloo IA
WXAN(FM) Ava IL
*WTSG(FM) Carlinville IL
WEIC(AM) Charleston IL
WGRB(AM) Chicago IL
*WSSD(FM) Chicago IL
WFFX(AM) East St. Louis IL
WBGX(AM) Harvey IL
WWHN(AM) Joliet IL
WGGH(AM) Marion IL
WVEL(AM) Pekin IL
WINU(AM) Shelbyville IL
WBCP(AM) Urbana IL
WYGS(FM) Columbus IN
WTLC(AM) Indianapolis IN
WMYJ(AM) Martinsville IN
WLHN(AM) Muncie IN
WFIA-FM New Albany IN
*WMYJ-FM Oolitic IN
WSLM(AM) Salem IN
WFRI(FM) Winamac IN
WYWY(AM) Barbourville KY
*WAAJ(FM) Benton KY
*WVHM(FM) Benton KY
WEKT(AM) Elkton KY
WIOK(FM) Falmouth KY
WFLE(AM) Flemingsburg KY
WFUL(AM) Fulton KY
WOKE(FM) Garrison KY
WLGC(AM) Greenup KY
WFSR(AM) Harlan KY
WHBN(AM) Harrodsburg KY
WEKG(AM) Jackson KY
WKVG(AM) Jenkins KY
*WVCT(FM) Keavy KY
WGWM(AM) London KY
WLLV(AM) Louisville KY
WLOU(AM) Louisville KY
WMIK(AM) Middlesboro KY
WFLW(AM) Monticello KY
WRVK(AM) Mt. Vernon KY
WLOC(AM) Munfordville KY

WCGW(AM) Nicholasville KY
WSIP-FM Paintsville KY
WYGH(AM) Paris KY
WDOC(AM) Prestonsburg KY
WCBR(AM) Richmond KY
WIDS(AM) Russell Springs KY
WBFC(AM) Stanton KY
*WJCR-FM Upton KY
KWDF(AM) Ball LA
WXOK(AM) Baton Rouge LA
WIKC(AM) Bogalusa LA
KDLA(AM) De Ridder LA
KGGM(FM) Delhi LA
*KGRM(FM) Grambling LA
KKNO(AM) Gretna LA
KJCB(AM) Lafayette LA
KRJO(AM) Monroe LA
WBOK(AM) New Orleans LA
WYLD(AM) New Orleans LA
KTTP(AM) Pineville LA
WPRF(FM) Reserve LA
KRUS(AM) Ruston LA
KIOU(AM) Shreveport LA
KOKA(AM) Shreveport LA
KSYB(AM) Shreveport LA
KBYO(AM) Tallulah LA
KTJZ(FM) Tallulah LA
KNCB(AM) Vivian LA
KVCL(AM) Winnfield LA
WBGR(AM) Baltimore MD
WCAO(AM) Baltimore MD
WWIN(AM) Baltimore MD
*WLIC(FM) Frostburg MD
WFBR(AM) Glen Burnie MD
*WAIJ(FM) Grantsville MD
WPGC(AM) Morningside MD
WJDY(AM) Salisbury MD
WPRS-FM Waldorf MD
*WHCF(FM) Bangor ME
*WFST(AM) Caribou ME
WFLT(AM) Flint MI
WDRJ(AM) Inkster MI
*WUNN(AM) Mason MI
WEXL(AM) Royal Oak MI
WWSJ(AM) Saint Johns MI
KRJJ(AM) Brooklyn Park MN
WWJC(AM) Duluth MN
KNOF(FM) Saint Paul MN
*KGNA-FM Arnold MO
KOMC(AM) Branson MO
*KNLH(FM) Cedar Hill MO
*KYRV(FM) Concordia MO
*KGNN-FM Cuba MO
*KNLQ(FM) Cuba MO
KAUL(FM) Ellington MO
KGGN(AM) Gladstone MO
*KJIR(FM) Hannibal MO
KPRT(AM) Kansas City MO
KMHM(FM) Lutesville MO
KMRF(AM) Marshfield MO
*KJAB-FM Mexico MO
*KNLG(FM) New Bloomfield MO
KPWB(AM) Piedmont MO
*KOKS(FM) Poplar Bluff MO
*KNLP(FM) Potosi MO
KATZ(AM) Saint Louis MO
KSTL(AM) Saint Louis MO
*KBIY(FM) Van Buren MO
*KGNV(FM) Washington MO
WFCA(FM) Ackerman MS
WIZK(AM) Bay Springs MS
WBYP(FM) Belzoni MS
WELZ(AM) Belzoni MS
WCHJ(AM) Brookhaven MS
WAJV(FM) Brooksville MS
WMGO(AM) Canton MS
WONG(AM) Canton MS
WCLD(AM) Cleveland MS
WFFF(AM) Columbia MS
WTWG(AM) Columbus MS
*WAUM(FM) Duck Hill MS
WTYJ(FM) Fayette MS
WJIW(FM) Greenville MS
WGRM(AM) Greenwood MS
WGRM-FM Greenwood MS
WKXG(AM) Greenwood MS

WQFX(AM) Gulfport MS
WORV(AM) Hattiesburg MS
WOEG(AM) Hazlehurst MS
WHAL-FM Horn Lake MS
WCPC(AM) Houston MS
WNLA(AM) Indianola MS
*WVSD(FM) Itta Bena MS
WHLH(FM) Jackson MS
*WMPR(FM) Jackson MS
WOAD(AM) Jackson MS
WZRX(AM) Jackson MS
*WJTA(FM) Kosciusko MS
WAML(AM) Laurel MS
WESY(AM) Leland MS
WXTN(AM) Lexington MS
WRBE(AM) Lucedale MS
WAPF(AM) McComb MS
WMER(AM) Meridian MS
WNBN(AM) Meridian MS
WMIS(AM) Natchez MS
WOSM(FM) Ocean Springs MS
WRJW(AM) Picayune MS
WSEL(AM) Pontotoc MS
WSEL-FM Pontotoc MS
WRPM(AM) Poplarville MS
WXHB(FM) Richton MS
WSAO(AM) Senatobia MS
WAVN(AM) Southaven MS
*WAQB(FM) Tupelo MS
WFCG(FM) Tylertown MS
WROB(AM) West Point MS
*KPGB(FM) Pryor MT
WRCS(AM) Ahoskie NC
*WTJY(FM) Asheboro NC
WZOO(AM) Asheboro NC
WSKY(AM) Asheville NC
WPYB(AM) Benson NC
WVOE(AM) Chadbourn NC
WRNA(AM) China Grove NC
WYZD(AM) Dobson NC
WCKB(AM) Dunn NC
WDUR(AM) Durham NC
WCLW(AM) Eden NC
WBXB(FM) Edenton NC
WGQR(FM) Elizabethtown NC
WBOB-FM Enfield NC
WFMO(AM) Fairmont NC
WSTS(FM) Fairmont NC
WIDU(AM) Fayetteville NC
WWOL(AM) Forest City NC
WNNL(FM) Fuquay-Varina NC
WFMC(AM) Goldsboro NC
WYCV(AM) Granite Falls NC
WEAL(AM) Greensboro NC
WKEW(AM) Greensboro NC
*WNAA(FM) Greensboro NC
WPET(AM) Greensboro NC
WKDX(AM) Hamlet NC
WHNC(AM) Henderson NC
*WPIR(FM) Hickory NC
WJCV(AM) Jacksonville NC
WRKB(AM) Kannapolis NC
WKTE(AM) King NC
WELS(AM) Kinston NC
WELS-FM Kinston NC
WEWO(AM) Laurinburg NC
WAGR(AM) Lumberton NC
WHBK(AM) Marshall NC
WDSL(AM) Mocksville NC
WDEX(AM) Monroe NC
WIXE(AM) Monroe NC
WCIS(AM) Morganton NC
WSYD(AM) Mount Airy NC
WWDR(AM) Murfreesboro NC
WCVP(AM) Murphy NC
WWNB(AM) New Bern NC
WAUG(AM) New Hope NC
WCBQ(AM) Oxford NC
WMFA(AM) Raeford NC
WTEL(AM) Red Springs NC
WEGG(AM) Rose Hill NC
*WOGR-FM Salisbury NC
WXKL(AM) Sanford NC
WYAL(AM) Scotland Neck NC
WVCB(AM) Shallotte NC
WMPM(AM) Smithfield NC

WGMA(AM) Spindale NC
WTAB(AM) Tabor City NC
WCPS(AM) Tarboro NC
WTLK(AM) Taylorsville NC
WJRM(AM) Troy NC
WZKB(FM) Wallace NC
WOBX(AM) Wanchese NC
WARR(AM) Warrenton NC
WENC(AM) Whiteville NC
WWWC(AM) Wilkesboro NC
WIAM(AM) Williamston NC
WLSG(AM) Wilmington NC
WWIL(AM) Wilmington NC
WGTM(AM) Wilson NC
WLLY(AM) Wilson NC
WBTE(AM) Windsor NC
WGTI(FM) Windsor NC
WNBR-FM Windsor NC
WPOL(AM) Winston-Salem NC
WSMX(AM) Winston-Salem NC
*WXRI(FM) Winston-Salem NC
WECU(AM) Winterville NC
WYNC(AM) Yanceyville NC
KHRT(AM) Minot ND
KTFJ(AM) Dakota City NE
*KJLT-FM North Platte NE
WIMG(AM) Ewing NJ
*WXXY-FM Port Republic NJ
WFAI(AM) Salem NJ
WNJC(AM) Washington Township NJ
*KUPR(FM) Alamogordo NM
*KYCM(FM) Alamogordo NM
*KHII(FM) Cloudcroft NM
WUFO(AM) Amherst NY
WBBF(AM) Buffalo NY
WTHE(AM) Mineola NY
WLIB(AM) New York NY
WHLD(AM) Niagara Falls NY
WGFT(AM) Campbell OH
WINW(AM) Canton OH
WJMO(AM) Cleveland OH
WVKO(AM) Columbus OH
*WBIE(FM) Delphos OH
WCNW(AM) Fairfield OH
WJYD(FM) London OH
WRTK(AM) Niles OH
WJTB(AM) North Ridgeville OH
WABQ(AM) Painesville OH
WXIC(AM) Waverly OH
WRAC(FM) West Union OH
*WCSU-FM Wilberforce OH
WLRD(FM) Willard OH
*KAJT(FM) Ada OK
*KTGS(FM) Ada OK
*KSSX(FM) Chickasha OK
*KVAZ(FM) Henryetta OK
KKRX(AM) Lawton OK
KVIS(AM) Miami OK
KTLV(AM) Midwest City OK
*KLRB(FM) Stuart OK
*KFXT(FM) Sulphur OK
KTLQ(AM) Tahlequah OK
*KAZC(FM) Tishomingo OK
KIMY(FM) Watonga OK
WPWA(AM) Chester PA
WADV(AM) Lebanon PA
WJSM-FM Martinsburg PA
*WRIJ(FM) Masontown PA
WPGR(AM) Monroeville PA
*WPEL(AM) Montrose PA
*WRWJ(FM) Murrysville PA
WGBN(AM) New Kensington PA
WNAP(AM) Norristown PA
WANS(AM) Anderson SC
WVGB(AM) Beaufort SC
WBSC(AM) Bennettsville SC
WSPX(FM) Bowman SC
WEAF(AM) Camden SC
WGCV(AM) Cayce SC
WXTC(AM) Charleston SC
WGCD(AM) Chester SC
WWRK(AM) Darlington SC
WDSC(AM) Dillon SC
WEGX(FM) Dillon SC
WYNN(AM) Florence SC
WNOW-FM Gaffney SC

WLMC(AM) Georgetown SC
*WTBI-FM Greenville SC
WCZZ(AM) Greenwood SC
WPJM(AM) Greer SC
WHGS(AM) Hampton SC
WJDJ(AM) Hartsville SC
*WLGI(FM) Hemingway SC
WJBS(AM) Holly Hill SC
WPDT(FM) Johnsonville SC
WGSS(FM) Kingstree SC
*WKCL(FM) Ladson SC
WAGL(AM) Lancaster SC
WJAY(AM) Mullins SC
WKZK(AM) North Augusta SC
WPJK(AM) Orangeburg SC
*WSSB-FM Orangeburg SC
WPOG(AM) Saint Matthews SC
WTUA(FM) Saint Stephen SC
WCSZ(AM) Sans Souci SC
WQMC(AM) Sumter SC
WALD(AM) Walterboro SC
WAAW(FM) Williston SC
WWGM(FM) Alamo TN
WOJG(FM) Bolivar TN
WIGN(AM) Bristol TN
WJOC(AM) Chattanooga TN
WMCH(AM) Church Hill TN
WQZQ(AM) Clarksville TN
WCLE(AM) Cleveland TN
WSGM(FM) Coalmont TN
WMRB(AM) Columbia TN
WHUB(AM) Cookeville TN
WENR(AM) Englewood TN
WEMB(AM) Erwin TN
WEKR(AM) Fayetteville TN
WJAK(AM) Jackson TN
WDEB-FM Jamestown TN
WWAM(AM) Jasper TN
WJJT(AM) Jellico TN
WETB(AM) Johnson City TN
WBBX(AM) Kingston TN
WKGN(AM) Knoxville TN
WKXV(AM) Knoxville TN
WQLA(AM) La Follette TN
WEEN(AM) Lafayette TN
WDXL(AM) Lexington TN
WFLI(AM) Lookout Mountain TN
WBMC(AM) McMinnville TN
WBBP(AM) Memphis TN
WLOK(AM) Memphis TN
WXRQ(AM) Mount Pleasant TN
WNAH(AM) Nashville TN
WUAT(AM) Pikeville TN
WMXV(FM) Saint Joseph TN
WSIB(FM) Selmer TN
WJBZ-FM Seymour TN
WLIJ(AM) Shelbyville TN
WSMT(AM) Sparta TN
WDEH(AM) Sweetwater TN
WJIG(AM) Tullahoma TN
WECO(AM) Wartburg TN
WVRY(FM) Waverly TN
WQSE(AM) White Bluff TN
WBOZ(FM) Woodbury TN
KLVQ(AM) Athens TX
KNRB(FM) Atlanta TX
*KAZI-FM Austin TX
KWWJ(AM) Baytown TX
KZZB(AM) Beaumont TX
*KHVT(FM) Bloomington TX
KGAS(AM) Carthage TX
KDET(AM) Center TX
KCOM(AM) Comanche TX
KYOK(AM) Conroe TX
KNGR(AM) Daingerfield TX
*KDKR(FM) Decatur TX
KNNK(FM) Dimmitt TX
KHVN(AM) Fort Worth TX
KKGM(AM) Fort Worth TX
KWRD(AM) Henderson TX
KTXJ-FM Jasper TX
KJTX(FM) Jefferson TX
KRMY(AM) Killeen TX
KZZN(AM) Littlefield TX
KFIT(AM) Lockhart TX
*KFTG(FM) Pasadena TX

*KPVU(FM) Prairie View TX
KCHL(AM) San Antonio TX
KFIT EXP STN San Antonio TX
KROI(FM) Seabrook TX
KZEE(AM) Weatherford TX
KWNS(FM) Winnsboro TX
KLLB(AM) West Jordan UT
WWDW(FM) Alberta VA
WTTX-FM Appomattox VA
WBTX(AM) Broadway-Timberville VA
WKBY(AM) Chatham VA
WCPK(AM) Chesapeake VA
WGGM(AM) Chester VA
WFIC(AM) Collinsville VA
WPZZ(FM) Crewe VA
WDVA(AM) Danville VA
WILA(AM) Danville VA
WGFC(AM) Floyd VA
WLQM(AM) Franklin VA
WWWJ(AM) Galax VA
WGAT(AM) Gate City VA
WNRG(AM) Grundy VA
WHHV(AM) Hillsville VA
WHAP(AM) Hopewell VA
WHFD(FM) Lawrenceville VA
WLRV(AM) Lebanon VA
WLLL(AM) Lynchburg VA
WMEV(AM) Marion VA
WTJZ(AM) Newport News VA
WGPL(AM) Portsmouth VA
WGTH(AM) Richlands VA
WGTH-FM Richlands VA
WFTH(AM) Richmond VA
WWWR(AM) Roanoke VA
WYTI(AM) Rocky Mount VA
WXMY(AM) Saltville VA
*WTRM(FM) Winchester VA
WGOD(AM) Charlotte Amalie VI
KGSG(FM) Pasco WA
KZIZ(AM) Sumner WA
WGLB(AM) Elm Grove WI
WJLS(AM) Beckley WV
*WKJL(FM) Clarksburg WV
WPDX(AM) Clarksburg WV
*WFGH(FM) Fort Gay WV
WEMM-FM Huntington WV
WOTR(FM) Lost Creek WV
WHJC(AM) Matewan WV
WWYO(AM) Pineville WV
WBGS(AM) Point Pleasant WV
*WPCN(FM) Point Pleasant WV
WAEY(AM) Princeton WV
WRRL(AM) Rainelle WV
WYKM(AM) Rupert WV
WVRC(AM) Spencer WV
*WMLJ(FM) Summersville WV
WAFD(FM) Webster Springs WV

Greek

WXYB(AM) Indian Rocks Beach FL
WPSO(AM) New Port Richey FL
WNTN(AM) Newton MA

Inspirational

*KUDU(FM) Tok AK
*WOCG(FM) Huntsville AL
*KFLR-FM Phoenix AZ
*KNAQ(FM) Prescott AZ
*KFLT(AM) Tucson AZ
*KYCJ(FM) Camino CA
*KAXL(FM) Green Acres CA
*KTLW(FM) Lancaster CA
*KCJH(FM) Livingston CA
*KHCS(FM) Palm Desert CA
*KVIP(AM) Redding CA
*KVIP-FM Redding CA
*KYCC(FM) Stockton CA
*KARM(FM) Visalia CA
KLLV(AM) Breen CO
*KTPL(FM) Pueblo CO
*KTAD(FM) Sterling CO
*WKZM(FM) Sarasota FL
WTJH(AM) East Point GA

*KSDA-FM Agat GU
*KCMR(FM) Mason City IA
KBGN(AM) Caldwell ID
WRMS(AM) Beardstown IL
*WZRS(FM) Pana IL
*WLWJ(FM) Petersburg IL
*WIHM(AM) Taylorville IL
*WGNR-FM Anderson IN
*KXJH(FM) Linton IN
*WUBS(FM) South Bend IN
*WATI(FM) Vincennes IN
*KFLO-FM Blanchard LA
*KBAN(FM) De Ridder LA
*KYLC(FM) Lake Charles LA
*WHCF(FM) Bangor ME
*WUFN(FM) Albion MI
*WNFA(FM) Port Huron MI
*WNFR(FM) Sandusky MI
*WTHN(FM) Sault Ste. Marie MI
*KTGG(AM) Spring Arbor MI
*KDNI(FM) Duluth MN
*KTIS-FM Minneapolis MN
*KBPG(FM) Montevideo MN
*KBHZ(FM) Willmar MN
*KGNA-FM Arnold MO
*KGNN-FM Cuba MO
WBVV(FM) Guntown MS
*WURC(FM) Holly Springs MS
*KBLW(FM) Billings MT
KMCJ(FM) Colstrip MT
*KGFC(FM) Great Falls MT
*KXEI(FM) Havre MT
*KVCM(FM) Helena MT
WPZS(FM) Albemarle NC
WCLN-FM Clinton NC
*WGPS(FM) Elizabeth City NC
WNNL(FM) Fuquay-Varina NC
*KAYA(FM) Hubbard NE
*KPNO(FM) Norfolk NE
*KJLT-FM North Platte NE
*KGRD(FM) Orchard NE
WIBG(AM) Ocean City NJ
*KYCM(FM) Alamogordo NM
*KFLQ(FM) Albuquerque NM
*WNGN(FM) Argyle NY
*WCII(FM) Spencer NY
*WCRF(FM) Cleveland OH
*WVML(FM) Millersburg OH
*WVMS(FM) Sandusky OH
*KSYE(FM) Frederick OK
KGLC(FM) Miami OK
*KLVV(FM) Ponca City OK
*KMWR(FM) Brookings OR
WDBA(FM) DuBois PA
*WAWN(FM) Franklin PA
*WVME(FM) Meadville PA
*WVMN(FM) New Castle PA
WLMC(AM) Georgetown SC
WJNI(FM) Ladson SC
WFMV(FM) South Congaree SC
WLJI(FM) Summerton SC
*KLMP(FM) Rapid City SD
WLRM(AM) Millington TN
KPEZ(FM) Austin TX
*KCBI(FM) Dallas TX
KCRN(AM) San Angelo TX
KCRN-FM San Angelo TX
*KHTA(FM) Wake Village TX
*WAUQ(FM) Charles City VA
*WARN(FM) Culpeper VA
WTJZ(AM) Newport News VA
WRIS(AM) Roanoke VA
*WCMD-FM Barre VT
*WGLY-FM Bolton VT
*WCKJ(FM) Saint Johnsbury VT
*WGLV(FM) Woodstock VT
KCIS(AM) Edmonds WA
*KCSH(FM) Ellensburg WA
*KAYB(FM) Sunnyside WA
WRLB(FM) Rainelle WV
*KOHR(FM) Sheridan WY

Italian

WEST(AM) Easton PA

Japanese

KZOO(AM) Honolulu HI

Jazz

WAUD(AM) Auburn AL
*WVSU-FM Birmingham AL
*WJAB(FM) Huntsville AL
*WQPR(FM) Muscle Shoals AL
*WAPR(FM) Selma AL
*WUAL-FM Tuscaloosa AL
*KBSA(FM) El Dorado AR
*KUAF(FM) Fayetteville AR
*KASU(FM) Jonesboro AR
*KABF(FM) Little Rock AR
*KUAR(FM) Little Rock AR
*KXRJ(FM) Russellville AR
*KJZA(FM) Drake AZ
*KJZZ(FM) Phoenix AZ
KYOT-FM Phoenix AZ
*KUAZ(AM) Tucson AZ
*KUAZ-FM Tucson AZ
KWRK(FM) Window Rock AZ
*KAWC-FM Yuma AZ
*KNCA(FM) Burney CA
KRML(AM) Carmel CA
*KCHO(FM) Chico CA
*KSPC(FM) Claremont CA
KRVR(FM) Copperopolis CA
*KECG(FM) El Cerrito CA
*KFSR(FM) Fresno CA
KMGQ(FM) Goleta CA
*KKJZ(FM) Long Beach CA
KTWV(FM) Los Angeles CA
*KSBR(FM) Mission Viejo CA
*KNSQ(FM) Mount Shasta CA
*KQNC(FM) Quincy CA
*KXJZ(FM) Sacramento CA
*KSDS(FM) San Diego CA
KKSF(FM) San Francisco CA
KJZN(FM) San Joaquin CA
*KSJS(FM) San Jose CA
*KCBX(FM) San Luis Obispo CA
*KCSM(FM) San Mateo CA
KRUZ(FM) Santa Barbara CA
*KSBX(FM) Santa Barbara CA
*KCRW(FM) Santa Monica CA
*KXJS(FM) Sutter CA
*KCLU(FM) Thousand Oaks CA
*KRZA(FM) Alamosa CO
*KAJX(FM) Aspen CO
*KCJX(FM) Carbondale CO
*KUVO(FM) Denver CO
*KCME(FM) Manitou Springs CO
KTNI-FM Strasburg CO
*WDJW(FM) Somers CT
*WECS(FM) Willimantic CT
*WPFW(FM) Washington DC
*WRTX(FM) Dover DE
WJSJ(FM) Fernandina Beach FL
*WGCU-FM Fort Myers FL
*WUFT-FM Gainesville FL
*WJUF(FM) Inverness FL
WSJT(FM) Lakeland FL
*WMKO(FM) Marco FL
*WDNA(FM) Miami FL
WLVE(FM) Miami Beach FL
*WUCF-FM Orlando FL
WASJ(FM) Panama City Beach FL
WSJF(FM) Saint Augustine Beach FL
*WUSF(FM) Tampa FL
WGYL(FM) Vero Beach FL
WLOQ(FM) Winter Park FL
*WCLK(FM) Atlanta GA
WJZZ-FM Roswell GA
*WSVH(FM) Savannah GA
*KIPO(FM) Honolulu HI
*WOI-FM Ames IA
*KCCK-FM Cedar Rapids IA
*KALA(FM) Davenport IA
*KJMC(FM) Des Moines IA
*KTPR(FM) Fort Dodge IA
*KBBG(FM) Waterloo IA
*KBSU(AM) Boise ID
*KIBX(FM) Bonners Ferry ID

*KBSK(FM) McCall ID
*KISU-FM Pocatello ID
KYZK(FM) Sun Valley ID
*KEZJ(AM) Twin Falls ID
*WBEZ(FM) Chicago IL
*WHPK-FM Chicago IL
WNUA(FM) Chicago IL
*WNIJ(FM) De Kalb IL
*WSIE(FM) Edwardsville IL
*WNUR-FM Evanston IL
*WDCB(FM) Glen Ellyn IL
*WBEQ(FM) Morris IL
*WGLT(FM) Normal IL
*WCBU(FM) Peoria IL
*WIPA(FM) Pittsfield IL
*WQUB(FM) Quincy IL
*WUIS(FM) Springfield IL
WBCP(AM) Urbana IL
*WFIU(FM) Bloomington IN
*WBEW(FM) Chesterton IN
*WVPE(FM) Elkhart IN
*WUEV(FM) Evansville IN
*WBOI(FM) Fort Wayne IN
*WICR(FM) Indianapolis IN
*WBAA(AM) West Lafayette IN
*KANH(FM) Emporia KS
KKCI(FM) Goodland KS
*KANU(FM) Lawrence KS
*KJHK(FM) Lawrence KS
*KANV(FM) Olsburg KS
*KRPS(FM) Pittsburg KS
*KMUW(FM) Wichita KS
*KLSA(FM) Alexandria LA
*WBRH(FM) Baton Rouge LA
*WWNO(FM) New Orleans LA
*WWOZ(FM) New Orleans LA
*KDAQ(FM) Shreveport LA
*KTLN(FM) Thibodaux LA
*WFCR(FM) Amherst MA
*WGBH(FM) Boston MA
WHRB(FM) Cambridge MA
*WICN(FM) Worcester MA
*WEAA(FM) Baltimore MD
WSMJ(FM) Baltimore MD
*WYPR(FM) Baltimore MD
*WFWM(FM) Frostburg MD
WQJZ(FM) Ocean Pines MD
*WESM(FM) Princess Anne MD
WBQI(FM) Bar Harbor ME
*WYAR(FM) Yarmouth ME
*WGVU-FM Allendale MI
*WCML-FM Alpena MI
*WUCX-FM Bay City MI
*WDET-FM Detroit MI
WGPR(FM) Detroit MI
*WBLU-FM Grand Rapids MI
*WCMW-FM Harbor Springs MI
*WMUK(FM) Kalamazoo MI
*WLNZ(FM) Lansing MI
*WNMU-FM Marquette MI
*WCMU-FM Mount Pleasant MI
*WCMB-FM Oscoda MI
*WSGR-FM Port Huron MI
*WCMZ-FM Sault Ste. Marie MI
WWCM(FM) Standish MI
*WNMC-FM Traverse City MI
*WBLV(FM) Twin Lake MI
WGVS-FM Whitehall MI
*WEMU(FM) Ypsilanti MI
*KBEM-FM Minneapolis MN
*WMCN(FM) Saint Paul MN
*KQAL(FM) Winona MN
*KRCU(FM) Cape Girardeau MO
*KRNW(FM) Chillicothe MO
*KWWC-FM Columbia MO
*KSEF(FM) Farmington MO
*KJLU(FM) Jefferson City MO
KRLI(FM) Malta Bend MO
*KXCV(FM) Maryville MO
*KCOZ(FM) Point Lookout MO
*KCLC(FM) Saint Charles MO
*WMAH-FM Biloxi MS
*WMAE-FM Booneville MS
*WMAU-FM Bude MS
*WMAO-FM Greenwood MS
*WURC(FM) Holly Springs MS
*WVSD(FM) Itta Bena MS
*WJSU(FM) Jackson MS
*WMPN-FM Jackson MS
*WMAW-FM Meridian MS
*WMAB-FM Mississippi State MS
*WMAV-FM Oxford MS
WKBB(FM) West Point MS
*KEMC(FM) Billings MT
*KBMC(FM) Bozeman MT
*KAPC(FM) Butte MT
*KGPR(FM) Great Falls MT
*KUFN(FM) Hamilton MT
*KNMC(FM) Havre MT
*KUHM(FM) Helena MT
*KUKL(FM) Kalispell MT
*KUFM(FM) Missoula MT
*WCQS(FM) Asheville NC
WVOE(AM) Chadbourn NC
*WNCU(FM) Durham NC
*WFSS(FM) Fayetteville NC
*WFQS(FM) Franklin NC
*WNAA(FM) Greensboro NC
*WSHA(FM) Raleigh NC
*WDCC(FM) Sanford NC
*WSNC(FM) Winston-Salem NC
*KCND(FM) Bismarck ND
*KDPR(FM) Dickinson ND
*KFJM(FM) Grand Forks ND
*KPRJ(FM) Jamestown ND
*KMPR(FM) Minot ND
*KPPR(FM) Williston ND
*KLPR(FM) Kearney NE
*KZUM(FM) Lincoln NE
*KIOS-FM Omaha NE
*WNEC-FM Henniker NH
WEZS(AM) Laconia NH
*WBGO(FM) Newark NJ
*WRTQ(FM) Ocean City NJ
WPRB(FM) Princeton NJ
*KGLP(FM) Gallup NM
*KRWG(FM) Las Cruces NM
*KSFR(FM) Santa Fe NM
*KRNM(FM) Chalan Kanoa-Saipan NP
*KNCC(FM) Elko NV
*KBSJ(FM) Jackpot NV
*KUNV(FM) Las Vegas NV
*KUNR(FM) Reno NV
KJZS(FM) Sparks NV
*WSQX-FM Binghamton NY
*WCWP(FM) Brookville NY
*WBFO(FM) Buffalo NY
*WEOS(FM) Geneva NY
*WGMC(FM) Greece NY
*WRCU-FM Hamilton NY
*WVHC(FM) Herkimer NY
*WSQA(FM) Hornell NY
*WUBJ(FM) Jamestown NY
*WHCR-FM New York NY
*WKCR-FM New York NY
*WOLN(FM) Olean NY
WJZR(FM) Rochester NY
*WLIU(FM) Southampton NY
*WAER(FM) Syracuse NY
*WBGU(FM) Bowling Green OH
*WCPN(FM) Cleveland OH
*WDPS(FM) Dayton OH
WDSJ(FM) Greenville OH
WJZA(FM) Lancaster OH
*WMRT(FM) Marietta OH
*WMUB(FM) Oxford OH
WJZK(FM) Richwood OH
*WXTS-FM Toledo OH
*WCSU-FM Wilberforce OH
*KALU(FM) Langston OK
*KGOU(FM) Norman OK
*KROU(FM) Spencer OK
*KSMF(FM) Ashland OR
*KSBA(FM) Coos Bay OR
*KBVR(FM) Corvallis OR
*KMHD(FM) Gresham OR
*KSKF(FM) Klamath Falls OR
*KLCO(FM) Newport OR
*WBUQ(FM) Bloomsburg PA
*WRTL(FM) Ephrata PA
*WQLN-FM Erie PA
*WRTY(FM) Jackson Township PA
*WRTI(FM) Philadelphia PA
*WDUQ(FM) Pittsburgh PA
*WXAC(FM) Reading PA
*WUSR(FM) Scranton PA
*WVIA-FM Scranton PA
*WJAZ(FM) Summerdale PA
*WRKC(FM) Wilkes-Barre PA
*WPTC(FM) Williamsport PA
*WVYA(FM) Williamsport PA
WIBS(AM) Guayama PR
*WRUO(FM) Mayaguez PR
*WRTU(FM) San Juan PR
WYAS(FM) Vieques PR
*WELH(FM) Providence RI
*WSCI(FM) Charleston SC
*WSSB-FM Orangeburg SC
*WNSC-FM Rock Hill SC
WAZS(AM) Summerville SC
*WRJA-FM Sumter SC
KPSD(FM) Faith SD
*KQSD-FM Lowry SD
*KZSD-FM Martin SD
*KDSD-FM Pierpont SD
*KBHE-FM Rapid City SD
*KTSD-FM Reliance SD
*KAUR(FM) Sioux Falls SD
*KUSD(FM) Vermillion SD
WBOL(AM) Bolivar TN
*WUTC(FM) Chattanooga TN
*WFHU(FM) Henderson TN
*WUOT(FM) Knoxville TN
*WFMQ(FM) Lebanon TN
*WOEZ(FM) Maynardville TN
*WUMR(FM) Memphis TN
*WMOT(FM) Murfreesboro TN
*WRVU(FM) Nashville TN
*KVLU(FM) Beaumont TX
*KHPU(FM) Brownwood TX
*KAMU-FM College Station TX
*KETR(FM) Commerce TX
*KEDT-FM Corpus Christi TX
*KTEP(FM) El Paso TX
*KTCU-FM Fort Worth TX
*KMBH-FM Harlingen TX
KHJZ-FM Houston TX
*KTSU(FM) Houston TX
*KSHU(FM) Huntsville TX
*KLDN(FM) Lufkin TX
*KHID(FM) McAllen TX
*KNTU(FM) McKinney TX
*KSAU(FM) Nacogdoches TX
*KWLD(FM) Plainview TX
*KRTU(FM) San Antonio TX
KZSP(FM) South Padre Island TX
*KVRT(FM) Victoria TX
*KMCU(FM) Wichita Falls TX
*KUSR(FM) Logan UT
KBZN(FM) Ogden UT
*KUER(FM) Salt Lake City UT
*WVTU(FM) Charlottesville VA
*WVTW(FM) Charlottesville VA
*WVTR(FM) Marion VA
*WHRV(FM) Norfolk VA
*WVST-FM Petersburg VA
*WVRU(FM) Radford VA
*WVTF(FM) Roanoke VA
*WRUV(FM) Burlington VT
*WVPS(FM) Burlington VT
*WNCH(FM) Norwich VT
*WRVT(FM) Rutland VT
*WVPA(FM) Saint Johnsbury VT
*WVPR(FM) Windsor VT
*KBCS(FM) Bellevue WA
*KZAZ(FM) Bellingham WA
*KEWU-FM Cheney WA
KOHO-FM Leavenworth WA
*KPLI(FM) Olympia WA
*KVIX(FM) Port Angeles WA
*KZUU(FM) Pullman WA
*KPBX-FM Spokane WA
*KPLU-FM Tacoma WA
KYNR(AM) Toppenish WA
*WUEC(FM) Eau Claire WI
*WLSU(FM) La Crosse WI
*WWSP(FM) Stevens Point WI
*WSUW(FM) Whitewater WI
*WVPB(FM) Beckley WV
*WVPW(FM) Buckhannon WV
*WVPN(FM) Charleston WV
*WVWV(FM) Huntington WV
*WVEP(FM) Martinsburg WV
*WVPM(FM) Morgantown WV
*WVPG(FM) Parkersburg WV
*WVNP(FM) Wheeling WV
*KCWC-FM Riverton WY
*KPRQ(FM) Sheridan WY

Korean

KXMX(AM) Anaheim CA
KMPC(AM) Los Angeles CA
KYPA(AM) Los Angeles CA
KFOX(AM) Torrance CA
KHRA(AM) Honolulu HI
KREA(AM) Honolulu HI
WKTA(AM) Evanston IL
WWRU(AM) Jersey City NJ
WDCT(AM) Fairfax VA
KWYZ(AM) Everett WA
KSUH(AM) Puyallup WA

Light Rock

WRSA-FM Decatur AL
WTID(FM) Repton AL
KSMJ(FM) Shafter CA
WLRQ-FM Cocoa FL
WMEZ(FM) Pensacola FL
WEGC(FM) Sasser GA
WTGA(AM) Thomaston GA
WTGA-FM Thomaston GA
KDAT(FM) Cedar Rapids IA
KLLT(FM) Spencer IA
WAVJ(FM) Princeton KY
WCDV(FM) Hammond LA
WSAG(FM) Linwood MI
WGER(FM) Saginaw MI
KBEK(FM) Mora MN
KRVY-FM Starbuck MN
KHME(FM) Winona MN
*WCCE(FM) Buie's Creek NC
WCNG(FM) Murphy NC
KZPR(FM) Minot ND
KSFT-FM South Sioux City NE
WFPG(FM) Atlantic City NJ
WBTA(AM) Batavia NY
WWFS(FM) New York NY
WDOH(FM) Delphos OH
KDPM(FM) Cottage Grove OR
KDEP(FM) Garibaldi OR
WLTJ(FM) Pittsburgh PA
WSBG(FM) Stroudsburg PA
KOLY(AM) Mobridge SD
KVIL(FM) Highland Park-Dallas TX
KLFX(FM) Nolanville TX
KODM(FM) Odessa TX
*KYQX(FM) Weatherford TX
KRPX(FM) Wellington UT
WJDV(FM) Broadway VA
*KTCV(FM) Kennewick WA
WLDB(FM) Milwaukee WI
WLJY(FM) Whiting WI

MOR

KGTL(AM) Homer AK
KIFW(AM) Sitka AK
*WSTF(FM) Andalusia AL
*WDYF(FM) Dothan AL
WHEP(AM) Foley AL
WLVV(AM) Mobile AL
*WLBF(FM) Montgomery AL
WNZZ(AM) Montgomery AL
WTLM(AM) Opelika AL
KFFB(FM) Fairfield Bay AR
KOOU(FM) Hardy AR
KVRE(FM) Hot Springs Village AR
KTLO-FM Mountain Home AR
KBHC(AM) Nashville AR
KFLG(AM) Bullhead City AZ
KYBC(AM) Cottonwood AZ
KSAZ(AM) Marana AZ
KNOT(AM) Prescott AZ
KWXY-FM Cathedral City CA
KNTI(FM) Lakeport CA
KXBX(AM) Lakeport CA
*KADV(FM) Modesto CA
KESP(AM) Modesto CA
KIDD(AM) Monterey CA
KEZW(AM) Aurora CO
*KTSG(FM) Sidney CO
KRDZ(AM) Wray CO
WWFL(AM) Clermont FL
WOSN(FM) Indian River Shores FL
WONN(AM) Lakeland FL
*WJTF(FM) Panama City FL
WITS(AM) Sebring FL
WGHC(AM) Clayton GA
WCON(AM) Cornelia GA
WNGA(AM) Elberton GA
WLRR(FM) Milledgeville GA
WMNZ(AM) Montezuma GA
WNEG(AM) Toccoa GA
*WRAF-FM Toccoa Falls GA
WLOV(AM) Washington GA
*WYFW(FM) Winder GA
KUAM(AM) Hagatna GU
KJAN(AM) Atlantic IA
KBUR(AM) Burlington IA
KCHA(AM) Charles City IA
KDEC(AM) Decorah IA
KRNT(AM) Des Moines IA
KMAQ-FM Maquoketa IA
KPTO(AM) Pocatello ID
WFRL(AM) Freeport IL
WAIK(AM) Galesburg IL
WHHL(FM) Jerseyville IL
WLRB(AM) Macomb IL
WLBH(AM) Mattoon IL
WCSJ(AM) Morris IL
WCSJ-FM Morris IL
WOAM(AM) Peoria IL
WNTA(AM) Rockford IL
WBNL(AM) Boonville IN
WLOI(AM) La Porte IN
KABI(AM) Abilene KS
*KONQ(FM) Dodge City KS
KIND(AM) Independence KS
*WCVK(FM) Bowling Green KY
WCTT(AM) Corbin KY
WSON(AM) Henderson KY
WYMC(AM) Mayfield KY
WEMD(AM) Easton MD
*WYPF(FM) Frederick MD
WVAE(AM) Biddeford ME
XETRA(AM) Tijuana MEX
WAAM(AM) Ann Arbor MI
WMHG(AM) Muskegon MI
WMJH(AM) Rockford MI
WSAM(AM) Saginaw MI
WCSY-FM South Haven MI
KATE(AM) Albert Lea MN
*KMSK(FM) Austin MN
KROX(AM) Crookston MN
KMRS(AM) Morris MN
WCMP(AM) Pine City MN
KTRF(AM) Thief River Falls MN
KZYM(AM) Joplin MO
KCXL(AM) Liberty MO
KBFL(AM) Springfield MO
*KDWG(FM) Dillon MT
KLCY(AM) East Missoula MT
KXGF(AM) Great Falls MT
KTNY(FM) Libby MT
KMMR(FM) Malta MT
WSQL(AM) Brevard NC
WBAG(AM) Burlington-Graham NC
WAZZ(AM) Fayetteville NC
WTZQ(AM) Hendersonville NC
WCVP(AM) Murphy NC
WNOS(AM) New Bern NC
WIOZ(AM) Pinehurst NC
WSAT(AM) Salisbury NC
KBRB(AM) Ainsworth NE
WWNH(AM) Madbury NH
WFEA(AM) Manchester NH
WOF(AM) Andover NJ

WVNJ(AM) Oakland NJ
WBUD(AM) Trenton NJ
WCMC(AM) Wildwood NJ
KATK(AM) Carlsbad NM
KKOB Exp Stn Santa Fe NM
WKLI-FM Albany NY
WINR(AM) Binghamton NY
WCGR(AM) Canandaigua NY
*WITR(FM) Henrietta NY
WHUC(AM) Hudson NY
WVIP(FM) New Rochelle NY
WOEN(AM) Olean NY
*WPSA(FM) Paul Smiths NY
WLNA(AM) Peekskill NY
WKIP(AM) Poughkeepsie NY
WTLB(AM) Utica NY
WFAS(AM) White Plains NY
WAKR(AM) Akron OH
WNCO(AM) Ashland OH
WATH(AM) Athens OH
WBNO-FM Bryan OH
WBCO(AM) Bucyrus OH
WILE-FM Byesville OH
WMNI(AM) Columbus OH
WOHI(AM) East Liverpool OH
WCHO(AM) Washington Court House OH
WYPC(AM) Wellston OH
WHIZ(AM) Zanesville OH
KSHO(AM) Lebanon OR
KBCH(AM) Lincoln City OR
KTIL-FM Tillamook OR
WLNP(FM) Carbondale PA
WEST(AM) Easton PA
WFRA(AM) Franklin PA
WQMU(FM) Indiana PA
WCRO(AM) Johnstown PA
WLSH(AM) Lansford PA
WNAK(AM) Nanticoke PA
WEGH(FM) Northumberland PA
WOYL(AM) Oil City PA
WJAS(AM) Pittsburgh PA
WISL(AM) Shamokin PA
WNBT(AM) Wellsboro PA
WMSW(AM) Hatillo PR
WISA(AM) Isabela PR
WPPC(AM) Penuelas PR
WBMJ(AM) San Juan PR
WIVV(AM) Vieques PR
WXEW(AM) Yabucoa PR
*WEPC(FM) Belton SC
WCRS(AM) Greenwood SC
*WKCL(FM) Ladson SC
WSNW(AM) Seneca SC
KWAT(AM) Watertown SD
WAMB(AM) Nashville TN
KAAM(AM) Garland TX
KMVL(AM) Madisonville TX
KNBO(AM) New Boston TX
KAHL(AM) San Antonio TX
KDAE(AM) Sinton TX
KLGN(AM) Logan UT
KOGN(AM) Ogden UT
KENT(AM) Parowan UT
KNFL(AM) Tremonton UT
WAMV(AM) Amherst VA
WPYA(FM) Chesapeake VA
WFVA(AM) Fredericksburg VA
WMXH-FM Luray VA
WRVQ(FM) Richmond VA
WTON-FM Staunton VA
KKAD(AM) Vancouver WA
*WLBL(AM) Auburndale WI
WFCL(AM) Clintonville WI
*WJTY(FM) Lancaster WI
WOMT(AM) Manitowoc WI
WRJC(AM) Mauston WI
WJMT(AM) Merrill WI
WRIT-FM Milwaukee WI
WJUB(AM) Plymouth WI
WRCO(AM) Richland Center WI
WXCX(FM) Siren WI
WLSL(FM) Three Lakes WI
WKLP(AM) Keyser WV
*WHFI(FM) Lindside WV
WWYO(AM) Pineville WV
WSLW(AM) White Sulphur Springs WV
KWYO(AM) Sheridan WY

Native American

*KNBA(FM) Anchorage AK
*KUYI(FM) Hotevilla AZ
*KRMH(FM) Red Mesa AZ
*KOHN(FM) Sells AZ
*KGHR(FM) Tuba City AZ
*KSUT(FM) Ignacio CO
WKAM(AM) Goshen IN
*KINI(FM) Crookston NE
KGAK(AM) Gallup NM

New Age

*KFJC(FM) Los Altos CA
KEST(AM) San Francisco CA
KSFF(FM) San Francisco CA
*KBSM(FM) McCall ID
*WNUR-FM Evanston IL
*KHCT(FM) Great Bend KS
*KHCC-FM Hutchinson KS
*KHCD(FM) Salina KS
*WYAJ(FM) Sudbury MA
*WMTB-FM Emmittsburg MD
*WMHW-FM Mount Pleasant MI
WELY-FM Ely MN
*KCOZ(FM) Point Lookout MO
*KLPR(FM) Kearney NE
*WNEC-FM Henniker NH
WBUZ(FM) La Vergne TN
*KSAU(FM) Nacogdoches TX
KBZN(FM) Ogden UT
KHQN(AM) Spanish Fork UT
*KGHP(FM) Gig Harbor WA
*KCWC-FM Riverton WY

News

*KSKA(FM) Anchorage AK
KLAM(AM) Cordova AK
KFBX(AM) Fairbanks AK
*KTOO(FM) Juneau AK
*KDLL(FM) Kenai AK
*KMXT(FM) Kodiak AK
KAKN(FM) Naknek AK
*KFSK(FM) Petersburg AK
*KCAW(FM) Sitka AK
WNSI-FM Atmore AL
*WBHM(FM) Birmingham AL
WHOS(AM) Decatur AL
*WRWA(FM) Dothan AL
WULA(AM) Eufaula AL
WDLT(AM) Fairhope AL
WHEP(AM) Foley AL
*WSGN(FM) Gadsden AL
WJBB(AM) Haleyville AL
*WLRH(FM) Huntsville AL
WNZZ(AM) Montgomery AL
*WTSU(FM) Montgomery-Troy AL
*WQPR(FM) Muscle Shoals AL
*WAPR(FM) Selma AL
KEWI(AM) Benton AR
*KUCA(FM) Conway AR
KCAB(AM) Dardanelle AR
*KBSA(FM) El Dorado AR
*KUAF(FM) Fayetteville AR
KXJK(AM) Forrest City AR
KAFN(FM) Gould AR
*KASU(FM) Jonesboro AR
KBOK(AM) Malvern AR
KARN-FM Sheridan AR
KWAK(AM) Stuttgart AR
KFNX(AM) Cave Creek AZ
*KNAU(FM) Flagstaff AZ
*KPUB(FM) Flagstaff AZ
*KNAG(FM) Grand Canyon AZ
*KNAD(FM) Page AZ
*KBAQ-FM Phoenix AZ
KIDR(AM) Phoenix AZ
*KJZZ(FM) Phoenix AZ
*KNAQ(FM) Prescott AZ
*KNAA(FM) Show Low AZ
KQTH(FM) Tucson AZ
*KUAZ(AM) Tucson AZ
KVOI(AM) Tucson AZ
*KAWC-FM Yuma AZ
*KHSU-FM Arcata CA
*KPRX(FM) Bakersfield CA
KSZL(AM) Barstow CA
KBLX-FM Berkeley CA
*KNCA(FM) Burney CA
*KCHO(FM) Chico CA
*KHSR(FM) Crescent City CA
*KVPR(FM) Fresno CA
*KFRN(AM) Long Beach CA
KFWB(AM) Los Angeles CA
KNX(AM) Los Angeles CA
*KCRY(FM) Mojave CA
*KNSQ(FM) Mount Shasta CA
KTOX(AM) Needles CA
*KCRU(FM) Oxnard CA
*KAZU(FM) Pacific Grove CA
KPSI(AM) Palm Springs CA
KNTS(AM) Palo Alto CA
*KZYX(FM) Philo CA
KAHZ(AM) Pomona CA
*KQNC(FM) Quincy CA
KQMS(AM) Redding CA
*KNHT(FM) Rio Dell CA
*KVCR(FM) San Bernardino CA
KCBS(AM) San Francisco CA
KGO(AM) San Francisco CA
KSFO(AM) San Francisco CA
KLIV(AM) San Jose CA
*KCBX(FM) San Luis Obispo CA
*KCRW(FM) Santa Monica CA
KJPR(AM) Shasta Lake City CA
KTHO(AM) South Lake Tahoe CA
*KUOP(FM) Stockton CA
*KXJS(FM) Sutter CA
*KKTO(FM) Tahoe City CA
*KZYZ(FM) Willits CA
*KNYR(FM) Yreka CA
*KRZA(FM) Alamosa CO
KNFO(FM) Basalt CO
KCFC(AM) Boulder CO
*KCJX(FM) Carbondale CO
*KRCC(FM) Colorado Springs CO
*KPYR(FM) Craig CO
*KCFR(AM) Denver CO
*KDNK(FM) Glenwood Springs CO
*KPRN(FM) Grand Junction CO
*KUNC(FM) Greeley CO
KPKE(AM) Gunnison CO
*KRLJ(FM) La Junta CO
*KPRH(FM) Montrose CO
*KVMT(FM) Montrose CO
*KVNF(FM) Paonia CO
*KCFP(FM) Pueblo CO
KKPC(AM) Pueblo CO
*KRNC(FM) Steamboat Springs CO
*KPRE(FM) Vail CO
WGCH(AM) Greenwich CT
WQUN(AM) Hamden CT
WXLM(FM) Stonington CT
*WETA(FM) Washington DC
WTOP-FM Washington DC
WILM(AM) Wilmington DE
WTAN(AM) Clearwater FL
*WGCU-FM Fort Myers FL
*WQCS(FM) Fort Pierce FL
WTOT-FM Graceville FL
WBXY(FM) La Crosse FL
*WMKO(FM) Marco FL
*WFIT(FM) Melbourne FL
WEBY(AM) Milton FL
WCFI(AM) Ocala FL
*WMFE-FM Orlando FL
*WFSW(FM) Panama City FL
*WKGC-FM Panama City FL
WPNN(AM) Pensacola FL
*WUWF(FM) Pensacola FL
*WANM(FM) Tallahassee FL
*WUSF(FM) Tampa FL
*WXEL(FM) West Palm Beach FL
V6AK(AM) Truk FM
*V6AI(AM) Yap FM
*WUNV(FM) Albany GA
*WUGA(FM) Athens GA
*WACG-FM Augusta GA
WBBQ-FM Augusta GA
WNRR(AM) Augusta GA
*WWIO-FM Brunswick GA
*WUWG(FM) Carrollton GA
WBHF(AM) Cartersville GA
WYXC(AM) Cartersville GA
*WMUM-FM Cochran GA
WDAK(AM) Columbus GA
*WTJB(FM) Columbus GA
*WNGU(FM) Dahlonega GA
*WPPR(FM) Demorest GA
WUFF-FM Eastman GA
*WJWV(FM) Fort Gaines GA
WKEU(AM) Griffin GA
WQCH(AM) La Fayette GA
WMVG(AM) Milledgeville GA
WCNN(AM) North Atlanta GA
WGPB(FM) Rome GA
WWIO(AM) Saint Mary's GA
*WSVH(FM) Savannah GA
WJAT(AM) Swainsboro GA
*WABR-FM Tifton GA
WNEG(AM) Toccoa GA
*WWET(FM) Valdosta GA
WVOP(AM) Vidalia GA
*WJSP-FM Warm Springs GA
*WXVS(FM) Waycross GA
*KPRG(FM) Hagatna GU
KRTR(AM) Honolulu HI
KJAN(AM) Atlantic IA
*KWOI(FM) Carroll IA
*KUNI(FM) Cedar Falls IA
*KLNI(FM) Decorah IA
KILR-FM Estherville IA
KIOW(FM) Forest City IA
*KTPR(FM) Fort Dodge IA
KVFD(AM) Fort Dodge IA
KXIC(AM) Iowa City IA
KIFG-FM Iowa Falls IA
*KOWI(FM) Lamoni IA
KLEM(AM) Le Mars IA
*KRNI(AM) Mason City IA
*KUNY(FM) Mason City IA
*KDMR(FM) Mitchellville IA
KCOB(AM) Newton IA
KAYL-FM Storm Lake IA
KCII-FM Washington IA
*KBSU-FM Boise ID
*KBSX(FM) Boise ID
*KIBX(FM) Bonners Ferry ID
*KBSY(FM) Burley ID
KVNI(AM) Coeur d'Alene ID
*KNWO(FM) Cottonwood ID
*KBSM(FM) McCall ID
*KBSQ(FM) McCall ID
*KRFA-FM Moscow ID
KIDO(AM) Nampa ID
*KBYI(FM) Rexburg ID
*KBSS(FM) Sun Valley ID
WRMJ(FM) Aledo IL
*WSIU(FM) Carbondale IL
WBBM(AM) Chicago IL
*WCRX(FM) Chicago IL
*WNIJ(FM) De Kalb IL
WIXN(AM) Dixon IL
*WNIE(FM) Freeport IL
*WDCB(FM) Glen Ellyn IL
WKIF(FM) Kankakee IL
*WNIW(FM) La Salle IL
*WIUM(FM) Macomb IL
*WVSI(FM) Mount Vernon IL
*WUSI(FM) Olney IL
WPRS(AM) Paris IL
*WIPA(FM) Pittsfield IL
*WQUB(FM) Quincy IL
*WVIK(FM) Rock Island IL
WNTA(AM) Rockford IL
WJBD-FM Salem IL
WCCI(FM) Savanna IL
*WUIS(FM) Springfield IL
WSDR(AM) Sterling IL
WSQR(AM) Sycamore IL
*WIUW(FM) Warsaw IL
WFRX(AM) West Frankfort IL
*WBSB(FM) Anderson IN
WBIW(AM) Bedford IN
WZBD(FM) Berne IN
*WFHB(FM) Bloomington IN
*WFIU(FM) Bloomington IN
WREB(FM) Greencastle IN
*WBSH(FM) Hagerstown IN
WXGO(AM) Madison IN
*WBSW(FM) Marion IN
*WBST(FM) Muncie IN
WLBC-FM Muncie IN
*WBSJ(FM) Portland IN
WRAY(AM) Princeton IN
WSLM-FM Salem IN
WZZB(AM) Seymour IN
*WBAA-FM West Lafayette IN
KDCC(AM) Dodge City KS
KGNO(AM) Dodge City KS
KVOE(AM) Emporia KS
*KHCT(FM) Great Bend KS
KVGB(AM) Great Bend KS
*KHCC-FM Hutchinson KS
KWBW(AM) Hutchinson KS
KKAN(AM) Phillipsburg KS
*KRPS(FM) Pittsburg KS
*KHCD(FM) Salina KS
KMZA(FM) Seneca KS
*KMUW(FM) Wichita KS
WBRT(AM) Bardstown KY
*WKYU-FM Bowling Green KY
*WEKF(FM) Corbin KY
WCPM(AM) Cumberland KY
*WKUE(FM) Elizabethtown KY
*WEKH(FM) Hazard KY
*WKPB(FM) Henderson KY
*WNKU(FM) Highland Heights KY
WLAP(AM) Lexington KY
*WUKY(FM) Lexington KY
WKJK(AM) Louisville KY
WTTL(AM) Madisonville KY
*WKMS-FM Murray KY
WNBS(AM) Murray KY
*WEKU(FM) Richmond KY
*WDCL-FM Somerset KY
WMSK-FM Sturgis KY
WEZJ(AM) Williamsburg KY
*KLSA(FM) Alexandria LA
WJBO(AM) Baton Rouge LA
WWL-FM Kenner LA
KVOL(AM) Lafayette LA
*WRBH(FM) New Orleans LA
*WWNO(FM) New Orleans LA
*KDAQ(FM) Shreveport LA
KRMD(AM) Shreveport LA
*WHAB(FM) Acton MA
*WFCR(FM) Amherst MA
*WBUR-FM Boston MA
*WGBH(FM) Boston MA
WHMQ(AM) Greenfield MA
WCAP(AM) Lowell MA
WBIX(AM) Natick MA
WESO(AM) Southbridge MA
WTLP(FM) Braddock Heights MD
WTWT(AM) Frederick MD
*WGMS(FM) Hagerstown MD
WKHZ(AM) Ocean City MD
*WSCL(FM) Salisbury MD
WFED(AM) Silver Spring MD
*WMEH(FM) Bangor ME
*WMED(FM) Calais ME
WCXU(FM) Caribou ME
WCXX(FM) Madawaska ME
WHXR(FM) North Windham ME
WEGP(AM) Presque Isle ME
WTME(AM) Rumford ME
WKTQ(AM) South Paris ME
WNZS(AM) Veazie ME
*WGVU-FM Allendale MI
*WCML-FM Alpena MI
WMAX(AM) Bay City MI
*WUCX-FM Bay City MI
WHFB(AM) Benton Harbor-St. Joseph MI
WNEM(AM) Bridgeport MI
WNZK(AM) Dearborn Heights MI
*WDET-FM Detroit MI

WJR(AM) Detroit MI
WWJ(AM) Detroit MI
*WKAR-FM East Lansing MI
WSHN(AM) Fremont MI
WMJZ-FM Gaylord MI
*WBLU-FM Grand Rapids MI
WSCG(AM) Greenville MI
*WCMW-FM Harbor Springs MI
*WGGL-FM Houghton MI
*WIAA(FM) Interlochen MI
*WMUK(FM) Kalamazoo MI
WJNL(AM) Kingsley MI
WJIM(AM) Lansing MI
*WIAB(FM) Mackinaw City MI
*WNMU-FM Marquette MI
*WCMU-FM Mount Pleasant MI
WKBZ(AM) Muskegon MI
*WCMB-FM Oscoda MI
*WCMZ-FM Sault Ste. Marie MI
WWCM(FM) Standish MI
WMSH(AM) Sturgis MI
*WICA(FM) Traverse City MI
*WBLV(FM) Twin Lake MI
WGVS-FM Whitehall MI
*WEMU(FM) Ypsilanti MI
WPNW(AM) Zeeland MI
KASM(AM) Albany MN
*KNCM(FM) Appleton MN
*KNSE(FM) Austin MN
*KNBJ(FM) Bemidji MN
*KBPN(FM) Brainerd MN
*WIRN(FM) Buhl MN
*WSCN(FM) Cloquet MN
*KNSR(FM) Collegeville MN
KSUM(AM) Fairmont MN
KDHL(AM) Faribault MN
*KNWF(FM) Fergus Falls MN
*WLSN(FM) Grand Marais MN
KDWA(AM) Hastings MN
KDUZ(AM) Hutchinson MN
*KXLC(FM) La Crescent MN
*KMSU(FM) Mankato MN
KYSM(AM) Mankato MN
*KTIS(AM) Minneapolis MN
*KNOW-FM Minneapolis-St. Paul MN
*KCCD(FM) Moorhead MN
WCMP(AM) Pine City MN
WCMP-FM Pine City MN
KLOH(AM) Pipestone MN
*KZSE(FM) Rochester MN
*KNGA(FM) Saint Peter MN
*KNTN(FM) Thief River Falls MN
KTRF(AM) Thief River Falls MN
KOWZ(AM) Waseca MN
KSGF-FM Ash Grove MO
KAAN(AM) Bethany MO
KAAN-FM Bethany MO
KYOO(AM) Bolivar MO
KAPE(AM) Cape Girardeau MO
*KRCU(FM) Cape Girardeau MO
KCHI(AM) Chillicothe MO
KCHI-FM Chillicothe MO
*KRNW(FM) Chillicothe MO
KDKD-FM Clinton MO
*KBIA(FM) Columbia MO
*KSEF(FM) Farmington MO
*KCUR-FM Kansas City MO
KMBZ(AM) Kansas City MO
*KKTR(FM) Kirksville MO
KNIM(AM) Maryville MO
*KXCV(FM) Maryville MO
KBDZ(FM) Perryville MO
*KSMS-FM Point Lookout MO
KMIS(AM) Portageville MO
KYRO(AM) Potosi MO
*KMST(FM) Rolla MO
*KWMU(FM) Saint Louis MO
KLFJ(AM) Springfield MO
*KSMU(FM) Springfield MO
KSAR(FM) Thayer MO
KTTN-FM Trenton MO
*KSMW(FM) West Plains MO
WWZQ(AM) Aberdeen MS
WHSY(AM) Hattiesburg MS
*WURC(FM) Holly Springs MS
*WJSU(FM) Jackson MS

*WMPN-FM Jackson MS
*WMAB-FM Mississippi State MS
*WMAV-FM Oxford MS
KBUL(AM) Billings MT
*KEMC(FM) Billings MT
*KBMC(FM) Bozeman MT
*KAPC(FM) Butte MT
*KGPR(FM) Great Falls MT
*KUFN(FM) Hamilton MT
KHDN(AM) Hardin MT
*KUHM(FM) Helena MT
*KUKL(FM) Kalispell MT
*WCQS(FM) Asheville NC
WGCR(AM) Brevard NC
WLOE(AM) Eden NC
*WFSS(FM) Fayetteville NC
*WFQS(FM) Franklin NC
WTRU(AM) Kernersville NC
*WKNS(FM) Kinston NC
WMYN(AM) Mayodan NC
WCVP(AM) Murphy NC
*WTEB(FM) New Bern NC
*WZNB(FM) New Bern NC
*WZRN(FM) Norlina NC
WAYN(AM) Rockingham NC
*WNCW(FM) Spindale NC
WCIE(AM) Spring Lake NC
WSIC(AM) Statesville NC
WWTB(FM) Topsail Beach NC
*WHQR(FM) Wilmington NC
*WFDD-FM Winston-Salem NC
*KCND(FM) Bismarck ND
KDLR(AM) Devils Lake ND
*KDPR(FM) Dickinson ND
*KFBN(FM) Fargo ND
KKXL(AM) Grand Forks ND
*KPRJ(FM) Jamestown ND
*KMPR(FM) Minot ND
KDDR(AM) Oakes ND
KOVC(AM) Valley City ND
*KPPR(FM) Williston ND
*KTNE-FM Alliance NE
KNCY-FM Auburn NE
*KMNE-FM Bassett NE
*KCNE-FM Chadron NE
KGMT(AM) Fairbury NE
*KHNE-FM Hastings NE
KRVN(AM) Lexington NE
KRVN-FM Lexington NE
*KUCV(FM) Lincoln NE
*KRNE-FM Merriman NE
*KXNE-FM Norfolk NE
*KPNE-FM North Platte NE
KOLT(AM) Scottsbluff NE
WKBK(AM) Keene NH
WASR(AM) Wolfeboro NH
WSUS(FM) Franklin NJ
*WBJB-FM Lincroft NJ
KDEF(AM) Albuquerque NM
*KGLP(FM) Gallup NM
KRSN(AM) Los Alamos NM
*KMTH(FM) Maljamar NM
*KTDB(FM) Ramah NM
KZXQ(FM) Reserve NM
KRUI(AM) Ruidoso Downs NM
KCHS(AM) Truth or Consequences NM
*KNCC(FM) Elko NV
KTSN(AM) Elko NV
*KNPR(FM) Las Vegas NV
KRLV(AM) Las Vegas NV
*KWPR(FM) Lund NV
*KLNR(FM) Panaca NV
*KUNR(FM) Reno NV
*KTPH(FM) Tonopah NV
*WAMC-FM Albany NY
WVTL(AM) Amsterdam NY
WYSL(AM) Avon NY
*WSKG-FM Binghamton NY
*WSQX-FM Binghamton NY
*WCWP(FM) Brookville NY
*WBFO(FM) Buffalo NY
*WSQE(FM) Corning NY
*WGSU(FM) Geneseo NY
*WSQA(FM) Hornell NY
*WSQG-FM Ithaca NY
WJTN(AM) Jamestown NY

*WUBJ(FM) Jamestown NY
WLLG(FM) Lowville NY
*WOSR(FM) Middletown NY
WBBR(AM) New York NY
WCBS(AM) New York NY
WINS(AM) New York NY
*WNYC-FM New York NY
*WOLN(FM) Olean NY
*WSQC-FM Oneonta NY
*WDFH(FM) Ossining NY
WEBO(AM) Owego NY
*WCEL(FM) Plattsburgh NY
WRCR(AM) Spring Valley NY
*WAER(FM) Syracuse NY
*WANC(FM) Ticonderoga NY
WTNY(AM) Watertown NY
WAKR(AM) Akron OH
WFUN(AM) Ashtabula OH
WYBL(FM) Ashtabula OH
*WCVV(FM) Belpre OH
*WGBE(FM) Bryan OH
WAIS(AM) Buchtel OH
WCSM-FM Celina OH
*WCPN(FM) Cleveland OH
*WCBE(FM) Columbus OH
*WGDE(FM) Defiance OH
*WKSU-FM Kent OH
*WFCO(FM) Lancaster OH
*WGLE(FM) Lima OH
WLTP(AM) Marietta OH
WMOA(AM) Marietta OH
*WOSB(FM) Marion OH
*WKRJ(FM) New Philadelphia OH
*WNRK(FM) Norwalk OH
*WKSV(FM) Thompson OH
*WGTE-FM Toledo OH
*WKRW(FM) Wooster OH
*WYSU(FM) Youngstown OH
*KOCU(FM) Altus OK
*KYCU(FM) Clinton OK
KCRC(AM) Enid OK
KIHN(AM) Hugo OK
*KOSN(FM) Ketchum OK
*KCCU(FM) Lawton OK
KPGM(AM) Pawhuska OK
*KOSU(FM) Stillwater OK
*KWGS(FM) Tulsa OK
*KSMF(FM) Ashland OR
*KSOR(FM) Ashland OR
*KOAB-FM Bend OR
KZZR(AM) Burns OR
*KSBA(FM) Coos Bay OR
*KLFO(FM) Florence OR
*KAGI(AM) Grants Pass OR
*KLMF(FM) Klamath Falls OR
*KSKF(FM) Klamath Falls OR
*KTVR-FM La Grande OR
*KOAP(FM) Lakeview OR
KGAL(AM) Lebanon OR
*KOOZ(FM) Myrtle Point OR
*KLCO(FM) Newport OR
*KRBM(FM) Pendleton OR
KUMA(AM) Pendleton OR
KCMX(AM) Phoenix OR
KEX(AM) Portland OR
*KOPB-FM Portland OR
*KSRS(FM) Roseburg OR
*KTBR(AM) Roseburg OR
*KSJK(AM) Talent OR
*WDIY(FM) Allentown PA
WBVP(AM) Beaver Falls PA
WZUM(AM) Carnegie PA
*WFSE(FM) Edinboro PA
*WQLN-FM Erie PA
WFRA(AM) Franklin PA
WGET(AM) Gettysburg PA
WJSM(AM) Martinsburg PA
WKQW(AM) Oil City PA
WKQW-FM Oil City PA
KYW(AM) Philadelphia PA
*WHYY-FM Philadelphia PA
KQV(AM) Pittsburgh PA
*WDUQ(FM) Pittsburgh PA
WECZ(AM) Punxsutawney PA
*WVIA-FM Scranton PA
WQRM(FM) Smethport PA

*WPSU(FM) State College PA
WLIH(FM) Whitneyville PA
*WVYA(FM) Williamsport PA
WORA(AM) Mayaguez PR
*WRUO(FM) Mayaguez PR
WEKO(AM) Morovis PR
WEXS(AM) Patillas PR
WSOL(AM) San German PR
WCNX(AM) Hope Valley RI
WXNI(AM) Westerly RI
*WLJK(FM) Aiken SC
*WJWJ-FM Beaufort SC
WQNT(AM) Charleston SC
*WSCI(FM) Charleston SC
*WLTR(FM) Columbia SC
*WHMC-FM Conway SC
WJMX(AM) Florence SC
*WEPR(FM) Greenville SC
WFXH(AM) Hilton Head Island SC
WRHM(FM) Lancaster SC
WSNW(AM) Seneca SC
WBCU(AM) Union SC
KGIM(AM) Aberdeen SD
KBFS(AM) Belle Fourche SD
*KESD(FM) Brookings SD
KDSJ(AM) Deadwood SD
KPSD(FM) Faith SD
KOKK(AM) Huron SD
*KQSD-FM Lowry SD
KJAM-FM Madison SD
*KZSD-FM Martin SD
*KDSD-FM Pierpont SD
*KTSD-FM Reliance SD
*KNWC(AM) Sioux Falls SD
*KNWC-FM Sioux Falls SD
*KRSD(FM) Sioux Falls SD
*KUSD(FM) Vermillion SD
KJJQ(AM) Volga SD
KWAT(AM) Watertown SD
WCTA(AM) Alamo TN
WJZM(AM) Clarksville TN
*WSMC-FM Collegedale TN
*WHRS(FM) Cookeville TN
WDXI(AM) Jackson TN
*WKNP(FM) Jackson TN
*WUOT(FM) Knoxville TN
WLIV(AM) Livingston TN
*WKNO-FM Memphis TN
*WYPL(FM) Memphis TN
*WPLN-FM Nashville TN
WNTT(AM) Tazewell TN
*WTML(FM) Tullahoma TN
*KACU(FM) Abilene TX
*KUT(FM) Austin TX
KSKY(AM) Balch Springs TX
*KVLU(FM) Beaumont TX
KBST-FM Big Spring TX
KRHC(AM) Burnet TX
*KTXP(FM) Bushland TX
*KAMU-FM College Station TX
*KETR(FM) Commerce TX
*KEDT-FM Corpus Christi TX
KAND(AM) Corsicana TX
KRLD(AM) Dallas TX
*KTEP(FM) El Paso TX
*KMBH-FM Harlingen TX
*KPFT(FM) Houston TX
KTRH(AM) Houston TX
*KUHF(FM) Houston TX
*KSHU(FM) Huntsville TX
*KTXI(FM) Ingram TX
KAML(AM) Kenedy-Karnes City TX
*KLDN(FM) Lufkin TX
KCUL(AM) Marshall TX
*KHID(FM) McAllen TX
KWEL(AM) Midland TX
KGNB(AM) New Braunfels TX
*KOCV(FM) Odessa TX
KOGT(AM) Orange TX
KBUS(FM) Paris TX
KGKL(AM) San Angelo TX
*KUTX(FM) San Angelo TX
*KSTX(FM) San Antonio TX
KJIM(AM) Sherman TX
*KTOT(FM) Spearman TX
*KQXS(FM) Stephenville TX

*KVRT(FM) Victoria TX
*KMCU(FM) Wichita Falls TX
KMTI(AM) Manti UT
KOGN(AM) Ogden UT
*KPCW(FM) Park City UT
*KUER(FM) Salt Lake City UT
KNFL(AM) Tremonton UT
WKDE(AM) Altavista VA
WOWZ(AM) Appomattox VA
WDIC-FM Clinchco VA
*WOTC(FM) Edinburg VA
*WEMC(FM) Harrisonburg VA
*WMRA(FM) Harrisonburg VA
WAGE(AM) Leesburg VA
*WMRL(FM) Lexington VA
WLVA(AM) Lynchburg VA
WLEE(AM) Richmond VA
*WVTF(FM) Roanoke VA
WVAB(AM) Virginia Beach VA
*WISE-FM Wise VA
WYVE(AM) Wytheville VA
WJOY(AM) Burlington VT
*WVPS(FM) Burlington VT
*WNCH(FM) Norwich VT
*WRVT(FM) Rutland VT
*WVPA(FM) Saint Johnsbury VT
WDEV(AM) Waterbury VT
*WVPR(FM) Windsor VT
KWLE(AM) Anacortes WA
*KASB(FM) Bellevue WA
*KZAZ(FM) Bellingham WA
KOZI-FM Chelan WA
*KNWV(FM) Clarkston WA
*KGHP(FM) Gig Harbor WA
KEDO(AM) Longview WA
*KLWS(FM) Moses Lake WA
*KPLI(FM) Olympia WA
*KQWS(FM) Omak WA
*KNWP(FM) Port Angeles WA
*KVIX(FM) Port Angeles WA
*KWSU(AM) Pullman WA
KWNC(AM) Quincy WA
KOMO(AM) Seattle WA
KTTH(AM) Seattle WA
*KUOW-FM Seattle WA
KBBO(AM) Selah WA
*KPBX-FM Spokane WA
KSBN(AM) Spokane WA
*KSFC(FM) Spokane WA
*KPLU-FM Tacoma WA
KYNR(AM) Toppenish WA
KUOW(AM) Tumwater WA
*KNWY(FM) Yakima WA
WATW(AM) Ashland WI
*WUEC(FM) Eau Claire WI
*WPNE-FM Green Bay WI
*WGTD(FM) Kenosha WI
*WLSU(FM) La Crosse WI
*WERN(FM) Madison WI
*WHA(AM) Madison WI
WISN(AM) Milwaukee WI
*WUWM(FM) Milwaukee WI
*WSSW(FM) Platteville WI
WRDB(AM) Reedsburg WI
WRCO-FM Richland Center WI
WEVR-FM River Falls WI
*WHND(FM) Sister Bay WI
WJJQ-FM Tomahawk WI
WXCO(AM) Wausau WI
*WXPW(FM) Wausau WI
*WVPB(FM) Beckley WV
*WVPW(FM) Buckhannon WV
*WVPN(FM) Charleston WV
*WVWV(FM) Huntington WV
*WVEP(FM) Martinsburg WV
*WVPM(FM) Morgantown WV
*WVPG(FM) Parkersburg WV
*WAUA(FM) Petersburg WV
WRON(AM) Ronceverte WV
*WVNP(FM) Wheeling WV
KRSV(AM) Afton WY
KRSV-FM Afton WY
*KBUW(FM) Buffalo WY
*KUWC(FM) Casper WY
*KDUW(FM) Douglas WY
KIML(AM) Gillette WY

*KUWG(FM) Gillette WY
*KUWJ(FM) Jackson WY
*KUWR(FM) Laramie WY
*KUWN(FM) Newcastle WY
*KUWX(FM) Pinedale WY
KPOW(AM) Powell WY
*KUWZ(FM) Rock Springs WY
*KPRQ(FM) Sheridan WY
*KSUW(FM) Sheridan WY
*KUWD(FM) Sundance WY
KYDT(FM) Sundance WY
KUWT(FM) Thermopolis WY

News/talk

KBYR(AM) Anchorage AK
KFQD(AM) Anchorage AK
KFAR(AM) Fairbanks AK
*KUAC(FM) Fairbanks AK
KTKN(AM) Ketchikan AK
*KXKM(FM) McCarthy AK
KIAM(AM) Nenana AK
*KNOM(AM) Nome AK
*KNOM-FM Nome AK
KIFW(AM) Sitka AK
KSRM(AM) Soldotna AK
*KTNA(FM) Talkeetna AK
WDNG(AM) Anniston AL
WVNN(AM) Athens AL
WAPI(AM) Birmingham AL
WERC(AM) Birmingham AL
WCMA(AM) Daleville AL
WXAL(AM) Demopolis AL
WWNT(AM) Dothan AL
WBCF(AM) Florence AL
WAAX(AM) Gadsden AL
WGEA(AM) Geneva AL
WRJM-FM Geneva AL
WGYV(AM) Greenville AL
WGSV(AM) Guntersville AL
WABB(AM) Mobile AL
WNTM(AM) Mobile AL
WACV(AM) Montgomery AL
WLWI(AM) Montgomery AL
WANI(AM) Opelika AL
WHBB(AM) Selma AL
WFEB(AM) Sylacauga AL
WVNN-FM Trinity AL
WTBC(AM) Tuscaloosa AL
*WUAL-FM Tuscaloosa AL
WVNA(AM) Tuscumbia AL
WAPZ(AM) Wetumpka AL
KAPZ(AM) Bald Knob AR
KFPW-FM Barling AR
KLCN(AM) Blytheville AR
KJMT(FM) Calico Rock AR
KFAY(AM) Farmington AR
KBJT(AM) Fordyce AR
KWHN(AM) Fort Smith AR
KYHN(AM) Fort Smith AR
KELD-FM Hampton AR
KAWW(AM) Heber Springs AR
KZNG(AM) Hot Springs AR
KBTM(AM) Jonesboro AR
KARN(AM) Little Rock AR
*KUAR(FM) Little Rock AR
KVOM(AM) Morrilton AR
KNBY(AM) Newport AR
KARV-FM Ola AR
KSMD(FM) Pangburn AR
KCLA(AM) Pine Bluff AR
KOTN(AM) Pine Bluff AR
KARV(AM) Russellville AR
KWYN(AM) Wynne AR
KFNX(AM) Cave Creek AZ
KAPR(AM) Douglas AZ
KVNA(AM) Flagstaff AZ
KTAR-FM Glendale AZ
KJAA(AM) Globe AZ
KDJI(AM) Holbrook AZ
KAAA(AM) Kingman AZ
KNTR(AM) Lake Havasu City AZ
KFNN(AM) Mesa AZ
KLPZ(AM) Parker AZ
KFYI(AM) Phoenix AZ
KKNT(AM) Phoenix AZ
KYCA(AM) Prescott AZ
KQNA(AM) Prescott Valley AZ
KATO(AM) Safford AZ
KAZM(AM) Sedona AZ
KVWM(AM) Show Low AZ
KTAN(AM) Sierra Vista AZ
KJLL(AM) South Tucson AZ
KNUV(AM) Tolleson AZ
KNST(AM) Tucson AZ
*KUAZ-FM Tucson AZ
KYET(AM) Williams AZ
KBLU(AM) Yuma AZ
KJOK(AM) Yuma AZ
KCFJ(AM) Alturas CA
KERN(AM) Bakersfield CA
KNZR(AM) Bakersfield CA
*KQVO(FM) Calexico CA
KPAY(AM) Chico CA
*KZFR(FM) Chico CA
KGOE(AM) Eureka CA
KINS(AM) Eureka CA
KMJ(AM) Fresno CA
KRLA(AM) Glendale CA
KNCO(AM) Grass Valley CA
KQAB(AM) Lake Isabella CA
KCAA(AM) Loma Linda CA
*KPFK(FM) Los Angeles CA
KTNQ(AM) Los Angeles CA
KWKW(AM) Los Angeles CA
*KPMO(AM) Mendocino CA
KTIQ(AM) Merced CA
KYOS(AM) Merced CA
KFIV(AM) Modesto CA
KNRY(AM) Monterey CA
KXTY(FM) Morro Bay CA
KMJC(AM) Mount Shasta CA
KVON(AM) Napa CA
*KQEI-FM North Highlands CA
KNEW(AM) Oakland CA
KLAA(AM) Orange CA
KGAM(AM) Palm Springs CA
KNWQ(AM) Palm Springs CA
KUTY(AM) Palmdale CA
KKXX(AM) Paradise CA
*KPCC(FM) Pasadena CA
KPRL(AM) Paso Robles CA
KWKU(AM) Pomona CA
KVTA(AM) Port Hueneme CA
KTIP(AM) Porterville CA
KHGQ(FM) Quincy CA
KPCO(AM) Quincy CA
KFBK(AM) Sacramento CA
KSAC(AM) Sacramento CA
*KXJZ(FM) Sacramento CA
KION(AM) Salinas CA
KTIE(AM) San Bernardino CA
KCBQ(AM) San Diego CA
KFMB(AM) San Diego CA
KOGO(AM) San Diego CA
*KPBS-FM San Diego CA
KIQI(AM) San Francisco CA
*KQED-FM San Francisco CA
KZSF(AM) San Jose CA
KVEC(AM) San Luis Obispo CA
KYNS(AM) San Luis Obispo CA
KZSB(AM) Santa Barbara CA
KSCO(AM) Santa Cruz CA
KSMA(AM) Santa Maria CA
*KRCB-FM Santa Rosa CA
KSRO(AM) Santa Rosa CA
KIRN(AM) Simi Valley CA
KVML(AM) Sonora CA
KOWL(AM) South Lake Tahoe CA
KWSX(AM) Stockton CA
KSUE(AM) Susanville CA
*KCLU(FM) Thousand Oaks CA
KNWH(AM) Twentynine Palms CA
KNTK(FM) Weed CA
KUBA(AM) Yuba City CA
*KAJX(FM) Aspen CO
KRLN(AM) Canon City CO
KRDO(AM) Colorado Springs CO
KVOR(AM) Colorado Springs CO
KZNT(AM) Colorado Springs CO
KVFC(AM) Cortez CO
*KBUT(FM) Crested Butte CO
KBJD(AM) Denver CO
KBNO(AM) Denver CO
KNUS(AM) Denver CO
KOA(AM) Denver CO
KDGO(AM) Durango CO
KNRV(AM) Englewood CO
KFTM(AM) Fort Morgan CO
KGLN(AM) Glenwood Springs CO
KNZZ(AM) Grand Junction CO
KFKA(AM) Greeley CO
KHNC(AM) Johnstown CO
KWUF(AM) Pagosa Springs CO
KCSJ(AM) Pueblo CO
KGFT(FM) Pueblo CO
KRDO-FM Security CO
KCOL(AM) Wellington CO
WPRX(AM) Bristol CT
*WQAQ(FM) Hamden CT
WDRC(AM) Hartford CT
WTIC(AM) Hartford CT
WZBG(FM) Litchfield CT
WMMW(AM) Meriden CT
*WPKT(FM) Meriden CT
WELI(AM) New Haven CT
WNLK(AM) Norwalk CT
*WNPR(FM) Norwich CT
*WEDW-FM Stamford CT
WSTC(AM) Stamford CT
WATR(AM) Waterbury CT
WWCO(AM) Waterbury CT
*WSHU(AM) Westport CT
WILI(AM) Willimantic CT
*WAMU(FM) Washington DC
WMAL(AM) Washington DC
WOL(AM) Washington DC
*WPFW(FM) Washington DC
WTWP(AM) Washington DC
WDOV(AM) Dover DE
WGMD(FM) Rehoboth Beach DE
WDEL(AM) Wilmington DE
WORL(AM) Altamonte Springs FL
WFLN(AM) Arcadia FL
WTWB(AM) Auburndale FL
WWJB(AM) Brooksville FL
WKFL(AM) Bushnell FL
WMMV(AM) Cocoa FL
WJTK(FM) Columbia City FL
WRHC(AM) Coral Gables FL
WTPS(AM) Coral Gables FL
WNDB(AM) Daytona Beach FL
WZEP(AM) De Funiak Springs FL
WTJV(AM) De Land FL
WYND(AM) De Land FL
WNWF(AM) Destin FL
WGUL(AM) Dunedin FL
WENG(AM) Englewood FL
*WAFG(FM) Fort Lauderdale FL
WINK(AM) Fort Myers FL
WJNX(AM) Fort Pierce FL
WFTW(AM) Fort Walton Beach FL
WRUF(AM) Gainesville FL
WXYB(AM) Indian Rocks Beach FL
WBOB(AM) Jacksonville FL
*WJCT-FM Jacksonville FL
WOKV(AM) Jacksonville FL
WJBW(AM) Jupiter FL
WKIZ(AM) Key West FL
WLKF(AM) Lakeland FL
WPBR(AM) Lantana FL
WFFG(AM) Marathon FL
*WHWY(FM) Marathon FL
WGUF(FM) Marco FL
WMEL(AM) Melbourne FL
WAQI(AM) Miami FL
WIOD(AM) Miami FL
*WLRN-FM Miami FL
WOCN(AM) Miami FL
WQBA(AM) Miami FL
WSUA(AM) Miami FL
WWFE(AM) Miami FL
WSKY-FM Micanopy FL
WNOG(AM) Naples FL
WPSO(AM) New Port Richey FL
WKAT(AM) North Miami FL
WOCA(AM) Ocala FL
WDBO(AM) Orlando FL
WELE(AM) Ormond Beach FL
WLTG(AM) Panama City FL
WCOA(AM) Pensacola FL
WAMT(AM) Pine Castle-Sky Lake FL
WFLF(AM) Pine Hills FL
WWBA(AM) Pinellas Park FL
WPSL(AM) Port St. Lucie FL
WCCF(AM) Punta Gorda FL
WFOY(AM) Saint Augustine FL
WSDO(AM) Sanford FL
WLSS(AM) Sarasota FL
WSRQ(AM) Sarasota FL
WSTU(AM) Stuart FL
*WFSU-FM Tallahassee FL
WTAL(AM) Tallahassee FL
WUTL(FM) Tallahassee FL
WFLA(AM) Tampa FL
WHNZ(AM) Tampa FL
WFTL(AM) West Palm Beach FL
WJNO(AM) West Palm Beach FL
WPRD(AM) Winter Park FL
WDDQ(FM) Adel GA
WALG(AM) Albany GA
WLTA(AM) Alpharetta GA
WGAU(AM) Athens GA
*WABE(FM) Atlanta GA
WAOK(AM) Atlanta GA
WGKA(AM) Atlanta GA
WGST(AM) Atlanta GA
WSB(AM) Atlanta GA
WGAC(AM) Augusta GA
WRDW(AM) Augusta GA
WGIG(AM) Brunswick GA
WMOG(AM) Brunswick GA
WGRA(AM) Cairo GA
WJTH(AM) Calhoun GA
WLBB(AM) Carrollton GA
WRCG(AM) Columbus GA
WSRM(FM) Coosa GA
WBLJ(AM) Dalton GA
WDUN(AM) Gainesville GA
WHIE(AM) Griffin GA
WRPG(FM) Hawkinsville GA
WVCC(AM) Hogansville GA
WLOP(AM) Jesup GA
WVGA(FM) Lakeland GA
WMAC(AM) Macon GA
WLAQ(AM) Rome GA
WRGA(AM) Rome GA
WBMQ(AM) Savannah GA
WWNS(AM) Statesboro GA
WKWN(AM) Trenton GA
WRFV(AM) Valdosta GA
WGAC-FM Warrenton GA
WWGA(AM) Waycross GA
WCGA(AM) Woodbine GA
KGUM(AM) Hagatna GU
KPUA(AM) Hilo HI
KHBZ(AM) Honolulu HI
KHNR-FM Honolulu HI
*KHPR(FM) Honolulu HI
KHVH(AM) Honolulu HI
*KIPO(FM) Honolulu HI
KWAI(AM) Honolulu HI
KNUI(AM) Kahului HI
KAOI(AM) Kihei HI
KQNG(AM) Lihue HI
*KKUA(FM) Wailuku HI
KASI(AM) Ames IA
*WOI(AM) Ames IA
KWBG(AM) Boone IA
KBUR(AM) Burlington IA
WMT(AM) Cedar Rapids IA
WOC(AM) Davenport IA
WHO(AM) Des Moines IA
*KDUB(FM) Dubuque IA
WDBQ(AM) Dubuque IA
KILR(AM) Estherville IA
KMCD(AM) Fairfield IA
KBKB(AM) Fort Madison IA
*WSUI(AM) Iowa City IA
KOKX(AM) Keokuk IA
KFJB(AM) Marshalltown IA
KOEL(AM) Oelwein IA
*KOJI(FM) Okoboji IA
KBOE-FM Oskaloosa IA
KLEE(AM) Ottumwa IA
KIWA(AM) Sheldon IA
KMA(AM) Shenandoah IA
KMNS(AM) Sioux City IA
KSCJ(AM) Sioux City IA
*KWIT(FM) Sioux City IA
KXEL(AM) Waterloo IA
KQWC-FM Webster City IA
KBOI(AM) Boise ID
*KBSU(AM) Boise ID
KBFI(AM) Bonners Ferry ID
KID(AM) Idaho Falls ID
KRLC(AM) Lewiston ID
*KISU-FM Pocatello ID
KSEI(AM) Pocatello ID
KWIK(AM) Pocatello ID
KSPT(AM) Sandpoint ID
*KEZJ(AM) Twin Falls ID
KLIX(AM) Twin Falls ID
KWEI(AM) Weiser ID
WBGZ(AM) Alton IL
WBIG(AM) Aurora IL
WCIL(AM) Carbondale IL
WDWS(AM) Champaign IL
KSGM(AM) Chester IL
*WBEZ(FM) Chicago IL
WGN(AM) Chicago IL
WIND(AM) Chicago IL
WLS(AM) Chicago IL
WHOW(AM) Clinton IL
WDAN(AM) Danville IL
WLBK(AM) De Kalb IL
WSOY(AM) Decatur IL
WCRA(AM) Effingham IL
WRMN(AM) Elgin IL
WJJG(AM) Elmhurst IL
WFIW(AM) Fairfield IL
*WGNN(FM) Fisher IL
WGIL(AM) Galesburg IL
WGEN(AM) Geneseo IL
WKYX-FM Golconda IL
WJPF(AM) Herrin IL
WJIL(AM) Jacksonville IL
WLDS(AM) Jacksonville IL
WKEI(AM) Kewanee IL
WLPO(AM) La Salle IL
WSMI(AM) Litchfield IL
WLBH(AM) Mattoon IL
*WBEQ(FM) Morris IL
WCSJ(AM) Morris IL
WCSJ-FM Morris IL
WINI(AM) Murphysboro IL
WCMY(AM) Ottawa IL
WMBD(AM) Peoria IL
WZOE(AM) Princeton IL
WGEM-FM Quincy IL
WTAD(AM) Quincy IL
WRHL(AM) Rochelle IL
WROK(AM) Rockford IL
WHCO(AM) Sparta IL
WMAY(AM) Springfield IL
WTAX(AM) Springfield IL
*WNIQ(FM) Sterling IL
WSPL(AM) Streator IL
WTIM-FM Taylorville IL
*WILL(AM) Urbana IL
WKRS(AM) Waukegan IL
WHBU(AM) Anderson IN
WGCL(AM) Bloomington IN
WHON(AM) Centerville IN
*WBEW(FM) Chesterton IN
WCSI(AM) Columbus IN
WTRC(AM) Elkhart IN
*WVPE(FM) Elkhart IN
WGBF(AM) Evansville IN
*WNIN-FM Evansville IN
*WBOI(FM) Fort Wayne IN
WFCV(AM) Fort Wayne IN
WOWO(AM) Fort Wayne IN
WLTH(AM) Gary IN
WTRE(AM) Greensburg IN
WJOB(AM) Hammond IN
*WFYI-FM Indianapolis IN
WIBC(AM) Indianapolis IN
WXNT(AM) Indianapolis IN
WIOU(AM) Kokomo IN

WSAL(AM) Logansport IN
WBAT(AM) Marion IN
WKBV(AM) Richmond IN
WSBT(AM) South Bend IN
WAOV(AM) Vincennes IN
*WBAA(AM) West Lafayette IN
KKOY(AM) Chanute KS
KGGF(AM) Coffeyville KS
KIUL(AM) Garden City KS
KLOE(AM) Goodland KS
*KZAN(FM) Hays KS
*KZNA(FM) Hill City KS
KJCK(AM) Junction City KS
KLWN(AM) Lawrence KS
KSCB(AM) Liberal KS
KMAN(AM) Manhattan KS
KCCV(AM) Overland Park KS
KQMA-FM Phillipsburg KS
KINA(AM) Salina KS
KSAL(AM) Salina KS
KCVT(FM) Silver Lake KS
KMAJ(AM) Topeka KS
WIBW(AM) Topeka KS
KLEY(AM) Wellington KS
KNSS(AM) Wichita KS
KKLE(AM) Winfield KS
WAIA(AM) Beaver Dam KY
WKXO(AM) Berea KY
WKCT(AM) Bowling Green KY
WCTT(AM) Corbin KY
WKDP(AM) Corbin KY
WHIR(AM) Danville KY
WHOP(AM) Hopkinsville KY
WVLK(AM) Lexington KY
*WFPL(FM) Louisville KY
WGTK(AM) Louisville KY
WHAS(AM) Louisville KY
WNGO(AM) Mayfield KY
*WBFI(FM) McDaniels KY
WMST(AM) Mt. Sterling KY
WVKY(AM) Nicholasville KY
WOMI(AM) Owensboro KY
WKYX(AM) Paducah KY
WKYH(AM) Paintsville KY
WEKY(AM) Richmond KY
WRUS(AM) Russellville KY
WTLO(AM) Somerset KY
KPEL-FM Abbeville LA
WABL(AM) Amite LA
*WRKF(FM) Baton Rouge LA
WIKC(AM) Bogalusa LA
WASO(AM) Covington LA
KAOK(AM) Lake Charles LA
KMLB(AM) Monroe LA
KNOE(AM) Monroe LA
KNOC(AM) Natchitoches LA
WGSO(AM) New Orleans LA
WIST(AM) New Orleans LA
WRNO-FM New Orleans LA
WWL(AM) New Orleans LA
KEEL(AM) Shreveport LA
KNCB(AM) Vivian LA
WARL(AM) Attleboro MA
WBZ(AM) Boston MA
WILD(AM) Boston MA
WZAI(FM) Brewster MA
WXBR(AM) Brockton MA
WBNW(AM) Concord MA
WHNP(AM) East Longmeadow MA
WSAR(AM) Fall River MA
WEIM(AM) Fitchburg MA
WGAW(AM) Gardner MA
*WAMQ(FM) Great Barrington MA
*WMLN-FM Milton MA
*WNAN(FM) Nantucket MA
WBSM(AM) New Bedford MA
WDIS(AM) Norfolk MA
WHMP(AM) Northampton MA
WBEC(AM) Pittsfield MA
WESX(AM) Salem MA
WHYN(AM) Springfield MA
WPEP(AM) Taunton MA
WGTX(FM) Truro MA
*WBUR(AM) West Yarmouth MA
WXTK(FM) West Yarmouth MA
WNNZ(AM) Westfield MA

*WCAI(FM) Woods Hole MA
WTAG(AM) Worcester MA
WBAL(AM) Baltimore MD
*WEAA(FM) Baltimore MD
WOLB(AM) Baltimore MD
*WYPR(FM) Baltimore MD
WCBC(AM) Cumberland MD
WFMD(AM) Frederick MD
WFRB(AM) Frostburg MD
WHAG(AM) Halfway MD
WJSS(AM) Havre de Grace MD
WPTX(AM) Lexington Park MD
*WSDL(FM) Ocean City MD
WICO(AM) Salisbury MD
WQMR(FM) Snow Hill MD
WCME(FM) Boothbay Harbor ME
*WMEP(FM) Camden ME
*WMEF(FM) Fort Kent ME
WFAU(AM) Gardiner ME
WVOM(FM) Howland ME
WJCX(FM) Pittsfield ME
WGAN(AM) Portland ME
WLOB(AM) Portland ME
*WMEA(FM) Portland ME
*WMEM(FM) Presque Isle ME
WLOB-FM Rumford ME
WWNZ(AM) Veazie ME
*WMEW(FM) Waterville ME
WABJ(AM) Adrian MI
WAAM(AM) Ann Arbor MI
*WUOM(FM) Ann Arbor MI
WBCK(AM) Battle Creek MI
WBRN(AM) Big Rapids MI
WATT(AM) Cadillac MI
WSGW-FM Carrollton MI
WMKT(AM) Charlevoix MI
WDTK(AM) Detroit MI
*WKAR(AM) East Lansing MI
WCHT(AM) Escanaba MI
*WFUM-FM Flint MI
WOOD(AM) Grand Rapids MI
*WVGR(FM) Grand Rapids MI
WMPL(AM) Hancock MI
WBCH(AM) Hastings MI
WHTC(AM) Holland MI
WMIQ(AM) Iron Mountain MI
WIAN(AM) Ishpeming MI
WKHM(AM) Jackson MI
WKMI(AM) Kalamazoo MI
WKZO(AM) Kalamazoo MI
WKZO(AM) Kalamazoo MI
*WGVU(AM) Kentwood MI
WKLA(AM) Ludington MI
WMTE(AM) Manistee MI
WPIQ(FM) Manistique MI
WDMJ(AM) Marquette MI
WGVS(AM) Muskegon MI
WJML(AM) Petoskey MI
WPHM(AM) Port Huron MI
WSGW(AM) Saginaw MI
WSJM(AM) Saint Joseph MI
WMIC(AM) Sandusky MI
WKNW(AM) Sault Ste. Marie MI
WMMI(AM) Shepherd MI
WCHB(AM) Taylor MI
WTCM(AM) Traverse City MI
WMFN(AM) Zeeland MI
KATE(AM) Albert Lea MN
KXRA(AM) Alexandria MN
KAUS(AM) Austin MN
WWWI(AM) Baxter MN
KBEW(AM) Blue Earth MN
KRWC(AM) Buffalo MN
KDLM(AM) Detroit Lakes MN
KDAL(AM) Duluth MN
WEBC(AM) Duluth MN
KCNN(AM) East Grand Forks MN
KRBT(AM) Eveleth MN
KBRF(AM) Fergus Falls MN
KLTF(AM) Little Falls MN
WYRQ(FM) Little Falls MN
KMHL(AM) Marshall MN
KTLK-FM Minneapolis MN
WCCO(AM) Minneapolis MN
WWTC(AM) Minneapolis MN
KMRS(AM) Morris MN

WNMT(AM) Nashwauk MN
KCHK(AM) New Prague MN
KCUE(AM) Red Wing MN
KROC(AM) Rochester MN
KNSI(AM) Saint Cloud MN
WJON(AM) Saint Cloud MN
*KNGA(FM) Saint Peter MN
KRBI(AM) Saint Peter MN
KWLM(AM) Willmar MN
KDOM(AM) Windom MN
KDOM-FM Windom MN
KWNO(AM) Winona MN
KWOA(AM) Worthington MN
*KNSW(FM) Worthington-Marshall MN
*KGNA-FM Arnold MO
KSWM(AM) Aurora MO
KKOZ(AM) Ava MO
KKOZ-FM Ava MO
KBFL-FM Buffalo MO
KMRN(AM) Cameron MO
KZIM(AM) Cape Girardeau MO
KFRU(AM) Columbia MO
*KOPN(FM) Columbia MO
*KGNN-FM Cuba MO
KREI(AM) Farmington MO
KSSZ(FM) Fayette MO
KJFF(AM) Festus MO
KHMO(AM) Hannibal MO
KLIK(AM) Jefferson City MO
KWOS(AM) Jefferson City MO
KQYX(AM) Joplin MO
KZRG(AM) Joplin MO
*KKFI(FM) Kansas City MO
KLWT(AM) Lebanon MO
KMAL(AM) Malden MO
*KNEO(FM) Neosho MO
KRMS(AM) Osage Beach MO
KFMO(AM) Park Hills MO
*KCOZ(FM) Point Lookout MO
KWOC(AM) Poplar Bluff MO
KTTR(AM) Rolla MO
KTTR-FM Saint James MO
KFEQ(AM) Saint Joseph MO
KMOX(AM) Saint Louis MO
KTRS(AM) Saint Louis MO
KSMO(AM) Salem MO
KSIS(AM) Sedalia MO
KSIM(AM) Sikeston MO
KURM-FM South West City MO
KSGF(AM) Springfield MO
KWTO(AM) Springfield MO
KTUI(AM) Sullivan MO
KALM(AM) Thayer MO
*KGNV(FM) Washington MO
KOZQ(AM) Waynesville MO
KWPM(AM) West Plains MO
KUKU(AM) Willow Springs MO
WAMY(AM) Amory MS
*WMAH-FM Biloxi MS
WTNI(AM) Biloxi MS
*WMAU-FM Bude MS
WDSK(AM) Cleveland MS
WCJU(AM) Columbia MS
WJWF(AM) Columbus MS
WPBQ(AM) Flowood MS
*WMAO-FM Greenwood MS
WTCD(FM) Indianola MS
WKOZ(AM) Kosciusko MS
WMXI(FM) Laurel MS
WJZD(FM) Long Beach MS
WAKK(AM) McComb MS
WHNY(AM) McComb MS
*WMAW-FM Meridian MS
WMOX(AM) Meridian MS
WBUV(FM) Moss Point MS
WNAT(AM) Natchez MS
WJNT(AM) Pearl MS
WFMM(FM) Sumrall MS
WQBC(AM) Vicksburg MS
WVBG(AM) Vicksburg MS
WKBB(FM) West Point MS
KGVW(AM) Belgrade MT
KBLG(AM) Billings MT
KMMS(AM) Bozeman MT
*KGVA(FM) Fort Belknap Agency MT
KQDI(AM) Great Falls MT

KLYQ(AM) Hamilton MT
KBLL(AM) Helena MT
KCAP(AM) Helena MT
KGEZ(AM) Kalispell MT
KOFI(AM) Kalispell MT
KGVO(AM) Missoula MT
KJJR(AM) Whitefish MT
WQNX(AM) Aberdeen NC
WSPC(AM) Albemarle NC
WWNC(AM) Asheville NC
WTKF(FM) Atlantic NC
*WBJD(FM) Atlantic Beach NC
WXIT(AM) Blowing Rock NC
WATA(AM) Boone NC
*WBUX(FM) Buxton NC
WCHL(AM) Chapel Hill NC
*WUNC(FM) Chapel Hill NC
WBT(AM) Charlotte NC
*WFAE(FM) Charlotte NC
*WNCU(FM) Durham NC
WGAI(AM) Elizabeth City NC
WFNC(AM) Fayetteville NC
WIDU(AM) Fayetteville NC
WGBR(AM) Goldsboro NC
WSML(AM) Graham NC
*WFHE(FM) Hickory NC
WHKY(AM) Hickory NC
WMFR(AM) High Point NC
WJNC(AM) Jacksonville NC
WAAV(AM) Leland NC
WJRI(AM) Lenoir NC
WLXN(AM) Lexington NC
*WUND-FM Manteo NC
*WURI(FM) Manteo NC
WKRK(AM) Murphy NC
WAUG(AM) New Hope NC
WDOX(AM) Raleigh NC
WPTF(AM) Raleigh NC
*WRQM(FM) Rocky Mount NC
WCAB(AM) Rutherfordton NC
WSTP(AM) Salisbury NC
WLTT(FM) Shallotte NC
WNCA(AM) Siler City NC
WEEB(AM) Southern Pines NC
WTXY(AM) Whiteville NC
WSJS(AM) Winston-Salem NC
KFYR(AM) Bismarck ND
KLXX(AM) Bismarck-Mandan ND
KFGO(AM) Fargo ND
WDAY(AM) Fargo ND
KNOX(AM) Grand Forks ND
KQDJ(AM) Jamestown ND
KNDK(AM) Langdon ND
KQLX(AM) Lisbon ND
KHRT(AM) Minot ND
KEYZ(AM) Williston ND
KCOW(AM) Alliance NE
KWBE(AM) Beatrice NE
KJSK(AM) Columbus NE
KHUB(AM) Fremont NE
KRGI(AM) Grand Island NE
KGFW(AM) Kearney NE
KIMB(AM) Kimball NE
*KLNE-FM Lexington NE
KFOR(AM) Lincoln NE
KLIN(AM) Lincoln NE
WJAG(AM) Norfolk NE
KODY(AM) North Platte NE
KFAB(AM) Omaha NE
KKAR(AM) Omaha NE
KOTK(AM) Omaha NE
KNEB(AM) Scottsbluff NE
*WEVO(FM) Concord NH
WKXL(AM) Concord NH
WTSN(AM) Dover NH
WGIP(AM) Exeter NH
*WEVC(FM) Gorham NH
*WEVH(FM) Hanover NH
WTSL(AM) Hanover NH
WTPL(FM) Hillsboro NH
WEVJ(FM) Jackson NH
*WEVN(FM) Keene NH
WUVR(AM) Lebanon NH
WGIR(AM) Manchester NH
*WEVS(FM) Nashua NH
WSMN(AM) Nashua NH

WNTK-FM New London NH
WGIN(AM) Rochester NH
WCCM(AM) Salem NH
*WNJN-FM Atlantic City NJ
*WNJS-FM Berlin NJ
*WNJB(FM) Bridgeton NJ
*WNJZ(FM) Cape May Court House NJ
WRNJ(AM) Hackettstown NJ
WGYM(AM) Hammonton NJ
*WNJM(FM) Manahawkin NJ
WCTC(AM) New Brunswick NJ
WOND(AM) Pleasantville NJ
WHWH(AM) Princeton NJ
*WNJP(FM) Sussex NJ
*WNJT-FM Trenton NJ
KINN(AM) Alamogordo NM
KKOB(AM) Albuquerque NM
*KUNM(FM) Albuquerque NM
KENN(AM) Farmington NM
KYKK(AM) Hobbs NM
KOBE(AM) Las Cruces NM
KNMX(AM) Las Vegas NM
KTBL(AM) Los Ranchos de Albuquerque NM
*KENW-FM Portales NM
KSEL(AM) Portales NM
KBIM(AM) Roswell NM
KCKN(AM) Roswell NM
*KSFR(FM) Santa Fe NM
*KRNM(FM) Chalan Kanoa-Saipan NP
KCNM(AM) Garapan-Saipan NP
KKFT(FM) Gardnerville-Minden NV
KDWN(AM) Las Vegas NV
KXNT(AM) North Las Vegas NV
KNUU(AM) Paradise NV
KKOH(AM) Reno NV
KBZZ(AM) Sparks NV
KWNA(AM) Winnemucca NV
*WAMC(AM) Albany NY
WROW(AM) Albany NY
WCSS(AM) Amsterdam NY
WBTA(AM) Batavia NY
WINR(AM) Binghamton NY
WNBF(AM) Binghamton NY
WBEN(AM) Buffalo NY
*WNED(AM) Buffalo NY
*WCAN(FM) Canajoharie NY
WCGR(AM) Canandaigua NY
WENI(AM) Corning NY
WWLE(AM) Cornwall NY
WKRT(AM) Cortland NY
WFLR(AM) Dundee NY
WFLR-FM Dundee NY
WDOE(AM) Dunkirk NY
WENY(AM) Elmira NY
*WEOS(FM) Geneva NY
WGVA(AM) Geneva NY
WWSC(AM) Glens Falls NY
WLEA(AM) Hornell NY
WWLZ(AM) Horseheads NY
WHCU(AM) Ithaca NY
*WJFF(FM) Jeffersonville NY
*WAMK(FM) Kingston NY
WGHQ(AM) Kingston NY
WLVL(AM) Lockport NY
WYBG(AM) Massena NY
WALL(AM) Middletown NY
WVOX(AM) New Rochelle NY
WADO(AM) New York NY
*WBAI(FM) New York NY
*WNYC(AM) New York NY
WOR(AM) New York NY
WACK(AM) Newark NY
*WSUF(FM) Noyack NY
*WRVO(FM) Oswego NY
WHAM(AM) Rochester NY
WROC(AM) Rochester NY
*WXXI(AM) Rochester NY
WGY(AM) Schenectady NY
*WRLI-FM Southampton NY
*WUSB(FM) Stony Brook NY
*WRVD(FM) Syracuse NY
WSYR(AM) Syracuse NY
WIBX(AM) Utica NY
*WRUN(AM) Utica NY

*WRVN(FM) Utica NY
*WRVJ(FM) Watertown NY
WHLO(AM) Akron OH
WATH(AM) Athens OH
*WOUB(AM) Athens OH
WBLL(AM) Bellefontaine OH
*WOUC-FM Cambridge OH
WCER(AM) Canton OH
WBEX(AM) Chillicothe OH
*WOUH-FM Chillicothe OH
WKRC(AM) Cincinnati OH
WLW(AM) Cincinnati OH
*WVXU(FM) Cincinnati OH
*WCSB(FM) Cleveland OH
WTAM(AM) Cleveland OH
WERE(AM) Cleveland Heights OH
*WOSU(AM) Columbus OH
WTVN(AM) Columbus OH
*WOSE(FM) Coshocton OH
WHIO(AM) Dayton OH
WING(AM) Dayton OH
WONW(AM) Defiance OH
WEOL(AM) Elyria OH
WFIN(AM) Findlay OH
WIRO(AM) Ironton OH
*WOUL-FM Ironton OH
WIMA(AM) Lima OH
WMAN(AM) Mansfield OH
*WMRT(FM) Marietta OH
WMRN(AM) Marion OH
WMVO(AM) Mount Vernon OH
WCLT(AM) Newark OH
*WMUB(FM) Oxford OH
WHIO-FM Piqua OH
WULM(AM) Springfield OH
WCWA(AM) Toledo OH
WSPD(AM) Toledo OH
WBTC(AM) Uhrichsville OH
WANR(AM) Warren OH
*WYSO(FM) Yellow Springs OH
WKBN(AM) Youngstown OH
WHIZ(AM) Zanesville OH
*WOUZ(FM) Zanesville OH
KWHW(AM) Altus OK
KWON(AM) Bartlesville OK
KCLI(AM) Clinton OK
KUSH(AM) Cushing OK
KGWA(AM) Enid OK
KTJS(AM) Hobart OK
*KGOU(FM) Norman OK
KOKC(AM) Oklahoma City OK
KQCV(AM) Oklahoma City OK
KTOK(AM) Oklahoma City OK
*KROU(FM) Spencer OK
KSPI(AM) Stillwater OK
KFAQ(AM) Tulsa OK
KRMG(AM) Tulsa OK
KAST(AM) Astoria OR
KBKR(AM) Baker City OR
*KOBK(FM) Baker City OR
KWRO(AM) Coquille OR
KLOO(AM) Corvallis OR
*KOAC(AM) Corvallis OR
KNND(AM) Cottage Grove OR
KWVR(AM) Enterprise OR
*KLCC(FM) Eugene OR
KPNW(AM) Eugene OR
KUGN(AM) Eugene OR
KAJO(AM) Grants Pass OR
KUIK(AM) Hillsboro OR
KYKN(AM) Keizer OR
KAGO(AM) Klamath Falls OR
KFLS(AM) Klamath Falls OR
KLBM(AM) La Grande OR
*KDOV(FM) Medford OR
KMED(AM) Medford OR
KNPT(AM) Newport OR
KBBR(AM) North Bend OR
*KAPL(AM) Phoenix OR
KBNP(AM) Portland OR
KPOJ(AM) Portland OR
KXL(AM) Portland OR
KDUN(AM) Reedsport OR
*KLFR(FM) Reedsport OR
*KMPQ(FM) Roseburg OR
KQEN(AM) Roseburg OR
KACI(AM) The Dalles OR
KMBD(AM) Tillamook OR
KPAM(AM) Troutdale OR
KLWJ(AM) Umatilla OR
WAEB(AM) Allentown PA
WRTA(AM) Altoona PA
WILK-FM Avoca PA
WBFD(AM) Bedford PA
WBLF(AM) Bellefonte PA
WGPA(AM) Bethlehem PA
*WBUQ(FM) Bloomsburg PA
WISR(AM) Butler PA
WHYL(AM) Carlisle PA
WCHA(AM) Chambersburg PA
WWCH(AM) Clarion PA
WCOJ(AM) Coatesville PA
WFRM(AM) Coudersport PA
WCED(AM) DuBois PA
WRDD(AM) Ebensburg PA
WJET(AM) Erie PA
WPSE(AM) Erie PA
WZSK(AM) Everett PA
WHP(AM) Harrisburg PA
*WITF-FM Harrisburg PA
WRKK(AM) Hughesville PA
WNPV(AM) Lansdale PA
WLBR(AM) Lebanon PA
WIEZ(AM) Lewistown PA
WWGE(AM) Loretto PA
WJSM-FM Martinsburg PA
WPTT(AM) McKeesport PA
WMGW(AM) Meadville PA
WKST(AM) New Castle PA
WOYL(AM) Oil City PA
WNTP(AM) Philadelphia PA
WPHB(AM) Philipsburg PA
KDKA(AM) Pittsburgh PA
WDVE(FM) Pittsburgh PA
WPGB(FM) Pittsburgh PA
WPAZ(AM) Pottstown PA
*WYBF(FM) Radnor Township PA
WEEU(AM) Reading PA
WBZU(AM) Scranton PA
WPIC(AM) Sharon PA
WRSC(AM) State College PA
WKOK(AM) Sunbury PA
WTIV(AM) Titusville PA
WEMR(AM) Tunkhannock PA
*WNJR(FM) Washington PA
WKZN(AM) West Hazleton PA
WILK(AM) Wilkes-Barre PA
WRAK(AM) Williamsport PA
*WRLC(FM) Williamsport PA
WWPA(AM) Williamsport PA
WSBA(AM) York PA
WCMN(AM) Arecibo PR
WYAC(AM) Cabo Rojo PR
WNEL(AM) Caguas PR
WLEY(AM) Cayey PR
WOIZ(AM) Guayanilla PR
WMSW(AM) Hatillo PR
WBSG(AM) Lajas PR
WMNT(AM) Manati PR
WKJB(AM) Mayaguez PR
WDEP(AM) Ponce PR
WISO(AM) Ponce PR
WPAB(AM) Ponce PR
WPRP(AM) Ponce PR
WUKQ(AM) Ponce PR
WAPA(AM) San Juan PR
*WIPR(AM) San Juan PR
WKAQ(AM) San Juan PR
WOSO(AM) San Juan PR
WSKN(AM) San Juan PR
WUNO(AM) San Juan PR
WUPR(AM) Utuado PR
WENA(AM) Yauco PR
WKFE(AM) Yauco PR
WRNI-FM Narragansett Pier RI
WADK(AM) Newport RI
WHJJ(AM) Providence RI
WPRO(AM) Providence RI
WRNI(AM) Providence RI
WNRI(AM) Woonsocket RI
WAIM(AM) Anderson SC
WTMA(AM) Charleston SC
WBT-FM Chester SC
WVOC(AM) Columbia SC
WFIS(AM) Fountain Inn SC
WGTN(AM) Georgetown SC
WYRD(AM) Greenville SC
WCRS(AM) Greenwood SC
WRIX-FM Honea Path SC
WRNN(AM) Myrtle Beach SC
WRHI(AM) Rock Hill SC
WOLI(AM) Spartanburg SC
WORD(AM) Spartanburg SC
WSPG(AM) Spartanburg SC
WDXY(AM) Sumter SC
*WRJA-FM Sumter SC
KJAM(AM) Madison SD
KMSD(AM) Milbank SD
KORN(AM) Mitchell SD
KCCR(AM) Pierre SD
KOTA(AM) Rapid City SD
KELO(AM) Sioux Falls SD
KSOO(AM) Sioux Falls SD
WNAX(AM) Yankton SD
WBCR(AM) Alcoa TN
WBIN(AM) Benton TN
WTBG(FM) Brownsville TN
WDEF(AM) Chattanooga TN
WGOW(AM) Chattanooga TN
WBAC(AM) Cleveland TN
WCRV(AM) Collierville TN
WPTN(AM) Cookeville TN
*WMKW(FM) Crossville TN
WDNT(AM) Dayton TN
WDKN(AM) Dickson TN
WCPH(AM) Etowah TN
WAKM(AM) Franklin TN
WHEW(AM) Franklin TN
WNWS-FM Jackson TN
WTJS(AM) Jackson TN
*WETS(FM) Johnson City TN
WETR(AM) Knoxville TN
WNML(AM) Knoxville TN
WCOR(AM) Lebanon TN
WLOD(AM) Loudon TN
WPLN(AM) Madison TN
WRKQ(AM) Madisonville TN
WWTN(FM) Manchester TN
WCMT(AM) Martin TN
WAKI(AM) McMinnville TN
KWAM(AM) Memphis TN
WREC(AM) Memphis TN
WMTN(AM) Morristown TN
WGNS(AM) Murfreesboro TN
WLAC(AM) Nashville TN
WNOX(FM) Oak Ridge TN
*WFCM(AM) Smyrna TN
WGOW-FM Soddy-Daisy TN
WTZX(AM) Sparta TN
WXQK(AM) Spring City TN
KWKC(AM) Abilene TX
KGNC(AM) Amarillo TX
KIXZ(AM) Amarillo TX
KRGN(FM) Amarillo TX
KLBJ(AM) Austin TX
KLVI(AM) Beaumont TX
KRCM(AM) Beaumont TX
KWHI(AM) Brenham TX
KVNS(AM) Brownsville TX
KXYL-FM Brownwood TX
*KEOS(FM) College Station TX
WTAW(AM) College Station TX
KEYS(AM) Corpus Christi TX
KKTX(AM) Corpus Christi TX
KHER(FM) Crystal City TX
*KERA(FM) Dallas TX
KLIF(AM) Dallas TX
KRLD(AM) Dallas TX
KURV(AM) Edinburg TX
KULP(AM) El Campo TX
KROD(AM) El Paso TX
KTSM(AM) El Paso TX
KFLC(AM) Fort Worth TX
WBAP(AM) Fort Worth TX
KGAF(AM) Gainesville TX
KLAT(AM) Houston TX
KNTH(AM) Houston TX
KPRC(AM) Houston TX
KMNY(AM) Hurst TX
KFRO(AM) Longview TX
KFYO(AM) Lubbock TX
KJDL(AM) Lubbock TX
KJTV(AM) Lubbock TX
KRFE(AM) Lubbock TX
KCRS(AM) Midland TX
KSFA(AM) Nacogdoches TX
KWBC(AM) Navasota TX
KOKE(AM) Pflugerville TX
KOLE(AM) Port Arthur TX
KKSA(AM) San Angelo TX
KCOR(AM) San Antonio TX
KTSA(AM) San Antonio TX
WOAI(AM) San Antonio TX
*KTSW(FM) San Marcos TX
KWED(AM) Seguin TX
KBKH(FM) Shamrock TX
KTEM(AM) Temple TX
KLUP(AM) Terrell Hills TX
KTFS(AM) Texarkana TX
KSEV(AM) Tomball TX
KTBB(AM) Tyler TX
KBCT-FM Waco TX
KWTX(AM) Waco TX
KWFS(AM) Wichita Falls TX
KSUB(AM) Cedar City UT
*KUSU-FM Logan UT
KVNU(AM) Logan UT
KSL-FM Midvale UT
KOAL(AM) Price UT
*KBYU-FM Provo UT
KSVC(AM) Richfield UT
KDXU(AM) Saint George UT
KZNU(AM) Saint George UT
*KCPW-FM Salt Lake City UT
KNRS(AM) Salt Lake City UT
KSL(AM) Salt Lake City UT
KCPW(AM) Tooele UT
KVEL(AM) Vernal UT
WFHG-FM Abingdon VA
WBVA(AM) Bayside VA
WFNR(AM) Blacksburg VA
WCHV(AM) Charlottesville VA
WINA(AM) Charlottesville VA
WFNR-FM Christiansburg VA
WFLO(AM) Farmville VA
WSVA(AM) Harrisonburg VA
*WCNV(FM) Heathsville VA
WREL(AM) Lexington VA
WBRG(AM) Lynchburg VA
WLNI(FM) Lynchburg VA
*WHRV(FM) Norfolk VA
WNIS(AM) Norfolk VA
WROU(AM) Petersburg VA
*WCVE(FM) Richmond VA
WRVA(AM) Richmond VA
WRVA(AM) Richmond VA
WFIR(AM) Roanoke VA
WTZE(AM) Tazewell VA
WTWP-FM Warrenton VA
WKCI(AM) Waynesboro VA
WINC(AM) Winchester VA
WVGN(FM) Charlotte Amalie VI
WVWI(AM) Charlotte Amalie VI
WSNO(AM) Barre VT
WBTN(AM) Bennington VT
WKVT(AM) Brattleboro VT
WVMT(AM) Burlington VT
WTSJ(AM) Randolph VT
WSYB(AM) Rutland VT
WDEV-FM Warren VT
KBKW(AM) Aberdeen WA
KXRO(AM) Aberdeen WA
KGMI(AM) Bellingham WA
*KUGS(FM) Bellingham WA
KARI(AM) Blaine WA
KELA(AM) Centralia-Chehalis WA
KOZI(AM) Chelan WA
*KNWR(FM) Ellensburg WA
KXLE(AM) Ellensburg WA
KULE(AM) Ephrata WA
KONA(AM) Kennewick WA
KTCR(AM) Kennewick WA
KBSN(AM) Moses Lake WA
*KMWS(FM) Mount Vernon WA
*KSVR(FM) Mount Vernon WA
KONP(AM) Port Angeles WA
KZXR(AM) Prosser WA
KQQQ(AM) Pullman WA
*KFAE-FM Richland WA
KIRO(AM) Seattle WA
KKNW(AM) Seattle WA
KKOL(AM) Seattle WA
KGA(AM) Spokane WA
KQNT(AM) Spokane WA
KXLY(AM) Spokane WA
*KXOT(FM) Tacoma WA
KGDC(AM) Walla Walla WA
KUJ(AM) Walla Walla WA
*KWWS(FM) Walla Walla WA
KPQ(AM) Wenatchee WA
*KDNA(FM) Yakima WA
KIT(AM) Yakima WA
WXCE(AM) Amery WI
*WLBL(AM) Auburndale WI
WBEV(AM) Beaver Dam WI
WISS(AM) Berlin WI
*WHSA(FM) Brule WI
*WHAD(FM) Delafield WI
WAYY(AM) Eau Claire WI
KFIZ(AM) Fond du Lac WI
WFAW(AM) Fort Atkinson WI
WDUZ(AM) Green Bay WI
WTAQ(AM) Green Bay WI
WOGO(AM) Hallie WI
*WHHI(FM) Highland WI
WCLO(AM) Janesville WI
WHBY(AM) Kimberly WI
WIZM(AM) La Crosse WI
WLDY(AM) Ladysmith WI
WIBA(AM) Madison WI
WTDY(AM) Madison WI
WDLB(AM) Marshfield WI
WMEQ(AM) Menomonie WI
*WVSS(FM) Menomonie WI
WTMJ(AM) Milwaukee WI
WOSH(AM) Oshkosh WI
*WHBM-FM Park Falls WI
WPDR(AM) Portage WI
WRJN(AM) Racine WI
WJMC(AM) Rice Lake WI
WRPN(AM) Ripon WI
*WRPN-FM Ripon WI
WHBL(AM) Sheboygan WI
WSPT(AM) Stevens Point WI
*KUWS(FM) Superior WI
WFDL(AM) Waupun WI
WSAU(AM) Wausau WI
WFHR(AM) Wisconsin Rapids WI
WWNR(AM) Beckley WV
WHIS(AM) Bluefield WV
WCHS(AM) Charleston WV
WTCS(AM) Fairmont WV
WMTD(AM) Hinton WV
WVHU(AM) Huntington WV
WEPM(AM) Martinsburg WV
WRNR(AM) Martinsburg WV
WAJR(AM) Morgantown WV
WVLY(AM) Moundsville WV
WVNT(AM) Parkersburg WV
WRRL(AM) Rainelle WV
WAJR-FM Salem WV
WWVA(AM) Wheeling WV
*KUWA(FM) Afton WY
KBBS(AM) Buffalo WY
KFBC(AM) Cheyenne WY
KLEN(FM) Cheyenne WY
KODI(AM) Cody WY
KUGR(AM) Green River WY
KOWB(AM) Laramie WY
KGAB(AM) Orchard Valley WY
KROE(AM) Sheridan WY

Nostalgia

WGMZ(FM) Glencoe AL
KVRC(AM) Arkadelphia AR
KHBM(AM) Monticello AR
KFTT(FM) Bagdad AZ
KFLG(AM) Bullhead City AZ
*KCEA(FM) Atherton CA

KXBX(AM) Lakeport CA
KZPO(FM) Lindsay CA
KIDD(AM) Monterey CA
KBLF(AM) Red Bluff CA
KXDZ(FM) Templeton CA
KEZW(AM) Aurora CO
WNLC(FM) East Lyme CT
WFCT(FM) Apalachicola FL
WLQH(AM) Chiefland FL
WROD(AM) Daytona Beach FL
WTKE(AM) Fort Walton Beach FL
WDIZ(AM) Panama City FL
WMOG(AM) Brunswick GA
WCGO(AM) Chicago Heights IL
WAIK(AM) Galesburg IL
WMIX(AM) Mount Vernon IL
WCVL(AM) Crawfordsville IN
WILO(AM) Frankfort IN
WMRI(AM) Marion IN
KTOP(AM) Topeka KS
WIVY(FM) Morehead KY
WVJS(AM) Owensboro KY
WIZZ(AM) Greenfield MA
WCAP(AM) Lowell MA
WTBO(AM) Cumberland MD
WJZN(AM) Augusta ME
WABI(AM) Bangor ME
WDEA(AM) Ellsworth ME
WBAE(AM) Portland ME
*WYAR(FM) Yarmouth ME
WCBY(AM) Cheboygan MI
WXLA(AM) Dimondale MI
WDBC(AM) Escanaba MI
WFNT(AM) Flint MI
WGRY(AM) Grayling MI
WHIT(AM) Hudsonville MI
WLXT(FM) Petoskey MI
WMJH(AM) Rockford MI
KKIN(AM) Aitkin MN
KBMO(AM) Benson MN
KDKK-FM Park Rapids MN
KBFL-FM Buffalo MO
KBFL(AM) Springfield MO
WJBI(AM) Batesville MS
WIIN(AM) Ridgeland MS
*KFBN(FM) Fargo ND
WMYF(AM) Portsmouth NH
WDNY(AM) Dansville NY
WELG(AM) Ellenville NY
WTLA(AM) North Syracuse NY
*WRVO(FM) Oswego NY
WSGO(AM) Oswego NY
WBZA(FM) Rochester NY
*WRVD(FM) Syracuse NY
*WRVN(FM) Utica NY
*WRVJ(FM) Watertown NY
WCDK(FM) Cadiz OH
WCHI(AM) Chillicothe OH
WJEH(AM) Gallipolis OH
*WMKV(FM) Reading OH
WSOM(AM) Salem OH
WNIO(AM) Youngstown OH
KTIL-FM Tillamook OR
WLNP(FM) Carbondale PA
WNAK(AM) Nanticoke PA
WJAS(AM) Pittsburgh PA
WLKW(AM) West Warwick RI
WCAM(AM) Camden SC
WQMV(AM) Waverly TN
KSLI(AM) Abilene TX
KBST(AM) Big Spring TX
KRHC(AM) Burnet TX
WSNZ(FM) Appomattox VA
WCVA(AM) Culpeper VA
WZEZ(FM) Goochland VA
WRAD(AM) Radford VA
KBRD(AM) Lacey WA
KAST-FM Long Beach WA
KOMW(AM) Omak WA
KTRW(AM) Opportunity WA
WCCN(AM) Neillsville WI
WETZ(AM) New Martinsville WV
WETZ-FM New Martinsville WV

Oldies

KEAG(FM) Anchorage AK
KHAR(AM) Anchorage AK
*KMJG(FM) Homer AK
KBBO-FM Houston AK
*KWJG(FM) Kasilof AK
KFMJ(FM) Ketchikan AK
KXBA(FM) Nikiski AK
KIFW(AM) Sitka AK
WAFN-FM Arab AL
WATV(AM) Birmingham AL
WEBJ(AM) Brewton AL
WTBF-FM Brundidge AL
WACQ(AM) Carrville AL
WAVH(FM) Daphne AL
WZNJ(FM) Demopolis AL
WEEL(AM) Dothan AL
WGAD(AM) Gadsden AL
WDGM(FM) Greensboro AL
WGYV(AM) Greenville AL
WBPT(FM) Homewood AL
WHPH(FM) Jemison AL
WLAY-FM Littleville AL
WMFC-FM Monroeville AL
WOPP(AM) Opp AL
WVOK(AM) Oxford AL
WPID(AM) Piedmont AL
WMRK(AM) Selma AL
WJDB(AM) Thomasville AL
WRLD-FM Valley AL
WKXM-FM Winfield AL
KEWI(AM) Benton AR
KQBK(FM) Booneville AR
KFCM(FM) Cherokee Village AR
KGFL(AM) Clinton AR
KTRQ(FM) Colt AR
KBKG(FM) Corning AR
KFLI(FM) Des Arc AR
KFPW(AM) Fort Smith AR
KLSZ-FM Fort Smith AR
KBYB(FM) Hope AR
KKIK(FM) Horseshoe Bend AR
KOLL-FM Lonoke AR
KQOR(FM) Mena AR
KNAS(FM) Nashville AR
KQDD(FM) Osceola AR
KAMO-FM Rogers AR
KFFK(AM) Rogers AR
KWAK-FM Stuttgart AR
KRLW(AM) Walnut Ridge AR
KCTT-FM Yellville AR
KXFF(FM) Colorado City AZ
KCEE(AM) Cortaro AZ
KKLD(FM) Cottonwood AZ
KRDE(FM) Globe AZ
KGVY(AM) Green Valley AZ
KRCY-FM Lake Havasu City AZ
KIKO(AM) Miami AZ
KGMG(FM) Oracle AZ
KOOL-FM Phoenix AZ
KOY(AM) Phoenix AZ
KPHX(AM) Phoenix AZ
KAZG(AM) Scottsdale AZ
KZKE(FM) Seligman AZ
KSNX(FM) Show Low AZ
KVSL(AM) Show Low AZ
KKYZ(FM) Sierra Vista AZ
KTKT(AM) Tucson AZ
KWFM(AM) Tucson AZ
KJOK(AM) Yuma AZ
KIQO(FM) Atascadero CA
KBRE(FM) Atwater CA
KHYL(FM) Auburn CA
KKBB(FM) Bakersfield CA
KBOV(AM) Bishop CA
KODS(FM) Carnelian Bay CA
KZRO(FM) Dunsmuir CA
KXO(AM) El Centro CA
KKHB(FM) Eureka CA
*KYAF(FM) Firebaugh CA
KMGV(FM) Fresno CA
KAZA(AM) Gilroy CA
KRAK(AM) Hesperia CA
KZLA(FM) Huron CA
KOKO-FM Kerman CA
KOMY(AM) La Selva Beach CA
KVLI-FM Lake Isabella CA

KBOX(FM) Lompoc CA
KRTO(FM) Lompoc CA
KRTH(FM) Los Angeles CA
KCEZ(FM) Los Molinos CA
KABX-FM Merced CA
KNAH(FM) Merced CA
KNKK(FM) Needles CA
KHOP(FM) Oakdale CA
KOCN(FM) Pacific Grove CA
KDES-FM Palm Springs CA
KNWQ(AM) Palm Springs CA
KCCL(FM) Placerville CA
KLOA(AM) Ridgecrest CA
KOLA(FM) San Bernardino CA
KURS(AM) San Diego CA
KISQ(FM) San Francisco CA
KRDG(FM) Shingletown CA
KQOD(FM) Stockton CA
KSTN(AM) Stockton CA
KFSO-FM Visalia CA
KRCV(FM) West Covina CA
KRSX-FM Yermo CA
KUBA(AM) Yuba City CA
KCMN(AM) Colorado Springs CO
KXKL-FM Denver CO
KBKL(FM) Grand Junction CO
KEJJ(FM) Gunnison CO
KZMV(FM) Kremmling CO
KBLJ(AM) La Junta CO
KLIM(AM) Limon CO
KJEB(FM) New Castle CO
KDZA-FM Pueblo CO
KVRH(AM) Salida CO
KSTC(AM) Sterling CO
WQUN(AM) Hamden CT
WDRC-FM Hartford CT
WKNL(FM) New London CT
WREF(AM) Ridgefield CT
WQQQ(FM) Sharon CT
WATR(AM) Waterbury CT
WBIG-FM Washington DC
WLBW(FM) Fenwick Island DE
WNCL(FM) Milford DE
*WMHS(FM) Pike Creek DE
WFHT(AM) Avon Park FL
WWBF(AM) Bartow FL
WRGO(FM) Cedar Key FL
WPFL(FM) Century FL
WMMV(AM) Cocoa FL
WSRZ-FM Coral Cove FL
WZEP(AM) De Funiak Springs FL
WOLZ(FM) Fort Myers FL
WOLL(FM) Hobe Sound FL
WAFZ(AM) Immokalee FL
WJGL(FM) Jacksonville FL
WMXQ(FM) Jacksonville FL
WEGT(FM) Lafayette FL
WMGG(AM) Largo FL
WQHL(AM) Live Oak FL
WWWK(FM) Marathon FL
WINT(AM) Melbourne FL
WEBZ(FM) Mexico Beach FL
WECM(AM) Milton FL
WSBB(AM) New Smyrna Beach FL
WMFQ(FM) Ocala FL
WEUS(AM) Orlovista FL
WYCL(FM) Pensacola FL
WMXJ(FM) Pompano Beach FL
WPOI(FM) Saint Petersburg FL
WWAV-FM Santa Rosa Beach FL
WJCM(AM) Sebring FL
WJQB(FM) Spring Hill FL
WRBQ-FM Tampa FL
WQOL(FM) Vero Beach FL
WTTB(AM) Vero Beach FL
WISK(AM) Americus GA
WMGR(AM) Bainbridge GA
WEBS(AM) Calhoun GA
WBHF(AM) Cartersville GA
WCLA(AM) Claxton GA
WGFS(AM) Covington GA
WCUG(AM) Cuthbert GA
WMRZ(FM) Dawson GA
WKVQ(AM) Eatonton GA
WDDK(FM) Greensboro GA
WKEU(AM) Griffin GA

WNGM(AM) Hiawassee GA
WTRP(AM) La Grange GA
WAYS(AM) Macon GA
WYTH(AM) Madison GA
WYIS(AM) McRae GA
WYSC(FM) McRae GA
WGCO(FM) Midway GA
WMNZ(AM) Montezuma GA
WMTM-FM Moultrie GA
WQXZ(FM) Pinehurst GA
WSFB(AM) Quitman GA
WUUS(AM) Rossville GA
WJFL(FM) Tennille GA
WPAX(AM) Thomasville GA
WATG(FM) Trion GA
WVOP(AM) Vidalia GA
WWUF(FM) Waycross GA
WRLA(AM) West Point GA
KGMZ-FM Aiea HI
KQMQ-FM Honolulu HI
KTOH(FM) Kalaheo HI
KBGX(FM) Keaau HI
KONI(FM) Lanai City HI
KLBA-FM Albia IA
KASI(AM) Ames IA
KCIM(AM) Carroll IA
KCHE(AM) Cherokee IA
KCLN(AM) Clinton IA
KIOA(FM) Des Moines IA
*KJMC(FM) Des Moines IA
KGRR(FM) Epworth IA
KIOW(FM) Forest City IA
KVFD(AM) Fort Dodge IA
KLMJ(FM) Hampton IA
KNOD(FM) Harlan IA
KKMA(FM) Le Mars IA
KDAO(AM) Marshalltown IA
KRIB(AM) Mason City IA
KWPC(AM) Muscatine IA
*KIGC(FM) Oskaloosa IA
KBIZ(AM) Ottumwa IA
KIYX(FM) Sageville IA
KTLB(FM) Twin Lakes IA
KRQN(FM) Vinton IA
KCII(AM) Washington IA
KCII-FM Washington IA
KOKZ(FM) Waterloo IA
KHPP(AM) Waukon IA
KAWO(FM) Boise ID
KGEM(AM) Boise ID
KBAR(AM) Burley ID
KCID(AM) Caldwell ID
KVNI(AM) Coeur d'Alene ID
KAYN(FM) Gooding ID
KRPL(AM) Moscow ID
KACH(AM) Preston ID
KGTM(FM) Rexburg ID
KLIX-FM Twin Falls ID
KTFI(AM) Twin Falls ID
WQRL(FM) Benton IL
WBYS(AM) Canton IL
WROY(AM) Carmi IL
WILY(AM) Centralia IL
WJMK(FM) Chicago IL
WZZN(FM) Chicago IL
WPMJ(FM) Chillicothe IL
WMQZ(FM) Colchester IL
WIXN(AM) Dixon IL
WWYW(FM) Dundee IL
KUUL(FM) East Moline IL
*WRSE(FM) Elmhurst IL
WDBQ-FM Galena IL
WMCW(AM) Harvard IL
WJBM(AM) Jerseyville IL
WEAI(FM) Lynnville IL
WDKR(FM) Maroa IL
WPWQ(FM) Mount Sterling IL
WHQQ(FM) Neoga IL
WSEI(FM) Olney IL
WSEY(FM) Oregon IL
WTRX-FM Pontiac IL
WZOE-FM Princeton IL
WTRH(FM) Ramsey IL
WKXQ(FM) Rushville IL
WEJT(FM) Shelbyville IL
WQQL(FM) Springfield IL

WYYS(FM) Streator IL
WCFF(FM) Urbana IL
WYKT(FM) Wilmington IL
WASK-FM Battle Ground IN
WQRK(FM) Bedford IN
WUZR(FM) Bicknell IN
WKLU(FM) Brownsburg IN
WLME(FM) Cannelton IN
WIFE(AM) Connersville IN
WOCC(AM) Corydon IN
WIBN(FM) Earl Park IN
WURK(FM) Elwood IN
WJLT(FM) Evansville IN
WLDE(FM) Fort Wayne IN
WIAU(FM) Franklin IN
WTLC-FM Greenwood IN
WBZQ(AM) Huntington IN
*WJPR(FM) Jasper IN
WAWK(AM) Kendallville IN
*WJEF(FM) Lafayette IN
WQTY(FM) Linton IN
WXGO(AM) Madison IN
WBAT(AM) Marion IN
WEFM(FM) Michigan City IN
WERK(FM) Muncie IN
WJJK(FM) Noblesville IN
WSEZ(AM) Paoli IN
WTCA(AM) Plymouth IN
WZOC(FM) Plymouth IN
WSJD(FM) Princeton IN
WRIN(AM) Rensselaer IN
WROI(FM) Rochester IN
WAXI(FM) Rockville IN
WZZB(AM) Seymour IN
WTCJ(AM) Tell City IN
WSKL(FM) Veedersburg IN
WJOT(AM) Wabash IN
WJOT-FM Wabash IN
WRSW-FM Warsaw IN
KSAJ-FM Abilene KS
KVOE(AM) Emporia KS
KMDO(AM) Fort Scott KS
KOMB(FM) Fort Scott KS
KCAR-FM Galena KS
KAYS(AM) Hays KS
KWHK(FM) Hutchinson KS
KALN(AM) Iola KS
KBGL(FM) Larned KS
KGTR(FM) Larned KS
KNGL(AM) McPherson KS
KLKC-FM Parsons KS
KWLS(AM) Pratt KS
KSKL(FM) Scott City KS
KWME(FM) Wellington KS
KEYN-FM Wichita KS
KFTI(AM) Wichita KS
WMDJ-FM Allen KY
WANV(FM) Annville KY
WCBL-FM Benton KY
WKDZ(AM) Cadiz KY
WBVX(FM) Carlisle KY
WAIN(AM) Columbia KY
WCTT(AM) Corbin KY
WCYN(AM) Cynthiana KY
WEKB(AM) Elkhorn City KY
WDJO(AM) Florence KY
WFKY(AM) Frankfort KY
WOVO(FM) Glasgow KY
WUGO(FM) Grayson KY
WKYA(FM) Greenville KY
WXMZ(FM) Hartford KY
WKIC(AM) Hazard KY
WQXY(AM) Hazard KY
WHVO(AM) Hopkinsville KY
WIRV(AM) Irvine KY
WLBN(AM) Lebanon KY
WLSK(FM) Lebanon KY
WWXL(AM) Manchester KY
WMJL-FM Marion KY
WFTM(AM) Maysville KY
WMKJ(FM) Mt. Sterling KY
WVJS(AM) Owensboro KY
WSIP(AM) Paintsville KY
WSIP(AM) Paintsville KY
WPKE(AM) Pikeville KY
WANO(AM) Pineville KY

WXKZ-FM Prestonsburg KY
WWKY(FM) Providence KY
WAKY(FM) Radcliff KY
WCND(AM) Shelbyville KY
WKEQ(FM) Somerset KY
WTLO(AM) Somerset KY
WLKS(AM) West Liberty KY
WXKQ-FM Whitesburg KY
WGKY(FM) Wickliffe KY
KDBS(AM) Alexandria LA
WTGG(FM) Amite LA
KQIS(FM) Basile LA
KEZP(FM) Bunkie LA
KSIG(AM) Crowley LA
KNOU(FM) Empire LA
KFNV-FM Ferriday LA
WTIX-FM Galliano LA
KLEB(AM) Golden Meadow LA
KHLA(FM) Jennings LA
KLLA(AM) Leesville LA
KASO(AM) Minden LA
KLKL(FM) Minden LA
KLIL(FM) Moreauville LA
KMYO-FM Morgan City LA
KZBL(FM) Natchitoches LA
KRDJ(FM) New Iberia LA
KWCL-FM Oak Grove LA
KBEB-FM Rayne LA
KPCH(FM) Ruston LA
KVPI-FM Ville Platte LA
WODS(FM) Boston MA
WCIB(FM) Falmouth MA
WATD-FM Marshfield MA
WUPE-FM North Adams MA
WJOE(AM) Orange-Athol MA
WUPE(AM) Pittsfield MA
WJDA(AM) Quincy MA
WARE(AM) Ware MA
WORC-FM Webster MA
WAMD(AM) Aberdeen MD
WCMD(AM) Cumberland MD
WTBO(AM) Cumberland MD
WTDK(FM) Federalsburg MD
WFRB(AM) Frostburg MD
WARK(AM) Hagerstown MD
WTTR(AM) Westminster MD
WICL(FM) Williamsport MD
WJTO(AM) Bath ME
WCXU(FM) Caribou ME
WGUY(FM) Dexter ME
WABK-FM Gardiner ME
WBYA(FM) Islesboro ME
WYNZ(FM) Westbrook ME
*WYAR(FM) Yarmouth ME
XHRM-FM Tijuana MEX
WHNN(FM) Bay City MI
WMRX-FM Beaverton MI
WCZW(FM) Charlevoix MI
WTVB(AM) Coldwater MI
WOBE(FM) Crystal Falls MI
WOMC(FM) Detroit MI
WGKL(FM) Gladstone MI
WFGR(FM) Grand Rapids MI
WKAD(FM) Harrietta MI
WHIT-FM Hartford MI
*WHPR(FM) Highland Park MI
*WKHN(FM) Hubbard Lake MI
WHTO(FM) Iron Mountain MI
WIKB(AM) Iron River MI
WIKB-FM Iron River MI
*WJCQ(FM) Jackson MI
*WJKQ(FM) Jackson MI
WMTE-FM Manistee MI
WTIQ(AM) Manistique MI
WHLX(AM) Marine City MI
WAGN(AM) Menominee MI
WQXO(AM) Munising MI
*WPQZ(FM) Muskegon MI
WNGE(FM) Negaunee MI
WNBY-FM Newberry MI
WNIL(AM) Niles MI
WLCS(FM) North Muskegon MI
*WAQQ(FM) Onsted MI
WAKV(AM) Otsego MI
WQXC-FM Otsego MI
WLXT(FM) Petoskey MI

WHLS(AM) Port Huron MI
WDEE-FM Reed City MI
WHAK-FM Rogers City MI
*WAAQ(FM) Rogers Heights MI
WQTX(FM) Saint Johns MI
WYVN(FM) Saugatuck MI
*WSHJ(FM) Southfield MI
WMSH-FM Sturgis MI
WCCW-FM Traverse City MI
WPON(AM) Walled Lake MI
WEFG-FM Whitehall MI
KULO(FM) Alexandria MN
KQQL(FM) Anoka MN
KAUS(AM) Austin MN
KSCR-FM Benson MN
KBEW(AM) Blue Earth MN
KUAL-FM Brainerd MN
KRWC(AM) Buffalo MN
WGVY(FM) Cambridge MN
WKLK(AM) Cloquet MN
KDAL-FM Duluth MN
KLDJ(FM) Duluth MN
KQDS(AM) Duluth MN
WGVZ(FM) Eden Prairie MN
KJJK(AM) Fergus Falls MN
KKCQ(AM) Fosston MN
KKRC(FM) Granite Falls MN
WMFG-FM Hibbing MN
KDUZ(AM) Hutchinson MN
KGHS(AM) International Falls MN
KRAQ(FM) Jackson MN
KQEG(FM) La Crescent MN
WGVX(FM) Lakeville MN
KLQP(FM) Madison MN
KRJM(FM) Mahnomen MN
WMOZ(FM) Moose Lake MN
KRDS-FM New Prague MN
KRFO(AM) Owatonna MN
KISD(FM) Pipestone MN
KWNG(FM) Red Wing MN
KLGR-FM Redwood Falls MN
KNXR(FM) Rochester MN
KLCX(FM) Saint Charles MN
KXAC(FM) Saint James MN
KZJK(FM) Saint Louis Park MN
KVGO(FM) Spring Valley MN
KLBB(AM) Stillwater MN
KAKK(AM) Walker MN
KRUE(FM) Waseca MN
KDJS(AM) Willmar MN
KWNO(AM) Winona MN
KOZX(FM) Cabool MO
KCRV-FM Caruthersville MO
KWKZ(FM) Charleston MO
KCHI(AM) Chillicothe MO
KCHI-FM Chillicothe MO
KDKD(AM) Clinton MO
*KWWC-FM Columbia MO
KDFN(AM) Doniphan MO
KYMO-FM East Prairie MO
KESM(AM) El Dorado Springs MO
KYLS(AM) Fredericktown MO
KKCA(FM) Fulton MO
KCMO-FM Kansas City MO
KBOA(AM) Kennett MO
KXOQ(FM) Kennett MO
KOMC-FM Kimberling City MO
KIRX(AM) Kirksville MO
KQUL(FM) Lake Ozark MO
KJMO(FM) Linn MO
KRLI(FM) Malta Bend MO
KYRX(FM) Marble Hill MO
KZZT(FM) Moberly MO
KXMO-FM Owensville MO
KLID(AM) Poplar Bluff MO
KSFT(AM) Saint Joseph MO
KOKO(AM) Warrensburg MO
KWMO(AM) Washington MO
KOSP(FM) Willard MO
KUKU-FM Willow Springs MO
WAFM(FM) Amory MS
WIZK(AM) Bay Springs MS
WVIM-FM Coldwater MS
WHKL(FM) Crenshaw MS
WNIX(AM) Greenville MS
WOHT(FM) Grenada MS

WDXO(FM) Hazlehurst MS
WQJQ(FM) Kosciusko MS
WAGR-FM Lexington MS
WAZA(FM) Liberty MS
WNAU(AM) New Albany MS
WHTU(FM) Newton MS
WMTI(FM) Picayune MS
WOAD-FM Pickens MS
WCJU-FM Prentiss MS
WVBG-FM Redwood MS
WYAB(FM) Yazoo City MS
KANA(AM) Anaconda MT
KKBR(FM) Billings MT
KOBB-FM Bozeman MT
KXTL(AM) Butte MT
KKGR(AM) East Helena MT
KIKC(AM) Forsyth MT
KXGN(AM) Glendive MT
KEIN(AM) Great Falls MT
KLFM(FM) Great Falls MT
KMTX(AM) Helena MT
KOFI(AM) Kalispell MT
KTNY(FM) Libby MT
KYLT(AM) Missoula MT
KBQQ(FM) Pinesdale MT
KSEN(AM) Shelby MT
KWYS(AM) West Yellowstone MT
KWOL-FM Whitefish MT
KVCK(AM) Wolf Point MT
WZKY(AM) Albemarle NC
WOXL-FM Biltmore Forest NC
WZGM(AM) Black Mountain NC
WBHN(AM) Bryson City NC
WPCM(AM) Burlington NC
WCSL(AM) Cherryville NC
WCLN(AM) Clinton NC
WERX-FM Columbia NC
WEGO(AM) Concord NC
WBLA(AM) Elizabethtown NC
WODR(FM) Fair Bluff NC
WSRC(AM) Fair Bluff NC
WFSC(AM) Franklin NC
WTRG(FM) Gaston NC
WGNC(AM) Gastonia NC
WWNF(FM) Goldsboro NC
WNCT-FM Greenville NC
WFMZ(FM) Hertford NC
WAIZ(AM) Hickory NC
WXRC(FM) Hickory NC
WCCG(FM) Hope Mills NC
WILT(FM) Jacksonville NC
WTHZ(FM) Lexington NC
WLON(AM) Lincolnton NC
WFNC-FM Lumberton NC
WHIP(AM) Mooresville NC
WCDG(FM) Moyock NC
WWMY(FM) Raleigh NC
WLWL(AM) Rockingham NC
WFJA(FM) Sanford NC
WOHS(AM) Shelby NC
WFVL(FM) Southern Pines NC
WTOE(AM) Spruce Pine NC
WACB(AM) Taylorsville NC
WSVM(AM) Valdese NC
WMXF(AM) Waynesville NC
*KEYA(FM) Belcourt ND
KACL(FM) Bismarck ND
KXGT(FM) Carrington ND
KDVL(FM) Devils Lake ND
KDIX(AM) Dickinson ND
KAUJ(FM) Grafton ND
KKXL(AM) Grand Forks ND
KRRZ(AM) Minot ND
KEGK(FM) Wahpeton ND
KCOW(AM) Alliance NE
KGMT(AM) Fairbury NE
KTNC(AM) Falls City NE
KFMT-FM Fremont NE
KSDZ(FM) Gordon NE
KUVR(AM) Holdrege NE
KADL(FM) Imperial NE
KKPR-FM Kearney NE
KYOY(FM) Kimball NE
KLNC(FM) Lincoln NE
KBRL(AM) McCook NE
KOOQ(AM) North Platte NE

KOGA(AM) Ogallala NE
KGOR(FM) Omaha NE
KRFS(AM) Superior NE
KOAQ(AM) Terrytown NE
KCTY(FM) Wayne NE
KAWL(AM) York NE
WMOU(AM) Berlin NH
WMEX(FM) Farmington NH
WNNH(FM) Henniker NH
WZBK(AM) Keene NH
WLTN(AM) Littleton NH
WQSO(FM) Rochester NH
WPLY-FM Walpole NH
WLKZ(FM) Wolfeboro NH
WMID(AM) Atlantic City NJ
WILW(FM) Avalon NJ
WRNJ(AM) Hackettstown NJ
WOBM(AM) Lakewood NJ
WJRZ-FM Manahawkin NJ
WMTR(AM) Morristown NJ
WNNJ(AM) Newton NJ
WTKU-FM Ocean City NJ
WTAA(AM) Pleasantville NJ
WGHT(AM) Pompton Lakes NJ
KDEF(AM) Albuquerque NM
KRKE(AM) Albuquerque NM
KKFG(FM) Bloomfield NM
KCCC(AM) Carlsbad NM
KWKA(AM) Clovis NM
KDSK(FM) Grants NM
KYVA-FM Grants NM
KVLC(FM) Hatch NM
KHOB(AM) Hobbs NM
*KEDP(FM) Las Vegas NM
KABG(FM) Los Alamos NM
KLEA-FM Lovington NM
KRTN-FM Raton NM
*KRDR(FM) Red River NM
KBCQ(AM) Roswell NM
KCRX(AM) Roswell NM
KBUY(AM) Ruidoso NM
KCHS(AM) Truth or Consequences NM
KSTJ(FM) Boulder City NV
KELY(AM) Ely NV
KNYE(FM) Pahrump NV
WSEN(AM) Baldwinsville NY
WSEN-FM Baldwinsville NY
WABH(AM) Bath NY
WBZO(FM) Bay Shore NY
WENI-FM Big Flats NY
WHTT-FM Buffalo NY
WPTR(FM) Clifton Park NY
WIYN(FM) Deposit NY
WDOE(AM) Dunkirk NY
WALK(AM) East Patchogue NY
WFKL(FM) Fairport NY
WCQL(FM) Glens Falls NY
WGIX-FM Gouverneur NY
WXUR(FM) Herkimer NY
WXHC(FM) Homer NY
WLEA(AM) Hornell NY
WZCR(FM) Hudson NY
WNYQ(FM) Hudson Falls NY
WHVW(AM) Hyde Park NY
WKSN(AM) Jamestown NY
WCKM-FM Lake George NY
WICY(AM) Malone NY
WCBS-FM New York NY
WGNY(AM) Newburgh NY
WPAC(FM) Ogdensburg NY
WHDL(AM) Olean NY
WMCR(AM) Oneida NY
WMCR-FM Oneida NY
WZOZ(FM) Oneonta NY
WKOL(FM) Plattsburgh NY
WVTK(FM) Port Henry NY
WDLC(AM) Port Jervis NY
WRIV(AM) Riverhead NY
WODZ-FM Rome NY
WTRY-FM Rotterdam NY
WLNG(FM) Sag Harbor NY
WVKZ(AM) Schenectady NY
WCDO-FM Sidney NY
WIPS(AM) Ticonderoga NY
WDLA(AM) Walton NY
WTBQ(AM) Warwick NY

WJQZ(FM) Wellsville NY
WNYV(FM) Whitehall NY
WZAD(FM) Wurtsboro NY
*WRMU(FM) Alliance OH
WOHF(FM) Bellevue OH
WRQN(FM) Bowling Green OH
WQCT(AM) Bryan OH
WHBC(AM) Canton OH
WMJI(FM) Cleveland OH
WBVB(FM) Coal Grove OH
WYBZ(FM) Crooksville OH
WJER(AM) Dover-New Philadelphia OH
WZRX-FM Fort Shawnee OH
WGRR(FM) Hamilton OH
WRBP(FM) Hubbard OH
WDLW(AM) Lorain OH
WAGX(FM) Manchester OH
WMRN(AM) Marion OH
*WLMH(FM) Morrow OH
WIOI(AM) New Boston OH
WBBG(FM) Niles OH
WBUK(FM) Ottawa OH
WPTW(AM) Piqua OH
WMLX(FM) Saint Mary's OH
WSWR(FM) Shelby OH
WULM(AM) Springfield OH
WDIG(AM) Steubenville OH
WWWM-FM Sylvania OH
WTTF(AM) Tiffin OH
WYNT(FM) Upper Sandusky OH
WKVX(AM) Wooster OH
KALV(AM) Alva OK
KACO(FM) Apache OK
KVSO(AM) Ardmore OK
KRVT(AM) Claremore OK
KTAT(AM) Frederick OK
KTSO(FM) Glenpool OK
KVRW(FM) Lawton OK
KTMC(AM) McAlester OK
KOMA(FM) Oklahoma City OK
KOKL(AM) Okmulgee OK
KVLH(AM) Pauls Valley OK
KOSB(FM) Perry OK
KLOR-FM Ponca City OK
KMUR(AM) Pryor OK
KVRO(FM) Stillwater OK
KSLE(FM) Wewoka OK
KRAT(FM) Altamont OR
KKBC-FM Baker City OR
KQAK(FM) Bend OR
KURY(AM) Brookings OR
KCNA(FM) Cave Junction OR
KDCQ(FM) Coos Bay OR
KRJT(FM) Elgin OR
KKNX(AM) Eugene OR
KODZ(FM) Eugene OR
KCST-FM Florence OR
KLTH(FM) Lake Oswego OR
KLYC(AM) McMinnville OR
KLDZ(FM) Medford OR
KVAN-FM Pilot Rock OR
KBZY(AM) Salem OR
KSWB(AM) Seaside OR
KAVJ(FM) Sutherlin OR
KACI-FM The Dalles OR
KCUP(AM) Toledo OR
KPPT-FM Toledo OR
WFBG(AM) Altoona PA
WLZS(FM) Beaver Springs PA
WALY(FM) Bellwood PA
WFBS(AM) Berwick PA
WHLM-FM Berwick PA
WKQL(FM) Brookville PA
WASP(AM) Brownsville PA
WCCL(FM) Central City PA
WCPA(AM) Clearfield PA
WYJK(AM) Connellsville PA
WKPL(FM) Ellwood City PA
WFNN(AM) Erie PA
WHPA(FM) Gallitzin PA
WHKF(FM) Harrisburg PA
WTKT(AM) Harrisburg PA
WYCY(FM) Hawley PA
WAZL(AM) Hazleton PA
WDAD(AM) Indiana PA

WTYM(AM) Kittanning PA
WKVA(AM) Lewistown PA
WMVL(FM) Linesville PA
WBPZ(AM) Lock Haven PA
WQZS(FM) Meyersdale PA
WPLY(AM) Mount Pocono PA
WBZD-FM Muncy PA
WJST(AM) New Castle PA
WYNE(AM) North East PA
WNCC(AM) Northern Cambria PA
WOGL(FM) Philadelphia PA
WWSW-FM Pittsburgh PA
WSOX(FM) Red Lion PA
WUZZ(FM) Saegertown PA
WKBI(AM) Saint Marys PA
WLSW(FM) Scottdale PA
WARM(AM) Scranton PA
WYFM(FM) Sharon PA
WPHD(FM) South Waverly PA
WCDW(FM) Susquehanna PA
WTTC(AM) Towanda PA
WTTC-FM Towanda PA
WPKL(FM) Uniontown PA
WOWY(FM) University Park PA
WRRN(FM) Warren PA
WJPA(AM) Washington PA
WJPA-FM Washington PA
WLUZ(AM) Bayamon PR
WLUZ(FM) Bayamon PR
WCMA-FM Fajardo PR
WOIZ(AM) Guayanilla PR
WTIL(AM) Mayaguez PR
WYEL(AM) Mayaguez PR
WLEO(AM) Ponce PR
WZAR(FM) Ponce PR
WIDI(FM) Quebradillas PR
WKVM(AM) San Juan PR
WRSS(AM) San Sebastian PR
WWBB(FM) Providence RI
WZLA-FM Abbeville SC
WZLA-FM Abbeville SC
WKSP(FM) Aiken SC
WBSC(AM) Bennettsville SC
WGFG(FM) Branchville SC
WPUB-FM Camden SC
WLTQ(AM) Charleston SC
WCRE(AM) Cheraw SC
WAHT(AM) Clemson SC
WOMG(FM) Columbia SC
WYNN(AM) Florence SC
WSYN(FM) Georgetown SC
WPCI(AM) Greenville SC
WOLT(FM) Greer SC
WWNU(FM) Irmo SC
WKSX-FM Johnston SC
WKSC(AM) Kershaw SC
WCOO(FM) Kiawah Island SC
WAGL(AM) Lancaster SC
WGUS-FM New Ellenton SC
WKDK(AM) Newberry SC
WXLY(FM) North Charleston SC
WGZO(FM) Parris Island SC
WLOW(FM) Port Royal SC
*WRBK(FM) Richburg SC
WJES(AM) Saluda SC
WICI(FM) Sumter SC
WWOF(AM) Walhalla SC
KAWK(FM) Custer SD
KDSJ(AM) Deadwood SD
KSQB-FM Dell Rapids SD
KBJM(AM) Lemmon SD
KKSD(FM) Milbank SD
KMSD(AM) Milbank SD
KCCR(AM) Pierre SD
KKLS(AM) Rapid City SD
KLQT(FM) Wessington Springs SD
WVOL(AM) Berry Hill TN
WBOL(AM) Bolivar TN
WOPI(AM) Bristol TN
WDOD(AM) Chattanooga TN
WKOM(FM) Columbia TN
WZYX(AM) Cowan TN
WCDZ(FM) Dresden TN
WTRO(AM) Dyersburg TN
*WVCP(FM) Gallatin TN
WRQQ(FM) Goodlettsville TN
WSMG(AM) Greeneville TN
WIRJ(AM) Humboldt TN
WMXX-FM Jackson TN
WKTP(AM) Jonesborough TN
WHGG(AM) Kingsport TN
WKOS(FM) Kingsport TN
WKPT(AM) Kingsport TN
*WKCS(FM) Knoxville TN
WMSR(AM) Manchester TN
WCMT(AM) Martin TN
WHDM(AM) McKenzie TN
WTPR-FM McKinnon TN
WXMX(FM) Millington TN
WCRK(AM) Morristown TN
WLIK(AM) Newport TN
WATO(AM) Oak Ridge TN
WOKI(FM) Oliver Springs TN
WTPR(AM) Paris TN
WQKR(AM) Portland TN
WKSR(AM) Pulaski TN
WKSR-FM Pulaski TN
WORM(AM) Savannah TN
WLOD-FM Sweetwater TN
WNTT(AM) Tazewell TN
WENK(AM) Union City TN
KULL(FM) Abilene TX
KBYG(AM) Big Spring TX
KNTX(AM) Bowie TX
KNEL(AM) Brady TX
KROO(AM) Breckenridge TX
KRVA-FM Campbell TX
KPUR-FM Canyon TX
KCTX(AM) Childress TX
KEFH(FM) Clarendon TX
KGAP(FM) Clarksville TX
KCCT(AM) Corpus Christi TX
KMXR(FM) Corpus Christi TX
KIVY(AM) Crockett TX
KXIT-FM Dalhart TX
KBFB(FM) Dallas TX
KLUV(FM) Dallas TX
KWMC(AM) Del Rio TX
KDDD-FM Dumas TX
KINL(FM) Eagle Pass TX
KAMA(AM) El Paso TX
KOFX(FM) El Paso TX
KINV(FM) Georgetown TX
KFRO-FM Gilmer TX
KKDA(AM) Grand Prairie TX
KCOL-FM Groves TX
KONO-FM Helotes TX
KTBZ-FM Houston TX
KHVL(AM) Huntsville TX
KRVF(FM) Kerens TX
KHTC(FM) Lake Jackson TX
KSHN-FM Liberty TX
KITY(FM) Llano TX
KKCL(FM) Lorenzo TX
KDAV(AM) Lubbock TX
KQXX-FM Mission TX
KCKM(AM) Monahans TX
KEWL-FM New Boston TX
KZRB(FM) New Boston TX
KMCM(FM) Odessa TX
KREH(AM) Pecan Grove TX
KEYE-FM Perryton TX
KREW(AM) Plainview TX
KITE(FM) Port Lavaca TX
KROB(AM) Robstown TX
KTLU(AM) Rusk TX
KWRW(FM) Rusk TX
KONO(AM) San Antonio TX
KBPA(FM) San Marcos TX
KBKH(FM) Shamrock TX
KJIM(AM) Sherman TX
*KSQX(FM) Springtown TX
*KQXS(FM) Stephenville TX
KSST(AM) Sulphur Springs TX
KKTK(AM) Texarkana TX
KDOK(FM) Tyler TX
KGLD(AM) Tyler TX
KVOU(AM) Uvalde TX
KVWC(AM) Vernon TX
KVWC-FM Vernon TX
KNAL(AM) Victoria TX
KBGO(FM) Waco TX
KVLL-FM Wells TX
KWUD(AM) Woodville TX
KBDX(FM) Blanding UT
KXOL(AM) Brigham City UT
KXBN(FM) Cedar City UT
KLGN(AM) Logan UT
KODJ(FM) Salt Lake City UT
KGNT(FM) Smithfield UT
WLEQ(FM) Bedford VA
WODI(AM) Brookneal VA
WBOP(FM) Buffalo Gap VA
*WFOS(FM) Chesapeake VA
WCTG(FM) Chincoteague VA
WDIC-FM Clinchco VA
WGRQ(FM) Colonial Beach VA
WKEY(AM) Covington VA
WSUH(FM) Crozet VA
WBTM(AM) Danville VA
WILA(AM) Danville VA
WYTT(FM) Emporia VA
WZRV(FM) Front Royal VA
WHAP(AM) Hopewell VA
WZFM(FM) Narrows VA
WVCV(AM) Orange VA
WARV-FM Petersburg VA
WBBT-FM Powhatan VA
WBTJ(FM) Richmond VA
WRXL(FM) Richmond VA
WXBX(FM) Rural Retreat VA
WSVO(FM) Staunton VA
WKQY(FM) Tazewell VA
WMBG(AM) Williamsburg VA
WSTA(AM) Charlotte Amalie VI
WMNG(FM) Christiansted VI
WWOD(FM) Hartford VT
WSKI(AM) Montpelier VT
WVNR(AM) Poultney VT
WCFR(AM) Springfield VT
KWLE(AM) Anacortes WA
KBRO(AM) Bremerton WA
KRQT(FM) Castle Rock WA
KITI(AM) Chehalis-Centralia WA
KEYF-FM Cheney WA
KEYF(AM) Dishman WA
KQBE(FM) Ellensburg WA
KLCK(AM) Goldendale WA
KEYG-FM Grand Coulee WA
KARY-FM Grandview WA
KNTB(AM) Lakewood WA
KEDO(AM) Longview WA
KBRC(AM) Mount Vernon WA
KWDB(AM) Oak Harbor WA
KRIZ(AM) Renton WA
KBBD(FM) Spokane WA
KBSG-FM Tacoma WA
KYNR(AM) Toppenish WA
KCSY(FM) Twisp WA
KTEL(AM) Walla Walla WA
WDKM(FM) Adams WI
WBDK(FM) Algoma WI
WATW(AM) Ashland WI
WOLX-FM Baraboo WI
WAQE-FM Barron WI
WGEZ(AM) Beloit WI
WWIS(AM) Black River Falls WI
WHLK(FM) De Forest WI
WRJO(FM) Eagle River WI
WRKU(FM) Forestville WI
WMIN(AM) Hudson WI
WHRY(AM) Hurley WI
WNXR(FM) Iron River WI
WOGB(FM) Kaukauna WI
WJBL(FM) Ladysmith WI
WLTU(FM) Manitowoc WI
WMDC(FM) Mayville WI
WSFQ(FM) Peshtigo WI
WPVL-FM Platteville WI
WPRE(AM) Prairie du Chien WI
WRDB(AM) Reedsburg WI
WRIG(AM) Schofield WI
WSBW(FM) Sister Bay WI
WBOG(AM) Tomah WI
WJJQ-FM Tomahawk WI
WGBW(AM) Two Rivers WI
WVRQ(AM) Viroqua WI
WTTN(AM) Watertown WI
WHTL-FM Whitehall WI
WKCH(FM) Whitewater WI
WVBO(FM) Winneconne WI
WFGM-FM Barrackville WV
WIWS(AM) Beckley WV
WUKL(FM) Bethlehem WV
WMRE(AM) Charles Town WV
WELD(AM) Fisher WV
*WFGH(FM) Fort Gay WV
WMTD(AM) Hinton WV
WMTD-FM Hinton WV
WTCR(AM) Kenova WV
WCBC-FM Keyser WV
WFSP-FM Kingwood WV
WKAZ-FM Miami WV
WCLG(AM) Morgantown WV
WAXS(FM) Oak Hill WV
WADC(AM) Parkersburg WV
WRON-FM Ronceverte WV
WDMX(FM) Vienna WV
WHAW(AM) Weston WV
WBBD(AM) Wheeling WV
WBTH(AM) Williamson WV
KBBS(AM) Buffalo WY
KMLD(FM) Casper WY
KRRR(FM) Cheyenne WY
KKTY(AM) Douglas WY
KZWB(FM) Green River WY
KMER(AM) Kemmerer WY
KARS-FM Laramie WY
KREO(FM) Pine Bluffs WY
KPIN(FM) Pinedale WY
KVOW(AM) Riverton WY
KRKK(AM) Rock Springs WY
KTHE(AM) Thermopolis WY
KWOR(AM) Worland WY

Other

*KBBI(AM) Homer AK
*KBBI(AM) Homer AK
WUHT(FM) Birmingham AL
WOZK(AM) Ozark AL
WTBF(AM) Troy AL
KVLD(FM) Atkins AR
KVLD(FM) Atkins AR
KXHT(FM) Marion AR
KWAK(AM) Stuttgart AR
KQPN(AM) West Memphis AR
KVVA-FM Apache Junction AZ
*KJTA(FM) Flagstaff AZ
KLNZ(FM) Glendale AZ
KTZR-FM Green Valley AZ
KSZR(FM) Oro Valley AZ
KAZM(AM) Sedona AZ
*KOHN(FM) Sells AZ
KHOV-FM Wickenburg AZ
KCIK(AM) Blue Lake CA
KOND(FM) Clovis CA
KDAC(AM) Fort Bragg CA
KNCR(AM) Fortuna CA
KGBA-FM Holtville CA
*KCRY(FM) Mojave CA
KLLE(FM) North Fork CA
*KKSM(AM) Oceanside CA
KMAK(FM) Orange Cove CA
*KCRU(FM) Oxnard CA
KXTZ(FM) Pismo Beach CA
KGBY(FM) Sacramento CA
KJAY(AM) Sacramento CA
KLOK(AM) San Jose CA
KWIZ(FM) Santa Ana CA
KKXX-FM Shafter CA
KYAA(AM) Soquel CA
KGEN(AM) Tulare CA
*KBDG(FM) Turlock CA
KSFM(FM) Woodland CA
KZYR(FM) Avon CO
KKKK(AM) Colorado Springs CO
KLVZ(AM) Denver CO
KLZ(AM) Denver CO
*KDNK(FM) Glenwood Springs CO
KKCH(FM) Glenwood Springs CO
KVLE-FM Gunnison CO
KIDN-FM Hayden CO
*KUTE(FM) Ignacio CO
KQKS(FM) Lakewood CO
KAYW(FM) Meeker CO
*KVMT(FM) Montrose CO
*KPGS(FM) Pagosa Springs CO
*KVNF(FM) Paonia CO
*KTSC-FM Pueblo CO
*KTSC-FM Pueblo CO
KPHT(FM) Rocky Ford CO
KJJD(AM) Windsor CO
WCUM(AM) Bridgeport CT
WZMX(FM) Hartford CT
WFNW(AM) Naugatuck CT
WTMP(AM) Egypt Lake FL
WTZB(FM) Englewood FL
WMIB(FM) Fort Lauderdale FL
WMIB(FM) Fort Lauderdale FL
WRNE(AM) Gulf Breeze FL
WLLD(FM) Holmes Beach FL
WJAX(AM) Jacksonville FL
WMFM(FM) Key West FL
WWAB(AM) Lakeland FL
WLBE(AM) Leesburg FL
WPYO(FM) Maitland FL
WFLC(FM) Miami FL
WFLA-FM Midway FL
WONQ(AM) Oviedo FL
WONQ(AM) Oviedo FL
WPLK(AM) Palatka FL
WMNE(AM) Riviera Beach FL
WSIR(AM) Winter Haven FL
WLVX(FM) Elberton GA
WPLO(AM) Grayson GA
WTYB(FM) Springfield GA
WHLJ(FM) Statenville GA
WHGH(AM) Thomasville GA
WLET(AM) Toccoa GA
KUAI(AM) Eleele HI
KIPA(AM) Hilo HI
KSSK(AM) Honolulu HI
KDMU(FM) Bloomfield IA
*KJMC(FM) Des Moines IA
*KSUI(FM) Iowa City IA
KRIB(AM) Mason City IA
*KBBG(FM) Waterloo IA
WKKD(AM) Aurora IL
WSDZ(AM) Belleville IL
WFUN-FM Bethalto IL
WGCI-FM Chicago IL
WWHP(FM) Farmer City IL
*WKCC(FM) Kankakee IL
*WVJC(FM) Mount Carmel IL
WBCP(AM) Urbana IL
WEJK(FM) Boonville IN
*WSWI(AM) Evansville IN
WSYW(AM) Indianapolis IN
WLOI(AM) La Porte IN
WLQI(FM) Rensselaer IN
*KMUW(FM) Wichita KS
KSJM(FM) Winfield KS
WXBC(FM) Hardinsburg KY
KMEZ(FM) Belle Chasse LA
KSYR(FM) Benton LA
KQLQ(FM) Columbia LA
*KRVS(FM) Lafayette LA
KLRZ(FM) Larose LA
KMRC(AM) Morgan City LA
KAGY(AM) Port Sulphur LA
WMKI(AM) Boston MA
WVBF(AM) Middleborough Center MA
WNTN(AM) Newton MA
WORC(AM) Worcester MA
*WESM(FM) Princess Anne MD
XHRM-FM Tijuana MEX
WFDF(AM) Farmington Hills MI
WBCH-FM Hastings MI
WFAT(FM) Portage MI
*WSHJ(FM) Southfield MI
WMFN(AM) Zeeland MI
*KMSU(FM) Mankato MN
*KCCM-FM Moorhead MN
*KRLX(FM) Northfield MN
*KGAC(FM) Saint Peter MN
KRBI-FM Saint Peter MN
*KSMR(FM) Winona MN
*KSMR(FM) Winona MN
KBXR(FM) Columbia MO

KUGT(AM) Jackson MO
WKCU(AM) Corinth MS
*WURC(FM) Holly Springs MS
KBSR(AM) Laurel MT
WZJS(FM) Banner Elk NC
WKXB(FM) Burgaw NC
WYNA(FM) Calabash NC
WQNQ(FM) Fletcher NC
*WDCC(FM) Sanford NC
WOHS(AM) Shelby NC
*KABU(FM) Fort Totten ND
KSJZ(FM) Jamestown ND
WFTN(AM) Franklin NH
WNTK(AM) Newport NH
WHTG(AM) Eatontown NJ
*WYRS(FM) Manahawkin NJ
*WYRS(FM) Manahawkin NJ
*KUUT(FM) Farmington NM
*KUSW(FM) Flora Vista NM
*KGLP(FM) Gallup NM
KRSN(AM) Los Alamos NM
KLEA(AM) Lovington NM
KSFX(FM) Roswell NM
KSHP(AM) North Las Vegas NV
WDDY(AM) Albany NY
WCSS(AM) Amsterdam NY
WBNR(AM) Beacon NY
WEHH(AM) Elmira Heights-Horseheads NY
WBEA(FM) Southold NY
WTOR(AM) Youngstown NY
*KWYA(FM) Astoria OR
KQFM(FM) Hermiston OR
KDZR(AM) Lake Oswego OR
KRXF(FM) Sunriver OR
WBRR(FM) Bradford PA
WLER-FM Butler PA
WZUM(AM) Carnegie PA
WBCB(AM) Levittown-Fairless Hills PA
WEDO(AM) McKeesport PA
WPHI-FM Media PA
WVPO(AM) Stroudsburg PA
*WRKC(FM) Wilkes-Barre PA
WGIT(AM) Canovanas PR
WJZG(FM) Culebra PR
WGDL(AM) Lares PR
WZMT(FM) Ponce PR
WIDI(FM) Quebradillas PR
WGTN-FM Andrews SC
WVCO(FM) Loris SC
WXBT(FM) West Columbia SC
KBRK(AM) Brookings SD
WMPS(AM) Bartlett TN
WXIS(FM) Erwin TN
WOWW(AM) Germantown TN
WQJK(FM) Maryville TN
*WOEZ(FM) Maynardville TN
WUUS-FM South Pittsburg TN
*KACV-FM Amarillo TX
KBZD(FM) Amarillo TX
KOOC(FM) Belton TX
KQFX(FM) Borger TX
KNEL(AM) Brady TX
KBSO(FM) Corpus Christi TX
KBXX(FM) Houston TX
KDAV(AM) Lubbock TX
KIKK(AM) Pasadena TX
KMJR(FM) Portland TX
KROM(FM) San Antonio TX
*KSYM-FM San Antonio TX
*KSYM-FM San Antonio TX
KPYK(AM) Terrell TX
KBTE(FM) Tulia TX
KBTE(FM) Tulia TX
KTXN-FM Victoria TX
KWUD(AM) Woodville TX
*KZMU(FM) Moab UT
*KRDC-FM Saint George UT
KMRI(AM) West Valley City UT
WMXH-FM Luray VA
*WHRO-FM Norfolk VA
WJOI(AM) Norfolk VA
WHKT(AM) Portsmouth VA
*WWPV-FM Colchester VT
WDEV-FM Warren VT
KXAA(FM) Cle Elum WA
KSBN(AM) Spokane WA
WERL(AM) Eagle River WI
WHSM-FM Hayward WI
WTUX(AM) Madison WI
*WMSE(FM) Milwaukee WI
WOKY(AM) Milwaukee WI
WKFX(FM) Rice Lake WI
WKSH(AM) Sussex WI
WMMM-FM Verona WI
*WXPW(FM) Wausau WI
WDZN(FM) Romney WV

Polish

WNWI(AM) Oak Lawn IL
WPNA(AM) Oak Park IL
WNVR(AM) Vernon Hills IL
WRKL(AM) New City NY
WLIM(AM) Patchogue NY

Polka

*WOES(FM) Ovid-Elsie MI
KASM(AM) Albany MN
WKTX(AM) Cortland OH
WTKM(AM) Hartford WI
WTKM-FM Hartford WI

Portugese

KLBS(AM) Los Banos CA
*WFAR(FM) Danbury CT
*WFHL(FM) New Bedford MA

Progressive

*KRUA(FM) Anchorage AK
*KSUA(FM) Fairbanks AK
*KXCI(FM) Tucson AZ
*KSPB(FM) Pebble Beach CA
*KRCB-FM Santa Rosa CA
*KZSU(FM) Stanford CA
*KCSU-FM Fort Collins CO
*WVOF(FM) Fairfield CT
*WDJW(FM) Somers CT
*WVUD(FM) Newark DE
*WKPX(FM) Sunrise FL
*WPRK(FM) Winter Park FL
*WREK(FM) Atlanta GA
*WVGS(FM) Statesboro GA
*KIWR(FM) Council Bluffs IA
*KALA(FM) Davenport IA
*KWLC(AM) Decorah IA
*KRUI-FM Iowa City IA
*KRNL-FM Mount Vernon IA
*WJMU(FM) Decatur IL
*WNUR-FM Evanston IL
*WLCA(FM) Godfrey IL
*WMXM(FM) Lake Forest IL
*WIUS(FM) Macomb IL
*WMHD-FM Terre Haute IN
*KSDB-FM Manhattan KS
*WWHR(FM) Bowling Green KY
*WTUL(FM) New Orleans LA
*WIQH(FM) Concord MA
*WUML(FM) Lowell MA
*WJJW(FM) North Adams MA
*WUMD(FM) North Dartmouth MA
*WZLY(FM) Wellesley MA
*WCHC(FM) Worcester MA
WRNR-FM Grasonville MD
*WUMF-FM Farmington ME
*WMEB-FM Orono ME
*WMEW(FM) Waterville ME
*WYAR(FM) Yarmouth ME
WGRD-FM Grand Rapids MI
*WIDR(FM) Kalamazoo MI
*WMHW-FM Mount Pleasant MI
*WSDP(FM) Plymouth MI
*KUOM(AM) Minneapolis MN
*WMCN(FM) Saint Paul MN
*KWUR(FM) Clayton MO
*KCOU(FM) Columbia MO
*WUAG(FM) Greensboro NC
*WDCC(FM) Sanford NC
*WSIF(FM) Wilkesboro NC
*KDNE(FM) Crete NE
KIBZ(FM) Crete NE
*WUNH(FM) Durham NH
*WNEC-FM Henniker NH
*WKNH(FM) Keene NH
*WPCR-FM Plymouth NH
*WRPR(FM) Mahwah NJ
*WVPH(FM) Piscataway NJ
WPRB(FM) Princeton NJ
*WTSR(FM) Trenton NJ
*WKNJ-FM Union Township NJ
WNJC(AM) Washington Township NJ
WRRB(FM) Arlington NY
*WHCL-FM Clinton NY
WEHN(FM) East Hampton NY
*WCVF-FM Fredonia NY
*WEOS(FM) Geneva NY
*WGFR(FM) Glens Falls NY
*WRCU-FM Hamilton NY
*WRHO(FM) Oneonta NY
*WIRQ(FM) Rochester NY
*WFNP(FM) Rosendale NY
*WRUC(FM) Schenectady NY
WEHM(FM) Southampton NY
*WUSB(FM) Stony Brook NY
WCLX(FM) Westport NY
WLKK(FM) Wethersfield Township NY
WDST(FM) Woodstock NY
*WOUB(AM) Athens OH
*WUFM(FM) Columbus OH
*WDUB(FM) Granville OH
*WMCO(FM) New Concord OH
*WUSO(FM) Springfield OH
*WJCU(FM) University Heights OH
*WOBN(FM) Westerville OH
*WYSO(FM) Yellow Springs OH
*KWVA(FM) Eugene OR
*KEOL(FM) La Grande OR
*WWEC(FM) Elizabethtown PA
*WZBT(FM) Gettysburg PA
*WTGP(FM) Greenville PA
*WKVR-FM Huntingdon PA
*WARC(FM) Meadville PA
*WMSS(FM) Middletown PA
*WIXQ(FM) Millersville PA
*WKDU(FM) Philadelphia PA
*WPTS-FM Pittsburgh PA
*WXAC(FM) Reading PA
*WSRU(FM) Slippery Rock PA
*WRLC(FM) Williamsport PA
*WVYC(FM) York PA
*WDOM(FM) Providence RI
*WDOM(FM) Providence RI
*WSBF-FM Clemson SC
*KAOR(FM) Vermillion SD
*WNRZ(FM) Dickson TN
*WRVU(FM) Nashville TN
*WUTS(FM) Sewanee TN
*KERA(FM) Dallas TX
*KSAU(FM) Nacogdoches TX
KHQN(AM) Spanish Fork UT
WVAX(AM) Charlottesville VA
*WEHC(FM) Emory VA
*WMLU(FM) Farmville VA
*WFFC(FM) Ferrum VA
*WWHS-FM Hampden-Sydney VA
*WXJM(FM) Harrisonburg VA
*WDCE(FM) Richmond VA
*WCWM(FM) Williamsburg VA
*WIUV(FM) Castleton VT
*WRMC-FM Middlebury VT
*KUGS(FM) Bellingham WA
*KSVR(FM) Mount Vernon WA
*KEXP-FM Seattle WA
*KWRS(FM) Spokane WA
*KUPS(FM) Tacoma WA
*WBSD(FM) Burlington WI
*WWSP(FM) Stevens Point WI
*WVBC(FM) Bethany WV
*WVWC(FM) Buckhannon WV
*WWVU-FM Morgantown WV
*KUWA(FM) Afton WY
*KDUW(FM) Douglas WY
*KUWG(FM) Gillette WY
*KUWJ(FM) Jackson WY
*KUWR(FM) Laramie WY
*KUWN(FM) Newcastle WY
*KUWX(FM) Pinedale WY
*KCWC-FM Riverton WY
*KUWZ(FM) Rock Springs WY
*KSUW(FM) Sheridan WY
*KUWD(FM) Sundance WY
KUWT(FM) Thermopolis WY

Public Affairs

*KYUK(AM) Bethel AK
*KFSK(FM) Petersburg AK
KIAL(AM) Unalaska AK
*KCHU(AM) Valdez AK
KFNN(AM) Mesa AZ
*KNAI(FM) Phoenix AZ
*KPFA(FM) Berkeley CA
*KPFB(FM) Berkeley CA
KLTX(AM) Long Beach CA
*KSJD(FM) Cortez CO
*KAFM(FM) Grand Junction CO
*KVMT(FM) Montrose CO
*KVNF(FM) Paonia CO
WQUN(AM) Hamden CT
*WCSP-FM Washington DC
*WETA(FM) Washington DC
WTJV(AM) De Land FL
*WUFT-FM Gainesville FL
*WJUF(FM) Inverness FL
WPBR(AM) Lantana FL
*WKFA(FM) Saint Catherine FL
*WPIO(FM) Titusville FL
*WGLT(FM) Normal IL
*WFHB(FM) Bloomington IN
WILO(AM) Frankfort IN
*WGVE(FM) Gary IN
*WJCF(FM) Morristown IN
WKCT(AM) Bowling Green KY
*WRPS(FM) Rockland MA
*WGMS(FM) Hagerstown MD
*WDIH(FM) Salisbury MD
*WMEH(FM) Bangor ME
*WMED(FM) Calais ME
*WMEP(FM) Camden ME
*WMEF(FM) Fort Kent ME
*WMEA(FM) Portland ME
*WMEM(FM) Presque Isle ME
WMPL(AM) Hancock MI
*KMSK(FM) Austin MN
*KMSU(FM) Mankato MN
KAUL(FM) Ellington MO
*KCUR-FM Kansas City MO
*KBIY(FM) Van Buren MO
*WJSU(FM) Jackson MS
*WRVS-FM Elizabeth City NC
WSMX(AM) Winston-Salem NC
*KABU(FM) Fort Totten ND
KSWV(AM) Santa Fe NM
*WNYC(AM) New York NY
WOR(AM) New York NY
*WPSA(FM) Paul Smiths NY
*WGBE(FM) Bryan OH
*WOSU(AM) Columbus OH
*WGDE(FM) Defiance OH
*WGLE(FM) Lima OH
*WGTE-FM Toledo OH
KIXR(FM) Ponca City OK
*KTCB(FM) Tillamook OR
*WDIY(FM) Allentown PA
*WDUQ(FM) Pittsburgh PA
WIAC(AM) San Juan PR
*WJWJ-FM Beaufort SC
*KQSD-FM Lowry SD
*KUSD(FM) Vermillion SD
*WQOX(FM) Memphis TN
*WYPL(FM) Memphis TN
*KUSR(FM) Logan UT
WFHR(AM) Wisconsin Rapids WI

Reggae

WJKC(FM) Christiansted VI
WSTX-FM Christiansted VI

Religious

*KATB(FM) Anchorage AK
*KNOM(AM) Nome AK
*KNOM-FM Nome AK
*KJNP(AM) North Pole AK
KRSA(AM) Petersburg AK
*KUDU(FM) Tok AK
*WSTF(FM) Andalusia AL
WRAB(AM) Arab AL
WATV(AM) Birmingham AL
*WBFR(FM) Birmingham AL
*WLJR(FM) Birmingham AL
WBSA(AM) Boaz AL
WYMR(AM) Bridgeport AL
*WYFD(FM) Decatur AL
*WMBV(FM) Dixons Mills AL
*WDYF(FM) Dothan AL
*WGTF(FM) Dothan AL
*WFIX(FM) Florence AL
*WTBB(FM) Gadsden AL
WGEA(AM) Geneva AL
WBXR(AM) Hazel Green AL
*WBHY-FM Mobile AL
WMOB(AM) Mobile AL
*WLBF(FM) Montgomery AL
*WTBJ(FM) Oxford AL
*WJCK(FM) Piedmont AL
WBTG(AM) Sheffield AL
WYEA(AM) Sylacauga AL
WAPZ(AM) Wetumpka AL
*KVMN(FM) Cave City AR
*KBDO(FM) Des Arc AR
KTCN(FM) Eureka Springs AR
*KARH(FM) Forrest City AR
*KAOW(FM) Fort Smith AR
*KEAF(FM) Fort Smith AR
*KLRO(FM) Hot Springs AR
*KAOG(FM) Jonesboro AR
KUUZ(FM) Lake Village AR
KAAY(AM) Little Rock AR
KWLR(FM) Maumelle AR
*KCMH(FM) Mountain Home AR
*KANX(FM) Sheridan AR
KMTL(AM) Sherwood AR
*KLRC(FM) Siloam Springs AR
*KRMB(FM) Bisbee AZ
*KWRB(FM) Bisbee AZ
KPXQ(AM) Glendale AZ
*KNLB(FM) Lake Havasu City AZ
KRIT(FM) Parker AZ
*KWFH(FM) Parker AZ
KASA(AM) Phoenix AZ
*KPHF(FM) Phoenix AZ
*KGCB(FM) Prescott AZ
KXXT(AM) Tolleson AZ
KTBA(AM) Tuba City AZ
KWIM(FM) Window Rock AZ
*KYRM(FM) Yuma AZ
*KNDL(FM) Angwin CA
KLFF(AM) Arroyo Grande CA
KBRT(AM) Avalon CA
*KFRB(FM) Bakersfield CA
KLHC(AM) Bakersfield CA
*KWTH(FM) Barstow CA
*KWTW(FM) Bishop CA
*KIBC(FM) Burney CA
*KMRO(FM) Camarillo CA
KFIA(AM) Carmichael CA
*KHAP(FM) Chico CA
*KFRJ(FM) China Lake CA
*KBXO(FM) Coachella CA
*KPSH(FM) Coachella CA
*KFRP(FM) Coalinga CA
KRDU(AM) Dinuba CA
*KECR(AM) El Cajon CA
*KASK(FM) Fairfield CA
KJPG(AM) Frazier Park CA
*KEYQ(FM) Fresno CA
*KFNO(FM) Fresno CA
KYNO(AM) Fresno CA
KKMC(AM) Gonzales CA
*KPJP(FM) Greenville CA
KESQ(AM) Indio CA
KTYM(AM) Inglewood CA
*KJCU(FM) Laytonville CA

*KEFR(FM) Le Grand CA
KJOP(AM) Lemoore CA
*KLWG(FM) Lompoc CA
KLTX(AM) Long Beach CA
KHOT(AM) Madera CA
*KADV(FM) Modesto CA
KKXX(AM) Paradise CA
KCBC(AM) Riverbank CA
KPRO(AM) Riverside CA
*KSGN(FM) Riverside CA
*KEBR(AM) Rocklin CA
*KEAR-FM Sacramento CA
KEZY(AM) San Bernardino CA
KWVE(FM) San Clemente CA
KSDO(AM) San Diego CA
KEAR(AM) San Francisco CA
KFAX(AM) San Francisco CA
KJDJ(AM) San Luis Obispo CA
*KHFR(FM) Santa Maria CA
*KGZO(FM) Shafter CA
*KFRS(FM) Soledad CA
*KWG(AM) Stockton CA
*KPRA(FM) Ukiah CA
KDIA(AM) Vallejo CA
KRSX(AM) Victorville CA
KERI(AM) Wasco-Greenacres CA
KSMH(AM) West Sacramento CA
KRKS-FM Boulder CO
KLTT(AM) Commerce City CO
KRKS(AM) Denver CO
*KCIC(FM) Grand Junction CO
KFEL(AM) Pueblo CO
KGFT(FM) Pueblo CO
WADS(AM) Ansonia CT
*WFAR(FM) Danbury CT
*WJMJ(FM) Hartford CT
WFIF(AM) Milford CT
*WCTF(AM) Vernon CT
WHIM(AM) Apopka FL
WQOP(AM) Atlantic Beach FL
WSWN(AM) Belle Glade FL
*WYFZ(FM) Belleview FL
WLVJ(AM) Boynton Beach FL
*WRMB(FM) Boynton Beach FL
WTCL(AM) Chattahoochee FL
*WMIE(FM) Cocoa FL
WWBC(AM) Cocoa FL
WMFJ(AM) Daytona Beach FL
WYND(AM) De Land FL
*WJED(FM) Dogwood Lakes Estate FL
*WKTO(FM) Edgewater FL
*WMFL(FM) Florida City FL
*WAFG(FM) Fort Lauderdale FL
*WJYO(FM) Fort Myers FL
WMYR(AM) Fort Myers FL
*WJFP(FM) Fort Pierce FL
*WYFB(FM) Gainesville FL
WACC(AM) Hialeah FL
WOIR(AM) Homestead FL
WCGL(AM) Jacksonville FL
*WJFR(FM) Jacksonville FL
*WJIR(FM) Key West FL
WOTS(AM) Kissimmee FL
*WKES(FM) Lakeland FL
*WYFO(FM) Lakeland FL
*WAPB(FM) Madison FL
*WJNF(FM) Marianna FL
*WGSG(FM) Mayo FL
WCIF(FM) Melbourne FL
*WMCU(FM) Miami FL
*WFRF-FM Monticello FL
*WJLU(FM) New Smyrna Beach FL
*WJTF(FM) Panama City FL
WDWR(AM) Pensacola FL
*WPCS(FM) Pensacola FL
*WVIJ(FM) Port Charlotte FL
*WKFA(FM) Saint Catherine FL
*WUJC(FM) Saint Marks FL
*WFTI-FM Saint Petersburg FL
WDEO-FM San Carlos Park FL
*WTLG(FM) Starke FL
*WWFR(FM) Stuart FL
WTAL(AM) Tallahassee FL
WTIS(AM) Tampa FL
*WYFE(FM) Tarpon Springs FL
*WPIO(FM) Titusville FL

*WWXC(FM) Albany GA
WLTA(AM) Alpharetta GA
*WFRP(FM) Americus GA
WAEC(AM) Atlanta GA
WGUN(AM) Atlanta GA
WFAM(AM) Augusta GA
WBAF(AM) Barnesville GA
WUFE(AM) Baxley GA
WGMI(AM) Bremen GA
*WCCV(FM) Cartersville GA
*WFRC(FM) Columbus GA
*WYFK(FM) Columbus GA
*WWEV-FM Cumming GA
WDPC(AM) Dallas GA
WDCY(AM) Douglasville GA
*WMVV(FM) Griffin GA
WWWE(AM) Hapeville GA
WGML(AM) Hinesville GA
*WOAK(FM) La Grange GA
WBML(AM) Macon GA
*WYFS(FM) Savannah GA
*WRAF-FM Toccoa Falls GA
WAFT(FM) Valdosta GA
WGOV(AM) Valdosta GA
WYFA(FM) Waynesboro GA
*WYFW(FM) Winder GA
*KHMG(FM) Barrigada GU
KTWG(AM) Hagatna GU
*KCIF(FM) Hilo HI
KHNR(AM) Honolulu HI
KLHT(AM) Honolulu HI
KFFF(AM) Boone IA
KFFF-FM Boone IA
*KDFR(FM) Des Moines IA
KWKY(AM) Des Moines IA
*KEGR(FM) Fort Dodge IA
*KDCR(FM) Sioux Center IA
KTFC(FM) Sioux City IA
KTFG(FM) Sioux Rapids IA
*KNWS(AM) Waterloo IA
*KNWS-FM Waterloo IA
KNJY(AM) Boise ID
KBXL(FM) Caldwell ID
*KWJT(FM) Rathdrum ID
*KBYR-FM Rexburg ID
*KAWZ(FM) Twin Falls ID
*KEFX(FM) Twin Falls ID
*KTWD(FM) Wallace ID
*WBJW(FM) Albion IL
WXAN(FM) Ava IL
WRMS(AM) Beardstown IL
*WOLG(FM) Carlinville IL
*WTSG(FM) Carlinville IL
*WBGL(FM) Champaign IL
*WMBI-FM Chicago IL
WYLL(AM) Chicago IL
*WDLM(AM) East Moline IL
*WDLM-FM East Moline IL
*WCBW-FM East St. Louis IL
*WRYT(AM) Edwardsville IL
*WGNN(FM) Fisher IL
WBGX(AM) Harvey IL
*WJCH(FM) Joliet IL
*WGSL(FM) Loves Park IL
WGGH(AM) Marion IL
*WOTW(FM) Monee IL
*WBNH(FM) Pekin IL
*WCIC(FM) Pekin IL
WVEL(AM) Pekin IL
WPEO(AM) Peoria IL
*WGNJ(FM) Saint Joseph IL
WAUR(AM) Sandwich IL
*WSOG(FM) Spring Valley IL
*WIHM(AM) Taylorville IL
WAIT(AM) Willow Springs IL
WGLL(AM) Auburn IN
*WSPM(FM) Cloverdale IN
*WFOF(FM) Covington IN
WVHI(AM) Evansville IN
*WBGW(FM) Fort Branch IN
WLYV(AM) Fort Wayne IN
*WHLP(FM) Hanna IN
WBRI(AM) Indianapolis IN
*WIWC(FM) Kokomo IN
*WYTJ(FM) Linton IN
*WBHW(FM) Loogootee IN

*WTMK(FM) Lowell IN
*WMBL(FM) Mitchell IN
WNDZ(AM) Portage IN
WHLY(AM) South Bend IN
*WCRT-FM Terre Haute IN
*WHOJ(FM) Terre Haute IN
*WKRY(FM) Versailles IN
*WATI(FM) Vincennes IN
*WHPL(FM) West Lafayette IN
KAIR(AM) Atchison KS
*KBJQ(FM) Bronson KS
KCLY(FM) Clay Center KS
*KHYM(FM) Copeland KS
*KJIL(FM) Copeland KS
KAHS(AM) El Dorado KS
*KPOR(FM) Emporia KS
*KVCY(FM) Fort Scott KS
*KGCR(FM) Goodland KS
*KPRD(FM) Hays KS
KREJ(FM) Medicine Lodge KS
KJRG(AM) Newton KS
KCCV-FM Olathe KS
KCCV(AM) Overland Park KS
*KAKA(FM) Salina KS
KYUL(AM) Scott City KS
KCVT(FM) Silver Lake KS
*KBUZ(FM) Topeka KS
*KJTY(FM) Topeka KS
KSGL(AM) Wichita KS
*KYWA(FM) Wichita KS
WYWY(AM) Barbourville KY
WLJC(FM) Beattyville KY
*WCVK(FM) Bowling Green KY
WKDP(AM) Corbin KY
WCPM(AM) Cumberland KY
WWLK(AM) Eddyville KY
WIOK(FM) Falmouth KY
WJMD(FM) Hazard KY
WKVG(AM) Jenkins KY
WDFB(AM) Junction City KY
WJMM-FM Keene KY
WYGE(FM) London KY
*WSOF-FM Madisonville KY
*WBFI(FM) McDaniels KY
WMSK(AM) Morganfield KY
*WAXG(FM) Mt. Sterling KY
WLCR(AM) Mt. Washington KY
WVSG(AM) Neon KY
WNOP(AM) Newport KY
*WKWC(FM) Owensboro KY
*WPTJ(FM) Paris KY
*WJSO(FM) Pikeville KY
WLCK(AM) Scottsville KY
*WTHL(FM) Somerset KY
WMTC(AM) Vancleve KY
WMTC-FM Vancleve KY
WBCE(AM) Wickliffe KY
WEKC(AM) Williamsburg KY
KJMJ(AM) Alexandria LA
KBZE(FM) Berwick LA
WIKC(AM) Bogalusa LA
KAJN-FM Crowley LA
*KVDP(FM) Dry Prong LA
*KPAE(FM) Erwinville LA
*WCKW(AM) Garyville LA
KKNO(AM) Gretna LA
*KAYT(FM) Jena LA
*KOJO(FM) Lake Charles LA
*KBIO(FM) Natchitoches LA
WLNO(AM) New Orleans LA
WPFC(AM) Port Allen LA
KTKC(FM) Springhill LA
WSRO(AM) Ashland MA
WEZE(AM) Boston MA
WACE(AM) Chicopee MA
*WJWT(FM) Gardner MA
WVNE(AM) Leicester MA
WCMX(AM) Leominster MA
*WSMA(FM) Scituate MA
WNEB(AM) Worcester MA
*WFSI(FM) Annapolis MD
WBMD(AM) Baltimore MD
*WOEL-FM Elkton MD
WFRB(AM) Frostburg MD
*WLIC(FM) Frostburg MD
*WAIJ(FM) Grantsville MD

WOLC(FM) Princess Anne MD
*WGTS(FM) Takoma Park MD
WWPN(FM) Westernport MD
*WCRH(FM) Williamsport MD
*WFST(AM) Caribou ME
*WYFP(FM) Harpswell ME
WTME(AM) Rumford ME
*WSEW(FM) Sanford ME
WKTQ(AM) South Paris ME
WMAX(AM) Bay City MI
WLJW(AM) Cadillac MI
*WOLW(FM) Cadillac MI
*WTCK(FM) Charlevoix MI
WLCM(AM) Charlotte MI
WLQV(AM) Detroit MI
*WLJN(AM) Elmwood Township MI
*WPHN(FM) Gaylord MI
WGDN(AM) Gladwin MI
WFUR(AM) Grand Rapids MI
WFUR-FM Grand Rapids MI
WKPR(AM) Kalamazoo MI
*WEUL(FM) Kingsford MI
*WMPC(AM) Lapeer MI
*WHWL(FM) Marquette MI
WRDT(AM) Monroe MI
WOAP(AM) Owosso MI
WMKD(FM) Pickford MI
*WPCJ(FM) Pittsford MI
*WNFA(FM) Port Huron MI
*WMSD(FM) Rose Township MI
*WNFR(FM) Sandusky MI
*WOFR(FM) Schoolcraft MI
*KTGG(AM) Spring Arbor MI
*WUFL(AM) Sterling Heights MI
WHST(FM) Tawas City MI
*WLJN-FM Traverse City MI
*WHWG(FM) Trout Lake MI
*WGNB(FM) Zeeland MI
KJLY(FM) Blue Earth MN
KYCR(AM) Golden Valley MN
WCTS(AM) Maplewood MN
*KTIS(AM) Minneapolis MN
KVOX(AM) Moorhead MN
KCGN-FM Ortonville MN
KCUE(AM) Red Wing MN
KCWJ(AM) Blue Springs MO
KOMC(AM) Branson MO
*KOZO(FM) Branson MO
KMFC(FM) Centralia MO
*KFUO(AM) Clayton MO
*KJCV(FM) Country Club MO
KDJR(FM) De Soto MO
*KEXS(AM) Excelsior Springs MO
*KMCV(FM) High Point MO
KBCV(AM) Hollister MO
KUGT(AM) Jackson MO
*KHGN(FM) Kirksville MO
KLEX(AM) Lexington MO
KMRF(AM) Marshfield MO
*KLUH(FM) Poplar Bluff MO
*KOKS(FM) Poplar Bluff MO
*KSIV-FM Saint Louis MO
KXEN(AM) Saint Louis MO
KRHW(AM) Sikeston MO
*KWFC(FM) Springfield MO
*KNLN(FM) Vienna MO
*WPAE(FM) Centreville MS
WTGY(FM) Charleston MS
WWUN-FM Friar's Point MS
*WAII(FM) Hattiesburg MS
WAML(AM) Laurel MS
*WATP(FM) Laurel MS
WESY(AM) Leland MS
WAKK(AM) McComb MS
*WAQL(FM) McComb MS
*WASM(FM) Natchez MS
*WATU(FM) Port Gibson MS
*WJZB(FM) Starkville MS
*WZKM(FM) Waynesboro MS
WJNS-FM Yazoo City MS
KGVW(AM) Belgrade MT
KURL(AM) Billings MT
KGLE(AM) Glendive MT
*KSPL(FM) Kalispell MT
*KPLG(FM) Plains MT
WCGC(AM) Belmont NC

WGCR(AM) Brevard NC
*WCCE(FM) Buie's Creek NC
WPTL(AM) Canton NC
WMYT(AM) Carolina Beach NC
WHVN(AM) Charlotte NC
WCKB(AM) Dunn NC
WRJD(AM) Durham NC
WLOE(AM) Eden NC
WFMO(AM) Fairmont NC
WWOL(AM) Forest City NC
WFMC(AM) Goldsboro NC
WYCV(AM) Granite Falls NC
WKEW(AM) Greensboro NC
WYFL(FM) Henderson NC
*WHPE-FM High Point NC
WMYN(AM) Mayodan NC
*WOTJ(FM) Morehead City NC
WDJS(AM) Mount Olive NC
WKRK(AM) Murphy NC
*WAAE(FM) New Bern NC
WAUG(AM) New Hope NC
WCLY(AM) Raleigh NC
WEED(AM) Rocky Mount NC
WEGG(AM) Rose Hill NC
WVCB(AM) Shallotte NC
*WAGO(FM) Snow Hill NC
WCOK(AM) Sparta NC
WGMA(AM) Spindale NC
*WYFQ-FM Wadesboro NC
WOBX(AM) Wanchese NC
WSMY(AM) Weldon NC
WIAM(AM) Williamston NC
WLSG(AM) Wilmington NC
WBTE(AM) Windsor NC
WPOL(AM) Winston-Salem NC
*WSNC(FM) Winston-Salem NC
*KBFR(FM) Bismarck ND
KHRT(AM) Minot ND
*KFNW(AM) West Fargo ND
KPNY(FM) Alliance NE
*KNBE(FM) Beatrice NE
*KTLX(FM) Columbus NE
*KNFA(FM) Grand Island NE
*KNHA(FM) Grand Isle NE
*KAYA(FM) Hubbard NE
*KLCV(FM) Lincoln NE
*KNGN(AM) McCook NE
*KPNO(FM) Norfolk NE
*KVSS(FM) Omaha NE
KCMI(FM) Terrytown NE
WDER(AM) Derry NH
*WVFA(FM) Lebanon NH
*WYGG(FM) Asbury Park NJ
*WKDN-FM Camden NJ
WTMR(AM) Camden NJ
WCHR(AM) Flemington NJ
WIFI(AM) Florence NJ
*WDDM(FM) Hazlet NJ
WXMC(AM) Parsippany-Troy Hills NJ
*WJPH(FM) Woodbine NJ
*KFLQ(FM) Albuquerque NM
KNKT(FM) Armijo, Albuquerque NM
*KZPI(FM) Deming NM
*KRUC(FM) Las Cruces NM
*KXXQ(FM) Milan NM
KCKN(AM) Roswell NM
*KWFL(FM) Roswell NM
KHAC(AM) Tse Bonito NM
KKVV(AM) Las Vegas NV
*KIHM(AM) Reno NV
KXTO(AM) Reno NV
WLOF(FM) Attica NY
WDCX(FM) Buffalo NY
*WMHI(FM) Cape Vincent NY
WCHP(AM) Champlain NY
WJIV(FM) Cherry Valley NY
WSIV(AM) East Syracuse NY
WLNL(AM) Horseheads NY
*WFRH(FM) Kingston NY
WBAR-FM Lake Luzerne NY
WVOA-FM Mexico NY
WTHE(AM) Mineola NY
WJUX(FM) Monticello NY
*WWRV(AM) New York NY
*WFSO(FM) Olivebridge NY
WZXV(FM) Palmyra NY

WHIC(AM) Rochester NY
WYFY(AM) Rome NY
*WFRS(FM) Smithtown NY
*WFRW(FM) Webster NY
*WOFN(FM) Beach City OH
WJYM(AM) Bowling Green OH
WILB(AM) Canton OH
WGGN(FM) Castalia OH
*WCDR-FM Cedarville OH
*WOHC(FM) Chillicothe OH
WCVX(AM) Cincinnati OH
*WCRF(FM) Cleveland OH
*WHVT(FM) Clyde OH
WRFD(AM) Columbus-Worthington OH
WWOW(AM) Conneaut OH
*WCUE(AM) Cuyahoga Falls OH
*WJJE(FM) Delaware OH
WGNZ(AM) Fairborn OH
WNLT(FM) Harrison OH
*WFOT(FM) Lexington OH
*WVML(FM) Millersburg OH
WCCD(AM) Parma OH
*WOHP(FM) Portsmouth OH
*WLRY(FM) Rushville OH
*WVMS(FM) Sandusky OH
*WCVZ(FM) South Zanesville OH
*WEEC(FM) Springfield OH
*WOTL(FM) Toledo OH
*WXML(FM) Upper Sandusky OH
WASN(AM) Youngstown OH
*WYTN(FM) Youngstown OH
*WJIC(FM) Zanesville OH
*KKVO(FM) Altus OK
*KNYD(FM) Broken Arrow OK
KEOR(AM) Catoosa OK
*KDIM(FM) Coweta OK
*KKRD(FM) Enid OK
KMFS(AM) Guthrie OK
*KALU(FM) Langston OK
*KMSI(FM) Moore OK
KQCV(AM) Oklahoma City OK
KXOJ-FM Sapulpa OK
KCFO(AM) Tulsa OK
*KORM(FM) Astoria OR
*KDJC(FM) Baker City OR
*KJCH(FM) Coos Bay OR
*KDOV(FM) Medford OR
*KLRF(FM) Milton-Freewater OR
*KBVM(FM) Portland OR
*KAJC(FM) Salem OR
*KQFE(FM) Springfield OR
KLWJ(AM) Umatilla OR
KGRV(AM) Winston OR
*WITX(FM) Beaver Falls PA
WHJB(AM) Bedford PA
*WFKJ(AM) Cashtown PA
*WZXQ(FM) Chambersburg PA
WPWA(AM) Chester PA
WVCH(AM) Chester PA
*WPGM(AM) Danville PA
*WPGM-FM Danville PA
WISP(AM) Doylestown PA
*WEFR(FM) Erie PA
WPPZ-FM Jenkintown PA
WJSA(AM) Jersey Shore PA
WJSA-FM Jersey Shore PA
WJSM(AM) Martinsburg PA
WJSM-FM Martinsburg PA
*WRIJ(FM) Masontown PA
*WPEL-FM Montrose PA
*WRWJ(FM) Murrysville PA
*WBGM(FM) New Berlin PA
*WVMN(FM) New Castle PA
WPCL(FM) Northern Cambria PA
WQOR(AM) Olyphant PA
WFIL(AM) Philadelphia PA
WPHE(AM) Phoenixville PA
WITK(AM) Pittston PA
WGLD(AM) Red Lion PA
*WBYO(FM) Sellersville PA
*WBYX(FM) Stroudsburg PA
*WRGN(FM) Sweet Valley PA
*WBMR(FM) Telford PA
WLIH(FM) Whitneyville PA
WFAB(AM) Ceiba PR
*WJVP(FM) Culebra PR

WCGB(AM) Juana Diaz PR
WRRE(AM) Juncos PR
*WPLI(FM) Levittown PR
WZOL(FM) Luquillo PR
WCHQ(AM) Quebradillas PR
WBMJ(AM) San Juan PR
WKVM(AM) San Juan PR
WIVV(AM) Vieques PR
WARV(AM) Warwick RI
WMIR(AM) Atlantic Beach SC
WVCD(AM) Bamberg-Denmark SC
*WAGP(FM) Beaufort SC
WVGB(AM) Beaufort SC
*WEPC(FM) Belton SC
*WYFV(FM) Cayce SC
*WFCH(FM) Charleston SC
*WMHK(FM) Columbia SC
WQXL(AM) Columbia SC
WELP(AM) Easley SC
WFGN(AM) Gaffney SC
*WYFG(FM) Gaffney SC
WMUU(AM) Greenville SC
*WTBI-FM Greenville SC
WBZF(FM) Hartsville SC
WRIX(AM) Homeland Park SC
WKZK(AM) North Augusta SC
*WYFH(FM) North Charleston SC
*WKVC(FM) North Myrtle Beach SC
WNMB(AM) North Myrtle Beach SC
WPJK(AM) Orangeburg SC
WAVO(AM) Rock Hill SC
WQIZ(AM) Saint George SC
WDRF(AM) Woodruff SC
*KKAA(AM) Aberdeen SD
*KVCF(FM) Freeman SD
*KVCX(FM) Gregory SD
*KVFL(FM) Pierre SD
*KQFR(FM) Rapid City SD
*KQKD(AM) Redfield SD
*KNWC(AM) Sioux Falls SD
*KJBB(FM) Watertown SD
WWGM(FM) Alamo TN
WBIN(AM) Benton TN
*WDYN-FM Chattanooga TN
WLMR(AM) Chattanooga TN
*WMBW(FM) Chattanooga TN
*WYFC(FM) Clinton TN
WSGM(FM) Coalmont TN
WCRV(AM) Collierville TN
*WAYM(FM) Columbia TN
WMRB(AM) Columbia TN
*WWOG(FM) Cookeville TN
*WMKW(FM) Crossville TN
WYXE(AM) Gallatin TN
WCLC(AM) Jamestown TN
WCLC-FM Jamestown TN
WDEB(AM) Jamestown TN
WKXV(AM) Knoxville TN
WLAF(AM) La Follette TN
WMQM(AM) Lakeland TN
*WAWI(FM) Lawrenceburg TN
WBLC(AM) Lenoir City TN
WFLI(AM) Lookout Mountain TN
WBBP(AM) Memphis TN
WMCT(AM) Mountain City TN
*WFCM-FM Murfreesboro TN
WENO(AM) Nashville TN
WLAC(AM) Nashville TN
WNQM(AM) Nashville TN
WYFN(AM) Nashville TN
*WDNX(FM) Olive Hill TN
WKJQ(AM) Parsons TN
*WAUV(FM) Ripley TN
*WAZD(FM) Savannah TN
*WFCM(AM) Smyrna TN
WSTN(AM) Somerville TN
*WZYZ(FM) Spencer TN
*WAUT-FM Tullahoma TN
KDRY(AM) Alamo Heights TX
*KJRT(FM) Amarillo TX
KRGN(FM) Amarillo TX
KWWJ(AM) Baytown TX
KPDB(FM) Big Lake TX
KMZZ(FM) Bishop TX
*KASV(FM) Borger TX
*KBNR(FM) Brownsville TX

KBEN(AM) Carrizo Springs TX
KJON(AM) Carrollton TX
*KBNJ(FM) Corpus Christi TX
KCTA(AM) Corpus Christi TX
KGGR(AM) Dallas TX
*KDKR(FM) Decatur TX
KVOZ(AM) Del Mar Hills TX
*KYFB(FM) Denison TX
KDHN(AM) Dimmitt TX
*KOIR(FM) Edinburg TX
*KVER(FM) El Paso TX
KVIV(AM) El Paso TX
KIJN(AM) Farwell TX
KIJN-FM Farwell TX
*KJMA(FM) Floresville TX
KFST(AM) Fort Stockton TX
KGBC(AM) Galveston TX
*KHCB(AM) Galveston TX
*KBJS(FM) Jacksonville TX
*KJCR(FM) Keene TX
KRMY(AM) Killeen TX
KINE(AM) Kingsville TX
*KBNL(FM) Laredo TX
KYMI(FM) Los Ybanez TX
*KAMY(FM) Lubbock TX
*KBIB(AM) Marion TX
KRIO(AM) McAllen TX
*KCAS(FM) McCook TX
KJBC(AM) Midland TX
*KFLB(AM) Odessa TX
*KFLB-FM Odessa TX
*KYFP(FM) Palestine TX
*KPMB(FM) Plainview TX
*KDEI(AM) Port Arthur TX
KMFM(FM) Premont TX
KBIC(FM) Raymondville TX
KCRN(AM) San Angelo TX
KCRN-FM San Angelo TX
KTUE(AM) Tulia TX
*KGLY(FM) Tyler TX
*KVNE(FM) Tyler TX
*KXBJ(FM) Victoria TX
KANI(AM) Wharton TX
*KPDR(FM) Wheeler TX
KNAK(AM) Delta UT
KHQN(AM) Spanish Fork UT
WABN(AM) Abingdon VA
WYFJ(FM) Ashland VA
WLSD(AM) Big Stone Gap VA
WZAP(AM) Bristol VA
WNLR(AM) Churchville VA
WKTR(AM) Earlysville VA
*WOTC(FM) Edinburg VA
WKNV(AM) Fairlawn VA
WFAX(AM) Falls Church VA
WPAK(AM) Farmville VA
WYFT(FM) Luray VA
WKPA(AM) Lynchburg VA
*WRVL(FM) Lynchburg VA
WNRV(AM) Narrows-Pearisburg VA
WYFI(FM) Norfolk VA
WYRM(AM) Norfolk VA
WSWV(AM) Pennington Gap VA
WPCE(AM) Portsmouth VA
WGTH-FM Richlands VA
WREJ(AM) Richmond VA
WRIS(AM) Roanoke VA
WXLZ(AM) Saint Paul VA
WKGM(AM) Smithfield VA
WKBA(AM) Vinton VA
*WYCS(FM) Yorktown VA
WGOD-FM Charlotte Amalie VI
*WIVH(FM) Christiansted VI
WTWN(AM) Wells River VT
KARI(AM) Blaine WA
*KACS(FM) Chehalis WA
KSPO(FM) Dishman WA
*KCSH(FM) Ellensburg WA
KTAC(FM) Ephrata WA
KTBI(AM) Ephrata WA
*KARR(AM) Kirkland WA
KWPZ(FM) Lynden WA
*KMLW(FM) Moses Lake WA
*KOLU(FM) Pasco WA
*KBLE(AM) Seattle WA
*KMBI(AM) Spokane WA

*KMBI-FM Spokane WA
KTTO(AM) Spokane WA
KHSS(FM) Walla Walla WA
*KSOH(FM) Wapato WA
*KPLW(FM) Wenatchee WA
*KBNO-FM White Salmon WA
*WEMI(FM) Appleton WI
*WOVM(FM) Appleton WI
WYNW(FM) Birnamwood WI
*WDVM(AM) Eau Claire WI
*WVCF(FM) Eau Claire WI
*WVFL(FM) Fond du Lac WI
*WEMY(FM) Green Bay WI
*WJTY(FM) Lancaster WI
*WJWD(FM) Marshall WI
*WVCY-FM Milwaukee WI
WMMA(FM) Nekoosa WI
*WVCY(AM) Oshkosh WI
WHFA(AM) Poynette WI
*WNWC(AM) Sun Prairie WI
*WVCX(FM) Tomah WI
*WEGZ(FM) Washburn WI
*WVRN(FM) Wittenberg WI
WJLS(AM) Beckley WV
*WKJL(FM) Clarksburg WV
WEMM(AM) Huntington WV
WFSP(AM) Kingwood WV
WOTR(FM) Lost Creek WV
WMON(AM) Montgomery WV
WOAY(AM) Oak Hill WV
WELD-FM Petersburg WV
WBGS(AM) Point Pleasant WV
WJYP(AM) Saint Albans WV
WMXE(FM) South Charleston WV
WSCW(AM) South Charleston WV
KUYO(AM) Evansville WY
*KWCF(FM) Sheridan WY

Rock/AOR

KWHL(FM) Anchorage AK
*KCUK(FM) Chevak AK
*KCDS(FM) Deadhorse AK
*KDLG(AM) Dillingham AK
KKED(FM) Fairbanks AK
KKED(FM) Fairbanks AK
KXLW(FM) Houston AK
KZND-FM Houston AK
KSUP(FM) Juneau AK
WYSF(FM) Birmingham AL
WRTR(FM) Brookwood AL
WESP(FM) Dothan AL
WENN(FM) Hoover AL
WRTT-FM Huntsville AL
WQNR(FM) Tallassee AL
*WVUA-FM Tuscaloosa AL
WTGZ(FM) Tuskegee AL
KCCB(AM) Corning AR
KKEG(FM) Fayetteville AR
KTRN(FM) White Hall AR
KDKB(FM) Mesa AZ
KUPD-FM Tempe AZ
KMXZ-FM Tucson AZ
KWCX(FM) Willcox AZ
KWMX(FM) Williams AZ
*KAWC(AM) Yuma AZ
KBRE(FM) Atwater CA
KRRX(FM) Burney CA
KWTY(FM) Cartago CA
KFMF(FM) Chico CA
KSSI(FM) China Lake CA
KCLB-FM Coachella CA
*KVHS(FM) Concord CA
*KDVS(FM) Davis CA
KSOF(FM) Dinuba CA
KRAB(FM) Green Acres CA
KURQ(FM) Grover Beach CA
KRZR(FM) Hanford CA
KCXX(FM) Lake Arrowhead CA
*KRQZ(FM) Lompoc CA
KFFG(FM) Los Altos CA
KLOS(FM) Los Angeles CA
*KXLU(FM) Los Angeles CA
KIXA(FM) Lucerne Valley CA
KRCK-FM Mecca CA
*KAKX(FM) Mendocino CA

KMFB(FM) Mendocino CA
KHKK(FM) Modesto CA
KJEE(FM) Montecito CA
*KSFH(FM) Mountain View CA
KRQR(FM) Orland CA
KMRJ(FM) Rancho Mirage CA
KCAL-FM Redlands CA
KRXQ(FM) Sacramento CA
KIOZ(FM) San Diego CA
KYXY(FM) San Diego CA
KFOG(FM) San Francisco CA
KITS(FM) San Francisco CA
*KSJS(FM) San Jose CA
KZOZ(FM) San Luis Obispo CA
*KSCU(FM) Santa Clara CA
KXFX(FM) Santa Rosa CA
KMKX(FM) Willits CA
KILO(FM) Colorado Springs CO
KKDC(FM) Dolores CO
KEPL(AM) Estes Park CO
KRKV(FM) Las Animas CO
*KDRE(FM) Sterling CO
*WERB(FM) Berlin CT
*WQAQ(FM) Hamden CT
WCCC-FM Hartford CT
WHCN(FM) Hartford CT
WPLR(FM) New Haven CT
WPHH(FM) Waterbury CT
*WECS(FM) Willimantic CT
WWDC-FM Washington DC
WOSC(FM) Bethany Beach DE
WZBH(FM) Georgetown DE
WYYX(FM) Bonifay FL
WNDN(FM) Chiefland FL
WXTB(FM) Clearwater FL
WJRR(FM) Cocoa Beach FL
WJBX(FM) Fort Myers Beach FL
WJBX(FM) Fort Myers Beach FL
WKSM(FM) Fort Walton Beach FL
WRUF-FM Gainesville FL
WHDR(FM) Miami FL
WHHZ(FM) Newberry FL
WMMO(FM) Orlando FL
WHOG-FM Ormond-by-the-Sea FL
WFKZ(FM) Plantation Key FL
WXSR(FM) Quincy FL
WSJZ-FM Sebastian FL
WGLF(FM) Tallahassee FL
WLTQ-FM Venice FL
*WRAS(FM) Atlanta GA
WHFX(FM) Darien GA
WCHZ(FM) Harlem GA
WJAD(FM) Leesburg GA
WRHQ(FM) Richmond Hill GA
WRXR-FM Rossville GA
WWRQ-FM Valdosta GA
KUMU-FM Honolulu HI
KQCS(FM) Bettendorf IA
KBOB-FM De Witt IA
KAZR(FM) Pella IA
KSEZ(FM) Sioux City IA
KCVI(FM) Blackfoot ID
KZDX(FM) Burley ID
KHTQ(FM) Hayden ID
KOZE-FM Lewiston ID
KQXR(FM) Payette ID
KSNA(FM) Rexburg ID
KIRQ(FM) Twin Falls ID
WXRX(FM) Belvidere IL
WRXX(FM) Centralia IL
WWGO(FM) Charleston IL
WKQX(FM) Chicago IL
WLUP-FM Chicago IL
WXRT-FM Chicago IL
WXLT(FM) Christopher IL
*WRSE(FM) Elmhurst IL
WKTA(AM) Evanston IL
*WHFH(FM) Flossmoor IL
WLSR(FM) Galesburg IL
*WLCA(FM) Godfrey IL
*WCSF(FM) Joliet IL
*WLTL(FM) La Grange IL
WDQZ(FM) Lexington IL
*WLNX(FM) Lincoln IL
WNLF(FM) Macomb IL
*WLKL(FM) Mattoon IL

WTAO(FM) Murphysboro IL
*WONC(FM) Naperville IL
WIHN(FM) Normal IL
WIXO(FM) Peoria IL
WMYE(FM) Rantoul IL
WQLZ(FM) Taylorville IL
WEBX(FM) Tuscola IL
WPGU(FM) Urbana IL
WDML(FM) Woodlawn IL
WSDM-FM Brazil IN
*WDSO(FM) Chesterton IN
WTFX-FM Clarksville IN
*WRGF(FM) Greenfield IN
*WBDG(FM) Indianapolis IN
WRZX(FM) Indianapolis IN
WYXB(FM) Indianapolis IN
WKHY(FM) Lafayette IN
*WCYT(FM) Lafayette Township IN
WWWY(FM) North Vernon IN
WBTO-FM Petersburg IN
*WPUM(FM) Rensselaer IN
WWSY(FM) Seelyville IN
WRBR-FM South Bend IN
*WISU(FM) Terre Haute IN
KJML(FM) Columbus KS
KMDO(AM) Fort Scott KS
*KJHK(FM) Lawrence KS
KQRC-FM Leavenworth KS
KMKF(FM) Manhattan KS
*KSDB-FM Manhattan KS
KFIX(FM) Plainville KS
KICT-FM Wichita KS
*KSWC(FM) Winfield KS
WRVC-FM Catlettsburg KY
WPKE-FM Coal Run KY
WXZZ(FM) Georgetown KY
WGBF-FM Henderson KY
WKCB-FM Hindman KY
WIFX-FM Jenkins KY
WKTG(FM) Madisonville KY
WGKS(FM) Paris KY
WZZL(FM) Reidland KY
WRZI(FM) Vine Grove KY
WXCM(FM) Whitesville KY
KFTE(FM) Breaux Bridge LA
KNXX(FM) Donaldsonville LA
KYRK(FM) Houma LA
KXRR(FM) Monroe LA
*KXUL(FM) Monroe LA
KKND(FM) Port Sulphur LA
KXOR-FM Thibodaux LA
WMJX(FM) Boston MA
*WBMT(FM) Boxford MA
WHRB(FM) Cambridge MA
*WIQH(FM) Concord MA
WPXC(FM) Hyannis MA
WLZX(FM) Northampton MA
*WMHC(FM) South Hadley MA
WWFX(FM) Southbridge MA
*WZLY(FM) Wellesley MA
WYRX(FM) Lexington Park MD
WXMD(FM) Pocomoke City MD
WCYY(FM) Biddeford ME
*WUMF-FM Farmington ME
*WRBC(FM) Lewiston ME
*WMEB-FM Orono ME
WTOS-FM Skowhegan ME
*WQAC-FM Alma MI
*WCHW-FM Bay City MI
WRIF(FM) Detroit MI
WBNZ(FM) Frankfort MI
WGFN(FM) Glen Arbor MI
WJZJ(FM) Glen Arbor MI
WKLQ(FM) Greenville MI
WGLI(FM) Hancock MI
WKLT(FM) Kalkaska MI
WQUS(FM) Lapeer MI
WLJZ(FM) Mackinaw City MI
WKQZ(FM) Midland MI
WAOR(FM) Niles MI
*WBLD(FM) Orchard Lake MI
WKLZ-FM Petoskey MI
WRKR(FM) Portage MI
WIRX(FM) Saint Joseph MI
*WNMC-FM Traverse City MI
WWBN(FM) Tuscola MI

KAOD(FM) Babbitt MN
KQHT(FM) Crookston MN
KBAJ(FM) Deer River MN
KZCR(FM) Fergus Falls MN
KXXR(FM) Minneapolis MN
KQWB-FM Moorhead MN
KPHR(FM) Ortonville MN
*KDXL(FM) Saint Louis Park MN
WHMH-FM Sauk Rapids MN
KZIO(FM) Two Harbors MN
*KQAL(FM) Winona MN
*KSMR(FM) Winona MN
KDBB(FM) Bonne Terre MO
KFEB(FM) Campbell MO
*KCOU(FM) Columbia MO
KCGQ-FM Gordonville MO
KBBM(FM) Jefferson City MO
KRBZ(FM) Kansas City MO
KYYS(FM) Kansas City MO
*KWJC(FM) Liberty MO
KNIM-FM Maryville MO
KBHI(FM) Miner MO
KMYK(FM) Osage Beach MO
KJEZ(FM) Poplar Bluff MO
KFLW(FM) Saint Robert MO
KPNT(FM) Sainte Genevieve MO
WCPR-FM D'Iberville MS
WUSW(FM) Hattiesburg MS
WRXW(FM) Pearl MS
KQBL(FM) Billings MT
KRZN(FM) Billings MT
*KDWG(FM) Dillon MT
KQDI-FM Great Falls MT
KMHK(FM) Hardin MT
KBBZ(FM) Kalispell MT
KOZB(FM) Livingston MT
*KBGA(FM) Missoula MT
*WWCU(FM) Cullowhee NC
WRCQ(FM) Dunn NC
WTPT(FM) Forest City NC
WXNR(FM) Grifton NC
WVBZ(FM) High Point NC
WXQR(FM) Jacksonville NC
WRFX-FM Kannapolis NC
WOBX-FM Manteo NC
WMGV(FM) Newport NC
WBBB(FM) Raleigh NC
*WKNC-FM Raleigh NC
WEND(FM) Salisbury NC
*WDCC(FM) Sanford NC
WAZO(FM) Southport NC
WTMT(FM) Weaverville NC
*KEYA(FM) Belcourt ND
KQZZ(FM) Devils Lake ND
KZRX(FM) Dickinson ND
KDSR(FM) Williston ND
KRVX(FM) Wimbledon ND
KHUS(FM) Bennington NE
*KINI(FM) Crookston NE
KLZA(FM) Falls City NE
KOZY-FM Gering NE
*KFKX(FM) Hastings NE
*KLPR(FM) Kearney NE
KLMY(FM) Lincoln NE
KZMC(FM) McCook NE
KOGA-FM Ogallala NE
KEZO-FM Omaha NE
KKCD(FM) Omaha NE
KKJK(FM) Ravenna NE
*KWSC(FM) Wayne NE
WDCR(AM) Hanover NH
WFRD(FM) Hanover NH
WGIR-FM Manchester NH
WVRR(FM) Newport NH
*WPCR-FM Plymouth NH
WHEB(FM) Portsmouth NH
WWYY(FM) Belvidere NJ
WDHA-FM Dover NJ
*WNTI(FM) Hackettstown NJ
*WJSV(FM) Morristown NJ
WJSE(FM) Petersburg NJ
WRAT(FM) Point Pleasant NJ
*WSOU(FM) South Orange NJ
*WMCX(FM) West Long Branch NJ
KRWN(FM) Farmington NM
KFMQ(FM) Gallup NM

KXPZ(FM) Las Cruces NM
*KEDP(FM) Las Vegas NM
KPSA-FM Lordsburg NM
KQBA(FM) Los Alamos NM
KQNM(AM) Milan NM
*KRDR(FM) Red River NM
KSFX(FM) Roswell NM
KRSI(FM) Garapan-Saipan NP
KOYT(FM) Elko NV
KRNO(FM) Incline Village NV
KOMP(FM) Las Vegas NV
KDOT(FM) Reno NV
KRZQ-FM Sparks NV
*WCDB(FM) Albany NY
*WETD(FM) Alfred NY
WZMR(FM) Altamont NY
WRRB(FM) Arlington NY
WBAB(FM) Babylon NY
*WGCC-FM Batavia NY
WRCD(FM) Canton NY
*WHCL-FM Clinton NY
WQBJ(FM) Cobleskill NY
WKGB-FM Conklin NY
*WCEB(FM) Corning NY
*WSUC-FM Cortland NY
WEHN(FM) East Hampton NY
WKLL(FM) Frankfort NY
*WGFR(FM) Glens Falls NY
WOTT(FM) Henderson NY
WFXF(FM) Honeoye Falls NY
*WICB(FM) Ithaca NY
WVBR-FM Ithaca NY
WPDA(FM) Jeffersonville NY
WAQX-FM Manlius NY
WRRV(FM) Middletown NY
WKRH(FM) Minetto NY
WXRK(FM) New York NY
WKRL-FM North Syracuse NY
*WRHO(FM) Oneonta NY
*WDFH(FM) Ossining NY
*WNYO(FM) Oswego NY
*WPOB-FM Plainview NY
WRCN-FM Riverhead NY
*WSBU(FM) Saint Bonaventure NY
WEHM(FM) Southampton NY
WHFM(FM) Southampton NY
WOUR(FM) Utica NY
*WPNR-FM Utica NY
WDST(FM) Woodstock NY
WONE-FM Akron OH
*WZIP(FM) Akron OH
*WRMU(FM) Alliance OH
*WRDL(FM) Ashland OH
*WBWC(FM) Berea OH
WMMS(FM) Cleveland OH
WBZX(FM) Columbus OH
*WUFM(FM) Columbus OH
WTUE(FM) Dayton OH
WSWD(FM) Fairfield OH
WKXA-FM Findlay OH
*WLFC(FM) Findlay OH
WZRX-FM Fort Shawnee OH
WWCD(FM) Grove City OH
*WHSS(FM) Hamilton OH
*WKET(FM) Kettering OH
*WCMO(FM) Marietta OH
*WYSZ(FM) Maumee OH
WJZE(FM) Oak Harbor OH
*WUSO(FM) Springfield OH
*WSTB(FM) Streetsboro OH
WIOT(FM) Toledo OH
*WYSA(FM) Wauseon OH
*WLHS(FM) West Chester OH
*WOBN(FM) Westerville OH
*WCWS(FM) Wooster OH
WNCD(FM) Youngstown OH
KACO(FM) Apache OK
KKBS(FM) Guymon OK
KZCD(FM) Lawton OK
KMYZ-FM Pryor OK
KCXR(FM) Taft OK
KMOD-FM Tulsa OK
KZZE(FM) Eagle Point OR
KUFO-FM Portland OR
WZZO(FM) Bethlehem PA
*WBUQ(FM) Bloomsburg PA

WRKW(FM) Ebensburg PA
*WFSE(FM) Edinboro PA
*WVMM(FM) Grantham PA
WEXC(FM) Greenville PA
WRVV(FM) Harrisburg PA
WBSX(FM) Hazleton PA
WRKY-FM Hollidaysburg PA
*WKVR-FM Huntingdon PA
*WVBU-FM Lewisburg PA
*WNTE(FM) Mansfield PA
WEEO-FM McConnellsburg PA
WTPA(FM) Mechanicsburg PA
WWIZ(FM) Mercer PA
*WKDU(FM) Philadelphia PA
WRFF(FM) Philadelphia PA
WYSP(FM) Philadelphia PA
WJOW(FM) Philipsburg PA
WDVE(FM) Pittsburgh PA
WPAM(AM) Pottsville PA
*WXAC(FM) Reading PA
*WUSR(FM) Scranton PA
*WSYC-FM Shippensburg PA
WZXR(FM) South Williamsport PA
WMTT(FM) Tioga PA
WFYN(FM) Waynesboro PA
*WRKC(FM) Wilkes-Barre PA
*WPTC(FM) Williamsport PA
*WVYC(FM) York PA
WMEG(FM) Guayama PR
WCAD(FM) San Juan PR
WIAC-FM San Juan PR
*WQRI(FM) Bristol RI
WHJY(FM) Providence RI
WSUY(FM) Charleston SC
WARQ(FM) Columbia SC
WYBB(FM) Folly Beach SC
WKZQ-FM Myrtle Beach SC
KVAR(FM) Pine Ridge SD
*KAUR(FM) Sioux Falls SD
KDDX(FM) Spearfish SD
*KAOR(FM) Vermillion SD
WBXE(FM) Baxter TN
WDEF-FM Chattanooga TN
WDOD-FM Chattanooga TN
WSKZ(FM) Chattanooga TN
WRZK(FM) Colonial Heights TN
WZDQ(FM) Humboldt TN
*WUTK-FM Knoxville TN
WKXD(FM) Monterey TN
*KACC(FM) Alvin TX
KLBJ-FM Austin TX
KROX-FM Buda TX
KJXJ(FM) Cameron TX
*KWTS(FM) Canyon TX
KZRK-FM Canyon TX
KTUX(FM) Carthage TX
KLAQ(FM) El Paso TX
*KTCU-FM Fort Worth TX
KHFI-FM Georgetown TX
KHLT(AM) Hallettsville TX
KFRQ(FM) Harlingen TX
*KTAI(FM) Kingsville TX
KIOL(FM) La Porte TX
KFMX-FM Lubbock TX
KZRC(FM) Markham TX
KMJR(FM) Portland TX
KMDX(FM) San Angelo TX
KISS-FM San Antonio TX
*KVRK(FM) Sanger TX
KNCN(FM) Sinton TX
KLYD(FM) Snyder TX
KXRV(FM) Centerville UT
KBER(FM) Ogden UT
WBRW(FM) Blacksburg VA
WCJZ(FM) Charlottesville VA
WDYL(FM) Chester VA
WACL(FM) Elkton VA
WROX-FM Exmore VA
WFQX(FM) Front Royal VA
WZZU(FM) Lynchburg VA
WROV-FM Martinsville VA
WNOR(FM) Norfolk VA
WRXL(FM) Richmond VA
WYSK-FM Spotsylvania VA
*WNRS-FM Sweet Briar VA
WZZI(FM) Vinton VA

WEXP(FM) Brandon VT
*WWLR(FM) Lyndonville VT
*WNUB-FM Northfield VT
*WVTC(FM) Randolph Center VT
WIZN(FM) Vergennes VT
WMXR(FM) Woodstock VT
*KGRG-FM Auburn WA
KNRK(FM) Camas WA
*KCWU(FM) Ellensburg WA
KBIS(AM) Forks WA
*KZUU(FM) Pullman WA
KEGX(FM) Richland WA
KISW(FM) Seattle WA
*KWRS(FM) Spokane WA
*KWCW(FM) Walla Walla WA
KXRX(FM) Walla Walla WA
KATS(FM) Yakima WA
WAPL-FM Appleton WI
WDRK(FM) Cornell WI
WWHG(FM) Evansville WI
WIIL(FM) Kenosha WI
WRQT(FM) La Crosse WI
WIBA-FM Madison WI
WMZK(FM) Merrill WI
WHQG(FM) Milwaukee WI
WLUM-FM Milwaukee WI
WZOR(FM) Mishicot WI
WCCN-FM Neillsville WI
WWWX(FM) Oshkosh WI
*WSUP(FM) Platteville WI
*WSHS(FM) Sheboygan WI
WHBZ(FM) Sheboygan Falls WI
*WRGX(FM) Sturgeon Bay WI
WZBY(FM) Sturgeon Bay WI
KRBR(FM) Superior WI
WJJO(FM) Watertown WI
WBCV(FM) Wausau WI
*WSUW(FM) Whitewater WI
WAMX(FM) Milton WV
WHBR-FM Parkersburg WV
WKLC-FM Saint Albans WV
WEGW(FM) Wheeling WV
KAML-FM Gillette WY
KMTN(FM) Jackson WY

Russian

KICY(AM) Nome AK
WKTA(AM) Evanston IL

Smooth Jazz

KNIK-FM Anchorage AK
*WVAS(FM) Montgomery AL
*KUAP(FM) Pine Bluff AR
KOAS(FM) Dolan Springs AZ
KMET(AM) Banning CA
KSSJ(FM) Fair Oaks CA
KJJZ(FM) Indio CA
KIFM(FM) San Diego CA
KJZY(FM) Sebastopol CA
KKXS(FM) Shingletown CA
KMYT(FM) Temecula CA
KJCD(FM) Longmont CO
WXJZ(FM) Gainesville FL
*WKTZ-FM Jacksonville FL
WSBZ(FM) Miramar Beach FL
WZJZ(FM) Port Charlotte FL
WJZT(FM) Woodville FL
WSSJ(FM) Rincon GA
WARH(FM) Granite City IL
WUBU(FM) South Bend IN
WYJZ(FM) Speedway IN
WKYL(FM) Lawrenceburg KY
WJSH(FM) Folsom LA
KKSJ(FM) Maurice LA
WEIB(FM) Northampton MA
WJZQ(FM) Cadillac MI
WJZL(FM) Charlotte MI
WVMV(FM) Detroit MI
KTLK-FM Minneapolis MN
KZWV(FM) Eldon MO
WDAF-FM Liberty MO
*WCCE(FM) Buie's Creek NC
WJJZ(FM) Burlington NJ

*WLFR(FM) Pomona NJ
KQBT(FM) Rio Rancho NM
KABQ-FM Santa Fe NM
WQCD(FM) New York NY
*WRMU(FM) Alliance OH
WNWV(FM) Elyria OH
KUJJ(FM) Weston OR
WSJW(FM) Starview PA
*WFSK-FM Nashville TN
KMVK(FM) Fort Worth TX
KERV(AM) Kerrville TX
*KPVU(FM) Prairie View TX
WCJZ(FM) Charlottesville VA
WJCD(FM) Windsor VA
WJZW(FM) Woodbridge VA
KZAL(FM) Manson WA
KWJZ(FM) Seattle WA
KIJZ(FM) Vancouver WA
WBJZ(FM) Berlin WI
WJZX(FM) Brookfield WI
WAUN(FM) Kewaunee WI

Soul

WZZA(AM) Tuscumbia AL
KOKY(FM) Sherwood AR
WAMJ(FM) Mableton GA
WTYB(FM) Springfield GA
WFUN-FM Bethalto IL
KMXH(FM) Alexandria LA
KXZZ(AM) Lake Charles LA
WRBO(FM) Como MS
WQMG-FM Greensboro NC
WRKS(FM) New York NY
WAMO(AM) Millvale PA

Spanish

WQCR(AM) Alabaster AL
WWGC(AM) Albertville AL
WZGX(AM) Bessemer AL
WPSB(AM) Birmingham AL
WCOC(AM) Dora AL
WYAM(AM) Hartselle AL
WRLM(AM) Irondale AL
WJHX(AM) Lexington AL
WCRL(AM) Oneonta AL
WHAL(AM) Phenix City AL
WKAX(AM) Russellville AL
WQRX(AM) Valley Head AL
KSEC(FM) Bentonville AR
KZTD(AM) Cabot AR
KDQN(AM) De Queen AR
KPZA(AM) Hot Springs AR
KAKS(FM) Huntsville AR
KTUV(AM) Little Rock AR
KYNG(AM) Springdale AR
KOAI(AM) Van Buren AR
KVVA-FM Apache Junction AZ
KKMR(FM) Arizona City AZ
*KRMB(FM) Bisbee AZ
KMIA(AM) Black Canyon City AZ
KCKY(AM) Coolidge AZ
KDAP(AM) Douglas AZ
*KRMC(FM) Douglas AZ
KLNZ(FM) Glendale AZ
KQMR(FM) Globe AZ
KZLZ(FM) Kearny AZ
KRRN(FM) Kingman AZ
*KNOG(FM) Nogales AZ
KOFH(FM) Nogales AZ
KCMT(FM) Oro Valley AZ
KHOT-FM Paradise Valley AZ
KRIT(FM) Parker AZ
KIDR(AM) Phoenix AZ
*KNAI(FM) Phoenix AZ
KSUN(AM) Phoenix AZ
KEVT(AM) Sahuarita AZ
KOMR(FM) Sun City AZ
KVIB(FM) Sun City West AZ
KNUV(AM) Tolleson AZ
KTKT(AM) Tucson AZ
KCEC-FM Wellton AZ
*KYRM(FM) Yuma AZ
KWRN(AM) Apple Valley CA
KSSE(FM) Arcadia CA
KBYN(FM) Arnold CA
*KCFA(FM) Arnold CA
KMYX-FM Arvin CA
KLBN(FM) Auberry CA
KAFY(AM) Bakersfield CA
KLHC(AM) Bakersfield CA
KPSL-FM Bakersfield CA
*KTQX(FM) Bakersfield CA
KWAC(AM) Bakersfield CA
KIQQ(AM) Barstow CA
*KODV(FM) Barstow CA
KXXZ(FM) Barstow CA
KAEH(FM) Beaumont CA
KXSB(FM) Big Bear Lake CA
KSEH(FM) Brawley CA
KGBA(AM) Calexico CA
*KUBO(FM) Calexico CA
KCEL(FM) California City CA
KXTS(FM) Calistoga CA
*KMRO(FM) Camarillo CA
KNTO(FM) Chowchilla CA
*KHDC(FM) Chualar CA
KCVR-FM Columbia CA
KBLO(FM) Corcoran CA
KFVR(AM) Crescent City CA
KXSE(FM) Davis CA
KCHJ(AM) Delano CA
KCHJ(AM) Delano CA
KLAX-FM East Los Angeles CA
KMQA(FM) East Porterville CA
KMLA(FM) El Rio CA
KSPE-FM Ellwood CA
KTTA(FM) Esparto CA
KSSD(FM) Fallbrook CA
KXZM(FM) Felton CA
*KEYQ(AM) Fresno CA
KGST(AM) Fresno CA
KIRV(AM) Fresno CA
KOQO-FM Fresno CA
*KSJV(FM) Fresno CA
KWRU(AM) Fresno CA
KYNO(AM) Fresno CA
KEBN(FM) Garden Grove CA
KAZA(AM) Gilroy CA
KSCA(FM) Glendale CA
KBAA(FM) Grass Valley CA
KLOK-FM Greenfield CA
KSEA(FM) Greenfield CA
KIDI(FM) Guadalupe CA
KGEN-FM Hanford CA
KRDA(FM) Hanford CA
KXRS(FM) Hemet CA
KMPG(AM) Hollister CA
KXSM(FM) Hollister CA
KMXX(FM) Imperial CA
KESQ(AM) Indio CA
KRCD(FM) Inglewood CA
KLMG(FM) Jackson CA
KEDD(FM) Johannesburg CA
KBHH(FM) Kerman CA
KEXA(FM) King City CA
KSRN(FM) Kings Beach CA
KUNA-FM La Quinta CA
KWJL(AM) Lancaster CA
*KVUH(FM) Laytonville CA
KBTW(FM) Lenwood CA
KSKD(FM) Livingston CA
KCVR(AM) Lodi CA
KRQK(FM) Lompoc CA
KSMY(FM) Lompoc CA
KBUE(FM) Long Beach CA
KLTX(AM) Long Beach CA
KHJ(AM) Los Angeles CA
KLVE(FM) Los Angeles CA
KTNQ(AM) Los Angeles CA
KWKW(AM) Los Angeles CA
KXOL-FM Los Angeles CA
KQLB(FM) Los Banos CA
KMMM(FM) Madera CA
KMRQ(FM) Manteca CA
KRCX-FM Marysville CA
KIWI(FM) McFarland CA
KMEN(FM) Mendota CA
KTIQ(AM) Merced CA
KBBU(FM) Modesto CA
KHPY(AM) Moreno Valley CA
KLMM(FM) Morro Bay CA
KIQQ-FM Newberry Springs CA
KAAT(FM) Oakhurst CA
KEWE(AM) Oroville CA
KHHZ(FM) Oroville CA
KOXR(AM) Oxnard CA
KXLM(FM) Oxnard CA
KLUN(FM) Paso Robles CA
KTSE-FM Patterson CA
KTOB(AM) Petaluma CA
KATD(AM) Pittsburg CA
KWKU(AM) Pomona CA
KCAL(AM) Redlands CA
KDIF(AM) Riverside CA
KFSG(AM) Roseville CA
KDBV(AM) Salinas CA
KPRC-FM Salinas CA
KRAY-FM Salinas CA
KTGE(AM) Salinas CA
KEZY(AM) San Bernardino CA
KLNV(FM) San Diego CA
KLQV(FM) San Diego CA
KSDO(AM) San Diego CA
KBUA(FM) San Fernando CA
KIQI(AM) San Francisco CA
KSOL(FM) San Francisco CA
KRQB(FM) San Jacinto CA
KBRG(FM) San Jose CA
KSJO(FM) San Jose CA
KZSF(AM) San Jose CA
KVVZ(FM) San Rafael CA
KWIZ(FM) Santa Ana CA
KBKO(AM) Santa Barbara CA
KIST-FM Santa Barbara CA
KZER(AM) Santa Barbara CA
KVVF(FM) Santa Clara CA
KSQL(FM) Santa Cruz CA
KSBQ(AM) Santa Maria CA
KTAP(AM) Santa Maria CA
KBLA(AM) Santa Monica CA
KLJR-FM Santa Paula CA
KUNX(AM) Santa Paula CA
*KBBF(FM) Santa Rosa CA
KRRS(AM) Santa Rosa CA
KSES-FM Seaside CA
*KGZO(FM) Shafter CA
KMBX(AM) Soledad CA
KMJV(FM) Soledad CA
KSTN-FM Stockton CA
KMIX(FM) Tracy CA
KGEN(AM) Tulare CA
KLOC(AM) Turlock CA
KUKI(AM) Ukiah CA
KSSC(FM) Ventura CA
KRSX(AM) Victorville CA
KALI(AM) West Covina CA
KLLK(AM) Willits CA
KLOQ-FM Winton CA
KOBO(AM) Yuba City CA
KPVW(FM) Aspen CO
KMXA(AM) Aurora CO
KJMN(FM) Castle Rock CO
KGDQ(FM) Colorado Springs CO
KBNO(AM) Denver CO
KNRV(AM) Englewood CO
KXPK(FM) Evergreen CO
KFTM(AM) Fort Morgan CO
KEXO(AM) Grand Junction CO
KGRE(AM) Greeley CO
KFVR-FM La Junta CO
KXRE(AM) Manitou Springs CO
KAVA(AM) Pueblo CO
KNKN(FM) Pueblo CO
KRMX(AM) Pueblo CO
KJJD(AM) Windsor CO
WADS(AM) Ansonia CT
WCUM(AM) Bridgeport CT
WPRX(AM) Bristol CT
WSUB(AM) Groton CT
WNEZ(AM) Manchester CT
WFNW(AM) Naugatuck CT
WLAT(AM) New Britain CT
WRYM(AM) New Britain CT
WXCT(AM) Southington CT
WJWL(AM) Georgetown DE
WYUS(AM) Milford DE
WNWK(AM) Newark DE
WNTF(AM) Bithlo FL
WLCC(AM) Brandon FL
WEWC(AM) Callahan FL
WAFC(AM) Clewiston FL
WTIR(AM) Cocoa Beach FL
WRHC(AM) Coral Gables FL
WDCF(AM) Dade City FL
WRLZ(AM) Eatonville FL
WVOJ(AM) Fernandina Beach FL
WRMA(FM) Fort Lauderdale FL
WCRM(AM) Fort Myers FL
WJNX(AM) Fort Pierce FL
WRTO-FM Goulds FL
WACC(AM) Hialeah FL
WCMQ-FM Hialeah FL
WOIR(AM) Homestead FL
WAFZ-FM Immokalee FL
*WGES-FM Key Largo FL
WZMQ(FM) Key Largo FL
*WLAZ(FM) Kissimmee FL
WOTS(AM) Kissimmee FL
*WBIY(FM) La Belle FL
WIPC(AM) Lake Wales FL
WWRF(AM) Lake Worth FL
WMGG(AM) Largo FL
WQBQ(AM) Leesburg FL
WWCL(AM) Lehigh Acres FL
WAMR-FM Miami FL
WAQI(AM) Miami FL
*WDNA(FM) Miami FL
WOCN(AM) Miami FL
WQBA(AM) Miami FL
WSUA(AM) Miami FL
WWFE(AM) Miami FL
WNMA(AM) Miami Springs FL
WZSP(FM) Nocatee FL
WXDJ(FM) North Miami Beach FL
WUNA(AM) Ocoee FL
WRUM(FM) Orlando FL
*WIRP(FM) Pennsuco FL
WPTK(AM) Pine Island Center FL
WTLQ-FM Punta Rassa FL
WPSP(AM) Royal Palm Beach FL
WYUU(FM) Safety Harbor FL
WIWA(AM) Saint Cloud FL
WGES(AM) Saint Petersburg FL
WSDO(AM) Sanford FL
WPIK(FM) Summerland Key FL
WAMA(AM) Tampa FL
WQBN(AM) Temple Terrace FL
WNUE-FM Titusville FL
WAUC(AM) Wauchula FL
WLAA(AM) Winter Garden FL
WPRD(AM) Winter Park FL
WAOS(AM) Austell GA
WBZY(FM) Bowdon GA
WLKQ-FM Buford GA
WXEM(AM) Buford GA
WWVA-FM Canton GA
WPBS(AM) Conyers GA
WDAL(AM) Dalton GA
WFJO(FM) Folkston GA
WLBA(AM) Gainesville GA
WPLO(AM) Grayson GA
WWWE(AM) Hapeville GA
WFTD(AM) Marietta GA
WKTM(FM) Soperton GA
*WJDS(FM) Sparta GA
WGTA(AM) Summerville GA
WNSY(FM) Talking Rock GA
KDNZ(AM) Cedar Falls IA
KBGG(AM) Des Moines IA
KXLQ(AM) Indianola IA
*KOJI(FM) Okoboji IA
KDLS-FM Perry IA
*KWIT(FM) Sioux City IA
KWSL(AM) Sioux City IA
KSPZ(AM) Ammon ID
KDBI(FM) Emmett ID
KWEI-FM Fruitland ID
KRXR(AM) Gooding ID
KFTA(AM) Rupert ID
KIGO(AM) Saint Anthony ID
KWEI(AM) Weiser ID
WNTD(AM) Chicago IL
WRTO(AM) Chicago IL
WPPN(FM) Des Plaines IL
WOJO(FM) Evanston IL
WONX(AM) Evanston IL
WVIV-FM Highland Park IL
WGBJ(FM) Auburn IN
WNTS(AM) Beech Grove IN
WEDJ(FM) Danville IN
WRSW-FM Warsaw IN
KYQQ(FM) Arkansas City KS
KANR(FM) Belle Plaine KS
KDCC(AM) Dodge City KS
KFXX-FM Hugoton KS
KSSA(FM) Ingalls KS
KDTD(AM) Kansas City KS
KKHK(AM) Kansas City KS
KTCM(FM) Kingman KS
KYUU(AM) Liberal KS
KCZZ(AM) Mission KS
*KYWA(FM) Wichita KS
WCVG(AM) Covington KY
WNTC(FM) Drakesboro KY
WTUV-FM Eminence KY
WXRA(AM) Georgetown KY
KGLA(AM) Gretna LA
WFNO(AM) Norco LA
WMSX(AM) Brockton MA
WUNR(AM) Brookline MA
WXKS(AM) Everett MA
WKOX(AM) Framingham MA
WCEC(AM) Haverhill MA
WNNW(AM) Lawrence MA
*WFHL(FM) New Bedford MA
WSPR(AM) Springfield MA
WRCA(AM) Waltham MA
WACM(AM) West Springfield MA
WLZL(FM) Annapolis MD
WTRI(AM) Brunswick MD
WYRE(AM) Essex MD
WWGB(AM) Indian Head MD
WILC(AM) Laurel MD
WCTN(AM) Potomac-Cabin John MD
WLXE(AM) Rockville MD
WACA(AM) Wheaton MD
WNWZ(AM) Grand Rapids MI
WYGR(AM) Wyoming MI
KMNV(AM) Saint Paul MN
KZGX(AM) Watertown MN
KQMO(FM) Shell Knob MO
WMYT(AM) Carolina Beach NC
WLLQ(AM) Chapel Hill NC
WGSP(AM) Charlotte NC
WCXN(AM) Claremont NC
WRRZ(AM) Clinton NC
WTIK(AM) Durham NC
WGBT(FM) Eden NC
WZBO(AM) Edenton NC
WCNC(AM) Elizabeth City NC
WFAY(AM) Fayetteville NC
WRTG(AM) Garner NC
WYMY(FM) Goldsboro NC
WWBG(AM) Greensboro NC
WGOS(AM) High Point NC
WSRP(AM) Jacksonville NC
WLNR(AM) Kinston NC
WSGH(AM) Lewisville NC
*WLLN(AM) Lillington NC
WGSB(AM) Mebane NC
WNOW(AM) Mint Hill NC
WREV(AM) Reidsville NC
WZUP(FM) Rose Hill NC
WNCA(AM) Siler City NC
WCIE(AM) Spring Lake NC
WETC(AM) Wendell-Zebulon NC
WTOB(AM) Winston-Salem NC
KYDZ(AM) Bellevue NE
KBBX-FM Nebraska City NE
WADB(AM) Asbury Park NJ
WEMG(AM) Camden NJ
WJDM(AM) Elizabeth NJ
WCAA(FM) Newark NJ
WNSW(AM) Newark NJ
WXMC(AM) Parsippany-Troy Hills NJ
WPAT(AM) Paterson NJ
WPAT-FM Paterson NJ

WMIZ(AM) Vineland NJ
*KANW(FM) Albuquerque NM
KKRG(FM) Albuquerque NM
KRZY(AM) Albuquerque NM
KLVO(FM) Belen NM
KPZE-FM Carlsbad NM
KKNS(AM) Corrales NM
*KZPI(FM) Deming NM
KDCE(AM) Espanola NM
KYBR(FM) Espanola NM
KRZE(AM) Farmington NM
KLMA(FM) Hobbs NM
KPZA-FM Jal NM
*KRUC(FM) Las Cruces NM
*KRWG(FM) Las Cruces NM
KFUN(AM) Las Vegas NM
KNMX(AM) Las Vegas NM
KZNM(FM) Los Alamos NM
KRDD(AM) Roswell NM
KNUW(FM) Santa Clara NM
KJFA(FM) Santa Fe NM
KRZY-FM Santa Fe NM
KSWV(AM) Santa Fe NM
*KNLK(FM) Santa Rosa NM
KSSR(AM) Santa Rosa NM
KXMT(FM) Taos NM
KKVS(FM) Truth or Consequences NM
KDOX(AM) Henderson NV
KRGT(FM) Indian Springs NV
KISF(FM) Las Vegas NV
KKVV(AM) Las Vegas NV
KQRT(FM) Las Vegas NV
KRLV(AM) Las Vegas NV
KWID(FM) Las Vegas NV
KRNV-FM Reno NV
KXEQ(AM) Reno NV
KXTO(AM) Reno NV
KBDB(AM) Sparks NV
KQLO(AM) Sun Valley NV
KLSQ(AM) Whitney NV
WQBU-FM Garden City NY
WALL(AM) Middletown NY
WADO(AM) New York NY
*WHCR-FM New York NY
WKDM(AM) New York NY
WSKQ-FM New York NY
*WWRV(AM) New York NY
WEOK(AM) Poughkeepsie NY
WDLR(AM) Delaware OH
WVKO-FM Johnstown OH
*WNZN(FM) Lorain OH
KIZS(FM) Collinsville OK
KZUE(AM) El Reno OK
KTAT(AM) Frederick OK
KINB(FM) Kingfisher OK
KTUZ-FM Okarche OK
KREU(FM) Roland OK
KJMU(AM) Sand Springs OK
KXTD(AM) Wagoner OK
KTMT(AM) Ashland OR
*KORM(FM) Astoria OR
KDPM(FM) Cottage Grove OR
KRYP(FM) Gladstone OR
KMUZ(AM) Gresham OR
KUIK(AM) Hillsboro OR
KXOR(AM) Junction City OR
KRTA(AM) Medford OR
KZTB(FM) Milton-Freewater OR
KSZN(AM) Milwaukie OR
KGDD(AM) Oregon City OR
KXMG(AM) Portland OR
KRDM(AM) Redmond OR
*KXPD(AM) Tigard OR
KWBY(AM) Woodburn OR
WHOL(AM) Allentown PA
*WLCH(FM) Lancaster PA
WUBA(AM) Philadelphia PA
WPHE(AM) Phoenixville PA
*WXAC(FM) Reading PA
WOQI(AM) Adjuntas PR
WABA(AM) Aguadilla PR
WIVA-FM Aguadilla PR
WTPM(FM) Aguadilla PR
WWNA(AM) Aguadilla PR
WCMN(AM) Arecibo PR
WCMN-FM Arecibo PR
WMIA(AM) Arecibo PR
WNIK(AM) Arecibo PR
WBQN(AM) Barceloneta-Manati PR
WOLA(AM) Barranquitas PR
WLUZ(AM) Bayamon PR
WODA(FM) Bayamon PR
WRSJ(AM) Bayamon PR
WYAC(AM) Cabo Rojo PR
WNEL(AM) Caguas PR
WVJP(AM) Caguas PR
WVJP-FM Caguas PR
WDIN(FM) Camuy PR
WGIT(AM) Canovanas PR
WIDA(AM) Carolina PR
WBRQ(FM) Cidra PR
WJZG(FM) Culebra PR
WIBS(AM) Guayama PR
WZET(FM) Hormigueros PR
WALO(AM) Humacao PR
WCGB(AM) Juana Diaz PR
WBSG(AM) Lajas PR
WGDL(AM) Lares PR
WMNT(AM) Manati PR
WAEL-FM Maricao PR
WIOB(FM) Mayaguez PR
WPRA(AM) Mayaguez PR
WTIL(AM) Mayaguez PR
WEKO(AM) Morovis PR
WYQE(FM) Naguabo PR
WPPC(AM) Penuelas PR
WDEP(AM) Ponce PR
WPAB(AM) Ponce PR
WPRP(AM) Ponce PR
WRIO(FM) Ponce PR
WZAR(FM) Ponce PR
WZMT(FM) Ponce PR
WCHQ(AM) Quebradillas PR
WIDI(FM) Quebradillas PR
WHOY(AM) Salinas PR
WSOL(AM) San German PR
WAPA(AM) San Juan PR
WIAC(AM) San Juan PR
WIOA(FM) San Juan PR
WQBS(AM) San Juan PR
*WRTU(FM) San Juan PR
WSKN(AM) San Juan PR
WZNT(FM) San Juan PR
WRSS(AM) San Sebastian PR
WRSS(AM) San Sebastian PR
WERR(FM) Utuado PR
WUPR(AM) Utuado PR
WIVV(AM) Vieques PR
WIVV(AM) Vieques PR
WIVV(AM) Vieques PR
WXEW(AM) Yabucoa PR
WKFE(AM) Yauco PR
WALE(AM) Greenville RI
WKKB(FM) Middletown RI
*WELH(FM) Providence RI
WPMZ(AM) Providence RI
WBLR(AM) Batesburg SC
WCEO(AM) Columbia SC
WGVL(AM) Greenville SC
WQVA(AM) Lexington SC
WAZS-FM McClellanville SC
WZJY(AM) Mt. Pleasant SC
WKMG(AM) Newberry SC
WRML(FM) Pageland SC
WNFO(AM) Ridgeland SC
WBZK(AM) York SC
WSAA(FM) Benton TN
WNWS(AM) Brownsville TN
WHEW(AM) Franklin TN
WYXE(AM) Gallatin TN
WKCE(AM) Maryville TN
WGSF(AM) Memphis TN
WMXK(FM) Morristown TN
WMGC(AM) Murfreesboro TN
*WMTS-FM Murfreesboro TN
WNQM(AM) Nashville TN
WNVL(AM) Nashville TN
KJAV(FM) Alamo TX
KTCY(FM) Azle TX
KXGJ(FM) Bay City TX
KQQK(FM) Beaumont TX
KIBL(AM) Beeville TX
KDXX(FM) Benbrook TX
KPDB(FM) Big Lake TX
KBYG(AM) Big Spring TX
KBRN(AM) Boerne TX
KKUB(AM) Brownfield TX
*KPBB(FM) Brownfield TX
*KBNR(FM) Brownsville TX
KKPS(FM) Brownsville TX
*KPBE(FM) Brownwood TX
KXYL(AM) Brownwood TX
KTAM(AM) Bryan TX
KTAE(AM) Cameron TX
KAYG(FM) Camp Wood TX
KBEN(AM) Carrizo Springs TX
KCZO(FM) Carrizo Springs TX
KJON(AM) Carrollton TX
KDET(AM) Center TX
KWOW(FM) Clifton TX
KCOR-FM Comfort TX
KUNO(AM) Corpus Christi TX
KMMZ(FM) Crane TX
KXOI(AM) Crane TX
KHER(FM) Crystal City TX
KNIT(AM) Dallas TX
KFZO(FM) Denton TX
KSML(AM) Diboll TX
KLMO-FM Dilley TX
KDHN(AM) Dimmitt TX
KXBT(FM) Dripping Springs TX
KEPS(AM) Eagle Pass TX
*KEPX(FM) Eagle Pass TX
*KOIR(FM) Edinburg TX
KVLY(FM) Edinburg TX
KAMA(AM) El Paso TX
KBNA(AM) El Paso TX
KBNA-FM El Paso TX
KINT-FM El Paso TX
KSVE(AM) El Paso TX
*KVER(FM) El Paso TX
KVIV(AM) El Paso TX
KXPL(AM) El Paso TX
KYSE(FM) El Paso TX
KXXS(FM) Elgin TX
KDFM(FM) Falfurrias TX
KPSO-FM Falfurrias TX
KIJN(AM) Farwell TX
KMUL(AM) Farwell TX
KDFT(AM) Ferris TX
KFST-FM Fort Stockton TX
KEGL(FM) Fort Worth TX
KFJZ(AM) Fort Worth TX
KFLC(AM) Fort Worth TX
KLNO(FM) Fort Worth TX
KBRZ(AM) Freeport TX
KJOJ-FM Freeport TX
*KPBN(FM) Freer TX
KGRW(FM) Friona TX
KATH(AM) Frisco TX
*KHCB(AM) Galveston TX
KOVE-FM Galveston TX
KINV(FM) Georgetown TX
KHMC(FM) Goliad TX
KGBT(AM) Harlingen TX
KVJM(FM) Hearne TX
KEKO(FM) Hebronville TX
KJNZ(FM) Hereford TX
KEYH(AM) Houston TX
KLAT(AM) Houston TX
KLOL(FM) Houston TX
KLTN(FM) Houston TX
KQUE(AM) Houston TX
KXYZ(AM) Houston TX
*KHCH(AM) Huntsville TX
KLEY-FM Jourdanton TX
KERB(AM) Kermit TX
KERB-FM Kermit TX
*KHKV(FM) Kerrville TX
KINE(AM) Kingsville TX
KNOR(FM) Krum TX
*KBNL(FM) Laredo TX
*KHOY(FM) Laredo TX
KLAR(AM) Laredo TX
KLNT(AM) Laredo TX
KNEX(FM) Laredo TX
KHHL(FM) Leander TX
KESS-FM Lewisville TX
KYMI(FM) Los Ybanez TX
KBZO(AM) Lubbock TX
KXTQ-FM Lubbock TX
KELG(AM) Manor TX
*KBIB(AM) Marion TX
KCUL-FM Marshall TX
KGBT-FM McAllen TX
KRIO(AM) McAllen TX
KLTO-FM McQueeney TX
*KPBJ(FM) Midland TX
KBDR(FM) Mirando City TX
KIRT(AM) Mission TX
KIMP(AM) Mount Pleasant TX
KLHB(FM) Odem TX
KOZA(AM) Odessa TX
KQLM(FM) Odessa TX
KLVL(AM) Pasadena TX
KRIO-FM Pearsall TX
KOKE(AM) Pflugerville TX
KDVE(FM) Pittsburg TX
*KGWP(FM) Pittsburg TX
*KPMB(FM) Plainview TX
KREW(AM) Plainview TX
KRIA(FM) Plainview TX
KWMF(AM) Pleasanton TX
KQBU-FM Port Arthur TX
KTJM(FM) Port Arthur TX
KZPL(FM) Port Isabel TX
KMFM(FM) Premont TX
KCLR(AM) Ralls TX
KBIC(FM) Raymondville TX
KBUC(FM) Raymondville TX
KQBO(FM) Rio Grande City TX
KHCK-FM Robinson TX
KSAB(FM) Robstown TX
KRTX(AM) Rosenberg-Richmond TX
*KTER(FM) Rudolph TX
KSJT-FM San Angelo TX
KCOR(AM) San Antonio TX
KROM(FM) San Antonio TX
KXTN-FM San Antonio TX
KZDC(AM) San Antonio TX
KQSI(FM) San Augustine TX
KUKA(FM) San Diego TX
KUBR(AM) San Juan TX
KUOL(AM) San Marcos TX
KMAT(FM) Seadrift TX
KIKZ(AM) Seminole TX
KSEY(AM) Seymour TX
KDAE(AM) Sinton TX
KSTV(AM) Stephenville TX
KAMZ(FM) Tahoka TX
KXAL-FM Tatum TX
KLQB(FM) Taylor TX
KWNX(AM) Taylor TX
KYST(AM) Texas City TX
KTUE(AM) Tulia TX
KSAH(AM) Universal City TX
KTNO(AM) University Park TX
KZMP(AM) University Park TX
KUVA(FM) Uvalde TX
KRGE(AM) Weslaco TX
KTXZ(AM) West Lake Hills TX
KAIQ(FM) Wolfforth TX
KBAW(FM) Zapata TX
KXTA(AM) Centerville UT
KSVN(AM) Ogden UT
KDUT(FM) Randolph UT
KBJA(AM) Sandy UT
WZHF(AM) Arlington VA
WPWC(AM) Dumfries-Triangle VA
WTOX(AM) Glen Allen VA
WKDV(AM) Manassas VA
WVXX(AM) Norfolk VA
WBTK(AM) Richmond VA
WVNZ(AM) Richmond VA
WKCW(AM) Warrenton VA
WAMM(AM) Woodstock VA
WMYP(FM) Frederiksted VI
KBKW(AM) Aberdeen WA
KWMG(AM) Auburn-Federal Way WA
KXPA(AM) Bellevue WA
KMMG(FM) Benton City WA
*KCED(FM) Centralia WA
KZHR(FM) Dayton WA
KDDS-FM Elma WA
KMNA(FM) Mabton WA
*KSVR(FM) Mount Vernon WA
KQSN(FM) Naches WA
KZTA(FM) Naches WA
KRSC(AM) Othello WA
KLES(FM) Prosser WA
KZML(FM) Quincy WA
KRCW(FM) Royal City WA
KDOW(AM) Seattle WA
KZTS(AM) Sunnyside WA
KKMO(AM) Tacoma WA
KYXE(AM) Union Gap WA
KWWX(AM) Wenatchee WA
*KBNO-FM White Salmon WA
KWLN(FM) Wilson Creek WA
*KDNA(FM) Yakima WA
WDGY(AM) Hudson WI
WLMV(AM) Madison WI
WJTI(AM) Racine WI
WDDW(FM) Sturtevant WI
KBMG(FM) Evanston WY
KOLT-FM Warren AFB WY

Sports

KTZN(AM) Anchorage AK
KCBF(AM) Fairbanks AK
WANA(AM) Anniston AL
WNSI-FM Atmore AL
WAUD(AM) Auburn AL
WNSP(FM) Bay Minette AL
WSPZ(AM) Birmingham AL
WFMH(AM) Cullman AL
WKUL(FM) Cullman AL
WWTM(AM) Decatur AL
WZNJ(FM) Demopolis AL
WOOF(AM) Dothan AL
WSBM(AM) Florence AL
WHEP(AM) Foley AL
WTKI(AM) Huntsville AL
WUMP(AM) Madison AL
WNTM(AM) Mobile AL
WACV(AM) Montgomery AL
WMSP(AM) Montgomery AL
WLAY(AM) Muscle Shoals AL
WJOX(FM) Northport AL
WIQR(AM) Prattville AL
WELR(AM) Roanoke AL
WNSI(AM) Robertsdale AL
WYTK(FM) Rogersville AL
WWIC(AM) Scottsboro AL
WFEB(AM) Sylacauga AL
WACT(AM) Tuscaloosa AL
WTBC(AM) Tuscaloosa AL
WVNA(AM) Tuscumbia AL
WVSA(AM) Vernon AL
WAPZ(AM) Wetumpka AL
WKXM(AM) Winfield AL
KBTA(AM) Batesville AR
KEWI(AM) Benton AR
KREB(AM) Bentonville-Bella Vista AR
KASR(FM) Conway AR
KXXA(AM) Conway AR
KCAB(AM) Dardanelle AR
KELD(AM) El Dorado AR
KFFA-FM Helena AR
KTTG(FM) Mena AR
KVOM(AM) Morrilton AR
KDXE(AM) North Little Rock AR
KDRS(AM) Paragould AR
KOTN(AM) Pine Bluff AR
KARV(AM) Russellville AR
KHGG(AM) Van Buren AR
KHGG-FM Waldron AR
KMIA(AM) Black Canyon City AZ
KVNA(AM) Flagstaff AZ
KIKO(AM) Miami AZ
KGME(AM) Phoenix AZ
KIDR(AM) Phoenix AZ
KMVP(AM) Phoenix AZ
KTAR(AM) Phoenix AZ
KQNA(AM) Prescott Valley AZ
KATO(AM) Safford AZ
KAZM(AM) Sedona AZ
KTAN(AM) Sierra Vista AZ
KDUS(AM) Tempe AZ

KNUV(AM) Tolleson AZ
KCUB(AM) Tucson AZ
KFFN(AM) Tucson AZ
KNST(AM) Tucson AZ
KJOK(AM) Yuma AZ
KATA(AM) Arcata CA
KXTK(AM) Arroyo Grande CA
KBFP(AM) Bakersfield CA
KGEO(AM) Bakersfield CA
KMET(AM) Banning CA
KFPT(AM) Clovis CA
KWRM(AM) Corona CA
KCBL(AM) Fresno CA
KFIG(AM) Fresno CA
KGST(AM) Fresno CA
KXEX(AM) Fresno CA
KRER(FM) Hamilton City CA
KAVL(AM) Lancaster CA
KLAC(AM) Los Angeles CA
KSPN(AM) Los Angeles CA
KWKW(AM) Los Angeles CA
KMFB(FM) Mendocino CA
KBKY(FM) Merced CA
KESP(AM) Modesto CA
KXTY(FM) Morro Bay CA
KLAA(AM) Orange CA
KNTS(AM) Palo Alto CA
KPRL(AM) Paso Robles CA
KWKU(AM) Pomona CA
KNLF(FM) Quincy CA
KNRO(AM) Redding CA
KCTC(AM) Sacramento CA
KHTK(AM) Sacramento CA
KSAC(AM) Sacramento CA
KNBR(AM) San Francisco CA
KKJL(AM) San Luis Obispo CA
KTCT(AM) San Mateo CA
KIST(AM) Santa Barbara CA
KCNR(AM) Shasta CA
KIRN(AM) Simi Valley CA
KOWL(AM) South Lake Tahoe CA
KSUE(AM) Susanville CA
KXPS(AM) Thousand Palms CA
KEZL(AM) Visalia CA
KSLK(FM) Visalia CA
KNFO(FM) Basalt CO
KSIR(AM) Brush CO
KSKE(AM) Buena Vista CO
KKKK(AM) Colorado Springs CO
KKML(AM) Colorado Springs CO
KVOR(AM) Colorado Springs CO
KBNO(AM) Denver CO
KKFN(AM) Denver CO
KOA(AM) Denver CO
KIUP(AM) Durango CO
KIIX(AM) Fort Collins CO
KTMM(AM) Grand Junction CO
KEPN(AM) Lakewood CO
KRCN(AM) Longmont/Denver CO
KSXT(AM) Loveland CO
KWUF(AM) Pagosa Springs CO
KJMP(AM) Pierce CO
KGHF(AM) Pueblo CO
KRGS(AM) Rifle CO
KBCR(AM) Steamboat Springs CO
KVLE(AM) Vail CO
KSPK(FM) Walsenburg CO
WINE(AM) Brookfield CT
WPOP(AM) Hartford CT
WMMW(AM) Meriden CT
WAVZ(AM) New Haven CT
WXLM(FM) Stonington CT
WTEM(AM) Washington DC
WWTX(AM) Wilmington DE
WBGF(FM) Belle Glade FL
WWJB(AM) Brooksville FL
WKFL(AM) Bushnell FL
WRHC(AM) Coral Gables FL
WNDB(AM) Daytona Beach FL
WTJV(AM) De Land FL
WHBO(AM) Dunedin FL
WFLL(AM) Fort Lauderdale FL
WGGG(AM) Gainesville FL
WRUF(AM) Gainesville FL
WTKE-FM Holt FL
WFXJ(AM) Jacksonville FL

WNNR(AM) Jacksonville FL
WZNZ(AM) Jacksonville FL
WJXL(AM) Jacksonville Beach FL
WKWF(AM) Key West FL
WHOO(AM) Kissimmee FL
WBXY(FM) La Crosse FL
WFFG(AM) Marathon FL
WMEL(AM) Melbourne FL
WQAM(AM) Miami FL
WNMA(AM) Miami Springs FL
WWCN(AM) North Fort Myers FL
WMOP(AM) Ocala FL
WQTM(AM) Orlando FL
WELE(AM) Ormond Beach FL
WLTG(AM) Panama City FL
WPCF(AM) Panama City Beach FL
WPSL(AM) Port St. Lucie FL
WAOC(AM) Saint Augustine FL
WFOY(AM) Saint Augustine FL
WDAE(AM) Saint Petersburg FL
WSRQ(AM) Sarasota FL
WQYK(AM) Seffner FL
WAXY(AM) South Miami FL
WSTU(AM) Stuart FL
*WANM(FM) Tallahassee FL
WHBT(AM) Tallahassee FL
WNLS(AM) Tallahassee FL
WHNZ(AM) Tampa FL
WEFL(AM) Tequesta FL
WIXC(AM) Titusville FL
WFSH(AM) Valparaiso-Niceville FL
WJNO(AM) West Palm Beach FL
WGPC(AM) Albany GA
WSRA(AM) Albany GA
WRFC(AM) Athens GA
WQXI(AM) Atlanta GA
WGAC(AM) Augusta GA
WNRR(AM) Augusta GA
WRDW(AM) Augusta GA
WSGF(AM) Augusta GA
WFNS(AM) Blackshear GA
WPPL(FM) Blue Ridge GA
WMOG(AM) Brunswick GA
WSFN(AM) Brunswick GA
WBHF(AM) Cartersville GA
WYXC(AM) Cartersville GA
WEAM(AM) Columbus GA
WRCG(AM) Columbus GA
WSHE(AM) Columbus GA
WSHE(AM) Columbus GA
WDMG(AM) Douglas GA
WGGA(AM) Gainesville GA
WHIE(AM) Griffin GA
WLOP(AM) Jesup GA
WLAG(AM) La Grange GA
WIFN(FM) Macon GA
WMVG(AM) Milledgeville GA
WCNN(AM) North Atlanta GA
WLAQ(AM) Rome GA
WSNT(AM) Sandersville GA
WJLG(AM) Savannah GA
WTKS(AM) Savannah GA
WPTB(AM) Statesboro GA
WWNS(AM) Statesboro GA
WJAT(AM) Swainsboro GA
WVLD(AM) Valdosta GA
WVOP(AM) Vidalia GA
KHLO(AM) Hilo HI
KPUA(AM) Hilo HI
KKEA(AM) Honolulu HI
KKON(AM) Kealakekua HI
KAOI(AM) Kihei HI
KQNG(AM) Lihue HI
KUPA(AM) Pearl City HI
KMVI(AM) Wailuku HI
KCNZ(AM) Cedar Falls IA
KGYM(AM) Cedar Rapids IA
KMJM(AM) Cedar Rapids IA
KJOC(AM) Davenport IA
KWKY(AM) Des Moines IA
KXNO(AM) Des Moines IA
WDBQ(AM) Dubuque IA
KILR-FM Estherville IA
KVFD(AM) Fort Dodge IA
KIFG-FM Iowa Falls IA
KOKX(AM) Keokuk IA

KLEM(AM) Le Mars IA
KOEL(AM) Oelwein IA
KSCJ(AM) Sioux City IA
KAYL-FM Storm Lake IA
KBFI(AM) Bonners Ferry ID
KRLC(AM) Lewiston ID
KTIK(AM) Nampa ID
KIOV(AM) Payette ID
KWIK(AM) Pocatello ID
KRXK(AM) Rexburg ID
KSPT(AM) Sandpoint ID
WBIG(AM) Aurora IL
WCIL(AM) Carbondale IL
WDWS(AM) Champaign IL
*WCRX(FM) Chicago IL
WGN(AM) Chicago IL
WMVP(AM) Chicago IL
WSCR(AM) Chicago IL
WHOW-FM Clinton IL
WDAN(AM) Danville IL
WSOY(AM) Decatur IL
WZPN(FM) Farmington IL
WGIL(AM) Galesburg IL
*WGBK(FM) Glenview IL
WJPF(AM) Herrin IL
WDDD(AM) Johnston City IL
WLPO(AM) La Salle IL
WLUV(AM) Loves Park IL
WZZT(FM) Morrison IL
WVMC(AM) Mount Carmel IL
WVLN(AM) Olney IL
WPRS(AM) Paris IL
WZOE(AM) Princeton IL
WGEM(AM) Quincy IL
WKJR(AM) Rantoul IL
WTJK(AM) South Beloit IL
WHCO(AM) Sparta IL
WFMB(AM) Springfield IL
WTAX(AM) Springfield IL
WSDR(AM) Sterling IL
WSPL(AM) Streator IL
WVIL(FM) Virginia IL
KFNS(AM) Wood River IL
WHBU(AM) Anderson IN
WBIW(AM) Bedford IN
WSDX(AM) Brazil IN
WCSI(AM) Columbus IN
WKJG(AM) Fort Wayne IN
WOWO(AM) Fort Wayne IN
WLTH(AM) Gary IN
WREB(FM) Greencastle IN
WIBC(AM) Indianapolis IN
WNDE(AM) Indianapolis IN
WXLW(AM) Indianapolis IN
WIOU(AM) Kokomo IN
WASK(AM) Lafayette IN
WBAT(AM) Marion IN
WYFX(FM) Mount Vernon IN
WLBC-FM Muncie IN
WXFN(AM) Muncie IN
WWSZ(AM) New Albany IN
WJCP(AM) North Vernon IN
WZZB(AM) Seymour IN
WDND(AM) South Bend IN
WSBT(AM) South Bend IN
WWLV(AM) South Bend IN
WBOW(AM) Terre Haute IN
WAOV(AM) Vincennes IN
WRSW(AM) Warsaw IN
KFH-FM Clearwater KS
KGNO(AM) Dodge City KS
KIUL(AM) Garden City KS
KKCI(FM) Goodland KS
KVGB(AM) Great Bend KS
KNNS(AM) Larned KS
KLWN(AM) Lawrence KS
KMAN(AM) Manhattan KS
KLKC(AM) Parsons KS
KSEK(AM) Pittsburg KS
KINA(AM) Salina KS
KMAJ(AM) Topeka KS
WIBW(AM) Topeka KS
KGSO(AM) Wichita KS
KKLE(AM) Winfield KS
WAIA(AM) Beaver Dam KY
WCBL(AM) Benton KY

WBGN(AM) Bowling Green KY
WXAM(AM) Buffalo KY
WTCO(AM) Campbellsville KY
WNES(AM) Central City KY
WIEL(AM) Elizabethtown KY
WTSZ(AM) Eminence KY
WCDS(AM) Glasgow KY
WLGC(AM) Greenup KY
WLXG(AM) Lexington KY
WVLK(AM) Lexington KY
WKRD(AM) Louisville KY
WTTL(AM) Madisonville KY
WNBS(AM) Murray KY
WPAD(AM) Paducah KY
WKYH(AM) Paintsville KY
WPKY(AM) Princeton KY
WVLK-FM Richmond KY
WKRD-FM Shelbyville KY
WMSK-FM Sturgis KY
WIBR(AM) Baton Rouge LA
WJBO(AM) Baton Rouge LA
KBZE(FM) Berwick LA
KRRP(AM) Coushatta LA
WSKR(AM) Denham Springs LA
KNBB(FM) Dubach LA
KEUN(AM) Eunice LA
KJIN(AM) Houma LA
KFXZ(AM) Lafayette LA
KPEL(AM) Lafayette LA
KNOE(AM) Monroe LA
WODT(AM) New Orleans LA
WWL(AM) New Orleans LA
WWWL(AM) New Orleans LA
KRMD(AM) Shreveport LA
WSLA(AM) Slidell LA
KEZM(AM) Sulphur LA
WARL(AM) Attleboro MA
WEEI(AM) Boston MA
WWZN(AM) Boston MA
WXBR(AM) Brockton MA
WAMG(AM) Dedham MA
WVEI-FM Easthampton MA
WSAR(AM) Fall River MA
WEIM(AM) Fitchburg MA
WCAP(AM) Lowell MA
WLLH(AM) Lowell MA
WBSM(AM) New Bedford MA
WHMP(AM) Northampton MA
WBEC(AM) Pittsfield MA
WESX(AM) Salem MA
WESO(AM) Southbridge MA
WXTK(FM) West Yarmouth MA
WVEI(AM) Worcester MA
WBAL(AM) Baltimore MD
WJFK(AM) Baltimore MD
WCEM(AM) Cambridge MD
WSRY(AM) Elkton MD
WFMD(AM) Frederick MD
WWXT(FM) Prince Frederick MD
WICO(AM) Salisbury MD
WTGM(AM) Salisbury MD
WQMR(FM) Snow Hill MD
WNST(AM) Towson MD
WABI(AM) Bangor ME
WZON(AM) Bangor ME
WBQI(FM) Bar Harbor ME
WJJB(AM) Brunswick ME
WLVP(AM) Gorham ME
WLAM(AM) Lewiston ME
WIGY(FM) Madison ME
WTBM(FM) Mexico ME
WSYY(AM) Millinocket ME
WOXO-FM Norway ME
WRKD(AM) Rockland ME
WPHX(AM) Sanford ME
WSKW(AM) Skowhegan ME
WJJB-FM Topsham ME
WJAE(AM) Westbrook ME
WFYC(AM) Alma MI
WTKA(AM) Ann Arbor MI
WBFN(AM) Battle Creek MI
WBRN(AM) Big Rapids MI
WDTW(AM) Dearborn MI
WDFN(AM) Detroit MI
WXYT(AM) Detroit MI
WDOW(AM) Dowagiac MI

WVFN(AM) East Lansing MI
WTRX(AM) Flint MI
WMJZ-FM Gaylord MI
WBBL(AM) Grand Rapids MI
WTKG(AM) Grand Rapids MI
WSCG(AM) Greenville MI
WMPL(AM) Hancock MI
WMIQ(AM) Iron Mountain MI
WZAM(AM) Ishpeming MI
WIBM(AM) Jackson MI
WQLR(AM) Kalamazoo MI
WJNL(AM) Kingsley MI
WIDG(AM) Saint Ignace MI
WSJM(AM) Saint Joseph MI
WLBY(AM) Saline MI
WKNW(AM) Sault Ste. Marie MI
WMSH(AM) Sturgis MI
WCCW(AM) Traverse City MI
WMFN(AM) Zeeland MI
KNFX(AM) Austin MN
KBUN(AM) Bemidji MN
KLIZ(AM) Brainerd MN
KVBR(AM) Brainerd MN
KDLM(AM) Detroit Lakes MN
WEBC(AM) Duluth MN
KSUM(AM) Fairmont MN
KDHL(AM) Faribault MN
KDWA(AM) Hastings MN
WMFG(AM) Hibbing MN
KDUZ(AM) Hutchinson MN
KYSM(AM) Mankato MN
KFAN(AM) Minneapolis MN
KFXN(AM) Minneapolis MN
WCMP-FM Pine City MN
KOLM(AM) Rochester MN
KWEB(AM) Rochester MN
KRBI(AM) Saint Peter MN
WBHR(AM) Sauk Rapids MN
KXSS(AM) Waite Park MN
KDOM(AM) Windom MN
KWNO(AM) Winona MN
KBFL-FM Buffalo MO
KAPE(AM) Cape Girardeau MO
KGIR(AM) Cape Girardeau MO
KDKD-FM Clinton MO
KTGR(AM) Columbia MO
KRFT(AM) De Soto MO
KHMO(AM) Hannibal MO
KCTE(AM) Independence MO
KCSP(AM) Kansas City MO
WHB(AM) Kansas City MO
KLWT(AM) Lebanon MO
KCXM(FM) Lee's Summit MO
KNIM(AM) Maryville MO
KFMO(AM) Park Hills MO
KLID(AM) Poplar Bluff MO
KMIS(AM) Portageville MO
KTTR(AM) Rolla MO
KFEQ(AM) Saint Joseph MO
KMOX(AM) Saint Louis MO
KSLG(AM) Saint Louis MO
KTRS(AM) Saint Louis MO
KSMO(AM) Salem MO
KLTK(AM) South West City MO
KGMY(AM) Springfield MO
KWTO(AM) Springfield MO
KWTO-FM Springfield MO
KTUI-FM Sullivan MO
KTTN-FM Trenton MO
KFNS-FM Troy MO
KOKO(AM) Warrensburg MO
WWZQ(AM) Aberdeen MS
WAMY(AM) Amory MS
WXBD(AM) Biloxi MS
WCJU(AM) Columbia MS
WJWF(AM) Columbus MS
WPBQ(AM) Flowood MS
WFOR(AM) Hattiesburg MS
WHSY(AM) Hattiesburg MS
WJDX(AM) Jackson MS
WSFZ(AM) Jackson MS
WHNY(AM) McComb MS
WMOX(AM) Meridian MS
WYHL(AM) Meridian MS
WMLC(AM) Monticello MS
WNAT(AM) Natchez MS

WQMS(AM) Quitman MS
WKOR(AM) Starkville MS
WSSO(AM) Starkville MS
WTUP(AM) Tupelo MS
WQBC(AM) Vicksburg MS
KBLG(AM) Billings MT
KMMS(AM) Bozeman MT
KBOW(AM) Butte MT
KGEZ(AM) Kalispell MT
KGRZ(AM) Missoula MT
WISE(AM) Asheville NC
WTKF(FM) Atlantic NC
WATA(AM) Boone NC
WYSE(AM) Canton NC
WFNA(AM) Charlotte NC
WFNZ(AM) Charlotte NC
*WWCU(FM) Cullowhee NC
WDNC(AM) Durham NC
WGAI(AM) Elizabeth City NC
WGHB(AM) Farmville NC
WGNC(AM) Gastonia NC
WNCT(AM) Greenville NC
WYSR(AM) High Point NC
WJNC(AM) Jacksonville NC
WLXN(AM) Lexington NC
WLON(AM) Lincolnton NC
WRBZ(AM) Raleigh NC
WCBT(AM) Roanoke Rapids NC
WRMT(AM) Rocky Mount NC
WCAB(AM) Rutherfordton NC
WTSB(AM) Selma NC
WOHS(AM) Shelby NC
WNCA(AM) Siler City NC
WSIC(AM) Statesville NC
WBLO(AM) Thomasville NC
WMFD(AM) Wilmington NC
WVOT(AM) Wilson NC
KDKT(AM) Beulah ND
KXMR(AM) Bismarck ND
KLXX(AM) Bismarck-Mandan ND
KKAG(AM) Fargo ND
WDAY(AM) Fargo ND
KQDJ(AM) Jamestown ND
KOVC(AM) Valley City ND
KOZN(AM) Bellevue NE
KICS(AM) Hastings NE
KXPN(AM) Kearney NE
KLMS(AM) Lincoln NE
KSWN(FM) McCook NE
KXSP(AM) Omaha NE
KOLT(AM) Scottsbluff NE
WTSV(AM) Claremont NH
WTSN(AM) Dover NH
WGIP(AM) Exeter NH
WTSL(AM) Hanover NH
WTPL(FM) Hillsboro NH
WLTN(AM) Littleton NH
WGIR(AM) Manchester NH
WKBR(AM) Manchester NH
WGAM(AM) Nashua NH
WSMN(AM) Nashua NH
WGIN(AM) Rochester NH
WADB(AM) Asbury Park NJ
WPAT(AM) Paterson NJ
WPHY(AM) Trenton NJ
KRSY(AM) Alamogordo NM
KDEF(AM) Albuquerque NM
KNML(AM) Albuquerque NM
KCQL(AM) Aztec NM
KNFT(AM) Bayard NM
KCLV(AM) Clovis NM
KENN(AM) Farmington NM
KYKK(AM) Hobbs NM
KOBE(AM) Las Cruces NM
KRSN(AM) Los Alamos NM
KRUI(AM) Ruidoso Downs NM
KVSF(AM) Santa Fe NM
KTSN(AM) Elko NV
KBAD(AM) Las Vegas NV
KENO(AM) Las Vegas NV
KLAV(AM) Las Vegas NV
KRLV(AM) Las Vegas NV
KSHP(AM) North Las Vegas NV
KPLY(AM) Reno NV
KBZZ(AM) Sparks NV
WVTL(AM) Amsterdam NY

WYSL(AM) Avon NY
WYOS(AM) Binghamton NY
WPUT(AM) Brewster NY
WBEN(AM) Buffalo NY
WGR(AM) Buffalo NY
WCBA(AM) Corning NY
WELM(AM) Elmira NY
WENE(AM) Endicott NY
WMML(AM) Glens Falls NY
WWSC(AM) Glens Falls NY
WNRS(AM) Herkimer NY
WHCU(AM) Ithaca NY
WJTN(AM) Jamestown NY
WIRD(AM) Lake Placid NY
WIXT(AM) Little Falls NY
WLVL(AM) Lockport NY
WTMM-FM Mechanicville NY
WADO(AM) New York NY
WEPN(AM) New York NY
WFAN(AM) New York NY
WACK(AM) Newark NY
WLNA(AM) Peekskill NY
WEAV(AM) Plattsburgh NY
WADR(AM) Remsen NY
WTMM(AM) Rensselaer NY
WRNY(AM) Rome NY
WGGO(AM) Salamanca NY
WSPQ(AM) Springville NY
*WAER(FM) Syracuse NY
WHEN(AM) Syracuse NY
WNSS(AM) Syracuse NY
WOFX(AM) Troy NY
WPIE(AM) Trumansburg NY
WIBX(AM) Utica NY
WUTQ(AM) Utica NY
WNER(AM) Watertown NY
WARF(AM) Akron OH
WFUN(AM) Ashtabula OH
WATH(AM) Athens OH
WOMP(AM) Bellaire OH
WBLL(AM) Bellefontaine OH
WQEL(FM) Bucyrus OH
WILE(AM) Cambridge OH
WCER(AM) Canton OH
WCSM-FM Celina OH
WCKY(AM) Cincinnati OH
WSAI(AM) Cincinnati OH
WKNR(AM) Cleveland OH
WTAM(AM) Cleveland OH
WWGK(AM) Cleveland OH
WBNS(AM) Columbus OH
WING(AM) Dayton OH
WONE(AM) Dayton OH
WONW(AM) Defiance OH
WEOL(AM) Elyria OH
WMOH(AM) Hamilton OH
WJMP(AM) Kent OH
WIMA(AM) Lima OH
WMOA(AM) Marietta OH
WTIG(AM) Massillon OH
WJAW-FM McConnelsville OH
WMPO(AM) Middleport-Pomeroy OH
WLKR(AM) Norwalk OH
WLKR-FM Norwalk OH
WRGM(AM) Ontario OH
WNXT(AM) Portsmouth OH
WLEC(AM) Sandusky OH
WIZE(AM) Springfield OH
WSTV(AM) Steubenville OH
WLQR(AM) Toledo OH
WBTC(AM) Uhrichsville OH
WELW(AM) Willoughby-Eastlake OH
WQKT(FM) Wooster OH
WBBW(AM) Youngstown OH
WKBN(AM) Youngstown OH
KADA(AM) Ada OK
KOKB(AM) Blackwell OK
KUSH(AM) Cushing OK
KPNS(AM) Duncan OK
WWLS-FM Edmond OK
KADS(AM) Elk City OK
KFXY(AM) Enid OK
KBEL(AM) Idabel OK
KXCA(AM) Lawton OK
KEBC(AM) Midwest City OK
WWLS(AM) Moore OK

KBIX(AM) Muskogee OK
KYAL-FM Muskogee OK
KREF(AM) Norman OK
WKY(AM) Oklahoma City OK
KOKP(AM) Perry OK
KYAL(AM) Sapulpa OK
KSPI(AM) Stillwater OK
KEOK(FM) Tahlequah OK
KTLQ(AM) Tahlequah OK
KAKC(AM) Tulsa OK
KCFO(AM) Tulsa OK
KTBZ(AM) Tulsa OK
KSIW(AM) Woodward OK
KAST(AM) Astoria OR
KKEE(AM) Astoria OR
KBND(AM) Bend OR
KICE(AM) Bend OR
KHSN(AM) Coos Bay OR
KLOO(AM) Corvallis OR
KSCR(AM) Eugene OR
KUIK(AM) Hillsboro OR
KFLS(AM) Klamath Falls OR
KLAD(AM) Klamath Falls OR
KGAL(AM) Lebanon OR
KNPT(AM) Newport OR
KTIX(AM) Pendleton OR
KFXX(AM) Portland OR
KQEN(AM) Roseburg OR
KOHI(AM) Saint Helens OR
KKSN(AM) Salem OR
KMBD(AM) Tillamook OR
WSAN(AM) Allentown PA
WTKZ(AM) Allentown PA
WVAM(AM) Altoona PA
WMBA(AM) Ambridge PA
WBVP(AM) Beaver Falls PA
*WBUQ(FM) Bloomsburg PA
WISR(AM) Butler PA
WWCB(AM) Corry PA
*WESS(FM) East Stroudsburg PA
WEEX(AM) Easton PA
WPDC(AM) Elizabethtown PA
WPSE(AM) Erie PA
WRIE(AM) Erie PA
WFRA(AM) Franklin PA
WGET(AM) Gettysburg PA
WTKT(AM) Harrisburg PA
WPSN(AM) Honesdale PA
WPRR(AM) Johnstown PA
WTYM(AM) Kittanning PA
WLPA(AM) Lancaster PA
WWGE(AM) Loretto PA
WMGW(AM) Meadville PA
WJUN(AM) Mexico PA
WKST(AM) New Castle PA
WIP(AM) Philadelphia PA
WPEN(AM) Philadelphia PA
WPHB(AM) Philipsburg PA
WBGG(AM) Pittsburgh PA
WEAE(AM) Pittsburgh PA
WPGB(FM) Pittsburgh PA
WYCK(AM) Plains PA
WEEU(AM) Reading PA
WIOV(AM) Reading PA
WKBI(AM) Saint Marys PA
WEJL(AM) Scranton PA
WICK(AM) Scranton PA
WQRM(FM) Smethport PA
WBHV(AM) Somerset PA
WMAJ(AM) State College PA
WKOK(AM) Sunbury PA
WTZN(AM) Troy PA
WBAX(AM) Wilkes-Barre PA
WLYC(AM) Williamsport PA
WRAK(AM) Williamsport PA
WOYK(AM) York PA
WGIT(AM) Canovanas PR
WXRF(AM) Guayama PR
WMNT(AM) Manati PR
WYEL(AM) Mayaguez PR
WVOZ(AM) San Juan PR
WADK(AM) Newport RI
WPRO(AM) Providence RI
WSKO(AM) Providence RI
WSKO-FM Wakefield-Peacedale RI
WEEI-FM Westerly RI

WZMJ(FM) Batesburg SC
WCCP-FM Clemson SC
WPCC(AM) Clinton SC
WCOS(AM) Columbia SC
WOIC(AM) Columbia SC
WVOC(AM) Columbia SC
WIQB(AM) Conway SC
WJXY-FM Conway SC
WTMZ(AM) Dorchester Terrace-Brentwood SC
WOLS(AM) Florence SC
WFIS(AM) Fountain Inn SC
WXJY(FM) Georgetown SC
WFXH(AM) Hilton Head Island SC
WWFN-FM Lake City SC
WRHM(FM) Lancaster SC
WJKB(AM) Moncks Corner SC
WRNN(AM) Myrtle Beach SC
WRHI(AM) Rock Hill SC
WOLI(AM) Spartanburg SC
WSPG(AM) Spartanburg SC
WBCU(AM) Union SC
WALI(FM) Walterboro SC
KGIM(AM) Aberdeen SD
KBFS(AM) Belle Fourche SD
KIJV(AM) Huron SD
KORN(AM) Mitchell SD
KSOO(AM) Sioux Falls SD
KSQB(AM) Sioux Falls SD
KWSN(AM) Sioux Falls SD
KVTK(AM) Vermillion SD
WNFN(FM) Belle Meade TN
WXSM(AM) Blountville TN
WNSR(AM) Brentwood TN
WRKM(AM) Carthage TN
WDEF(AM) Chattanooga TN
WJZM(AM) Clarksville TN
WCSV(AM) Crossville TN
WEMB(AM) Erwin TN
WCPH(AM) Etowah TN
WEKR(AM) Fayetteville TN
WHEW(AM) Franklin TN
WGFX(FM) Gallatin TN
WMLR(AM) Hohenwald TN
WGOC(AM) Kingsport TN
WNML(AM) Knoxville TN
WVLZ(AM) Knoxville TN
WQLA-FM La Follette TN
WCOR(AM) Lebanon TN
WLIV(AM) Livingston TN
WNML-FM Loudon TN
WMSR(AM) Manchester TN
WWTN(FM) Manchester TN
WKCE(AM) Maryville TN
WHBQ(AM) Memphis TN
WREC(AM) Memphis TN
*WUMR(FM) Memphis TN
WLIV-FM Monterey TN
WGNS(AM) Murfreesboro TN
WBFG(FM) Parker's Crossroads TN
WQBB(AM) Powell TN
WTNE-FM Trenton TN
KZQQ(AM) Abilene TX
KESN(FM) Allen TX
KGNC(AM) Amarillo TX
KPUR(AM) Amarillo TX
KVET(AM) Austin TX
KRUN(AM) Ballinger TX
KFNC(FM) Beaumont TX
KIKR(AM) Beaumont TX
KGOW(AM) Bellaire TX
KBST(AM) Big Spring TX
KZRK(AM) Canyon TX
KZNE(AM) College Station TX
KEYS(AM) Corpus Christi TX
KSIX(AM) Corpus Christi TX
KZNX(AM) Creedmoor TX
KNIT(AM) Dallas TX
KRLD(AM) Dallas TX
KTCK(AM) Dallas TX
KURV(AM) Edinburg TX
KULP(AM) El Campo TX
KHEY(AM) El Paso TX
KROD(AM) El Paso TX
KFLC(AM) Fort Worth TX
KKGM(AM) Fort Worth TX

KATH(AM) Frisco TX
KBME(AM) Houston TX
KILT(AM) Houston TX
KTRH(AM) Houston TX
*KSHU(FM) Huntsville TX
KAML(AM) Kenedy-Karnes City TX
KFRO(AM) Longview TX
KKAM(AM) Lubbock TX
KMND(AM) Midland TX
KBED(AM) Nederland TX
KGNB(AM) New Braunfels TX
KRIL(AM) Odessa TX
KOGT(AM) Orange TX
KBPO(AM) Port Neches TX
KSOX(AM) Raymondville TX
KGKL(AM) San Angelo TX
KKSA(AM) San Angelo TX
KTKR(AM) San Antonio TX
*KTSW(FM) San Marcos TX
KTDK(FM) Sanger TX
KJIM(AM) Sherman TX
KSET(AM) Silsbee TX
KWNX(AM) Taylor TX
KTEM(AM) Temple TX
KCMC(AM) Texarkana TX
KTBB(AM) Tyler TX
KYZS(AM) Tyler TX
KRZI(AM) Waco TX
KSL-FM Midvale UT
KJQS(AM) Murray UT
KALL(AM) North Salt Lake City UT
KOAL(AM) Price UT
KOVO(AM) Provo UT
KSVC(AM) Richfield UT
KFNZ(AM) Salt Lake City UT
KSL(AM) Salt Lake City UT
KZNS(AM) Salt Lake City UT
KVEL(AM) Vernal UT
KUNF(AM) Washington UT
WXTR(AM) Alexandria VA
WCBX(AM) Bassett VA
WBLT(AM) Bedford VA
WKEX(AM) Blacksburg VA
WKLV(AM) Blackstone VA
WBDY(AM) Bluefield VA
WFHG(AM) Bristol VA
WINA(AM) Charlottesville VA
WKAV(AM) Charlottesville VA
WDIC-FM Clinchco VA
WPIN(AM) Dublin VA
WODY(AM) Fieldale VA
WGAT(AM) Gate City VA
WLRT(AM) Hampton VA
WHBG(AM) Harrisonburg VA
WREL(AM) Lexington VA
WBRG(AM) Lynchburg VA
WLNI(FM) Lynchburg VA
WVGM(AM) Lynchburg VA
*WWMC(FM) Lynchburg VA
WJFK-FM Manassas VA
WNRV(AM) Narrows-Pearisburg VA
WCMS(AM) Newport News VA
WTAR(AM) Norfolk VA
WRAD(AM) Radford VA
WRNL(AM) Richmond VA
WXGI(AM) Richmond VA
WGMN(AM) Roanoke VA
WTON(AM) Staunton VA
WXTG(FM) Virginia Beach VA
WWXX(FM) Warrenton VA
WINC(AM) Winchester VA
WYVE(AM) Wytheville VA
WVWI(AM) Charlotte Amalie VI
WSNO(AM) Barre VT
WTSA(AM) Brattleboro VT
WCAT(AM) Burlington VT
WVMT(AM) Burlington VT
WFAD(AM) Middlebury VT
WDEV-FM Warren VT
WDEV(AM) Waterbury VT
WNHV(AM) White River Junction VT
KXLX(AM) Airway Heights WA
KWLE(AM) Anacortes WA
KCLK(AM) Asotin WA
KPUG(AM) Bellingham WA
KELA(AM) Centralia-Chehalis WA

KXLE(AM) Ellensburg WA
KULE(AM) Ephrata WA
KRKO(AM) Everett WA
KWOK(AM) Hoquiam WA
KBSN(AM) Moses Lake WA
KWIQ(AM) Moses Lake North WA
KRSC(AM) Othello WA
KFLD(AM) Pasco WA
KZXR(AM) Prosser WA
KALE(AM) Richland WA
KIRO(AM) Seattle WA
KJR(AM) Seattle WA
KBBO(AM) Selah WA
KJRB(AM) Spokane WA
KHHO(AM) Tacoma WA
KYNR(AM) Toppenish WA
KUJ(AM) Walla Walla WA
KKRT(AM) Wenatchee WA
WSCO(AM) Appleton WI
WISS(AM) Berlin WI
WDUZ-FM Brillion WI
WMBE(AM) Chilton WI
WEAQ(AM) Chippewa Falls WI
WTLX(FM) Columbus WI
WBIZ(AM) Eau Claire WI
KFIZ(AM) Fond du Lac WI
WFAW(AM) Fort Atkinson WI
WNFL(AM) Green Bay WI
WKTY(AM) La Crosse WI
WTSO(AM) Madison WI
WMAM(AM) Marinette WI
WDLB(AM) Marshfield WI
WIGM(AM) Medford WI
WMEQ(AM) Menomonie WI
WTMJ(AM) Milwaukee WI
WLKD(AM) Minocqua WI
WOSH(AM) Oshkosh WI
WNBI(AM) Park Falls WI
WPVL(AM) Platteville WI
WOBT(AM) Rhinelander WI
WRCO-FM Richland Center WI
WEVR(AM) River Falls WI
WEVR-FM River Falls WI
WCLB(AM) Sheboygan WI
WKLJ(AM) Sparta WI
WDOR-FM Sturgeon Bay WI
WJJQ(AM) Tomahawk WI
WJJQ-FM Tomahawk WI
WAUK(AM) Waukesha WI
WKEZ(AM) Bluefield WV
WCHS(AM) Charleston WV
WSWW(AM) Charleston WV
WXKX(AM) Clarksburg WV
WBES(AM) Dunbar WV
WMMN(AM) Fairmont WV
WAMN(AM) Green Valley WV
WRVC(AM) Huntington WV
WEPM(AM) Martinsburg WV
WRNR(AM) Martinsburg WV
WVLY(AM) Moundsville WV
WJAW(AM) Saint Marys WV
WKKX(AM) Wheeling WV
KRSV(AM) Afton WY
KRSV-FM Afton WY
KBBS(AM) Buffalo WY
KKTL(AM) Casper WY
KVOC(AM) Casper WY
KFBC(AM) Cheyenne WY
KRAE(AM) Cheyenne WY
KODI(AM) Cody WY
KIML(AM) Gillette WY
KHAT(AM) Laramie WY
KOWB(AM) Laramie WY
KYDT(FM) Sundance WY

Talk

KENI(AM) Anchorage AK
KUDO(AM) Anchorage AK
*KYUK(AM) Bethel AK
KRUP(FM) Dillingham AK
KFBX(AM) Fairbanks AK
KJNO(AM) Juneau AK
KVOK(AM) Kodiak AK
*KUDU(FM) Tok AK
KVAK(AM) Valdez AK

WASG(AM) Atmore AL
WXJC(AM) Birmingham AL
WYMR(AM) Bridgeport AL
WKUL(FM) Cullman AL
WDPT(AM) Decatur AL
WOOF(AM) Dothan AL
WULA(AM) Eufaula AL
WJLD(AM) Fairfield AL
WABF(AM) Fairhope AL
WHEP(AM) Foley AL
WBHP(AM) Huntsville AL
WTKI(AM) Huntsville AL
WNSI(AM) Robertsdale AL
WBTG(AM) Sheffield AL
WTBF(AM) Troy AL
WVSA(AM) Vernon AL
KEWI(AM) Benton AR
KREB(AM) Bentonville-Bella Vista AR
KVDW(AM) England AR
KXJK(AM) Forrest City AR
KHOZ(AM) Harrison AR
KXAR(AM) Hope AR
KABZ(FM) Little Rock AR
KTTG(FM) Mena AR
KFFK(AM) Rogers AR
KWCK(AM) Searcy AR
KARN-FM Sheridan AR
KHGG-FM Waldron AR
KZZZ(AM) Bullhead City AZ
KFNX(AM) Cave Creek AZ
KXAM(AM) Mesa AZ
KGME(AM) Phoenix AZ
KIDR(AM) Phoenix AZ
KZON(FM) Phoenix AZ
KNXN(AM) Sierra Vista AZ
KRVZ(AM) Springerville-Eagar AZ
KVOI(AM) Tucson AZ
KXMX(AM) Anaheim CA
KIXW(AM) Apple Valley CA
KBRT(AM) Avalon CA
KAFY(AM) Bakersfield CA
KGEO(AM) Bakersfield CA
KHTY(AM) Bakersfield CA
KRXA(AM) Carmel Valley CA
KPTR(AM) Cathedral City CA
KNWZ(AM) Coachella CA
KTKZ-FM Dunnigan CA
*KMUE(FM) Eureka CA
KWSW(AM) Eureka CA
KIRV(AM) Fresno CA
*KMUD(FM) Garberville CA
KGBA-FM Holtville CA
KABC(AM) Los Angeles CA
KFI(AM) Los Angeles CA
KLSX(FM) Los Angeles CA
KTLK(AM) Los Angeles CA
KMYC(AM) Marysville CA
KTOX(AM) Needles CA
KKGN(AM) Oakland CA
KGDP(AM) Oildale CA
KDAR(FM) Oxnard CA
KGAM(AM) Palm Springs CA
KPSI(AM) Palm Springs CA
KNTS(AM) Palo Alto CA
*KZYX(FM) Philo CA
KAHZ(AM) Pomona CA
KNLF(FM) Quincy CA
KSTE(AM) Rancho Cordova CA
KQMS(AM) Redding CA
*KVIP(AM) Redding CA
KTKZ(AM) Sacramento CA
*KVCR(FM) San Bernardino CA
KLSD(AM) San Diego CA
KEST(AM) San Francisco CA
KFAX(AM) San Francisco CA
KGO(AM) San Francisco CA
KSFO(AM) San Francisco CA
KTRB(AM) San Francisco CA
KYCY(AM) San Francisco CA
KPRZ(AM) San Marcos-Poway CA
KIST(AM) Santa Barbara CA
KTMS(AM) Santa Barbara CA
KCNR(AM) Shasta CA
KFUT(AM) Thousand Palms CA
KXPS(AM) Thousand Palms CA
KCEO(AM) Vista CA

KERI(AM) Wasco-Greenacres CA
*KZYZ(FM) Willits CA
KNFO(FM) Basalt CO
KSIR(AM) Brush CO
KSKE(AM) Buena Vista CO
KKKK(AM) Colorado Springs CO
KLTT(AM) Commerce City CO
KHOW(AM) Denver CO
KKFN(AM) Denver CO
KJOL(AM) Grand Junction CO
KRCN(AM) Longmont/Denver CO
KKZN(AM) Thornton CO
KVLE(AM) Vail CO
WICC(AM) Bridgeport CT
WGCH(AM) Greenwich CT
WMMW(AM) Meriden CT
WMRD(AM) Middletown CT
WLIS(AM) Old Saybrook CT
WHDD(AM) Sharon CT
WXLM(FM) Stonington CT
WSNG(AM) Torrington CT
WKND(AM) Windsor CT
WTEM(AM) Washington DC
WWRC(AM) Washington DC
WILM(AM) Wilmington DE
WTMC(AM) Wilmington DE
WHIM(AM) Apopka FL
WQOP(AM) Atlantic Beach FL
WSBR(AM) Boca Raton FL
WWPR(AM) Bradenton FL
WTAN(AM) Clearwater FL
WWBC(AM) Cocoa FL
WTKS-FM Cocoa Beach FL
WDJA(AM) Delray Beach FL
WTOT-FM Graceville FL
WACC(AM) Hialeah FL
WTKE-FM Holt FL
WYMM(AM) Jacksonville FL
WBXY(FM) La Crosse FL
WWTK(AM) Lake Placid FL
WWAB(AM) Lakeland FL
*WCNZ(AM) Marco Island FL
WINZ(AM) Miami FL
WMBM(AM) Miami Beach FL
WNMA(AM) Miami Springs FL
WFLA-FM Midway FL
WEBY(AM) Milton FL
*WEGS(FM) Milton FL
WPGS(AM) Mims FL
*WSOR(FM) Naples FL
WWCN(AM) North Fort Myers FL
WMOP(AM) Ocala FL
WQTM(AM) Orlando FL
WTLN(AM) Orlando FL
*WFSW(FM) Panama City FL
*WJTF(FM) Panama City FL
WTBN(AM) Pinellas Park FL
WHSR(AM) Pompano Beach FL
WWNN(AM) Pompano Beach FL
WZZR(FM) Riviera Beach FL
WMEN(AM) Royal Palm Beach FL
WTMY(AM) Sarasota FL
WPUL(AM) South Daytona FL
WYOO(FM) Springfield FL
WCZR(FM) Vero Beach FL
WTTB(AM) Vero Beach FL
WZTA(AM) Vero Beach FL
WBZT(AM) West Palm Beach FL
WZHR(AM) Zephyrhills FL
WRFC(AM) Athens GA
WALR(AM) Atlanta GA
WGUN(AM) Atlanta GA
WNIV(AM) Atlanta GA
WQXI(AM) Atlanta GA
WNRR(AM) Augusta GA
WBBK(AM) Blakely GA
WYXC(AM) Cartersville GA
WGHC(AM) Clayton GA
WJJC(AM) Commerce GA
WSEM(AM) Donalsonville GA
WCFO(AM) East Point GA
WNGA(AM) Elberton GA
WDDK(FM) Greensboro GA
WKLY(AM) Hartwell GA
WBKZ(AM) Jefferson GA
WFOM(AM) Marietta GA

WTNL(AM) Reidsville GA
WJAT(AM) Swainsboro GA
WZQZ(AM) Trion GA
*WJSP-FM Warm Springs GA
WPLV(AM) West Point GA
WIMO(AM) Winder GA
KKEA(AM) Honolulu HI
KORL(AM) Honolulu HI
KRTR(AM) Honolulu HI
KUMU(AM) Honolulu HI
KWQW(FM) Boone IA
KCPS(AM) Burlington IA
KCNZ(AM) Cedar Falls IA
*KUNI(FM) Cedar Falls IA
KJOC(AM) Davenport IA
KPSZ(AM) Des Moines IA
KWKY(AM) Des Moines IA
KCJJ(AM) Iowa City IA
KGLO(AM) Mason City IA
*KDMR(FM) Mitchellville IA
KICD(AM) Spencer IA
*KNWS(AM) Waterloo IA
KBLI(AM) Blackfoot ID
KFXD(AM) Boise ID
KSPD(AM) Boise ID
KBAR(AM) Burley ID
KBGN(AM) Caldwell ID
KBLY(AM) Idaho Falls ID
KOZE(AM) Lewiston ID
KTMB(FM) Mountain Home ID
KIDO(AM) Nampa ID
KTIK(AM) Nampa ID
*KBSW(FM) Twin Falls ID
WVON(AM) Berwyn IL
WTHQ(AM) Brookport IL
WCAZ(AM) Carthage IL
WMVP(AM) Chicago IL
*WSSD(FM) Chicago IL
WYLL(AM) Chicago IL
WHOW-FM Clinton IL
WCPT(AM) Crystal Lake IL
WCKG(FM) Elmwood Park IL
WGNU(AM) Granite City IL
WXOZ(AM) Highland IL
WDDD(AM) Johnston City IL
WJOL(AM) Joliet IL
WKAN(AM) Kankakee IL
WGGH(AM) Marion IL
WFXN(AM) Moline IL
WMIX(AM) Mount Vernon IL
*WPTH(FM) Olney IL
WPRS(AM) Paris IL
WPEO(AM) Peoria IL
WLIQ(AM) Quincy IL
WTRH(FM) Ramsey IL
WNTA(AM) Rockford IL
*WGNJ(FM) Saint Joseph IL
WAUR(AM) Sandwich IL
WFMB(AM) Springfield IL
WSDR(AM) Sterling IL
*WGNR(AM) Anderson IN
WBIW(AM) Bedford IN
WFRN(AM) Elkhart IN
WWFT(FM) Fishers IN
WWCA(AM) Gary IN
WNDE(AM) Indianapolis IN
WTLC(AM) Indianapolis IN
WIMS(AM) Michigan City IN
WMRS(FM) Monticello IN
WYFX(FM) Mount Vernon IN
WFIA-FM New Albany IN
WRAY(AM) Princeton IN
WSLM-FM Salem IN
*WHOJ(FM) Terre Haute IN
KFH-FM Clearwater KS
KGNO(AM) Dodge City KS
KCNW(AM) Fairway KS
KVGB(AM) Great Bend KS
KBUF(AM) Holcomb KS
KWBW(AM) Hutchinson KS
KCVW(FM) Kingman KS
KCCV-FM Olathe KS
KLKC(AM) Parsons KS
KFRM(AM) Salina KS
KFH(AM) Wichita KS
*KYWA(FM) Wichita KS

KKLE(AM) Winfield KS
WCMI(AM) Ashland KY
WCBL(AM) Benton KY
WSFE(AM) Burnside KY
WNES(AM) Central City KY
WLGC(AM) Greenup KY
WFTG(AM) London KY
WFIA(AM) Louisville KY
WKJK(AM) Louisville KY
WTTL(AM) Madisonville KY
WNBS(AM) Murray KY
WLSI(AM) Pikeville KY
WPRT(AM) Prestonsburg KY
WSFC(AM) Somerset KY
WLXO(FM) Stamping Ground KY
WEZJ(AM) Williamsburg KY
KJMJ(AM) Alexandria LA
KSYL(AM) Alexandria LA
WJBO(AM) Baton Rouge LA
WPYR(AM) Baton Rouge LA
KBCL(AM) Bossier City LA
KEUN(AM) Eunice LA
WWL-FM Kenner LA
KFXZ(AM) Lafayette LA
KVOL(AM) Lafayette LA
*KOJO(FM) Lake Charles LA
KWLA(AM) Many LA
KLIC(AM) Monroe LA
*KBIO(FM) Natchitoches LA
WSHO(AM) New Orleans LA
WWWL(AM) New Orleans LA
KRMD(AM) Shreveport LA
WSRO(AM) Ashland MA
WNSH(AM) Beverly MA
*WBUR-FM Boston MA
WEZE(AM) Boston MA
WRKO(AM) Boston MA
WTKK(FM) Boston MA
WTTT(AM) Boston MA
WACE(AM) Chicopee MA
WHTB(AM) Fall River MA
*WCCT-FM Harwich MA
WCAP(AM) Lowell MA
WBIX(AM) Natick MA
WPLM(AM) Plymouth MA
WESO(AM) Southbridge MA
*WSKB(FM) Westfield MA
WCRN(AM) Worcester MA
WBIS(AM) Annapolis MD
WCBM(AM) Baltimore MD
WJFK(AM) Baltimore MD
WTNT(AM) Bethesda MD
WHFS(FM) Catonsville MD
WKDI(AM) Denton MD
*WYPF(FM) Frederick MD
WARK(AM) Hagerstown MD
WVIE(AM) Pikesville MD
WDMV(AM) Walkersville MD
WZON(AM) Bangor ME
WBCI(FM) Bath ME
WJJB(AM) Brunswick ME
WZAN(AM) Portland ME
WEGP(AM) Presque Isle ME
WFMX(FM) Skowhegan ME
WJJB-FM Topsham ME
WJAE(AM) Westbrook ME
WATZ(AM) Alpena MI
WMAX(AM) Bay City MI
WHFB(AM) Benton Harbor-St. Joseph MI
WDTW(AM) Dearborn MI
WNZK(AM) Dearborn Heights MI
WDFN(AM) Detroit MI
WKRK-FM Detroit MI
WVFN(AM) East Lansing MI
*WLJN(AM) Elmwood Township MI
WSNL(AM) Flint MI
WTRX(AM) Flint MI
WSHN(AM) Fremont MI
WTKG(AM) Grand Rapids MI
WSCG(AM) Greenville MI
*WHPR(FM) Highland Park MI
WJMS(AM) Ironwood MI
WJNL(AM) Kingsley MI
WILS(AM) Lansing MI
WJIM(AM) Lansing MI

WKBZ(AM) Muskegon MI
WHAK(AM) Rogers City MI
WSDS(AM) Salem Township MI
WLBY(AM) Saline MI
WIOS(AM) Tawas City MI
*WICA(FM) Traverse City MI
*WLJN-FM Traverse City MI
WPON(AM) Walled Lake MI
WODJ(AM) Whitehall MI
WDEO(AM) Ypsilanti MI
WPNW(AM) Zeeland MI
KNFX(AM) Austin MN
KKBJ(AM) Bemidji MN
WFMP(FM) Coon Rapids MN
KROX(AM) Crookston MN
*KDNI(FM) Duluth MN
WWJC(AM) Duluth MN
KKCQ(AM) Fosston MN
KYCR(AM) Golden Valley MN
KDWA(AM) Hastings MN
WMFG(AM) Hibbing MN
KTOE(AM) Mankato MN
KFAN(AM) Minneapolis MN
KFXN(AM) Minneapolis MN
KLOH(AM) Pipestone MN
KKMS(AM) Richfield MN
KWEB(AM) Rochester MN
KTNF(AM) Saint Louis Park MN
KSTP(AM) Saint Paul MN
KOWZ(AM) Waseca MN
KSGF-FM Ash Grove MO
KAPE(AM) Cape Girardeau MO
*KNLH(FM) Cedar Hill MO
KCHR(AM) Charleston MO
*KFUO(AM) Clayton MO
KCTO(AM) Cleveland MO
KFTK(FM) Florissant MO
KCMO(AM) Kansas City MO
*KKTR(FM) Kirksville MO
KBNN(AM) Lebanon MO
KCXL(AM) Liberty MO
KTRI-FM Mansfield MO
KWIX(AM) Moberly MO
KQYS(AM) Neosho MO
KLID(AM) Poplar Bluff MO
*KNLP(FM) Potosi MO
KJSL(AM) Saint Louis MO
KLTK(AM) South West City MO
KADI(AM) Springfield MO
KLFJ(AM) Springfield MO
*KSCV(FM) Springfield MO
KWTO-FM Springfield MO
KSAR(FM) Thayer MO
KLPW(AM) Union MO
KJPW(AM) Waynesville MO
WWZQ(AM) Aberdeen MS
WBSL(AM) Bay St. Louis MS
WXRZ(FM) Corinth MS
WFMN(FM) Flora MS
WABG(AM) Greenwood MS
WFOR(AM) Hattiesburg MS
WHSY(AM) Hattiesburg MS
WJDX(AM) Jackson MS
WALT(AM) Meridian MS
WRQO(FM) Monticello MS
WPMP(AM) Pascagoula-Moss Point MS
WHOC(AM) Philadelphia MS
WKMQ(AM) Tupelo MS
WTNM(FM) Water Valley MS
KURL(AM) Billings MT
KBOZ(AM) Bozeman MT
KGRZ(AM) Missoula MT
*KUFM(FM) Missoula MT
WCGC(AM) Belmont NC
WFGW(AM) Black Mountain NC
WATA(AM) Boone NC
WSQL(AM) Brevard NC
WZTK(FM) Burlington NC
WBAG(AM) Burlington-Graham NC
WOGR(AM) Charlotte NC
WZRH(AM) Dallas NC
WLOE(AM) Eden NC
WPEK(AM) Fairview NC
WGHB(AM) Farmville NC
WNCT(AM) Greenville NC

WGOS(AM) High Point NC
WYSR(AM) High Point NC
WDYT(AM) Kings Mountain NC
WKXU(FM) Louisburg NC
WYRN(AM) Louisburg NC
WIXE(AM) Monroe NC
WXNC(AM) Monroe NC
*WAAE(FM) New Bern NC
*WZRN(FM) Norlina NC
WFMI(FM) Southern Shores NC
WBLO(AM) Thomasville NC
WWTB(FM) Topsail Beach NC
WDLX(AM) Washington NC
WBFJ(AM) Winston-Salem NC
KQWB(AM) West Fargo ND
KTTT(AM) Columbus NE
KOLT(AM) Scottsbluff NE
KAWL(AM) York NE
WDER(AM) Derry NH
WKBK(AM) Keene NH
WEMJ(AM) Laconia NH
WENJ(AM) Atlantic City NJ
WTMR(AM) Camden NJ
WWDJ(AM) Hackensack NJ
WOBM(AM) Lakewood NJ
WXKW(FM) Millville NJ
WVNJ(AM) Oakland NJ
*WVPH(FM) Piscataway NJ
WGHT(AM) Pompton Lakes NJ
WKMB(AM) Stirling NJ
WKXW(FM) Trenton NJ
WNJC(AM) Washington Township NJ
KRSY(AM) Alamogordo NM
KABQ(AM) Albuquerque NM
KBZU(FM) Albuquerque NM
KKIM(AM) Albuquerque NM
KSVP(AM) Artesia NM
KNFT(AM) Bayard NM
KICA(AM) Clovis NM
*KNMI(FM) Farmington NM
KZXQ(FM) Reserve NM
KRUI(AM) Ruidoso Downs NM
KTRC(AM) Santa Fe NM
KVOT(AM) Taos NM
*KNIS(FM) Carson City NV
KTSN(AM) Elko NV
KKVV(AM) Las Vegas NV
KLAV(AM) Las Vegas NV
KSFN(AM) North Las Vegas NV
KXTE(FM) Pahrump NV
KBZB(FM) Pioche NV
KJFK(AM) Reno NV
KRZQ-FM Sparks NV
*WAMC-FM Albany NY
WVTL(AM) Amsterdam NY
WAUB(AM) Auburn NY
WBUF(FM) Buffalo NY
WWKB(AM) Buffalo NY
WCKL(AM) Catskill NY
WCHP(AM) Champlain NY
WJIV(FM) Cherry Valley NY
WSDE(AM) Cobleskill NY
WENE(AM) Endicott NY
WHHO(AM) Hornell NY
WLIE(AM) Islip NY
WNYY(AM) Ithaca NY
WJTN(AM) Jamestown NY
*WOSR(FM) Middletown NY
WABC(AM) New York NY
WEPN(AM) New York NY
WFAN(AM) New York NY
WMCA(AM) New York NY
WWRL(AM) New York NY
WSLB(AM) Ogdensburg NY
WEBO(AM) Owego NY
*WCEL(FM) Plattsburgh NY
WEAV(AM) Plattsburgh NY
WTWK(AM) Plattsburgh NY
*WAIH(FM) Potsdam NY
WHTK(AM) Rochester NY
WSFW(AM) Seneca Falls NY
WFBL(AM) Syracuse NY
WNSS(AM) Syracuse NY
*WANC(FM) Ticonderoga NY
WTBQ(AM) Warwick NY
WATN(AM) Watertown NY

WRCI(FM) Webster NY
WBZB(FM) Westhampton NY
WXZO(FM) Willsboro NY
WFUN(AM) Ashtabula OH
WAIS(AM) Buchtel OH
WGFT(AM) Campbell OH
WCIN(AM) Cincinnati OH
WCVX(AM) Cincinnati OH
WDBZ(AM) Cincinnati OH
WHK(AM) Cleveland OH
WHKW(AM) Cleveland OH
WYTS(AM) Columbus OH
WWOW(AM) Conneaut OH
WFOB(AM) Fostoria OH
WHTH(AM) Heath OH
WNIR(FM) Kent OH
WLOH(AM) Lancaster OH
WFTK(FM) Lebanon OH
WLTP(AM) Marietta OH
WUCO(AM) Marysville OH
WCCD(AM) Parma OH
WNXT(AM) Portsmouth OH
WPAY(AM) Portsmouth OH
*WLRY(FM) Rushville OH
*WCVZ(FM) South Zanesville OH
WTDA(FM) Westerville OH
WELW(AM) Willoughby-Eastlake OH
WASN(AM) Youngstown OH
KOKB(AM) Blackwell OK
KPNS(AM) Duncan OK
KXCA(AM) Lawton OK
KTLR(AM) Oklahoma City OK
KCFO(AM) Tulsa OK
KWFX(FM) Woodward OK
KZZR(AM) Burns OR
KEJO(AM) Corvallis OR
KLZS(AM) Eugene OR
KOPT(AM) Eugene OR
*KRVM(AM) Eugene OR
KGAL(AM) Lebanon OR
*KEZX(AM) Medford OR
KUMA(AM) Pendleton OR
KCMX(AM) Phoenix OR
KEX(AM) Portland OR
KKPZ(AM) Portland OR
KPDQ(AM) Portland OR
KPDQ-FM Portland OR
KOHI(AM) Saint Helens OR
WMBA(AM) Ambridge PA
WBVP(AM) Beaver Falls PA
*WBUQ(FM) Bloomsburg PA
WLFP(AM) Braddock PA
WFYL(AM) King of Prussia PA
WDAC(FM) Lancaster PA
WJSM(AM) Martinsburg PA
WEDO(AM) McKeesport PA
WMLP(AM) Milton PA
WFIL(AM) Philadelphia PA
WHAT(AM) Philadelphia PA
*WHYY-FM Philadelphia PA
WPHT(AM) Philadelphia PA
WURD(AM) Philadelphia PA
WWDB(AM) Philadelphia PA
WBGG(AM) Pittsburgh PA
WEAE(AM) Pittsburgh PA
WTZN-FM Pittsburgh PA
WWNL(AM) Pittsburgh PA
WECZ(AM) Punxsutawney PA
*WNJR(FM) Washington PA
WCHE(AM) West Chester PA
WPYT(AM) Wilkinsburg PA
WYYC(AM) York PA
WLUZ(AM) Bayamon PR
WPRA(AM) Mayaguez PR
WTIL(AM) Mayaguez PR
WBMJ(AM) San Juan PR
WQII(AM) San Juan PR
WRSS(AM) San Sebastian PR
WIVV(AM) Vieques PR
WXEW(AM) Yabucoa PR
WEEI-FM Westerly RI
WXNI(AM) Westerly RI
*WLJK(FM) Aiken SC
WQSC(AM) Charleston SC
WISW(AM) Columbia SC
WJMX(AM) Florence SC

WOLS(AM) Florence SC
WSCC-FM Goose Creek SC
WCKI(AM) Greer SC
*WMBJ(FM) Murrell's Inlet SC
WAVO(AM) Rock Hill SC
WRNN-FM Socastee SC
WQMC(AM) Sumter SC
WWOF(AM) Walhalla SC
KSDN(AM) Aberdeen SD
KBFS(AM) Belle Fourche SD
KIJV(AM) Huron SD
KORN(AM) Mitchell SD
KSQP(AM) Pierre SD
KTOQ(AM) Rapid City SD
KSDR(AM) Watertown SD
WCTA(AM) Alamo TN
WYXI(AM) Athens TN
*WHCB(FM) Bristol TN
WNKX(AM) Centerville TN
WJZM(AM) Clarksville TN
WQZQ(AM) Clarksville TN
WPWT(AM) Colonial Heights TN
WZYX(AM) Cowan TN
WAEW(AM) Crossville TN
WRWB(AM) Harrogate TN
WIRJ(AM) Humboldt TN
WJCW(AM) Johnson City TN
WKVL(AM) Knoxville TN
WRJZ(AM) Knoxville TN
WNML-FM Loudon TN
WMSR(AM) Manchester TN
WLOK(AM) Memphis TN
WSMB(AM) Memphis TN
*WFSK-FM Nashville TN
WQBB(AM) Powell TN
WZNG(AM) Shelbyville TN
KZQQ(AM) Abilene TX
KPUR(AM) Amarillo TX
KZIP(AM) Amarillo TX
KVET(AM) Austin TX
KSKY(AM) Balch Springs TX
KBST(AM) Big Spring TX
KBYG(AM) Big Spring TX
KZNX(AM) Creedmoor TX
KGGR(AM) Dallas TX
KLLI(FM) Dallas TX
KTCK(AM) Dallas TX
KIXL(AM) Del Valle TX
KATX(FM) Eastland TX
KEAS(AM) Eastland TX
KHRO(AM) El Paso TX
KFCD(AM) Farmersville TX
KNAF(AM) Fredericksburg TX
KEES(AM) Gladewater TX
KVCE(AM) Highland Park TX
KCOH(AM) Houston TX
KILT(AM) Houston TX
KXYZ(AM) Houston TX
KTXX(FM) Karnes City TX
KERV(AM) Kerrville TX
*KAVX(FM) Lufkin TX
KWEL(AM) Midland TX
KVOP(AM) Plainview TX
KKLF(AM) Richardson TX
KJCE(AM) Rollingwood TX
KGKL(AM) San Angelo TX
KSLR(AM) San Antonio TX
*KSQX(FM) Springtown TX
KLGO(FM) Thorndale TX
KTNO(AM) University Park TX
KALL(AM) North Salt Lake City UT
KLO(AM) Ogden UT
KTKK(AM) Sandy UT
KCPX(AM) Spanish Valley UT
WAVA(AM) Arlington VA
WAVA-FM Arlington VA
WHAN(AM) Ashland VA
WLES(AM) Bon Air VA
WODI(AM) Brookneal VA
WWVT(AM) Christiansburg VA
WEVA(AM) Emporia VA
WFVA(AM) Fredericksburg VA
WMNA-FM Gretna VA
WKCY(AM) Harrisonburg VA
WAGE(AM) Leesburg VA
WJFK-FM Manassas VA

WHEE(AM) Martinsville VA
WCMS(AM) Newport News VA
WESR(AM) Onley-Onancock VA
WVCV(AM) Orange VA
*WRAR(AM) Tappahannock VA
WDHP(AM) Frederiksted VI
WRSA(AM) Saint Albans VT
KCLK(AM) Asotin WA
KPUG(AM) Bellingham WA
KGNW(AM) Burien-Seattle WA
KOZI-FM Chelan WA
KMAX(AM) Colfax WA
KSPO(FM) Dishman WA
KTAC(FM) Ephrata WA
KTBI(AM) Ephrata WA
KRKO(AM) Everett WA
KLAY(AM) Lakewood WA
KGTK(AM) Olympia WA
KOMW(AM) Omak WA
KTRW(AM) Opportunity WA
KRSC(AM) Othello WA
KPTK(AM) Seattle WA
*KUOW-FM Seattle WA
KVI(AM) Seattle WA
KBBO(AM) Selah WA
KITZ(AM) Silverdale WA
KJRB(AM) Spokane WA
KPTQ(AM) Spokane WA
KSBN(AM) Spokane WA
KUOW(AM) Tumwater WA
KBMS(AM) Vancouver WA
KTRO(AM) Vancouver WA
WEAQ(AM) Chippewa Falls WI
WTLX(FM) Columbus WI
*WDVM(AM) Eau Claire WI
*WHID(FM) Green Bay WI
WMCS(AM) Greenfield WI
WTKM(AM) Hartford WI
WTKM-FM Hartford WI
WKBH(AM) Holmen WI
*WHLA(FM) La Crosse WI
WKTY(AM) La Crosse WI
WMAM(AM) Marinette WI
*WHWC(FM) Menomonie WI
WJMT(AM) Merrill WI
WISN(AM) Milwaukee WI
WCSW(AM) Shell Lake WI
*WHDI(FM) Sister Bay WI
*WNWC(AM) Sun Prairie WI
WXXM(FM) Sun Prairie WI
WDSM(AM) Superior WI
*WVCX(FM) Tomah WI
WJJQ(AM) Tomahawk WI
*WEGZ(FM) Washburn WI
*WLBL-FM Wausau WI
WBUC(AM) Buckhannon WV
WVTS(AM) Charleston WV
WBES(AM) Dunbar WV
WRKP(FM) Moundsville WV
WRON(AM) Ronceverte WV
WEIR(AM) Weirton WV
KTWO(AM) Casper WY
KUYO(AM) Evansville WY
KIML(AM) Gillette WY
KYOD(FM) Glendo WY
KPOW(AM) Powell WY

Tejano

KXEW(AM) South Tucson AZ
KOQO-FM Fresno CA
KKHR(FM) Abilene TX
KOPY-FM Alice TX
KFON(AM) Austin TX
KKPS(FM) Brownsville TX
KTJK(AM) Del Rio TX
KFZO(FM) Denton TX
KTMR(AM) Edna TX
KXXS(FM) Elgin TX
KPSO-FM Falfurrias TX
KTFM(FM) Floresville TX
KGRW(FM) Friona TX
KHMC(FM) Goliad TX
KBTQ(FM) Harlingen TX
KJBZ(FM) Laredo TX
KEJS(FM) Lubbock TX

KHCK-FM Robinson TX
KMIQ(FM) Robstown TX
KSAB(FM) Robstown TX
KRTX(AM) Rosenberg-Richmond TX
KCLL(FM) San Angelo TX
KXTN-FM San Antonio TX
KUVA(FM) Uvalde TX

Top-40

WJDB-FM Thomasville AL
KLBQ(FM) El Dorado AR
KOFH(FM) Nogales AZ
KQST(FM) Sedona AZ
KWRN(AM) Apple Valley CA
KSME(FM) Greeley CO
KKMG(FM) Pueblo CO
WKCI-FM Hamden CT
WYKS(FM) Gainesville FL
WBVD(FM) Melbourne FL
WFKS(FM) Neptune Beach FL
WMTX(FM) Tampa FL
WWWQ(FM) College Park GA
WBTS(FM) Doraville GA
WBTY(FM) Homerville GA
KSHK(FM) Kekaha HI
KSAS-FM Caldwell ID
WAZY-FM Lafayette IN
WSVX(AM) Shelbyville IN
KVSV(AM) Beloit KS
KJCK-FM Junction City KS
KLZR(FM) Lawrence KS
KCHZ(FM) Ottawa KS
WLTO(FM) Nicholasville KY
WKKS-FM Vanceburg KY
KSMB(FM) Lafayette LA
WJMN(FM) Boston MA
WRZE(FM) Nantucket MA
WDKZ(FM) Salisbury MD
WKQI(FM) Detroit MI
KCAJ-FM Roseau MN
KOQL(FM) Ashland MO
WHGO(FM) Pascagoula MS
KBUD(FM) Sardis MS
WFLB(FM) Laurinburg NC
WDAY-FM Fargo ND
KQKY(FM) Kearney NE
WERZ(FM) Exeter NH
WWJZ(AM) Mount Holly NJ
WHTZ(FM) Newark NJ
KHOD(FM) Des Moines NM
WDRE(FM) Calverton-Roanoke NY
*WECW(FM) Elmira NY
WMCR(AM) Oneida NY
WMCR-FM Oneida NY
WLNG(FM) Sag Harbor NY
WAKS(FM) Akron OH
WMTR-FM Archbold OH
WODB(FM) Delaware OH
WDKF(FM) Englewood OH
KQLL-FM Owasso OK
KOSG(FM) Pawhuska OK
*KAYE-FM Tonkawa OK
KLKY(FM) Stanfield OR
WFKB(FM) Boyertown PA
WXYX(FM) Bayamon PR
WNOD(FM) Mayaguez PR
WKAQ-FM San Juan PR
WSSX-FM Charleston SC
WYMB(AM) Manning SC
KDSJ(AM) Deadwood SD
WHBQ-FM Germantown TN
WBMC(AM) McMinnville TN
WYDL(FM) Middleton TN
WIHG(FM) Rockwood TN
KNDE(FM) College Station TX
KYSE(FM) El Paso TX
KMKI(AM) Plano TX
WNRJ(FM) Poquoson VA
KAZZ(FM) Deer Park WA
KQQB-FM Newport WA
KJET(FM) Raymond WA
KRXY(FM) Shelton WA
WKSZ(FM) De Pere WI
WIZM-FM La Crosse WI
WELK(FM) Elkins WV
*WQAB(FM) Philippi WV
KLQQ(FM) Clearmont WY

Triple A

*KXLL(FM) Juneau AK
*KGHR(FM) Tuba City AZ
KWMT-FM Tucson AZ
KBHR(FM) Big Bear City CA
*KNCA(FM) Burney CA
KPYG(FM) Cambria CA
KPRI(FM) Encinitas CA
KOZT(FM) Fort Bragg CA
KPIG-FM Freedom CA
KHUM(FM) Garberville CA
KRSH(FM) Healdsburg CA
*KNSQ(FM) Mount Shasta CA
KZAP(FM) Paradise CA
KPIG(AM) Piedmont CA
KTKE(FM) Truckee CA
KWCA(FM) Weaverville CA
KSPN-FM Aspen CO
KBCO-FM Boulder CO
KSMT(FM) Breckenridge CO
KRKY-FM Estes Park CO
KYSL(FM) Frisco CO
KCUV(FM) Greenwood Village CO
*KUTE(FM) Ignacio CO
KFMU-FM Oak Creek CO
*KPGS(FM) Pagosa Springs CO
KSNO-FM Snowmass Village CO
WKZE-FM Salisbury CT
WOCM(FM) Selbyville DE
*WFIT(FM) Melbourne FL
*WUWF(FM) Pensacola FL
WPUP(FM) Royston GA
KPTL(FM) Ankeny IA
*KUNI(FM) Cedar Falls IA
KDEC-FM Decorah IA
*KRNI(AM) Mason City IA
*KUNY(FM) Mason City IA
*KDMR(FM) Mitchellville IA
*KISU-FM Pocatello ID
KPND(FM) Sandpoint ID
WTTS(FM) Bloomington IN
*WEEM-FM Pendleton IN
KACY(FM) Arkansas City KS
*WNKU(FM) Highland Heights KY
*WUKY(FM) Lexington KY
KROK(FM) South Fort Polk LA
WRNX(FM) Amherst MA
WXRV(FM) Haverhill MA
WRSI(FM) Turners Falls MA
WCYI(FM) Lewiston ME
*WLNZ(FM) Lansing MI
*KUMD-FM Duluth MN
*WTIP(FM) Grand Marais MN
*KCMP(FM) Northfield MN
*KMSE(FM) Rochester MN
*KSRQ(FM) Thief River Falls MN
*KTBG(FM) Warrensburg MO
*WUSM-FM Hattiesburg MS
KMMS-FM Bozeman MT
KDTR(FM) Florence MT
*WGWG(FM) Boiling Springs NC
WUIN(FM) Carolina Beach NC
*WSGE(FM) Dallas NC
WVOD(FM) Manteo NC
*WNCW(FM) Spindale NC
WLKC(FM) Campton NH
WMWV(FM) Conway NH
*WBJB-FM Lincroft NJ
*KUUT(FM) Farmington NM
*KUSW(FM) Flora Vista NM
KTAO(FM) Taos NM
*WGFR(FM) Glens Falls NY
*WFUV(FM) New York NY
*WAPS(FM) Akron OH
*WDPS(FM) Dayton OH
*WYSO(FM) Yellow Springs OH
*KRSC-FM Claremore OK
*KSMF(FM) Ashland OR
*KRVM-FM Eugene OR
*KSKF(FM) Klamath Falls OR
KINK(FM) Portland OR
*KSYD(FM) Reedsport OR
*WVMM(FM) Grantham PA
*WYEP-FM Pittsburgh PA
*KSDJ(FM) Brookings SD
KSQY(FM) Deadwood SD
*WUTC(FM) Chattanooga TN
WRLT(FM) Franklin TN
WFIV-FM Loudon TN
KEEP(FM) Bandera TX
KFAN-FM Johnson City TX
*KSYM-FM San Antonio TX
*WNRN(FM) Charlottesville VA
WTYD(FM) Deltaville VA
WCNR(FM) Keswick VA
*WVRU(FM) Radford VA
WIVI(FM) Charlotte Amalie VI
WDOT(FM) Danville VT
WEBK(FM) Killington VT
WRSY(FM) Marlboro VT
WNCS(FM) Montpelier VT
*WNUB-FM Northfield VT
*KGHP(FM) Gig Harbor WA
KMTT(FM) Tacoma WA
*WUWM(FM) Milwaukee WI
*WYMS(FM) Milwaukee WI
*KBUW(FM) Buffalo WY
*KUWC(FM) Casper WY

Underground

*WZBC(FM) Newton MA

Urban Contemporary

WDLT-FM Chickasaw AL
WJJN(FM) Columbia AL
WQZZ(FM) Eutaw AL
WAGH(FM) Fort Mitchell AL
WMGJ(AM) Gadsden AL
WKXN(FM) Greenville AL
WXQW(FM) Gurley AL
WHOG(AM) Hobson City AL
WLOR(AM) Huntsville AL
WJUS(AM) Marion AL
WBHJ(FM) Midfield AL
WBLX-FM Mobile AL
WEUP-FM Moulton AL
WJAM-FM Orrville AL
WZMG(AM) Pepperell AL
WKXK(FM) Pine Hill AL
WBFZ(FM) Selma AL
WBFA(FM) Smiths AL
WZHT(FM) Troy AL
WBHK(FM) Warrior AL
WJWZ(FM) Wetumpka AL
*KSWH(FM) Arkadelphia AR
KMGC(FM) Camden AR
KMLK(FM) El Dorado AR
KAMJ-FM Gosnell AR
KAKJ(FM) Marianna AR
KIPR(FM) Pine Bluff AR
KZYP(FM) Pine Bluff AR
KTOY(FM) Texarkana AR
KCLT(FM) West Helena AR
KKFR(FM) Mayer AZ
KTUC(AM) Tucson AZ
KNGY(FM) Alameda CA
KVIN(AM) Ceres CA
KJLH-FM Compton CA
*KCRH(FM) Hayward CA
KKUU(FM) Indio CA
KRAJ(FM) Johannesburg CA
KHHT(FM) Los Angeles CA
KRBV(FM) Los Angeles CA
*KSFH(FM) Mountain View CA
KWIE(FM) Ontario CA
KPAT(FM) Orcutt CA
KDAY(FM) Redondo Beach CA
KBMB(FM) Sacramento CA
KDON-FM Salinas CA
KYLD(FM) San Francisco CA
*KSJS(FM) San Jose CA
KKJL(AM) San Luis Obispo CA
KVYB(FM) Santa Barbara CA
KWYL(FM) South Lake Tahoe CA
*KSAK(FM) Walnut CA
KIBT(FM) Fountain CO
*WQTQ(FM) Hartford CT
WYBC-FM New Haven CT
*WECS(FM) Willimantic CT
WKND(AM) Windsor CT
WHUR-FM Washington DC
WKYS(FM) Washington DC
WHJX(FM) Baldwin FL
WBTP(FM) Clearwater FL
*WJCB(FM) Clewiston FL
WHQT(FM) Coral Gables FL
WHNR(AM) Cypress Gardens FL
WCFB(FM) Daytona Beach FL
*WJFP(FM) Fort Pierce FL
WJBT(FM) Green Cove Springs FL
WRNE(AM) Gulf Breeze FL
WRRX(FM) Gulf Breeze FL
WNEW(FM) Jupiter FL
WPBH(FM) Port St. Joe FL
WRXB(AM) Saint Petersburg Beach FL
WHBX(FM) Tallahassee FL
WTMG(FM) Williston FL
*WPRK(FM) Winter Park FL
WJIZ-FM Albany GA
WVEE(FM) Atlanta GA
WFXA-FM Augusta GA
WBBK-FM Blakely GA
WSOL-FM Brunswick GA
WWLD(FM) Cairo GA
WZBN(FM) Camilla GA
WFXE(FM) Columbus GA
WMRZ(FM) Dawson GA
WQMJ(FM) Forsyth GA
WIBB-FM Fort Valley GA
WFXM(FM) Gordon GA
WHTA(FM) Hampton GA
WVKX(FM) Irwinton GA
WALR-FM La Grange GA
WLZN(FM) Macon GA
WPRW-FM Martinez GA
WHBS(AM) Moultrie GA
WMPZ(FM) Ringgold GA
WBGA(FM) Saint Simons Island GA
WQBT(FM) Savannah GA
WEAS-FM Springfield GA
WHLJ(FM) Statenville GA
WNUQ(FM) Sylvester GA
WGOV(AM) Valdosta GA
WLYX(FM) Valdosta GA
*WVVS(FM) Valdosta GA
WRBV(FM) Warner Robins GA
WAKB(FM) Wrens GA
KUAM-FM Hagatna GU
KZGZ(FM) Hagatna GU
*KALA(FM) Davenport IA
*KJMC(FM) Des Moines IA
KATZ-FM Alton IL
WKRO(AM) Cairo IL
*WPCD(FM) Champaign IL
*WKKC(FM) Chicago IL
*WMBI(AM) Chicago IL
WDZ(AM) Decatur IL
*WIUS(FM) Macomb IL
WCZQ(FM) Monticello IL
WMOJ-FM Connersville IN
WEOA(AM) Evansville IN
WPWX(FM) Hammond IN
*WBDG(FM) Indianapolis IN
WHHH(FM) Indianapolis IN
WGZB-FM Lanesville IN
*WISU(FM) Terre Haute IN
*KSDB-FM Manhattan KS
*KYWA(FM) Wichita KS
KSJM(FM) Winfield KS
WIZF(FM) Erlanger KY
WMJM(FM) Jeffersontown KY
WBTF(FM) Midway KY
WDXR(AM) Paducah KY
WUBT(FM) Russellville KY
KEDG(FM) Alexandria LA
KJMG(FM) Bastrop LA
KRVV(FM) Bastrop LA
KBZE(FM) Berwick LA
KDKS-FM Blanchard LA
KBCE(FM) Boyce LA
*KGRM(FM) Grambling LA
KBTT(FM) Haughton LA
KTGV(FM) Jonesville LA
WEMX(FM) Kentwood LA
KJCB(AM) Lafayette LA
KRRQ(FM) Lafayette LA
KJMH(FM) Lake Arthur LA
KZWA(FM) Moss Bluff LA
WQUE-FM New Orleans LA
*WWOZ(FM) New Orleans LA
WYLD-FM New Orleans LA
KQXL-FM New Roads LA
KKST(FM) Oakdale LA
KMJJ-FM Shreveport LA
KVMA-FM Shreveport LA
KBSF(AM) Springhill LA
KTJZ(FM) Tallulah LA
KNEK(AM) Washington LA
WKAF(FM) Brockton MA
*WOZQ(FM) Northampton MA
*WMHC(FM) South Hadley MA
WERQ-FM Baltimore MD
WOCQ(FM) Berlin MD
WMMJ(FM) Bethesda MD
WWIN-FM Glen Burnie MD
WSBY-FM Salisbury MD
WRED(FM) Saco ME
WDMK(FM) Detroit MI
WJLB(FM) Detroit MI
WMXD(FM) Detroit MI
WQHH(FM) Dewitt MI
WDZZ-FM Flint MI
WVIB(FM) Holton MI
WJNZ(AM) Kentwood MI
WHTD(FM) Mount Clemens MI
WSNX-FM Muskegon MI
WNWN(AM) Portage MI
WTLZ(FM) Saginaw MI
WOWE(FM) Vassar MI
*KMOJ(FM) Minneapolis MN
KDMO(AM) Carthage MO
WMBH(AM) Joplin MO
KPRS(FM) Kansas City MO
KMJK(FM) Lexington MO
WACR-FM Aberdeen MS
WESE(FM) Baldwyn MS
WRBJ-FM Brandon MS
WAJV(FM) Brooksville MS
WMGO(AM) Canton MS
WCLD-FM Cleveland MS
WWKZ(FM) Columbus MS
WJXM(FM) De Kalb MS
WJKX(FM) Ellisville MS
WJMG(FM) Hattiesburg MS
WJMI(FM) Jackson MS
*WMPR(FM) Jackson MS
WBAD(FM) Leland MS
WJZD(FM) Long Beach MS
WKXI-FM Magee MS
WZLD(FM) Petal MS
WMSU(FM) Starkville MS
WMXU(FM) Starkville MS
WGNG(FM) Tchula MS
WZKS(FM) Union MS
WJXN-FM Utica MS
KLSK(FM) Great Falls MT
WVOE(AM) Chadbourn NC
WMKS(FM) Clemmons NC
WPEG(FM) Concord NC
WFXC(FM) Durham NC
*WRVS-FM Elizabeth City NC
WKXS-FM Leland NC
WIKS(FM) New Bern NC
WPWZ(FM) Pinetops NC
WJMH(FM) Reidsville NC
WRSV(FM) Rocky Mount NC
WUKS(FM) Saint Pauls NC
WFXK(FM) Tarboro NC
WSMY(AM) Weldon NC
WENC(AM) Whiteville NC
WZFX(FM) Whiteville NC
WMNX(FM) Wilmington NC
*KZUM(FM) Lincoln NE
KOPW(FM) Plattsmouth NE
*WSPS(FM) Concord NH
*WNEC-FM Henniker NH
WJKS(FM) Canton NJ

WIMG(AM) Ewing NJ
WTTH(FM) Margate City NJ
WRNB(FM) Pennsauken NJ
WEZW(FM) Wildwood Crest NJ
*KCEP(FM) Las Vegas NV
KVGS(FM) Laughlin NV
*WCDB(FM) Albany NY
WPHR-FM Auburn NY
*WCWP(FM) Brookville NY
WWWS(AM) Buffalo NY
WBLK(FM) Depew NY
*WRCU-FM Hamilton NY
*WVHC(FM) Herkimer NY
*WICB(FM) Ithaca NY
WBLS(FM) New York NY
WWPR-FM New York NY
*WOSS(FM) Ossining NY
*WNYO(FM) Oswego NY
WDKX(FM) Rochester NY
*WFNP(FM) Rosendale NY
*WPNR-FM Utica NY
WENZ(FM) Cleveland OH
WZAK(FM) Cleveland OH
WCKX(FM) Columbus OH
WIMX(FM) Gibsonburg OH
WRBP(FM) Hubbard OH
WZOQ(AM) Lima OH
WRTK(AM) Niles OH
WJTB(AM) North Ridgeville OH
WDHT(FM) Springfield OH
WDIG(AM) Steubenville OH
WJUC(FM) Swanton OH
WROU-FM West Carrollton OH
*WCSU-FM Wilberforce OH
KVSP(FM) Anadarko OK
KJMM(FM) Bixby OK
KJMZ(FM) Cache OK
*KALU(FM) Langston OK
KRMP(AM) Oklahoma City OK
KGTO(AM) Tulsa OK
*KBVR(FM) Corvallis OR
*KWVA(FM) Eugene OR
WAMO-FM Beaver Falls PA
WTCY(AM) Harrisburg PA
WLAN(AM) Lancaster PA
*WWLU(FM) Lincoln University PA
WHAT(AM) Philadelphia PA
WUSL(FM) Philadelphia PA
*WCLH(FM) Wilkes-Barre PA
*WRLC(FM) Williamsport PA
WBRU(FM) Providence RI
*WJMF(FM) Smithfield RI
WJMZ-FM Anderson SC
WIIZ(FM) Blackville SC
WYNN-FM Florence SC
WLVH(FM) Hardeeville SC
*WLGI(FM) Hemingway SC
WXST(FM) Hollywood SC
WJNI(FM) Ladson SC
WCMG(FM) Latta SC
WLXC(FM) Lexington SC
WHXT(FM) Orangeburg SC
WPJK(AM) Orangeburg SC
*WSSB-FM Orangeburg SC
WDAI(FM) Pawley's Island SC
WPAL-FM Ridgeville SC
WZTF(FM) Scranton SC
WASC(AM) Spartanburg SC
WWWZ(FM) Summerville SC
WWDM(FM) Sumter SC
WXBT(FM) West Columbia SC
WNOO(AM) Chattanooga TN
WFKX(FM) Henderson TN
WQQK(FM) Hendersonville TN
WKGN(AM) Knoxville TN
KJMS(FM) Memphis TN
WDIA(AM) Memphis TN
WHRK(FM) Memphis TN
*WQOX(FM) Memphis TN
WEUZ(FM) Minor Hill TN
WJTT(FM) Red Bank TN
WHRP(FM) Tullahoma TN
*KAZI-FM Austin TX
KTCX(FM) Beaumont TX
KHPT(FM) Conroe TX
KSSM(FM) Copperas Cove TX
KPTI(FM) Crystal Beach TX
KBFB(FM) Dallas TX
KKDA-FM Dallas TX
KEPS(AM) Eagle Pass TX
KSOC(FM) Gainesville TX
KCOH(AM) Houston TX
KMJQ(FM) Houston TX
KIIZ-FM Killeen TX
*KBWC(FM) Marshall TX
KPTY(FM) Missouri City TX
KZZA(FM) Muenster TX
KZRB(FM) New Boston TX
KPWT(FM) Terrell Hills TX
KZEY(AM) Tyler TX
KBLZ(FM) Winona TX
*KRDC-FM Saint George UT
*WNRN(FM) Charlottesville VA
WUVA(FM) Charlottesville VA
WVBE-FM Lynchburg VA
WCDX(FM) Mechanicsville VA
WKUS(FM) Norfolk VA
*WNSB(FM) Norfolk VA
WOWI(FM) Norfolk VA
WKJM(FM) Petersburg VA
WBTJ(FM) Richmond VA
WKJS(FM) Richmond VA
WVBE(AM) Roanoke VA
WTOY(AM) Salem VA
WQOK(FM) South Boston VA
WXEZ(FM) Yorktown VA
WSTA(AM) Charlotte Amalie VI
WVJZ(FM) Charlotte Amalie VI
WJKC(FM) Christiansted VI
WWKS(FM) Cruz Bay VI
*WVTC(FM) Randolph Center VT
KYIZ(AM) Renton WA
KYNR(AM) Toppenish WA
KBMS(AM) Vancouver WA
KHHK(FM) Yakima WA
WJMR-FM Menomonee Falls WI
WNOV(AM) Milwaukee WI
WJQM(FM) Mount Horeb WI

Variety/Diverse

*KBRW(AM) Barrow AK
*KYUK(AM) Bethel AK
*KCUK(FM) Chevak AK
KZPA(AM) Fort Yukon AK
*KIYU(AM) Galena AK
*KEUL(FM) Girdwood AK
*KHNS(FM) Haines AK
*KMJG(FM) Homer AK
*KWJG(FM) Kasilof AK
*KDLL(FM) Kenai AK
*KOTZ(AM) Kotzebue AK
*KUHB-FM Saint Paul AK
*KCAW(FM) Sitka AK
KNSA(AM) Unalakleet AK
WKAC(AM) Athens AL
WRVX(FM) Eufaula AL
*WLRH(FM) Huntsville AL
*WLJS-FM Jacksonville AL
KBVA(FM) Bella Vista AR
KAVH(FM) Eudora AR
KURM(AM) Rogers AR
*KXRJ(FM) Russellville AR
KUOA(AM) Siloam Springs AR
*KRMH(FM) Red Mesa AZ
KSED(FM) Sedona AZ
*KXCI(FM) Tucson AZ
KBSZ(AM) Wickenburg AZ
*KAWC(AM) Yuma AZ
*KHSU-FM Arcata CA
KAHI(AM) Auburn CA
*KISL(FM) Avalon CA
KWRM(AM) Corona CA
*KHSR(FM) Crescent City CA
*KKUP(FM) Cupertino CA
*KDVS(FM) Davis CA
*KFCF(FM) Fresno CA
KTDE(FM) Gualala CA
*KCRH(FM) Hayward CA
KTYM(AM) Inglewood CA
*KFJC(FM) Los Altos CA
*KAKX(FM) Mendocino CA
*KVMR(FM) Nevada City CA
*KCSN(FM) Northridge CA
*KZYX(FM) Philo CA
*KWMR(FM) Point Reyes Station CA
*KFPR(FM) Redding CA
*KYDS(FM) Sacramento CA
*KSRH(FM) San Rafael CA
*KCSB-FM Santa Barbara CA
*KFER(FM) Santa Cruz CA
*KSVY(FM) Sonoma CA
*KRZA(FM) Alamosa CO
KSYY-FM Bennett CO
*KGNU-FM Boulder CO
*KEPC(FM) Colorado Springs CO
*KRFC(FM) Fort Collins CO
KJME(AM) Fountain CO
*KWSB-FM Gunnison CO
KCEG(AM) Pueblo CO
*KOTO(FM) Telluride CO
KJAC(FM) Timnath CO
*WVOF(FM) Fairfield CT
*WESU(FM) Middletown CT
*WCNI(FM) New London CT
*WBVC(FM) Pomfret CT
*WAPJ(FM) Torrington CT
*WWEB(FM) Wallingford CT
*WWUH(FM) West Hartford CT
*WWPT(FM) Westport CT
WJWK(AM) Seaford DE
WBGC(AM) Chipley FL
*WRGP(FM) Homestead FL
WWFE(AM) Miami FL
*WEJF(FM) Palm Bay FL
WPIK(FM) Summerland Key FL
WVLG(AM) Wildwood FL
*WRFG(FM) Atlanta GA
WMLB(AM) Avondale Estates GA
WMJE(FM) Clarkesville GA
*WBCX(FM) Gainesville GA
*WHCJ(FM) Savannah GA
WZAT(FM) Savannah GA
*KURE(FM) Ames IA
*KWLC(AM) Decorah IA
*KHOE(FM) Fairfield IA
KOKX-FM Keokuk IA
*KRNL-FM Mount Vernon IA
KZSR(FM) Onawa IA
KDLS(AM) Perry IA
*KLCZ(FM) Lewiston ID
*KUOI-FM Moscow ID
WKIE(FM) Arlington Heights IL
*WEFT(FM) Champaign IL
*WEIU(FM) Charleston IL
*WIIT(FM) Chicago IL
*WKKC(FM) Chicago IL
*WLUW(FM) Chicago IL
*WRTE(FM) Chicago IL
WSBC(AM) Chicago IL
*WZRD(FM) Chicago IL
WCFJ(AM) Chicago Heights IL
*WVKC(FM) Galesburg IL
WDUK(FM) Havana IL
*WHSD(FM) Hinsdale IL
*WLTL(FM) La Grange IL
*WAES(FM) Lincolnshire IL
WKOT(FM) Marseilles IL
*WMTH(FM) Park Ridge IL
*WILL-FM Urbana IL
*WFWR(FM) Attica IN
*WNDY(FM) Crawfordsville IN
*WPSR(FM) Evansville IN
WTRE(AM) Greensburg IN
*WHCI(FM) Hartford City IN
WGL-FM Huntington IN
*WJEL(FM) Indianapolis IN
WAWK(AM) Kendallville IN
*WBRO(FM) Marengo IN
*WECI(FM) Richmond IN
*KVCO(FM) Concordia KS
*KONQ(FM) Dodge City KS
*KBCU(FM) North Newton KS
KQMA-FM Phillipsburg KS
*WRFL(FM) Lexington KY
*WMKY(FM) Morehead KY
*WKMS-FM Murray KY
WRKA(FM) Saint Matthews KY
*KLSU(FM) Baton Rouge LA
*KRVS(FM) Lafayette LA
*KEDM(FM) Monroe LA
*KNWD(FM) Natchitoches LA
KKAY(AM) White Castle LA
*WAMH(FM) Amherst MA
*WMUA(FM) Amherst MA
WBMX(FM) Boston MA
*WERS(FM) Boston MA
*WRBB(FM) Boston MA
*WBIM-FM Bridgewater MA
*WMBR(FM) Cambridge MA
*WGAJ(FM) Deerfield MA
*WXPL(FM) Fitchburg MA
*WHHB(FM) Holliston MA
*WCCH(FM) Holyoke MA
WMKK(FM) Lawrence MA
*WWTA(FM) Marion MA
*WAVM(FM) Maynard MA
*WMFO(FM) Medford MA
WMRC(AM) Milford MA
*WNCK(FM) Nantucket MA
WNBP(AM) Newburyport MA
WNTN(AM) Newton MA
*WOMR(FM) Provincetown MA
*WMHC(FM) South Hadley MA
*WSCB(FM) Springfield MA
*WZLY(FM) Wellesley MA
WQSR(FM) Baltimore MD
*WHFC(FM) Bel Air MD
*WMUC-FM College Park MD
WMET(AM) Gaithersburg MD
*WUMF-FM Farmington ME
*WXOU(FM) Auburn Hills MI
*WHFR(FM) Dearborn MI
*WRCJ-FM Detroit MI
WMJO(FM) Essexville MI
*WMTU-FM Houghton MI
*WIDR(FM) Kalamazoo MI
*WYDM(FM) Monroe MI
*WOAS(FM) Ontonagon MI
*WBLD(FM) Orchard Lake MI
*WLSO(FM) Sault Ste. Marie MI
*WYCE(FM) Wyoming MI
WYGR(AM) Wyoming MI
WELY(AM) Ely MN
WELY-FM Ely MN
*WTIP(FM) Grand Marais MN
*KAXE(FM) Grand Rapids MN
*KVSC(FM) Saint Cloud MN
*KSMR(FM) Winona MN
*KYMC(FM) Ballwin MO
KCXL(AM) Liberty MO
*KGSP(FM) Parkville MO
*WUSM-FM Hattiesburg MS
*WPRL(FM) Lorman MS
WAPF(AM) McComb MS
*KGVA(FM) Fort Belknap Agency MT
*KGPR(FM) Great Falls MT
KINX(FM) Great Falls MT
*WSGE(FM) Dallas NC
*WRVS-FM Elizabeth City NC
*WUAW(FM) Erwin NC
*WQFS(FM) Greensboro NC
WHKP(AM) Hendersonville NC
*WYQS(FM) Mars Hill NC
WAME(AM) Statesville NC
*WHYC(FM) Swanquarter NC
*KDSU(FM) Fargo ND
*KDCV-FM Blair NE
KNCY(AM) Nebraska City NE
*WPEA(FM) Exeter NH
WSNJ(AM) Bridgeton NJ
*WRRC(FM) Lawrenceville NJ
WMVB(AM) Millville NJ
*WRSU-FM New Brunswick NJ
*WLFR(FM) Pomona NJ
*WWPH(FM) Princeton Junction NJ
*WFDU(FM) Teaneck NJ
KDAZ(AM) Albuquerque NM
*KRUX(FM) Las Cruces NM
KVSF-FM Pecos NM
KRSI(FM) Garapan-Saipan NP
*WALF(FM) Alfred NY
*WHRW(FM) Binghamton NY
*WXLH(FM) Blue Mountain Lake NY
*WSLU(FM) Canton NY
*WHCL-FM Clinton NY
*WSUC-FM Cortland NY
*WXXE(FM) Fenner NY
WGBB(AM) Freeport NY
*WHPC(FM) Garden City NY
*WRHU(FM) Hempstead NY
*WJFF(FM) Jeffersonville NY
*WSHR(FM) Lake Ronkonkoma NY
*WVCR-FM Loudonville NY
*WSLO(FM) Malone NY
*WBAI(FM) New York NY
*WKCR-FM New York NY
*WXLG(FM) North Creek NY
*WXLU(FM) Peru NY
*WAIH(FM) Potsdam NY
*WSLL(FM) Saranac Lake NY
*WSPN(FM) Saratoga Springs NY
WSPQ(AM) Springville NY
*WRPI(FM) Troy NY
*WSLJ(FM) Watertown NY
*WAPS(FM) Akron OH
*WOBO(FM) Batavia OH
*WBGU(FM) Bowling Green OH
*WRUW-FM Cleveland OH
*WCBE(FM) Columbus OH
WKTX(AM) Cortland OH
*WUDR(FM) Dayton OH
*WWSU(FM) Dayton OH
*WSLN(FM) Delaware OH
*WKCO(FM) Gambier OH
*WDUB(FM) Granville OH
WFGA(FM) Hicksville OH
WMVO(AM) Mount Vernon OH
KQOB(FM) Enid OK
KIHN(AM) Hugo OK
*KSRG(FM) Ashland OR
KMHS(AM) Coos Bay OR
*KLFO(FM) Florence OR
*KEOL(FM) La Grande OR
*WJCS(FM) Allentown PA
WGPA(AM) Bethlehem PA
*WLVR(FM) Bethlehem PA
WHLM(AM) Bloomsburg PA
*WLBS(FM) Bristol PA
WZUM(AM) Carnegie PA
*WDNR(FM) Chester PA
*WJRH(FM) Easton PA
*WERG(FM) Erie PA
WGSM(FM) Greensburg PA
*WIUP-FM Indiana PA
WKHB(AM) Irwin PA
WKFB(AM) Jeannette PA
*WFNM(FM) Lancaster PA
WEDO(AM) McKeesport PA
*WKDU(FM) Philadelphia PA
WNWR(AM) Philadelphia PA
*WKPS(FM) State College PA
*WPSU(FM) State College PA
WMBS(AM) Uniontown PA
*WRDV(FM) Warminster PA
WOQI(AM) Adjuntas PR
WCGB(AM) Juana Diaz PR
WDEP(AM) Ponce PR
WJIT(AM) Sabana PR
WEGA(AM) Vega Baja PR
*WUSC-FM Columbia SC
WLBG(AM) Laurens SC
*KLND(FM) Little Eagle SD
*KILI(FM) Porcupine SD
WQSV(AM) Ashland City TN
*WAPX-FM Clarksville TN
WHHM-FM Henderson TN
*WCSK(FM) Kingsport TN
WYGO(FM) Madisonville TN
*WEVL(FM) Memphis TN
*WUTS(FM) Sewanee TN
WCMT-FM South Fulton TN
WSGI(AM) Springfield TN
WQAK(FM) Union City TN
KVLF(AM) Alpine TX
*KNON(FM) Dallas TX
*KOOP(FM) Hornsby TX
KRBA(AM) Lufkin TX
*KEOM(FM) Mesquite TX
KNBT(FM) New Braunfels TX

KREH(AM) Pecan Grove TX
KRTX(AM) Rosenberg-Richmond TX
*KZMU(FM) Moab UT
*KPGR(FM) Pleasant Grove UT
KLGL(FM) Richfield UT
*WTJU(FM) Charlottesville VA
*WVTW(FM) Charlottesville VA
*WHOV(FM) Hampton VA
WCLM(AM) Highland Springs VA
WHAP(AM) Hopewell VA
*WWMC(FM) Lynchburg VA
*WVTR(FM) Marion VA
WWLB(FM) Midlothian VA
WSTX(AM) Christiansted VI
*WIUV(FM) Castleton VT
WWMP(FM) Waterbury VT
*KUBS(FM) Newport WA
KSUH(AM) Puyallup WA
*WBCR-FM Beloit WI
*WORT(FM) Madison WI
WOCO(AM) Oconto WI
*WXPR(FM) Rhinelander WI
*WRFW(FM) River Falls WI
*WCCX(FM) Waukesha WI
WQWV(FM) Fisher WV
*WRSG(FM) Middlebourne WV
WKWK-FM Wheeling WV
WVVV(FM) Williamstown WV
*KWRR(FM) Ethete WY
KYOD(FM) Glendo WY
KJJL(AM) Pine Bluffs WY

Vietnamese

KZSJ(AM) San Martin CA
KVNR(AM) Santa Ana CA
KVVN(AM) Santa Clara CA
KJOJ(AM) Conroe TX
KYND(AM) Cypress TX

Women

WNSH(AM) Beverly MA
WDSS(AM) Ada MI

Programming on Radio Stations in Canada

Adult Contemp

CFIT-FM Airdrie AB
CIXF-FM Brooks AB
CHFM-FM Calgary AB
CIQX-FM Calgary AB
CKCE-FM Calgary AB
CKMX(AM) Calgary AB
CHMN-FM Canmore AB
CFXE-FM Edson AB
*CKRP-FM Falher AB
CJUV-FM Lacombe AB
CFRV-FM Lethbridge AB
CJOC-FM Lethbridge AB
CFMY-FM Medicine Hat AB
CKKX-FM Peace River AB
CHUB-FM Red Deer AB
CIZZ-FM Red Deer AB
CFMG-FM Saint Albert AB
CHSL-FM Slave Lake AB
CJBZ-FM Taber AB
CKWY-FM Wainwright AB
CKGO-FM-1 Boston Bar BC
CFLD(AM) Burns Lake BC
CJSU-FM Duncan BC
CIEG-FM Egmont BC
CISC-FM Gibsons BC
CFSR-FM Hope BC
CHNL(AM) Kamloops BC
CKBZ-FM Kamloops BC
CHSU-FM Kelowna BC
CILK-FM Kelowna BC
CJNL(AM) Merritt BC
CKWV-FM Nanaimo BC
CKKC-FM Nelson BC
CKZX-FM New Denver BC
CJOR(AM) Osoyoos BC
CIBH-FM Parksville BC
CISP-FM Pemberton BC
CIPN-FM Pender Harbour BC
CIGV-FM Penticton BC
CJAV-FM Port Alberni BC
CFNI(AM) Port Hardy BC
CKKN-FM Prince George BC
CIOR(AM) Princeton BC
CKXR-FM Salmon Arm BC
CKAY-FM Sechelt BC
CFBV(AM) Smithers BC
CISQ-FM Squamish BC
CJAT-FM Trail BC
CHQM-FM Vancouver BC
CKLG-FM Vancouver BC
CKZZ-FM Vancouver BC
CICF-FM Vernon BC
CHBE-FM Victoria BC
CIOC-FM Victoria BC
CISW-FM Whistler BC
CKLF-FM Brandon MB
CKDM(AM) Dauphin MB
CFAR(AM) Flin Flon MB
CILT-FM Steinbach MB
CJAR(AM) The Pas MB
CHTM(AM) Thompson MB
CJEL-FM Winkler MB
CFWM-FM Winnipeg MB
CKY-FM Winnipeg MB
CKBC-FM Bathurst NB
CKNB(AM) Campbellton NB
CJVA(AM) Caraquet NB
CJEM-FM Edmundston NB
*CJPN-FM Fredericton NB
CIKX-FM Grand Falls NB
CKMV-FM Grand Falls NB
CFAN-FM Miramichi City NB
CFQM-FM Moncton NB
CIOK-FM Saint John NB
CJCW(AM) Sussex NB
CJCJ-FM Woodstock NB
CFOZ-FM Argentia NF
CJOZ-FM Bonavista Bay NF
CJKK-FM Clarenville NF
CKOZ-FM Corner Brook NF
CFLN(AM) Goose Bay NF
CIFX-FM Lewisporte NF
CIOZ-FM Marystown NF
CHOS-FM Rattling Brook NF
CKSS-FM Red Rocks NF
CHOZ-FM Saint John's NF
CKSJ-FM Saint John's NF
VOCM(AM) Saint John's NF
CIOS-FM Stephenville NF
CJLS-FM-1 Barrington NS
CKBW-FM Bridgewater NS
CIOO-FM Halifax NS
CKWM-FM Kentville NS
CKBW-FM-1 Liverpool NS
CKEC(AM) New Glasgow NS
CJLS-FM-2 New Tusket NS
CIGO-FM Port Hawkesbury NS
CKBW-FM-2 Shelburne NS
CBI-FM Sydney NS
CKTO-FM Truro NS
CJLS-FM Yarmouth NS
CJCD-FM-1 Hay River NT
CJCD-FM Yellowknife NT
CBQR-FM Rankin Inlet NU
CHMS-FM Bancroft ON
CHAY-FM Barrie ON
CIQB-FM Barrie ON
CIGL-FM Belleville ON
CJOJ-FM Belleville ON
CFBG-FM Bracebridge ON
CKPC-FM Brantford ON
CFJR-FM Brockville ON
CJPT-FM Brockville ON
CJDV-FM Cambridge ON
CKSY-FM Chatham ON
CHUC-FM Cobourg ON
CKSG-FM Cobourg ON
CHPB-FM Cochrane ON
CKCB-FM Collingwood ON
CFLG-FM Cornwall ON
CKDR-FM Dryden ON
CJBB-FM Englehart ON
CFOB-FM Fort Frances ON
CHWC-FM Goderich ON
CIMJ-FM Guelph ON
CKLH-FM Hamilton ON
CHPR-FM Hawkesbury ON
CHYK-FM-3 Hearst ON
*CINN-FM Hearst ON
CFBK-FM Huntsville ON
CFQK-FM Kaministiquia ON
CJRL-FM Kenora ON
CIYN-FM Kincardine ON
CFLY-FM Kingston ON
CJKL-FM Kirkland Lake ON
CKWR-FM Kitchener ON
CHYR-FM Leamington ON
CKLY-FM Lindsay (city of Kawartha Lakes) ON
CFPL(AM) London ON
CIQM-FM London ON
CFNO-FM Marathon ON
CKDX-FM Newmarket ON
CFLZ-FM Niagara Falls ON
CHUR-FM North Bay ON
CICX-FM Orillia ON
CISS-FM Ottawa ON
CJMJ-FM Ottawa ON
CJWL-FM Ottawa ON
CKLP-FM Parry Sound ON
CIMY-FM Pembroke ON
*CFRH-FM Penetanguishene ON
CHLK-FM Perth ON
CFPS-FM Port Elgin ON
CHMY-FM Renfrew ON
CHSC(AM) Saint Catharines ON
CFGX-FM Sarnia ON
CHAS-FM Sault Ste. Marie ON
CHCD-FM Simcoe ON
CHGK-FM Stratford ON
CJMI-FM Strathroy ON
CFSF-FM Sturgeon Falls ON
CHYC-FM Sudbury ON
CJMX-FM Sudbury ON
CJUK-FM Thunder Bay ON
CJUK-FM Thunder Bay ON
CKPR-FM Thunder Bay ON
CHYK-FM Timmins ON
CHFI-FM Toronto ON
CHUM-FM Toronto ON
CJEZ-FM Toronto ON
CKQV-FM Vermillion Bay ON
CHGB-FM Wasaga Beach ON
CJWA-FM Wawa ON
CIXL-FM Welland ON
CIDR-FM Windsor ON
CKWW(AM) Windsor ON
CKNX-FM Wingham ON
CIHR-FM Woodstock ON
CHLQ-FM Charlottetown PE
CFVM-FM Amqui PQ
CJAN-FM Asbestos PQ
CHLC-FM Baie Comeau PQ
CIEU-FM Carleton PQ
CHAI-FM Chateauguay PQ
CBJ-FM Chicoutimi PQ
CFVD-FM Degelis PQ
CHRD-FM Drummondville PQ
CJDM-FM Drummondville PQ
CFMF-FM Fermont PQ
CFRP(AM) Forestville PQ
*CFTH-FM-1 Harrington Harbour PQ
*CKRK-FM Kahnawake PQ
CFLM(AM) La Tuque PQ
CJLA-FM Lachute PQ
CFGL-FM Laval PQ
CFOM-FM Levis PQ
CHAA-FM Longueuil PQ
CHGA-FM Maniwaki PQ
CFLO-FM Mont-Laurier PQ
CFEL-FM Montmagny PQ
CITE-FM Montreal PQ
CJFM-FM Montreal PQ
CHNC(AM) New Carlisle PQ
CHIK-FM Quebec PQ
*CION-FM Quebec PQ
CITF-FM Quebec PQ
CJEC-FM Quebec PQ
CJMF-FM Quebec PQ
*CKRL-FM Quebec PQ
CJBR-FM Rimouski PQ
CKMN-FM Rimouski-Mont Joli PQ
CIEL-FM Riviere du Loup PQ
CHOA-FM Rouyn-Noranda PQ
CJMM-FM Rouyn-Noranda PQ
CJAB-FM Saguenay PQ
CJAS-FM Saint Augustin PQ
CKRB-FM Saint Georges-de-Beauce PQ
CIME-FM Saint Jerome PQ
CHEQ-FM Sainte-Marie-de-Beauce PQ
CKCN-FM Sept-Iles PQ
CKSM(AM) Shawinigan PQ
CFGE-FM Sherbrooke PQ
CITE-FM-1 Sherbrooke PQ
CKLD-FM Thetford Mines PQ
CJEB-FM Trois Rivieres PQ
CKOD-FM Valleyfield PQ
CFDA-FM Victoriaville PQ
CKVM-FM Ville-Marie PQ
CFYM(AM) Kindersley SK
CFNK-FM Pinehouse Lake SK
CKBI(AM) Prince Albert SK
CHMX-FM Regina SK
CIZL-FM Regina SK
CFMC-FM Saskatoon SK
CJMK-FM Saskatoon SK
CKFI-FM Swift Current SK
CFGW-FM Yorkton SK
CIAY-FM Whitehorse YT

Agriculture

CFAM(AM) Altona MB
CJRB(AM) Boissevain MB
CKRM(AM) Regina SK

Album-Oriented Rock

CKLZ-FM Kelowna BC
CJAR(AM) The Pas MB
CHRE-FM Saint Catharines ON
CJMX-FM Sudbury ON
CJQQ-FM Timmins ON
CIZL-FM Regina SK

Alternative

CFEX-FM Calgary AB
*CJSW-FM Calgary AB
CJSR-FM Edmonton AB
CFEQ-FM Winnipeg MB
*CHMR-FM Saint John's NF
*CIOI-FM Hamilton ON
*CHRW-FM London ON
*CKDJ-FM Ottawa ON
CHRY-FM Toronto ON
*CKLN-FM Toronto ON
CIMI-FM Charlesbourg PQ
*CISM-FM Montreal PQ
CHOI-FM Quebec PQ

American Indian

CHDH-FM Siksika AB
CHFN-FM Cape Croker (Neyaashiinigmiing) ON
CFNK-FM Pinehouse Lake SK

Arabic

CHOU(AM) Montreal PQ

Beautiful Music

CBX-FM Edmonton AB
*CION-FM Quebec PQ

Big Band

CFEP-FM Eastern Passage NS

Black

CHRY-FM Toronto ON

Blues

*CBR-FM Calgary AB
CHMC-FM Edmonton AB
*CHRW-FM London ON
*CKRZ-FM Ohsweken ON
CKLX-FM Montreal PQ

Chinese

CKER-FM Edmonton AB
CJVB(AM) Richmond BC
CHMB(AM) Vancouver BC
CHKT(AM) Toronto ON

Christian

CJSI-FM Calgary AB
CJCA(AM) Edmonton AB
CKVN-FM Lethbridge AB
CJLT-FM Medicine Hat AB
CIAJ-FM Prince Rupert BC
CHVN-FM Winnipeg MB
CKJS(AM) Winnipeg MB
CJFY-FM Blackville NB
CIXN-FM Fredericton NB
CJRI-FM Fredericton NB
CITA-FM Moncton NB
CKOE-FM Moncton NB
CINB-FM Saint John NB
CJLU-FM Halifax NS
CINU-FM Truro NS
CJLF-FM Barrie ON
CKJJ-FM Belleville ON
CFWC-FM Brantford ON
CKGW-FM Chatham ON
*CJIV-FM Dryden ON
CJTW-FM Kitchener-Waterloo ON
CHJX-FM London ON
CJYE(AM) Oakville ON
CHRI-FM Ottawa ON
CJTK-FM Sudbury ON
CKSO-FM Sudbury ON
*CJOA-FM Thunder Bay ON
*CHIM-FM Timmins ON
CJFH-FM Woodstock ON
CHIC-FM Rouyn-Noranda PQ
CFAQ-FM Blucher SK
CIOT-FM Nipawin SK
CJJC-FM Yorkton SK
CIAY-FM Whitehorse YT

Classic Rock

CFGQ-FM Calgary AB
CKIS-FM Calgary AB
CFBR-FM Edmonton AB
CIRK-FM Edmonton AB
CJRY-FM Edmonton AB
CKYX-FM Fort McMurray AB
CJXK-FM Grand Centre (Cold Lake) AB
CFRI-FM Grande Prairie AB
CFRV-FM Lethbridge AB
CFDV-FM Red Deer AB
CFXW-FM Whitecourt AB
CKQR-FM Castlegar BC
CKNL-FM Fort St. John BC
CKGF-FM-2 Greenwood BC
CFMI-FM New Westminster BC
CKDV-FM Prince George BC
CFNR-FM Terrace BC
CKKQ-FM Victoria BC
CKX-FM Brandon MB
CHTM(AM) Thompson MB
CITI-FM Winnipeg MB
CFAI-FM Edmundston NB
CFRK-FM Fredericton NB
CIBX-FM Fredericton NB
CJMO-FM Moncton NB
CJYC-FM Saint John NB
CFOZ-FM Argentia NF
CJOZ-FM Bonavista Bay NF
CJKK-FM Clarenville NF
CKOZ-FM Corner Brook NF
CKXD-FM Gander NF
CKXG-FM Grand Falls-Windsor NF
CIOZ-FM Marystown NF
CHOS-FM Rattling Brook NF
CKSS-FM Red Rocks NF
CHOZ-FM Saint John's NF
VOCM-FM Saint John's NF
CIOS-FM Stephenville NF
CFRQ-FM Dartmouth NS
CHNS-FM Halifax NS
CKEC(AM) New Glasgow NS
CHER-FM Sydney NS
CJIJ-FM Sydney NS
CKIQ-FM Iqaluit NU
CFJB-FM Barrie ON
CJXY-FM Burlington ON
CFZN-FM Haliburton ON

CKTI-FM Kettle Point ON
CFCA-FM Kitchener ON
CFRM-FM Little Current ON
CILQ-FM North York ON
*CKRZ-FM Ohsweken ON
CKGE-FM Oshawa ON
CHEZ-FM Ottawa ON
CKQB-FM Ottawa ON
CKWF-FM Peterborough ON
CKTG-FM Thunder Bay ON
CKDK-FM Woodstock ON
CKQK-FM Charlottetown PE
CFVM-FM Amqui PQ
CKAJ-FM Jonquiere PQ
CFOM-FM Levis PQ
CJMF-FM Quebec PQ
CKIA-FM Quebec PQ
CFJO-FM Thetford Mines PQ
CHGO-FM Val d'Or PQ
CJNE-FM Nipawin SK
CFMM-FM Prince Albert SK
CFWF-FM Regina SK
CIZL-FM Regina SK
CKCK-FM Regina SK
CFET-FM Tagish YT
CHON-FM Whitehorse YT

Classical

*CBR-FM Calgary AB
CBX-FM Edmonton AB
CKUA(AM) Edmonton AB
*CKUA-FM Edmonton AB
CBW-FM Winnipeg MB
*CBZ-FM Fredericton NB
CBA-FM Moncton NB
*CBN-FM Saint John's NF
*VOWR(AM) Saint John's NF
*CBH-FM Halifax NS
CFMZ-FM Cobourg ON
*CBOQ-FM Ottawa ON
*CBOX-FM Ottawa ON
CIXK-FM Owen Sound ON
CBL-FM Toronto ON
CFMZ-FM-1 Toronto ON
CJBC-FM Toronto ON
*CBE-FM Windsor ON
*CBJX-FM Chicoutimi PQ
CHLX-FM Gatineau PQ
CBF-FM Montreal PQ
CBM-FM Montreal PQ
CJPX-FM Montreal PQ
*CBVX-FM Quebec PQ
CJSQ-FM Quebec PQ
*CKRL-FM Quebec PQ
*CBRX-FM Rimouski PQ
CBF-FM-1 Trois Rivieres PQ
*CBK-FM Regina SK
*CBKS-FM Saskatoon SK

Comedy

CJEF-FM Saint John NB

Contemporary Hit/Top-40

CJEG-FM Bonnyville AB
CIBQ(AM) Brooks AB
CIBK-FM Calgary AB
CHBN-FM Edmonton AB
CKNG-FM Edmonton AB
CFXE-FM Edson AB
CFGP-FM Grande Prairie AB
CFRV-FM Lethbridge AB
CIZZ-FM Red Deer AB
CJBZ-FM Taber AB
CFCP-FM Courtenay BC
CKRX-FM Fort Nelson BC
CKRV-FM Kamloops BC
CKTK-FM Kitimat BC
CHTK(AM) Prince Rupert BC
CKDM(AM) Dauphin MB
CJPG-FM Portage la Prairie MB
CICY-FM Selkirk MB
CHIQ-FM Winnipeg MB
CKMM-FM Winnipeg MB
CKLE-FM Bathurst NB
CKNB(AM) Campbellton NB
CJEM-FM Edmundston NB
CKMV-FM Grand Falls NB
CHWV-FM Saint John NB
CFOZ-FM Argentia NF
CJOZ-FM Bonavista Bay NF
CJKK-FM Clarenville NF
CKOZ-FM Corner Brook NF
CIFX-FM Lewisporte NF
CIOZ-FM Marystown NF
CHOS-FM Rattling Brook NF
CKSS-FM Red Rocks NF
CHOZ-FM Saint John's NF
CKIX-FM Saint John's NF
CIOS-FM Stephenville NF
CJFX-FM Antigonish NS
CJIJ-FM Sydney NS
CKPE-FM Sydney NS
CKMB-FM Barrie ON
CJOJ-FM Belleville ON
CKPC(AM) Brantford ON
CKEY-FM Fort Erie ON
CKAP-FM Kapuskasing ON
CKBT-FM Kitchener-Waterloo ON
CHST-FM London ON
*CHMO(AM) Moosonee ON
CIDC-FM Orangeville ON
CHRI-FM Ottawa ON
CIHT-FM Ottawa ON
CIXK-FM Owen Sound ON
CKDR-5(AM) Red Lake ON
CKDR-2(AM) Sioux Lookout ON
CFHK-FM St. Thomas ON
CHYC-FM Sudbury ON
CKYK-FM Alma PQ
CFVM-FM Amqui PQ
CIEU-FM Carleton PQ
CHAI-FM Chateauguay PQ
CFVD-FM Degelis PQ
CFTX-FM Gatineau PQ
*CKRK-FM Kahnawake PQ
CHOX-FM La Pocatiere PQ
CFLM(AM) La Tuque PQ
CBGA-FM Matane PQ
CKMF-FM Montreal PQ
CFVD-FM-2 Pohenegamook PQ
CIPC-FM Port-Cartier PQ
CKMN-FM Rimouski-Mont Joli PQ
CHLN(AM) Trois Rivieres PQ
CIGB-FM Trois Rivieres PQ
CKOI-FM Verdun PQ
CKJH(AM) Melfort SK
CFMM-FM Prince Albert SK
CKRW(AM) Whitehorse YT

Country

CKBA(AM) Athabasca AB
CJPR-FM Blairmore AB
CIBQ(AM) Brooks AB
CKRY-FM Calgary AB
CFCW(AM) Camrose AB
CIBW-FM Drayton Valley AB
CKDQ(AM) Drumheller AB
CISN-FM Edmonton AB
CKRA-FM Edmonton AB
CJOK-FM Fort McMurray AB
CJXX-FM Grande Prairie AB
CKHL-FM High Level AB
CKVH(AM) High Prairie AB
CHRB(AM) High River AB
CFWE-FM Lac La Biche AB
CHLB-FM Lethbridge AB
CKSA-FM Lloydminster AB
CHAT-FM Medicine Hat AB
CKLJ-FM Olds AB
CKYL(AM) Peace River AB
CKGY-FM Red Deer AB
CHBW-FM Rocky Mountain House AB
CHLW(AM) Saint Paul AB
CKSQ(AM) Stettler AB
CKKY(AM) Wainwright AB
CFOK(AM) Westlock AB
CIHS-FM Wetaskiwin AB
CKJR(AM) Wetaskiwin AB
CIXM-FM Whitecourt AB
CKBX(AM) 100 Mile House BC
CKQC-FM Abbotsford BC
CFWB(AM) Campbell River BC
CHBZ-FM Cranbrook BC
CJDC(AM) Dawson Creek BC
CILZ-FM Greenville BC
CJKC-FM Kamloops BC
CIGV-FM Penticton BC
CHQB(AM) Powell River BC
CJCI-FM Prince George BC
CKCQ-FM Quesnel BC
CJFW-FM Terrace BC
CHMZ-FM Tofino BC
CJJR-FM Vancouver BC
CIVH(AM) Vanderhoof BC
CKWL(AM) Williams Lake BC
CKLQ(AM) Brandon MB
CKXA-FM Brandon MB
CFRY(AM) Portage la Prairie MB
CFRY-FM Portage la Prairie MB
CFPX-FM Pukatawagan MB
CFQX-FM Selkirk MB
CICY-FM Selkirk MB
CJAR(AM) The Pas MB
CHTM(AM) Thompson MB
CKMW(AM) Winkler-Morden MB
CHNK-FM Winnipeg MB
CKNB(AM) Campbellton NB
CFXY-FM Fredericton NB
CKHJ(AM) Fredericton NB
CJXL-FM Moncton NB
CHSJ-FM Saint John NB
CHTD-FM Saint Stephen NB
CHVO(AM) Carbonear NF
CFLC-FM Churchill Falls NF
CKVO(AM) Clarenville NF
CFCB(AM) Corner Brook NF
CKGA(AM) Gander NF
CKCM(AM) Grand Falls NF
CHCM(AM) Marystown NF
CFNW(AM) Port au Choix NF
CFCV-FM Saint Andrews NF
CJYQ(AM) Saint John's NF
VOCM(AM) Saint John's NF
*VOWR(AM) Saint John's NF
CFSX(AM) Stephenville NF
CFLW(AM) Wabush NF
*CKJM-FM Cheticamp NS
CFDR(AM) Dartmouth NS
CKDY(AM) Digby NS
CHFX-FM Halifax NS
CKEN-FM Kentville NS
CKAD(AM) Middleton NS
CKEC(AM) New Glasgow NS
CJCB(AM) Sydney NS
CKTY-FM Truro NS
CFAB(AM) Windsor NS
CKLB-FM Yellowknife NT
CJKX-FM Ajax ON
CHCQ-FM Belleville ON
CJBQ(AM) Belleville ON
CHAM(AM) Hamilton ON
CING-FM Hamilton ON
CKTI-FM Kettle Point ON
CIKZ-FM Kitchener-Waterloo ON
CJSP-FM Leamington ON
CJBX-FM London ON
CICZ-FM Midland ON
*CHMO(AM) Moosonee ON
CFXN-FM North Bay ON
CKAT(AM) North Bay ON
*CKRZ-FM Ohsweken ON
CKYC-FM Owen Sound ON
CHVR-FM Pembroke ON
CKQM-FM Peterborough ON
CJHR-FM Renfrew ON
CJQM-FM Sault Ste. Marie ON
CKBY-FM Smiths Falls ON
CIGM(AM) Sudbury ON
CKOT(AM) Tillsonburg ON
CKNX(AM) Wingham ON
CFCY-FM Charlottetown PE
CJRW-FM Summerside PE
CFGT(AM) Alma PQ
CFVD-FM Degelis PQ
CHIP-FM Fort Coulonge PQ
*CFTH-FM-1 Harrington Harbour PQ
CKAJ-FM Jonquiere PQ
*CKRK-FM Kahnawake PQ
*CFIN-FM Lac-Etchemin PQ
CFIC-FM Listuguj PQ
CHRG-FM Maria (Reserve) PQ
CHNC(AM) New Carlisle PQ
CKIA-FM Quebec PQ
CKMN-FM Rimouski-Mont Joli PQ
CJMS(AM) Saint Constant PQ
CJSL(AM) Estevan SK
CFMQ-FM Hudson Bay SK
CKVX-FM Kindersley SK
CJLR-FM La Ronge SK
CFDM-FM Meadow Lake SK
CJNS(AM) Meadow Lake SK
CJNS-FM Meadow Lake SK
CJVR-FM Melfort SK
CILG-FM Moose Jaw SK
CJNB(AM) North Battleford SK
CKRM(AM) Regina SK
CJWW(AM) Saskatoon SK
CJSN(AM) Shaunavon SK
CKSW(AM) Swift Current SK
CFSL(AM) Weyburn SK
CJGX(AM) Yorkton SK
CHON-FM Whitehorse YT

Diversified

*CBR(AM) Calgary AB
*CHFA(AM) Edmonton AB
CKUA(AM) Edmonton AB
*CKUA-FM Edmonton AB
*CJSF-FM Burnaby BC
CHET-FM Chetwynd BC
CHAD-FM Dawson Creek BC
CFBX-FM Kamloops BC
CHLY-FM Nanaimo BC
*CJLY-FM Nelson BC
*CFUR-FM Prince George BC
CBUF-FM Vancouver BC
*CITR-FM Vancouver BC
CKAV-FM-2 Vancouver BC
*CFUV-FM Victoria BC
*CKSB(AM) Saint Boniface MB
CKXL-FM Saint Boniface MB
CICY-FM Selkirk MB
*CBWK-FM Thompson MB
CBW(AM) Winnipeg MB
CBW-FM Winnipeg MB
*CKUW-FM Winnipeg MB
CJVA(AM) Caraquet NB
CBZF-FM Fredericton NB
CHSR-FM Fredericton NB
*CFJU-FM Kedgwick NB
*CHMA-FM Sackville NB
CJRM-FM Labrador City NF
*CBN(AM) Saint John's NF
*CHMR-FM Saint John's NF
*VOWR(AM) Saint John's NF
*CIFA-FM Comeauville NS
CBHA-FM Halifax NS
*CBH-FM Halifax NS
*CKDU-FM Halifax NS
CFCT(AM) Tuktoyaktuk NT
*CFYK(AM) Yellowknife NT
CKON-FM Akwesasne ON
CKOL-FM Campbellford ON
CKUN-FM Christian Island ON
CFRU-FM Guelph ON
CKJN-FM Haldimand County ON
*CFMU-FM Hamilton ON
CFBW-FM Hanover ON
CHCR-FM Killaloe ON
CBBK-FM Kingston ON
*CFRC-FM Kingston ON
*CKVI-FM Kingston ON
CKWR-FM Kitchener ON
CJIQ-FM Kitchener/Paris ON
CBBL-FM London ON
CJBC-FM-4 London ON
*CKRZ-FM Ohsweken ON
*CBOQ-FM Ottawa ON
*CBOX-FM Ottawa ON
CKAV-FM-9 Ottawa ON
CKCU-FM Ottawa ON
CKQB-FM Ottawa ON
*CFFF-FM Peterborough ON
CHRY-FM Toronto ON
*CKMS-FM Waterloo ON
*CBE(AM) Windsor ON
*CBEF(AM) Windsor ON
*CBE-FM Windsor ON
CFMF-FM Fermont PQ
CFIM-FM Iles-de-la-Madeleine PQ
CKAJ-FM Jonquiere PQ
CBOF-1(AM) Maniwaki PQ
CBGA-FM Matane PQ
CIRA-FM Montreal PQ
*CKUT-FM Montreal PQ
CFVD-FM-2 Pohenegamook PQ
CBVE-FM Quebec PQ
CKIA-FM Quebec PQ
CHRL-FM Roberval PQ
*CBSI-FM Sept-Iles PQ
CFLX-FM Sherbrooke PQ
CFOU-FM Trois Rivieres PQ
CIAX-FM Windsor PQ
*CBKF-1(AM) Gravelbourg SK
CFMQ-FM Hudson Bay SK
CJLR-FM La Ronge SK
*CBK(AM) Regina SK
CBKF-FM Regina SK
*CJTR-FM Regina SK
CBKF-2(AM) Saskatoon SK
*CBKS-FM Saskatoon SK
CIDD-FM White Bear Lake Resort SK

Drama/Literature

CBF-FM-1 Trois Rivieres PQ

Easy Listening

CJRB(AM) Boissevain MB
CFEP-FM Eastern Passage NS
CHAL-FM Halifax NS
CJHR-FM Renfrew ON
CKOT-FM Tillsonburg ON
CFMQ-FM Hudson Bay SK

Educational

CHDH-FM Siksika AB
*CBG(AM) Gander NF
CBT(AM) Grand Falls-Windsor NF
*CFGB-FM Happy Valley NF
*CBCS-FM Sudbury ON
CJBC(AM) Toronto ON
*CHYZ-FM Sainte Foy PQ
*CBSI-FM Sept-Iles PQ

Eskimo

CKQN-FM Baker Lake NU

Ethnic

CHKF-FM Calgary AB
CKER-FM Edmonton AB
CFWE-FM Lac La Biche AB
CHKG-FM Richmond BC
CJVB(AM) Richmond BC
CJRJ(AM) Vancouver BC
CKYE-FM Vancouver BC
CKJS(AM) Winnipeg MB
CFRU-FM Guelph ON
*CFMU-FM Hamilton ON
CJMR(AM) Mississauga ON
CJLL-FM Ottawa ON
CHHA(AM) Toronto ON
CHKT(AM) Toronto ON
CHTO(AM) Toronto ON
CJSA-FM Toronto ON
*CJAM-FM Windsor ON
CHUK-FM Mashteuiatsh (Pointe-Bleue) PQ

CFMB(AM) Montreal PQ
CHOU(AM) Montreal PQ
CJWI(AM) Montreal PQ
CKDG-FM Montreal PQ
CJLR-FM La Ronge SK

Filipino

CJSA-FM Toronto ON

Folk

*VOWR(AM) Saint John's NF

Foreign/Ethnic

CHMB(AM) Vancouver BC
CKQN-FM Baker Lake NU
CIAO(AM) Brampton ON
CHIN(AM) Toronto ON
CHIN-FM Toronto ON
CIRV-FM Toronto ON
CFMB(AM) Montreal PQ
CINQ-FM Montreal PQ
CJCF-FM Cumberland House SK

French

CBCX-FM Calgary AB
CBRF-FM Calgary AB
*CKRP-FM Falher AB
CKXU-FM Lethbridge AB
CBUX-FM Vancouver BC
CILS-FM Victoria BC
*CKSB(AM) Saint Boniface MB
CKXL-FM Saint Boniface MB
*CFJU-FM Kedgwick NB
CHOY-FM Moncton NB
CKUM-FM Moncton NB
CKRO-FM Pokemouche NB
CHQC-FM Saint John NB
CJRM-FM Labrador City NF
*CKJM-FM Cheticamp NS
CKRH-FM Halifax NS
CIVR-FM Yellowknife NT
CFRT-FM Iqaluit NU
CHOD-FM Cornwall ON
CHYK-FM-3 Hearst ON
CKGN-FM Kapuskasing ON
CJBC-FM-4 London ON
*CHUO-FM Ottawa ON
*CFRH-FM Penetanguishene ON
CFSF-FM Sturgeon Falls ON
CHYK-FM Timmins ON
CHOQ-FM Toronto ON
CFID-FM Acton Vale PQ
CFMV-FM Chandler PQ
CBJ-FM Chicoutimi PQ
CKII-FM Dolbeau-Mistassini PQ
CJDM-FM Drummondville PQ
CFMF-FM Fermont PQ
CHIP-FM Fort Coulonge PQ
CFTX-FM Gatineau PQ
CHLX-FM Gatineau PQ
CHME-FM Les Escoumins PQ
CHHO-FM Louiseville PQ
CHUK-FM Mashteuiatsh (Pointe-Bleue) PQ
CHEF-FM Matagami PQ
CFLO-FM Mont-Laurier PQ
CINQ-FM Montreal PQ
CHOI-FM Quebec PQ
CHIC-FM Rouyn-Noranda PQ
*CHYZ-FM Sainte Foy PQ
CFUT-FM Shawinigan PQ
CFAK-FM Sherbrooke PQ
CFLX-FM Sherbrooke PQ
CIGR-FM Sherbrooke PQ
CJEB-FM Trois Rivieres PQ
CFRG-FM Gravelbourg SK

German

CHPD-FM Aylmer ON

Golden Oldies

CKDO(AM) Oshawa ON
CHAB(AM) Moose Jaw SK
CJNE-FM Nipawin SK

Gospel

CIHS-FM Wetaskiwin AB
CJRI-FM Fredericton NB
*VOAR(AM) Mount Pearl NF

Greek

CHTO(AM) Toronto ON

Hindi

CJSA-FM Toronto ON

Inspirational

CKER-FM Edmonton AB
CJRB(AM) Boissevain MB

Jazz

*CJSW-FM Calgary AB
CKUA(AM) Edmonton AB
*CKUA-FM Edmonton AB
CBAX-FM Halifax NS
CBHA-FM Halifax NS
CBBL-FM London ON
*CHRW-FM London ON
CBBS-FM Sudbury ON
*CKLU-FM Sudbury ON
*CJRT-FM Toronto ON
*CBJX-FM Chicoutimi PQ
CHLX-FM Gatineau PQ
CKLX-FM Montreal PQ
*CKRL-FM Quebec PQ
*CBRX-FM Rimouski PQ
*CBK-FM Regina SK
*CBKS-FM Saskatoon SK

Jewish

CJRS(AM) Montreal PQ

Light Rock

CHMC-FM Edmonton AB
CKFU-FM Fort St. John BC
CIOC-FM Victoria BC
CKXL-FM Saint Boniface MB
CKDH(AM) Amherst NS
CKNR-FM Elliot Lake ON
CJTN-FM Quinte West ON
CKGB-FM Timmins ON
CIMF-FM Gatineau PQ
CHOE-FM Matane PQ
CFQR-FM Montreal PQ
CHEY-FM Trois Rivieres PQ
CHSN-FM Estevan SK

MOR

*CHFA(AM) Edmonton AB
CHET-FM Chetwynd BC
CKCL-FM Chilliwack BC
CFKC(AM) Creston BC
CHAD-FM Dawson Creek BC
CFTK(AM) Terrace BC
CKBD(AM) Vancouver BC
CFAM(AM) Altona MB
CHSM(AM) Steinbach MB
CJVA(AM) Caraquet NB
CKRO-FM Pokemouche NB
CKEC(AM) New Glasgow NS
CKHR-FM Hay River NT
CFCO(AM) Chatham ON
CHOD-FM Cornwall ON
CJMI-FM Strathroy ON
CHWO(AM) Toronto ON

CJAN-FM Asbestos PQ
CHLC-FM Baie Comeau PQ
CFIX-FM Chicoutimi PQ
CHVD-FM Dolbeau-Mistassini PQ
CHRD-FM Drummondville PQ
CFRP(AM) Forestville PQ
CJRG-FM Gaspe PQ
CILE-FM Havre-Saint-Pierre PQ
CJLM-FM Joliette PQ
CJIT-FM Lac Megantic PQ
*CFIN-FM Lac-Etchemin PQ
CHEF-FM Matagami PQ
CHRM-FM Matane PQ
CKNA-FM Natashquan PQ
CKYQ-FM Plessisville PQ
CJBR-FM Rimouski PQ
CJRE-FM Riviere au Renard PQ
CFNJ-FM Saint Gabriel-de-Brandon PQ
CIHO-FM Saint Hilarion PQ
CJDS-FM Saint Pamphile PQ
CJMC-FM Sainte Anne des Monts PQ
CHEQ-FM Sainte-Marie-de-Beauce PQ
CHLN(AM) Trois Rivieres PQ
CKBI(AM) Prince Albert SK
CJSN(AM) Shaunavon SK
CIMG-FM Swift Current SK

Native American

CICU-FM Eskasoni Indian Reserve NS
CKAV-FM Toronto ON
CKUJ-FM Kuujjuaq PQ

News

CFFR(AM) Calgary AB
*CJSW-FM Calgary AB
CBX-FM Edmonton AB
CFML-FM Burnaby BC
CBTE-FM Crawford Bay BC
*CBTK-FM Kelowna BC
CJNL(AM) Merritt BC
*CFPR(AM) Prince Rupert BC
CKWX(AM) Vancouver BC
CKIZ-FM Vernon BC
CBA(AM) Moncton NB
CBD-FM Saint John NB
CBGY(AM) Bonavista Bay NF
*CIFA-FM Comeauville NS
CBHA-FM Halifax NS
CBQX-FM Kenora ON
CKGE-FM Oshawa ON
CHSC(AM) Saint Catharines ON
CBEG-FM Sarnia ON
*CBQT-FM Thunder Bay ON
CFTR(AM) Toronto ON
CBMI-FM Baie Comeau PQ
CHRD-FM Drummondville PQ
*CBV-FM-6 La Malbaie PQ
CFBS-FM Lourdes-de-Blanc-Sablon PQ
CKWE-FM Maniwaki (Kitigan Zibi Anishinabeg Reserve) PQ
CBME-FM Montreal PQ
CINW(AM) Montreal PQ
CINF(AM) Verdun PQ
CFMM-FM Prince Albert SK
*CBK(AM) Regina SK

News/talk

*CBR(AM) Calgary AB
CHQR(AM) Calgary AB
CBX(AM) Edmonton AB
*CHFA(AM) Edmonton AB
CHDH-FM Siksika AB
CKOV(AM) Kelowna BC
CKNW(AM) New Westminster BC
*CBU(AM) Vancouver BC
*CFRO-FM Vancouver BC
CFAX(AM) Victoria BC
CJOB(AM) Winnipeg MB
*CKUW-FM Winnipeg MB
*CBZ-FM Fredericton NB

*CBAF-FM Moncton NB
CKNI-FM Moncton NB
CHNI-FM Saint John NB
*CBY(AM) Corner Brook NF
*CBG(AM) Gander NF
*CBN-FM Saint John's NF
VOCM(AM) Saint John's NF
CJNI-FM Halifax NS
*CHAK(AM) Inuvik NT
*CFYK(AM) Yellowknife NT
CFFB(AM) Iqaluit NU
CHML(AM) Hamilton ON
CKGL(AM) Kitchener ON
CFPL(AM) London ON
CJBK(AM) London ON
CFRA(AM) Ottawa ON
CFOS(AM) Owen Sound ON
CKTB(AM) Saint Catharines ON
*CBCS-FM Sudbury ON
*CKLU-FM Sudbury ON
*CBQ-FM Thunder Bay ON
CFMJ(AM) Toronto ON
CFRB(AM) Toronto ON
*CBE(AM) Windsor ON
CKLW(AM) Windsor ON
CBJ-FM Chicoutimi PQ
CKNU-FM Donnacona PQ
CJRC-FM Gatineau PQ
CFIM-FM Iles-de-la-Madeleine PQ
CBGA-FM Matane PQ
CJAD(AM) Montreal PQ
CBVE-FM Quebec PQ
*CBV-FM Quebec PQ
CJBR-FM Rimouski PQ
CKRS(AM) Saguenay PQ
CBGN(AM) Sainte Anne des Monts PQ
CKCN-FM Sept-Iles PQ
CKSM(AM) Shawinigan PQ
CFLX-FM Sherbrooke PQ
CHLT(AM) Sherbrooke PQ
CHLN(AM) Trois Rivieres PQ
CJME(AM) Regina SK
CJWW(AM) Saskatoon SK
CKOM(AM) Saskatoon SK

Nostalgia

CJNU-FM Winnipeg MB
CJCH(AM) Halifax NS
CKPT(AM) Peterborough ON
*CFIN-FM Lac-Etchemin PQ
CFAV(AM) Laval PQ

Oldies

CHQT(AM) Edmonton AB
CFXH-FM Hinton AB
CKLR-FM Courtenay BC
CHAD-FM Dawson Creek BC
CHNL(AM) Kamloops BC
CKFR(AM) Kelowna BC
CJNL(AM) Merritt BC
CIBH-FM Parksville BC
CKOR(AM) Penticton BC
CISL(AM) Richmond BC
CKKS-FM Sechelt BC
CHOR(AM) Summerland BC
CFRW(AM) Winnipeg MB
CFBC(AM) Saint John NB
CFCV-FM Saint Andrews NF
*VOWR(AM) Saint John's NF
CBI-FM Sydney NS
CKPC(AM) Brantford ON
CJUL(AM) Cornwall ON
CJOY(AM) Guelph ON
CKOC(AM) Hamilton ON
CFFX(AM) Kingston ON
CFMK-FM Kingston ON
CKLC(AM) Kingston ON
CKKW(AM) Kitchener ON
CKSL(AM) London ON
CIWW(AM) Ottawa ON
CJMJ-FM Ottawa ON
CKKL-FM Ottawa ON
CFOS(AM) Owen Sound ON

CKRU(AM) Peterborough ON
CHOK(AM) Sarnia ON
CKDR-2(AM) Sioux Lookout ON
CJCS(AM) Stratford ON
CFMX-FM Toronto ON
CHUM(AM) Toronto ON
CHTN-FM Charlottetown PE
*CFTH-FM-1 Harrington Harbour PQ
CFOM-FM Levis PQ
CHRG-FM Maria (Reserve) PQ
CKRS(AM) Saguenay PQ
CFEI-FM Saint Hyacinthe PQ
CFZZ-FM Saint Jean-Iberville PQ
CFYM(AM) Kindersley SK
CFNK-FM Pinehouse Lake SK
CKBI(AM) Prince Albert SK
CJYM(AM) Rosetown SK

Other

*CBR(AM) Calgary AB
CFCW-FM Camrose AB
CHET-FM Chetwynd BC
CKFU-FM Fort St. John BC
*CFPR(AM) Prince Rupert BC
CKIZ-FM Vernon BC
*CBCV-FM Victoria BC
CHTT-FM Victoria BC
CKCW-FM Moncton NB
CJFX-FM Antigonish NS
CKUL-FM Halifax NS
CFFB(AM) Iqaluit NU
CKGN-FM Kapuskasing ON
CJTT-FM New Liskeard ON
CBQS-FM Sioux Narrows ON
CJET-FM Smiths Falls ON
CHIP-FM Fort Coulonge PQ
*CBV-FM-6 La Malbaie PQ
CHUT-FM Lac-Simon (Louvicourt) PQ
CINQ-FM Montreal PQ

Progressive

CKCU-FM Ottawa ON
CFNY-FM Toronto ON
*CJAM-FM Windsor ON

Public Affairs

*CKRP-FM Falher AB
CBTE-FM Crawford Bay BC
*CBTK-FM Kelowna BC
CBYG-FM Prince George BC
*CFRO-FM Vancouver BC
CFNC(AM) Cross Lake MB
CBA(AM) Moncton NB
CBGY(AM) Bonavista Bay NF
CBI(AM) Sydney NS
CJLX-FM Belleville ON
*CFMU-FM Hamilton ON
CBQX-FM Kenora ON
CKCU-FM Ottawa ON
CBQL-FM Savant Lake ON
*CBQ-FM Thunder Bay ON
CBLA-FM Toronto ON
CBL-FM Toronto ON
*CKMS-FM Waterloo ON
CBMI-FM Baie Comeau PQ
*CHOC-FM Saint Remi PQ
*CBKA-FM La Ronge SK

Religious

CHRB(AM) High River AB
*VOAR(AM) Mount Pearl NF
CHSB-FM Bedford NS
CKKK-FM Peterborough ON
CJTK-FM Sudbury ON
CIRA-FM Montreal PQ
CJRS(AM) Montreal PQ
*CION-FM Quebec PQ
CKSW(AM) Swift Current SK

Rock/AOR

CKLM-FM-1 Bonnyville AB
CJAY-FM Calgary AB
CFBR-FM Edmonton AB
CHDI-FM Edmonton AB
CJRX-FM Lethbridge AB
CKLM-FM Lloydminster AB
CKBX(AM) 100 Mile House BC
CKSR-FM Chilliwack BC
CHDR-FM Cranbrook BC
CJDR-FM Fernie BC
CHRX-FM Fort St. John BC
CKGR(AM) Golden BC
CKIR(AM) Invermere BC
CIFM-FM Kamloops BC
CHWF-FM Nanaimo BC
CHNV-FM Nelson BC
CJMG-FM Penticton BC
CFNI(AM) Port Hardy BC
CIRX-FM Prince George BC
CJCI-FM Prince George BC
CKCQ-FM Quesnel BC
CKCR(AM) Revelstoke BC
CHMZ-FM Tofino BC
CFOX-FM Vancouver BC
CIVH(AM) Vanderhoof BC
CJZN-FM Victoria BC
CFFM-FM Williams Lake BC
CKWL(AM) Williams Lake BC
CFPX-FM Pukatawagan MB
CJKR-FM Winnipeg MB
*CKUW-FM Winnipeg MB
CFXY-FM Fredericton NB
CJEF-FM Saint John NB
CKIM(AM) Baie Verte NF
CKXX-FM Corner Brook NF
CKOA-FM Glace Bay NS
CKTO-FM Truro NS
CFJB-FM Barrie ON
CJLX-FM Belleville ON
CKUE-FM Chatham ON
CJSS-FM Cornwall ON
CJBB-FM Englehart ON
CIKR-FM Kingston ON
CHYM-FM Kitchener ON
CFPL-FM London ON
CKFX-FM North Bay ON
CKQB-FM Ottawa ON
CHTZ-FM Saint Catharines ON
CHKS-FM Sarnia ON
CJRQ-FM Sudbury ON
CJSD-FM Thunder Bay ON
CFNY-FM Toronto ON
CIMX-FM Windsor ON
CIBU-FM Wingham ON
CFMV-FM Chandler PQ
CKTF-FM Gatineau PQ
CFOR-FM Maniwaki PQ
CHOM-FM Montreal PQ
*CKRL-FM Quebec PQ
CIKI-FM Rimouski PQ
CIBM-FM Riviere du Loup PQ
CHJM-FM Saint Georges PQ
CIGR-FM Sherbrooke PQ
CJSO-FM Sorel PQ
CFJO-FM Thetford Mines PQ
CJCQ-FM North Battleford SK
CHQX-FM Prince Albert SK
CJDJ-FM Saskatoon SK
CKRC-FM Weyburn SK

Smooth Jazz

CIQX-FM Calgary AB
CHMC-FM Edmonton AB
CJZZ-FM Winnipeg MB
CIWV-FM Hamilton ON

Spanish

CHHA(AM) Toronto ON

Sports

CFAC(AM) Calgary AB
CHQR(AM) Calgary AB
CFRN(AM) Edmonton AB
CHED(AM) Edmonton AB
CKNW(AM) New Westminster BC
CKST(AM) Vancouver BC
CHML(AM) Hamilton ON
CFPL(AM) London ON
CFGO(AM) Ottawa ON
CIGM(AM) Sudbury ON
CJCL(AM) Toronto ON
CKAC(AM) Montreal PQ
CKGM(AM) Montreal PQ
CHRC(AM) Quebec PQ
CKSM(AM) Shawinigan PQ
CHLN(AM) Trois Rivieres PQ

Talk

CHED(AM) Edmonton AB
CFUN(AM) Vancouver BC
CHMJ(AM) Vancouver BC
*CBWK-FM Thompson MB
CKTP-FM Fredericton Centre NB
CBD-FM Saint John NB
CBDQ-FM Labrador City NF
CIFX-FM Lewisporte NF
CBQR-FM Rankin Inlet NU
CFGO(AM) Ottawa ON
CBLA-FM Toronto ON
CIRR-FM Toronto ON
CJCL(AM) Toronto ON
*CJAM-FM Windsor ON
*CBCT-FM Charlottetown PE
CFGT(AM) Alma PQ
CBJE-FM Chicoutimi PQ
CIDI-FM Lac-Brome PQ
CHMP-FM Longueuil PQ
CKWE-FM Maniwaki (Kitigan Zibi Anishinabeg Reserve) PQ
*CIBL-FM Montreal PQ
*CBRX-FM Rimouski PQ
*CBSI-FM Sept-Iles PQ
CHLT-FM Sherbrooke PQ
*CBK(AM) Regina SK
*CFWH(AM) Whitehorse YT

Top-40

CFBT-FM Vancouver BC
CKHZ-FM Halifax NS
CFIF-FM Iroquois Falls ON
CIDC-FM Orangeville ON
CISS-FM Ottawa ON
CFSF-FM Sturgeon Falls ON
CHNO-FM Sudbury ON
CHMT-FM Timmins ON
CIRR-FM Toronto ON
CJAQ-FM Toronto ON
CKTF-FM Gatineau PQ
CJIT-FM Lac Megantic PQ
CIMO-FM Magog PQ
CJMV-FM Val d'Or PQ
CFDM-FM Meadow Lake SK

Triple A

CFUL-FM Calgary AB
CFML-FM Burnaby BC

Urban Contemporary

CHPQ-FM Parksville BC
CJWV-FM Winnipeg MB
*CKUW-FM Winnipeg MB
CJEF-FM Saint John NB
*CIXX-FM London ON
CFXJ-FM Toronto ON

Variety/Diverse

CIAM-FM Fort Vermilion AB
CKXU-FM Lethbridge AB
CIVL-FM Abbotsford BC
CFCH-FM Chase BC
CIDO-FM Creston BC
*CBTK-FM Kelowna BC
CHLS-FM Lillooet BC
CHMM-FM MacKenzie BC
CJMP-FM Powell River BC
CIMM-FM Ucluelet BC
*CBU(AM) Vancouver BC
CBUX-FM Vancouver BC
CJJJ-FM Brandon MB
CHFC(AM) Churchill MB
CJSB-FM Swan River MB
CINC-FM Thompson MB
CJUM-FM Winnipeg MB
CKIC-FM Winnipeg MB
CIMS-FM Balmoral NB
*CBAF-FM Moncton NB
*CBAL-FM Moncton NB
CKUM-FM Moncton NB
CFMH-FM Saint John NB
CJSE-FM Shediac NB
*CBY(AM) Corner Brook NF
CFXU-FM Antigonish NS
*CKJM-FM Cheticamp NS
CBAX-FM Halifax NS
CKHR-FM Hay River NT
CBQR-FM Rankin Inlet NU
CHES-FM Erin ON
CFGI-FM Georgina Island ON
CKHA-FM Haliburton ON
CKGN-FM Kapuskasing ON
*CBOF-FM Ottawa ON
*CBO-FM Ottawa ON
CFBU-FM Saint Catharines ON
CJAI-FM Stella ON
CBBX-FM Sudbury ON
*CBON-FM Sudbury ON
*CKLU-FM Sudbury ON
CILU-FM Thunder Bay ON
CHOQ-FM Toronto ON
*CIUT-FM Toronto ON
CJBC(AM) Toronto ON
CKHC-FM Toronto ON
CFWP-FM Wahta Mohawk Territory near Bala ON
CIWS-FM Whitchurch-Stouffville ON
CFID-FM Acton Vale PQ
CFTH-FM-2 La Tabatiere PQ
CFLM(AM) La Tuque PQ
CIDI-FM Lac-Brome PQ
CHHO-FM Louiseville PQ
CKAU-FM Maliotenam PQ
CHGA-FM Maniwaki PQ
CKWE-FM Maniwaki (Kitigan Zibi Anishinabeg Reserve) PQ
CHRG-FM Maria (Reserve) PQ
CBFX-FM Montreal PQ
CJLO(AM) Montreal PQ
CJBE-FM Port-Menier PQ
*CBV-FM Quebec PQ
CIAU-FM Radisson PQ
*CHYZ-FM Sainte Foy PQ
CIBO-FM Senneterre PQ
CJMQ-FM Sherbrooke PQ
CHXL-FM Okanese Indian Reserve SK
*CBK-FM Regina SK
CFCR-FM Saskatoon SK
CKZP-FM Zenon Park SK
CFET-FM Tagish YT

Special Programming on Radio Stations in the U.S.

Adult Contemp

KNTI(FM) Lakeport CA 3 hrs
KOLV(FM) Olivia MN
WKBO(AM) Harrisburg PA
WRDW-FM Philadelphia PA 11 hrs

Agriculture

WKAC(AM) Athens AL 5 hrs
WVNN(AM) Athens AL 2 hrs
WNSI-FM Atmore AL 5 hrs
WACQ(AM) Carrville AL 1 hr
WZTQ(AM) Centre AL 1 hr
WKUL(FM) Cullman AL 15 hrs
WTVY-FM Dothan AL 5 hrs
WULA(AM) Eufaula AL 2 hrs
WABF(AM) Fairhope AL 1 hr
WKWL(AM) Florala AL 1 hr
WHEP(AM) Foley AL 2 hrs
WZOB(AM) Fort Payne AL 2 hrs
WJBB(AM) Haleyville AL 3 hrs
WERH(AM) Hamilton AL
WINL(FM) Linden AL 10 hrs
WACV(AM) Montgomery AL 5 hrs
WOPP(AM) Opp AL 2 hrs
WKEA-FM Scottsboro AL 1 hr
WHBB(AM) Selma AL 10 hrs
WBTG-FM Sheffield AL 1 hr
WTLS(AM) Tallassee AL 6 hrs
WTBF(AM) Troy AL 17 hrs
KAPZ(AM) Bald Knob AR 3 hrs
KAAB(AM) Batesville AR 5 hrs
KEWI(AM) Benton AR 4 hrs
KTHS(AM) Berryville AR 15 hrs
KTHS-FM Berryville AR 15 hrs
KXXA(AM) Conway AR 6 hrs
KHTE-FM England AR 5 hrs
KVDW(AM) England AR 5 hrs
KXJK(AM) Forrest City AR 16 hrs
KFFA(AM) Helena AR 16 hrs
KFIN(FM) Jonesboro AR 13 hrs
KNEA(AM) Jonesboro AR 6 hrs
KVMA(AM) Magnolia AR 2 hrs
KVSA(AM) McGehee AR 5 hrs
KBOA-FM Piggott AR 5 hrs
KPBQ-FM Pine Bluff AR 2 hrs
KPOC(AM) Pocahontas AR 10 hrs
KPOC-FM Pocahontas AR 10 hrs
KURM(AM) Rogers AR 10 hrs
KARV(AM) Russellville AR 5 hrs
KWCK(AM) Searcy AR 10 hrs
KWCK-FM Searcy AR 3 hrs
KUOA(AM) Siloam Springs AR 1 hr
KWAK(AM) Stuttgart AR 6 hrs
KFYX(FM) Texarkana AR 2 hrs
KOSE(AM) Wilson AR 5 hrs
KWYN(AM) Wynne AR 6 hrs
KDJI(AM) Holbrook AZ 8 hrs
KLPZ(AM) Parker AZ 1 hr
KVSL(AM) Show Low AZ 2 hrs
KCFJ(AM) Alturas CA 1 hr
KERN(AM) Bakersfield CA 1 hr
KISV(FM) Bakersfield CA 1 hr
KJMB(FM) Blythe CA 5 hrs
KXO(AM) El Centro CA 7 hrs
KRKC(AM) King City CA 10 hrs
KUBB(FM) Mariposa CA 2 hrs
KYOS(AM) Merced CA 5 hrs
KESP(AM) Modesto CA 3 hrs
KBLF(AM) Red Bluff CA 5 hrs
KHTK(AM) Sacramento CA 4 hrs
KSTN(AM) Stockton CA 3 hrs
KJUG(AM) Tulare CA 5 hrs
KJUG-FM Tulare CA 5 hrs
KWNE(FM) Ukiah CA 1 hr
KUBA(AM) Yuba City CA 4 hrs
KGIW(AM) Alamosa CO 6 hrs
KRAI(AM) Craig CO 1 hr
KFTM(AM) Fort Morgan CO 6 hrs
KRKY(AM) Granby CO 1 hr
KFKA(AM) Greeley CO 15 hrs
KRCN(AM) Longmont/Denver CO 1 hr
KSLV(AM) Monte Vista CO 1 hr
KSTC(AM) Sterling CO 15 hrs
KCRT(AM) Trinidad CO 1 hr
KCOL(AM) Wellington CO 2 hrs
KRDZ(AM) Wray CO 10 hrs
WGMD(FM) Rehoboth Beach DE 2 hrs
WGBG(FM) Seaford DE 1 hr
WBGF(FM) Belle Glade FL 5 hrs
WLBE(AM) Leesburg FL 3 hrs
WTYS(AM) Marianna FL 1 hr
WDVH-FM Trenton FL 2 hrs
WZZS(FM) Zolfo Springs FL 1 hr
V6AH(AM) Pohnpei FM 20 hrs
WAJQ(AM) Alma GA 2 hrs
WAJQ-FM Alma GA 2 hrs
WDEC-FM Americus GA 1 hr
WGAC(AM) Augusta GA 4 hrs
WMGR(AM) Bainbridge GA 3 hrs
WJTH(AM) Calhoun GA 1 hr
WDCO(AM) Cochran GA 3 hrs
WDXQ-FM Cochran GA 3 hrs
WRCG(AM) Columbus GA 3 hrs
WJJC(AM) Commerce GA 2 hrs
WCUG(AM) Cuthbert GA 8 hrs
WDMG(AM) Douglas GA 4 hrs
WMLT(AM) Dublin GA 6 hrs
WCEH(AM) Hawkinsville GA 5 hrs
WVOH(AM) Hazlehurst GA 2 hrs
WIFO-FM Jesup GA 5 hrs
WLOP(AM) Jesup GA 10 hrs
WQCH(AM) La Fayette GA 2 hrs
WMCG(FM) Milan GA 1 hr
WHKN(FM) Millen GA 10 hrs
WMTM(AM) Moultrie GA 16 hrs
WALH(AM) Mountain City GA 2 hrs
WSTI-FM Quitman GA 5 hrs
WWNS(AM) Statesboro GA 12 hrs
WJAT(AM) Swainsboro GA 5 hrs
WSYL(AM) Sylvania GA 5 hrs
WTHO-FM Thomson GA 3 hrs
WTIF(AM) Tifton GA 5 hrs
KLGA(AM) Algona IA
KLGA-FM Algona IA
KJAN(AM) Atlantic IA 12 hrs
KWBG(AM) Boone IA 15 hrs
KBUR(AM) Burlington IA 19 hrs
KCPS(AM) Burlington IA 10 hrs
WMT(AM) Cedar Rapids IA 19 hrs
KCHA-FM Charles City IA 12 hrs
KCHE(AM) Cherokee IA 12 hrs
KCLN(AM) Clinton IA 10 hrs
KROS(AM) Clinton IA 5 hrs
KCZQ(FM) Cresco IA 12 hrs
WOC(AM) Davenport IA 10 hrs
KDSN(AM) Denison IA 12 hrs
KDSN-FM Denison IA 7 hrs
WHO(AM) Des Moines IA 15 hrs
KDTH(AM) Dubuque IA 17 hrs
KDST(FM) Dyersville IA
KILR(AM) Estherville IA 9 hrs
KILR-FM Estherville IA 4 hrs
KMCD(AM) Fairfield IA 10 hrs
KIOW(FM) Forest City IA 15 hrs
KWMT(AM) Fort Dodge IA
KCTN(FM) Garnavillo IA
KGRN(AM) Grinnell IA 12 hrs
KLMJ(FM) Hampton IA 8 hrs
KNOD(FM) Harlan IA 3 hrs
KHBT(FM) Humboldt IA 10 hrs
KKIA(FM) Ida Grove IA 10 hrs
KIFG(AM) Iowa Falls IA 5 hrs
KOKX(AM) Keokuk IA 6 hrs
KOKX-FM Keokuk IA 3 hrs
KLEM(AM) Le Mars IA 18 hrs
KMCH(FM) Manchester IA 7 hrs
KMAQ(AM) Maquoketa IA 10 hrs
KGLO(AM) Mason City IA 15 hrs
KCZE(FM) New Hampton IA 12 hrs
KCOB(AM) Newton IA 2 hrs
KOEL(AM) Oelwein IA 16 hrs
KBIZ(AM) Ottumwa IA 12 hrs
KMA(AM) Shenandoah IA
*KDCR(FM) Sioux Center IA 2 hrs
KMNS(AM) Sioux City IA 20 hrs
KSCJ(AM) Sioux City IA 5 hrs
KTFC(FM) Sioux City IA 1 hr
KKRF(FM) Stuart IA 5 hrs
KTLB(FM) Twin Lakes IA 15 hrs
KCII-FM Washington IA
KQWC(AM) Webster City IA 8 hrs
KNJY(AM) Boise ID 5 hrs
KCHQ(FM) Driggs ID 1 hr
KORT(AM) Grangeville ID 2 hrs
KID(AM) Idaho Falls ID 18 hrs
KOZE(AM) Lewiston ID 2 hrs
KRLC(AM) Lewiston ID 5 hrs
KVSI(AM) Montpelier ID 2 hrs
KRPL(AM) Moscow ID 4 hrs
KWIK(AM) Pocatello ID 3 hrs
KACH(AM) Preston ID 3 hrs
KSRA(AM) Salmon ID 4 hrs
KSRA-FM Salmon ID 4 hrs
KLIX(AM) Twin Falls ID 2 hrs
KTFI(AM) Twin Falls ID 3 hrs
WQRL(FM) Benton IL 3 hrs
WBNQ(FM) Bloomington IL 1 hr
WJBC(AM) Bloomington IL 13 hrs
WLMD(FM) Bushnell IL 3 hrs
WKRO(AM) Cairo IL 12 hrs
WBYS(AM) Canton IL 10 hrs
WCIL(AM) Carbondale IL 1 hr
WROY(AM) Carmi IL 6 hrs
WDWS(AM) Champaign IL 10 hrs
WEIC(AM) Charleston IL 8 hrs
KSGM(AM) Chester IL 2 hrs
WDAN(AM) Danville IL 20 hrs
WLBK(AM) De Kalb IL 12 hrs
WSOY(AM) Decatur IL 10 hrs
WIXN(AM) Dixon IL 11 hrs
WDQN(AM) Du Quoin IL 3 hrs
KUUL(FM) East Moline IL 1 hr
WFIW(AM) Fairfield IL 16 hrs
WFIW-FM Fairfield IL 12 hrs
WFRL(AM) Freeport IL 15 hrs
WAAG(FM) Galesburg IL 5 hrs
WGIL(AM) Galesburg IL 10 hrs
WGEL(FM) Greenville IL 19 hrs
WEBQ(AM) Harrisburg IL 6 hrs
WDUK(FM) Havana IL 8 hrs
WRVY-FM Henry IL 4 hrs
WIJR(AM) Highland IL 3 hrs
WJIL(AM) Jacksonville IL 8 hrs
WLDS(AM) Jacksonville IL 20 hrs
WJBM(AM) Jerseyville IL 18 hrs
WJOL(AM) Joliet IL 3 hrs
WKAN(AM) Kankakee IL 10 hrs
WKEI(AM) Kewanee IL 20 hrs
WSMI(AM) Litchfield IL 18 hrs
WSMI-FM Litchfield IL 18 hrs
WLUV(AM) Loves Park IL 6 hrs
WJEQ(FM) Macomb IL 1 hr
WLRB(AM) Macomb IL 1.25 hrs
WZUS(FM) Macon IL 5 hrs
WLBH-FM Mattoon IL 9 hrs
WMCI(FM) Mattoon IL 5 hrs
WFXN(AM) Moline IL 1 hr
WRAM(AM) Monmouth IL 18 hrs
WCZQ(FM) Monticello IL 11 hrs
WCSJ-FM Morris IL 10 hrs
WVMC(AM) Mount Carmel IL 5 hrs
WMIX(AM) Mount Vernon IL 18 hrs
WMIX-FM Mount Vernon IL 12 hrs
WHQQ(FM) Neoga IL 7 hrs
WCMY(AM) Ottawa IL 9 hrs
WPXN(FM) Paxton IL 10 hrs
WMBD(AM) Peoria IL 15 hrs
WBBA-FM Pittsfield IL 10 hrs
WSPY-FM Plano IL 18 hrs
WZOE(AM) Princeton IL 15 hrs
WTAY(AM) Robinson IL 3 hrs
WTYE(FM) Robinson IL 6 hrs
WKXQ(FM) Rushville IL 6 hrs
WJBD(AM) Salem IL 4 hrs
WJBD-FM Salem IL 4 hrs
WAUR(AM) Sandwich IL 15 hrs
WHCO(AM) Sparta IL 20 hrs
WTAX(AM) Springfield IL 16 hrs
WSDR(AM) Sterling IL 16 hrs
WSQR(AM) Sycamore IL 6 hrs
WTIM-FM Taylorville IL 20 hrs
WRAN(FM) Tower Hill IL 6 hrs
*WILL(AM) Urbana IL 7 hrs
WPMB(AM) Vandalia IL 4 hrs
WGFA-FM Watseka IL 18 hrs
WSCH(FM) Aurora IN 3 hrs
WRBI(FM) Batesville IN 5 hrs
WBIW(AM) Bedford IN 3 hrs
WCVL(AM) Crawfordsville IN 5 hrs
WADM(AM) Decatur IN 2 hrs
WIKY-FM Evansville IN 17 hrs
WILO(AM) Frankfort IN 12 hrs
WSHW(FM) Frankfort IN 8 hrs
WFLQ(FM) French Lick IN 3 hrs
WREB(FM) Greencastle IN 5 hrs
WTRE(AM) Greensburg IN 10 hrs
WBDC(FM) Huntingburg IN 5 hrs
WAWK(AM) Kendallville IN 1 hr
WLOI(AM) La Porte IN 8 hrs
WASK(AM) Lafayette IN 2 hrs
WKOA(FM) Lafayette IN 3 hrs
WSAL(AM) Logansport IN 10 hrs
WRZR(FM) Loogootee IN 2 hrs
WORX-FM Madison IN 3 hrs
WXGO(AM) Madison IN 3 hrs
WCBK-FM Martinsville IN 1 hr
WMYJ(AM) Martinsville IN 1 hr
WEFM(FM) Michigan City IN
WRCY(AM) Mount Vernon IN 5 hrs
WMDH(AM) New Castle IN 2 hrs
WMDH-FM New Castle IN 1 hr
WSEZ(AM) Paoli IN 7 hrs
WTCA(AM) Plymouth IN 11 hrs
WLQI(FM) Rensselaer IN 15 hrs
WRIN(AM) Rensselaer IN 12 hrs
WKBV(AM) Richmond IN 4 hrs
WROI(FM) Rochester IN 10 hrs
WIFE-FM Rushville IN 18 hrs
WAXL(FM) Santa Claus IN 3 hrs
WZZB(AM) Seymour IN 2 hrs
WNDI(AM) Sullivan IN 6 hrs
WKUZ(FM) Wabash IN 5 hrs
WRSW(AM) Warsaw IN 11 hrs
WRSW-FM Warsaw IN 2 hrs
WWBL(FM) Washington IN 15 hrs
KSOK(AM) Arkansas City KS 10 hrs
KAIR(AM) Atchison KS 7 hrs
KVSV(AM) Beloit KS 9 hrs
KSNP(FM) Burlington KS 8 hrs
KGNO(AM) Dodge City KS 15 hrs
KDNS(FM) Downs KS 6 hrs
KIUL(AM) Garden City KS 5 hrs
*KGCR(FM) Goodland KS 2 hrs
KVGB(AM) Great Bend KS 7 hrs
KHAZ(FM) Hays KS 10 hrs
KNZA(FM) Hiawatha KS 14 hrs
KBUF(AM) Holcomb KS 15 hrs
KNNS(AM) Larned KS 10 hrs
KSCB(AM) Liberal KS 6 hrs
KSLS(FM) Liberal KS 12 hrs
KOFO(AM) Ottawa KS 2 hrs
KLKC(AM) Parsons KS 2 hrs
KKAN(AM) Phillipsburg KS 10 hrs
KRSL(AM) Russell KS 2 hrs
KFRM(AM) Salina KS 5 hrs
KSAL(AM) Salina KS 3 hrs
KYUL(AM) Scott City KS 5 hrs
KMZA(FM) Seneca KS 7 hrs
KULY(AM) Ulysses KS 12 hrs
KLEY(AM) Wellington KS 10 hrs
WANY(AM) Albany KY 2 hrs
WBRT(AM) Bardstown KY 10 hrs
WKDZ(AM) Cadiz KY 2 hrs
WNES(AM) Central City KY 7 hrs
WAIN(AM) Columbia KY 2 hrs
WCPM(AM) Cumberland KY 1 hr
WHSX(FM) Edmonton KY 15 hrs
WFKN(AM) Franklin KY 6 hrs
WGRK(AM) Greensburg KY 5 hrs
WKCM(AM) Hawesville KY 3 hrs
WSON(AM) Henderson KY 2 hrs
WKMO(FM) Hodgenville KY 2 hrs
WHOP-FM Hopkinsville KY 15 hrs
WMJL-FM Marion KY 3 hrs
WQQR(FM) Mayfield KY 4 hrs
WFTM(AM) Maysville KY 6 hrs
WFTM-FM Maysville KY 6 hrs
WMIK(AM) Middlesboro KY 1 hr
WFLW(AM) Monticello KY 5 hrs
WMST(AM) Mt. Sterling KY 2 hrs
WBKR(FM) Owensboro KY 2 hrs
WVJS(AM) Owensboro KY 1 hr
WBIO(FM) Philpot KY 5 hrs
WVLE(FM) Scottsville KY 3 hrs
WKEQ(FM) Somerset KY 1 hr
WTLO(AM) Somerset KY 1 hr
WYSB(FM) Springfield KY 10 hrs
WRSL(AM) Stanford KY 4 hrs
WMTC(AM) Vancleve KY 1 hr
WMTC-FM Vancleve KY 2 hrs
WLKS(AM) West Liberty KY 5 hrs
KQLQ(FM) Columbia LA 3 hrs
KSIG(AM) Crowley LA 5 hrs
WFPR(AM) Hammond LA 1 hr
WTRI(AM) Brunswick MD 2 hrs
WSRY(AM) Elkton MD 1 hr
WJEJ(AM) Hagerstown MD 1 hr
WICO(AM) Salisbury MD 1 hr
WTTR(AM) Westminster MD 4 hrs
WABJ(AM) Adrian MI 7 hrs
WFYC(AM) Alma MI 4 hrs
WQBX(FM) Alma MI 4 hrs
WATZ(AM) Alpena MI 3 hrs
WLEW(AM) Bad Axe MI 4 hrs
WBCM(FM) Boyne City MI 1 hr
WKYO(AM) Caro MI 18 hrs
WTVB(AM) Coldwater MI 6 hrs
WDOW(AM) Dowagiac MI 7 hrs
WCHT(AM) Escanaba MI 1 hr
WGHN(AM) Grand Haven MI 5 hrs
WCSR(AM) Hillsdale MI 3 hrs
WHIT(AM) Hudsonville MI 3 hrs
WKZO(AM) Kalamazoo MI 10 hrs
WSGW(AM) Saginaw MI 10 hrs
WMLM(AM) Saint Louis MI 5 hrs
WMIC(AM) Sandusky MI 12 hrs
WTCM(AM) Traverse City MI 5 hrs
WPNW(AM) Zeeland MI 1 hr
KASM(AM) Albany MN 6 hrs
KATE(AM) Albert Lea MN 18 hrs
KXRA(AM) Alexandria MN 5 hrs
KKCQ-FM Bagley MN 10 hrs
KBMO(AM) Benson MN 10 hrs
KSCR-FM Benson MN 10 hrs
KBEW(AM) Blue Earth MN 15 hrs
KJLY(FM) Blue Earth MN 5 hrs
WGVY(FM) Cambridge MN 8 hrs
WFMP(FM) Coon Rapids MN
KROX(AM) Crookston MN 10 hrs
KDLM(AM) Detroit Lakes MN 1 hr
KCNN(AM) East Grand Forks MN 6 hrs
KFMC(FM) Fairmont MN 6 hrs
KBRF(AM) Fergus Falls MN 15 hrs
KKCQ(AM) Fosston MN 5 hrs
KDUZ(AM) Hutchinson MN 18 hrs
KKOJ(AM) Jackson MN 15 hrs
KLFD(AM) Litchfield MN 20 hrs
KLTF(AM) Little Falls MN 8 hrs
KEYL(AM) Long Prairie MN 5 hrs

KLQL(FM) Luverne MN 10 hrs
KLQP(FM) Madison MN 5 hrs
KDMA(AM) Montevideo MN 7 hrs
KMGM(FM) Montevideo MN 1 hr
KYMN(AM) Northfield MN 6 hrs
KDIO(AM) Ortonville MN 18 hrs
WCMP(AM) Pine City MN 6 hrs
WQPM(AM) Princeton MN 2 hrs
KROC(AM) Rochester MN 12 hrs
KIKV-FM Sauk Centre MN 20 hrs
KNSG(FM) Springfield MN 15 hrs
KNSP(AM) Staples MN 10 hrs
KSNR(FM) Thief River Falls MN 1 hr
KTRF(AM) Thief River Falls MN 12 hrs
KARL(FM) Tracy MN 15 hrs
KWAD(AM) Wadena MN 10 hrs
KDJS(AM) Willmar MN 5 hrs
KDJS-FM Willmar MN
KWLM(AM) Willmar MN 8 hrs
KAGE(AM) Winona MN
KWOA(AM) Worthington MN
KAAN(AM) Bethany MO 10 hrs
KWRT(AM) Boonville MO 5 hrs
KMAM(AM) Butler MO 15 hrs
KMOE(FM) Butler MO 15 hrs
KOZX(FM) Cabool MO 2 hrs
KATI(FM) California MO 5 hrs
KRLL(AM) California MO 5 hrs
KMRN(AM) Cameron MO 12 hrs
KDMO(AM) Carthage MO 12 hrs
KWKZ(FM) Charleston MO 1 hr
KDFN(AM) Doniphan MO 5 hrs
KREI(AM) Farmington MO 5 hrs
KTJJ(FM) Farmington MO 6 hrs
KUNQ(FM) Houston MO 2 hrs
KWOS(AM) Jefferson City MO 12 hrs
KBOA(AM) Kennett MO 5 hrs
KIRX(AM) Kirksville MO 10 hrs
KBNN(AM) Lebanon MO 8 hrs
KNIM(AM) Maryville MO 5 hrs
KMEM-FM Memphis MO 8 hrs
KWWR(FM) Mexico MO 4 hrs
KXEO(AM) Mexico MO 3 hrs
KMCR(FM) Montgomery City MO 2 hrs
KQYS(AM) Neosho MO 6 hrs
KNEM(AM) Nevada MO 1 hr
KNMO(FM) Nevada MO 1 hr
KBDZ(FM) Perryville MO 2 hrs
KZNN(FM) Rolla MO 3 hrs
KFEQ(AM) Saint Joseph MO 20 hrs
KSMO(AM) Salem MO 18 hrs
KDRO(AM) Sedalia MO 6 hrs
KBXB(FM) Sikeston MO 3 hrs
KRHW(AM) Sikeston MO 6 hrs
KSIM(AM) Sikeston MO 12 hrs
KTTS-FM Springfield MO 5 hrs
KWTO(AM) Springfield MO 20 hrs
KRWP(FM) Stockton MO 8 hrs
KSAR(FM) Thayer MO 4 hrs
KTKS(FM) Versailles MO 2 hrs
KWRE(AM) Warrenton MO 1 hr
KUKU-FM Willow Springs MO 6 hrs
WCKK(FM) Carthage MS
WBAQ(FM) Greenville MS 1 hr
WROA(AM) Gulfport MS 1 hr
WTCD(FM) Indianola MS 5 hrs
WJDX(AM) Jackson MS 2 hrs
WMSI(FM) Jackson MS 1 hr
WIQQ(FM) Leland MS 6 hrs
WRQO(FM) Monticello MS 1 hr
WWMS(FM) Oxford MS 2 hrs
WHOC(AM) Philadelphia MS 2 hrs
WRJW(AM) Picayune MS 6 hrs
WELO(AM) Tupelo MS 1 hr
WTYL(AM) Tylertown MS 6 hrs
WJNS-FM Yazoo City MS 16 hrs
KFLN(AM) Baker MT 10 hrs
KBOW(AM) Butte MT 5 hrs
KBCK(AM) Deer Lodge MT 2 hrs
KXGN(AM) Glendive MT 2 hrs
KXGF(AM) Great Falls MT 2 hrs
KOJM(AM) Havre MT 4 hrs
KPQX(FM) Havre MT 10 hrs
*KXEI(FM) Havre MT 1 hr
KXLO(AM) Lewistown MT
KYUS-FM Miles City MT 1 hr
KATQ(AM) Plentywood MT 5 hrs
KATQ-FM Plentywood MT 5 hrs
KCGM(FM) Scobey MT 6 hrs
KSEN(AM) Shelby MT 8 hrs
KZIN-FM Shelby MT 4 hrs
KVCK(AM) Wolf Point MT 6 hrs
KVCK-FM Wolf Point MT 6 hrs
WQDK(FM) Ahoskie NC 7 hrs
WKXR(AM) Asheboro NC 1 hr
WWNC(AM) Asheville NC 1 hr
WGAI(AM) Elizabeth City NC 3 hrs
WFMO(AM) Fairmont NC 5 hrs
WCXL(FM) Kill Devil Hills NC 2 hrs
WKTE(AM) King NC 2 hrs
WHBK(AM) Marshall NC 3 hrs
WPAQ(AM) Mount Airy NC 1 hr
WWDR(AM) Murfreesboro NC 10 hrs
WCVP(AM) Murphy NC 3 hrs
WCBQ(AM) Oxford NC
WPTF(AM) Raleigh NC 10 hrs
WTEL(AM) Red Springs NC 5 hrs
WPTM(FM) Roanoke Rapids NC 15 hrs
WEGG(AM) Rose Hill NC 9 hrs
WKRX(FM) Roxboro NC 5 hrs
WRXO(AM) Roxboro NC 5 hrs
WWGP(AM) Sanford NC 7 hrs
WYAL(AM) Scotland Neck NC 2 hrs
WMPM(AM) Smithfield NC 2 hrs
WTAB(AM) Tabor City NC 12 hrs
WADE(AM) Wadesboro NC 1 hr
WNMX-FM Waxhaw NC
WKSK(AM) West Jefferson NC 3 hrs
WENC(AM) Whiteville NC 5 hrs
WTXY(AM) Whiteville NC 2 hrs
KBMR(AM) Bismarck ND 4 hrs
KPOK(AM) Bowman ND 2 hrs
*KMHA(FM) Four Bears ND 1 hr
KAUJ(FM) Grafton ND 18 hrs
KXPO(AM) Grafton ND 18 hrs
KNDC(AM) Hettinger ND
KQDJ(AM) Jamestown ND 6 hrs
KSJB(AM) Jamestown ND 12 hrs
KNDK(AM) Langdon ND 12 hrs
KQLX(AM) Lisbon ND 14 hrs
KCJB(AM) Minot ND 4 hrs
KHRT(AM) Minot ND 1 hr
KZPR(FM) Minot ND 4 hrs
KDDR(AM) Oakes ND 15 hrs
KUSO(FM) Albion NE 10 hrs
KAAQ(FM) Alliance NE 4 hrs
KCOW(AM) Alliance NE 18 hrs
KWBE(AM) Beatrice NE 14 hrs
KCNI(AM) Broken Bow NE
KZEN(FM) Central City NE 20 hrs
KCSR(AM) Chadron NE 6 hrs
KQSK(FM) Chadron NE 4 hrs
KKOT(FM) Columbus NE 8 hrs
KTTT(AM) Columbus NE 5 hrs
KGMT(AM) Fairbury NE 18 hrs
KHUB(AM) Fremont NE 6 hrs
KHAS(AM) Hastings NE 2 hrs
KMTY(FM) Holdrege NE 5 hrs
KUVR(AM) Holdrege NE 5 hrs
KGFW(AM) Kearney NE 8 hrs
KRVN-FM Lexington NE 10 hrs
KICX-FM McCook NE 6 hrs
KIOD(FM) McCook NE 6 hrs
KNCY(AM) Nebraska City NE 3 hrs
KNEN(FM) Norfolk NE 10 hrs
KMCX(FM) Ogallala NE 2 hrs
KOGA(AM) Ogallala NE 10 hrs
KFAB(AM) Omaha NE 5 hrs
KBRX(AM) O'Neill NE 12 hrs
KNEB(AM) Scottsbluff NE 18 hrs
KNEB-FM Scottsbluff NE 18 hrs
KOLT(AM) Scottsbluff NE 5 hrs
KSID(AM) Sidney NE 5 hrs
KRFS(AM) Superior NE 5 hrs
KRFS-FM Superior NE
KOAQ(AM) Terrytown NE 5 hrs
KTCH(AM) Wayne NE 20 hrs
KTIC-FM West Point NE 10 hrs
KAWL(AM) York NE 7 hrs
*WNEC-FM Henniker NH 4 hrs
WSNJ(AM) Bridgeton NJ 10 hrs
WFAI(AM) Salem NJ 8 hrs
KRSY(AM) Alamogordo NM 1 hr
KICA(AM) Clovis NM 5 hrs
KOTS(AM) Deming NM 5 hrs
KSEL-FM Portales NM 4 hrs
KMXQ(FM) Socorro NM 2 hrs
KWNA(AM) Winnemucca NV 2 hrs
WABH(AM) Bath NY 1 hr
WVIN-FM Bath NY 1 hr
WBRV-FM Boonville NY 3 hrs
WKGB-FM Conklin NY 1 hr
WWLE(AM) Cornwall NY 1 hr
WRWD-FM Highland NY 1 hr
WHHO(AM) Hornell NY
WJTN(AM) Jamestown NY 1 hr
WIXT(AM) Little Falls NY 1 hr
WLVL(AM) Lockport NY 6 hrs
WLLG(FM) Lowville NY 3 hrs
WICY(AM) Malone NY 1 hr
WYBG(AM) Massena NY
WALL(AM) Middletown NY 1 hr
WACK(AM) Newark NY 5 hrs
WEOK(AM) Poughkeepsie NY 2 hrs
WRIV(AM) Riverhead NY 8 hrs
WUAM(AM) Saratoga Springs NY 1 hr
WSPQ(AM) Springville NY 5 hrs
WIPS(AM) Ticonderoga NY 6 hrs
WIBX(AM) Utica NY 14 hrs
WDLA(AM) Walton NY 2 hrs
WCJW(AM) Warsaw NY 11 hrs
WTNY(AM) Watertown NY 3 hrs
WNCO(AM) Ashland OH 3 hrs
WAIS(AM) Buchtel OH 5 hrs
WBCO(AM) Bucyrus OH 4 hrs
WCER(AM) Canton OH 6 hrs
WHBC(AM) Canton OH 1 hr
WDOH(FM) Delphos OH 8 hrs
WEDI(AM) Eaton OH 2 hrs
WFIN(AM) Findlay OH 7 hrs
WAXZ(FM) Georgetown OH 10 hrs
WKTN(FM) Kenton OH 2 hrs
WLOH(AM) Lancaster OH 1 hr
WIMA(AM) Lima OH 5 hrs
WIMT(FM) Lima OH 5 hrs
WMOA(AM) Marietta OH 1 hr
WMRN(AM) Marion OH 5 hrs
WMPO(AM) Middleport-Pomeroy OH 1 hr
WSEO(FM) Nelsonville OH
WHOF(FM) North Canton OH 2 hrs
WLKR-FM Norwalk OH 3 hrs
WOBL(AM) Oberlin OH 5 hrs
WBUK(FM) Ottawa OH 5 hrs
WPTW(AM) Piqua OH 6 hrs
WNTO(FM) Racine OH 1 hr
WSOM(AM) Salem OH 2 hrs
WMVR-FM Sidney OH 10 hrs
*WEEC(FM) Springfield OH 1 hr
WCKY-FM Tiffin OH 3 hrs
WTTF(AM) Tiffin OH 3 hrs
WSPD(AM) Toledo OH 3 hrs
WTUZ(FM) Uhrichsville OH 2 hrs
WYNT(FM) Upper Sandusky OH 3 hrs
WCHO(AM) Washington Court House OH 5 hrs
WRAC(FM) West Union OH 10 hrs
WKFI(AM) Wilmington OH 2 hrs
WQKT(FM) Wooster OH 2 hrs
WBZI(AM) Xenia OH 2 hrs
WHIZ(AM) Zanesville OH 2 hrs
KEYB(FM) Altus OK 2 hrs
KKBI(FM) Broken Bow OK 5 hrs
KZUE(AM) El Reno OK 7 hrs
KGWA(AM) Enid OK 3 hrs
KTAT(AM) Frederick OK 2 hrs
KIHN(AM) Hugo OK 1 hr
KMAD(AM) Madill OK 2 hrs
KYAL-FM Muskogee OK 5 hrs
KPNC-FM Ponca City OK 5 hrs
KCXR(FM) Taft OK 1 hr
KFAQ(AM) Tulsa OK 5 hrs
KBKR(AM) Baker City OR 2 hrs
KZZR(AM) Burns OR 6 hrs
KWVR(AM) Enterprise OR 4 hrs
KWVR-FM Enterprise OR 4 hrs
KAGO(AM) Klamath Falls OR 3 hrs
KLAD(AM) Klamath Falls OR 1 hr
KLBM(AM) La Grande OR 2 hrs
KSRV(AM) Ontario OR 15 hrs
KUMA(AM) Pendleton OR 10 hrs
KCKX(AM) Stayton OR 15 hrs
KACI(AM) The Dalles OR 1 hr
KODL(AM) The Dalles OR 4 hrs
KLWJ(AM) Umatilla OR 1 hr
WHLM(AM) Bloomsburg PA 1 hr
WCOJ(AM) Coatesville PA 2 hrs
WFRM(AM) Coudersport PA 2 hrs
WDAC(FM) Lancaster PA 4 hrs
WJSM(AM) Martinsburg PA 1 hr
*WPEL(AM) Montrose PA 1 hr
WATS(AM) Sayre PA 1 hr
WSBA(AM) York PA 4 hrs
WSOL(AM) San German PR 2 hrs
WVCD(AM) Bamberg-Denmark SC 1 hr
WOLS(AM) Florence SC 1 hr
WBZF(FM) Hartsville SC 3 hrs
WHSC(AM) Hartsville SC 3 hrs
WJBS(AM) Holly Hill SC 2 hrs
WJAY(AM) Mullins SC 10 hrs
WQMC(AM) Sumter SC 2 hrs
KGIM(AM) Aberdeen SD 12 hrs
KSDN(AM) Aberdeen SD 15 hrs
KBFS(AM) Belle Fourche SD 20 hrs
KBRK(AM) Brookings SD 9 hrs
KZMX(AM) Hot Springs SD 6 hrs
KJAM(AM) Madison SD 11 hrs
KMSD(AM) Milbank SD 6 hrs
KMIT(FM) Mitchell SD 18 hrs
KORN(AM) Mitchell SD 10 hrs
KOLY(AM) Mobridge SD
KIMM(AM) Rapid City SD 1 hr
KTOQ(AM) Rapid City SD 2 hrs
KPLO-FM Reliance SD 5 hrs
KSDR-FM Watertown SD 8 hrs
KKYA(FM) Yankton SD 5 hrs
KYNT(AM) Yankton SD 5 hrs
WNAX(AM) Yankton SD 35 hrs
WLAR(AM) Athens TN 2 hrs
*WHCB(FM) Bristol TN 1 hr
WNKX(AM) Centerville TN 1 hr
WNKX-FM Centerville TN
WMCH(AM) Church Hill TN 1 hr
WMCP(AM) Columbia TN 4 hrs
WZYX(AM) Cowan TN 2 hrs
WEKR(AM) Fayetteville TN 1 hr
WHIN(AM) Gallatin TN 5 hrs
WMYL(FM) Halls Crossroads TN 2 hrs
WDXI(AM) Jackson TN 12 hrs
WCLC(AM) Jamestown TN 2 hrs
WDEB(AM) Jamestown TN 3 hrs
WJFC(AM) Jefferson City TN 1 hr
WBUZ(FM) La Vergne TN 1 hr
WEEN(AM) Lafayette TN 5 hrs
WKZX-FM Lenoir City TN 4 hrs
WLIL(AM) Lenoir City TN 4 hrs
WLIV(AM) Livingston TN 2 hrs
WMSR(AM) Manchester TN
WAKI(AM) McMinnville TN 2 hrs
WBMC(AM) McMinnville TN 5 hrs
WREC(AM) Memphis TN 3 hrs
WSM(AM) Nashville TN 6 hrs
*WDNX(FM) Olive Hill TN 1 hr
WMUF(AM) Paris TN 2 hrs
WUAT(AM) Pikeville TN 5 hrs
WLIJ(AM) Shelbyville TN 3 hrs
WDBL(AM) Springfield TN 10 hrs
WCDT(AM) Winchester TN 15 hrs
KYYW(AM) Abilene TX 3 hrs
KGNC(AM) Amarillo TX 11 hrs
KFYN(AM) Bonham TX 6 hrs
KNEL-FM Brady TX 4 hrs
KWHI(AM) Brenham TX 3 hrs
KOXE(FM) Brownwood TX 3 hrs
KCAR(AM) Clarksville TX 2 hrs
KQBZ(FM) Coleman TX 14 hrs
KSTA(AM) Coleman TX 14 hrs
KZNE(AM) College Station TX 10 hrs
KVMC(AM) Colorado City TX 8 hrs
KXIT(AM) Dalhart TX 7 hrs
KDDD-FM Dumas TX 5 hrs
KURV(AM) Edinburg TX 10 hrs
KNES(FM) Fairfield TX 3 hrs
KMUL(AM) Farwell TX 3 hrs
WBAP(AM) Fort Worth TX 6 hrs
KNAF(AM) Fredericksburg TX 5 hrs
KGAF(AM) Gainesville TX 2 hrs
KCTI(AM) Gonzales TX 5 hrs
KPIR(AM) Granbury TX 1 hr
KGVL(AM) Greenville TX 5 hrs
KHLT(AM) Hallettsville TX 5 hrs
KVRP-FM Haskell TX 5 hrs
KWRD(AM) Henderson TX 5 hrs
KPAN(AM) Hereford TX 12 hrs
KEBE(AM) Jacksonville TX 9 hrs
KOOI-FM Jacksonville TX 9 hrs
KMBL(AM) Junction TX 6 hrs
KVLG(AM) La Grange TX 4 hrs
KCYL(AM) Lampasas TX 5 hrs
KZZN(AM) Littlefield TX 7 hrs
KFYO(AM) Lubbock TX 15 hrs
KCUL(AM) Marshall TX 3 hrs
KRQX(AM) Mexia TX 12 hrs
KSFA(AM) Nacogdoches TX 7 hrs
KNET(AM) Palestine TX 6 hrs
KOMX(FM) Pampa TX 10 hrs
KBUS(FM) Paris TX 6 hrs
KEYE(AM) Perryton TX 2 hrs
KEYE-FM Perryton TX 2 hrs
KKYN-FM Plainview TX 12 hrs
KVOP(AM) Plainview TX 12 hrs
KITE(FM) Port Lavaca TX 1 hr
KWFB(FM) Quanah TX 3 hrs
KGKL(AM) San Angelo TX 6 hrs
KBAL(AM) San Saba TX 2 hrs
KBAL-FM San Saba TX 2 hrs
KWED(AM) Seguin TX 6 hrs
KIKZ(AM) Seminole TX 5 hrs
KSEM-FM Seminole TX 5 hrs
KBKH(FM) Shamrock TX 25 hrs
KDAE(AM) Sinton TX 6 hrs
KSTV(AM) Stephenville TX 5 hrs
KXOX(AM) Sweetwater TX 5 hrs
KVOU(AM) Uvalde TX 12 hrs
KBEC(AM) Waxahachie TX 3 hrs
KWUD(AM) Woodville TX 02 hrs
KSUB(AM) Cedar City UT 6 hrs
KNAK(AM) Delta UT 5 hrs
KVNU(AM) Logan UT 2 hrs
KMTI(AM) Manti UT 5 hrs
KOAL(AM) Price UT 5 hrs
KSVC(AM) Richfield UT 1 hr
KHQN(AM) Spanish Fork UT 4 hrs
KVEL(AM) Vernal UT 2 hrs
WBTX(AM) Broadway-Timberville VA 1 hr
WDIC(AM) Clinchco VA 1 hr
WPZZ(FM) Crewe VA 10 hrs
WEVA(AM) Emporia VA one hrs
WPAK(AM) Farmville VA 1 hr
WWWJ(AM) Galax VA 1 hr
WXGM(AM) Gloucester VA 2 hrs
WXGM-FM Gloucester VA 2 hrs
WMNA(AM) Gretna VA 8 hrs
WNRG(AM) Grundy VA 5 hrs
WSVA(AM) Harrisonburg VA 8 hrs
WHHV(AM) Hillsville VA 1 hr
WKWI(FM) Kilmarnock VA 2 hrs
WOJL(FM) Louisa VA 2 hrs
WRVA(AM) Richmond VA 3 hrs
WXLZ(AM) Saint Paul VA 2 hrs
WKGM(AM) Smithfield VA 02 hrs
WSBV(AM) South Boston VA 1 hr
WKDW(AM) Staunton VA 1 hr
WHEO(AM) Stuart VA 4 hrs
WKCW(AM) Warrenton VA 1 hr
WKCI(AM) Waynesboro VA 2 hrs
WZLF(FM) Bellows Falls VT 1 hr
KNBQ(FM) Centralia WA 1 hr
KOZI(AM) Chelan WA 3 hrs
KCLX(AM) Colfax WA 5 hrs
KYSN(FM) East Wenatchee WA 1 hr
KTBI(AM) Ephrata WA 15 hrs
KULE(AM) Ephrata WA 1 hr
KULE-FM Ephrata WA 5 hrs
KONA(AM) Kennewick WA 3 hrs
KBRC(AM) Mount Vernon WA 5 hrs
KNCW(FM) Omak WA 1 hr

KOMW(AM) Omak WA 2 hrs
KQQQ(AM) Pullman WA 3 hrs
KWNC(AM) Quincy WA 5 hrs
KGA(AM) Spokane WA 1 hr
KQNT(AM) Spokane WA 6 hrs
KTEL(AM) Walla Walla WA 5 hrs
KPQ(AM) Wenatchee WA 3 hrs
KIT(AM) Yakima WA 6 hrs
WBEV(AM) Beaver Dam WI 8 hrs
WXRO(FM) Beaver Dam WI 6 hrs
WFCL(AM) Clintonville WI 8 hrs
WAXX(FM) Eau Claire WI 15 hrs
WAYY(AM) Eau Claire WI 5 hrs
KFIZ(AM) Fond du Lac WI 10 hrs
WTAQ(AM) Green Bay WI 5 hrs
WAUN(FM) Kewaunee WI 10 hrs
WKTY(AM) La Crosse WI 5 hrs
WGLR(AM) Lancaster WI 10 hrs
WTSO(AM) Madison WI 20 hrs
WMAM(AM) Marinette WI 3 hrs
WDLB(AM) Marshfield WI 14 hrs
WMEQ(AM) Menomonie WI 15 hrs
WJMT(AM) Merrill WI 7 hrs
WCCN(AM) Neillsville WI 19 hrs
WPKR(FM) Omro WI 1 hr
WPVL(AM) Platteville WI 12 hrs
WJUB(AM) Plymouth WI 5 hrs
WPDR(AM) Portage WI 6 hrs
WNFM(FM) Reedsburg WI 18 hrs
WRCO(AM) Richland Center WI 2 hrs
WRCO-FM Richland Center WI 18 hrs
WEVR(AM) River Falls WI 18 hrs
*WRFW(FM) River Falls WI 10 hrs
WTCH(AM) Shawano WI 21 hrs
WHBL(AM) Sheboygan WI 10 hrs
WCSW(AM) Shell Lake WI 3 hrs
WOSQ(FM) Spencer WI 8 hrs
WDOR-FM Sturgeon Bay WI 5 hrs
WVRQ(AM) Viroqua WI 1 hr
WVRQ-FM Viroqua WI 1 hr
WDUX(AM) Waupaca WI 6 hrs
WFDL(AM) Waupun WI 5 hrs
WDEZ(FM) Wausau WI 4 hrs
WSAU(AM) Wausau WI 5 hrs
WHTL-FM Whitehall WI 5 hrs
WELD(AM) Fisher WV 3 hrs
*WVMR(AM) Frost WV 5 hrs
WKCJ(FM) Lewisburg WV 3 hrs
WGGE(FM) Parkersburg WV 2 hrs
WWVA(AM) Wheeling WV 2 hrs
KFBC(AM) Cheyenne WY 1 hr
KMER(AM) Kemmerer WY 3 hrs
KCGY(FM) Laramie WY 1 hr
KHAT(AM) Laramie WY 1 hr
KASL(AM) Newcastle WY 5 hrs
KPOW(AM) Powell WY 19 hrs
KTAK(FM) Riverton WY 5 hrs
KVOW(AM) Riverton WY 5 hrs
KYCN(AM) Wheatland WY 3 hrs
KZEW(FM) Wheatland WY 7 hrs

Alternative

KWKM(FM) Saint Johns AZ 3 hrs
*KALW(FM) San Francisco CA 4 hrs
KMBY-FM Seaside CA 4 hrs
KFMU-FM Oak Creek CO 4 hrs
WRUF-FM Gainesville FL 6 hrs
WXXL(FM) Tavares FL 6 hrs
KECH-FM Sun Valley ID 5 hrs
*WQUB(FM) Quincy IL 12 hrs
KIND-FM Independence KS 4 hrs
WFRD(FM) Hanover NH 1 hr
KAGM(FM) Los Lunas NM 18 hrs
KRNG(FM) Fallon/ Reno NV
*WGCC-FM Batavia NY
KSPI-FM Stillwater OK
WFBC-FM Greenville SC 2 hrs
WRLT(FM) Franklin TN 1 hr
KWKQ(FM) Graham TX 10 hrs
*KRLF(FM) Pullman WA 1 hr
*KYVT(FM) Yakima WA 3 hrs
*KUWS(FM) Superior WI 16 hrs

American Indian

*KDLG(AM) Dillingham AK 1 hr
*KIYU(AM) Galena AK 2 hrs
KCAM(AM) Glennallen AK 1 hr
*KTOO(FM) Juneau AK 1 hr
*KDLL(FM) Kenai AK 10 hrs
KIAM(AM) Nenana AK 3 hrs
*KJNP(AM) North Pole AK 2 hrs
WNSI-FM Atmore AL 1 hr
*KABF(FM) Little Rock AR 2 hrs
KRDE(FM) Globe AZ 6 hrs
*KUAZ(AM) Tucson AZ 1 hr
*KXCI(FM) Tucson AZ 2 hrs
*KNNB(FM) Whiteriver AZ 8 hrs
KTNN(AM) Window Rock AZ
*KZFR(FM) Chico CA 2 hrs
*KFCF(FM) Fresno CA 2 hrs
*KMUD(FM) Garberville CA 1 hr
*KIDE(FM) Hoopa CA 20 hrs
*KCSB-FM Santa Barbara CA 3 hrs
KRTZ(FM) Cortez CO 1 hr
*KSUT(FM) Ignacio CO 7 hrs
*WUCF-FM Orlando FL 2 hrs
WKHC(FM) Dahlonega GA 1 hr
*WBCX(FM) Gainesville GA 4 hrs
KWIK(AM) Pocatello ID 1 hr
*WUEV(FM) Evansville IN 1 hr
*WHFC(FM) Bel Air MD 3 hrs
*WUPI(FM) Presque Isle ME 2 hrs
WCUP(FM) L'Anse MI 2 hrs
*WLNZ(FM) Lansing MI 3. hrs
*WNMC-FM Traverse City MI 1 hr
*KBSB(FM) Bemidji MN 3 hrs
*KKFI(FM) Kansas City MO 2 hrs
*WWCU(FM) Cullowhee NC 5. hrs
*KEYA(FM) Belcourt ND 9 hrs
*KCND(FM) Bismarck ND 2 hrs
*KDPR(FM) Dickinson ND 2 hrs
*KABU(FM) Fort Totten ND 19 hrs
*KMHA(FM) Four Bears ND 4 hrs
*KPRJ(FM) Jamestown ND 2 hrs
*KMPR(FM) Minot ND 2 hrs
*KPPR(FM) Williston ND 2 hrs
*KINI(FM) Crookston NE 15 hrs
*WNEC-FM Henniker NH 18 hrs
*KABR(AM) Alamo Community NM 10 hrs
KPCL(FM) Farmington NM 7 hrs
*KGLP(FM) Gallup NM 10 hrs
KGLX(FM) Gallup NM 3 hrs
KYVA-FM Grants NM 10 hrs
*KTDB(FM) Ramah NM 8 hrs
*KSFR(FM) Santa Fe NM 4 hrs
*KSHI(FM) Zuni NM 20 hrs
WYBG(AM) Massena NY
*WBAI(FM) New York NY 1 hr
WNTO(FM) Racine OH 1 hr
KVSP(FM) Anadarko OK 1 hr
KZUE(AM) El Reno OK 1 hr
*KOSN(FM) Ketchum OK 1 hr
KIRC(FM) Seminole OK 1 hr
*KOSU(FM) Stillwater OK 1 hr
KWSH(AM) Wewoka OK 1 hr
*KMUN(FM) Astoria OR 2 hrs
*KTEC(FM) Klamath Falls OR 1 hr
KOLY(AM) Mobridge SD
KOLY-FM Mobridge SD 1 hr
*KBHE-FM Rapid City SD 1 hr
*KTSD-FM Reliance SD 1 hr
KBHB(AM) Sturgis SD 1 hr
*KAOR(FM) Vermillion SD 2 hrs
*KUSD(FM) Vermillion SD 1 hr
WLIL(AM) Lenoir City TN 1 hr
*WRVU(FM) Nashville TN 2 hrs
*KOOP(FM) Hornsby TX 1 hr
*KZMU(FM) Moab UT 5 hrs
*KRCL(FM) Salt Lake City UT 4 hrs
*WUVT-FM Blacksburg VA 2 hrs
*KSER(FM) Everett WA 2 hrs
*KAOS(FM) Olympia WA 3 hrs
*KSFC(FM) Spokane WA 5 hrs
KYNR(AM) Toppenish WA 20 hrs
*WPNE-FM Green Bay WI 2. hrs
*WOJB(FM) Reserve WI 15 hrs
*WHND(FM) Sister Bay WI 2 hrs

Arabic

WPNA(AM) Oak Park IL 2 hrs
KLAV(AM) Las Vegas NV 7 hrs
*WDIY(FM) Allentown PA 1 hr
*WMUH(FM) Allentown PA 2 hrs
KARI(AM) Blaine WA 1 hr

Armenian

KTYM(AM) Inglewood CA 2 hrs
*KUSF(FM) San Francisco CA 1 hr
*WJCU(FM) University Heights OH 2 hrs

Beautiful Music

*KNOG(FM) Nogales AZ 5 hrs
WMDJ-FM Allen KY
WMRX-FM Beaverton MI 2 hrs
KHND(AM) Harvey ND 3 hrs
WALK-FM Patchogue NY
KNNK(FM) Dimmitt TX 20 hrs
*KZAZ(FM) Bellingham WA 7 hrs

Big Band

*KASU(FM) Jonesboro AR 2 hrs
*WGRS(FM) Guilford CT 8 hrs
WQUN(AM) Hamden CT 4 hrs
*WMNR(FM) Monroe CT 8 hrs
*WRXC(FM) Shelton CT 8 hrs
*WGSK(FM) South Kent CT 8 hrs
WTAN(AM) Clearwater FL 40 hrs
KROS(AM) Clinton IA 3 hrs
*WSIU(FM) Carbondale IL 4 hrs
WLBK(AM) De Kalb IL 5 hrs
WHPO(FM) Hoopeston IL 2 hrs
WTAY(AM) Robinson IL 10 hrs
WTYE(FM) Robinson IL 10 hrs
WMAY(AM) Springfield IL 5 hrs
WXNT(AM) Indianapolis IN 2 hrs
WAWK(AM) Kendallville IN 2 hrs
KIND(AM) Independence KS 2 hrs
KLKC(AM) Parsons KS 3 hrs
KVCL(AM) Winnfield LA 4 hrs
*WESM(FM) Princess Anne MD 10 hrs
*WKHS(FM) Worton MD 2 hrs
*WHFR(FM) Dearborn MI 6 hrs
WBNZ(FM) Frankfort MI 2 hrs
*WLNZ(FM) Lansing MI 3 hrs
WIOS(AM) Tawas City MI 6 hrs
KYMN(AM) Northfield MN 3 hrs
KOLV(FM) Olivia MN
*KXMS(FM) Joplin MO 2 hrs
KPRK(AM) Livingston MT 4 hrs
WXIT(AM) Blowing Rock NC 4 hrs
*WCCE(FM) Buie's Creek NC 4 hrs
*WZRU(FM) Roanoke Rapids NC 4 hrs
KMTY(FM) Holdrege NE 5 hrs
KUVR(AM) Holdrege NE 5 hrs
WSNJ(AM) Bridgeton NJ
*WNTI(FM) Hackettstown NJ 4 hrs
KRSY(AM) Alamogordo NM 6 hrs
KRSN(AM) Los Alamos NM 8 hrs
WDNY(AM) Dansville NY 3 hrs
WDNY-FM Dansville NY 3 hrs
*WNYC(AM) New York NY 2 hrs
WDOS(AM) Oneonta NY 7 hrs
WRGR(FM) Tupper Lake NY 2 hrs
WNYV(FM) Whitehall NY 3 hrs
WATH(AM) Athens OH 15 hrs
WOHI(AM) East Liverpool OH 4 hrs
WBBZ(AM) Ponca City OK 3 hrs
KBEZ(FM) Tulsa OK 5 hrs
WNPV(AM) Lansdale PA 3 hrs
WLSH(AM) Lansford PA 4 hrs
WPHB(AM) Philipsburg PA 5 hrs
WCRI(FM) Block Island RI 4 hrs
WJMX(AM) Florence SC 3 hrs
WBZF(FM) Hartsville SC 3 hrs
WHSC(AM) Hartsville SC 3 hrs
WMYB(FM) Myrtle Beach SC
*KNCT-FM Killeen TX 6 hrs
WKSI-FM Stephens City VA 2 hrs
WVNR(AM) Poultney VT 3 hrs
KELA(AM) Centralia-Chehalis WA 3 hrs
KEYG(AM) Grand Coulee WA 4 hrs
*WVMR(AM) Frost WV 3 hrs
*WWVU-FM Morgantown WV 2 hrs

Black

*KSUA(FM) Fairbanks AK 8 hrs
KIAL(AM) Unalaska AK 04 hrs
WHMA-FM Ashland AL 6 hrs
WNSI-FM Atmore AL 1 hr
WULA(AM) Eufaula AL 3 hrs
WGYV(AM) Greenville AL 6 hrs
*WJAB(FM) Huntsville AL 3 hrs
WMGY(AM) Montgomery AL 15 hrs
*WVAS(FM) Montgomery AL 5 hrs
WOPP(AM) Opp AL 4 hrs
WKAX(AM) Russellville AL 4 hrs
WHBB(AM) Selma AL 18 hrs
WTRB-FM Sylacauga AL 6 hrs
KAMD-FM Camden AR 5 hrs
*KUAF(FM) Fayetteville AR 5 hrs
KFFA(AM) Helena AR 10 hrs
KOSE(AM) Wilson AR 6 hrs
KDVA(FM) Buckeye AZ 6 hrs
KTKT(AM) Tucson AZ 1 hr
*KXCI(FM) Tucson AZ 4 hrs
*KPFA(FM) Berkeley CA 18 hrs
*KMUD(FM) Garberville CA 3 hrs
*KXLU(FM) Los Angeles CA 10 hrs
*KMPO(FM) Modesto CA 3 hrs
*KVMR(FM) Nevada City CA 4 hrs
*KSPB(FM) Pebble Beach CA 18 hrs
*KZYX(FM) Philo CA 6 hrs
*KUCR(FM) Riverside CA 18 hrs
*KYCC(FM) Stockton CA 6 hrs
KDIA(AM) Vallejo CA 2 hrs
KDYA(AM) Vallejo CA 2 hrs
*KGNU-FM Boulder CO 7 hrs
KLDC(AM) Brighton CO 2 hrs
*KCSU-FM Fort Collins CO 3 hrs
*KAFM(FM) Grand Junction CO 3 hrs
*KMSA(FM) Grand Junction CO 6 hrs
*WPKN(FM) Bridgeport CT 4 hrs
WFIF(AM) Milford CT 6 hrs
*WCNI(FM) New London CT 3 hrs
*WVUD(FM) Newark DE 10 hrs
WPHK(FM) Blountstown FL 12 hrs
*WVUM(FM) Coral Gables FL 7 hrs
WAVS(AM) Davie FL
WRUF(AM) Gainesville FL 4 hrs
*WUFT-FM Gainesville FL 4 hrs
WGWD(FM) Gretna FL 20 hrs
WLBE(AM) Leesburg FL 3 hrs
WMEL(AM) Melbourne FL 2 hrs
WOCA(AM) Ocala FL 2 hrs
*WKGC-FM Panama City FL 6 hrs
WPRY(AM) Perry FL 2 hrs
WMEN(AM) Royal Palm Beach FL 2 hrs
*WKPX(FM) Sunrise FL 3 hrs
*WVFS(FM) Tallahassee FL 8 hrs
*WBVM(FM) Tampa FL 4 hrs
WDEC-FM Americus GA 5 hrs
WRFC(AM) Athens GA 15 hrs
WMOG(AM) Brunswick GA 8 hrs
WGRA(AM) Cairo GA 6 hrs
WEBS(AM) Calhoun GA 2 hrs
WBTR-FM Carrollton GA 5 hrs
WDCO(AM) Cochran GA 6 hrs
WDXQ-FM Cochran GA 6 hrs
WCUG(AM) Cuthbert GA 4 hrs
WUFF(AM) Eastman GA 5 hrs
*WBCX(FM) Gainesville GA 12 hrs
WYIS(AM) McRae GA 4 hrs
WMVG(AM) Milledgeville GA 4 hrs
WJEP(AM) Ochlocknee GA 2 hrs
WTGA(AM) Thomaston GA 4 hrs
WLET(AM) Toccoa GA
KLNG(AM) Council Bluffs IA 6 hrs
*KUOI-FM Moscow ID 3 hrs
KATZ-FM Alton IL
*WESN(FM) Bloomington IL 18 hrs
*WEFT(FM) Champaign IL 8 hrs
*WVKC(FM) Galesburg IL 6 hrs
*WCSF(FM) Joliet IL 2 hrs
WJOL(AM) Joliet IL 1 hr
*WMXM(FM) Lake Forest IL 6 hrs
*WLRA(FM) Lockport IL 15 hrs
*WQNA(FM) Springfield IL 9 hrs
*WFYI-FM Indianapolis IN 5 hrs
WXFN(AM) Muncie IN 3 hrs
WWVR(FM) West Terre Haute IN 6 hrs
*KONQ(FM) Dodge City KS 10 hrs
KWBW(AM) Hutchinson KS 2 hrs
*KSDB-FM Manhattan KS 4 hrs
*WCVK(FM) Bowling Green KY 2 hrs
*WWHR(FM) Bowling Green KY 2 hrs
WCPM(AM) Cumberland KY 3 hrs
*WNKJ(FM) Hopkinsville KY 5.5 hrs
WFXY(AM) Middlesboro KY 2 hrs
WEKY(AM) Richmond KY 12 hrs
*KLSP(FM) Angola LA 10 hrs
KAJN-FM Crowley LA 2 hrs
KVPI-FM Ville Platte LA 10 hrs
KVCL(AM) Winnfield LA 3 hrs
KVCL-FM Winnfield LA 20 hrs
*WERS(FM) Boston MA 15 hrs
*WOMR(FM) Provincetown MA 6 hrs
*WBSL-FM Sheffield MA 2 hrs
WMAS(AM) Springfield MA 1 hr
*WYAJ(FM) Sudbury MA 6 hrs
*WCHC(FM) Worcester MA 8 hrs
WKHI(FM) Fruitland MD 5 hrs
*WUPI(FM) Presque Isle ME 2 hrs
*WMHB(FM) Waterville ME 8 hrs
WLQV(AM) Detroit MI 10 hrs
*WUPX(FM) Marquette MI 6 hrs
WHLS(AM) Port Huron MI 1 hr
WSJM(AM) Saint Joseph MI 5 hrs
*WNMC-FM Traverse City MI 20 hrs
WMFN(AM) Zeeland MI 2hrs hrs
*KFAI(FM) Minneapolis MN 10 hrs
KNOF(FM) Saint Paul MN 5 hrs
KMFC(FM) Centralia MO 3 hrs
*KOPN(FM) Columbia MO 10 hrs
*KCFV(FM) Ferguson MO 4 hrs
KLSC(FM) Malden MO 6 hrs
*KMVC(FM) Marshall MO 10 hrs
KLID(AM) Poplar Bluff MO 2 hrs
KDRO(AM) Sedalia MO 1 hr
WBLE(FM) Batesville MS 3 hrs
WELZ(AM) Belzoni MS 20 hrs
WCJU(AM) Columbia MS 112 hrs
WKCU(AM) Corinth MS 2 hrs
WABG(AM) Greenwood MS 4 hrs
WGRM(AM) Greenwood MS 2 hrs
WNBN(AM) Meridian MS
WRJW(AM) Picayune MS 8 hrs
WJDR(FM) Prentiss MS 5 hrs
WSSO(AM) Starkville MS 12 hrs
*KGLT(FM) Bozeman MT 1 hr
WFGW(AM) Black Mountain NC 1 hr
WRRZ(AM) Clinton NC 5 hrs
*WNCU(FM) Durham NC
WGAI(AM) Elizabeth City NC 4 hrs
WSML(AM) Graham NC 18 hrs
WYRN(AM) Louisburg NC
WHIP(AM) Mooresville NC 6 hrs
WDJS(AM) Mount Olive NC 5 hrs
WNNC(AM) Newton NC 2 hrs
WRXO(AM) Roxboro NC 4 hrs
WTSB(AM) Selma NC 6 hrs
*KRNU(FM) Lincoln NE 2 hrs
*KZUM(FM) Lincoln NE 13 hrs
*KIOS-FM Omaha NE 2 hrs
*KWSC(FM) Wayne NE 4 hrs
*WUNH(FM) Durham NH 4 hrs
*WGLS-FM Glassboro NJ 10 hrs
*WRRC(FM) Lawrenceville NJ 10 hrs
*WKNJ-FM Union Township NJ 2 hrs
*WMSC(FM) Upper Montclair NJ 4 hrs
KKIM(AM) Albuquerque NM 2 hrs
*WXBA(FM) Brentwood NY 5 hrs
*WBSU(FM) Brockport NY 1 hr
*WBNY(FM) Buffalo NY 12 hrs
*WSLU(FM) Canton NY
WSIV(AM) East Syracuse NY 20 hrs
*WRCU-FM Hamilton NY 10 hrs
WHLI(AM) Hempstead NY 1 hr

WKJY(FM) Hempstead NY 1 hr
WLNL(AM) Horseheads NY 1 hr
WVOX(AM) New Rochelle NY 1 hr
*WBAI(FM) New York NY 10 hrs
*WKCR-FM New York NY 12 hrs
*WNYU-FM New York NY 5 hrs
WJJL(AM) Niagara Falls NY 2 hrs
*WNYK(FM) Nyack NY 6 hrs
*WQKE(FM) Plattsburgh NY 9 hrs
WRNY(AM) Rome NY 3 hrs
*WFNP(FM) Rosendale NY 14 hrs
*WUSB(FM) Stony Brook NY 12 hrs
*WJPZ-FM Syracuse NY 12 hrs
*WOUB(AM) Athens OH 8 hrs
*WCDR-FM Cedarville OH 2 hrs
*WOHC(FM) Chillicothe OH 2 hrs
*WOSU(AM) Columbus OH 1 hr
*WWSU(FM) Dayton OH 12 hrs
*WDUB(FM) Granville OH 9 hrs
WRBP(FM) Hubbard OH
WFCJ(FM) Miamisburg OH 3 hrs
WNPQ(FM) New Philadelphia OH 4 hrs
WCCD(AM) Parma OH 3 hrs
*WOHP(FM) Portsmouth OH 2 hrs
WNTO(FM) Racine OH 1 hr
*WEEC(FM) Springfield OH 1 hr
*WXUT(FM) Toledo OH 8 hrs
*WOBN(FM) Westerville OH 2 hrs
KZBB(FM) Poteau OK 2 hrs
*KMUN(FM) Astoria OR 2 hrs
KKNX(AM) Eugene OR 3 hrs
*KLCC(FM) Eugene OR 3 hrs
*KRVM-FM Eugene OR 2 hrs
*KWVA(FM) Eugene OR 4 hrs
*KTEC(FM) Klamath Falls OR 3 hrs
*KEOL(FM) La Grande OR 12 hrs
*KSLC(FM) McMinnville OR 2 hrs
*KLCO(FM) Newport OR 3 hrs
*KRRC(FM) Portland OR 10 hrs
KXMG(AM) Portland OR 1 hr
KAVJ(FM) Sutherlin OR 1 hr
*WLVR(FM) Bethlehem PA 12 hrs
*WCAL(FM) California PA 4 hrs
*WCUC-FM Clarion PA 6 hrs
*WFSE(FM) Edinboro PA 15 hrs
*WKVR-FM Huntingdon PA 10 hrs
*WIUP-FM Indiana PA 14 hrs
*WFNM(FM) Lancaster PA 6 hrs
*WNTE(FM) Mansfield PA 5 hrs
*WARC(FM) Meadville PA 8 hrs
WJST(AM) New Castle PA 1 hr
*WKDU(FM) Philadelphia PA 8 hrs
*WRCT(FM) Pittsburgh PA 12 hrs
*WVMW-FM Scranton PA 2 hrs
*WSYC-FM Shippensburg PA 9 hrs
*WXVU(FM) Villanova PA 10 hrs
WSBA(AM) York PA 3 hrs
WBRU(FM) Providence RI 20 hrs
WARV(AM) Warwick RI 2 hrs
WOON(AM) Woonsocket RI 1 hr
WLUA(AM) Belton SC 8 hrs
WBSC(AM) Bennettsville SC 15 hrs
*WSCI(FM) Charleston SC 5 hrs
WCRE(AM) Cheraw SC 5 hrs
WFIS(AM) Fountain Inn SC 4 hrs
WJBS(AM) Holly Hill SC 17 hrs
WRIX(AM) Homeland Park SC 7 hrs
*KSDJ(FM) Brookings SD 8 hrs
WYXI(AM) Athens TN 1 hr
*WHCB(FM) Bristol TN 1 hr
*WMBW(FM) Chattanooga TN 1 hr
*WAPX-FM Clarksville TN 6 hrs
WKBL(AM) Covington TN 12 hrs
WHIN(AM) Gallatin TN 2 hrs
*WVCP(FM) Gallatin TN 8 hrs
WITA(AM) Knoxville TN 8 hrs
WKZX-FM Lenoir City TN 1 hr
WLIL(AM) Lenoir City TN 1 hr
WDXL(AM) Lexington TN 4 hrs
*WEVL(FM) Memphis TN 10 hrs
WXRQ(AM) Mount Pleasant TN 4 hrs
WGNS(AM) Murfreesboro TN 9 hrs
WLAC(AM) Nashville TN 20 hrs
*WRVU(FM) Nashville TN 3 hrs
*WUTS(FM) Sewanee TN 2 hrs
WLIJ(AM) Shelbyville TN 1 hr
*KGNZ(FM) Abilene TX 2 hrs
KLVQ(AM) Athens TX 1 hr
KAGC(AM) Bryan TX 2 hrs
*KWTS(FM) Canyon TX 3 hrs
KTMR(AM) Edna TX 3 hrs
KNES(FM) Fairfield TX 3 hrs
KGVL(AM) Greenville TX 1 hr
*KOOP(FM) Hornsby TX 2 hrs
*KPFT(FM) Houston TX 15 hrs
KHVL(AM) Huntsville TX 5 hrs
*KBJS(FM) Jacksonville TX 1 hr
*KTAI(FM) Kingsville TX 8 hrs
KVLG(AM) La Grange TX 1 hr
KSHN-FM Liberty TX 3 hrs
KFRO(AM) Longview TX 3 hrs
KHKZ(FM) Mercedes TX 3 hrs
KLVL(AM) Pasadena TX 4 hrs
*KPVU(FM) Prairie View TX 6 hrs
KPYK(AM) Terrell TX 8 hrs
KLGO(FM) Thorndale TX 6 hrs
KQRL(AM) Waco TX 4 hrs
*KZMU(FM) Moab UT 3 hrs
*KRCL(FM) Salt Lake City UT 20 hrs
*WTJU(FM) Charlottesville VA 8 hrs
WKEY(AM) Covington VA 1 hr
WFAX(AM) Falls Church VA 15 hrs
WMNA(AM) Gretna VA 2 hrs
WKWI(FM) Kilmarnock VA 6 hrs
*WVRU(FM) Radford VA 6 hrs
WKBA(AM) Vinton VA 10 hrs
WKCI(AM) Waynesboro VA 3 hrs
*KUGS(FM) Bellingham WA 10 hrs
*KZUU(FM) Pullman WA 12 hrs
*KNHC(FM) Seattle WA
*KSFC(FM) Spokane WA 2 hrs
*KWRS(FM) Spokane WA 12 hrs
*KUPS(FM) Tacoma WA 18 hrs
KJOX(AM) Yakima WA 2 hrs
*WORT(FM) Madison WI 3 hrs
*WMSE(FM) Milwaukee WI 13 hrs
*KUWS(FM) Superior WI 4 hrs
*WCCX(FM) Waukesha WI 3 hrs
*WVWC(FM) Buckhannon WV 4 hrs
*WWVU-FM Morgantown WV 9 hrs
*WQAB(FM) Philippi WV 2 hrs
*WPHP(FM) Wheeling WV 4 hrs

Bluegrass

*WQPR(FM) Muscle Shoals AL
WNUZ(AM) Talladega AL 6 hrs
*WUAL-FM Tuscaloosa AL
*KABF(FM) Little Rock AR 6 hrs
*KFJC(FM) Los Altos CA 8 hrs
*KCSN(FM) Northridge CA 5 hrs
*KAJX(FM) Aspen CO 2 hrs
*KDUR(FM) Durango CO 6 hrs
WWOJ(FM) Avon Park FL 2 hrs
*WUCF-FM Orlando FL 3 hrs
WTUF(FM) Boston GA 5 hrs
*WUWG(FM) Carrollton GA 2 hrs
WJJC(AM) Commerce GA 13 hrs
*WQNA(FM) Springfield IL 6 hrs
*WUIS(FM) Springfield IL 2 hrs
WAWK(AM) Kendallville IN 2 hrs
WMRS(FM) Monticello IN
*WPUM(FM) Rensselaer IN 3 hrs
*WECI(FM) Richmond IN 19 hrs
*WMHD-FM Terre Haute IN 1 hr
*KANU(FM) Lawrence KS 4 hrs
WWAG(FM) McKee KY 9 hrs
*WKMS-FM Murray KY 3 hrs
WKWY(FM) Tompkinsville KY 6 hrs
WTKY-FM Tompkinsville KY 6 hrs
WKHW(FM) Pocomoke City MD 4 hrs
*WDET-FM Detroit MI 3 hrs
*WMUK(FM) Kalamazoo MI 4 hrs
*KBEM-FM Minneapolis MN 4 hrs
*KOPN(FM) Columbia MO 6 hrs
KTJJ(FM) Farmington MO 2 hrs
*KMST(FM) Rolla MO 5 hrs
*KCLC(FM) Saint Charles MO 12 hrs
WKZU(FM) Ripley MS 2 hrs
WECR-FM Beech Mountain NC
*WCCE(FM) Buie's Creek NC 3 hrs
WECR(AM) Newland NC 2 hrs
WQDR(FM) Raleigh NC
WLHC(FM) Robbins NC 5 hrs
WEGG(AM) Rose Hill NC 10 hrs
WTQR(FM) Winston-Salem NC 2 hrs
*WDVR(FM) Delaware Township NJ 6 hrs
*WBJB-FM Lincroft NJ 3 hrs
*WBZC(FM) Pemberton NJ 4 hrs
*KGLP(FM) Gallup NM 12 hrs
*KRWG(FM) Las Cruces NM 8 hrs
*WBFO(FM) Buffalo NY 3 hrs
*WSQG-FM Ithaca NY 5 hrs
*WUBJ(FM) Jamestown NY 3 hrs
*WOLN(FM) Olean NY 3 hrs
*WCNY-FM Syracuse NY 3 hrs
*WUNY(FM) Utica NY 3 hrs
*WJNY(FM) Watertown NY 3 hrs
*WOSU(AM) Columbus OH 12 hrs
*WYSO(FM) Yellow Springs OH 6 hrs
KVSP(FM) Anadarko OK 2 hrs
WWSM(AM) Annville-Cleona PA 3 hrs
WSKE(FM) Everett PA 3 hrs
*WVMM(FM) Grantham PA 2 hrs
WLMI(FM) Kane PA 1 hr
WPHB(AM) Philipsburg PA 4 hrs
*WYEP-FM Pittsburgh PA 4 hrs
*WQSU(FM) Selinsgrove PA 7 hrs
*WBYO(FM) Sellersville PA 2 hrs
*WBYX(FM) Stroudsburg PA 3. hrs
*WZZD(FM) Warwick PA 3 hrs
WCRI(FM) Block Island RI 4 hrs
WVFB(FM) Celina TN 6 hrs
*WHRS(FM) Cookeville TN 1 hr
WEMB(AM) Erwin TN 2 hrs
*WVCP(FM) Gallatin TN 2 hrs
WCLC(AM) Jamestown TN 3 hrs
WWAM(AM) Jasper TN 1 hr
WLAF(AM) La Follette TN 7 hrs
*WRVU(FM) Nashville TN 3 hrs
*WUTS(FM) Sewanee TN 2 hrs
WSBI(AM) Static TN 1 hr
WNTT(AM) Tazewell TN
*KETR(FM) Commerce TX 3 hrs
KSWA(AM) Graham TX 2 hrs
KSHN-FM Liberty TX 2 hrs
*KOCV(FM) Odessa TX 2 hrs
KZHN(AM) Paris TX
WKDE-FM Altavista VA 10 hrs
WMNA(AM) Gretna VA 20 hrs
WOJL(FM) Louisa VA 14 hrs
WSIG(FM) Mount Jackson VA 6 hrs
WKDW(AM) Staunton VA 1 hr
WKCW(AM) Warrenton VA
*KBCS(FM) Bellevue WA
*WOJB(FM) Reserve WI 2 hrs
WVRQ-FM Viroqua WI 2 hrs
*WVMR(AM) Frost WV 5 hrs
*WWVU-FM Morgantown WV 1 hr
WMOV(AM) Ravenswood WV 10 hrs
WHAW(AM) Weston WV 8 hrs hrs
KKTY-FM Douglas WY 1 hr

Blues

*KUAC(FM) Fairbanks AK 4 hrs
*KTNA(FM) Talkeetna AK 5 hrs
WDLT-FM Chickasaw AL 18 hrs
*WVAS(FM) Montgomery AL 6 hrs
*WQPR(FM) Muscle Shoals AL
*WUAL-FM Tuscaloosa AL
*WVUA-FM Tuscaloosa AL 3 hrs
KFFA(AM) Helena AR 8 hrs
*KASU(FM) Jonesboro AR
KERX(FM) Paris AR 3 hrs
KWKM(FM) Saint Johns AZ 1 hr
*KNCA(FM) Burney CA 6 hrs
*KSPC(FM) Claremont CA 4 hrs
*KFSR(FM) Fresno CA 6 hrs
*KKJZ(FM) Long Beach CA 15 hrs
*KSBR(FM) Mission Viejo CA 3 hrs
*KNSQ(FM) Mount Shasta CA 6 hrs
*KVMR(FM) Nevada City CA 7 hrs
KHOP(FM) Oakdale CA 2 hrs
*KZYX(FM) Philo CA 3 hrs
KRXQ(FM) Sacramento CA 1 hr
*KXJZ(FM) Sacramento CA 7 hrs
*KCPR(FM) San Luis Obispo CA 3 hrs
*KCSM(FM) San Mateo CA 5 hrs
*KSCU(FM) Santa Clara CA 3 hrs
*KCLU(FM) Thousand Oaks CA 5 hrs
*KRCC(FM) Colorado Springs CO 5 hrs
*KDUR(FM) Durango CO 6 hrs
KIBT(FM) Fountain CO 3 hrs
*KWSB-FM Gunnison CO 3 hrs
KWUF-FM Pagosa Springs CO 10 hrs
*KVNF(FM) Paonia CO 3 hrs
*KOTO(FM) Telluride CO 7 hrs
*WESU(FM) Middletown CT 10 hrs
*WFCS(FM) New Britain CT 12 hrs
*WUCF-FM Orlando FL 2 hrs
*WFCF(FM) Saint Augustine FL 4 hrs
*WKPX(FM) Sunrise FL 3 hrs
*WCLK(FM) Atlanta GA 3 hrs
WZBN(FM) Camilla GA 5 hrs
*WOI(AM) Ames IA 3 hrs
*KUNI(FM) Cedar Falls IA 5 hrs
KROS(AM) Clinton IA 1 hr
*KDUB(FM) Dubuque IA 5 hrs
*KRNI(AM) Mason City IA 5 hrs
*KUNY(FM) Mason City IA 5 hrs
*KDMR(FM) Mitchellville IA 5 hrs
*KOJI(FM) Okoboji IA 4 hrs
*KWIT(FM) Sioux City IA 2 hrs
KECH-FM Sun Valley ID 8. hrs
*WEFT(FM) Champaign IL 10 hrs
WMKB(FM) Earlville IL 5 hrs
WMVN(FM) East St. Louis IL 1 hr
*WIUM(FM) Macomb IL 7 hrs
*WIUS(FM) Macomb IL 4 hrs
*WQUB(FM) Quincy IL 2 hrs
*WNIU(FM) Rockford IL 4 hrs
*WQNA(FM) Springfield IL 3 hrs
WDML(FM) Woodlawn IL 1 hr
WTTS(FM) Bloomington IN 2 hrs
*WVPE(FM) Elkhart IN 15 hrs
*WFYI-FM Indianapolis IN 4 hrs
*WCYT(FM) Lafayette Township IN 2 hrs
*KTCC(FM) Colby KS 3 hrs
*KANU(FM) Lawrence KS 4 hrs
*KJHK(FM) Lawrence KS 2 hrs
*WKMS-FM Murray KY 2 hrs
KJMG(FM) Bastrop LA 12 hrs
*WERS(FM) Boston MA 15 hrs
*WGBH(FM) Boston MA 8 hrs
*WESM(FM) Princess Anne MD 5 hrs
WBQI(FM) Bar Harbor ME 15 hrs
*WQAC-FM Alma MI 7 hrs
*WCBN-FM Ann Arbor MI 3 hrs hrs
WGTO(AM) Cassopolis MI 4 hrs
*WHFR(FM) Dearborn MI 12 hrs
*WDET-FM Detroit MI 10 hrs
*WDBM(FM) East Lansing MI 4 hrs
WKLT(FM) Kalkaska MI 2 hrs
WNWN(AM) Portage MI 3 hrs
WRKR(FM) Portage MI 4 hrs
*WPHS(FM) Warren MI 3 hrs
*WTIP(FM) Grand Marais MN 15 hrs
*KOPN(FM) Columbia MO 13 hrs
*KKFI(FM) Kansas City MO 9 hrs
KPOW-FM La Monte MO 6 hrs
*KGSP(FM) Parkville MO 12 hrs
*KCLC(FM) Saint Charles MO 7 hrs
KBFL(AM) Springfield MO 4 hrs
WESE(FM) Baldwyn MS 6 hrs
*WJSU(FM) Jackson MS 2 hrs
WJZD(FM) Long Beach MS 20 hrs
*WASU-FM Boone NC 2 hrs
*WNAA(FM) Greensboro NC 3 hrs
WVOD(FM) Manteo NC 2 hrs
*WSHA(FM) Raleigh NC 8 hrs
*WNCW(FM) Spindale NC 4 hrs
*KCND(FM) Bismarck ND 2 hrs
*KFJM(FM) Grand Forks ND 3 hrs
*KRNU(FM) Lincoln NE 2 hrs
*KZUM(FM) Lincoln NE 13 hrs
KKCD(FM) Omaha NE 1 hr
*KWSC(FM) Wayne NE 2 hrs
WHDQ(FM) Claremont NH 3 hrs
*WUNH(FM) Durham NH 3 hrs
WFRD(FM) Hanover NH 1 hr
*WNEC-FM Henniker NH 4 hrs
*WKNH(FM) Keene NH 3 hrs
*WPCR-FM Plymouth NH 3 hrs
*WNTI(FM) Hackettstown NJ 11 hrs
*WBJB-FM Lincroft NJ 4 hrs
KRSI(FM) Garapan-Saipan NP 6 hrs
*WBFO(FM) Buffalo NY 8 hrs
*WGMC(FM) Greece NY 3 hrs
*WICB(FM) Ithaca NY 2 hrs
WVBR-FM Ithaca NY 5 hrs
*WUBJ(FM) Jamestown NY 8 hrs
*WOLN(FM) Olean NY 8 hrs
WZOZ(FM) Oneonta NY 2 hrs
WTKV(FM) Oswego NY 1 hr
WPDH(FM) Poughkeepsie NY
*WSPN(FM) Saratoga Springs NY 9 hrs
*WUSB(FM) Stony Brook NY 10 hrs
*WAER(FM) Syracuse NY 3 hrs
*WCBE(FM) Columbus OH 3 hrs
WJZE(FM) Oak Harbor OH
*WUSO(FM) Springfield OH 3 hrs
WJUC(FM) Swanton OH 5 hrs wkly hrs
*WXTS-FM Toledo OH 5 hrs
*WXUT(FM) Toledo OH 2 hrs
*WYSO(FM) Yellow Springs OH 4 hrs
*KRSC-FM Claremore OK 6 hrs
*KGOU(FM) Norman OK 8 hrs
*KROU(FM) Spencer OK 8 hrs
*KSMF(FM) Ashland OR 6 hrs
*KSBA(FM) Coos Bay OR 6 hrs
KUJZ(FM) Creswell OR 2 hrs
*KLCC(FM) Eugene OR 3 hrs
*KMHD(FM) Gresham OR 15 hrs
*KSKF(FM) Klamath Falls OR 6 hrs
*KLCO(FM) Newport OR 4 hrs
KYTE(FM) Newport OR 2 hrs
KMCQ(FM) The Dalles OR 4 hrs
*WDCV-FM Carlisle PA 6 hrs
*WDNR(FM) Chester PA 2 hrs
*WPSX(FM) Kane PA 3 hrs
*WKDU(FM) Philadelphia PA 3 hrs
*WYEP-FM Pittsburgh PA 7 hrs
*WSYC-FM Shippensburg PA 2 hrs
WWII(AM) Shiremanstown PA 4 hrs
*WPSU(FM) State College PA 3 hrs
*WRDV(FM) Warminster PA 3 hrs
*WRLC(FM) Williamsport PA 3 hrs
*WRIU(FM) Kingston RI 3 hrs
*WSSB-FM Orangeburg SC 2 hrs
*KAUR(FM) Sioux Falls SD 3 hrs
*WTTU(FM) Cookeville TN 3 hrs
WRLT(FM) Franklin TN 2 hrs
*WETS(FM) Johnson City TN 12 hrs
*WRVU(FM) Nashville TN 3 hrs
*WUTS(FM) Sewanee TN 3 hrs
*KAZI-FM Austin TX 6 hrs
*KUT(FM) Austin TX 6 hrs
KLUB(FM) Bloomington TX 1 hr
*KSAU(FM) Nacogdoches TX 2 hrs
*KOCV(FM) Odessa TX 4 hrs
*KUTX(FM) San Angelo TX 6 hrs
*KSTX(FM) San Antonio TX 6 hrs
KNCN(FM) Sinton TX 2 hrs
*KSUU(FM) Cedar City UT 4 hrs
*KZMU(FM) Moab UT 19 hrs
*WMRY(FM) Crozet VA 4 hrs
*WWHS-FM Hampden-Sydney VA 2 hrs
*WHOV(FM) Hampton VA 3 hrs
*WMRA(FM) Harrisonburg VA 4 hrs
*WMRL(FM) Lexington VA 4 hrs
*WVRU(FM) Radford VA 2 hrs
*WCVE(FM) Richmond VA 3 hrs
*WCWM(FM) Williamsburg VA 3 hrs
*WRMC-FM Middlebury VT 10 hrs
WIZN(FM) Vergennes VT 3 hrs
*KSER(FM) Everett WA 6 hrs
KKZX(FM) Spokane WA 2 hrs
*KUPS(FM) Tacoma WA 6 hrs
*WBSD(FM) Burlington WI 3 hrs
*WUEC(FM) Eau Claire WI 3 hrs
WMCS(AM) Greenfield WI 6 hrs
*WHND(FM) Sister Bay WI 2 hrs

*WWSP(FM) Stevens Point WI 4 hrs
*WVWC(FM) Buckhannon WV 2 hrs

Children

KLEF(FM) Anchorage AK 1 hr
*KTOO(FM) Juneau AK 1 hr
KRSA(AM) Petersburg AK 10 hrs
*WMBV(FM) Dixons Mills AL 3 hrs
*WRWA(FM) Dothan AL 1 hr
*WTSU(FM) Montgomery-Troy AL 1 hr
KFMM(FM) Thatcher AZ 2 hrs
*KCFY(FM) Yuma AZ 7 hrs wkly hrs
*KYRM(FM) Yuma AZ 6 hrs
KZRO(FM) Dunsmuir CA 2 hrs
KGBA-FM Holtville CA 4 hrs
*KPFK(FM) Los Angeles CA 1 hr
KSPN(AM) Los Angeles CA 24 hrs
*KXLU(FM) Los Angeles CA 1 hr
KWVE(FM) San Clemente CA 3 hrs
KFAX(AM) San Francisco CA 1 hr
*WIHS(FM) Middletown CT 9 hrs
*WAFG(FM) Fort Lauderdale FL 4 hrs
*WJYO(FM) Fort Myers FL 5 hrs
*WJLF(FM) Gainesville FL 1 hr
WMEL(AM) Melbourne FL one hrs
WZSP(FM) Nocatee FL
*WIRP(FM) Pennsuco FL 6 hrs
*WBVM(FM) Tampa FL 4 hrs
WDWD(AM) Atlanta GA
*WTJB(FM) Columbus GA 1 hr
*KHMG(FM) Barrigada GU 2 hrs
*KHOE(FM) Fairfield IA 3 hrs
KTFC(FM) Sioux City IA 5 hrs
KTFG(FM) Sioux Rapids IA 5 hrs
*KCIR(FM) Twin Falls ID 2 hrs
*WUEV(FM) Evansville IN 5 hrs
*KJTY(FM) Topeka KS 5 hrs
WFCC-FM Chatham MA 1 hr
*WKHS(FM) Worton MD 5 hrs
*WMDR(AM) Augusta ME
*WHCF(FM) Bangor ME 5 1/2 hrs
WBQX(FM) Thomaston ME 1 hr
WDBC(AM) Escanaba MI 1 hr
*WLNZ(FM) Lansing MI 1 hr
KJLY(FM) Blue Earth MN 4 hrs
*WTIP(FM) Grand Marais MN
*KYMC(FM) Ballwin MO 1 hr
*KGNN-FM Cuba MO 7 hrs
KKBL(FM) Monett MO 2 hrs
*KGNV(FM) Washington MO 6 hrs
*WPAE(FM) Centreville MS 5 hrs
*WMBU(FM) Forest MS 5 hrs
WLRC(AM) Walnut MS
*KABU(FM) Fort Totten ND 12 hrs
WMVB(AM) Millville NJ 3 hrs
*WMHI(FM) Cape Vincent NY 11 hrs
WYBG(AM) Massena NY
*WRHO(FM) Oneonta NY 2 hrs
*WMHR(FM) Syracuse NY 11 hrs
*WMHN(FM) Webster NY 11 hrs
WFCJ(FM) Miamisburg OH 2 hrs
KIHN(AM) Hugo OK 1 hr
*KMUN(FM) Astoria OR 6 hrs
*KWVA(FM) Eugene OR 4 hrs
*WFKJ(AM) Cashtown PA 12 hrs
*WDNR(FM) Chester PA 2 hrs
WCOJ(AM) Coatesville PA 2 hrs
WDBA(FM) DuBois PA 2 hrs
*WXPN(FM) Philadelphia PA 5 hrs
WCTL(FM) Union City PA 1 hr
*WTMV(FM) Youngsville PA 10 hrs
*KLND(FM) Little Eagle SD 4 hrs
*WHCB(FM) Bristol TN 10 hrs
KESN(FM) Allen TX 2 hrs
KBYG(AM) Big Spring TX 1 hr
KPSM(FM) Brownwood TX 3 hrs
WRR(FM) Dallas TX 2 hrs
*KNLE-FM Round Rock TX 4 hrs
*KTER(FM) Rudolph TX 4 hrs
*KVNE(FM) Tyler TX 4 hrs
*WTJU(FM) Charlottesville VA 2 hrs
WPRZ(AM) Warrenton VA 7 hrs
*WVPS(FM) Burlington VT .5 hrs
WWOD(FM) Hartford VT 3 hrs
WCVT(FM) Stowe VT 1 hr
WWMP(FM) Waterbury VT 3 hrs
WTWN(AM) Wells River VT
*KRLF(FM) Pullman WA 5 hrs
*KDNA(FM) Yakima WA 5. hrs
*WGNV(FM) Milladore WI 4 hrs
*WVPG(FM) Parkersburg WV 1 hr
*WQAB(FM) Philippi WV 2 hrs

Chinese

KGBA-FM Holtville CA 14 hrs
KKLA-FM Los Angeles CA 2 hrs
KEST(AM) San Francisco CA
*KUSF(FM) San Francisco CA 9 hrs
KSJX(AM) San Jose CA 10 hrs
WJDA(AM) Quincy MA 3 hrs
*WDBM(FM) East Lansing MI 4 hrs
WSDS(AM) Salem Township MI 4 hrs
*KRNM(FM) Chalan Kanoa-Saipan NP 1 hr
*WUSB(FM) Stony Brook NY 1 hr
*WJCU(FM) University Heights OH 1 hr
WBZK(AM) York SC 10 hrs
*KHCB(AM) Galveston TX 13 hrs
*KHCB-FM Houston TX 1 hr
*KKER(FM) Kerrville TX 1 hr
*WUVT-FM Blacksburg VA 2 hrs

Christian

KCAM(AM) Glennallen AK 8 hrs
KJLH-FM Compton CA 6 hrs
KNCO(AM) Grass Valley CA 4 hrs
KFTM(AM) Fort Morgan CO 6 hrs
WEBY(AM) Milton FL 7 hrs
WMGF(FM) Mount Dora FL 20 hrs
WROM(AM) Rome GA
WDML(FM) Woodlawn IL 3 hrs
*WMHD-FM Terre Haute IN 2 hrs
KIND-FM Independence KS 2 hrs
*WHFC(FM) Bel Air MD 6 hrs
*WPHS(FM) Warren MI 4 hrs
*KYMC(FM) Ballwin MO 4 hrs
*KWJC(FM) Liberty MO 10 hrs
KNEM(AM) Nevada MO 5 hrs
KNMO(FM) Nevada MO 5 hrs
KAYX(FM) Richmond MO
KKDY(FM) West Plains MO 3 hrs
*WASU-FM Boone NC 3 hrs
*WYQS(FM) Mars Hill NC 15 hrs
WLHC(FM) Robbins NC 5 hrs
KTNC(AM) Falls City NE 1 hr
*WKNH(FM) Keene NH 3 hrs
*WITR(FM) Henrietta NY 10 hrs
WRIP(FM) Windham NY 2 hrs
*WRDL(FM) Ashland OH 7 hrs
KGFY(FM) Stillwater OK 4 hrs
KLDR(FM) Harbeck-Fruitdale OR 2 hrs wkly hrs
*WCYJ-FM Waynesburg PA 3 hrs
WFIS(AM) Fountain Inn SC 3 hrs
KSLT(FM) Spearfish SD 10 hrs
*WKTS(FM) Kingston TN 2 hrs wkly hrs
WNRQ(FM) Nashville TN 6 hrs
KRUN(AM) Ballinger TX 4 hrs
KQTY(AM) Borger TX 5 hrs
KQTY-FM Borger TX 5 hrs
*KVTT(FM) Dallas TX 6 hrs
KWEL(AM) Midland TX 2.5 hrs
*KSAU(FM) Nacogdoches TX 2 hrs
*KRTU(FM) San Antonio TX 2 hrs
KMAS(AM) Shelton WA 4 hrs

Classic Rock

KMMT(FM) Mammoth Lakes CA 4 hrs
KVAY(FM) Lamar CO 4 hrs
*WXCI(FM) Danbury CT 3 hrs
WSGL(FM) Naples FL 5 hrs
*WCSF(FM) Joliet IL 4 hrs
*WRRG(FM) River Grove IL 2 hrs
*WBKE-FM North Manchester IN 6 hrs
*WECI(FM) Richmond IN 16 hrs
*WVUR-FM Valparaiso IN 3 hrs
*KTCC(FM) Colby KS 3 hrs
*WRBC(FM) Lewiston ME 11 hrs
WKLT(FM) Kalkaska MI
*KMVC(FM) Marshall MO 4 hrs
WERX-FM Columbia NC
*WPSC-FM Wayne NJ 12 hrs
*WCWP(FM) Brookville NY 2 hrs
*WQKE(FM) Plattsburgh NY 12 hrs
WMKX(FM) Brookville PA 6 hrs
*WJRH(FM) Easton PA 4 hrs
*WCYJ-FM Waynesburg PA 3 hrs
*WDOM(FM) Providence RI 3 hrs
*KSAU(FM) Nacogdoches TX 14 hrs
KNCN(FM) Sinton TX 2 hrs
*WVBC(FM) Bethany WV 8 hrs
KKTY-FM Douglas WY 3 hrs

Classical

*KBRW(AM) Barrow AK 2 hrs
*KYUK(AM) Bethel AK 4 hrs
KCAM(AM) Glennallen AK 10 hrs
*KRBD(FM) Ketchikan AK 11 hrs
KIAM(AM) Nenana AK 1 hr
KICY-FM Nome AK 2 hrs
*KNOM(AM) Nome AK 5 hrs
*KNOM-FM Nome AK 5 hrs
KRSA(AM) Petersburg AK 5 hrs
*KCAW(FM) Sitka AK 15 hrs
*KSTK(FM) Wrangell AK 4 hrs
KQSM-FM Bentonville AR 2 hrs
KTCN(FM) Eureka Springs AR 5 hrs
*KSMC(FM) Moraga CA 4 hrs
*KAZU(FM) Pacific Grove CA 3 hrs
*KZYX(FM) Philo CA 14 hrs
*KUCR(FM) Riverside CA 14 hrs
KHJQ(FM) Susanville CA 5 hrs
*KCSS(FM) Turlock CA 9 hrs
*KGNU-FM Boulder CO 13 hrs
*KDUR(FM) Durango CO 6 hrs
*KCIC(FM) Grand Junction CO 14 hrs
*KSUT(FM) Ignacio CO 8 hrs
*KVNF(FM) Paonia CO 15 hrs
*KOTO(FM) Telluride CO 9 hrs
*WPKN(FM) Bridgeport CT 2 hrs
*WRTC-FM Hartford CT 4 hrs
*WCNI(FM) New London CT 6 hrs
*WGSK(FM) South Kent CT 1 hr
*WWEB(FM) Wallingford CT 2 hrs
*WNHU(FM) West Haven CT 9 hrs
*WVUD(FM) Newark DE 10 hrs
WKEY-FM Key West FL 4 hrs
*WRAS(FM) Atlanta GA 3 hrs
*WREK(FM) Atlanta GA 15 hrs
WMOG(AM) Brunswick GA 1 hr
*WBCX(FM) Gainesville GA 20 hrs
*KDFR(FM) Des Moines IA 2 hrs
*KCMR(FM) Mason City IA 5 hrs
KSAS-FM Caldwell ID 2 hrs
KSRA(AM) Salmon ID 1 hr
KSRA-FM Salmon ID 1 hr
*WESN(FM) Bloomington IL 6 hrs
*WHPK-FM Chicago IL 10 hrs
*WEPS(FM) Elgin IL 3 hrs
*WVKC(FM) Galesburg IL 18 hrs
*WJCH(FM) Joliet IL 2 hrs
*WMXM(FM) Lake Forest IL 3 hrs
*WLRA(FM) Lockport IL 6 hrs
*WDSO(FM) Chesterton IN 1 hr
*WBKE-FM North Manchester IN 10 hrs
*WPUM(FM) Rensselaer IN 3 hrs
*WMHD-FM Terre Haute IN 4 hrs
*WVUR-FM Valparaiso IN 3 hrs
*WVUB(FM) Vincennes IN 6 hrs
KIND(AM) Independence KS 3 hrs
*WOMR(FM) Provincetown MA 16 hrs
*WYAJ(FM) Sudbury MA 3 hrs
WMVY(FM) Tisbury MA 4 hrs
*WCHC(FM) Worcester MA 6 hrs
*WHFC(FM) Bel Air MD 18 hrs
*WSJB-FM Standish ME 2 hrs
*WOES(FM) Ovid-Elsie MI 1 hr
WRSR(FM) Owosso MI 1 hr
KNXR(FM) Rochester MN 4 hrs
*WMCN(FM) Saint Paul MN 4 hrs
*KQAL(FM) Winona MN 14 hrs
KIXQ(FM) Joplin MO 3 hrs
*KWJC(FM) Liberty MO 10 hrs
*KGNV(FM) Washington MO 5 hrs
*KGLT(FM) Bozeman MT 11 hrs
*KMSM-FM Butte MT 2 hrs
*KNMC(FM) Havre MT 10 hrs
KALS(FM) Kalispell MT 1 hr
*WFSS(FM) Fayetteville NC 5 hrs
WVOD(FM) Manteo NC 6 hrs
*WYQS(FM) Mars Hill NC 7 hrs
WCVP(AM) Murphy NC 20 hrs
*WSNC(FM) Winston-Salem NC 4 hrs
KHAS(AM) Hastings NE 2 hrs
*KLPR(FM) Kearney NE 18 hrs
KCMI(FM) Terrytown NE 4 hrs
*WKNH(FM) Keene NH 4 hrs
*WPCR-FM Plymouth NH 3 hrs
*WKDN-FM Camden NJ 2 hrs
WSJQ(FM) North Cape May NJ 3 hrs
*WLFR(FM) Pomona NJ 4 hrs
*KUNM(FM) Albuquerque NM 12 hrs
KPCL(FM) Farmington NM 1 hr
*KGLP(FM) Gallup NM 8 hrs
*WXLH(FM) Blue Mountain Lake NY
*WBFO(FM) Buffalo NY 1 hr
*WSLU(FM) Canton NY
*WHCL-FM Clinton NY 9 hrs
*WRCU-FM Hamilton NY 4 hrs
*WJSL(FM) Houghton NY 5 hrs
WHVW(AM) Hyde Park NY 16 hrs
*WBAI(FM) New York NY 5 hrs
WSRK(FM) Oneonta NY 2 hrs
WLIM(AM) Patchogue NY 7 hrs
*WUSB(FM) Stony Brook NY 14 hrs
*WKWZ(FM) Syosset NY 6 hrs
*WPNR-FM Utica NY 10 hrs
*WFRW(FM) Webster NY 2 hrs
*WBGU(FM) Bowling Green OH 3 hrs
*WCUE(AM) Cuyahoga Falls OH 2 hrs
*WLFC(FM) Findlay OH 3 hrs
*WMCO(FM) New Concord OH 4 hrs
*WUSO(FM) Springfield OH 3 hrs
*WYTN(FM) Youngstown OH 2 hrs
WBBZ(AM) Ponca City OK 5 hrs
*KBVR(FM) Corvallis OR 4 hrs
*KEOL(FM) La Grande OR 4 hrs
*KRRC(FM) Portland OR 4 hrs
*WLVR(FM) Bethlehem PA 8 hrs
*WESS(FM) East Stroudsburg PA 4 hrs
*WZBT(FM) Gettysburg PA 3 hrs
*WIUP-FM Indiana PA 15 hrs
WJSA(AM) Jersey Shore PA 1 hr
WJSA-FM Jersey Shore PA 1 hr
*WFNM(FM) Lancaster PA 2 hrs
*WARC(FM) Meadville PA 10 hrs
*WPEL(AM) Montrose PA 1 hr
WJST(AM) New Castle PA 1 hr
*WHYY-FM Philadelphia PA 4 hrs
*WRCT(FM) Pittsburgh PA 3 hrs
*WXLV(FM) Schnecksville PA 13 hrs
*WUSR(FM) Scranton PA 5 hrs
*WVMW-FM Scranton PA 7 hrs
*WSYC-FM Shippensburg PA 2 hrs
*WRLC(FM) Williamsport PA 1 hr
*WVYC(FM) York PA 8 hrs
*WTMV(FM) Youngsville PA 2.5 hrs
WTPM(FM) Aguadilla PR 7 hrs
WMUU-FM Greenville SC 14 hrs
WMYB(FM) Myrtle Beach SC
WQMC(AM) Sumter SC 15 hrs
*WHCB(FM) Bristol TN 1 hr
*WFHU(FM) Henderson TN 7 hrs
*WFMQ(FM) Lebanon TN 6 hrs
*WRVU(FM) Nashville TN 6 hrs
*WDNX(FM) Olive Hill TN 5 hrs
*WUTS(FM) Sewanee TN 4 hrs
KYKR(FM) Beaumont TX 4 hrs. hrs
*KWTS(FM) Canyon TX 4 hrs
*KTRU(FM) Houston TX 8 hrs
*KNTU(FM) McKinney TX 6 hrs
*KSUU(FM) Cedar City UT 3 hrs
*KPCW(FM) Park City UT 17 hrs
*KRDC-FM Saint George UT 15 hrs
*WWHS-FM Hampden-Sydney VA 2 hrs
*WVRU(FM) Radford VA 15 hrs
*WDCE(FM) Richmond VA 3 hrs
*WCWM(FM) Williamsburg VA 11 hrs
*WIUJ(FM) Charlotte Amalie VI 5 hrs
*WIUV(FM) Castleton VT 3 hrs
*WJSC-FM Johnson VT 3 hrs
*WWLR(FM) Lyndonville VT 2 hrs
*WRMC-FM Middlebury VT 15 hrs
WDEV(AM) Waterbury VT 1 hr
KELA(AM) Centralia-Chehalis WA 2 hrs
KULE-FM Ephrata WA 5 hrs
KEYG(AM) Grand Coulee WA 4 hrs
*KMLW(FM) Moses Lake WA 1 hr
*KAGU(FM) Spokane WA 2 hrs
*WMSE(FM) Milwaukee WI 3 hrs
*WSUP(FM) Platteville WI 4 hrs
WRJN(AM) Racine WI 2 hrs
*WVBC(FM) Bethany WV 2 hrs
*WVWC(FM) Buckhannon WV 2 hrs
WZJO(FM) Dunbar WV 2 hrs
*WWVU-FM Morgantown WV 4 hrs
KMTN(FM) Jackson WY

Comedy

KTCL(FM) Wheat Ridge CO 1 hr
KEYN-FM Wichita KS 2 hrs
*WVSD(FM) Itta Bena MS 2 hrs
*WPCR-FM Plymouth NH 3 hrs
*WITR(FM) Henrietta NY 2 hrs
WLUZ(AM) Bayamon PR 19 hrs

Contemporary Hit/Top-40

*KNOM(AM) Nome AK 12 hrs
*KNOM-FM Nome AK 12 hrs
*KCSS(FM) Turlock CA 10 hrs
*WXCI(FM) Danbury CT 3 hrs
WZHR(AM) Zephyrhills FL 10 hrs
KIOW(FM) Forest City IA 19 hrs
*WBKE-FM North Manchester IN 10 hrs
WTNM(FM) Water Valley MS 4 hrs
KATQ-FM Plentywood MT
*WIRQ(FM) Rochester NY 3 hrs
WQIO(FM) Mount Vernon OH 4 hrs
*WKVR-FM Huntingdon PA 15 hrs
WVOZ(AM) San Juan PR 15 hrs
KYKR(FM) Beaumont TX 4 hrs. hrs
*KRLF(FM) Pullman WA 1 hr
*KVTI(FM) Tacoma WA 3 hrs

Country

*KBRW(AM) Barrow AK 7 hrs
*KYUK(AM) Bethel AK 4 hrs
*KRBD(FM) Ketchikan AK 14 hrs
KIAL(AM) Unalaska AK 04 hrs
*KSTK(FM) Wrangell AK 16 hrs
WKAC(AM) Athens AL 13 hrs
*KPFA(FM) Berkeley CA 18 hrs
*KFJC(FM) Los Altos CA 8 hrs
*KVMR(FM) Nevada City CA 7 hrs
*KAZU(FM) Pacific Grove CA 6 hrs
*KVNF(FM) Paonia CO 5 hrs
*KOTO(FM) Telluride CO 12 hrs
*WWEB(FM) Wallingford CT 2 hrs
*WAMU(FM) Washington DC 4 hrs
WBTS(FM) Doraville GA 5 hrs
KSTO(FM) Hagatna GU 12 hrs
KROS(AM) Clinton IA 6 hrs
KGRN(AM) Grinnell IA 12 hrs
KCHQ(FM) Driggs ID 3 hrs
WTAY(AM) Robinson IL 12 hrs
WTYE(FM) Robinson IL 12 hrs
*WEEM-FM Pendleton IN 4 hrs
*WPUM(FM) Rensselaer IN 3 hrs
WLVK(FM) Fort Knox KY
*KLSP(FM) Angola LA 6 hrs
*WKHS(FM) Worton MD 2 hrs
WBPW(FM) Presque Isle ME

*WSJB-FM Standish ME 3 hrs
*WCBN-FM Ann Arbor MI 3 hrs hrs
*WDBM(FM) East Lansing MI 4 hrs
*WPHS(FM) Warren MI 4 hrs
*WMCN(FM) Saint Paul MN 2 hrs
WLRC(AM) Walnut MS
*KXEI(FM) Havre MT 1 hr
*WASU-FM Boone NC 8 hrs
WCVP(AM) Murphy NC 12 hrs
*KMHA(FM) Four Bears ND 8 hrs
KCNI(AM) Broken Bow NE
*WNEC-FM Henniker NH 3 hrs
*WDVR(FM) Delaware Township NJ 12 hrs hrs
WLNL(AM) Horseheads NY 1 hr
WVBR-FM Ithaca NY 4 hrs
*WKCR-FM New York NY 6 hrs
WGGO(AM) Salamanca NY 5 hrs
*WKWZ(FM) Syosset NY 6 hrs
*WBGU(FM) Bowling Green OH 4 hrs
*KRSC-FM Claremore OK 5 hrs
*KRVM-FM Eugene OR 1 hr
*KWVA(FM) Eugene OR 3 hrs
*KRRC(FM) Portland OR 2 hrs
*WFKJ(AM) Cashtown PA 4 hrs
*WCUC-FM Clarion PA 6 hrs
WSKE(FM) Everett PA 2 hrs
*WRCT(FM) Pittsburgh PA 3 hrs
*WQSU(FM) Selinsgrove PA 6 hrs
*WRDV(FM) Warminster PA 4 hrs
*WCYJ-FM Waynesburg PA 3 hrs
WCHE(AM) West Chester PA 2 hrs
*WDOM(FM) Providence RI 2 hrs
WWLX(AM) Lawrenceburg TN 8 hrs
*WEVL(FM) Memphis TN 15 hrs
WSM(AM) Nashville TN 12 hrs
*WUTS(FM) Sewanee TN 2 hrs
KQTY(AM) Borger TX 10 hrs
*KPCW(FM) Park City UT 18 hrs
WCLM(AM) Highland Springs VA
*WJSC-FM Johnson VT 3 hrs
*WBSD(FM) Burlington WI 4 hrs
*WOJB(FM) Reserve WI 15 hrs

Croation

WELW(AM) Willoughby-Eastlake OH 12 hrs
WKBN(AM) Youngstown OH 2 hrs
WFGI(AM) Charleroi PA 1 hr
WOGI(FM) Charleroi PA 1 hr
WEDO(AM) McKeesport PA 1 hr
WKZV(AM) Washington PA 1 hr
WRRD(AM) Jackson WI 2 hrs

Czech

KMRY(AM) Cedar Rapids IA 3 hrs hrs
*WOES(FM) Ovid-Elsie MI 1 hr
WAAL(FM) Binghamton NY 1 hr
WOMP(AM) Bellaire OH 2 hrs
KAGC(AM) Bryan TX 2 hrs
KTAE(AM) Cameron TX 8 hrs
KULP(AM) El Campo TX 5 hrs
KHLT(AM) Hallettsville TX 5 hrs
KHBR(AM) Hillsboro TX 2 hrs
KVLG(AM) La Grange TX 6 hrs
KTEM(AM) Temple TX 3 hrs
WAUN(FM) Kewaunee WI 1 hr

Disco

*WVBU-FM Lewisburg PA 6 hrs

Discussion

WNDN(FM) Chiefland FL

Diversified

*KALW(FM) San Francisco CA 1 1/2 hrs
WUMS(FM) University MS 2 hrs
*WNAA(FM) Greensboro NC 7 hrs

Drama/Literature

*KUAR(FM) Little Rock AR 2 hrs
*KZYX(FM) Philo CA 3 hrs
*KOTO(FM) Telluride CO 3 hrs
*WGLT(FM) Normal IL 2 hrs
*WOES(FM) Ovid-Elsie MI 1 hr
*KMSK(FM) Austin MN 3 hrs
*KMSU(FM) Mankato MN 3 hrs
*WNCW(FM) Spindale NC 3 hrs
*WNYC-FM New York NY 5 hrs
KPYK(AM) Terrell TX 7 hrs
*KAGU(FM) Spokane WA 2 hrs
*WQAB(FM) Philippi WV 2 hrs

Easy Listening

KRBB(FM) Wichita KS 18 hrs
KCHR(AM) Charleston MO

Educational

*KXRJ(FM) Russellville AR
*KCRH(FM) Hayward CA 1 hr
KVEC(AM) San Luis Obispo CA
*KYCC(FM) Stockton CA 1 hr
WGCH(AM) Greenwich CT 2 hrs
*WJMJ(FM) Hartford CT
*WHIF(FM) Palatka FL 10 hrs
*WKGC(AM) Panama City Beach FL 8 hrs
KGUM(AM) Hagatna GU 1 hr
*WEPS(FM) Elgin IL 13 hrs
*WDCB(FM) Glen Ellyn IL 12 hrs
*WEEM-FM Pendleton IN 10 hrs
*WOMR(FM) Provincetown MA 10 hrs
*WFWM(FM) Frostburg MD 5 hrs
*WNMU-FM Marquette MI
*WBAI(FM) New York NY
*WCWS(FM) Wooster OH 8 hrs
KBNP(AM) Portland OR
*WESS(FM) East Stroudsburg PA 7 hrs
*WPUC-FM Ponce PR
WPWT(AM) Colonial Heights TN 1 hr
*KMFA(FM) Austin TX 2 hrs
WRVA(AM) Richmond VA 2 hrs
WWYO(AM) Pineville WV 2 hrs

Eskimo

KAGV(AM) Big Lake AK 3 hrs
*KNOM(AM) Nome AK 6 hrs
*KNOM-FM Nome AK 6 hrs

Ethnic

*KRBD(FM) Ketchikan AK 5 hrs
*KALW(FM) San Francisco CA 1 hr
*KALW(FM) San Francisco CA 4 hrs
*WJFP(FM) Fort Pierce FL 8 hrs
WPSO(AM) New Port Richey FL
WSB(AM) Atlanta GA 1 hr
*WUIS(FM) Springfield IL 1 hr
KVPI(AM) Ville Platte LA 12 hrs
WCAR(AM) Livonia MI
*KGVA(FM) Fort Belknap Agency MT 15 hrs
*WSOU(FM) South Orange NJ 10 hrs
*KRNM(FM) Chalan Kanoa-Saipan NP 2 hrs
KRSI(FM) Garapan-Saipan NP 6 hrs
WVBR-FM Ithaca NY 5 hrs
WLIM(AM) Patchogue NY 1 hr
*WJCU(FM) University Heights OH 2 hrs
*KWVA(FM) Eugene OR 4 hrs
*WDIY(FM) Allentown PA 1 hr
WEDO(AM) McKeesport PA 1 hr
WVWI(AM) Charlotte Amalie VI 2 hrs
WWKS(FM) Cruz Bay VI 25 hrs
WIZM(AM) La Crosse WI 2 hrs

Farsi

WUST(AM) Washington DC 5 hrs

Filipino

*KBRW(AM) Barrow AK 2 hrs
KCHJ(AM) Delano CA 3 hrs
*KECG(FM) El Cerrito CA 2 hrs
KKLA-FM Los Angeles CA 1 hr
*KMPO(FM) Modesto CA 1 hr
KSJX(AM) San Jose CA 2 hrs
WPSO(AM) New Port Richey FL 1 hr
*V6AI(AM) Yap FM 5 hrs
KWAI(AM) Honolulu HI 7 hrs
KLAV(AM) Las Vegas NV 5 hrs
KBRO(AM) Bremerton WA 4 hrs

Finnish

*KUSF(FM) San Francisco CA 1 hr
*WFHB(FM) Bloomington IN 3 hrs
KRBT(AM) Eveleth MN 1 hr
*WBXL(FM) Baldwinsville NY 2 hrs

Folk

*KUAC(FM) Fairbanks AK 10 hrs
*KRBD(FM) Ketchikan AK 10 hrs
*WSGN(FM) Gadsden AL 2 hrs
*WQPR(FM) Muscle Shoals AL 5 hrs
*WUAL-FM Tuscaloosa AL 5 hrs
*KUAF(FM) Fayetteville AR 5 hrs
*KASU(FM) Jonesboro AR 4 hrs
*KABF(FM) Little Rock AR 10 hrs
*KUAR(FM) Little Rock AR 3 hrs
KCTT-FM Yellville AR 10 hrs
KVNA(AM) Flagstaff AZ 4hrs hrs
*KXCI(FM) Tucson AZ 2 hrs
*KPFA(FM) Berkeley CA 10 hrs
*KNCA(FM) Burney CA 3 hrs
*KFSR(FM) Fresno CA 3 hrs
*KXLU(FM) Los Angeles CA 1 hr
*KSBR(FM) Mission Viejo CA 2 hrs
*KMPO(FM) Modesto CA 4 hrs
*KNSQ(FM) Mount Shasta CA 3 hrs
*KVMR(FM) Nevada City CA 13 hrs
*KAZU(FM) Pacific Grove CA 6 hrs
*KZYX(FM) Philo CA 8 hrs
KVYN(FM) Saint Helena CA 2 hrs
*KCBX(FM) San Luis Obispo CA 15 hrs
*KRCB-FM Santa Rosa CA 6 hrs
*KGNU-FM Boulder CO 20 hrs
*KAFM(FM) Grand Junction CO 10 hrs
*KMSA(FM) Grand Junction CO 2 hrs
*WSHU(FM) Fairfield CT 5 hrs
*WGRS(FM) Guilford CT 2 hrs
*WMNR(FM) Monroe CT 2 hrs
WYBC-FM New Haven CT 3 hrs
*WCNI(FM) New London CT 9 hrs
*WRXC(FM) Shelton CT 2 hrs
*WGSK(FM) South Kent CT 2 hrs
*WNHU(FM) West Haven CT 6 hrs
*WVUD(FM) Newark DE 15 hrs
*WUFT-FM Gainesville FL 1 hr
*WKGC(AM) Panama City Beach FL 2 hrs
*WFCF(FM) Saint Augustine FL 3 hrs
*WVFS(FM) Tallahassee FL 3 hrs
V6AH(AM) Pohnpei FM 20 hrs
*WUGA(FM) Athens GA 4 hrs
*WUWG(FM) Carrollton GA 2 hrs
*KUNI(FM) Cedar Falls IA 4 hrs
KROS(AM) Clinton IA 2 hrs
*KIWR(FM) Council Bluffs IA 2 hrs
*KDUB(FM) Dubuque IA 4 hrs
*KHOE(FM) Fairfield IA 6 hrs
*KRNI(AM) Mason City IA 4 hrs
*KUNY(FM) Mason City IA 4 hrs
*KDMR(FM) Mitchellville IA 4 hrs
*KRNL-FM Mount Vernon IA 2 hrs
*KWAR(FM) Waverly IA 3 hrs
*KBSU(AM) Boise ID
*KRFA-FM Moscow ID
*KUOI-FM Moscow ID 3 hrs
*WSIU(FM) Carbondale IL 3 hrs
*WEFT(FM) Champaign IL 10 hrs
*WEIU(FM) Charleston IL 4 hrs
WFMT(FM) Chicago IL 4 hrs
*WNIJ(FM) De Kalb IL 4 hrs
*WNUR-FM Evanston IL 3 hrs
*WDCB(FM) Glen Ellyn IL 12 hrs
*WIUM(FM) Macomb IL 7 hrs
*WGLT(FM) Normal IL 4 hrs
*WQUB(FM) Quincy IL 2 hrs
*WQNA(FM) Springfield IL 8 hrs
*WFHB(FM) Bloomington IN 10 hrs
*WVPE(FM) Elkhart IN 9 hrs
*WICR(FM) Indianapolis IN 1 hr
*KANZ(FM) Garden City KS 6 hrs
*KZNA(FM) Hill City KS 6 hrs
*KRPS(FM) Pittsburg KS 3 hrs
*KMUW(FM) Wichita KS 6 hrs
*WKUE(FM) Elizabethtown KY 5 hrs
*WKMS-FM Murray KY 3 hrs
*WDCL-FM Somerset KY 5 hrs
*WFCR(FM) Amherst MA 4 hrs
*WGBH(FM) Boston MA 10 hrs
*WAMQ(FM) Great Barrington MA 7 hrs
*WOMR(FM) Provincetown MA 19 hrs
*WBSL-FM Sheffield MA 2 hrs
*WMTB-FM Emmittsburg MD 1 hr
*WRBC(FM) Lewiston ME 4 hrs
*WMHB(FM) Waterville ME 12 hrs
*WCBN-FM Ann Arbor MI 3 hrs
*WDET-FM Detroit MI 3 hrs
WBNZ(FM) Frankfort MI 2 hrs
*WBLU-FM Grand Rapids MI 5 hrs
*WCMW-FM Harbor Springs MI 3 hrs
*WLNZ(FM) Lansing MI 3 hrs
*WNMC-FM Traverse City MI 11 hrs
*WBLV(FM) Twin Lake MI 5 hrs
*WYCE(FM) Wyoming MI 1 hr
*KMSK(FM) Austin MN 5 hrs
*KBSB(FM) Bemidji MN 3 hrs
WMFG(AM) Hibbing MN
*KMSU(FM) Mankato MN 5 hrs
*KFAI(FM) Minneapolis MN 6 hrs
*WMCN(FM) Saint Paul MN 2 hrs
*KGAC(FM) Saint Peter MN 9 hrs
*KOPN(FM) Columbia MO 2 hrs
*KKFI(FM) Kansas City MO 4 hrs
*KCOZ(FM) Point Lookout MO 10 hrs
*KMST(FM) Rolla MO 5 hrs
*KEMC(FM) Billings MT 5 hrs
*KGLT(FM) Bozeman MT 12 hrs
*WCQS(FM) Asheville NC 9 hrs
*WBUX(FM) Buxton NC 20 hrs
*WUNC(FM) Chapel Hill NC 20 hrs
*WWCU(FM) Cullowhee NC 4. hrs
*WFSS(FM) Fayetteville NC 3 hrs
*WFQS(FM) Franklin NC 9 hrs
*WUND-FM Manteo NC 20 hrs
*WURI(FM) Manteo NC 20 hrs
*WZRU(FM) Roanoke Rapids NC 5 hrs
WLHC(FM) Robbins NC 3 hrs
*WNCW(FM) Spindale NC 12 hrs
*KEYA(FM) Belcourt ND 4 hrs
*KCND(FM) Bismarck ND 6 hrs
*KDPR(FM) Dickinson ND 6 hrs
*KPRJ(FM) Jamestown ND 6 hrs
*KMPR(FM) Minot ND 6 hrs
*KPPR(FM) Williston ND 6 hrs
*KRNU(FM) Lincoln NE 2 hrs
*KZUM(FM) Lincoln NE 8 hrs
*WEVO(FM) Concord NH 3 hrs
*WUNH(FM) Durham NH 4 hrs
*WEVC(FM) Gorham NH 3 hrs
*WEVH(FM) Hanover NH 3 hrs
*WNEC-FM Henniker NH 18 hrs
WEVJ(FM) Jackson NH 3 hrs wkly hrs
*WEVN(FM) Keene NH 3 hrs
*WKNH(FM) Keene NH 6 hrs
*WDVR(FM) Delaware Township NJ 6 hrs
*WBZC(FM) Pemberton NJ 4 hrs
*WLFR(FM) Pomona NJ 3 hrs
*WTSR(FM) Trenton NJ 4 hrs
*KSJE(FM) Farmington NM 15 hrs
*KGLP(FM) Gallup NM 10 hrs
*KRWG(FM) Las Cruces NM 8 hrs
*KUNR(FM) Reno NV 2 hrs
*WAMC-FM Albany NY 7 hrs
*WSKG-FM Binghamton NY 5 hrs
*WXLH(FM) Blue Mountain Lake NY
*WBNY(FM) Buffalo NY 3 hrs
*WCAN(FM) Canajoharie NY 7 hrs
*WSLU(FM) Canton NY
*WSQE(FM) Corning NY 5 hrs
*WCVF-FM Fredonia NY 4 hrs
*WICB(FM) Ithaca NY 2 hrs
*WSQG-FM Ithaca NY 5 hrs
WVBR-FM Ithaca NY 8 hrs
*WAMK(FM) Kingston NY 7 hrs
WYBG(AM) Massena NY
*WOSR(FM) Middletown NY 7 hrs
*WBAI(FM) New York NY 2 hrs
*WSUF(FM) Noyack NY 5 hrs
*WRHO(FM) Oneonta NY 5 hrs
*WSQC-FM Oneonta NY 5 hrs
WTKV(FM) Oswego NY 3 hrs
*WCEL(FM) Plattsburgh NY 7 hrs
*WSPN(FM) Saratoga Springs NY 6 hrs
*WUSB(FM) Stony Brook NY 15 hrs
*WANC(FM) Ticonderoga NY 6 hrs
WTBQ(AM) Warwick NY
*WBGU(FM) Bowling Green OH 4 hrs
*WLFC(FM) Findlay OH 3 hrs
*WKSU-FM Kent OH 12 hrs
*WOBC-FM Oberlin OH 12 hrs
*WMUB(FM) Oxford OH 5 hrs
WNTO(FM) Racine OH 2 hrs
*WKRW(FM) Wooster OH 12 hrs
*WYSO(FM) Yellow Springs OH 2 hrs
*WYSU(FM) Youngstown OH 3 hrs
*KRSC-FM Claremore OK 5 hrs
*KSMF(FM) Ashland OR 3 hrs
*KMUN(FM) Astoria OR 18 hrs
*KSBA(FM) Coos Bay OR 3 hrs
*KBVR(FM) Corvallis OR 4 hrs
*KLCC(FM) Eugene OR 12 hrs
*KRVM-FM Eugene OR 3 hrs
*KSKF(FM) Klamath Falls OR 3 hrs
*KTEC(FM) Klamath Falls OR 3 hrs
*KLCO(FM) Newport OR 12 hrs
*WDIY(FM) Allentown PA 12 hrs
WWCS(AM) Canonsburg PA 2 hrs
*WZBT(FM) Gettysburg PA 6 hrs
*WIUP-FM Indiana PA 4 hrs
*WPSX(FM) Kane PA 10 hrs
*WHYY-FM Philadelphia PA 4 hrs
*WXPN(FM) Philadelphia PA 5 hrs
*WRCT(FM) Pittsburgh PA 3 hrs
*WYEP-FM Pittsburgh PA 9 hrs
WEEU(AM) Reading PA 3 hrs
*WPSU(FM) State College PA 12 hrs
*WRDV(FM) Warminster PA 4 hrs
*WRIU(FM) Kingston RI 15 hrs
*WJMF(FM) Smithfield RI 4 hrs
*KAUR(FM) Sioux Falls SD 2 hrs
*KCSD(FM) Sioux Falls SD 5 hrs
*WHCB(FM) Bristol TN 1 hr
*WTTU(FM) Cookeville TN 3 hrs
*WRVU(FM) Nashville TN 3 hrs
*KUT(FM) Austin TX 4 hrs
*KAMU-FM College Station TX 3 hrs
*KEOS(FM) College Station TX 10 hrs
*KTEP(FM) El Paso TX 3 hrs
*KOOP(FM) Hornsby TX 7 hrs
*KTRU(FM) Houston TX 3 hrs
*KOCV(FM) Odessa TX 4 hrs
*KUTX(FM) San Angelo TX 4 hrs
*KSTX(FM) San Antonio TX 5 hrs
*KZMU(FM) Moab UT 6 hrs
*WNRN(FM) Charlottesville VA 19 hrs
*WTJU(FM) Charlottesville VA 20 hrs
*WMRY(FM) Crozet VA 8 hrs
*WMRA(FM) Harrisonburg VA 8 hrs
*WMRL(FM) Lexington VA 8 hrs
*WHRV(FM) Norfolk VA 7 hrs
*WVRU(FM) Radford VA 1 hr
*WCVE(FM) Richmond VA 6 hrs
*WVPS(FM) Burlington VT 4 hrs
*WIUV(FM) Castleton VT 4 hrs

*WRMC-FM Middlebury VT 15 hrs
WNCS(FM) Montpelier VT 4 hrs
*WRVT(FM) Rutland VT 4 hrs
*WVPR(FM) Windsor VT 6 hrs
*KUGS(FM) Bellingham WA 6 hrs wkly hrs
*KZAZ(FM) Bellingham WA 8 hrs
*KNWR(FM) Ellensburg WA
*KSER(FM) Everett WA 2 hrs
*KAOS(FM) Olympia WA 16 hrs
*KZUU(FM) Pullman WA 4 hrs
*KFAE-FM Richland WA
*KAGU(FM) Spokane WA 2 hrs
*KPBX-FM Spokane WA
*KWRS(FM) Spokane WA 8 hrs
*WHSA(FM) Brule WI 3 hrs
*WBSD(FM) Burlington WI 5 hrs
*WUEC(FM) Eau Claire WI 3 hrs
*WPNE-FM Green Bay WI 8 hrs
*WVSS(FM) Menomonie WI 6 hrs
*WHND(FM) Sister Bay WI 3 hrs
*WLBL-FM Wausau WI 3 hrs
*WVBC(FM) Bethany WV 2 hrs
WWYO(AM) Pineville WV 1 hr
WMOV(AM) Ravenswood WV 2 hrs
WHAW(AM) Weston WV 4 hrs
*KUWA(FM) Afton WY 5 hrs
*KUWJ(FM) Jackson WY 10 hrs
*KUWR(FM) Laramie WY 10 hrs
*KUWZ(FM) Rock Springs WY 10 hrs

Foreign/Ethnic

*KJHA(FM) Houston AK 3 hrs
*KJNP(AM) North Pole AK 1 hr
*KCAW(FM) Sitka AK 3 hrs
KSAZ(AM) Marana AZ 2 hrs
*KKUP(FM) Cupertino CA 19 hrs
KBIF(AM) Fresno CA 56 hrs
*KFCF(FM) Fresno CA 1 hr
*KMUD(FM) Garberville CA 1 hr
KMYC(AM) Marysville CA 2 hrs
*KVMR(FM) Nevada City CA 20 hrs
*KAZU(FM) Pacific Grove CA 6 hrs
*KXJZ(FM) Sacramento CA 2 hrs
KEST(AM) San Francisco CA
*KUSF(FM) San Francisco CA 2 hrs
KMRB(AM) San Gabriel CA 4 hrs
*KCSB-FM Santa Barbara CA 2 hrs
KOBO(AM) Yuba City CA 3 hrs
*KRZA(FM) Alamosa CO 3 hrs
*WFAR(FM) Danbury CT 1 hr
*WJMJ(FM) Hartford CT 1 hr
*WWUH(FM) West Hartford CT 14 hrs
WHUR-FM Washington DC 6 hrs
WYUS(AM) Milford DE 3 hrs
WXYB(AM) Indian Rocks Beach FL 5 hrs
WHOO(AM) Kissimmee FL 2 hrs
*WDNA(FM) Miami FL 10 hrs
*WLRN-FM Miami FL 3 hrs
WTMY(AM) Sarasota FL 4 hrs
*WRFG(FM) Atlanta GA 9 hrs
WWWE(AM) Hapeville GA 6 hrs
*WWET(FM) Valdosta GA 3 hrs
KUAM(AM) Hagatna GU 10 hrs
KUAI(AM) Eleele HI 5 hrs
KWAI(AM) Honolulu HI 2 hrs
KKON(AM) Kealakekua HI
*KDCR(FM) Sioux Center IA 1 hr
*WHPK-FM Chicago IL 4 hrs
WONX(AM) Evanston IL 28 hrs
WVIV-FM Highland Park IL
*WQNA(FM) Springfield IL 3 hrs
*WFHB(FM) Bloomington IN 3 hrs
WNDZ(AM) Portage IN 4 hrs
*WCUW(FM) Worcester MA 10 hrs
*WEAA(FM) Baltimore MD 7 hrs
WMET(AM) Gaithersburg MD 4 hrs
*WUPI(FM) Presque Isle ME 2 hrs
*WMHB(FM) Waterville ME 10 hrs
*WHFR(FM) Dearborn MI 2 hrs
KNOF(FM) Saint Paul MN 1 hr
*WMCN(FM) Saint Paul MN 16 hrs
*WSHA(FM) Raleigh NC 7 hrs
KMTY(FM) Holdrege NE 1 hr
*WFMU(FM) East Orange NJ 15 hrs
*WBJB-FM Lincroft NJ 3 hrs
WPRB(FM) Princeton NJ 6 hrs
*KUNM(FM) Albuquerque NM 9 hrs
KTAO(FM) Taos NM 5 hrs
*KUNR(FM) Reno NV 9 hrs
*WGMC(FM) Greece NY 1 hr
*WICB(FM) Ithaca NY 2 hrs
WJTN(AM) Jamestown NY 1 hr
WKSN(AM) Jamestown NY 1 hr
*WRHO(FM) Oneonta NY 2 hrs
*WRHV(FM) Poughkeepsie NY 1 hr
*WSPN(FM) Saratoga Springs NY 3 hrs
*WAER(FM) Syracuse NY 4 hrs
*WAPS(FM) Akron OH 3 hrs
WIMX(FM) Gibsonburg OH 8 hrs
*WDUB(FM) Granville OH 2 hrs
*KLCO(FM) Newport OR 3 hrs
*KBOO(FM) Portland OR 4 hrs
KXMG(AM) Portland OR 1 hr
WWCS(AM) Canonsburg PA 2 hrs
WZUM(AM) Carnegie PA 2 hrs
*WKDU(FM) Philadelphia PA 12 hrs
WTPM(FM) Aguadilla PR 1 hr
KEDA(AM) San Antonio TX 4 hrs
*KZMU(FM) Moab UT 1 hr
*KRCL(FM) Salt Lake City UT 5 hrs
*WUVT-FM Blacksburg VA 8 hrs
*KAOS(FM) Olympia WA 3 hrs
*KPBX-FM Spokane WA
*WHID(FM) Green Bay WI 2. hrs
WEKZ(AM) Monroe WI 3 hrs
KBBS(AM) Buffalo WY 1 hr

French

*KTOO(FM) Juneau AK 2 hrs
*KUSF(FM) San Francisco CA 2 hrs
*KSRH(FM) San Rafael CA 1 hr
*WPKN(FM) Bridgeport CT 2 hrs
WUST(AM) Washington DC 15 hrs
WOKB(AM) Ocoee FL 7 hrs
*WUCF-FM Orlando FL 1 hr
KSIG(AM) Crowley LA 18 hrs
KVPI-FM Ville Platte LA 18 hrs
WJIB(AM) Cambridge MA 10 hrs
WHTB(AM) Fall River MA 1 hr
*WCUW(FM) Worcester MA 2 hrs
*WRBC(FM) Lewiston ME 2 hrs
*WCBN-FM Ann Arbor MI 1 hr
*KFAI(FM) Minneapolis MN 2 hrs
WMOU(AM) Berlin NH 3 hrs
WFEA(AM) Manchester NH 3 hrs
WCHP(AM) Champlain NY
*WOBC-FM Oberlin OH 1 hr
*KRRC(FM) Portland OR 2 hrs
WWCS(AM) Canonsburg PA 2 hrs
WNRI(AM) Woonsocket RI 4 hrs
WOON(AM) Woonsocket RI 1 hr
*WEVL(FM) Memphis TN 1 hr
*WRVU(FM) Nashville TN 1 hr
*WUTS(FM) Sewanee TN 2 hrs
*WIUJ(FM) Charlotte Amalie VI 4 hrs

Full Service

WTMY(AM) Sarasota FL 2 hrs

German

*KCSN(FM) Northridge CA 3 hrs
KFKA(AM) Greeley CO 1 hr
WUST(AM) Washington DC 7 hrs
*WKTO(FM) Edgewater FL 1.5 hrs
KSKB(FM) Brooklyn IA 1 hr
*KRNL-FM Mount Vernon IA 2 hrs
*WIIT(FM) Chicago IL 3 hrs
WKTA(AM) Evanston IL 5 hrs
WGNU(AM) Granite City IL 2 hrs
WIJR(AM) Highland IL 1 hr
WVIV-FM Highland Park IL
*WICR(FM) Indianapolis IN 1 hr
WNDZ(AM) Portage IN 2 hrs
*WCUW(FM) Worcester MA 2 hrs
WBMD(AM) Baltimore MD 1 hr
WATZ(AM) Alpena MI 2 hrs
WKCQ(FM) Saginaw MI 3 hrs
KASM(AM) Albany MN 2 hrs
WEW(AM) Saint Louis MO 3 hrs
KTTT(AM) Columbus NE 5 hrs
KBRX(AM) O'Neill NE 6 hrs
*KUNV(FM) Las Vegas NV 1 hr
*WBXL(FM) Baldwinsville NY 3 hrs
WHVW(AM) Hyde Park NY 1 hr
WKNY(AM) Kingston NY 1 hr
WXRL(AM) Lancaster NY 1 hr
WVOA-FM Mexico NY 2 hrs
WTLA(AM) North Syracuse NY 2 hrs
WSGO(AM) Oswego NY 2 hrs
*WAPS(FM) Akron OH 2 hrs
*WOBO(FM) Batavia OH 5 hrs
*WCPN(FM) Cleveland OH 1 hr
WKTX(AM) Cortland OH 5 hrs
*WQRP(FM) Dayton OH 3 hrs
WONW(AM) Defiance OH 4 hrs
WDLW(AM) Lorain OH one hrs
WCWA(AM) Toledo OH 1 hr
WELW(AM) Willoughby-Eastlake OH 1 hr
KKRX(AM) Lawton OK 1 hr
*WMUH(FM) Allentown PA 2 hrs
WGPA(AM) Bethlehem PA 2 hrs
WWCS(AM) Canonsburg PA 2 hrs
*WDCV-FM Carlisle PA 1 hr
*WMCE(FM) Erie PA 4 hrs
WKST(AM) New Castle PA 1 hr
WEEU(AM) Reading PA 2 hrs
*WCLH(FM) Wilkes-Barre PA 3 hrs
*WRVU(FM) Nashville TN 1 hr
KHLT(AM) Hallettsville TX 5 hrs
*KOOP(FM) Hornsby TX .5 hrs
KVLG(AM) La Grange TX 1 hr
KIKZ(AM) Seminole TX 1 hr
KSEM-FM Seminole TX 1 hr
WKGM(AM) Smithfield VA 1 hr
KARI(AM) Blaine WA 2 hrs
WSSP(AM) Milwaukee WI 8 hrs
WEKZ(AM) Monroe WI 3 hrs
WXER(FM) Plymouth WI 3 hrs

Golden Oldies

KGFT(FM) Pueblo CO 11 hrs
WSYY-FM Millinocket ME

Gospel

KFAR(AM) Fairbanks AK 2 hrs
*KSDP(AM) Sand Point AK 3 hrs
KIAL(AM) Unalaska AK 02 hrs
WVNN(AM) Athens AL 3 hrs
WNSI-FM Atmore AL 12 hrs
WSPZ(AM) Birmingham AL 5 hrs
WBSA(AM) Boaz AL
WAOQ(FM) Brantley AL 14. hrs
WKNU(FM) Brewton AL 2 hrs
WACQ(AM) Carrville AL 5 hrs
WGZZ(FM) Dadeville AL 3 hrs
WOOF(AM) Dothan AL 17 hrs
WTVY-FM Dothan AL 4 hrs
WELB(AM) Elba AL 12 hrs
*WFIX(FM) Florence AL 6 hrs
WZOB(AM) Fort Payne AL 5 hrs
WGAD(AM) Gadsden AL 5 hrs
WDJL(AM) Huntsville AL 6 hrs
*WJAB(FM) Huntsville AL 5 hrs
WCKA(AM) Jacksonville AL 2 hrs
WIXI(AM) Jasper AL 6 hrs
WINL(FM) Linden AL 5 hrs
WLWI-FM Montgomery AL 4 hrs
*WVAS(FM) Montgomery AL 5 hrs
WAMI(AM) Opp AL 15 hrs
WAMI-FM Opp AL 15 hrs
WOPP(AM) Opp AL 19 hrs
WJRL-FM Ozark AL 8 hrs
WGOL(AM) Russellville AL 4 hrs
WFEB(AM) Sylacauga AL 6 hrs
WNUZ(AM) Talladega AL 7 hrs
*WEBT(FM) Valley AL
KEWI(AM) Benton AR 10 hrs
KAMD-FM Camden AR 19 hrs
KBJT(AM) Fordyce AR 11 hrs
KHOZ(AM) Harrison AR 10 hrs
KFFA(AM) Helena AR 4 hrs
KBOK(AM) Malvern AR 8 hrs
KUOA(AM) Siloam Springs AR 2 hrs
KZHE(FM) Stamps AR 8 hrs
KFYX(FM) Texarkana AR 2 hrs
KWRF(AM) Warren AR 8 hrs
KWRF-FM Warren AR 8 hrs
KCLT(FM) West Helena AR 15 hrs
KDVA(FM) Buckeye AZ 7 hrs
*KXCI(FM) Tucson AZ 2 hrs
KXMX(AM) Anaheim CA
KCEL(FM) California City CA 3 hrs
KRML(AM) Carmel CA 6 hrs
KJLH-FM Compton CA 6 hrs
*KECG(FM) El Cerrito CA 5 hrs
KTDE(FM) Gualala CA 1 hr
KGBA-FM Holtville CA 7 hrs
KSRN(FM) Kings Beach CA 1 hr
*KPFK(FM) Los Angeles CA 2 hrs
KYOS(AM) Merced CA 1 hr
*KAZU(FM) Pacific Grove CA 4 hrs
*KZYX(FM) Philo CA 2 hrs
KEST(AM) San Francisco CA
KISQ(FM) San Francisco CA 3 hrs
KDIA(AM) Vallejo CA 7 hrs
KDYA(AM) Vallejo CA 7 hrs
KUBA(AM) Yuba City CA 2 hrs
*KASF(FM) Alamosa CO 4 hrs
KRTZ(FM) Cortez CO 1 hr
KVAY(FM) Lamar CO 4 hrs
KSLV(AM) Monte Vista CO 4 hrs
*KVNF(FM) Paonia CO 3 hrs
KGFT(FM) Pueblo CO 3 3 hrs
*WQTQ(FM) Hartford CT 12 hrs
*WRTC-FM Hartford CT 6 hrs
*WESU(FM) Middletown CT 6 hrs
WYBC-FM New Haven CT 8 hrs
*WCNI(FM) New London CT 3 hrs
*WNHU(FM) West Haven CT 4 hrs
WKND(AM) Windsor CT 5 hrs
WHUR-FM Washington DC 14 hrs
WAFL(FM) Milford DE 2 hrs
WYBT(AM) Blountstown FL 15 hrs
WWPR(AM) Bradenton FL 6 hrs
WTMP-FM Dade City FL 20 hrs
WZEP(AM) De Funiak Springs FL 10 hrs
WTMP(AM) Egypt Lake FL 20 hrs
WENG(AM) Englewood FL 2 hrs
*WUFT-FM Gainesville FL 1 hr
WRNE(AM) Gulf Breeze FL
*WJUF(FM) Inverness FL 1 hr
WWAB(AM) Lakeland FL 12 hrs
WLBE(AM) Leesburg FL 4 hrs
WQHL(AM) Live Oak FL 7 hrs
WQHL-FM Live Oak FL 7 hrs
WMAF(AM) Madison FL
WTYS(AM) Marianna FL 11 hrs
WOKB(AM) Ocoee FL 12 hrs
WLTG(AM) Panama City FL 7 hrs
WTMY(AM) Sarasota FL 6 hrs
WPUL(AM) South Daytona FL
*WANM(FM) Tallahassee FL 18 hrs hrs
WQLC(FM) Watertown FL 4 hrs
WFLM(FM) White City FL 20 hrs
WZZS(FM) Zolfo Springs FL 2 hrs
V6AH(AM) Pohnpei FM 2 hrs
WFSH-FM Athens GA 2 hrs
*WCLK(FM) Atlanta GA 17 hrs
WTUF(FM) Boston GA 7 hrs
WJTH(AM) Calhoun GA 2 hrs
WZBN(FM) Camilla GA 16 hrs
WDCO(AM) Cochran GA 6 hrs
WDXQ-FM Cochran GA 6 hrs
WCON-FM Cornelia GA 10 hrs
WOKA-FM Douglas GA 4 hrs
WQZY(FM) Dublin GA 3 hrs
WRDO(FM) Fitzgerald GA 6 hrs
WFJO(FM) Folkston GA 5 hrs
*WBCX(FM) Gainesville GA 6 hrs
WJGA-FM Jackson GA 15 hrs
WLOP(AM) Jesup GA 10 hrs
WHCG(AM) Metter GA 4 hrs
WNEA(AM) Newnan GA 15 hrs
WJEP(AM) Ochlocknee GA 3 hrs
WRBX(FM) Reidsville GA 161 hrs
WPTB(AM) Statesboro GA 6 hrs
WTHO-FM Thomson GA 1 hr
WLET(AM) Toccoa GA
WNGC(FM) Toccoa GA
WGOV(AM) Valdosta GA 14 hrs
WVLD(AM) Valdosta GA 2 hrs
KSTO(FM) Hagatna GU 6 hrs
KELR-FM Chariton IA 5 hrs
KROS(AM) Clinton IA 1 hr
*KALA(FM) Davenport IA 13 hrs
*KHOE(FM) Fairfield IA 3 hrs
KYTC(FM) Northwood IA 1 hr
KBOE(AM) Oskaloosa IA 9 hrs
*KIGC(FM) Oskaloosa IA 12 hrs
KLEE(AM) Ottumwa IA 6 hrs
KIHK(FM) Rock Valley IA 3 hrs
KTLB(FM) Twin Lakes IA 2 hrs
*KBBG(FM) Waterloo IA
KWIK(AM) Pocatello ID 2 hrs
WBGZ(AM) Alton IL 4 hrs
WKRO(AM) Cairo IL 12 hrs
*WSSD(FM) Chicago IL
WWHP(FM) Farmer City IL 3 hrs
WJRE(FM) Galva IL 1 hr
*WDCB(FM) Glen Ellyn IL 2 hrs
WHPO(FM) Hoopeston IL 7 hrs
WJOL(AM) Joliet IL 1 hr
*WMXM(FM) Lake Forest IL 3 hrs
WPNA(AM) Oak Park IL 2 hrs
WVAZ(FM) Oak Park IL 4 hrs
WBBA-FM Pittsfield IL 3 hrs
WNTA(AM) Rockford IL 20 hrs
WPMB(AM) Vandalia IL 3 hrs
WYKT(FM) Wilmington IL 4 hrs
WBNL(AM) Boonville IN 5 hrs
WURK(FM) Elwood IN 4 hrs
WFLQ(FM) French Lick IN 4 hrs
WKAM(AM) Goshen IN 6 hrs
WXLW(AM) Indianapolis IN 5 hrs
WYXB(FM) Indianapolis IN 10 hrs
*WKPW(FM) Knightstown IN
WMRS(FM) Monticello IN 5 hrs
WNDZ(AM) Portage IN 2 hrs
WRIN(AM) Rensselaer IN 2 hrs
WAXI(FM) Rockville IN 2 hrs
WTCJ(AM) Tell City IN 6 hrs
WWVR(FM) West Terre Haute IN
KDNS(FM) Downs KS 5 hrs
KHAZ(FM) Hays KS 3 hrs
KINZ(FM) Humboldt KS 3 hrs
KHUT(FM) Hutchinson KS 5 hrs
KWBW(AM) Hutchinson KS 11 hrs
KNNS(AM) Larned KS 6 hrs
*KSDB-FM Manhattan KS 3 hrs
KFNF(FM) Oberlin KS 3 hrs
KKAN(AM) Phillipsburg KS 12 hrs
KFRM(AM) Salina KS 5 hrs
KSKG(FM) Salina KS 3 hrs
*KMUW(FM) Wichita KS 3 hrs
WANY(AM) Albany KY 6 hrs
WMDJ-FM Allen KY
WLFX(FM) Berea KY 12 hrs
WAIN-FM Columbia KY 15 hrs
WJQI(AM) Fort Campbell KY 10 hrs
WHVO(AM) Hopkinsville KY 3 hrs
WLBN(AM) Lebanon KY 5 hrs
WFTM(AM) Maysville KY 5 hrs
WFTM-FM Maysville KY 5 hrs
WFXY(AM) Middlesboro KY 3 hrs
WMIK(AM) Middlesboro KY 2 hrs
WKYQ(FM) Paducah KY 2 hrs
WRLV-FM Salyersville KY 4 hrs
WKWY(FM) Tompkinsville KY 7 hrs
WTKY-FM Tompkinsville KY 8 hrs
KRVV(FM) Bastrop LA 4 hrs
WFPR(AM) Hammond LA 12 hrs
WHMD(FM) Hammond LA 4 hrs
KJLO-FM Monroe LA 4 hrs
KBSF(AM) Springhill LA 15 hrs
KVCL-FM Winnfield LA 20 hrs
WJIB(AM) Cambridge MA 4 hrs
*WAIC(FM) Springfield MA
WPEP(AM) Taunton MA 12 hrs

*WEAA(FM) Baltimore MD 13 hrs
*WMTB-FM Emmittsburg MD 1 hr
WKHI(FM) Fruitland MD 5 hrs
WAAI(FM) Hurlock MD 3 hrs
WKHW(FM) Pocomoke City MD 5 hrs
*WESM(FM) Princess Anne MD 20 hrs
WQTE(FM) Adrian MI 2 hrs
*WCBN-FM Ann Arbor MI 1 hr
WGTO(AM) Cassopolis MI 10 hrs
WLCM(AM) Charlotte MI 3 hrs
WDZZ-FM Flint MI 8 hrs
WJNZ(AM) Kentwood MI 4 hrs
WTLZ(FM) Saginaw MI 6 hrs
WMLM(AM) Saint Louis MI 2 hrs
KDUZ(AM) Hutchinson MN 3 hrs
KLQL(FM) Luverne MN 4 hrs
KYOO(AM) Bolivar MO 2 hrs
KBFL-FM Buffalo MO 3 hrs
KATI(FM) California MO 3 hrs
KRLL(AM) California MO 3 hrs
KMFC(FM) Centralia MO 2 hrs
KCHR(AM) Charleston MO 10 hrs
KWKZ(FM) Charleston MO 6 hrs
*KOPN(FM) Columbia MO 3 hrs
KUNQ(FM) Houston MO 10 hrs
*KTTK(FM) Lebanon MO 7 hrs
*KTTK(FM) Lebanon MO 80 hrs
*KJAB-FM Mexico MO 20 hrs
KWBZ(FM) Monroe City MO 4 hrs
*KGSP(FM) Parkville MO 3 hrs
*KCLC(FM) Saint Charles MO 9 hrs
KDRO(AM) Sedalia MO 6 hrs
*KWND(FM) Springfield MO 3 hrs
KTTN-FM Trenton MO 6 hrs
WWZQ(AM) Aberdeen MS 8 hrs
WAMY(AM) Amory MS 6 hrs
WESE(FM) Baldwyn MS 6 hrs
WJBI(AM) Batesville MS
WBSL(AM) Bay St. Louis MS 13 hrs
WBKN(FM) Brookhaven MS 3 hrs
*WPAE(FM) Centreville MS 15 hrs
WKRA(AM) Holly Springs MS 12 hrs
*WJSU(FM) Jackson MS 18 hrs
WNAT(AM) Natchez MS 18 hrs
WNAU(AM) New Albany MS
WWZD-FM New Albany MS 4 hrs
WOXD(FM) Oxford MS 12 hrs
WKZU(FM) Ripley MS 6 hrs
WSAO(AM) Senatobia MS
WAVN(AM) Southaven MS 19 hrs
WLRC(AM) Walnut MS
WROB(AM) West Point MS 2 hrs
WIGG(AM) Wiggins MS
WKXR(AM) Asheboro NC 10 hrs
WWNC(AM) Asheville NC 3 hrs
*WGWG(FM) Boiling Springs NC 15 hrs
WATA(AM) Boone NC 5 hrs
WGCR(AM) Brevard NC
WSQL(AM) Brevard NC 8 hrs
*WCCE(FM) Buie's Creek NC 11 hrs
WKYK(AM) Burnsville NC 12 hrs
WCLN(AM) Clinton NC 7.5 hrs
WPEG(FM) Concord NC 6 hrs
WFXC(FM) Durham NC 4 hrs
*WNCU(FM) Durham NC
*WRVS-FM Elizabeth City NC 20 hrs
WBLA(AM) Elizabethtown NC 8 hrs
*WFSS(FM) Fayetteville NC 2 hrs
WYCV(AM) Granite Falls NC
WLNC(AM) Laurinburg NC 4 hrs
WLON(AM) Lincolnton NC 5 hrs
WBRM(AM) Marion NC 5 hrs
WDJS(AM) Mount Olive NC 5 hrs
WIKS(FM) New Bern NC 4 hrs
WECR(AM) Newland NC 10 hrs
WPJL(AM) Raleigh NC
*WSHA(FM) Raleigh NC 20 hrs
*WZRU(FM) Roanoke Rapids NC 6 hrs
WRSV(FM) Rocky Mount NC 20 hrs
WKRX(FM) Roxboro NC 4 hrs
WKRX(FM) Roxboro NC 5 hrs
WRXO(AM) Roxboro NC 5 hrs
WNCA(AM) Siler City NC 10 hrs
WEEB(AM) Southern Pines NC 6 hrs
*WNCW(FM) Spindale NC 2 hrs
WAME(AM) Statesville NC 5 hrs
WFXK(FM) Tarboro NC 3 hrs
WACB(AM) Taylorsville NC 12 hrs
WSVM(AM) Valdese NC 4 hrs
WKSK(AM) West Jefferson NC 5 hrs
WYNC(AM) Yanceyville NC
*KABU(FM) Fort Totten ND 7 hrs
KAUJ(FM) Grafton ND 4 hrs
KXPO(AM) Grafton ND 5 hrs
KQLX(AM) Lisbon ND 6 hrs
KTGO(AM) Tioga ND 11 hrs
*KINI(FM) Crookston NE 6 hrs
KMTY(FM) Holdrege NE 5 hrs
KUVR(AM) Holdrege NE 5 hrs
*KRNU(FM) Lincoln NE 2 hrs
KRFS(AM) Superior NE 3 hrs
WTMR(AM) Camden NJ 20 hrs
*WJPG(FM) Cape May Court House NJ 1 hr
WTTH(FM) Margate City NJ 5 hrs
WMVB(AM) Millville NJ 8 hrs
*WTSR(FM) Trenton NJ 6 hrs
*WMSC(FM) Upper Montclair NJ 2 hrs
*WMCX(FM) West Long Branch NJ 6 hrs
*WJPH(FM) Woodbine NJ 1 hr
KRSY(AM) Alamogordo NM 8 hrs
KATK(AM) Carlsbad NM 2 hrs
*KCEP(FM) Las Vegas NV 19 hrs
*WCDB(FM) Albany NY 3 hrs
WROW(AM) Albany NY 3 hrs
*WXLH(FM) Blue Mountain Lake NY
*WSLU(FM) Canton NY
WSIV(AM) East Syracuse NY 1 hr
*WEOS(FM) Geneva NY 3 hrs
*WITR(FM) Henrietta NY 8 hrs
WHVW(AM) Hyde Park NY 1.5 hrs
*WVCR-FM Loudonville NY 3 hrs
WVOX(AM) New Rochelle NY 1 hr
WJJL(AM) Niagara Falls NY 1 hr
WLIM(AM) Patchogue NY 2 hrs
WDKX(FM) Rochester NY 7 hrs
WMYY(FM) Schoharie NY
*WAER(FM) Syracuse NY 3 hrs
*WONB(FM) Ada OH 3 hrs
*WRMU(FM) Alliance OH 2 hrs
WAIS(AM) Buchtel OH 3 hrs
WCER(AM) Canton OH 11 hrs
WCVX(AM) Cincinnati OH 15 hrs
*WWSU(FM) Dayton OH 3 hrs
WOHI(AM) East Liverpool OH 1 hr
WEDI(AM) Eaton OH 5 hrs
*WCVO(FM) Gahanna OH 6 hrs
WPOS-FM Holland OH 20 hrs
WJYD(FM) London OH 10 hrs
WTIG(AM) Massillon OH 6 hrs
WMPO(AM) Middleport-Pomeroy OH 18 hrs
WQIO(FM) Mount Vernon OH 2 hrs
WNPQ(FM) New Philadelphia OH 4 hrs
WPAY(AM) Portsmouth OH 6 hrs
WNTO(FM) Racine OH 7 hrs
WULM(AM) Springfield OH 6 hrs
WDIG(AM) Steubenville OH
WJUC(FM) Swanton OH 12 hrs wkly hrs
WERT(AM) Van Wert OH 3 hrs
WKFI(AM) Wilmington OH 5 hrs
WBZI(AM) Xenia OH 5 hrs
KADA(AM) Ada OK 5 hrs
KADA-FM Ada OK 5 hrs
KYFM(FM) Bartlesville OK 4 hrs
KOKB(AM) Blackwell OK 11 hrs
KKBI(FM) Broken Bow OK 4 hrs
*KRSC-FM Claremore OK 4 hrs
KDDQ(FM) Comanche OK 2 hrs
KTNT(FM) Eufaula OK 3 hrs
KPRV-FM Heavener OK 12 hrs
KIHN(AM) Hugo OK 5 hrs
KFXI(FM) Marlow OK 8 hrs
KTMC(AM) McAlester OK 5 hrs
KTMC-FM McAlester OK 2 hrs
KKNG-FM Newcastle OK 8 hrs
KRIG-FM Nowata OK 4 hrs
KPRV(AM) Poteau OK 12 hrs
KFAQ(AM) Tulsa OK 2 hrs
KAJO(AM) Grants Pass OR 1 hr
KYKN(AM) Keizer OR 6 hrs
KWBY(AM) Woodburn OR 3 hrs
WWSM(AM) Annville-Cleona PA 3 hrs
WBVP(AM) Beaver Falls PA 2 hrs
WCHA(AM) Chambersburg PA 2 hrs
WCOJ(AM) Coatesville PA 2 hrs
*WERG(FM) Erie PA 3 hrs
WSKE(FM) Everett PA 1 hr
*WZBT(FM) Gettysburg PA 1 hr
*WVMM(FM) Grantham PA 2 hrs
WTCY(AM) Harrisburg PA
WTKT(AM) Harrisburg PA 2 hrs
*WIUP-FM Indiana PA 1 hr
WJSA(AM) Jersey Shore PA 2 hrs
WJSA-FM Jersey Shore PA 2 hrs
WQZS(FM) Meyersdale PA 5 hrs
WWBE(FM) Mifflinburg PA 2 hrs
WHAT(AM) Philadelphia PA 12 hrs
*WKDU(FM) Philadelphia PA 4 hrs
WURD(AM) Philadelphia PA 9 hrs
WUSL(FM) Philadelphia PA 4 hrs
WPHB(AM) Philipsburg PA 6 hrs
WPAM(AM) Pottsville PA
*WXLV(FM) Schnecksville PA 4 hrs
*WBYO(FM) Sellersville PA 2 hrs
WWII(AM) Shiremanstown PA 2 hrs
*WZZD(FM) Warwick PA 3 hrs
WKZV(AM) Washington PA 1 hr
*WRLC(FM) Williamsport PA 6 hrs
*WTMV(FM) Youngsville PA 2 hrs
*WJMF(FM) Smithfield RI 2 hrs
WOON(AM) Woonsocket RI 1 hr
WZLA-FM Abbeville SC 8 hrs
WBT-FM Chester SC 6 hrs
WOLS(AM) Florence SC 4 hrs
WFIS(AM) Fountain Inn SC 4 hrs
WBZF(FM) Hartsville SC 3 hrs
WHSC(AM) Hartsville SC 3 hrs
WKMG(AM) Newberry SC 3 hrs
WALI(FM) Walterboro SC 5 hrs
*KLND(FM) Little Eagle SD 3 hrs
KBHB(AM) Sturgis SD 3 hrs
WATX(AM) Algood TN
WVOL(AM) Berry Hill TN 6 hrs
*WHCB(FM) Bristol TN 15 hrs
WFWL(AM) Camden TN 8 hrs
WRJB(FM) Camden TN 4 hrs
WVFB(FM) Celina TN 7 hrs
WNKX(AM) Centerville TN 5 hrs
WNKX-FM Centerville TN
WMSR-FM Collinwood TN
WHUB(AM) Cookeville TN 11 hrs
WZYX(AM) Cowan TN 10 hrs
WCDZ(FM) Dresden TN 1 hr
WSDQ(AM) Dunlap TN 7 hrs
WEMB(AM) Erwin TN 10 hrs
*WVCP(FM) Gallatin TN 6 hrs
WMYL(FM) Halls Crossroads TN 2 hrs
WTNK(AM) Hartsville TN 4 hrs
*WFHU(FM) Henderson TN 9 hrs
WFKX(FM) Henderson TN 3 hrs
WHHM-FM Henderson TN 10 hrs
WQQK(FM) Hendersonville TN 6 hrs
WMLR(AM) Hohenwald TN 4 hrs
WDXI(AM) Jackson TN 16 hrs
WNRX(FM) Jefferson City TN 4 hrs
WKZX-FM Lenoir City TN 18 hrs
WLIL(AM) Lenoir City TN 18 hrs
WAXO(AM) Lewisburg TN 12 hrs
WDXL(AM) Lexington TN 10 hrs
WLIV(AM) Livingston TN 18 hrs
WDIA(AM) Memphis TN
WTRB(AM) Ripley TN 6 hrs
WJLE(AM) Smithville TN 15 hrs
WJLE-FM Smithville TN 15 hrs
WEPG(AM) South Pittsburg TN 10 hrs
WDBL(AM) Springfield TN 10 hrs
WSBI(AM) Static TN 10 hrs
WNTT(AM) Tazewell TN
*KGNZ(FM) Abilene TX 2 hrs
KIXZ(AM) Amarillo TX 6 hrs
*KAZI-FM Austin TX 18 hrs
KORQ(FM) Baird TX 3 hrs
KYKR(FM) Beaumont TX 4 hrs. hrs
KBYG(AM) Big Spring TX 6 hrs
KQTY(AM) Borger TX 3 hrs
KQTY-FM Borger TX 3 hrs
KNTX(AM) Bowie TX 4 hrs
KLTR(FM) Brenham TX 10 hrs
KPSM(FM) Brownwood TX 2 hrs
KTAE(AM) Cameron TX 6 hrs
KCAR(AM) Clarksville TX 6 hrs
KQBZ(FM) Coleman TX 7 hrs
KSTA(AM) Coleman TX 7 hrs
*KEOS(FM) College Station TX 3 hrs
KCOM(AM) Comanche TX 5 hrs
KSSM(FM) Copperas Cove TX
KBHT(FM) Crockett TX 6 hrs
KDDD-FM Dumas TX 5 hrs
KATX(FM) Eastland TX 5 hrs
KEAS(AM) Eastland TX 5 hrs
*KTEP(FM) El Paso TX 4 hrs
KNES(FM) Fairfield TX 3 hrs
KSWA(AM) Graham TX 2 hrs
KHBR(AM) Hillsboro TX 6 hrs
KOOK(FM) Junction TX 2 hrs
KRNH(FM) Kerrville TX 3 hrs
KRVL(FM) Kerrville TX 1 hr
*KTAI(FM) Kingsville TX 6 hrs
KYMI(FM) Los Ybanez TX 16 hrs
KZRC(FM) Markham TX 4 hrs
KHKZ(FM) Mercedes TX 2 hrs
KRQX(AM) Mexia TX 3 hrs
KCKM(AM) Monahans TX 3 hrs
KJCS(FM) Nacogdoches TX 3 hrs
KMBV(FM) Navasota TX 4 hrs
KPLT(AM) Paris TX 4 hrs
KZHN(AM) Paris TX
KRXT(FM) Rockdale TX 2 hrs hrs
KBAL(AM) San Saba TX 7 hrs
KBAL-FM San Saba TX 7 hrs
KBKH(FM) Shamrock TX 6 hrs
KXOX(AM) Sweetwater TX 4 hrs
KTFS(AM) Texarkana TX 2 hrs
KTBB(AM) Tyler TX 5 hrs
*KVNE(FM) Tyler TX 4 hrs
KZEY(AM) Tyler TX 16 hrs
KVWC(AM) Vernon TX 16 hrs
KALK(FM) Winfield TX 1 hr
KWUD(AM) Woodville TX 6 hrs
KBLQ-FM Logan UT 8 hrs
*KWCR-FM Ogden UT 3 hrs
*KUER(FM) Salt Lake City UT 3 hrs
WKDE(AM) Altavista VA 5 hrs
WKDE-FM Altavista VA 5 hrs
WKEX(AM) Blacksburg VA 5 hrs
*WTJU(FM) Charlottesville VA 1 hr
WKEY(AM) Covington VA 2 hrs
WEVA(AM) Emporia VA 3 hrs
WPAK(AM) Farmville VA 12 hrs
WLQM(AM) Franklin VA 12 hrs
WGAT(AM) Gate City VA 20 hrs
WMNA(AM) Gretna VA 15 hrs
WCLM(AM) Highland Springs VA 5 wkly hrs
WKWI(FM) Kilmarnock VA 5 hrs
WOJL(FM) Louisa VA 3 hrs
WMEV(AM) Marion VA 2 hrs
WSIG(FM) Mount Jackson VA 4 hrs
WNVA(AM) Norton VA 15 hrs
WVCV(AM) Orange VA 2 hrs
*WVST-FM Petersburg VA 13 hrs
WXLZ(AM) Saint Paul VA 15 hrs
WKGM(AM) Smithfield VA 5 hrs
WQOK(FM) South Boston VA 9 hrs
WTZE(AM) Tazewell VA 5 hrs
WKCW(AM) Warrenton VA 6 hrs
WMBG(AM) Williamsburg VA 5 hrs
WYVE(AM) Wytheville VA 4 hrs
WWOD(FM) Hartford VT 2 hrs
WMNV(FM) Rupert VT
WTWN(AM) Wells River VT
KBDB-FM Forks WA 4 hrs
KBIS(AM) Forks WA 6.5 hrs
KWDB(AM) Oak Harbor WA 6 hrs
*KNHC(FM) Seattle WA 6 hrs
KITZ(AM) Silverdale WA 2 hrs
KJOX(AM) Yakima WA 2 hrs
WATK(AM) Antigo WI 2 hrs
WMCS(AM) Greenfield WI 5 hrs
WKKV-FM Racine WI 6 hrs
WRCO-FM Richland Center WI 6 hrs
WKCJ(FM) Lewisburg WV 6 hrs
WVNT(AM) Parkersburg WV 4 hrs
WELD-FM Petersburg WV 3 hrs
WCWV(FM) Summersville WV 15 hrs
WHAW(AM) Weston WV 18 hrs
WXCC(FM) Williamson WV 8 hrs
KJUA(AM) Cheyenne WY 1 hr
KKTY-FM Douglas WY 1 hr

Greek

WONX(AM) Evanston IL 2 hrs
WVIV-FM Highland Park IL
WJOB(AM) Hammond IN 1 hr
WLYN(AM) Lynn MA 2 hrs
WGFP(AM) Webster MA 1 hr
WBMD(AM) Baltimore MD 2 hrs
WSDS(AM) Salem Township MI 4 hrs
WKBR(AM) Manchester NH 8 hrs
WAAL(FM) Binghamton NY 1 hr
WKTX(AM) Cortland OH 2 hrs
WCCD(AM) Parma OH 2 hrs
WEDO(AM) McKeesport PA 1 hr
WKST(AM) New Castle PA 1 hr
WBZK(AM) York SC 10 hrs
*WUVT-FM Blacksburg VA 2 hrs
WEIR(AM) Weirton WV 1 hr

Hardcore

*WITR(FM) Henrietta NY 2 hrs
*WNRN(FM) Charlottesville VA 4 hrs
*WWHS-FM Hampden-Sydney VA 4 hrs

Hebrew

*WHPK-FM Chicago IL 1 hr
KLAV(AM) Las Vegas NV 1 hr
*WKDU(FM) Philadelphia PA 3 hrs

Hindi

KLAV(AM) Las Vegas NV 1 hr
*KBOO(FM) Portland OR 1 hr
KXMG(AM) Portland OR 1 hr
WWII(AM) Shiremanstown PA 1 hr
*WEVL(FM) Memphis TN 1 hr
KTEK(AM) Alvin TX 3 hrs
KGOL(AM) Humble TX 15 hrs
*KTAI(FM) Kingsville TX 3 hrs

Hungarian

KTYM(AM) Inglewood CA 1 hr
*WAPS(FM) Akron OH 1 hr
*WCPN(FM) Cleveland OH 1 hr
WKTX(AM) Cortland OH 7 hrs
*WQRP(FM) Dayton OH 3 hrs
*WJCU(FM) University Heights OH 3 hrs
WEDO(AM) McKeesport PA 1 hr

Inspirational

*WGTS(FM) Takoma Park MD
WNBN(AM) Meridian MS

Irish

*KUSF(FM) San Francisco CA 1 hr
*KRCC(FM) Colorado Springs CO 5 hrs
WQUN(AM) Hamden CT 2 hrs
WMRD(AM) Middletown CT 1 hr
*WNHU(FM) West Haven CT 5 hrs
*WUCF-FM Orlando FL 1 hr
*WHPK-FM Chicago IL 1 hr
WPNA(AM) Oak Park IL 7 hrs
*KANU(FM) Lawrence KS 2 hrs
*WGBH(FM) Boston MA 2 hrs

WACE(AM) Chicopee MA 2 hrs
WNBP(AM) Newburyport MA 4 hrs
WNTN(AM) Newton MA 6 hrs
WBRK(AM) Pittsfield MA 1 hr
*WUNH(FM) Durham NH 2 hrs
WAAL(FM) Binghamton NY 2 hrs
WHVW(AM) Hyde Park NY 1 hr
WKNY(AM) Kingston NY 1 hr
*WVCR-FM Loudonville NY 3 hrs
WVOX(AM) New Rochelle NY 1 hr
*WFUV(FM) New York NY 10 hrs
WLIM(AM) Patchogue NY 1 hr
WLLW(FM) Seneca Falls NY 2 hrs
WSFW(AM) Seneca Falls NY 2 hrs
WTBQ(AM) Warwick NY 1 hr hrs
WTBQ(AM) Warwick NY 2 hrs
*WCBE(FM) Columbus OH 4 hrs
WGBN(AM) New Kensington PA 2 hrs
*WYEP-FM Pittsburgh PA 2 hrs
WADK(AM) Newport RI 2 hrs
*WEVL(FM) Memphis TN 4 hrs

Italian

KTYM(AM) Inglewood CA 3 hrs
*KUSF(FM) San Francisco CA 1 hr
WICC(AM) Bridgeport CT 5 hrs
*WFAR(FM) Danbury CT 1 hr
WGCH(AM) Greenwich CT 1 hr
WMRD(AM) Middletown CT 2 hrs
WRYM(AM) New Britain CT 2 hrs
WATR(AM) Waterbury CT 3 hrs
*WWUH(FM) West Hartford CT 3 hrs
WXYB(AM) Indian Rocks Beach FL 2 hrs
WMEL(AM) Melbourne FL 2 hrs
WDUV(FM) New Port Richey FL 1 hr
WPSO(AM) New Port Richey FL 2 hrs
*WUCF-FM Orlando FL 1 hr
WNTN(AM) Newton MA 2 hrs
*WHRW(FM) Binghamton NY 3 hrs
WAMF(AM) Fulton NY 2 hrs
WHVW(AM) Hyde Park NY 1 hr
WJTN(AM) Jamestown NY 1 hr
WIZR(AM) Johnstown NY 1 hr
WLVL(AM) Lockport NY 2 hrs
WVOA-FM Mexico NY 2 hrs
WVIP(FM) New Rochelle NY 6 hrs
WJJL(AM) Niagara Falls NY 4 hrs
WLIM(AM) Patchogue NY 4 hrs
*WRUC(FM) Schenectady NY 1 hr
WLLW(FM) Seneca Falls NY 2 hrs
WSFW(AM) Seneca Falls NY 2 hrs
WUTQ(AM) Utica NY 2 hrs
*WAPS(FM) Akron OH 2 hrs
WRTK(AM) Niles OH 1 hr
*WJCU(FM) University Heights OH 2 hrs
WELW(AM) Willoughby-Eastlake OH 1 hr
WNIO(AM) Youngstown OH 3 hrs
KXMG(AM) Portland OR 1 hr
*WMUH(FM) Allentown PA 2 hrs
WWCS(AM) Canonsburg PA 2 hrs
*WERG(FM) Erie PA 3 hrs
WEDO(AM) McKeesport PA 1 hr
WGBN(AM) New Kensington PA 2 hrs
WURD(AM) Philadelphia PA 3 hrs
WPIC(AM) Sharon PA 2 hrs
WFAX(AM) Falls Church VA 1 hr
*WMSE(FM) Milwaukee WI 3 hrs
WRJN(AM) Racine WI 2.5 hrs
WRLF(FM) Fairmont WV 3 hrs
WTCS(AM) Fairmont WV 3 hrs
WEIR(AM) Weirton WV 3 hrs

Japanese

KTYM(AM) Inglewood CA 1 hr
KEST(AM) San Francisco CA
*KCSB-FM Santa Barbara CA 1 hr
*V6AI(AM) Yap FM 5 hrs
KUAM(AM) Hagatna GU 3 hrs
KPUA(AM) Hilo HI 6 hrs
*WBXL(FM) Baldwinsville NY 1 hr
*KWVA(FM) Eugene OR 2 hrs

*WRVU(FM) Nashville TN 2 hrs

Jazz

*KBRW(AM) Barrow AK 6 hrs
*KUAC(FM) Fairbanks AK 15 hrs
*KIYU(AM) Galena AK 4 hrs
*KBBI(AM) Homer AK 5 hrs
*KTOO(FM) Juneau AK 14 hrs
*KRBD(FM) Ketchikan AK 10 hrs
*KSTK(FM) Wrangell AK 8 hrs
WDLT-FM Chickasaw AL 5 hrs
WZEW(FM) Fairhope AL 6 hrs
*WFIX(FM) Florence AL 6 hrs
WJRL-FM Ozark AL 4 hrs
WWPG(AM) Tuscaloosa AL 2 hrs
KEZA(FM) Fayetteville AR
*KXRJ(FM) Russellville AR 15 hrs
KNOT(AM) Prescott AZ 2 hrs
*KXCI(FM) Tucson AZ 2 hrs
KHOV-FM Wickenburg AZ 2 hrs
*KHSU-FM Arcata CA 10 hrs
*KPFA(FM) Berkeley CA 15 hrs
KRML(AM) Carmel CA 140 hrs
*KHSR(FM) Crescent City CA 10 hrs
*KMUD(FM) Garberville CA 6 hrs
KFGY(FM) Healdsburg CA 5 hrs
*KFJC(FM) Los Altos CA 7 hrs
*KPFK(FM) Los Angeles CA 5 hrs
KMMT(FM) Mammoth Lakes CA 2 hrs
*KSMC(FM) Moraga CA 5 hrs
KHOP(FM) Oakdale CA 2 hrs
*KZYX(FM) Philo CA 11 hrs
*KUCR(FM) Riverside CA 6 hrs
KYCY(AM) San Francisco CA
*KRCB-FM Santa Rosa CA 6 hrs
KNNN(FM) Shasta Lake City CA 3 hrs
KZSQ-FM Sonora CA 3 hrs
KTHO(AM) South Lake Tahoe CA
*KCSS(FM) Turlock CA 4 hrs
*KASF(FM) Alamosa CO 6 hrs
*KGNU-FM Boulder CO 15 hrs
KVCU(AM) Boulder CO 3 hrs
*KRCC(FM) Colorado Springs CO 15 hrs
*KDUR(FM) Durango CO 9 hrs
KIQX(FM) Durango CO 7 hrs
*KCSU-FM Fort Collins CO 3 hrs
*KMSA(FM) Grand Junction CO 12 hrs
*KWSB-FM Gunnison CO 3 hrs
*KSUT(FM) Ignacio CO 15 hrs
KFMU-FM Oak Creek CO 4 hrs
KWUF-FM Pagosa Springs CO 10 hrs
*KVNF(FM) Paonia CO 17 hrs
*KOTO(FM) Telluride CO 9 hrs
*WPKN(FM) Bridgeport CT 16 hrs
*WXCI(FM) Danbury CT 3 hrs
*WQTQ(FM) Hartford CT 12 hrs
WZBG(FM) Litchfield CT 2 hrs
WMRD(AM) Middletown CT 4 hrs. weekly hrs
WYBC-FM New Haven CT 8 hrs
*WCNI(FM) New London CT 9 hrs
WLIS(AM) Old Saybrook CT 4 hrs
WXLM(FM) Stonington CT 2 hrs
*WNHU(FM) West Haven CT 12 hrs
*WECS(FM) Willimantic CT 16 hrs
WKND(AM) Windsor CT 3 hrs
*WAMU(FM) Washington DC 3 hrs
*WVUD(FM) Newark DE 15 hrs
WGMD(FM) Rehoboth Beach DE 2 hrs
*WKTO(FM) Edgewater FL 4 hrs
*WJLF(FM) Gainesville FL 2 hrs
WXCV(FM) Homosassa Springs FL 7 hrs
WAVV(FM) Marco FL 3 hrs
WRXB(AM) Saint Petersburg Beach FL 15 hrs
*WANM(FM) Tallahassee FL 15 hrs hrs
*WFSU-FM Tallahassee FL 8 hrs
WFLM(FM) White City FL 4 hrs
*WPRK(FM) Winter Park FL 3 hrs
*WUGA(FM) Athens GA 4 hrs
*WABE(FM) Atlanta GA 7 hrs
*WREK(FM) Atlanta GA 15 hrs
*WACG-FM Augusta GA 18 hrs

*WMUM-FM Cochran GA 4 hrs
*WNGU(FM) Dahlonega GA 4 hrs
*WPPR(FM) Demorest GA 4 hrs
*WJWV(FM) Fort Gaines GA 4 hrs
WFXM(FM) Gordon GA 3 hrs
*WSVH(FM) Savannah GA 4 hrs
*WABR-FM Tifton GA 4 hrs
*WWET(FM) Valdosta GA 4 hrs
WWSN(FM) Waycross GA 5 hrs
*WXVS(FM) Waycross GA 4 hrs
KUAI(AM) Eleele HI 4 hrs
*WOI(AM) Ames IA 7 hrs
KMXG(FM) Clinton IA 3 hrs
KROS(AM) Clinton IA 1 hr
*KHOE(FM) Fairfield IA 2 hrs
*KRNL-FM Mount Vernon IA 2 hrs
*KOJI(FM) Okoboji IA 17 hrs
*KIGC(FM) Oskaloosa IA 12 hrs
*KWIT(FM) Sioux City IA 17 hrs
KSAS-FM Caldwell ID 4 hrs
*KBSM(FM) McCall ID
*KRFA-FM Moscow ID
*KUOI-FM Moscow ID 4 hrs
KECH-FM Sun Valley ID 6. hrs
*WESN(FM) Bloomington IL 6 hrs
*WEIU(FM) Charleston IL 4 hrs
WFMT(FM) Chicago IL 5 hrs
*WIIT(FM) Chicago IL 9 hrs
*WSSD(FM) Chicago IL
*WEPS(FM) Elgin IL 6 hrs
*WVKC(FM) Galesburg IL 15 hrs
WYMG(FM) Jacksonville IL 2 hrs
*WCSF(FM) Joliet IL 2 hrs
*WMXM(FM) Lake Forest IL 6 hrs
*WLRA(FM) Lockport IL 15 hrs
*WIUM(FM) Macomb IL 5 hrs
*WIUS(FM) Macomb IL 2 hrs
WPNA(AM) Oak Park IL 4 hrs
WOAM(AM) Peoria IL 2 hrs
*WRRG(FM) River Grove IL 5 hrs
*WVIK(FM) Rock Island IL 9 hrs
WNNS(FM) Springfield IL 6 hrs
*WGRE(FM) Greencastle IN 3 hrs
WNTR(FM) Indianapolis IN 6 hrs
*WMHD-FM Terre Haute IN 2 hrs
*WVUR-FM Valparaiso IN 3 hrs
*KANZ(FM) Garden City KS 15 hrs
*KZNA(FM) Hill City KS 15 hrs
*KSDB-FM Manhattan KS 3 hrs
KRBB(FM) Wichita KS 2 hrs
*WKUE(FM) Elizabethtown KY 15 hrs
WRNZ(FM) Lancaster KY 2 hrs
*WDCL-FM Somerset KY 15 hrs
*KLSP(FM) Angola LA 7 hrs
*WERS(FM) Boston MA 15 hrs
*WAMQ(FM) Great Barrington MA 13 hrs
WFNX(FM) Lynn MA 8 hrs
*WBSL-FM Sheffield MA 15 hrs
*WYAJ(FM) Sudbury MA 5 hrs
WMVY(FM) Tisbury MA 4 hrs
*WCHC(FM) Worcester MA 6 hrs
WLIF(FM) Baltimore MD 8 hrs
*WHFC(FM) Bel Air MD 18 hrs
*WKHS(FM) Worton MD 2 hrs
*WUMF-FM Farmington ME 15 hrs
*WRBC(FM) Lewiston ME 4 hrs
WPHX-FM Sanford ME 6 hrs
WBQW(FM) Scarborough ME 5 hrs
*WSJB-FM Standish ME 3 hrs
WBQX(FM) Thomaston ME 2 hrs
*WMHB(FM) Waterville ME 8 hrs
*WQAC-FM Alma MI 2 hrs
*WCBN-FM Ann Arbor MI 17 hrs
*WHFR(FM) Dearborn MI 16 hrs
*WDBM(FM) East Lansing MI 5 hrs
*WKAR-FM East Lansing MI 7 hrs
*WTHS(FM) Holland MI 6 hrs
WJNZ(AM) Kentwood MI 6 hrs
*WUPX(FM) Marquette MI 2 hrs
WRKR(FM) Portage MI
*KFAI(FM) Minneapolis MN 12 hrs
*WMCN(FM) Saint Paul MN 4 hrs
*KYMC(FM) Ballwin MO 6 hrs
*KOPN(FM) Columbia MO 4 hrs
*KCFV(FM) Ferguson MO 4 hrs

*KKFI(FM) Kansas City MO 10 hrs
*KGSP(FM) Parkville MO 14 hrs
*KMNR(FM) Rolla MO 3 hrs
*KMST(FM) Rolla MO 3 hrs
KMOX(AM) Saint Louis MO 4 hrs
KBFL(AM) Springfield MO 12 hrs
*KSMU(FM) Springfield MO 10 hrs
WGNL(FM) Greenwood MS 6 hrs
*KMSM-FM Butte MT 5 hrs
WSQL(AM) Brevard NC 10 hrs
*WXDU(FM) Durham NC 18 hrs
*WRVS-FM Elizabeth City NC 6 hrs
*WKNS(FM) Kinston NC 6 hrs
*WYQS(FM) Mars Hill NC 7 hrs
WIKS(FM) New Bern NC 2 hrs
WNOS(AM) New Bern NC 12 hrs
*WTEB(FM) New Bern NC 4 hrs
WNNC(AM) Newton NC 3 hrs
*WZRU(FM) Roanoke Rapids NC 6 hrs
WLHC(FM) Robbins NC 2 hrs
*WNCW(FM) Spindale NC 5 hrs
WFXK(FM) Tarboro NC 4 hrs
*WFDD-FM Winston-Salem NC 16 hrs
*KFKX(FM) Hastings NE 3 hrs
*KRNU(FM) Lincoln NE 2 hrs
KEZO-FM Omaha NE 3 hrs
*KIOS-FM Omaha NE 10 hrs
KKCD(FM) Omaha NE 4 hrs
*KWSC(FM) Wayne NE 2 hrs
*WUNH(FM) Durham NH 5 hrs
*WKNH(FM) Keene NH 3 hrs
WFEX(FM) Peterborough NH 6 hrs
*WPCR-FM Plymouth NH 3 hrs
*WDVR(FM) Delaware Township NJ 11 hrs
*WNTI(FM) Hackettstown NJ 9 hrs
*WBZC(FM) Pemberton NJ 4 hrs
*WLFR(FM) Pomona NJ 10 hrs
*WTSR(FM) Trenton NJ 4 hrs
*WKNJ-FM Union Township NJ 8 hrs
*WMSC(FM) Upper Montclair NJ 2 hrs
*WPSC-FM Wayne NJ 12 hrs
*WMCX(FM) West Long Branch NJ 3 hrs
KKTC(FM) Angel Fire NM 4 hrs
*KSJE(FM) Farmington NM 15 hrs
KRSN(AM) Los Alamos NM 2 hrs
KTAO(FM) Taos NM 5 hrs
KRSI(FM) Garapan-Saipan NP 1 hr
*KCEP(FM) Las Vegas NV 12 hrs
KNEV(FM) Reno NV 2 hrs
*WAMC-FM Albany NY 18 hrs
*WCDB(FM) Albany NY 10 hrs
WVIN-FM Bath NY 2 hrs
*WHRW(FM) Binghamton NY 9 hrs
*WSKG-FM Binghamton NY
*WXLH(FM) Blue Mountain Lake NY
*WBNY(FM) Buffalo NY 3 hrs
*WCAN(FM) Canajoharie NY 17 hrs
*WSLU(FM) Canton NY
*WHCL-FM Clinton NY 9 hrs
WKGB-FM Conklin NY 2 hrs
*WSQE(FM) Corning NY
*WRCU-FM Hamilton NY 12 hrs
*WITR(FM) Henrietta NY 8 hrs
*WICB(FM) Ithaca NY 13 hrs
*WSQG-FM Ithaca NY 7 hrs
*WAMK(FM) Kingston NY 13 hrs
*WOSR(FM) Middletown NY 13 hrs
*WBAI(FM) New York NY 5 hrs
*WNYC-FM New York NY 4 hrs
WQXR-FM New York NY 2 hrs
*WRHO(FM) Oneonta NY 4 hrs
*WSQC-FM Oneonta NY
WZOZ(FM) Oneonta NY 2 hrs
WLIM(AM) Patchogue NY 4 hrs
*WCEL(FM) Plattsburgh NY 13 hrs
*WRHV(FM) Poughkeepsie NY 2 hrs
WDKX(FM) Rochester NY 4 hrs
*WFNP(FM) Rosendale NY 4 hrs
*WMHT-FM Schenectady NY 1 hr
*WRUC(FM) Schenectady NY 15 hrs
WLLW(FM) Seneca Falls NY 1 hr
WSFW(AM) Seneca Falls NY 1 hr
*WKWZ(FM) Syosset NY 12 hrs

*WCNY-FM Syracuse NY 7 hrs
*WANC(FM) Ticonderoga NY 17 hrs
*WPNR-FM Utica NY 14 hrs
*WUNY(FM) Utica NY 7 hrs
*WJNY(FM) Watertown NY 5 hrs
WRIP(FM) Windham NY 3 hrs
*WRDL(FM) Ashland OH 5 hrs
*WGBE(FM) Bryan OH
*WCBE(FM) Columbus OH 4 hrs
*WWSU(FM) Dayton OH 3 hrs
*WGDE(FM) Defiance OH 16 hrs
WBVI(FM) Fostoria OH 3 hrs
WWCD(FM) Grove City OH 3 hrs
WRBP(FM) Hubbard OH 12 hrs
WVKO-FM Johnstown OH 10 hrs
*WGLE(FM) Lima OH 16 hrs
WZOQ(AM) Lima OH 2 hrs
WCLV(FM) Lorain OH 5 hrs
*WMCO(FM) New Concord OH 10 hrs
*WOBC-FM Oberlin OH 15 hrs
*WUSO(FM) Springfield OH 6 hrs
*WGTE-FM Toledo OH 16 hrs
WRVF(FM) Toledo OH 6 hrs
*WOBN(FM) Westerville OH 1 hr
*WYSO(FM) Yellow Springs OH 12 hrs
*KRSC-FM Claremore OK 5 hrs
KBZQ(FM) Lawton OK 4 hrs
*KCCU(FM) Lawton OK
KZBB(FM) Poteau OK 2 hrs
*KOAC(AM) Corvallis OR 12 hrs
*KWVA(FM) Eugene OR 6 hrs
*KEOL(FM) La Grande OR 6 hrs
KYTE(FM) Newport OR 4 hrs
*KRRC(FM) Portland OR 10 hrs
KLRR(FM) Redmond OR 5 hrs
KAVJ(FM) Sutherlin OR 2 hrs
*WDIY(FM) Allentown PA 10 hrs
*WLVR(FM) Bethlehem PA 12 hrs
WMKX(FM) Brookville PA 3 hrs
*WDCV-FM Carlisle PA 6 hrs
*WDNR(FM) Chester PA 2 hrs
*WCUC-FM Clarion PA 3 hrs
*WESS(FM) East Stroudsburg PA 6 hrs
*WJRH(FM) Easton PA 9 hrs
*WMCE(FM) Erie PA 4 hrs
*WZBT(FM) Gettysburg PA 4 hrs
*WVMM(FM) Grantham PA 4 hrs
*WKVR-FM Huntingdon PA 3 hrs
*WIUP-FM Indiana PA 15 hrs
*WPSX(FM) Kane PA 3 hrs
*WFNM(FM) Lancaster PA 8 hrs
*WVBU-FM Lewisburg PA 3 hrs
*WNTE(FM) Mansfield PA 2 hrs
*WARC(FM) Meadville PA 4 hrs
*WIXQ(FM) Millersville PA 2 hrs
*WHYY-FM Philadelphia PA 4 hrs
*WRCT(FM) Pittsburgh PA 18 hrs
*WXLV(FM) Schnecksville PA 6 hrs
*WVMW-FM Scranton PA 10 hrs
*WSYC-FM Shippensburg PA 3 hrs
*WPSU(FM) State College PA 4 hrs
*WRDV(FM) Warminster PA 3 hrs
*WNJR(FM) Washington PA 3 hrs
*WRLC(FM) Williamsport PA 8 hrs
*WVYC(FM) York PA 8 hrs
WWNA(AM) Aguadilla PR 3 hrs
WOLA(AM) Barranquitas PR 5 hrs
WUKQ-FM Mayaguez PR 6 hrs
WADK(AM) Newport RI 7 hrs
WBRU(FM) Providence RI 18 hrs
WEGX(FM) Dillon SC 1 hr
WOLS(AM) Florence SC 4 hrs
WYNN(AM) Florence SC
*WLGI(FM) Hemingway SC 18 hrs
*WSSB-FM Orangeburg SC 10 hrs
WSPA-FM Spartanburg SC 6 hrs
*KSDJ(FM) Brookings SD 2 hrs
*KCSD(FM) Sioux Falls SD 10 hrs
*WAPX-FM Clarksville TN 6 hrs
*WTTU(FM) Cookeville TN 3 hrs
WRLT(FM) Franklin TN 2 hrs
*WFHU(FM) Henderson TN 45 hrs
*WEVL(FM) Memphis TN 15 hrs
*WMTS-FM Murfreesboro TN 4 hrs
*WRVU(FM) Nashville TN 18 hrs

*WUTS(FM) Sewanee TN 4 hrs
*KACU(FM) Abilene TX 3 hrs
*KACV-FM Amarillo TX 12 hrs
KGSR(FM) Bastrop TX 6 hrs
*KWTS(FM) Canyon TX 3 hrs
*KEOS(FM) College Station TX 3 hrs
KFRO-FM Gilmer TX 5 hrs
KODA(FM) Houston TX 4 hrs
*KTRU(FM) Houston TX 14 hrs
KFAN-FM Johnson City TX
*KNCT-FM Killeen TX 15 hrs
KMND(AM) Midland TX 1 hr
*KOCV(FM) Odessa TX 4 hrs
KQXT(FM) San Antonio TX 4 hrs
*KSTX(FM) San Antonio TX 6 hrs
*KTXK(FM) Texarkana TX 15 hrs
*KWBU-FM Waco TX 10 hrs
KBLQ-FM Logan UT 4 hrs
*KPCW(FM) Park City UT 12 hrs
*KRDC-FM Saint George UT 10 hrs
*WWHS-FM Hampden-Sydney VA 6 hrs
*WXJM(FM) Harrisonburg VA 14 hrs
*WVRU(FM) Radford VA 19 hrs
*WCVE(FM) Richmond VA 18 hrs
*WDCE(FM) Richmond VA 9 hrs
*WCWM(FM) Williamsburg VA 13 hrs
*WISE-FM Wise VA 9 hrs
*WIUJ(FM) Charlotte Amalie VI 6 hrs
*WIUV(FM) Castleton VT 10 hrs
*WWLR(FM) Lyndonville VT 3 hrs
WEQX(FM) Manchester VT 4 hrs
*WRMC-FM Middlebury VT 10 hrs
WNCS(FM) Montpelier VT 5 hrs
WXLF(FM) White River Junction VT 4 hrs
*KUGS(FM) Bellingham WA 10 hrs wkly hrs
*KNWR(FM) Ellensburg WA
*KSER(FM) Everett WA 2 hrs
KBRD(AM) Lacey WA 1 hr
*KWSU(AM) Pullman WA 14 hrs
*KZUU(FM) Pullman WA 12 hrs
*KFAE-FM Richland WA 15 hrs
*KUOW-FM Seattle WA 5 hrs
*KAGU(FM) Spokane WA 2 hrs
*KPBX-FM Spokane WA
*KUPS(FM) Tacoma WA 12 hrs
*WHSA(FM) Brule WI 6 hrs
*WBSD(FM) Burlington WI 4 hrs
*WPNE-FM Green Bay WI 10 hrs
*WORT(FM) Madison WI 15 hrs
*WHWC(FM) Menomonie WI 3 hrs
*WVSS(FM) Menomonie WI 5 hrs
*WMSE(FM) Milwaukee WI 15 hrs
*WSUP(FM) Platteville WI 3 hrs
*WOJB(FM) Reserve WI 10 hrs
*WXPR(FM) Rhinelander WI 8 hrs
*WHND(FM) Sister Bay WI 10 hrs
*KUWS(FM) Superior WI 15 hrs
*WSUW(FM) Whitewater WI 6 hrs
*WVWC(FM) Buckhannon WV 8 hrs
WQZK-FM Keyser WV 2 hrs
*WWVU-FM Morgantown WV 9 hrs
*WQAB(FM) Philippi WV 4 hrs
WMOV(AM) Ravenswood WV 2 hrs
*WPHP(FM) Wheeling WV 1 hr
*KUWA(FM) Afton WY 5 hrs
KMTN(FM) Jackson WY
*KUWJ(FM) Jackson WY 6 hrs
*KUWR(FM) Laramie WY 6 hrs
*KUWZ(FM) Rock Springs WY 6 hrs

Jewish

*KCSN(FM) Northridge CA 3 hrs
WMRD(AM) Middletown CT 1 hr
WILI(AM) Willimantic CT 1 hr
WPBR(AM) Lantana FL 3 hrs
WMEL(AM) Melbourne FL 2 hrs
*WCBN-FM Ann Arbor MI 1 hr hrs
WVIP(FM) New Rochelle NY 1 hr
WVOX(AM) New Rochelle NY 2 hrs
WMCA(AM) New York NY 9 hrs
*WDIY(FM) Allentown PA 1 hr
*WHCB(FM) Bristol TN 1 hr
*KEOS(FM) College Station TX 2 hrs

Korean

WONX(AM) Evanston IL 20 hrs
*WNKJ(FM) Hopkinsville KY 1 hr
*KRNM(FM) Chalan Kanoa-Saipan NP 1 hr
*WUSB(FM) Stony Brook NY 1 hr

Light Rock

*KTNA(FM) Talkeetna AK 5 hrs

Lithuanian

WONX(AM) Evanston IL 1 hr
WNDZ(AM) Portage IN 1 hr
WBMD(AM) Baltimore MD 1 hr
*WGMC(FM) Greece NY 1 hr
*WCPN(FM) Cleveland OH 1 hr
*WJCU(FM) University Heights OH 2 hrs
WEDO(AM) McKeesport PA 1 hr

MOR

WSNJ(AM) Bridgeton NJ
WBUK(FM) Ottawa OH 4 hrs

Native American

*KDUR(FM) Durango CO 3 hrs
KIXR(FM) Ponca City OK 3 hrs
*KRVM-FM Eugene OR 2 hrs

New Age

*KUAC(FM) Fairbanks AK 3 hrs
*WBHM(FM) Birmingham AL 8 hrs
*WSGN(FM) Gadsden AL 10 hrs
*WQPR(FM) Muscle Shoals AL 20 hrs
*WUAL-FM Tuscaloosa AL 20 hrs
*KASU(FM) Jonesboro AR
KEST(AM) San Francisco CA
*KVNF(FM) Paonia CO 6 hrs
*WXCI(FM) Danbury CT 3 hrs
*WSHU(FM) Fairfield CT 6 hrs
*WGRS(FM) Guilford CT 1 hr
*WMNR(FM) Monroe CT 1 hr
*WRXC(FM) Shelton CT 1 hr
*WGSK(FM) South Kent CT 1 hr
*WMFE-FM Orlando FL 4 hrs
*WFCF(FM) Saint Augustine FL 4 hrs
*WRAS(FM) Atlanta GA 3 hrs
*WUWG(FM) Carrollton GA 3 hrs
*KCCK-FM Cedar Rapids IA 7 hrs
*KTPR(FM) Fort Dodge IA 10 hrs
*WSIU(FM) Carbondale IL 4 hrs
*WSIE(FM) Edwardsville IL 10 hrs
*WNIU(FM) Rockford IL 2 hrs
*WKMS-FM Murray KY 3 hrs
*KMSK(FM) Austin MN 5 hrs
*KMSU(FM) Mankato MN 5 hrs
*KCOZ(FM) Point Lookout MO 10 hrs
*WYQS(FM) Mars Hill NC 6 hrs
*WZRU(FM) Roanoke Rapids NC 10 hrs
*KUND-FM Grand Forks ND 4 hrs
*KZUM(FM) Lincoln NE 8 hrs
*WKNH(FM) Keene NH 4 hrs
*WKNJ-FM Union Township NJ 8 hrs
*WCVF-FM Fredonia NY 4 hrs
*WSUF(FM) Noyack NY 6 hrs
*WAER(FM) Syracuse NY 3 hrs
*WGDE(FM) Defiance OH 4 hrs
*WGLE(FM) Lima OH
*WGTE-FM Toledo OH 4 hrs
*WYSO(FM) Yellow Springs OH 4 hrs
*KGOU(FM) Norman OK 4 hrs
*KROU(FM) Spencer OK 4 hrs
*WQLN-FM Erie PA 2 hrs
*WIUP-FM Indiana PA 4 hrs
*WVBU-FM Lewisburg PA 1 hr
*WKDU(FM) Philadelphia PA 2 hrs
*WRDV(FM) Warminster PA 3 hrs
WCRI(FM) Block Island RI 4 hrs
*WUTS(FM) Sewanee TN 4 hrs
*KAMU-FM College Station TX 5 hrs
*KTRU(FM) Houston TX 1 hr
*KNTU(FM) McKinney TX 3 hrs
WSNZ(FM) Appomattox VA 8 hrs
*WVRU(FM) Radford VA 3 hrs
*KPBX-FM Spokane WA
*WWVU-FM Morgantown WV 6 hrs
KMTN(FM) Jackson WY

New Wave

*WVUA-FM Tuscaloosa AL 4 hrs

News

KIAL(AM) Unalaska AK 14 hrs
WBCF(AM) Florence AL 5 hrs
KWBF-FM North Little Rock AR 15 hrs
KBHR(FM) Big Bear City CA 8 hrs
KWXY-FM Cathedral City CA 2 hrs
KNTS(AM) Palo Alto CA
*KRZA(FM) Alamosa CO 4 hrs
WILI-FM Willimantic CT 1 hr
WRZN(AM) Hernando FL 4 hrs
*WRGP(FM) Homestead FL 3 hrs
WCNK(FM) Key West FL 1 hr
WAZX(AM) Smyrna GA
KLGA(AM) Algona IA
KLGA-FM Algona IA
KTFC(FM) Sioux City IA 10 hrs
WXET(FM) Arcola IL
*WPCD(FM) Champaign IL 8 hrs
WUZR(FM) Bicknell IN
*WGCS(FM) Goshen IN 8 hrs
WMJL-FM Marion KY
WKIK(AM) La Plata MD
WSMD-FM Mechanicsville MD
*WPHS(FM) Warren MI 5 hrs
KBNN(AM) Lebanon MO
KCXL(AM) Liberty MO 4 hrs
KSIM(AM) Sikeston MO
KRWP(FM) Stockton MO
WKXU(FM) Louisburg NC
*KZUM(FM) Lincoln NE 11 hrs
KDSS(FM) Ely NV
WBAZ(FM) Bridgehampton NY
WFXF(FM) Honeoye Falls NY 1 hr
*WKCR-FM New York NY 3 hrs
WJZR(FM) Rochester NY 1 hr
WBEA(FM) Southold NY
*WCII(FM) Spencer NY 14 hrs
WCLV(FM) Lorain OH 1 hr
KSPI(AM) Stillwater OK
*KMHD(FM) Gresham OR 5 hrs
KMUZ(AM) Gresham OR 3 hrs
*WRTU(FM) San Juan PR 7 hrs
WIVV(AM) Vieques PR 7 hrs
WHYM(AM) Lake City SC
KCYL(AM) Lampasas TX 14 hrs
*KSUU(FM) Cedar City UT
WEVA(AM) Emporia VA 14 hrs
WVCV(AM) Orange VA 18 hrs
WHEO(AM) Stuart VA 10 hrs
*KASB(FM) Bellevue WA 3 hrs
KQQQ(AM) Pullman WA 10 hrs
WOMT(AM) Manitowoc WI 18 hrs
*KUWJ(FM) Jackson WY 3 hrs
*KUWR(FM) Laramie WY 3 hrs

News/talk

*WVAS(FM) Montgomery AL 4 hrs
KWKM(FM) Saint Johns AZ 2 hrs
WCCC-FM Hartford CT 10 hrs
WILM(AM) Wilmington DE
WWWE(AM) Hapeville GA 14 hrs
*WHFH(FM) Flossmoor IL 1 hr
WHHL(FM) Jerseyville IL 3 hrs
*WLTL(FM) La Grange IL 10 hrs
*WLRA(FM) Lockport IL 10 hrs
WITZ-FM Jasper IN 3 hrs
WWVR(FM) West Terre Haute IN
*KCFN(FM) Wichita KS
WFLW(AM) Monticello KY 10 hrs
WCBR(AM) Richmond KY
KUGT(AM) Jackson MO
WQNQ(FM) Fletcher NC 5 hrs
WCBQ(AM) Oxford NC
WLHC(FM) Robbins NC 1 hr
WMPM(AM) Smithfield NC 12 hrs
KLIQ(FM) Hastings NE
WSNJ(AM) Bridgeton NJ
*WJSV(FM) Morristown NJ 3 hrs
KSNE-FM Las Vegas NV 1 hr
*WCDB(FM) Albany NY 1 hr
*WITC(FM) Cazenovia NY 3 hrs
WCKM-FM Lake George NY
WJJL(AM) Niagara Falls NY 5 hrs
*WNYO(FM) Oswego NY 4 hrs
*WFNP(FM) Rosendale NY 5 hrs
*WRMU(FM) Alliance OH 5 hrs
WRBP(FM) Hubbard OH
*WESS(FM) East Stroudsburg PA 6 hrs
*WFSE(FM) Edinboro PA 3 hrs
*WQLN-FM Erie PA 3 hrs
WXTU(FM) Philadelphia PA 1/2 hrs
WBCU(AM) Union SC 10 hrs
*KLND(FM) Little Eagle SD 5 hrs
KYMI(FM) Los Ybanez TX 10 hrs
KRFE(AM) Lubbock TX 15 hrs
KTFS(AM) Texarkana TX
KMER(AM) Kemmerer WY 7 hrs
*KUWZ(FM) Rock Springs WY 3 hrs

Nostalgia

KLXR(AM) Redding CA 10 hrs. hrs
KNTK(FM) Weed CA 2 hrs
*WAMU(FM) Washington DC 4 hrs
*KCMR(FM) Mason City IA 10 hrs
WAAM(AM) Ann Arbor MI 6 hrs
KRDS-FM New Prague MN 18 hrs
WECR-FM Beech Mountain NC
WDOS(AM) Oneonta NY 2 hrs
WZKL(FM) Alliance OH
*WFSE(FM) Edinboro PA 2 hrs
WLXC(FM) Lexington SC
KLSR-FM Memphis TX 10 hrs
KEYG(AM) Grand Coulee WA 4 hrs
KARS-FM Laramie WY 6 hrs

Oldies

KRSA(AM) Petersburg AK 5 hrs
*WJAB(FM) Huntsville AL 3 hrs
WJRL-FM Ozark AL 6 hrs
KTCN(FM) Eureka Springs AR 2 hrs
KEZA(FM) Fayetteville AR
*KAZU(FM) Pacific Grove CA 5 hrs
*KSPB(FM) Pebble Beach CA 4 hrs
KRAI-FM Craig CO 9 hrs
*KWSB-FM Gunnison CO 3 hrs
*WPFW(FM) Washington DC 3 hrs
*WVUM(FM) Coral Gables FL 5 hrs
WXCV(FM) Homosassa Springs FL 6 hrs
WMAF(AM) Madison FL
WGOV(AM) Valdosta GA 10 hrs
*WQUB(FM) Quincy IL 2 hrs
*WRRG(FM) River Grove IL 11 hrs
WJVO(FM) South Jacksonville IL 5 hrs
WBIW(AM) Bedford IN
*WCYT(FM) Lafayette Township IN 2 hrs
WXLO(FM) Fitchburg MA 5 hrs
WHAI(FM) Greenfield MA 13 hrs
*WOMR(FM) Provincetown MA 9 hrs
*WEAA(FM) Baltimore MD 5 hrs
*WKHS(FM) Worton MD 6 hrs
WHKB(FM) Houghton MI 16 hrs
WMPX(AM) Midland MI 2 hrs
KASM(AM) Albany MN
KEYL(AM) Long Prairie MN 3 hrs
KNUJ(AM) New Ulm MN 8 hrs
KXLP(FM) New Ulm MN 15 hrs
KOLV(FM) Olivia MN
KKAQ(AM) Thief River Falls MN 6 hrs
KKDQ(FM) Thief River Falls MN 6 hrs
KBFL(AM) Springfield MO 4 hrs
*WVSD(FM) Itta Bena MS 10 hrs
WUMS(FM) University MS 2 hrs
KBAZ(FM) Hamilton MT 3 hrs
KPRK(AM) Livingston MT 5 hrs
*WNAA(FM) Greensboro NC 5 hrs
WIXE(AM) Monroe NC 5 hrs
*WZRU(FM) Roanoke Rapids NC 4 hrs
WLTT(FM) Shallotte NC 6 hrs
KAUJ(FM) Grafton ND 4 hrs
KLIR(FM) Columbus NE 12 hrs
*WDVR(FM) Delaware Township NJ 13 hrs
*WNTI(FM) Hackettstown NJ 3 hrs
*WTSR(FM) Trenton NJ 6 hrs
*WGCC-FM Batavia NY
*WCEB(FM) Corning NY 6 hrs
WVBR-FM Ithaca NY 5 hrs
*WNYU-FM New York NY 3 hrs
WZOZ(FM) Oneonta NY 2 hrs
WHUD(FM) Peekskill NY 5 hrs
WPDH(FM) Poughkeepsie NY 4 hrs
WSFW(AM) Seneca Falls NY 3 hrs
*WRDL(FM) Ashland OH 4 hrs
KBZQ(FM) Lawton OK 6 hrs
KKNG-FM Newcastle OK 12 hrs
WMBA(AM) Ambridge PA 3 hrs
WMKX(FM) Brookville PA 8 hrs
WWCS(AM) Canonsburg PA 4 hrs
*WDNR(FM) Chester PA 2 hrs
WOKW(FM) Curwensville PA 2 hrs
*WESS(FM) East Stroudsburg PA 8 hrs
*WVMM(FM) Grantham PA 4 hrs
WCCS(AM) Homer City PA 9 hrs
WLSH(AM) Lansford PA 3 hrs
WPTT(AM) McKeesport PA 9 hrs
WMGH-FM Tamaqua PA 14 hrs
*WNJR(FM) Washington PA 3 hrs
*WCYJ-FM Waynesburg PA 3 hrs
WYKZ(FM) Beaufort SC 5 hrs
WHSC(AM) Hartsville SC 6 hrs
WSPA-FM Spartanburg SC 10 hrs
WMXX-FM Jackson TN 15 hrs
WWLX(AM) Lawrenceburg TN 8 hrs
KFGL(FM) Abilene TX 2 hrs
KFYN(AM) Bonham TX 6 hrs
*KTRU(FM) Houston TX 3 hrs
KJAS(FM) Jasper TX 3 hrs
*WVRU(FM) Radford VA 5 hrs
WTSA-FM Brattleboro VT 16 hrs
WIZN(FM) Vergennes VT 3 hrs
WLKG(FM) Lake Geneva WI 10 hrs
*WFGH(FM) Fort Gay WV 19 hrs
*WWVU-FM Morgantown WV 8 hrs

Other

*KSUA(FM) Fairbanks AK 19 hrs
*KTOO(FM) Juneau AK 1 hr
*KNOM(AM) Nome AK 14 hrs
*KNOM-FM Nome AK 14 hrs
KIFW(AM) Sitka AK 3 hrs
*WMBV(FM) Dixons Mills AL 3 hrs
WABF(AM) Fairhope AL 6 hrs
WBCF(AM) Florence AL 5 hrs
*WVUA-FM Tuscaloosa AL 3 hrs
*KUAR(FM) Little Rock AR 4 hrs
*KNAU(FM) Flagstaff AZ
*KHSU-FM Arcata CA 14 hrs
KIXF(FM) Baker CA
*KSPC(FM) Claremont CA 6 hrs
*KHSR(FM) Crescent City CA 14 hrs
*KFSR(FM) Fresno CA 1 hr
*KFSR(FM) Fresno CA 9 hrs
*KSBR(FM) Mission Viejo CA 9 hrs
KNTS(AM) Palo Alto CA
*KZYX(FM) Philo CA 8 hrs
KTIP(AM) Porterville CA
KTIP(AM) Porterville CA 1 hr
KLXR(AM) Redding CA 1 hr. hrs
KVEC(AM) San Luis Obispo CA
*KSCU(FM) Santa Clara CA 1 hr
*KSCU(FM) Santa Clara CA 6 hrs

KLCA(FM) Tahoe City CA 2 hrs
KTCL(FM) Wheat Ridge CO 1 hr
KRDZ(AM) Wray CO 3 hrs
*WGRS(FM) Guilford CT 1 hr
WQUN(AM) Hamden CT 2 hrs
WCCC-FM Hartford CT 10 hrs
*WQTQ(FM) Hartford CT 18 hrs
*WRTC-FM Hartford CT 6 hrs
*WESU(FM) Middletown CT 5 hrs
*WMNR(FM) Monroe CT 1 hr
*WRXC(FM) Shelton CT 1 hr
*WPFW(FM) Washington DC 1 hr
WWUS(FM) Big Pine Key FL 4 hrs
WTKE(AM) Fort Walton Beach FL 1 hr
*WJLF(FM) Gainesville FL 5 hrs
WXYB(AM) Indian Rocks Beach FL
WIIS(FM) Key West FL 4 hrs
WPBR(AM) Lantana FL 17 hrs
*WHWY(FM) Marathon FL 3 hrs
WOCN(AM) Miami FL 84 hrs
WPSO(AM) New Port Richey FL
WCTH(FM) Plantation Key FL 3 hrs
WHKR(FM) Rockledge FL 1 hr
*WFCF(FM) Saint Augustine FL 4 hrs
*WCLK(FM) Atlanta GA 12 hrs
*WRAS(FM) Atlanta GA 3 hrs
*WRAS(FM) Atlanta GA 6 hrs
*WREK(FM) Atlanta GA 18 hrs
WGAC(AM) Augusta GA 3 hrs
WBTS(FM) Doraville GA 2 hrs
*WBCX(FM) Gainesville GA 6 hrs
KGUM(AM) Hagatna GU 7 hrs
KHBC(AM) Hilo HI
KWXX-FM Hilo HI 20 hrs
KNDI(AM) Honolulu HI 2 hrs
KNDI(AM) Honolulu HI 7 hrs
KWAI(AM) Honolulu HI 14 hrs
KWAI(AM) Honolulu HI 2 hrs
*KKUA(FM) Wailuku HI 3 hrs
KLGA(AM) Algona IA
KLGA-FM Algona IA
*KIWR(FM) Council Bluffs IA 4 hrs
*KMSC(FM) Sioux City IA 4 hrs
*KWAR(FM) Waverly IA 3 hrs
*KWAR(FM) Waverly IA 5 hrs
KRLC(AM) Lewiston ID 2 hrs
WILY(AM) Centralia IL
*WRSE(FM) Elmhurst IL 3 hrs
*WRSE(FM) Elmhurst IL 6 hrs
WGNU(AM) Granite City IL 2 hrs
WYMG(FM) Jacksonville IL 1 hr
*WLRA(FM) Lockport IL 15 hrs
*WUIS(FM) Springfield IL 19 hrs
WGFA-FM Watseka IL 2 hrs
*WETN(FM) Wheaton IL 2 hrs
*WGRE(FM) Greencastle IN 2 hrs
WNDV-FM South Bend IN 4 hrs
*WWHR(FM) Bowling Green KY 2 hrs
WVLC(FM) Mannsville KY 5 hrs
*WKMS-FM Murray KY 1 hr
WRLV-FM Salyersville KY 6 1/2 hrs
WYSB(FM) Springfield KY 12 hrs
WTKY-FM Tompkinsville KY 1 hr
*KLSP(FM) Angola LA 4 hrs
KEUN-FM Eunice LA 12 hrs
*WERS(FM) Boston MA 4 hrs
*WGBH(FM) Boston MA 3 hrs
WHTB(AM) Fall River MA 11 hrs
WFNX(FM) Lynn MA 2 hrs
WNTN(AM) Newton MA 2 hrs
WESX(AM) Salem MA 8 hrs
*WCHC(FM) Worcester MA 9 hrs
*WEAA(FM) Baltimore MD 5 hrs
*WHFC(FM) Bel Air MD 15 hrs
*WUMF-FM Farmington ME 15 hrs
*WMPG(FM) Gorham ME 14 hrs
*WYFP(FM) Harpswell ME 3 hrs
*WDBM(FM) East Lansing MI 4 hrs
WCHT(AM) Escanaba MI 1 hr
WSRT(FM) Gaylord MI 5. hrs
WMFN(AM) Zeeland MI 1 hr
WLKX-FM Forest Lake MN 9 hrs
KLTF(AM) Little Falls MN 5 hrs
KCHK(AM) New Prague MN 40 hrs
*WMCN(FM) Saint Paul MN 4 hrs
KYOO(AM) Bolivar MO 3 hrs
*KCFV(FM) Ferguson MO 4 hrs
*KCFV(FM) Ferguson MO 8 hrs
KFAL(AM) Fulton MO 6 hrs
KCXL(AM) Liberty MO 12 hrs
KRWP(FM) Stockton MO
*KGNV(FM) Washington MO 5 hrs
WLRC(AM) Walnut MS
WJNS-FM Yazoo City MS 16 hrs
KXGN(AM) Glendive MT 5 hrs
WERX-FM Columbia NC
*WWCU(FM) Cullowhee NC 5. hrs
WCKB(AM) Dunn NC 9 hrs
*WFSS(FM) Fayetteville NC 2 hrs
WBAV-FM Gastonia NC 8 hrs
*WUAG(FM) Greensboro NC 14 hrs
WKGX(AM) Lenoir NC 16 hrs
WPAQ(AM) Mount Airy NC 15 hrs
WTAB(AM) Tabor City NC
KBTO(FM) Bottineau ND 5 hrs wkly hrs
WMOU(AM) Berlin NH 3 hrs
WFRD(FM) Hanover NH 2 hrs
*WKNH(FM) Keene NH 7 hrs
WWNH(AM) Madbury NH 24 hrs
WFEX(FM) Peterborough NH 2 hrs
WPNH(AM) Plymouth NH 6 hrs
*WPSC-FM Wayne NJ 18 hrs
*WPSC-FM Wayne NJ 3 hrs
KCNM(AM) Garapan-Saipan NP
KDSS(FM) Ely NV
KELY(AM) Ely NV 8 hrs
KLAV(AM) Las Vegas NV 4 hrs
*KUNV(FM) Las Vegas NV 2 hrs
*WCDB(FM) Albany NY 10 hrs
*WCDB(FM) Albany NY 3 hrs
WCSS(AM) Amsterdam NY 12 hrs
*WBNY(FM) Buffalo NY 3 hrs
WDNY(AM) Dansville NY 4 hrs
*WCVF-FM Fredonia NY 4 hrs
*WEOS(FM) Geneva NY 10 hrs
*WITR(FM) Henrietta NY 2 hrs
WVBR-FM Ithaca NY 6 hrs
*WKCR-FM New York NY 6 hrs
*WKCR-FM New York NY 8 hrs
*WNYU-FM New York NY 13 hrs
WEOK(AM) Poughkeepsie NY 2 hrs
WTMM(AM) Rensselaer NY 5 hrs
WBEA(FM) Southold NY
*WOBO(FM) Batavia OH 3 hrs
WXXR(FM) Fredericktown OH 2hrs hrs
WFXN-FM Galion OH 2hrs hrs
WWCD(FM) Grove City OH 4 hrs
WXXF(FM) Loudonville OH 2hrs hrs
WPFB(AM) Middletown OH 2 hrs
*WOBC-FM Oberlin OH 20 hrs
*WOBC-FM Oberlin OH 8 hrs
WSTV(AM) Steubenville OH 2 hrs
*WOBN(FM) Westerville OH 2 hrs
*WYSO(FM) Yellow Springs OH 3 hrs
KKBS(FM) Guymon OK 5 hrs
KSPI(AM) Stillwater OK
KUJZ(FM) Creswell OR 4 hrs
*KLCC(FM) Eugene OR 9 hrs
*KTEC(FM) Klamath Falls OR 15 hrs
*KLCO(FM) Newport OR 6 hrs
KACI(AM) The Dalles OR 3 hrs
KMCQ(FM) The Dalles OR 3 hrs
WBVP(AM) Beaver Falls PA 2 hrs
*WBUQ(FM) Bloomsburg PA 10 hrs
*WDCV-FM Carlisle PA 1 hr
WCCR(FM) Clarion PA
*WVBU-FM Lewisburg PA 2 hrs
WRKT(FM) North East PA 1 hr
*WKDU(FM) Philadelphia PA 4 hrs
WJAS(AM) Pittsburgh PA 2 hrs
*WRCT(FM) Pittsburgh PA 12 hrs
*WXLV(FM) Schnecksville PA 6 hrs
*WUSR(FM) Scranton PA 10 hrs
WPIC(AM) Sharon PA 12 hrs
WMBS(AM) Uniontown PA 3 hrs
*WCYJ-FM Waynesburg PA 3 hrs
WMNT(AM) Manati PR
WUKQ(AM) Ponce PR
*WRIU(FM) Kingston RI 6 hrs
WADK(AM) Newport RI 1 hr
WADK(AM) Newport RI 2 hrs
WVGB(AM) Beaufort SC
WJDJ(AM) Hartsville SC 2 hrs
WJBS(AM) Holly Hill SC 2 hrs
WMXT(FM) Pamplico SC 5 hrs
KGIM(AM) Aberdeen SD
*KLND(FM) Little Eagle SD 2 hrs
*KAUR(FM) Sioux Falls SD 6 hrs
WNAX(AM) Yankton SD 15 hrs
*WHCB(FM) Bristol TN 2 hrs
*WHRS(FM) Cookeville TN 1 hr
WWAM(AM) Jasper TN 1 hr
WKGN(AM) Knoxville TN 1 hr
WEGR(FM) Memphis TN
*WMTS-FM Murfreesboro TN 2 hrs
*WRVU(FM) Nashville TN 11 hrs
*KACV-FM Amarillo TX 6 hrs
KBYG(AM) Big Spring TX 12 hrs
KPSM(FM) Brownwood TX 5 hrs
*KWTS(FM) Canyon TX 11 hrs
*KAMU-FM College Station TX 5 hrs
KTSM(AM) El Paso TX 1 hr
KSWA(AM) Graham TX 2 hrs
KPRC(AM) Houston TX 16 hrs
KRVL(FM) Kerrville TX 1 hr
KHKZ(FM) Mercedes TX 3 hrs
KWEL(AM) Midland TX 1 hr
*KOCV(FM) Odessa TX 4 hrs
KRXT(FM) Rockdale TX 10 hrs hrs
KNCN(FM) Sinton TX 1 hr
WKEX(AM) Blacksburg VA 3 hrs
*WNRN(FM) Charlottesville VA 6 hrs
WEVA(AM) Emporia VA 3 hrs
WRVA(AM) Richmond VA 2 hrs
WVBW(FM) Suffolk VA 1 hr
*WISE-FM Wise VA 2 hrs
*WVPS(FM) Burlington VT 8 hrs
WEQX(FM) Manchester VT 5 hrs
WIKE(AM) Newport VT 5 hrs
WVNR(AM) Poultney VT 3 hrs
*WRVT(FM) Rutland VT 3 hrs
*WVPR(FM) Windsor VT 3 hrs
*KGRG-FM Auburn WA 3 hrs
*KUGS(FM) Bellingham WA 17 hrs wkly hrs
*KUGS(FM) Bellingham WA 2 hrs wkly hrs
KBRO(AM) Bremerton WA 4 hrs
KGNW(AM) Burien-Seattle WA
*KSER(FM) Everett WA 2 hrs
*KZUU(FM) Pullman WA 2 hrs
KLFE(AM) Seattle WA 2 hrs
KLFE(AM) Seattle WA 4 hrs
*KNHC(FM) Seattle WA 6 hrs
*KPBX-FM Spokane WA
*KUPS(FM) Tacoma WA 4 hrs
*WBSD(FM) Burlington WI 5 hrs
*WWSP(FM) Stevens Point WI 1 hr
*WCCX(FM) Waukesha WI 3 hrs
WDLS(AM) Wisconsin Dells WI 5 hrs
WGGE(FM) Parkersburg WV 5 hrs
*WVPG(FM) Parkersburg WV 2 hrs
*WVSB(FM) Romney WV

Polish

KXMX(AM) Anaheim CA
*KSPC(FM) Claremont CA 3 hrs
*KUSF(FM) San Francisco CA 1 hr
WPRX(AM) Bristol CT 2 hrs
WGCH(AM) Greenwich CT 1 hr
*WRTC-FM Hartford CT 3 hrs
WMMW(AM) Meriden CT 1 hr
WMRD(AM) Middletown CT 2 hrs
WRYM(AM) New Britain CT 5 hrs
*WCNI(FM) New London CT 3 hrs
WICH(AM) Norwich CT 1 hr
WATR(AM) Waterbury CT 2 hrs
*WWUH(FM) West Hartford CT 3 hrs
*WKTO(FM) Edgewater FL 1.5 hrs
WXYB(AM) Indian Rocks Beach FL 2 hrs
WLBE(AM) Leesburg FL 2 hrs
WPSO(AM) New Port Richey FL 1 hr
WSBB(AM) New Smyrna Beach FL 1 hr
WTMY(AM) Sarasota FL 1 hr
KSKB(FM) Brooklyn IA 1 hr
WVIV-FM Highland Park IL
WJOL(AM) Joliet IL 1 hr
WHCO(AM) Sparta IL 2 hrs
WJOB(AM) Hammond IN 2 hrs
WIMS(AM) Michigan City IN 3 hrs
WACE(AM) Chicopee MA 1 hr
WHTB(AM) Fall River MA 1 hr
WNBH(AM) New Bedford MA 2 hrs
WHMP(AM) Northampton MA 3 hrs
WBRK(AM) Pittsfield MA 2 hrs
WESX(AM) Salem MA 2 hrs
*WBSL-FM Sheffield MA 1 hr
WESO(AM) Southbridge MA 3 hrs
WARE(AM) Ware MA 4 hrs
WGFP(AM) Webster MA 1 hr
*WCUW(FM) Worcester MA 6 hrs
WORC(AM) Worcester MA 4 hrs
WBMD(AM) Baltimore MD 2 hrs
WATZ(AM) Alpena MI 2 hrs
WATZ-FM Alpena MI 2 hrs
WLEW(AM) Bad Axe MI 2 hrs
WIBM(AM) Jackson MI 1.5 hrs
WMTE-FM Manistee MI 6 hrs
*WOES(FM) Ovid-Elsie MI 1 hr
WMIC(AM) Sandusky MI 5 hrs
*WPHS(FM) Warren MI 2 hrs
WBMI(FM) West Branch MI 6 hrs
WNMT(AM) Nashwauk MN 2 hrs
KRDS-FM New Prague MN 18 hrs
WEW(AM) Saint Louis MO 2 hrs
KJSK(AM) Columbus NE 4 hrs
KTTT(AM) Columbus NE 5 hrs
*WUNH(FM) Durham NH 2 hrs
*WSOU(FM) South Orange NJ 2 hrs
WAAL(FM) Binghamton NY 3 hrs
*WHRW(FM) Binghamton NY 3 hrs
*WBFO(FM) Buffalo NY 3 hrs
WECK(AM) Cheektowaga NY 2 hrs
WDOE(AM) Dunkirk NY 6 hrs
WAMF(AM) Fulton NY 5 hrs
*WGMC(FM) Greece NY 2 hrs
*WUBJ(FM) Jamestown NY 3 hrs
WIZR(AM) Johnstown NY 1 hr
WKNY(AM) Kingston NY 1 hr
WXRL(AM) Lancaster NY 19 hrs
WLVL(AM) Lockport NY 1 hr
*WVCR-FM Loudonville NY 3 hrs
WVOA-FM Mexico NY 4 hrs
WJJL(AM) Niagara Falls NY 2 hrs
WTLA(AM) North Syracuse NY 2 hrs
*WOLN(FM) Olean NY 3 hrs
WSGO(AM) Oswego NY 2 hrs
WEOK(AM) Poughkeepsie NY 1 hr
WRIV(AM) Riverhead NY 4 hrs
WGGO(AM) Salamanca NY 1 hr
*WSPN(FM) Saratoga Springs NY 3 hrs
WLLW(FM) Seneca Falls NY 2 hrs
WSFW(AM) Seneca Falls NY 2 hrs
*WUSB(FM) Stony Brook NY 1 hr
WIBX(AM) Utica NY 3 hrs
WUTQ(AM) Utica NY 4 hrs
WTBQ(AM) Warwick NY
WNYV(FM) Whitehall NY 1 hr
*WOBO(FM) Batavia OH 3 hrs
WOMP(AM) Bellaire OH 2 hrs
*WCPN(FM) Cleveland OH 1 hr
WKTX(AM) Cortland OH 1 hr
WDLW(AM) Lorain OH 3 hrs
WRTK(AM) Niles OH 1 hr
WSTV(AM) Steubenville OH 2 hrs
WCWA(AM) Toledo OH 1 hr
WTOD(AM) Toledo OH 4 hrs
*WJCU(FM) University Heights OH 2 hrs
WELW(AM) Willoughby-Eastlake OH 1 hr
*WMUH(FM) Allentown PA 2 hrs
WVAM(AM) Altoona PA 1 hr
WMBA(AM) Ambridge PA 2 hrs
WWSM(AM) Annville-Cleona PA 2 hrs
WWCS(AM) Canonsburg PA 2 hrs
WFGI(AM) Charleroi PA 2 hrs
WOGI(FM) Charleroi PA 2 hrs
*WERG(FM) Erie PA 3 hrs
*WMCE(FM) Erie PA 3 hrs
WCCS(AM) Homer City PA 3 hrs
WNTJ(AM) Johnstown PA 5 hrs
WEDO(AM) McKeesport PA 1 hr
WQFM(FM) Nanticoke PA 3 hrs
WKST(AM) New Castle PA 1 hr
WGBN(AM) New Kensington PA 3 hrs
WYCK(AM) Plains PA 3 hrs
WPAZ(AM) Pottstown PA 1 hr
WECZ(AM) Punxsutawney PA 3 hrs
*WXLV(FM) Schnecksville PA 6 hrs
WICK(AM) Scranton PA 2 hrs
WPIC(AM) Sharon PA 3 hrs
WCDW(FM) Susquehanna PA 5 hrs
WKZV(AM) Washington PA 3 hrs
WLKW(AM) West Warwick RI 2 hrs
WNRI(AM) Woonsocket RI 2 hrs
WOON(AM) Woonsocket RI 3 hrs
KYNT(AM) Yankton SD 1 hr
*KOOP(FM) Hornsby TX .5 hrs
KVLG(AM) La Grange TX 6 hrs
KBEC(AM) Waxahachie TX 2 hrs
KANI(AM) Wharton TX 6 hrs
WVNR(AM) Poultney VT 1 hr
WJMT(AM) Merrill WI 3 hrs
WRPN(AM) Ripon WI 2 hrs
*WRPN-FM Ripon WI 2 hrs
WSPT(AM) Stevens Point WI 1 hr
WSAU(AM) Wausau WI 3 hrs
WMOV(AM) Ravenswood WV 1 hr
WBBD(AM) Wheeling WV 2 hrs

Polka

KZAT-FM Belle Plaine IA 2 hrs
KDSN(AM) Denison IA 4 hrs
KMAQ(AM) Maquoketa IA 3 hrs
KLEE(AM) Ottumwa IA 1 hr
WLUV(AM) Loves Park IL 6 hrs
WPNA(AM) Oak Park IL 15 hrs
WTAY(AM) Robinson IL 3 hrs
KCAY(FM) Russell KS
KRSL(AM) Russell KS 4 hrs
WNBY(AM) Newberry MI 2 hrs
WUPY(FM) Ontonagon MI 1 hr
WYGR(AM) Wyoming MI 3 hrs
KRBT(AM) Eveleth MN 3 hrs
WMFG(AM) Hibbing MN 4 hrs
KDUZ(AM) Hutchinson MN 8 hrs
KLTF(AM) Little Falls MN 2 hrs
KWNO(AM) Winona MN 5 hrs
KHND(AM) Harvey ND 3 hrs
KTTT(AM) Columbus NE
WGHT(AM) Pompton Lakes NJ 1 hr
WAAL(FM) Binghamton NY 4 hrs
*WZIP(FM) Akron OH 4 hrs
WNDH(FM) Napoleon OH 2 hrs
WELW(AM) Willoughby-Eastlake OH 15 hrs
WKBN(AM) Youngstown OH 2 hrs
WMBA(AM) Ambridge PA 2 hrs
WGPA(AM) Bethlehem PA 12 hrs
WHYL(AM) Carlisle PA 2 hrs
WLMI(FM) Kane PA 1 hr
WPTT(AM) McKeesport PA 2 hrs
WPHB(AM) Philipsburg PA 6 hrs
WARM(AM) Scranton PA 4 hrs
WWII(AM) Shiremanstown PA 7 hrs
WMGH-FM Tamaqua PA 3 hrs
WKZV(AM) Washington PA 2 hrs
KYNT(AM) Yankton SD 1 hr
*WMTS-FM Murfreesboro TN 2 hrs
KWHI(AM) Brenham TX 2 hrs
KULM(FM) Columbus TX 12 hrs
KNAF(AM) Fredericksburg TX 4.5 hrs
KCTI(AM) Gonzales TX 5 hrs
KRXT(FM) Rockdale TX 2 hrs hrs
KYKM(FM) Yoakum TX 9 hrs
WDKM(FM) Adams WI 14 hrs
WLDY(AM) Ladysmith WI 3 hrs
WCCN(AM) Neillsville WI 2 hrs
WIZD(FM) Rudolph WI 3 hrs
WVRQ(AM) Viroqua WI 6 hrs

Portugese

KSTN-FM Stockton CA 4 hrs
*WRTC-FM Hartford CT 8 hrs
*WWUH(FM) West Hartford CT 3 hrs
WACE(AM) Chicopee MA 1 hr
WSAR(AM) Fall River MA 3 hrs
WPEP(AM) Taunton MA 10 hrs
WPHE(AM) Phoenixville PA 3 hrs
WNRI(AM) Woonsocket RI 2 hrs
*WRVU(FM) Nashville TN 2 hrs

Progressive

*KFJC(FM) Los Altos CA 4 hrs
*WONC(FM) Naperville IL 14 hrs
WTTS(FM) Bloomington IN 4 hrs
*WDSO(FM) Chesterton IN 8 hrs
*WBKE-FM North Manchester IN 7 hrs
*WECI(FM) Richmond IN 18 hrs
*KMVC(FM) Marshall MO 10 hrs
*KCLC(FM) Saint Charles MO 7 hrs
*KZUM(FM) Lincoln NE 15 hrs
WRRV(FM) Middletown NY 2 hrs
*WIRQ(FM) Rochester NY 3 hrs
WIOT(FM) Toledo OH 2 hrs
*KRSC-FM Claremore OK 12 hrs
*WCAL(FM) California PA 6 hrs
*WCUC-FM Clarion PA 9 hrs
*WHRV(FM) Norfolk VA 14 hrs
WIZN(FM) Vergennes VT 1 hr
*KZUU(FM) Pullman WA 10 hrs
*WSUP(FM) Platteville WI 6 hrs

Public Affairs

WDLT-FM Chickasaw AL 4 hrs
KMLE(FM) Chandler AZ 1 hr
*KGHR(FM) Tuba City AZ 5 hrs
*KPFA(FM) Berkeley CA 18 hrs
*KNCA(FM) Burney CA 7 hrs
*KSPC(FM) Claremont CA 3 hrs
*KCRH(FM) Hayward CA 5 hrs
KYSR(FM) Los Angeles CA 2 hrs
*KNSQ(FM) Mount Shasta CA 7 hrs
KTYD(FM) Santa Barbara CA 1 hr
KIMN(FM) Denver CO 2 hrs
KJJD(AM) Windsor CO
WSTW(FM) Wilmington DE 1 hr
WXTB(FM) Clearwater FL 4 hrs
WIRA(AM) Fort Pierce FL 1 hr
WVOP(AM) Vidalia GA 2 hrs
*WEFT(FM) Champaign IL 5 hrs
*WDGC-FM Downers Grove IL 6 hrs
KUUL(FM) East Moline IL 6 hrs
*WEPS(FM) Elgin IL 3 hrs
WFXN(AM) Moline IL 4 hrs
WPNA(AM) Oak Park IL 5 hrs
WVAZ(FM) Oak Park IL 2 hrs
WYKT(FM) Wilmington IL 4 hrs
KNCK(AM) Concordia KS 2 hrs
WMJL-FM Marion KY 1 hr
WSNE-FM Taunton MA 4 hrs
*WCBN-FM Ann Arbor MI 5 hrs
WDMK(FM) Detroit MI 2 hrs
WCCY(AM) Houghton MI 1 hr
*KQAL(FM) Winona MN 10 hrs
KCMO(AM) Kansas City MO 1 1/2 hrs
KCMO-FM Kansas City MO 1 1/2 hrs
WCLN(AM) Clinton NC
WPAQ(AM) Mount Airy NC 1 hr
WKRK(AM) Murphy NC 3 hrs
*KABU(FM) Fort Totten ND 5 hrs
KLIQ(FM) Hastings NE
*KZUM(FM) Lincoln NE 11 hrs
*WBJB-FM Lincroft NJ 5 hrs
*WRPR(FM) Mahwah NJ 12 hrs
WMTR(AM) Morristown NJ 5 hrs
WNNJ(AM) Newton NJ 1 hr
*WSOU(FM) South Orange NJ 3 hrs
*WTSR(FM) Trenton NJ 8 hrs
*KUNV(FM) Las Vegas NV 4 hrs
*WXLH(FM) Blue Mountain Lake NY
*WBSU(FM) Brockport NY 8 hrs
*WCWP(FM) Brookville NY 5 hrs
WJYE(FM) Buffalo NY 2 hrs
*WNED(AM) Buffalo NY
*WSLU(FM) Canton NY
WAQX-FM Manlius NY 1 hr
*WDFH(FM) Ossining NY 20 hrs
WUMX(FM) Rome NY 1 hr
WQAR(FM) Stillwater NY
*WJPZ-FM Syracuse NY 13 hrs
*WARY(FM) Valhalla NY 10 hrs
WNYV(FM) Whitehall NY 5 hrs
*WZIP(FM) Akron OH 11 hrs
WRQN(FM) Bowling Green OH 1 hr
WKKY(FM) Geneva OH 2 hrs
KJSR(FM) Tulsa OK 2 hrs
*KSMF(FM) Ashland OR 7 hrs
*KSOR(FM) Ashland OR 7 hrs
*KSBA(FM) Coos Bay OR 7 hrs
*KSKF(FM) Klamath Falls OR 7 hrs
KBNP(AM) Portland OR
KACI(AM) The Dalles OR 1 hr
WWCH(AM) Clarion PA
WSJR(FM) Dallas PA 1 hr
*WQLN-FM Erie PA 5 hrs
WRTS(FM) Erie PA 1 hr
WTKT(AM) Harrisburg PA 2 hrs
WIOQ(FM) Philadelphia PA
WSHH(FM) Pittsburgh PA 1 hr
*WSRU(FM) Slippery Rock PA 1 hr
*WRLC(FM) Williamsport PA 2 hrs
WAGS(AM) Bishopville SC 2 hrs
WSAA(FM) Benton TN 2 hrs
WKHT(FM) Knoxville TN 2 hrs
WQBB(AM) Powell TN 2 hrs
KDHT(FM) Cedar Park TX 2 hrs
KBFM(FM) Edinburg TX
KTSM(AM) El Paso TX 1 hr
KTBZ-FM Houston TX 1 hr
*KSWP(FM) Lufkin TX 2 hrs
KAMX(FM) Luling TX 2 hrs
*KNTU(FM) McKinney TX 2 hrs
KCYY(FM) San Antonio TX 2 hrs
KKYX(AM) San Antonio TX 2 hrs
KQXT(FM) San Antonio TX 1 hr
KNCN(FM) Sinton TX 1 hr
KVEL(AM) Vernal UT 1 hr
*WVRU(FM) Radford VA 6 hrs
WMOO(FM) Derby Center VT 8 hrs
WJJR(FM) Rutland VT 1 hr
KEDO(AM) Longview WA 2 hrs
WNBI(AM) Park Falls WI 5 hrs
*WWSP(FM) Stevens Point WI 5 hrs
WWYO(AM) Pineville WV 8 hrs

Reggae

*KBBI(AM) Homer AK 3 hrs
*WJAB(FM) Huntsville AL 4 hrs
*WVUA-FM Tuscaloosa AL 3 hrs
*KABF(FM) Little Rock AR 4 hrs
*KSPC(FM) Claremont CA 8 hrs
*KFSR(FM) Fresno CA 6 hrs
*KSBR(FM) Mission Viejo CA 3 hrs
*KSPB(FM) Pebble Beach CA 2 hrs
*KZYX(FM) Philo CA 5 hrs
*KCSB-FM Santa Barbara CA 6 hrs
*KCSS(FM) Turlock CA
*KASF(FM) Alamosa CO 5 hrs
KSMT(FM) Breckenridge CO 2 hrs
*KRCC(FM) Colorado Springs CO 6 hrs
*KWSB-FM Gunnison CO 6 hrs
KTCL(FM) Wheat Ridge CO 2 hrs
*WXCI(FM) Danbury CT 2 hrs
*WQTQ(FM) Hartford CT 4 hrs
*WESU(FM) Middletown CT 10 hrs
WMRD(AM) Middletown CT 1 hr
*WRGP(FM) Homestead FL 3 hrs
WIIS(FM) Key West FL 4 hrs
*WFCF(FM) Saint Augustine FL 4 hrs
*WANM(FM) Tallahassee FL 3 hrs
WFLM(FM) White City FL 4 hrs
*WCLK(FM) Atlanta GA 3 hrs
*WRAS(FM) Atlanta GA 4 hrs
KWXX-FM Hilo HI 20 hrs
KAOY(FM) Kealakekua HI 4 hrs
*WNUR-FM Evanston IL 4 hrs
*KJHK(FM) Lawrence KS 3 hrs
*WESM(FM) Princess Anne MD 2 hrs
*WCBN-FM Ann Arbor MI 4 hrs hrs
*WDET-FM Detroit MI 2 hrs
*WLNZ(FM) Lansing MI 4 hrs
*WVSD(FM) Itta Bena MS 3 hrs
*WJSU(FM) Jackson MS 2 hrs
*WNAA(FM) Greensboro NC 7 hrs
WVOD(FM) Manteo NC 2 hrs
KKCD(FM) Omaha NE 1 hr
*WKNH(FM) Keene NH 3 hrs
*WPCR-FM Plymouth NH 3 hrs
*WNTI(FM) Hackettstown NJ 3 hrs
*WBZC(FM) Pemberton NJ 4 hrs
KRSI(FM) Garapan-Saipan NP 22 hrs
*WBNY(FM) Buffalo NY 3 hrs
*WHCL-FM Clinton NY 2 hrs
*WCVF-FM Fredonia NY 4 hrs
*WEOS(FM) Geneva NY 3 hrs
*WITR(FM) Henrietta NY 5 hrs
*WICB(FM) Ithaca NY 2 hrs
*WNYU-FM New York NY 2 hrs
*WPNR-FM Utica NY 5 hrs
*WDUB(FM) Granville OH 2 hrs
WVKO-FM Johnstown OH 5 hrs
*KEOL(FM) La Grande OR 7 hrs
*KBOO(FM) Portland OR 10 hrs
*WLVR(FM) Bethlehem PA 6 hrs
WWCS(AM) Canonsburg PA 2 hrs
*WJRH(FM) Easton PA 6 hrs
*WERG(FM) Erie PA 4 hrs
*WKVR-FM Huntingdon PA 3 hrs
*WRIU(FM) Kingston RI 7 hrs
*WSSB-FM Orangeburg SC 4 hrs
*WRVU(FM) Nashville TN 3 hrs
*WUTS(FM) Sewanee TN 5 hrs
*KAZI-FM Austin TX 6 hrs
*KTRU(FM) Houston TX 6 hrs
*KTSU(FM) Houston TX 8 hrs
*KSAU(FM) Nacogdoches TX 2 hrs
*WWHS-FM Hampden-Sydney VA 4 hrs
*WHOV(FM) Hampton VA 4 hrs
*WCWM(FM) Williamsburg VA 6 hrs
WIZN(FM) Vergennes VT 1 hr
*KGRG-FM Auburn WA 4 hrs
*KSER(FM) Everett WA 2 hrs
*KUPS(FM) Tacoma WA 6 hrs
*WBSD(FM) Burlington WI 3 hrs
*WWVU-FM Morgantown WV 4 hrs

Religious

KYMG(FM) Anchorage AK 1 hr
*KBRW(AM) Barrow AK 1 hr
*KNOM(AM) Nome AK 20 hrs
*KNOM-FM Nome AK 20 hrs
KIAL(AM) Unalaska AK 04 hrs
WKNU(FM) Brewton AL 2 hrs
WTVY-FM Dothan AL 3 hrs
WABF(AM) Fairhope AL 6 hrs
WKWL(AM) Florala AL 12 hrs
WAGH(FM) Fort Mitchell AL 6 hrs
WZOB(AM) Fort Payne AL 5 hrs
WTKI(AM) Huntsville AL 4 hrs
*WLJS-FM Jacksonville AL 3 hrs
WEUP-FM Moulton AL 1 hr
WNSI(AM) Robertsdale AL 6 hrs
WGOL(AM) Russellville AL 10 hrs
WKEA-FM Scottsboro AL 4 hrs
WTBC(AM) Tuscaloosa AL 3 hrs
*WVUA-FM Tuscaloosa AL 3 hrs
KMJI(FM) Ashdown AR 4 hrs
KEWI(AM) Benton AR 5 hrs
KLYR(AM) Clarksville AR 8 hrs
KAVV(FM) Benson AZ 3 hrs
KCUZ(AM) Clifton AZ 2 hrs
KFYI(AM) Phoenix AZ 2 hrs
KTKT(AM) Tucson AZ 2 hrs
KHOV-FM Wickenburg AZ 1 hr
KXMX(AM) Anaheim CA
KERN(AM) Bakersfield CA 1 hr
KISV(FM) Bakersfield CA 1 hr
KMET(AM) Banning CA 4 hrs
KJLH-FM Compton CA 7 hrs
KCNQ(FM) Kernville CA 1 hr
KLBS(AM) Los Banos CA 8 hrs
KDUQ(FM) Ludlow CA 1 hr
*KSMC(FM) Moraga CA 2 hrs
KAAT(FM) Oakhurst CA 2 hrs
KWKU(AM) Pomona CA 15 hrs
KVML(AM) Sonora CA 3 hrs
KSTN(AM) Stockton CA 5 hrs
KSUE(AM) Susanville CA 3 hrs
KXPS(AM) Thousand Palms CA 17 hrs
KDIA(AM) Vallejo CA 5 hrs
KDYA(AM) Vallejo CA 5 hrs
KUBA(AM) Yuba City CA 2 hrs
KCMN(AM) Colorado Springs CO 3 hrs
*KSJD(FM) Cortez CO 1 hr
KFKA(AM) Greeley CO 4 hrs
KUBC(AM) Montrose CO 3 hrs
KCRT(AM) Trinidad CO 5 hrs
KSPK(FM) Walsenburg CO 2 hrs
KCOL(AM) Wellington CO 1 hr
WGCH(AM) Greenwich CT 3 hrs
WMMW(AM) Meriden CT 4 hrs
*WECS(FM) Willimantic CT 3 hrs
WILI(AM) Willimantic CT 2 hrs
WJWL(AM) Georgetown DE 6 hrs
WYUS(AM) Milford DE 10 hrs
WGMD(FM) Rehoboth Beach DE 2 hrs
WSTW(FM) Wilmington DE 1 hr
WYBT(AM) Blountstown FL
WWPR(AM) Bradenton FL 6 hrs
WLQH(AM) Chiefland FL 9 hrs
*WVUM(FM) Coral Gables FL 6 hrs
WYNY(AM) Cross City FL 5 hrs
WHNR(AM) Cypress Gardens FL 10 hrs
WNDB(AM) Daytona Beach FL 5 hrs
WHBO(AM) Dunedin FL 2 hrs
WIRA(AM) Fort Pierce FL 1 hr
WXYB(AM) Indian Rocks Beach FL 8 hrs
WPLA(AM) Jacksonville FL 2 hrs
WWRF(AM) Lake Worth FL 4 hrs
WONN(AM) Lakeland FL 2 hrs
WQHL(AM) Live Oak FL 6 hrs
WMEL(AM) Melbourne FL 6 hrs
WQAM(AM) Miami FL 4 hrs
WARO(FM) Naples FL 2 hrs
WPSO(AM) New Port Richey FL 8 hrs
WSBB(AM) New Smyrna Beach FL 5 hrs
*WHIF(FM) Palatka FL 10 hrs
WIYD(AM) Palatka FL 5 hrs
WFBX(FM) Parker FL 5 hrs
WCOA(AM) Pensacola FL
WPSL(AM) Port St. Lucie FL 6 hrs
WSDO(AM) Sanford FL 3 hrs
WKII(AM) Solana FL 4 hrs
WIXC(AM) Titusville FL 6 hrs
WSIR(AM) Winter Haven FL 14 hrs
V6AH(AM) Pohnpei FM 2 hrs
*V6AI(AM) Yap FM 4 hrs
WSB(AM) Atlanta GA 3 hrs
WMGR(AM) Bainbridge GA 12 hrs
WJTH(AM) Calhoun GA 16 hrs
WBLJ(AM) Dalton GA
WDMG(AM) Douglas GA 6 hrs
WWWE(AM) Hapeville GA 12 hrs
WCEH(AM) Hawkinsville GA 1 hr
WMAC(AM) Macon GA 4 hrs
WHKN(FM) Millen GA 2 hrs
WROM(AM) Rome GA
WGTA(AM) Summerville GA 12 hrs
WTHO-FM Thomson GA 6 hrs
WLET(AM) Toccoa GA
WVOP(AM) Vidalia GA 8 hrs
KCHE-FM Cherokee IA 4 hrs
KROS(AM) Clinton IA 4 hrs
KCQQ(FM) Davenport IA 1 hr
KILR(AM) Estherville IA 11 hrs
KILR-FM Estherville IA 11 hrs
KNOD(FM) Harlan IA 2 hrs
KNIA(AM) Knoxville IA 18 hrs
KMCH(FM) Manchester IA 4 hrs
KRIB(AM) Mason City IA 5 hrs
KYTC(FM) Northwood IA 2 hrs
KBIZ(AM) Ottumwa IA 4 hrs
KLLT(FM) Spencer IA 2 hrs
KTLB(FM) Twin Lakes IA 2 hrs
KXEL(AM) Waterloo IA 20 hrs
KVNI(AM) Coeur d'Alene ID 3 hrs
KMCL-FM McCall ID 1 hr
KVSI(AM) Montpelier ID 2 hrs
KRPL(AM) Moscow ID 2 hrs
KWIK(AM) Pocatello ID 1 hr
KSPT(AM) Sandpoint ID 2 hrs
KTFI(AM) Twin Falls ID 5 hrs
WRMJ(FM) Aledo IL 3 hrs
KATZ-FM Alton IL 2 hrs
WBGZ(AM) Alton IL 3 hrs
WBIG(AM) Aurora IL 6 hrs
WDWS(AM) Champaign IL 4 hrs
KSGM(AM) Chester IL 6 hrs
*WIIT(FM) Chicago IL 2 hrs
WDKB(FM) De Kalb IL 1 hr
WLBK(AM) De Kalb IL 6 hrs
WDQN(AM) Du Quoin IL 5 hrs
WAIK(AM) Galesburg IL 6 hrs
WGIL(AM) Galesburg IL 4 hrs
WLSR(FM) Galesburg IL 2 hrs
WKYX-FM Golconda IL 1 hr
WJBM(AM) Jerseyville IL 3 hrs
WKEI(AM) Kewanee IL 6 hrs
WLBH(AM) Mattoon IL 5 hrs
WMOK(AM) Metropolis IL 5 hrs
WRAM(AM) Monmouth IL 3 hrs
WDQX(FM) Morton IL 1 hr
WINI(AM) Murphysboro IL 6 hrs
*WONC(FM) Naperville IL 4 hrs
WPNA(AM) Oak Park IL 4 hrs
WLCE(FM) Petersburg IL 5 hrs
WBBA-FM Pittsfield IL 3 hrs
WKXQ(FM) Rushville IL 6 hrs
WJBD(AM) Salem IL 6 hrs
WHCO(AM) Sparta IL 10 hrs
WTIM-FM Taylorville IL 4 hrs
WRAN(FM) Tower Hill IL 3 hrs
*WETN(FM) Wheaton IL 3 hrs
WZSR(FM) Woodstock IL 1 hr
WQME(FM) Anderson IN 9 hrs
WNUY(FM) Bluffton IN 4 hrs
WIFE(AM) Connersville IN 12 hrs
WBYT(FM) Elkhart IN 2 hrs
WWFT(FM) Fishers IN 3 hrs
WFLQ(FM) French Lick IN 6 hrs
*WGRE(FM) Greencastle IN 2 hrs
WTRE(AM) Greensburg IN 3 hrs
WJOB(AM) Hammond IN 2 hrs
WBDC(FM) Huntingburg IN 5 hrs
WXGO(AM) Madison IN 6 hrs
WEFM(FM) Michigan City IN 4 hrs
WMDH(AM) New Castle IN 2 hrs
WRIN(AM) Rensselaer IN 10 hrs
WROI(FM) Rochester IN 6 hrs
WQKC(FM) Seymour IN 3 hrs
WZZB(AM) Seymour IN 6 hrs
WSBT(AM) South Bend IN 2 hrs
WCLS(FM) Spencer IN 6 hrs
WAWC(FM) Syracuse IN 4 hrs
WCJC(FM) Van Buren IN 3 hrs
KABI(AM) Abilene KS 4 hrs
KSOK(AM) Arkansas City KS 5 hrs
KSNP(FM) Burlington KS 3 hrs
KDCC(AM) Dodge City KS
KVGB(AM) Great Bend KS 2 hrs
KFBZ(FM) Haysville KS 2 hrs
KHOK(FM) Hoisington KS 2 hrs
KNNS(AM) Larned KS 5 hrs
KLWN(AM) Lawrence KS 4 hrs
KNGL(AM) McPherson KS 5 hrs
KLKC(AM) Parsons KS 2 hrs
WIBW(AM) Topeka KS 6 hrs
KLEY(AM) Wellington KS 4 hrs
*KCFN(FM) Wichita KS
WMMG(AM) Brandenburg KY 8 hrs
WKDP-FM Corbin KY 5 hrs
WJQI(AM) Fort Campbell KY 4 hrs
WFKN(AM) Franklin KY
WHVO(AM) Hopkinsville KY 6 hrs
WRNZ(FM) Lancaster KY 4 hrs
WKYL(FM) Lawrenceburg KY 2 hrs
WQQR(FM) Mayfield KY 10 hrs
WFTM(AM) Maysville KY 5 hrs
WFXY(AM) Middlesboro KY 3 hrs

WCBR(AM) Richmond KY 35 hrs
WEKY(AM) Richmond KY 6 hrs
WTLO(AM) Somerset KY 4 hrs
*WJCR-FM Upton KY
KVVP(FM) Leesville LA 9 hrs
KVCL(AM) Winnfield LA 15 hrs
*WERS(FM) Boston MA 1 hr
*WGAO(FM) Franklin MA 8 hrs
WBEC(AM) Pittsfield MA 3 hrs
WBRK(AM) Pittsfield MA 2 hrs
WMAS(AM) Springfield MA 2 hrs
WOLB(AM) Baltimore MD 2 hrs
WCEM(AM) Cambridge MD 5 hrs
WSRY(AM) Elkton MD 3 hrs
*WMTB-FM Emmittsburg MD 4 hrs
WAYZ(FM) Hagerstown MD 3 hrs
WXCY(FM) Havre de Grace MD 2 hrs
WVAE(AM) Biddeford ME 2 hrs
WABJ(AM) Adrian MI 3 hrs
*WQAC-FM Alma MI 2 hrs
WATZ(AM) Alpena MI 2 hrs
WATZ-FM Alpena MI 1 hr
WAAM(AM) Ann Arbor MI 4 hrs
*WAUS(FM) Berrien Springs MI 10 hrs
WYBR(FM) Big Rapids MI 1 hr
WDBC(AM) Escanaba MI 4 hrs
WDZZ-FM Flint MI 1 hr
WWKR(FM) Hart MI 3 hrs
WCSR(AM) Hillsdale MI 10 hrs
*WTHS(FM) Holland MI 14 hrs
WCCY(AM) Houghton MI 1 hr
WKHM(AM) Jackson MI 4 hrs
WKZO(AM) Kalamazoo MI 5 hrs
WFCX(FM) Leland MI 1 hr
WTWR-FM Luna Pier MI 3 hrs
WMTE-FM Manistee MI 2 hrs
WMPX(AM) Midland MI 3 hrs
WNIL(AM) Niles MI 6 hrs
WUPY(FM) Ontonagon MI 3 hrs
WRSR(FM) Owosso MI 1 hr
WJML(AM) Petoskey MI 8 hrs
KXRA(AM) Alexandria MN 2 hrs
KXRZ(FM) Alexandria MN 3 hrs
KKCQ-FM Bagley MN 9 hrs
KDLM(AM) Detroit Lakes MN 8 hrs
KBRF(AM) Fergus Falls MN 7 hrs
KJJK(AM) Fergus Falls MN 3 hrs
WLKX-FM Forest Lake MN 6 hrs
KKCQ(AM) Fosston MN 4 hrs
WMFG(AM) Hibbing MN 4 hrs
KLFD(AM) Litchfield MN 3 hrs
KFML(FM) Little Falls MN 6 hrs
KLTF(AM) Little Falls MN 4 hrs
KEYL(AM) Long Prairie MN 4 hrs
KQAD(AM) Luverne MN 5 hrs
KMHL(AM) Marshall MN 7 hrs
KYMN(AM) Northfield MN 2 hrs
KDIO(AM) Ortonville MN 5 hrs
KLOH(AM) Pipestone MN 5 hrs
KQIC(FM) Willmar MN 2 hrs
KAGE(AM) Winona MN
KAAN(AM) Bethany MO 1 hr
KMRN(AM) Cameron MO 4 hrs
KCRV(AM) Caruthersville MO 20 hrs
KCXL(AM) Liberty MO 3 hrs
*KMVC(FM) Marshall MO 16 hrs
KMEM-FM Memphis MO 4 hrs
*KJAB-FM Mexico MO 20 hrs
KMCR(FM) Montgomery City MO 2 hrs
KELE-FM Mountain Grove MO 3 hrs
KBDZ(FM) Perryville MO 5 hrs
KLID(AM) Poplar Bluff MO 2 hrs
KMIS(AM) Portageville MO 5 hrs
*KCLC(FM) Saint Charles MO 7 hrs
KMOX(AM) Saint Louis MO 1 hr
KDRO(AM) Sedalia MO 3 hrs
KWTO(AM) Springfield MO 1 hr
KLPW(AM) Union MO 6 hrs
KTKS(FM) Versailles MO 3 hrs
KJPW(AM) Waynesville MO 3 hrs
WAFM(FM) Amory MS 2 hrs
WAMY(AM) Amory MS 6 hrs
WHJT(FM) Clinton MS 6 hrs
WBAQ(FM) Greenville MS 4 hrs
WDMS(FM) Greenville MS 1 hr
WFOR(AM) Hattiesburg MS 8 hrs

WTCD(FM) Indianola MS 11 hrs
WIQQ(FM) Leland MS 6 hrs
WMOX(AM) Meridian MS 8 hrs
WRQO(FM) Monticello MS 10 hrs
WRJW(AM) Picayune MS 16 hrs
WKZU(FM) Ripley MS 2 hrs
KBOW(AM) Butte MT 2 hrs
*KMSM-FM Butte MT 3 hrs
KXTL(AM) Butte MT 1 hr
KMSO(FM) Missoula MT 1 hr
KATQ(AM) Plentywood MT 6 hrs
KATQ-FM Plentywood MT 6 hrs
WWNC(AM) Asheville NC 3 hrs
WXIT(AM) Blowing Rock NC 8 hrs
WSQL(AM) Brevard NC 4 hrs
WBAG(AM) Burlington-Graham NC 5 hrs
WRRZ(AM) Clinton NC 6 hrs
WGAI(AM) Elizabeth City NC 4 hrs
WQNQ(FM) Fletcher NC 2 hrs
WBRM(AM) Marion NC 7 hrs
WHIP(AM) Mooresville NC 6 hrs
WECR(AM) Newland NC 5 hrs
WPTM(FM) Roanoke Rapids NC 3 hrs
WNCA(AM) Siler City NC 15 hrs
WMPM(AM) Smithfield NC 8 hrs
WTOE(AM) Spruce Pine NC 8 hrs
WMXF(AM) Waynesville NC 3 hrs
WENC(AM) Whiteville NC 4 hrs
WTXY(AM) Whiteville NC 10 hrs
*KEYA(FM) Belcourt ND 10 hrs
KAUJ(FM) Grafton ND 3 hrs
KXPO(AM) Grafton ND 3 hrs
KNDK(AM) Langdon ND 4 hrs
KDDR(AM) Oakes ND 3 hrs
KEYZ(AM) Williston ND 5 hrs
KZEN(FM) Central City NE 5 hrs
KJSK(AM) Columbus NE 20 hrs
KLIR(FM) Columbus NE 4 hrs
*KINI(FM) Crookston NE 6 hrs
*KFKX(FM) Hastings NE 3 hrs
KRVN(AM) Lexington NE 12 hrs
KRFS(AM) Superior NE 3 hrs
WHCY(FM) Blairstown NJ 1 hr
*WDVR(FM) Delaware Township NJ 6 hrs
WSJO(FM) Egg Harbor City NJ 4 hrs
*WRDR(FM) Freehold Township NJ 6 hrs
*WNTI(FM) Hackettstown NJ 4 hrs
WTTH(FM) Margate City NJ 1 hr
WNNJ(AM) Newton NJ 1 hr
WGHT(AM) Pompton Lakes NJ 2 hrs
*WSOU(FM) South Orange NJ 5 hrs
WNJC(AM) Washington Township NJ 6 hrs
KRSY(AM) Alamogordo NM 4 hrs
KKTC(FM) Angel Fire NM 8 hrs
KARS(AM) Belen NM
KSIL(FM) Hurley NM
KLEA(AM) Lovington NM 3 hrs
KLAV(AM) Las Vegas NV 3 hrs
KSNE-FM Las Vegas NV 1 hr
KNEV(FM) Reno NV 1 hr
WYSL(AM) Avon NY 4 hrs
WBNR(AM) Beacon NY 5 hrs
WAAL(FM) Binghamton NY 3 hrs
*WHRW(FM) Binghamton NY 6 hrs
WBRV-FM Boonville NY 3 hrs
*WCWP(FM) Brookville NY 4 hrs
*WHCL-FM Clinton NY 2 hrs
WDNY(AM) Dansville NY 1 hr
WDNY-FM Dansville NY 1 hr
WFLR(AM) Dundee NY 5 hrs
WFLR-FM Dundee NY 5 hrs
WELM(AM) Elmira NY 1 hr
WGBB(AM) Freeport NY 6 hrs
WMML(AM) Glens Falls NY 3 hrs
WLIE(AM) Islip NY 3 hrs
WKSN(AM) Jamestown NY 2 hrs
WGHQ(AM) Kingston NY 3 hrs
WIXT(AM) Little Falls NY 1 hr
WLVL(AM) Lockport NY 3 hrs
WLLG(FM) Lowville NY 2 hrs
WVOX(AM) New Rochelle NY 3 hrs
WOR(AM) New York NY 4 hrs

WGNY(AM) Newburgh NY 4 hrs
WTLA(AM) North Syracuse NY 2 hrs
*WNYK(FM) Nyack NY 2 hrs
WDOS(AM) Oneonta NY 7 hrs
WEBO(AM) Owego NY 6 hrs
*WQKE(FM) Plattsburgh NY 3 hrs
WEOK(AM) Poughkeepsie NY 2 hrs
WMYY(FM) Schoharie NY
WSPQ(AM) Springville NY 2 hrs
WRGR(FM) Tupper Lake NY 1 hr
WAJZ(FM) Voorheesville NY 1 hr
WTBQ(AM) Warwick NY
WNYV(FM) Whitehall NY 2 hrs
*WONB(FM) Ada OH 1 hr
WBLL(AM) Bellefontaine OH 6 hrs
WBCO(AM) Bucyrus OH 4 hrs
WQEL(FM) Bucyrus OH 5 hrs
WCER(AM) Canton OH 5 hrs
WKKI(FM) Celina OH 2 hrs
WMNI(AM) Columbus OH 4 hrs
WHIO(AM) Dayton OH 2 hrs
*WWSU(FM) Dayton OH 11 hrs
WDFM(FM) Defiance OH 3 hrs
WZOM(FM) Defiance OH 6 hrs
*WLFC(FM) Findlay OH 3 hrs
WRBP(FM) Hubbard OH 6 hrs
WBKS(FM) Ironton OH 2 hrs
WIRO(AM) Ironton OH 7 hrs
WVKO-FM Johnstown OH 13 hrs
WZOQ(AM) Lima OH 11 hrs
WMOA(AM) Marietta OH 1 hr
WYVK(FM) Middleport OH 6 hrs
WMPO(AM) Middleport-Pomeroy OH 6 hrs
WMVO(AM) Mount Vernon OH 7 hrs
*WMCO(FM) New Concord OH 2 hrs
WOBL(AM) Oberlin OH 1 hr
WNTO(FM) Racine OH 7 hrs
WDMN(AM) Rossford OH
WCWA(AM) Toledo OH 3 hrs
WSPD(AM) Toledo OH 5 hrs
WBTC(AM) Uhrichsville OH 2 hrs
WTUZ(FM) Uhrichsville OH 4 hrs
*WOBN(FM) Westerville OH 4 hrs
WHIZ-FM Zanesville OH 1 hr
KVSP(FM) Anadarko OK 5 hrs
KWON(AM) Bartlesville OK 5 hrs
KGYN(AM) Guymon OK 8 hrs
KICM(FM) Healdton OK 6 hrs
KTJS(AM) Hobart OK 15 hrs
KTMC-FM McAlester OK 5 hrs
WBBZ(AM) Ponca City OK 5 hrs
KZBB(FM) Poteau OK 1 hr
KGFF(AM) Shawnee OK 4 hrs
KWIL(AM) Albany OR 168 hrs
KNND(AM) Cottage Grove OR 3 hrs
KAJO(AM) Grants Pass OR 8 hrs
KMUZ(AM) Gresham OR 1 hr
KUIK(AM) Hillsboro OR 2 hrs
KQIK(AM) Lakeview OR 2 hrs
KQIK-FM Lakeview OR 2 hrs
*KSLC(FM) McMinnville OR 2 hrs
KNPT(AM) Newport OR 2 hrs
KOHI(AM) Saint Helens OR
KACI(AM) The Dalles OR 1 hr
KWBY(AM) Woodburn OR 5 hrs
WBVP(AM) Beaver Falls PA 2 hrs
WHLM(AM) Bloomsburg PA 2 hrs
WISR(AM) Butler PA 6 hrs
*WCAL(FM) California PA 6 hrs
WWCS(AM) Canonsburg PA 2 hrs
WIOO(AM) Carlisle PA 5 hrs
WCCL(FM) Central City PA 4 hrs
WCHA(AM) Chambersburg PA 8 hrs
WHGT(AM) Chambersburg PA 1 hr
WCCR(FM) Clarion PA 4 hrs
WWCH(AM) Clarion PA 8 hrs
WCOJ(AM) Coatesville PA 8 hrs
*WFSE(FM) Edinboro PA 4 hrs
*WWEC(FM) Elizabethtown PA 4 hrs
*WTGP(FM) Greenville PA 1 hr
WHUN(AM) Huntingdon PA 2 hrs
*WKVR-FM Huntingdon PA 3 hrs
WDAD(AM) Indiana PA 2 hrs
WTYM(AM) Kittanning PA 4 hrs
WNPV(AM) Lansdale PA 5 hrs

WCNS(AM) Latrobe PA 2 hrs
WEDO(AM) McKeesport PA 1 hr
WJUN(AM) Mexico PA 2 hrs
*WMSS(FM) Middletown PA 8 hrs
WPGR(AM) Monroeville PA 6 hrs
WNAK(AM) Nanticoke PA 8 hrs
WGBN(AM) New Kensington PA 2 hrs
*WWNW(FM) New Wilmington PA 3 hrs
WWKL(FM) Palmyra PA 6 hrs
WHKS(FM) Port Allegany PA 1 hr
WPAZ(AM) Pottstown PA 12 hrs
WKBI-FM Saint Marys PA 2 hrs
WBZU(AM) Scranton PA 1 hr
WICK(AM) Scranton PA 3 hrs
*WUSR(FM) Scranton PA 4 hrs
WPIC(AM) Sharon PA 2 hrs
*WSRU(FM) Slippery Rock PA 1 hr
WQRM(FM) Smethport PA 2 hrs
WTRN(AM) Tyrone PA 4 hrs
*WXVU(FM) Villanova PA 2 hrs
WKZV(AM) Washington PA 1 hr
WCHE(AM) West Chester PA 8 hrs
WKZN(AM) West Hazleton PA 1 hr
WILK(AM) Wilkes-Barre PA 1 hr
*WRLC(FM) Williamsport PA 3 hrs
WALO(AM) Humacao PR 2 hrs
WEXS(AM) Patillas PR 2 hrs
WKVM(AM) San Juan PR 40 hrs wkly hrs
WCRI(FM) Block Island RI 2 hrs
*WJMF(FM) Smithfield RI 2 hrs
WAGS(AM) Bishopville SC 7 hrs
WQNT(AM) Charleston SC 1 hr
WQSC(AM) Charleston SC 1 hr
WMUU-FM Greenville SC 20 hrs
WBHC-FM Hampton SC 11 hrs
WBZF(FM) Hartsville SC 6 hrs
WHSC(AM) Hartsville SC 6 hrs
WJDJ(AM) Hartsville SC 7 hrs
WKMG(AM) Newberry SC 2 hrs
WQKI-FM Orangeburg SC 6 hrs
WOLI(AM) Spartanburg SC 5 hrs
WSPA-FM Spartanburg SC 3 hrs
KBFS(AM) Belle Fourche SD 2 hrs
WNAX(AM) Yankton SD 16 hrs
WCTA(AM) Alamo TN 9 hrs
WYXI(AM) Athens TN 8 hrs
WMPS(AM) Bartlett TN 7 hrs
WJZM(AM) Clarksville TN
WCLE(AM) Cleveland TN 6 hrs
WYSH(AM) Clinton TN 15 hrs
WKRM(AM) Columbia TN 4 hrs
WZYX(AM) Cowan TN 12 hrs
WDKN(AM) Dickson TN 12 hrs
WAKM(AM) Franklin TN 6 hrs
WMYL(FM) Halls Crossroads TN 2 hrs
WJFC(AM) Jefferson City TN 4 hrs
WNRX(FM) Jefferson City TN 2 hrs
WJCW(AM) Johnson City TN 4 hrs
WKGN(AM) Knoxville TN 5 hrs
WLIV(AM) Livingston TN 15 hrs
WREC(AM) Memphis TN 4 hrs
WMGC(AM) Murfreesboro TN 4 hrs
WLIK(AM) Newport TN 18 hrs
WUAT(AM) Pikeville TN 15 hrs
WJTT(FM) Red Bank TN 4 hrs
WLIJ(AM) Shelbyville TN 11 hrs
WZNG(AM) Shelbyville TN 5 hrs
WSMT(AM) Sparta TN 10 hrs
WCDT(AM) Winchester TN 6 hrs
KYYW(AM) Abilene TX 5 hrs
KGNC(AM) Amarillo TX 4 hrs
KQIZ-FM Amarillo TX 2 hrs
KLVQ(AM) Athens TX 8 hrs
KFYN(AM) Bonham TX 6 hrs
KQTY(AM) Borger TX 1 hr
KQTY-FM Borger TX 1 hr
KLXK(FM) Breckenridge TX 3 hrs
KWHI(AM) Brenham TX 3 hrs
*KWTS(FM) Canyon TX 3 hrs
KGAS(AM) Carthage TX 10 hrs
KGAS-FM Carthage TX 10 hrs
KCTX-FM Childress TX 5 hrs
KHER(FM) Crystal City TX 2 hrs
KTDR(FM) Del Rio TX 3 hrs

KWMC(AM) Del Rio TX 5 hrs
KTSM(AM) El Paso TX 3 hrs
KGVL(AM) Greenville TX 6 hrs
KIKT(FM) Greenville TX 6 hrs
KCLW(AM) Hamilton TX 6 hrs
KVRP-FM Haskell TX 6 hrs
KTBZ-FM Houston TX 1 hr
KVLG(AM) La Grange TX 5 hrs
KCYL(AM) Lampasas TX 10 hrs
KSHN-FM Liberty TX 5 hrs
KZZN(AM) Littlefield TX 5 hrs
KFYO(AM) Lubbock TX 6 hrs
KCKL(FM) Malakoff TX 7 hrs
KLSR-FM Memphis TX 5 hrs
KBED(AM) Nederland TX 4 hrs
KNBT(FM) New Braunfels TX 3 hrs
KNET(AM) Palestine TX 7 hrs
KZHN(AM) Paris TX
KEYE(AM) Perryton TX 3 hrs
KEYE-FM Perryton TX 3 hrs
KWFB(FM) Quanah TX 2 hrs
KMIQ(FM) Robstown TX 6 hrs
KDCD(FM) San Angelo TX 2 hrs
KELI(FM) San Angelo TX 6 hrs
KQXT(FM) San Antonio TX 2 hrs
KIKZ(AM) Seminole TX 6 hrs
KSEM-FM Seminole TX 6 hrs
KSTV(AM) Stephenville TX 6 hrs
KPYK(AM) Terrell TX 13 hrs
KBCY(FM) Tye TX 5 hrs
KTXZ(AM) West Lake Hills TX 6 hrs
KWUD(AM) Woodville TX 03 hrs
KSUB(AM) Cedar City UT
KVNU(AM) Logan UT 2 hrs
*KWCR-FM Ogden UT 3 hrs
*KBYU-FM Provo UT 2 hrs
KXRQ(FM) Roosevelt UT 8 hrs
KDXU(AM) Saint George UT 4 hrs
KVEL(AM) Vernal UT
WKEX(AM) Blacksburg VA 3 hrs
WDIC(AM) Clinchco VA 12 hrs
WBNN-FM Dillwyn VA
WFLO(AM) Farmville VA 10 hrs
WFLO-FM Farmville VA 1 hr
WBQB(FM) Fredericksburg VA 2 hrs
WXGM(AM) Gloucester VA 2 hrs
WXGM-FM Gloucester VA 2 hrs
WKCY(AM) Harrisonburg VA 2 hrs
*WXJM(FM) Harrisonburg VA 2 hrs
WOJL(FM) Louisa VA 3 hrs
WMXH-FM Luray VA 5 hrs
WRAA(AM) Luray VA 7 hrs
WZZU(FM) Lynchburg VA 1 hr
WMEV(AM) Marion VA 8 hrs
WNVA(AM) Norton VA 2 hrs
WVCV(AM) Orange VA 1 hr
*WDCE(FM) Richmond VA 3 hrs
WRVA(AM) Richmond VA 10 hrs
WYTI(AM) Rocky Mount VA 10 hrs
WHEO(AM) Stuart VA 10 hrs
WVBW(FM) Suffolk VA 2 hrs
WKCW(AM) Warrenton VA 6 hrs
WVWI(AM) Charlotte Amalie VI 6 hrs
WYAC-FM Christiansted VI 3 hrs
*WRMC-FM Middlebury VT 1 hr
WVNR(AM) Poultney VT 2 hrs
WMNV(FM) Rupert VT
KWLE(AM) Anacortes WA 1 hr
KYSN(FM) East Wenatchee WA 1 hr
KARY-FM Grandview WA 8 hrs
KRIZ(AM) Renton WA 18 hrs
*KWRS(FM) Spokane WA 6 hrs
KZIZ(AM) Sumner WA 18 hrs
KHSS(FM) Walla Walla WA 3 hrs
*KDNA(FM) Yakima WA 4. hrs
WATW(AM) Ashland WI 5 hrs
WRPQ(AM) Baraboo WI 4 hrs
WCFW(FM) Chippewa Falls WI 2 hrs
WPCK(FM) Denmark WI 3 hrs
WMCS(AM) Greenfield WI 3.25 hrs
WJOK(AM) Kaukauna WI 2 hrs
WAUN(FM) Kewaunee WI 3 hrs
WJMT(AM) Merrill WI 3 hrs
WMYX(FM) Milwaukee WI 1 hr
WSSP(AM) Milwaukee WI 3 hrs
WTMJ(AM) Milwaukee WI 2 hrs

WLKD(AM) Minocqua WI 3 hrs
WMQA-FM Minocqua WI 3 hrs
WNBI(AM) Park Falls WI 1 hr
WVRQ(AM) Viroqua WI 6 hrs
WTTN(AM) Watertown WI 4 hrs
WSAU(AM) Wausau WI 3 hrs
*WVBC(FM) Bethany WV 4 hrs
WKEZ(AM) Bluefield WV 5 hrs
WBUC(AM) Buckhannon WV 7 hrs
*WVWC(FM) Buckhannon WV 4 hrs
WMRE(AM) Charles Town WV
WELD(AM) Fisher WV 10 hrs
*WVMR(AM) Frost WV 10 hrs
WRVC(AM) Huntington WV 3 hrs
WKCJ(FM) Lewisburg WV 4 hrs
WEPM(AM) Martinsburg WV 6 hrs
WADC(AM) Parkersburg WV 3 hrs
WRON(AM) Ronceverte WV 5 hrs
WCWV(FM) Summersville WV 18 hrs
WELC(AM) Welch WV 15 hrs
WZST(FM) Westover WV 2 hrs
KLGT(FM) Buffalo WY 1 hr
KYOD(FM) Glendo WY 2 hrs wkly hrs
KASL(AM) Newcastle WY 2 hrs

Rock/AOR

KIAL(AM) Unalaska AK 08 hrs
*KSPB(FM) Pebble Beach CA 2 hrs
*KZYX(FM) Philo CA 6 hrs
*WXCI(FM) Danbury CT 3 hrs
*WNHU(FM) West Haven CT 9 hrs
*KMSC(FM) Sioux City IA 4 hrs
*WBKE-FM North Manchester IN 6 hrs
*WVUR-FM Valparaiso IN 3 hrs
*KTCC(FM) Colby KS 7 hrs
*WYAJ(FM) Sudbury MA 3 hrs
*WYFP(FM) Harpswell ME 4 hrs
*WQAC-FM Alma MI 2 hrs
*WQAC(FM) Alma MI 5 hrs
*WSGR-FM Port Huron MI 12 hrs
KMGK(FM) Glenwood MN 8 hrs
*WTIP(FM) Grand Marais MN 15 hrs
*KOPN(FM) Columbia MO 6 hrs
*WKNC-FM Raleigh NC 12 hrs
*KWSC(FM) Wayne NE 2 hrs
*WNTI(FM) Hackettstown NJ 6 hrs
*WRRC(FM) Lawrenceville NJ 10 hrs
*WGCC-FM Batavia NY
*WEOS(FM) Geneva NY 6 hrs
*WQKE(FM) Plattsburgh NY 10 hrs
*WFNP(FM) Rosendale NY 7 hrs
WIOT(FM) Toledo OH 2 hrs
*WXUT(FM) Toledo OH 4 hrs
*KSLC(FM) McMinnville OR 7 hrs
*WCAL(FM) California PA 5 hrs
*WJRH(FM) Easton PA 6 hrs
*WUSR(FM) Scranton PA 8 hrs
WSBG(FM) Stroudsburg PA 3 hrs
*WNJR(FM) Washington PA 10 hrs
*WPTC(FM) Williamsport PA 5 hrs
*WDOM(FM) Providence RI 6 hrs
*WTTU(FM) Cookeville TN 3 hrs
*WVCP(FM) Gallatin TN 15 hrs
WHHM-FM Henderson TN 6 hrs
*KRTU(FM) San Antonio TX 2 hrs
KNCN(FM) Sinton TX 2 hrs
*KSUU(FM) Cedar City UT 4 hrs
*KGRG-FM Auburn WA 2 hrs
*KRLF(FM) Pullman WA 1 hr
*KUPS(FM) Tacoma WA 8 hrs
*WSUP(FM) Platteville WI 6 hrs
*WSUW(FM) Whitewater WI 14 hrs
*WWVU-FM Morgantown WV 3 hrs
KYOD(FM) Glendo WY 2 hrs

Russian

*KJNP(AM) North Pole AK 11 hrs
KTYM(AM) Inglewood CA 1 hr
WUST(AM) Washington DC 5 hrs
WAZN(AM) Watertown MA 10 hrs
*WUPI(FM) Presque Isle ME 2 hrs
*WDCV-FM Carlisle PA 1 hr
*WRVU(FM) Nashville TN 1 hr

*KUGS(FM) Bellingham WA 2 hrs wkly hrs
KKNW(AM) Seattle WA 5 hrs
KLFE(AM) Seattle WA 12 hrs

Sacred

WHOU-FM Houlton ME 1 hr

Scottish

*WOBO(FM) Batavia OH 1 hr
*KBCS(FM) Bellevue WA 5 hrs

Serbian

WNDZ(AM) Portage IN 2 hrs
WEDO(AM) McKeesport PA 1 hr
WRJN(AM) Racine WI 2 hrs

Slovak

*WWPT(FM) Westport CT 3 hrs
WPNA(AM) Oak Park IL
*WCPN(FM) Cleveland OH 1 hr
WEDO(AM) McKeesport PA 1 hr

Slovenian

*WAPS(FM) Akron OH 2 hrs
WKTX(AM) Cortland OH 2 hrs
WEDO(AM) McKeesport PA 1 hr

Smooth Jazz

KLXR(AM) Redding CA 3 hrs. hrs
*WONB(FM) Ada OH 18 hrs
WVLY-FM Milton PA 6 hrs

Soul

*WYEP-FM Pittsburgh PA 3 hrs

Spanish

*KSUA(FM) Fairbanks AK 3 hrs
*KTOO(FM) Juneau AK 2 hrs
WKAC(AM) Athens AL 24 hrs
*WTBB(FM) Gadsden AL 1 hr
*WJAB(FM) Huntsville AL 2 hrs
*WTBJ(FM) Oxford AL 1 hr
KFPW(AM) Fort Smith AR 6 hrs
KLSZ-FM Fort Smith AR 6 hrs
KAAY(AM) Little Rock AR 2 hrs
*KABF(FM) Little Rock AR 10 hrs
KVNA(AM) Flagstaff AZ 3 hrs
KXEG(AM) Phoenix AZ 5 hrs
*KUAZ(AM) Tucson AZ 5 hrs
*KUAZ-FM Tucson AZ 3 hrs
*KXCI(FM) Tucson AZ 4 hrs
KINO(AM) Winslow AZ 3 hrs
*KAWC(AM) Yuma AZ 15 hrs
KXMX(AM) Anaheim CA
*KZFR(FM) Chico CA 6 hrs
KKCY(FM) Colusa CA 1 hr
KTHU(FM) Corning CA 1 hr
*KKUP(FM) Cupertino CA 3 hrs
*KECG(FM) El Cerrito CA 3 hrs
KBIF(AM) Fresno CA 6 hrs
*KFCF(FM) Fresno CA 5 hrs
*KMUD(FM) Garberville CA 2 hrs
KAZA(AM) Gilroy CA 133 hrs
KKMC(AM) Gonzales CA 5 hrs
KTYM(AM) Inglewood CA 1 hr
KNTI(FM) Lakeport CA 3 hrs
KXBX(AM) Lakeport CA 3 hrs
*KPFK(FM) Los Angeles CA 15 hrs
*KADV(FM) Modesto CA 1 hr
*KSMC(FM) Moraga CA 3 hrs
KVON(AM) Napa CA 5 hrs
KBLF(AM) Red Bluff CA 4 hrs
*KFPR(FM) Redding CA 4 hrs
KDIF(AM) Riverside CA 168 hrs
*KCPR(FM) San Luis Obispo CA 3 hrs

KPRZ(AM) San Marcos-Poway CA 22 hrs
*KCSB-FM Santa Barbara CA 12 hrs
KSTN-FM Stockton CA
KWNE(FM) Ukiah CA 4 hrs
KGIW(AM) Alamosa CO 6 hrs
*KRZA(FM) Alamosa CO 14 hrs
*KGNU-FM Boulder CO 3 hrs
KSMT(FM) Breckenridge CO 2 hrs
*KUVO(FM) Denver CO 15 hrs
KRKY(AM) Granby CO 1 hr
*KAFM(FM) Grand Junction CO 3 hrs
*KLFV(FM) Grand Junction CO 2 hrs
*KWSB-FM Gunnison CO 1 hr
KRCN(AM) Longmont/Denver CO 1 hr
KSLV(AM) Monte Vista CO 10 hrs
*KVNF(FM) Paonia CO 2 hrs
*WPKN(FM) Bridgeport CT 4 hrs
*WRTC-FM Hartford CT 6 hrs
*WFCS(FM) New Britain CT 2 hrs
*WCNI(FM) New London CT 3 hrs
*WHUS(FM) Storrs CT 3 hrs
*WWUH(FM) West Hartford CT 3 hrs
*WECS(FM) Willimantic CT 9 hrs
WILI(AM) Willimantic CT 1 hr
WUST(AM) Washington DC 15 hrs
*WVUD(FM) Newark DE 2 hrs
WDEL(AM) Wilmington DE 2 hrs
WWPR(AM) Bradenton FL 40 hrs
*WVUM(FM) Coral Gables FL 2 hrs
*WKTO(FM) Edgewater FL 2 hrs
*WJFP(FM) Fort Pierce FL 2 hrs
*WAPN(FM) Holly Hill FL 4 hrs
*WRGP(FM) Homestead FL 3 hrs
WXYB(AM) Indian Rocks Beach FL 2 hrs
WKEY-FM Key West FL 3 hrs
*WJNF(FM) Marianna FL 1 1/4 hrs
WMNE(AM) Riviera Beach FL
WMEN(AM) Royal Palm Beach FL 2 hrs
*WFCF(FM) Saint Augustine FL 4 hrs
*WVFS(FM) Tallahassee FL 2 hrs
*WBVM(FM) Tampa FL 4 hrs
WTIS(AM) Tampa FL 1 hr
WSIR(AM) Winter Haven FL 10 hrs
*WRFG(FM) Atlanta GA 5 hrs
WTHV(AM) Hahira GA 5 hrs
WGML(AM) Hinesville GA 1 hr
KWAI(AM) Honolulu HI 3 hrs
KZAT-FM Belle Plaine IA 2 hrs
KDMU(FM) Bloomfield IA 1wkly hrs
KCHE(AM) Cherokee IA 1 hr
KLNG(AM) Council Bluffs IA 10 hrs
*KALA(FM) Davenport IA 15 hrs
KDSN(AM) Denison IA 4 hrs
*KHOE(FM) Fairfield IA 2 hrs
*KRNL-FM Mount Vernon IA 1 hr
KBGN(AM) Caldwell ID 5 hrs
KLLP(FM) Chubbuck ID 4 hrs
KART(AM) Jerome ID 9 hrs
KMHI(AM) Mountain Home ID 7 hrs
KFTA(AM) Rupert ID
*WEFT(FM) Champaign IL 3 hrs
*WPCD(FM) Champaign IL 4 hrs
*WIIT(FM) Chicago IL 5 hrs
*WMBI(AM) Chicago IL 12 hrs
WRMN(AM) Elgin IL 10 hrs
WWHN(AM) Joliet IL 1 hr
*WIUS(FM) Macomb IL 3 hrs
*WFEN(FM) Rockford IL 1 hr
WSDR(AM) Sterling IL 4 hrs
WLYV(AM) Fort Wayne IN 1 hr
*WGCS(FM) Goshen IN 8 hrs
WBDC(FM) Huntingburg IN .5 hrs
*WBKE-FM North Manchester IN 1 hr
WRSW(AM) Warsaw IN 2 hrs
WRSW-FM Warsaw IN 2 hrs
KDCC(AM) Dodge City KS
KFFX(FM) Emporia KS 2 hrs
KVOE(AM) Emporia KS 3 hrs
*KANZ(FM) Garden City KS 6 hrs
KKCI(FM) Goodland KS 1 hr
*KZNA(FM) Hill City KS 6 hrs
KWKR(FM) Leoti KS 3 hrs
KSLS(FM) Liberal KS 5 hrs

*KBCU(FM) North Newton KS 2 hrs
KULY(AM) Ulysses KS 3 hrs
*KMUW(FM) Wichita KS 2 hrs
KRBB(FM) Wichita KS 3 hrs
*WNKJ(FM) Hopkinsville KY 1 hr
WYGH(AM) Paris KY
*WFCR(FM) Amherst MA 4 hrs
*WBSL-FM Sheffield MA 2 hrs
WORC-FM Webster MA 1 hr
*WBUR(AM) West Yarmouth MA 5 hrs
*WCUW(FM) Worcester MA 19 hrs
WOLB(AM) Baltimore MD 2 hrs
*WMPG(FM) Gorham ME 4 hrs
WLEN(FM) Adrian MI 4 hrs
WQTE(FM) Adrian MI 3 hrs
*WCBN-FM Ann Arbor MI 3 hrs
*WKAR(AM) East Lansing MI 3 hrs
WHTC(AM) Holland MI 3 hrs
*WTHS(FM) Holland MI 8 hrs
WIBM(AM) Jackson MI 2 hrs
*WLNZ(FM) Lansing MI 4 hrs
WHLS(AM) Port Huron MI 1 hr
*WNMC-FM Traverse City MI 2 hrs
*WYCE(FM) Wyoming MI 10 hrs
WPNW(AM) Zeeland MI 2 hrs
KATE(AM) Albert Lea MN 2 hrs
KYCR(AM) Golden Valley MN 14 hrs
KDUZ(AM) Hutchinson MN 1 hr
KYSM-FM Mankato MN 1 hr
*KBEM-FM Minneapolis MN 4 hrs
*KFAI(FM) Minneapolis MN 8 hrs
KCHK(AM) New Prague MN 6 hrs
KYMN(AM) Northfield MN 2 hrs
KRFO(AM) Owatonna MN 2 hrs
KLOH(AM) Pipestone MN 2 hrs
KRRW(FM) Saint James MN 1 hr
KNOF(FM) Saint Paul MN 1 hr
*KSRQ(FM) Thief River Falls MN 1 hr
KAOL(AM) Carrollton MO 3 hrs
KMZU(FM) Carrollton MO 3 hrs
KDMO(AM) Carthage MO 6 hrs
KMFC(FM) Centralia MO 1 hr
*KCUR-FM Kansas City MO 2 hrs
*KKFI(FM) Kansas City MO 16 hrs
WHB(AM) Kansas City MO 2 hrs
KCXL(AM) Liberty MO 5 hrs
WEW(AM) Saint Louis MO 10 hrs
WRRZ(AM) Clinton NC 5 hrs
WCLW(AM) Eden NC 6 hrs
WDJS(AM) Mount Olive NC 5 hrs
WMFA(AM) Raeford NC 6 hrs
*WSHA(FM) Raleigh NC 3 hrs
WTSB(AM) Selma NC 5 hrs
WNCA(AM) Siler City NC 25 hrs
WKSK(AM) West Jefferson NC 1 hr
KAUJ(FM) Grafton ND 2 hrs
KJSK(AM) Columbus NE 6 hrs
*KRNU(FM) Lincoln NE 2 hrs
*KZUM(FM) Lincoln NE 4 hrs
*KJLT(AM) North Platte NE 1 hr
KNEB(AM) Scottsbluff NE 5 hrs
WFEA(AM) Manchester NH 2 hrs
WOF(AM) Andover NJ 3 hrs
*WBJB-FM Lincroft NJ 4 hrs
WMVB(AM) Millville NJ 2 hrs
*WFDU(FM) Teaneck NJ 3 hrs
*WMSC(FM) Upper Montclair NJ 2 hrs
KSVA(AM) Albuquerque NM 4 hrs
*KUNM(FM) Albuquerque NM 9 hrs
KCQL(AM) Aztec NM 6 hrs
KATK-FM Carlsbad NM
KLMX(AM) Clayton NM 2 hrs hrs
KOTS(AM) Deming NM 8 hrs
*KGLP(FM) Gallup NM 10 hrs
KYVA-FM Grants NM 4 hrs
KVLC(FM) Hatch NM 6 hrs
*KRWG(FM) Las Cruces NM 10 hrs
KBUY(AM) Ruidoso NM 4 hrs
*KSFR(FM) Santa Fe NM 4 hrs
KTNM(AM) Tucumcari NM 18 hrs
KKVV(AM) Las Vegas NV 20 hrs
*KUNV(FM) Las Vegas NV 5 hrs
KKOH(AM) Reno NV 2 hrs
*WCDB(FM) Albany NY 3 hrs
*WHRW(FM) Binghamton NY 9 hrs
*WCWP(FM) Brookville NY 1 hr

WCHP(AM) Champlain NY
WDOE(AM) Dunkirk NY 2 hrs
*WCVF-FM Fredonia NY 4 hrs
WGBB(AM) Freeport NY 2 hrs
*WGMC(FM) Greece NY 10 hrs
WJTN(AM) Jamestown NY 1 hr
WIZR(AM) Johnstown NY 1 hr
*WVCR-FM Loudonville NY 3 hrs
WVOA-FM Mexico NY 15 hrs
*WBAI(FM) New York NY 3 hrs
*WKCR-FM New York NY 10 hrs
*WNYU-FM New York NY 2 hrs
WGNY(AM) Newburgh NY 1 hr
*WRHO(FM) Oneonta NY 2 hrs
*WNYO(FM) Oswego NY 6 hrs
*WFNP(FM) Rosendale NY 3 hrs
*WSPN(FM) Saratoga Springs NY 3 hrs
*WRUC(FM) Schenectady NY 3 hrs
*WUSB(FM) Stony Brook NY 3 hrs
WTBQ(AM) Warwick NY
*WBGU(FM) Bowling Green OH 4 hrs
WCVX(AM) Cincinnati OH 3 hrs
*WWSU(FM) Dayton OH 3 hrs
WEOL(AM) Elyria OH 2 hrs
*WLFC(FM) Findlay OH 3 hrs
WFOB(AM) Fostoria OH 3 hrs
*WDUB(FM) Granville OH 2 hrs
WRBP(FM) Hubbard OH 2 hrs
WDLW(AM) Lorain OH 7 hrs
WXKR(FM) Port Clinton OH 1 hr
*WJCU(FM) University Heights OH 4 hrs
WERT(AM) Van Wert OH 1 hr
KWHW(AM) Altus OK 16 hrs
KGYN(AM) Guymon OK 8 hrs
KBZQ(FM) Lawton OK 4 hrs
KTLR(AM) Oklahoma City OK 3 hrs
*KMUN(FM) Astoria OR 3 hrs
*KBVR(FM) Corvallis OR 4 hrs
*KLCC(FM) Eugene OR 5 hrs
*KWVA(FM) Eugene OR 6 hrs
*KAGI(AM) Grants Pass OR 6 hrs
KOHU(AM) Hermiston OR 6 hrs
KUIK(AM) Hillsboro OR 21 hrs
KCGB-FM Hood River OR 4 hrs
KIHR(AM) Hood River OR 14 hrs
KAGO(AM) Klamath Falls OR 5 hrs
*KTEC(FM) Klamath Falls OR 3 hrs
KLYC(AM) McMinnville OR 8 hrs
*KLCO(FM) Newport OR 5 hrs
*KBOO(FM) Portland OR 10 hrs
*KBPS(AM) Portland OR 1 hr
*KBVM(FM) Portland OR 14 hrs
KKPZ(AM) Portland OR 20 hrs
*KRRC(FM) Portland OR 2 hrs
KACI-FM The Dalles OR 1 hr
KODL(AM) The Dalles OR 2 hrs
KLWJ(AM) Umatilla OR 1 hr
*WDIY(FM) Allentown PA 3 hrs
*WMUH(FM) Allentown PA 4 hrs
WGPA(AM) Bethlehem PA 2-4 hrs
WWCS(AM) Canonsburg PA 2 hrs
*WDCV-FM Carlisle PA 1 hr
*WJRH(FM) Easton PA 6 hrs
*WERG(FM) Erie PA 3 hrs
*WMCE(FM) Erie PA 3 hrs
*WQLN-FM Erie PA 1 hr
*WZBT(FM) Gettysburg PA 4 hrs
WEDO(AM) McKeesport PA 1 hr
WWKL(FM) Palmyra PA 14 hrs
WNTP(AM) Philadelphia PA 2 hrs
*WCLH(FM) Wilkes-Barre PA 3 hrs
*WVYC(FM) York PA 1 hr
WKJB(AM) Mayaguez PR 1 hr
*WRIU(FM) Kingston RI 3 hrs
WKMG(AM) Newberry SC 10 hrs
*KLND(FM) Little Eagle SD 1 hr
*WHCB(FM) Bristol TN 1 hr
*WETS(FM) Johnson City TN 1 hr
*WYPL(FM) Memphis TN 1 hr
WNQM(AM) Nashville TN
*WRVU(FM) Nashville TN 2 hrs
KVLF(AM) Alpine TX 10 hrs
KKCN(FM) Ballinger TX 4 hrs
*KVLU(FM) Beaumont TX 5 hrs

KNEL-FM Brady TX 6 hrs
KXYL(AM) Brownwood TX 36 hrs
*KWTS(FM) Canyon TX 3 hrs
KDHT(FM) Cedar Park TX 2 hrs
KUZN(FM) Centerville TX 6 hrs hrs
KQBZ(FM) Coleman TX 5 hrs
KSTA(AM) Coleman TX 5 hrs
KCTA(AM) Corpus Christi TX 6 hrs
*KEDT-FM Corpus Christi TX 4 hrs
KDHN(AM) Dimmitt TX 17 hrs
KTMR(AM) Edna TX 12 hrs
KULP(AM) El Campo TX 14 hrs
KELP(AM) El Paso TX 12 hrs
KCTI(AM) Gonzales TX 6 hrs
KCLW(AM) Hamilton TX 12 hrs
*KMBH-FM Harlingen TX 3 hrs
*KOOP(FM) Hornsby TX 10 hrs
*KHCB-FM Houston TX 10 hrs
*KSHU(FM) Huntsville TX 4 hrs
*KBJS(FM) Jacksonville TX 1 hr
*KKER(FM) Kerrville TX 6 hrs
KRNH(FM) Kerrville TX 4 hrs
*KTAI(FM) Kingsville TX 3 hrs
*KNTU(FM) McKinney TX 6 hrs
KLSR-FM Memphis TX 6 hrs
KIMP(AM) Mount Pleasant TX 25 hrs
*KLUX(FM) Robstown TX 3 hrs
KTLU(AM) Rusk TX 10 hrs
KWRW(FM) Rusk TX 10 hrs
KSLR(AM) San Antonio TX 18 hrs
KXOX(AM) Sweetwater TX 8 hrs
*KVNE(FM) Tyler TX 2 hrs
KQRL(AM) Waco TX 14 hrs
KANI(AM) Wharton TX 3 hrs
*KPDR(FM) Wheeler TX 5 hrs
KWFS(AM) Wichita Falls TX 4 hrs
*KWCR-FM Ogden UT 16 hrs
*KEYY(AM) Provo UT 5 hrs
*KRCL(FM) Salt Lake City UT 9 hrs
KVEL(AM) Vernal UT
*WTJU(FM) Charlottesville VA 2 hrs
WWWJ(AM) Galax VA 20 hrs
*WHOV(FM) Hampton VA 12 hrs
*WXJM(FM) Harrisonburg VA 3 hrs
WKGM(AM) Smithfield VA 01 hrs
*WIUJ(FM) Charlotte Amalie VI 4 hrs
*WIUV(FM) Castleton VT 1 hr
*WRMC-FM Middlebury VT 1 hr
KWLE(AM) Anacortes WA 6 hrs
KOZI(AM) Chelan WA 5 hrs
KBSN(AM) Moses Lake WA 9 hrs
KBRC(AM) Mount Vernon WA 3 hrs
*KAOS(FM) Olympia WA 6 hrs
KOMW(AM) Omak WA 2 hrs
*KZUU(FM) Pullman WA 3 hrs
*KUOW-FM Seattle WA 2 hrs
KMAS(AM) Shelton WA 2.5 hrs
*KWRS(FM) Spokane WA 2 hrs
*KDNA(FM) Yakima WA 106 hrs
*WEMI(FM) Appleton WI 1 hr
WDUZ-FM Brillion WI 2 hrs
*WEMY(FM) Green Bay WI 1 hr
*WHID(FM) Green Bay WI 3 hrs
WMCS(AM) Greenfield WI 5 hrs
WRRD(AM) Jackson WI 2 hrs
*WMSE(FM) Milwaukee WI 3 hrs
WSSP(AM) Milwaukee WI 3 hrs
*WSHS(FM) Sheboygan WI 3. hrs
*WCCX(FM) Waukesha WI 8 hrs
*WMLJ(FM) Summersville WV 1 hr
KRAE(AM) Cheyenne WY 2 hrs
KUGR(AM) Green River WY 5 hrs

Sports

*KRUA(FM) Anchorage AK 1 hr
*WVSU-FM Birmingham AL
*WMBV(FM) Dixons Mills AL 1 hr
*WFIX(FM) Florence AL 6 hrs
WUMP(AM) Madison AL
WMGY(AM) Montgomery AL 6 hrs
WKLD(FM) Oneonta AL
WZCT(AM) Scottsboro AL 19 hrs
WKXM-FM Winfield AL 10 hrs
KYEL(FM) Danville AR
KFFA(AM) Helena AR 15 hrs
KHGG(AM) Van Buren AR
*KLVA(FM) Casa Grande AZ 7 hrs
KDJI(AM) Holbrook AZ 10 hrs
KAAA(AM) Kingman AZ
KNOT(AM) Prescott AZ 8 hrs
KMET(AM) Banning CA 3 hrs
KRKC(AM) King City CA 9 hrs
KLAA(AM) Orange CA
KTIP(AM) Porterville CA 6 hrs
KLIV(AM) San Jose CA
KKJL(AM) San Luis Obispo CA
KVEC(AM) San Luis Obispo CA
KVML(AM) Sonora CA 15 hrs
KFTM(AM) Fort Morgan CO 10 hrs
*KAFM(FM) Grand Junction CO 3 hrs
KUBC(AM) Montrose CO 5 hrs
KCOL(AM) Wellington CO 7 hrs
WGCH(AM) Greenwich CT 6 hrs
WXJN(FM) Lewes DE
WWBF(AM) Bartow FL
WSWN(AM) Belle Glade FL
*WVUM(FM) Coral Gables FL 5 hrs
WNDB(AM) Daytona Beach FL
WDSP(AM) De Funiak Springs FL
WACC(AM) Hialeah FL
WFLF(AM) Pine Hills FL
WTBN(AM) Pinellas Park FL varies hrs
WAOC(AM) Saint Augustine FL
WKII(AM) Solana FL
WFSH(AM) Valparaiso-Niceville FL
WGAU(AM) Athens GA
WMGR(AM) Bainbridge GA 15 hrs
WFNS(AM) Blackshear GA 10 hrs
WBTS(FM) Doraville GA
WCEH(AM) Hawkinsville GA
WFOM(AM) Marietta GA fall hrs
WAZX(AM) Smyrna GA
KPUA(AM) Hilo HI
KCPS(AM) Burlington IA 10 hrs
KLMJ(FM) Hampton IA
KIKD(FM) Lake City IA
KMCH(FM) Manchester IA 7 hrs
*KWDM(FM) West Des Moines IA 3 hrs
KMCL-FM McCall ID 1 hr
KTFI(AM) Twin Falls ID 3 hrs
WXET(FM) Arcola IL
WRXX(FM) Centralia IL
WGN(AM) Chicago IL
WXEF(FM) Effingham IL
WWHP(FM) Farmer City IL 5 hrs
*WHFH(FM) Flossmoor IL 4 hrs
WAIK(AM) Galesburg IL 10 hrs
WGIL(AM) Galesburg IL 15 hrs
*WGBK(FM) Glenview IL 9 hrs
WJBM(AM) Jerseyville IL 13 hrs
WVLI(FM) Kankakee IL
*WLTL(FM) La Grange IL 5 hrs
WFXN(AM) Moline IL 8 hrs
WBBA-FM Pittsfield IL 6 hrs
WTJK(AM) South Beloit IL
WGFA-FM Watseka IL 18 hrs
*WETN(FM) Wheaton IL 5 hrs
WYKT(FM) Wilmington IL 12 hrs
WBIW(AM) Bedford IN
WQRK(FM) Bedford IN
WUZR(FM) Bicknell IN
WLME(FM) Cannelton IN 6 hrs
*WGCS(FM) Goshen IN 10 hrs
WJOB(AM) Hammond IN
WBDC(FM) Huntingburg IN 6 hrs
WIVR(FM) Kentland IN
*WJEF(FM) Lafayette IN
*WEEM-FM Pendleton IN 10 hrs
WFMG(FM) Richmond IN 8 hrs
WAXI(FM) Rockville IN
WNDV-FM South Bend IN
*KTCC(FM) Colby KS 3 hrs
*KONQ(FM) Dodge City KS 5 hrs
KAYS(AM) Hays KS
KSEK(AM) Pittsburg KS
WKCM(AM) Hawesville KY 6 hrs
WCBR(AM) Richmond KY
WMSK-FM Sturgis KY
WEZJ-FM Williamsburg KY
WORC(AM) Worcester MA 6 hrs
WTAG(AM) Worcester MA
WNAV(AM) Annapolis MD
WTRI(AM) Brunswick MD 5 hrs
WXCY(FM) Havre de Grace MD 6 hrs
WKIK(AM) La Plata MD
WBPW(FM) Presque Isle ME
WMCM(FM) Rockland ME
*WBFH(FM) Bloomfield Hills MI 6 hrs
WDMK(FM) Detroit MI 2 hrs
WKMI(AM) Kalamazoo MI
WAGN(AM) Menominee MI 12 hrs
WRSR(FM) Owosso MI 4 hrs
WJML(AM) Petoskey MI
WBJI(FM) Blackduck MN 4 hrs
KXDL(FM) Browerville MN 2 hrs
*WTIP(FM) Grand Marais MN 2 hrs
KEYL(AM) Long Prairie MN 10 hrs
KZYM(AM) Joplin MO
KBDZ(FM) Perryville MO 5 hrs
KSIM(AM) Sikeston MO
KURM-FM South West City MO
KTXR(FM) Springfield MO
KRWP(FM) Stockton MO
WZLD(FM) Petal MS 3 hrs
WRJW(AM) Picayune MS 4 hrs
WIGG(AM) Wiggins MS
KMON(AM) Great Falls MT 5 hrs
WECR-FM Beech Mountain NC
WCGC(AM) Belmont NC
WCSL(AM) Cherryville NC 3 hrs
WAZZ(AM) Fayetteville NC
WGOS(AM) High Point NC 10 hrs
WELS-FM Kinston NC
WRNS-FM Kinston NC 6 hrs
WFLB(FM) Laurinburg NC
WKXU(FM) Louisburg NC
WECR(AM) Newland NC
WCBQ(AM) Oxford NC
WQDR(FM) Raleigh NC
WNCA(AM) Siler City NC 6 hrs
WEEB(AM) Southern Pines NC
WAME(AM) Statesville NC
WSVM(AM) Valdese NC 15 hrs
WNMX-FM Waxhaw NC
WTQR(FM) Winston-Salem NC
KLXX(AM) Bismarck-Mandan ND 6 hrs
KDLR(AM) Devils Lake ND
KDIX(AM) Dickinson ND
KMAV(AM) Mayville ND
KMAV-FM Mayville ND
KRRZ(AM) Minot ND
KGFW(AM) Kearney NE 10 hrs
*KRNU(FM) Lincoln NE 2 hrs
KIOD(FM) McCook NE 10 hrs
KNCY(AM) Nebraska City NE 6 hrs
WHTG(AM) Eatontown NJ 29 hrs
*WJSV(FM) Morristown NJ 3 hrs
WCTC(AM) New Brunswick NJ
WGHT(AM) Pompton Lakes NJ 2 hrs
WHWH(AM) Princeton NJ
*WMSC(FM) Upper Montclair NJ 3 hrs
*WMCX(FM) West Long Branch NJ 11 hrs
KICA(AM) Clovis NM 4 hrs
KBIM-FM Roswell NM
KKVS(FM) Truth or Consequences NM
KKOH(AM) Reno NV
WBTA(AM) Batavia NY 9 hrs
WBAZ(FM) Bridgehampton NY
*WCWP(FM) Brookville NY 3 hrs
WBEN(AM) Buffalo NY
*WHCL-FM Clinton NY 6 hrs
WDNY-FM Dansville NY 4 hrs
*WKCR-FM New York NY 3 hrs
WEBO(AM) Owego NY 16 hrs
WODZ-FM Rome NY
*WRUC(FM) Schenectady NY 4 hrs
WUUF(FM) Sodus NY 5 hrs
WFAS(AM) White Plains NY 8 hrs
*WZIP(FM) Akron OH 3 hrs
WBNO-FM Bryan OH
WCDK(FM) Cadiz OH
WBVB(FM) Coal Grove OH
WFIN(AM) Findlay OH 12 hrs
WLOH(AM) Lancaster OH 8 hrs
WFGF(FM) Lima OH
WMOA(AM) Marietta OH 15 hrs
WKLM(FM) Millersburg OH
WBUK(FM) Ottawa OH 2 hrs
WPTW(AM) Piqua OH 8 hrs
WCWA(AM) Toledo OH 15 hrs
WHIZ-FM Zanesville OH 3 hrs
KADA-FM Ada OK 3 hrs
KRVT(AM) Claremore OK 4 hrs
KOCY(AM) Del City OK
KXXY-FM Oklahoma City OK
KOKL(AM) Okmulgee OK 10 hrs
KSPI(AM) Stillwater OK
KURY-FM Brookings OR
KPNW(AM) Eugene OR
KYKN(AM) Keizer OR
KRTA(AM) Medford OR
KBBR(AM) North Bend OR
KEX(AM) Portland OR
KQEN(AM) Roseburg OR
KOHI(AM) Saint Helens OR
KCKX(AM) Stayton OR
WRTA(AM) Altoona PA
WBVE(FM) Bedford PA
WZWW(FM) Bellefonte PA 3 hrs
*WFSE(FM) Edinboro PA
WTCY(AM) Harrisburg PA
WNTJ(AM) Johnstown PA 20 hrs
*WFNM(FM) Lancaster PA 2 hrs
WNPV(AM) Lansdale PA 6 hrs
WBPZ(AM) Lock Haven PA
*WMSS(FM) Middletown PA 5 hrs
WQLV(FM) Millersburg PA 6 hrs
WWKL(FM) Palmyra PA
WNTP(AM) Philadelphia PA 5 hrs
KQV(AM) Pittsburgh PA
WEAE(AM) Pittsburgh PA
WDMT(FM) Pittston PA
WARM(AM) Scranton PA
*WQSU(FM) Selinsgrove PA 4 hrs
*WSRU(FM) Slippery Rock PA 1 hr
WQRM(FM) Smethport PA 3 hrs
WMBS(AM) Uniontown PA
WKZV(AM) Washington PA 1 hr
WBAX(AM) Wilkes-Barre PA
*WPTC(FM) Williamsport PA 2 hrs
WEXS(AM) Patillas PR 6 hrs
WLEO(AM) Ponce PR
*WDOM(FM) Providence RI 2 hrs
WVGB(AM) Beaufort SC
WGTR(FM) Bucksport SC 8 hrs
WTMA(AM) Charleston SC
WPCC(AM) Clinton SC
WISW(AM) Columbia SC
WNOW-FM Gaffney SC
WAVO(AM) Rock Hill SC
WOLI(AM) Spartanburg SC 12 hrs
WBCU(AM) Union SC 10 hrs
KZKK(FM) Huron SD
KKSD(FM) Milbank SD 5 hrs
KIMM(AM) Rapid City SD
KSDR-FM Watertown SD 8 hrs
WNAX(AM) Yankton SD 10 hrs
WSAA(FM) Benton TN
WHUB(AM) Cookeville TN 10 hrs
WAKM(AM) Franklin TN 6 hrs
WAEZ(FM) Greeneville TN
WDXI(AM) Jackson TN 16 hrs
WFLI(AM) Lookout Mountain TN
WMSR(AM) Manchester TN
WTNE-FM Trenton TN 20 hrs
WBOZ(FM) Woodbury TN 5 hrs
KSKY(AM) Balch Springs TX
KCAR(AM) Clarksville TX 10 hrs
KFXR(AM) Dallas TX 6 hrs
KRLD(AM) Dallas TX
KROD(AM) El Paso TX
WBAP(AM) Fort Worth TX
KGAF(AM) Gainesville TX 3 hrs
KCOH(AM) Houston TX
KCYL(AM) Lampasas TX 8 hrs
KJTV(AM) Lubbock TX
KSFA(AM) Nacogdoches TX
KJAK(FM) Slaton TX 5 hrs
KBTE(FM) Tulia TX
KGLD(AM) Tyler TX
KVOU-FM Uvalde TX
*KPGR(FM) Pleasant Grove UT 4 hrs
WKEX(AM) Blacksburg VA 4 hrs
WLQM-FM Franklin VA
WFVA(AM) Fredericksburg VA
WMNA(AM) Gretna VA 10 hrs
*WRVL(FM) Lynchburg VA
WYTI(AM) Rocky Mount VA
WTFX(AM) Winchester VA
WKVT-FM Brattleboro VT 6 hrs
WVNR(AM) Poultney VT 10 hrs
WJEN(FM) Rutland VT 4 hrs
*KASB(FM) Bellevue WA 6 hrs
*KCED(FM) Centralia WA 4 hrs
KCLX(AM) Colfax WA 7 hrs
KBDB-FM Forks WA 20 hrs
KBIS(AM) Forks WA 20 hrs
KONA(AM) Kennewick WA 8 hrs
KTCR(AM) Kennewick WA
KKNW(AM) Seattle WA 15 hrs
KXLY(AM) Spokane WA 20 hrs
KKAD(AM) Vancouver WA 18 hrs
WBEV(AM) Beaver Dam WI 18 hrs
WNFL(AM) Green Bay WI 120 hrs
WLKG(FM) Lake Geneva WI 2 hrs
WMAM(AM) Marinette WI
WPVL(AM) Platteville WI 15 hrs
WRPN(AM) Ripon WI 15 hrs
*WRPN-FM Ripon WI 15 hrs
WCOW-FM Sparta WI
WOSQ(FM) Spencer WI
*WWSP(FM) Stevens Point WI 3 hrs
*KUWS(FM) Superior WI 6 hrs
WAUK(AM) Waukesha WI
WDUX-FM Waupaca WI 15 hrs
WJLS(AM) Beckley WV 3 hrs
WXKX(AM) Clarksburg WV
WTNJ(FM) Mount Hope WV
WWYO(AM) Pineville WV 18 hrs

Talk

WATV(AM) Birmingham AL 8 hrs. hrs
WBCF(AM) Florence AL 12 hrs
WNUZ(AM) Talladega AL 5 hrs
KHTE-FM England AR 10 hrs
KVDW(AM) England AR 10 hrs
KBOK(AM) Malvern AR 6 hrs
KCGS(AM) Marshall AR 7 hrs
KCUZ(AM) Clifton AZ 8 hrs
KJLH-FM Compton CA 8.5 hrs
KXBX(AM) Lakeport CA 5 hrs
KTOX(AM) Needles CA 22 hrs
KXTZ(FM) Pismo Beach CA 1 hr
*KASF(FM) Alamosa CO 6 hrs
KFTM(AM) Fort Morgan CO 5 hrs
KGRE(AM) Greeley CO 1 hr
WICH(AM) Norwich CT 15 hrs
WINY(AM) Putnam CT 10 hrs
WEBE(FM) Westport CT 1 hr
WWPR(AM) Bradenton FL 15 hrs
WRNE(AM) Gulf Breeze FL
WOTS(AM) Kissimmee FL 20 hrs
WFOY(AM) Saint Augustine FL
*WWFR(FM) Stuart FL 1 hr
WBAF(AM) Barnesville GA 5 hrs
*WFRC(FM) Columbus GA 8 hrs
WSDZ(AM) Belleville IL 1/2 hrs
*WSSD(FM) Chicago IL 6 hrs
*WHFH(FM) Flossmoor IL 1 hr
WAIK(AM) Galesburg IL 10 hrs
*WCSF(FM) Joliet IL 4 hrs
WGFA(AM) Watseka IL 5 hrs
WKLU(FM) Brownsburg IN
WWFT(FM) Fishers IN 1 hr
*WPUM(FM) Rensselaer IN 1 hr
WLBN(AM) Lebanon KY 5 hrs
WTBK(FM) Manchester KY 8 hrs
WTRI(AM) Brunswick MD 2 hrs
WJEJ(AM) Hagerstown MD 8 hrs
WPHX-FM Sanford ME 2 hrs
WDZZ-FM Flint MI 1 hr
KNXR(FM) Rochester MN 2 hrs
KAYX(FM) Richmond MO
KSIM(AM) Sikeston MO
WTCD(FM) Indianola MS 10 hrs
WATA(AM) Boone NC 20 hrs

WRFX-FM Kannapolis NC 3 hrs
WENC(AM) Whiteville NC 5 hrs
KHND(AM) Harvey ND 8 hrs
WMOU(AM) Berlin NH 2 hrs
WRNJ(AM) Hackettstown NJ 10 hrs
WOBM(AM) Lakewood NJ 12 hrs
KSNM(AM) Las Cruces NM 20 hrs
KBCQ(AM) Roswell NM 15 hrs
KMXQ(FM) Socorro NM 1 hr
WUFO(AM) Amherst NY 8 hrs
*WSQX-FM Binghamton NY 10 hrs
WENT(AM) Gloversville NY 1 hr
*WNYC(AM) New York NY 3 hrs
WEOK(AM) Poughkeepsie NY 5 hrs
WBEA(FM) Southold NY
WONW(AM) Defiance OH 3 hrs
WEGE(FM) Lima OH 5 hrs
*WLRY(FM) Rushville OH 16 hrs
KGFF(AM) Shawnee OK
KWIP(AM) Dallas OR 5 hrs
KWVR-FM Enterprise OR 15 hrs
KGAL(AM) Lebanon OR 5 hrs
KQEN(AM) Roseburg OR
*KSJK(AM) Talent OR 6 hrs
KACI(AM) The Dalles OR 10 hrs
*WBUQ(FM) Bloomsburg PA 10 hrs
WBHD(FM) Olyphant PA
WPAM(AM) Pottsville PA
WMBS(AM) Uniontown PA
WZAR(FM) Ponce PR 15 hrs
*WRTU(FM) San Juan PR 2 hrs
WOLS(AM) Florence SC 16 hrs
WNOW-FM Gaffney SC 10 hrs
WLMC(AM) Georgetown SC 4 hrs
WRIX-FM Honea Path SC 20 hrs
WSAA(FM) Benton TN
WZYX(AM) Cowan TN
KIXZ(AM) Amarillo TX 2 hrs
*KAZI-FM Austin TX 10 hrs
KNES(FM) Fairfield TX 15 hrs
KMJQ(FM) Houston TX 3 hrs
KTBZ-FM Houston TX 2 hrs
KKCL(FM) Lorenzo TX 17 hrs
*KSTX(FM) San Antonio TX 6 hrs
KSUB(AM) Cedar City UT
WFVA(AM) Fredericksburg VA
WRVA(AM) Richmond VA 1 hr
KBKW(AM) Aberdeen WA
KGNW(AM) Burien-Seattle WA
KXLY(AM) Spokane WA 15 hrs
KXLY(AM) Spokane WA 5 hrs
*KVTI(FM) Tacoma WA 4 hrs
WMOV(AM) Ravenswood WV 2 hrs
KOVE(AM) Lander WY 15 hrs

Tejano

KPAN(AM) Hereford TX 15 hrs

Top-40

WQHR(FM) Presque Isle ME
KOLV(FM) Olivia MN
KATQ(AM) Plentywood MT
WGFY(AM) Charlotte NC
WCLM(AM) Highland Springs VA
KMPS-FM Seattle WA

Triple A

*KOJI(FM) Okoboji IA 18 hrs
*KWIT(FM) Sioux City IA 12 hrs
*KMUW(FM) Wichita KS 14 hrs
*WKMS-FM Murray KY 3 hrs
*KOPN(FM) Columbia MO 10 hrs
*WEOS(FM) Geneva NY 12 hrs
WEQX(FM) Manchester VT 4 hrs

Ukranian

WILI(AM) Willimantic CT 1 hr
WPNA(AM) Oak Park IL
WCCD(AM) Parma OH 1 hr
KARI(AM) Blaine WA 1 hr

Underground

*WMTS-FM Murfreesboro TN 10 hrs

Urban Contemporary

*KSCU(FM) Santa Clara CA 15 hrs
*KCSU-FM Fort Collins CO 3 hrs
*WQTQ(FM) Hartford CT 19 hrs
*WRGP(FM) Homestead FL 12 hrs
*KICB(FM) Fort Dodge IA 2 hrs
*KMSC(FM) Sioux City IA 4 hrs
*WQUB(FM) Quincy IL 2 hrs
*WVUR-FM Valparaiso IN 3 hrs
*KTCC(FM) Colby KS 4 hrs
*WKMS-FM Murray KY 3 hrs
*WKHS(FM) Worton MD 2 hrs
*WSGR-FM Port Huron MI 6 hrs
*KSRQ(FM) Thief River Falls MN 5 hrs
*KMVC(FM) Marshall MO 20 hrs
*WASU-FM Boone NC 6 hrs
*WXDU(FM) Durham NC 12 hrs
*WKNC-FM Raleigh NC 12 hrs
*KFKX(FM) Hastings NE 8 hrs
*WKNH(FM) Keene NH 4 hrs
*WPSC-FM Wayne NJ 18 hrs
KRNG(FM) Fallon/ Reno NV
*WIRQ(FM) Rochester NY 3 hrs
*WUSO(FM) Springfield OH 9 hrs
*WBUQ(FM) Bloomsburg PA 10 hrs
*WCAL(FM) California PA 12 hrs
*WDCV-FM Carlisle PA 15 hrs
*WUSR(FM) Scranton PA 6 hrs
*WPTC(FM) Williamsport PA 5 hrs
WBRU(FM) Providence RI 20 hrs
*WDOM(FM) Providence RI 16 hrs
*KCFS(FM) Sioux Falls SD 6 hrs
*WTTU(FM) Cookeville TN 3 hrs
*KSAU(FM) Nacogdoches TX 4 hrs
*WNRN(FM) Charlottesville VA 16 hrs
*WIUV(FM) Castleton VT 5 hrs
*WRMC-FM Middlebury VT 12 hrs
*KGRG-FM Auburn WA 3 hrs
*WSUP(FM) Platteville WI 4 hrs
*WSUW(FM) Whitewater WI 14 hrs

Variety/Diverse

*KRUA(FM) Anchorage AK 20 hrs
*WEGL(FM) Auburn AL
KBHR(FM) Big Bear City CA 2 hrs
WAZX(AM) Smyrna GA
*KIWR(FM) Council Bluffs IA 16 hrs
*WNUR-FM Evanston IL 12 hrs
*WDCB(FM) Glen Ellyn IL 7 hrs
WVEZ(FM) Louisville KY 5 hrs
*KGAC(FM) Saint Peter MN 8 hrs
WPEG(FM) Concord NC 8 hrs
WHKP(AM) Hendersonville NC 18 hrs
*WSHA(FM) Raleigh NC 4 hrs
*WITC(FM) Cazenovia NY 10 hrs
*WBAI(FM) New York NY
WJZA(FM) Lancaster OH 15 hrs
WJZK(FM) Richwood OH 15v hrs
WEZY(FM) Racine WI

Vietnamese

KXMX(AM) Anaheim CA 2 hrs
*KHCB(AM) Galveston TX 4 hrs
*WSHS(FM) Sheboygan WI 3 hrs

Women

*KPFA(FM) Berkeley CA 10 hrs
*KAZU(FM) Pacific Grove CA 6 hrs
*WCNI(FM) New London CT 3 hrs
*WPFW(FM) Washington DC 3 hrs
KROS(AM) Clinton IA 5 hrs
*KMSC(FM) Sioux City IA 5 hrs
KNSG(FM) Springfield MN 3 hrs
WNBN(AM) Meridian MS
WUMS(FM) University MS 1 hr
KAWL(AM) York NE 3 hrs
*WXUT(FM) Toledo OH 2 hrs
KUJZ(FM) Creswell OR 2 hrs
KGNW(AM) Burien-Seattle WA

Special Programming on Radio Stations in Canada

Agriculture

CFAC(AM) Calgary, AB 10 hrs
CFCW(AM) Camrose, AB 5 hrs
CKDQ(AM) Drumheller, AB 8 hrs
CFXE-FM Edson, AB 2 hrs
CHRB(AM) High River, AB 5 hrs
CHLW(AM) Saint Paul, AB 5 hrs
CKKY(AM) Wainwright, AB 10 hrs
CFOK(AM) Westlock, AB 5 hrs
CKSR-FM Chilliwack, BC 2 hrs
CIGV-FM Penticton, BC 1 hr
CKLQ(AM) Brandon, MB 18 hrs
CFRY(AM) Portage la Prairie, MB 4 hrs
CHSM(AM) Steinbach, MB
CBW(AM) Winnipeg, MB 6 hrs
CKDH(AM) Amherst, NS 2 hrs
CKDY(AM) Digby, NS 7 hrs
CKEN-FM Kentville, NS 5 hrs
CKAD(AM) Middleton, NS 3 hrs
CJBQ(AM) Belleville, ON 3 hrs
CFCO(AM) Chatham, ON 3 hrs
CHCD-FM Simcoe, ON 5 hrs
CFRB(AM) Toronto, ON 2 hrs
CJSL(AM) Estevan, SK 4 hrs
CFYM(AM) Kindersley, SK
CJNB(AM) North Battleford, SK 7 hrs
CKBI(AM) Prince Albert, SK 2 hrs
*CBK(AM) Regina, SK 5 hrs
CKSW(AM) Swift Current, SK 5 hrs
CFSL(AM) Weyburn, SK 3 hrs

Album-Oriented Rock

CFMQ-FM Hudson Bay, SK

American Indian

CJSR-FM Edmonton, AB 2 hrs
CFNR-FM Terrace, BC 9 hrs
CHMB(AM) Vancouver, BC 1 hr
*CFUV-FM Victoria, BC 1 hr
CHTM(AM) Thompson, MB 10 hrs
CHSR-FM Fredericton, NB 1 hr
CKON-FM Akwesasne, ON
*CHMO(AM) Moosonee, ON 5 hrs
CKCU-FM Ottawa, ON 2 hrs
CKLP-FM Parry Sound, ON 1 hr
CFBU-FM Saint Catharines, ON 1 hr
CBQS-FM Sioux Narrows, ON 1 hr
CHNO-FM Sudbury, ON 1 hr
*CKRK-FM Kahnawake, PQ 10 hrs
*CBKA-FM La Ronge, SK 20 hrs
*CJTR-FM Regina, SK 4 hrs

Arabic

*CKRL-FM Quebec, PQ 3 hrs

Big Band

CJLX-FM Belleville, ON 2 hrs
CFBG-FM Bracebridge, ON 1 hr

Black

CJSR-FM Edmonton, AB 7 hrs
*CFRO-FM Vancouver, BC 14 hrs
*CITR-FM Vancouver, BC 18 hrs
CHSR-FM Fredericton, NB 3 hrs
CKLB-FM Yellowknife, NT 1 hr
*CFMU-FM Hamilton, ON 4 hrs
CKCU-FM Ottawa, ON 12 hrs
CHRY-FM Toronto, ON
*CJAM-FM Windsor, ON 10 hrs
*CIBL-FM Montreal, PQ 13 hrs
*CKUT-FM Montreal, PQ 20 hrs
*CKRL-FM Quebec, PQ 6 hrs
CFNJ-FM Saint Gabriel-de-Brandon, PQ 1 hr
*CJTR-FM Regina, SK 2 hrs

Bluegrass

CKOL-FM Campbellford, ON 3 hrs
CHCR-FM Killaloe, ON 6 hrs

Blues

CKXL-FM Saint Boniface, MB 2 hrs
*CHMR-FM Saint John's, NF 6 hrs
CJLX-FM Belleville, ON 1 hr
*CFMU-FM Hamilton, ON 5 hrs
CFBW-FM Hanover, ON 2 hrs
CFRM-FM Little Current, ON 2 hrs
CFBU-FM Saint Catharines, ON 2 hrs
CFMX-FM Toronto, ON 1 hr
CIEU-FM Carleton, PQ 5 hrs

Children

CHVN-FM Winnipeg, MB 1 hr
*CKUW-FM Winnipeg, MB 2 hrs
CHRI-FM Ottawa, ON 2 hrs

Chinese

*CFRO-FM Vancouver, BC 2 hrs
CHSR-FM Fredericton, NB 3 hrs
*CHUO-FM Ottawa, ON 2 hrs
CFBU-FM Saint Catharines, ON 2 hrs
CHRY-FM Toronto, ON
CINQ-FM Montreal, PQ 5 hrs
*CJTR-FM Regina, SK 1 hr

Christian

*CIXX-FM London, ON 3 hrs

Classical

CHQT(AM) Edmonton, AB 12 hrs
CJSR-FM Edmonton, AB 2 hrs
CKBX(AM) 100 Mile House, BC 1 hr
CIGV-FM Penticton, BC 2 hrs
CHOR(AM) Summerland, BC 3 hrs
CIOC-FM Victoria, BC 5 hrs
CFAM(AM) Altona, MB 15 hrs
CBW(AM) Winnipeg, MB 8 hrs
*CKUW-FM Winnipeg, MB 4 hrs
CHSR-FM Fredericton, NB 6 hrs
*CHMR-FM Saint John's, NF 10 hrs
CJLX-FM Belleville, ON 2 hrs
CHOD-FM Cornwall, ON 4 hrs
*CFMU-FM Hamilton, ON 5 hrs
CFBU-FM Saint Catharines, ON 1 hr
CHAS-FM Sault Ste. Marie, ON 5 hrs
CFRB(AM) Toronto, ON 7 hrs
*CBE(AM) Windsor, ON 4 hrs
*CJAM-FM Windsor, ON 4 hrs
CIEU-FM Carleton, PQ 3 hrs
CHIP-FM Fort Coulonge, PQ 2 hrs
CJRG-FM Gaspe, PQ 2 hrs
*CFIN-FM Lac-Etchemin, PQ 4 hrs
CHGA-FM Maniwaki, PQ 1 hr
*CIBL-FM Montreal, PQ 4 hrs
CKIA-FM Quebec, PQ 3 hrs
CKMN-FM Rimouski-Mont Joli, PQ 3 hrs
CFNJ-FM Saint Gabriel-de-Brandon, PQ 2 hrs
CIHO-FM Saint Hilarion, PQ 2 hrs
CJMC-FM Sainte Anne des Monts, PQ 2 hrs
CFLX-FM Sherbrooke, PQ 7 hrs
CJSO-FM Sorel, PQ 2 hrs
CFMC-FM Saskatoon, SK 2 hrs

Country

CKXL-FM Saint Boniface, MB 2 hrs wkly hrs
CBW(AM) Winnipeg, MB 1 hr
CJVA(AM) Caraquet, NB 15 hrs
*CFJU-FM Kedgwick, NB 12 hrs
CKDH(AM) Amherst, NS 11 hrs
CFRM-FM Little Current, ON 5 hrs
CFNO-FM Marathon, ON 12 hrs
CFMF-FM Fermont, PQ 4 hrs
*CFIN-FM Lac-Etchemin, PQ 6 hrs
CHGA-FM Maniwaki, PQ 8 hrs
CFNJ-FM Saint Gabriel-de-Brandon, PQ 4 hrs
CJMC-FM Sainte Anne des Monts, PQ 2 hrs
CKCN-FM Sept-Iles, PQ 8 hrs
CFDA-FM Victoriaville, PQ 3 hrs

Disco

CKXL-FM Saint Boniface, MB 4 hrs
CKMF-FM Montreal, PQ

Educational

CKUA(AM) Edmonton, AB
CBW-FM Winnipeg, MB 5 hrs
CJLX-FM Belleville, ON 4 hrs
*CIXX-FM London, ON 4 hrs
CHRY-FM Toronto, ON

Ethnic

CKER-FM Edmonton, AB 10 hrs
*CITR-FM Vancouver, BC 2 hrs
*CIOI-FM Hamilton, ON 2 hrs
*CHUO-FM Ottawa, ON 6 hrs

Filipino

CKJS(AM) Winnipeg, MB 20 hrs

Finnish

CKTG-FM Thunder Bay, ON 1 hr

Folk

CJSR-FM Edmonton, AB 16 hrs
CKXL-FM Saint Boniface, MB 2 hrs
*CKUW-FM Winnipeg, MB 10 hrs
CFAN-FM Miramichi City, NB 2 hrs
*CHMR-FM Saint John's, NF 6 hrs
*VOWR(AM) Saint John's, NF 15 hrs
CJLX-FM Belleville, ON 1 hr
CKCU-FM Ottawa, ON 12 hrs
*CJAM-FM Windsor, ON 4 hrs
CIEU-FM Carleton, PQ 3 hrs
CHGA-FM Maniwaki, PQ 5 hrs
*CKUT-FM Montreal, PQ 3 hrs

Foreign/Ethnic

*CJSF-FM Burnaby, BC 2 hrs
CFKC(AM) Creston, BC 2 hrs
CJAT-FM Trail, BC 2 hrs
*CFRO-FM Vancouver, BC 8 hrs
CHSR-FM Fredericton, NB 5 hrs
*CKJM-FM Cheticamp, NS 1 hr
*CHAK(AM) Inuvik, NT 18 hrs
CFCT(AM) Tuktoyaktuk, NT 5 hrs
*CFYK(AM) Yellowknife, NT 16 hrs
CBQR-FM Rankin Inlet, NU 10 hrs
CJLX-FM Belleville, ON 2 hrs
CKWR-FM Kitchener, ON 4 hrs
*CBQT-FM Thunder Bay, ON 1 hr
CKOT(AM) Tillsonburg, ON 2 hrs
CHRY-FM Toronto, ON
CINQ-FM Montreal, PQ 18 hrs
CKIA-FM Quebec, PQ 3 hrs
CHON-FM Whitehorse, YT 15 hrs

French

*CJSW-FM Calgary, AB 1 hr
CJSR-FM Edmonton, AB 1 hr
*CFUV-FM Victoria, BC 2 hrs
CKXL-FM Saint Boniface, MB 120 hrs wkly hrs
CKBC-FM Bathurst, NB 9 hrs
CKNB(AM) Campbellton, NB 18 hrs
CHSR-FM Fredericton, NB 2 hrs
CKHJ(AM) Fredericton, NB 1 hr
*CHMR-FM Saint John's, NF 2 hrs
*CFMU-FM Hamilton, ON 1 hr
CHCR-FM Killaloe, ON 8 hrs
CKCU-FM Ottawa, ON 2 hrs
*CKLU-FM Sudbury, ON 19 hrs
CHRY-FM Toronto, ON
*CIUT-FM Toronto, ON 2 hrs
CKAJ-FM Jonquiere, PQ
*CFIN-FM Lac-Etchemin, PQ
CHAA-FM Longueuil, PQ 18 hrs
*CKUT-FM Montreal, PQ 9 hrs
CJBR-FM Rimouski, PQ
CFCR-FM Saskatoon, SK 1 hr
*CFWH(AM) Whitehorse, YT 1 hr

German

*CJSW-FM Calgary, AB 2 hrs
CKJS(AM) Winnipeg, MB 6 hrs
CKWR-FM Kitchener, ON 3 hrs
*CHUO-FM Ottawa, ON 2 hrs
*CKLU-FM Sudbury, ON 1 hr
CKOT(AM) Tillsonburg, ON 1 hr
CFCR-FM Saskatoon, SK 2 hrs
CKSW(AM) Swift Current, SK 1 hr

Gospel

CJSR-FM Edmonton, AB 2 hrs
CILK-FM Kelowna, BC 3 hrs
CHVN-FM Winnipeg, MB 4 hrs
CKOL-FM Campbellford, ON 3 hrs
CFCO(AM) Chatham, ON 2 hrs
CKSY-FM Chatham, ON 2 hrs
CFBW-FM Hanover, ON 6 hrs
CFRM-FM Little Current, ON 3 hrs
CKLP-FM Parry Sound, ON 1 hr
CKOT-FM Tillsonburg, ON 1 hr
CHRY-FM Toronto, ON
CHIP-FM Fort Coulonge, PQ 7 hrs
*CKUT-FM Montreal, PQ 2 hrs
CJWW(AM) Saskatoon, SK 3 hrs

Greek

CKJR(AM) Wetaskiwin, AB 2 hrs
*CFRO-FM Vancouver, BC 1 hr
CHMB(AM) Vancouver, BC 1/2 hrs
*CITR-FM Vancouver, BC 1 hr
CKJS(AM) Winnipeg, MB 1 hr
CJLX-FM Belleville, ON 1 hr
CKWR-FM Kitchener, ON 2 hrs
CHAA-FM Longueuil, PQ 5 hrs
CINQ-FM Montreal, PQ 13 hrs

Hebrew

CHRY-FM Toronto, ON

Hindi

*CHMR-FM Saint John's, NF 1 hr

Hungarian

CKOT(AM) Tillsonburg, ON 1 hr

Irish

CHVO(AM) Carbonear, NF 8 hrs
CKXD-FM Gander, NF 12 hrs
CIGO-FM Port Hawkesbury, NS 1 hr
CHWO(AM) Toronto, ON 1 hr hrs

Italian

*CJSW-FM Calgary, AB 1 hr
CKER-FM Edmonton, AB 3 hrs
CHMB(AM) Vancouver, BC 1 hr
*CFUV-FM Victoria, BC 2 hrs
CKJS(AM) Winnipeg, MB 5 hrs
CFRU-FM Guelph, ON 1 hr
*CFMU-FM Hamilton, ON 1 hr
CHYR-FM Leamington, ON 3 hrs
CKCU-FM Ottawa, ON 1 hr
CHAS-FM Sault Ste. Marie, ON 2 hrs
CJQM-FM Sault Ste. Marie, ON 4 hrs
*CKLU-FM Sudbury, ON 1 hr
CKTG-FM Thunder Bay, ON
*CKRL-FM Quebec, PQ 2 hrs
*CJTR-FM Regina, SK 1 hr
CFCR-FM Saskatoon, SK 1 hr

Japanese

CHMB(AM) Vancouver, BC 7 hrs

Jazz

CKMX(AM) Calgary, AB 5 hrs
CHOR(AM) Summerland, BC 3 hrs
CFOX-FM Vancouver, BC 2 hrs
*CFRO-FM Vancouver, BC 16 hrs
CKXL-FM Saint Boniface, MB 4 hrs
CBW-FM Winnipeg, MB 3 hrs
*CKUW-FM Winnipeg, MB 10 hrs
CKLE-FM Bathurst, NB 2 hrs
CHSR-FM Fredericton, NB 6 hrs
CJMO-FM Moncton, NB 2 hrs
CKUM-FM Moncton, NB 4 hrs
*CHMR-FM Saint John's, NF 6 hrs
*CKJM-FM Cheticamp, NS 3 hrs
CFRQ-FM Dartmouth, NS 3 hrs
CIVR-FM Yellowknife, NT 4 hrs
CJLX-FM Belleville, ON 3 hrs
CFBG-FM Bracebridge, ON 2 hrs
CHOD-FM Cornwall, ON 4 hrs
*CFMU-FM Hamilton, ON 10 hrs
CKGE-FM Oshawa, ON 2 hrs
*CHUO-FM Ottawa, ON 5 hrs
CKCU-FM Ottawa, ON 15 hrs
CKWF-FM Peterborough, ON 12 hrs
CFBU-FM Saint Catharines, ON 4 hrs
CHAS-FM Sault Ste. Marie, ON 2 hrs
CHRY-FM Toronto, ON
*CBE(AM) Windsor, ON 2 hrs
*CJAM-FM Windsor, ON 6 hrs
CIEU-FM Carleton, PQ 3 hrs
CFMF-FM Fermont, PQ 1 hr
CJRG-FM Gaspe, PQ 2 hrs
*CFIN-FM Lac-Etchemin, PQ 6 hrs
CHGA-FM Maniwaki, PQ 3 hrs
*CIBL-FM Montreal, PQ 14 hrs
*CBVX-FM Quebec, PQ 16 hrs
CKIA-FM Quebec, PQ 4 hrs
CFNJ-FM Saint Gabriel-de-Brandon, PQ 2 hrs
CIHO-FM Saint Hilarion, PQ 2 hrs
CFLX-FM Sherbrooke, PQ 8 hrs

CFMC-FM Saskatoon, SK 2 hrs

Korean

CHMB(AM) Vancouver, BC 1 hr

New Age

CFGX-FM Sarnia, ON 7 hrs

News

*CFPR(AM) Prince Rupert, BC
*CBVX-FM Quebec, PQ 7 hrs

News/talk

CKPR-FM Thunder Bay, ON 10 hrs
*CFTH-FM-1 Harrington Harbour, PQ

Nostalgia

CFRN(AM) Edmonton, AB

Oldies

CFRN(AM) Edmonton, AB
CJSU-FM Duncan, BC 12 hrs
CKXX-FM Corner Brook, NF 3 hrs
CHIP-FM Fort Coulonge, PQ
CJLM-FM Joliette, PQ 6 hrs
CHAA-FM Longueuil, PQ 9 hrs
CKMN-FM Rimouski-Mont Joli, PQ 3 hrs
CFDA-FM Victoriaville, PQ 3 hrs

Other

CKUA(AM) Edmonton, AB
CISC-FM Gibsons, BC 3 hrs
CKFR(AM) Kelowna, BC 2 hrs
CKFR(AM) Kelowna, BC 4 hrs
CHLS-FM Lillooet, BC 16 hrs
CKKS-FM Sechelt, BC 3 hrs
CHMB(AM) Vancouver, BC 1 hr
CHMB(AM) Vancouver, BC 1/2 hrs
CKXL-FM Saint Boniface, MB 4 hrs
CJAR(AM) The Pas, MB 5 hrs
CHVN-FM Winnipeg, MB 6 hrs
CIOK-FM Saint John, NB 4 hrs
*CBN(AM) Saint John's, NF 3 hrs
CIVR-FM Yellowknife, NT 3 hrs
CJLX-FM Belleville, ON 1 hr
CFBG-FM Bracebridge, ON 1 hr
CFMJ(AM) Toronto, ON
CHWO(AM) Toronto, ON 1 hr hrs
*CIUT-FM Toronto, ON 5 hrs
*CIBL-FM Montreal, PQ 8 hrs
CINQ-FM Montreal, PQ 16 hrs

Polish

CJSR-FM Edmonton, AB 2 hrs
CKER-FM Edmonton, AB 6 hrs
*CFRO-FM Vancouver, BC 5 hrs
*CFUV-FM Victoria, BC 1 hr
CKJS(AM) Winnipeg, MB 7 hrs
CHCR-FM Killaloe, ON 1 hr
CKWR-FM Kitchener, ON 4 hrs
CKCU-FM Ottawa, ON 1 hr
*CKLU-FM Sudbury, ON 1 hr
*CJAM-FM Windsor, ON 1 hr
CFCR-FM Saskatoon, SK 1 hr

Portugese

CKER-FM Edmonton, AB 2 hrs
CFOK(AM) Westlock, AB 5 hrs
*CJSF-FM Burnaby, BC 2 hrs
CJOR(AM) Osoyoos, BC 3 hrs

CHMB(AM) Vancouver, BC 1 hr
CKJS(AM) Winnipeg, MB 8 hrs
CJDV-FM Cambridge, ON 2 hrs
CKWR-FM Kitchener, ON 5 hrs
CFBU-FM Saint Catharines, ON 2 hrs
CINQ-FM Montreal, PQ 12 hrs
*CJTR-FM Regina, SK 1 hr

Public Affairs

*CFPR(AM) Prince Rupert, BC
CHLN(AM) Trois Rivieres, PQ 8 hrs

Reggae

CKXL-FM Saint Boniface, MB 2 hrs
*CHMR-FM Saint John's, NF 2 hrs
*CFMU-FM Hamilton, ON 4 hrs
CFMX-FM Toronto, ON 1 hr
*CIBL-FM Montreal, PQ 4 hrs

Religious

CKDQ(AM) Drumheller, AB 2 hrs
CHLW(AM) Saint Paul, AB 5 hrs
CFOK(AM) Westlock, AB 6 hrs
CIVH(AM) Vanderhoof, BC 5 hrs
CHTM(AM) Thompson, MB 12 hrs
CFAN-FM Miramichi City, NB 4 hrs
CJCW(AM) Sussex, NB 4 hrs
*CHMR-FM Saint John's, NF 4 hrs
*VOWR(AM) Saint John's, NF 10 hrs
CFSX(AM) Stephenville, NF 1 hr
CHMS-FM Bancroft, ON 1 hr
CJSS-FM Cornwall, ON 1 hr
CFRU-FM Guelph, ON 1 hr
CHYR-FM Leamington, ON 3 hrs
CKDO(AM) Oshawa, ON 1 hr
CKGE-FM Oshawa, ON 1 hr
*CHUO-FM Ottawa, ON 2 hrs
CKCU-FM Ottawa, ON 3 hrs
CKTG-FM Thunder Bay, ON
CKNX(AM) Wingham, ON 6 hrs

CHRD-FM Drummondville, PQ 1 hr
*CFIN-FM Lac-Etchemin, PQ 1 hr
CJSL(AM) Estevan, SK 10 hrs
CKJH(AM) Melfort, SK 9 hrs
CJNB(AM) North Battleford, SK 10 hrs
CJDJ-FM Saskatoon, SK 6 hrs
CFSL(AM) Weyburn, SK 9 hrs

Rock/AOR

CJDC(AM) Dawson Creek, BC 4 hrs
CKXL-FM Saint Boniface, MB 10 hrs wkly hrs
CHIP-FM Fort Coulonge, PQ

Scottish

CKEC(AM) New Glasgow, NS
CIGO-FM Port Hawkesbury, NS 2 hrs
CFBW-FM Hanover, ON 2 hrs
CJTN-FM Quinte West, ON 1 hr
CHWO(AM) Toronto, ON 2 hrs hrs

Spanish

*CJSW-FM Calgary, AB 1 hr
CJSR-FM Edmonton, AB 2 hrs
CKER-FM Edmonton, AB 8 hrs
*CJSF-FM Burnaby, BC 4 hrs
*CITR-FM Vancouver, BC 2 hrs
*CFUV-FM Victoria, BC 2 hrs
CKXL-FM Saint Boniface, MB 2 hrs
CKJS(AM) Winnipeg, MB 3 hrs
CFRU-FM Guelph, ON
*CFMU-FM Hamilton, ON 1 hr
*CIOI-FM Hamilton, ON 3 hrs
CKWR-FM Kitchener, ON 4 hrs
*CHUO-FM Ottawa, ON 3 hrs
CFBU-FM Saint Catharines, ON 4 hrs
*CKLU-FM Sudbury, ON 1 hr
CHRY-FM Toronto, ON
*CIUT-FM Toronto, ON 4 hrs
*CJAM-FM Windsor, ON 1 hr

*CFIN-FM Lac-Etchemin, PQ
CHAA-FM Longueuil, PQ 3 hrs
CINQ-FM Montreal, PQ 16 hrs
*CKUT-FM Montreal, PQ 6 hrs
CKIA-FM Quebec, PQ 4 hrs
*CKRL-FM Quebec, PQ 2 hrs
CFLX-FM Sherbrooke, PQ 3 hrs
CFCR-FM Saskatoon, SK 2 hrs

Sports

CFFR(AM) Calgary, AB 15 hrs
CFCW(AM) Camrose, AB
CKFR(AM) Kelowna, BC 10 hrs
CHMS-FM Bancroft, ON 2 hrs
CFFX(AM) Kingston, ON
CJBX-FM London, ON
CHOK(AM) Sarnia, ON
CFMJ(AM) Toronto, ON

Talk

CISQ-FM Squamish, BC 5 hrs
CHMS-FM Bancroft, ON 5 hrs
CFMQ-FM Hudson Bay, SK

Ukranian

CKER-FM Edmonton, AB 10 hrs
CHMB(AM) Vancouver, BC 1 hr
CJRL-FM Kenora, ON 1 hr

Urban Contemporary

*CHYZ-FM Sainte Foy, PQ 15 hrs

Vietnamese

CHMB(AM) Vancouver, BC 2 hrs
CKCU-FM Ottawa, ON 1 hr
CHAA-FM Longueuil, PQ 5 hrs

U.S. Radio Markets: Arbitron Metro Survey Area Ranking

This chart ranks the 300 radio markets by Metro Survey Area population. Figures include all persons aged 12 or older and are based on 2000 U.S. Bureau of Census estimates updated and projected to January 1, 2008. Data reflects the Spring 2007 Arbitron market definitions. A single asterisk (*) following a market name indicates that the market is embedded in a larger nearby market. See U.S. Radio Markets beginning on pg. D-36 for details.

Market	Population
1.New York, NY	15,291,100
2.Los Angeles, CA	10,826,600
3.Chicago, IL	7,738,000
4.San Francisco, CA	5,891,900
5.Dallas-Ft. Worth, TX	4,838,600
6.Houston-Galveston, TX	4,469,900
7.Philadelphia, PA	4,360,200
8.Washington, DC	4,176,300
9.Atlanta, GA	4,085,000
10.Detroit, MI	3,888,300
11.Boston, MA	3,838,800
12.Miami-Ft. Lauderdale-Hollywood, FL	3,533,000
13.Puerto Rico	3,296,800
14.Seattle-Tacoma, WA	3,257,200
15.Phoenix, AZ	3,058,000
16.Minneapolis-St. Paul, MN	2,662,100
17.San Diego, CA	2,497,000
18.Nassau-Suffolk, NY (Long Island)	2,373,900
19.Tampa-St. Petersburg-Clearwater, FL	2,314,300
20.St. Louis, MO	2,282,700
21.Baltimore, MD	2,257,900
22.Denver-Boulder, CO	2,194,800
23.Portland, OR	2,001,600
24.Pittsburgh, PA	1,998,800
25.Riverside-San Bernardino, CA	1,806,800
26.Cleveland, OH	1,794,200
27.Sacramento, CA	1,785,400
28.Cincinnati, OH	1,721,200
29.San Antonio, TX	1,586,000
30.Kansas City, MO-KS	1,575,300
31.Salt Lake City-Ogden-Provo, UT	1,554,200
32.Las Vegas, NV	1,484,400
33.Charlotte-Gastonia-Rock Hill, NC-SC	1,456,600
34.Orlando, FL	1,448,600
35.San Jose, CA	1,436,400
36.Milwaukee-Racine, WI	1,433,300
37.Columbus, OH	1,422,700
38.Providence-Warwick-Pawtucket, RI	1,393,500
39.Middlesex-Somerset-Union, NJ	1,382,800
40.Indianapolis, IN	1,328,100
41.Norfolk-VA Beach-Newport News, VA	1,327,600
42.Austin, TX	1,252,400
43.Raleigh-Durham, NC	1,184,200
44.Nashville, TN	1,158,800
45.Greensboro-Winston Salem-High Point, NC	1,131,200
46.West Palm Beach-Boca Raton, FL	1,116,800
47.Jacksonville, FL	1,083,700
48.Oklahoma City, OK	1,075,700
49.Memphis, TN	1,060,700
50.Hartford-New Britain-Middletown, CT	1,047,700
51.Monmouth-Ocean, NJ	1,020,500
52.Buffalo-Niagara Falls, NY	979,600
53.Rochester, NY	936,000
54.Louisville, KY	930,600
55.Richmond, VA	916,400
56.Birmingham, AL	875,300
57.New Orleans, LA	864,100
58.McAllen-Brownsville-Harlingen, TX	838,400
59.Dayton, OH	835,500
60.Greenville-Spartanburg, SC	824,900
61.Tucson, AZ	803,300
62.Ft. Myers-Naples-Marco Island, FL	783,100
63.Albany-Schenectady-Troy, NY	778,800
64.Honolulu, HI	768,300
65.Tulsa, OK	732,000
66.Fresno, CA	723,400
67.Grand Rapids, MI	708,400
68.Allentown-Bethlehem, PA	690,600
69.Wilkes Barre-Scranton, PA	684,200
70.Albuquerque, NM	672,000
71.Knoxville, TN	644,100
72.Omaha-Council Bluffs, NE-IA	617,600
73.Sarasota-Bradenton, FL	610,100
74.Akron, OH	599,000
75.Wilmington, DE	590,300
76.El Paso, TX	580,900
77.Baton Rouge, LA	568,700
78.Bakersfield, CA	567,800
79.Harrisburg-Lebanon-Carlisle, PA	560,800
80.Monterey-Salinas-Santa Cruz, CA	556,200
81.Stockton, CA	555,500
82.Syracuse, NY	555,100
83.Gainesville-Ocala, FL	530,600
84.Springfield, MA	529,500
85.Little Rock, AR	523,200
86.Daytona Beach, FL	518,500
87.Toledo, OH	517,200
88.Charleston, SC	511,500
89.Greenville-New Bern-Jacksonville, NC	504,900
90.Mobile, AL	501,300
91.Columbia, SC	494,600
92.Des Moines, IA	493,600
93.Spokane, WA	491,800
94.Melbourne-Titusville-Cocoa, FL	476,400
95.Madison, WI	468,800
96.Lakeland-Winter Haven, FL	468,500
97.Colorado Springs, CO	467,900
98.Wichita, KS	466,700
99.Ft. Pierce-Stuart-Vero Beach, FL	465,700
100.Visalia-Tulare-Hanford, CA	455,700
101.Johnson City-Kingsport-Bristol, TN-VA	455,600
102.Lafayette, LA	447,200
103.York, PA	445,500
104.Lexington-Fayette, KY	440,200
105.Boise, ID	435,100
106.Ft. Wayne, IN	429,100
107.Chattanooga, TN	423,600
108.Modesto, CA	421,800
109.New Haven, CT	419,200
109.Augusta, GA	419,200
111.Worcester, MA	417,100
112.Morristown, NJ	416,500
113.Huntsville, AL	415,900
114.Lancaster, PA	413,500
115.Roanoke-Lynchburg, VA	412,300
116.Portsmouth-Dover-Rochester, NH	412,200
117.Youngstown-Warren, OH	407,200
118.Jackson, MS	405,500
119.Santa Rosa, CA	403,400
120.Oxnard-Ventura, CA	402,600
121.Bridgeport, CT	395,900
122.Lansing-East Lansing, MI	389,400
123.Reno, NV	384,900
124.Pensacola, FL	379,800
124.Ft. Collins-Greeley, CO	379,800
126.Victor Valley, CA	376,600
127.Flint, MI	369,400
128.Canton, OH	350,100
129.Fayetteville, NC	348,800
130.Reading, PA	342,800
131.Saginaw-Bay City-Midland, MI	339,200
132.Shreveport, LA	336,200
133.Beaumont-Port Arthur, TX	324,500
134.Appleton-Oshkosh, WI	323,300
135.Fayetteville (NW AR)	319,100
136.Corpus Christi, TX	319,000
137.Palm Springs, CA	318,800
138.Burlington-Plattsburg, VT-NY	317,200
139.Atlantic City-Cape May, NJ	317,100
140.Newburgh-Middletown (Mid-Hudson Valley), NY	314,800

U.S. Radio Markets: Arbitron Metro Survey Area Ranking

141.Trenton, NJ ... 312,600
142.Springfield, MO ... 307,500
143.Quad Cities (Davenport-Rock Island-Moline), IA-IL ... 304,300
144.Biloxi-Gulfport-Pascagoula, MS ... 303,400
145.Stamford-Norwalk, CT ... 302,000
145.Salisbury-Ocean City, MD ... 302,000
147.Ann Arbor, MI ... 297,100
148.Tyler-Longview, TX ... 296,800
149.Peoria, IL ... 295,800
150.Eugene-Springfield, OR ... 294,800

151.Flagstaff-Prescott, AZ ... 290,700
151.Montgomery, AL ... 290,700
153.Fredericksburg, VA ... 288,400
154.Rockford, IL ... 287,300
155.Macon, GA ... 280,100
156.Killeen-Temple, TX ... 273,100
157.Huntington-Ashland, WV-KY ... 270,300
158.Savannah, GA ... 265,300
159.Asheville, NC ... 258,800
160.Utica-Rome, NY ... 257,600

161.Myrtle Beach, SC ... 257,200
162.Evansville, IN ... 257,100
163.Poughkeepsie, NY ... 255,100
164.Tallahassee, FL ... 253,300
165.Hagerstown-Chambersburg-Waynesboro, MD-PA ... 243,600
166.Wilmington, NC ... 241,800
167.Portland, ME ... 240,600
168.Erie, PA ... 238,000
169.Concord, NH (Lake Regions) ... 236,400
170.Wausau-Stevens Pt., WI (Central Wisconsin) ... 236,000

171.Anchorage, AK ... 229,700
172.San Luis Obispo, CA ... 229,500
173.New London, CT ... 228,600
174.Lincoln, NE ... 227,700
175.Morgantown-Clarksburg-Fairmont, WV ... 224,700
176.Ft. Smith, AR ... 224,600
177.New Bedford-Fall River, MA ... 222,600
178.South Bend, IN ... 220,500
179.Lebanon-Rutland-White River Junction, NH-VT ... 215,900
180.Merced, CA ... 214,500

181.Binghamton, NY ... 214,200
182.Charleston, WV ... 213,000
183.Lubbock, TX ... 211,600
184.Kalamazoo, MI ... 204,500
185.Green Bay, WI ... 203,500
186.Columbus, GA ... 203,000
187.Odessa-Midland, TX ... 202,600
188.Tupelo, MS ... 201,500
189.Cape Cod, MA ... 201,000
190.Manchester, NH ... 198,700

191.Johnstown, PA ... 198,300
192.Traverse City-Petoskey, MI ... 197,600
193.Dothan, AL ... 196,500
194.Topeka, KS ... 194,600
195.Amarillo, TX ... 192,600
196.Danbury, CT ... 189,400
197.Frederick, MD ... 188,500
198.Chico, CA ... 187,700
199.Tri-Cities, WA (Richland-Kennewick-Pasco) ... 187,500
200.Yakima, WA ... 186,900

201.Waco, TX ... 186,200
202.Rocky Mount-Wilson, NC ... 185,800
203.Clarksville-Hopkinsville, TN-KY ... 177,800
204.Duluth-Superior, MN-WI ... 175,500
205.Laredo, TX ... 174,900
206.Terre Haute, IN ... 173,900
207.Santa Maria-Lompoc, CA ... 173,100
208.Bowling Green, KY ... 172,400
209.Laurel-Hattiesburg, MS ... 172,300
210.Medford-Ashland, OR ... 171,000

211.Santa Barbara, CA ... 170,900
212.Muncie-Marion, IN ... 170,500
213.Cedar Rapids, IA ... 168,700
214.Sunbury-Selinsgrove-Lewisburg, PA ... 167,700
215.Olean, NY ... 167,600
216.Florence, SC ... 167,200
217.Bend, OR ... 166,100
218.St. Cloud, MN ... 164,800
219.Hilton Head, SC ... 164,500
220.Bangor, ME ... 163,500

221.Alexandria, LA ... 162,000
222Champaign, IL ... 161,300
223.Fargo-Moorhead, ND-MN ... 160,300
224.Elmira-Corning, NY ... 160,300
225.Winchester, VA ... 159,800
226.Ft. Walton Beach, FL ... 159,600
227.Las Cruces, NM ... 159,200
228.Redding, CA ... 158,100
229.Lake Charles, LA ... 156,900
230.La Crosse, WI ... 155,300

231.Charlottesville, VA ... 152,500
232.Rochester, MN ... 151,400
233.Muskegon, MI ... 148,000
234.Tuscaloosa, AL ... 145,500
235.Twin Falls (Sun Valley), ID ... 142,700
236.Dubuque, IA ... 142,200
237.Santa Fe, NM ... 140,700
238.Panama City, FL ... 140,600
239.Joplin, MO ... 140,500
240.Marion-Carbondale, IL (Southern Illinois) ... 140,400

241.Bryan-College Station, TX ... 138,400
241.Bloomington, IL ... 137,000
243.Pittsburgh, KS (Southeast Kansas) ... 136,300
244.Abilene, TX ... 136,100
245.Eau Claire, WI ... 135,500
246.Lafayette, IN ... 134,100
247.LaSalle-Peru, IL ... 131,700
248.Sussex, NJ ... 131,500
249.Wheeling, WV ... 129,500
250.Parkersburg-Marietta, WV-OH ... 128,500

251.Lima, OH ... 128,400
252.Waterloo-Cedar Falls, IA ... 128,300
253.Lufkin-Nacogdoches, TX ... 127,800
254.Pueblo, CO ... 127,700
255.State College, PA ... 126,400
256.Columbia, MO ... 124,700
257.Monroe, LA ... 123,900
258.Florence-Muscle Shoals, AL ... 122,900
259.Billings, MT ... 118,500
260.Hamptons-Riverhead, NY ... 117,200

261.Battle Creek, MI ... 116,900
261.Kalispell-Flathead Valley, MT ... 115,500
263.Texarkana, TX-AR ... 114,400
264.Grand Junction, CO ... 113,200
265.Wichita Falls, TX ... 111,400
266.Montpelier-Barre-St. Johnsbury, VT ... 109,300
267.Altoona, PA ... 109,100
268.Augusta-Waterville, ME ... 106,800
269.Valdosta, GA ... 105,600
270.Albany, GA ... 104,700

271.Williamsport, PA ... 102,600
272.Elkins-Buckhannon-Weston, WV ... 101,900
273.Columbus-Starkville-West Point, MS ... 101,700
274.Mankato-New Ulm-St. Peter, MN ... 100,900
275.Sioux City, IA ... 100,300
276.Rapid City, SD ... 99,500
277.Harrisonburg, VA ... 99,300
278.Sheboygan, WI ... 99,000
279.Watertown, NY ... 98,300
280.Lewiston-Auburn, ME ... 94,100

281.Decatur, IL ... 92,200
282.Lawton, OK ... 91,800
283.Bluefield, WV ... 91,700
284.Ithaca, NY ... 90,100
285.Bismarck, ND ... 87,500
286.San Angelo, TX ... 86,000
287.Cookeville, TN ... 85,700
288.Sebring, FL ... 84,400
289.Grand Forks, ND-MN ... 83,300
290.Hot Springs, AR ... 82,600

291.Jackson, TN ... 80,200
292.Jonesboro, AR ... 74,200
293.Cheyenne, WY ... 72,500
294.The Florida Keys, FL ... 69,900
295.Beckley, WV ... 69,000
296.Mason City, IA ... 68,700
297.Meridian, MS ... 66,600
298.Brunswick, GA ... 61,400
299.Casper, WY ... 60,300
300.Aspen, CO ... 55,800

U.S. Radio Markets

Abilene, TX: Rank 244
MSA: 136,100
American Family Radio: KAQD(FM)
Canfin Enterprises Inc.: KKHR-FM, KWKC, KZQQ(AM)
Cumulus Media Inc.: KBCY(FM), KCDD(FM), KHXS-FM, KTLT(FM)
EMF Broadcasting: KAGT(FM)
GAP Broadcasting LLC: KEAN-FM, KEYJ-FM, KFGL(FM), KSLI(AM), KULL-FM, KYYW(AM)
Weston Entertainment L.P.: KVRP, KVRP-FM

Akron, OH: Rank 74
MSA: 599,00
Clear Channel Communications Inc.: WARF(AM), WHLO
Family Stations Inc.: WCUE
Rubber City Radio Group Inc.: WAKR, WQMX-FM

Albany, GA: Rank 270
MSA: 104,700
Clear Channel Communications Inc.: WJIZ-FM, WJYZ, WOBB-FM, WRAK-FM
Cumulus Media Inc.: WALG, WEGC-FM, WGPC, WJAD-FM, WQVE(FM), WZBN(FM)
EMF Broadcasting: WHKV(FM)
Good News Network: WZIQ-FM

Albany-Schenectady-Troy, NY: Rank 63
MSA: 778,800
ABC Inc.: WDDY(AM)
Anastos Media Group Inc.: WABY(AM), WQAR-FM, WUAM(AM), WVKZ
Capital Media Corp.: WHAZ, WHAZ-FM, WMYY-FM
Clear Channel Communications Inc.: WGY, WHRL-FM, WKKF(FM), WOFX(AM), WPYX-FM, WRVE-FM, WTRY-FM
Crawford Broadcasting Co.: WDCD(AM), WPTR(FM)
EMF Broadcasting: WYAI(FM), WYKV(FM)
Pamal Broadcasting Ltd.: WAJZ(FM), WENU(AM), WFLY-FM, WIZR, WKLI-FM, WROW(AM), WYJB(FM), WZMR-FM
Regent Communications Inc.: WBZZ(FM), WGNA-FM, WQBJ-FM, WQBK-FM, WTMM(AM), WTMM-FM
Vox Radio Group L.P.: WUPE(AM)
WAMC/Northeast Public Radio: WAMC(AM), WAMC-FM, WCAN-FM

Albuquerque, NM: Rank 70
MSA: 672,000
KBZU(FM), KKNS(AM)
ABC Inc.: KALY
American General Media: KABG(FM), KAGM(FM), KARS, KHFM(FM), KKIM, KLVO-FM
Citadel Broadcasting Corp.: KDRF(FM), KKOB(AM), KKOB-FM, KNML(AM), KRST-FM, KTBL(AM)
Clear Channel Communications Inc.: KABQ, KABQ-FM, KBQI(FM), KPEK-FM, KTEG(FM), KZRR-FM
Entravision Communications Corp.: KRZY, KRZY-FM
Family Life Communications Inc.: KFLQ(FM)
Univision Radio: KIOT-FM, KJFA(FM), KKRG(FM), KKSS-FM, KQBT(FM)
Wilkins Communications Network Inc.: KXKS(AM)

Alexandria, LA: Rank 221
MSA: 162,000
KWDF
American Family Radio: KAPM(FM)
Cenla Broadcasting Co. Inc.: KDBS, KKST-FM, KQID-FM, KRRV-FM, KSYL, KZMZ-FM
EMF Broadcasting: KLXA-FM
Opus Media Holdings LLC: KBKK(FM), KEZP-FM, KLAA-FM
Radio Maria Inc.: KJMJ(AM)
The Radio Group: KAPB-FM

Allentown-Bethlehem, PA: Rank 68
MSA: 690,600
Citadel Broadcasting Corp.: WCTO-FM, WLEV-FM
Clear Channel Communications Inc.: WAEB, WSAN(AM), WZZO-FM
J-Systems Franchising Corp.: WLSH
Nassau Broadcasting Partners L.P.: WBYN(AM), WEEX, WTKZ, WWYY-FM

Altoona, PA: Rank 267
MSA: 109,100
WKMC
Allegheny Mountain Network Stations: WTRN
Forever Broadcasting: WALY-FM, WFBG, WRKY-FM, WVAM, WWOT(FM)
Vernal Enterprises Inc.: WHPA(FM)

Amarillo, TX: Rank 195
MSA: 192,600
American Family Radio: KAVW(FM)
Cumulus Media Inc.: KARX(FM), KPUR, KPUR-FM, KQIZ-FM, KZRK-FM
EMF Broadcasting: KXLV(FM), KXRI(FM)
Family Life Communications Inc.: KRGN-FM
GAP Broadcasting LLC: KATP(FM), KIXZ, KMML-FM, KMXJ-FM, KPRF(FM)
Morris Radio LLC: KGNC
Tejas Broadcasting Ltd. LLP: KBZD(FM), KQFX-FM, KTNZ

Anchorage, AK: Rank 171
MSA: 229,700
KAXX, KLEF-FM, KUDO(AM), KZND-FM
Clear Channel Communications Inc.: KASH-FM, KBFX(FM), KENI, KGOT-FM, KTZN, KYMG-FM
EMF Broadcasting: KAKL(FM)
Morris Radio LLC: KBRJ(FM), KEAG-FM, KFQD, KHAR, KMXS-FM, KWHL-FM
New Northwest Broadcasters LLC: KBBO-FM, KDBZ(FM), KFAT(FM)

Ann Arbor, MI: Rank 147
MSA: 297,100
WDEO(AM)
Birach Broadcasting Corp.: WSDS(AM)
Clear Channel Communications Inc.: WLBY(AM), WTKA, WWWW-FM
First Broadcasting Operating Inc.: WAAM

Appleton-Oshkosh, WI: Rank 134
MSA: 323,300
Cumulus Media Inc.: WNAM, WOGB-FM, WOSH, WPKR-FM, WWWX-FM
Evangel Ministries Inc.: WEMI(FM)
Midwest Communications Inc.: WNCY-FM, WOZZ-FM, WROE-FM
Mountain Dog Media: WFON(FM)
Relevant Radio: WJOK, WOVM(FM)
Results Broadcasting: WFCL, WJMQ(FM)
VCY America Inc.: WVCY
Woodward Communications Inc.: WAPL-FM, WHBY(AM), WSCO(AM)

Asheville, NC: Rank 159
MSA: 258,800
Clear Channel Communications Inc.: WPEK(AM), WQNQ(FM), WWNC
HRN Broadcasting Inc.: WZGM(AM)
Saga Communications Inc.: WTMT(FM), WYSE(AM)
Wilkins Communications Network Inc.: WSKY(AM)

Atlanta, GA: Rank 9
MSA: 4,085,000
ABC Inc.: WDWD
Beasley Broadcast Group Inc.: WAEC, WWWE(AM)
CBS Radio: WAOK, WZGC-FM
Citadel Broadcasting Corp.: WKHX-FM, WYAY-FM
Clear Channel Communications Inc.: WBZY(FM), WCOH, WGST, WKLS(FM), WLTM(FM), WUBL(FM), WWVA-FM
Cox Radio Inc.: WALR-FM, WBTS(FM), WSB, WSB-FM, WSRV(FM)
Cumulus Media Partners LLC: WNNX-FM, WWWQ(FM)
Davis Broadcasting Inc.: WCHK, WLKQ-FM
Dickey Broadcasting Co.: WALR, WCNN, WFOM
GHB Radio Group: WYZE
Jacobs Media Corp.: WGGA
La Favorita Inc.: WAOS(AM), WXEM
Lincoln Financial Media: WQXI, WSTR-FM
Multicultural Radio Broadcasting Inc.: WGFS
Radio One Inc.: WAMJ(FM), WHTA(FM), WPZE(FM)
Salem Communications Corp.: WAFS(AM), WFSH-FM, WGKA(AM), WLTA, WNIV
Sheridan Broadcasting Corp.: WIGO(AM)
Willis Broadcasting Corp.: WTJH

Atlantic City-Cape May, NJ: Rank 139
MSA: 317,100
Access.1 Communications Corp.: WGYM(AM), WJSE-FM, WMGM(FM), WOND(AM), WTAA(AM), WTKU-FM
Equity Communications LP: WAIV(FM), WAYV(FM), WCMC(AM), WEZW(FM), WGBZ(FM), WMID(AM), WTTH(FM), WZBZ(FM), WZXL(FM)
Millennium Radio Group LLC: WENJ(AM), WFPG(FM), WPUR-FM, WSJO(FM), WXKW(FM)
Press Communications L.L.C.: WBBO(FM)

Augusta, GA: Rank 109
MSA: 419,200
WKSX-FM, WNRR(AM)
Beasley Broadcast Group Inc.: WCHZ-FM, WDRR(FM), WGAC, WGAC-FM, WGUS(AM), WGUS-FM, WHHD(FM), WKXC-FM, WRDW(AM)
Bible Broadcasting Network: WYFA-FM
Clear Channel Communications Inc.: WBBQ-FM, WEKL(FM), WIBL(FM), WKSP(FM), WPRW-FM, WYNF(AM)
Good News Network: WLPE(FM)
Radio One Inc.: WAEG-FM, WAKB-FM, WFXA-FM, WTHB-FM
Wilkins Communications Network Inc.: WFAM(AM)

Augusta-Waterville, ME: Rank 268
MSA: 106,800
Atlantic Coast Radio L.L.C.: WLOB-FM
Citadel Broadcasting Corp.: WEBB-FM, WJZN(AM), WMME-FM, WTVL(AM)
Clear Channel Communications Inc.: WFAU, WKCG-FM, WTOS-FM
Mountain Wireless Inc.: WCTB-FM, WFMX(FM), WSKW

Austin, TX: Rank 42
MSA: 1,252,400
KHHL(FM)
Border Media Partners LLC: KELG(AM), KFON, KTXZ, KXBT(FM), KXXS(FM)
CBS Radio: KAMX(FM), KJCE, KKMJ-FM, KLQB(FM)
Clear Channel Communications Inc.: KASE-FM, KFMK-FM, KHFI-FM, KPEZ-FM, KVET-FM
EMF Broadcasting: KYLR(FM)
Emmis Communications Corp.: KBPA(FM), KDHT(FM), KGSR-FM, KLBJ, KROX-FM
Houston Christian Broadcasters Inc.: KHIB(FM)
Relevant Radio: KIXL(AM)
Simmons Media Group: KWNX(AM), KZNX(AM)
Univision Radio: KINV(FM)

Bakersfield, CA: Rank 78
MSA: 567,800
American General Media: KEBT(FM), KERI(AM), KERN, KGEO, KISV(FM), KKXX-FM
Buck Owens Productions Inc.: KCWR(FM), KUZZ, KUZZ-FM
Buckley Broadcasting Corp.: KKBB-FM, KLLY-FM, KNZR, KSMJ(FM)
Clear Channel Communications Inc.: KBFP(AM), KBFP-FM, KBKO-FM, KDFO-FM, KHTY(AM), KRAB-FM
Family Stations Inc.: KFRB-FM
Gore-Overgaard Broadcasting Inc.: KLHC(AM)
IHR Educational Broadcasting: KJPG(AM)
Lotus Communications Corp.: KCHJ, KIWI(FM), KPSL-FM, KWAC

Baltimore, MD: Rank 21
MSA: 2,257,900
WFBR(AM), WRBS(AM), WTTR(AM)
CBS Radio: WHFS(FM), WJFK, WQSR(FM), WWMX-FM
Clear Channel Communications Inc.: WCAO, WPOC-FM, WSMJ(FM)
Family Stations Inc.: WBGR, WBMD
First Broadcasting Operating Inc.: WAMD
Radio One Inc.: WERQ-FM, WWIN, WWIN-FM
Shamrock Communications Inc.: WZBA(FM)

Bangor, ME: Rank 220
MSA: 163,500
WNSX(FM), WNZS(AM), WWNZ(AM)
Brantley Broadcast Associates LLC: WFGO(AM)
Clear Channel Communications Inc.: WABI, WBFB-FM, WGUY(FM), WKSQ-FM, WVOM-FM, WWBX-FM
Cumulus Media Inc.: WBZN-FM, WEZQ-FM, WQCB-FM
The Zone Corp.: WKIT-FM, WZON

Baton Rouge, LA: Rank 77
MSA: 568,700
Citadel Broadcasting Corp.: KQXL-FM, KRDJ(FM), WCDV(FM), WEMX-FM, WIBR, WXOK
Clear Channel Communications Inc.: KRVE-FM, WJBO, WSKR, WYNK-FM
Communications Capital Managers LLC: WUBR(AM)
EMF Broadcasting: WBKL(FM)
Family Worship Center Church Inc.: WJFM-FM
Guaranty Broadcasting Co. of Baton Rouge, LLC: KNXX(FM), WDGL-FM, WNXX(FM), WTGE(FM), WYPY(FM)
Pamal Broadcasting Ltd.: WPYR(AM)

Battle Creek, MI: Rank 261
MSA: 116,900
WOLY
Clear Channel Communications Inc.: WBCK, WBFN(AM), WRCC(FM)
Family Life Communications Inc.: WUFN(FM)

Beaumont-Port Arthur, TX: Rank 133
MSA: 324,500
Clear Channel Communications Inc.: KCOL-FM, KIOC-FM, KKMY-FM, KLVI, KYKR-FM
Cumulus Media Inc.: KAYD-FM, KBED(AM), KIKR, KQXY-FM, KTCX-FM
Family Stations Inc.: KTXB-FM
Martin Broadcasting Inc.: KZZB

Radio Maria Inc.: KDEI(AM)

Beckley, WV: Rank 295
MSA: 69,000
First Media Radio LLC: WJLS, WJLS-FM
Southern Communications Corp.: WAXS-FM, WCIR-FM, WIWS, WMTD, WMTD-FM, WTNJ-FM, WWNR

Bend, OR: Rank 217
MSA: 166,100
Horizon Broadcasting Group.: KLTW-FM, KQAK(FM), KRCO(AM), KWLZ-FM, KWPK-FM

Billings, MT: Rank 259
MSA: 118.500
Chaparral Communications: KWMY(FM)
Cherry Creek Radio LLC: KBLG, KRKX(FM), KRZN(FM), KYYA-FM
Clear Channel Communications Inc.: KBBB(FM), KBUL, KCTR-FM, KKBR-FM, KMHK-FM
Connoisseur Media LLC: KPBR(FM), KPLN(FM)
EMF Broadcasting: KBIL(FM), KLRV(FM)
New Northwest Broadcasters LLC: KGHL, KGHL-FM, KQBL(FM), KRPM(FM), KRSQ-FM
Sun Mountain Inc.: KBSR

Biloxi-Gulfport-Pascagoula, MS: Rank 144
MSA: 303,400
American Family Radio: WAOY-FM
Clear Channel Communications Inc.: WBUV(FM), WKNN-FM, WMJY-FM, WQYZ-FM
Triad Broadcasting Co. L.L.C.: WCPR-FM, WHGO(FM), WTNI(AM), WUJM(FM), WXBD, WXYK-FM
Walking by Faith Ministries Inc.: WQFX

Binghamton, NY: Rank 181
MSA: 214,200
Citadel Broadcasting Corp.: WNBF, WWYL(FM), WYOS(AM)
Clear Channel Communications Inc.: WBBI(FM), WENE(AM), WINR, WKGB-FM, WMRV-FM, WMXW-FM
CSN International: WIFF(FM)
Double O Radio L.L.C.: WIYN-FM

Birmingham, AL: Rank 56
MSA: 875,300
Citadel Broadcasting Corp.: WAPI, WFFN-FM, WSPZ(AM), WUHT(FM), WYSF-FM, WZRR-FM
Clear Channel Communications Inc.: WDXB(FM), WENN(FM), WMJJ-FM, WQEN(FM)
Cox Radio Inc.: WAGG(AM), WBHJ(FM), WBHK-FM, WBPT(FM), WNCB(FM), WPSB(AM), WZZK-FM
Crawford Broadcasting Co.: WDJC-FM, WXJC(AM), WXJC-FM, WYDE(AM), WYDE-FM
Davidson Media Group LLC: WAYE, WRLM(AM)
Family Stations Inc.: WBFR-FM
Joy Christian Communications Inc.: WLYJ(AM)
Sheridan Broadcasting Corp.: WATV(AM)

Bismarck, ND: Rank 285
MSA: 87,500
Clear Channel Communications Inc.: KBMR, KFYR, KQDY(FM), KSSS-FM, KXMR
Cumulus Media Inc.: KACL(FM), KBYZ(FM), KKCT-FM, KLXX, KUSB(FM)
EMF Broadcasting: KNRI(FM)
Family Stations Inc.: KBFR(FM)

Bloomington, IL: Rank 242
MSA:
Connoisseur Media LLC: WIHN-FM
Great Plains Media Inc.: WYST(FM)
Regent Communications Inc.: WBWN-FM, WJBC

Bluefield, WV: Rank 283
MSA: 91,700
Baker Family Stations: WAMN
Peggy Sue Broadcasting Corp.: WRIC-FM
Triad Broadcasting Co. L.L.C.: WBDY, WHIS, WHQX-FM, WKEZ, WKOY-FM, WTZE

Boise, ID: Rank 105
MSA: 435,100
Citadel Broadcasting Corp.: KBOI, KIZN-FM, KKGL-FM, KQFC-FM, KTIK, KZMG(FM)
FM Idaho Co. LLC: KAYN(FM), KSRV-FM, KTMB(FM)
Journal Communications Inc.: KCID(AM), KGEM(AM), KJOT(FM), KQXR-FM, KRVB(FM), KTHI(FM)
KSPD Inc.: KBXL(FM), KSPD
Peak Broadcasting LLC: KAWO(FM), KCIX(FM), KFXD(AM), KIDO(AM), KSAS-FM, KXLT-FM

Boston: Rank 11
MSA: 3,838,800
WAMG(AM), WLLH
ABC Inc.: WMKI(AM)
Beasley Broadcast Group Inc.: WRCA
Bob Bittner Broadcasting Inc.: WJIB
CBS Radio: WBCN-FM, WBMX-FM, WBZ, WODS-FM, WZLX-FM
Clear Channel Communications Inc.: WJMN-FM, WKOX, WXKS
Costa-Eagle Radio Ventures L.P.: WCEC(AM), WNNW(AM)
Entercom Communications Corp.: WAAF(FM), WEEI, WKAF(FM), WMKK(FM), WRKO
Greater Media Inc.: WBOS-FM, WKLB-FM, WMJX-FM, WROR-FM, WTKK(FM)
Langer Broadcasting Group L.L.C.: WBIX(AM), WSRO(AM)
Multicultural Radio Broadcasting Inc.: WAZN(AM), WLYN
Nassau Broadcasting Partners L.P.: WCRB(FM)
Northeast Broadcasting Company Inc.: WXRV-FM
Phoenix Media Communications Group: WFEX(FM), WFNX(FM)
Radio One Inc.: WILD
Rose City Radio Corp.: WWZN(AM)
Salem Communications Corp.: WEZE, WROL, WTTT(AM)

Bowling Green, KY: Rank 208
MSA: 172,400
Commonwealth Broadcasting Corp.: WCDS(AM), WHHT(FM), WOVO(FM), WPTQ(FM)
Forever Communications Inc.: WBGN(AM), WBVR-FM, WLYE-FM, WUHU(FM)

Bridgeport, CT: Rank 121
MSA: 395,900
Blount Communications Group: WFIF
Cox Radio Inc.: WEZN-FM
Cumulus Media Inc.: WICC

Brunswick, GA: Rank 298
MSA: 61,400
MarMac Communications LLC: WSFN
Qantum Communications Corp.: WBGA(FM), WGIG(AM), WHFX(FM), WMOG, WWSN-FM, WYNR(FM)

Bryan-College Station, TX: Rank 241
MSA: 138,400
American Family Radio: KLGS(FM)
Brazos Valley Communications Ltd.: KJXJ(FM), KORA-FM, KTAM, KZTR-FM
Bryan Broadcasting Corp.: KNDE(FM), KZNE(AM), WTAW(AM)
Clear Channel Communications Inc.: KAGG(FM), KKYS-FM, KNFX-FM
Fort Bend Broadcasting Co.: KMBV(FM)

Buffalo-Niagara Falls, NY: Rank 52
MSA: 976,600
Citadel Broadcasting Corp.: WBBF(AM), WEDG-FM, WGRF-FM, WHLD, WHTT-FM
Corus Entertainment Inc.: CFNY-FM
Crawford Broadcasting Co.: WDCX-FM
Entercom Communications Corp.: WBEN, WGR, WKSE-FM, WLKK(FM), WWKB, WWWS
Family Stations Inc.: WFBF-FM
Regent Communications Inc.: WBLK-FM, WBUF(FM), WECK, WJYE-FM, WYRK-FM
Sheridan Broadcasting Corp.: WUFO

Burlington-Plattsburgh, VT-NY: Rank 138
MSA: 317,200
Clear Channel Communications Inc.: WCPV-FM, WEAV, WEZF-FM, WVTK(FM), WXZO(FM)
Hall Communications Inc.: WBTZ-FM, WIZN-FM, WJOY(AM), WKOL-FM, WOKO-FM
Northeast Broadcasting Company Inc.: WCAT(AM), WTWK(AM), WWMP(FM)
Radio Vermont Group Inc.: WCVT-FM, WDEV-FM

Canton, OH: Rank 128
MSA: 350,100
Clear Channel Communications Inc.: WHOF(FM), WKDD(FM)
Cumulus Media Inc.: WRQK-FM
NextMedia Group Inc.: WHBC

Cape Cod, MA: Rank 189
MSA: 201,000
Nassau Broadcasting Partners L.P.: WFQR(FM), WFRQ(FM), WPXC-FM
Qantum Communications Corp.: WCIB-FM, WCOD-FM, WRZE(FM), WXTK-FM
Sandab Communications L.P. II: WFCC-FM, WKPE-FM, WOCN-FM, WQRC(FM)

Casper, WY: Rank 299
MSA: 60,300
Clear Channel Communications Inc.: KKTL, KRVK(FM), KTRS-FM, KTWO, KWYY-FM

Cochise Broadcasting LLC: KWYX(FM)
Mt. Rushmore Broadcasting Inc.: KASS(FM), KHOC-FM, KMLD(FM), KQLT-FM, KVOC

Cedar Rapids, IA: Rank 213
MSA: 168,700
KGYM(AM)
Clear Channel Communications Inc.: KKRQ-FM, KKSY(FM), KMJM(AM), KXIC, WMT, WMT-FM
Cumulus Media Inc.: KDAT(FM), KHAK-FM, KRNA-FM

Champaign, IL: Rank 222
MSA: 161,300
Illinois Bible Institute Inc.: WBGL-FM
RadioStar Inc.: WEBX-FM, WGKC-FM, WMYE(FM), WQQB(FM)
Saga Communications Inc.: WCFF(FM), WIXY-FM, WLRW-FM, WXTT(FM)

Charleston, SC: Rank 88
MSA: 511,500
WMGL-FM
Apex Broadcasting Inc.: WAVF-FM, WIHB(FM), WXST(FM)
Bible Broadcasting Network: WYFH-FM
Citadel Broadcasting Corp.: WNKT-FM, WSSX-FM, WSUY-FM, WTMA, WWWZ-FM, WXTC
Clear Channel Communications Inc.: WALC-FM, WEZL-FM, WLTQ(AM), WRFQ-FM, WSCC-FM, WXLY-FM
Family Stations Inc.: WFCH-FM
Glory Communications Inc.: WTUA-FM
Jabar Communications Inc.: WAZS(AM), WAZS-FM
Kirkman Broadcasting Inc.: WJKB(AM), WQNT(AM), WQSC, WTMZ
L M Communications Inc.: WCOO(FM), WYBB-FM

Charleston, WV: Rank 182
MSA: 213,000
Baker Family Stations: WOKU
Bristol Broadcasting Co. Inc.: WBES(AM), WVTS(AM), WZJO(FM)
L M Communications Inc.: WJYP(AM), WKLC-FM, WMXE(FM), WSCW
West Virginia Radio Corp.: WCHS, WKAZ(AM), WKAZ-FM, WKWS-FM, WRVZ-FM, WSWW, WVAF-FM

Charlotte-Gastonia-Rock Hill, NC-SC: Rank 33
MSA: 1,456,600
WRKB, WRNA
ABC Inc.: WGFY
Bible Broadcasting Network: WYFQ(AM)
CBS Radio: WBAV-FM, WFNA(AM), WFNZ, WKQC(FM), WNKS-FM, WPEG-FM, WSOC-FM
Clear Channel Communications Inc.: WEND-FM, WIBT(FM), WKKT-FM, WLYT-FM, WRFX-FM
Davidson Media Group LLC: WBZK, WNOW
GHB Radio Group: WAVO, WCGC, WEGO, WHVN, WNMX-FM
HRN Broadcasting Inc.: WCSL, WGNC, WLON, WOHS
Lincoln Financial Media: WBT, WBT-FM, WLNK(FM)
Neely Enterprises: WGIV(AM)
Norsan Consulting and Management Inc.: WGSP, WXNC(AM)
Our Three Sons Broadcasting L.L.P.: WRHI(AM)
Radio One Inc.: WQNC(FM)
Truth Broadcasting Corp.: WZRH(AM)

Charlottesville, VA: Rank 231
MSA: 152,500
Baker Family Stations: WKTR
Clear Channel Communications Inc.: WCHV, WCJZ(FM), WCYK-FM, WHTE-FM, WKAV, WSUH(FM)
Piedmont Communications Inc.: WOJL(FM)
Saga Communications Inc.: WCNR(FM), WINA, WQMZ(FM), WVAX(AM), WWWV-FM

Chattanooga, TN: Rank 107
MSA: 423,600
WNOO
3 Daughters Media Inc.: WUUS(AM), WUUS-FM
Bahakel Communications: WDEF-FM, WDOD
Brewer Broadcasting Corp.: WJTT-FM, WMPZ-FM
Citadel Broadcasting Corp.: WGOW, WGOW-FM, WOGT-FM
Clear Channel Communications Inc.: WLND(FM), WRXR-FM, WUSY-FM
North Georgia Radio Group L.P.: WOCE(FM)
The Moody Bible Institute of Chicago: WMBW-FM
Wilkins Communications Network Inc.: WLMR(AM)
Willis Broadcasting Corp.: WSDT

Cheyenne, WY: Rank 293
MSA: 72,500
Clear Channel Communications Inc.: KCGY(FM), KGAB, KIGN(FM), KLEN-FM, KQMY(FM)
EMF Broadcasting: KAIX(FM)
Northeast Broadcasting Company Inc.: KHAT(AM), KRAE, KZDR(FM)
Regent Communications Inc.: KARS-FM, KKPL(FM)

Chicago: Rank 3
MSA: 7,738,000
ABC Inc.: WMVP, WRDZ
Birach Broadcasting Corp.: WNWI
Bonneville International Corporation: WDRV(FM), WILV(FM), WTMX-FM, WWDV(FM)
CBS Radio: WBBM, WBBM-FM, WCKG-FM, WJMK-FM, WSCR(AM), WUSN-FM, WXRT-FM
Citadel Broadcasting Corp.: WLS, WZZN(FM)
Clear Channel Communications Inc.: WGCI-FM, WGRB(AM), WKSC-FM, WLIT-FM, WNUA-FM, WVAZ-FM, WVON(AM)
Crawford Broadcasting Co.: WPWX(FM), WSRB(FM), WYCA(FM)
EMF Broadcasting: WCLR(FM), WJKL(FM), WSRI(FM)
Emmis Communications Corp.: WKQX-FM, WLUP-FM
Family Stations Inc.: WJCH-FM
Kovas Communications of Indiana Inc.: WCGO, WKKD
McNaughton-Jakle Stations: WBIG, WRMN
Multicultural Radio Broadcasting Inc.: WNTD
Newsweb Corp.: WAIT(AM), WCFJ, WCPT(AM), WKIE-FM, WNDZ, WRZA(FM), WSBC(AM)
NextMedia Group Inc.: WERV-FM, WJOL, WKRS, WLIP, WRXQ(FM), WWYW(FM), WXLC-FM, WZSR-FM
Polnet Communications Ltd.: WEEF, WKTA, WNVR, WPJX(AM)
Porter County Broadcasting Corp.: WAKE, WLJE-FM, WXRD-FM, WZVN-FM
Relevant Radio: WAUR, WWCA
Salem Communications Corp.: WIND, WYLL(AM)
Spanish Broadcasting System Inc.: WLEY-FM
STARadio Corp.: WYKT-FM
The Moody Bible Institute of Chicago: WMBI(AM), WMBI-FM
Three Eagles Communications: WCCQ-FM
Tribune Broadcasting Co.: WGN(AM)
Univision Radio: WOJO-FM, WPPN(FM), WRTO(AM), WVIV-FM, WVIX(FM)

Chico, CA: Rank 198
MSA: 187,700
Deer Creek Broadcasting LLC: KEWE(AM), KHHZ(FM), KHSL-FM, KMXI-FM, KPAY
Family Stations Inc.: KHAP(FM)
Fritz Communications Inc.: KBQB(FM), KCEZ(FM), KKCY-FM, KMJE-FM, KRQR-FM, KTHU(FM)
Mapleton Communications LLC: KALF(FM), KFMF-FM, KQPT(FM), KZAP-FM

Cincinnati, OH: Rank 28
MSA: 1,721,200
CBS Radio: WGRR-FM, WKRQ-FM, WUBE-FM, WYGY(FM)
Christian Broadcasting System Ltd.: WCVX(AM), WDJO(AM)
Clear Channel Communications Inc.: WCKY(AM), WKFS-FM, WKRC, WLW, WSAI(AM), WVMX-FM
Cumulus Media Partners LLC: WFTK(FM), WRRM-FM, WSWD(FM)
EMF Broadcasting: WORI(FM)
First Broadcasting Operating Inc.: WAOL-FM, WOXY-FM
Gateway Radio Works Inc.: WAXZ-FM
Pillar of Fire Inc.: WAKW-FM
Plessinger Radio Group: WCVG
Radio One Inc.: WDBZ(AM), WIZF-FM
Vernon R Baldwin Inc.: WCNW, WMOH, WNLT-FM

Clarksville-Hopkinsville, TN-KY: Rank 203
MSA: 177,800
Key Broadcasting Inc.: WHOP(AM), WHOP-FM
Saga Communications Inc.: WEGI(FM), WVVR(FM) 2SUBHEAD1Cleveland, OH: Rank 26
MSA: 1,794,200
WCCD(AM)
ABC Inc.: WWMK
CBS Radio: WDOK-FM, WKRI(FM), WNCX(FM), WQAL-FM
Clear Channel Communications Inc.: WGAR-FM, WMJI-FM, WMMS-FM, WMVX-FM, WTAM
Elyria-Lorain Broadcasting Co.: WEOL, WNWV-FM
Good Karma Broadcasting L.L.C.: WKNR(AM), WWGK(AM)
Radio One Inc.: WENZ-FM, WERE(AM), WJMO(AM), WZAK-FM
Salem Communications Corp.: WFHM-FM, WHK(AM), WHKW(AM)
The Moody Bible Institute of Chicago: WCRF-FM

Colorado Springs, CO: Rank 97
MSA: 467,900
Bahakel Communications: KILO-FM
Bustos Media LLC: KGDQ(FM)
Citadel Broadcasting Corp.: KATC-FM, KKFM-FM, KKMG-FM, KKML(AM), KKPK(FM), KVOR
Clear Channel Communications Inc.: KIBT(FM), KKLI-FM, KVUU-FM
Crawford Broadcasting Co.: KCBR, KCMN
Latino Communications LLC: KXRE(AM)
News-Press & Gazette Co.: KRDO(AM)
Salem Communications Corp.: KBIQ(FM), KGFT-FM, KZNT(AM)

Columbia, MO: Rank 256
MSA: 124,700
KATI(FM), KFAL, KTXY-FM, KWRT(AM)
Best Broadcast Group: KZZT-FM
Cumulus Media Inc.: KBXR(FM), KFRU, KOQL-FM, KPLA(FM)
The Curators of the University of Missouri: KBIA(FM)

Columbia, SC: Rank 91
MSA: 494,600
WPOG(AM)
Bible Broadcasting Network: WYFV-FM
Citadel Broadcasting Corp.: WISW, WLXC-FM, WTCB-FM
Clear Channel Communications Inc.: WCOS, WLTY-FM, WNOK(FM), WVOC, WXBT(FM)
Double O Radio L.L.C.: WWNQ(FM), WWNU(FM)
Glory Communications Inc.: WFMV(FM), WGCV(AM)
Good News Network: WBLR
Inner City Broadcasting: WARQ(FM), WHXT-FM, WMFX(FM), WOIC(AM), WWDM(FM), WZMJ-FM
Miller Communications Inc.: WIGL(FM)
Norsan Consulting and Management Inc.: WCEO(AM)

Columbus, GA: Rank 186
MSA: 203,000
Archway Broadcasting Group: WCGQ-FM, WKCN-FM, WRCG, WRLD-FM
Bible Broadcasting Network: WYFK-FM
Clear Channel Communications Inc.: WAGH-FM, WDAK, WGSY-FM, WHAL(AM), WSHE(AM), WSTH-FM, WVRK-FM
Davis Broadcasting Inc.: WEAM, WEAM-FM, WIOL(FM), WKZJ(FM), WOKS
Family Stations Inc.: WFRC-FM

Columbus, OH: Rank 37
MSA: 1,422,700
American Family Radio: WWGV(FM)
Clear Channel Communications Inc.: WBWR(FM), WCOL-FM, WKKJ(FM), WLZT(FM), WNCI-FM, WQIO-FM, WTVN(AM), WYTS(AM)
Dispatch Broadcast Group: WBNS, WBNS-FM
North American Broadcasting Co. Inc.: WBZX-FM, WMNI, WTDA(FM)
Radio One Inc.: WCKX-FM, WJYD(FM), WXMG-FM
Saga Communications Inc.: WJZA-FM, WJZK(FM), WODB(FM), WSNY(FM)
Salem Communications Corp.: WRFD(AM)
Wilks Broadcast Group LLC: WHOK-FM, WLVQ(FM), WNKK(FM)

Columbus-Starkville-West Point, MS: Rank 273
MSA: 101,700
WTWG(AM)
Air South Radio Inc.: WLZA(FM)
American Family Radio: WCSO(FM), WJZB(FM)
Cumulus Media Inc.: WJWF(AM), WKOR(AM), WKOR-FM, WMBC(FM), WSMS(FM), WSSO(AM)
TeleSouth Communications Inc.: WROB(AM)
Urban Radio Licenses LLC: WACR-FM, WMSU(FM)

Concord (Lake Regions), NH: Rank 169
MSA: 236,400
Citadel Broadcasting Corp.: WPKQ(FM)

Cookeville, TN: Rank 287
MSA: 85,700
Clear Channel Communications Inc.: WGIC-FM, WGSQ-FM, WHUB, WPTN
JWC Broadcasting: WKXD-FM

Corpus Christi, TX: Rank 136
MSA: 319,000
Clear Channel Communications Inc.: KMXR-FM, KNCN-FM, KRYS-FM, KSAB-FM, KUNO
Convergent Broadcasting LLC: KJKE(FM), KKPN(FM), KPUS(FM)
EMF Broadcasting: KKLM(FM)
Gerald Benavides Stns: KMZZ(FM)
Malkan Broadcast Associates: KEYS, KKBA(FM), KZFM(FM)
Tejas Broadcasting Ltd. LLP: KLHB(FM), KLTG-FM, KMJR(FM), KOUL-FM
World Radio Network Inc.: KBNJ(FM)

Dallas-Fort Worth: Rank 5
MSA: 4,838,600
KATH(AM), KJON(AM), KVCE(AM)
ABC Inc.: KESN(FM), KMKI
Border Media Partners LLC: KFJZ
CBS Radio: KJKK(FM), KLLI(FM), KLUV(FM), KMVK(FM), KRLD, KVIL(FM)
Citadel Broadcasting Corp.: KSCS-FM, KTYS(FM), WBAP
Clear Channel Communications Inc.: KDGE(FM), KDMX(FM), KEGL-FM, KFXR(AM), KHKS-FM, KZPS-FM
Crawford Broadcasting Co.: KAAM(AM)
Criswell Communications: KCBI(FM)
Cumulus Media Partners LLC: KDBN(FM), KKLF(AM), KLIF(AM), KPLX(FM), KTCK, KTDK(FM)
James Crystal Inc.: KNIT(AM)
Liberman Broadcasting Inc.: KBOC(FM), KTCY(FM), KZMP(AM), KZMP-FM, KZZA(FM)
LKCM Radio Group L.P.: KFWR(FM), KRVA-FM, KRVF(FM), KTFW-FM
M&M Broadcasters Ltd.: KCLE(AM), KHFX(AM), KJSA
Mortenson Broadcasting Co.: KGGR, KHVN, KKGM(AM), KRVA, KTNO(AM)
Multicultural Radio Broadcasting Inc.: KDFT, KMNY(AM)
NextMedia Group Inc.: KLAK(FM)
Radio One Inc.: KBFB(FM), KSOC(FM)
Salem Communications Corp.: KLTY(FM), KSKY(AM)
Service Broadcasting Group LLC: KKDA, KKDA-FM, KRNB-FM
Univision Radio: KDXX(FM), KFLC(AM), KFZO(FM), KLNO(FM)

Danbury, CT: Rank 196
MSA: 189,400
Berkshire Broadcasting Corp.: WLAD, WREF
Cumulus Media Inc.: WDBY(FM), WINE

Dayton, OH: Rank 59
MSA: 835,500
Brewer Broadcasting Corp.: WQLK-FM
Clear Channel Communications Inc.: WDKF(FM), WIZE, WLQT-FM, WMMX-FM, WONE, WTUE-FM, WXEG-FM
Cox Radio Inc.: WHIO, WHIO-FM, WHKO-FM, WZLR(FM)
EMF Broadcasting: WOKL(FM)
Radio One Inc.: WDHT(FM), WGTZ-FM, WING, WKSW-FM, WROU-FM
Radio Stations WPAY/WPFB Inc.: WPFB, WPFB-FM
Town and Country Broadcasting Inc.: WBZI, WEDI(AM)

Daytona Beach, FL: Rank 86
MSA: 518,500
Black Crow Media Group LLC: WHOG-FM, WKRO-FM, WNDB, WVYB-FM
Buddy Tucker Association Inc.: WYND
Gore-Overgaard Broadcasting Inc.: WROD, WSBB
J&V Communications Inc.: WTJV(AM)
Mega Communications LLC: WNUE-FM
Renda Broadcasting Corp.: WGNE-FM

Decatur, IL: Rank 281
MSA: 92,200
NextMedia Group Inc.: WDZ(AM), WSOY-FM
The Cromwell Group Inc.: WZNX(FM), WZUS(FM)

Denver-Boulder, CO: Rank 22
MSA: 2,194,800
KNRV(AM), KTNI-FM
CBS Radio: KIMN-FM, KWLI(FM), KXKL-FM
Clear Channel Communications Inc.: KBCO-FM, KBPI(FM), KHOW, KKZN(AM), KOA, KPTT(FM), KRFX-FM, KTCL(FM)
Crawford Broadcasting Co.: KLDC, KLTT, KLVZ(AM), KLZ
EMF Broadcasting: KLDV(FM)
Entercom Communications Corp.: KALC(FM), KEZW, KOSI-FM, KQMT(FM)
Entravision Communications Corp.: KJMN-FM, KMXA, KXPK-FM
Latino Communications LLC: KBNO(AM)
Lincoln Financial Media: KEPN(AM), KJCD(FM), KKFN, KQKS-FM, KYGO-FM
NRC Broadcasting Inc.: KCKK(AM), KCUV(FM)
Pillar of Fire Inc.: KPOF
Salem Communications Corp.: KBJD, KNUS, KRKS(AM), KRKS-FM

Des Moines, IA: Rank 92
MSA: 493,600
KDLS-FM
Birch Broadcasting Corp.: KXLQ
Citadel Broadcasting Corp.: KGGO-FM, KHKI-FM, KJJY(FM), KWQW(FM)
Clear Channel Communications Inc.: KASI, KCCQ(FM), KDRB(FM), KKDM-FM, KPTL(FM), KXNO(AM), WHO(AM)
Connoisseur Media LLC: KZLN(FM)
Coon Valley Communications Inc.: KDLS
Family Stations Inc.: KDFR(FM)
Northwestern College & Radio: KNWI(FM), KNWM(FM)
Saga Communications Inc.: KAZR(FM), KIOA(FM), KLTI-FM, KRNT, KSTZ(FM)

Detroit: Rank 10
MSA: 3,888,300
ABC Inc.: WFDF(AM)
Birach Broadcasting Corp.: WNZK, WPON
CBS Radio: WOMC-FM, WVMV-FM, WWJ, WXYT, WYCD-FM
CHUM Ltd.: CIDR-FM, CIMX-FM, CKLW, CKWW
Citadel Broadcasting Corp.: WDRQ-FM, WDVD(FM), WJR
Clear Channel Communications Inc.: WDFN, WDTW(AM), WDTW-FM, WJLB-FM, WKQI-FM, WMXD-FM, WNIC-FM
Crawford Broadcasting Co.: WEXL, WMUZ-FM, WRDT(AM)
Davidson Media Group LLC: WDRJ(AM)
Family Life Communications Inc.: WUFL
Greater Media Inc.: WCSX-FM, WMGC-FM, WRIF(FM)
Liggett Communications L.L.C.: WHLX(AM), WPHM
Radio One Inc.: WCHB, WDMK(FM), WHTD(FM)
Salem Communications Corp.: WDTK(AM), WLQV

Dothan, AL: Rank 193
MSA: 196,500
WELB, WQLS
Magic Broadcasting LLC: WJRL-FM, WKMX-FM, WLDA(FM), WTVY-FM
Wilson Broadcasting Inc.: WAGF, WAGF-FM, WJJN-FM

Dubuque, IA: Rank 236
MSA: 142,200
American Family Radio: KIAD(FM)
Cumulus Media Inc.: KLYV-FM, KXGE-FM, WDBQ, WDBQ-FM, WJOD-FM
Family Life Communications Inc.: WJTY-FM
Morgan Murphy Stations (Evening Telegram Co): WGLR, WPVL, WPVL-FM
Radio Dubuque Inc.: KATF(FM), KDTH, KGRR-FM

Duluth-Superior, MN-WI: Rank 204
MSA: 175,500
Clear Channel Communications Inc.: KBMX(FM), KKCB-FM, KLDJ-FM, WEBC
Heartland Communications Group LLC: WNXR(FM)
Midwest Communications Inc.: KDAL, KRBR-FM, KTCO-FM, WDSM, WGEE(AM), WUSZ-FM
Northwestern College & Radio: KDNI(FM), KDNW(FM)
Red Rock Radio Corp.: KQDS, KQDS-FM, KZIO-FM, WWAX-FM

Eau Claire, WI: Rank 245
MSA: 135,500
Clear Channel Communications Inc.: WATQ(FM), WBIZ, WISM-FM, WMEQ(AM), WQRB-FM
Maverick Media LLC: WAXX-FM, WAYY(AM), WDRK(FM), WEAQ, WECL-FM, WIAL-FM
Relevant Radio: WDVM(AM)
VCY America Inc.: WVCF-FM

El Paso, TX: Rank 76
MSA: 580,900
Bravo Mic Communications LLC: KXPZ(FM)
Clear Channel Communications Inc.: KHEY(AM), KPRR-FM, KTSM(AM)
EMF Broadcasting: KKLY(FM)
Entravision Communications Corp.: KHRO(AM), KINT-FM, KOFX-FM, KSVE, KYSE(FM)
Regent Communications Inc.: KROD, KSII-FM
Univision Radio: KAMA, KBNA(AM), KBNA-FM
World Radio Network Inc.: KVER-FM

Elizabeth City-Nags Head, NC: Rank 0
MSA:
CapSan Media LLC: WFMZ(FM), WVOD(FM), WYND-FM, WZPR(FM)
CSN International: WGPS(FM)
East Carolina Radio Group: WCNC(AM), WERX-FM, WKJX(FM), WOBR-FM, WOBX(AM), WRSF(FM), WZBO(AM)
MAX Media L.L.C.: WCMS-FM, WCXL(FM), WGAI(AM)
Willis Broadcasting Corp.: WBXB(FM)

Elkins-Buckhannon-Weston, WV: Rank 272
MSA: 101,900

Elmira-Corning, NY: Rank 223
MSA: 160,300
Backyard Broadcasting LLC: WNKI-FM, WPGI-FM, WWLZ
CSN International: WREQ-FM
Family Life Network: WCIH-FM, WCIK-FM
Pembrook Pines Media Group: WABH, WEHH, WELM, WOKN-FM
Route 81 Radio LLC: WCBA, WENI(AM), WENI-FM, WENY, WENY-FM, WGMM(FM)

Erie, PA: Rank 168
MSA: 238,000
Citadel Broadcasting Corp.: WQHZ(FM), WRIE, WXTA-FM
Connoisseur Media LLC: WFNN(AM), WJET(AM), WRKT-FM, WRTS-FM, WXBB(FM)
Family Stations Inc.: WEFR-FM

Eugene-Springfield, OR: Rank 150
MSA: 294,800
Bicoastal Media L.L.C.: KDUK-FM, KODZ-FM, KPNW
Churchill Communications LLC: KLZS(AM), KOPT(AM), KXOR(AM)
Cumulus Media Inc.: KEHK-FM, KSCR(AM), KUGN, KUJZ(FM), KZEL-FM
Family Stations Inc.: KQFE-FM
McKenzie River Broadcasting Company, Inc.: KEUG(FM), KKNU(FM), KMGE-FM

Evansville, IN: Rank 162
MSA: 257,100
WEZG(FM)
Regent Communications Inc.: WDKS-FM, WGBF, WGBF-FM, WJLT(FM), WKDQ-FM
South Central Communications Corp.: WABX-FM, WEJK(FM), WEOA, WIKY-FM, WLFW(FM), WSTO-FM
The Original Company Inc.: WRCY(AM), WYFX(FM)
Withers Broadcasting Co.: WYNG(FM)
Word Broadcasting Network Inc.: WVHI

Fargo-Moorhead, ND-MN: Rank 223
MSA: 160,300
KVOX
Brantley Broadcast Associates LLC: WZFN(AM)
Forum Communications Co.: WDAY
Leighton Enterprises Inc.: KBOQ(FM)
Northwestern College & Radio: KFNL(FM), KFNW(AM)
Radio Fargo-Moorhead Inc.: KBVB(FM), KFGO, KKAG(AM), KRWK(FM), WDAY-FM
Triad Broadcasting Co. L.L.C.: KLTA(FM), KPFX(FM), KQWB-FM, KVOX-FM

Fayetteville (Northwest Arkansas), AR: Rank 135
MSA: 319,100
American Family Radio: KBNV(FM)
Clear Channel Communications Inc.: KEZA-FM, KIGL(FM), KKIX-FM, KMXF-FM
Cumulus Media Inc.: KAMO-FM, KFAY, KKEG-FM, KMCK-FM, KQSM-FM, KYNF(FM), KYNG(AM)
Davidson Media Group LLC: KAKS(FM)
KERM Inc.: KURM

Fayetteville, NC: Rank 129
MSA: 348,800
WFMO
Beasley Broadcast Group Inc.: WAZZ(AM), WFLB-FM, WKML-FM, WTEL(AM), WUKS(FM), WZFX-FM
Bible Broadcasting Network: WYBH(FM)
Cumulus Media Inc.: WFNC, WFNC-FM, WFVL(FM), WRCQ-FM
Davidson Media Group LLC: WSTS-FM
Norsan Consulting and Management Inc.: WFAY(AM)

Flagstaff-Prescott, AZ: Rank 151
MSA: 290,700
KAHM(FM), KYCA(AM), KZGL(FM)
LKCM Radio Group L.P.: KFSZ(FM)
Prescott Valley Broadcasting Co. Inc.: KPPV(FM), KQNA(AM)
Yavapai Broadcasting Corp.: KKLD(FM), KVNA(AM), KVNA-FM, KVRD-FM, KYBC(AM)

Flint, MI: Rank 127
MSA: 369,400
Birach Broadcasting Corp.: WCXI(AM)
Christian Broadcasting System Ltd.: WSNL(AM)
Citadel Broadcasting Corp.: WFBE-FM, WTRX
Cumulus Media Inc.: WDZZ-FM, WRSR-FM, WWCK
EMF Broadcasting: WAKL(FM)
Regent Communications Inc.: WFNT, WLCO(AM), WQUS(FM), WWBN-FM
Superior Communications: WTAC(FM)

Florence, SC: Rank 216
MSA: 167,200
American Family Radio: WDLL(FM)
Cumulus Media Inc.: WCMG-FM, WHLZ(FM), WHSC, WMXT-FM, WWFN-FM, WYNN
GHB Radio Group: WHYM(AM), WOLS(AM)
Glory Communications Inc.: WPDT-FM
Miller Communications Inc.: WSIM(FM)
Qantum Communications Corp.: WDAR-FM, WJMX, WJMX-FM, WWRK(AM), WZTF(FM)

Florence-Muscle Shoals, AL: Rank258
MSA: 122,900
Big River Broadcasting Corp.: WQLT-FM, WSBM(AM), WXFL(FM)
Urban Radio Licenses LLC: WLAY-FM

Fort Collins-Greeley, CO: Rank 124
MSA: 379,800
Clear Channel Communications Inc.: KIIX(AM), KSME(FM)
EMF Broadcasting: KLHV(FM)
NRC Broadcasting Inc.: KJAC(FM), KRKY-FM

Fort Myers-Naples-Marco Island, FL: Rank 62
MSA: 783,100
Beasley Broadcast Group Inc.: WJBX-FM, WJPT(FM), WRXK-FM, WWCN, WXKB-FM
Clear Channel Communications Inc.: WBTT(FM), WCKT(FM), WOLZ-FM, WZJZ(FM)
Fort Myers Broadcasting Co.: WINK(AM), WINK-FM, WPTK(AM), WTLQ-FM
Meridian Broadcasting Inc.: WARO-FM, WNOG, WTLT-FM, WUSV(FM)
Relevant Radio: WCNZ(AM), WMYR, WVOI(AM)
Renda Broadcasting Corp.: WGUF-FM, WJGO(FM), WSGL-FM, WWGR-FM
WAY-FM Media Group Inc.: WAYJ-FM

Fort Pierce-Stuart-Vero Beach, FL: Rank 99
MSA: 465,700
Black Media Works Inc.: WJFP-FM
Clear Channel Communications Inc.: WAVW(FM), WCZR(FM), WQOL-FM, WSYR-FM, WZTA(AM)
Vero Beach Broadcasters LLC: WGYL-FM, WJKD(FM), WOSN-FM, WTTB

Fort Smith, AR: Rank 176
MSA: 224,600
American Family Radio: KAOW(FM)
Clear Channel Communications Inc.: KKBD(FM), KMAG-FM, KWHN(AM), KZBB-FM
Cumulus Media Inc.: KBBQ-FM, KLSZ-FM, KOAI(AM), KOMS-FM
Family Stations Inc.: KEAF(FM)
Pearson Broadcasting: KERX(FM), KTTG-FM
Pharis Broadcasting Inc.: KFPW, KFPW-FM, KHGG(AM), KQBK(FM)

Fort Walton Beach, FL: Rank 226
MSA: 159,600
Cumulus Media Inc.: WFTW, WNCV(FM), WYZB-FM, WZNS-FM
Qantum Communications Corp.: WMXZ-FM, WWAV-FM
Star Broadcasting Inc.: WRKN(FM), WTKE(AM), WTKE-FM

Fort Wayne, IN: Rank 106
MSA: 429,100
WBNI-FM, WGBJ(FM), WLYV
Artistic Media Partners Inc.: WVBB(FM)
Bott Radio Network: WFCV
Federated Media: WBYR-FM, WFWI-FM, WKJG(AM), WMEE-FM, WOWO, WQHK-FM
Independence Media Holdings LLC: WNUY(FM)
Sarkes Tarzian Inc.: WAJI(FM), WLDE-FM
Summit City Radio Group: WGL, WGL-FM, WNHT(FM), WXKE(FM)

Frederick, MD: Rank 197
MSA: 188,500
Clear Channel Communications Inc.: WFMD(AM), WFRE(FM)

Fredericksburg, VA: Rank 153
MSA: 288,400

Fresno, CA: Rank 66
MSA: 723,400
KSXE(FM)
Clear Channel Communications Inc.: KALZ(FM), KBOS-FM, KCBL, KFSO-FM, KHGE(FM), KRDU, KRZR-FM
Family Stations Inc.: KFNO-FM
Gore-Overgaard Broadcasting Inc.: KBIF, KIRV
Lotus Communications Corp.: KGST, KLBN-FM, KMMM(FM)
Moon Broadcasting: KAAT(FM)
Multicultural Radio Broadcasting Inc.: KWRU(AM)
Peak Broadcasting LLC: KFJK(FM), KFPT(AM), KMGV-FM, KMJ, KOQO-FM, KSKS-FM, KWYE(FM)
Univision Radio: KLLE(FM), KOND(FM), KRDA(FM)
Wilks Broadcast Group LLC: KFRR-FM, KJFX-FM, KJZN(FM)

Gainesville-Ocala, FL: Rank 83
MSA: 530,600
Asterisk Inc.: WMFQ-FM, WXJZ(FM), WYGC(FM)
Bible Broadcasting Network: WYFB-FM, WYFZ(FM)
Entercom Communications Corp.: WKTK-FM, WSKY-FM
Pamal Broadcasting Ltd.: WDVH(AM), WDVH-FM, WHHZ(FM), WKZY(FM), WTMG-FM, WTMN(AM)
Wooster Republican Printing Co.: WNDD-FM, WNDT(FM), WOGK(FM)

Grand Forks, ND-MN: Rank 289
MSA: 83,300
Clear Channel Communications Inc.: KJKJ-FM, KKXL, KQHT-FM, KSNR-FM
Leighton Enterprises Inc.: KCNN, KNOX, KNOX-FM, KYCK-FM, KZLT-FM

Grand Junction, CO: Rank 264
MSA: 113,200
Cherry Creek Radio LLC: KKXK-FM, KUBC(AM)
Cumulus Media Inc.: KBKL(FM), KEKB-FM, KEXO, KKNN-FM, KMXY-FM
EMF Broadcasting: KLFV(FM)
MBC Grand Broadcasting Inc.: KJYE-FM, KMGJ(FM), KMOZ-FM, KNZZ(AM), KSTR-FM, KTMM(AM)
United Ministries: KJOL(AM)

Grand Rapids, MI: Rank 67
MSA: 708,400
Birach Broadcasting Corp.: WMFN, WMJH
Citadel Broadcasting Corp.: WHTS(FM), WKLQ(FM), WLAV-FM, WTNR(FM)
Clear Channel Communications Inc.: WBCT-FM, WBFX(FM), WMAX-FM, WMRR-FM, WOOD, WOOD-FM, WSNX-FM, WTKG
Kuiper Stns: WFUR
Midwest Communications Inc.: WHTC(AM)
Regent Communications Inc.: WFGR-FM, WGRD-FM, WNWZ, WTRV-FM
The Mid-West Family Broadcast Group: WHIT(AM)

Great Falls, MT: Rank 259
MSA:
American Family Radio: KAFH(FM)
Cherry Creek Radio LLC: KAAK(FM), KLFM-FM, KMON, KMON-FM, KVVR(FM)
College Creek Media LLC: KEAU(FM), KUUS(FM), KZUS(FM)
EMF Broadcasting: KGFA(FM)
Family Stations Inc.: KFRW(FM)
Fisher Communications Inc.: KINX(FM), KQDI, KQDI-FM, KXGF

Green Bay, WI: Rank 185
MSA: 203,500
Cumulus Media Inc.: WDUZ(AM), WDUZ-FM, WPCK(FM), WQLH-FM, WZNN(FM)
Evangel Ministries Inc.: WEMY-FM
Midwest Communications Inc.: WIXX-FM, WNFL, WTAQ(AM), WZBY(FM)
Results Broadcasting: WOWN-FM, WTCH
Woodward Communications Inc.: WKSZ(FM), WZOR(FM)

Greensboro-Winston Salem-High Point, NC: Rank 45
MSA: 1,131,200
WCOG, WYSR(AM)
Clear Channel Communications Inc.: WGBT(FM), WMAG(FM), WMKS(FM), WTQR-FM, WVBZ(FM)
Curtis Media Group: WMFR, WPCM, WSJS, WSML, WZTK(FM)
Davidson Media Group LLC: WSGH, WTOB, WWBG
Entercom Communications Corp.: WJMH-FM, WPAW(FM), WPET, WQMG-FM, WSMW(FM)
GHB Radio Group: WBLO(AM), WIST-FM, WTIX(AM)
Positive Alternative Radio Inc.: WXRI-FM
Truth Broadcasting Corp.: WKEW, WPOL, WTRU(AM)

Greenville-New Bern-Jacksonville, NC: Rank 89
MSA: 504,900
WDLX
American Family Radio: WAAE-FM, WJKA(FM)
Archway Broadcasting Group: WLGT(FM)
Beasley Broadcast Group Inc.: WIKS-FM, WMGV-FM, WNCT, WNCT-FM, WSFL-FM, WXNR-FM
CTC Media Group Inc.: WECU(AM), WNOS, WSME(AM), WWNB
Curtis Media Group: WKIX(FM)
Inner Banks Media LLC: WRHD(FM), WRHT-FM, WWHA(FM), WWNK(FM)
NextMedia Group Inc.: WANG, WERO(FM), WILT(FM), WQSL-FM, WQZL(FM), WRNS-FM, WSSM(FM), WXQR-FM
Radio La Grande: WLNR, WSRP(AM)

Greenville-Spartanburg, SC: Rank 60
MSA: 824,900
Clear Channel Communications Inc.: WESC-FM, WGVL, WLFJ(AM), WMYI(FM)
Cox Radio Inc.: WHZT(FM), WJMZ-FM
Davidson Media Group LLC: WOLI(AM), WOLI-FM, WOLT-FM
Entercom Communications Corp.: WGVC(FM), WORD(AM), WROQ-FM, WSPA-FM, WTPT-FM, WYRD
Wilkins Communications Network Inc.: WELP(AM)

Hagerstown-Chambersburg-Waynesboro, MD-PA: Rank 165
MSA: 243,600
Allegheny Mountain Network Stations: WEEO-FM
Main Line Broadcasting LLC: WCHA, WDLD(FM), WHAG, WIKZ-FM, WQCM(FM)
Nassau Broadcasting Partners L.P.: WARK, WWEG(FM)
Prettyman Broadcasting Co.: WICL(FM), WLTF(FM)
VerStandig Broadcasting: WAYZ(FM), WCBG(AM), WFYN(FM), WPPT(FM)

Hamptons-Riverhead, NY: Rank 260
MSA: 117,200

Harrisburg-Lebanon-Carlisle, PA: Rank 79
MSA: 660,800
Citadel Broadcasting Corp.: WCAT-FM, WMHX(FM)
Clear Channel Communications Inc.: WHP, WKBO, WRBT-FM, WTKT(AM)
Cumulus Media Inc.: WTCY, WTPA-FM, WWKL(FM)
MAX Media L.L.C.: WYGL-FM
Route 81 Radio LLC: WHYL

Harrisonburg, VA: Rank 277
MSA: 99,300
Clear Channel Communications Inc.: WACL-FM, WKCY
VerStandig Broadcasting: WBHB-FM, WHBG, WJDV(FM), WQPO-FM, WSVA

Hartford-New Britain-Middletown, CT: Rank 50
MSA: 1,047,700
WCCC(AM), WCCC-FM
ABC Inc.: WDZK
Buckley Broadcasting Corp.: WDRC(AM), WDRC-FM
CBS Radio: WRCH-FM, WTIC, WZMX-FM
Clear Channel Communications Inc.: WHCN-FM, WKSS-FM, WPHH(FM), WWYZ-FM
Davidson Media Group LLC: WXCT(AM)
Family Stations Inc.: WCTF
Freedom Communications of Connecticut Inc.: WKND(AM), WLAT(AM), WNEZ(AM)

Hilton Head, SC: Rank 219
MSA: 164,500

Honolulu, HI: Rank 64
MSA: 768,300
KORL(AM)
Clear Channel Communications Inc.: KDNN(FM), KHBZ(AM), KHVH, KSSK(AM), KSSK-FM, KUCD(FM)
Cox Radio Inc.: KCCN-FM, KINE-FM, KKNE(AM), KPHW(FM), KRTR(AM), KRTR-FM
Salem Communications Corp.: KAIM-FM, KGMZ-FM, KGU, KHNR(AM), KHNR-FM, KHUI(FM)
Visionary Related Entertainment L.L.C.: KDDB(FM), KPOI-FM, KQMQ-FM, KUMU, KUMU-FM

Houston-Galveston: Rank 6
MSA: 4,469,900
KGOW(AM)
ABC Inc.: KMIC(AM)
CBS Radio: KHJZ-FM, KIKK, KILT
Clear Channel Communications Inc.: KBME, KHMX-FM, KKRW-FM, KODA-FM, KPRC, KTBZ-FM, KTRH
Cox Radio Inc.: KHPT(FM), KHTC(FM), KKBQ-FM, KTHT(FM)
Cumulus Media Inc.: KFNC(FM), KIOL(FM), KSTB-FM
Cumulus Media Partners LLC: KRBE(FM)
Entravision Communications Corp.: KGOL
Houston Christian Broadcasters Inc.: KHCB, KHCB-FM
Liberman Broadcasting Inc.: KEYH, KJOJ, KJOJ-FM, KQQK(FM), KQUE, KSEV, KTJM-FM
Martin Broadcasting Inc.: KYOK(AM)
Multicultural Radio Broadcasting Inc.: KXYZ
Pacifica Foundation Inc.: KPFT-FM
Radio One Inc.: KBXX(FM), KMJQ-FM, KROI(FM)
Salem Communications Corp.: KKHT-FM, KNTH(AM), KTEK
SIGA Broadcasting Corp.: KGBC, KLVL
Univision Radio: KLAT, KLTN(FM), KOVE-FM, KPTI(FM), KPTY(FM), KQBU-FM, KRTX

Huntington-Ashland, WV-KY: Rank 157
MSA: 270,300
WLGC, WLGC-FM
Baker Family Stations: WOKT
Clear Channel Communications Inc.: WAMX-FM, WBKS(FM), WBVB-FM, WIRO, WTCR, WTCR-FM, WVHU(AM), WZZW
Connoisseur Media LLC: WMGA(FM), WRYV-FM
Kindred Communications Inc.: WCMI, WDGG-FM, WRVC, WRVC-FM
Mortenson Broadcasting Co.: WEMM(AM), WEMM-FM
Positive Alternative Radio Inc.: WKAO(FM)

Huntsville, AL: Rank 113
MSA: 415,900
Black Crow Media Group LLC: WAHR-FM, WLOR, WRTT-FM
Clear Channel Communications Inc.: WBHP, WDRM-FM, WHOS, WTAK-FM
Cumulus Media Inc.: WHRP(FM), WUMP, WVNN, WVNN-FM, WXQW(FM), WZYP-FM
Wilkins Communications Network Inc.: WBXR(AM)

Indianapolis, IN: Rank 40
MSA: 1,328,100
ABC Inc.: WRDZ-FM
Clear Channel Communications Inc.: WFBQ-FM, WNDE, WRZX-FM
Cumulus Media Partners LLC: WFMS-FM, WJJK(FM), WWFT(FM)
Davidson Media Group LLC: WNTS
Emmis Communications Corp.: WIBC(AM), WLHK(FM), WNOU(FM), WYXB(FM)
Entercom Communications Corp.: WNTR(FM), WXNT(AM), WZPL-FM
Mid-America Radio Group Inc.: WCBK-FM, WMYJ(AM)
Radio One Inc.: WHHH-FM, WTLC, WTLC-FM, WYJZ(FM)
Rodgers Broadcasting Corp.: WIFE(AM)
Sarkes Tarzian Inc.: WTTS-FM
The Moody Bible Institute of Chicago: WGNR-FM
Wilkins Communications Network Inc.: WBRI(AM)

Ithaca, NY: Rank 284
MSA: 90,100
Citadel Broadcasting Corp.: WIII(FM), WKRT(AM)
Pembrook Pines Media Group: WPIE
Saga Communications Inc.: WHCU, WNYY(AM), WQNY-FM, WYXL-FM

Jackson, MS: Rank 118
MSA: 405,500
WRBJ-FM
Backyard Broadcasting LLC: WRXW(FM), WWJK(FM)
Clear Channel Communications Inc.: WHLH(FM), WJDX, WMSI-FM, WQJQ-FM, WSTZ-FM, WZRX
Inner City Broadcasting: WJMI-FM, WKXI, WKXI-FM, WOAD, WOAD-FM
New South Communications Inc.: WIIN, WJKK-FM, WUSJ(FM), WYOY-FM
TeleSouth Communications Inc.: WFMN-FM

Jackson, TN: Rank 291
MSA: 80,200
Black Crow Media Group LLC: WFKX(FM), WHHM-FM, WWYN(FM), WZDQ(FM)
Forever Communications Inc.: WLSZ(FM), WTJS, WYNU-FM
Grace Broadcasting Services Inc.: WWGM(FM)

Jacksonville, FL: Rank 47
MSA: 1,083,700
WFOY(AM)
ABC Inc.: WBWL
Chesapeake-Portsmouth Broadcasting Corp.: WBOB(AM), WZNZ
Clear Channel Communications Inc.: WFKS(FM), WFXJ(AM), WJBT-FM, WPLA(FM), WQIK-FM, WROO(FM), WSOL-FM
Cox Radio Inc.: WAPE-FM, WFYV-FM, WJGL(FM), WMXQ-FM, WOKV, WOKV-FM
Norsan Consulting and Management Inc.: WNNR(AM), WSOS(AM), WVOJ(AM)
Renda Broadcasting Corp.: WEJZ-FM, WMUV(FM), WSOS-FM
Salem Communications Corp.: WZAZ
Tama Broadcasting Inc.: WHJX(FM), WJSJ(FM), WSJF(FM)
Word Broadcasting Network Inc.: WYMM(AM)

Johnson City-Kingsport-Bristol, TN-VA: Rank 101
MSA: 455,600
Bristol Broadcasting Co. Inc.: WAEZ(FM), WFHG(AM), WFHG-FM, WTZR(FM), WXBQ-FM
Citadel Broadcasting Corp.: WGOC(AM), WJCW, WKOS-FM, WQUT-FM, WXSM(AM)
Glenwood Communications Corp.: WKPT, WKTP, WMEV, WMEV-FM, WOPI, WTFM-FM
Information Communications Corp.: WABN, WHGG(AM), WPWT(AM)
Positive Alternative Radio Inc.: WCQR-FM

Johnstown, PA: Rank 191
MSA: 198,300
2510 Licenses LLC: WBHV(AM), WCCL(FM), WLKH(FM), WLKJ(FM), WPRR(AM)
Family Stations Inc.: WFRJ-FM
Forever Broadcasting: WFGI-FM, WJHT(FM), WKYE(FM), WNTJ(AM), WNTW(AM), WRKW(FM)
He's Alive Inc.: WPCL(FM)
Renda Broadcasting Corp.: WKQL(FM)
Vernal Enterprises Inc.: WNCC(AM), WRDD

Jonesboro, AR: Rank 292
MSA: 74,200
American Family Radio: KAOG(FM)
Clear Channel Communications Inc.: KFIN(FM), KIYS(FM), KNEA(AM)
EMF Broadcasting: KJLV(FM)
Saga Communications Inc.: KEGI(FM), KJBX(FM)

Joplin, MO: Rank 239
MSA: 140,500
KHST(FM), KOBC-FM, KWXD-FM, WMBH(AM)
Textron Financial Corp.: KCAR-FM, KJML-FM, KMOQ-FM, KQYS(AM), KQYX(AM)
Zimmer Radio Inc.: KIXQ(FM), KJMK-FM, KSYN-FM, KXDG-FM, KZRG(AM), KZYM(AM)

Kalamazoo, MI: Rank 184
MSA: 204,500
Cumulus Media Inc.: WKFR-FM, WKMI, WRKR-FM
Forum Communications Co.: WZUU-FM
Kuiper Stns: WKPR
Midwest Communications Inc.: WKZO(AM), WNWN, WNWN-FM, WQLR(AM), WVFM(FM)

Kansas City, MO-KS: Rank 30
MSA: 1,575,300
KCHZ(FM)
ABC Inc.: KPHN
Bick Broadcasting Co.: KSIS(AM)
Bott Radio Network: KAYX(FM), KCCV, KCCV-FM, KLEX
Carter Broadcast Group Inc.: KPRS-FM, KPRT
Cumulus Media Inc.: KMJK(FM)
Cumulus Media Partners LLC: KCFX(FM), KCJK(FM), KCMO, KCMO-FM
Davidson Media Group LLC: KCZZ(AM), KDTD(AM)
Entercom Communications Corp.: KCSP(AM), KKHK(AM), KMBZ, KQRC-FM, KRBZ(FM), KUDL-FM, KXTR(AM), KYYS-FM, WDAF-FM
The Curators of the University of Missouri: KCUR-FM
Wilkins Communications Network Inc.: KCNW(AM)
Wilks Broadcast Group LLC: KBEQ-FM, KCKC(FM), KFKF-FM, KMXV-FM

Killeen-Temple, TX: Rank 156
MSA: 273,100
KTON
Clear Channel Communications Inc.: KIIZ-FM, KLFX-FM
Cumulus Media Inc.: KLTD-FM, KOOC(FM), KSSM(FM), KTEM, KUSJ(FM)
EMF Broadcasting: KYAR(FM)
Martin Broadcasting Inc.: KRMY

Knoxville, TN: Rank 71
MSA: 644,100
WIFA(AM), WIHG(FM), WIJV(FM), WITA
Citadel Broadcasting Corp.: WIVK-FM, WNML(AM), WNML-FM, WOKI(FM)
East Tennessee Radio Group L.P.: WSEV, WSEV-FM
Horne Radio Group: WATO, WFIV-FM, WGAP, WKVL(AM), WLOD, WMTY(AM)
Journal Communications Inc.: WKHT(FM), WMYU(FM), WQBB, WWST(FM)
Norsan Consulting and Management Inc.: WKGN
Peg Broadcasting Crossville LLC: WPBX(FM)
South Central Communications Corp.: WIMZ-FM, WJXB-FM, WQJK(FM)

La Crosse, WI: Rank 230
MSA: 155,300

La Salle-Peru, IL: Rank 247
MSA: 131,700

Lafayette, IN: Rank 246
MSA: 134,100
WBAA(AM), WBAA-FM
American Family Radio: WQSG(FM)
Artistic Media Partners Inc.: WAZY-FM, WLFF(FM), WSHP(FM), WSHY(AM)
Kaspar Broadcasting Group: WSHW-FM
RadioWorks Inc.: WKHY(FM)
Schurz Communications Inc.: WASK, WASK-FM, WKOA-FM, WXXB(FM)
The Moody Bible Institute of Chicago: WHPL-FM

Lafayette, LA: Rank 102
MSA: 447,200
KEUN, KEUN-FM, KNEK-FM
Citadel Broadcasting Corp.: KNEK, KRRQ-FM, KSMB-FM, KXKC(FM)
EMF Broadcasting: KIKL(FM)
Pittman Broadcasting Services LLC: KFXZ(AM), KFXZ-FM, KKSJ(FM), KVOL
Radio Maria Inc.: KNIR
Regent Communications Inc.: KFTE-FM, KMDL-FM, KPEL, KPEL-FM, KRKA(FM), KROF, KTDY-FM

Lake Charles, LA: Rank 229
MSA: 156,900
American Family Radio: KYLC(FM)
Apex Broadcasting Inc.: KLCL, KNGT(FM), KTSR(FM)
Cumulus Media Inc.: KAOK, KBIU(FM), KKGB-FM, KQLK(FM), KXZZ, KYKZ-FM
EMF Broadcasting: KRLR(FM)
Radio Maria Inc.: KOJO-FM

Lakeland-Winter Haven, FL: Rank 96
MSA: 468,500
Bible Broadcasting Network: WYFO-FM
CBS Radio: WSJT-FM
Hall Communications Inc.: WLKF, WONN, WPCV-FM, WWRZ-FM
The Moody Bible Institute of Chicago: WKES-FM

Lancaster, PA: Rank 114
MSA: 413,500
Citadel Broadcasting Corp.: WIOV-FM
Clear Channel Communications Inc.: WLAN, WLAN-FM
Hall Communications Inc.: WLPA, WROZ-FM

Lansing-East Lansing, MI: Rank 122
MSA: 389,400
Christian Broadcasting System Ltd.: WLCM
Citadel Broadcasting Corp.: WFMK(FM), WITL-FM, WJIM(AM), WVFN(AM)
Family Life Communications Inc.: WUNN(AM)
MacDonald Broadcasting Co.: WHZZ-FM, WILS, WQHH-FM, WXLA
Rubber City Radio Group Inc.: WJXQ-FM, WJZL(FM), WQTX(FM), WVIC(FM)

Laredo, TX: Rank 205
MSA: 174,900
Border Media Partners LLC: KBDR(FM), KLNT, KNEX-FM

Las Vegas, NV: Rank 32
MSA: 1,484,400
Beasley Broadcast Group Inc.: KCYE(FM), KDWN, KKLZ-FM, KSTJ-FM
CBS Radio: KKJJ(FM), KLUC-FM, KMXB(FM), KXNT, KXTE-FM
Clear Channel Communications Inc.: KPLV(FM), KSNE-FM, KWID(FM), KWNR-FM
Entravision Communications Corp.: KQRT(FM)
Faith Communications Corp.: KSOS(FM)
Kemp Communications Inc.: KONV(FM), KVEG(FM)
Lotus Communications Corp.: KBAD, KENO, KOMP-FM, KWWN(AM), KXPT-FM
McNaughton-Jakle Stations: KSHP
Riviera Broadcast Group LLC: KOAS(FM), KVGS(FM)
Univision Radio: KISF-FM, KLSQ(AM)

Laurel-Hattiesburg, MS: Rank 209
MSA: 172,300
American Family Radio: WAII-FM, WATP-FM
Blakeney Communications Inc.: WBBN(FM), WKZW-FM, WXHB(FM), WXRR-FM
Clear Channel Communications Inc.: WEEZ, WFOR, WHER-FM, WJKX-FM, WNSL-FM, WUSW-FM, WZLD(FM)
EMF Broadcasting: WKNZ-FM
Walking by Faith Ministries Inc.: WAML

Lawton, OK: Rank 282
MSA: 91,800
American Family Radio: KVRS-FM
GAP Broadcasting LLC: KLAW(FM), KVRW-FM, KZCD(FM)
Monarch Broadcasting Inc.: KQTZ-FM
Perry Publishing & Broadcasting Co.: KJMZ(FM), KKRX, KVSP(FM), KXCA(AM)

Lebanon-Rutland-White River Junction, NH-VT: Rank 179
MSA: 215,900

Lewiston-Auburn, ME: Rank 280
MSA: 94,100
Gleason Radio Group: WEZR(AM)
Nassau Broadcasting Partners L.P.: WLAM(AM)

Lexington-Fayette, KY: Rank 104
MSA: 440,200
Christian Broadcasting System Ltd.: WCGW, WVKY(AM)
Clear Channel Communications Inc.: WBUL-FM, WKQQ-FM, WLKT-FM, WMXL-FM, WXRA(AM)
Cumulus Media Inc.: WLTO-FM, WVLK, WVLK-FM, WXZZ-FM
EMF Broadcasting: WRVG(FM)
L M Communications Inc.: WBTF-FM, WBVX(FM), WCDA-FM, WGKS-FM, WLXG
Vernon R Baldwin Inc.: WVRB-FM
Wallingford Broadcasting Co.: WEKY, WKXO, WLFX(FM)

Lima, OH: Rank 251
MSA: 128,400
Clear Channel Communications Inc.: WIMA, WIMT(FM), WZRX-FM
Maverick Media LLC: WDOH-FM, WEGE(FM), WFGF(FM), WWSR(FM), WZOQ(AM)

Lincoln, NE: Rank 174
MSA: 227,700
Bott Radio Network: KLCV-FM
Clear Channel Communications Inc.: KIBZ(FM), KLMY(FM), KTGL-FM, KZKX-FM
Family Worship Center Church Inc.: KNBE(FM)
Three Eagles Communications: KFRX(FM), KRKR(FM)
Triad Broadcasting Co. L.L.C.: KLIN, KLNC(FM)

Little Rock, AR: Rank 85
MSA: 523,600
KARN-FM, KASR(FM), KPZK-FM, KZTD(AM)
ABC Inc.: KDIS-FM
Archway Broadcasting Group: KOLL-FM
Citadel Broadcasting Corp.: KAAY, KARN, KIPR-FM, KLAL-FM, KOKY-FM, KPZK(AM), KURB-FM
Clear Channel Communications Inc.: KDJE(FM), KHKN(FM), KHLR(FM), KMJX-FM, KSSN-FM
Metropolitan Radio Group Inc.: KGHT
Noalmark Broadcasting Corp.: KBOK(AM)
Simmons Media Group: KDXE(AM)
US Stations LLC: KQUS-FM

Los Angeles: Rank 2
MSA: 10,826,600
KMPC(AM)
ABC Inc.: KDIS(AM), KSPN(AM)
CBS Radio: KFWB, KLSX-FM, KNX, KROQ-FM, KRTH-FM, KTWV-FM
Citadel Broadcasting Corp.: KABC, KLOS-FM
Clear Channel Communications Inc.: KAVL, KBIG-FM, KFI, KHHT(FM), KIIS-FM, KLAC, KOSS-FM, KOST-FM, KTLK(AM), KTPI-FM, KYSR-FM
Crawford Broadcasting Co.: KBRT
Emmis Communications Corp.: KMVN(FM), KPWR-FM
Entravision Communications Corp.: KDLD(FM), KDLE(FM), KLYY(FM), KSSE(FM)
Family Stations Inc.: KFRN
Hi-Favor Broadcasting LLC: KLTX
High Desert Broadcasting LLC: KGMX-FM, KUTY, KWJL(AM)
Liberman Broadcasting Inc.: KBUA(FM), KBUE(FM), KEBN(FM), KHJ(AM), KWIZ-FM
Lotus Communications Corp.: KIRN(AM), KWKU(AM), KWKW
Magic Broadcasting LLC: KDAY(FM)
Multicultural Radio Broadcasting Inc.: KAHZ(AM), KALI, KAZN, KBLA, KYPA
Pacifica Foundation Inc.: KPFK-FM
Radio One Inc.: KRBV(FM)
Salem Communications Corp.: KFSH-FM, KKLA-FM, KRLA(AM), KXMX(AM)
Spanish Broadcasting System Inc.: KLAX-FM, KXOL-FM
Univision Radio: KLVE-FM, KRCD(FM), KRCV(FM), KSCA-FM, KTNQ

Louisville, KY: Rank 54
MSA: 930,600
WWSZ(AM)
ABC Inc.: WDRD(AM)
Clear Channel Communications Inc.: WAMZ(FM), WCND, WHAS, WKJK, WKRD(AM), WKRD-FM, WLUE(FM), WQMF-FM, WTFX-FM, WZKF(FM)
Commonwealth Broadcasting Corp.: WTSZ(AM)
Cox Radio Inc.: WPTI(FM), WRKA-FM, WSFR(FM), WVEZ(FM)
Cumulus Media Partners LLC: WAVG(AM)
Davidson Media Group LLC: WLLV, WLOU, WTUV(AM), WTUV-FM
EMF Broadcasting: WARA(FM)
Radio One Inc.: WDJX-FM, WGZB-FM, WLRS(FM), WLRX(FM), WMJM-FM, WXMA(FM)
Salem Communications Corp.: WFIA, WFIA-FM, WGTK(AM), WRVI-FM

Lubbock, TX: Rank 183
MSA: 211,600
KTXT-FM
EMF Broadcasting: KKLU(FM)
Entravision Communications Corp.: KAIQ(FM), KBZO
Family Life Communications Inc.: KAMY(FM)
GAP Broadcasting LLC: KFMX-FM, KFYO, KKAM, KKCL-FM, KQBR(FM), KZII-FM
Ramar Communications II Ltd.: KJTV(AM), KLZK(FM), KXTQ-FM
Wilks Broadcast Group LLC: KLLL-FM, KMMX(FM), KONE(FM)

Macon, GA: Rank 155
MSA: 280,100
WFXM(FM), WQMJ(FM), WXJO(AM), WXKO
American Family Radio: WBKG-FM
Clear Channel Communications Inc.: WEBL(FM), WIBB-FM, WLCG, WPCH(FM), WQBZ-FM, WRBV-FM, WVVM(AM)
Cumulus Media Inc.: WAYS(AM), WDDO, WDEN-FM, WIFN(FM), WLZN(FM), WMAC, WMGB(FM), WPEZ(FM)
Family Life Communications Inc.: WJTG(FM)
Georgia Eagle Broadcasting Inc.: WNNG(AM)
Rodgers Broadcasting Corp.: WBML

Madison, WI: Rank 95
MSA: 468,800
Clear Channel Communications Inc.: WIBA, WMAD(FM), WTSO, WXXM(FM)
Entercom Communications Corp.: WCHY(FM), WMMM-FM, WOLX-FM
Family Stations Inc.: WJJO-FM
Northwestern College & Radio: WNWC(AM), WNWC-FM
NRG Media LLC: WSJY-FM
Relevant Radio: WHFA(AM)
The Mid-West Family Broadcast Group: WLMV(AM), WTDY, WTUX(AM), WWQM-FM

Manchester, NH: Rank 190
MSA: 198,700
Blount Communications Group: WDER(AM)
Northeast Broadcasting Company Inc.: WKBR
Saga Communications Inc.: WFEA, WMLL(FM), WZID(FM)

Mankato-New Ulm-St. Peter, MN: Rank 274
MSA: 100,900
Ingstad Brothers Broadcasting LLC: KNUJ(AM)
Linder Broadcasting Group: KTOE(AM), KXLP(FM)
Three Eagles Communications: KEEZ-FM, KRBI(AM)

Marion-Carbondale (Southern Illinois): Rank 240
MSA: 140,400
American Family Radio: WAWJ(FM)
Clear Channel Communications Inc.: WDDD, WDDD-FM, WFRX, WQUL-FM, WTAO-FM, WVZA-FM
MAX Media L.L.C.: WCIL, WJPF, WOOZ-FM, WUEZ(FM), WXLT(FM)

Mason City, IA: Rank 296
MSA: 68,700
American Family Radio: KBDC(FM)
Clear Channel Communications Inc.: KGLO(AM), KIAI(FM), KLKK(FM), KSMA-FM
Three Eagles Communications: KRIB(AM)

McAllen-Brownsville-Harlingen, TX: Rank 58
MSA: 838,400
Border Media Partners LLC: KBUC(FM), KESO-FM, KJAV-FM, KSOX, KURV, KVJY, KZSP-FM
Clear Channel Communications Inc.: KBFM(FM), KHKZ(FM), KQXX-FM, KTEX-FM, KVNS(AM)
Entravision Communications Corp.: KFRQ-FM, KKPS-FM, KVLY-FM, KZPL(FM)
Univision Radio: KBTQ(FM), KGBT, KGBT-FM
World Radio Network Inc.: KVMV-FM

Meadville-Franklin, PA: Rank 0
MSA:

Medford-Ashland, OR: Rank 210
MSA: 171,000
Bicoastal Media L.L.C.: KIFS(FM), KLDZ(FM), KMED, KRWQ-FM, KZZE-FM
Mapleton Communications LLC: KAKT(FM), KBOY-FM, KCMX, KCMX-FM, KTMT
Opus Broadcasting Systems Inc.: KCNA(FM), KEZX(AM), KROG-FM, KRTA, KRVC(FM)

Melbourne-Titusville-Cocoa, FL: Rank 94
MSA: 476,400
Clear Channel Communications Inc.: WMMB, WMMV
Cumulus Media Inc.: WHKR-FM, WINT(AM), WSJZ-FM
Genesis Communications Inc.: WIXC(AM)
Rama Communications Inc.: WTIR(AM)

Memphis, TN: Rank 49
MSA: 1,060,700
Bott Radio Network: WCRV
CBS Radio: WMC(AM), WMC-FM, WMFS(FM)
Citadel Broadcasting Corp.: WGKX-FM, WRBO(FM), WXMX(FM)
Clear Channel Communications Inc.: KJMS(FM), WDIA, WEGR(FM), WHAL-FM, WHRK-FM, WREC(AM)
EMF Broadcasting: KKLV(FM), WPLX(AM)
Entercom Communications Corp.: WRVR(FM), WSMB(AM), WSNA(FM)
F W Robbert Broadcasting Co. Inc.: WMQM(AM)
First Broadcasting Operating Inc.: WVIM-FM
Simmons Media Group: KQPN(AM)
Sudbury Services Inc.: KHLS-FM, KLCN, KOSE, KQDD(FM)

Merced, CA: Rank 180
MSA: 214,500
Bott Radio Network: KCIV(FM)
Buckley Broadcasting Corp.: KHTN(FM), KUBB-FM
Citadel Broadcasting Corp.: KDJK(FM)
EMF Broadcasting: KLVN(FM)
Mapleton Communications LLC: KLOQ-FM, KNAH(FM), KTIQ(AM), KYOS

Meridian, MS: Rank 297
MSA: 66,600
Clear Channel Communications Inc.: WJDQ(FM), WMSO(FM), WYHL(AM)
Mississippi Broadcasters L.L.C.: WJXM(FM), WKZB(FM), WUCL(FM)
New South Communications Inc.: WALT

Miami-Fort Lauderdale-Hollywood, FL: Rank 12
MSA: 3,533,000
WSRF
ABC Inc.: WMYM(AM)
Beasley Broadcast Group Inc.: WHSR, WKIS-FM, WPOW-FM, WQAM, WWNN
Clear Channel Communications Inc.: WBGG-FM, WHYI-FM, WINZ(AM), WIOD, WLVE-FM, WMGE(FM), WMIB(FM)
Cox Radio Inc.: WEDR-FM, WFLC-FM, WHDR(FM), WHQT-FM
Entravision Communications Corp.: WLQY
Independence Media Holdings LLC: WOCN
James Crystal Inc.: WFLL(AM)
Lincoln Financial Media: WAXY, WLYF(FM), WMXJ-FM
Multicultural Radio Broadcasting Inc.: WEXY, WNMA
Radio One Inc.: WTPS(AM)
Salem Communications Corp.: WKAT
Spanish Broadcasting System Inc.: WCMQ-FM, WRMA-FM, WXDJ-FM
Univision Radio: WAMR-FM, WAQI, WQBA, WRTO-FM

Middlesex-Somerset-Union, NJ: Rank 39
MSA: 1,382,800

Milwaukee-Racine, WI: Rank 36
MSA: 1,433,300
WTKM(AM), WTKM-FM
ABC Inc.: WKSH(AM)
Bliss Communications Inc.: WBKV, WBWI-FM, WRJN
Bustos Media LLC: WDDW(FM)
Clear Channel Communications Inc.: WISN, WKKV-FM, WMIL-FM, WOKY, WQBW(FM), WRIT-FM
Entercom Communications Corp.: WMYX-FM, WSSP(AM), WXSS-FM
Family Stations Inc.: WMWK-FM
Good Karma Broadcasting L.L.C.: WAUK
Journal Communications Inc.: WTMJ
Milwaukee Radio Alliance L.L.C.: WLDB(FM), WLUM-FM, WMCS
Relevant Radio: WPJP(FM), WZRK(AM)
Saga Communications Inc.: WHQG(FM), WJMR-FM, WJYI, WKLH-FM
Salem Communications Corp.: WFZH(FM), WRRD(AM)
VCY America Inc.: WVCY-FM

Minneapolis-St. Paul, MN: Rank 16
MSA: 2,662,100
KQSP(AM), KTNF(AM)
ABC Inc.: KDIZ
CBS Radio: KZJK(FM), WCCO
Citadel Broadcasting Corp.: KQRS-FM, KXXR(FM), WGVY(FM), WGVZ(FM)
Clear Channel Communications Inc.: KDWB-FM, KEEY-FM, KFAN, KFXN, KQQL-FM, KTCZ-FM, KTLK-FM
Davidson Media Group LLC: KMNV(AM), KRJJ(AM)
EMF Broadcasting: KMKL(FM)
Hubbard Broadcasting Inc.: KSTP(AM), KSTP-FM, WFMP(FM), WIXK(AM)
Ingstad Brothers Broadcasting LLC: KCHK
Northwestern College & Radio: KTIS, KTIS-FM
Relevant Radio: WLOL(AM)
Salem Communications Corp.: KKMS, KYCR, WWTC

Mobile, AL: Rank 90
MSA: 501,300
WABB, WABB-FM, WABF(AM), WCSN-FM
Brantley Broadcast Associates LLC: WHOA(AM)
Buddy Tucker Association Inc.: WMOB, WTOF(AM)
Clear Channel Communications Inc.: WKSJ-FM, WMXC-FM, WNTM(AM), WRKH-FM
Cumulus Media Inc.: WBLX-FM, WDLT, WDLT-FM, WGOK, WYOK-FM
Family Worship Center Church Inc.: WQUA-FM
Goforth Media Inc.: WBHY-FM
Martin Broadcasting Inc.: WLVV

Modesto, CA: Rank 108
MSA: 421,800
Bustos Media LLC: KBBU(FM)
Citadel Broadcasting Corp.: KATM(FM), KESP(AM), KHKK-FM, KHOP-FM, KWNN-FM
Clear Channel Communications Inc.: KFIV, KJSN-FM, KMRQ(FM), KOSO-FM
Entravision Communications Corp.: KTSE-FM
Pappas Telecasting Companies: KMPH(AM)

Monmouth-Ocean, NJ: Rank 51
MSA: 1,020,500

Monroe, LA: Rank 257
MSA: 123,900
Bible Broadcasting Network: KYFL-FM
Communications Capital Managers LLC: KNBB(FM), KXKZ-FM
Holladay Broadcasting of Louisiana LLC: KJMG-FM, KLIP-FM, KMLB, KRJO(AM), KRVV-FM
New South Communications Inc.: KJLO-FM
Opus Media Holdings LLC: KMYY(FM), KXRR(FM), KZRZ(FM)

Monterey-Salinas-Santa Cruz, CA: Rank 80
MSA: 556,200
KNRY, KRXA(AM), KYAA(AM)
Buckley Broadcasting Corp.: KWAV-FM, KYZZ(FM)
Bustos Media LLC: KZSJ(AM)
Clear Channel Communications Inc.: KDON-FM, KION(AM), KOCN-FM, KPRC-FM, KTOM-FM
Entravision Communications Corp.: KMBX(AM), KSES-FM
Lazer Broadcasting Corp.: KXSM(FM)
Mapleton Communications LLC: KBOQ(FM), KCDU(FM), KHIP(FM), KMBY-FM, KPIG-FM
Univision Radio: KSQL(FM)
Wolfhouse Radio Group Inc.: KEXA(FM), KMJV(FM), KRAY-FM, KTGE

Montgomery, AL: Rank 151
MSA: 290,700
Clear Channel Communications Inc.: WHLW(FM), WWMG(FM), WZHT-FM
Cumulus Media Inc.: WHHY-FM, WMSP, WNZZ, WXFX-FM
GHB Radio Group: WMGY
Tiger Communications Inc.: WTGZ(FM)

Montpelier-Barre-St. Johnsbury, VT: Rank 266
MSA: 109,300

Morgantown-Clarksburg-Fairmont, WV: Rank 175
MSA: 224,700
WFGM-FM
Burbach Broadcasting Group: WGIE(FM), WGYE(FM), WOBG, WOBG-FM, WXKX(AM)
McGraw/Elliott Group Stations: WBRB(FM)
Tschudy Broadcast Group: WPDX(AM), WPDX-FM, WZST-FM
West Virginia Radio Corp.: WAJR, WAJR-FM, WVAQ-FM, WWLW(FM)

Morristown, NJ: Rank 112
MSA: 416,500
Greater Media Inc.: WMTR(AM)

Muncie-Marion, IN: Rank 212
MSA: 170,500
Citadel Broadcasting Corp.: WMDH-FM

Muskegon, MI: Rank 233
MSA: 148,000
American Family Radio: WMCQ(FM)
Citadel Broadcasting Corp.: WVIB(FM)
Clear Channel Communications Inc.: WKBZ(AM), WMUS(FM), WSHZ(FM)
Cumulus Media Inc.: WMHG(AM)

Myrtle Beach, SC: Rank 161
MSA: 257,200
Cumulus Media Inc.: WDAI-FM, WIQB(AM), WSEA(FM), WSYN-FM, WXJY-FM, WYAK-FM
NextMedia Group Inc.: WKZQ-FM, WMYB(FM), WRNN(AM), WRNN-FM, WYAV(FM)
Qantum Communications Corp.: WGTR-FM, WQSD(FM), WWXM(FM)

Nashville, TN: Rank 44
MSA: 1,158,800
Bible Broadcasting Network: WYFN
Bott Radio Network: WCRT(AM)
Citadel Broadcasting Corp.: WGFX-FM, WKDF-FM
Clear Channel Communications Inc.: WLAC, WRVW-FM, WSIX-FM, WUBT(FM)
Cumulus Media Inc.: WQQK-FM, WRQQ(FM), WSM-FM, WWTN-FM
Davidson Media Group LLC: WNSG(AM), WNVL(AM)
F W Robbert Broadcasting Co. Inc.: WNQM
Grace Broadcasting Services Inc.: WFGZ-FM
Salem Communications Corp.: WBOZ(FM), WFFH(FM), WFFI(FM), WVRY-FM
South Central Communications Corp.: WCJK(FM), WJXA-FM
Southern Wabash Communications Corp.: WMGC, WNSR
The Cromwell Group Inc.: WBUZ(FM), WVNS-FM
WAY-FM Media Group Inc.: WAYM(FM)

Nassau-Suffolk, NY (Long Island): Rank 18
MSA: 2,373,900

New Bedford-Fall River, MA: Rank 177
MSA: 222,600

New Haven, CT: Rank 109
MSA: 419,200
Buckley Broadcasting Corp.: WMMW
Clear Channel Communications Inc.: WAVZ, WELI, WKCI-FM
Cox Radio Inc.: WPLR-FM

New London, CT: Rank 173
MSA: 228,600
Citadel Broadcasting Corp.: WSUB, WXLM(FM)
Hall Communications Inc.: WCTY-FM, WICH, WILI-FM, WKNL(FM), WNLC-FM

New Orleans, LA: Rank 57
MSA: 864,100
KLRZ-FM, WBOK, WPRF(FM), WYLK(FM)
ABC Inc.: WBYU
Citadel Broadcasting Corp.: KKND-FM, KMEZ-FM, WDVW(FM), WMTI(FM)
Clear Channel Communications Inc.: KYRK(FM), WNOE-FM, WODT, WQUE-FM, WRNO-FM, WYLD, WYLD-FM
Covenant Network: WCKW(AM)
Davidson Media Group LLC: WFNO
Entercom Communications Corp.: WEZB-FM, WKBU(FM), WLMG(FM), WWL(AM), WWL-FM, WWWL(AM)
F W Robbert Broadcasting Co. Inc.: WVOG
GHB Radio Group: WIST(AM)
Southwest Broadcasting Inc.: WJSH(FM)
Spotlight Broadcasting LLC: KAGY
Sunburst Media-Louisiana LLC: KCIL(FM)

New York: Rank 1
MSA: 15,291,100
WQXR-FM, WVOX
ABC Inc.: WEPN(AM)
Access.1 Communications Corp.: WWRL
Back Bay Broadcasters LLC: WBAZ(FM), WBEA(FM), WEHN(FM)
Barnstable Corporation: WBZO-FM, WHLI, WKJY-FM, WMJC(FM), WRCN-FM
Buckley Broadcasting Corp.: WOR
CBS Radio: WCBS, WCBS-FM, WFAN, WINS, WWFS(FM), WXRK(FM)
Citadel Broadcasting Corp.: WABC(AM), WMOS(FM), WPLJ(FM)

Clear Channel Communications Inc.: WALK(AM), WAXQ-FM, WHTZ-FM, WKTU(FM), WLTW-FM, WWPR-FM
Cox Radio Inc.: WBAB(FM), WBLI-FM, WCTZ(FM), WFOX(FM), WGBB, WHFM-FM, WNLK(AM), WSTC(AM)
Cumulus Media Inc.: WEBE-FM, WFAS-FM
Emmis Communications Corp.: WQCD(FM), WQHT-FM, WRKS(FM)
Family Stations Inc.: WFME-FM
Greater Media Inc.: WCTC(AM), WDHA-FM, WMGQ(FM), WWTR(AM)
Inner City Broadcasting: WBLS-FM, WLIB
Millennium Radio Group LLC: WADB, WJLK(FM)
Multicultural Radio Broadcasting Inc.: WJDM, WKDM(AM), WNSW, WNYG, WPAT, WZRC
Pacifica Foundation Inc.: WBAI-FM
Pamal Broadcasting Ltd.: WLNA(AM), WXPK(FM)
Pillar of Fire Inc.: WAWZ(FM)
Polnet Communications Ltd.: WLIM, WRKL
Press Communications L.L.C.: WHTG, WHTG-FM, WWZY-FM
Rose City Radio Corp.: WSNR(AM)
Salem Communications Corp.: WMCA, WWDJ
Spanish Broadcasting System Inc.: WPAT-FM, WSKQ-FM
The Morey Organization Inc.: WBZB(FM), WDRE(FM), WLIR-FM
Universal Broadcasting of New York Inc.: WTHE(AM), WVNJ(AM)
Univision Radio: WADO, WCAA-FM, WQBU-FM

Newburgh-Middletown, NY (Mid-Hudson Valley): Rank 140
MSA: 314,800
Clear Channel Communications Inc.: WRWC(FM)
Cumulus Media Inc.: WALL, WZAD-FM
Sunrise Broadcasting Corp.: WGNY, WGNY-FM
WAMC/Northeast Public Radio: WOSR-FM

Norfolk-Virginia Beach-Newport News, VA: Rank 41
MSA: 1,327,600
ABC Inc.: WHKT, WRJR(AM)
Baker Family Stations: WKGM
Bible Broadcasting Network: WYFI-FM
Chesapeake-Portsmouth Broadcasting Corp.: WTJZ
Clear Channel Communications Inc.: WCDG(FM), WJCD(FM), WKUS(FM), WOWI-FM
Davidson Media Group LLC: WVXX(AM)
Entercom Communications Corp.: WNVZ-FM, WPTE-FM, WVKL-FM, WWDE-FM
MAX Media L.L.C.: WCMS(AM), WGH-FM, WVBW(FM), WXEZ-FM, WXMM(FM)
Red Zebra Holdings LLC: WXTG(FM)
Saga Communications Inc.: WAFX-FM, WJOI
Sinclair Communications Inc.: WNIS, WNRJ(FM), WPYA(FM), WROX-FM, WTAR(AM)
Willis Broadcasting Corp.: WCPK, WGPL, WPCE
Word Broadcasting Network Inc.: WYRM(AM)

Odessa-Midland, TX: Rank 187
MSA: 202,600
American Family Radio: KBMM(FM)
Cumulus Media Inc.: KBAT(FM), KMND, KODM-FM, KRIL, KZBT(FM)
Double O Radio L.L.C.: KHKX(FM), KMCM-FM, KQRX-FM
EMF Broadcasting: KLVW(FM)
Family Life Communications Inc.: KFLB(AM), KFLB-FM
GAP Broadcasting LLC: KCHX(FM), KCRS(AM), KCRS-FM, KFZX(FM), KMRK-FM
La Promesa Foundation: KJBC(AM) Oklahoma City, OK: Rank 48
MSA: 1,075,700
KKWD(FM)
Bott Radio Network: KQCV, KQCV-FM
Citadel Broadcasting Corp.: KATT-FM, KYIS-FM, WKY, WWLS, WWLS-FM
Clear Channel Communications Inc.: KHBZ-FM, KTOK(AM), KTST-FM, KXXY-FM
EMF Broadcasting: KOKF-FM, KYLV-FM
Family Worship Center Church Inc.: KMFS(AM)
One Ten Broadcast Group Inc.: KIRC-FM
Perry Publishing & Broadcasting Co.: KRMP(AM)
Renda Broadcasting Corp.: KMGL-FM, KOKC(AM), KOMA(FM), KRXO-FM
Tyler Media Broadcasting Corp.: KOCY(AM), KOJK(FM), KTLR(AM), KTUZ-FM

Olean, NY: Rank 215
MSA: 167,600

Omaha-Council Bluffs, NE-IA: Rank 72
MSA: 617.600
Clear Channel Communications Inc.: KFAB, KGOR-FM, KHUS(FM), KQBW(FM), KXKT(FM)
Cochise Broadcasting LLC: KOMJ(AM)
EMF Broadcasting: KMLV(FM)
Journal Communications Inc.: KEZO-FM, KKCD-FM, KQCH(FM), KSRZ-FM, KXSP(AM)
Nebraska Rural Radio Association: KTIC-FM
Salem Communications Corp.: KCRO, KGBI-FM, KOTK(AM)
Waitt Omaha LLC: KKAR, KLTQ(FM), KOIL(AM), KOPW(FM), KOZN(AM), KQKQ-FM, KYDZ(AM)
Wilkins Communications Network Inc.: KLNG(AM)

Orlando, FL: Rank 34
MSA: 1,448,600
ABC Inc.: WDYZ(AM)
CBS Radio: WJHM-FM, WOCL-FM, WOMX-FM
Clear Channel Communications Inc.: WFLF(AM), WJRR-FM, WMGF(FM), WQTM(AM), WRUM(FM), WTKS-FM, WXXL-FM
Cox Radio Inc.: WCFB-FM, WDBO, WHTQ-FM, WMMO-FM, WPYO(FM)
Genesis Communications Inc.: WAMT(AM), WHOO(AM)
J&V Communications Inc.: WOTS, WPRD, WSDO(AM)
Rama Communications Inc.: WLAA(AM), WNTF, WOKB(AM)
Salem Communications Corp.: WHIM, WTLN

Oxnard-Ventura, CA: Rank 120
MSA: 402,600
Cumulus Media Inc.: KBBY-FM, KHAY-FM, KVEN
Entravision Communications Corp.: KSSC(FM)
Gold Coast Broadcasting LLC: KCAQ(FM), KFYV(FM), KKZZ(AM), KOCP-FM, KUNX(AM), KVTA
Lazer Broadcasting Corp.: KLJR-FM, KOXR
Salem Communications Corp.: KDAR(FM)

Palm Springs, CA: Rank 137
MSA: 318,800
KWXY-FM
CBS Radio: KEZN-FM
Entravision Communications Corp.: KLOB-FM
Morris Radio LLC: KDGL(FM), KFUT(AM), KKUU(FM), KNWQ(AM), KNWZ(AM), KXPS(AM)
News-Press & Gazette Co.: KESQ, KUNA-FM
RR Broadcasting: KDES-FM, KGAM, KPSI, KPSI-FM, KPTR(AM)

Panama City, FL: Rank 238
MSA: 140,600
Clear Channel Communications Inc.: WDIZ, WFBX(FM), WPAP-FM, WPBH(FM)
Double O Radio L.L.C.: WAKT-FM, WASJ(FM), WPFM-FM, WRBA(FM)
Family Life Communications Inc.: WJTF-FM
Family Worship Center Church Inc.: WFFL(FM)
Magic Broadcasting LLC: WILN-FM, WPCF(AM), WVVE(FM), WYOO-FM, WYYX-FM
Williams Communications Inc.: WLTG

Parkersburg-Marietta, WV-OH: Rank 250
MSA: 128,500
Burbach Broadcasting Group: WADC, WGGE(FM), WHBR-FM, WRZZ-FM, WVNT(AM), WXIL-FM
Clear Channel Communications Inc.: WDMX-FM, WHNK(AM), WLTP(AM), WNUS-FM, WRVB-FM

Pensacola, FL: Rank 124
MSA: 379,800
Clear Channel Communications Inc.: WTKX-FM, WYCL-FM
Cumulus Media Inc.: WCOA
Pamal Broadcasting Ltd.: WMEZ-FM, WXBM-FM
Wilkins Communications Network Inc.: WVTJ(AM)

Peoria, IL: Rank 149
MSA: 295,800
Illinois Bible Institute Inc.: WCIC-FM
Independence Media Holdings LLC: WPIA(FM), WWCT(FM), WXMP(FM)
Regent Communications Inc.: WFYR-FM, WGLO-FM, WIXO(FM), WVEL, WZPW(FM)
Triad Broadcasting Co. L.L.C.: WDQX(FM), WIRL(AM), WMBD, WPBG-FM, WSWT-FM, WXCL(FM)

Philadelphia: Rank 7
MSA: 4,360,200
WURD(AM)
ABC Inc.: WWJZ
Beasley Broadcast Group Inc.: WRDW-FM, WTMR, WWDB(AM), WXTU-FM
CBS Radio: KYW, WIP, WOGL(FM), WPHT
Clear Channel Communications Inc.: WDAS-FM, WIOQ-FM, WISX(FM), WRFF(FM), WUBA(AM), WUSL-FM
Davidson Media Group LLC: WEMG(AM)
EMF Broadcasting: WKVP(FM)
Family Stations Inc.: WKDN-FM
GHB Radio Group: WNAP
Great Scott Broadcasting: WPAZ
Greater Media Inc.: WBEN-FM, WMMR-FM, WPEN
Inner City Broadcasting: WHAT
Langer Broadcasting Group L.L.C.: WFYL(AM)
Mountain Broadcasting Corp.: WPWA
Multicultural Radio Broadcasting Inc.: WTTM(AM)
Radio One Inc.: WPHI-FM, WPPZ-FM, WRNB(FM)
Route 81 Radio LLC: WCOJ(AM)
Salem Communications Corp.: WFIL(AM), WNTP(AM)

Phoenix, AZ: Rank 15
MSA: 3,058,000
KASA, KMVA(FM)
ABC Inc.: KMIK
Bonneville International Corporation: KMVP, KPKX(FM), KTAR(AM), KTAR-FM
CBS Radio: KMLE-FM, KOOL-FM, KZON(FM)
Clear Channel Communications Inc.: KESZ-FM, KGME(AM), KMXP-FM, KNIX-FM, KOY, KYOT-FM, KZZP-FM
Communicom Broadcasting LLC: KXEG(AM), KXXT(AM)
EMF Broadcasting: KLVK(FM)
Entravision Communications Corp.: KDVA(FM), KLNZ-FM, KMIA(AM), KVVA-FM
Family Life Communications Inc.: KFLR-FM
Family Stations Inc.: KPHF-FM
Multicultural Radio Broadcasting Inc.: KIDR
Riviera Broadcast Group LLC: KEDJ(FM), KKFR(FM)
Salem Communications Corp.: KKNT(AM), KPXQ(AM)
Sandusky Radio: KAZG(AM), KDKB(FM), KDUS
Univision Radio: KHOT-FM, KHOV-FM, KOMR(FM), KQMR(FM)

Pittsburg, KS (Southeast Kansas): Rank 243
MSA: 136,300

Pittsburgh, PA: Rank 24
MSA: 1,998,800
ABC Inc.: WEAE
Birach Broadcasting Corp.: WWCS
Broadcast Communications Inc.: WKFB(AM), WKHB(AM)
Butler County Radio Network Inc.: WBUT, WISR, WLER-FM
CBS Radio: KDKA, WDSY-FM, WTZN-FM, WZPT-FM
Clear Channel Communications Inc.: WBGG(AM), WDVE-FM, WKST-FM, WPGB(FM), WXDX-FM
Inner City Broadcasting: WLFP(AM)
Keymarket Communications LLC: WASP, WFGI(AM), WOGF(FM), WOGG(FM), WPKL(FM), WYJK(AM)
Langer Broadcasting Group L.L.C.: WPYT(AM)
Renda Broadcasting Corp.: WGSM(FM), WJAS
Salem Communications Corp.: WORD-FM
Sheridan Broadcasting Corp.: WAMO(AM), WAMO-FM, WPGR(AM)
Wilkins Communications Network Inc.: WWNL(AM)

Portland, ME: Rank 167
MSA: 240,600
WCLZ(FM), WCYI(FM)
Atlantic Coast Radio L.L.C.: WJAE, WJJB(AM), WLOB, WRED-FM
Blount Communications Group: WBCI-FM
Bob Bittner Broadcasting Inc.: WJTO
Citadel Broadcasting Corp.: WBLM-FM, WCYY-FM, WHOM-FM, WJBQ-FM
Gleason Radio Group: WOXO-FM, WTBM(FM)
Nassau Broadcasting Partners L.P.: WBQW-FM, WFNK(FM), WHXR(FM), WLVP(AM), WTHT(FM)
Saga Communications Inc.: WBAE, WGAN(AM), WMGX(FM), WPOR(FM), WYNZ-FM, WZAN

Portland, OR: Rank 23
MSA: 2,001,600
ABC Inc.: KDZR(AM)
Bustos Media LLC: KGDD(AM), KMUZ, KSZN(AM), KXMG(AM)
CBS Radio: KCMD(AM), KINK(FM), KLTH(FM), KUFO-FM, KUPL-FM, KVMX(FM)
Churchill Communications LLC: KXPD(AM)
Clear Channel Communications Inc.: KEX, KIJZ(FM), KKCW-FM, KPOJ(AM)
Crawford Broadcasting Co.: KKPZ(AM)
EMF Broadcasting: KLVP-FM, KZRI(FM)
Entercom Communications Corp.: KFXX(AM), KGON-FM, KKSN(AM), KNRK-FM, KRSK(FM), KTRO(AM), KYCH-FM
Pamplin Broadcasting: KKAD(AM), KPAM
Rose City Radio Corp.: KXJM-FM, KXL
Salem Communications Corp.: KPDQ-FM, KRYP(FM)

Portsmouth-Dover-Rochester, NH: Rank 116
MSA: 412,200
Citadel Broadcasting Corp.: WOKQ-FM, WSAK(FM), WSHK(FM)
Clear Channel Communications Inc.: WERZ-FM, WGIP, WHEB-FM, WQSO-FM, WUBB-FM
Costa-Eagle Radio Ventures L.P.: WCCM(AM)
Phoenix Media Communications Group: WPHX-FM

Poughkeepsie, NY: Rank 163
MSA: 255,100
Clear Channel Communications Inc.: WBWZ-FM, WKIP, WPKF(FM), WRNQ-FM, WRWD-FM
Cumulus Media Inc.: WCZX-FM, WEOK, WRRB(FM)
Pamal Broadcasting Ltd.: WBNR, WBPM(FM), WSPK-FM

Providence-Warwick-Pawtucket, RI: Rank 38
MSA: 1,393,500
WNRI, WRNI-FM
ABC Inc.: WDDZ(AM)
Anastos Media Group Inc.: WPEP
Astro Tele-Communications Corp. Rhode Island: WADK, WJZS(FM), WKFD(AM)
Blount Communications Group: WARV
Citadel Broadcasting Corp.: WBSM, WFHN-FM, WPRO, WSKO, WSKO-FM, WWKX-FM
Clear Channel Communications Inc.: WHJJ, WHJY-FM, WSNE-FM, WWBB-FM

Davidson Media Group LLC: WKKB(FM)
Entercom Communications Corp.: WEEI-FM
Hall Communications Inc.: WCTK(FM), WLKW(AM), WNBH

Pueblo, CO: Rank 254
MSA: 127,700
KNKN-FM, KRLN(AM), KRMX(AM)
Bahakel Communications: KYZX(FM)
Clear Channel Communications Inc.: KCCY(FM), KCSJ, KDZA-FM, KGHF
Family Stations Inc.: KFRY(FM)
Latino Communications LLC: KAVA
Timothy C. Cutforth Stns: KCEG(AM)

Puerto Rico: Rank 13
MSA: 3,296,800
WNIK, WNIK-FM, WQBS
Bestov Broadcasting Inc.: WIAC, WIAC-FM, WYAC(AM)
Calvary Evangelistic Mission Inc.: WBMJ, WCGB, WIVV
International Broadcasting Corp.: WCHQ(AM), WEKO(AM), WIBS, WRSJ(AM), WTIL, WVOZ-FM, WXRF
Media Power Group Inc.: WDEP(AM), WKFE, WLEY, WSKN(AM)
Spanish Broadcasting System Inc.: WCMA-FM, WEGM(FM), WIOA(FM), WIOB-FM, WIOC-FM, WMEG-FM, WNOD(FM), WODA(FM), WZET(FM), WZMT-FM, WZNT-FM
Univision Radio: WKAQ, WUKQ(AM), WUKQ-FM, WYEL(AM)
Uno Radio Group: WCMN, WFDT(FM), WFID-FM, WIVA-FM, WLEO(AM), WMIO(FM), WNEL, WORA, WPRM-FM, WPRP, WRIO-FM, WUNO(AM)

Quad Cities, IA-IL (Davenport-Rock Island-Moline): Rank 143
MSA: 304,300
WKBF
Clear Channel Communications Inc.: KCQQ(FM), KMXG(FM), KUUL(FM), WFXN(AM), WLLR-FM, WOC
Cumulus Media Inc.: KBEA-FM, KBOB-FM, KJOC, KQCS(FM), WXLP-FM
EMF Broadcasting: WLKU(FM)
Miller Media Group: WGEN, WKEI(AM), WYEC(FM)

Raleigh-Durham, NC: Rank 43
MSA: 1,184,200
WDUR
Capitol Broadcasting Co. Inc.: WCMC-FM, WRAL(FM)
Clear Channel Communications Inc.: WDCG-FM, WKSL(FM), WRDU-FM, WRVA-FM
Curtis Media Group: WBBB-FM, WCLY, WDNC(AM), WDOX(AM), WFMC, WKXU(FM), WPTF, WQDR-FM, WWMY(FM), WYMY(FM), WYRN(AM)
Davidson Media Group LLC: WRJD(AM), WTIK
Radio La Grande: WGSB, WLLQ(AM), WRTG
Radio One Inc.: WFXC-FM, WFXK-FM, WNNL-FM, WQOK-FM
Truth Broadcasting Corp.: WDRU(AM)

Rapid City, SD: Rank 276
MSA: 99,500
American Family Radio: KASD(FM)
Bethesda Christian Broadcasting: KLMP(FM), KSLT-FM, KTPT(FM)
CSN International: KWRC(FM)
Duhamel Broadcasting Enterprises: KDDX(FM), KOTA
Family Stations Inc.: KQFR(FM)
Haugo Broadcasting Inc.: KSQY-FM, KTOQ
Schurz Communications Inc.: KBHB, KFXS-FM, KKLS, KKMK-FM, KOUT-FM, KRCS-FM

Reading, PA: Rank 130
MSA: 342,800
Citadel Broadcasting Corp.: WIOV
Clear Channel Communications Inc.: WRAW(AM), WRFY-FM

Redding, CA: Rank 228
MSA: 158,100
EMF Broadcasting: KKRO(FM), KLVB(FM)
Fritz Communications Inc.: KESR(FM), KEWB-FM, KHRD(FM), KNCQ-FM
Mapleton Communications LLC: KNNN(FM), KNRO(AM), KQMS, KRRX-FM, KSHA-FM

Reno, NV: Rank 123
MSA: 384,900
Americom: KBZZ(AM), KJFK(AM), KLCA-FM, KRNO(FM), KZTQ(FM)
Azteca Broadcasting Corp.: KXEQ
Citadel Broadcasting Corp.: KBUL-FM, KKOH, KNEV-FM
Entravision Communications Corp.: KRNV-FM
IHR Educational Broadcasting: KIHM(AM)
Lotus Communications Corp.: KDOT-FM, KHIT, KOZZ-FM, KPLY(AM), KUUB(FM)
Wilks Broadcast Group LLC: KJZS(FM), KRZQ-FM, KTHX-FM, KURK(FM)

Richmond, VA: Rank 55
MSA: 916,400
WHAP
ABC Inc.: WDZY
American Family Radio: WRIH(FM)
Chesapeake-Portsmouth Broadcasting Corp.: WLES(AM)
Clear Channel Communications Inc.: WBTJ(FM), WRNL, WRVA, WRVQ-FM, WRXL-FM, WTVR-FM
Cox Radio Inc.: WDYL-FM, WKHK-FM, WKLR-FM, WMXB-FM
Davidson Media Group LLC: WLEE(AM), WREJ, WVNZ(AM)
Main Line Broadcasting LLC: WARV-FM, WBBT-FM, WLFV(FM)
Mountain Broadcasting Corp.: WBTK(AM)
Radio One Inc.: WCDX-FM, WKJM(FM), WKJS(FM), WPZZ(FM), WROU(AM)
Red Zebra Holdings LLC: WXGI

Riverside-San Bernardino, CA: Rank 25
MSA: 1,806,800
Anaheim Broadcasting Corp.: KCAL-FM, KOLA(FM)
CBS Radio: KFRG-FM
Clear Channel Communications Inc.: KDIF, KGGI-FM, KKDD, KTDD(AM)
Hi-Favor Broadcasting LLC: KEZY(AM)
Lazer Broadcasting Corp.: KCAL, KXRS-FM, KXSB-FM
Magic Broadcasting LLC: KRQB(FM)
Salem Communications Corp.: KTIE(AM)

Roanoke-Lynchburg, VA: Rank 115
MSA: 412,300
3 Daughters Media Inc.: WGMN, WVGM
Centennial Broadcasting LLC: WLEQ(FM), WLNI-FM, WZZI-FM, WZZU(FM)
Clear Channel Communications Inc.: WJJS-FM, WJJX-FM, WROV-FM, WSNV(FM), WSNZ(FM), WYYD-FM, WZBL(FM)
Mel Wheeler Inc.: WFIR(AM), WSLC-FM, WSLQ-FM, WVBE(AM), WVBE-FM, WXLK-FM
Perception Media Group Inc.: WCQV(AM), WWWR
Positive Alternative Radio Inc.: WPAR-FM, WRXT-FM, WTTX-FM
Truth Broadcasting Corp.: WLVA

Rochester, MN: Rank 232
MSA: 151,400
Clear Channel Communications Inc.: KRCH(FM), KWEB(AM)
Cumulus Media Inc.: KOLM(AM), KROC(AM), KROC-FM, KWWK(FM)

Rochester, NY: Rank 53
MSA: 936,000
WACK(AM), WASB
CBS Radio: WCMF-FM, WPXY-FM, WRMM-FM
Clear Channel Communications Inc.: WDVI(FM), WFXF(FM), WHAM, WHTK, WKGS-FM, WVOR(FM)
Crawford Broadcasting Co.: WLGZ(AM), WRCI(FM)
Entercom Communications Corp.: WBEE-FM, WBZA(FM), WFKL(FM)
Holy Family Communications: WHIC(AM)
M.B. Communications: WFLK-FM, WYLF(AM)

Rockford, IL: Rank 154
MSA: 287,300
Cumulus Media Inc.: WKGL-FM, WROK, WXXQ-FM, WZOK-FM
Good Karma Broadcasting L.L.C.: WTJK
Maverick Media LLC: WGFB(FM), WNTA, WRTB(FM), WXRX-FM

Rocky Mount-Wilson, NC: Rank 202
MSA: 185,800

Sacramento, CA: Rank 27
MSA: 1,785,400
KSAC(AM)
ABC Inc.: KIID(AM)
Bustos Media LLC: KTTA-FM
CBS Radio: KHTK, KNCI-FM, KQJK(FM), KSFM-FM, KYMX-FM, KZZO-FM
Cherry Creek Radio LLC: KRLT-FM
Citadel Broadcasting Corp.: KWYL(FM)
Clear Channel Communications Inc.: KFBK, KGBY-FM, KHYL-FM, KSTE
EMF Broadcasting: KLRS(FM), KLVS(FM)
Entercom Communications Corp.: KCTC, KDND(FM), KRXQ-FM, KSEG-FM, KSSJ-FM, KWOD-FM
Entravision Communications Corp.: KBMB(FM), KNTY(FM), KRCX-FM, KXSE(FM)
Family Stations Inc.: KEAR-FM
First Broadcasting Operating Inc.: KCCL(FM)
IHR Educational Broadcasting: KAHI, KSMH(AM)
Lazer Broadcasting Corp.: KSRN(FM)
Salem Communications Corp.: KFIA, KTKZ

Saginaw-Bay City-Midland, MI: Rank 131
MSA: 339,200
WYLZ-FM
Citadel Broadcasting Corp.: WHNN-FM, WILZ-FM, WIOG(FM), WKQZ-FM
EMF Broadcasting: WLKB(FM)
MacDonald Broadcasting Co.: WKCQ-FM, WMJO(FM), WSAG(FM), WSAM
Meredith Broadcasting Group, Meredith Corp.: WNEM(AM)
NextMedia Group Inc.: WCEN-FM, WSGW, WSGW-FM, WTLZ-FM

Salisbury-Ocean City, MD: Rank 145
MSA: 302,000
Birach Broadcasting Corp.: WGOP(AM)
Clear Channel Communications Inc.: WDKZ(FM), WLBW-FM, WOSC-FM, WSBY-FM, WTGM, WWFG-FM
Delmarva Broadcasting Co.: WAFL-FM, WICO(AM), WICO-FM, WNCL(FM), WQJZ-FM, WXJN-FM, WXMD(FM), WYUS
Great Scott Broadcasting: WGBG-FM, WJWL, WKDB(FM), WKHW-FM, WOCQ-FM, WZBH-FM, WZEB(FM)

Salt Lake City-Ogden-Provo, UT: Rank 31
MSA: 1,554,200
KALL(AM)
3 Point Media: KHTB(FM), KOAY(FM)
ABC Inc.: KWDZ(AM)
Azteca Broadcasting Corp.: KSVN
Bible Broadcasting Network: KYFO-FM
Bonneville International Corporation: KRSP-FM, KSFI-FM, KSL, KSL-FM, KUTR(AM)
Bustos Media LLC: KDUT(FM), KXTA(AM)
Carlson Communications International: KDYL(AM)
Citadel Broadcasting Corp.: KBEE(FM), KBER(FM), KENZ(FM), KJQS(AM), KKAT(AM), KKAT-FM
Clear Channel Communications Inc.: KJMY(FM), KNRS, KODJ-FM, KOSY-FM, KXRV(FM), KZHT(FM)
Faith Communications Corp.: KANN
Family Stations Inc.: KUFR-FM
Legacy Communications Corporation: KOGN(AM)
Millcreek Broadcasting L.L.C.: KUDD(FM), KUUU(FM)
Simmons Media Group: KEGH(FM), KOVO, KXOL, KXRK-FM, KYMV(FM), KZNS(AM)

San Angelo, TX: Rank 286
MSA: 86,000
Criswell Communications: KCRN, KCRN-FM
Double O Radio L.L.C.: KELI-FM, KGKL, KGKL-FM, KKCN(FM), KNRX(FM)
EMF Broadcasting: KLRW(FM), KLTP(FM), KNAR(FM)
Foster Communications Co. Inc.: KCLL(FM), KIXY-FM, KKSA, KWFR-FM

San Antonio, TX: Rank 29
MSA: 1,586,000
KYTY(AM)
ABC Inc.: KRDY(AM)
Bible Broadcasting Network: KYFS-FM
Border Media Partners LLC: KJXK(FM), KSAH(AM), KTFM(FM), KTSA, KZDC
Clear Channel Communications Inc.: KAJA(FM), KQXT-FM, KTKR, KXXM-FM, WOAI
Cox Radio Inc.: KCYY(FM), KISS-FM, KKYX, KONO, KONO-FM, KPWT(FM), KSMG(FM)
GAP Broadcasting LLC: KIXS-FM
La Promesa Foundation: KJMA(FM), KWMF(AM)
Lotus Communications Corp.: KZEP-FM
Martin Broadcasting Inc.: KCHL
Salem Communications Corp.: KLUP(AM), KSLR
SIGA Broadcasting Corp.: KTMR
Univision Radio: KAHL(AM), KBBT(FM), KCOR(AM), KCOR-FM, KLTO-FM, KXTN-FM
Victoria RadioWorks Ltd.: KBAR-FM, KNAL(AM)

San Diego, CA: Rank 17
MSA: 2,497,000
KFMB, XETRA
Astor Broadcast Group: KCEO, KFSD(AM)
CBS Radio: KSCF(FM), KYXY-FM
Clear Channel Communications Inc.: KGB-FM, KHTS-FM, KIOZ-FM, KLSD(AM), KMYI(FM), KOGO(AM), KUSS(FM)
EMF Broadcasting: KLVJ-FM
Entravision Communications Corp.: KSSD(FM)
Hi-Favor Broadcasting LLC: KSDO
Lincoln Financial Media: KBZT(FM), KIFM-FM, KSON, KSON-FM, KSOQ-FM
Salem Communications Corp.: KCBQ, KPRZ
Univision Radio: KLNV-FM, KLQV-FM

San Francisco: Rank 4
MSA: 5,891,900
ABC Inc.: KMKY
Bicoastal Media L.L.C.: KLLK
Bonneville International Corporation: KBWF(FM), KDFC-FM, KOIT, KOIT-FM
CBS Radio: KCBS, KFRC-FM, KITS-FM, KLLC-FM, KMVQ-FM, KYCY
Citadel Broadcasting Corp.: KGO, KSFO
Clear Channel Communications Inc.: KCNL(FM), KIOI-FM, KISQ-FM, KKGN(AM), KKSF-FM, KMEL-FM, KNEW, KSJO-FM, KYLD-FM
Coast Radio Company Inc.: KKDV(FM), KKIQ-FM, KUIC-FM
Cumulus Media Partners LLC: KFFG-FM, KFOG-FM, KNBR, KSAN-FM, KTCT
EMF Broadcasting: KLVR-FM
Family Stations Inc.: KEAR(AM)
Inner City Broadcasting: KBLX-FM, KVTO, KVVN
Mapleton Communications LLC: KPIG(AM)
Maverick Media LLC: KMHX(FM), KSRO, KVRV(FM), KXFX-FM
Moon Broadcasting: KRRS, KTOB(AM)

Multicultural Radio Broadcasting Inc.: KATD, KEST, KIQI
NextMedia Group Inc.: KBAY(FM), KEZR(FM)
Pacifica Foundation Inc.: KPFA-FM
Pappas Telecasting Companies: KTRB(AM)
Salem Communications Corp.: KFAX, KNTS(AM)
Sinclair Communications Inc.: KSXY(FM)
Spanish Broadcasting System Inc.: KRZZ(FM)
Univision Radio: KBRG(FM), KLOK, KSOL(FM), KVVF(FM), KVVZ(FM)

San Jose, CA: Rank 35
MSA: 1,436,400
Multicultural Radio Broadcasting Inc.: KSJX(AM)

San Luis Obispo, CA: Rank 172
MSA: 229,500
American General Media: KIQO-FM, KKAL(FM), KKJG-FM, KZOZ-FM
Clear Channel Communications Inc.: KSLY-FM, KSTT-FM, KURQ(FM), KVEC
EMF Broadcasting: KLVH(FM)
Lazer Broadcasting Corp.: KLMM(FM)
Mapleton Communications LLC: KPYG(FM), KXTZ-FM, KYNS(AM)

Santa Barbara, CA: Rank 211
MSA: 170,900
KIST(AM), KTMS
Cumulus Media Inc.: KMGQ(FM), KRUZ(FM), KVYB(FM)
Lazer Broadcasting Corp.: KZER(AM)
Point Broadcasting Company: KBKO, KIST-FM, KSBL-FM, KSPE-FM, KTYD-FM

Santa Fe, NM: Rank 237
MSA: 140,700
KVSF(AM)
American General Media: KKIM-FM, KTRC(AM), KZNM(FM)
EMF Broadcasting: KBAC(FM), KSFQ-FM

Santa Maria-Lompoc, CA: Rank 207
MSA: 173,100
American General Media: KBOX(FM), KPAT-FM, KRQK-FM
Clear Channel Communications Inc.: KSMA, KSMY(FM), KSNI-FM, KXFM-FM
Emerald Wave Media: KIDI-FM, KRTO(FM), KTAP
Family Stations Inc.: KHFR(FM)
Knight Broadcasting Inc.: KINF(AM), KSYV-FM, KUHL(AM)
Lazer Broadcasting Corp.: KSBQ

Santa Rosa, CA: Rank 119
MSA: 403,400
Ace Radio Corp.: KGRP(FM)

Sarasota-Bradenton, FL: Rank 73
MSA: 610,100
CBS Radio: WLLD-FM
Clear Channel Communications Inc.: WCTQ(FM), WDDV(AM), WLTQ-FM, WSDV(AM), WSRZ-FM, WTZB(FM)
Cox Radio Inc.: WHPT(FM)
Metropolitan Radio Group Inc.: WBRD, WTMY
Northwestern College & Radio: WSMR-FM
Salem Communications Corp.: WLSS(AM)
The Moody Bible Institute of Chicago: WKZM-FM
Viper Communications Broadcast Group: WENG

Savannah, GA: Rank 158
MSA: 265,300
WRHQ-FM, WWJN(FM)
Bible Broadcasting Network: WYFS-FM
Clear Channel Communications Inc.: WAEV-FM, WLVH-FM, WSOK, WTKS(AM), WYKZ(FM)
Cumulus Media Inc.: WBMQ, WEAS-FM, WIXV-FM, WJCL-FM, WJLG, WTYB(FM), WZAT-FM
MarMac Communications LLC: WSEG(AM)
Tama Broadcasting Inc.: WSSJ(FM), WTHG(FM)
Triad Broadcasting Co. L.L.C.: WFXH(AM), WGCO(FM), WGZR(FM), WLOW(FM)

Seattle-Tacoma, WA: Rank 14
MSA: 3,257,200
ABC Inc.: KKDZ
Bustos Media LLC: KWMG(AM)
CBS Radio: KBKS-FM, KJAQ(FM), KMPS-FM, KPTK(AM), KZOK-FM
Clear Channel Communications Inc.: KFNK(FM), KHHO, KJR, KJR-FM, KUBE-FM
CRISTA Broadcasting: KCIS(AM), KCMS(FM)
Entercom Communications Corp.: KBSG-FM, KIRO, KISW-FM, KKWF(FM), KMTT-FM, KNDD-FM, KTTH(AM)
Family Stations Inc.: KARR
Fisher Communications Inc.: KOMO(AM), KPLZ(FM), KVI(AM)
Multicultural Radio Broadcasting Inc.: KXPA
Premier Broadcasters: KRXY-FM
Salem Communications Corp.: KDOW(AM), KGNW, KKMO, KKOL, KLFE

Sandusky Radio: KIXI, KQMV(FM), KRWM-FM, KWJZ(FM)
Seattle Streaming Radio LLC: KBRO, KNTB

Sebring, FL: Rank 288
MSA: 84,400
Cohan Radio Group Inc.: WITS(AM), WJCM(AM), WWLL(FM), WWOJ(FM), WWTK(AM)

Sheboygan, WI : Rank 278
MSA: 99,000
Midwest Communications Inc.: WBFM(FM), WHBL(AM), WHBZ(FM), WXER(FM)
Mountain Dog Media: WCLB(AM)

Shreveport, LA: Rank 132
MSA: 336,200
Access.1 Communications Corp.: KBTT(FM), KLKL(FM), KOKA, KSYR(FM), KTAL-FM
Amistad Communications Inc.: KASO, KSYB(AM)
Cumulus Media Inc.: KMJJ-FM, KQHN(FM), KRMD-FM, KVMA-FM
GAP Broadcasting LLC: KEEL(AM), KRUF(FM), KTUX-FM, KVKI-FM, KWKH(AM), KXKS-FM
Metropolitan Radio Group Inc.: KIOU, KORI(FM), KTKC-FM

Sioux City, IA: Rank 275
MSA: 100,300
Clear Channel Communications Inc.: KGLI-FM, KMNS, KSEZ-FM, KSFT-FM, KWSL
Powell Broadcasting Co. Inc.: KKMA(FM), KKYY(FM), KSCJ, KSUX(FM)

South Bend, IN: Rank 178
MSA: 220,500
WHLY(AM)
Artistic Media Partners Inc.: WDND(AM), WNDV-FM, WWLV(AM), WZOW(FM)
Federated Media: WAOR-FM, WBYT-FM
Le Sea Broadcasting: WHME-FM
Progressive Broadcasting System Inc.: WFRN-FM
Schurz Communications Inc.: WNSN-FM, WSBT

Spokane, WA: Rank 93
MSA: 491,800
KAZZ(FM), KTRW(AM)
Citadel Broadcasting Corp.: KBBD(FM), KDRK-FM, KEYF(AM), KEYF-FM, KGA, KJRB, KZBD(FM)
Clear Channel Communications Inc.: KCDA(FM), KIXZ-FM, KKZX-FM, KPTQ(AM), KQNT(AM)
Morgan Murphy Stations (Evening Telegram Co): KEZE-FM, KXLY, KZZU-FM
Morris Radio LLC: KWIQ(AM)
Pamplin Broadcasting: KTSL-FM
The Moody Bible Institute of Chicago: KMBI-FM

Springfield, MA: Rank 84
MSA: 529,500
Citadel Broadcasting Corp.: WMAS
Clear Channel Communications Inc.: WHYN, WHYN-FM, WNNZ, WPKX-FM, WRNX-FM
Davidson Media Group LLC: WACM, WSPR
Pamal Broadcasting Ltd.: WPNI(AM)
Saga Communications Inc.: WAQY-FM, WHMP, WHNP(AM), WLZX(FM), WRSI(FM)
Vox Radio Group L.P.: WVEI-FM

Springfield, MO: Rank 142
MSA: 307,500
Bott Radio Network: KSCV(FM)
Clear Channel Communications Inc.: KGMY, KSWF(FM), KTOZ-FM, KXUS-FM
Journal Communications Inc.: KSGF(AM), KSGF-FM, KSPW(FM)
Meyer Communications Inc.: KBFL(AM), KTXR-FM, KWTO, KWTO-FM
The Mid-West Family Broadcast Group: KKLH(FM), KOMG(FM), KOSP(FM), KQRA(FM)

St. Cloud, MN: Rank 218
MSA: 164,800
Leighton Enterprises Inc.: KCML(FM), KNSI, KZPK-FM
Regent Communications Inc.: KKSR(FM), KLZZ-FM, KMXK-FM, KXSS(AM), WJON
Tri-County Broadcasting Inc.: WBHR, WHMH-FM

St. Louis, MO: Rank 20
MSA: 2,282,700
ABC Inc.: WSDZ
Big League Broadcasting LLC: KFNS(AM), KFNS-FM, KRFT(AM)
Birach Broadcasting Corp.: WEW, WIJR(AM)
Bonneville International Corporation: WARH(FM), WIL(AM), WIL-FM, WMVN(FM)
Bott Radio Network: KSIV(AM), KSIV-FM
CBS Radio: KEZK-FM, KMOX, KYKY-FM

Clear Channel Communications Inc.: KATZ(AM), KATZ-FM, KLOU(FM), KMJM-FM, KSD(FM), KSLZ(FM)
Covenant Network: KHOJ(AM), WRYT(AM)
Crawford Broadcasting Co.: KJSL, KSTL
Emmis Communications Corp.: KFTK(FM), KIHT-FM, KPNT-FM, KSHE(FM)
Family Worship Center Church Inc.: KDJR(FM)
Kaspar Broadcasting Group: KFAV-FM, KWRE(AM)
Missouri River Christian Broadcasting Inc.: KGNA-FM
Radio One Inc.: WFUN-FM, WHHL(FM)
Shepherd Group: KJFF(AM)
Simmons Media Group: KSLG(AM), WFFX(AM)
The Curators of the University of Missouri: KWMU-FM

Stamford-Norwalk, CT: Rank 145
MSA: 302,000

State College, PA: Rank 255
MSA: 126,400
2510 Licenses LLC: WBHV-FM, WOWY(FM)
Allegheny Mountain Network Stations: WGMR(FM)
Family Stations Inc.: WXFR(FM)
First Media Radio LLC: WZWW-FM
Forever Broadcasting: WBUS(FM), WLTS(FM), WMAJ, WQWK(FM), WRSC, WSGY(FM)
Magnum Broadcasting Inc.: WBLF, WJOW(FM), WPHB

Stockton, CA: Rank 81
MSA: 555,500
Citadel Broadcasting Corp.: KJOY-FM, KWIN-FM
Clear Channel Communications Inc.: KQOD-FM, KWSX(AM)
Entravision Communications Corp.: KCVR, KMIX-FM
IHR Educational Broadcasting: KWG(AM)

Sunbury-Selinsgrove-Lewisburg, PA: Rank 214
MSA: 167,700
Sunbury Broadcasting Corp.: WMLP(AM), WVLY-FM

Sussex, NJ: Rank 248
MSA: 131,500
Clear Channel Communications Inc.: WNNJ, WNNJ-FM, WSUS-FM

Syracuse, NY: Rank 82
MSA: 555,100
Buckley Broadcasting Corp.: WFBL(AM), WSEN(AM)
Citadel Broadcasting Corp.: WAQX-FM, WLTI(FM), WNTQ(FM)
Clear Channel Communications Inc.: WBBS-FM, WHEN, WPHR-FM, WSYR, WWDG(FM), WWHT-FM, WYYY-FM
Galaxy Communications L.P.: WKRH-FM, WKRL-FM, WSCP, WSGO, WTKV-FM, WTKW-FM, WTLA, WZUN(FM)
Mars Hill Network: WMHR-FM
WOLF Radio Inc.: WOLF, WOLF-FM

Tallahassee, FL: Rank 164
MSA: 253,300
Clear Channel Communications Inc.: WFLA-FM, WTLY(FM), WTNT-FM, WXSR-FM
Cumulus Media Inc.: WGLF(FM), WHBT(AM), WHBX(FM), WWLD(FM)
Opus Media Holdings LLC: WAIB-FM, WEGT(FM), WHTF-FM, WUTL(FM)

Tampa-St. Petersburg-Clearwater, FL: Rank 19
MSA: 2,314,300
ABC Inc.: WWMI
Bible Broadcasting Network: WYFE-FM
CBS Radio: WQYK(AM), WQYK-FM, WRBQ-FM, WYUU(FM)
Clear Channel Communications Inc.: WBTP(FM), WFLA(AM), WFUS(FM), WHNZ(AM), WMTX(FM), WXTB-FM
Cox Radio Inc.: WDUV-FM, WPOI(FM), WSUN-FM, WWRM(FM), WXGL(FM)
Family Stations Inc.: WFTI-FM
Genesis Communications Inc.: WHBO(AM), WWBA
Mega Communications LLC: WLCC, WMGG(AM)
Metropolitan Radio Group Inc.: WRXB
Salem Communications Corp.: WGUL(AM), WTBN(AM), WTWD(AM)
Tama Broadcasting Inc.: WTMP, WTMP-FM
Wagenvoord Advertising Group Inc.: WDCF, WTAN(AM), WZHR(AM)

Terre Haute, IN: Rank 206
MSA: 173,900
Covenant Network: WHOJ(FM)
Crossroads Communications Inc.: WAXI-FM, WBOW(AM), WSDM-FM, WSDX(AM)
Emmis Communications Corp.: WTHI-FM, WWVR-FM
Illinois Bible Institute Inc.: WCRT-FM
Midwest Communications Inc.: WINH(FM), WMGI-FM, WPRS
The Cromwell Group Inc.: WCBH-FM
The Original Company Inc.: WQTY-FM

Texarkana, TX-AR: Rank 263
MSA: 114,400

Arklatex LLC: KBYB(FM), KCMC, KFYX(FM), KTFS(AM), KTOY(FM)
EMF Broadcasting: KKLT(FM)
Family Worship Center Church Inc.: KNRB(FM)
GAP Broadcasting LLC: KKYR-FM, KOSY(AM), KPWW-FM, KYGL-FM
Textron Financial Corp.: KEWL-FM, KKTK(AM), KPGG(FM)
Tower Investment Trust Inc.: KTTY(FM)

The Florida Keys: Rank 294
MSA: 69,900
Clear Channel Communications Inc.: WAIL(FM), WCTH(FM), WEOW(FM), WFKZ(FM), WKEY-FM, WKEZ-FM
Vox Radio Group L.P.: WCNK(FM), WWUS(FM)

Toledo, OH: Rank 87
MSA: 517,200
Clear Channel Communications Inc.: WCWA, WIOT(FM), WPFX-FM, WRVF(FM), WSPD, WVKS-FM
Cumulus Media Inc.: WLQR, WRQN-FM, WRWK(FM), WTOD, WTWR-FM, WWWM-FM, WXKR-FM
Family Stations Inc.: WOTL-FM
Family Worship Center Church Inc.: WJYM
Lake Cities Broadcasting Corp.: WLZZ(FM)
Urban Radio Licenses LLC: WIMX-FM, WJZE-FM

Topeka, KS: Rank 194
MSA: 194,600
KMAJ-FM
American Family Radio: KBUZ(FM)
Bott Radio Network: KCVT(FM)
Cumulus Media Inc.: KDVV-FM, KMAJ, KTOP, KWIC-FM
Family Life Communications Inc.: KJTY-FM
Great Plains Media Inc.: KLZR(FM), KMXN(FM)
Morris Radio LLC: WIBW

Traverse City-Petoskey, MI: Rank 192
MSA: 197,600
Fort Bend Broadcasting Co.: WBNZ(FM), WLDR(AM), WLDR-FM, WOUF(FM)
Good News Media Inc.: WLJN-FM
MacDonald Garber Broadcasting Co.: WKHQ-FM, WLXV(FM), WMBN(AM)
Midwestern Broadcasting Co.: WBCM(FM), WCCW(AM), WJZQ(FM), WTCM(AM)
Northern Star Broadcasting L.L.C.: WGFM(FM), WGFN(FM), WJZJ(FM), WLJZ(FM), WMKC(FM)

Trenton, NJ: Rank 141
MSA: 312,600
Greater Media Inc.: WJJZ(FM)
Millennium Radio Group LLC: WBUD, WKXW(FM)
Multicultural Radio Broadcasting Inc.: WHWH
Nassau Broadcasting Partners L.P.: WPHY(AM), WPST(FM)

Tri-Cities, WA (Richland-Kennewick-Pasco): Rank 199
MSA: 187,500
KVAN(AM)
Bustos Media LLC: KMMG(FM), KZTB(FM)
Cherry Creek Radio LLC: KONA, KONA-FM
Clear Channel Communications Inc.: KEYW-FM, KFLD
CSN International: KBLD(FM)
EMF Broadcasting: KRKL(FM)
Moon Broadcasting: KLES(FM), KZXR
New Northwest Broadcasters LLC: KALE, KEGX(FM), KIOK-FM, KNLT-FM, KTCR, KUJ-FM

Tucson, AZ: Rank 61
MSA: 803,300
KEVT(AM)
Citadel Broadcasting Corp.: KCUB, KHYT-FM, KIIM-FM, KSZR(FM), KTUC
Clear Channel Communications Inc.: KNST, KOHT-FM, KRQQ-FM, KTZR-FM, KWFM(AM), KWMT-FM, KXEW
EMF Broadcasting: KAIC(FM)
Family Life Communications Inc.: KFLT, KFLT-FM
Good News Communications Inc.: KGMS(AM), KVOI
Journal Communications Inc.: KGMG-FM, KMXZ-FM, KQTH(FM)
Lotus Communications Corp.: KFMA-FM, KLPX-FM, KTKT

Tulsa, OK: Rank 65
MSA: 732,000
KMUR(AM), KRVT(AM)
ABC Inc.: KMUS(AM)
Adonai Radio Group: KCXR(FM), KEMX-FM, KXOJ-FM, KYAL(AM), KYAL-FM
Clear Channel Communications Inc.: KAKC(AM), KIZS(FM), KMOD-FM, KQLL-FM, KTBT(FM), KTBZ(AM)
Cox Radio Inc.: KJSR-FM, KKCM(FM), KRAV-FM, KRMG, KWEN-FM
Davidson Media Group LLC: KJMU(AM)
Journal Communications Inc.: KFAQ(AM), KXBL(FM)
K95.5 Inc.: KTFX-FM
KCD Enterprises Inc.: KPGM(AM)
M&M Broadcasters Ltd.: KEOR(AM)
Perry Publishing & Broadcasting Co.: KGTO, KJMM-FM

Renda Broadcasting Corp.: KBEZ(FM), KHTT-FM
Shamrock Communications Inc.: KMYZ-FM, KTSO(FM)

Tupelo, MS: Rank 188
MSA: 201,500
Air South Radio Inc.: WFTA-FM
American Family Radio: WAQB-FM
Clear Channel Communications Inc.: WBVV(FM), WESE-FM, WKMQ(AM), WTUP, WWKZ(FM), WWZD-FM
Stanford Communications Inc.: WAFM-FM, WAMY, WWZQ

Tuscaloosa, AL: Rank 234
MSA: 145,500
Citadel Broadcasting Corp.: WBEI(FM), WJOX(FM), WJRD(AM), WTSK, WTUG-FM
Clear Channel Communications Inc.: WACT, WRTR(FM), WTXT-FM, WZBQ-FM
The Moody Bible Institute of Chicago: WMFT(FM)

Tyler-Longview, TX: Rank 148
MSA: 296,800
Access.1 Communications Corp.: KFRO, KKUS(FM), KOOI-FM, KYKX-FM
American Family Radio: KATG(FM)
Bible Broadcasting Network: KYFP(FM)
Bott Radio Network: KTAA(FM)
EMF Broadcasting: KZLO(FM)
GAP Broadcasting LLC: KBGE(AM), KISX-FM, KKTX-FM, KNUE(FM), KTYL-FM
Gleiser Communications LLC: KDOK-FM, KEES, KTBB, KYZS
Reynolds Radio Inc.: KAJK(FM)
Salem Communications Corp.: KPXI-FM
Waller Broadcasting: KEBE, KFRO-FM, KLJT-FM, KXAL-FM

Utica-Rome, NY: Rank 160
MSA: 257,600
Bible Broadcasting Network: WYFY
Clear Channel Communications Inc.: WADR, WIXT(AM), WOKR(FM), WOUR-FM, WRNY, WSKS(FM), WSKU(FM), WUMX(FM), WUTQ
EMF Broadcasting: WKVU(FM)
Galaxy Communications L.P.: WKLL-FM, WRCK-FM, WTLB
Regent Communications Inc.: WFRG-FM, WIBX, WODZ-FM
WAMC/Northeast Public Radio: WRUN(AM)

Valdosta, GA: Rank 269
MSA: 105,600
Black Crow Media Group LLC: WQPW(FM), WSTI-FM, WVGA(FM), WVLD(AM), WWRQ-FM, WXHT(FM)
Dee Rivers Radio Group: WAAC(FM), WGOV(AM), WLYX(FM)
EMF Broadcasting: WVDA(FM)
Rama Communications Inc.: WRFV(AM)
Three Trees Communications Inc.: WJYF(FM)

Victor Valley, CA: Rank 126
MSA: 376,600

Visalia-Tulare-Hanford, CA: Rank 100
MSA: 455,700
Azteca Broadcasting Corp.: KGEN, KGEN-FM
Buckley Broadcasting Corp.: KIOO-FM, KSEQ-FM
Clear Channel Communications Inc.: KEZL(AM)
IHR Educational Broadcasting: KJOP
Moon Broadcasting: KMQA-FM

Waco, TX: Rank 201
MSA: 186,200
American Family Radio: KBDE(FM), KSUR(FM)
Border Media Partners LLC: KWOW-FM
Clear Channel Communications Inc.: KBGO(FM), KBRQ(FM), KWTX, KWTX-FM, WACO-FM
Simmons Media Group: KLRK(FM), KQRL(AM), KRZI(AM)
Univision Radio: KHCK-FM
William W. McCutchen III Stns: KDRW(FM)

Washington, DC: Rank 8
MSA: 4,176,300
WKDV
Bonneville International Corporation: WFED(AM), WPRS-FM, WTLP(FM), WTOP-FM, WTWP(AM), WTWP-FM, WTWT(AM)
CBS Radio: WJFK-FM, WLZL(FM), WPGC, WPGC-FM, WTGB-FM
Citadel Broadcasting Corp.: WJZW-FM, WMAL, WRQX-FM
Clear Channel Communications Inc.: WASH-FM, WBIG-FM, WIHT(FM), WMZQ-FM, WTEM, WTNT(AM), WWDC-FM, WWRC(AM)
Entravision Communications Corp.: WACA
Mountain Broadcasting Corp.: WWGB
Multicultural Radio Broadcasting Inc.: WLXE(AM), WZHF
Nassau Broadcasting Partners L.P.: WAFY-FM
Pacifica Foundation Inc.: WPFW-FM
Radio One Inc.: WKYS-FM, WMMJ-FM, WYCB
Red Zebra Holdings LLC: WWXT(FM), WWXX(FM), WXTR(AM)
Salem Communications Corp.: WAVA(AM), WAVA-FM
Somar Communications Inc.: WKIK, WYRX(FM)

Waterloo-Cedar Falls, IA: Rank 252
MSA: 128,300
Ace Radio Corp.: KCOO(FM)
Bahakel Communications: KFMW-FM, KOKZ-FM, KWLO(AM), KXEL
Cumulus Media Inc.: KCRR(FM), KKHQ-FM, KOEL-FM
KM Communications Inc.: KQMG, KQMG-FM
Northwestern College & Radio: KNWS

Watertown, NY: Rank 279
MSA: 98,300
Clancy-Mance Communications: WBDR(FM)
Community Broadcasters LLC: WATN, WBDI(FM), WOTT-FM, WTOJ(FM)
EMF Broadcasting: WKWV(FM)
Mars Hill Network: WMHI-FM
Regent Communications Inc.: WNER(AM), WTNY

Wausau-Stevens Point, WI (Central Wisconsin): Rank 170
MSA: 236,000
Evangel Ministries Inc.: WGNV(FM)
Midwest Communications Inc.: WDEZ-FM, WIZD-FM, WOFM-FM, WRIG, WSAU
Muzzy Broadcasting L.L.C.: WKQH(FM), WSPT, WSPT-FM
NRG Media LLC: WBCV(FM), WGLX-FM, WLJY(FM), WYTE(FM)
Seehafer Broadcasting Corp.: WDLB, WFHR, WOSQ-FM, WXCO

West Palm Beach-Boca Raton, FL: Rank 46
MSA: 1,116,800
WDJA(AM), WJBW(AM)
ABC Inc.: WMNE(AM)
Beasley Broadcast Group Inc.: WSBR
CBS Radio: WEAT-FM, WIRK-FM, WMBX-FM, WNEW(FM), WPBZ-FM
Clear Channel Communications Inc.: WBZT(AM), WJNO(AM), WKGR-FM, WLDI-FM, WOLL-FM, WRLX-FM, WZZR(FM)
Communicom Broadcasting LLC: WLVJ(AM)
James Crystal Inc.: WFTL(AM), WMEN(AM)
WAY-FM Media Group Inc.: WAYF(FM)

Wheeling, WV: Rank 249
MSA: 129,500
Clear Channel Communications Inc.: WBBD, WEGW-FM, WKWK-FM, WVKF(FM), WWVA(AM)
Keymarket Communications LLC: WOMP, WUKL(FM), WYJK-FM

Wichita Falls, TX: Rank 265
MSA: 111,400
Cumulus Media Inc.: KLUR-FM, KOLI-FM, KQXC-FM, KYYI-FM
EMF Broadcasting: KZKL(FM)
GAP Broadcasting LLC: KBZS(FM), KNIN-FM, KWFS, KWFS-FM
James Falcon Stns: KZAM(FM)
Tower Investment Trust Inc.: KXXN(FM)

Wichita, KS: Rank 98
MSA: 466,700
KAHS(AM)
ABC Inc.: KQAM
American Family Radio: KCFN(FM)
Bible Broadcasting Network: KYFW-FM
Bott Radio Network: KJRG
Clear Channel Communications Inc.: KRBB-FM, KTHR(FM), KZCH(FM), KZSN(FM)
Connoisseur Media LLC: KIBB(FM)
EMF Broadcasting: KTLI(FM)
Entercom Communications Corp.: KDGS(FM), KEYN-FM, KFBZ(FM), KFH(AM), KFH-FM, KNSS(AM)
Journal Communications Inc.: KFTI(AM), KFTI-FM, KFXJ(FM), KICT-FM, KYQQ(FM)
Robert Ingstad Broadcast Properties: KBUF(AM)
Steckline Communications Inc.: KGSO(AM)
WAY-FM Media Group Inc.: KYWA(FM)

Wilkes Barre-Scranton, PA: Rank 69
MSA: 684,200
WEMR, WITK(AM)
Bold Gold Media Group LP: WFBS(AM), WICK, WWRR(FM), WYCK
Citadel Broadcasting Corp.: WARM, WBHD(FM), WBHT-FM, WBSX(FM), WMGS-FM, WSJR(FM)
Clear Channel Communications Inc.: WHCY-FM
Entercom Communications Corp.: WBZU(AM), WDMT(FM), WGGI-FM, WGGY-FM, WILK, WILK-FM, WKRF(FM), WKRZ-FM, WKZN(AM)
Holy Family Communications: WQOR(AM)
MAX Media L.L.C.: WFYY(FM)
Nassau Broadcasting Partners L.P.: WVPO
Route 81 Radio LLC: WAZL, WCDL(AM), WLNP(FM), WNAK
Shamrock Communications Inc.: WBAX, WEJL, WEZX-FM, WQFM-FM, WQFN(FM)

Williamsport, PA: Rank 271
MSA: 102,600
Backyard Broadcasting LLC: WBZD-FM, WILQ-FM, WRVH(FM), WWPA
Clear Channel Communications Inc.: WBYL(FM), WRAK, WRKK, WVRT(FM)

Wilmington, DE: Rank 75
MSA: 590,300
Clear Channel Communications Inc.: WILM, WWTX(AM)
Delmarva Broadcasting Co.: WDEL, WSTW-FM, WXCY(FM)
NextMedia Group Inc.: WJBR-FM
Priority Radio Inc.: WSRY(AM)

Wilmington, NC: Rank 166
MSA: 241,800
Carolina Christian Radio: WDVV-FM, WLSG(AM), WMYT, WWIL, WWIL-FM, WZDG(FM)
Cumulus Media Inc.: WAAV, WGNI-FM, WMNX-FM, WWQQ-FM
NextMedia Group Inc.: WAZO(FM), WKXB-FM, WMFD, WRQR-FM
Sea-Comm Inc.: WBNE(FM), WLTT(FM), WNTB(FM)

Winchester, VA: Rank 225
MSA: 159,800
WSVG(AM)

Clear Channel Communications Inc.: WAZR-FM, WFQX-FM, WTFX(AM)
Mid Atlantic Network: WINC, WWRE(FM), WWRT(FM)
Vox Radio Group L.P.: WSIG(FM)

Worcester, MA: Rank 111
MSA: 417,100
WORC
Blount Communications Group: WNEB, WVNE(AM)
Citadel Broadcasting Corp.: WORC-FM, WWFX(FM), WXLO-FM
Clear Channel Communications Inc.: WSRS-FM, WTAG
Entercom Communications Corp.: WVEI(AM)
Northeast Broadcasting Company Inc.: WJOE(AM), WNYN-FM

Yakima, WA: Rank 200
MSA: 186,900
Bustos Media LLC: KYXE(AM), KZTA-FM, KZTS(AM)
Clear Channel Communications Inc.: KDBL(FM), KIT, KUTI(AM)
New Northwest Broadcasters LLC: KARY-FM, KBBO(AM), KHHK-FM, KJOX(AM)

York, PA: Rank 103
MSA: 445,500
Citadel Broadcasting Corp.: WQXA-FM
Cumulus Media Partners LLC: WARM-FM, WGLD(AM), WSBA, WSOX-FM
Hall Communications Inc.: WSJW(FM)
Wilkins Communications Network Inc.: WYYC(AM)

Youngstown-Warren, OH: Rank 117
MSA: 407,200
Beacon Broadcasting Inc.: WANR, WLOA(AM), WRTK(AM)
Clear Channel Communications Inc.: WAKZ(FM), WBBG(FM), WKBN, WNIO(AM)
Cumulus Media Inc.: WBBW, WLLF-FM, WPIC, WSOM, WWIZ-FM
Family Stations Inc.: WYTN-FM
Forever Broadcasting: WWGY(FM)
Salem Communications Corp.: WHKZ(AM)

Section E

Programming

Major Broadcast TV Networks

ABC

ABC Inc., 500 S. Buena Vista St., Burbank, CA 91521. Phone: 818-460-7477; Web site: abc.go.com.

77 W. 66th St., New York, NY 10023. Phone: 212-456-7777.

Anne Sweeney, pres, Disney-ABC Television Group; Julia Franz, sr VP, Comedy Series; Michael Shaw, pres, sls and mktg, Touchstone Television; Stephen McPherson, pres, ABC Primetime Entertainment; John Rouse, sr VP/affil rel, ABC Television; Brain Frons, pres, ABC Daytime; George Bodenheimer, pres, ABC Sports and ESPN Inc.; Walter Liss, pres, ABC Owned Television.

Ownership: The Walt Disney Co., 500 S. Buena Vista St., Burbank, CA 91521-9722. Phone: 818-560-1000; Web site: www.disney.com.

George Mitchell, chmn; Robert Iger, pres/CEO; Thomas O. Staggs, sr exec VP/CFO; Christine M. McCarthy, exec VP, corporate finance; Alan Braveman, sr exec VP/gen counsel; Zenia Mucha, exec VP corporate communication; Preston Padden, exec VP/govt rel.

CBS

51 W. 52nd St., New York, NY 10019. Phone: 212-975-4321; Web site: www.cbs.com.

7800 Beverly Blvd., Los Angeles, CA 90036. Phone: 323-575-2200.

Leslie Moonves, chmn, pres/CEO; Louis Briskman, exec VP/gen counsel; Gil Schwartz, exec VP, corporate communications, CBS Corporation; Nancy Tellem, pres; Nina Tassler, exec VP/drama series dev; Jodi Roth, VP/special, CBS Entertainment.

Ownership: CBS Corporation (formerly known as Viacom Inc.), 1515 Broadway, New York, NY 10036. Phone: 212-258-6000; Web site: www.viacom.com.

Sumner Redstone, chmn/CEO; Leslie Moonves, pres/CEO; Martin Shea, exec VP pres, investor rel; Frederic Reynolds, exec VP/CFO; Richard Jones, sr VP/gen tax counsel.

The CW Television Network

The CW is a joint venture between CBS Corporation and Warner Bros. Entertainment, a subsidiary of Time Warner.

4000 Warner Blvd., Burbank, CA 91522. Phone: 818-977-5000; Web site: www.cwtv.com.

Dawn Ostroff, pres, entertainment; John Maatta, COO; Michael Roberts, exec VP of current progmg; Betsy McGowen sr VP and gen mgr of Kids WB!; Rick Mater, sr VP, bcst standards (from The WB); Eric Cardinal, sr VP, research; Elizabeth Tumulty, sr VP, network distribution; Bill Morningstar, exec VP natl sls; Kim Fleary, exec VP comedy dev; Rick Haskins, exec VP mktg brand strategy; Paul McGuire, exec VP network communications; Thorn Sherman, exec VP drama dev; Lori Openden, sr VP talent and casting.

Ownership: CBS Corporation; Warner Bros. Entertainment Inc. (Time Warner).

FOX

Twentieth Century Fox, 10201 W. Pico Blvd., Los Angeles, CA 90064. Phone: 310-369-1000; Web site: www.fox.com.

Peter Liguori, pres; Marcy Ross, sr VP, Fox Broadcasting Co.; Anthony Vinciquera, pres/CEO, Fox Networks Group; Preston Beckman, exec VP, strategic program planning; David Hill, chmn/CEO, Fox Group; Roger Ailes, chmn/pres Fox News channel; Jack Abernethy, CEO, Fox Television stns; Dennis Swanson, pres stns opns.

Ownership: News Corp., 1211 Ave. of the Americas, New York, NY 10036. Phone: 212-852-7000; Web site: wwww.newscorp.com.

K. Rupert Murdoch, chmn/CEO; Peter Chernin, pres/COO; David DeVoe, CFO.

i

ION Media Networks (Independent Television)

601 Clearwater Park Rd., West Palm Beach, FL 33401. Phone: 888-467-2988; Web site: www.ionline.tv.

Brandon Burgess, pres/CEO; Richard Garcia, sr VP/CFO; Adam K. Weinstein, sr VP, sec/chief legal officer; Steven J. Friedman, pres cable; Stephen Appel, pres, sls/mktg; David A. Glenn, pres engrg; Doug Barker, pres bcst distribution.

Ownership: Paxson Communications Corp. (dba ION Media Networks), 601 Clearwater Park Rd., West Palm Beach, FL 33401. Phone: 888-467-2988.

Lowell Paxson, chmn Emeritus; Independent Directors: W. Lawence Patrick, chnm; Henry Brandon; Brandon Burges; Raymond S. Rajewski; William A. Roskin; Lucille S. Salhany; Frederick M.R. Smith.

NBC

30 Rockefeller Plaza, 25th Fl., New York, NY 10112. Phone: 212-664-4444; Web site: www.nbc.com.

3000 W. Alameda Ave., Los Angeles, CA 91505.

Jeff Zucker, pres/CEO, NBC Universal; Lynn Calpeter, exec VP/CFO; Richard Cotton, exec VP/gen counsel; John Eck, pres Media Works/CIO/NBC Universal; Jay Ireland, pres, NBC Universal; Cory Shields, exec VP/communications; Marc Chini, exec VP human resources .

Ownership: NBC Universal, a subsidiary of General Electric; NBC Universal, 30 Rockefeller Plaza, New York, NY 10112. Phone: 212-664-4444; Web site: www.nbcuni.com.

General Electric, 3135 Eastern Tpke., Fairfield, CT 06828. Phone: 203-373-2211; Web site: www.ge.com.

Jeffrey Immelt, chmn/CEO GE; Brackett Denniston III, sr VP/gen counsel; Pamela Daley, sr VP/corporate business dev; Dan Henson, chief mktg officer; Kathryn A. Cassidy, VP/treas; John Lynch, sr VP/corporate human resources.

Major TV Program Syndicators/Distributors

Buena Vista Television

Head Office: 500 S. Buena Vista St., Burbank, CA 91521; Tel: 818-560-6212.

Officers: Janice Marinelli, pres; Tom Malanga, sr VP finance; John Bryan, exec VP/gen sls mgr; Dan Cohen, sr VP/gen mgr; Mary Kellogg, exec VP/current progmg; Lloyd Komesar, exec VP/strategic rsch; Sal Sardo, exec VP/mktg.

New York Office: Advertising and Sales, 7 W. 66th St., New York, NY 10023; Tel: 212-456-1756; Fax: 212-456-0396.

Management: Howard Levy, exec VP/sls.

Major first-run programming: Live with Regis and Kelly, Unsolved Mysteries, Ebert & Roeper.

Major off-network programming: Alias, According to Jim, Home Improvement, My Wife and Kids.

King World Productions

Head Office: 2401 Colorado Ave., Suite 110, Santa Monica, CA 90404; Tel: 310-264-3300.

Management: Roger King, CEO, CBS Enterprises and King World Productions, Inc.; Robert Madden, COO; Delilah Loud, sr VP, adv/promotion; Arthur Sando, sr VP/communications; Armando Nunez, exec VP Mike Stornello, sr VP, dev.

New York Office: 1700 Broadway, 32nd & 33rd Fl., New York, NY 10019; Tel: 212-315-4000; Fax: 212-582-9255.

Management: Joe DiSalvo, pres, domestic television sls; Steve Hirsch, pres; Michael Auerbach, sr VP, King World media sls; Jonathan Birkhahn, sr VP/business legal affrs.

New Jersey Office: 830 Morris Tpke., Short Hills, NJ 07078; Tel: 973-376-1313; Fax: 973-376-7787.

Management: Steve LoCascio, exec VP/CFO; Moira Coffey, sr VP/rsch.

Major first-run programming: The Oprah Winfrey Show, Dr. Phil, Wheel of Fortune, Jeopardy!, Inside Edition, Mr. Food, CBS MarketWatch and Bob Vila's Home Again.

Major off-network programming: Everybody Loves Raymond, CSI: Crime Scene Investigation and CSI: Miami.

NBC Universal Television Distribution

Head Office: NBC Enterprises. 3400 W. Olive Ave., 6th Fl., Burbank, CA 91505; Tel: 818-526-6900.

Management: Ed Wilson, pres; Jon Hookstratten, exec VP, current progmg; Jerry Petry, exec VP; Linda Finnell, sr VP progmg; Mary Beth McAdaragh, VP mktg; Mara Jacobberger, dir mktg.

Domestic syndication programming: Martha Stewart, Access Hollywood, Blind Date, Jerry Springer Show, Maury and Starting Over.

First-run weekly syndicated programming: The Chris Matthews Show, The George Michael Sports Machine, Rebecca's Garden, The Wall Street Journal Report with Maria Bartiromo and Your Total Health.

Off-Network Distribution: Crossing Jordan, Fear Factor, Providence, and the Law & Order franchise.

International Television Distribution Featured Shows: Bionic Women, Life, Lipstick Jungle and The it Crowd.

Paramount Worldwide Television Distribution

Head Office: 5555 Melrose Ave., Hollywood, CA 90038-3197; Tel: 323-956-5000.

Management: Joel Berman, pres; John Nogawski, pres; Greg Meidel, pres progmg; Mark Dvornik, exec VP/gen sls mgr, Paramount Domestic Television; Michael Mischler, exec VP/mktg; John Wentworth, exec VP, mktg/media rel; Dawn Abel, sr VP/rsch; Dennis Emerson, sr VP/off-net sls mgr; John Kohler, sr VP/creative affrs; Phil Murphy, sr VP/opns; Bill Weber, sr VP/TV systems.

New York Office: 1515 Broadway, 33rd Fl., New York, NY 10036; Tel: (212) 258-6000.

Scott Koondel, SVP, natl sls mgr.

CBS Paramount International Television

Armando Nunez, president, CBS Paramount International Television

Major first-run programming: Entertainment Tonight, ET Weekend, The Insider, Insider Weekend, Judge Joe Brown, Judge Judy, Maximum Exposure, The Montel Williams Show, Unexplained Mysteries.

Major off-network programming: Becker, Cheers, Frasier, Star Trek Next Generation, Girlfriends, The Parkers.

Sony Pictures Television

Head Office: 10202 W. Washington Blvd., Culver City, CA 90232; Tel: 310-244-4000.

Management: Steve Mosko, pres; Russ Krasnoff, pres, progmg/production; John Weiser, pres distribution; Jeannie Bradley, exec VP, current progmg; Gregory Boone, exec VP, legal affrs; Richard Frankie, exec VP/business opns; Ed Lammi, exec VP/production; Don Loughery, exec VP/business affrs; David Mumford, exec VP, planning/opns; Robert Oswaks, exec VP/mktg; Helen Verno, exec VP/movies; Wayne Goldstein, VP, new-media rsch/sls planning.

New York Office: 550 Madison Ave., New York, NY 10022; Tel: 212-833-8500.

Management: Barbara "Bo" Argentino, exec VP/adv sls; John J. Rohrs, exec VP/cable sls.

Daytime dramas: The Young and the Restless and Days of our Lives.

Major cable network: Heist, Huff, Rescue Me, The Shield, Strong Medicine and Walker, Texas Ranger.

Major first-run syndication: Life & Style, Pat Croce: Moving In, Wheel of Fortune, Jeopardy!.

Major off-network comedies: Just Shoot Me!, Seinfeld, The King of Queens, Mad About You and Married with Children.

Twentieth Television

Head Office: 2121 Avenues of the Stars, 21st Fl., Los Angeles, CA 90067. Tel: 310-369-1000; Fax: 310-369-3899. Web site: www.fox.com.

Office: Fox Television PO Box 900, Beverly Hills, CA. Tel: 310-369-2731.

Management: Roger Ailes, chmn; Les Eisner, VP, media rel; Bob Cook, pres/COO; Robb Dalton, pres, progmg/dev; Paul Franklin, exec VP/gen sls mgr; Marisa Fermin, exec VP, business/legal affrs; Joanne Burns, sr VP mktg, rsch/new media; Steve MacDonald, sr VP, gen sls mgr/basic cable; Matthew Rodriguez, sr VP, mktg/creative; Elaine Bauer-Brooks, sr VP, progmg/dev; Daniel Tibbets, VP/production.

New York Office: 1211 Ave. of the Americas, 3rd Fl., New York, NY 10036; Tel: 212-556-2400.

Management: Andrew Butcher, VP corporate affrs/communication; Bob Cesa, exec VP, adv sls/basic cable sls.

Major first-run programming: Geraldo at Large, hosted by legend Geraldo Rivera, Divorce Court and Judge Alex.

Major off-network programming: The Simpsons, Malcolm in the Middle, Yes, Dear, "24", Reba, The Bernie Mac Show, Dharma & Greg and King of the Hill.

Warner Bros. Domestic Cable Distribution

Head Office: 4000 Warner Blvd., Bldg. 160, Burbank, CA 91522; Tel: 818-977-4340; Fax: 818-977-4474.

Management: Eric Frankel, pres; Gus Lucas, exec VP; Ron Sunderland, sr VP, legal/business affrs; Linda Abrams, sr VP, mktg; Maury Litner, VP; Mark DeVitre, VP/sls; Donald Putrimas, VP/finance; Pam Ritchie, VP, sls planning/program inventory; Mike Russo, VP, sls; Timothy Ramirez, sls dir.

Major Cable Network: ER, The West Wing, The Gilmore Girls, Smallville, Cold Case, Without a Trace, Nip/Tuck and Two and a Half Men.

Warner Bros. Domestic Television Distribution

Head Office: 4001 Olive Ave., 4th Fl., Burbank, CA 91522; Tel: 818-954-5652; Fax: 818-954-5697.

Management: Dick Robertson, pres; Jim Paratore, exec VP; Bruce K. Rosenblum, exec VP/media rsch; Bill Marcus, sr VP/sls; Rick Meril, sr VP/sls; Liz Huszarik, sr VP, media rsch; Peter Roth, pres, Warner Bros. Television Production; Len Goldstein, exec VP/dev; Scott Rowe, VP/corporate communications, Warner Bros. Entertainment.

New York Office: 1325 Ave. of the Americas, 31st Fl., New York, NY 10019; Tel: 212-636-5300; Fax: 212-636-5380.

Management: Michael Teicher, exec VP/media sls; Roseann Cacciola, sr VP, gen sls mgr/media sls.

Major first-run programming: The Ellen DeGeneres Show, The Tyra Banks Show, The People's Court, Judge Mathis and Showtime at the Apollo.

Major off-network programming: Friends, This Old House, Smallville, Will & Grace and Without a Trace.

Regional Broadcast TV Networks

ALIN-TV, 149 Madison Ave., Suite 602, New York, NY, 10016. Phone: (212) 889-1327. Fax: (212) 213-6968.

Alan Cohen, pres.

ALIN-TV offers locally originated progmg on a line-up of leading ind stns providing natl participation on a daily basis. Specific networks are provided to zero in on target audience progmg: prime, prime access, teen/young adult, late night entertainment, daytime, weekend entertainment, news & kids.

American Public Television, 55 Summer St. 4th Fl., Boston, MA, 02110. Phone: (617) 338-4455. Fax: (617) 338-5369. E-mail: info@aptonline.org Web Site: www.aptonline.org.

Cynthia Fennerman, pres; Chris Funkhauser, VP/exch & distr svs.

Comprises WETA-TV, WHMM, both Washington, DC; WEDW(TV) Bridgeport, WEDY(TV) New Haven, WEDH(TV) Hartford, WEDN(TV) Norwich, all Connecticut; WCBB(TV) Augusta, WMED-TV Calais, WMEB-TV Orono, WMEM-TV Presque Isle, all Maine; WGBH-TV Boston, WGBX-TV Boston, WGBY-TV Springfield, all Massachusetts; WENH-TV Durham, WEKW-TV Keene, WLED-TV Littleton, all New Hampshire; WNJT-TV Trenton, New Jersey; WSKG(TV) Binghamton, WNED-TV Buffalo, WNET(TV) New York, WLIW-TV Plainview, WXXI(TV) Rochester, WMHT(TV) Schenectady, WCNY-TV Syracuse, WNPE-TV Watertown, all New York; WCET-TV Cincinnati, WVIZ-TV Cleveland, WPTD-TV Dayton, all Ohio; WLVT-TV Bethlehem-Allentown, WPSX-TV Clearfield, WITF-TV Harrisburg, WHYY-TV Philadelphia, WQED(TV) Pittsburgh, WVIA(TV) Scranton-Wilkes Barre, WQLN Erie, all Pennsylvania; WSBE-TV Providence, Rhode Island; WMPT-TV Annapolis, WMPB(TV) Baltimore, WWPB(TV) Hagerstown, WCPB-TV Salisbury, all Maryland. Service virtually all Public Television stations in the U.S.

California Farm Network, 2300 River Plaza Dr., Sacramento, CA, 95833. Phone: (916) 561-5550. Fax: (916) 561-5695. E-mail: cfbf@cfbf.com Web Site: www.cfbf.com.

Bob Krauter, exec dir; Ron Miller, opns mgr; Doug Mosebar, pres.

Comprises KUVI-TV Bakersfield, KAEF-TV Eureka, KSEE-TV Fresno, KIXE-TV Redding, KRCR-TV Redding, KSBW-TV Salinas, KSBY-TV San Luis Obispo, K26AY Lakeport, KXTV-TV Sacramento, KPXN-TV Los Angeles, KFTY-TV Santa Rosa, KBHAA-TV San Francisco, all California; KYMA(TV) Yuma, Arizona; RFD-TV dish net direct TV.

California-Oregon Broadcasting Inc., Box 1489, Medford, OR, 97501. Phone: (541) 779-5555. Fax: (541) 779-1151. E-mail: kobi@kobi5.com Web Site: www.localnewscomesfirst.com.

Patricia C. Smullin, pres.

Eugene, OR 97408. KEVU-KISR-TV, 2940 Chad Dr. Phone: (541) 682-2525. Mark Mitzger, gen mgr.

Medford, OR 91501. KOBI-TV, 125 S.Fir St. Phone: (541) 779-5555. Fax: (541) 779-1151. Pat Sumllin, owner.

Comprises KLSR-TV & KEVU-TV Eugene, KOTI(TV) Klamath Falls, KOBI(TV) Medford, all Oregon. Represented by John Blair & Co., Northwest.

4X Network, Box 1686, 3425 S. Broadway, Minot, ND, 58701. Phone: (701) 852-2104. Fax: (701) 838-9360. E-mail: webmaster@kxmcnews.com Web Site: www.kxmc.com.

David Reiten, gen mgr.

Comprises KXMB-TV Bismarck, KXMA-TV Dickinson, KXMC-TV Minot, KXMD-TV Williston, all North Dakota. Represented by Katz Continental.

KMWB, (Sinclair Communication Inc.). 1640 Como Ave., St. Paul, MN, 55108. Phone: (651) 646-2300. Fax: (651) 646-1220. E-mail: sales@kmwb23.com Web Site: www.kmwb23.com.

Art Lanham, pres & gen mgr; Miles Kennedy, VP & controller; Bob Weinstein, natl sls mgr; Jeff Ongstad, creative svcs dir.

Minnesota, Wisconsin. Represented by Millenium.

KSN Television Group, Box 333, 833 N. Main St., Wichita, KS, 67201. Phone: (316) 265-3333. Fax: (316) 292-1197. Web Site: www.ksn.com.

Comprises KSNG Garden City, KSNC Great Bend, KSNT Topeka, KSNW Wichita, KSNK Oberlin, all Kansas. Represented by TeleRep. Above TV stns affiliated with NBC Television Network. Owned by SJL of KS.

KWCH-TV, Schurz Communications, Inc., Box 12, Wichita, KS, 67201. Phone: (316) 838-1212. Fax: (316) 831-6198. Web Site: www.kwch.com.

Joan Barrett, VP/gen mgr; Brian McDonough, gen sls mgr.

Comprises KBSD-TV Ensign-Dodge City, KBSL-TV Goodland, KBSH-TV Hays, KWCH-TV Wichita-Hutchinson, all Kansas. Represented by HRP.

Kansas Television Network, 1500 N. West St., Wichita, KS, 67203. Phone: (316) 943-4221. Fax: (316) 943-5493. Web Site: www.kake.com.

Terry Cole, gen mgr.

Comprises KLBY-TV Colby, KUPK-TV Garden City, KAKE-TV Wichita, all Kansas. Represented by Katz.

Keloland TV Young Broadcasting of Sioux Falls Inc., KELO TV Bldg., 501 S. Phillips Ave., Sioux Falls, SD, 57104. Phone: (605) 336-1100. Fax: (605) 334-3447. Fax: (605) 336-0202. Web Site: www.keloland.com.

Vincent Young, chmn; Mark Millage, news dir.

Comprises KDLO-TV Florence, KPLO-TV Reliance, KELO-TV Sioux Falls, KCLO-TV Rapid City, all South Dakota. Represented by Adam Young Inc.

Montecito Broadcast Group LLC, (formerly SJL Broadcast Management Corp.). 559 San Ysidro Rd., Suite I, Montecito, CA, 93108. Phone: (805) 969-9278. Fax: (805) 969-2399.

George Lilly, CEO.

Montecito, CA 93108, 633 Picacho Ln. Phone: (805) 969-9278. Fax: (805) 969-2399. George D. Lilly, pres.

Comprises KHON-TV, Honolu, HI; KSNT-TV, Topeka & KSNW-TV, Witchita both KS; KOIN-TV, Portland, OR.

National Educational Telecommunications Association, Box 50008, Columbia, SC, 29250. Phone: (803) 799-5517. Fax: (803) 771-4831. E-mail: skip@netaonline.org Web Site: www.netaonline.org.

Skip Hinton, pres.

Comprises Alabama PTV Birmingham, Alabama; KUAC Fairbanks, KYUK Bethel, Alaska; KUAT Tucson, Arizona; Arkansas ETV Conway, Arkansas; KOCE Huntington Beach, KLCS Los Angeles, both California; WBCC Cocoa, WCEU Daytona Beach, WFSU Tallahassee, WGCU Fort Myers, WLRN Miami, WSRE Pensacola, WUFT Gainesville, WUSF Tampa, WXEL West Palm Beach, all Florida; Georgia Public Broadcasting Atlanta, WPBA Atlanta, both Georgia; Idaho PTV Boise, Idaho; WNIT, Elkhart, WYIN Merrillville, both Indiana; Iowa PTV Johnston, Iowa; KOOD Bunker Hill, Kansas; Kentucky ETV Lexington, WKYU Bowling Green, both Kentucky; LA Public Broadcasting Baton Rouge, WLAE New Orleans, both Louisiana; Maryland Public Broadcasting Owings Mills, Maryland; WKAR, E. Lansing, WGVU Grand Rapids, Michigan; Minnesota, Twin Cities PTV St. Paul, Minnesota; Mississippi EB Jackson, Mississippi; KCPT Kansas City, KETC St. Louis, KMOS Warrensburg, KOZK Ozarks Public TV Springfield, all Missouri; Montana PTV Bozeman, Montana; Nebraska ETV Lincoln, Nebraska; KLVX Las Vegas, Nevada; New Hampshire PTV Durham, New Hampshire; NJN Trenton, New Jersey; KENW Portales, KNME Albuquerque, KRWG Las Cruces, all New Mexico; WLIW Long Island, WMHT Schenectady, WNET New York, WPBS Watertown, all New York; UNC-TV Research Triangle Park, North Carolina; Prairie Public Television Fargo, North Dakota; WOUB Athens, WOSU, Columbus, WPTD Dayton, WNEO, Kent, all Ohio; KRSC Claremore, Oklahoma ETV Oklahoma City, both Oklahoma; Oregon Public Broadcasting Portland, Oregon; WLVT Allentown, WPSU, University Park, WYBE Philadelphia, all Pennsylvania; WSBE Providence, Rhode Island; South Carolina ETV Columbia, South Carolina; South Dakota Public Television Vermillion, South Dakota, WCTE Cookeville, WNPT Nashville, WKNO Memphis, WLJT Martin, WSJK Knoxville, WTCI Chattanooga, all Tennessee; KAMU College Station, KWBU Waco, KEDT Corpus Christi, KERA Dallas, KLRU Austin, KMBH Harlingen, KNCT Killeen, KOCV Odessa, KTXT Lubbock, all Texas; KBYU Provo, KUED Salt Lake City, Utah; Vermont PTV Colchester, Vermont; WTJX St. Thomas, Virgin Islands; WBRA Roanoke, WCVE Richmond, WHRO Norfolk, WVPT Harrisonburg, all Virginia; KSPS Spokane, Washington; West Virginia Public Broadcasting Charleston, West Virginia; Wisconsin PTV Milwaukee , ETV Madison, both Wisconsin; Wyoming PTV Riverton, Wyoming.

Nebraska Television Network (NTV), Box 220, Kearney, NE, 68848. Phone: (308) 743-2494. Fax: (308) 743-2644. E-mail: news@nebraskatv.net Web Site: www.nebraska.tv.

Janet Noll, gen mgr.

Comprises KTVG(TV) Grand Island; KHGI-TV13, Kearney, Hastings, Grand Island; KWNB-TV North Platte & KSNB-TV Superior and the translators of K02HB, K17CI, K11KV, K12KW, K13OM, K13NP, K13VO, K06EY. Represented by Petry.

North Dakota Television, 200 N. 4th St., Bismarck, ND, 58501. Phone: (701) 255-5757. Fax: (701) 255-8220. Web Site: www.kfyrtv.com. E-mail: kfyrtv@kfyrtv.com

Julie Jensen, natl sls mgr; Barry Shumaier, rgnl sls mgr; Dick Heidt, gen mgr.

Comprises KFYR-TV Bismarck, KQCD-TV Dickinson, KMOT-TV Minot, KUMV-TV Williston, all North Dakota; KVLY-TV, serving North Dakota, South Dakota & Montana. Represented by Blair.

Ohio Educational Telecommunications Network Commission, 2470 North Star Rd., Columbus, OH, 43221. Phone: (614) 644-1714. Fax: (614) 644-3112. E-mail: christofi@oet.state.oh.us Web Site: www.oet.edu.

Denos Christofi, exec dir.

Comprises WEAO Akron, WNEO-TV Alliance, WOUB-TV Athens, WBGU-TV Bowling Green, WOUC-TV Cambridge, WCET Cincinnati, WVIZ-TV Cleveland, WOSU-TV Columbus, WPTD Dayton, WPTO Oxford, WPBO-TV Portsmouth, WGTE-TV Toledo, all Ohio.

Pennsylvania Public Television Network, 24 Northeast Dr., Hershey, PA, 17033. Phone: (717) 533-6011. Fax: (717) 533-4236. E-mail: sstrobel@state.pa.us Web Site: www.pptn.pa.us.

Sylvia L. Strobel, pres.

PPTN provides leadership & acountablity in guiding, supporting & advocating public telecommunications to educate, enlighten, inspire & connect the citizens of PA. Public stns operating across PA : WQLN-TV Erie; WITF-TV Harrisburg; WHYY-TV & WYBE-TV Philadelphia; WQED-TV Pittsburgh; WVIA-TV Scranton; WPSU-TV University Park.

Wisconsin Educational Communications Board, 3319 W. Beltline Hwy., Madison, WI, 53713-4296. Phone: (608) 264-9600. Fax: (608) 264-9664. Web Site: www.ecb.org.

Wendy Wink, exec dir.

Comprises WPNE(TV) Green Bay, WHLA-TV La Crosse, WHWC-TV Menomonie/Eau Claire, WLEF-TV Park Falls, WHRM-TV Wausau, all Wisconsin. Affils: Wisconsin: WMVS(TV) Milwaukee; WDSE-TV Duluth, Minnesota.

National Cable Networks

A&E Network, 235 E. 45th St., New York, NY, 10017. Phone: (212) 210-1400. Fax: (212) 210-9755. Web Site: www.aetv.com.

Abbe Raven, pres/CEO; Whitney Goit II, exec VP; David Zagin, sr VP; Mel Berning, exec VP; Bob DeBitetto, gen mgr.

A&E Network offers discerning viewers a unique blend of original progmg featuring its signature series BIOGRAPHY, original movies, dramas, series & documentaries. A&E Network also administers The History Channel (see listing).

Serving 88 million subs in the United States & Canada. On 10,000 cable systems. Satellite: Galaxy V, transponder 23.

ABC Family Channel, 500 S. Buena Vista St., Burbank, CA, 91521. Phone: (818) 560-1000. Web Site: www.abcfamily.com.

Ben Pyne, sr VP; Laura Nathanson, exec VP; Nicole Nochols, VP; Anne Sweeney, chmn.

Newport Beach, CA 92660, 660 Newport Center Dr, Suite 770. Phone: (714) 759-7685. Fax: (714) 759-9491. Janice Slipp, dir western rgn.

Englewood, CO 80111, 5445 DCT Pkwy, Suite 525. Phone: (303) 220-8901. Fax: (303) 220-9102. Tracy Jenkins, J.D., dir Rocky Mountain rgn.

Atlanta, GA 30349, Box 492347. Phone: (770) 461-4929. Fax: (770) 461-8678. Russell A. Breault, VP eastern division.

Oakbrook, IL 60523, 1301 W. 22nd St, Suite 902. Phone: (630) 990-0437. Fax: (630) 990-0463. Ralph Trentadue, natl dir lcl adv sls; Shirley Hill, VP western division.

New York, NY 10036, 1133 Ave. of the Americas, 36th Fl. Phone: (212) 782-1860. Fax: (212) 782-1865. Steve Israelsky, VP Northeast rgn.

Lewisville, TX 75067, 1422 W. Main St, Suite 201. Phone: (972) 436-2217. Fax: (972) 436-0209. Mark Solow, dir Southwest rgn.

Virginia Beach, VA 23450-2050, Box 2050, 2877 Guardian Ln. Phone: (757) 459-6281. Fax: (757) 459-6429. Craig Sherwood, sr VP/mgng dir. (Hqtrs affil sls & rel).

Basic cable network available in over 87 million homes nationwide 24 hours; delivers a dynamic mix of quality entertainment with original series & movies, classics fron Disney. Satellite: Galaxy V, transponder 11.

ABS-CBN (The Filipino Channel), 859 Cowan Rd., Burlingame, CA, 94010. Phone: (650) 697-3700. Fax: (650) 697-3500. Web Site: www.abs-bn.com.

Rafael Lopez, mgng dir; Eugenio Lopezill, chmn; Augusto Almeda-Lopez, vice chmn.

A 24-hour all Filipino premium svc ch delivered via satellite from the Philippines.

Serving 35,000 subs on 11 systems. Satellite: Galaxy 11, transponder 24.

ANA Television Network, 1510 H St. N.W., Suite 400, Washington, DC, 20005. Phone: (202) 898-8222. Fax: (202) 898-8088.

Angelyn Adams, CFO.

Arabic-language TV net bcstg to the Arab-American community 24 hours via cable, wireless cable, satellite.

Satellite: DIRECTV Plus. Satellite: Galaxy V, transponder 11.

ART (Arab Radio & Television), 315 Arden Ave., Suite 26, Glendale, 91203. Phone: (818) 243-0278. Fax: (818) 243-9278. Web Site: www.art.tv.net.

Michael Scott, VP.

ART's foundation is based on the largest gen entertainment library in the Middle East. Available 24 hours a day in North America, has progmg targeted to second generation Arab Americans. Satellite: Galaxy 11, transponder 24.

AYM Sports, Avenida Chapultepac 405, Colonia Juarez, Delegacion Cuauhtemoc, Mexico, 06600. Web Site: www.aymsports.tv.

Benjamin Hinojosa, pres; Carlos Carrillo, progmg VP.

Available 24 hours a day, 100% Mexican network consisting of soccer, basketball, rodeo charreadas, horse racing, boxing, kick boxing, jujitsu, karate, tae kwon do, swimming, driving, truck series, rallies and much more. Satellite: Telestar 7, transponder 12.

AZN Television, 4100 E. Dry Creek Rd., Centennial, CO, 800122. Phone: (303) 712-5400. Fax: (303) 712-5401. Web Site: azntv.com.

Rod Shanks, gen mgr; Bill Georges, sr VP affiliate & adv sls; Scott Wheeler, sr VP net dev; Victor Perez, VP technology.

Los Angeles, CA 90064, 12100 W. Olympic Blvd, Suite 200. Phone: (310) 979-2246.

New York, NY 10036, 1114 Ave. of the Americas, 22nd Fl. Phone: (917) 934-1558.

AZN Telvision is the nt for Asian America. The channel's progmg targets the fast-growing, affluent & multi-generational Asian American community, as well as a broader American audience interested in the Asian experience. Geres include the most popular Asian films, dramas, documentaries, anime, & news, as well as original progmg. Progmg on AZN Television is either in English or subtitled in English. Satellite: Galaxy 11. 24.

14BS-CBN International, (ABS-CBN International). 150 Shoreline Dr., Redwood City, CA, 94065-1400. Phone: (650) 508-6000. Fax: (650) 551-1062. E-mail: tfc@abs-cbni Web Site: www.abs-cbni.com.

Rafael Lopez, COO & sr VP; Jun del Rosario, production mgr.

A 24-hour svc for Filipinos worldwide. Progmg originates at ABS-CBN, the Philippines top-rated net. Time-shifted for North America.

Satellite: Galaxy 11, PAS-2 for Pacific delivery, Telstar 5 (ku) transponder 22 for North America. On 144 cable systems. 100,000 subs.

Access Television Network Inc., 2600 Michelson Dr., Suite 1650, Irvine, CA, 92612. Phone: (949) 263-9900. Fax: (949) 757-1526. Web Site: www.accesstv.com. E-mail: info@accesstv.com

George Henry, CEO; Robert T. Tyler, CFO; Mark R. Russo, sr VP opns.

Denver, CO 80202, 1020 15th St, Unit 30.

Organized natl marketplace for paid progmg on loc cable systems. Galaxy 11. On 650 cable systems. 40 million subs. transponder 9.

African Independent Television (AIT), One AIT Rd., PMB 1309, Apapa, Alagbado-Lagos Nigeria. Phone: (212) 213-2070. Web Site: www.aittv.com.

Ladi Lawal, COO.

AIT is a Pan-African gen entertainment ch offering news, talk show, soap opera, sports, Afician culture & music 24 hours a day. Areas: United States & Afica. Satellite: Telstar 5, transponder 5.

AMC, (Rainbow Programming Service Holdings Inc.). 200 Jericho Quadrangle, Jericho, NY, 11753. Phone: (516) 803-3000. Fax: (516) 803-4426. Web Site: www.amctv.com.

David Sehring, sr VP; Kathleen Dore, pres.

AMC is a 24 hour, movie-based mix or original series, documentaries & specials. 85 million subs. Satellite: Satcom C-4, transponder 1.

America's Collectibles Network (ACNTV), 10001 Kingston Park, Suite 57, Knoxville, 37922. Phone: (865) 693-8471. Fax: (865) 560-3298. Web Site: www.acntv.com.

F. Robert Hall, pres; Harris Bagley, VP.

ACNT offers a wide var of jewelry & gemstones at reduced prices, 24 hours a day home shopping, in the United States & Canada. Serving more than 63 million subs. Satellite: Telstar7, Transponder 21, Galaxy 11, Transponder 19.

America's Store, (A Division of Home Shopping Network). One HSN Dr., St. Petersburg, FL, 33729. Phone: (727) 872-1000. Fax: (727) 872-7356. Web Site: www.americasstore.com.

Peter Ruben, exec VP; Tom Mclnerney, CEO; Barry Diller, chmn/CEO; Will Keller, VP/gen mgr; Mike McMahon, opns VP.

America's Store offers live, 24 hour video retail net featuring merchandise with proven appeal in an upbeat, spontaneous style. 18 million subs. Satellite: Satcom C-3, transponder 10.

Animal Planet, One Discovery Pl., Silver Spring, MD, 20910. Phone: (240) 662-2000. Fax: (240) 662-1854. E-mail: first_last@discovery.com Web Site: www.animalplanet.com.

Marjorie Kaplan, gen mgr & pres.

Animal Planet TV network, 24 hrs broadband ch, Animal Planet Beyond & PetsIncredible, a producer, distributor of pet-training videos & web svc & other media platforms, a robust VOD svc, mobile content & merchandising extensions. Satellite: AMC 10. 90.5 million subs. Transponder 14.

Anime Network, 10114 West Sam Houston Pkwy. S., Suite 200, Houston, 77099. Phone: (713) 341-7200. Fax: (713) 341-7199. Web Site: www.theanimenetwork.com.

John Ledford, chmn/CEO; Kevin Corcoran, pres.

Anime is an exploration of Western pop culture. Anime reaches males 18-35 demographic with four different genres, including martial arts, comedy, science fiction & drama 24 hours a day, basic ad-supported. Serving more than 84 million subs.

The Anti-Aging Network, Inc., National Cable & Telecommunications Association, 1724 Massachusetts Ave., Washington, DC, 20036. Phone: (202) 775-3550. Fax: (805) 373-6595. Web Site: www.anti-agingnetwork.tv.

Progmg is informational, educ, & motivational, designed to empower viewer's to "live better longer" & "get an edge on aging." Slated to launch as a cable net in fall 2004.

Automotive Networks, Corp., WheelsTV. 289 Great Rd., Acton, MA, 01720. Phone: (978) 264-4333. Fax: (978) 264-9547. E-mail: jimbar@wheelstv.net Web Site: www.wheelstv.net.

Automotive entertainment & info via video on demand, worldwide web & dir mktg.

Serving 9.1 million subs.

BBC America, 747 3rd Ave., Fl. 6, New York, NY, 10017-2871. Phone: (301) 347-2222. Phone: (212) 705-9387. Web Site: www.bbcamerica.com.

Bill Hilary, pres/CEO; Kathryn Mitchell, gen mgr.

BBC America is a 24 hour award-winning TV featuring razor-sharp comedies, provocative dramas & life changing makeovers. Digital, Analog & DBS. Satellite: Satcom C3. 37 million. Satellite: Galaxy VII, transponder 22.

BET (Black Entertainment Television), 2020 Pennsylvania Ave. N.W., Washington, DC, 20006. Phone: (202) 608-2000. Fax: (202) 608-2631. E-mail: bet-tv_bizdev@bet.com Web Site: www.bet.com.

Scott Mills, pres/COO.

Burbank, CA 91505. BET/Los Angeles - Production, 2801 W. Olive Ave. Phone: (818) 566-9940. Fax: (818) 566-1655.

Washington, DC 20018. Network Operations, 1899 Ninth St. N.E. Phone: (202) 608-2800.

Washington, DC 20018. BET Film Production Facility, 2000 West Pl. N.E. Phone: (202) 608-2800. Fax: (202) 608-2629.

Chicago, IL 60607. BET/Chicago, 180 N. Stetson Ct, Suite 4350. Phone: (312) 819-8600.

New York, NY 10017. BET/New York, 380 Madison Ave, 20th Fl. Phone: (212) 716-5600. Fax: (212) 697-2050.

BET is the nation's leading TV network providing 24 hour for African-American audience in the United States, Canada & the Caribbean. BET Digital Networks-BET jazz, BET Gospel & BET Hip-Hop. 78 million subs. Satellite: Galaxy V, transponder 20.

BET Jazz: The Jazz Channel, 1235 W St. N. E., Washington, DC, 20018. Phone: (202) 608-2000. Phone: (800) 711-1630. Fax: (202) 608-2631. Web Site: www.bet.com.

Robert L. Johnson, chmn/CEO; Debra L. Lee, pres/COO.

The Jazz Channel is a 24-hour TV progmg svc dedicated exclusively to jazz through in-studio performances, documentaries,concert coverage & celebrity interviews. 9 million subs. Satellite: Galaxy VII, transponder 21.

BabyFirstTV, Box 25639, Los Angeles, CA, 90025. Phone: (888) 251-2229. Web Site: www.babyfirsttv.com.

Guy Oranim, CEO; Sharon Rechter, exec VP; Arik Kerman, sr VP.

Bandamax, 5999 Center Dr., Los Angeles, 90045. Phone: (301) 348-3371. Web Site: www.tutv.tv.

Mark Feldman, pres/CEO; Carlos Madrazo, CFO; Chris Fager, exec VP; Ariela Nerobay, VP sls.

Bandamax features a 24-hour a day country music video, including best artists in Tex-Mex, Norteno, Banda & Manachi genres. De Pelicula Sp-language films, De Pelicula Clasico films of Mexico's golden era.

Black Belt TV, The Martial Arts Network, 880 Calle Primavera, San Dimas, CA, 91773. Phone: (909) 971-9300. Fax: (909) 854-9329. Fax: (909) 394-0791. Web Site: www.blackbelttv.com.

Erik D. Jones, chmn/CEO.

Based on the popularity & participation in martial arts, we have created a 24 hours/7 days a week cable TV net that targets demographics highly desired by advertisers. Our appeal focuses on all income level individuals & families. BLACK BELT TV provides progmg for advertisers seeking to attract martial arts practitioners/enthusiasts, health/fitness-minded individuals, as well as sports enthusiasts, suppliers of exercise equipment, & other companies & product manufacturers that can directly reach their target audience. Our progmg includes martial arts movies, martial arts training/self-defense, self-improvement programs, sports, women, & children-oriented programs, martial arts news, & much more.

Bloomberg Television, 499 Park Ave., New York, 10022. Phone: (212) 893-3331. Fax: (202) 522-2400. Web Site: www.bloomberg.com/media/tv.

Kenneth Kohn, editor; Betsy Alekman, mktg.

A sophisticated 24-hour business & financial news ch. Serving over 200 million subs worldwide, United States, Canada, Central & South America, Europe,& Asia/Pacific. Satellite: Galaxy 11, HITS, C3.

BlueHighways TV, 111 Shivel Dr., Hendersonville, TN, 37075. Phone: (615) 264-3292. Fax: (615) 264-3308. Web Site: www.bluehighwaystv.com.

Stan Hitchcock, chmn/CEO.

The Boating Channel (TBC), 60 Bay St., Sag Harbor, NY, 11963-2022. Phone: (631) 725-4440. Fax: (631) 725-0748. Web Site: www.boatingchannel.com.

Barbara-Jo London, CEO; Daniel E. London, COO; Phillip J. Kassel, Esq & legal counsel; Gregory Hahn, exec producer.

Marine news & weather, entertainment, info & educ for the recreational, professional boater & cruise vacationer.

Boston Kids & Family TV, 43 Hawkins St., Suite 1B, Boston, 02114. Phone: (617) 635-3112. Fax: (617) 635-4475. E-mail: cable@ci.boston.ma.us Web Site: www.cityofboston.gov/cable.

Michael Lynch, dir; David Burt, stn mgr.

A partnership between the City of Boston & WGBH. Available 24 hour a day , edu TV progmg. PBS Kids from WGBH in Boston.

Bravo, (NBC Cable Network). 30 Rockefeller Plaza, 14th Fl. East, New York, NY, 10112. Phone: (212) 664-4444. E-mail: support@bravotv.com Web Site: www.bravotv.com.

Lauren Zalaznick, pres.

Bravo offers innovative arts & entertainment progmg with a unique point of view featuring original series, theater, dance, music & documentaries 24 hour.

Serving 75 million subs. Satellite: Satcom C-4, transponder 7.

Buzztime Entertainment, Inc., 5966 La Place Ct., Suite 100, Carlsbad, 92008. Phone: (760) 476-1976. Fax: (760) 438-3505. Web Site: www.buzztime.com.

Stanley Kinsey, chmn/CEO; Tyrone Lam, pres/COO; Dan Sweeney, sr VP; Pat Ruble, VP.

Buzztime is the only 24-hour interactive entertainment bcst created exclusively for TV audiences. Featuring play-along trivia games for players of all interests & ability levels.

CCTV 4, 11 Fuxing Rd., Beijing, 100859. China. Phone: (310) 414-2110. Phone: 011-86-10-6-850-6517. Fax: (310) 141-2101. Fax: 011-86-10-6-851-4993. Web Site: www.cctv-4.com.

Michael Scott, VP.

China Central TV (CCTV) China's only national bcstg network. Provide Info about China's politics, economy, society, culture, science, edu & history, also Chinese viewers, living outside of China, 24 hours a day. Serving more than 100 million subs. Satellite: Galaxy 11, transponder 24.

CMT: Country Music Television, 330 Commerce St., Nashville, TN, 37201. Phone: (615) 335-8400. Fax: (615) 335-8615. Web Site: www.cmt.com.

Brian Philips, sr VP/gen mgr.

CMT, America's # one country music network, 24 hours a day. CMT, owned & operated by MTV networks. Serving 73.5 million subs. Satellite: Satcom C-4, transponder 24, Satcom C3, transponder 18 west coast.

CNBC, 900 Sylvan Ave., Englewood Cliffs, NJ, 07632. Phone: (201) 735-2622. Web Site: www.cnbc.com.

Mark Hoffman, pres.

CNBC set the standard for up-to-the-minute business news & incisive analysis of global financial markets. During primetime, the network presents broad-base news, talk, interview & entertainment progmg.

Serving 86 million subs. Satellite: Galaxy 5, transponder 13.

CNN-Cable News Network, 1 CNN Ctr., Atlanta, GA, 30303. Phone: (404) 827-2300. Web Site: www.cnn.com.

David Payne, sr VP & gen mgr; Mich Gelman, sr VP & exec producer; Susan Grant, sr. VP news svcs.

CNN provides coverage of major breaking stories, business, weather, sports & special reports, worldwide audience, 24 hour.

Serving over 75.1 million subs. Satellite: Galaxy 5, transponder 5.

CNN en Espanol, One CNN Ctr., Atlanta, GA, 30303. Phone: (404) 878-1515. Phone: (404) 878-1555. Fax: (404) 878-0050. Web Site: www.cnnenespanol.com.

Christopher Crommett, sr VP; Chris Cramer, pres.

A 24-hour Sp-language news network in the United States & Latin America. The network keeps its loyal viewers connected with the events, issues trends that matter most to them & their families.

Serving more than 13 million subs.

CNN Headline News, One CNN Ctr., Atlanta, GA, 30348-5366. Phone: (404) 827-1500. Web Site: www.cnn.com/hln.

Roland Santo, exec VP & gen mgr.

Provides viewers with a 30-minute news, 24 hours a day. Each half-hour covers major news stories as well as business, sports, medicine, entertainment, weather & human interest topics. Areas: United States, Canada, Mexico & Caribbean.

Serving over 86 million subs. Satellite: Galaxy 5, transponder 22.

CNNI (CNN International), One CNN Ctr., 3rd Flr., Atlanta, GA, 30348. Phone: (404) 827-1500. Web Site: www.cnn.com/cnni.

Rena Golden, sr VP; Eric Ludgood, VP; Chris Cramer, mgng dir; Debra Kocker, VP.

A 24-hour global news & info, with live, breaking world news, sports, features & weather. Serving 170 million worldwide.

c/net: the computer network, 235 Second St., San Francisco, CA, 94105. Phone: (415) 344-2000. Fax: (415) 274-3750. Web Site: www.cnet.com.

Shelby Bonnie, CEO.

The current line up is CNET news.com, airs weekly on CNBC- every Sat. & Sun., 4 pm(eastern) & CNET tv.com airs on syndication.

C-SPAN (Cable Satellite Public Affairs Network), 400 N. Capitol St. N.W., Suite 650, Washington, DC, 20001. Phone: (202) 737-3220. Fax: (202) 737-3323. Web Site: www.c-span.org.

Brian P. Lamb, chmn/CEO; Rob Kennedy, exec VP & COO; Susan Swain, exec VP & COO; Bruce Collins, exec VP & corporate counsel.

C-SPAN progmg includes live coverage of the House of Representatives, National Pres Club speeches & congressional hearings. C-SPAN 2 live coverage of the U.S. Senate & C-SPAN 3 pub affrs TV. 87 million subs. Satellite: Satcom C-3, transponder 7.

CSTV: College Sports Television, 85 10th Ave., 3rd Fl., New York, NY, 10011. Phone: (212) 342-8700. Fax: (212) 342-8899. Web Site: www.cstv.com.

Brian T. Bedol, CEO.

The network televises regular season & championship events coverage from every major collegiate athletic conference & televises nine NCAA Championships available 24-hour, ad-supported. 15 million subs. Satellite: Galaxy IR, transponder 22.

CTI Zhong Tian Channel, 1255 Corporate Center Dr., Suite 212, Monterey Park, 91754. Phone: (323) 415-0068.

Andy Chung, gen mgr.

A 24-hour Mandarin-Chinese ch, consists of progmg derived from Chinese TV Int'l reputable Zhong Tian news. Satelite: Galaxy 11, transponder 24.

The California Channel, 1121 L St., Suite 110, Sacramento, CA, 95814. Phone: (916) 444-9792. Fax: (916) 444-9812. E-mail: contactus@calchannel.com Web Site: www.calchannel.com.

John Hancock, pres.

Televised coverage of California state legislature & govt agency proceedings. M-F, 9 AM-3:30 PM. Satellite: Galaxy 15. On 114 cable systems. 5.8 million subs. transponder 3 C.

Canal 24 Horas, 1100 Ponce de Leon Blvd., Coral Gables, 33134. Phone: (305) 444-4402. Fax: (305) 444-6301. E-mail: aragon@tveamerica.com Web Site: www.rtve.es.

Mariano Aragon, mgr.

A 24 hour news network from TVE which offers a Headline News fromat with 30 minute blocks. The network also produces 17 different 30 minutes daily & wkly news magazines. Areas: United States, Mexico, Caribbean, Central & South America & Europe. 250,000 subs. Satellite: Telstar 5, transponder 1.

Cartoon Network, 1050 Techwood Dr. N.W., Altanta, GA, 30318. Phone: (404) 827-4700. Fax: (404) 827-1700. Web Site: www.cartoonnetwork.com.

Jim Samples, gen mgr; Dennis Adamovich, mktg VP; Gary Albright, sr VP; Bob Higgins, progmg VP.

A Turner Broadcasting System. Inc.'s 24-hour Cartoon Network offers the best in animated entertainment. Drawing from the world's largest cartoon library, also showcases unique original ventures such as "Johnny Bravo,' "Cow and Chicken," "Dexter's Laboratory," "Ed, Edd n Eddy," and "Cartoon Cartoon." Serving 86 million subs around the world. Satellite: Galaxy 1, transponder 15 (West). Satellite: Galaxy IR, transponder 8 (East).

Cartoon Network Latin America, 1030 Techwood Dr., N.W., Atlanta, GA, 3031815264. Phone: (404) 885-2434. Phone: (404)827-1700. Fax: (404) 885-2157. E-mail: ruthie.stephenson@turner.com Web Site: www.cartoonnetworkla.com.

The first global 24-hour cable ch programmed entirely with cartoons. Available in Sp, Portuguese or English.

Serving more than 10.9 million subs. Satellite: PanAmSat 3R.

Celticvision: The Irish Channel, 95 Wexford St., Needham, MA, 02494. Phone: (781) 444-2080. Fax: (781) 449-7074. E-mail: cvbarnwell@aol.com

Material drawn from RTE, UTV & the BBC, with original material creating a 24-hour ch of alt English-language progmg with a distinctly Irish flavor. Serving 200,000 subs on 3 multiple systems.

Channel One Russia Worldwide Network, 20 Frunzenskaya Daberezhmaya, Suite D, Moscow, 119146. Phone: (310) 414-2110. Fax: (310) 414-2101. Web Site: www.firstchannel.tv.

David Quinn, dir.

A 24-hour Russian language ch for Russian communities throughout the United States. The ch consists of dramas, movies, news, children progmg, sports, talk show and more. Satellite: Galaxy 11, transponder 24.

Chronicle DTV, 53 W. 36th St., Suite 203, New York, 10018. Phone: (212) 337-9700 ext.105. Fax: (212) 352-1190. Web Site: www.chronicledtv.com.

David Peipers, chmn; Richard Blume, pres/CEO.

A 24-hour digital TV progmg network offering diverse selection of feature length non-fiction & documentary programs. Chronicle is seen in 35 cities in the United States, Los Angeles, Miami, West Palm Beach & Orlando. one million subs. Satellite: Telstar 5, transponder 22.

Cinemax, (Home Box Office). 1100 Ave. of the Americas, New York, NY, 10036. Phone: (212) 512-1000. Fax: (212) 512-5637. Web Site: www.cinemax.com.

Cinemax is a 24-hour digital pay-TV svc designed to provide viewers with the most movies & fewest repeats. Multiplex chs: Cinemax, MoreMAX, ActionMAX, ThrillerMAX, WMAX, @MAX, 5StarMAX, OuterMAX. Cinemax is seen in the United States & Puerto Rico. 39 million subs. Satellite: Galaxy IR, transponder 23.

Classic Arts Showcase, Box 828, Burbank, CA, 91503. Phone: (323) 878-0283. Fax: (323) 878-0329. E-mail: casmail@sbglobal.net Web Site: www.classicartsshowcase.org.

James Rigler, pres & dir progmg; Charlie Mount, gen mgr.

CAS is a non-profit arts progmg svc that include 16 art disciplines. The svc also features classic video from independent producers with the right to show clips. We require copies from masters on BetaCam-SP Tap. Available 24 hour.

Serving 60 million subs. Satellite: Galaxy 1R, transponder 5.

College Entertainment Network, 6255 Sunset Blvd., Suite 611, Hollywood, 90028. Phone: (323) 465-9880. Fax: (323) 465-9881. Web Site: www.collegeentertainment.com.

Robert Artura, pres/CEO; Georgina Montalvan, VP mktg.

College Entertainment Network is a world of college TV stn on one network, it features block of programs, from extreme sports to entertainment. 10 million subs.

Comedy Central, 1775 Broadway, New York, NY, 10019. Phone: (212) 767-8600. Fax: (212) 767-8592. Web Site: www.comedycentral.com.

Michele Ganeless, exec VP & gen mgr.

Los Angeles, CA 90067, 2049 Century Park E, Suite 4250. Phone: (310) 201-9500. Fax: (310) 201-9488.

A 24-hour all comedy TV network that covers stand-up, sketch comedy, movies, talk shows, sitcoms, specials & classics TV shows.

Serving 84.8 million subs. Satellites: Satcom C-3, transponder 21 (east), Galaxy 1R, transponder 1 (west).

Courtroom Television Network (Court TV), 600 3rd Ave., 2nd Fl., New York, NY, 10016. Phone: (212) 973-2800. Fax: (212) 973-3210. Web Site: www.courttv.com.

Robert Rose, exec VP; Marc Juris, gen mgr & progmg & mktg; Ira Fields, exec VP & CFO; Art Bell, pres/COO.

Court TV telecast trails day by day & high profile original programs 24-hour. Serving 80 million subs.

The Crime Channel, 78206 Varner Rd., Suite D131, Palm Desert, CA, 92211. Phone: (760) 360-6151. Fax: (760) 360-3258. E-mail: crimechannel@dcrr.com

Arnie Frank, pres.

The Crime Channel offers series, movies, documentaries, original productions, on the spot crime news & foreign programs. Satellite: Satcom C-1, transponder 11.

Daystar Television Network, 4201 Pool Rd., Colleyville, 76034. Phone: (817) 571-1229. Fax: (817) 571-7458. E-mail: comments@daystar.com Web Site: www.daystar.com.

Janice Smith, VP progmg; David Troxel, VP.

Our progmg is multi-ch & interdenominational. Christian TV net, is available from DirectTV, Dish Network 24-hours a day.

De Pelicula, 5999 Center Dr., Los Angeles, 90045. Phone: (301) 348-3371. Web Site: www.tutv.tv.

Mark Feldman, pres/CEO; Carlos Madrazo, CFO.

A 24-hour contemporary & classic movie ch featuring the best Sp language films.

Multiplex ch: De Pelicula Clasico is a 24-hour movie ch featuring the best films of Mexico's golden era.

TeleHit is a young, hip & cutting-edge, trend setting lite-style & music ch. 750,000 subs.

Deep Dish TV, 339 Lafayette St., New York, NY, 10012. Phone: (212) 473-8933. Fax: (212) 420-8223. E-mail: deepdish@igc.org Web Site: www.deepdishtv.org.

Ron Davis, chmn; Tom Pool, exec dir; Victoria Macdonado, VP.

Educational progmg (one hour a wk) distributed to PBS & pub access chs. Satellite: Galaxy IR, transponder 15.

Discovery Channel, 4 Choke Cherry Rd., Rockville, MD, 20850. Phone: (301) 354-2000. Fax: (240) 662-1854. E-mail: first_last@discovery.com Web Site: www.discovery.com.

Billy Campbell, pres; Joe Abruzzese, pres & adv sls; Billy Goodwyn, pres & affil sls & mktg; Jane Root, exec VP & gen mgr.

Discovery Channel, the United States' largest cable TV net, is the nation's premier provider of real-world entertainment, offering a signature mix of compelling, high-end production values & vivid cinematography that consistently represents qulity for viewers. Primetime progmg features science a& technology, exploration, adventure, history and in-depth, behind-the scenes glimpses at the people, places & organizations that shape & share our world, & is dedicated to creating the highest quality TV & media to inspire audiences by delivering knowledge about the world in an energizing way; evolving a timeless brand for a changing world. 74.9 million subs. Satellite: Satcom C-4, Transponder 13.

Discovery en Espanol, One Discovery Place, Silver Spring, MD, 20910. Phone: (240) 662-2000. Fax: (240) 662-1854. E-mail: first_last@discovery.com Web Site: www.discovery.com.

Billy Campbell, pres; Joe Abruzzese, pres & adv sls; Bill Goodwyn, pres & affil sls & mktg; Luis Silberwasser, exec VP & gen mgr.

Discovery en Espanol brings U.S. Sp-speaking viewers the world's best entertainment. Science & technology. People, places & worlds cultures. Nature progmg. All in the Discovery tradition of quality production, spectabular cinematography & fantastick storytelling. And today, Discovery en Espanol features even more of the very best programs from the family of Discovery Networks than ever before.

Discovery HD Theater, One Discovery Pl., Silver Spring, MD, 20910. Phone: (240) 662-2000. Fax: (240) 662-1854. E-mail: first_last@discovery.com Web Site: www.discovery.com.

Billy Campbell, pres; Joe Abruzzese, pres & adv sls; Patrick Younge, exec VP & gen mgr; Bill Goodwyn, pres & affil sls & mktg.

Discovery HD Theater is a 24-hour high-definition(HD) network from the creators of the Discovery Channel. Offering consistent, real-world entertainment in virtually all categories of entertainment created by Discovery Networks, Discovery HD Theater showcases progmg about adventure, nature, wildlife, science & technology, world culture & more - all designed to provide viewers with the hightest-quality TV experience available.

Discovery Health Channel, One Discover Pl., Silver Spring, MD, 20910. Phone: (240) 662-2000. Fax: (240) 662-1854. E-mail: first_last@discovery.com Web Site: www.discovery.com.

Eileen O'Neill, exec VP & gen mgr; Billy Campbell, pres; Joe Abruzzese, pres & adv sls; Bill Goodwyn, pres & affil sls & mktg.

Discovery Health Channel takes viewers inside the fascinating and informative world of health and medicine to experience firsthand, compelling, real life stories of medical breakthroughs and human triumphs.

Discovery Home Channel, One Discovery Pl., Silver Spring, MD, 20910. Phone: (240) 662-2000. Fax: (240) 662-1854. E-mail: first_last@discovery.com Web Site: www.discovery.com.

Bill Campbell, pres; Joe Abruzzese, pres & adv sls ; Bill Goodwyn, pres & affil sls & mktg; David Abraham, exec VP & gen mgr.

Discovery Home Channel is the only 24-hour progmg destination for every aspect of the home. On the net, viewers can find progmg about design, home improvement & food-all under one roof. Relaunched in Spring 2004, Discovery Home Channel offers progmg that informs, motivates & appeals to viewers" sense of creativity, style & fun, whether they're preparing anew dinner recipe, designing a room of landscaping the backyard.

Discovery Kids Channel, (Discovery Communications). One Discovery Pl., Silver Spring, MD, 20910. Phone: (240) 662-2000. Fax: (240) 662-1854. E-mail: first_last@discovery.com Web Site: www.discovery.com.

Billy Campbell, pres; Joe Abruzzese, pres & adv sls; Bill Goodwyn, pres & affil sls & mktg; Marjorie Kaplan, exec VP & gen mgr.

Discovery Kids, nominated for a record 14 Daytime Emmys in 2006, is the only place on TV devoted to helping kids explore the real world - from the depths of the ocean to outer space to self-awareness. From the Emmy-winning DISCOVERY KIDS ON NBC progmg block to READY, SET, LEARN! for preschoolers & the Discovery Kids Channel, Discovery Kids offers the best in real-world entertainment for kids.

Discovery Kids en Espanol, One Discovery Pl., Silver Spring, MD, 20910. Phone: (240) 662-2000. Fax: (240) 662-1854. E-mail: first_last@discovery.com Web Site: www.discovery.com.

Billy Campbell, pres; Joe Abruzzese, pres & adv sls ; Bill Goodwyn, pres & affil sls & mktg; Luis Silberwasser, exec VP & gen mgr.

Discovery Kids en Espanol is the first U.S. Sp-language network that appeals to everyone in the household, targeting preschoolers, tweens & families. It's the perfect place to stimulate & inspire kids' thirst for knowledge, while challenging them to explore their world.

Discovery Times Channel, One Discovery Pl., Silver Spring, MD, 20910. Phone: (240) 662-2000. Fax: (240) 662-1854. E-mail: first_last@discovery.com Web Site: www.discovery.com.

Billy Campbell, pres; Joe Abruzzese, pres & adv sls; Bill Goodwyn, pres & affil sls & mktg; Vivian Schiller, exec VP & gen mgr.

The Discovery Times Channel combines the journalistic authority of The New York Times with the progmg excellence of Discovery Communications to bring viewers provocative, engaging and relevant documentary series & specials about the events & ideas shaping our times. The network has won mulitple industry honors for its breakthrough progmg, including three Emmys, an Overseas Press Club Award, three Natl Headliner Awards & a BANFF Rockie Award. In addition the network has had films premiere at various film festivals, including the Sundance Film Festival, Tribeca Film Festival, San Francisco International Film Festival & the Tokyo Video Festival. Available to over 39 million suscribers in the United States, the Discovery Times Channel is a joint venture between the New York Times Co. (NYSE-NYT) & Discovery Communications, Inc., the leading global real-world media & entertainment Co.

Discovery Travel & Living, One Discover Pl., Silver Spring, MD, 20910. Phone: (240) 662-2000. Fax: (240) 662-1854. E-mail: first_last@discovery.com Web Site: www.discovery.com.

Billy Campbell, pres; Joe Abruzzese, opns VP; Bill Goodwyn, pres & affil sls & mktg; Luis Silberwasser, opns VP.

Discovery Travel & Living (Viajar y Vivir) is a getaway for adults who want to experience the best the world has to offer-in their native language. The net provides an eclectic mix of progmg on travel, well-being, food, design a& decor & encourages Hispanic viewers to live life well & make the most of their free time.

Disney Channel, 3800 W. Alameda Ave., Burbank, CA, 91505. Phone: (818) 569-7500. Fax: (818) 566-1358. Web Site: www.disneychannel.com.

Rich Ross, pres.

A 24-hour gen entertainment network for kids & families through original series, movies & contempary acquired progmg. Serving over 87 million subs. Satellite: Galaxy 5, transponder 1 east, Galaxy 1-R, transponder 7 west.

The Dream Network, 9300 Georgia Ave., Suite 206, Silver Springs, 20910. Phone: (301) 587-0000. Fax: (301) 587-7464. Web Site: www.thedreamnetwork.com.

Alvin Augustus Jones, pres/CEO.

The Dream Network is the urban family choice for news, talk, sports & gospel music. Seen in the United States, Canada, Caribbean, Europe, Africa & Asia. Serving 25 million subs on 10 million cable systems. Satellite: DirecTV.

E! Entertainment Television, 5750 Wilshire Blvd., Los Angeles, CA, 90036-3709. Phone: (323) 954-2400. Fax: (323) 954-2500. Web Site: www.eonline.com.

Neil Baker, sr VP & adv sls; Ted Harbert, pres/CEO; Ken Bettsteller, CEO.

New York, NY 10036, 11 W. 42nd St. Phone: (212) 852-5100. (212) 852-5151. Dave Cassaro, exec VP.

A 24-hours progmg net covering celebrities, entertainment news, gossip & pop-culture, feature behind the scenes with today's biggest stars. 84 million subs. Satellite: Satcom C-3, transponder 23.

ESPN, ESPN Plaza, Bristol, CT, 06010-9454. Phone: (860) 766-2000. Web Site: www.espn.com.

Edward Erhardt, pres; Ed Durso, exec VP & admin; Lee Ann Daly, exec VP mktg; Chris Driessen, exec VP & CFO.

New York, NY 10158-0180, 605 3rd Ave. Phone: (212) 916-9200.

A 24-hour svc covering sports events, news, info, & lifestyle progmg.

ESPN CLASSIC is a 24-hour, all sports network devoted to telecasting the greatest games, stories, heroes & memories in the history of sports. Serving 50 millon subs. Satellite: Galaxy 10R, transponder 20.

ESPN DEPORTE offers a wide var of domestic & intl sports progmg 24-hour. Serving 14 million subs. Satellite: Galaxy 10R, transponder 20.

ESPN HD offers a 24-hour high definition TV svc from ESPN, features high profile telecast. Satellite: Galaxy 10R, transponder 22.

ESPNEWS , the nation's only 24-hour TV sports news svc, provides an expanded window for news & highlights,as well as live coverage of beaking news. Serving 50 million subs. Satellite: Galaxy 10R, transponder 21. 88 million subs. Satellite: Galaxy V, transponder 9.

ESPN2, ESPN Plaza, Bristol, CT, 06010-9454. Phone: (860) 766-2000. Web Site: www.espn.com.

A 24-hour sports network features a progmg line-up on par with ESPN. 86 million subs. Satellite: Galaxy V, transponder 14.

ESPNU, c/o ESPN Regional Television, 11001 Rushmore Dr., Charlotte, NC, 28277. Phone: (704) 973-5000. Fax: (877) 378-3776. Web Site: www.espn.com.

Burke Magnus, VP/gen mgr.

EWTN, The Global Catholic Network. 5817 Old Leeds Rd., Irondale, AL, 35210. Phone: (205) 271-2900. Fax: (205) 271-2925. Web Site: www.ewtn.com.

R. William Steltemeier, chmn/CEO; Michael P. Warsaw, pres; Chris Wegemer, VP mktg; Scott Hults, dir; Doug Keck, VP.

America's largest relg cable network offers coml-free family-oriented progmg in English & Sp. EWTN features documentaries, music, drama, live talk shows, animated children shows & special church events from around the world. 53 million subs. Satellite: Galaxy IR, transponder 11.

Ecology Communications, 9171 Victoria Dr., Ellicott City, 21042. Phone: (410) 465-0480. Fax: (410) 461-5152. Web Site: www.ecology.com.

Shelley Duvall, chmn/CEO; Eric McLamb, pres.

Ecology Communications focuse on ecology & enviroment in a var of entertainment-driven formats wkly. 10 million subs.

Encore Media Corp., 8900 Liberty Cir., Englewood, CO, 80112. Phone: (720) 852-7700. Fax: (720) 852-7710. Web Site: www.starz.com/appmanager/seg/s.

Bill Myers, exec VP & CFO; Jerry Maglio, exec VP & mktg; Robert B. Clasen, pres/CEO.

Los Angeles, CA 90025. International Channel, 11766 Wilshire Blvd, Suite 710. Phone: (310) 477-9922. Fax: (310) 477-4544. Victoria Kent, rgnl VP.

Englewood, CO 80111. Founders, 5445 DTC Pkwy, Suite 600. Phone: (303) 771-7700. Fax: (303) 267-4001. Robin Feller, rgnl VP.

Englewood, CO 80111. New Media, 5445 DTC Pkwy, Suite 600. Phone: (303) 771-7700. Fax: (303) 267-4098. Leslie Nittler, VP sls & mktg.

Atlanta, GA 30342. southeast region, 5775 Peachtree Dunwoody Rd, Suite D-620. Phone: (404) 531-7060. Fax: (404) 531-7075. Cindy Feinberg, rgnl VP.

Chicago, IL 60601. central region, 111 E. Wacker Dr, Suite 1300. Phone: (312) 938-8900. Fax: (312) 938-8902. Susan LeVarsky, rgnl VP.

Hoboken, NJ 07030. eastern region, 70 Hudson St. Phone: (201) 239-9020. Fax: (201) 239-2290. Shanita Evans, rgnl VP.

Carrollton, TX 75006. Time Warner & Central region office, 2340 E. Trinity Mills Rd, Suite 300. Phone: (972) 417-2866. Fax: (972) 417-2806. Paige Holmes.

Encore Media Corp. offers the following premium cable progmg svcs:

ENCORE presents hit movies from the 1960s, 1970s & 1980s, coml-free, 24-hours per day. Serves 6 millions subs. Satellite: Galaxy 1-R, transponder, 3.

STARZ!-encore 8 features first-run releases from leading distributors 24-hours per day. Serves 2 million subs. Satellite: Satcom C-4, transponder 2.

MOVIEPLEX is a STARZ! ch, never shows R-rated progmg & is offered as an analog feed only. Serves 5.4 million subs. Satellite: Satcom C 4, transponder 5 east & west.

MYSTERY features great coml-free mystery movies, 24-hours a day.

STARZ ON DEMAND is an enchancement to STARZ!. Satellite: Galaxy 1R, transponder 13 east & west.

STARZ! CINEMA features movies for everyone who is passionate about movies. Satellite: Galaxy 1R, transponder 13 east, Satcom C4, transponder 5 west.

STARZ! FAMILY is the first & only coml-free ch showing family movies. Satellite: Galaxy 1R, transponder 13 east, Satcom C4, transponder 5 west.

STARZ! SUPER PAK features new hit movies. Serves 12 million subs. Satellite: Galaxy 1R, transponder 13 east & west.

STARZ KIDS is a 24-hour ch for kids ages 2-8. Satellite: Galaxy 1R, transponder 13 east, Satcom C4, transponder 5 west.

STARZ! THEATER is programme like a movie theater. Satellite: Galaxy 1R, transponder 13 east, Satcom C4, transponder 5 west.

TRUE STORIES tell is like it is, presenting real life dramas. Satellite: Galaxy 1R, transponder 13 east & west.

WESTERN ch features Hollywood's most popular past & contemporary western films.

ENCORE THEMATIC MULTIPLEX offers six, genre-specific movie svcs 24-hours per day.

Satellite: Galaxy 1-R, transponder 13.

ENCORE THEME BY DAY offers hit movies from the 1960s, 1970s & 1980s. Genres are determined by the day of the week.

ENCORE HD is the high definition version of the flagship Encore movies svc, 24-hour. Serves over 63,699 subs. Satellite: Galaxy 13, transponder 7 west, transponder 9 east, Digital Digicipher II.

ACTION offers non-stop excitement, action movies featured nightly.

LOVE STORIES is the place for movie romance. Satellite: Galaxy 1R, transponder 3, east & west. Starz Entertainment Group is a premium movie svc provider in the US offering movie chs & airs over 1,000 movies a month.

eSignal, 3955 Point Eden Way, Hayward, CA, 94545. Phone: (510) 266-6000. Fax: (510) 266-6100. Web Site: www.esignal.com.

Chuck Thompson, pres; Grant Mader, VP net ops & telecommunications.

Real-time stock, option, commodity quotation service & sports service delivered via cable TV, FM frequency & bcst VBI to end user PC.

Serving more than 20,000 subs on more than 850 systems. Satellites: Satcom F3, transponder 11; Galaxy 3, transponder 24.

EUROCINEMA, European Movies on TV. 387 Park Ave., 3rd Fl., New York, NY, 10016. Phone: (212) 763-5533. Fax: (212) 685-7636. E-mail: eurocinema@envrocinema.com Web Site: www.eurocinema.com.

Sebastiaen Perioche, chmn/CEO.

Non-Hollywood movies service for Broadband & Digital TV

FX Networks Inc., (A subsidiary of Fox, Inc.). 1440 S. Sepulveda Blvd., Los Angeles, CA, 90025. Phone: (310) 444-8777. Fax: (310) 444-8667. Web Site: www.fxnetworks.com.

John Landgraf, pres; Chuck Saftler, VP progmg; Christy Dees, dev dir; Michael Sakin, sr VP; Mark DeVitre, sr VP; Chris Carlisle, sr VP; Lindsay Gardner, exec VP; Steve LeBlang, rsch dir; John Solberg, VP; Steve Webster, VP; Eric Shiu, VP.

An entertainment basic cable net from Fox Television involving hit series, daily films, original programs & sports.

Serving more than 53 million subs on 3,293 cable systems. Satellite: Hughes Communications Galaxy 7, transponder 4 & 5.

FamilyNet Television & Radio, 6350 West Fwy., Fort Worth, TX, 76116. Phone: (800) 832-6638. Phone: (817)737-4011. Fax: (817) 298-3388. E-mail: info@familynet.com Web Site: www.familynet.com.

Martin Coleman, COO; R. Chip Turner, VP mktg; Randy Singer, pres/CEO; Ray Raley, engrg VP.

FamilyNet TV is a full-time cable network, including original values-based programs & operates Christian talk ch 161 on sirius statellite radio. Reliable, safe TV for today's family. Satellite Intersat Americas 13, transponder 20.

FitTV, One Discovery Pl., Silver Spring, MD, 20910. Phone: (240) 662-2000. Fax: (240) 662-1854. E-mail: first_last@discovery.com Web Site: www.discovery.com.

Billy Campbell, pres; Joe Abruzzese, pres & adv sls; Bill Goodwyn, pres & affil sls & mktg; Eileen O'Neill, exec VP & gen mgr.

FitTV is the premier, interactive fitness brand that inspires consumers to improve their fitness & well-being-on their terms. FitTV features approachable experts & entertaining shows that help real people learn how to incorporate fitness in to their busy lives.

Food Network, 1180 6th Ave., 12th Fl., New York, NY, 10036. Phone: (212) 398-8836. Fax: (212) 736-7716. Web Site: www.foodnetwork.com.

Brooke Johnson, pres; Adam Rockmore, VP mktg.

A 24-hour network committed to exploring new, different & interesting ways to approach food. 82 million subs. Satellite: Galaxy IR, transponder 4.

Fox Movie Channel, (FX Networks Inc.). Box 900, Beverly Hills, CA, 90213. Phone: (310) 369-0923. Fax: (310) 969-4687. Web Site: www.thefoxmoviechannel.com.

Chuck Saftler, gen mgr.

The network is dedicated to preserving Hollywood history through original series & specials. Satellite: Galaxy GE7, transponder 7.

FX (FOX BASIC CABLE) is a flagship gen entertainment basic cable network from Fox, 19 hours a day. Serving 84 million subs.

Fox Net, 10201 W. Pico Blvd., Bldg. 100, Los Angeles, CA, 90064. Phone: (310) 369-1000. Fax: (310) 969-1360. Web Site: www.fox.com.

Susan Kiel, VP; Dwayne Bright, progmg dir; Wendy Chambers, dir; Betty Wang, dir mktg; Keith Goldberg, dir; Mark Handwerger, natl sls mgr; Julie Allen, natl sls mgr; Steve Nazar, opns mgr.

Los Angeles, CA 90035. FOX Broadcasting Co, 10201 W. Pico Blvd. Phone: (310) 369-5153. Fax: (310) 969-0316. Susan Kiel, VP, Network Distribution & Cable Operations.

A 24-hour basic cable affil to Fox Broadcasting Co.

Serving 2 million subs on 1,350 cable systems. Satellite: Satcom C-1, transponder 19.

Fox News Channel

See listing in Major National TV News Organizations, this section.

Free Speech TV, Box 44099, Denver, CO, 80201. Phone: (303) 442-8445. Fax: (303) 442-6472. E-mail: jon@freespeech.org Web Site: www.freespeech.org.

John Schwartz, pres; Jon Stout, gen mgr.

FSTV airs primarily social, political, cultural & environmental documentaries & news progms, 24-hours a day. Serving 25 million subs on cable & DISH Network. 163. 15,500,000.

Fuel, 1440 S. Sepulveda Blvd., Suite 1900, Los Angeles, 90025. Fax: (310) 444-8559. E-mail: hookup@fuel.tv Web Site: www.fuel.tv.

Cj Olivares, VP progmg; David Sternberg, gen mgr; Kelsey Martinez, dir.

A 24-hour sports network featuring snowboarding, wakeboarding, surfing, BMX, motorcross & skateboading. Serving 5 million subs. Satellite: Galaxy II, transponder 5.

Fuse, 11 Penn Plaza, 15th Fl., New York, NY, 10001. Phone: (212) 324-3400. Fax: (212) 324-3445. Web Site: www.fuse.tv.

Eric Sherman, pres; Norman Schoenfeld, VP progmg; Michael Goldstein, VP; Kim Martin, exec VP; Theano Apostolou, VP.

Santa Monica, CA 90404. The Water Garden, 2425 W. Olympic Blvd, Suite 5050. Phone: (310) 998-9300. John Pezzini, VP.

Chicago, IL 60601. Chicago Office, 205 N. Michigan, Suite 803A. Phone: (312) 938-4222. Joseph Glennon, Sr. VP.

Fuse is the only all-music, viewer-influenced TV network, featuring music videos, exclusive artist interviews, live concerts & specials. Serving 35 million subs. Satellite; Loral Skynet Telstar 7, transponder 14.

Galavision, 605 3rd Ave., 12th Fl., New York, NY, 10158. Phone: (212) 455-5300. Fax: (212) 986-4731. Web Site: www.univisionnetworks.com.

Tim Krass, exec VP affil affrs; Timothy Spillane, VP.

Los Angeles, CA 90045, 6701 Center Dr. W, Suite 650. Phone: (310) 348-3600.

Chicago, IL 60611, 541 N. Fairbanks Ct, Suite 1240. Phone: (312) 494-5101.

Dallas, TX 75201, 2323 Bryan St, Suite 1900. Phone: (214) 758-2300.

A 24-hr Sp language cable network for united states hispanics in distribution & viewership. 37 million subs. Satellite: Satcom C-4, transponder 4.

Game Show Network, 2150 Colorado Ave., Santa Monica, CA, 90404. Phone: (310) 255-6800. Fax: (310) 255-6810. E-mail: distribution@gsn.com Web Site: www.gameshownetwork.com.

Chicago, IL 60610, 515 N. State St. Suite 2120. Phone: (312) 261-4500. Fax: (312) 261-4521.

New York, NY 10019, 680 Fifth Ave. 11th Fl. Phone: (212) 333-2510. Fax: (646) 557-2996.

Dallas, TX 75201, The Republic Center, 325 N. St. Paul St. Suite 1500. Phone: (214) 965-8500. Fax: (214) 965-8576.

Game Show Network (GSN) is the only U.S. television network dedicated to game progmg and interactive game playing, featuring over 65 hours per week of original progmg & enhanced classics.

Reaching 50 million subs.

The Golf Channel, 7580 Commerce Center Dr., Orlando, FL, 32819. Phone: (407) 363-4653. Fax: (407) 363-7976. Web Site: www.thegolfchannel.com.

Page H. Thompson, pres.

A 24-hour ch offering a blend of tournament coverage from the PGA, LPGA, Sr Tour, Nike, EPGA Tours, as well as instruction, interactive talk, news, profiles, classics, travel & more.

Serving 30 million subs on 2,200 cable systems. Satellite: Galaxy VI, Transponder 7.

GoodLife TV Network, 650 Massachusetts Ave. N.W., Washington, DC, 20001. Phone: (202) 289-6633. Fax: (202) 289-6632. Web Site: www.goodtv.com.

Mark Ringwald, VP progmg; Lawrence R. Meli, pres/CEO.

GoodLife TV Network is the nation's only full-time cable ch dedicated to improving the quality of American life through info & entertainment programs. Satellite: Galaxy IR, transponder 22.

Great American Country, 9697 E. Mineral Ave., Centennial, CO, 80112. Phone: (303) 792-3111. Fax: (303) 784-8518. Web Site: www.countrystars.com.

Jeffrey Wayne, pres/CEO; Kenneth O. Street, sr VP; Anthony M. Aiello, VP; Scott Durand, VP mktg; Glenn Jones, chmn.

GAC is a country music video network, features national & loc advertising, also featuring a broad var of videos programs for ages 25-54, 24 hours a day. 24 million subs. Satellite: Satcom C-3, transponder 20.

Guthy-Renker Television, 41-550 Eclectic St., Suite 200, Palm Desert, CA, 92260. Phone: (760) 773-9022. Fax: (310) 234-3601. Web Site: www.guthy-renker.com.

Direct-response TV

HBO (Home Box Office), 1100 Ave. of the Americas, New York, NY, 10036. Phone: (212) 512-1000. Fax: (212) 512-1166. Web Site: www.hbo.com. E-mail: info@hob.com

Bill Nelson, chmn/CEO.

Features 24-hour var progmg including theatrical films, original movies, specials, documentaries, sports, & series.

Multiplex chs. HBO, HBO 2, HBO Latino, HBO Signature, HBO Family, HBO Comedy, HBO Zone - known collectively as HBO The Works. MoreMAX.

Serving 39 million subs. Satellites: Analog, Galaxy 1R, transponder 23; Galaxy 1R, transponder 18; Analog: Galaxy 5, transponder 8; Galaxy 3R, transponder 16; Galaxy 3R, transponder 20; Galaxy 3R, transponder 19.

HDNET, 2400 N. Ulster St., Denver, 80238. Phone: (303) 388-8500. Fax: (303) 388-9600. E-mail: info@hd.net Web Site: www.hd.net.

Mark Cuban, chmn/pres; Philip Garvin, gen mgr.

HDNET, the leader in high-indefinition bcstg, produces & televises 24-hour a day. Satellite: Galaxy 9, transponder 19C.

HDNET Movies is a 24-hour coml-free schedule of full-length feature films. Satellite: Galaxy 9, transponder 19C.

HSN, The Home Shopping Network, One HSN Dr., St. Petersburg, FL, 33729. Phone: (727) 872-1000. Fax: (727) 872-7356. Web Site: www.hsn.com.

Bob Rosenblatt, pres; Peter Ruben, exec VP; Tom McInerney, CEO.

HSN offers live, 24-hour video retailing. 81 million subs. Satellite: Satcom C-4, transponder 10.

Hallmark Channel, 12700 Ventura Blvd., Suite 200, Studio City, CA, 91604-2463. Phone: (818) 755-2400. Fax: (818) 755-2564. Web Site: www.hallmarkchannel.com.

David Kenin, VP progmg.

A 24-hour basic cable ch that provides high quality entertainment progmg to a national audience. 56 million subs. Satellite: Satcom C-3, transponder 5.

Home & Garden Television Network (HGTV), 9721 Sherrill Blvd., Knoxville, TN, 37932. Phone: (865) 694-2700. Fax: (865) 531-1588. Web Site: www.hgtv.com.

Mike Boyd, VP mktg; Judy Girard, pres.

HGTV is a 24-hour network that provide practical info & creative ideas to help the viewers to make the most of their lives at home & is designed to appeal to all ages & lifestyles. Satellite: Galaxy IR, transponder 4.

Home Improvement Television Network, 3441 Baker St., San Diego, CA, 92117. Phone: (858) 273-0572. Fax: (858) 273-8410. E-mail: homefix@hometvnet.com Web Site: www.hometvnet.com.

Bruce Lamb, pres.

Providers of home improvement progmg & 90-second video vignettes.

ICF Film, 11 Penn Plaza, New York, NY, 10001. Phone: (516) 803-4500. Fax: (516) 803-4506. Web Site: www.ifctv.com. E-mail: help@ifctv.com

Jonathan Sehring, pres. Satellite: Galaxy VII.

I.M.A.G.E. LLC, (Interactive Meet and Greet Entertainment). Box 702, 67 River St., Hudson, MA, 01749-0702. Phone: (508) 788-5474. E-mail: wexrex@aol.com Web Site: www.meetandgreet.tv.

Gary Sohmers, chmn/pres; Chip Sohmers, CEO.

Basic svc, 24-hour progmg targeted to collectors of various merchandise & memorabilia. Educ & entertaining progmg with a shopping element, with patent pending interactive applications.

The Idea Channel, 2002 Filmore Ave., Suite 1, Erie, PA, 16506-6913. Phone: (814) 464-9068. Fax: (814) 833-7415. E-mail: info@iclearchannel.com Web Site: www.ideachannel.com.

Bob Chitester, pres/founder; Rick Platt, VP.

Educ progmg.

The Independent Film Channel (IFC), (A division of Rainbow Media Programming Holdings). 200 Jericho Quadrangle, Jericho, NY, 11753. Phone: (516) 803-3000. Fax: (516) 803-4616. E-mail: webmaster@ifctv.com Web Site: www.ifctv.com.

Jonathan D. Sehring, pres.

A 24-hour uncut coml-free ch, capturing the true spirit of independent film, original series, live events & enchanced new media progmg. Serving 29 million subs. Satellite: Galaxy 7, transponder 14.

The Inspiration Network INSP, 7910 Crescent Executive Dr., 5th Fl., Charlotte, NC, 28217. Phone: (704) 525-9800. Web Site: www.insp.com.

David Cerullo, pres/CEO; Rod Tapp, exec VP; Tom Hohman, sr VP; Larry Simms, VP sls; Ron Shuping, VP progmg.

INSP blends ministry programs with family-oriented movies, dramas, music & children's programs along with concerts & specials. 21 million subs. Satellite: Galaxy IR, transponder 17.

Inspirational Life Television, (I-Lifetv). 7910 Crescent Executive Dr., 5th Fl., Charlotte, NC, 28217. Phone: (704) 525-9800. Fax: (704) 527-9899.

David Cerullo, pres/CEO; Tom Hohman, sr VP; Larry Sims, VP sls.

Ilifetv is a 24-hour a day programmer & digital TV network which distributes life-enriching, edu entertainment progmg. 6 million subs. Satellite: Galaxy IR, transponder 7.

International Channel, 8900 Liberty Circle, Englewood, CO, 80112. Phone: (720) 853-2933. Fax: (720) 853-2901. Web Site: www.internationalchannel.com.

Bill Georges, sr VP; Rod Shanks, gen mgr.

International Channel is the only cable ch serving multiple ethnic audiences 24-hour a day nationwide. On 1400. 12.5 million subs. Satellite: Galaxy IX, transponder 24.

International Networks, 4100 E. Dry Creek Rd., Suite A 300, Centennial, CO, 80122. Phone: (303) 712-5400. Fax: (303) 712-5401. Web Site: internationalnetworks.com.

Rod Shanks, mgng dir; Scott Wheeler, sr VP net dev; Michael Scott, VP distribution; Victor Perez, VP technology.

New York, NY 10036, 1114 Ave. of the Americas, 22nd Fl. Phone: (917) 934-1558.

Los Angeles, CA 90064, 12100 W. Olympic Blvd, Suite 200. Phone: (310) 979-2246.

Internet Networks is the leading expert in providing & mktg global prgmg to the growing multi-ethnic, multi-cultural & multi-lingual communities in the US. International Neworks distributes 16 networks that provide Asian, European & Middle Eastern progmg from a variety of international sources. Arabic Radio & TV (Arabic); China Central TV-4 (Mandarin; Channel One Russia Worldwide (Russian); CTI Zhong Tian Channel (Mandarin); Munhwa Bcstg Corp. (Korean); ProSiebenSat.1 Welt (German); RAI International (Italian); Russian TV Net (Russian); Saigon Bcstg TV Net (Vietnamese); The Filipino Channel (Filipino); TV Asia (S. Asian); TV JAPAN (Japanese); tvK (Korean); tvK2 (Korean); TV5MONDE (French); TV Polonia (Polish). Satellite: Galaxy 11. 14 Million. 24.

Jewelry Television By ACN, 10001 Kingston Pike, Suite 57, Knoxville, TN, 37922. Phone: (865) 692-6000. Fax: (865) 693-3688. Web Site: www.jewelrytelevision.com.

Harry Bagley, sr VP affil rel.

ACN is the only network that focuses exclusively on the sls of fine jewelry & gemstones, 24-hours a day. On 11 million cable systems. 32 milion subs. Satellite: Telstar 5, transponder 19.

Jewish Television Network, 13743 Ventura Blvd., Suite 200, Sherman Oaks, CA, 91423. Phone: (818) 789-5891. E-mail: jewishtv@earthlink.net Web Site: www.jewishtvnetwork.com.

Jay Sanderson, CEO.

Production & cablecasting of net-quality Jewish progmg in news, pub affrs, education, arts, PBS & entertainment.

Serving 6 million subs on 15 cable systems.

The Jones Companies, 9697 E. Mineral Ave., Englewood, CO, 80112. Phone: (303) 784-8232. Fax: (303) 784-8549. E-mail: publicrelations@ones.com Web Site: www.jones.com.

Glenn R. Jones, pres/CEO; Timothy J. Burke, VP.

A 24-hour revenue providing svc that features full-length product demonstrations & introduction of new products in a hands-on demonstration format.

Serving 26 million subs. Satellite: GE Satcom C3, transponder 20.

KTLA, United Video (a company of the United Video Satellite Group). 5800 Sunset Blvd., Los Angeles, CA, 90028. Phone: (323) 460-5500. Fax: (323) 460-5333. Web Site: ktla.trb.com.

John Reardon, VP/gen mgr.

Los Angeles WB stn offers movies, news, specials & live sporting events, featuring the Los Angeles Clippers basketball . Services offered: Ind satellite carrier serving CATV, SMATV & MMDS distributing WGN, WPIX, KTLA & WFMT (FM) Network Svcs.

Satellite: Spacenet 6E. On 220 cable systems. 1 million subs. transponder 15.

The Learning Channel (TLC), (Discovery Communications). One Discovery Pl., Silver Spring, MD, 20910. Phone: (240) 662-2000. Fax: (240) 662-1854. E-mail: first_last@discovery.com Web Site: www.discovery.com.

Billy Campbell, pres; Joe Abruzzese, pres & adv sls; Bill Goodwyn, pres & affil sls & mktg; David Abraham, exec VP & gen mgr.

TLC is the only TV network dedicated to lifelong learning for viewers who want to grow up, not old. Featuring progmg that explores like's key transitions & turning points. TLC presents high-quality, relatable & authentic personal stories. TLC connects more than 97 million homes in North America to the human experience with life's lessons you can't learn from books.

Lifetime, (A&E Television Networks). 309 West 49th St., 17th Fl., New York, NY, 10019. Phone: (212) 424-7183. Fax: (212) 907-9409. E-mail: morgenst@lifetimetv.com

Nickolas Davatzes, pres/CEO; Dan Davids, VP/gen mgr; Charles Maday, sr VP.

A 24-hour progmg svc featuring historical documentaries, specials & mini-series. 86 million subs. Satellite: Satcom C-3, transponder 12.

Lifetime Movie Network, 309 W. 49th St., New York, NY, 10019. Phone: (212) 424-7000. Web Site: www.lifetimetv.com.

Andrea Wong, pres/CEO; James Wesley, CFO; Rick Haskins, VP/gen mgr.

Lifetime is committed to offering the highest quality entertainment & info progmg 24-hours a day. Serving more than 36 million subs. Satellite: Telstar 7, transponder 14, Galaxy XR, transponder 20.

LIFETIME Television, 309 W. 49th St., New York, NY, 10019. Phone: (212) 424-7000. Web Site: www.lifetimetv.com.

Andrea Wong, pres/CEO; Barbara Fisher, exec VP; James Wesley, CFO; Lynn Picard, exec VP; Louise Henry Bryson, exec VP; Patricia Langer, VP business affrs; Meredith Wagner, exec VP; Tim Brooks, VP rsch; Richard Basso, sr VP; Gwynne McConkey, opns VP.

Lifetime is committed to offering the highest quality entertainment, info progmg & advocating a wide range of issues affecting women & their families, 24 hours a day nationwide . Launched in 1984. Serving over 87 million subs. Lifetime Movie Network in 1998 & a second sister svc, Lifetime Real Women, launched in August 2001. On the web, Lifetime Online (www.lifetimetv.com) features informational resources & interactive entertainment. All four svcs, Lifetime TV, Lifetime Movie Network, Lifetime Real Women Online, are part of Lifetime Entertainment svcs, a 50/50 joint venture of The Hearst Corp & The Walt Disney Co. Satellite: GAlaxy V, transponder 21 east, Satcom, C3, tansponder 4 west. Satellite: Galaxy V, transponder 21.

The Locomotion Channel, 420 Lincoln Rd., Suite 235, Miami Beach, FL, 33139. Phone: (786) 276-1140. Fax: (786) 276-1141.

Rodrigo Piza, gen mgr.

The Locomotion features the best intl productions created specifically for viewers ages 18-35, combining electronic music & digital culture 24-hours a day. On 1.2 milion cable systems. Satellite: Pan-Am Satcom 5.

MSNBC, One MSNBC Plaza, Secaucus, NJ, 07094. Phone: (201) 583-5000. Fax: (201) 583-5453. Web Site: www.msnbc.com.

Val Nicholas, VP mktg; Dan Abrams, gen mgr.

MSNBC is an all news network 24-hours a day. 82 million subs. Satellite: Galaxy IR, transponder 10.

MTV Latino, 1111 Lincoln Rd., 6th Fl., Miami Beach, FL, 33139. Phone: (305) 535-3700. Fax: (305) 672-5204. Web Site: www.mtvla.com.

MTV's sixth global network, 24-hour progmg, advertiser supported and available in Latin America & the United States.

Serving over 8.5 million subs. Satellite: Satcom C3, transponder 19 (USA); PanAm Sat 3, transponder 5C & 6C (Latin America).

MTV: Music Television, MTV Networks Inc. 1515 Broadway, New York, NY, 10036. Phone: (212) 258-8000. Fax: (212) 258-8100. Web Site: www.mtv.com.

Brian Garden, pres; Judy McGrath, pres.

A 24-hour music video ch in stereo.

Serving 88.7 million subs. Satellites: Satcom C-4, transponder 17 east, Satcom C-3, transponder 16 west.

Multiplex chs: MTV Espanol is a 24-hour music network featuring ground-breaking Latin pop, rock en Espanol, & Latin alternative music for the new Latino generation in the united states. Serving 13 million subs. Satellite: GE 3, transponder 15.

MTV Hits offers the best in pop, rock & hip hop music videos for ages 12-24. Serving 18 million subs. Satellite: GE 3, transponder 15.

MTV Jam is the home of hip hop, R&B, soul & everything urban. Serves under 10 million subs. Satellite: GE 3, transponder 15. MTV2 is a full svc music ch. Serves 55 million subs. Satellite: GE 3, transponder 15.

Nickelodeon Gas-Games & Sports For Kids, is part of MTV Network's Digital Suite. Serves 15.6 million subs. Satellite: C-3, transponder 15.

MTV Networks, 1515 Broadway, New York, NY, 10036. Phone: (212) 846-8000. Fax: (212)528-8100. Web Site: www.mtvn.com.

David Cohn, gen mgr.

MTV2 is the premier destination to find the hottest mix of music videos, long form music programs, exclusive access to their favorite bands, and ground-breaking music before it hits mainstream.

Military Channel, One Discovery Pl., Silver Spring, MD, 20910. Phone: (240) 662-2000. Fax: (240) 662-1854. E-mail: first_last@discovery.com Web Site: www.discovery.com.

Billy Campbell, pres; Joe Abruzzese, pres & adv sls; Bill Goodwyn, pres & affil sls & mktg; Jane Root, exec VP & gen mgr.

The Military Channel brings viewers compelling, real-world stories of heroism, military strategy, technological breakthroughs & turning points in history. The network takes viewers "behind the lines" to hear the personal stories of servicemen & women & offfers in-depth explorations of military technology, battlefield strategy, aviation & history. As the only cable network devoted to military subjects, it also provides unique access to this world, allowing viewers to experience & understand a world full of human drama, courage, innovation & long-held military traditions.

The Movie Channel (TMC), (Showtime Networks Inc.). 1633 Broadway, New York, NY, 10019. Phone: (212) 708-1699. Fax: (212) 708-1530. Web Site: www.showtimeonline.com.

TMC features daily movie marathons, overnight & double vision weekends, 24-hours a day. Serving 34.8 million subs. Satellites: Satcom C-3, transponder 19 east; Satcom G-9, transponder 5 west.

Movie Channel Xtra, The offers more of viewers favorite movies 24-hour.

NFL Network, 10950 Washington Blvd., Culver City, CA, 90232. Phone: (310) 840-4635. Fax: (310) 280-1132. E-mail: palanskys@nfl.com Web Site: www.nfl.com/nflnetwork/home.

Steve Bornstein, pres/CEO; Adam Shaw, sr VP.

A national cable & satellite ch telecasting NFL content 24 hrs a day. Satellites: Galaxy 11, Transponder 9. Serving 30 million subs on more than 60 cable systems.

National Geographic Channel, 1145 17th St. N.W., Washington, DC, 20036-4688. Phone: (202) 912-6500. Fax: (202) 912-6603. E-mail: comments@natgeochannel.com Web Site: www.nationalgeographic.com/channel.

John Ford, exec VP progmg; Kiera Hynninen, sr VP mktg; David Haslingden, CEO.

NGC provides viewers with direct connection to adventures all over the world & unique access to the most repected scientists, journalists & filmmakers, 24-hours a day. 47 million subs. Satellite: Satcom C-3, transponder 1.

National Jewish Television Network, Box 480, Wilton, CT, 06897. Phone: (203) 834-3799. E-mail: nj@jewishmail.com

Joel A. Levitch, pres/CEO.

NJT offers documentaries, children's programs, news magazines, Info, cultural & relg progmg for the Jewish community, presented 3 hours every Sunday.

Serving 10 million subs. Satellite: C-Band.

Nationality Broadcasting Network, (Radio-WKTX 830 AM/NBN TV). 11906 Madison Ave., Lakewood, OH, 44107. Phone: (216) 221-0330. Fax: (216) 221-3638.

Jim Georgiades, opns mgr.

Provides internationally & locally-produced nationality TV & radio progmg, special programs & comls; also full-svc production house.

Serving one cable system and the scola tv network.

Newsworld International, (North American Television Inc.). 1230 Ave. of the America, New York, NY, 10020. Phone: (212) 413-5000. Fax: (212) 413-6546. E-mail: nwifeedback @indtvholdings.com Web Site: www.nwitv.com.

Lou Cooper, VP progmg; John Bernbach, chmn; Patrick Vien, pres/CEO; Doug Halloway, pres network distribution.

Intl progmg covering top stories from around the world plus current affrs, documentaries, the latest business, financial & sports news, 24-hours a day.

Serving 18 million subs. Satellite: Galaxy 1R, GE3, transponder 22, MPEG2.

NICK at NITE

See Nickelodeon.

Nickelodeon, (MTV Networks Inc.). 1515 Broadway, New York, NY, 10036. Phone: (212) 258-8000. Web Site: www.nickelodeon.com.

Tom Freston, chmn/CEO; Cyma Zarghami, VP/gen mgr; Jeff Dunn, CEO.

Nickelodeon cable network targets kids. NICK at NITE provides entertainment svc for the TV generation.

Serving 87.9 million subs. Satellite: Satcom C-4, transponder 3 east. NICK at NITE is on 4,381 cable systems. Satellites: Satcom C-3, transponder 18 west. NICKTOONS is the 24-hour digital destination for the next generation of animation. Serving 13 million subs. Satellite: C-3, transponder 15. NICK2 gives viewers the convenience of watching their favorite Nickelodeon & NICK at NITE shows at different times of the day.

Noah's World International, 11448 Kanapali Ln., Boynton Beach, 33437. Phone: (561) 732-2108. Fax: (516) 732-5108. Web Site: www.noahsworldtv.com.

Ken Klein, pres; Ilan Klein, exec VP.

Noah's World Internation profiles & present enlightened countries around the globe, 24-hours a day.

Noggin/The N, 1633 Broadway, 7th Fl., New York, NY, 10019. Phone: (212) 654-7707. Web Site: www.noggin.com.

Tom Ascheim, VP/gen mgr; Kenny Miller, VP progmg.

The N, the nighttime network for teens, 24-hours a day. Satellite: Satcom C-3, transponder 15.

OASIS TV, 1875 Century Park E., Suite 600, Los Angeles, CA, 90067. Phone: (310) 553-4300. Fax: (310) 553-4300. E-mail: service@oasistv.com Web Site: www.oasistv.com.

Robert Schnitzer, CEO; Azim Khamisa, chmn.

Oasis TV is a global TV programmer of cutting-edge body-mind-spirit news, inspiration & entertainment, 24-hours. On 16 cable systems. 1.8 million subs.

Open TV, 350 5th Ave., Fl. 59, New York, NY, 10118-5999. Phone: (212) 497-7000. Fax: (212) 497-7001. E-mail: info@actv.com Web Site: www.actv.com.

David Reese, pres/CEO.

Branchburg, NJ 08876, 3040 Rte 22 W., Suite 210. Phone: (908) 252-3800. David Reese.

Individualized TV progmg for educ & entertainment.

The Outdoor Channel, 43445 Business Park Dr., Suite 103, Temecula, CA, 92590. Phone: (800) 770-5750. Fax: (909) 699-6313. Web Site: www.outdoorchannel.com.

Jake Hartwidk, exec VP; Andrew Dale, pres/CEO; Wade Sherman, sr VP; Amy Hendrickson, VP mktg; Greg Harrigan, sr VP.

Outdoor Channel offers progmg such as fishing, hunting, hiking , competitive shooting & motor sports, 24-hours a day.

Serving 26 million subs. Satellite: Galaxy 10R, Transponder 24, Digital.

OVATION-The Arts Network, 5801 Duke St., Suite D-112, Alexandria, VA, 22304. Phone: (703) 813-6310. Fax: (703) 813-6336. E-mail: info@ovationtv.com Web Site: www.ovationtv.com.

Edward J. Mathias, chmn; Harold E. Morse, pres/CEO; Susan Wittenberg, VP progmg; Lee J. Lindbloom, VP network opns.

The network covers arts news from around the world & children's arts programs, 20 hours a day. 6.6 million subs. Satellite: Galaxy VII, transponder 13.

Oxygen Media Inc., 75 9th Ave., 7th Fl., New York, NY, 10011. Phone: (212) 651-2070. Fax: (212) 651-2099. E-mail: feedback@oxygen.com Web Site: www.oxygen.com.

Geraldine B. Laybourne, chmn/CEO; Lisa G. Hall, pres/COO; Mary G. Murano, exec VP; Daniel H. Taitz, gen counsel.

Oxygen Media is a 24-hour cable network TV for women. Serving 49 million subs. Satellite: Satcom C3, transponder east Analog, Galaxy II, transponder 13 west Digital.

Pennsylvania Cable Network (PCN), 401 Fallowfield Rd, Camp Hill, PA, 17011. Phone: (717) 730-6000. Fax: (717) 730-6005. Fax: (717) 441-4540. E-mail: pcntv@pcntv.com Web Site: www.pcntv.com.

Brian Lockman, pres/CEO; William J. Bova, VP progmg; Debra Kohr Sheppard, opns VP; Rick Cochran, VP mktg.

Philadelphia, PA 19101, 400 N. Broad St. Phone: (215) 854-4455. Corey Clarke, bureau chief.

Pittsburgh, PA 15222, Pittsburgh Post Gazette Bldg, 34 Blvd. Phone: (412) 263-1300. Doug Sicchitano, bureau chief.

The nation's preeminent state pub affrs net, with live & same-day coverage of the Pennsylvania General Assembly. PCN also covers significant state events, such as high school sports finals.

Serving 3.3 million subs on 150 cable systems. Satellite: AMC-6, transponder 15.

Pentagon Channel, 601 North Fairfax St., Alexandria, 22314. Phone: (703) 428-0265. Fax: (703) 428-0466. E-mail: pentagonchannel@hq.afis.osd.mil

Gene Brink, gen mgr.

Pentagon Channel is a gov owned TV of the Department of Defense, providing internal communications to svc members, families, the National Guard, the Reserve & military retirees, available 24-hours. 5.3 million subs.

Plato Learning Inc., 10801 Nesbitt Ave. South, Bloomington, MN, 55437. Phone: (800) 447-5286. E-mail: info@plato.com Web Site: www.plato.com.

Terri Reden, dir communications.

Interactive TV progmg for children.

Playboy TV, (Playboy Entertainment Group). 2706 Media Center Dr., Los Angeles, CA, 90065. Phone: (323) 276-4000. Fax: (323) 276-4500. Web Site: www.playboytv.com.

James English, pres; Sol Weisel, exec VP; Jeff Jenest, exec VP; Craig Simon, sr VP.

Entertainment targeted to adults. Schedule consists of nearly 100% original Playboy programs with the balance comprised of acquired programs & feature films. Serving 4.5 under PPV svc. Satellite: Galaxy 5, transponder 2.

Hot Zone is a 24-hour Pay-Per-View ch for adults. Satellite: Telstar 7, transponder 5.

Hot Networks is a 24-hour Pay-Per-View ch for adults. Satellite: Telstar 7, transponder 5.

Praise Television, 28059 US Hwy. 19 N., Suite 300, Clearwater, 33761. Phone: (800) 921-9692. Fax: (727) 530-0671.

Dustin Rubeck, pres.

A 24-hours, Christian music for family entertainment. Satellite: GE-1, transponder 7.

Product Information Network (PIN), 9697 East Mineral Ave., Englewood, 80155-3309. Phone: (303) 784-8321. Fax: (303) 784-8549.

Jon Shaver, COO; Richard Steele, VP sls; Tom Cahill, VP mktg.

A 24-hours info net, ad-supported. 35 million subs. Satellite: Satcom C-3, transponder 20.

Puma TV, 2029 S.W. 105th Ct., Miami, 33165-7937. Phone: (305) 554-1876. Fax: (305) 554-6776.

Jose Luis Rodriguez, pres; Osvaldo Rodriguez, VP.

Puma TV offers a wide var of info programs on fashion, modeling & entertainment, music ch by Hispanic for Hispanic, 24-hours. 2.7 million subs. Satellite: Telstar 5.

QVC, Studio Park, West Chester, PA, 19380. Phone: (484) 701-1000. E-mail: webmaster@qvc.com Web Site: www.qvc.com.

Douglas S. Briggs, pres; Tim Megaw, sr VP.

QVC, Inc. is the world's preeminent electronic retailer mktg a wide var of brand name products, categories as home furnishings, licensed products, fashions, beauty, electronics & fine jewelry. The company's other divisions/subsidiaries include, Loc, on Q Music, QVC International. 85.4 million subs. Satellite: Satcom C-4, transponder 9.

The Real Estate Network-TREN, 325 Sharon Park Dr., Suite 512, Menlo Park, CA, 94025. Phone: (650) 361-1000. Fax: (650) 361-1880. E-mail: kevintren@yahoo.com

Kevin L. Keithley, CEO; Ronald D. Keithley, COO.

Niche advertiser-supported entertainment progmg & interactive svc focusing on merchandising real estate listings & related products & svcs throughout America.

Recovery Network, 1411 5th St., Suite 250, Santa Monica, CA, 90401. Phone: (310) 393-3979. Fax: (310) 393-5749. E-mail: info@recoverynetwork.com Web Site: www.recoverynetwork.org.

Progmg addresses behavioral & alternative health care issues & treatments for eating disorders, addictions, depression, sexual addictions, substance abuse, etc.

Serving 5 million subs on 36 systems. Satellite: Galaxy 7. Launched Spring 1996.

SCOLA, 21557 270th St., McClelland, IA, 51548-0619. Phone: (712) 566-2202. Fax: (712) 566-2502. E-mail: scora@scora.org Web Site: www.scola.org.

Francis Lajba, pres; John Millar, VP.

Foreign language news & educ progmg, 24-hours a day. Satellite: Telstar 5.

STARNET, 1332 Enterprise Dr., Suite 200, West Chester, PA, 19380. Phone: (610) 692-5900. Fax: (610) 692-6487.

Jerry Lenfest, pres/CEO; Joy Tartar, CFO.

Automatic cross-ch tune-in promotion svc for basic, pay & pay-per-view delivered via satellite. Nu-Star - automatic cross ch tune-in promotion svc for basic pay & pay-per-view delivered via satellite. The Promoter-individualized tune-in promotion for PPV delivered via satellite.

Serving 23 million subs on 960 cable systems for STARNET. Satellite: Satcom C-4, transponder 12.

Sci-Fi Channel, (USA Networks). 30 Rockefeller Plaza, New York, NY, 10112. Phone: (212) 644-4444. Fax: (212) 413-6509. Web Site: www.scifi.com.

David Howe, gen mgr.

Dedicated to a broad range of science fiction & fact, fantasy, & horror programs, 24-hours a day.

Serving 83 million subs. Satellite: Galaxy 5, transponder 4 east, Galaxy 1R, transponder 24 west.

Scripps Works, 9721 Sherrill Blvd., Knoxville, TN, 37932. Phone: (865) 694-2700. Fax: (865) 690-9281. Web Site: www.scrippsnetworks.com.

Bob Baskerville, pres; Robyn Ulrich, VP mktg; Jeff Sears, VP.

DIY cable TV network operated by Scripps Networks, providing in depth demonstrations & tips for categories such as home improvement, home bldg, tools, products, gardening, landscaping, automotive, boating, decorating, design, arts, crafts, cooking, hobby & recreations, 24-hour a day. 25 million subs. Satellite: Galaxy IR, transponder 4.

Shop at Home Network, 5388 Hickory Hollow Pkwy., Nashville, TN, 37013. Phone: (615) 263-8000. Fax: (615) 263-8084. Web Site: www.shopathometv.com.

Tim Engle, pres.

ShopNBC, ValueVision Media Inc., 6740 Shady Oak Rd., Eden Prairie, MN, 55344. Phone: (952) 943-6000. Fax: (952) 943-6011. Web Site: www.shopnbc.com.

Will Lansing, CEO; Frank Elsenbast, CFO.

ValueVision Media (Nasdaq: VVTV) operates in the rapidly growing converged world of TV, the Internet & e-commerce. The company flagship media property, ShopNBC, the nation's fasting growing shoppng network, is bcst into 70 million homes 24 hrs a day. GE Equity & NBC own approximately 27% of ValueVision Media. Satellite used and transponder: Galaxy 15, transponder 12. 1,900.

Short TV, 580 Broadway, Suite 1104, New York, 10012. Phone: (212) 226-6258. Fax: (212) 925-5802. E-mail: info@shorttv.com Web Site: www.shorttv.com.

Roland Dib, pres.

Short films, available 24-hours, ad-supported. 2.5 million subs.

Showtime Networks Inc., 1633 Broadway, New York, NY, 10019. Phone: (212) 708-1699. Fax: (212) 708-1530. Web Site: www.showtimeonline.com.

Matthew C. Blank, chmn/CEO.

Showtime Networks Inc. (SNI), which is a wholly owned subsidiary of Viacom Inc., owns the premium TV nets SHOWTIME, THE MOVIE CHANNEL & FLIX. SNI also operates & manages the premium TV net SUNDANCE CHANNEL which is owned by SNI, Robert Redford & Polygram Filmed Entertainment. SHOWTIME on Espanol, a separate audio feed of SHOWTIME, is available for the Spanish-speaking audience. SNI also markets & distributes sports & entertainment events for exhibition to subscribers on a pay-per-view basis. Multiplex chs: SHOWTIME BEYOND features sci-fi, horror & fantasy. SHOWTIME PAY-PER-VIEW. SHOWTIME EXTREME is comprised exclusively of action movies. SHOWTIME FAMILY ZONE is a ch for the entire family. SHOWTIME NEXT is SHOWTIME'S interactive playground targeting gen young adults 18-24. SHOWTIME SHOWCASE offers great progmg with Showtime originals airing. SHOWTIME TOO is a multi-feed ch. SHOWTIME WOMEN highlights women who are instrumental in moving the story ahead, in front of & behind the camera. THE MOVIE CHANNEL HD is a all movie svc., 34.8 million subs. Satellite: Satcom C-3, transponder 19 (east).

Si TV, 3030 Andrita St., Bldg. A, Los Angeles, 90065. Phone: (323) 256-8900. Fax: (323) 256-9888. Web Site: www.sitv.com.

Jeff Valdez, chmn/CEO; Leo Perez, COO; Rita Morales, VP progmg.

SiTV is a Latino-themed network in English that features original progmg, comedy, drama, var shows, talk-format strips, music & style shows, available 24-hours.

SingleVision Entertainment Television Network, 760 Skipper Dr., Atlanta, GA, 30318. Phone: (678) 596-5909. E-mail: singlevisiontv@netscape.net Web Site: www.singlevisiontv.com.

Michael Wilson, CEO.

Entertainment & info for unmarried people. Progmg targeted to Generation X , includes interactive televised personals, advertiser supported & IPTV offering. 4 million subs. Satellite: Galaxy VI.

Skyview World Media, Two Executive Dr., Suite 6000, Fort Lee, 07024. Phone: (201) 242-3000. Fax: (201) 944-5961. E-mail: info@kskyviewmedia.com Web Site: www.skyviewmedia.com.

John Lunsford, pres/CEO; James Helfott, sr VP.

Skyview World Media is North America's leading provider of foreign ethnic progmg. 100,000 subs.

Sleuth, NBC Universal Cable, 900 Sylvan Ave., 1 CNBC Plaza, Englewood Cliffs, NJ, 07632. Phone: (201) 735-3604. Web Site: www.sleuthchannel.com.

Soapnet, 3800 West Alameda Ave., Burbank, 91505. Phone: (818) 569-7500. Fax: (818) 566-1358. Web Site: www.soapnet.com.

Deborah Blackwell, sr VP; Mary Ellen DiPrisco, VP progmg; Sherri York, VP mktg.

Soapnet features today's soaps tonight, classic soaps, news & info from the world of soaps, 24-hours. Serving 35.7 million subs. Satellite: Galaxy 10 R.

Sorpresa, 6125 Airport Fwy., Suite 200, Fort Worth, 76117. Phone: (817) 222-1234. Fax: (817) 222-9809. Web Site: www.sorpresatv.net.

Leonard Firestone, chmn/CEO; Michael Fletcher, pres.

The nation's first network dedicated to America's Hispanic children, available 24-hours, ad-supported, Digital Premium. Satellite: Telstar 5, transponder 24.

SourceSuite, LLC, 5601 MacArthur Blvd., Suite 201, Irving, 75038. Phone: (469) 524-0116. Fax: (469) 417-0314. Web Site: www.intchan.com.

Charlie Barnes, gen mgr.

SourceSuite is a leading provider of interactive TV products, available 24-hours. 400,000 subs.

SPEED, 9711 Southern Pines Blvd., Charlotte, NC, 28273. Phone: (704) 731-2222. Web Site: www.speedtv.com.

Rick Miner, sr VP & exec producer; Bill Osborn, sr VP mktg; Bob Ecker, progmg VP; Kevin Annison, VP opns & business dev; Kevin Wilson, VP business & legal; Chris Long, VP/exec producer; Francois McGillicuddy, VP finance & admin; Hunter Nickell, exec VP & gen mgr.

SPEED celebrates its 10th Anniversary in 2006 as the premier destination for everything automotive, providing insight & action. 57.1 million subs. Satellite: Satcom C-4, transponder 11.

Spice 1, 2706 Media Center Dr., Los Angeles, 90065. Phone: (323) 276-4000. Fax: (323) 276-4500. Web Site: www.spicetv.com.

James English, pres.

An erotic adult-theme movie net, Premium, Pay-Per-View, available 24-hours. Multiplex ch: Spice 2. 11,000 subs. Satellite: Telstar 5.

Spike TV, (division of MTV Networks). 1775Broadway, 37th Fl., New York, NY, 10019. Phone: (212) 846-8705. Fax: (212) 846-2560. Web Site: www.spiketv.com.

Kevin Kay, gen mgr.

Nashville, TN 37214, 2806 Opryland Dr. Phone: (615) 457-7230.

Network for men. 87.2 million subs. Satellite: Satcom C-3, transponder 18.

The Sportsman Channel, 2855 S. James Dr., Suite 101, New Berlin, WI, 53151. Phone: (262) 432-9100. Fax: (262) 432-9101. Web Site: www.thesportsmanchannel.com.

C. Michael Cooley, pres/CEO; Todd D. Hansen, sr VP; Jim Seley, progmg dir; Darrell Lake, VP sls.

The Sportsman Channel provides continuos hunting & fishing progmg 24-hours a day. 10.7 million subs. Satellite: Telstar 5, Transponder 1.

Starz!, See Encore Media Corp.

The Style Network, 5750 Wilshire Blvd., Los Angeles, CA, 90036-3709. Phone: (323) 954-2400. Fax: (323) 954-2500. Web Site: www.eonline.com.

Style Network covers the gamut of the lifestyle genre. 34 million subs.

Sun TV, 2245 Godby Rd., Atlanta, 30349. Phone: (404) 766-9197. Fax: (404) 767-5264.

David C. Simon, pres/CEO; Virgil Scott, VP.

Sun TV network features news, sports, entertainment, sitcoms, soaps, talk shows from 6:30 AM-midnight Saturday-Sunday (EST). Original progmg from the Caribbean, Central & South America. Serving 700,000 subs. Satellite: Galaxy 4-13.

Sundance Channel, 1633 Broadway, 8th Fl., New York, NY, 10019. Phone: (212) 654-1500. Fax: (212) 654-4738. Web Site: www.sundancechannel.com.

Larry Aidem, pres/CEO.

Sundance Channel is a 24-hours a day ch, featuring uncut coml-free programs, providing TV viewers daring & engaging feature films, short, documentaries, world cinema & animation. 17 million subs. Satellite: Satcom C-4, transponder 20.

TBN-Trinity Broadcasting Network, (TBN Cable Network). 2823 W. Irving Blvd., Irving, TX, 75061. Phone: (972) 313-9500. Phone: (800) 735-5542. Fax: (972) 313-1010. Web Site: www.tbn.org.

Paul Crouch, pres; Robert Higley, VP mktg.

TBN is America's most watched relg network, offering 24-hours of coml-free inspiritional original programs, that appeal to viewers in many denominations. Progmg includes Nashville gospel concerts, health & fitness, talk shows & svcs from America's largest Churches. Multiplex ch: TBN Enlace USA is a 24-hour multi-faith Hispanic ch from Trinity Broadcasting Network. The Church Channel is a new digital network from TBN features church svc program from Protestant, Catholic & Jewish faith groups, 24-hours. On 43.4 million cable systems. 49 million subs. Satellite: Galaxy V, transponder 3.

TBS Superstation, 1050 Techwood Dr. N.W., Atlanta, GA, 30318. Phone: (404) 885-4339. Fax: (404) 885-4319. Web Site: www.tbssuperstation.com.

Ken Schwab, sr VP progmg & TNT & TBS; Mark Lazarus, pres of Turner Entertainment Group; Steve Koonin, exec VP & COO &; Tricia Melton, mktg VP.

TBS is television's "very funny" network, serving as home to such hot contemp comedies as Sex and the City, Everybody Loves Raymond, Family Guy, Seinfield and Friends; high-profile original reality series; blockbuster movies; and hosted movie showcases. 90.9 million subs. Satellite: Galaxy V, transponder 6.

TEN-The Erotic Network, 7007 Winchester Cir., Suite 200, Boulder, 80301. Phone: (303) 786-8700. Fax: (303) 938-8388. Web Site: www.noof.com.

Ken Boenish, pres; Michael Weiner, CEO; William Mossa, VP mktg.

TEN is a network that uses the "un-inhibited" editing standard, 24-hours a day. Multiplex chs: TEN On Demand, TENBlox, TENClips & TENXtsy, PLEASURE. Serving 13.7 million subs. Satellite: Telstar 7-24, TUN, G10R-7.

TNT Latin America, 1030 Techwood Dr., Atlanta, GA, 30318. Phone: (404) 885-2434. Fax: (404) 885-2157. Web Site: www.tntla.turner.com.

Rick Perez, VP/gen mgr.

24-hours cable net bcst in Sp, Portuguese, English featuring contemp, original movies, NBA coverage & exclusive premieres.

Serving more than 8.5 million subs in 39 countries in the rgn. Satellite: PanAmSat 1, transponder 3.

TNT (Turner Network Television), 1050 Techwood Dr. N.W., Atlanta, GA, 30318. Phone: (404) 885-4339. Fax: (404) 885-4319. E-mail: karen.cassell@turner.com Web Site: www.tnt.tv.

Mark Lazarus, pres; Steve Koonin, exec VP.

Turner Network Television (TNT), television's destination for drama and one of cable's top-rated networks, offers original movies & series, including the detective drama series "The Closer," starring Golden Globe & Screen actors Guild Awards nominee Kyra Sedgwick; Saved, a new character-driven drama starring Tom Everett Scott; and this summer's eagerly anticipated anthology series Nightmares & Deamscapes: From the Stories of Stephen King. TNT is also home to powerful one-hour dramas, such as Law & Order, Without a Trace, Las Vegas, Cold Case, ER, NYPD Blue, Charmed and Judging Amy; bcst premiere movies; compelling prime-time specials, such as the Screen Actors Guild Awards; & championship sports coverage, including NASCAR &the NBA. TNT is also available in high definition. 90.3 million subs. Satellite: Galaxy V, transponder 17.

TR!O, 1230 Ave. of the Americas, New York, 10020. Phone: (212) 413-5000. Fax: (212) 413-6552. Web Site: www.triotv.com.

Lauren Zalaznick, pres.

TR!O is an entertainment cable TV ch reflecting pop

culture, 24-hours a day. Serving 20 million subs. Satellite: Galaxy 1R, transponder 24, Satcom C3, transponder 8.

TV Asia, (Asian Star Broadcasting Network Inc.). 76 National Rd., Edison, NJ, 08817. Phone: (732) 650-1100. Fax: (732) 650-1112. E-mail: into@tvasiausa.com Web Site: www.tvasiausa.com.

TV Asia provides a wide range of prgmg produced for South Asian Americans, 24-hours.

Satellite: Galaxy II, transponder 24.

TV Games (TVG) Network, 6701 Center Dr. W., Los Angeles, 90045. Phone: (310) 242-9500. Web Site: www.tvgnetwork.com.

Ryan O'Hara, COO. 12 million subs. Satellite: GE-1, transponder 23.

TV Guide Channel, 6922 Hollywood Blvd., Los Angeles, CA, 90028. Phone: (323) 817-4600. Web Site: www.tvguide.com.

Ray Hopkins, exec VP.

New York, NY 10017, 708 Third Ave., 21st Fl. Phone: (212) 370-1799. Fax: (212) 370-7575. Chris Manning, eastern sls mgr.

The network combines original etertaining long-form progmg with comprehensive listing info 24-hours a day. 70 million subs. Satellite: Satcom C-4.

TV Japan, (Japan Network Group Inc.). 100 Broadway, 15th Fl., New York, NY, 10005. Phone: (212) 262-3377. Fax: (212) 262-5577. E-mail: takeucki@tvjapan.net Web Site: www.tvjapan.net.

Mitsuo Sekino, pres/CEO; Masao Watari, exec VP; Koki Matsumoto, exec VP.

TV Japan is a Japanese language ch, available 24-hours a day. Satellite: Galaxy II, transponder 24.

TV Land, 1515 Broadway, New York, NY, 10036. Phone: (212) 258-8000. Fax: (212) 846-1775.

Larry W. Jones, VP/gen mgr.

TV Land is the only network dedicated to the best of everything TV from the past 50 years, available 24-hours. 82.1 million subs. Satellite: Satcom C-3, transponder 18.

Talk Internetwork, Inc., 9662 E. Volture Dr., Scottsdale, AZ, 85260. Phone: (480) 551-9774. E-mail: info@talkinternet@aol.com Web Site: www.talkinternet.com.

Edwin Cooperstein, pres/CEO; Pat McMahon, progmg dir.

A 24-hour, live, all-talk progmg internet site.

Talkline Communications Television Network, Box 20108, Park West Stn., New York, NY, 10025-1510. Phone: (212) 769-1925. Fax: (212) 799-4195.

Jewish programs with newsmaker guests, celebrity interviews as well as informational progmg. Presented Sundays 11 AM-6 PM EST nationally & Sundays 2-5 PM EST & 9-11 PM EST in the New York area.

Serving more than 18 million subs on 825 cable systems.

TechTV, 650 Townsend St., 3rd Fl., San Francisco, 94103. Phone: (415) 355-4000. Fax: (415) 355-4670. E-mail: techtvinfo@techtv.com

Joseph Gillespie, exec VP; Greg Brannan, sr VP; Peter Gochis, VP sls.

TechTV intrigues viewers with everything from help & info to cutting-edge factual progmg to outrageous late-night fun. 43 milion subs. Satellite: Satcom C-4, transponder 12.

Telemundo, 2290 W. 8th Ave., Hialeah, 10019. Phone: (305) 889-7200. Fax: (305) 889-7205.

Jim McNamara, pres/CEO; Don Brown, CFO; Ramon Escobar, exec VP.

Telemundo, a United States Sp-language TV network, available 24-hours. Multiplex ch: Telemundo Internacional. Serving 32 million subs. Satellite: Satcom, transponder 20 east, AMC-4, transponder 8 west.

The Tennis Channel Inc., 2850 Ocean Park Blvd., Suite 150, Santa Monica, CA, 90405. Phone: (310) 656-9400. Fax: (310) 656-9433. Web Site: www.thetennischannel.com.

Ken Soloman, CEO; John Brady, CFO; Steve Bellamy, pres & founder; Randy Brown, sr VP.

The Tennis Channel, 2850 Ocean Park Blvd., Santa Monica, 90405. Phone: (310) 314-9400. Fax: (310) 314-9433. Web Site: www.thetennischannel.com.

Steve Bellamy, pres; Ken Solomon, chmn/CEO; John Brady, CFO; Keith Manasco, VP opns.

The Tennis Channel is the 24-hour cable TV network devoted to tennis & other racquet sports, ad-supported. 3 million subs. Satellite: Telstar 5, transponder 15.

The Irish Channel (CelticVision), 179 Amory St., Brookline, MA, 02446. Phone: (617) 566-4844. Fax: (617) 566-7536. E-mail: celticty@aol.com

Serving 140,000 subs on 2 cable systems.

The Science Channel, One Discovery Pl., Silver Spring, MD, 20910. Phone: (240) 662-2000. Fax: (240) 662-1854. E-mail: first_last@discovery.com Web Site: www.discovery.com.

Billy Campbell, pres; Joe Abruzzese, pres & adv sls; Bill Goodwyn, pres & affil sls & mktg; Jane Root, exec VP & gen mgr.

The Science Channel is the only net dedicated to making science progmg accessible, relevant, substantive & entertaining. Now reaching over 40 million homes in the United States, The Science Channel explores science's past, present & future, from uncovering lost worlds to exploring the latest in scientific discoveries in today's headlines.

The Theatre Channel, Box 2676, Venice, CA, 90294. Phone: (310) 823-6508. Fax: (310) 823-3431. E-mail: info@theatrechannel.com Web Site: www.theatrechannel.com.

Cheryl Beach, CEO.

Traditional & alternative live theatre dance, opera, children's theatre in a videotape format.

Time Warner Cable, One Time Warner Ctr., New York, NY, 10019. Phone: (212) 598-7200. Web Site: www.timewarner.com.

Richard D. Parsons, CEO; Jeffrey L. Bewkes, pres/COO.

Flushing, NY 11355, 41-61 Kissena Blvd.

Cable service.

Satellite: 1.3 million subs.

Toon Disney, 3800 W. Alameda Ave., Burbank, CA, 91505. Phone: (818) 569-7500. Fax: (818) 566-1358. Web Site: www.toondisney.com.

Ann Sweeney, pres.

Toon Disney is a var of acquired animated programs, 24-hours a day. Serving more than 43 million subs. Satellite: Galaxy 10R.

Travel Channel, One Discovery Pl., Silver Spring, MD, 20910. Phone: (240) 662-2000. Fax: (240) 662-1854. E-mail: first_last@discovery.com Web Site: www.discovery.com.

Billy Campbell, pres; Joe Abruzzese, pres & adv sls; Bill Goodwyn, pres & affil sls & mktg; Patrick Younge, exec VP & gen mgr.

Launched in Feb of 1987, the Travel Channel is the only TV net devoted exclusively to travel entertainment. Capturing the fascination, freedom & fun of travel, Travel Channel delivers insightful stories from the world's most popular destinations & inspiring diversions.

Tribune Media Services, TMS TV Listings. 435 N. Michigan Ave., Suite 1500, Chicago, IL, 60611. Phone: (312) 222-8654. Fax: (312) 222-1360. E-mail: jfennel@tribune.com Web Site: www.tribunemedia entertainment.com.

Cameron Young, exec dir sls; Kathleen Tolstrip, gen mgr.

Glens Falls, NY 12801, One Apollo Dr. Phone: (800) 424-4747. Brian Ward, opns dir.

Content leader in print TV program guides, program guide/loc community ch video products, TV program schedules/cable lineups for digital & advanced analog IPG set-top box applications, movie showtimes.

Turner Classic Movies (TCM), 1050 Techwood Dr. N.W., Atlanta, GA, 30318. Phone: (404) 885-5535. Web Site: www.turnerclassicmovies.com.

Tom Karsch, VP/gen mgr; Mark Lazarus, pres.

Features Hollywood's greatest movies of all time, presented 24-hours, coml-free. 63.9 million subs.

USA Network, 30 Rockefeller Plaza, c/o NBC, New York, NY, 10112. Phone: (212) 664-4444. Web Site: www.wnbc.com.

Gen entertainment network featuring movies, original series, sports specials, teen & children's progmg, 24-hours a day.

Serving 89.7 million subs. Satellites: Galaxy 5 transponder 19 east, Galaxy 1R, transponder 21 west.

Univision Television Group, 5999 Center Dr., Los Angeles, CA, 90045. Phone: (310) 348-4865. Fax: (310) 348-3643. E-mail: tkrass@univision.net Web Site: www.univisionnetworks.com.

New York, NY 10158. East, 605 Third Ave, 26 Fl. Phone: (212) 455-5342. Fax: (212) 986-4731. John Heffron, VP affil rels.

Dallas, TX 75201. Central, 2323 Bryan St, Suite 1900. Phone: (214) 758-2405. Fax: (214) 758-2395. Deanna Andaverde, VP affil rel.

A 24-hour most watched TV network (Sp or English) among the nation's growing Hispanic population, featuring movies, novellas, sports, children progmg, musical special, national & loc newscasts.

Serving 34.5 million subs. Satellite: Galaxy 1R.

Telefutura is the newest 24-hour gen-interest Sp language bcstg network. Serves 7.2 million subs. Satellite: Galaxy 1. Satellite: Galaxy IR.

Urban Television Network Corp., 27075 S. Cooper, Suite 119, Arlington, TX, 76015. Phone: (817) 303-7449. Fax: (817) 459-2942. E-mail: info@uatvn.com Web Site: www.uatvn.com.

Jacob R. Miles III, CEO; Randy Moseley, exec VP/CFO.

VH1 (Music First), (MTV Networks Inc.). 1515 Broadway, New York, NY, 10036. Phone: (212) 846-7840. Web Site: www.vh1.com.

VH1 is a 24-hour ch that features new, current & classic music video, for viewers ages 18-49 who grew up with music videos.

Multiplex chs: VH1 Classic, VH1 Country, VH1 Megahits, VH1 Soul & VH Uno. 86.3 million subs. Satellite: Satcom C-4, transponder 23.

Versus, 281 Tresser Blvd., 9th Fl., Stamford, CT, 06901. Phone: (203) 406-2500. Fax: (203) 406-2534. E-mail: feedback@olntv.com Web Site: www.olntv.com.

Los Angeles, CA 90064, 11835 W. Olympic Blvd, Suite 980. Phone: (310) 473-5404. Fax: (310) 473-6525.

New York, NY 10016, 90 Park Ave., 2nd Fl. Phone: (212) 883-4000. Fax: (212) 687-1819.

Versus provides viewers with non-stop action, 24-hours a day.

Serving 58 million subs. Satellite: Galaxy II, transponder 21. On 4,893 cable systems.

Viewer's Choice, 909 3rd Ave., 21st Fl., New York, NY, 10022. Phone: (212) 486-6600. Fax: (212) 688-9497. E-mail: webmaster@ppv.com Web Site: www.ppv.com.

Los Angeles, CA 90067, 1888 Century Park E. Phone: (310) 785-9094. (310) 785-9194. Fax: (310) 785-9195. (310) 785-9769.

Atlanta, GA 30338, 1117 Perimeter Ctr. W, Suite 500 E. Phone: (404) 399-3119. Fax: (404) 399-3014.

Southfield, MI 48034, 26677 W. Twelve Mile Rd. Phone: (810) 354-3375. Fax: (810) 358-9693.

Leading pay-per-view network, offering top box office titles, sports, & entertainment events through 35 plus digital chs of NVOD svc.

Serving 28 million addressable households & 100 million sub units on 1,700 cable systems. Satellites: Satcom C-3, transponder 3; Satcom C-4, transponder 18.

WSBK-TV, (Boscom). 1170 Soldiers Field Rd., Boston, MA, 02134. Phone: (617) 787-7000.

Betsy Bianco, dir mktg.

WSBK-TV is a 24-hour ind ch from Boston featuring sports, movies, news & specials. Satellite: GE-3, Transponder 3.

The Weather Channel, 300 Interstate North Pkwy., Atlanta, GA, 30339. Phone: (770) 226-0000. Fax: (770) 226-2950. Web Site: www.weather.com.

Patrick Scott, pres; Terry Connelly, VP/gen mgr; Lyn Andrews, pres.

New York, NY 10022, 845 Third Ave, 11th Floor. Phone: (212) 893-2245. Lyn Andrews, Pres. TWC Media Solutions.

All-weather progmg 24-hours a day; natl, international, rgnl & loc weather forecasts & features. Multiplex ch: WeatherScan.

Serving 87.5 million subs. Satellite: GE Satcom C-3, transponder 13.

WE-Women's Entertainment, 200 Jericho Quadrangle, Jericho, NY, 11753. Phone: (516) 803-4400. Fax: (516) 803-4398. E-mail: hckaras@rainbow-media.com Web Site: www.we-womenscenterentertainment.com.

Helen Karas, sr VP.

A 24-hour cable network featuring classic movies & TV progmg devoted entirely to romance.

Serving 51.4 million subs. Satellite: Galaxy 7, transponder 12.

Worship Network, Box 428, Safe Harbor, FL, 34695-0365. Phone: (727) 536-0036. Fax: (727) 530-0671. Web Site: www.worship.net.

Bruce Kobish, pres/CEO.

Worship Network features, scenery from around the world with words of wisdom from scriptures & inspirational music. Available 24 hours a day. 66 million subs. transponder 7.

Regional Cable News Networks

Allbritton Communications, 1000 Wilson Blvd., Suite 2700, Arlington, VA, 22209. Phone: (703) 647-8745. Fax: (703) 647-8746. E-mail: jkillen@allbrittontv.com Web Site: www.newschannel8.net.

James Killen, sls VP.

There are 8 stns: WJLA. Newschannel 8. WHTM, WSET, WCIU, WBMA, KATV, KTUL.

The Arizona News Channel, 5555 N 7th Ave., Phoenix, AZ, 85013. Phone: (602) 207-3762. Fax: (602) 379-2459. E-mail: advertising@azfamily.com Web Site: www.azfamily.com.

Arizona's first & only 24-hour loc news svc built on a unique partnership; live loc breaking coverage gives viewers the latest news from around the Valley; NewsChannel 3's "Good Morning Arizona," "Good Day Arizona," "Good Evening Arizona" & "The News Show" replay throughout the day on the AZ News Channel, Cox Cable Channel 14; progmg also includes loc productions exclusive to cable, such as "Project Parenting."

Bay News 9, 700 Carillon Pkwy., Ste. 9, St. Petersburg, FL, 33716. Phone: (727) 329-2300. Fax: (727) 329-2434. E-mail: viewer@baynews9.com Web Site: www.baynews9.com.

Elliott Wiser, VP/gen mgr; Mike Gautreau, news dir.

Bay News 9 is a 24-hour ch owned & operated by Bright House Network. The ch serves over 1 million cable customers in Tampa Bay. Also programs a 24-hour Sp & sports ch.

Bay TV, 1001 Van Ness Ave., San Francisco, CA, 94109. Phone: (415) 441-4444. Fax: (415) 561-8745. E-mail: 4listens@klon4.com Web Site: www.baytv.com.

Paul Dinovitz, gen mgr.

A 24-hour loc news, sports & info cable ch.

CLTV News (ChicagoLand Television News), 2000 York Rd., Suite 114, Oak Brook, IL, 60523. Phone: (630) 368-4000. Fax: (630) 571-0489. Web Site: www.cltv.com.

Steve Farber, gen mgr.

Covers Chicago loc & rgnl news, sports, news, weather & traf info, serving 1.9 million subs.

CN8 - The Comcast Network, 1500 Market St., 28th Fl., W. Tower, Philadelphia, PA, 19102. Phone: (215) 981-7750. Fax: (215) 981-8420. Web Site: www.cn8.com.

Michael A. Doyle, pres; Melissa Kennedy, dir mktg.

CN8, The Comcast Network, is an award-winning, 24-hours news, talk, sports & entertainment cable net created by Comcast Cable Communications, that has steadily won viewers, awards & accolades since its inception in 1996. CN8 provides quality locally-produced progmg in four main areas-live, interactive television; rgnl news; entertainment; coverage of high school, college & professional sports. CN8 continues to expand its compelling mix of news, talk, sports & entertainment progmg through the eastern seaboard, from Washington DC, to the new England area, broadcasting to 6.2 million viewers everyday.

CablePulse (CP 24), 299 Queen St. W., Toronto, ON, M5V 2Z5. Canada. Phone: (416) 591-5757. Fax: (416) 593-6397. E-mail: info@cp24.com Web Site: www.pulse24.com.

Stephen Hurlbut, VP/gen mgr; Dan Hamilton, VP sls; Tina Cortese, dir of news progmg; Keith Wilson, opns dir; Karen Reid, news dir; David Kirkwood, VP; Jenny Norush, adv dir; Bev Nenson, dir of publicity; Allan Schwebel, VP mktg & VP sls.

Rgnl 24-hour a day English language news & information channel.

The California Channel, 1121 L St, Suite 110, Sacramento, CA, 95814. Phone: (916) 444-9792. Fax: (916) 444-9812. Web Site: www.calchannel.com.

John Hancock, pres.

The California Channel is an independent, nonprofit, public affairs cable television network. Programming includes coverage of California Assembly and Senate floor sessions and committee meetings, capitol press conferences, and proceedings of regulatory boards and state commissions. 33.5 hours/week, serving 5,800,000 subs.

CBS News 4, 8900 N.W. 18th Terrace, Miami, FL, 33172. Phone: (305) 591-4444. Web Site: www.cbs4news.com.

Central Florida News 13, 20 N. Orange Ave., Suite 13, Orlando, FL, 32801. Phone: (407) 513-1300. Fax: (407) 513-1310. E-mail: newsdesk@cfnews13.com Web Site: www.cfnews13.com.

Robin A. Smythe, gen mgr & VP.

24 hour cable news channel serving Orlando & the Central Florida Region.

Country Television Network San Diego, 1600 Pacific Hwy., Rm. 208, San Diego, CA, 92101-2422. Phone: (619) 595-4600. Fax: (619)557-4027. E-mail: ctn@sdcountry.ca.gov Web Site: www.ctn.org.

Michael Workman, dir; Janice McGee, owner; Barry Fraser, cable franchise admin.

Country Television Network San Diego makes country govt more accessible & understandable to the citizens of San Diego County through informational progmg focusing on the svcs, programs & current issues of county gov, 24 hr a day, serving 708,700 subs.

Florida's News Channel, 1801 Halstead Blvd., Tallahassee, FL, 32309. Phone: (850) 222-6397. Fax: (850) 894-5202.

Bob Brillante, CEO; Frank Watson, VP/gen mgr.

Las Vegas One, 3228 Channel 8 Drive, Las Vegas, NV, 89109. Phone: (702) 792-8888. Fax: (702) 696-7222. Web Site: www.klas-tv.com.

Robert Stoldal, VP opns; Emily Neilson, pres & gen mgr; Bob Comings, news dir.

Las Vegas One is a 24-hours loc news ch serving the Las Vegas area.

Michigan Government Television, 111 S. Capitol Ave., 4th Fl., Romney Bldg., Lansing, MI, 48909. Phone: (517) 373-4250. Fax: (517) 335-7342. E-mail: mgtv@mgtv.org Web Site: www.mgtv.org.

Bill Trevarthen, exec dir.

Cable network covering all branches of Michigan's state government.

Neighborhood News 12, 111 New South Rd., Hicksville, NY, 11801. Phone: (516) 393-3378. Fax: (516) 393-0021. Web Site: www.news12.com.

Barry J. Romanski, gen mgr.

Neighborhood News 12 is a 24-hour news channel for individual communities, as defined by a single zip code. Literally "Neighborhood News," the format is sometimes described as an electronic community newspaper combining traditional video packages, voiceovers, music tracks, animation and still digital images. 24/7 serving 209,000 subs.

New England Cable News, 160 Wells Ave., Newton, MA, 02459. Phone: (617) 630-5000. Fax: (617) 630-5057. Fax: (617) 630-5055. Web Site: www.necnews.com.

Philip Balboni, pres; Charles Kravetz, stn mgr & news dir.

A 24-hour rgnl news net.

New York 1 News, 75 9th Ave., 6th Fl., New York, NY, 10011. Phone: (212) 691-6397. Fax: (212) 563-7154. E-mail: ny/news@ny1.com Web Site: www.ny1.com.

Steve Paulus, sr VP; Brad Shapiro, dir; Marc Nathanson, exec producer; Bernie Han, news dir; Kevin Dugan, producer.

A 24-hour, all-news cable ch devoted primarily to coverage of New York City & its neighborhoods.

News 12 Bronx, 930 Soundview Ave., Bronx, NY, 10473. Phone: (718) 861-6800. E-mail: customerservice@news12.com Web Site: www.news12.com.

News 12 Bronx is a 24-hours rgnl news progmg svc (a News 12 Regional Network). 24 hours serving 250,000 subs.

News 14 Carolina, 316 East Morehead St., Suite 100, Charlotte, NC, 28202. Phone: (704) 973-5800. Fax: (704) 973-2770. E-mail: feedback@news14.com Web Site: www.news14.com.

New 14 Carolina offers 24-hours loc news & weather every ten minutes on the Ones.

News Channel 8, 1100 Wilson Blvd., 6th Fl., Arlington, VA, 22209. Phone: (703) 236-9555. Fax: (703) 236-2345. Web Site: www.news8.net.

A 24-hour news svc offered in the Washington, DC metropolitan area over all cable svcs. It is available to cable subs in Alexandria, Arlington County, Fairfax City, Fairfax County & Loudoun County in Virginia, Montgomery & Prince George's Counties in Maryland & Washington, DC.

News Now 53, 777 Northwest Grand Blvd., Suite 600, Oklahoma City, OK, 73118. Phone: (405) 600-6600. Fax: (405) 600-0670. Web Site: www.kotv.com.

News On One, 3501 Farnam, Omaha, NE, 68131. Phone: (402) 346-6666. Fax: (402) 233-7888. E-mail: sixonlin@wowt.com Web Site: www.wowt.com.

Dir of prom & production & tech opns.

News 10 Now, 815 Erie Blvd. E., Syracuse, NY, 13210. Phone: (315) 234-1000. Fax: (315) 234-0635. E-mail: info@news10now.com Web Site: www.news10now.com.

Ron Lombard, gen mgr & news dir.

A 24-hour loc/rgnl news ch serving 560,000 Time Warner cable subscribers throughout central/upstate New York.

News 12 Connecticut, 28 Cross St., Norwalk, CT, 06851. Phone: (203) 849-1321. Fax: (203) 849-1327. E-mail: news12.ct@news12.com Web Site: www.news12.com.

Tom Appleby, news dir; Carmela Williams, news coord.

24 hour, 7 day week reg news ch featuring hyper-loc news coverage including sports & weather.

News 12 Long Island, One Media Crossways, Woodbury, NY, 11797. Phone: (516) 393-1200. Fax: (516) 393-1456. Web Site: www.news12.com. E-mail: new12li@news12.com

Patrick Dolan, news dir.

A 24-hours rgnl news svc.

News 12 New Jersey, 450 Raritan Ctr. Pkwy., Suite H, Edison, NJ, 08837-3994. Phone: (732) 346-3200. Fax: (732) 417-5155. E-mail: news12nj@news12.com Web Site: www.news12.com.

Larry B. Meyrowitz, dir opns; Laura Johnson, sls dir; Randal W. Stanley, news dir & gen mgr; Patrick O. Young, dir mktg & prom dir.

A cable exclusive svc, owned & operated by Cablevision Systems Corporation, reaching 1.8 million households in 14 northern & central counties in New Jersey. Available on Cablevision, Comcast, Service Electirc & Time Warner cable systems. Corporate office & studio is located in Edison, New Jersey, with additional newsrooms located in Oakland, Madison, Newark, Belmar & Trenton.

News 12 Westchester, 6 Executive Plaza, Yonkers, NY, 10701. Phone: (914) 378-8916. Fax: (914) 378-8938. E-mail: news12wc@news12.com Web Site: www.news12.com.

Marguerite Tolliver, mgr.

24-hour news organization covering Westchester County.

NewsChannel 5+, 474 James Robertson Pkwy., Nashville, TN, 37219. Phone: (615) 248-5371. Fax: (615) 248-5394. E-mail: mbonnett@newschannel5.com Web Site: www.newschannel5.com.

Michelle Bonnett, exec dir.

NewsChannel 5+ is a loc news & info stn, serving 550,000 homes in Middle Tennessee & Southern Kentucky.

Ohio News Network, 770 Twin Rivers Dr., Columbus, OH, 43215. Phone: (614) 280-3700. Fax: (614) 280-6305. Web Site: www.ohionewsnow.com.

Tom Griesdorn, VP/gen mgr; Frank Willson, mktg dir; Barb Geller, affil rel mgr; Greg Fisher, news dir; Jason Pheister, progmg dir; Vince Jones, opns mgr; Chuck DeVendra, sls dir.

A 24-hour cable news ch featuring loc news, weather & sports for the people of Ohio. Currently seen in over 1.5 million homes.

Orange County Newschannel, 625 North Grand Avenue, Santa Ana, CA, 92701. Phone: (714) 565-3850. Fax: (714) 565-3869. Web Site: www.rtnda.org.

Mike Sweeney, gen mgr; Don Engelhardt, chief engr; Mya Bulwa, exec producer; Susanne Lysak, news dir.

Orange County Newschannel features exclusive coverage of loc news, sports, weather & traf in southern California's Orange County, 24 hours serving over 575,000 subs.

Pennsylvania Cable Network, 401 Fallowfield Rd., Camp Hill, PA, 17011. Phone: (717) 730-6000. Fax: (717) 730-6005. E-mail: pcntv@pcntv.com Web Site: www.pcntv.com.

Brian Lockman, pres/CEO; William J. Bova, VP progmg; Debra Kohr Sheppard, VP opns; Richard Cochran, VP mktg.

PCN is the nation's pre-eminent state pub affairs network, with live & same day coverage of the Pennsylvania Senate/House & other govt activities. PCN televises significant state events (such as high school sports championships), tours museums & mfg facilities in the state & distributes educ progmg.

Pittsburgh Cable News Channel (PCNC), 11 Television Hill, Pittsburgh, PA, 15214. Phone: (412) 237-1190. Fax: (412) 237-1286. Web Site: www.realpittsburgh.com. E-mail: burgnews@wpix.com

John Howell, VP; Mark W. Barash, stn mgr; Jennifer Rigby, news dir.

Loc & rgnl news, talk & info.

R News/Time Warner Communications, 71 Mt. Hope Ave., Rochester, NY, 14620. Phone: (585) 756-2424. Fax: (585) 756-1673. Web Site: www.rnews.com. E-mail: assignment@rnews.com

Ed Buttaccio, news dir.

Loc news 24-hours 7 days per week. Interactive daily call-in show. Nightly loc Sp newscast.

Regional News Network (RNN), 721 Broadway, Kingston, NY, 12401. Phone: (914) 339-6200. Fax: (914) 339-6264. Web Site: www.rtnda.org/resources/nonstopnews/regionalnews.html.

Richard French, gen mgr; James Sweeny, news dir.

RNN is a 24 hour provider of news & pub info targeted to suburban New York, Connecticut & and New Jersey serving over 250,000 subs.

Rhode Island News Channel, 10 Orms St., Providence, RI, 02904. Phone: (401) 453-8000. Fax: (401) 331-4431. E-mail: radeszkoza@abc6.com Web Site: www.abc6.com.

Ronald Adeszko, gen mgr.

Rhode Island News Channel is a simulcast & rebroadcast of WLNE newscasts for Rhode Island.

SNN News 6, 1741 Main St., Sarasota, FL, 34236. Phone: (941) 361-4600. Web Site: www.snn6.com.

Linda DesMarais, gen mgr.

A 24-hour cable news ch with focus on loc news & info. The Sarasota Herald -Tribune newspaper owned by the New York Times Co.

San Diego's Newschannel 15, Box 85347, San Diego, CA, 92186. Phone: (619) 237-1010. Fax: (619) 527-0369. Web Site: www.10news.com.

San Diego's Newchannel 15 provides original newscasts, repeats of KGTV-10 newscasts & live break-ins 24 hours.

Texas Cable News, 570 Young St., Dallas, TX, 75202. Phone: (214) 977-4500. Fax: (214) 977-4610. Web Site: www.txcn.com.

James T. Aitken, VP/gen mgr; Steve Ackermann, exec news dir.

24-hour loc/rgnl news ch covering the state of Texas.

Tri-State Media News (TSM news), 2215 DuPont Pkwy., New Castle, DE, 19770. Phone: (877) TSM-NEWS. Web Site: www.tsmnews.com.

Stanley H. Green, pres/CEO.

WJLA-TV/Newschannel 8, 1100 Wilson Blvd., 6th Fl., Arlington, VA, 22209. Phone: (703) 647-8745. Fax: (703) 647-8746. Web Site: www.newschannel8.net. E-mail: jkillen@allbrittontv.com

James Killen, sls VP.

A 24-hours rgnl news svc for Washington, DC, suburban Maryland & northern Virginia. On 15 cable systems serving 1,125,000 subs.

Regional Cable Sports Networks

Big Ten Network, 444 N. Michigan Ave., Suite 1200, Chicago, IL, 60611. Phone: (312) 665-0700. Fax: (312) 665-0740. Web Site: www.bigten.org.

Mark Silverman, pres.

Channel 4 San Diego, 350 10th Ave., Suite 500, San Diego, CA, 92101. Phone: (619) 683-1900. Fax: (619) 876-4993. Web Site: 4sd.com.

Craig Nichols, VP/gen mgr.

San Diego Padres baseball, etc.

College Sports Television, Chelsea Piers, Pier 62, Suite 316, New York, NY, 10011. Phone: (212) 342-8700. Fax: (212) 342-8899. E-mail: customerservice@website.cstv.com Web Site: www.cstv.com.

Eric Krasnoo, VP.

Comcast SportsNet, 3601 South Broad St., Philadelphia, PA, 19148. Phone: (215) 336-3500. Fax: (215) 952-5996. E-mail: askcsn@comcastsportsnetwork.com Web Site: www.comcastsportsnet.com.

Jack Williams, pres.

Rgnl TV progmg svcs includes live coverage of Philadelphia Flyers ice hockey, Philadelphia 76ers basketball, Philadelphia Phillies baseball, pro boxing, college basketball, football, indoor lacross, ABL, loc sports news & sports talk programs.

Serving 3 million subs on MSOS(16).

Comcast SportsNet Mid-Atlantic, 7700 Wisconsin Ave., Suite 200, Bethesda, MD, 20814. Phone: (301) 718-3200. Fax: (301) 718-3300. E-mail: viewmail@comcastsportsnet.com Web Site: midatlantic.comcastsportsnet.com.

Rgnl sports net serving mid-Atlantic. Progmg includes Orioles baseball, Capitals, hockey, Wizards, basketball, ACC & CAA.

Serving 4.4 million subs on over 200 cable systems.

Satellite: Spacenet III, transponder 12-H, ch 23 (scrambled).

Cox Sports AZ, 20401 N. 29th Ave., Phoenix, AZ, 85027. Phone: (623) 322-8001. Fax: (623) 322-7424. Web Site: phoenix.cox.net.

Phoenix Suns basketball, sports specials, high school sports & high school championships, etc. Phoenix metropolitan area serving over 500,000 subs. Loc microwave/fiber distributed regionally to additional operators.

Cox Sports Television, 2121 Airline Dr., 4th Fl - Central, Metairie, LA, 70001. Phone: 504-304-8134. Fax: 504-304-2243. E-mail: coxsportstv@cox.com Web Site: www.coxsportstv.com.

Cox Sports TV is an innovated, 24 hours net providing compelling & rgnl sports progmg. 1.3 million.

CSS - Comcast/Charter Sports Southeast, 2995 Courtyards Dr., Norcross, GA, 30071. Phone: (770) 559-7800. Phone: (770) 559-2742 (Jeff Miller). Fax: (770) 559-2329. E-mail: css@csssports.com Web Site: www.csssports.com.

ESPN Inc., ESPN Plaza, 545 Middle St., Bristol, CT, 06010. Phone: (860) 766-2000. Fax: (860) 766-2400. Web Site: www.espn.com.

ESPN offers a var of professional & amateur sports, including NFL, college basketball, NHL major league baseball, the woman's NCAA tournament.

On 28,000 affiliating cable systems serving over 77 million subs.

ESPNews, ESPN Plaza, 545 Middle St., Bristol, CT, 06010-9454. Phone: (860) 585-2000. Fax: (860) 766-2400. Web Site: www.espn.com.

ESPN Classic, ESPN Plaza, 545 Middle St., Bristol, CT, 06010. Phone: (860) 766-2000. Fax: (860) 766-2400. Web Site: www.espn.com.

Classic sporting events, sports series, documentaries & movies; home shopping for sports merchandise & interactive sports games.

Serving 20 million subs on 400 plus cable systems.

Satellite: Galaxy 7, transponder 13 (compressed).

FSN Arizona, 2 North Central, Suite 1700, One Renaissance Sq., Phoenix, AZ, 85004. Phone: (602) 257-9500. Fax: (602) 257-0848. Web Site: www.foxsports.com/arizona.

Mike Connelly, VP/gen mgr; Amy Serafin, progmg mgr; Brett Hansen, dir mktg; Jen Baker, account exec; Michael Bardess, producer.

Provides rgnl coverage of loc interest sports progmg. Serving 2.3 million subs in Arizona & Mexico. Satellite: C-1/16.

FSN Chicago, 205 N. Michigan Ave., Chicago, IL, 60601. Phone: (312) 729-9300. Web Site: fsnchicago.com.

Rgnl all-sports cable net serving Illinois, Indiana & Iowa, featuring Chicago Bulls, Blackhawks, Cubs/White Sox games, collegiate & high school sports.

Serving 3.8 million subs on over 160 cable systems.

Satellite: GE-1, transponder 13.

FSN Florida, 1550 Sawgrass Corporate Pkwy., Suite 350, Sunrise, FL, 33323. Phone: (954) 845-9994. Fax: (954) 845-0923. Web Site: www.fsnflorida.com.

Brad Heard, VP progmg; Jeff Genthner, VP/gen mgr; Larry Hoepfner, VP mktg.

FSN Florida progmg includes Major League Baseball's Marlins, Tampa Bay Devil Rays & National Hockey League's Florida Panthers. Serving 5 million subs on 150 cable systems.

Satellite: GE 1, transponder 4, channel 230.

FSN Pittsburgh, 2 Allegheny Ctr., Suite 1000, Pittsburgh, PA, 15212. Phone: (412) 322-9500. Fax: (412) 237-8439. Web Site: www.foxsports.com.

Rgnl sports network available to cable companies in Pennsylvania, Ohio, West Virginia & Maryland. Progmg includes Pittsburgh Pirates, Penguins & collegiate sports featuring Pitt men's basketball, West Virginia, Penn State, Mid America, conference events, nightly show "Sauran on Sports Beat" & Pittsburgh Sports Tonight, Pittsburgh Steeler wkly press conference, loc high school football championships.

Serving 2.3 million subs on 64 cable systems.

Satellite: Galaxy XI.

FSN South, 1175 Peachtree St. N.E., Bldg. 100, Suite 200, Atlanta, GA, 30361. Phone: (404) 230-7300. Fax: (404) 230-7399. Web Site: www.foxsports.com.

Chris Killebrew, VP sls; Bill Irish, VP progmg & VP progmg; Cheryl Raiford, controller; Jamie Kimbrough, dir; Brian Hogan, gen sls mgr; Steve Craddock, exec producer & VP; Jeff Genthner, VP/gen mgr.

NCAA sports, Atlanta Hawks, Memphis Grizzlies basketball, Atlanta Braves, Baltimore Orioles, Cincinnati Reds, St. Louis Cardinals baseball, Carolina Hurricanes, Nashville Predators hockey, NASCAR, golf, tennis & much more.

Serving 11.3 million subs on more than 1,100 cable systems.

Satellite: Galaxy 11, transponder 4, ch 2.

FSN West, 1100 S. Flower St. #2200, Los Angeles, CA, 90015. Phone: (213) 743-7800. Fax: (213) 743-7841. Web Site: msn.foxsports.com/regional/west.

Steve Simpson, VP/gen mgr.

Los Angeles Lakers basketball, Kings hockey, Lazers indoor soccer, Strings tennis, San Diego Soccers soccer & collegiate sports, etc.

FSN West 2, 1100 S. Flower St., #2200, Los Angeles, CA, 90015. Phone: (213) 743-7800. Fax: (213) 743-7841. Web Site: msn.foxsports.com/regional/west.

LaVada Heath, sls VP; Lisa Laky, VP sls; Alex Tevllin, progmg mgr; Greg Dowling, news dir; Steve Simpson, VP/gen mgr; Amy Wilson, dir mktg.

Rgnl sports net featuring the Los Angeles Dodgers, Los Angeles Clippers, Los Angeles Galaxy, Mighty Ducks of Anaheim, USC & UCLA athletic events & other sports.

Serving 3 million subs.

Satellite: G1, transponder 21. 4,800,000.

Fox Soccer Channel, 1440 S. Sepulveda Blvd., 2nd Fl., Los Angeles, CA, 90025. Phone: (310) 444-8642. Fax: (310) 444-8445. Web Site: www.foxsoccer.com. E-mail: ben.alkaly@fox.com

Frank Uddo, VP progmg; Dermot McQuarrie, sr VP; Sean Riley, sr VP; David Sternberg, exec VP; Raul de Quesada, sr VP; David Stenberg, gen mgr; Ed Derse, VP & interactive media; Raul Palma, VP & production; Fausto Ceballos, VP & on-air promotions; Mike Petruzzi, natl adv sls mgr; Veronica Alvarez, dir of mktg & promotions.

Fox Soccer Channel is the nation's leading TV destination for young, passionate & affluent soccer fans. The best in exclusive coverage of professional, college & youth soccer. Satellite: Galaxy 11, transponder 8. 550+. over 30 million.

Fox Sports en Espanol, 1440 S. Sepulveda Blvd., Los Angeles, CA, 90025. Phone: (310) 444-8658. Fax: (310) 444-8445. E-mail: patrick.ilabaca@fox.com Web Site: www.fse.tv.

New York, NY 10036, 1211 Ave. of the Americas. Phone: (212) 822-7000. Tom Maney, VP adv sls.

Live, exclusive coverage in Sp of the Copa Toyota Libertadores & Major League Baseball's All-Star game & World Series postseason, boxing & nightly sports news. Serving 7.9 million subs on 1,321 cable systems. Satellite: Satcom C1, Transponder 1, ch 8 (SA Power VU IRD D9225).

Fox Sports Net, 10201 W. Pico Blvd., FNC/Bldg. 101, 5th Fl., Los Angeles, CA, 90035. Phone: (310) 369-6000. Fax: (310) 969-6700. Web Site: www.foxsports.com.

Tracy Dolgin, exec VP; Arthur Smith, VP progmg; Jim Martin, exec VP; Jeff Shell, CFO; Randy Freer, pres.

A natl, rgnl & loc supplier of sports progmg.

Serves 68 million subs through 22 rgnl sports nets.

Fox Sports Net Bay Area, 77 Geary St., 5th Fl., San Francisco, CA, 94108. Phone: (415) 296-8900. Fax: (415) 296-9198. E-mail: fsnbayinfo@fsnbayarea.com Web Site: www.fsnbayarea.com.

Jeff Krolik, VP/gen mgr; Chris Geer, VP sls; Ted Griggs, VP progmg; Michael McCright, gen sls mgr.

Programing: San Francisco Giants, Oakland Athletics, Golden State Warriors, San Jose Sharks, San Jose Saber Cats & San Jose Stealth.

Serving more than 4 million households in Northern California & Northern Nevada.

Satellite: Compressed, AMC 1, T 18 Channel 110.

Fox Sports Net Detroit, 26555 Evergreen Rd., Suite 90, Southfield, MI, 480276. Phone: (248) 226-9700. Fax: (248) 226-9725. Web Site: www.foxsports.com/detroit. E-mail: detroit@foxsports.net

Lisa Giles, progmg dir; Greg Hammaren, VP/gen mgr; John Tuohey, exec producer.

Cable sports net featuring Detroit Pistons, Red Wings, Tigers, Fury, Shock, CCHA hockey & Michigan High School Association championship contests.

Serving 3.2 million subs on more than 70 cable systems.

Fox Sports Net Midwest, 700 St. Louis Union Station, Suite 300, St. Louis, MO Phone: (314) 206-7020. Fax: (314) 206-7070. E-mail: midwest@foxsports.net Web Site: www.foxsports.com.

Jack Donovan, VP/gen mgr.

Indianapolis, IN 46204. FSN Indiana, 135 N. Pennsylvania St., Suite 720.

Fox Sports Net Midwest reaches more than 5.4 million cable & satellite TV homes in six Midwest states. It telecasts more that 2,000 hours of loc progmg each year, including coverage of St. Louis Cardinals baseball, St. Louis Blues hockey, Indiana Pacers basketball, Indiana Fever basketball, Kansas City Royals baseball, Cincinnati Reds baseball, Big 12: football, women's basketball & showcase, Univ. of Missouri athletics, Kansas State Univ. athletics, Univ. of Nebraska Basketball, Missouri Valley Conference basketball, championship events, Gateway Conference football, & loc high school sports programs, collegiate coaches shows.

Fox Sports Net New England, 42 3rd Ave., Burlington, MA, 01803-4414. Phone: (781) 270-7200. Web Site: www.foxsportsnewengland.com.

Boston Celtics basketball, New York Mets (Connecticut only), college basketball, golf, football, hockey, professional tennis, soccer & auto racing.

On 215 cable systems serving 2.9 million subs.

Satellite: GE1, transponder 14.

Fox Sports Net New York, Two Penn Plaza, 4th Fl., New York, NY, 10001. Phone: (212) 465-6000. Fax: (212) 465-6024. Web Site: msn.foxsports.com.

A two-ch rgnl sports network that delivers approximately 300 live games of the New York Islanders, Mets, New Jersey Nets & Devils, in addition to horse racing, college football, basketball & variety of sports specials.

On 128 affil cable systems serving more than 2.7 million subs.

Satellite: GE SpaceNet 2, transponders 1.

Fox Sports Net North, 1 Main St. S.E., # 600, Minneapolis, MN, 55414-1036. Phone: (612) 330-2468. Fax: (612) 330-9010. Web Site: www.foxcable.com.

Rgnl Sports Network: Minnesota, Iowa, Wisconsin, South Dakota & North Dakota. MLB & Brewers, NBA Timberwolves & Bucks, University of Minnesota hockey, & women's athletics, University of Wisconsin men's & women's athletics, Marquette University athletics.

Serving 3 million subs.

Satellite: GE 3, transponder 6.

Fox Sports Net Northwest, 3626 156th Ave. S.E., Bellevue, WA, 98006. Phone: (425) 641-0104. Fax: (425) 641-9811. Web Site: www.foxsports.com/northwest.

Mark Shuken, VP/gen mgr; Amy Affeld, progmg dir; Mike Smith, controller; Liz Serrette, opns mgr; Julie McCormack, mgr; Brett Bibby, gen sls mgr.

Coverage of PAC-10, Big Sky, other collegiate conference athletic events; Mariners, SuperSonics & other professional & high school events in the Pacific Northwest rgn.

Serving 2.4 million subs on 100 cable systems.

Satellite: G7, transponder 4.

Fox Sports Net Ohio, 9200 S. Hills Blvd., Suite 200, Broadview Heights, OH, 44147. Phone: (440) 746-8000. Fax: (440) 746-9480. Web Site: www.foxsports.com.

Steve Pawlowski, communications dir.

Cincinnati, OH 45242, 11311 Cornell Park Dr, Suite 406. Phone: (513) 469-2006. Fax: (513) 469-2007. Web Site: www.foxsports.net.

Live sports progmg: Cleveland Indians, Cleveland Cavaliers, Cincinnati Reds, Columbus Blue Jackets, college football, basketball & sports news.

Serving 4.5 million subs on 206 cable systems.

Satellite: Satcom GE1, transponder T4. Alternate: GE1 T17.

Fox Sports Net Rocky Mountain, 2300 15th St., Suite 300, Denver, CO, 80202. Phone: (720) 898-2700. Fax: (720) 898-2735. Web Site: www.foxsports.com.

Steven Gravlin, dir; Tim Griggs, gen mgr; Amy Turner, dir.

Rgnl sports net serving 8 states. Progmg includes Denver Nuggets, Utah Jazz, Colorado Avalanche, Colorado Rockies, Univ of Denver & Big 12 conference. Serving 2.2 million subs on 300 cable systems.

Satellites: G7.

Fox Sports Net Southwest, 100 E. Royal Ln., Suite 200, Irving, TX, 75039. Phone: (972) 868-1800. Fax: (972) 868-1678. Web Site: www.foxsports.com.

Jon Heidtke, VP/gen mgr; Mike Anastassiou, exec producer; Mike Ibanez, affil sls dir.

Rgnl sports net serving Texas, Oklahoma, Arkansas, Louisiana & parts of New Mexico. Serving 9 million subs on 1,300 cable and satellite systems. Satellite: Galaxy 11, transponder 4 (digitally compressed).

Madison Square Garden Network, Two Penn Plaza, 4th Fl., New York, NY, 10001. Phone: (212) 465-6000. Fax: (212) 465-6024. E-mail: mscnetpr@msgnetwork.com Web Site: www.msgnetwork.com.

New York Knicks, Rangers & Yankees; college football & basketball games; boxing. Exclusive Garden events as well as original series progmg.

Serving more than 6.1 million subs on more than 250 cable systems.

Satellite: Satcom 4, transponder 6.

New England Sports Network (NESN), 480 Arsenal St. #1, Watertown, MA, 02472-2805. Phone: (617) 536-9233. Fax: (617) 536-7814. Web Site: www.boston.com/sports.nesn.

NESN is a cable sports svc that delivers Boston Bruins, Red Sox, New England college sports as well as boxing, tennis, fishing, bowling & wrestling.

Serving 3.5 million subs on 28 cable systems.

Satellites: Satcom F-4, transponder 13; GE C-3, transponder 14.

The Sports Network, 2200 Byberry Rd., Hatboro, PA, 19040. Phone: (215) 441-8444. Fax: (215) 441-5767. E-mail: kzajac@sportsnetwork.com Web Site: www.sportsnetwork.com.

International real-time sports wire svc providing content, branded web pages, satellite and/or computer feeds directly to broadcasters (radio & TV), print, Internet sites, wireless with state of the art technology.

SportsNet New York, 75 Rockefeller Plaza, 29th Fl., New York, NY, 10019. Phone: (212) 485-4800. Fax: (212) 485-4802. Web Site: www.sny.tv.

Jon Litner, pres.

Sun Sports, 1000 Legion Place, Suite1600, Orlando, FL, 32801-1060. Phone: (407) 648-1150. Fax: (407) 245-2571. E-mail: askus@foxsports.net Web Site: www.sunsportstv.com.

Cathy Weeden, VP/gen mgr.

Sunrise, FL 33303, 1550 Sawgrass Corp. Pkwy, Suite 350.

Rgnl sports cable net. Progmg includes Orlando Magic & Miami Heat NBA basketball, Tampa Bay Lightning NHL hockey, Florida State, Univ. of Florida, SEC, & FHSAA, athletics, as well as a wide var of loc & rgnl sports events plus Chevy Tailgate Saturday, In My Own Words, Chevy FL Fishing Report. Serving 6.3 million subs.

Satellites: Galaxy II, transponder 4.

Victory Sports One, 60 S. 6th St., Suite 3700, Minneapolis, MN, 55403. Phone: (612) 661-3778. E-mail: info@victorysports.com

VideoSeat Pay-Per-View, (A division of Host Communications Inc.). 546 E. Main St., Lexington, KY, 40508. Phone: (859) 226-4678. Fax: (859) 226-4391. E-mail: dossd@hcionline.com Web Site: www.hostcommunications.com.

Lawthon Logan, sr exec VP sls.

VideoSeat handles turnkey pay-per-view syndication of several top schools in college football, including: Kentucky, Mississippi State, South Carolina, & Tennessee. Systems in Kentucky, Georgia, Mississippi, South Carolina & Tennessee.

Yankees Entertainment and Sports Network LLC, The Chrysler Bldg., 405 Lexington Ave., 36th Fl., New York, NY, 10174-3699. Phone: (646) 487-3600. Fax: (646) 487-3612. E-mail: info@yesnetwork.com Web Site: www.yesnetwork.com.

Michael Wach, exec VP.

Cable Audio Services

CRN Digital Talk Radio, 10487 Sunland Blvd., Sunland, CA, 91040. Phone: (818) 352-7152. Fax: (818) 352-3229. E-mail: info@crni.net Web Site: www.crni.net.

Michael Horn, pres/CEO; Jennifer Horn, VP sls & mktg VP; Paul Stern, VP opns.

Premier provider for radio syndication. CRN develops the hottest new, unique talk talent & distributes it worldwide via cable TV audio, radio, satellite audio & the internet. ON 125 cable systems nationwide serving 26 million subs. Satellite: IA 13, transponder 15, virtual ch 521-526. CRN 1-6 with 6 talk networks.

The Classical Station, WCPE, Box 897, Wake Forest, NC, 27588. Phone: (919) 556-5178. Fax: (919) 556-9273. E-mail: wcpe@wcpe.org Web Site: theclassicalstation.org.

Deborah S. Proctor, chief engr.

Free 24-hour classical music progmg with live announcers for radio, cable, other distributors. Weekly request programs, opera and features. Satellite: Galaxy 14, Transponder 8, Vert, 6.30/6. 48 MHz, or 4DTV G5 959.

DMX, Inc., 600 Congress Ave., Fl. 14, Austin, TX, 78701. Phone: (512) 380-8500. Fax: (512) 380-8501. Web Site: www.dmx.com.

Steve Hicks, chmn; John Cullins, pres; Paul Stove, COO; Kim Shipman, CFO.

Offers uninterrupted premium digital audio music progmg via satellite & cable to residential & coml subs.

Moody Broadcasting Network, 820 N. LaSalle Blvd., Chicago, IL, 60610. Phone: (800) 621-7031. Phone: (312) 329-4433. Fax: (312) 329-4339. E-mail: mbn@moody.edu Web Site: www.mbn.org.

Doug Hastings, opns mgr & progmg mgr; Denny Nugent, progmg dir; Wayne Pederson, VP.

Provides 24-hours format of relg & educ progmg; music, drama, talk, news & pub affrs.

On 426 radio stns nationwide.

Satellites: AMC-3, transponder 17H (DVB-stereo digital) on AMC-8 (SCPC, digital stereo).

Music Choice, 110 Gibraltar Rd., Suite 200, Horsham, PA, 19044. Phone: (215) 784-5840. Fax: (215) 784-5869. Web Site: www.musicchoice.com.

David J. Del Beccaro, pres/CEO; Damon Williams, sr VP.

Music Choice is the premier music television network, reaching U.S .households through digital cable and satellite television. Music Choice programs interruption-free music for homes and businesses and distributes televised concerts and music shows. The Music Choice music channels reach 33 million households and the Music Choice Concert Series airs in 44 million homes nationally. Music Choice is a partnership among subsidiaries of Microsoft Corporation, Motorola, Inc., Sony Corporation of America, Warner Music Group, Inc., EMI Music and several leading U.S. cable providers: Adelphia Cable Communications, Comcast Cable Communications, Cox Communications, and Time Warner Cable.

WFMT Radio Network, 5400 N. St. Louis Ave., Chicago, IL, 60625. Phone: (773) 279-2112/2114. Fax: (773) 279-2199. E-mail: tmedia@wfmt.com Web Site: www.wfmt.com.

Steve Robinson, gen mgr & sr VP.

Classical music, spoken arts & fine arts program series & specials. Satellite- & tape-delivered. Major symphony orchestras, opera, jazz, exclusive BBC & Radio Deutsche Welle progmg, WFMT-produced archival & spoken-word progmg, live studio performances, folk music. Since 1976.

Serving more than 900 radio outlets worldwide.

Satellite: Galaxy 4, digital frequeney B72.0.

"The Weather Center", (a broadcast service of Aviation Weather Inc.). 701 Gervais St., Suite 150-224, Columbia, SC, 29201. Phone: (803) 739-2827. E-mail: wxcenter@aviationweatherinc.com Web Site: www.aviationweatherinc.com.

Liam Richard Ferguson, pres.

"Regional Radio Broadcast/Weathercast Network" across the Carolinas and Georgia in over 20 bcst markets. Weather forecasting, site-specific bcst svc for stns all across America. 100% barter.

Yesterday U.S.A., 2001 Plymouth Rock, Richardson, TX, 75081. Phone: (972) 889-8255. Fax: (972) 889-2329. E-mail: yesterdayusa@mail.com Web Site: www.yesterdayusa.com.

William J. Bragg, founder.

A 24-hour natl radio voice of the National Museum of Communication of Irving, TX. Presenting public domain old-time radio shows & vintage music free of charge & without comls.

Satellites: Galaxy 11, transponder 18, Ku verticel, frequency 12060, audio PID 1620 left ch .

Major National TV News Organizations

ABC News

Ownership: Walt Disney Company

147 Columbus Ave., New York, NY 10023; Tel: 212-456-1000

Executives: David Westin, pres, opns/admin; Paul Mason, sr VP; Paul Slavin, sr VP producer; Phyllis McGrady, exec producer/special progmg; Bob Murphy, sr VP/multimedia; Chris Isham, sr producer; Amy Entelis, sr VP, talent recruitment/business affrs; Kerry Marash, VP/editorial quality; Dawn Porter, dir/news practices; Barbara Fedida, dir/news practices; Andrea Cohen, VP/business affrs; Jeffrey Schneider, VP/news media; Derek Medina, VP/business dev; Dick Wald, sr VP/consultant; Jacqueline Shire, news consultant; Roger Goodman, VP/special projects.

Domestic Bureau

Atlanta: 2580 Cumberland Pkwy. S.E., Suite 160, Atlanta, GA 30339; Tel: 770-431-2380; Fax: 770-431-7800

Kate O'Brian, bureau chief.

Chicago: 190 N. State St., Chicago, IL 60601; Tel: 312-899-4015; Fax: 312-899-4050

Ron Schofield, Midwest bureau chief.

Los Angeles: 4151 Prospect Ave., Los Angeles, CA 90027; Tel: 323-671-5261, Fax: 323-671-5210

David Eaton, bureau chief; Charlie Herman, deputy bureau chief; Michael Ray Gammon; Derick Yanehiro & Marilyn Heck, assignment editors; Roger Scott & Chris Cahan, assignment editors, weekend.

New York: 47 W. 66th St., 3rd Fl., New York, NY 10023; Tel: 212-456-2700, 212-456-7777; Fax: 212-456-2214

Mimi Gurbst, VP, news coverage; Kris Sebastian, VP, Northeast bureau chief; Chuck Lustig, dir of foreign news; Patrick Sullivan, assignment mgr; Assignment editors: Justin Anderson; Ed Bailey; Wendy Fisher; Barbara Chen; Michael Kreisel; Barbara Garci; Clem Lane & Eva Price; Ursula Fahy, International domestic/assignment editor.

Washington: 1717 DeSales St., N.W., Washington, DC 20036; Tel: 202-222-7300; Fax: 202-222-7684

Robin Sproul, bureau chief; Dennis Dunleavey, deputy bureau chief; Assignment editors: Diane Boozer; Dee Carden; George Sanchez; Zack Wolf & Julianne Donofrio.

Dallas: 606 Young St., Dallas, TX 75202; Tel: 214-748-9631

Denver: 123 Speer Blvd., Denver, CO 80203; Tel: 303-832-7777

Miami: 1320 South Dixie Hwy., Coral Gables, FL 33146; Tel: 305-662-2116

Seattle: 140 4th Ave., Seattle, WA 98109; Tel: 206-404-4000

International Bureaus

Beijing: 4-1-71 Jian Guo Men Wai Compound, Beijing, China; Tel: 861 06532 2671; Fax: 861 06532 2668

Josh Gerstein; Chito Romana.

Havana: Tel: 537 8793 650; Fax: 537 8730 087

Mara Valdes.

Hong Kong: 21/F Shell Tower, Times Square, One Matheson St., Causeway Bay, Hong Kong; Tel: 852 2203 2000; Fax: 852 2203 1400

Mark Litke; Andrew Morse.

Jerusalem: 206 Jaffa Rd., Jerusalem, Israel; Tel: 9722 500 5911; Fax: 9722 500 2051.

John Yang; Simon McGregor-Wood; Bruno Nota.

Kenya: Tel: 2542 522 624

Martin Seemungal.

London: 3 Queen Caroline St., London W6 9PE, United Kingdom; Tel: 44 208 222 5500 / 8309-5500; Fax: 44 208 222 5020

Marcus Wilford; Robin Wiener.

Mexico City: Reforma 350 Piso 12, Colonia Juarez, C. P. 06600 Mexico D.F.; Tel: 525 55 511 2790; Fax: 525 55 511 2785

Jose Cohen.

Moscow: Bolshoi Afanieveskay, Per 7, Moscow, CIS, Russia; Tel: 7095 232 3737 / 7095 291 1987 / 8590-2500; Fax: 7095 202 5827

Tomek Rolski.

Paris: 155 Rue du Faubourg St. Honore, 75008 Paris, France; Tel: 33 1 58 56 35 00; Fax: 33 1 43 59 38 28

Bruno Silvestre.

Rome: Piazza Grazioli, 5, Rome 00186, Italy; Tel: 3906 679 7715; Fax: 3906 679 7704

Phoebe Natanson.

Tokyo: NHK Hoso Center East Bldg., 7F, Jinnan 2-2-1, Shibuya-Ku, Tokyo, 150-8001, Japan; Tel: 813 3485 2631; Fax: 813 3485 2641

Chika Nakayama.

Primetime Shows Executive Producers: 20/20: David Sloan; Good Morning America: Shelley Ross; Nightline: Leroy Sievers; Primetime Thursday: David Doss; This Week with George Stephanopoulos: Tom Bettag; World News Tonight with Charles Gibson: Jon Banner; World News Tonight Saturday/Sunday: Craig Bengtson.

Affiliate News Service: ABSAT, 47 W. 66th St., New York, NY 10023; Tel: 212-456-4134 Mike Huitt, dir; Chris Myers, opns mgr.

CBS News

Ownership: Viacom

51 W. 52nd St., New York, NY 10019; Tel: 212-975-4321; Fax: 212-975-3285; national desk: 212-975-4114; Fax: 212-975-1893; foreign desk: 212-975-3019; Fax: 212-245-7560. Web site: www.cbs.com.

Executives: Marcy McGinnis, sr VP/news coverage; John Frazee, VP/news svcs; Sandra Genelius, spokeswoman CBS; Frank Governale, VP/news opns; Linda Mason, sr VP, standards/special projects; James McKenna, VP, finance/admin; Janet Leissner, VP/Washington, DC bureau cheif; John Paxson, VP, Europe/London bureau chief.

Domestic Bureaus

Atlanta: 260 14th St. NE, Atlanta, GA 30318; Tel: 404-872-8301

Dallas: 1011 N. Central Expwy, Dallas, TX 75231; Tel: 214-739-1199; Fax: 214-696-9011

Los Angeles: 7800 Beverly Blvd., Los Angeles, CA 90036; Tel: 323-575-2345

Jennifer Siebens, bureau chief.

Miami: 4770 Biscayne Blvd., Miami, FL 33101; Tel: 305-571-4400

San Francisco: 825 Battery St., San Francisco, CA 94111; Tel: 415-362-8177

Washington: 2020 M St., N.W., Washington, DC 20036; Tel: 202-457-4321; Fax: 202-331-1765 (newsroom)

Janet Leissner, bureau chief.

International Bureaus

Amman, Jordan; Baghdad, Iraq; Beijing, China; Bonn, Germany; Hong Kong; Johannesburg, South Africa; London, England; Moscow, Russia; Paris, France; Rome, Italy; Tel Aviv, Israel; Tokyo, Japan.

Primetime Shows Executive Producers: CBS Evening News: Bob Schieffer, anchor; Up to the Minute: Karen Sacks; CBS Morning News: Michael Bass; 60 Minutes (Sunday Edition): Jeff Fager, 60 Minutes (Weekday Edition): Josh Howard; 48 Hours Mystery: Susan Zirinsky & CBS News Sunday Morning: Rand Morrison; The Saturday Early Show/The Early Show: Michael Bass, sr exec producer.

Affiliate News Service: CBS News Service, 524 W. 57th St., New York, NY 10019; Tel: 212-975-5641 John Frazee, sr VP.

CNBC

CNBC, Inc., 900 Sylvan Ave., Englewood Cliffs, NJ 07632; Tel: 201-735-2622; Fax: 201-735-3200. Web site: www.cnbc.com.

Executives: Mark Hoffman, pres and CEO; Lilach Asofsky, sr VP, mktg and rsch; Judith H. Dobrzynski, mgng editor, business news; Lauren Donovan, sr VP; Robert Foothorap, digital products; Scott Drake, VP, mgr technology; David Friend, sr VP, business news; Cheryl Gould, VP primetime and weekend progmg; Nikki Gonzalez, VP, human resources; Bob Meyers, gen mgr, CNBC Ventures; Kevin Egan, VP and CFO; Steve Fastook, VP, tech and coml opns; Amy Zelvin, VP, PR.

Domestic Bureaus

Washington: 1025 Connecticut Ave., N.W., Suite 800, Washington, DC 20036; Tel: 202-467-5400; Fax: 202-737-4985

Alan Murray, bureau chief/anchor.

Los Angeles: 3000 W. Alameda Ave., Burbank, CA 91523; Tel: 818-840-3214; Fax: 818-840-4181

Heather Allen, bureau chief.

International Bureaus: London; Singapore.

Primetime Shows and Executive Producers: Wake Up Call: Gary Kanofsky; Squawk Box: Bill

McCandless; Morning Call: Rich Fisherman; Power Lunch: Bob Fasbender; Street Signs: Andy Hoffman; Closing Bell: Alex Crippen; Kudlow & Cramer: Matt Quayle; Capital Report: Steve Lewis; Special Report: Diane Galligan; Dennis Miller: Eddie Feldman; Suze Orman: Amy Fellar.

CNN

News Desk, One CNN Center, Atlanta, GA 30303; Tel: 404-827-1700

Executives: Jim Walton, pres/CNN worldwide; Greg D'Alba, exec sr VP, news COO/adv sls and mktg; Brad Ferrer, exec VP finance/admin.

Domestic Bureaus

New York: One Time Warner Center, New York, NY 10019; Tel: 212-275-7800

Karen Curry, bureau chief.

Boston: 637 Washington St., Suite 208, Brookline, MA 02446; Tel: 617-264-9905

Dan Lothian, bureau chief.

Chicago: 435 N. Michigan Ave., Chicago, IL 60611; Tel: 312-645-8555

Ron Hess, deputy bureau chief.

Los Angeles: 6430 W. Sunset Blvd., Los Angeles, CA 90028; Tel: 323-993-5000

Miami: 12000 Biscayne Blvd. N., Miami, FL 33181; Tel: (America) 305-892-5100; (Espanol) 305-895-4885

John Zarrella, bureau chief.

Washington: 820 First St. N.E., Washington, DC 20022; Tel: 202-898-7900

David Bohrman, bureau chief.

International Bureaus

Hong Kong: 30 F Oxford House, 979 Kings Rd., Taikoo Pl., Quarry Bay, Hong Kong

London: CNN, Turner House, 16 Great Marlborough, London WIS7HS, United Kingdom

CNN TV Programs: Anderson Cooper 360: Terry Baker, sr exec producer; Lou Dobbs Tonight: Bill Dorman, exec producer; American Morning with Soledad O'Brien and Bill Hemmer: Wil Surrat, exec producer; Larry King Live; Interview Debate late edition; Reliable Sources; The Situation Room; This What.

Fox News Channel

Ownership: News Corp.

1211 Ave. of the Americas, New York, NY 10036; Tel: 212 852-7002; Fax: 212-301-5888

Executives: Roger Ailes, chmn/CEO; Kevin Magee, sr VP/radio; Mark Kranz, CFO; John Moody, sr VP, news/editorial; Paul Rittenberg, sr VP/adv sls; Bill Shine, VP/progmg; News Corp.; Rupert Murdoch, chmn/CEO, News Corp.; Peter Chernin, pres/COO, News Corp.; Martin Pompadur, exec VP; David DeVoe, sr VP/CFO; Andrew Butcher, VP, corporate affrs/communications; Dianne Brandi, VP, legal/business affrs; Kim Hume, VP, DC bureau chief; Brian Lewis, exec sr VP, corporate/communications.

Domestic Bureaus

Atlanta: 260 14th St., N.W., Atlanta, GA 30318; Tel: 404-685-2280

Todd Ciganek, bureau chief.

Denver: Total Bldg N. Tower, 999 18th St., Suite 1665, Denver, CO 80202; Tel: 303-383-1170; Fax: 303-383-1171

Dallas: 301 N. Market St., Suite 450, Dallas, TX 75202: Tel: 214-742-5005: Fax: 214-742-1067

Russ Cosby, bureau chief.

Los Angeles: 2044 Armocast Ave., Los Angeles, CA 90025; Tel: 310-571-2000/5/7; Fax: 310-571-2009

John Brady, bureau chief.

Miami: 1440 79th St. Causeway, Suite 208, North Bay Village, FL 33141; Tel: 305-866-8007; Fax: 305-866-5444

Nancy Harmeyer, bureau chief.

San Francisco: 901 Battery St., Suite 210, San Francisco, CA 94111; Tel: 415-951-8550

Washington: 2201 C St., N.W., Washington, DC 20520; Tel: 202-496-0109; Fax: 202-824-6426

International Bureaus

Hong Kong: One Harbourfront, 18 Tak Fung St., 15th Fl. Hunghan, Kkowloon, Hong Kong; Tel: 011-852-2821-8853; Fax: 011-852-2621-8658

Vivian McGrath, bureau chief.

Jerusalem: 206 Jaffa Rd., PO Box 13172, Jerusalem 91131, Israel; Tel: 011-972-2500-1421; Fax: 011-972-3642-2226

London: Grant Way, Isleworth, Middlesex TW7 500, United Kingdom; Tel: 011-44-171-805-7146; Fax: 011-44-171-805-7140

Moscow: 3 Gruzinsky Pereulok, KV. 311-312, Moskow 123056, Russia; Tel: 011-7502-221-3221; Fax: 011-7095-254-5828

Primetime Shows: Special Report with Brit Hume; The O'Reilly Factor; Hannity and Colmes; On the Record with Greta Van Susteren; Your World with Neil Cavuto; The Big Story with John Gibson.

MSNBC

1 MSNBC Plaza, Seacaucus, NJ 07094; Tel: 201-583-5000; Fax: 201-583-5453

Executives: Rick Kaplan, pres/gen mgr; Phil Griffin, exec in charge; Mark Effron, VP, news/daytime progmg; Robin Garfield, VP, strategic opns; Val Nicholas, VP, ad/promotion; Jeremy Gaines, VP, communications; Nick Tzanis, VP, technical opns.

Primetime Shows: The Abrams Report: Meghan Shaefer, exec producer; Hardball: Tammy Haddad, exec producer; Lester Holt Live; Countdown with Keith Olbermann: Izzy Povich, exec producer; MSNBC Investigates; Headlines and Legends; Scarborough Country: Lia Macko, exec producer.

NBC News

Ownership: NBC Universal

30 Rockefeller Plaza, New York, NY 10112; Tel: 212-664-4444

Executives: Bob Wright, chm/CEO; Lynn Calpeter, exec VP/CFO; David Overbeeke, exec VP/chief info off; John Eck, Media Work pres/chief info off; Anna Perez, exec VP/communications; Rick Cotton, exec VP/gen counsel; Lisa Hsia, VP; Elena Nachmanoff, VP/talent dev; David McCormick, chief, standards/practices; Lloyd Siegel, dir/news partnerships; Jocelyn Cordova, dir, talent recruitment/dev.

Domestic Bureaus

Midwest Bureau: 454 N. Columbus Dr., 1st Fl., Chicago, IL 60611

Stewart Dan, bureau chief.

Primetime Shows and Executive Producers: Nightly News: Steve Capus; TODAY: Jim Bell; Dateline: David Corvo, exec producer; Meet the Press: Betsy Fischer; Weekend Today: Alex Wallace; Weekend Nightly News: Bob Epstein; Chris Matthews Show: Nancy Nathan.

Affiliate News Service: NBC News Channel, 925 Woodridge Center Dr., Charlotte, NC 28217, Tel: 704-329-8741 Bob Horner, pres; Sharon Houston, exec producer.

Television News Services

ABC News

See ABC listing in Major National TV News Organizations, this section.

APTN Productions, The Interchange, Oval Rd., Camden Lock, London, NW1 7DZ. United Kingdom. Phone: (0) 20 7482 7400. Fax: (0) 20 7413 8312. Web Site: www.aptn.com.

Eric Braun, mgng dir; Nigel Baker, exec dir.

New York, NY 10023, 1995 Broadway. Phone: (212) 362-4440.

International TV svcs company, daily satellite news feeds to bcstrs worldwide, tech facilities, camera crew hire worldwide.

Serves TV.

AccuWeather Inc., 385 Science Park Rd., State College, PA, 16803. Phone: (814) 235-8600. Fax: (814) 235-8609. E-mail: sales@accuweather.com Web Site: www.accuweather.com.

Dr. Joel N. Myers, pres; Evan Myers, sr VP/dir opns.

TV, radio, weather progmg & systems, plus turnkey solutions to take your loc news to the mobile web.

Agence France-Presse, 1500 K St., Suite 600, Washington, DC, 20005. Phone: (202) 289-0700. Fax: (202) 414-0635. Web Site: www.afp.com. E-mail: afp-usa@afp.com

Peter Mackler, editor-in-chief English; Philipe Raater, editor-in-chief & Fr edition.

Produces a variety of international news svcs for radio & TV, including text wires in six languages, photo wires, graphics & financial wires (plus multimedia svcs).

All Africa Global Media, 920 M Street S.E., Washington, DC, 20002. Phone: (202) 546-0777. Fax: (202) 546-0676. Web Site: www.allafrica.com. E-mail: newsdesk@allafrica.com

Reed Kramer, CEO; Amadou Mahtar Ba, pres.

A news & info service on African affrs for TV, radio & print news svcs.

American Academy of Dermatology, Communications Dept., American Academy of Dermatology. Box 4014, Schaumburg, IL, 60168-4014. Phone: (847) 330-0230. Fax: (847) 330-8907. Web Site: www.aad.org.

Washington, DC 20005-3319, 1350 I Street NW, Suite 870. Phone: (202) 842-3555. Fax: (202) 842-4355.

Expert physicians available for TV & radio interviews, audio & video tapes on skin cancer detection, as well as info on skin, hair & nail conditions.

American Heart Association National Center, 7272 Greenville Ave., Dallas, TX, 75231. Phone: (214) 706-1330. Phone: (800) 242-8721. Fax: (214) 706-5243. Web Site: www.americanheart.org/news.

Julie Del Barto, bcst mgr.

Periodic satellite news feeds of medical rsch stories.

The Associated Press, AP Broadcast News Center, 1825 K St. N.W., Suite 800, Washington, DC, 20006-1202. Phone: (202) 736-1100. Phone: (800) 821-4747. Fax: (202) 736-1124. Web Site: www.apbroadcast.com.

James R. Williams III, VP.

AP Services for TV: Video: APTN Video News. Wires: APTV Wire, AP News Tickers, AP NewsPower, AP Data Stream. AP Alert Graphics: AP GraphicsBank. Software: AP NewsCenter; AP NewsDesk; AP NewsDesk (LAN), ENPS, SNAPfeed Satellite Delivery: AP Express. Elections: ENPS Stats, AP Politics, AP Election Wire. Online content: CustomNews, AP Online, Online Video Network, AP Spanish Online. Photos: Photo Archive, Photo Stream.

Audio-Video News, 3622 Stanford Cir., Falls Church, VA, 22041. Phone: (703) 354-6795. E-mail: connielawn@aol.com Web Site: dcski.com.

Covers major natl, international & specialty stories for radio & TV stns in the U.S. & around the world. Also do live talk-back features. Serves radio, TV & write ski reports.

Bloomberg L.P., 731 Lexington Ave., New York, NY, 10022. Phone: (212) 318-2200, EXT. 2201. Fax: (917) 369-5000. Web Site: www.bloomberg.com.

John Meehan, chief of bcstg.

24-hour TV ch. Business & news reports for radio & TV stns. Full news service.

British Information Services, 845 3rd Ave., New York, NY, 10022. Phone: (212) 745-0277. Fax: (212) 745-0359.

Mark Hopkinson, Head Radio/TV div; Sarah Kendall, mktg.

Assists radio & TV crews visiting the United Kingdom.

Serves radio & TV.

Broadcast Interview Source, 2233 Wisconsin Ave. N.W., Washington, DC, 20007. Phone: (202) 333-5000. Phone: (202) 333-5000. Fax: (202) 342-5411. E-mail: editor@yearbook.com Web Site: www.expertclick.com.

Mitchell P. Davis, editor.

Free source of interview contacts

Broadcast News Ltd., 36 King St. E., Toronto, ON, M5C 2L9. Canada. Phone: (416) 364-3172. Fax: (416) 364-1325. E-mail: tscott@broadcastnews.ca Web Site: www.cp.org.

Wayne Waldroff, gen mgr; David Ross, CFO/dir opns; Terry Scott, news dir.

Full wire & audio svcs (news agency), satellite delivery for radio program syndicators.

Serves radio & TV.

CBS News

See CBS listing in Major National TV News Organizations, this section.

CNN and CNN Headline News

See listing in Major National TV News Organizations, this section.

Camera Planet, 253 Fifth Ave., New York, NY, 10016. Phone: (212) 779-0500. E-mail: archive@cameraplanet.com Web Site: www.cameraplanet.com.

Steve Carlis, pres/COO; Steve Rosenbaum, pres/CEO.

Canada NewsWire Ltd., 1500, 20 Bay St., WaterPark Pl., Toronto, ON, M5J 2N8. Canada. Phone: (416) 863-9350. Phone: (866) 805-9530. Fax: (416) 863-9429. E-mail: cnwtor@newswire.ca Web Site: www.newswire.ca.

Sylvia Kavanagh, mgr; Tim Griffin, bcst mgr; Carolyn McGill, dir mktg.

Calgary, AB T2P 3C5 Canada, Gulf Canada Sq, 401 Ninth Ave. S.W., Suite 835. Phone: (403) 269-7605. Fax: (403) 263-7888. TWX: 03-824872. E-mail: Michle.dauphine@newswire.ca. Krista Wightman, mgr.

Vancouver, BC V6B 4NB Canada, 650 West Georgia St, Suite 1103. Phone: (604) 669-7764. Fax: (604) 669-4356. TWX: 04-508529. Larry Cardy, VP western Canada.

Halifax, NS B4A 1E6 Canada, Sun Tower, 1550 Bedford Hwy., Suite 410. Phone: (902) 422-1411. Fax: (902) 422-3507. TWX: 019-21534. E-mail: jgallant@newswire.ca. Robert Moffatt, mgr Atlantic Canada.

Ottawa, ON K1P 6A9 Canada, 255 Albert St, Suite 460. Phone: (613) 563-4465. Fax: (613) 563-0548. TWX: 053-3292. Hugh Johnson, VP natl capital rgn.

Montreal, PQ H3B 2J6 Canada, 1155 Rene Levesque Blvd. W, Suite 3310. Phone: (514) 878-2520. Fax: (514) 878-4451. TWX: 055-60936. E-mail: scmtl@newswire.ca. Elaire Carr, VP Quebec.

Offers a range of industry leading communication products & svcs for companies looking to maximize the strength of their news. Whether you are an investor rel off, or a specialist in PR, CNW offers the right tools for your communications.

Capital Television News Service (CTNS), 1629 S. St., Sacramento, CA, 95814. Phone: (916) 446-7890. Fax: (916) 446-7893. E-mail: pacsat@pacsat.com

Video wire svc providing daily news coverage, via satellite, of California's capitol for subscribing TV stns throughout the state.

The Church of Jesus Christ of Latter-day Saints (Mormons), 50 East North Temple St., Salt Lake City, UT, 84150. Phone: (801) 240-2205. Fax: (801) 240-1167. E-mail: purdyrm@ldschurch.org Web Site: www.lds.org.

Michael Purdy, mgr; Dale Bills, mgr; Kim Farah, mgr.

Offers free pub affrs, news & feature progmg; also guests for talk shows. Pub affrs progmg is not church-oriented.

Serves radio & TV.

Compu-Weather Inc., 2566 Rt. 52, Hopewell Junction, NY, 12533. Phone: (800) 825-4445. Fax: (800) 825-4441. E-mail: sales@compu-weather.com Web Site: www.compu-weather.com.

Jeff Wimmer, pres.

TV & radio svc providing weather forecasts, features, info & actualities.

Congressional Quarterly Inc., 1255 22nd St. NW, Washington, DC, 20037. Phone: (202) 419-8500. Phone: (800) 432-2250. Fax: (800) 380-3810. E-mail: kwhite@cq.com Web Site: www.cq.com.

Keith White, opns mgr; David Rapp, editor-in-chief.

Congressional Quarterly Weekly Report & News Service, editorial rsch reports, newsletters, seminars, rsch, reference volumes, paperbacks; daily & wkly congressional info publications.

Connecticut Weather Center Inc., 18 Woodside Ave., Danbury, CT, 06810-7123. Phone: (203) 730-2899. Fax: (203) 730-2839. E-mail: weatherlab@ctweather.com Web Site: www.ctweather.com.

William Jacquemin, pres.

Weather forecasts for all media. Custom intros/outros/lives. Accurate forecasts. Barter or cash arrangement available.

Feature Story News, 1730 Rhode Island Ave., Suite 405, Washington, DC, 20036. Phone: (202) 296-9012. Fax: (202) 296-9205. E-mail: markss@featurestory.com Web Site: www.featurestory.com.

Simon Marks, pres.

Orlando, FL 32801, 1103 Palmer St. Phone: (407) 898-1929. Steve Mort, correspondent.

New York, NY 10036, 226 W. 47th St, 2nd floor. Phone: (212) 764-5848. Nathan King, correspondent.

Ind supplier of radio & TV news to English-language bcstrs worldwide. Bureaus in Washington, Moscow, London, New York, Orlando & Beijing.

Fox News Channel

See listing in Major National TV News Organizations, this section.

Golden Lamb Productions, Box 47, Schoolhouse Rd., Nassau, NY, 12123. Phone: (866) 457-2739. Fax: (518) 766-4558. Web Site: www.glpvideoproduction.com.

Dow Haynor, pres.

ENG, EFP crews, HD and SD; SNG available. Serves the Northeast; 24-hour call; packages, live remotes, camera crane news & sports.

Hollywood News Service, 13636 Ventura Blvd., Suite 303, Sherman Oaks, CA, 91423. Phone: (818) 986-8168. Phone: (818) 990-5945. Fax: (818) 789-8047. E-mail: editor@newscalender.com Web Site: www.newscalendar.com.

A wire service to the entertainment media. Publisher of Hollywood News Calendar in Los Angeles; Entertainment News Calendar in New York.

Independent Television News of London Ltd., 400 N. Capital St., Suite 899, Washington, DC, 20001. Phone: (202) 429-9080. Fax: (202) 429-8948. E-mail: michael.herrod@itn.co.uk Web Site: www.itn.co.uk.

London WC1X 8XV, ITN House, 200 Grays Inn Rd. Phone: 011-441-637-2424. 017-833-3000. Stewart Purvis, editor.

Other branches: South Africa, Moscow, London. Hong Kong. British TV news, Washington bureau.

Israel Broadcasting Service, 800 2nd Ave., New York, NY, 10017. Phone: (212) 499-5402. Fax: (212) 499-5425. E-mail: yonih@newyork.mfa,gov.il Web Site: www.israel.org.

Yoni Heilman, dir community & interrelg affrs.

Free radio & TV programs & footage about Israel.

Kyodo News New York Bureau, 747 Third Ave., Suite 1801, New York, NY, 10017. Phone: (212) 508-5460. Fax: (212) 508-5461. Web Site: www.kyodo.co.jp.

Japan's leading newsgathering organization serving virtually all media in all parts of the world. The combined circulation of newspaper subscribers is about 50 million.

Medialink, 708 Third Ave., New York, NY, 10017. Phone: (212) 682-8300. Phone: (800) 843-0677. Fax: (212) 682-5260. Web Site: www.medialink.com. E-mail: info@medialink.com

Michele Wallace, sr VP; Larry Thomas, COO.

London W1P 5AH, 7 Fitzroy Sq. Phone: 44-207 554 2700. Fax: 44-207 554 2710.

Los Angeles, CA 90028, 6430 Sunset Blvd, Suite 1100. Phone: (323) 465-0111. Fax: (323) 465-9230.

San Francisco, CA 94111, One Maritime Plaza, Suite 1670. Phone: (415) 296-8877. Fax: (415) 296-9929.

Washington, DC 20045, Natl. Press Bldg., 529 14th St. N.W., Suite 1230-A. Phone: (202) 628-3800. Fax: (202) 628-2377.

Chicago, IL 60611, The Time & Life Bldg, 541 N.

Fairbanks Ct, Suite 1910. Phone: (312) 222-9850. Fax: (312) 222-9810.

Dallas, TX 75254, 5000 Quorum, Suite 450. Phone: (972) 774-0200. Fax: (972) 774-0222.

International video & audio PR, satellite feed & news advisory service. Accessible by computer/newswire in U.S. & European newsrooms.

Serves radio & TV.

MediaOne Services, 901 Battery St., Suite 220, San Francisco, CA, 94111. Phone: (415) 693-5000. Fax: (415) 693-5005. E-mail: info@mediaoneservices.com Web Site: www.mediaoneservices.com.

Benjamin Schick, pres/COO; Nelson Ferreira, opns mgr.

Satellite uplinking & fiber-optic transmission capabilities. 20'x 40'& 30' x 30' studios for production of cable progmg, teleconferences, live interviews, satellite press tours.

Serves radio & TV.

Metro Weather Service Inc., 71 S. Central Ave., Valley Stream, NY, 11580. Phone: (800) 488-7866. Fax: (516) 568-8853. E-mail: metrowx@aol.com Web Site: www.metrowx.com.

Pat Pagano, pres.

Tailored weather forecasts for TV & briefings to weathercasters.

Serves radio & TV.

Miami News Net, 2641 S.W. 27th St., Miami, FL, 33133. Phone: (305) 285-0044. Fax: (305) 285-0074. E-mail: mnn@bellsouth.net Web Site: www.miaminewsnet.com.

Catherine A. Scull, pres.

A 24-hour TV news, sports & entertainment svc that provides crews, video archive, avid, beta edit & feed facilities. Live talkback studio facilities, dual path digital KU uplink trunk.

Mountain News Corporation, 50 Vashell Way, Suite 200, Orinda, CA, 94563. Phone: (925) 254-4456. Fax: (925) 254-7923. E-mail: info@mountainnews.com Web Site: www.mountainnews.com.

Rob Brown, producer.

Mountain News Corporation, formally AMI News, is the largest & oldest producer of winter & summer progmg for media. We deliver the most accurate & timing news & info covering mountain activities.

NBC News

See NBC listing in Major National TV News Organizations, this section.

NOAA/National Weather Service Headquarters, 1325 East-West Hwy., Silver Spring, MD, 20910. Phone: (301) 713-0700. Fax: (301) 713-1598. Web Site: www.nws.noaa.gov.

D.L. Johnson, dir.

Anchorage, AK 99513-7575, 222 W. Seventh Ave, 23, Rm. 517. Phone: (907) 271-5136.

Honolulu, HI 96813, Grosvenor Ctr. Mauka Tower, 737 Bishop St., Suite 2200. Phone: (808) 532-6416. Fax: (808) 532-5569. James Weyman, dir Pacific rgn.

Kansas City, MO 64153-2371, 7220 N.W. 101 Terr. Phone: (816) 891-8914. Sandy Boyse, dir central rgn.

Bohemia, NY 11716-2626, 630 Johnson Ave. Phone: (516) 244-0101. Dean Gulezian, dir eastern rgn.

Fort Worth, TX 76102-6171, 819 Taylor St, Rm. 10A03. Phone: (817) 978-1000. Erma Nations, dir southern rgn.

Salt Lake City, UT 84147-1102, Federal Bldg, 125 S. State St., Rm. 1311. Phone: (801) 524-5122. Vickie L. Nadolski, dir western rgn.

Weather & flood warnings, forecasts & related info for the media & gen public.

The Nasdaq Stock Market, 1 Liberty Plaza, New York, NY, 10006. Phone: (212) 858-5211. Fax: (646) 625-6548. E-mail: petersos@nasdaq.com Web Site: www.nasdaq.com.

Robert Greifeld, pres/CEO.

Customized loc data for the stock market.

Serves radio & TV.

Nemo News Service, 7179 Via Maria, San Jose, CA, 95139. Phone: (408) 226-6339. Fax: (408) 226-6339. E-mail: broadcast@reizner.com

Richard Reizner, pres.

On-assignment coverage of news & sporting events for radio & TV stns worldwide.

Serves radio & TV.

Nielsen Entertainment News Wire, 101 Federal St., Suite 600, Boston, MA, 02110. Phone: (617) 478-5500. Fax: (617) 478-5501. E-mail: donald.gallagher@nielsen.com Web Site: www.nielsenenw.com.

Donald Gallagher, mgng editor.

Advance news from Nielsen-owned publications serving radio, TV, nwsprs & online.

Nippon TV Network Corp., 645 5th Ave., Ste. 303, New York, NY, 10022. Phone: (212) 660-6900. Fax: (212) 489-8395. Fax: (212) 265-8495. E-mail: motoko@ntvic.com Web Site: www.ntv.co.jp.

Motoko Hasegawa, editor-in-chief; Jusaburo Hayashi, pres.

Localized TV news svc.

NorthStar Studios Inc., 3201 Dickerson Pike, Nashville, TN, 37207. Phone: (615) 650-6000 ext. 6031 (Grant Barbre). Phone: (615) 650-6000. Fax: (615) 650-6300. E-mail: grant.barbre@northstarstudios.tv Web Site: www.northstarstudios.tv.

Grant Barbre, pres.

Nashville, TN 37207, 3201 Dickerson Pike. Phone: (615) 650-6000. Grant Barbre, pres.

Complete TV production svcs: 7 stages, mobile production/uplink trucks, network origination, transmissions, digital archiving, Avid/DS, linear editing, graphics/animations & ENG crews.

Potomac Television - News & Video Service, 1510 H. St. N.W., Suite 202B, Washington, DC, 20005. Phone: (202) 783-6464. Fax: (202) 783-1132. E-mail: jnorins@potomactv.com Web Site: www.ptpngroup.com.

Jamie Norins, news bureau chief.

Washington, DC, news coverage, studios, editing facilities, live shots, crews, satellite capability & duplications.

Presson Perspectives, 600 Druid Rd. E., Clearwater, FL, 33756. Phone: (727) 461-1885. Phone: (800) 249-4521. Fax: (727) 443-1984. E-mail: gpresson@tampabayrr.com

Gina Presson, producer & pres.

Specializing in TV news & documentary production & electronic publishing. Svcs include rsch, field production, videography, postproduction, & satellite feeds for radio & TV.

Serves radio & TV.

Radio Press News Services, 8633 Arbor Dr., El Cerrito, CA, 94530-2728. Phone: (510) 524-9559. E-mail: jag4jl@aol.com

Robert Miles Master, editor-in-chief.

Natl coverage, with special unit for northern California, Bay Area of California & adjacent states, photographer on staff. Multimedia news, Travellands & Vacationland. Special features, articles, transcriptions & video features. TV assignment accepted.

Reuters America, 3 Times Sq., New York, NY, 10036. Phone: (646) 223-4000. Web Site: www.reuters.com. E-mail: boblagrassa@reuters.com

Bob Lagrassa, dir tv opns.

Worldwide TV news production & transmission svcs for loc TV stns/producers. Camera crews, production facilities, news bureaus, satellite svcs, video/slide archives.

The Seattle Video Bureau, Box 99218, Seattle, WA, 98199. Phone: (206) 448-2500. Fax: (206) 378-1700. E-mail: dave@seattlevideo.com Web Site: www.seattlevideo.com.

David Oglevie, pres.

ENG/EFP crews with BETACAM SP kits. Net experienced.

Skywatch Weather Center, 347 Prestley Rd., Bridgeville, PA, 15017. Phone: (800) 759-9282. Phone: (412) 221-6000. Fax: (412) 221-3160. E-mail: airsci@skywatchweather.com Web Site: www.skywatchweather.com.

Dr. Stanley J. Penkala, pres; Daniel Krzywiecki, VP; Stanley Bostjancic, treas; Harry Green, sec.

Weather forecasts targeted to the lstng area, & comprehensive briefings for on-air talent.

Serves radio & TV.

Skyways Communications, L.L.C., 89 Access Rd., Suite 20, Norwood, MA, 02062. Phone: (781) 551-9960. Fax: (781) 551-5956. E-mail: scott@skyways.net Web Site: www.skyways.net.

LuAnn Reeb, pres; Scott Hess, VP/chief tech off.

Custom TV news gathering, producing & mobile satellite uplinking svcs. Provides bcst-experienced crews, producers, reporters & technicians for breaking news, live-event coverage & webcasting.

The Sports Network, 2200 Byberry Rd., Suite 200, Hatboro, PA, 19040. Phone: (215) 441-8444. Fax: (215) 441-5767. E-mail: kzajac@sportsnetwork.com Web Site: www.sportsnetwork.com.

Mickey Charles, pres/CEO; Ken Zajac, sls dir.

International real-time sports wire svc providing content, branded web pages, satellite and/or computer feeds directly to broadcasters (radio & TV), print, Internet sites, wireless with state of the art technology.

Tankersley Productions, Inc., St. Louis, 858 Hanley Industrial Ct., St. Louis, MO, 63144. Phone: (314) 725-0116. E-mail: randy@tankersleyproductions.com Web Site: www.tankersleyproductions.com.

Randy Tankersley, pres.

Full-bcst svcs. ENG/EFP Beta SP crews with/without producers, avid nonlinear editing.

U.S. Conference of Catholic Bishops, Department of Communication, Film/TV Review Svcs. Office for Film & Broadcasting, 1011 First Ave., New York, NY, 10022. Phone: (212) 644-1880. Fax: (212) 644-1886. E-mail: ofb@msn.com Web Site: www.usccb.org.

Harry Forbes, dir.

Publishes wkly reviews of movies, TV with artistic & moral observations.

WB11.com WPIX-TV New York, 220 E. 42nd St., New York, NY, 10017. Phone: (212) 949-1100. Fax: (212) 210-2591. Web Site: wb11.com. E-mail: jziegler@tribune.com

Betty Ellen Berlamino, gen mgr; John Ziegler, mktg dir.

WSI (Weather Services International), 400 Minuteman Rd., Andover, MA, 01810. Phone: (978) 983-6300. Fax: (978) 983-6400. Web Site: www.wsi.com.

Mark Gildersleeve, pres; Linda Maynard, VP mktg.

WSI is the leading source of professional on-air weather systems, solutions & forecasting svc for TV, including TrueView, the most innovative weather storytelling tool available.

The Washington Bureau, 400 N. Capitol St. N.W., Suite 775, Washington, DC, 20001. Phone: (202) 347-6396. Fax: (202) 628-6295. E-mail: rtillery@twbnews.com Web Site: www.twbnews.com.

Richard Tillery, Bureau chief; Julia Rockler, CEO.

Custom TV news coverage: ENG crews, producers & talent. Prod svcs: editing, studio, remote & satellite capabilities. Two live studios. Teleconference capability. Fiber Optic connectivity with Capital, White House & other locations.

Washington News Network, 400 N. Capitol St. N.W., Suite G-50, Washington, DC, 20001. Phone: (202) 628-4000. Fax: (202) 628-4015.

David Oziel, bureau chief.

Washington news bureau for more than 100 TV stns, rgnl nets & news programs nationwide. Provides reporter packages & vo/sots, crew hires & hearing video transcripts. Full editing, studio & satellite facilities on Capitol Hill.

"The Weather Center", (a broadcast service of Aviation Weather Inc.). 701 Gervais St., Suite 150-224, Columbia, SC, 29201. Phone: (803) 739-2827. E-mail: wxcenter@aviationweatherinc.com Web Site: www.aviationweatherinc.com.

Liam Richard Ferguson, pres.

"Regional Radio Broadcast/Weathercast Network" across the Carolinas and Georgia in over 20 bcst markets. Weather forecasting, site-specific bcst svc for stns all across America. 100% barter.

WeatherData Inc., 245 N. Waco St., Suite 310, Wichita, KS, 67202. Phone: (316) 265-9127. Fax: (316) 265-1949. Web Site: www.weatherdata.com. E-mail: ceo@weatherdata.com

Mike Smith, CEO.

Forecasts for radio & TV, meteorology training, slides & videotape of weather & related phenomena. Nexrad radar interpretation seminar; distributor of Nexrad weather display systems. Meteorologist 24/7, storm monitoring & customer svc.

WeatherVision Inc., 916 Foley St., Jackson, MS, 39202-3406. Phone: (601) 948-7018. Fax: (601) 948-6052. E-mail: edward@weathervision.com Web Site: www.weathervision.com.

Edward Saint-Pe', pres; Jason McCleave, VP.

Customized, localized TV weathercasts with or without meteorologists. Barter/cash via Ku-band satellite. Complete studio teleport for use by news media on site. Avid editing available on site. Serves radio, TV, 3-D branding animation svcs.

National Radio Programming Services

ABC Radio Networks

ABC Inc. Executives: David Westin, pres; Alan Braverman, exec VP/gen counsel; John Hare, pres ABC Radio Division; Larry Hyams, VP primetime audience analysis; John E. McConnell, progmg/comments.

ABC Radio Networks Executives: John Hare, pres ABC Radio Division; Darryl Brown, exec VP/gen mgr; Jean-Paul Colaco, pres/gen mgr, Radio Disney; Mitch Dolan, pres ABC Radio Station Group 1; Anne Gatoff, VP/dir finance; Kevin Miller, sr VP/ad sls.

ABC 24-Hour Formats: ABC AC, Classic R & B (Urban Oldies),Today's Hits & Yesterday's Favorites (Best of the 70s through today), Classic Rock (Classic AOR), Country Coast-to-Coast (Contemporary Country), Hot AC (Young AC), Memories (Adult Soft Oldies), Oldies Radio, Real Country, Rejoice! (Gospel), Stardust (MOR), The Touch (Urban AC).

Music & Sports: *American Country Countdown with Bob Kingsley; American Gold; ESPN Radio; Flashback; Radio Disney; Rock & Roll's Greatest Hits with Dick Bartley; The Ride.*

Urban Programming: *DeDe McGuire's Word on the Street; The Doug Banks Morning Show; The Smiley Report with Tavis Smiley; The Tom Joyner Morning Show.*

News & Talk Programming: *ABC News; ABC Sports; America's Most Wanted; Focus on the Family; MoneyTalk with Bob Brinker; Paul Harvey; Satellite Sisters; Mark Levin Show; Sean Hannity Show.*

East Region

77 W. 66th St., New York, NY 10023. (212) 456-7777.

Departments: Advertising Sales, Affiliate Marketing East, Finance, Research, MIS.

Executives: Geoff Rich, exec VP/progmg.

125 West End Ave., New York 10023. (212) 735-1700.

Departments: ABC News Radio, Engineering, Network Programming, International.

Executives: Robert Donnelly, VP engrg; Chris Berry, VP news radio.

West Region

13725 Montfort Dr., Dallas, TX 75240. (972) 991-9200.

Departments: Affiliate Marketing West; Entertainment Programming; Marketing & Promotion; ABC 24-Hour Formats; Advertising Sales (Southwest); Engineering; Finance; MIS; Research; Clearance; International.

Executives: James Robinson, pres; Michael Connolly, sr VP/ad sls; John Rosso, sr VP affil rels/business admin.

American Urban Radio Networks

Executive Headquarters: 655 Third Ave., 24th Fl., New York, NY 10017. (212) 883-2100. Fax: (212) 297-2571. Web site: www.aurnol.com. E-mail: information@aurnol.com.

Officers: Vernon Wright, exec VP/client dev; Howard Eisen, exec/sls; Basil Murrair, VP/promotion; Barry Feldman, exec dir mktg.

Program Headquarters: 960 Penn Ave., Suite 200, Pittsburgh, PA 15222-3811. (412) 456-4000. Fax: (412) 456-4040.

Officers: Jerry Lopes, pres, progmg opns/affiliations; Glenn Bryant, sr VP/opns; Kathy Gersna, VP human resources/corporate opns; Adele Lawhead, VP/controller; Tene Croom, dir news; Dian Sirko, dir/corporate traffic; Bob Sharkey, chief engr; Ty Miller, dir sports; Jay Silvers, production dir; Laurene Gaines, exec producer, The Bev Smith Show.

Chicago Office: 30 N. Michigan Ave., Suite 1218, Chicago, IL. (312) 558-9090. Fax: (312) 558-9280. Contact: Jon Krongard, VP/sls Western Region.

Detroit Office: 1133 Whittier Rd., Grosse Pointe, MI 48230. (313) 885-4243. Fax: (313) 885-2192. Contact: J.D. MacKay, exec sls dir.

AP Radio Networks

The AP Radio Networks are **AP All News Radio (ANR)** and **AP Network News (APNN)**, both administered by the Radio Division of the AP Broadcast organization. ANR is a live news network that taps AP's worldwide resources to deliver the latest audio news from around the globe 24 hours a day. APNN provides newscasts, sportscasts, business reports, entertainment reports and features plus actuality feeds. APNN provides regularly scheduled progmg on the Main Channel and live, long-form coverage of special events and major breaking news on the Hotline channel.

AP (Associated Press)

AP is a not-for-profit cooperative with more than 4,000 employees working in more than 240 worldwide bureau owned by its U.S. daily nwspr members. Any nwspr, radio or TV stn can become a member.

International Headquarters: 450 W. 33rd St., New York, NY 10001. (212) 621-1500. General/National Desk: (212) 621-1600. International Desk: (212) 621-1750. Fax: (212) 621-5469, arts and entertainment; (212) 621-1587 business news. Web site: www.ap.org. E-mail: info@ap.org (no attachments).

AP Broadcast

AP Broadcast News Center: 1825 K St. N.W., Suite 800, Washington, DC 20006-1253. (800) 821-4747 (TV). Fax: (202) 736-1199, news; (202) 736-1124, admin; (202) 736-1107, VP office. Web site: www.apbroadcast.com.

Advisory Board: President: Ed Christian, CEO SAGA Communications, Grosse Pointe Farms, MI; VP for Radio: Jim Farley, VP news & progmg WTOP AM/FM, Washington, DC; Jim Williams, VP; Greg Groce, dir.

Advisory Board for Radio: Clark Brown, pres radio division Jefferson-Pilot Communications, Atlanta, GA; John Dickey, exec VP Cumulus Media Inc., Atlanta, GA; Dick Ferguson, co-COO Cox Radio, Milford, CT; Dan Halyburton, sr VP/gen mgr Susquehanna Radio Corp., Dallas, TX; Gabe Hobbs, natl dir news/talk Clear Channel Communications, Tampa, FL; Zemira Jones, pres/gen mgr ABC Radio - Chicago, Chicago, IL; Laura Morris, VP/mkt mgr Infinity Broadcasting Inc., Houston, TX; Jim Russell, sr VP Minnesota Public Radio, Los Angeles, CA; Thomas Callahan, gen mgr.

Senior Management: James R. Williams III, VP/dir bcst svcs; Thomas Callahan, gen mgr/AP radio; Greg Groce, dir business opns/dev; Brad Kalbfeld, deputy dir/mgr editor; Roger Lockhart, dir mktg comms; Lee Perryman, deputy dir bcst svcs/dir bcst tech; John Phillips, dir financial planning; Montrese Garner-Sampson, dir human resources.

Sales Management: Bill Burke, product mgr for bcst tech; Carol Robinson, dir group sls; Susan Spaulding, dir Radio Groups/Internet sls.

Newsroom Management: Denise Vance, international mgr of the Americas for APTN; Wally Hindes, asst mgng editor/radio; Ed Tobias, asst mgng editor bcst news; Barbara Worth, asst mgng editor news.

AP Broadcast: Radio Division

(800) 527-7234. E-mail: apradio@ap.org.

Executives: Thomas Callahan, gen mgr (202)736-1105, tcallahan@ap.org; Susan Spaulding, dir Radio Group and Internet Sales (202) 736-9622, sspaulding@ap.org; Dave Herring, dir radio network sls; Cushmeer Singleton, Radio, Division II mgr.

CBS

Headquarters: CBS Television Network 51 W. 52nd St., New York, NY 10019; (212) 975-4321; Fax: (212) 975-4516.

CBS

Leslie Moonves, pres/chmn/CEO; Sean McManus, pres CBS Sports; Bill Korn, exec VP strategic planning, business dev/bcst opns; Fred Reynolds, exec VP/CFO; Jack Bergen, sr VP corporate rel; Martin Franks, exec VP CBS/pres CBS Foundation; Helene Blieberg, VP/exec dir CBS Foundation; Dean Daniels, VP/gen mgr CBS New Media; Matthew Margo, VP program practices East Coast; Carol Altieri, VP program practices West Coast; Gil Schwartz, exec VP communications; Peter K. Schruth, sr VP/gen mgr affil rel; Jay Gold, VP finance; Kenneth Cooper, VP facilities opns; Bruce Taub, exec VP/CFO; Tom Gentile, VP strategic planning; Derek Reisfeld, VP business dev.

David Zemelman, sr VP/human resources corporate; Ed Yergeau, sr VP/industrial rel; Leon Schulzinger, VP industrial labor rel East Coast; John McLean, VP industrial rel West Coast; Joseph A. Flaherty, sr VP technology; Brent Stranathan, VP bcst distribution; Robert Seidel, VP engrg/advanced technology; Michael Klausman, pres/studio opns; Steve Schifrin, VP program production svcs; Harvey Holt, VP stage opns; Barry Zegal, VP production; David Zink, VP chief info officer; Michael Vinyard, VP info systems West Coast; Thomas Maile, VP telecommunications; Dennis D'Oca, VP risk mgmt; Elliott Matz, VP/dir real estate; Ellen Kaden, exec VP, gen counsel/sec; Martin P. Messinger, sr VP/deputy gen counsel; Susan J. Holliday, sr VP/deputy gen counsel; Howard F. Jaeckel, VP assoc/gen counsel; Susanna M. Lowy, assoc gen counsel litigation; Mark W. Engstrom, assoc gen counsel labor; Sanford I. Kryle, assoc gen counsel contracts/rights dev; Mark W. Johnson, assoc gen counsel Washington DC; Derk Zimmerman, sr VP, new ventures/business dev.

CBS News

Andrew Heyward, pres; Linda Mason, sr VP, Standards/special projects; Lane Venardos, VP, hard news/special events; Al Ortiz, exec producer/dir special events; Scott Herman, sr VP/news radio; Harvey Nagler, VP/gen mgr radio; Marcy McGinnis, sr VP/news coverage; John Frazee, sr VP/news svcs; Frank Governale, VP opns; James McKenna, VP finance/admin; Josie Thomas, sr VP diversity.

CBS Sports

SportsLine.com. Inc.: 2200 W. Cypress Creek Rd., Ft. Lauderdale, FL 33309. (954) 489-4000 ext 5026. Fax: (954) 771-2807. Web Site: www.sportsline.com

Management: Michael Levy, pres; Mark J. Mariani, pres sls/mktg; Kenneth W. Sanders, exec VP/strategic financial planning; Stephen E. Snyder, COO/VP; Tom Arrix, sr VP sls/client svc.

Board of Directors: Michael Levy, chmn; Thomas Cullen; Gerry Hogan; Richard B. Horrow; Peter Glusker;

Sean McManus; Andrew Nibley; Michael P. Schulhof.

CBS Affiliate Relations

Peter K. Schruth, exec VP/gen mgr affil rel; Preston Farr, VP/dir affil rel; Jeffrey McIntyre, VP mktg/affil rel.

CBS Enterprises

Roger King, chmn/CEO; Armando Nunez, pres/bcst international; Vec Russo, VP/controller.

East Coast

1700 Broadway, 32nd & 33rd Fl., New York, NY 10019. (212) 315-4000; Fax: (212) 582-9255.

CBS Station Group

Farid Suleman, sr VP/CFO; Wes Spencer, VP/controller; Stephen A. Hildebrandt, VP/gen counsel; Scott Herman, exec VP news.

CBS Owned Radio Stations

WVEE-FM Atlanta; WZGC-FM Atlanta; WCAO Baltimore; WLIF-FM Baltimore; WBCN-FM Boston; WBZ/WODS-FM Boston; WZLX-FM Boston; WBBM/WBBM-FM Chicago; WCKG-FM Chicago (Elmwood Park, IL); WJMK-FM Chicago; WMAQ Chicago; WSCR/WXRT-FM Chicago; KHVN Dallas; KOAI-FM Dallas; KLUV-FM Dallas; KRLD Dallas; KVIL-FM Dallas; WOMC-FM & WVMV-FM Detroit; WWJ/WKRK-FM, WXYT Detroit; WYCD-FM Detroit; KILT-AM-FM Houston; KIKK-AM-FM Houston (Pasadena, TX); KXYZ Houston; KNX Los Angeles; KCBS-FM Los Angeles; KFWB Los Angeles; KRLA Los Angeles (Pasadena); KLSX-FM Los Angeles; KROQ-FM Los Angeles (Pasadena); KRTH-FM Los Angeles; KTWV-FM Los Angeles; WCCO Minneapolis; WLTE-FM Minneapolis; WCBS/WCBS-FM New York; WFAN New York; WINS New York; WNEW-FM New York; WXRK-FM/WZRC New York; KYW Philadelphia; WPHT Philadelphia; WIP Philadelphia; WOGL-FM Philadelphia; KDKA Pittsburgh; KOME-FM San Jose; KCBS San Francisco; KFRC-AM-FM/KYCY-AM-FM; KITS-FM & KLLC-FM San Francisco; KPIX-AM-FM San Francisco; KMOX; Lee Clear, KLOU-FM St. Louis; WQYK-AM-FM Tampa (Seffner/St. Petersburg); WARW-FM Washington, DC (Bethesda, MD); WHFS-FM Washington, DC (Annapolis); WJFK-AM-FM Washington, DC (Baltimore); WPGC-AM-FM Washington, DC (Morningside, MD).

CNN Radio Networks

Headquarters: One CNN Ctr. N.W., Atlanta, GA 30303-2762. (404) 827-2750. Web Site: www.cnnradionet.com.

CNN Radio: Natl & International radio news network, features web accessible sound bites.

Parent Company: Time Warner

Principal Executives: Jim Walton, pres CNN USA; Rick Davis, exec VP, news standards/practices; Princell Hair, exec VP, gen mgr/CNN/US; Nancy Lane, VP/exec dir CNN Newsgathering; Cindy Patrick, exec VP/opns; Jack Womack, sr VP/domestic news opns; Robert Garcia, VP CNN Radio; Harley Hotchkiss, dir opns CNN Radio; Richard Benson, exec producer CNN Radio.

Bureaus: Atlanta; Boston; Chicago; Dallas; Denver; Los Angeles; Miami; New York; San Francisco; Seattle; Washington, DC; Baghdad; Bangkok; Beijing; Beirut; Berlin; Buenos Aires, Cairo; Dubai, Frankfurt; Havana; Hong Kong, Islamabad; Istanbul; Jakarta; Jerusalem; Johannesburg; Lagos; London; Madrid; Mexico City; Moscow; Nairobi; New Delhi; Paris; Rome; Tokyo; Seoul; Sidney.

Eastern Region Public Media (Eastern Public Radio)

Mailing Address: Georgette Bronfman, exec dir, Eastern Public Radio, Box 615, Kensington, MD 20895. Phone: (301) 943-2930. Web Site: http://www.easternpublicradio.org.

Executive Committee: Georgette Bronfman, exec dir; Lee Ferraro (co-chmn) gen mgr, WYEP Pittsburgh, PA; Jeanne Fisher (co-vice-chair) VP of radio WXXI-FM Rochester, NY; Michael Black (sec) gen mgr, WEOS Geneva, NY; John Kraus (treas) gen mgr, WRVO Oswego, NY; Quyen Shanahan (at-large) assoc gen mgr, WXPN Philadelphia, PA; Rob Gordon (at-large) gen mgr, WPLN Nashville, TN; Maxie Jackson (at-large) sr dir prgram dev, WNYC New York, NY; Dave Spizale (co-chmn) gen mgr, KRVS Lafayette, LA; Kate Lochte (at-large) gen mgr, WKMS Murray, KY; Earl Johnson (co-vice-chair) gen mgr, WABE Atlanta, GA.

Family Stations Inc.

Headquarters: 290 Hegenberger Rd., Oakland, CA 94621. (510) 568-6200. E-Mail: info@familyradio.com; Web site: www.familyradio.com.

Executives: Harold Camping, pres/gen mgr; David Hoff, progmg mgr; Dan Elyea, engrg mgr.

Family Radio is a nondenominational, noncommercial, nonprofit, listener-supported, 24-hour, Christian ministry. Conservative Christian music & progmg. Some talk, limited news six days per week.

Jones Radio Networks

Headquarters: 8200 S. Akron St., Suite 103, Centennial, CO 80112 (303) 784-8700; (800) 609-5663 (Denver); (800) 426-9082 (Seattle); (800) 611-5663 (Washington, DC); (888) 644-8255. Fax: (303) 784-8612. Web site: www.jonesradio.com.

Executives: Glenn R. Jones, chmn; Jeffrey C. Wayne, pres/COO, Jones International Networks; Phil Barry, VP/gen mgr; Frank De Santis, VP, gen mgr news/talk; James LaMarca, exec VP/COO; Amy Bolton, VP, gen mgr news/talk; Susan Stephens, VP/gen mgr Seattle.

Sales & Marketing: Patrick Crocker, sls dir; Kim Ketchel, ad sls/mktg.

24-Hour Formats: Country, CD Country, Classic Hit Country, Adult Contemporary, Adult Hit Radio, Rock Classics, Good Time Oldies, Music of Your Life, Smooth Jazz, Branding Power.

News & Talk Programming: Long Form: Neal Boortz, Stephanie Miller, Ed Schultz, Bill Press & Midnight Radio. Short Form: Fight Back with David Horowitz, The Clark Howard Minute, Wall Street Wake-Up with Chris Byron, Something You Should Know. Weekends: Newsweek On-Air, Best of Stephanie Miller, Best of Neal Boortz. Newsweek on Air, The Ed Schultz Show.

Personalities: Bill Cody Classic Country Weekend, Lia, Danny Wright All Night.

Research & Prep: American Comedy Network, Jimmy Carter Entertainment Report, Gossip To Go With Flo, Jones Research Network, Jones Prep Country, AC, CHR, Rock and Oldies stns.

Moody Broadcasting Network

Headquarters: 820 N. LaSalle Blvd., Chicago, IL 60610. (800) 621-7031; (312) 329-4271. Fax: (312) 329-4368. Web Site: http://www.mbn.org; E-mail: mbn@moody.edu.

Executives: Doug Hastings, progmg/opns division mgr; David Woodworth, sr network rep; David Trout, sr network rep; Peter Straw, network rep.

Relg & educ stereo audio progmg; music, talk, news & pub affrs 24 hours a day. Services 370 radio affils in 50 states, Washington, DC, Puerto Rico & the Virgin Islands & on 7 cable systems serving over 66,220 subs. Also provides ACCUWatch, an automatic transmitter monitoring svcs to radio stns for unattended opns. Satellites: AMC-3, transponder 17 (digital FM-quad, aka DVB, stereo audio), & AMC-8 (Aurora III), transponder 10 (SCPC mono audio).

National Public Radio (NPR)

Headquarters: 635 Massachusetts Ave. N.W., Washington, DC 20001. (202) 513-2000. Fax: (202) 513-3329. Web Site: http://www.npr.org.

Corporate Officers: Kevin Klose, pres/CEO; Ken Stern CEO; Ellen Weiss, VP news; Audi Sporkin, VP communications; Jim Elder, CFO/treas, VP finance; Barbara Hall, VP, dev/exec dir; Kathleen Jackson, VP/human resources; Neal Jackson, VP/legal affrs; Jay Kernis, sr VP/progmg; Peter Loewenstein, VP/distribution; Jackie Nixon, dir, audience/corporate rsch; Dana Davis Rehm, sr VP, member/program svcs; Margaret Low Smith, VP/progmg; Mike Starling, CTO/exec dir, NPR Labs; Maria C. Thomas, VP/gen mgr, NPR Digital Media; Walt Swanston, dir diversity mgmt.

Board of Directors: Tim Eby (chmn of bd); JoAnn Urofsky (vice-chmn of bd); John A. Herrmann Jr. (chair) NPR Foundation; Carol Cartwright (member), pres, Kent State University; Howard H. Stevenson (member), Sarofirm-Rock professor business/admin, Harvard University; Lyle Logan (member), sr VP/personal financal svcs; Judith Winston (member) principal Winston Withers & Associates, LLC.

Member Managers: Rob Gordon, pres/gen mgr, WPLN; Cephas Bowles, gen mgr, WBGO-FM; Tom Rieland, radio mgr, WOSU-FM; John Stark, gen mgr, KNAU; Scott Hanley, dir/gen mgr, WDUQ-FM; Michael Lazar, pres/gen mgr, Capital Public Radio Inc.; Ellen Rocco, stn mgr, North Country Radio; JoAnn Urofsky, gen mgr, WUSF; Mark Vogelzang, pres/gen mgr, Vermont Public Radio.

This noncommercial, satellite-delivered radio system serves a growing audience of more than 15 million Americans each week via 620 public radio stns and the Internet. NPR also serves: Europe, Asia, Australia and Africa via NPR Worldwide; military installations overseas via American Forces Network; and Japan via cable.

NPR provides member stations with progmg, professional dev, promotional support, program distribution/representation in Washington on issues affecting bcstg. Programs include: All Things Considered, Morning Edition, Weekend Edition, Talk of the Nation, Fresh Air, The Motley Fool Radio Show, Day To Day, Tell Me More, World Cafe and Rough Cuts. & *The Thistle & Shamrock.*

Public Radio International

Headquarters: 100 N. 6th St., Suite 900A, Minneapolis, MN 55403. (612) 338-5000. Fax: (612) 330-9222. Web Site: http://www.pri.org.

Management Staff: Alisa Miller, pres/CEO; Timothy J. Engel, sr VP/CFO; Melinda Ward, sr VP production; Eleanor Harris, sr VP, head mktg/distribution; Elinor Gould Zimmerman, VP/resource dev; Dan Jensen, dir.

Background Information: PRI is a Minneapolis-based pub radio network and audio publisher that provides over 400 hours each week of original progmg bcst by over 715 pub radio stn affiliates. Its progmg also is available on locally branded pub radio stn Web sites, internationally through the World Radio Network, and nationwide via Sirius Satellite Radio. PRI was founded in 1983 as American Public Radio by five leading public radio stations to dev distinctive radio programs and to div the pub radio offerings available to American listeners. News programs includes BBC World Service, Living on Earth, Open Source, PRI's The World. Entertainment programs includes Bob Edwards Weekends, Fair Game from PRI with Faith Salie, This America Life, The Travis Smiley Show and Michael Feldmam's What d'Ya Know?. Classical Music Progams includes Classical 24, Schickel Mix Major symphony orchestras.

Superadio Network

Headquarters MA: 1661 Worcester Rd., Suite 205, Framingham, MA 01701. (508) 620-0006. Fax: (508) 628-1590. Web Site: http://www.superadio.com; E-mail: mixes@superadio.com.

Executives: Rich O'Brien, sr VP progmg/affil rels; Joan Brooks, business mgr; Alexis Coble, distribution mgr; Sheila Pellegrini, network affil coord; Dianne Cook, admin asst.

Headquarters NY: 11 Penn Plaza, 16th Fl., New York, New York. (212) 714-1000. Fax: (212) 714-1563.

Executives: Jack Bryant, COO; Eric Faison, VP affil rel; John Campanario, VP/dir ubran progmg; Pat Gillen.

Programs: CHR, weeknights 5 hrs. Host: Matthew Reid of Z95.7 San Francisco. Deliv: interactive digital satellite. Listener requests, mixing, contests and celebrity gossip. Interfaces with local production system.

Behind the Scenes Gospel: Gospel, wkly 2 hrs. Hosts: Eric Faison; Tracy Foye-Green. Deliv: CD. "Praise Party" with contemporary and classic gospel; "The Gos-Pill" with minister Dr. Val.

City Jam: Dance, wkly 4 hrs. Deliv: CD. Dance and rhythm classics.

Classic Jam: Urban, wkly 4 hrs. Deliv: CD. 80s, 90s Hip-Hop and R&B.

Classic Jam Mini-Mixx: Urban, wkly 7.5 mins. Deliv: CD. Core artists include Run-DMC, The Notorious BIG, Zhane, Mary J. Blige, Soul II Soul, New Edition, Snoop Dogg, LL Cool J, Salt 'n Pepa, En Vogue, and Guy, etc.

Elvis Only: Host: Jay Gordon, Retro Country USA Host: Ken Cooper.

Hip-Hop Comedy Kut: Urban, 10 min daily. Mixed by DJ Kut and DJ Rated R. Deliv: CD. Comedy bits mixed in with song hooks.

Hit AC Mix: Hot AC, 4 hrs wkly. Mixed by Aaron Scofield. Deliv: CD, MP2, MP3. Incorporates the playlist of the station, remixed and then beat-mixed.

Howie Carr Show: Talk, 3 hrs daily. Host: Howie Carr. Unique talk with his own take on the world.

Inspiration Jam: Gospel, 2 hrs wkly. Deliv: CD. Gospel and inspirational secular mus mixed, beat-to-beat.

Jump Off: The takeove mix, 3 hrs wkly mix show for urban and rythmic radio spotlight some of the hottest mixes.

Kool Jam: Urban, 4 hrs wkly. Deliv: CD. Funky old school R&B classics with no rap.

Lost in the 80's: Urban AC, 2 hrs wkly. Hosts: Derrick Jonzun; Stephanie Williams. R&B hits from the 80s Club Jam," "One-Hit Wonder," "80's Timeline," and actualities and bytes from artists.

New Skool Mini-Mixx: Urban, 5 min wkly. Deliv: CD. Core artists include Missy Elliot, Snoop Dogg, LL Cool J, 112, P-Diddy, Jagged Edge, Dr. Dre, Ashanti, Mary J. Blige, Ludacris, etc.

Old Skool Mini-Mixx: Urban AC, 7.5 min wkly. Deliv: CD. Core artists include Parliament, Stevie Wonder, Prince, Zapp, Kool & The Gang, Rick James, Teena Marie, Shalamar, Michael Jackson, The Commodores, Gap Band, etc.

Oldies Jam: R&B hits of the 70s and 80s. Core artists include Earth Wind & Fire, Barry White, Prince, Rick James, Parliament, Tavares, and Kool & The Gang, Luther Vandross, The SOS Band, Jody Watley and Cameo.

On the Air: R&B, 3 hrs wkly. Host: Russ Parr. Features celebrity interviews, "Horror-Scopes," "The Wrong Songs Mix" by DJ 6th Sense, and "The Fat Five," counting down the top five songs on the Urban charts.

Open House Party: CHR, S-Su 5 hrs. Hosted by John Garabedian. Deliv: satellite. Live request call-ins, live celebrity interviews, great giveaways.

Paul Oakenfold Presents: Alternative rock, 2 or 3 hrs wkly. Mixed by Paul Oakenfold.

Pecos Pero Locos: Latin, 2 hrs wkly. Deliv: CD. Hip-Hop served up by Khool Aid and Johnny Cuervo.

Rap Jam: Urban, 4 hrs wkly. Deliv: CD. Aggressive, cutting-edge, all-rap/Hip-Hop mix.

Retro Country USA: Country, 2 hrs wkly. Host: Ken Cooper. Deliv: CD, MP2, MP3. Greatest country hits 1980-1989.

Slam Jam: Urban, 4 hrs wkly. Deliv: CD. Today's hottest Hip-Hop and R&B mixed by high-profile major-market radio DJ's.

Smooth Jam: Urban, 4 hrs wkly. Deliv: CD. Contemporary and classic R&B with no rap.

The Soul Lounge: Urban, 2 hrs wkly. Host: Terry Bello. Deliv: CD. Urban Vibe/neo-soul format and lifestyle. Features include "The Happs" entertainment report, "Tongue & Groove," a poem/spoken word segment, Artist of the Week, artist interview, and "The Vibe Session," an hour mix.

Supermixx 80's: 80s, 4 hrs wkly. Deliv: CD. Beat-mixed retro.

Supermixx Mainstream: CHR, 4 hrs wkly. Supermixx blends the best, the hits and Beat-mixed Top 40.

Supermixx Rhythmic: CHR rhythmic, 4 hrs wkly. Deliv: CD. Beat-mixed Urban.

Supermixx Rock: Alternative, 4 hrs wkly. Deliv: CD. Beat-mixed Alternative (or Modern) Rock.

United Press International

Headquarters: 1510 H St. N.W., Washington, DC 20005. (202) 898-8000. Fax: (202) 898-8057. E-mail: tips@upi.com (news); sales@upi.com (general); support@upi.com (cust svc). Web site: www.upi.com.

Leadership Team: Nicholas Chiaia, COO; Christopher Ching, VP of finance; Adriana Avakian, dir/mktg and licensing.

Executives Editors: Arnand de Borchgrave (editor at large); Michael Marshall (editor in chief); Dr. Chung Hwan Kwak, chmn/pres; Dalal Saoud; middle east bureau chief; Martin Walker, editor English language opns. Editorial Staffs: Alejandra Aguirre; Mohamad Assaf; Larry Moffitt; Phil Berardelli; John Hendel; T.K. Maloy; Martin Sieff; Steve Mitchell; Shaun Waterman.

Broadcast History: UPI is a global opn hqtr in Washington, DC with offices in Seoul, Korea; United Kingdom; Beirut Lebanon; Tokyo, Japan; Santiago, Chile and Hong Kong, China. UPI was founded in 1907 by E.W. Scripps as the United Press (UP). It became known as UPI when the UP merged with the International News Service in 1958, which was founded in 1909 by William Randolph Hearst. UPI is owned by News World Communications, a global multi-media company. In 1935, UPI became the first news svc to supply news to bcstrs. Ten years later, UPI started the first sports wire. In 1958, UPI began the first wire svc radio network, providing radio stns with voice reports from correspondents all over the globe.

USA Radio Networks, Inc.

Headquarters: 2290 Springlake Rd., Suite 107, Dallas, TX 75234. (972) 484-3900; (800) 829-8111. Web site: www.usaradio.com

Executives: Mark Maddoux, pres/CEO; Tim Maddoux, VP/COO; David Maddoux, IS technology.

Affiliate Relations: Tim Lee, Robert Jimenez, affil sls; Tracy Maddoux, affil rel.

Advertiser Sales Division: Buddy Vaughn; Tiffany Forney.

Foreign Correspondents: Andrew Adams, Tokyo; Anya Ardayeva and Guy Chazan, Moscow; Susan Lackey, London; Ellen Ratner, DC; Connie Lawn, Washington; Laurence Frost, Paris; Nathan Morley, Cyprus; Ronnie Nathaniels, Manila; David Bendo, Jerusalem.

News Directors: Bob Morrison; Judy Hydock, asst news.

Editors/Producers (weekday): Charlie Butts, relg editor; Andy McCall; Judy Siegel; Melanie Smith; Curt Lewis; Richard Curtis, Lindsay Hooker.

Anchors: Russ Rossman, sr anchor; John Scott; Jason Walker; Allen Stone.

Weekend News Anchors: Cynthia King; Jack Dereat.

Sports News: Ray Canevari.

Broadcast Operations & Engineering: Tom King, engr; Andrew Hydock, opns.

Service available via Satcom C-5 & through other nets & outlets. The USA Radio Network includes over 1,500 affiliated radio stns.

Programs

Daily:

DayBreak USA: weekdays, ET, "Magazine " style show, focuses on National news and sports, dollars & sense financial info, entertainment, movies music & TV shows. Tips on relationships, getting in shape and adding some Fun to Life and much more!

Point of View: weekdays, ET. Marlin Maddoux hosts America's original daily political roundtable with top-name guests & solid values also Insight of todays event from a conservative perpective.

The Roth Show: weekdays 1 pm-2 pm, Host Laurie Roth is a talk show standing on principles and leading the charge for the legions of Americans who are saying "Enough Already!"

The Judicial Watch Report: From behind the scenes and inside the beltway in Washington, DC. Can be described as "no holds barred" radio.

Wize Trade Live: A "live" call-in-talk show that determines with accuracy that action of an investor's stock is based on supply and demand.

International News Hour: A "Live" from Jerusalem bureau brings the latest news from around the world.

Weekends:

Outdoors, The Weekend: Host, Alex Langer covers virtually every aspect of outdoor and recreational topics.

Cruise Control: Enjoy working on your own automobile? Perfect for the weekend mechanic as well as the seasoned professional.

ApParently: Radio for Parents with Susan Sierra is for and about Parents and Parenting from infancy to the teenage years.

Golden Age of Radio Theater: Classic radio program from the early days of radio, Hosted by Vic Ives.

The Ron Seggi Show Live from Universal Studios Florida: The show is two hours of the hottest stars, entertainers and celebrities.

Mick William's Cyber Line: Mick Williams answers listeners' questions on a wide range of topics from the internet, software, hardware, and what's new in technology.

News Products

USA News: five-minute top-of-the-hour, two-minute bottom-of-the-hour NewsBreaks; special reports; coverage of national & world news via USA's worldwide team of news professionals.

USA Sports: delivered on the 45 minutes mark with breaking news, pro coverage, & all college scores in a fast-paced delivery.

USA Business Reports: latest stock closings, corporate news, as well as developments in personal finance to help listeners stretch their dollars.

The Wall Street Journal Radio Network

Headquarters: 1155 Avenue of the Americas, 8th Fl., New York, NY 10036. (212) 416-2375, (609) 520-4777. Fax: (800) 828-6397. E-Mail: wsjradio@dowjones.com. Web site: www.wsjradio.com.

Executives: Paul Bell, exec VP (212) 597-5606; Nancy Abramson, dir affil rels (914) 244-0655; Debra Adamski, mgr/affil radio sls (212) 597-5605; Bryan Mitchell, Eastern rgnl sls mgr (212) 597-5934; Ken Alandt, Mid-West-Western sls mgr (313) 226-1226; Ken Martin, affil mkt rep (212) 597-5610; Janie Edwards, traf dir (212) 597-5609; Jay Colon, network coord affidavits/commercials (212) 597-5608; Patrice Sikora, mgng editor (609) 520-4477; Pat O'Neill, afternoon newsroom editor/news inquiries (609) 520-4356; Jeff Bellinger, morning news editor/news inquiries (609) 520-4389; Chuck Fishman, producer/The Wall Street Journal This Morning (609) 520-7904.

Provides *The Wall Street Journal Report,* 18 two-minute hourly business & financial newscasts each weekday & six weekend reports. *The Dow Jones Money Report* provides 16 one-minute newsbriefs focusing on money news & consumer trends. The net also provides six weekend reports. Progmg is satellite-delivered on Satcom C5, Transponder 23. Business news script svcs also available.

Westwood One

Headquarters: 40 W. 57th St. 5th Fl., New York, NY 10019. (212) 641-2000. Fax: (212) 641-2185. Web site: www.westwoodone.com.

Principal Executives: Peter Kosann, pres/CEO; Gary Yusko, CFO; Roby Wiener, chief/ mktg off; Paul Gregrey, exec VP/dir sls; Dennis Green, sr VP/affil sls; James Starace, VP, affil info/compliance; David Hillmant, chief admin officer/exec VP/gen counsel; Carolyn Jones, VP/human resources; Paul Bronstein, VP rsch; Luis Rodriguez, VP/info technology; Conrad Trautman, sr VP, opn/engrg.

Westwood One Programming

News

News Networks: *CBS Radio News, CNBC Business Radio, CNN Radio News, Marketwatch.com, NBC News Radio, Westwood One News.*

Features Programs: *ET Radio Minute, CBS Healthwatch, Dave Ross, Meet The Press, In the Marketplace, Late Night on Jimmy Kimmel Live, Osgood File, Raising our Kids, What's in the News, World News Roundup, Harry Smith Reporting.*

Talk

Talk Programs: *America in the Morning, America this Week, The Jim Bonhannon Show, The Don & Mike Show, First Light, The Adam Carolla Show, Phil Valentine Show, Dennis Miller, The Lars Larson Show, Larry King Live, Loveline, The Tom Leykis Show, On the Garden Line with Jerry Baker, The Radio Factor with Bill O'Reilly, Troubleshooter Tom Martino, The Week in Review.*

Sports

Football: *Monday Night Football, Sunday Night Football, NFL Playoffs, NFL Championships, NFL Super Bowl, NFL Pro Bowl, NFL Sunday Doubleheaders, NCAA Football, NFL Insider, NFL Preview, The NFL Today.*

Basketball: *NCAA Basketball, March To Madness, National Invitational Tournament.*

Golf: *British Open Championship, Masters, PGA Championship, US Open.*

Other Sports: *Focus on Racing Radio, Wimbledon, HBO Boxing.*

Sports Features: *John Madden Sports Quiz, The Madden Minute, Scoreboard, Sports Central USA, Sports Time, Sports World Roundup, Sports Feed, Today in Auto Racing, Today in Golf, Today in Sports, Westwood One Sports Report, 3rd & Long with Howie Long.*

Entertainment

Features: *Daily Show with John Stewart, ET Radio Minute, Late Show with David Letterman, Late Late Show with Craig Kilborn, Randy Jackson Hit List.*

Music

Alt/Modern Rock: *Absolutely Live, The Fax Prep Service, Loveline, MTV Radio Network, Out of Order.*

Contemporary/CHR: *The E! Radio Network Prep Service, MTV Radio Network, MTV's Total Request Live Weekend Countdown, Night Flight, Saturday Night All Request 80s, VH1 Behind the Music, Storytellers and Concerts, VH1 Radio Network.*

Classic Rock: *The Beatle Brunch, The Beatle Years, Off the Record, Superstar Concert Series, VH1 Behind the Music, Storytellers and Concerts, VH1 Radio Network.*

Country: *CMT Radio Network, Country's Cutting Edge, Country Gold, Country's Inside Trak, Country Six Pack, CMT's Country Countdown USA with Lon Helton, Grand Ole Opry, Stars of Country, The Weekly Country Music Countdown, Young and Verna.*

Oldies: *The Beatle Brunch, The Beatle Years, Doo Wop Heaven, The Motown Show, Oldies Six Pack.*

Urban/Hip Hop: *BET Radio Network, MTV Radio Network, MTV's TRL Weekend Countdown (Rhythmic).*

Urban/Hip Hop: *The Academy Of Country Music Awards, The BET Awards, Country Artist Album Premieres and Specials, The GRAMMY Awards, MTV Concerts and Specials, Music Events and Concerts, NFL Kickoff, VH1 Concerts and Specials.*

Prep Services

BET Radio Network, The CBS Morning Resource, The E! Radio Network, CMT Radio Morning Facts, MTV Radio Prep, VH1 Morning Prep, Westwood One Prep, BET Prep, Entertainment Newsfeed, Westwood One Celebrity Satellite Tours.

24-Hour Formats

Adult Rock & Roll, Adult Standards, Bright AC, CNN Headline News, Hot Country, Mainstream Country, The Oldies Channel, Soft AC.

WFMT Radio Network

Headquarters: 5400 N. St. Louis Ave., Chicago, IL 60625. (773) 279-2000. Fax: (773) 279-2199. Web Site: http://www.wfmt.com

Executives: Daniel Schmidt, pres/CEO; Steve Robinson, sr VP radio; Reese P. Marcusson, exec VP/CFO; Sandra P. Guthman, chmn; Farrell Frentress, exec VP/dev; V.J. McAleer, sr VP/production; Joanie Bayhack, sr VP, corporate communication/direct mktg; Donna L. Davies, sr VP/dev; Parke Richeson, sr VP, finance/business dev; sr VP; John Brennan, treas; Renee Crown (vice-chmn); Roger Plummer (vice-chmn); Don Mueller, opns mgr; Carol Martinez, stn rels mgr.

Satellite-delivered performing arts, jazz & spoken word progmg to over 1500 coml & pub radio stns domestically & 40 countries abroad. Among the feature programs are concerts by major symphony orchestras; productions by opera companies; concerts from Europe; concerts & spoken word from the BBC; music & verbal documentaries; folk music; *BSN Around the Clock,* a customized classical music format svc; the WFMT Jazz Satellite Network, a customized jazz format svc & periodic specials.

Regional Radio Programming Services

Agrinet News Network, 104 Radio Rd., Powells Point, NC, 27966. Phone: (252) 491-2414. Fax: (252) 491-2939. Web Site: www.agrinetradio.com. E-mail: info@agrinetradio.com

Bill Ray, pres; Lisa Ray, gen sls mgr.

Comprises 150 stns in Virginia, Maryland, Pennsylvania, New Jersey, West Virginia, North Carolina, South Carolina, Delaware, Alabama, Florida, Georgia, New York, Illinois, Indiana. Nationally distributed Agrinet news program.

Alaska Public Radio Network, 810 E. Ninth Ave., Anchorage, AK, 99501-3826. Phone: (907) 277-2776. Fax: (907) 263-7450. E-mail: aprn@alaska.net Web Site: www.aprn.org.

Bede Trantina, stn mgr; Duncan Moon, news dir.

Juneau, AK 99801. Juneau Alaska News Bureau, 530 Park St. Phone: (907) 586-6948. Dave Donaldson, state capitol bureau chief.

Washington, DC 20008. Washington, DC News Bureau, 2801 Quebec St. N.W, Suite 505. Phone: (202) 488-1961. Joel Southern, capitol bureau chief.

Satellite-delivered news/info programs to 26 member stn across Alaska from state-of-the-art studios, hqtr in Anchorage.

Allegheny Mountain Network, Box 247, Tyrone, PA, 16686. Phone: (814) 684-3200. Fax: (814) 684-1220. E-mail: amnnet@aol.com

Cary Simpson, pres; John F. Simpson, exec VP; Alfred Haper, VP sls.

Wellsboro, PA 16901, Box 98. Phone: (570) 724-1490. Albert Harper, VP sls.

Comprises 11 stns in Pennsylvania. Represented by Dome & Associates.

American Ag Network, 214 W. Pleasant Dr., Pierre, SD, 57501-2472. Phone: (605) 224-9911. Fax: (605) 224-8984. E-mail: markswendsen@amfmradio.biz

Mark Swendsen, pres.

Comprises 40 stns: 16 in South Dakota, 22 in North Dakota & 2 in Montana.

Arkansas Radio Network, 700 Wellington Hills Rd., Little Rock, AR, 72211. Phone: (501) 401-0228. Phone: (800) 839.4610. Fax: (501) 401-0367. E-mail: gordon.stephan@citcomm.com Web Site: www.arkansasradionetwork.com.

Comprises 55 interconnected stns, all in Arkansas, Texas & Mississippi. Represented by StateNets & McGauren Guild.

Beasley Broadcast Group, 3033 Riviera Dr., Suite 200, Naples, FL, 34103. Phone: (239) 263-5000. Fax: (239) 263-8191. E-mail: email@bbg.com Web Site: www.bbgi.com.

George G. Beasley, chmn/CEO; Bruce Beasley, pres.

Radio stns 44: 27 FMs & 17 AMs in 11 large, mid-sized markets.

Berkshire Broadcasting Co. Inc., 466 Curran Hwy., North Adams, MA, 01247. Phone: (413) 663-6567. Fax: (413) 662-2143. E-mail: wnaw@wnaw.com Web Site: www.wnaw.com.

Brownfield Network, (A division of Learfield Communications Inc.). 505 Hobbs Rd., Jefferson City, MO, 65109-6829. Phone: (573) 893-5700. Fax: (573) 893-8094. E-mail: jsteinman@learfield.com Web Site: www.brownfieldnetwork.com.

Bruce Beasley, pres; Joyce Steinman, adv.

Comprises 205 stns in Illinois, Iowa, Missouri, Nebraska, Indiana, South Dakota & Wisconsin. Represented by In-House.

Paul Bunyan Network, Paul Bunyan Bldg., 314 E. Front, Traverse City, MI, 49684. Phone: (231) 947-7675. Fax: (231) 929-3988. E-mail: wtcm@wtcmradio.com

Ross Biederman, pres & gen mgr; Jim Sofonia, chief engr; Jon Patrick, natl sls mgr; Jack O'Malley, progmg dir.

Comprises five stns in Michigan. Represented by Katz Radio.

CRN International, Inc., One Circular Ave., Hamden, CT, 06514. Phone: (203) 288-2002. Fax: (203) 281-3291. E-mail: info@crnradio.com Web Site: www.crnradio.com.

Barry Berman, pres; S. Richard Kalt, exec VP; Patrick Kane, sr VP; Steve Wakeen, VP.

Stragtegy & execution, retail mktg, lifestyle progmg, promotions, digital media & other non-traditional communication tactics, small business, collaborative mktg, weather-triggered media placement, Ethnic, Just -In-Time Marketing. Branch Office: Minneapolis, MN.

California News Radio, 14605 N. Airport Dr., Suite 370, Scottsdale, AZ, 85260. Phone: (480) 503-8700. Fax: (480) 503-8745. Web Site: www.skyviewsatellite.com.

Jeanne-Marie Condo, gen mgr; Ken Thiele, pres.

Compu-Weather Inc., 2566 Rt. 52, Hopewell Junction, NY, 12533. Phone: (800) 825-4445. Fax: (800) 825-4441. E-mail: sales@compu-weather.com Web Site: www.compu-weather.com.

Jeff Wimmer, pres.

Florida Public Radio Network, 1600 Red Barber Plaza, Tallahassee, FL, 32310. Phone: (850) 487-3194. Fax: (850) 487-3293. E-mail: fpr@wfsu.org Web Site: www.fsu.edu.

Carolina Austin, opns mgr; Tom Flanigan, news dir.

Serves 13 FM public radio stns in Florida.

Florida's Radio Networks, 2500 Maitland Ctr. Pkwy., Suite 407, Maitland, FL, 32751. Phone: (407) 916-7810. Phone: (407) 916-7800. Fax: (407) 916-7425. E-mail: rgreen@frn.com Web Site: www.frn.com.

Jim Poling, opns mgr; Rick Green, gen mgr; Jim Underwood, gen sls mgr.

Comprises 58 stns in Florida. Represented by StateNets Inc.

Georgia News Network, 1819 Peachtree Rd., Suite 700, Atlanta, GA, 30309. Phone: (404) 607-9045. Phone: (800) 776-4638. E-mail: robmaynard@clearchannel.com Web Site: www.georgianewsnetwork.com.

Linda Kent, sls; Rob Maynard, progmg.

Comprises 108 stns in Georgia. Represented by StateNts.

Hawkeye Network, 505 Hobbs Rd., Jefferson City, MO, 65109. Phone: (573) 893-7200. Fax: (573) 893-2321. Web Site: www.learfield.com.

Clyde G. Lear, pres; Greg Brown, VP; Keith Sampson, exec producer; Bob Agramonte, gen mgr.

Comprises 50 stns in Iowa.

Hispanic Communications Network, 1126 16th St. N.W., Suite 350, Washington, DC, 20036. Phone: (202) 637-8800. Fax: (202) 637-8801. E-mail: news@hrn.org Web Site: www.hrn.org.

Comprises 80 stns in Arizona, California, Florida, Georgia, Idaho, Illinois, Louisiana, Massachusetts, Missouri, Nevada, New Mexico, New York, Oklahoma, Oregon, Pennsylvania, Rhode Island, Texas, Utah, Washington, & Washington, DC.

Hometown Radio Network, 1100 Chester Ave., Suite 100, Cleveland, OH, 44115. Phone: (216) 781-0035. Fax: (216) 348-8408. E-mail: sjsharpe@regionalreps.com Web Site: www.regionalreps.com.

Stuart J. Sharpe, pres.

Comprises over 1000 affils in Delaware, Florida, Georgia, Iowa, Illinois, Indiana, Kansas, Kentucky, Maryland, Nebraska, North Carolina, Ohio, Oklahoma, Pennsylvania, South Carolina, Virginia, & West Virginia. Represented by: Rgnl Reps Corp.

ION Radio Network, Box 1223, Airport Rd., Morristown, NJ, 07960. Phone: (973) 983-8222. Fax: (973) 983-1390. E-mail: steve@ionweather.com Web Site: www.ionweather.com.

Stephen Pellettiere, pres.

Comprises 19 stns: five in New Jersey, six in New York, three in Pennsylvania, two in Maryland, two in Connecticut & one in Rhode Island.

Illinois Radio Network, 430 W. Erie, Suite 505, Chicago, IL, 60610. Phone: (312) 943-6363. Fax: (312) 943-5109. Web Site: www.illinoisradionetwork.com.

Dennis Mellott, gen mgr & pres.

A statewide satellite-delivered net providing news, sports, business & special progmg. IRN 67 affils. Representative: StateNets.

KEDA Radio, 510 S. Flores, San Antonio, TX, 78204. Phone: (210) 226-5254. Fax: (210) 227-7937. E-mail: kedakid@aol.com Web Site: www.kedaradio.com.

Comprises three stns in Texas, three affiliates. Represented by Caballero Spanish Media.

Kansas Agriculture Network, Box 1818, Topeka, KS, 66601-1818. Phone: (785) 272-2199. Fax: (785) 272-3536. Web Site: www.radionetworks.com.

Craig Colbach, gen mgr; Ed O'Donnell, opns mgr; Jason Weil, sls mgr.

Comprises 37 stns in Kansas. Represented by Katz Radio.

Kansas Information Network, Box 1818, Topeka, KS, 66601-1818. Phone: (785) 272-2199. Fax: (785) 272-3536. Web Site: www.radionetworks.com.

Craig Colbach, gen mgr; Ed O'Donnell, opns mgr; Liz Montano, news dir.

Comprises 45 stns in Kansas. Represented by StateNets.

Kansas State Sports Network, 1632 S. Maze Rd., Wichita, 67209. Phone: (316) 721-8484.

Kentucky News Network, (A subsidiary of Clear Channel Radio Inc.). One Radio Dr., Louisville, KY, 40218. Phone: (502) 479-2240. Fax: (502) 479-2229. Web Site: www.kentuckynewsnetwork.com.

Ed Huckleberry, editor; Doug Wethington, gen sls mgr; Price Allen, dir; Jack Crowner, dir.

Comprises 93 stns: 92 in Kentucky, one in West Virginia. Represented by StatesNets. Live via satellite.

Linder Farm Network, 255 Cedardale Dr., Owatonna, MN, 55060. Phone: (507) 444-9224. Fax: (507) 444-9080. E-mail: farm@linderradio.com Web Site: www.linderfarmnetwork.com.

Jeff Stewart, sls mgr; Lynn Ketelsen, gen mgr.

Comprises 22 stns in Minnesota. Represented by Katz Radio.

Louisiana Agri-News Network, 10500 Coursey Blvd., Ste. 104, Baton Rouge, LA, 70816. Phone: (225) 291-2727. Fax: (225) 297-7539. E-mail: jim@la-net.net Web Site: www.la-net.net.

Bill Rigell, pres/CEO; Jim Engster, gen mgr.

Comprises 66 stns in Louisiana & Mississippi. Represented by McGavren Guild.

Louisiana Network Inc., 10500 Coursey Blvd., Suite 104, Baton Rouge, LA, 70816. Phone: (225) 291-2727. Fax: (225) 297-7539. E-mail: bill.rigell@la-net.net Web Site: www.la-net.net.

Comprises 84 stns in Louisiana. Represented by news net: State Nets, Agri-News Network: McGavren Guild.

Michigan Farm Radio Network, 325 South Walnut, Lansing, MI, 48933. Phone: (517) 484-4888. Fax: (517) 484-5015. E-mail: kbuys@mfrn.com Web Site: www.mfrn.com.

Dennis Mellott, gen mgr & pres; Kirsten Buys, dir; Janelle Brose, dir.

Comprises 21 Michigan. Represented by J.L. Farmakis.

Michigan Radio Network, 325 S. Walnut, Lansing, MI, 48933. Phone: (517) 484-4888. Fax: (517) 484-1389. Web Site: www.michiganradionetwork.com.

Dennis Mellott, pres & gen mgr; Rob Baykian, news dir; Kirsten Buys, dir.

A statewide satellite-delivered net providing news, sports, business & special progmg. MRN 70 affils. Representative: StateNets.

Mid-America Ag Network, 1632 S. Maize Rd., Wichita, KS, 67209. Phone: (316) 721-8484. Fax: (316) 721-8276. Web Site: www.maanradio.com.

Rick Betzen, gen mgr; Greg Steckline, pres.

Comprises 44 stns in Colorado, Kansas, Nebraska & Oklahoma. Represented by Torbet Radio.

Mississippi Agri Network, 6311 Ridgewood Rd., Jackson, MS, 39211. Phone: (601) 957-1700. Fax: (601) 956-5228. E-mail: lsheldon@telesouth.net Web Site: www.telesouth.com; www.supertalkms.com.

Stacy Long, gen sls mgr.

Comprises 35 affils in Mississippi. Represented by McGavren/Guild.

Mississippi News Network, 6311 Ridgewood Rd., Jackson, MS, 39211. Phone: (601) 957-1700. Fax: (601) 956-5228. E-mail: lsheldon@telesouth.net Web Site: www.telesouth.com.

Stacy Long, gen sls mgr.

Comprises 82 affils in Mississippi. Represented by StateNets.

Mississippi State Basketball Network, 6311 Ridgewood Rd., Jackson, MS, 39211. Phone: (601) 957-1700. Fax: (601) 956-5228. E-mail: kdillon@telesouth.com

Kim Dillon, mktg dir.

Comprises 28 affils in Mississippi & one in Tennessee.

Mississippi State Football Network, 6311 Ridgewood Rd., Jackson, MS, 39211. Phone: (601) 957-1700. Fax: (601) 956-5228. E-mail: kdillon@telesouth.com Web Site: www.telesouth.com.

Kim Dillon, mktg dir; Steve Davenport, pres/CEO.

Comprises 30 affils in Mississippi & one in Alabama. Represented by Kim Dillon.

Missourinet, (A division of Learfield Communications Inc.). 505 Hobbs Rd., Jefferson City, MO, 65109. Phone: (573) 893-2829. Fax: (573) 893-8094. E-mail: info@missourinet.com Web Site: www.missourinet.com.

Bob Priddy, news dir.

Serves Missouri - 70 affils.

Mountain News Network, 50 Vashell Way, Suite 200, Orinda, CA, 94563. Phone: (925) 254-4456. Phone: (800) 736-0370 Eastern Bureau. Fax: (925) 254-6135. E-mail: news@mnn.net Web Site: www.mnn.net.

Comprises 1,537 stns nationwide.

NRG Media, LLC, Box 94, Fort Atkinson, WI, 53538. Phone: (920) 563-2667. Fax: (920) 563-0315. E-mail: jvriezen@nrgmedia.com Web Site: www.lite1073.com.

Jim Vriezen, gen mgr.

National Educational Telecommunications Association, Box 50008, Columbia, SC, 29250. Phone: (803) 799-5517. Fax: (803) 771-4831. E-mail: skip@netaonline.org Web Site: www.netaonline.org.

Skip Hinton, pres.

Eighty-four members in 38 states & the U.S. Virgin Islands.

New South Communications Inc., Box 5797, Meridian, MS, 39302. Phone: (601) 693-2661. Fax: (601) 483-0826.

Ed Holladau, pres.

Comprises 18 stns: five in Louisiana, ten in Mississippi & three in Alabama. Represented by McGavren Guild.

North Carolina News Network, 711 Hillsborough St., Raleigh, NC, 27603. Phone: (919) 890-6128. Fax: (919) 890-6024. E-mail: rhankin@ncnn.com Web Site: www.ncnn.com.

Ardie Gregory, VP/gen mgr.

Comprises 75 stns in North Carolina. Represented by StateNets.

North Dakota News Network, Box 1197, Pierre, SD, 57501. Phone: (605) 224-9911. Fax: (605) 224-8984. E-mail: markswendsen@amfmradio.biz

Mark Swendsen, pres.

Comprises 24 stns in North Dakota. Represented by StateNets.

Ohio Educational Telecommunications Network Commission, 2470 North Star Rd., Columbus, OH, 43221. Phone: (614) 644-1714. Fax: (614) 644-3112. Web Site: www.oet.edu.

Comprising 12 TV stns & 32 radio stns, 10 radio reading svcs, & eight educ technology.

Oklahoma News Network, (Oklahoma Agrinet). Box 1000, Oklahoma City, OK, 73101. Phone: (405) 858-1400 Ext. 278. Fax: (405) 840-5808.

Jerry Bohnen, news dir.

Comprises 55 stns in Oklahoma.

Pittsburgh Country Network, 1439 Denniston St., Pittsburgh, PA, 15217. Phone: (412) 421-2600. Fax: (412) 421-6001. E-mail: rafson@cmsradio.com Web Site: www.cmsradio.com.

Roger Rafson, pres.

Comprises three stns in Pennsylvania. Represented by Commercial Media Sales.

Radio Iowa, (A division of Learfield Communications Inc.). 2700 Grand Ave., Suite 103, Des Moines, IA, 50312. Phone: (515) 282-1984. Fax: (515) 282-1879. E-mail: radioiowa@learfield.com Web Site: www.radioiowa.com.

Clyde G. Lear, pres/CEO; Jennifer Shaefer, adv; O. Kay Henderson, news dir.

Comprises 55 affils in Iowa.

Radio Pennsylvania Network, (A Division of WITF Inc.). 1982 Locust Ln., Harrisburg, PA, 17109. Phone: (717) 232-8400. Phone: (717) 236-6000. Fax: (717) 232-7612. E-mail: craig_rhodes@radiopa.org Web Site: www.radiopa.com.

Craig Rhodes, opns mgr; Brad Christman, editor.

Bcsts state news, sports & features to affils in Pennsylvania. Comprises 80 stns. Represented by NASRN.

Saga Communications, Inc., Corporate Offices, Ste. 201, 73 Kercheval Ave., Grosse :Point Farms, MI, 55404-1009. Phone: (313) 886-7070. Fax: (313) 886-7150.

Timothy Shears, pres; Warren Lada, VP progmg; Edward hrstian, pres/CEO.

Comprises 93 stns in Minnesota, North Dakota, South Dakota, Iowa. Represented by State Networks Inc. (retail) & self represented (agricultural).

South Carolina News Network, (A division of Telesouth Communications Inc.). 3710 Landmark Dr., Suite 100, Columbia, SC, 29204. Phone: (803) 790-4300. Fax: (803) 790-4309.

Comprises 40 affils in South Carolina.

South Dakota News Network, Box 1197, Pierre, SD, 57501. Phone: (605) 224-9911. Fax: (605) 224-8984. E-mail: markswendsen@amfmradio.biz

Mark Swendsen, pres.

Comprises 20 stns in South Dakota. Represented by StateNets.

Southeast AgNet, 5053 N.W. Highway 225 A, Ocala, FL, 34482. Phone: (407) 436-1909. Fax: (407) 436-1364. E-mail: gary@southeastagnet.com Web Site: www.southeastagnet.com.

Gary Cooper, pres; Robin Loftin, VP; Randall Weiseman, dir opns.

Stns interconnected via Internet. Comprises 65 affils in Florida, Georgia & the Alabama rgn.

Southern Farm Network, 3012 Highwoods Blvd., Suite 200, Raleigh, NC, 27604. Phone: (919) 876-0674. Fax: (919) 790-8369. E-mail: bpprice@southernfarmnetwork.com Web Site: www.southernfarmnetwork.com.

Barbara G. Price, opns mgr & sls.

Comprises 20 affils in North Carolina & South Carolina.

Tennessee Agri-Net, (Subsidiary of Clear Channel Communications Inc.). 55 Music Square West, Nashville, TN, 37203. Phone: (615) 774-4785. Fax: (615) 687-9797. E-mail: janetpatterson@clearchannel.com Web Site: www.tennesseeradionetwork.com.

Comprises 48 stns in Tennessee.

Tennessee Radio Network, (A subsidiary of Clear Channel Broadcasting Inc.). 55 Music Sq. W., Nashville, TN, 37203. Phone: (615) 664-2400. Fax: (615) 687-9797. E-mail: janetpatterson@clearchannel.com Web Site: www.tennesseeradionetwork.com.

Chris Romer, opns mgr; Janet Patterson, gen sls mgr.

Comprises 76 stns in Tennessee. Representative Buddy Sadler, affil rel dir.

Texas State Network, 4131 N. Central Expwy., Suite 500, Dallas, TX, 75204. Phone: (214) 525-7400. Fax: (214) 525-7371. E-mail: dbell@cbs.com Web Site: www.tsnradio.com.

Jerry Bobo, VP & gen mgr; Dan Bell, gen sls mgr; Julis Graw, dir news & opns; Brian Purdy, VP/mktg mgr.

Austin, TX 78701, 502 E. 11th St, Suite 320. Phone: (512) 474-5275. Fax: 512 476 9232. Candy Schmidt: Regional Sales Director.

Provides newscasts, sportscasts, agriculture reports, longform programs to 165 stns in Texas & the Texas Rangers Radio Network. The oldest & largest state radio network owned by CBS Radio. Represented by StateNets.

Tiger Network, (A division of Learfield Communications Inc.). 505 Hobbs Rd., Jefferson City, MO, 65109. Phone: (314) 893-7200. Fax: (314) 893-2321. Web Site: www.learfield.com.

Clyde G. Lear, pres; Greg Brown, VP; Keith Sampson, exec producer; Bob Agramonte, gen mgr; Aaron Worsham, VP opns.

Comprises 55 stns in Missouri.

Tribune Radio Networks, 435 N. Michigan Ave., Chicago, IL, 60611. Phone: (312) 222-3342. Fax: (312) 222-4876. E-mail: bpabst@tribune.com Web Site: www.tribuneradio.com.

Barbra Pabst, opns; Kurt Vanderan, opns mgr.

Network comprised of: Chicago Cubs Network (50 stns), National Farm Report (260 stns), Farming America (200 stns), Agri-Voice Network (95 stns), Samuelson's Soapbox (150 stns). Represented by Eastman.

University of Mississippi Baseball Network, 6311 Ridgewood Rd., Jackson, MS, 39211. Phone: (601) 957-1700. Fax: (601) 956-5228. Web Site: www. telesouth.com.

Comprises 30 affils in Mississippi & one in Tennessee. Represented by Kim Dillon.

University of Mississippi Basketball Network, 6311 Ridgewood Rd., Jackson, MS, 39211. Phone: (601) 957-1700. Fax: (601) 956-5228. Web Site: www.telesouth.com.

Steve Davenport, pres; Stacy Long, gen sls mgr.

Comprises 30 affils in Mississippi & one in Tennessee. Represented by Kim Dillon.

University of Mississippi Football Network, 6311 Ridgewood Rd., Jackson, MS, 39211. Phone: (601) 957-1700. Fax: (601) 956-5228. Web Site: www.telesouth.com.

Steve Davenport, pres; Stacy Long, gen sls mgr.

Comprises 30 affils in Mississippi & one in Tennessee. Represented by Tim Fritts.

Univision Radio, 3102 Oak Lawn Ave., Suite 215, Dallas, TX, 75219. Phone: (214) 525-7700. Fax: (214) 525-7750. Web Site: www.univision.net.

Alan F. Horn, pres/COO.

Comprises 42 stns: seven in California, four in Florida, three in Illinois, two in Nevada, two in New York, 22 in Texas & two in Arizona. Represented by Katz Radio.

WRTI-FM, 1509 Cecil B. Moore, 3rd Fl., Philadelphia, PA, 19121-3410. Phone: (215) 204-8405. Fax: (215) 204-7027. Web Site: www.wrti.org.

David S. Conant, exec dir; Patty Prevost, dev dir; Tobias Poole, opns.

Comprises four stns in Pennsylvania, one in New Jersey & one in Delaware.

WV Radio Corp. and Metronews Radio Network, Greer Bldg., 1251 Earl L. Core Rd., Morgantown, WV, 26505. Phone: (304) 296-0029. Fax: (304) 296-3876. Web Site: www.wvmetronews.com. E-mail: dmiller@wvradio.com

Dale B. Miller, pres; Hoppy Kercheval, VP opns.

Charleston, WV 25301, 1111 Virginia St. E. Phone: (304) 342-8131.

Comprises West Virginia News (58 stns in West Virginia) & Mountaineer Sports Network (72 stns in West Virginia).

"The Weather Center", (A broadcast service of Aviation Weather Inc.). 701 Gervais St., Suite 150-224, Columbia, SC, 29201. Phone: (803) 739-2827. E-mail: wxcenter@aviationweatherinc.com Web Site: www.aviationweatherinc.com.

"Regional Radio Broadcast/Weathercast Network" across the Carolinas & Georgia in over 20 broadcast markets. Weather forecasting, site-specific bcst svc for stns all across America. 100% barter.

Western Agri-Radio Networks, (dba California Agri-Radio Network & Southwest Agri-Radio Network). 1700 S. 1st Ave., Suite 214, Yuma, AZ, 85364. Phone: (928) 782-1440. Fax: (928) 782-1474. E-mail: ggatley@sprynet.com Web Site: www.westernagri-radio.net.

George G. Gatley, pres.

Yuma, AZ 85364. Southwest Agri-Radio Network, 1700 S. 1st Ave, Suite 214. Phone: (800) 944-6077. Fax: (520) 782-1474. E-mail: ggatley@sprynet.com. Web Site: www.home.com/sprynet/ggatley.

Yuma, AZ 85364. California Agri-Radio Network, 1700 S. 1st Ave, Suite 214. Phone: (800) 944-6077. (520) 782-1440. Fax: (520) 782-1474. E-mail: ggatley@sprynet.com. Web Site: www.home.com/sprynet/ggatley.

Fifteen radio stns in California & two in Arizona. Represented by J.L. Famakis.

Wisconsin Radio Network, 222 State St., Suite 401, Madison, WI, 53703. Phone: (608) 251-3900. Fax: (608) 251-7233. E-mail: info@wrn.com Web Site: www.wrn.com.

Joyce Steinman, adv; Bob Hague, news dir.

Statewide satellite-delivered net providing Wisconsin news & sports.

Yancey AG Network, Box 1000, Oklahoma City, OK, 73101. Phone: (405) 858-10297. E-mail: ashliacker @clearchannel.com Web Site: www.oklahomaagrinet.net.

Ron Hays, progmg dir; Ben Buckland, network mgr.

Represents 17 stns.

Radio News Services

ABC News Radio, 125 West End Ave., New York, NY, 10023. Phone: (212) 456-5100. Fax: (212) 456-5150. E-mail: customerservice@abc.com Web Site: www.abcradionetworks.com.

Steve Jones, VP; Andrew Kalb, exec dir & news progmg; Michael Rizzo, exec dir & news & sports coverage; Robert Garcia, exec dir & Washington bureau chief.

Serves 2,500 affiliates & 92.5 million wkly listeners nationwide, with bcst facilities in New York City & Washington, DC.

AMI News, 50 Vashell Way, Suite 200, Orinda, CA, 94563. Phone: (925) 254-4456. Fax: (925) 254-6135. E-mail: info@mountainviewnews.com Web Site: www.theamigroup.com.

Chad Dyer, VP/internet news svcs., Eastern Bureau. Phone: (800) 736-0370.

MP3 Wave phone- & tape-supplied features focusing on skiing, fishing, camping, travel & beach conditions with related news & information. Offered seasonally. Available on the Internet.

AccuWeather Inc., 385 Science Park Rd., State College, PA, 16803. Phone: (814) 235-8600. Fax: (814) 235-8609. E-mail: info@accuweather.com Web Site: www.accuweather.com.

Gary Kemp, sls VP; Dr. Joel N. Myers, pres/CEO.

World's most accurate svc provides products & svcs for TV, radio, internet, mobile web & all new media platforms.

Africa News Service Inc., 920 M St. S.E., Washington, DC, 20003. Phone: (202) 546-0777. Fax: (202) 546-0676. Web Site: www.allafrica.com.

Reed Kramer, CEO; Amrada Mahtar Ba, pres.

News & info svc on African affrs.

Agence France-Presse, 1015 15th St. N.W., Washington, DC, 20005. Phone: (202) 289-0700. Fax: (202) 414-0525. E-mail: afp-usa@afp.com Web Site: www.afp.com.

Pierre Louette, chief exec; Denis Hiault, dir; Jean-Pierre Vignolle, dir.

Produces a var of international news svcs, including text wires in six languages, photo wires, graphics & financial wires.

Alaska Public Radio Network, 810 E. Ninth Ave., Anchorage, AK, 99501-3826. Phone: (907) 277-2776. Fax: (907) 263-7450. E-mail: aprn@alaska.net Web Site: www.aprn.org.

Bede Trantina, stn mgr; Duncan Moon, news dir.

Juneau, AK 99801. Juneau Alaska News Bureau, 530 Park St. Phone: (907) 586-6948. Dave Donaldson, state capitol bureau chief.

Washington, DC 20008. Washington, DC News Bureau, 2801 Quebec St. N.W, Suite 505. Phone: (202) 488-1961. Joel Southern, capitol bureau chief.

Satellite-delivered news/info programs to 26 member stn across Alaska from state-of-the-art studios, hqtr in Anchorage.

American Academy of Dermatology, Communications Dept., American Academy of Dermatology. Box 4014, Schaumburg, IL, 60168-4014. Phone: (847) 330-0230. Fax: (847) 330-8907. Web Site: www.aad.org.

Washington, DC 20005-3319, 1350 I Street NW, Suite 870. Phone: (202) 842-3555. Fax: (202) 842-4355.

Expert physicians available for TV & radio interviews, audio & video tapes on skin cancer detection, as well as info on skin, hair & nail conditions.

American Heart Association, 7272 Greenville Ave., Dallas, TX, 75231-4596. Phone: (214) 373-6300. Fax: (214) 706-5243. Web Site: www.americanheart.org.

M. Cass Wheeler, CEO.

Rsch & lifestyle reports, distributed by tape & live copy.

American Urban Radio Networks, 960 Penn Ave., Suite 200, Pittsburgh, PA, 15222. Phone: (412) 456-4000. Fax: (412) 456-4040. E-mail: jlopes@aurn.com Web Site: www.aurn.com.

Jerry Lopes, opns.

New York, NY 10016, 432 Park Ave. S., 14th Fl. Phone: (212) 883-2100. Fax: (212) 297-2571. E-mail: jay@aurn.com. E.J. "Jay" Williams, pres.

Info, news, sports & entertainment of special interest to Blacks & other minorities.

Associated Press Broadcast Services, 1825 K St. N.W., Suite 800, Washington, DC, 20006-1202. Phone: (202) 736-1100. Fax: (202) 736-1199 (news). Fax: (202) 736-1124 (admin). Web Site: www.apbroadcast.com.

Audio: PrimeCuts, Sound Bank. On Air: Radio News. On Line: News Tickers, Online Video Network; Custom News, Sp Online. Prep: Power Prep. Software: News Desk, APENPS. Sound Desk, News Center. Text Headlines, Sp, Sports Power, News Power, Images & Multimedia: asap, all AP.

Associated Press Network News

See Associated Press listing in Major National Radio Networks, this section.

Audio-Video News, 3622 Stanford Cir., Falls Church, VA, 22041. Phone: (703) 354-6795. E-mail: connielawn@aol.com Web Site: www.dcski.com.

Connie Lawn, pres.

Covers major natl, international & ski stories for radio & TV stns in the United States & around the world. Also live "inserts" into radio & TV shows.

The Berns Bureau, Box 2939, Washington, DC, 20013-2939. Phone: (202) 314-5165. Fax: (202) 628-1432.

Matt Kaye,.

Complete "localized" coverage of Washington, DC. Satellite ISDN & telephone transmission. Audio news releases. Govt, politics, farm, relg & other progmg for radio. Audio svc for TV.

Black Radio Network Inc. (BRN), 166 Madison Ave., New York, NY, 10016. Phone: (212) 686-6850. Fax: (212) 686-7308. E-mail: news@blackradionetwork.com

Roy Thompson, VP.

Provides a daily actuality news service emphasizing minority-oriented items.

British Information Services, 845 Third Ave., New York, NY, 10022. Phone: (212) 745-0395. Fax: (212) 758-5395.

Provides daily audio news feed svc filed by digital line from London at no cost to stns. Assists radio & TV crews visiting the United Kingdom.

Broadcast News Ltd., 36 King St. E., Toronto, ON, M5C 2L9. Canada. Phone: (416) 364-3172. Fax: (416) 364-8896. Web Site: www.broadcastnews.ca. E-mail: info@broadcastnews.ca

Full wire & audio svcs (news agency), satellite delivery for radio program syndicators.

CBS News

See CBS listing in National Radio Programming Services, this section.

CNN Radio News, One CNN Ctr., Atlanta, GA, 30303. Phone: (404) 878-2276. Fax: (404) 827-1995. Web Site: www.westwoodone.com.

Fred Bennett, sr VP.

Top- & bottom-of-the-hour radio newscasts 24-hours a day plus business, sports & lifestyle updates.

Canada NewsWire Ltd., 1500, 20 Bay St., WaterPark Pl., Toronto, ON, M5J 2N8. Canada. Phone: (416) 863-9350. Phone: (866) 805-9530. Fax: (416) 863-9429. E-mail: cnwtor@newswire.ca Web Site: www.newswire.ca.

Sylvia Kavanagh, mgr; Tim Griffin, bcst mgr; Carolyn McGill, dir mktg.

Calgary, AB T2P 3C5 Canada, Gulf Canada Sq, 401 Ninth Ave. S.W., Suite 835. Phone: (403) 269-7605. Fax: (403) 263-7888. TWX: 03-824872. E-mail: Michle.dauphine@newswire.ca. Krista Wightman, mgr.

Vancouver, BC V6B 4NB Canada, 650 West Georgia St, Suite 1103. Phone: (604) 669-7764. Fax: (604) 669-4356. TWX: 04-508529. Larry Cardy, VP western Canada.

Halifax, NS B4A 1E6 Canada, Sun Tower, 1550 Bedford Hwy., Suite 410. Phone: (902) 422-1411. Fax: (902) 422-3507. TWX: 019-21534. E-mail: jgallant@newswire.ca. Robert Moffatt, mgr Atlantic Canada.

Ottawa, ON K1P 6A9 Canada, 255 Albert St, Suite 460. Phone: (613) 563-4465. Fax: (613) 563-0548. TWX: 053-3292. Hugh Johnson, VP natl capital rgn.

Montreal, PQ H3B 2J6 Canada, 1155 Rene Levesque Blvd. W, Suite 3310. Phone: (514) 878-2520. Fax: (514) 878-4451. TWX: 055-60936. E-mail: scmtl@newswire.ca. Elaire Carr, VP Quebec.

Offers a range of industry leading communication products & svcs for companies looking to maximize the strength of their news. Whether you are an investor rel off, or a specialist in PR, CNW offers the right tools for your communications.

The Church of Jesus Christ of Latter-day Saints (Mormons), 50 East North Temple, Salt Lake City, UT, 84150. Phone: (801) 240-4612. Fax: (801) 240-5449. E-mail: russelldg@chq.byu.edu Web Site: www.lds.org.

Gordon Hinckley, pres; Donald G. Russell, media rel.

Offers free pub affrs, news & feature progmg for TV & radio; also guests for talk shows. Pub affrs progmg is not church-oriented.

Compu-Weather Inc., 2566 Rt. 52, Hopewell Junction, NY, 12533. Phone: (800) 284-7246. Fax: (845) 226-1918. E-mail: sales@compu-weather.com Web Site: www.compu-weather.com.

Jeff Wimmer, pres.

Weather forecasts, features, info & actualities for TV & radio.

Congressional Quarterly Inc., 1255 22nd St. N.W., Washington, DC, 20037. Phone: (202) 419-8500. Fax: (202) 728-1863. E-mail: customerservice@cq.com Web Site: www.cq.com.

Keith A. White, VP/gen mgr; David Rapp, editor.

Print & Web-based info products & svcs on govt, politics & current interest topics. Daily & wkly publications, reference books & newsletters.

Connecticut Weather Center Inc., 18 Woodside Ave., Danbury, CT, 06810-7123. Phone: (203) 730-2899. Fax: (203) 730-2839. E-mail: weatherlab@ctweather.com Web Site: www.ctweather.com.

William Jacquemin, pres.

Weather forecasts for all media. Custom intros/outros/lives. Accurate forecasts. Barter or cash arrangement available.

Corus Radio Network, 700 W. Georgia St., Suite 2000, Vancouver, BC, V7Y 1K9. Canada. Phone: (604) 331-2830. Fax: (604) 331-2722. E-mail: akrueger@cknw.com Web Site: www.corusent.com.

Allan Krueger, opns mgr; John P. Hayes, pres.

Live & pre-recorded info & entertainment program production & satellite delivery to rgnl & natl Canadian radio stns.

Dairyline Radio, 1843 Front St., Suite A, Lynden, WA, 98264. Phone: (360) 354-5596, EXT. 101. Fax: (360) 354-7517. E-mail: bbaker@dairyline.com Web Site: www.dairyline.com.

Lee Mielke, pres; Bill Baker, mktg dir.

Wilmington, NC 28403-7224. DairyBusiness Communications, 7225 Wrightsville Ave, 204.

Lynden, WA 98264

Five minute & 9 1/2 minute Dairy Report- weekdays. Daily updates of news affecting the dairy industry.

Earth Reports Environmental News Science, 18600 Queen Anne Rd, Upper Marlboro, MD, 20774. Phone: (301) 249-8200. Fax: (301) 249-3613. E-mail: info@earthreports.org

Fred Tutman, pres.

Environmental news svc offering actualities & specials.

Entertainment News Calendar, 250 W. 57th St., Suite #1431, New York, NY, 10107. Phone: (212) 421-1370. Fax: (212) 563-3488. E-mail: editor@newscalendar.com Web Site: www.newscalendar.com.

Evelyn Heyward, editor.

Sherman Oaks, CA 91403. Hollywood News Calender, 15030 Ventura Blvd., Ste 742. Phone: (818) 990-5945. Carolyn Fox, publisher.

Daily entertainment news svc.

Fairchild Broadcast News, 405 E. 42nd St., Suite 310, New York, NY, 10012. Phone: (212) 593-3294. Fax: (212) 686-7308. Web Site: www.fairchildgroup.com/news.

Gathers & disseminates news around the world.

Feature Story News, 1730 Rhode Island Ave., Suite 405, Washington, DC, 20036. Phone: (202) 296-9012. Fax: (202) 296-9205. E-mail: markss@featurestory.com Web Site: www.featurestory.com.

Simon Marks, pres.

Orlando, FL 32801, 1103 Palmer St. Phone: (407) 898-1929. Steve Mort, correspondent.

New York, NY 10036, 226 W. 47th St, 2nd floor. Phone: (212) 764-5848. Nathan King, correspondent.

Ind supplier of radio & TV news to English-language bcstrs worldwide. Bureaus in Washington, Moscow, London, New York, Orlando & Beijing.

Hollywood News Calendar, 13636 Ventura Blvd., #303, Sherman Oaks, CA, 91423. Phone: (818) 990-5945. Phone: (818) 986-8186. Fax: (818) 789-8047. E-mail: editor@newscalendar.com Web Site: www.newscalendar.com.

Carolyn Fox, publisher.

New York, NY 10107. Entertainment News Calender, 250 W. 57th St, #1431. Phone: (212) 421-1370. Fax: (212) 563-3488. Carolyn Fox, publisher.

Daily entertainment news svc. Publisher of Hollywood News Calendar in Los Angeles & Entertainment News Calendar in New York.

Israel Broadcasting Service, 800 Second Ave., New York, NY, 10017. Phone: (212) 499-5402. Fax: (212) 499-5425. E-mail: newyork@israel.org

Free radio & TV programs, features from & about Israel.

Medialink, 708 3rd Ave., 9th Fl., New York, NY, 10017. Phone: (212) 682-8300. Fax: (212) 682-5260. Web Site: www.medialink.com.

Mary C. Buhhay, sr VP; Laurence Moskowitz, pres; Larry Thomas, COO.

London W1R 3AA, 37/38 Golden Sq. Phone: 44-71-240-3923. Jim Gold.

Los Angeles, CA 90028, 6430 Sunset Blvd, Suite 1100. Phone: (323) 465-0111.

San Francisco, CA 94111, One Maritime Plaza. Phone: (415) 296-8877. Fax: (415) 296-9929.

Washington, DC 20005, 1401 New York Ave. N.W, Suite 520. Phone: (202) 628-3800.

Atlanta, GA 30326, 3340 Peachtree Rd. N.E, Suite 1520. Phone: (404) 848-7500.

Chicago, IL 60611, The Time & Life Bldg., 541 N. Fairbanks Ct., Suite 1910. Phone: (312) 222-9850.

Dallas, TX 75244, 4851 LBJ Fwy, Suite 605. Phone: (972) 774-0200.

International video & audio PR, satellite feed & news advisory svc. Advisories accessible by computer/newswire in United States & Europe.

Metro Networks/Shadow Broadcast Services, a Westwood One Co., 2800 Post Oak, Suite 4000, Houston, TX, 77056. Phone: (713) 407-6000. Fax: (713) 407-6049. Web Site: www.westwoodone.com.

Chuck Bortnick, pres; Peter Kosann, pres/CEO.

Provider of traf reporting svcs & leading supplier of loc news, sports, weather & video news svcs to the TV & radio bcst industries.

Metro Weather Service Inc., 71 So. Central Ave., Ste. 102, Valley Stream, NY, 11580. Phone: (516) 568-8844. Fax: (516) 568-8853. E-mail: metrowx@aol.com Web Site: www.metroweather.com.

Pat Pagano, pres.

Tailored weather forecasts for radio & TV. Feature reports farming, marine, ski, long-range forecasts via phone, computer, fax, switched 56.

NOAA/National Weather Service, 1325 East-West Hwy., Silver Spring, MD, 20910. Phone: (301) 713-0622. Fax: (301) 713-1292. Web Site: www.noaa.gov.

Jack F. Kelly, dir; Curtis Carey, pub affrs dir; George Hernandez, editor.

Weather & flood warnings, forecasts & related info for the media & general pub.

Nemo News Service, 7179 Via Maria, San Jose, CA, 95139. Phone: (408) 226-6339. Phone: (800) 243-8433. Fax: (408) 226-6403. E-mail: dickreizner@worldnet.att.net

Dick Reizner,.

On-assignment coverage of news & sporting events for radio & TV stns worldwide.

News Broadcast Network, 451 Park Ave S., New York, NY, 10016. Phone: (212) 684-8919. Phone: (800) 920-6397. Fax: (212) 684-9650. E-mail: info@newbroadcastnetwork.com Web Site: www.newsbroadcastnetwork.com.

Robert Hill, exec producer; Michael J. Hill, pres.

Washington Phone: (703) 893-4577. (202) 638-1603. Fax: (703) 893-6967. (202) 638-1607.

Los Angeles, CA Phone: (909) 621-6903. Fax: (909) 621-9492.

Chicago, IL Phone: (603) 963-4455. Fax: (603) 963-4487.

Seattle, WA Phone: (206) 624-7505. Fax: (206) 624-7556.

Milwaukee, WI Phone: (414) 321-6210. Fax: (414) 321-3608.

Production & distribution of electronic news releases, actualities & pub affrs programs distributed by satellite, telephone & tape.

North American Network, 7910 Woodmont Ave., Suite 1400, Bethesda, MD, 20814. Phone: (301) 654-9810. Fax: (301) 654-9828. Web Site: www.radiospace.com.

Tom Sweeney, pres.

Audio news releases & talk show interviews. On-site coverage for corps, govt agencies & assns.

PA-SportsTicker, 989 6th Ave., 2nd Fl., New York, NY, 10018. Phone: (212) 738-5611. Fax: (212) 695-8560. E-mail: newsroom@sportsticker.com Web Site: www.sportsticker.com.

Jim Morganthaler, gen mgr; Jay Imus, sls dir.

Boston, MA 02210, Boston Fish Pier, West Bldg. #1, Suite 302. Phone: (617) 951-0070. Fax: (617) 737-9960.

New York, NY 10158, 19 E 34th St. Phone: (212) 515-1000. Fax: (212) 515-1211.

Provides 24-hr sports news, info, instant scores & complete sports news coverage on all professional & major college events.

Radio America, 1100 N. Glebe Rd., Suite 900, Arlington, VA, 22201. Phone: (703) 302-1000. Phone: (800) 807-4703. Fax: (571) 480-4140. E-mail: radio@radioamerica.org Web Site: www.radioamerica.org.

Michael Paradiso, COO; Rich McFadden, news dir.

Short & long-form programs, special series, documentaries & daily one hr news show.

Radio Press News Services, 8633 Arbor Dr., El Cerrito, CA, 94530-2728. Phone: (510) 524-9559. E-mail: jag4jl@aol.com

Robert Miles Master, editor-in-chief.

Natl coverage, with special unit for northern California, Bay Area of California & adjacent states, photographer on staff. Multimedia news, Travellands & Vacationland. Special features, articles, transcriptions & video features. TV assignment accepted.

Radio Pulsebeat News, Box 418, Hewlett, NY, 11557. Phone: (212) 686-6850. Fax: (212) 686-7308.

Jay R. Levy, VP.

Gen actuality news service.

RadioTour.com, 2233 Wisconsin Ave. N.W., Washington, DC, 20007. Phone: (202) 333-4904. Fax: (202) 342-5411. Web Site: www.expertclick.com.

Publisher of free *Yearbook of Experts, Authorities & Spokespersons.*

Reuters America Inc., 1333 H St. N.W., Washington, DC, 20005. Phone: (202) 898-8300. Fax: (202) 898-8383. Web Site: www.reuters.com.

Steve Ginsburg, editor-in-charge (bcst svcs & online); Rob Doherty, bureau chief; Mitch Koppelman, VP; David Weissler, assignment editor.

The Reuter Broadcast Report & Reuter Broadcast PLUS, features natl & international news, sports, business news, entertainment & weather.

Skywatch Weather Center, 347 Prestley Rd., Bridgeville, PA, 15017. Phone: (800) SKY-WATCH. Fax: (412) 221-3160. E-mail: airsci@skywatchweather.com Web Site: www.skywatchweather.com.

Dr. Stanley Penkala, pres.

Taped, live & MP3 weathercasts targeted to the listing area, in stn-specified formats. Featuring accuracy, clarity & mature voices.

The Sports Network, 2200 Byberry Rd., Hatboro, PA, 19040. Phone: (215) 441-8444. Fax: (215) 441-5767. Web Site: www.sportsnetwork.com. E-mail: jschonewolf@sportshetwork.com

International real-time sports wire svc providing content, branded web pages, satellite and/or computer feeds directly to broadcasters (radio & TV), print, Internet sites, wireless with state of the art technology.

Studio M Productions, 4032 Wilshire Blvd., Ste 403, Los Angeles, CA, 90010. Phone: (888) 389-7372. Fax: (213) 389-3299. E-mail: senator@sound4film-tv.com Web Site: www.sound4film-tv.com.

Mike Michaels, owner.

Honolulu, HI 96830, 8715 Waikiki Stn. Phone: (888) 389-7372. Fax: (213) 389-3299. E-mail: senator@sound4film-tv.com. Web Site: www.sound4film-tv.com.

Stringers, crew news, sports, features, remote bcsts, engrs, announcers, reporters, equipment for radio, TV, film & video.

Texas State Networks, 4131 N. Central Expwy., Suite 500, Dallas, TX, 75204-2175. Phone: (214) 525-7400. Fax: (214) 525-7372. E-mail: tsnnews@cbs.com Web Site: www.tsnradio.com.

Julius Graw, dir news & opns; Dan Bell, gen sls mgr; Brian Purdy, gen mgr.

Austin, TX 78701. Austin News Bureau, 502 E. 11th, Suite 320. Phone: (512) 474-5264 (NEWS). (512) 474-5275 (SALES). Robert Wood, dir.

News svc of the Texas State Networks. Provides Texas news, sports, agriculture, business & weather, Texas Rangers Radio Network & special features & long form programs.

The Nasdaq Stock Market, 9513 Key West Ave., Rockville, MD, 20850. Phone: (202) 728-8884. Fax: (202) 728-6993. Web Site: www.nasdaqnews.com. E-mail: nasdaqnew@nasdaq.com

Bethany Sherman, sr VP.

Free daily stock market reports tailored for loc & rgnl audiences. Voicers with Nasdaq market data & financial analysis.

Trans World Communications Inc., Box 418, Hewlett, NY, 11557. Phone: (212) 686-6850. Fax: (212) 686-7308.

Jay Levy, VP.

Produces audio news svcs for radio & TV bcstg.

United Press International Inc., 1510 H St. N.W., Washington, DC, 20005. Phone: (202) 898-8000. Fax: (202) 898-8057. E-mail: editorforms@upi.com Web Site: www.upi.com.

Nicholas Chiaia, COO; Christopher Ching, VP finance.

Full global text, audio, photo news , info svcs, morning drive progmg, world, natl news, sports, weather, features & financial reports 24 hours.

VNU Entertainment News Wire, 100 Boylston St., Suite 210, Boston, MA, 02116. Phone: (617) 482-9447. Fax: (617) 482-9562. Web Site: www.vnuenw.com.

Advance news from VNU owned publications.

Views and People in the News (VP News), 1212 Fifth Ave., New York, NY, 10029. Phone: (212) 876-6503. Fax: (212) 876-6503.

Other offices located in Chicago, Los Angeles & San Francisco. News svcs to radio stns currently serving more than 100,000 subs; sells syndicated less than half-hour radio shows.

WINGS: Women's International News Gathering Service, Box 95090, Vancouver, BC, BC V5T 4T8. Canada. Phone: (604) 876-6994. Web Site: www.wings.org. E-mail: wings@wings.org

Frieda Werden, producer.

Austin, TX 78764, Box 33220. Stacy Pettigrew, bureau mgr.

Syndicate audio news & current affrs program, both produced in-house & acquired, focus on women & hard news. Distribution CD, satellite, & FTP.

The Wall Street Journal Radio Network, 335 Madison Ave., 18th Fl., New York, NY, 10017. Phone: (800) 828-6397. E-mail: wsjradio@dowjones.com Web Site: www.wsjradio.com.

Nancy Abramson, exec dir.

Hourly business & financial news reports transmitted live via satellite 18 times daily from the Journal's New York newsroom. Dow Jones Money Report also transmitted 18 times daily.

"The Weather Center", (a broadcast service of Aviation Weather Inc.). 701 Gervais St., Suite 150-224, Columbia, SC, 29201. Phone: (803) 739-2827. E-mail: wxcenter@aviationweatherinc.com Web Site: www.aviationweatherinc.com.

Liam Richard Ferguson, pres.

"Regional Radio Broadcast/Weathercast Network" across the Carolinas and Georgia in over 20 bcst markets. Weather forecasting, site-specific bcst svc for stns all across America. 100% barter.

Weather-One, (A wholly-owned division of Liberty Hill Broadcasting). 31800 Northwestern Hwy., Suite 100, Farmington Hills, MI, 48334. Phone: (248) 737-3000. Fax: (248) 737-3555. E-mail: barryzate@aol.com

Barry Zate, sr VP.

Provides weather forecasting svcs, advanced storm warnings, agricultural & ski info to radio stns.

WeatherData Inc., 245 N. Waco, Suite 310, Wichita, KS, 67202. Phone: (316) 265-9127. Fax: (316) 265-1949. Web Site: www.weatherdata.com. E-mail: ceo@weatherdata.com

Mike Smith, CEO.

Weather radar, graphic & info display systems, training, on air forecast & storm warning svcs. Select Warn, Storm Hawk, 24/7 storm monitoring, & customer svc. Complete system integration & training.

Evan Weiner Productions, Box 1656, Mount Vernon, NY, 10552. Phone: (914) 667-9070. Phone: (203) 288-2597 (producer). Fax: (914) 667-3043. E-mail: evan4256@aol.com Web Site: www.bickley.com/evan_weiner.html.

Evan Weiner, exec producer.

Sports commentaries & reporting. Current program: The Business of Sports, commentaries on Metro Source.

Westwood One, Radio New Services. 40 W. 57th St., New York, NY, 10019. Phone: (212) 641-2000. Fax: (212) 641-2185. Web Site: www.westwoodone.com.

Producer & distributor of radio progmg including CNN, NBC, Mutual, CNBC Business Radio, 24-hours music formats, long-short-form talk, music & news programs.

World Radio Network, Box 1212, London, SW8 2ZF. United Kingdom. Phone: 44-20-7896-9000. Fax: 44-20-7896-9007. E-mail: contactus@wrn.org Web Site: www.wrn.org.

Karl Miosga, chmn; Jeff Cohen, dev dir; Gary Edgerton, mgng dir.

WRN Network One (via Galaxy 5) news & features ch, comprises live progmg segments in English & languages from more than 20 international bcstrs. Also supplies many customized progmg feeds to radio stns.

Radio Format Providers

ABC Radio Networks, 13725 Montfort Dr., Dallas, TX, 75240. Phone: (972) 991-9200. Fax: (972) 448-3378. Web Site: www.abcradionetwork.com.

James Robinson, pres; Julie Atherton, dir mktg.

Bcsts five full-service line nets, Paul Harvey News & Comment, ESPN Radio, long-form progmg, 24-hour formats, ABC News, ABC Sports, & d/wkly features.

Alternative Programming, 4215 Brendenwood Rd., Rockford, IL, 61107. Phone: (815) 229-3995. Fax: (815) 229-5043. E-mail: altprog@sbcglobal.net

Gary A. Knoll, owner.

Complete music formats for radio - current music for various formats - custom CD svc.

American Blues Network, Box 6216, Gulfport, MS, 39506. Phone: (800) 896-5307, ext 117. Fax: (228) 896-5703. E-mail: rip@americianbluesnetwork.com Web Site: www.americianbluesnetwork.com.

Stan Daniels, CEO.

American Comedy Network & Onion Radio News, 91 River St., Milford, CT, 06460. Phone: (203) 877-8210. Fax: (203) 877-8242. E-mail: acn@americancomedynetwork.com Web Site: www.americancomedynetwork.com.

Adrienne Munos, sls; Kurt Luchs, gen mgr; Ben Churchill, producer.

Comedy svc providing daily topical audio sound bites, song parodies, fake comls. Comedy CD, e-mail prep & gold library. Onion Radio News 10 features every week via web and more.

Toby Arnold & Associates, 3234 Commander Dr., Carrollton, TX, 75006. Phone: (972) 661-8201. Phone: (800) 527-5335. Fax: (972) 250-6014. E-mail: toby@taamusic.com Web Site: www.taamusic.com.

Toby Arnold, pres/CEO; Dolly Arnold, VP & COO; Lawrence Mangiameli, VP & dir.

Audio Production libraries for radio. Station Imaging, Morning show promo sweeper, stager packages for all formats. Cash or Barter.

The Beethoven Satellite Network, (BSN Around the Clock). 5400 N. St. Louis Ave., Chicago, IL, 60625. Phone: (773) 279-2112. Fax: (773) 279-2199. E-mail: cmartinez@wfmt.com Web Site: www.wfmt.com.

Steve Robinson, VP/gen mgr.

Satellite-delivered classical mus format svc 24-hours daily serving over 300 outlets nationwide. Produced by WFMT-FM Chicago. Since 1986.

Satellite: Galaxy 6, Digital frequency B72.4.

CBS Radio Networks, 524 West 57th St., New York, NY, 10019. Phone: (212) 975-2044. Fax: (212) 974-0615. E-mail: barobinson@cbs.com Web Site: www.westwoodone.com.

Beth Robinson, VP progmg; Peter Kosann, pres/CEO; Andrew Zaref, CFO.

This division currently offers NFL Football, NCAA Basketball & College Football. Also syndicates *David Letterman's Top Ten List.*

CRN International Inc., One Circular Ave., Hamden, CT, 06514. Phone: (203) 288-2002. Fax: (203) 281-3291. E-mail: info@crnradio.com Web Site: www.crnradio.com.

Barry Berman, pres; S. Richard Kalt, exec VP; Patrick Kane, sr VP; Doug Harris, dir.

Features include *Ski Watch®*, a 60-second, daily ski conditions update. Summer-oriented progmg includes *Beach Watch®* & *Summer Watch.®* Small business programs include the *Small Business Report* & *Small Business Profile.* All programs available on a barter basis.

Christmas Music Networks, 11000 W. 96th Terr., Overland Park, KS, 66214-2258. Fax: (913) 492-7941. Web Site: www.christmasradionetwork.com.

Ross Reagan, pres; John Jessup, VP.

"The best Christmas music progmg available." 36-hours of digitally produced Christmas Eve & Christmas Day progmg. Four different formats: Adult contemp, news/talk, oldies & country.

The Classical Station, WCPE, Box 897, Wake Forest, NC, 27588. Phone: (919) 556-5178. Fax: (919) 556-9273. E-mail: wcpe@wcpe.org Web Site: theclassicalstation.org.

Deborah Proctor, gen mgr; Dick Storck, progmg dir; Rae Weaver, dev dir; Curtis Brothers, dir.

Free 24-hour classical music progmg with live announcers for radio, cable, other distributors. Wkly request programs, opera & features.

Creative Radio Network, Box 7749, Thousand Oaks, CA, 91359. Phone: (818) 991-3892. Fax: (818) 991-3894.

Darwin Lamm, pres/CEO.

Radio music program for A/C—country & modern. Elvis international forum magazine.

Dialogue, One Woodrow Wilson Plaza, 1300 Pennsylvania Ave., N.W., Washington, DC, 20004-3027. Phone: (202) 691-4146. Fax: (202) 691-4141. E-mail: dialogue@wwic.si.edu Web Site: www.wilsoncenter.org/dialogue.

Rachel Edmonds, producer; John Tyler, dir; George Liston Seay, exec producer.

Wkly half-hour program of conversations on natl, internatl affrs, history & culture. Available to pub & coml stns free of charge on CD. Progmg produced by the Woodrow Wilson International Center for Scholars.

Eagle Media Productions Ltd., Box 580, Northford, CT, 06472. Phone: (203) 294-1190. Fax: (203) 294-9512. E-mail: lou.adler@sbcglobal.net Web Site: www.louadler.com.

Louis Adler, pres; Thalia Adler, VP.

Offers *Medical Journal*, 90-second feature-barter; CD delivery.

Excelsior Radio Networks, 220 W. 42nd St., New York, NY, 10036-7202. Phone: (212) 679-3200. Fax: (212) 681-1952. E-mail: robscolaro@aol.com

Michael R. Ewing, pres/COO; Jonathan Goldman, exec VP.

Las Vegas, NV 89119, 1445 E. Tropicana Ave. Phil Hall, gen mgr.

Sports talk 24-hours a day. Sports analysis & commentary on AOL & the internet.

Executive Broadcast Services, 30 Mobray Ct., Colorado Springs, CO, 80906. Phone: (719) 579-6676. Fax: (719) 579-6664. E-mail: skip@executivebroadcast.com

Skip Joeckel, pres.

Markets & sells a select line of programs, products & svcs to U.S. radio stns.

Far West Communications Inc., 2401 Rockdell St., La Crescenta, CA, 91214-1738. Phone: (818) 248-2400. Fax: (818) 248-2596. E-mail: farwestinc@aol.com

Paul J. Ward, gen mgr; Skip Joeckel, mktg dir; Paul J. Ward, pres; Ron Blassnig, chief engr.

Formats: *Gold Plus, 30 Plus, True Country, True Country II, Modern MOR.* Svcs: MASTERDISC, custom song libraries on CD; delivery on analog tape, CD, DAT.

Fischer Broadcast Services, 10841 Bittersweet Lane, Fishers, IN, 46038-2203. Phone: (317) 514-5757. Fax: (317) 578-3884. Web Site: www.superfisch.com.

Scott Fischer, pres.

SUPERFISCH—THE PROMO VOICE SUPERHERO. Scott Fischer, signature voice artist for DirecTV Sports, Fox Sports Net Networks, and other fine affiliates. Myriad reads—always right! ISDN/MP3.

Ghostwriters/Radio Mall, 2412 Unity Ave. N., Dept BR, Minneapolis, MN, 55422-3450. Phone: (800) 759-4561. Fax: (763) 522-6256. E-mail: info@radiomall.com Web Site: www.radiomall.com.

David Dworkin, owner.

Over 21 years experience with products sold to more than 7,200 radio stns worldwide as well as TV stns, audio-video producers & cable operators. If you have a finished product that you'd like to mkt to radio stns, contact us. We offer a 60-day money-back guarantee.

Hispanic Communications Network, 1126 16th St., N.W., Suite 350, Washington, DC, 20036. Phone: (202) 637-8800. Fax: (202) 637-8801. E-mail: info@hcnmedia.com Web Site: www.hcnmedia.org.

Jeff Kline, chmn/CEO.

Produces six daily Spanish radio programs & distributes them to Hispanic Radio net affiliates.

J.N. Productions, 902-1790 Bayshore Dr., Vancouver, BC, V6G 3G5. Canada. Phone: 604-331-0690. E-mail: jnproductions@telus.net Web Site: www.jnproductions.bc.ca.

Jakob Nortman, pres.

For all your voice-over needs, including narration, corporate videos, on-hold telephone messages & announcements for GPS systems. Radio production facilities available.

Jameson Broadcast Inc., 1644 Hawthorne St., Sarasota, FL, 34239. Phone: (941) 906-8800. Fax: (941) 906-8801. E-mail: jamie@jamesonbcast.com Web Site: www.jamesonbroadcast.com/radio.

Specializes in short-form entertainment, info programs & promotions.

Jones MediaAmerica Inc., 1133 Ave. of the Americas, 11th Fl., New York, NY, 10036. Phone: (212) 302-1100. Fax: (212) 556-9575. E-mail: dbrown@mediaamerica.com Web Site: www.mediaamerica.com.

Catherine Csukas; Gary Schonfeld, pres; Susan Love, VP sls.

Troy, MI 48084, 100 W. Big Beaver Rd, Suite 300. Phone: (248) 526-3002. Ken Alandt, mgng dir, midwest region.

Sherman Oaks, CA 91403, 15233 Ventura Blvd, Penthouse 8. Phone: (818) 986-8500. Brenda Holland, mgng dir western rgn; Daniel Depercin, natl account mgr.

Marathon Shores, FL 33050, 11399 Overseas Hwy. Phone: (305) 289-4524. Judy Langley, natl account mgr S.E.

Chicago, IL 60611, 401 N. Michigan Ave, Suite 1200. Phone: (312) 840-8260. Dave Simon, natl acc mgr.

Nashville, TN 37203, 49 Music Sq. W, Suite 301. Phone: (615) 327-7539. John Alexander, dir mus mktg.

MediaAmerica is the largest independent marketer of natl radio programs in the United States. MAI represents over 70 radio programs servicing the following formats: AOR, modern rock, CHR, urban, country, AC, news info, class, jazz/new age, talk, 24-hour satellite delivered & sports progmg.

Jones Radio Networks, (A Jones Media Networks Company). 8200 S. Akron St., Suite 103, Centennial, CO, 80112. Phone: (303) 784-8700. Fax: (303) 784-8612. E-mail: pcrocker@joneradio.net Web Site: www.jonesradio.com.

Phil Barry, VP/gen mgr; Patrick Crocker, VP; Amy Bolton, VP.

Washington, DC 20003, 418 10th St. S.E. Phone: (800) 611-5663. Fax: (202) 546-8435. Amy Bolton, VP/gen mgr-news/talk.

Centennial, CO 80112, 8200 S. Akron St, Suite 103. Phone: (303) 784-8700. Fax: (303) 784-8612. Phil Barry, Group VP/gen mgr.

New York, NY 10036, 1133 Avenue of the Americas. Phone: (888) 644-8255. Fax: (212) 556-9529. Gary Schonfeld, president.

Seattle, WA 98121, 3131 Elliot Avenue, #770. Phone: (800) 426-9082. Fax: (206) 441-6582. Susan Stephens, VP/gen gr.

Premier producer of 24-hr format progmg, daypart personality, talk programs & various music svcs to radio stns.

Launch Radio Networks, (a division of United Stations Radio Network). 1065 Ave. of the Americas, 3rd Fl., New York, NY, 10018. Phone: (212) 536-3600. Fax: (212) 536-3601. Web Site: www.launchradionetworks.com. E-mail: ccolombo@launchradionetworks.com

Dave Ankers, dir opns; Charlie Colombo, VP/gen mgr.

Launch Radio Networks produces, distributes music, entertainment news & svcs for radio stns as well as other media worldwide.

MRN Radio (Motor Racing Network), 1801 W. International Speedway Blvd., Daytona Beach, FL, 32114. Phone: (386) 947-6400. Fax: (386) 947-6716. E-mail: mrn-radio@mrnnet.com Web Site: www.mrnradio.com.

Live bcsts of NASCAR stock car racing & related programs via satellite.

J J McKay Productions Inc., 800 Penn St., Denver, CO, 80227. Phone: (303) 980-1948.

J.J. McKay, pres.

Maximum-impact production and versatile voice-over talent. All formats. Choose the voice for today AND tomorrow! Delivered via analog tape, DAT, ISDN/Zephyr, MP3, DCI.

Miller Broadcast Management, 616 W. Fulton St., Suite 516, Chicago, IL, 60661. Phone: (312) 454-1111. Fax: (312) 454-0044. Web Site: info@millerbroadcast.com.

Lisa Miller, pres; Matt Miller, VP.

Musical Starstreams, Box 12685, LaJolla, CA, 92039-2685. Phone: (619) 276-8989. E-mail: forest@starstreams.com Web Site: www.starstreams.com.

Musical Starstreams is a wkly two-hour program of "exotic electronica" targeted to adults age 25-54.

Orange Productions, 523 Righters Ferry Rd., 1st Fl., Bala Cynwyd, PA, 19004. Phone: (610) 667-8620. Fax: (610) 667-8939. E-mail: orange@snip.net Web Site: www.soundsofsinatra.com.

Sid Mark, pres; Jon Harmelin, VP/gen mgr.

Production & distribution of a wkly two-hour program *Sounds of Sinatra*.

Premiere Radio Networks Inc., 15260 Ventura Blvd., 4th Fl., Sherman Oaks, CA, 91403-5339. Phone: (818) 377-5300. Fax: (818) 377-5333. E-mail: webmaster@premradio.com Web Site: www.premrad.com.

Kraig T. Kitchin, pres/COO.

Sherman Oaks, CA 91403, 15260 Ventura Blvd., 5th Fl. Phone: (818) 377-5300. Theresa Gage, sr VP/western sls.

Atlanta, GA 30305, 3405 Piedmont Rd, Suite 500. Phone: (404) 870-5070. Alexandra Fenech, VP/southern sls.

Chicago, IL 60611, 875 N. Michigan Ave, Suite 1450. Phone: (312) 266-3870. Susan McDonald , VP/Chicago rgnl sls mgr.

Royal Oak, MI 48067, 306 S. Washington Ave, Suite 214.

New York, NY 10020, 1270 Ave. of the Americas. Phone: (212) 445-3900. Catherine Mongarella, sr VP/eastern sls.

Dallas, TX 75240, 14001 N. Dallas Pkwy, Suite 500. Phone: (972) 239-6220. Jeff Steele. sr VP/natl music syndication.

Premiere Radio features the following personalities: Rush Limbaugh, Delilah Ryan Seacrest, Steve Harvey, Jim Rome, Glenn Beck, Ty Pennington, Blair Garner, Whoopi Goldberg, Dr. Laura Schlessinger, Maria Bartiromo, George Noory, Casey Kasem, Ben & Brian, Bill Handel, Bob (Kevoian) & Tom (Griswold), Jeff Foxworthy, Jay Leno, Hohn Boy & Billy, T.D. Jakes, Matt Drudge, Big Tigger, Art Bell, Big D & Bubba, & Dr. Dean Edell.

RPM Radio Programming and Management Inc., 1133 West Long Lake Rd., Bloomfield Hills, MI, 48302. Phone: (800) 521-2537. Phone: (248) 647-1068. Fax: (248) 647-3936. E-mail: info@tophitsusa.com Web Site: www.tophitsusa.com.

Thomas M. Krikorian, pres.

Top Hits USA wkly CD svc & CD libraries including Solid Gold, Spectrum A/C & Country One. Classic rock, CD Christmas library.

Radio Center for People with Disabilities (RCPD), 230 E. Ohio St., Suite 101, Chicago, IL, 60611. Phone: (312) 640-5000. Fax: (312) 640-5010. E-mail: rc4pd@aol.com Web Site: www.rcpd.org.

Brad Saul, CEO.

PCPD is a non-profit agency founded to recruit, train & place people with disabilities in paying, off-air jobs in the radio business.

Radio Express Inc., 1415 W. Magnolia Blvd., Suite 201, Burbank, CA, 91506. Phone: (818) 295-5800. Fax: (818) 295-5801. E-mail: radioinfo@radioexpress.com Web Site: www.radioexpress.com.

Tom Rounds, CEO; John Fleck, pres.

Distributors outside the United States: *The World Chart Show, Rick Dees Weekly Top 40, Country Countdown, Hot Mix, Hitdisc, Golddisc, Supercharger Production tool kit* & *Production libraries by firstcom music*. Programs available by cash or barter, libraries & products cash only.

Radio Spirits, Box 3107, Cedar Knolls, NJ, 07927. Phone: (973) 539-7557. Fax: (847) 524-8245. Web Site: www.radiospirits.com.

Hakan Lindskog, pres.

Radio producers of the nationally-syndicated old time radio program *When Radio Was*. Complete digital recording studio features Sonic Solutions Digital Work Station with No-Noise.

SFX Radio Network, Clear Channel Entertainment, 220 W. 42nd St., New York, NY, 10036. Phone: (917) 421-4000. Web Site: www.sfxnet.com.

Offers both wkly & mthy shows featuring classic rock, country, urban contemp & live concerts in addition to stn prep svcs.

Salem Music Network Inc., (A division of Salem Music Networks/Salem Radio Network). 402 BNA Dr., Suite 400, Nashville, TN, 37217. Phone: (615) 367-2210. Fax: (615) 367-0758. E-mail: info@salemmusicnetwork.com Web Site: www.salemmusicnetwork.com.

Michael S. Miller, gen mgr.

Provider of three different 24-hour Christian music formats via digital satellite to 230+ radio stns throughout the US, Canada, and operator of two greater Nashville (TN) radio stations.

Sheridan Broadcasting Corp., 960 Penn Ave., Suite 200, Pittsburgh, PA, 15222. Phone: (412) 456-4008. Phone: (800) 456-4211. Fax: (412) 456-4040 (progmg). Fax: (412) 457-4077 (admin).

Ronald Davenport, chmn.

Provides hourly news & sports, longform talk & mus progmg as well as *USA Music Magazine*. Other alternative progmg includes *Coming Soon* movie review, *Straight Up* with Bev Smith & *White House Report* with White House correspondent April Ryan.

Sirius Satellite Radio, 1221 Ave. of the Americas, 36th Fl., New York, NY, 10020. Phone: (212) 584-5100. Fax: (212) 584-5200. E-mail: rshnall@siriusradio.com Web Site: www.sirius.com.

Rebecca Schnall, media rel.

Southcott Productions, Box 33185, Granada Hills, CA, 91343. Phone: (818) 368-4938. Fax: (818) 368-4938. E-mail: info@musicofyourlife.com Web Site: www.musicofyourlife.com.

Chuck Southcott, owner.

North Hollywood, CA 91602, 4605 Lankershim Blvd, Suite 702. Phone: (818) 755-9952. Chuck Southcott, program dir.

Adult pop standards, progmg hqtrs for *Music of Your Life Radio Network*. Talent includes: Gary Owens, Wink Martindale, Peter Marshall, Pat Boone, Johnny Magnus, Ken Young & Les Brown Jr.

TM Century Inc., 2002 Academy, Dallas, TX, 75234. Phone: (972) 406-6800. Fax: (972) 406-6890. E-mail: tmci@tmcentury.com Web Site: www.tmcentury.com.

David Graupner, pres/CEO; Eve Mayer Orsburn, VP Sales & Marketing; Erik Hastings, exec producer.

GoldDisc music libraries, HitDisc wkly mus svc, music on hard drive, jingles, mus libraries production, special programs, CD-ROM.

Talkline Communications Network, Box 20108, Park West Station, New York, NY, 10025-1510. Phone: (212) 769-1925. Fax: (212) 799-4195. E-mail: tcntalk@aol.com Web Site: www.talkline communication.com.

Zev J. Brenner, pres.

National Jewish radio net featuring news, interviews with newsmaker guests & celebrities; live call-in format; live segments from Israel; satellite delivered. Available on barter.

24 Karat Productions Inc., 1717 W. Sanderling Ln., Port St. Lucie, FL, 34982. Phone: (772) 465-5511. Fax: (772) 465-5511. Web Site: www.marg101z@bellsouth.net.

Mark Prichard, pres/CEO; Gloria Prichard, exec VP.

Music 1, classic pop, over 10,000 stereo selections w/mthy updates for radio, offices; on tape or CD, consultant svcs.

United Press International, 1510 H St. N.W., Washington, DC, 20005. Phone: (202) 898-8111. Phone: (202) 898-8100. Fax: (202) 898-8057. E-mail: editorforms@upi.com Web Site: www.upi.com.

Tobin C. Beck, news dir; Michael Marshall, editor.

Full-svc company offers a number of short-form info features to its affil radio stns.

Virtual Radio, Box 579, 4521 Campus Dr., Irvine, CA, 92612. Phone: (949) 752-9237. Fax: (949)752-9456. E-mail: joel@virtualradio.com Web Site: www.vradio.com.

Joel Easton, mgng dir.

Virtual Radio is the oldest music website providing a new radio format, content & internet expertise to bcstrs world wide.

WFMT Fine Arts Radio, 5400 N. St. Louis Ave., Chicago, IL, 60625. Phone: (773) 279-2000. Fax: (773) 279-2199. E-mail: guide@networkchicago.com Web Site: www.wfmt.com.

Daniel Schmidt, pres/CEO; Peter Whorf, progmg dir.

Classical, opera & folk mus, news & fine art progmg 24-hours per day through United Video Inc. Serving 200 cable systems in 30 states with 850,000 subs.

WFMT Radio Network/BSN Around the Clock, 5400 N. St. Louis Ave., Chicago, IL, 60625. Phone: (773) 279-2000. Fax: (773) 279-2199. Web Site: www.wfmt.com.

Steve Robinson, sr VP.

Provides 24-hour-a-week classical music program format svc via satellite to more than 500 stns; programs in one-hour modules with program host & loc sound.

"The Weather Center", (a broadcast service of Aviation Weather Inc.). 701 Gervais St., Suite 150-224, Columbia, SC, 29201. Phone: (803) 739-2827. E-mail: wxcenter@aviationweatherinc.com Web Site: www.aviationweatherinc.com.

Liam Richard Ferguson, pres.

"Regional Radio Broadcast/Weathercast Network" across the Carolinas and Georgia in over 20 bcst markets. Weather forecasting, site-specific bcst svc for stns all across America. 100% barter.

Westwood One, Radio New Services. 40 W. 57th St., New York, NY, 10019. Phone: (212) 641-2000. Fax: (212) 641-2185. Web Site: www.westwoodone.com.

Producer & distributor of radio progmg including CNN, NBC, Mutual, CNBC Business Radio, 24-hours music formats, long-short-form talk, music & news programs.

World Radio Network, Box 1212, London, SW8 2ZF. United Kingdom. Phone: 44-20-7896-9000. Fax: 44-20-7896-9007. E-mail: contactus@wrn.org Web Site: www.wrn.org.

Karl Miosga, chmn; Jeff Cohen, dev dir; Gary Edgerton, mgng dir.

WRN Network One (via Galaxy 5) news & features ch, comprises live progmg segments in English & languages from more than 20 international bcstrs. Also supplies many customized progmg feeds to radio stns.

Music Licensing

APM/Associated Production Music, 6255 Sunset Blvd., Suite 820, Hollywood, CA, 90028. Phone: (323) 461-3211. Phone: (800) 543-4276. Fax: (323) 461-9102. E-mail: accountservices@apmmusic.com Web Site: www.apmmusic.com.

Connie Red, sls dir.

New York, NY 10016, 240 Madison Ave., 11th Fl. Phone: (800) 276-6874. (212) 856-9800. Fax: (212) 856-9807. George Macisa, natl sls mgr.

Sixteen libraries: KPM, Bruton, Sonoton, Carlin, Castle, NFL. Over 3,000 CDs, personalized packages, music search, 15-20 new CD releases mthy.

American Society of Composers, Authors & Publishers (ASCAP), One Lincoln Plaza, New York, NY, 10023. Phone: (212) 621-6000. Fax: (212) 724-9064. E-mail: info@ascap.com Web Site: www.ascap.com.

Marilyn Bergman, chmn/pres.

London W1X 1PB. ASCAP - London, 8 Cork St. Phone: 01-44-207-439-0909. Fax: 001-44-207-434-0073.

Los Angeles, CA 90046. ASCAP - Los Angeles, 7920 W. Sunset Blvd., 3rd Fl. Phone: (323) 883-1000. Fax: (323) 883-1049.

Miami Beach, FL 33139. ASCAP - Miami, 420 Lincoln Rd, Suite 385. Phone: (305) 673-3446. Fax: (305) 673-2446.

Atlanta, GA 30318. ASCAP - Atlanta, PMB 400, 541 Tenth St. N.W. Phone: (404) 351-1224. Fax: (404) 351-1252. (Not a membership office).

Chicago, IL 60647. ASCAP - Midwest, 1608 N. Milwaukee Ave, Suite 1007. Phone: (773) 394-4286. Fax: (773) 394-5639.

Nashville, TN 37203. ASCAP - Nashville, 2 Music Sq. W. Phone: (615) 742-5000. Fax: (615) 742-5020.

A membership assn of more than 275,000 composers, lyricists, & music publishers, ASCAP licenses the pub performances of its members' works. ASCAP has reciprocal agreements with foreign societies representing virtually every country that has laws protecting copyright.

BMI-Broadcast Music Inc., 320 W. 57th St., New York, NY, 10019. Phone: (212) 586-2000. Fax: (212) 489-2368. Web Site: www.bmi.com.

Del Bryant, pres/CEO; John E. Cody, COO.

Miami, FL 33126, 5201 Blue Lagoon Dr, Suite 310. Phone: (305) 266-3636.

Atlanta, GA 30326, Tower Pl. 100, 3340 Peachtree Rd. N.E., Suite 570. Phone: (404) 261-5151.

Hato Rey, PR 00917, Bank Trust Plaza, 255 Ponce de leon Ave, East Wing, Suite A-262. Phone: (787) 754-6490.

London, NO NWISHN United Kingdom, 84 Harley House, Marlebone Rd. Phone: 0114420 7486 2036.

Los Angeles, CA 90069, 8730 Sunset Blvd. Phone: (310) 659-9109.

Nashville, TN 37203, 10 Music Sq. E. Phone: (615) 401-2000.

Licenses the pub performance rights of musical compositions for more than 300,000 songwriters, composers & music publishers; maintains reciprocal arrangements with more than 40 licensing organizations worldwide.

European American Music Distributors L.L.C., 35 E. 21st St., Fl. 8, New York, NY, 10010. Phone: (212) 358-4999. Fax: (212) 871-0237. E-mail: info@eamdllc.com Web Site: www.eamdllc.com.

Music publisher & distributor.

The Harry Fox Agency Inc., 711 3rd Ave., New York, NY, 10017. Phone: (212) 370-5330. Fax: (646) 487-6779. Web Site: www.harryfox.com. E-mail: press@harryfox.com

Music licensing.

SESAC Inc., 55 Music Sq. E., Nashville, TN, 37203. Phone: (615) 320-0055. Fax: (615) 329-9627. E-mail: dhoughton@sesac.com Web Site: www.sesac.com.

London W1H 3FF, 67 Upper Berkeley St. Phone: 020-7616-9284. Fax: 020-7563-7029. Wayne Bickerton.

New York, NY 10019, 152 W. 57th St., 57th Fl. Phone: (212) 586-3450. Fax: (212) 489-5699. Deb Houghton, VP bcst licensing.

Performing rights organization representing a diversity of copyrighted music.

Society of Composers, Authors & Music Publishers of Canada (SOCAN), Societe Canadienne des auteurs, compositeurs et editeurs de musique. 41 Valleybrook Dr., Toronto, ON, M3B 2S6. Canada. Phone: (416) 445-8700. Phone: (866) 307-6226. Fax: (416) 445-7108. E-mail: socan@socan.ca Web Site: www.socan.ca.

Dartmouth B2Y 2N6, Queen Sq., 45 Alderney Dr., Suite 802. Phone: (902) 464-7000. Fax: (902) 464-9696.

Edmonton T6H 5P9, 1145 Weber Centre, 5555 Calgary Tr. Phone: (780) 439-9049. Fax: (780) 432-1555.

Montreal H3A 3J2, 600, boul. de Maisonneuve Ouest, Bureau 500. Phone: (514) 844-8377. Fax: (514) 849-8446.

Vancouver V6E 2V2, 1201 W. Pender St, Suite 400. Phone: (604) 669-5569. Fax: (604) 688-1142.

SOCAN licenses the public performance of music in Canada & distributes performance royalties to copyright holders worldwide.

Warner Bros. Publications, 15800 N.W. 48th Ave., Miami, FL, 33014. Phone: (305) 620-1500. Fax: (305) 621-4869. Fax: (305) 621-1094. Web Site: www.warnerbros.com.

Full-line music publishers of popular, standard & educ music as well as instructional videos from influential musicians. International market.

Canadian Broadcast Networks

Astral Television Networks Inc.

Head Office: Astral Media Inc., 2100, rue Sainte-Catherine Ouest, Bureau 1000, Montreal PQ H3H 2T3. (514) 939-5000. Fax: (514) 939-1515. Web site: www.astralmedia.com.

Television Division (English): BCE Place, 181 Bay Street, Box 787, Suite 100, Toronto, ON M5J 2T3. (416) 956-2010. Fax: (416) 956-2018.

Television Division (French): 2100 rue Sainte-Catherine Ouest, Bureau 900, Montreal, PQ H3H 2T3. (514) 939-5090. Fax: (514) 939-5098.

Management: John Riley, pres; Domenic Vivolo, sr VP sls/mktg; Kevin Wright, sr VP progmg; Deborah Wilson, VP communications; Chris Bell, VP technology; John Pow, VP finance admin; Eric Malette, VP human resources; Megan O'Neal, dir business/legal affrs; Alicia Barin, VP strategic planning; Stephen Green, VP sls/affiliate mktg; Bill Custers, dir opns; Marni Goldman, mgng producer; Kira Murdock, rsch advisor; Roopa Shah, communications mgr.

Television: The Movie Network; Speciality TV; Super Ecran; Mpix; Canal Vie; VRAK.TV; Z; Canal D; Family; Viewer's Choice; TELETOON; Canal Indigo; Canada Cinepop.

CTV Inc., a division of Bell Globemedia

Head Office: 9 Channel Nine Ct., Scarborough, ON, Canada M1S 4B5. (416) 332-5000. Fax: (416) 332-5022. Web site: www.ctv.ca.

CTV Executives: Ivan Fecan, CEO, CTV Inc.; Rick Brace, pres; Paul Sparke, sr VP corporate/pub affrs; Scott Henderson, sr dir communication; Andre Serero, legal/corporate sec, Bell Globemedia.

CTV Sales Offices

Montreal, PQ H2K 4R2: 1010 Sherbrooke Street, Suite 1803. (514) 529-2105. Fax: (514) 521-0102.

Vancouver, BC V6Z 1X5: 750 Burrard Street, (604) 608-2868. Fax: (604) 609-5796.

New York, NY 10017: Telerep, One Dag Hammarskjold Plaza, 25th Fl. (212) 759-8787. Fax: (212) 486-8746.

Canadian Broadcasting Corp.

The Canadian Broadcasting Corp. (CBC) is a publicly owned corporation established by the Broadcasting Act (1936) of the Canadian Parliament to provide the natl bcstg svc in Canada in the two official languages English & French. Under this legislation, the CBC is subject to regulations of the Canadian Radio-Television & Telecommunications Commission (CRTC).

Program Services: The progmg on CBC networks is nearly all Canadian and virtually free of coml adv. Newsworld is a 24-hour natl satellite to cable English-language news & info svc. Le Réeseau de l'information (RDI) is a 24-hour natl satellite-to-cable French-language news & info svc. The heart of CBC's natl distribution system is Canada's Anik E2 satellite, carrying progmg through six different time zones.

Head Office: 181 Queen Street, Box 3220, Station C, Ottawa, ON K1P 1K9. (613) 288-6000. TDD: (613) 288-6455. E-mail: liaison@cbc.ca.

CBC Board of Directors: Robert Rabinovitch, pres/CEO; Guy Fournier, C.M., chmn/bd of dir; Bernd Christman, CEO.

Principal Officers: Robert Rabinovitch, pres/CEO; Richard Stursberg, exec VP, CBC Television; Sylvain Lafrance, exec VP, French svcs; Jane Chalmers, VP, CBC Radio; Pierre Nollet, VP, gen counsel/corporate sec; George Smith, sr VP, human resources/organization; Johanne Charbonneau, VP/CFO; Raymond Carnovale, VP/chief technology officer; Michel Tremblay, VP, strategy/business dev; William Chambers, VP/communications; Michel Saint-Cyr, pres/real estate division.

CBC Ombudsmen: English svcs, Canadian Broadcasting Corporation, Box 500, Station A, Toronto, ON M5W 1E6; E-mail: ombudsman@cbc.ca. Web Site: www.cbc.ca/obudsman. Renaud Gilbert, French svcs, 1400 Rene-Levesque Boulevard East, Box 6000, Montreal, PQ H3C 3A8; E-mail: ombudsman@radio-canada.ca. Web Site: www.radio-canada-ca/obudsman.

English Services: 250 Front Street West, Box 500, Station A, Toronto, ON M5W 1E6. (416) 205-3311. TDD: (416) 205-6688; E-mail: cbcinput@cbc.ca.

French Services: 1400 Rene-Levesque Boulevard East, Box 6000, Montreal, PQ H3C 3A8. (514) 597-6000. TDD: (514) 597-6013; E-mail: auditoire@radio-canada.ca.

Newfoundland Region (English svcs): Radio bldg.: 25 Henry Street, Television bldg.: 95 University Ave., Box 12010, Station A, St. John's, NF A1B 3T8. (709) 576-5000.

Maritime Region (English svcs): Radio bldg.: 5600 Sackville Street, Television bldg.: 1840 Bell Street, Box 3000, Halifax, NS B3J 3E9. (902) 420-8311.

Atlantic Provinces (French svcs): 250 Universite Ave., Box 950, Moncton, NB E1C 8N8. (506) 853-6666.

Quebec Region (English svcs): 1400 Rene-Levesque Boulevard East, Box 6000, Montreal, PQ H3C 3A8. (514) 597-6000.

Quebec City & Eastern Quebec Region (French svcs): 888 St-Jean Street, Box 18800, QC City, QC G1K 9L4. (418) 654-1341.

Ontario Region (English svcs): 250 Front Street West, Box 500, Station A, Toronto, ON M5W 1E6. (416) 205-3311. TDD: (866) 220-6045.

Ontario Region (French svcs): 181 Queen Street, Box 3220, Station C, Ottawa, ON K1P 1K9. (613) 724-1200. TDD: (613) 288-6455.

Manitoba Region (English & French svcs): 541 Portage Ave., Box 160, Winnipeg, MB R3C 2H1. (204) 788-3222.

Saskatchewan Region (English & French svcs): 2440 Broad Street, Box 540, Regina, SK S4P 4A1. (306) 347-9540.

Alberta Region (English & French svcs): 10062-102nd Ave., Room 123, Edmonton City Centre, Box 555, Edmonton, AB T5J 2P4. (780) 468-7500.

British Columbia Region (English & French svcs): Box 4600, Vancouver, BC V6B 4A2. (604) 662-6000.

CBC North: 5129 49th Street, Box 160, Yellowknife, NT X1A 1P8. (867) 920-5400.

Global Television Network

Head Office: 81 Barber Greene Rd., Toronto, ON M3C 2A2. (416) 446-5311. Fax: (416) 446-5449. Web site: www.canada.com/globaltv.

CanWest Global Executive Management: Derek Burney, O.C., chmn; Leonard Asper, pres/CEO; Peter Viner, pres/CEO, Canadian opns; David Asper, exec VP/chmn National Post; John Mcguire, CFO; Thomas Strike, pres, CanWest MediaWorks International; Richard Leipsic, sr VP/gen counsel; Gail Asper, corporate sec; John Culligan, sr VP/corporate dev; Grace Palombo, sr VP/human resources.

TVA

Head Office: Groupe TVA Inc., 1600 de Maisonneuve Boulevard. E., Montreal, PQ H2L 4P2. (514) 526-2951; Fax: (514) 598-6086. Web site: www.tva.ca.

Executive: Pierre Dion, pres/CEO.

TVA Owned: CFCM-Quebec City; CHLT-Sherbrooke; CHEM-Trois-Rivieres; CFER-Rimouski; CJPM-Saguenay.

Regional Affiliates: CHAU-Carleton; CIMT-Riviere-du-Loup; CFEM-Rouyn; CHOT-Gatineau.

Television Quatre Saisons

Head Office: Corp Owner: TQS Inc., 612 rue Saint-Jacques, Montreal, PQ H3C 5R1. (514) 390-6035. Web site: www.tqs.ca.

Principal Officers: M. Rene Guimond, pres/CEO; Mme. Marie Carrier, dir corporate communications; Mme. Therese David, VP communications; Mme. Monique Lacharite, VP finance/admin; M. Guy Meuier, VP sls; M. Richard Roy, gen mgr operations.

Affiliates: CFJP Montreal, PQ; CFAP Quebec City, PQ; CFRS Saguenay, PQ; CFKS Sherbrooke, PQ; CFKM Trois Rivieres, PQ; CFTF Riviere Du Loup, PQ; CFVS Val d'Or, PQ; CFGS Gatineau, PQ.

Canadian Cable Networks

ARTV, 1400 boul. Rene-Levesque Est, Bureau A-53-1, Montreal, PQ, H2L 2M2. Canada. Phone: (514) 597-3636. Fax: (514) 597-3633. Web Site: www.artv.ca. Marie Cote, gen mgr; Jacinthe Brisebois, production mgr; Catherine Dupont, opns mgr; Luc Leblanc, Artistic Dir.; Gilbert Morin, comptroller; Gilles Desjardins, Head & Distribution; Marc Pichette, Head & Communications.

ARTV is a French-language channel dedicated entirely to arts and culture. Twenty-four hours a day of great performances, films, documentaries, dramas & design. The pleasure of capturing the art and culture of Quebec, Canada & the whole world.

Aboriginal Peoples Television Network, 339 Portage Ave., Winnipeg, MB, R3B 2C3. Canada. Phone: (204) 947-9331. Fax: (204) 947-9307. E-mail: info@aptn.ca Web Site: www.aptn.ca. Jean LaRose, CEO.

Alliance Broadcasting, (dba Showcase TV & History TV). 121 Bloor St. E., Toronto, ON, M4W 3M5. Canada. Phone: (416) 967-1174. Fax: (416) 960-0971. Web Site: www.allianceatlantis.com. Phyllis Yaffe, pres/CEO.

Best of Canadian & international TV series & movies. Serving 5 million subs on 100 cable systems. Satellite: ANIK-E2.

Atlantic Satellite Network (ASN), 2885 Robie St., Halifax, NS, B3K 5Z4. Canada. Phone: (902) 453-4000. Fax: (902) 454-3302. E-mail: bt@ctv.ca Web Site: www.ctv.ca. Rich Marchand, gen sls mgr; L. Wartman, opns dir.

Movies & news, educ programs weekend mornings.

Serves 51 cable systems. Satellite: Anik C-1.

BookTelevision: The Channel, 299 Queen St. W., Toronto, ON, M5V 2Z5. Canada. Phone: (416) 591-5757. E-mail: info@booktelevision.com Web Site: www.booktelevision.com.

BookTelevision: The Channel spotlights all the writing that informs & entertains us in our daily lives.

Bravo!, 299 Queen St. W., Toronto, ON, M5V 2Z5. Canada. Phone: (416) 591-5757. Fax: (416) 591-8497. E-mail: bravomail@bravo.ca Web Site: www.bravo.ca.

Bravo! NewStyle Arts Channel is dedicated to entertaining, stimulating & enlightening veiwers who have a taste for more complex TV. Bravo! delivers a wide array of fine arts progmg, balancing longer-form structured shows & shorter pieces that appear in a more random way as "flow" to create a fluid mix of distinctive music, dance, opera, drama, literature, cinema, visual art, the art of TV & the art of talk. Serving 5.8 million subs on 700 cable systems.

CBC Newsworld, Box 500, Station A, Toronto, ON, M5W 1E6. Canada. Phone: (416) 205-2409. Fax: (416) 205-8684. Web Site: www.newsworld.com. E-mail: maria_mirowicz@cbc.ca Maria Mirowicz, program dir.

Live 24-hours news & info net on basic cable, satellite & wireless in Canada.

On 1500 cable systems serving 8 million subs. Satellite: Anik E2 (Ku-band).

CPAC-Cable Public Affairs Channel, (A subsidiary of Consortium of Canadian Cable Companies). 1750-45 O'Connor St., Ottawa, ON, K1P 1A4. Canada. Phone: (613) 567-2722. Fax: (613) 567-2741. E-mail: comments@cpac.ca Web Site: www.cpac.ca.

Uncut, unfiltered coverage of Canadian pub affrs issues including LIVE bcsts of the House of Commons & its Standing Committees. Serving 7.2 million subs.

CTV Newsnet, Box 9, Station "O", Toronto, ON, M4A 2M9. Canada. Phone: (416) 332-5000. Fax: (416) 291-5337. E-mail: news@ctv.ca Web Site: www.ctv.ca. Jana Juginovic, progmg dir & news dir.

Continually updated headline news, business, sports, weather & entertainment, every 15 minutes.

Canadian Learning Television (CLT), 299 Queen St. W., Toronto, ON Canada. Phone: (416) 591-5757. Phone: (780) 440-7777. Web Site: www.clt.ca.

Adult education stn offering access to accredited opportunities, including university & college credit courses, personal dev training & job opportunities.

Canadian Satellite Communications Inc. (CANCOM), 2055 Flavelle Blvd., Mississauga, ON, L5K 1Z8. Canada. Phone: (905) 403-2020. Fax: (905) 403-2022. Web Site: www.cancom.ca. Don Fletcher, VP.

Expert in evaluating, selecting, integrating & implementing satellite-based solutions for business. Cancom operates in four main lines of business: broadcast solutions, tracking solutions, learning solutions & data solutions.

Le Canal Nouvelles, 1600 boul. de Maisonneuve est, Montreal, PQ, H2L 4P2. Canada. Phone: (514) 598-2869. Fax: (514) 598-6037. Web Site: www.tva.canoe.ca. Martin Cloutier, gen mgr.

Les Chaines Tele Astral, Les Chaines Tele Astral, Une division d'Astra Media, 2100, Ste-Catherine St. W., Rm. 700, Montreal, PQ, H3H 2T3. Canada. Phone: (514) 939-3150. Fax: (514) 939-3151. Web Site: www.astral.com. Pierre Roy, pres; Johanne Saint-Laurent, VP.

Progmg includes Super cran, the Fr pay-TV stn; Canal Famille, children's progmg stn devoted to children from ages 3 to 14; Canal D, a specialty ch featuring mainly documentaries.

Serving 245,000 subs (Super Ecran); 2,110,000 subs (Canal Famille), &1,705,000 subs (Canal D).

Serving 370 cable systems.

Satellite: Anik E-2, transponder 11-A.

The Comedy Network, Box 1000, Station "O", Toronto, ON, M4A 2W3. Canada. Phone: (416) 332-5300. Fax: (416) 332-5283. Rick Brace, pres; Brent Haynes, program dir.

A 24-hour service featuring Canadian & international programs devoted exclusively to comedy sketches, standup comedy, & ongoing comedy series. Coverage area: national.

Country Music Television (Canada), 64 Jefferson Ave, Unit 18, Toronto, ON, M6K 3H4. Canada. Phone: (416) 534-1191. Fax: (416) 530-2215. E-mail: info@cmt.ca Web Site: www.cmtcanada.ca. Michael Harris, CEO.

A 24-hour mus & entertainment net that combines mus videos with programs and features that focus on the artists and their mus.

Serving 7 million subs on 1,487 cable systems in Canada.

Satellite: Anik E2 (Ku-Band), transponder T4.

Court TV Canada, 10212 Jasper Ave., Edmonton, AB, T5J 5A3. Canada. Phone: (780) 440-7777. Fax: (780) 440-8899. E-mail: info@courttvcanada.ca Web Site: www.courttvcanada.ca. Jill Bonenfant, news dir & progmg dir.

Court TV Canada, in partnership with the U.S. based Court TV, combines Court TV's compelling daytime live trial coverage, legal analysis from inside U.S. courts with legal & police dramas, movies, documentaries & series from Canada & abroad.

DMX Music-Canada, 7260 12th St. S.E., Suite 120, Calgary, AB, T2H 2S5. Canada. Phone: (403) 640-8527. Fax: (403) 253-2788. E-mail: brad.trumble@dmxmusic.com Web Site: www.dmx.ca. Brad Trumble, VP.

Formerly a residential svc, now a coml svc exculsively. Considering a return to the Canadian market. DMX commercial audio svc; 102 formats digital audio.

Serving 8000 subs. Satellites: C3 Bank, TBA (Ku-band) delivered by satellite ant.

Discovery Channel, #9 Channel 9 Ct., Toronto, ON, M1S 4B5. Canada. Phone: (416) 332-5000. Web Site: www.ctv.ca. Ivan Fecan, pres/CEO; Corrie Coe, program dir.

Non-fiction documentary TV progmg focusing on the themes of nature, science & technology, adventure.

On 385 cable systems serving 5.6 million subs.

Satellite: Anik E2, Channel 210.

Drive-In Classics, 299 Queen St. W., Toronto, ON, M5V 2Z5. Canada. Phone: (416) 591-5757. E-mail: driveinclassics @driveinclassics.ca Web Site: www.driveinclassics.ca.

Drive-In classics is a movie ch that celebrates the funny, entertaining & sometimes thought-provoking drive-in movies of the 50s, 60s & 70s.

Fairchild Television Ltd., #3300-415, Hazelbridge Way, Aberdeen Centre, Richmond, BC, V6X 4J7. Canada. Phone: (604) 295-1313. Fax: (604) 295-1300. E-mail: info@fairchildtv.com Web Site: www.fairchildtv.com. Joseph Chan, pres.

The only Chinese language specialty TV across Canada. Serving 360,000 subs on 8 cable systems & DTH.

The Family Channel Inc., Box 787, BCE Place, 181 Bay St., Toronto, ON, M5J 2T3. Canada. Phone: (416) 956-2030. Fax: (416) 956-2035. E-mail: info@family.ca Web Site: www.family.ca. Kevin Wright, VP progmg; Barbara Bailie, dir.

Premium TV net offering family entertainment based on 60% from the Disney Channel, 25% Canadian & 15% international progmg.

Serving 5.4 million subs, on 300 cable systems, transponder T20.

FashionTelevisionChannel, 299 Queen St. W., Toronto, ON, M5V 2Z5. Canada. Phone: (416) 591-5757. E-mail: infoft@ftchannel.com Web Site: www.ftchannel.com. Marcia Martin, VP/gen mgr; David Kirkwood, VP sls; Ellen Baine, VP progmg; Scott Greig, creative dir; Jay Levine, production supvr; Bev Nenson, dir publicity; Allan Schwebel, VP-affil sls & mktg.

Canada's first & only 24-hour English language fashion channel dedicated to the world of art, architecture, photgraphy & designb with a celebration of style. Designer TV 24-hours a day.

Food Network Canada, 121 Bloor St. E., Toronto, ON, M4W 3M5. Canada. Phone: (416) 967-1174. E-mail: info@allianceatlantis.com Web Site: www.foodtv.ca. Bill Kossman, program dir.

HGTV Canada, 121 Bloor St. E., Suite 200, Toronto, ON, M4W 3M5. Canada. Phone: (866) 967-4488 (viewer relations). Phone: (416) 967-0022 (main reception). Fax: (416) 960-0971. E-mail: feedback@hgtv.ca Web Site: www.hgtv.ca. Norm Bolen, exec VP.

A 24-hour Canadian home & garden progmg resource.

Serving 5.2 million subs.

Satellite: F1, transponder T19.

History Television, 121 Bloor St. E., Suite B1, Toronto, ON, M4W 3M5. Canada. Phone: (416) 967-1174. Fax: (416) 960-0971. Web Site: www.allianceatlantis.com. Marc Etkind, progmg dir.

A 24-hour program svc featuring current & world history told in documentaries, mini-series & feature films.

Satellite: Launching September 1997. On 350 cable systems serving 4 million subs.

Satellite: Anik e-2, transponder 19.

Life Network, 121 Bloor St. E., Suite 200, Toronto, ON, M4W 3M5. Canada. Phone: (416) 967-0022. Fax: (416) 960-0971. E-mail: info@lifenetwork.ca Web Site: www.lifenetwork.ca. Kirstine Layfield, exec dir.

Offers lifestyle entertainment progmg about the people, places & experiences that make the journey of life worthwhile & interesting.

Serving 26 million English & Fr subs on 100 cable systems.

Satellite: Anik E2 (Ku-band), transponder T19 (horizontal).

La Magnetotheque, 1055 Rene Levesque E., Suite 501, Montreal, PQ, H2L 4S5. Canada. Phone: (514) 282-1999. Fax: (514) 282-1676. Web Site: www.lamagnetotheque.qc.ca. E-mail: info@lamagnetotheque.qc.ca Majorie Theodore, gen mgr.

French-language reading svc for persons who are blind, visually impaired, or print-handicapped.

Movie Central, 5324 Calgary Tr., Suite 200, Edmonton, AB, T6H 4J8. Canada. Phone: (780) 430-2800. Fax: (780) 437-3188. Web Site: www.moviecentral.com. Sandy Perkins, progmg dir; Andrew Eddy, VP/gen mgr.

Coml-free premium pay TV svc including movies, mus & comedy specials, major sports events & boxing (Superchannel, Movie Max!, Viewers Choice, Pay-Per-View).

Serving 300,000 subs on 170 cable systems.

Satellite: Anik E2.

MuchLOUD, 299 Queen St. W., Toronto, ON, M5V 2Z5. Canada. Phone: (416) 591-5757. E-mail: muchloud@muchmusic.com Web Site: www.muchloud.com.

For fans of hard music everywhere - MuchLOUD delivers. Alternative, metal and punk music videos, featured alongside exclusive artist interviews, specials, classic archival material and up-to-the-minute concert info.

MuchMoreMusic, 299 Queen St. W., Toronto, ON, M5V 2Z5. Canada. Phone: (416) 591-5757. Fax: (416) 926-4026. E-mail: muchmoremail@muchmoremusic.com Web Site: www.muchmoremusic.com. David Kines, VP/gen mgr.

Brings music fans Hot AC MusicVideo, top international specials, documentaries, movies and a growing roster of exclusive, original programming they can't find anywhere else.

MuchMoreRetro, 299 Queen St. W., Toronto, ON, M5V 2Z5. Canada. Phone: (416) 591-5757. Fax: (416) 926-4026. E-mail: request@muchmoreretro.com Web Site: www.muchmoreretro.com. David Kines, VP/gen mgr.

Source for 24/7 classic videoflow from artists including The Police, Madonna, Bon Jovi, Corey Hart, Prince, Aerosmith, Duran Duran, Janet Jackson, Rush, Nirvana and Alanis Morissette and more.

MuchMusic, 299 Queen St. W., Toronto, ON, M5V 2Z5. Canada. Phone: (416) 591-5757. E-mail: muchmail@muchmusic.com Web Site: www.muchmusic.com.

Live to air approximately 8 hours daily from streetfront headquarters in downtown Toronto, with videoflow showcasing live performance & interviews from musical artists & celebrity guests.

Serving 7,063,468 subs.

Satellite: Anik F1, transponder 17, L-Band Frequency 977.75 mhz.

MuchVibe, 299 Queen St. W., Toronto, ON, M5V 2Z5. Canada. Phone: (416) 591-5757. E-mail: muchvibe@muchmusic.com Web Site: www.muchvibe.ca.

The source for top music videos, interviews, concert specials, concert listings and classic clips from the CHUM music video archive. Hip Hop, Rap, R&B, Old School, Reggae and more.

MusiMax & MusiquePlus, 355 rue Ste- Catherine O., Montreal, PQ, H3B 1A5. Canada. Phone: (514) 284-7587. Fax: (514) 284-1889. Web Site: www.musiqueplus.com. Pierre Marchand, VP/gen mgr.

Musimax is a French-language speciality svc owned equally by Astral Media Inc. of Montreal and CHUM Ltd. of Toronto. MusiquePlus is MuchMusic's French-language counterpart in Quebec. Serving 2.078 million subs on approximately 120 cable systems . Satellite: Anik F1, transponder 9B.

OLN, Outdoor Life Network. 9 Channel Nine Crt., Scarborough, ON, M1S 4B5. Canada. Phone: (416) 332-5000. Fax: (416) 332-5861. Web Site: www.tsn.ca/oln. Anna Stamboic, dir.

Canada's destination for adventurous entertainment. Going beyond the comforts of home, OLN's progmg reveals the onsatible human drive for adventure.

Prime TV, 2100 One Lombard Pl., Winnipeg, MB, R3B-OX3. Canada. Phone: (204) 926-4800. Web Site: www.globaltv.com. Tim Schellenberg, gen mgr.

The best of TV. Classy & classic entertainment & informational progmg for those moving on from youth-skewed traditional TV fare. 5 million subs.

RDI-Le Reseau de l'information de Radio - Canada, 1400 Blvd. Rene-Levesque E., Montreal, PQ, H2L 2M2. Canada. Phone: (514) 597-7224. Fax: (514) 597-5226. E-mail: gilles_desjardins@radio-canada.ca Web Site: www.radio-canada.ca/rdi/distribution. Martin Cloutier, exec dir; Robert Nadeau, progmg.

RDI-Le Reseau de l'information is Canada's French-language news network. RDI provide live of coverage major events, newscasts every half hour, sports, financial news, as well as info programs on a wide range of topics. 9.2 million subs.

Report on Business Television, 720 King St. W., 10th Fl., Toronto, ON, M5V 2T3. Canada. Phone: (416) 957-8100. Fax: (416) 957-8180. Web Site: www.robtv.com. Jack Fleischmann, gen mgr.

Le Reseau des sports (RDS), 1755 Blvd. Rene-Levesque Est, Suite 300, Montreal, PQ, H2K 4P6. Canada. Phone: (514) 599-2244. Fax: (514) 599-2299. E-mail: webmaster@rds.ca Web Site: www.rds.ca. Jerry Frappier, gen mgr.

Provides 24-hour sports TV in Fr.

Satellite: ANIK E-2, transponder T-18.

The Score Television Network, 370 King St. W., Suite 304, Toronto, ON, M5V 1J9. Canada. Phone: (416) 977-6787. Fax: (416) 977-0238. E-mail: info@thescore.ca Web Site: www.thescore.ca. John Levy, CEO; David Errington, VP/gen mgr.

Delivers the most comprehensive svc of professional & amateur sports news & info from Canada & around the world & is in every major Canadian cable market. Available in more than 5 million cable homes.

Serving 5.4 million subs on 370 cable systems.

Satellite: Anik F1, transponder 19.

Sex TV: The Channel, 299 Queen St. W., Toronto, ON, M5V 2Z5. Canada. Phone: (416) 591-5757. E-mail: sextvchannel@cum.com Web Site: www.sextvthechannel.com.

The Shopping Channel, 59 Ambassador Dr., Mississauga, ON, L5T 2P9. Canada. Phone: (905) 565-3500. Phone: (905) 565-2600 (voicemail attendant). Fax: (905) 565-2641. Web Site: www.theshoppingchannel.ca. Ted Starkman, VP/gen mgr.

Live, shop-at-home televised retail svc, offering a var of consumer products.

Serving 5.7 million subs across Canada via cable & satellite.

Satellite: Anik E2, transponder 5.

SPACE: The Imagination Station, 299 Queen St. W., Toronto, ON, M5V 2Z5. Canada. Phone: (416) 591-5757. E-mail: space@spacecast.com Web Site: www.spacecast.com.

Cable-delivered, national, 24 hour, English-language Science Fiction, Science Fact, Speculation and Fantasy channel. The program mix includes memorable sci-fi classics and current popular series, plus feature films, documentaries, specials and daily original productions with a tilt to information and new age speculation.

Star! The Entertainment Information Station, 299 Queen St. W., Toronto, ON, M5V 2Z5. Canada. Phone: (416) 591-7400. E-mail: info@star-tv.com Web Site: www.star-tv.com.

Canada's only 24-hour national specialty service dedicated to the world of showbiz news and information. Programming includes in-depth specials and events, detailed behind-the scene features on major movies, exclusive interviews with the world's biggest celebrities and extensive live coverage of award shows, premieres and galas.

TMN—The Movie Network/MOVIEPIX, Box 787, BCE Place, Suite 100, 181 Bay St., Toronto, ON, M5J 2T3. Canada. Phone: (416) 956-2010. Fax: (416) 956-2018. E-mail: kwright@tv.astral.com Web Site: www.movienetwork.ca. Kevin Wright, VP progmg.

Two English-language, gen interest, pay TV nets featuring recent movie titles on the multi-channeled TMN, & new classics on MOVIEPIX.

Serving 350,000 subs on 200 cable systems.

Satellite: Anik E1 (Ku-band), transponder T31 (TMN); Anik E2 (Ku-band), transponder T27 (MOVIEPIX).

TSN—The Sports Network, 9 Channel Nine Crt., Toronto, ON, M1S 4B5. Canada. Phone: (416) 332-5000. Fax: (416) 332-7656. Web Site: www.tsn.ca. Adam Ashton, VP mktg; Rick Chisholm, sr VP; Andrea Goldstein, dir; Phil King, pres; Kim McKenney, dir; Judy Needham, dir; Nikki Moffat, dir.

TSN's flagship news program, SportsCentre, NHL & first three rounds of the Stanley Cups Playoffs, Toronto Maple Leafs hockey, International Hockey including the IIHF World Junior Championship, the Olympic Games through 2012. CFL, NFL, PGA Tour & all four golf Majors, Season of Champions Curling, NASCAR. A 24-hour sports ch distributed on cable in Canada. Covers all major professional & amateur sports.

Serving 8.8 million subs on more than 2,000 cable systems.

Satellite: Anik E1, transponder 18 KU-H.

TVOntario, 2180 Yonge St., Toronto, ON, M4T 2T1. Canada. Phone: (416) 484-2600. Fax: (416) 484-6285. E-mail: jjavet@tvontario.org Web Site: www.tvontario.org. Lee Robock, gen mgr; Ray Newell, dir opns.

Provides educ progmg in English & Fr off air & via cable systems throughout Ontario.

TVO network (English) serves 98% of Ontario households. (Fr) serves 75% of Ontario households & 300,000 households in Quebec. Together the nets are on 327 cable systems.

Satellites: Anik F1, transponder 21.

Talk TV, Box 9, Station "O", Toronto, ON, M4A 2M9. Canada. Phone: (416) 332-5030. Fax: (416) 332-5283. Web Site: www.talktv.ca. Ed Robinson, gen mgr; Patrick Patterson, sls dir.

TELETOON, Box 787, 181 Bay St., Toronto, ON, M5J 2T3. Canada. Phone: (416) 956-2060. Fax: (416) 956-2070. E-mail: info@teletoon.com Web Site: www.teletoon.com. Darrell Atherley, mktg VP & sls VP; Leslie Kruger, dir mktg & prom dir.

This specialty net shows the best in animation from Canada & around the planet.

Serving 6 million subs on 1,000 cable systems.

Satellite: Anik E2, transponder 20. 1000. 6 Million.

Treehouse TV, 64 Jefferson Ave., Unit 18, Toronto, ON, M6K-3H4. Canada. Phone: (416) 534-1191. Web Site: www.treehousetv.ca. Phil Piazza, VP progmg; Susan Ross, VP/gen mgr.

Treehouse TV is a specialty net dedicated to providing a variety of imaginative, stimulating and coml-free progmg for preschoolers from morning until bedtime.

Serving 4 million subs on 180 cable systems.

Satellite: Anik E-2, transponder 5.

Viewer's Choice Canada, Box 787, BCE Place, Suite 100, 181 Bay St., Toronto, ON, M5J 2T3. Canada. Phone: (416) 956-2010. Fax: (416) 956-2055. Web Site: www.viewerschoice.com. John Riley, pres/CEO.

Eastern Canada's pay-per-view network.

On 50 cable systems serving 600,000 addressable subs.

Satellites: Anik E1; Anik E2.

Vision TV: (Canada's Multi Faith Network), 80 Bond St., Toronto, ON, M5B 1X2. Canada. Phone: (416) 368-3194. Fax: (416) 368-9774. Web Site: www.visiontv.ca. E-mail: estella@visiontv.ca Bill Roberts, pres/CEO; Mark Prasuhn, COO.

Programs presented by 30 plus faith groups, British comedies, movies dramas, documentaries, pub affrs, music & performance.

Serving 7.8 million subs on 12 cable systems. Satellite: Anik F1, transponder 5.

VoicePrint(TM), (A division of The National Broadcast Reading Service Inc.). 1090 Don Mills Rd., Suite 303, Toronto, ON, M3C 3R6. Canada. Phone: (416) 422-4222. Fax: (416) 422-1633. E-mail: nbrs@nbrscanada.com Web Site: www.voiceprintcanada.com. Robert S. Trimbee, pres; Mike Hanson, mgng dir.

Read published news in audio format for blind, vision-restricted & sr Canadians.

W Network, 64 Jefferson Ave., Unit 18, Toronto, ON, M6K 3H4. Canada. Phone: (416) 534-1191. Web Site: www.wnetwork.com.

The Weather Network/MeteoMedia Inc., (A division of Pelmorex Communications Inc.). 1755 Rene-Levesque Blvd. E., Suite 251, Montreal, PQ, H2K 4P6. Canada. Phone: (514) 597-1700. Fax: (514) 597-2981. Web Site: www.theweathernetwork.com. Pierre L. Morrissette, pres/CEO; Luc Perreault, VP.

Natl satellite-to-cable TV network bcstg in Fr (MétéoMédia) & English (The Weather Network) offering weather & environmental info 24-hours a day, 7 days a week.

Serving 8.2 million subs on 752 headends.

Satellite: Anik E2, transponder 1A.

YTV Canada Inc., 64 Jefferson Ave., Unit 18, Toronto, ON, M6K 3H4. Canada. Phone: (416) 534-1191. Fax: (416) 533-0346. E-mail: info@ytv.ca Web Site: www.ytv.ca. Susan Schaefer, mktg VP; Phil Piazza, VP progmg.

English language basic cable specialty svc dedicated to children, teens & their families.

On approximately 1,200 cable systems serving an estimated 8.1 million subs.

Satellite: ANIK E1 East/West-DVC, transponder 7 (nationwide), 111 degrees (Ku-band), vert polarization, 11900 MHZ.

Canadian Radio Networks and Services

Astral Radio and Énergie

Head Office: Astral Media Inc., Bureaux de la direction, 2100, rue Sainte-Catherine, Bureau 1000, Montreal, PQ H3H 2T3. (514) 939-5000. Fax: (514) 939-1515. Web site: www.astralmedia.com.

Radio Division: 1717 boul. Rene-Levesque E. St., Bureau 200, Montreal, PQ H2L 4T9. (514) 529-3229. FAX: (514) 529-9308. Web site: www.radioenergie.com.

Principal Officers: Ian Greenberg, pres/CEO; Andre Bureau, chmn of bd; Sidney Greenberg, VP; Louis Ryan, VP strategic planning; Alain Bergaron, VP Brand Management/corporate communications; Sophie Emond, VP reg/govt affrs; Rachel Yates, dir/corporate communications; Louis Marcotte, asst VP finance; Brigitte Catellier, VP legal affrs; Michael Arpin, sr advisor reg/govt affrs; Claude Gagnon, sr VP/CFO; Arnold Chiasson, VP humam resource; Louis Tasse, asst VP human resource.

Affiliates: FM Stations: 94.3 (Montreal); 98.9 (Quebec); 99.1 (Rouyn-Noranda); 102.3 (Mauricie); 102.7 (Val d'Or); 104.1 (Outaouais); 106.1 (Estrie); CITF (Quebec); CHEY (Trois-Rivieres); CIKX (Grand Falls, N-B); CJCJ (Woodstock, N-B); CFXY (Fredericton, N-B); CIBX (Fredericton, N-B); CKTO (Truro, Nova Scotia); CKTY (Truro, Nova Scotia); AM Stations: CKHJ (Fredericton, New-Brunswick); CKBC (Bathurst, New-Brunswick).

Canadian Broadcasting Corp.

The Canadian Broadcasting Corp. (CBC) is a publicly owned corporation established by the Broadcasting Act (1936) of the Canadian Parliament to provide the natl bcstg svc in Canada in the two official languages, English & French. Under this legislation, the CBC is governed by the 1991 Broadcasting Act and subject to regulations of the Canadian Radio-Television and Telecommunications Commission (CRTC).

Program Services: The CBC operates English and French AM & FM stereo networks. The progmg on these networks is nearly all Canadian and virtually free of coml adv. CBC North bcsts radio programs to Canada's north in English, French and eight native languages, serving the special needs of native & non-native groups in the Yukon, the Northwest Territories and northern Quebec. Radio Canada International is Canada's voice abroad. Bcstg on shortwave in seven languages, RCI's progmg reflects Canada's political, economic, social & cultural spectrum to an international audience. Newsworld is a 24-hour natl satellite to cable English-language news & info svc. Le Rèeseau de l'information (RDI) is a 24-hour natl satellite to cable French-language news & info svc. The heart of CBC's natl distribution system is Canada's Anik E2 satellite, carrying progmg through six different time zones. CBC's progmg is bcst over 684 AM & FM stns.

Head Office: 181 Queen St., Box 3220, Station C, Ottawa, ON K1P 1K9. (613) 288-6000. TDD: (613) 288-6455. Web site: www.cbc.ca; E-mail: liaison@cbc.ca. Box 5000 Stn. A, Toronto, ON M5W 1E6. Phone: (416) 205-7264.

CBC Board of Directors: Timothy W. Casgrain, chair; Helene Fortin; Marie Giguere; Roy L. Heenan; Jane Heffelfinger; Clarence LeBreton; Howard McNutt; L. Richard O'Hagan; James S. Palmer.

Principal Officers: Robert Rabinovitch, pres/CEO; Jane Chalmers, VP CBC Radio; Sylvain Lafrance, exec VP French Radio and New Media; Pierre Nollet, VP/gen counsel/corporate sec; George C.B. Smith, sr VP human resources and organization; Johanne Charbonneau, VP/CFO; Raymond Carnovale, VP/CTO; Michel Tremblay, VP strategy/business dev; William B. Chambers, VP communications; Michel Saint-Cyr, pres Real Estate div.

CBC Ombudsman: David Bazay, William Morgan & Vince Carlin, English svcs, Box 500, Station A, Toronto, ON M5W 1E6; E-mail: ombudsman@cbc.ca. Web site: www.cbc.ca/ombudsman. Julie Miville-Dechene, French svcs, Box 6000, Montreal, PQ H3C 3A8; E-mail ombudsman@radio-canada.ca. Web Site: www.radio-canda-ca/ombudsman.

Ombudsman English Services: Canadian Broadcasting Corporation, Box 500, Station A, Toronto, ON M5W 1E6. E-mail: ombudsman@cbc.ca. Web site: www.cbc.ca/ombudsman.

Ombudsman French Services: Box 6000, Montreal, PQ H3C 3A8. E-mail: ombudsman@radio-canada.ca. Web site: www.radio-canada.ca/ombudsman.

Communications-English Networks: 250 Front Street West, Box 500, Station A, Toronto, ON M5W 1E6. (866) 306-4636. TDD: (416) 205-6688; E-mail: cbcinput@cbc.ca.

Communications-French Services: Box 6000, Montreal, PQ H3C 3A8. (514) 597-6000. TDD: (514) 597-6013; E-mail: auditoire@radio-canada.ca.

Newfoundland Region (English): Radio bldg.: 25 Henry Street, Television bldg.: 95 University Ave., Box 12010, Station A, St. John's, NF A1B 3T8. (709) 576-5000.

Maritime Region (English): Radio bldg.: 56 Sackville Street, Television bldg.: 1840 Bell St., Box 3000, Halifax, NS B3J 3E9. (902) 420-8311.

Atlantic Provinces (French): 250 Universite Ave., Box 950, Moncton, NB E1C 8N8. (506) 853-6666.

Quebec Region (English): 1400 Rene-Levesque Boulevard East, Box 6000, Montreal, PQ H3C 3A8. (514) 597-6000.

Quebec City & Eastern Quebec Region (French): 88 St-Jean Street, Box 18800, Ste. Foy, PQ G1V 9L4. (418) 654-1341.

Ontario Region (English): 250 Front Street West, Box 500, Station A, Toronto, ON M5W 1E6. (416) 205-3311.

Ontario Region (French): 181 Queen Street, Box 3220, Station C, Ottawa, ON K1Y 1E4. (613) 724-1200.

Manitoba Region (English & French): 541 Portage Ave., Box 160, Winnipeg, MB R3C 2H1. (204) 788-3222.

Saskatchewan Region (English & French): 2440 Broad St., Box 540, Regina, SK S4P 4A1. (306) 347-9540.

Alberta Region (English & French): 10062-102nd Ave., Room 123, Edmonton City Centre, Box 555, Edmonton, AB T5J 2P4. (780) 468-7500.

British Columbia Region (English & French): 700 Hamilton Road, Box 4600, Vancouver, BC V6B 4A2. (604) 662-6000.

CBC North: 5129 49th Street, Box 160, Yellowknife, NT X1A 1P8. (867) 920-5400.

Producers, Distributors, and Production Services Alphabetical Index

A

ABC Family Channel, 500 S. Buena Vista St., Burbank, CA, 91521. Phone: (818) 560-1000. Web Site: www.abcfamily.com.

TV-CATV only.

ABC Family features quality, contemporary entertainment for all members of the family including original series, movies & specials. Available in over 87 million homes via basic cable.

ABC Radio Networks, 444 Madison Ave., 9th Fl., New York, NY, 10022. Phone: (212) 735-1700. Fax: (212) 735-1799. Web Site: www.abcradio.com.

John McConnell, sr VP; Tom Powell, VP; Dave Kaufman, VP; Dan Formento, VP.

Radio Only.

Producers of natl & international radio features, such as *Flashback, Flashback Pop Quiz* & *Rock Slides.*

ABC Radio Networks, 13725 Montfort Dr., Dallas, TX, 75240. Phone: (972) 991-9200. Fax: (972) 991-9890. Web Site: www.abcradio.com.

Traug Keller, pres; Darryl Brown, exec VP; Robert Hall, progmg VP; John Russo, VP affil rel west; T.J. Lambert, VP sports.

Radio Only.

Live 24-hour-a-day premium progmg available featuring 10 radio formats. Also includes SMN PRIZM rsch clustering.

ACC Entertainment, Bavariafilmplatz 7, 82031 Grünwald, Munich Phone: 49-89 64981-332. Phone: 49-89 64981-232. E-mail: accficm@acc.com

TV-CATV only.

Film & TV producers, distributors.

ACTV Inc., 233 Park Ave., 10th Floor, New York, NY, 10020. Phone: (212) 497-7000. Fax: (212) 459-9548. E-mail: info@actv.com

Christopher Cline, CFO; David Reese, chmn/CEO.

TV-CATV only.

Interactive TV progmg for educ & entertainment.

ADM—International Film & TV Distribution, Drienerwolde House, Drienerwoldeweg, Hengelo, IL, 7552 PC. Netherlands. Phone: 31 74 250 6843. Fax: 31 74 250 1874.

Carole K. Hodson, mgng dir; Herman Melzer, chmn acquisitions; Sarah J. Mydlak, dir sales & mktg.

TV-CATV only.

International distributor of film & TV programs including features, classics, documentaries, children's, plus much more.

ALIN TV, 149 Madison Ave., Suite 602, New York, NY, 10016. Phone: (212) 889-1327. Fax: (212) 213-6968. Web Site: www.alintv.com.

Alan Cohen, pres.

TV-CATV only.

Unwired TV natl network, syndication, digital media sls & mktg.

ANA Television Network, 1510 H St. N.W., Suite 400, Washington, DC, 20005. Phone: (202) 898-8222. Fax: (202) 898-8088.

Angelyn Adams, CFO.

TV-CATV only.

Arabic-language TV net bcstg to the Arab-American community 24 hours via cable, wireless cable, satellite.

Satellite: DIRECTV Plus.

APA International Film Distributors Inc., 14260 S.W. 136th, Suite 16, Miami, FL, 33186. Phone: (305) 666-0020. Fax: (305) 234-7565. E-mail: apafilm@bellsouth.net

TV-CATV only.

TV program production & distribution.

APM/Associated Production Music, 6255 Sunset Blvd., Suite 820, Hollywood, CA, 90028. Phone: (323) 461-3211. Phone: (800) 543-4276. Fax: (323) 461-9102. E-mail: sales@apmmusic.com Web Site: www.apmmusic.com.

New York, NY 10173, 342 Madison Ave, Suite 1200. Phone: (800) 276-6874. Craig Giummarra, sls mgr.

TV-CATV only.

Sixteen Libraries: KPM, Bruton, Sonoton, Carlin, Castle, NFL. Over 3,000 CDs, Personalized Packages, Music Search, 15-20 New CD releases mthy.

ATA Trading Corp., Box 307, Massapequa Park, NY, 11762. Phone: (516) 541-5336. Fax: (516) 541-5336. E-mail: atat@verizon.net

Harold G. Lewis, pres; Susan Lewis, VP.

TV-CATV only.

Worldwide distributors for ind producers in all areas of feature films, made-for-TV productions, series, documentaries & children's programs.

Academy Entertainment, 59 Westminster Ave., Bergenfield, NJ, 07621. Phone: (201) 385-8139. Phone: (201) 394-1849. Fax: (201) 385-8196. E-mail: mlrfilms@aol.com

Alan Miller, pres; Al Leifer, co-pres.

TV-CATV only.

Distribution of film, TV & video progmg worldwide.

Accuracy in Media Inc., 4455 Connecticut Ave. N.W., Suite 330, Washington, DC, 20008. Phone: (202) 364-4401. Fax: (202) 364-4098. E-mail: info@aim.org Web Site: www.aim.org.

Don Irvine, chmn.

Radio Only.

Nationwide media monitoring organization produces documentary TV films, one-minute weekday radioo commentaries, bi-monthly publications, and programs that critique media coverage.

Acme, 9976 W. Wanda Dr., Beverly Hills, CA, 90210. Phone: (310) 276-5509. Fax: (310) 276-1183.

Bradley Friedman, pres; David Temianka, dir; Fred Wietzchz, CEO.

TV-CATV only.

Feature film & TV production, music videos, childrens progmg, commercials, robotics, scripting, tin ton props, rock music stock footage, space stock footage.

Advanced Digital Services, Inc., 948 N. Cahuenga Blvd., Hollywood, CA, 90038. Phone: (323) 468-2200. Fax: (323) 468-2211. Web Site: www.adshollywood.com.

Andrew McIntyre, pres; Kevin Yates, COO.

TV-CATV only.

Video duplication, standard conversion.

Adventist Media, 101 W. Cochran, Simi Valley, CA, 93065. Phone: (805) 955-7777. Fax: (805) 522-1082. E-mail: info@faithfortoday.tv

Marshall Chase, gen mgr.

TV-CATV only.

TV program production & distribution.

African Family Film Foundation, Box 630, Santa Cruz, CA, 95061-0630. Phone: (831) 426-3133. E-mail: taale@africanfamily.org Web Site: www.africanfamily.org.

Taale Laafi Rosellini, dir.

TV-CATV only.

Production & distribution of films & videotapes promoting African family life & culture.

Agency for Instructional Technology (AIT), Box A, Bloomington, IN, 47402-0120. Phone: (800) 457-4509. Phone: (812) 339-2203. Fax: (812) 333-4218. E-mail: info@ait.net Web Site: www.ait.net.

Bloomington, IN 47404, 1800 N. Stonelake Dr. (Shipping address).

TV-CATV only.

Produces, acquires & distributes technology-based learning resources—including video, videodisc, software & print—for all K-12 curricular areas, vocational educ/tech prep, early childhood, & professional dev.

Agora TV, 195 Hicks Dr. S.E., Marietta, GA, 30060. Phone: (404) 226-4503. Fax: (678) 581-3750. E-mail: joe@agoratv.tv Web Site: www.agoratv.tv.

Joseph Gora, pres.

TV-CATV only.

TV production & equipment rental.

Agrinet Farm Radio Network, 104 Radio Rd., Powell's Point, NC, 27966. Phone: (252) 491-2414. Fax: (252) 491-2939. Web Site: www.agrinetradio.com.

Bill Ray, dir; Gary Gross, dir opns; Lisa Ray, natl sls mgr.

TV-CATV only.

State, rgnl & natl agricultural news, mkts & weather.

Airwaves Audio Inc., 150 Mutual St., Toronto, ON, M5B 2M1. Canada. Phone: (416) 977-1098.

James Kennedy, pres.

TV-CATV only.

Audiovisual & industrial postproduction. Audio recording & mixing for radio & TV.

Alden Films, Box 449, Clarksburg, NJ, 08510. Phone: (732) 462-3522. Fax: (732) 294-0330. E-mail: info@aldenfilms.com Web Site: www.aldenfilms.com.

Paul Weinberg, pres; Fran Fried, admin asst.

TV-CATV only.

Distributes nearly 600 films & videos on Israel & Judaica. Official distributor for state of Israel.

All Media Productions Inc., 12261 Cleveland Ave. Ste F, Nunica, MI, 49448-9309. Phone: (616) 837-0776. Fax: (616) 837-0897. E-mail: linda@allmediaproductions.com Web Site: www.allmediaproductions.com.

Linda Langs, pres.

TV-CATV only.

Web dev, internet mktg & film distribution.

All My Features Inc., 9190 Clearstream Terr., Mechanicsville, VA, 23111. Phone: (804) 730-1534. Fax: (804) 559-4809.

TV-CATV only.

Provides daily entertainment news, entertainment-related features via audio & computer feeds.

All Productions, 7025 Regner Rd., Suite 5, San Diego, CA, 92119. Phone: (619) 284-2566. Fax: (619) 460-6160. E-mail: mikeall@eudoramail.com Web Site: www.allproductions.com.

Michael J. All, CEO; Stephen A. All, CFO; Jean M. All, pres.

San Diego, CA 92119-1941, 7025 Regner Rd. Phone: (619) 460-4837. (619) 286-7733. Fax: (619) 460-6160.

TV-CATV-Radio.

TV & radio program production, distribution; cable-ready TV progmg; promotion film production, production svcs; TV, radio spots & coml announcers.

Allegro Productions Inc., 1000 Clint Moore Rd., Suite 211, Boca Raton, FL, 33487. Phone: (800) 275-4636. Fax: (888) 329-3737. E-mail: allegro@ssrvideo.com Web Site: www.ssrvideo.com.

Jerome G. Forman, pres.

TV-CATV only.

Educational & corporate progmg, including documentary/bcst. From concept to completion, offering full service video post production, CD-ROM/DVD authoring, multi-format duplication, 3D animation & effects.

Allied Production and Distribution Services, 135 W. Hancock St., Decatur, GA, 30030. Phone: (404) 373-1227. Fax: (404) 373-1227.

Edwin Clark, pres.

TV-CATV only.

TV program production & distribution.

Aloha Productions, Box 33648, San Diego, CA, 92163. Phone: (619) 275-7357. Phone: (800) 223-2564. Fax: (619) 296-5909. E-mail: jhal@alohajingles.com

Hal Hodgson, exec producer.

TV-CATV-Radio.

Original coml music production, scoring, jingles, long-form; movie & TV scores.

Alternative Programming, 4215 Brendenwood Rd., Rockford, IL, 61107. Phone: (815) 229-3995. Fax: (815) 229-5043. E-mail: altprog@sbcglobal.net

Gary A. Knoll, owner.

Radio Only.

Complete music formats for radio - current music for various formats - custom CD svc.

Altman Productions, 3401 Macomb St. N.W., Washington, DC, 20016. Phone: (202) 362-3088. Fax: (202) 362-0234. E-mail: itsacademicquiz@aol.com

Sophie B. Altman, exec producer; Susan Altman, producer; Susan Lechner, editor.

TV-CATV only.

TV & radio program production. Producers of It's Academic, the high school quiz program.

Altruist Media, 2601A Wilson Blvd., Arlington, VA, 22201. Phone: (703) 812-8813. Fax: (703) 812-9710. E-mail: info@altruistmedia.com Web Site: altruistmedia.com.

Jan Dearth, pres.

TV-CATV only.

A full-svc visual communications firm. Staff producers, writers & dirs provide full creative direction & project mgmt from concept dev, treatment, scripting & graphics design to production & delivery. Offers videotape, live event production, consulting svcs for organizational communications. Provides comprehensive production svcs for videotape, special event & live business TV-video conference progmg. Also offers media training, VNR production, video press tours, consulting for private networks, new media svcs including distributed multimedia, WWW design & CD-ROM dev.

Americ Disc, 11 Oval Dr., Islandia, NY, 11788. Phone: (631) 234-0200. Fax: (631) 232-4430.

TV-CATV only.

Film processing, film-to-tape transfers, video editing, videocassette, audiocassette duplication, CD-ROM & CD-Audio duplication packaging & fulfillment.

America On The Road, 4038 Exultant Dr., Rancho Palos Verdes, CA, 90275. Phone: (310) 265-9873. Fax: (310) 544-4318. E-mail: aotrradio@cox.net Web Site: www.americaontheroad.com.

Ed Yelin, producer; Al Herskovitz, ptnr.

TV-CATV only.

One hour wkly, 2.5-minute daily automotive consumer show.

America One Television Network, 6125 Airport Freeway, Suite 100, Ft. Worth, TX, 76117. Phone: (817) 546-1400. Fax: (682) 647-0756. Web Site: www.americaone.com.

Matt Reiff, pres; Preston Bornman, sr VP.

TV-CATV only.

24 hour gen entertainment bcst network.

American Blues Network, Box 6216, Gulfport, MS, 39506. Phone: (800) 896-5307, ext 117. Fax: (228) 896-5703. E-mail: rip@americianbluesnetwork.com Web Site: www.americianbluesnetwork.com.

Stan Daniels, CEO.

Radio Only.

American Chiropractic Association Inc., 1701 Clarendon Blvd., Arlington, VA, 22209. Phone: (703) 276-8800. Fax: (703) 243-2593. Web Site: www.acatoday.com.

TV-CATV only.

Professional membership organization.

American Farm Bureau Inc., 1501 E. Woodfield Rd., Suite 300W, Schaumburg, IL, 60173. Phone: (847) 685-8752. Fax: (847) 685-8950. E-mail: stut@fb.org Web Site: www.fb.org.

Stewart Truelsen, dir.

TV-CATV only.

AGFeed, mthy video feed of news stories about food & agriculture, *Newsline* radio svc, *Focus on Agriculture* commentary, stock footage.

American Foundation for the Blind, 11 Penn Plaza, Suite 300, New York, NY, 10001. Phone: (212) 502-7600. Phone: (800) 232-5463. Fax: (212) 502-7777. E-mail: afbinfo@afb.net Web Site: www.afb.org.

Carl Augusto, pres; Liz Greco-Rocks, dir.

TV-CATV only.

Provides consultation & referrals, social & technological rsch, publications, info svcs, public educ, govt rel & talking books.

American Heart Association, 7272 Greenville Ave., Dallas, TX, 75231-4596. Phone: (214) 706-1330. Fax: (214) 706-5243. Web Site: americanheart.org.

Julie Del Barto, communications mgr bcstg.

TV-CATV only.

Video news releases, stock footage related to heart & disease for news programs.

American Public Television, 55 Summer St., Boston, MA, 02110. Phone: (617) 338-4455. Fax: (617) 338-5369. Web Site: www.aptonline.org.

Cynthia Fenneman, pres/CEO.

TV-CATV only.

Major distributor of high quality TV programs to all U.S. public TV stns. Also distributor of programs to international media.

American Stock Exchange, 86 Trinity Pl., New York, NY, 10006. Phone: (212) 306-1229. Fax: (212) 306-5489. E-mail: kenneth.meyer@amex.com Web Site: www.amex.com.

Kimberly Zapien, dir.

TV-CATV only.

TV studio location on a trading floor, teleprompters, access to industry analysts, production and postproduction svcs.

American TelNet, 855 SW 78th Ave., Plantation, FL, 33324. Phone: (954) 453-7000. Fax: (954) 453-7809. E-mail: success@americantelnet.com

An 800/900 Interactive svc bureau offering a wide var of turnkey pay-per-call entertainment & business programs.

AmericaNurse TV Productions, Box 7717, Romeoville, IL, 60446. Phone: (815) 773-4497. Web Site: www.americanurse.com.

Karon Gibson, R.N., producer.

TV-CATV only.

Consumer educ shows on health, safety & other self-help titles. Entertaining introduction to optional alternative & mainstream medicine & Rx.

America's Most Wanted, 5151 Wisconsin Ave. N.W., Washington, DC, 20016. Phone: (202) 205-2600. Fax: (202) 204-2604.

TV-CATV only.

Wkly reality-based program for Fox TV.

Anderson Productions Ltd. (APL), 37 W. 20th St., Loft #904, New York, NY, 10011. Phone: (212) 414-9220. Fax: (212) 206-0279. E-mail: andersontv@aol.com

Steven C. F. Anderson, exec producer & pres; Brian Peter Falk, dir.

TV-CATV only.

Reality-based TV program production for bcst & cable TV.

Angel Films Co., 967 Hwy. 40, New Franklin, MO, 65274-9778. Phone: (573) 698-3900. Fax: (573) 698-3900. E-mail: phoeenix@phoeenix.org

William H. Hoehne Jr., chmn; Joyce L. Chow, CEO; Arlene Hulse, pres; Leana Le Gee, VP mktg & adv VP; Matthew P. Eastman, VP production.

TV-CATV only.

Production, distribution, syndication of progmg for adults & children.

Animated Production Services, 321 W. 44th St., New York, NY, 10036. Phone: (212) 265-2942. Fax: (212) 265-2944. Web Site: www.digitaltofilm.com.

TV-CATV only.

TV program, coml, promotional film production, distribution & production svcs, digital film.

Antenne 2 - French TV 2, 1290 Ave. of the Americas, Suite 3410, New York, NY, 10104. Phone: (212) 581-1771. Fax: (212) 541-4309.

TV-CATV only.

TV program production.

The Arabic Channel, 366 86 St., 1st Fl., Brooklyn, NY, 11209-5002. Phone: (718) 238-2450. Fax: (718) 238-2465.

Gamil Tawfol, pres/CEO; Marguerite M. Moore, VP; Dr. Saleh El-Ahwal, VP.

TV-CATV only.

Arabic language progmg bcst Time Warner/Comcast Cablevision.

Archive Films/Archive Photos, 75 Varick St., 5th Floor, New York, NY, 10013. Phone: (646) 613-4000. Phone: (800) 876-5115. Fax: (646) 613-4140. E-mail: sales@archivefilms.com Web Site: www.archivefilms.com.

111 45 Stockholm. Archive Films/Archive Photos Scandinavia, Birger Jarlsgatan 55. Phone: 46 8 20 89 20. Fax: 46 8 20 89 33. Contact: Lennert Karlsson.

Cologne 50969. Archive Films GMBH, Bremstrasse 12. Phone: 49 221 936 4080. Fax: 49 221 360 4112. Contact: Craig Burns.

London W1P 6EE. Archive Films/Archive Photos, 17 Conway St. Phone: 44 171 312 0300. Fax: 44 171 391 9123. Contact: Chris Blakeston.

Milan 20123. Archive Films/Archive Photos Italy, Via Terraggio 17. Phone: 39 2 874 693. Fax: 39 2 805 7739. Contact: Guido Rossi.

Paris Ducaud 75009. Archive Films/Archive Photos, 4 Boulevard Poissonniere. Phone: 33 1 55 77 00 00. Fax: 33 1 55 77 00 66. Contact: Sylvie Ducaud.

TV-CATV only.

Stock footage/photo library providing all types of historical footage & photos for use in products for TV/CATV.

Ardustry Home Entertainment LLC, 21250 Califa St., Woodland Hills, CA, 91367. Phone: (818) 712-9070. Fax: (818) 712-9000. Web Site: www.ardustry.com.

Cheryl Freeman, CEO.

TV-CATV only.

Distribution, dev, production documentaries, TV series, kids, specials.

Arkadia Entertainment Corp., 34 E. 23rd St., 3rd Fl., New York, 10010. Phone: (212) 533-0007. Fax: (212) 979-0266. E-mail: arkadian@aol.com Web Site: www.arkadiarecords.com.

Bob Karcy, CEO.

Radio Only.

CD, DVD, video production & distribution worldwide. A broad range of exclusive progmg.

Armedia Communications, 307-3219 Young St., Toronto, ON, M4N 2L3. Canada. Phone: (905) 889-0076. Fax: (905) 889-0078.

David Mazmanian, owner.

TV-CATV only.

Audio, video, music production & bcst svcs.

J. Arnold Productions, 363 Massachusetts Ave., Lexington, MA, 02420. Phone: (781) 674-2277. Fax: (781) 674-0272. E-mail: jarpro@aol.com

James Arnold, pres; Lori Arnold, production mgr; Eric Fisher, production mgr.

Charlotte, NC 28117, 147 Cove Creek Rd. Phone: (704) 663-4444. Fax: (704) 663-6696. James Arnold, pres.

TV-CATV only.

Full-svc on location video production. ENG-EFP crews & Betacam equipment packages.

The Kay Arnold Group, 34 Kramer Dr., Paramus, NJ, 07652. Phone: (201) 652-6037. Fax: (201) 612-8578.

Kay Arnold, pres.

TV-CATV only.

Production & distribution of film & tape programs for TV, satellite, cable, home video & non-theatrical.

Toby Arnold and Associates Inc., 3234 Commander Dr., Carrollton, TX, 75006. Phone: (800) 527-5335. Phone: (972) 661-8200. Fax: (972) 250-6014. E-mail: toby@taamusic.com Web Site: www.taamusic.com.

Toby Arnold, pres; Dolly Arnold, VP/COO; Lawrence Mangiameli, VP/creative dir.

TV-CATV only.

Audio production libraries for TV & radio. Station Imaging, Morning show promo sweeper, stager packages for all formats.

Artisan PictureWorks Ltd., 800 Forrest St. N.W., Atlanta, GA, 30318. Phone: (404) 355-3398. Fax: (404) 350-0302. E-mail: info@artisanpicture.com Web Site: www.artisanpictureworks.com.

Bryan Gartman, pres; Amy Thompson, production mgr; Dan Valdes, mgr.

TV-CATV only.

Studio facilities feature Ultimatte, fiber optics to satellite uplink. Live multicam specialists. Remote & in-house production facilities.

Artist View Entertainment Inc., 4425 Irvine Ave., Studio City, CA, 91502-1919. Phone: (818) 752-2480. Fax: (818) 752-9339. E-mail: artistview@earthlink.net Web Site: artistviewent.com.

Scott J. Jones, pres; Jay E. Joyce, VP.

TV-CATV only.

Worldwide distribution in all media specializing in feature films.

Ascent Entertainment Group Inc., 1225 17th St., Suite 1800, Denver, CO, 80202. Phone: (303) 308-7000. Fax: (303) 308-0485.

TV-CATV-Radio.

Hotel entertainment & info svcs, video conferencing, satellite bcst distribution, & net construction & maintenance.

Ascent Media Management East, 235 Pegasus Ave., Northvale, NJ, 07647. Phone: (201) 767-3800. Fax: (201) 784-2769. E-mail: agavin@apvi.com Web Site: www.apvi.com.

Don Buck, pres; Al Gavin, sls VP; Tony Beswick, VP/gen mgr.

Burbank, CA 91502. Audio Plus Video - West, 200 S. Flower St. Phone: (818) 841-7100. Larry Kingen, VP/gen mgr.

TV-CATV only.

Standards & aspect ratio conversion, international duplication, PAL/NTSC editing, film-to-tape transfers, 16 X 9 audio layback, restoration & satellite svcs, dud authoring, compression, streaming.

Ascent Media Network Services, 2901 W. Alameda Ave., Burbank, CA, 91505. Phone: (818) 840-7174. Fax: (818) 567-1131. Web Site: www.ascentmedia.com.

Sharon Pyne, opns dir; Lennis Schwartz, VP opns; Jodynne Wood, sls dir.

London, NO WIT 2NS United Kingdom, 48 Charlotte St.

TV-CATV only.

AM NS powers the bcst-cable nets around the world. We distribute progmg content over our integrated fiber & satellite net.

Ascent Media Network Services, 250 Harbor Dr., Stamford, CT, 06902. Phone: (203) 965-6000. Fax: (203) 965-6405. Web Site: www.ascentmedia.com.

Francis G. Luperella, sr VP; Matt Armstrong, VP mktg.

Singapore. Asia Bcst Centre Phone: (612) 330-2639. E-mail: vhendra@abc.gwns.com. Vincent Helseth.

Minneapolis, MN 55038. GWNS Minneapolis, 6845 20th Ave. Phone: (612) 330-2639. Joel Helseth, sls & mktg.

TV-CATV only.

Video transmission, origination; technical consulting, new media products; private networks, post production, studio, graphics; bcst event svcs & satellite svcs.

Asia Pacific Productions, 19698 S.E. Cottonwood St., Portland, OR, 97267. Phone: (503) 723-6456. Fax: (503) 723-6456. E-mail: info@approd.com Web Site: approd.com.

Thomas F. Hopkins, pres; Miyuki Shigeji, VP.

Kobe 653-086 Japan, 3-17 Higashi Maruyama Cho. Phone: 81 78 691 2450 Telephone/Fax.

TV-CATV only.

Provides news, documentary & program production; coml production; business/promotional film & video production; production svcs for TV/CATV.

Associated Press Television News, 1825 K St., N.W., Suite 800, Washington, DC, 20036. Phone: (202) 736-9595. Fax: (202) 736-9619. Web Site: www.ap.org.

TV-CATV only.

TV news production, news library, video editing & ENG production.

Associated Television International, 4401 Wilshire Blvd., Los Angeles, CA, 90010. Phone: (323) 556-5600. Fax: (323) 556-5610. Fax: www.associatedtelevision.com. E-mail: atiwest@aol.com

TV-CATV only.

Full-svc production, distribution & syndication company in business for over 20 years.

Association of Islamic Charitable Projects, 4431 Walnut St., Philadelphia, PA, 19104. Phone: (215) 387-8888. Fax: (215) 387-3815. Web Site: www.aicp.org.

TV-CATV only.

Islamic progmg. Educational micro-bcst net svc for metro Philadelphia.

At a Glance, 6350 W. Freeway, Fort Worth, TX, 76116. Phone: (817) 570-1400. Phone: (800) 266-1837. Fax: (817) 737-9436. E-mail: info@familynet.com Web Site: www.familynetradio.com.

Lisa Bratton, radio mktg & distribution; Donna Senn, radio Distribution; Chuck Ries, producer.

Radio Only.

Variety of topics: health, fitness, character, parenting, etc. 60 second spots, 10 per month, on CD.

Atlantic Video Inc., 650 Massachusetts Ave. N.W., Washington, DC, 20001. Phone: (202) 408-0900. Fax: (202) 408-8496. Web Site: www.atlanticvideo.com.

Doug Moon Joo, pres; John Sommers, VP/gen mgr; Amy Schwab, mktg dir.

Alexandria, VA 22304, 150 S. Gordon St. Phone: (703) 823-2800.

TV-CATV only.

Soundstages, postproduction, graphics, duplication, remote, satellite uplink, videoconferencing, film-to-tape, D-2, digital 110 pathways & audio sweetening.

Auburn Television, (A division of Telecommunications). Auburn University, Admin. Bldg., Corner of Samford & Donahue, Auburn, AL, 36849-5423. Phone: (334) 844-5707. Fax: (334) 844-5708. Web Site: www.auburn.edu.

Richard Burnett, exec dir; John Gober, engr; Deborah Howard, opns mgr.

Montgomery, AL 36104. Broadview Media, 401 Adams St, Box 7. Phone: (344) 223-5708. Rich Michaelson, dir.

TV-CATV-Radio.

TV/studio/remote production, Tape/CD/DVD production, satellite uplink/downlink & avid edit suite, CATV

The Audio Department Inc., 119 W. 57th St., 4th Fl., New York, NY, 10019. Phone: (212) 586-3503. Fax: (212) 245-1675. Web Site: www.theaudiodepartment.com.

Aimee Mitchaud, mgr; Lola Norarevian, mgr.

TV-CATV only.

Audio & audio for video, adv & media promotion.

Audio Production Services, University of Colorado, Campus Box 379, 312 Stadium Bldg., Boulder, CO, 80309. Phone: (303) 492-2675. Fax: (303) 492-7017.

Radio Only.

Radio program production.

Auritt Communications Group, 729 Seventh Ave., 5th Fl., New York, NY, 10019. Phone: (212) 302-6230. Fax: (212) 302-2969. Web Site: www.auritt.com.

Joan Auritt, pres.

TV-CATV only.

Satellite media tours, event coverage, video news releases, B-roll packages, radio tours, audio new releases, sls/corporate videos, web casting, print tours

Australian Tourist Commission, 2049 Century Park E., Fl. 19, Los Angeles, CA, 90067-3121. Phone: (310) 229-4871. Fax: (310) 552-1215. E-mail: rmonfrini@atc.australia.com Web Site: www.australia.com.

Robert Monfrini, dir.

TV-CATV only.

TV program distribution.

Avid Technology Inc., Avid Technology Park, One Park W., Tewksbury, MA, 01876. Phone: (800) 949-2843. Phone: (978) 640-6789. Fax: (978) 640-1366. E-mail: info@avid.com Web Site: www.avid.com.

TV-CATV only.

Avid Technology is a leading supplier of newsroom computer, editing, playback & effects systems. Implemented as stand-alone or networked systems, Avid solutions provide speed, creativity & operating efficiencies throughout the newsroom.

Axcess Broadcast Services Inc., 4801 Spring Valley, Suite 105-B, Dallas, TX, 75244. Phone: (972) 386-6847. Fax: (972) 386-5207.

TV-CATV only.

Sls consulting for new businesses in the top 100 markets. CD production library, radio, TV promotions & IDs.

B

BBC Worldwide Americas Inc., 747 3rd Ave., 7th Fl., New York, NY, 10017. Phone: (212) 705-9300. Fax: (212) 888-0576.

TV-CATV only.

TV program production & distribution, home video, library sls, licensing.

BBC Worldwide Television Ltd., 80 Woodlands, London, W12 0TT. Fax: (181) 749-0538. Fax: (181) 576-2000. E-mail: webguide@bbc.co.uk Web Site: www.bbc.co.uk.

TV-CATV-Radio.

Program licensing to international bcstrs & generation of co-production business. Dev of BBC branded satellite & cable channels worldwide.

S. Banks Group Inc., 174 Johnston Ave., Toronto, ON, M2N 1H3. Canada. Phone: (416) 224-0296. Fax: (416) 224-8542.

Sydney Banks, pres.

TV-CATV only.

Feature film and TV program production.

Bardel Entertainment Inc., 548 Beatty St., Vancouver, BC, V6B 2L3. Canada. Phone: (604) 669-5589. Fax: (604) 669-9079. E-mail: bardel@bardelanimation.com Web Site: www.bardelentertainment.com or www.bardelanimation.com.

Barry Ward, pres; Delna Bhesania, CEO; Cathy Schoch, producer.

TV-CATV only.

High quality 3D, Maya & hybrids of digital & traditional animation svcs for feature film, television, interactive media, internet & commercials.

Bavaria Film GmbH, Bavariafilmplatz 7, 82031Geiselgasteig/Munich Phone: 49 89 6499 0. Fax: 49 89 6492 507. E-mail: presse@bavaria-film.de Web Site: www.bavaria-film.de.

Dieter Frank, pres; Thilo Kleine, pres; Peter Kussius, sls mgr.

TV-CATV only.

Dubbing, film laboratories, film & tape transfers, film & TV production, production svcs, multimedia svcs.

Bayliss, (Formerly Gene Bayliss). 208 Good Hill Rd., Weston, CT, 06883-2326. Phone: (203) 227-7521. Fax: (203) 454-1032. Web Site: www.genebayliss.com.

Gene Bayliss, producer & consultant.

Produces, directs video conferences, videotapes for corporations & industries, meetings & special events.

Beckmann International, Meadow Ct., West St., Ramsey, Isle of Man, IM8 1AE. Phone: 44 01624 816585. Fax: 44 01624 816589. E-mail: beckmann@enterprise.net Web Site: www.beckmanngroup.co.uk.

TV-CATV only.

International sls distributor specializing in non-fiction progmg.

Beethoven Satellite Network, (Classical Music Format Service). c/o WFMT Fine Arts Radio, 5400 N. St. Louis Ave., Chicago, IL, 60625. Phone: (773) 279-2000. Phone: (800) USA-WFMT. Fax: (773) 279-2199. Web Site: www.wfmt.com.

Steve Robinson, VP; Peter Vandegraaff, progmg dir; Carol Martinez, stn mgr.

Radio Only.

Program production & distribution; 168 hour-a-week classical music format with program hosts in one-hour modules, loc sound included.

Dave Bell Associates Inc., 3211 Cahuenga Blvd. W., Hollywood, CA, 90068. Phone: (323) 851-7801. Fax: (323) 851-9349. E-mail: dbmovies@aol.com

Dave Bell, pres; Ted Weiant, VP; Fred Putman, VP; Kitty Stallings, Associate.

TV-CATV only.

Dev & production of TV movies, reality series, feature films, documentaries & game shows.

Bell Foto Art Productions, 375 Josephine St., Suite C, Denver, CO, 80206. Phone: (303) 377-4606. Fax: (303) 322-2443. E-mail: bellfoto@att.net Web Site: www.bellfoto.tv.

Chris Bell, owner.

TV-CATV only.

Award winning video production: VNR, corporate, news sports, medical, training & legal. Story tellers in Beta SP, DV-cam, 24 & High Definition-Varicam.

Bellon Entertainment, 250 W. 57th St., Suite 1414, New York, NY, 10107. Phone: (212) 265-1222. Fax: (212) 265-7318. E-mail: bellonent@aol.com

Gregory P. Bellon, pres.

TV-CATV only.

Represent and develop TV formats for worldwide distribution.

Best Film & Video Corp., 157 Fairview Ave., East Meadow, NY, 11554. Phone: (516) 931-6969. Fax: (516) 931-5959.

Roy B. Winnick, pres; Dana Miller, dir mktg.

Beverly Hills, CA 90210, 242 N. Canon Dr. Phone: (310) 274-9944. Fax: (310) 274-9960.

TV program production & distribution of home video.

Black Audio Devices, Box 106, Ventura, CA, 93002-0106. Phone: (805) 653-5557. Fax: (805) 653-5557. Web Site: www.blackaudio.com.

TV-CATV only.

Blackbird Productions, 535 King's Rd., Suite 115, The Plaza, London, SWIO 0SZ. Phone: +44 (0171) 352-4882. Fax: +44 (0171) 351-3728.

TV-CATV only.

Program production & distribution.

Blackstone Stock Footage, 509 Upsall Drive, Antioch, TN, 37013. Phone: (615) 731-5310. Fax: (615) 731-5232. E-mail: g.clifford@worldnet.att.net Web Site: www.blackstonestockfootage.com.

Glenda Clifford, pres.

TV-CATV only.

We offer: Archival newsreel footage, medical, extreme sports, landmarks from around the world, food, people, animals, underwater, timelapse cities & nature.

Blanc Communications Corp., 171 Pier Ave., Suite 517, Santa Monica, CA, 90405. Phone: (310) 278-2600. Fax: (310) 396-8434.

TV program production; TV & radio coml production & distribution.

The Chuck Blore Co., 17428 Tarzana St., Encino, CA, 91316. Phone: (818) 784-5104. Fax: (818) 986-1196. E-mail: bloregroup@aol.com Web Site: www.chuckblore.com.

Chuck Blore, CEO.

TV-CATV only.

TV programs & coml production svcs. Radio coml production svcs. TV programs consultation.

Blue Canyon Productions, Box 6622, Santa Fe, NM, 87502. Phone: (505) 989-9298. Web Site: www.bluecanyonproductions.com.

Jim Terr, pres.

TV-CATV only.

Award-winning, nationally-bcst jingle, PSA & radio spot production, voice-overs, video production, as well as music production, scoring & scripting.

Blue Heaven Productions, 11 Glenwood Rd., Toms River, NJ, 08753-4117. Phone: (732) 349-8569. E-mail: raynorman3@juno.com

Ray Norman, pres.

Radio Only.

Nostalgia music production library, CD masters made

Blue Sky Studios, 44 S. Broadway, White Plains, NY, 10601. Phone: (914) 259-6500. Fax: (914) 259-6499. E-mail: query@blueskystudios.com Web Site: www.blueskystudios.com.

Brian Keane, gen mgr.

TV-CATV only.

Dev & production of CG animated films.

Blue Star Media, Dallas Cowboys Broadcasting. Dallas Cowboys Football Club, One Cowboys Pkwy., Irving, TX, 75063. Phone: (972) 556-9345. Fax: (972) 556-9339. Web Site: dallascowboys.com.

Scott Purcel, dir bcstg.

TV-CATV-Radio.

Radio play by play, wkly sports TV for NFL Dallas Cowboys & show for Dallas Cowboys.

Robert L. Bocchino, 264 Montgomery Ave., Haverford, PA, 19041-1531. Phone: (610) 649-0993. Fax: (610) 649-0895.

Robert L. Bocchino, owner.

TV-CATV only.

Voice over artist, coml spokesperson.

Bonneville Communications, 5 Triad Ctr., Suite 700, Salt Lake City, UT, 84180-1121. Phone: (801) 237-2600. Fax: (801) 237-2614. E-mail: bonneville@bonneville.com Web Site: www.bonneville.com.

Gregg D. Garber, gen mgr; Marc Lee, dir; Paul Yates, controller & VP.

TV-CATV only.

A values-driven adv agency engaged in communications for quality life.

Boston Symphony Orchestra, Symphony Hall, 301 Massachusetts Ave., Boston, MA, 02115. Phone: (617) 266-1492. Fax: (617) 638-9367. Web Site: www.bso.org.

Mark Volpe, mgng dir.

TV-CATV only.

Evening at Pops TV series & other special TV productions. Originates regular radio bcst of BSO concerts.

Dick Brescia Associates, 164 Garfield St., Haworth, NJ, 07641. Phone: (201) 385-6566. Fax: (201) 385-6449. E-mail: dbasyndicators@prodigy.net Web Site: www.ictx.com/dba.

Radio Only.

Radio shows: *When Radio Was*, *Stan Freberg Here*. Radio movie classics, radio super heroes.

Brillig Productions Inc., 770 Amalfi Dr., Pacific Palisades, CA, 90272. Phone: (310) 459-4450. Fax: (310) 459-4456. E-mail: brilligprod@cooliwk.net

Barry Brown, pres; Joy Brown, VP.

TV-CATV only.

Feature films, TV features, TV comls, documentaries.

British Broadcasting Corp., (Fine Arts Programs.). c/o WFMT Radio Network, 5400 N. St. Louis Ave., Chicago, IL, 60625. Phone: (773) 279-2000. Fax: (773) 279-2199. E-mail: finearts@wfmt.com Web Site: www.wfmt.com.

Carol Martinez, mgr; Steve Robinson, sr VP.

Radio Only.

Distribute wkly series *My Music*, & *My Word* for coml & public stns in the United States by WFMT Fine Arts Network.

Broadcast News Service, Box 919, Norwood, MA, 02062-0919. Phone: (781) 344-6988. Fax: (781 344-8928.

P. J. Romano, dir.

TV-CATV only.

Radio, TV features & productions, audio news & features.

Broadcast Programming, 2211 5th Ave., Seattle, WA, 98121. Phone: (206) 728-2741. Phone: (800) 426-9082. Fax: (206) 441-6582. E-mail: experts@jmseattle.com Web Site: www.jonesradio.com.

Radio Only.

Daypart personality progmg, music log & consulting services.

Broadview Media, 4455 W. 77th St., Minneapolis, MN, 55435. Phone: (952) 835-4455. Fax: (952) 835-0971. E-mail: michaels@broadviewmedia.com

Michael Smith, VP progmg.

TV-CATV only.

Full-svc production, postproduction & creative svcs for the production of TV programs.

Himan Brown-Radio Drama Network, 285 Central Park W., New York, NY, 10024. Phone: (212) 724-4333.

Himan Brown, owner.

Radio Only.

Radio program production, TV, film program production & CD-ROM audio dramas.

Bruder Releasing Inc. (BRI), 2020 Broadway, Santa Monica, CA, 90404. Phone: (310) 829-2222. Fax: (310) 829-0202. E-mail: bruder@brivideo.net Web Site: www.46ri.net.

Marc Bruder, pres.

TV-CATV only.

Supplies ind films to pay-per-view, cable, bcst & video markets worldwide.

Bulbeck & Mas SL, Quinones, 2, 28015 Madrid Phone: 34 91 594 2709. Fax: 34 91 445 7212. E-mail: bymfilms@bulbeckymas.com

TV-CATV only.

Specialists in libraries of Spanish features.

Burrud Productions Inc., 16351 Gothard St., Unit D, Huntington Beach, CA, 92647. Phone: (714) 842-8422. Fax: (714) 842-0433. E-mail: burrudprod@aol.com Web Site: www.burrud.com.

John Burrund, pres/CEO; Linda Karabin, VP; Drew Horton, VP; Valerie Chow, VP; Shannon Mead, exec dir & CEO.

TV-CATV only.

Feature film & TV production of reality, wildlife, oceanic, human adventure, documentary & world exploration progmg.

Buzzco Associates Inc., 33 Bleecker St., Suite 5A, New York, NY, 10012. Phone: (212) 473-8800. Fax: (212) 473-8891. E-mail: info@buzzzco.com Web Site: www.buzzzco.com.

Marilyn Kraemer, sec/treas.

TV-CATV only.

A full range of animation from traditional to innovative computer 2-D.

C

CABLEready Corp., 98 East Ave., Norwalk, CT, 06851-5029. Phone: (203) 855-7979. Fax: (203) 855-8370. E-mail: info@cableready.net Web Site: www.cableready.net.

Gary Lico, pres; Lou Occhicone, VP progmg; Sabrina Sanchez, dir; Kerry Novick, VP sls.

TV-CATV only.

Dev & sls of programs to U.S. cable TV networks & systems & all international telecasters.

CA Media Development, Box 1141, 1415 Hooper Ave., Suite 203, Toms River, NJ, 08754. Phone: (732) 797-1965. Fax: (732) 797-1260. E-mail: ca.media@comcast.net

Greg Koziar, pres.

TV-CATV only.

Full svc adv agency, as well as coml production, for radio & cable TV.

CBC International Sales, Box 500, Stn. A, Toronto, ON, M5W 1E6. Canada. Phone: (416) 205-3500. Fax: (416) 205-3482. E-mail: cbcis@toronto.cbc.ca

Christina Criss Hajek; Susan Hewitt, head international sls (London) & new business dev; Sandra Sarciada-Naughton,.

London W1P 8DD, 43/51 Great Titchfield St.

Los Angeles, CA 90025, 1950 Sawtelle Blvd, Suite 333.

TV-CATV only.

CBC is Canada's natl bcstr. Produces & distributes TV progmg in both English & French.

CBS Paramount International Television, 5555 Melrose Ave., Los Angeles, CA, 90038. Phone: (323) 956-5000. Fax: (323) 862-2217. E-mail: jennifer.weingroff@cbsparamount.com Web Site: www.cbscorporation.com.

Armando Nunez Jr., pres.

TV-CATV only.

International television distribution, co-production, local production formats, channel management.

CBS Television Distribution, 2401 Colorado Ave., Suite 110, Santa Monica, CA, 90404. Phone: (310) 264-3300. Fax: (310) 264-3301. Web Site: www.cbscorporation.com.

Roger King, CEO; Robert Madden, pres/COO; John Nogawski, pres/COO; Terry Wood, pres & creative affrs & dev.

TV-CATV only.

TV program production, distribution, mktg to domestic syndication & other TV venues.

CCI Entertainment Ltd., 18 Dupont St., Toronto, ON, MSR 1V2. Canada. Phone: (416) 964-8750. Fax: (416) 964-1980.

Arnie Zipursky, pres/CEO; Annette Frymer, COO.

TV-CATV only.

Distributor & co-producer, producer

CCM Media Services, 104 Woodmont Blvd., Suite 300, Nashville, TN, 37205. Phone: (615) 386-3011. Fax: (615) 312-4266. E-mail: bengland@ccmcom.com Web Site: www.ccmmagazine.com.

Radio Only.

Nationally syndicated radio programs, as well as spot and radio special production. "ccm radio magazine" is the flagship show

CDC United Network, 40 Rue Souveraine, 1050 Brussels Phone: (322) 502-6640. Fax: (322) 502-6656. E-mail: alexandre@cdc.skynet.be MOBILE: 3275713057

TV-CATV only.

TV distribution & merchandising in Latin America.

CDR Communications Inc., 9310-B Old Keene Mill Rd., Burke, VA, 22015. Phone: (703) 569-3400. Fax: (703) 569-3448. E-mail: chris@cdrcommunications.com Web Site: www.communications.com.

Christopher D. Rogers, pres; Nancy B. Rogers, VP.

TV-CATV-Radio.

Film, TV, video & radio production: teleconferences, documentaries, adv campaigns, PSAs; graphics, animation, syndication, promotion, publishing, distribution, postproduction & mktg.

CFP Video Productions, Box 86, Caldwell, NJ, 07006-0086. Phone: (973) 226-2481. Fax: (973) 226-2480. E-mail: don.spitzmiller@verizon.net

Donald Spitzmiller, pres.

TV-CATV-Radio.

Full video, audio svcs for TV & industrial productions. Avid edit svc/post production.

CIFEX International Inc., One Peconic Hills Ct., Southampton, NY, 11968-1618. Phone: (631) 283-9454. Fax: (631) 283-4210. E-mail: cifex@prodigy.net

Gerald J. Rappoport, pres; Beulah Rappoport, VP business affrs; Shirley Clarke, VP mktg.

TV-CATV only.

Distributor of foreign-language feature films, animated & live-action short films & documentaries.

CMT, 330 Commerce St., Nashville, TN, 37201. Phone: (615) 335-8400. Fax: (615) 335-8615. Web Site: www.cmt.com.

Brian Philips, sr VP; Judy McGrath, pres; Jama Bowen, VP; Martin Clayton, VP/Gen Mgr & CMT.com; Mary Beth Cunin, VP/Program Planning & scheduling; James Hitchcock, VP/ Creative & Marketing; Nick Loria, VP/ Ad Sls; Chris Parr, VP/ Music & Talent.

TV-CATV only.

America's #1 country music net, provides original progmg, live concerts, events, music videos by established & cutting edge artists, news & info.

CNBC Syndication, 900 Sylvan Ave., Englewood Cliffs, NJ, 07632. Phone: (201) 735-2622. Fax: (201) 585-3365.

Howard Homonoff, VP/gen mgr; Steve Blechman, mgr; Margaret Agsteribbe, mgr; Pamela Thomas Graham, CEO.

TV-CATV only.

Syndicated TV program, *Wall Street Journal Report*, business events.

CNDP—Centre National de Documentation Pedagogique, 29, rue D'Ulm, 75005 Paris Phone: 330146349310. Phone: 33146329308. Fax: 330-14-40-72-789. Web Site: www.cndp.fr.

TV-CATV-Radio.

Production & distribution of educational TV & multimedia programs.

CN8, The Comcast Network, Penns Landing Studio, 1351 S. Columbus Blvd., Philadelphia, PA, 19147. Phone: (215) 468-2222. Fax: (215) 468-3812. Web Site: www.cn8.tv.

Jonathan Gorchow; David Shane, dir of progmg & CN8; Cheryl Flamini, VP of business dev & Eastern division; Peggy Giordano, mgr & CN8 progmg; Denise Pettyford, dir of network adv sls & CN8; Larry Watzman, creative svcs dir & CN8; Alex Soumbenioits, mktg & PR mgr & CN8; Brian McLendon, dir of network productions & CN8; Scott Clark, dir of engrg & CN8; Stephanie Millagranna, admin coord; Mark Dudzinski, stn mgr; Rich Frantz, mgr engr; Jon Gurevitch, VP sports; Buck Dopp, VP.

New Castle, DE 19720. New Castle Studio, 2215 N. Dupont Hwy. Phone: (302) 661-4202. Fax: (302) 661-4201. Assignment desk: (302) 661-4290. Fax: (302) 661-4291.

TV-CATV only.

CN8, The Comcast Network, is a rgnl cable net offering news, sports, & entertainment progmg to 3.9 million cable homes.

CNN Newsource Sales, Inc., One CNN Center, 12 North, Atlanta, GA, 30303. Phone: (404) 827-5475. Fax: (404) 827-4466. E-mail: cnn.newsource@turner.com Web Site: newsource.cnn.com.

Chicago, IL 60601, 180 North Stetson Ave., Suite 2700. Phone: (312) 729-5925. Gary Butterfield, VP sls & affil rel.

New York, NY 10019, One Time Warner Ctr., 20th Fl. Phone: (212) 275-6734. (508) 627-6437. Joe Middleburg, VP sls & affil rel; Doug Jones, VP sls & affil rel.

TV-CATV only.

Provider of news & info content to the loc bcst news industry.

C N R Radio, Box 22246, Minneapolis, MN, 55422-0246. Phone: (763) 537-5868. E-mail: CNRadio@comcast.net Web Site: http://CNRadio.notlong.net.

George Carden, pres & producer.

Radio Only.

News interviews, soundbites & features with newsmakers for primarily Christian radio stns & nets.

CONUS Archive, 3415 University Ave., St. Paul, MN, 55114. Phone: (651) 642-4576. Fax: (651) 642-4669. Web Site: www.conus.com.

Chris Bridson, sr sls exec; Jim Richter, VP.

TV-CATV only.

Video archive svcs to natl & international program producers.

CRM Learning, 2215 Faraday Ave., Carlsbad, CA, 92008-7295. Phone: (800) 421-0833. Fax: (760) 931-5792. Web Site: www.crmlearning.com.

Peter J. Jordan, pres/CEO.

TV-CATV only.

Production & distribution of business training films.

CRN International, One Circular Ave., Hamden, CT, 06514. Phone: (203) 288-2002. Fax: (203) 281-3291. Web Site: www.crnradio.com.

Barry Berman, pres; S. Richard Kalt, exec VP.

Radio Only.

Short-form customized radio progms & promotions; *SkiWatch® BeachWatch* & small business reports.

CS Associates, 200 Dexter Ave, Watertown, MA, 02472-4236. Phone: (617) 923-0077. Fax: (617) 923-0025. E-mail: programs@csassociates.com

Charles Schuerhoff, pres; Brian Gilbert, aquisitions; Lisa Carey, VP intl sales; Jason Redmond, Mgr of Acquisitions.

TV-CATV only.

Program distribution, specializing in documentaries, foreign & domestic TV & cable; broker co-productions.

CTVC Hillside Studios, Merry Hill Rd., Bushey, Watford, Herts, WD23 1DR. Phone: 020 8950 4426. Fax: 020 8950 1437. E-mail: barrie.allcott@ctvc.co.uk Web Site: www.ctvc.co.uk.

Barrie Allcott, mgng dir; Ray Bruce, producer.

TV-CATV only.

Producers of programs with humanitarian values, especially relg. Also full bcst facilities available for hire.

CTV Television Inc., Box 9, Stn. O, Toronto, ON, M4A 2M9. Canada. Phone: (416) 332-5000. Fax: (416) 332-5065. Web Site: www.ctv.ca.

Susanne Boyce, pres CTV progmg.

TV-CATV only.

TV bcstg, program production & distribution.

C 2 Productions Inc., 15430 Catalpa Cove Ln., Fort Myers, FL, 33908. Phone: (239) 437-4222. Fax: (239) 437-2042. E-mail: C2Productions@earthlink.net Web Site: chriscorley.com.

Chris Corley, pres.

TV-CATV-Radio.

Voice-overs delivered digitally or in person.

Cable Films & Video, Box 7171, Country Club Station, Kansas City, MO, 64113. Phone: (913) 362-2804. Phone: (800) 514-2804. Fax: (913) 362-2804. E-mail: cablefilms@msn.com Web Site: www.onlineworld.com/movies.

Herbert Miller, CEO.

TV-CATV only.

Classic films, all formats: one inch BETA SP, CD-ROM, U-Matic, PAL, NTSC, SECAM, DVD. Over 300 motion pictures & classic cartoons, clips available.

Call For Action Inc., 5272 River Rd., Suite 300, Bethesda, MD, 20816. Phone: (301) 657-8260. Fax: (301) 657-2914. Web Site: www.callforaction.org.

Shirley L. Rooker, pres.

TV-CATV only.

International hotline svc, affiliated with the bcst media, that provides info, assistance to individuals & small businesses with consumer problems.

Camera Group, 3920 N. 29th Ave., Hollywood, FL, 33022. Phone: (305) 945-2020. Fax: (305) 945-1117. E-mail: cameragrp@aol.com Web Site: www.cameragroup.com.

Eileen Garcia-Di Rosa, pres.

TV-CATV only.

Rental, sls, svc & maintenance of motion picture, TV & video production equipment.

CamMate Studios, 425 E. Comstock, Chandler, AZ, 85225. Phone: (480) 813-9500. Fax: (480) 813-9292. E-mail: cammate@cammate.com Web Site: www.cammate.com.

Ron Mitchell, CEO.

TV-CATV only.

Camera cranes, telescopic cranes & mini cranes.

Campbell-Ewald Advertising, 30400 Van Dyke, Warren, MI, 48093. Phone: (586) 574-3400. Fax: (586) 558-5891. Web Site: www.campbell-ewald.com.

Anthony J. Hopp, CEO/chmn/pres; S.H. Gilbert, exec VP/CFO; D.A. Karnowsky, VP; W.J. Ludwig, VP; L.M. Schultz, VP; J.T. Seregny, VP.

Los Angeles, CA 90025, 11100 Santa Monica Blvd, 6th Fl. Phone: (213) 914-2200.

Chicago, IL 60611, One Magnificent Mile, 930 N. Michigan Ave, Suite 1060. Phone: (312) 587-2650.

New York, NY 10017, One Dag Hammarskjold Plaza. Phone: (212) 605-8000.

TV-CATV only.

TV programs, TV radio coml, promotion film production.

Canamedia Productions Ltd., 381 Richmond St. E., Suite 200, Toronto, ON, M5A 1P6. Canada. Phone: (416) 483-7446. Fax: (416) 483-7529. E-mail: canamed@canamedia.com Web Site: www.canamedia.com.

Les Harris, pres; Andrea Stokes, intl sls & acquisitions mgr.

TV-CATV only.

Canamedia offers production & international distribution svcs. It also exclusively represents in Canada the ITN source Archive & Natural History New Zealand Archives as well as the PUMP audio music archive.

CanLib Inc., 4819 Galendo St., Woodland Hills, CA, 91364-4326. Phone: (818) 888-6005. Fax: (818) 888-2505. E-mail: canlibinc@adelphia.net

Gene Accas, pres; Carol Stevens, exec VP & sec/treas.

TV-CATV only.

Bcstg & media consulting: rsch for producers, distributors, advertisers, agencies & law firms (legal expert witness).

Cannell Studios, 7083 Hollywood Blvd., Suite 600, Hollywood, CA, 90028. Phone: (323) 465-5800. Fax: (323) 856-7390. Web Site: www.cannell.com.

Stephen J. Cannell, chmn/CEO.

TV-CATV only.

Capital Communications, 2357-3 South Tamiami Trl., Venice, FL, 34293. Phone: (941) 492-4688. Fax: (941) 492-4923. E-mail: cap5678@isp.com Web Site: www.isp.com.

James Springer, pres/COO.

TV-CATV only.

International distributor of pre-packaged TV programs.

Carden & Cherry Syndication Inc., 1220 McGavock St., Nashville, TN, 37203. Phone: (615) 255-6694. Fax: (615) 255-8345.

TV-CATV only.

TV & radio coml production & distribution; production svcs.

Careco Television Productions, 5717 N.W. Pkwy., Suite 104, San Antonio, TX, 78249. Phone: (800) 668-8081. Fax: (210) 697-0150. Web Site: www.outdooraction.com.

Charles Goodloe, pres; Lavonne Kacalek, VP.

TV-CATV only.

Producer of *American Outdoors* & *Fishing Texas*, weekly half hour series.

Caridi Entertainment, 250 W. 57th St., Suite 1326, New York, NY, 10107. Phone: (212) 581-2277. Fax: (212) 581-2278. E-mail: c.caridi@att.net

TV-CATV only.

Full-svc international distributor & production company.

Carleton Productions International Inc., 1500 Merivale Rd., 5th Fl., Nepean, ON, K2E 6Z5. Canada. Phone: (613) 224-9666. Fax: (613) 224-9074. E-mail: cpi@magi.com Web Site: www.carletonproductions.com.

Mark Ross, pres.

TV & radio programs, coml production & distribution & production svcs.

George Carlson & Associates, 323 First Ave. W., Seattle, WA, 98119. Phone: (206) 213-0562. Fax: (206) 213-0562.

George Carlson, producer.

TV-CATV only.

Producers/distributors of 1/2-hour color, true life, travel adventure series to all parts of the world called *The Traveler & Northwest Traveler.*

Carlton International Media Inc., 11145 N.W. 1st Pl., Coral Springs, FL, 33071. Phone: (954) 345-1620. Fax: (954) 345-1490. E-mail: clarea@msn.com Web Site: www.carltonint.co.uk.

Claire Alter, VP; Rupert Dillnot-Cooper, CEO; Louise Pedersen, mgr.

Studio City, CA 91604 Phone: (818) 753-6363. Jeri Sacks, VP/US sls.

TV-CATV only.

British TV distributor, licenses a wide range of programs worldwide.

Carpel Video Inc., 429 E. Patrick St., Frederick, MD, 21701. Phone: (800) 238-4300. Phone: (301) 694-3500. Fax: (301) 694-9510. Web Site: www.carpelvideo.com.

Andy Carpel, pres.

TV-CATV only.

Videotape recyclers; production svcs, video tape to DVD duplication.

Carriage House Studios, 119 Westhill Rd., Stamford, CT, 06902. Phone: (203) 358-0065. Fax: (203) 964-4988. E-mail: chstudios@aol.com

John Montagnese, pres & Studio mgr.

TV-CATV only.

Recording studio.

Carsey-Werner Distribution, 12001 Ventura Pl., Suite 600, Studio City, CA, 91604. Phone: (818) 299-9600. Fax: (818) 299-9650. Web Site: www.carseywerner.com.

Bob Dubelko, pres/COO; Herbert Lazarus, pres; James Kraus, pres.

TV-CATV only.

TV program distribution.

Sandra Carter Global, Inc., 230 W. 79th St., Suite 102, New York, NY, 10024. Phone: (212) 875-1811. Fax: (212) 875-0088. E-mail: sales@sandra-carter.com Web Site: www.sandra-carter.com.

Sandra Carter, pres; Ettore Botta, VP sls.

TV-CATV only.

Distribution to all media, co-production deals, production, principle product in factual series.

Castle Hill Productions Inc., 36 W. 25th St., 2nd Fl., New York, NY, 10010. Phone: (212) 242-1500. Fax: (212) 414-5737. E-mail: mm@castlehillproductions.com Web Site: www.castlehillproductions.com.

Julian Schlossberg, chmn; Mel Maron, pres; Barbara Karmel, VP TV sls.

Boca Raton, FL 33431, 2385 Executive Center Dr, Suite 100.

TV-CATV only.

Movie distribution for theater, TV, cable, and video.

Catholic Communication Campaign, 3211 4th St. N.E., Washington, DC, 20017. Phone: (202) 541-3204. Fax: (202) 541-3129. E-mail: pgarcia@usccb.org Web Site: www.usccb.org/ccc.

Pat Ryan Garcia, dir.

TV-CATV-Radio.

TV & radio pub svc programs.

Catholic Communications Corp., 65 Elliot St., Springfield, MA, 01101. Phone: (413) 452-0648. Fax: (413) 747-0273. E-mail: m.graziano@diospringfield.org

Mark Duport, CEO.

TV-CATV-Radio.

TV & radio production svcs.

Catholic Television Network, Box 430, 9531 Akron-Canfield Rd., Canfield, OH, 44406-0430. Phone: (330) 533-2243. Fax: (330) 533-1907. E-mail: judyctny@aol.com Web Site: www.doy.org.

Bob Gavalier, gen mgr.

TV-CATV only.

24-hour ecumenical TV ch.

Celebrities Productions, 230 S. Bemiston Ave., Suite 1400, St. Louis, MO, 63105. Phone: (314) 862-7800. Fax: (314) 721-5171.

I.J. Davis, pres; David Dovich, VP; Walt Williams, VP.

TV-CATV-Radio.

Creation, production of radio, TV spots, programs, audio visuals; arrangement for celebrity talent, music, syndication & video conference production.

CelebrityFootage, 320 South Almont Dr., Beverly Hills, CA, 90211. Phone: (310) 360-9600. Fax: (310) 360-9696. E-mail: michael@celebrityfootage.com Web Site: www.celebrityfootage.com.

Michael Goldberg, pres.

TV-CATV only.

Provides broadcasters with celebrity entertainment news from the Los Angeles area, including movie premieres, award shows & charity benefits.

Celluloid Dreams, 24 rue Lamartine 75009, Paris Phone: (33) 1 49 70 83 20. Fax: (33) 1 49 70 03 71. Web Site: www.celluloid-dreams.com.

TV-CATV only.

International distribution of ind features, documentaries & animation films.

Center City Film & Video, 1503-05 Walnut St., Philadelphia, PA, 19102. Phone: (215) 568-4134. Fax: (215) 568-6011. E-mail: centercity@ccfv.com Web Site: www.ccfv.com.

Jordan M. Schwartz, chmn/pres; Brian Isely, VP/gen mgr; John Gillespie, exec producer.

TV-CATV only.

Award winning production staff, video; film production, studio; remote camera packages including ultimatte, digital audio suite, flint, complete post production.

Central City Productions, Inc., 212 East Ohio St. #3, Chicago, IL, 60611. Phone: (312) 654-1100. Fax: (312) 321-9921. Web Site: www.ccptv.com.

Don Jackson, chmn/CEO; Rosemary Jackson, VP; Erma Gray Davis, pres/COO; Jennifer J. Jackson, gen mgr; Heather Davis, sls VP.

TV-CATV only.

Production & mktg of bcst & cable TV progmg targeted towards minority viewers.

Central Park Media, 250 W. 57th St., Suite 1723, New York, NY, 10107. Phone: (646) 957-8301 (X-201). Fax: (646) 957-8316. E-mail: jod@teamcpm.com Web Site: www.centralparkmedia.com.

John O'Donnell, mgng dir.

TV-CATV only.

Over 200 Japanese Anime titles available for TV & cable.

Century III at Universal Studios Florida, 2000 Universal Studios Plaza, Orlando, FL, 32819-7606. Phone: (407) 354-1000. Fax: (407) 352-8662. E-mail: rcibella@century3.com Web Site: www.century3.com.

TV-CATV-Radio.

Full-svc production & postproduction facility, audio department, custom graphics, digital editing capabilities, film transfers, interactive department, satellite uplink svcs.

Channel Four Television, 124 Horseferry Rd., London, SW1 2TX. Phone: 44 20 7396 4444. E-mail: righttoreply@channel.4.com Web Site: www.channel4.com.

TV-CATV only.

UK bcstr.

Chicago Radio Syndicate Inc., 15003 Lemay St., Van Nuys, CA, 91405. Phone: (800) 621-6949. Fax: (818) 376-8529. Web Site: www.sandyorkin-crs.com. E-mail: sandyo@earthlink.net

Sandy Orkin, pres.

Radio Only.

Syndication of Dick Orkin comedy features—*Chickenman, Tooth Fairy & Mini-People* & commercial camgaigns.

Children's Media Productions, Box 40400, Pasadena, CA, 91114-7400. Phone: (626) 797-5462. E-mail: childrensmedia@yahoo.com Web Site: www.childrensmedia.com.

C. Ray Carlson, pres; Joy Carlson, PR.

TV-CATV only.

Producer & distributor of children's progmg, videos & feature films, worldwide.

Christian Children's Associates Inc., Box 446, Toms River, NJ, 08754. Phone: (732) 240-3003. Fax: (732) 286-4244. E-mail: adventurepals@juno.com Web Site: www.adventurepals.com.

Jean Donaldson, pres; Frank Troilo, VP; Reverend William Cook, dir.

TV-CATV-Radio.

Production, distribution of radio & TV progmg for children.

Christian Media Network, Box 448, Jacksonville, OR, 97530. Phone: (541) 899-8888. Web Site: www.christianmedianetwork.com.

TV-CATV-Radio.

Christian Science Sentinel - Radio Edition, One Norway St., C4-20, Boston, MA, 02115-3122. Phone: (617) 450-2000. Fax: (617) 450-3997. E-mail: sentinelradio@csps.com Web Site: www.sentinelradio.com.

Susan Kerr, producer.

Radio Only.

Religious radio programs.

Christian TV Services of Ellicottville Inc., P. O. Box 209, Ellicottville, NY, 14731-0209. Phone: (716) 699-2549. Fax: (716) 699-2590. E-mail: geothayer@yahoo.com Web Site: www.christiantvservices.com.

Rev. George A. Thayer, pres/CEO.

TV-CATV only.

Christian media consultants "Ministering to ministries around the world"; locally linked area worship places, internet listing places; svcs , inc. 1974 internet.

The Christophers Inc., 12 E. 48th St., New York, NY, 10017. Phone: (212) 759-4050. Fax: (212) 838-5073. E-mail: mail@christophers.org Web Site: www.christophers.org.

Tony Rossi, producer; Lisa Mantineo, producer.

TV-CATV only.

TV & radio production & distribution.

Chrysalis Distribution, 13 Bramley Rd., London, WI0 6SP. Phone: (44) 207 4674. Fax: (44) 207 221 6286. E-mail: distribution@chrysalis.co.uk

Christina Willoughby, mgng dir.

TV-CATV only.

International sale of TV programs to all media worldwide.

Cimarron Group, 6855 Santa Monica Blvd., Hollywood, CA, 90038. Phone: (323) 337-0300. Fax: (323) 337-0333. Web Site: www.cimarrongroup.com.

Cheryl Savala, sr art dir; Bob Farina, owner.

TV-CATV only.

TV promotions, spec shoots, graphics, sls presentations, trade & consumer print design, title treatment & image campaigns.

Cinecraft Productions Inc., 2515 Franklin Blvd., Cleveland, OH, 44113. Phone: (216) 781-2300. Fax: (216) 781-1067. E-mail: info@cinecraft.com Web Site: www.cinecraft.com.

Neil G. McCormick, chmn; Neil G. MCCormick, mgr; Maria E. Keckan, pres.

TV-CATV only.

Betacam field production; 60' x 70' sound stage with hard cyc; AVID MC1000NT; Animation with Soft Image; interactive DVD & CD-R dev.

CineFilm/CineTransfer, 2156 Faulkner Rd. N.E., Atlanta, GA, 30324. Phone: (404) 633-1448. Phone: (800) 633-1448. Fax: (404) 633-3867. E-mail: csr@cinefilmlab.com Web Site: www.cinefilmlab.com.

William G. Thorton, pres; Jim Ogburn, gen mgr.

TV-CATV only.

16mm, super 16mm, 35mm color negative processing & printing. Dailies thru release prints. State-of-the-art video dailies & scene-to-scene transfers. Spirit Data Cini, all HD Formats

CineGroupe, 1151 Alexandre-DeSeve St., Montreal, PQ, H2L 2T7. Canada. Phone: (514) 524-7567. Fax: (514) 849-5001. E-mail: distribution@cinegroupe.ca Web Site: www.cinegroupe.com.

Jacques Pettigrew, pres/CEO; Michel Lemire, exec VP; Marie-Christine DuFour, exec VP.

TV-CATV only.

Animation, TV production, postproduction, paint & trace studio, distribution of animation TV series for children.

Cinema Concepts Animation Studio, 2030 Powers Ferry Rd., #214, Atlanta, GA, 30339. Phone: (770) 956-7460. Fax: (770) 956-8358. E-mail: info@cinemaconcepts.com Web Site: www.cinemaconcepts.com.

Stewart D. Harnell, CEO; Sharron A. Harnell, VP; John Price, studio dir; Theresa Dickey, gen mgr.

TV-CATV-Radio.

Animated corporate IDs, presentation/policy trailers for TV, cable & motion picture theatres, theatrical trailer fulfillment.

The Cinema Guild Inc., 130 Madison Ave., 2nd Fl., New York, NY, 10016-7038. Phone: (212) 685-6242. Fax: (212) 685-4717. E-mail: info@cinemaguild.com Web Site: www.cinemaguild.com.

Gary Crowdus, gen mgr; Philip Hobel, chmn/CEO; Mary Ann Hobel, co-chmn.

TV-CATV only.

Film & video distribution to theatrical, non-theatrical, TV & home video mkts, worldwide.

Cinema Sound Ltd., 311 W. 75th St., New York, NY, 10023. Phone: (212) 799-4800. Fax: (212) 799-2057.

Joan S. Franklin, pres; John S. Rockwell, production dir.

Radio Only.

State of the art recording studio.

Circle Oak Productions Inc., 33 N. Birch Hill Rd., Patterson, NY, 12563. Phone: (845) 878-9017. Fax: (845) 878-9018.

Educ film production.

Tim Cissell Music, 1120 Grassmere Dr., Richardson, TX, 75080-2909. Phone: (972) 680-0817. Fax: (972) 680-0866. E-mail: tcissell@wt.net Web Site: www.web.wt.net/~tcissell.

Tim Cissell, owner.

TV-CATV-Radio.

Offers music composition & production for all media (TV/CATV & radio)—jingles, IDs, film & video.

The Dick Clark Productions, 9200 W Sunset Blvd., Loa Angeles, CA, 90069-3502. Phone: (818) 841-3003. Fax: (818) 954-8609. Web Site: www.dickclarkproductions.com.

Dick Clark, chmn/CEO; Francis C. La Maina, pres/COO; Bill Simon, CFO.

TV-CATV only.

TV production for networks, cable & syndication. Produces series, specials & movies for TV.

Classic Media, 860 Broadway, 6th Fl., New York, NY, 10003. Phone: (212) 659-3011. Fax: (212) 659-1958.

Douglas Schwalbe, head of international; Bob Higgins, head of creative affrs & production.

Beverly Hills, CA 90211, 8640 Wilshire Blvd. Phone: (310) 659-6004. Fax: (310) 659-4599. Leslie Levine, Dorothy Schecter.

TV-CATV only.

Classic Media is a New York-based entertainment company that manages some of the most recognizable family oriented properties across all media including feature film, television, home video & consumer products.

The Classical Station, WCPE, Box 897, Wake Forest, NC, 27588. Phone: (919) 556-5178. Fax: (919) 556-9273. E-mail: wcpe@wcpe.org Web Site: theclassicalstation.org.

Deborah Proctor, gen mgr; Dick Storck, progmg dir; Rae Weaver, dev dir; Curtis Brothers, dir.

TV-CATV only.

Free 24-hour classical music progmg with live announcers for radio, cable, other distributors. Wkly request programs, opera & features.

Clausen Communications Inc., PMB 150, 235 W. Brandon Blvd., Brandon, FL, 33511-5103. Phone: (407) 696-9095.

Chris Clausen, pres.

TV-CATV only.

Network quality voice-overs for TV & radio stns. Same day service on internet, switched 56/ISDN, CD, DAT or analog delivery.

Clayton-Davis & Associates Inc., 230 S. Bemiston Ave., Suite 1400, St. Louis, MO, 63105. Phone: (314) 862-7800. Fax: (314) 721-5171. Web Site: www.claytondavis.com.

Jennifer Jermak, pres; Steve Pezold, VP.

TV-CATV-Radio.

Program production, syndication & barter.

Clear Channel Broadcasting Inc., 55 Music Sq. West, Nashville, TN, 37203. Phone: (615) 664-2400. Fax: (615) 664-2457. E-mail: davealpert@clearchannel.com

Dave Alpert, pres; Kevin Moore, VP sls; Tom Stevens, opns dir.

TV-CATV-Radio.

State radio networking, collegiate radio & TV networking.

Clear Channel Entertainment Television, 220 W. 42nd St., 9th Fl., New York, NY, 10036. Phone: (917) 421-5206. Fax: (917) 421-5239. Web Site: www.clearchannelentertainment.com.

Steve Stern, exec producer; Joe Townley, pres; Marc Forest, progmg VP; Dawn Olejar, opns VP.

TV-CATV only.

Produces TV programs & promotion films, documentary film production, sports event TV production.

Clever Cleaver Productions, 4718 N. Placita Ventana Del Rio, Tucson, AZ, 85750-6271. Phone: (520) 615-1582. Fax: (520) 615-1586. E-mail: cleverccook@comcast.net Web Site: www.clevercleaver.com.

Lee N. Gerovitz, pres; Steve Cassarino, VP.

TV-CATV only.

Offers 260 3-minute cooking vignettes & 27 30-minute cooking shows, 90-second high-definition cooking vignettes (cash or barter) & 2-minute tailgate cooking vignettes (free licensing fee).

Coe Film Associates Inc., 70 E. 96th St., New York, NY, 10128. Phone: (212) 831-5355. Fax: (212) 996-6728. E-mail: cfainc@juno.com

TV-CATV only.

TV program distribution.

Colon & Associates Inc., 7100 Blvd. East, Guttenberg, NJ, 07093. Phone: (201) 869-4615. Fax: (201) 869-6217. E-mail: rei616@aol.com Web Site: www.saptv.com.

Reinaldo Colon, pres & CEO.

TV-CATV-Radio.

Program distribution, production, Sp language dubbing.

ComBridges, 70 Irwin St., San Rafael, CA, 94901. Phone: (415) 454-5505. Fax: (415) 454-1941. Web Site: www.combridges.com. E-mail: info@combridges.com

Jon Leland, pres & creative dir; Donna Nieddu, mgr; John Kraus, exec producer.

TV-CATV only.

Source of videos, seminars & interactive media. Producer, websites.

Combs Music, 421 Cedar Trail, Winston-Salem, NC, 27104. Phone: (336) 760-3905. Fax: (336) 760-3855. E-mail: dave@combsmusic.com Web Site: www.combsmusic.com.

TV-CATV only.

Composes, produces, publishes & distributes easy lstng instrumental music, e.g., *Rachel's Song*.

Comcast Media Center, 4100 E. Dry Creek Rd., Littleton, CO, 80122. Phone: (303) 486-3800. Fax: (303) 486-3891. Web Site: www.comcast.com.

TV-CATV only.

Network origination, production, postproduction, uplinking, compression, remote production, audio production.

Command Productions, Box 3000, Sausalito, CA, 94966-3000. Phone: (415) 332-3161. Fax: (415) 332-1901. E-mail: audio@commandproductions.com Web Site: www.commandproductions.com.

Warren Weagant, pres; Kitt Weagant, VP.

TV-CATV-Radio.

Radio-TV, CATV audio voice identification and promotion production.

Communications III Inc., 921 Eastwind Dr., Suite 104, Westerville, OH, 43081. Phone: (614) 901-7720. Fax: (614) 901-7721. E-mail: shalliday@comiii.com Web Site: www.comIII.com.

Scott Halliday, pres.

TV-CATV only.

Central Ohio C-band satellite & Teleport svc, with access to studio, edit suites, at Ohio State University & various downtown locations.

CompuWeather Inc., 2566 Rt. 52, Hopewell Junction, NY, 12533. Phone: (800) 825-4445. Fax: (800) 825-4441. E-mail: forecasts@compuweather.com Web Site: www.compuweather.com.

Jeff Wimmer, principal; Todd Gross, principal.

TV-CATV-Radio.

Weather, environmental features, actualities, forecasts, info, worldwide weather consulting svc, forecast, outcodes, studies, advice, site specific, 24/7 & 31 years experience.

Concept Videos, 5371 Punta Alta, Apt.1E, Laguna Hills, CA, 92653. Phone: (800) 333-8252. Fax: (877) 523-5592. E-mail: wjconnell@preschoolpower.com Web Site: www.preschoolpower.com.

William Connell, pres.

TV-CATV only.

Gold medal winning children's series, *Pre-School Power*, (13 x 30) recently telecast on 190 public TV stns.

Consolidated Film Industries (CFI), 959 N. Seward St., Hollywood, CA, 90038. Phone: (323) 960-7444. Fax: (323) 960-7573. Web Site: www.technicolor.com.

TV-CATV only.

Film processing; titles & opticals; videotape transfers; office rentals.

Continental Recordings Inc., 23 Mirimichi Street, Plainville, MA, 02762-1710. Phone: (508) 699- 0003. Fax: (617) 699-0005. E-mail: danf31@earthlink.net

L. Daniel Flynn, pres.

TV-CATV only.

Coml jingles, stn IDs, original music creation & production, cassette duplication & bcstg adv consultation, CD duplicator.

William F. Cooke Television Programs, 890 Yonge St., Suite 800, Toronto, ON, M4W 3P4. Canada. Phone: (416) 967-6141. Fax: (416) 967-5133. E-mail: cooketv@canada.com

William F. Cooke, pres/CEO; Alex McWilliams, pres distribution.

TV-CATV only.

TV program production & distribution.

Cookie Jar Group, 1055 Rene Levesque E., 9th Floor, Montreal, PQ, H2L 4S5. Canada. Phone: (514) 843-7070. Fax: (514) 843-6773. E-mail: info@cinar.com Web Site: www.cinar.com.

Steward Syder, CEO; Peter Moss, pres; David Ferguson, pres; Steve Carson, Cinar Education.

TV-CATV only.

Dev, production, postproduction & distribution of non-violent, quality kids & family live action animated progmg, & educational products.

Coote Communications, 568 Carver Hill, Milton, ON, LQT 5K5. Canada. Phone: (905) 203-0065. E-mail: amcoote@hotmail.com

Morgan Coote, pres; Donald Coote, VP.

TV-CATV only.

Complete film & videotape production from script to screen.

Cornell University Educational Television Center, 126 CCCGarden Avenue, Ithaca, NY, 14853-6601. Phone: (607) 255-8162. Fax: (607) 255-1563. E-mail: grp2@cornell.edu Web Site: www.DLS.cornell.edu.

David O. Watkins; Glen Palmer, mgr business/production svcs.

TV-CATV only.

Satellite uplinks, Betacam DvcPro video production & postproduction, audio production

Country Crossroads, 6350 W. Freeway, Fort Worth, TX, 76116. Phone: (817) 570-1491. Phone: (800) 292-2287. Fax: (817) 737-9436. E-mail: cries@familynetradio.com Web Site: www.countrycrossroadsradio.com.

Chuck Ries, producer; Kirk Teegarden, producer.

Radio Only.

Country music with interviews. Program hosted by Brother Jon Rivers, 30 minute wkly on CD or download.

Cramer Productions Center, 425 University Ave., Norwood, MA, 02062. Phone: (781) 278-2300. Fax: (781) 255-0721. E-mail: info@crameronline.com Web Site: www.crameronline.com.

Tom Martin, pres; Rich Sturchio, VP.

TV-CATV only.

Film & video production svcs from design to presentation; staging svcs; video duplications. Web-casting, interactive media.

Thomas Craven Film Corp., 5 W. 19th St., New York, NY, 10011-4216. Phone: (212) 463-7190. Fax: (212) 627-4761. E-mail: michael@cravenfilms.com Web Site: www.cravenfilms.com.

Michael Craven, pres; Ernest Barbieri, VP.

TV-CATV-Radio.

Complete film & video production svcs from scripting through shooting & editing to distribution.

Crawford Communications, 3845 Pleasantdale Rd., Atlanta, GA, 30340. Phone: (404) 876-7149. Fax: (678) 421-6717. Web Site: www.crawford.com.

Jesse C. Crawford, owner; Paul Hansel, pres; Bill Thompson, sls dir; Jessica Moore, mktg dir.

TV-CATV-Radio.

Computer graphics, animation, production & postproduction svcs domestic & international teleport.

Creative International Activities Ltd., 372 Central Park W., New York, NY, 10025. Phone: (212) 663-8944. Fax: (212) 865-8486. E-mail: ciaklaus@aol.com

Klaud J. Lehmann, pres.

International TV program syndication & consultation.

Creative Marketing & Communications Corp., 7633 Athenia Dr., Cincinnati, OH, 45244. Phone: (513) 624-8301. Phone: (800) 845-8477. Fax: (513) 624-8302. Web Site: www.cmcideas.com.

Terry Dean, pres; Susan Dean, VP.

Radio Only.

Syndicated 30-60-second coml wraparounds.

Creative Radio Network, Box 7749, Thousand Oaks, CA, 91359. Phone: (818) 991-3892. Fax: (818) 991-3894.

Darwin Lamm, pres.

Radio Only.

Radio program production & syndication—all formats; holiday & artist specials.

Crest National Digital Media Complex, 1000 N. Highland Ave., Hollywood, CA, 90038. Phone: (323) 466-0624. Fax: (323) 461-8901. Web Site: www.crestnational.com. E-mail: info@crestnational.com

Ron Stein, pres; John Walker, exec sls mgr.

TV-CATV only.

Full videotape postproduction including film transfer, processing, sweetening & duplication.

The Crime Channel, PMB 371, 42335 Washington St., Suite F, Palm Desert, CA, 92211. Phone: (760) 360-6151, EXT. 4. Fax: (760) 360-3258.

Arnold Frank, pres.

TV-CATV-Radio.

TV program distribution & bcstg.

Critical Mass Releasing Inc., 77 Mowat Ave., Suite 110, Toronto, ON, M6K 3E3. Canada. Phone: (416) 538-2535. Fax: (416) 538-3367. E-mail: cmass@netcom.ca

William Alexander, pres; Lisa-Marie Doorey, Dir of International sls.

TV-CATV only.

International TV & domestic distribution, TV & film, film & series production, theatrical & video releasing.

Ben Cromer Communications, Box 526, Round Hill, VA, 20142. Phone: (540) 338-5486. Fax: (540) 338-5486. E-mail: info@bencromer.com

Ben Cromer, pres.

TV-CATV-Radio.

Feature writing & script preparation for print & bcst media; specializing in the music & entertainment industry; telecommunications; business/economics & travel/history.

Crossroads Christian Communications Inc., Box 5100, Burlington, ON, L7R 4M2. Canada. Phone: (905) 335-7100.
TV-CATV only.
Produces 100 Huntley Street program.

Crown International Pictures Inc., 8701 Wilshire Blvd., Beverly Hills, CA, 90211. Phone: (310) 657-6700. Fax: (310) 657-4489. E-mail: crown@crownintlpictures.com Web Site: www.crownintlpictures.com.
Scott E. Schwimer, sr VP; Mark Tenser, pres/CEO; Lisa Agay, dir publ & adv.
TV-CATV only.
Film production & distribution.

Crystal Pictures Inc., 9 Pack Sq., #204, Asheville, NC, 28801. Phone: (828) 285-9995. Fax: (828) 285-9997. E-mail: cryspic@aol.com
Joshua Tager, pres; Jane Anne Rolston, gen sls mgr.
Ashville, NC 28801, 9 S.W. Pack Sq. Phone: (828) 285-9996. Fax: (828) 285-9997.
TV-CATV only.
Distribution of feature film & svcs to all TV outlets in United States & abroad.

Cube International, 1863 Pamela Ct., Suite 50, Simi Valley, CA, 9365. Phone: (661) 255-1945. Fax: (805) 527-1160. E-mail: phill@cubeinternational.com Web Site: www.cubeinternational.com.
Olivier de Courson, mgng dir; Phillip G. Catherall, mgng dir.
TV-CATV only.
Film & TV distribution to all worldwide markets & media.

Curb Entertainment International Corp., 3907 W. Alameda Ave., Burbank, CA, 91505. Phone: (818) 843-8580. Fax: (818) 566-1719. E-mail: info@curbentertainment.com Web Site: www.curbentertainment.com.
Carole Curb, pres; Mike Curb, chmn; Ilda Toth, exec dir; Aaron Rogers, dir mktg.
TV-CATV only.
International production & distribution co.

Custom Productions Inc., 1334 3rd St. Promenade, Suite 300, Santa Monica, CA, 90401. Phone: (310) 393-4144. Fax: (310) 393-1143. Web Site: www.customproductions.TV.
Steve Stockman, pres.
TV-CATV only.
Creation & production of custom TV campaigns for radio stns and TV news stns in the top 25 markets.

D

DC Audio, (Dryden Clarke Audio & Daily Feed). 1783 Lanier Pl. N.W., Suite B, Washington, DC, 20009. Phone: (202) 667-1234. Fax: (202) 667-5578. E-mail: dfeed@dailyfeed.com Web Site: www.dailyfeed.com.
John Dryden, pres.
Radio Only.
Produces The Daily Feed, a 90-second political, social satire radio commentary. Markets cash and bartered radio inventory to 18 + demos.

DG Systems, 750 W. John Carpenter Fwy., Irving, TX, 75039. Phone: (972) 581-2000. Fax: (972) 581-2001. Web Site: www.dgsystems.com.
Marty Melody, VP sls.
TV-CATV-Radio.
Duplication & distribution of corporate training & educ, TV & radio programs. Distribution of syndicated TV programs via satellite & videotape.

DIC Entertainment, 4100 W. Alameda Ave., Burbank, CA, 91505. Phone: (818) 955-5400. Fax: (818) 955-5696. Web Site: www.dicentertainment.com.
Andy Heyward, chmn/CEO; Brad Brooks, pres; Jedd Gold, mktg VP.
TV-CATV only.
DIC Entertainment, a leading children's entertainment company, is a full-service studio dedicated to creating, developing, producing, distributing, mktg & mdsg children's & family-based intellectual properties.

DLT Entertainment Ltd., 124 E. 55th St., New York, NY, 10022-4501. Phone: (212) 245-4680. Fax: (212) 315-1132. Web Site: www.dltentertainment.com.
Donald L. Taffner, owner; John Fitzgerald, CEO; Donald Taffner Jr., VP; Jeff Cotugno, VP.
TV-CATV only.
TV program production & distribution.

DMX Music, 900 E. Pine St., Seattle, WA, 98122. Phone: (800) 831-8001. Fax: (206) 329-9952. Web Site: www.dmxmusic.com.
Liberty Media, owner.
TV-CATV-Radio.
Programmer & supplier of satellite-delivered music svcs for business & cable TV. Available satellite direct or through FM subcarrier.

D-Squared Media, 30 E. 20th St., 7th Fl., New York, NY, 10003. Phone: (212) 478-1005. Fax: (212) 254-3489. E-mail: solutions@dsquaredmedia.com Web Site: www.dsquaredmedia.com.
Adriana E. Davis, exec producer; Bryan S. Durr, dir/videographer.
TV-CATV only.
Film, video, & radio production svcs from script to screen.

D-V-X International, (A division of Demo-Vox Sound Studio Inc.). 1038 Bay Ridge Ave., Brooklyn, NY, 11219. Phone: (718) 680-7234. Fax: (718) 680-7234.
TV-CATV only.
Video recording, creative production & postproduction svcs.

DWJ Television, One Robinson Ln., Ridgewood, NJ, 07450. Phone: (201) 445-1711. Fax: (201) 445-8352. E-mail: dwjinfo@dwjtv.com Web Site: www.dwjtv.com.
Daniel G. Johnson, pres; Michael L. Friedman, exec VP; Cynthia Boseski, sr VP.
TV-CATV-Radio.
Provides TV, radio progmg, production; promotional video production & production svcs.

Daley Video, 4095 Hitchcock Rd., Concord, CA, 94518. Phone: (925) 676-7260. Fax: (413) 541-8354. E-mail: gadaley@aol.com
Greg Daley, owner/CEO.
TV-CATV only.
Betacam SP, D1 digital betacom editing & 3/4-inch postproduction; TV production.

Dandelion Distribution Ltd., Unit 5 Churchill Court, Station Rd., North Harrow, Middlesex, HA2 7SA. Fax: 44(0)181 863 0463. Phone: 44(0)181 863 1888.
TV-CATV only.
International distribution & production company producing drama, documentaries, animation & movies.

Dargaud-Marina, 15-27 rue Moussorgski, Paris, 75018. France. Phone: 331-5326-3100. Fax: 331-5326-3113. E-mail: sales@dargaudmarina.fr
Gaspard De Chavagnac, mgng dir; Claude De Saint Vincent, pres; Patrick Desiev, VP finance; G. Guillot, sls.
TV-CATV only.
Distribution & production company specialized in children's progmg, mainly animation.

Darino Films/Library of Special Effects, 222 Park Ave. S., New York, NY, 10003. Phone: (212) 228-4024. E-mail: edarino@hotmail.com
Ed Darino, owner.
TV-CATV only.
Distributors of TV programs, video, CD, animation, educationals, effects libraries, stock footage libraries, production CD & DVD.

Daro Film Distribution, Le Victoria, 13 Blvd. Princess Charlotte, MC, 98000. Phone: (377) 979-1600. Fax: (377) 979-1590. E-mail: daro@meditnet.com
TV-CATV only.
International distribution, co-production, co-financing of TV programs & films.

Jeff Davis Productions Inc., 6166 Mulholland Hwy., Los Angeles, CA, 90068. Phone: (323) 464-3500. Fax: (323) 464-1414. E-mail: jeffdavies@jeffdavies.com Web Site: www.jeffdavis.com.
Jeff Davies, CEO.
TV-CATV-Radio.
Voiceover, production, TV & radio.

DaviSound, Box 521, 1504 Sunset, Newberry, SC, 29108. Phone: (803) 276-0639. E-mail: davisound@hotmail.com Web Site: www.davisound.com.
Hayne Davis; Annette Davis, opns mgr.
TV-CATV only.
Coml & promotional writing & producing for radio, jingles & program production & distribution. Also provides DaviSound "Tool Boxes," custom fabricated pro audio equipment.

De Wolfe Music Library Inc., 25 W. 45th St., New York, NY, 10036. Phone: (212) 382-0220. Fax: (212) 382-0278. E-mail: info@dewolfmusic.com Web Site: www.dewolfemusic.com.
Andrew M. Jacobs, pres; Jamie Gillespie, Mgr Music Sls.
TV-CATV-Radio.
The largest independant production music library in the world!

DeLuxe Laboratories, 1377 N. Serrano Ave., Hollywood, CA, 90027. Phone: (323) 462-6171. Fax: (323) 461-0608. Web Site: www.bydeluxe.com.
Cyril Drabinsky, pres; Steve Van Anda, sls VP.
TV-CATV only.
Full svc motion picture processing lab with labs in Toronto, London, Rome.

Design Partners Inc., 1438 N. Gower St., Hollywood, CA, 90028. Phone: (323) 856-9191. Fax: (323) 856-9258. E-mail: designpartners@dpi-ld.com Web Site: www.dpi-ld.com.
Greg Brunton, pres.
TV-CATV only.
TV, lighting design, industrial production & TV production & tech supervision svcs.

Devillier Donegan Enterprises L.P., 4401 Connecticut Ave. N.W., 6th Floor, Washington, DC, 20008. Phone: (202) 686-3980. Fax: (202) 686-3999. Web Site: www.ddegroup.com.
Ronald J. Devillier, pres/CEO; Brian Donegan, exec VP; Joan Lanigan, VP business legal affrs; Linda Ekizian, VP mktg; Gregory Diefenbach, production & dev; John Esteban, VP finance/admin.
TV-CATV only.
Worldwide distribution of progmg: international & ind documentaries, Hollywood profiles, science series, drama, natural history progmg & the performing arts.

Devlin Design Group Inc., 12526 High Bluff, # 300, San Diego, CA, 92130. Phone: (760) 634-6515. Fax: (760) 634-6929. E-mail: creative@ddgtv.com Web Site: www.ddgtv.com.
Dan Devlin,.
TV-CATV only.
Designs, builds & installs news sets & newsrooms. Facility planning, broadcast consulting, tech & lighting direction. Virtual Reality Rsch & Dev Ctr. Virtual sets, soft sets.

Digital Brewery L.L.C., 3820 Packard, Suite 150, Ann Arbor, MI, 48108. Phone: (800) 572-0098. Phone: (734) 975-8880. Fax: (734) 975-8915. Web Site: www.digitalbrewery.com.
Terry Dollhoff, co-pres; Sal Calabrese, co-pres.
TV-CATV only.
Packaged animations include backgrounds, holidays, corporate, adv & globes, maps & flags.

Digital Force, 149 Madison Ave., 12th Fl., New York, NY, 10016. Phone: (212) 252-9300. Fax: (212) 252-7377. E-mail: info@digitalforce.com Web Site: www.digitalforce.com.
Jerome Bunke, pres; Vanessa Towle-Mullin, production mgr; Arthur Crumlish, sr VP.
TV-CATV only.
Compact disc, CD-ROM & DVDproduction service to meet the needs of bcstrs, cable networks, & small labels/ind artists. Clients include the National Football League (NFL) & National Hockey League (NHL) on Fox TV as well as PSAs & promotional discs from bcstrs nationwide, ABC-TV & NBC, Westwood One & CBS.

Dimension 3 Corp, 5240 Medina Rd., Woodland Hills, CA, 91364-1913. Phone: (818) 592-0999. Fax: (818) 592-0987. E-mail: info@d3.com Web Site: www.d3.com.
Daniel L. Symmes, pres.
TV-CATV only.
Supplies 3-D bcst TV process & 3-D film; equipment, consultation & 3-D glasses.

Disney Channel, 3800 W. Alameda Ave., Burbank, CA, 91505. Phone: (818) 569-7700. Fax: (818) 845-8249. Web Site: www.disneychannel.com.
Anne M. Sweeney, pres.
TV-CATV only.
Original TV program production & distribution.

Walt Disney Company, 500 S. Buena Vista St., Burbank, CA, 91521-0990. Phone: (818) 560-1000. Fax: (818) 560-1930.
Zenia Mucha, corporate comm.
TV-CATV only.
The Walt Disney Company subsidiary; devs & syndicates

first-run adult & children's progmg, off-net progmg & feature film packages.

The Walt Disney Company, 500 S. Buena Vista St., Burbank, CA, 91521. Phone: (818) 560-1000. Fax: (818) 560-1930. Web Site: www.disney.com.

Michael Eisner, chmn/CEO.

TV-CATV only.

Diversified Communications Inc., 2000 M St. N.W., Suite 340, Washington, DC, 20036. Phone: (202) 775-4300. Fax: (202) 775-4363. Web Site: www.dciteleport.com.

Al Levin, pres; Nelson Crumling, VP.

TV-CATV only.

Complete mobile facilities, Ku-band uplink trucks, extensive loc & global connectivity; internationally compliant, fully redundant Ku-band air transportable uplink.

D'Ocon Films Productions, C/Calaf.3 Bajos, Barcelona, 08021. Spain. Phone: 34-93-240-41-22. Fax: 34-93-240-41-24. E-mail: docon@docon.es

TV-CATV only.

Principally an animation company offering full range of pre-production, production & postproduction svcs either developing our own concepts or co-producing.

The Dolmatch Group Ltd., Box 3298, 19697 Glen Brae Dr., Saratoga, CA, 95070. Phone: (408) 741-8620. Fax: (408) 741-8620. E-mail: tdgdolmatch@yahoo.com

Murray Dolmatch, pres; Sandra Dolmatch, VP.

TV-CATV only.

Represents producers in the United States, United Kingdom, Germany, France & Italy; distributes feature films, documentaries & animation; active in co-production & co-financing.

Domain Communications L.L.C., 289 S. Main Pl., Carol Stream, IL, 60188-2425. Phone: (630) 668-5300. Fax: (630) 668-0158. E-mail: dmorris@domaincommuications.com Web Site: www.domaincommunications.com.

David Morris, pres; Jim Draper, VP sls.

Radio Only.

Recording audio studios, production, CD replication, high-speed cassette duplicating, fulfillment.

Dome Productions, 1 Blue Jays Way, Suite #3400, Toronto, ON, M5V1J3. Canada. Phone: (416) 341-2001. Phone: (514) 731-3663. Fax: (416) 341-2020. Fax: (514) 731-4646. E-mail: mcarlyle@domeprod.com Web Site: www.domeproductions.com.

Mary Ellen Carlyle, VP/gen mgr.

Mont-Royal, PQ Canada, 5647 Ferrier.

TV-CATV only.

Mobile production trucks/airpacks (High Definition, Digital, Analog), telecommunications (Fibre/Satellite transmission, satellite media tours, playouts), Host bcst (Design, production, engrg, opons).

Donnelly & Associates, 7507 Sunset Blvd., Suite 202, Los Angeles, CA, 90046. Phone: (323) 850-5861. Fax: (323) 850-5866. E-mail: wpdonnelly@earthlink.net

W.P. Donnelly, pres.

TV-CATV only.

Mktg & licensing films to pay-TV, network & syndication packages.

Dorling Kindersley Vision, 80 Strand, London, WC2R 0RL. Phone: 0044 207 010 3000. Fax: 0044 207 010 6636. E-mail: dkvision@dk-uk.com Web Site: www.dk-uk.com.

TV-CATV only.

Produces programs for the international TV & video markets, incorporating visual design with universally appealing subjects.

John Driscoll/VoiceOver America, Box 744, Mill Valley, CA, 94941. Phone: (415) 388-8701. Phone: (888) 766-2049. Fax: (415) 388-8719. E-mail: johndriscoll @voiceoveramerica.com Web Site: www.johndriscoll.com.

John Moore, pres/CEO. VoiceOverPeople.com, c/o AT&A. Phone: (818) 760-6688.

TV-CATV only.

The new millenium's premiere voice talent for stn branding available instantly via ISDN & web download.

Mark Druck Productions Inc., 300 E. 40th St., New York, NY, 10016. Phone: (212) 682-5980. Fax: (212) 682-5981. E-mail: markdruck@aol.com

Mark Druck, pres; Lisa Dodenhoff, producer.

TV-CATV only.

TV & video tape industrial progmg, prom film production & distribution.

Duke International, Box 46, Douglas, Isle of Man, IM99 1DD. Phone: (+44) 1624 640020. Fax: (+44) 1624 640001. E-mail: info@dukesales.com Web Site: www.dukesales.com.

Jon Quayle, sls dir.

TV-CATV only.

A wide range of powersport progmg, documentaries, clips; also production & editing facilities.

The D.L. Dykes Jr. Foundation, 305 E. Capitol St., Jackson, MS, 39201. Phone: (601) 354-0767.

Joe Todaro, production mgr; David R. Dyker, CEO.

TV-CATV only.

Program production, Collage® non-linear post suite, digital audio capable. Nonprofit.

E

E! Entertainment Television, 5750 Wilshire Blvd., Los Angeles, CA, 90036-3709. Phone: (323) 954-2400. Fax: (323) 954-2500. Web Site: www.eonline.com.

Neil Baker, sr VP & adv sls; Ted Harbert, pres/CEO; Ken Bettsteller, CEO.

New York, NY 10036, 11 W. 42nd St. Phone: (212) 852-5100. (212) 852-5151. Dave Cassaro, exec VP.

TV-CATV only.

A 24-hours progmg net covering celebrities, entertainment news, gossip & pop-culture, feature behind the scenes with today's biggest stars.

ESPI Video, 4801 Spring Valley Rd., Suite 116, Dallas, TX, 75244. Phone: (214) 522-6699. Fax: (214) 522-7699. E-mail: gsleeper@espivideo.com Web Site: www.espivideo.com.

Gary Sleeper, pres.

TV-CATV only.

Full-svc production company specializing in corporate video production. Postproduction facilities & on-location svcs also available.

ESPN Radio Network, ESPN Plaza, Bristol, CT, 06010. Phone: (860) 766-2661. Fax: (860) 860-5523. Web Site: www.espnradio.com.

John A. Walsh, exec editor; Len Weiner, progmg dir; John Martin, exec producer.

Radio Only.

NBA On ESPN Radio; College Game Day (Sat); ESPN Radio weekends; Brent Musburger afternoon drive sportscasts; AM & PM drive commentaries; NFL Gameday (Sun)

ESPN Regional Television, 11001 Rushmore Dr., Charlotte, NC, 28277. Phone: (704) 973-5000. Fax: (704) 973-5090. Web Site: www.espn.com.

Chuck Gerber, exec VP/gen mgr.

TV-CATV only.

Producer & distributor of TV sports events including college & professional basketball, boxing & auto racing for over-the-air & cable.

ETN—Educational Telecommunications Network, (An ETV service of the Los Angeles County Office of Education). 9300 Imperial Hwy., Rm. 126, Los Angeles County Office of Education, Downey, CA, 90242-2890. Phone: (562) 401-5622. Fax: (562) 922-8841. E-mail: etn@lacoe.edu Web Site: www.lacoe.edu.

Richard Quinones PhD., division dir.

TV-CATV only.

ETN develops & transmits educ programs in the major K-12 curriculum areas as well as adult educ & parent educ programs (via satellite over Ku-band) to schools, homes & offices nationwide.

EUE Screen Gems Studios, 222 E. 44th St., New York, NY, 10017. Phone: (212) 450-1600. Fax: (212) 450-1610.

Bill Vassar, VP; Larry Coatto, technical officer.

Wilmington, NC 28405, 1223 N. 23rd St. Phone: (910) 343-3500.

TV program, coml production & distribution.

Eagle Eye Film Company, 824 N. Victory Blvd., Burbank, CA, 91502. Phone: (818) 506-6100. Fax: (818) 506-4313. Web Site: www.eagleyepost.com.

Chuck Spatariu, pres; Joel Minnich, opns coord.

TV-CATV only.

Editing facility, editing rentals & RAID storage solutions.

Eagle Media Productions Ltd., Box 580, Northford, CT, 06472. Phone: (203) 294-1190. Fax: (203) 294-9512. E-mail: louadler@sbcglobal.net

Louis C. Adler, pres; Thalia Adler, VP.

TV-CATV only.

Radio program syndication, program & news consultant. Producers of *Medical Journal.*

Earthwatch Radio, 10 Science Hall, 550 N. Park St., Madison, WI, 53706. Phone: (608) 263-3063. Fax: (608) 262-2273. E-mail: spomplun@wise.edu Web Site: ewradio.org.

Steve Pomplun, producer; Richard Hoops, producer.

Radio Only.

Daily two-minute radio feature on environment & science. Ten progms distributed mthy on compact disc. Programs distruted mthy on compact disc.

Eaton Films Ltd., 10 Holbein Mews, London, SW1W 8NN. Phone: (44) 207-823-6173. Fax: (44) 207-823-6017. E-mail: eaton.films@talk21.com

Judith Bland, dir; Liz Cook, dir internatonal sls.

TV & video distribution.

Ebbets Field Productions Ltd., Box 42, Wykagyl Stn, New Rochelle, NY, 10804. Phone: (914) 636-1281.

David Saperstein, pres.

TV-CATV only.

Writers, dirs & producers of film (features, TV, cable) & video.

Echo Radio Productions Inc., 44895 Hwy. 82, Aspen, CO, 81611. Phone: (800) 385-4612. Phone: (970) 925-2640. Fax: (970) 925-9369. E-mail: kayla@echoradio.com Web Site: www.echoradio.com.

Kayla Hoffman-Cook, VP; Rodney H. Jacobs, CEO.

Radio Only.

Syndicator & producer of radio vignette progmg.

Ecumedia News Service, Box 358, Ridgefield, CT, 06877. Phone: (203) 431-6092. Fax: (212) 870-2030. E-mail: roy.lloyd@ecunet.org

Roy T. Lloyd, dir.

Radio Only.

News stories, features & actualities about ethics & relg produced for radio.

Ecumenical Communications, 48 Eastview Rd., Terryville, CT, 06786. Phone: (860) 585-5090. E-mail: info@ecucomm.ro Web Site: www.ecucomm.ro.

Robert J. Geckler, owner.

Radio Only.

Radio, podcast program production, distribution; production, restoration svcs, internet, website & podcast svcs.

Ellis Entertainment, 1300 Yonge St., Suite 300, Toronto, ON, M4T 1X3. Canada. Phone: (416) 924-2186. Fax: (416) 924-6115. E-mail: sales@ellisent.com

Caroline Godin, natl sls mgr; Stephen Ellis, pres; Kip Spidell, production mgr.

TV-CATV only.

Producers and distributors of 1000+ hrs of non-fiction and family entertainment for TV and video for four decades.

Empire Burbank Studio, 1845 Empire Ave., Burbank, CA, 91504. Phone: (818) 840-1400. Fax: (818) 567-1062.

TV-CATV only.

Sound stage studio rental, audience rated TV studios, full-svc production facilities & equipment, ultimate stage.

Encore Video Productions Inc., 811 Main St., Myrtle Beach, SC, 29577. Phone: (843) 448-9900. Fax: (843) 448-9235. E-mail: frank@encorevideo.biz Web Site: www.encorevideo.biz.

Rik Dickinson, pres; Frank Payne, VP.

TV-CATV only.

Location & studio production, specializing in EFP/ENG 1-Camera productions, full script to screen svc, VNR, EPK satellite media tours, teleconferences, & magazine TV production. Betagami SP, non-linear editing.

Enoki Films U.S.A. Inc., 16430 Ventura Blvd., Suite 308, Encino, CA, 91436. Phone: (818) 907-6503. Fax: (818) 907-6506. E-mail: info@enokifilmsusa.com Web Site: www.enokifilmsusa.com.

Yoshi Enoki, pres; Ricki Ames, VP worldwide distribution; Madoka Koike, distribution coord.

TV-CATV only.

Producer & distributor of children's animation for TV & video.

Envoy Productions, 660 Mason RIdge Ctr. Dr., St. Louis, MO, 63141-8557. Phone: (314) 317-4216. Fax: (314) 317-4299. E-mail: sandi.clement@lhm.org Web Site: www.envoyproductions.com.

Sandi Clement, Manager; Kurt R. Klaus, pres.

TV-CATV only.

TV, radio production, distribution (English & Sp). Syndicates

30-minute wkly radio shows, The Lutheran Hour & TV holiday specials.

Episcopal Church Center, 815 2nd Ave., New York, NY, 10017. Phone: (212) 716-6102. Fax: (212) 949-8059. Web Site: www.episcopalchurch.org.

Spokespersons for church & society issues.

Essence Television Productions Inc., 135 W. 50th St., Frnt 4, New York, NY, 10020-1201. Phone: (212) 642-0600. Fax: (212) 921-5173. Web Site: www.essence.com.

Edward Lewis, chmn/CEO.

TV-CATV only.

TV program production.

Ethnic-American Broadcasting Co., Two Executive Drive, Fort Lee, NJ, 07024. Phone: (201) 242-3000. Fax: (201) 944-5961. E-mail: info@skyview

Fort Lee, NJ 07024, 2 Executive Dr, Suite 600. Phone: (201) 242-3000.

TV-CATV-Radio.

Provides ethnic radio & TV language svcs via DBS & through cable systems throughout North America.

Eurocine, 33 Ave. Des Champs Elysees, Paris, 75008. Phone: 33.1.42.25.6492. Fax: 33.1.42.25.7338. E-mail: eurocine@club-internet.fr Web Site: www.eurocine.net.

TV-CATV only.

Production & distribution in all media.

Europe Images International, 1 Rond-Point Victor Hugo, F-92130, Issy-Les-Moulineaux France. Phone: (33) 1 55 95 58 00. Fax: (33) 1 55 95 58 10. E-mail: Europe-Images @europeimages.com Web Site: www.europeimages.com.

John Rouilly, CEO.

TV-CATV only.

Acquires, distributes & invests in international TV progmg. Catalog close to 5 hours broken into three categories, drama, children's documentaries.

Evangelical Lutheran Church in America, 8765 W. Higgins Rd., Chicago, IL, 60631. Phone: (773) 380-2941. Fax: (773) 380-2406. E-mail: ava.martin@elca.org Web Site: www.elca.org.

Ava Martin, dir.

TV-CATV-Radio.

TV & radio progmg, promotional film production, distribution, production svcs & news.

Evergreen Entertainment Group, 1825 Ponce De Leon Blvd., Suite 450, Coral Gables, FL, 33134-3626. Phone: (305) 460-4448. E-mail: evergreenenter@juno.com

Migdalia Inocencio, pres.

TV-CATV only.

Worldwide programs distribution; international co-production liaison; mktg & progmg cable/satellite.

Expand Images, 7 Rue Taylor, Paris, 75010. Phone: (33) 0148-0305-44. Fax: (33) 0148-0345-04. E-mail: communication@expand.fr Web Site: www.expand.fr.

TV-CATV only.

Production & distribution company.

Eye in the Woods, Box 89, Brewton, AL, 36427. Phone: (251) 809-1909. Fax: (251) 809-0729. Web Site: www.eyeinthewoods.com.

Dale Faust, pres/exec producer.

TV-CATV only.

Produce a weekly show viewed on The Outdoor Channel.

Eyewitness Kids News, LLC, 182 Sound Beach Ave., Old Greenwich, CT, 06870-0116. Phone: (203) 637-0044. Fax: (203) 698-0812. E-mail: primonews@aol.com Web Site: www.educationtelevisionfund.org.

Albert T. Primo, pres.

TV-CATV only.

Coaching of TV news, program talent, strategic planning & production svcs.

F

FTC/Orlando, 324 DeSota Cir., Orlando, FL, 32804. Phone: (407) 422-8246. Fax: (407) 843-0738. E-mail: ftcorlando@aol.com Web Site: www.ftcorlando.com.

A.J. Foresta, pres.

TV-CATV only.

Full-svc film & TV production company, specializing in coml & feature production. Area specialty: steadicam.

Faith for Today, 101 W. Cochran St., Simi Valley, CA, 93065. Phone: (888) 940-0062. Fax: (805) 522-2114. E-mail: info@faithfortoday.tv Web Site: www.faithfortoday.tv.

Michael Tucker, Speaker/Director.

TV-CATV only.

Producer & distributor of *Lifestyle Magazine, McDougall M.D.* & *The Evidence.*

Family Stations Inc., 290 Hegenberger Rd., Oakland, CA, 94621. Phone: (510) 568-6200. Phone: (800) 543-1495. Fax: (510) 633-7983. E-mail: famradio@familyradio.com Web Site: www.familyradio.com.

Harold E. Camping, pres/gen mgr; Rick Prime, tech dir; W. Craig Hulsebos, progmg mgr; William Thornton, VP.

Radio Only.

Radio program production & distribution.

FamilyNet, 6350 W. Freeway, Fort Worth, TX, 76116. Phone: (817) 737-4011. Fax: (817) 377-4372. E-mail: ddavis@familynet.com Web Site: www.familynet.com.

David Clark, pres; Glenn McEowen, VP tech opns; Martin Coleman, VP production team; Chip Turner, VP mktg; Darin Davis, VP sls & traffic.

TV-CATV only.

FamilyNet is a 24/7 cable net. In addition, FamilyNet produces 5 syndicated radio progms. Values based, family oriented progmg.

FamilyNet Radio, 6350 W. Freeway, Fort Worth, TX, 76116-4511. Phone: (817) 570-1416. Phone: (800) 266-1837. Fax: (817) 735-1790. E-mail: smiller@familynet.com Web Site: www.familynetradio.com.

Scott Miller, progmg dir.

Radio Only.

FamilyNet Radio produces radio programs: Powerline, Country Crossroads, Strength for Living, Christian Talk ch for SIRIUS & At a Glance PSAs.

Faraone Communications Inc., 75 West End Ave., R-9A, New York, NY, 10023. Phone: (212) 489-1313. Fax: (212) 489-8978. E-mail: tedfaraone@verizon.net Web Site: www.worldwidepublicrelations.com.

Ted Faraone, principal; Randolph Nader, VP.

Valley Village, CA 91607. Valley Village, 4804 Laurel Canyon Blvd., Ste 516.

TV-CATV-Radio.

Media rel svcs to producers, distributors of radio, TV programs, talent & home video.

Federal Citizen Information Center, 1800 F St. N.W., Rm. G-142, Washington, DC, 20405. Phone: (202) 501-1794. Fax: (202) 501-4281. E-mail: nancy.tyler@gsa.gov Web Site: www.pueblo.gsa.gov.

Teresa Nasif, dir; Nancy Gregory Tyler, bcst mgr.

TV-CATV-Radio.

TV & radio PSAs promoting FirstGov.gov, the official web portal of the federal government.

Festival de Television de Monte-Carlo, 4, Boulevard du Jardin Exotique, Monte Carlo, 98000. Phone: 377 93 10 40 60. Fax: 377 93 50 70 14. E-mail: info@tvfestival.com Web Site: www.tvfestival.com.

TV-CATV only.

Competition of TV films & miniseries; news programs; producers. Conferences, panels & other market-related activities.

Film House Inc., 810 Dominican Dr., Nashville, TN, 37228. Phone: (615) 255-4000. Fax: (615) 255-4111. E-mail: results@filmhouse.com

Curt Hahn, CEO; Ron Routson, pres/COO; Wayne Campbell, VP mktg; Andy Cohen, CFO; Edith Johnson, VP.

TV-CATV only.

Creates & produces TV mktg campaigns for radio & TV stns worldwide.

Film Roman Inc., 12020 Chandler Blvd., Suite 200, North Hollywood, CA, 91607. Phone: (818) 761-2544. Fax: (818) 985-2973. Web Site: www.filmroman.com.

John Hyde, pres/CEO.

TV-CATV only.

Animation production studio.

Filmoption International Inc., 3401 St. Antoine St., Westmount, PQ, H3Z 1X1. Canada. Phone: (416) 598-1557. Fax: (416) 593-0013. E-mail: mrosilo@filmoption.com Web Site: www.filmoption.com.

Maryse Rouillard, pres; Lizanne Rouillard, VP; Muriel Rosilio, sr exec sls & co-productions; Evangelia Ozek, sls exec.

Toronto, ON M5J 2L7 Canada, 144 Front St. West, Suite 760. Phone: (416) 598-1557. Fax: (416) 593-0013. E-mail: mrosilio@filmoption.com. Muriel rosilio, sr exec sls co-productions.

Westmount, PQ H3Z 1X1 Canada, 3401 St-Antoine. Phone: (215) 931-6180. Fax: (514) 939-2034. E-mail: mrouilla@filmoption.com. Maryse Rouillard, pres.

TV-CATV only.

International distribution of TV programs.

Films Five Inc., 42 Overlook Rd., Great Neck, NY, 11020. Phone: (516) 487-5865.

Walter Bergman, pres.

TV-CATV only.

Pre- & postproduction; film & video comls, documentaries, sls films.

Films Media Group, Box 2053, Princeton, NJ, 08543-2053. Phone: (609) 671-0266. Phone: (800) 257-5126. Fax: (609) 671-5772. E-mail: custserv@films.com Web Site: www.films.com.

Amy Bevilacqua, exec VP/ gen mgr; Diane Bilello, VP sls.

TV-CATV only.

Distributes programs for bcst & cable industries to non-theatrical, educ, institutional, home video & business markets.

Films of the Nations, Box 449, Clarksburg, NJ, 08510. Phone: (732) 462-3522. Fax: (732) 294-0330. E-mail: aldfilms@bellatlantic.net Web Site: www.aldenfilms.com.

Paul Weinberg, pres.

TV-CATV only.

TV program, promotion & educ film distribution.

Financial Media Services, Inc., Box 870928, Stone Mountain, GA, 30087. Phone: (770) 413-2258. Fax: (770) 465-0180. E-mail: charles@charlesross.com

TV-CATV-Radio.

Produces & syndicates nationally syndicated radio show *Your Personal Finance.*

David Finch Distribution Ltd., Box 264, Walton-on-Thames, KT12 3YR. United Kingdom. Phone: 44-1932-882733. Fax: 44-1932-882108. E-mail: sales@david-finch.com

David Finch, chief exec.

TV-CATV only.

Supply of programs for home video & TV worldwide. Acquisition for United Kingdom home video.

Finger Lakes Productions International, 119 S. Cayuga St., Ithaca, NY, 14850. Phone: (607) 275-9400. Fax: (607) 277-0961. E-mail: world@flpradio.com Web Site: www.flpradio.com.

Paul Bartishevich, pres/CEO.

Radio Only.

Full-svc radio mktg, production & syndication of short-form radio features. International marketing, sales & consulting.

First Marketing, 3300 Gateway Dr., Pompano Beach, FL, 33069. Phone: (954) 979-0700. Phone: (800) 641-9251. Fax: (954) 971-4707. Web Site: www.first-marketing.com.

Ronald Drenning, pres; Neil Rosenblum, VP/business dev.

TV-CATV only.

First Marketing offers 30 yrs of experience partnering with marketing professionals to dev custom communications programs designed to enhance custom relationships & profitability.

1st Miracle Productions, 3439 W. Cahuenga Blvd., Hollywood, CA, 90068. Phone: (323) 874-6000. Fax: (323) 874-4252. E-mail: sales@1stmiracleproductions.com Web Site: www.1stmiracleproductions.com.

Moshe Bibiyan, CEO; Simon Bibiyan, pres.

TV-CATV only.

International distribution, co-production, postproduction finance.

First Run/Icarus Films, 32 Court St., 21st Fl., Brooklyn, NY, 10201. Phone: (718) 488-8900. Fax: (718) 488-8642. E-mail: info@frif.com Web Site: www.frif.com.

Jonathan Miller, pres.

TV-CATV only.

International TV program distribution: documentaries, current affrs, music, arts, cultural programs.

FirstCom Music, 1325 Capital Pkwy., Suite 109, Carrollton, TX, 75006. Phone: (800) 858-8880. Fax: (972) 242-6526. E-mail: info@firstcom.com Web Site: www.firstcom.com.

Ken Nelson, exec producer; Carol Riffert, exec VP.

TV-CATV-Radio.

FirstCom Music offers the highest quality, most professionally produced, best sounding, and easiest-to-use music libraries in the industry. Guaranteed.

Fischer Broadcast Services, 10841 Bittersweet Lane, Fishers, IN, 46038-2203. Phone: (317) 514-5757. Fax: (317) 578-3884. Web Site: www.superfisch.com.

Scott Fischer, pres.

TV-CATV-Radio.

SUPERFISCH—THE PROMO VOICE SUPERHERO. Scott Fischer, signature voice artist for DirecTV Sports, Fox Sports Net Networks, and other fine affiliates. Myriad reads—always right! ISDN/MP3.

Forde Motion Picture Labs, 1001 Lenora St., Seattle, WA, 98121-2706. Phone: (206) 682-2510. Phone: (800) 682-2510. Fax: (206) 682-2560. Web Site: www.fordelabs.com.

Richard E. Vedvick, pres.

TV-CATV only.

Overnight processing of 35mm/16mm Eastman color negative dailies; release printing.

Four Star Media, 201 E. 15th St., New York, NY, 10003. Phone: (212) 533-5994.

Radio Only.

Radio progmg production, distribution & mktg.

Fox Digital, Fox Network Ctr., 10201 W. Pico Blvd., Los Angeles, CA, 90035. Phone: (310) 369-6622. Fax: (310) 969-6125. Web Site: www.fox.com.

TV-CATV only.

Videotape production facilities, stages & equipment.

Fox 17 Studio Productions, 631 Mainstream Dr., Nashville, TN, 37228. Phone: (615) 244-1717. Fax: (615) 259-3962. Web Site: www.wztv.com. E-mail: production@fox17.com

Bill Zuckerman, prom dir.

TV-CATV only.

Full range video & film production facility; 25 x 40 studio & 60 x 60 studio soundstage; betacams & a var of tape formats for TV/CATV.

Fox Sports West, 1100 S. Flower St., Los Angeles, CA, 90015. Phone: (213) 743-7800. Fax: (213) 743-7835. Web Site: www.foxsports.com.

Steve Simpson, gen mgr; Dennis Johnson, public relations dir.

TV-CATV only.

TV program production & distribution.

Fox 29 WUTV Sinclair, 951 Whitehaven Rd., Grand Island, NY, 14072. Phone: (716) 773-7531. Fax: (716) 773-5753. Web Site: www.wutv.com.

Don Moran, gen mgr; Jon May, progmg coord.

TV-CATV only.

TV coml production; U.S. rep, Katz; Canadian rep, Airtime.

France TV Distribution, Immeuble Le Barjac, 1, blvd. Victor, Paris, 75015. Phone: 01-44-2501-18. Fax: 01-44-2501-01. Web Site: www.francetv.com.

TV-CATV only.

Marketing & sls of French progmg.

Sandy Frank Entertainment Inc., 954 Lexington Ave., Suite 255, New York, NY, 10021. Phone: (212) 772-1889. Fax: (212) 772-2297. E-mail: filmsfe@aol.com Web Site: www.sandyfrankent.com.

Sandy Frank, chmn/CEO; Damaso V. Santana, exec VP; Rosalie Perrone, controller; Sandi Spidell, VP opns; Maury Shields, VP business affrs; Barbara Kalicinska, sls; Sophia Evans, sls; Susan Piscitello, sls.

TV-CATV only.

TV production & syndication.

Free Speech TV (FSTV), Box 44099, Denver, CO, 80201. Phone: (303) 442-8445. Fax: (303) 442-6472. Web Site: www.freespeech.org. E-mail: viewercomments@fstv.org

John Schwartz, bd pres; Jon Stout, gen mgr; Nathaniel Reeder, opns dir; Eric Galatas, progmg dir; Jason McKain, dir outreach/member svcs.

TV-CATV only.

Acquires works from activists, independent film/video artists & community based media; providing exposure to progressive ideas.

Freewheelin' Films Ltd., 44895 Hwy. 82, Aspen, CO, 81611. Phone: (970) 925-2640. Fax: (970) 925-9369. Web Site: www.fwf.com. E-mail: kalyla@fwf.com

Rodney H. Jacobs, CEO; Kayla Hoffman-Cook, VP.

TV-CATV only.

25-yr old production company specializing in entertainment, sports & lifestyle specials.

The Fremantle Corp., 660 Madison Ave., 21 Floor, New York, NY, 10021. Phone: (212) 421-4530. Fax: (212) 207-8357. Web Site: www.fremantlecorp.com.

Paul Talbot, pres; Diane Tripp, co-managing dir; Blanca Oca Pertierra, VP Latin America & home video; Keith Talbot, co-managing dir.

TV-CATV only.

International TV program distribution & co-production.

Fremantle Media Ltd., 1 Stephen St., London, W1T 1AL. United Kingdom. Phone: 44 (0)20 7691-6000. Fax: 44 (0)20 7691-6100. E-mail: feedback@freemantlemedia.com Web Site: www.pearsontv.com.

Greg Dyke, CEO; Tony Cohen, mgng dir; James Bennet, CEO.

TV production & distribution.

FremantleMedia North America Inc., 2700 Colorado Ave., Suite 450, Santa Monica, CA, 90404. Phone: (310) 255-4700. Fax: (310) 255-4800. Web Site: www.fremantlemedia.com.

David Lyle, pres; Cecile Frot Coutaz, COO.

New York, NY 10036, 1540 Broadway. Catherine V, MacKay, Deputy CEO.

TV-CATV only.

TV production.

Chuck Fries Productions Inc., 6922 Hollywood Blvd., 12th Fl., Hollywood, CA, 90028. Phone: (310) 203-9520. Fax: (323) 466-2266. E-mail: chuckfries@aol.com

TV-CATV only.

Domestic & international TV, home video, & feature film production & distribution.

G

GLL TV Enterprises Inc., 8009 Via Fiore, Sarasota, FL, 34238. Phone: (941) 925-4339. Fax: (941) 925-3976. E-mail: glltv@pobox.com

Gunther L. Less, pres; Ellen G. Less, sec/treas.

TV-CATV only.

TV program, coml, promotional film production & distribution. Journey to Adventure, the longest-running syndicated travel show on TV.

GMI Media L.L.C., 2211 5th Ave., Seattle, WA, 98121. Phone: (206) 374-8889. Fax: (206) 374-2150. E-mail: moreinfo@gmimedia.com Web Site: www.gmimedia.com.

Ron Erak, pres; Richard Germaine, VP/gen mgr.

TV-CATV only.

Custom ID jingle packages for all radio & TV formats. CD production libraries. Voice overs & production for promotions & spots.

GRB Entertainment, 13400 Riverside Dr., 3rd Fl., Studio City, CA, 91423. Phone: (818) 728-7697. Fax: (818) 728-7601. E-mail: gbenz@grbtv.com Web Site: www.grbtv.com.

Gary R. Benz, pres/CEO.

TV-CATV only.

Production & distribution (TV).

GTN, 13320 Northend Ave., Oak Park, MI, 48237. Phone: (248) 548-2500. Fax: (248) 548-1916. Web Site: www.gtninc.com.

Doug Cheek, pres.

TV-CATV only.

Studios, remote equipment, multi-format editing, film transfer, audio & duplication svcs, on-site satellite svcs & graphics.

GVI, 1775 K St. N.W., Suite 220, Washington, DC, 20006. Phone: (202) 293-4488. Fax: (202) 293-3293. E-mail: andy@g-v-i.com Web Site: www.g-v-i.com.

Andy Hemmindinger, pres; Bob Burnett, VP.

TV-CATV only.

Full creative script-to-screen production, Beta SP fieldcrews, computer graphics. Avid editing, equipment rental & DVD authoring.

Galavision, 605 Third Ave., 12th Fl., New York, NY, 10158-0180. Phone: (212) 455-5300. Fax: (212) 953-0198. Web Site: www.univision.com.

Ray Rodriguez, pres/COO; Joanne Lynch, gen mgr; Cesar Conde, opns mgr.

Miami, FL 33178, 9405 N.W. 41st St. Phone: (305) 471-4022. Fax: (305) 471-3977.

Chicago, IL 60611, 541 N. Fairbanks Ct., 12th Fl., Suite 1240. Phone: (312) 494-5100. Fax: (312) 494-5115.

Los Angeles, CA 90045, 5999 Center Dr. Phone: (310) 348-3621. Fax: (310) 348-3619.

Dallas, TX 75201, 2323 Bryan St, Suite 1900. Phone: (214) 758-2420. Fax: (214) 758-2430.

TV-CATV only.

Spanish TV program distribution, cable.

Gedeon Programmes, 44-50 av du Capitaine Glarner, Saint-Quen, 93585. Fax: 33 01 49 48 65 03. Phone: 33 01 49 48 65 00. Web Site: www.gedeonprogrammes.com.

TV-CATV only.

Films, TV movies, TV series, interactive fiction.

General Broadcasting Co. Inc., 8 N. Bothwell St., Suite 103, Palatine, IL, 60067. Phone: (847) 202-8804. Fax: (847) 202-8834.

Robert E. Potter, pres; Charles E. Maples, VP; Dean Mulchaey, production mgr; Eric Edgerton, gen mgr.

Background music, environmental music progmg.

Georgia Film, Video & Music Office, Georgia Department of Economic Dev, 75 Fifth St. N.W., Atlanta, GA, 30308. Phone: (404) 962-4052. Fax: (404) 962-4053. E-mail: film@georgia.org Web Site: www.georgia.org.

Bill Thompson, dir.

TV-CATV only.

Location scouting & preproduction svcs provided to feature film, TV movie, coml & multimedia production companies.

Getty Images, 601 N. 34th St., Seattle, WA, 98103. Phone: (206) 925-5000. Fax: (206) 925-5001. E-mail: sales@gettyimages.com Web Site: www.gettyimages.com.

TV-CATV only.

Getty Images is an imagery company creating and providing still and moving images to communications professionals around the globe.

Ghostwriters/Radio Mall, 2412 Unity Ave. N., Dept BR, Minneapolis, MN, 55422-3450. Phone: (800) 759-4561. Fax: (763) 522-6256. E-mail: info@radiomall.com Web Site: www.radiomall.com.

David Dworkin, owner.

TV-CATV only.

Over 21 years experience with products sold to more than 7,200 radio stns worldwide as well as TV stns, audio-video producers & cable operators. If you have a finished product that you'd like to mkt to radio stns, contact us. We offer a 60-day money-back guarantee.

Lon Gibby Productions, Inc./ Gibby Media Group, 1213 S Pines Rd., Ste C, Spokane Valley, WA, 99206-5485. Phone: (509) 467-1113. Phone: (800) 200-1113. Fax: (509) 467-4763. E-mail: lon@longibby.com Web Site: www.longibby.com.

Lon Gibby, pres/CEO.

Multimedia productions, video, CD-Rom, CD-I, producers of bcst TV programs, comls, infomercials, corporate videos & webcasting

Gladney Communications Ltd., 101 Reni Rd., Manhasset, NY, 11030. Phone: (516) 627-3016. Fax: (516) 767-1957.

Norman Gladney, pres; Marion Gladney, exec VP.

TV-CATV only.

TV & radio production & distribution.

Glenray Productions Inc., Box 40400, Pasadena, CA, 91114-7400. Phone: (626) 797-5462. E-mail: glenray@pacbell.net Web Site: www.familymedia.net.

C. Ray Carlson, pres.

TV-CATV only.

Films, TV series, video distribution & production, primarily for family & children.

Global Entertainment Media, 1200 N.W. 78th Ave., Suite 104, Miami, FL, 33126. Phone: (786) 206-4873. Fax: (786) 206-4889. E-mail: sales@gem-media.com Web Site: www.gem-media.com.

Alexander A. Fiore, CEO; Mercedes M. Fiore, pres.

TV-CATV only.

Production & distribution company.

Global Telemedia Inc., 98 East Ave., Norwalk, CT, 06851. Phone: (203) 854-9985. Fax: (203) 855-8370. E-mail: gt@globaltelemedia.com Web Site: www.globaltelemedia.com.

Greg Kimmelman, pres/CEO; Anne Gulledge, gen sls mgr.

TV-CATV only.

Production & distribution of bcst TV & DVD progmg worldwide.

GlobeCast, 1270 Avenue of the Americas, Ste 2800, New York, NY, 10020. Phone: (212) 373-5140. Fax: (212) 399-1949. E-mail: info@globecastna.com Web Site: www.globecast.com.

David Sprechman, pres/CEO; Mary Frost, Sr sls VP; Jonathan Feldman, VP business affrs; Keven Cahoon, VP mktg.

Culver City, CA 90232. GlobeCast Los Angeles/California, 10525 W. Washington Blvd. Phone: 1-310-845-3900. Fax: 1-310-845-3904.

Washington, DC 20005. GlobeCast Washington/DC, 1120 G Street, NW - 2nd Floor. Phone: 1-202-383-2745. Fax: 1-202-393-4914.

Miami, FL 33166. GlobeCast Miami Headquarters America/Florida, 7291 NW 74th St. Phone: 1-310-687-1600. Fax: 1-305-341-4424.

Salt Lake City, UT 84119. GlobeCast Salt Lake City/Utah, 1193 West 2400 South, Suite A. Phone: 1-801-908-1100. Fax: 1-801-954-0991.

TV-CATV only.

GlobeCast Audio division supports a comprehensive package of audio transmission svcs:

ABC/Keystone Ventures provides Satcom C5 DATS/SEDAT distribution svcs to 7,000 radio stns.

3D2, a high-quality digital net designed to svc the entertainment industry, connections post-production facilities, recording studios & voice-over talent worldwide.

A/FX Network utilizing Telos Zephyr code located at venues for sports backhauls & special events.

"Hybrid" bridging svcs to simplify a digital world full of different flavored audio codecs.

Remote Production Packages for single or multi-stn remote bcsts.

GlobeCast North America, 10525 W. Washington Blvd., Culver City, CA, 90230. Phone: (310) 845-3900. Fax: (310) 845-3904. Web Site: www.globecast.com.

Ken Drake, dir opns.

TV-CATV only.

In the center of Hollywood, GlobeCast North America's Sunset facility provides studio production second audio progmg (SAP), & related client facility svcs on an as-scheduled or contractual basis. GlobeCast's studio is completely integrated w/GlobeCast's network of global, end-to-end connectivity via satellite, fiber optics & microwave.

GlobeCast North America, provides the bcstg industry with a unique combination of both recognized expertise & extensive inter-continental svcs. Through GlobeCast's vast global infrastructure of over 100 transponders, 30 teleports & interconnect facilities, the company provides instant access to the world's major media markets. GlobeCast North America is part of France Telecom, one of the worlds largest telecommunications companies.

Jeff Gold Productions Inc., 13900 Panay Way, M-307, Marina del Rey, CA, 90292. Phone: (310) 827-9165.

Jeff Gold, dir.

TV-CATV only.

Production svcs, film & coml production.

Golden Gate Studios, (KTLN TV 68). 400 Tamal Plaza, Suite 428, Corte Madera, CA, 94925. Phone: (415) 945-7500. Fax: (415) 924-0264. Web Site: www.goldengatestudios.com.

TV-CATV only.

2 Studios, Green Screen Options, Full Production Packages

The Samuel Goldwyn Films, 9570 W. Pico Blvd., Suite 400, Los Angeles, CA, 90035-6405. Phone: (310) 860-3100. Fax: (310) 860-3195.

Samuel Goldwyn Jr., chmn/CEO; Meyer Gottlieb, pres/COO.

Movie acquisition & distribution.

Good Life Associates, Box 82808, Lincoln, NE, 68501-2808. Phone: (402) 464-6440. Fax: (402) 464-6880. E-mail: martinj@backtothebible.org Web Site: www.goodlifeassociates.org.

Thomas C. Schindler, pres; Martin Jones, dir.

Radio Only.

Radio program, coml production & distribution svcs.

Good News Broadcasting Association Inc., 6400 Cornhusker Hwy., Lincoln, NE, 68501. Phone: (402) 464-7200. Fax: (402) 464-7474. E-mail: info@backtothebible /backtothebible.com Web Site: www.backtothebible.org.

Woodrow Kroll, pres.

TV-CATV only.

Radio & TV program production & distribution.

Gordon Productions, Box 640549, San Francisco, CA, 94164. Phone: (415) 776-7484. Fax: (415) 776-7822. E-mail: john@gpvideo.com Web Site: www.gpvideo.com.

John Gordon, pres; Les Lieurance, VP; Jerry Gordon, CEO.

TV-CATV only.

Production/distribution of video news releases, TV pub svc announcements, audio news releases. Distribution via satellite path fire & cassette.

Gould Entertainment Corp., 10 Chicken St., Wilton, CT, 06897-1807. Phone: (203) 762-3525.

Michael J. Gould, pres.

TV-CATV only.

Consultants, packagers, distributors of progmg; specialists in mktg foreign programs.

Billy Graham Evangelistic Association, Radio Department, Box 1270, Charlotte, NC, 28201. Phone: (704) 401-2432. Fax: (704) 401-3028. E-mail: had@bgea.org

Franklin Graham, pres; Roger Flessing, dir.

Radio Only.

Granada America, 15303 Ventura Blvd., Suite 800, Sherman Oaks, CA, 91403. Phone: (818) 455-4600. Fax: (818) 455-4700. Web Site: www.itv.com.

David Gyngell, CEO; Sam Zoda, exec VP; Emily Brecher, CFO.

TV-CATV only.

Worldwide distribution, production of TV series & made-for-TV movies.

Great Chefs Television/Publishing, (A division of G.C.I., Inc.). 747 Magazine St., New Orleans, LA, 70156. Phone: (504) 581-5000. Fax: (504) 581-1188. E-mail: info@greatchefs.com Web Site: www.greatchefs.com.

John Shoup, pres/CEO; Linda Nix, dir.

TV-CATV only.

Production, distribution of cooking, jazz TV programs, videos, CDs, CD-ROMs & books.

Great North Productions, 3720-76 Ave., Edmonton, AB, T6B 2N9. Canada. Phone: (780) 440-2022. Fax: (403) 440-3400. Web Site: www.greatnorth.ab.ca.

Penny Ritco, VP.

TV-CATV only.

A full-svc international production & distribution company, providing worldwide distribution for bcst home video, non-theatrical markets & co-production opportunities.

Great Plains National (GPN), Box 80669, Lincoln, NE, 68501. Phone: (800) 228-4630. Fax: (800) 306-2330. E-mail: gpn@unl.edu Web Site: www.gpn.unl.edu.

Stephen C. Lenzen, dir.

TV-CATV only.

Acquires, produces, promotes & distributes videotaped instructional videos for bcst, cablecast & audiovisual use.

The Griffin Group, 130 S. El Camino Dr., Beverly Hills, CA, 90212. Phone: (310) 385-2700. Fax: (310) 358-2701. Web Site: www.merv.com.

Larry Cohen, pres/CEO.

TV-CATV only.

Full-svc dev & production company; slate includes series, specials & films. Also real estate & hotel ownership.

Grinberg Film Libraries Inc., 21011 Itasca St., Unit D, Chatworth, CA, 91311. Phone: (818) 709-2450. Fax: (818) 709-8540. Web Site: www.grinberg.com.

W. "Bill" Brewington, CEO.

TV-CATV only.

Stock footage & news library.

Groove Addicts, 12211 West Washington Blvd., Los Angeles, CA, 90066. Phone: (310) 572-4646. Fax: (310) 572-4647. E-mail: info@grooveaddicts.com Web Site: www.grooveaddicts.com.

Dain Eric Blair, CEO; Bill Stolier, VP/gen mgr.

Chicago, IL 60610. Groove Addicts - Chicago, 108 W. Hubbard St. Phone: (312) 467-7047. Fax: (312) 467-7049.

TV-CATV only.

Custom, syndicated news, image, promotion music for bcst & entertainment projects worldwide. Standout jingle & I.D. packages. Barter & Cash opportunities.

Groove Addicts Production Music Catalog, 12211 West Washington Blvd., Los Angeles, CA, 90066. Phone: (310) 572-4644. Fax: (310) 572-4647. E-mail: info@groveaddicts.com Web Site: www.grooveaddicts.com.

Bill Stolier, VP/gen mgr.

TV-CATV only.

Our catalog is growing! We now have 16 libraries, all continuously updated, sound design & SFX. Annual blanket license & custom music packages.

Grove Television Enterprises Inc., 46216 Dry Creek Dr., Badger, CA, 93603. Phone: (559) 337-2595. Fax: (559) 337-1334.

John W. Hyde, pres/CEO.

TV-CATV only.

TV production & distribution international & domestic.

H

HAVE Inc., 350 Power Ave., Hudson, NY, 12534-2448. Phone: (518) 828-2000. Phone: (800) 999-4283. Fax: (518) 828-2008. E-mail: have@haveinc.com Web Site: www.haveinc.com.

Nancy Gordon, pres; Paul Swedenburg, VP.

TV-CATV only.

Distribution of audio, videotape, equipment, accessories & supplies, featuring BELDEN, CANARE & MOGAMI cable. Duplication, postproduction svcs & international standards conversion svcs. CD-Audio & CD-ROM, DVD replication & duplication, DVD authoring.

HEA Productions, 313 Gahbauer Rd., Hudson, NY, 12534. Phone: (518) 822-1717. Fax: (518) 822-1042. E-mail: susan@susanhamilton.com

Susan Hamilton, pres.

TV-CATV only.

Radio & TV music production.

HIT Entertainment P.L.C., Maple House, 149-150 Tottenham Ct. Rd., 5th Fl., London, W1T 7NF. United Kingdom. Phone: 20-7554-2500. Fax: 20-7388-9321. E-mail: contactus@hitentertainment.com Web Site: www.hitentertainment.com.

Rob Lawes, CEO; Peter Orton, chmn; Charles Caminada, sls dir; Steve Ruffini, CFO.

Beverly Hills, CA 90212, 9300 Wilshire Blvd., 2nd Fl. Phone: (301) 724-8979.

Allen, TX 7500-3320, 830 Greenville Ave. Phone: (972) 390-6000.

TV-CATV only.

Distributor, co-producer & financier of quality animiation, children's & natural history progmg.

Alfred Haber Distribution Inc., 111 Grand Ave., Palisades Park, NJ, 07650. Phone: (201) 224-8000. Fax: (201) 947-4500. E-mail: info@haberinc.com Web Site: www.alfredhaber.com.

Robert Kennedy, exec VP; Alfred Haber, pres.

TV-CATV only.

TV program distribution.

Halland Broadcast Services, 2412 Unity Ave. N., Minneapolis, MN, 55422. Phone: (763) 522-6256. Fax: (763) 522-6256. E-mail: info@h-b-s.com Web Site: www.h-b-s.com.

Dave Dworkin, mgr.

Radio Only.

Rock 'n' Roll Graffiti oldies library on compact disc, *The Eighties Plus* AC/CHR library on compact disc & *The Seventies* AC/CHR gold library on compact disc. Country music libraries on compact disc. Also available on hard drive.

Hamilton Productions Inc., 7732 Georgetown Pike, McLean, VA, 22102. Phone: (703) 734-5444. Fax: (703) 734-5449. E-mail: jah@dgsys.net

John Hamilton, pres; Jay Hamilton, VP; Anne H. Deger, VP.

TV-CATV only.

Ind TV production firm.

Hanna-Barbera Productions Inc., 15303 Ventura Blvd., Suite 1400, Sherman Oaks, CA, 91403. Phone: (818) 977-7500. Fax: (818) 977-7510. Web Site: www.hanna-barbera.com.

William Hanna, co-chmn; Joseph Barbera, co-chmn.

TV program production.

Happi Associates, Box 110892, Nashville, TN, 37222. Phone: (615) 220-6050. Phone: (615) 604-1981. E-mail: doddrace@aol.com

Skeeter Dodd, gen mgr.

Radio Only.

Radio program & mgmt; country formats; motivational speaking, jingles ID & coml, production music, features, customized productions.

Larry Harmon Pictures Corp., 7080 Hollywood Blvd., Suite 202, Hollywood, CA, 90028. Phone: (323) 463-2331. Fax: (323) 463-7219. E-mail: tellbozo@aol.com Web Site: www.bozo.com.

Larry Harmon, pres; Susan Harmon, exec VP; Marci Breth, VP corporate affrs.

TV-CATV only.

Owner & distributor of *Bozo* cartoons & live show franchise, & *Laurel & Hardy* cartoons.

Harmony Gold U.S.A. Inc., 7655 Sunset Blvd., Los Angeles, CA, 90046. Phone: (323) 851-4900. Fax: (323) 851-5599. E-mail: sales@harmonygold.com Web Site: www.harmonygold.com.

Frank Agrama, chmn/CEO; Melissa Wohl, VP sls & acquisition.

TV-CATV only.

TV production, international TV distribution.

Harpo Productions, 110 N. Carpenter St., Chicago, IL, 60607. Phone: (312) 633-1000. Web Site: www.oprah.com.

Produces "The Oprah Winfrey Show."

Health Net Productions & Pet Talk, 185 N. New Ballas Rd., St. Louis, MO, 63141. Phone: (314) 997-5422. Fax: (314) 997-5422. E-mail: judyleven@aol.com

Judy Leventhal, pres; Chuck LeRoi, VP.

TV-CATV only.

Distributes 90-second pharmacy vignettes, 45 to 60-second pet care vignettes & 90-second sports medicine vignettes.

Hearst Entertainment, Inc., 300 W. 57th St., 15th Fl., New York, NY, 10019. Phone: (212) 969-7553. Fax: (646) 280-1553. Web Site: www.hearstent.com.

Bruce L. Paisner, pres; Stacey Valenza, VP sls.

Woodland Hills, CA 91364. Hearst Entertainment Productions, 20335 Ventura Blvd., Suite 300. Phone: (818) 444-5010. Fax: (818) 444-5011.

TV-CATV only.

Leading producer & distributor of made-for-television movies, first-run entertainment, animated series, reality & documentary progmg for the global marketplace.

Hearts of Space Inc., 454 Las Gallinas #333, San Rafael, CA, 94903. Phone: (415) 499-9901. Fax: (415) 499-9903. E-mail: help@hos.com Web Site: timelessmedia.net/hos.

Stephen M. Hill, pres & producer; Leyla Rael Hill, VP/gen mgr.

Radio Only.

Syndicated one-hour progmg of ambient, electronic, multi-cultural & contemplative spacemusic via NPR satellite transmission.

Heil Enterprises, Box 1372, Lancaster, PA, 17608-1372. Phone: (717) 898-9100. Fax: (717) 898-6600. E-mail: info@thegospelgreats.com Web Site: www.thegospelgreats.com.

Paul Heil, owner.

Radio Only.

Radio program, production & distribution.

Arthur Henley Productions, 101 W. 23rd St., #2462, New York, NY, 10011. Phone: (718) 263-0136. E-mail: ah55@webtv.net

Arthur Henley, pres.

TV-CATV only.

TV & radio program production; radio program distribution.

Henninger Media Services, Inc., 2601-A Wilson Blvd., Arlington, VA, 22201. Phone: (703) 243-3444. Phone: (888) 243-3444. Fax: (703) 243-5697. Fax: (703) 243-4023. Web Site: www.henninger.com.

Rob Henninger, CEO; Doug Miller, sls dir & dir mktg; Brian J. Kelly, gen mgr; Adam Kranitz, mktg mgr.

Los Angeles, CA 90404. Tribeca Henninger Editing Tools-CA (T.H.E. Tools), 3000 W. Olympic Blvd, Ste. 1350. Phone: (310) 264-4192. Fax: (310) 264-4194. Jon Tronowski, facility mgr.

Washington, DC 20007. Henninger Capitol, 2121 Wisconsin Ave. N.W. Phone: (202) 965-7800. Fax: (202) 965-7815. Bobby Wright, gen mgr.

Washington, DC 20036. Henninger 1150 Post, 1150 17th St, Suite 401. Phone: (202) 833-3444. Fax: (202) 833-3995. Peggy Polito, facility mgr.

New York, NY 10013. Tribeca Henninger Editing Tools-NY (T.H.E. Tools), Tribeca Media Center, 65 North Moore St. Phone: (212) 226-7770. Fax: (212) 226-8157. Jon Miles, gen mgr.

Nashville, TN 37228. Henninger Elite, Metro Center, 50 Vantage Way, Suite 100. Phone: (615) 256-7678. Fax: (615) 255-7212. Roy Giorgio, gen mgr.

Arlington, VA 22201. Henninger Media Development, 2601-A Wilson Blvd. Phone: (703) 243-3444. Fax: (703) 243-5697. Steven Schupak, gen mgr.

Richmond, VA 23220. Commonwealth Film Labs, 1500 Brook Rd. Phone: (804) 649-8611. Fax: (804) 648-7715. Roger Robison, gen mgr.

Richmond, VA 23223. Henninger Richmond, 1901 E. Franklin St, Suite 103. Phone: (804) 644-5006. Fax: (804) 783-0820. Scott Witthaus, gen mgr.

TV-CATV only.

TV postproduction, film & video, film processing, 2-D & 3-D graphics, TV progmg dev, distribution; multi-media & DVDs.

Heritage/Baruch Television Distribution, 1025 Connecticut Ave. N.W., Suite 1012, Washington, DC, 20036-5417. Phone: (202) 833-1777. Fax: (202) 496-0162.

Ed Baruch, pres; Steve Smallwood, VP; Valerie Cooley-Elliott, dir mktg.

TV-CATV only.

Mktg, syndication & production/distribution of progmg to network syndication international.

Highland Laboratories, Administration Bldg., Pier 96, San Francisco, CA, 94124. Phone: (415) 981-5010. Fax: (415) 981-5019. Web Site: www.highlandlab.com.

B.J. Brose, pres.

TV-CATV only.

Video, audio, film duplication, film transfers: D-2, Betacam, 2 inches, 1 inch, 3/4 inch, 1/2 inch.

Jack Hilton Inc., 230 Park Ave., Suite 1530, New York, NY, 10169. Phone: (212) 687-2002. Fax: (212) 697-9008.

TV-CATV only.

TV & video productions.

The History Makers, 1900 S. Michigan Ave., Chicago, IL, 60616. Phone: (312) 674-1900. Fax: (312) 674-1915. Web Site: www.thehistorymakers.com.

Julieanna L. Richardson,.

TV-CATV only.

Video production.

Holigan Investment Group Ltd., 15950 N. Dallas Pkwy., Suite 750, Dallas, TX, 75248. Phone: (972) 387-7999. Fax: (972) 387-1685. Web Site: www.michael.holigan.com.

Michael Holigan, exec producer; Tim Dickey, exec producer.

Dallas, TX 75240, 6029 Beltline, Suite 110.

TV-CATV only.

Production & syndication of YOUR NEW HOUSE and THE REALITY OF SPEED; 30 minute TV programs.

Home Improvement Television Network, 3441 Baker St., San Diego, CA, 92117. Phone: (858) 273-0572. Fax: (858) 273-8410. E-mail: homefix@hometvnet.com Web Site: www.hometvnet.com.

Bruce Lamb, pres.

TV-CATV only.

Providers of home improvement progmg & 90-second video vignettes.

Hometown Illinois Radio Network, Box 169, 918 E. Park, Taylorville, IL, 62568-0169. Phone: (217) 824-3395. Fax: (217) 824-3301. Web Site: www.hometownillinoisradio.com.

Randal J. Miller, pres.

Radio Only.

Wired network providing loc reports from Illinois State Fair, Illinois Farm Bureau Convention & Commodity Classic.

Hope Channel, Box 4000, Silver Spring, MD, 20914. Phone: (301) 680-6689. Fax: (301) 680-6312. E-mail: info@hopetv.org Web Site: www.hopetv.org.

Brad Thorp, dir; Gary Gibbs, VP.

TV-CATV only.

Family friendly TV progmg, 24/7.

Horizon Audio Creations, Box 486, 74 Chemin De Lanse, Hudson Heights-Rigaud, PQ, J0P 1J0. Canada. Phone: (450) 451-4549. Fax: (450) 451-4549. E-mail: reachcraigcutler@mac.com

Craig W. Cutler, pres; Marguerite Blais, progmg mgr; Mary-Lou Dodd, opns mgr.

Radio Only.

Radio program, coml production; production svcs; inflight audio progmg & adv.

Horizons Television Inc., 9305 Monalaine Ct., Great Falls, VA, 22066. Phone: (703) 759-7500. Fax: (703) 759-1620. E-mail: admin@horizonstv.com Web Site: www.horizonstv.com.

Timothy E. Donner, exec dir.

TV-CATV only.

Creative dev & full-svc production of reality-based TV programs & commissioned videos for diverse organizations & assns. Avid media composer.

Thomas Horton Associates Inc., 408 Bryant Cir., Suite K, Ojai, CA, 93023. Phone: (805) 646-7866. Fax: (805) 646-3600. E-mail: tha@sharktv.com Web Site: www.sharktv.com.

Thomas F. Horton, pres; Jean Horton Garner, sr VP; Garry R. Garner, dir mktg.

TV-CATV only.

TV program full-svc production, postproduction, international & domestic distribution, specializing in award-winning documentaries.

Host Communications Inc., 546 E. Main St., Lexington, KY, 40508-2300. Phone: (859) 226-4678. Fax: (859) 226-4419. Web Site: www.hostcommunications.com.

James Host, CEO; Gordon Whitner, pres/COO.

TV-CATV only.

TV & radio production & syndication.

Hot Box Digital, 367 N. Hwy. 101, Solana Beach, CA, 92075. Phone: (858) 292-8520. Fax: (858) 292-8520.

Cam MacMillan, exec producer.

TV-CATV only.

Design & production of bcst 3-D computer graphics. Logo animation, stn packages. All tape formats supported. Productions of Subito Studio Video Graphics Volumes.

Marie Hoy Film & TV, 18 Bruton Pl. Berkeley Sq., Mayfair, London, W1X 7AA. Phone: 020-7851-6666. Fax: 017-1493-3997. E-mail: mariehoy@cocoon.co.uk

TV-CATV only.

Co-production Funding & Financial Packaging.

Huntridge Video Productions Inc., Box 3813, Greenville, SC, 29608-3813. Phone: (864) 271-3348. Fax: (864) 232-4462. E-mail: mat@huntridge.com Web Site: www.huntridge.com.

TV-CATV only.

TV production & postproduction.

I

ISL Television Ltd., Seymour News House, Seymour News, London, W1H 9PE. Phone: (44) 171 616 11 11. Fax: (44) 171 616-1110. Web Site: www.islworld.com.

TV-CATV only.

TV program sls & distribution, bcst sponsorship, events & TV program production & TV consultancy.

The Idea Channel, 2002 Filmore Ave., Suite 1, Erie, PA, 16506. Phone: (814) 464-9068. Fax: (814) 464-9069. E-mail: info@ideachannel.com Web Site: www.ideachannel.com.

Bob Chitester, pres/CEO; Rick Platt, VP/COO.

TV-CATV only.

Discussions 20-40 minutes in length, featuring two or three leading scholars on a wide variety of subjects. Internet offerings also.

The Image Generators, 18156 Darnell Dr., Olney, MD, 20832. Phone: (301) 924-5700. Fax: (301) 570-8916. E-mail: mweiner@imagegenerators.com Web Site: www.imagegenerators.com.

Michael J. Weiner, pres/CEO.

TV-CATV only.

Voice-overs, radio spot & program production; media training, progmg concept to completion; ISDN-equipped (TELOS).

Imagers Inc., 1575 Northside Dr., Suite 490, Atlanta, GA, 30318. Phone: (404) 351-5800. Fax: (404) 351-9020. Web Site: www.imagers.com.

TV coml production & distribution; production svcs.

Images Communication Arts Corp., 366 N. Broadway, Suite 410, Jericho, NY, 11753. Phone: (516) 939-2990.

Robert Braverman, pres; Paul Fritz-Nemeth, sls.

TV-CATV only.

Syndication & co-production of TV & radio program series & specials. Syndication of "old-time radio" dramas.

In-Motion Pictures, 5 Percy St., London, W1T 1DG. United Kingdom. Phone: (207) 467-6880. Fax: (207) 467-6890. E-mail: Sales@jment.com Web Site: www.jment.com.

Dr. Hilmar Siebert, chmn; Julian Freeston, CFO.

Beverly Hills, CA 90212, 412 S. Beverly Dr., 5th Fl. Phone: (301) 789-4500.

TV-CATV only.

Film & TV production & distribution.

Independent Edge Films, 719 52nd St. N., St. Petersburg, FL, 33710. Phone: (727) 321-2898. E-mail: michaelfox@indi-edge.com Web Site: www.indi-edge.com.

Michael D. Fox, dir.

TV-CATV only.

Ind motion picture/TV/web production & distribution.

Integrity Media, 401 E. Corpoorate Dr., #222, Lewisville, TX, 75057. Phone: (214) 222-7878. Fax: (214) 222-7838. E-mail: schalupka@integritymedia.net

Douglas Neece, pres; Sandy Chalupka, VP.

TV-CATV only.

TV & radio program distribution, time buying & media planning.

International Broadcasting Network, Box 691111, 5206 FM 1960 W., Suite 105, Houston, TX, 77269. Phone: (281) 587-8900. Fax: (281) 774-9923. E-mail: ibn@ev1.net

Paul Broyles, pres.

TV-CATV only.

Network of ten low power stns; including loc produced progmg.

International Program Consultants Inc., 52 E. End Ave., New York, NY, 10028. Phone: (212) 734-9096. Fax: (212) 734-6495.

Russell J. Kagan, mgng dir.

TV-CATV only.

International TV distribution, TV progmg & home video acquisition consultation, co-production consultation.

International Tele-Film, 41 Horner Ave., Unit #3, Toronto, ON, M8Z 4X4. Canada. Phone: (416) 252-1173. Fax: (416) 252-1676. E-mail: info@itf.ca Web Site: www.itf.ca.

TV-CATV only.

Distributor for documentaries, features, series & specials, in Canada & worldwide.

International Television Broadcasting Inc., 36-01 36th Ave., 2nd Fl., Long Island City, NY, 11106. Phone: (718) 784-8555. Fax: (718) 784-8901. E-mail: info@itvgold.com Web Site: www.itvgold.com.

Dr. Sathya Viswanath, pres.

TV-CATV only.

Full time Indian TV program for cable & bcst TV.

International Television Corp., 4380 N.W. 128th St., Miami, FL, 33054. Phone: (305) 688-7475. Fax: (305) 685-5697. Web Site: www.coralintl.com.

Jose Escalante, VP/gen mgr; Guadalupe D'Agostino, VP international sls.

TV-CATV only.

TV progmg distribution & production. Worldwide distribution, co-productions.

Ion Weather Network, 13 B East Main St., Denville, NJ, 07834. Phone: (973) 983-8222. Fax: (973) 983-1390. Web Site: www.ionweather.com. E-mail: steve@ionweather.com

Stephen Pellettiere Sr., pres; Stephen Pellettiere Jr., consultant.

Radio Only.

Gen weather forecasts, science info.

Irving Productions Inc., 3202 E. 21st, Tulsa, OK, 74114. Phone: (918) 744-1221. Fax: (918) 744-1223. E-mail: irving@irvingproductions.com Web Site: www.irvingproductions.com.

Dick Schmitz, pres.

TV-CATV only.

Audio recording & production svcs for all media.

It Is Written Television, Box O, Thousand Oaks, CA, 91360. Phone: (805) 955-7733. Fax: (805) 955-7734. E-mail: iiw@iiw.org Web Site: www.iiw.org.

Mark Finley, dir; Shawn Boonstra, assoc speaker.

TV-CATV only.

TV & radio program production & distribution; internet.

Italtoons Corp., 32 W. 40th St., New York, NY, 10018. Phone: (212) 730-0280. Fax: (212) 730-0313. E-mail: salesinfo@italtoons.com Web Site: www.italtoons.com.

Giuliana Nicodemi, pres; Luisa Rivosecchi, sls; Ken Priester, gen mgr.

TV-CATV only.

TV program production & distribution. Children's animation.

Ivanhoe Broadcast News Inc., 2745 W. Fairbanks Ave., Winter Park, FL, 32789. Phone: (407) 740-0789. Fax: (407) 740-5320. E-mail: jcherry@ivanhoe.com Web Site: www.ivanhoe.com.

Majorie BeKaert Thomas, pres; Bette Bon Fleur, CEO; John Cherry, pres/sls.

TV-CATV only.

Producer & syndicator of targeted new series. Medical breakthrough Inside Science, Prescription Health & Smart Woman.

J

JAM Creative Productions Inc., 5454 Parkdale Dr., Dallas, TX, 75227. Phone: (214) 388-5454. Fax: (214) 381-4647. E-mail: sales@jingles.com Web Site: www.jingles.com.

Jonathan M. Wolfert, pres; Mary Lyn Wolfert, sr VP; Tom Parma, sls; Cary Bass, sls; Randy Bell, sls.

TV-CATV only.

ID jingle & coml production for radio & TV, custom music & production svcs.

J&H Music Programming, 5814 Fleming Terr. Rd., Greensboro, NC, 27410. Phone: (336) 218-8052. Fax: (336) 218-8052.

Joseph V. Gelo, pres; Helen J. Gelo, VP.

Radio Only.

Radio program distribution.

JC Productions Inc., 1851 Murray Hill Station, New York, NY, 10016. Phone: (212) 213-0455. Fax: (212) 532-2820. E-mail: jcpro@bellatlantic.net Web Site: www.jc-productions.com.

Joe Conforti, dir.

TV-CATV only.

Live action coml production; CD-ROM multimedia production; 3D animation & graphic design.

JGT Media Productions, 12408 86th Pl., N.E., Kirkland, WA, 98034-2601. Phone: (425) 820-4523. Fax: (425) 820-4523.

J. Graley Taylor,.

TV program, promotion film production & distribution; production svcs; rgnl award program, ARBY Awards; film & video production.

J.N. Productions, 902-1790 Bayshore Dr., Vancouver, BC, V6G 3G5. Canada. Phone: 604-331-0690. E-mail: jnproductions@telus.net Web Site: www.jnproductions.bc.ca.

Jakob Nortman, pres.

Radio Only.

For all your voice-over needs, including narration, corporate videos, on-hold telephone messages & announcements for GPS systems. Radio production facilities available.

Jameson Broadcast Inc., 1644 Hawthorne St., Sarasota, FL, 34239. Phone: (941) 906-8800. Fax: (941) 906-8801. E-mail: radio@jamesonbcast.com Web Site: www.jamesonbroadcast.com.

Jamie G. Jameson, pres; Trulee C. Jameson, VP.

Radio Only.

Radio program production, syndication & special projects.

Janson Media, 88 Semmens Rd., Harrington Park, NJ, 07640. Phone: (201) 784-8488. Fax: (201) 784-3993. E-mail: info@janson.com Web Site: www.janson.com.

Stephen Janson, pres; Zara Janson, VP; Betsy Van Ost, dir; Lynne Warshavsky, office mgr.

TV-CATV only.

International TV & video/ DVD program distribution & production; video/DVD publishing.

Jefferson-Pilot Sports, 1900 W. Morehead St., Charlotte, NC, 28208. Phone: (704) 374-3669. Fax: (704) 374-3859. E-mail: jweber@jpsports.com Web Site: www.jpsports.com.

Edward M. Hull, pres & gen mgr; Pam Hawthorne, gen sls mgr & VP; Powell Kidd, VP opns; Jimmy Rayburn, opns mgr, VP & opns mgr; Jeff Tennant, VP sls & VP mktg.

Atlanta, GA 30326, 3390 Peachtree Rd, NE, Suite 1000. Phone: (404) 364-6556. Fax: (404) 364-6557. Jim Weilbaecher, dir sls.

Rutherford, NJ 07070, Meadows Office Complex, 201 Rt. 17 N, Suite 300. Phone: (201) 438-2088. Fax: (201) 939-0224. Adam J. Moore, dir sls.

TV-CATV only.

TV & CATV sports production & syndication.

Jerusalem Radio Productions, 545 W. 111th St., Suite 8-I, New York, NY, 10025. Phone: (212) 666-2144.

Radio program production & distribution; production svcs.

The Johnson Group, 6800 Fleetwood Rd, Suite 100, McLean, VA, 22101. Phone: (703) 356-4004. Fax: (703) 356-6969. E-mail: rmjcameron@aol.com Web Site: www.thejgroup.com.

Robert M. Johnson, pres; Joe Fab, VP.

Radio Only.

Video, film, multimedia & radio creative svcs & production.

Joe Jones Productions, 10556 Arnwood Rd., Lake View Terrace, CA, 91342. Phone: (818) 899-4457. Fax: (818) 899-4457. E-mail: jojones@jojonesnetcom.com

Joe Jones, exec producer; Marion Jones, VP opns.

TV-CATV only.

TV, radio coml & program production; jingle production & production svcs.

Jones Radio Networks, (a Jones Media Networks company). 8200 S. Akron St., Suite 103, Centennial, CO, 80112. Phone: (303) 784-8700. Fax: (303) 784-8612. E-mail: betterservice@jonesradio.net Web Site: www.jonesradio.com.

Bob Hampton, pres; Phil Barry, gen mgr; Susan Stephens, dir; Patrick Crocker, dir; Amy Onsager, VP talk progmg; Kim Ketchel, VP mktg.

Centennial, CO 80112, 8200 S. Akron St, Suite 103. Phone: (303) 784-8700.

New York, NY 10036, 1133 Avenue of the Americas. Phone: (888) 644-8255.

Seattle, WA 98121, 3131 Elliott Ave, #770. Phone: (800) 426-9082.

Radio Only.

Premier producer of 24-hr. format progmg, daypart personality programs & various music svcs to radio stns.

Tom Jones Recording Studios, 1620 Greenview Dr. S.W., Rochester, MN, 55902-1034. Phone: (507) 288-7711. Fax: (507) 288-4531.

Thomas H. Jones, pres; Aaron Manthei, chief engr.

TV-CATV-Radio.

Recording studio, compact disc duplication recording svcs. Radio program & coml production.

Jordan Klein Film & Video, 10197 S.E. 144th Pl., Summerfield, FL, 34491. Phone: (352) 288-3999. Fax: (352) 288-5538. E-mail: jkfv01@gate.net Web Site: www.jordy.com.

Jordan Klein Jr., pres.

TV-CATV only.

Underwater, on-the-water production; rental film, video housing & crews; Bahamas specialist.

Nicole Jouve, 54 Avenue du Roule, Neuilly Sur Seine, 92200. France. Phone: 33 1 47 22 43 27. Fax: 33 1 47 22 43 27. E-mail: interamany@aol.com

Nicole Jouve, pres.

TV-CATV only.

Distribution of French films (non-theatrical, TV & video), documentaries (Jean Rouch) childrens programs.

Juravic Entertainment, 620 Glenridge Dr., Glenview, IL, 60025. Phone: (847) 998-5998. Fax: (847) 998-6013. E-mail: dljuravic@aol.com

TV-CATV only.

TV program syndication.

K

KCRA-TV, (Hearst-Argyle Television Inc). 3 Television Cir., Sacramento, CA, 95814-0794. Phone: (916) 446-3333. Fax: (916) 325-3731. Web Site: www.thekcrachannel.com.

Elliott Troshinsky, pres & gen mgr.

TV-CATV only.

TV program & coml production; production svcs.

KCSN 88.5 FM, California State University, Northridge, 18111 Nordhoff St., Northridge, CA, 91330-8312. Phone: (818) 677-3090. E-mail: frederick.d.johnson@csun.edu Web Site: www.kcsn.org.

Michael Worrall, chief engr; Fred Johnson, gen mgr; Martin Perlich, progmg dir; Laura Kelly, dev dir.

Radio Only.

Public radio serving parts of Los Angeles, CA—classical

weekdays, eclectic weeknights & weekends. PRI, AP affil. The best of public radio.

KJD Teleproductions, 30 Whyte Dr., Voorhees, NJ, 08043. Phone: (856) 751-3500. Fax: (856) 751-7729. E-mail: mactoday@earthlink.net Web Site: www.kjdteleproductions.com.

Larry Scott, pres/CEO.

TV-CATV only.

TV, radio program, coml, promotion film production & distribution; production svcs; TV processing lab.

KPTS-TV, 320 West 21st St. N., Wichita, KS, 67203-2499. Phone: (316) 838-3090. Fax: (316) 838-8586. E-mail: dchecots@kpts.org Web Site: www.kpts.org.

Don Checots, gen mgr; Dave McClintock, dir opns & engrg dir.

TV-CATV only.

Industrial video production for corporate training & mktg communications.

KTOO-TV & Radio Station, 360 Egan Dr., Juneau, AK, 99801. Phone: (907) 586-1670. Fax: (907) 586-3612. E-mail: info@ktoo.org Web Site: www.ktoo.org.

Jeff Brown, dir.

TV-CATV only.

A radio program highlighting diversified music & stories for children.

KUSA Television, 500 Speer Blvd., Denver, CO, 80203. Phone: (303) 871-9999. Fax: (303) 698-4700. E-mail: kusa@9news.com Web Site: www.9news.com.

Asa Darrow, production mgr.

TV-CATV only.

News production only.

David Kaye Productions Inc., 1361 Paseo Redondo Ave., Burbank, CA, 91501. Phone: (800) 843-3933. Phone: (310) 403-1714. Fax: (604) 921-1926. E-mail: info@davidkaye.com Web Site: www.davidkaye.com.

David Kaye, pres; Stephan J. Sisk, opns.

TV-CATV-Radio.

Full svc voice-over production company, providing radio & TV imaging & branding around the world.

Kazmark Entertainment Group, 14320 Ventura Blvd., Suite 601, Sherman Oaks, CA, 91423. Phone: (818) 981-4410. Fax: (818) 501-2211. E-mail: jkaz@earthlink.net Web Site: paulae@earthlink.net.

TV-CATV only.

Pay-per-view productions & distributors for cable TV.

The Kenwood Group, 75 Varney Pl., San Francisco, CA, 94107-1922. Phone: (415) 957-5333. Fax: (415) 957-5311. Web Site: www.kenwoodgroup.com.

Christina Crowley, pres.

Creative svcs & production of comls & corporate communications, film, video, multimedia, meetings & events.

Killer Tracks, 8750 Wilshire Blvd., FL 2, Beverly Hills, CA, 90211-2715. Phone: (323) 957-4455. Fax: (323) 957-4470. E-mail: sales@killertracks.com Web Site: www.killertracks.com.

TV-CATV only.

Provides production music library & sound effects.

Kipany Productions Ltd., 32 E. 39th St., New York, NY, 10016. Phone: (212) 883-8300. Fax: (212) 883-0409. E-mail: share82308@aol.com

T. Hendry, pres; K. Colligan, CEO.

TV-CATV only.

Video production, bcst progmg, mktg, sls & communications experts/web designers specializing in Telcom, event mgmt.

Klein &, 20530 Pacific Coast Hwy., Malibu, CA, 90265. Phone: (310) 317-9599. Fax: (310) 456-7701. E-mail: imagedoctor@kleinand.com Web Site: www.kleinand.com.

Bob Klein, pres.

TV-CATV only.

Mktg consulting, creative svcs for bcst, cable & Internet companies.

Knowledge In A Nutshell Inc., 1420 Centre Ave., Suite 2213, Pittsburgh, PA, 15219. Phone: (800) 688-7435. Phone: (412) 765-2020. Fax: (412) 765-3672. E-mail: audrey@knowledgeinanutshell.com Web Site: www.knowledgeinanutshell.com.

Charles Reichblum, pres.

TV-CATV only.

Syndicates radio/TV program, Knowledge in a Nutshell & Knowledge minute.

Kultur International Films, 195 Hwy. 36, West Long Branch, NJ, 07764. Phone: (732) 229-2343. Fax: (732) 229-0066. E-mail: info@kultur.com Web Site: www.kultur.com.

Dennis M. Hedlund, chmn; Pearl Lee, VP; Ronald Davis, mgng dir.

TV-CATV only.

Suppliers of programs on DVD in North America. Selection includes documentaries, opera, ballet, classical music, profiles, theater, comedy, fitness, country music, rock & roll.

L

Lakeside TV Co., 300 Highpoint Dr., Suite 712, Hartsdale, NY, 10530. Phone: (914) 946-7806. Fax: (914) 946-7806 (fax/phone). E-mail: bernshu@msn.com

Bernard Schulman, pres; Diane Ross, VP.

TV-CATV only.

TV program distribution, production & syndication.

Lambert Television, 100 N. Crescent Dr., 2nd Flr., Beverly Hills, CA, 90210. Phone: (310) 551-1900. Fax: (310) 385-4004. E-mail: jones@lamberttv.com Web Site: www.lamberttv.com.

Michael Jones, exec VP.

TV-CATV only.

Lambert Television owns and operates a television station group.

Lapco Communications, 437 E. Beil Ave., Nazareth, PA, 18064. Phone: (610) 759-9444. Fax: (610) 759-8589. E-mail: sales@lapcocom.com Web Site: www.lapcocom.com.

P. Pagilaro, pres; L. Van Winkle, VP.

TV-CATV only.

TV progmg & production, computer graphics, producers of original progmg, TV comls, TV promotion production, production mgmt, computer stock background library.

Launch Radio Networks, (a division of United Stations Radio Network). 1065 Ave. of the Americas, 3rd Fl., New York, NY, 10018. Phone: (212) 536-3600. Fax: (212) 536-3601. Web Site: www.launchradionetworks.com. E-mail: ccolombo@launchradionetworks.com

Dave Ankers, dir opns; Charlie Colombo, VP/gen mgr.

Radio Only.

Launch Radio Networks produces, distributes music, entertainment news & svcs for radio stns as well as other media worldwide.

Leadem to Water Production Inc., Box 279, Oregon City, OR, 97045. Phone: (503) 631-7661. Fax: (503) 631-7672. E-mail: info@horsemansworld.com Web Site: www.horsemansworld.com.

Jeff Tracy,.

TV-CATV only.

Produces syndicated radio show *Horseman's World*, coml video narrations, produces World bcst for A.Q.H.A. on 250 + stn, cable TV houses northwest & cowboy cooking.

John Lemmon Films, 1325 Rock Point Rd., Charlotte, NC, 28270. Phone: (704) 532-1944. Fax: (704) 566-1984. E-mail: jlemmon@jlf.com Web Site: www.jlf.com.

Mike Rosinski, head animator.

TV-CATV only.

Clay, cel & stop-motion animation for TV specials, comls, program openings & on-air IDs.

Leo Productions, 1 Rond Point Victor Hugo, 92130-Issy-Les-Moulineaux France. Phone: (331) 55 95 57 00. Fax: (331) 55 95 57 01. E-mail: leo@leoproductions.com Web Site: www.leoproductions.com.

Jean-Louis Brugat, mgng dir.

TV-CATV only.

TV production, progmg consultants, live bcsts.

Leukemia & Lymphoma Society, 1311 Mamaroneck Ave., White Plains, NY, 10605. Phone: (914) 949-5213. Fax: (914) 949-6691. E-mail: lanereg@lls.org Web Site: www.leukemia-lymphoma.org.

Nancy Klein, VP mktg; Geralyn Laneve, VP.

TV-CATV only.

Produces & distributes educational ideas to inform & educate viewers about leukemia & related diseases & available treatment.

Liberty Studios Inc., 238 E. 26th St., New York, NY, 10010. Phone: (212) 532-1865. Fax: (212) 779-2207. E-mail: email@libertystudios.us

Anthony Lover, pres; John Sawyer, VP.

TV-CATV only.

TV program, coml, film & video production.

Lifestyle Magazine/The Evidence, 101 W. Cochran St., Simi Valley, CA, 93065. Phone: (888) 940-0062. Fax: (805) 522-2114. E-mail: info@ffttv.org Web Site: www.faithfortoday.tv.

Michael Tucker, dir.

TV-CATV only.

TV program production & distribution.

Lightbridge Production & Distribution, 1051 Broadway, Sonoma, CA, 95476. Phone: (707) 939-4920. Fax: (707) 939-4919.

Roy Walkenhorst, CEO; Judy Brooks, founder.

TV-CATV only.

TV & video progmg.

Lighthouse Productions, 118 S. Main St., Goshen, 46526. Phone: (574) 533-1400. Fax: (661) 760-8775. E-mail: audicolabels@audicolabels.com Web Site: www.audicolabels.com.

Bill Landow, owner.

TV-CATV only.

Audio, video full production svc, documentaries, training, audio & video recording svcs.

Lightyear Entertainment L.P., 434 Ave of the Americas, 6th flr., New York, NY, 10111. Phone: (212) 353-5084. Fax: (212) 353-5083. E-mail: mail@lightyear.com Web Site: www.lightyear.com.

Arnold Holland, pres/CEO.

TV-CATV only.

TV program production & distribution. Audio & video distribution.

Limelight Communications Inc., 2532 W. Meredith Dr., Vienna, VA, 22181. Phone: (703) 242-4596. Phone: Cell (703) 626-3167. Fax: (703) 242-0324. E-mail: moreinfo@limelight.com Web Site: www.limelight.com.

Kenneth Reff, pres; Linda Falkerson, VP.

TV-CATV only.

Writing, producing & editing svcs for bcst & industrial clients. Available as sub-contractors for specific svcs, or to fully produce complete shows.

Lindberg Productions Inc., 24 Mulford Ave., East Hampton, NY, 11937. Phone: (212) 599-1239. Phone: (917) 696-1826. E-mail: ctimany@aol.com

Larry Lindberg, pres; Erika Shapeero, producer.

TV-CATV only.

Video production for business & industry.

Lion and Fox Recording Studios, 9517 Baltimore Ave., College Park, MD, 20740-1321. Phone: (301) 982-4431. E-mail: mail@lionfox.com Web Site: www.lionfox.com.

James Fox, pres.

TV-CATV-Radio.

Digital audio production for TV & radio; music & EFX libraries; CD and CD-ROM & cassette duplication; location audio.

Lions Gate Entertainment, (A division of Trimark Holdings). 2700 Colorado Ave., Suite 200, Santa Monica, CA, 90404. Phone: (310) 449-9200. Fax: (310) 392-0252. Web Site: www.lionsgatefilms.com.

Don Feltheimer, pres/CEO.

TV-CATV only.

Domestic & international TV, film, video distribution & adv sls firm. Builds, manages, invests in domestic & foreign bcst networks.

Litton Syndications Inc., 790 Johnnie Dodds Blvd., Suite 201, Mount Pleasant, SC, 29464. Phone: (843) 883-5060. Fax: (843) 883-9957. E-mail: tim@litton.tv Web Site: www.litton.tv.

Tim Voit, VP/dir sls.

TV-CATV only.

TV distribution (TV program sls & mktg).

London Weekend Television International, South Bank TV Ctr., Upper ground, London, SE1 9LT. United Kingdom. Phone: (020) 7620-1620. E-mail: images@lwt.co.uk Web Site: www.lwt.co.uk.

Charles Allen, mgng dir; Steve Morrison, head admin.

New York, NY 10110, 500 Fifth Ave, Suite 1710. Phone: (212) 682-3055. Fax: (212) 869-3693. Ellis Bell, CPA.

TV-CATV only.

TV program production & distribution.

Longhorn Radio Network, 1 University Station, (A0704) University of Texas, Austin, TX, 78712-1090. Phone: (512) 471-1631. Fax: (512) 471-3700. Web Site: www.kut.org.

J. Stewart Vanderwilt, dir; Hawk Mendenhall, progmg dir.

Radio Only.

Loral Skynet, (A subsidiary of Loral Cyberstar). 2440 Research Blvd. #200, Rockville, MD, 20850. Phone: (301) 258-8101. Fax: (301) 258-8119. Web Site: www.loralskynet.com.

Terry Hart, pres/CEO; Gayle Armstrong, VP mktg.

London W1P 8AE. Loral Cyberstar-Europe, Inc., 131-151 Great Titchfield St. Phone: +44 171 892 3700.

TV-CATV-Radio.

International satellite communications company that leases capacity for video transmissions for TV & other program distributors. Also provides Internet access & private net svcs directly to Internet Service Providers & multinational businesses worldwide. Svcs include data networking, voice, video, teleconferencing & news distribution to multiple points worldwide.

M

MAKWDE Productions, 10556 Arnwood Rd., Lake View Terrace, CA, 91342. Phone: (818) 899-4457. Fax: (818) 890-4050. E-mail: feetrell@ix.netcom.com

Marion Jones, pres/CEO; Dwayne Jones, dir sls & mktg; Keith Jones, adv mgr; Detra Jones, opns mgr; Joe Jones, international.

TV-CATV only.

TV coml producers, radio programs, coml & jingle producers.

MAN QC Creations, 123 E. Dania Beach Blvd., Dania, FL, 33004. Phone: (954) 921-1111.

TV-CATV only.

Worldwide on-location production; complete digital production & postproduction svcs; studio facility; TV progmg.

MGC The Multimedia Group of Canada, 415-A Mount Pleasant, Montreal Westmount, PQ, H3Y 3G9. Canada. Phone: (514) 844-3636. Fax: (514) 844-4990. E-mail: mgc@the-mgc.com Web Site: www.the-mgc.com.

Jacques Bouchard, pres/CEO; Roselyne Brovillet, sls; David Seeler, dev VP.

TV-CATV only.

Participates in the dev & distribution of progmg in the intl mkt.

MGM Inc., 2500 Broadway, Santa Monica, CA, 90404. Phone: (310) 449-3000. Fax: (310) 264-1244. Web Site: www.mgm.com.

TV-CATV only.

TV program distribution.

MGM TV Canada, 20 Queen St. West #3500, Toronto, ON, H5H 3R3. Canada. Phone: (416) 260-9680. Fax: (416) 260-9993. Web Site: www.mgm.com.

TV-CATV only.

Film distribution for all UA, Polygram & Orion film library (features, series & animated).

MG/Perin Inc., 110 Green St. Suite 304, New York, NY, 10012. Phone: (212) 941-9750. Fax: (212) 941-9122. E-mail: mgperin@aol.com

Richard Perin, pres.

TV-CATV only.

TV program production & distribution.

MPL Media, 621 Mainstream Dr., Ste 260, Nashville, TN, 37228. Phone: (615) 256-1675. Fax: (615) 256-0757. E-mail: daviddeeb@mplmedia.com Web Site: www.mplmedia.com.

Peggy Shedlock, VP; David Deeb, dir mktg & sls dir.

TV-CATV only.

16/35 color negative processing, rank cintel & ursagold transfer, video edit, graphics, DVD authoring & fulfillment, HD video editing.

MRC Films, Box 697, Plainview, NY, 11803. Phone: (516) 796-7568. E-mail: jlmollot3@verizon.net

Larry Mollot, exec producer.

TV-CATV only.

Producers of original progmg for bcst, cablecast, TV comls & PSAs.

MRN Radio, 1801 International Speedway Blvd., Daytona Beach, FL, 32114. Phone: (386) 947-6400. Fax: (386) 947-6716. Web Site: www.mrnradio.com.

Cheryl Knight, dir affil; Steve Harrison, natl sls mgr.

Radio Only.

Live coverage of NASCAR stock car racing plus *NASCAR LIVE* wkly telephone talk, *NASCAR Today* daily news program, via satellite.

MSE, 540 Toby Hill Rd., Westbrook, CT, 06498. Phone: (860) 399-0191. Fax: (860) 399-0196. E-mail: marcia@mseusa.com Web Site: www.mseusa.com.

Marcia Simon, pres.

TV-CATV-Radio.

Bcst & promotional writing & production; specialists in health & medical progmg.

MTI The Image Group, 885 2nd Ave., Level C, New York, NY, 10017. Phone: (212) 548-7700. Fax: (212) 759-7465. E-mail: jromano@image-group.com Web Site: www.image-group.com.

New York, NY 10019, 727 11th Ave. Phone: (212) 649-6333. Jerry Romano, dir business dev.

New York, NY 10016, 401 Fifth Ave. Phone: (212) 592-0600.

TV-CATV only.

Nine studios, 25 Digital online suites, 10 Avids, four infernos, 63 D platforms, NIT/MAC platforms, URSA Diamond/c-Reality, Digital Sound mixing Duplication.

MTM Entertainment Inc., 12700 Ventura Blvd., Suite 200, Studio City, CA, 91604. Phone: (310) 235-9700.

TV-CATV only.

TV program production & distribution.

MVI Post, 6320 Castle Pl., Falls Church, VA, 22044. Phone: (703) 536-7678. Fax: (703) 536-9490. E-mail: mailbox@mvipost.com Web Site: www.mvipost.com.

Frank Maniglia Jr., pres; Craig Maniglia, VP.

TV-CATV-Radio.

Full-svc video, audio & graphics postproduction; features screensound digital audio system 601 component digital video suite, high definition digital postproduction facility.

MacNeil/Lehrer Productions, 2700 S. Quincy St., Suite 250, Arlington, VA, 22206. Phone: (703) 998-2170. Fax: (703) 998-5707. Web Site: www.pbs.org/newshour. E-mail: pbs@newshour.org

Dan Werner, pres; David Sit, VP; Harold Crawford, controller; Susan Mills, dir program dev; Pam Wyatt, dir admin.

TV-CATV only.

Produces news & info programs for public TV & other coml & cable networks. Production of *The News Hour with Jim Lehrer.*

Madison Square Garden Network, 4 Penn Plaza, 4th Floor, New York, NY, 10001. Phone: (212) 465-5926. Fax: (212) 465-6024. E-mail: msgnetpr@thegarden.com Web Site: www.msgnetwork.com.

Mike McCarthy, exec SP MSG; Neil Davis, exec VP adv sls; Leon Schwier, exec production sports design.

TV-CATV only.

NY Knicks, NY Rangers & NY Mets, NY MetroStars, NY Power, NY Islanders & NY Liberty; boxing, college football & basketball; exclusive Garden events; original series.

Magno Sound & Video, 729 7th Ave., New York, NY, 10019. Phone: (212) 302-2505. Fax: (212) 819-1282. E-mail: david@magnosound.com Web Site: www.magnosound.com.

Robert Friedman, pres; David Friedman, VP.

TV-CATV only.

Complete film, TV & radio production & postproduction svcs for agency, feature, network, corporate & industrial clients.

Make It Happen Productions Inc., 5925 Troost Ave., No. Hollywood, CA, 91601. Phone: (323) 851-6444. Fax: (323) 851-6465. E-mail: bfrank@mihpitv

Billy Frank, pres.

TV-CATV only.

Ind & co-productions, dev of projects, package projects, production svcs, post production.

Man From Mars Productions, 159 Orange St., Manchester, NH, 03104-4217. Phone: (603) 668-0652. Fax: (603) 666-4878. E-mail: brouder@juno.com Web Site: www.manfrommars.com.

Ed Brouder, owner.

Radio Only.

Aircheck sls for radio collectors; coml production.

Manhattan Production Music, 355 W. 52nd St., 6th Fl., New York, NY, 10019. Phone: (212) 333-5766. Phone: (800) 227-1954. Fax: (212) 262-0814. E-mail: info@mpmmusic.com Web Site: www.mpmmusic.com.

TV-CATV only.

Five libraries containing over 300 CDs, including the Audiophile Sound effects series, and the Chesky Classical Library.

Manhattan Transfer Miami, 2850 Tiger Tail, Coconut Grove, FL, 33133. Phone: (800) 826-8864. Phone: (305) 857-0350. Fax: (305) 857-0175. Web Site: www.bvinet.com.

George O'Neil, VP.

Miami, FL 33179, 2028 N.E. 15th Ct. Phone: (305) 652-5762.

TV-CATV-Radio.

Full-svc video post-production including film transfer, on-line, off-line editing, audio, graphics, progmg, voice dubbing translation, sound stage & new media dev.

Ben Manilla Productions, 3361 20th St., San Francisco, CA, 94110-2627. Phone: (415) 970-8020. Fax: (415) 970-8024. E-mail: info@bmpaudio.com Web Site: www.bmpaudio.com.

J. Ben Manilla, pres; Jim Swenson, producer.

Radio Only.

Audio production & progmg for a var of formats.

Mar Vista Entertainment, 12519 Venice Blvd., Los Angeles, CA, 90066. Phone: (310) 737-0950. Fax: (310) 737-9115. E-mail: info@marvista.net Web Site: www.marvista.com.

Ferdando Szew, COO; Michael Jacobs, pres; George Port, exec VP.

Sharon, MA 02067, 210 N. Main St. Phone: (781) 784-2480.

TV-CATV only.

Domestic & international distribution of children's animation & live action features & series, films & documentaries.

Marathon International, 74 rue Bonaparte, 75006 Paris Phone: 331-53-1091-00. Fax: 331-43-2504-66. E-mail: marathon@marathon.fr Web Site: www.marathon.fr.

TV-CATV only.

Distributor of TV programs worldwide; series, documentaries, animation, wildlife, TV movies.

Maryknoll Productions, 75 Ryder Rd., Maryknoll, NY, 10545-0308. Fax: (914) 762-6567. E-mail: nkeel@maryknoll.com Web Site: www.maryknollmall.org.

Lawrence M. Rich, exec producer.

TV-CATV only.

Offers a library of video & film productions featuring Third World countries; radio & TV programs also available.

Maryland Public Television, 11767 Owings Mills Blvd., Owings Mills, MD, 21117. Phone: (410) 356-5600. Fax: (410) 581-4338. Web Site: www.mpt.org.

Robert Shuman, pres/CEO; Larry Unger, exec VP; Eric Eggleton, sr VP.

TV-CATV only.

TV program production & distribution.

Masai Films Inc., 6922 Hollywood Blvd., Suite 401, Hollywood, CA, 90028. Phone: (323) 466-5451. Fax: (323) 466-2440.

Fritz Goode,.

TV-CATV only.

TV & radio program & coml producers; production svcs.

Maslow Media Group Inc., 2134 Wisconsin Ave., N.W., Washington, DC, 20007. Phone: (202) 965-1100. Fax: (202) 965-6171. E-mail: lmaslow@maslowmedia.com Web Site: www.maslowmedia.com.

Linda Maslow, CEO.

TV-CATV only.

Staffing & recruitment (freelance & fulltime), for best camera crews, cable, film, multimedia, payroll & paymaster svcs nationwide for bcst film & new media industries.

Mason Video, 9632 N. 34th St., Omaha, NE, 68112. Phone: (402) 455-9422. Fax: (402) 455-0707. E-mail: melemason@aol.com Web Site: www.masonvideo.com.

Mele Mason, owner.

TV-CATV only.

Offers bcst video productions. Equipment includes Ikigami HLV55 Betacam, DVcam & HDV.

MasterControl FamilyNet Radio, The Broadcast Communications Group, NAMB. 6350 W. Freeway, Fort Worth, TX, 76116. Phone: (817) 570-1400. Fax: (817) 737-9436. E-mail: lbratton@familynet.com Web Site: www.familynetradio.com.

Lisa Young, radio mktg & distribution; Donna Senn, radio

distribution; Chuck Ries, producer.
Radio Only.
Total health program featuring interviews with experts on physical, mental, financial, and spiritual health, hosts Ralph Baker & Terri Barrett.

Matchframe Video, 610 N. Hollywood Way, Suite 101, Burbank, CA, 91505. Phone: (818) 840-6800. Fax: (818) 840-2726.
George Francisco, CEO; Rand Gladden, pres; Pam Hollander, exec VP; Michael Levy, VP/gen mgr.
TV-CATV only.
In-house editing suites; audio sweetening; portable on-line/off-line (AVID/HD and SD) editing systems; graphics; telecine; tape to tape color connection.

William Mauldin Productions Inc., 1010 Canonero Dr., Greensboro, NC, 27410-3804. Phone: (336) 632-9801. Fax: (540) 301-0399. E-mail: productions@mauldin.net Web Site: www.mauldin.net.
William D. Mauldin, pres/CEO.
TV-CATV-Radio.
Major market talent for narrations & voice overs, documentaries, station IDs, program intros for small market radio. Digital audio & video.

Maximum Marketing Services Inc., 833 W. Jackson, Ste 300, Chicago, IL, 60607. Phone: (312) 226-4111. Fax: (312) 226-5765. Web Site: www.maxmarketing.com.
John McGowan, pres.
TV-CATV only.
TV & radio program, production & distribution; public relation services.

Maysles Films, Inc., 250 W. 54th St., New York, NY, 10019. Phone: (212) 582-6050. Fax: (212) 586-2057. E-mail: info@mayslesfilms.com Web Site: www.mayslesfilms.com.
Albert Maysles, pres.
TV-CATV only.
Full production svcs for theatrical & TV non-fiction films; adv comls; industrial films, including pre- & postproduction.

Lynn McAfee Photography, 11324 1/2 Hatteras St., North Hollywood, CA, 91601. Phone: (818) 761-1317.
Lynn McAfee, owner/photographer.
TV-CATV-Radio.
Unit production stills, special photography & all photographic svcs.

McClain Enterprises Inc., 4405-B Belmont Park Terrace, Nashville, TN, 37215-3609. Phone: (615) 269-6517. Fax: (615) 269-6648. E-mail: carolyn@mcclaintv.com Web Site: www.mcclaintv.com.
Carolyn McClain, pres.
TV-CATV only.
Custom & syndicated TV mktg for radio, sls consulting, sls & promotional projects for radio & TV stns.

Media Access Group at WGBH, One Quest St., Boston, MA, 02135. Phone: (617) 300-3600. Fax: (617) 300-1020. E-mail: access@wgbh.org Web Site: access.wgbh.org.
Larry Goldberg, dir.
Burbank, CA 91502, 300 E. Magnolia Blvd., 2nd Fl. Phone: (818) 562-3344.
TV-CATV only.
Provides real-time & off-line captioning, subtitling, descriptive narration & consulting.

The Media Group of Connecticut Inc., 17 Maple St., Weston, CT, 06883-1026. Phone: (203) 544-0018. E-mail: mediagr@aol.com
Harvey F. Bellin, pres.
TV-CATV only.
TV, video production, writing, directing & editing; digital animation; dramatization & documentary; TV, corporate & govt svcs offered.

Media Planning Group (MPG), 195 Broadway, 12th Fl., New York, NY, 10007. Phone: (646) 587-5000. Fax: (646) 587-5005. Web Site: www.mpgsite.com.
Bob Riordan, mgr.
TV-CATV-Radio.
Producers, distributors & TV program packagers; videocassette producer/distributor. Full svc media buying & planning company.

Media Visions, 8430 Terminal Rd., Newington, VA, 22079. Phone: (703) 550-1500. Phone: (800) 628-3556. Fax: (703) 550-9711. E-mail: mrock@mediavisions.net Web Site: www.mediavisions.net.
Mike Rock,.
Provides full-svc video duplication, packaging, warehousing & complete order fulfillment. CD, DVD.

Medialink, 708 3rd Ave., 8th Fl., New York, NY, 10017. Phone: (212) 682-8300. Fax: (212) 682-2370. E-mail: mwallace@medialink.com Web Site: www.medialink.com.
Mary Buhay, sr VP; Monica Jennings, sr VP.
TV-CATV only.
Video & audio news release distributor & producer to TV & radio stns throughout the United States & Europe.

MediaTracks Inc., 2250 E. Devon Ave., Suite 150, Des Plaines, IL, 60018-4507. Phone: (847) 299-9500. Fax: (847) 299-9501. E-mail: slustig@mediatracks.com Web Site: www.mediatracks.com.
Shel Lustig, pres; Reed Pence, VP.
Radio Only.
Produce, syndicate & distribute radio progmg, news, comls & PSAs. Specialists in health & medicine, news & pub affrs.

Medstar Television Inc., 5920 Hamilton Blvd., Allentown, PA, 18106. Phone: (610) 395-1300. Fax: (610) 391-1556. Web Site: www.medstar.com.
William P. Ferretti, CEO; Paul Dowling, pres; Ron Petrovich, VP.
TV-CATV only.
Health & medical news progmg includes one-hour specials, Health Matters TV series, MedstarSource & MedstarAdvances news svcs.

Megatrax Production Music Inc., 7629 Fulton Ave., North Hollywood, CA, 91605. Phone: (818) 255-7100. Phone: (888) 634-2555. Fax: (818) 255-7199. E-mail: megatrax@megatrax.com Web Site: www.megatrax.com.
John Dwyer, owner; Ron Mendelsohn, owner.
TV-CATV-Radio.
Production music for bcst promotion & adv. Custom scoring & news music packages available.

Bill Melendez Productions Inc., 13400 Riverside Dr., Suite 201, Sherman Oaks, CA, 91423. Phone: (818) 382-7382. Fax: (818) 382-7377. E-mail: bmpi@aol.com
Bill Melendez, pres.
TV program, coml animation production.

Message on Hold, Box 747, Hendersonville, NC, 28793-0747. Phone: (828) 692-7200. Phone: (800) 223-1930. Fax: (828) 692-9147. E-mail: molton.ad@bellsouth.net
J. Ellis Molton, pres; Randy Molton, opns mgr.
Radio Only.
Producers of high-quality comls for telephone "hold" lines. Specializing in automotive, financial, medical & pharmacies.

Metro Music Productions Inc., 37 W. 20th St., Suite 906, New York, NY, 10011. Phone: (212) 229-1700. Phone: (800) 697-7392. Fax: (212) 229-9063. E-mail: info@metromusicinc.com Web Site: www.metromusicinc.com.
Mitch Coodley, pres; Katrina Haskell, office mgr.
TV-CATV only.
Original music scoring for TV progmg, prom, news, sports, comls. Production music library geared toward bcst.

Metro Networks, A Westwood One Co. 300 Bridge St., New Cumberland, PA, 17070. Phone: (717) 774-8150. Fax: (717) 774-8160. E-mail: elaine.konkle@metronetworks.com Web Site: www.metronetworks.com.
Elaine Konkle, gen mgr.
TV-CATV-Radio.
TV & radio net; radio production/distribution.

Metro Weather Service Inc., 571 So. Central Ave., Ste 102, Valley Stream, NY, 11580. Phone: (516) 568-8844. Phone: (800) 488-7866. Fax: (516) 568-8853. Fax: (800) 768-7998. E-mail: metrowx@aol.com Web Site: www.metrowx.com.
Pat Pagano, pres.
TV-CATV only.
Provides accurate & understandable weather forecasts. Serves any part of the nation; live consultations.

Robert Michelson Inc., 508 3rd Ave., San Francisco, CA, 94118. Phone: (415) 386-6862. Fax: (415) 386-2714. E-mail: rm@rmitv.com Web Site: www.rmitv.com.
Robert Michelson, pres; David Alexander, dir.
TV-CATV-Radio.
TV coml production; custom & syndicated TV spots for radio stns. Leading producer of TV spots for rock radio.

Midwest Video Communications Inc., Box 11627, Omaha, NE, 68111. Phone: (402) 991-2981. Fax: (402) 933-8990. E-mail: midwestvideo@msn.com
John S. Turner, pres; John Lott, producer/dir; Artes Johnson, dir.
TV-CATV only.
TV program & distribution; coml production, satellite teleconference productions; business & promotion film productions; production svcs, TV news features production.

Miller Broadcast Management, 616 W. Fulton St., Suite 516, Chicago, IL, 60661. Phone: (312) 454-1111. Fax: (312) 454-0044. E-mail: info@millerbroadcast.com Web Site: www.millerbroadcast.com.
Matt Miller, VP; Lisa Miller, pres.
Radio Only.
Produces & syndicates natl progmg including *KidsRadio*.

Robin Miller, Filmaker Inc., 606 W. Broad St., Bethlehem, PA, 18018. Phone: (610) 691-0900. Fax: (610) 691-0952. E-mail: mail@filmaker.com Web Site: www.filmaker.com.
TV-CATV only.
TV program, promotion film production; production svcs.

Warren Miller Entertainment, 2540 Frontier Ave., Suite 104, Boulder, CO, 80301. Phone: (303) 442-3430. Fax: (303) 442-3402. Web Site: www.warrenmiller.com.
Josh Haskins, producer; Tim Malone, dir.
TV-CATV only.
Second unit feature, coml, TV program, promotion; film production & distribution specializing in snow & outdoor adventure sports.

Miss Universe, L.P., 4111 W. Alameda, Suite 605, Burbank, CA, 91505. Phone: (818) 972-9202. Fax: (818) 972-9001. Web Site: www.missuniverse.com.
Paula M. Shugart, pres; Tony Santomauro, VP business dev.
TV-CATV only.
TV program production.

Mobile Video Services Ltd., 1620 I St. N.W., Washington, DC, 20006. Phone: (202) 331-8882. Fax: (202) 331-9064. E-mail: bookfeed@mobilevideo.net Web Site: www.mobilevideo.net.
Lawrence J. VanderVeen, pres; Christine Baber, opns mgr.
TV-CATV only.
Bcst production svcs, best "Official Washington", live shot, remote crews, studio svcs, editing, graphics suites, satellite & fiber transmission svcs.

Modern Entertainment, Box 8075, Van Nuys, CA, 91409-8075. Phone: (818) 386-0444. Fax: (818) 728-3677.
Michael Weiser, pres/CEO; Ken Du Bow, sr VP worldwide sls; Allyson Hall, VP international sls.
TV-CATV only.
Distribution of movies world wide, CD, DVD.

Modern Sound Pictures Inc., 1402 Howard St., Omaha, NE, 68102. Phone: (402) 341-8476. Fax: (402) 341-8487. E-mail: mspi1@att.net Web Site: www.modernsoundpictures.com.
Sandra L. Smith, pres.
Non-theatrical 16mm film, video rental library & retail audiovisual equipment for rental & sls.

Molton Advertising Inc., Box 747, Hendersonville, NC, 28793-0747. Phone: (828) 692-7200. Phone: (800) 223-1930. Fax: (828) 692-9147. E-mail: molton@bellsouth.net Web Site: www.moltonad.com.
Ellis Molton, pres; Randy Molton, opns mgr.
Radio Only.
Syndicated radio & nwspr series for loc use.

Mondo TV, Via G. Gatti 8/A, 00162 Rome Italy. Phone: 39-6-86320364. Phone: 39-06-86323293. Fax: 39-06-86209836. E-mail: mondotv@mondotv.it
Orlando Corradi, pres/CEO; Gian Claudio Galatoli, dir; Roberto Farina, head international sls.
TV-CATV only.
Animated TV series.

Montgomery Community Television Inc., 7548 Standish Pl., Rockville, MD, 20855. Phone: (301) 424-1730. Fax: (301) 294-7476. Web Site: www.montgomerycommunitytv.com.
Don Katzen, mktg dir.
TV-CATV only.
Full-svc video production & postproduction, 2,400 sq ft. studio including complete control room, GVG200 switcher,

DVE, on- & off-line editing. Also, opn of two cable chs reaching 220,000 subs.

Moody Broadcasting Network, 820 N. LaSalle Blvd., Chicago, IL, 60610-3284. Phone: (312) 329-4433. Phone: (800) 621-7031. Fax: (312) 329-4339. E-mail: mbn@moody.edu Web Site: www.mbn.org.

Douglas Hastings, opns mgr & progmg mgr; David Woodworth, administrator tech dev; Robert Neff, VP; Tony Rufo, satellite dept mgr; Wayne Shepherd, progmg mgr.

Radio Only.

Radio program production, full-svc relg digital stereo audio progmg via, satellite internet file download & syndicated tape distribution & ACCUWatch radio transmitter monitoring.

Moonstone Entertainment, Box 7400, Studio City, CA, 91614-7400. Phone: (818) 985-3003. Fax: (818) 985-3009.

Ernst "Etchie" Stroh, CEO; Yael Stroh, pres.

TV-CATV only.

International distribution and production.

The Charles Morrow Associates Company LLC, 307 7th Ave., Suite 1402, New York, NY, 10001. Phone: (212) 989-2400. Fax: (212) 989-2697. E-mail: cmorrow@cmorrow.com Web Site: www.cmorrow.com.

Charlie Morrow, pres.

Radio Only.

Sound design, audio production, music production, audio/visual service, music composition, multimedia producers, surround sound studio, public service announcements.

MotorNet, Box 69, Farmingdale, NJ, 07727-0069. Phone: (732) 751-1020. Fax: (732) 751-1038. E-mail: motornet@iop.com Web Site: www.motorsportsreport.com.

Charlie Roberts, pres; Ken Stout, producer; Jack Schultz, dir mktg.

Radio Only.

Radio program production & distribution.

Mountain News Corporation, 50 Vashell Way, Suite 200, Orinda, CA, 94563-3020. Phone: (925) 254-4456. Fax: (925) 254-7923. E-mail: admin@aminews.com Web Site: www.theamigroup.com.

Rob Brown, pres; Chad Dyen, VP. Eastern Bureau Phone: (800) 736-0370.

Radio Only.

Produce & package outdoor recreation reports of 30, 60 & 90-seconds in length. Also deliver ski reports via phone, computer, facsimilie for on-air, phone lines & web sites.

Munhwa Broadcasting Corp. (MBC), National Press Bldg., Suite 1131, 529 14th St. N.W., Washington, DC, 20045. Phone: (202) 347-0078. Fax: (202) 347-0079. E-mail: mgchoi@imbc.com Web Site: www.imbc.com.

Chang-Young Choi, bureau chief.

TV-CATV only.

Korean natl TV net news.

Munz, 2470 W. 8th Ave., Hialeah, FL, 33010. Phone: (305) 884-8200. Fax: (305) 889-7212. Web Site: www.mun2television.com.

Don Browne, COO; Yolanda Foster, progmg VP; Tracy McDonough, Dir of Business Operations; Joe Bernard, sls dir; Alvaro Krupkin, Dir of Creative Services+; Maria Acosta, dir opns; Laura Dergal, dir mktg.

TV-CATV only.

English language cable network targeting young U.S. Hispanics.

Music of Your Life, 6525 Babcock St., SE, Malabar, FL, 32950-5002. Phone: (321) 725-0014. Fax: 3(21) 725-0098. E-mail: info@musicofyourlife.com Web Site: www.musicofyourlife.com.

Kerry Fink, CEO.

Radio Only.

Music of Your Life satellite radio network for coml radio affil stns, featuring adult standard format

Musical Starstreams, Box 12685, La Jolla, CA, 92039-2685. Phone: (619) 276-8989. Fax: (619) 276-0918. E-mail: forest@starstreams.com Web Site: www.starstreams.com.

Radio Only.

Two-hour wkly syndicated exotic electronica music progmg also available as a full-time format; radio adv production.

Musivision, 185 E. 85th St., New York, NY, 10028. Phone: (212) 860-4420.

TV-CATV-Radio.

Design & production of computer graphics & animation for stn IDs, promotions, show openings & coml applications.

Muzak, 3318 Lakemont Blvd., Fort Mill, SC, 29708. Phone: (803) 396-3000. Phone: (800) 331-3340. Fax: (803) 396-3136. Web Site: www.muzak.com. E-mail: feedback@muzak.com

Bill Boyd, CEO.

Bcsts 60 chs of business mus, adv parting audio messages, ZNET data bcstg & video via direct bcst satellite.

Myriad Pictures, 3015 Main St., Suite 400, Santa Monica, CA, 90405. Phone: (310) 279-4000. Fax: (310) 279-4001. E-mail: info@myriadpictures.com Web Site: www.myriadpictures.com.

TV-CATV only.

An ind TV co-production & distribution company specializing in music series, features, documentaries & drama for the international market.

N

NAHB Production Group, (National Association of Home Builders). 1201 15th St. N.W., 5th Fl., Washington, DC, 20005. Phone: (202) 822-0200 ext 8543. Fax: (202) 266-8054. E-mail: cgoldweber@nahb.com

Cary Goldweber, exec producer; Joyce Pearson, sr supevrg producer.

Radio Only.

Complete video production/editing facility with large stock library. Productions include weekly HGTV series, political spots, PSA, & instructional/mktg progmgs.

NASA Broadcast & Imaging Branch, NASA Headquarters (PMD), 300 E. St. S.W., Rm. CL78, Washington, DC, 20546. Phone: (202) 358-0000. Fax: (202) 358-4333.

Mike Crnkovic, chief printing & design.

TV-CATV-Radio.

Aeronautics & Space Report, an hour magazine quarterly (Betacam SP) to media & producers only.

NBD Television Ltd., 2, Royalty Studios, 105 Lancaster Rd., London, W11 1QF. United Kingdom. Phone: 44 (0) 20 7243 3646. Fax: 44 (0) 7243 3656. E-mail: distribution@nbdtv.com Web Site: www.nbdtv.com.

Nicky Davies Williams, CEO; Andrew Winter, gen sls mgr.

International TV progmg sls distribution.

NCAA, Box 6222, Indianapolis, IN, 46206-6222. Phone: (317) 917-6222. Fax: (317) 917-6807. Fax: (317) 917-6856. E-mail: jrinebold@ncaa.org Web Site: www.ncaasports.com.

Chris Farrow, dir; Ron Schwartz, dir; Jeramy Michiaels, mgr; Frank Rhodes, mgr; Greg Weitekamp, mgr.

New York, NY 10019, 19 West 57th St. Phone: (212) 541-8840. Fax: (212) 262-4647. Ron Schwartz, dir NCAA TV new svcs.

TV-CATV only.

NCAA championship progmg, distribution & footage requests.

NDR Media, Rothenbaumchaussee 159+161, Hamburg, 20149. Germany. Phone: (040) 44 1920. E-mail: info@ndrtv.de Web Site: www.ndrtv.de.

Horst Bennit, mgng dir; Hans-Stefan Heyne, head international acquisition; Ulla Lamas-Torres, sls.

TV-CATV only.

Distribution of TV plays, dramas, wildlife, educ, children's & documentary programs.

NEP Studios, 1 Dag Hammarskjold Plaza, Concourse Level, New York, NY, 10017-2201. Phone: (212) 548-7700. Fax: (212) 355-0523. E-mail: wsheehy@nepstudios.com Web Site: www.nepinc.com.

William Sheehy, VP.

TV-CATV-Radio.

Shooting stage, Filmor tape, duplication all standards editing & DVD-R duplication.

NFL Films, 1 NFL Plaza, Mt. Laurel, NJ, 08054. Phone: (856) 222-3500. Fax: (856) 722-6779. Web Site: www.nflfilms.com.

Steve Sabol, pres; William Driber, VP production; Rick Angeli, dir facilities; Barry Wolper, COO; Phil Tuckett, VP entertainment videos; Jeff Howard, VP video opns.

TV-CATV only.

Teleproduction facility: digital editing suites, 16mm & 35mm film processing, film-to-tape transfer, studio & remote production, sound studios, animation & flame.

NHK Japan Broadcasting Corp., 2030 M St. N.W., Suite 706, Washington, DC, 20036. Phone: (202) 828-5180. Fax: (202) 828-4571. Web Site: www.nhk.or.jp/englishtop/.

Ryuichi Teshima, bureau chief.

TV-CATV only.

TV & radio program distribution.

NRS Group PTY Ltd., 9-13 Lawry Pl., Macquarie, Canberra, Act 2614. Australia. Phone: (61) 2-6251-6333. Fax: 61 2-6251-6240. E-mail: grahampatrick@nrsgroup.com.au

Graham Patrick, mgng dir.

TV-CATV only.

Produces range of quality TV series & documentaries, full TV & audio production facility with qualified personnel, international program distributor.

NTN Communications Inc., 5966 La Place Ct., Suite 100, Carlsbad, CA, 92008-8830. Phone: (888) PLAYNTN. Fax: (760) 438-3505. Web Site: www.ntn.com.

Stanley B. Kinsey, chmn/CEO; Mark deGorter, pres/COO; James B. Frakes, CFO; Tyrone Lam, pres Buzztime Entertainment Inc.

TV-CATV only.

NTN Communications, Inc.®, a leading producer & distributor of live interactive TV entertainment, bcsts exciting multi-player games to hospitality venues.

NTV International Corp., 50 Rockefeller Plaza, Suite 940, New York, NY, 10020. Phone: (212) 489-8390. Fax: (212) 489-8395. Web Site: www.ntvic.com.

Yoichi Shimada, pres.

TV-CATV only.

Complete video production & postproduction facility; satellite transmission capabilities worldwide; ENG package international TV coord, program sls & acquisitions.

NVC Arts, The Forum, 74-80 Camden St., London, NW1 0EG. United Kingdom. Phone: 44 (0) 7388 3833. Fax: 44 (0)20 7388 7174. E-mail: mia_fjox-non@nvcarts.com Web Site: www.nvcarts-tv.com.

John Kelleher, mgng dir; Elfyn Morris, stn mgr.

TV-CATV only.

Producers & distributors of opera, ballet & performing arts programs for world TV.

N W Media, 106 SE 11th Avenue, Portland, OR, 97214. Phone: (503) 223-5010. Fax: (503) 223-4737. Web Site: www.nwmedia.com. E-mail: info@nwmedia.com

Jeanne Alldredge, pres; Mitchell Harris, sls dir.

TV-CATV-Radio.

Audio/videotape duplication; CD/DVD Duplication, multimedia authoring, graphic design, mastering svcs & messaging.

National Church Broadcasting, Box 8263, Haledon, NJ, 07508. Phone: (973) 956-2900. Fax: (973) 956-0600.

Samuel Cummings, pres; June Young, VP.

TV-CATV only.

Distributor of church & children's religious progmg to radio stns.

National Collegiate Athletic Association, Box 6222, Indianapolis, IN, 46206-6222. Phone: (317) 917-6222. Fax: (317) 917-6888. Web Site: www.ncaa.org.

Miles Brand, pres; JoJo H. Rinebold, dir.

TV-CATV only.

Producers of selected NCAA championships.

National Council of Churches Communications Unit, 475 Riverside Dr., Rm. 850, New York, NY, 10115. Phone: (212) 870-2227. Fax: (212) 870-2030. E-mail: news@ncccusa.org Web Site: www.ncccusa.org.

Shirley . Struchen, dir media resources; Wesley T. Pettillo, dir communications; Carol J. Fouke, dir news svcs.

TV-CATV only.

Bcst production, distribution & assistance to reporters, networks, stns; prepared radio reports & actualities without charge from bcst news professionals.

National Film Board of Canada, 350 Fifth Ave., Suite 4820, New York, NY, 10118. Phone: (212) 629-8890. Fax: (212) 629-8502. E-mail: newyork@nfb.ca Web Site: www.nfb.ca.

Christina Rogers, US mktg mgr TV.

TV-CATV only.

TV program distribution.

National Mobile Television, 2740 California St., Torrance, CA, 90503. Phone: (310) 782-9945. Phone: (800) 242-0642. Fax: (310) 782-9949. Web Site: www.nmtv.com.

Kevin Sublette, gen mgr; Stephanie Hampton, opns dir.

Seattle, WA 98168, 12698 Gateway Dr. Phone: (206)-242-0642.

TV-CATV only.

Mobile TV facilities for remote production of multi-camera events.

National Public Radio, 635 Massachusetts Ave. N.W., Washington, DC, 20001-3753. Phone: (513) 414-2000. Fax: (513) 414-3329. Web Site: www.npr.org.

Bill Davis; Kevin Klose, pres/CEO; Peter J. Loewenstein, VP distribution; Barbara Hall, dev VP; Kathleen D. Jackson, VP human resources; Mike Starling, engrg VP; Jim Elder, VP/CFO; Celeste James, VP communications; Ken Stern, exec VP; Jeffrey Dvorkin, omsbudman; Jay Kernis, sr VP progmg; Margaret Low-Smith, VP progmg; Maria Thomas, VP online; Dana Davis Rehm, member/progm svcs.

Radio Only.

Radio program production & distribution.

Native American Public Telecommunications Inc., Box 83111, Lincoln, NE, 68501. Phone: (402) 472-3522. Fax: (402) 472-8675. Web Site: www.nativetelecom.org.

Frank Blythe, exec dir; Mary Ann Koehler, business mgr; Carol Cornsilk, progm/production dir.

TV-CATV only.

Producing & developing educ telecommunication programs for all media including TV & pub radio.

Nemo News, 7179 Via Maria, San Jose, CA, 95139. Phone: (408) 226-6339. E-mail: dick@reizner.net

Dick Reizner, owner.

TV-CATV only.

ENG unit (all formats) covering assignments worldwide.

Network Music, 8750 Wilshire Blvd., Beverly Hills, CA, 90211. Phone: (800) 854-2075. Fax: (310) 358-4311. E-mail: sales@networkmusic.com Web Site: www.networkmusic.com.

Gary Gross, pres; Chuck Ansel, VP opns; Dennis Dunn, VP sls; Todd Kern, dir mktg; Carl Peel, dir.

TV-CATV-Radio.

Produces music, sound effects & production elements libraries.

New City Releasing Inc., 20700 Ventura Blvd., Suite 350, Woodland Hills, CA, 91364. Phone: (818) 348-2500. Fax: (818) 348-3022. Web Site: www.newcityreleasing.com.

Alan B. Burnsteen, pres; David Bursteen, VP.

TV-CATV only.

Producer & distributor of motion pictures to cable TV.

New Dimensions Radio, Box 569, Ukiah, CA, 95482. Phone: (800) 935-8273. Phone: (707) 468-5215. E-mail: info@newdimensions.org Web Site: www.newdimensions.org.

Michael A. Toms, co-pres; Justine Toms, co-pres.

Radio Only.

Radio program production & distribution.

New Films International, 8484 Wilshire Blvd., Suite 510, Beverly Hills, CA, 90211. Phone: (323) 655-1050. Fax: (323) 655-1070. E-mail: newfilms@newfilmsint.com Web Site: www.newfilmsint.com.

Nesin Hason, pres; Sezin Sonar, VP.

TV-CATV-Radio.

U.S.-based distribution company specialized in Romania, Bulgaria, & Turkey.

New Line Television, 888 Seventh Ave., 20th Fl., New York, NY, 10106. Phone: (212) 649-4900. Fax: (212) 956-1936. Web Site: www.newline.com.

Jim Rosenthal, pres; David Spiegelman, exec VP; Robin Seidner, sr VP.

Los Angeles, CA 90048, 116 N. Robetson.

TV-CATV only.

TV program production & distribution.

New Visions Syndication Inc., 44895 Hwy. 82, Aspen, CO, 81611. Phone: (970) 925-2640. Fax: (970) 925-9369. Web Site: www.newvisionssyndication.com. E-mail: kayla@nvs.com

Kayla Hoffman-Cook, VP.

TV-CATV only.

International & domestic syndicator of specials & series; sports, lifestyle & entertainment.

New Zoo Revue, 6399 Wilshire Blvd., Suite 816, Los Angeles, CA, 90048. Phone: (323) 782-3525. Fax: (323) 782-3530. E-mail: newzoo@aol.com Web Site: www.newzoorevue.com.

Barbara Atlas, pres.

TV-CATV only.

TV program production & distribution.

News Broadcast Network, 451 Park Ave. S., 7th Fl., New York, NY, 10016. Phone: (212) 684-8910. Fax: (212) 684-9650. Web Site: www.newsbroadcastnetwork.com.

Mike Hill, pres; Jill Hill, sr chmn.

TV-CATV-Radio.

Produce & distribute news & feature material to TV & radio stns.

Nightingale-Conant Corp., 6245 W. Howard St., Niles, IL, 60714. Phone: (847) 647-0300. Fax: (847) 647-7145. Web Site: www.nightingale.com.

Vic Conant, pres.

Radio Only.

Radio program production & distribution.

Nine Network Australia, 6255 Sunset Blvd., Suite 1500, Los Angeles, CA, 90028. Phone: (323) 461-3853. Fax: (323) 462-4849.

TV-CATV only.

Studio production facility, standard conversion & all format tape facilities.

No Soap Productions, 936 Broadway, 4th Fl., New York, NY, 10010. Phone: (212) 581-5572. Fax: (212) 586-0045. E-mail: dan@nosoap.net Web Site: www.nosoap.net.

Dan Aron, pres.

TV-CATV-Radio.

Radio comls, sound design for radio & TV, voicecasting, production studio-digital.

North American Network, Inc., 7910 Woodmont Ave., Suite 1400, Bethesda, MD, 20814. Phone: (301) 654-9810. Fax: (301) 654-9828. E-mail: info@nan.com Web Site: www.radiospace.com.

Thomas P. Sweeney, pres; Tammy Lemley, VP.

Radio Only.

Full-svc radio P.R. providing progmg, news PSAs, promotional campaigns & sls opportunities to stns nationwide in English & Spanish.

North by Northwest Productions, 903 W. Broadway, Spokane, WA, 99201. Phone: (509) 324-2949. Fax: (509) 324-2959. E-mail: marcdahlstrom@nxnw.net Web Site: www.nxnw.net.

Marc Dahlstrom, mgng ptnr.

Boise, ID 83702, 601 W. Broad St. Phone: (208) 345-7870. Shane Jibben, opns mgr.

TV-CATV only.

High end video & film production; D1/D2 postproduction; Paint/3D animation; sound design/production; HD ediiting.

North Shore Productions, Box 1308, Detroit Lakes, MN, 56502. Phone: (218) 846-1936. Fax: (218) 846-1936. E-mail: nspsfrm@tekstar.com Web Site: www.agriculture.com/sfradio.

Radio Only.

Produces & distributes "The Career Clinic®" & the "Successful Farming® Radio Magazine", two-minute features with affil sharing natl revenues.

North Star Music, 22 London St., East Greenwich, RI, 02818. Phone: (401) 886-8888. Fax: (401) 886-8886. E-mail: info@northstarmusic.com Web Site: www.northstarmusic.com.

Richard R. Waterman, pres.

Radio Only.

Production, distribution, mktg & promotion of recorded music.

Northwest Imaging & FX, 2339 Columbia St., Suite 100, Vancouver, BC, V5Y 3Y3. Canada. Phone: (604) 873-9330. Fax: (604) 873-9339. E-mail: nwfx@nwfx.com

Alex Tkach, VP.

TV-CATV only.

Shooting, visual effects, animation, digital editing suites, audio sweetening, Digital Betacam Sp. D-1, 1 inch, duplication Cintel Diamond Film transfer.

O

O. Atlas Enterprises Inc., (New Zoo Review). 327 North Palm Dr., Beverly Hills, CA, 90210. Phone: (323) 782-3525. Fax: (323) 782-3530. E-mail: newzoo@aol.com Web Site: www.newzoorevue.com.

TV-CATV only.

International licensing, distribution of progmg for TV, all media.

O'Grady & Associates, 8431 Sabal Palm Ct., Vero Beach, FL, 32963-4296. Phone: (772) 234-4177. Fax: (772) 231-9819. E-mail: jfogjr@juno.com

James F. O'Grady Jr., pres.

TV-CATV-Radio.

Brokerage/consulting.

OGM Production Music, 6464 Sunset Blvd., Suite 790, Hollywood, CA, 90028. Phone: (323) 461-2701. Phone: (800) 421-4163 (Sales). Fax: (323) 461-1543. E-mail: ogmmusic@ogmmusic.com Web Site: www.ogmmusic.com.

Ole Georg, pres.

TV-CATV only.

Video, cable, films, CD-ROM, bcst, multimedia, infomercials, satellite program, interactive TV, electronic publishing, theatrical features.

Oasis International, 6 Pardee Ave., Suite 103, Toronto, ON, M6K 3H5. Canada. Phone: (416) 588-6821. Fax: (416) 588-7276. E-mail: info@oasisinternational.com Web Site: www.oasisinternational.com.

Peter Emerson, pres; Valerie Cabrera, exec VP; Ben Bishop, sls & acquistion exec; Prentiss Holman, dir internaational sls.

TV-CATV only.

Worldwide TV & format distribution.

Oasis TV Inc., 9887 Santa Monica Blvd., Suite 200, Beverly Hills, CA, 90212. Phone: (310) 553-4300. Fax: (310) 553-1159. Web Site: www.oasistv.com.

TV-CATV only.

24-hours cable/satellite net providing a broad, well-branded var of new age/human potential progmg.

Omnimusic, 52 Main St., Port Washington, NY, 11050. Phone: (800) 828-6664. Phone: (516) 883-0121. Fax: (516) 883-0271. E-mail: bring@omnimusic.com Web Site: www.omnimusic.com.

Doug Wood, pres; Patti Wood, VP; Barbara Ring, mktg dir.

TV-CATV-Radio.

Creative production music designed for maximum impact. The best, easiest online search & download available. Digitally produced music & sound effect libraries for TV, radio & cable featuring real instruments in various styles & orchestrations.

On Track, 6350 W. Freeway, Fort Worth, TX, 76116. Phone: (817) 570-1400. Phone: (800) 266-1837. Fax: (817) 737-9436. E-mail: info@familynet.com Web Site: www.familynetradio.com.

Lisa Bratton, radio mktg & distribution; Donna Senn, radio distribution; Chuck Ries, producer.

Radio Only.

Contemporary Christian music with artist interviews, 30 minutes wkly, on CD.

One Hundred Biblemen & Women of the U.S.A., Box 8263, Haledon, NJ, 07508. Phone: (973) 956-2900. Fax: (973) 956-0600.

Sam Cummings, pres.

TV-CATV only.

Low-cost bcstg & buying net for churches.

Oppix Productions Inc., 353111 Laurel Leaf Ln., Fairfax, VA, 22031-3212. Phone: (703) 280-8200. Fax: (703) 280-9292. E-mail: info@oppix.com Web Site: www.oppix.com.

James Oppenheimer, pres.

TV-CATV-Radio.

Full-svc video internet, DVD production & post, serving bcst, corporate assns, nonprofit & govt. All formats.

Orange Productions Inc., 523 Righters Ferry Rd., 1st floor, Bala Cynwyd, PA, 19004. Phone: (610) 667-8620. Fax: (610) 667-8939. E-mail: orange@snip.net Web Site: www.soundsofsinatra.com.

Sid Mark, pres; Brian Mark, opns mgr.

Radio Only.

Production & distribution of a wkly two-hour program *Sounds of Sinatra.*

Dick Orkin's Amazing Radio, 15003 Lemay St., Van Nuys, CA, 91405. Phone: (800) 621-6949. Fax: (818) 376-8529. E-mail: sandyo@earthlink.net Web Site: www.sandyorkin-crs.com.

Sandy Orkin, pres.

Radio Only.

Syndicated packages of Dick Orkin comls customized for loc advertisers. Available in six categories.

Outdoor Media Group, Box 2151, Lake Oswego, OR, 97035. Phone: (503) 675-7345. Fax: (503) 675-7820. E-mail: omg@aojtv.com Web Site: www.aojtv.com.

Russell Cameron, pres.

TV-CATV only.

TV program, coml production & distribution; production svcs.

Jim Owens Entertainment, 624 Grassmere Park, Suite 16, Nashville, TN, 37211. Phone: (615) 256-7700. Fax: (615) 242-9735. Web Site: www.crookandchase.com.

Jim Owens, pres; Jennifer Anderson, producer.

TV-CATV only.

Radio program production company for syndication. TV production for cable & home video.

P

PACSAT, 1629 S St., Sacramento, CA, 95814. Phone: (916) 446-7890. Fax: (916) 446-7893. E-mail: pacsat@pacsat.com Web Site: www.pacsat.com.

Steve Mallory, pres; Marcia Calvin, opns mgr.

TV-CATV-Radio.

Video, audio & satellite professionals. Ku-HD satellite trucks. ENG crews. Fly pack with CCU'S. Post production & graphics.

PAULAR Entertainment L.L.C., 13700 Marina Pointe Dr., Suite 901, Marina del Rey, CA, 90292. Phone: (310) 821-0430. Fax: (310) 821-3793. E-mail: thirdwayv@aol.com

Larry Friedricks, ptnr; Paula Fierman, ptnr.

TV-CATV only.

World-wide distribution of feature films, video & TV including movies, series & mini-series.

PBS Video, 1320 Braddock Pl., Alexandria, VA, 22314. Phone: (703) 739-5000, EXT. 8614. Fax: (703) 739-8487. Web Site: www.pbs.org.

Pat Mitchell, pres/CEO.

Handles videocassette, DVD's sls, rental & licensing of selected PBS programs to schools, colleges, libraries, hospitals & other institutions.

PMTV Producers Management Television, 681 Moore Rd., Suite 100, King of Prussia, PA, 19406. Phone: (610) 768-1770. Fax: (610) 768-1773. E-mail: mailto.pmtv@pmtv.com Web Site: www.pmtv.com.

TV-CATV only.

Full-svc mobile TV production company, providing mobile units, crews, satellite svcs, lighting, staging, etc. for sports, entertainment & teleconferences worldwide.

PPM Multimedia, Brezo 4 - URB Los Robles, Torrelodones, Madrid, 28250. Spain. Phone: (34) 91 859 1913. Fax: (34) 91 859 0932. E-mail: multimedia@ppmm.es Web Site: www.ppmm.es.

Paco Rodriguez, mgng dir.

TV-CATV only.

Distribution of animation, feature films, documentaries, co-production setting.

(PSSI) Production & Satellite Services Inc., 11860 Mississippi Ave., Los Angeles, CA, 90025. Phone: (310) 575-4400. Fax: (310) 575-4451. E-mail: pssi@pssi-usa.com Web Site: www.pssi-usa.com.

Robert C. Lamb, pres; Brian Nelles, sr VP; Melissa D. Meek, mktg.

TV-CATV only.

Full service production & satellite transmission company with 19 fully redundant C- & Ku-band satellite trucks located nationwide. Available for news, sports, corporate, entertainment events, media tours, video conferencing, webcasting & downlinks. Additional svcs include Standard, High Definition digital transmission, encryption, multiple camera productions, event coordination & a C/Ku Flyaway system for international transmission. skIP Broadband also available for internet, voice connectivity, & webcasting via satellite, ideal in remote areas with little or no connectivity.

Palace Digital Studios, 29 N. Main St., South Norwalk, CT, 06854. Phone: (203) 853-1740. Fax: (203) 855-9608. E-mail: wendy@palacedigital.com Web Site: www.palacedigital.com.

Wendy Lambert, pres.

New York, NY 10036. Servi Digital Studios, 35 W. 45th St. Phone: (212) 921-0555. Carol McCoy, VP opns.

TV-CATV only.

On-air promotion design & animation; program opns; identity packages; sls tape creation longform editorial, CD-ROM, & Website design.

Shelly Palmer Productions, P. O. Box 1877, New York, NY, 10156-1877. Phone: (212) 532-3880. E-mail: info@shellypalmer.com Web Site: www.shellypalmer.com.

Shelly Palmer, pres.

TV-CATV only.

Music, video, TV, film production & creative svcs. Adv & mktg. Music libraries, sound design, sls videos & post-production.

Pan American Video, 3144 Broadway, Suite 4, Eureka, CA, 95501. Phone: (707) 822-3800. Fax: (707) 822-0800. E-mail: panam@panamvideo.com Web Site: www.panamvideo.com.

Teri Lane, pres; Sheila McQuillen, VP.

TV-CATV only.

Public domain movies & TV shows, bcst quality & stock footage.

Pantomime Pictures Inc., 12144 Riverside Dr., North Hollywood, CA, 91607. Phone: (818) 980-5555. Fax: (818) 984-3470.

Fred Crippen, dir; Matt Crippen, producer.

TV-CATV only.

Animation production & design for TV comls, educ & industrial use. Animation camera for 35 mm & 16 mm.

Parrot Communications International Inc., 2917 N. Ontario St., Burbank, CA, 91504. Phone: (818) 567-4700. Fax: (818) 567-4600. E-mail: info@parrotmedia.com Web Site: www.parrotmedia.com.

Robert W. Mertz, pres/CEO; Rae Ann Mertz, exec VP; Michael Norris, VP.

TV-CATV only.

Database mgmt, direct mail svcs, promotional fulfillment, warehousing, contest fulfillment, bcst faxing, high speed duplication & videotape duplication.

Pathe International, 21 rue Francois 1 er, 75008 Paris France. Fax: 33-1-40-76-9194. Fax: 33-1-40-76-9169. E-mail: christine.hayet@pathe.com Web Site: www.pathe.fr.

Jerome Seydoux; Eduardo Malone, co-chmn/CEO; Emma Rami, VP finance; Michel Crepon, COO.

TV-CATV only.

Production & distribution of TV films & documentaries. Production of multimedia programs.

Paulist Media Works, 3055 4th St. N.E., Washington, DC, 20017. Phone: (202) 269-6064. Fax: (202) 269-4304. E-mail: info@paulist.org Web Site: www.paulist.org/pmw.

Sue Donovan, pres.

TV-CATV only.

Support for non-profit organization in internet svcs/website/web design distribution of relg radio programs, video/documentary production

Paulist Productions, Box 1057, Pacific Palisades, CA, 90272. Phone: (310) 454-0688. Fax: (310) 459-6549. E-mail: paulistmail@paulistproductions.org Web Site: www.paulistproductions.org.

Frank Desiderio, CSP, pres; Enid Sevilla, gen mgr & finacial off; Barbara Gangi, producer; Joseph Kim, VP business affrs.

TV-CATV only.

TV program production.

Peckham Productions, 50 S. Buckhout St., Irvington, NY, 10533. Phone: (914) 591-4140. Fax: (914) 591-4149. Web Site: www.peckhampix.com.

Peter H. Peckham, pres; Waldine Peckham, producer; Peter Peckham, producer; Russell Peckham, dir & producer.

TV-CATV only.

Full-svc film & video producer of TV comls, TV programs & TV net promos.

Perception Media Group, 1848 Clay St., Roanoke, VA, 24013. Phone: (540) 563-5225. Fax: (540) 563-0117.

Ben Peyton, pres.

TV-CATV only.

TV & radio coml program production, distribution & jingles. Religious programs & distribution.

Peters Communications, 1555 Berenda Pl., El Cajon, CA, 92020. Phone: (858) 565-8511. Fax: (619) 440-1481. E-mail: ppiep@cox.net

Edward J. Peters, pres.

TV-CATV only.

Media mktg consultants, providing rsch, concept, mktg plan, music, graphics & animation.

Philadelphia Flyers Hockey Club, 3601 S. Broad St., Philadelphia, PA, 19148. Phone: (215) 465-4500. Fax: (215) 952-4103. E-mail: rryan@comcast-spectacor.com Web Site: www.philadelphiaflyers.com.

Ed Snider, chmn; Bob Clarke, pres; Shawn Tilger, VP mktg.

TV-CATV only.

TV & radio program production.

Phoebus Communications Inc., 10905 Ft. Washington Rd., Suite 300, Fort Washington, MD, 20744. Phone: (301) 292-9800. Fax: (301) 292-0829. E-mail: phoebuscom@aol.com Web Site: www.phoebusinc.net.

Gail C. Arnall, Ph.D., pres.

TV-CATV only.

Full scale distance learning net mgmt.

Phoenix Communications Group, 3 Empire Blvd., South Hackensack, NJ, 07606. Phone: (201) 807-0888. Fax: (201) 807-0272. Web Site: www.phoenixcomm.com.

Joe Podesta, chmn; Jim Holland, pres; Rich Domich, sls VP & mktg; Geoff Belinfante, exec producer & sr VP; Trish Ferreri, mgr.

TV-CATV only.

Major League Sports Newsatellite; TV & video production & distribution; stock footage licensing.

Pied Piper Films Ltd., 825941 Mel-Nott TL, R. R. 2, Shelburne, ON, L0N 1S6. Canada. Phone: (519) 925-6558. Fax: (519) 925-6558.

Allan Wargon; Lee Israelski, producer.

TV-CATV only.

Motion picture production.

Pike Productions Inc., Box 300, 11 Clarke St., Newport, RI, 02840. Phone: (401) 846-8890. Fax: (401) 847-0070. E-mail: info@pikefilmtrailers.com

James A. Pike, pres; Cornelia M. Pike, sls mgr.

Custom ads & special announcement trailers produced & distributed in all formats—35mm, 185x1, Scope, stereo, 70mm stereo.

Planet Pictures Ltd., 4764 Park Granada, Suite 208, Calabasas, CA, 91302. Phone: (818) 222-9000. Fax: (818) 222-4370. E-mail: info@planetpictures.com Web Site: www.planetpictures.com.

Jim Hayden, pres; Jennifer Hayden, mgng dir; Peter Torvik, business affrs.

TV-CATV only.

Production & distribution for documentary, informational, reality-based TV programs.

Playboy Entertainment Group Inc., 2706 Media Center Dr., Los Angeles, CA, 90065. Phone: (323) 276-4000. Fax: (323) 276-4500.

Jim English, pres; James Griffith, sr VP.

Los Angeles, CA 90065. Andrita Studios, 3030 Andrita St. Sol Weisel, VP production.

TV-CATV only.

TV program production, distribution & video.

Playhouse Pictures, Box 2089, Los Angeles, CA, 90078-2089. Phone: (323) 851-2112. Fax: (323) 851-2117. E-mail: playpix@aol.com

Ted Woolery, producer; Gerry Woolery, dir; Todd Shalter, dir.

TV-CATV only.

Animated TV coml production & short films.

Point 360, 1133 North Hollywood Way, Burbank, CA, 91505. Phone: (818) 556-5700. Fax: (818) 556-5753. Web Site: www.point360.com.

Dennis Imbler, gen mgr; Paul Ponzio, VP sls/producer.

Hollywood, CA 90038, 1220 N. Highland Ave. Phone: (323) 957-5500. Rich Appel, gen mgr; Brian Grant, VP business dev.

Hollywood, CA 90038, 712 N. Seward St. Phone: (323) 462-5330. Yvonne Parker, gen mgr; Brian Grant, VP business dev.

Hollywood, CA 90038, 1025 N. McCadden Pl. Phone: (323) 461-8383. Fabian Sanchez, gen mgr; Brian Grant, VP business dev.

Los Angeles, CA 90064, 12421 W. Olympic Blvd. Phone: (310) 207-7079. Carl Segal, gen mgr; Brian Grant, VP business dev.

Postproduction & duplication, film-to-tape transfers, audio svcs. Digital editing, distribution & syndication.

PorchLight Entertainment Inc., 11777 Mississippi Ave., Los Angeles, CA, 90025. Phone: (310) 477-8400. Fax: (310) 477-5555. Web Site: www.porchlight.com.

Bruce D. Johnson, pres/CEO; William T. Baumann, exec VP & CFO.

TV-CATV only.

Produces & distributes family entertainment progmg including animation, TV movies & interactive multimedia progmg, licensing & merchandising.

Ports of Paradise, Box 33648, San Diego, CA, 92163. Phone: (619) 275-7357. Phone: (800) 223-2564. Fax: (619) 296-5909. E-mail: aloharn@portparadise.com Web Site: www.portparadise.com.

J. Hal Hodgson, exec producer.

Radio Only.

Hour-long radio program with Hawaiian music & info. Available on a barter basis.

PostWorks, New York, 100 Ave. of the Americas, New York, NY, 10013. Phone: (212) 557-4949. Fax: (212) 983-4083. Web Site: www.pwny.com.

TV-CATV only.

Film-to-tape or data in standard or Hi-Definition; editing: digital, Hi-Definition, non-linear; duplication, conversion, satellite & fibre transmissions.

Potomac TV/Communications, 15110 H St. N.W., Suite 202, Washington, DC, 20005. Phone: (202) 783-8000. Fax: (202) 783-1132. E-mail: www.sgreenaway@potomactv.com Web Site: www.potomactv.com.

Nick Chiaia, pres.

TV-CATV only.

New content & video svcs production facility. C- & Ku-band satellite transmitters/receive svcs.

Power Play Music Video L.L.C., 223-225 Washington St., Newark, NJ, 07102. Phone: (973) 642-5132. Fax: (973) 642-5747. Web Site: www.powerplay-mvtv.com. E-mail: powerplaytv@cs.com

Greg Ferguson, pres.

TV-CATV only.

TV & radio program; coml production & distribution; production svcs.

Powerline, 6350 W. Fwy., Fort Worth, TX, 76116. Phone: (817) 570-1491. Phone: (800) 292-2287. Fax: (817) 737-9436. E-mail: cries@familynetradio.com Web Site: www.powerlineradio.com.

Chuck Ries, producer; Kirk Teegarden, producer.

Radio Only.

Adult contemp music blended with brief commentaries about life by host Brother Jon Rivers, 30 min wkly on CD or download.

Powersports/Millenium International, 14242 Ventura Blvd., Ste 300, Sherman Oaks, CA, 91423-2757. Phone: (818) 708-9995. Fax: (818) 708-0598. E-mail: intl@ps-mill.com Web Site: www.ps-mill.com.

William McAbian, pres; Tal Dean McAbian, exec VP.

TV-CATV only.

Production & distribution of special interest & documentaries for TV, cable & home video.

Prairie Dog Entertainment, Box 28700, San Diego, CA, 92198. Phone: (800) 448-7664. Web Site: www.buckhowdy.com.

Steve Vaus, CEO.

Radio Only.

Exclusively offering Buck Howdy's Cow Pie Radio, the fastest growing wkly kids radio program.

Praxis Media Inc., 9 Twilight Pl., South Norwalk, CT, 06854. Phone: (203) 866-6666. Fax: (203) 853-8299. E-mail: praxiscc@aol.com

Christopher Campbell, pres & dir; Deborah Weingrad, VP & dir.

TV-CATV only.

TV program, coml, promotional film production; production svcs.

Premiere Radio Networks Inc., 15260 Ventura Blvd., 5th Fl., Sherman Oaks, CA, 91403-5339. Phone: (818) 377-5300. Fax: (818) 377-5333. E-mail: webmaster@premrad.com Web Site: www.premrad.com.

Kraig Kitchin, pres; Dan Yukelson, VP finance; Eileen Thorgusen, VP; Rich Meyer, pres; Nancy Deitemeyer, VP opns.

Radio Only.

Line-up includes Bcst Results Group (BRG) features, Olympia, Premiere Comedy Networks Formats, Long-form features, Mediabase rsch formats, music svcs, online, plain-wrap formats, prep & short-form. All features & svcs offered on a barter basis are available via satellite, disc, tape, script or phone, depending on the program.

Presbyterian Church (U.S.A.), 100 Witherspoon St., Louisville, KY, 40202-1396. Phone: (502) 569-5211. Phone: (502) 569-5493. Fax: (502) 569-8845.

Jerry Van Marter, news dir.

TV-CATV only.

Radio & TV production, video, audio production, distribution, mktg & Internet svcs.

Presson Perspectives, 600 Druid Rd. E., Clearwater, FL, 33756. Phone: (727) 461-1885. Fax: (727) 443-1984. E-mail: gpresson@tampabay.rr.com Web Site: www.medforum.com.

Gina Presson, pres & exec producer.

TV-CATV only.

News, documentary & internet production ranging from turnkey pieces to any segment.

Prime Cut Productions Inc., 11909 E. Trail, San Fernando, CA, 91342. Phone: (818) 897-7321. Fax: (818) 834-9889. E-mail: janicekaplan@primecutproductions.com Web Site: www.primecutproductions.com.

Edward Flaherty, VP; Jan Kaplan, pres.

TV-CATV-Radio.

NTSC Betacam European producers for American TV; bilingual location professionals; distribution, co-production, stock footage, TV script consultant/writer & editing.

Primedia Workplace Learning, 4101 International Pkwy., Carrollton, TX, 75007. Phone: (800) 848-1717. Fax: (972) 309-5666. Web Site: www.pwpl.com.

Josh Karin, pres/CEO; Gina Valencia, dir mktg.

TV-CATV only.

Video-based, interactive tech training programs for industry, utilities, govt relating to maintenance, opns & safety.

Primo Newservice Inc., Box 116, 182 Sound Beach Ave., Old Greenwich, CT, 06870-0116. Phone: (203) 637-0044. Fax: (203) 698-0812. E-mail: primonews@aol.com Web Site: www.teenkidsnews.com.

Albert T. Primo, pres/CEO.

TV-CATV-Radio.

TV news consulting, strategic news positioning talent & mgmt, coaching. Cable news training; Internet broadband svc.

Pro Video, 2904-A Colorado Ave., Santa Monica, CA, 90404. Phone: (310) 828-2292. E-mail: provideo1@earthlink.net

Joel Webb, pres.

TV-CATV only.

Commercials mastered to DVD & 3/4 inches.

Producers Group, Ltd., 713 S. Pacific Coast Hwy., Suite B, Redondo Beach, CA, 90277-4233. Phone: (310) 316-0481. Fax: (310) 316-1482. E-mail: lee.gluckman@producers-group.tv

Lee Gluckman Jr., pres.

TV-CATV only.

Dev & production of theatrical, TV films & series.

Production Garden Music Libraries, 510 E. Ramsey Rd., Suite 4, San Antonio, TX, 78216. Phone: (800) 247-5317. Fax: (210) 530-5230. E-mail: sales@productiongarden.com Web Site: www.productiongarden.com.

TV-CATV-Radio.

Ten distinct music libraries featuring production music, including production elements, sound effects; both lease & buy-out options available.

Productions La Fete, 387 St. Paul W., Montreal, PQ, H2Y 2A7. Canada. Phone: (514) 848-0417. Fax: (514) 848-0064. E-mail: info@lafete.com

Rock Demers, pres; Xiao Juan Zhou, VP distribution; Daniel Proulx, VP finance.

TV-CATV only.

Production of children/family feature films, drama series, documentaries, multimedia, etc.

The Program Exchange, 375 Hudson St., New York, NY, 10014. Phone: (212) 463-3500. Fax: (212) 463-2662. E-mail: info@programexchange.com Web Site: www.programexchange.com.

Allen Banks, pres; Chris Hallowell, sr VP/mgmg dir.

TV-CATV only.

TV program distribution.

Promark Television, 500 S. Palm Canyon Dr., Suite 220, Palm Springs, CA, 92264. Phone: (760) 322-7776. Fax: (760) 322-5149. E-mail: hdlevine@promarktv.com Web Site: www.promarktv.com.

David Levine, pres/CEO.

TV-CATV only.

TV program production & distribution.

Promusic, 941-A Clint Moore Rd., Boca Raton, FL, 33487. Phone: (561) 995-0331. Phone: (800) 322-7879. Fax: (561) 995-8434. E-mail: mail@promusiclibrary.com Web Site: www.promusiclibrary.com.

Alain Leroux, pres; Mike Spitz, sls dir.

TV-CATV only.

Production music for film, TV & more. Vast CD catalog to choose from with extensive classical & opera.

Protestant Hour Inc., 644 W. Peachtree St., Suite 300, Atlanta, GA, 30309-1925. Phone: (404) 815-9110. Fax: (404) 815-0258. E-mail: info@day1.net Web Site: www.day1.net.

Peter Wallace, pres & exec producer.

Radio Only.

Radio program production & distribution; relg ecumenical media.

Public Media Incorporated, 1560 Sherman Ave. Ste 440, Evanston, IL, 60201-4803. Phone: (773) 878-2600. Fax: (773) 878-8406. E-mail: webmasters@homevision.com Web Site: www.homevision.net.

Adrianne Furniss, pres/CEO; Carole Little, VP.

TV-CATV only.

Sls to all TV outlets in North America, foreign & classic cinema distribution.

Q

Quality Film & Video, 232 Cockeysville Rd., Hunt Valley, MD, 21030. Phone: (410) 785-1920. E-mail: qfv@qualityfilmvideo.com Web Site: www.qualityfilmvideo.com.

Peter A. Garey, pres; Guy G. Garey, VP.

TV-CATV only.

Video production, postproduction svcs, videotape, CD-ROM & DVD duplication.

Questar, 680 N. Lake Shore Dr., Suite 900, Chicago, IL, 60611. Phone: (312) 266-9400. Fax: (312) 266-9523. Web Site: www.questar1.com.

Jason Nader, pres.

TV-CATV only.

Producer, distributor of travel documentaries, children's cultural historical & natural history.

R

RAD Marketing & Cabletowns, 167 Crary-on-the-Park, Mount Vernon, NY, 10550. Phone: (914) 668-3563. Fax: (914) 668-4247. E-mail: cabletown@verizon.net

Bob Dadarria, pres; Adele Dadarria, account exec.

TV-CATV-Radio.

Consumer list, individuals & household lifestyles. Svcs include: cable TV subs with addressable box direct TV buyers with Tel # scrubbed mailing lists, rgnl area mkts, natl area mkts, sweepstakes entrants list, children lists, donors, travel.

RAI Corp., (Italian Radio TV System). 32 Avenue of the Americas Bldg 1, New York, NY, 10013-2473. Phone: (212) 468-2500. Fax: (212) 765-1956. Web Site: www.raicorp.net.

Mario Bona, CEO; Guido Corso, pres.

TV-CATV only.

Italian natl radio & TV.

RBC Ministries, Box 2222, Grand Rapids, MI, 49501-2222. Phone: (616) 942-6770. Fax: (616) 957-5741. E-mail: rbc@rbc.org Web Site: www.rbc.net.

Martin De Haan II, pres.

TV-CATV only.

TV & radio production & distribution. Programs: TV Day of Discovery, Radio-Discover the Word, Words To Live By, Our Daily Bread, Sports Spectrum, My Utmost for His Highest & Walk In The Wood.

RBC Ministries/Midwest Media Managers, Box 2606, Grand Rapids, MI, 49501-2606. Phone: (877) 245-0550. Phone: (616) 942-6360. Fax: (616) 957-5741. E-mail: mmm@rbc.net Web Site: www.rbc.net/mmm.

Martin De Haan II, pres; John Nasby, dir; Rod McNany, dir.

TV-CATV only.

In-house agency for RBC Ministries; providing TV/Radio placement & promotional support of RBC resources including the devotional, Our Daily Bread.

RDF Media, 48-49 Princes Pl., London, W11 4QA. United Kingdom. Phone: 44 (0)20 7908-1200. Fax: 44 (0)20 7908-1234. E-mail: sales@rdfmedia.com Web Site: www.rdfmedia.com.

David Frank, chief exec; Joely Fether, production dir; Stephen Lambert, progmg dir.

TV-CATV only.

Production & distribution of TV & radio programs.

RMD & Assoc. Inc., 534 Rosemary Cir., Media, PA, 19063. Phone: (610) 566-3799. Fax: (610) 566-3799. E-mail: rmdassociates@yahoo.com

Dick D'Anjolell, pres & exec producer; Hank Shaw, tech support svcs; Celeste Walsh, production mgr.

TV-CATV only.

Program creative svcs, production & distribution, specializing in promotion & business info.

RPM Media Enterprises-The Relic Rack Review, 108 Holmes Oval, New Providence, NJ, 07974-1425. Phone: (908) 464-2222. E-mail: richardjlorenzo@relicrack.com Web Site: www.relic-rack.com.

Richard J. Lorenzo, pres/CEO; Margaret P. Lorenzo, CFO; Jack Kratoville, VP.

Radio Only.

Four-hour wkly rock & roll oldies syndicated entertainment program called the Relic Rack Review, that includes music, news & nostalgia entertainment. Prep svcs including Relic Rack Fast Facts. Multi format radio program consulting svcs.

RPM-Radio Programming & Management Inc., 1133 W. Long Lake Rd., Ste 200, Bloomfield Hills, MI, 48302. Phone: (248) 647-1068. Phone: (800) 521-2537. Fax: (888) 776-0006. E-mail: rpmorlk@aol.com Web Site: www.tophitsusa.com.

Thomas M. Kirkorian, pres.

Radio Only.

Wkly CD svc top hits U.S. & CD music libraries. Full CD format svcs & progmg consultation.

The Radio Almanac, 107 Jensen Cir., West Springfield, MA, 01089-4451. Phone: (413) 737-7600. Fax: (413) 737-7600. E-mail: cspencer@mail.map.com

Charles G. Spencer, publisher.

TV-CATV-Radio.

Lifestyle info; special rsch; writing, voice & production projects. Monthly filler material publication, featuring events, sports, biographical sketches & other info.

Radio America, 1030 15th St. N.W., Suite 1040, Washington, DC, 20005. Phone: (202) 408-0944. Fax: (202) 408-1087. E-mail: radioa@radioamerica.org Web Site: www.radioamerica.org.

James C. Roberts, pres; Mike Paradiso, COO; Rich McFadden, producer; Greg Corombos, producer.

Radio Only.

News & feature svc providing daily, wkly & special programs (90 seconds to one hr) & multi-part documentaries.

Radio & TV Roundup Productions, 653 Sunhaven Dr., Clayton, NJ, 08312-1955. Phone: (856) 881-2570. Fax: (856) 307-9506. E-mail: delfon@att.net Web Site: www.nanes.com.

Bill Bertenshaw, CEO; Bobbi Cherrelle, exec producer; B.C. Slachofsky, mgr; Richard Nanes, dir; Kathy Hanley, production coord.

Cape May, NJ 08204-0108, Box 108. Bobbi Cherrelle, exec producer.

TV-CATV-Radio.

Production, placement of TV & radio progmg including comls & PSAs. Distribute free classical CDs & TV progmg.

Radio Canada International/Canadian Broadcasting Corp., Box 6000, Montreal, PQ, H3C 3A8. Canada. Phone: (514) 597-7656. Fax: (514) 597-6607. E-mail: rci@montreal.src.ca Web Site: www.rcinet.ca.

Jean Larin, exec dir.

Radio Only.

Daily shortwave & internet, seven languages, 24-hour eutelsat F6 Europe, intelsat 707 Africa, asiasat 2. Recorded & live program placement on foreign stns.

Radio City Entertainement, 2 Penn Plaza, New York, NY, 10021. Phone: (212) 465-6000. Phone: (212) 485-7000. Web Site: www.thegarden.com.

Katie Schroeder, dir pub affrs.

Los Angeles, CA 90067, 2049 Century Park E, Suite 1200. Phone: (310) 551-2721.

TV-CATV only.

TV program producers; production svcs.

Radio Express Inc., 1415 W. Magnolia Blvd., Burbank, CA, 91506. Phone: (818) 295-5800. Fax: (818) 295-5801. E-mail: radioinfo@radioexpress.com Web Site: www.radioexpress.com.

Tom Rounds, CEO; Jessica D'Agostin, VP sls; Anita Antonio, gen mgr; Christopher DiMatteo, VP mktg; Christian Jones, VP production.

Radio Only.

Radio Express exports Radioplay, the best music svc for radio, production music & progmg to enhance any music format, worldwide.

Radio Production Services Inc., 201 Lena Dr., Easley, SC, 29640-9647. Phone: (864) 855-7191. Fax: (864) 855-7191, EXT 2. E-mail: kenroy2@aol.com

R. Kenneth Rogers, pres; W.L. Ames, CFO; E.C. Rogers, VP; Ron Rackley, engrg dir; Nan Cohen, mgr; Jason Gold, sls.

Salem, MA 01970. Radio Production Services Inc., 27 Congress Street. Phone: (978)-740-0990. Fax: (978)-740-0660.

Cold Spring Harbor, NY 11724. Radio Production Services Inc., Box 26. Phone: (516) 246-9182. Alex Silvers, A&R dir.

Nashville, TN 37201. Radio Production Services Inc., 206 Union St, Ext. Phone: (615) 227-1947. Fred James, VP.

TV-CATV-Radio.

Radio program production specializing in early rhythm and blues & oldies formats. Radio, TV coml production & distribution. Stn ID packages, jingles, consultation svcs in sls & progmg.

Radio Sound Network, 770 Twin Rivers Dr., Columbus, OH, 43215. Phone: (614) 621-5600. Fax: (614) 621-5620. E-mail: steve.clawson@radiohio.com Web Site: www.radiohio.com.

Steve Clawson, engrg dir; Tony Miller, dir.

Columbus, OH 43215, 605 S. Front St. Phone: (614) 460-3850.

Radio Only.

Full-svc digital satellite audio & data distribution, including affil rel & net bldg. for new & existing sports, news/talk, music & specialty networks.

Radio Spirits, 2 Ridgedale Ave., Cedar Knolls, NJ, 07927. Phone: (800) 359-0570, ext.236. Phone: (973) 539-7557. Fax: (973) 539-1273. Web Site: www.radiospirits.com. E-mail: wholesale@radiospirits.com

Hakan Lindskog, pres; David Carroll, VP sls.

Radio Only.

Syndicated radio production specializing in "Golden Age of Radio." Production of *When Radio Was* with Stan Freberg, bartered to 300 affls.

Radio Television Espanola (RTVE), Edificio Prado Del Rey/ Desp. 3/023, Prado Del Rey, Madrid, 28223. Phone: (34 91) 581-54 91. Fax: (34 91) 581-77 41. E-mail: contratos_canales_inter.ep@rtve.es Web Site: www.rtve.es.

TV-CATV-Radio.

Production & distribution of its own productions as well as some 250 feature films in co-production with independent Sp & Latin American film producers.

Radioguide People Inc., (Vuolo Video). Box 880, Novi, MI, 48376. Phone: (248) 926-1234. E-mail: artvuolo@aol.com Web Site: www.vuolovideo.com.

Arthur R. Vuolo Jr., pres.

Radio Only.

Publishers of radio stn guides for the gen public, co-sponsored by loc stns & natl advertisers. Produces videos of radio stns, radio events for educational & entertainment purposes.

The Radio-Studio Network, Box 683, Times Square Station, New York, NY, 10108-0683. Phone: (800) 827-1722. Fax: (212) 868-5663. E-mail: programming@radio-studio.net Web Site: www.radio-studio.net.

Steve Warren, pres.

Radio Only.

Multi-Format network progmg by MP3 download. Music, news, talk, Infomercials.

Rampion Visual Productions L.L.C., 125 Walnut St., Watertown, MA, 02472. Phone: (617) 972-1777. Fax: (617) 972-9157. E-mail: info@rampion.com Web Site: www.rampion.com.

Michael R. Garneau, ptnr; Steven Tringali, dir & mgng ptnr.

TV-CATV only.

Full Digital Component Editing (BetaSP & DigiBeta), Computer Graphics (2D & 3D), Green Screen Studio, Digital Compositing & DVD creation.

Ray Sports Network, 104 Radio Rd., Powell's Point, NC, 27966. Phone: (252) 491-2414. Fax: (252) 491-2939.

Bill Ray, pres; Jody O'Donnell, sports dir.

Radio Only.

TV & radio program & coml production, distribution. Sports syndication.

Raycom Sports, 2815 Coliseum Ctr. Dr., Suite 200, Charlotte, NC, 28217. Phone: (704) 378-4400. Fax: (704) 378-4461. E-mail: khaines@raycomsports.com Web Site: www.raycomsports.com.

Ken Haines, pres/CEO; Colin Smith, VP station relations; De Cordell, VP sls; Peter Rolfe, Production.

Mobile, AL 36602. Mobile, 200 Government St., Ste 302. Phone: (251) 432-9340. Steve Haraleson, Dir.

Sacramento, CA 95814. Sacramento, 1107 Second St., Ste 210. Phone: (916) 443-2503. Brian Flojola, VP.

TV-CATV only.

Grant mgmt, sls distribution, produce, market, distribute sports, entertainment progmg nationally & internationally.

The Real Estate Network, 19925 Stevens Creek Blvd., Cupertino, CA, 95014. Phone: (408) 725-7530.

TV-CATV only.

Multimedia computer, interactive TV networks for mdse, mktg real estate listings, related goods & svcs.

Reel Media International Inc., 4516 Lovers Ln., # 178, Dallas, TX, 75225-6925. Phone: (214) 521-3301. Fax: (214) 522-3448. E-mail: reelmedia@aol.com Web Site: www.reelmediaintl.com.

Tom T. Moore, pres.

TV-CATV only.

Worldwide distributor of motion pictures, Public Libray of 2,000 movies, series & documentaries. Servicing all rights.

Reid/Land Productions Inc., 425 E. 58th St., Suite 46H, New York, NY, 10022. Phone: (212) 754-3348. Fax: (212) 754-7034. E-mail: reidland@rcn.com

Allen Reid, pres; Mady Land, exec VP.

TV-CATV only.

Packaging, creation & production of TV programs— var, entertainment, music, games, sports how-to & cooking.

Russ Reid Company, 2 N. Lake Ave., Suite 600, Pasadena, CA, 91101. Phone: (626) 449-6100. Fax: (626) 449-4756. Web Site: www.russreid.com.

Carol Philips, VP.

TV-CATV only.

Specializes in fund-raising, direct response adv, direct mail, TV, radio program production, placement & PR for nonprofit organizations.

Reizner & Reizner Film & Video, 7179 Via Maria, San Jose, CA, 95139. Phone: (408) 226-6339. Fax: (408) 226-6403. E-mail: dick@reizner.com

Dick Reizner, owner.

TV-CATV only.

Bcst & industrial production in all formats. Certified Legal Video Specialist. Gyrozoom rental.

Reliance Audio Visual Corp., 575 Lexington Ave., New York, NY, 10022. Phone: (212) 586-5000. Fax: (914) 237-1004. E-mail: rav@aol.com

Gil M. Meyer, pres; Norma E. Matthews, exec VP.

TV-CATV only.

Audio & video permanent installations, design, video & teleconferencing, consultation, rentals, staging multimedia, video projection, dealerships, leasing & display design.

Response Reward Systems L.C., 1850 Bay Rd., 2-C, Vero Beach, FL, 32963. Phone: (772) 234-5449. Fax: (772) 234-5949. E-mail: marcyuk@mpinet.net

Henry Von Kohorn, MBA, Ph.D., CEO.

TV-CATV only.

Patented technology enablling TV viewers in the United States to legally bet, cost-free and risk-free, on the outcome of sports events, from their homes via the Internet.

Reuters Media, 3 Times Sq., 18th Fl., New York, NY, 10036. Phone: (646) 223-4300. Fax: (646) 223-4390. Fax: (646) 223-4370. Web Site: www.reuters.com.

Richard Sabreen, exec VP.

TV-CATV only.

International news for interactive multimedia news archive for CD-ROM & on-demand applications. Stock photos & film footage.

Reuters Television, 1333 H St. N.W., Washington, DC, 20005. Phone: (202) 898-0056. Fax: (202) 898-1236. Web Site: www.reuters.com.

John Clarke, editor.

TV-CATV only.

Reuters news & sports svcs include 128 Reuters bureaus, camera crews & a comprehensive satellite network. It serves more than 200 bctrs & their affils in 84 countries. Satellite & news production svcs feature satellite delivery networks. International bcst centers in Moscow, Washington, DC & London. Offers live positions, studio facilities & direct access to the satellite network. Library & program packages cover major news & sporting events.

Rex Post, 610 S.W. 17th Ave., Portland, OR, 97205. Phone: (503) 238-4525. Fax: (503) 236-8347. E-mail: info@rexpost.com Web Site: www.rexpost.com.

Russell Gorsline, gen mgr; Marc Kennedy, sls; Lee Rooklin, sls.

TV-CATV-Radio.

TV & radio production, audio recording, video production, CD-ROM & DVD, website. Audio/video production & post, ISDN digital patch, DVD authoring & web design.

Richter Productions Inc., 330 W. 42nd St., Suite 2410, New York, NY, 10036. Phone: (212) 947-1395. Fax: (212) 643-1208. E-mail: richter330@aol.com Web Site: www.richtervideos.com.

Robert Richter, pres; Amy Kessler, production mgr.

TV-CATV only.

TV program, promotion film production, distribution, film & video.

Riden International Inc., 6024 Paseo Palmilla, Goleta, CA, 93117. Phone: (805) 964-7041. Fax: (805) 964-1338. E-mail: rideninc@aol.com Web Site: www.rideninc.com.

Richard Dennison, pres.

TV-CATV only.

TV program production & distribution.

Rigel Entertainment, 4201 Wilshire Blvd., Suite 555, Los Angeles, CA, 90010. Phone: (323) 954-8555. Fax: (323) 954-8592. E-mail: info@rigel.tv Web Site: www.rigel.tv.

John Laing, pres/CEO; Kristie Smith, sls; Sherrie Guerrero, sls; Bryan Hambleton, sls.

TV-CATV only.

Offers international TV, video rights to TV series, MOWs, specials & feature films.

River City Video Productions, Box 310601, New Braunfels, TX, 78131-0601. Phone: (830) 625-3474. Fax: (830) 625-3710. Web Site: www.fishingandoutdoor.com.

Deborah J. Dougherty, pres.

TV-CATV only.

Mktg, instructional & promotional videos & TV commercials.

Roberts Communications Network Inc., 4175 Cameron St., Suite B-10, Las Vegas, NV, 89103. Phone: (702) 227-7500. Fax: (702) 227-7501.

Tommy Roberts, chmn; Todd Roberts, pres/CEO.

TV-CATV only.

C-Bond satellite transponder capacity, uplinking, encoding & decoding. 50 Ch "Direct To Home" Platform

Rockey Hill and Knowlton, (A division of the Rockey Co.). 221 Yale Ave. N., Ste 530, Seattle, WA, 98109-5490. Phone: (206) 728-1100. Fax: (206) 728-1106. Web Site: www.rockey-seattle.com.

Corporate & financial film production. Public relations.

Peter Rodgers Organization, 1800 N. Highland Ave., #412, Hollywood, CA, 90028. Phone: (323) 962-1778. Fax: (323) 962-7174. E-mail: profilms@ixpres.com Web Site: www.profilms.com.

Stephen Rodgers, CEO; Teresa Rouse, CFO; Roaul Peter Mongilardi, dir opns.

TV-CATV only.

Celebrating its 26th year representing over 30 productions, companies & independent producers. Consultants, distributors, reps. Over 2,000 hours of progmg.

Romano & Associates Inc., 5094 Dorsey Hall Dr., Suite 104, Ellicott City, MD, 21042. Phone: (410) 730-4133. Fax: (410) 730-2219. Web Site: www.racommunications.com.

Jim Carroll, VP; Neil Romano, dir & producer.

TV-CATV only.

Full-svc production company specializing in issue-oriented short-feature films & documentaries, PSAs; professional of children's educ videos & comls.

Rose Entertainment, 5529 McLennan Ave., Encino, CA, 91436. Phone: (818) 817-7554. Fax: (818) 817-7585. E-mail: rosenter@pacbell.net

Rosamaria Gonzalez, pres; Flory Quiroa, opns mgr.

TV-CATV only.

TV progmg distribution company for Latin America.

Rosler Creative, 88 Howard St., #2307, San Francisco, CA, 94105. Phone: (415) 896-1414. Fax: (415) 896-1616. E-mail: Peter@RoslerCreative.com Web Site: www.roslercreative.com.

Peter Rosler, owner.

Radio Only.

TV comls, creative dev & production.

TV-CATV only.

Rosnay International, 6 Rue Robert Estienne, Paris, 75008. France. Phone: 01.42.89.18.54. Fax: 01.42.25.34.39. E-mail: fronet3038@aol.com

TV-CATV only.

Distribution & production company.

Steve Rotfeld Productions Inc., 610 Old Lancaster Rd., Suite 210, Bryn Mawr, PA, 19010. Phone: (610) 520-0671. Fax: (610) 520-0681. Web Site: www.rotfeldproductions.com.

Steve Rotfeld, pres/exec producer.

TV-CATV only.

A TV production company that produces & syndicates TV shows.

Jack Rourke Productions, Box 1705, Burbank, CA, 91507. Phone: (818) 843-4839.

TV-CATV only.

TV & radio program production.

ROZON, 2101 Blvd St. Laurent, Montreal, H2X 2T5. Phone: (514) 845-3155. Fax: (514) 845-4140. Web Site: www.hahaha.com.

Nathalie Bourdon, dir tv sls; Bruce Hills, COO; Gilbert Rozon, pres; Isabelle Begin, dir international TV; Christos Sourligas, international TV sls, publicity, mktg & tv sls.

TV-CATV only.

Producer/distributor of comedy programs, standup comedy & nonverbal light entertainment.

S

SFP Productions, 2 Ave. de L' Europe, 94360 Bry-Sur-Marne Cedex France. Phone: 3316 9833 602. Fax: 0033 11498 33604. E-mail: distribution@sfr.fr Web Site: www.sfp.fr.

Roland Fiszel, CEO; Sophie Vtueite, sls dir.

Distribution worldwide rights (TV movies, series, mini-series documentaries).

SPI International, 55 White St., Suite 1A, New York, NY, 10013. Phone: (212) 673-5103. Fax: (212) 673-5183. Web Site: www.spiintl.com.

Loni Farhi, pres; Stacey Sobel, VP.

TV-CATV only.

SPI International is an independent distribution company, supplying a wide variety of quality progmg worldwide. SPI specializes in programs with broad global appeal: light entertainment, game shows & animation.

Sak Entertainment, 398 W. Amelia St., Orlando, FL, 32801. Phone: (407) 648-0001. Fax: (407) 648-1333. E-mail: info@sak.com Web Site: www.sak.com.

David Russell, mgng & artistic dir.

TV-CATV only.

Professional comedy actors, dirs, producers & writers. Entertainment consultants for WDW, Universal Studio, Harrahs Corp. & Busch Gardens.

Sanctuary Records Group Ltd., Sanctuary House, 45-53 Sinclair Rd., London, W14 0NS. United Kingdom. Phone: 44-020-7602-6351. Fax: 44-020-7603-5941. E-mail: info@sanctuarygroup.com Web Site: www.sanctuarygroup.com.

TV-CATV only.

Program production & sls.

Edward Sarson Productions, 30 Duke St., Suite 511, Kitchener, ON, N2H 3W5. Canada. Phone: (519) 576-1824. Fax: (519) 740-6766.

George Sarson, pres.

Producers for the *Toad Patrol* TV series.

The Saturday Evening Post Television Department, (Benjamin Franklin Literary & Medical Society). 1100 Waterway Blvd., Indianapolis, IN, 46202. Phone: (317) 634-1100. Fax: (317) 637-0126. E-mail: q.lee@satevepost.com Web Site: www.satevepost.org.

Quinton Lee, dir TV.

TV-CATV only.

Producers & distributors of health shows for radio & TV; coml production.

SB Management, 890 Monterey, Box 12837, San Luis Obispo, CA, 93406. Phone: (805) 543-9214. Fax: (805) 543-9243. E-mail: michael@mikehesser.com Web Site: www.mikehesser.com.

Mike Hesser, pres.

Radio Only.

Consulting & coaching for sls & mgmt.

SCOLA, 21557 270th St, McClelland, IA, 51548. Phone: (712) 566-2202. Fax: (712) 566-2502. E-mail: scola@scola.org Web Site: www.scola.org.

Francis Lajba, pres; John Millar, VP.

TV-CATV-Radio.

SCOLA progmg from more than 90 countries in more than 80 languages. These programs are available via internet, satellite, cable to learners of languages study, ethnic communities & anyone seeking a global perspective. Mission is to help the people of the world learn about one another.

Scottish Television Ltd., 200 Renfield St., Glasgow, G2 3PR. Phone: 011-44-141-300-3000. Fax: 011-44-141-300-3030. Web Site: www.scottishtv.co.uk.

TV-CATV only.

Bcstr & production company.

Seattle Video Crew, (Formerly The Seattle Video Bureau). Box 99218, Seattle, WA, 98199. Phone: (206) 448-2500. Fax: (206) 378-1700. E-mail: crew@seattlevideo.com Web Site: www.seattlevideo.com.

David Oglevie, pres.

TV-CATV only.

Location video production for bcst news, corporate & industrial, mktg, & medical. BETACAM SP, & DVCAM, NTSC or PAL formats.

Semaphore Entertainment Group, 32 E. 57th St., New York, NY, 10022. Phone: (212) 371-8650. Fax: (212) 888-8650. Web Site: www.seg.com.

TV-CATV-Radio.

TV program production, radio program production & distribution.

SeniorVision Productions, Inc., 418 North Central St., East Bridgewater, MA, 02333. Phone: (508) 350-9700. Web Site: www.seniorvision.com.

Steve Brown, co-owner; Noah Brookoff, co-owner.

TV-CATV only.

Full-svc video production company offering all aspects of program dev, creation & implementation bcst, S.I.V., corporate.

Seraphim Communications Inc., 1568 Eustis St., St. Paul, MN, 55108. Phone: (651) 645-9173. Fax: (651) 645-3515. E-mail: info@seracomm.com Web Site: www.seracomm.com.

Hal Dragseth, pres; Kristin Wiersma, VP.

TV-CATV only.

Full-svc video & AV capabilities. Emphasis on video production from concept through final product. In-house grahics, interactive Web, CD-ROM/DVD.

SESAC Inc., 55 Music Sq. E., Nashville, TN, 37203. Phone: (615) 320-0055. Fax: (615) 329-9627. Web Site: www.sesac.com.

Pat Colliuns, COO.

London W1H 3FF, 67 Upper Berkeley St. Phone: 020 7616 9284. Fax: 020 7563 7029. Dr. Wayne Bickerton.

Santa Monica, CA 90401, 501 Santa Monica Blvd, Suite 450. Phone: (310) 393-9671. Fax: (310) 393-6497. Pat Rogers, Sr VP - Writer /Publisher Relations.

New York, NY 10019, 152 W. 57th St., 5th Fl. Phone: (212) 484-0600. Fax: (212) 489-5699. Stephen Swid, Freddie Gershon, Ira Smith, Co-Chmn.

Music rights organization representing the performance rights of affil composers, authors & publishers.

Sesame Workshop, One Lincoln Plaza, New York, NY, 10023. Phone: (212) 595-3456. Fax: (212) 875-6111. Web Site: www.sesameworkshop.org.

Gary Knell, pres.

TV-CATV only.

TV program production.

Seven Network Australia Inc., 10100 Santa Monica Blvd., Suite 1750, Los Angeles, CA, 90067. Phone: (310)-553-3345. Fax: (310) 553-4812. Web Site: www.seven.com.au.

Zane Bair, gen mgr; Mike Amor, bureau chief.

TV-CATV only.

U.S. office & news bureau of ch 7, Australia—a major coml TV network of Australia.

1776 Productions, 5 Sparrow Dr., Livingston, NJ, 07039. Phone: (973) 533-0762. Fax: (973) 992-1010. E-mail: nce@rcn.com

Ralph Weisinger, pres.

Promotional film production.

Seville Pictures, (division of Behaviour Entertainment Inc.). 150 Eglington Ave. E., Suite 804, Toronto, ON, M4P 1E8. Canada. Phone: (416) 480-0453. Fax: (416) 480-0501. E-mail: 1ufor@seville.com Web Site: www.seville.com.

Pierre Brousseau, pres; Andrew Austin, sr VP; David Reckziegel, pres.

Seville pictures is involved in production and distribution of multimedia content for film, TV, video & on-line communication marketplace.

Sam Shad Productions, Box 10853, Reno, NV, 89510. Phone: (775) 857-2244. Fax: (775) 857-2272. E-mail: sam@shad.reno.nv.us Web Site: www.bestofreno.tv.

Sam Shad, pres; Bonnie McCorkle, program dev.

TV-CATV only.

Radio progmg & coml production, TV progmg & coml production, TV & radio progmg concepts dev from start to finish, internet design & adv, public relations.

Harvey Sheldon Productions, 7855 E. Horizon View Dr., Anaheim Hills, CA, 92808. Phone: (714) 281-5929. Fax: (714) 281-5929.

Harvey Sheldon, pres.

TV-CATV only.

Daily or wkly classic rock/swing video format for TV & cable stns. On Century Cable available for syndication serving 150,000 TV/cable households in Los Angeles/Orange county.

Shield Productions Inc., 11964 N Lake Dr., Boynton Beach, FL, 33436. Phone: (561) 734-5599. Fax: (561) 734-8176.

James C. Dolan, pres.

TV-CATV only.

Creation & production of radio & TV comls.

Shukovsky English Entertainment, 4605 Lankershim Blvd., Suite 510, North Hollywood, CA, 91602. Phone: (818) 763-9191. Fax: (818) 763-9878.

Joel Shukovsky, pres; Diane English, producer.

TV-CATV only.

Producer & distributor of TV progmg, especially half-hour comedy.

Sidewater Enterprises Inc., 2647 Laurel Pass, Los Angeles, CA, 90046. Phone: (310) 358-4960. Fax: (323) 656-7853. E-mail: fredericsid@aol.com

Frederic M. Sidewater, pres; Rosalyn G. Sidewater, exec VP.

TV-CATV only.

Consulting regarding financing & distribution for independent productions.

Silverline Pictures, 22837 Ventura Blvd., Ste. 205, Woodland Hills, CA, 91364. Phone: (818) 225-9032. Fax: (818) 225-9053. E-mail: silverline@earthlink.net Web Site: www.silverlinepictures.com.

Leman Cetiner, CEO; Axel Munch, pres.

TV-CATV only.

Full-svc production & distribution company producing theatrical, TV & kids series.

Silverman Productions Inc., 106 E. Cary St., Richmond, VA, 23219. Phone: (804) 343-1934. Fax: (804) 343-1938. E-mail: donald@silvermstockfootage.com Web Site: www.silvermanstockfootage.com.

Donald Silverman, pres.

TV-CATV-Radio.

Cityscase, arch val, nature & stock footage.

Sing For Joy, Saint Olaf College, 1520 Saint Olaf Ave, Northfield, MN, 55057. Phone: (507) 786-8596. Fax: (507) 786-8608. E-mail: singforjoy@stolaf.edu Web Site: www.singforjoy.org.

Jeff O'Donnell, producer & music dir; Jennifer Rowe, communication coord.

Radio Only.

Sacred choral works with host comments relating the texts to the current scriptural lessons of the ecumenical church year.

Skywatch Weather Center, 347 Prestley Rd., Bridgeville, PA, 15017. Phone: (800)-SKYWATCH. Phone: (412) 221-6000. Fax: (412) 221-3160. E-mail: airsci@skyweather.com Web Site: www.skywatchweather.com.

Stanley J. Penkala, pres; Daniel Krzywiecki, VP.

TV-CATV-Radio.

Specially formatted weathercasts produced in the Skywatch Weather Center®.

Smith/Lee Productions, Inc., 7420 Manchester Rd., St. Louis, MO, 63143. Phone: (314) 647-3900. Fax: (314) 647-3959. E-mail: global@smithlee.com Web Site: www.smithlee.com.

David Smith, pres; Barry Lee, VP.

TV-CATV only.

TV & radio coml promotional film production; production svcs; studio specializing in Audio Post; music production, voice recording & multimedia.

P. Allen Smith Gardens, Box 7347, Little Rock, AR, 72217. Phone: (501) 376-1894. Fax: (501) 376-1896. Web Site: www.pallensmith.com.

P. Allen Smith, pres.

TV-CATV only.

Nationally syndicated gardening & lifestyle news inserts reported by professional garden designer Allen Smith.

Soldiers Radio & Television, U.S. Army Public Affairs, Box 31, 2511 Jefferson Davis Hwy., Arlington, VA, 22202. Phone: (703) 602-4675. Fax: (703) 602-5220. E-mail: armynewswatch@smc.army.mil Web Site: www.army.mil/srtv.

George McNamara, dir; Paul Schultz, opns mgr; Gene Gunderson, chief engr; Jim Ryan, opns mgr; Melody Day, mktg.

TV-CATV only.

Radio, TV news bureau, soldiers radio network & Army Newswatch a biweekly TV newscast.

Solid Gospel Network (Reach Satellite Network, Inc.), 402 BNA Drive, Suite 400, Nashville, TN, 37217. Phone: (615) 367-2210. E-mail: info@SalemMusicNetwork.com Web Site: www.solidgospel.com.

Michael S. Miller, gen mgr; Don Burns, progmg dir; Wade Schoenemann, opns mgr; Ed Evensen, Local Traffic Manager; Jim Black, director of Affiliate Relations; Rick Shelton, mgr.

Radio Only.

24-hour, satellite delivered, Christian country & southern gospel network, featuring artists like Bill Gaither, The Isaacs, Gold City, Jeff & Sheri Easter & The Martins, live from Nashville, the Christian music capital of the world.

Sony Pictures, 10202 West Washington Blvd., Culver City, CA, 90232. Phone: (310) 244-4000. Fax: (310) 244-2626. Web Site: www.spe.sony.com.

Michael Lynton, CEO.

Atlanta, GA 30339, 2859 Paces Ferry Rd, Suite 1130. Phone: (770) 434-5400.

Chicago, IL 60611, 455 N. Cityfront Plaza Dr, Suite 2520. Phone: (312) 644-0770.

New York, NY 10022, 550 Madison Ave. Phone: (212) 883-8500.

Dallas, TX 75219, 3500 Maple Ave, Suite 205. Phone: (214) 520-7070.

TV-CATV only.

SoperSound Music Library, Box 869, Ashland, OR, 97520. Phone: (800) 227-9980. Fax: (541) 552-0832. E-mail: info@sopersound.com Web Site: www.sopersound.com.

Dennis Reed, pres.

TV-CATV only.

Contemp music library for all production needs. Available on CDs & direct digital down load online.

Sound*Bytes, 1425 Hopkins St. N.W., Suite 401, Washington, DC, 20036. Phone: (202) 296-2022. E-mail: press@soundbytesradio.com Web Site: www.soundbytesradio.com.

Jan Ziff, pres.

Radio Only.

Computer audio show production & syndication.

Sound Idea Productions, 417 Nursery St., Nevada City, CA, 95959. Phone: (510) 832-5178. Fax: (510) 832-4829. E-mail: soundidea@oro.net

Glenn Davidson, pres.

Oakland, CA 94612, 405 14th St, Suite 612. Phone: (530) 478-9770.

Radio Only.

Archival Service. Radio production & syndication.

Sound Ideas, 105 W. Beaver Creek Rd., Suite 4, Richmond Hill, ON, L4B 1C6. Canada. Phone: (905) 886-5000. Phone: (800) 387-3030. Fax: (905) 886-6800. E-mail: info@sound-ideas.com Web Site: www.sound-ideas.com.

Brian Nimens, pres/CEO; Mike Bell, VP.

TV-CATV-Radio.

Royalty free sound effects, imaging elements, music for the professional audio industry; including bcst, cable, film, multimedia & internet applications.

Sound of Birmingham Productions, 3625 5th Ave. S., Birmingham, AL, 35222. Phone: (205) 595-8497. E-mail: don@soundofbirmingham.com Web Site: soundofbirmingham.com.

Don Mosley, pres; Betty Mosley, VP/off mgr.

TV-CATV only.

Radio & TV voice-overs, jingles, video sweetening, custom music, recording studios, full-svc studios, ISDN.

Sound Source Networks, 2 St. Clair Ave. W., 11th Fl., Toronto, ON, M4V 1L6. Canada. Phone: (416) 922-1290. Fax: (416) 323-6819. E-mail: info@soundsource.ca Web Site: www.soundsource.ca.

Jean Marie Heimrath, pres & gen mgr; Lesley Soldat, VP opns.

Radio Only.

Radio net that produces, markets & distributes radio programs nationally.

Soundshop Recording Studio LLC, 1307 Division St., Nashville, TN, 37203. Phone: (615) 244-4149. Fax: (615) 242-8759. E-mail: soundshopstudio@aol.com

Mike Bradley, owner; Don Cook, owner.

TV-CATV only.

Music production, recording studios, 2-48 track digital or 24 track analog studios.

Soundtrack, 162 Columbus Ave., Boston, MA, 02116-5222. Phone: (617) 303-7500. Fax: (617) 303-7555. Web Site: www.soundtrackboston.com.

Jeanne Priest, mgr.

New York, NY 10010, 936 Broadway. Phone: (212) 420-6010. Chris Rich, opns mgr.

Production, postproduction & custom music of all kinds; specializing in sound designs.

Southcott Productions, Box 33185, Granada Hills, CA, 91394. Phone: (818) 368-4938. Fax: (818) 368-4938. E-mail: csouthcott@aol.com Web Site: www.musicofyourlife.com.

Chuck Southscott, owner.

Radio Only.

Radio program, coml production & distribution.

Southern STAR, Phone: 612 9519 2677. Fax: 612 9517 2530. E-mail: info@duplitek.com.au Web Site: www.southern-star.com.au.

Neil Balnaves; Errol Sullivan, chief exec entertainment; Robyn Watts, chief exec sls.

TV-CATV only.

Southern Star is an integrated film & TV production distribution & manufacturing group. Southern Star is a public listed company.

Spanish Broadcasting System, 26 W. 56th St., New York, NY, 10019. Phone: (212) 541-9200. Fax: (212) 541-6904. Web Site: www.lamusica.com.

Raul Alarcon, CEO; Carey Davis, VP/gen mgr; Joseph A. Garcia, CFO.

Radio Only.

Spanish progmg syndication.

Spelling Television Inc., 5700 Wilshire Blvd., Suite 575, Los Angeles, CA, 90036. Phone: (323) 965-5700. Fax: (323) 965-5895.

Aaron Spelling, chmn of bd; E. Duke Vincent, Vice-chmn; Jonathan Levin, pres.

TV-CATV only.

TV program production.

Charlie Spencer Productions, 107 Jensen Cir., West Springfield, MA, 01089-4451. Phone: (413) 737-7600. Fax: (413) 737-7600 (Voice First). E-mail: cspencer@mail.map.com

Charlie Spencer, producer & dir.

TV-CATV only.

Specializes in nature & environmental progmg, as well as outdoor recreation & nature travel. Svcs include consulting, producing, directing, rsch, writing & narration. Progms include *Gardening with Wildflowers, The Urban Canoeist* & *American Waste.*

Sport International Inc., Villa del Mar East, 14 J Ave. Isla Verde, Carolina, PR, 00979. Phone: (787) 268-8751. Fax: (787) 726-7683. E-mail: hector@hjfsport.com Web Site: www.hjfsport.com.

Hector Figueroa, chmn of bd; Juliet Giamartino, co-chmn; Jennifer Marin, VP event sls & production.

TV-CATV only.

Global mktg, distribution, bcstg, sports events, Pay-Per-View & special entertainment progmg. Current properties include: *This Day in Sports, Wide World of Bloopers*, live buying & buying documentaries.

Sports Byline U.S.A., 300 Broadway, Suite 8, San Francisco, CA, 94133. Phone: (415) 434-8300. Fax: (415) 391-2569. E-mail: webmaster@sportsbyline.com Web Site: www.sportsbyline.com.

Ron Barr, chmn; Darren Peck, pres; Doug McCormack, exec producer.

Radio Only.

Satellite-delivered nationwide radio sports talk net, listener 800 number, 7days (West Coast), barter.

StarDate/Universo, University of Texas at Austin, 2609 University Ave., Suite 3118, Austin, TX, 78712. Phone: (512) 471-5285. Fax: (512) 471-5060. E-mail: perez@stardate.org Web Site: www.stardate.org.

Sandra Preston, exec producer; Damond Benningfield, producer.

Radio Only.

Syndicated two-minute radio programs on stars & planets visible in the night sky. Each program is date-specific. English/Spanish.

Charles H Stern Agency Inc., 1999 Ave. of the Stars, Suite 1400, Los Angeles, CA, 90267. Phone: (310) 788-4570.

TV-CATV-Radio.

Radio coml production & distribution. Represents a variety of TV & radio personalities.

Steven B. Stevens, 7400 Sweetwater Branch, West Chester, OH, 45069. Phone: (513) 755-7300. Fax: (513) 755-7507. E-mail: sbstevens@aol.com

Steven B. Stevens, pres.

TV-CATV-Radio.

Accomplished narrator/voice talent with deep warm authoratative voice.

Marty Stouffer Productions Ltd., 44190 Hwy. 82 E., Aspen, CO, 81611. Phone: (970) 925-5536. Fax: (970) 920-3820. E-mail: mary@stoufferoffice.com

TV-CATV only.

TV program production.

Strand Media Group Inc., 12240 Venice Blvd., Suite 23, Los Angeles, CA, 90066. Phone: (310) 390-2248. Fax: (310) 390-2857. E-mail: strandmg@aol.com Web Site: www.somethingyoushouldknow.net.

Mike Carruthers, pres.

Radio Only.

Radio programs, coml production & distribution; media buying agent.

Strength for Living, 6350 W. Freeway, Fort Worth, TX, 76116. Phone: (817) 570-1400. Phone: (800) 266-1837. Fax: (817) 737-9436. E-mail: info@familynet.org Web Site: www.familynetradio.com.

Lisa Bratton, radio mkktg & distribution; Donna Senn, radio distribution; Chuck Ries, producer.

Radio Only.

Offers audiences relevant messages of hope & spiritual encouragement in their search to find strength for living. Host Bob Reccord.

M C Stuart & Associates Pty Ltd., 2/34 Power St., Balwyn Victoria, 03103. Phone: 61 3 9888 5835. Phone: 61 409 885 831 (mobile). Fax: 61 3 9888 5831.

Max Stuart, mgng dir & sec.

TV-CATV only.

Worldwide TV & feature film distributors.

Studio Babelsberg GmbH, August-Bebel-St. 26-53, 14482 Potsdam Germany. Phone: 49 (0) 331-72 -13151. Fax: 49 (0) 331-72-12525. E-mail: info@studiobabelsberg.com Web Site: www.studiobabelsberg.de.

Gerhard Bergfried, CEO.

TV-CATV only.

Studio area, studio technology, set design & construction, film laboratory, postproduction, dubbing theatres for films, TV & video.

Studio Center Corp., 200 W. 22nd St., Norfolk, VA, 23517. Phone: (757) 622-2111. Fax: (757) 623-5512. E-mail: info@studiocenter.com Web Site: www.studiocenter.com.

William Prettyman, CEO & exec producer.

Las Vegas, CA 89103, 3875 S. Jones Blvd. Phone: (702) 248-2777. Fax: (702) 248-5400. Curt Kroeger, gen mgr.

Memphis, TN 38112, 2693 Union Ave. Ext. Phone: (901)323-7060. Fax: (901) 320-0003. Sheldon Borgelt.

TV-CATV-Radio.

Radio comls, stn promotions, TV voice-over svcs, original music, on hold & wedsite audio.

Studio M Productions Unlimited, 4032 Wilshire Blvd., Suite 403, Los Angeles, CA, 90010. Phone: (213) 389-7372. Fax: (213) 389-3299. E-mail: mixer@sound4film-tv.com Web Site: www.mandy.com/stu001.html.

Mike Michaels, owner/sr engr/producer; Hugo Buehring, progmg mgr.

TV-CATV-Radio.

TV & radio program production, tape, film & live production svcs. Live remotes for TV, radio, news & sports. (888) 389-7372.

Suite Audio, 21 Stone Wall Ln., Clinton, CT, 06413. Phone: (860) 664-9499. E-mail: info@suiteaudio.com Web Site: www.suiteaudio.com.

Bob Nary, owner/pres.

Clinton, CT 06413. Clinton, CT, 21 Stonewall Ln. Bob Nary, owner.

TV-CATV only.

Facility: ProTools HDII/TDM, Waves restoration & Platinum, Sonic Solutions. Commercial - Program - surround production. CD mastering - authoring - duplication. Cassette duplication.

Sullivan Entertainment Inc./Sullivan Entertainment International, 110 Davenport Rd., Toronto, ON, M5R 3R3. Canada. Phone: (416) 921-7177. Fax: (416) 921-7538. Web Site: www.sullivan-ent.com.

London United Kingdom. Sullivan Entertainment Europe Ltd, Savant House, 63-65 Camden High St. Phone: (207) 383-5192. Muriel Thomas, sr VP intl sls.

TV-CATV only.

Production, distribution, home video, dev series & feature film.

Sullivan Video Services Inc., 7 Pickman Dr., Bedford, MA, 01730. Phone: (781) 271-1720. Fax: (781) 271-1740. E-mail: johnsul@sull Web Site: www.sullivanvideo.com.

John M. Sullivan, pres.

TV-CATV only.

Full-svc video production company. Provides ENG/EFP news crews & Ku-band satellite svcs, Fiber Optic Studio Tape feeding capabilities.

The Summit Media Group Inc. (Sub 4 Kids Entertainment), 1414 Ave. of the Americas, New York, NY, 10019. Phone: (212) 754-4900. Fax: (212) 754-5480. E-mail: shirsch@4kidsent.com

Alfred Kahn, chmn.

TV-CATV only.

Distribution of programs in the United States with emphasis on programs for children.

Sunbow Entertainment, 100 5th Ave., Fl. 3, New York, NY, 10011. Phone: (212) 893-1600. Fax: (212) 893-1630. Web Site: www.tvloonland.com.

Suzanne Berman, dir creative affrs; Rebecca Gallivan, opns dir.

TV-CATV only.

Developers, producers & distributors of quality children's & family programs, licensors & merchandise.

Sundial Productions, 275 Huyler St., S. Hackensack, NJ, 07606. Phone: (201) 525-5100. Fax: (201) 525-5111.

Jack Kreismer, Principal; David Lapidus, Principal.

Radio Only.

Voice production company; radio & telephone feature production.

SunGard Output Solutions, 350 Automation Way, Irondale, AL, 35210. Phone: (205) 307-6713. Fax: (205) 307-6813. E-mail: jamey.vella@sungard.com Web Site: www.sungard.com.

Joseph Harper, pres; Jamey Vella, natl sls exec.

TV-CATV-Radio.

SunGard's document distribution output channels include: paper, electronic (ebill/email), optical and/ or magnetic media.

Swell Pictures Inc., 455 N. City Front Plaza, 18th Fl., Chicago, IL, 60611. Phone: (312) 464-8000. Fax: (312) 464-8020. Web Site: www.swellinc.com.

Michael Topel, pres; Joe Flores, sr VP/chief engr; Brian Clark, exec VP; Dave Mueller, VP client svcs; Radi Akel, opns mgr.

TV-CATV only.

Film-to-tape transfer, videotape editing, digital sound editing & original music. Avid, 3-D graphics & compositing, audio & new media.

System TV, 45/47 rue Paul Bert, 92100 Boulogne, Paris France. Phone: (33)-1-55-38-2020. Fax: (33)-1-55-38-20-30. E-mail: daniel@systemtv.fr Web Site: www.systemtv.fr.

Daniel Renoug, pres.

TV-CATV only.

TV program production, press agency, produce weather & info svcs.

T

TM Century Inc., 2002 Academy Ln., Dallas, TX, 75234. Phone: (972) 406-6800. Fax: (972) 406-6890. E-mail: tmci@tmcentury.com Web Site: www.tmcentury.com.

David Graupner, pres/CEO.

TV-CATV only.

TMC creates, produces & distributes music-based products for bcst, including compilation libraries, production music, I.D. packages & coml jingles.

TRF Production Music Libraries, 747 Chestnut Ridge Rd., Chestnut Ridge, NY, 10977. Phone: (800) 899-MUSIC. Phone: (845) 356-0800. Fax: (845) 356-0895. E-mail: info@trfmusic.com Web Site: www.trfmusic.com.

Michael Nurko, pres/CEO.

TV-CATV-Radio.

More than 50,000 selections (4,000 discs) of both contemporary & traditional music. Includes Dennis Music, MP 2000, PAN, PowerSound, Pyramid, Supraphon, TRF Alpha, Cobra, Kool Kat, Spain is Music, Adrenalin, Bravo & Stock Music libraries. Also includes complete classical series, sound effects & authentic International Ethnic series.

TR Productions, 209 W. Central St., Suite 108, Natick, MA, 01760. Phone: (508) 650-3400. Fax: (508) 650-3455. Web Site: www.trprod.com.

Cary M. Benjamin, principal.

TV-CATV only.

Production svcs.

TS James & Associates, 83 Christopher St., New York, NY, 10014-4246. Phone: (212) 331-0186. Fax: (212) 505-0959. E-mail: tom@jandasound.com

Thomas S. James, pres.

TV-CATV only.

Original scoring for film & TV.

Talco Productions, 279 E. 44th St., New York, NY, 10017. Phone: (212) 697-4015. Fax: (212) 697-4827. E-mail: alaw1@mindspring.com

Alan Lawrence, pres; Marty Holberton, VP.

TV-CATV-Radio.

TV, radio progmg, documentaries, ind, educ production, PR consultation & production.

Talk America Radio Networks, 520 Broad St., Newark, NJ, 07102. Phone: (973) 438-3026. Fax: (973) 438-1637. Web Site: www.talkamerica.com.

Trang Nguyen, COO; Maurice Bortz, opns VP.

Radio Only.

24-hour talk seven days a week offering live call-in programs. All programs barter & every minute is covered.

Talk Radio Network, (a subsidiary of Premiere Radio Networks). Box 3755, Central Point, OR, 97502. Phone: (541) 664-8827. Fax: (541) 664-6250. Web Site: www.talkradionetwork.com.

Mark Masters, pres.

Radio Only.

Full-svc live talk radio format 24 hours daily. Serving more than 400 affils nationwide.

Talkline Communications Radio Network, Box 20108, Park West Station, New York, NY, 10025-1510. Phone: (212) 769-1925. Fax: (212) 799-4195. E-mail: tcntalk@aol.com Web Site: www.talklinecommunications.com.

Zev J. Brenner, exec producer.

Radio Only.

Natl Jewish radio net. Carried in over 2,500 markets. Contemp Jewish progmg; interview & call-in format with newsmaker guests from politics, entertainment & Israel. Live segments from Israel. Available on barter. Satellite delivered.

Tamouz Media, 37 W. 20th, Suite 1007, New York, NY, 10011. Phone: (212) 463-7437. Fax: (212) 463-7409. E-mail: tamouzmedia@aol.com Web Site: www.tamouz.com.

Ilan Ziv, producer.

TV-CATV only.

TV program & documentary production.

Tankersley Productions, Inc., St. Louis, 858 Hanley Industrial Ct., St. Louis, MO, 63144. Phone: (314) 725-0116. E-mail: randy@tankersleyproductions.com Web Site: www.tankersleyproductions.com.

Randy Tankersley, pres.

Full-bcst svcs. ENG/EFP Beta SP crews with/without producers, avid nonlinear editing.

Tapestry International, Ltd., 11 Hanover Sq., 14th Fl., New York, NY, 10005. Phone: (212) 505-2288. Fax: (212) 505-5059. E-mail: tunein@tapestry.tv

Nancy L. Walzog, pres; Karen Carlson, dir; Laura Ibanez, sls exec.

TV-CATV only.

Production & distribution company working in the international TV marketplace.

Technisonic Studios, 500 S. Ewing Ave., Suite G, St. Louis, MO, 63103. Phone: (314) 533-1777. Fax: (314) 533-6527. E-mail: mstroot@technisonic.com Web Site: www.technisonic.com.

Mike Stroot,.

TV-CATV only.

16 & 35 mm film production, full-sound studios, video production, film & video editing.

Tel-A-Cast Productions, 1016 Everee Inn Rd., Griffin, GA, 30224. Phone: (770) 233-4200. Fax: (770) 233-4247. E-mail: mrenew@osmose.com

Michael Renew, production mgr.

TV-CATV only.

From concept to execution, produces TV & radio comls, training videos, mktg videos & other projects quickly & efficiently.

Tel-Air Interests Inc., 2040 Sherman St., Hollywood, FL, 33020. Phone: (954) 924-4949. Fax: (954) 924-4980. E-mail: telair@aol.com Web Site: www.telairint.com.

Grant H. Gravitt Jr., pres; M. L. Gravitt, sec/treas.

TV-CATV only.

Syndicated TV specials (sports & music), contract production, theatrical short subjects & documentary TV & films, infomercials. Digital recording studio, digital & linear editing.

TeleCom Productions Inc., 5875 Peachtree Industrial Blvd., Suite 150, Norcross, GA, 30092. Phone: (770) 455-3569. Fax: (770) 455-3938. Web Site: www.tcpatlanta.com.

Budd O. Libby, pres; Roger B. Clark, co-chmn; Dan Bower, co-chmn.

TV-CATV-Radio.

Production & syndicator of *Let's Go to the Races, Free Cash Lotto, Daily Race Game* & *Post Time* retail prize promotions; random animated digital drawing systems (RADDS).

Telegenic Programs Inc., 161 Forest Hill Rd., Toronto, ON, M5P 2N3. Canada. Phone: (416) 484-8000. Fax: (416) 484-8001. E-mail: telegenic@aol.com

H. Lawrence Fein, chmn/CEO; Ronda Taylor, dir opns.

TV-CATV only.

One of Canada's leading TV distribution companies now in its 30th successful year.

Telenium Studios, 525 Mildred Ave., Primos, PA, 19018. Phone: (610) 626-6500. Fax: (610) 626-2638. Web Site: www.telenium.com.

Todd Strine, CEO; Peter Hayes, mgr; Cathie Hunt, mobile sls.

Philadelphia, PA 19123. Telenium Post, 520 N. Columbus Blvd, Suite 204. Phone: (215) 629-2000. Fax: (219) 625-8353. Mark Reidenaver, dir of post prod.

TV-CATV only.

Studio & location production & postproduction for networks, syndication & cable. Full facilities, production mgmt & creative svcs.

Telepros, Box 1116, Belmont, CA, 94002. Phone: (650) 345-0505. E-mail: telepros@comcast.net

Niels Melo, pres.

TV-CATV only.

Producers of live entertainment, sports, news, performing arts, corporate videos and webcasts.

Televents Ltd., 2450 Virginia Ave. N.W., Washington, DC, 20037. Phone: (202) 296-0541. Fax: (202) 296-0541.

James Avis, mgr.

TV-CATV only.

TV program production & distribution.

Television & Radio Features Inc., Box 237, Lincolnshire, IL, 60069-0327. Phone: (847) 541-7600. Fax: (847) 541-2600.

Morton A. Small, pres.

TV-CATV only.

Promotions, prize svcs; mktg.

Television Representatives Inc., 9720 Wilshire Blvd., Suite 202, Beverly Hills, CA, 90212-2006. Phone: (310) 278-4050. Fax: (310) 278-3350.

Alan Silverbach, pres.

TV-CATV only.

TV program distribution, United States & international.

The Television Syndication Company, Inc., 501 Sabal Lake Dr., Suite 105, Longwood, FL, 32779. Phone: (407) 788-6407. Fax: (407) 788-4397. E-mail: tvsco@prodigy.net Web Site: www.tvsco.com.

Cassie M. Yde, pres; Robert E. Yde, dir mktg.

TV-CATV only.

A full-svc TV syndication & distribution organization offering TV progmg to bcstrs worldwide.

Televix Entertainment Inc., 449 S. Beverly Dr., 3rd Fl., Suite 300, Beverly Hills, CA, 90212. Phone: (310) 788-5500. Fax: (310) 286-0207. E-mail: postmaster@televix.com Web Site: www.televix.com.

Hugo Rose, CEO; Pamela Popp, sr VP.

TV-CATV only.

Distribution of TV programs in the Latin American & Spanish U.S. markets.

Telfax Inc., 3305 Pleasant Valley Ln., Arlington, TX, 76015. Phone: (817) 468-0070. Fax: (817) 468-0111. E-mail: ts4telfax@aol.com

Tony Symanovich, pres.

TV-CATV only.

Remote TV production svcs.

Tepuy, 2745 Ponce de Leon Blvd., Coral Gables, FL, 33134. Phone: (305) 774-0033. Fax: (305) 774-7372. Web Site: www.tepuy.com.

Marcos Santana, chmn/CEO; Esperanza Garay, VP; Fernando Espejo, dir; Igancio Barrera, exec VP.

Pozuelo de Alarcon 28224 Spain, Via Dos Castillas 9C-P2, 2B. Phone: (3491) 351-7107. Fernando Espeto.

Caracas 1060 Venezuela, Ave. Libertador Torre E, EXA PH-1. Phone: (58212) 953-3363.

TV-CATV only.

Program distribution.

Danny Thomas Productions, 10100 Santa Monica Blvd., Suite 950, Los Angeles, CA, 90067. Phone: (310) 277-4866. Fax: (310) 286-1963. E-mail: anita@dethomasbobo.com

Anita De Thomas, pres.

TV-CATV only.

TV programs, features, coml production.

Thompson Creative, 4631 Insurance Ln., Dallas, TX, 75205. Phone: (214) 559-4000. Fax: (214) 521-8578. E-mail: info@thompsoncreative.com Web Site: www.thompsoncreative.com.

J. Larry Thompson, CEO; Susan Price Thompson, pres.

TV-CATV only.

Contemp radio ID jingles for all formats.

Time Capsule Inc., 124 Cottonwood Ln., Centerville, MA, 02632. Phone: (800) 822-7785. Fax: (508) 778-5590. E-mail: tc@tcapsule.com Web Site: www.tcapsule.com.

Richard T. Teimer, pres & sls dir; Nancy Q. Proctor, VP affl rel; Bill Stephens, VP special projects.

Radio Only.

System to bring 1,000 more cash ads per year; daily quizzes fit all formats.

Today Video, 475 10th Ave. 10th Fl., New York, NY, 11024. Phone: (212) 239-3999. Fax: (212) 239-2999.

David Seeger, CEO.

TV-CATV only.

Production svcs.

Toes Production Inc., 22 Hickory Dr., Maplewood, NJ, 07040. Phone: (973) 793-5440. Fax: (973) 363-7798. E-mail: phonepatch@aol.com

Brad Abelle, pres.

TV-CATV-Radio.

Brad is a radio talk show host and voice artist.

Tomwil Inc., 4905 Gentry Ave., Valley Village, CA, 91607. Phone: (818) 769-0883. Fax: (818) 769-0887. E-mail: tomwil@earthlink.net

James R. Rokos, pres; Wilda A. Rokos, VP.

TV-CATV only.

Distributors of features, light entertainment, sports, series & documentaries to all media in the world market.

Toucan Productions, 60 Fiddlers Elbow Rd., Margaretville, NY, 12455. Phone: (212) 580-4882.

TV-CATV only.

Corporate video productions from conception to completion.

Traffic Pulse Networks (A Unit of Mobility Technologies), Mobility Technology - Traffic Pluse Network, 851 Duportail Rd., Suite 220, Wayne, PA, 19087. Phone: (610) 725-9700. Fax: (610) 725-0530. E-mail: info@mobility technologies.com Web Site: www.mobilitytechnologies.com.

Doug Alexander, CEO; Jim Brown, VP; Al McGowan, sr VP; Robert Pollan, COO/CFO.

TV-CATV only.

Traffic Pulse Networks provides digital & traditional traffic data to radio, TV & CATV. It also provides an inventory rep service.

Traffic Scan Network, Inc, (Formerly Shadow Broadcast Services). 7707 Waco Ave., Baton Rouge, LA, 70806-1440. Phone: (225) 926-7152. Fax: (225) 923-0704. E-mail: johnny@trafficscannetwork.com Web Site: www.trafficscannetwork.com.

Johnny Ahysian, pres/gen mgr.

TV-CATV-Radio.

Traf, news, sports, weather reports for radio, TV & cable systems.

The Transcription Company, 4100 W. Burbank Blvd., 3rd Flr., Burbank, CA, 91505. Phone: (818) 848-6500. Fax: (818) 556-4150. Web Site: www.transcripts.net. E-mail: customerservice@transcripts.net

Michele Bartmon, dir of admin; Samantha Somers, mgr entertainment transcription.

Newport Beach, CA 92660. Rapidtext Inc., 1801 Dove St, Suite 101. Phone: (949) 399-9200. Glory Johnson, COO & VP.

TV-CATV only.

Transcribe all media: TV shows, films, news, sports, documentaries, meetings interviews. Provide closed captioning. Provide translations. Sell ABC News transcripts.

The Transfer Zone, 13251 Northend, Oak Park, MI, 48237-3261. Phone: (248) 548-7580. Fax: (248) 548-0924. E-mail: transferzone@juno.com Web Site: thetransferzone.com.

Roxane B. Newhouse, dir mktg.

TV-CATV only.

International video standard conversions, duplication, film/slide transfers: A-B roll editing, video slide/film. Video to CD & DVD & duplication.

Triage Entertainment Inc., 6701 Center Dr. W., Suite 1111, Los Angeles, CA, 90045-1552. Phone: (818) 386-6800. Fax: (818) 386-9889. Web Site: www.triageinc.com.

Stu Schreiberg, pres; Chris Greenleaf, dev VP; John Bravakis, business mgr; Steve Kroonpnick, exec dev.

TV-CATV only.

Independent, full-svc production company with extensive experience in the production of TV & film; post-production facilities & motion control/graphics department.

Tribune Entertainment Co., 5800 Sunset Blvd., Los Angeles, CA, 90028. Phone: (323) 460-5800. Fax: (323) 460-3858. Web Site: www.tribtv.com.

David Berson, sr VP business affairs; Karen Corbin, sr VP progmg & dev; Larry Hutchings, VP; Richard Inouye, VP; Henry Urick, VP mktg; Gina Brittle-Mackey, dir; Taylor Fuller, dir; Seth Howard, dir; Debra McCormick, dir mktg; George NeJame, dir; Natalie Sackin, creative dir & mktg; Jon Krobot, account exec.

Roswell, GA 30076, 1580 Warsaw Rd, Suite 210. Phone: (770) 643-4504. Fax: (770) 643-2549. Samuel K. Fuller, dir, SE rgnl sls.

Chicago, IL 60611, 435 N. Michigan Ave, Suite 1800. Phone: (312) 222-4000. Fax: (312) 222-3815. Michael Adinamis, VP bcst opns; Dick Bailey, dir midwest adviser sls; Jennifer Dreyer, acc exec midwest rgnl sls; Jeff McElheney, acc exec midwest rgnl sls.

New York, NY 10017, 220 E. 42nd St, Suite 400. Phone: (212) 210-1000. Fax: (212) 210-1056. Liz Koman, sr VP adv sls; Steve Mulderrig, sr VP/gen sls mgr.

TV-CATV only.

Acquires, develops, produces & distributes progmg for TV including "Gene Roddenberry's Andromeda," "Gene Roddenberry's Earth: Final Conflict," "BeastMaster," Malibu, CA," "Soul Train," "U.S. Farm Report," "Soul Train Music Awards," "Soul Train Lady of Soul Awards," "Soul Train Christmas Starfest," & "Live from the Academy Awards."

Tribune Radio Networks, 435 N. Michigan Ave., Chicago, IL, 60611. Phone: (312) 222-3342. Fax: (312) 222-4876. Web Site: www.tribuneradio.com.

Barbara Pabst, opns mgr; Tom Langmyer, gen mgr & VP; Wendi Power, mgr.

Radio Only.

Tape & internet digital delivered farm, sports & specialty programs including: *Chicago Cubs Network, Agri-Voice, National Farm Report, Farming America, Samuelson's Sez.*

TRI-COMM Productions, (A First Vision Group Co). 11 Palmetto Pkwy., Suite 201, Hilton Head Island, SC, 29926-3703. Phone: (843) 681-5000. Fax: (843) 681-2945. Web Site: www.tri-comm.tv.

William J. Robinson, pres/CEO.

TV-CATV only.

A full-svc production company specializing in film HD & digital betacam production, sound design, graphics/animation. Also offers sugar sand beaches & the best golf courses around.

Trident Releasing, 8401 Melrose Pl., 2nd Fl., Los Angeles, CA, 90069. Phone: (323) 655-8818. Fax: (323) 655-0515. E-mail: tridents@aol.com

Jean Ovrum, chmn; Victoria Plummer, pres; Kristi Mailing, sls VP.

TV-CATV only.

Acquire feature films in the postproduction & completed stages.

Troma Entertainment, Inc., 733 Ninth Ave., New York, NY, 10019. Phone: (212) 757-4555. Fax: (212) 399-9885. Web Site: www.troma.com.

TV-CATV only.

Troma is one of the oldest ind film companies in the world. We produce & distribute films & offer stock footage.

Turner Entertainment Co., 1888 Century Park E., 10th Fl., Los Angeles, CA, 90067. Phone: (310) 788-6801. Fax: (310) 788-6810. E-mail: roger.mayer@turner.com

Robert L. Mayer, pres/COO.

TV-CATV only.

Sls, licensing & servicing of major film & TV library.

TVOntario, (TVO network, TFO network). Box 200, Stn Q, Toronto, ON, M4T 2T1. Canada. Phone: (416) 484-2600. Fax: (416) 484-2662. Web Site: www.tvo.org.

Lee Robock, COO; Yuonne Carey-Le, dir.

Educ production, educ bcst international program sls, coproduction.

20th Century Fox/Incendo Television Distribution Ltd., 101 Bloor St. W., Suite 400, Toronto, ON, M5S 2Z7. Canada. Phone: (416) 643-3897. Fax: (416) 643-3898.

Michael Murphy, sr VP; Kimberley Ball, dir mktg; David Heaph, sls dir.

TV-CATV only.

TV program distribution.

Twentieth Century Fox Television Distribution, Box 900, Beverly Hills, CA, 90213-0900. Phone: (310) 369-1000. Fax: (310) 369-8892. Web Site: www.foxnow.com.

Mark Kaner, pres; Peter Levinshon, pres; Marion Edwards, exec VP.

London WID 3AP. Twentieth Century Fox Television Distribution, 31-32 Soho Square.. Phone: 44-20-7437-7766. Fax: 44-20-7439-1806.

Moore Park, NO 1363 Australia. Fox Studios Austrailia, Driver Avenue. Phone: 61-8353-2200. Fax: 61-2-835-2205.

Sao Paulo 04543-121 Brazil. Fox Film Do Brasil Ltda., Rua Dr. EDuardo De Souza Arrrrranha, 387-3o Andar. Phone: 5511-3365-5205. Fax: (5511) 3365-5177.

Toronto, ON M5S 2Z7 Canada. Fox/Incendo, 101 Bloor Street West, Suite #400. Phone: 416-643-3897. Fax: 416-643-3907.

Paris 75008 France. Twentieth Century Fox France, Inc., TV Division, 21 bis rue Lord Byron. Phone: 33-1-5393-9398. Fax: 33-1-5393-9397.

Pembroke Pines, FL 33028-2867. Twentieth Century Fox Television Distribution, 2000 N.W. 150th Avenue, Suite 1110. Phone: 954-322-5000. Fax: 954-322-5275.

TV-CATV only.

Production & distribution.

Twentieth Television, 2121 Ave. of the Stars, Suite 2100, Los Angeles, CA, 90067. Phone: (310) 369-3924. Fax: (310) 369-1506. Web Site: www.fox.com.

Bob Cook, pres/COO; David Shall, exec VP; Mark Kaner, pres; Robb Dalton, pres; Bob Cesa, exec VP.

New York, NY 10036. Twentieth Television, 1211 Avenue of the Americas16th Fl. Bob Cesa, exec VP adv sls & cable progmg sls.

TV-CATV only.

Produces, distributes progmg for net TV, domestic & international TV markets.

Two Oceans Entertainment Group, 2017 Lemoyne St., Suite 800, Los Angeles, CA, 90026. Phone: (818) 501-6550. Fax: (818) 501-6558. E-mail: twoceans@aol.com

Meryl Marshall, pres; Susan Whittaker, dev VP.

TV-CATV only.

Domestic & international TV production. Most recent—*When Danger Follows You Home, Baby Monitor, Sound of Fear, Happily Ever After, Fairy Tales For Every Child.*

U

UBC Radio, 230 Ohio St., Suite 101, Chicago, IL, 60611. Phone: (312) 640-5000. Phone: (312) 751-0135. Fax: (312) 640-5010.

Bradley Saul, CEO; Ron Gleason, pres.

Radio Only.

Talk radio long form progmg plus short form features.

U.S. Plan B Inc., 466 Orange St., Suite 280, Redlands, CA, 92374. Phone: (818) 998-8833. Phone: (888) 877-5262. Fax: (702) 926-2532. E-mail: office@usplanb.com Web Site: www.usplanb.com.

Marc Curtis, producer.

TV-CATV only.

TV news gathering & production crews, stock footage, rsch.

Ukrainian Melody Hour, Box 2257, Washington, DC, 20013. Phone: (202) 529-7606. Phone: (202) 269-1824. Fax: (202) 638-5995.

Roman V. Marynowych, producer & dir; Odile E. Marynowych, exec sec.

TV-CATV only.

Ukrainian radio, TV & cable program productions.

United Learning Co., 1560 Sherman Ave., Suite 100, Evanston, IL, 60201. Phone: (800) 323-9084. Fax: (847) 328-6706. E-mail: joel.altschul@unitedlearning.com Web Site: www.unitedlearning.com.

Joel Altschul, chmn.

TV-CATV only.

Educ programs, series & documentaries, videos & curriculums, streaming.

United Methodist Communications, 810 - 12th Ave. S., Nashville, TN, 37203. Phone: (615) 742-5400. Fax: (615) 742-5125. E-mail: lalexander@umcom.org Web Site: www.umcom.org.

Larry Hollon, sec; Jeneane Jones, dir TV progmg.

TV-CATV only.

TV & radio production & distribution.

United Sound Systems Inc., 15849 Wyoming St., Detroit, MI, 42838. Phone: (313) 340-4200. Fax: (313) 832-5666.

TV-CATV-Radio.

Audio & duplicating recording, postproduction audio facilities & svcs, & music production.

U.S. Air Force Recruiting Service, Randolph AFB, 550 D Street W., Suite 1, Universal City, TX, 78150-5421. Phone: (210) 652-3937. Fax: (210) 652-4892. E-mail: rspsa@rs.af.mil Web Site: www.airforce.com.

Gary Quesenberry, supt bcstg; Ted Northrup, producer & dir.

TV-CATV only.

Custom production of radio & TV PSAs for Air Force recruiting for local communities.

United Stations Radio Network, 25 W. 45 St., New York, NY, 10036. Phone: (212) 869-1111. Fax: (212) 869-1115. E-mail: info@unitedstations.com Web Site: www.unitedstations.com.

Nick Verbitsky, CEO.

Los Angeles, CA 90064, 11400 W. Olympic Blvd., # 200. Phone: (310) 914-0188. Anne Martinez, mgr.

Chicago, IL 60606, 333 W. Weeker Dr, # 700. Phone: (312) 444-2034. Rich Baum, VP midwest.

Dallas, TX Phone: (972) 506-8776. Rob Ellis, mgr.

Radio Only.

Entertainment & comedy. Progmg for radio stns, news, business & weather features.

University of Colorado Television, Campus Box 379, Boulder, CO, 80309. Phone: (303) 492-1857. Fax: (303) 492-7017. E-mail: kathleen.albers@colorado.edu

Kate Albers, producer & dir.

TV-CATV only.

TV & radio program production & distribution; production svcs.

University of Detroit Mercy, 4001 W. McNichols, Communication Studies Dept., Detroit, MI, 48221-3038. Phone: (313) 578-0311. Phone: (313) 993-2005. Fax: (313) 993-1166. Web Site: www.udmercy.edu.

Michael Jayson, engr.

TV-CATV only.

Radio program syndication.

University of Kentucky Public Relations & Radio-TV News Bureau, Mathews Bldg., Rm. 4, Lexington, KY, 40506. Phone: (859) 257-1754. Fax: (859) 257-4017. E-mail: cnath1@email.uky.edu Web Site: www.uky.edu.

Carl Nathe, dir radio/TV news bureau.

TV-CATV only.

TV & radio program & coml, promotional film production & distribution.

Univision Communications Inc., 5999 Center Drive, Los Angeles, CA, 90045. Phone: (310) 556-7676. Fax: (310) 556-7615. Web Site: www.univision.net.

A. Jerrold Perenchio, chmn/CEO; Ray Rodriguez, pres/COO.

TV-CATV only.

Sp language bcst & cable TV net radio stns, music record labels & an internet destination.

V

VA-Tech Video/Broadcast Services, 285 Whittemore Hall, Blacksburg, VA, 24061. Phone: (540) 231-5930. Fax: (540) 231-4622. Web Site: www.vbs.vt.edu.

Mark Harden, mgr.

TV-CATV only.

Complete video & audio production & postproduction svcs. Uplink, downlink & CATV opn. Telephone & data communication.

VRI (Video Rentals Inc.), 100 Stonehurst Ct., Northvale, NJ, 07647. Phone: (800) 255-2874. Phone: (201) 750-3200. Fax: (201) 784-2795. E-mail: info@rentvri.com Web Site: www.rentvri.com.

Tom Canavan, VP/gen mgr.

TV-CATV only.

Full-svc rental facility with a complete inventory of bcst & industrial video equipment.

VTTV Videothek Electronic TV-Production GmbH + Co Kopier KG, Havelchaussee 161, Berlin, 14055. Germany. Phone: (030) 300 95-3. Fax: (030) 300 95-500. E-mail: info@vttv.de Web Site: www.vttv.de.

Paul Bielicki, pres; Friedel Lux, head dev/rsch; Herbert Bauermeister, head creative dept.

TV-CATV only.

Complete production service: studios (600/225 m2), professional equipment for production & postproduction, high definition production units. Commercials, documentaries, features, pop promos, TV.

Valentino Music & Sound Effect Libraries, 500 Executive Blvd., Elmsford, NY, 10523. Phone: (914) 347-7878. Phone: (800) 223-6278. Fax: (914) 347-4764. E-mail: tvmusic@ibm.net Web Site: www.tvmusic.com.

Thomas J. Valention, pres; Francis T. Valentino, VP.

TV-CATV only.

Compact disc music & sound effects libraries.

Van Vliet Media, 420 E. 55th St., Suite 6L, New York, NY, 10022. Phone: (212) 486-6577. Fax: (212) 980-9826. E-mail: vanvlietmedia@att.net Web Site: vanvlietmedia.com.

Rochelle Bebell, pres; Harlan B. DeBell, VP.

TV-CATV only.

Digital transfers, dup cardiac angiography.

Venevision International, 550 Biltmore Way, Suite 1180, Coral Gables, FL, 33134. Phone: (305) 442-3411. Fax: (305) 448-4762. E-mail: info@venevisionintl.com Web Site: www.venevisionintl.com.

Luis A. Villanueva, pres; Benjamin F. Perez, exec VP; Cristobal Ponte, VP; Mario Castro, mktg & creative svc dir/sls dir Central America; Miguel Somoza, sls.

TV-CATV only.

Distribution of progmg.

Venice Media Services, 299 W. Houston, 10th Fl., New York, NY, 10014-3620. Phone: (212) 859-5100. Fax: (212) 727-9495.

Peggy Green, pres/Natl Bcstg of Venice U.S.

TV-CATV only.

TV program distribution.

Viacom Inc., 1515 Broadway, New York, NY, 10036. Phone: (212) 258-6000. Fax: (212) 258-6465. Web Site: www.viacom.com.

Summer M. Redstone, chmn/CEO; Matthew Blank, CEO; Martin Shea, sr VP; Al Weber, pres; Jonathan L. Dolgen, chmn; John Antioco, chmn/CEO; Carl D. Folta, sr VP; Thomas E. Freston, chmn/CEO; Michael D. Fricklas, exec VP; Sherry Lansing, chmn; Herb Scannell, pres; Jack Ramanos, pres; William A. Roskin, sr VP; Carol Melton, sr VP; Mel Karmazin, pres/COO; Richard Bressler, sr VP/CFO.

Hollywood, CA 90038. Paramount Pictures, 5555 Melrose Ave.

Charlotte, NC 28217. Paramount Parks, 8720 Red Oak Blvd, Suite 375.

New York, NY 10019. Showtime Networks Inc., 1633 Broadway.

New York, NY 10020. Simon & Schuster, 1230 Ave. of the Americas.

New York, NY 10036. MTV Networks, 1515 Broadway.

Dallas, TX 75270. Blockbuster Entertainment, 1201 Elm St.

TV-CATV only.

TV program production & distribution, motion pictures production & distribution, book publishing, video production & distribution, theme parks.

Viacom Video Services, 524 W. 57th St., New York, NY, 10019-2924. Phone: (212) 975-8139. Fax: (212) 975-7272. E-mail: mjeffers@cbs.com Web Site: www.viacom.com.

Mark Jeffers, dir distribution & opns.

Los Angeles, CA 90024, 10877 Wilshire Blvd. Phone: (310) 446-6051. Fax: (310) 446-6066. Lee Salas, VP & natl sls mgr.

TV-CATV only.

Duplicate & distribute syndicated TV progmg via tape & satellite, coml integration, international standards conversion, uplink/downlink, tape duplication, space segment; high definition.

Video-Cinema Films Inc., 510 E. 86th St., New York, NY, 10028. Phone: (212) 734-1632. Fax: (212) 734-1632.

Larry Stern, pres.

TV-CATV only.

Distribution/licensor of motion pictures & excerpts for all forms of TV & allied media in US and worldwide.

Video Enterprises Inc., 575 29th St., Manhattan Beach, CA, 90266-3430. Phone: (310) 796-5555. Fax: (310) 546-2921. E-mail: hambrose@earthlink.net

Heidi Lane-Ambrose, pres; Renska Somers, office mgr.

TV-CATV-Radio.

Natl placement of ten-second promotional spots on game shows, talk, var & sports programs.

Video I-D Teleproductions Inc., 105 Muller Rd., Washington, IL, 61571. Phone: (800) 333-9123. Fax: (309) 444-4333. E-mail: videoid@videoid.com Web Site: www.videoid.com.

Sam B. Wagner, pres; Gwen Wagner, mktg mgr; Larry Strantz, sls consultant.

TV-CATV only.

Full teleproduction svcs, DVD & CD-ROM capabilities, location production, linear & nonlinear editing, 3D graphics, specializing in corporate image, safety, training, sls & mktg.

Video/Media Distribution Inc., 1050 N. State St., Chicago, IL, 60610. Phone: (312) 944-4700. Fax: (312) 944-1582. E-mail: shelvm@aol.com

Shel Beugen, pres.

TV-CATV only.

Program sls & syndication svcs to bcst, cable stns & networks.

Video One Inc., 4952 Nagle Ave., Sherman Oaks, CA, 91423. Phone: (818) 781-9824. Fax: (818) 753-4704.

Robert G. Kaufmann, pres; Kevin E. Hamburger, VP.

Remote TV production facilities.

Video Services, 1033 Elm Hill Pike, Nashville, TN, 37210. Phone: (615) 248-1010. Fax: (615) 244-5712. E-mail: info@siffordvideoservices.com Web Site: www.siffordvideoservices.com.

Joel Covington, owner.

TV-CATV only.

CD, DVD, videotape duplication, standard conversions.

Video Techniques Inc., Box 9649, Bradenton, FL, 34206-9649. Phone: (941) 758-3077. Fax: (941) 758-4896. E-mail: vti@videotechniques.com Web Site: www.videotechniques.com.

Bob Lorentzen, pres.

TV-CATV only.

Beta SP & MII field & postproduction facilities. CD, DVD authoring & duplication, VHS duplication, streaming video hosting.

VideoActive Productions, 1560 Broadway, Suite 610, New York, NY, 10036. Phone: (212) 541-6592. Web Site: www.videoactiveprod.com.

Steven Garrin, pres.

TV-CATV-Radio.

Radio, TV, film production/post production.

Videographic West, Box 1093, 30 Benchmark Rd., Suite 203, Avon, CO, 81620. Phone: (970) 949-5593. Fax: (970) 949-6331. E-mail: video@colorado.net Web Site: www.skitv.com.

Michael Billingsley, principal; Stephanie Billingsley, principal.

TV-CATV only.

Specializing in cable sports progmg & distribution, extensive stock footage of action sports, mountain lifestyle, sking, golf, family & travel. Full-svc production house, field crews, two SD/component pinnacle Liquid uncompressed NLE workstations, complete multimedia suite with DVD authoring/encoding, 2D animation suite with talented artists.

Videomedia, C/Jose Isbert, 2, Ciudad de la Imagen, Pozuelo De Alarcon, Madrid, 28223. Spain. Phone: 34-91-512.8000. Fax: 34-91-518.8017. E-mail: videomedia@videomedia.es Web Site: www.videomedia.es.

Jorge Arque, CEO & pres; Mireia Acosta, fiction mgr; Daniel Acuna, entertainment mgr; Fernanda Montoro, intnl div.

TV-CATV only.

Independent production company. Entertainment formats & programs, documentaries, fiction.

Videosmith Inc., 100 Spring Garden St., Philadelphia, PA, 19123. Phone: (215) 238-5070. Fax: (215) 238-5075. E-mail: info@videosmith.com Web Site: www.videosmith.com.

Steven T. Smith, pres.

TV-CATV only.

TV progmg, equipment rentals & production mgmt.

VIEW Video Inc., 34 E. 23rd St., New York, NY, 10010. Phone: (212) 674-5550. Fax: (212) 979-0266. E-mail: viewvid@aol.com Web Site: www.view.com.

Bob Karcy, pres; Stephen R. Kates, dir mktg.

TV-CATV only.

International home video production & distribution of special interest progms in the areas of art, jazz, pop music, opera, dance, children's interactive sports & modern lifestyle progms.

Virginia Tech, CNS (Communications Network Services), 1770 Forecast Dr., Blacksburg, VA, 24061-0506. Phone: (540) 231-6460. Fax: (540) 231-8418. Web Site: www.cns.vt.edu.

Judy Lilly, dir communications network svcs.

TV-CATV only.

Video & Audio Service to campus - faculty staff & students.

Vision Broadcasting - KVBA TV 19, 1017 New York Ave., Alamogordo, NM, 88310-6921. Phone: (505) 437-1919. E-mail: kvba@kvbatv.com

William J. Oechsner Jr., gen mgr & pres.

TV-CATV only.

Bcst of Christian, loc TV, sports & public interest TV.

Vuolo Video Air-Chex, Box 880, Novi, MI, 48376. Phone: (248) 926-1234. E-mail: artvuolo@aol.com Web Site: www.vuolovideo.com.

Arthur Vuolo, Jr., producer.

Radio Only.

Video air checks of American radio stns & An Inside Look.

Vyvx, One Technology Center, Tulsa, OK, 74103. Phone: (800) 364-0807. Fax: (918) 547-2989. Web Site: www.vyvx.com.

Derek Smith, VP/gen mgr.

TV-CATV only.

Distributor of TV comls/traf via satellite by remote VTR control to 600 stns; radio coml distribution; production, postproduction. Multi-format duplication.

W

WKMG Productions, 4466 N. John Young Pkwy., Orlando, FL, 32804. Phone: (407) 291-6000. Fax: (407) 521-1204. Web Site: www.local6.com.

Henry Maldonado, gen mgr; Jim Murphy, production.

TV program, coml, promotional film production; production svcs, post production & field production. WKMG production resources primarily dedicated to stn use.

WCTD AM 1620, 244 Post Rd., Westerly, RI, 02891. Phone: (401) 322-1743. Phone: (401) 322-9091. Fax: (401) 322-1645. Web Site: www.wblq.org/htm.wctd.

J.J. MacDade Nunez, gen mgr; Chris DiPaola, pres.

Radio Only.

Traveler information.

WFMT Radio Network, 5400 N. St. Louis Ave., Chicago, IL, 60625. Phone: (773) 279-2112. Phone: (773) 279-2114. Fax: (773) 279-2119. Web Site: www.wfmt.com.

Steve Robinson, sr VP.

Radio Only.

Produces wkly series including Chicago Symphony retrospective & the New York Philharmonic this week for coml & pub stns. Broad range of symphonic, opera & classical music documentary progmg, jazz & folk music, including exclusive features from BBC & Radio Deutsche Welle, Germany.

WGN Television, 2501 W. Bradley Pl., Chicago, IL, 60618. Phone: (773) 528-2311. Fax: (773) 528-6857. Web Site: www.wgntv.com. E-mail: wgnttvnews@tribune.com

John Vitanovec, VP/gen mgr; Dominick Mancuso, stn mgr; Merri Dee, community rel dir; Marty Wilke, sls dir; JoAnn Stern, svcs dir; Rob Salerno, dir finance & admin; Greg Caputo, news dir.

TV-CATV only.

TV programs & comls, promotional film production & distribution, production svcs.

WMAQ-TV, NBC Tower, 454 N. Columbus Dr., Chicago, IL, 60611-5555. Phone: (312) 836-5555. Fax: (312) 527-4290. Web Site: www.nbc5.com.

Larry Wert, pres/gen mgr; Patricia Golden, VP sls.

TV-CATV only.

TV program production; production svcs.

WNN Health and Wealth Motivation, 6699 N. Federal Hwy., Boca Raton, FL, 33487. Phone: (561) 997-0074. Fax: (561) 997-0476. Web Site: www.wnnhealthtalkradio.com.

Robert Morency, VP/gen mgr.

Radio Only.

Worldwide 24-hour format of motivational speakers & self-help info.

The WPA Film Library, 16101 S. 108th Ave., Orland Park, IL, 60467. Phone: (708) 460-0555. Fax: (708) 460-0187. E-mail: sales@wpafilmlibrary.com Web Site: www.wpafilmlibrary.com.

Diane Paradiso, sls.

TV-CATV only.

One of the largest stock footage libraries in the U.S. WPA offers holdings in newsreels, music, pop culture & stock shots.

WQED Multimedia, (The Metropolitan Pittsburgh Public Broadcasting Station). 4802 5th Ave., Pittsburgh, PA, 15213. Phone: (412) 622-1300. Fax: (412) 622-6413. E-mail: info@wqed.org Web Site: www.wqed.org.

George Miles, pres/CEO; Steven Reubi, controller & treas; Rosemary Martinelli, exec dir; Deborah Acklin, VP/gen mgr; Patricia Walker, VP finance; Susan Johnson Radlo, exec dir; Lilli Mosco, dev VP.

TV-CATV-Radio.

TV & radio program/production, distribution; production svcs & web, publishing.

WQXR, 122 5th Ave., 3rd Fl., New York, NY, 10011. Phone: (212) 633-7600. Fax: (212) 633-7666. E-mail: listener.mail@wqxr.com Web Site: www.wqxr.com.

Penny Gaffney, sls VP; Thomas Bartunek, gen mgr.

Radio Only.

Radio program production. The classical Radio Station of the New York Times.

WRS/Channel One, (A division of WRS Inc.). 1000 Napor Blvd., Pittsburgh, PA, 15205. Phone: (412) 937-7700. Fax: (412) 922-1020. E-mail: jackn@wrslabs.com Web Site: www.wrslabs.com.

F. Jack Napor, pres.

TV-CATV only.

Complete syndication & distribution svcs via satellite & tape. Integration, AMOL encoding, duplication, uplink/downlink fulfillment, audio, film-to-tape transfer; replication of CD/DVD.

WSI Corp., 400 Minuteman Rd., Andover, MA, 01810. Phone: (978) 983-6300. Fax: (978) 983-6400. Web Site: www.wsi.com.

TV-CATV only.

World leader in providing real-time weather data, imagery, forecasting & weather progmg for bcst stns, cable operators & cable nets.

WTOB Channel 2—Public/Government Access TV, 300 S. Main St., Blacksburg, VA, 24062. Phone: (540) 961-1199. Fax: (540) 961-1875. E-mail: wtob@blacksburg.gov Web Site: www.blacksburg.gov.

Carlton Herman, stn mgr.

TV-CATV only.

Locally originated progmg, live town meetings, equipment training & loan svcs in inch, S-VHS & VHS formats.

WWE Entertainment Inc., Titan Tower, 1241 E. Main St., Stamford, CT, 06902. Phone: (203) 352-8600. Fax: (203) 352-8699. Web Site: www.wwe.com.

Vincent K. McMahon, chmn; Linda E. McMahon, CEO; James Rothschild, sr VP; Donna Goldsmith, sr VP/licensing & merchandising; Phil Livingston, CFO.

TV-CATV-Radio.

Exclusive worldwide distributor of WWF events, TV programs (bcst network/syndication, PPV & basic cable) & other sports/entertainment properties.

Wade Productions Inc., 493 High Cliffe Lane, Tarrytown, NY, 10591. Phone: (212) 286-9111. E-mail: wade@1cj@aol.com

Carolyn J. Wade, pres.

TV-CATV only.

Meetings, video, entertainment, staging & teleconferencing for corporations & assns.

Warner Bros. International Distribution Inc. (Canada), 4576 Yonge St., 2nd Fl., North York, ON, M2N 6P1. Canada. Phone: (416) 250-8384. Fax: (416) 250-8598. Web Site: www.wbitv.com.

Robert Blair, VP/gen mgr; Leslie Hibbins, prom dir & publicity.

TV-CATV only.

TV progm distribution & promotions for Canada.

Warner Bros., Bldg. 140, 300 Television Plaza, Burbank, CA, 91505. Phone: (818) 954-7500. Fax: (818) 954-7322.

Peter Ross, pres.

TV-CATV only.

TV program production & distribution.

Warner Bros. Animation, 15301 Ventura Blvd., Suite 115301, Unit E, Sherman Oaks, CA, 91403. Phone: (818) 977-8700. Fax: (818) 382-6056. Web Site: www.warnerbros.com.

Sander Schwartz, pres; Andy Lewis, VP/gen mgr.

TV-CATV only.

Dev & produces animated progmg for TV, video & other media.

Warner Bros. Domestic Television Distribution, 4000 Warner Blvd., Burbank, CA, 91522. Phone: (818) 954-5877. Fax: (818) 954-5820. Web Site: www.warnerbros.com.

Dick Robertson, pres; Jim Paratore, exec VP; Jacqueline Hartley, VP; Jeff Hufford, VP; Mark O'Brien, sr VP.

Print Department—Burbank: Bud Rowe, foreign TV print administrator. (818) 954-3731.TV program syndication.

Warner Bros. International Television, 4000 Warner Blvd., Burbank, CA, 91522. Phone: (818) 954-6000. Fax: (818) 954-4040.

Jeffrey R. Schlesinger, pres; Mauro Sardi, sls VP; James P. Marrinan, sr VP; Josh Berger, VP; David Camp, VP finance; Lisa Gregorian, VP; Ron Miele, VP; John Whitesell, VP; Catherine Malatesta, sr VP; Susan Kroll, sr VP; Malcolm Dudley-Smith, sr VP; Marsha Armstrong, VP; Monica Dodi, VP; Sal LoCurto, VP mktg; Kelley Nichols, VP; Robert Nitkin, VP; Matthew Robinson, VP.

Acapulco 37 06140. Warner Bros. (Mexico) S.A., Colonea Codesa. Phone: 011-525-211-3353-211-0293. 211-0298-211-0466. Fax: 011-525-553-2822-553-2002. Jorge Sanchez, VP Latin America.

London W1V 4AP. Warner Bros. International TV, 135 Wardour St. Phone: 011-44-171-494-3710. Fax: 011-44-171-287-9086. Richard Milnes, VP UK territories, Turkey, Israel.

Madrid 28033. Warner Bros. International TV, Artuto Soria, 336 1. Phone: 011-34-1-384-06-40. Fax: 011-34-1-384-06-41. Jose Abad, sls exec Sp territories.

Minato-ku, Tokyo 105. Time Warner Entertainment Japan, 1-2-4 Hamamatsu-cho. Phone: 011-81-3-5472-8341. Fax: 011-81-3-5472-6343. Teruji Mochimaru, mgng dir Japan.

North Sydney NSW 2060. Warner Bros. PTY. Ltd., 8-20 Napier St. Phone: 011-61-2-9957-3899. Fax: 011-61-2-9956-7788. Wayne Broun, VP mgng dir Australia, Asia Pacific; Greg Robertson, VP Asia.

Paris 75017. Warner Bros. International TV, 67 Avenue Dewagram. Phone: 011-33-1-5537-5933. Fax: 011-33-1-5537-4968. Michel Lecourt, VP Fr territories.

Rome 00195. Warner Bros. Italia S.R.L., Via Giuseppe Avezzana, 51. Phone: 011-39-6-321-7779. Fax: 011-39-6-321-7278. Rosario Ponzio, mgng dir It-speaking Europe.

North York, ON M2N 6P1 Canada. Warner Bros. International TV, 4576 Yonge St, 2nd Fl. Phone: (416)250-8384. Fax: 416-250-8598. Kevin Byles, VP & gen mgr, Canadian opns.

TV program production & distribution.

Warner Bros. Television Production, 4000 Warner Blvd., Burbank, CA, 91522. Phone: (818) 954-6000. Fax: (818) 954-7048. Web Site: www.warnerbros.com.

Greg Maday, sr VP movies & mini-series; Mary Buck, sr VP talent & casting; Steve Pearlman, sr VP current programs; Robert Rosenbaum, sr VP net prodction; David Sacks, sr VP current progmg; Paul Stager, sr VP studio gen counsel; Julie Waxman, sr VP business affrs.

TV-CATV only.

TV program production.

Warren Only Media Group, Box 2372, Times Square Station, New York, NY, 10036. Phone: (856) 507-9368. Fax: (856) 507-9368. E-mail: warrenonly@90.com Web Site: www.warrenonly.com.

TV-CATV only.

Satellite program distribution, syndication, playout, back hauling, film & video production.

Washington Korean Broadcasting Co., 7004-K Little River Tpke, Annandale, VA, 22003. Phone: (703) 354-4900. Fax: (703) 354-6078. E-mail: wkbc@wkbc.biz Web Site: www.wkbc.biz.

Yong C. Pak, pres; Yong S. Lee, gen mgr.

Radio Only.

All ethnic progmg in Korean language, news, music, drama & talk show.

Wawatay Native Communication Society, Box 1180, 16 Fifth Ave., Sioux Lookout, ON, P8T 1B7. Canada. Phone: (807) 737-2951. Fax: (807) 737-3224. Web Site: www.wawatay.on.ca.

Mike Metatawabin, pres; Christine Chisel, exec dir.

TV-CATV only.

Radio & TV (Cree, Ojibway & English) net, bilingual nwspr, aboriginal language translations, multi-track audio recording.

Wax Music, Sound Design & Mix, 18 W. 21st St., 10th Fl., New York, NY, 10010-6903. Phone: (212) 989-9292. Fax: (212) 989-5195. E-mail: chris@waxnyc.com Web Site: www.waxnyc.com.

James Wolcott, composer/sound designer; Chris Arbisi, chief engr.

TV-CATV-Radio.

Original music, sound design & mixing for all media. Three digital studios.

We the People to BNK Kids, 41 Madison Ave., New York, NY, 10010. Phone: (212) 213-2700. Fax: (212) 685-8332. Web Site: www.amazin.com.

TV-CATV only.

Entertainment & media company specializing in the youth market. Hqtrs in New York & offices in Chicago, Los Angeles & Paris, France.

"The Weather Center", (a broadcast service of Aviation Weather Inc.). 701 Gervais St., Suite 150-224, Columbia, SC, 29201. Phone: (803) 739-2827. E-mail: wxcenter@aviationweatherinc.com Web Site: www.aviationweatherinc.com.

Liam Richard Ferguson, pres.

TV-CATV only.

"Regional Radio Broadcast/Weathercast Network" across the Carolinas and Georgia in over 20 bcst markets. Weather forecasting, site-specific bcst svc for stns all across America. 100% barter.

WeatherVision Inc., 916 Foley St., Jackson, MS, 39202-3406. Phone: (601) 948-7018. Fax: (601) 948-6052. E-mail: edward@weathervision.com Web Site: www.weathervision.com.

Edward Saint-Pe', pres; Jason McCleave, VP.

TV-CATV only.

Customized, localized TV weathercasts with or without meteorologists. Barter/cash via Ku-band satellite. Complete studio teleport for use by news media on site. Avid editing available on site. Serves radio, TV, 3-D branding animation svcs.

Alan Weiss Productions, 355 W. 52nd St., New York, NY, 10019. Phone: (212) 974-0606. Fax: (212) 974-0976. E-mail: myacoub@awptv.com Web Site: www.awptv.com.

Alan J. Weiss, pres; Tania Wilk, VP.

TV-CATV-Radio.

Fourteen Emmys for video production. We handle any broadcast, PR or corporate from concept to distribution.

Wellspring, (formerly Fox Lorber Associates, Inc.). 419 Park Ave. S., 20th Fl., New York, NY, 10016. Phone: (212) 686-6777. Fax: (212) 685-2625. Web Site: www.wellspring.com.

Sheri Levine, sr VP international distribution; Marie Therese Guirgis, dir acquisitions.

TV-CATV only.

Worldwide distributors of film & video properties for home video, standard & non-standard TV.

Welwood International Film Production, 160 Washington S.E., Suite 138, Albuquerque, NM, 87108-2731. Phone: (505) 265-1899. E-mail: welwoodint@aol.com

Bill Swortwood, pres/creative dir; Barbara Ferrel, VP/CEO.

TV-CATV only.

Works with consultants, rsch companies & client stns to create effective TV campaigns since 1986.

Westar Music, 105 West Beaver Creek Rd. Suite 3, Richmond Hill, ON, L4B 1C6. Canada. Phone: (905) 886-3100. Fax: (905) 886-6800. E-mail: info@westarmusic.com Web Site: www.westarmusic.com.

Brian Nimens, pres/CEO.

TV-CATV-Radio.

High caliber production music in a wide var of categories for bcst, cable, film, corporate video & multimedia applications.

Western International Syndication, 12100 Wilshire Blvd., Suite 150, Los Angeles, CA, 90025. Phone: (310) 820-8485. Fax: (310) 820-8376. Web Site: www.wistelevision.com. E-mail: info@wistelevision.com

Chris Lancey, pres/CEO; Danielle Valdivia, rsch dir.

TV-CATV only.

Distributes wkly & special progmg nationwide as well as internationally.

Westwood One, Radio New Services. 40 W. 57th St., New York, NY, 10019. Phone: (212) 641-2000. Fax: (212) 641-2185. Web Site: www.westwoodone.com.

Producer & distributor of radio progmg including CNN, NBC, Mutual, CNBC Business Radio, 24-hours music formats, long-short-form talk, music & news programs.

White Rabbit Productions, 1587 S. Main St., Salt Lake City, UT, 84115. Phone: (800) 549-3115. Phone: (801) 463-9292. Fax: (801) 463-7226. E-mail: info@whiterabbitproductions.com Web Site: www.whiterabbitproductions.com.

Sam Prigg, pres/dir photography.

TV-CATV only.

Complete film & video production svcs; two Ikegami HL-V55 Beta Sp camera packages, digital video point-of-view cam, non-linear editing.

Wide Eye Productions, Inc., 686 N. 9th, Boise, ID, 83702. Phone: (208) 336-0391. Fax: (208) 336-6644. E-mail: info@wideeye.tv Web Site: www.wideeye.tv.

Tom Hadzor, dir; Jennifer Isenhart, producer.

TV-CATV only.

Full-service bcst & industrial video production. ENG/EFP. High definition & Sony B-600.

Daniel Wilson Productions Inc., 300 W. 55th St. Ste 8V, New York, NY, 10019. Phone: (212) 765-7148. Fax: (212) 765-7916. E-mail: wilprod@verizon.net

Daniel Wilson, pres.

TV & theatrical film production & distribution.

Witt/Thomas Productions, 11901 Santa Monica Blvd., Suite 596, W. Los Angeles, CA, 90025. Phone: (310) 472-6004. Fax: (310) 476-5015.

Paul Junger Witt; Tony Thomas, partner; Susan Harris, partner.

TV-CATV only.

TV & film production company.

Robert Wold Co., 88 Three Vines Ct., Ladera Ranch, CA, 92694. Phone: (949) 363-0993. E-mail: robertnwold@cox.net

Robert N. Wold, owner.

TV-CATV only.

Special-interest & entertainment progmg for bcst & cable TV. Production, mktg & distribution of syndicated programs.

Fred Wolf Films, 4222 W. Burbank Blvd., Burbank, CA, 91505. Phone: (818) 846-0611. Fax: (818) 846-0979. E-mail: administration@fredwolffilms.com Web Site: www.fredwolffilms.com.

Fred Wolf, pres.

TV-CATV only.

TV program production & distribution.

Work Edit, 270 W. 39th St., 11th Floor, New York, NY, 10018. Phone: (212) 719-4577. Fax: (212) 719-4380. Web Site: www.workedit.com.

Dalton Helms, owner; Ken Sackheim, owner.

TV-CATV only.

Post production svcs, DVD authoring.

World Events Productions Ltd., One Memorial Dr., St. Louis, MO, 63102. Phone: (314) 345-1000. Fax: (314) 345-1091. E-mail: wep@wep.com Web Site: www.wep.com.

Edward J. Koplar, pres; Tiffany Ilardi, mgng dir.

TV-CATV only.

TV program production & distribution.

World Radio Network, Box 1212, London, SW8 2ZF. United Kingdom. Phone: 44-20-7896-9000. Fax: 44-20-7896-9007. E-mail: contactus@wrn.org Web Site: www.wrn.org.

Karl Miosga, chmn; Jeff Cohen, dev dir; Gary Edgerton, mgng dir.

WRN Network One (via Galaxy 5) news & features ch, comprises live progmg segments in English & languages from more than 20 international bcstrs. Also supplies many customized progmg feeds to radio stns.

Worldview Entertainment Inc., The Killiam Collection. 145 W. 55th St., Suite 7-D, New York, NY, 10019. Phone: (212) 582-6997. Fax: (212) 925-2314. E-mail: birnhardt@aol.com

Sandra J. Birnhak, CEO; Glenn E. Shealey, pres.

TV-CATV only.

Archival stock footage library, distribution to international bcstrs.

Worldvision NY, 143 W. 29th St., New York, NY, 10001. Phone: (212) 736-2997. Fax: (212) 736-9755. Web Site: www.worldvision.org.

John Claus, exec dir.

London SW1X OAE, Worldvision Enterprises U.K. Ltd, 54 Pont St. Phone: 011-441-71-584-5357. Fax: 011-441-71-581-3483. Bill Peck, Janice Wilson, Zsuzsanna Jung.

Paris 75008, Worldvision Enterprises S.A.R.L, 28, Rue Bayard. Phone: 011-33-1-4723-3995. Fax: 011-33-1-4070-9269. Mary Jane Fourniel, Catherine Molinier, John Hernan.

Rio de Janeiro CEP 22270, 22270 Rua Voluntarios Da Patria N, Gr.604. Phone: 011-55-21-539-2992. Fax: 011-55-21-266-4737. Raymundo Rodriguez, Maria Alice Freire.

Rome 00186, Adalia Anstalt, Via del Corso, 22/Int 10. Phone: 011-39-6-322-5190. Fax: 011-39-6-322-6450. Michael Kiwe, Dorothy Shaw.

Sydney, Milsons Point 02061. Worldvision Enterprises of Australia PTY Ltd., 5-13 Northcliff St. Phone: 011-61-2-9922-4722. Fax: 011-61-2-9955-8207. TWX: (790) 70474. Brian Rhys-Jones, Paul Stuart, Karen Zylstra.

Tokyo 104, Tsukiji Hamarikyu Bldg, 7th Fl, 5-3-3 Tsukiji, Chou-ku. Phone: 011-81-3-3545-3978. Fax: 011-81-3-5550-8316. Mie Horasawa, Yukie Kumagai.

Toronto, ON M5R 2A5 Canada, Worldvision Enterprises of Canada, 1200 Bay St, Suite 802. Phone: (416) 967-1200. Fax: (416) 967-0521. Bruce Swanson, Kathy Fraser.

Los Angeles, CA 90036, 5700 Wilshire Blvd, 5th Fl. Phone: (213) 965-5910. Fax: (213) 965-5915. David McNaney, VP western division.

Coconut Grove, FL 33133. Tele-UNO, Grand Bay Plaza, 2665 S. Bayshore Dr. Phone: (305) 285-4307. Fax: (305) 285-4308. Hilary Hattler, John McDonald.

Atlanta, GA 30346. Worldvision Enterprises Inc. Latin America, 400 Perimeter Ctr. Terr, Suite 185. Phone: (770) 394-3967. Fax: (770) 394-9002. Mary Ann Pasante, Leticia Estrada, Carla Araya.

Atlanta, GA 30346, 400 Perimeter Center Terr, Suite 150. Phone: (404) 394-3967. Fax: (404) 394-9002. John Barrett, VP southern division.

Chicago, IL 60610, 515 N. State St, Suite 2305. Phone: (312) 527-0461. Fax: (312) 527-0688. Tony Bauer, VP central division.

New York, NY 10019, 1700 Broadway. Phone: (212) 261-2700. Fax: (212) 261-2724. Bill Baffi, VP cable & new technologies; Frank L. Browne, VP eastern division.

TV-CATV only.

TV program distribution for ind productions.

The Worship Network, 28059 U.S. Hwy. 19 N., Suite 300, Clearwater, FL, 33761. Phone: (727) 536-0036. Fax: (727) 530-0671. E-mail: ken@worship.net Web Site: www.worship.net.

Bruce Koblish, pres/CEO; Bob Shreffler, VP finance; Tim Brown, VP.

TV-CATV only.

Inspirational music set to nature scenes, overlaid with scripture 24 hours a day. Progmg is interspersed with short devotional teachings.

Larry John Wright Inc., 1045 E. University Dr., Suite 1, Mesa, AZ, 85203. Phone: (480) 833-8111. Fax: (480) 969-2895. E-mail: jessica@ljohnw.com Web Site: www.larryjohnwright.com.

Larry F. John, CEO; John N. Wright, pres.

TV-CATV only.

Own & operate production studios; produce film & video comls, radio comls & jingles, TV shows & industrial videos.

The Wyland Group, 101 W. Cochran St., Simi Valley, CA, 93065. Phone: (805) 955-7680. Fax: (805) 522-1082. Web Site: www.lifestyle.org.

Chauncey Smith, account exec; Linda Walter, dir.

TV-CATV only.

Production of health related, family values progmg.

X

XL Media Solutions, 110 N. Ditmar St., Oceanside, CA, 92054. Phone: (760) 722-8284. Fax: (888) 722-8234. E-mail: staff@exxelaudio.com Web Site: www.exxelaudio.com.

William Kottcamp, mgr.

Radio Only.

Radio program, coml production & distribution.

Y

Yada/Levine Video Productions, 606 N. Larchmont Blvd., Suite 100, Los Angeles, CA, 90004. Phone: (323) 461-1616. Fax: (323) 461-2288. E-mail: video@yadalevine.com Web Site: www.yadalevine.com.

Michael Yada, pres/CEO.

TV-CATV only.

Betacam SP, Digital Betacam and HD production. Camera crews with equipment package. Avid editing.

Yale Video Inc., 2441 W. La Palma Ave., Suite 530, Anaheim, CA, 92801. Phone: (714) 693-5300. Fax: (714) 693-5395. E-mail: burty@webcastingtv.com Web Site: www.webcastingtv.com.

Burton A. Yale, CEO.

Offers the latest in editing technology; from D-2 to Hi8.

Yorkshire Television, (A division of Granada Media Group). Television Centre, 104 Kirkstall Rd., Leeds, LS3 IJS. United Kingdom. Phone: 0113-243-8283. Fax: 0113-244-5107. E-mail: communications@granadamedia.com Web Site: www.granada.co.uk.

David M.B. Croft; Charles Allen CBE, chmn.

TV-CATV only.

Independent TV program maker & bcstr.

Z

ZBS Foundation, 174 N. River Rd., Fort Edward, NY, 12828. Phone: (518) 695-6406. Fax: (518) 695-4041. E-mail: info@zbs.org Web Site: www.zbs.org.

Thomas Lopez, pres.

Radio Only.

Producer of audio drama.

Zachry Associates, 500 Chestnut, Suite 2000, Abilene, TX, 79602. Phone: (325) 677-1342. Fax: (325) 672-2001.

H.C. Zachry, pres.

TV-CATV only.

Produces & distributes TV & radio programs & comls as well as promotional films & production svcs.

Sandy Zimmerman Productions, 4800 Black Bear Rd., Suite 204, Las Vegas, NV, 89149. Phone: (702) 731-6491. E-mail: sandyzimm@go.com

Sandy Zimmerman, producer & owner; Robert Gonzales, production mgr.

TV-CATV only.

Develops, produces & distributes TV programs, documentaries, infomercials, travel specials, TV comls, & industrial & corporate videos. Syndicates one-two-five -minute program fillers.

Producers, Distributors, Production Services Subject Index

3-D Films

Blue Sky Studios
Dimension 3 Corp
WeatherVision Inc.

3-D TV Systems

Dimension 3 Corp
Pro Video

Agricultural Programming, Radio

Agrinet Farm Radio Network
American Farm Bureau Inc.
Clear Channel Broadcasting Inc.
Cookie Jar Group
Leadem to Water Production Inc.
Montgomery Community Television Inc.
North Shore Productions
Tribune Entertainment Co.
Tribune Radio Networks

Animation

Angel Films Co.
The Kay Arnold Group
Bardel Entertainment Inc.
Blue Sky Studios
Buzzco Associates Inc.
Central Park Media
CineGroupe
Cinema Concepts Animation Studio
Classic Media
Clayton-Davis & Associates Inc.
Crawford Communications
Dargaud-Marina
Walt Disney Company
Enoki Films U.S.A. Inc.
Film House Inc.
Hearst Entertainment, Inc.
Italtoons Corp.
JC Productions Inc.
John Lemmon Films
Magno Sound & Video
Mar Vista Entertainment
The Media Group of Connecticut Inc.
Bill Melendez Productions Inc.
Mondo TV
North by Northwest Productions
Northwest Imaging & FX
PPM Multimedia
Pantomime Pictures Inc.
Pike Productions Inc.
Playhouse Pictures
PorchLight Entertainment Inc.
PostWorks, New York
The Program Exchange
Rampion Visual Productions L.L.C.
Romano & Associates Inc.
Edward Sarson Productions
Silverline Pictures
Southern STAR
The Summit Media Group Inc. (Sub 4 Kids Entertainment)
Sunbow Entertainment
Two Oceans Entertainment Group
Videographic West
Warner Bros. Animation
WeatherVision Inc.
Fred Wolf Films
World Events Productions Ltd.

Audio Production

American TelNet
Armedia Communications
Broadcast News Service
Himan Brown-Radio Drama Network
CA Media Development
CCM Media Services
C N R Radio
CRN International
Clayton-Davis & Associates Inc.
Command Productions
Continental Recordings Inc.
Creative Marketing & Communications Corp.
Jeff Davis Productions Inc.
DaviSound
Digital Force
Domain Communications L.L.C.
John Driscoll/VoiceOver America
Ecumenical Communications
Finger Lakes Productions International
GMI Media L.L.C.
Good Life Associates
Good News Broadcasting Association Inc.
Heil Enterprises
Horizon Audio Creations
Host Communications Inc.
Irving Productions Inc.
J.N. Productions
The Johnson Group
KJD Teleproductions
David Kaye Productions Inc.
Lion and Fox Recording Studios
Man From Mars Productions
Manhattan Transfer Miami
Ben Manilla Productions
William Mauldin Productions Inc.
MediaTracks Inc.
Metro Networks
MotorNet
N W Media
National Public Radio
New Dimensions Radio
No Soap Productions
North Star Music
Dick Orkin's Amazing Radio
Paulist Media Works
Perception Media Group
Presbyterian Church (U.S.A.)
Protestant Hour Inc.
Radio Spirits
Shield Productions Inc.
Smith/Lee Productions, Inc.
Sound of Birmingham Productions
Soundshop Recording Studio LLC
Soundtrack
Studio M Productions Unlimited
Suite Audio
TR Productions
Talco Productions
Technisonic Studios
Thompson Creative
University of Colorado Television
University of Detroit Mercy
University of Kentucky Public Relations & Radio-TV News Bureau
VideoActive Productions
WQXR
ZBS Foundation

Audio Production Library

American TelNet
Blue Heaven Productions
Himan Brown-Radio Drama Network
Clayton-Davis & Associates Inc.
Ghostwriters/Radio Mall
Groove Addicts Production Music Catalog
J&H Music Programming
OGM Production Music
Radio America
Sound Ideas
TM Century Inc.

Audio Recording Services

American TelNet
The Audio Department Inc.
Bruder Releasing Inc. (BRI)
C 2 Productions Inc.
Cinema Sound Ltd.
Command Productions
Continental Recordings Inc.
Jeff Davis Productions Inc.
Domain Communications L.L.C.
John Driscoll/VoiceOver America
Horizon Audio Creations
The Image Generators
Irving Productions Inc.
Tom Jones Recording Studios
Lion and Fox Recording Studios
MVI Post
Matchframe Video
Media Access Group at WGBH
Metro Networks
MotorNet
New Dimensions Radio
No Soap Productions
Omnimusic
Paulist Media Works
Protestant Hour Inc.
Radio Production Services Inc.
Radio Spirits
Reizner & Reizner Film & Video
Rex Post
Sound of Birmingham Productions
Soundshop Recording Studio LLC
Studio M Productions Unlimited
Suite Audio
UBC Radio
VideoActive Productions
WQXR
WRS/Channel One
Washington Korean Broadcasting Co.

Audio/Visual Services

American TelNet
Capital Communications
Coote Communications
DG Systems
Great Plains National (GPN)
Host Communications Inc.
Kipany Productions Ltd.
Modern Sound Pictures Inc.
Reliance Audio Visual Corp.
TR Productions
Talco Productions
Videosmith Inc.
Wade Productions Inc.

Background Music

American TelNet
FirstCom Music
Groove Addicts Production Music Catalog
Horizon Audio Creations
Joe Jones Productions
Message on Hold
Muzak
OGM Production Music
Omnimusic
Promusic
RPM-Radio Programming & Management Inc.
Sound Ideas
Westar Music

Camera Operators

J. Arnold Productions
Asia Pacific Productions
Bell Foto Art Productions
CamMate Studios
D-V-X International
Daley Video
FTC/Orlando
Maslow Media Group Inc.
Nemo News
PACSAT
PMTV Producers Management Television
Reizner & Reizner Film & Video
Tankersley Productions, Inc., St. Louis
Telepros
Videosmith Inc.
WKMG Productions
WTOB Channel 2—Public/Government Access TV
White Rabbit Productions
Yada/Levine Video Productions

Cassette Duplicating

Command Productions
Continental Recordings Inc.
Domain Communications L.L.C.
Jeff Gold Productions Inc.
Man From Mars Productions
N W Media

CD Production Library

Children's Programming, Radio

Children's Programming, Radio & TV

Children's Programming, TV

Commercial Distribution, Radio

Commercial Distribution, Radio & TV

Commercial Distribution, TV

Commercial Production, Radio

Commercial Production, Radio & TV

Commercial Production, TV

Jordan Klein Film & Video
The Kenwood Group
Lapco Communications
MAKWDE Productions
MRC Films
Mason Video
Maysles Films, Inc.
McClain Enterprises Inc.
Midwest Video Communications Inc.
Warren Miller Entertainment
Mobile Video Services Ltd.
Pantomime Pictures Inc.
Peckham Productions
Pied Piper Films Ltd.
Playhouse Pictures
Prime Cut Productions Inc.
Producers Group, Ltd.
Reuters Television
River City Video Productions
Rosler Creative
ROZON
Sanctuary Records Group Ltd.
Sport International Inc.
Sullivan Video Services Inc.
Technisonic Studios
Tel-A-Cast Productions
Triage Entertainment Inc.
Video Techniques Inc.
Videosmith Inc.
WKMG Productions
Warner Bros. Animation
Welwood International Film Production
White Rabbit Productions
Fred Wolf Films
Larry John Wright Inc.
Yada/Levine Video Productions

Computer Graphics

CA Media Development
Center City Film & Video
CineGroupe
Crest National Digital Media Complex
Daley Video
Darino Films/Library of Special Effects
ETN—Educational Telecommunications Network
Film House Inc.
Henninger Media Services, Inc.
The History Makers
JC Productions Inc.
Kipany Productions Ltd.
Magno Sound & Video
Warren Miller Entertainment
Northwest Imaging & FX
PACSAT
Palace Digital Studios
Pike Productions Inc.
Playhouse Pictures
PostWorks, New York
RMD & Assoc. Inc.
Rampion Visual Productions L.L.C.
Rex Post
TR Productions
Traffic Pulse Networks (A Unit of Mobility Technologies)
VA-Tech Video/Broadcast Services
WRS/Channel One

Creative Services

Aloha Productions
American TelNet
Ascent Media Network Services
Bayliss
The Chuck Blore Co.
Campbell-Ewald Advertising
Center City Film & Video
Cimarron Group
ComBridges
DWJ Television
Jeff Davis Productions Inc.
Devlin Design Group Inc.
ESPI Video
ETN—Educational Telecommunications Network
Ebbets Field Productions Ltd.
Faraone Communications Inc.
Film House Inc.
First Marketing
Fox 29 WUTV Sinclair
Jeff Gold Productions Inc.
JC Productions Inc.
The Johnson Group
The Kenwood Group
Kipany Productions Ltd.
Klein &
Make It Happen Productions Inc.
N W Media
No Soap Productions
Oppix Productions Inc.
Peters Communications
Praxis Media Inc.
RMD & Assoc. Inc.
Radioguide People Inc.
Rosler Creative
Sak Entertainment
Edward Sarson Productions
SeniorVision Productions, Inc.
Strand Media Group Inc.
Talco Productions
TeleCom Productions Inc.
Video I-D Teleproductions Inc.
Wade Productions Inc.
Warren Only Media Group
Larry John Wright Inc.
Sandy Zimmerman Productions

Development, Films

Ardustry Home Entertainment LLC
CIFEX International Inc.
Curb Entertainment International Corp.
The Walt Disney Company
The Dolmatch Group Ltd.
ETN—Educational Telecommunications Network
Ebbets Field Productions Ltd.
Faraone Communications Inc.
1st Miracle Productions
JC Productions Inc.
Lions Gate Entertainment
Make It Happen Productions Inc.
Richter Productions Inc.
Rosler Creative
Studio Babelsberg GmbH
Tamouz Media
Triage Entertainment Inc.

Development, Films, TV Series & Video

Angel Films Co.
Burrud Productions Inc.
CABLEready Corp.
CDR Communications Inc.
Celebrities Productions
Children's Media Productions
Clayton-Davis & Associates Inc.
ComBridges
Critical Mass Releasing Inc.
DLT Entertainment Ltd.
Devillier Donegan Enterprises L.P.
Duke International
ETN—Educational Telecommunications Network
Film Roman Inc.
Jeff Gold Productions Inc.
Independent Edge Films
Make It Happen Productions Inc.
Oasis International
Pied Piper Films Ltd.
Producers Group, Ltd.
Reel Media International Inc.
Sak Entertainment
SeniorVision Productions, Inc.
Silverline Pictures
Two Oceans Entertainment Group
Warren Only Media Group

Development, TV Films, Series

Dave Bell Associates Inc.
The Chuck Blore Co.
Canamedia Productions Ltd.
CanLib Inc.
Walt Disney Company
The Walt Disney Company
The Dolmatch Group Ltd.
ETN—Educational Telecommunications Network
Ebbets Field Productions Ltd.
Essence Television Productions Inc.
Eurocine
1st Miracle Productions
Freewheelin' Films Ltd.
The Fremantle Corp.
FremantleMedia North America Inc.
Glenray Productions Inc.
The Griffin Group
Thomas Horton Associates Inc.
JC Productions Inc.
Nicole Jouve
MacNeil/Lehrer Productions
Make It Happen Productions Inc.
New Line Television
Pantomime Pictures Inc.
Planet Pictures Ltd.
Questar
Reuters Media
Richter Productions Inc.
Peter Rodgers Organization
SPI International
Sullivan Entertainment Inc./Sullivan Entertainment International
Sunbow Entertainment
Tamouz Media
Tapestry International, Ltd.
Triage Entertainment Inc.
Twentieth Television
Warner Bros. Animation

Development, Video

AmericaNurse TV Productions
Clayton-Davis & Associates Inc.
ComBridges
The Walt Disney Company
The Dolmatch Group Ltd.
ETN—Educational Telecommunications Network
Essence Television Productions Inc.
Eurocine
Glenray Productions Inc.
Global Telemedia Inc.
The Griffin Group
JC Productions Inc.
Kipany Productions Ltd.
Lions Gate Entertainment
Make It Happen Productions Inc.
Midwest Video Communications Inc.
Questar
Richter Productions Inc.
Tapestry International, Ltd.
Triage Entertainment Inc.
VIEW Video Inc.

Distribution, Audio

DaviSound
Digital Force
Domain Communications L.L.C.
Good Life Associates
The Image Generators
Irving Productions Inc.
North Star Music
Dick Orkin's Amazing Radio
Paulist Media Works

Distribution, Cable

ATA Trading Corp.
Academy Entertainment
The Kay Arnold Group
Bruder Releasing Inc. (BRI)
Castle Hill Productions Inc.
Central Park Media
DLT Entertainment Ltd.
The Walt Disney Company
FamilyNet
GLL TV Enterprises Inc.
Galavision
Granada America
Great Plains National (GPN)
Grove Television Enterprises Inc.
HAVE Inc.
Janson Media
Kazmark Entertainment Group
Loral Skynet
Mar Vista Entertainment
Media Planning Group (MPG)
Modern Entertainment
Moonstone Entertainment
New City Releasing Inc.
O. Atlas Enterprises Inc.
Oasis International
Playboy Entertainment Group Inc.
Powersports/Millenium International
Rigel Entertainment
Peter Rodgers Organization
Harvey Sheldon Productions
Video-Cinema Films Inc.
Worldview Entertainment Inc.

Distribution, Cartoons

ATA Trading Corp.
Academy Entertainment
Central Park Media
CineGroupe
Classic Media
Dargaud-Marina
Walt Disney Company
The Dolmatch Group Ltd.
Film Roman Inc.
The Fremantle Corp.
Larry Harmon Pictures Corp.
Italtoons Corp.
Mondo TV
Pan American Video
The Summit Media Group Inc. (Sub 4 Kids Entertainment)
World Events Productions Ltd.

Distribution, Film and Video

ADM—International Film & TV Distribution
ATA Trading Corp.
Academy Entertainment
Accuracy in Media Inc.
Angel Films Co.
Ardustry Home Entertainment LLC
Arkadia Entertainment Corp.
Ascent Media Network Services
Ascent Media Network Services
CDR Communications Inc.
CIFEX International Inc.
CRM Learning
Castle Hill Productions Inc.
Children's Media Productions
Cinema Concepts Animation Studio
The Cinema Guild Inc.
Crystal Pictures Inc.
Cube International
Curb Entertainment International Corp.
Donnelly & Associates
Duke International
Enoki Films U.S.A. Inc.
Filmoption International Inc.
Films Media Group
Films of the Nations
David Finch Distribution Ltd.
1st Miracle Productions
First Run/Icarus Films
Free Speech TV (FSTV)
Glenray Productions Inc.
Granada America
Great Plains National (GPN)
Grove Television Enterprises Inc.
Alfred Haber Distribution Inc.
Independent Edge Films
International Tele-Film
JGT Media Productions
Nicole Jouve
Kazmark Entertainment Group
Kultur International Films
Lions Gate Entertainment
Mar Vista Entertainment
Maryknoll Productions
Maysles Films, Inc.
Modern Entertainment
Modern Sound Pictures Inc.
Moonstone Entertainment
Myriad Pictures
New City Releasing Inc.
New Line Television
O. Atlas Enterprises Inc.
PAULAR Entertainment L.L.C.
PPM Multimedia
Pike Productions Inc.
Planet Pictures Ltd.
Playboy Entertainment Group Inc.
Producers Group, Ltd.
Public Media Incorporated
RMD & Assoc. Inc.
Reel Media International Inc.
Richter Productions Inc.
Rigel Entertainment
Peter Rodgers Organization
Sanctuary Records Group Ltd.
SeniorVision Productions, Inc.
Seville Pictures
Silverline Pictures
Southern STAR
Marty Stouffer Productions Ltd.
M C Stuart & Associates Pty Ltd.
Television Representatives Inc.
Trident Releasing
Video/Media Distribution Inc.
VIEW Video Inc.
Worldview Entertainment Inc.

Distribution, Music

ADM—International Film & TV Distribution
Arkadia Entertainment Corp.
Sandra Carter Global, Inc.
Cube International
Joe Jones Productions
Kultur International Films
Loral Skynet
North Star Music

Distribution, Radio & TV Programming

Accuracy in Media Inc.
AmericaNurse TV Productions
CDR Communications Inc.
Catholic Communication Campaign
Christian Children's Associates Inc.
The Christophers Inc.
The Crime Channel
DG Systems
Evangelical Lutheran Church in America
Evergreen Entertainment Group
FamilyNet
Gordon Productions
International Television Broadcasting Inc.
Medialink
National Council of Churches Communications Unit
New Visions Syndication Inc.
News Broadcast Network
Outdoor Media Group
Prairie Dog Entertainment
RBC Ministries/Midwest Media Managers
The Radio-Studio Network
Viacom Video Services
Warren Only Media Group
The Worship Network

Distribution, Radio Programming

Agrinet Farm Radio Network
At a Glance
The Classical Station, WCPE
Clear Channel Broadcasting Inc.
Country Crossroads
The Walt Disney Company
Domain Communications L.L.C.
Good Life Associates
Happi Associates
It Is Written Television
J&H Music Programming
Jameson Broadcast Inc.
The Johnson Group
Jones Radio Networks
Knowledge In A Nutshell Inc.
Leadem to Water Production Inc.
Longhorn Radio Network
Loral Skynet
MAKWDE Productions
MRN Radio
MasterControl FamilyNet Radio
Media Planning Group (MPG)
Metro Networks
Moody Broadcasting Network
MotorNet
Musical Starstreams
National Public Radio
New Dimensions Radio
North American Network, Inc.
North Shore Productions
On Track
Ports of Paradise
Powerline
Premiere Radio Networks Inc.
Protestant Hour Inc.
RPM-Radio Programming & Management Inc.
Radio Canada International/Canadian Broadcasting Corp.
Radio Express Inc.
Radio Spirits
Ray Sports Network
Solid Gospel Network (Reach Satellite Network, Inc.)
Sound Source Networks
Sports Byline U.S.A.
Talk America Radio Networks
Talk Radio Network
UBC Radio
United Stations Radio Network
WFMT Radio Network
Washington Korean Broadcasting Co.

Distribution, TV Programming

ADM—International Film & TV Distribution
ATA Trading Corp.
Academy Entertainment
American Public Television
The Kay Arnold Group
Ascent Media Management East
Ascent Media Network Services
Bellon Entertainment
CABLEready Corp.
CBS Paramount International Television
CBS Television Distribution
CCI Entertainment Ltd.
CNBC Syndication
CS Associates
CTV Television Inc.
Cable Films & Video
Canamedia Productions Ltd.
CanLib Inc.
Carleton Productions International Inc.
Carsey-Werner Distribution
Sandra Carter Global, Inc.
Castle Hill Productions Inc.
Central Park Media
Chrysalis Distribution
Classic Media
Clever Cleaver Productions
Concept Videos
William F. Cooke Television Programs
The Crime Channel
Critical Mass Releasing Inc.
Cube International
DLT Entertainment Ltd.
Dargaud-Marina
Devillier Donegan Enterprises L.P.
Walt Disney Company
The Walt Disney Company
The Dolmatch Group Ltd.
Donnelly & Associates
Dorling Kindersley Vision
Duke International
Eaton Films Ltd.
Ellis Entertainment
Europe Images International
FamilyNet
Film Roman Inc.
Filmoption International Inc.
1st Miracle Productions
First Run/Icarus Films
Fox Sports West
Free Speech TV (FSTV)
The Fremantle Corp.
GRB Entertainment
Galavision
Global Entertainment Media
Granada America
Grove Television Enterprises Inc.
Alfred Haber Distribution Inc.
Harmony Gold U.S.A. Inc.
Hearst Entertainment, Inc.
Henninger Media Services, Inc.
Holigan Investment Group Ltd.
Hope Channel
Thomas Horton Associates Inc.
International Tele-Film
International Television Corp.
It Is Written Television
Italtoons Corp.
Ivanhoe Broadcast News Inc.
Janson Media
Jefferson-Pilot Sports
Nicole Jouve
Juravic Entertainment
Kazmark Entertainment Group
Knowledge In A Nutshell Inc.
Kultur International Films
Lakeside TV Co.
Lambert Television
Litton Syndications Inc.
Loral Skynet
MAKWDE Productions
MGC The Multimedia Group of Canada
MGM TV Canada
MG/Perin Inc.
Mar Vista Entertainment
Media Planning Group (MPG)
Metro Networks
Mobile Video Services Ltd.
Modern Entertainment

Dubbing Services

Duplication Services

Editing Services

Educational Programming, Radio

Educational Programming, Radio & TV

Educational Programming, TV

Entertainment Programming, Radio

ABC Radio Networks
Agrinet Farm Radio Network
Broadcast Programming
C N R Radio
CRN International
The Dick Clark Productions
Creative Marketing & Communications Corp.
DC Audio
The Walt Disney Company
Ecumenical Communications
Essence Television Productions Inc.
Jameson Broadcast Inc.
Jones Radio Networks
Launch Radio Networks
Leadem to Water Production Inc.
Ben Manilla Productions
MasterControl FamilyNet Radio
Media Planning Group (MPG)
North American Network, Inc.
On Track
Powerline
Prairie Dog Entertainment
Premiere Radio Networks Inc.
Radio America
Radio Spirits
Sam Shad Productions
StarDate/Universo
Talk Radio Network
UBC Radio
United Stations Radio Network
WCTD AM 1620
"The Weather Center"
Larry John Wright Inc.
ZBS Foundation

Entertainment Programming, Radio & TV

Arkadia Entertainment Corp.
The Chuck Blore Co.
Arthur Henley Productions
KJD Teleproductions
Bill Melendez Productions Inc.
New Visions Syndication Inc.
Jim Owens Entertainment
SCOLA
Ukrainian Melody Hour
University of Colorado Television
WQED Multimedia
WWE Entertainment Inc.

Entertainment Programming, TV

ATA Trading Corp.
America One Television Network
J. Arnold Productions
Burrud Productions Inc.
Buzzco Associates Inc.
CBS Television Distribution
CN8, The Comcast Network
Cable Films & Video
CanLib Inc.
Carsey-Werner Distribution
Sandra Carter Global, Inc.
Central City Productions, Inc.
Chrysalis Distribution
The Dick Clark Productions
Clear Channel Broadcasting Inc.
Clever Cleaver Productions
Country Crossroads
The Crime Channel
Crystal Pictures Inc.
Cube International
The Walt Disney Company
E! Entertainment Television
Ellis Entertainment
Essence Television Productions Inc.
Evergreen Entertainment Group
First Run/Icarus Films
GLL TV Enterprises Inc.
GRB Entertainment
Glenray Productions Inc.
Global Entertainment Media
Global Telemedia Inc.
Gould Entertainment Corp.
Great North Productions
The Griffin Group
Alfred Haber Distribution Inc.
Hearst Entertainment, Inc.
International Tele-Film
International Television Corp.
Janson Media
Jefferson-Pilot Sports
Juravic Entertainment
KPTS-TV
KUSA Television
Kultur International Films
Lakeside TV Co.
Lapco Communications
Litton Syndications Inc.
MGC The Multimedia Group of Canada
Media Planning Group (MPG)
Myriad Pictures
NTN Communications Inc.
New Zoo Revue
Planet Pictures Ltd.
PorchLight Entertainment Inc.
Powersports/Millenium International
Prime Cut Productions Inc.
Promark Television
Questar
Raycom Sports
The Real Estate Network
Reid/Land Productions Inc.
ROZON
Sam Shad Productions
Harvey Sheldon Productions
Sullivan Entertainment Inc./Sullivan Entertainment International
Tel-Air Interests Inc.
Telepros
The Television Syndication Company, Inc.
Tomwil Inc.
Triage Entertainment Inc.
Turner Entertainment Co.
Twentieth Television
Viacom Inc.
Video-Cinema Films Inc.
Videographic West
Vision Broadcasting - KVBA TV 19
Warner Bros.
Warner Bros. Animation
Wide Eye Productions, Inc.
Daniel Wilson Productions Inc.
World Events Productions Ltd.
The Wyland Group
Yorkshire Television

Film and Tape Transfers (Film-to-Tape)

Ascent Media Management East
Carpel Video Inc.
Crawford Communications
Crest National Digital Media Complex
Highland Laboratories
MPL Media
Magno Sound & Video
Manhattan Transfer Miami
Matchframe Video
Northwest Imaging & FX
Point 360
PostWorks, New York
Quality Film & Video
The Transfer Zone
VTTV Videothek Electronic TV-Production GmbH + Co Kopier KG
Van Vliet Media

Film Laboratories

Crawford Communications
Crest National Digital Media Complex
DeLuxe Laboratories
Forde Motion Picture Labs
Henninger Media Services, Inc.
MPL Media
Media Visions
Point 360
Studio Babelsberg GmbH

Film Preservation/Restoration

Worldview Entertainment Inc.

Graphic Effects Library

Darino Films/Library of Special Effects
WeatherVision Inc.

Graphics

Cimarron Group
Devlin Design Group Inc.
MVI Post
Manhattan Transfer Miami
Matchframe Video
Peters Communications
Pro Video
Radioguide People Inc.
River City Video Productions
Technisonic Studios
TRI-COMM Productions

Industrial Films

CRM Learning
DWJ Television
ESPI Video
Encore Video Productions Inc.
The Kenwood Group
Limelight Communications Inc.
Lindberg Productions Inc.
MRC Films
Make It Happen Productions Inc.
Pantomime Pictures Inc.
Peckham Productions
Primedia Workplace Learning
Tel-Air Interests Inc.
Telepros
Video/Media Distribution Inc.
Wade Productions Inc.
Work Edit

Inflight Audio Progamming

Horizon Audio Creations
Ports of Paradise
Prairie Dog Entertainment
RDF Media
Time Capsule Inc.

Interactive Television

Ascent Media Network Services
NTN Communications Inc.
Parrot Communications International Inc.
PostWorks, New York
The Real Estate Network
SPI International
Video I-D Teleproductions Inc.

Interactive Television Programming

Eurocine
Ivanhoe Broadcast News Inc.
NTN Communications Inc.
Phoebus Communications Inc.
Presson Perspectives
Producers Group, Ltd.
The Real Estate Network
Reuters Media
Sunbow Entertainment
University of Colorado Television

Jingles

Aloha Productions
Blue Heaven Productions
CA Media Development
Continental Recordings Inc.
GMI Media L.L.C.
Groove Addicts
JAM Creative Productions Inc.
Joe Jones Productions
MAKWDE Productions
Network Music
Oppix Productions Inc.
Radio Express Inc.
Radio Production Services Inc.
Shield Productions Inc.
Sound of Birmingham Productions
TM Century Inc.
Thompson Creative

Libraries, Film

Getty Images
Granada America
Grinberg Film Libraries Inc.
Grove Television Enterprises Inc.
MGM TV Canada
Modern Entertainment
Modern Sound Pictures Inc.
Pan American Video
Reel Media International Inc.
Peter Rodgers Organization
Televix Entertainment Inc.
Turner Entertainment Co.
Video-Cinema Films Inc.
The WPA Film Library
Worldview Entertainment Inc.

Libraries, TV

ADM—International Film & TV Distribution
CBS Paramount International Television
CBS Television Distribution
CelebrityFootage
Cube International
Film Roman Inc.
Getty Images
Granada America
Grinberg Film Libraries Inc.
Grove Television Enterprises Inc.
Hearst Entertainment, Inc.
Modern Entertainment
NRS Group PTY Ltd.
New Films International
New Zoo Revue
Pan American Video
Sanctuary Records Group Ltd.
Televix Entertainment Inc.
Turner Entertainment Co.
Twentieth Century Fox Television Distribution
Viacom Video Services
WWE Entertainment Inc.

Libraries, Video

CelebrityFootage
Children's Media Productions
The Dick Clark Productions
First Run/Icarus Films
Getty Images
Grinberg Film Libraries Inc.
Grove Television Enterprises Inc.
The Idea Channel
National Collegiate Athletic Association
New Zoo Revue
Public Media Incorporated
Reuters Television
River City Video Productions
WWE Entertainment Inc.

Licensing Services

Classic Media
Film Roman Inc.
Getty Images
Grinberg Film Libraries Inc.
Larry Harmon Pictures Corp.
Harmony Gold U.S.A. Inc.
Southern STAR
The Summit Media Group Inc. (Sub 4 Kids Entertainment)
Sunbow Entertainment
TRF Production Music Libraries
Televix Entertainment Inc.
Turner Entertainment Co.
WWE Entertainment Inc.

Location Services

American Stock Exchange
CompuWeather Inc.
Georgia Film, Video & Music Office
Lion and Fox Recording Studios
Maslow Media Group Inc.
U.S. Plan B Inc.

Medical Programming, Radio

AmericaNurse TV Productions
Creative Marketing & Communications Corp.
Eagle Media Productions Ltd.
Essence Television Productions Inc.
Jameson Broadcast Inc.
MSE
MasterControl FamilyNet Radio
MediaTracks Inc.
Powerline
RPM-Radio Programming & Management Inc.
Radio Production Services Inc.
The Saturday Evening Post Television Department
Talk America Radio Networks
Talk Radio Network

Medical Programming, Radio & TV

AmericaNurse TV Productions
CDR Communications Inc.
Faith for Today
Arthur Henley Productions
Lifestyle Magazine/The Evidence
MSE
Presson Perspectives
Radio & TV Roundup Productions

Medical Programming, TV

J. Arnold Productions
Bell Foto Art Productions
CABLEready Corp.
The D.L. Dykes Jr. Foundation
Hamilton Productions Inc.
Ivanhoe Broadcast News Inc.
Limelight Communications Inc.
Mason Video
Medstar Television Inc.
The Saturday Evening Post Television Department
The Wyland Group
Sandy Zimmerman Productions

Mobile Production Units

CMT
Cinecraft Productions Inc.
Dome Productions
KUSA Television
Nemo News
PMTV Producers Management Television
(PSSI) Production & Satellite Services Inc.
Telenium Studios
Telfax Inc.
VTTV Videothek Electronic TV-Production GmbH + Co Kopier KG
Videosmith Inc.
WKMG Productions

Music and Sound Effects

Blue Heaven Productions
FirstCom Music
GMI Media L.L.C.
Ghostwriters/Radio Mall
Groove Addicts Production Music Catalog
Manhattan Production Music
Megatrax Production Music Inc.
OGM Production Music
Omnimusic
Production Garden Music Libraries
Promusic
SoperSound Music Library
Sound Ideas
TRF Production Music Libraries
Westar Music

Music Composition

Carriage House Studios
Tim Cissell Music
Groove Addicts
Megatrax Production Music Inc.
Metro Music Productions Inc.
Peters Communications
SoperSound Music Library
Valentino Music & Sound Effect Libraries
Wax Music, Sound Design & Mix

Music Libraries

Alternative Programming
Broadcast Programming
FirstCom Music
Ghostwriters/Radio Mall
Groove Addicts Production Music Catalog
Halland Broadcast Services
Killer Tracks
MVI Post
Megatrax Production Music Inc.
Metro Music Productions Inc.
Network Music
New Zoo Revue
OGM Production Music
Omnimusic
Production Garden Music Libraries
Promusic
RPM Media Enterprises-The Relic Rack Review
RPM-Radio Programming & Management Inc.
Radio & TV Roundup Productions
Radio Express Inc.
Smith/Lee Productions, Inc.
Sound Ideas
TM Century Inc.
TRF Production Music Libraries
Valentino Music & Sound Effect Libraries
Westar Music

Music Lyrics

Wax Music, Sound Design & Mix

Music Production

Aloha Productions
Arkadia Entertainment Corp.
Armedia Communications
Carriage House Studios
Tim Cissell Music
Digital Force
Groove Addicts
Joe Jones Productions
Megatrax Production Music Inc.
North Star Music
Shelly Palmer Productions
Peters Communications
Ports of Paradise
Prairie Dog Entertainment
Smith/Lee Productions, Inc.
Soundshop Recording Studio LLC
Valentino Music & Sound Effect Libraries

Music Production, Film and Video

Carriage House Studios
Tim Cissell Music
TS James & Associates
Valentino Music & Sound Effect Libraries
Warren Only Media Group
Wax Music, Sound Design & Mix

Music Production, Radio

ABC Radio Networks
Toby Arnold and Associates Inc.
Blue Heaven Productions
Carriage House Studios
Digital Force
Halland Broadcast Services
Happi Associates
Network Music
RPM Media Enterprises-The Relic Rack Review
Spanish Broadcasting System
Valentino Music & Sound Effect Libraries
WCTD AM 1620

Music Production, Radio & TV

Aloha Productions
Tim Cissell Music
Continental Recordings Inc.
JAM Creative Productions Inc.
Shield Productions Inc.
SoperSound Music Library
Sound of Birmingham Productions
TM Century Inc.
Ukrainian Melody Hour
Valentino Music & Sound Effect Libraries
Warren Only Media Group
Wax Music, Sound Design & Mix

Postproduction Services

Processing Labs

Producers, Documentaries

Producers, Film

Producers, Multimedia

Producers, Radio Programming

Radio Spirits
Ray Sports Network
Sam Shad Productions
Sound Source Networks
Sports Byline U.S.A.
StarDate/Universo
Talco Productions
Talk Radio Network
Time Capsule Inc.
United Stations Radio Network
WFMT Radio Network
WQXR
Wawatay Native Communication Society
"The Weather Center"

Producers, TV Programming

Accuracy in Media Inc.
Agora TV
AmericaNurse TV Productions
Anderson Productions Ltd. (APL)
Angel Films Co.
Armedia Communications
Ascent Media Network Services
S. Banks Group Inc.
Bardel Entertainment Inc.
Dave Bell Associates Inc.
Bell Foto Art Productions
Bellon Entertainment
Brillig Productions Inc.
Broadcast News Service
Himan Brown-Radio Drama Network
Bruder Releasing Inc. (BRI)
Buzzco Associates Inc.
CBS Paramount International Television
CBS Television Distribution
CCI Entertainment Ltd.
CDR Communications Inc.
CFP Video Productions
Call For Action Inc.
Canamedia Productions Ltd.
CanLib Inc.
George Carlson & Associates
Carlton International Media Inc.
Sandra Carter Global, Inc.
Catholic Communication Campaign
Catholic Communications Corp.
Central City Productions, Inc.
Children's Media Productions
Christian Children's Associates Inc.
The Christophers Inc.
The Dick Clark Productions
Classic Media
Clever Cleaver Productions
William F. Cooke Television Programs
Cookie Jar Group
Cornell University Educational Television Center
Critical Mass Releasing Inc.
Crystal Pictures Inc.
Curb Entertainment International Corp.
DLT Entertainment Ltd.
D-Squared Media
Dargaud-Marina
Walt Disney Company
Dorling Kindersley Vision
E! Entertainment Television
Ebbets Field Productions Ltd.
Ellis Entertainment
Envoy Productions
Faith for Today
FamilyNet
Film Roman Inc.
David Finch Distribution Ltd.
Fox 17 Studio Productions
Fox Sports West
Fox 29 WUTV Sinclair
Sandy Frank Entertainment Inc.
Free Speech TV (FSTV)
Freewheelin' Films Ltd.
The Fremantle Corp.
GLL TV Enterprises Inc.
GRB Entertainment
Lon Gibby Productions, Inc./ Gibby Media Group
Glenray Productions Inc.
Global Telemedia Inc.
Great Chefs Television/Publishing
Great North Productions
Great Plains National (GPN)
The Griffin Group
Hamilton Productions Inc.
Larry Harmon Pictures Corp.
Health Net Productions & Pet Talk
Heritage/Baruch Television Distribution
Jack Hilton Inc.
The History Makers
Holigan Investment Group Ltd.
Home Improvement Television Network
Hope Channel
The Idea Channel
Independent Edge Films
International Broadcasting Network
International Television Broadcasting Inc.
It Is Written Television
Italtoons Corp.
Ivanhoe Broadcast News Inc.
KPTS-TV
KUSA Television
Lakeside TV Co.
Lapco Communications
Lifestyle Magazine/The Evidence
Lightbridge Production & Distribution
MAKWDE Productions
MG/Perin Inc.
MacNeil/Lehrer Productions
Madison Square Garden Network
Maryknoll Productions
Maryland Public Television
Maysles Films, Inc.
The Media Group of Connecticut Inc.
Media Planning Group (MPG)
Montgomery Community Television Inc.
NAHB Production Group
NRS Group PTY Ltd.
National Collegiate Athletic Association
National Council of Churches Communications Unit
News Broadcast Network
North by Northwest Productions
O. Atlas Enterprises Inc.
O'Grady & Associates
Jim Owens Entertainment
PAULAR Entertainment L.L.C.
Shelly Palmer Productions
Paulist Productions
Peckham Productions
Pied Piper Films Ltd.
PorchLight Entertainment Inc.
Power Play Music Video L.L.C.
Powersports/Millenium International
Presbyterian Church (U.S.A.)
Prime Cut Productions Inc.
Questar
RBC Ministries
Rampion Visual Productions L.L.C.
Raycom Sports
The Real Estate Network
Reel Media International Inc.
Reid/Land Productions Inc.
Reuters Media
Rigel Entertainment
Steve Rotfeld Productions Inc.
ROZON
Sanctuary Records Group Ltd.
The Saturday Evening Post Television Department
SeniorVision Productions, Inc.
Seville Pictures
Sam Shad Productions
Silverline Pictures
Southern STAR
Charlie Spencer Productions
Sport International Inc.
Marty Stouffer Productions Ltd.
Sullivan Entertainment Inc./Sullivan Entertainment International
Sunbow Entertainment
System TV
Talco Productions
Tamouz Media
Tapestry International, Ltd.
Tel-A-Cast Productions
Tel-Air Interests Inc.
The Television Syndication Company, Inc.
Triage Entertainment Inc.
Tribune Entertainment Co.
Twentieth Century Fox Television Distribution
Twentieth Television
Two Oceans Entertainment Group
Viacom Inc.
Videographic West
Vision Broadcasting - KVBA TV 19
WQED Multimedia
WTOB Channel 2—Public/Government Access TV
Warner Bros. Animation
Warren Only Media Group
Wawatay Native Communication Society
Wide Eye Productions, Inc.
Daniel Wilson Productions Inc.
The Wyland Group

Producers, Video

Altruist Media
American Farm Bureau Inc.
Armedia Communications
J. Arnold Productions
Asia Pacific Productions
Bayliss
Bell Foto Art Productions
Himan Brown-Radio Drama Network
Bruder Releasing Inc. (BRI)
CFP Video Productions
CRM Learning
Catholic Communications Corp.
Children's Media Productions
Cinecraft Productions Inc.
Clever Cleaver Productions
Concept Videos
Coote Communications
Cramer Productions Center
Thomas Craven Film Corp.
Custom Productions Inc.
D-Squared Media
D-V-X International
DWJ Television
ESPI Video
Ebbets Field Productions Ltd.
Encore Video Productions Inc.
Envoy Productions
FTC/Orlando
Faith for Today
Films Five Inc.
David Finch Distribution Ltd.
GLL TV Enterprises Inc.
GVI
Lon Gibby Productions, Inc./ Gibby Media Group
Good News Broadcasting Association Inc.
Great Chefs Television/Publishing
Health Net Productions & Pet Talk
Jack Hilton Inc.
The History Makers
Home Improvement Television Network
Horizons Television Inc.
The Idea Channel
JGT Media Productions
The Johnson Group
Jordan Klein Film & Video
Kipany Productions Ltd.
Lapco Communications
Lifestyle Magazine/The Evidence
Lightbridge Production & Distribution
Limelight Communications Inc.
Lindberg Productions Inc.
MRC Films
MSE
Maslow Media Group Inc.
Mason Video
William Mauldin Productions Inc.
McClain Enterprises Inc.
Medialink
Montgomery Community Television Inc.
NAHB Production Group
Nemo News
News Broadcast Network
North by Northwest Productions
O. Atlas Enterprises Inc.
O'Grady & Associates
Shelly Palmer Productions
Peckham Productions
Pike Productions Inc.
Playboy Entertainment Group Inc.
Potomac TV/Communications
Powersports/Millenium International
Praxis Media Inc.
Prime Cut Productions Inc.
Primedia Workplace Learning
Quality Film & Video
RMD & Assoc. Inc.
Reizner & Reizner Film & Video
Rex Post
River City Video Productions
Rosler Creative
Steve Rotfeld Productions Inc.
Sak Entertainment
SeniorVision Productions, Inc.
Seraphim Communications Inc.
Charlie Spencer Productions
Marty Stouffer Productions Ltd.
Sullivan Video Services Inc.
TR Productions
Tankersley Productions, Inc., St. Louis
Tel-A-Cast Productions
Telepros
Toucan Productions
Triage Entertainment Inc.

Production Music Libraries

Production Services

Promotion Design

Promotion Film Distribution/Production

Promotion Production, Radio

Promotion Production, Radio & TV

Promotion Production, TV

Public Service Announcements

Publishing, Video and Print

Recording Studios

Religious Programming, Radio

Religious Programming, Radio & TV

Religious Programming, TV

Remote Facilities

Satellite Uplink Services

Scriptwriters

Set Design

Sound Design

Sound Effects/Sound Effect Libraries

Sound Recording

Sound Stages

Special Effect Libraries

Special Effects

Sports Programming, Radio

Sports Programming, Radio & TV

Sports Programming, TV

Stage and Studio Rental

Standards Conversion

Stock Footage/Tape

American Farm Bureau Inc.
Burrud Productions Inc.
Carpel Video Inc.
CelebrityFootage
Cinema Sound Ltd.
Darino Films/Library of Special Effects
Freewheelin' Films Ltd.
Lon Gibby Productions, Inc./ Gibby Media Group
Thomas Horton Associates Inc.
Media Planning Group (MPG)
Warren Miller Entertainment
National Collegiate Athletic Association
Pan American Video
System TV
The WPA Film Library
Worldview Entertainment Inc.

Studio Facilities

ABC Radio Networks
Ascent Media Network Services
Bardel Entertainment Inc.
CamMate Studios
Continental Recordings Inc.
D-V-X International
Devlin Design Group Inc.
Dome Productions
ETN—Educational Telecommunications Network
Fox Digital
Fox 17 Studio Productions
Fox 29 WUTV Sinclair
Golden Gate Studios
J.N. Productions
MTI The Image Group
Man From Mars Productions
Maryland Public Television
Mobile Video Services Ltd.
NEP Studios
National Public Radio
PACSAT
The Radio-Studio Network
Reuters Television
Soundshop Recording Studio LLC
Suite Audio
Telenium Studios
University of Colorado Television
University of Detroit Mercy
VTTV Videothek Electronic TV-Production GmbH + Co Kopier KG
WQED Multimedia

Syndication, Cable

ALIN TV
Accuracy in Media Inc.
CABLEready Corp.
CBS Television Distribution
Cable Films & Video
Carsey-Werner Distribution
Crystal Pictures Inc.
Evergreen Entertainment Group
Sandy Frank Entertainment Inc.
GLL TV Enterprises Inc.
Gould Entertainment Corp.
Great Chefs Television/Publishing
Images Communication Arts Corp.
Jefferson-Pilot Sports
Lakeside TV Co.
Lifestyle Magazine/The Evidence
Lions Gate Entertainment
Media Planning Group (MPG)
Parrot Communications International Inc.
Playboy Entertainment Group Inc.
Power Play Music Video L.L.C.
Raycom Sports
Harvey Sheldon Productions
The Summit Media Group Inc. (Sub 4 Kids Entertainment)
Twentieth Television
University of Detroit Mercy
Video/Media Distribution Inc.

Syndication, Radio

ABC Radio Networks
Accuracy in Media Inc.
Armedia Communications
At a Glance
Country Crossroads
ESPN Radio Network
Eagle Media Productions Ltd.
Envoy Productions
Good News Broadcasting Association Inc.
Happi Associates
Hometown Illinois Radio Network
Images Communication Arts Corp.
Jameson Broadcast Inc.
Jones Radio Networks
Knowledge In A Nutshell Inc.
Longhorn Radio Network
MRN Radio
MasterControl FamilyNet Radio
MediaTracks Inc.
Miller Broadcast Management
Molton Advertising Inc.
Music of Your Life
Musical Starstreams
NCAA
New Dimensions Radio
North Shore Productions
On Track
Orange Productions Inc.
Dick Orkin's Amazing Radio
Jim Owens Entertainment
Ports of Paradise
Powerline
Premiere Radio Networks Inc.
RPM Media Enterprises-The Relic Rack Review
RPM-Radio Programming & Management Inc.
Radio America
Ray Sports Network
Sing For Joy
Solid Gospel Network (Reach Satellite Network, Inc.)
Sound Source Networks
Sports Byline U.S.A.
Strand Media Group Inc.
Strength for Living
Talk America Radio Networks
Talk Radio Network
Time Capsule Inc.
Tribune Radio Networks
UBC Radio
United Stations Radio Network
University of Detroit Mercy
WFMT Radio Network
"The Weather Center"
Larry John Wright Inc.
ZBS Foundation

Syndication, TV

ALIN TV
Angel Films Co.
CBS Television Distribution
CNBC Syndication
Cable Films & Video
CanLib Inc.
George Carlson & Associates
Carsey-Werner Distribution
William F. Cooke Television Programs
DG Systems
DLT Entertainment Ltd.
Envoy Productions
Eyewitness Kids News, LLC
Sandy Frank Entertainment Inc.
GRB Entertainment
Golden Gate Studios
Gould Entertainment Corp.
Great Chefs Television/Publishing
Hearst Entertainment, Inc.
Heritage/Baruch Television Distribution
Holigan Investment Group Ltd.
Jefferson-Pilot Sports
Juravic Entertainment
KJD Teleproductions
Lakeside TV Co.
Lifestyle Magazine/The Evidence
Lions Gate Entertainment
MGM TV Canada
MG/Perin Inc.
McClain Enterprises Inc.
Music of Your Life
NCAA
National Collegiate Athletic Association
New Visions Syndication Inc.
Pan American Video
Parrot Communications International Inc.
Planet Pictures Ltd.
The Program Exchange
Promark Television
Peter Rodgers Organization
Steve Rotfeld Productions Inc.
Harvey Sheldon Productions
P. Allen Smith Gardens
TeleCom Productions Inc.
Television Representatives Inc.
The Television Syndication Company, Inc.
The Transcription Company
Tribune Entertainment Co.
Twentieth Century Fox Television Distribution
Twentieth Television
Viacom Inc.
Video/Media Distribution Inc.
Warren Only Media Group
Welwood International Film Production
Western International Syndication
Robert Wold Co.
Worldvision NY
The Wyland Group

Teleconferences

Ascent Media Network Services
Broadview Media
Celebrities Productions
Dome Productions
Encore Video Productions Inc.
Episcopal Church Center
KPTS-TV
National Mobile Television
PMTV Producers Management Television
(PSSI) Production & Satellite Services Inc.
Reizner & Reizner Film & Video
Vyvx
Wade Productions Inc.

Traffic Reporting

Metro Networks
Traffic Pulse Networks (A Unit of Mobility Technologies)
Traffic Scan Network, Inc

Training Film Productions

Cinecraft Productions Inc.
ComBridges
Encore Video Productions Inc.
Gordon Productions
Mason Video
National Collegiate Athletic Association
News Broadcast Network
Primedia Workplace Learning
Strand Media Group Inc.
Toucan Productions

Training Films

Bell Foto Art Productions
CRM Learning
Coote Communications
Thomas Craven Film Corp.
D-Squared Media
Films Media Group
International Tele-Film
Primedia Workplace Learning
Toucan Productions

Travel Programming, TV

Burrud Productions Inc.
Canamedia Productions Ltd.
Capital Communications
George Carlson & Associates
Janson Media
William Mauldin Productions Inc.
NRS Group PTY Ltd.
Promark Television
Questar
RAD Marketing & Cabletowns
M C Stuart & Associates Pty Ltd.
System TV
TRI-COMM Productions
U.S. Plan B Inc.
Videosmith Inc.

Travelogues

George Carlson & Associates
Promark Television
RAD Marketing & Cabletowns
Sandy Zimmerman Productions

Video Conferences

Videotape Editing

Voice-Overs

Weather Programming

Weather Programming, Radio

Section F

Technology

Equipment Manufacturers and Distributors Alphabetical Index

A

A & S Case Co. Inc., 5260 Vineland Ave., N. Hollywood, CA, 91601. Phone: (818) 509-5920. Fax: (818) 509-1397. E-mail: ascase@earthlink.net Web Site: www.ascase.com. Kenneth E. Berry, pres.

Reusable shipping cases; computer, musical instrument cases.

A.C.C. Electronix, Inc., 1845 W. Hovey Ave., Normal, IL, 61761-4315. Phone: (309) 888-9990. Fax: (309) 452-0893. E-mail: acc@accelectronix.com Web Site: www.accelectronix.com. Andrew M. Rector, pres.

Repair cartridge tape recorders & reproducers.

ADC, Box 1101, Minneapolis, MN, 55440-1101. Phone: (952) 938-8080. Fax: (952) 917-1717. Web Site: www.adc.com. Mike Day, VP; Gokul Hemmady, VP & CFO; Robert E. Switz, pres & chmn.

ADC provides the connections for wireline, wireless, cable, bcst & enterprise nets around the world. ADC's equipment & svcs enable high-speed Internet, data, video & voice svcs.

ADCOUR, 623 Main St., Woburn, MA, 01801. Phone: (781) 935-9944. Fax: (781) 937-3499. Richard Jacobs, pres.

Batteries, chargers, power supplies & power conditioning.

ADSCO Line Products Inc., 3500 Washington Ave., Houston, TX, 77007. Phone: (713) 880-2424. Fax: (713) 880-2456. Linda Schmuck, pres.

Outside plant line hardware for CATV: guy strand, messengers, lashing wire/rods, formed grips/dead-ends & related line hardware. Stainless steel poleline hardware.

ADTEC Inc., 408 Russell St., Nashville, TN, 37206. Phone: (615) 256-6619. Fax: (615) 256-6593. E-mail: sales@adtecinc.com Web Site: www.adtecinc.com. Ron Johnson, VP.

Jacksonville, FL 32216, 2231 Corporate Square Blvd. Phone: (904) 720-2003. Kevin Ancelin, pres.

Products offered: Loc origination controllers & systems, coml insertion controllers & systems, network delay recording.

AKG Acoustics, U.S., 914 Airpark Center Dr., Nashville, TN, 37217. Phone: (615) 620-3800. Fax: (615) 620-3875. E-mail: akgusa@harman.com Web Site: www.akg-acoustics.com. Doug Mac Callum, VP/gen mgr.

Microphones, headphones, wireless microphones, in-ear monitoring systems, wireless loudspeakers, conferencing products.

AMCO Engineering Co., 3801 Rose St., Schiller Park, IL, 60176. Phone: (847) 671-6670. Fax: (847) 671-9469. Web Site: www.amcoengineering.com. Thomas Anderson, pres; Tom Ligman, natl sls mgr; James Walenda, dir mktg.

Data/Communications, monitoring & EMI cabinets, single or multiple bay. Inline or curved configurations, standard or custom.

AMS Neve Inc., 100 Ave. of the Americas, 5th Fl., New York, NY, 10013. Phone: (212) 965-1400. Fax: (212) 965-9306. Web Site: www.amsneve.com. John Hart, pres.

North Hollywood, CA 91602, 4220 Lankershim Blvd., 2nd Fl. Phone: (818) 753-8789. John Hart.

Distributors of AMS-NEVE analog & digital audio equipment for the bcst, video postproduction, film, music recording & mastering industries.

APM/Associated Production Music, 6255 Sunset Blvd., Suite 820, Hollywood, CA, 90028. Phone: (323) 461-3211. Phone: (800) 543-4276. Fax: (323) 461-9102. E-mail: sales@apmmusic.com Web Site: www.apmmusic.com. Connie Red, account exec; Tia Sommer, sls dir.

New York, NY 10016, 240 Madison Ave. Phone: (800) 276-6874. (212) 856-9800. Fax: (212) 856-9807. George Macias, natl sls mgr. (East Coast).

Sixteen production music libraries, over 3,000 CDs, personalized packages, music search svc, 15-20 new CD releases mthy.

APW Mayville/Stantron, 403 Degner Ave., Mayville, WI, 53050-0028. Phone: (800) 558-7297. E-mail: customerservice @mayvilleproducts.com Web Site: www.apwmayville.com. Daniel Eder, pres; Rich Runnels, sls dir.

Stantron racks, cabinets, enclosures, & related accessories for bcst integrators & professional audio video installations.

ARRI, Inc., 617 Rt. 303, Blauvelt, NY, 10913-1109. Phone: (845) 353-1400. Fax: (845) 425-1250. E-mail: arriflex@arri.com Web Site: www.arri.com. Juergen Schwinzer, VP; John Gresch, VP; Volker Bahnemann, pres.

Burbank, CA 91502, 600 N. Victory Blvd. Phone: (818) 841-7070. Bill Russell, VP, western opns.

Manufacturer of professional motion picture film cameras & accessories, lighting equipment & post-production tools.

A R T Applied Research and Technology, 215 Tremont St., Rochester, NY, 14608. Phone: (585) 436-2720. Fax: (585) 436-3942. Web Site: www.artproaudoio.com. Philip Betette, pres.

Digital audio signal processors & enhancement devices.

ATCI/Antenna Technology Communications Inc., 450 N. McKemy, Chandler, AZ, 85226. Phone: (480) 844-8501. Fax: (480) 898-7667. Web Site: www.atci.com. Gary Hatch, CEO; Ron Kahle, COO/CFO.

Simpson, PA 18407, 289 Atlas St. Phone: (570) 282-3590. William Pryle, area mgr.

Simulsat multibeam earth stns; parabolic antennas from 1.8 m to 32 m. Headend electronics, design & maintenance, used/refurbished equipment.

AVAB America Inc., 434 Payran Street, Petaluma, CA, 94952. Phone: (707)778-8990. E-mail: sales@avab.com Web Site: www.avab.com. Hans J. Lau, pres.

Manufacturer of studio & theatrical lighting equipment; lighting controllers, dimmers, fixtures.

AVCOM of Virginia Inc., 7730 Whitehead Rd., Richmond, VA, 23237. Phone: (804) 794-2500. Fax: (804) 794-8284. Web Site: www.avcomofva.com. Jay T. Evans, pres.

Manufacturer of portable spectrum analyzers & accessories.

AVI Systems, 6271 Bury Dr., Eden Prairie, MN, 55346. Phone: (952) 949-3700. Fax: (952) 949-6000. E-mail: info@avisys.com Web Site: www.avisystems.com. Joe Stoebner, CEO.

Cameras, consoles, monitors, speakers, microphones, remote control, transmitters, exciters, video-conferencing systems & TV bcstg equipment.

AVS Graphics & Media Inc., 963 Autumn Ave, Salt Lake City, UT, 84116-2243. Phone: (801) 975-9799. Fax: (801) 975-0970. E-mail: sales@avsgmedia.com Web Site: www.avsgmedia.com. Gavin Hunter, CEO.

Bcst & production character generators & Still stores.

AVX Corp., 3900 Electronics Dr., Raleigh, NC, 27604. Phone: (919) 878-6200. Fax: (919) 878-6470. Web Site: www.avxcorp.com. Jimmy White, sls dir; Craig Hunter, mktg dir.

Electronic component.

AZCAR U.S.A. Inc., 121 Hillpointe Dr., Suite 700, Canonsburgh, PA, 15317. Phone: (724) 873-0800. Fax: (724) 873-4770. E-mail: info@azcar.com Web Site: www.azcar.com. Stephen Pumple, pres/CEO; John Luff, sr VP business dev; Karl Paulsen, sr VP engrg; Marv Nolan, dire business dev.

Markham, ON L3R 0H3 Canada, 3235 14th Ave. Phone: (905) 470-2545. Fax: (905)470-2556. S. Pumple, pres/CEO.

Cambridge CB1 3HD United Kingdom, #1 College Business Park, Coldhams Lane. Phone: 44 1223 414101. Fax: 44 123 414102. F. Jarvis, dir.

Bcst engrg, systems integration.

Abroyd Communications Ltd., 50 Goebel Unit 5, Cambridge, ON, N3C 1Z1. Canada. Phone: (519) 220-0420. Fax: (519) 658-1094. Web Site: www.abroyd.com.

Designers, manufacturers & installers of communication towers; manufacturer for Lightning Dissipation Arrays Chem-Rod from LEC.

Access Intelligence LLC, 4 Choke Cherry Rd., 2nd Fl., Rockville, MD, 20850. Phone: (301) 354-2000. Web Site: www.accessintel.com. Donald Pazour, pres/CEO; Ed Pinedo, exec VP & CFO; Macy Fecto, exec VP.

Magazines, trade shows & seminars.

Accom Inc., 1490 O'Brien Dr., Menlo Park, CA, 94025. Phone: (650) 328-3818. Fax: (650) 327-2511. E-mail: info@accom.com Web Site: www.accom.com. Junaid Sheikh, chmn/CEO; Phil Bennett, exec VP; Bill Ludwig, VP sls.

Accom designs, manufactures, sell, support a complete line of digital video production, disk recording & editing tools for use in the worldwide professional TV marketplace-encompassing the production, post production, bcstg & computer video markets.

Accurate Sound Corp., 3475A Edison Way, Menlo Park, CA, 94025. Phone: (650) 365-2843. Fax: (650) 365-3057. E-mail: ron@accuratesound.com Web Site: www.accuratesound.com. Ronald M. Newdoll, pres.

High-speed tape duplicating & recording equipment, digital audio logging recorders, audio & videotape conditioners, audio recorders. CD-R recorders for audio & ROM.

AccuWeather Inc., 385 Science Park Rd., State College, PA, 16803-2215. Phone: (814) 235-8600. Fax: (814) 235-8609. E-mail: sales@accuux.com Web Site: www.accuweather.com. Dr. Joel N. Myers, founder & pres; Evan Myers, sr VP/COO; Barry Lee Myers, exec VP/ gen counsel; Elliot Abrams, sr VP/ chief meteorologist; Dr. Joe Sobel, sr VP/dir forensics.

AccuWeather, Inc. offers a broad new menu of powerful integrated, muturally supporting weather content & weather brand-building solutions.

Acme Electric Corp., Aerospace Division, 528 W. 21st St., Tempe, AZ, 85282. Phone: (480) 894-6864. Fax: (480) 921-0470. Web Site: www.acme-electric.com/aerospace. Kevin Kriegel, dir mktg.

Sealed fiber nickel-cadmium batteries, battery chargers, battery control units, & AC/DC & DC/OC converters.

Acoustic Systems/ETS-Lindgren, 1301 Arrow Point Dr., Cedar Park, TX, 78613. Phone: (512) 531-6400. Fax: (512) 531-6500. E-mail: info@acousticsystems.com Web Site: www.acousticsystems.com.

Acoustical Solutions Inc., 2852 East Parham Rd., Richmond, VA, 23228. Phone: (800) 782-5742. Fax: (804) 346-8808. E-mail: info@accousticalsolutions.com Web Site: www.acousticalsolutions.com. Michael Binns, pres; Don Strahle, sales & mktg.

Sound & noise control materials including products for the bcst/recording industry, telecommunications industry, architectural acoustics & industrial noise control.

Acrodyne Industries Inc., 10706 Beaver Dam Rd., Cockeysville, PA, 21030. Phone: (410) 568-2105. Web Site: www.acrodyne.com. Nat Ostroff, chmn; Mark Polovick, natl sls mgr; Ellen Rainey, mktg mgr.

Phoenixville, PA 19460, 200 Schell Ln. Phone: (610) 917-1300. Fax: (610) 917-8148.

Television transmitters - IOT equipped Quantum Line; solid-state Rohde & Schwarz medium-high power, supported by Acrodyne in North America.

Acterna, Cable Networks Division, 5808 Churchman Bypass, Indianapolis, IN, 46203. Phone: (317) 788-9351. Fax: (317) 614-8308. Web Site: www.acterna.com.

Test equipment for video nets, including broadband RF & fiber optics. (SLMs, system analyzers, leakage, sweeps & OTDRs) software.

Adcom, 310 Judson St., Unit 5, Toronto, ON, M8Z 5T6. Canada. Phone: (416) 251-3355. Fax: (416) 251-3977.

"Night Suite" DI, non-linear editing systems, bcst control systems, video conferencing. Room control systems "1 room".

Adrienne Electronics Corp., 7225 Bermuda Rd., Unit G, Las Vegas, NV, 89119. Phone: (702) 896-1858. Fax: (702) 896-3034. E-mail: info@adrielec.com Web Site: www.adrielec.com.

Small routing switchers, time code products, machine control products.

Advance Products Co. Inc., 1199 E. Central, Wichita, KS, 67214. Phone: (316) 263-4231. Fax: (316) 263-4245. Harold Knapp, gen mgr.

Mobile projector, TV & video, tables & cabinets, wall & ceiling mount brackets.

Advanced Designs Corp., 1169 W. 2nd St., Bloomington, IN, 47403. Phone: (812) 333-1922. Fax: (812) 333-2030. E-mail: adc@doprad.com Web Site: www.doprad.com. Matt McGrath, pres; James Sawtelle, mktg mgr.

DOPRAD® 32 doppler radar system, weather data display system, storm path analyzer, street-level maps, lightning, low-cost remoting & composite live doppler.

Advanced Media Inc., Boc 599, Bohemia, NY, 11716-0599. Phone: (631) 244-1616. Fax: (631) 244-1415. E-mail: team@advancedmedia.com Web Site: www.advancemedia.com.

Kiosk-interactive technology.

Advanced Media Technologies, Inc., 720 S. Powerline Rd., Suite G, Deerfield Beach, FL, 33442-8156. Phone: (888) 293-5856. Fax: (954) 427-9688. E-mail: sales@advancedmediatech.com Web Site: www.advancedmediatech.com. Ken Mosca, pres.

AMT supplies high performance products from well-known manufacturers including R, fiber distribution, video, data & IP.

Advent Communications Ltd., Nashleigh Hill, Chesham, Bucks, HP5 3HE. United Kingdom. Phone: 44 1494 774400. Fax: 44 1494 791127. E-mail: sales@adventcomms.com Web Site: www.adventcomms.com. Stephen Rudd, mgng dir; George Koumblis, sls dir.

Provides satellite communication solutions for bcst, telecommunications, military & coml applications-design, manufacture & integrating a complete range of digital SNG flyaway & vehicle mounted terminals, a complete range of subsystems upconverters, downconverters, DVB modulators, MPEG II Video Exciters, equalizers & remote control systems.

Aeroflex, 35 South Service Rd., Plainview, NY, 11803. Phone: (516) 694-6700. Fax: (516) 694-2562. Web Site: www.aeroflex.com. Jeff Bloomer, pres/CEO.

Trophy Club, TX 76262, 49 Trophy Club Rd. Phone: (817) 430-5842. Carlos Blanco, sls mgr - South America.

Test & measurement instrumentation.

AheadTek, 6410 Via Del Oro, San Jose, CA, 95119. Phone: (408) 226-9800. Phone: (408) 226-9991. Fax: (408) 226-9195. Fax: (408) 226-9194. E-mail: patj@drs-ahead.com Web Site: www.aheadtek.com. Art Hoegger, pres; Pat Johnston, product mgr.

Cost effective solutions for your specialty magnetic head applications.

Peter Albrecht Company Inc., 6250 Industrial Ct., Greendale, WI, 53129-2432. Phone: (414) 421-6630. Fax: (414) 421-9091. E-mail: sales@peteralbrecht.com Web Site: www.peteralbrecht.com. T.C. Ziolkowski, pres.

Motorized studio battens, plaks & other rigging systems. Cyclorama track & curtain systems designed & installed.

Alesis, 300 Corporate Pointe, Suite 500, Culver City, CA, 90230-8721. Phone: (310) 821-5000. Fax: (310) 306-2650. Web Site: www.alesis.com.

Digital tape recording system, mixing consoles, digital & analogue signal processing, amplification, drum machines, keyboards.

Alexander Technology, 1938 University, Lisle, IL, 60532. Phone: (800) 247-1821. Phone: (641) 423-8955. Fax: (641) 423-1644. E-mail: info@alexandertechnology.com Web Site: www.alexandertechnology.com. John Casey, pres/CEO.

Rechargeable nicad in-board, on-board & battery belts; nicad battery chargers & analyzer/conditioners; portable radio & pager batteries.

Alias/WaveFront Inc., 210 King St. E., Toronto, ON, M5A 1J7. Canada. Phone: (416) 362-9181. Fax: (416) 369-6140. Web Site: www.aliaswavefront.com.

2D & 3D computer graphic imaging & animation software for professionals in entertainment & industrial markets.

All Mobile Video Inc., 221 W. 26th St., New York, NY, 10001. Phone: (212) 727-1234. Fax: (212) 255-6644. Web Site: www.allmobilevideo.com. Anton Duke, CEO; Eric Duke, pres.

San Diego, CA 92123, 9670 Aero Dr. Phone: (619) 569-8451. N. Tabkum, dir West Coast opns.

Saint Petersburg, FL 33742, 10490 Gandy Blvd. Phone: (813) 579-8902. Bary Spencer, dir teleport opns.

Bcst video equipment rental including truck remotes & total carry-in packages, complete studio facilities.

Allen & Heath USA, Agoura Business Center, E. 5304 Derry Ave., Suite C, Agoura Hills, CA, 91301. Web Site: www.allen-heath.com. Al Nickols, sls dir.

Audio mixing consoles for recording & live sound applications including automated consoles.

Allen Avionics, Inc., 224 E. Second St., Mineola, NY, 11501. Phone: (516) 248-8080. Fax: (516) 747-6724. E-mail: jim@allenavionics.com Web Site: www.allenavionics.com. Jim Lyons, VP.

All passive components, video filters & delay lines, audio & video hum eliminators, LC filters & electromagnetic delay lines. SDI hum eliminators.

Allied Electronics Inc., 7410 Pebble Dr., Fort Worth, TX, 76118. Fax: (817) 595-6444. Web Site: www.alliedelec.com. Lee Davidson, VP; Robert Pfleg, pres; Bob Whetson, sls dir; Rob Birse, dir.

Broad line distributor of electronic components.

Allied Tower Co. Inc., 4646 Mandale, Alvin, TX, 77511. Phone: (281) 331-9627. Fax: (281) 331-9822. Max Bowen, CEO; Jeff Bowen, pres; Doug W. Moore, VP.

Design, fabrication & erection of FM, AM, TV & communication towers.

Allison Payments Systems L.L.C., 2200 Production Dr., Indianapolis, IN, 46241. Phone: (317) 808-2400. Fax: (317) 808-2477. Web Site: www.apsllc.com.

Coupon payment billing systems.

Allsop Inc., Box 23, Bellingham, WA, 98227. Phone: (360) 734-9090. Fax: (360) 734-9858. Web Site: www.allsop.com. Jim Allsop; Mike Allsop, co-pres.

Cleaning accessories for audio & video, record care products & compact discs, computer accessories.

Alpack Associates, Inc., 10 Commerce Rd., Unit D, Fairfield, NJ, 07004. Phone: (973) 244-4414. Fax: (973) 244-4483. E-mail: info@alpack-pic.com Les Weinstock, pres.

Standard & custom carrying & shipping cases for all bcst equipment. Both hard & soft case styles.

Alpha Technologies Inc., 3767 Alpha Way, Bellingham, WA, 98226. Phone: (360) 647-2360. Fax: (360) 671-4936. E-mail: alpha@alpha.com Web Site: www.alpha.com. Paul Humphreys, VP mktg; Warren Johnson, pres/COO.

Manufacturer of power systems for coaxial & fiber optic networks. UPS systems, DC products, batteries & surge suppression.

Alpha Video & Electronics (AVEC), 200 Mingo Church Rd., Finleyville, PA, 15332. Phone: (412) 429-2000. Fax: (724) 348-8600. E-mail: henry@aveceng.com Web Site: www.aveceng.com. Henry Lassige, Sr., pres.

O.B. vans, eng vans, DSNG vans, ENG mast safety device, turnkey systems, camera transporter.

Alpine Optics Inc., 9913 N.W. 20th St., Coral Springs, FL, 33071. Phone: (954) 344-9871. Fax: (954) 344-3665. E-mail: toalpine_optics@bellsouth.net Web Site: www.alpine-optics.com. Horst Stahl, pres.

Repair, maintenance of all Canon, Fujinon, Nikon, Schneider & JVC lenses.

Altronic Research Inc., Box 249, Yellville, AR, 72687. Phone: (800) 482-5623. Fax: (870) 449-6000. E-mail: altronic@mtnhome.com Web Site: www.altronic.com. John Dyess, pres.

Omegaline RF coaxial load resistors (dummy loads).

Aluma Tower Company Inc., Box 2806-BC, Vero Beach, FL, 32961-2806. Phone: (772) 567-3423. Fax: (772) 567-3432. E-mail: atc@alumatower.com Web Site: www.alumatower.com. Theodore E. Gottry, VP.

Aluminum telescoping towers combined with trailers & optional shelters provides mobile units. Vehicle mounted towers for installation on customer's vehicle.

Amek U.S.A., 8500 Balboa Blvd., Northridge, CA, 91329. Phone: (818) 920-3212. Fax: (818) 920-3208. E-mail: amekusa@harman.com

Amek, TAC (Total Audio Concepts) & Langley audio consoles for production, postproduction, audio recording, sound reinforcement & Medici signal processing equipment.

American Antenna Inc., 4707 Roosevelt St., Glen Park, IN, 46408. Phone: (219) 985-4000. Fax: (219) 985-4001. E-mail: sales@americanantenna.com Web Site: www.americanantenna.com. Nick Michels, pres; Chuck Forsyth, VP sls; Rick Gard, sec.

Distributor, installer of Earth Station Antennas from .45cm to 6.1m motorized actuators, receivers, controllers, LNB's & accessories.

American Eurocopter Corp., 2701 Forum Dr., Grand Prairie, TX, 75052-7099. Phone: (972) 641-0000. Fax: (972) 641-3550. Web Site: www.eurocopterusa.com. Marc Paganini, pres; Brenda Revland, dir.

Servicing North American market; manufactures & sells complete line of single- & twin-engine turbine helicopters.

Ampex Data Systems Corp., 1228 Douglas Ave., Redwood City, CA, 94063-3199. Phone: (650) 367-2011. Fax: (650) 367-2444. E-mail: info@ampexdata.com Web Site: www.ampexdata.com. Ed Bramson, pres/CEO; Bob Atchison, VP; Joel Talcott, VP.

Data recorders, data systems, mass data storage, instrumentation recorder products; 19 mm scanning recorders, library systems (DST & DIS products), related tape, after-market parts & video recorder support.

Amplivox Portable Sound Systems, 3149 MacArthur Blvd., Northbrook, IL, 60062. Phone: (847) 498-9000. Fax: (800) 267-5489. E-mail: droth@ampli.com Web Site: www.ampli.com. Don Roth, CEO.

Portable sound systems/lecterns/wireless/indoor-outdoor, made in USA, UL, CSA, CE, 6 years warranty.

Amtel Network, 431 Myrtle St., Suite 6, Glendale, CA, 91203. Phone: (818) 842-8088. Fax: (818) 551-4999. E-mail: amtel@amtel.com Web Site: www.amtelsystems.com. Mike Takamatsu, pres.

Text-visual intercom system.

Analog Digital International Inc., 20 E. 49th St., 2nd Fl., New York, NY, 10017-1023. Phone: (212) 688-5110. Fax: (212) 688-5405. E-mail: info@analogdigitalinc.com Web Site: www.analogdigitalinc.com.

Sls & Rentals of professional /bcst NTSC/PA equipment, post production, DVD authoring, AVID editing , final cut pro editing & training.

Anchor Audio Inc., 2565 W. 237th St., Torrance, CA, 90505. Phone: (310) 784-2300. Phone: (888) 444-6077. Fax: (310) 784-0066. E-mail: sales@anchoraudio.com Web Site: www.anchoraudio.com. David Jacobs, pres.

Bcst intercom equipment, 2-ch high performance & low cost.

Andrew Corp., 10500 W. 153rd St., Orland Park, IL, 60462. Phone: (708) 349-3300. Phone: (800) 255-1479. Fax: (708) 349-5943. Web Site: www.andrew.com. Paul Cox, pres; Barry Cohen, sls dir; George Tong, mgr.

Orland Park, IL 60462, 10500 W. 153rd St. Phone: (708) 349-3300. Paul Cox, group pres; Bary Cohen, sls; George Tong, govt Antennas & ESAs.

VHF & UHF-TV transmitting, microwave & ESA's; coaxial cable; waveguides; towers; equipment shelters; instal svcs, combiners & pressurization equipment.

Antenna Concepts Inc., 6626 Merchandise Way, Diamond Springs, CA, 95619. Phone: (530) 621-2015. Fax: (530) 622-3274. E-mail: sales@antennaconcepts.com Web Site: www.antennaconcepts.com. Mark A. Cunningham, pres/CEO.

Custom & standard low-, medium- and high-power omni or directional digital & analog UHF, VHF, FM, & MMDS bcst antennas. Full power Broadcast antennas. TV: UHF/VHF analog/digital antennas including full-UHF band CP panel. Fm: Ultra Tracker single-lobe.

Anton/Bauer Inc., 14 Progress Dr., Shelton, CT, 06484. Phone: (203) 929-1100. Fax: (203) 929-9935. Web Site: www.antonbauer.com. Alex DeSorbo, pres.

NiCad & NIMH cameras batteries, chargers, lighting & diagnostic accessories for the professional video industry.

Anvil Cases, 15730 Salt Lake Ave., City of Industry, CA, 91745. Phone: (626) 968-4100. Fax: (626) 968-1703. Web Site: www.anvilcase.com. E-mail: info@anvilcase.com

Heavy-duty reuseable, custom, standard shipping cases & containers for all bcst equipment.

Aphex Systems Ltd., 11068 Randall St., Sun Valley, CA, 91352. Phone: (818) 767-2929. Fax: (818) 767-2641. E-mail: sales@aphex.com Web Site: www.aphex.com. Marvin Caesar, pres; Wayne La Farr, product specialist.

Model 2020 MKIII, Compellor-intelligent AGC, Dominator II precision multi-band peak limiter, Aural Exciter, Expressor, remote controlled mic preams, TVGS MIC/instrument, preamplifiers, analog to digital converters.

Argraph Corp., 111 Asia Pl., Carlstadt, NJ, 07072. Phone: (201) 939-7722. Fax: (201) 939-7782. Mark Roth, pres; Martin Lipton, sls.

Hayward, CA 94545. Argraph West, 2710 McCone. Phone: (510) 298-0575.

Anti-stat cleaning cloths, samigron video tripods.

Aries Industries Inc., Corporate Office, 550 Elizabeth St., Waukesha, WI, 53186. Phone: (262) 896-7205. Phone: (800) 234-7205. Fax: (262) 246-7099. E-mail: sales@ariesind.com Web Site: www.ariesind.com.

Fresno, CA 93727, 5748 E. Shields. Phone: (800) 671-0383. Fax: (559) 291-0463. J. Lenahan, CEO.

Manufacture pipeline inspection televising test & seal equipment.

Arista Information Systems, 2150 Boggs Rd., Suite 430, Duluth, GA, 30096. Phone: (678) 473-1885. Fax: (678) 473-1051. E-mail: sales@aristainfo.com Web Site: www.aristainfo.com. Scott Ford, sls dir.

Cable TV subscriber & statement printing.

Arrakis Systems Inc., 6604 Powell St., Loveland, CO, 80538. Phone: (970) 461-0730. Fax: (970) 663-1010. E-mail: sales@arrakis-systems.com Web Site: www.arrakis-systems.com. Michael C. Palmer, pres; Jon Young, VP sls; Roderic M. Graham, VP.

Audio consoles, digital audio, satellite, live-assist, hard drive automation & production systems, studio furniture.

Arri Canada Ltd., 415 Horner Ave., Unit 11, Etobicoke, ON, M8W 4W3. Canada. Phone: (416) 255-3335. Fax: (416) 255-3399. E-mail: david@arrican.com Web Site: www.arri.com. David Rosengarten, pres.

ARRI camera, lightning equipment & all professional accessories, sales & service.

Arris, Corporate Headquarters, 3871 Lakefield Dr., Suwanee, GA, 30024. Phone: (800) 469-6569 (in US). Phone: (770) 622-8400. Fax: (770) 622-8770. Web Site: www.arrisi.com.

CMTS, cable modems, telephony voice ports & modems, oss/provisioning systems, HFC infrastructure products.

Artel Video Systems, 330 Codman Hill Rd., Boxborough, MA, 01719. Phone: (978) 263-5775. Fax: (978) 263-9755. E-mail: info@artel.com Web Site: www.artel.com. Richard Dellacanonica, pres/CEO.

A pioneer, global leader in the dev, mfg, support of bcst quality video transport hardware for video svc providers, cable TV operators & bcstrs.

Artesia Technologies, 700 King Farm Blvd., Suite 400, Rockville, MD, 20850. Phone: 301-548-7850. Fax: 301-548-4015. Web Site: www.artesia.com. Brian Hedquist, dir mktg.

Did you know that Digital Asset Management typically has a return on investment of 9 to 18 months? Learn more at Artesia's Seminar Series presented with The Gartner Group & Frank Gilbane.

Ascent Media Management Services, 2901 W. Alameda Ave., Burbank, CA, 91505. Phone: (818) 840-7000. Fax: (818) 840-7129. Web Site: www.4mc.com. William Humphrey, pres; Beth Simon, sls VP & sr VP; Andre Macaluso, opns VP.

Northvale, NJ 07647, 235 Pegasus Ave. Phone: (201) 767-3800. Fax: (201) 767-4568. Beth Simon, sr VP sls/mktg.

Postproduction video & film svcs: editing, telecine, sound, duplication, satellite svcs, film lab, standard conversion tape to film transfers & digital asset mgmt.

Ascent Media Services, 6344 Fountain Ave., Hollywood, CA, 90028. Phone: (323) 988-6520. Web Site: www.ascentmedia.com. Edward Olson, VP engrg & techhnology; Lou DiMauro, dir video sls; David Lostracco, dir opns & client svcs., Asia Broadcast Centre. Phone: (65) 548-0388. Jim Crowe, mgng dir.

Minneapolis, MN 55403. GWNS Phone: (612) 330-2639. Joel Helseth, dir sls & mktg.

Video transmission, origination, tech consulting, new media products, private networks, post-production, studio, graphics, bcst event svcs and satellite svcs.

Ascent Media Services, (Formerly Waterfront Communications Corp.). 545 5th Ave., New York, NY, 10017. Phone: (212) 907-1208. Fax: (212) 599-4172.

A transmission company specializing in video switching, quality control, last mile connections, remote transmissions, production & audiovisual svcs to the bcst, cable & corporate TV industries.

Associated Press Broadcast Services, 1825 K St. N.W., Suite 800, Washington, DC, 20006-1202. Phone: (202) 736-1100. Fax: (202) 736-1124. Fax: (202) 736-1199. Web Site: www.apbroadcasting.com. James R. Williams, VP; Lee Perryman, dir.

AP NewsDesk: Newsroom computer software program for mgng TV, radio news & info resources.

Atlantic Inc., 10018 Santa Fe Springs Rd., Sante Fe Springs, CA, 90670-2922. Phone: (562) 903-9550. Fax: (562) 903-9053. E-mail: atlantic@atlantic-inc.com Web Site: www.atlantic-inc.com. Leo Dardashti, pres; Don Dolliver, VP sls.

Manufacturer of metal storage systems for DVDs, CDs, & VHS.

Atlantic Sound Systems, R.R. 2, New Glasgow, Pictou County, NS, B2H 5C5. Canada. Phone: (902) 752-8527. E-mail: plann@wisic.com

Professional bcstg, sound & lighting equipment. Rental & PA Installations.

Atlantic Video Inc., 650 Massachusetts Ave. N.W., Washington, DC, 20001. Phone: (202) 408-0900. Fax: (202) 408-8496. E-mail: aschwab@atlanticvideo.com Web Site: www.atlanticvideo.com. Gary DeMoss, dev VP; Ed Milligan, pres; Ted Nelson, dir.

Atlantic video is a full svc bcst production svc facility, from studio, remote, post & transmission.

Atlas Case Corp., 1380 So. Cherokee St., Denver, CO, 80223. Phone: (888) 325-2199. Fax: (877) 525-2339. Web Site: www.atlascases.com. Randy Sabey, pres.

Airline-approved shipping & carrying cases. Local transport cases, custom or from stock.

Atlas Sound, 4545 E. Baseline Rd., Phoenix, AZ, 85042. Phone: (800) 876-3332. Phone: (602) 438-4545. Fax: (800) 765-3435. E-mail: atlascustser@atlassound.com Web Site: www.atlassound.com. Kent Meske, sls VP; Manny Kitagawa, gen sls mgr; Ken Peck, rgnl sls mgr; Steve Young, mktg VP.

Ennis, TX 75119. Atlas Sound Manufacturing, 1601 Jack McKay.

Atlas Sound brand microphone & equipment stands, accessories; equipment consoles, racks & cabinets; loudspeaker systems; a/v monitoring devices.

Audico Labels, 118 South Main St., Goshen, IN, 46526. Phone: (574) 533-6688. Phone: (800) 252-5667. Fax: (661) 760-8775. E-mail: audicolabels@audicolabels.com Web Site: www.audicolabels.com. Bill Landow, owner; Claudia Landow, owner.

Media pressure sensitive labels.

Audio Accessories Inc., 25 Mill St., Marlow, NH, 03456. Phone: (603) 446-3335. Fax: (603) 446-7543. E-mail: audioacc@patchbays.com Web Site: www.patchbays.com. M.B. Hall, pres; T.J. Symonds, opns mgr.

Jack panels, (audio & video patchbays) patch cords, telephone jacks & plugs, pre-wired jack panels (miniature & full-size) & video panels.

Audio Implements/GKC, 1703 Pearl St., Waukesha, WI, 53186-5626. Phone: (262) 524-2424. Fax: (262) 524-7898. E-mail: info@audioimplements.com Web Site: www.audioimplements.com. Walter L. Kolb, owner; Anita Brown, sec.

Acoustic coiled earpiece, receivers & cords, microphone line & monitor amplifiers, used in conjunction with IFB system.

Audio Precision Inc., 5750 S.W. Arctic Dr., Beaverton, OR, 97005. Phone: (503) 627-0832. Fax: (503) 641-8906. E-mail: sales@audioprecision.com Web Site: www.audioprecision.com. David Solomon, sls dir; Al Miksch, CEO.

2700 Series, Portable One & ATS-1, ATS-2 audio test sets for bcst & satellite use.

Audio Processing Technology Ltd./APT, Unit 6 Edgewater Rd., Belfast, BT3 9JQ. Ireland. Phone: 44 0 28 9037 1110. Fax: 44 0 28 9037 1137. E-mail: aptmarketing@aptx.com Web Site: www.aptx.com.

Los Angeles, CA 90028, 6255 Sunset Blvd, Suite 1025. Phone: (323) 463-2963. Fax: (323) 463-8878. Simav Factor.

Digital (apt-X) audio compression system for professional applications such as storage & transmission of audio over low capacity digital circuits such as ISDN.

Audio-Technica U.S., Inc., 1221 Commerce Dr., Stow, OH, 44224. Phone: (330) 686-2600. Fax: (330) 688-3752. E-mail: pro@atus.com Web Site: www.audio-technica.com. Phil Cajka, pres/CEO; Fred Nichols, sr VP; Richard Strungle, VP opns; Jackie Green, VP; Steven Lefkowitz, CFO.

Microphones, wireless microphones, headphones, automatic microphone mixers, phono cartridges, turntables, audio & video accessories.

Audio Technologies Inc. (ATI), 154 Cooper Rd., West Berlin, NJ, 08091. Phone: (856) 719-9900. Fax: (856) 719-9903. Web Site: www.atiaudio.com. David V. Day, pres.

Bcst audio, mic, line, distribution, interface & turntable amplifiers, power amps, on-air consoles & audio processors.

Audio-Video Engineering Co., One Pineapple Ln., Stuart, FL, 34996. Phone: (772) 219-3623. Fax: (772) 219-3624. Olga M. Drucker, pres.

Video hum stop coil (hum bucker).

Audioarts Engineering, 600 Industrial Dr., New Bern, NC, 28562. Phone: (252) 638-7000. Fax: (252) 635-4857. E-mail: sales@wheatstone.com Web Site: www.audioarts.net. Gary C. Snow, pres; Andrew Calvanese, VP; Jay Tyler, sls dir.

Manufacturer of digital, analog bcst audio mixing consoles & processing equipment.

Audiolab Electronics Inc., 620 Commerce Dr., Suite C, Roseville, CA, 95678. Phone: (916) 784-0200. Fax: (916) 784-1425. E-mail: info@audiolabelectronics.com Web Site: www.audiolabelectronics.com. Ronald A. Stofan, pres/CEO.

Professional line of bulk tape degaussers for all formats of tape including: Beta SP, DAT 2" reels up to 16" diameters, hard drives, DLT media & degaussing svc.

Auernheimer Labs Corp., 4561 E. Florence Ave., Fresno, CA, 93725. Phone: (559) 442-1048. Curley Auernheimer, pres; Warren Auernheimer, VP; Dwayne Auernheimer, sec/treas.

Loudspeaker systems; loudspeaker repair, studio, monitor, control room & auditorium.

Austin Insulators Inc., 7510 Airport Rd., Mississauga, ON, L4T 2H5. Canada. Phone: (905) 405-1144. Fax: (905) 405-1150. E-mail: sales@austin-insulators.com Web Site: www.austin-insulators.com. Patrick Warr, pres; Beverly O'Brien, exec VP.

Base/guyline insulators, static drain devices, tower lighting transformers, replacements for obsolete insulators & LED lighting for AM(MW), LW antennas.

Autodesk, 10 Duke St., Old Montreal, PQ, H3C 2L7. Canada. Phone: (514) 393-1616. Fax: (514) 393-0110. E-mail: med-ent@autodesk.com Web Site: www.autodesk.com. Martin Vann, VP; Marc Petit, VP product dev.

London W1V 5FJ, 22 Soho Sq. Phone: 44 71 734 4224. Simon Shaw, gen mgr.

Santa Monica, CA 90405, 2110 Main St, Suite 207. Phone: (310) 396-1167. Brian Gaffney, gen mgr.

Autodesk solutions for creating, mgng & distributing digital content, so artists can create once and use anywhere.

Autogram Corp., Box 456, 1500 Capital Ave., Plano, TX, 75074-8113. Phone: (972) 424-8585. Phone: (800) 327-6901. Fax: (972) 423-6334. E-mail: info@autogramcorp.com Web Site: www.autogramcorp.com. Ernest T. Ankele Jr., pres/CEO; Delores Ankele, comptroller; John A. Stanley Jr., dir mktg.

Pacemaker IIk audio consoles, Pacemaker 6, 8, 10 Slide Pot, Mini-Mix 8 & Mini-Mix 12 Economy Consoles.Solution 20 audio systems, CYA-4 emergency switchers. Autoclock clock/timer/thermometer.

Automatic Devices Company, 2121 S. 12th St., Allentown, PA, 18103. Phone: (610) 797-6000. Fax: (610) 797-4088. Web Site: www.automaticdevices.com.

Cyclorama tracks, lighting tracks, lift & draw machines, electronic limit switches.

Autoscript, (formerly BDL-Autoscript). 391 Meadow St., Fairfield, CT, 06824. Phone: (203) 338-8356. Fax: (203) 338-8359. George Andros, consultant; Michael Accardi, pres.

London E14 9RL United Kingdom. Autoscript (UK), 10 Preston Rd. Phone: 44 0 20 7538 1427. Fax: 44 0 20 7515 9529. E-mail: b.larter@autoscript.tv. Brian Larter, mgng dir.

Design & manufacture of digital teleprompting systems, maintain a high level of new product dev.

AVerMedia Technologies Inc., 423 Dixon Landing Rd., Milpitas, CA, 95035. Phone: (800) 863-2332. Phone: (408) 263-3828. Web Site: www.aver.com. Arthur Pait, pres.

Aside from TV Turner/Desktop TV Personal Video Recorder products, AVerMeida also provides digital camera picture TV display devices, Document Camera & PC-to-TV Converters.

Avid Broadcast, 1925 Andover St., Tewksbury, MA, 01876. Phone: (978) 640-6789. Fax: (978) 640-1366. Web Site: www.avid.com. David Schleifer, dir bcstg & workgroups; Adam Taylor, VP sls & customer svc.

Burbank, CA 91502, 115 N. First St. Phone: (818) 557-2520. Fax: (818) 557-2558. John Steinhauer, dir bcst group sls.

New York, NY 10022, 575 Lexington Ave., 14th Fl. Phone: (212) 983-2424. Fax: (212) 983-8718. Michael Wright, dir sls.

Madison, WI 53719, 6400 Enterprise Ln, Suite 200. Phone: (608) 274-8686. Fax: (608) 273-5876. Robert Long, CTO.

Automated bcst newsroom systems. Non-linear video editing systems, media storage & networking systems, video server, content mgmt systems & asset mgmt systems.

Avid Technology Inc., Metropolitan Technology Park, One Park W., Tewksbury, MA, 01876. Phone: (800) 949-AVID. Phone: (978) 640-3669. Fax: (978) 851-0418. Fax: (978) 640-1366. E-mail: info@avid.com Web Site: www.avid.com. David Krall, pres/CEO; Paul Milbury, CFO; Joe Bentivegna, VP; Chas Smith, VP sls.

Burbank, CA 91502, 115 N. 1st St, Suite 100. Phone: (818) 557-2520. John Steinhauer, dir strategic sls America.

New York, NY 10017, 317 Madison Ave., Suite 521, 5th Fl. Phone: (212) 983-2424. Michael Wright, dir strategic sls.Avid's networked bcst news productions are designed to facilitate the process of digital news gathering (DNG).

Avtech Systems Inc., 141 Ayers Ct., Teaneck, NJ, 07666. Phone: (201) 833-8777. Fax: (201) 833-4995. E-mail: disamuel@aol.com Fred M. Samuel, pres; David Samuel, sls VP.

Closed circuit video equipment, components & accessories security video systems.

Axcera, Box 525, 103 Freedom Dr., Lawrence, PA, 15055. Phone: (724) 873-8100. Fax: (724) 873-8105. E-mail: info@axcera.com Web Site: www.axcera.com. Mike Rosso, sr VP; Richard Schwartz, VP mktg.

High and low power digital, analog, UHF and VHF transmitters and translators; exciter retofits; MMDS systems and Broadband wireless access technology.

B

B&B Systems, 1840 Flower St., Glendale, CA, 91201. Phone: (818) 551-5871. Fax: (818) 247-3487. E-mail: info@b-bsystems.com Web Site: www.b-bsystems.com.

Design & instal of production & postproduction systems, vans & mobile units, manufacturer of audio monitoring products.

BBE Sound Inc., 5381 Production Dr., Huntington Beach, CA, 92649. Phone: (714) 897-6766. Fax: (714) 896-0736. Web Site: www.bbesound.com. Rob Rizzuto, VP sls.

Audio/video signal processors to eliminate phase & amplitude distortion.

BEI Duncan Electronics, (BEI Technologies, Inc.). 170 Technology Dr., Irvine, CA, 92618-2401. Phone: (949) 341-9500. Fax: (714) 258-8120. E-mail: sales@beiducan.com Web Site: www.beiduncan.com. Dr. Demetris A. Agrotis, VP/gen mgr.

BEXT Inc., 1045 10th Ave., San Diego, CA, 92101. Phone: (619) 239-8462. Fax: (619) 239-8474. E-mail: bext@bext.com Web Site: www.bext.com. Dennis Pieri, CEO; Claudio Tilesi, CFO.

Radio & Digital TV Transmitters, Antennas, Amplifiers, Receivers, Boosters, STLs, RF Combiners, RF Filters, Stereo Generators, FmExtra Digital Radio Encoders & Receivers.

BGW Systems, Amplifier Technologies, Inc., 1749 Chapin Rd., Montebello, CA, 90640. Phone: (323) 278-0001. Fax: (310) 323-0083. E-mail: sales@bgw.com Web Site: www.ati-amp.com. Morris Kessler, pres; Angie Scott, dir opns.

Professional, bcst, coml audio power amplifiers & self-powered subwoofer systems.

BHP Inc., 4700 Chase Ave., Lincolnwood, IL, 60712. Phone: (847) 677-3000. Fax: (847) 677-1311. E-mail: sales@bhpinc.com Web Site: www.bhpinc.com. Jonathan Banks, pres.

Motion picture laboratory equipment, film printers & accessories.

BMG, 1540 Broadway, New York, NY, 10036. Phone: (212) 930-4000. Fax: (212) 930-4015. Web Site: www.bmg.com. Rolf Schmidt Hotz, chmn.

Produce, market & distribute recorded music.

Bald Mountain Laboratory, 222 Bellevue Rd., Troy, NY, 12180. Phone: (518) 279-9753. E-mail: hambob@highstream.net Robert S. Henry, owner.

Frequency readings.

Band Pro Film & Digital Inc., 3403 W. Pacific Ave., Burbank, CA, 91505. Phone: (818) 841-9655. Fax: (818) 841-7649. E-mail: sales@bandpro.com Web Site: www.bandpro.com. Amnon Band, owner; Renee Contreras, exec VP.

Dornach 85909 Germany. Band Pro Munich GmbH, Karl-Hammerschmidt Str. 38. Phone: 49 89-945 48 490. Gerhard Baieir, mgmg dir.

Tel Aviv 67897 Israel. Band Pro Israel, Hasolelim 3. Phone: (972) 3-562-1631. Fax: (972) 3-562-1632. Ofer Menashe, mgng dir.

Band Pro Film and Digital, Home of HD, offers cinemtographers the highest level of expertise & finest equipment available.

Barco Visual Solutions, LLC, 3059 Premiere Pkwy., Duluth, GA, 30097. Phone: (770) 218-3200. Fax: (770) 218-3250. E-mail: bpsmarketing@barco.com Web Site: www.barco.com/projection_systems. Larry Steelman, sls dir; Tom Ray, exec VP & gen mgr; Jim Durant, mktg mgr; Ellyce Kelly, mgr.

BARCO offers complete monitoring solutions for control rooms in telecom traffic, surveillance, public utilities, process control & financing.

Baron Telecom, 2355 Industrial Park Blvd., Cumming, GA, 30041. Phone: (678) 455-6298. Ran Bukshpan, CEO; Ron Raviv, CFO; Ross Kruchten, pres.

Frederick, MD 21703, 4640 Wedgewood Blvd. Phone: (301) 663-9300. Fax: (301) 663-9584. Tom Cureton.

Houston, TX 77070, 10430 Rogers Rd. Phone: (713) 973-6904. Fax: (713) 973-0205. Ross Kruchten.

Kirkland, WA 98033, 11112 117th Pl. N.E. Phone: (425) 739-9342. Fax: (425) 739-9314. Russ Stromberg.

Project mgmt & turn-key construction of communications towers, including erection, maintenance & inspection of tall towers.

Russ Bassett, 8189 Byron Rd., Whittier, CA, 90606. Phone: (562) 945-2445. Fax: (562) 698-8972. E-mail: E-mail@russbassett.com Web Site: www.russbassett.com. Maurice Heaton, VP mktg; Joe Malerba, VP sls.

High density storage cabinets for video tape, audio tape & CDs.

Battery Pros Inc., 5659 BlackJack Rd., Flowery Branch, GA, 30542-5402. Phone: (770) 271-8801. Phone: (800) 451-7171. Fax: (770) 271-9714. E-mail: sales@batteryprosinc.com Web Site: www.batteryprosinc.com. Patti Novak, pres; Maria Arce, sec.

Battery recelling/rebuilding for Bricks & Belts, primary & secondary batteries, custom battery pack design & manufacture.

Bauer Transmitters, 10870 Pellicano, #252, El Paso, TX, 79935. Phone: (915) 595-1048. Fax: (915) 595-1840. E-mail: paul @bauertx.com Web Site: www.bauertx.com. Paul E. Gregg, pres.

Remanufactured Bauer-Sparta & Elcom Bauer AM/FM transmitters, AM combining and antenna coupling equipment.

Belar Electronics Laboratory Inc., Box 76, 119 Lancaster Ave., Devon, PA, 19333. Phone: (610) 687-5550. Fax: (610) 687-2686. E-mail: sales@belar.com Web Site: www.belar.com. Arno Meyer, pres.

AM, FM, FM stereo, SCA, RDS/RBDS, shortwave, TV, TV stereo modulation & frequency monitors.

Belden Electronics Divison, Box 1980, Richmond, IN, 47375. Phone: (765) 983-5200. Fax: (765) 983-5294. E-mail: info@belden.com Web Site: www.belden.com. John Stroup, pres/CEO; Peter Sheehan, pres; Mark Myrick, VP sls; Ian O'Connell, mktg VP.

Precision video coaxial, triaxial cables, professional music cables, ENG cables, audio snakes, RGB cables, 50 ohm transmission cables.

Bencher Inc., 241 Depot St., Antioch, IL, 60002. Phone: (847) 838-3195. Fax: (847) 838-3479. E-mail: bencher@bencher.com Web Site: www.bencher.com.

Photographic & video vertical camera copystands & accessories, including Motion Picture Maker movable copy stage.

Benchmark Media Systems Inc., 5925 Court St. Rd., Syracuse, NY, 13206-1707. Phone: (315) 437-6300. Fax: (315) 437-8119. E-mail: info@benchmarkmedia.com Web Site: www.benchmarkmedia.com. Allen H. Burdick, pres; R. Rory Rall, sls mgr.

Audio processing & distribution systems, VU/PPM meters, interface/headphone amplifiers, microphone pre-amplifiers; digital to analog & analog to digital converters.

Bend-A-Lite Flexible Neon, 905 G St., Hampton, VA, 23661. Phone: (757) 245-7675. Fax: (757) 244-4819. Web Site: www.bendalite.com. E-mail: info@bendalite.com Hugh Jones, pres; Ron Koppel, mktg dir.

Flexible neon that can be cut with scissors, cut section can be re-electrified. 110v, 12v, 220v, 24v, indoor/outdoor. Lengths up to 300 ft., brilliant neon colors.

Benner-Nawman Inc., 3450 Sabin Brown Rd., Wickenburg, AZ, 85390. Phone: (800) 992-3833. Fax: (928) 684-7041. E-mail: mail@bnproducts.com Web Site: www.bnproducts.com. Edward R. Kientz, pres.

Specialty tools for CATV, cable termination & distribution boxes (cabinets).

Bexel Corp., 2701 N. Ontario St., Burbank, CA, 91504. Phone: (818) 841-5051. Fax: (818) 841-1572. E-mail: burbank@bexel.com Web Site: www.bexel.com. Andy Crist, CEO; Justin Paxton, dir opns & engrg dir.

Irvine, CA 92614, 1821 Kaiser Ave. Phone: (949) 955-2222. Fax: (949) 955-2294. Debbie White, rental mgr.

Miami, FL 33179, 20239 N.E. 15th Ct. Phone: (305) 653-5051. Fax: (305) 655-6209.

Norcross, GA 30093, 5555 Oak Brook Pkwy, Suite 160. Phone: (770) 448-3000. Fax: (770) 449-5747. Frank Zamor, rental mgr.

New York, NY 10019, 625 W. 55th St. Phone: (212) 246-5051. Fax: (212) 246-6373. Mke King.

Irving, TX 75061, 1001 N. Union Bower, Suite 130. Phone: (214) 946-5051. Fax: (972) 831-9860. Jim Barrett, contact.

Rental of video equipment & ancillary items to the video & audio production community. Used equipment sls.

Bexel, Inc., 5555 Oak Brook Pkwy., Suite 160, Norcross, GA, 30093. Phone: (770) 448-3000. E-mail: info@charterabs.com Web Site: www.bexel.com.

Orlando, FL 32811, 4201 Vineland Rd, Suite I-12. Phone: (407) 872-0054. Joan Hagle, natl sls mgr.

Chicago, IL 60622, 870 W. Division St, Unit E. Russell Roberts, lead rental coord.

Irving, TX 75038, 3251 W. Story Rd. Phone: (972) 869-9100. Alan McDonald, natl sls mgr.

Leading provider of rental equipment to the bcst & corporate video markets. Cameras, VTRs, digital edit systems, specialty gear.

Beyerdynamic, 56 Central Ave., Farmingdale, NY, 11735. Phone: (631) 293-3200. Phone: (800) 293-4463. Fax: (631) 293-3288. E-mail: salesusa@beyerdynamic.com Web Site: www.beyerdynamic.com. Nel Keinz, mgr; Bob Lowig, sls; Alan Feckanin, natl sls mgr.

Microphones, headsets, monitor headphones, studio & on-location UHF & VHF wireless systems.

Bird Electronic Corp., 30303 Aurora Rd., Solon, OH, 44139-2794. Phone: (866) 695-4569. Fax: (866) 546-4306. E-mail: sales@bird-technologies.com Web Site: www.bird-electronic.com.

Bird Technologies Group is a knowledgeable & innovative supplier providing solutions for RF Measurement & Mgmt in your world.

Birns and Sawyer Inc., 6381 De Longpre Ave., Los Angeles, CA, 90028. Phone: (323) 466-8211. Fax: (323) 466-1868. E-mail: info@birnsandsawyer.com Web Site: www.birnsandsawyer.com. William Meurer, pres.

Bitcentral Inc., (formerly Miralite Communications Inc.). 18872 Bardeen Ave., Irvine, CA, 92612. Phone: (949) 253-9003. Fax: (949) 474-1885. E-mail: sales@bitcentral.com Web Site: www.bitcentral.com. Fred Fourcher, CEO.

Bicentral is in the business of providing innovative solutions that transform the mgmt, production & distribution of the news.

Black Audio, Box 106, Ventura, CA, 93002. Phone: (805) 653-5557. E-mail: sales@blackaudio.com Web Site: www.blackaudio.com. Bruce Black, pres.

We provide parts, tools & accessories to all areas of pro audio.

Blimpy Floating Signs/Bend-A-Lite, 905 G St., Hampton, VA, 23661. Phone: (757) 245-7675. Fax: (757) 244-4819. E-mail: info@blimpy.com Web Site: www.blimpy.com. Hugh Jones, pres; Ron Koppel, mktg dir.

Giant blimps, hot air balloons & rooftop balloons. Complete custom department for any shape or size, flexible neon in seven brilliant colors.

Blonder Tongue Laboratories Inc., Box 1000, One Jake Brown Rd., Old Bridge, NJ, 08857-1000. Phone: (732) 679-4000. Fax: (732) 679-4353. Web Site: www.blondertongue.com.

Manufacturer of private cable equipment, including satellite receivers, modulators, processors, amplifiers, combiners, passives.

Bogen Communications Inc., Box 575, 50 Spring St., Ramsey, NJ, 07446. Phone: (201) 934-8500. Fax: (201) 934-9832. E-mail: info@bogen.com Web Site: www.bogen.com. Michael Fleischer, pres; David Chambers, sls VP; Maureen Flotard, CFO.

Audio amplifiers, mixer-preamplifiers, power amplifiers; FM/AM tuners & receivers; intercom systems; public address & sound reinforcement systems; digital repeater products; speakers.

Bogen Imaging Inc., 565 E. Crescent Ave., Ramsey, NJ, 07446. Phone: (201) 818-9500. Fax: (201) 818-9177. E-mail: info@bogenimaging.com Web Site: www.bogenimaging.us. Paul Wagner, sls dir; Mark Bender, sls dir.

Professional video products including tripods, fluid heads, dollies, stands & accessories. Grip equipment & lighting filters.

Boonton Electronics Corp., Box 465, 25 Eastmans Rd., Parsippany, NJ, 07054-0465. Phone: (973) 386-9696. Fax: (973) 386-1053. E-mail: boonton@boonton.com Edward Garcia, pres/CEO; John Kenneally, VP sls; Marc Wolfsohn, CFO; Brent Hessen-Schmidt, mktg dir; Richard Blackwell, engrg VP.

Electronic test & measuring equipment: microwave/RF power, RF voltmeters, capacitance/inductance & modulation meters.

Bradley Broadcast Sales, 7313 Grove Rd., Frederick, MD, 21704. Phone: (800) 732-7665. Phone: (301) 682-8700. Fax: (301) 682-8377. E-mail: info@bradleybroadcast.com Web Site: www.bradleybroadcast. Art Reed, gen mgr; Joellen Reed, mktg mgr.

Radio control room & transmission equipment, professional sound equipment, telephone interface devices.

Broadcast Data Consultants, 51 S. Main Ave., Suite 312, Clearwater, FL, 33765. Phone: (727) 442-5566. Phone: (800) 275-6204. E-mail: bdc@broadcastdata.com Web Site: www.broadcastdata.com. Neil Edwards, VP; Scott Wachtler, pres.

The Traffic C.O.P. for windows traffic billing progm. Free CD Rom demo available.

Broadcast Electronic Services, 4825 Trawler Ct., Jacksonville, FL, 32225. Phone: (904) 646-1630. Fax: (904) 641-1443.

Betabox/GPI net 410, video-editing interface products for E.N.G. & postproduction; T.B.C. remote devices.

Broadcast Electronics Inc., 4100 N. 24th St., Quincy, IL, 62305. Phone: (217) 224-9600. Fax: (217) 224-9607. E-mail: bdcast@bdcast.com Web Site: www.bdcast.com. Neil Glassman, VP mktg; Ray Miklius, VP; Tim Bealor, VP.

Radio bcst equipment including digital studio systems, AM, FM transmitters, RPUs & STLs.

Broadcast Engineering, 9800 Metcalf Ave., Overland Park, KS, 66212. Phone: (913) 341-1300. Fax: (913) 967-1905. E-mail: jchalon@prismbzb.com Web Site: www.broadcastengineering.com. Brad Dick, editor; Jon Chalon, publisher.

Banbury, Oxon OX16 8YJ, Box 250. Phone: +44-129-527-8407. Fax: +44-129-527-8408. Richard Woolley.

Shinjuku-ku, Tokyo 162-0822. Orient Echo Inc., 1101 Grand Maison, Shimomiyabi-cho 2-18. Phone: (03) 3235-5961. Fax: (03) 3235-5852. Mashy Yoshikawa.

Brooklyn, NY 11231, 335 Court St., 9. Phone: (718) 802-0488. Fax: (718) 522-4751. Josh Gordon.

Broadcast Engineering: Published for mgmt & engrg personnel working in bcst, production, postproduction, cable facilities in North America.

Broadcast Equipment Surplus Inc., Box 1300, Raymond, MS, 39154. Phone: (601) 857-8573. Fax: (601) 857-2346. E-mail: corkren@netdoor.com Jeffrey Corkren, VP.

Represents bcst equipment manufacturers; sls, svc, instal, turnkey designs, engrg; new & used equipment.

Broadcast International Group, 10458 N.W. 31st Terr., Miami, FL, 33172. Phone: (305) 599-2112. Fax: (305) 599-1133. E-mail: anamaria@bigmiami.com Web Site: www.bigmiami.com. Ana Maria Sagastegui, pres.

Bcst TV equipment.

Broadcast Microwave Services Inc., 12367 Crosthwaite Cir., Dock 10, Poway, CA, 92064. Phone: (858) 391-3050. Phone: (800) 669-9667. Fax: (858) 391-3049. E-mail: dept111@bms-inc.com Web Site: www.bms-inc.com. Graham Bunney, pres & gen mgr.

Los Angeles, CA 93065, 293 Sycamore Grove. Phone: (805) 581-4566. Fax: (805) 527-8263. Jim Kubit , sls engr.

Waynesboro, VA 22980, 105A Lew Dewitt Blvd, #278. Phone: (540) 932-3660. Fax: (858) 391-3049. Russell Murphy, sls engr.

COFDM wireless microwave, transmitters, receivers & antenna systems for ENG vehicles, helicopters, autotrackers, central receive sites.

Broadcast Sports Technologies, 1360 Blair Dr., Suite A, Odenton, MD, 21113. Phone: (410) 672-3900. Fax: (410) 672-3906. Peter Larsson, gen mgr.

Supply microwave, camera & cable equipment for large sporting events. Supply remote control cameras & communication systems.

Broadcast Store Inc., 1840 Flower, Glendale, CA, 91201. Phone: (818) 551-5858. Fax: (818) 551-0686. E-mail: bcssales@broadcaststore.com Web Site: www.broadcaststore.com. Lou Claude, pres.

Miami, FL 33179, 1031 Ives Dairy Rd. Phone: 305-266-2112. Fax: 305-266-2113.

New York, NY 10018, 500 W. 37th St. Phone: (212) 268-8800. Fax: (212) 268-1858.

Buy, sell, consign new & preowned Audio/Video bcst equiptment for production & post-production needs.

Broadcast Supply Worldwide, 7012 27th St. W., Tacoma, WA, 98466. Phone: (800) 426-8434. Fax: (800) 231-7055. E-mail: sales@bswusa.com Web Site: www.bswusa.com. Irv Law, chmn; Tim Schwieger, pres.

Audio bcst equipment distributor. Representing over 200 manufacturers worldwide.

Broadcast Video Systems Corp., 10 Woltner Way, Markham, ON, L3R 4R4. Canada. Phone: (905) 305-0565. Fax: (416) 946-1964. E-mail: bvs@bvs.ca Web Site: www.bvs.ca. Bert Verwey, pres.

SDI, analog video keyers, chroma keyers, closed captioning, encoders/decoders, positioner, bridge, V-chip, data transmission, encoders & transcoders.

Broadcasters General Store Inc., 2480 S.E. 52nd St., Ocala, FL, 34480. Phone: (352) 622-7700. Fax: (352) 629-7000. E-mail: info@bgs.cc Web Site: www.bgs.cc.

Professional audio, video & RF equipment. Telco interfaces, digital codecs, 400 vendor line card.

Bryston Ltd., Box 2170, 677 Neal Dr., Peterborough, ON, K9J 7Y4. Canada. Phone: (705) 742-5325. Fax: (705) 742-0882. Web Site: www.bryston.ca.

Audio amplifiers, pre-amplifiers, crossovers, & microphone pre-amps.

Bud Industries Inc., 4605 E. 355 St., Willoughby, OH, 44094. Phone: (440) 946-3200. Fax: (440) 951-4015. E-mail: saleseast@budind.com Web Site: www.budind.com. Blair K. Haas, VP mktg.

Phoenix, AZ 85080, Box 41190. Phone: (623) 516-9494.

Open & welded racks; cabinets & accessories.

Burk Technology, 7 Beaver Brook Rd., Littleton, MA, 01460. Phone: (978) 486-0086. Fax: (978) 486-0081. E-mail: sales@burk.com Web Site: www.burk.com. Peter C. Burk, pres; Anita Russell, gen mgr.

Transmitter remote control systems including multi-site, unattended units & automatic transmitter control system. LX-1 six input stereo selector.

Burle Industries Inc., 1000 New Holland Ave., Lancaster, PA, 17601-5688. Phone: (717) 295-6000. Fax: (717) 295-6096. E-mail: burlesls@burle.com Web Site: www.burle.com. E. Burlefinger, pres/CEO; Carl Rintz, exec VP; Kirk Jenne, gen counsel.

VHF/FM power tubes, photomultipliers & imaging devices.

Burlington A/V Recording Media Inc., 106 Mott St., Oceanside, NY, 11572. Phone: (516) 678-4414. Phone: (800) 331-3191. Fax: (516) 678-8959. E-mail: sales@burlington-av.com Web Site: www.burlington-av.com. Ruth Schwartz, VP; Jan Alan, pres.

Wholesale distributor for all formats of recording media, blank audio/video tape, CD-R, DVD-RR, diskettes, data media, A/V recording equipment.

Burst Electronics Inc., Box 65972, Albuquerque, NM, 87193. Phone: (505) 898-1455. Fax: (505) 898-0159. E-mail: sales@burstelectronics.com Web Site: www.burstelectronics.com. Brad Hamlin, pres.

CG, DA's, HD/analog video switchers, video mixers, decoders, logo generator, video generators, TBC & GPI converters.

C

CADCO Systems Inc., 2363 Merritt Dr., Garland, TX, 75041. Phone: (972) 271-3651. Phone: (800) 877-2288. Fax: (972) 271-3654. E-mail: carmen@cadcosystems.com Web Site: www.cadcosystems.com. Steven G. Johnson, chmn/CEO; Carmen Howard, stn mgr.

Manufacturer of CATV & broadband communication products such as modulators, demodulators, signal processors, channel converters, translators & special application headend equipment, fixed-channel & frequency agile.

CATV Services Inc., 12099 N.W. 98th Ave., Hialeah Gardens, FL, 33010-2927. Phone: (305) 512-5601. Phone: (800) 227-1200. Fax: (305) 512-5606. E-mail: ryan.richard@catvservices.com Web Site: www.catvservices.com. Richard C. Richmond, pres.

Excess inventory professionals, buy & sell.

CBT Systems, 10115 Carroll Canyon Rd., San Diego, CA, 92131. Phone: (858) 536-2927. Fax: (858) 536-2354. Darrell Wendhardt, pres.

TV bcst studio systems & mobile unit design, engrg & integration.

CCI, 101 Merritt Ave., Iron Mountain, MI, 49801. Phone: (906) 774-1755. Fax: (906) 774-6117. E-mail: info@ccinc,us Web Site: www.cciinc.us. John P. Jamar, pres.

CCI is an employee owned, full service turnkey provider of engineering, mapping, construction, and integration services for converging Video, Voice, Data, Transportation and Utility networks.

C-COR, 60 Decibel Rd., State College, PA, 16801. Phone: (814) 238-2461. Fax: (814) 238-4065. Web Site: www.c-cor.net. David Woodle, chmn/CEO.

Pleasanton, CA 94588. Broadband Management Soultions, Software Divison Headquarters, 5673 Gibraltar Dr., Suite 100. Phone: (925) 251-3000. Fax: (925) 467-0600. Douglas W. Engerman.

Lakewood, CO 80228. Broadband Network Services, Services Division, 300 Union Blvd., Suite 515. Phone: (303) 980-8058. Paul E. Janson.

Meriden, CT 06450. Broadband Communication Products, Product Division Headquarters, 999 Research Pkwy. Phone: (203) 630-5700. Fax: (203) 630-5701. John O. Caezza.

Globally-tailored fiber optic, RF & digital video transport telecommunications products, OSS mgmt solutions & high-end tech field svcs for broadband networks.

CEA-Computer Engineering Associates, 7526 Connelley Dr., Suite H, Hanover, MD, 21076. Phone: (410) 987-7003. Fax: (410) 987-6710. E-mail: ceanews@erols.com Web Site: www.ceanews.com. Paul Keys, pres.

Bridgeport, NJ 08014, 600 Heron Dr. Phone: (800) 888-3922. Steve McKemy, VP.

Gibsonia, PA 15044, 5465 Rt. 8. Phone: (412) 443-2600. Bill Interthal, mgr.

CEA newsroom system—complete automation systems for radio & TV newsrooms.

CECO International Corp., 440 W. 15th St., New York, NY, 10011. Phone: (212) 206-8280. Fax: (212) 727-2144. Web Site: www.cecostudios.com. Donald Kline, owner & pres.

Motion picture & TV equipment; sound stages; location trucks with generators.

CED, 3590 N.W. 34th St., Miami, FL, 33142. Phone: (305) 635-5361. Fax: (305) 635-5366. David Levy, mgr.

MMDS, wireless cable, UHF, VHF quality transmission systems. Electrical Distributors.

C I S Inc., 3360 Martin Farm Rd., Suwanee, GA, 30024. Phone: (678) 482-2000. Fax: (678) 482-2007. E-mail: sales@cisfocus.com Web Site: www.cisfocus.com. Jeffery Eichler, pres; Lynn Hamlin, VP sls.

Integrated broadband & fiber design software products. Software solutions for network mapping, planning, design, & management of the outside plant.

CMP Enclosures Inc., 3901 Grove Ave., Gurnee, IL, 60031. Phone: (847) 244-3230. Fax: (847) 244-3257. E-mail: cmpencl@aol.com Web Site: www.enclosures.com. Mike Gober, pres; Wendi Lee, VP.

Manufacturers of electronic enclosures for rack mounting equipment.

COASTCOM, 1151 Harbor Bay Pkwy., Alameda, CA, 94502. Phone: (510) 523-6000. Fax: (510) 523-6150. E-mail: info@coastcom.com Web Site: www.coastcom.com. E. M. "Ted" Buttner, pres.

Alameda, CA 94502, 1151 Harbor Bay Pkwy. Phone: (510) 523-6000. Mark Packwood, rgnl sls mgr.

Charlotte, NC 28210, 5000 Sharonwoods Ln. Phone: (704) 643-7221. Duane Hardie, rgnl sls mgr.

Lewisville, TX 75067, 562 Continental Dr. Phone: (972) 316-3611. Mike Walsh, rgnl sls mgr.

Pearl River, WY 10965, 19 Harding St. Phone: (914) 980-3703. Tom McCafferty, rgnl sls mgr.

Manufacturer of T1 voice data network systems specializing in T3 cross connecting, T1 multiplexing & digital program channels for audio bcstg.

COMTEK Inc., 357 W. 2700 S., Salt Lake City, UT, 84115. Phone: (801) 466-3463. Phone: (800) 496-3463. Fax: (801) 484-6906. E-mail: sales@comtek.com Web Site: www.comtek.com. Ralph Belgique, chief engr; Laurel Robertson, sls dir; Jon Belgique, Communication Director.

COMTEK manufactures synthesized & fixed frequency wireless communication equipment & accessories, including cuing systems (IFB) & wireless microphones.

CONTEC Corp., 1023 State St., Schenectady, NY, 12307-1511. Phone: (518) 382-8000. Fax: (518) 382-8452. Richard Kielb, sr VP; Steve Knuth, pres/CEO.

Phoenix, AZ 85040, 4114 E. Wood St, Suite 103. Phone: (602) 437-2890. Paul Hagert, mgr.

Tampa, FL 33610, 5906 Breckenridge Rd, Suite A. Phone: (813) 623-1721. Dick Lawton, mgr.

Bloomington, IN 47404, 2480 N. Curry Pike. Phone: (812) 330-8727. Jeff Van Horne, mgr.

Seattle, WA 98148, 1250 S. 192nd St. Phone: (206) 244-5770. Bob Vick, mgr.

Motorala , Pace, Phillips & Scientific Atlanta Authorized Warranty Digital Repair Svc Ct. Manufactor of universal remote controls for digital terminals.

CORPLEX Inc., 203 Northfield Rd., Northfield, IL, 60093-3311. Phone: (847) 784-9700. Fax: (847) 784-9701. E-mail: carter730@aol.com Web Site: www.corplextv.com. Carter Ruehrdaz, pres.

Video production equipment, postproduction equipment, rental & mobile TV.

CPC-Computer Prompting & Captioning Co., 1010 Rockville Pike, Suite 306, Rockville, MD, 20852. Phone: (301) 738-8487. Phone: (800) 977-6678. Fax: (301) 738-8488. Web Site: www.cpcweb.com. Dr. Dilip Som, pres; Sidney Hoffman, VP.

Closed captioning, subtitling, DVD, V-Chip, teleprompting systems & svcs, Crossover Links for WebTV.

CS Communications Inc., 9825 Bridleridge Ct., Vienna, VA, 22181. Phone: (703) 938-5365. Fax: (703) 938-5823. E-mail: chazsamp@aol.com Charles E. Sampson, pres.

Engrg & consulting svcs for wireless and satellite systems. System design, feasibility & economic analysis.

CSG Systems, 7887 E. Belleview, Suite 1000, Englewood, CO, 80111. Phone: (303) 796-2850. Fax: (303) 804-4088. Web Site: www.csgsystems.com. Jack Pogge, pres/COO; Neal Hansen, CEO; Peter Kalan, CFO; Randall Cardinal; Kurt Silverman, CTO; William Fisher, pres GSS; Ed Nafus, pres BSD; Ed Mangold, sr VP global sls; Sally Else, sr VP product mgmt; Liz Bauer, sr VP investor rel & corp communications; Darren Walsh, sr VP global professional svcs; Alan Michels, VP/gen mgr., MA 038986 Singapore, 6 Temasek Blvd. Phone: 65 6883 1900. 65 6883 1990.

London WC2N6HT United Kingdom, 1-11 John Adams St. Phone: 44 20 7004 1840. 44 20 7004 1841.

Miami, FL 33126, 6303 Blue Lagoon Dr. Phone: (305) 421-8900. (305) 421-8934.

Complete sub info mgmt & data processing systems for the cable TV & telephone industries.

CSI-Camera Support International, Box 681, Woodland Hills, CA, 91365. Phone: (818) 224-4850. Fax: (818) 887-5727. E-mail: markintash@aol.com Web Site: www.csitripods.com.

Camera support dollies, tripods, pan/tilt heads & accessories ENG EFP & studio application for bcst & industrial application.

Cable Leakage Technologies, 903 N. Bowser, Suite 150, Richardson, TX, 75081-2375. Phone: (972) 907-8100. Fax: (972) 907-2950. Web Site: www.wavetracker.com. Perry Havens, pres.

Digital RF tracking/mapping system used in CLI monitoring.

Cable Prep, (Ben Hughes Communication Products Co.). Box 373, 207 Middlesex Ave, Chester, CT, 06412-0373. Phone: (860) 526-4337. Fax: (860) 526-2291. Web Site: www.cableprep.com. E-mail: toolmaker@cableprep.com Deborah Morrow, pres; David Morrow, VP.

Cable Prep® TerminX, hex crimp, coring & stripping, drop wire stripping, jacket strippers, messenger removal & tools.

Cable Serv Inc., 4560 Eastgate Pkwy., Mississauga, ON, L4W 3W6. Canada. Phone: (905) 629-1111. Fax: (905) 629-1115. Web Site: www.cableserv.com.

TV Exciters, 5-10-20 watt LPTV trans & transmitters, TV modulators, demodulators, processors, & satellite receivers.

Cable Services Company Inc., 2113 Marydale Ave., Williamsport, PA, 17701. Phone: (570) 323-8518. Fax: (570) 322-5373. Web Site: www.cable-services.com.

Turnkey fiber-optic & coaxial construction; distributor of CATV products.

Cable Technologies International, 460 Oakdale Ave., Hatboro, PA, 19040. Phone: (215) 672-5400. Fax: (215) 672-0440. E-mail: sales@cabletechnologies.com Web Site: www.cabletechnologies.com.

Hand held remotes, converter parts, cosmetic & electronic; test equipment, headend & linegear.

Cable Yellow Pages, 20917 Higging Court, Torrance, CA, 90501. Phone: (800) 777-4320. Fax: (310) 212-5392.

Phone directory for cable TV systems.

CablePro, (A division of ICM Corp.). 6260 Downing St., Denver, CO, 80216. Phone: (303) 288-8107. Fax: (303) 288-4769. E-mail: sales@icmcorp.net Web Site: www.icmcorp.net. Randy Holiday, pres; Gary Williams, sls VP.

CablePro's attention to design, material & workmanship produces the highest quality for instal tools.

CableReady Inc., (A division of ICM Corp). 6260 Downing St., Denver, CO, 80216. Phone: (303) 288-8107. Fax: (303) 288-4769. E-mail: sales@icmcorp.net Web Site: www.icmcorp.net. Randy Holliday, pres; Gary Williams, VP sls.

Painted galvolume molding with custom fittings backed by a 15-year warranty, U.L. listed and Class A fire rated.

CableTek Wiring Products Inc., 1150 Taylor St., Elyria, OH, 44035. Phone: (440) 365-2487. Fax: (440) 322-0321. E-mail: treilly@apk.net Web Site: www.cable-tek.com. Tim Reilly, gen mgr.

Interior & exterior surface wiring products; terminal enclosures, residential enclosures, security products.

Cablynx, Inc., (Formerly Nova Systems/Shintron). 500 W. Cummings Park, Suite 1150, Woburn, MA, 01801. Phone: (781) 933-2000. Fax: (781) 933-4641. E-mail: sales@nova-sys.com Sam Asano, pres.

Routing switchers, distribution amplifiers, time code, component video, PC accessories, compugraphics to video, frame synchronizer.

Calculated Industries Inc., 4840 Hytech Dr., Carson City, NV, 89706. Phone: (775) 885-4900. Fax: (775) 885-4949. E-mail: info@calculated.com Web Site: www.calculated.com. Mark Paulsen,.

Time code calculators work in & convert between all time formats; drop/non-drop, multiple EPS rates for all SMPTE/PAL equations.

California Amplifier, 1401 N. Rice Ave., Oxnard, CA, 93030. Phone: (805) 987-9000. Fax: (805) 987-8359. Web Site: www.calamp.com. E-mail: sales@calcamp.som Tom Prochnow, sls VP; Rick Wheeler, VP; Philip Cox, VP.

Manufacturer of mesh & offset satellite antennas ranging in size from 18" to 16'.

Calumet Photographic, 900 W. Bliss St., Chicago, IL, 60622. Phone: (630) 860-7458. Phone: (312) 944-2774. Fax: (312) 944-4035. Web Site: www.calumetphoto.com. Peter Biasotti, pres.

Calzone Case Co., 225 Black Rock Ave., Bridgeport, CT, 06605. Phone: (203) 367-5766. Fax: (203) 336-4406. E-mail: vin.calzone@calzonecase.com Web Site: www.calzonecase.com. Joseph E. Calzone, pres; Vincent J. Calzone, sls VP.

City of Industry, CA 91745, 15730 Salt Lake Ave. Phone: (626) 968-4100. Fax: (626) 968-1703. Mike Herman, VP admin/sls.

Carrollton, TX 75007, 75006 Luna Rd, Suite 126. Phone: (972) 241-3900. Fax: (972) 241-3998. Tom Mackno, VP.

Manufacturers of custom & standard shipping cases for all industries featuring Escort, LD-ATA, Military, X series, Titan

Camera Service Center, (A division of Arri, Inc.). 619 W. 54th St., New York, NY, 10019. Phone: (212) 757-0906. Fax: (212) 713-0075. Web Site: www.cameraservice.com. Hardwrick Johnson, VP opns; Simon Broad, mktg VP.

Fort Lauderdale, FL 33312, 2385 Stirling Rd. Phone: (954) 322-4545. Fax: (954) 322-4188. Ed Stamm.

The largest full-svc film equipment rental company, carrying a complete line of camera & lighting products.

CamMate Studios/Systems, 425 E. Comstock, Chandler, AZ, 85225. Phone: (480) 813-9500. Fax: (480) 813-9292. E-mail: cammate@cammate.com Web Site: www.cammate.com. Ron Mitchell, CEO.

Exclusive sls & rental of the CamMate, a single operator remote camera crane in various configurations for video & film.

Camplex Corporation, 3302 W. 6th Ave., Emporia, KS, 66801. Phone: (620) 342-7743. Fax: (620) 342-7405. E-mail: jtwebb@camplex.com Web Site: www.camplex.com. J. Thomas Webb, CEO; C. Duane Woodmas, pres.

CAMPLEX is a universally adaptable video/audio signals multiplexing system for ENG/EFP/SNG Prosamer cameras, camcorders used in remote applications.

Canare Corp., 45 Commerce Way, Totowa, NJ, 07512. Phone: (201) 944-3433. Fax: (201) 944-2290. E-mail: info5@canare.com Web Site: www.canare.com. Larry Cano, sls; Kazuo Urata, pres/CEO.

Professional Audio & Video Cable, 75 ohm connectors, patchbays, snake systems, assemblies, strip & crimp tools, fiber optical products.

Canon U.S.A. Inc., (Broadcast Equipment Division Headquarters). 65 Challenger Rd., Ridgefield Park, NJ, 07660. Phone: (800) 321-4388. Fax: (201) 807-3333. E-mail: bctv@cusa.canon.com Web Site: www.canonbroadcast.com. Rich Eiles, sls; Patrick Breheny, sls; John Rose, sls.

Irvine, CA 92618, 15955 Alton Pkwy. Phone: (949) 753-4330. Fax: (949) 753-4337. Tom Bender, sls; Joe Patton, sls; Stephanie Franz, sls.

Norcross, GA 30093, 5625 Oakbrook Pkwy. Phone: (770) 849-7890. Fax: (770) 849-7888. Mack Mc Crary, sls; Jim Dobbins, sls.

Itasca, IL 60143, 100 Park Blvd. Phone: (630) 250-6236. Fax: (630) 250-0399. Lou Bobroff, sls; Jon Giunchedi, sls.

Irvine, TX 75063, 3200 Regent Blvd. Phone: (972) 409-8871. Fax: (972) 409-8869. Mark Parks, sls.

The Canon Broadcast & Communications (BCTV) division a part of the larger Canon U.S.A. Inc. The BCTV lens products is squarely based on the highly advanced optical, mechanical & digital technologies for which Canon became legendary. Studio, field & ENG lenses & svc, (HDTV/SDTV) video, audio, data optical beam transmission, remote control P/T/Z camera system.

Capstone Communications Inc., 163 Grandview Ln., Mahwah, NJ, 07430. Phone: (905) 472-2330. Web Site: www.capstonecomm.com.

Brokerage, rsch consultation & bcst equipment brokerge.

Carpel Video Inc., 429 E. Patrick St., Frederick, MD, 21701. Phone: (800) 238-4300. Phone: (301) 694-3500. Fax: (301) 694-9510. Web Site: www.carpelvideo.com. Andy Carpel, pres.

Videotape wholesalers. Mail order post production in MD; store: DVD production and duplication. Lowest prices on 6 blank video tapes. 800-238-4300.

Celco, 8660 Red Oak Ave., Rancho Cucamonga, CA, 91730. Phone: (909) 481-4648. Fax: (909) 481-6899. E-mail: info@celco.com Web Site: www.celco.com.

Design & manufacture of motion picture film recorders.

Center City Film & Video, 1503-05 Walnut St., Philadelphia, PA, 19102. Phone: (215) 568-4134. Fax: (215) 568-6011. Web Site: www.ccfv.com. Jordan Schwartz, chmn; Brian Tsely, VP/gen mgr.

Studio/remote/postproduction D-2, D-3, 1" - Beta - 3/4" - ADO - Paint Box/Abekas 62/GV300 with E-Mem; film/tape DaVinci color correction, ADO repositioning & interactive motion control; D-s, D-3; AVID, Digital Betacam.

Century Precision Optics, 7701 Haskell Ave., Van Nuys, CA, 91406. Phone: (818) 766-3715. Fax: (818) 505-9865. E-mail: info@centuryoptics.com Web Site: www.centuryoptics.com. Bill Turner, VP.

Wide angle & telephoto lens for video & motion picture cameras; lens accessories; lens service; schneider filters.

Channel Master L.L.C., 1315 Industrial Park Dr., Smithfield, NC, 27577. Phone: (919) 934-9711. Fax: (919) 989-2200. Web Site: www.channelmaster.com. Bill Currer, pres; George Jusaites, dir.

Manufacturer of Satellite Broadband, DBS & Off-Air Antenna sysstems, Electronics & Accessories, and Point-to-Point Microwave equipment.

Channel One Lighting Systems Inc., 1522 E. 6th St., Tulsa, OK, 74120-4026. Phone: (918) 587-2663. W. Blair Powell, pres.

Complete line of lighting equipment for TV, theatre & industrial applications; specializes in the design & manufacture of electrical distribution, grid & cyclorama systems, curtain & track.

Channell Commercial Corp., Box 9022, 26040 Ynez Rd., Temecula, CA, 92589-9022. Phone: (909) 719-2600. Fax: (909) 296-2322. E-mail: info@channellcorp.com Web Site: www.channellcomm.com. William H. Channell Jr., pres; Andrew M. Zogby, VP global mktg; John Kaiser, VP N America sls.

Global designer & manufacturer of equipment, offers a complete line of enclosures for CATV & telecommunication

Charles Industries Ltd., 5600 Apollo Dr., Rolling Meadows, IL, 60008. Phone: (847) 806-6300. Fax: (847) 806-6231. Web Site: www.charlesindustries.com. Joseph T. Charles, pres.

Pedestals, custom security boxes, amplifier & TAP brackets-hardware, splicing vaults, taps, splitters & couplers.

Cheetah International, 8120 Sheridan Blvd., Suite C-206, Westminister, CO, 80003. Phone: (520) 751-8681. Fax: (520) 722-1699. E-mail: sales@caption.com Web Site: www.caption.com. Donald Miller, pres.

Closed captioning software on-line & postproduction & related hardware.

Chicago Condenser Corp., (A division of Capacitor Industries). 6455 N. Avondale Ave., Chicago, IL, 60631. Phone: (773) 774-6666. Fax: (773) 774-6690. E-mail: info@capacitorindustries.com Web Site: www.capacitorindustries.com. Terry Noone, pres.

High Voltage Filter Capacitors for radio & TV bcst transmission.

Chrono-Log Corp., 2 W. Park Rd., Havertown, PA, 19083. Phone: (610) 853-1130. Fax: (610) 853-3972. E-mail: chronlog@chronolog.com Web Site: www.chronolog.com. Paula Freilich, pres.

GPS Receiver (time only), WWV synchronizer, digital clocks & time display systems, time code generators.

Chyron Corp., 5 Hub Dr., Melville, NY, 11747. Phone: (631) 845-2000. Fax: (631) 845-3895. Web Site: www.chyron.com. Alec Shapiro, Vp sls & mktg; Ed Grebow, pres/CEO; Steve Sloane, VP international sls; Patricia Lampe, VP/treas; David Buckler, client dev.

Cupertino, CA 95014. Chryon Corp. West, 10121 Miller Ave, Suite 201. Phone: (408) 873-3800. Fax: (408) 986-0452. Denise Gallant, product mgr.

Atlanta, GA 30303, One CNN Ctr., South Towers, Suite 558. Phone: (404) 880-9004. Fax: (404) 880-9104. Ryad Kahale, Chyron rgnl sls.

A leading provider of broadcast hardware, software & services spanning television & the Internet. Provides a broad range of leading edge hardware & software products, including paint & animation systems, character generators, master control switches, & bcst, automation & media mgmt packages.

Cine 60 Inc., 630 9th Ave., New York, NY, 10036. Phone: (212) 586-8782. Fax: (212) 459-9556. E-mail: info@cine60newyork.com Web Site: www.cine60newyork.com. Paul Wildum, pres; Vidal Ortiz, gen mgr; Richard Ortiz, mgr.

Nickel-Cadmium battery belts, battery packs, chargers, sun-guns, kits, dir chair, dir viewfinders, slates, cables & snaplocks.

Cintel Inc., 25020 Ave. Stanford, Suite 190, Valencia, CA, 91355. Phone: (661) 294-2310. Fax: (661) 294-1019. E-mail: sales@cintelinc.com Web Site: www.cintelinc.com. Adam Welsh, mgng dir; Curtis Christianson, opns mgr; David Saville, sls dir.

Chestnut Ridge, NY 10977, 80 Red Schoolhouse Rd, Suite 103. Phone: (914) 371-7220. Fax: (914) 371-6896. David Saville, sls dir.

Flying spot telecines, DVE system, keycode system, high-resolution scanner, color correctors.

Circuit Research Labs Inc. (CRL Systems, Inc.), 7970 S. Kyrene Rd., Tempe, AZ, 85284-2199. Phone: (480) 403-8300. Fax: (480) 403-8301. E-mail: crl@crlsystems.com Web Site: www.crlsystems.com. Robert McMartin, CEO; Jay Brentlinger, pres; Robert Orban, VP; Greg Ogonowski, VP new prod dev; Gary Clarkson, VP/sec; Phillip Zeni, COO/VP.

Multiband audio AGCs, compressors & limiters for AM/FM; MTS processors, stereo generators, shortwave, AES/EBU digital audio tester.

Clark Wire & Cable Co. Inc., 1355 Armour Blvd., Mundelein, IL, 60060-4401. Phone: (847) 949-9944. Fax: (847) 949-9595. E-mail: sales@clarkwire.com Web Site: www.clarkwire.com. Susan Clark, owner.

Audio, video, camera & speciality cable products for bcst industry, available in bulk or assembled harnesses, connectors, panels, reels, & boxes.

Clear-Com Communication Systems, 4065 Hollis St., Emeryville, CA, 94608-3505. Phone: (510) 496-6666. Fax: (510) 496-6699. E-mail: sales@clearcom.com Web Site: www.clearcom.com. Michael Wang, gen mgr; Ed Fitzgerald, natl sls mgr. Eastleigh, England, Eastleigh, England. Phone: 44-23-8090-7000. Patrick Woolcocks, EMEA Direct Sales.

Walnut Creek, CA 94596, Box 302. Phone: (925) 932-8134. Peter Giddings, Asia/Pacific dir of sls. (Export division office).

Single & multi-ch hardwire intercom systems for use in teleproduction. Wired & wireless partyline & digital matrix intercom systems.

Clearone Communications Corp., 1825 Research Way, Salt Lake City, UT, 84119. Phone: (801) 975-7200. Phone: (800) 945-7730. Fax: (801) 977-0087. Web Site: www.clearone.com. Fran Flood, CEO; Randy Wichinski, CFO.

Professional audio & teleconferencing.

CoarcVideo, Box 2, Rt. 217, Mellenville, NY, 12544. Phone: (800) 888-4451. Phone: (518) 672-4451. Fax: (518) 672-4048. E-mail: coarc@aol.com Bob Spiewak, dir mktg; Alva Stalker, production mgr.

Used by bcstrs, cable systems, duplicating houses, production companies for environmentally-designed videotape reloaded products & standard video tape products; provides Umatic & Betacam VHS tape, program fulfillment svcs. CoarcVideo is part of the Coarc organization, which trains employees & provides various programs for the disabled.

Coaxial Dynamics, 6800 Lake Abram Dr., Middleburg Hts., OH, 44130. Phone: (440) 243-1100. Fax: (440) 243-1101. E-mail: coaxial@apk.net Web Site: www.coaxial.com. Joe Kluha, gen mgr.

RF wattmeters, terminations, RF load resistors, RF couplers & accessories.

Cohu Inc., (Electronics Division). Box 85623, San Diego, CA, 92186-5623. Phone: (858) 277-6700. Fax: (858) 277-0221. E-mail: info@cohu.com Web Site: www.cohu.com/cctv. Joe Olmstead Jr., natl sls mgr; Jeff Tyler, mktg mgr.

CCTV cameras & camera control systems, color, CCD, B/W.

Colorado Video Inc., Box 928, Boulder, CO, 80306. Phone: (303) 530-9580. Fax: (303) 530-9569. E-mail: kirk@colorado-video.com Web Site: www.colorado-video.com. Kirk Fowler, pres.

Image transmission for UBI system; time-division video multiplexing/demultiplexing system.

Comex Worldwide Corp., Box 8, Aldie, VA, 20105-0008. Phone: (703) 327-1520. Fax: (703) 327-1540. E-mail: cwcmmds@hotmail.com Web Site: www.comexworldwide.com. Jack A. Rickel, pres/CEO; Susan Rose, gen mgr.

CWC develops bcst & pay TV systems, VHF/UHF bcsts, satellite communications, MMDS & cable systems & turnkey communication systems.

Comm Scope Inc., 1100 Comm Scope Pl. S.E., Hickory, NC, 28602. Phone: (800) 982-1708. Phone: (828) 324-2200. Fax: (828) 328-3400. Web Site: www.commscope.com. Frank Drendel, CEO.

Coaxial & fiber-optic cables including CRD & NEC approved drop cables, QR, P3 & CableGuard.

Commercial Electronics Ltd., 1335 Burrard St., Vancouver, BC, V6Z 1Z7. Canada. Phone: (604) 669-5525. Fax: (604) 669-6347. E-mail: pro@cemail.ca Web Site: www.commercialelectronics.ca. H.H. von Tiesenhausen, pres.

Nanaimo, BC V9X-1A5 Canada, 1678 Extension Rd, Unit 2. Phone: (250) 754-7612. Mark Arleft, video sls.

Audio video equipment, systems designs

Commercial Radio Monitoring Co., 103 S.W. Market St., Lee's Summit, MO, 64063. Phone: (816) 524-3777. Fax: (816) 524-3777. Web Site: www.commercialradio.us. W. R. Thorsen, pres; Ronald Thorsen, VP.

Frequency measurements & equipment calibration.

Communication & Power Industries, 811 Hansen Way, Palo Alto, CA, 94304. Phone: (650) 846-2800. Fax: (650) 846-3706. Web Site: www.cpii.com. Joseph Caldarelli, CEO.

Georgetown, ON L7G-2J4 Canada, 45 River Dr. Phone: (905) 877-0161. Joseph Caldarelli, pres communications & medical products.

Palo Alto, CA 94304, 607 Hansen Way. Phone: (415) 846-2900. Al Ferriera, pres traveling wave tube products.

Palo Alto, CA 94304, 811 Hansen Way. Phone: (415) 846-2800. (415) 846-3700. Armand Staprans, pres microwave products; Jim Commendatore, pres Satcom.

San Carlos, CA 94070, 301 Industrial Way. Phone: (415) 592-1221. H. Frederick Koehler, pres Eimac Div.

Beverly, MA 01915. Beverly Microwave Division, 150 Sohier Rd. Phone: (978) 922-6000. Dennis Gleason, division pres.

Manufactures a complete line of power grid tubes, klystrons & klystrode IOTs, traveling wave tubes, satellite communication transmitters, microwave components.

Communication & Power Industries, EIMAC Division, 301 Industrial Rd., San Carlos, CA, 94070. Phone: (650) 592-1221. Phone: (800) 414-8823. Fax: (650) 592-9988. E-mail: powergrid@eimac.cpii.com Web Site: www.eimac.com. Michael Chen, pres; John Allan, VP mktg.

Power grid tubes, cavity amplifiers, IOT (UHF TV).

Communication Graphics, Inc., 1765 N. Juniper, Broken Arrow, OK, 74012. Phone: (800) 331-4438. Phone: (918) 258-6502 (Okla). Fax: (918) 251-8223. Web Site: www.cgilink.com. Dave Cleveland, pres.

Choose the company MORE radio stations have selected for printing decals, event stickers, statics, concert patches, magnets, media kits and more!

Communications General Corp., 2685 Alta Vista Dr., Fallbrook, CA, 92028-9739. Phone: (760) 723-2700. E-mail: r.gonsett@ieee.org Robert F. Gonsett, pres.

Monthly AM, FM & TV frequency measurements in the Southern California area & spectral measurements.

Communications Specialties Inc., 55 Cabot Ct., Hauppauge, NY, 11788. Phone: (631) 273-0404. Fax: (631) 273-1638. E-mail: info@commspecial.com Web Site: www.commspecial.com. Paul Seiden, sls dir.

Shaw Tower 189702 Singapore, 100 Bencoolen Rd., # 22-09. Phone: (+665) 656 391-8790. Fax: (+656) 656 396-0138. Jeohan Tohkingkeo, rgnl mgr Asia Pacific.

Manufacturer of fiber-optic transmission sytems, including the Pure Digital Fiberlink line for professional quality video, audio and data.

Communications Structures & Services, 645 C. E. Renfro St., Burleson, TX, 76028. Phone: (817) 295-8183. Fax: (817) 295-8075. Keith Cendrick, pres.

Tower mf, Erection, maintenance, true turn-key installation, foundations, emergency svcs, antenna & transmission line replacement, site acquistion.

Comprehensive Video Group, 55 Ruta Ct., South Hackensack, NJ, 07606. Phone: (201) 229-0025. Phone: (800) 526-0242. Fax: (201) 814-0510. Web Site: www.compvideo.com. Scott Schaefer, VP.

Digital HDTV UpConverter, High Resolution bulk cable, Video/Audio Multi media & Data Cable assemblies (lifetime warranty), connectors, adaptors, wallplates, distribution amps, switches, convertors, etc.

Comprompter Inc., 1601 Caledonia St., Suite E, La Crosse, WI, 54603-3606. Phone: (800) 785-7766. Fax: (608) 784-5013. E-mail: enrnews@enrnews.com Web Site: www.enrnews.com. Ralph King, pres.

Offers PC-compatible prompting & networked computerized newsroom & newsroom automation systems for radio, TV, corporate & industrial use.

Computer Concepts Corp., 13375 Stemmons Fwy., Suite 400, Dallas, TX, 75234. Phone: (800) 255-6350. Phone: (913) 541-0900. Fax: (913) 541-0169. Web Site: www.ccc-dcs.com. Greg L. Dean, chmn.

Total digital integration improves sound, progmg production & scheduling. Business software includes traf & billing for radio.

Computer Resolutions, 35 Benham Ave., Bridgeport, CT, 06605. Phone: (203) 384-0742. Fax: (203) 384-0473. Web Site: www.cri1.com. Carl Palmieri, CEO.

PC- & mainframe-based traf systems; both offer multistation capability.

Comrex Corp., 19 Pine Rd., Devens, MA, 01434. Phone: (978) 784-1776. Fax: (978) 784-1717. E-mail: info@comrex.com Web Site: www.comrex.com. Lynn Cheney, pres; Kris Bobo, VP.

Comsearch, 19700 Janelia Farm Blvd., Ashburn, VA, 20147. Phone: (703) 726-5500. Fax: (703) 726-5600. E-mail: info@comsearch.com Web Site: www.comsearch.com.

Communication engrg svcs for mobile, microwave & satellite systems, including frequency, propagation & integrations svcs.

ComSonics Inc., 1350 Port Republic Rd., Harrisonburg, VA, 22801. Phone: (540) 434-5965. Fax: (540) 432-9794. Web Site: www.comsonics.com. Dennis A. Zimmerman, pres/CEO; Dale Lann, CFO; Donn E. Meyerhoeffer, COO; Donald J. Sommerville, dir sls/mktg.

Manufacture RF signal level meter & RF leakage detector, CATV repair facility.

Comtech Antenna Systems Inc., 3100 Communications Rd., St. Cloud, FL, 34769. Phone: (407) 892-6111. Fax: (407) 892-0994. E-mail: info@comtechantenna.com Web Site: www.comtechantenna.com. Thomas Christy, pres; William Parker, dir mktg.

Satellite antenna systems, sizes 1.8-7.3 meters; Offsat(tm), 2 degree spacing antenna; 3.8, 5.0m & Offsat(tm) transportables.

Concerto Software, 4450 River Green Pkwy., Suite 100, Duluth, GA, 30096-8326. Phone: (770) 446-7800. Fax: (770) 239-4725. Web Site: www.concerto.com. Andrew Philipowski, pres.

Newport Beach, CA 92660, 1300 Bristol St. N, Suite 100. Phone: (714) 261-9330. Roy Rich, sls rep.

Great Neck, NY 11021, 1010 Northern Blvd, Suite 208. Phone: (516) 829-0390. Rich Bogner, sls rep.

Dallas, TX 75244, 5001 LBJ Fwy, Suite 727. Phone: (214) 387-5210. Nick Pollard, western rgnl sls mgr.

Dallas, TX 75238, 3778 Realty Rd. Phone: (609) 235-1771. Randy Pugh, natl acct exec.

Automated telephone call processing products for inbound & outbound call ctrs.

Condor D C Power Supplies Inc., (A subsidiary of SL Industries). 2311 Statham Pkwy., Oxnard, CA, 93033. Phone: (805) 486-4565. Phone: (800) 235-5929. Fax: (805) 487-8911. Web Site: www.condorpower.com. Owen Farren, owner; Sal Ronchetti, pres.

Multiple outlet strips, surge & noise suppressors, & uninterruptible power supplies.

Condux International, Box 247, 154 Kingswood Rd., Mankato, MN, 56001. Phone: (800) 533-2077. Phone: (507) 387-6576. Fax: (507) 387-1442. E-mail: cndxinfo@condux.com Web Site: www.condux.com. Brad Radichel, pres.

Underground & aerial construction tools & equipment for coaxial cable, telephone & fiber.

Connectronics Corp., Box 3355, 2745 Avondale Ave., Toledo, OH, 43607. Phone: (419) 537-0020. Fax: (419) 537-0007. Tom Ricketts, pres; Al Mocek, VP.

Morgan Hill, CA 95037. California, Box 2047. Phone: (408) 779-8888. Fax: (408) 778-0722.

Audio wire & cable, special wire & cable assys. Interconnect products for audio, video, data & telephone.

Conrac Systems Inc., 5124 Commerce Dr., Baldwin Park, CA, 91706. Phone: (626) 480-0095. Fax: (626) 480-0077. Web Site: www.conrac.com. Bill Moeller, pres.

Manufacturer of a var of color & monochrome video monitors for bcst & computer graphic display.

Control Concepts Corp., (A subsidiary of Liebert Corp). Box 1380, 328 Water St., Binghamton, NY, 13902-1380. Phone: (607) 724-2484. Phone: (800) 288-6169. Fax: (607) 722-8713. E-mail: info@control-concepts.com Web Site: www.control-concepts.com. Bill Fierle, pres; Sarah Beadle, dir mktg.

Power protection products for transmitters, studios & CATVs from transients & lightning induced voltages.

Convergent Media Systems Corp., One Convergent Center, 190 Bluegrass Valley Pkwy., Alpharetta, GA, 30005. Phone: (800) 877-7804. Fax: (707) 369-9100. E-mail: convergent@convergent.com Web Site: www.convergent.com. Murray Holland, owner; Bryan Allen, pres.

Transportable satellite uplinking & downlinking svcs. Includes facilities & transponder time for Ku- & C-band applications.

Convergys Inc., 1551 Sawgrass Corporate Pkwy., Suite 300, Sunrise, FL, 33323. Phone: (954) 851-9200. Fax: (954) 851-9224. Web Site: www.convergys.com. Mike Bauza, VP; Kurt Champion, product mgmt.

Atlanta, GA 30319, 4170 Ashford Dunwoody Rd, Suite 525. Phone: (404) 845-4400.

"Cablemaster/Icoms" customer mgmt & billing system running on IBM as/400 platform; solution for the convergent cable TV/Telephone industry.

Cooper Sound Systems Inc., 645 Main St., Suite C, Morro Bay, CA, 93442-2273. Phone: (805) 772-1007. Fax: (805) 772-1098. E-mail: coopersound@charterinternet.com Web Site: www.coopersound.com. Andrew Cooper, pres; Janet Cooper, VP.

Film & video location mixers & accessories, microphone preamplifiers & time code resolvers.

Copperweld Fayetteville Division, 254 Cotton Mill Rd., Fayetteville, TN, 37334. Phone: (931) 433-7177. Fax: (931) 433-0419. John D. Turner, pres/CEO; Steve Levy, VP mktg & sls.

Copper-clad aluminium wire, copper-clad steel wire, & aluminum-clad steel wire.

Coptervision, 7625 Hayvenhurst Ave., #36, Van Nuys, CA, 91406. Phone: (818) 781-3003. Fax: (818) 782-4070. E-mail: info@coptervision.com Web Site: www.coptervision.com. Sarita Spiwak, pres/CEO.

CVG-high tech , Rollvision 3-axes camera system, Remotevision a multifaceted radio contral, Photovision a new unmaned helicopter for aerial still photography, Flexvision & CVG-A.

Corning Cable Systems, 800 17th St. N.W., Hickory, NC, 28601. Phone: (828) 901-5000. Fax: (828) 901-5488. Web Site: www.corning.com/cablesystems. Larry Aiello, pres; Mike Genovese, sr VP/mgng dir.

Manufacturer of optical fiber cables & accessories for video, data, voice communications applications.

Corning Gilbert Inc., 5310 W. Camelback, Glendale, AZ, 85301. Phone: (623) 845-5613. Fax: (623) 845-5160. Kathy Murphy, CEO.

Trunk, distribution & "F" connectors for CATV.

Corning Inc., (Telecommunications Products Division). One Riverfront Plaza, Corning, NY, 14831. Phone: (800) 525-2524/539-3632 (US & Canada). Phone: (607) 986-8125/3344 (International). E-mail: cofic@corning.com Web Site: www.corning.com/opticalfiber. Eric Musser, gen mgr; Eric musser, VP.

Single-mode & multimode optical fibers including: InfiniCor laser-optimized multimode fibers, NexCor fiber, SMF-28e fiber, MetroCor fiber, LEAF fiber, & Vascade submarine fibers.

Cortana Corp., Box 2548, Farmington, NM, 87499-2548. Phone: (888) 325-5336. Fax: (505) 326-2337. E-mail: cortana@cyberport.com Evelyn Nott, pres; Henry Bond, VP.

Stati-Cat Lightning Prevention System.

Cortland Cable Co. Inc., Box 330, 44 River St., Cortland, NY, 13045-0330. Phone: (607) 753-8276. Fax: (607) 753-3183. E-mail: cortlandcable@cortlandcable.com Web Site: www.cortlandcable.com. John Stidd, pres; Rick Nye, VP.

Kevlar fiber antenna guys & ropes, including eye splice end terminations-potted sockets.

Costume Armour Inc./Christo Vac, 2 Mill St., Box 85, Cornwall, NY, 12518. Phone: (845) 534-9120. Fax: (845) 534-8602. Web Site: www.costumearmour.com. Nino Novellino, pres.

Period armor & weapons, vacuum-formed background panels, custom made props & sculpture.

Countryman Associates Inc., 417 Stanford Ave., Redwood City, CA, 94063. Phone: (800) 669-1422. Phone: (650) 364-9988. Fax: (650) 364-2794. E-mail: sales@countryman.com Carl Countryman, res/chief engr.

Very small precision electret condenser microphones for wide applications & the Type-85 Direct Box.

Crouse-Kimzey Co., 1320 Post & Paddock Rd., Suite 200, Grand Prairie, TX, 75050. Phone: (800) 433-2105. Fax: (972) 623-2800. E-mail: sales@proaudio.com Web Site: www.proaudio.com. John Paul Kimzey, pres.

Rockaway Beach, MO 65740. Crouse-Kimzey of Missouri, 381 Molly Ln. Phone: (417) 561-1050. (800) 955-6800. Fax: (417) 561-1052. Bill Wallace.

Colorado Springs, CO 80911. Crouse-Kimzey of Colorado, 4125 Novia Dr. Phone: (800) 257-6233. Fax: (719) 392-8876. Lee Edwards, gen mgr.

Lynn, IN 47355. Crouse Kimzey/Mid-America, 9170 South U.S. Hwy. 27. Phone: (877) 223-2221. Fax: (765) 874-2540. Barry Pike, gen mgr.

Bcst equipment & pro audio equipment sls.

Crown Broadcast IREC, (division of Crown International Inc). Box 2000, 25166 Leer Dr., Elkhart, IN, 46515-2000. Phone: (574) 262-8900. Fax: (574) 262-5399. E-mail: fmsaes@irecl.com Web Site: www.crownbroadcast.com. Steve Burns, pres/CEO.

Bcst RF equipment, FM radio transmitters. Supplier to the Natl weather svc for emergency weather radio transmitters.

Cygnal Technology, (formerly Normex Telecom Incorp.). 70 Valleywood Dr., Markham, ON, L3R 4T5. Canada. Phone: (905) 944-6500. Fax: (905) 944-6520. E-mail: normex@normex.com Web Site: www.cygnal.ca.

Mgmt, instal & maintenance svcs for studios, radio-TV transmitters, satellite systems & CATV.

D

DBX Professional Products, 8760 S. Sandy Pkwy., Sandy, UT, 84070. Phone: (801) 568-7660. Fax: (801) 568-7662. E-mail: customer@dbxpro.com Web Site: www.dbxpro.com. Robert Benson, sls VP.

Audio signal processing devices: compressor/limiters, De-essers, equalizers, gates & noise reduction.

DEDOTEC USA Inc., 84 Sheffield Business Park, Ashley Falls, MA, 01222. Phone: (413) 229-2550. Fax: (413) 229-2556. Web Site: www.dedolight.com. E-mail: info@dedeolight.com

Dedolight low voltage, high intensity lighting fixtures for film, video, ENG & EFP applications. Battery or AC operation.

D.H. Satellite, Box 239, 600 N. Marquette Rd., Prairie du Chien, WI, 53821. Phone: (608) 326-6041. Fax: (608) 326-4233. E-mail: mdoll@mhtc.net Web Site: www.dhsatellite.com. Mike Doll, VP.

Manufacturer of solid spun aluminum antennas & mounts. Antennas range from .6m (24") to 5m (16') with various mounting options. Delivery & instal is available from DH for all of our antenna equipment.

DPA Microphones, Inc., 2432 Main St., Longmont, CO, 80501. Phone: (303) 485-1025. Fax: (303) 485-6470. E-mail: info-usa@dpamicrophones.com Web Site: www.dpamicrophones.com. Bruce Myers, pres.

DPA Microphones features a complete line of cardioid & omnidirectional microphones & accessories for all applictions.

DRS Technologies, 128 S. Industrial Blvd., Enterprise, AL, 36330. Phone: (334) 347-3478. Fax: (334) 393-4556. Web Site: www.drs.com. Larry Sabourin, pres; Frank Sloan, dir mktg; Gary Bruce, sls dir.

Doppler weather radar systems (rain & wind measurements) with PC-based graphics display & control.

DSC Laboratories, 3565 Nashua Dr., Mississauga, ON, L4V 1R1. Canada. Phone: (905) 673-3211. Fax: (905) 673-0929. E-mail: dsc@dsclabs.com Web Site: www.dsclabs.com. D. Corley, pres; S. Corley, mktg.

Combi Optical Signal Generators (OSGs) & CamAlign chip charts for camera alignment & matching-deal for studio, shop & stadium.

DST Innovis, 1104 Investment Blvd., Eldorado Hills, CA, 95762. Phone: (800) 835-8389. Fax: (916) 934-7054. Web Site: www.dstinnovis.com. Michael McGrail, pres; Anthony Piniella, mgr.

North Sydney NSW 2061. CableData (Asia Pacific), Level 4, 44 Miller St, Suite 404. Phone: +61 29.460.2250. Fax: +61 29.460.2238.

Sao Paulo 04571-010. CableData (Latin America), Andar, Suite 81, Ave. Eng Luis Carlos Berrini, 1297. Phone: +55 11.5505.6799. Fax: +55 11.5505.8691.

Customer mgmt & billing solutions for communications & utilities industries. Clients include providers of CATV, telephony, DBS, wireless, electricity, water, gas, waste mgmt, utility & multi-svcs in over 20 countries.

Dage-MTI Inc., 701 N. Roeske Ave., Michigan City, IN, 46360. Phone: (219) 872-5514. Fax: (219) 872-5559. E-mail: dagemti@dagemti.com Web Site: www.dagemti.com. Arthur D. Sterling, pres; Peggy Moore, dir mktg.

Closed circuit TV cameras & accessories.

Peter W. Dahl Co. Inc., 5869 Waycross, El Paso, TX, 79924. Phone: (915) 751-2300. Fax: (915) 751-0768. E-mail: pwdco@pwdahl.com Web Site: www.pwdahl.com. Peter W. Dahl, pres; Gary L. Komassa, VP.

Heavy duty plate, power, filament, modulation transformers & reactor; single- & three-phase rectifiers, vacuum & oil filled capacitors.

Daily Electronics Corp., Box 822437, Vancouver, WA, 98682-0053. Phone: (360) 896-8856. Phone: (800) 346-6667. Fax: (360) 896-5476. E-mail: daily@worldaccessnet.com Web Site: www.dailyelectronics.net. Jim Grimes, pres.

Produces vacuum tubes—transmitting, camera, industrial & receiving. Tube rebuilding.

Dalet Digital Media Systems, 50 Broadway, Suite 1500, New York, NY, 10004. Phone: (212) 825-3322. Fax: (212) 825-0182. E-mail: sales@dalet.com Web Site: www.dalet.com. Stephane Guez, CEO; Ken Tankel, gen mgr; Fred Roux, opns mgr.

London SW179SH, Trident Business Centre, 89, Bickersteth Rd. Phone: +44 181 516 7750.

Madrid 28036, Calle Dr. Fleming 16. Phone: +34 914 581 988.

Paris 75010, 251, rue du Fbg St. Martin. Phone: +33 1 40 38 01 39.

Singapore 088449, FBC Singapore, 89 Neil Rd. Phone: +65 3260 672.

Software for radio & TV. Newsroom computer systems, acquisition, cataloging, producing, sharing, archiving, and distribution of video & audio assets.

Data Security Inc., 729 Q St., Lincoln, NE, 68508. Phone: (800) 225-7554. Phone: (402) 434-5959. Fax: (402) 434-3291. E-mail: eschafer@telesis-inc.com Web Site: www.datasecurityinc.com. Brian Boles, CEO; Eric Schafer, VP.

Tape Enhancement Series features bulk tape deguassers & videotape cleaner/evaluators.

Delta Electronics Inc., Box 11268, 5730 General Washington Dr., Alexandria, VA, 22312. Phone: (703) 354-3350. Fax: (703) 354-0216. Web Site: www.deltaelectronics.com. John Wright, pres; William R. Fox, VP engrg; Joseph S. Novak, VP mktg.

RF instrumentation including ammeters, operating impedance bridges, receiver generators, AM stereo exciters, monitors & audio processors.

DeSisti Lighting, 1109 Grand Ave., North Bergen, NJ, 07047. Phone: (908) 317-0020. Fax: (201) 319-1104.

Rome, NO Italy. World Headquarters (Desisti Lighting, Spa), Via Cancelliera 10/A, 00040 Cecchina, Albano Laziale. Phone: 01139-06934991. Fax: 01139-069343489. Fabio Desisti, gen mgr.

Complete professional lighting equipment & svcs. Quartz fresnels, softlights & cyc lights; HMI fresnels, softlights & sunguns; motorized studio lighting.

Devlin Design Group Inc., 625 Broadway, Ste. 1101, Box 5208, Frisco, CO, 80443-5208. Phone: (970) 688-2772. Fax: (970) 688-2772. E-mail: ddgemail@ddgtv.com Web Site: www.ddgtv.com. Dan Devlin, CEO; Judy Parker, dir mktg; Kristina Jones, Media Dir.

Specializes in bcst news productions. News sets, newsrooms, turnkey & design only. Set design, virtual sets, hard set construction, consultation.

Dialogic Communications Corp., 730 Cool Springs Blvd., Franklin, TN, 37067. Phone: (615) 790-2882. Fax: (615) 790-1329. Gene Kirby, pres; Charles Smith, VP engrg.

Interactive audio response voice processing equipment & software for pay-per-view, appointment confirmation, outage reporting, etc.

Dictaphone Corp., 3191 Broadbridge Ave., Stratford, CT, 06614. Phone: (203) 381-7000. Fax: (203) 386-8597. Web Site: www.dictaphone.com. Rob Schwager, chmn/pres.

Multi-ch voice communications tape recorders (loggers).

Dielectric Communications, Box 949, 22 Tower Rd., Raymond, ME, 04071. Phone: (207) 655-8100. Fax: (207) 655-8177. E-mail: dcsales@dielectric.spx.com Web Site: www.dielectric.com. David Wilson, pres; Jay Martin, VP sls; Kerry Cozad, VP adv.

Antennas, inside equipment, waveguide, transmission line, switches, loads, pressurization, lighting, TV, Radio, Wireless towers, combiners.

DiGi Co. Ltd., (Formerly Soundtracs, P.L.C.). Box 2260, Keller, TN, 76244. Phone: (877) 292-1623. Web Site: www.digiconsoles.com.

Digital audio mixing consoles

Digidesign, 2001 Junipero Serra Blvd., Daly City, CA, 94014-3886. Phone: (650) 731-6300. Fax: (650) 731-6399. E-mail: prodinfo@digidesign.com Web Site: www.digidesign.com. David Lebolt, gen mgr; Christopher Bock, VP sls. France. France Office, 44 Ave. Georges Pompidou, 92300 Levallois-Perret. Phone: 33 1 41 49 40 10. Fax: 33 1 47 49 40 10. France. France Office, 44 Ave. Georges Pompidou, 92300 Levallois-Perret.

Tokyo 107-0052 Japan. Japan Office, 4F ATT Bldg, 2-11-7 Akasaka, Minato-ku. Phone: 81-3-3505-7963. Fax: 81-3-3505-3417.

Iver Heath, Bucks SLO ONH United Kingdom. UK Office, West Complex, Pinewood Studios, Pinewood Rd. Phone: 44 1753 653322. Fax: 44 1753 658501.

New York, NY 10019. New York Office, 1650 Broadway, Suite 1113. Phone: (212)664-7627.

Digitel Corp., 2719 Piedmont Ave., Duluth, MN, 55811. Phone: (218) 727-0202. Jeffrey Stromquist, owner.

Bcst remote control & automation control equipment.

Dimension 3, 5240 Medina Rd., Woodland Hills, CA, 91364. Phone: (818) 592-0999. Web Site: www.d3.com.

Provides 3-D bcst TV processes. Supplies equipment, consultation & 3-D glasses.

Direct Broadcast Services Inc., 612 Corporate Way, Suite 8, Valley Cottage, NY, 10989. Phone: (845) 267-2800. Fax: (845) 267-2123. E-mail: lrosengerg@directbroadcast.com Web Site: www.directbroadcast.com. Leo Rosenberg, pres.

Transmission svcs & rentals: Ku-band uplinking/downlinking, portable microwave, fiber-optic systems, newsvan opns.

The Display & Exhibit Source, 4715 McEwen St., Dallas, TX, 75244. Phone: (972) 239-0061. Fax: (972) 239-0089. E-mail: sales@displaysource.com Web Site: www.displaysource.com. Dan South, sls.

Designs, manufactures modular, portable backdrops, displays, signal & graphic systems.

Display Devices Inc., 5880 N. Sheridan Blvd., Arvada, CO, 80003. Phone: (303) 412-0399. Fax: (303) 412-9346. Web Site: www.displaydevices.com. Merv Perkins, pres; Ruth Perkins, VP.

CRT, LCD, slide projector motorized lifts & stationary mounts. Custom applications.

Display Systems International Inc., 2214 Hanselman Ave., Saskatoon, SK, S7L 6A4. Canada. Phone: (306) 934-6884. Fax: (306) 934-6447. E-mail: sales@displaysystemsintl.com Web Site: www.displaysystemsintl.com. Dale Lemke, pres.

Electronic progmg guide to display on-screen scrolling TV listing. Also info display software & systems for bulletin boards & adv.

Ditch Witch, Box 66, 1959 W. Firr Ave., Perry, OK, 73077. Phone: (580) 336-4402. Fax: (580) 572-3523. E-mail: info@ditchwitch.com Web Site: www.ditchwitch.com.

Manufacturer of trenching, vibratory plow & trenchless technology equipment, electronic locating & tracking equipment, mini-skid steers, excavators & escavator tool-carriers.

DMT USA, Inc., 1224 Forest Pkwy., Unit 140, West Deptford, NJ, 08066. Phone: (856) 423-0010. Fax: (856) 423-7002. E-mail: sales@dmtonline.us Web Site: www.dmtonline.us. Alberto Giorgini, exec VP; Tom Newman, dir mktg.

TV transmitters, translators, DAB, FM, microwaves, stems, antennas, system integration & scientific applications.

Dolby Laboratories Inc., 100 Potrero Ave., San Francisco, CA, 94103. Phone: (415) 558-0200. Fax: (415) 863-1373. Web Site: www.dolby.com. Ray M. Dolby, chmn; Bill Jasper, pres.

Wootton Bassett, Wiltshire SN4 8QJ Phone: 1793 842 100. Tony Spath, bcst projects mgr.

Audio noise reduction & signal processing equipment; digital audio coding for ISDN, cable, satellite & other applications; dolby surround equipment.

Dorrough Electronics, 5221 Collier Pl., Woodland Hills, CA, 91364. Phone: (818) 998-2824. Fax: (818) 998-1507. E-mail: dorroughel@aol.com Web Site: www.dorrough.com. Mike Dorrough, owner.

Chatsworth, CA 91311, 20434 Corisco St. Phone: (818) 998-4886.

Dorrough Electronics manufactures Audio Loudness Meters featuring Peak & Average signals ballistically set for a highly accurate reading.

Doty-Moore Tower Services, 1570 W. Beltline Rd., Cedar Hill, TX, 75104. Phone: (972) 637-5000. Fax: (972) 293-1255. E-mail: services@stainlessllc.com Web Site: www.stainlessllc.com. J. Patrick Moore, pres; Donald T. Doty, VP; Les Kutasi, Sls mgr; Tom Hoenninger, Chief Engr/VP-opns.Full spectrum of tower maintenance, costruction & inspections. RF svcs include RF mapping of tower & facilities. 24 hrs emergency svcs.

Dove Systems, 3563 Sueldo St., Suite E, San Luis Obispo, CA, 93401-7590. Phone: (805) 541-8292. Fax: (805) 541-8293. E-mail: dove@dovesystems.com Web Site: www.dovesystems.com. Gary Dove, owner.

Studio & stage lighting control equipment.

Dow-Key Microwave Corp., 4822 McGrath, Ventura, CA, 93003. Phone: (805) 650-0260. Fax: (805) 650-1734. E-mail: askdk@dowkey.com Web Site: www.dowkey.com. Mark Mandrell, pres.

Microwave switches, coaxial RF relays & switches, 75 ohm & 50 ohm styles available.

R.L. Drake Co., 230 Industrial Dr., Franklin, OH, 45005-4496. Phone: (937) 746-4556. Fax: (937) 806-1510. E-mail: bcyearbook@rldrake.com Web Site: www.rldrake.com. Ron Wysong, pres/CEO; Mike Brubakerr, sls VP; Andy Ruffin, sls dir.

Peterborough, ON K9J 7M1 Canada, 655 The Queensway. Phone: (705) 742-3122. Steve Roe, sls mgr.

Analog, digital cable headend equipment including receivers, modulators, processors, accessories for reception & distribution of progmg.

DRS Broadcast Technology, 4212 S. Buckner Blvd., Dallas, TX, 75227. Phone: (214) 381-7161. Fax: (214) 381-3250. E-mail: info@drs-bt.com Web Site: www.contelec.com. Adil Mina, VP Business Development; John Uvodich, VP/gen mgr; Bret Brewer, mktg mgr.

Birmingham, AL 35226, 2280 Rockcreek Trail. Phone: (205) 822-1078. Dave Hultsman, Regional Sales Manager.

Specialists in AM, FM & SW transmitters, antenna systems & other RF equipment.

Dubner International Inc., 13 Westervelt Pl., Westwood, NJ, 07675. Phone: (201) 664-6434. Fax: (201) 358-9377. E-mail: rdubner@compuserve.com Web Site: www.dubner.com. Robert Dubner, pres.

Manufacturers PC bcst & production equipment. Products include VideoALERT for on-air signal monitoring & response & SCENE STEALER for videotape logging & archiving.

M. Ducommun Co., 58 Main St., Warwick, NY, 10990. Phone: (845) 986-5757. Fax: (845) 986-7720. M. Ducommun Jr., pres.

Stopwatches for radio, TV, sls & svc.

Dynamic Solutions 2000, 527 Carey Ave., Wilkes-Barre, PA, 18702. Phone: (570) 824-7626. Fax: (570) 824-0556. E-mail: jpgibbons@prodigy.net Web Site: www.ds2000.net. John P. Gibbons, pres.

Convergent billing solutions for the communications & utility industries.

E

e2v technologies Inc., 4 Westchester Plaza, Elmsford, NY, 10523. Phone: (914) 592-6050. Fax: (914) 592-5148. E-mail: enquiries@e2v.com Web Site: www.e2v.com. Mike Kirk, VP.

Mississauga, ON L5A 4H2 Canada, Box 29667. Phone: (905) 848-6430. Fax: (905) 848-9343. Ann Au-Yong.

Buffalo, NY 14221, 80 Post Rd. Phone: (716) 626-9055. Fax: (716) 631-5117. Rick Bossert, natl sls mgr.

Manufacturer of Digital & Analog IOTs, ESCiors, Klystrons for UHF TV transmitters; Stellar range of satellite uplink amplifiers.

Electronics Corp., 7130 National Parks Hwy., Carlsbad, NM, 88220. Phone: (800) 854-0259. Fax: (505) 887-6880. E-mail: sales@edcorusa.com Web Site: www.edcorusa.com.

Audio mic/line mixers, line amplifiers, audio mixers, audio transformers & custom transformers.

EDX Wireless LLC, Box 1547, Eugene, OR, 97440-1547. Phone: (541) 345-0019. Fax: (541) 345-8145. E-mail: info@edx.com Web Site: www.edx.com.

Engrg software & svcs for AM, TV, FM bcst & communication svcs.

EEG Enterprises Inc., 586 Main St., Farmingdale, NY, 11735. Phone: (516) 293-7472. Fax: (516) 293-7417. E-mail: sales@eegent.com Web Site: www.eegent.com. Philip McLaughlin, pres; Eric McErlain, sls dir.

TV closed captioning technology; HDTV & SDTV, closes caption encorders, decorders; V-chip encorders, decorders & systems; affil communications.

EFI Electronics Corp., 1751 S. 4800 W., Salt Lake City, UT, 84104. Phone: (800) 877-1174. Fax: (801) 977-0200. Web Site: www.efinet.com. Terry O'Neal, pres/CEO; Steve Wallace, CFO.

Manufactures a complete line of power protection systems for industrial, coml & computer applications.

ENCO Systems Inc., 29444 Northwestern Hwy., Southfield, MI, 48034. Phone: (248) 827-4440. Fax: (248) 827-4441. E-mail: sales@enco.com Web Site: www.enco.com. Gene Novacek, pres; Don Backus, sls VP.

DADpro32 digital audio delivery systems, custom software engrg for the bcst industry.

E-N-G Mobile Systems Inc., 2245 Via De Mercados, Concord, CA, 94520. Phone: (925) 798-4060. Fax: (925) 798-0152. E-mail: sales@e-n-g.com Web Site: www.e-n-g.com. Dick A. Glass, pres; Ted Kendrick, VP; Ray Iddon, sls.

West Grove, PA 19390, 119 Lloyd Rd. Phone: (610) 659-2640. John Watkins, opns mgr.

Custom-designed ENG & EFP vehicles, rack-ready & turnkey systems. Other mobile electronic systems. ENG system components.

EON Corp., 360 Herndon Pkwy., Herndon, VA, 20170. Phone: (703) 467-0230.

Dev & mfg of wireless two-way interactive technology for consumers & businesses which operate via radio frequency.

ERI-Installations Inc., 7777 Gardner Rd., Chandler, IN, 47610. Phone: (812) 925-6000. Fax: (812) 925-4030. E-mail: sbeeler@eriinc.com Web Site: www.eriinc.com. Thomas B. Silliman, pres; Bart Wenderoth, mgr; Todd Forbes, controller.

Antenna, tower instal svc; antenna rebuilding, stand-by antennas, tower up-grades, reinforcing; fully bonded & insured.

ESE, 142 Sierra St., El Segundo, CA, 90245. Phone: (310) 322-2136. Fax: (310) 322-8127. E-mail: ese@ese-web.com Web Site: www.ese-web.com. Brian Way, VP; William Kaiser, pres.

Master clocks, digital clocks, programmable timers, time code generators & readers, distribution amplifiers, programmable clocks.

ETS-Lindgren, 1301 Arrow Point Dr., Cedar Park, TX, 78613. Phone: (512) 531-6400. Fax: (512) 531-6500. E-mail: info@ets-lindgren.com Web Site: www.ets-lindgren.com. Dave Baron, sls; Bruce Butler, pres; Mark Mawdsley, VP sls.

Non-ionizing radiation test equipment; low frequency survey meters; RF/microwave broadband field strength meters; calibration svcs, software & training.

E-Z Trench Manufacturing Co. Inc., 2315 S. Hwy. 701, Loris, SC, 29569. Phone: (843) 756-6444. Fax: (843) 756-6442. Web Site: www.eztrench.com. Roger Porter, pres.

Lightweight trenchers—digs trench, lays cables & covers all in one pass.

Eagle Comtronics Inc., 7665 Henry Clay Blvd., Liverpool, NY, 13008. Phone: (315) 622-3402. Phone: (800) 448-7474. Fax: (315) 622-3800. E-mail: sales@eaglecomtronics.com Web Site: www.eaglecomtronics.com.

CATV manufacturer & designer of security traps, decoders, & tier traps. Custom OEM filter designs.

Eastman Kodak Co., 343 State St., Rochester, NY, 14650. Phone: (585) 724-4000. Fax: (585) 724-0663. Web Site: www.kodak.com. Daniel Carp, chmn/CEO; Michael Morley, exec VP.

Cameras, projectors, graphic & motion picture products.

Echostar Communications Corp., 9601 S. Meridian Blvd., Englewood, CO, 80211. Phone: (303) 723-1000. Fax: (303) 723-1046. Web Site: www.echostar.com. Charles Ergen, CEO; Michael Dugan, pres/COO; Mark Jackson, sr VP.

Littleton, CO 80120, 5701 S. Santa Fe Rd. Phone: (303) 723-1000. Fax: (303) 723-1099. Charles Ergen, CEO/chman of bd; Brent Gale, dir bcst engrg.

Satellite TV reception systems.

Eddie Egan & Associates, 6136 W. Washington Blvd., Culver City, CA, 90048. Phone: (310) 278-0370. Fax: (310) 275-6412. E-mail: eganfloor@aol.com Daniel Egan, pres; Armand Egan, VP.

Floor coverings for video stages including wood, vinyl & carpeting.

Eigen, 13366 Grass Valley Ave., Grass Valley, CA, 95945. Phone: (530) 265-2020. Fax: (530) 265-2792.

Digital image processors with optional floppy or Winchester disc storage. High-resolution video disc recorders.

Elan Enterprises Ltd., 506 E. St. Charles Rd., Carol Stream, IL, 60188. Phone: (800) 331-8382. Fax: (630) 690-6618. E-mail: eeljim8720@aol.com Web Site: www.generator-inverter.com. Jim Johnsen, pres & sls dir; Joe Johnsen Jr., VP.

Redi-line electric generators. Tripp Lite inverters & sure power isolators.

Elcom Systems Inc., PMB 255, 20423 State Rd. 7 #F6, Boca Raton, FL, 33498-6797. Phone: (561) 883-1945. Fax: (561) 883-1945. E-mail: sales@elcomsystems.com Web Site: www.elcomsystems.com. Leonard Pollachek, pres.

RF coaxial attenuators, terminations, couplers, double balanced mixers, detectors, DC-4.2 Ghz, impedance transformers.

Electro Impulse Laboratory Inc., Box 278, 1805 Rt. 33, Neptune, NJ, 07754-0278. Phone: (732) 776-5800. Fax: (732) 776-6793. E-mail: sales@electroimpulse.com Web Site: www.electroimpulse.com. Mark Rubin, pres.

Manufacturer of dry, forced, air-cooled FM dummy loads & RF calorimeters.

Electro Rent Corp., (Instrument Rental Division). 6060 Sepulveda Blvd., Van Nuys, CA, 91411. Phone: (818) 787-2100. Fax: (818) 787-4354. Web Site: www.electrorent.com. Craig Birgi, rgnl mgr.

Duluth, GA 30096, 3500 Corporate Way. Phone: (770) 813-7000. (800) 688-1111. Rich Curry, eastern rgnl sls mgr.

Test rental equipment including CATV sweep analyzers, signal level meters, video generators/monitors, & cable fault locators.

Electroline Equipment Inc., 8265 Blvd. St. Michel, Montreal, PQ, H1Z 3E4. Canada. Phone: (514) 374 6335. Fax: (514) 374-9370. E-mail: info@electroline.com Web Site: www.electroline.com. John Vincent, pres/CEO; Winston Rodrigues, VP opns; Jay Staiger, VP mktg; Alain Servant, VP.

Cable TV equipment, off-premises addressable systems, passive devices, filters, amplifiers, headend RF signal mgnt equipment & transponders.

Electronic Script Prompting, 6129 Western Ave., Clarendon Hills, IL, 60514. Phone: (630) 887-0346. Fax: (630) 887-0389. Web Site: www.prompting.com.

Teleprompting rental & sale.

Electronic Theatre Controls Inc., 3031 Pleasant View Rd., Middleton, WI, 53562. Phone: (608) 831-4116. Fax: (608) 836-1736. Web Site: www.etcconnect.com. Fred Foster, pres.

Orlando, FL 32811, 4201 Vineland Rd, Suite I-1. Phone: (407) 843-7770. Rob Raff, southeast rgnl mgr. (Southeast rgnl office).

New York, NY 10036, Film Center Bldg., 630 Ninth Ave., Suite 1001. Phone: (212) 397-8080. Joe DiNardo, northeast rgnl mgr. (Northeast rgnl office).

Entertainment lighting systems, including control consoles, dimming equipment & interface products.

Electronology, Inc., 508 Lakeland Blvd., Mattoon, IL, 61938. Phone: (800) 278-2050. Fax: (217) 258-5558. E-mail: info@einc.com Web Site: www.einc.com. Jay Martin, mgr; John Sullivan, consultant; Jim Renkel, consultant.

Distributor of ITC digicenter digital audio systems, ITC cart machines & products, other audio equipment to bcstrs.

EMCOR Enclosures, 1600 4th Ave. N.W., Rochester, MN, 55901. Phone: (507) 287-3535. Fax: (507) 287-3405. E-mail: sales@emcorenclosures.com Web Site: www.emcorenclosures.com. Tom Ryan, sls dir; Karla Grunewaldd, mktg mgr.

Conventional & Flat Panel Display Consoles, modification /custom capabilities, EMI/RFI shielded & Seismic qualified enclosures, a full range of component accessories.

Emerson Network Power-Viewsonics, (formerly Viewsonics Inc.). 3103 N. Andrews Ave. Ext, Pompano Beach, FL, 33064-2118. Phone: (507) 833-8822. Fax: (507) 833-6287. Web Site: www.emersonnetworkpower.com/connectivity.

One GHz amplifiers, security systems, apartment boxes, combiners, LAN, CATV, one GHz splitters, taps, custom design systems & products, head end signal coupler/splitter system.

Encoda Systems Inc., (formerly Columbine JDS, Enterprise and DAL/Brake Automation). 1999 Broadway, Suite 4000, Denver, CO, 80202-3050. Phone: (303) 237-4000. Fax: (303) 237-0085. E-mail: info@encodasystems.com Web Site: www.encodasystems.com. Barry Goldsmith, CEO; Rob McConnell, COO.

Encoda is the authority in seamless automation for the business of media. Encoda is the only company offering end-to-end technological solutions to buyers & sellers of adv time within the electronic media marketplace (bcst, cable, wireless, & DBS).

Energy-Onix Broadcast Equipment Co. Inc., Box 801, 1306 River St., Valatie, NY, 12184. Phone: (518) 758-1690. Fax: (518) 758-1476. E-mail: energy-onix@energy-onix.com Web Site: www.energy-onix.com. Bernard Wise, pres.

Transmitters: FM solid state to 10 kw, grounded grid triode to 50 kw & AM & SW to 100 kw. STL, Translator & remote pick up.

Enghouse Systems Limited, 80 Tiverton Ct., Suite 800, Markham, ON, L3R 0G4. Canada. Phone: (905) 946-3200. Fax: (905) 946-3201. E-mail: info@enghouse.com Web Site: www.enghouse.com. Michael Ford, CEO; Tony Murphy, pres; Andrew Nellestyn, pres; Jerry Diakow, VP sls.

CableCad®—automated mapping/facilities mgmt software with integrated design capabilities for Cable TV companies.

Ensemble Designs, Box 993, Grass Valley, CA, 95945. Phone: (530) 478-1830. Fax: (530) 478-1832. E-mail: info@ensembledesigns.com Web Site: www.ensembledesigns.com. David S. Wood, pres; Cindy Zuelsdorf, mktg mgr; Mondae Hott, gen sls mgr.

Video, audio conversion distribution, HD/down conversion, fiber satellite & desktop video applications.

Entertainment Communications Network (ECN), 4370 Tujunga Ave., Studio City, CA, 91604. Phone: (818) 752-1400. Fax: (818) 752-1443. E-mail: csd@ecnmedia.com Web Site: www.ecnmedia.com. Barry Weintraub, chmn; Dennis Fitch, chmn.

Wall, NJ 07719, 1628 Dubac Rd. Phone: (732) 280-7107.

Bcst faxing to entertainment data bases, online resources, E-mail networks, digital graphics-delivery.

Equipment Technology Inc., 341 N.W. 122nd, Oklahoma City, OK, 73114. Phone: (405) 748-3841. Fax: (405) 755-6829. Web Site: www.eti1.com. James Neuberger, pres.

Oklahoma City, OK 73114, 341 N.W. 122 No. Glenn Smith, VP mktg/sls.

Aerial buckets: articulating & telescoping; truck & van mounted; working height ranges 33 to 43 ft.

Euphonix Inc., 220 Portage Ave., Palo Alto, CA, 94306. Phone: (650) 855-0400. Fax: (650) 855-0410. Web Site: www.euphonix.com. James Dobbie, dir.

North Hollywood, CA 91602. Euphonix Sales & Marketing, 10647B Riverside Dr. Phone: (818) 766-1666. Fax: (818) 766-3401. Andy Wild, VP sls & mktg.

Manufactures the Euphonix CSII digitally-controlled analog audio mixing system.

Even Technologies, Formerly Vertigo Technology Inc. 1255 W. Pender St., Vancouver, BC, V6E 2V1. Canada. Phone: (604) 684-2113. Fax: (604) 684-2108.

3D plug-ins for Photoshop & Illustrator.

Eventide Inc., One Alsan Way, Little Ferry, NJ, 07643. Phone: (201) 641-1200. Fax: (201) 641-1640. E-mail: audio@eventide.com Web Site: www.eventide.com. Gordon Moore, gen mgr; Richard Factor, pres; Tony Agnello, CTO; Ray Maxwell, VP.

Audio delay lines, time compression/expansion, pitch change effects, digital reverb, effects processor & digital audio logger.

Evertz Microsystems Ltd., 5288 John Lucas Dr., Burlington, ON, L7L 5Z9. Canada. Phone: (905) 335-3700. Fax: (905) 335-3573. E-mail: sales@evertz.com Web Site: www.evertz.com. Orest Holyk, sls dir; Joe Cirincione, dir.

Reading, NO RG6 1AZ United Kingdom, 59 Suttons Business Park.

Burbank, CA 91505, 212 N. Evergreen St.

Manassas, VA 20110, 9250 Mosby St, Suite 201.

Evertz provides the most comprehensive line of Fiber Optic Transport equipment, the most advanced line of Multi-Image Display, Monitoring Systems, SDTV & HDTV conversion, synchronization products for use in satellite, cable & bcst applications.

The Express Group, 3360 Thorn St., San Diego, CA, 92104. Phone: (619) 280-9061. Fax: (619) 280-9030. E-mail: egmail@theexpressgroup.com Web Site: www.theexpressgroup.com. Byron Andrus, pres; George Andrus, consultant.

Design, fabrication, lighting of custom news sets, newsrooms, interview sets; custom & modular radio cabinetry.

F

F&F Productions, L.L.C., 14333 Myerlake Cir., Clearwater, FL, 33760. Phone: (727) 535-6776. Fax: (727) 577-5011. George Orgera, pres; Connie Vizaro, VP sls; Bill McKechney, engrg VP.

Remote production svcs & TV mobile units.

F-Conn Industries, (A division of ICM Corp). 6260 Downing St., Denver, CO, 80216. Phone: (303) 288-8107. Fax: (303) 288-4769. E-mail: sales@icmcorp.net Web Site: www.icmcorp.net. Randy Holliday, pres; Susan Stockstill, sls VP.

Get the performance of a high end connector at a standard F-fitting cost.

FM Atlas—Publishing and Electronics, Box 336, Esko, MN, 55733-0336. Phone: (218) 879-7676. Fax: (218) 879-7676. E-mail: fmatlas@aol.com Web Site: www.user.aol.com/fmatlas. Bruce Elving, owner.

Tunable FM/SCS & SAP-modified TV audio radios, adaptor kits with LED display. Brailled radios for the blind.

FM Systems Inc., 3877 S. Main St., Santa Ana, CA, 92707. Phone: (714) 979-3355. Phone: (800) 235-6960. Fax: (714) 979-0913. Web Site: www.fmsystems-inc.com. E-mail: fmsystemsinc@worldnet.att.net Frank McClatchie, pres; Don McClatchie, COO.

Stereo performance meter, audio level masters, digital video volt meters, video & audio modulation meters, multichannel subcarriers & video gain controls (VM771).

FOR.A Corp. of America, 11125 Knott Ave., Suite A, Cypress, CA, 90630. Phone: (714) 894-3311. Fax: (714) 894-5399. Web Site: www.for-a.com. Robert Browne, gen sls mgr.

Video & audio bcst & postproduction equipment; TBCs, color correctors, production switchers, de/encoders, complete video editing systems, virtual stoid, multiviewing.

FWT Inc., Box 8597, 5750 E. I-20, Fort Worth, TX, 76124. Phone: (817) 255-3060. Fax: (817) 255-2957. Web Site: www.fwtinc.com.

Towers, communications bldgs, standby power systems, mobile communications bldgs, fiber optics, & splicing trailers.

Faroudja Laboratories, (A Division of Genesis Microchip). 180 Baytech Dr., Ste.110, San Jose, CA, 95134. Phone: (408) 635-4200. Fax: (408) 957-0364. Web Site: www.faroudja.com. Yves Faroudja, Contact.

NTSC Encoder; NTSC & PAL/NTSC Decoder (RGB or D1 output); Bidirectional Transcoder; NTSC & PAL/NTSC Line Doublers & Line Quadruplers.

Farrtronics Ltd., 39 Kent Ave., Kitchener, ON, N2G 3R2. Canada. Phone: (519) 741-1010. Fax: (519) 578-2044.

Audio & video patchfields; intercom systems, IFB systems, audio distribution amplifiers, beltpack party line systems, monitor packages.

Fast Forward Video, 18200-B W. McDurmott, Irvine, CA, 92614. Phone: (949) 852-8404. Fax: (949) 852-1226. Web Site: www.ffv.com. E-mail: mplaydon@ffv.com Paul Dekeyser, pres; Mark Playdon, dir channel sls; Dennis Mallon, opns mgr.

Extrodinary recorders-miniature DVR, bcst DVR, solid state DVR, on-body or on-bd, rugged & reliable, digital video recorders.

Feldmar Watch and Clock Center, 9000 W. Pico Blvd., Los Angeles, CA, 90035. Phone: (310) 274-8016. Fax: (310) 274-2081. E-mail: sales@feldmarwatch.com Web Site: www.feldmarwatch.com. Sol Meller, pres.

Stopwatches, clocks, watches, timers, sls & repairs.

Fermont ACTS Co., (subsidiary of ESSI). 141 N. Ave., Bridgeport, CT, 06606. Phone: (203) 366-5211. Fax: (203) 367-3642. Web Site: www.engineeredsupport.com. Thomas Santoro, pres.

Emergency standby power & diesel generator sets. Co-generation plant modules.

Ferno-Washington Inc., 70 Weil Way, Wilmington, OH, 45177-9371. Phone: (937) 382-1451. Fax: (937) 382-1191. Web Site: www.ferno.com. Joe Bourgraf, CEO; Tim Schroeder, mktg mgr.

Carts designed to aid in the movement of heavy & bulky equipment.

Fiber Options, 4575 Research Way, Suite 250, Corvallis, OR, 97333. Phone: (800) 469-1676. Phone: (541) 754-9134. Fax: (541) 752-9097. E-mail: cvovideosales@ge.com John Collins, pres; Fred Scott, sls VP; Vic Milani, VP.

West Yorkshire L527 OLQ. Fiber Options Europe Ltd., Unit 7, Cliff Pk, Morley, Leeds. Phone: 44-1132-3816668. Fax: 44-1132-2588121. Steve Clarke, mgng dir.

Manufactures fiber-optic video, data, and audio transmission systems for security, bcst, educ, teleconferencing, ITS, and industrial markets.

Film/Video Equipment Service Co. Inc., 800 S. Jason St., Denver, CO, 80223. Phone: (303) 778-8616. Fax: (303) 778-8657. E-mail: fvesco@fvesco.com Web Site: www.fvesco.com. Dean D. Schneider, pres; Kay Baker, sls/mktg mgr.

Film & video equipment rental & sls-cameras, lenses, lighting, grip, pro audio, camera support, specialty gear for quality production.

FitzCo. Inc., 2600 W. Wall, Midland, TX, 79701. Phone: (432) 684-0861. Fax: (432) 682-9978. Michael Fitz-Gerald, pres.

Speakers, recorders, amplifiers, mixers, tapes, microphones, & headphones; sound reinforcement & bcst equipment.

Flash Technology Corporation of America, 332 Nichol Mill Ln., Franklin, TN, 37067. Phone: (615) 261-2000. Fax: (615) 261-2600. E-mail: info@flashtechnology.com Web Site: www.flashtechnology.com. Mark Joss, VP.

Nashua, NH 03060, 55 Lake St. Phone: (603) 883-6500.

Aviation high-intensity obstruction lights for tall structures & medium intensity for structures up to 500 ft.

FloriCal Systems Inc., 4581 N.W. 6th St., Gainesville, FL, 32609. Phone: (352) 372-8326. Fax: (352) 375-0859. E-mail: sales@florical.com Web Site: www.florical.com. Sunda Scanlon, VP/gen mgr.

TV Automation, on-air presentation, assset mgmt, material aquisition, meida prep, hi/lo resbrowsing, multi-ch monitoring/alarms & electronic progmg guides.

Fluke Corp., Box 9090, Everett, WA, 98206-9090. Phone: (800) 443-5853. Fax: (206) 446-5116. E-mail: fluke-info @fluke.com Web Site: www.fluke.com. George Sherman, pres/CEO; H. Lawrence Culp Jr., pres; Jim Lico, pres.

Electronic test, measurement & control instrumentation.

Focus Enhancements, 1370 Dell Ave., Campbell, CA, 95008. Phone: (408) 866-8300. Fax: (408) 866-4859. E-mail: info@focusinfo.com Web Site: www.focusinfo.com. Bret Moyer, CEO.

World class manufacturer of ASICs for scan conversion, internet & interavtive TV applications. Manufactures PC to TV scan converters, scalers, line quadruplers & DV video production equipment & effects generators.

Fostex America, (former name Fostex Corp. of America). 13701 Cimarron Ave., Gardena, CA, 90249. Phone: (310) 329-2960. Fax: (310) 329-1230. E-mail: info@fostex.com Web Site: www.fostex.com. Bob Schmidt, pres; Budd Johnson, mktg dir.

Manufacturer & marketer of innovative digital recorders, editors, mixers, performance products for creative musicians, producers, as well as personal & professional studios.

Four Seasons Solar Products Corp., 5005 Veterans Memorial Hwy., Holbrook, NY, 11741. Phone: (631) 563-4000. Fax: (631) 563-4010. Web Site: www. oikos.com. David Ewing, pres.

Freeland Products Inc., 75412 Hwy. 25, Covington, LA, 70435. Phone: (985) 893-1243. Phone: (800) 624-7626. Fax: (985) 892-7323. E-mail: freeland-inc.com @freeland-inc.com Web Site: www.freeland-inc.com. Joel H. Freeland, pres.

Rebuilding of TV & radio transmitter tubes.

Frequency Measuring Service Inc., Box 353, Commerce City, CO, 80037. Phone: (303) 288-1482. Fax: (303) 289-8006. Howard S. Eldridge, pres & dir.

Frequency measurements, modulation calibration, field intensity measurements, spectrum analysis.

Frezzolini Electronics Inc., 5-7 Valley St., Hawthorne, NJ, 07506. Phone: (973) 427-1160. Fax: (973) 427-0934. E-mail: infoi@frezzi.com Web Site: www.frezzi.com. James J. Crawford, pres.

High-capacity NICAD/NIMH batteries for all professional cameras & camcorders; advanced power supplies; microcomputer control chargers; ENG lighting & accessories.

Frontline Communications, 12770 44th St. N., Clearwater, FL, 33762. Phone: (727) 573-0400. Fax: (727) 571-3295. E-mail: dmckay@frontlinecomm.com Web Site: www.frontlinecomm.com. Jonathan Sherr, gen mgr; Doug McKay, sls mgr; Bob King, international sls mgr.

Manufacturer of custom bcst vehicles for ENG microwave, analog & digital satellite uplink & remote field production applications.

Fuji Film USA, Inc., 200 Summit Lake Dr., Valhalla, NY, 10595-1356. Phone: (914) 789-8100. Fax: (914) 789-8530. Web Site: www.fujifilmusa.com. Terry Takahashi, gen mgr; Terry Takahshi, exec VP; Tom Daily, dir mktg; Tom Volpicella, VP sls; Jim Hegadorn, mgr.

Cypress, CA 90630, 6200 Phyllis Dr. Phone: (714) 372-4200.

Norcross, GA 30092, 250 Scientific Dr. Phone: (770) 813-5100.

Hanover Park, IL 60133, 850 Central Ave. Phone: (312) 924-5800.

Edison, NJ 08837, 1100 King George Post Rd. Phone: (732) 857-3000.

Irving, TX 75063, 4100 West Royal Ln, Suite 175. Phone: (972) 852-5500.

Professional videotape, data media for bcst, production, cinematography & industrial applications.

Fujinon Inc., 10 High Point Dr., Wayne, NJ, 07470. Phone: (973) 633-5600. Fax: (973) 633-5216. T. Nakamura, pres; John Newton, VP; Tom Calabro, natl sls mgr.

Redondo, CA 90278, West Bay Business Park, 2621A Manhattan Beach Blvd. Phone: (310) 536-0800. Miles Shozuya, West Coast sls mgr; Chuck Lee, mktg mgr.

Hollywood, FL 33021, 4101 N. 48th Terr. Phone: (954) 966-0484. Kelly Nelson, Southeast rgnl sls mgr.

Addison, TX 75001, 4951 Airport Pkwy, Suite 802A. Phone: (972) 385-8902. David Waddell, mktg mgr.

CTV, ENG, EFP lenses, optical systems, accessories.

Full Compass Systems Ltd., 8001 Terrace Ave., Middleton, WI, 53562-3194. Phone: (800) 356-5844. Phone: (608) 831-7330. Fax: (608) 831-6330. E-mail: webmaster@fullcompass.com Web Site: www.fullcompass.com. Jonathan Lipp, CEO.

Over 300 product lines for broadcast recording, entertainment, video & sound reinforcement industries.

Fuller Manufacturing, 695 S. Glenwood Pl., Burbank, CA, 91506. Phone: (818) 500-0116. Fax: (818) 238-9959. Ron Fuller, engrg mgr.

IFB for news & satellite trunks.

Furman Sound Inc., 1997 S. McDowell Blvd., Petaluma, CA, 94954. Phone: (707) 763-1010. Fax: (707) 763-1310. E-mail: info@furmansound.com Web Site: www.furmansound.com. Joe Desmond, VP sls & mktg VP; John Humphrey, CFO.

Analog, digital, video monitor systems; power conditioning/distribution, mixers, equalizers, compressors, crossovers, patch bays,voltage regulators, headphone amplifliers & distribution sustems.

Future Productions Inc., 100 Industrial Ave., Little Ferry, NJ, 07643-1913. Phone: (201) 727-0903. Fax: (201) 727-0908. Web Site: members.aol.com/futureprd/video.html.

Manufactures & svcs audio/video distribution amplifiers, video duplication control systems, bcst camera control systems, computer graphic systems; & video duplication svc.

G

GAMPRODUCTS Inc., 4975 W. Pico Blvd., Los Angeles, CA, 90019. Phone: (323) 935-4975. Fax: (323) 935-2002. Web Site: www.gamonline.com. Joseph N. Tawil, owner/gen mgr; Harry Beard, opns mgr; Pascal Zandt, sls.

Lighting equipment, portable, studio special effects, projections, control console, dimming, color, correction, diffusion filters & patterns (gobos).

G Prime Ltd., Radio City Sn., Box 1525, New York, NY, 10101. Phone: (212) 765-3415. Fax: (212) 581-8938. E-mail: info@gprime.com Web Site: www.gprime.com. Russ O. Hamm, pres.

Importer & distributor of European professional audio equipment for the bcst & recording industries.

Gala, (A division of Paco Corp). 3185 First Street, St. Hubert, PQ, J3Y 8Y6. Canada. Phone: (450) 678-7226. Fax: (450) 678-4060. E-mail: Info@pacocorp.com Web Site: www.pacocorp.com.

Es Condido, CA 92025, 655 Calle Ladra. Phone: (760) 738-5555. Richard R. Haller.

Theatrical rigging, revolving stages & orchestra lifts.

Galaxy Audio Inc., (dba Valley Audio). Box 16285, Wichita, KS, 67216-0285. Phone: (316) 263-2852. Fax: (316) 263-0642. Web Site: www.galaxyaudio.com. Brock Jabara, CEO; Yule Jabara, natl sls mgr.

Microphone preamplifiers, loudspeaker, powered amplifiers, combiners, splitters & test equipment.

Garner Products, (division of Audiolab Electronics Inc.). 620 Commerce Dr., Suite C, Roseville, 95678. Phone: (926) 784- 0200. Fax: (916) 784-1425. E-mail: info@garner-products.com Web Site: www.garner-products.com. Ronald A. Stofan, pres.

Professional line of bulk tape degaussers for all formats of tape including: Beta SP, DAT 2" reels up to 16" diameters, hard drives, DLT media & degaussing svc.

Geac Libra, 462 Bearcat Dr., Suite C, Salt Lake City, UT, 04115-2520. Phone: (800) 453-3827. Fax: (801) 974-1900. E-mail: info@geac.com Web Site: www.gcs.geac.com. Eric Schlor, sls.

Accounting software for radio, including billing affidavits & sls analysis.

Gefen Inc., 6265 Variel Ave., Woodland Hills, CA, 91367. Phone: (800) 545-6900. Phone: (818) 884-6294. Fax: (818) 884-3108. E-mail: gsinfo@gefen.com Web Site: www.gefen.com. Hagai Gefen, pres; John Guzman, tech & sls.

Manufacturer of computer peripherals that enhance the studio environment. HDTV, KVM, USB, CAT-5 DVI, ADC extenders, switchers, converters, adapters, distribution amplifiers & cable.

General Atomics, 4949 Greencraig Ln., San Diego, CA, 92123. Phone: (858) 522-8300. Fax: (858) 522-8301. Web Site: www.ga-esi.com. Phil Arneson, pres.

Terminal automation products, radiation, monitoring system, triqq and manufacturer of Maxwell high voltage capactiors & power supplies.

General Cable, 4 Tesseneer Dr., Highland Heights, KY, 41076. Phone: (859) 572-8000. Fax: (859) 572-8458. Web Site: www.generalcable.com.

Copper, aluminum and fiberoptic wire, cable products for communications, energy & electrical markets.

General Electric Co., 3135 Easton Tpke., Fairfield, CT, 06431. Phone: (800) 626-2004. Phone: (203)373-2039. Fax: (203) 373-3198. Web Site: www.ge.com. E-mail: geinfo@www.ge.com

Fort Lee, NJ 07024. CNBC & MSNBC, 2200 Fletcher Ave. Bill Bolste, pres.

New York, NY 10020. NBC, 30 Rockefeller Plaza. Phone: (212) 664-4444. Robert C. Wright, pres.

Cleveland, OH 44112, 4338 Nela Park. Phone: (216) 362-5600. Fax: (216) 266-2310. Keith T.S. Ward, quartz-stage studio production mgr.

NBC bcstg; CNBC & MSNBC; lighting products; Americom satellite; electrical distribution & control; intercast; MSNBC desktop video.

General Electrodynamics Corp., 8000 Calendar Rd., Arlington, TX, 76001. Phone: (817) 572-0366. Fax: (817) 572-0373. Web Site: www.gecscales.com. Dick Davis, pres.

Tubes, TV cameras, electronics, aircraft weighing equipment, contract weighing svcs, truck scales, load scales.

Geneva Aviation Inc., 20021 80th Ave. S., Kent, WA, 98032. Phone: (253) 395-9105. Fax: (253) 395-9150. E-mail: info@genevaaviation.com Web Site: www.genevaaviation.com. Gary Hasson, pres.

Design, manufacture & instal of E.N.G. & microwave equipment for news helicopters.

Gennum Corp., Box 489, Station A, Burlington, ON, L7R 3Y3. Canada. Phone: (905) 632-2996. Fax: (905) 632-2055. Web Site: www.gennum.com.

High performance integrated circuits, including switches & processing functions, for analog & digital video applications.

Gepco International Inc., 1770 Birchwood Ave., Des Plaines, IL, 60018. Phone: (847) 795-9555. Fax: (847) 795-8770. E-mail: gepco@gepco.com Web Site: www.gepco.com. Gary R. Geppert, pres; David Mecklenburger, CFO.

Burbank, CA 91502, 826 N. Lake St. Phone: (818) 894-3446. Fax: (818) 569-5226.

Audio cable & video cable in bulk or cut to length. Assemblies, boxes, connectors, patchbays. ADC, Kings Neutrik & switchcraft

Glentronix, 90 Nolan Ct., Unit 7, Markham, ON, L3R 4L9. Canada. Phone: (905) 475-8494. Fax: (905) 475-0955. Web Site: www.glentronix.com.

Studio video equipment, audio jackfield & test equipment.

Global Microwave Systems Inc., 1916 Palomar Oaks Way, Suite 100, Carlsbad, CA, 92008. Phone: (760) 496-0055. Fax: (760) 496-0057. E-mail: gms@gmsinc.com Web Site: www.gmsinc.com. Sam Nasiri, pres; Wayne Rogers, sls.

The latest in microwave communications equipment.

Globecomm Systems Inc., 45 Oser Ave., Hauppauge, NY, 11788-3816. Phone: (631) 231-9800. Fax: (631) 231-1557. Web Site: www.globecommsystems.com. David Hershberg, CEO; Kenneth Miller, pres; F. Dugourd, Contact.

Earth stn ground segements. Video Broadcasting Service & Content Delivery Service.

Alan Gordon Enterprises Inc., 5625 Melrose Ave., Hollywood, CA, 90038. Phone: (323) 466-3561. Fax: (323) 871-2193. E-mail: info@alangordon.com Web Site: www.alangordon.com. Grant Loucks, pres; Wayne Loucks, gen mgr; Don Sahlein, exec VP.

Professional motion picture & video equipment sls mf & rental, & video equipment.

Gorman-Redlich Manufacturing Co., 257 W. Union St., Athens, OH, 45701. Phone: (740) 593-3150. Fax: (740) 592-3898. Web Site: www.gorman-redlich.com. E-mail: jimg@gorman-redlich.com James T. Gorman, owner.

EAS encoders, decoders, encoder-decoders & receivers, digital antennas monitors, NOAA weather radios. With same decoding. All 7 frequencies.

Graham-Patten Systems Inc., The ISIS Group, 119 E. McKnight Way, Unit A, Grass Valley, CA, 95949-9503. Phone: (888) 622-4747. Phone: (530) 477-2984. Fax: (530) 477-2986. Web Site: www.gpsys.com.

Digital audio mixers, digital audio systemization products.

Grant Tower, Inc., 13064 Wisner Ave., Grant, MI, 49327. Phone: (231) 834-5665. Fax: (231) 834-7870. Terry L. Sharp Jr., pres; Walter Knoch, office mgr.

Bcst tower erection & maintenance service.

Gray Engineering Laboratories Inc., 504 W. Chapman Ave., Suite O, Orange, CA, 92668. Phone: (714) 997-4151. Fax: (714) 997-1939. Web Site: www.grayengineeringlabs.com. Scott R. Gray, pres.

SMPTE time-code generators & readers, safe area generators, video-assisted film editing components.

Great Lakes Data Systems, Inc., 5954 Priestly Dr., Carlsbad, CA, 92008. Phone: (760) 753-1024. Fax: (760) 753-2538. E-mail: sales@cablebilling.com Web Site: www.cablebilling.com. J. Alonzo Rosado, pres; Laura Rosado, VP sls/mktg.

Beaver Dam, WI 53916, Box 295. Phone: (920) 887-7651. Fax: (920) 887-7653.

Affordable PC/Network billing & subscriber mgmt systems. Addressable interface, PPV, ARU, ANI, Hotel PPV. Training, data conversion & toll-free support.

Greenberg Teleprompting, 115 S. Olive St., Orange, CA, 92866. Phone: (818) 838-4437. Fax: (818) 838-0447. E-mail: info@greenprompt.com Web Site: www.greenprompt.com. Jim Estochin, owner.

Cameral mounted teleprompting, speech prompting, nationwide clients.

Group One Ltd., 70 Sea Ln., Farmingdale, NY, 11735. Phone: (561) 249-1399. Fax: (516) 249-8870. E-mail: jackk@g1limited.com Web Site: www.g1limited.com.

Exclusive distributor for a number of prominent audio & lighting products including: MC2, Celestion, XTA Electronics, Elektralite, Pulsar, Blue Sky.

Gyrocam Systems, 8100 15th St. E., Sarasota, FL, 34243. Phone: (941) 355-3206. Fax: (941) 355-3417. E-mail: info@gyrocamsystems.com Web Site: www.gyrocamsystems.com. Ken Sanborn, pres/CEO; Joe Stark, VP sls; Stefanie Kowitt, exec dir.

Manufacturer of the Gyrocam-gyrostablized camera systems for aircraft, boats or vehicles. High Definition Cameras & V700 watt searchlight also available.

H

HM Electronics Inc., 14110 Stowe Dr., Poway, CA, 92064-7147. Phone: (858) 535-6000. Fax: (858) 452-7207. Web Site: www.hme.com. Mike Hughes, VP/gen mgr; John Kowalski, Sls. Mgr.; Rick Molina, production mgr.

Wireless Intercoms

Hardigg Cases, (A division of Hardigg Industries Inc.). Box 201, 147 N. Main St., South Deerfield, MA, 01373-0201. Phone: (413) 665-2163. Fax: (413) 665-8330. E-mail: cases@hardigg.com Web Site: www.hardigg.com. James S. Hardigg, pres.

Reusable shipping cases. Rugged 19 rack, shock-mounted enclosures. Bcst equipment cases.

Harman International Industries Inc., (A subsidiary of JBL Professional.). 8500 Balboa Blvd., Northridge, CA, 91329. Phone: (818) 893-8411. Fax: (818) 892-9590. Web Site: www.harman.com. Dr. Sidney Harman, pres; Bernard Girod, CEO.

Manufacturer of audio signal processing equipment designed for sound reinforcement, recording & bcstg.

Harmonic Inc., 549 Baltic Way, Sunnyvale, CA, 94089. Phone: (408) 542-2500. Fax: (408) 542-2510. Web Site: www.harmonicinc.com. Anthony J. Ley, pres.

Fiber-optic & digital transmission systems for cable TV, including transmitters, receivers, return path equipment & net mgmt hardware & software.

Harris Automation Solutions, 1134 E. Arques Ave., Sunnyvale, CA, 595-8250. Phone: (408) 990-8200. Fax: (408) 990-8250. E-mail: sales@harris.com Web Site: www.harris.com. Jim Wood, mgr.

Wantagh, NY 11793, Bos 3200. Phone: (516) 783-6022. Martin Frange, dir sls, northeast & southeast rgn.

Issaquah, WA 98027, 700 NW Gilman Blvd, 133-227. Phone: (425) 837-3799. Brian Lay, dir sls, western & central rgn.

Louth is a supplier of media mgmt, automation system: for bcst & cable TV.

Harris Corp., Broadcast Communications Division, 4393 Digital Way, Mason, OH, 45040. Phone: (513) 459-3400. Phone: (513) 459-3547 (Jackie Broo, mktg). Fax: (513) 701-5315. E-mail: broadcast@harris.com Web Site: www.harris.com. Timothy E. Thorsteinson, pres.

Sunnyvale, CA 94086, 1134 E. Arques Ave. Phone: (408) 990-8200.

Quincy, IL 62305. Harris Corp., Broadcast Communications Division, Box 4290, 3200 Wismann Ln. Phone: (217)

222-8200. Fax: (217) 222-7041.

Supplier of products, systems, svc & automation solutions for the bcst industry.

Harris Corp., Broadcast Division, Box 4290, 3200 Wismann Ln., Quincy, IL, 62305-4290. Phone: (217) 222-8200. Fax: (217) 221-7085. Web Site: www.harris.com. Bob Weirather, dir TV product line; Jack O'Dear, dir international sls; Gaylen C. Evans, dir N.American field sls.

South Glens Falls, NY 12803, Box 1179, 10373 Saratoga Rd. Phone: (518) 793-2181. Fax: (518) 793-7423. Rich Redmond, sls mgr. (Northeast U.S. radio sls).

Federal Way, WA 98003, 33430 13th Pl. S, Suite 205A. Phone: (206) 874-7444. Fax: (206) 874-8866. Cal Vandegrift, sls rep. (Northwest U.S. radio sls).

Digital radio & TV transmission equipment, svc, tower studies, training, turnkey RF systems.

Harris Corporation Co., 107 Gilbert Rd., Saratoga Springs, NY, 12866. Phone: (518) 226-0918. Fax: (518) 226-0741. Web Site: www.harris.com. Brian Szewczyk, mgr.

Plano, TX 75023, Box 867717. Phone: (214) 612-2053. (800) 729-0494. Fax: (214) 612-2145. "Doc" Masoomian.

Complete AM & FM bcst systems—equipment sls, svc & instal. Bcst systems intergrators.

Harris-Farinon, 350 Twin Dolphin Dr., Redwood Shores, CA, 94065. Phone: (650) 594-3000. Fax: (650) 594-3110. Web Site: www.harris.com.

Microwave for intercity relay & STLs.

Harrison Consoles, (Formerly Harrison by GLW). 1024 Firestone Pkwy., LaVergne, TN, 37086. Phone: (615) 641-7200. Fax: (615) 641-7224. E-mail: info@harrisonconsoles.com Web Site: www.harrisonconsoles.com. William B. Owen, pres; Gary Thielman, advanced product mgr; Claude Hill, dir mktg & sls dir.

Analog & digital audio mixing consoles for on-air bcst, production, video, film sound postproduction, live sound & music recording.

HAVE Inc., 350 Power Ave., Hudson, NY, 12534-2448. Phone: (518) 828-2000. Phone: (800) 999-4283. Fax: (518) 828-2008. E-mail: have@haveinc.com Web Site: www.haveinc.com. Nancy Gordon, pres; Paul Swedenburg, VP.

Canare, Belden, Gepco, Flexygy, Mogami Cable; Network connectors & adaptors. Professional blank media, equipment & accessories. DVD, CD duplication & postproduction svcs.

Henry Engineering, 503 Key Vista Dr., Sierra Madre, CA, 91024. Phone: (626) 355-3656. Fax: (626) 355-0077. E-mail: info@henryeng.com Web Site: www.henryeng.com. Hank Landsberg, pres.

The Matchbox, other audio interface, control interface & digital audio storage devices.

Hessler Enterprises Inc., 106 Susan Dr., #1, Elkins Park, PA, 19027. Phone: (215) 379-2300. Fax: (215) 663-8839. Web Site: www.hessler.com. Ed Hessler, pres; Brian Hessler, VP.

Produces bcstg forms including script sets, contracts, program logs, invoices, labels, A/R statements & computer stock paper.

Hewlett Packard Co., 3000 Hanover St., Palo Alto, CA, 94304. Phone: (650) 857-1501. Fax: (650) 857-5518. Web Site: www.hewlettpackard.com.

High Tech Industries, 298 N. Smith Ave., Corona, CA, 91720. Phone: (888) 747-9817. Fax: (951) 279-5773. Web Site: www.customstudio.com. Douglas J. Kanczuzewski, gen mgr.

Bcst TV & radio equipment consoles, rack, cabinetry both standard & custom for edit suites, control rooms & machine rooms.

Highway Information Systems, Inc., 4021 Stirrup Creek Dr., Suite 100, Durham, NC, 27703. Phone: (919) 361-2479. Phone: (800) 849-4447. Fax: (800) 849-2947. E-mail: sales@high Web Site: www.highwayinfo.com. Bruce Reimer, gen mgr; Mike Corbett, gen sls mgr.

Manufacturer of travelers info stns & hwy advisory bcst systems on low-power AM radio for motorists.

Hignite Tower Service, 9945 Arkansas St., Bellflower, CA, 90706. Phone: (562) 925-1951. Fax: (562) 925-6171. John Hignite, owner; Jackie Hignite, office mgr.

Tower engrg, erection, fabrication, maintenance & painting.

Hipotronics Inc., Box 414, 1650 Rt. 22, Brewster, NY, 10509. Phone: (845) 279-8091. Fax: (845) 279-2467. Web Site: www.hipotronics.com. Gary Amato, gen mgr.

High-voltage DC power supplies & industrial grade voltage regulators for medium-to-high-power applications.

Hitachi Denshi America, Ltd., 150 Crossway Park Dr., Woodbury, NY, 11797. Phone: (516) 921-7200. Fax: (516) 496-3718. E-mail: info@hdac.com Web Site: www.hdal.com. M. Matsuhashi, pres; J. Breitenbucher, VP; S. Moran, rgnl sls mgr.

Torrance, CA 90501, 371 Van Ness Way. Phone: (310) 328-6116. David Morris, rgnl sls mgr.

Bcst, professional & industrial cameras, monitor, test & measurement equipment.

Hoagland Instrument, Inc., 78 Stone Pl., Melrose, MA, 02176-0004. Phone: (781) 665-4428. Fax: (781) 665-3855. Web Site: www.hoagland-instrument.com. Debbie D'Ambrosio, pres.

Thermal & electronic time delay relays.

Hoffend & Sons Inc., 66 School St., Victor, NY, 14564. Phone: (585) 924-5000. Fax: (585) 924-0545. Web Site: www.hoffend.net. Donald A. Hoffend, CEO; Peter Hoffend, pres.

Engrg, mfg & instal of studio rigging systems, motorized hoists, tracks, turntables & controls.

Hogg & Davis Inc., Box 405, 3800 Eagle Loop, Odell, OR, 97044. Phone: (541) 354-1001. Fax: (541) 354-1080. E-mail: info@hoggdavis.com Web Site: www.hoggdavis.com. F. Neil Hogg, pres.

Cable reels, cable reel trailers, pole tongs, cable sheaves, break-away reels, 36" & 52" tensioners underground puller, 4 drum puller.

Hollywood Rentals Production Services, 19731 Nordhoff St., North Ridge, CA, 91324. Phone: (818) 407-7800. Fax: (818) 407-7875. Anil Sharma, pres.

Charlotte, NC 28216, 9100-C Perimeter Woods Dr. Phone: (740) 597-1308. Jeff Pentek.

Production equipment & vehicles for film & video (rental); sale of new equipment & expendable items.

Hollywood Vaults Inc., 742 N. Seward St., Hollywood, CA, 90038. Phone: (323) 461-6464. Phone: (800) 569-5336. Fax: (323) 461-6479. Web Site: www.hollywoodvaults.com. David Wexler, pres; Julianna Wexler, VP.

Santa Barbara, CA 93103. (Corporate office of Hollywood Vaults Inc.), 1780 Prospect Ave. Phone: (805) 569-5336. Chris Robinson, exec admin.

State-of-the-art film & tape storage vault. Secure, climate-controlled, 24-hours self svc access.

Homalite, 11 Brookside Dr., Wilmington, DE, 19804. Phone: (302) 652-3686. Fax: (302) 652-4578. Web Site: www.homalite.com. Robert Cahill, pres.

Manufactures low-reflectance, contrast enhancement filters for use on CRTs, LEDs & other forms of info display.

Honeywell Airport Systems, 2121 Union Pl., Simi Valley, CA, 93065. Phone: (805) 581-5591. Fax: (805) 581-5032. Ed Wheeler, VP opns.

Tower, obstruction lighting & controls.

Hoodman Corp., 20445 Gramercy Pl., Suite 201, Torrance, CA, 90501. Phone: (310) 222-8608. Phone: (800) 818-3946 (US). Fax: (310) 222-8623. E-mail: lou@hoodmanusa.com Web Site: www.hoodmanusa.com. Mike Schmidt, pres; Louis Schmidt, VP mktg.

TV sun shades/monitor hoods for glare-free outdoor viewing & video carts.

Horita, Box 3993, Mission Viejo, CA, 92690. Phone: (949) 489-0240. Fax: (949) 489-0242. E-mail: horita@horita.com Web Site: www.horita.com. Gerald Hester, pres; Christopher Lovallo, sls.

SMPTE time code readers, generators, inserters, PC tape logging software; color bar, black, sync generators; titler, distribution amplifiers, audio meter, matte generator.

Hotbox Digital, 367 N. Hwy. 101, Solana Beach, CA, 92075. Phone: (858) 292-8520. Fax: (858) 292-1812. Cam MacMillan, owner/exec producer.

Design & production of bcst 3-D computer graphics. Logo animation, stn packages. All tape formats supported. Producers of Subito Studio Video Graphic Library.

Hotronic Inc., 1875 S. Winchester Blvd., Campbell, CA, 95008. Phone: (408) 378-3883. Fax: (408) 378-3888. Web Site: www.hotronics.com. Andy Ho, pres; Linda Chang, sls & mktg mgr.

Time base corrector & frame synchronizer with freeze frame/field, digital effects, & 8x2 Asynchronized Router etc.

I

ICM (International Crystal Mfg.), 10 N. Lee Ave., Box 1768, Oklahoma City, OK, 73101. Phone: (405) 236-3741. Fax: (405) 235-1904. E-mail: freeland@icmfg.com Web Site: www.icmfg.com. Beth Freeland, pres; Royden Freeland, chief engr.

Precision electronic crystals, crystal filters, clock oscillators, TCXO's, VCXO's.

ICX Global, 8136 S. Grantway, Littleton, CO, 80122-2726. Phone: (720) 873-8400. Phone: (303) 706-1275. Web Site: www.icxglobal.com. Johhny Iverson, gen mgr.

Amherst, NY 14226, 3960 Harlem Rd. Phone: (716) 839-3803.

ICX Global designs, manufactures & markets a wide range of remote control products for bcst, satellite, cable, & consumer electronics devices.

IMS (Interactive Market Systems Inc.), 770 Broadway, 15th Fl., New York, NY, 10003. Phone: (646) 654-5900. Fax: (646) 654-5901. E-mail: sales@imsusa.com Web Site: www.imsms.com. Lisa Finn, VP sls; Sherry Orr, dir media/agency sls.

IMS is the leading international provider of info systems & solutions for the media industry. IMS systems & software form an integral part of media & mktg decisions around the world. Media professionals trust IMS for innovative technologies, an unparalled global perspective & valuable insights.

IPITEK, 2330 Faraday Ave., Carlsbad, CA, 92008. Phone: (760) 438-1010. Fax: (760) 438-2462. E-mail: sales@ipetk.com Web Site: www.ipitek.com. Michael M. Salour, chmn/CEO; Horace Tsiang, VP sls.

IRIS Technologies Inc., R.R. 12, Box 36, Westmoreland Industrial Pk., Greensburg, PA, 15601. Phone: (724) 832-9855. Fax: (724) 832-8999. Jerry Salandro, pres/CEO.

Bountiful, VT 84010, 563 W 500 South. Phone: (801) 296-8250. Fax: (801) 296-8248. (Engineering).

Video Commander icon based routing, iNED & SmartPort product lines which allow complete headend control from anywhere in the world.

ITI Electronics Inc., 32 Stonewall Dr., Livingston, NJ, 07039-1822. Phone: (973) 890-7888. Fax: (973) 992-0459. E-mail: itielect@aol.com Robert A Stein, pres.

Connectorized & Prewired jackfields; patch panels; telephone line amplifiers; other series 400 & 10 line cards.

ITT Cannon Electric, 666 E. Dyer Rd., Santa Ana, CA, 92707. Phone: (714) 557-4700. Fax: (714) 628-2142. Web Site: www.ittcannon.com.

Electronic connectors & interconnect systems & info card technology suppliers to a var of industries, including bcst & data communications companies.

Identix, 5600 Rowland Road, Minnetonka, MN, 55343. Phone: (952) 932-0888. Fax: (952) 932-7181. Web Site: www.identix.com. Bob MacCashin, CEO.

Identix is a leading biometrics solutions provider with proven & cost-effective verification security for applications including banking, healthcare, government, & access control.

Ikegami Electronics (U.S.A.) Inc., 37 Brook Ave., Maywood, NJ, 07607. Phone: (201) 368-9171. Fax: (201) 569-1626. E-mail: sales@ikegami.com Web Site: www.ikegami.com. Alan Keil, engrg VP; Teri Zastrow, mktg dir & sls dir.

Ft. Lauderdale, FL 33309, 5200 N.W. 33rd Ave, Suite 111.

Manhattan Beach, CA 90273, 2631 Manhattan Beach Blvd. Phone: (310) 297-1900. Fax: (310) 536-9550.

Elmhurst, IL 60126, 747 Church Rd, Unit C1.

Waxahachie, TX 75167, 773 Bearden.

Bcst/professional video cameras, monitors, microwave equipment.

Illbruck Inc., 3800 Washington Ave. N., Minneapolis, MN, 55412. Phone: (612) 520-3620. Phone: (800) 662-0032. Fax: (612) 521-5639. E-mail: sales@illbruck-sonex.com Web Site: www.illbruck-sonex.com.

Sonex® accoustical products including wall panels & ceiling tiles.

Illumination Dynamics Inc., (A division of Arri, Inc.). 3823 Barringer Dr., Charlotte, NC, 28217. Phone: (704) 679-9400. Fax: (704) 679-9420. Web Site: www.illuminationdynamics.com. Steve Hipsley, VP; Carly Barber, pres; Jeff Pentek, COO; Maria Carpenter, dir mktg.

Pacoima, CA 91331, 10232 Glenoaks Blvd. Phone: (818) 686-6400. Fax: (818) 686-6776. Craig Chiapuzio, Dir. of Operations.

Complete line of lighting, grip power distributors, generators for feature film, commercials, bcst & special events.

The Image Group Post, LLC., (Formerly MTI/The Image Group, Inc.). 885 2nd Ave., New York, NY, 10017. Phone: (212) 548-7700. Fax: (212) 355-0523. Web Site: www.image-group.com. Charles Pontillo, chmn; Willie Sheehy, pres.

New York, NY 10017, 305 E. 46th St. Phone: (212) 548-4400.

Production & postproduction svcs, including remotes, computer animation, scenic svcs, satellite transmissions & networking.

Image Logic Corp., 6807 Brennon Ln., Chevy Chase, MD, 20815. Phone: (301) 907-8891. Fax: (301) 652-6584. E-mail: info@imagelogic.com Web Site: www.imagelogic.com. Woodrow Landay, pres.

Log producer, automated videotape logging system: Autocaption, automated desktop closed-captioning & sub-titling system.

Image Video, (A division of 1077541 Ontario Ltd). 1620 Midland Ave., Toronto, ON, M1P 3C2. Canada. Phone: (416) 750-8872. Fax: (416) 750-8015. E-mail: sales@imagevideo.com Web Site: www.imagevideo.com. Andy A. Vanags, pres; Dave Russell, VP.

Under monitor tally display systems, tally mappers, multi-video display systems & alarm systems.

Imagine Products Inc., 1052 Summit Dr., Carmel, IN, 46032. Phone: (317) 843-0706. Fax: (317) 843-0807. E-mail: sales@imagineproducts.com Web Site: www.imagineproducts.com. Dan Montgomery, pres/CEO; M. Jane Montgomery, VP.

PC software/hardware for logging, video library, EDL transfers & tape libraries. Timecode readers & VTR controls.

Imaging Automation, 269 Concord Rd., 3rd Fl., Billerica, MA, 01821. Phone: (978) 932-2200. Fax: (978) 932-2225. E-mail: info@viisage.com Web Site: www.imagingauto.com. Bernard C. Bailey, pres/CEO; Bradley T. Miller, CFO & sr. VP.

Supplier of optical-based systems for storing & retrieving documents & images.

Industrial Acoustics Co., Inc., 1160 Commerce Ave., Bronx, NY, 10462. Phone: (718) 931-8000. Fax: (718) 863-1138. Web Site: www.industrialacoustics.com. Kenneth DeLasho, VP.

Staines, Middlesex TW18 4XB, Walton House, Central Trading Estate. Simon White, dir mktg.

Wanchai, Hopewell Centre, 183 Queen's Rd. E, Rm. 2501, 25/F. Phone: 557-8633. Alvin Leung Jr.

Complete accu-tone II acoustical environments for bcst industry plus noise-lock sound control doors, windows, walls & silencers.

Industrial Equipment Representatives (IER), 1685 Precision Park Ln., Suite E, San Diego, CA, 92173. Phone: (619) 428-2261. Phone: (619) 428-2262. Fax: (619) 428-3483. Web Site: www.ier-broadcast.com. Alex Rodriguez, gen sls mgr; Juan Biosca, gen mgr.

Bcst, TV & recording studios equipment & supplies.

Innovision Optics Inc., 1719 21st St., Santa Monica, CA, 90404. Phone: (310) 453-4866. Fax: (310) 453-4677. E-mail: innovision@innovision-optics.com Web Site: www.innovisionoptics.com. Mark Centkowski, pres.

Remote-controlledTracking Systems, Floor Cams, Vertical Cams, Vertical Cams, Tower Cams, Birds Eye, HD Probe Lens, HD Cine SpeedCam.

Inovonics Inc., 1305 Fair Ave., Santa Cruz, CA, 95060. Phone: (831) 458-0552. Fax: (831) 458-0554. E-mail: info@inovon.com Web Site: www.inovon.com. James B. Wood,.

Manufacturers of bcst audio signal processing, encoding/decoding, sound recording & instrumentation equipment.

Inscriber Technology Corporation, 26 Peppler St., Waterloo, ON, N2J 3C4. Canada. Phone: (519) 570-9111. Fax: (519) 570-9140. E-mail: info@inscriber.com Web Site: www.inscriber.com. Dan Mance, pres; Mike Bernhardt, sls dir; Randy Fowlie, COO.

1431 EE Aalsmeer. Inscriber Technology-European Rep Office, Zijdsraat 72. Phone: +31-297-362030. Fax: +31-297-380939. David Hughes, dir European oper.

Chiyoda-ky, Tokyo 100-0005. Inscriber Technology-Asian Rep Office, Level 9, AIG Bldg, 1-1-3 Marunouchi. Phone: 81-3-5288-5237. Fax: 81-3-5288-5111. Doug Strable, dir oper - Asia Pacific.

Software for desktop & bcst video markets, including Character generators, digital stores & Video Server/Sequencers

Insulated Wire Inc. Microwave Products Division, 20 E. Franklin St., Danbury, CT, 06810. Phone: (203) 791-1999. Fax: (203) 748-5217. Saverio T. Bruno, pres.

High-frequency, low-loss microwave cable & cable assemblies featuring IW's Tuf-Flex Series to 60 GHz.

Integrys Holdings L.L.C., 770 Pelham Rd., Greenville, SC, 29615. Phone: (864) 297-9290. Fax: (864) 297-9213. Web Site: www.integryslic.com. William C. Cox, pres/CEO.

Subscriber mgmt and billing system.

Intelligent Media Technology, 9460 Delegates Dr., Suite 108, Orlando, FL, 32837. Phone: (407) 855-8181. Fax: (407) 855-1653. Bob Proctor, engrg dir.

Digital audio snakes, A/D conversion & transmission, D/A conversion receiver/repeaters, multimedia fiber optic transmission systems (audio/video/voice/data).

Intelliprompt, 9039 Lucerne Ave., Culver City, CA, 90232. Phone: (310) 837-0389. Fax: (310) 837-0806. E-mail: la@intelliprompt.com Web Site: www.intelliprompt.com. Tony Finetti, W. coast opns; Ernest Boyden, pres.

New York, NY 10036, 630 9th Ave., Suite. 907, (corner of44th & 45th). Phone: (212) 765-0555. Fax: (888) 504-5047. E-mail: prompt@intelliprompt.com. Trish Devine.

Toronto, ON M5V 2R8 Canada, 44 Tecumseth St. Phone: (888) 504-9535. Fax: (888) 504-5047. E-mail: prompt@intelliprompt.com. Trish Devine.

Computerized teleprompting svcs.

Interface Media Group, 1233 20th St. N.W., Washington, DC, 20036. Phone: (202) 861-0500. Fax: (202) 296-4492. E-mail: info@interfacevideo.com Web Site: www.interfacevideo.com. Tom Angell, pres; Adam Hurst, VP.

FACILITY: film transfer/location/studio/motion control, Avid/interformat digital edit, audio, graphics, dubs, Vyvx/3D2/DGS/satellite, standards conversion.

Interlogix, 280 Huyler St., South Hackensack, NJ, 07606. Phone: (201) 489-9595. Fax: (201) 489-0111.

Closed circuit TV cameras, monitors & accessories, specializing in covert surveillance cameras.

International Cinema Equipment, Division of Magna-Tech Electronic Co. 5600 N.W. 32nd Ave., Miami, FL, 33142. Phone: (305) 573-7339. Fax: (305) 573-8101. E-mail: iceco@aol.com Web Site: www.iceco.com. Steve Krams, pres; Dara Reusch, VP.

16mm, 35mm, 70mm film projection equipment, film-to-tape transfer equipment, sound systems, editing equipment.

International Datacasting Corp., 2680 Queensview Dr., Ottawa, ON, K2B 8H6. Canada. Phone: (613) 596-4120. Fax: (613) 596-4863. Fax: (613) 596-9208. E-mail: corporate@intldata.ca Web Site: www.intldata.ca. Ron W. Clifton, pres/CEO; Denzil Doyle, chmn.

Rsch, dev, manufacture & mktg of value added high speed digital data transmission network & svcs.

International Electro-Magnetics, 350 N. Eric Dr., Palatine, IL, 60067. Phone: (847) 358-4622. Fax: (847) 358-4623. E-mail: mail@iemmag.com Web Site: www.iemmag.com. Anthony Pretto, pres.

Standard replacement & custom recording heads for audio, video & film.

Intersil Corp. Headquarters, 1001 Murphy Ranch Rd., Milpitas, CA, 95035. Phone: (408) 432-8888. Fax: (408) 432-0640. Web Site: www.intersil.com. Rich Beyer, pres/CEO; Dan Heneghan, CFO.

Tsimshatsui, Kowlon, NO Hongkong, The Gateway, 9 Canton Rd., Suite 1506, 15F Tower 6. Phone: +852 2709 7600.

Yokohama, NO 220-5820 Japan, Queen Tower A, 12F 2-3-1, Minato-Mirai, Nishi-ku. Phone: +81 45 682 5820. Fax: +81 45 682 5821.

Palm Bay, FL 32905, 2401 Palm Bay Rd. Phone: (321) 724-7000. (888) 486-3774. E-mail: investor@intersil.com. Web Site: www.intersil.com.

ICs for wireless networking, high performance analog-flat panel displays, optical storage (CD,DVD recordaable) & power mgmt.

Isaia & Co., 4650 Lankershim Blvd., North Hollywood, CA, 91602. Phone: (818) 752-3104. Fax: (818) 752-3105. Web Site: www.isaia.com. Roy N. Isaia, pres.

Remote camera heads "Power-Pod," "Runford Baker" tripods & heads, camera cranes "Egripment"—"Cinerent carbon fiber," used cameras.

J

J and R Moviola Inc., 1135 N. Mansfield Ave., Los Angeles, CA, 90038. Phone: (323) 467-3107. Fax: (213) 466-2201. Web Site: www.movieola.com. Joe Paskal, pres; Randy Paskal, exec VP.

Denver, CO 80238, 8000 E. 40th Ave. Phone: (303) 321-1099. Randy Urlik, exec VP.

Chicago, IL 60610, 416 W. Ontario. Phone: (312) 787-0622. Jeff McNeir, VP.

New York, NY 10036, 636 11th Ave. Phone: (212) 247-0972. Bob Herman, VP.

Film editing equipment, film & video shipping & storage, film-to-video transfer machine.

JBL Professional, Box 2200, 8500 Balboa Blvd., Northridge, CA, 91329. Phone: (818) 894-8850. Fax: (818) 830-1220. Web Site: www.jblpro.com. E-mail: info@jblpro.com Mark Gander, VP.

Manufacturers of loudspeaker systems for bcstg, recording studios, theaters, concerts, stadiums & other applications.

JC Sound Stages, 1160 N. Las Palmas Ave., Los Angeles, CA, 90038. Phone: (323) 467-7870. Fax: (323) 467-7832. Web Site: www.jcband.com/jcsoundstages.html. J.C. Belanger, owner.

Cable-controlled camera booms equipped for film or video, rehearsal pre-production recording all in classiest vibe avail anywhere.

The J-Lab Co., Box 6530, Malibu, CA, 90264. Phone: (310) 457-4090. Fax: (310) 457-4494. Web Site: www.j-lab.com. Jerry LaBarbera, pres.

Component accessories, battery-operated video, audio DAs, camera controls, variable speed shutter devices & portable switchers.

JNJ Industries Inc., 290 Beaver St., Suite 303, Franklin, MA, 02038. Phone: (508) 553-0529. Fax: (508) 553-9973. E-mail: sales@jnj-industries.com Web Site: www.jnj-industries.com. Jack Volpe, pres; Gail Howe, VP; Bob Enterkin, dir mktg.

CFC & HCFC free solvents, presaturated cloth wipes, spray bottles, dry cloth wipes, lens wipes; aqueous chemistries & ultra-sonic cleaning machines.

JOA Cartridge Service, 448 E. Hancock St., Lansdale, PA, 19446. Phone: (215) 362-8796. Fax: (215) 368-2336. E-mail: mmolyneaux_joa@yahoo.com Web Site: www.joaonlone.com. Mark P. Molyneaux, owner.

Bcst audiotape cartridges, audio, videotape & cassettes, tape accessories, DAT tape & cassettes, data storage diskettes & cassettes, optical disks, recordable CDs & reloading svc.

JSB Service Co., 204 S. Bayard Ave., Waynesboro, VA, 22980. Phone: (540) 949-5899. Fax: (540) 949-5863. Web Site: www.jsbservice.com. Joseph S. Brumbelow, pres/CEO.

Repair, resale of microwave communication devices, receivers, transmitters, solid state sources, amplifiers. Manufacturer of microwave components & modules, miniature dielectric resonant oscillators & VCO.

JVC Professional Products Company, 1700 Valley Rd., Wayne, NJ, 07470. Phone: (973) 317-5000. Fax: (973) 317-5030. Web Site: www.jvc.com/pro. E-mail: proinfo@jvc.com Kirk Hirota, pres; Bob Mueller, CEO & exec VP.

Cypress, CA 90630, 5665 Corporate Ave. Phone: (714) 229-8024. Eric Rosenberg, rgnl sls mgr.

Aurora, IL 60504-8149, 705 Enterprise St. Phone: (630) 851-7809. Chris Dalaly, branch mgr.

Wayne, NJ 07470, 1700 Valley Rd. Phone: (973) 317-5000. Paul Kasparian, branch mgr.

Plasmas; full line of professional video equipment including digital VTRs cameras, monitors & projectors.

Jampro Antennas/RF Systems Inc., 6340 Skycreek Dr., Sacramento, CA, 95828. Phone: (916) 383-1177. Fax: (916) 383-1182. E-mail: jampro@jampro.com Web Site: www.jampro.com. Alex Perchevitch, pres; Doug McCabe, VP.

Manufacturers of TV & FM bcst antennas, combiners, filters & a complete line of rigid coaxial transmission line.

Jennings Technology Co., 970 McLaughlin Ave., San Jose, CA, 95122. Phone: (408) 292-4025. Fax: (408) 286-1789. E-mail: sales@jenningstech.com Web Site: www.jenningstech.com. Steve Randazzo, pres; J. Horton, controller.

High-voltage vacuum & gas capacitors; relays, switches, single- & three-phase contactors & instruments.

Jensen Tools, 7815 S. 46th St., Phoenix, AZ, 85044. Phone: (602) 453-3169. Phone: (800) 366-9662. Fax: (602) 438-1690. E-mail: jensen@stanleyworks.com Web Site: www.jensentools.com. Bridget Marnocha, pres.

Electronic tool kits & cases, tools, test equipment.

Jensen Transformers Inc., 9304 Deering Ave., Chatsworth, CA, 91311-5857. Phone: (818) 374-5857. Fax: (818) 374-5856. E-mail: info@jensen-transformers.com Web Site: www.jensen transformers.com. Bill Whitlock, pres.

Audio transformers, ISO-MAX audio & video ground isolation boxes

E.F. Johnson Co., (A division of Transcript International). 123 State St. N., Waseca, MN, 56093. Web Site: www.efjohnson.com. Michael Jalbert, pres/CEO.

K

K&H Products Ltd (Porta-Brace), Box 249, North Bennington, VT, 05257. Phone: (802) 442-8171. Fax: (802) 442-9118. E-mail: info@portabrace.com Web Site: www.portabrace.com. Robert Howe, pres; John Fairley, plant mgr.

Soft carrying cases for professional portable video/audio equipment.

Kahn Communications Inc., 338 Westbury Ave., Suite 2, Carle Place, NY, 111514. Phone: (516) 222-2221. Leonard R. Kahn, pres.

New York, NY 10017, 767 3rd Ave. Phone: (212) 983-6765. Leonard R. Kahn, pres.

Kalun Communications Inc., 44 Larkfield Dr., Toronto, ON, M3B 2H1. Canada. Phone: (416) 410-4138. Fax: (416) 410-4138. E-mail: postmaster@kalun.4t.com Web Site: www.kalun.4t.com. Paul Wong, engrg dir.

RF test equipment including wideband sweep generators, sweep comparator, switched attenautor, return loss bridge, detector & headend equipment for ATSC

Kangaroo Products Inc., 10845 Wheatlands Ave., Suite C, Santee, CA, 92071-2856. Phone: (619) 562-9696. Fax: (619) 449-7244. E-mail: sales@kangarooproducts.com Steve Leiserson, pres; Nancy Byrd, VP.

Custom contract carrying cases.

Kathrein Inc., Scala Division, Box 4580, Medford, OR, 97501. Phone: (541) 779-6500. Fax: (541) 779-6575. E-mail: broadcast@kathrein.com Web Site: www.kathrein-scala.com. Manfred Muenzel, pres; Judy Young, sls; Michael Bach, sls engr; Mike Johnson, sls engr.

Antennas & filters, low to full power, includes STL/TSL, LPTV, CATV, RPU, translator & FM/TV monitoring. Custom patterns our specialty.

Kay Industries Inc., 604 N. Hill St., South Bend, IN, 46617. Phone: (574) 236-6220. Fax: (574) 289-5932. E-mail: phasemaster@kayind.com Web Site: www.kayind.com. Larry Katz, natl sls mgr.

Rotary phase converters for single phase to three phase power.

Key West Technology, 14563 W. 96th Terr., Lenexa, KS, 66215. Phone: (800) 331-2019. Phone: (913) 492-4666. Fax: (913) 322-1864. E-mail: sales@keywesttechnology.com Web Site: www.keywesttechnology.com.

Manufacturer of titlers, character generators, Logo & ID inserters, automated display & control systems; graphic display generators; playback systems & emergency alert systems.

Kidde-Fenwal Inc., 400 Main St., Ashland, MA, 01721. Phone: (508) 881-2000. Fax: (508) 881-6729. Web Site: www.fenwalcontrols.com. John Sullivan, pres; Kathleen Schoonmaker, VP.

High-speed fire protection systems.

Kings Electronics Co. Inc., 1685 Overview Dr., Rock Hill, SC, 29730. Phone: (803) 909-5000. Fax: (803) 909-5092. Web Site: www.kingselectronics.com.

Video patch panels, patch cords, coaxial connectors, triaxial connectors, twinaxial connectors.

Kintronic Labs Inc., Box 845, Bristol, TN, 37621-0845. Phone: (423) 878-3141. Fax: (423) 878-4224. E-mail: ktl@kintronic.com Web Site: www.kintronic.com. Louis A. King, CEO; Gwen King, VP; Tom King, pres.

AM matching & directional antenna phasing systems, AM multiplexers, transmitter combiners, AM dummy loads, isocouplers, passive KF components & transmission lines.

Kline Towers, 828 Williams Street, West Columbia, SC, 29169. Phone: (803) 251-8000. Fax: (803) 251-6200. E-mail: raywhite@klinetowers.com Web Site: www.klinetowers.com. J.C. Kline, pres; R.C. White, VP; Anthony J. Fronseca, sls.

Designers, fabricators & erectors of TV, FM & other bcst towers & specialty structures.

Knox Video, 8547 Grovemont Cir., Gaithersburg, MD, 20877. Phone: (301) 840-5805. Fax: (301) 840-2946. Web Site: www.knoxvideo.com. Philip Edwards, pres; Stefan Seigel, production mgr; Roland Blood, exec VP.

Electronic bulletin bd for video messages. VCR control units. Full matrix routing switches.

Konica Minolta Corp., 725 Darlington Ave., Mahwah, NJ, 07430. Phone: (201) 529-6060. Fax: (201) 529-6070. E-mail: isddisplay@minolta.com Web Site: www.minoltausa.com.

CRT & LCD color analyzing instrumentation.

Kuhnel Co. Inc., 155 Harmony Rd., Mickleton, NJ, 08056. Phone: (856) 423-4277. Fax: (856) 423-5105. Mary Kuhnel, pres.

Instal & maintenance of antennas & towers.

L

L-3 Communications Telemetry East, Box 729, Bristol, PA, 19007-0729. Phone: (267) 545-7000. Fax: (267) 545-0100. Web Site: www.l-3com.com. Rod Oren, gen mgr & sr VP; William Wargo, VP business affrs.

Manufacturer of satellite receiving systems, antennas, telemetry receiving systems, ancillary equipment, communications for aerospace & defense.

LARCAN, 228 Ambassador Dr., Mississauga, ON, L5T 2J2. Canada. Phone: (905) 564-9222. Fax: (905) 564-9244. E-mail: sales@larcan.com Web Site: www.larcan.com. Jim Adamson, sr VP & sls & mktg; Sheryl Richmond, US sls opns mgr.

LARCAN innovates, designs, and manufactures superior analog and digital television transmitters for wireless and broadcast markets worldwide. LARCAN specializes in custom network planning and RF synergies for broadcast and mobile video/DVB-H. "End to End' engineering solutions in solid state VHF, UHF, high power IOT transmitters, as well as low power transmitters/translators and FM solutions. Broadcast innovations and transmitters from 1 w to 100 kw strong.

LARCAN USA, 1390 Overlook Dr., Lafayette, CO, 80026. Phone: (303) 665-8000. Fax: (303) 673-9900. Web Site: www.larcan.com. David Hale, chmn & VP sls; Jim Adamson, pres.

Repair & sls of high power UHF TV transmitters, low power TV transmitters & translators, FM transmitters & translators & AC line surge protectors.

LBA Technology Inc., Box 8026, 3400 Tupper Dr., Greenville, NC, 27835-8026. Phone: (800) 522-4464. Phone: (252) 757-0279. Fax: (252) 752-9155. E-mail: lbatech@lbagroup.com Web Site: www.lbagroup.com. Lawrence Behr, CEO; Marcian Bouchard, pres; Javier Castillo, VP.

Design & manufacture medium wave antenna systems marketed worldwide, including folded unipole antennas, tuning units, transmitter combiners, diplexers, triplexers, RF components & collocation equiptment.

LEA International, 6520 Harney Rd., Tampa, FL, 33610. Phone: (813) 621-1324. Fax: (813) 621-8980. Web Site: www.leaintl.com. Mike Everson, VP sls; Shawn Thompson, mgng dir.

Hayden Lake, ID 83835, 10701 Airport Dr. Phone: (800) 881-8506. Fax: (208) 762-6099.

Manufacturers of transient voltage surge suppression & power conditioning equipment.

LINK Electronics Inc., 2137 Rust Ave., Cape Girardeau, MO, 63703. Phone: (573) 334-4433. Fax: (573) 334-9255. E-mail: link@linkelectronics.com Web Site: www.linkelectronics.com. Bob Henson, pres; Ellen Henson, exec VP; James Timberlake, VP opns; Dave Aufdenberg, customer svc.

San Pedro, CA 90731. LINK Electronics Inc.-Western Rgnl Sls, 1035 W. 20th St. Phone: (310) 548-3925. Fax: (310) 548-3350. E-mail: philipburnslink@wmconnect.com. Phil Burns, rgnl mgr-Western rgn.

Lawrenceville, NJ 08648. LINK Electronics Inc.-Northeast & Southeast Rgnl Sls, 2 W. Laurelwood Dr. Phone: (609) 561-9551. E-mail: raybouchard@aol.com. Ray Bouchard, rgnl mgr-N.E. & S.W. rgns.

Manufacturer of Sync Generators, system timing, audio & video DAs, power amps, encoders, decoders, video processing, test equipment & video presence detectors, closed caption encoders, decoders, video switchers, digital distribution & conversion.

LTM Corp. of America, 7755 Haskell Ave., Van Nuys, CA, 91406. Phone: (818) 780-9828. Fax: (818) 780-9848. E-mail: info@ltmlighting.com Web Site: www.ltmlighting.com. Dennis Knopf, exec VP.

HMI & quartz lighting fixtures for film & video production; fresnels, open face, fiber optic, soft lights & fluorescents from 18w to 18,000 w. Also complete line of microphone poles, windscreens & muffs.

Laser Diode Inc., Fiber Optic Business Unit Tyco/Electronics. 2 Olsen Ave., Edison, NJ, 08820. Phone: (732) 549-9001. Fax: (732) 906-1559. E-mail: sales@laserdiode.com Web Site: www.laserdiode.com. Rollin Ball, dir; Peggy Scarillo, sls; Scott Grayman, production mgr; Steve Lerner, production mgr.

Manufacture FP, high power, & CW lasers along with high sensitivity detectors, FDDI/SONET modules for short/long haul transmission, test DWDM & military, and commercial fiber optic systems. Also offer Hi-Reliability custom packaging svcs.

The Laumic Rental Co., 432 W. 45th St., New York, NY, 10036. Phone: (212) 586-6161. Fax: (212) 245-0974. Stuart Mann, gen mgr.

Sls, rental, svc, training for bcst, industrial equipment; systems designed & installed.

Leader Instruments Corp., 6484 Commerce Dr., Cypress, CA, 90630. Phone: (714) 527-9300. Fax: (714) 527-7490. E-mail: lopez@leaderusa.com Web Site: www.leaderusa.com. M. Sawa, pres.

Cypress, CA 90630, 6484 Commerce Dr. Phone: (714) 527-9300.

Electronic test equipment for video, audio, RF, microwave, oscilloscopes & gen use.

Leaming Industries, 3972 Barranca Pkwy., J 608, Irvine, CA, 92606. Phone: (949) 743-5233. Fax: (949) 743-5233. E-mail: sales@leaming.com Web Site: www.leaming.com. Robert F. Leaming, pres; Keith G. Rauch, sr engr.

BTSC Stereo/SAP encoders, modulators

The Leather Specialty Co., 2690 W. Airport Blvd., Sanford, FL, 32771. Phone: (407) 323-1830. Fax: (407) 330-1317.

Transit/shipping cases, custom manufactured to specifications & tool cases.

Lectrosonics Inc., 581 Laser Rd. NE, Rio Rancho, NM, 87124. Phone: (505) 892-4501. Fax: (505) 892-6243. E-mail: sales@lectrosonics.com Web Site: www.lectrosonics.com. Larry E. Fisher, pres; Bruce C. Jones, mktg VP; Gordon Moore, sls VP; Bob Cunnings, engrg VP.

Wireless microphone systems for bcst, motion picture & tele-product applications. Automatic sound system mixing & control.

Leightronix Inc., 2330 Jarco Dr., Holt, MI, 48842. Phone: (800) 243-5589. Fax: (517) 694-1600. E-mail: sales@leightronix.com Web Site: www.leightronix.com. Jeff Possanza,.

Video servers & TV automation.

Leitch Inc., 4400 Vanowen Street, Burbank, CA, 91505. Phone: (757) 548-2300. Phone: (800) 231-9673. Fax: (757) 548-0019. E-mail: leitch@leitch.com Web Site: www.leitch.com. Paula Moore, gen mgr; Tom Jordan, sls VP; Don Thompson, dir mktg.

Toronto, ON M3C 3E5 Canada. Leitch Technology Corp., 150 Ferrand Dr, Suite 700. Phone: (416) 445-9640. (800) 387-0233. John Nielson, South Central rgnl sls mgr.

Audio & video distribution amplifiers, sync generators, clock systems & timers, synchronizers, test equipment, still storage, scramblers & descramblers. Audio, video, digital & data routing switchers, terminations, serial digital products.

Lemco Tool Corp., 1850 Metzger Ave., Cogan Station, PA, 17728. Phone: (570) 494-0620. Fax: (570) 494-0860. E-mail: toolinfo@lemco-tool.com Web Site: www.lemco-tool.com. Glenn G. Miller, pres.

Designers & manufacturers of mechanical tools, equipment & materials for the construction & maintenance of CATV systems.

LEMO USA Inc., Box 2408, Rohnert Park, CA, 94927-2408. Phone: (707) 578-8811. Fax: (707) 578-0869. E-mail: info@lemousa.com Web Site: www.lemo.com. Jim Hassett, gen mgr; Carol Taylor, natl sls & mktg mgr; Julie Carlson, mktg mgr.

LEMO designs & manufactures precision custom connection solutions. LEMO developed the 3K.93C series connector which is now the SMPTE standard for natl & international bcst companies.

Leviton NSI Colortran, (A division of NSI Corp.). 20497 S.W. Teton, Tualatin, OR, 97062. Phone: (503) 404-5500. Phone: (800) 576-6060. Fax: (503) 404-5600. E-mail: pauls@leviton.com Harold Leviton, pres; Paul Sherbo, VP sls & mktg.

Lighting fixtures & control devices for theater, TV & architectural applications.

Lightning Eliminators & Consultants Inc., 6687 Arapahoe Rd., Boulder, CO, 80303. Phone: (303) 447-2828 ext.100. Fax: (303) 447-8122. E-mail: info@lecglobal.com Web Site: www.lecglobal.com. Roy B. Carpenter Jr., chief technologist; Peter A. Carpenter, exec VP; Jerry V. Dollar, pres/CEO.

Designers & manufacturers of lightning strike prevention, grounding & power conditioning systems.

Lightning Master Corp., Box 6017, 1351 N. Arcturas Ave., Clearwater, FL, 33765. Phone: (727) 447-6800. Fax: (727) 461-3177. E-mail: bak@lightningmaster.com Web Site: www.lightningmaster.com. Bruce A. Kaiser, pres; Dave Anderson, opns VP; Mark Shubin, sls VP.

Gwangju City 500-827 Korea, Democratic People's Republic Of, 3F 12 1-2 Sinan-Dong, Buk-Gu. Phone: 826 252 162 16. Sang Su Lee, gen mgr.

Structural lightning protection equipment, transient voltage surge suppression, bonding & grounding products, consulting svcs; site survey, analysis & training.

Lightning Prevention Systems, 154 Cooper Rd., Suite 1201, West Berlin, NJ, 08091-9116. Phone: (856) 767-7806. Phone: (888) 667-8745. Fax: (856) 767-7547. E-mail: info@lpsnet.com Web Site: www.lpsnet.com. Ian E. Fawthrop, pres.

Manufactures equipment utilizing point discharge technology to remove the lightning attractive static charge on towers or structures that they're on, preventing lightning strikes.

Lindsay Broadband Inc., 2035 Fisher Dr., R.R. #5, Peterborough, ON, K9J 6X6. Canada. Phone: (705) 742-1350. Fax: (705) 742-7669. E-mail: sales@lindsaybroadbandinc.com Web Site: www.lindsaybroadbandinc.com. David Atman, pres; David Hayford, VP opns; Linda Curtin, VP finance; Chris Skarica, VP engrg.

Full line of CATV equipment & supplies, hybrid modules, tap-offs & power supplies, distribution amplifiers, equalizers, fiber optic transmission systems, line surge protectors & status monitoring.

Linear Acoustic Inc., 354 N. Prince St., Lancaster, PA, 17603. Phone: (717) 735-3611. Fax: (717) 735-3612. E-mail: tim@linearacoustic.com Web Site: www.linearacoustic.com. Tim Carroll, pres; Steven Strassberg, VP sls & mktg.

Lipsner Smith Co., 4700 Chase Ave., Lincolnwood, IL, 60712. Phone: (847) 677-3000. Phone: (800) 323-7520. Fax: (847) 677-1311. Fax: (800) 784-6733. E-mail: sales@lipsner.com Web Site: www.lipsner.com.

Motion Picture Film Laboratory Equipment.

Listec Video Corp., 2001 Palm Beach Lakes Blvd., Suite 411, West Palm Beach, FL, 33409. Phone: (561) 683-3002. Fax: (561) 683-7336. E-mail: sales@listec.com Web Site: www.listec.com. Joanne Camarda, pres.

Hauppauge, NY 11788, 40-3 Oser Ave. Phone: (631) 273-3029. Fax: (631) 435-4544. Bob Lorello, sls & product support.

Fully professional range of flat-panel, monitor prompters for studio, field & conferencing applications complemented by Windows prompting software.

Location Sound Corp., 10639 Riverside Dr., North Hollywood, CA, 91602. Phone: (818) 980-9891. Fax: (818) 980-9911. E-mail: information@locationsound.com Web Site: www.locationsound.com. Steve Joachim, gen sls mgr; Robert Noone, rental mgr.

Dealer of professional audio & communications solutions for film, video, bcst, business, institutional & recording applications. Over 30 years experience.

Logica Inc., 655 3rd Ave., Suite 700, New York, NY, 10017. Phone: (212) 682-7411. Fax: (212) 682-0715.

Consulting.

Logitek, 5622 Edgemoor Dr., Houston, TX, 77081. Phone: (713) 664-4470. Phone: (800) 231-5870. Fax: (713) 664-4479. E-mail: info@logitekaudio.com Web Site: www.logitekaudio.com. Tag Borland, pres; Frank Grondstein, sls dir.

Digital audio consoles, digital audio routers & audio level indicators (meters).

Lowel-Light Manufacturing Inc., 140 58th St., Brooklyn, NY, 11220. Phone: (718) 921-0600. Fax: (718) 921-0303. E-mail: info@lowel.com Web Site: www.lowel.com. Don Youngberg, midwest sls; Dale Marks, sls rep; Toni Pearl, dealer liaison; Eric Drucker, eastern sls mgr.

Lights, controls, mounts & kits for imaging professionals, innovatively designed & built for rugged dependable use, ease of operation and portability.

Luxor, 2245 Delany Rd., Waukegan, IL, 60087. Phone: (847) 244-1800. Fax: (800) 327-1698. E-mail: luxorfurn@ameritech.net Web Site: www.luxorfurn.com. Robert T. Raw, gen mgr; Robert White, sls & mktg mgr.

Computer stands, A/V equipment stands, conference room furniture, TV stands, library & office furniture.

M

M/A-COM, (A division of AMP Incorporated). 1011 Pawtucket Blvd., Lowell, MA, 01853. Phone: (978) 442-5000. Phone: (800) 366-2266. Fax: (978) 442-5350. Web Site: www.macom.com. Rick P. Hess, pres/CEO; Tim Emery, dir.

RF microwave & mm wave components & subsystems.

MATCO Inc., 15000 Stetson Rd., Los Gatos, CA, 95033-9770. Phone: (408) 353-2670. Phone: (800) 348-1843. Fax: (408) 353-8781. E-mail: sales@matco.video.com Web Site: www.matco-video.com. David Harbert, pres; Rita Harbert, gen mgr; William Meyer, dir.

Playback automation, coml insertion, & machine control systems for bcst, cable & coml, industrial & medical. MPEG 2 video servers with automation software options.

MCG Surge Protection, 12 Burt Dr., Deer Park, NY, 11729. Phone: (631) 586-5125. Fax: (631) 586-5120. E-mail: info1@mcgsurge.com Web Site: www.mcgsurge.com. Christine Jelley, CEO; Diane Lanciotti, CFO; Sue Baron, gen sls mgr.

Surge protectors for AC power lines, telephone/signal & data lines. Protecting industry since 1967.

MCL Inc., 501 S. Woodcreek Rd., Bolingbrook, IL, 60440-4999. Phone: (630) 759-9500. Fax: (630) 759-5018. E-mail: sales@mcl.com Web Site: www.mcl.com. Frank P. Morgan, VP mktg; Art Faverio, pres/CEO; Frank Morgan, VP sls.

Satellite communication fixed & mobile High Power Amplifiers in C-band, X-Band, Ku-band, DBS, V-Band, Ka-Band & Multi-Band.

MGE UPS SYSTEMS Inc., 1660 Scenic Ave., Costa Mesa, CA, 92626. Phone: (714) 557-1636. Fax: (714) 557-9788. E-mail: info@mgeups.com Web Site: www.mgeups.com. Mike Chmura, VP; Ray Prince, pres.

Hoffman Estates, IL 60195, 2895 Greenspoint Pkwy. #350. Phone: (847) 585-1113. Fax: (847) 585 1125. Mike Chmura, VP sls & mktg.

New York, NY 10018, 520 8th Ave., 21st Fl. Phone: (212) 594-9333. Fax: (212) 594-3691.

Manufacturers of uninterruptible power systems (UPS) power conditioners & inverters that protect equipment from power related problems.

MODCOMP Inc., 1500 S. Powerline Rd., Suite A, Deerfield Beach, FL, 33442. Phone: (954) 571-4600. Fax: (954) 571-4700. E-mail: info@modcomp.com Web Site: www.modcomp.com. Alex Lupinetti, pres; Ron Cook, opns VP.

Minicomputer systems, hardware & software for ground stn monitoring & control. SCADA applications & website enabling software.

MRPP Inc., 201 W. Chatham St., Suite 202, Cary, NC, 27511. Phone: (919) 468-1000. Fax: (919) 468-1956. Web Site: www.mrppinc.com. Sheila Ogle, CEO; Sue Toth, pres.

Procuring, servicing, instal, & sale of bcstg & satellite equipment. Leasing plans available.

M2 America, 470 Riverside St., Portland, ME, 04103. Phone: (207) 797-2600. Fax: (207) 797-2604. E-mail: info@m2america.com Web Site: www.m2america.com.

CD, CD-R, DVD-R, optical disk duplicators.

MUSICAM U.S.A., Bldg. 4, 670 N. Beers St., Holmdel, NJ, 07733. Phone: (732) 739-5600. Fax: (732) 739-1818. E-mail: sales@musicamusa.com Web Site: www.musicamusa.com. Jill A. Fitzpatrick, mktg.

Digital Audio codecs for remote bcstg with Bandwidth up to 20 khz for ISDN, POTS or IP.

MYAT Inc., 360 Franklin Tpke., Mahwah, NJ, 07430. Phone: (201) 684-0100. Fax: (201) 684-0104. Web Site: www.myat.com. E-mail: sales@myat.com Philip Cindrich, pres; Derek Small, dir filter products; Dennis Heymans, sls mgr.

Falmouth, ME 04105, 60 Gray Rd. Phone: (207) 878-7807. (207) 767-7806.

Transmision line systems, filters, combiners, UHF & L-band antenna.

Macrovision Corp., 2830 De La Cruz Blvd., Santa Clara, CA, 95050-2619. Phone: (408) 562-8400. Fax: (408) 567-1800. Web Site: www.macrovision.com. E-mail: info@macrovision.com John Ryan, chmn; Bill Krepick, pres/CEO.

Tokyo 150-0001 Japan. Macrovision Japan K.K., Takaba Bldg. 2F, 6-18-5, Jingumae, Shibuya-Ku. Phone: 81-35-774-6253. Masao Kumei.

Beeshire SL 61 BR United Kingdom. Macrovision UK Ltd., 14-18 Bell St, Maiden Head. Phone: (44) 870 871 1111. Martin Brooker.

Copy protection & rights mgmt for videocassettes, pay-per-view cable, satellite TV, & video conferencing.

Magna-Tech Electronic Co. Inc., 5600 N.W. 32nd Ave., Miami, FL, 33142. Phone: (305) 573-7339. Fax: (305) 573-8101. E-mail: magnatech@iceco.com Web Site: www.magna-tech.com. Steven Krams, pres; Barnet Kaufman, VP.

Manufactures professional motion picture sound recording, reproducing & projection equipment, film recorders & reproducers; 16 & 35mm recorders, telecine followers, counters, pre amps, dubbers & looping systems.readers.

Magni Systems Inc., 22965 N.W. Evergreen Pkwy., Hillsboro, OR, 97124. Phone: (503) 615-1900. Fax: (503) 615-1999. E-mail: sales@magnisystems.com Web Site: www.magnisystems.com. Victor L. Kong, CEO; Chuck Barrows, VP sls.

Video Test Equipment and Scan Converters. Automated video test & monitoring equipment, waveform monitors, vectorscopes, test signal generators, VIT inserter, PC graphics to video encoders & video overlay scan converters.

Magnum Towers Inc., 9370 Elder Creek Rd., Sacramento, CA, 95829. Phone: (916) 381-5053. Fax: (916) 381-2144. E-mail: office@magnumtowers.com Web Site: magnumtowers.com. Lawrence Smith, pres; Lori Morris, office mgr.

Radio, TV & microwave towers.

Marathon Norco Aerospace, Inc., 8301 Imperial Dr, Waco, TX, 76712. Phone: (254) 776-0650. Fax: (254) 776-6558. Web Site: www.mptc.com. Al Rodriquez, pres.

CASP universal battery support systems & AC/DC power supplies.

Marcom, 540 Hauer Apple Way, Aptos, CA, 95003-9501. Phone: (831) 768-8668. Fax: (831) 768-7810. E-mail: marty@mar-com.com Web Site: www.mar-com.com. Martin Jackson, pres.

FM, AM, TV & microwave transmitting equipment; sls engrg, instal & maintenance.

Marconi Communications, 4350 Weaver Pkwy., Warrenville, IL, 60555. Phone: (630) 579-5000. Fax: (630) 579-5050. Web Site: www.marconi.com. Dusty Becker, VP; David Smith, VP.

London WIK2HD United Kingdom, 34 Grosvenor. Phone: 44 (0) 20 7493 8484. Fax: 44 (0) 20 7493 1974.

CATV enclosures, security enclosures, protection devices, pole line hardware & connectors.

Marietta Design Group, 82 Plantation Point, Suite 200, Fairhope, AL, 36532. Phone: (251) 990-3558. Fax: (360) 838-9046. E-mail: support@mariettadesign.com Web Site: www.mariettadesign.com.

AccuPrompt— & QuickPrompt— teleprompting software for MacIntosh.

QuickPrompt 1.7.2—for professional video prompting.

Maritz Inc., 1375 North Highway Dr., Fenton, MO, 63099. Phone: (877) 462-7489. Web Site: www.maritz.com.

Communications, film/video training, business meetings, mktg.

Marketron International, 700 Airport Blvd., Suite 130, Burlingame, CA, 94010-2001. Phone: (800) 788-9245. Fax: (650) 548-2295. E-mail: lcarpenter@marketron.com Web Site: www.marketron.com. Mike Jackson, CEO.

Toronto, ON M2N 6C6 Canada, 5075 Yonge St, Suite 404. Phone: (416) 221-9944. Les Bridgen, gen mgr.

Birmingham, AL 35244, 3000 Riverchase Galleria, 8th Fl. Phone: (205) 987-7456. Fax: (205) 733-4535. E-mail: tvsales@marketron.com. Michael Hunter, gen mgr.

Hailey, ID 83333, 101 Empty Saddle Trail. Phone: (208) 788-6272. Gary Coats, gen mgr.

Software applications for radio, TV, networks, syndicators, traf, accounting, mgmt, demand pricing, inventory control, rsch & proposals.

Marshall Electronics, 1910 E. Maple Ave., El Segundo, CA, 90245. Phone: (310) 333-0606. Phone: (800) 800-6608. Fax: (310) 333-0688. E-mail: sales@lcdracks.com Web Site: www.marshall-usa.com. Nathan Mordukhay, exec VP; Leonard Marshall, CEO.

Wire cable & connectors—Mogami superflex wire and cable, Tajimi, connectors & LCD bcst monitors. Marshall provides the highest quality products to the bcst, vidio & music recording mkts. Marshall specializes in manufacturing optics, LCD panels, microphones & mogami wire.

Marti Electronics, 4100 N. 24th St., Quincy, IL, 62305. Phone: (217) 224-9600. Fax: (217) 224-9607. E-mail: sales@martielectronics.com Web Site: www.martielectronics.com. John Lackness, sls.

Composite, dual mono & digital STL systems, remote pickup systems, telemetry links, studio to transmitter links, FM exciters, transmitters & pots remote pickup systems.

Martinsound Inc., 1151 W. Valley Blvd., Alhambra, CA, 91803-2440. Phone: (626) 281-3555. Fax: (626) 284-3092. E-mail: info@martinsound.com Web Site: www.martinsound.com. Joe Martinson, pres; Doug Osborne, dir & sls & mktg.

Complete line of audio control consoles for music recording, bcst, & video postproduction applications. MultiMax surround monitor control system, flying faders console automation, Martech MSS-10 precision microphone preamplifier.

Masterclock Inc., 2484 W. Clay St., St. Charles, MO, 63301. Phone: (800) 940-2248. Fax: (636) 724-3776. Web Site: www.massterclock.com. William J. Clark, pres.

Masterclock systems, clock displays & time code products mfg & distribution.

Matrox Video Products Grp, 1055 St. Regis Blvd., Dorval, PQ, H9P 2T4. Canada. Phone: (514) 685-2630, EXT. 2636. Fax: (514) 685-2853. E-mail: video.info@matrox.com Web Site: www.matrox.com/video. Lorne Trottier, pres; Spiro Plagakis, VP sls & mktg.

Emmy award-winning technology & mktg leader in the field of digital video hardware for reeltime editing, DVD authoring & web streaming.

Matthews Studio Equipment Inc. (MSE), 2405 Empire Ave., Burbank, CA, 91504-3399. Phone: (818) 843-6715. Fax: (323) 849-1525. E-mail: info@msegrip.com Web Site: www.msegrip.com. Robert Kulesh, VP sls & mktg.

TV camera support dollies, land tripods, studio pedestals, pan/tilt heads, cases.

Maxell Corp. of America, 2208 Rt. 208, Fairlawn, NJ, 07410. Phone: (201) 794-5900. Fax: (201) 796-8790. Web Site: www.maxellpromedia.com.

Blank audio & video recording tape for professional bcstrs & duplicators.

Maze Corporation, 3855 Rock Ridge Rd., Birmingham, AL, 35210-3797. Phone: (205) 706-2080. Fax: (205) 956-5027. E-mail: maze@mazecorp.com Web Site: www.mazecorp.com. Vira J. Maze, pres.

Remarketers of TV & video equipment.

McCurdy Radio Ltd., 30 Kelfield St., Toronto, ON, M9W 5A2. Canada. Phone: (416) 248-6155. Fax: (416) 248-6755. E-mail: sales88@mcradio.com Web Site: www.mcradio.com. Paul Hudson, pres; Stuart Hobbs, gen mgr.

Buffalo, NY 14206, 1051 Clinton St.

Audio monitors & meters.

Media Computing Inc., Box 4169, Cave Creek, AZ, 85327-4169. Phone: (480) 575-7281. E-mail: info@mediacomputing.com Web Site: www.mediacomputing.com. Michael Rich, CEO; Kathryn A. Hulka, treas; Larry L. Baum, mgr.

ANGIS-PC-based software automatically updates displays on characters generators & web pages with real-time data like elections, news tickers, closing.

Media Concepts Inc., 200 Spring Garden, Unit B, Philadelphia, PA, 19123. Phone: (215) 923-2545. Fax: (215) 928-0750. E-mail: mediacon@libertynet.org Bob Weissman, pres.

Video duplication, international video standards conversion, CD-Rom duplication, DVD duplication authoring. Macrovision copy-protection.

Mediasoft Inc., 7200 N. Broadway Ext., Oklahoma City, OK, 73116. Phone: (405) 607-2000. Fax: (405) 607-2071. E-mail: info@mediaofusa.com Web Site: www.mediasoftusa.com. Bob Alfson, pres.

Microcomputer products & svcs.

Mega Hertz, 4100 International Plaza, Suite 150, Fort Worth, TX, 76109. Phone: (800) 883-8839. Fax: (817) 529-0745. E-mail: sales@megahz.com Web Site: www.megahz.com. Doug Sherar, gen sls mgr.

Mega Hertz is a Value-Added-Reseller of "Unique" Multi-Vendor System Solutions that support the deployment of advanced technologies in hybrid Fiber/Coax Braodband Networks.

Megastar Inc., 4709 Compass Bow Ln., Las Vegas, NV, 89130. Phone: (702) 386-2844. Fax: (702) 388-1250. Nigel Macrae, pres.

Design & instal of integrated satellite networks. Reseller, earth stns, earth stn equip & microwave equipment & all support equipment.

MEGGER, 2621 Van Buren Ave., Norristown, PA, 19403. Phone: (610) 676-8500. Fax: (610) 676-8610. E-mail: sales@megger.com Web Site: www.megger.com.

Cable fault-locating equipment & other electrical testing instruments.

Memorex Products Inc., 17777 Center Court Dr., Suite 800, Cerritos, CA, 90703. Phone: (562) 653-2800. Fax: (562) 653-2900. E-mail: generaling@memorex.com Web Site: www.memorex.com. Scott Stroup, VP.

Memorex is a manufacture, marketer of consumer media & computer products.

Meridian Design Associates, Architects, 1140 Broadway, New York, NY, 10001. Phone: (212) 431-8643. Fax: (212) 431-8775. E-mail: info@meridiandesign.com Web Site: www.meridiandesign.com.

Miami, FL 33144, 907 S.W. 79th Ave. Phone: (305) 262-7663. Fax: (305) 262-7675. Antonio Argibay.

Architectural firm specializing in the design of bcst & media facilities.

Merlin Engineering Works Inc., 1888 Embarcadero Rd., Palo Alto, CA, 94303. Phone: (650) 856-0900. Phone: (800) 227-1980. Fax: (650) 858-2302. E-mail: sales@merlineng.com Web Site: www.merlineng.com. Debbie Dirickson, dir.

Bcst VTRs, custom VTRs & accessories, VTR automation systems, stereo audio encoders, standards converters.

Metz Engineering, 15684 Old Mormon Bridge Rd., Crescent, IA, 51526-4138. Phone: (712) 545-3222. Fax: (712) 545-9111. Joanne M. Metz, owner; John P. Metz III, dir.

Machine & welding shop plus construction.

Michael Stevens & Partners Ltd., Invicta Works, Elliott Rd., Bromley, Kent, BR2 9NT. United Kingdom. Phone: 44 0 020 8460 7299. Fax: 44 0 020 8460 0499. E-mail: simon@michael-stevens.com Web Site: www.michael-stevens.com.

Kingston Springs, TN 37082, 149 Dillard Ct, Suite E. Phone: (615) 952-2345. Fax: (615) 952-2342. Megan McCullough, sls office mgr.

Micro Communications Inc., Box 4365, 438 Kelley Ave., Grenier Field, Manchester, NH, 03108-4365. Phone: (603) 624-4351. Phone: (800) 545-0608. Fax: (603) 624-4822. E-mail: frank.malanga@mcibroadcast.com Web Site: www.mcibroadcast.com. Al Kula, sls engr; Sam Matthews, mktg mgr; Paul Smith, CEO.

Waveguide & coaxial transmission line; complete RF system packages for UHF, VHF, FM & LPTV panel antennas; antennas for UHF, VHF & FM.

Micro Technology Unlimited, 6900 Six Forks Rd., Raleigh, NC, 27615. Phone: (919) 870-0344. Fax: (919) 870-7163. E-mail: info@mtu.com Web Site: www.mtu.com. David B. Cox, pres.

Karaoke software products & pro workstations.

MicrQlog Corp., 20270 Goldenrod Ln., Germantown, MD, 20876. Phone: (301) 540-5500. Fax: (301) 540-5557. E-mail: info@mlog.com Web Site: www.mlog.com. Joe Brookman, pres/CEO.

Automated outbound/inbound voice messaging systems & service bureau; interactive voice response to mainframe computer.

Micron Audio Products Ltd., 216 Little Falls Rd., Cedar Grove, NJ, 07009. Phone: (973) 857-8150. Fax: (973) 857-3756. E-mail: micronaudio@cs.com Paul Tepper, pres.

TRAM lavalier microphones, sls & svc.

Microspace Communications Corp., 3100 Highwoods Blvd., Suite 120, Raleigh, NC, 27604. Phone: (919) 850-4500. Fax: (919) 850-4518. E-mail: uplink@microspace.com Web Site: www.microspace.com. Joseph Amor III, VP/gen mgr; Greg Hurt, dir; Carolyn Newey, dir.

Providing video, data & audio transmission svcs designed for antennas as small as 30 inches. Operates on domestic satellites for coverage of North America. Also providing fixed C- & Ku-band uplink svcs for video transmissions supporting applications such as news, sports, program origination (live or taped); & business TV. Remote & studio production available. Turnaround svc to & from domestic & international satellites.

Microwave Filter Co. Inc., 6743 Kinne St., East Syracuse, NY, 13057. Phone: (315) 438-4700. Phone: (800) 448-1666. Fax: (315) 463-1467. E-mail: mfcsales@microwavefilter.com Web Site: www.microwavefilter.com. Carl Fahrenkrug, pres; Scott Parsells, sls VP.

Filters, traps, combiners & custom networks for TV, radio, CATV, wireless cable, LAN & mobile radio.

Milestek Corp., 1506 I-35W, Denton, TX, 76207-2402. Phone: (940) 484-9400. Phone: (800) 524-7444. Fax: (940) 484-9402. E-mail: salesinfo@milestek.com Web Site: www.milestek.com. Brett Powers, pres.

Connectors including both 50 ohm & 75 ohm BNCs, cabling, patching & tools for coaxial cable.

Milestone Technologies Inc., Box 37145, Raleigh, NC, 27627. Phone: (919) 773-1772. E-mail: info@milestonetechnologies.com Web Site: www.milestonetechnologies.com. Miles Beam, pres.

Data bcstg file transfer software (SATX). Bcst binary files over one-way data nets (DBS, VSAT, TV, FM, VBI, RDS, MPEG2, etc.). Consulting & system integration svcs.

Miller Camera Support, L.L.C., 216 Little Falls Rd., Cedar Grove, NJ, 07009. Phone: (973) 857-8300. Fax: (973) 857-8188. E-mail: info@millertripods.us Web Site: www.millertripods.com. Gus Harilaou, gen sls mgr.

Pan & tilt fluid heads, tripods & camera support systems & accessories for DV, ENG & EFP (OB).

Miranda Technologies Inc., 3499 Douglas B. Floreani, Montreal, PQ, H4S 2C6. Canada. Phone: (514) 333-1772. Phone: (800) 224-7882. Fax: (514) 333-9828. Web Site: www.miranda.com. Strath Goodship, CEO; Spiro Plagakis, sr VP sls & mktg.

Beijing, NO 100037 China. Miranda China, Rm. 2402, Sichuan Bldg., E. Tower, 1 Fuchengmenwai St., Xicheng District. Phone: +86 10-68364818. Fax: +81 10-68364817. E-mail: chinasales@miranda.com.

Montreuil 93100 France. Miranda France, 216, rue de Rosny, 931000 Montreuil. Phone: +33 (0) 1 55 86 87 88. Fax: +33 (0) 1 55 86 00 29. E-mail: francesales@miranda.com.

Wanchai Hongkong. Miranda Asia, Unit 1706, Tai Tung Bldg., 8 Fleming Rd. Phone: +852-2539-6987. Fax: +852-2539-0804. E-mail: asiasales@miranda.com.

Tokyo 103-0013 Japan. Miranda Japan, 3-1-17 Nihombashi Ningyacho, Ishii Bldg. 2F, Cjuo-ku. Phone: +81 (0) 3-5644-7533. Fax: +81 (0) 3-3662-7555.

Oxfordshire OX10 9DG United Kingdom. Miranda Europe, Hithercroft Rd, Wallingford. Phone: +44 (0) 1491 820 000. Fax: +44 (0) 1491 820 001. E-mail: europesales@miranda.com.

Springfield, NJ 07081. Miranda USA, 195 Mountain Ave. Phone: (973) 379-0089. Fax: (973) 379-1953. E-mail: ussales@miranda.com.

Digital video interface products for bcstg & postproduction: serializers, digital-to-analog converters, NTSC encoders, computer video interfaces.

Mitsubishi Digital Electronics America Inc., 9351 Jeronimo Rd., Irvine, CA, 92618. Phone: (949) 465-6000. Fax: (949) 465-6046. Web Site: www.mitsubishi-tv.com.

Portable videotape recorder systems, consumer VCR's & audio visual big screen TV's.

Mobile Video Services Ltd., 1620 Eye St. N.W., Washington, DC, 20006. Phone: (202) 331-8882. Fax: (202) 331-9064. E-mail: Bookfeed@mobilevideo.net Web Site: www.mobilevideo.net. Lawrence VanderVeen, pres; Christine Baber, opns mgr.

Complete EFP & ENG svcs, multi-camera remote packages, editing teleco & satellite transmission svcs available. CBS & CNN news feeds available.

Modulation Sciences Inc., 12A World's Fair Dr., Somerset, NJ, 08873. Phone: (732) 320-3090. Fax: (732) 302-0206. E-mail: sales@modsci.com Web Site: www.modsci.com. Eric Small, CEO; Judy Mueller, pres.

With 20 + years experience in the bcst industry, we manufacture full line of FM & TV equipment including: composite clipper, STL's distribution amplifiers, SteroMaxx—Spatial image englarger, modulation monitors,

SCA & Data SCA equipment, TV stereo reference decoder, SAP & PRO generators, PRO ch receivers, SAP receivers, NTSC precision video demodulators.

Mohawk, 9 Mohawk Dr., Leominster, MA, 01453. Phone: (978) 537-9961. Fax: (978) 537-4358. E-mail: info@mohawk-cdt.com Web Site: www.mohawk-cable.com. Jeff Miller, VP mktg & VP sls.

Mohawk offers an end to end solution for your HDTV cabling needs. We use LEMO stainless connectors & have many var of SMPTE cable. Mohawk is the OEM for fiber & copper camera cable assemblies for all of the major camera manufacturers.

/ole-Richardson Co., 937 N. Sycamore Ave., Hollywood, CA, 90038-2384. Phone: (323) 851-0111. Fax: (323) 851-5593. E-mail: info@mole.com Web Site: www.mole.com. Michael C. Parker, pres; Don Phillips, VP sls; Larry Mole Parker, exec VP.

Lighting equipment for the motion picture, TV, video & still photographic industries.

Moseley Associates Inc., 111 Castilian Dr., Santa Barbara, CA, 93117-3093. Phone: (805) 968-9621. Fax: (805) 685-9638. E-mail: info@moseleysb.com Web Site: www.moseleysb.com. Jamal Hamdani, pres/CEO; Bruce Tarr, CFO.

Remote control systems, AM & FM stereo STLs, aural RPLs, data transmission systems & telecommunications, digital transmission system.

Motion Picture Enterprises Inc., Box 276, Tarrytown, NY, 10591-0276. Phone: (212) 245-0969. Fax: (212) 245-0974. E-mail: mpeny@aol.com Web Site: www.mpe.net. Neal R. Pilzer, pres.

Shipping cases, cabinets & cans for film & tape; custom made fibre cases, film & video equipment, supplies, sls, rental & repairs.

Motor Capacitors Inc., 6455 Avondale Ave., Chicago, IL, 60631. Phone: (773) 774-6666. Fax: (773) 774-6690. E-mail: info@capacitorindustries.com Web Site: www.capacitorindustries.com.

Motor-run, motor-start, metalized, oil-filtered, high voltage, film, electrolytic & power capacitors, & R.C. networks.

Motorola Broadband Communications Sector, 101 Tournament Dr., Horsham, PA, 19044. Phone: (215) 323-1000. Fax: (215) 323-0242. E-mail: broadband@motorola.com Web Site: www.motorola.com/broadband. Daniel M. Moloney, exec VP & pres/CEO.

Englewood, CO 80111, 6400 S. Fiddler-Green Cir. Phone: (303) 740-6118. Pete Wornski, VP.

Lewisville, TX 75057, 1330 Capital Pkwy. Phone: (972) 323-4100. Tim Roberti, rgnl mgr.

CATV headend & distribution equipment; sub terminals, addressable systems & interactive products.

Motorola Digital Media Systems, 55 Las Colinas Ln., San Jose, CA, 95119. Phone: (408) 362-4800. Fax: (408) 362-4825. Web Site: www.motorola.com. Doug Means, VP/gen mgr.

San Jose, CA 95119. Motorola Digital Media Systems, 55 Las Colinas LN. Phone: (408) 362-4800. Fax: (408) 362-4851. (408) 362-4825. Web Site: www.motorola.com.

Fiber-optic video links for bcst & CATV use. 1550nm AM systems & AM return systems.

Moviola, (Formerly Videotape Distributors Inc.). 545 W. 45th St., New York, NY, 10036. Phone: (212) 581-7111. Phone: (800) 327-3724. Fax: (212) 581-7977. Robert Schoenberg, VP/gen mgr.

Full-svc supplier of videotape, accessories & digital data storage products.

Multi-Image Network, 312 Otterson Dr., Suite F, Chico, CA, 95928. Phone: (530) 345-4211. Fax: (530) 345-7737. Katheryn Schifferle, pres/CEO.

Multimedia production systems for cable TV, bcst, PE G training, education, & corporate TV. Pre- & post-launch consulting, training seminars, repair svcs.

Murphy Studio Furniture, 4153 N. Bonita St., Spring Valley, CA, 91977. Phone: (619) 698-4658. Fax: (619) 698-1268. E-mail: dennismurphy@cox.net Web Site: www.murphystudiofurniture.com. Dennis W. Murphy, pres.

Design/construction of studio furniture for radio, TV & production facilities. Five modular lines. Custom designs.

Murray Co., 1807 Park 270 Dr., Suite 460, St. Louis, MO, 63146. Phone: (314) 576-2818. Fax: (314) 434-5780. Web Site: www.murray-company.com. John O'Hara, principal.

Kansas City, KS 66210, 7300 College, Suite 210. Phone: (913) 451 1884. Fax: (913) 451-3761.

Gen construction, design, program mgmt, space planning, project budgeting & consolidation planning.

Murry Rosenblum Sound Assoc., Inc., Audio Limited U.S.A. 21-36 33rd Rd., Long Island City, NY, 11106. Phone: (718) 728-2654. Fax: (718) 728-2654. E-mail: murryrosenblum2@nyc.rr.com Murry Rosenblum, pres.

Audio limited wireless microphones—two switchable frequencies—small UHF standard or diversity receiver—pocket transmitter or handhold transmitter.

Musco Mobile Lighting Ltd., Box 808, 100 First Ave. W., Oskaloosa, IA, 52577. Phone: (641) 673-0411. Fax: (641) 672-1996. Web Site: www.musco.com. Jerome Fynaardt, gen sls mgr.

Mobile location lighting utilizing 6K HMIs; remote control of pan, tilt & focus.

N

NEC America Inc., (Broadcast Equipment Dept.). 6555 N. State Hwy. 161, Irving, TX, 75039. Phone: (214) 262-2000. Phone: (214) 262-6299. Fax: (972) 751-7001. Web Site: www.nec.com.

VUES on-line digital editing system (video).

NSI, 9050 Red Branch Rd., Columbia, MD, 21045. Phone: (410) 964-8400. Fax: (410) 964-9661. Web Site: www.nsystems.com. E-mail: sales@nsystem.com Stephen Neuberth, pres.

Microwave antennas & remote controls for ENG applications.

NUCOMM Inc., 101 Bilby Rd., Hackettstown, NJ, 07840. Phone: (908) 852-3700. Fax: (908) 813-0399. Web Site: www.nucomm.com.

Microwave transmitters, receivers including digital video microwave systems & accessories for both portable & fixed line of sight applications. Modulators/demodulators & color bar generators.

NVISION Products, 125 Crown Point Ct., Grass Valley, CA, 95945. Phone: (530) 265-1000. Fax: (530) 265-1021. E-mail: nvsales@nvision1.com Web Site: www.nvision.tv. Charles S. Meyer, pres; Jay Kuca, dir.

Digital audio & data distribution, conversion, routing & transmission equipment for production/postproduction applications for bcstg industry.

NWL Capacitors, Box 10416, Riviera Beach, FL, 33419-0416. Phone: (561) 848-9009. Fax: (561) 848-9011. Robert Scitz, gen mgr; Linda Nixon, VP.

Manufacturers.

Nady Systems Inc., 6701 Shellmound St., Emeryville, CA, 94608. Phone: (510) 652-2411. Fax: (510) 652-5075. E-mail: ussales@nady.com Web Site: www.nady.com. John Nady, pres/CEO; Scott Wunschel, sls dir.

Wireless VMP & AMF products for bcst, film, video, stage, fixed instals. Consumer audio & communication equipment.

Nalpak, 1267 Vernon Way, El Cajon, CA, 92020-1838. Phone: (619) 258-1200. Fax: (619) 258-0925. E-mail: nalpaak@nalpak.com Web Site: www.nalpakcom.com. Robert S. Kaplan, pres; Debra S. Kaplan, VP.

Packaging & Material Handling Products; teffpak,Torm, Magliner, Leatherman, Gerber, Buck, Surefire, Steamlight.

Narda Satellite Networks, 435 Moreland Road, Hauppauge, NY, 11788. Phone: (631) 231-1700. Fax: (631) 272-5500. E-mail: sn.mktg@l-3com.com Web Site: www.l-3com.com /satellitenetworks. Greg Federline, gen mgr.

Turnkey satellite earth stns & networks, SNG & Fly Away electronics, ground communications equipment & M&C systems. Manufactures & implements a full line of earth stn network monitors & control systems.

Nardal - An L-3 Communications Co., 435 Moreland Rd., Hauppauge, NY, 11788. Phone: (631) 231-1700. Fax: (631) 231-1711. E-mail: nardaeast@l-3com.com Web Site: nardamicrowave.com. John Mega, pres; Michael Sanatore, opns VP.

Portable RF/microwave test instruments, power density meters, coaxial power monitors & meters.

Narragansett Imaging, 51 Industrial Dr., North Smithfield, RI, 02896. Phone: (401) 762-3800. Fax: (401) 767-4407. Web Site: www.nimaging.com. Donald Borwne, VP; Michael Halloran, VP.

Camera tubes, CCD camera modules.

National Audio Co. Inc., Box 3657, Glenstone Station, Springfield, MO, 65808. Phone: (417) 863-1925. Fax: (417) 863-7825. E-mail: nac@nactape.com Web Site: www.national-audiocompany.com. Steve Stepp, pres; Maxine Bass, sec/treas.

Quantegy audio & videotapes, Audio Pro-custom-loaded audio cassettes, Recordex Duplicators & Video Pro-professional video cassettes, Recordex equipment, CDL & DVD.

National Mobile Television, 2740 California St., Torrence, CA, 90503. Phone: (800) 242-0642. Phone: (206) 782-9945. Fax: (206) 782-9949. Web Site: www.nmtv.com. Mark Howorth, chmn/CEO.

Equipment includes: 40' & 34' trailers. Philips LDK-26 cameras with 44X lenses, Ikegami HL-79EAL cameras, Grass Valley switchers, Sony BVH-2000 & BVH-3100 1" VTR's, Chyron 4100 EXB with CCM, Abekas A-53D, Yamaha audio mixers, RTS intercom & IFB, Sony BVE-900 edit on bd.

National Mobile TV - Houston, 10 Greenway Plaza, Houston, TX, 77046. Phone: (713) 627-9270. Fax: (713) 871-9617. Bob Robinson, dir; Tim Jopplin, opns mgr.

Irving, TX 75039, 6 Communications Complex, 6221 N. O'Connor, Suite 117. Phone: (972) 556-1816. Fax: (972) 556-2543. (Dallas office).

Multi-camera location production company with 4-48' & 1-36' location production trucks. Eng Package & full-service offices in Houston & Dallas. Additional permanent facilities at the Summit in Houston.

National Steel Erectors Corp., Box 709, Muskogee, OK, 74402. Phone: (918) 683-6511. Fax: (918) 683-0888. B.R. Bayless, pres; Neal Bayless, exec VP.

Erection of radio, TV & microwave towers, including turnkey construction, from design to completion.

National Video Services Inc., 18 Commerce Rd., Newtown, CT, 06470. Phone: (203) 270-0677. Fax: (203) 270-9619. E-mail: sales@intermedvideo.com Web Site: www.intermedvideo.com. Harry Davies, sec/treas.

Distribution of video equipment for corporate & industrial use; design & install of TV studios; mfg of video equipment; rsch & engrg.

National Video Tape Co. Inc., 6800 Sierra Ct., Suite D, Dublin, CA, 94568. Phone: (925) 803-1440. Fax: (925) 803-0227. E-mail: jlittlefield@nationalvideotape.com Web Site: nationalvideotape.com. Jack E. Dixon, pres; Josh Littlefield, sls.

Seattle, WA 98199, 1471 Elliott Ave. W. Phone: (206) 284-3340. Mari Scimeca.

Custom length VHS cassettes; Sony, Fuji, Maxell Panasonic video & data media products. CD/DVD printing & duplication.

Nationwide Tower Company Inc., 414 N. Ingram St., Henderson, KY, 42420. Phone: (270) 869-8000. Fax: (270) 869-8500. E-mail: hjohnston@nationwidetower.com Web Site: www.globalcommnet.com. Kevin Roth, sls VP; Diane Pruitt, gen sls mgr.

Tower inspections, painting, repair re-guy, lighting, antennas, feedlines, analysis, erect, dismantle, line sweeping, site monitoring, & tower tracker svcs.

Nautel Ltd., 10089 Peggy's Cove Rd., Hackett's Cove, NS, B3Z 3J4. Canada. Phone: (902) 823-2233. Fax: (902) 823-3183. E-mail: info@nautel.com Web Site: www.nautel.com. Jorgen Jensen, mgr sls & mktg.

Bangor, ME 04401. Nautel Maine Inc., 201 Target Industrial Cir.

Solid state AM/FM bcst transmitters.

Navitar Inc., Buhl Optical Div. 200 Commerce Dr., Rochester, NY, 14623. Phone: (585) 359-4000. Fax: (585) 359-4999. E-mail: info@navitar.com Web Site: www.navitar.com.

Projection lenses, LCD, slide & overhead projectors.

L.E. Nelson Sales Corp., (Thorn-EMI Studio & Theatre Lamps). 4800 W. University Ave., Las Vegas, NV, 89103. Phone: (702) 367-3656. Fax: (702) 367-7058. L.E. Nelson, pres; H.F. Nelson, VP western rgn; D.R. Imfeld, VP eastern rgn.

Fair Lawn, NJ 07410, 18-02 River Rd. Phone: (201) 794-6700. Dan Imfeld, VP eastern rgn.

Studio lamps, quartz (tungsten-halogen) from 25 w to 10 kw & projection lamps. Exculsive importer of Thorn Lamps.

Nemal Electronics International Inc., 12240 N.E. 14th Ave., North Miami, FL, 33161. Phone: (305) 899-0900. Fax: (305) 895-8178. E-mail: info@nemal.com Web Site: www.nemal.com. Benjamin L. Nemser, pres.

Sao Paulo, Av. Morumbi 7948. Phone: 011-5535-2368. Carlos Heckmann Jr., gen mgr.

Manufacturer of electronic cable, connectors, assemblies, & interconnect products for use in bcst applications.

Noise Control Corp., Box 81774, Bakersfield, CA, 93380. Phone: (800) 606-6473. E-mail: ncc@noisecontrol.com Web Site: www.noisecontrol.com.

Acoustical noise control products.

Neumade Products Corp., 30-40 Pecks Ln., Newtown, CT, 06470. Phone: (203) 270-1100. Fax: (203) 270-7778. E-mail: neumadels@aol.com Web Site: www.neumade.com. R.N. Jones, CEO; Gregory Jones, VP.

Film handling & editing equipment; storage facilities for film, slides, videotape, overhead & opaque projectors, motion picture projection systems.

Neutrik U.S.A. Inc., 195 Lehigh Ave., Lakewood, NJ, 08701. Phone: (732) 901-9488. Fax: (732) 901-9608. E-mail: info@neutrikusa.com Web Site: www.neutrikusa.com. James E. Cowan, pres; Julie Applegate, mgr.

Audio connectors, plugs & jacks, patch panels, patch cord assemblies, circular, industrial connectors & accessories, knobs, BNC jacks & plugs, RJ45, 3-5 mm plugs.

New York City Lites, 242 W. 27th St., 6th Floor, New York, NY, 10001. Phone: (212) 366-9800. Fax: (212) 366-5040. E-mail: nycl@nycl.tv Deke Hazirjian, pres.

Lighting design for video & TV.

Newark Electronics, (A Premier Co.). 4801 N. Ravenswood Ave., Chicago, IL, 60640. Phone: (773) 784-5100. Fax: (888) 551-4801. Web Site: www.newark.com. Mike Ruprich, CEO.

Distributor of bcst cable, assemblies, connectors voice/data networking & electronic component parts. Branches throughout the U.S., Canada, U.K. & Germany.

Nigel B. Furniture/Marketec, 4417 W. Magnolia Blvd., Burbank, CA, 91505. Phone: (818) 557-2661. Fax: (818) 557-2665. E-mail: info@marketec.com Web Site: www.nigelb.com. Penny Russell, owner.

Everything for rack mounting equipment, from vertical racks to desks, consoles, & workcenters for bcst & cable, post production, audio, & multimedia applications.

Norlight Telecommunications Inc., 13935 Bishops Dr., Brookfield, WI, 53005. Phone: (262) 792-9700. Fax: (262) 792-7793. Web Site: www.norlight.com. James Ditter, pres; Robert Rogers, VP.

Skokie, IL 60076, 3617 Oakton St. Phone: (847) 674-7476. Dave Pritchard, dir.

Fiber-optic & microwave transmission of bcst level video.

Norpak Corporation, 10 Hearst Way, Kanata, ON, K2L 2P4. Canada. Phone: (613) 592-4164. Fax: (613) 592-6560. E-mail: sales@nordak.ca Web Site: www.norpak.ca. James Carruthers, pres.

TV Data Broadcast; Interactive TV; Financial, News, Weather Radar Information Broadcast; HDTV Data Encoding; Closed Captioning; V-Chip; NABTS

Norsat International Inc., 300-4401 Still Creek Dr., Burnaby, BC, V5C 6G-9. Canada. Phone: (604) 292-9000. Phone: (800) 644-4562. Fax: (604) 292-9100. Web Site: www.norsat.com.

Beijing 100029P.R. Beijing Broadcasting Institute, 1704-A Union Plaza, 20 Chao Wai Plaza. Phone: (011) 86 10 65871281. Fax: (011) 86 10 65871081.

South Carlton, Lincoln LN1 2RL, The Old School. Phone: (011) 44 1522 730 800. Fax: 011- 44 1522 730 927. E-mail: smullery@noisat.com. Stan Mullery.

High speed, reliable data transmission products & networks, microware products & worldwide installations of opns STDs, DVB & SAT networks.

Nortel Networks, 8200 Dixie Rd., Suite 100, Brampton, ON, L6T 5P6. Canada. Phone: (905) 863-0000. Web Site: www.northelworks.com. Frank A. Dunn, pres/CEO; Nicholas J. De Roma, chief legal off; Doug Beatty, CFO; Greg Munford, chief tech off; Chahram Bolouri, pres; William J. Donavan, sr VP; Albert Hitchock, CIO; Masood Tarig, pres.

Supplier of telecommunications equipment. Provides voice over packets, multimedia svcs & applications, wireless data, and broadband networking to public network carriers, wireless operators and multi-svc operators.

North American Cable Equipment Inc., 1085 Andrew Dr., Suite A, West Chester, PA, 19380. E-mail: sales@northamericancable.com Web Site: www.northamericancable.com. Aaron Starr, pres; Kirk Davies, gen sls mgr.

Manufacturer of CATV, RF modulators, demodulators and processors.

North Dakota Television L.L.C., (Formerly Sunrise Television). 200 N. Fourth St., Bismark, ND, 58501. Phone: (701) 255-5757. Fax: (701) 255-8220. Web Site: www.kfyr.com. Jim Sande, progmg dir.

35' multi-camera bcst truck, Betacam SP VTRs, DVE, A/B roll edit, 24 ch audio bd, wireless IFB, 7kw generator, intercom systems.

North Hills Signal Processing, a PORTA Systems Co., 6851 Jericho Tpke., Suite 170, Syosset, NY, 11791. Phone: (516) 682-7740. Fax: (516) 682-7704. E-mail: info@northills-sp.com Web Site: www.northhills-sp.com.

Manufactures of MIL-STDT553 data bus products & wideband/video transformer. Our standard & custom product, offer unmatched performance & reliability for a wide range of applications in the military, aerospace & industrial OEM mkts.

Northeast Towers Inc., 199 Brickyard Rd., Farmington, CT, 06032. Phone: (860) 677-1999. Fax: (860) 677-1300. E-mail: netowers@ctl.nai.net Stephen Savino Jr., pres.

HDTV, TV, Cellular, PCS, AM, FM, CATV & microwave towers; ground systems; maintenance, materials, turnkey instals, specialty coatings, & strobes.

Northeastern Communications Concepts Inc., 40 Benford Dr., Princeton Junction, NJ, 08550. Phone: (609) 936-0006. E-mail: webmaster@nccnewyork.com Web Site: www.nccnewyork.com. Alfred W. D'Alessio, pres.

Bcst design svcs, studio furniture, custom audio equipment, custom data systems/components, cabinets, racks, panels, recording studios construction & prefab.

Northern Magnetics Inc., 25026 Anza Dr., Santa Clarita, CA, 91355-3413. Phone: (805) 257-0216. Fax: (805) 257-2037. Robert R. Rocheleau, VP.

Sls & svc of magnetic recording heads.

Northern Power Systems, 182 Mad River Park, Waitsfield, VT, 05673. Phone: (802) 496-2955. Fax: (802) 496-2953. E-mail: info@northernpower.com Web Site: www.nothernpower.com. Clint Coleman, pres.

Remote power systems based on renewable energy inputs (wind/solar); hybrid power systems.

Northern Technologies Inc., Box 610, 23123 E. Mission Ave., Liberty Lake, WA, 99019. Phone: (509) 927-0401. Fax: (509) 927-0435. E-mail: webmaster@northern-tech.com Web Site: www.nothern-tech.com. Jarrod Goodwin, dir N. American sls.

Full line of transient control systems for AC, dataline & telephone, including UPS systems & regulators.

Northrup Grumman, 1840 Century Park E., Los Angeles, CA, 90067. Phone: (310) 553-6262. Fax: (310) 553-2076. Kent Kresa, pres.

Film & video cameras for military; video-to-film recorders; optics & optical systems.

Northwest Monitoring Service, Box 70144, Eugene, OR, 97401. Phone: (541) 345-2236. James C. Bradley, owner.

Mthy frequency measurements for AM-FM-TV. Mobile svc includes California, Oregon, Washington, Idaho & Nevada.

NTV International Corporation, 645 5th Ave., Suite 303, New York, NY, 10022. Phone: (212) 660-6900. Fax: (212) 660-6998. Yoishi Shimada, pres.

Fred A. Nudd Corp., Box 577, 1743, Rt. 104, Ontario, NY, 14519. Phone: (315) 524-2531. Fax: (315) 524-4249. Web Site: www.nuddtowers.com. Fred Nudd, VP; Tom Nudd, pres.

Design, manufacture, instal, maintenance & analysis of communication towers.

O

O'Connor Professional Camera Support Systems, 100 Kalmus Dr., Costa Mesa, CA, 92626. Phone: (714) 979-3993. Fax: (714) 957-8138. E-mail: sales@ocon.com Web Site: www.ocon.com. Joel Johnson, VP/gen mgr; Robert Low, VP sls.

Manufacturer of camera support equipment including fluid heads, tripods & accessories.

Olesen, (A division of Entertainment Resources Inc.). 19731 Nordhoff St., North Ridge, CA, 91324. Phone: (818) 407-7800. Fax: (818) 407-7868. Web Site: www.hollywoodrentals.com. Carlos DeMattos, CEO.

All production supplies, equipment for TV, theater, both live & taped.

Omicron Video, 22251 Roscoe Blvd., West Hills, CA, 91304. Phone: (818) 704-0704. Fax: (818) 704-0475. E-mail: sales@omicronvideo.com Web Site: www.omicronvideo.com. Kimiharu Akiyama, pres.

Video/audio distribution equipment. Computer graphics/HDTV distribution equipment.

Omnimount Systems, 8201 S. 48th St., Phoenix, AZ, 85044. Phone: (480) 829-8000. Fax: (480) 756-9000. E-mail: info@omnimount.com Web Site: www.omnimount.com. Garrett Weyand, CEO; Alexander Cyrell, pres.

Loudspeaker mounts-omnidirectional adjustability supporting ounces to hundreds of pounds. Also, flexible, refined mounting systems for TV's/computer monitors & peripherals.

180 Connect, 6365 N.W. 6th Way, Suite 200, Ft. Lauderdale, FL, 33309. Phone: (800) 683-0253. Fax: (954) 671-8619. Dalia Rodborne, mgr.

Have been providing Turkey residential/commercial inside premise wiring & outside plant construction svcs for 20 yrs.

Opamp Labs Inc., 1033 N. Sycamore Ave., Los Angeles, CA, 90038. Phone: (323) 934-3566. Fax: (323) 462-6490. E-mail: bel@opamplabs.com Web Site: www.opamplabs.com. B. Losmandy, chief engr & pres.

Amplifiers: audio, video, microphone, line & power. Audio oscillators & transformers. Power supplies, network audio/video feed boxes, audio/video routing switches.

Optical Disc Corp., 12150 Mora Dr., Sante Fe Springs, CA, 90670. Phone: (562) 946-3050. Fax: (562) 946-6030. Web Site: www.optical-disc.com. Richard Wilkinson, pres; Ken Shrimplin, sr VP; John Brown, VP.

Recordable laser video discs, videodisc recording systems & other auxiliary equipment. Compact disc & videodisc mastering systems.

Orban, (A Harman International Co.). 1525 Alvarado St., San Leandro, CA, 94577. Phone: (510) 351-3500. Fax: (510) 351-0500. E-mail: info@orban.com Web Site: www.orban.com. Jay Brentlinger, pres/CEO.

Orban manufacturers bcst audio equipment for radio & TV including processors for TV, FM, AM & HF & the Audicy digital audio workstation, & the Airtime digital audio delivery system.

Ortel, 2015 W. Chestnut St., Alhambra, CA, 91803. Phone: (626) 293-3400. Fax: (626) 293-3428. E-mail: docmaster@agere.com Web Site: www.emcore.com. Gyo Shinozaki, mgr.

Signal transmission products, specializing in opto electronics & RF electronics technologies.

Allen Osborne Associates Inc., 756 Lakefield Rd., Westlake Village, CA, 91361. Phone: (805) 495-8420. Fax: (805) 373-6067. E-mail: j_osborne@aoa-gps.com Web Site: www.aoa-gps.com. Jim Osborne, VP.

Pneumatic masts systems for remote E.N.G., fixed or mobile radio communications, etc.

Otari USA Sales Inc., 9420 Lurline Ave., Unit C, Chatsworth, CA, 91311. Phone: (818) 734-1785. Fax: (818) 594-7208. Fax: (818) 734-1786. Web Site: www.otari.com.

Manufacturer of audio & video cassette loaders & duplicators. Manufacturer of audio mixing consoles, hard disk audio recorders, tape recorders, DAT recorders, minidisc recorders & players, CD changers, digital audio format converters.

P

PC& E, 2235 Defoor Hills Rd., Atlanta, GA, 30318. Phone: (404) 609-9001. Fax: (404) 609-9926. Web Site: www.pce-atlanta.com. Doug Smith, pres; Matt Timmons, gen mgr.

Equipment rental, grip & process trailers; rental & sls for lighting, expendables, camera.

PESA Switching Systems, Inc., 330 A Wynn Dr., Huntsville, AL, 35805. Phone: (256) 726-9200. Fax: (256) 726-9271. Web Site: www.pesa.com. Dave Gass, CFO.

Melville, NY 11747, 35 Pinelawn Rd, Suite 99E. Phone: (800) 328-1008. Robert McAlpine.

Routing switchers.

PMTV Producers Management Television, 681 Moore Rd., Suite 100, King of Prussia, PA, 19406. Phone: (610) 768-1770. Fax: (610) 768-1773. E-mail: mailto.pmtv@pmtv.com Web Site: www.pmtv.com.

Full-svc mobile TV production company, providing mobile units, crews, satellite svcs, lighting, staging, etc. for sports, entertainment & teleconferences worldwide.

POA/Paul Olivier & Associates Ltd., Box 410, 626 Forest View Way, Palmer Lake, CO, 80133. Phone: (719) 488-2270. Fax: (719) 488-2648. Paul Oliver, pres; Howard Phillips, engrg dir; Rick Brandon, controller.

TV bcst equipment, design engrg, fabrication, turnkey TV systems, bcst & cable consulting svcs, mobile units.

Pace Micro Technology P.L.C., 3701 FAU Blvd., Suite 200, Baca Raton, FL, 33431. Phone: (561) 995-6000. Fax: (561) 995-6001. E-mail: info@pacemirco.com Web Site: www.pacemirco.com. Mike McTighe, chmn; Neil Gaydon, CEO; David McKinney, COO.

First DVB MPEG-2 set-top boxes, the first to integrate DOCSIS into a digital cable set-top box & launching the first ever H.264 DVB-S2 high definition set-top box.

Packaged Lighting Systems Inc., P.O. Box 285, 29 Grant St., Walden, NY, 12586. Phone: (845) 778-3515. Phone: (800) 836-1024 (orders). Fax: (845) 778-1286. E-mail: info@packagedlighting.com Web Site: www.packagedlighting.com. Hy Hilzen, pres.

Factory prewired, self-contained TV studio systems complete with lighting/dimming/grid/power distribution.

Panasonic Broadcast & Television Systems Co., One Panasonic Way, Panazip 2E-7, Secaucus, NJ, 07094. Phone: (201) 348-5300. Fax: (201) 348-5318. Web Site: www.panasonic.com/broadcast. Andy Takani, pres.

Los Angeles, CA 90068, 3330 Cahuenga Blvd. W. Phone: (323) 436-3500.

Secaucus, NJ 07094, One Panasonic Way, 4E-7. Phone: (201) 348-7621.

MII VCR, D3 digital VCRs, digital processed cameras, Carts (MARC) analog & digital, tapes, DVC pro, D5, Post Box, RAMJA products, monitors, projectors.

Panavision New York, 540 W. 36th St., New York, NY, 10018. Phone: (212) 606-0700. Fax: (212) 244-4457. Web Site: www.panavisionnewyork.com. Peter Schnitzler, pres; Ira Goodman, VP.

16mm & 35mm motion picture & video equipment, lighting & grip equipment, generators, trucks, dollies & cranes.

Panel Authority Inc., 411 New Ave., Lockport, IL, 60441. Phone: (815) 838-0488. Fax: (815) 838-7852. E-mail: preston@panelauthority.com Web Site: www.panelauthority.com. Preston Wakeland, pres.

Custom made engraved aluminum connector panels & enclosures.

Parsons Audio, 192 Worcester St., Wellesley Hills, MA, 02481. Phone: (781) 431-8708. Fax: (781) 431-8783. E-mail: sales@paudio.com Web Site: www.paudio.com. Mark Parsons, owner; Les Arnold, sls; Rick Scott, sls.

Equipment sls for on-air & production Sony, Yamaha, Dolby, Digidesign, Lexicon, Denon, Tascam, etc.

Parsons Manufacturing Corp., 1055 O'Brien Dr., Menlo Park, CA, 94025. Phone: (650) 324-4726. Fax: (650) 324-3051. E-mail: pmccase@aol.com Web Site: www.pmccases.com. Alan R. Parsons, CEO; Alan Hall, controller.

Instrument carrying cases, shipping cases molded plastic, retracting wheels & recessed hardware.

Paulmar Industries Inc., Box 638, Antioch, IL, 60002. Phone: (847) 395-2080. Fax: (847) 589-2070. Web Site: www.paulmar.com. E-mail: sales@paulmar.com Robert F. Menary, pres.

Automatic film, video inspection machines, film & video supplies, DVD repair & rejuvenation equipment.

Peavey Electronics, 711 A St., Meridian, MS, 39301-2898. Phone: (601) 483-5365. Fax: (601) 486-1278. Web Site: www.peavey.com. Hartley Peavey, CEO.

Recording & audio products, SMPTE/MIDI synchronization signal processing, reference monitors, microphones & production mixing consoles.

Peerless Industries Inc., 3215 W. North Ave., Melrose Park, IL, 60160. Phone: (708) 865-8870. Fax: (708) 865-0760. Web Site: www.peerlessindustries.com. E-mail: info@peerlessindustries.com Ken Dillon, CEO.

Video Mounting hardware including stands, carts & brackets for floor, furniture, wall & ceiling applications.

Penn Elcom Inc., 12691 Monarch St., Garden Grove, CA, 92841. Phone: (714) 230-6200. Fax: (714) 230-6222. E-mail: california@penn-elcom.com Web Site: www.penn-elcom.com. Frank McCourt, pres; Phil Strafford, dir.

Hardware & accessories for flightcases, racks, speaker cabinets, stagelights & trussing.

Penny & Giles Inc., 5875 Obispo Ave., Long Beach, CA, 90805. Phone: (562) 531- 6500. Fax: (562) 531-4020. E-mail: u.s.sales@pennyandgiles.com Web Site: www.pgcontrols.com.

Cwmfelinfach, Gwent NP1 7HZ Phone: (44) 1495-202024.

Studio faders; joystick controllers; T-Bar controllers for video effects generators; MIDI mgr & D.A.W. interface.

Penta Laboratories, 9740 Cozycroft Ave., Chatsworth, CA, 91311. Phone: (818) 882-3872. Phone: (800) 421-4219. Fax: (818) 882-3968. Web Site: www.pentalabs.com. Steve Sanett, pres; Peter Russell, VP; Stacey Romm, CFO; Veronica Calderon, mgr.

Electron tubes distribution & mfg.

Pentax Imaging Co., 600 12th St., Suite 300, Golden, CO, 80401. Phone: (303) 799-8000. Fax: (303) 728-0226. Web Site: www.pentaxusa.com. Robert Bender, pres.

Manufacture camera lens.

Performance Power Technologies, Box 947, Roswell, GA, 30077. Phone: (770) 475-3192. E-mail: poweringcatv@yahoo.com Web Site: www.performance-power.com. Jud Williams, pres.

Standby power supplies, AC power supplies & battery testers.

PerkinElmer, 35 Congress St., Salem, MA, 01970. Phone: (978) 745-3200. Fax: (978) 745-0894. Web Site: www.perkinelmer.com. John Pautler, opns mgr.

High-medium-intensity aviation obstruction lighting & beacons. FAA-approved; StrobeGuard & FlashGuard.

Phasetek Inc., 550 California Rd., Unit 11, Quakertown, PA, 18951. Phone: (215) 536-6648. Fax: (215) 536-7180. E-mail: phasetekinc1@earthlink.net Web Site: www.phasetekinc.com. Kurt Gorman, pres; David Gorman, mktg VP.

Manufactures AM/MW antenna, phasing equipment, antenna tuning units, diplexers, dummy loads, RF inducters & components.

Philip-Cooke Co., 132 N. 11th St., Allentown, PA, 18102. Phone: (800) 887-0950. Phone: (610) 437-2251. Fax: (610) 437-6164. E-mail: kentk@philipcooke.com Web Site: www.philipcooke.com. Kent Kjellgren, pres.

Distribute video cassette duplications equipment & CDs.

Phillystran Inc., 151 Commerce Dr., Montgomeryville, PA, 18936. Phone: (215) 368-6611. Fax: (215) 362-7956. E-mail: info@phillystran.com Web Site: www.phillystran.com. Wynne Wister III, pres; Kenneth A. Knight, sls.

Phillystran HPTG; electrically transparent, maintenance free tower guy system; specially designed systems for high-power applications.

Phoenix E N G, Inc., 6832 Foxhill Ln., Cincinnati, OH, 45236. Phone: (513) 891-1444. Fax: (513) 891-3453. E-mail: engphoenix@aol.com Jennifer Braun; Kevin Jordan, pres; Bob Braun, VP mktg.

"One man band" live trucks, vans, 4-wheel-drive. On-location radio vehicles & production trucks.

Photo Research, 9731 Topanga Canyon Pl., Chatsworth, CA, 91311. Phone: (818) 341-5151. Fax: (818) 341-7070. Web Site: www.photoresearch.com. Francis Dominic, pres.

Brightness photometers, footcandle meters, telephotometers, spectroradiometers, spectral & spatial scanners.

Photomart Cine-Video Inc., 6327 S. Orange Ave., Orlando, FL, 32809. Phone: (407) 851-2780. Phone: (800) 443-2901. Fax: (407) 851-2553. E-mail: info@photomartusa.com Web Site: www.photomartusa.com. Cloyd Taylor, pres.

Sls, svc, of professional support equipment, supplies for video, film & still photography.

Pinnacle Systems Inc., 280 N. Bernardo Ave., Mountain View, CA, 94043. Phone: (650) 526-1600. Fax: (650) 526-1601. Web Site: www.pinnaclesys.com. E-mail: sales@pinnaclesys.com Mark L. Sanders, chmn; Charles J. Vaughan, pres/CEO; Ajay Chopra, dir.

Manufacturer digital video effects & graphics work stns. Digital video effects & graphics.

Pinzone Engineering Group Inc., 10142 Fairmount Rd., Newbury, OH, 44065. Phone: (304) 368-7950. Fax: (440) 729-5591. E-mail: systemsengineering@pinzone.com Web Site: www.pinzone.com. Basil F. Pinzone Jr., pres.

Turnkey satellite uplink/downlink systems, networks & Pinzone CORUM AM Anti-Skywave antenna.

Pioneer New Media Technologies Inc., 2265 E. 220th St., Long Beach, CA, 90810. Phone: (310) 952-2000. Fax: (310) 952-2100. Web Site: www.pioneerbroadband.com. Paul Dempsey, pres.

Columbus, OH 43228, 2200 Dividend Dr. Phone: (614) 876-0771. John Unverzagt, dir engrg. (Engrg office).

Set-top & remote-controlled tunable converters; one- & two way addressable converters & control systems.

Pirod Inc., Box 128, 1545 Pidco Dr., Plymouth, IN, 46563. Phone: (574) 936-4221. Fax: (574) 936-6796. E-mail: pirod@pirod.com Web Site: www.pirod.com. Myron C. Noble, CEO; Hillary Asher, sls VP.

Solid-rod towers, monopoles & tower accessories for cellular, PCs, bdcst, microwave & two-way communication.

Pixel Instruments Corp., 160-B Albright Way, Los Gatos, CA, 95032. Phone: (408) 871-1975. Fax: (408) 871-1976. E-mail: info@pixelinstruments.tv Web Site: www.pixelinstruments.tv. Mirko Vojnovic, pres.

Audio & video signals processing equipment including audio synchronizers, video frame synchronizers, audio delays & video delay detectors.

Plastic Reel Corp. of America, Box 296, Park Ridge, NJ, 07656. Fax: (201) 933-9468. Benjamin Zuk, pres; Pat Baccarella, exec VP; Carole Pinker, pres.

North Hollywood, CA 91605, 8140 Webb Ave. Phone: (818) 504-0400. Carole Pinker, VP.

Chicago, IL 60644, 5410 W. Roosevelt Rd. Phone: (800) 929-0356. Edwin Santani, office mgr.

Videotape, audiotape reels, boxes, video cassette mailing, storage boxes, video supplies, recording media, video & audio.

Polyline Corp., 1401 Estes Ave., Elk Grove Village, IL, 60007-5405. Phone: (800) 701-7689. Phone: (847) 357-1266. Fax: (800) 816-3330. Web Site: www.polylinecorp.com. Ed Kaiser, pres.

Irwindale, CA 91702, 16018 Adelante, Unit C. Phone: (800) 701-7689.

Recording, duplicating & packaging supplies for audio, video & CD from stock.

Potomac Instruments, Inc., 932 Philadelphia Ave., Silver Spring, MD, 20910. Phone: (301) 589-2662. Fax: (301) 589-2665. E-mail: sales@pi-usa.com Web Site: www.pi-usa.com. David G. Harry, COO; Guy E. Berry, mgr special project.

Antenna monitors, field strength meters, audio test equipment.

Power & Telephone Supply Co., 2673 Yale Ave., Memphis, TN, 38112. Phone: (901) 324-6116. Fax: (901) 320-3082. Web Site: www.ptsupply.com. Jim Pentecost, pres; Laburn Dye, VP; Larry Smith, VP.

Los Angeles, CA 90670, 12314 Bell Ranch Rd. Phone: (310) 903-1701. Fax: (310) 903-1705. Sonny Dickinson.

Miami, FL 33166, 7535 N.W. 52nd St. Phone: (305) 597-0091; 597-0262. Tommy Browder.

Des Moines, IA 50321, 3107 S.W. 61st St, Bldg. D. Phone: (515) 244-4375. Fax: (515) 244-4757. Doug McPhee.

Lexington, NC 27292, Box 1856, 2950 Greensboro St. Phone: (704) 249-0256. Fax: (704) 249-7475. Don Skinner.

Tigard, OR 97224, 16666 S.W. 72nd, Bldg. 12. Phone: (503) 620-4909. Fax: (503) 620-9074. Andy Baker. (Portland branch).

Reamstown, PA 17567, Box 244, Rt. 272. Phone: (215) 267-4991. Fax: (215) 267-4367. Don Skinner.

Memphis, TN 38112, Box 12383, 2673 Yale Ave. Phone: (901) 324-6116. Fax: (901) 320-3082. Dale Stevenson.

Dallas, TX 76063, 1456 S. 2nd Ave. Phone: (817) 477-1556. Fax: (817) 477-1557. Ray Morrison.

Neenah, WI 54956, 987 Ehlers Rd. Phone: (414) 725-5454. Fax: (414) 725-6162. Roger Rademacher.

Full-line supplier of communication products, including telecom, data & cable TV.

Powr-Ups Corp., One Roned Rd., Shirley, NY, 11967. Phone: (631) 345-5700. Fax: (631) 345-0060. Steven E. Summer, pres.

DC-motor controls.

Precision Microproducts of America, #1 Comac Loop, Unit 13, Ronkonkoma, NY, 11779. Phone: (631) 580-3456. Fax: (631) 580-3003. E-mail: sales@p-m-a.com Web Site: www.p-m-a.com. Jerry Wasserman, pres.

Photographic processing machines & accessories.

Prime Image, Inc., 662 Giguere Ct., Suite C, San Jose, CA, 95133-1742. Phone: (408) 867-6519. Fax: (408) 926-7294. E-mail: ssales@primeimageinc.com Web Site: www.primeimageinc.com. Bob Kelly, sls VP; Elizabeth Fjeldheim, mktg VP; Rodney Hampton, opns VP.

Provides digital progmg time reduction/editingequipment; audio & video delays; transcoding time base correctors; synchronizers; digital standards converters; computer video products.

Prisma Packaging, 15787 W. Ryerson Rd., West Berlin, WI, 53151-3617. Phone: (414) 342-6464. Phone: (800) 325-1089. Fax: (414) 342-0932. E-mail: info@prismapkg.com Web Site: www.prismapkg.com. Richard Schmaelzle, pres.

Printer manufacturer specializing in presentation folders, media kits, sls kits & videocassette packaging.

Pro Video & Film Equipment Co. Inc., 11425 Mathis Ave., Dallas, TX, 75234. Phone: (972) 869-9990. Phone: (888) 869-9998. Fax: (972) 869-0145. E-mail: providfilm @aol.com Web Site: www.provideofilm.com. Bill Reiter, pres.

Used equipment dealer specializing in video, bcst, film, lighting, audio. Consignment, sales, leasing & appraisal svcs available. Service & repairs.

Production Intercom Inc., Box 3247, Barrington, IL, 60011-3247. Phone: (800) 562-5872. Fax: (847) 381-4360. E-mail: info@beltpack.com Web Site: www.beltpack.com. Glenn mullis, pres; Sibbelina Mullis, sec.

Unique talent receiver (IFB), small to large intercom systems, headsets for cameras & new half-duplex wireless system.

Products International Inc., 9893 Brewers Ct., Laurel, MD, 20723. Phone: (240) 568-3940. Fax: (240) 568-3948. E-mail: sales@prodintl.com Web Site: www.prodintl.com.

Equipment, instruments, tools, supplies for electronic production, maintenance & svc.

Professional Communications Systems, (A division of Media General Broadcasting, Inc). 5426 Beaumont Center Blvd., Suite 350, Tampa, FL, 33634. Phone: (800) 447-4714. Fax: (813) 886-9477. Web Site: www.pcomsys.com. Ray A. Stephens, pres.

Pensacola, FL 32507, 2001 Augusta Ave. Phone: (850) 455-9800. Hardy Morris.

Jacksonville, FL 32216, 4110 Southpoint Blvd, Suite 129. Phone: (904) 281-0650. Fax: (904) 281-0309. Ed Kothera.

Miami, FL 33186, 11921 S.W. 144th St. Phone: (305) 253-4900. Fax: (305) 253-2551. Lloyd Hicks.

Plantation, FL 32792, 7860 Peters Rd, F-104. Phone: (954) 472-9400. Fax: (954) 424-6065. Charles Ross.

Winter Park, FL 32792, 7051 University Blvd, Suite 310. Phone: (407) 657-6421. Fax: (407) 657-0475.

Albany, GA 31701, 210 N. Monroe St. Phone: (229) 439-4954. Fax: (229) 434-0719. Karl Laster.

Consulting, design, procurement, integration, training & support.

Professional Sound Corp., 28085 Smyth Dr., Valencia, CA, 91355. Phone: (661) 295-9395. Fax: (661) 295-8398. E-mail: sales@professionalsound.com Web Site: www.professionalsound.com. Ron Meyer, pres.

Design, manufacture of portable sound recording products for film & video industries

Professional Sound Services Inc., 311 W. 43rd St., Suite 1100, New York, NY, 10036. Phone: (212) 586-1033. Fax: (212) 586-0970. Web Site: www.pro-sound.com. Rich Topham, pres.

Wireless microphones, wireless, wired intercoms, IFB, telephone interfaces, analog, digital recorders, mixers, lavaliers, boompoles.Sls, rentals & svc.

Prophet Systems Innovations, 111 W. Third St., Ogallala, NE, 69153. Phone: (877) 774-1010. Fax: (308) 284-4181. E-mail: prophetsales@prophetsys.com Web Site: www.prophetsys.com. Kevin Lockhart, pres.

Protech Audio Corp., 192 Cedar River Rd., Indian Lake, NY, 12842. Phone: (518) 648-6410. Fax: (518) 648-6395. E-mail: sales@protechaudio.com Bill Murphy, pres.

Professional audio preamps, power amps & signal processors, Dugan Automatic Mixing Controllers.

Prysmian Communications Cables and Systems, 700 Industrial Dr., Lexington, SC, 29072-3799. Phone: (803) 951-4800. Fax: (803) 951-4898. Web Site: www.prysmianusa.com. Brian DiLascia, VP/gen mgr; Gale Thrall, natl sls dir; Martin Hanchard, pres/CEO.

ISO 9001-registered manufacturer of fiber-optic cables & supplier of optical amplifier products for the cable TV industry.

Q

QEI Corporation, Box 805, Williamstown, NJ, 08094. Phone: (856) 728-2020. Fax: (856) 629-1751. E-mail: qeisales@qei-broadcast.com Web Site: www.qei-broadcast.com. Charles H. Haubrich, pres.

QSC Audio Products Inc., 1675 MacArthur Blvd., Costa Mesa, CA, 92626-1440. Phone: (714) 754-6175. Fax: (714) 754-6174. Web Site: www.qscaudio.com. Barry Andrews, CEO; Pat N. Quilter, VP engrg; John Andrews, COO; Pete Kalmen, natl sls mgr.

Professional power amplifiers, dual monaural power amplifiers, plug-in accessory products, integrated amplifiers, music & paging system.

QTV, 208 Harbor Dr., Stamford, CT, 06902. Phone: (203) 406-1400. Fax: (203) 323-3394. E-mail: sales@qtv.com Web Site: www.qtv.com. Michael Accardi, VP sls.

New York, NY 10010, 19 W. 21st St. Phone: (212) 929-7755. Fax: (212) 929-2105. Steve Carofalo, gen mgr.

Los Angeles, CA 90036, 5919 W. 3rd St. Phone: (213) 936-6195. Steve Hulkower, gen mgr.

Computer prompter software. 9", 12" & 15" on-camera prompters. Lightweight flat panel prompters.

Qintar Technologies Inc., 31352 Via Colinas, Suite 104, West Lake Village, CA, 91362. Phone: (818) 991-7300. Fax: (818) 889-7400. E-mail: sales@qintar.com Web Site: www.qintar.com. Randall Tishkoff, pres.

Active & passive devices for CATV, amplifiers, filters, connectors, wall plates & wiring products.

Quality Tower Erectors Inc., 2280 10th St. S.E., Largo, FL, 33771. Phone: (727) 585-6176. Fax: (727) 581-3277. Robert F. Diamond, pres.

QTE offers a full line of tower svcs in addition to our other offerings. Now more than ever, QTE is the complete solution for your site & asset needs. nance, erection, antenna systems, microwave, cellular, painting, turnkey service & tower site rental.

Quantel Inc., 1950 Old Gallows Rd., Suite 101, Vienna, VA, 22182. Phone: (203)972-3199. Fax: (203) 972-3189. Web Site: www.quantel.com. Ken Ellis, CEO.

Toronto, ON M5E 1E5 Canada, 1Yonge St, Suite 1100. Phone: (416) 362-9522. Mark Northeast.

Los Angeles, CA 90212, 8501 Wilshire Blvd, Suite 340. Phone: (310) 652-9227. Fax: (310) 657-8869. Mark Grasso.

San Francisco, CA 94104, 100 Bush St, Suite 1910. Phone: (650) 225-9036. (415) 263-1300. Fax: (650) 225-9091. Tom McGowan.

Atlanta, GA 30338, 5 Concourse Pkwy, Suite 330. Phone: (770) 804-5470. Fax: (770) 804-5479. Dan Wingard.

Chicago, IL 60611, 541 N. Fairbanks, Suite 1225. Phone: (312) 755-1766. Fax: (312) 755-1767.

New York, NY 10019, 111 W. 57th St., 10th Fl. Phone: (212) 977-4877. Fax: (212) 977-6539. Dave Saadatmandi.

Irving, TX 75038, 1425 Greenway Dr, Suite 470. Phone: (972) 751-1818. Fax: (752) 756-0006. Mike Rucker.

Quantel is the world's leading designer & manufacturer of digital image processing & manipulation products for video, film & print.

Quick-Set International Inc., 3650 Woodhead Dr., Northbrook, IL, 60062-1895. Phone: (847) 498-0700. Fax: (847) 498-1258. E-mail: sales@tripods.com Web Site: www.tripods.com. Jim Fenning, VP sls.

Instrument positioning equipment. Tripods, pan & tilts.

R

R-Columbia Products Co. Inc., 2008 St. Johns Ave., Highland Park, IL, 60035. Phone: (847) 432-7915. Fax: (847) 432-9181. E-mail: sales@rcolumbia.com Web Site: www.rcolumbia.com. I. Rozak, pres; Ed Hill, sls.

Headphones with & without microphone; cameraman headphones, wired & wireless intercom systems, ultralight headphones & IFB/ENG telephones.

RF Specialties Group, (RF Specialties of Missouri). 22406 N.E. 159th St., Kearney, MO, 64060. Phone: (800) 467-7373. Fax: (816) 628-4508. Web Site: www.rfspec.com. E-mail: rfmo@uniteone.net Patricia Kreger, chmn/pres; John Sims, sls; Chris Kreger, VP & sec.

Makati City, Metro Manila, NO 15239 Philippines. RF Specialties of Asia Corporation, 4958 Guerrero St, Poblacion. Phone: +63-2-412-4327. Fax: +63-2-895-6509. E-mail: eedmiston@rfsasia.com. Ed Edmiston.

Santa Barbara, CA 93105. RF Specialties of California, 3463 State St, Suite 229. Phone: (805) 682-9429. (800) 346-6434. Fax: (805) 682-5170. E-mail: rfsca@aol.com. Sam Lane.

Crestview, FL 32539. RF Specialties of Florida, 4706 Young Rd. Phone: (850) 423-7335. (800) 476-8943. Fax: (850) 423-7331. E-mail: rfoffl@aol.com. William Hoisington. Cell (850) 621-3680.

Richmond, IN 47374-1501. RF Specialties of Missouri, Inc., 1651 Capri Lane. Phone: 888-966-1990. Fax: 800-859-5481. E-mail: rf@insightbb.com. Rick Funk. Cell: 765-914-7778.

Kearney, MO 64060. RF Specialties of Missouri, Inc., 22406 N.E. 159th St. Phone: (800) 467-7373. (816) 628-5959. Fax: (816) 628-4508. E-mail: rfmo@uniteone.net. Chris Kreger; John Sims. Chris Cell: 816-506-7473.

New Ipswich, NH 03071. RF Specialties of Pennsylvania, Inc., 40 Settlement Hill. Phone: (603) 878-0618. (800) 485-8684. Fax: (603) 878-1527. E-mail: sam_on_the_hill@Monad.net. S.A. Matthews. Cell: (603) 801- 8466.

Las Vegas, NV 89129. RF Specialties of California, 3416 Lacebark Pine Street. Phone: 888-737-7321. Fax: 866-737-7321. E-mail: newbro@ix.netcom.com. Bill Newbrough.

Ebensburg, PA 15931. RF Specialties of Pennsylvania, Inc., 619 Industrial Park Road, Ste 200. Phone: (814) 472-2000. Fax: (814) 472-2230. E-mail: rfofpa@aol.com. Dave Edmiston. Cell: (814) 659-6575.

Monroeville, PA 15146-0002. RF Specialties of Pennsylvania, Inc., Box 2. Phone: (866) 412-7373. Fax: (412) 291-1135. E-mail: edrfofpa@nb.net. Pittsburgh office - Ed Young.

Southampton, PA 18966. RF Specialties of Pennsylvania Inc., Box 477. Phone: (888) 260-9298. (215) 322-2410. Fax: (215) 322-4585. E-mail: harrynlarkin@cs.com. Harry Larkin. (Philadelphia Office).

Amarillo, TX 79114. RF Specialties of Texas, Box 7630. Phone: (800) 537-1801. (806) 372-4518. Fax: (806) 373-8036. E-mail: rfstx@swbell.net. Don Jones. Cell (817) 312-7489.

Fort Worth, TX 76119. RF Specialties of Texas (Fort Worth Sales Office), 3528 Fairfax. Phone: (888) 839-7373. (817) 535-1979. Fax: (817) 535-0784. E-mail: rfstxftw@charter.net. Wray Reed.

Mukilteo, WA 98275-2226. RF Specialties of Washington, Inc., 885 18th Street. Phone: 425-210-9196. Fax: 925-476-7886. E-mail: waltlowery@msn.com. Walt Lowery.

Vancouver, WA 98687. RF Specialties of Washington Inc., Box 87571. Phone: (800) 735-7051. (360) 828-5992. Fax: (360) 883-4940. E-mail: rfswa@bobtheitguy.com. Bob Trimble.

Full-line radio bcst equipment suppliers. AM & FM transmitters, towers, lines, antenna systems, studios, microwave & digital systems.

RF Technologies Corp., 12 Foss Rd., Lewiston, ME, 04240. Phone: (207) 777-7778. Fax: (207) 777-7784. Web Site: www.rftechnologies.net. George M. Harris, CEO.

Designs & manufactures high-power bcst RF nets, components for FM & TV bcstrs. Products include antennas, diplexers, combiners, filters, switches, coax, waveguides & coaxal.

RTS Systems Telex Communications Inc, /. 2550 N. Hollywood Way, Suite 207, Burbank, CA, 91505-1055. Phone: (818) 566-6700. Fax: (818) 843-7953. Web Site: www.telex.com. Ralph Strader, VP; Murray Porteous, natl sls mgr; Dave Richardson, rgnl sls mgr.

Destin, FL 32541, 311 Stillwater Cove. Phone: (850) 654-4058. Rick Fisher, sls regl mgr.

Butler, NJ 07405, Box 866, 10 Park Pl. Bldg. 1. Phone: (973) 283-6200. Ken Smalley, rgnl sls.

Milford, PA 18337, 3807 Sunrise Lakes. Phone: (570) 686-5444. Chuck Roberts, tech support/engrg.

Terrell, TX 75160, 10927 FM 1565. Phone: (972) 524-6047. Britt Bowers, rgnl sls.

Centerville, VA 20120, 15463 Waters Creek. Phone: (703) 867-8333. Michael Brown, sls regnl.

Intercommunication systems, IFB systems, pro-audio amplifiers, microphones & phono preamplifiers.

Radian Communication Services Inc, 461 Cornwall Rd., Box 880, Oakville, ON, L6J 5C5. Canada. Phone: (905) 844-1242. Fax: (905) 844-8837. E-mail: info@radiancorp.com Web Site: www.radiancorp.com.

Design, supply, instal of bcst transmitters, antennas & towers.

Radio Aids Inc., 313 Kintzele Rd., Michigan City, IN, 46350. Phone: (219) 879-2215. Fax: (219) 874-8239. John M. Carpenter, pres.

Measurement of occupied bandwidth, TV aural & visual, radio carriers, subcarriers, pilots, STL/TSL links.

Radio Computing Services (RCS), 12 Water St., White Plains, NY, 10601. Phone: (914) 428-4600. Fax: (914) 428-5922. E-mail: info@rcsworks.com Web Site: www.rcsworks.com. Philippe Generali, pres; Mike Powell, VP; Richard Darr, sls VP.

Frankfurt 60388, Borsigallee 37. Phone: 49-610-973-4450. Fax: 49-610-973-4499. E-mail: info@rcseurope.de. Karl

Kessler, gen mgr.

Richmond, BC V6X 3R9 Canada, Box 32060, 410 #5 Rd. Phone: (604) 986-4468. Fax: (604) 986-4469. Ross Langbell.

Paris 75011 France, 83 Ave. Philippe Auguste. Phone: 33-1-53-27-36-36. Fax: 33- 1-53- 27- 36-60. Eric Vanryckeghem.

Bandra Mumboi (West), NO 400050 India, 262 Hart Niwas, 30th Rd. Phone: +91 22 697 1600. Fax: +91 22 695 5760. Elliot Stechman.

Christchurch, NO 8005 New Zealand. RCS (NZ) Ltd., 33 Sir William Pickering Dr. Phone: +64.3.358.4333. Fax: +64.3.358.4330. E-mail: info@rcs.co.nz. Web Site: www.rsc.co.nz. Ian Campbell.

Singapore 079903 Singapore, 10 Anson Rd, 10-10 International Plaza. Phone: +65 6324 6658. Fax: 65-6324-6659. E-mail: cfawell@attglobel.net. Colin Fawell, gen mgr.

Bergbron, Johannesburg, NO 1709 South Africa. RCS Africa, Leephy Studios, 11 Jonkershoek Rd. Phone: +27.11.477.1229. Fax: +27.11.673.3948. E-mail: hayden@rscafrica.co.za. Hayden Beetar.

Malmo SE-211 35 Sweden. RCS Scandinavia, Kalendgatan 26. Phone: +46.40.66.55.880. Fax: +46.40.66.55.888. E-mail: info@rcs.se. Web Site: www.rcs.se. Sven Andrae.

London, NO W1N 5FD United Kingdom. RCS United Kingdom, 167-169 Great Portland St. Phone: 44.20.7636.9636. Fax: 44.20.7636.7766. E-mail: info@rcsuk.com. Web Site: www.rcsuk.com. Sebastian Holmes.

Live Oak, CA 95953, 6018 Madden Ave. Phone: (530) 695-3997. Fax: (530) 674-5780. E-mail: hshaw@rcsworks.com. Dean Cull, western sls; Jennifer Cull, govt.

Miami, FL 33145, 1385 Carol Way, #202. Phone: (305) 860-5870. Fax: (305) 860-5832. Candice Castillo.

Digital studio automation & digital audio ripping/analysis, music scheduling, traf, sls, newsroom & talk show software/hardware, internet/streaming tools.

Radio Design Labs. (RDL), Box 1286, Carpinteria, CA, 93014. Phone: (805) 684-5415. Fax: (805) 684-9316. E-mail: sales@rdlnet.com Web Site: www.rdlnet.com. Joel Bump, pres; Jerry Clements, VP.

Full line of microphone & line level amplifiers, mixers, DAs & processors.

Radio Detection/Riser Bond, 154 Portland Rd., Bridgton, ME, 04009. Phone: (207) 647-9495. Fax: (207) 647-9496. E-mail: bridgton@radiodetection.spx.com Web Site: www.riserbond.com. Jim Walton, sls VP.

Electronic test equipment; cable fault locators; time domain reflectometer.

Radio Engineering Industries Inc., 6534 L St., Omaha, NE, 68117. Phone: (402) 339-2200. Fax: (402) 339-1704. E-mail: sales@radioeng.com Web Site: www.radioeng.com. Terry Jukes, chmn.

Sls, svc of bcst equipment, amplifiers, paging systems, SCA & coml sound equipment.

Radio Frequency Systems, 200 Pondview Dr., Meriden, CT, 06450. Phone: (203) 630-3311. Fax: (203) 634-2272. E-mail: sales.americas@rfsworld.com Bill Bayne, pres.

Rigid coaxial line (7/8" to 9 3/16"), FM antennas, FM, VHF/UHF IFTS, MMDS, TV antennas, dehydrators, instal accesories, RF, microwave antenna subsystems, instal & field svc.

Radio Photo Antennas Inc., 48 Mountain Rd., Farmington, CT, 06032. Phone: (860) 676-0051. Fax: (860) 677-9639. E-mail: cfaricher@snet.net Robert E. Richer, pres; Professor Maurice Hately, chief tech off; Alec Thomas, head engr.

Hansworth, Middlesex, NO TW13 7 DW United Kingdom, 97 Foxwood Close. Phone: (1) 44 0797 085. Fax: 8175. Alex Thomas, head engrg.

Company mkts medium wave & long wave antenna.

Radio Research Instrument Co. Inc., 584 N. Main St., Waterbury, CT, 06704. Phone: (203) 753-5840. Fax: (203) 754-2567. E-mail: radiores@prodigy.net Web Site: www.radioresearch.thomasregister.com. P. J. Plishner, pres; E. B. Doyle, exec VP.

Provides radar systems, threat emitters & spare parts; complete maintenance facility for repair.

Radio Systems Inc., 601 Heron Dr., Logan Township, NJ, 08085-1741. Phone: (856) 467-8000. Fax: (856) 467-3044. E-mail: sales@radiosystems.com Web Site: www.radiosystems.com. Daniel Braverman, pres; Gerrett Conover, VP.

Audio consoles, distribution amplifiers, low-power AM transmitters, clock, timer systems, telephone hybrids & studio wiring systems.

Radiogear Inc., 8746 Gerst Ave., Perry Hall, MD, 21128. Phone: (410) 933-8445. Fax: (410) 933-8352. E-mail: email@radiogearinc.com Web Site: www.radiogearinc.com. Charles Spencer, pres.

Radio bcst equipment, audio supplies (RF, audio) & quality pre-owned radio bcst equipment.

Radyne ComStream Corp., 3138 E. Elwood St., Phoenix, AZ, 85034. Phone: (602) 437-9620. Fax: (602) 437-4811. E-mail: sales@radynecomstream.com Web Site: www.radynecomstream.com. Brian Duggan, pres; Robert C. Fitting, CEO; Steve Eymann, VP engrg; David Koblanski, gen mgr.

Complete digital audio bcst net for compressed CD quality audio over satellite.

Digital data distribution products for the financial industry.

Railway Systems Design, Inc., Valley Forge Corporate Center, 1010 Adams Ave., Audubon, PA, 19403-2402. Phone: (610) 650-7730. Fax: (610) 650-8190. Web Site: www.rsdconsulting.com. Walter J. Clarke, ptnr; Terry A. Shantz, ptnr.

Consulting engrs, tower engrg, design & construction mgmt.

Ram Broadcast Systems, Box 277, Wauconda, IL, 60084-0277. Phone: (800) 779-7575. Phone: (847) 487-7575. Fax: (847) 487-2440. Web Site: www.ramsyscom.com. Ron Mitchell, pres.

Switchers (audio & video) mixers, intercom systems, audio/video DAs, systems engrg & custom cabinetry.

Rangertone Research Inc., 40 Entin Rd., Clifton, NJ, 07014. Phone: (973) 594-8722. Fax: (973) 594-8724. George P. Zazzali, pres; Daniel J. Zazzali, VP.

Audiovisual equipment.

Raven Screen Corp., 112 Spring St., Monroe, NY, 10950. Phone: (212) 534-8408. Fax: (845) 782-1840. Martin Soss, pres.

Manual, motorized & custom projection screens & materials.

Record/Play Tek Inc., Box 790, 110 E. Vistula St., Bristol, IN, 46507-0790. Phone: (574) 848-5233. Fax: (574) 848-5333. E-mail: stoll@recordplaytek.com Web Site: www.recordplaytek.com. Michael Stoll, CEO.

Voice logging recorders "911," cassette, reel-to-reel, VHS, computer CDR & DVD RAM.

Recortec Inc., 1620-A Berryessa Rd., San Jose, CA, 95133-1026. Phone: (408) 928-1480. Fax: (408) 729-3661. E-mail: sales@recortec.com Web Site: www.recortec.com. Dr. Lester H. Lee, pres.

Manufacturer of coml disc players & LCD players.

Reel-O-Matic Inc., Box 95309, Oklahoma City, OK, 73143. Phone: (405) 672-0000. Fax: (405) 672-7200. Web Site: www.reel-o-matic.com. Terry Simmons, pres.

Equipment to re-spool, coil, measure & distribute cable.

Rees Associates Inc., 9211 Lake Hefner Pkwy., Oklahoma City, OK, 73120. Phone: (405) 942-7337. Fax: (405) 948-1261. Web Site: www.rees-associates.com. Frank W. Rees Jr., pres; C. Leroy James, exec VP; William H. Yost, VP mktg.

Dallas, TX 75219-4341, 3102 Oak Lawn, Suite 200. Phone: (214) 522-7337. Frank W. Rees Jr., pres.

Bcst & production facility design; architectural svcs; studio design; equipment planning; facility business plans; interior design; consulting.

Register Data Systems, 1691 Forsyth St., Macon, GA, 31201. Phone: (478) 745-5500. Fax: (478) 745-0500. E-mail: sales@registerdata.com Web Site: www.registerdata.com. Lowell L. Register, pres; Ricky Lockerman, sls.

Digital audio automation systems for live assist, satellite, traf & billing software packages for radio & TV.

Renkus-Heinz Inc., 19201 Cook St., Foothill Ranch, CA, 92610-3510. Phone: (949) 588-9997. Fax: (949) 588-9514. E-mail: sales@renkus-heinz.com Web Site: www.renkus-heinz.com. Harro K. Heinz, pres; Carl Dorwaldt, mktg mgr.

Reference point arrays, powered network loudspeakers, R-control remote supervision network. Reference point arrays, powered network loudspeakers.

Research Technology International Inc., 4700 Chase Ave., Lincolnwood, IL, 60712-1689. Phone: (847) 677-3000. Phone: (800) 323-7520. Fax: (847) 677-1311. E-mail: sales@rtico.com Web Site: www.rtico.com. Ray L. Short Jr., pres; Thomas W. Boyle, sr VP; Bill Wolavka, sls VP.

Videotape evaluator/cleaners; degaussers; storage & care, supplies, film cleaners. CD/DVD cleaners-restorers inspectors.

Reuters American Inc., 3 Times Square, New York, NY, 10036. Phone: (646) 223-4000. Web Site: www.reuters.com.

Ottawa, ON K1P 5P8 Canada, 165 Sparks St., Booth Bldg. Phone: (613) 235-6745. Antony Parry, financial correspondent; John Rogers, gen news.

Toronto, ON M5H 3T9 Canada, Standard Life Centre, 121 King St. W., 20th Fl. Phone: (416) 869-3600. Peter Thomas, mgr Canada.

Montreal, PQ H3A 2A5 Canada, 2020 Rue Universite, Suite 1020. Phone: (514) 282-0705. William Miller.

Los Angeles, CA 90071, 445 S. Figueroa, Suite 2100. Phone: (213) 380-2014. Ronald Clarke, chief.

Washington, DC 20005, 1333 H St. N.W, Suite 410. Phone: (202) 898-8300. Bruce Russell, chief.

Chicago, IL 60606, 311 S. Wacker Dr, Suite 1100. Phone: (312) 922-6038. Geoffrey Atkins, chief.

Kansas City, MO 64112, 4800 Main. Phone: (816) 561-8671. Bob Martin.

Supplier of natl, world, business news, info to media & professionals.

Richardson Electronics, (A division of Broadcast Richardson). PO Box 393, 40W267 Keslinger Rd., LaFox, IL, 60147. Phone: (630) 208-2200. Fax: (630) 208-2662. E-mail: broadcast@rell.com Web Site: broadcast.rell.com. Edward Richardson, CEO; Dario Sacomani, CFO; Robert Prince, VP.

LaFox, IL 60147-0393, Box 393, 40W267 Keslinger Rd. Phone: (630) 208-2200. (800) 882-3872. Fax: (630) 208-2550. E-mail: broadcast@rell.com. Web Site: broadcast.rell.com.

Global provider of power tubes, TV, radio transmitters, IP, digital satellite systems, NLE video systems & studio pakages.

Richmond Sound Design Ltd., Box 19523, Vancouver, BC, V5T 4E7. Canada. Phone: (604) 715-9441. Fax: (250) 414-5205. E-mail: sales@richmondsounddesign.com Web Site: www.richmondsounddesign.com. C.B. Richmond, pres; M. Williams, mgr.

London, ON EC2A 3PB United Kingdom, 23 Charlotte Rd. Phone: +44 20 7613 3305. John Leonard.

Automated multichannel audio for live shows & theatre; show controllers.

Ripley Company, 46 Nooks Hill Rd., Cromwell, CT, 06416. Phone: (860) 635-2200. Phone: (800) 528-8665. Fax: (860) 635-3631. E-mail: info@ripley-tools.com Web Site: www.ripley-tools.com. Keith D'Amato, sls dir; Tom Lindenmuth, gen mgr; Ken Grey, dir mktg.

Ripley's Cablematic, Miller & Utility tool lines offer manufacturers cable preparation tools for CATV telecomm data & electric utiliy.

Rodelco Electronics Corp., 111 Haynes Ct., Ronkonkoma, NY, 11779. Phone: (631) 981-0900. Fax: (631) 981-1792. E-mail: rodelco@erols.com Joseph M. Rodgers, gen mgr.

TV translators, VHF & UHF.

Rohn Industries Inc., 6718 W. Plank Rd., Peoria, IL, 61604. Phone: (309) 697-4400. Fax: (309) 697-5612. E-mail: mail@rohnnet.com Web Site: www.rohnnet.com. Horace Ward, pres/CEO; Craig Ahlstrom, VP sls.

Towers (up to 2,000 feet) monopoles, antenna mounts for communication industry. Turnkey construction & installation avaible worldwide.

Roland Corp. U.S., Box 910921, 5100 S. Eastern Ave., Los Angeles, CA, 90091-0921. Phone: (323) 890-3700. Fax: (323) 890-3701. Web Site: www.rolandus.com. Dennis Houlihan, pres; Mark Malbon, exec VP.

Electronic musical instruments, signal processors, sound reinforcement, hard disk editors, noise eliminators, bcst production equipment & post production equipment.

Rosco Laboratories Inc., 52 Harbor View Ave., Stamford, CT, 06902. Phone: (203) 708-8900. Fax: (203) 708-8919. E-mail: info@rosco.com Web Site: www.rosco.com. Stan Miller, pres; Stan Schwartz, exec VP.

Hollywood, CA 90038, 1120 N. Citrus Ave. Phone: (323) 462-2233. Fax: (323) 462-3338. Jim Meyer, mgr.

Lighting filters & diffusers, studio floor covering, connectors & digital (or rental & custom) backdrops.

Roscor Corp., 1061 Feehanville Dr., Mount Prospect, IL, 60056. Phone: (847) 299-8080. Fax: (847) 299-4206. Fax: (847) 803-8089. E-mail: sales@roscor.com Web Site: www.roscor.com. Paul Roston, pres; Mitch Roston, exec VP; Tom Voights, sls VP; Edward Jones, VP finance.

Cincinnati, OH 45241, 2868 E. Kemper Rd. Phone: (513) 772-3393. Tim Navaro, branch mgr.

Farmington Hills, MI 48331, 27280 Haggerty Rd, Suite

C2. Phone: (248) 489-0090. Paul Niehaus, branch mgr.
Milwaukee, WI 53204, 600 W. Virginia St. Phone: (414) 223-2600. Steve Olson, branch mgr.

Professional audio/video/RF/presentation equipment. Turnkey engrg & instal svcs.

Ross Video Ltd., Box 220, 8 John St., Iroquois, ON, K0E 1K0. Canada. Phone: (613) 652-4886. Fax: (613) 652-4425. E-mail: solutions@rossvideo.com Web Site: www.rossvideo.com.

Video production switchers, analog & digital terminal gear, video keyers, encoders & decoders, mini master control & telecine switchers.

Royal Consumer Information Products, 379 Campus Dr., Somerset, NJ, 08875. Phone: (732) 627-9977. Web Site: www.olivettiofficeusa.com. Salomon Suwalsky, pres; Todd Althoff, VP mktg.

S

SAIC, 1710 Saic Dr., Mclean, VA, 22102. Phone: (703) 821-4300. Web Site: www.saic.com.

Eidophor large screen projectors. Rental source & sole North American distributor.

S&L Plastics Inc., 2860 Bath Pike, Nazareth, PA, 18064. Phone: (610) 759-0280. Fax: (610) 759-0650. Web Site: www.slpinc.cc. John Bungert, pres.

Thermo plastic products.

SES Americom, 4 Research Way, Princeton, NJ, 08540-6684. Phone: (609) 987-4000. Fax: (609) 987-4495. Web Site: www.ses-americom.com. Bryan mcGurik, pres.

Satellite distribution svcs for coml bcst & cable TV; prog syndicators, SNG & bcst radio distribution svcs.

S W R Inc., (Systems with Reliability.). 619 Industrial Park Rd., Ebensburg, PA, 15931. Phone: (814) 472-5436. Fax: (814) 472-5552. Web Site: www.swr-rf.com. Edward J. Edmiston, pres; David K. Edmiston, gen sls mgr.

Timog, Quezon City, 31-E Scout Bayoran. Phone: 011-632-411-0068. Edward J. Edmiston, pres.

Manufacturers of TV & FM transmit antennas, rigid coax, waveguide & associated accessories.

Sabine Inc., 13301 Hwy. 441, Alachua, FL, 32615. Phone: (386) 418-2000. Fax: (386) 418-2001. E-mail: sabine@sabine.com Web Site: www.sabine.com. Doran Oster, pres.

Manufacturers of digital signal processing equipment for sound systems. Makers of the patented FBX Feedback Exterminator & True MobilityTM wireless microphones.

Sachtler Corp. of America, 709 Executive Blvd., Valley Cottage, NY, 10989. Phone: (845) 268-2113. Fax: (845) 268-9324. E-mail: sales@sachtler.com Web Site: www.sachtler.com. Eric Falkenberg, pres.

Burbank, CA 91505, 3316 W. Victory Blvd. Phone: (818) 854-4446.

Complete line of camera support equipment for ENG, EFP, O.B. & the new generation of studio cameras. Lighting for news, production & studio open-face technology & fresnel.

Sacramento Theatrical Lighting (STL), 950 Richards Blvd., Sacramento, CA, 95814. Phone: (800) 283-2785. Fax: (916) 447-5012. E-mail: saclight@aol.com Steve Odehnal, mgr.

Specialists in studio & location lighting, grip equipment, draperies, rigging & grid work. Consultation & production svcs. Sls, rentals & svcs.

Sadelco Inc., 75 W. Forest Ave., Englewood, NJ, 07631. Phone: (201) 569-3323. Fax: (201) 569-6285. E-mail: sadelco@aol.com Web Site: www.sadelco.com. Les Kaplan, pres.

Signal level meters, calibrators & leakage detectors.

Samson Technologies Corp., Box 9031, Syosset, NY, 11791. Phone: (516) 364-2244. Fax: (516) 364-3888. E-mail: sales@samsontech.com Web Site: www.samsontech.com. Douglas Bryant, pres; Scott Goodman, CEO; Jack Knight, VP opns; Bob Caputo, VP sls; Pete Moe, VP mktg.

Manufacturer of wireless microphones, mixing consoles, power amplifiers & audio products. Behringer audio processing, Hartke speakers & Zoom effects processors.

Sanyo Fisher Co., 21605 Plummer St., Chatsworth, CA, 91311. Phone: (818) 998-7322. Fax: (818) 998-3533. Web Site: www.sanyo.com. Paul W. D'Arcy, VP.

Audio amplifiers & receivers, CD players, audiotape recorders, turntables, dictation machines, cordless telephones, TVs, VTRs & LCD projectors.

Sarnoff Corp., (A Subsidiary of SRI International). 201 Washington Rd., Princeton, NJ, 08543-5300. Phone: (609) 734-2000. Fax: (609) 734-2221. Web Site: www.sarnoff.com. Satyam C. Cherukuri, pres.

Contract rsch & dev facility for electronic, biomedical, & info technologies, specializing in digital video.

Sascom Marketing Group, 34 Nelson St., Oakville, ON, L6L 3H6. Canada. Phone: (905) 469-8080. Fax: (905) 469-1129. E-mail: c.smith@sascom.com Web Site: www.sascom.com. Curt Smith, pres.

Sascom Represents: Adgil, Audio Cube, Digital Audio Denmark, Doremi, Msoft Server Sound & Cube-Tec.

Satellite Systems Corp., 101 Malibu Dr., Virginia Beach, VA, 23452. Phone: (757) 463-3553. Fax: (757) 463-3891. Web Site: www.satsyscorp.com. Bob Kite, pres.

SCPC & video subcarrier satellite systems for radio, SNG & data bcst networks.

Schafer International, 220 Surrey Dr., Bonita, CA, 91902. Phone: (619) 267-9000. Fax: (619) 267-9003. E-mail: patty@schaferinternational.com Web Site: www.schaferinternational.com. Paul C. Schafer, pres.

Equipment & parts, for radio & TV stns, primarily in Mexico.

Schafer World Communications Corp., Box 1047, Marion, VA, 24354-1047. Phone: (276) 783-2000. Fax: (276) 783-2064. Bob Dix, pres; Ann Dix, VP; Kevin Soos, mktg.

Schafer offers two levels of sophistication in hard disk audio systems, "GENESIS" Digital Studio: touch screen & remote control interface; excellent live assist & full automation capability (complete music scheduling software included); cut-to-cut mixing on hard disk including editing; simultaneous record/playback from hard disk; can connect to external machines & CD multi-players.

Schneider Optics Inc., 285 Oser Ave., Hauppauge, NY, 11788. Phone: (631) 761-5000. Fax: (631) 761-5090. E-mail: info@schneideroptics.com Web Site: www.schneideroptics.com. Ron Leven, sr VP; Dwight Lindsay, sr VP.

Manufacturer/distributor of high quality optical filters for video, still photography & motion picture. Product line also includes a wide range of lenses for CCTV, large format photography, darkroom enlarging & slide & film projection.

Scientific Atlanta, 5030 Sugarloaf Pkwy., Lawrenceville, GA, 30044. Phone: (770) 236-5000. Fax: (770) 902-2591. E-mail: gregg.echols@sciatl.com Web Site: www.sciatl.com. James McDonald, chmn/CEO; Patrick Tylka, pres; Dwight Duke, pres; Michael Harney, pres.

A complete line of cable TV & broadband communications systems, products. and professional services.

Scientific Atlanta Canada Inc. Nexus Division, Satellite TV Networks, 100 Middlefield Rd. Unit 1, Scarborough, ON, M1S 4M6. Canada. Phone: (416) 299-6888. Fax: (416) 299-7145. Web Site: www.scientificatlanta.com. Michael Harney, pres; Dwight Duke, pres; Patrick Tylka, pres.

TV RF signal processing equipment, transmission products & cable TV amplifiers.

Scott Studios Corp., 13375 Stemmons Fwy., Suite 400, Dallas, TX, 75234. Phone: (972) 620-0070. Web Site: www.scottsstudios.com. David Scott, pres.

Radio automation systems; digital audio.

ScreenLight & Grip, 502 Sprague St., Dedham, MA, 02026. Phone: (781) 326-5088. Fax: (781) 326-4751. E-mail: lightsne@aol.com Web Site: www.screenlightandgrip.com. Guy Holt, pres.

Location lighting & production svcs, equipment rental, trucks, vans, etc.

Second Chance Body Armor Inc., 7915 Cameron St., Central Lake, MI, 49622-0573. Phone: (231) 544-5721. Phone: (800) 253-7090. Fax: (231) 544-9824. E-mail: email@secondchance.com Web Site: www.secondchance.com. Paul J. Banducci, VP/gen mgr.

Leading body armor manufacturer now offering ballistic protection for news media reporters & photographers.

Seger Electronics, 97 Libbey Pkwy, Weymouth, MA, 02189. Phone: (781) 682-4844. Web Site: www.seger.com. Frank Flynn, pres; Ray Norton, CEO.

Largest inventory of electromechanical components. No mimimums. Liberal sampling. 24-hours order check. Custom assembly, engraving & printing.

Selco Products Co., 605 S. East Street, Anaheim, CA, 92805. Phone: (714) 717-1333. Fax: (714) 917-1355. E-mail: sales@selcoproducts.com Web Site: www.selcoproducts.com. Tim Wilkinson, pres; Michelle Blakeslee, mktg.

A full range of product lines are offered by selco including thermal products, control knobs, electronic controls & digital panel meters.

Sencore Inc./AAVS, 3200 Sencore Dr., Sioux Falls, SD, 57107. Phone: (605) 339-0100. Fax: (605) 339-0317. Fax: (605) 335-6379. Web Site: www.sencore.com. Doug Bowden, VP.

Electronic test equipment for servicing & performance testing of consumer electronics & CATV/MATV equipment.

Senior Aerospace, Ketema Division. 790 Greenfield Dr., El Cajon, CA, 92021. Phone: (619) 442-3451. Fax: (619) 440-1456. Ron Case, gen mgr.

Design build-to-print aerospace products, cryogenic lines, valves, burst discs, electric motors, actuators.

Sennheiser Electronic Corp., One Enterprise Dr., Old Lyme, CT, 06371. Phone: (860) 434-9190. Fax: (860) 434-1759. E-mail: info@sennheiserusa.com Web Site: www.sennheiserusa.com. John Falcone, pres/CEO; Scott Schumer, VP sls.

Col. Del Valle, D.F. Mexico 03100, Av. Xola No. 613 PH6. Phone: (525) 639-0956. Fax: (525) 639-9482.

Burbank, CA 91505, 4116 W. Magnolia Blvd, Suite 100. Phone: (818) 845-7366. Fax: (818) 845-7140.

Microphones, headphones, boomsets, wireless microphones & infrared products as well as DAS audio loudspeakers & Chevin rsch amplifiers.

Servoreeler Systems, (Xedit Corp.). 218-31 97th Ave., Queens Village, NY, 11429. Phone: (718) 464-9400. Fax: (718) 464-9435. E-mail: srsystems@servoreelers.com Web Site: www.servoreelers.com. Claude M. Karczmer, pres; Eileen Karczmer, sls dir.

SUSPENDED MICROPHONE SERVOREELERS - deploy, retract and position suspended microphones by remote pushbutton or computer control. ex-teleconferencing, corp board rooms, churches, concert halls, sports arenas & universities.

Sescom Inc., 608 Main St., wellsville, KS, 66092. Phone: (785) 883-3009. Fax: (785) 883-4422. E-mail: sescom@sescom.com Web Site: www.sescom.com.

Audio interfacing equipment, audio transformers & modules.

Setcom Corp., 1400 N. Shoreline Blvd., Mountain View, CA, 94043-1385. E-mail: rvb@setcomcorp.com Web Site: www.setcomcorp.com. L. Kent Schwartzman, pres; Bob Von Buelow, mktg mgr.

Portable & fixed position intercom systems, bcst intercom & portable radio headsets.

Seton Identification Products, Box 819, Branford, CT, 06405. Phone: (203) 488-8059. Fax: (203) 488-7259. Web Site: www.seton.com. E-mail: comments@seton.com Richard L. Fisk, pres.

Signs, tags, labels, pipe markers, valve tags, & nameplates to meet OSHA/ANSI specifications.

Shallco Inc., Box 1089, 308 Components Dr., Smithfield, NC, 27577. Phone: (800) 876-3135 (USA only). Phone: (919) 934-3298 (outside USA). Fax: (919) 934-3135 (outside USA only). E-mail: sales@shallco.com Web Site: www.shallco.com. John Shallcross Sr., chmn; Jason S. Shallcross, pres.

Variable & fixed audio attenuators.

Sharp Electronics Corp., CCD Products Div., (LCD Products Group). Sharp Plaza, Mail Stop One, Mahwah, NJ, 07430-2135. Phone: (201) 529-8200. Phone: (866) 4-VISUAL. Fax: (201) 529-9636. E-mail: ProLCD@SharpSEC.com Web Site: www.SharpLCD.com. Ron Colgan, VP; Fred Krazeisze, Dir. Strategic Marketing; Bruce Pollack, Assoc. Dir. Marketing; Bob Soucy, sls dir.

Data/video projection systems for portable and permanent installation applications; LCD video monitors, TVs, VCRs, TV/VCRs and Viewcam Camcorders.

Shively Labs, Box 389, Bridgton, ME, 04009. Phone: (207) 647-3327. Fax: (207) 647-8273. E-mail: sales@shively.com Web Site: www.shively.com. David G. Allen, sls; Joe Rohrer, sls; Edd Forke, sls; Angela Gillespie, sls.

FM antennas, FM translators, branched & balanced combiners, coax, patch panels, filters, compressor dehydrators, & related RF equipment, pattern work & field svcs.

Shook Mobile Technology, LP, 7451 FM 3009, Schertz, TX, 78154. Phone: (210) 651-5700. Fax: (210) 651-5220. E-mail: shook@shook-usa.com Web Site: www.shook-usa.com. John Heaney, CEO; Ronald Crockett, pres & dir mktg.

Mobile TV production, ENG, SNV vehicles. Rack ready or turnkey delivery. HD/SD Systems integration.

Shure Inc., 5800 W. Touhy Ave., Niles, IL, 60714. Phone: (847) 600-2000. Fax: (847) 600-1212. Web Site: www.shure.com. E-mail: info@shur.com R.L. Shure, chmn; S. LaMantia, pres.

World-standard microphones, wireless audio systems, phonograph cartridges, mixers, digital signal processors, & personal monitors.

Siemens Dematic Limited, 167 Hunt St., Ajax, ON, L1S 1P6. Canada. Phone: (905) 683-8200. Fax: (905) 683-0186. Web Site: www.siemens.ca.

Solid state FM transmitters to 5 kw, automatic coaxial changeover units, shortwave transmitters.

Sierra Automated Systems & Engineering Corp., 2625 N. San Fernando Blvd., Burbank, CA, 91504. Phone: (818) 840-6749. Fax: (818) 840-6751. Web Site: www.sasaudio.com. Edward O. Fritz, pres; Al Salci, VP; Giovanni Morales, gen mgr.

Audio switching & mixing systems maunufacturer. Mix-Minus/IFB, satellite distribution/switching, automated switching & distribution, studio intercom, on-air routing, teleconferencing.

Sifford Video Services, 1033 Elm Hill Pike, Nashville, TN, 37210. Phone: (615) 248-1010. Fax: (615) 244-5712. Joel Covington, pres.

Sigma Electronics Inc., 1027 Commercial Ave., Box 448, East Petersburg, PA, 17520-0448. Phone: (717) 569-2681. Fax: (717) 569-4056. E-mail: sales@sigmaelectronics.com Web Site: www.sigmaelectronics.com. Billy Smilley, CEO.

Santa Rosa, CA. Western rgnl office Phone: (707)539-5314. Randy Smith, western rgnl sls mgr.

Routing switchers for audio & video; distribution amplifiers; sync & test signal generators; encoders, decoders, transcoders, converters.

Signal Monitoring Service, 773 Upper Fredricktown Rd., Mt. Vernon, OH, 43050. Phone: (888) 449-5643. Fax: (740) 397-2769. Robert (Bob) Bowman, owner.

AM/FM/TV frequency & modulation documentation - NRSC proof for AM.

Sinar Bron Inc., 17 Progress St., Edison, NJ, 08820. Phone: (908) 754-5800. Fax: (908) 754-5807. Web Site: www.sinarbron.com. James Bellina, pres; William D. Andrews, VP.

Pro-Cyc prefabricated coves for infiniti walls in video & photo studios; & studio lighting/HMI.

Sitco Antenna Company, Box 20456, 10330 N.E. Marx St., Portland, OR, 97220-1139. Phone: (503) 253-2000. Fax: (503) 253-2009. E-mail: sitco@simplicitytool.com Web Site: www.simplicitytool.com. Gustave Berliner, CEO; Markus Burcker, pres.

CATV, MATV antennas.

SiteSafe Inc., 200 N. Glebe Rd., Suite 1000, Arlington, VA, 22203. Phone: (703) 276-1100. Fax: (703) 276-1169. Web Site: www.sitesafe.com. Wesley O. McGee, pres.

Engrg software & wireless telecom engrg consulting svcs.

Skotel Corp., 92094 boul Taschereau, bureau 7400, Longueuil, PQ, J4W 3K8. Canada. Phone: (514) 806-2340. Fax: (514) 221-2338. E-mail: stephenscott@videotron.ca Stephen Scott, pres.

Switching Equipment: Distribution switchers, Video equipment: Time code generators & time code readers.

Skytec, Inc., 23 Inland Farm Rd., Windam, ME, 04062. Phone: (207) 893-1700. Fax: (207) 893-1717. E-mail: skytecinc@aol.com Web Site: www.skytecinc.com. Rick Sullivan, pres.

Manufacturer of skystrobe obstruction lighting systems - ETL certified, FAA approved. Parts, sls, svc & training seminars for obstruction lighting.

Snell & Wilcox Inc., 2225-I Martin Ave., Santa Clara, CA, 95050. Phone: (408) 260-1000. Phone: (800) 827-4544. Fax: (408) 260-2800. E-mail: snellcal@aol.com Web Site: www.snellwilcox.com. Dick Crippa, pres.

Havant, Harts P09 2PE. Smith & Wilcox Ltd., Southleigh Park House, Eastleigh Rd.

Petersfield, Hampshire GU33 5AZ. Snell & Wilcox Ltd., Durford Mill. Phone: 44-0-730-821-188. 44-0-730-821-199. David Youlton, chmn.

Snell & Wilcox is one of the world's largest manufacturers of bcst electronics. The complete product family includes a full range of video & audio processing equipment consisting of Decoding, Encoding, High Definition Format Conversion, MPEG Compression & Pre-processing, Display, Noise Reduction, Post Production Switchers (both SDTV & HDTV), Standards Conversion, Synchronization, Test & Measurement & IQ Modular products.

Solid State Logic Inc., 320 W. 46th St., New York, NY, 10036. Phone: (212) 315-1111. Fax: (212) 315-0251. E-mail: nysales@solid-state-logic.com Web Site: www.solid-state-logic.com. Steve Zaretsky, VP.

Los Angeles, CA 90036, 5757 Wilshire Blvd. Phone: (323) 549-9090. Phil Wagner, VP/western opns.

SSL C100 digital bcst console is perfect for on-air, live to tape productors. Its cost effective, reliable & friendly.

Solutec Ltd. (HA), 4360 D'Iberville, Montreal, PQ, H2H 2LB. Canada. Phone: (514) 522-8960. E-mail: gilles.fortin@sympatico.ca Gilles Fortin, pres.

Closed caption encoders (analog & SDI) & software.

Sound Designers Studio, 424 W. 45th St., New York, NY, 10036-3565. Phone: (212) 757-5679. Fax: (212) 265-1250.

Electronic equipment racking systems, console automation systems, digital recording facilities.

Soundcraft U.S.A., 8500 Balboa Blvd., Northridge, CA, 91329. Phone: (818) 920-3212. Fax: (818) 920-3208. E-mail: soundcraft-usa@harman.com Web Site: www.soundcraft.com. Dave Neal, mktg; Tom Der, natl sls mgr.

Audio mixing consoles for recording, theater, concert sound reinforcement & bcstg.

Southern Broadcast Services, 80 Commerce Dr., Suite B, Pelham, AL, 35124. Phone: (800) 256-9235. Fax: (205) 663-7108. Web Site: www.southernbroadcastservices.com. Jim Coleman, pres.

Tower erection, antenna instal, maintenance svcs.

Spacenet Services Inc., 1750 Old Meadow Rd., McLean, VA, 22102. Phone: (703) 848-1000. Fax: (703) 848-1010. Web Site: www.spacenet.com. David Shiff, VP mktg.

Satellite-based interactive data, bcst data & bcst video for coml companies worldwide.

Specialized Communications, 20940 Twin Springs Dr., Smithsburg, MD, 21783-1510. Phone: (800) 359-1858. Fax: (301) 790-0173. E-mail: service@spec-comm.com Web Site: www.spec-comm.com. David Linetsky, pres; Beth A. Linetsky, mktg dir; Andrew Hoffman, opns mgr.

Factory Authorized Service Center providing repair & maintenance of bcst video equipment. Factory Integrator & Dealer for distinctive industry brands.

Spectra Sonics, 3750 Airport Rd., Ogden, UT, 84405. Phone: (801) 392-7531. Fax: (801) 392-7531. Jean Dilley, controller; Gregory D. Dilley, pres.

Professional audio production, including power amps, compressor/limiters, portable speaker system, mixers, & line/distribution amps.

Spotcat Software, 1734 Green Valley Rd., Havertown, PA, 19083. Phone: (610) 446-1515. E-mail: support@spotcat.com

Sprague Magnetics Inc., 12806 Bradley Ave., Sylmar, CA, 91342. Phone: (818) 364-1800. Phone: (800) 553-8712. Fax: (818) 364-1810. E-mail: smiav@spraguemagnetics.com Web Site: www.spraguemagnetics.com. Dorothy Sprague, pres; John Austin, mgr.

Long-wearing cart, film, reel-to-reel tape heads, refurbishment svcs, replacement parts, alignment tapes, accessories.

Stage Equipment & Lighting Inc., 12250 N.E. 13th Ct., North Miami, FL, 33161. Phone: (305) 891-2010. Fax: (305) 893-2828. Fax: (800) 597-2010. E-mail: mail@seal-fla.com Web Site: www.seal-fla.com. Vivian Gill, pres; Michael Grosz, VP; Rick Rudolph, VP.

Orlando, FL 32811, 4600 S.W. 36th St. Phone: (407) 425-2010. Fax: (407) 648-2604. Mike Collins, tech consultant.

Tampa, FL 33619, 9207 Palm River Rd, Suite 108. Phone: (813) 626-8500. Fax: (813) 620-1404.

Film, video & theatrical lighting & grip, & related support equipment.

Stahl, A Scott & Fetzer Co., 3201 Old Lincoln Way, Wooster, OH, 44691. Phone: (330) 264-7441. Fax: (330) 264-3319. Web Site: www.stahl.cc. Bob McBride, pres; Tom Cole, sls dir.

Merced, CA 95340 Phone: (209) 383-4336.

Cardington, OH 43315 Phone: (419) 864-6871. Eric McNaly, plant mgr.

Durant, OK 74701 Phone: (405) 924-5575. Steve Shepard, plant mgr.

Stainless LLC, 1140 Welsh Rd., Suite 250, North Wales, PA, 19454. Phone: (215) 631-1400. Phone: (800) 486-3333. Fax: (215) 631-1427. E-mail: sales@stainlessllc.com Web Site: www.stainlessllc.com. Donald T. Doty, pres; J. Patrick Moore, VP; Tom Hoenninger, Chief Engr/VP-opns; Les Kutasi, Sls mgr.McKinney, TX 75069. Stainless Doty Moore, 213 E. Louisiana St., Ste. 250. Phone: (214) 585-0117.Cedar Hill, TX 75104. Doty Moore Stainless, 1570 W. Beltline Rd. Phone: (972) 637-5000.Design, engrg, fabrication of communications & bcst towers. Existing tower engrg studies & DTV analysis.

Stancil Corp., 2644 S. Croddy Way, Santa Ana, CA, 92704. Phone: (714) 546-2002 ext. 4316. Phone: (800) 782-6245. Fax: (714) 546-2092. E-mail: guy.churchouse@stancilcorp.com Web Site: www.stancilcorp.com. Michael Custer, CEO.

Voice logging recorders, multichannel, 4-144 channels, 24-hour recording time; digital format, instant recall recorders, windows 2000 voiceXP.

Standard Communications Corp., 1111 Knox Street, Torrance, CA, 90502. Phone: (310) 532-5300. Fax: (310) 532-0397. E-mail: SatcommSales@stdcom.com Web Site: www.standardcom.com. Ron Blanchard, pres/CEO.

Broadband TV receivers & cable headend products for broadcast and CATV.

Stanton Group, 3000 S.W. 42nd St., Ft. Lauderdale, FL, 33312. Phone: (954) 689-8833. Fax: (954) 689-8460. Web Site: www.stantonmagnetics.com. E-mail: info@stantonmagnetics.com Henri Cohen, sls dir.

Huntington, CA 92649. KRK, 5242 Business Dr. Phone: (714) 373-4600. (714) 373-0421.

Simi Valley, CA 93065. Kerwin Vega, 555 Fast Easy Street.

Turntables, professional cartridges, CD players, final scratch, monitors, speakers.

Star Case Manufacturing Co. Inc., 648 Superior Ave., Munster, IN, 46321. Phone: (219) 922-4440. Phone: (800) 822-STAR. Fax: (219) 922-4442. E-mail: starcase@starcase.com Web Site: www.starcase.com. Dennis Toma, pres; Ralph G. Hoopes, VP.

Flight cases (protective casement)—Carry Star, ATA Star, Super Star, Ultra Star, Star Light.

StarGuide Digital Networks Inc., 750 W. John Carpenter Fwy., Suite 700, Irving, TX, 75039. Phone: (972) 581-2000. Fax: (972) 581-2001. Web Site: www.starguidedigital.com. E-mail: hq@starguidedigital.com

High speed internet networking of digital audio, video & web. Software, satellite, terrestrial & DSL systems.

Storeel Corp., Box 80523, Atlanta, GA, 30366. Phone: (770) 458-3280. Fax: (770) 457-5585. E-mail: reely@mindspring.com Carolyn S. Galvin, pres; Michael Valerio, gen sls mgr; Elizabeth Galvin, VP sls.

Space-efficient storage for all formats of tape & film; double-drive systems for longer lengths; set-up trucks; CD storage.

Peter Storer & Associates Inc., 1361 W. Towne Square Rd., Mequon, WI, 53092. Phone: (262) 241-9005. Fax: (262) 241-9036. E-mail: storer@storertv.com Web Site: www.storertv.com. Peter Storer, pres; Doug Knight, sls VP.

SIMS-Multi-channel, multi-user, PC-based TV program schedule, amortization & liability system. Manage VOD as well as linear schedules.

Strand Lighting Inc., 6603 Darin Way, Cypress, CA, 90630. Phone: (714) 230-8200. Fax: (714) 899-0042. E-mail: sales@strandlight.com Web Site: www.strandlight.com. Bill King, pres.

New York, NY, 928 Broadway. Phone: (212) 242-1042.

Studio & remote lighting, & control equipment.

Strata Marketing Inc., 30 W. Monroe, Suite 1900, Chicago, IL, 60603. Phone: (312) 222-1555. Fax: (312) 222-2510. Web Site: www.stratag.com. John Shelton, pres.

Computer software for quantitative & qualitative radio, TV & nwspr media, cable.

Strata Marketing Inc., 30 West Monroe, Suite 1900, Chicago, IL, 60603. Phone: (312) 222-1555. Fax: (312) 222-2510. E-mail: rsparks@stratag.com Web Site: www.stratag.com. Bruce W. Johnson, pres.

TV, radio & media ratings analysis. Microsoft Windows-based software systems for cable systems & bcst TV stns.

Strong International, c/o Ballantyne of Omaha Inc., 4350 McKinley St., Omaha, NE, 68112. Phone: (402) 453-4444. Fax: (402) 453-7238. Web Site: www.ballantyne-omaha.com. John P. Wilmers, pres; Ray Boegner, sr VP.

35/70mm projection equipment, Xenon lamphouse systems, platters, Xenon bulbs, follow spotlights.

Structural System Technology Inc., 6867 Elm St., Suite 200, McLean, VA, 22101. Phone: (703) 356-9765. Fax: (703) 448-0979. E-mail: contact@sst-towers.com Web Site: www.sst-towers.com. Fred Purdy, pres; Kaveh Mehrnama, VP engrg; Greg Pleinka, natl sls mgr.

Structural engrg studies, analysis, design, modifications, inspections, fabrication, erection of towers & antenna.

Studio Technologies Inc., 5520 W. Touhy Ave., Skokie, IL, 60077. Phone: (847) 676-9177. Fax: (847) 982-0747. E-mail: stisales@studio-tech.com Web Site: www.studio-tech.com. Gordon Kapes, pres; Carrie Loving, mgr.

Microphone pre-amplifiers, stereo simulators & recognition units, telephone & hard-wired IFB communications systems on-air announcer's consoles. Accessories for digital audio workstations.

Studio Technology, 529 Rosedale Rd., Suite #103, Kenneth Square, PA, 19348. Phone: (610) 925-2785. Fax: (610) 925-2787. E-mail: sales@studiotechnology.com Web Site: www.studiotechnology.com. Vince Fiola, owner.

Bcst furniture, design & instal svcs.

Summit Software Systems Inc., 555 Camino del Rio, Unit A1, Durango, CO, 81303. Phone: (970) 385-4411. Fax: (970) 385-4734. E-mail: sales@summitsoft Web Site: www.summitsoftware.com. Paul Adams, pres.

PC-based traf, sls, billing, accounts receivable, accounts payable, payroll & gen ledger for single or multi-stns & single or multi-users.

Sundance Digital Inc., 545 E. John Carpenter Fwy., Suite 200, Irving, TX, 75062. Phone: (972) 444-8442. Fax: (972) 444-8450. E-mail: sales@sundig.com Web Site: www.sundancedigital.com. Jacque Durocher, gen mgr; Steve Krant, Vp sls & mktg.

Sundance Digital a part of Avid is an award-winning leader in TV automation solutions for individual & multistation bcstrs.

Superior Satellite Engineers Inc., 1743 Middle Rd., Columbia Falls, MT, 59912. Phone: (406) 257-9590. Fax: (406) 257-9599. E-mail: superior@superiorsatellliteusa.com Web Site: www.superiorsatelliteusa.com. Steve Catlett, pres; Jackie Williams, dir.

Navigator steerable & fixed antenna systems; multiple satellite feed systems & LNBs for cable & bcst TV.

Superior Tower Services Inc., 5757 FM 1696, Iola, TX, 77861. Phone: (936) 394-9925. Phone: (800) 306-4504. Fax: (936) 394-4020. Edward Carter, pres.

For all your tower & antenna needs: antenna, transmission line analysis, emergency repairs, two way, microwave, cellular, AM/FM, installations, tower erections, inspections, & maintenance.

Superscope Technologies Professionals, 2640 White Oaks Circle, Suite A, Aurora, IL, 60504. Phone: (630) 820-4800. Fax: (630) 820-8103. Web Site: www.superscopetechnologies.com. Fred Hackendahl, pres.

Products include portable cassette recorders, single & dual cassette recorders, CD players, multi-track recorders, compact recorders, portable, & rackmount.

Swager Communications Inc., Box 656, Fremont, IN, 46737. Phone: (260) 495-2515. Fax: (260) 495-4205. E-mail: bswager@dmci.net Web Site: www.swager.com. Dan J. Swager, pres; Lee Swager, VP; Tim Swager, sec/treas.

Designs, fabricates, installs & maintains AM/FM, TV/CATV & microwave communication towers.

Swintek Enterprises Inc., 965 Shulman Ave., Santa Clara, CA, 95050. Phone: (408) 727-4889. Phone: (408) 727-7544. Fax: (408) 727-3025. E-mail: ssales@swintek.com Web Site: www.swintek.com. William P. Swintek, pres.

18 ch wireless intercom, 1 w IFB with wireless EAR piece receiver, complete linear headsets.

Switchcraft Inc., 5555 N. Elston Ave., Chicago, IL, 60630. Phone: (773) 792-2700. Fax: (773) 792-2129. Web Site: www.switchcraft.com. Keith A. Bandolik, pres.

Offers a variety of products including audio patchbays, connectors, adapters, jacks & plugs, and video patchbays.

Symetrix Inc., 6408 216th St. S.W., Mountlake Terrace, WA, 98043. Phone: (425) 778-7728. Fax: (425) 778-7727. E-mail: symetrix@symetrixaudio.com Web Site: www.symetrixaudio.com. Dane Butcher, pres; Jim Latimer, sls dir.

Digital & analog audio signal processing.

Symmetricom, 3750 Westwind Blvd, Santa Rosa, CA, 95403. Phone: (707) 528-1230. Fax: (707) 527-6640. Web Site: www.symmetricom.com.

Time & frequency receivers traceable to NIST & USNO. Complete line of time code instrumentation.

Symmetricom, 2300 Orchard Pkwy., San Jose, CA, 95131. Phone: (949) 598-7500. Fax: (949) 598-7524. Erik Van derKay, pres; Bob Krist, VP.

Irvine, CA 92718. Datum-Irvine, 3 Parker. Phone: (714) 770-5000. Heinz Badura, pres.

Austin, TX 78761. Datum-Austin, Box 14766. Phone: (512) 251-2341. Jack Rice, pres.

Time code generators, readers, displays, encoders, search systems, distribution amplifiers, transmitters & receivers; design & manufacture of precision frequency products & timing instruments.

Syntellect Inc., 16610 N. Black Canyon Hwy., Suite 100, Phoenix, AZ, 85053. Phone: (770) 587-0700. Phone: (800) 347-9907. Fax: (770) 587-0589. Web Site: www.syntellect.com.

Phoenix, AZ 85027, 20401 N. 29th Ave. Phone: (602) 789-2800. Scott Coleman, pres.

ARUs for automated customer svc, ANI svcs for PPV order processing, predictive dialing systems for telemarketing & collections.

SyntheSys Research Inc., 3475-D Edison Way, Menlo Park, CA, 94025. Phone: (650) 364-1853. Fax: (650) 364-5716. E-mail: info@synthesysresearch.com Web Site: www.synthesysresearch.com. Jim Waschura, pres; John Ryan, gen sls mgr.

SyntheSys is a leading manufacturer of test & measurement specializing in serial digital video analyzers for SDI & high definition.

System Associates, Box 5925, Glendale, AZ, 85312. Phone: (866) 937-0209. Fax: (866) 435-0160. E-mail: video@systemsassociates.com Web Site: www.systemsassociates.tv. Mike Ferguson, gen mgr.

Used bcst TV equipment, to buy and sell, appraisals, auctions

Systems Wireless Ltd., 555 Herndon Pkwy., Ste. 135, Herndon, VA, 20170. Phone: (800) 542-3332. Fax: (703) 437-1107. E-mail: sales@swl.com Web Site: www.swl.com. Bill Sien, sls.

Sls, service & rental of wireless microphones, wireless intercom, wireless listening devices, wireless video & Clear Com cabled intercom systems.

T

TAI Audio, 5828 Old Winter Garden Rd., Orlando, FL, 32835. Phone: (407) 296-9959. Fax: (407) 648-1352. E-mail: sales@taiaudio.com Web Site: www.taiaudio.com. Joseph Guzzi, pres.

Rental, sls & svc of professional audio for film, video, TV & postproduction. Specializes in wireless communication equipment.

TALX Corp., 1850 Borman Ct., St. Louis, MO, 63146. Phone: (314) 214-7000. Fax: (314) 214-7588. Web Site: www.talx.com. William W. Canfield, pres; Michael E. Smith, VP.

Interactive communications; more specific svc; Interactive voice response, Interactive web, employment verification (work # for everyone) & Outsource svc.

TC Electronic, 5706 Corsa Ave., Suite 107, Westlake Village, CA, 91362. Phone: (818) 665-4900. Fax: (818) 665-4901. E-mail: infous@tcelectronic.com Web Site: www.tcelectronic.com. John Maier, CEO; Ed Simeone, chmn.

Digital compressor/limiter expander, digital signal processors, DTV audio processors & high-resolution digital delays.

T-C Specialties Co., Box 192, Coudersport, PA, 16915. Phone: (814) 274-8060. Phone: (800) 458-6074. Fax: (814) 274-0690. E-mail: tcsmail@adelphia.net Web Site: www.tcspecialties.com. Daniel C. Major, pres; Bill Crown, VP production; Judi Tucker, sec; Mike Harris, VP opns.

Coupon billing systems & related forms; large volume dir mail inkjetting, presort/barcoding mailing

TDK Electronics Corp., 3190 E. Miraloma Ave., Anaheim, CA, 92806. Phone: (714) 238-7900. Fax: (714) 632-1868. Web Site: www.tdk.com. Hajime Sawabe, pres.

TEAC America Inc., 7733 Telegraph Rd., Montebello, CA, 90640. Phone: (323) 726-0303. Fax: (323) 727-7656. Hajime Yamaguchi, pres; Gary S. Beckerman, exec VP.

Consumer audio/video & professional recording equipment, airborne video recorder, instrumentation data recorders, computer peripherals/floppy disks, tape backup & industrial optical disk recorders & playback.

TFT Inc., 1953 Concourse Dr., San Jose, CA, 95131-1708. Phone: (408) 943-9323. Fax: (408) 432-9218. E-mail: info@tftinc.com Web Site: www.tftinc.com. Darryl E. Parker, VP.

Digital, analog STLs, Reciters, synchronous boosters, modulation monitors & emergency alert systems(EAS).

TOA Electronics Inc., 601 Gateway Blvd., Suite 300, South San Francisco, CA, 94080. Phone: (650) 588-2538. Fax: (650) 588-3349. E-mail: info@toaelectronics.com Allan Lamberti, sls dir; Hisayuki Okvoka, pres/CEO.

Sound, communication equipment for coml sound & audio/video industries. Manufacturer, distributor of high quality, reliable audio & security products.

TTE Inc., 11652 W. Olympic Blvd., Los Angeles, CA, 90064. Phone: (310) 478-8224. Fax: (800) 473-2791. E-mail: sls@tte.com Web Site: www.tte.com. Stephen J. Sodaro, sls VP & mktg VP.

LC filters to 18 GHz, balun, matching transformers, combiners, active filters to 1 MHz. RF, microwave filters DC-18ghz & video splitters.

T.T. Technologies Inc., 2020 E. New York St., Aurora, IL, 60504. Phone: (800) 533-2078. Phone: (630) 851-8200. Fax: (630) 851-8299. E-mail: info@tttechnologies.com Web Site: www.tttechnologies.com. Chris Brahler, pres; Dave Holcomb, VP.

Ocala, FL 34479, 3701 N.E. 36th Ave, Suite C. Phone: (352) 622-2077. Tom Garner, gen mgr.

Grundomat pneumatic piercing tools, Grundoram pipe ramming system & Grundocrack Preumatic pipe bursting systems, Grundoburst Static Pipe Bursting.

TWR Lighting, (Division of 02 Wireless Solutions). 4300 Windfern, Suite 100, Houston, TX, 77041-0943. Phone: (713) 973-6905. Fax: (713) 973-9352. Web Site: www.twrlighting.com. Ken Meador, pres/CEO; Raymond Kraemer, VP sls & mktg; Jeff Huchlefeld, VP opns.

Aviation obstruction lighting manufacturer, sls & svc of low, medium, LED products & High Intensity systems.

Talk-A-Phone Co., 5013 N. Kedzie Ave., Chicago, IL, 60625. Phone: (773) 539-1100. Fax: (773) 539-1241. E-mail: info@talkaphone.com Web Site: www.talkaphone.com. S. Shanes, exec VP; Robert Shanes, VP sls.

Intercommunication systems, ADA compliant emergency phones, ADA areas of rescue, apartment access systems.

Tamron U.S.A. Inc., 10 Austin Blvd., Commack, NY, 11725. Phone: (631) 858-8400. Fax: (631) 543-5666. Web Site: www.tamron.com. E-mail: feedback@tamron.com Tak Inoue, pres; Bert Krank, western sls mgr; John VanSteenberg, sls; Stacie Errera, chief mktg off; Gregg Maniaci, eastern sls mgr.

Lenses for 35mm SLR digital & film cameras, & CCTV lenses.

Tandberg Television Inc., 4500 River Green Pkwy., Suite 110, Duluth, GA, 30095-2580. Phone: (678) 812-6300. Fax: (678) 812-6400. Web Site: www.tandbergtv.com. Eric Cooney, pres/CEO; Fraser Park, CFO.

Tannoy North America Inc., 335 Gage Ave., Suite 1, Kitchener, ON, N2M 5E1. Canada. Phone: (519) 745-1158. Fax: (519) 745-2364. E-mail: inquiries@tannoyna.com Web Site: www.tannoy.com. Marc Bertrand, mgng dir.

Tannoy is a leading innovator of premium audio solutions utilizing cutting edge acoustic, electronic & digital experise.

Tapeswitch Corp., 100 Schmitt Blvd., Farmingdale, NY, 11735. Phone: (631) 630-0442. Fax: (631) 630-0454. E-mail: sales@tapes.com Web Site: www.tapeswitch.com.

Mission, CA 92692 Phone: (949) 588-9387. Jeff Johnson, rgnl sls mgr.

Fishers, IN 46038 Phone: (317) 570-6178. Tom Bertellotti, rgnl sls mgr.

New Bern, NC 28562 Phone: (252) 637-7728. Vinnie Colucci, rgnl sls mgr.

Franklin, TN 37064 Phone: (615) 591 7399. Tim DePeri, rgnl sls mgr.

Safety light curtains, sensing mats, edges, ribbon switches, electronic zone controllers, sensing bumpers, safety & protection equipment.

J.A. Taylor & Associates, Box 331, Boyertown, PA, 19512-0331. Phone: (610) 754-6800. Fax: (610) 754-9766. E-mail: jataylor@broadcastassociates.com Web Site: www.broadcastassociates.com.

Appraisers & brokers of TV production equipment. Serves video production companies, TV stns & financial institutions.

Teatronics/Entertainment Lighting Control, P.O. Box 508, Santa Mangarita, CA, 93453. Phone: (805) 438-4000. Fax: (805) 438-5400. E-mail: sales@teatronics.com Web Site: www.teatronics.com.

Lighting control & power distribution systems for stage, studio & remote applications.

Tech Laboratories Inc., 955 Belmont Ave., North Haledon, NJ, 07508. Phone: (973) 427-5333. Fax: (973) 427-5455. E-mail: corporate@techlabs.com Web Site: www.techlabsinc.com. Bernard M. Ciongoli, pres; Earl M. Bjorndal, VP.

Rotary switches; electrical/electronic subcontract, attenuators, transformers, pcb assembly & infrared security systems.

TECH-SA-PORT, Box 5372, 120 S. Whitfield St., Pittsburgh, PA, 15206-0372. Phone: (412) 661-1620. Phone: (800) 543-2233. E-mail: tech-sa-port@juno.com Web Site: www.tech-sa-port.com. Lewis J. Scheinman, pres.

Computer & electronic equipment cleaning supplies, including lint-free wipers, contamination-free chemicals, & spray dusters. All types of wiping materials.

Technet Systems Group, (A Division of Steve Vanni Associates Inc). Box 422, Auburn, NH, 03032. Phone: (603) 483-5365. Fax: (603) 483-0512. E-mail: svanni@technetsystems.com Web Site: www.technetsystems.com. Steve Vanni, pres.

Bcst equipment supplier & distributor for radio & TV, specializing in complete turnkey packages including planning, design, equipment, instal, towers & FCC licensing.

Techni-Tool Inc., 1547 N. Trooper Rd., Worcester, PA, 19490-1117. Phone: (800) 832-4866. Phone: (610) 825-4990. Fax: (610) 828-5623. Fax: (800) 854-8665. E-mail: sales@techni-tool.com Web Site: www.techni-tool.com. Paul Weiss, pres; David Weitner, mktg dir; Stuart Weiss, VP; Michael T. Ryan, sls dir.

Master distributor of hand tools/kits & shipping cases, solder/desolder equipment, ESD products, cleanroom products, test equipment & telecommunications products for the bcst Industry.

Technologies for Worship, 3891 Holburn Rd., Queensville, ON, L0G 1R0. Canada. Phone: (905) 473-9822. Fax: (905) 473-9928. E-mail: bc@tfwm.com Web Site: www.tfwm.com. Shelagh Rogers, pres; Barry Cobus, VP; Kevin Rogers-Cobus, exec editor.

Trade magazine & tech directory for houses of worship involved in audio, AV, bcst, computers, film, video & music. Owner of "Inspiration Conferences & Expositions."

Tekskil Industries Inc., #102-998 Harbourside Dr., North Vancouver, BC, V7P 3T2. Canada. Phone: (604) 985-2250. Fax: (604) 985-2248. E-mail: inquiries@tekskil.com Web Site: www.tekskil.com. John Veenstra, pres.

Manufacturer of video, speech & computer prompting equipments.

Tektronix Inc., 14200 S.W. Karl Braun Dr., Beaverton, OR, 97077. Phone: (800) 835-9433. Phone: (503) 627-7111. Fax: (503) 627-6108. Web Site: www.tektronix.com. Richard Willis, pres/CEO.

Telcom Research, 3375 N. Service Rd., A7, Burlington, ON, L7N 3G2. Canada. Phone: (905) 336-2450. Fax: (905) 336-1487. E-mail: dougl@tecomresearch.com Web Site: www.timecode.com. Tom Banting, pres.

SMPTE/EBU time code generators, readers; character inserters, LTC-VITC & VITC-LTC trans. Logging/offline/EDL software.

Tele-Measurements Inc., 145 Main Ave., Clifton, NJ, 07014-1078. Phone: (973) 473-8822. Fax: (973) 473-0521. E-mail: tmcorp@aol.com Web Site: www.tele-measurements.com. William E. Endres, pres; W. Chris Endres, gen mgr.

Bcst video equipment, tapes TV systems, teleconferencing, maintenance support, CCTV & rentals, distance learning.

Telecast Fiber Systems Inc., 102 Grove St., Worcester, MA, 01605. Phone: (508) 754-4858. Fax: (508) 752-1520. E-mail: jcommare@telecast-fiber.com Web Site: www.telecast-fiber.com. Richard A. Cerny, pres; Eugene E. Baker, VP; Joseph Commare, VP mktg & intl sls.

Mill Valley, CA 94941, 835 Autumn Ln. Phone: (415) 383-5388. Fax: (650) 745-3711. James Hurwitz, mgr Western U.S.

Saco, ME 04072, 280 Ferry Rd. Phone: (207) 282-9772. Fax: (207) 282-8666.

Fuquay-Varina, NC 27526, 3009 Bentwillow Dr. Phone: (919) 557-6059. Fax: (919) 557-5206. Bryan Keen, sr sls engr.

Dallas, TX 75231, 8712 Lacrosse Dr. Phone: (214) 553-1366. Fax: (241) 553-1367.

Fiber-optic video & audio systems for TV bcst production.

Telecrafter Products, 12687 W. Cedar Dr., Suite 100, Lakewood, CO, 80228-2031. Phone: (303) 986-0086. Fax: (303) 986-1042. E-mail: mail@telecrafter.com Web Site: www.telecrafter.com.

Drop installation products for broadband telecommunications delivery svc, including cable clips, cablemakers, cable guard, house boxes, fitting savers, & more.

Telemetrics Inc., 6 Leighton Pl., Mahwah, NJ, 07430. Phone: (201) 848-9818. Fax: (201) 848-9819. Web Site: www.telemetricsinc.com. Anthony C. Cuomo, pres; Anthony E. Cuomo, mgr.

Camera pan, tilt systems & triax camera control systems.

Telenium Mobile, 525 Mildred Ave., Primos, PA, 19018. Phone: (610) 626-6500. Fax: (610) 626-2638. Web Site: www.telenium.com. Tod Strine, pres.

48-foot mobile unit with Ikegami cameras; Chyron 4100EXB; Chyron Infinity A42 & A53 & Sony 1" VTRs; & BVW 75's. Available in United States & Canada.

Teleplex Inc., (Alford Division). 4801 Industrial Pkwy., Indianapolis, IN, 46226. Phone: (317) 895-8800. Fax: (317) 895-2900. Tom L. Fitch, pres/CEO; Lois E. Clark, VP.

Produces precision electronics, components, FM stn combiners, custom FM radio antenna arrays, TV antenna systems, & HDTV monitor antennas. Supports all Alford products & antenna products.

Telescript Inc., 445 Livingston St., Norwood, NJ, 07648. Phone: (201) 767-6733. Fax: (201) 784-0323. E-mail: info@telescript.com Web Site: www.telescript.com. John McGrath, mgng dir.

Austin, TX 78752. Telescript West, 7801 N. Lamar Blvd. Phone: (512) 302-0766. Jim Stringer, mgr.

IBM & compatibles prompting programs & equipment. Lightweight, high-resolution 12" & 17" monitor prompters. Flat panel prompters, window based prompting software for bcst & video productions applications, comprehensive line of LCD prompters.

Teletech Inc., 38235 Executive Drive, Westland, MI, 48185. Phone: (734) 641-2300. Fax: (734) 641-2323. Web Site: www.teletech-inc.com. Keith Johnson, VP; Todd Osment, field svc mgr; Richard Humphrey, sr engr; Kevin Beltramo, field engr.

Facility construction, antenna instal for AM, FM, TV, LPTV & microwave; antenna site mgmt; FCC applicacations; EME studies; Aeronautical studies.

teletech.ca, 211 Telson Rd., Unit 3, Markham, ON, L3R 1E7. Canada. Phone: (905) 475-5646. Fax: (905) 475-5684. E-mail: jacksrwld@rogers.com Web Site: www.teletech.ca. Jack Kirkpatrick, pres.

London, ON N5Z 3M7 Canada, 931 Leathorne St. Brad Rose, mgr.

Total bcst, postproduction, audio & video sls, service, & rentals of equipment & supplies.

Televideo San Diego, 4783 Ruffner St., San Diego, CA, 92111. Phone: (858) 268-1100. Fax: (858) 268-1790. Web Site: www.televideosd.com. Linda Stepp, CFO; David Stepp, pres.

Televideo designs, installs, svcs video production, training, distance learning & videocon ferencing systems.

Television Engineering Corp., 2647 Rock Hill Industrial Ct., Saint Louis, MO, 63144. Phone: (314) 961-2800. Fax: (314) 961-2808. Web Site: www.tvengineering.com. Jack Vines Jr., gen sls mgr; Jack Vines Sr., design engr.

Manufacturer of news vans, satellite vehicles, Eagle Eye camera, IFB controller & turnkey systems.

Television Equipment Assoc. Inc./Matthey, Box 404, Brewster, NY, 10509-0404. Phone: (845) 278-0960. Fax: (845) 278-0964. Bill Pegler, pres; Joseph Tocidlowski, mgr.

Serial digital interface products, NTSC/PAL Decoders, analog & digital DAs, video & pulse delays, video filters, A/D & D/A converters, headsets, serial digital/Fiberoptic links, Routing switches for digital video and radio.

Telex Communications Inc., 12000 Portland Ave. S., Burnsville, MN, 55337. Phone: (952) 884-4051. Fax: (952) 884-0043. E-mail: prosound@telex.com Web Site: www.telex.com. Ned Jackson, CEO.

Burbank, CA 91505, 2550 Hollywood Way, Suite 207. Phone: (818) 566-6700.

Wired & wireless microphones; headphones/headsets; wired & wireless intercoms; audio duplicators/copiers.

TELLABS, 1415 W. Diehl Rd., Naperville, IL, 60563. Phone: (630) 798-8800. Fax: (630) 798-2000. Web Site: www.tellabs.com.

Abingdon, Oxfordshire 0X14 3Y3, 29 The Quadrant. Phone: 011-44-235-524-400.

Hauppauge, NY 11788, 60 Commerce Dr. Phone: (516) 231-1550.

Teleconferencing systems, digital echo cancellers, data over voice multiplexers, signaling systems, video conferencing, audio systems.

Telos Systems, 2101 Superior Ave., Cleveland, OH, 44114. Phone: (216) 241-7225. Fax: (216) 241-4103. Web Site: www.telos-systems.com. Steve Church, pres/CEO; Frank Foti, pres; Michael Dosch, mgng dir.

Manufacturer of MP 3 ISDN codecs, digital network, telephone interfaces (e.g. hybrids), MP 3 internet hardware & software for webcasting.

Teltron Technologies Inc., 2 Riga Ln., Birdsboro, PA, 19508. Phone: (610) 582-9450. Phone: (800) 835-8766. Fax: (610) 582-0851. E-mail: teltron@ptdprolog.net Web Site: www.teltrontech.com.

Camera tubes for monochrome, color, special purpose applications & view finder CRTs.

Tenco Tower Co., 9647 Folsom Blvd., Sacramento, CA, 95827-1326. Phone: (916) 638-8833. Fax: (916) 366-7383. E-mail: donald.tenn@towerguys.com Donald Joseph Tenn, pres/CEO.

Instal, maintenance, sls of towers, antennas, hardware for the bcst, cable & communications industry.

Tentel, 333 Industrial Dr. #4, Placerville, CA, 95667. Phone: (530) 344-0183. Fax: (530) 344-0186. Web Site: www.tentel.com. E-mail: info@tentel.com John Chavers, gen mgr.

Texas Electronics Inc., Box 7225, Dallas, TX, 75209. Phone: (214) 631-2490. Fax: (214) 631-4218. E-mail: info@texaselectronics.com Web Site: www.texaselectronics.com. Carol Westlund, pres; Jane Hansen, VP; Jason Burson, sls.

Manufacturer of meteorological instruments & controls.

Texscan MSI, 2210 W. Alexander Street Ste. A, Salt Lake City, UT, 84119. Phone: (801) 956-0000. Fax: (801) 956-0750. Web Site: www.texscan.com. Leonard J. Fabiano, pres.

Character generators, digital, analog commercial insertion systems, audio/video playback systems, weather data svc, multimedia graphics production systems & VCR controllers.

Thales Broadcast & Multimedia Inc., 104 Feeding Hills Rd., Southwick, MA, 01077. Phone: (413) 998-1100. Fax: (413) 569-0679. E-mail: joe.turbolski@us.thales.bm.com Web Site: www.hales-bm.com. Mark Kearns, CEO; Richard E. Fiore Jr., sr VP.

Coral Gables, FL 33146, Gables One Tower, Suite 780. Phone: (305) 665-0067. Perry Priestley, sls dir Canada & Latin America.

Thales B&M desugbs devekios, manufactures & markets equipment sytems and solutions in the fields of terrestrial transmission, digital videl processing & multimedia distribution.

Thales Broadcast & Multimedia S.A., 1 rue de l'Hautil, Z.A. les Bountries, Conflans Sainte Honorine Cedex, 78702. France. Phone: 1 34 90 31 00. Fax: 1 34 90 30 00. Web Site: www.thomcast.thanson-csf.con.

Thales Components Corp., 40 G Commerce Way, Totowa, NJ, 07511. Phone: (973) 812-9000. Fax: (973) 812-9050. Web Site: www.tccus.com. S. Shpock, CEO.

Power grid triodes & tetrodes, cavities, klystrons travel wave tubes & I.O.T.s.

Theatre Service & Supply Corp., 1792 Union Ave., Baltimore, MD, 21211. Phone: (410) 467-1225. Fax: (410) 467-1289. E-mail: sales@stage-n-studio.com Web Site: www.stage-n-studio.com. Richard A. Antisdel, pres; Jacauelin Keleman, sls.

Manufacturer of studio, theatrical curtains, track systems, distributor of lighting & theatrical hardware.

Theatrical Services Inc., 128 S. Washington, Wichita, KS, 67202. Phone: (316) 263-4415. Fax: (316) 263-9927. E-mail: tsi@theatricalservices.com Web Site: www.theatricalservices.com. Stephen A. Wolf, pres.

Manufacturers & distributors of studio lighting & control equipment, studio cycloramas, curtains & track.

Thermodyne Cases, 1841 Business Pkwy., Ontario, CA, 91761. Phone: (909) 923-9945. Fax: (909) 923-7505. Web Site: www.thermodyne-online.com. Gary S. Ackerman, pres.

Reusable shipping cases; rack-mounted operating cases. Provides protection for all electronic equipment during transit.

Thomas & Betts, 745 Avoca Ave., Dorval, PQ, H9P 1G4. Canada. Phone: (514) 636-6560. Fax: (514) 631-4306. Web Site: www.t&b.com.

Manufacturer of quality products for aerial construction & subscriber instal hardware for the Cable TV & telephone industry.

Thomas & Betts Corp., 8155 T&B Blvd., Memphis, TN, 38120. Phone: (800) 920-0328. Fax: (800) 283-6756.

Manufacturer of Poleline hardware aerial & drop systems, fiber-optic hand holes, MMDS antenna mounting hardware.

James Thomas Engineering, 10240 Caneel Dr., Knoxville, TN, 37931. Phone: (865) 692-3060. Fax: (865) 692-9020. Web Site: www.jthomaseng.com. Mike Garl, pres.

Worcestershire WR10 2DB, Station Approach, Pershore Trading Estate, NR Pershore. Fax: (0386) 553002. Graham Thomas, pres.

Manufactures & distributes spun aluminum PAR fixtures, modular aluminum trussing, ground support, pre-wired lamp bars, spot banks & lighting accessories. Distributes Socapex & VEAM multipin connectors & cable. Assembles custom multicables, breakouts & distributes boxes to specifications. Distributor of olflex cables.

Thomson Broadcast & Media Solutions, Box 599000, Nevada City, CA, 95959-5900. Phone: (800) 824-5127. Fax: (530) 478-3166. Web Site: www.thomsongrassvalley.com. Tim Thorsteinson, pres/CEO; Russ Johnson, the Americas sls; Stephen Wong, Pacific rgn sls.

Video servers/disk recorders, media platforms, video production centers (switchers), signal mgmt systems (routers, modular), DVEs & HDTV equipment.

Thomson, Inc (RCA/GE), 10330 N. Meridian St., Indianapolis, IN, 46290. Phone: (317) 587-3000. Fax: (317) 587-6708. E-mail: dave.arland@thomson.net Web Site: www.rca.com. Michael D. O'Hara, exec VP.

Boulogne, Cedex 92045. Thomson S.A. (parent co.), 46 Quai Alphonse Le Gallo. Phone: 3301-418650.

Manufactures mkt audio, video communications & accessories products.

Thomson Multi Media, 2300 S. Decker Lake Blvd., Salt Lake City, UT, 84119. Phone: (801) 972-8000. Fax: (801) 972-6304. Web Site: www.thomsonmultimedia.com. Martin Fry, CEO.

Wyomissing, PA 19610, 13 Kevin Ct. Phone: (215) 678-8711. Fax: (215) 678-8784. Jeff Rosica, sls mgr.

Thorn-EMI Studio Lamps/L.C., 4800 W. Univ. Ave., Las Vegas, NV, 89103. Phone: (702) 367-3656. Fax: (702) 367-7058. L.E. Nelson, pres; H.F. Dowd, VP; D.R. Imfeld, VP.

Thorn-EMI studio lamps & lamps from: G.E., Philips, Osram & Ushio.

360 Systems, 31355 Agoura Rd., Westlake Village, CA, 91361-4610. Phone: (818) 991-0360. Fax: (818) 991-1360. E-mail: info@360systems.com Web Site: www.360systems.com. Robert Easton, pres.

Image Server Video & Graphics servers, Progm Time Delays, Content Mirroring Systems, DigiCart, Instant Replay and ShortCut audio editors and players.

3M, 3M Product Information Center, St. Paul, MN, 55144-1000. Phone: (651) 737-6501. Phone: (800) 3m-helps. Fax: (800) 713-6329. E-mail: innovation@mmm.com Web Site: www.3m.com.

Fault locators; cable & cabling equipment & supplies; Post-it notes & flags; Telephony: copper & fiber optic networks; Vikuiti display enhancement films; splicing kts; volition fiber optics; electrical tubing, tapes, terminations & connectors.

The Tiffen Company, 90 Oser Ave., Hauppauge, NY, 11788. Phone: (631) 273-2500. Phone: (800) 645-2522. Fax: (631) 273-2557. Steve Tiffen, pres/CEO; Hilary Araujo, VP mktg; Jeff Cohen, VP; Michael Cannatta, COO.

Photographic filters, lens accessories for motion picture, still photography, digital video, Davis & Sanford tripods, Domke Bags, support systems; steadicam camera stabilizing systems.

Time Logic Inc., 1914 Palomar Oaks Way, Suite 150, Carlsbad, CA, 92008. Phone: (760) 517-0445. Fax: (760) 431-1351. E-mail: jiml@timelogic.com Web Site: www.timelogic.com. Jim Lindelien, pres.

Automation systems for TV bcstrs & radio stns. Custom software dev for Tektronix Profile disks, HDTV time delay systems, using disk or tape.

Time Manufacturing Co., Box 20368, Waco, TX, 76702-0368. Phone: (254) 399-2100. Fax: (254) 399-2651. E-mail: renees@timemfg.com Web Site: www.timemfg.com.

Truck mounted aerial lifts ranging from 29' to 210' in height.

Times Fiber Communications Inc., 358 Hall Ave., Wallingford, CT, 06492. Phone: (203) 265-8500. Fax: (203) 265-8422. Web Site: www.timesfiber.com. Timothy F. Cohane, pres/COO; Stan VonFeldt, VP sls; Chris Huffman, dir mktg.

Renfrew, ON Canada, Box 430. Phone: (613) 432-8557.

Phoenix, AZ 85063, Box 14975. Phone: (602) 278-5576. Les Judd.

Chatham, VA 24531, Box 119A, Rt. 2. Phone: (804) 432-1800.

Coaxial, twisted pair composite cables for broadband, cellular/PCS applications, semiflex, svc entry, drop cables & connectors.

Tinsley Laboratory Inc., (A division of Silicon Valley Group). 4040 Lakeside Dr., Richmond, CA, 94806. Phone: (510) 222-8110. Fax: (510) 223-4534. Dan Desmond, pres.

Gyrozoom image stabilizing lens, GX3 integrated CCD camera/stabilizing system.

Toner Cable Equipment Inc., 969 Horsham Rd., Horsham, PA, 19044. Phone: (215) 675-2053. Phone: (800) 523-5947. Fax: (215) 675-7543. E-mail: info@tonercable.com Web Site: www.tonercable.com. Robert L. Toner, pres; B.J. Toner, VP.

International distributor & manufacturer of a complete line of cable TV & wireless cable equipment.

Torpey Time, 98-2220 Midland Ave., Scarborough, ON, M1P 3E6. Canada. Phone: (416) 298-7788. Phone: (800) 387-6141. Fax: (416) 298-7789. E-mail: sales@torpeytime.com Web Site: www.torpeytime.com. Bob Torpey, pres.

Master clock systems, digital & analog slave clocks, timers, video time & temperature equipment.

Toshiba America Consumer Products, 1420 Toshiba Dr., Lebanon, TN, 37087. Phone: (615) 444-8501. Fax: (615) 443-3810. Web Site: www.toshiba.com. Robert Arnett, sr VP.

Wayne, NJ 07470, 82 Totowa Rd. Phone: (973) 628-8000. Fax: (973) 628-1875. (New Jersey office).

HDTV products: HD-VCR (Analog-UniHi), HD monitor (projection & CRT), NTSC to HDTV upconverter, HD-CCD color camera, HD horizon system.

Tower Innovations, (A Dielectric Company). 2855 Hwy. 261, Newburgh, IN, 47630. Phone: (812) 853-0595. Fax: (812) 853-6652. E-mail: towers@towerinnovations.net Web Site: www.towerinnovations.net. David Nicholson, exec VP & COO; Tim Ryan, VP sls/mktg.

Manufacture towers for bcst, cellular & PCS Communications Innovative engrg solutions, fabrication, construction planning, tower erection & turnkey systems.

Tower Inspection Inc., Box 709, Muskogee, OK, 74402-0709. Phone: (918) 683-8915. Fax: (918) 683-0888. E-mail: sales@towerinspection.com Web Site: www.towerinspection.com. Barry R. Bayless, pres; Gary G. Lehman, VP.

Inspection svcs during construction; maintenance inspection, painting, repairs of radio, microwave & TV towers.

Tower Network Services, 2009 Ranch Rd. 620 N., Suite 710, Austin, TX, 78734. Phone: (512) 266-6200. Fax: (512) 266-6210. Web Site: www.towernetwork.com.

Svc tower, antenna. RF testing & tower structural analysis. Tower elevator repair and upgrading.

Tower Structures Inc., 2567 Business Pkwy., Minden, NV, 89423. Phone: (888) 219-0299. Fax: (775) 267-1308. E-mail: sales@towerstructures.com Web Site: www.towerstructures.com. Steven Hopkins, pres; Donald G. Weirauch, VP.

Vista, CA 92083, 430 Olive. Phone: (760) 631-4577. Tony Sierra, rgnl mgr.

Towers, design & construction of transmitter bldgs, instal of bcst, microwave & satellite antennas.

Transcom Corp., Box 26744, Elkins Park, PA, 19027. Phone: (215) 938-7304. Fax: (215) 938-7361. E-mail: transcom@fmamtv.com Web Site: www.fmamtv.com. Martin Cooper, pres.

Huntington Valley, PA 19006, 2655 Philmont Ave, Suite 200. Phone: (215) 938-7304. (800) 441-8454. Fax: (215) 938-7361. Martin Cooper, pres.

New digital (8VSB/DVB-T/H), angle TV transmitters, microwave links, antenna, cable & select used TV, AM, FM transmitters.

Transcrypt International, 3900 N.W. 12th St., Suite 200, Lincoln, NE, 68521. Phone: (402) 474-4800. Mike Kelley, gen mgr.

Transtector Systems Inc., 10701 N. Airport Rd., Hayden, ID, 83835. Phone: (800) 882-9110. Phone: (208) 772-8515. Fax: (208) 762-6133. E-mail: sales@transtector.com Web Site: www.transtector.com. Shawn Thompson, mgng dir.

Transient overvoltage protective devices, power quality consulting svcs; college-accredited, power-quality assurance education courses.

Transvision/Vision Accomplished Inc., 550 Maulhardt Ave., Oxnard, CA, 93030. Phone: (805) 981-8740. Fax: (805) 981-8738. E-mail: info@txvision.com Web Site: www.txvision.com. Kimithy Vaughan, pres.

Transportable, flyaway satellite transmission; mobile/TVROs; studio/remote production & transmission mgmt.

Tribune Media Services, 333 Glen St., Glens Falls, NY, 12801. Phone: (800) 833-9581. Phone: (518) 792-9914. Fax: (518) 761-7118. E-mail: tvdata@tvdata.com Web Site: www.tvdata.com. James McCormick, VP opns; Kathleen Tolstrup, VP sls; Lanna Langlois, VP finance.

International source for TV info. Clients include interactive on-screen & on-line guides, nwsprs, print publications, cable companies, telephone companies, rsch organizations, producers, syndicators of TV programs & advertisers.

Trident Media Group/ Spector Entertainment Group Inc., 2441 Impala Dr., Carlsbad, CA, 92008. Phone: (760) 438-9080. Fax: (760) 438-0968. Eric M. Spector, exec VP; Evan M. Spector, pres.

International audio, video, data & telephone communications svcs between United States/Canada & Mexico/Latin America; private TV networks; 12 C-band uplinks; teleports; satellite space segment; encryption; organization svcs; closed circuit TV & security systems.

Trilogy Communications Inc., 2910 Hwy. 80 E., Pearl, MS, 39208. Phone: (601) 932-4461. Phone: (800) 874-5649. Fax: (601) 939-6637. E-mail: info@trilogycoax.com Web Site: www.trilogycoax.com. Shinn Lee, chmn/pres/CEO; John Kaye, vice chmn; Grace Lee, exec VP; Bill Lee, sr VP; Jim Oldham, sls dir; Fei Wei, Mkt. Asso.

World leading manufacturer of advanced technology coaxial cables for wireless networks for cellular, paging, PCS, SMR and in- building networking applications application, ISO-9001 certified.

Trimm Inc., 407 Railroad St., Butner, NC, 27509. Phone: (847) 362-3700. Fax: (847) 680-3888. E-mail: trimminc@frontiernet.net Web Site: www.trimminc.com.

Manufacturer of fuse panel s& terminal blocks.

Triplett Corp., One Triplett Dr., Bluffton, OH, 45817. Phone: (419) 358-5015. Phone: (800) 874-7538. Fax: (419) 358-7956. Fax: (888) trip-fax. E-mail: wjh@triplett.com Web Site: www.triplett.com. Warren Hess, pres/CEO.

Panel instruments & test equipment. Electrical, electronic, telecommunication & railroad testers.

Trompeter Semflex, 55550 E. McDowell Rd., Mesa, AZ, 85215. Phone: (480) 985-9000. Fax: (480) 985-0334. E-mail: sales@trompeter.com Web Site: www.trompeter.com. Joe Norwood, pres.

DS3 interconnection & DSX products for central office.

Tulsat/An Addvantage Technologies Co., 1221 E. Houston St., Broken Arrow, OK, 74012. Phone: (800) 331-5997. Fax: (918) 251-1138. E-mail: tulsat@tulsat.com Web Site: www.tulsat.com. David Chymiak, pres; Ken Chymick, VP.

Turner Studios Field Operations, 1020 Techwood Dr., Atlanta, GA, 30318. Phone: (404) 885-4746. Fax: (404) 885-2175. E-mail: charli.whitfield@turner.com Bob McGee, dir; Scott Marks, VP; Charli Whitfield, opns mgr.

Two 53 ft expandable mobile units with or without crews; 12 cams, 9 VTRs, EVS, DVEous, Infinit & Deko; FFV Omega DDRs.

Tyco Electronics, 300 Constitution Dr., Menlo Park, CA, 94025. Phone: (650) 361-3333. Fax: (650) 361-2288. Web Site: www.tycoelectronics.com. Jackie Heisse, CEO.

Coaxial connectors, environmental sealing products & antenna de-icers.

U

U.S. Tape & Label Corp., 2092 Westport Ctr. Dr., St. Louis, MO, 63146. Phone: (314) 824-4444. Fax: (314) 824-4400. Web Site: www.ustl.com. Jim Eiseman, pres.

Custom printed bumper strips & window labels for the bcst industry. Industrial labels, direct mail printing & label-aire equipment.

U.S. Traffic & Display Solutions, 9603 John St., Santa Fe Springs, CA, 90670. Phone: (562) 923-9600. Fax: (562) 923-7555. E-mail: dsi@dsiusa.com Web Site: www.displaysolutionsinc.com.

Changeable outdoor electronic adv displays.

Ultimate Precision/GKM, 200 Finn Ct., Farmingdale, NY, 11735. Phone: (631) 249-7816. Fax: (631) 777-1828. E-mail: lmallia@afcosystems.com Web Site: www.ultimateprecision.com. Michael Mallia, pres; Larry Mallia, VP/gen mgr.

Complete precision sheet metal design & mfg servicing leading companies in high technology, military, telecommunications & medial equipment.

Ultimate Support Systems Inc., Box 470, Fort Collins, CO, 80522. Phone: (970) 493-4488. Fax: (970) 221-2274. E-mail: custserv@ultimatesupport.com Jim Dismore, chmn/CEO.

Strong, lightweight speaker & lighting tripods. Microphone stands for nearly any application.

Ultimatte Corp., 20945 Plummer St., Chatsworth, CA, 91311. Phone: (818) 993-8007. Fax: (818) 993-3762. Web Site: www.ultimatte.com. Reid Baker, bussiness dev; Alan Dadourian, chief engr; Lynne Sauve, pres.

Video compositing devices for comls, live bcst, production, postproduction & computerized tripod head.

Uni-Set Corp., 449 Ave. A, Rochester, NY, 14621. Phone: (585) 544-3820. Fax: (585) 544-1110. E-mail: info@unisetcorp.com Web Site: www.unisetcorp.com. Ronald D. Kniffin, pres.

Modular studio staging systems for studio settings; news setting; 7 top talent tables, & UNI-CYC anywhere cyclorama.

Union Connector Co., 40 Dale St., West Babylon, NY, 11704-1104. Phone: (631) 753-9550. Fax: (631) 753-9560. Web Site: www.unionconnector.com. Richard A. Wolpert, pres; Alan T. Wolpert, VP.

Electrical connectors, power distribution systems, portable power cabinets, custom switchgear, cases, carts & grip equipment.

Unique Business Systems, 2901 Ocean Park Blvd., Suite 215, Santa Monica, CA, 90405. Phone: (310) 396-3929. Fax: (310) 396-6114. E-mail: pbatra@unibiz.com Web Site: www.unibiz.com. Pradeep Batra, pres.

Langhorne, PA 19047, 3000 Cabot Blvd. W, Suite 220E. Phone: (215) 702-3530.

RentTrace-asset mgmt software to track rental equipment. Handles quotes, reservation, contracts, inventory control, invoicing & accounts receivable. Completely barcode compatible.

Unisys Corp., Unisys Way, Blue Bell, PA, 19424. Phone: (215) 986-4011. Web Site: www.unisys.com. George Gazerwitz, pres.

Cable info business systems. Unisys hardware: A1, A4, A6, A10, A12, A17 & IBM PC compatibles.

United Media Inc., 4771 E. Hunter, Anaheim, CA, 92807. Phone: (714) 777-4510. Fax: (714) 777-2434. E-mail: umi@unitedmediainc.com Web Site: www.unitedmediainc.com.

United Media Inc is a developer & manufacturer which recognizes the ongoing need for high quality, affordable professional video equipment, developer of the On-Line Express non-linear editing system for Windows NT and multicom. The On-Line Express offers uncompromised digital editing, compositing, digital audio editing, titling, 2D & 3D realtime effects & sophisticated media mgmt.

United States Broadcast, 1371 Production Dr., Burlington, KY, 41005. Phone: (859) 282-1802. Fax: (859) 282-1804. E-mail: genmgr@usbroadcast.com

New, used TV, audio equipment, bcst batteries & chargers.

UniVision Inc., 2801 S. Russell, Missoula, MT, 59801. Phone: (406) 721-8876. Fax: (406) 721-0810. E-mail: sales@univision-computers.com Web Site: www.univision-computers.com. Jim Green, pres.

Sell & repair computers. Program software, fiberoptics & wiring.

Utah Scientific Inc., 4750 Wiley Post Way, Suite 150, Salt Lake City, UT, 84116. Phone: (801) 575-8801. Fax: (801) 537-3099. E-mail: sales @utahscientific.com Tom Harmon, CEO; David Burland, COO.

Routing, master control switches & control systems. Industry's best 10 years warranty.

Utility Tower Company, Box 12369, 3200 N.W. 38th, Oklahoma City, OK, 73157. Phone: (405) 946-5551. Fax: (405) 947-8466. E-mail: utctower@aol.com Gloria Nelson, pres; Ron Nelson Jr., VP; Joe James, production mgr.

Tower structures, accessories for bcstg & wireless applications; tower design, engrg analysis, turnkey instals; modifications, maintenance & inspections.

V

VCI-Video Communications Inc., 146 Chestnut St., Springfield, MA, 01103. Phone: (413) 272-7200. Fax: (413) 272-7201. E-mail: sales@vcisolutions.com Web Site: www.vcisolutions.com. W. Lowell Putnam, CEO; Jay Batista, VP sls & VP mktg.

Sales, traffic & automation solutions for the media industry. Improves work flow & increases ROI.

VRI (Video Rentals Inc.), 100 Stonehurst Ct., Northvale, NJ, 07647. Phone: (800) 255-2874. Phone: (201) 750-3206. Fax: (201) 784-2795. E-mail: info@renturi.com Alan Schneider, VP/gen mgr.

Burbank, CA 91502, 200 S Flower St. Phone: (877)

736-8874. Fax: (818) 729-0073.

New York, NY 10019, 423 W. 55th St. Phone: (212) 582-4400.

Full-svc rental facility with a complete inventory of bcst & industrial video equipment.

V-Soft Communications, 721 W. 1st St., Suite A, Cedar Falls, IA, 50613. Phone: (319) 266-8402. Fax: (319) 266-9212. E-mail: info@v-soft.com Web Site: www.v-soft.com. Doug Vernier, pres; Adam Puls, tech support & mktg; Kate Michler, tech consultant.

Bcst engrg software for AM, FM, TV & gen communications. Signal propagation, allocation work, path profiles, custom mapping & more.

VTECH Communications, 9590 S.W. Gemini Dr., Suite 120, Beaverton, OR, 97008-7109. Phone: (503) 643-8981. Fax: (503) 644-9887. Web Site: www.vtechphones.com.

900 mhz cordless analog digital telephones.

Valmont Communications Inc., (Formerly Microflect Co. Inc.). 3575 25th St. S.E., Salem, OR, 97302-1190. Phone: (503) 363-9267. Phone: (800) 547-2151. Fax: (503) 363-4613. E-mail: custoinfo@microflect.com Web Site: www.valmont.com. Doug Kochenderfer, pres.

Towers, microwave passive repeaters, waveguide support systems & tech svcs.

Van Nostrand Radio Engineering Service, 256 Strickland Pasture Rd., Jackson, GA, 30233-3928. Phone: (770) 775-7575. E-mail: vnres@yahoo.com W.L. Van Nostrand,.

Melbourne Beach, FL 32951-0458, Box 510458. Phone: (321) 723-1250. Samuel B. Boor, co-owner.

Frequency measurements up to 26 ghz.

Vantage Lighting Inc., 175 Paul Dr., San Rafael, CA, 94903. Phone: (800) 445-2677. Phone: (415) 507-0402. Fax: (415) 507-0502. E-mail: eight@vanltg.com Web Site: www.vanltg.com. Marc Allsman, pres; Peter Allsman, sec/treas.

Replacement lamps including stage, studio, projection audiovisual, HMI, Xenon, 3D video & laser system. Electronic ballasts for HID lighting.

Veetronix Inc., Box 480, 1311 W. Pacific, Lexington, NE, 68850. Phone: (308) 324-6661. Fax: (308) 324-4985. E-mail: sales@veetronix.com Web Site: www.veetronix.com. Roger Teeters, gen sls mgr.

Keyboard & panel reed switches & keycaps with in-house tooling.

Vega, (A Telex Company). 8601 Cornhusker E. Hwy., Lincoln, NE, 68505-5321. Phone: (402) 467-5324. Fax: (402) 467-3279. E-mail: vega@telex.com Web Site: www.vega-signaling.com. Don Poysa, sls.

Dispatch control consoles, amplifiers & monitoring products.

Veriad, 650 Columbia St., Brea, CA, 92821. Phone: (800) 423-4643. Phone: (714) 990-2700. Fax: (800) 962-0658. Web Site: www.veriad.com. E-mail: info@veriad.com

Videotape & audiotape format labels, CD & DVD labels tape mgmt labels, labeling software, packaging products & production supplies.

Vermeer Manufacturing Co., 1210 Vermeer Rd. E., Pella, IA, 50219. Phone: (515) 628-3141. Phone: (888) 837-6337. Fax: (515) 621-7734. E-mail: salesinfo@vermeermfg.com Web Site: www.vermeer.com. Robert Vermeer, chmn/CEO; Mary Andringa, pres/COO.

Cable plows, trenchers, backhoes, stump cutters & hyraulic boring equipment.

Vertex Communications Corp., 2600 N. Longview St., Kilgore, TX, 75662-6842. Phone: (903) 984-0555. Fax: (903) 984-1826. E-mail: info@vertexcomm.com Gary Kanipe, VP.

Beijing 100005. Vertex Beijing Office, COFCO Plaza, Suite 411, Tower B, N. 8 Jian Guo Men Nei Ave. Phone: (+86-10) 6528-7258. Fax: (+86-10) 6528-7261. E-mail: info@vertex.com.cn.

Burntisland, Fife KY3 9EA. Vertex International Ltd., 37 Kinghorn Rd. Phone: 44-1592-873-956. E-mail: vertes@mbox3.singnet.com.sg.

Duisburg D-47198. Vertex Antennentechnik GmbH, Baumstr. 50. Phone: 49-2066-20960. Fax: 49-2066-209611. E-mail: info@vertexant.com.

Singapore 038987. Vertex Asia (Singapore Representative Office), 21-03 Suntec Tower One, 7 Temasek Blvd. Phone: 65-430-9524. Fax: 65-430-9516. E-mail: vertex@pacific.net.sg.

Santa Clara, CA 95054. Vertex Antenna Systems LLC, 2211 Lawson Ln. Phone: (408) 654-5600. Fax: (408) 654-5613/5614. E-mail: bernard@tiw.com.

Torrance, CA 90505. Vertex Microwave Products Inc., 3111 Fujita St. Phone: (310) 539-6704. Fax: (310) 539-7463. E-mail: info@vertexmpi.com.

West Melbourne, FL 32904. Vertex Florida Office, 255 East Dr, Suite F. Phone: (407) 956-8999. Fax: (407) 956-7999. E-mail: vertexflorida@ibm.net.

Albuquerque, NM 87121. Vertex-New Mexico Inc., 1255 Old Coors Rd. S.W. Phone: (505) 242-5251. Fax: (505)243-5630. E-mail: vnm@abq.com.

State College, PA 16803. Vertex Electronic Products Inc., 2120 Old Gatesburg Rd. Phone: (814) 238-2700. Fax: (814) 238-6589. E-mail: info@vertexepi.com.

Kilgore, TX 75662-6842. Vertex Antenna Products Division, 2600 N. Longview St. Phone: (903) 984-0555. (903) 984-1826. E-mail: vapdmktg@vertexcomm.com.

Longview, TX 75604. Vertex Control Systems Division, 1915 Harrison Rd. Phone: (903) 295-1480. Fax: (903) 295-1479. E-mail: sales@vcsd.com.

Richardson, TX 75081. Vertex Special Projects Division, 101 W. Buckingham Rd. Phone: (972) 643-1869. Fax: (972) 643-1870. E-mail: info@vertexdallas.com.

Antennas, control systems, passive microwave devices, field svcs, satcom net equipment, custom engrg solutions, SSPAs, LNAs, RF components/subsystems.

Vicon Industries Inc., 89 Arkay Dr., Hauppauge, NY, 11788. Phone: (631) 952-2288. Fax: (631) 951-2288. Web Site: www.vicon-cctv.com. Ken Darby, pres/CEO; John Badke, CFO; Peter Horn, VP opns; Yacov Pshtissky, VP tech & dev; Bret McGowan, VP sls & mktg.

Closed circuit TV equipment, systems for the security & surveillance industry.

VidCAD Documentation Programs (VDP Inc.), 755 South Telshor Blvd., Suite D9, Las Cruces, NM, 88011. Phone: (505) 522-0003. Fax: (505) 522-0009. E-mail: sales@vidcad.com Web Site: www.vidcad.com. Dr. Walter Black, CEO.

VidCAD software connects your idea from diagram design & rack planning to installation, maintenance & rebuilds-on time & on budget.

Video Accessory Corp., 2450 Central Ave., Suite G, Boulder, CO, 80301. Phone: (800) 821-0426. Fax: (303) 440-8878. E-mail: vac@vac-brick.com Web Site: www.vac-brick.com. Richard Frey, chief technical officer; Amy Barnes Frey, dir; Frank S. Barnes, pres.

Black Burst generators, video & audio distribution amplifiers, video line isolators, video & audio switches.

Video International Development Corp., 65-15 Brook Ave., Deer Park, NY, 11729. Phone: (631) 243-5414. Fax: (631) 243-4314. E-mail: info@videointernational.com Web Site: www.videointernational.com. Bernard Bressel, pres.

Barsinghausen 30890, Ulmenweg II. Phone: (05105) 81144. S. Freitag, VP.

Digital TV standards converters with four-field/four-line interpolation for bcst & industrial use as well as noise reducers, analog to digital converters, audio delay lines & transcoders.

Videomagnetics Inc., 3970 Clearview Frontage Rd., Colorado Springs, CO, 80911. Phone: (719) 390-1313. Fax: (719) 390-1316. E-mail: vmi@csprings.com Web Site: www.videomagnetics.com. Tony B. Korte, pres; Jane C. Pennie, sls VP.

Full service specialists in betacam camera & recorders. Refurbished video heads & scanners.

Videotek, A Division of Leitch Technology. 243 Shoemaker Rd., Pottstown, PA, 19464-6433. Phone: (800) 800-5719. Fax: (610) 327-9295. E-mail: sales@videotek.com Web Site: www.videotek.com. Philip Steyaert, pres; Richard R. Hollowbush, VP; Bob Landingham, VP sls; Jeff Viola, dev VP.

Manufacturer of test/measurement equipment, video demodulators, routing switchers, color correctors/processors, related equipment for professional video/TV bcst markets.

Videotron Ltee, 300 Viger Ave. E., Montreal, PQ, H28 3W4. Canada. Phone: (514) 281-1711. Fax: (514) 985-8652.

Montreal, PQ H2X 3W4 Canada. LeGroupe Videotron Ltee., 300 ave Viger est.

Cable TV, digital TV, interactive TV and telecommunications.

Videssence L.L.C., 10768 Lower Azusa Rd., El Monte, CA, 91731. Phone: (626) 579-0943. Fax: (626) 579-6803. E-mail: contact@videssence.tv Web Site: www.videssence.tv. Toni Swarens, pres; Lauri Maines, VP.

Energy-efficient floorescent, studio lighting products for TV, film, stage & industrial communications applications.

Viking Cases, 10480 Oak St. N.E., St. Petersburg, FL, 33716. Phone: (800) 237-8560. Fax: (727) 577-2082. E-mail: sales@vikingcases.com Web Site: www.vikingcases.com. Arthur W. Stemler, CEO; Bruce S. Stemler, pres; Reese Autry, VP.

Heavy-duty reusable shipping cases, lightweight carrying cases & EIA rack cases.

Vinten, 709 Executive Blvd., Ste. A, Valley Cottage, NY, 10989-2024. Phone: (845) 268-0100. Fax: (845) 268-0113. Web Site: www.vinten.com. Bob Carr, pres.

Best quality studio camera robotics, Parliamentary camera robotics & virtual sets.

Vinten Inc., 709 Executive Blvd., Valley Cottage, NY, 10989. Phone: (845) 268-0100. Fax: (845) 268-0113. E-mail: mike.denicola@vinten.com Web Site: www.vinten.com. Michael DeNicola, pres.

Toronto M4CC 2B1 Canada, 50 Moberly Ave. Phone: (416) 693-8578. Fax: (416) 693-9489. Sam Duncan, Canadian rgnl sls mgr.

Burbank, CA 91506, G 95 S. Glenwood, Suite B. Phone: (818) 843-5244. Fax: (818) 843-5176. Mark Playdon, west rgn sls mgr.

Sunrise, FL 33351, 10208 N.W. 47th St. Phone: (945) 572-4344. Fax: (945) 572-4565. Joseph Lantowski, southern rgnl sls mgr.

Shamong, NJ 08088, 4 Birch Ct. Phone: (609) 268-2405. Fax: 609-268-3204. Len Donovan, northeast rgnl sls mgr.

Remote control camera systems. Pneumatic studio pedestals, pan & tilt heads, lightweight tripods & heads.

Vision Database Systems, 1095 Jupiter Park Dr., Suite 3, Jupiter, FL, 33458. Phone: (561) 748-0711. Fax: (561) 748-0712. Web Site: www.visiondatabase.com. Emil Bonaduce, pres.

Photo image software.

Visual Sound Inc., 485 Pkwy. S., Broomall, PA, 19008. Phone: (610) 544-8700. Phone: (800) 523-7525. Fax: (610) 544-3385. Web Site: www.visualsound.com. Karen Bogosian, pres.

Halethorpe, MD 21227. Beltsville, 3919 Vero Rd. # J. Phone: (410) 242-4216.

Camp Hill, PA 17011, 490 S. St. John's Church Rd. Phone: (717) 730-6651. (800) 382-1301. Fax: (717) 761-0874.

Audio-video sls, svcs, installation, maintainance of teleconferencing rooms; ENG, production vans, studios, events production & rentals.

W

WIREMAX Ltd., Box 3336, 705 Wamba Ave., Toledo, 43607. Phone: (800) 843-9479. Fax: (419) 531-9503. Al Mocek, pres; Mark Robinson, gen mgr; Tom Ricketts, VP.

Manufacturer of high temp, wire, cable, high voltage wire & cable.

WOIO & WUAB TV, 1717 E. 12th St., Cleveland, OH, 44114. Phone: (216) 771-1943. Fax: (216) 515-7152. Web Site: www.hometeam19.com. Jim Stunek, mgr; Sharon Ohlson, product coord.

Full studio facilities; GVG 300 switcher, CMX 3100B edit, 1-inch, 3/4-inch, Beta formats; Abekas digital F/X; Artstar graphics; remote packages.

Ward-Beck Systems Ltd., Unit 10, 455 Milner Ave., Toronto, ON, M1B 2K4. Canada. Phone: (416) 335-5999. Fax: (416) 335-5202. E-mail: request@ward-beck.com Web Site: ward-beck.com. Eugene L. Johnson, mgng dir; Michael Jordan, sls dir; Doug Bascombe, engrg dir.

Distribution metering, monitoring, conversion of AES & analog audio, video & serial digital bcstg signals. Radio consoles.

Wearguard, 141 Longwater Dr., Norwell, MA, 02061. Fax: (800) 867-7160. Web Site: www.wearguard.com. David Gold, pres.

Offers a comprehensive line of work clothing & identity apparel serving the cable industry.

WeatherBank Inc., 1015 Waterwood Pkwy., Suite J, Edmond, OK, 73034. Phone: (405) 359-0773. Phone: (800) 687-3562. Fax: (405) 341-0115. E-mail: sroot@weatherbank.com Web Site: www.weatherbank.com. Steven A. Root, pres/CEO.

Satellite-delivered weather info, audio forecasting svcs & consulting to all industries.

Wegener Communications Inc., 11350 Technology Cir., Duluth, GA, 30097. Phone: (770) 814-4000. Fax: (770) 623-0698. E-mail: info@wegener.com Web Site: www.wegener.com. Robert Placek, CEO; Ned Mountain, exec VP.

Provider of digital solutions for IP data, video, audio nets, bcst TV, cable TV, radio nets, dist. ed., compel net control for regionalized progmg & coml insertion.

Wescam Inc., 649 N. Service Rd. W., Burlington, ON, L7P 5B9. Canada. Phone: (905) 633-4000. Fax: (905) 633-4100. Web Site: www.wescam.com.

Van Nuys, CA 91406-3823, 7150 Hayvenhurst Ave. Phone: (818) 785-9282. Fax: (818) 785-9787. Chris White.

Featuring Wescam Helicopter film, video, HD system, new & ultimately stable XR for all group applications.

Weschler Instruments, Weshler Instruments Division of Hughes Corporation. 16900 Foltz Pkwy., Cleveland, OH, 44149. Phone: (440) 238-2550. Fax: (440) 238-0660. E-mail: sales@weschler.com Web Site: www.weschler.com. Jerry Lucak, gen sls mgr.

Digital & analog panel meters, RF AM meters, process indicators, digital multimeters, circuit tracers, power analyzers & test equipment.

Westcott, 1447 Summit St., Toledo, OH, 43603. Phone: (419) 243-7311. Fax: (419) 243-8401. Web Site: www.fjwestcott.com. E-mail: info@fjwestcott.com Thomas A. Waltz, pres.

Lightweight, portable & collapsible light control equipment: silks & solids, scrims, Illuminator® reflectors, umbrellas, light modifiers, & Scrim Jim modular light panels.

Westlake Audio, Professional Products Manufacturing Group, 2696 Lavery Ct., Unit 18, Newbury Park, CA, 91320. Phone: (805) 499-3686. Fax: (805) 498-2571. Web Site: www.westakeaudio.com. E-mail: kenc@westlakeaudio.com Glenn Phoenix, pres; Ken Centrofante, mktg.

Audio monitors & accessories.

Westlake Audio, Professional Sales Group, 7265 Santa Monica Blvd., Los Angeles, CA, 90046. Phone: (323) 851-9800. Fax: (323) 851-0182. Web Site: www.westlakeaudio.com. Deborah Rally, gen mgr; David Logan, gen sls mgr; Steve Burdick, VP.

Los Angeles, CA 90048, 8447 Beverly Blvd. Phone: (323) 654-2155. Fax: (323) 655-0478.

Newbury Park, CA 91320, 2696 Lavery Ct, Unit 18. Phone: (805) 499-3686. Fax: (805) 498-2571. Glenn Phoenix, pres.

Professional audio equipment repairs, professional recording equipment sales, rentals, studio design. Westlake can provide all your Pro Audio needs, from Pro Audio equipment sls to full tracking & mixing.

Wheatstone Corp., 600 Industrial Dr., New Bern, NC, 28562. Phone: (252) 638-7000. Fax: (252) 635-4857. E-mail: sales@wheatstone.com Web Site: www.wheatstone.com. Gary C. Snow, pres; Andrew Calvanese, VP; Brad Harrison, sls dir.

Manufacturer of analog, digital bcst audio mixing consoles, processing equipment, radio & TV products since 1976.

Wheelit Inc., 440 Arco Dr, Toledo, OH, 43635-2800. Phone: (419) 531-4900. Fax: (419) 531-6415. E-mail: wheelit@solarstop.net Web Site: www.wheelitinc.com. John M. Skilliter, pres.

Video production carts (folding & non-folding), video display stands & computer carts.

Whirlwind, 99 Ling Rd., Rochester, NY, 14612. Phone: (585) 663-8820. Fax: (585) 865-8930. Fax: (888) 733-4396. E-mail: sales@whirlwindusa.com Web Site: www.whirlwindusa.com. Michael Laiacona, pres.

Mix-6 audio mixers, presspower 2 active pressbox, active splitters; P-12 & power amplifiers, MD-1 MIC/line driver.

Wicks Broadcast Solutions, Box 3078, 508 S. 7th St., Opelika, AL, 36803. Phone: (334) 749-5641. Fax: (334) 749-5666. E-mail: sales@datacount.com Web Site: www.datacount.com. Bill Price, gen mgr.

Datacount produces a bcst traf system for radio, called Darts. In addition to single user, Darts is available for multi-user as well as multi-stn groups. Datacount is one of the largest supplier of bcstg software in the world. PC & window based system encompassing all aspects of logging, traf, co-op, billing, accts receivable & sls mgmt.

Wicks Broadcast Solutions L.L.C., Box 67, 1950 Winchester Ave., Reedsport, OR, 97467. Phone: (800) 547-3930. Phone: (541) 271-3681. Fax: (541) 271-5721. Web Site: www.wicksbroadcastsolutions.com. Bob Richardson, pres; Jeffrey Kimmel, sls dir; Bob Leighton, dev dir.

Windows-based software for radio operations, traf, billing, sls, analysis functions. Digital Universe system for uncompressed audio in live assist/automation applications.

Wil-Can Electronics Ltd., 8560 Torbram Rd., Unit #35, Brampton, ON, L6T 5C9. Canada. Phone: (888) 596-2020. Fax: (888) 866-7775. E-mail: wilcan@lightningtvss.com Web Site: www.powersurges.com. William J. Black, pres; Gregory J. Black, gen mgr.

Buffalo, NY 14216, 2316 Delaware Ave, Suite 285. Phone: (888) 596-2020. Fax: (888) 866-7775. Gregory J. Black, gen mgr/sec treas.

Designers, manufacturers & consultants. Lightning & high energy transient control including protection for telephone, signal & data lines.

WILL-BURT Co., Box 900, Orrville, OH, 44667. Phone: (330) 682-7015. Fax: (330) 684-1190. Web Site: www.willburt.com. E-mail: contact_us@willburt.com Jeff Evans, CEO; Steven Pinkley, gen sls mgr.

Telescoping mast used to position antennas, lights & cameras to heights of 20 to 134 ft.

Wiltronix Inc., Box 364, 16850 Oakmont Ave., Washington Grove, MD, 20880. Phone: (301) 258-7676. Fax: (301) 963-8624. E-mail: equipsales@wiltronix.com Web Site: www.wiltronix.com. Dwight Wilcox, pres; Ellen Packard, gen sls mgr.

Manufacturers rep for digital HD, SD, video, audio, signal processing, transmission for bcst production & govt video facilities.

Winsted Corp., 10901 Hampshire Ave. S., Minneapolis, MN, 55438. Phone: (952) 944-9050. Fax: (952) 944-1546. E-mail: info@winsted.com Web Site: www.winsted.com. G.R. Hoska, chmn; Randy Smith, pres; Stephen Hoska, CEO.

Editing & production consoles, space saving tape & data storage systems. Multimedia & Lan/Wan server workstations.

Wireless Accessories Group, 1840 County Line Rd., Suite 301, Huntingdon Valley, PA, 19006. Phone: (888) 233-0202. Phone: (215) 322-4600. Fax: (215) 322-4606. Web Site: www.wirexgroupl.com.

Two-way radio equipment.

WireReady NSI, 56 Hudson St., Northboro, MA, 01532. Phone: (800) 833-4459. Phone: (508) 393-0200. Fax: (508) 393-0255. E-mail: sales@wireready.com Web Site: www.wireready.com. David Gerstmann, pres.

NewsReady 32, CartReady, ControlReady, NewsReady, StormReady & SalesReady software. Cart replacement, satellite, music on HD, newsrooms & sls automation. All windows/pc.

Wireworks Corp., 380 Hillside Ave., Hillside, NJ, 07205. Phone: (908) 686-7400. Phone: (800) 642-9473. Fax: (908) 686-0483. E-mail: sales@wireworks.com Web Site: www.wireworks.com. Gerald J. Krulewicz, pres; Larry J. Williams, controller.

Audio, video, & audio/video combination cabling assemblies for bcst market; cable testers, transformer isolated mic splitters; perfect custom panels.

Wohler Technologies Inc., 31055 Huntwood Ave., Hayward, CA, 94544. Phone: (510) 870-0810. Fax: (510) 870-0811. Web Site: www.wohler.com. Will C. Wohler, pres.

Wolf Coach Inc., 7 B St., Auburn Industrial Park, Auburn, MA, 01501. Phone: (508) 791-1950. Fax: (508) 799-2384. E-mail: sales@wolfcoach.com Web Site: www.wolfcoach.com. Richard Wolf, VP; Mark A. Leonard, natl sls reps; Thomas P. Jennings, natl sls reps; Emeric Feldmar, mgr.

Salt Lake City, UT 84115, 2451 South 600 W, 200. Phone: (801) 977-9533. Rex A. Reed, systems mgr/engr, Wesley K. Gordon, chief engr.

News vans, satellite vehicles, production trailers, vehicle based microwave, satellite uplink, digital SNG, audio/video systems & turnkey systems.

World Tower Co. Inc., Box 508, 1213 Compressor Dr., Mayfield, KY, 42066. Phone: (270) 247-3642. Fax: (270) 247-0909. E-mail: worldtow@ldd.net Web Site: www.worldtower.com. Doug Walker, pres.

Manufactures & erects bcst & CATV towers, microwave & cellular.

World Video Sales Co., Box 331, Boyertown, PA, 19512. Phone: (610) 754-6800. Fax: (610) 754-9766. E-mail: sales@mivs.com John A. Taylor, pres.

Manufacturers of: video timers/titlers, screen splitters, pattern generators, routing systems, distribution amplifiers & other special purpose video equipments.

Xintekvideo Inc., 56 W. Broad St., Stamford, CT, 06902. Phone: (203) 348-9229. Fax: (203) 348-9266. E-mail: john@xintekvideo.com Web Site: www.xintekvideo.com. John Rossi, pres.

Video processing equipment, including transcoders, color correctors, image enhancers, noise reducers, co-channel filters, ghost removers, impulse noise eliminators.

Yamaha Corp. of America, 6600 Orangethorpe Ave., Buena Park, CA, 90620. Phone: (714) 522-9109. Fax: (714) 739-2680. Web Site: www.yamaha.com. Tom Sumner, VP/gen mgr.

Portable keyboards, synthesizers & drums. Manufactures a complete line of professional audio products targeted to the project studio, commercial studio, postproduction, bcst & sound reinforcement markets.

Yanchar Design & Consulting Group, 26741 Portola Pkwy., Suite 1E, Foothill Ranch, CA, 92610-1713. Phone: (949) 770-6601. Fax: (949) 770-6575. E-mail: info@yanchardesign.com Web Site: www.yanchardesign.com. Carl J. Yanchar, pres.

Acoustical design, facility design, consultation, systems design, studio construction & instal.

Z

Zack Electronics Inc., 1070 Hamilton Rd., Duarte, CA, 91010. Phone: (626) 303-0655. Fax: (626) 303-8694. E-mail: jlomas@zackinc.com Web Site: www.zackelectronics.com. Judi Lomas, mgr; Dennis Awad, pres.

Cable, connectors, & comprehensive core products for the bcst industry, custom audio/video/data cable assemblies. Featuring: Belden, Neutrik, Switchcraft, Pomona & avbcable.com.

Zenith Electronics Corp., 2000 Millbrook Dr., Lincolnshire, IL, 60069. Phone: (847) 391-7000. Fax: (847) 941-9200. Web Site: www.zenith.com. T.J. Lee, pres/CEO.

Full line of CATV converters; MMDS systems; cable & pay TV systems for PAL & SECAM international markets; PC-based system controllers; accessories.

Ziehl Electronic Service, 8611 Dale Rd., Gasport, NY, 14067. Phone: (716) 772-7800. Fax: (716) 772-7985. Richard F. Ziehl, owner.

Frequency measurement svc.

Zomax Inc., 800 Corporate Way, Fremont, CA, 94539. Phone: (510) 657-8425. Fax: (510) 657-8427. Web Site: www.zomax.com.

CD manufacturer specializing in printing fulfillment svcs, & custom work in CD entertainment & CD-ROM.

Equipment Manufacturers and Technical Services Subject Index

Audio Accessories

AKG Acoustics, U.S.
Allsop Inc.
Amek U.S.A.
Aphex Systems Ltd.
Audico Labels
Audio Implements/GKC
Black Audio
Bogen Communications Inc.
Electronics Corp.
HAVE Inc.
Henry Engineering
The J-Lab Co.
JOA Cartridge Service
Jensen Transformers Inc.
Peavey Electronics
Polyline Corp.
Professional Sound Corp.
Radio Design Labs. (RDL)
Seger Electronics
Sescom Inc.
Sprague Magnetics Inc.
Star Case Manufacturing Co. Inc.
TAI Audio
Trimm Inc.
Video Accessory Corp.
Westlake Audio, Professional Products Manufacturing Group
Wireworks Corp.
Zack Electronics Inc.

Audio Amps, AGC & Limiters

Professional Sound Corp.
Renkus-Heinz Inc.
Samson Technologies Corp.
Sescom Inc.
Spectra Sonics

Audio Cartridge Library Labels

Veriad

Audio Cartridges

Spectra Sonics
Stanton Group

Audio Compressors

Alesis
Circuit Research Labs Inc. (CRL Systems, Inc.)
Furman Sound Inc.
Sascom Marketing Group
Sescom Inc.
Spectra Sonics
TC Electronic

Audio Consoles

Arrakis Systems Inc.
Autogram Corp.
Harris Corp., Broadcast Communications Division
Harrison Consoles
Logitek
North American Cable Equipment Inc.
Ward-Beck Systems Ltd.
Wheatstone Corp.

Audio Equipment

AMS Neve Inc.
AVI Systems
Access Intelligence LLC
Audio Implements/GKC
Audio Precision Inc.
BBE Sound Inc.
Black Audio
Bogen Communications Inc.
Broadcast Store Inc.
Burlington A/V Recording Media Inc.
Comprehensive Video Group
Comrex Corp.
Countryman Associates Inc.
Crouse-Kimzey Co.
DBX Professional Products
Film/Video Equipment Service Co. Inc.
FitzCo. Inc.
Fostex America
Full Compass Systems Ltd.
Galaxy Audio Inc.
Geneva Aviation Inc.
Group One Ltd.
Harris Corporation Co.
Industrial Equipment Representatives (IER)
Jensen Transformers Inc.
Lectrosonics Inc.
Memorex Products Inc.
Nady Systems Inc.
National Video Services Inc.
Opamp Labs Inc.
Peavey Electronics
Photomart Cine-Video Inc.
Professional Sound Corp.
Protech Audio Corp.
QSC Audio Products Inc.
Ram Broadcast Systems
Rangertone Research Inc.
Samson Technologies Corp.
Sanyo Fisher Co.
Sascom Marketing Group
Sennheiser Electronic Corp.
Spectra Sonics
Systems Wireless Ltd.
TAI Audio
TEAC America Inc.
Tannoy North America Inc.
Thomson, Inc, (RCA/GE)
Visual Sound Inc.
Whirlwind

Audio Jackfields, Pre-Wired

Audio Accessories Inc.
Farrtronics Ltd.
Fostex America
Furman Sound Inc.
Glentronix
Milestek Corp.
Penny & Giles Inc.
Seger Electronics
Switchcraft Inc.

Audio Limiters

Amek U.S.A.
Circuit Research Labs Inc. (CRL Systems, Inc.)
Harman International Industries Inc.
Spectra Sonics
Symetrix Inc.

Audio Mixers and Recorders

Alesis
Allen & Heath USA
Amek U.S.A.
Audio-Technica U.S., Inc.
Audio Technologies Inc. (ATI)
Cooper Sound Systems Inc.
Electronics Corp.
Graham-Patten Systems Inc.
Harrison Consoles
The Image Group Post, LLC.
Location Sound Corp.
Martinsound Inc.
Micro Technology Unlimited
Nady Systems Inc.
Otari USA Sales Inc.
Penny & Giles Inc.
Professional Sound Corp.
Professional Sound Services Inc.
Protech Audio Corp.
Radio Design Labs. (RDL)
Shure Inc.
Superscope Technologies Professionals
United States Broadcast
Yamaha Corp. of America

Audio Monitoring Systems

AKG Acoustics, U.S.
B&B Systems
Dorrough Electronics
Furman Sound Inc.
IRIS Technologies Inc.
Martinsound Inc.
McCurdy Radio Ltd.
Renkus-Heinz Inc.
Westlake Audio, Professional Products Manufacturing Group

Audio Noise Reduction Systems

Allen Avionics, Inc.
Digidesign
Dolby Laboratories Inc.
Image Video
Noise Control Corp.
Symetrix Inc.

Audio Processors

Audioarts Engineering
BBE Sound Inc.
Delta Electronics Inc.
Inovonics Inc.
Linear Acoustic Inc.
Modulation Sciences Inc.
Nady Systems Inc.
Peavey Electronics
Protech Audio Corp.
Samson Technologies Corp.
Sascom Marketing Group
Symetrix Inc.

Audio Replacement Heads

International Electro-Magnetics
Northern Magnetics Inc.
Sprague Magnetics Inc.

Audio Routing Switches

Burk Technology
Burst Electronics Inc.
Electronology, Inc.
Harrison Consoles
Logitek
NVISION Products
NTV International Corporation
Omicron Video
Radio Design Labs. (RDL)
Richmond Sound Design Ltd.
Sierra Automated Systems & Engineering Corp.
Sigma Electronics Inc.

Audio Signal Processing Systems

Alesis
Aphex Systems Ltd.
BBE Sound Inc.
Broadcasters General Store Inc.
Circuit Research Labs Inc. (CRL Systems, Inc.)
Communications Specialties Inc.
Dolby Laboratories Inc.
G Prime Ltd.
Graham-Patten Systems Inc.
Group One Ltd.
International Datacasting Corp.
Lectrosonics Inc.
Pixel Instruments Corp.
Sabine Inc.
Shure Inc.
Symetrix Inc.
TC Electronic

Audio Systems and Components

Audio Implements/GKC
Commercial Electronics Ltd.
Gefen Inc.
Parsons Audio
Protech Audio Corp.
VidCAD Documentation Programs (VDP Inc.)
Visual Sound Inc.

Audio Test Tapes, Gauges & Equipment

Audio Precision Inc.
Leader Instruments Corp.
Tentel

Audio Transmission Equipment

Artel Video Systems
Audio Processing Technology Ltd./APT
Fiber Options
Intelligent Media Technology
International Datacasting Corp.
Leaming Industries

MUSICAM U.S.A.
Neutrik U.S.A. Inc.
TFT Inc.

Audio/Video Cartridges

CoarcVideo
Memorex Products Inc.

Audiotape

BMG
Burlington A/V Recording Media Inc.
CoarcVideo
Maxell Corp. of America
Memorex Products Inc.
Moviola
National Audio Co. Inc.
National Video Services Inc.

Audiotape Cartridge Machines

A.C.C. Electronix, Inc.

Automated Newsroom Systems

Avid Broadcast
CEA-Computer Engineering Associates
Comprompter Inc.
Computer Concepts Corp.
Dalet Digital Media Systems
FloriCal Systems Inc.
Harris Automation Solutions
Reuters American Inc.
Scott Studios Corp.
WireReady NSI

Automated Radio

Dalet Digital Media Systems
ENCO Systems Inc.
Encoda Systems Inc.
Marketron International
Radio Computing Services (RCS)
Register Data Systems
Scott Studios Corp.
Spotcat Software
Wicks Broadcast Solutions L.L.C.
WireReady NSI

Automated Tape Winders

Otari USA Sales Inc.

Automated Telephone & Voice Mail

Concerto Software
Microlog Corp.
TALX Corp.

Automatic Cassette Loaders

Otari USA Sales Inc.

Automatic Transmission Systems

Encoda Systems Inc.
Harris Automation Solutions

Automation Systems

ADTEC Inc.
Arrakis Systems Inc.
Broadcast Electronics Inc.
CEA-Computer Engineering Associates
Computer Concepts Corp.
ENCO Systems Inc.
Encoda Systems Inc.
Harris Automation Solutions
Harris Corp., Broadcast Communications Division
Leightronix Inc.
MATCO Inc.
Marketron International
Media Computing Inc.
Orban
Radio Computing Services (RCS)
Register Data Systems
Scott Studios Corp.
Spotcat Software
Sundance Digital Inc.
Thomson Multi Media
Time Logic Inc.
Wicks Broadcast Solutions L.L.C.
WireReady NSI

Automation, Switching and Control

ADTEC Inc.
Digitel Corp.
FloriCal Systems Inc.
Harris Automation Solutions
Leightronix Inc.
MATCO Inc.
Register Data Systems
Richmond Sound Design Ltd.
Sierra Automated Systems & Engineering Corp.
Sundance Digital Inc.
Texscan MSI
Thomson Multi Media

Automation, TV Station

Avid Broadcast
Dalet Digital Media Systems
Encoda Systems Inc.
FloriCal Systems Inc.
Harris Automation Solutions
Leightronix Inc.
MATCO Inc.
Marketron International
Media Computing Inc.
Peter Storer & Associates Inc.
Sundance Digital Inc.
Thomson Multi Media
Time Logic Inc.
VCI-Video Communications Inc.
Videotron Ltee
WireReady NSI

Batteries and Accessories

ADCOUR
Acme Electric Corp., Aerospace Division
Alexander Technology
Alpha Technologies Inc.
Anton/Bauer Inc.
Arri Canada Ltd.
Battery Pros Inc.
Burlington A/V Recording Media Inc.
Cine 60 Inc.
Frezzolini Electronics Inc.
Alan Gordon Enterprises Inc.
Marathon Norco Aerospace, Inc.
North American Cable Equipment Inc.
Performance Power Technologies
The Tiffen Company

Blimps

Blimpy Floating Signs/Bend-A-Lite

Blowers and Fans

APW Mayville/Stantron
Allied Electronics Inc.
Bud Industries Inc.
CMP Enclosures Inc.
Condux International
EMCOR Enclosures
Seger Electronics

Booms and Cameras

Alan Gordon Enterprises Inc.

Boosters, TV

Axcera

Broadcast & Program Logging Recorders

Dictaphone Corp.
Wicks Broadcast Solutions

Broadcast Audio Products

A.C.C. Electronix, Inc.
ADC
AKG Acoustics, U.S.
AMS Neve Inc.
AheadTek
Aphex Systems Ltd.
Audio Implements/GKC
Audio Processing Technology Ltd./APT
Autogram Corp.
Belden Electronics Divison
Benchmark Media Systems Inc.
Comrex Corp.
DBX Professional Products
Dolby Laboratories Inc.
Ensemble Designs
FM Systems Inc.
Farrtronics Ltd.
Fiber Options
Full Compass Systems Ltd.
Furman Sound Inc.
Galaxy Audio Inc.
Gefen Inc.
Harrison Consoles
JOA Cartridge Service
Lectrosonics Inc.
Logitek
MUSICAM U.S.A.
Orban
Parsons Audio
Polyline Corp.
Sennheiser Electronic Corp.
TAI Audio
teletech.ca
Telex Communications Inc.
360 Systems
Wegener Communications Inc.
Wheatstone Corp.
Wicks Broadcast Solutions L.L.C.

Broadcast Equipment

Advanced Media Technologies, Inc.
Allison Payments Systems L.L.C.
Analog Digital International Inc.
Atlantic Sound Systems
Audio Implements/GKC
Autodesk
BEXT Inc.
Bexel, Inc.
Broadcast Store Inc.
Broadcast Supply Worldwide
C-COR
Crouse-Kimzey Co.
Digitel Corp.
DMT USA, Inc.
Energy-Onix Broadcast Equipment Co. Inc.
Ensemble Designs
FitzCo. Inc.
Harris Corporation Co.
Ikegami Electronics (U.S.A.) Inc.
Industrial Equipment Representatives (IER)
Inscriber Technology Corporation
JVC Professional Products Company
Kathrein Inc., Scala Division
Kay Industries Inc.
Konica Minolta Corp.
The Laumic Rental Co.
MRPP Inc.
MUSICAM U.S.A.
Marshall Electronics
Merlin Engineering Works Inc.
Miranda Technologies Inc.
Modulation Sciences Inc.
NVISION Products
National Video Services Inc.
Nautel Ltd.
Allen Osborne Associates Inc.
POA/Paul Olivier & Associates Ltd.
Panasonic Broadcast & Television Systems Co.
Phasetek Inc.
RF Specialties Group
Schafer International
Scientific Atlanta Canada Inc. Nexus Division
Specialized Communications
J.A. Taylor & Associates
Technet Systems Group
Thales Broadcast & Multimedia Inc.
Transcom Corp.
Westcott

Broadcast Radio Equipment

Arrakis Systems Inc.
BEXT Inc.
Belar Electronics Laboratory Inc.
Bradley Broadcast Sales
Broadcast Electronics Inc.
Broadcasters General Store Inc.
Comrex Corp.
DBX Professional Products
Dalet Digital Media Systems
DRS Broadcast Technology
Highway Information Systems, Inc.
Industrial Equipment Representatives (IER)
Inovonics Inc.
Jampro Antennas/RF Systems Inc.
Nautel Ltd.
Radiogear Inc.
Schafer International
Shively Labs
Swintek Enterprises Inc.
TFT Inc.
Tenco Tower Co.
Wicks Broadcast Solutions L.L.C.

Broadcast RF Equipment

Allen Avionics, Inc.
Antenna Concepts Inc.
Bauer Transmitters
Belar Electronics Laboratory Inc.
Bird Electronic Corp.
Broadcast Microwave Services Inc.
CED
Comex Worldwide Corp.
Communication & Power Industries
Communication & Power Industries, EIMAC Division
Crown Broadcast IREC
Dielectric Communications
DMT USA, Inc.
DRS Broadcast Technology
e2v technologies Inc.
Industrial Equipment Representatives (IER)
JSB Service Co.
Jampro Antennas/RF Systems Inc.
Kathrein Inc., Scala Division
Kintronic Labs Inc.
LARCAN USA
Micro Communications Inc.
Nautel Ltd.
Radiogear Inc.
Shively Labs
Swintek Enterprises Inc.
TFT Inc.
Teleplex Inc.
Televideo San Diego
V-Soft Communications
WILL-BURT Co.

Broadcast Studio Construction, Prefab

Devlin Design Group Inc.
Industrial Acoustics Co., Inc.
Northeastern Communications Concepts Inc.

Broadcast Studio Equipment

A.C.C. Electronix, Inc.
ADC
AKG Acoustics, U.S.
Audioarts Engineering
Barco Visual Solutions, LLC
Bradley Broadcast Sales
Broadcast Electronic Services
Broadcasters General Store Inc.
Canon U.S.A. Inc.
Digidesign
ENCO Systems Inc.
Furman Sound Inc.
Harris Corp., Broadcast Division
Inscriber Technology Corporation
MUSICAM U.S.A.
Marshall Electronics
/ole-Richardson Co.
O'Connor Professional Camera Support Systems
Parsons Audio
Radiogear Inc.
Richardson Electronics
Rosco Laboratories Inc.
Telecast Fiber Systems Inc.
Thales Broadcast & Multimedia Inc.
Theatre Service & Supply Corp.
Transcom Corp.
United States Broadcast
Videotron Ltee
Vinten

Broadcast TV Equipment

AKG Acoustics, U.S.
AVS Graphics & Media Inc.
AZCAR U.S.A. Inc.
Accom Inc.
AccuWeather Inc.
All Mobile Video Inc.
Alpine Optics Inc.
BEXT Inc.
Belar Electronics Laboratory Inc.
Bexel Corp.
Bexel, Inc.
Bitcentral Inc.
Broadcast Electronic Services
Broadcast International Group
Canon U.S.A. Inc.
Comtech Antenna Systems Inc.
Dalet Digital Media Systems
DeSisti Lighting
DMT USA, Inc.
Ensemble Designs
Fiber Options
Freeland Products Inc.
Full Compass Systems Ltd.
Hitachi Denshi America, Ltd.
Jampro Antennas/RF Systems Inc.
LARCAN
Maze Corporation
/ole-Richardson Co.
Allen Osborne Associates Inc.
POA/Paul Olivier & Associates Ltd.
Schafer International
Snell & Wilcox Inc.
Specialized Communications
Standard Communications Corp.
J.A. Taylor & Associates
Telecast Fiber Systems Inc.
Thales Broadcast & Multimedia Inc.
Toshiba America Consumer Products
VCI-Video Communications Inc.
Wegener Communications Inc.
Wescam Inc.
WILL-BURT Co.
Wiltronix Inc.

Broadcast Video Products

ADC
Accom Inc.
Alpine Optics Inc.
Analog Digital International Inc.
Battery Pros Inc.
Belden Electronics Divison
Bitcentral Inc.
Broadcast Electronic Services
Canon U.S.A. Inc.
Fast Forward Video
Glentronix
LINK Electronics Inc.
MATCO Inc.
Marshall Electronics
Matrox Video Products Grp
Maxell Corp. of America
Roscor Corp.
Snell & Wilcox Inc.
Telemetrics Inc.
teletech.ca
360 Systems
The Tiffen Company
Videomagnetics Inc.
Wegener Communications Inc.

Bulktape DeGausser

Audiolab Electronics Inc.
Data Security Inc.
Garner Products
Glentronix
Paulmar Industries Inc.
Sprague Magnetics Inc.

Bulktape, Audio Cassette

CoarcVideo

Cabinets, Racks, Panels

A & S Case Co. Inc.
ADTEC Inc.
AMCO Engineering Co.
APW Mayville/Stantron
Allied Electronics Inc.
Atlas Sound
Benner-Nawman Inc.
Bud Industries Inc.
CMP Enclosures Inc.
Calzone Case Co.
EMCOR Enclosures
Gepco International Inc.
Hardigg Cases
High Tech Industries
Murphy Studio Furniture
National Video Services Inc.
Neumade Products Corp.
Newark Electronics
North American Cable Equipment Inc.
Northern Technologies Inc.
Parsons Manufacturing Corp.
Penn Elcom Inc.
Performance Power Technologies
Ram Broadcast Systems
Seger Electronics
Seton Identification Products
Stahl, A Scott & Fetzer Co.
Star Case Manufacturing Co. Inc.
Storeel Corp.
Thermodyne Cases
Winsted Corp.
Zack Electronics Inc.

Cable and Accessories

Allied Electronics Inc.
Belden Electronics Divison
Canare Corp.
Clark Wire & Cable Co. Inc.
Communications Specialties Inc.
Comprehensive Video Group
Corning Cable Systems
Cortland Cable Co. Inc.
EMCOR Enclosures
Gefen Inc.
General Cable
Gepco International Inc.
HAVE Inc.
Insulated Wire Inc. Microwave Products Division
Jensen Tools
LEMO USA Inc.
M/A-COM
Marshall Electronics
Milestek Corp.
Mohawk
Nemal Electronics International Inc.
Newark Electronics
North American Cable Equipment Inc.
Phillystran Inc.
Prysmian Communications Cables and Systems
Radio Frequency Systems
Servoreeler Systems
Stage Equipment & Lighting Inc.
James Thomas Engineering
Times Fiber Communications Inc.
Trilogy Communications Inc.
Trompeter Semflex
WIREMAX Ltd.
Whirlwind
Wireworks Corp.
Zack Electronics Inc.

Cable Security Systems

Electroline Equipment Inc.
Emerson Network Power-Viewsonics
Marshall Electronics
North American Cable Equipment Inc.

Cable Termination Equipment, A/V

Clark Wire & Cable Co. Inc.
North American Cable Equipment Inc.
Videotron Ltee
Zack Electronics Inc.

Calibrators, TV Cameras/Monitors

Commercial Radio Monitoring Co.
Dage-MTI Inc.

Camera Mounts

Camera Pan/Tilt Heads

Camera Tubes

Cameras, Projectors & Accessories

Capacitors

Captioning Equipment

Cartridge Automatic Tape

Cartridge Storage Racks

Cases

Cassette Duplication, Audio/Video

Cassette, Audiotape Equip. & Access.

Cassette, Videotape Equip. & Access.

Cassettes

CATV Equipment and Supplies

CATV Hybrid Modules

CATV Power Supplies

CD Players

Cellular Mobile Telephones

Character Generators

Chroma Keyers

Chronometers, Clocks

Solid State Logic Inc.
Soundcraft U.S.A.
Vega
Wheatstone Corp.
Winsted Corp.

Consoles, On-Air

AMCO Engineering Co.
Audioarts Engineering
Autogram Corp.
Broadcast Equipment Surplus Inc.
Logitek
North American Cable Equipment Inc.
Radio Systems Inc.
Soundcraft U.S.A.
Wheatstone Corp.

Construction Services

CCI
Cygnal Technology
180 Connect
Quality Tower Erectors Inc.
T.T. Technologies Inc.
Teletech Inc.
Tenco Tower Co.

Control Systems

Identix
Image Logic Corp.
Knox Video
MATCO Inc.
NSI
Richmond Sound Design Ltd.
Sacramento Theatrical Lighting (STL)
Vega
Vertex Communications Corp.
Vicon Industries Inc.

Converters and Switchers (CATV)

CONTEC Corp.
Gefen Inc.
Sascom Marketing Group
Utah Scientific Inc.

Converters, Standards

CATV Services Inc.
Ikegami Electronics (U.S.A.) Inc.
The Image Group Post, LLC.
Kay Industries Inc.
Merlin Engineering Works Inc.
Philip-Cooke Co.
Snell & Wilcox Inc.
Video International Development Corp.

Converters, TV

Kay Industries Inc.
Pioneer New Media Technologies Inc.
Scientific Atlanta
Wiltronix Inc.

Copy Stands

Murphy Studio Furniture

Costumes and Properties

Costume Armour Inc./Christo Vac

Crystal Units

ICM (International Crystal Mfg.)

Cue Systems

COMTEK Inc.
Greenberg Teleprompting
Studio Technologies Inc.

Custom Consoles

APW Mayville/Stantron
CMP Enclosures Inc.
Calzone Case Co.
High Tech Industries
Northeastern Communications Concepts Inc.
Studio Technology
Winsted Corp.

Custom Studios

Channel One Lighting Systems Inc.
High Tech Industries
Industrial Acoustics Co., Inc.
Murphy Studio Furniture
POA/Paul Olivier & Associates Ltd.
Studio Technology
Yanchar Design & Consulting Group

Cyclorama Tracks

Peter Albrecht Company Inc.
Automatic Devices Company
Channel One Lighting Systems Inc.
Olesen
Theatre Service & Supply Corp.
Theatrical Services Inc.
Uni-Set Corp.

Data Communications Systems

Canon U.S.A. Inc.
Dynamic Solutions 2000
EEG Enterprises Inc.
Entertainment Communications Network (ECN)
FM Systems Inc.
Great Lakes Data Systems, Inc.
MCG Surge Protection
MODCOMP Inc.
Microspace Communications Corp.
Milestone Technologies Inc.
Modulation Sciences Inc.
Norpak Corporation
180 Connect
Spacenet Services Inc.
Trident Media Group/ Spector Entertainment Group Inc.

Data Transmission Equipment

Broadcast Video Systems Corp.
Canon U.S.A. Inc.
Fiber Options
IPITEK
Inovonics Inc.
Intelligent Media Technology
International Datacasting Corp.
MCL Inc.
Moseley Associates Inc.
Norpak Corporation
Nortel Networks
Radyne ComStream Corp.
Spacenet Services Inc.
TELLABS
Wegener Communications Inc.

Decals

Seton Identification Products

Decoders

Audio Processing Technology Ltd./APT
Belar Electronics Laboratory Inc.
EEG Enterprises Inc.
Faroudja Laboratories
Macrovision Corp.
Mega Hertz
Pixel Instruments Corp.
Sarnoff Corp.
Snell & Wilcox Inc.
Television Equipment Assoc. Inc./Matthey
Trident Media Group/ Spector Entertainment Group Inc.
Vega
Video International Development Corp.

Dehydrators and Accessories

Dielectric Communications
Radio Frequency Systems
Shively Labs

Demodulators and Modulators

CADCO Systems Inc.
Cable Serv Inc.
FM Systems Inc.
General Atomics
L-3 Communications Telemetry East
NUCOMM Inc.
North American Cable Equipment Inc.
Videotek
Videotron Ltee

Descramblers, Pay TV

CONTEC Corp.
Macrovision Corp.
Motorola Broadband Communications Sector
North American Cable Equipment Inc.

Design Services, Broadcast

Cygnal Technology
DeSisti Lighting
Meridian Design Associates, Architects
Murphy Studio Furniture
Northeastern Communications Concepts Inc.
Professional Communications Systems
Quality Tower Erectors Inc.
Rees Associates Inc.
Studio Technology
VidCAD Documentation Programs (VDP Inc.)
Yanchar Design & Consulting Group

Designers, Production Facilities

CBT Systems
Four Seasons Solar Products Corp.
High Tech Industries
Meridian Design Associates, Architects
Northeastern Communications Concepts Inc.
POA/Paul Olivier & Associates Ltd.
Rees Associates Inc.
Yanchar Design & Consulting Group

Digital Audio Processing Equipment

Aphex Systems Ltd.
Audio Processing Technology Ltd./APT
Benchmark Media Systems Inc.
Broadcast Supply Worldwide
COASTCOM
Circuit Research Labs Inc. (CRL Systems, Inc.)
Dialogic Communications Corp.
Eventide Inc.
G Prime Ltd.
Graham-Patten Systems Inc.
Linear Acoustic Inc.
Logitek
Micro Technology Unlimited
NVISION Products
Nortel Networks
Orban
Penny & Giles Inc.
Radio Computing Services (RCS)
Roland Corp. U.S.
Sabine Inc.
Sascom Marketing Group
Symetrix Inc.
TC Electronic
Ward-Beck Systems Ltd.

Digital Audio Recorders

Alesis
Electronology, Inc.
Fostex America
Micro Technology Unlimited
Roland Corp. U.S.
Scott Studios Corp.
Stancil Corp.
Superscope Technologies Professionals
360 Systems
Yamaha Corp. of America

Digital Audio Recording & Editing Station

AMS Neve Inc.
Avid Broadcast
Computer Concepts Corp.
Digidesign
ENCO Systems Inc.
Electronology, Inc.
Micro Technology Unlimited
Orban
Parsons Audio
Penny & Giles Inc.
Radio Computing Services (RCS)
Scott Studios Corp.
360 Systems
Visual Sound Inc.
WireReady NSI

Digital Broadcast Equipment

ADTEC Inc.
AMS Neve Inc.
ATCI/Antenna Technology Communications Inc.
Accom Inc.
Acterna
Arrakis Systems Inc.
Audio Technologies Inc. (ATI)
Axcera
BEXT Inc.
Belar Electronics Laboratory Inc.
Bexel Corp.
Bexel, Inc.
Bitcentral Inc.
Broadcast Electronics Inc.
Broadcast Store Inc.
Broadcast Supply Worldwide
COASTCOM
CORPLEX Inc.
Computer Concepts Corp.
Crouse-Kimzey Co.
Dalet Digital Media Systems
DiGi Co. Ltd.
R.L. Drake Co.
Dubner International Inc.
Fast Forward Video
General Atomics
Harris Corp., Broadcast Communications Division
Harris Corp., Broadcast Division
Hitachi Denshi America, Ltd.
IPITEK
International Datacasting Corp.
JVC Professional Products Company
MATCO Inc.
MUSICAM U.S.A.
Marshall Electronics
Mega Hertz
Mohawk
Panasonic Broadcast & Television Systems Co.
Pinnacle Systems Inc.
Radyne ComStream Corp.
Register Data Systems
Telos Systems
Utah Scientific Inc.
VCI-Video Communications Inc.
Visual Sound Inc.

Digital Image Processors

Barco Visual Solutions, LLC
Eigen
Multi-Image Network
Xintekvideo Inc.

Digital Special Effects Systems

AccuWeather Inc.
Alias/WaveFront Inc.
Autodesk
Chyron Corp.
Cintel Inc.
Eastman Kodak Co.
FOR.A Corp. of America
Focus Enhancements
Pinnacle Systems Inc.
Quantel Inc.
The Tiffen Company
Toshiba America Consumer Products

Digital Video Graphics and Animation

AccuWeather Inc.
Alias/WaveFront Inc.
Autodesk
Commercial Electronics Ltd.
Even Technologies
Key West Technology
Matrox Video Products Grp
Multi-Image Network
Quantel Inc.

Digital Video Processing Equipment

Dubner International Inc.
Ensemble Designs
FOR.A Corp. of America
Focus Enhancements
Gennum Corp.
LINK Electronics Inc.
Leitch Inc.
Mega Hertz
NVISION Products
NTV International Corporation
Scientific Atlanta
Television Equipment Assoc. Inc./Matthey
Visual Sound Inc.
Ward-Beck Systems Ltd.
Wiltronix Inc.
Xintekvideo Inc.

Digital Video Production Systems

Bitcentral Inc.
Commercial Electronics Ltd.
Matrox Video Products Grp
Multi-Image Network
NVISION Products
Specialized Communications
Thomson Broadcast & Media Solutions

Distortion Analyzers

Audio Precision Inc.
Boonton Electronics Corp.
Electro Rent Corp.
SyntheSys Research Inc.

Distribution Amplifiers

Audio Technologies Inc. (ATI)
Autogram Corp.
Benchmark Media Systems Inc.
Burst Electronics Inc.
Cablynx, Inc.
ESE
Farrtronics Ltd.
Henry Engineering
Horita
Intersil Corp. Headquarters
Leitch Inc.
North American Cable Equipment Inc.
Omicron Video
Philip-Cooke Co.
Qintar Technologies Inc.
Radio Systems Inc.
Ross Video Ltd.
Sigma Electronics Inc.
Studio Technologies Inc.
Symmetricom
Symmetricom
Television Equipment Assoc. Inc./Matthey
Video Accessory Corp.
Videotek
Ward-Beck Systems Ltd.

Distribution Systems

Enghouse Systems Limited
IPITEK
Motorola Digital Media Systems
North American Cable Equipment Inc.
Philip-Cooke Co.
Scientific Atlanta
StarGuide Digital Networks Inc.

Dollies, Instrument Carts, Etc.

Ferno-Washington Inc.
Hogg & Davis Inc.
Matthews Studio Equipment Inc. (MSE)
Panavision New York
Wheelit Inc.

Dummy Loads

Altronic Research Inc.
Bird Electronic Corp.
Electro Impulse Laboratory Inc.
Kintronic Labs Inc.
Nardal - An L-3 Communications Co.
Phasetek Inc.
Sencore Inc./AAVS

Duplicators

Accurate Sound Corp.
Ascent Media Management Services
Interface Media Group
M2 America
National Audio Co. Inc.

Earth Stations

Ascent Media Services
D.H. Satellite
General Electric Co.
Maze Corporation
Narda Satellite Networks
Pinzone Engineering Group Inc.
TDK Electronics Corp.
Trident Media Group/ Spector Entertainment Group Inc.
Vertex Communications Corp.

Editing Equipment, Sales-Rental-Service

Adcom
Commercial Electronics Ltd.
Dubner International Inc.
International Cinema Equipment
Neumade Products Corp.
Plastic Reel Corp. of America
Specialized Communications
TDK Electronics Corp.
United Media Inc.
VRI (Video Rentals Inc.)

Editing Film and Tape

Ascent Media Management Services
Avid Broadcast
Center City Film & Video
Chyron Corp.
The Image Group Post, LLC.
J and R Moviola Inc.
Paulmar Industries Inc.
Servoreeler Systems
Skotel Corp.
TDK Electronics Corp.

EFP (Electronic Field Production)

Bexel Corp.
Mobile Video Services Ltd.
PMTV Producers Management Television
Panasonic Broadcast & Television Systems Co.
Shook Mobile Technology, LP
TDK Electronics Corp.

Electronic Advertising Displays

Display Systems International Inc.
U.S. Traffic & Display Solutions

Electronic Components

ADCOUR
Communication & Power Industries, EIMAC Division
Dow-Key Microwave Corp.
Jennings Technology Co.
Kay Industries Inc.
Laser Diode Inc.
M/A-COM

Electronic Equipment

Electronic Protection Equipment

Emergency Alerting Systems

Emergency Broadcast Equipment

EMI Cabinets

Encoders

ENG Equipment and Accessories

ENG Vans

Engineering Systems

Equalizers

Equipment Maintenance and Repair

Erasers, Magnetic Tape

Exciters

Facilities Planning

Fiber Optic Cable and Accessories

Fiber Optic Transmission Systems

Field Strength Meters

Film Equipment

Film Printers, Motion Pictures

BHP Inc.
Eastman Kodak Co.

Film Processors

Eastman Kodak Co.
Lipsner Smith Co.
Precision Microproducts of America

Film Scanners

ARRI, Inc.
Cintel Inc.

Film-to-Tape Transfer Equipment

International Cinema Equipment
J and R Moviola Inc.
Lipsner Smith Co.

Filters and Delay Lines

Allen Avionics, Inc.
Electroline Equipment Inc.
Microwave Filter Co. Inc.
Motorola Broadband Communications Sector
Schneider Optics Inc.
TTE Inc.
Television Equipment Assoc. Inc./Matthey

Fire Detection System

Kidde-Fenwal Inc.

Floor Covering Stages

Eddie Egan & Associates
Rosco Laboratories Inc.

Frame Synchronizers

Cablynx, Inc.
Hotronic Inc.
Leitch Inc.
Pixel Instruments Corp.
Videotek
World Video Sales Co.

Frequency Measuring Services

Antenna Concepts Inc.
Commercial Radio Monitoring Co.
Communications General Corp.
Frequency Measuring Service Inc.
Northwest Monitoring Service
Radio Aids Inc.
Sencore Inc./AAVS
Signal Monitoring Service
Symmetricom
Van Nostrand Radio Engineering Service
Ziehl Electronic Service

Frequency Monitors

Symmetricom

Generators, Electric

CECO International Corp.
Elan Enterprises Ltd.
FWT Inc.
Hollywood Rentals Production Services
PMTV Producers Management Television
Panavision New York
ScreenLight & Grip

Generators, Signal

Kalun Communications Inc.
Magni Systems Inc.
Sencore Inc./AAVS
Video Accessory Corp.

Graphics

AccuWeather Inc.
Alias/WaveFront Inc.
All Mobile Video Inc.
Atlantic Video Inc.
Chyron Corp.
Even Technologies
The Image Group Post, LLC.
Inscriber Technology Corporation
Interface Media Group
Multi-Image Network

HDTV Equipment

Accom Inc.
Acrodyne Industries Inc.
Antenna Concepts Inc.
Autodesk
Bexel Corp.
Camplex Corporation
Canon U.S.A. Inc.
Communication & Power Industries
Dielectric Communications
Enghouse Systems Limited
Frequency Measuring Service Inc.
Harris Corp., Broadcast Division
Hitachi Denshi America, Ltd.
JVC Professional Products Company
Kalun Communications Inc.
Kathrein Inc., Scala Division
LEMO USA Inc.
Marshall Electronics
Milestone Technologies Inc.
Mohawk
O'Connor Professional Camera Support Systems
Panasonic Broadcast & Television Systems Co.
Professional Communications Systems
Radyne ComStream Corp.
Sarnoff Corp.
Sharp Electronics Corp., CCD Products Div.
SyntheSys Research Inc.
Teleplex Inc.
Thomson Broadcast & Media Solutions
Toshiba America Consumer Products
Utah Scientific Inc.
Ward-Beck Systems Ltd.
Wiltronix Inc.
Winsted Corp.

Headend Systems

ADC
Alpha Technologies Inc.
Blonder Tongue Laboratories Inc.
CCI
C-COR
ComSonics Inc.
D.H. Satellite
General Atomics
Kalun Communications Inc.
Mega Hertz
Norsat International Inc.
North American Cable Equipment Inc.
Ortel
Scientific Atlanta
Sitco Antenna Company
Standard Communications Corp.
Toner Cable Equipment Inc.

Heads, Magnetic Film & Tape, Disk

AheadTek
International Electro-Magnetics
Northern Magnetics Inc.
Sprague Magnetics Inc.

Heads, Refurbishing

International Electro-Magnetics
Northern Magnetics Inc.
Sprague Magnetics Inc.
Videomagnetics Inc.

Headset Amplifiers

Fostex America
Henry Engineering
R-Columbia Products Co. Inc.

Headsets, Headphones

AKG Acoustics, U.S.
Anchor Audio Inc.
Audio-Technica U.S., Inc.
COMTEK Inc.
Clear-Com Communication Systems
Fostex America
Production Intercom Inc.
R-Columbia Products Co. Inc.
RTS Systems Telex Communications Inc
Sacramento Theatrical Lighting (STL)
Sennheiser Electronic Corp.
Setcom Corp.
Stanton Group
Telex Communications Inc.

Helicopters

American Eurocopter Corp.
Geneva Aviation Inc.
Gyrocam Systems
Wescam Inc.

High Definition Television (HDTV)

Cintel Inc.
LARCAN USA
Leitch Inc.
Miranda Technologies Inc.
Sarnoff Corp.
Telecast Fiber Systems Inc.
Toshiba America Consumer Products

Image Enhancers, TV

Colorado Video Inc.
Xintekvideo Inc.

Infrared Transmission Systems

Sennheiser Electronic Corp.
Sound Designers Studio

Installation Services

American Antenna Inc.
B&B Systems
Broadcast Store Inc.
CBT Systems
Harris Corporation Co.
180 Connect
Professional Communications Systems
Quality Tower Erectors Inc.
Superior Tower Services Inc.
Teletech Inc.
VidCAD Documentation Programs (VDP Inc.)

Instruments Cases

A & S Case Co. Inc.
Atlas Case Corp.
Calzone Case Co.
Hardigg Cases
Jensen Tools
Penn Elcom Inc.
Star Case Manufacturing Co. Inc.
Thermodyne Cases

Interactive Television

Avid Broadcast
Tribune Media Services

Intercom Systems

Anchor Audio Inc.
Clear-Com Communication Systems
Farrtronics Ltd.
Fuller Manufacturing
HM Electronics Inc.
ITI Electronics Inc.
Olesen
Production Intercom Inc.
R-Columbia Products Co. Inc.
RTS Systems Telex Communications Inc
Ram Broadcast Systems
Sacramento Theatrical Lighting (STL)
Setcom Corp.
Sierra Automated Systems & Engineering Corp.

Studio Technologies Inc.
Swintek Enterprises Inc.
Systems Wireless Ltd.
Talk-A-Phone Co.
Telex Communications Inc.
Wiltronix Inc.

ISO Couplers (AM & FM)

Phasetek Inc.

Jack Panels and Accessories

ADC
Audio Accessories Inc.
Auernheimer Labs Corp.
Clark Wire & Cable Co. Inc.
Gepco International Inc.
ITI Electronics Inc.
Kings Electronics Co. Inc.
NVISION Products
Penn Elcom Inc.
Penny & Giles Inc.
Trimm Inc.

Klystron Amplifiers/Lead Oxide Vidicon

e2v technologies Inc.
Penta Laboratories

Klystrons

Communication & Power Industries
Daily Electronics Corp.
e2v technologies Inc.
LARCAN USA
MRPP Inc.
Penta Laboratories
Thales Components Corp.

Labels

Audico Labels
Memorex Products Inc.
National Audio Co. Inc.
Seton Identification Products
Telecrafter Products
U.S. Tape & Label Corp.
Veriad

LED, VU and S Panel Meters

Audio Technologies Inc. (ATI)
Dorrough Electronics
Logitek
Sescom Inc.
Weschler Instruments

Lenses, Optical and Camera

Alpine Optics Inc.
Band Pro Film & Digital Inc.
Canon U.S.A. Inc.
Century Precision Optics
Dimension 3
Film/Video Equipment Service Co. Inc.
Fujinon Inc.
Homalite
Innovision Optics Inc.
Marshall Electronics
Navitar Inc.
Pentax Imaging Co.
Schneider Optics Inc.
Tamron U.S.A. Inc.

Library Storage Systems

Hollywood Vaults Inc.
Imagine Products Inc.
Neumade Products Corp.
Paulmar Industries Inc.

Lighting Design

DeSisti Lighting
Devlin Design Group Inc.
The Express Group
L.E. Nelson Sales Corp.
New York City Lites
Packaged Lighting Systems Inc.
Sachtler Corp. of America
Thorn-EMI Studio Lamps/L.C.
Videssence L.L.C.

Lighting Equipment

ARRI, Inc.
AVAB America Inc.
Peter Albrecht Company Inc.
Anton/Bauer Inc.
Arri Canada Ltd.
Atlantic Sound Systems
Automatic Devices Company
Band Pro Film & Digital Inc.
Bend-A-Lite Flexible Neon
CECO International Corp.
Camera Service Center
Channel One Lighting Systems Inc.
Comprehensive Video Group
Control Concepts Corp.
DEDOTEC USA Inc.
DeSisti Lighting
Dove Systems
Electronic Theatre Controls Inc.
Film/Video Equipment Service Co. Inc.
Flash Technology Corporation of America
Frezzolini Electronics Inc.
Full Compass Systems Ltd.
GAMPRODUCTS Inc.
Group One Ltd.
Honeywell Airport Systems
LTM Corp. of America
Leviton NSI Colortran
Lowel-Light Manufacturing Inc.
Matthews Studio Equipment Inc. (MSE)
/ole-Richardson Co.
Musco Mobile Lighting Ltd.
L.E. Nelson Sales Corp.
Neutrik U.S.A. Inc.
New York City Lites
Olesen
Allen Osborne Associates Inc.
PC& E
Packaged Lighting Systems Inc.
Panavision New York
Photomart Cine-Video Inc.
Pro Video & Film Equipment Co. Inc.
Rosco Laboratories Inc.
Sachtler Corp. of America
ScreenLight & Grip
Sinar Bron Inc.
Skytec, Inc.
Stage Equipment & Lighting Inc.
Strand Lighting Inc.
Strong International
Teatronics/Entertainment Lighting Control
Theatre Service & Supply Corp.
Theatrical Services Inc.
James Thomas Engineering
Thorn-EMI Studio Lamps/L.C.
Ultimate Support Systems Inc.
Vantage Lighting Inc.
Videssence L.L.C.
Westcott

Lightning Protection Equip. & Systems

Abroyd Communications Ltd.
Cortana Corp.
EFI Electronics Corp.
LBA Technology Inc.
LEA International
Lightning Eliminators & Consultants Inc.
Lightning Master Corp.
Lightning Prevention Systems
MCG Surge Protection
/ole-Richardson Co.
Neumade Products Corp.
Northern Technologies Inc.
Stage Equipment & Lighting Inc.
Thorn-EMI Studio Lamps/L.C.
Transtector Systems Inc.
Wil-Can Electronics Ltd.

Lights, On-Air

Electronic Theatre Controls Inc.
General Electric Co.
Sachtler Corp. of America
Sinar Bron Inc.

Lights, Recording

Electronic Theatre Controls Inc.
General Electric Co.
Sinar Bron Inc.

Lights, Stage

AVAB America Inc.
Bend-A-Lite Flexible Neon
Electronic Theatre Controls Inc.
GAMPRODUCTS Inc.
General Electric Co.
Hollywood Rentals Production Services
/ole-Richardson Co.
L.E. Nelson Sales Corp.
Olesen
Packaged Lighting Systems Inc.
Sacramento Theatrical Lighting (STL)
Stage Equipment & Lighting Inc.
Strong International
Theatre Service & Supply Corp.
James Thomas Engineering
Thorn-EMI Studio Lamps/L.C.
Vantage Lighting Inc.

Line Conditioning

ADCOUR
Control Concepts Corp.
Furman Sound Inc.
LEA International
MCG Surge Protection
MGE UPS SYSTEMS Inc.
Newark Electronics
Nigel B. Furniture/Marketec
Northern Technologies Inc.
Wil-Can Electronics Ltd.

Line Surge Protectors

ADCOUR
Control Concepts Corp.
EFI Electronics Corp.
Henry Engineering
LEA International
Lightning Master Corp.
MCG Surge Protection
MGE UPS SYSTEMS Inc.
Northern Technologies Inc.
Transtector Systems Inc.
Wil-Can Electronics Ltd.

Locks

CableTek Wiring Products Inc.
Penn Elcom Inc.

Logging System

Accurate Sound Corp.
Dubner International Inc.
Geac Libra
Horita
Image Logic Corp.
Imagine Products Inc.
Spotcat Software
Telcom Research

Loudspeakers and Accessories

Amplivox Portable Sound Systems
Atlas Sound
Auernheimer Labs Corp.
JBL Professional
Sennheiser Electronic Corp.
Ultimate Support Systems Inc.
Yanchar Design & Consulting Group

Machine Control Systems

ADTEC Inc.
Peter Albrecht Company Inc.
Focus Enhancements
General Electric Co.
Leightronix Inc.
MATCO Inc.
Micro Technology Unlimited
Philip-Cooke Co.
Tapeswitch Corp.

Master Control Switches

Evertz Microsystems Ltd.
Leitch Inc.
Miranda Technologies Inc.
Tapeswitch Corp.
Utah Scientific Inc.

Meters

Aeroflex
B&B Systems
Benchmark Media Systems Inc.
ComSonics Inc.
Dorrough Electronics
Electro Rent Corp.
G Prime Ltd.
Jennings Technology Co.
Konica Minolta Corp.
McCurdy Radio Ltd.
Photo Research
Radio Detection/Riser Bond
Techni-Tool Inc.
Tentel
Triplett Corp.
Weschler Instruments

Microphones and Accessories

AKG Acoustics, U.S.
AVI Systems
Audio-Technica U.S., Inc.
Black Audio
Bogen Communications Inc.
COMTEK Inc.
Countryman Associates Inc.
DPA Microphones, Inc.
FitzCo. Inc.
G Prime Ltd.
Group One Ltd.
LTM Corp. of America
Location Sound Corp.
Marshall Electronics
Micron Audio Products Ltd.
Nady Systems Inc.
Professional Sound Corp.
R-Columbia Products Co. Inc.
RTS Systems Telex Communications Inc
Sennheiser Electronic Corp.
Servoreeler Systems
Sescom Inc.
Shure Inc.
Systems Wireless Ltd.
Telex Communications Inc.
Ultimate Support Systems Inc.

Microwave

Ascent Media Services
Broadcast Microwave Services Inc.
Broadcast Sports Technologies
Comex Worldwide Corp.
Comsearch
Direct Broadcast Services Inc.
Global Microwave Systems Inc.
Harris-Farinon
M/A-COM
Marcom
NSI
PMTV Producers Management Television
RF Specialties Group
TFT Inc.

Microwave Amplifiers

Comex Worldwide Corp.
Harris-Farinon
NUCOMM Inc.

Microwave Antennas

Broadcast Microwave Services Inc.
Broadcast Sports Technologies
Comex Worldwide Corp.
D.H. Satellite
NSI
NUCOMM Inc.
Radio Frequency Systems
Radio Research Instrument Co. Inc.
Southern Broadcast Services

Microwave Equipment

AVCOM of Virginia Inc.
Blonder Tongue Laboratories Inc.
Broadcast Equipment Surplus Inc.
Broadcast Microwave Services Inc.
Broadcast Sports Technologies
Comex Worldwide Corp.
Dow-Key Microwave Corp.
E-N-G Mobile Systems Inc.
Frontline Communications
Geneva Aviation Inc.
Hitachi Denshi America, Ltd.
JSB Service Co.
Marti Electronics
Moseley Associates Inc.
NUCOMM Inc.
Norsat International Inc.
Ortel
Radio Research Instrument Co. Inc.

Microwave Transmitters

Broadcast Microwave Services Inc.
Broadcast Sports Technologies
CED
Comex Worldwide Corp.
Global Microwave Systems Inc.
JSB Service Co.
NUCOMM Inc.
Norlight Telecommunications Inc.

Mobile Communications

Aluma Tower Company Inc.
CSG Systems
DST Innovis
Geneva Aviation Inc.
Hardigg Cases
ICM (International Crystal Mfg.)
MCL Inc.
Thermodyne Cases

Mobile Studio Equipment

Calzone Case Co.
Camplex Corporation
F&F Productions, L.L.C.
Ferno-Washington Inc.
Harrison Consoles
Hollywood Rentals Production Services
National Mobile Television
National Mobile TV - Houston
North Dakota Television L.L.C.
Packaged Lighting Systems Inc.
Telos Systems
United States Broadcast

Mobile Units-Sales/Rental

All Mobile Video Inc.
Frontline Communications
PMTV Producers Management Television
J.A. Taylor & Associates
Telenium Mobile
Turner Studios Field Operations

Mobile Vans

Alpha Video & Electronics (AVEC)
E-N-G Mobile Systems Inc.
Frontline Communications
North Dakota Television L.L.C.
Panavision New York
Phoenix E N G, Inc.
Roscor Corp.
ScreenLight & Grip
Shook Mobile Technology, LP
Television Engineering Corp.

Modular Set Design System

The Display & Exhibit Source
Uni-Set Corp.

Modulators

Advanced Media Technologies, Inc.
Blonder Tongue Laboratories Inc.
CADCO Systems Inc.
Cable Serv Inc.
Cable Technologies International
R.L. Drake Co.
Leaming Industries
Radyne ComStream Corp.
Standard Communications Corp.
Toner Cable Equipment Inc.

Moldings

CablePro
CableReady Inc.
The Display & Exhibit Source

Monitor Amplifiers

Alesis
Audio Technologies Inc. (ATI)
Ross Video Ltd.

Monitor Speakers

Group One Ltd.
Harman International Industries Inc.
JBL Professional
Stanton Group
Tannoy North America Inc.
Westlake Audio, Professional Sales Group
Yamaha Corp. of America
Yanchar Design & Consulting Group

Monitors, Audio and Video

AVI Systems
Band Pro Film & Digital Inc.
Conrac Systems Inc.
Delta Electronics Inc.
Dorrough Electronics
Farrtronics Ltd.
Furman Sound Inc.
Hoodman Corp.
Ikegami Electronics (U.S.A.) Inc.
Image Video
Interlogix
JVC Professional Products Company
Magni Systems Inc.
Marshall Electronics
Sharp Electronics Corp., CCD Products Div.

Monitors, Frequency, Modulation Phase

Belar Electronics Laboratory Inc.
Commercial Radio Monitoring Co.
Frequency Measuring Service Inc.
Inovonics Inc.
Nardal - An L-3 Communications Co.
TFT Inc.

Monopoles

FWT Inc.
Fred A. Nudd Corp.
Pirod Inc.
Rohn Industries Inc.

Motion Control Equipment

Innovision Optics Inc.
Richmond Sound Design Ltd.
Tapeswitch Corp.
Vinten Inc.

Motion Picture Equipment

ARRI, Inc.
BHP Inc.
CECO International Corp.
Celco
Dimension 3
Eastman Kodak Co.
Alan Gordon Enterprises Inc.
Hollywood Rentals Production Services
/ole-Richardson Co.
Motion Picture Enterprises Inc.
O'Connor Professional Camera Support Systems
PC& E
Panavision New York
Pro Video & Film Equipment Co. Inc.

Motors

Powr-Ups Corp.
Senior Aerospace

Mounting Products

Advance Products Co. Inc.
California Amplifier
Omnimount Systems
Peerless Industries Inc.
Penn Elcom Inc.

Multiplexers

COASTCOM
Kintronic Labs Inc.
3M

Multiplexing

Camplex Corporation
Colorado Video Inc.
L-3 Communications Telemetry East

Multistandard TV, VCR Camcorders

Analog Digital International Inc.

Music Equipment

Whirlwind
Yamaha Corp. of America

Name Plates

Seton Identification Products
3M

Neon Lighting

Bend-A-Lite Flexible Neon

Network Delay Systems

StarGuide Digital Networks Inc.

Noise Reduction Systems, Video

Noise Control Corp.
North Hills Signal Processing, a PORTA Systems Co.
Video International Development Corp.
Xintekvideo Inc.

Office Equipment

JOA Cartridge Service
Luxor
Royal Consumer Information Products

Office Supplies and Forms

Hessler Enterprises Inc.
3M
Veriad

Optical Disk Systems

Identix
Imaging Automation
TEAC America Inc.

Oscillators

ICM (International Crystal Mfg.)
JSB Service Co.
Opamp Labs Inc.
Potomac Instruments, Inc.

Panels

APW Mayville/Stantron
Benner-Nawman Inc.
Bud Industries Inc.
Kings Electronics Co. Inc.
Panel Authority Inc.
Servoreeler Systems
Wireworks Corp.

Passive Components

Allen Avionics, Inc.
Connectronics Corp.
Peter W. Dahl Co. Inc.
Elcom Systems Inc.
IPITEK
Kintronic Labs Inc.
MYAT Inc.
Micro Communications Inc.
Microwave Filter Co. Inc.
Power & Telephone Supply Co.
Qintar Technologies Inc.
S W R Inc.
Thales Components Corp.

Patch Cords

Audio Accessories Inc.
Canare Corp.
Kings Electronics Co. Inc.
Switchcraft Inc.
Trompeter Semflex

Pay TV Equipment and Services

DST Innovis
Electroline Equipment Inc.
Great Lakes Data Systems, Inc.
Macrovision Corp.
Syntellect Inc.
T-C Specialties Co.
Zenith Electronics Corp.

Pedestals

Argraph Corp.
CSI-Camera Support International
Channell Commercial Corp.
Charles Industries Ltd.
Matthews Studio Equipment Inc. (MSE)
Miller Camera Support, L.L.C.
Sachtler Corp. of America
Thomas & Betts Corp.
Vinten Inc.

Phasing Equipment

Intersil Corp. Headquarters
Kay Industries Inc.
Kintronic Labs Inc.
Phasetek Inc.

Phono Equipment and E Micro-Trak

Audio-Technica U.S., Inc.

Photographic Equipment

Bencher Inc.
Calumet Photographic
Precision Microproducts of America
Sinar Bron Inc.
Tamron U.S.A. Inc.
The Tiffen Company
Tinsley Laboratory Inc.

Westcott

Photographic Processing Machines

Precision Microproducts of America

Plastics and Injection Molders

S&L Plastics Inc.

Plugs and Connectors

Canare Corp.
Condor D C Power Supplies Inc.
Connectronics Corp.
Corning Cable Systems
Kings Electronics Co. Inc.
Marconi Communications
Olesen
Selco Products Co.
Switchcraft Inc.

Pole Line Hardware

Condux International
Hogg & Davis Inc.
Marconi Communications
Power & Telephone Supply Co.
Thomas & Betts Corp.

Portable Power Supplies

Fermont ACTS Co.
Frezzolini Electronics Inc.
Hollywood Rentals Production Services
Northern Power Systems

Portable Video Tape Recorder Systems

Mitsubishi Digital Electronics America Inc.

Postproduction Systems

Accom Inc.
Adcom
Analog Digital International Inc.
Ascent Media Services
Atlantic Video Inc.
B&B Systems
CORPLEX Inc.
Cheetah International
Interface Media Group

Power Meters

Bird Electronic Corp.
Boonton Electronics Corp.
Coaxial Dynamics
Dorrough Electronics
Products International Inc.
Weschler Instruments

Power Supplies and Accessories

ADCOUR
Alpha Technologies Inc.
CableReady Inc.
JSB Service Co.
Kay Industries Inc.
Leader Instruments Corp.
MGE UPS SYSTEMS Inc.
/ole-Richardson Co.
Northern Power Systems
Northern Technologies Inc.
Products International Inc.
Transtector Systems Inc.

Pre-Amps, Microphone

Benchmark Media Systems Inc.
Beyerdynamic
Bryston Ltd.
Electronics Corp.
Magna-Tech Electronic Co. Inc.
Martinsound Inc.
Stanton Group
Studio Technologies Inc.

Symetrix Inc.

Pressurizing Equipment and Accessories

Andrew Corp.
Radio Frequency Systems

Professional Audio Equipment

BBE Sound Inc.
Countryman Associates Inc.
DBX Professional Products
Dolby Laboratories Inc.
Electronology, Inc.
Henry Engineering
JOA Cartridge Service
Jensen Transformers Inc.
Lectrosonics Inc.
Location Sound Corp.
Parsons Audio
Professional Sound Services Inc.
QSC Audio Products Inc.
Radio Design Labs. (RDL)
Roland Corp. U.S.
Roscor Corp.
Sabine Inc.
Shure Inc.
Sound Designers Studio
TAI Audio
TC Electronic
Tannoy North America Inc.
Thermodyne Cases
Westlake Audio, Professional Sales Group
Yamaha Corp. of America

Professional Recording Equipment

Aphex Systems Ltd.
Digidesign
Harman International Industries Inc.
Professional Sound Services Inc.
Roland Corp. U.S.
Soundcraft U.S.A.
Superscope Technologies Professionals
TEAC America Inc.
Westlake Audio, Professional Sales Group

Professional Sound Equipment

A R T Applied Research and Technology
BBE Sound Inc.
Bradley Broadcast Sales
Broadcasters General Store Inc.
Cooper Sound Systems Inc.
Countryman Associates Inc.
JBL Professional
Professional Sound Services Inc.
Sound Designers Studio

Professional Video Equipment

AVS Graphics & Media Inc.
Alpine Optics Inc.
Bexel, Inc.
Bogen Imaging Inc.
Dubner International Inc.
Fast Forward Video
Gray Engineering Laboratories Inc.
MATCO Inc.
Merlin Engineering Works Inc.
Pinnacle Systems Inc.
Roscor Corp.
Televideo San Diego
Thermodyne Cases
Videomagnetics Inc.

Projectors, Projection Systems

Barco Visual Solutions, LLC
International Cinema Equipment
Motion Picture Enterprises Inc.
Navitar Inc.
Raven Screen Corp.
SAIC
Sharp Electronics Corp., CCD Products Div.
Strong International
Theatre Service & Supply Corp.

Promotion Products

Blimpy Floating Signs/Bend-A-Lite
Communication Graphics, Inc.
Prisma Packaging

Prompting Equipment

COMTEK Inc.
CPC-Computer Prompting & Captioning Co.
Comprompter Inc.
Electronic Script Prompting
Intelliprompt
Listec Video Corp.
Marietta Design Group
QTV
Tekskil Industries Inc.

Protective Apparel

Second Chance Body Armor Inc.

Public Address Systems

Amplivox Portable Sound Systems
Atlantic Sound Systems
Greenberg Teleprompting

Publications, Broadcast

Broadcast Engineering
Cable Yellow Pages
FM Atlas—Publishing and Electronics
Tribune Media Services

Racks

A & S Case Co. Inc.
AMCO Engineering Co.
APW Mayville/Stantron
Atlas Case Corp.
Atlas Sound
Avtech Systems Inc.
Benner-Nawman Inc.
Bud Industries Inc.
CMP Enclosures Inc.
Gepco International Inc.
Nigel B. Furniture/Marketec
Packaged Lighting Systems Inc.
Panel Authority Inc.
Parsons Manufacturing Corp.
Penn Elcom Inc.
Sound Designers Studio
Star Case Manufacturing Co. Inc.
Storeel Corp.
Thermodyne Cases
Ultimate Precision/GKM
Viking Cases
Winsted Corp.

Radar Systems

Advanced Designs Corp.
Radio Research Instrument Co. Inc.

Radio Control Equipment

Motor Capacitors Inc.

Radio Equipment

Motor Capacitors Inc.
RF Specialties Group
Schafer International
Wicks Broadcast Solutions L.L.C.
Wireless Accessories Group

Radio Equipment, 2-Way

ICM (International Crystal Mfg.)
Location Sound Corp.
Nady Systems Inc.
R-Columbia Products Co. Inc.
Systems Wireless Ltd.
Wireless Accessories Group

Receivers, Shortwave, AM-FM-TV, Multiplex

R.L. Drake Co.
FM Atlas—Publishing and Electronics
Freeland Products Inc.
Samson Technologies Corp.
Symmetricom

Recorders, Accessories

Electronology, Inc.
Record/Play Tek Inc.
Stancil Corp.
360 Systems

Recorders, Audio

Bradley Broadcast Sales
National Audio Co. Inc.
Otari USA Sales Inc.
Record/Play Tek Inc.
Sanyo Fisher Co.
Stancil Corp.
Superscope Technologies Professionals

Recorders, Cassette Audio Logging

Stancil Corp.

Recorders, Video

Accom Inc.
Eigen
Fast Forward Video
JVC Professional Products Company
MATCO Inc.
Merlin Engineering Works Inc.
Northrup Grumman
Optical Disc Corp.
Sanyo Fisher Co.
TEAC America Inc.
Thomson, Inc, (RCA/GE)
Videomagnetics Inc.

Recording Studios

JC Sound Stages
Sound Designers Studio
Westlake Audio, Professional Sales Group
Yanchar Design & Consulting Group

Recording Studios Construction, Prefab

Industrial Acoustics Co., Inc.
Northeastern Communications Concepts Inc.

Reels, Magnetic Tape

Motion Picture Enterprises Inc.
Polyline Corp.
Stancil Corp.

Remote Broadcast Equipment

Aluma Tower Company Inc.
Broadcast Supply Worldwide
Camplex Corporation
Comrex Corp.
Digitel Corp.
LBA Technology Inc.
MUSICAM U.S.A.
Marti Electronics
PMTV Producers Management Television
StarGuide Digital Networks Inc.
Telecast Fiber Systems Inc.
Telos Systems
Vinten
Wescam Inc.

Remote Control

AVI Systems
Broadcast Equipment Surplus Inc.
Broadcast Sports Technologies
Burk Technology
CONTEC Corp.

Innovision Optics Inc.
Moseley Associates Inc.
Motorola Broadband Communications Sector
NSI
Nigel B. Furniture/Marketec
Quick-Set International Inc.
Solutec Ltd. (HA)
Vinten
Vinten Inc.
Wescam Inc.

Remote Control for VTRs

Imagine Products Inc.

Remote Control Switching Systems

Evertz Microsystems Ltd.
Vicon Industries Inc.

Remote Control Systems for TV Cameras

Interlogix
Quick-Set International Inc.
Telemetrics Inc.
Vinten

Rental Equipment (Broadcast and Cable)

Analog Digital International Inc.
Bexel Corp.
Bexel, Inc.
CORPLEX Inc.
Direct Broadcast Services Inc.
Greenberg Teleprompting
The Laumic Rental Co.
PC& E
ScreenLight & Grip
Systems Wireless Ltd.
Tele-Measurements Inc.
Unique Business Systems
VRI (Video Rentals Inc.)

Repair

CONTEC Corp.
ComSonics Inc.
Feldmar Watch and Clock Center
Flash Technology Corporation of America
Multi-Image Network
Skytec, Inc.
Swintek Enterprises Inc.
Tele-Measurements Inc.
Tentel

Research and Equipment

Future Productions Inc.
IMS (Interactive Market Systems Inc.)
National Video Services Inc.
Sarnoff Corp.
Tech Laboratories Inc.
Teleplex Inc.
Tribune Media Services

Reverberation Chambers

Industrial Acoustics Co., Inc.

RF Bridging Equipment

Lindsay Broadband Inc.
Potomac Instruments, Inc.

RF Coaxial Load Resistors

Altronic Research Inc.
Bird Electronic Corp.
Coaxial Dynamics
LARCAN USA
Trompeter Semflex

RF Instrumentation & Components

Bird Electronic Corp.
Boonton Electronics Corp.
Coaxial Dynamics
Dow-Key Microwave Corp.
Jennings Technology Co.
LBA Technology Inc.
MYAT Inc.
Phasetek Inc.
Potomac Instruments, Inc.
RF Specialties Group
Sadelco Inc.
Shively Labs
TTE Inc.

RF Power Attenuators

Bird Electronic Corp.
Kalun Communications Inc.

Rigging Systems

Peter Albrecht Company Inc.
DeSisti Lighting
Gala
Hoffend & Sons Inc.
Hollywood Rentals Production Services
Olesen
Packaged Lighting Systems Inc.
Sacramento Theatrical Lighting (STL)
Superior Tower Services Inc.
James Thomas Engineering

Robotics

CSI-Camera Support International
DeSisti Lighting
Frezzolini Electronics Inc.
Fujinon Inc.
O'Connor Professional Camera Support Systems
Vinten
Vinten Inc.

Routing Switchers

Adrienne Electronics Corp.
Amek U.S.A.
Burst Electronics Inc.
Cablynx, Inc.
Evertz Microsystems Ltd.
General Atomics
IRIS Technologies Inc.
Image Video
Knox Video
Leitch Inc.
NTV International Corporation
PESA Switching Systems, Inc.
Richmond Sound Design Ltd.
Sierra Automated Systems & Engineering Corp.
Sigma Electronics Inc.
Thomson Broadcast & Media Solutions
Utah Scientific Inc.
Video Accessory Corp.
Wheatstone Corp.
Wiltronix Inc.
World Video Sales Co.

Satellite Audio Systems

Arrakis Systems Inc.
International Datacasting Corp.
Register Data Systems
Standard Communications Corp.
StarGuide Digital Networks Inc.

Satellite Communications Systems

ATCI/Antenna Technology Communications Inc.
Andrew Corp.
Ascent Media Services
Channel Master L.L.C.
Communication & Power Industries
L-3 Communications Telemetry East
MCL Inc.
MRPP Inc.
Maze Corporation
Megastar Inc.
Microspace Communications Corp.
Milestone Technologies Inc.
Narda Satellite Networks
Norsat International Inc.
Pinzone Engineering Group Inc.
Richardson Electronics
Rodelco Electronics Corp.
Roscor Corp.
SES Americom
Satellite Systems Corp.
Spacenet Services Inc.
Superior Satellite Engineers Inc.

Satellite News Vehicle

Direct Broadcast Services Inc.
E-N-G Mobile Systems Inc.
Frontline Communications
Shook Mobile Technology, LP
J.A. Taylor & Associates
Television Engineering Corp.
Wolf Coach Inc.

Satellite Receiver

Advanced Media Technologies, Inc.
Blonder Tongue Laboratories Inc.
Cable Serv Inc.
Chrono-Log Corp.
R.L. Drake Co.
International Datacasting Corp.
L-3 Communications Telemetry East
Standard Communications Corp.

Satellite Resale and Common Carriers

Ascent Media Services
Megastar Inc.
Microspace Communications Corp.
Norlight Telecommunications Inc.
SES Americom
Trident Media Group/ Spector Entertainment Group Inc.

Satellite Services

Ascent Media Management Services
Ascent Media Services
Ascent Media Services
Comsearch
Direct Broadcast Services Inc.
Megastar Inc.
Microspace Communications Corp.
Milestone Technologies Inc.
Mobile Video Services Ltd.
PMTV Producers Management Television
SES Americom
Trident Media Group/ Spector Entertainment Group Inc.

Satellite Terminals

Channel Master L.L.C.

Scan Converters

AVerMedia Technologies Inc.
Communications Specialties Inc.
Faroudja Laboratories
Focus Enhancements
Magni Systems Inc.

Scramblers, Pay TV

CED
Macrovision Corp.

Security Surveillance

AVCOM of Virginia Inc.
Avtech Systems Inc.
Cablynx, Inc.
Colorado Video Inc.
EMCOR Enclosures
HM Electronics Inc.
Macrovision Corp.
TOA Electronics Inc.
Tele-Measurements Inc.
Vicon Industries Inc.

Security Systems

Avtech Systems Inc.
Imaging Automation
TOA Electronics Inc.
Tech Laboratories Inc.

Service, Repair and Maintenance

Alpine Optics Inc.
Arri Canada Ltd.
C-COR
ComSonics Inc.
Cygnal Technology
Flash Technology Corporation of America
HM Electronics Inc.
JSB Service Co.
Signal Monitoring Service
Skytec, Inc.
United States Broadcast
Videomagnetics Inc.
Westlake Audio, Professional Sales Group

Signal Generators

Burst Electronics Inc.
Condux International
Leader Instruments Corp.
Magni Systems Inc.
Techni-Tool Inc.
Video Accessory Corp.
World Video Sales Co.

Signal Processor

A R T Applied Research and Technology
BBE Sound Inc.
CADCO Systems Inc.
Ensemble Designs
Eventide Inc.
Evertz Microsystems Ltd.
G Prime Ltd.
LINK Electronics Inc.
Miranda Technologies Inc.
Renkus-Heinz Inc.
Roland Corp. U.S.
Ross Video Ltd.
Sabine Inc.
Scientific Atlanta Canada Inc. Nexus Division
TC Electronic

Signs, Engraved, Plastic

Blimpy Floating Signs/Bend-A-Lite
The Display & Exhibit Source
Seton Identification Products

Sound Editing

Ascent Media Management Services
Sascom Marketing Group

Sound Equipment

Atlantic Sound Systems
Atlas Sound
Cooper Sound Systems Inc.
Crouse-Kimzey Co.
International Cinema Equipment
Radio Design Labs. (RDL)
Renkus-Heinz Inc.
TOA Electronics Inc.

Sound Mixers

Cooper Sound Systems Inc.
Harrison Consoles
Professional Sound Corp.

Sound Recording Equipment

A R T Applied Research and Technology
Atlantic Sound Systems
Bradley Broadcast Sales
Cooper Sound Systems Inc.
Countryman Associates Inc.
Digidesign
Roland Corp. U.S.
TAI Audio

Sound Systems

Amplivox Portable Sound Systems
Full Compass Systems Ltd.
JBL Professional
Renkus-Heinz Inc.

Speakers

Amplivox Portable Sound Systems
Atlas Sound
Auernheimer Labs Corp.
FitzCo. Inc.
Galaxy Audio Inc.
JBL Professional
QSC Audio Products Inc.
Renkus-Heinz Inc.
Stanton Group
TOA Electronics Inc.

Special Effects Generators

Alias/WaveFront Inc.

Special Effects, Audiovisual

Ampex Data Systems Corp.
Autodesk
Chyron Corp.
Dimension 3
Richmond Sound Design Ltd.

Spectrum Analyzers

AVCOM of Virginia Inc.
Frequency Measuring Service Inc.
Products International Inc.
SyntheSys Research Inc.

Splicing Equipment, Film, Tape

Amek U.S.A.
Neumade Products Corp.
Servoreeler Systems

Standards Converters

Glentronix
Interface Media Group
Merlin Engineering Works Inc.
Snell & Wilcox Inc.
Video International Development Corp.

Standby Power

ADCOUR
Alpha Technologies Inc.
LEA International
MGE UPS SYSTEMS Inc.
Northern Technologies Inc.
Performance Power Technologies

Stands, Computer, A/V, TV Etc.

Advance Products Co. Inc.
Calzone Case Co.
Luxor
Matthews Studio Equipment Inc. (MSE)
Peerless Industries Inc.
Wheelit Inc.

Stands, Microphone

Atlas Sound
Ultimate Support Systems Inc.

Station Automation

CEA-Computer Engineering Associates
Encoda Systems Inc.
FloriCal Systems Inc.
Key West Technology
Leightronix Inc.
Professional Communications Systems
Scott Studios Corp.
Spotcat Software
VCI-Video Communications Inc.
WireReady NSI

Status Monitoring

Acterna
C I S Inc.
Flash Technology Corporation of America
Image Video

Stereo Equipment, Audio

Advanced Media Technologies, Inc.
Leaming Industries

Stereo Generation Equipment

Aphex Systems Ltd.
Circuit Research Labs Inc. (CRL Systems, Inc.)
Inovonics Inc.
Leaming Industries
Modulation Sciences Inc.

Stereo Simulation

Studio Technologies Inc.

Still Stores

AVS Graphics & Media Inc.
Inscriber Technology Corporation

Still Stores-Digital

AVS Graphics & Media Inc.
Multi-Image Network

Stopwatches

M. Ducommun Co.
Feldmar Watch and Clock Center

Storage Facilities

Atlantic Inc.
Hollywood Vaults Inc.
Research Technology International Inc.

Studio Equipment

Peter Albrecht Company Inc.
Atlantic Video Inc.
Bradley Broadcast Sales
DBX Professional Products
ENCO Systems Inc.
Ferno-Washington Inc.
Matthews Studio Equipment Inc. (MSE)
PC& E
Parsons Audio
Shure Inc.
Stage Equipment & Lighting Inc.
Theatre Service & Supply Corp.

Studio Facilities

All Mobile Video Inc.
Ascent Media Services
CECO International Corp.
Center City Film & Video
Devlin Design Group Inc.
The Image Group Post, LLC.
Interface Media Group
Technet Systems Group

Studio Furniture

Arrakis Systems Inc.
Devlin Design Group Inc.
The Express Group
High Tech Industries
Murphy Studio Furniture
Nigel B. Furniture/Marketec
Northeastern Communications Concepts Inc.
Studio Technology
Wheatstone Corp.
Winsted Corp.

Studio Sets, Custom

Devlin Design Group Inc.
The Express Group
Studio Technology
Uni-Set Corp.

Sub-Carrier Generators

Marti Electronics

Switches and Accessories

Communications Specialties Inc.
Dow-Key Microwave Corp.
IRIS Technologies Inc.
Nortel Networks
Qintar Technologies Inc.
Seger Electronics
Selco Products Co.
Tapeswitch Corp.
Tech Laboratories Inc.
Veetronix Inc.

Switching Equipment, Audio

Autogram Corp.
Broadcast Electronic Services
IRIS Technologies Inc.
NTV International Corporation
Ram Broadcast Systems
Sierra Automated Systems & Engineering Corp.
Utah Scientific Inc.
Veetronix Inc.
Video Accessory Corp.

Switching Equipment, Video

Ampex Data Systems Corp.
Broadcast Electronic Services
Gennum Corp.
Hotronic Inc.
IRIS Technologies Inc.
Ikegami Electronics (U.S.A.) Inc.
Ross Video Ltd.
Utah Scientific Inc.
Veetronix Inc.
Vicon Industries Inc.
Video Accessory Corp.

Tape Conditioner, Audio/Video

Data Security Inc.

Tape Duplicators

Accurate Sound Corp.
Ascent Media Management Services
Future Productions Inc.
IRIS Technologies Inc.
M2 America

Tape Equipment and Accessories

Audiolab Electronics Inc.
Plastic Reel Corp. of America
Record/Play Tek Inc.
Research Technology International Inc.
Tentel

Tape Heads, Audio

AheadTek
International Electro-Magnetics
Northern Magnetics Inc.
Record/Play Tek Inc.

Tape Recorders

Access Intelligence LLC
National Audio Co. Inc.
Northrup Grumman
Record/Play Tek Inc.
Sanyo Fisher Co.
Superscope Technologies Professionals
TEAC America Inc.
Thomson, Inc, (RCA/GE)

Tape Synchronizers

Adcom

Tape, Audio and Video

BMG
Fuji Film USA, Inc.
HAVE Inc.
Maxell Corp. of America
Memorex Products Inc.
Motion Picture Enterprises Inc.
Moviola
National Video Tape Co. Inc.
Panasonic Broadcast & Television Systems Co.
Plastic Reel Corp. of America
Polyline Corp.
Record/Play Tek Inc.

Tap-Offs (CATV)

Emerson Network Power-Viewsonics

Telecine

Accom Inc.
Ascent Media Management Services
B&B Systems
Cintel Inc.

Telecommunications Products

Aluma Tower Company Inc.
Audio Processing Technology Ltd./APT
CSG Systems
Cable Technologies International
Concerto Software
Corning Cable Systems
DST Innovis
Dolby Laboratories Inc.
EDX Wireless LLC
Enghouse Systems Limited
Hogg & Davis Inc.
ICM (International Crystal Mfg.)
IPITEK
ITI Electronics Inc.
Intersil Corp. Headquarters
MODCOMP Inc.
McCurdy Radio Ltd.
MEGGER
Milestone Technologies Inc.
Newark Electronics
Nortel Networks
Northern Power Systems
Ortel
Power & Telephone Supply Co.
Prysmian Communications Cables and Systems
Sierra Automated Systems & Engineering Corp.
SiteSafe Inc.
Syntellect Inc.
T-C Specialties Co.
Techni-Tool Inc.
TELLABS
Telos Systems
Thomas & Betts Corp.
V-Soft Communications
VTECH Communications

Teleconferencing

Artesia Technologies
Canon U.S.A. Inc.
Lectrosonics Inc.
MODCOMP Inc.
Servoreeler Systems
Tamron U.S.A. Inc.
Tele-Measurements Inc.
Televideo San Diego
TELLABS
VidCAD Documentation Programs (VDP Inc.)

Telemetry Receiving Systems

L-3 Communications Telemetry East

Telemetry Transmission Links

Burk Technology
L-3 Communications Telemetry East
Marti Electronics

Telephone Control Systems

Inovonics Inc.

Telephone Interface Equipment

Broadcasters General Store Inc.
Comrex Corp.
Radio Systems Inc.
Syntellect Inc.
Telos Systems

Telephone Line Amplifiers

ITI Electronics Inc.
Prysmian Communications Cables and Systems

Teleprompters

Autoscript
CPC-Computer Prompting & Captioning Co.
Electronic Script Prompting
Greenberg Teleprompting
Intellipropmt
QTV
Telescript Inc.

Test Equipment

Acterna
Aeroflex
Audio Precision Inc.
Bird Electronic Corp.
Boonton Electronics Corp.
Broadcast Video Systems Corp.
Cable Leakage Technologies
Coaxial Dynamics
Comprehensive Video Group
ComSonics Inc.
Crouse-Kimzey Co.
DSC Laboratories
Dorrough Electronics
ETS-Lindgren
Electro Rent Corp.
FM Systems Inc.
Fluke Corp.
Galaxy Audio Inc.
Horita
Jennings Technology Co.
Jensen Tools
Kalun Communications Inc.
L-3 Communications Telemetry East
Leader Instruments Corp.
Magni Systems Inc.
MEGGER
Modulation Sciences Inc.
Products International Inc.
Radio Detection/Riser Bond
Sadelco Inc.
Sarnoff Corp.
Shallco Inc.
Techni-Tool Inc.
Tentel
Thales Broadcast & Multimedia Inc.
3M
Triplett Corp.
Videotek
Weschler Instruments
Wireworks Corp.
Zack Electronics Inc.

Time Base Correctors

Broadcast Electronic Services
Burst Electronics Inc.
FOR.A Corp. of America
Glentronix
Hotronic Inc.
Prime Image, Inc.
Symmetricom

Time Code Equipment

Adrienne Electronics Corp.
Analog Digital International Inc.
Calculated Industries Inc.
Chrono-Log Corp.
Digidesign
ESE
Electronic Theatre Controls Inc.
Gray Engineering Laboratories Inc.

Time Delay Units, Audio

Timers

Tone Generator and Detectors

Tools and Accessories

Tower Design, Manufacture

Tower Erection

Tower Inspections

Tower Maintenance

Tower Obstruction Lighting and Controls

Tower Structural Analysis

Towers, Accessories and Service

Towers, Used

Traffic Advisory System

Transcoders

Transformers

Translators and Accessories

Axcera
CADCO Systems Inc.
Energy-Onix Broadcast Equipment Co. Inc.
Rodelco Electronics Corp.

Transmission Lines

Andrew Corp.
Ascent Media Services
Atlantic Video Inc.
Belden Electronics Divison
Dielectric Communications
Industrial Equipment Representatives (IER)
MYAT Inc.
S W R Inc.
Transcom Corp.

Transmitter Building

Allied Tower Co. Inc.
Rees Associates Inc.

Transmitter Systems

Acrodyne Industries Inc.
Axcera
B&B Systems
Broadcasters General Store Inc.
Colorado Video Inc.
DRS Broadcast Technology
Energy-Onix Broadcast Equipment Co. Inc.
Marcom
Nautel Ltd.
RF Specialties Group
Scientific Atlanta Canada Inc. Nexus Division
Siemens Dematic Limited
Swintek Enterprises Inc.
Technet Systems Group

Transmitters

Acrodyne Industries Inc.
Broadcast Supply Worldwide
CCI
CED
Colorado Video Inc.
DMT USA, Inc.
DRS Broadcast Technology
Energy-Onix Broadcast Equipment Co. Inc.
L-3 Communications Telemetry East
Laser Diode Inc.
Norsat International Inc.
Samson Technologies Corp.
Scientific Atlanta Canada Inc. Nexus Division
Thales Broadcast & Multimedia Inc.

Transmitters, Radio

Bauer Transmitters
Broadcast Electronics Inc.
Broadcast Equipment Surplus Inc.
DRS Broadcast Technology
Energy-Onix Broadcast Equipment Co. Inc.
Harris Corp., Broadcast Communications Division
Harris Corp., Broadcast Division
Industrial Equipment Representatives (IER)
Marcom
Nautel Ltd.
Richardson Electronics
TFT Inc.
Transcom Corp.

Transmitters, TV

Acrodyne Industries Inc.
Axcera
CED
DMT USA, Inc.
Harris Corp., Broadcast Communications Division
Harris Corp., Broadcast Division
Kalun Communications Inc.
LARCAN
MRPP Inc.
Marcom
Radian Communication Services Inc
Scientific Atlanta Canada Inc. Nexus Division
Transcom Corp.

Traveling Wave Tubes

Communication & Power Industries
Daily Electronics Corp.
e2v technologies Inc.
Penta Laboratories
Thales Components Corp.

Tripods, Pedestals and Accessories

Argraph Corp.
Band Pro Film & Digital Inc.
Bogen Imaging Inc.
Isaia & Co.
Miller Camera Support, L.L.C.
O'Connor Professional Camera Support Systems
Photomart Cine-Video Inc.
Pro Video & Film Equipment Co. Inc.
Quick-Set International Inc.
Sachtler Corp. of America
Sinar Bron Inc.
Stage Equipment & Lighting Inc.
Telemetrics Inc.
Thomas & Betts Corp.
The Tiffen Company
Ultimate Support Systems Inc.
Vinten Inc.

Tubes and Tube Rebuilding

Auernheimer Labs Corp.
Burle Industries Inc.
Communication & Power Industries, EIMAC Division
Daily Electronics Corp.
Freeland Products Inc.
General Electrodynamics Corp.
Narragansett Imaging
Penta Laboratories
Richardson Electronics
Thales Components Corp.

Tuning and Phasing Units

LBA Technology Inc.

Turnkey Studio Systems

Peter Albrecht Company Inc.
Alpha Video & Electronics (AVEC)
CBT Systems
Cygnal Technology
Harris Corp., Broadcast Communications Division
POA/Paul Olivier & Associates Ltd.
Radio Systems Inc.
Technet Systems Group

TV Equipment

AVerMedia Technologies Inc.
Broadcast Electronic Services
CECO International Corp.
CED
Cable Serv Inc.
Freeland Products Inc.
Hitachi Denshi America, Ltd.
LARCAN
POA/Paul Olivier & Associates Ltd.
Schafer International
Sharp Electronics Corp., CCD Products Div.
Zenith Electronics Corp.

TV Standards Converter

Prime Image, Inc.
Snell & Wilcox Inc.
Video International Development Corp.

TV Terminal Equipment

Hoodman Corp.

Used Broadcast Equipment

Intersil Corp. Headquarters
Maze Corporation
Pro Video & Film Equipment Co. Inc.
Radiogear Inc.
Solutec Ltd. (HA)
System Associates
J.A. Taylor & Associates
United States Broadcast

Vacuum Capacitors

Jennings Technology Co.
Penta Laboratories
Richardson Electronics

Vans and Mobile Units

Alpha Video & Electronics (AVEC)
CBT Systems
E-N-G Mobile Systems Inc.
F&F Productions, L.L.C.
Frontline Communications
Maze Corporation
Panavision New York
Phoenix E N G, Inc.
Shook Mobile Technology, LP
J.A. Taylor & Associates
Television Engineering Corp.
WOIO & WUAB TV

VBI Equipment

Broadcast Video Systems Corp.
CPC-Computer Prompting & Captioning Co.
Milestone Technologies Inc.
Norpak Corporation

Video Accessories

ARRI, Inc.
Adrienne Electronics Corp.
Audiolab Electronics Inc.
AVerMedia Technologies Inc.
Battery Pros Inc.
Bencher Inc.
Broadcast Electronic Services
The J-Lab Co.
Jensen Transformers Inc.
Kings Electronics Co. Inc.
North Hills Signal Processing, a PORTA Systems Co.
Peerless Industries Inc.
Plastic Reel Corp. of America
Polyline Corp.
Research Technology International Inc.
Star Case Manufacturing Co. Inc.
Switchcraft Inc.
Thermodyne Cases
The Tiffen Company
Veriad
Video Accessory Corp.
Viking Cases
Westcott
Zack Electronics Inc.

Video Character Generators

Burst Electronics Inc.
Display Systems International Inc.
EEG Enterprises Inc.
FOR.A Corp. of America
Horita
Hotbox Digital
Inscriber Technology Corporation
Key West Technology

Video Delay Lines

Allen Avionics, Inc.
Television Equipment Assoc. Inc./Matthey

Video Delay Subsystems

Television Equipment Assoc. Inc./Matthey

Video Disc Recorders

Accom Inc.
Leightronix Inc.
MATCO Inc.
Optical Disc Corp.
Pinnacle Systems Inc.

Video Editing Equipment

Access Intelligence LLC
Accom Inc.
Avid Broadcast
Bitcentral Inc.
Broadcast Electronic Services
CORPLEX Inc.
Commercial Electronics Ltd.
Dubner International Inc.
Focus Enhancements
The Image Group Post, LLC.
Imagine Products Inc.
Matrox Video Products Grp
NEC America Inc.
Paulmar Industries Inc.
Pinnacle Systems Inc.
Richardson Electronics
Tele-Measurements Inc.
United Media Inc.
United States Broadcast

Video Effects Generators and Accessories

Cintel Inc.
FOR.A Corp. of America
Utah Scientific Inc.

Video Equipment

Access Intelligence LLC
All Mobile Video Inc.
AVerMedia Technologies Inc.
Avtech Systems Inc.
Band Pro Film & Digital Inc.
Bencher Inc.
Bexel, Inc.
Broadcast Store Inc.
CORPLEX Inc.
CamMate Studios/Systems
Dage-MTI Inc.
Dimension 3
Ensemble Designs
Fiber Options
Film/Video Equipment Service Co. Inc.
Full Compass Systems Ltd.
Geneva Aviation Inc.
HAVE Inc.
Horita
Innovision Optics Inc.
MATCO Inc.
Maze Corporation
Mohawk
National Mobile Television
Photomart Cine-Video Inc.
Pro Video & Film Equipment Co. Inc.
Professional Communications Systems
Switchcraft Inc.
Telcom Research
Televideo San Diego
Thermodyne Cases
United Media Inc.
Vicon Industries Inc.
Westcott

Video Graphics Systems

Accom Inc.
Alias/WaveFront Inc.
Chyron Corp.
Commercial Electronics Ltd.
Even Technologies
Hotbox Digital
Miranda Technologies Inc.

Video Processing Equipment

FM Systems Inc.
Pixel Instruments Corp.
Thomson Broadcast & Media Solutions
Xintekvideo Inc.

Video Production Switchers

FOR.A Corp. of America
Focus Enhancements
Ross Video Ltd.
Thomson Broadcast & Media Solutions
Videotek

Video Projection System, Large Screen

Barco Visual Solutions, LLC
SAIC
Sharp Electronics Corp., CCD Products Div.

Video Routing Switchers

Adrienne Electronics Corp.
Broadcast Electronic Services
MATCO Inc.
NTV International Corporation
Omicron Video
Sigma Electronics Inc.
Videotek

Video Special Effects Systems

Access Intelligence LLC
Accom Inc.
Chyron Corp.
Even Technologies
Image Logic Corp.
Matrox Video Products Grp
Ultimatte Corp.

Videotape Recorders

Fast Forward Video
Northrup Grumman
Sanyo Fisher Co.
Sharp Electronics Corp., CCD Products Div.
Thomson, Inc, (RCA/GE)
Toshiba America Consumer Products

Videotape Recorders, Portable

Fast Forward Video

Videotape Suppliers

BMG
Carpel Video Inc.
Fuji Film USA, Inc.
JOA Cartridge Service
Maxell Corp. of America
Memorex Products Inc.
Moviola
National Video Tape Co. Inc.
Plastic Reel Corp. of America
Tele-Measurements Inc.

Voltage Regulators

Furman Sound Inc.
Hipotronics Inc.
LEA International

Watt Meters

Coaxial Dynamics

Wave Guide Support Systems

MYAT Inc.

Wave Guides

Dielectric Communications
MYAT Inc.
Radio Research Instrument Co. Inc.
S W R Inc.

Weather Data Display Systems (WDDS)

AccuWeather Inc.
Advanced Designs Corp.
DRS Technologies
Key West Technology
Norpak Corporation
Texscan MSI
Ultimatte Corp.
WeatherBank Inc.

Weather Forecasting

AccuWeather Inc.
DRS Technologies
WeatherBank Inc.

Weather Instruments

AccuWeather Inc.
DRS Technologies
Gorman-Redlich Manufacturing Co.
Texas Electronics Inc.
Texscan MSI
WeatherBank Inc.

Weather Radar & Graphic Displays, Color

Advanced Designs Corp.
DRS Technologies
Norpak Corporation
Radio Research Instrument Co. Inc.
WeatherBank Inc.

Weighing Equipment

General Electrodynamics Corp.

Wireless Microphones

Audio-Technica U.S., Inc.
Beyerdynamic
COMTEK Inc.
Countryman Associates Inc.
Lectrosonics Inc.
Location Sound Corp.
Murry Rosenblum Sound Assoc., Inc.
Nady Systems Inc.
Professional Sound Services Inc.
R-Columbia Products Co. Inc.
Radio Engineering Industries Inc.
Sabine Inc.
Shure Inc.
Systems Wireless Ltd.
TAI Audio
TOA Electronics Inc.
Telex Communications Inc.
Vega

Wiring Products

AKG Acoustics, U.S.
Belden Electronics Divison
CableTek Wiring Products Inc.
Clark Wire & Cable Co. Inc.
Condux International
Connectronics Corp.
Copperweld Fayetteville Division
General Cable
Gepco International Inc.
Radio Systems Inc.
Ripley Company
Techni-Tool Inc.
3M
WIREMAX Ltd.
Wireworks Corp.

Satellite and Transmission Services

ABC Family Worldwide, (a subsidiary of International Family Entertainment Inc.). 500 S. Buena Vista, Burbank, CA, 91521-0001. Phone: (818) 560-1000.

Haim Saban, chmn/CEO; Maureen Smith, exec VP.

Studio City, CA 91604, 12700 Ventura Blvd. Phone: (818) 755-2400. Tony Thomopoulos, CEO MTM Entertainment.

New York, NY 10036, 1133 Ave. of the Americas, 37th Fl. Phone: (212) 782-0600. Rick Sirvaitis, pres adv sls; Barbara Bekkedahl, exec VP/ adv sls.

Basic cable network available in over 76 million homes nationwide; delivers a dynamic mix of quality entertainment with original movies, specials & series in prime time & a fun-filled daytime lineup of newly produced and classic series for kids.

AMV Gateway, Williams Communications Group. Box 420, 27 Randolph St., Carteret, NJ, 07008. Phone: (732) 969-3191. Fax: (732) 541-2007. E-mail: chelsea@amvtraffic Web Site: www.allmobilevideo.com.

Michael Carberry, dir.

Operates teleport facilities serving New York City. Videotape svcs; satellite transmission for the bcst, CATV & videoconferencing industries. International wideband voice, data & videoconferencing svcs overseas.

AT&T Alascom, 505 E. Bluff Dr., Anchorage, AK, 99501-1100. Phone: (907) 264-7274. Fax: (907) 274-5029. Web Site: www.attalascom.com.

Mike Felix, pres; Kay Witt, opns mgr.

Telecommunications, long-distance telephone carrier for the state of Alaska offering bcst, voice, data, WATS, Alaskanet & dedicated private line long-distance svcs.

Agri Net Ray Communications Inc., 104 Radio Rd., Powells Point, NC, 27966-9601. Phone: (252) 491-2414. Fax: (252) 491-2939. E-mail: info@agrinetradio.com Web Site: www.agrinetradio.com.

William S. Ray, pres/gen mgr; Lisa Ray, gen sls mgr; Bob Yanacek, opns mgr.

Turnkey satellite transmission service for radio news, sports, syndicated program distribution, audio conferencing; transportable bcst studio/uplinks for remote bcsts & domestic back hauling; network coord svcs & space segment bookings available.

American Microwave & Communications, (A division of Western Tele-Communications). 4616 N. Grand River Ave., Suite D, Lansing, MI, 48906-2576. Phone: (517) 327-3000. Fax: (517) 327-4706. E-mail: albronson@sbcglobal.net

Microwave delivery of distant signals to cable systems; net TV & radio service to bcst stns; networking among TV stations.

Arqiva Limited, Crawley Ct., Winchester, Hampshire, S021 2QA. United Kingdom. Phone: 01962 823434. Fax: 01962 822378. Web Site: www.kingstoninmedia.com. E-mail: inmedia@kcom.com

Peter Douglas, exec chmn; Joe Barry, dir opns.

Satellite svcs company providing data bcst, business TV, international VSAT nets & uplink facilities.

Ascent Media Network Services, 250 Harbor Dr., Stamford, CT, 06904. Phone: (203) 965-6000. Fax: (203) 965-6320. Web Site: www.ascentmedia.com.

Peter Brickman, mgng dir.

Minneapolis, MN 55403. Teleport Minnesota, 90 S. 11th St. Phone: (612) 330-2433. Fax: (612) 330-2603. Mark Durenberger, gen mgr.

C- & Ku-band domestic & international transmission service; fiber-optic connectivity to metropolitan New York. Cable origination, bcst, business TV, transponder availability, studio & postproduction.

BAF Satellite & Technology Corp., 65 E. NASA Blvd., Melbourne, FL, 32901. Phone: (800) 966-3822. Phone: (800) 223-1860 (24 hr). Fax: (800) 486-5983. Fax: (800) 223-1866 (24 hr). E-mail: info@bafsat.com Web Site: www.bafsat.com.

James Vautrot, pres/CEO; George Sperring, dev VP.

Long-term, short-term & occasional analog & digital KU & C band satellite space, full or partial transponder or transponders for domestic or international video, voice or data transmissions.

BT North America, Broadcast Services, 2025 M St. N.W., Washington, DC, 20036. Phone: (202) 721-8598. Fax: (202) 721-8595. E-mail: andy.cassels@bt.com Web Site: www.broadcast.bt.com.

Andrew Cassells, mgr bcstg svcs.

Operates international bcst center; 24-hour satellite rooftop teleport, digital compression & encryption svcs.

CATV Services Inc., (Penn Service Microwave Co. Inc.). 115 Mill St., Danville, PA, 17821. Phone: (570) 275-1431. Fax: (570) 275-3888. Web Site: www.catvservice.com.

Samuel Haulman, gen mgr.

Video distribution of TV signals to various CATV companies in Pennsylvania.

Communications III, 921 Eastwind Dr., Suite 104, Westerville, OH, 43081. Phone: (614) 901-4420. Fax: (614) 901-7721. E-mail: shalliday@comiii.com Web Site: www.comiii.com.

Scott Halliday, pres.

Common carrier/C-band uplink svcs. ISDN, IP, & Satellite videoconferencing, Polycom & First Virtual.

Crawford Satellite Services, 3845 Pleasantdale Rd., Atlanta, GA, 30340. Phone: (404) 876-7149. Fax: (678) 421-6717. E-mail: info@crawford.com Web Site: www.crawford.com.

Jesse Crawford, pres; Greg West, CFO.

Domestic/international satellite trasmission, network origination/playback, satellite uplink trucks, IP-media, fiber ooptics, studio & remote video production, HD/SD post production, DVD.

DCT Transmission L.L.C., 10040 E. Happy Valley Rd., Unit 454, Scottsdale, AR, 85255. Phone: (480) 515-0913. Fax: (480) 515-4632. E-mail: wiesenberg@spacedata.net

Jim Wiesenberg, dir.

Broadband microwave provider for metro-wide analog, digital, data voice or video delivery to end users or as interconnect —2-38 ghz.

DIRECTV Latin America, 1211 Ave. of the Americas, 6th Fl., New York, NY, 10036. Phone: (212) 462-5000. Fax: (212) 462-5081. E-mail: webmaster@directvla.com Web Site: www.directvla.com.

Bruce Churchill, pres.

Provides direct TV for Latin America.

GlobeCast America, 10525 Washington Blvd., Culver City, CA, 90232. Phone: (310) 845-3900. Fax: (310) 845-3904. Web Site: www.globecastamerica.com.

Paul Rush, dir opns.

Miami, FL 33166. GlobeCast Hero Productions, 7291 N.W. 74 St. Phone: (305) 887-1600. Fax: (305) 391-4424. (Corporate hqtrs).

GlobeCast America provides the bcstg industry with a unique combination of both recognized expertise & extensive inter-continental svcs. Through GlobeCast's vast global infrastructure of over 100 transponders, 30 teleports & interconnect facilities, the company provides instant access to the world's major media markets. GlobeCast America is part of France Telecom, one of the world's largest telecommunications companies.

Home Shopping Network, 1 HSN Dr., St. Petersburg, FL, 33729. Phone: (727) 872-1000. Fax: (727) 872-6615. E-mail: grossmanm@hsn.met Web Site: www.hsn.com.

Mindy Grossman, CEO; Rob Solomon, engrg VP.

Svcs include C- & Ku-band transmissions from Tampa, FL, C-band transmissions from New York, NY, & Los Angeles, CA, & postproduction.

International Telecommunications Satellite Organization (INTELSAT), 3400 International Dr. N.W., Washington, DC, 20008-3098. Phone: (202) 944-6800. Fax: (202) 944-8125. Web Site: www.intelsat.com.

John Romm, pres & media svcs.

Provider of international & domestic satellite telecommunications svcs serving more than 130 nations. Operates a global net of more than 20 satellites.

Kaufman Broadcast Services, 3655 Olive St., St. Louis, MO, 63108. Phone: (314) 533-6633. Fax: (314) 533-1113. E-mail: info@kaufmanbroadcast.com Web Site: www.kaufmanbroadcast.com.

Bill Kaufman, pres.

Transmission svcs via satellite or fiber optics, production & editing facilities.

Loral Skynet, (A subsidiary of Loral Space & Communications). 500 Hills Dr., Bedminster, NJ, 07921. Phone: (908) 470-2300. Fax: (908) 470-2459. E-mail: info@loralskynet.com Web Site: www.loralskynet.com.

Patrick K. Brandt, pres.

Miami, FL 33126, 7820 N.W. LeJeune Rd. Phone: (305) 476-0503. Fax: (305) 476-0722. (Parent Company).

Full- & part-time C- & Ku-band transponder svcs on Telstar 4, 5, 6 & 7. Provides service on both C-band & Ku-band transponders on Telstar satellites & other satellites, in addition to tech consulting & tracking, telemetry & control of satellite fleets for other customers. Applications include: coml & pub TV bcst, syndication, satellite news gathering, distance learning, videoconferencing, data networking, & occasional use. The company operates two state-of-the-art satellite earth stns in Hawley, PA, & Three Peaks, CA. The main Satellite Operations Center & the Transponder Booking Center reside at Hawley.

MCI Tampa, 3608 Queen Palm Dr., Tampa, FL, 33619-1311. Phone: (813) 829-0011. Web Site: www.mci.com.

Michael Capellas, pres/CEO.

Serves Pennsylvania Public Television Network, Pennsylvania, Ohio, New Jersey, Vermont, New Hampshire, New York & Massachusetts. CATV systems with a variety of svcs.

MCI, 500 Clinton Ctr. Dr., Clinton, MS, 39056. Phone: (601) 460-5600. Phone: (800) 644-news. Web Site: www.mci.com.

Tammy McLean, dir communications.

Data, internet, international long distance & telecom svcs throughout the United States.

Megastar Inc., 4709 Compass Bow Ln., Las Vegas, NV, 89130. Phone: (702) 386-2844. Fax: (702) 388-1250.

Nigel Macrae, pres.

Houston, TX Teleport svcs, C-band & Ku-band, fiber-optic & microwave interconnect. 11M Intelsat antenna, PanAmSat, etc. Other svcs available.

Microwave Networks Inc., 4000 Greenbriar St. #100A, Stafford, TX, 77477-3921. Phone: (281) 263-6500. Fax: (281) 263-6400. Web Site: www.microwavenetworks.com.

Jim Gordon, gen mgr.

Microwave Service Co., 1359 Rd. 681, Saltillo, MS, 38866. Phone: (662) 842-7620. Fax: (662) 844-7061. E-mail: lharris@wtba.com

Jane Spain, CEO; Larry Harris, gen sls mgr.

Point-to-point transmission of video by microwave.

Multicomm Sciences International Inc. (MSI), 266 W. Main St., Denville, NJ, 07834. Phone: (973) 627-7400. Fax: (973) 215-2168. E-mail: mail@multicommsciences.com Web Site: www.multicommsciences.com.

Victor J. Nexon, Jr., pres.

Serves satellite, microwave, lightwave, radio & cable industries. Market info, field surveys, system design, feasibility studies, frequency coordination & project mgmt.

NPR Satellite Services, 635 Massachusetts Ave. N.W., Washington, DC, 20001. Phone: (202) 513-2626. Fax: (202) 513-3035. E-mail: linkup@npr.org Web Site: www.nprss.org.

George Gimourginas, dir.

Comprehensive satellite solutions for audio & video distribution: space segment, equipment, systems design, engrg support 24/7 customer svc.

Norlight Telecommunications, 13935 Bishops Drive, Brookfield, WI, 53005. Phone: (888) 210-3100. Fax: (262) 792-7660. E-mail: mjt@norlight.com Web Site: www.norlight.com.

James Ditter, pres; Michael J. Turnbull, mgr new business

dev.

Design, installation & maintenance of custom network solutions, including HDTV transport, on-site data center - secure co-location svc, off-site video file servers & managed router svc.

Novanet Communications Ltd., 725 Westney Rd. S., Suite 4, Ajax, ON, L1J 7J7. Canada. Phone: (905) 686-6666. Fax: (905) 619-1053. E-mail: getinfo@novanetcomm.com Web Site: www.novanetcomm.com.

Joseph Uyede, pres/CEO; Debbie MacLeod, VP sls & mktg; Stewart Sheriff, VP opns.

Provides info distribution svcs; audio svcs from 3.5 khz to 20 khz in analog or digital formats, also a line of data bcst offerings ranging in speed from 75 baud to T-1; & "Satpac," a packet-switched data net using receiver technology.

PMTV Producers Management Television, 681 Moore Rd., Suite 100, King of Prussia, PA, 19406. Phone: (610) 768-1770. Fax: (610) 768-1773. E-mail: mailto.pmtv@pmtv.com Web Site: www.pmtv.com.

Full-svc mobile TV production company, providing mobile units, crews, satellite svcs, lighting, staging, etc. for sports, entertainment & teleconferences worldwide.

Pacific Microwave, 926 S. Grape St., Medford, OR, 97501. Phone: (541) 773-4171.

Common carrier microwave svc for cable systems & bcst TV.

PanAmSat, 20 Westport Rd., Suite 270, Wilton, CT, 06897. Phone: (203) 210-8000. Fax: (203) 210-8001. Web Site: www.panamsat.com. E-mail: corpcomm@panamsat.com

Mike Antonovich; Joseph R. Wright, Jr., CEO.

Long Beach, CA 90810, 1600 Forbes Way. Phone: (310) 525-5500. Fax: (310) 525-5505. Kurt Riegelman. (North America office).

Coral Gables, FL 33134, One Alhambra Plaza, Suite 100. Phone: (305) 445-5536. Fax: (305) 445-5315. Estevao Ghizoni, sr dir. (Latin America office).

Owns & operates private global network of communications satellites providing bcst, business communications, telephony & data svcs to customers worldwide.

Potomac Television-News & Video Services, 1510 H St. N.W., Suite 202B, Washington, DC, 20005. Phone: (202) 783-6464. Fax: (202) 783-1132. E-mail: jnorins@potomactv.com Web Site: www.ptpngroup.com.

Jamie Norins, News Bureau Chief.

TV production facilities include studios, on-line & non-linear editing & transmission facilities to Ku-band & C-band satellites. Interconnect to major news-gathering points in Washington, DC, at the National Press Bldg.

Production & Satellite Services Inc. (PSSI), 11860 Mississippi Ave., Los Angeles, CA, 90025. Phone: (310) 575-4400. Fax: (310) 575-4451. E-mail: pssi@800satlink.com Web Site: www.pssi-usa.com.

Joseph A. Kittrell, VP bcst sls; Robert Lamb, pres.

Full service production & satellite transmission company specializing in coordination, production & transmission of international live-event progmg. Own & operate 12 fully redundant transportable upling/production trucks which are maintained in Los Angeles, San Francisco, Seattle, Las Vegas, Denver, Phoenix & Chicago. Entire Western region of the United States is covered including Albuquerque, Salt Lake City & Portland. Also subcontract with vendors throughout U.S. to provide uplink, downlink & production svcs and production svcs for teleconferences & other events.

Radio Sound Network, 770 Twin Rivers Dr., Columbus, OH, 43215. Phone: (614) 621-5600. Fax: (614) 621-5620. E-mail: steve.clawson@radiohio.com Web Site: www.radiohio.com.

Steve Clawson, engrg dir; Tony Miller, dir.

Columbus, OH 43215, 605 S. Front St. Phone: (614) 460-3850.

Full-svc digital satellite audio & data distribution, including affil rel & net bldg. for new & existing sports, news/talk, music & specialty networks.

Reuters Television Internationale, 3 Times Sq., 4th Fl., New York, NY, 10036. Phone: (646) 223-6600. Fax: (646) 223-6615. Web Site: www.reuters.com.

Bob LaGrasso, dir client svcs.

Satellite svcs for news departments. Satellite production & communication departments handle satellite feed requirements from anywhere in the world. Satellite coordinates, video crews & major studios on six continents. Standards conversion.

SES Americom, 4 Research Way, Princeton, NJ, 08540. Phone: (609) 987-4000. Fax: (609) 987-4517. Web Site: www.ses-americom.com.

Bryan A. McGuirk, pres & media solutions.

Operates GE-1-3, SATCOM, GSTAR & SPACENET domestic satellites (C-band & six Ku-band). The fleet svcs the cable TV, bcst, radio & educ market & govt businesses. Supports net of earth stns, central terminal offices & TT&C facilities.

SpaceCom Systems, A TV Guide Company. 1950 E. 71st St., Tulsa, OK, 74136-5422. Phone: (800) 950-6690. Fax: (918) 477-6861. Web Site: www.spacecom.com.

David Pollack, pres/CEO; Ruth Ann Odom, mktg coordinator.

Satellite transmission svcs, equipment & space segment on C- & Ku-Band for point-to-multipoint applications. Also two-way high-speed Satellite Broadband for remote locations, quick connects, disaster recovery, Internet.

Spacenet Services Inc., 1750 Old Meadow Rd., McLean, VA, 22102. Phone: (703) 848-1000. Web Site: www.spacenet.com.

Jim Norton, VP network opns; Glenn Katz, COO.

Comprehensive range of satellite-based communication svcs for video & data; digital video networks for business applications.

StarNet, 3417 N. First St., Abilene, TX, 79603. Phone: (888) 828-7352. Phone: (325) 672-9618. E-mail: info@usdlc.org Web Site: www.usdlc.org.

Glenda Mathis, owner.

Ku-band distance learning, Ku-band transponder time, & program/production svcs.

Teleglobe, 1555 Rue Carrie-Derick, Montreal, PQ, H3C 6W2. Canada. Phone: (514) 868-7272. Fax: (514) 868-7234. Web Site: www.teleglobe.com. E-mail: jean-louis-houde@vsnlinternational.com

Vinod Kumar, pres; Jean-Louis Houde, VP opns.

Hong Kong 02508, Two Pacifco Place, 88 Queensway. Phone: 852-2530-8500. Fax: 852-2-537-7417. Andrew Kwok, dir.

International satellite transmission svcs from Lawrentides/Lake Cowichan earth stns in Canada. Signatory on Intelsat, Immarsat. Full range of svcs to world satellite systems.

Telemundo Network, 2470 W. 8th Ave, Hialeah, FL, 33010. Phone: (305) 884-8200. Web Site: www.telemundo.com.

James McNammara, pres.

Production capabilities & svcs. Uplink & video transmission. C- & Ku-band uplink & downlink. Fiber & microwave also available.

Telenor Satellite Services, 1101 Wooton Pkwy, Rockville, MD, 20852. Phone: (301) 838-7800. Fax: (301) 838-7801. E-mail: customer.care@telenor.com Web Site: www.telenor.com/satellite.

Bob Baker, pres; Dave Farmer, VP mktg.

Rockville, MD 20852. Telenor Satellite Services, Inc., 1101 Wootton Pkwy., 4th Fl. Phone: (301) 838-7800. Fax: (301) 838-7701.

Telenor is a global provider of satellite svcs, inmarsat, intelsat, new skies, satmex, digital networking svcs & technology.

Teleport Chicago, 3617 Oakton St.-rear, Skokie, IL, 60076. Phone: (800) 875-4657. Phone: (888) 255-8755. Fax: (847) 674-7485. E-mail: sales@norlight.com Web Site: www.norlight.com.

Dave Pritchard, dir.

A full-svc teleport serving the upper Midwest via the Norlight Telecommunications microwave & fiber-optic transmission system.

Telesat Canada, 1601 Telesat Ct., Gloucester, ON, K1B 5P4. Canada. Phone: (613) 748-0123. Fax: (613) 748-8712. E-mail: info@telesat.ca Web Site: www.telesat.ca.

Calgary, AB T2E 0A6 Canada, 1780 Centre Ave. N.E. Phone: (403) 235-5751. Fax: (403) 273-3337. (Calgary rgnl office).

Montreal, PQ H2K 4R5 Canada, 1200 Papineau Ave, Suite 140. Phone: (514) 521-7862. Fax: (514) 527-6429. (Montreal rgnl office).

Communications via satellite, consulting, & satellite earth stn nets.

Time Warner Cable, 290 Harbor Dr., Stamford, CT, 06902. Phone: (800) 479-0624. Web Site: www.timewarnercable.com.

Glenn A. Britt, pres/CEO.

New York, NY 10019, 75 Rockerfeller Plaza. Phone: (212) 484-8000.

Time Warner Cable is committed to in-home entertainment, communications, info, customer care & quality products that create the best possible customer experience.

Transvision Inc., 550 Maulhardt Ave., Oxnard, CA, 93030. Phone: (805) 981-8740. Fax: (805) 981-8738. E-mail: kvvaughn@txvision.com Web Site: www.txvision.com.

Kimithy Vaughan, sls/mktg dir; Vince Waterson, engr.

Twelve transportable & satellite transmission facilities (video, audio, voice, data); flypack production & SNG svcs; digital compression; domestic & international.

TV Guide Inc., One TV Guide Plaza, 7140 S. Lewis Ave., Tulsa, OK, 74136-5422. Phone: (918) 488-4000. Fax: (918) 488-4979. Web Site: www.tvguide.com.

Josh Axelrod, dir progmg mgmt; Mike Burks, sr database admin.

Diversified communications company serving cable, home satellite TV, radio/data networks, private businesses; operating companies: UVTV, Prevue Networks, Superstar Satellite Entertainment, SpaceCom Systems.

Verestar, 2010 Corporate RDG, Suite 600, McClean, VA, 22102. Phone: (703) 610-1000. E-mail: info@verestar.com Web Site: www.verestar.com.

Raymond J. O'Brien, pres/COO.

Verestar, a global service provider, delivers integrated satellite, terrestrial, and Internet solutions to customers around the globe. With an unparalleled infrastructure including teleports in the US and Europe, more than 150 antennas and extensive space and terrestrial capacity, Verestar seamlessly manages customers' communications with end-to-end solutions and flexible service options.

Videocom Media Services, LLC, Box 212, Boston, MA, 02137. Phone: (781) 329-4080. Fax: (781) 329-8534. E-mail: traffic@videocom.com Web Site: www.videocom.com.

Daniel V. Swartz, mgr.

Videocom Teleport Boston. Steerable C- & Ku-band antennas interconnected via private microwave & telco loops co-located with single- & multiple-camera program origination facility. Receives & transmits all formats of videotape; transportable uplinks; bcst news distribution; link with Canadian satellites. Digital video transmission & net mgmt svcs. Videocam Teleport Rio de Janerio & Videocam Teleport Sao Paulo.

Vision Accomplished/Hawaii Overseas Teleport, 91-110 Hanua St., Suite 31, Barbers Point, HI, 96862. Phone: (808) 682-0032.

Craig Landis, pres.

Portable uplink & downlink systems plus full-svc teleport facility on Honolulu. Studio & tape svcs plus fiber connections.

WGVU-TV, 301 W. Fulton, Grand Rapids, MI, 49504-6492. Phone: (616) 331-6666. Fax: (616) 331-6625. E-mail: wgvu@gvsu.edu Web Site: www.wgvu.org.

Michael Walenta, gen mgr; Ken Kolbe, stn mgr; Bob Lumbert, engrg dir.

Provides multiple studio post production, teleconferencing & satellite uplink svcs.

WHYY Inc., Independence Mall W., 150 N. 6th St., Philadelphia, PA, 19106. Phone: (215) 351-1200. Fax: (215) 351-0398. E-mail: pgluck@whyy.org Web Site: www.whyy.org.

Paul Gluck, VP/stn mgr.

Occasional video, encryption svcs, transponder time available. Interconnect with Telco, on-site production facilities, & teleconferencing for up to 1,000 people. Transportable Ku-band earth stn; C- or Ku-band uplinking.

WSAC, 9601 South Meridian, Englewood, CO, 80112. Phone: (800) 367-3193. Fax: (303) 723-3006. E-mail: pcocommercial@echostar.com

Charles Ergen, pres; Sean Payton, opns mgr.

The premiere provider of digital satellite progmg, mktg support & informational svcs to the private cable & wireless cable industry serving multi dwelling units (MDVs).

WTVS-Analog 56 & Digital 43 Detroit Public Television, 7441 Second Ave., Detroit, MI, 48202. Phone: (313) 873-7200. Fax: (313) 876-8118. E-mail: email@dptv.org Web Site: www.detroitpublictv.org.

Daniel Alpert, gen mgr.

Offers Ku-band uplinking to any satellite; uplink, fiberlink, downlink, production, postproduction & teleconference svcs available.

Warren Only Media Group, 19 West Almond Street, Vineland, NJ, 08360. Phone: (856) 507-9368. Fax: (856) 507-9368. E-mail: warren@warrenonly.com Web Site: www.warrenonly.com.

Warren Only, pres.

New York, NY 10036, Box 2372. Phone: (917) 339-0036.

Program playout, turnarounds, uplink services.

Williams Communications, 111 E. 1st St., Vyvx Services, Tulsa, OK, 74103. Phone: (918) 547-5760.

Jeff Storey, CEO.

Long Beach, CA 90802, 200 Oceangate, Suite 570. Phone: (800) 747-7074.

Atlanta, GA 30329, 1802 Briarcliff Rd. Phone: (800) 648-3333.

Park Ridge, NJ 07656, One Maynard Dr. Phone: (800) 746-3019.

Provider of integrated fiber-optic, satellite & teleport multimedia and data gathering, management and transmission services.

Wiltel Communications - VYVX, One Technology Center, Tulsa, OK, 74103. Phone: (866) 945-8351. E-mail: info@vyvx.com Web Site: www.wiltel.com.

F/T K2 lease, time available. Two SNV units. C- & Ku-band earth stns, turnaround. Complete production facility, multi-format duplication.

Teleports

Alexandria, VA—Verestar, 2010 Corporate RDG, Suite 600, McClean, VA, 22102. Phone: (703) 610-1000. E-mail: info@verestar.com Web Site: www.verestar.com.

Raymond J. O'Brien, pres/COO.

Verestar, a global service provider, delivers integrated satellite, terrestrial, and Internet solutions to customers around the globe. With an unparalleled infrastructure including teleports in the US and Europe, more than 150 antennas and extensive space and terrestrial capacity, Verestar seamlessly manages customers' communications with end-to-end solutions and flexible service options.

Atlanta, GA—Atlanta International Teleport, 3530 Bomar Rd., Douglasville, GA, 30135. Phone: (770) 949-6600. Fax: (770) 942-6653. E-mail: sales@atlantateleport.com Web Site: www.atlantateleport.com.

Adam Grow, III, chief engr.

Internet, private business networks, VSAT Hub, C/Ku Up/down, audio, video, data, telephony, TDMA, SCPC, VCII+, fiber, standards conversion, Intelsat-B Station.

Atlanta, GA—Crawford Satellite Services, 3845 Pleasantdale Rd., Atlanta, GA, 30340. Phone: (404) 876-7149. Phone: (800) 831-8027. Fax: 678) 421-6717. E-mail: pweber@crawford.com Web Site: www.crawford.com.

Pat Weber, satillite account exec; Jesse Crawford, chmn; Jim Schuster, sr VP satellite svcs.

On-line, net origination, transportables, & transponder brokerage.

Atlanta, GA—Turner Teleport Inc., Box 105366, One CNN Ctr., Atlanta, GA, 30348-5366. Phone: (404) 827-1500.

Satellite uplink & downlink svcs for Turner Broadcasting Services.

Atlanta, GA—VYVX Teleport Atlanta, 1802 Briarcliff Rd., Atlanta, GA, 30329. Phone: (404) 325-0818. Fax: (404) 325-3949. E-mail: jeff.huffman@level3.com Web Site: www.level3.com.

Jeff Huffman, gen mgr.

VYVX's UpSouth Teleport offers domestic U.S. & international uplink, downlink & transponder service for video (analog or compressed), data, & voice telecommunication. Fiber-optic & microwave links to points of presence for major bcst & telco locations in the Atlanta, GA, metropolitan area. Direct access to all C- & Ku-band satellites in domestic U.S. arc & AOR. Svcs include program origination; tape playback, recording & editing; standards conversion; encryption; & turnarounds.

Boston, MA—Videocom Media Services, LLC, Box 212, Boston, MA, 02137. Phone: (781) 329-4080. Fax: (781) 329-8534. E-mail: traffic@videocom.com Web Site: www.videocom.com.

Daniel V. Swartz, mgr. Ownership: Private.

Videocom Teleport Boston. Steerable C- & Ku-band antennas interconnected via private microwave & telco loops co-located with single- & multiple-camera program origination facility. Receives & transmits all formats of videotape; transportable uplinks; bcst news distribution; link with Canadian satellites. Digital video transmission & net mgmt svcs. Videocam Teleport Rio de Janerio & Videocam Teleport Sao Paulo.

Catawissa, PA—Roaring Creek International Teleport/AT&T, 311 Earth Station Rd., Catawissa, PA, 17820. Phone: (570) 799-1000. Phone: (570) 799-1012. Fax: (570) 799-1042. E-mail: erubenstein@att.com Web Site: www.att.com.

Doug Rubenstein, mgr.

Ant 1, 2, 3, 4, 5. G/T 40.91 international voice, data, video to primary path, major path 1 & 2, international satellites, Fuji fiber, connectivity. Facilities are available for transportable interconnect or permanent transmission space. All C-band ants are Intelesat standard meters & all Ku-band ants are standard C meters.

Chicago, IL—Echostar, 6723 W. Steger Rd., Monee, IL, 60449. Phone: (708) 534-2400. Fax: (708) 534-0060. E-mail: lawrence.baer@echostar.com

Larry Baer, mgr. Ownership: Chicago, IL.

C- & Ku-band satellite transmission svcs. Audio, video, data, uplink/downlink communication.

Chicago, IL—Norlight Teleport Chicago, 3617 Oakton St., Skokie, IL, 60076. Phone: (847) 568-7195. Fax: (847) 674-7485. E-mail: drp@norlight.com Web Site: www.norlight.com.

Dave Pritchard, dir.

Brookfield, WI 53045, 275 N. Corporate.

Full-svc teleport, domestic & international, satellite uplink, downlink, turnaround, encryption, rgn access on Norlight Telecommunications Interstate Network, VYVX ant AT&T fiber access.

Cincinnati, OH—WLWT-TV, 1700 Young St., Cincinnati, OH, 45202. Phone: (513) 412-5000. Fax: (513) 412-6100. E-mail: rdyer@hearst.com Web Site: www.channelcincinnati.com.

Richard Dyer, gen mgr; Mark Diangelo, gen sls mgr.

On-line. Provides occasional C-band uplinking svcs.

Culver City, CA—GlobeCast North America, 3872 Keystone Ave. S., Culver City, CA, 90232. Phone: (310) 845-3900 (sales). Phone: (310) 845-3939. Fax: (310) 845-3904. E-mail: robert.marking@globecastna.com Web Site: www.globecastna.com.

Mary Frost, CEO.

Washington, DC 20006, 1825 K St. N.W., 9th Fl. Phone: (202) 861-0894. Fax: (202) 861-3107.

Washington, DC 20001, 400 North Capitol St. N.W, Suite 880. Phone: (202) 737-4440. Fax: (202) 737-1476. (International Sales).

Miami, FL 33166. GlobeCast Hero Productions, 7291 N.W. 74 St. Phone: (305) 887-1600. Fax: (305) 887-7076. (Corporate Headquarters).

New York, NY 10017, 110 E. 42nd St., 11th Fl. Phone: (212) 885-8700. Fax: (212) 885-8701. (Eastern Region office).

Keystone's Los Angeles International Teleport, in Culver City, CA & Sylmar, offers a transmission network of domestic & international satellite transponders providing video & audio origination svcs for over 1,000 united states & international clients, including news, sports, program distribution & business TV clientele. The company utilizes its extensive worldwide network of space & terrestrial facilities in the united states, as well as through its international parternships.

Dallas, TX—Megastar Inc., 4709 Compass Bow Ln., Las Vegas, NV, 89130. Phone: (702) 386-2844. Fax: (702) 388-1250.

Nigel Macrae, pres.

Denver, CO—VYVX Teleport Denver, 9174 S. Jamaica St., Englewood, CO, 80112. Phone: (303) 397-4100. Fax: (303) 799-8325.

Theran Davis, opns mgr. Ownership: Denver.

Domestic & international uplink, downlink & transponder service for video & date communications. Regional Fiber & microwave interconnectivity to broadcast affiliates, PoPs & sports venues

Edmonton, AB—Edmonton Teleport. Telesat Canada, 5311 Allard Way, Edmonton, AB, T6H 5B8. Canada. Phone: (780) 437-6167. Fax: (780) 436-5667. Web Site: www.telesat.ca.

Major bcst teleport offering full North American arc at C-band, & Anik E1 & E2 at Ku-band for occasional use needs.

Jackson, MS—Jackson Teleport Inc., 916 Foley St., Jackson, MS, 39202. Phone: (800) 353-9177. Phone: (601) 352-6673. Fax: (601) 948-6052. E-mail: edward@weathervision.com Web Site: www.weathervision.com.

Edward Saint Pe, pres; Jason McCleave, VP.

Fixed 7-meter earth stn on site. Video satellite transmission & reception. Videoconferencing, business TV, news, sports & weathercast feed, origination, program distribution & syndication svcs.

Los Angeles, CA—Williams Services/VYVX Steele Valley Teleports, 20021 Santa Rosa Mine Rd., Perris, CA, 92570. Phone: (909) 943-5399. Fax: (909) 943-3459. E-mail: gene.brookhart@wcg.com

Gene Brookhart, gen mgr.

Englewood, CO 80112, 58 Inverness Dr. E. Phone: (303) 397-4100. (800) 424-9757 (full-time svcs).

Steele Valley Teleports offers domestic U.S. & international uplink, downlink & transponder service for video (analog or compressed), image, data & voice telecommunication. Fiber-optic & microwave links to points of presence for major bcst & telco locations in the Los Angeles metropolitan area. Direct access to all C- & Ku-band satellites in domestic U.S. arc & POR. Svcs include program origination; tape playback, recording & editing; standards conversion; encryption; & turnarounds. Williams Services/VYVX Steele Valley Teleports coast-to-coast teleport locations, which include Los Angeles, CA; New York, NY; Denver, CO; & Atlanta, GA, are interconnected via fiber-optic backbone.

Montreal, PQ—Montreal Teleport. Telesat Canada, 1601 Telesat Ct., Gloucester, ON, K1B 5P4. Canada. Phone: (613) 748-0123. Fax: (613) 748-8712. E-mail: info@telesat.ca Web Site: www.telesat.ca. Ownership: Montreal.

Access to all Telesat Anik and Nimiq satellites & most U.S. domestic satellites.

New Orleans, LA—Network Teleports Inc., 3200 Chartres St., New Orleans, LA, 70117. Phone: (504) 942-9200. Fax: (504) 942-9204. E-mail: uplink@networkteleports.com Web Site: www.networkteleports.com.

Barbara Lamont, pres; Ludwig Gelobter, VP; C.E. Feltner, chmn of bd; Nolly Paul, VP opns. Ownership: New Orleans.

C-band voice/video & data, Ku-band data/voice, B-MAC encryption, newsfeeds, production, fiber-optic links, audio-subcarrier & C- & Ku-band 5CPC, satellite telephones, IP multicasting, webcasting.

New York, NY—GlobeCast North America, 5 Teleport Dr., Staten Island, NY, 10311. Phone: (718) 983-2600. Fax: (718) 983-2615. E-mail: robert.marking @globecastna.com Web Site: www.globecast.com.

Mary Frost, CEO.

Keystone's New York Teleport in Staten Island, NY, offers a transmission network of leased domestic & international satellite transponders providing radio & TV origination svcs for more than 1,000 U.S. & international clients every year, including news, sports, program distribution & business TV clientele. Bob Bechar, pres.

Oakland, CA—ICG Telecom Group, 161 Inverness Dr., West Englewood, CO, 80112.

Alternative Access Carrier. Bay Area Teleport provides communications svcs at DS0, T1 or T3 levels for primary or alternative access applications. Bay Area Teleport's system connects 12 counties in northern California. Includes a fiber-optic network in San Francisco, CA & across to Oakland, CA, as well as access to satellite svcs through its earth stn complex in Niles Canyon.

Pittsburgh, PA—Pittsburgh Telecommunications, 201 Browntown Rd., New Kensington, PA, 15068. Phone: (724) 337-1888. Fax: (724) 337-1754. E-mail: info@pitcomm.com Web Site: www.pitcomm.com.

Bill Sciolla, mgr video opns; Al Stem, pres.

Fixed & remote uplinking & downlinking, turnarounds, space segment, microwave & fiber-optic interconnects, point-to-point & VSAT svcs.

Stamford, CT—Ascent Media Network Services, 250 Harbor Dr., Stamford, CT, 06902. Phone: (888) 826-4016. Fax: (203) 965-6405. E-mail: networkservices @ascentmedia.com Web Site: www.ascentmedia.com.

Peter Brickman, mgng dir. Ownership: Stamford

C- & Ku-band domestic & international transmission service; connectivity to metro New York via proprietary fiber origination, bcst, business TV, transponder availability; studio & postproduction svcs.

Toronto, ON—Toronto Teleport. Telesat Canada, 1601 Telesat Ct., Gloucester, ON, K1B 5P4. Canada. Phone: (613) 748-0123. Fax: (613) 748-8712. E-mail: info@telesat.ca Web Site: www.telesat.ca. Ownership: Toronto.

Access to all Telesat Anik & Nimiq satellites & most united states domestic satellites.

Vancouver, BC—Vancouver Teleport. Telesat Canada, 1601 Telesat Ct., Gloucester, ON, K1B 5P4. Canada. Phone: (613) 748-0123. Fax: (613) 748-8712. E-mail: info@telesat.ca Web Site: www.telesat.ca. Ownership: Vancouver.

Access to all Telesat Anik & Nimiq satellites & most North American satellites.

Washington, DC—Potomac Television-News & Video Services, 1510 H St. N.W., Suite 202B, Washington, DC, 20005. Phone: (202) 783-6464. Fax: (202) 783-1132. E-mail: jnorins@potomactv.com Web Site: www.ptpngroup.com.

Jamie Norins, News Bureau Chief.

TV production facilities include studios, on-line & non-linear editing & transmission facilities to Ku-band & C-band satellites. Interconnect to major news-gathering points in Washington, DC, at the National Press Bldg.

Woodbury, NY—Rainbow Network Communications, 620 Hicksville Rd., Bethpage, NY, 11714. Phone: (516) 803-0355. Fax: (516) 803-4924. E-mail: togreco@rainbow-media.com Web Site: www.rncnetwork.com.

Steve Pontillo, pres; Thomas A. Greco, VP business & dev.

Multiple 11-meter & 9-meter uplinking antennas, multiple downlinking ants, both servicing the entire satellite arc. Connectivity in & out of New York & metropolitan area, Ku-band transportable, origination/editing svcs, & full, longterm & occasional transponder leasing. Compression svcs.

Section G

Professional Services

Station and Cable System Brokers

American Media Services L.L.C., Box 20696, Charleston, SC, 29413. Phone: (843) 972-2200. Fax: (843) 881-4436. E-mail: ams@ams.fm Web Site: www.americanmediaservices.com.

Edward F. Seeger, pres/CEO.

Bonham, TX 75418. Dallas office:, 9208 Timbercreek Dr. Phone: (903) 640-5857. Fax: (903) 640-5859. David Reeder, rgnl broker.

Forest Lake, IL 60047. Chicago office:, 24180 N. Forest Dr. Phone: (847) 540-5410. Frank McCoy, exec VP/engrg.

Marble Falls, TX 78654. Austin office:, 303 Avenue Q. Phone: (877) 267-2636. Patrick McNamara, rgnl broker.

Developers & brokers of radio properties. Also appraisals, search svcs (buyer's agent), upgrade studies.

Associated Broadcasters, Inc., Box 42566, Cincinnati, OH, 45242. Phone: (513) 791-5982. Fax: (513) 891-5727.

Irv Schwartz, mgng dir & pres.

Legal & filing svcs in turnkey packages; consulting & appraisal svcs available.

Barger Broadcast Brokerage, Ltd., 8023 Vantage Dr., Suite 840, San Antonio, TX, 78230. Phone: (210) 340-7080. Fax: (210) 341-1777. E-mail: jwbarger@sbcglobal.net

John W. Barger, pres.

Media brokerage, loc mktg agreements, financial placements, appraisals & mgmt/financial consulting.

Blackburn & Co. Inc., 201 N. Union St., Suite 340, Alexandria, VA, 22314. Phone: (703) 519-3703. Fax: (703) 519-9756.

James W. Blackburn Jr., chmn.

Radio & TV stations and communications tower brokerage, financing and appraisals.

Frank Boyle & Co., L.L.C., 2001 W. Main St., Suite 280, Stamford, CT, 06902. Phone: (203) 969-2020. Fax: (203) 316-0800. E-mail: fboylebrkr@aol.com

Frank Boyle, pres.

Radio & TV media brokerage, mergers & acquisitions/appraisals.

Broadcast Media Associates, Box 1233, Santa Maria, CA, 93456. Phone: (805) 937-1553. Fax: (805) 937 1553. E-mail: cliffhunter@cliffhunter.com Web Site: broadcastmediabroker.com.

Clifford M. Hunter, pres.

Radio/TV/cable brokerage in the western states. Confidential mktg for radio, TV & LPTV properties; valuation packages & financial analysis; consultants to sellers & buyers.

Broadcast Media Partners, 8216 N. 54th St., Scottsdale, AZ, 85253. Phone: (480) 951-1897. Fax: (480) 596-8385. E-mail: ted@nicholsoncompany.com

Ted Nicholson, ptnr.

Murrieta, CA 92562, 3855 Lochinvar Ct. Phone: (909) 698-1131. Fax: (909) 696-0998. F. Patrick Nugent, ptnr. (Los Angeles office).

Brokerage svcs, valuations, due diligence, Far East consulting, court-approved takeover specialists.

BroadcastStations4Sale.com, 905 N. Gilbreath St., Graham, NC, 27253. Phone: (336) 570-9133. Fax: (336) 570-3464. E-mail: tedjgray@netzero.net Web Site: www.broadcaststations4sale.com.

Ted J. Gray, pres.

Buy or sell radio stns or TV stns. List to sell yourself or let BroadcastStations4Sale.com sell it for you.

Broadcasting Asset Management Corp., 1323 Forest Glen, Winnetka, IL, 60093. Phone: (847) 446-8882. Fax: (847) 446-4855.

Jack Minkow, pres.

Radio stn & group stn brokerage, mergers, acquisition analysis, appraisals, feasibility studies; subdebt & equity placement.

Bulkley Capital, L.P., 5949 Sherry Ln., #1370, Dallas, TX, 75225. Phone: (214) 692-5476. Fax: (214) 692-9309. E-mail: info@bulkleycapital.com Web Site: www.bulkleycapital.com.

G. Bradford Bulkey, pres/CEO; Oliver Cone, VP; Lisa Bulkley, VP.

Investment banking: mergers, acquisitions & private placements of debt & equity capital.

Business Broker Associates, Box 4757, Chattanooga, TN, 37405-0757. Phone: (423) 756-7635. Fax: (423) 870-9036. E-mail: bba@cdc.net Web Site: www.cdc.net/~bba.

Alfred C. Dick, broker.

Media broker & consultant for radio, TV & cable systems.

Chaisson & Company Inc., 154 Indian Waters Dr., New Canaan, CT, 06840. Phone: (203) 966-6333. Fax: (203) 966-1298. E-mail: rchaisco@aol.com

Robert A. Chaisson, pres.

Brokerage of radio/TV sls & acquisitions. Robert A. Chaisson, pres.

S.R. Chanen & Co. Inc./Media Technology Capital Corp., 3300 N. 3rd Ave., Phoenix, AZ, 85013. Phone: (602) 234-1411. Fax: (602) 285-9268.

Steven R. Chanen, chmn; Donald E. New, pres acquisitions & investments.

Investment banking, brokerage & financial advisory svcs for the communications & entertainment industries.

Chapin Enterprises, 1248 "O" St., Lincoln, NE, 68508. Phone: (402) 475-5285. Fax: (402) 475-5293. E-mail: dchapin@inetnebr.com

R.W. Chapin, pres.

Stn sls, consulting, financial counseling, per location & receiver. Media broker, consulting.

CobbCorp, LLC, 800 Laurel Oak Dr., Suite 210, Naples, FL, 34108. Phone: (212) 812-5020. Fax: (239) 596-0660. E-mail: briancobb@cobbcorp.com Web Site: www.cobbcorp.com.

Brian Cobb, pres; Denis LeClair, VP; Daniel Graves, mgng dir; Jack Higgins, mgng dir.

New York, NY 10022, 445 Park Ave., 9th Fl. Phone: (212) 812-5020.

Mergers, acquisitions, investment banking & merchant banking.

Communication Resources Media Brokers, 5343 E. 22nd St., Tulsa, OK, 74114. Phone: (918) 743-8300. Fax: (918) 749-3348. E-mail: tbelc@cox.net

Tom Belcher, pres.

Brokerage svc to radio & TV stns, cable TV companies, & individual telephone companies.

Communications Equity Associates, 101 E. Kennedy Blvd., Suite 3300, Tampa, FL, 33602. Phone: (813) 226-8844. Fax: (813) 225-1513. Web Site: www.ceaworldwide.com.

J. Patrick Michaels Jr., chmn/CEO; Robert Berger, mgng dir; Carsten Philipson, mgng dir; Ken Jones, sr VP & gen counsel; Donald Russell, mgng dir; Ming Jung Sr., mgng dir/CFO-CEA.

Westport, CT 06880, 191 Post Rd. W. Phone: (203) 221-2662. Fax: (203) 221-2663. Dave Moyer, mgng dir.

Miami, FL 33131, 150 S.E. 2nd Ave., Suite 609. Phone: (305) 810-2740. Fax: (305) 810-2741. Jose Rodriguez, mgng dir-Latin Americian Communication Ptnr.

New York, NY 10020, 1270 Avenue of the Americias, Suite 1818. Phone: (212) 218-5085. Fax: (212) 218-5099. Alexander Rossi, Bob Ennis, Waldo Glasman, Paul Miller, Evan Blum, mgng directors, Jason Donnell, dir.

Investment banking, corporate finance & private equity firm specializing in the cable, bcstg, telecommunications, media, & entertainment industries.

The Connelly Co., 17909 Holly Brook Dr., Tampa, FL, 33647-2245. Phone: (813) 991-9494. Fax: (813) 991-9444. E-mail: connellyradiotv@verizon.net

Robert J. Connelly, pres.

New Market, NH 03857, 198 S. Main St. Phone: (603) 659-3648.

South Effingham, NH 03882. Effingham (Summer only), Bailey Rd. Phone: (603) 522-6462. Fax: (603) 522-6348.

Brokers, consultants & recovery unit to assist banks & financial institutions.

Cox & Cox L.L.C., 2454 Shiva Ct., St. Louis, MO, 63011. Phone: (636) 458-4780. Fax: (636) 273-1312. E-mail: bc@coxandcoxllc.com Web Site: www.coxandcoxllc.com.

Robert Cox, pres.

Mergers & acquisitions, appraisals, consulting & expert testimony.

Diversified Investment Services, Inc., 17146 S.E. 23rd Dr., Suite 58, Vancouver, WA, 98683. Phone: (503) 221-1122. Phone: (800) 635-1772. Fax: (360) 882-4760. E-mail: a.stilli@yahoo.com

Armand J. Santilli, pres.

Media brokers/finders, real estate investment bankers & brokers.

Earl Reilly Enterprises, 550 Aloha St., Suite 404, Seattle, WA, 98019. Phone: (206) 282-6914. Fax: (206) 285-2765. E-mail: earlcant@comcast.net Web Site: www.tvspotnet.com.

Earl F. Reilly, pres.

Freeland, WA 98249, Box 1300. Phone: (206) 331-7223. Fax: (206) 331-7223.

Bcst rep, representing U.S. TV stns in Canada. Also licensed bcst stn brokers.

Edwin Tornberg & Co. Inc., 8917 Cherbourg Dr., Potomac, MD, 20854. Phone: (301) 983-8700. Fax: (301) 299-2297.

Edwin Tornberg, pres.

Appraisals, brokerage, financial & mgmt consulting for radio, TV & cable.

EnVest Media, LLC, 6802 Patterson Ave., Richmond, VA, 23226. Phone: (804) 282-5561. Fax: (804) 282-5703. E-mail: mitt@envestmedia.com Web Site: www.envestmedia.com.

Mitt Younts, mgng member.

Alpharetta, GA 30004, 11770 Haynes Bridge Rd, #205. Phone: (770) 753-9650. Fax: (770) 753-0089. E-mail: jswnet@aol.com. Jesse Weatherby, member, mgr.

Nationwide radio, TV acquisition, valuation, financing & consulting firm. The company provides brokerage svcs to stn transaction, appraisal svcs to stn owners & financial institutions. The group secures debt & equity acquisition financing, offers consulting & asset mgmt svcs, acting as court appointed receivers or trustees for bcst stns.

The Exline Company, 4340 Redwood Hwy., Suite F-230, San Rafael, CA, 94903. Phone: (415) 479-3484. Fax: (415) 479-1574. E-mail: exline@pacbell.net Web Site: www.exline.com.

Andrew P. McClure, pres; Erick Stunberg, assoc.

Complete brokerage, consulting & appraisal svcs for radio & TV properties.

Explorer Communications Inc., 3615 W. Treyburn Path, Lecanto, FL, 34461. Phone: (352) 746-7121. Fax: (352) 746-4255. E-mail: jfhoff@tampabay.rr.com

Jim Hoffman, pres.

Bcst media brokerage svcs. Specialists in medium & small market entrepreneurial transactions.

Norman Fischer & Associates Inc., Box 5308, Austin, TX, 78763-5308. Phone: (512) 476-9457. Fax: (512) 476-0540. E-mail: terrill@nfainc.com Web Site: www.nfainc.com.

Terrill Fischer, pres.

Brokerage in radio, TV & cable. Consultation in mgmt & opns, appraisals, feasibility studies, expert testimony, financial planning & assistance.

Richard A. Foreman Associates, Inc., 330 Emery Dr. E., Stamford, CT, 06902-2210. Phone: (203) 327-2800. Fax: (203) 967-9393. E-mail: ra@rafamedia.com Web Site: www.rafamedia.com.

Richard A. Foreman, pres.

Specializing in cash-positive radio & TV stns in major growth mkts.

Michael Fox International, Inc., 11425 Cronhill Dr., Owings Mills, MD, 21117. Phone: (410) 654-7500. Fax: (410) 654-5876. E-mail: info@michaelfox.com Web Site: www.michaelfox.com.

William Z. Fox, chmn; David S. Fox, CEO.

Beverly Hills, CA 90212, 9454 Wilshire Blvd, Penthouse. Phone: (310) 248-2821. Adam Reich, co-pres.

Baltimore, MD 21117, 11425 Cronhill Dr. Phone: (410) 654-7500. Gilbert Schwartzman, sr VP.

New York, NY 10601, 75 S. Broadway, 4th Fl. Phone: (914) 723-7600. Jonathan Reich, pres.

Auction sls of bcst properties.

Fugatt Media Services, 9214 Butternut Dr., Crystal Lake, IL, 60014. Phone: (815) 546-1470. E-mail: fugattmediaservices@comcast.net

Michael L. Fugatt, pres.

Media brokerage firm specialing in radio & cable svcs including appraisals.

Gammon Media Brokers L.L.C., 5219 N. Casa Blanca Dr., Suite 16, Paradise Valley, AZ, 85253. Phone: (480) 614-6612. Fax: (480) 614-6613. E-mail: cmiller@gmbi.com Web Site: www.gmbi.com.

Christopher D. Miller, pres/CEO; James A. Gammon, chmn.

Chevy Chase, MD 20815, 5600 Wisconsin Ave, Suite 308. Phone: (301) 332-0940. James A. Gammon, chmn.

Brokerage & strategy advice to sellers & buyers of radio stns & TV stns; nwspr and cable TV systems.

Clifton Gardiner & Company, L.L.C., 2437 S. Chase Ln., Lakewood, CO, 80227. Phone: (303) 758-6900. Fax: (303) 479-9210. E-mail: cliff@cliftongardiner.com Web Site: www.cliftongardner.com.

Clifton H. Gardiner, pres.

Brokerage & financial svcs for the bcst & cable industries.

Dave Garland Media Brokerage, 1110 Hackney St., Houston, TX, 77023. Phone: (713) 921-9603. Fax: (713) 926-2694. E-mail: garland@radiobroker.com Web Site: www.radiobroker.com.

David Garland, owner.

Broker of radio stn properties in Texas & surrounding states.

HPC Puckett & Co., Box 9063, Rancho Santa Fe, CA, 92130. Phone: (858) 756-4915. Fax: (858) 756-4534.

Thomas F. Puckett, CEO.

Topeka, KS 66614, 2921 S.W. Wanamaker Dr, Suite 104. Phone: (913) 273-0017.

Communications brokerage & investment banking.

Hadden & Associates, Media Brokers - Orlando. 147 Eastpark Dr., Celebration, FL, 34447. Phone: (321) 939-3141. Fax: (321) 939-3142. E-mail: haddenws@aol.com Web Site: www.haddenonline.com.

Doyle Hadden, pres/CEO; Ryan P. Hadden, VP.

Communications broker, acquisitions, divestitures; financial assistance and appraisal to the broadcasting industry.

Margret Haney Media Brokerage & Consultants, 2995 Woodside Rd., Bldg. 400, Woodside, CA, 94062-2446. Phone: (408) 557-8887. Fax: (408) 557-0808. E-mail: www.hhmargret@yahoo.com Web Site: www.margret@margrethaney.com.

Margret Haney, CEO.

Media brokerage/consulting, adv, media buyers.

Hardesty & Associates, 500 East Balboa, Newport Beach, CA, 92661. Phone: (949) 723-2230. Fax: (949) 723-2240. E-mail: info@hardestyassociates.com Web Site: www.hardestyassociates.com.

Bill Hardesty, pres; Lillian Barnhart, office mgr.

Acquisitions, divestitures, appraisals, financing & consulting.

Hawkeye Radio Properties Inc., 3325 Conservancy Ln., Middleton, WI, 53562. Phone: (608) 831-8708. Fax: (608) 831-6100. E-mail: dganske@charter.net

Dale A. Ganske, pres.

Complete radio/TV brokering, consulting, FCC rules & regulations.

Henson Media Inc., 455 S. Fourth Ave., Suite 427, Louisville, KY, 40202-2508. Phone: (502) 589-0060. Fax: (502) 589-0058. E-mail: edhenson1@cs.com Web Site: www.hensonmedia.com.

Ed Henson, pres.

Campbellsville, KY 42718, 7811 Saloma Rd. Phone: (270) 789-9513. Bryan McForland.

Radio & TV brokers.

The Ted Hepburn Co., 325 Garden Rd., Palm Beach, FL, 33480. Phone: (561) 863-8995. Fax: (561) 863-8997. E-mail: tedhep2@mac.com

Ted Hepburn, mgng dir.

Radio, TV & cable brokerage; appraisals.

R. Miller Hicks & Co., 1011 W. 11th St., Austin, TX, 78703. Phone: (512) 477-7000. Fax: (512) 477-9697. E-mail: millerhicks@rmhicks.com

R. Miller Hicks, pres.

Business dev & consulting svcs since 1957.

Holt Media Group, 2178 Industrial Dr., Suite 914, Bethlehem, PA, 18017. Phone: (610) 814-2821. Fax: (610) 814-2826. E-mail: artholt@holtmedia.com Web Site: www.holtmedia.com.

Christine E. Borger, exec VP.

Macon, GA 31220-5260, 6826 Bay Point Dr. Phone: (478) 474-8161. Fax: (610) 814-2826. Carl Strandell, assoc.

Brokerage, consulting, appraisals.

Bruce Houston Associates, Inc., 2251 Hunter Mill Rd., Vienna, VA, 22181. Phone: (703) 938-1016. Fax: (703) 938-6078. Web Site: brucehouston@aol.com.

Bruce Houston, pres.

Media brokers for radio & TV.

International Media Consulting, 48 Mountain Rd., Farmington, CT, 06032-2341. Phone: (860) 677-9688. Fax: (860) 677-9639. E-mail: robert.richer@snet.net

Robert E. Richer, owner.

Broker specializing in the sale of overseas media properties.

Johnson Communication Properties Inc., 880 Old Crystal Bay Rd., Minneapolis, MN, 55391. Phone: (952) 4041-1104. Fax: (952) 404-1102. E-mail: johncomm88@aol.com

Jerry Johnson, pres.

Radio & TV broker for 25 years.

Jorgenson Broadcast Brokerage Inc., 426 S. River Rd., Tryon, NC, 28782-7879. Phone: (828) 859-6982. Fax: (828) 859-6831. E-mail: goradiotv@aol.com

Mark W. Jorgenson, pres.

Cupertino, CA 95014-2358, 19995 Stevens Creek Blvd. Phone: (408) 973-7292. Fax: (408) 516-9526. Peter Mieuli.

Confidential, nationwide brokerage of bcst properties.

Kalil & Co. Inc., 3444 N. Country Club, Suite 200, Tucson, AZ, 85716. Phone: (520) 795-1050. Fax: (520) 322-0584. E-mail: kalil@kalilco.com

Media brokerage firm dealing in radio, TV & cable. Handles exclusive listings, confidential searches.

Kempff Communications Co., 3301 Bayshore Blvd., Suite 1407, Tampa, FL, 33629. Phone: (813) 258-3433. Fax: (813) 902-1360. E-mail: kempffcc@aol.com Web Site: www.kepffbarr.com.

Ron Kempff, pres; Aurelia Serna, VP.

Broker & consultant also offering financial & mgmt svcs, court ordered sale of stns.

Kepper, Tupper & Company, 112 High Ridge Ave., Ridgefield, CT, 06877. Phone: (203) 431-3366. Fax: (203) 431-3864. E-mail: keppertup@aol.com Web Site: kepper-tupper.com.

John B. Tupper, pres.

Brokerage & investment banking svcs for the cable & bcst TV industries. Please visit our website kepper-tupper.com.

Knowles Media Brokerage Services, Box 9698, Bakersfield, CA, 93389. Phone: (661) 833-3834. Fax: (661) 833-3845. E-mail: gregg.knowles@netzero.net Web Site: www.media-broker.com.

Gregg K. Knowles, pres.

Daily & wkly nwsprs, print publications/sls, consultation. Sls, mergers, acquistions, appraisals.

The Kompas Group, Box 250813, Milwaukee, WI, 53225-6513. Phone: (262) 781-0188. Fax: (262) 781-5313. E-mail: kompasgroup@toast.net Web Site: www.thelptvstore.com.

John Kompas, pres; Burt Sherwood, ptnr.

Sarasota, FL 34242, 5053 Ocean Blvd., Suite 14. Phone: (941) 349-2165. Fax: (941) 312-0974.

Financial & mktg svcs for LPTV; audience demographic reports; stn coverage maps; appraisals/brokerage; financial & strategic planning.

Kozacko Media Services, Box 948, Elmira, NY, 14902. Phone: (607) 733-7138. Fax: (607) 733-1212. E-mail: rkozacko@stny.rr.com Web Site: www.kozackomediaservices.com.

Richard L. Kozacko, pres.

Tucson, AZ 85750, 6890 E. Sunrise Dr, Box120-40. Phone: (520) 299-4869. Fax: (520) 299-1786. George W. Kimble, assoc.

Keswick, VA 22947, 1071 Club Dr. Phone: (434) 244-2653. Fax: (434) 244-2666. W. Donald Roberts, Jr., assoc.

Appraisals & current market evaluations of radio & TV stns; bcst stn acquisition brokers.

H.B. LaRue, Media Brokers, 9454 Wilshire Blvd., Suite 628, Beverly Hills, CA, 90212. Phone: (310) 275-9266. Fax: (310) 274-4076. E-mail: hblarue@sbcglobal.net

New York, NY 10021, 500 E. 77th St, Suite 1909. Phone: (212) 288-0737. Hugh Ben LaRue, pres; Joy Thomas, VP.

Media brokerage; TV, radio & CATV. Appraisals & feasibility studies.

Lattice Communications, L.L.C., 441 Vine St., Suite 3900, Cincinnati, OH, 45202. Phone: (513) 381-7775. Fax: (513) 381-8808. Web Site: www.latticecommunications.com.

R. Dean Meiszer, pres/CEO.

Acquisition, merger, appraisal & financial svcs to radio, TV, nwspr, cable & other media-related industries.

Lazard L.L.C., 30 Rockefeller Plaza, New York, NY, 10020. Phone: (212) 632-6000. Web Site: www.lazard.com.

Michael Castellano, CFO; Robert Hougie, mgng dir; Bruce Wasserstein, CEO.

Lazard's broad range of svcs includes: gen financial advice; domestic, cross-border mergers & acquisitions; divestitures; privatizations; special committee assignments; takeover defenses; corporate restructurings; strategic partnerships/joint ventures; & debt/equity underwriting.

Legacy Securities Corp., 4684 Rosewell Rd., N.E., Atlanta, GA, 30342. Phone: (404) 965-2420. Fax: (404) 965-2421. Web Site: www.legacysecurities.com.

Michael D. Easterly, chmn/CEO; Chris Battel, pres.

Joe M. Leonard Jr. & Associates Inc., Box 222, Gainesville, TX, 76241. Phone: (940) 665-4076. E-mail: lin45@ntin.net Web Site: www.rockabillyhall.com/joeleonard.html.

Joe M. Leonard Jr., pres.

Brokerage of radio & TV.

Jack Maloney Inc., 28 Shore Dr., Huntington, NY, 11743. Phone: (631) 549-2656. Fax: (631) 549-2656.

Jack Maloney, pres.

Provides confidential svcs to buyers & sellers in the radio & TV business.

Mayo Communications Inc., Box 82784, Tampa, FL, 33682. Phone: (813) 264-5050. Fax: (813) 264-5353.

Lincoln A. Mayo, pres.

Media brokerage, & appraisals for bcst & print. Consulting concerning media sls & acquisitions.

R.E. Meador & Associates, Inc., Box 36, Lexington, MO, 64067. Phone: (660) 259-2544. Fax: (660) 259-6424. E-mail: remeador@earthlink.net

Ralph E. Meador, pres.

Acquisitions, sls, mktg studies & appraisal svcs in central & midwestern states.

Media Services Group Inc., 3948 S. Third St. #191, Jacksonville Beach, FL, 32250. Phone: (904) 285-3239. Fax: (904) 285-5618. E-mail: reedmsconsulting@cs.com Web Site: www.mediaservicesgroup.com.

George R. Reed, mgng dir.

Morristown, NJ 07960. New York Metro, 45 Park Pl. S. #146. Phone: (973) 631-6612. Fax: (973) 631-6613. E-mail: rtmck2515@aol.com. Tom McKinley, dir.

Placerville, CA 95667, 2020 Carson Rd. Phone: (415) 828-9576. Fax: (415) 933-3362. E-mail: mckinley55@msn.com. Michael McKinley, assoc.

Colorado Springs, CO 80906-1073. Colorado Springs, CO, 2910 Electra Drive. Phone: (719) 630-3111. Fax: (719) 630-1871. E-mail: jbmccoy@adelphia.net. Jody McCoy, dir.

St. Simons Island, GA 31522. St. Simons Island, GA, 205 Marina Dr. Phone: (912) 634-6575. Fax: (912) 634-5770. E-mail: edwesser@adlphia.net. Eddie Esserman, dir.

Overland Park, KS 66209. Kansas City, KS, 5225 W. 122nd Street. Phone: (913) 498-0040. Fax: (913) 498-0041. E-mail: 75767.3151@compuserve.com. Bill Lytle, dir; Mike Lytle, assoc.

Providence, RI 02903. Providence, RI, 170 Westminster St., Suite 701. Phone: (401) 454-3130. Fax: (401) 454-3131. E-mail: rmaccini@ cox.net; scs@scsloan.com. Robert J. Maccini, dir; Stephan Sloan, assoc; Ted Clark, analyst.

Richardson, TX 75080. Dallas, TX, 1131 Rockingham Dr, Suite 209. Phone: (972) 231-4500. Fax: (972) 231-4509. E-mail: whitelytx@cs.com. Bill Whitley, dir.

Logan, UT 84341. Salt Lake City, 1289 North 1500 E. Phone: (435) 753-8090. Fax: (435) 753-2980. E-mail: ggm@cache.net. Greg Merrill, dir.

One of the nation's leading full svc media brokerage, valuation & consulting firms with in-depth industry knowledge & market expertise.

Media Venture Partners, 2 Jackson St., Suite 100, San Francisco, CA, 94111. Phone: (415) 391-4877. Fax: (415) 391-4912. Web Site: www.mediaventurepartners.com.

Elliot B. Evers, mgng dir; Brian Pryor, VP; Adam Altsuler, VP.

Radio & TV brokerage svcs; mergers & acquisitions; telecom; investment banking.

Mitchell & Associates, 8319 DL Tally Dr., Shreveport, LA, 71115. Phone: (318) 798-7816. Fax: (318) 798-6084. Fax: (318) 797-5987.

John Mitchell, pres.

Media brokers, appraisers, consultants.

Montcalm, Box 4608, Rolling Bay, WA, 98061-0608. Phone: (206) 780-1700. Fax: (206) 842-7151. E-mail: jerden@aol.com

Jerry Dennon, pres.

Radio & TV brokerage svcs, mergers & acquisitions, & investment banking.

George Moore & Associates Inc., 6918 Wildglen Dr., Suite 100 W, Dallas, TX, 75230. Phone: (214) 369-5665. Fax: (214) 369-5667. Web Site: www.gmaservices.com.

W. James Moore, pres.

Brokerage of radio, TV, asset & market appraisals prepared for owners, buyers & lenders.

Gordon P. Moul & Associates Inc., Box 42, York Haven, PA, 17370. Phone: (717) 266-4212. Fax: (717) 266-0780. E-mail: gpmassociatesinc@netscape.com

Gordon Moul, pres; Joyce Moul, VP & sec.

Bcst broker & consultant. Specializing in East Coast AM-FM & TV.

MyMediaBroker.com, 407 Broadmoor Acres, Portales, NM, 88130. Phone: (505) 356-2000. Fax: (505) 356-2003. E-mail: sandi@mymediabroker.com Web Site: www.mymediabroker.com.

Sandi Bergman, pres/CEO.

Full svc media brokerage firm.

New England Media L.L.C., Box 594, Willimantic, CT, 06226. Phone: (860) 456-1111. Fax: (860) 456-9501. E-mail: support@expage.com Web Site: www.expage.com/pagenemedia.

Michael C. Rice, mgng ptnr.

Mansfield Center, CT 06250, 50 Kaya Lane. Phone: (860) 455-1414. Michael Rice, mgng ptnr.

Media brokers, consultants & appraisers specializing in radio in the Northeast.

O'Grady & Associates, 8431 Sabal Palm Ct., Vero Beach, FL, 32963-4296. Phone: (772) 234-4177. Fax: (772) 231-9819. E-mail: jfogjr@juno.com

James F. O'Grady, Jr., pres; Jane A. O'Grady, VP.

Confidential media brokerage svcs, consultants & appraisals.

Patrick Communications L.L.C., 6805 Douglas Legum Dr., Suite 100, Elkridge, MD, 21075. Phone: (410) 799-1740. Fax: (410) 799-1705. E-mail: patrick@patcomm.com Web Site: www.patcomm.com.

Larry Patrick, pres; Susan Patrick, exec VP; Greg Guy, VP.

Stn brokerage, investment banking, mgmt consulting svcs, appraisals & opns consulting.

John Pierce & Company L.L.C., 11 Spiral Dr., Suite 3, Florence, KY, 41042. Phone: (859) 647-0101. Fax: (859) 647-2616. E-mail: jpierce@johnpierceco.com Web Site: www.johnpierceco.com.

Rebecca Grizovic, finance dir.

Radio, TV, & cable sls & appraisals.

Questcom Media Services Inc., 10925 David Taylor Dr., Suite 100, Charlotte, NC, 28262. Phone: (704) 948-9800. Fax: (704) 948-9888. E-mail: drbussell@aol.om

Donald R. Bussell, pres.

Radio & TV stn brokerage specialists concentrating in top 150 markets; offering asst with mergers & consolidations. Registered FDIC & RTC broker/appraiser.

RBC Daniels, L.P., 3200 Cherry Creek Dr. S., Suite 500, Denver, CO, 80209. Phone: (303) 778-5555. Fax: (303) 778-5599. Web Site: www.rbcdaniels.com. E-mail: info@rbcdaniels.com

Brian Deevy, chmn/CEO.

New York, NY 10022, 711 Fifth Ave, Suite 405. Phone: (212) 935-5900. Fax: (212) 863-4859. Greg Ainsworth, sr mgng dir ; David Tolliver, mgng dir.

Provides mergers, acquisitions, corporate finance & financial advisory svcs to the cable, telecommunications media & technology industries.

Stan Raymond & Associates Inc., Box 8231, Longboat Key, FL, 34228. Phone: (941) 383-9404. Fax: (941) 383-9132. E-mail: stnray@aol.com Web Site: stanraymond.com.

Stan Raymond, pres.

Financial svcs, media brokers, appraisers & consultants specializing in the Southeast.

Gordon Rice Associates, Box 20398, Charleston, SC, 29413. Phone: (843) 884-3590. Fax: (843) 881-0358. E-mail: gordon@gordonriceassociates.com

Gordon Rice, broker.

Brokerage svcs, appraisals & investment analysis for radio & TV.

Riley Representatives, 240 Lakeland Dr., Highland Village, TX, 75077. Phone: (972) 966-1715. Fax: (972) 966-1015.

Jack Riley, pres.

Roehling Broadcast Services Ltd., 7340 Oak Knoll Dr., Indianapolis, IN, 46217. Phone: (317) 887-1945. Fax: (317) 887-1947. E-mail: edradiobr@aol.com Web Site: roehlingbroadcast.com.

Edward W. Roehling, pres; Sandra Roehling, sec/treas.

Bcst appraisers, brokers, consultants, also financing, sls, mgmt consultation.

Ray H. Rosenblum, Media Broker / Appraiser / Consultant, Box 38296, Pittsburgh, PA, 15238. Phone: (412) 362-6311. Fax: (412) 362-6317.

Ray H. Rosenblum, broker.

Media brokering, appraising, financing & consulting for radio & TV stns in 50 states plus territories.

Rumbaut & Company, 888 Brickell Bay Dr., Miami Beach, FL, 33141-1024. Phone: (305) 868-0000. Fax: (305) 577-8222. E-mail: julio@rumbaut.com Web Site: www.rumbaut.com.

Julio Rumbaut, pres.

Media brokers & consultants in all facets of the TV & radio industries.

Sailors & Associates, 150 Executive Center Dr., Suite 4, Box 67, Greenville, SC, 29615. Phone: (864) 297-0530. Fax: (864) 297-0595.

Don F. Sailors, owner.

Media brokerage & specializing in radio, TV & cable.

SalesGroup, 41 Herbert Rd., Boston, MA, 02184-5507. Phone: (781) 848-4201. Fax: (781) 848-4715. E-mail: salesgroup@beld.net

Abbe Raven, pres.

Stn brokers, Northeast.

Satterfield & Perry Inc., 7211 Fourth Ave. S., St. Petersburg, FL, 33707. Phone: (727) 345-7338. Fax: (727) 345-3809. E-mail: eraust@prodigy.net

Robert Austin, pres; John Willis, sec/treas.

Wetumpka, AL 36093, 169 Mountain Meadows La. Phone: (334) 514-2241. Fax: (334) 514-2291. Ken Hawkins, VP.

Centennial, CO 80016, 20456 E. Orchard Pl. Phone: (303) 400-5150. Fax: (303) 400-5063. Jim Birschbach, VP.

Denver, CO 80210, 2020 S. Monroe St. #302. Phone: (303) 758-1876. Fax: (303) 756-1865. Al Perry, chmn emeritus.

Littleton, CO 80162, Box 620308-B. Phone: (303) 948-2200. Fax: (303) 948-3468. Joe Benkert, VP.

Coos Bay, OR 97420, PO Box 362. Phone: (541) 751-0043. Fax: (541) 751-0043. Dick McMahon, VP.

Overland Park, KS 66207, 4918 W. 101st Terr. Phone: (913) 649-5103. Fax: (913) 649-5103. Douglas Stephens, sr VP.

Aiken, SC 29803, 131 Inwood Dr. Phone: (803) 649-0031. Fax: (803) 649-7786. John Willis, VP.

Buffton, SC 29909, 14 Bellereve Dr. Phone: (843) 706-9737. Fax: (843) 706-9737. Tony Rizza, VP.

Radio & TV broker, mgmt & sls consultant, FDIC-approved appraiser & expert witness.

John W. Saunders, Media Broker, 1207 Woodhollow Dr., Suite 3101, Houston, TX, 77057. Phone: (713) 789-4222. Fax: (713) 789-4322. E-mail: theradiobroker@aol.com Web Site: www.theradiobroker.com.

John W. Saunders, owner.

Nationwide radio brokerage & appraisals. Buyers or sellers represented on a confidential, professional & personal basis. Top 10 to small markets.

Serafin Bros. Inc., Box 262888, Tampa, FL, 33685. Phone: (813) 885-6060. Fax: (813) 885-6857. E-mail: gserafin@tampabay.rr.com

Glenn Serafin, pres.

Tampa, FL 33615, 4212 Deepwater Ln.

Bcst brokerage, finance & valuation svcs.

Burt Sherwood & Associates Inc., 6415 Midnight Pass Rd., Suite 206, Sarasota, FL, 34242. Phone: (941) 349-2165. Fax: (941) 312-0974. E-mail: bohica1@comcast.net

Burt Sherwood, pres.

Brokerage radio, TV & LPTV; appraisals.

Barry Skidelsky, Esq., 185 E. 85th St., 23D, New York, NY, 10028. Phone: (212) 832-4800.

Barry Skidelsky, owner.

Acquisitions, divestitures, mergers, time brokerage, financing.

Snowden Associates, Box 1966, One Commerce Sq., Suite 200, Washington, NC, 27889. Phone: (252) 940-1680. Fax: (252) 940-1682. E-mail: zophsnowden@earthlink.net

Zoph Potts, chmn; Ray Bergevin, pres.

Brokers, consultants & appraisers to the bcst industry in the Southeast.

Howard E. Stark, 575 Madison Ave., 10th Fl., New York, NY, 10022. Phone: (212) 355-0405.

Howard E. Stark, pres.

Media broker; mergers & acquisitions in the communications field.

Gary Stevens & Co., 49 Locust Ave., Suite 107, New Canaan, CT, 06840. Phone: (203) 966-6465. Fax: (203) 966-6522. E-mail: deelmakur@aol.com

Gary Stevens, pres.

Bcst mergers, acquisitions & investment banking svcs.

Stonemark Inc., Box 21305, Seattle, WA, 98111. Phone: (206) 343-7777. Fax: (206) 628-0839. E-mail: info@stonemark.net Web Site: www.stonemark.net.

William V. May, pres.

Specializing in business brokerage, acquisition searches, capital arrangement & debt placement for bcst properties as well as other industries.

The Thorburn Co., 6625 Hwy. 53 E., Suite 410-72, Dawsonville, GA, 30534. Phone: (678) 513-1363. Fax: (678) 513-1615. E-mail: thorburnco@aol.com Web Site: www.thorburncompany.com.

Robert M. Thorburn, pres.

Appraisals, brokerage, financial & mgmt consulting for radio, TV & cable.

Van Huss Media Services Inc., 4239 Heyward Pl., Indianapolis, IN, 46250. Phone: (317) 813-0106. Fax: (317) 813-0107. E-mail: vanhussmediaservices@aol.com

William "Bill" Van Huss, pres.

Brokerage, financial consulting & appraisals for radio, TV & cable.

The Venture Group, 415 Discovery Rd., Virginia Beach, VA, 23451. Phone: (757) 491-5444. Fax: (757) 422-0727. E-mail: hhurst@hrfn.net

Herbert M. Hurst, pres.

Intermediaries in the sale, merger & acquisition of radio & TV stns.

Ed Walters & Associates, 1170 Clearwater Ct., Palatine, IL, 60067. Phone: (847) 359-6117. Fax: (847) 359-6167. E-mail: radiobroker@msn.com Web Site: www.edwaltersandassoc.com.

Ed Walters, pres; Michael Walters, VP; Karrol Walters, sec.

Waukesha, WI 53186, 1801 Coral D. Phone: (414) 544-6800. Fax: (414) 544-1705. Ed Walters, pres.

Nationwide brokers, specializing in radio, TV & cable systems, acquisition searches with confidentiality ensured.

The Whittle Agency, 12716 Lindley Dr., Raleigh, NC, 27614. Phone: (919) 848-3596. Fax: (919) 848-0519. E-mail: thewhittleagency@nc.rr.com

Gary L. Whittle, pres.

Total media brokerage svcs, including sls/appraisals of radio stns in the Carolinas, Virginia & Southeast.

Willis Broadcasting, 645 Church St., Suite 400, Norfolk, VA, 23510. Phone: (757) 624-6500. Fax: (757) 624-6515. E-mail: willisbroadcasting@yahoo.com

L. E. Willis Sr., pres.

Wood & Co. Inc., 431 Ohio Pike, Suite 200 N., Cincinnati, OH, 45255. Phone: (513) 528-7373. Fax: (513) 528-7374.

Larry C. Wood, pres.

Nationwide brokerage service to buyers & sellers of TV & radio properties.

Management and Marketing Consultants

AVI Communications Inc., 517 Huffines Blvd., Lewisville, TX, 75056. Phone: (214) 637-5464. Phone: (800) 221-2842. Fax: (214) 637-6285. E-mail: info@avi-communications.com Web Site: www.avi-communications.com. Patrick Shaughnessy, pres/CEO.

TV sls training, new business dev svcs & Butch Harmon golf tips for TV & radio.

Abt Associates Inc., 55 Wheeler St., Cambridge, MA, 02138. Phone: (617) 492-7100. Fax: (617) 492-5219. E-mail: webmaster@abtassociated.com Web Site: www.abtassoc.com. Peg Laplan, pres/CEO; John Shane, chmn.

Washington, DC 20005, 1110 Vermont Ave. N.W. Phone: (202) 263-1800. (202) 263-1801.

Chicago, IL 60610, 640 N. LaSalle. Phone: (312) 867-4000. Fax: (312) 867-4200.

Bethesda, MD 20814, 4800 Montgomery Ln, Suite 600. Phone: (301) 913-0500. Fax: (301) 652-3618.

Mktg rsch, strategic planning, mgmt consulting, audience rsch & segmentation; customer satisfaction programs, quality of svc programs, social science survey rsch, publ policy rsch, economical rsch.

John P. Allen Airspace Consultants Inc., 290 Marsh Lakes Dr., Fernandina Beach, FL, 32034. Phone: (904) 261-6523. Fax: (904) 277-3651. E-mail: maryjpa@bellsouth.net Web Site: johnpallenairspace.com. Mary C. Lowe, pres.

Conducts FAA aeronautical evaluations as specified in Subpart C of Part 77 of the Federal Aviation Regulations.

Anderson Productions Ltd., 37 W. 20th St., Suite 904, New York, NY, 10011. Phone: (212) 414-9220. Fax: (212) 206-0279. E-mail: andersontv@aol.com Steven C.F. Anderson, pres & exec producer.

TV program production & consulting firm specializing in info progmg for cable & bcstg TV.

Nick Anthony & Associates Inc., 1795 W. Market St., Akron, OH, 44313. Phone: (330) 864-2268. Fax: (330) 864-2261. E-mail: nick@nickanthony.com Web Site: www.nickanthony.com. Nick Anthony, pres.

Mktg, progmg & motivational consultant.

The Aspen Institute Communications & Society Program, 1 Dupont Cir. N.W., Suite 700, Washington, DC, 20036. Phone: (202) 736-5818. Fax: (202) 467-0790. E-mail: firestone@aspeninstitute.org Web Site: www.aspeninstitute.org/c&s. Charles M. Firestone, exec dir.

Pub policy seminars & reports.

Associated Broadcasters, Inc., Box 42566, Cincinnati, OH, 45242. Phone: (513) 791-5982. Fax: (513) 891-5727. Irv Schwartz, mgng dir & pres.

Legal & filing svcs in turnkey packages; consulting & appraisal svcs available.

Audience Research & Development (AR&D), 8828 Stemmons Fwy., Suite 500, Dallas, TX, 75247. Phone: (214) 630-5097. Fax: (214) 630-4951. E-mail: jgumbert@ar-d.com Web Site: www.ar-d.com. Jerry Gumbert, CEO & pres.

Rsch-based, full-svc, new media consulting firm serving TV stns, cable systems, internet companies, nwsprs & program syndicators.

The Austin Company, 6095 Parkland Blvd., Cleveland, OH, 44124. Phone: (440) 544-2684. Fax: (440) 544-2690. E-mail: austininfo@theaustin.com Web Site: www.theaustin.com. Patrick B. Flanagan, pres/COO; Dennis Raymond, sr VP.

Other Branches: Atlanta, GA; Cleveland, OH; Kansas City, MO; Irvine, CA.

Consulting, architectural design, engrg & construction svcs for TV, cable & radio bcstg facilities.

AZCAR, 121 Hillpointe Dr., Suite 700, Canonsburg, PA, 15317. Phone: (724) 873-0800. Fax: (724) 873-4770. E-mail: info@azcar.com Web Site: www.azcar.com. Richard Bisignano, pres; Mary Nahra, VP.

Video & audio system consultation, systems integration, design, instal & training; serving cable systems, cable mfg, corporate, bcst & teleproduction facilities & engng.

BIA Financial Network, 15120 Enterprise Ct., Suite 100, Chantilly, VA, 20151. Phone: (703) 818-2425. Fax: (703) 803-3299. E-mail: info@bia.com Web Site: www.bia.com. Thomas J. Buono, chmn/CEO.

Financial & strategic consultants to communications industries offering fair market valuations, expert tax appraisals, due diligence, acquisition consulting, business plans, internal operational audits, litigation support, investment banking, venture funding, capital, industry rsch & analysis publications & software.

BTMI (Broadcast Trustee Management Inc.), 1090 Vermont Ave. N.W., Suite 800, Washington, DC, 20005. Phone: (202) 408-7036. Fax: (202) 408-1590. E-mail: Probinson@aol.cin Web Site: www.btmi.com.

Financial-asset mgmt, valuation, mktg, restructuring & recovery consultation svcs.

Bayliss Broadcast Foundation, Box 51126, Pacific Grove, CA, 93950. Phone: (831) 655-5229. Fax: (831) 655-5228. E-mail: khfranke@baylissfoundation.org Web Site: www.baylissfoundation.org. Kit Hunter Franke, exec dir; Carl Butrum, pres.

Bayliss Radio Intern program & scholarships for college students studying for a career in radio are primary focus of Foundation.

The Benchmark Co., 907 S. Congress, Suite 207, Austin, TX, 78704-1700. Phone: (512) 707-7500. E-mail: Benchmark@aol.com Web Site: www.Benchmarkresearch.com. Rob Balon, pres.

Full-svc bcst consulting & rsch company featuring benchmark perceptual phone surveys & the Focus 100 system, which replaces focus groups.

The Benton Group, Box 5076, Vancouver, WA, 98668. Phone: (360) 574-7369. Fax: (360) 576-6866. Web Site: www.thebentongroup.net. Donald Benton, CEO.

Yellow-page & nwspr experts, specialized sls training programs & seminars.

Beveridge Institute of Sales & Sales Management, 113 North Grant St., Barringto, IL, 60010. Phone: (847) 381-7797. Phone: (800) 227-4332. Fax: (847) 381-7301. E-mail: info@beveridgeinc.com Web Site: www.beveridgeinc.com. Dick Beveridge, pres.

Sls & sls mgmt performance & productivity improvement programs; training workshops.

Big Blue Dot, 63 Pleasant St., Watertown, MA, 02472. Phone: (617) 923-2583. Fax: (617) 923-8014. E-mail: bigbluedot@bigblue.com Web Site: www.bigblue.com. Jan Craige Singer, pres.

Trend tracking resources for the kids' market, consulting, & newsletter via e-mail; creative svcs.

Blackburn & Co. Inc., 201 N. Union St., Suite 340, Alexandria, VA, 22314. Phone: (703) 519-3703. Fax: (703) 519-9756. James W. Blackburn, chmn; Richard F. Blackburn, pres; Tony Rizzo, broker.

Acquisition svcs of all kinds including appraisals, brokerage & financing for radio & TV stations and communications towers.

Blair Productions, Box 42513, Washington, DC, 20015. Phone: (301) 652-9040.

Full-svc mgmt & financial counseling to the bcst industry with emphasis on radio turnarounds, problems & start-ups; investment banking svc.

Mark Blinoff Inc., 1837 S.E. Harold St., Portland, OR, 97202-4932. Phone: (503) 232-9787. Phone: (800) 929-5119. Fax: (503) 232-9787. Fax: (800) 929-5119. E-mail: acmrl@myexcel.com Eric Norberg, pres.

Consulting firm for radio bcstrs, including sls dev, mgmt training & progmg.

Block Communications Group Inc., 2910 Neilson Way, Suite 503, Santa Monica, CA, 90405-5368. Phone: (310) 452-3355. Fax: (310) 452-4077. E-mail: dblock@earthlink.net Web Site: www.blockcommunicationsgroup.com. Richard C. Block, pres.

Consultants specializing in new cable svcs, bcstg stns, syndicated progmg, distribution & mktg serving U.S. & international clients since 1974.

Bond & Pecaro Inc., 1920 N St. N.W., Suite 350, Washington, DC, 20036. Phone: (202) 775-8870. Fax: (202) 775-0175. E-mail: bp@bondpecaro.com Web Site: www.bondpecaro.com. Timothy Pecaro, principals.

Economic & financial consulting, valuation studies, asset allocations, appraisals, feasibility studies, fairness opinions, Internet valuations & expert testimony.

Bortz Media & Sports Group, 4582 S. Ulster St., Ste 1450, Denver, CO, 80237. Phone: (303) 893-9902. Fax: (303) 893-9913. E-mail: info@bortz.com Web Site: www.bortz.com. James M. Trautman, mgng dir; Arthur Steiker, mgng dir.

TV stn mgmt consulting, cable financial & market analysis; corporate strategic planning.

Bowman Valuation Services, 706 Duke St., Alexandria, VA, 22314. Phone: (703) 549-5681. Fax: (703) 549-5682. E-mail: bowman@bowmanvaluation.com Web Site: www.bowmanvaluation.com. Peter Bowman, ptnr; Chip Snyder, ptnr.

Appraisals, asset allocations, specialized studies.

Frank Boyle & Co., L.L.C., 2001 W. Main St., Suite 280, Stamford, CT, 06902. Phone: (203) 969-2020. Fax: (203) 316-0800. E-mail: fboylebrkr@aol.com Frank Boyle, pres.

Radio & TV media brokerage, mergers & acquisitions/appraisals.

Broadcast Media Associates, Box 1233, Santa Maria, CA, 93456. Phone: (805) 937-1553. Fax: (805) 937-1553. E-mail: cliffhunter@cliffhunter.com Web Site: broadcastmediabroker.com. Clifford M. Hunter, pres.

Mgmt consulting, mktg studies, bcst investment analysis.

Broadcast Services Inc., Box 6418, Brattleboro, VT, 05302-6418. Phone: (802) 258-3000. Phone: (802) 258-4500 svc. Fax: (802) 258-2500. E-mail: mh@markhutchins.com Web Site: www.markhutchins.com. Mark F. Hutchins, pres.

Predicted-coverage mapping, signal-improvement studies, interference mitigation, satellite & microwave facilities inter-connection, RF radiation safety/compliance.

Broadcasting Asset Management Corp., 1323 Forest Glen Dr. N., Winnetka, IL, 60093. Phone: (847) 446-8882. Fax: (847) 446-4855. Jack Minkow, pres.

Merger & feasibility studies; acquisition analyses; capital structuring, brokerage, mgmt procurement & consulting; sr, subordinated & equity placement.

Broadcasting Unlimited Inc., 35 Main St., Wayland, MA, 01778. Phone: (508) 653-7200. E-mail: jwilliams@dmrinteractive.com Jay Williams, CEO.

Strategic planning & execution of direct mktg, telemarketing, & e-marketing promotion svcs for radio.

Broward Alliance, 110 E. Broward Blvd., #1990, Fort Lauderdale, FL, 33301-2248. Phone: (954) 524-3113 ext 220. Phone: (800) 741-1420. Fax: (954) 524-3167. E-mail: info@browardalliance.org Web Site: www.browardalliance.org/film.

Resource for film, TV & print industry production and business relocation.

Howard Burkat Communications, 16 Drake Rd., Scarsdale, NY, 10583. Phone: (914) 723-9043. E-mail: hburkat@burkat.com

Sls, mktg & mgmt consulting for TV, cable, radio, domestic & international. Planning analysis, rsch & recruiting.

Kent Burkhart's Office Inc., 133 East End Dr., Key Biscayne, FL, 33149. Phone: (305) 439-8871. Fax: (305) 361-0650. E-mail: radiokent@aol.com Kent Burkhart, chmn.

Media consultant to radio stns & networks, the Internet, cable TV, & audio & product mktg. LMA liaisons & negotiators.

Alan Burns & Associates, 17357 Perdido Key Dr., Unit 7E, Pensacola, FL, 32507-7866. Phone: (703) 648-0000. Phone: (850) 49R-ADIO. Fax: (703) 264-1710. E-mail: alan@burnsradio.com Web Site: www.burnsradio.com.

Progmg & mktg consultants.

Cable Audit Associates Inc., 5340 So. Quebec St., Suite 100, Greenwood Village, CO, 80111. Phone: (303) 694-0444. Fax: (303) 694-2559. E-mail: blazarus@cableaudit.com Web Site: www.cableaudit.com. Bruce N. Lazarus, CEO; Mitch Walker, sr VP.

Progmg license fee audits of cable operators, MMDS, SMATVs & TVRO middlemen.

Carolina Media Professionals Inc., Box 3325, Spartanburg, SC, 29304. Phone: (864) 597-1301. Fax: (864) 596-7539. E-mail: rachel@upstate.net Web Site: www.carolinamedia.com. Rachel Greene, pres.

Specializes in the placement of natl adv through radio, TV for products, svs and programs. Print: natl adv. Production: audio & video.

The Center for Sales Strategy, Inc., 610 W. DeLeon St., Tampa, FL, 33606-2720. Phone: (813) 254-2222. Fax: (813) 254-9222. Web Site: www.csscenter.com. Steve Marx, CEO; John Henley, exec VP; Jim Hopes, pres.

Comprehensive consulting & training svcs for radio & TV, cable, and nwspr, in sls, mktg & mgmt, exclusively on a long-term, multi-year basis.

Chenevert Songy Rodi Soderberg, (An Engrg/Architectural Corp.). 6767 Perkins Rd., Suite 200, Baton Rouge, LA, 70808. Phone: (225) 769-0546. Fax: (225) 767-0060. E-mail: csrs@csrsonline.com Web Site: www.csrsonline.com.

Architects, planners & tech designers specializing in new & renovated bcst/cable production facilities.

Christian TV Services, 18 Elizabeth St., P.O. Box 209, Ellicottville, NY, 14731-0209. Phone: (716) 699-2549. Fax: (716) 699-2590. E-mail: george@christianservices.com Web Site: www.christiantvservices.com. Russell A. Thayer, dir; Roger A. Thayer, progmg dir; Randall A. Thayer, prom dir; George A. Thayer, pres/CEO; Joyce E. Thayer, admin asst.

TVRO consultant for Christian media started: 1974; affil: TBN/TCT/CTS Ministering to Ministries, listing media organizations & Internet places of worship. (800) 982-8823.

Christian Television Services, 9775 S.W. 87th Ave., Miami Beach, FL, 33141. Phone: (305) 592-7642. Fax: (305) 596-4564. E-mail: webmaster@citv.com Web Site: www.citv.com. Russell Thorne, CEO.

Relg media buying.

Chubb Group of Insurance Companies, 15 Mountain View Rd., Warren, NJ, 07059. Phone: (908) 903-2000. Fax: (908) 903-2027. E-mail: info@chubb.com Web Site: www.chubb.com.

Endorsed multi-natl property/casualty carrier by the Bcst Financial Mgrs Assns.

Branches in more than 115 offices in 30 countries.

Claritas Inc., 1525 Wilson Blvd., Suite 1000, Arlington, VA, 22209. Phone: (703) 812-2700. Fax: (703) 812-2701. Web Site: www.claritas.com. E-mail: info@claritas.com

Chicago, IL 60604-4302, 332 S. Michigan Ave, Suite 200. Phone: (312) 986-2650. Margie Lymperis, VP electronic media.

Mktg data, software & consulting designed for cable, TV, radio & other new media companies.

Clark-Mann & Associates, 203 Columbus Ave., San Francisco, CA, 94133. Phone: (415) 421-0220. Fax: (415) 421-0417. E-mail: wclarkmann2@msn.com William Clark, CEO.

Full-svc adv agency & PR firm with background in bcst & print.

Clear Channel Satellite, 76 Inverness Dr. E., Suite B, Englewood, CO, 80112. Phone: (303) 925-1708. Fax: (303) 925-1714. E-mail: sales@clearchannelsatellite.com Web Site: www.clearchannelsatellite.com. Don Harms, pres; Monty Dent, gen sls mgr.

Satellite space-time; audio distribution via satellite; WAN protection svcs; DSNG svcs; satellite equipment sls; Satellite installation svcs.

Colorado Springs Film Commission, 515 S. Cascade Ave., Colorado Springs, CO, 80903. Phone: (719) 635-7506 x127. Fax: (719) 635-4968. E-mail: kgriffis@experiencecoloradosprings.com Web Site: www.filmcoloradosprings.com.

Free location svcs, Production Resource Guide, location guide available for the Colorado Springs, CO area; help with crews, hotels & permits.

Coltrin & Associates Inc., 1212 Ave. of the Americas, 10th Fl., New York, NY, 10036. Phone: (212) 221-1616. Fax: (212) 221-7718. E-mail: steve_coltrin@coltrin.com Web Site: www.coltrin.com.

Yardley, PA 19067, 801 Floral Vale Blvd. Phone: (215) 497-3188. Gwen Coltrin, COO.

Singapore 048623 Indonesia, 50 Raffles Pl., 37th Fl. Phone: +65 6829 7149. Chan Chee Pong, VP.

London W1J0DW United Kingdom, 35 Picadilly, 3rd Fl. Phone: +44 20 7494 4748.

Burlingame, CA 94010, 433 Airport Blvd, Suite 414. Phone: (650) 373-2005. Benoit Rungeard, dir. (San Francisco office).

Celebration, FL 344747, 215 Celebration Pl, Suite 500. Phone: (321) 559-1112.

Salt Lake City, UT 84111, 215 S. State St, Suite 675. Phone: (801) 350-9412.

Consultant svcs to bcst mgmt, mktg sls, promotional

rsch; New York, NY, & Washington, DC representation in corporate, govt & PR.

Columbia Management Advisors, One S. Wacker Dr., Chicago, IL, 60606. Phone: (312) 443-4000. Fax: (312) 855-2552. William Rarkin, pres.

Investment counsel & mutual fund mgmt. Offices in Chicago, IL; Cleveland, OH; New York, NY; San Francisco, CA; & Puerto Rico.

ComBridges, 70 Irwin, San Rafael, CA, 94901. Phone: (415) 454-5505. Fax: (888) 530-5505. E-mail: info@combridges.com Web Site: www.combridges.com. Jon Leland, pres.

Complete creative & production svcs including: animation, special effects, video production & video streaming, web site design, & computer-based production system design. Experience with creative svc departments & corporate communications.

CommNOW, 15 Random Farm Rd., Chappaqua, NY, 10514. Phone: (203) 440-3636. E-mail: research@commnow.com Web Site: www.commnow.com. Robert Daigle, VP mktg.

Telecommunications, high-technology, mktg rsch.

Communication Trends Inc., 6120 Powers Ferry Rd. N.W., Suite 140, Atlanta, GA, 30339. Phone: (404) 843-8717. Fax: (404) 843-6869.

Mktg & adv for cable, direct bcst, bcst communications industries & related technologies.

Communications Design Associates Inc., 437 Turnpike St., Canton, MA, 02021-2702. Phone: (339) 502-6551. Fax: (339) 502-6595. E-mail: srandall@cdaconsultants.com Web Site: www.cdaconsultants.com.

Ind consultants to radio, TV, corp & govt clients. Designers of studios, production, presentation & multi-media facilities.

Communications Equity Associates, LLC, 101 E. Kennedy Blvd., Suite 3300, Tampa, FL, 33602. Phone: (813) 226-8844. Fax: (813) 225-1513. E-mail: spark@ceaworldwide.com Web Site: www.ceaworldwide.com. J. Patrick Michaels, Jr., chmn/CEO.

Miami, FL 33131, 150 S.E. 2nd Ave. Phone: (305) 810-2740. Fax: (305) 810-2741. Jose Rodriguez, mgng dir.

New York, NY 10020, 1270 Ave. of the Americas, Suite 1818. Phone: (212) 218-5085. Fax: (212) 218-5099. Alex Rossi.

Prague, 110 00 Czech Republic, Melantrichova 17. Phone: 420-2-216-32451. Fax: 420-2-242 29 412. Valclav Matatko, mgng dir CEA Prague.

Paris 75008 France, 9, Rue Royale. Phone: 33 153 308607. Fax: 33 142 42 3214. Frank Portais, mgng dir.

Munich D-80538 Germany, Prinzregentenstrasse 56. Phone: 49-(0)89-290 725 0. Fax: 49-(0)89-290 725 200. Dr. Stephan Goetz, mgng dir.

Madrid 28006 Spain, Serrano 93, 6E. Phone: 34-91-745 13 13. Fax: 34-91-563 38 62. Jose Cabera-Kabana, mgng dir.

London W1k 5DL United Kingdom, 32 Brock St., 3rd Fl. Phone: 44-207-647-7700. Fax: 44-207-647-7710. Hal Young, exec dir.

Westport, CT 06880, 191 Post Rd W. Phone: (203) 221-2662. Fax: (203) 221-2663. E-mail: dmoyer@ceaworldwide.com. Dave Moyer, pres-CEA principal advisors.

Tampa, FL 33602, 101 E. Kennedy Blvd, Suite 1818. Phone: (813) 226-8844. Fax: (813) 225-1513. Bob Berger, sr. VP.

Provides investment banking, brokerage, regulatory affrs & mgmt svcs for bcst, CATV & related communications industries.

Comsearch, 19700 Janelia Farm Blvd., Ashburn, VA, 20147. Phone: (703) 726-5500. Fax: (703) 726-5600. E-mail: info@comsearch.com Web Site: www.comsearch.com. Douglass R. Hall, pres.

Provides frequency coord, site selection, RFI measurements, path surveys, protection for satellite earth stn dishes & terrestrial microwave facilities & wireless engrg svcs & data.

Conley & Associates, LLC, 1459 Interstate Loop, Bismarck, ND, 58503-5560. Phone: (701) 222-3902. Fax: (701) 222-4815. E-mail: info@conleyassociates.net Web Site: www.conleyassociates.net. Christopher J. Conley, gen ptnr; Candace Christianson, gen ptnr.

Consultants in the areas of strategic planning, appraisals, finance, opns, engrg, mktg, human resources & pub/govt rel.

Connecticut Film Video & Media Office, 805 Brook St., Bldg 4, Rocky Hill, CT, 06067 3405. Phone: (800) 392-2122. Phone: (860) 571-7130. Fax: (860) 721-7088. E-mail: info@ctfilm.com Web Site: www.CTfilm.com. Guy Ortoleva, exec dir; Heidi Hamilton, dir.

A film commission eager to respond to any situation or need.

Connelly Co. Inc., 17909 Holly Brook Dr., Tampa, FL, 33647-2245. Phone: (813) 991-9494. Fax: (813) 991-9494. E-mail: connellyradiotv@verizon.net Robert J. Connelly, pres.

New Market, NH 03857, 198 S. Main St. Phone: (603) 659-3648. Fax: (603) 659-3681. Rob Connelly.

South Effingham, NH 03882. Connelly Co. Inc. (Summer only), Bailey Rd. Phone: (603) 522-6462. Fax: (603) 522-6348. R.J. Connelly, pres. (Summer only).

Brokers, consultants & recovery units to assist banks & financial institutions.

Consolidated Communications Consultants, 1837 S.E. Harold St., Portland, OR, 97202-4932. Phone: (503) 232-9787. Phone: (800) 929-5119. Fax: (503) 232-9787. Fax: (800) 929-5119. E-mail: acmrl@myexcel.com Web Site: www.acmusicresearch.com. Eric Norberg, gen mgr; Jane A. Kennedy, rsch dir.

Provide progmg, sls & mktg assistance for radio stns (AM mass-appeal, A/C stns a specialty).

Contemporary Communications, 9408 Grand Gate St., Las Vegas, NV, 89143. Phone: (702) 898-4669. Fax: (208) 567-6865. E-mail: larryfuss@cox.net Larry G. Fuss, pres.

Progmg, opns & mgmt consulting for small- & medium-market radio stns; tech svcs; FCC compliance; computer software svcs.

Convergent Media Systems, 190 Bluegrass Valley Pky., Suite 800, Alpharetta, GA, 30305. Phone: (770) 369-9000. Fax: (770) 369-9100. E-mail: convergent@convergent.com Web Site: www.convergent.com. Bryan Allen, pres/CEO; William Wheless Jr., CFO.

Provider of video & data technologies to support the communication & training needs of companies. Svcs include consultation, design, instal, net & systems mgmt; systems integration in the following areas: special event TV, business TV, desktop video, video production, videoconferencing, & interactive multimedia.

Cox & Cox L.L.C., 2454 Shiva Ct., St. Louis, MO, 63011. Phone: (636) 458-4780. Fax: (636) 273-1312. E-mail: bc@coxandcoxllc.com Web Site: www.coxandcoxllc.com. Robert Cox, pres.

Mergers & acquisitions, appraisals, consulting & expert testimony.

Cross-Country Communications Inc., Box 535, Suffern, NY, 10901. Phone: (845) 368-1720. Fax: (208) 692-2181. E-mail: cccomm@aol.com Web Site: www.cross-country.com. Joe Capobianco, pres.

Strategic planning, business development and new product launches in media/entertainment, production for Radio-TV-Web.

DDS Sales Training, 6904 W. Sagamore Cir., Sioux Falls, SD, 57106. Phone: (605) 361-9923. Fax: (605) 361-1828. E-mail: ddssales@sio.mideo.net Darrell Solberg, pres.

Radio sls training & consulting; mgmt training & consulting; mktg/adv seminars for businesses.

DIS Consulting Corp., 10 Waterside Plaza, #33D, New York, NY, 10010-2608. Phone: (212) 213-6872. Fax: (212) 213-6876. E-mail: dougsheer@aol.com Web Site: www.disconsultingcorporation.com. Douglas I. Sheer, CEO.

Mktg consultants to 1,000 equipment manufacturers since 1982. Mktg consultation, business plan writing, financial & market rsch, mktg & distribution plans.

DST Innovis, 1104 Investment Blvd., Eldorado Hills, CA, 95762. Phone: (800) 835-8389. Fax: (916) 934-7054. Web Site: www.dstinnovis.com. Michael McGrail, pres; Anthony Piniella, mgr.

North Sydney NSW 2061. CableData (Asia Pacific), Level 4, 44 Miller St, Suite 404. Phone: +61 29.460.2250. Fax: +61 29.460.2238.

Sao Paulo 04571-010. CableData (Latin America), Andar, Suite 81, Ave. Eng Luis Carlos Berrini, 1297. Phone: +55 11.5505.6799. Fax: +55 11.5505.8691.

Customer mgmt & billing solutions for communications & utilities industries. Clients include providers of CATV, telephony, DBS, wireless, electricity, water, gas, waste mgmt, utility & multi-svcs in over 20 countries.

David Tait Appraisal, 1848 Laurel Canyon Rd, Los Angeles, CA, 90046-2029. Phone: (323) 654-8420. Fax: (323) 656-1854. E-mail: dta87@earthlink.net David M. Tait, BCBA, owner; Mel Fineberg, assoc.

Fair market value appraisals of radio/TV/CATV for purchase allocation, finance, estate planning, ESOPs, bankruptcy.

E. Alvin Davis & Associates Inc., 35 Hampton Ln., Cincinnati, OH, 45208. Phone: (513) 325-5600. Fax: (513) 272-2303. E-mail: ealvin@ealvin.com Web Site: www.ealvin.com. E. Alvin Davis, pres; Ted McAllister, VP.

Provides expert counsel to oldies stns.

Direct Mail Express Inc., 2441 Bellevue Ave., Daytona Beach, FL, 32114. Phone: (386) 257-2500. Fax: (386) 271-3001. E-mail: tpanaggio@dmenet.com Web Site: www.dmenet.com. Mike Panaggio, CEO; Mike Waither, pres.

High-impact, direct-mail campaigns, data base mgmt, audience rsch via cluster-targeted mktg, market exclusive.

Ditingo Media Enterprises, 100 Park Ave., 6th Fl., New York, NY, 10017. Phone: (212) 308-8810. Fax: (212) 916-0772. E-mail: vditingo@aol.com Vincent M. Ditingo, pres.

Media & mktg consulting, training, corporate writing, books, profiles, articles, speeches, news announcements, brochures, web site content & newsletters for bcstg & cable.

Electronicast Corp., 800 S. Claremont St., Suite 105, San Mateo, CA, 94402. Phone: (650) 343-1398. Fax: (650) 343-1698. E-mail: massaf@electronicast.com Web Site: www.electronicast.com.

Market forecast consulting concern for the fiber-optic, optoelectronic, telecommunication & CATV industries. Multi-client & custom reports available.

Enterprise Appraisal Co., 489 Devon Park Dr., Suite 320, Wayne, PA, 19087. Phone: (610) 687-5855. Fax: (610) 971-0760. Web Site: www.enterpriseappraisal.com.

Washington, DC Phone: (202) 887-0948.

New York, NY Phone: (212) 517-8037.

Evaluates communications-oriented assets, such as equipment & real estate, for TV, CATV, radio, cellular systems, & satellites.

EnVest Media, LLC, 6802 Patterson Ave., Richmond, VA, 23226. Phone: (804) 282-5561. Fax: (804) 282-5703. E-mail: mitt@envestmedia.com Web Site: www.envestmedia.com. Mitt Younts, mgng member.

Alpharetta, GA 30004, 11770 Haynes Bridge Rd, #205. Phone: (770) 753-9650. Fax: (770) 753-0089. E-mail: jswnet@aol.com. Jesse Weatherby, member, mgr.

Nationwide radio, TV acquisition, valuation, financing & consulting firm. The company provides brokerage svcs to stn transaction, appraisal svcs to stn owners & financial institutions. The group secures debt & equity acquisition financing, offers consulting & asset mgmt svcs, acting as court appointed receivers or trustees for bcst stns.

Equidata, 724 Thimble Shoals Blvd., Newport News, VA, 23606. Phone: (757) 873-3395. Phone: (757) 873-0519. Fax: (800) 873-9752. Fax: (757) 873-1224. Web Site: www.equidata.net. Thomas E. Cucuel, pres; Mary Emmett, exec VP.

Nationwide collection agency/credit reporting agency.

Evalueserve, Inc., Box 2037, Saratoga, CA, 95070-0037. Phone: (408) 872-1078. Fax: (720) 294-0943. E-mail: alok.aggarwal@evalueserve.com Web Site: evalueserve.com. Alok Aggarwal, chmn.

Telecommunications, high-technology, businessrsch, data anaylsis.

Executive Broadcast Services, 30 MoBray Ct., Colorado Springs, CO, 80906. Phone: (719) 579-6676. Fax: (719) 579-6664. E-mail: skip@executivebroadcast.com Web Site: www.executivebroadcast.com. Skip Joeckel, pres.

Offers talk progmg & sports guides.

Executive Decision Systems Inc., 6421 W. Weaver Dr., Littleton, CO, 80123-3815. Phone: (303) 795-9090. E-mail: dlenoble@comcast.net Web Site: www.retailinsights.com.

Provide sls & mgmt training, focusing on generating long-term, loc direct revenues. Provides academic approach to media mktg to stns around the United States & abroad. System 21 is guaranteed to return 12 times the revenues within 150 days or your money is refunded.

Executive Media Services, 138 E. Waterford Dr., Seneca, SC, 299672. Phone: (864) 985-1133. Fax: (864) 985-1137. E-mail: excomm@bellsouth.net Web Site: www.executivecomm.com.

Loc sls consulting for TV & radio stns, cable TV systems. Provides source for sls strategy & sls support, sls seminars; start-up & turnaround specialists. New revenue dev & sls, training, using the internet, the home of one-on-one sls & training.

The Exline Company, 4340 Redwood Hwy., Suite F-230, San Rafael, CA, 94903. Phone: (415) 479-3484. Fax: (415) 479-1574. E-mail: exline@pacbell.net Web Site: www.exlinecompany.com. Andy McClure, pres; Erick Steinberg, broker.

Mgmt, financial rsch, appraisal, receiverships & bankruptcies.

The Express Group, 2231 Westland Ave., San Diego, CA, 92104. Phone: (619) 280-9061. Fax: (619) 280-9030. E-mail: egmail@theexpressgroup.com Web Site: www.theexpressgroup.com. Byron Andrus, pres; Roberta Andrus, VP.

Design, fabrication, instal & lighting of news environments, newsrooms, interview & talkshow sets.

Eyewitness Newservice Inc., Box 116, 182 Sound Beach Ave., Old Greenwich, CT, 06870-0116. Phone: (203) 637-0044. Fax: (203) 698-0812. E-mail: primonews@aol.com Web Site: www.teenkidsnews.tv. Albert T. Primo, pres/CEO.

New York, NY 10019, 355 W. 52nd St. Phone: (212) 974-0606. Al Primo, exec producer.

TV news strategic planning, focus group rsch talent & mgmt, coaching. Cable news training; Internet Broadband Svc; TV production.

FM Atlas - Publishing and Electronics, Box 336, Esko, MN, 55733-0336. Phone: (218) 879-7676. Fax: (218) 879-7676. E-mail: fmatlas@aol.com Web Site: users.aol.com/fmatlas. Bruce F. Elving, owner.

FM radio directory, rsch on FM-SCS & FM trans, FM-SCS receivers & newsletter.

Faraone Communications Inc./dba Worldwide Public Relations, 75 West End Ave., Suite R-9A, New York, NY, 10023. Phone: (212) 489-1313. Fax: (212) 489-8978. E-mail: ted.faraone@verizon.net Web Site: www.worldwidepublicrelations.com. Ted Faraone, chmn.

PR svc to bcst, cable, radio, TV & other entertainment properties & media companies.

Faries & Associates, 67 Central Ave., Los Gatos, CA, 95030. Phone: (408) 354-7308. Fax: (408) 395-6670. Web Site: www.fariesinc.com. David A. Faries, gen ptnr.

Business forcasting & analysis.

Federal Engineering Inc., Redwood Plaza II, 10600 Arrowhead Dr., Fairfax, VA, 22030. Phone: (703) 359-8200. Fax: (703) 359-8204. E-mail: info@fedeng.com Web Site: www.fedeng.com. Ronald F. Bosco, pres; John E. Murray, sr VP.

Strategic planning, coverage analysis, new product definition, market rsch, competitive analysis, rates & tariffs, bcst stn design, mergers & acquisitions, expert testimony, regulatory support.

Ferraro Communications Inc., 39 Byron Rd., Weston, MA, 02493. Phone: (781) 235-5556. Fax: (781) 235-5558. Tom Ferraro, pres.

Concept, script & production/direction for coml, radio, film & videotape productions.

Norman Fischer & Associates Inc., Box 5308, Austin, TX, 78763. Phone: (512) 476-9457. Fax: (512) 476-0540. E-mail: terrill@nfainc.com Web Site: www.nfainc.com. Terrill Fischer, pres.

Brokerage in radio, TV & cable, consultation in mgmt & opns, appraisals, feasibility studies, expert testimony, financial planning & assistance.

William Fleming & Associates, 176 N. Beacon St., Hartford, CT, 06105. Phone: (860) 236-4453. E-mail: wlfleming@comcast.net William L. Fleming, pres.

Investment banking & financial consultant to bcstrs; assists in structuring mergers, acquisitions & refinancings; arranges equity, debt or other funds needed.

Florical Systems, 4581 N. W. 6th St., Gainesville, FL, 32609. Phone: (352) 372-8326. Fax: (352) 375-0859. E-mail: sales@florical.com Web Site: www.florical.com. Jim Berry, rgnl sls mgr.

Manufacturer of TV automations, controls & effects.

Focal Press, 30 Corporate Dr., Suite 400, Burlington, MA, 01803. Phone: (781) 212-2212. Fax: (781) 313-4880. E-mail: j.tracy@elsevier.com Web Site: www.focalpress.com. Chris Mebegon, mktg mgr; Joanne Tracy, dir.

Oxford OX2 8DP, Linacre House, Jordan Hill. Phone: 011-44-1-865-310366. Jennifer Welham.

Publishes professional tech books in bcstg, film, video, multimedia, theatre & photography.

Ford Foundation, Media, Arts, & Culture, 320 E. 43rd St., New York, NY, 10017. Phone: (212) 573-5000. Fax: (212) 351-3649. Web Site: www.fordfound.org.

Richard A. Foreman Associates Inc., 330 Emery Dr. E., Stamford, CT, 06902-2210. Phone: (203) 327-2800. Fax: (203) 967-9393. E-mail: raf@rafamedia.com Web Site: www.rafamedia.com. Richard A. Foreman, pres.

Fair market evaluations & asset appraisals, media brokerage, stn financing, & mgmt/production consultation.

Franey, Muha & Alliant Inc., 9901 Business Pkwy., Suite B, Lanham, MD, 20706. Phone: (301) 459-0055. Fax: (301) 459-5405. Web Site: www.franeymuhaalliant.com. Bill Franey, CEO.

Herndon, VA 20171, 13921 Park Central Rd, Suite160. Phone: (703) 397-0977. Fax: (703) 397-0995. John Muha.

Insurance, bonding & benefits admin.

Clifton Gardiner & Company, L.L.C., 2437 S. Chase Ln., Denver, CO, 80227. Phone: (303) 758-6900. Fax: (303) 479-9210. E-mail: cliff@cliftongardiner.com Web Site: www.cliftongardiner.com. Clifton Gardiner, pres.

Brokerage, consulting & financial svcs for the bcst & cable TV industries.

Georgia Film, Video & Music Office, Georgia Department of Economic Dev, 75 Fifth St. N.W., Atlanta, GA, 30308. Phone: (404) 962-4052. Fax: (404) 962-4053. E-mail: film@georgia.org Web Site: www.georgia.org. Bill Thompson, dir.

Location scouting & preproduction svcs provided to feature film, TV movie, coml & multimedia production companies.

Getty Images, 601 N. 34th St., Seattle, WA, 98103. Phone: (206) 925-5000. Fax: (206) 925-5001. E-mail: feedback@gettyimages.com Web Site: www.gettyimages.com. Jonathan Klein, CEO.

Business & mgmt consulting, training, long-term strategic planning, organizations analysis, mktg positioning; seminars on goal setting, leadership, mgmt skills, sls training. Retail training.

Dave Gifford International, 1142 Tano Del Este, Santa Fe, NM, 87506. Phone: (505) 989-7007. Fax: (505) 988-1991. E-mail: giff@talkgiff.com

Sls & sls management training, sls turnarounds & troubleshooting. Sls, mgmt & adv seminars. New account sls & client dev., creator of graduate school of sls.

Gilbert Communications, 4101 Legends Way, Maryville, TN, 37801. Phone: (865) 982-2889. Fax: (865) 977-6633. E-mail: rgilbert63090@mindspring.com Robert W. Gilbert, consultant.

Full-svc radio/TV news consulting, writing seminars, staff motivation, news policy formulation, profit center strategy, *Broadcast News Handbook*.

Greenwood Performance Systems LLC, 907 S. Detroit, Suite 720, Tulsa, OK, 74120. Phone: (800) 331-9115. Phone: (918) 665-7252. Fax: (918) 665-7233. Fax: (800) 378-2544. E-mail: info@greenwoodperformance.com Web Site: www.greenwoodperformance.com. Jim Rhea, CEO.

Bcst-specific sls & mgmt training, seminars & courses including sls mgmt consultation, strategic planning, compensation, selection & evaluation.

Guidestar Corp., 10600 Arrowhead Dr., Fairfax, VA, 22030. Phone: (703) 352-5700. Fax: (703) 359-8204. E-mail: info@fedeng.com Web Site: www.fedeng.com. Ronald F. Bosco, dir.

Mktg communications & PR specifically tailored to serve the telecommunications & info processing marketplaces.

Halper & Associates, 304 Newbury St., # 506, Boston, MA, 02115. Phone: (617) 786-0666. Fax: (617) 786-1809. E-mail: dlh@donnahalper.com Web Site: www.donnahalper.com. Donna L. Halper, pres.

Radio progmg & mgmt consulting, market studies, format changes, music library software. Staff training, motivation. Specialize in small & medium markets, new owners, turnarounds. Also bcst historian.

Margret Haney Media Brokerage & Consultants, 2995 Woodside Rd., Bldg. 400, Woodside, CA, 94062-2446. Phone: (408) 557-8887. Fax: (408) 557-0808. E-mail: www.hhmargret@yahoo.com Web Site: www.margret @margrethaney.com. Margret Haney, CEO.

Media brokerage/consulting, adv, media buyers.

Bill Hennes & Associates, 5009 Crosswinds Dr., Wilmington, NC, 28409. Phone: (910) 313-2491. Fax: (910) 313-0228. E-mail: bhennes105@aol.com Web Site: www.allaboutcountry.com.

Progmg & mgmt consulting.

The Ted Hepburn Co., 325 Garden Rd., Palm Beach, FL, 33480. Phone: (561) 863-8995. Fax: (561) 863-8997. E-mail: tedhep2@mac.com Ted Hepburn, mgng dir.

Radio, TV & cable brokerage; appraisals.

R. Miller Hicks & Co., 1011 W. 11th St., Austin, TX, 78703. Phone: (512) 477-7000. Fax: (512) 477-9697. E-mail: millerhicks@rmhicks.com Web Site: www.rmhicks.com. R. Miller Hicks, pres.

Brokerage, financing, mgmt consulting.

Hoffman Schutz Media Capital Inc., 2044 West California St., San Diego, CA, 92110. Phone: (619) 291-7070. Web Site: www.hs-media.com. Anthony M. Hoffman, pres; David E. Schutz, VP.

Strategic planning for lender & investor appraisals & litigation support.

Host Communications Inc., 546 E. Main St., Lexington, KY, 40508. Phone: (859) 226-4678. Fax: (859) 226-4391. Web Site: www.hostcommunications.com.

College sports bcstg & TV syndications; publishing & sports mktg.

The Howard-Sloan-Koller Group, 300 E. 42nd St., New York, NY, 10016. Phone: (212) 661-5250. Fax: (212) 557-9178. E-mail: hsk@hsksearch.com Web Site: www.hsksearch.com. Edward R. Koller Jr., pres/CEO.

Exec search & consulting in the cable, infotechnology, entertainment, new media & publishing industries.

The Image Generators, (A division of Voicelines Inc.). 18156 Darnell Dr., Olney, MD, 20832. Phone: (301) 924-5700. Fax: (301) 570-8916. E-mail: mweiner@imagegenerators.com Web Site: www.imagegenerators.com. Michael J. Weiner, CEO.

Mktg & mgmt issues; talent training workshops & coaching.

International Media Consulting, Inc., 48 Mountain Rd., Farmington, CT, 06032-2341. Phone: (860) 677-9688. Fax: (860) 677-9639. E-mail: robert.richer@snet.net Robert E. Richer, owner.

International brokerage firm dealing exclusively with the buying & selling of media properties located outside of the United States.

IPI Report, The International Journalism Magazine. 320 Lee Hills Hall, Columbia, MO, 65211. Phone: (573) 884-7542. Fax: (573) 884-1870. E-mail: ipi-report@missouri.edu Shawn Donnelly, editor.

Columbia, MO 65211, 132A Neff Annex. Phone: (573) 884-7542. Stuart H. Loory, editor.

IPI Report defends, celebrates, relects & explores the international media & freedom of expression.

Jones TM, Inc., 2002 Academy, Dallas, TX, 75234. Phone: (972) 406-6800. Fax: (972) 406-6890. E-mail: jtm@jonestm.com Web Site: www.jonestm.com. David Graupner, CEO; Jay Noble, VP sls.

The world's leading supplier of jingles, production & imaging libraries, wkly music svc & music libraries on hard drive.

KSL 5 Television, 55 N. 300 W., Salt Lake City, UT, 84180. Phone: (801) 575-5555. Fax: (801) 575-5830. Web Site: www.ksl.com. Bruce Christensen, pres; Greg James, sr VP; Con Psarras, VP news; Mark West, VP sls; Steve Poulsen, VP mktg; Michelle Kettle, progmg dir; Brent Robinson, chief engr.

An advertiser-supported NABTS news & info svc available through TV decoders or personal computers equipped with modems. Modem number is (801) 575-5911.

Kagan Media Appraisals, a division of Media Central/Primedia, One Lower Ragsdale Dr., Building One, Suite 130, Monterey, CA, 93940. Phone: (831) 624-1536. Fax: (831) 625-3225. E-mail: info@kagan.com Web Site: www.Kagan.com. Robin Flynn, VP bcstg.

Specializes in the valuation & appraisal of media & communications properties. As part of the Media Central Group of Companies, we maintain the industry's most comprehensive data base of stn values, so we know what yesterday's stns sold for, what buyers are paying today & what they are likely to pay tomorrow. Svcs include: Fair market valuations, expert witness testimony, asset appraisals, ESOP valuations, fairness opinions, minority interest valuations, financial feasibility studies, strategic planning, custom rsch & reports & consulting.

Kagan World Media/Primedia, Inc., One Lower Ragsdale Dr., Building One, Suite 130, Monterey, CA, 93940. Phone: (831) 624-1536. Fax: (831) 625-3225. Web Site: www.kagan.com. Larry Gerbrandt, COO; Sandie Borthwick, dir opns.

Strategic conferences on media & communications topics, including interactive, multimedia, telecommunications, entertainment deals & financing.

Kalba International Inc., 116 McKinley Ave., New Haven, CT, 06515. Phone: (203) 397-2199. Fax: (781) 240-2657. E-mail: kalba@comcast.net Web Site: www.kalbainternational.com. Kas Kalba, pres; Pat Kalba, VP; F. Roberts, VP.

Consulting & advisory svcs on telecommunications, bcstg & cable TV, including international ventures, due diligence, litigation support.

Kane Reece Associates Inc., 822 South Ave. W., Westfield, NJ, 07090-1460. Phone: (908) 317-5757. Fax: (908) 317-4434. E-mail: info@kanereece.com Web Site: www.kanereece.com. John "Jack" Kane, principal; Norval D. Reece, principal.

Asset appraisals, business valuations, due diligence, expert testimony, property tax compliance & control, system mgmt & mgmt/engrg consulting.

Kempff Communications Co., 3301 Bayshore Blvd., Suite 1407, Tampa, FL, 33629. Phone: (813) 258-3433. Fax: (813) 902-1360. E-mail: kempffcc@aol.com Web Site: www.kepffbarr.com. Ron Kempff, pres; Aurelia Serna, VP.

Broker & consultant also offering financial & mgmt svcs, court ordered sale of stns.

The Kompas Group, Box 250813, Milwaukee, WI, 53225-6513. Phone: (262) 781-0188. Fax: (262) 781-5313. E-mail: kompasgroup@toast.net Web Site: www.thelptvstore.com. John Kompas, pres; Burt Sherwood, ptnr.

Sarasota, FL 34242, 5053 Ocean Blvd., Suite 14. Phone: (941) 349-2165. Fax: (941) 312-0974.

Financial & mktg svcs for LPTV; audience demographic reports; stn coverage maps; appraisals/brokerage; financial & strategic planning.

Kovsky & Miller Research, 37 Sawmill River Rd., Hawthorne, NY, 10532. Phone: (914) 347-3606. Fax: (914) 347-3976. E-mail: hkkmrresearch@aol.com Web Site: kovskymiller.com. Harry Kovsky, pres.

Specialists in content & format analysis of loc & network TV news & entertainment programs, promotional analysis, ratings analysis, audience promotional rsch & audience survey rsch.

Kozacko Media Services, Box 948, Elmira, NY, 14902. Phone: (607) 733-7138. Fax: (607) 733-1212. E-mail: rkozacko@stny.rr.com Web Site: www.kozackomediaservices.com. Richard L. Kozacko, pres.

Tucson, AZ 85750, 6890 E. Sunrise Dr, Box 120-40. Phone: (520) 299-4869. Fax: (520) 299-1786. George W. Kimble, assoc.

Elmira, NY 14902, Box 948. Phone: (607) 733-7138. Fax: (607) 733-1212. Jack Clancy, assoc.

Keswick, VA 22947, 1071 Club Dr. Phone: (434) 244-2653. Fax: (434) 244-2666. W. Donald Roberts Jr., assoc.

Appraisals & current market evaluations of radio & TV stns. Bcst acquisition planning.

Lawson & Associates Architects, 7939 Norfolk Ave., Suite 200, Bethesda, MD, 20814. Phone: (301) 654-1600. Fax: (301) 654-1601. E-mail: blawson@lawsonarch.com Web Site: www.lawsonarch.com.

Consulting architectural design & construction mgmt svcs for the TV & cable industry; facility planning, design & coordination of construction svcs.

Liberty Hill Corp., Box 253003, West Bloomfield, MI, 48325. Phone: (248) 737-3000. Fax: (248) 737-3555. E-mail: barryzate@aol.com Barry Zate, sr VP; Ronald Zate, sr VP.

Consulting for radio progmg, talent & admin svcs; on-site seminars, problem targeting & complete strategic planning for the communications industry.

Lipson & Co., 1900 Ave. of the Stars, Suite 2810, Los Angeles, CA, 90067. Phone: (310) 277-4646. Fax: (310) 277-8585. E-mail: inquries@lipsonco.com Web Site: www.lipsonco.com. Howard R. Lipson, pres; Harriet L. Lipson, VP.

Specializes in recruiting for international & domestic bcstg, cable, entertainment, adv, mktg, finance, mdse & licensing.

Locations Tasmania Pty. Ltd., Box 537, Sandy Bay, Tasmania, 07005. Phone: 61-362-243578. Fax: 61-362-24248211. E-mail: wildangels@bigpond.com

Coordination svcs for international film & TV production. Second unit svcs stock footage library.

Loral Skynet, 2400 Research Blvd., Suite 200, Rockville, MD, 20850. Phone: (301) 258-8101. Fax: (301) 258-8119. E-mail: info@loralskynet.com Web Site: www.loralskynet.com. Terry Hart, pres/CEO.

International telecommunication svcs in Asia Pacific rgn, specializing in private networks via satellite.

Lund Consultants to Broadcast Management Inc., 840 Hinckley Rd., Suite 123, Burlingame, CA, 94010-1505. Phone: (650) 692-7777. Fax: (650) 692-7799. E-mail: lundradio@aol.com Web Site: www.lundradio.com. John C. Lund, pres; Dan R. Spice, VP.

Experts in progmg consulting; multiopoly strategy. Adult contemp, country, top 40, rock, classic rock, oldies, news-talk. Music, formatics, proms, talent dev, perceptual rsch.

Frank N. Magid Associates Inc., One Research Ctr., Marion, IA, 52302. Phone: (319) 377-7345. Fax: (319) 377-5861. E-mail: mailia@magid.com Web Site: www.magid.com. Brent Magid, pres/CEO; Steve Ridge, exec VP.

Sherman Oaks, CA 91403, 15260 Ventura Blvd, Suite 2130. Phone: (818) 263-3300. Fax: (818) 263-3311. Jack MacKenzie, exec VP.

New York, NY 10019, 1775 Broadway, Suite 1401. Phone: (212) 974-2310. Fax: (212) 515-4540. Vicki Cohen, exec VP.

Specialists in rsch-driven consultation to traditional & new media firms; svcs include strategic planning, web site evaluation & dev, program evaluation, talent search, coaching, TMI, & the Magid Network.

Mahlum Architects, 71 Columbia, Suite 400, Seattle, WA, 98104. Phone: (206) 441-4151. Fax: (206) 441-0478. E-mail: info@mahlum.com Web Site: www.mahlum.com.

Design; tech consulting; feasibility & facilities studies; cost analysis & construction admin for TV/radio stns, production & equipment storage facilities, & film studios.

Marketing & Creative Services, (Division of Frank N. Magid Associates Inc.). One Research Ctr., Marion, IA, 52302. Phone: (319) 377-7345. Fax: (319) 377-5861. E-mail: mailia@magid.com Web Site: www.magid.com. Steve Ridge, exec VP; Brent Magid, pres/CEO; Bill Hague, sr VP.

Sherman Oaks, CA 91403, 15260 Ventura Blvd, Suite 2130. Phone: (818) 263-3300. Fax: (818) 263-3311. Jack MacKenzie, exec VP.

New York, NY 10019, 1775 Broadway, Suite 1401. Phone: (201) 974-2310. Fax: (212) 515-4540. Vicki Cohen, exec VP.

Rsch & consultation.

Marshall & Stevens Inc., 355 S. Grand Ave, Ste 1759, ste. 1750, Los Angeles, CA, 90017. Phone: (213) 612-8000. Fax: (213) 612-8010. E-mail: info@marshall-stevens.com Web Site: www.marshall-stevens.com. Fred Thomas, VP.

St. Louis, MO 63101, 701 Market Street, # 370. Phone: (800) 325-7337. (800)-325-7337. Raymond Essma, VP Central div.

New York, NY 10036, 1156 Ave. of the Americas, #703. Phone: (212) 425-4300. Wiley Scott, VP sls.

Natl appraisal firm with extensive bcstg client base. Value real estate, equipment, intangible assets & overall business valuations. Assist in financing sls, purchase price allocation, & cast segregation.

Maxagrid, 3939 Belt Line Rd., Suite 250, Addison, TX, 75001. Phone: (972) 241-2110. Fax: (972) 241-2174. E-mail: maxagrid@maxagrid.com Web Site: www.maxagrid.com. Jim Tiller, pres/CEO; Karen Brian, mgr.

Yield mgmt systems for bcst in USA, Australia, & Canada. Systems & strategies that help mgrs improve yields on adv revenues.

Maxwell Media Group, 6053 Bunker Hill, Pittsburgh, PA, 15206. Phone: (412) 441-2020. Fax: (412) 661-9377. Bill Maxwell, pres.

Consultant.

Mayo Communications Inc., Box 82784, Tampa, FL, 33682. Phone: (813) 264-5050. Fax: (813) 264-5353. Lincoln A. Mayo, pres.

Media brokerage, appraisals for bcst & print. Consulting, concerning media sls & acquisitions.

Mazer & Associates, 3452 Grayton Rd., Detroit, MI, 48224. Phone: (313) 885-5686. John Mazer, J.D. Jr., pres.

Radio & TV progmg svcs, market rsch, format design, talent evaluation, license renewal preparation, labor rel, mgmt & admin consulting.

McNulty Consultants, 1926 E. 34th Ave., Spokane, WA, 99203. Phone: (509) 535-5168. Wayne F. McNulty, pres.

Scottsdale, AZ 95258, 7819 Via Rio. Phone: (480) 922-9546. Wayne F. McNulty, pres.

Mgmt & financial assistance in purchasing & operating TV & radio stns. Effective bottom-line mgmt, all phases of stn & group opns.

McVay Media, 2001 Crocker Rd., Suite 260, Cleveland, OH, 44145. Phone: (440) 892-1910. E-mail: mcvaymedia@aol.com Web Site: www.mcvaymedia.com. Mike McVay, consultant; Doris McVay, gen mgr.

Atlanta, GA 30101. Atlanta, 628 Braidwood Dr. Fax: (770) 795-1022.

Consultant radio stns in progmg various formats.

Media & Marketing, 4245 Sarah St., Burbank, CA, 91505. Phone: (818) 753-9510. E-mail: mel.lambert @mediaandmarketing.com Web Site: www.mediaandmarketing.com. Mel Lambert, creative director.

Consulting service for the audio & multimedia industries.

Media Communications Group Inc., Box 335, 11 Spiral Dr., Suite 3, Florence, KY, 41042. Phone: (859) 647-0055. Fax: (859) 647-2611. E-mail: jpierce@paragoncomm.com Web Site: www.paragoncomm.com. Dan Hubbard, sr VP; Rebecca Neal, chief of opns; John L. Pierce, pres/CEO.

Representing, consulting & mgmt svcs to radio stns nationwide.

Media Economics, 69 N. Sheridan Ave., Bethpage, NY, 11714. Phone: (516) 931-0248. E-mail: beconomist@aol.com Layton W. Franko PhD., pres.

Market analysis, forecasting, pricing, planning, sports economics & business rsch for TV, cable & radio industries.

Media Perspectives, 127 Greensward Ln., Cherry Hill, NJ, 08002. Phone: (856) 482-7979. Fax: (856) 482-0957. Steven G. Apel, pres.

Progmg & mktg counseling through applied audience & advertiser rsch.

Media Sales Management, Potsdamer Strasse 31, Berlin, 10783. Phone: 49-30-215-30300. Fax: 49-30-215-1182.

Hamburg D-22397, Sthamerstrasse 58. Phone: 49-40-605-2114. Fax: 49-40-605-1762. Norbert Schmidt, pres.

Offers a complete package of bcst svcs, finance, acquisition, progmg sls mgmt, rsch & valuations. German & European specialists.

The Mediacenter, 1500 Harbor Blvd., Weehawken, NJ, 07086. Phone: (866) 412-0866. Fax: (201) 348-1761. E-mail: info@mediacenteronline.com Web Site: www.mediacenteronline.com. Barbara Zeiger, pres; Russell Sands, gen mgr.

Promote increased mktg professionalism among TV execs & mgrs; provide sls support tools that identify & dev new adv budgets.

Mercer Capital Management Inc., 5860 Ridgeway Center Pkwy., 4th Fl., Memphis, TN, 38120. Phone: (901) 685-2120. Fax: (901) 685-2199. E-mail: mcm@mercercapital.com Web Site: www.mercercapital.com.

Louisville, KY 40202, 206 Kentuckey Towers. Phone: (502) 585-6340. James E. Graves, VP.

Mercer Capital provides high-quality independent business appraisals & other financial advisory svcs for all types of media including radio.

Metro Orlando Film & Television Commission, 301 E. Pine St., Suite 900, Orlando, FL, 32801. Phone: (407) 422-7159. Fax: (407) 841-9069. E-mail: suzy@filmorlando.com Web Site: www.filmorlando.com. Suzy Allen, VP.

One stop permitting, locations library, location scouting, community familiarization tours, Filmbook with complete listings of crews, technicians & production support vendors.

J.M. Miller, Box 190, Ashburn, VA, 20146. Phone: 703-729-7745. Fax: 703-729-7745. E-mail: broadcastappraisal@yahoo.com Jan M. Miller, owner.

Bcst TV, DTV, AM, FM, production, satellite, microwave facility & equipment inspection, asset appraisal reports, engrg evaluation overviews & consulting svcs. For valuation, acquisition, finance, purchase price allocation, ad valorem tax, insurance, leasing, liquidation & litigation.

Jay Mitchell Associates Inc., 4 Ventana, Aliso Viejo, CA, 92656. Phone: (949) 533-4912. Fax: (949) 666-5045. E-mail: mitchell @jaymitchell.com Web Site: www.jaymitchell.com. Jay Mitchell, pres.

Mgmt, mktg, progmg, promotions consulting; market analysis, web site design & consulting.

George Moore & Associates Inc., 6918 Wildglen Dr., Suite 100 W, Dallas, TX, 75230. Phone: (214) 369-5665. Phone: (800) 220-3287. Fax: (214) 369-5667. W. James Moore, pres.

Brokerage of radio, TV & CATV properties; asset & market appraisals; introduction to institutional financing sources.

Multimedia Research Group Inc. (MRG, Inc.), 1754 Technology Dr., Suite 132, San Jose, CA, 95110. Phone: (408) 453-5553. Fax: (408) 453-5559. E-mail: info@mrgco.com Web Site: www.mrgco.com. Gary Schultz, pres.

Provides strategic consulting & published market intelligence on content dev, content distribution, channels & networks.

Nashville Mayor's Office of Film, 222 2nd Ave., Suite 418, Nashville, TN, 37201. Phone: (615) 880-1827. Fax: (615) 862-6025. E-mail: andyvr@nashville.org Web Site: www.filmnashville.com. Tessa Atkins, dir.

Location scouting, permits, produce annual production directory, assist with all logistics of TV/film/video projects, liaison to media & govt.

National Strategies Inc., 1100 H St. N.W., Suite 1200, Washington, DC, 20005. Phone: (202) 349-7001. Fax: (202) 783-1041. E-mail: daylward@natstrat.com Web Site: www.nationalstrategiesinc.com. Al Gordon, CFO.

Milan 20121 Italy, Via San Senatore 10. Phone: +39 02 720 94266. Fax: +39 02 720 94759.

New York, NY 10022, 14 E. 60th St, Suite 1002. Phone: (212) 758-0690. (212) 750-6518.

Pub policy strategies & implementation, business & investment dev.

Navigant International, 945 Hornet Drive,Ste.101, Hazelwood, MO, 63042. Phone: (314) 592-3800. Fax: (314) 592-3900. Mike Million, dir opns.

Corporate & leisure travel, specializes in promotional packages & incentive programs.

New England Media L.L.C., Box 594, Willimantic, CT, 06226. Phone: (860) 456-7400. Fax: (860) 456-5688. E-mail: mcrice@prodigy.net Michael Rice, ptnr.

Mansfield Center, CT 06250, 50 Kaya Lane. Phone: (860) 455-1414. Michael Rice, mgng ptnr.

Media brokers, consultants & appraisers specializing in radio in the Northeast.

Newbrough Associates Inc., Box 1822, Des Moines, IA, 50305-1822. Phone: (515) 244-8909. Fax: (515) 244-8909. E-mail: wbn@att.net Bill Newbrough, pres.

Provides alternative advisory consultaions worldwide, focusing on communications. Forty years of mass media & mgmt experiences. WISDOM+RESEARCH=SUCCESS.

NEWSDirections, 1521 Rocky Knoll Ln., Dacula, GA, 30019-6754. Phone: (770) 569- 2277. E-mail: tony@news-direction.com Web Site: www.news-directions.com. Tony Windsor, pres.

Professional dev & career mktg for TV news reporters, anchors & producers.

Nathan M. Nickolaus, 320 E. McCarthy St., Jefferson City, MO, 65101-3115. Phone: (573) 634-6313. Fax: (573) 634-6504. E-mail: nnickolaus@jeffcity.mo.org Web Site: nnickolaus@jeffcitymo.org. Nathan M. Nickolaus, consultant.

Noll & Associates, 475 Gate Five Rd., Suite 211, Sausalito, CA, 94965. Phone: (415) 332-2254. Fax: (415) 332-5519. E-mail: kennen@nollmedia.com Web Site: www.nollmedia.com. Kennen Williams, pres.

Bcst mktg/sls training, new business dev & organizational dev.

Northwest Broadcasting Co., Box 332, Dallastown, PA, 17313-0332. Phone: (815 308-7613. E-mail: mkrafcisin@hotmail.com Michael H. Krafcisin, pres.

Mgmt, progmg, production, scriptwriting, talent, voice-over svcs, opns & engrg consultation

The Omnia Group, 601 South Blvd., Tampa, FL, 33606. Phone: (800) 525-7117. Fax: (813) 254-8558. E-mail: mcleveland@omniagroup.com Web Site: www.omniagroup.com.

Same-day response on best industry-validated selection tools that help hire the right person the first time.

Ott & Associates, 9225 Chatham Grove Ln., Suite D, Richmond, VA, 23236. Phone: (804) 276-7202. Fax: (804) 745-7778. E-mail: rick@rickott.com Web Site: www.rickott.com. Rick Ott, pres.

Problem solving, consultation in complete confidentiality. Mgmt consulting.

PMA Marketing Inc., 4359 S. Howell, Suite 106, Milwaukee, WI, 53207. Phone: (414) 482-2638. Fax: (414) 483-1980. E-mail: patrick@amfmtv.com Web Site: www.amfmtv.com. Patrick Martin, pres.

Buy & sell new & used bcst equipment, radio stn start-ups & turnarounds, problem solving for difficult bcst situations.

PR/PR, 775 S. Kirkman Rd., Suite 104, Orlando, FL, 32811. Phone: (407) 299-6128. Fax: (407) 299-2166. E-mail: pam@prpr.net Web Site: www.prpr.net. Pam Lontos, pres; Rick Dudnick, VP.

Publicity in Radio, TV, print for speakers & authors.

Palazzo Intercreative, 308 Occidental Ave. South, Suite 200, Seattle, WA, 98104. Phone: (206) 328-5555. Fax: (206) 324-4348. E-mail: palazzo@palazzo.com Web Site: www.palazzo.com. Richard Roberts, pres & dir.

Stn identity design & consultation svcs, including on-air, print, outdoor, graphics, syndicated animation packages, movie & news opns, & radio spots.

Paragon Media Strategies, 12345 W. Alameda Pkwy., Suite 325, Denver, CO, 80228. Phone: (303) 922-5600. Fax: (303) 922-1589. E-mail: info@paragonmediastrategies.com Web Site: www.paragonmediastrategies.com. Mike Henry, CEO; John Stevens, pres; Bob Harper, sr VP.

Media rsch & consulting.

Patrick Communications L.L.C., 6805 Douglas Legum Dr., Suite 100, Elkridge, MD, 21075. Phone: (410) 799-1740. Fax: (410) 799-1705. E-mail: patrick@patcomm.com Web Site: www.patcomm.com. Larry Patrick, pres; Susan Patrick, exec VP; Greg Guy, VP.

Stn brokerage, investment banking, mgmt consulting svcs, appraisals & opns consulting.

Donald A Perry & Associates Inc., Box 1275, Newport News, VA, 23601. Phone: (757) 877-4367. Fax: (757) 693-2885. E-mail: dperry@cablefirst.net Donald A. Perry, pres.

Mgmt, brokerage & appraisal svcs to the cable TV industry.

Peters Communications, 1555 Berenda Pl., El Cajon, CA, 92020. Phone: (858) 565-8511. Fax: (619) 440-1481. E-mail: ppiep@cox.net Edward J. Peters, pres.

Media mktg consultants, providing rsch, concept, mktg plan, music, graphics & animation.

Point Broadcasting Company, Point 3G. 715 Broadway, Suite 320, Santa Monica, CA, 90401. Phone: (310) 451-4430. Fax: (310) 451-1423. John Hearne, mgr.

Operating, technical & financial mgmt.

Pollack Media Group Inc., 860 Via De La Paz, Suite D2, Pacific Palisades, CA, 90272. Phone: (310) 459-8556. Fax: (310) 454-5046. E-mail: hq@pollackmedia.com Web Site: www.pollackmedia.com. Jeff Pollack, chmn/CEO; Tommy Hadges, pres; Dave Brewer, exec VP; Jim Kerr, VP news.

Worldwide bcst progmg advisory firm, all facets of progmg, positioning, mktg, adv, rsch, music. All formats.

Poorman & Group, 143-147 E. Main St., Suite 2C, Lock Haven, PA, 17745. Phone: (570) 748-7000. Fax: (570) 748-7700. Web Site: www.lockhaven.com. Stephen P. Poorman, pres.

Pennsylvania & Texas-based mgmt consulting firm offers "no-charge" interviews to radio & TV stns relating to business & real estate issues. Specializes in organizing & mgng financially distressed businesses.

Price Waterhouse Coopers, 1 North Wacker Drive, Chicago, IL, 60606. Phone: (312) 298-2000. Fax: (312) 298-2001. Web Site: www.pwc.com.

Provides valuation consulting svcs for acquisitions, swaps, estate planning & litigation.

W.L. Pritchard & Co. L.C., 4405 E.W. Hwy., Suite 501, Bethesda, MD, 20814. Phone: (301) 654-1144. Fax: (301) 654-1814. E-mail: wlpritchard-co@verizon.net Web Site: www.wlpco.com. Ellen Hoff, pres; Jack Dicks, VP engrg.

Professional engrg and business problem solving in telecommunications, competitor analysis, satellite communications, earth stns, & launch vehicles.

RBC Daniels L.P., 3200 Cherry Creek S. Dr., Suite 500, Denver, CO, 80209. Phone: (303) 778-5555. Fax: (303) 778-5599. E-mail: info@rbcdaniels.com Web Site: www.rbcdaniels.com. Brian Deevy, chmn/CEO; David Tolliver, mgng dir.

New York, NY 10022, 711 5th Ave, Suite 405. Phone: (212) 935-5900. Fax: (212) 863-4859. Greg Ainsworth, sr mgng dir; David Tolliver, mgng dir, media bcst group.

Provides both mergers & acquisitions, corporate financial svcs to the cable telecommunications, media & technology industries.

The R Corp., 2477 Stickney Point Rd., Suite 201 B, Sarasota, FL, 34231. Phone: (941) 924-2400. Fax: (941) 924-1650. E-mail: rcorp@comcast.net

Consulting to electronic media, cable, wireless, & bcstg.

R.F. Technologies Corp., 12 Foss Rd., Lewiston, ME, 04240. Phone: (207) 777-7778. Fax: (207) 777-7784. E-mail: info@rftechnologies.net Web Site: www.rftechnologies.net. George M. Harris, pres.

Provides file engrg & svc for TV & FM antennas,

transmission lines, diplexers & combiners.

RPM Radio Programming & Management, 1133 W. Longlake Rd., Suite 200, Bloomfield Hills, MI, 48302. Phone: (888) 776-0006. Fax: (248) 647-2663. E-mail: info@tophitusa.com Web Site: www.tophitsusa.com. Thomas M. Krikorian, pres.

Top Hits U.S. wkly CD svc & CD libraries including Solid Gold, Spectrum A/C & Country One; CD Christmas library.

The Radio-Studio Network, Box 683, Times Square Station, New York, NY, 10108-0683. Phone: (800) 827-1722. Fax: (212) 868-5663. E-mail: programming@radio-studio.net Web Site: www.radio-studio.net. Steve Warren, pres.

Multi-Format network progmg by MP3 download. Music, news, talk, Infomercials.

Rattigan Resources, 3409 Wilshire Rd., Portsmouth, VA, 23703. Phone: (757) 484-3017. E-mail: rattiganjack@aol.com Jack M. Rattigan CRMC, CEO.

Specializes in "mktg to the 50+ demographic" (baby boomers & beyond). Customized sls training & seminars for stns programmed to the mature & wealthiest audience.

Rees Associates Inc., Rees Plaza at East Wharf, 9211 Lake Hefner Pkwy, Suite 300, Oklahoma City, OK, 73120. Phone: (405) 942-7337. Phone: (888) 942-7337. Fax: (405) 948-1261. E-mail: rees@rees.com Web Site: www.rees.com. Frank W. Rees Jr., pres; Leroy James, exec VP; Kristina Dover, mktg.

Atlanta, GA 30309 United Kingdom. The Metropolis Bldg., 951 Peachtree St. N.E. Phone: (404) 351-6869. Fax: (404) 351-8343.

Dallas, TX 75202-1711, 1801 N. Lamar St, Suite 600. Phone: (214) 522-7337. Fax: (214) 522-0444. Frank Rees Jr., pres.

Bcst & production facility design; architectural svcs; facility business plans; interior design; studio design; equipment planning, consulting.

Restivo Communications/Starstruck Entertainment Company, 73 Widdicombe Hill Blvd., Suite 1515, Toronto, ON, M9R 4B3. Canada. Phone: (416) 242-7009. Web Site: www.prmediaconnection.com. Peter J. Restivo, pres.

Bcst & media consultants; media training, MOW development.

George Rodman Associates, 100 Christwood Blvd., Apt. 119, Covington, LA, 70433-4601. Phone: (831) 626-1630. Fax: (831) 626-8662. E-mail: grodman@viaworldwide.com George T. Rodman, pres; Sally M. Rodman, VP.

Provides stns, networks, groups & program suppliers with mktg counseling & promotional materials, including adv campaigns, logos, on-air design, TV spots & animation.

Roehling Broadcast Services Ltd., 7340 Oak Knoll Dr., Indianapolis, IN, 46217. Phone: (317) 887-1945. Fax: (317) 887-1947. E-mail: edradiobr@aol.com Web Site: roehlingbroadcast.com. Edward W. Roehling, pres; Sandra Roehling, sec/treas.

Bcst appraisers, brokers, consultants, also financing, sls, mgmt consultation.

Ray H. Rosenblum Sales & Management Consultant, Box 38296, Pittsburgh, PA, 15238. Phone: (412) 362-6311. Fax: (412) 362-6317. E-mail: rayhrosenblum@hotmail.com Ray H. Rosenblum, consultant.

Consultant & appraiser for radio & TV stns & political candidates, with focus on mgmt, sls, proms, news & PR.

Rumbaut & Co., 1331 Brickell Bay Dr., Suite 3401, Miami Beach, FL, 33141-1024. Phone: (305) 868-0000. Fax: (305) 577-8222. E-mail: julio@rumbaut.com Web Site: www.rumbaut.com. Julio Rumbaut, pres.

Media brokers & consultants in all facets of the TV & radio industries.

William Russell & Associates Inc., 305 W. Masonic View Ave., Alexandria, VA, 22301. Phone: (703) 739-6277. Fax: (703) 797-7584. E-mail: russell@williamrussellassociates.com Web Site: www.wmrussellassociates.com. William A. Russell Jr., pres.

Govt rel & public rel consultants specializing in telecommunications & international trade issues.

SB Management, 890 Monterey St., Suite G, San Luis Obispo, CA, 93401. Phone: (805) 543-9214. Fax: (805) 543-9243. E-mail: Michael@mikehesser.com Web Site: www.mikehesser.com. Michael Hesser, pres.

Assist in finding, evaluating, financing & structuring acquisitions. Also, consult mgmt & sls.

S C Research International Inc., 1317 Third Ave., Suite 100, New York, NY, 10021. Phone: (212) 867-6060. Fax: (212) 867-6579. E-mail: info@scri.com Web Site: www.scri.com. Desmond C. Chaskelson, dir.

Syndicated reports & custom rsch for manufacturers & investors in bcstg, professional video & audio; publishers of *Broadcast Equipment Marketplace (BEM), Professional Video Marketplace (PFM), Professional Multi-Media Marketplace (PMM), European Telemedia Marketplace (ETM), Asian Telemedia Marketplace (ATM).*

SRCS/Markits, 17 Royal Rd., Bangor, ME, 04401. Phone: (207) 942-5548. Fax: (207) 942-9164. Web Site: www.markits99.com. Steve Robbins, pres.

Client-directed mktg program for coml bcst properties (radio & TV).

San Antonio Film Commission, 203 S. St. Mary's St., 2nd Fl., San Antonio, TX, 78298. Phone: (210) 207-6730. Fax: (210) 207-6843. E-mail: filmsa@filmsanantonio.com Web Site: www.filmsanantonio.com. Drew Mayer-Oakes, dir.

City film commission. Photo Library. Liaison with all city offices. Filming permits. Parking assistance.

SATMAGAZINE.COM, 800 Siesta Way, Sonoma, CA, 95476. Phone: (707) 939-9306. Fax: (707) 939-9235. E-mail: design@satnews.com Web Site: www.satnews.com.

Publishers of the mthy Satmagazine online magazine on coml satellite systems. Also available on a CD-ROM & through the web at http://www.satnews.com.

Satterfield & Perry Inc., 7211 Fourth Ave. S., St. Petersburg, FL, 33707. Phone: (727) 345-7338. Fax: (727) 345-3809. E-mail: eraus@prodigy.net Robert Austin, pres; John Willis, sec/treas.

Wetumpa, AL 36093, 169 Mountain Meadows Ln. Phone: (334) 514-2241. (334) 514-2291. Ken Hawkins, VP.

Denver, CO 80210, 2020 S. Monroe St., Suite 302. Al Perry, pres emeritus; Joe Benkert, VP; Jim Birschbach, VP.

Overland Park, KS 66207, 4918 W. 101st Terr. Phone: (913) 649-5103. Fax: (913) 649-5103. Doug Stephens, VP.

Coos Bay, OR 97420, Box 362. Phone: (541) 751-0043. (541) 256-5553. Dick McMahon, VP.

Aiken, SC 29803, 131 Inwood Dr, Suite 302. Phone: (803) 649-0031. (803) 649-7786. John Willis, VP.

Radio, TV, broker, mgmt & sls consultant, FDIC-approved appraiser & expert witness.

Seabrook Travel Consultants, 4225 Sawgrass Dr., Summerville, SC, 29420. Phone: (843) 552-0702. Fax: (843) 552-3717. E-mail: larry@thekirbycompanies.com Larry Kirby, CEO.

Bcst incentive trips worldwide; all major sporting events; owned & operated by bcstrs.

Shane Media Services, 2500 Tanglewilde, Suite 106, Houston, TX, 77063. Phone: (713) 952-9221. Fax: (713) 952-1207. E-mail: smsofc@shanemedia.com Web Site: www.shanemedia.com. Ed Shane, CEO; Renee Revett, consultant.

Radio progmg and mgmt consultation, custom designed perceptual and qualitative rsch for electronic media outlets.

Shotmakers, Inc., One Horizon Rd., Fort Lee, NJ, 07024. Phone: (201) 886-0287. Fax: (201) 886-0287. E-mail: info@shotmakers.org Web Site: www.shotmakers.org. Dan Robinson, pres.

Media consultant, original film & TV productions; novelist. Represent historic photos & film of New York City & Atlanta.

Barry Skidelsky, Esq., 185 East 85th St., 23 D, New York, NY, 10028. Phone: (212) 832-4800. E-mail: bskidelsky@mindspring.com Barry Skidelsky, consultant.

Consults lenders, investors, owners & mgmt on M&A, strategy & opns, FCC ownership, bankruptcy, trustee, arbitrator & expert witness.

Barry Skidelsky, Esq., 185 East 85th St., 23 D, New York, NY, 10028. Phone: (212) 832-4800. E-mail: bskidelsky@mindspring.com

Skywatch Weather Center, 347 Prestley Rd., Bridgeville, PA, 15017. Phone: (412) 221-6000. Phone: (800) SKY-WATCH. Fax: (412) 221-3160. E-mail: airsci@skywatchweather.com Web Site: www.skywatchweather.com. Dr. Stanley J. Penkala, pres; Daniel Krywiecki, VP.

Specially formatted weathercasts from the Skywatch Weather Center.®

Bill Slatter & Associates, 423 Main St., Natchez, MS, 39120. Phone: (601) 442-1828. E-mail: slatterb@natchez.net

Talent coaching.

Smart Target Marketing, 6800 Southwest 40th St., #304, Miami, FL, 33155. Phone: (305) 667-6665. Fax: (305) 667-3508. E-mail: contact@smarttarget.com Web Site: www.smarttarget.com.

Custom strategic direct-mktg programs; complete promotional & adv svcs including direct mail/targeted mailing lists, telemarketing, data base, custom publishing, sls training, & interactive phone/prom, smart targets & prizm targeting svcs.

Soundtrack, 162 Columbus Ave., Boston, MA, 02116-5222. Phone: (617) 303-7500. Fax: (617) 303-7555. Web Site: www.soundtrackgroup.com. Amy Blankenship, opns.

New York, NY 10010, 936 Broadway. Phone: (212) 420-6010. Chris Rich, opns mgr.

Production, postproduction & custom music of all kinds; specializing in sound designs.

Southern Surveys, 1551 Olde Mill Pl., Marietta, GA, 30066. Phone: (770) 924-3584. Phone: (678) 467-8650. Fax: (770) 924-3584. E-mail: rick@streetlevelviews.com Web Site: www.streetlevelviews.com. Rick Phillips, pres.

Natl qualitative moderator for one-on-one rsch. Nationwide recruiting for auditorium music tests since 1986. Video market rsch.

Wayne A. Stacey & Assoc. Ltd., 2145 Hubbard Cr., Ottawa, ON, K1J 6L3. Canada. Phone: (613) 745-9151. E-mail: wstacey@stacey.ca Wayne A. Stacey, pres.

Govt rels, CRTC/IC applications, bcst rsch, bcst consulting, engrg svcs, demographic studies.

Gary Stevens & Co., 49 Locust Ave., Suite 107, New Canaan, CT, 06840. Phone: (203) 966-6465. Fax: (203) 966-6522. E-mail: deelmakur@aol.com Gary Stevens, pres.

Bcst mergers, acquistions & investment banking svcs.

Stonick Recruitment Inc., 1230 Lake Deeson Pointe, Lakeland, FL, 33805. Phone: (863) 680-1379. Fax: (863) 680-1397. E-mail: stonick@gate.net Web Site: www.stonickrecruitment.com. Chris Stonick, pres.

A natl radio sls consulting firm bringing "recruitment adv" to radio (strictly new business dev).

Peter Storer & Associates Inc., 1361 W. Towne Sq. Rd., Mequon, WI, 53092. Phone: (262) 241-9005. Fax: (262) 241-9036. E-mail: doug@storertv.com Web Site: www.storertv.com. Doug Knight, sls VP; Peter Storer Jr., pres.

Storer Information System (SIMS): Multi-ch, multi-user, PC based TV program schedule, amortization & liability system.

Structural System Technology Inc., 6867 Elm St., McLean, VA, 22101. Phone: (703) 356-9765. Fax: (703) 448-0979. E-mail: fred.purdy@sst-towers.com Web Site: www.sst-towers.com. Fred Purdy, P.E., pres; Kaveh Mehrnama, P.E., VP.

Structural engrg studies, analysis, design, modifications, inspections, fabrication & erection of towers & antenna structures.

Joe Sullivan Executive Search & Recruiting, Box 178, 1202 Lexington Ave., New York, NY, 10028. Phone: (212) 734-7890. Fax: (212) 734-0631. E-mail: jsa6@aol.com Web Site: www.joesullivanexecutivesearch.com. Joseph J. Sullivan Jr., pres; Shane P. Sullivan, VP; Barbara Sullivan, CFO.

Retain exec recruitment in media & academia for positions in middle & sr mgmt with incomes in excess of $120,000 per year.

Synovate, 8600 N.W. 17th St., Suite 100, Miami, FL, 33126. Phone: (305) 716-6800. Fax: (305) 716-6756. Web Site: www.synovate.com. Richard Tobin, pres.

Laguna Hills, CA 92653, 23151 Alde Dr, Suite C4. Phone: (714) 598-9055. Dave Thomas, VP/gen mgr. (Los Angeles office).

Full-svc mktg rsch company, specializing in the Hispanic market; focus group testing, awareness usage tracking studies, media ratings.

Szabo Associates Inc., Media Collection Professionals, 3355 Lenox Rd. NE, 9th Fl., Atlanta, GA, 30326. Phone: (404) 266-2464. Fax: (404) 266-2165. E-mail: info@szabo.com Web Site: www.szabo.com. C. Robin Szabo, VP.

Experts in creditor & debtor rights; consulting media properties in the accounts receivable process; domestic & international collections.

TalentTrainers, 10807 Waring Pl., Charlotte, NC, 28277. Phone: (704) 541-0892. Fax: 1-800-787-4284. E-mail: brice@talenttrainers.com Web Site: www.talenttrainers.com. Shirley Brice, pres.

Talent coaching for TV stns & newspapers. Media trainer for corporate executives. One-on-one sessions, small workshops & individual critiques. Weekly "live" coaching chat on www.talent trainers.com.

J.A. Taylor & Associates, Box 331, Boyertown, PA, 19512-0331. Phone: (610) 754-6800. Fax: (610) 754-9766. E-mail: jataylor@broadcastassociates.com Web Site: www.broadcastassociates.com.

Appraisers & brokers of TV production equipment. Serves video production companies, TV stns & financial institutions.

Tele-Measurements Inc., 145 Main Ave., Clifton, NJ, 07014-1078. Phone: (973) 473-8822. Fax: (973) 473-0521. E-mail: contact@telemeasurements.com Web Site: www.tele-measurements.com. William E. Endres, pres; Douglas W. Cook, VP sls.

Bcst, professional video equipment, videotape, TV systems, teleconferencing, ongoing maintenance support, CCTV, equipment rentals.

Teletech Inc., Box 85567, Westland, MI, 48185. Phone: (734) 641-2300. Fax: (734) 641-2323. Web Site: www.teletech-inc.com. Keith Johnson, VP; Todd Osment, field svc mgr.

Antenna site mgmt; tower, studio & antenna construction & maintenance; frequency searches; FCC application preparation; EMI, radiation & microwave studies.

Television by Design Inc., 3277 Roswell Rd., Suite 714, Atlanta, GA, 30305. Phone: (404) 873-3277. Fax: (404) 873-7900. E-mail: jay@tvbd.com Web Site: www.tvbd.com. Jay Antzakas, pres.

Creators of electronic graphic design; consultants on visual design, equipment & opns for TV stns.

Tenner & Associates Inc., 121 Quail Run Rd., Henderson, NV, 89014. Phone: (702) 792-9430. Fax: (702) 792-5748. Web Site: www.tennerandassoc.com. Lisa Tenner, pres.

Event & conference producers for the entertainment industry.

3-H Cable Communications Consultants, 502 E. Main St., Auburn, WA, 98002-5502. Phone: (253) 833-8380. Fax: (253) 833-8430.

Cable franchise admin, negotiation, renewal, tech evaluation, community needs assessment & franchise fee audits.

Tony Lease Incentive Tours, 500 S. Palm Canyon Dr., Suite 215, Palm Springs, CA, 92264-7454. Phone: (760) 325-9799. Fax: (760) 325-9755. E-mail: info@tonylease Web Site: www.tonylease.com.

Laguna Niguel, CA 92677. Laguna Niguel, Box 7531. Phone: (949) 249-6867. Becky Cerato, exec asst mgr.

Five star sls incentive tours for media. Also has Media Sports Tours for Superbowl, Final Four OLympic packages. Forty years experience; world wide contacts.

Edwin Tornberg & Co. Inc., 8917 Cherbourg Dr., Potomac, MD, 20854. Phone: (301) 983-8700. Fax: (301) 299-2297. Edwin Tornberg, pres.

Negotiators for purchase & sale of radio, TV stns & CATV systems; appraisers & financial advisers; mgmt consultants.

Transcomm Inc., Box 2845, Fairfax, VA, 22031. Phone: (703) 323-5150. Fax: (703) 426-4527. E-mail: transcommusa@msn.com Web Site: www.transcommusa.com. Dr. Norman C. Lerner, engr.

Financial/economic analysis, market rsch, pricing studies & regulatory economics.

24 Karat Productions Inc., 1717 W. Sanderling Ln., Ft. Pierce, FL, 34982. Phone: (772) 465-5511. E-mail: www.marg101z@bellsouth.net Mark Prichard, pres/CEO; Gloria Prichard, exec VP.

Music-1 Mix: over 10,000 key stereo selections plus monthly updates, complete consultant svcs delivered on CD or tape.

Jon Ulmer & Associates, 2176 Highpoint Rd., Snellville, GA, 30078. Phone: (770) 979-3031. Fax: (770) 979-3789. E-mail: julmercpa@comcast.net John R. Ulmer, owner.

Accounting, computer & financial mgmt consulting.

VIP Research Inc., 5700 Broadmoor St., Suite 200, Mission, KS, 66202. Phone: (888) 384-9494. Fax: (913) 677-2727. E-mail: clark@vipresearch.net Web Site: www.vipresearch.net. Mike Heydman, rsch dir; Clark Roberts, CFO & sls dir.

Provides hook-tape production, listener screening, fielding & tabulation for all music testing, perceptual studies, focus groups & promotional telemarketing.

Vanguard Media Corp., 310 N. Westlake Blvd., Suite 230, Westlake Village, CA, 91362. Phone: (805) 446-4100. Fax: (805) 446-4111. E-mail: info@vanmedia.com Web Site: www.vanmedia.com. Rick Newberger, pres.

Advisory firm to media companies, specializing in planning, & dev of new progmg svcs (e.g., The Golf Channel) & distribution systems (e.g., DBS).

Veronis, Suhler, 350 Park Ave., New York, NY, 10022. Phone: (212) 935-4990. Fax: (212) 381-8168. Web Site: www.veronissuhler.com. John J. Veronis, chmn/CEO; John S. Suhler, pres/CEO; Jeffrey T. Stevenson, gen ptnr; James P. Rutherfurd, exec VP; Vernois Suhler Stevenson, gen ptnr.

Merchant bankers to media, communications & info industries, with focus on mergers & acquisitions, valuations, joint ventures & private equity.

A.G. Visk, 2973 Evans Oaks Ct., Atlanta, GA, 30340. Phone: (770) 939-5657. E-mail: tvisk@yahoo.com

Promotional concepts, scripts & publications for the bcstg & entertainment industries.

WW Associates, 10040 East Happy Valley Rd. unit 454, Scottsdale, AZ, 85255. Phone: (480) 515-0913. Fax: (480) 515-4632. E-mail: jwiesenberg@mba1977.hbs.edu

International new media dev, strategic planning & acquisition assistance, cable & wireless MMDS expertise, PPV event & movie studio liaison.

Warren Only Media Group, 19 W. Almond St., Vineland, NJ, 08360. Phone: (856) 507-9368. Fax: (856) 507-9368. E-mail: sales@warrenonly.com Web Site: www.warrenonly.com. Warren Only, consultant.

Bcstg radio & TV consultants, telecommunications, FCC applications.

Washington Information Group Ltd., 1655 N. Ft. Meyer Dr., Suite 800, Arlington, VA, 22209. Phone: (202) 463-7334. Fax: (703 527-4586. Web Site: www.winfogroup.com. Douglas House, pres.

Customized business rsch. Competitive intelligence, pub & private company rsch, industry & market studies. Strictly confidential; free consultation.

"The Weather Center", (A bcst svc of Aviation Weather Inc.). 701 Gervais St., Suite 150-224, Columbia, SC, 29201. Phone: (803) 739-2827. E-mail: wxcenter@aviationweatherinc.com Web Site: www.aviationweatherinc.com. L.R. Ferguson, pres.

"Rgnl Radio Bcst/Weathercast Network" across the Carolinas and Georgia in over 20 bcst markets. Weather forecasting, site-specific bcst svcs for stns all across America. 100% barter.

The Wexler Group, 1317 F St. N.W., Suite 600, Washington, DC, 20004. Phone: (202) 638-2121. Fax: (202) 638-7045. Web Site: www.wexlergroup.com. Anne Wexler, chmn; Dale Snape, gen mgr.

Consulting firm, specializing in govt rel & pub affrs with strong emphasis on mass media, telecommunications, copyright, trade.

Wind River Broadcast Center, 117 E. 11th St., Loveland, CO, 80537. Phone: (800) 669-3993. Phone: (970) 669-3442. Fax: (970) 663-6081. E-mail: jim@windriverbroadcast.com Web Site: www.windriverbroadcast.com. Jim McDonald, CEO.

TV/FM/AM tech & regulatory consulting; publishers of *The Bigbook Project,* a radio/TV stn tech & regulatory workbook system. FCC applications, exhibits.

Wishnow Group Inc., 82 Bubier Rd., Marblehead, MA, 01945-3640. Phone: (781) 631-2444. E-mail: jwishnow@yahoo.com Jerrold D. Wishnow, pres.

Position bcst clients as community service leaders.

Wolfe Media, 10755-F Scripps Poway Pkwy., #612, San Diego, CA, 92131. Phone: (858) 530-8787. Phone: (888) 965-3226. Fax: (858) 530-9974. E-mail: dw@wolfemedia.com Web Site: www.wolfemedia.com. David Wolfe, pres.

Media consultants.

Walter Wulff & Associates, FAA Consultants. Box 914, Point Clear, AL, 36564. Phone: (251) 990-2502. Fax: (334) 990-2503. E-mail: wulff@zebra.net Walter H. Wulff, CEO.

Conducts FAA tower studies & EMI evaluations.

Station Financing Services

ABN AMRO, Park Ave. Plaza 55 E. 52nd St., New York, NY, 10055. Phone: (212) 409-1000. Phone: (212) 251-3524. Fax: (212) 409-7291. Web Site: www.abnamro.com.

Joost Kuiper, chmn; Patrick Phalon, dir corporate communications.

Allied Capital Corp., 1919 Pennsylvania Ave. N.W., 3rd Fl., Washington, DC, 20006. Phone: (202) 331-1112. Fax: (202) 659-2053. E-mail: wrichardson@alliedcaptial.com Web Site: www.alliedcapital.com.

William Walton, CEO.

Subordinated debt, deal size $3 million to $8 million, natl & international.

Alta Communications, 200 Clarendon St., 51st Fl., Boston, MA, 02116. Phone: (617) 262-7770. Fax: (617) 262-9779. Web Site: www.altacomm.com. E-mail: clarason@altacomm.com

Brian McNeill, mgng gen ptnr; Kim Dibble, mgng gen ptnr.

Provide equity & subordinated debt for acquisitions, buyouts, recapitalizations, etc. for companies in radio, TV, cable TV & related industries.

BIA Capital Corp., 15120 Enterprise Ct., Suite 100, Chantilly, VA, 20151. Phone: (703) 818-8115. Fax: (703) 803-3299. E-mail: cwiebe@bia.com Web Site: www.bia.com.

Thomas Buono, CEO.

Investment banking svc, including placement of debt & equity, advice in capital structure & merger, & acquisition issues.

BIA Digital Partners, L.P., 15120 Enterprise Ct., Suite 100, Chantilly, VA, 20151. Phone: (703) 227-9600. Fax: (703) 803-3299. E-mail: gjohnson@bia.com Web Site: www.bia.com.

Greg Johnson, mgng principal.

Provides subordinated debt & preferred equity for communications companies in amounts from 4-25 million.

BIA Financial Network, 15120 Enterprise Ct., Suite 100, Chantilly, VA, 20151-1102. Phone: (703) 818-2425. Fax: (703) 803-3299. E-mail: info@bia.com Web Site: www.bia.com.

Thomas J. Buono, CEO.

Financial consultants to the communications industry; fair market valuations, tax appraisals, acquisition consulting, business plans, internal operational audits, litigation support, investment, publications, & database software, venture funding, capital.

BMO Nesbitt Burns (Bank of Montreal), 3 Times Sq., New York, NY, 10036. Phone: (212) 605-1424. Fax: (212) 605-1648. E-mail: yvonne.bos@bmo.com Web Site: www.bmo.com.

Provides lending & other capital raising svcs, derivatives, & cash mgmt to the bcst & cable industries.

Bank of America Illinois, 231 S. LaSalle St., Chicago, IL, 60697. Phone: (312) 828-2345. Fax: (312) 987-7148. E-mail: pat.mendrik@bankofamerica.com Web Site: www.bankofamerica.com.

John Brennan, pres.

The Barclays Group, 200 Park Ave., New York, NY, 10166. Phone: (212) 412-4000. Fax: (212) 412-7300. Web Site: www.barcap.com.

Thomas L. Kalaris, CEO.

San Francisco, CA 94111, 100 California St. Phone: (415) 765-4700. Fax: (415) 765-4760. Andrew Wynns, mgr.

Berkery, Noyes & Co., 1 Liberty Plaza, New York, NY, 10006. Phone: (212) 668-3022. Fax: (212) 747-9092. E-mail: cathy@berkerynoyes.com Web Site: www.berkerynoyes.com.

Joseph Berkery, pres.

San Francisco, CA 94104, 580 California St., 5th Fl. Phone: (415) 440-5001.

Newton, MA 02460, 40 Kirkstall Rd. Phone: (617) 969-7935. Marlowe G. Teig, mgng dir.

Assists with mergers, acquisitions, divestitures; financial analysis & counsel; debt or equity financing through private or pub chs, including LBOs, ESOPs & valuations.

Blackburn & Co. Capital Markets Group, 201 N. Union St., Suite 340, Alexandria, VA, 22314. Phone: (703) 519-3703. Fax: (703) 519-9756.

James Blackburn, Jr., chmn.

Brokerage of radio & TV stns. James Blackburn Jr., pres.

Bulkley Capital L.P., 5949 Sherry Ln., Suite 1370, Dallas, TX, 75225. Phone: (214) 692-5476. Fax: (214) 692-9309. E-mail: info@bulkleycapital.com Web Site: www.bulkleycapital.com.

Lisa Bulkley, investment off.

Investment banking; mergers, acquisitions, private placements of debt & equity capital.

CEA Inc., 1270 Ave. of the Americas, Suite 1818, New York, NY, 10020. Phone: (212) 218-5085. Fax: (212) 218-5099. Web Site: www.ceaworldwide.com.

Investment banking & brokerage.

CIBC World Markets, 300 Madison Ave., New York, NY, 10017. Phone: (212) 856-4000. Fax: (212) 856-3996. Web Site: www.cibcwm.com.

Gary W. Brown, pres/CEO.

Investment banking & asset mgmt.

CIT Equipment Rental and Finance, 900 Ashwood Pkwy., Suite 170, Atlanta, GA, 30338. Phone: (770) 551-7847. Fax: (770) 206-9295.

CIT Equipment Rental & Finance provides sr debt capital for equipment, expansion & modernization.

Chaisson & Company Inc., 154 Indian Waters Dr., New Canaan, CT, 06840. Phone: (203) 966-6333. Fax: (203) 966-1298. E-mail: rchaisco@aol.com

Robert A. Chaisson, pres.

Brokerage of radio/TV sls & acquisitions. Robert A. Chaisson, pres.

Communications Equity Associates, 101 E. Kennedy Blvd., Suite 3300, Tampa, FL, 33602. Phone: (813) 226-8844. Fax: (813) 225-1513. Web Site: www.ceaworldwide.com. E-mail: rmichaels@ccceaworldwide.com

J. Patrick Michaels Jr., chmn/CEO; John Turner, COO; Ming Jung Sr., mgng dir/CFO-CEA; Donald Russell, mgng dir; Carsten Philipson, mgng dir; Ken Jones, Sr VP & gen counsel.

London W1K 5DL, 3rd Fl., 32 Brook St. Phone: 44 207 647 7700. Fax: 44 207 647 7710. Hak Yeung, exc dir.

Muenchen 80538, Prinzregentenstrasse 56. Phone: 49-(0)89 290 7250. Fax: 49 (0)89 290 725 200. Dr. Stephan Goetz, Stefan Sanktjohanser, mgng ptnrs, Dr. Gemot Wunderle, mgng dir.

Paris, NO 75008 France, 19, avenue George V. Phone: 33 (0) 1 70 72 55 01. Fax: 33 (0) 1 70 72 55 00. Franck Portais, mgng dir.

Madrid 28010 Spain, Glorieta Ruben Dario 3, 3rd Fl. Phone: 34 91 745-1313. Fax: 34 91 310-1675. Jose Cabrera-Kabana, mgng dir.

Westport, CT 06880, 191 Post Rd. W. Phone: (203) 221-2662. Fax: (203) 221-2663. Dave Mover, mktg dir.

Miami, FL 33131, 150 S.E. 2nd Ave., Suite 609. Phone: (305) 810-2740. Fax: (305 810-2741. Jose Rodriguez, mgng dir.

New York, NY 10020, 1270 Avenue of the Americas, Suite 1818. Phone: (212) 218-5085. Fax: (212) 218-5099. Alexander Rossi, Bob Ennis, Waldo Glasman, Paul Miller, Evan Blum, mgng directors, Jason Donnell, dir.

Investment banking, corporate finance & private equity, firm specializing in cable, bcstg, new media & entertainment industries.

Cox & Cox L.L.C., 2454 Shiva Ct., St. Louis, MO, 63011. Phone: (636) 458-4780. Fax: (636) 273-1312. E-mail: bc@coxandcoxllc.com Web Site: www.coxandcoxllc.com.

Robert Cox, pres.

Mergers & acquisitions, appraisals, consulting & expert testimony.

Daniels & Associates, 3200 Cherry Creek S. Dr., Suite 500, Denver, CO, 80209. Phone: (303) 778-5555. Fax: (303) 778-5599. Web Site: www.danielsonline.com. E-mail: info@danielsonline.com

Brian Deevy, chmn/CEO; Brad Busse, pres/COO.

New York, NY 10022, 711 5th Ave, Suite 405. Phone: (212) 935-5900. Fax: (212) 8634859. David Tolliver, VP.

Provides mergers, acquisitions, corporate finance & financial advisory svcs to the tcable, telecom, media & internet industries.

The Deer River Group, 888 16th St., N.W., Suite 400, Washington, DC, 20006. Phone: (202) 939-9090. Fax: (202) 939-9091. E-mail: robin.martin@earthlink.net

Robin B. Martin, pres/CEO.

Assist bcst execs in acquiring stns, securing financing; financial consultant to single stns & group owners; investment banking svcs.

Dresdner Kleinwort, 1301 Ave. of the Americas, New York, NY, 10019. Phone: (212) 969-2700. E-mail: laurafazio@dresdnerkleinwort.com

Laura Fazio, mgng dir.

Integrated investment bank offering premier mergers & acquisitions advising as well as private equity investment & placement, structuring, underwriting & advisory for global debt & equity.

EnVest Media, LLC, 6802 Patterson Ave., Richmond, VA, 23226. Phone: (804) 282-5561. Fax: (804) 282-5703. E-mail: mitt@envestmedia.com Web Site: www.envestmedia.com.

Mitt Younts, mgng member.

Alpharetta, GA 30004, 11770 Haynes Bridge Rd, #205. Phone: (770) 753-9650. Fax: (770) 753-0089. E-mail: jswnet@aol.com. Jesse Weatherby, member, mgr.

Nationwide radio, TV acquisition, valuation, financing & consulting firm. The company provides brokerage svcs to stn transaction, appraisal svcs to stn owners & financial institutions. The group secures debt & equity acquisition financing, offers consulting & asset mgmt svcs, acting as court appointed receivers or trustees for bcst stns.

Norman Fischer & Associates Inc., Box 5308, Austin, TX, 78763. Phone: (512) 476-9457. Fax: (512) 476-0504. E-mail: terrill@nfainc.com Web Site: www.nfainc.com.

Terrill Fischer, pres.

Bethlehem, PA 18017, 1330 Biafore Ave. Phone: (610) 317-2424. Bernhard Fuhrmann, East Coast/Atlantic assoc.

Brokerage in radio, TV & cable; consultation in mgmt & opns; appraisals; feasibility studies; expert testimony; financial planning & assistance.

William Fleming & Associates, 176 N. Beacon St., Hartford, CT, 06105. Phone: (860) 236-4453. E-mail: wlfleming@comcast.net

William L. Fleming, pres.

Investment banking & financial consultant to bcstrs; assists in structuring mergers, acquisitions & refinancings; arranges equity, debt or other funds needed.

Richard A. Foreman Associates Inc., 330 Emery Dr. E., Stamford, CT, 06902-2210. Phone: (203) 327-2800. Fax: (203) 967-9393. E-mail: rafamedia@compuserve.com Web Site: www.rafamedia.com.

Richard A. Foreman, pres.

Debt & equity placement for radio & TV stn acquisitions in major growth markets.

GE Capital Inc., 500 W. Monroe St., Chicago, IL, 60661. Phone: (312) 441-7000. Fax: (312) 441-6728. Web Site: www.gecapital.com.

Kim Gutierrez, dir mktg.

Coml financial svcs.

GE Commercial Finance Global Media and Communications, 201 Merrit 7, 4th fl, Norwalk, CT, 06851. Phone: (203) 956-4000. Fax: (203) 956-4528. Web Site: www.geglobalmediacomm.com.

Robert V. Stefanowski, mgng dir.

Leading provider of capital to the bcstg, cable, entertainment, movie theater, outdoor adv, publishing, technology, towers, wireless & wireline industries. Locations in Atlanta, Chicago, Delhi, London, New York, Norwalk, & San Francisco.

Clifton Gardiner & Company L.L.C., 2437 S. Chase Ln., Denver, CO, 80227. Phone: (303) 758-6900. Fax: (303) 479-9210. E-mail: cliff@cliftongardiner.com Web Site: www.cliftongardiner.com.

Clifton H. Gardiner, pres.

Consulting svcs on debt & equity placements.

Gleacher Partners, 660 Madison Ave., New York, NY, 10021. Phone: (212) 418-4200. Fax: (212) 752-2711. Web Site: www.gleacher.com.

Eric Gleacher, chmn.

Provide advice & capital to companies in the media & telecommunications industries.

Great Hill Partners, One Liberty Sq., Boston, MA, 02109. Phone: (617) 790-9400. Fax: (617) 790-9401. Web Site: www.greathillpartners.com.

Private equity for media & communications companies.

HSBC, One HSBC Center., Buffalo, NY, 14203. Phone: (716) 841-7212. Fax: (716) 854-2751. Web Site: www.us.hsbc.com.

Postproduction & radio/TV equipment financing.

R. Miller Hicks & Co., 1011 W. 11th St., Austin, TX, 78703. Phone: (512) 477-7000. Fax: (512) 477-9697. E-mail: millerhicks@rmhicks.com

Business consultant & dev firm.

Hoffman Schutz Media Capital Inc., 2044 W. California St., San Diego, CA, 92110. Phone: (619) 291-7070. E-mail: dave@hs-media.com Web Site: www.hs-media.com.

David E. Schutz, pres.

Peru, VT 05152, Rock Bottom Ln. Phone: (802) 824-6544. Anthony Hoffman.

Appraisals, restructurings & litigation support.

Hungerford, Aldrin, Nichols & Carter, CPAs, 2910 Lucerne Dr. S.E., Grand Rapids, MI, 49546. Phone: (616) 949-3200. Fax: (616) 949-7720. E-mail: charper@hanc.com Web Site: www.hanc.com.

Jerry Nichols, pres.

Confidential radio & TV market revenue share reports for the bcst industry.

J P Morgan and Co. Inc., 60 Wall St., New York, NY, 10260. Phone: (212) 483-2323. Web Site: www.jpmorgan.com.

Jamie Dimon, CEO.

Los Angeles, CA 90071, 333 S. Hope St, 35th Fl. Phone: (213) 437-9300.

San Francisco, CA 94111, 101 California St, 38th Fl. Phone: (415) 954-3200.

Chicago, IL 60606, 227 W. Monroe St, Suite 2800m. Phone: (312) 541-3300.

Mergers & acquisitions; debt & equity capital raising; swaps & derivatives; credit arrangement & loan syndication; securities sls & trading; asset mgmt.

KeyBanc Capital Markets, 127 Public Sq., 6th Fl., Cleveland, OH, 44114-1306. Phone: (216) 689-5787. Fax: (216) 689-4666. E-mail: kmayher@keybanccm.com Web Site: www.key.com/media.

Kathleen Mayher, exec VP & mgr.

Financing for media—TV, radio, cable, nwsprs, bcstg & telecommunications.

M/C Venture Partners, 75 State St., Suite 2500, Boston, MA, 02109. Phone: (617) 345-7200. Fax: (617) 345-7201. E-mail: jwade@mcventurepartners.com Web Site: www.mcventurepartners.com.

Edward Keith, CFO.

Provides equity financing & strategic guidance to entrepreneurial ventures in the media & telecommunications industries.

The MFR Group, Box 1184, Rancho Mirage, CA, 92270. Phone: (760) 324-1516. Fax: (760) 324-8255. E-mail: bfenmore@mfrgroup.net Web Site: www.mfrgroup.net.

Bart Fenmore, pres.

Accounts receivable funding for radio/TV stns.

Media Capital Inc., 890 Monterey St., Suite G, San Luis Obispo, CA, 93401. Phone: (805) 543-9214. Fax: (805) 543-9243. E-mail: michael@mikehesser.com Web Site: www.mikehesser.com.

Michael B. Hesser, pres.

Assist in finding, evaluating, financing & structuring acquisitions. Also, consult mgmt & sls.

Multimedia Broadcast Investment Corp., 3101 South St. N.W., Washington, DC, 20007. Phone: (202) 293-1166. Fax: (202) 293-1181. E-mail: wthreadgill@comcast.net

Walter Threadgill, pres.

Provides sr, subordinated debt, equity for telecommunications & bcst ventures.

National Broadcast Finance Corp., Box 3167, 27 Harrison St., New Haven, CT, 06515-0267. Phone: (203) 389-6000. Fax: (203) 389-6020. E-mail: cherhoniak@sbcglobal.net

David C. Cherhoniak, pres.

Specialized investment banking & financial consulting, including raising debt & equity for bcst acquisitions & refinancings; also brokerage for acquisitions & divestitures.

Nautic Partners, 50 Kennedy Plaza, Providence, RI, 02903. Phone: (401) 278-6770. Fax: (401) 278-6387. Web Site: www.nauticpartners.com. E-mail: bwheeler@nautic.com

Cynthia Balasco, CFO.

Source of equity capital to well-managed, positive cash flowing companies.

PK World Media, 126 Clock Tower Pl., Carmel, CA, 93923-8734. Phone: (831) 624-5100. Fax: (831) 625-3225. E-mail: info@kagan.com

Paul Kagan, CEO.

Specializing in financial & investment rsch. Media financial newsletters & database reports. Strategic consulting, seminars & conferences.

BNP Paribas, 919 Third Ave., New York, NY, 10022-3901. Phone: (212) 471-6500. Fax: (212) 669-9606.

Leading underwriting & syndicating time-sensitive, non-investment grade debt financing; often requiring complex & creative capital structures; also financing & equity co-investments.

Branch offices located in Los Angeles, London, New York, & Paris.

Patrick Communications L.L.C., 6805 Douglas Legum Dr., Suite 100, Elkridge, MD, 21075. Phone: (410) 799-1740. Fax: (410) 799-1705. E-mail: patrick@patcomm.com Web Site: www.patcomm.com.

Larry Patrick, pres; Susan Patrick, exec VP; Greg Guy, VP.

Stn brokerage, investment banking, mgmt consulting svcs, appraisals & opns consulting.

Phoenix Cable Inc., 17 S. Franklin Tpke., Ramsey, NJ, 07446. Phone: (201) 825-9090. Fax: (201) 825-8794.

San Rafael, CA 94901, 2401 Kerner Blvd. Phone: (415) 485-4500. Gus Constantin, pres.

Cable TV systems ownership, system mgmt svcs, lease & debt financing svcs.

Premier Capital Group, 1308 8th St., Suite 5, West Des Moines, IA, 50265. Phone: (515) 698-9600. Fax: (515) 698-9699.

Equipment financing for business owners or financing of customers of a broker or distributor.

Rodgers Broadcasting, Box 1646, Richmond, IN, 47375. Phone: (765) 962-6533. Fax: (765) 966-1499.

David A. Rodgers, pres.

Financeing particulary for small operators.

Schroder Investment Management, Equitable Ctr., 875 Third Ave., New York, NY, 10022. Phone: (212) 641-3800. Fax: (212) 641-3985. Web Site: www.schroders.com.

Jim Foster, exec VP.

Silicon Valley Bank, 185 Berry St., Lobby 1, Suite 3000, San Francisco, CA, 94107. Phone: (415) 512-4227. Phone: (415) 512-4200. Fax: (415) 348-0258. Fax: (415) 856-0810. E-mail: moleary@svbank.com Web Site: www.svb.com.

Ken Wilcox, pres; Meghan O'Leary, dir public rel.

Comprehensive finance svcs for middle market bcst & cable operators.

Barry Skidelsky, Esq., 185 E. 85th St., 23 D, New York, NY, 10028. Phone: (212) 832-4800.

Barry Skidelsky, owner.

Full svc assistance to investors, owners, mgmt, aquisition, divestiture, start-up, improvement, trustee (bankruptcy & FCC ownership), expert witness & arbitrator.

Syndicated Communications Inc. (SYNCOM), 8401 Colesville Rd., Suite 300, Silver Spring, MD, 20910. Phone: (301) 608-3203. Fax: (301) 608-3307. Web Site: www.syncomfunds.com.

Terry L. Jones, dir.

Edwin Tornberg & Co. Inc., 8917 Cherbourg Dr., Potomac, MD, 20854. Phone: (301) 299-6661. Fax: (301) 299-2297.

Edwin Tornberg, pres.

Veronis Suhler Stevenson, 350 Park Ave, New York, NY, 10022. Phone: (212) 935-4990. Fax: (212) 381-8168. E-mail: stevenson@veronissuhler.com Web Site: www.veronissuhler.com.

Jeffrey T. Stevenson, mgng ptnr & co-chief exec off.; James P. Rutherfurd, chief exec VP

London SW1 Y4J0 United Kingdom. London, St. James Square, Buchanan House, 8th Fl. Phone: 44-207-484-1440. Fax: 44-207-484-1415. Nigel Stapleton, chmn Veloms Schler International.

Private equity/media buyout affil of Veronis, Suhler, established in 1987 & investing in companies across the spectrum of communications industry segments.

Wachovia Securities, Wachovia Securities, 301 S. College St., Charlotte, NC, 28288. Phone: (704) 348-9500. Fax: (704) 715-1997. E-mail: info@wachovia.com Web Site: www.wachovia.com.

Ken Thompson, CEO; Ben Jenkins, sr exec VP.

Secured financing for acquisition &/or recapitalization of bcst properties.

Waller Capital Corp., 30 Rockefeller Plaza, Suite 4350, New York, NY, 10112. Phone: (212) 632-3600. Fax: (212) 632-3607. E-mail: info@wallercc.com Web Site: www.wallercc.com.

John Waller, III, chmn; Gregory J. Attorri, pres/CEO.

Financing & investment svcs to cable TV industry, specializing in cable TV mergers & acquisitions, buyout financing, raising debt & equity.

Wells Fargo Equipment Finance Inc., 530 Fifth Ave., 15th Fl., New York, NY, 10036. Phone: (212) 805-1000. Fax: (212) 805-1050. E-mail: wfefi@wellsfargo.com Web Site: www.wellsfargo.com.

John M. McQueen, pres/CEO.

Tustin, CA 92780, 14081 Yorba St, Suite 205. Phone: (714) 544-4190. Deborah Anderson, VP.

Danbury, CT 06811, 100 Mill Plain Rd., 3rd Fl. Phone: (203) 791-3944. Brian Rodden, VP.

Leading provider of equipment leasing & financing, intermediate term lending & specialty finance products to the bcst industry.

Wood & Co. Inc., 431 Ohio Pike, Suite 200, Cincinnati, OH, 45255. Phone: (513) 528-7373. Fax: (513) 528-7374.

Larry C. Wood, pres.

Research Services

A & A Research, 690 Sunset Blvd., Kalispell, MT, 59901. Phone: (406) 752-7857. Fax: (406) 752-0194. E-mail: fireowl@in-tch.com

Judith Doonan, pres; Dr. E. B. Eiselein, rsch dir.

Qualitative & quantitative audience surveys for small- & medium-market radio stns, TV stns, cable. Experience with minority stns.

The ARS Group, 110 Walnut St., Evansville, IN, 47708. Phone: (812) 425-4562. Fax: (812) 425-2844. E-mail: info@arsrsc.com Web Site: www.ars-group.com.

Dr. Margaret Blair, CEO; John Walling, pres.

Ad mgmt & dev tools: ARS Copytest, Firststep Proposition test, Outlook Advertising planner, StarBoard Story Board reviewer. WOWWW Competitive Intelligence system.

Abt Associates Inc., 55 Wheeler St., Cambridge, MA, 02138. Phone: (617) 492-7100. Fax: (617) 492-5219. E-mail: webmaster@abtassociated.com Web Site: www.abtassoc.com.

Peg Laplan, pres/CEO; John Shane, chmn.

Washington, DC 20005, 1110 Vermont Ave. N.W. Phone: (202) 263-1800. (202) 263-1801.

Chicago, IL 60610, 640 N. LaSalle. Phone: (312) 867-4000. Fax: (312) 867-4200.

Bethesda, MD 20814, 4800 Montgomery Ln, Suite 600. Phone: (301) 913-0500. Fax: (301) 652-3618.

Mktg rsch, strategic planning, mgmt consulting, audience rsch & segmentation; customer satisfaction programs, quality of svc programs, social science survey rsch, publ policy rsch, economical rsch.

ADcom Information Services Inc., 8400 N.W. 52nd St., Suite 101, Doral, FL, 33166-5309. Phone: (954) 481-8380. Fax: (954) 427-8950. E-mail: dickspooner@cableratings.com Web Site: www.cableratings.com.

Bill Livek, pres/CEO; Dick Spooner, exec VP sls.

New York, NY 10003, 230 Park Ave. South. Phone: (212) 598-5400. Alan Trugman, VP agency svcs.

Viewership rsch systems designed for multi-ch systems, with emphasis on cable ratings & qualitative data.

Admar Group, Inc., Box 1098, 87 Ruckman Road, Alpine, NJ, 07620-1098. Phone: (201) 767-8000. Fax: (201) 767-8006.

Henry D. Ostberg, chmn.

Adv rsch, concept evaluation, product testing & tracking studies; specializes in rsch for legal purposes.

The Adult Contemporary Music Research Letter, 1837 S.E. Harold St., Portland, OR, 97202-4932. Phone: (503) 232-9787. Phone: (800) 929-5119. Fax: (503) 232-9787. Fax: (800) 929-5119. E-mail: acmrl@myexcel.com Web Site: www.acmusicresearch.com.

Eric Norberg, editor & publisher.

Rsch audience appeal of current adult contemp mus, reported in wkly newsletter. Book of oldies rsch also available.

Arbitron Inc., 142 W. 57th St., New York, NY, 10019. Phone: (212) 887-1300. Fax: (212) 887-1401. Web Site: www.arbitron.com.

Pierre Bouvard, pres.

Los Angeles, CA 90024, 10877 Wilshire Blvd. Phone: (310) 824-6600. Fax: (310) 824-6651. Tony Belzer-Western Div Mgr.- radio stn svcs; John Hegelmeyer, adv & Agency svcs. (Western office).

Atlanta, GA 30328, 9000 Central Pkwy. Phone: (770) 668-5400. Fax: (770) 688-5417. Jim Remeny, radio stn svcs; Dan Griffin, mgr adv/agency svcs. (Southeastern office).

Chicago, IL 60606, 222 Riverside Plaza. Phone: (312) 542-1900. Fax: (312) 542-1901. John Nolan-Radio stn svcs, James Tobolski-Agency & adv svcs. (Midwestern office).

Columbia, MD 21046, 9705 Patuxent Woods Dr. Phone: (410) 312-8000. Tom O'Sullivan RSS eastern div mgr-Rad Sta svcs; Julie Ellis- agency & adv svcs. (Eastern office).

New York, NY 10019, 142 W. 57th St., 12th Fl. Phone: (212) 887-1300. Fax: (212) 887-1401. Tom O'Sullivan-RSS Eastern Div Mgr-Rad Sta Servs; Julie Ellis-agency & adv svcs.

Dallas, TX 75240, One Galleria Tower. Phone: (972) 385-5388. Fax: (972) 385-5377. Harry Clark, radio stn svcs; Becky Burkett, adv agency svcs. (Southwestern office).

Loc radio audience measurement in 286 markets, qualitative svc through RetailDirect, Scarborough & the Qualitative Diary svc in 260+ markets; Arbitron NewMedia; rsch based info svcs for the new electronic media.

Audience Research & Development (AR&D), 2440 Lofton Terrace, Ft. Worth, TX, 76109. Phone: (817) 924-6922. Fax: (817) 924-7539. Web Site: www.ar-d.com. E-mail: jgumbert@ar-d.com

Jerry Gumbert, CEO; Jerry Florence, pres/rsch.

Rsch & consultation in loc & natl TV progmg, specializing in news. Program dev, prom, strategic planning, mktg & sls rsch for bcst TV.

BBM Canada, 1500 Don Mills Rd., Suite 305, Toronto, ON, M3B 3L7. Canada. Phone: (416) 445-9800. Fax: (416) 445-8644. Web Site: www.bbm.ca. E-mail: staffing@bbm.ca

Jim MacLeod, pres/CEO; Mark Johnston, CFO; Don Easter, VP; Ron Bremner, VP; Tom Saint, VP; Kathy Carson, dir; Jeff Osborne, pres.

Richmond, BC V6X 3C6 Canada, 10991 Shellbridge Way, 2nd Fl. Phone: (604) 249-3500. Fax: (604) 214-9648. Catherine Kelly, VP Western svcs.

Moncton, NB E1C 817 Canada, 1234 Main St, Suite 3000. Phone: (506) 859-7700. Fax: (506) 852-4445.

Montreal, PQ H3A 1V4 Canada, 2055 Peel St., 11th Fl. Phone: (514) 878-9711. Fax: (514) 878-4210. Robert Langlois, VP Quebec svcs.

Media rsch for radio & TV, & custom rsch through ComQuest Division.

BDS Radio, 6255 Sunset Blvd., 19th Fl., Hollywood, CA, 90028. Phone: (323) 817-1543. Fax: (323) 817-1511. E-mail: catriona.mcginn@nielsen.com Web Site: www.bdsradio.com.

Catriona McGinn, gen mgr.

White Plains, NY 10601, One N. Lexington Ave. Phone: (914) 684-5578. Catriona McGinn, gen mgr.

Leaders in off-the-air music monitoring for radio & record industry. Pattern recognition technology identifies songs on stns across west America.

BIA Financial Network, 15120 Enterprise Ct., Suite 100, Chantilly, VA, 20151-1102. Phone: (703) 818-2425. Fax: (703) 803-3299. E-mail: info@bia.com Web Site: www.bia.com.

Thomas J. Buono, CEO.

Financial consultants to the communications industry; fair market valuations, tax appraisals, acquisition consulting, business plans, internal operational audits, litigation support, investment, publications, & database software, venture funding, capital.

The Benchmark Co., 907 S. Congress, Suite 7, Austin, TX, 78704. Phone: (512) 707-7500. Fax: (512) 707-7757. E-mail: benchmarkr@aol.com Web Site: www.thebenchmarkcompany.net.

Dr. Robert E. Balon, pres/CEO; Holly Brown, rsch dir.

Rsch & mktg for the bcst industry.

Berry Best Services Ltd., 1990 M St. N.W., Suite 740, Washington, DC, 20036. Phone: (202) 293-4964. Fax: (202) 293-0287. E-mail: admin@berrybest.com Web Site: www.berrybest.com.

Thomas L. Berry, VP.

FCC rsch, hard-copy & full electronic distribution of FCC news releases, pub notices & texts; web based FCC data bases.

Big Blue Dot, 63 Pleasant St., Watertown, MA, 02472. Phone: (617) 923-2583. Fax: (617) 923-8014. E-mail: bigbluedot@bigblue.com Web Site: www.bigblue.com.

Jan Craige Singer, pres.

Trend tracking resources for the kids' market, consulting, & newsletter via e-mail; creative svcs.

Bolton Research Corporation, 2709 S.W. 22nd Ave., Miami, FL, 33133. Phone: (305) 854-3887. Fax: (305) 854-3807. E-mail: brct@aol.com Web Site: www.boltonresearch.com.

Ted Bolton, pres.

Perceptual rsch studies; marketplace positioning; music rsch; format opportunity studies; focus groups.

Broadcast News Service, Box 919, Norwood, MA, 02062-0919. Phone: (781) 344-6988. Fax: (781) 344-8928. E-mail: pjrbroadcasting@aol.com

P.J. Romano, dir.

Recording & bcst svcs.

Broadcast Research & Consulting Inc., Box 728, Port Washington, NY, 11050. Phone: (516) 883-8486. Fax: (516) 883-3090. E-mail: jaltman752@aol.com

Herbert Altman, pres.

Syndicated svcs include news & entertainment talent search, net anchor index. Natl & loc market rsch studies & consultation for bcstrs covering programs, movies, news, prom, stn image & new electronic media. Determined the nominees & winners for the annual American Music Awards for 31 consecutive years.

CRI Research / Center for Radio Information, 204 Charlotte Dr., Dugspur, VA, 24325-3869. Phone: (845) 265-4459. Phone: (800) 359-9898. Fax: (845) 265-2715. E-mail: info@the-cri.com Web Site: www.the-cri.com.

Scott Webster, pres.

Radioscan, market/probe, radio/link, mailing labels, phone lists, mktg file, net analysis, group owner, cp reports, radio/TV data bases, available calls.

Mark Clements Research Inc., 60 E. 42nd St., New York, NY, 10165. Phone: (212) 221-2470. Fax: (212) 221-7628. E-mail: mcresearch@aol.com

Mark Clements, pres; E.L. Reiter, exec VP.

Mktg, mgmt & product rsch; economic & progmg studies for TV & radio.

Coleman Research Inc., 4020 Aerial Center Pkwy., Research Triangle Park, NC, 27709. Phone: (919) 571-0000. Fax: (919) 571-9999. E-mail: coleman@colemanres.com Web Site: www.colemaninsights.com.

Jon Coleman, pres; Chris Ackerman, sls VP.

Perceptual rsch, including progmg, mktg & sls studies; continuing consultation.

Comsearch, 19700 Janelia Farm Blvd., Ashburn, VA, 20147. Phone: (703) 726-5500. Fax: (703) 726-5600. Web Site: www.comsearch.com.

Doug Hall, chmn/pres.

Computerized allocation studies, ch analysis, transinterference analysis, system design, detailed coverage prediction, FCC application preparation, site location & mktg rsch.

Core Research, 2161 N.W. Military Hwy., Suite 202, San Antonio, TX, 78213. Phone: (210) 366-4210. Fax: (210) 366-4323. Web Site: www.coreresearch.biz.

Susan Korbel, owner.

Critical Mass Media, 3857 Ivanhoe Ave., Cincinnati, OH, 45212. Phone: (513) 631-4266. Fax: (513) 631-4329. E-mail: help@criticalmassmedia.com Web Site: www.criticalmassmedia.com.

Carolyn Gilbert, pres.

Rsch, telemarketing, direct mail, data mgmt & strategic planning for the radio & TV industry.

Dataworld Inc., 15120 Enterprise Ct., Chantilly, VA, 20151-1217. Phone: (703) 227-9680. Fax: (301) 656-5341. E-mail: dataworldinfo@bia.com Web Site: www.dataworld.com.

David J. Doherty, pres; Rob Farbman, sr VP.

Internet access to feasibility studies & subscription svcs for Flag & DataXpert; FM Explorer; visual on-line allocation tool; custom mapping; bcst facilities database.

Eastlan Resources, Box 3500-404, Sisters, OR, 97759-3500. Phone: (877) 886-3320. Fax: (541) 318-4646. E-mail: info@eastlan.com Web Site: www.eastlan.com.

Mike Gould, pres/CEO; Dave Hastings, dir.

The second largest radio audience measurement company in the US.

Edison Media Research, 6 W. Cliff St., Somerville, NJ, 08876. Phone: (908) 707-4707. Fax: (908) 707-4740. E-mail: rfarbman@edisonresearch.com Web Site: www.edisonresearch.com.

Larry Rosin, pres; Joe Lenski, exec VP; Rob Farbman, sr VP.

Complete market surveys with fast turn-around. Telephone surveys, music testing, focus groups, exit polling.

Entertainment Partners, 2835 N. Naomi, Burbank, CA, 91504. Phone: (818) 955-6000. Fax: (818) 845-6507. Web Site: www.entertainmentpartners.com.

Mark Goldstein, pres/CEO; John Minton, exec VP; Joe Giarrusso, VP mktg.

Orlando, FL 32819, 2000 Universal Studios Plaza. Phone: (407) 354-5900. Joy Ellis.

New York, NY 10001, 875 6th Ave., 15th Fl. Phone: (646) 473-9000. Myfa Cirinna, VP/gen mgr.

Production payroll software, residual, commercials, music & casting svcs, The Paymaster Industry Guide. Branches in New York, Los Angeles, Florida, London, Toronto, Vancouver, Australia, Japan.

FM Atlas Publishing, PO Box 336, Esko, MN, 55733-0336. Phone: (218) 879-7676. Fax: (218) 879-8333. E-mail: fmatlas@aol.com

Carol J. Elving, office mgr.

FM radio directory & FM media, rsch on utilization of FM/SCA & FM translators.

FMR Associates Inc., 6045 E. Grant Rd., Tucson, AZ, 85712. Phone: (520) 886-5548. Fax: (520) 886-9307. Web Site: www.FMRassociates.com.

Bruce Fohr, pres; Andy Wellik, rsch mgr.

Perceptual progmg studies, wkly callout, EARS Music Studies, format opportunity & vulnerability positioning studies; TV coml testing. Electronic progmg simulation tests.

First Amendment Center, (An operating program of The Freedom Forum). 1207 18th Ave. S., Nashville, TN, 37212. Phone: (615) 727-1600. Fax: (615) 727-1309. E-mail: info@fac.org Web Site: www.firstamendmentcenter.org.

Gene Policinski, exec dir; Brian Buchanan, editor; Tiffany Villager, dir; John Seigenthaler, founder.

Arlington, VA 22209, 1101 Wilson Blvd. Phone: (703) 528-0800. Gene Policinski.

Rsch & commentary on media issues. The center is devoted to improving the understanding of media issues by the press & the public.

GFK Custom Research North America, 1060 State Rd., Princeton, NJ, 08540. Phone: (609) 683-6100. Fax: (609) 683-6211. E-mail: info@gfkamerica.com Web Site: www.gfkamerica.com.

Bruce Barr, VP; Jim Timony, sr VP & media svc.

Custom-designed surveys for bcst industry on loc, rgnl & natl basis, loc & network radio & TV, cable/pay TV/DBS, natl global omnibus studies, viewer engagement & ad sls adhoc.

Gallup Organization, 1001 Gallup Drive, Omaha, NE, 68102. Phone: (402) 951-2003. Web Site: www.gallup.com.

James K. Clifton, CEO; Jane Miller, COO.

Mgmt consulting.

GeoMart, 516 Villanova Ct., Fort Collins, CO, 80525. Phone: (970) 416-8340. Fax: (970) 416-8345. E-mail: sales@geomart.com Web Site: www.geomart.com.

Chuck Cotherman, owner.

All USGS & DMA digital & paper maps. All NOS/NOAA charts, international topographic series, aerial photography, raised relief maps, digital products, business & mktg maps, travel maps, globes, etc.

Global Research Institute, 747 Wire Rd., Auburn, AL, 36832. Phone: (334) 826-0390. Fax: (334) 826-0390.

H.D. Norman, pres; Enrico Valdez, VP; Cherry Foster, gen mgr.

Mktg, data & media studies. International radio & TV audience measurement & progmg consultants.

Hagen Media Research, Box 40542, Washington, DC, 20016-0542. Phone: (703) 534-3003. Fax: (703) 534-3073. E-mail: donhagen@aol.com Web Site: www.cmmtv.com.

Don Hagen, pres.

Perceptual rsch for radio. Designs, conducts & analyzes qualitative & quantitative studies. Focus groups, telephone surveys, one-on-one sessions, music tests.

Hamilton Beattie & Staff Inc., 4201 Connecticut Ave. N.W., Suite 212, Washington, DC, 20008. Phone: (202) 686-5900. Fax: (202) 686-7080. E-mail: dave@hbstaff.com Web Site: www.hbstaff.com.

David Beattie, pres; Bryan Dooley, VP.

Fernandina Beach, FL 32034, 102 South 10th St. Phone: (904) 491-0591. Fax: (904) 491-0594. David Beattie, pres.

News rsch, mktg rsch & pub opinion surveys relating to program evaluation, licensing, new products, & high-tech telecommunications.

Peter D. Hart Research Associates, 1724 Connecticut Ave. N.W., Washington, DC, 20009. Phone: (202) 234-5570. Fax: (202) 232-8134. E-mail: info@hartresearch.com Web Site: www.hartresearch.com.

Geoffrey Garin, pres; Peter D. Hart, CEO; Frederick Yang, sr VP.

Audience rsch; polling for on-air use; bcst, cable, ETV, radio rsch svcs, including mkt surveys, political communication & cable referenda rsch.

Norman Hecht Research Inc., 33 Queens St., 3rd Fl., Syosset, NY, 11791. Phone: (516) 496-8866. Fax: (516) 496-8165. E-mail: nhr@normanhechtresearch.com Web Site: www.normanhechtresearch.com.

Laura Greenberg, co-pres/COO; Dan Greenberg, co-pres.

Providing custom market rsch, insight & understanding to clients in bcst, cable, technology, coml svcs & communications industries.

Kenneth Hollander Associates Inc., Box 49625, Atlanta, GA, 30359. Phone: (404) 231-4077. Fax: (770) 729-9375. Web Site: www.kharesearch.com/legal.

Kenneth Hollander, pres.

Mendocino, CA 95460, 45431 Greenling Cir. Phone: (707) 962-1648. (707) 962-1635.

Hungerford, Aldrin, Nichols & Carter, CPAs, 2910 Lucerne Dr. S.E., Grand Rapids, MI, 49546. Phone: (616) 949-3200. Fax: (616) 949-7720. E-mail: caldrin@hanc.com Web Site: www.hanc.com.

Clifford A. Aldrin, ptnr.

Hungerford radio & TV revenue report preparation & year-end accounting/auditing.

Innovative Audience Research, 119 LaColima, Pismo Beach, CA, 93449. Phone: (805) 556-0772. Fax: (805) 556-0772. E-mail: lavinc@charter.net

Mike Silverstein, pres; Susan B. Silverstein, exec VP.

Strategic planning in news, progmg & prom, impacting sweep book ratings in TV metered markets.

Insite Media Research, 31510 Anacapa View Dr., Mailbu, CA, 90265. Phone: (310) 589-0223. E-mail: scott@tvsurveys.com Web Site: www.tvsurveys.com.

Scott V. Tallel, pres.

Full-svc audience rsch firm specializing in voice capture telephone surveys online interviewing and program testing focus groups for television & the internet

Institute for Research on Public Policy, 1470 Peel St., Suite 200, Montreal, PQ, H3A 1T1. Canada. Phone: (514) 985-2461. Fax: (514) 985-2559. E-mail: irpp@irpp.org Web Site: www.irpp.org.

Social policy, public finance, governance, city-rgns, educ, structural change, public security labor market.

Intermedia Analyses Inc., 8 Shadow Rd., Upper Saddle River, NJ, 07458. Phone: (201) 327-6223. Fax: (201) 934-0192.

Carol G. Mayberry, pres.

Strategies/analyses for buying and/or selling radio, optimizing revenue or investment. Broker bcst equipment.

Kagan World Media, One Lower Ragsdale Dr., Building One, Ste.130, Monterey, CA, 93940. Phone: (831) 624-1536. Fax: (831) 625-3225. E-mail: info@kagan.com Web Site: www.kagan.com.

Specializing in financial & investment rsch. Media financial newsletters & database reports.

Mark Kassof & Co., 527 E. Liberty St., Suite 202, Ann Arbor, MI, 48104. Phone: (734) 662-5700. Fax: (734) 662-3255. E-mail: contact@kassof.com Web Site: www.kassof.com.

Mark Kassof, pres.

Strategic audience rsch to pinpoint a radio stn's most profitable format strategy; focus groups, auditorium music testing & promotional testing.

KIDSNET, 2506 Campbell Pl., Kensington, MD, 20895. Phone: (202) 291-1400. E-mail: kidsnet@kidsnet.org Web Site: www.kidsnet.org.

Natl resource of TV, radio, audio & video for children & educ multimedia; program-related study guides; mthy print & electronic publications.

Knowledge Networks/SRI, 570 South Ave. E, Cranford, NJ, 07016. Fax: (908) 497-8001. Web Site: www.knowledgenetworks.com. E-mail: info@knowledgenetworks.com

Gale D. Metzger, gen mgr; Burton E. Michaels, VP.

Specialists in cross-media allocation strategies, consumer media technologies. How people use media.

The Kompas Group, Box 250813, Milwaukee, WI, 53225-6513. Phone: (262) 781-0188. Fax: (262) 781-5313. E-mail: kompasgroup@toast.net Web Site: www.thelptvstore.com.

John Kompas, pres; Burt Sherwood, ptnr.

Sarasota, FL 34242, 5053 Ocean Blvd., Suite 14. Phone: (941) 349-2165. Fax: (941) 312-0974.

Financial & mktg svcs for LPTV; audience demographic reports; stn coverage maps; appraisals/brokerage; financial & strategic planning.

Lincoln Property Company, 101 Constitution Ave. N.W., Suite 600 E., Washington, DC, 20001. Phone: (202) 513-6700. Fax: (202) 898-2001. E-mail: jconnelly@lpc.com Web Site: www.lincolnproperty.com.

James M. Connelly, VP.

Consulting coml real estate svcs, tenant/client representation, construction/financial analysis, turnkey lease/sale assumptions; roof & building zoning surveys.

Lund Media Research, 840 Hinckley Rd., Suite 123, Burlingame, CA, 94010-1505. Phone: (650) 692-7777. Fax: (650) 692-7799. E-mail: lundradio@aol.com Web Site: www.lundradio.com.

John C. Lund, pres; Dan R. Spice, VP.

Perceptual, focus group, mus rsch; customized Radio Marketing Ascertainment evaluates stn & competitive progmg; implementation & progmg consultation; format design & multiopoly strategy.

Magazine Publishers of America, 810 7th Ave., 24th Fl., New York, NY, 10019. Phone: (212) 872-3700. Fax: (212) 888-4217. E-mail: mpar@magazine.org Web Site: www.magazine.org/home.

Nina B. Link, pres/CEO.

Educ seminars; surveys members on various topics; extensive library on magazine publishing; monitors issues in the magazine industry. Publications: *Newsletter of Research, Newsletter of International Publishing, Washington Newsletter* & *Magazine Newsletter*.

Frank N. Magid Associates Inc., One Research Ctr., Marion, IA, 52302. Phone: (319) 377-7345. Fax: (319) 377-5861. E-mail: mailia@magid.com Web Site: www.magid.com.

Brent Magid, pres/CEO; Steve Ridge, exec VP; Bill Hague, sr VP; Richard Haynes, VP rsch.

Sherman Oaks, CA 91403, 15260 Ventura Blvd, Suite 2130. Phone: (818) 263-3300. Fax: (818) 263-3311. Jack MacKenzie, exec VP entertainment.

New York, NY 10019, 1775 Broadway, Suite 1401. Phone: (212) 974-2310. Fax: (212) 515-4540. Vicki Cohen, exec VP entertainment.

Strategic rsch applications for traditional & new media companies, including Executive Telefocus, Magid Media Futures, INstant-STUDIO, Magid Performance Predictor, Magid Revenue Enhancer.

Marketing Evaluations Inc. The Q Scores Company, (A division of Marketing Evaluations Inc.). 1615 Northern Blvd., Manhasset, NY, 11030. Phone: (516) 365-7979. Fax: (516) 365-9351. E-mail: info@qscores.com Web Site: www.qscores.com.

Steven Levitt, pres; Henry Schafer, exec VP.

Syndicated market rsch surveys measuring familiarity & appeal of TV programs, cable programs, performers, characters, company & brand names, sports personalities.

Marketron International, 700 Airport Blvd., Suite 130, Burlingame, CA, 94010-2001. Phone: (800) 788-9245. Fax: (650) 548-2295. E-mail: lcarpenter@marketron.com Web Site: www.marketron.com.

Mike Jackson, CEO.

Toronto, ON M2N 6C6 Canada, 5075 Yonge St, Suite 404. Phone: (416) 221-9944. Les Bridgen, gen mgr.

Birmingham, AL 35244, 3000 Riverchase Galleria, 8th Fl. Phone: (205) 987-7456. Fax: (205) 733-4535. E-mail: tvsales@marketron.com. Michael Hunter, gen mgr.

Hailey, ID 83333, 101 Empty Saddle Trail. Phone: (208) 788-6272. Gary Coats, gen mgr.

Software applications for radio, TV, networks, syndicators, traf, accounting, mgmt, demand pricing, inventory control, rsch & proposals.

MarketVision Research, 10300 Alliance Rd., Cincinnati, OH, 45242-5617. Phone: (513) 791-3100. Fax: (513) 794-3500. E-mail: tmcmullen@mv-research.com Web Site: www.mv-research.com.

Jon Pinnell, pres; Tyler McMullen, sr VP.

Charlotte, NC 28210, 6805-A Fairview Rd. Phone: (704) 442-0444. Ronald Miller, exec VP.

Secauces, NJ 07094, 6000 Meadowlands Pkwy, Suite 250A. Phone: (201) 865-4040. Fax: (201) 865-1922.

Full-svc mktg rsch firm. Focused on customized high-value rsch.

Marquest Meda & Entertainment Research, 314 Orange St., Beaufort, NC, 28516-1821. Phone: (252) 728-4047. Fax: (915) 200-1850. E-mail: paul.rule@marquest.net Web Site: www.marquest.net.

Paul Rule, pres.

Loc & natl cable, bcst & print surveys. Sydicated on demand menu planner studies.

Marshall Marketing & Communications Inc., 2600 Boyce Plaza Rd, Suite 210, Pittsburgh, PA, 15241-3949. Phone: (412) 914-0970. Fax: (412) 914-0971. E-mail: info@mm-c.com Web Site: www.mm-c.com.

Craig A. Marshall, chmn/CEO; Richard Kinzler, pres/COO.

Orlando, FL 32835, 1445 Saddleridge Dr. Phone: (407) 299-3510. Bruce Hahn, rsch & sls consultant.

Cary, NC 27513, 102 Silver Lining Lane. Phone: (919) 388-7622. James Filippi, rsch & sls consultant.

Knoxville, TN 37931, 6729 Heatherbrook Dr. Phone: (865) 938-7988. Cynthia Bridges, rsch & sls consultant.

Langley, WA 98260, 5280 Lakeside Dr. Phone: (360) 321-2139. Lori West, rsch & sls consultant.

Spokane, WA 99223, 3021 E. 62nd Ave. Phone: (509) 443-1362. Rick Hamm, rsch & sls consultant.

Sls dev rsch; custom-designed, consumer market-specific data; demographics; media usage & Risc American program. Leading-Edge software: exclusive & non-exclusive mktg programs.

Media Market Resources, Box 442, Littleton, NH, 03561. Phone: (603) 444-5720. Fax: (603) 444-2872.

Cathy Devine, VP/gen mgr.

Produces multimedia profile reports, printed & electronic bcst directories & software.

Media Monitors Inc., 8253 Thoroughbred Run, Suite 102, Indianapolis, IN, 46256-4320. Phone: (317) 547-1362. Fax: (317) 549-0331. E-mail: jselig@mediamonitors.com Web Site: www.mediamonitors.com.

John L. Selig, pres; Tom Zarecki, dir.

Radio, nwspr & magazine monitoring Co. Creative review for natl & loc advertisers' radio spots.

Media Perspectives, 127 Greensward Ln., Cherry Hill, NJ, 08002. Phone: (856) 482-7979. Fax: (856) 482-0957. E-mail: steve@apels.net

Steven G. Apels, pres.

Rschs audience tastes & perceptions, employs advanced analytical techniques to guide bcstrs in constructing strategic progmg & mktg plans.

Media Rating Council, Inc., 370 Lexington Ave., Suite 902, New York, NY, 10017. Phone: (212) 972-0300. Fax: (212) 972-2786.

George W. Ivie, exec dir.

Determines criteria & standards & administrators an audit system for accreditation of audience measurement svcs to assure conformance with criteria, standards & procedures developed.

Mediamark Research Inc., 75 Ninth Avenue, 5th Floor, New York, NY, 10011. Phone: (212) 884-9200. Fax: (212) 884-9339. Web Site: www.mediamark.com.

Alain J. Tessier, chmn; Kathi Love, pres/CEO.

Los Angeles, CA 90068, 3575 Cahuenga Blvd. W, Suite 223. Phone: (232) 882-6325. Chetan Shah, western rgnl mgr.

Chicago, IL 60611, 444 N. Michigan Ave, Suite 2050. Phone: (312) 329-0901. Scott Turner, VP.

Syndicated rsch—product, demographics for TV, radio, cable, new media & print. Custom recontact surveys.

Miller, Kaplan, Arase & Co., 4123 Lankershim Blvd., North Hollywood, CA, 91602. Phone: (818) 769-2010. Fax: (818) 769-3100. Web Site: www.millerkaplan.com.

George Nadel Rivin, C.P.A., ptnr.

San Francisco, CA 94104, 180 Montgomery St, Suite 1840. Phone: (415) 956-3600. Catherine C. Gardner, C.P.A., ptnr.

Market revenue reports revenue forecast software and market x-ray including newspaper, TV & radio expenditures by adv.

MORPACE International, 31700 Middlebelt, Suite 200, Farmington Hills, MI, 48334. Phone: (248) 737-5300. Fax: (248) 737-5326. E-mail: information@morpace.com Web Site: www.morpace.com.

Jim Leiman, media studies.

Market rsch, strategic planning, viewer satisfaction, economic modeling, audience analysis, data base mapping for TV & radio stns, cable TV franchises.

Multimedia Research Group Inc. (MRG, Inc.), 1754 Technology Dr., Suite 132, San Jose, CA, 95110. Phone: (408) 453-5553. Fax: (408) 453-5559. E-mail: info@mrgco.com Web Site: www.mrgco.com.

Gary Schultz, pres.

Provides strategic consulting & published market intelligence on content dev, content distribution, channels & networks.

Jack Myers LLC, 20 E. 68 St., Suite 7B, New York, NY, 10021. Phone: (212) 794-4926. Fax: (212) 794-5160. E-mail: jack@jackmyers.com Web Site: www.jackmyers.com.

Jack Myers, editor.

Syndicated & proprietary rsch, consulting & evaluation for media companies, advertisers, & adv agencies. Publishes *Jack Myers Report* industry newsletter.

National Broadcast Finance Corp., Box 3167, 27 Harrison St., New Haven, CT, 06515-0267. Phone: (203) 389-6000. Fax: (203) 389-6020. E-mail: cheerhoniak@sbcglobal.net

David C. Cherhoniak, pres.

Financial advisory & investment banking svcs to bcstg industry & brokerage.

National Economic Research Associates Inc. (NERA), 50 Main St., 14th Fl., White Plains, NY, 10606. Phone: (914) 448-4000. Fax: (914) 448-4040. Web Site: www.nera.com.

Richard Rapp, pres.

San Francisco, CA 94111, One Front St. Phone: (415) 291-1000. Gregory Duncan, sr VP.

Sydney NSW 2000 Australia, Level 6, 50 Bridge St. Fax: 2-8272-6500. Greg Houston, dir.

Brussels, NO B-1040 Belgium, rue de la Loi, 23 Wetstraat. Phone: 32-2282-4340. Mark Williams, dir.

Madrid 28046 Spain, Paseo de la Castellana, 13. Phone: 91-212-6400. David Robinson, dir.

London W1C 1BE United Kingdom, 15 Stratford Pl. Phone: 20-7659-8500. John Rhys, dir.

Los Angeles, CA 90017, 777 S. Figueroa St. Phone: (213) 346-3000. Gary Dorman, sr VP.

Washington, DC 20037, 1255 23rd St. N.W. Phone: (202) 466-3510. Andrew Joskow, VP.

Chicago, IL 60611, 875 N. Michigan Ave. Phone: (312) 573-2800. David Evans, sr VP.

Cambridge, MA 02142, One Main St. Phone: (617) 621-0444. William Taylor, sr VP.

Ithaca, NY 14850, 308 N. Cayuga St. Phone: (607) 277-3007. Alfred E. Kahn, special consultant.

New York, NY 10036, 1166 Ave. of the Americas, 31st Fl. Phone: (212) 345-3000. Linda McLaughlin, sr VP.

Philadelphia, PA 19103, Two Logan Sq. Phone: (215) 864-3880. Eugene Ericksen, special consultant.

Economic consultant to bcst & cable TV companies on economic, pub policy & business strategy issues.

Nielsen Media Research, 770 Broadway, New York, NY, 10003. Phone: (646) 654-8300. Fax: (646) 654-8990. E-mail: info@nielsenmedia.com Web Site: www.nielsenmedia.com.

Susan D. Whiting, pres/CEO; Paul Donato, chief rsch off; Bob Luff, chief technology off.

Los Angeles, CA 90028, 6255 Sunset Blvd. Phone: (323) 817-1200. Rob Hebenstreit, VP, NSI rgnl mgr.

San Francisco, CA 94111, 2 Embarcadero Ctr., Suite 1017. Phone: (415) 249-6000. Colleen Hall, VP, NSI rgnl mgr.

Alpharetta, GA 30004, 1145 Sanctuary Pkwy., Suite 100. Phone: (770) 777-4260. Lisa Schmidt, VP, Southeastern NSI rgnl mgr.

Chicago, IL 60606, 200 W. Jackson Blvd., Suite 2700. Phone: (312) 385-6500. Jane Ryan, sr VP, sls & mktg NSI.

Dallas, TX 75201, 1717 Main St. Phone: (214) 290-9200. Lucinda Nobles, VP, sls & mktg NSI midwest mgr.

Nielsen offers TV audience measurement for network, loc, syndication, cable, Sp-language networks & stns; metered market & individual loc market reports in all markets; natl sydicated program ratings; loc syndicated program ratings; daily, wkly, & mthly network program ratings; telephone coincidentals; & competitive adv intelligence.

Paragon Media Strategies, 12345 W. Alameda Pkwy., Suite 325, Denver, CO, 80228. Phone: (303) 922-5600. Fax: (303) 922-1589. E-mail: info@paragonmediastrategies.com Web Site: www.paragonmediastrategies.com.

Mike Henry, CEO; John Stevens, pres; Bob Harper, sr VP.

Media rsch & consulting.

Peters Communications, 1555 Berenda Pl., El Cajon, CA, 92020. Phone: (858) 565-8511. Fax: (619) 440-1481. E-mail: ppiep@cox.net

Edward J. Peters, pres.

Media mktg consultants, providing rsch, concept, mktg plan, music, graphics & animation.

Pike & Fischer Inc., 1010 Wayne Ave., Suite 1400, Silver Spring, MD, 20910. Phone: (301) 562-1530. Fax: (301) 562-1521. E-mail: customercare@.com Web Site: www.pf.com.

Meg Hargreaves, pres.

Publishers of communications print and eletronic info svcs including; Bcst Rules Svc, Cable TV Rules Svc, Communications Regulation and U.S. Spectrum Report.

RAD Marketing & CableTowns, 167 Crary-on-the-Park, Mount Vernon, NY, 10550-0572. Phone: (914) 668-3563. Fax: (914) 668-4247.

Robert Dadarria, pres.

Cable subs, or TV buyers, MOB buyers- Mailing lists provided to marketers largest most ideal cable TV H/H audience available. Phone numbers; data base by age, gender, MOBs, and direct marketing responses.

Radio Computing Services Inc., 12 Water St., White Plains, NY, 10601. Phone: (914) 428-4600. Fax: (914) 428-5922. Web Site: www.rcsworks.com.

Andrew M. Economos, chmn; Philippe Generali, pres; Ted Nygreen, gen mgr.

Computer software for bcstrs, mus scheduling, rsch, data bases, yield mgmt, audio logging, digital audio systems.

The Radio Journal, PO Box 442, Littleton, NH, 03561. Phone: (603) 444-5720. Fax: (603) 444-2872. E-mail: genemckay@insideradio.com Web Site: www.theradiojournal.com.

Summary of FCC data & format changes. Publishes the *The Radio Journal* & *M The Radio Book*, & Inside Radio Daily Fax.

Research Communications Ltd., 95 Washington St., 402-357, Canton, MA, 02021-4009. Phone: (781) 341-1190. Fax: (781) 341-1191. E-mail: cranev@aol.com Web Site: www.researchcommunications.com.

Valerie Crane, Ph.D., CEO; Susan Crohan, PhD., pres.

Full-svc media rsch company specializing in quantitative/qualitative news mktg, & progmg rsch for bcst, cable, radio & adv.

Research International U.S.A., 222 Merchandise Mart Plaza, Suite 275, Suite 2511, Chicago, IL, 606654-1308. Phone: (312) 787-4060. Fax: (312) 787-4156. E-mail: info@riusa.com Web Site: www.riusa.com.

Mary Vallender, exec VP; Diane Frederick, exec VP; Mark Willard, exec VP; Bruce E. Lervoog, exec VP.

Phoenix, AZ 85021-4258. Phoenix, 8800 North 22nd Ave. Phone: (602) 735-8800. Fax: (602) 735-3270. Diane Frederick, exec VP.

San Francisco, CA 94107. San Francisco, 303 2nd St., 9th Fl. S. Phone: (415) 281-2760. Fax: (415) 281-2799. Bruce Lervoog, exec VP.

Stamford, CT 06901, 3 Landmark Sq. Phone: (203) 358-0900. Fax: (203) 353-0883. Mark Willard, exec VP.

Cambridge, MA 02139. Cambridge, 955 Massachusetts Ave. Phone: (617) 661-0110. Fax: (617) 661-3575. Mark Willard, exec VP.

Full-svc custom rsch design through analysis. Facilities include 300-position WATS phones with CRTs & complete computer capabilities.

Rules Service, (a division of Pike & Fischer Inc.). c/o Pike & Fischer Inc., 1010 Wayne Ave., Suite 1400, Silver Spring, MD, 20910. Phone: (301) 562-1530. Fax: (301) 562-1521. E-mail: ruletwo@starpower.net Web Site: www.ruleserv.com.

Zachary Wheat, group publisher; Meg Hargreaves, pres.

FCC rules & regulations updated in loose-leaf & disk svcs, includes 0, 1, 2, 5, 11, 13, 15, 17, 18, 19, 20, 21, 22, 24, 25, 27, 73, 74, 76, 78, 79, 80, 87, 90, 95, 97, 100 & 101.

SatNews Publishers, 800 Siesta Way, Sonoma, CA, 95476. Phone: (707) 939-9306. Fax: (707) 939-9235. E-mail: design@satnews.com Web Site: www.satnews.com.

Publishes the *International Satellite Directory*, the complete guide to the satellite communications industry. Also available on a CD-Rom & through the web at www.satnews.com

Scarborough Research, 770 Broadway, 13th Fl., New York, NY, 10003-9595. Phone: (646) 654-8400. Fax: (646) 654-8450. Fax: (646) 654-8440. E-mail: info@scarborough.com Web Site: www.scarborough.com.

Robert Cohen, Ph.D, pres/CEO.

Chicago, IL 60606, 200 W. Jackson Blvd, Suite 1822. Phone: (312) 385-6700. Howard Goldberg, sr VP Radio/Sports mktg.

Specializes in loc market & multimedia consumer studies; hundreds of consumer categories & media behavior, all specific to the individual market.

Shane Media Services, 2500 Tanglewilde, Suite 106, Houston, TX, 77063. Phone: (713) 952-9221. Fax: (713) 952-1207. E-mail: smsofc@shanemedia.com Web Site: www.shanemedia.com.

Ed Shane, CEO; Renee Revett, consultant.

Radio progmg and mgmt consultation, custom designed perceptual and qualitative rsch for electronic media outlets.

The Shosteck Group, 11002 Viers Mill Rd., Suite 709, Silver Springs, MD, 20902-2574. Phone: (301) 589-2259. Fax: (301) 588-3311. E-mail: jzweig@shosteck.com Web Site: www.shosteck.com.

Dr. Herschel Shosteck, chmn; Jane Zweig, CEO.

Telecommunication economics & market analysis emphasizing cellular. International consultation on demand, svc & equipment competition, distribution chns, economic & market effects of privatization, liberalization, competition, deregulation & market acceptance. Newly published studies on wireless and wireless internet.

Simmons Market Research Bureau, Inc., 29 Broadway, Fl. 30, New York, NY, 10006-3216. Phone: (212) 863-4500. Fax: (212) 863-4495. E-mail: kenw@smrb.com Web Site: www.smrb.com.

Christopher T. Wilson, pres/COO; William P. Livek, CEO; William E. Engel, CEO.

Deerfield Beach, FL 33441, 700 W. Hillsboro Blvd., Bldg 4, Suite 201. Phone: (954) 427-4104. Fax: (954) 427-4104. Bill Livek, CEO; Bill Engel, CEO.

Schaumburg, IL 60173, 955 American Ln. Phone: (224) 698-8142. Fax: (224) 698-4139. Mary Kay Petrella, dir of clients.

Study of media & markets; comprehensive measurement of media & product usage. Other studies include kids, teens, Hispanic & the Internet.

Sindlinger & Co. Inc., 405 Osborne St., Wallingford, PA, 19086. Phone: (610) 565-0247. Fax: (610) 565-7174. E-mail: nelsind@aol.com

Albert E. Sindlinger, chmn/pres.

Daily polling to determine consumer attitudes; microeconomic forecasting.

Spectrum Research, 5000 Boardwalk, #602, Ventnor City, NJ, 08406-2918. Phone: (609) 822-0056. Fax: (609) 385-2953. E-mail: peter@spectrumresearch.com Web Site: www.spectrumresearch.com.

Focus groups, lifestyle studies, & strategic market studies. Custom rsch for radio & TV.

SQAD Inc., 303 S. Broadway, Suite 108, Tarrytown, NY, 10591-5410. Phone: (914) 524-7600. Fax: (914) 524-7650. E-mail: info@sqad.com Web Site: www.sqad.com.

Neil Klar, pres/CEO; Larry Fried, VP natl sls.

Cost projections for National broadcast, cable and syndication (Net Costs), local CPP/CPM projections for Spot TV, Radio & Hispanic TV. SNAP software for Sweeps and Overnights.

Strata Marketing Inc., 30 West Monroe, Suite 1900, Chicago, IL, 60603. Phone: (312) 222-1555. Fax: (312) 222-2510. E-mail: rsparks@stratag.com Web Site: www.stratag.com.

Bruce W. Johnson, pres.

TV, radio & media ratings analysis. Microsoft Windows-based software systems for cable systems & bcst TV stns.

Synovate-Americas, 8600 NW 17th St., Suite 100, Miami, FL, 33126. Phone: (305) 716-6800. Fax: (305) 716-6756. Web Site: www.synovate.com.

Richard Tobin, pres.

Laguna Hills, CA 92653, 23151 Alcalbe Dr, Suite C-4. Phone: (949) 598-9055. Dave Thomas, VP/gen mgr. (Los Angeles office).

Radio/TV Hispanic audience ratings, natl & loc; market studies, stn profiles, focus groups, product usage/awareness, progmg/format rsch, copy testing. Specialization in ethnic rsch.

TNS Canadian Facts, 900-2 Bloor St. E., Toronto, ON, M4W 3H8. Canada. Phone: (416) 924-5751. Fax: (416) 923-7085. E-mail: info@nfocfgroup.com Web Site: www.tns-global.com.

Michael LoPresti, pres/CEO; David Stark, pub affrs dir.

Vancouver, BC V6E 4A4 Canada, 1130 W. Pender St, Suite 600. Phone: (604) 668-3344. Fax: (604) 668-3333. Diana Tindall, rsch dir.

Ottawa, ON K1P 5W6 Canada, Place de Ville Tower B, 112 Kent St, Suite 2010A. Phone: (613) 232-4408. Fax: (613) 232-7102. Bente Nielsen, VP.

Montreal, PQ H3G 2B3 Canada, 1250 Guy St, Suite 1030. Phone: (514) 935-7666. Fax: (514) 935-6770. Michel Gauvreau, VP.

Full range of custom-designed & syndicated market rsch svcs.

The Tarrance Group, 201 N. Union St., Suite 410, Alexandria, VA, 22314. Phone: (703) 684-6688. Fax: (703) 836-8256. E-mail: tarrance@tarrance.com Web Site: www.tarrance.com.

David Sackett, ptnr.

Audience rsch svcs for radio, TV & cable. News, progmg, positioning, survey & promotional rsch.

Teen-age Research Unlimited, 707 Skokie Blvd., 7th Flr., Northbrook, IL, 60062. Phone: (847) 564-3440. Fax: (847) 564-0825. E-mail: info@teenresearch.com Web Site: www.teenresearch.com.

Peter Zollo, pres.

Twice annually, syndicated study of the teen market with optional custom/proprietary questions. Custom rsch in teenage markets.

Telecommunications Research Inc., (A division of Nathan Associates Inc.). 2101 Wilson Blvd., Suite 1200, Arlington, VA, 22201. Phone: (703) 516-7800. Fax: (703) 351-6162. E-mail: info@triresearch.com Web Site: www.triresearch.com.

Economic & mgmt consultants specializing in mktg & survey rsch, business & property valuation & financial viability analysis.

VIP Research Inc., 5700 Broadmoor, Suite 710, Mission, KS, 66202. Phone: (913) 384-9494. Fax: (913) 677-2727. E-mail: mike@mjmresearch.com Web Site: www.mjmresearch.com.

Clark Roberts, CFO & sls dir; Mike Heydman, COO & rsch dir.

Provides hook-tape production, listener screening, fielding, & tabulation for all music testing, perceptual studies, focus groups & promotional telemarketing.

VNU Consumer Research Services, Inc., 12350 N.W. 39th St., Coral Springs, FL, 33065. Phone: (954) 753-6043. Fax: (954) 346-8869.

Kathy Pilhuj, VP; Mary Glover, VP; Mark Manders, VP.

Tuczon, AZ 85710, 6339 Speedway, Suite 200. Phone: (520) 751-2223. Barbara Garvin, mgr.

Sarasota, FL 34236, 1751 Mound St, Suite 205. Phone: (941) 955-9877. Kathleen Goodwin, mgr.

San Antonio, TX 78229, 4801 N.W. Loop 410, Suite 125. Phone: (210) 647-3198. Juan Hernandez, mgr.

Data collection for Survey Research, production of The Scarborough Report and The Market Audit Report.

Vallie-Richards Consulting Inc., Box 3510, Vallie Richards Donovan Consulting, Alpharetta, GA, 30023. Phone: (770) 346-0026. Fax: (770) 346-0028. E-mail: jimrvr@aol.com Web Site: www.vallierichards.com.

Jim Richards, pres.

America's premier contemp radio consultancy specializing in all variations of contemp radio balancing art & science to create ratings & revenue success.

Ventures in Media Inc., 78175 Estancia Dr., Palm Desert, CA, 92211. Phone: (760) 360-7303. Fax: (805) 338-9500.

Morrie Gelman, pres.

Tokyo f116-0002 Japan, 3-76-1-604 Arabawa-Ku. Phone: 11 81-3-3891-3622. Adam Gelman, purchasing VP.

Market rsch info packaging & consulting.

Video Monitoring Services, 330 W. 42nd St., New York, NY, 10036. Phone: (212) 736-2010. Fax: (212) 736-8206. E-mail: sales@vmsinfo.com Web Site: www.vmsinfo.com.

Jeff Tidyman, gen mgr.

Los Angeles, CA 90028, 6430 W. Sunset Blvd, Suite 400. Phone: (323) 993-0111. Fax: (323) 467-7540. Ashley Griffin.

Washington, DC 20045, 1066 National Press Bldg. Phone: (202) 393-7110. Fax: (202) 393-5451. Meredith Imwalle.

Chicago, IL 60610, 212 W. Superior St. Phone: (312) 649-1131. Fax: (312) 649-1527. Jack Monson.

Provides transcripts, digests & analysis of radio & TV news & commentary; surveys of program content; monitoring of comls; tape & cassette recordings; photoboards; film conversions.

Mona Wargo, 1600 N. Oak St., Suite 1401, Arlington, VA, 22209. Phone: (703) 243-9352. Fax: (703) 243-5795. E-mail: mwrsrch@erols.com

Mona Wargo, MS-IT/TS; rsch analyst.

Ind rsch analyst in bcst & telecommunications, FCC regulatory policy, legal & engig rsch, and consultant.

Washington Information Group Ltd., 1655 N. Fort Myer Dr., Suite 825, Arlington, VA, 22209. Phone: (703) 312-6004. Fax: (703) 527-4586.

Doug House, pres.

Customized business rsch on client-specified aspects of the bcstg industry, including competitor intelligence on companies, products, svcs & markets.

World Information Technologies Inc., 70 Carley Ave., Huntington, NY, 11743. Phone: (631) 549-3629. Fax: (631) 549-7527.

Amadee Bender, pres.

Telecommunications market rsch.

Your Personal Researcher, 22848 Mesa Way, Lake Forest, CA, 92630-4643. Phone: (949) 472-8538. E-mail: dlbraunstein@prodigy.net

Donna Lee Braunstein, owner.

Script & documentary rsch.

Engineering and Technical Consultants

AF Associated Inc./Ascent Media, 100 Stonehurst Ct., Northvale, NJ, 07647. Phone: (201) 750-1200. Fax: (201) 784-8637. E-mail: consulting@ascentmedia.com Web Site: www.afassoc.com.

Tom Canavan, pres; Jack Dawson, sr VP; Chris Summey, sr VP.

Mgmt & distribution of content to major motion picture studios, ind producers, bcst networks, cable channels, adv agencies & other companies that produce, own/or distribute entertainment, adv, news, sports, corporate, educ & industrial content.

AZCAR U.S.A. Inc., 121 Hillpointe Dr., Canonsburg, PA, 15317. Phone: (724) 873-0800. Fax: (724) 873-4770. E-mail: info@azcar.com Web Site: www.azcar.com.

Stephen Pumple, pres/CEO.

Bcst, video & audio system consultation, design, instal & training; serving cable systems, corporate & teleproduction facilities.

Advanced Technology Systems Inc., 7915 Jones Branch Dr., McLean, VA, 22102. Phone: (703) 506-0088. Fax: (703) 903-0415.

Claude Rumsey, owner; Doug Manning, VP opns; Penelope Parker, mktg dir.

Software dev, lifecycle mgmt, systems integration, testing, infrastructure mgmt, business process improvement, wireless svcs & specialized functional expertise.

John H. Battison, P.E. & Associates, Consulting Radio Engineers. 2684 State Rt. 60, Loudonville, OH, 44842. Phone: (419) 994-3849. Fax: (419) 994-5419. E-mail: batcom@bright.net

John H. Battison, P.E.; S. Bennett, VP.

All FCC svcs: AM, FM, TV, LPTV, applications, licensing, DA-proofs, ITFS & MMDS expert witness svcs.

Richard S. Becker & Associates, Chartered, 7128 Fair Fax Rd., Bethesda, MD, 20814. Phone: (301) 986-9005. Fax: (301) 986-8496. E-mail: beckereng@aol.com

Richard Becker, principal atty; Siamak Harandi, consulting engr; Christopher Fedeli, assoc.

Legal & engrg svc for bcstg, cable TV, cellular, paging, microwave & private radio.

Lawrence Behr Associates Inc., Box 8026, Greenville, NC, 27835-8026. Phone: (252) 757-0279. Fax: (252) 752-9155. E-mail: lbagrp@lbagroup.com Web Site: www.lbagroup.com.

Lawrence Behr, CEO; *Jerry E. Brown, VP.

Provides wireless svcs: site acquisition, construction mgmt, AM detuning, AM tower colocation, RF hazard mgmt, RF shielding, due diligence, facility mgmt, dev, maintenance, support svcs.

Serge Bergen, P.E., 7503 Amkin Ct., Clifton, VA, 22024. Phone: (703) 250-2691. E-mail: wtranavitch@cox.net

William V. Tranavitch Jr., ptnr.

Engrg svcs: AM, FM, TV, translators, LPTV.

Bernard Associates, 143 Palmers Hill Rd., Stamford, CT, 06902-2111. Phone: (203) 348-0804. Fax: (203) 921-1016.

Bernard Eishwald, P.E.

Broadcast Engineering & Equipment Maintenance Co., (BEEM Co.). 2322 S. 2nd Ave., Arcadia, CA, 91006. Phone: (626) 446-3468. Fax: (626) 445-8028. E-mail: joel@beemco.com Web Site: www.beemco.com.

Joel T. Saxberg, pres.

Site studies, applications, AM directional arrays, allocation studies, field work, mobile signal analysis, bcst consulting, radiofrequency electromagnetic field measurements.

Broadcast Services Inc., Box 6418, Brattleboro, VT, 05302-6418. Phone: (802) 258-3000. Phone: (802) 258-4500 Svc. Fax: (802) 258-2500. E-mail: mh@markhutchins.com Web Site: www.markhutchins.com.

Mark F. Hutchins, pres.

Field studies: Spectrum analysis, NRSC/RFR compliance. Propagation analysis, coverage maps, path profiles & shadowing studies.

Broadcast Signal Lab, LLP, 64 Richdale Ave., Cambridge, MA, 02140-2629. Phone: (617) 864-4298. Fax: (617) 661-1345. E-mail: information@broadcastsignallab.com Web Site: www.broadcastsignallab.com.

Rick Levy, ptnr; David P. Maxson, ptnr.

RF safety evaluation, expert testimony, coverage analysis, license engrg & applications, interference & spectrum analysis, frequency monitoring, tech due diligence.

Bromo Communications Inc., Box 191747, Atlanta, GA, 31119-1747. Phone: (404) 636-2257. Fax: (404) 636-2256. E-mail: bill@bromocom.com Web Site: www.bromocom.com.

Bill Brown, engr; Gil Moor, engr.

Washington, DC Phone: (202) 429-0600.

Consulting engrg for bcst stns. AM/FM & TV allocations, including field instals.

John F.X. Browne & Associates P.C., 38710 N. Woodward Ave., Suite 220, Bloomfield Hills, MI, 48304. Phone: (248) 642-6226. Fax: (248) 642-6027. E-mail: consultants@jfxb.com Web Site: www.jfxb.com.

*John F.X. Browne, P.E., pres; John Fleming, engr; Leonard W. Eden, engr.

Bcst consulting AM/FM/TV/DTV, MMDS/ITFS, PCS & satellite systems. FCC/FAA applications, filings & studies. Field measurement vehicle AM/FM/TV/DTV.

Richard W. Burden Associates, 20944 Sherman Way, Suite 213, Canoga Park, CA, 91303. Phone: (818) 340-4590. Fax: (818) 884-8840. E-mail: rwburden@pacbell.net

Richard W. Burden, engr.

Bcst tech svcs, facilities design, Traveller's Information Service (TIS) & Educational FM (EDFM) FCC applications, Part 15 AM & FM bcst systems engrg.

C.S.I. Telecommunications, 750 Battery St., Suite 350, San Francisco, CA, 94111-1555. Phone: (415) 751-8845. Fax: (415) 292-9981. E-mail: info@csitele.com Web Site: www.csitele.com.

Michael S. Newman, sr VP; Philip M. Kane, P.E./Esq, VP; William F. Ruck, engr; Thomas Croda, engr; Tim Pozar, engr; David Doom, P.E., engr.

Telecommunications, radio & microwave engrg, feasibility studies, FCC applications, systems engrg; equipment specifications, project mgmt & lab measurements.

Cavell, Mertz & Associates, Inc., 7839 Ashton Ave., Manassas, VA, 20109. Phone: (703) 392-9090. Fax: (703) 392-9559. E-mail: office@cavellmertz.com Web Site: www.cavellmertz.com.

Garrison C. Cavell, pres; Richard H. Mertz, VP/sec; Michael D. Rhodes, P.E., sr. engr.; Daniel G. Ryson, sr. engr.; Mark B. Peabody, sr. engr.; Robert J. Clinton, assoc. engr.; J.M. Perryman, assoc. engr.

Broadcast and Communications Consulting Engineers located in suburban Washington, DC. Experts in radio/TV channel searches, upgrades, FCC application support, transmission, coverage studies, interference evaluation, RF safety, and strategic planning, both broadcast (AM/FM radio, TV, digital TV, microwave/satellite) & industrial. Consultants in engineering, including transmitter and studio systems design & Integration; digital transition issues; litigation support and zoning hearings; tech-mgmt communications support; project implementation & mgmt svcs. Ability to assist with financial & techpresentations; can provide project design, development, organization & implementation. Able to assist with due diligence, technology translation & negotiations, equipment procurement (package purchases), bid dev & evaluation, & price negotiations. Available nationally & internationally.

Chenevert Architects LLC, 6767 Perkins Rd., Suite 100, Baton Rouge, LA, 70808. Phone: (225) 757-0955. Fax: (225) 757-0765. E-mail: chenevert@architects.com Web Site: www.chenevertarchitects.com.

Norman J. Chenevert, pres.

Architects, planners, & technical designers, specializing in new & renovated bcst/cable production facilities.

Chevalier Aviation Associates, LLC, 928 Via Panorama, Palos Verdes, CA, 90274. Phone: (310) 375-2979. Fax: (310) 791-7181. E-mail: jack.chevalier@verizon.net

Jack Chevalier L.L.B., assoc; L. Gene Garrett, dir.

Part 77 studies, FCC registrations, EMI analysis, legal assistance & representation before FAA, state & loc aeronautical & zoning agencies.

Clear Channel Communications, 1834 Lisenby Ave., Panama City, FL, 32401. Phone: (850) 769-1408. Fax: (850) 769-0659. E-mail: contactus@clearchannel.com Web Site: www.clearchannel.com.

Turnkey installations (AM, FM studios & transmitters), tech appraisals, emergency repairs, upgrades. International startups & upgrades.

Cohen, Dippell and Everist, P.C., 1300 L St. N.W., Suite 1100, Washington, DC, 20005. Phone: (202) 898-0111. Fax: (202) 898-0895. E-mail: cde@attgobal.net Web Site: http://www.broadcast-consulting-engineers.com.

*Donald G. Everist, P.E., pres.

Professional engrg svcs to the bcstg industry, United States & worldwide.

Commercial Radio Co., One Duttonsville School Dr., Cavendish, VT, 05142. Phone: (802) 226-7582. Fax: (802) 226-7738. Web Site: commercialradiocompany.com.

Daniel W. Churchill, PE, pres; Centura L. Churchill, VP; Andre S. LaPlante, sls.

Custom bcst engrg; AM, FM & shortwave bcst equipment sls & svce, specializing in transmitting components.

Communications Design Associates Inc., 437 Turnpike St., Canton, MA, 02021-2702. Phone: (339) 502-6551. Fax: (781) 551-8491. E-mail: information@cdaconsultants.com Web Site: www.cdaconsultants.com.

Robert Hemenway, ptnr; Stewart Randall, ptnr; Greg Vincent, ptnr.

Ind consultants to radio, TV, corporate & govt clients. Designers of studios, production, presentation & multimedia facilities.

Communications General Corp., 2685 Alta Vista Dr., Fallbrook, CA, 92028-9739. Phone: (760) 723-2700. E-mail: r.gonsett@ieee.org

Robert F. Gonsett, E.E., pres.

Bcst engrg consulting, AM/FM/TV applications & field engrg. Specializes in southwest United States, Hawaii & Mexico.

Communications Technologies Inc., Box 1130, 65 Country Club Ln., Marlton, NJ, 08053. Phone: (856) 985-0077. Fax: (856) 985-8124. E-mail: info@commtechrf.com Web Site: www.commtechrf.com.

Clarence M. Beverage, pres; Laura M. Mizrahi, VP; James W. Pollock, P.E., engr.

Bcst engrg consulting svcs with emphasis on AM, FM & TV RF systems design & FCC application preparation consistent with FCC rules & policies.

Comsearch, 19700 Janellia Farm Blvd., Ashburn, VA, 20147. Phone: (703) 726-5500. Fax: (703) 726-5600. Web Site: www.comsearch.com.

Doug Hall, pres.

A complete communications engrg svc organization, specializing in frequency mgmt & propagation engrg.

ComSonics Inc., Box 1106, 1350 Port Republic Rd., Harrisonburg, VA, 22801. Phone: (540) 434-5965. Fax: (540) 432-9794. E-mail: info@comsonics.com Web Site: www.comsonics.com.

Dennis A. Zimmerman, pres/CEO; Donn E. Meyerhoeffer, dir opns; Don J. Sommerville, dir mktg & sls.

Microprocessor controlled signal level meters, RF leakage detection & CATV repair facility.

Contemporary Communications, 9408 Grand Gate St., Las Vegas, NV, 89143. Phone: (702) 898-4669. Fax: (208) 567-6865. E-mail: lfuss@cox.net

Larry G. Fuss, pres.

FM, TV, STL & RPU applications; FM upgrades; computerized frequency searches; site selection assistance.

C.P. Crossno & Associates, Consulting Engineers. Box 180312, Dallas, TX, 75218. Phone: (214) 321-9140. Fax: (214) 321-9146. E-mail: c.crossno@ieee.org

Charles Paul Crossno, owner.

Aeronautical issues; antenna design.

Crown Castle, 2000 Corporate Dr., Cannonsburg, PA, 15317. Phone: (724) 416-2000. Fax: (724) 416-2200. E-mail: webmaster@crowncastle.com Web Site: www.crowncastle.com.

Communications engrg consultants & site/tower mgrs.

William Culpepper & Associates Inc., 900 Jefferson Dr., Charlotte, NC, 28270. Phone: (704) 365-9995. Fax: (704) 364-4823.

*William A. Culpepper, pres.

AM & FM applications & feasibility studies, specializing in applications for AM power increases & transmitter relocation.

DSI RF Systems Inc., 26H World's Fair Dr., Somerset, NJ, 08873. Phone: (732) 563-1144. Fax: (732) 563-1818. E-mail: tcarroll@dsirf.com Web Site: www.dsirf.com.

Joseph Giardina, chief engr; Herb Squire, VP engrg.

Radio & TV system design, transmitter & studio instal, microwave & satellite engrg & instal, remote control camera systems.

D.C. Williams Ph.D, P.E., Consulting Radio Engineer. Department of Computer Science & Engineering, The University of Nevada, Reno, Reno, NV, 89557. Phone: (775) 784-6974. Fax: (775) 885-8705. E-mail: drdcw@cs.unr.edu Web Site: www.cs.unr.edu/˜drdcw.

*D.C. Williams, P.E. PhD., pres.

Registered professional engr, specializing in allocations, antenna system design, applications, construction, evaluation & measurement of AM & FM directional facilities.

John J. Davis & Associates, Box 128, Sierra Madre, CA, 91025-0128. Phone: (626) 355-6909. Fax: (626) 355-4890. E-mail: johnjdavis@roadrunner.com Web Site: www.socaltowers.com.

*John J. Davis, pres.

Primary focus on FM & TV ch allocation studies & applications; facility upgrades, FM & TV translator applications, tower site mgmt.

Dettra Communications Inc., 7906 Fox Hound Rd., McLean, VA, 22102. Phone: (703) 790-1427. Fax: (703) 790-0497.

John E. Dettra Jr., pres.

Applications & hearing exhibits for bcst, paging, mobile telephone, cellular, microwave & private radio svcs; MMDS/ITFS; FCC rsch & consulting.

Devlin Design Group Inc., Box 5208, Frisco, CO, 80443. Phone: (760) 453-9360. E-mail: contacts@ddgtv.com Web Site: www.ddgtv.com.

Dan Devlin, CEO.

News sets, softset, virtual set environments, promotions, newsrooms, facility planning, lighting direction, consultation, Videssence Integration & softset (Virtual Reality Sets).

Diversified Systems Inc., 385 Market St., Kenilworth, NJ, 07033. Phone: (908) 245-4833. Fax: (908) 245-0011. E-mail: info@divsysinc.com Web Site: www.divsysinc.com.

Alfred D'Alessandro, pres/CEO.

Full-svc engrg, specializing in video & RF systems.

Doug Vernier Telecom Consultants, 721 W. 1st St., Suite A, Cedar Falls, IA, 50613. Phone: (319) 266-8402. Fax: (319) 266-9212. E-mail: consulting@v-soft.com Web Site: www.v-soft.com.

Doug Vernier, pres; Kate Michler, tech consultant.

Tech consulting for AM, FM, & TV. Coverage mapping, frequency searches, applications, site evaluations, stn watches, stn audits & more.

The Downtown Group, 236 W. 27th St., New York, NY, 10001. Phone: (212) 675-9506. Fax: (212) 675-3276. E-mail: info@downtowngroup.com Web Site: www.downtowngroup.com.

Peter Wilcox, ptnr; Mark Winkleman, ptnr.

Design of tech facilities: architecture, acoustics, engrg, testing. Typical projects include edit rooms, stages, recording studios & support facilities.

du Treil, Lundin & Rackley Inc., 201 Fletcher Ave., Sarasota, FL, 34237-6019. Phone: (941) 329-6000. Fax: (941) 329-6030. E-mail: bobjr@dlr.com Web Site: www.dlr.com.

Tech consulting for the communications industry.

ERI - Electronics Research Inc., 7777 Gardner Rd., Chandler, IN, 47610. Phone: (812) 925-6000. Fax: (812) 925-4030. Web Site: www.eriinc.com.

*Thomas B. Sillman, pres; Bill Elmer, sls; Ernest R. Jones, P.E./structural engrg; David Davies, mktg & B.S.M.E.; Robert Rose, rsch & dev & B.S.M.E.; Jim Kemman, B.S.M.E.; Dan Dowdle, B.S.M.E.; John Robinson, P.E.; Eric Wandel, engr.

Mf & instal FM antennas, towers, filters, combiners, lightning protection & grounding systems. Structural analysis also available.

Evans Associates Consulting, 210 S. Main St., Thiensville, WI, 53092. Phone: (262) 242-6000. Fax: (262) 242-6045. E-mail: info@evansassoc.com Web Site: www.evansassoc.com.

*Ralph E. Evans Sr., ptnr; *B. Benjamin Evans, P.E., ptnr.

Telecommunications consulting engrs, net design, FCC applications, digital bcstg strategic planning, fieldwork for AM, FM, TV, CATV, ITFS, microwave relay facilities & fiber, wireless, & PCS networks.

Federal Engineering Inc., Redwood Plaza II, 10600 Arrowhead Dr., Fairfax, VA, 22030. Phone: (703) 359-8200. Fax: (703) 359-8204. E-mail: info@fedeng.com Web Site: www.fedeng.com.

Ronald F. Bosco, pres; John E. Murray, sr VP.

Strategic planning, coverage analysis, new product definition, market rsch, competitive analysis, rates & tariffs, bcst stn design, mergers & acquisitions, expert testimony, regulatory support.

Charles S. Fitch, P.E., 45 Sarah Dr., Avon, CT, 06001. Phone: (860) 673-7260. Fax: (860) 673-7260. E-mail: fitchpe@comcast.net

*Charles S. Fitch, P.E., engr.

FCC allocations & applications, facility design, system design, construction supervision, field surveys, facility appraisals & inspections, computer progmg.

Paul Dean Ford, Broadcast Engineering Consultant. 18889 N.2350th St., Dennison, IL, 62423. Phone: (217) 826-9673. E-mail: wkzi@rr1.net

Paul Dean Ford, P.E., owner.

Engineering consultant.

Freedman, Mel, 2612 Portsmouth Ln., Modesto, CA, 95355. Phone: (209) 522-1180. Fax: (209) 522-1750. E-mail: melengr@sbcglobal.net

Mel Freeman, engr.

George M. Frese, P.E., 1011 Denis Ct., East Wenatchee, WA, 98802. Phone: (509) 884-4558. Fax: (509) 884-9170. E-mail: frese@genext.net

*George M. Frese, owner.

AM Antenna design. Short, DA & Multiplexing.

GeoMart, 516 Villanova Ct., Fort Collins, CO, 80525. Phone: (970) 416-8340. Fax: (970) 416-8345. E-mail: sales@geomart.com Web Site: www.geomart.com.

Chuck Cotherman, owner.

All USGS & DMA digital & paper maps. All NOS/NOAA charts, international topographic series, aerial photography, raised relief maps, digital products, business & mktg maps, travel maps, globes, etc.

Graphic Enterprises Inc., 3874 Highland Park N.W., North Canton, OH, 44720. Phone: (905) 624-9529. Fax: (905) 624-3715. Web Site: www.geiworldwide.com.

Large-format digital printing systems.

Hammett & Edison Inc., Box 280068, San Francisco, CA, 94128-0068. Phone: (707) 996-5200. Phone: (202) 396-5200 (DC). Fax: (707) 996-5280. E-mail: engr@h-e.com Web Site: www.h-e.com.

***William F. Hammett, P.E., pres; Dane E. Ericksen, P.E., engr; *Stanley Salek, P.E., engr; *Robert D. Weller, P.E., engr; *Mark D. Neumann, P.E., engr; Rajat Mathur, P.E., engr.**

Design & FCC filings: AM, FM, TV, STL, wireless cable. Specialties: computerized coverage studies, AM directionals / diplexers, RF radiation predictions / measurements / mitigations, field strength measurements, due diligence technical surveys, FAA EMI analysis.

Hatfield & Dawson, Consulting Engineers L.L.C., 9500 Greenwood Ave. N., Seattle, WA, 98103. Phone: (206) 783-9151. Fax: (206) 789-9834. E-mail: hatdaw@hatdaw.com Web Site: www.hatdaw.com.

B. F. Dawson III, pres.

Telecommunications & radio physics engrg, including bcst, electromagnetic compatibility, NIER measurement & analysis, antenna & propagation analysis, & design.

Charles A. Hecht & Associates Inc., 16 Doe Run, Pittstown, NJ, 08867. Phone: (908) 730-7959. Fax: (908) 730-7408. E-mail: hechtassoc@sprintmail.com

Charles A. Hecht, pres; William L. Smith Sr., engr; Charles J. Hecht, engr.

Bcst engrg svcs including FCC studies & applications, directional antenna design, fieldwork, tech litigation; specialists in AM studies for telecommunications companies.

Hilding Communications, Box 1700, Morgan Hill, CA, 95038-2222. Phone: (408) 842-2222. E-mail: eric@hilding.com

Eric Hilding, owner.

FM ch studies, complex FM substitution proposals, site locations, 301 applications engrg, gen bcst consulting.

HN Telecom Inc., 1160 Douglas Rd., ., Burnaby, BC, V5C 4Z6. Canada. Phone: (604) 294-3401. Fax: (604) 299-6712. E-mail: contact@hntelecom.com Web Site: www.hntelecom.com.

P. Hostinsky, principal; Bruce W. Grantholm, sr VP engrg/opns.

Telecommunications tech consulting svcs for AM, FM, TV bcst & CATV systems, studio-to-transmitter links & studio systems.

Doug Holland Inc., 1871 Sweet Briar Ln., Birmingham, AL, 35235-3357. Phone: (205) 229-5628. Fax: (205) 655-4092. E-mail: dougholland@att.net Web Site: www.dougholland.com.

Doug Holland, pres.

Turnkey opns, allocation studies, due diligence, transmitter installations, upgrades & coverage maps

R.L. Hoover Consulting Telecommunications, Consulting Telecommunications Engineer. 11704 Seven Locks Rd., Potomac, MD, 20854. Phone: (301) 983-0054.

*Robert Lloyd Hoover, P.E., owner.

Professional engrg consulting for AM, FM & TV applications & testimony. Radiation hazard analyses & testimony. Ex-owner AM & FM stns. Patent agent.

Independent Broadcast Consultants Inc., 110 County Rd. 146, Trumansburg, NY, 14886. Phone: (607) 273-2970. Fax: (607) 273-5125. E-mail: ibcengineering@juno.com Web Site: www.trumansburgchamber.com.

William J. Sitzman, pres; M.F. Sitzman, VP; George Soltysik, P.E., engr; N.L. Hollenback, engr; R.A. Lynch, engr.

AM, FM & SW applications, specializing in AM allocation studies & broadband AM directional antenna design & AM diplexer design.

J. Boyd Ingram & Associates, Box 1528, Batesville, MS, 38606. Phone: (662) 563-4007. Fax: (601) 563-9002. E-mail: jboyd@panola.com

J. Boyd Ingram, A.E., pres.

Tech consultation, facility construction & repair.

George Jacobs & Associates Inc., 3210 N. Leisure World Blvd., Suite 1001, Silver Spring, MD, 20908-0298. Phone: (301) 598-1282. Fax: (301) 598-7788. E-mail: broadcaster@gjainc.com Web Site: www.gjainc.com.

*George Jacobs, P.E., pres.

Specialists in conceptional design, application filing & frequency mgmt for FCC-licensed International Broadcast Stations (shortwave). Consultative liaison with foreign bcst stns & organizations.

Vir James, P.C., 965 S. Irving St., Denver, CO, 80219. Phone: (303) 937-1900. Fax: (303) 937-1902.

***Timothy C. Cutforth, P.E., pres.**

AM/FM/TV allocation studies & applications, AM directional ant design & tune-up, conductivity measurements.

Jenel Systems and Design Inc./Smalling Systems, 6700 Spokane, Plano, TX, 75023. Phone: (972) 491-1442. Fax: (972) 491-1442. E-mail: smalling@smallingsystems.com

Elmer Smalling III, pres/CEO; Howard Halcomb, sls VP; Erin Day Loyd, VP mktg.

Digital bcstg solutions; studio, post, acoustic & earth stn, design; trucks; TV system design from planning to turnkey construction.

Carl T. Jones Corp., 7901 Yarnwood Ct., Springfield, VA, 22153. Phone: (703) 569-7704. Fax: (703) 569-6417. E-mail: hhurst@ctjc.com Web Site: www.ctjc.com.

*C. Thomas Jones, Jr., pres; *Herman E. Hurst, Jr., mgr.

Consulting engrs specializing in bcstg & CATV tech design & regulatory filings. Maintain EMC/EMI testing laboratory.

KCI Technologies Inc., 4601 Six Forks Rd., Landmark Center II, Suite 220, Raleigh, NC, 27609. Phone: (919) 783-9214. Fax: (919) 783-9266. E-mail: employment@kci.com Web Site: www.kci.com.

Tom Donohue, P.E., sr VP; James Blake, engrg mgr.

Full engrg svcs to the communications industry, including tower analysis & remediation, design of standard & non-standard sites, "stealth" engrg, photo realistic renderings & turnkey construction.

Kessler & Gehman Associates Inc., 507 N.W. 60th St., Suite C, Gainesville, FL, 32607. Phone: (352) 332-3157. Fax: (352) 332-6392. E-mail: rwilhour@bellsouth.net Web Site: www.kga.bz.

*Robert Gehman, P.E. Jr., pres; William Kessler, P.E., VP; Jeffrey C. Gehman, engr; Ryan C. Wilhour, engr; William T. Godfrey, P.E., engr.

Studies, system design, FCC applications, bidding documents & contract monitoring for bcst, ITFS, wireless cable, microwave & mobile communications systems & digital TV.

Lightning Eliminators & Consultants Inc., 6687 Arapahoe Rd., Boulder, CO, 80303. Phone: (303) 447-2828 ext 107. Fax: (303) 447-8122. E-mail: info@lecglobal.com Web Site: www.lecglobal.com.

Roy B. Carpenter Jr, dir; Jerry Kerr, VP mktg & sls; Darwin N. Sletten, chief engr.

Consulting & engrg svcs in lightning prevention, grounding & power / signal / telephone / data line conditioning.

Lohnes and Culver, 8309 Cherry Ln., Laurel, MD, 20707-4830. Phone: (301) 776-4488. Fax: (301) 776-4499. E-mail: locul@locul.com

*Robert D. Culver, P.E., ptnr; Frederick D. Veihmeyer, ptnr.

Communication consulting engrg svc for bcst & related fields. Design, application, optimization, system evaluation & expert representation svcs.

Cecil Lynch Consulting Engineers, 2460 Illinois Ave., Modesto, CA, 95358. Phone: (209) 523-3955. Fax: (209) 522-5287.

Cecil Lynch, CEO.

Bcst engrg, stn appraisals, customized computer progmg svc, GPS surveying & RFR measurements.

Magnusson Klemencic, 1301 Fifth Ave., Suite 3200, Seattle, WA, 98101-2699. Phone: (206) 292-1200. Fax: (206) 292-1201. E-mail: resume@mka.com Web Site: www.mka.com.

Jon D. Magnusson, CEO; Brian McIntyre, sr VP.

Tower inspection, wind rating, tower retrofit, concept studies, design, contract documents & construction mgmt.

Mahlum Architects, 71 Columbia, Suite 400, Seattle, WA, 98104. Phone: (206) 441-4151. Fax: (206) 441-0478. E-mail: info@mahlum.com Web Site: www.mahlum.com.

John Mahlum, pres.

D.L. Markley & Associates Inc., 2104 W. Moss Ave., Peoria, IL, 61604. Phone: (309) 673-7511. Fax: (309) 673-8128. E-mail: dlm@dlmarkley.com Web Site: www.dlmarkley.com.

*Donald L. Markley, P.E., pres; Jeremy Ruck, engr; Rich Wood, engr; Todd Chiodini, engr.

AM, FM, TV & microwave applications, construction & measurements. Allocation studies, non-ionizing radiation measurements.

Marsand Inc., Box 485, 6100 IH-35W, Alvarado, TX, 76009. Phone: (817) 783-5566. Fax: (817) 783-5577. E-mail: tvcowboy@marsand.com Web Site: www.marsand.com.

*Matthew A. Sanderford, Jr., P.E., pres; David Sanderford, VP engrg.

Turnkey installation svcs, CAD-VIDCAD wiring documentation, FM & TV proof-of-performance, analog & digital, FCC consulting & applications, RF troubleshooting, installations & conversions.

Frank J. Maynard, 44683 Mansfield Dr., Novi, MI, 48375. Phone: (248) 344-2965. E-mail: info@fmaynard.com Web Site: www.fmaynard.com.

Frank J. Maynard, C.F.B.E., owner.

Radio & TV engrg svcs & applications.

McClanathan & Associates Inc., Box 939, Portland, OR, 97207. Phone: (503) 246-8080. Fax: (503) 246-6304.

*Robert A. McClanathan P.E., pres.

Professional electrical engrs for radio & TV FCC applications, computer svcs, field engrg & construction svcs.

Meintel, Sgrignoli, & Wallace, 1282 Smallwood Dr., Suite 372, Waldorf, MD, 20603. Phone: (202) 251-7589. Fax: (301) 645-1426. E-mail: wallacedtv@aol.com Web Site: www.mswdtv.com.

Dennis Wallace, ptnr; William Meintel, ptnr; Gary Sgrignoli, ptnr.

Specializing in digital & analog TV & Radio technical software, consumer electronics.

MidAmerica Electronics Service Inc., 410 Mt. Tabor Rd., New Albany, IN, 47150. Phone: (812) 945-1209. Fax: (812) 945-1859. E-mail: peterclb@aol.com

AM & FM field engrg svcs, antenna measurements, AM stereo instal & proof of performance, AM & FM spectrum analysis, NRSC compliance measurement, new construction & rebuilding.

Lawrence L. Morton Associates, 4105 Tenango Rd., Claremont, CA, 91711-2341. Phone: (323) 467-5010. Fax: (323) 467-5848. E-mail: larry@radiotv.biz

*Lawrence L. Morton, P.E., owner.

Telecommunications engrg consulting svcs for AM, FM, TV & LPTV. Computerized engrg svcs, field svcs, FCC applications.

Mueller Broadcast Design, 613 S. La Grange Rd., La Grange, IL, 60525. Phone: (708) 352-2166. Fax: (708) 352-2170. E-mail: mark@muellerbroadcastdesign.com Web Site: www.muellerbroadcastdesign.com.

Mark A. Mueller, owner.

AM/FM tech consultant, AM directional systems.

Mullaney Engineering Inc., 9049 Shady Grove Ct., Gaithersburg, MD, 20877. Phone: (301) 921-0115. Fax: (301) 590-9757. E-mail: jmullaney@mullengr.com Web Site: www.mullengr.com.

*John J. Mullaney, pres; *Alan E. Gearing, P.E., engr; Timothy Z. Sawyer, engr.

Consulting communications engrs providing: design & optimization of AM directional arrays; analysis for new allocation, site relocation & upgrades AM FM TV LPTV wireless cable (MDS/MMDS/ITFS/OFS); environmental radiation analysis; fieldwork; expert testimony.

Multicomm Sciences International Inc., 266 W. Main St., Denville, NJ, 07834. Phone: (973) 627-7400. Fax: (973) 625-1002. E-mail: mail@multicommsciences.com Web Site: www.multicommsciences.com.

Victor J. Nexon Jr., pres.

Frequency coord, site surveys, earth stn interference studies, FCC license, radiation hazard testing.

Munn-Reese Inc., Box 220, 385 Airport Dr., Coldwater, MI, 49036-0220. Phone: (517) 278-7339. Fax: (517) 278-6973. E-mail: wayne@munn-reese.com Web Site: www.munn-reese.com.

Christine Reese, VP; Wayne S. Reese, pres.

AM, FM, TV, low power TV & engrg consulting service, including applications, field tuning & problem solving.

Nationwide Tower Company Inc., 414 N. Ingram St., Henderson, KY, 42420. Phone: (270) 869-8000. Fax: (270) 869-8500. E-mail: hjohnston@nationwidetower.com Web Site: www.globalcommnet.com.

Kevin Roth, sls VP; Diane Pruitt, gen sls mgr.

Tower inspections, painting, repair re-guy, lighting, antennas, feedlines, analysis, erect, dismantle, line sweeping, site monitoring, & tower tracker svcs.

Newman-Kees Frequency Measurements, Engineering, & Installations, 8611 Slate Rd., Evansville, IN, 47720. Phone: (812) 963-3294. E-mail: nkeng@insightbb.com

Frank Hertel, owner.

RF & frequency measurements for AM-FM-TV coml users via air or on location. Audio/video svc & instals.

Owl Engineering & EMC Test Labs, Inc., 5844 Avenue N., Shoreview, MN, 55126. Phone: (651) 784-7445. Fax: (763) 784-4541. E-mail: info@owleng.com Web Site: www.owleng.com.

*Garrett G. Lysiak, P.E., pres; Diane Stewart Lysiak, ptnr.

Telecommunications consulting engrg svcs, applications, facilities specifications svcs, field engrg svcs, maintenance & FCC compliance svcs, EMC testing.

Pacific Radio Electronics, 969 N. La Brea, Los Angeles, CA, 90038. Phone: (323) 969-2035. Phone: (800) 634-9476. Fax: (323) 969-2053. E-mail: info@pacrad.com Web Site: www.pacrad.com.

Joseph Phillips, pres.

Distributor of racks, patch bays, cable, adaptors, connectors, handtools, outlet strips & many other products for the bcst industry.

William F. Pohts Telecommunications, 225 Denfield Dr., Alexandria, VA, 22309. Phone: (703) 360-7193. Fax: (703) 360-0309. E-mail: bill@pohts.com

*William F. Pohts, P.E., engr.

Consulting engr specializing in the emerging technologies in telecommunications & electronic systems.

Rimma Posin, 3712 Carmel Ave., Irvine, CA, 92606. Phone: (949) 857-9639. Fax: (949) 857-9639.

Rimma Posin, owner.

Consulting for cable; FCC applications.

W.L. Pritchard & Co. L.C., 4405 E.W. Hwy., Suite 501, Bethesda, MD, 20814. Phone: (301) 654-1144. Fax: (301) 654-1814. E-mail: wlpritchard-co@verizon.net Web Site: www.wlpco.com.

Ellen Hoff, pres; Jack Dicks, VP engrg.

Professional engrg and business problem solving in telecommunications, competitor analysis, satellite communications, earth stns, & launch vehicles.

RFK Engineering, LLC, 1229 19th St., N.W., Washington, DC, 20036. Phone: (202) 463-1565. Fax: (202) 463-0344. E-mail: prubin@satpar.com Web Site: www.rfkengineering.

*Philip A. Rubin, P.E., pres; Ted Kaplan, VP engrg; Jeffrey Freedman, CFO; Alex Latker, dir; Arnold Berman, PhD., engr; William Meeker, engr.

MSS, FSSS & BSS satellite experts, TV & radio cellular & other new media technologies. Experts in FCC rules & regulations. International experience, experts in ITU regulation. Software developers, simulation & modeling. In business over 20 years.

RF Technologies Corp., 12 Foss Rd., Lewiston, ME, 04240. Phone: (207) 777-7778. Fax: (207) 777-7784. E-mail: info@rftechnologies.net Web Site: www.rftechnologies.net.

Designs & manufactures high-power bcst RF networks & components for FM & TV bcstrs. Products include ants, diplexers, combiners, filters, switches, coaxial & waveguides.

Radio/TV Engineering Co., 1416 Hollister Ln., Los Osos, CA, 93402. Phone: (805) 528-1996. Fax: (805) 528-1982.

Norwood J. Patterson, pres; G. Dawn Patterson, exec sec/asst engrg.

AM, FM, FCC applications, directional ant design. Serving bcstrs for over 35 years.

Radiotechniques Engineering, LLC, Box 367, 402 10th Ave., Haddon Heights, NJ, 08035-0367. Phone: (856) 546-8008. Fax: (856) 546-1841. E-mail: ted@radiotechniques.com

*Edward A. Schober, P.E., VP.

AM, FM, TV, digital bcst, boosters, FCC, equipment, field, & systems engrg. RF, financial, opns, & acoustical design.

Rogers Cable Systems, 35-73 Wolfdale, Mississauga, ON, L5C 3T6. Canada. Phone: (905) 273-8000. Fax: (905) 273-9661. Web Site: www.rogers.com.

Consulting engrg svcs with emphasis on design, instal & testing of CATV systems, fiber-optic nets.

D.W. Sargent Broadcast Service Inc., 804 Richard Rd., Cherry Hill, NJ, 08034. Phone: (856) 667-8573. Fax: (856) 667-1409.

Dean W. Sargent, pres.

Ant system design & measurements for FM & TV. FM & TV master ant system design.

T.Z. Sawyer Technical Consultants, 9049 Shady Grove Ct., Gaithersburg, MD, 20877. Phone: (301) 921-0115. Fax: (301) 590-9757. E-mail: info@tzsawyer.com Web Site: www.sawyer.com.

Timothy Z. Sawyer, pres; Trisha E. Ford, admin asst.

FCC applications for AM, FM, TV, LPTV & aux svcs; AM directional ant design; AM, FM, & TV ant measurements; allocation studies; site surveys & inspections.

Sellmeyer Engineering, Box 356, McKinney, TX, 75070. Phone: (972) 542-2056. Fax: (214) 636-5940.

*J.S. Sellmeyer, P.E., owner.

AM, FM, TV applications, hearing support, directional ant design & adjustment; facilities planning & specialized equipment design.

SiteSafe Inc., 200 N. Glebe Rd., Suite 1000, Arlington, VA, 22203. Phone: (703) 276-1100. Fax: (703) 276-1169. E-mail: info@sitesafe.com Web Site: www.sitesafe.com.

Wesley McGee, pres; Bill Zlotnick, opns VP.

Bcst & land mobile & wireless engrg consulting svcs.

Smith and Fisher, 2237 Tackett's Mill Dr., Suite A, Woodbridge, VA, 22192. Phone: (703) 494-2101. Fax: (703) 494-2132. E-mail: kevin@smithandfisher.com Web Site: www.smithandfisher.com.

Neil M. Smith, ptnr; Kevin T. Fisher, ptnr.

Tech consultants to FM, TV, & LPTV stns; FCC applications; allocations studies; RF measurements; expert witness testimony.

Carl E. Smith Consulting Engineers, Box 807, 2324 N. Cleveland-Massilon Rd., Bath, OH, 44210-0807. Phone: (330) 659-4440. Fax: (330) 659-9234.

Al Warmus, pres; Brian M. Warmus, sec; Roy Stype III, VP.

AM, FM, TV & LPTV engrg, FCC applications, ant systems adjustments. Sls: towers, ants, transmission line, phasing equipment. Turnkey instal.

Frederick A. Smith Engineers, 1123 Old River Rd., Elloree, SC, 29047. Phone: (803) 897-2815. Fax: (803) 897-2816.

Frederick A. Smith, P.E., pres; Cameron E. Smith, P.E., VP.

Communications systems design, microwave path surveys, ant impedance measurements. United States & foreign.

Southern Broadcast Services, 80 Commerce Dr., Suite B, Pelham, AL, 35124. Phone: (205) 663-3709. Phone: (800) 256-9235. Fax: (205) 663-7108. E-mail: jwcoleman@southernbroadcastservices.com Web Site: www.southernbroadcastservices.com.

Jim Coleman, pres.

Tower erection, ant & line instal, cellular & maintenance svcs.

Standard Frequency Measuring Service, 2092 Arrowood Pl., Cincinnati, OH, 45231. Phone: (513) 851-4964. E-mail: lawilliams@alum.mit.edu

*Louis A. Williams Jr., P.E., engr.

Frequency measurements on AM, FM & TV.

J.M. Stitt & Associates Inc., 621 Mehring Way, Suite 1907, Cincinnati, OH, 45202. Phone: (513) 621-9292. Fax: (513) 651-9622. E-mail: towerjim@aol.com Web Site: www.jmstittassociates.com.

James Stitt, pres.

Engrg consultants, facility design & instal, contract engrg svcs, acoustical consultants, tower site mgmt.

Structural Systems Technology Inc., 6867 Elm St., McLean, VA, 22101. Phone: (703) 356-9765. Fax: (703) 448-0979.

Fred W. Purdy, P.E., pres; Keveh Mehrnama, P.E., engrg VP.

Structural engrg studies, analysis, design, modifications, inspections, fabrication & erection of towers & ant structures.

Superior Satellite Engineers, 1743 Middle Rd., Columbia Falls, MT, 59912. Phone: (406) 257-9590. Fax: (406) 257-9599. E-mail: superior@superiorsatelliteusa.com Web Site: www.superiorsatelliteusa.com.

Steve Catlett, engrg VP; Jackie Williams, dir.

Coml quality, cost effective satellite access solutions for IPTV, CATV, bcst TV including navigator steerable antenna.

Technet Systems Group, (A division of Steve Vanni Associates Inc). Box 422, Auburn, NH, 03032. Phone: (603) 483-5365. Fax: (603) 483-0512. E-mail: sales@technetsystems.com Web Site: www.technetsystems.com.

Steve Vanni, pres.

Bcst equipment supplier/distributor for radio & TV, specializing in complete "turnkeyed" packages including planning, design, equipment, instal, towers & FCC licensing.

Teletech Inc., Box 85567, Westland, MI, 48185. Phone: (734) 641-2300. Fax: (734) 641-2323. Web Site: www.teletech-inc.com.

Keith Johnson, VP; Todd Osment, field svcs mgr.

Scottsdale, AZ 85261, Box 4221. Phone: (480) 367-1500.

Engrg consultants: AM, FM, TV, LPTV; FCC applications /filings; FAA filings, aeronautical studies; tower erection, maintenance & inspections; antenna site dev & mgmt; directorial antenna design & proof of performance; contract engrg svcs.

Cullen B. Tendick Consulting Radio Engineer, 11753 N. Cassiopeia Dr., Tucson, AZ, 85737. Phone: (520) 575-8265. Fax: (520) 498-0156. E-mail: cullen3@juno.com

Cullen B. Tendick, consulting radio engr.

AM, FM & TV measurements & allocations.

TransVision, 550 Maulhardt Ave., Oxnard, CA, 93030. Phone: (805) 981-8740. Fax: (805) 981-8738. E-mail: info@txvision.com Web Site: www.txvision.com.

Kimithy Vaughn, mktg & sls.

Twelve transportable & satellite transmission facilities (video, audio, voice, data). Flypack production & SNG svcs.

V-Soft Communications, L.L.C., 721 W. 1st St., Suite A, Cedar Falls, IA, 50613. Phone: (319) 266-8402. Fax: (319) 266-9212. E-mail: info@v-soft.com Web Site: www.v-soft.com.

Doug Vernier, pres; John Gray, program dir; Adam Puls, mktg.

Bcst engrg software for signal propagation, frequency searching, interference analysis & custom mapping for AM, FM, TV, & LPTV.

Steve Vanni Associates Inc., Box 422, Auburn, NH, 03032. Phone: (603) 483-5365. Fax: (603) 483-0512. E-mail: svanni@techsystems.com

Steve Vanni, pres.

Tech consulting, systems design, project mgmt; complete turnkey svcs including equipment & towers through Technet Systems Group.

The Richard L. Vega Group Inc., 2527 E. Semoran Blvd., Apoka, FL, 32703-5835. Phone: (407) 814-8100. E-mail: vega@magicnet.net

Richard L. Vega Jr., chmn.

Tech consulting svcs for all telecommunications svcs including FCC applications, allocation studies, site acquisition.

Vernier, Doug, Telecommunications Consultants, 721 W. 1st St., Suite A, Cedar Falls, IA, 50613. Phone: (319) 266-8402. Fax: (319) 266-9212. E-mail: info@v-soft.com Web Site: www.v-soft.com.

Doug Vernier, pres; Kate Michler, consultant; Jake Vernier, assoc; Gayle Vernier, mgr.

Bcst engrg & consultation; ch searches, FCC applications, allocations, custom mapping, coverage analysis. V-Soft Communications bcst engr software.

Willoughby & Voss, LLC, Box 701190, San Antonio, TX, 78270-1190. Phone: (210) 490-2778. Phone: (210) 525-1111. Fax: (210) 490-2779. E-mail: willvoss@satx.rr.com

Lyndon H. Willoughby, owner.

AM, FM, TV, STL, trans applications, directional ant design, field svcs, allocations, site studies, system planning, frequency searches, facility inspection & non-ionized radiation studies.

Wireless Systems Engineering Inc./MLJ, (formerly JMS/MLJ Worldwide Inc.). 15713 Crabbs Branch Way, Suite 140, Rockville, MD, 20855. Phone: (301) 840-2030. Fax: (301) 840-2031. E-mail: info@wse-mlj.com

IT & software engrg, wireless engrg & internet svcs needs.

Walter Wulff & Associates, FAA Consultants. Box 914, Pt. Clear, AL, 36564. Phone: (251) 990-2502. Fax: (251) 990-2503. E-mail: wulff@zebra.net

Walter H. Wulff, pres.

Conducts FAA obstruction evaluation studies, FM electro magnetic interference analysis & communications consulting engrs.

Law Firms Active in Communications Law

Abacus Communications Company, 1801 Columbus Rd. N.W., Suite 101, Washington, DC, 20009-2031. Phone: (202) 462-3680. Fax: (202) 462-3781. E-mail: abacuscommco@covad.net Benjamin Perez.

Akerman & Senterfitt, Citrus Center, 255 South Orange Ave. 17th fl., Orlando, FL, 32801. Phone: (407) 843-7860. Fax: (407) 843-6610. Tom Cardwell.

Akin Gump Strauss Hauer & Feld LLP, Robert S. Strauss Bldg., 1333 New Hampshire Ave. N.W., Washington, DC, 20036. Phone: (202) 887-4000. Fax: (202) 887-4288. E-mail: washdcinfo@akinggump.com Web Site: akingump.com. Kathleen Q. Abernathy; Martina Bradford; Tom Davidson; Philip Marchesiello.

Anderson, Kill & Olick L.L.P., 2100 M St. N.W., Suite 650, Washington, DC, 20037. Phone: (202) 218-0040. Fax: (202) 218-0055. Web Site: www.andersonkill.com.

Ardi, Dennis, 340 N. Camden Dr., Third Fl., Beverly Hills, CA, 90210. Phone: (310) 271-6900. Fax: (310) 271-6963.

Arent & Fox, PLLC, 1050 Connecticut Ave. N.W., Washington, DC, 20036-5339. Phone: (202) 857-6000. Fax: (202) 857-6395. Web Site: www.arentfox.com. E-mail: delorey.denise@arentfox.com

Arnold & Porter LLP, 555 12th St. N.W., Washington, DC, 20004-1206. Phone: (202) 942-5000. Fax: (202) 942-5999. E-mail: norman_sinel@aporter.com Web Site: www.arnoldporter.com. Phillip W. Horton, Richard L. Rosen, Norman M. Sinel, Richard M. Firestone, Stephanie M. Phillipps, Patrick J. Grant, Marcia Cranbeg, William E. Cook, Theodore D. Frank, P. Scott Feira, Maureen Jeffreys, Peter Schildkraut, Michael Ryan, Donald T. Stepka, Emma Wright.

Asbury, Philip S., 309 S. Broad St., Philadelphia, PA, 19107-5813. Phone: (215) 985-0911. Fax: (215) 985-1195. E-mail: pasburypir@ol.com

Attorney At Law, 2154 Wisconsin Ave., N.W., Suite 250, Washington, DC, 20007-2280. Phone: (202) 223-3772. Fax: (202) 315-3587. E-mail: kenhardman@att.net Kenneth E. Hardman.

Ausley & McMullen, Box 391, 227 S. Calhoun St., Tallahassee, FL, 32302. Phone: (850) 224-9115. Fax: (850) 222-7560.

Baker & Hostetler LLP, 1050 Connecticut Ave. N.W., Suite 1100, Washington, DC, 20036. Phone: (202) 861-1500. Fax: (202) 861-1783. E-mail: khoward@bakerlaw.com Web Site: www.bakerlaw.com. Kenneth C. Howard Jr., Bruce W. Sanford, Mark I. Bailen, Bruce D. Brown, Laurie Babinski.

Law Offices of Ruth S. Baker-Battist, 5600 Wisconsin Ave., Chevy Chase, MD, 20815. Phone: (301) 718-0955. Fax: (301) 718-8867. E-mail: rbattist@aol.com

Baker Botts L.L.P., 1299 Pennsylvania Ave. N.W., Washington, DC, 20004. Phone: (202) 639-7700. Fax: (202) 639-7890. E-mail: laurie.spielman@bakerbotts.com Herbert J. Miller Jr., John Joseph Cassidy.

Baker, Ravenel & Bender, Box 8057, 1730 Main St., Columbia, SC, 20292. Phone: (803) 799-9091. Fax: (803) 779-3423. Charles E. Baker, Jay Bender, ptnrs.

Law Office of Martin J. Barab, 10250 Constellation Blvd., MGM Tower 3rd Fl., Century City, CA, 90067. Phone: (310) 843-0464. Fax: (310) 859-6650. E-mail: martin@bauermartinez.com Martin J. Barab, Sean M. Fawcett.

Barron & Newburger, P.C., 1212 Guadalupe St., Suite 104, Austin, TX, 78701. Phone: (512) 476-9103. Fax: (512) 476-9253. E-mail: bbarron@bnpdaw.com Barbara M. Barron.

Bass, Berry & Sims, 315 Deaderick St., South Center Suite 2700, Nashville, TN, 37238-0002. Phone: (615) 742-6200. Fax: (615) 742-6293.

Richard S. Becker & Associates, 7128 Fairfax Rd., Bethesda, MD, 20814. Phone: (301) 986-9005. Fax: (301) 986-8456. E-mail: Beckereng@aol.com Richard S. Becker, James S. Finerfrock.

Beitchman & Hudson, 215 14th St. N.W., Atlanta, GA, 30318. Phone: (404) 897-5252. Fax: (404) 874-4270. E-mail: leebeebee@aol.com Lee B. Beitchman.

Bell, Boyd & Lloyd, 1615 L St. N.W., Suite 1200, Washington, DC, 20036. Phone: (202) 466-6300. Fax: (202) 463-0678.

Law Offices of Jeff Berke, 12100 Wilshire Blvd., Suite 1500, Los Angeles, CA, 90025-6538. Phone: (310) 571-2808. Fax: (301) 571-0207. E-mail: Jeffberke@yahoo.com Jeff Berke. Esq.

Berkowitz, Trager & Trager, P.C., 8 Wright St., Westport, CT, 06880. Phone: (203) 226-1001. Fax: (203) 226-3801.

Law Offices of Lawrence Bernstein, 3510 Springland Ln., N.W., Washington, DC, 20008. Phone: (202) 296-1800. Fax: (202) 296-1800. E-mail: lawberns@verizon.net Lawrence Bernstein.

Bingham McCuthen, LLP, 2020 K St. N.W., Washington, DC, 20006. Phone: (202) 373-6033. Fax: (202) 373-6001. E-mail: andrew.lipman@bingham.com Web Site: www.bingham.com. Jean Kiddoo, Russell Blau, Catherine Wang.

Birch, Horton, Bittner & Cherot, 1155 Connecticut Ave. N.W., Suite 1200, Washington, DC, 20036. Phone: (202) 659-5800. Fax: (202) 659-1027. Web Site: www.birchhorton.com. Elisabeth H. Ross, Thomas L. Albert, Ronald Birch, William Horn.

Bishop, Payne, Harvard & Kaitcer, L.L.P., 500 W. Seventh St., Suite 1800, Fort Worth, TX, 76102-4782. Phone: (817) 335-4911. Fax: (817) 870-2631.

Blank, Rome, LLP, 405 Lexington Ave., 23rd Floor, New York, NY, 10174. Phone: (212) 885-5000. Fax: (212) 885-5001. Web Site: www.blankrome.com.

Bleiweiss, Irene, U.S. FCC, Audio Services Division, 445 12th St. S.W., Rm. 2B450, Washington, DC, 20554. Phone: (202) 418-2700. Fax: (202) 418-1411. E-mail: irene.bleiweiss@fcc.gov Web Site: www.fcc.gov/mb/audio. Irene Bleiweiss.

Blooston, Mordkofsky, Dickens, Duffy & Prendergast, LLP, 2120 L St. N.W., Suite 300, Washington, DC, 20037. Phone: (202) 659-0830. Fax: (202) 828-5568. Web Site: www.bloostonlaw.com. Harold Mordkofsky, Benjamin H. Dickens Jr., John A. Prendergast, Gerald J. Duffy.

Blumberg, Grace Ganz, UCLA School of Law, 405 Hilgard Ave., Los Angeles, CA, 90095. Phone: (310) 825-1334. Fax: (310) 206-6489. E-mail: blumberg@law-ucla.edu Web Site: www.law-ucla.edu.

Blume & Associates LLC, 10 Ellsworth Rd., Suite 209, West Hartford, CT, 06107. Phone: (860) 231-8777. Fax: (860) 231-8763. E-mail: db@blumelegal.net Web Site: www.blumelegal.net. Daniel Blume.

Boelter & Perry, 4640 Admiralty Way, Suite 500, Marina Del Rey, CA, 90292. Phone: (310) 496-5710. Fax: (310) 823-4325. E-mail: boltperr@comcast.net

Boies, Schiller & Flexner, LLP, 100 S.E. 2nd St., Suite 2800, Miami, FL, 33131. Phone: (305) 539-8400. Fax: (305) 539-1307. Web Site: www.bsfllp.com.

Bone McAllester Norton PLLC, 511 Union St., Suite 1600, Nashville, TN, 37219. Phone: (615) 238-6330. Fax: (615) 238-6301. Web Site: www.bonelaw.com. E-mail: mnorton@bonelaw.com C. Michael Norton.

Boose, Casey, Ciklin, Lubitz, Martens, McBane & O'Connell, Northbridge Tower, 515 N. Flagler Dr., Suite 1900, West Palm Beach, FL, 33401. Phone: (561) 832-5900. Fax: (561) 833-4209. E-mail: rcrump@boosecasey.com Patrick J. Casey.

Booth, Freret, Imlay & Tepper P.C., 14356 Cape May Rd., Silver Spring, MD, 20904-6011. Phone: (301) 384-5525. Fax: (301) 384-6384. E-mail: bfitpc@aol.com Christopher D. Imlay.

Bordelon, Hamlin & Theriot, 701 S. Peters St., Suite 100, New Orleans, LA, 70130-1661. Phone: (504) 524-5328. Fax: (504) 523-1071.

Borsari and Assoc., P.L.C., Box 100009, Arlington, VA, 22210. Phone: (703) 524-5800. Fax: (703) 524-4329. E-mail: John@borsari.com Web Site: www.borsari.com. John A. Borsari.

Borsari & Paxson, 4000 Albemarle St. N.W., Suite 100, Washington, DC, 20016. Phone: (202) 296-4800. Fax: (202) 296-4460. E-mail: bap@baplaw.com Web Site: www.baplaw.com. George R. Borsari Jr., Anne Thomas Paxson.

Boult, Cummings, Conners & Berry, PLC, 1600 Division St., Suite 700, Nashville, TN, 37219. Phone: (615) 244-2582. Fax: (615) 252-6380. Web Site: www.boultcummings.com.

Law Offices of Timothy K. Brady, Box 930, Johnson City, TN, 30605-0930. Phone: (423) 477-7619. E-mail: tkbrady@earthlink.net Timothy K. Brady.

Golob, Bragin & Sassoe, 1990 S. Bundy Dr., Suite 540, Los Angeles, CA, 90025-5245. Phone: (310) 979-0321. Fax: (310) 979-0366.

Bramson, Plutzik, Mahler, Birkhaeuser, LLP, 2125 Oak Grove Rd., Suite 120, Walnut Creek, CA, 94598. Phone: (925) 945-0200. Fax: (925) 945-8792. E-mail: rbramson@bramsonplutzik.com Robert M. Branson, Alan Plutzik.

Brann & Isaacson, Box 3070, 184 Main St., Lewiston, ME, 04243. Phone: (207) 786-3566. Fax: (207) 783-9325. E-mail: gisaacson@brannlaw.com Web Site: www.brannlaw.com. George Isaacson.

Brenner, Daniel L., National Cable Telecommunications Association, 1724 Massachusetts Ave. N.W., Washington, DC, 20036. Phone: (202) 775-3664. Fax: (202) 775-3603. E-mail: dbrenner@ncta.com Web Site: www.ncta.com. Daniel L. Brenner, sr VP.

Brickfield, Burchette, Ritts & Stone, 1025 Thomas Jefferson St. N.W., Suite 800 West Tower, 8th Fl., Washington, DC, 20007. Phone: (202) 342-0800. Fax: (202) 342-0807. Web Site: www.bbrslaw.com. Peter Mattheis.

Brighton , Runyon & Callahan, 45 Main St., Suite 22, Peterborough, NH, 03458-0674. Phone: (603) 924-7276. Fax: (603) 924-9764. L. Phillips Runyon III, sr ptnr.

Brooks, Pierce, McLendon, Humphrey & Leonard, Box 1800, Wachovia Capitol Center, 150 Fayetteville St. Mall Ste 1600, Raleigh, NC, 27602. Phone: (919) 839-0300. Fax: (919) 839-0304. E-mail: whargrove@brookspierce.com Web Site: www.brookspierce.com. Wade H. Hargrove, Mark J. Prak, Marcus W. Trathen, Ed Turlington, Kathy Thornton, David Kushner, Coe Ramsey, Charles Coble, Stephen Hartzell-Jordan, Charles Marshall.

Brown, Dean, Wiseman, Lisert, Proctor & Hart, L.L.P., 306 W. 7th St., Suite 200, Fort Worth, TX, 76102. Phone: (817) 332-1391. Fax: (817) 870-2427. Web Site: www.browndean.com. Beale Dean.

Frederic E. Brown, Attorney at Law, Box 71718, Fairbanks, AK, 99707. Phone: (907) 452-3452. Fax: (907) 452-3733. E-mail: fbrown@mosquitonet.com Frederic E. Brown.

Brown, Nietert, & Kaufman, Chartered, 1301 Connecticut Ave., Suite 450, Washington, DC, 20036. Phone: (202) 887-0600. Fax: (202) 223-8685. E-mail: david@bnkcomlaw.com David J. Kaufman, Lorretta Tobin.

Brown, Steven Ames, 69 Grand View Ave., San Francisco, CA, 94114-2741. Phone: (415) 647-7700. Fax: (415) 285-3048. E-mail: sabrown@entertainmentlaw.com

Bubar, James S., Attorney at Law, 1776 K St. N.W., Suite 800, Washington, DC, 20006. Phone: (202) 223-2060. Fax: (202) 223-2061. E-mail: jbubar@aol.com Web Site: www.lawyers.com/jamesbubar/. James S. Bubar.

Don Buchwald & Associates, 10 E. 44 St., 7th Fl., New York, NY, 10017-3606. Phone: (212) 867-1200. Fax: (212) 972-3209. E-mail: richard@buchwald.com Web Site: www.buchwald.com.

Law Offices of Robert J. Buenzle, 11710 Plaza America Dr., Suite 2000, Reston, VA, 20190. Phone: (703) 430-6751. Fax: (703) 430-4994. E-mail: buenzle@buenzlelaw.com Robert J. Buenzle.

Bullivant, Houser, & Bailey, 300 Pioneer Tower, 888 S.W. 5th Ave., Suite 300, Portland, OR, 97204. Phone: (503) 228-6351. Fax: (503) 295-0915. Web Site: www.bullivant.com.

Byelas & Neigher, 1804 Post Rd. E., Westport, CT, 06880. Phone: (203) 259-0599. Fax: (203) 255-2570.

Cades Schutte, 1000 Bishop St., Honolulu, HI, 96813. Phone: (808) 521-9221. Fax: (808) 540-5040. E-mail: jportnoy@cades.com Jeffrey S. Portnoy.

Cahill, Gordon & Reindel LLP, 1990 K St. N.W., Suite 950, Washington, DC, 20006. Phone: (202) 862-8900. Fax: (202) 862-8958. E-mail: mulvid@cgrdc.com

New York, NY 10005, 80 Pine St. Phone: (212) 701-3000. Floyd Abrams.

Calfee, Halter & Griswold, 800 Superior Ave., Suite 1400 MacDonald Investment Ctr., Cleveland, OH, 44114. Phone: (216) 622-8200. Fax: (216) 241-0816. E-mail: cbowers@calfee.com Web Site: www.calfee.com.

Callister, Nebeker & McCullough, Gateway Tower E., Suite 900, Salt Lake City, UT, 84133. Phone: (801) 530-7300. Fax: (801) 364-9127. Web Site: www.cnmlaw.com. Laurie S. Hart, Randall D. Benson, Jennifer Ward.

Cameron & Mittleman LLP, 56 Exchange Terr., Providence, RI, 02903-1766. Phone: (401) 331-5700. Fax: (401) 331-5787. Web Site: www.cm-law.com. Richard Mittleman, John W. Wolfe, Esq.

Caridi, Carmella, Esq., Caridi Video Inc., 250 W. 57th, New York, NY, 10107-1722. Phone: (212) 581-2277. Fax: (212) 581-2278.

Carr, Morris & Graeff, 1120 G St. N.W., Suite 930, Washington, DC, 20005. Phone: (202) 789-1000. Fax: (202) 628-3834.

Carter Ledyard & Milburn LLP, 1401 Eye St. N.W., Suite 300, Washington, DC, 20005. Phone: (202) 898-1515. Fax: (202) 898-1521. E-mail: info@clm.com Web Site: www.clm.com. Thomas F. Bardo, Mary S. Diemer, Bradley A. Farrell, Peter K. Killough, Timothy J. Fitzgibbon, Jennifer E. Wagman, Robert L. Hoegle.

Peter A. Casciato P.C., 335 Bryant St., Suite 410, San Francisco, CA, 94107. Phone: (415) 291-8661. Fax: (415) 291-8165. E-mail: pcasciato@sbcglobal.net Peter A. Casciato.

Cavallo, Robert M., 400 Park Ave., 21st Fl., New York, NY, 10022-4406. Phone: (212) 753-2224. Fax: (212) 753-7113. E-mail: rcavallo@jtjsys.com Robert Cavallo, Esq.

Bryan Cave L.L.P., 700 13th St. N.W., Suite 700, Washington, DC, 20005. Phone: (202) 508-6000. Fax: (202) 508-6200. John R. Wilner.

Kansas City, MO 64105, 3500 One Kansas City Pl. Phone: (816) 474-7400. Fax: (816) 374-3300. Web Site: www.bryancave.com. John R. Wilner.

New York, NY 10104, 1290 Ave. of the Americas. Phone: (212) 541-2000. Fax: (212) 541-4630. Web Site: www.bryancave.com. Jerome S. Boros, Renee E. Frost, Andrew Irving, Alan Pearce, Michael Rosen.

Edward de R. Cayia, P.A., 432 N.E. 3rd Ave., Fort Lauderdale, FL, 33301. Phone: (954) 765-1400. Fax: (954) 765-1421.

Chadbourne & Parke, 1200 New Hampshire Ave. N.W., Suite 300, Washington, DC, 20036. Phone: (202) 974-5600. Fax: (202) 974-5602. E-mail: info@chadbourne.com Web Site: www.chadbourne.com. Dana Frix, Michael Salsburg, David Plandes, Aaron Bartell, Hwan Kim, Kyunghoon Lee.

Los Angeles, CA 90017, 601 S. Figueroa St. Phone: (213) 892-1000.

New York, NY 10112, 30 Rockefeller Plaza. Phone: (212) 408-5100.

Chetkof, Gary H., Box 367, 293 Tinker St., Woodstock, NY, 12498. Phone: (845) 679-7600. Fax: (845) 679-5395. Web Site: www.wdst.com.

Clark Hill P.L.C., 500 Woodward Ave., Suite 3500, Detroit, MI, 48226-3435. Phone: (313) 965-8300. Fax: (313) 962-4348. Fax: (313) 965-8252. E-mail: dlee@clarkhill.com Web Site: www.clarkhill.com. David E. Nims III, Roderick S. Coy, Haran C. Rashes.

Richard N. Clarvit, P.A., 1313 N.E. 125th St., North Miami, FL, 33161. Phone: (305) 893-4135. Fax: (305) 893-4173. E-mail: richsongs@aol.com Richard N. Clarvit.

Clifford, Chance, LLP, 31 West 52 St., New York, NY, 10019-6131. Phone: (212) 878-8000. Fax: (212) 878-8375. Web Site: www.cliffordchance.com.

Cohn and Marks LLP, 1920 N St. N.W., Suite 300, Washington, DC, 20036-1622. Phone: (202) 293-3860. Fax: (202) 293-4827. E-mail: roy.russo@cohnmarks.com Web Site: www.cohnmarks.com. Lawrence N. Cohn, Richard A. Helmick, Robert B. Jacobi, Roy R. Russo, J. Brian DeBoice, Ellen M. Edmundson, Jerold L. Jacobs, Susan V. Sachs, Ronald A. Siegel.

Colby, Lauren A., Box 113, 10 E. Fourth St., Frederick, MD, 21705-0113. Phone: (301) 663-1086. Fax: (301) 695-8734. E-mail: lac@lcolby.com Web Site: www.lcolby.com. Lauren A. Colby.

Cole, Raywid & Braverman, L.L.P., 1919 Pennsylvania Ave. N.W., Suite 200, Washington, DC, 20006. Phone: (202) 659-9750. Fax: (202) 452-0067. E-mail: info@crblaw.com Web Site: www.crblaw.com.

Cooke, James R., 2821 Beachwood Cir., Arlington, VA, 22207. Phone: (703) 841-1001. Fax: (703) 841-1004. E-mail: jrcde@comcast.net James R. Cooke.

Hackensack, NJ 07601, 2 University Plaza, 2nd Fl. Phone: (201) 488-2200. Fax: (201) 342-6677. Vincenzo Taparo.

Albany, NY 12211, 20 Corporate Woods Blvd. Phone: (518) 427-9706. Fax: (518) 427-0235. Terence Burke.

Hamburg, NY 14075, One Grimsby Dr. Phone: (716) 646-5050. Fax: (716) 648-8204. Raymond Stapell, mgng ptnr.

Ithaca, NY 14850, 119 E. Seneca St. Phone: (607) 273-6444. Fax: (607) 273-6802. Mark Wheeler.

New York, NY 10036, 530 Fifth Ave. Phone: (212) 687-0100. Fax: (212) 997-7868. William O'Connor.

Rochester, NY 14604, 130 E. Main St. Phone: (716) 232-4440. Fax: (716) 232-1925. Gunther Buerman, mgng ptnr.

Syracuse, NY 13202, 300 S. State St., 4th Fl. Phone: (315) 423-7100. Fax: (315) 426-9331. Thomas E. Taylor.

Cooper, White & Cooper, L.L.P., 201 California St., 17th Fl., San Francisco, CA, 94111. Phone: (415) 433-1900. Fax: (415) 433-5530. E-mail: whansell@cwclaw.com Web Site: www.cwclaw.com. Walter W. Hansell, Mark P. Schreiber, Jed Solomon, Garth Black, Patrick M. Rosvall, Jamie Jie-Ming Chou.

Cooter, Mangold, Tompert & Wayson, 5301 Wisconsin Ave N.W., Suite 500, Washington, DC, 20015. Phone: (202) 537-0700. Fax: (202) 364-3664. E-mail: mterry@cootermangold.com

Corberlaw, Box 44212, Panorama City, CA, 91412-0212. Phone: (818) 786-7133. E-mail: corberlaw@aol.com Brian L. Corber, owner.

Law Offices of Bernard R. Corbett, 6312 Barrister Pl., Alexandria, VA, 22307-1214. Phone: (703) 549-4700. Fax: (703) 549-5290.

Corn-Revere, Robert, Davis Wright Tremaine LLP, 1919 Pennsylvania Ave. N.W., Washington, DC, 20006. Phone: (202) 973-4225. E-mail: bobcornrevere@dwt.com Web Site: www.dwt.com.

Couzens, Michael, Box 3642, Oakland, CA, 94609. Phone: (510) 658-7654. Fax: (510) 654-6741. Web Site: www.lptv.tv.

Covington & Burling, 1201 Pennsylvania Ave. N.W., Washington, DC, 20004. Phone: (202) 662-6000. Fax: (202) 662-6291. E-mail: jblake@cov.com Web Site: www.cov.com. Jonathan Blake, John Blevins, Evan R. Cox, Michael E. Cutler, Cathrine J. Dargan, Matthew S. DelNero, Ronald G. Drove Jr., Erin M. Egan, David Fink, William A. Fitz, Douglas G. Gibson, Ellen P. Goodman, Eric Dodson Greenberg, Darrin Hurwitz, Jennifer A. Johnson, Joan L. Kutcher, Genevieve Michaux, Gina L. Paik, Robert M. Sherman, Brian D. Smith, James C. Snipes, Philipp Tamussino, Lee J. Tiedrich, W. Jeffrey Voltmer, Gerard J. Waldron, Stephen A. Weiswasser, Bruce S. Wilson, Kurt A. Wimmer.

Craven Law Office, 1005 N. 7th St., Springfield, IL, 62702. Phone: (217) 544-1777. Fax: (217) 544-0713. E-mail: presslaw@aol.com Web Site: www.cravenlawoffice.com. Donald M. Craven.

Crowell & Moring, 1001 Pennsylvania Ave. N.W., Washington, DC, 20004-2595. Phone: (202) 624-2500. Fax: (202) 628-5116. E-mail: sthomas@crowell.com Web Site: www.crowell.com. John I. Stewart Jr., Robert M. Halperin, William D. Wallace.

Cuni, Ferguson, Levay & Bergmann, 10655 Springfield Pike, Cincinnati, OH, 45215. Phone: (513) 771-6768. Fax: (513) 771-6781. E-mail: pmusgrove@cfl-law.com Thomas Cuni.

DLA Piper Rudnick Gray Cary US LLP, 1200 19th St. N.W., Suite 700, Washington, DC, 20036. Phone: (202) 861-3913. Fax: (202) 689-7626. Web Site: www.dlapiper.com. Mark J. Tauber, E. Ashton Johnston.

New York, NY 10020-1104, 1251 Ave. of the Americas. Phone: (212) 835-6000. Fax: (212) 835-6001. E-mail: mmccabe@piperrudnick.com. Monica McCabe.

Law Offices of George E. Darby, Box 893010, Mililani, HI, 96789-3010. Phone: (808) 626-1300. Fax: (808) 626-1350. E-mail: darbylaw@teleport-asia.com Web Site: www.teleport-asia.com. George Darby.

Davis Wright Tremaine LLP, 1500 K St. N.W., Suite 450, Washington, DC, 20005. Phone: (202) 508-6600. Fax: (202) 508-6699. Web Site: www.dwt.com. James Blitz, Richard Cys, Laura Handman, Bob Corn-Revere.

Davis Wright Tremaine LLP, Phone: (206) 622-3150. Fax: (206) 628-7699. Web Site: www.dwt.com. E-mail: seattle@dwt.com

Los Angeles, CA 90017, 1000 Wilshire Blvd, Suite 600. Phone: (213) 633-6800. Fax: (213) 633-6899. Kelli L. Sager.

Washington, DC 20036, 1155 Connecticut Ave. N.W. Phone: (202) 508-6600. Fax: (202) 508-6699.

Boise, ID 83702, 999 Main St, Suite 911. Phone: (208) 338-8200. Fax: (208) 338-8299. Deborah Kristensen.

Portland, OR 97201, 2300 First Interstate Tower, 1300 S.W. 5th Ave. Phone: (503) 241-2300. Fax: (503) 778-5299. Duane A. Bosworth.

Day & Associates, 1812 Waterfront Plaza, 325 W. Main, Louisville, KY, 40202-4251. Phone: (502) 585-4131. Fax: (502) 581-1210. E-mail: dayandassociates@bellsouth.net Joe Day, owner.

Day, Berry & Howard L.L.P., CityPlace I, Hartford, CT, 06103-3499. Phone: (860) 275-0122. Fax: (860) 275-0343. E-mail: mwelsass@dbh.com Web Site: www.dbh.com. Robert P. Knickerbocker Jr., Paul N. Belval, Michael F. Halloran, William A. Hunter, Ross A. Pascal, David A. Swerdloff, Sabino Rodriguez, David T. Doot.

Debevoise & Plimpton LLP, 919 3rd Ave., New York, NY, 10022. Phone: (212) 909-6000. Fax: (212) 909-6836. E-mail: rdbohm@debevoise.com Web Site: www.debevoise.com. Richard D. Bohm, Bruce Keller.

Hong Kong, 13/F Entertainment Bldg, 30 Queen's Rd. Central. Phone: 852-2810-7918. Jeffrey S. Wood.

London EC2N 1HQ, The International Financial Centre, 25 Old Broad St. Phone: 44-171-786-9000. Robert R. Bruce.

Moscow 103104, Bolshoi Palashevsky Per 13/2. Phone: 7503-956-3858. Dmitri V. Nikiforov.

Paris 75008, 21 Ave. George V. Phone: 33-1-40-73-12-12. James A. Kiernan III, Antoine F. Kirry.

Washington, DC 20004, 555 13th St. N.W, Suite 1100-E. Phone: (202) 383-8000. Jeffrey P. Cunard.

Decker, Jones, McMackin, McClane, Hall & Bates, 801 Cherry St., Suite 2000, Unit 46, Fort Worth, TX, 76102. Phone: (817) 336-2400. Fax: (817) 332-3043. Web Site: www.deckerjones.com. Charles Milliken.

Del, Shaw, Moonves, Tanaka, Finkelstein & Lezcano, 2120 Colorado Ave., Suite 200, Santa Monica, CA, 90404. Phone: (310) 979-7900. Fax: (310) 979-7999.

Denechaud & Denechaud, 1010 Common St., Suite 3010 & 1207, New Orleans, LA, 70112-2483. Phone: (504) 522-4756. Fax: (504) 568-0783. E-mail: cidlaw@bellsouth.net

Robert A. DePont, Attorney at Law, 140 South St., Annapolis, MD, 21401. Phone: (410) 263-0632. Fax: (410) 280-8624. E-mail: robertade@msn.com Web Site: www.robertdeport.com. Robert A. DePont.

Devine & Millimet, Box 719, 111 Amherst St., Manchester, NH, 03101. Phone: (603) 669-1000. Fax: (603) 669-8547. E-mail: kmcginley@dm.com Web Site: www.devinemillimet.com.

DeWitt, Ross & Stevens, 2 E. Mifflin St., Suite 600, Madison, WI, 53703. Phone: (608) 255-8891. Fax: (608) 252-9243. E-mail: info@dewuttriss.com Web Site: www.dewittross.com.

Dickstein Shapiro Morin & Oshinsky LLP, 2101 L St., N.W., Washington, DC, 20037-1526. Phone: (202) 785-9700. Fax: (202) 887-0689. E-mail: info@dsmo.com Web Site: www.dicksteinshapiro.com. Walter J. Walvick, counsel; Robert F. Aldrich, Jacob S. Farber, Robert Felger, Valerie M. Furman, Allan C. Hubbard, Andrew S. Kersting, Adam Kirschenbaum, Albert H. Kramer, Gregory D. Kwan, Edward Modell, Lewis J. Paper.

New York, NY 10036-2714, 1177 Ave. of the Americas, 41st Fl.

Dieguez, Richard P., 192 Garden St., Suite 2, Roslyn Heights, NY, 11577-1012. Phone: (516) 621-6424. Fax: (516) 621-6508. E-mail: rpdieguez@rpdieguez.com Web Site: www.rpdieguez.com. Richard P. Dieguez.

Dorsey & Whitney, L L P, 50 S. 6th St., Suite 1500, Minneapolis, MN, 55402-1498. Phone: (612) 340-2873. Fax: (612) 340-2868. Web Site: www.dorsey.com. E-mail: cattanach.robert@dorsey.com Robert E. Cattanach, Karly Baraga, Shannon Heim, MN; Heather Grahame, AK; Tucker Trautman, R. Stephen Hall, CO; Charles Ferguson, CA; Stefan Lopatkiewicz, DC.

London EC2A INQ, Veritas House, 125 Finsbury Pavement. Phone: 011-44-171-588-0800. Fax: 011-44-171-588-0555.

Vancouver, BC V6C 3J8 Canada, 666 Burrard St., Suite 1300, Park Pl. Phone: (604) 687-5151. Fax: (604) 687-8504.

Anchorage, AK 99501, 1031 West 4th Ave, Suite 600. Phone: (907) 276-4557. Fax: (907) 276-4152.

Irvine, CA 92618-5310, 38 Technology Dr. Phone: (714) 424-5555. Fax: (714) 424-5554.

Denver, CO 80202-5644, Republic Plaza Bldg., Suite 4400, 370 Seventeenth St. Phone: (303) 629-3400. Fax: (303) 629-3450.

Washington, DC 20004, 1001 Pennsylvania Ave. N.W, Suite 200 Soutth. Phone: (202) 824-8800. Fax: (202) 824-8990.

Minneapolis, MN 55402-1498, 50 S. 6th St. Phone: (612) 340-2600. Fax: (612) 340-2868.

Great Falls, MO 59401, 507 Davidson Bldg., 8 Third St. Phone: (406) 727-3632. Fax: (406) 727-3638.

Missoula, MO 59802-4407, 125 Bank St, Suite 600. Phone: (406) 721-6025. Fax: (406) 543-0863.

Fargo, ND 58107-1344, Dakota Ctr., 51 N. Broadway, Suite 402. Phone: (701) 235-6000. Fax: (701) 235-9969.

New York, NY 10177, 250 Park Ave. Phone: (212) 415-9200. Fax: (212) 953-7201.

Salt Lake City, UT 84101, Wells Fargo Plaza, 170 S. Main St., Suite 925. Phone: (801) 350-3581. Fax: (801) 350-3585.

Seattle, WA 98101, US Bank Building Ctr., 1420 5th Ave., Suite 400. Phone: (206) 654-5400. Fax: (206) 654-5500.

Dow, Lohnes & Albertson, PLLC, 1200 New Hampshire Ave. N.W., Suite 800, Washington, DC, 20036. Phone: (202) 776-2000. Phone: (202) 776-2939. Fax: (202) 776-2222. Web Site: www.dowlohnes.com. E-mail: info@dowlohnes.com Michael D. Basile, Raymond G. Bender, James M. Burger, Christina H. Burrow, Peter H. Feinberg, John R. Feore Jr., Jeffrey L. Gee, Todd D. Gray, J.G. Harrington, Nam E. Kim, Kevin P. Latek, John S. Logan, Gary S. Lutzker, Melissa A. Marshall, Elizabeth A. McFadden, Margaret L. Miller, Edward J. Palmieri, Scott S. Patrick, Barry S. Persh, Jason E. Rademacher, Christopher J. Redding, Kevin F. Reed, Kenneth D. Salomon, M. Anne Swanson, To-Quyen T. Truong.

Atlanta, GA 30346, One Ravinia Dr, Suite 1600. Phone: (770) 901-8800.

Downs, Bertis E., 170 College Ave., Athens, GA, 30601. Phone: (706) 353-6689. Fax: (706) 546-6069.

Drinker Biddle & Reath L.L.P., 1500 K St. N.W., Suite 1100, Washington, DC, 20005-1209. Phone: (202) 842-8800. Fax: (202) 842-8465. E-mail: joe.edge@dbr.com Web Site: www.dbr.com. Joe Dixon Edge, Mark L. Pelesh, Richard M. Singer, Joaquin A. Marquez, Philip J. Mause, Timothy R. Hughes, Tina M. Pidgeon, John R. Przypyszny.

Duane Morris LLP, 30 S. 17th St., Philadelphia, PA, 19103. Phone: (215) 979-1000. Fax: (215) 979-1020. Web Site: www.duanemorris.com. Abraham Frumkin, ptnr.

Dunham, Corydon B. Counsel, Cahill Gordon & Reindel, 80 Pine St., New York, NY, 10005. Phone: (212) 701-3776. Fax: (212) 269-5420. Web Site: www.cahill.com. Corydon B. Dunham.

Joseph E. Dunne III, Attorney at Law, P.O. Box 9203, Durango, CO, 81302-9203. Phone: (970) 385-7312. Fax: (970) 385-7343. E-mail: lawman@animas.net Joseph E. Dunne III.

Ross Eatman, Box 102, Bedford, NY, 10506-0102. Phone: (914) 234-4748. Fax: (914) 234-4750. E-mail: emstalent@aol.com Ross Eatman.

Eaton, Peabody, Box 1210, 80 Excahange St., Bangor, ME, 04402-1210. Phone: (207) 947-0111. Fax: (207) 942-3040. E-mail: eaton@eatonpeabody.com Web Site: www.eatonpeabody.com.

Eckert, Seamans, Cherin & Mellott, 1515 Market St. 9th Fl., Philadelphia, PA, 19102. Phone: (215) 851-8400. Fax: (215) 851-8383. Web Site: www.escm.com.

Edelstein, Laird & Sobel, L.L.P., 9255 Sunset Blvd., Suite 800, Los Angeles, CA, 90069. Phone: (310) 274-6184. Fax: (310) 274-6185. E-mail: laird@elsentlaw.com Web Site: elsentlaw.com.

Edwards Angell Palmer & Dodge L.L.P., 111 Huntington Ave., Boston, MA, 02199. Phone: (617) 951-2233. Fax: (888) 325-9120. E-mail: smeredith@eapdlaw.com Web Site: www.eapdlaw.com.

Elam & Burke, P.A., 251 E. Front St., Suite 300, Boise, ID, 83702-7311. Phone: (208) 343-5454. Fax: (208) 384-5844. E-mail: mag@elamburke.com

Epstein, Levinsohn, Bodine, Hurwitz & Weinstein, P.C., 1790 Broadway, 10th Fl., New York, NY, 10019. Phone: (212) 262-1000. Fax: (212) 262-5022. Web Site: entlawfirm.com.

Ezor, A. Edward, 201 S. Lake Ave., Suite 505, Pasadena, CA, 91101. Phone: (626) 568-8098. Fax: (626) 568-8475.

Faegre & Benson, L.L.P., 801 Grand , Suite 3100, Des Moines, IA, 50309-8002. Phone: (515) 248-9000. Fax: (515) 248-9010. E-mail: mgiudicessi@faegre.com Web Site: www.faegre.com. Michael A. Giudicessi.

Farmer, Shirley Stewart, One Lincoln Plaza, Suite 19S, New York, NY, 10023-7149. Phone: (212) 787-6566. Fax: (212) 787-6567. E-mail: stewfar@rcn.com Web Site: www.shirleystewartfarmer.com.

Farrand, Cooper P.C., 235 Montgomery St., Suite 1035, San Francisco, CA, 94104. Phone: (415) 399-0600. Fax: (415) 677-2950. Web Site: www.fcblaw.com. Wayne B. Cooper, Stephen R. Farrand.

Federal Communications Comission Public Safety & Homeland Security Bureau, Public Safety & Homeland Security Bureau, 445 12th St., S.W., Washington, DC, 20554. Phone: (202) 418-0680. E-mail: mwilhelm@fcc.gov Web Site: www.fcc.gov. Michael J. Wilhelm.

Lindsey S. Feldman, Attorney at Law, 4551 Glencoe Ave., Suite 300, Marina del Rey, CA, 90292. Phone: (310) 823-1600. Fax: (310) 775-8775. E-mail: lfeldman@bergerkahn.com Lindsey S. Feldman.

Ferris & Britton, 401 West A St., Suite 1600, San Diego, CA, 92101. Phone: (619) 233-3131. Fax: (619) 232-9316. E-mail: aferris@ferrisbritton.com Web Site: www.ferrisbritton.com. Alfred G. Ferris, Christopher Q. Britton, Lee Austin, Michael Weinstein.

Fine & Associates, P.L.C., 335-337 Decatur St., Vieux Carre, New Orleans, LA, 70130-1023. Phone: (504) 581-5152. Fax: (504) 581-5152, EXT. 124. E-mail: dnfinelaw@aol.com

Fine & Block, 2060 Mt. Paran Rd. N.W., Suite 106, Atlanta, GA, 30327. Phone: (404) 261-6800. Fax: (404) 261-6960. Web Site: www.fineandblock.com. A. J. Block Jr.

Finkelstein, Thompson & Loughran, 1050 30th St. N.W., Washington, DC, 20007. Phone: (202) 337-8000. Phone: (866) 592-1960. Fax: (202) 337-8090. E-mail: lks@stllaw.com Web Site: www.ftllaw.com. Douglas G. Thompson Jr., L. Kendall Satterfield.

Fleischman & Walsh, L.L.P., 1919 Pennsylvania Ave. N.W., 6th Fl., Washington, DC, 20006. Phone: (202) 939-7900. Fax: (202) 745-0916. E-mail: fw@fw-law.com Web Site: www.fw-law.com.

Fletcher, Heald & Hildreth, P.L.C., 1300 N. 17th St., 11th Fl., Arlington, VA, 22209. Phone: (703) 812-0400. Fax: (703) 812-0486. E-mail: office@fhh-telecomlaw.com Web Site: www.fhh-telcomlaw.com. Vincent J. Curtis Jr., Frank R. Jazzo, James P. Riley, Howard M. Weiss, Paul J. Feldman, Kathleen Victory, Harry Martin, Mitchel Lazarus, Susan Marshall, Harry F. Cole, Scott M. Johnson, ptnrs; Ann Bavender, Anne G. Crump, sr. counsel; Donal J. Evans, Robert M. Gurss, Eugene M. Lawson, Francisco R. Montero, Edwards O'Neil, of counsel.

Foley & Lardner, 150 E. Gilman St., Madison, 53703. Phone: (608) 257-5035. Fax: (608) 258-4258. E-mail: dwalsh@foleylaw.com David G. Walsh.

Forrest, Herbert E., Federal Programs Br., Civil Division, Rm. 7112, U.S. Dept. of Justice, 20 Massachusetts Ave. N.W., Washington, DC, 20530. Phone: (202) 514-2809. Fax: (202) 616-8470. E-mail: herbert.forest@usdos.gov Herbert E. Forrest, trial atty.

Fowler, Measle & Bell, 300 W. Vine St., Suite 600, Lexington, KY, 40507-1660. Phone: (859) 252-6700. Fax: (859) 255-3735. E-mail: fmb@fmb.com Web Site: www.fmblaw.com.

Fox & Film Entertainment, 10201 W. Pico Blvd., Los Angeles, CA, 90035. Phone: (310) 369-1000. Fax: (310) 369-3333. Web Site: www.fox.com. Gregory Gelfan, Esq., exec VP.

Frost, Mark E., Box 153, Glens Falls, NY, 12801-0153. Phone: (518) 792-1126. Fax: (518) 793-1587. Web Site: mfrost@loneoak.com. Mark E. Frost.

Gammon & Grange, P.C., 8280 Greensboro Dr., 7th Fl., McLean, VA, 22102-3807. Phone: (703) 761-5000. Fax: (703) 761-5023. E-mail: awf@gglaw.com Web Site: www.gandglaw.com. A. Wray Fiitch III, Timothy R. Obitts, Stephen M. Clarke.

Ganz & Hollinger, 1394 3rd Ave., New York, NY, 10021. Phone: (212) 517-5500. Fax: (212) 772-2720. Web Site: www.ganzhollinger.com.

Gardere Wynne Sewell LLP, 1601 Elm St., Thanksgiving Tower, Suite 3000, Dallas, TX, 75201-4761. Phone: (214) 999-3000. Fax: (214) 999-4667. E-mail: mwebb@gardere.com Web Site: www.gardere.com.

Gardner, Carton & Douglas, 1301 K St. N.W., Suite 900 E. Tower, Washington, DC, 20005. Phone: (202) 230-5000. Fax: (202) 230-5300. Web Site: www.gcd.com. Francis E. Fletcher Jr., M. Scott Johnson, Thomas Dougherty, Laura Mow, Lee Petro, Jennifer Lewis.

Law Offices of Michael R. Gardner, P.C., 1150 Connecticut Ave. N.W., Suite 710, Washington, DC, 20036. Phone: (202) 785-2828. Fax: (202) 785-1504. E-mail: mrgpc@aol.com Michael R. Gardner, mgng ptnr.

Garvey, Schubert & Barer, 1191 2nd Ave., 18th Fl., Seattle, WA, 98101-2939. Phone: (206) 464-3939. Fax: (206) 464-0125. E-mail: kdavis@gsblaw.com Web Site: www.gsblaw.com.

Washington, DC 20007-3501, 1000 Potomac St. N.W., 5th Fl. Phone: (202) 965-7880. Fax: (202) 965-1729. E-mail: jking@gsblaw.com. Web Site: www.gsblaw. Matthew R. Schneider, D.C. mgng dir. Contact: John Wells King, counsel.

New York, NY 10012-3235, 599 Broadway, 10th Fl. Phone: (212) 431-8700. Fax: (212) 334-1278. Web Site: www.gsblaw.com. Matthew R. Schneider, NY mgng dir.

Portland, OR 97204-3141, 121 S. W. Morrison St. Phone: (503) 228-3939. Fax: (503) 226-0259. Web Site: www.gsblaw.com. Larry Brant, Steve Connolly, Bob Weaver, Portland mgmt comm.

Gibbs & Associates, P.C., 146 Central Park W., Suite 19E, New York, NY, 10023. Phone: (212) 787-2828. Fax: (212) 787-2886. E-mail: bhg1cg2@aol.com Bud H. Gibbs.

Gibson, Dunn & Crutcher, 333 S. Grand Ave., Suite 4600, Los Angeles, CA, 90071-3197. Phone: (213) 229-7000. Fax: (213) 229-7520. E-mail: agravit@gibsondunn.com Web Site: www.gdclaw.com.

Washington, DC 20036, 1050 Connecticut Ave. N.W, Suite 900. Phone: (202) 955-8500. Jill Sterner.

Gold & Pyle, 526 Superior Ave. E., 1140 Leader Bldg., Cleveland, OH, 44114. Phone: (216) 696-6122. Fax: (216) 696-3214.

Goldberg, Godles, Wiener & Wright, 1229 19th St. N.W., Washington, DC, 20036. Phone: (202) 429-4900. Fax: (202) 429-4912. E-mail: general@g2w2.com Web Site: g2w2.com. Henry Goldberg, Joseph A. Godles, Jonathan L. Wiener, Laura Stefani.

Golden & Golden, P.C., 10627 Jones St., Suite 101B, Fairfax, VA, 22030. Phone: (703) 691-0117. E-mail: k8los@aol.com Web Site: www.gglawva.com. Richard A. Golden.

Glenn A. Goldstein, Attorney at Law, 1650 Market St., Suite 4900, Philadelphia, PA, 19103. Phone: (215) 981-5922. Fax: (215) 981-5959. E-mail: glenn802@aol.com Glenn A. Goldstein, Esq.

Goodkind, Labaton, Rudoff & Sucharow, L.L.P., 100 Park Ave., New York, NY, 10017. Phone: (212) 907-0700. Fax: (212) 818-0477. E-mail: rrosenblum@labton.com Web Site: www.glrs.com.

Law Offices of Jeffrey L. Graubart, 350 W. Colorado Blvd., Suite 200, Pasadena, CA, 91105. Phone: (310) 788-2650. Fax: (310) 788-2657. E-mail: graubart@gte.net Web Site: www.lawyers.com. Jeffrey L. Graubart.

Gray & Robinson, 301 E. Pine St., Suite 1400, Orlando, FL, 32802. Phone: (407) 843-8880. Fax: (407) 244-5690. E-mail: llee@gray-robinson.com Web Site: www.gray-robinson.com. J. Charles Gray, Richard M. Robinson, founding ptnrs.

Greensfelder, Hemker & Gale, P.C., 2000 Equitable Bldg., 10 S. Broadway, St. Louis, MO, 63102-1774. Phone: (314) 516-2662. Fax: (314) 345-5499. E-mail: mlw@greensfelder.com Web Site: www.greensfelder.com. Sheldon K. Stock, Mary Ann L. Wymore, Jason L. Ross.

Greiter, Pegger, Kofler & Partner, Maria Theresien-Strasse 24, A-6020, Innsbruck Phone: 43 512-57-1811. Fax: 43 512-5849-25. Fax: 43 512-5711-52.

Groveman, Amy S., Cablevision Systems Corp., 111 Stewart Ave., Bethpage, NY, 11714. Phone: (516) 803-2300. Fax: (516) 803-2575. E-mail: agrovema@cablevision.com Web Site: www.cablevision.com.

Grubb, Jay G., 12 Forrest Edge Dr., Titusville, NJ, 08560. Phone: (410) 329-2108. Fax: (410) 329-2109. Jay G. Grubb.

Grubman, Indursky & Shire, P.C., 152 W. 57th St., 31st Fl., New York, NY, 10019. Phone: (212) 554-0400. Fax: (212) 554-0444.

Gullett, Sanford, Robinson & Martin, 315 Deadrick St., Suite 1100, Nashville, TN, 37238. Phone: (615) 244-4994. Fax: (615) 256-6339. John D. Lentz.

Hall, Dickler, Kent, Goldstein & Wood, 909 3rd Ave., 27th Fl., New York, NY, 10022. Phone: (212) 339-5409. Web Site: www.halldickler.com. Jeffrey S. Edelstein.

Handman, Stanley H., 10160 Cielo Dr., Beverly Hills, CA, 90210-2037. Phone: (310) 276-7503. Fax: (310) 276-1559. E-mail: stanhandman@yahoo.com Web Site: www.fleischerstudios.com. Stanley H. Handman.

Hansen, Jacobson, Teller, Hoberman, Newman, Warren & Sloan, L.L.P., 450 N. Roxbury Dr., 8th Fl., Beverly Hills, CA, 90210-4222. Phone: (310) 248-3105/248-3101. Fax: (310) 275-2329/550-5209. E-mail: sdecker@hjth.com Gretchen Bruggeman, John Farrell, Tom Hanson, Jason Hendler, Tom Hoberman, Craig Jacobson, Tom McGuire, Jeanne Newman, Ken Richman, Jason Sloane, Don Steele, Walter Teller, Steve Warren.

Law Offices of Douglas W. Harold Jr., 109 Southdown Cir., Stephens City, VA, 22655. Phone: (540) 869-0040. Fax: (540) 869-0041. E-mail: douglasharoldjr@yahoo.com Douglas W. Harold Jr.

Harris, Wiltshire & Grannis, L.L.P., 1200 Eighteenth St. N.W., Washington, DC, 20036. Phone: (202) 730-1300. Fax: (202) 730-1301. E-mail: sharris@harris Web Site: www.harriswiltshire.com. Mark A. Grannis, William M. Wiltshire, Kent D. Bressie, Jonathan B. Mirsky, John T. Nakahata.

Hasse / Molesky P.C., 526 Columbus Ave., 2nd Fl., San Francisco, CA, 94133. Phone: (415) 433-4380. Fax: (415) 433-6580. E-mail: jmolesky@molesky.com Web Site: www.entertainmentlaw.leadcounsel.com.

James A. Hatcher, Cox Communications Inc., 1400 Lake Hearn Dr. N.E., Atlanta, GA, 30319. Phone: (404) 843-5000. Fax: (404) 843-5845. James Hatcher, sr VP/gen counsel.

Law Offices of Richard J. Hayes, Box 200, Lincolnville, ME, 04849. Phone: (207) 336-3333. Fax: (202) 478-0048. E-mail: fcclaw@rjhayes.com Web Site: www.rjhayes.com. Richard J. Hayes, Jr.

Head, Johnson & Kachigian, 228 W. 17th Pl., Tulsa, OK, 74119. Phone: (918) 587-2000. Fax: (918) 584-1718. E-mail: hjk@law.com Web Site: www.hjklaw.com. Mark G. Kachigian.

Hearn, Edward R., 84 W. Santa Clara St., Suite 660, San Jose, CA, 95113. Phone: (408) 998-3400. Fax: (408) 297-1104. E-mail: nedhearnml@aol.com

John Hearne, 715 Broadway, Suite 320, Santa Monica, CA, 90401. Phone: (310) 451-4430. Fax: (310) 451-1423. John Hearne, owner.

Hebert, Spencer, Cusimano & Fry, LLP, 701 Laurel St., Baton Rouge, LA, 70802. Phone: (225) 344-2601. Fax: (225) 387-1714. E-mail: clsatty@aol.com Charles L. Spencer.

Heller Ehrman LLP, 275 Middlefield Rd., Menlo Park, CA, 94025. Phone: (650) 324-7000. Fax: (650) 324-0638. Web Site: www.hellerehrman.com. Daniel L. Appelman.

Hendrickson, Thomas, 203 Alderwood Dr., N. Potomac, MD, 20878. Phone: (301) 519-0085. E-mail: thomashendrickson@yahoo.com

Hewitt Katz Stepp and Wright Attorney at Law, 945 E. Paces Ferry Rd., Resurgens Plaza Ste. 2610, Atlanta, GA, 30326. Phone: (404) 240-0400. Fax: (404) 240-0401. E-mail: khewitt@atllawofc.com Web Site: www.robertnkatz.com.

Hill & Welch, 1330 New Hampshire Ave. N.W., Suite 113, Washington, DC, 20036. Phone: (202) 775-0070. Fax: (202) 775-9026. E-mail: welchlaw@earthlink.net

Hillman, Adria S., 41 E. 57th St., 15th Fl., New York, NY, 10022. Phone: (212) 593-5223. Fax: (212) 593-4633. E-mail: ahillman@adriashillman.com Adria S. Hillman.

Hinshaw & Culbertson, 3100 Campbell Mithun Tower, 222 S. 9th St., Minneapolis, MN, 55402. Phone: (612) 333-3434. Fax: (612) 334-8888. Web Site: www.hinshawculbertson.com. David Mylrea, ptnr.

Hogan & Hartson, Columbia Sq., 555 13th St. N.W., Washington, DC, 20004. Phone: (202) 637-5600. Fax: (202) 637-5910. Web Site: www.hhlaw.com. Ptnrs: Robert Corn-Revere, Marvin J. Diamond, Gardner F. Gillespie III, William S. Reyner Jr., Richard S. Rodin, Peter A. Rohrbach, Mace J. Rosenstein, David J. Saylor, Joel S. Winnik, Gerald E. Oberst, Linda Oliver, Marissa G. Repp, Edgar W. Holtz. Associates: Jacqueline P. Cleary, Karis A. Hastings.

1040 Brussels, Ave. Des Arts 41. Phone: (32.2) 505.09.11. Fax: (32.2) 505.09.96.

Budapest, Szabadsag ter 7 01944, Bank Center, Granite Tower, 9th Floor. Phone: 36-1-302-9050. Fax: 36-1-302-9060.

London EC4V 2AU, 21 Garlick Hill. Phone: (44 171) 815.1200. Fax: (44 171) 329.0299.

Moscow 119048, Bldg. 3, 33/2 Usacheva Street. Phone: 7095-245-5190. Fax: 7095-245-5192.

Paris 75002, 12, rue de la Paix. Phone: (33-1) 42.61.57.71. Fax: (33-1) 42.61.79.21.

Prague 1 110 00. Hogan & Hogan Praha, Opletalova 37. Phone: (40-2) 2411-7111. Fax: (40-2) 2421-5105.

Warsaw 00-854. Hogan & Hartson, Sp ZO.O., Atrium Tower, Al. Jana Pawla II 25. Phone: (48-22) 653 4200. Fax: (48-22) 653 4250.

Los Angeles, CA 90071, Biltmore Tower, 500 S. Grand Ave., Suite 1900. Phone: (213) 337-6700. Fax: (213) 337-6701.

Newport Beach, CA 92660, 46-75 MacArthur, Suite 670. Phone: (949) 250-4550. Fax: (949) 833-0976.

Colorado Springs, CO 80903, 2 N. Cascade Ave, Suite 1300. Phone: (719) 448-5900. Fax: (719) 448-5922.

Denver, CO 80202, One Tabor Ctr., 1200 17th St., Suite 1500. Phone: (303) 899-7300. Fax: (303) 899-7333.

Baltimore, MD 21202, 111 S. Calvert St. Phone: (410) 659-2700. Fax: (410) 539-6981.

McLean, VA 22102, 8300 Greensboro Dr. Phone: (703) 610-6100. Fax: (703) 610-6200.

Holland & Knight LLC, 131 S. Dearborn St., 30 Fl., Chicago, IL, 60603. Phone: (312) 263-3600. Fax: (312) 578-6666. Web Site: www.hklaw.com.

Holland & Knight LLP, 2099 Pennsylvania Ave. N.W., Suite 100, Washington, DC, 20006. Phone: (202) 955-3000. Fax: (202) 955-5564. Web Site: www.hklaw.com. Janet R. Studley, Edward W. Hummers Jr., Marvin Rosenberg, Charles Naftalin, Peter Connolly, George Wheeler, Alan Naftalin, Xiaohau Zhao.

Law Office of David Honig, 3636 16th St. N.W., Suite B-366, Washington, DC, 20010. Phone: (202) 332-7005. Fax: (202) 332-7511. E-mail: dhonig@crosslink.net David Honig.

Horgan, Michael Owen, 407 E. Robert Toomb Ave., Washington, GA, 30673. Phone: (706) 678-1987. Fax: (706) 678-1999. E-mail: mhorgan@nu-z.net Michael O. Horgan.

Ice Miller, One American Sq., Box 82001, Indianapolis, IN, 46282-0002. Phone: (317) 236-2100. Fax: (317) 236-2219. E-mail: info@icemiller.com Web Site: www.icemiller.com. Thomas H. Ristine.

Inghram & Inghram, 529 Hampshire, Suite 409, Bank of America Bldg., Quincy, IL, 62301. Phone: (217) 222-7420. Fax: (217) 222-1653. E-mail: inghram@inghramlaw.com John T. Inghram IV, James R. Inghram.

Irwin, Campbell & Tannenwald, P.C., 1730 Rhode Island Ave. N.W. #200, Washington, DC, 20036-3120. Phone: (202) 728-0400. Fax: (202) 728-0354. Web Site: www.ictpc.com. Peter Tannenwald, Alan C. Campbell, David A. Irwin, Richard F. wift, Kevin M. Walsh, Michelle A. McClure, Nathaniel J. Hardy, Jared B. Weaver.

Isaacman, Kaufman & Painter, 8484 Wilshire Blvd., Suite 850, Beverly Hills, CA, 90211. Phone: (323) 782-7700. Fax: (323) 782-7744. E-mail: zucker@ikplan.com Web Site: www.ikplan.com. Alan L. Isaacman, Esq., Andrew S. Zucker, Esq.

Jackson & Campbell, P.C., 1120 20th St. N.W., Suite 300-S, Washington, DC, 20036. Phone: (202) 457-1600. Fax: (202) 457-1678. E-mail: jmatteo@jackscamp.com Web Site: www.jacksoncampbell.com. James R. Michal, Esq.

Jacobs & Associates, 11 North Washington St., Suite 640, Rockville, MD, 20850. Phone: (301) 251-5470. Fax: (301) 251-5481. E-mail: jacobs@internet-law-firm.com

Jeffer, Mangels, Butler & Marmaro LLP, 1900 Ave. of the Stars, 7th Fl., Los Angeles, CA, 90067-5010. Phone: (310) 203-8080. Fax: (310) 203-0567. Web Site: www.jmbm.com.

Jenner & Block, 601 13th St. N.W., 12th Fl., Washington, DC, 20005. Phone: (202) 639-6000. Fax: (202) 639-6066. E-mail: mstull@jenner.com Web Site: www.jenner.com. Donald B. Verrilli Jr., Mark D. Schneider, Paul M. Smith, Jerome Epstein.

Johnson, Andrea L., 225 Cedar St., California Western School of Law, San Diego, CA, 92101. Phone: (800) 255-4252, EXT. 1474. Fax: (619) 696-9999. E-mail: ajohnson@cwsl.edu Web Site: www.cwsl.edu.

Johnston & Buchan LLP, 275 Slater St., Suite 1700, Ottawa, ON, K1P 5H9. Canada. Phone: (613) 236-3882. Fax: (613) 230-6423/230-6762. Fax: (613) 230-6762. Web Site: www.johnstonbuchan.com.

Jones, Day, 51 Louisiana Ave. N. W., Washington, DC, 20001. Phone: (202) 879-3939. Fax: (202) 626-1700. Web Site: www.jonesday.com. Stephen Brogan, ptnr.

Julian & Associates, 1038 N. LaSalle Dr., Chicago, IL, 60610. Phone: (312) 266-1500. Fax: (312) 337-1972.

Julien & Associates, 1501 Broadway, Suite 2600, New York, NY, 10036-5503. Phone: (212) 221-7575. Fax: (212) 221-7386. E-mail: jijulien@aol.com

Kass, Mitek & Kass, 1050 17th St. N.W., Suite 1100, Washington, DC, 20036. Phone: (202) 659-6500. Fax: (202) 293-2608. Web Site: www.kmklawyers.com.

Katten Muchin Rosenman LLP, 1025 Thomas Jefferson St. NW, 700 East Lobby, Washington, DC, 20007. Phone: (202) 625-3500. Fax: (202) 298-7570. E-mail: howard.braun@kattenlaw.com Web Site: www.kattenlaw.com. Lee W. Shubert, Shelley Sadowsky, Howard Braun.

Kay, Sheldon L., 30445 Northwestern Hwy., Suite 320, Farmington Hills, MI, 48334. Phone: (248) 539-1111. Fax: (248) 539-1114. E-mail: sllaw@hotmail.com Web Site: www.myspace.com/rnrlawyershow.

Kaye, Scholer, L.L.P., 901 15th St. N.W., Suite 1100, Washington, DC, 20005. Phone: (202) 682-3500. Fax: (202) 682-3580. E-mail: jshrinsky@kayescholer.com Web Site: kayescholer.com. Jason L. Shrinsky, Esq.; Bruce A. Eisen, Esq.; Allan G. Moskowitz, Esq.

Keller & Heckman, 1001 G St. N.W., Suite 500 W, Washington, DC, 20001. Phone: (202) 434-4100. Fax: (202) 434-4646. Web Site: www.khlaw.com. Wayne V. Black, Martin W. Bercovici, Michael F. Morrone, John B. Richards, C. Douglas Jarrett, Richard J. Leighton, Richard F. Mann, ptnrs.

Kelley, Drye Collier Shannon, 3050 K St. N.W., Washington, DC, 20007. Phone: (202) 342-8400. Fax: (202) 342-8451. Web Site: www.colliershannon.com.

Law Office of Dennis J. Kelly, Box 41177, Washington, DC, 20018. Phone: 888-FCC-LAW-1. Phone: (202) 293-2300. Fax: (410) 626-1794. E-mail: dkellyfcclaw1@comcast.net Dennis J. Kelly.

Law Office of Edward M. Kelman, 100 Park Ave., 20th Fl., New York, NY, 10017. Phone: (212) 371-9490. Fax: (212) 750-1356. E-mail: emknyc@aol.com

Kenkel & Associates, 9908 Sorrel Ave., Potomac, MD, 20854. Phone: (301) 299-6260. Fax: (301) 299-0720. E-mail: jngkenkel@aol.com John B. Kenkel.

Kilpatrick & Stockton L.L.P., 1100 Peach Tree St., Suite 2800, Atlanta, GA, 30309. Phone: (404) 815-6500. Phone: (404) 745-2492. Fax: (404) 815-6555. E-mail: rbuttram@kilpatrickstockton.com Web Site: www.kilpatrickstockton.com.

King & Ballow, 1100 Union St. Plaza, 315 Union St., Nashville, TN, 37201. Phone: (615) 259-3456. Fax: (615) 254-7907. E-mail: lawfirm@kingballow.com Web Site: www.kingballow.com. Douglas R. Pierce, Mark E. Hunt.

Kirkpatrick & Lockhart L.L.P., 75 State St., Boston, MA, 02109. Phone: (617) 261-3100. Phone: (617) 951-9230. Fax: (617) 261-3175. E-mail: bmorrissey@klng.com Web Site: www.klng.com. Stephen L. Palmer, Esq.

Kleinberg, Lopez, Lange, Cuddy, Edel & Klein L.L.P., 2049 Century Park E., Suite 3180, Los Angeles, CA, 90067-3205. Phone: (310) 286-9696. Fax: (310) 277-7145. Fax: (310) 286-6445. E-mail: lawyers@kllcek.com Web Site: www.kllcek.com. Kenneth Kleinberg.

Kletter, Matthew L., 183 Madison Ave., Penthouse, New York, NY, 10016. Phone: (212) 726-0090. Fax: (212) 447-6677.

Klitzman, Stephen, Office of Legislative & Inter-Govt Affairs. U.S. Federal Communications Commission, 445 12th St., SW, Office of General Counsel, Washington, DC, 20554. Phone: (202) 418-1763. Fax: (202) 418-7540. E-mail: steve.klitzma@fcc.gov Stephen Klitzman.

Koerner & Olender, P.C., 11913 Grey Hollow Ct., North Bethesda, MD, 20852. Phone: (301) 468-3336. Fax: (301) 468-3343. E-mail: bkofcclaw@erols.com James A. Koerner, Robert L. Olender.

Kraditor & Haber, P.C., 1212 Ave. of the Americas 3rd Fl., New York, NY, 10036. Phone: (212) 768-2100. Fax: (212) 768-2450. Web Site: www.fcc.gov.

Krech, David H., Wireless Telecommunications Bureau, FCC, 445 12th St. S.W., Rm. 4C216, Washington, DC, 20554. Phone: (202) 418-7240. Fax: (202) 418-7224. E-mail: dkrech@fcc.gov

Lang, Richert & Patch, 5200 N. Palm Ave., Suite 401, Fresno, CA, 93704-2225. Phone: (559) 228-6700. Fax: (559) 228-6727. Web Site: www.www.lrplaw.net.

Latham & Watkins, 555 11th St. N.W., Suite 1000, Washington, DC, 20004. Phone: (202) 637-2200. Fax: (202) 637-2201. E-mail: eric.bernthal@lw.com Web Site: www.lw.com. Eric Bernthal, Gary Epstein, Jim Barker, Teresa Baer, Kevin Boyle, Kevin Boyle, Karen Brinkmaner, Matt Brill, Ray Grochowski, John Janka, James Hanna, Richard Cameron, James Rogers, Brian Weimer, David Burns, Nia Mathis, Jeff Marks, Stefanie Alfonso-Frank, Rick Bress, Joe Sullivan, David Dantzic, Jessica Gibson, Elizabeth Park.

Law Office of Dan J. Alpert, 2120 N. 21st Rd., Arlington, VA, 22201. Phone: (703) 243-8690. Fax: (703) 243-8692. E-mail: dja@commlaw.tv Web Site: commlaw.tv. Dan J. Alpert; Washington, DC: (202) 371-7200 John C. Quale, Antoinette Cook Bush, Kenneth M. Kaufman, Lawrence Roberts, Ivan A. Schlager, Richard A. Hindman, Brian D. Weimer, David H. Pawlik, Margaret E. Lancaster, John M. Beahn, Malcolm J. Tuesley, Jared S. Sher; Chicago Office: Warren Lavey, David S. Prohofsky.

Law Offices of Henry W. Root, P.C., 1541 Ocean Ave., Suite 200, Santa Monica, CA, 90401-2104. Phone: (310) 395-6800. Fax: (310) 393-7777. E-mail: henry@grrlaw.com Henry W. Root Esq., Bruce Grakal Esq., Richard Rosenthal Esq.

Lawrence & Eason, 14000 Quail Spgs Pkwy, Suite 200, Oklahoma City, OK, 73134-2638. Phone: (405) 841-6000. Fax: (405) 841-6006.

LeBoeuf, Lamb, Greene & MacRae, 1875 Connecticut Ave. N.W., Suite 1200, Washington, DC, 20009. Phone: (202) 986-8000. Fax: (202) 986-8102. Web Site: www.llgm.com. David R. Poe, Lawrence G. Acker, Catherine P. McCarthy, Yvonne Coviello.

San Francisco, CA 94111, One Embarcadero Ctr. Phone: (415) 951-1100. Fax: (415) 951-1180. Thomas McDonald, R. Scott Puddy.

Boston, MA 02110, 260 Franklin St. Phone: (617) 439-9500. Fax: (617) 439-0341. Paul Connolly, Meab Purcell.

Newark, NJ 07102-5311, One Gateway Ctr, Suite 603. Phone: (201) 643-8000. Hon. Fredrick B. Lacey.

Albany, NY 12210, One Commerce Plaza, 99 Washington Ave. Phone: (518) 465-1500. Brian Fitzgerald.

New York, NY 10019, 125 W. 55th St. Phone: (212) 424-8000. Fax: (212) 424-8500. Vivian Polak.

Harrisburg, PA 17108, Box 12105, Strawberry Sq, 320 Market St, Suite E-400. Phone: (717) 232-8199. Jim Cawley.

Pittsburgh, PA 15219, 601 Grant St. Phone: (412) 594-2300. Fax: (412) 594-5237.

Leibowitz & Associates, P.A., One S.E. 3rd Ave., Suite 1450, Miami, FL, 33131-1715. Phone: (305) 530-1322. Fax: (305) 530-9417. E-mail: firm@broadlaw.com Matthew L. Leibowitz, Joseph A. Belisle, Ila L. Feld, Nicki J. Fernandez.

Leopold, Petrich & Smith, 2049 Century Park E., Suite 3110, Los Angeles, CA, 90067-3274. Phone: (310) 277-3333. Fax: (310) 277-7444. E-mail: dmayeda@lpsla.com Web Site: www.lpsla.com. Daniel M. Mayeda.

Leventhal Senter & Lerman PLLC, 2000 K St. N.W., Suite 600, Washington, DC, 20006-1809. Phone: (202) 429-8970. Fax: (202) 293-7783. Web Site: www.lsl-law.com.Norman P. Leventhal, Meredith S. Senter Jr., Steven Alman Lerman, Raul R. Rodriguez, Dennis P. Corbett, Barbara K. Gardner, Stephen D. Baruch, Sally A. Buckman, Brian M. Madden, David S. Keir, Nancy L. Wolf, Deborah R. Coleman, Nancy A. Ory, John D. Poutasse, Christopher J. Sova, Philip A. Bonomo, Howard A. Topel, Linda D. Feldmann, S. Jenell Trigg, Beth-Sherri Akyereko, Peter M. Gould, John W. Bagwell, Jessica L. Schneider, Suzanne E. Head, Louis J. Levy, Jennifer T. Miller, Jennifer M. Babin, Jean W. Benz, Katy Chang, Katrina C. Gkeber, Stephen A. Hidebrandt, Erin E. Kucerik

Lewis, Lewis & Ferraro, 28 N. Main St., West Hartford, CT, 06107-1928. Phone: (860) 521-1500. Fax: (860) 521-4500. E-mail: attorney@lewislewisferraro.com

Law Firm of Rosalind Lichter, Tribeca Film Ctr., 375 Greenwich St., New York, NY, 10013. Phone: (212) 941-4075. Fax: (212) 941-4076. Rosalind Lichter.

Loeb & Loeb L.L.P., 345 Park Ave., New York, NY, 10154. Phone: (212) 407-4000. Phone: (212) 407-4987. Fax: (212) 407-4990. E-mail: jmanton@loeb.com Web Site: www.loeb.com. Donald L. B. Baraf, Marc Chamlin.

Los Angeles, CA 90067-4164, 10100 Santa Monica Blvd. Phone: (310) 282-2475. Fax: (310) 282-2192. Mickey Mayerson.

Los Angeles, CA 90017-2475, 1000 Wilshire Blvd. Phone: (213) 688-3400. Fax: (213) 688-3460.

Loftus & Borgstrom, One Court St., Suite 320, Lebanon, NH, 03766. Phone: (603) 448-6420. Fax: (603) 448-6147. E-mail: wrlpc@valley.net William R. Loftus, Esq.; Karen J. Borgstrom, Esq.

Lommen Abdo, Cole, King & Stageberg, P. A., 2000 IDS Ctr., 80 S. 8th St., Minneapolis, MN, 55402. Phone: (612) 339-8131. Fax: (612) 339-8064. Web Site: www.Lommen.com.

London, Michael B., 10452 Oletha Ln., Los Angeles, CA, 90077-2420. Phone: (310) 474-0577. Fax: (310) 474-5413.

Lowndes, Drosdick, Doster, Kantor & Reed, P.A., Box 2809, 215 N. Eola Dr., Orlando, FL, 32801. Phone: (407) 843-4600. Fax: (407) 843-4444. E-mail: karen.plunkett @lowndes-law.com Julia L. Frey, Louis Frey Jr.

Lukas, Nace, Gutierrez & Sachs Chartered, 1650 Tysons Blvd., Suite 1500, McLean, VA, 22102. Phone: (703) 584-8678. Fax: (703) 584-8696. E-mail: rlucas@fcclaw.com Web Site: www.fcclaw.com. Russell D. Lukas, David L. Nace, Thomas Gutierrez, Elizabeth R. Sachs.

Law Offices of Patrice Lyons, Chartered, 910 17th St. N.W., Suite 800, Washington, DC, 20006. Phone: (202) 293-5990. Fax: (202) 293-5121. E-mail: palyons@bellatlantic.net Patrice Lyons.

David L. Maddox & Associates, P.C., 1207 17 Ave. S., Suite 300, Nashville, TN, 37212. Phone: (615) 329-0086. Fax: (615) 320-7150. E-mail: david@dmaddox.com

Madigan & Getzendanner, 30 N. LaSalle St., Suite 3906, Chicago, IL, 60602. Phone: (312) 346-4321. Fax: (312) 346-5619. Michael J. Madigan, Vincent J. Getzendanner.

Magee Law Firm, PLLC, 6845 Elm St., Suite 205, McLean, VA, 22101. Phone: (703) 356-7500. Fax: (703) 356-6863. E-mail: jmagee@mageelawfirm.com James E. Magee, Kristie S. Hassett, Jennifer A. Newberry.

Margolin Law Firm, 5502 High Dr., Mission Hills, KS, 66208-1121. Phone: (816) 753-3838. Fax: (816) 753-3842. James S. Margolin.

The Marshall Firm, 271 Madison Ave., 20th Fl., New York, NY, 10016. Phone: (212) 382-2044. Fax: (212) 382-3610. E-mail: tmf@marshallfirm.com Paul Marshall.

Donald E. Martin, P.C., Box 8433, Falls Church, VA, 22041. Phone: (703) 642-2344. Fax: (703) 642-2357. E-mail: dempc@prodigy.net Donald E. Martin.

McDonald, Hopkins, L.P.A., 2100 Bank One Ctr., 600 Superior Ave. E., Cleveland, OH, 44114-2653. Phone: (216) 348-5400. Fax: (216) 348-5474. E-mail: attorneys@mhbh.com Web Site: www.mhbh.com. Brian M. O'Neil.

Mary A. McReynolds, P.C., 1050 Connecticut Ave. N.W., Suite 1000, Washington, DC, 20036. Phone: (202) 429-1770. Fax: (202) 772-3101. Mary A. McReynolds, Esq.

Mensch, Linda Susan, 200 S. Michigan Ave., Suite 1240, Chicago, IL, 60604. Phone: (312) 922-2910. Fax: (312) 922-1865. E-mail: menschlaw@yahoo.com Web Site: menschlaw@yahoo.com. Linda Mensch.

Messerli & Kramer, 150 S. 5th St., Suite 1800, Minneapolis, MN, 55402-4246. Phone: (612) 672-3600. Fax: (612) 672-3777. E-mail: djohnson@mandklaw.com Web Site: www.messerlikramer.com. William F. Messerli.

Meyers & Meyers, 360 E. Randolph St., Suite 3104, Chicago, IL, 60601. Phone: (312) 616-1500. Fax: (312) 616-1737. E-mail: peterarbme@aol.com Web Site: www.petermeyers.net. Therese Zaller, Peter R. Meyers, Irving Meyers.

Midlen Law Center, 7618 Lynn, Chevy Chase, MD, 20815-6043. Phone: (301) 656-3000. Fax: (301) 656-8262. E-mail: john@midlen.com Web Site: www.midlen.com. John H. Midlen Jr.

Miller & Neely, P.C., 6900 Wisconsin Ave., Suite 704, Bethesda, MD, 20815. Phone: (301) 986-4160. Fax: (301) 986-4162. E-mail: millaw@netkonnect.net Jerrold D. Miller, John S. Neely.

Miller & Van Eaton, P.L.L.C., 1155 Connecticut Ave. N.W., Suite 1000, Washington, DC, 20036. Phone: (202) 785-0600. Fax: (202) 785-1234. E-mail: info2@millervaneaton.com Web Site: www.millervaneaton.com.

Miller, Canfield, Paddock & Stone, P.L.C., 150 W Jefferson Ave., Suite 2500, Detroit, MI, 48226. Phone: (313) 963-6420. Fax: (313) 496-7500. E-mail: houser@millercanfield.com Web Site: www.millercanfield.com. Tillman L. Lay.

Miller, Balis and O'Neil, 1140 19th St. N.W., Suite 700, Washington, DC, 20006. Phone: (202) 296-2960. Fax: (202) 296-0166. E-mail: mgrossman@mbolaw.com Milton J. Grossman.

Mintz, Levin, Cohn, Ferris, Glovsky & Popeo, P.C., 701 Pennsylvania Ave. N.W., Suite 900, Washington, DC, 20004. Phone: (202) 434-7300. Fax: (202) 434-7400. Web Site: www.mintzlevin.com. Charles D. Ferris, Frank W. Lloyd, Bruce D. Sokler, Howard J. Symons.

Boston, MA 02111, One Financial Center. Phone: (617) 542-6000. Fax: (617) 542-2241. Irwin Heller.

Mirowski & Associates, 757 W. Ivy St., San Diego, CA, 92101. Phone: (619) 702-5300. Fax: (619) 702-4666. E-mail: pmirowski@mirlaw.com Web Site: www.mirlaw.com. Paul J. Mirowski.

Mitchell, Charles D., 1601 N. Frontage Rd., Suite F, Vicksburg, MS, 39180. Phone: (601) 636-4545, EXT. 123. Fax: (601) 634-0897. E-mail: fysadm@vicksburgpost.com Web Site: www.vicksburgpost.com. Charles D. Mitchell.

Mitchell Silberberg & Knupp, 11377 W. Olympic Blvd., Los Angeles, CA, 90064. Phone: (310) 312-2000. Fax: (310) 312-3100. E-mail: info@msk.com Web Site: www.msk.com.

Mizrack & Gantt, Suite 850, 555 11th St. N.W., Washington, DC, 20004-1304. Phone: (202) 628-1717. Fax: (202) 628-1919. E-mail: jbgantt@att.global.net

Morris, Rathnau & De La Rosa, 39 La Salle St., Fl 500, Chicago, IL, 60603. Phone: (312) 606-0876. Fax: (312) 606-0879. E-mail: mdlrlawchicago@aol.com Joseph A. Morris, ptnr.

Morrison & Foerster L.L.P., 2000 Pennsylvania Ave. N.W., Suite 5500, Washington, DC, 20006. Phone: (202) 887-1500. Fax: (202) 887-0763. Web Site: www.mofo.com. Cheryl A.Tritt.

Moss & Barnett, A Professional Assn, 4800 Wells Fargo Ctr., 90 S. 7th St., Minneapolis, MN, 55402-4129. Phone: (612) 347-0300. Fax: (612) 339-6686. E-mail: weinstockd@moss-barnett.com Web Site: www.moss-barnett.com. Brian T. Grogan, Esq.

Todd W. Musburger, Ltd., 142 E. Ontario St., Suite 500, Chicago, IL, 60611. Phone: (312) 664-2600. Fax: (312) 664-4137. E-mail: todd@musburger.com Todd W. Musburger.

Myman, Abell, Fineman, Greenspan & Light, 11601 Wilshire Blvd., Suite 2200, Los Angeles, CA, 90025. Phone: (310) 820-7717. Fax: (310) 207-2680. Robert M. Myman.

NBC Universal Television Group, 100 Universal City Plaza, Bldg #1320, Suite 3E, Universal City, CA, 91608. Phone: (818) 777-6968. Fax: (818) 866-7597. E-mail: tracy.rich@nbcuni.com

Nadel, Mark S., U.S. Federal Communications Commission, 445 12th St. S.W., Rm. 5B 551, Washington, DC, 20554. Phone: (202) 418-7385. Fax: (202) 418-7361. E-mail: mnadel@fcc.gov Web Site: www.fcc.gov.

Naphtali, Ashirah S., 130-33 217 St., Suite B, Laurelton, NY, 11413-1230. Phone: (718) 481-7236. Fax: (718) 481-7236. E-mail: anaphml@aol.com Ashirah S. Naphtali, Esq., MBA.

National Exchange Carrier Association, N.E.C.A., 80 S. Jefferson Rd., Whippany, NJ, 07981-1009. Phone: (973) 884-8000. Phone: (800) 228-8597. Fax: (973) 884-8469. Web Site: www.neca.org.

Nemeth, Valerie A., Attorney at Law, 191 Calle Magadalena, Suite 270, Encinitas, CA, 92024-3750. Phone: (760) 944-4130. Phone: (310) 471-7648 (L.A.). Fax: (760) 944-3325. E-mail: vanemeth@cs.com Web Site: www.entlawyer.com. Valerie Nemeth.

Neuland, Nordberg, Andrews & Whitney, 22502 Avenida Empresa, Rancho Santa Margarita, CA, 92688. Phone: (949) 766-4700. Fax: (949) 766-4712. E-mail: dottieneuland@nnawlaw.com

Nathan M. Nickolaus, 320 E. McCarthy St., Jefferson City, MO, 65101-3115. Phone: (573) 634-6313. Fax: (573) 634-6504. E-mail: nnickolaus@jeffcity.mo.org Web Site: nnickolaus@jeffcitymo.org.

Nilsson, Kent R., U.S. Federal Communications Commission, 445 12th St. S.W., Washington, DC, 20554. Phone: (202) 418-2478. Fax: (202) 418-2345. Web Site: www.fcc.gov.

Nixon Peabody L.L.P., 401 9th St. NW, Suite 900, Washington, DC, 20004. Phone: (202) 585-8000. Fax: (202) 585-8080. E-mail: aprilsteffan@nixonpeabody.com Web Site: www.nixonpeabody.com. Veronica M. Ahern, William S. Andrews.

Rochester, NY 14603, Box 1051, Clinton Sq. Phone: (716) 263-1000. Richard D. Rochford Jr.

Nixon, Wilbert E. Jr., Wireless Telecommunications Bureau, Police & Rules Branch. Federal Communications Commission, 445 12th St. S.W. Rm. 4-A207, Washington, DC, 20554. Phone: (202) 418-7240. Fax: (202) 418-7447. E-mail: wnixon@fcc.gov Web Site: www.fcc.gov.

Nossaman, Guthner, Knox & Elliott, L.L.P., 50 California St., 34th Fl., San Francisco, CA, 94111. Phone: (415) 398-3600. Fax: (415) 398-2438. E-mail: mmattes@nossaman.com Web Site: www.nossaman.com. Martin A. Mattes, Jose E. Guzman Jr.

OPASTCO, 21 Dupont Cir. N.W., Suite 700, Washington, DC, 20036. Phone: (202) 659-5990. Fax: (202) 659-4619. E-mail: vlf@opastco.org Web Site: www.opastco.org.

O'Connell & Aronowitz PC, 54 State St., Albany, NY, 12207. Phone: (518) 462-5601. Fax: (518) 462-2670. E-mail: o'connell@albany.net Web Site: www.oalaw.com. Peter Danziger, Neil H. Rivchin.

O'Connell, Susan Lee, U.S. Federal Communications Commission, 445 12th St. S.W., Rm. 6A847, Washington, DC, 20554. Phone: (202) 418-1484. Fax: (202) 418-2824. E-mail: soconnell@fcc.gov

O'Connor & Hannan, 1666 K St. N.W., Suite 500, Washington, DC, 20006. Phone: (202) 887-1400. Fax: (202) 466-2198. E-mail: gadler@oconnorhannan.com Gary Adler.

Ogden Murphy Wallace, P.L.L.C., 1601 5th Ave., Suite 2100, Westlake Ctr. Tower, Seattle, WA, 98101-1686. Phone: (206) 447-7000. Fax: (206) 447-0215. E-mail: nparks@omwlaw.com Web Site: www.omwlaw.com.

O'Melveny & Myers, 1625 Eye St. N.W., Washington, DC, 20006. Phone: (202) 383-5300. Fax: (202) 383-5414. E-mail: jbeisner@omm.com Web Site: www.omm.com. John H. Beisner, Donald T. Bliss, John Rogovin, Jessica Davidson-Miller, Carl R. Schenker Jr., Charles Read, Paul McNamara, Martine Apollon, Todd Rosenberg.

O'Neil, Cannon, Hollman, Dejong, Chase Tower, 111 E. Wisconsin Ave., Suite 1400, Milwaukee, WI, 53202-4803. Phone: (414) 276-5000. Fax: (414) 276-6581. Carl Holborn.

O'Neill, Athy & Casey, P.C., 1310 19th St. N.W., Washington, DC, 20036. Phone: (202) 466-6555. Fax: (202) 466-6596. E-mail: aathy@oacpc.com Web Site: oacpc.com. Andrew Athy Jr., Christopher R. O'Neill, ptnr.

O'Reilly, Rancilio, Nitz, Andrews, Turnbull, & Scott P.C., 12900 Hall Rd., Suite 350, Sterling Heights, MI, 48313-1151. Phone: (586) 726-1000. Fax: (586) 726-1560. Web Site: www.orlaw.com. Neil J. Lehto, Donald DeNault.

Orr & Reno, Box 3550, One Eagle Sq., Concord, NH, 03302-3550. Phone: (603) 224-2381. Fax: (603) 224-2318. E-mail: mmclean@orr-reno.com Web Site: www.orr-reno.com. William L. Chapman.

Overton, John B., 14 Manzanita Pl., Mill Valley, CA, 94941. Phone: (415) 331-2889. E-mail: overton@plownet.com

Law Offices of James L. Oyster, 108 Oyster Ln., Castleton, VA, 22716. Phone: (540) 937-4800. Fax: (540) 937-2148. E-mail: joyster@crosslink.net James L. Oyster.

Pankopf, Arthur, 7819 Hampden Ln., Bethesda, MD, 20814-1108. Phone: (301) 657-8790. Fax: (301) 657-3296. E-mail: apankopf@worldnet.att.net Arthur Pankopf.

Pardo & Pardo P.A., Box 398646, Miami Beach, FL, 33239. Phone: (305) 673-1515. Fax: (305) 673-9359.

Patton Boggs L.L.P., 2550 M St. N.W., Suite 900, Washington, DC, 20037. Phone: (202) 457-6000. Fax: (202) 457-6315. Web Site: www.pattonboggs.com. Thomas H. Boggs Jr., Stephen Diaz Gavin, Paul C. Besozzi, J. Jeffrey Craven, John F. Fithian, Penelope S. Farthing, Janet Fitzpatrick, Jeffrey L. Ross. Dallas: Charles Miller, Jennifer Boudreau.

Pauker, Molly, Fox Television Stations Inc., 5151 Wisconsin Ave. N.W., Washington, DC, 20016-4124. Phone: (202) 895-3088. Fax: (202) 895-3222. E-mail: mollyp@fox.com

Paul, Hastings, Janofsky & Walker LLP, 1299 Pennsylvania Ave. N.W., 10th Fl., Washington, DC, 20004-2400. Phone: (202) 508-9500. Fax: (202) 508-9700. Web Site: www.paulhastings.com. Ralph B. Everett, Bruce D. Ryan, Michelle Cohen, Carl W. Northrop, John G. Johnson, William D. DeGrandis, David Burns, Christine Crowe, G. Hamilton Loeb.

Paul, Weiss, Rifkind, Wharton & Garrison, L.L.P., 1615 L St. N.W., Suite 1300, Washington, DC, 20036. Phone: (202) 223-7300. Fax: (202) 223-7427. Web Site: www.paulweiss.com. Phillip L. Spector, Jeffrey H. Olson, Patrick S. Campbell.

Pearce & Durick, Box 400, 314 E. Thayer Ave., Bismarck, ND, 58502-0400. Phone: (701) 223-2890. Fax: (701) 223-7865. E-mail: law.office@pearce-durick.com Web Site: www.pearce-durick.com. Patrick W. Durick, Larry L. Boschee, Jerome C. Kettleson, Gary R. Thune, Jonathan P. Sanstead.

John D. Pellegrin P.C., 10515 Dominion Valley Dr., Fairfax Station, VA, 22039. Phone: (703) 455-6101. Fax: (703) 455-6106. E-mail: jd@lawpell.com Web Site: www.pellegrin-law.com. John D. Pellegrin.

Pepper Hamilton LLP, 3000 Two Logan Sq., Eighteenth & Arch St., Philadelphia, PA, 19103-2799. Phone: (215) 981-4000. Fax: (215) 981-4750. E-mail: phinfo@pepperlaw.com Web Site: www.pepperlaw.com. Pepper Hamilton LLP, David A. Wormser.

Washington, DC 20005-2004, 600 14th St. N.W. Phone: (202) 220-1200. Fax: (202) 220-1665. David A. Wormser.

Perkins, Jr., Roy F., 1724 Whitewood Ln., Herndon, VA, 20170. Phone: (703) 435-9700. Fax: (703) 435-9701. Roy F. Perkins Jr.

Larry D. Perry, Attorney at Law, 11464 Saga Ln., Suite 110, Knoxville, TN, 37931-2819. Phone: (865) 927-8474. Fax: (865) 927-4912. E-mail: larryperry@worldnet.att.net Larry D. Perry, Esq.

Peterson Law Firm, 2033 Walnut St., Philadelphia, PA, 19103. Phone: (215) 557-9001. Fax: (215) 557-9108.

Phillips Nizer LLP, 666 5th Ave., 28th Fl., New York, NY, 10103-0084. Phone: (212) 977-9700. Fax: (212) 262-5152. E-mail: ssalmon@phillipsnizer.com Web Site: www.phillipsnizer.com.

Pierce, Robinson & Greene P A, 600 W. 4th St., North Little Rock, AR, 72114-5360. Phone: (501) 372-3131. Fax: (501) 372-3825. Web Site: www.prg-law.com. William Robinson.

Pillsbury, Winthrop Shaw Pittman LLP, 1540 Broadway, New York, NY, 10036-4039. Phone: (877) 323-4171. Fax: (415) 983-1200. Web Site: www.pillsburylaw.com.

Powell, Goldstein, Frazer & Murphy, 1201 W. Peachtree St. N.W., Fl. 14, Atlanta, GA, 30309. Phone: (404) 572-6600. Fax: (404) 572-6999. E-mail: wmoeling@pgfm.com Web Site: www.pgfm.com.

Washington, DC 20004, 1001 Pennsylvania Ave. N.W, 6th Fl. Phone: (202) 347-0066. Fax: (202) 624-7222. Jerome S. Breed.

Pratcher & Associates P.C., 1133 Kensington Ave., Buffalo, NY, 14215-1611. Phone: (716) 838-4612. Fax: (716) 838-4828. E-mail: frpratcher@pratcher.com Franklin Pratcher.

Buffalo, NY 14204, Town Gardens Plaza, 447 Williams St, Suite H. Phone: (716) 847-0145.

Preston, Gates, Ellis & Rouvelas Meeds L.L.P., 1735 New York Ave. N.W., Suite 500, Washington, DC, 20006. Phone: (202) 628-1700. Fax: (202) 331-1024. E-mail: robert@prestongates.com Web Site: www.prestongates.com. Martin L. Stern.

Proskauer Rose L.L.P., 1585 Broadway, New York, NY, 10036. Phone: (212) 969-3000. Fax: (212) 969-2900. E-mail: lbudish@proskauer.com Web Site: www.proskauer.com. Bertram A. Abrams, Lawrence H. Budish.

Provosty, Sadler Delaunay, Fiorenza & Sobel, Box 1791, Hibernia National Bank, 8th Fl., Alexandria, LA, 71309-1791. Phone: (318) 445-3631. Fax: (318) 445-9377. David Sobel.

Pulis, Gregory M., (Creative Artists Agency). 9830 Wilshire Blvd., Beverly Hills, CA, 90212. Phone: (310) 288-4545. Fax: (310) 288-4800. E-mail: gpulis@caa.com

Putbrese, Hunsaker & Trent, P.C., 200 S. Church St., Woodstock, VA, 22664. Phone: (540) 459-7646. Fax: (540) 459-7656. E-mail: phtlaw@mindspring.com John C. Trent.

Reddy, Begley & McCormick, L.L.P., 1156 15th St. N.W., Suite 610, Washington, DC, 20005-1770. Phone: (202) 659-5700. Fax: (202) 659-5711. E-mail: rbm@rbmfcclaw.com Web Site: www.rbmfcclaw.com. Dennis F. Begley, Matthew E. McCormick.

Rees, Broome & Diaz, 8133 Leesburg Pike, 9th Fl., Vienna, VA, 22182. Phone: (703) 790-1911. Fax: (703) 848-2530. E-mail: info@rbdlaw.com Web Site: www.rbdlaw.com. Peter S. Philbin.

The Law Offices of George Edward Regis, 121 W. 27th St., Suite 1001, New York, NY, 10001-6207. Phone: (212) 645-8800. Fax: (212) 645-7900. E-mail: georgeregis@yahoo.com George Edward Regis.

Renouf & Polivy, 1532 16th St. N.W., Washington, DC, 20036. Phone: (202) 265-1807. Fax: (202) 265-1810. E-mail: thamber@aol.com Katrina Renouf, Margot Polivy.

Resnick, Bernard Max, Esq, P.C., Two Bala Plaza, Suite 300, Bala Cynwyd, PA, 19004-1501. Phone: (610) 660-7774. Fax: (610) 668-0574. E-mail: bmresnick@aol.com Web Site: www.bernardresnick.com. Bernard Max Resnick, Priscilla J. Mattison.

Reynolds & Manning, P.A., Box 2809, Prince Frederick, MD, 20678. Phone: (410) 535-9220. Fax: (410) 535-9171. E-mail: calvertlawyer@comcast.net Web Site: www.lawyers.com/reynoldsandmanning. Christopher J. Reynolds.

Richards, Mary Beth, U.S. Federal Communications Commission, 445 12th St. S.W., Rm. 8-C750, Washington, DC, 20554. Phone: (202) 418-1000. Fax: (202) 418-2801. Web Site: www.marybeth.richards@fcc.gov.

Riezman & Berger, 7700 Bonhomme Ave., 7th Fl., St. Louis, MO, 63105. Phone: (314) 727-0101. Fax: (314) 727-6458. E-mail: jacobs@riezmamberger.com Web Site: www.riezmanberger.com. Bob Jacobs.

Riker Danzig Scherer Hyland & Perretti LLP, One Speedwell Ave., Headquarters Plaza, Morristown, NJ, 07962-1981. Phone: (973) 538-0800. Fax: (973) 538-1984. E-mail: info@riker.com Web Site: www.riker.com. Sidney M. Schreiber, Vincent J. Sharkey Jr., Edward K. DeHope, James C. Meyer, Michael A. Schmerling, Mark T. Pasko.

Trenton, NJ 08608-1220, 50 W. State St, Suite 1010.

Robins, Kaplan, Miller & Ciresi, 2800 LaSalle Plaza, 800 LaSalle Ave., Minneapolis, MN, 55402-2015. Phone: (612) 349-8500. Fax: (612) 339-4181. E-mail: kamarron@rkmc.com Web Site: www.rkmc.com. Timothy Block, Doug Boettge, John F. Gibbs, Todd Hartman, Lisa Heller, Rebecca Liethen, Edward Muramoto, Thomas A. Miller, Ed Muramoto, Sara A. Poulos, Steven Safvanski.

Gust Rosenfeld P.L.C., 201 E. Washington, Suite 800, Phoenix, AZ, 85004-2327. Phone: (602) 257-7422. Fax: (602) 254-4878. E-mail: chauncey@gustlaw.com Web Site: www.gustlaw.com. Tom Chauncey II.

Rosenfeld, Meyer & Susman L.L.P., 9601 Wilshire Blvd., Suite 710, Beverly Hills, CA, 90210-5288. Phone: (310) 858-7700. Fax: (310) 860-2430. Web Site: www.rmslaw.com.

Rourke, Gerald S., 76 Northwood Rd., Madison, CT, 06443. Phone: (203) 421-3424. Fax: (203) 421-8683. E-mail: gerald.rourke@comast.net Gerald S. Rourke.

Rubin, Winston, Diercks, Harris & Cooke, L.L.P., 1155 Connecticut Ave. N.W., 6th Fl., Washington, DC, 20036. Phone: (202) 861-0870. Fax: (202) 429-0657. E-mail: jwinston@rwdhc.com Web Site: www.rwdhc.com. James L. Winston, Steven J. Stone.

Ryan, Swanson & Cleveland P.L.L.C., 1201 3rd Ave., Suite 3400, Seattle, WA, 98101-3034. Phone: (206) 464-4224. Phone: (800) 458-5973. Fax: (206) 583-0359. E-mail: collette@ryanlaw.com Web Site: www.ryanlaw.com.

Law Offices of Lee Sacks, 23852 Pacific Coast Hwy., Suite 157, Malibu, CA, 90265-4879. Phone: (310) 451-3113. Fax: (310) 451-0089. Lee Sacks.

Sahl, Jack, School of Law, University of Akron, Akron, OH, 44325-2901. Phone: (330) 972-6753. Fax: (330) 258-2343. E-mail: jps@uakron.edu Web Site: www.uakron.edu/law.

Sanchez Law Firm, 2300 M. St., N.W., Suite 800, Washington, DC, 20037. Phone: (202) 237-2814. Fax: (202) 237-5614. E-mail: esanchez@bellatlantic.net Ernest T. Sanchez.

Sapronov & Associates, P. C., 3 Ravinia Dr., Suite 1455, Atlanta, GA, 30346. Phone: (770) 399-9100. Fax: (770) 395-0505. E-mail: info@wstelecomlaw.com Web Site: www.wstelecomlaw.com. Walt Sapronov, William D. Friend, Charles A. Hudak, Michael Stewart, Ronald Jackson, Timothy Geraghty.

Gary P. Schonman, 445 12th St. S.W., Rm. 3A660, Federal Communications Commision, Washington, DC, 20554. Phone: (202) 418-1795. Fax: (202) 418-2080. Web Site: www.fcc.gov. Gary P. Schonman.

Schuman, Felts, Chartered, 4804 Moorland Ln., Bethesda, MD, 20814. Phone: (301) 986-0200. Fax: (301) 986-7960. Sheldon Paul Schuman.

Schuster & Associates, 3594 Armourdale Ave., Long Beach, CA, 90808. Phone: (562) 596-5900. Fax: (562) 431-4540. E-mail: attorney@flightlaw.com Web Site: www.flightlaw.com.

Schwaninger & Associates, P.C., 1331 H St. N.W., Suite 500, Washington, DC, 20005. Phone: (202) 347-8580. Fax: (202) 347-8607/347-8643. E-mail: rschwaninger@sa-lawyers.net Web Site: www.sa-lawyers.net. Richard P. Hanno.

Schwartz, Woods & Miller, 1233 20th St., N.W., Suite 610, Washington, DC, 20036. Phone: (202) 833-1700. Fax: (202) 833-2351. E-mail: [\]@swmlaw.com Web Site: www.swmlaw.com. Lawrence M. Miller, Steven C. Schaffer, Malcolm G. Stevenson.

Law Offices of Philip L. Schwartz PA, 2000 Glades Rd., Suite 208, Boca Raton, FL, 33431. Phone: (954) 760-7770. Fax: (954) 524-4169. E-mail: phil@philipschwartz.com Philip L. Schwartz, Esq.

The Seale Law Firm, Bryton Tower, 1271 Poplar Ave., Suite 101, Memphis, TN, 38104-7231. Phone: (901) 722-8188. Fax: (901) 278-7126. E-mail: artslaw@bellsouth.net William B. Seale, Lisa A. Maniscalco, Cynthia Marrone, David Allen Outlaw, Harold C. Streibich.

Sell & Melton, Box 229, 577 Mulberry St., Suite 1400, Macon, GA, 31202-0229. Phone: (478) 746-8521. Fax: (478) 745-6426. E-mail: eds@sell-melton.com Web Site: www.sell-melton.com. Ed S. Sell III.

Severaid, Ronald H., 1805 Tribute Rd., Suite J, Sacramento, CA, 95815. Phone: (916) 929-8383. Fax: (916) 925-4763. E-mail: rhseveraid@earthlink.net

Seyfarth Shaw, 2029 Century Park E., 33rd Fl., Los Angeles, CA, 90067-3063. Phone: (310) 277-7200. Fax: (310) 201-5219. Web Site: www.seyfarth.com. Kenwood C. Youmans.

Seyfarth Shaw LLP, 55 E. Monroe St., Chicago, IL, 60603. Phone: (312) 781-8655. Fax: (312) 269-8869. E-mail: drowland@seyfarth.com Web Site: www.seyfarth.com. David J. Rowland, ptnr.

Law Offices of Thomas G. Shack Jr., 1150 Connecticut Ave, N.W., Suite 900, Washington, DC, 20036. Phone: (202) 293-5900. Fax: (202) 659-3493. Thomas G. Shack, Jr.

Shapiro, Burton J., 2147 N. Beachwood Dr., Los Angeles, CA, 90068-3462. Phone: (323) 469-9452. Fax: (603) 710-8109. E-mail: burtjay@mail.com Web Site: www.burtshapiro.com.

Shaw Pittman L.L.P., Communications Practice Group, 2300 N St. N.W., Washington, DC, 20037-1128. Phone: (202) 663-8000. Fax: (202) 663-8007. E-mail: info@shawpittman.com Web Site: www.shawpittman.com. Paul Cicelski, Ben C. Fisher, Richard R. Zaragoza, Clifford M. Harrington, Kathryn R. Schmeltzer, David D. Oxenford, Bruce D. Jacobs, Barry H. Gottfried, Glenn S. Richards, Scott R. Flick, Lauren Lynch Flick, Miles S. Mason, Carroll John Yung, Jane Sullivan Roberts, Dawn M. Sciarrino, Bryan T. McGinnis, Cynthia D. Greer, Susan M. Hafeli, Brendan Holland, David S. Konczal, Tony Lin, Veronica D. McLaughlin, Tina R. Reynolds, Christopher J. Sadowski, Katherine T. Suh, Amy L. Van de Kerckhove.

Shine and Hardin L.L.P., 2810 Beaver Ave., Fort Wayne, IN, 46807. Phone: (260) 745-1970. Fax: (260) 744-5411. E-mail: sshine@shineandhardin.com Steven R. Shine.

Shukat, Arrow, Hafer & Weber, L.L.P., 111 W. 57th St., Suite 1120, New York, NY, 10019-2211. Phone: (212) 245-4580. Fax: (212) 956-6471. E-mail: Peter@musiclaw.com Peter Shukat.

Shulman, Rogers, Gandal, Pordy & Ecker, P.A., 11921 Rockville Pike, 3rd Fl., Rockville, MD, 20852. Phone: (301) 230-5200. Fax: (301) 230-2891. E-mail: atilles@srgpe.com Web Site: www.shulmanrogers.com. Allan S. Tilles.

Siegal, Joel H., 703 Market St., San Francisco, CA, 94103. Phone: (415) 777-5547.

Siegel, Kelleher & Kahn, 420 Franklin St., Buffalo, NY, 14202. Phone: (716) 881-5800. Fax: (716) 885-3369. E-mail: info@skklaw.com Web Site: www.skklaw.com. Herbtert M. Siegler.

Law Offices of William D. Silva, 5335 Wisconsin Ave. N.W., Suite 400, Washington, DC, 20015-2003. Phone: (202) 362-1711. Fax: (202) 686-8282. E-mail: bill@luselaw.com Web Site: wmsilvalaw.com. William D. Silva.

Silver, Garvett & Henkel, 1110 Brickell Ave. Penthouse 1, Miami, FL, 33131. Phone: (305) 377-8802. Fax: (305) 377-8804.

Skadden, Arps, Slate, Meagher & Flom L.L.P., 1440 New York Ave. N.W., Washington, DC, 20005. Phone: (202) 371-7000. Fax: (202) 393-5760. Web Site: www.skadden.com. Richard A. Hindman, David H. Pawlik, Brian D. Weimer, John C. Quale, John M. Beahn, Antoinette C. Bush, Margaret E. Lancaster, Ivan A. Schlager, Jennifer B. Irvin, Lawrence Roberts, Jared S. Sher, Malcom J. Tuesley, (Washington DC); Warren G. Lavey, David S. Prohofsky, (Chicago).

Barry Skidelsky, Esq., 185 East 85th St., 23 D, New York, NY, 10028. Phone: (212) 832-4800. E-mail: bskidelsky@mindspring.com Barry Skidelsky (New York & Washington, DC).

Barry Skidelsky, Esq., 185 East 85th St., 23 D, New York, NY, 10028. Phone: (212) 832-4800. E-mail: bskidelsky@mindspring.com

Smith & Metalitz LLP, 1747 Pennsylvania Ave. N.W., Suite 825, Washington, DC, 20006. Phone: (202) 833-4198. Fax: (202) 872-0546. E-mail: info@smimetlaw.com Web Site: www.smimetlaw.com.

Reed Smith LLP, 3110 Fairview Park Dr., Suite 1400, Falls Church, VA, 22042-4503. Phone: (703) 641-4200. Fax: (703) 641-4340. E-mail: reedsmith@reedsmith.com Web Site: www.reedsmith.com.

Smithwick & Belendiuk, P.C., Suite 301, 5028 Wisconsin Ave. N.W., Washington, DC, 20016. Phone: (202) 363-4050. Fax: (202) 363-4266. E-mail: gsmithwick@fccworld.com Web Site: fccworld.com. Gary S, Smithwick, Arthur V. Belendiuk, Robert Lewis Thompson, William M. Bernard.

Sodos & Kafkas, 16985 Bluemount Rd., Suite 202, Brookfield, WI, 53005. Phone: (414) 785-5500. Fax: (414) 785-1100. E-mail: office@sodos.com

Solomon, Elise, Lifetime Television. Lifetime Television, 309 W. 49th St., New York, NY, 10019. Phone: (212) 424-7112. Fax: (212) 957-4447. E-mail: solomon@lifetimetv.com Web Site: www.lifetimetv.com. Elise Soloman.

Sommers, Schwartz, Silver & Schwartz, P.C., 2000 Town Ctr., Suite 900, Southfield, MI, 48075. Phone: (248) 355-0300. Fax: (248) 746-4001. E-mail: patrickmccauley@sommers.com Web Site: www.sommerspc.com. Patrick B. McCauley.

Sonneman & Sonneman, P.A., 111 Riverfront, Suite 202, Winona, MN, 55987. Phone: (507) 454-8885. Fax: (507) 454-8887. E-mail: sonneman@luminet.net Karl W. Sonneman, VP.

Sonnenschein, Nath & Rosenthal LLP, 8000 Sears Tower, 233 S. Wacker Dr., Chicago, IL, 60606. Phone: (312) 876-3114. Fax: (312) 876-7934. E-mail: sfifer@sonnenschein.com Web Site: www.sonnenschein.com. David W. Maher, Samuel Fifer.

Southmayd & Miller, 1220 19th St. N.W., Suite 400, Washington, DC, 20036. Phone: (202) 331-4100. Fax: (202) 331-4123. E-mail: jdsouthmayd@msn.com Jeffrey D. Southmayd, Michael R. Miller.

Spawn, Coy U., 1815 Bering Dr., Houston, TX, 77057-3109. Phone: (713) 782-2977. Fax: (713) 782-2977.

Spiegel & McDiarmid, 1333 New Hampshire Ave. N.W., Washington, DC, 20036. Phone: (202) 879-4000. Fax: (202) 393-2866. E-mail: jim.horwood@spiegelmcd.com Web Site: www.spiegelmcd.com. James Horwood, Tillman Lay, Scott Strauss, Peter Hopkins, Ruben Gomez, Gloria Tristani.

Springman, Braden, Wilson & Pontius, P.C., 1022 Bannock St., Denver, CO, 80204. Phone: (303) 685-4897/685-4633. Fax: (303) 685-4627. E-mail: sbwp@indra.com

Squire, Sanders & Dempsey, Box 407, 1201 Pennsylvania Ave. N.W., Washington, DC, 20044-0407. Phone: (202) 626-6600. Fax: (202) 626-6780. E-mail: jnadler@ssd.com Web Site: www.ssd.com. Thomas J. Ramsey, Herbert E. Marks, Joseph P. Markoski, Jonathan J. Nadler.

Cleveland, OH 44114-1304, 4900 Key Tower, 127 Public Sq. Phone: (216) 479-8500. Fax: (216) 479-8780. Terrence J. Clark, ptnr.

Stennett, Wilkinson & Peden, 401 Legacy Park, Ridgeland, MS, 39157. Phone: (601) 206-1816. Fax: (601) 206-9132. E-mail: swplaw@attorney.net Web Site: www.swplaw.com. Gene A. Wilkinson, James A. Peden Jr.

Stephens Media LLC, P.O. Box 70, Las Vegas, NV, 89125. Phone: (702) 477-3830. Fax: (702) 383-0230. Web Site: www.stephensmedia.com.

Steptoe & Johnson, 1330 Connecticut Ave. N.W., Washington, DC, 20036. Phone: (202) 429-3000. Fax: (202) 429-3902. E-mail: amamlet@steptoe.com Web Site: www.steptoe.com.

Stevens, Sally L., Box 41, Lumberville, PA, 18933. Phone: (215) 297-8245. Fax: (215) 297-5106.

Stewart & Irwin P.C., 251 E. Ohio St., Suite 1100, Indianapolis, IN, 46204. Phone: (317) 639-5454. Fax: (317) 632-1319. E-mail: raikman@silegal.com Web Site: www.stewart-irwin.com. Richard E. Aikman Jr.

Stewart, Estes & Donnell, 424 Church St. Suntrust Ctr., Suite 1401, Nashville, TN, 37219. Phone: (615) 244-6538. Fax: (615) 256-8386. Web Site: www.sedlaw.com. Stephen Heard.

Strichartz, James L., 201 Queen Anne Ave. N., Suite 400, Seattle, WA, 98109. Phone: (206) 282-8020. Fax: (206) 286-2050. E-mail: jim@condo-lawyers.com

Stroock, Stroock & Lavin, 180 Maiden Lane, New York, NY, 10038-4982. Phone: (212) 806-5400. Fax: (212) 806-6006. E-mail: kcummings@stroock.com Web Site: www.stroock.com.

Stryker, Tams & Dill LLP, 2 Penn Plaza E., Newark, NJ, 07105-2293. Phone: (973) 491-9500. Fax: (973) 491-9692. E-mail: dlinken@strykertams.com Web Site: www.strykertams.com. Dennis C. Linken, Richard P. De Angelis Jr.

Taylor, Jack, 1289 Lincoln Rd., Yuba City, CA, 95991. Phone: (530) 671-6800. Fax: (530) 671-6447.

Leslie Taylor Associates, 6800 Carlynn Ct., Bethesda, MD, 20817-4302. Phone: (301) 229-9410. Fax: (301) 229-3148. E-mail: ltaylor@lta.com Web Site: www.lta.com. Leslie Taylor.

Taylor, Ray L., 11608 Chayote St., Los Angeles, CA, 90049. Phone: (310) 476-6493. Fax: (310) 471-2763.

Technology Law Group L.L.C., 5335 Wisconsin Ave. N.W., Suite 440, Washington, DC, 20015. Phone: (202) 895-1707. Fax: (202) 244-8257. E-mail: mail@tlgdc.com Web Site: www.tlgdc.com. Alex Bouton.

Teitelbaum, Israel, 11301 Amherst Ave., Suite 202, Silver Spring, MD, 20902. Phone: (301) 933-3373. Fax: (301) 933-3651. Israel Teitelbaum.

Thelen Reid & Priest LLP, Washington, D.C. 701 Pennsylvania Ave. N.W., Suite 800, Washington, DC, 20004. Phone: (202) 508-4000. Fax: (202) 508-4321. Web Site: www.thelenreid.com.

San Francisco, CA 94105, 101 2nd St, Suite 1800. Phone: (415) 371-1200. Fax: (415) 371-1211.

Thiemann, Aitken & Vohra, 908 King St., Suite 300, Alexandria, VA, 22314. Phone: (703) 836-9400. Fax: (703) 836-9410. E-mail: ajthiemann@ttalaw.com Russell C. Powell, Robert Lewis Thompson.

Thomas, Ballenger, Vogelman & Turner, 124 S. Royal St., Alexandria, VA, 22314. Phone: (703) 836-3400. Fax: (703) 836-3549. John M. Ballenger.

Thompson Hine LLP, 1920 N St. N.W., Suite 800, Washington, DC, 20036. Phone: (202) 331-8800. Fax: (202) 331-8330. E-mail: barry.friedman@thompsonhine.com Web Site: www.thompsonhine.com. Barry A. Friedman, Stephen T. Lovelady, John C. Butcher.

Columbus, OH 43215, 10 W. Broad St, Suite 700. Phone: (614) 469-3200. Fax: (614) 469-3361. E-mail: tom.lodge@thompsonhine.com. Web Site: www.thompsonhine.com. Thomas E. Lodge.

Thrasher, Dinsmore & Dolan, 100 7th Ave., Suite 150, Chardon, OH, 44024-1079. Phone: (440) 285-2242. Fax: (440) 285-9423. E-mail: dmoore@dolan.law.pro Matthew Dolan.

Troutman Sanders L.L.P., 600 Peachtree St., Suite 5200, Atlanta, GA, 30308. Phone: (404) 885-3000. Fax: (404) 885-3900. Web Site: www.troutmansanders.com. Robert W. Webb, Richard H. Brody, Alan E. Serby.

Troy & Gould, PC, 1801 Century Park E., 16th Fl., Los Angeles, CA, 90067. Phone: (310) 553-4441. Fax: (310) 201-4746. Web Site: www.troygould.com.

Trugman, Richard S., 9200 Sunset Blvd., Suite 808, Los Angeles, CA, 90069. Phone: (310) 273-8834. Fax: (310) 273-8345. E-mail: trulaw@sbcglobal.net

Edmund W. Turnley III, 30 Music Square W., Suite 302, Nashville, TN, 37203. Phone: (615) 321-8600. Fax: (615) 321-8602. Edmund W. Turnley.

Turtle, Joel S., 55 Santa Clara Ave., Suite 120, Oakland, CA, 94610. Phone: (510) 763-7600. Fax: (510) 763-7894. E-mail: joelturtle@yahoo.com Web Site: www.riotmedia.com.

Umansky, Barry D., Irwin, Campbell & Tannerwald, 1730 Rhode Island Ave. N.W., Suite 200, Washington, DC, 20036-3101. Phone: (202) 728-0400. Fax: (202) 728-0354. Web Site: www.ictpc.com.

Van Cott, Bagley, Cornwall & McCarthy, 50 S. Main St., Suite 1600, Salt Lake City, UT, 84144-0103. Phone: (801) 532-3333. Fax: (801) 534-0058. E-mail: info@vancott.com Web Site: www.vancott.com. Steven D. Swidle, Robert M. Anderson, Jennifer K. Anderson.

A. Chavis Vanias, Attorney at Law, P.O. Box 9612, Columbia, SC, 29209. Phone: (803) 256-1244. Fax: (803) 753-0007. E-mail: acvanias@scbar.org

Varnum, Riddering, Schmidt & Howlett LLP, Box 352, Bridgewater Pl., Grand Rapids, MI, 49501-0352. Phone: (616) 336-6000. Fax: (616) 336-7000. E-mail: generalinfo@varnumlaw.com Web Site: www.varnumlaw.com. John W. Pestle, Timothy J. Lundgren.

Venable LLP, 1800 Mercantile Bank Bldg., 2 Hopkins Plaza, Baltimore, MD, 21201. Phone: (410) 244-7400. Fax: (410) 244-7742. E-mail: nmazmanian@venable.com

Washington, DC 20005, 1201 New York Ave. N.W, Suite 1000. Phone: (202) 962-4800. (202) 962-8300.

Veryl Miles/Professor at Law School, Cardinal Stn, Columbus School of Law/Catholic Univ. America, Washington, DC, 20064. Phone: (202) 319-5140. Fax: (202) 319-4459. Web Site: www.law.edu.

Vorys, Sater, Seymour and Pease LLP, Box 1008, 52 E. Gay St., Columbus, OH, 43216-1008. Phone: (614) 464-6400. Fax: (614) 464-6350. E-mail: jhgross@vssp.com Web Site: www.vssp.com. Sheldon A. Taft, M. Howard Petricoff, Stephen M. Howard, James H. Gross, William D. Kloss, C. William O'Neill, William S. Newcomb Jr., Benita A. Kahn, Robert N. Webner.

Washington, DC 20036-5109, 1828 L St. N.W, Suite 1111. Phone: (202) 467-8800. Fax: (202) 467-8900. Mark J. Palchick, Robert E. Levine.

WGBH Educational Foundation, 125 Western Ave., Boston, MA, 02134-1098. Phone: (617) 300-2000. Fax: (617) 300-1014. Web Site: www.wgbh.org. Eric A. Brass.

WTTW Channel 11/Chicago, 5400 N. St. Louis Ave., Chicago, IL, 60625. Phone: (773) 583-5000. Fax: (773) 583-3046. Web Site: www.wttw.com.

Wagner, Michael Francis, Mass Media Bureau, 445 12th St. S.W., Rm. 2-A523, Washington, DC, 20554. Phone: (202) 418-2700. Fax: (202) 418-1410. Web Site: www.fcc.gov/mb/audio.

Waller Lansden Dortch & Davis, PLLC, 511 Union St., Suite 2700, Nashville, TN, 37219. Phone: (615) 244-6380. Fax: (615) 244-6804. Web Site: www.wallerlaw.com. Robb S. Harvey.

Jon M. Waxman Associates, 302 W. 12th St., New York, NY, 10014. Phone: (212) 929-2562. Fax: (212) 229-1625. E-mail: jon@jonwaxman.com

Law Offices of Edward L. Weidenfeld, 888 17th St. N.W., Suite 900, Washington, DC, 20006. Phone: (202) 785-2143. Fax: (202) 452-8938. E-mail: julie@weidenfeldlaw.com

Weil, Gotshal & Manges, L.L.P., 1300 I St. N.W., Suite 900, Washington, DC, 20005. Phone: (202) 682-7000. Fax: (202) 857-0940. E-mail: bruce.turnbull@weil.com Web Site: www.weil.com. Bruce H. Turnbull, ptnr.

Law Offices of Joel Weisman, P.C., 1901 Raymond Dr., Suite 6, Northbrook, IL, 60062. Phone: (847) 400-5900. Fax: (847) 400-5534. Web Site: weismanmedia.com. Joel Weisman, Scott A. Weisman.

Weissmann, Wolff, Bergman, Coleman, Grodin & Evall, 9665 Wilshire Blvd., Suite 900, Beverly Hills, CA, 90212-2345. Phone: (310) 858-7888. Fax: (310) 550-7191. E-mail: wwbcsh@wwllp.com Eric Weissman.

Westervelt, Johnson, Nicholl & Keller, LLC, Associated Bank Bldg., 411 Hamilton Blvd., 14th Fl., Peoria, IL, 61602. Phone: (309) 671-3550. Fax: (309) 671-3588. E-mail: westervelt@westerveltlaw.com

Wheeler Wolf Law Firm, Box 2056, 220 N. 4th. St., Bismarck, ND, 58502-2056. Phone: (701) 223-5300. Fax: (701) 223-5366. E-mail: jackmcdonald@wheelerwolf.com Jack McDonald.

Wildman, Harrold, Allen & Dixon, 225 W. Wacker Dr., Suite 3000, Chicago, IL, 60606. Phone: (312) 201-2000. Fax: (312) 201-2555. Web Site: www.wildmanharrold.com.

Wiley Rein LLP, 1776 K St. N.W., Washington, DC, 20006. Phone: (202) 719-7000. Fax: (202) 719/7049/719-7207. E-mail: dcorini@wrf.com Web Site: www.wileyrein.com.

Wilkinson Barker Knauer, L.L.P., 2300 N St. N.W., Suite 700, Washington, DC, 20037. Phone: (202) 383-4141. Fax: (202) 783-5851. Web Site: www.wbklaw.com.Partners: Jonathan V. Cohen, Timothy J. Cooney, Christine M. Crowe, J. Wade Lindsay, F. Thomas Moran, Lawrence J. Movshin, Kenneth D. Patrich, Robert D. Primosch, Kenneth E. Satten, William J. Sill, Paul J. Sinderbrand, David H. Solomon, L. Andrew Tollin, Michael D. Sullivan, Bryan N. Tramont, Kathryn A. Zachem, Paige K. Fronabarger, Craig E. Gilmore, Russell P. Hanser, Brian W. Higgins, William R. Layton, Robert G. Morse, Mary N. O'Connor. Associates: Patricia Chuh, Mia Hayes, David K. Judelsohn, Nguyen T. Vu

Willcox & Savage, P.C., One Commercial Place, Suite 1800, Norfolk, VA, 23510. Phone: (757) 628-5500. Fax: (757) 628-5566. Web Site: www.willcoxsavage.com. E-mail: mshearon@wilsav.com

Virginia Beach, VA 23466-1888, Box 61888, One Columbus Ctr, Suite 1010. Fax: (757) 628-5659. Jeffrey H. Gray.

William Morris Agency, 151 El Camino Dr., Beverly Hills, CA, 90212-2704. Phone: (310) 274-7451. Fax: (310) 859-4176.

Willkie Farr & Gallagher LLP, 1875 K St. N.W., Washington, DC, 20006. Phone: (202) 303-1000. Fax: (202) 303-2000. Web Site: www.willkie.com. Stephen Bell, Frank Buono, James Casserly, Jonathan Friedman, Michael Hammer, Karen Henein, Michael Jones, Thomas Jones, Sophie Keefer, Grace Koh, Angie Kronenberg, Jonathan Lechter, Jennifer McCarthy, John McGrew, Nirali Patel, Stephanie Podey, McLean Sieverding, Pamela Strauss, Megan Stull, Rayn Wallach, Theodore Whitehouse, Philip Verveer.

Wilmer Cutler Pickering Hale and Dorr LLP, 1875 Pennsylvania Ave., N.W., Washington, DC, 20006-3642. Phone: (202) 663-6000. Fax: (202) 663-6363. E-mail: william.richardson@wilmerhale.com Lynn Charytan, Jonathan Frankel, John Harwood, Samir Jain, William Lake, Jonathan Nuechterlein, William R. Richardson, John Rogovin, Catherine Ronis, Jack Goodman, David Mendel, Josh Roland, Will DeVries, Meredith Halama, Aaron Hurowitz, Daniel McCuaig, Nathan Mitchler, Jonathan Siegelbaum, Alison Southall, Kenny Wright, Heather Zachary, Dileep Srihari.

Winkler, Bevacqua & Simmons, P.C., 60 Park Pl., 19th Fl., Newark, NJ, 07102. Phone: (973) 676-1200. Fax: (973) 624-5980. Maury R. Winkler.

Winston & Strawn, 35 W. Wacker Dr., Chicago, IL, 60601-9703. Phone: (312) 558-5600. Fax: (312) 558-5700. E-mail: jneis@winston.com Web Site: www.winston.com.

Washington, DC 20005, 1400 L St. N.W, 8th Fl. Phone: (202) 371-5700. Fax: (202) 371-5950. Deborah C. Costlow.

Wolf, Block, Schorr, & Solis-Cohen, 250 Park Ave., New York, NY, 10177-0030. Phone: (212) 986-1116. Fax: (212) 986-0604. Web Site: www.wolfblock.com. Stuart A. Shorenstein, David E. Bronston.

Womble, Carlyle, Sandridge & Rice, PLLC, 1401 I St. N.W., 7th Fl., Washington, DC, 20005. Phone: (202) 467-6900. Fax: (202) 467-6910. E-mail: ledwardshall@wcsr.com Web Site: www.wcsr.com. Howard J. Barr, Patricia M. Chuh, John F. Garziglia, Peter Gutmann, Vicent A. Pepper, Gregg P Skall, Joan D. Stewart, Mark Blackwell.

Wood, Maines & Nolan, Chartered, 1827 Jefferson Pl. N.W., Washington, DC, 20036. Phone: (202) 293-5333. Fax: (202) 293-9811. E-mail: wmb@legalcompass.com Web Site: legalcompass.com. Barry D. Wood, Ronald D. Maines, Stuart W. Nolan, Jr.

Wright & Talisman, P.C., 1200 G St. N.W., Suite 600, Washington, DC, 20005. Phone: (202) 393-1200. Fax: (202) 393-1240. E-mail: benna@wrightlaw.com Web Site: www.wright.com.

Young, Williams, Kirk & Stone PC, First Tennessee Plaza, Suite 2021, Box 550, Knoxville, TN, 37901-0550. Phone: (865) 637-1440. Fax: (865) 546-9808. E-mail: bob@tn-attorneys.com Web Site: www.lawyers.com/mclaw. Robert S. Stone.

Young, Clement & Rivers L.L.P., 5000 Thurmond Mall Dr., Suite 320, Columbia, SC, 29201. Phone: (843) 577-4000. Phone: (803) 254-2238. Fax: (843) 724-6600. E-mail: email@ycrlaw.com Web Site: www.ycrlaw.com.

Talent Agents and Managers

Abrams Artists Agency, 9200 Sunset Blvd., 11th Fl., Los Angeles, CA, 90069. Phone: (310) 859-0625. Fax: (310) 276-6193. E-mail: harry.abrams@abramsart.com

Harry Abrams, pres.

New York, NY 10001. Abrams Artists Agency, 275 7th Ave, 26th Fl. Phone: (646) 486-4600. Fax: (646) 486-0100. Neal Altman, sr VP.

Performing artists reps-talent agency.

N S Bienstock Inc., 1740 Broadway, 24th Fl., New York, NY, 10019. Phone: (212) 765-3040. Fax: (212) 757-6411. E-mail: nsb@nsbienstock.com Web Site: www.nsbienstock.com.

News & syndication talent specialists—loc & net—on & off camera. Packager of talk & reality progmg—MOWs.

Eatman Media Services Inc., 5901 N. Cicero Ave., Suite 307, Chicago, IL, 60646. Phone: (773) 777-5463. Fax: (773) 777-7106. E-mail: emstalent@aol.com

Pacific Palisades, CA 90272, Box 853. Phone: (310) 459-3728. Robert Eatman, pres.

Bedford, NY 10506, Box 102. Phone: (914) 234-4748. Ross Eatman, Esq.

Representation of TV newspersons, TV personalities, & radio talent in job placement & contract negotiation.

Ephraim & Associates, P.C., 108 W. Grand Ave., Chicago, IL, 60610-4206. Phone: (312) 321-9700. Fax: (312) 321-3655. E-mail: eliot@ephraim.com Web Site: www.ephraim.com.

Donald M. Ephraim, pres; Joseph F. Coyne, VP; Eliot S. Ephraim, VP; David M. Ephraim, VP.

Talent representation, including contract negotiation, legal & career consulation, tax, estate & pension planning. Eliot Ephraim, atty/agent.

Ken Fishkin & Associates, 50 Milk St., 20th Fl., Boston, MA, 02109-5002. Phone: (617) 423-5800. Fax: (617) 426-2674. E-mail: cindy@kfishkin.com

Kenneth R. Fishkin, pres.

Contract negotiation & job placement, TV & radio.

Goldstein Management Group, Inc., 1601 N. Sepulveda Blvd., Suite 357, Manhattan Beach, CA, 90266. Phone: (310) 545-8530. Fax: (310) 943-1569. E-mail: glenn802@aol.com

Glenn A. Goldstein, pres.

Full-svc representation of bcst talent & bcst journalists.

Reece Halsey Agency, 8733 Sunset Blvd., Suite 101, West Hollywood, CA, 90069. Phone: (310) 652-2409. Fax: (310) 652-7595.

Tiburon, CA 99920. Reece Halsey Agency, Box 704, 98 Main St. Phone: (415) 789-9191. Kimberly Cameron.

Shirley Hamilton Inc., 333 E. Ontario, Chicago, IL, 60611. Phone: (312) 787-4700. Fax: (312) 787-8456. E-mail: 1shamilton@att.net Web Site: www.shirleyhamilton.com.

Shirley Hamilton, pres; Lynne Hamilton, office mgr.

Representing talent for TV, PRINT, RADIO, ON CAMERA, FILM, LIVE, THEATRICAL, VOICEOVER, INDUSTRIALS. Audition facilities for OnCamera, Digital voiceover, Print.

The Image Generators, 18156 Darnell Dr., Olney, MD, 20832. Phone: (301) 924-4327. Fax: (301) 570-8916. E-mail: mweiner@imagegenerators.com Web Site: www.imagegenerators.com.

Michael J. Weiner, pres; Valle Bonhag, dir admin.

Voiceover talent, audition svc, online demos of pro voices.

International Creative Management Inc., 40 W. 57th St., 18th Fl., New York, NY, 10019. Phone: (212) 556-5600. Fax: (212) 556-5665. Web Site: www.ICMtalent.com.

Jeff Berg, pres/CEO.

London W1R 1RB, 76 Oxford St.-OAX. Phone: (011) 44-1-81-743-0558. Fax: (011) 44-1-81-743-6598.

Beverly Hills, CA 90211, 8942 Wilshire Blvd. Phone: (310) 550-4000. Fax: (310) 550-4100.

Miller Broadcast Management Inc., 616 W. Fulton St., Suite 516, Chicago, IL, 60661. Phone: (312) 454-1111. Fax: (312) 454-0044. E-mail: info@millerbroadcast.com Web Site: www.millerbroadcast.com.

Lisa Miller, pres.

Representing radio & TV personalities.

William Morris Agency Inc., 1325 Ave. of the Americas, New York, NY, 10019. Phone: (212) 586-5100. Fax: (212) 246-3583. Web Site: www.wma.com.

Wayne Kabak, COO; Cara Stein, COO; Jim Griffin, VP.

London W1V 5DG, 31/32 Soho Sq. Phone: (71) 434-2191. Steve Kenis, dir.

Beverly Hills, CA 90212, 151 El Camino Dr. Phone: (310) 859-4000. Fax: (310) 859-4440. E-mail: bgoodman@wma.com. Brad Goodman.

Nashville, TN 37203, 2100 West End Ave. Phone: (615) 385-0310. Rick Shipp.

Paradigm, 360 N. Crescent Dr., Beverly Hills, CA, 90210. Phone: (310) 288-8000. Fax: (310) 288-2000. E-mail: info@paradigm-agency.com

Sam Gores, pres.

Monterey, CA 93940, 509 Hartnell St. Phone: (615) 251-4400. Fax: (615) 251-44001.

New York, NY 10010, 360 Park Ave. S., 16th Fl.

Nashville, TN 93940, 124 12th Ave. S., Suite 410. Phone: (831) 375-4889. Fax: (831) 375-2623.

Rebel Entertainment, 5700 Wilshire Blvd., Ste 456, Los Angeles, CA, 90038. Phone: (323) 935-1700. Fax: (323) 932-9901. E-mail: rlawre8075@aol.com Web Site: www.arltalent.com.

Richard Lawrence, pres; Debra Goldfarb, VP.

Talent placement & TV show packaging.

Screen Children's Casting, 4000 Riverside Dr., Suite A, Burbank, CA, 91505. Phone: (818) 846-4300. Fax: (818) 846-3745.

Irene B. Gallagher, casting dir & owner.

Full-svc children's casting representing babies (especially twins), children & teenagers up to age 18, extra work.

Burt Shapiro Management, 2147 N. Beachwood Dr., Los Angeles, CA, 90068. Phone: (323) 469-9452. Fax: (603) 710-8019. E-mail: burtjay@mail.com Web Site: www.burtshapiro.com.

Burt Shapiro, pres.

Represents on-air talent including anchors, reporters, hosts, & sports anchor/reporters, as well as producers & news directors.

Barry Skidelsky, Esq., 185 East 85th St., 23D, New York, NY, 10028. Phone: (212) 832-4800.

Barry Skidelsky, Esq., pres; Eliot S. Ephraim, principal atty.

Personal mgmt & representation. Contract negotiations, counsel, etc.

The Voicecaster, 1832 W. Burbank Blvd., Burbank, CA, 91506. Phone: (818) 841-5300. Fax: (818) 841-2085. E-mail: casting@voicecaster.com Web Site: www.voicecaster.com.

Huck Liggett, CEO.

Voice casting for comls, films, animation, theme parks, audiovisual projects, etc.

Weisman Media, Joel Weisman, P.C., 1901 Raymond Dr., North Brook, IL, 60062. Phone: (847) 400-5900. Fax: (847) 400-5534. E-mail: joel@weismanmedia.com Web Site: www.weismanmedia.com.

Joel Weisman, pres.

Bcst contract drafting & negotiation; career & performance counseling; job search & market placement for on-air producers.

Employment and Executive Search Services

Bishop Partners, 708 Third Ave., Suite 2200, New York, NY, 10017. Phone: (212) 986-3419. Fax: (212) 986-3350. E-mail: info@bishoppartners.com Web Site: www.bishop partners.com.

Susan K. Bishop, pres/CEO.

A retained exec search firm specializing in cable, bcst, telecommunications, wireless, entertainment, publishing & multimedia.

California Broadcasters Association, 915 L St., Sacramento, CA, 95814. Phone: (916) 444-2237. Fax: (916) 444-2043. E-mail: cbaberry@aol.com Web Site: www.cabroadcasters.org.

Stan Statham, pres/CEO.

Lobbyist for coml radio & TV for the state of California & other legal issues.

Eatman Media Services Inc., 5901 N. Cicero Ave., Suite 307, Chicago, IL, 60646. Phone: (773) 777-5463. Fax: (773) 777-7106. E-mail: emstalent@aol.com

Pacific Palisades, CA 90272, Box 853. Phone: (310) 459-3728. Robert Eatman, pres.

Bedford, NY 10506, Box 102. Phone: (914) 234-4748. Ross Eatman, Esq.

Representation of TV newspersons, TV personalities, & radio talent in job placement & contract negotiation.

Entertainment Employment Journal (T.M.), 5632 Van Nuys Blvd., Suite 320, Van Nuys, CA, 91401. Phone: (800) 335-4335. E-mail: sales@eej.com Web Site: www.eej.com.

Bimonthly magazine providing career information & job listings with major & independent motion picture, TV & cable companies.

The Howard-Sloan-Koller Group, 300 E. 42nd St., New York, NY, 10017. Phone: (212) 661-5250. Fax: (212) 490-5322. E-mail: ekoller@hsksearch.com Web Site: www.hsksearch.com.

Edward R. Koller Jr., pres/CEO.

Beverly Hills, CA 90212, 9701 Wilshire Blvd. Phone: (310) 601-7114. Fax: (310) 601-7110. Edward R. Koller Jr., president.

Exec search & consulting in the cable, digital, entertainment & publishing industries.

JOBPHONE, Box 5048, Newport Beach, CA, 92662. Phone: (949) 721-9280. Fax: (949) 721-8478. E-mail: jobphone@aol.com Web Site: www.infoguru.com.

Keith Mueller, pres.

Natl TV/radio employment hotline.To hear job openings nationwide: (900) 726-5627 (JOBS), $1.99 per minute.

Keystone America, 38 Thomas St., Exeter Plaza, Exeter, PA, 18643. Phone: (570) 655-7143. Fax: (570) 654-5765. E-mail: usmail@keystoneamerica.com Web Site: keystoneamerica.com.

Alan "Al" Kornish, VP opns.

Ntl Employment svc for bcst employers & candidates, placement of engrs & technicians. Serving all USA states. See our website: keystoneamerica.com.

Korn/Ferry International, 1900 Avenue of the Stars, Suite 2600, Los Angeles, CA, 90067. Phone: (310) 552-1834. Web Site: www.kornferry.com.

William D. Simon, mgng dir entertainment; Roysi Erbes, sr assoc.

Worldwide sr level mgmt exec search firm servicing all sectors of the entertainment industry.

Lipson & Co., 1900 Ave. of the Stars, Suite 2810, Los Angeles, CA, 90067. Phone: (310) 277-4646. Fax: (310) 277-8585. E-mail: inquiries@lipsonco.com Web Site: www.lipsonco.com.

Howard R. Lipson, pres; Harriet Lipson, sr VP.

Specialists in international & domestic bcstg (TV & radio), cable & entertainment, & related financial, professional audio/video & electronic recruiting. Svcs also for TV & film production, merchandising, licensing, & computers.

Brad Marks International, 15233 Ventura Blvd., PH 16, Sherman Oaks, CA, 91403. Phone: (818) 382-6300. Fax: (818) 386-0050. E-mail: bodysnatcher@bradmarks.com Web Site: www.bradmarks.com.

Brad Marks, chmn/CEO.

Exec search at sr mgmt levels for communications, bcst, cable & multimedia companies. Areas include TV & film production, progmg, sls, mktg, news, gen mgmt, financial svcs, postproduction & adv/promotion.

Maslow Media Group Inc., 2134 Wisconsin Ave. N.W., Washington, DC, 20007. Phone: (202) 965-1100. Fax: (202) 965-6171. E-mail: lmaslow@maslowmedia.com Web Site: www.maslowmedia.com.

Linda Maslow, CEO; Carl Neubecker, VP.

Freelance & fulltime staffing, crewing & payroll svcs for bcst, corporate, & federal govt. Find a job @ www.tvgigsonline.com

Media Management Resources Inc., 6890 S. Tuscon Way, Englewood, CO, 80112. Phone: (303) 290-9800. Fax: (303) 290-9596. E-mail: bwein@mediamanagement.com

Michael S. Wein, pres; David Reiber, gen mgr.

Full-svc consulting practice providing business support & technology svcs to select media & technology companies.

Media Staffing Network, 150 E. Huron, Suite 900, Chicago, IL, 60611. Phone: (312) 944-9194. Fax: (312) 944-9195. E-mail: laurie@mediastaffingnetwork.com Web Site: www.mediastaffingnetwork.com.

Laurie Kahn, pres/CEO.

Media Staffing Network the only full-service staffing company that specializes in media adv sls & associated departments, offering both temporary & full-time positions nationwide. Clients include radio & TV stns, rep firms, Internet, cable systems, networks, syndication, magazines & adv agencies. Openings range from entry-level support to sr mgmt positions in sls, prom, buying, planning, traf, continuity, customer svc & rsch.

MediaLine, Box 51909, Pacific Grove, CA, 93950. Phone: (800) 237-8073. E-mail: medialine@medialine.com Web Site: www.medialine.com.

Adrienne Laurent, pres; Mark Shilstone, owner & mgr.

Job listings for TV news, production & promotions; streaming video of resume tapes on the Internet; daily eletronic newsletter.

Miller Broadcast Management Inc., 616 W. Fulton St., Suite 516, Chicago, IL, 60661. Phone: (312) 454-1111. Fax: (312) 454-0044. E-mail: info@millerbroadcast.com Web Site: www.millerbroadcasts.com.

Lisa Miller, pres; Matt Miller, VP.

Representing radio personalities.

Promotion Recruiters Inc., 16 Drake Rd., Scarsdale, NY, 10583. Phone: (914) 723-9043.

Howard Burkat, pres.

Search for stn, net & cable promotion, mktg execs & producers exclusively.

RTNDA Job Services, 1600 K St. N.W., Suite 700, Washington, DC, 20006. Phone: (202) 659-6510. Fax: (202) 223-4007. E-mail: rtnda@rtnda.org Web Site: www.rtnda.org.

Barbara Cochran, pres; Sarah Stump, editor.

Job bulletin.

Search Source Inc., Box 1161, Granite City, IL, 62040-1161. Phone: (618) 931-6060. Fax: (618) 876-6071. E-mail: search@norcom2000.com

James R. McKechan, pres.

Search & recruitment of bcst professionals.

Barry Skidelsky, Esq., 185 E. 85th St., 23 D, New York, NY, 10028. Phone: (212) 832-4800. E-mail: bskidelsky@mindspring.com

Barry Skidelsky, atty/consultant.

Employment contact; EEO compliance, training, audits, litigation & arbitration.

R.A. Stone & Associates, 5495 Belt Line Rd., Suite 103, Dallas, TX, 75254-7671. Phone: (972) 233-0483. Fax: (972) 991-4995. E-mail: stonesearch@aol.com

Robert Stone, pres.

Retainer based exec search svcs for the domestic & international TV, radio, cable, multimedia, production & related communications/entertainment industries.

Joe Sullivan Executive Search & Recruiting, Box 178, 1202 Lexington Ave., New York, NY, 10028. Phone: (212) 734-7890. Fax: (212) 734-0631. E-mail: jsa6@aol.com Web Site: www.joesullivanexecutivesearch.com.

Joseph J. Sullivan Jr., pres; Shane P. Sullivan, VP; Barbara Sullivan, CFO.

Retain exec recruitment in media & academia for positions in middle & sr mgmt with incomes in excess of $120,000 per year.

Ron Sunshine Associates, 2404 Clear Field Dr., Plano, TX, 75025. Phone: (972) 618-3670. Phone: (214) 509-3778. Fax: (972) 599-9583. E-mail: Ron@Ronsunshineassociates.com

Ron Sunshine, pres; Barbara Blake, VP.

Radio, TV & cable middle & upper mgmt.

Warren & Morris Ltd., 2190 Carmel Valley Rd., Del Mar, CA, 92014. Phone: (858) 481-3388. Fax: (858) 481-6221. E-mail: swarren@warrenmorrismltd.com Web Site: www.warrenmorrisltd.com.

Charles Morris, ptnr; Lynn Cason, ptnr; Chip Cossitt, ptnr; Amy McCoy, assoc; Scott Warren, ptnr.

Portsmouth, NH 03801, 132 Chapel St. Phone: (603) 431-7929. Fax: (603) 431-3460. Scott Warren. sr ptn; Arron Chaffee, ptn.

Natl & international exec/mgmt-level recruitment svcs in the cable TV, wireless communications & digital media industries.

Youngs, Walker & Co., 1605 Colonial Pkwy., Inverness, IL, 60067. Phone: (847) 991-6900. Fax: (847) 934-6607. E-mail: info@youngswalker.com Web Site: www.youngswalker.com.

Carl Youngs, pres.

Exec recruitment on a retained basis for TV & radio stn mgmt levels & corporate positions.

Professional Cards
Engineering & Technical Consultants

Section H

Associations, Events, Education, and Awards

Associations

Events

Education

Awards

Major National Associations

Academy of Television Arts & Sciences

Headquarters: 5220 Lankershim Blvd., North Hollywood, CA 91601-3109. (818) 754-2800. FAX: (818) 761-2827. Web Site: http://www.emmys.org.

Executives: Dick Askin, chmn & CEO; Steve Mosko, Academy Foundation chmn; Alan Perris, COO; Sheila Manning, sec; Donna Kanter, treas.

Mission Statement: The mission of the Academy is to promote creativity, diversity, innovation and excellence through recognition, education and leadership in the advancement of the telecommunications arts and sciences.

Cabletelevision Advertising Bureau Inc. (CAB)

Headquarters: 830 3rd Ave., New York, NY 10022. (212) 508-1200. FAX: (212) 832-3268. Web Site: www.onetvworld.org.

Executives: Sean Cunningham, pres/CEO; Jammie Spears, CPA, CFO; Charles (Chuck) Thompson, sr VP, sls & mktg; Danielle DeLauro, VP, sls & mktg; Dave Leitner, VP, sls & mktg; Christopher Jones, VP, corporate communications; Ira Sussman, VP, research & insights for CAB; Ken Anderson, creative dir.

Mission Statement: The CAB is dedicated to providing advertisers & agencies with the most current, complete actionable cable TV media insights at the national DMA & local levels.

Board of Directors: Jack Olson, VP media dev, Adelphia Communications; Louis Carr, exec VP bcst adv sls, BET; Joseph Abruzzese, pres ad sls, Discovery Networks; Robert Bakish, exec VP opns, Viacom, Inc.; Lawrence Fischer, pres media sls, Time Warner Cable Media Sales; Phil Kent, chmn/CEO, Turner Broadcasting Systems Inc.; Paul Iaffaldano, sr VP adv sls, The Weather Channel; Lynn Picard, exec VP sls, Lifetime Entertainment; Lou LaTorre, pres ad sls/corporate, Fox Channel Group; Sonja Farrand, groups VP/adv opns, On Media; David Cassaro, pres sls, Comcast Network Sales; David Kline, pres/COO, Rainbow Advertising; Edward R. Erhardt, pres ad sls, ESPN; William Farina, VP ad sls, Cox Communications; Jim Heneghan, corporate VP/adv, Charter Communications; Melvin Berning, exec VP adv sls, A&E Networks; Steven Gigliotti, exec VP ad sls & emerging media, Scripps Networks; Michael Bowker, VP adv, Cable ONE; Charlie Thurston, pres, Comcast Spotlight; Jeffrey Lucas, sr VP sls, Comedy Central.

Media Rating Council

Headquarters: 370 Lexington Ave., Suite 902, New York, NY 10017. (212) 972-0300. FAX: (212) 972-2786. E-mail: staff@mediarating.org.

Executives: George Ivie, exec dir/CEO MRC; Joan Gillman, dir; Barry Fischer, dir.

Mission: Secure for the media industry and related user audience measurement that is valid, reliable and effective.

National Association of Broadcasters (NAB)

Headquarters: 1771 N St. N.W., Washington, DC 20036. Phone: (202) 429-5300. FAX: (202) 429-4199. E-mail: nab@nab.org. Web Site: www.nab.org.

NAB Officers and Staff

NAB Executive Committee: John L. Sander (chair), sr advisor Belo Corp., Scottsdale, AZ; Bruce T. Reese (past chair), pres/CEO, Bonneville International Corp., Salt Lake City, UT; David K. Rehr, pres & CEO, NAB exec Offices, Washington, DC; David J. Barrett, pres/CEO Hearst-Argle Television, Inc., New York, NY; Alan W. Frank, pres/CEO, Post-Newsweek Stations, Inc., Detroit, MI; Lynn Beall, exec VP Gannett Broadcasting, St. Louis, MO; Jay Ireland, pres NBC Universal Television Stations & Network Operations, New York, NY; Steven W. Newberry, pres/CEO Commonwealth Broadcasting Corp., Glasgow, KY; Charles M. Warfield, pres/COO ICBC Broadcast Holdings, Inc., New York, NY.

NAB Radio Board of Directors: W. Russell Withers Jr. (chair), WMIX AM-FM Withers Broadcasting Companies, Mount Vernon, IL; Steven W. Newberry (first vice-chmn), Commonwealth Broadcasting Corp., Glasgow, KY; Charles M. Warfield (second vice-chair) ICBC Broadcast Holdings, Inc., New York, NY.

NAB District Representatives: Howard B. Anderson, KHWY, Inc., district: 24 (S.CA-GU-HI) Los Angeles, CA; John W. Barger, VictoriaRadio Works, Ltd., KVIC/KEPG/KITE/KRNX/KNAL, district: 18 (SO. TX), San Antonia, TX; Joseph M. Bilotta, Buckley Radio, district: 2 (NY-NJ), Greenwich, CT; Robert L. Bundgaard, KLKS, Lakes Broadcasting, district: 21 (MN-SD-ND), Breezy Point, MN; Bobby Caldwell, East Arkansas Broadcasters, district: 15 (TN-AR), Wynne, AR; Rodney P. Chambers, Sierra Broadcasting Corp., district: 23 (N.CA-AK) Susanville, CA; Paul G. Gardner, Elko Broadcasters Co., district: 22 (AZ-NV-NM-UT), Elko, NV; Bruce Goldsen, WKHM AM/FM, Jackson Radio Works, Inc., district: 13 (MI) Jackson, MI; Randy D. Gravley, WLJA-FM/WPGY-AM, Tri-State Communications, district: 9 (GA-AL), Jasper, GA; Bill Hendrich, Cox Radio, district: 7 (FL-PR-VI) Orlando, FL; James E. Janes, Bick Broadcasting/KHMO/KICK, district: 12 (MO-KS), Hannibal, MO; Rolland C. Johnson, Three Eagles Communications, district: 16 (CO-NE), Larkspur, CO; Daniel Savadove, Main Line Broadcasting, LLC, district: 3 (PA) West Conshohocken, PA; Mathew Mnich, North American Broadcasting Co. Inc., district: 11 (OH) Columbus, OH; Joel Oxley, Bonneville Washington Radio Group, district: 14 (DE-DC-MD-VA), Washington, DC; Mary Quass, NRG Media, LLC, district: 14 (IA-WI) Cedar Rapids, IA; Peter H. Smyth, Greater Media Boston, district: 1 (New England), Braintree, MA; Alex Snipe Jr., Glory Communications, Inc., district: 6 (NC-SC) Columbia, SC; Peter Benedetti, New Northwest Broadcasters, district: 25 (OR-WA), Seattle, WA; Kevin S. Perry, Perry Publishing & Broadcasting Co., district: 19 (OK-N. TX), Oklahoma, OK; Richard Cummings, Emmis Communications Corp., district: 10 (IN), Burbank, CA; ronald J. Davis, Butte Broadcasting, district: 20 (MT-ID-WY), Butte, CA; Bob Holladay, Holladay Broadcasting, district: 8 (LA-MS), Monroe, LA.

NAB Designated Board Seats: Edward K. Christian, Sage Communications, Grosse Pointe Farms, MI; Lew Dickey Jr., Cumulus Media, Inc., Atlanta, GA; Susan Davenport Austin, Sheridan Broadcasting Corp.; Alfred Liggins, Radio One Inc., Lanham, MD; Mark P. Mays, Clear Channel Worldwide, San Antonio, TX; Susan K. Patrick, Legend Communications, Ellicott City, MD; Caroline Beasley, Beasley Broadcast Group, Naples, FL; Amador Bustos, Bustos Media, LLC, Sacramento, CA; David J. Field, Entercom Communications Corp., Bala Cynwyd, PA.

NAB TV Board of Directors: Alan W. Frank (chair), pres & CEO Post-NewsweekStation, Inc., Detroit, MI; Lynn Beal, (first vice-chair), exec VP, Garnnett Broadcasting, St. Louis, MO; David J. Barrett (second vice-chair), pres/CEO Hearst-Argle Television, Inc., New York, NY.

NAB Elected Representatives: Elizabeth Murphy Burns, pres, Morgan Murphy Media; Jim Conschafter, sr VP/bcst stns, Media General Broadcast Croup, Richmond, VA; Paul Karpowicz, pres Meredith Copr. Broadcasting Group, rocky Hill, CT; John C. Kueneke, pres New-Press & Gazette Broadcasting, St. Louis, MO; Paul H. McTear, pres/CEO Raycom Media Inc., Montgomery, AL; Ralph M. Oakley, VP/COO QNI Broadcast Group, Quincy, IL; William B. Peterson, sr VP/Television Group, The E.W. Scripps Co., Cincinnati, OH; John E. Reardon, pres/CEO Tribune Broadcasting Co., Chicago, IL; Doreen Wade, pres Freedom Broadcasting, Inc., West Palm Beach, FL;orp., Portsmouth, VA; Doreen Wade, pres, Freedom Broadcasting, Inc., West Palm Beach, FL; K. James Yager, CEO Barrington Broadcasting Co., LLC, Hoffman Estates, IL.

NAB Designated Board Seats: Andrew S. Fisher, pres, Cox Television, Atlanta, GA; Michael J. Fiorile (vice-chmn) & CEO, Dispatch Broadcast Group, Columbus, OH; Lyle Banks, pres/CEO Banks Broadcasting, Inc., Winnetka, IL; Robert G. Lee, pres/gen mgr WDBJ Television Inc., Roanoke, VA; Scott Blumenthal exec VP/Television, LIN Television Corp., Providence, RI; David Woods, pres, WCOV-TV, Montgomery, AL.

NAB Network Representatives: C. Douglas Kranwinkle, exec VP/Law, Univision Communications Inc., Los Angeles, CA; Brandon Burgess, pres/CEO ION Media Networks, West Palm Beach, FL; Preston R. Padden, exec VP Worldwide Govt Relations, The Walt Disney Company, Washington, DC; Jay Iteland, pres NBC Universal Television Stations & Network Operations.

National Association of Farm Broadcasters

Headquarters: Box 500, Platte City, MO 64079. (816) 431-4032. (800) 294-6232. E-mail: info@nafb.com. Web Site: www.nafb.com.

Executive Director: Bill O'Neill; Jeremy Provenmire, members svcs mge; Stacia Cudd, editor; Susan Tally, office mgr.

National Association of Television Program Executives (NATPE)

Headquarters: 5757 Wilshire Blvd., Penthouse 10, Los Angeles, CA 90036-3681. (310) 453-4440. FAX: (310) 453-5258. Web Site: www.natpe.org.

Mission Statement: A global, non-profit organization dedicated to the creation, development & distribution of televised progmg in all forms across all mature and emerging media platforms.

Staff & Consultants: Rick Feldman, pres/CEO; Nick Orfanopoulos, sr VP conferences, opns/sls; Beth Braen, sr VP mktg; Jon Dobkin, CFO; Lew Klein, pres NATPE Educational Foundation.

Publications: *The NATPE Monthly*; *Programmer's Guide*; *Station Listening Guide*; *Reps, Groups, Distributors, Networks, Ad Agencies, Telco-DBS Guide, NATPE News & NATPE Daily Lead.*

International Representatives

European Director: Pam Smithard, mgng dir, 452 Oakleigh Rd. N., London, England, UK N2O 0RZ +(44) 20 8449 9333. FAX: (44 20) 8368-3824.

Korean Representative: Ko Mayoung, Act International Inc., +(813) 5770 5581. FAX: +(813) 5770 5583.

Pacific Rim Representative: Nick McMahon, Crawford Productions Group, 259 Middleborough Rd., Box Hill 3128, Victoria, Australia. +(61) 398 95 2211. FAX: +(61) 398 90 5732.

Japanese Representative: Akio Taniguchi, Act International, Inc., 2-6-2 Shinkawa, Ishibashi Bldg, #502, Chuo-Ku, Tokyo 104, Japan +(813) 5770 5581. FAX: +(813) 5770 5583.

Indian Representative: Bhuvan Lall, Empire Entertainment Private Ltd., B 157 Shivalik, New Delhi 110017, India +(91) 11 2669 3472. FAX: +(91) 11 2699 1078.

Chinese Consultant: Grace Ip, Neutral Distributing Services Inc., 347 W. Arbor Vitae St., Inglewood, CA 90301 (310) 330-7888. FAX: (310) 330-7889.

National Cable and Telecommunications Association (NCTA)

Headquarters: 25 Massachusetts Ave. N.W., Washington, DC 20036. (202) 222-2300. E-mail: webmaster@ncta.com. Web Site: www.ncta.com.

NCTA Officer

Kyle McSlarrow, pres/CEO.

NCTA Board of Directors

Nicholas Davatzes, co-chmn of bd; Robert Mironr, co-chmn of bd.

National Cable Television Cooperative Inc.

Headquarters: 11200 Corporate Ave., Lenexa, KS 66219-3292. (913) 599-5900. Web Site: www.cabletvcoop.org.

Mission: NCTC is a not-for-profit member-operating purchasing organization dedicated to reducing operating costs of its member cable companies. The Co-op negotiates and administers master affiliation agreements with cable TV progmg networks, cable hardware & equipment manufacturers & other svc providers on behalf of its member companies.

NCTC Executive: Jeff Abbas, pres/CEO

Radio Advertising Bureau

Headquarters: 125 W. 55th St., 21st Fl., New York, NY 10019. (800) 252-7234. Web Site: www.rab.com.

Services and Administration Center: 1320 Greenway Dr., Suite 500, Irving, TX 75038-2510. (800) 232-3131.

Chicago Office: 30 South Wacker Dr., 22nd Fl., Chicago, IL 60606. (312) 466-5639.

Los Angeles Office: 21900 Burbank Blvd., Suite 3038, Woodland Hills, CA 91367. (323) 904-4357.

Detroit Office: 28175 Haggerty Rd., Novi, MN. (248) 994-7678.

Officers

Joe Bilotta (chair); Herb McCord (chair); Peter Smyth (chair); Jeff Haley, pres/CEO; Scott Herman, (finance chair); Rick Cummings, exec committee; Jerry Kersting, exec committee; Kraig Kitchin, corporate sec/exec committee; Gunther Meisse, exec committee

Officers of the Corporation

Gary Fries, pres/CEO; Van Allen, ex VP/CFO; Mary Bennett, exec VP mktg; George Hyde, exec VP/training; Mike Mahone, exec VP/svcs; Ron Ruth, exec VP/stns; David Casper, sr VP/svcs; Mary Malone, sr VP svcs.

Board of Directors: Joe Bilotta, CEO, Buckley Broadcasting; Peter Smyth (RAB vice chair) pres/CEO, Greater Media; Gary Fries, pres/CEO Radio Advertising Bureau, Inc.; Weezie Kramer (chair RAB finance committee), regl VP Entercom; Bruce Beasley, pres, Beasley Broadcast Group; Charles (Chuck) Bortnick, co-COO, Westwood One; David Benjamin, CEO, Triad Bcstg; Don Benson, pres/radio div, Jefferson-Pilot Communications; Michael Carter, pres, Carter Broadcast Group; Rick Cummings, pres/radio, Emmis Communications; Lee Davis (RAB exec committee), pres/gen mgr, Cub Radio, Inc.; Tom Dobrez, pres, State Nets Radio; John Douglas, pres, AIM Broadcasting; Glenn Cherry, pres/CEO, Tama Broadcasting Inc.; David Field, pres/CEO, Entercom Communications Corp.; Carl Gardner, pres, Radio Journal Company; Steffen Mueller, managing dir, MOIR RundfunkGmbh; Laura Hagan, pres, Univision Radio National Sales; David Kantor, CEO, Reach Media; Marc Guild, pres/mktg, Interep Radio Sales; John Hare, pres, ABC Radio; Scott Knight, pres/CEO, Knight Quality Stations; Jerry Lee, pres, WBEB Radio; Kraig Kitchin (RAB exec/finance committee), pres, Premiere Radio Networks, Inc.; Gunther Meisse (RAB exec committee), pres, Communications Corp.; Warren Lada, sr VP/opns, Saga Communications; Marc Morgan, exec VP/COO, Cox Radio; Stuart (Stu) Olds (RAB exec committee), CEO, Katz Media Group, Inc.; Michael Osterhout, COO, Morris Radio LLC; Norman D. Rau, pres, Sandusky Radio; Bruce Reese, pres/CEO, Bonneville International; Jerry Kersting, sr VO/CFC, Clearchannel Communications; Herb McCord (RAB Corp. sec), Pres/CEO, Granum; Ginny Morris, pres, Radio Hubbard Broadcasting; John Pinch, COO, Cumulus Media; Charlie Rahilly, exec BP/opns, Clear Channel Communications; Steve Newberg, pres, Commonwealth Broadcasting; James Robinson, pres, ABC Radio Networks; Dean Sorenson, CRMC. Pres, Sorenson Broadcasting; Tony Renda, pres, Renda Broadcasting Corp.; Pierre (Pepe) Sutton (chmn), Inner City Broadcasting Corp.; Bayard (Bud) Walters (RAB exec committee), pres/owner, Cromwell Group, Inc.; Samuel (Skip) Weller, pres/COO, Radio Division NextMedia Group, Inc.; E.J. "Jay" Williams, Jr., pres, American Urban Radio Networks; Warren Lada, sr VP/opns, Saga Communications; George Pine, pres/COO, Interep; Tony Renda, pres, Renda Broadcasting Corp.; James Robinson, pres, ABC Radio Networks; Art Rowbotham, pres, Hall Communications Inc.; William L. Stakelin, pres/COO, Regent Communication; Gary Stone, sr VP/COO, Univision Radio; Roger Utnehmer, pres, Nicolet Broadcasting (WBDK Radio); Nancy Vaeth-Dubroff, pres/COO, Susquehanna Radio Corp; Charles Warfield, pres/COO, ICBC Broadcast Holdings, Inc.

Executive Committee: Lee Davis; Kraig Kitchin; Gunther Meisse; Stuart (Stu) Olds; Bayard "Bud" Walters.

Past Chairmen Advisory Committee: Richard Buckley; Arthur Carlson; Richard Chalin; David Crowl; John Dillie; Paul Fiddick; Skip Finley; David Kennedy; Glenn Mahone Esq; Jeffrey Smulyan.

Finance Committee: Weezie Kramer; Herb McCord.

Radio-Television News Directors Association Foundation

Headquarters: 1600 K St. N.W. Suite 700, Washington, DC 20006-2838. (202) 659-6510. FAX: (202) 223-4007. E-mail: rtnda@rtnda.org. Web Site: www.rtnda.org

Officers: Barbara Cochran, pres RTNDA, Washington, DC; Angie Kucharski (past-chairwomen), WBZ-TV/WSBK-TV Boston, MA; Ed Espositi (chmn-elect), WAKR/WONE/WQMX Akron, OH; Loren Tobia, treas AccuWeather, Syracuse, NY.

Directors-at-Large: Janice Gin, KTVU-TV Oakland, CA. E-mail: janice.gin@ktvu.com; Ed Tobias, Associated Press, Washington. E-mail: ed_tobias@ap.org; David Louie, KGO-TV San Francisco, CA. E-mail: david@davidlouie.com; Brian Trauring, WTVG-TV, Toledo, OH. E-mail: brian.trauning@abc.com.

Active Members: About 1,100 radio & TV news directors.

Television Bureau of Advertising (TVB)

Headquarters: 3 E. 54th St., New York, NY 10022-3108. (212) 486-1111. FAX: (212) 935-5631. E-mail: info@tvb.org. Web Site: www.tvb.org.

Staff Directory

Christopher Rohrs, pres; Abby Auerbach, exec VP; Carrie Hart, VP/rsch; Zyanya Chan, rsch intern; B.J. Park, rsch dir; Susan Cucinello, sr VP/rsch; Jack Poor, VP/mktg; Blanca McKenna, rsch dir; Pat Yancovitz, rsch dir; Hope Etheridge, VP/finance and admin; Janice Garjian, VP member svcs; Peter Schmid, sr VP mktg; Joseph C. Tirinato, sr VP membership; Gary Belis, VP communications.

Board of Directors: Bruce Baker, exec VP Cox Television; Jim Beloyianis, pres Katz Television Group, Katz Media Group, Inc.; Scott Blumenthal, VP/of TV, LIN Television Corp; Craig Broitman, pres Millenium Sales & Marketing; Frank Comerford, pres/gen mgr WNBC-TV, New York, NY; Tom Conway, VP/sls & mktg, Media General Broadcast Group; Cathy Egan, sr VP, new business/mktg, ABC National Television

Sales, Inc.; Michael Fiorile, pres/CEO, Distpatch Broadcast Group; Alan Frank, pres Post-Newsweek Stations Inc.; Jay Ireland, pres NBC Universal Television Stations; Tom Kane, pres/CEO, CBS Television Stations; Paul Karpowicz, pres Meredith Broadcasting; Kathleen Keefe, VP sls Hearst-Argyle Television; Doug Kiel, vice-chmn/CEO Journal Broadcast Group; Walter Liss, pres ABC Owned Television Stations, Inc.; Leo MacCourtney, pres/CEO Blair Television; Ibra Morales, pres Telemundo Group, Inc.; John Reardon, pres bcstg Tribune Co.; Dick Robertson, pres domestic TV distribution Warner Brothers, Burbank, CA; Louis Wall, pres Sagamore Hill Broadcasting; John Cottingham, sr VP/Broadcast Stations, Media General Broadcast Group; Jim Monahan, pres & gen mgr Tribune Co.; John Weiser, pres/distribution Sony Pictures Television Sales Inc.; Doreen Wade, pres Freedom Broadcasting Inc.; Robert Silva, VP/dir sls E.W. Scripps Co.; Leo MacCourtney, pres/CEO Blair Television; Julio Marenghi, pres/sls CBS Television Stations; Jim Monohan, pres, gen mgr TeleRep, Inc.; Christopher Rohrs, pres Television Bureau of Advertising; Raymond Schonbak, sr VP/opns Emmis Communications; Robert Silva, VP/dir sls E.W. Scripps Co.; Perry Sook, pres/CEO Nexstar Broadcasting Group; Paul Trelstad, sr VP Gannet Broadcasting; Doreen Wade, pres Freedom Broadcasting, Inc.; Louis Wall, pres Sagamore Hill Broadcasting; John Watkins, pres ABC National Television Sales, Inc., New York, NY; John Weiser, pres ABC National Television Sales, Inc.

National Associations

AFCEA, 4400 Fair Lakes Ct., Fairfax, VA, 22033-3899. Phone: (703) 631-6100. Phone: (800) 336-4583. Fax: (703) 631-6130. E-mail: promo@afcea.org Web Site: www.afcea.org.

ANEPA, C/Castelló, 59 bis, 28001, Madrid Spain. Phone: 91 575 53 81. Fax: 91 435 66 53. E-mail: anepa@anepa.net Web Site: www.anepa.net.

Academy of Canadian Cinema & Television, Natl Office, 172 King St. E., Main Fl., Toronto, ON, M5A 1J3. Canada. Phone: (416) 366-2227. Phone: (800) 644-5194. Fax: (416) 366-8454. Web Site: www.academy.ca. E-mail: info@academy.ca

Vancouver, BC V6B 5M9, 1385 Homer St. Phone: (604) 684-4528. Fax: (604) 684-4574. E-mail: judy_rink@telus.net.

Montreal, PQ HZW 1M5 Canada, 225, rue Roy E, bureau 106. Phone: (514) 849-7448. Fax: (514) 849-5069.

Acoustical Society of America, 2 Huntington Quadrayle, Suite 1N01, Melville, NY, 11747-4502. Phone: (516) 576-2360. Fax: (516) 576-2377. E-mail: asa@aip.org Web Site: asa.aip.org.

Advanced Television Systems Committee (ATSC), 1750 K St. N.W., Suite 1200, Washington, DC, 20006. Phone: (202) 872-9160. Fax: (202) 872-9161. E-mail: atsc@atsc.org Web Site: www.atsc.org.

More than 130 members representing TV networks, mf assns & others. International voluntary tech standards-setting organization for advanced TV.

The Advertising Council Inc., 261 Madison Ave., New York, NY, 10016-2303. Phone: (212) 922-1500. Fax: (212) 922-1676. E-mail: info@adcouncil.org Web Site: www.adcouncil.org.

Washington, DC 20036, 1203 19th St. N.W., 4th Fl. Phone: (202) 331-9153. Fax: (202) 331-9790.

Advertising Research Foundation Inc., 432 Park Ave. S., New York, NY, 10016. Phone: (212) 751-5656. Fax: (212) 319-5265. Web Site: www.thearf.org.

Alliance for Community Media, 666 11th St. N.W., Suite 740, Washington, DC, 20001-4542. Phone: (202) 393-2650. Fax: (202) 393-2653. E-mail: raiseeveryvoice@yahoo.com Web Site: www.alliancecm.org.

Alliance of Motion Picture and Television Producers, 15503 Ventura Blvd., Encino, CA, 91436. Phone: (818) 995-3600. Fax: (818) 382-1793. Web Site: www.amptp.org.

American Advertising Federation, 1101 Vermont Ave. N.W., Suite 500, Washington, DC, 20005-6306. Phone: (202) 898-0089. Fax: (202) 898-0159. E-mail: aaf@aaf.org Web Site: www.aaf.org.

American Association of Advertising Agencies (AAAA), 405 Lexington Ave., 18th Fl., New York, NY, 10174-1801. Phone: (212) 682-2500. Fax: (212) 682-8391. Web Site: www.aaaa.org.

American Center for Children and Media, 5400 N. St. Louis Ave., Chicago, IL, 60625. Phone: (773) 509-5510. Fax: (773) 509-5303. E-mail: dkleeman@atgonline.org Web Site: www.centerforchildrenandmedia.org.

American Cinema Editors Inc., 100 Universal City Plaza, Bldg. 2282, Rm. 234, Universal City, CA, 91608. Phone: (818) 777-2900. Fax: (818) 733-5023. Web Site: www.ace-filmeditors.org.

American Composers Alliance (ACA), 648 Broadway, Rm. 803, New York, NY, 10012-2301. Phone: (212) 362-8900. Phone: (212) 925-0458. Fax: (212) 925-6798. E-mail: info@composers.com Web Site: www.composers.com.

American Electronics Association, 5201 Great America Pkwy., Suite 520, Santa Clara, CA, 95054. Phone: (408) 987-4200. Fax: (408) 987-4298. Web Site: www.aeanet.org.

Washington, DC 20005, 601 Pennsylvania Ave. Phone: (202) 682-9110. Fax: (202) 682-9111. William Archey, pres.

American Marketing Association, 311 S. Wacker Dr., Suite 5800, Chicago, IL, 60606. Phone: (312) 542-9000. Phone: (800) 262-1150. Fax: (312) 542-9001. Web Site: www.marketingpower.com. E-mail: info@ama.org

American Meteorological Society, 45 Beacon St., Boston, MA, 02108-3693. Phone: (617) 227-2425. Fax: (617) 742-8718. E-mail: amsinfo@ametsoc.org Web Site: www.ametsoc.org/ams.

American Radio Relay League, 225 Main St., Newington, CT, 06111. Phone: (860) 594-0200. Fax: (860) 594-0259. E-mail: hg@arrl.org Web Site: www.arrl.org.

Joel Harrison, 1st VP.

American Society of Composers, Authors & Publishers (ASCAP), One Lincoln Plaza, New York, NY, 10023. Phone: (212) 621-6000. Fax: (212) 724-9064. E-mail: info@ascap.com Web Site: www.ascap.com.

Hato Rey, PR 00918. ASCAP - Puerto Rico, 654 Ave. Munoz Rivera, IBM Plaza Suite 1101 B. Phone: (787) 281-0782. Fax: (787) 767-2805.

London W1X1PB United Kingdom. ASCAP - London, 8 Cork St. Phone: 011 44-207-439-0909. Fax: 011 44-207-434-0073.

Los Angeles, CA 90046. ASAP - Los Angeles, 7920 W. Sunset Blvd., 3rd Fl. Phone: (323) 883-1000. Fax: (323) 883-1049.

Miami Beach, FL 33139. ASCAP - Miami, 420 Lincoln Rd., Suite 385. Phone: (305) 673-3446. Fax: (305) 673-2446.

Atlanta, GA 30318. ASCAP - Atlanta, 541 Tenth St. N.W, PMB400. Phone: (404) 351-1224. Fax: (404) 351-1252.

Chicago, IL 60647. ASCAP - Chicago, 1608 N. Milwaukee, Suite 1007. Phone: (773) 394-4286. Fax: (773) 394-5639.

Nashville, TN 37203. ASCAP - Nashville, Two Music Square W. Phone: (615) 742-5000. Fax: (615) 742-5020.

(See listing under Music Licensing, Section G.)

American Society of Media Photographers (ASMP), 150 N. 2nd St., Philadelphia, PA, 19106. Phone: (215) 451-2767. Fax: (215) 451-0880. E-mail: info@asmp.org Web Site: www.asmp.org.

Robert Wilex, 1st VP. Directors: Susan Carr, Judy Herrman & Clem Spalding.

American Society of TV Cameramen Inc., (U.S. affil of International Society of Videographers.). 2520 Lotus Hill Dr., Las Vegas, NV, 89134-7855. Phone: (702) 228-6704. Fax: (702) 228-6714. E-mail: ruzwe7@aol.com

Directors: Tom Jocelyn, Peter Basil, Gino Guarna & Sol Bress.

Sparkill, NY 10976, Box 296. Nicole Zweck-Spanos, Production Planning.

American Sportscasters Association, 225 Broadway, Suite 2030, New York, NY, 10007. Phone: (212) 227-8080. Fax: (212) 571-0556. E-mail: lschwa8918@aol.com Web Site: americansportscastersonline.com.

Directors: Lou Schwartz, Jon Miller, Jim Nantz, Dick Enberg & Bill Walton.

American Sportscasters Hall of Fame Trust, 225 Broadway, Suite 2030, New York, NY, 10007. Phone: (212) 227-8080. Fax: (212) 571-0556. E-mail: lschwa8918@aol.com Web Site: www.americansportscastersonline.com.

American Women in Radio and Television Inc., 8405 Greenboro Dr., Suite 800, McLean, VA, 22102. Phone: (703) 506-3290. Fax: (703) 506-3266. Web Site: www.awrt.org. E-mail: info@awrt.org

Association for Education in Journalism & Mass Communication (AEJMC), 234 Outlet Point Blvd., Suite A, Columbia, SC, 29210. Phone: (803) 798-0271. Fax: (803) 772-3509. E-mail: aejmc@aejmc.org Web Site: www.aejmc.org.

Association for Interactive Marketing, (AIM). 1120 Avenue of the Americas, New York, NY, 10036. Phone: (888) 337-0008. Fax: (212) 391-9233. Web Site: www.greenlight.co.uk.

Association for Maximum Service Television Inc., (MSTV, Inc.). Box 9897, 4100 Wisconsin Ave., N.W., Washington, DC, 20016. Phone: (202) 966-1956. Fax: (202) 966-9617. Web Site: www.mstv.org. E-mail: lmillory@mst.org

The Association for Women In Communications, 3337 Duke St., Alexandria, VA, 22314. Phone: (703) 370-7436. Fax: (703) 370-7437. E-mail: info@womcom.com Web Site: www.womcom.org.

Directors: Pamela Valenzvela.

Association of American Railroads, American Railroads Bldg., 50 F St. N.W., Washington, DC, 20001. Phone: (202) 639-2100. Fax: (202) 639-2558. E-mail: twhite@www.aar.org Web Site: www.aar.org.

Association of Cable Communicators, Box 75007, Washington, DC, 20013. Phone: (202) 222-2370. Fax: (202) 222-2371. E-mail: services@cablecommunicators.org Web Site: www.cablecommunicators.org.

Rob Stoddard, 1st VP; James Maiella Jr., 2nd VP; Libby O'Connell, Ph.D. Directors: Pete Abel, Bobby Amirshahi, Portia E. Badham, Janice Caluda, Sandra Colony, Annie Howell, Ellen Kroner, Margaret Lejuste, Jennifer Mooney, Misty Skedgell, Jean Margaret Smith & Thomas Southwick.

Association of Canadian Advertisers Inc., 175 Bloor St. E., South Tower, Suite 307, Toronto, ON, M4W 3R8. Canada. Phone: (416) 964-3805. Phone: (800) 565-0109 (CA). Fax: (416) 964-0771. E-mail: rlund@acaweb.com Web Site: www.aca-online.com.

Montreal, PQ H7A 3C6, 500 Sherbrooke St. W. Phone: (514) 842-6422. (800) 883-0422. Fax: (514) 842-6223. Roger Sirard, VP/member svcs.

Association of Federal Communications Consulting Engineers (AFCCE), Box 1933, Washington, DC, 20036-0333. Phone: (812) 925-6000 Ext 270. Phone: 2202) 898-0111. Fax: (812) 925-4030. Web Site: www.afcce.org. E-mail: president@afcce.org

Marnie K. Sarver, Esq. Directors: Glen Clark, Donald G. Everist, Carl T. Jones Jr. & David H. Layer.

Association of National Advertisers Inc. (ANA), 708 3rd Ave., New York, NY, 10017. Phone: (212) 697-5950. Fax: (212) 661-8057. Web Site: www.ana.net. E-mail: bduggan@ana.net

Directors: Donald F. Calhoon, Jocelyn Carter-Miller, J. Andrea Alstrup, Catherine D. Constable, Christopher Fraleigh, James J. Garrity, David B. Green, John D. Hayes, Stephen C. Jones, Dawn Hudson, David N. Iauco, Abby F. Kohnstamm, Ann Lewnes, Eric W. Leininger, Robert D. Liodice, Paula S. Sneed, Gary E. McCullough, James R. Stengel, James D. Speros, Allan H. Stefl, Stephen G. Sullivan, Joseph V. Tripodi, Rebecca Saeger, James L. McDowell, Nancy J. Wiese, Robert J. Gamgort & Robert C Lachky.

Washington, DC 20036. Washington Office, 1120 20th St. N.W, Suite 5206. Phone: (202) 296-1883. Fax: (202) 296-1430.

Association of Public Television Stations, 666 11th St. NW, Suite 200, Washington, DC, 20001. Phone: (202) 654-4200. Fax: (202) 654-4237. Web Site: www.apts.org. E-mail: john@lpts.org

Directors: Meegan White.

The Audio Engineering Society Inc., 60 E. 42nd St., Rm. 2520, New York, NY, 10165. Phone: (212) 661-8528. Fax: (212) 682-0477. E-mail: hq@aes.org Web Site: www.aes.org.

BMI Broadcast Music Inc., 320 W. 57th St., New York, NY, 10019-3790. Phone: (212) 586-2000. Fax: (212) 246-2163. E-mail: abooth@bmi.com Web Site: www.bmi.com.

Directors: Philip A. Jones, Frances W. Preston, James G. Babb, Harold C. Crump, N. John Douglas, Frank E. Melton, George V. Willoughby, K. James Yager, G. Neil Smith, David Sherman, Donald A. Thurston, Cecil L. Walker, Catherine L. Hughes, Craig A. Dubow, Amador Bustos & John L. Sander.

West Hollywood, CA 90069-2211, 8730 Sunset Blvd, 3rd Fl. W. Phone: (310) 659-9109.

Nashville, TN 37203-4399, 10 Music Sq. E. Phone: (615) 401-2000.

Broadcast Cable Credit Association Inc. (BCCA), 550 W. Frontage Rd., Suite 3600, Northfield, IL, 60093. Phone: (847) 881-8757. Fax: (847) 784-8059. E-mail: info@bccacredit.com Web Site: www.bccacredit.com.

Broadcast Cable Financial Management Association (BCFM), 550 W. Frontage Rd., Suite 3600, Northfield, IL, 60093-1243. Phone: (847) 716-7000. Fax: (847) 716-7004. E-mail: info@bcfm.com Web Site: www.bcfm.com.

Broadcast Education Association, 1771 N St. N.W., Washington, DC, 20036-2891. Phone: (202) 429-3935. Fax: (202) 775-2981. E-mail: beainfo@beaweb.org Web Site: www.beaweb.org.

Directors: Mary Alice Molgard, Thomas R. Berg, Rustin Greene, Joe Misiewicz, David Byland, Robert K. Avery, Gary Martin, Greg Luft, D'Artagnan Bebel, Stephen J. Cohen, Larry Patrick, Alan R. Albarran, Gary Corbitt, Steven Anderson, Norman Pattiz & Jannette L. Dates.

Broadcasters Foundation of America, 7 Lincoln Ave., Greenwich, CT, 06830. Phone: (203) 862-8577. Fax: (203) 629-5739. E-mail: ghhbcast@aol.com Web Site: broadcastersfoundation.org.

Broadcasters Hall of Fame, 1240 Ashford Ln. #1A, Akron, OH, 44313. Phone: (330) 836-4864.

Jeannette Camak, sec.

Akron, OH 044320, Box 8192. Henry Dunn, treas. (Broadcaster's Hall of Fame).

CTAM, Cable & Telecommunications Association. 201 N. Union St., Suite 440, Alexandria, VA, 22314. Phone: (703) 549-4200. Fax: (703) 684-1167. E-mail: info@ctam.com Web Site: www.ctam.com.

Directors: Shelly Good-Cook, Jessica Robinson, Dave Watson, Lisa Jackson, Patrick Dougherty, Sharon Radziewski, John Powers, Kelly Gordon, Monique Sadler-Powell & MariAnne Woehrle.

The Cable Center, 2000 Buchtel Blvd., Denver, CO, 80210. Phone: (303) 871-4885. Fax: (303) 871-4514. E-mail: info@cablecenter.org Web Site: www.cablecenter.org.

Cable in the Classroom, 1724 Massachusetts Ave. N.W., Washington, DC, 20036. Phone: (202) 775-1040. Fax: (202) 775-1047. Web Site: www.ciconline.org.

Cable Television Laboratories, 858 Coal Creek Cir., Louisville, CO, 80027-9750. Phone: (303) 661-9100. Fax: (303) 661-9199. Web Site: www.cablelabs.com. E-mail: m.schwartz@cablelabs.com

Cabletelevision Advertising Bureau Inc. (CAB),

See listing under Major National Associations, this section.

Can-West Media Sales, 333 King St. E., Toronto, ON, M5A 4R7. Canada. Phone: (416) 350-6002. Fax: (416) 442-2209. E-mail: queries@nationalpost.co Web Site: www.canada.com.

Canadian Association of Broadcast Consultants, 130 Cree Crescent, Winnipeg, MB, R3J 3W1. Canada. Phone: (204) 889-9202. Fax: (204) 831-6650. E-mail: kpelser@deema.mb.ca

Maurice Beausejour, pres; Kerry Pelser, sec/treas.

The Canadian Association of Broadcasters, Box 627, Station B, 306-350 Sparks St., Ottawa, ON, K1P 5S2. Canada. Phone: (613) 233-4035. Fax: (613) 233-6961. E-mail: cab@cab-acr.ca Web Site: www.cab-acr.ca.

Directors: Fawn-Dell Flanagan, Antoinette Mensour, Jim Patrick & Susan Wheeler.

Ottawa, ON K1R 7S8 Canada, 306-350 Sparks St. Phone: (613) 233-4035.

Canadian Association of Ethnic (Radio) Broadcasters, 622 College St., Toronto, ON, M6G 1B6. Canada. Phone: (416) 531-9991. Fax: (416) 531-5274. E-mail: info@chinradio.com Web Site: www.chinradio.com.

Canadian Film and Television Production Association (CFTPA), 160 John St., Toronto, ON, M5C 2E5. Canada. Phone: (416) 304-0280. Fax: (416) 304-0499. E-mail: toronto@cftpa.ca Web Site: www.cftpa.ca.

Directors: Cara Martin.

Ottawa, ON K1P 5H3, 151 Slater St, Suite 605. Phone: (613)-233-1444. Fax: (613)-233-0073.

Caribbean Broadcasting Union (CBU), Harbour Industrial Estate, Harbour Rd., Suite 1B, Bldg. 6 A, St. Michael, BB, 11145. Barbados. Phone: (246) 430-1006. Fax: (246) 228-9524. Web Site: www.caribunion.com.

Catholic Academy for Communication Arts Professionals, 1645 Brook Lynn Dr., Suite 2, Dayton, OH, 45432-1944. Phone: (937) 458-0265. Fax: (937) 458-0263. E-mail: admin@catholicacademy.org Web Site: www.catholicacademy.org.

Jeanean Merkel, 1st VP; Vicki Bedard, 2nd VP.

Center for Communication Inc., 561 Broadway, Suite 12 B, New York, NY, 10012. Phone: (212) 686-5005. Fax: (212) 504-2632. E-mail: info@cencom.org Web Site: www.cencom.org.

Edward Bleier, chmn; Frank Stanton, dir emeritus. Directors: Timothy Barry, William F. Baker, Robert M. Batscha, Patricia T. Carbine, Antoinette Cook Bush, John A. Dimling, David R. Drobis, Michael Eigner, Peter R. Ezersky, Charles B. Fruit, Ralph Guild, Andrew Heyward, Peter Jennings, Gerald M. Levin, Kate McEnroe, Martin Nisenholtz, Herbert Scannell, Alan Siegel, Alfred C. Sikes, Kenneth Stoddard, Howard Stringer, Alberto Vitale, Stephen A. Weiswasser, David Westin, Bob Wright, Lois Wyse, Mortimer B. Zuckerman, Simon Michael Bessie, Louis D. Boccardi, David W. Burke, Henry A. Grunwald, Irwin Segelstein, Burton B. Staniar & Loet A. Velmans.

Commonwealth Broadcasting Assn, CBA Secretariat, 17 Fleet St., London, EC4Y 1AA. United Kingdom. Phone: 011-44-171-5835550. Fax: 011-44-171-5835549. E-mail: cba@cba.org.uk Web Site: www.cba.org.uk.

Directors: George Valarino, Ronald Abraham, Roger Grant, Robert O'Rielly, Tombong Saidy, Sharon Crosbie & Cecilia Khuzwayo.

Community Broadcasters Association (CBA), 515 King St., Suite 420, Alexandria, VA, 22314. Phone: (703) 562-3588. Fax: (703) 684-6048. Web Site: www.communitybroadcasters.com.

Directors: Gary Co Cola, Larry Morton, Eleanor St. John, Warren Trumbly, Doug Williams, Sandra Woodworth & Lou Zanoni.

Produces major country radio convention; rgnl radio seminar; Country DJ Hall of Fame.

Council of Better Business Bureaus Inc., 4200 Wilson Blvd., 8th Fl., Arlington, VA, 22203-1838. Phone: (703) 276-0100. Fax: (703) 525-8277. E-mail: bbb@bbb.org Web Site: www.bbb.org.

Ottawa, ON K1N 7A2 Canada, 44 Byward Market Sq, Suite 220.

Country Music Association Inc., One Music Cir. S., Nashville, TN, 37203. Phone: (615) 244-2840. Fax: (615) 726-0314. Web Site: www.cmaworld.com.

Country Radio Broadcasters Inc., 819 18th Ave. S., Nashville, TN, 37203. Phone: (615) 327-4487. Fax: (615) 329-4492. Web Site: www.crb.org.

This non-profit organization is the only assn specificaly serving country radio. Holds annual Country Radio Seminar; rgnl conventions; trustee of Country DJ Hall of Fame & Country Radio Hall of Fame.

Electro Federation Canada, 5800 Explorer Dr., Suite 200, Mississauga, ON, L4W 5K9. Canada. Phone: (866) 602-8877. Fax: (905) 602-5686. E-mail: info@electrofed.com Web Site: www.electrofed.com.

Electronic Industries Alliance (EIA), 2500 Wilson Blvd., Arlington, VA, 22201. Phone: (703) 907-7500. Fax: (703) 907-7501. E-mail: dmccurdy@eia.org Web Site: www.eia.org.

Electronic Retailing Association (ERA), 200 N. 14th St., Suite 300, Arlington, VA, 22201. Phone: (703) 841-1751. Phone: (800) 987-6462. Fax: (703) 841-1751. E-mail: contact@retailing.org Web Site: www.retailing.org.

Stephen F. Breimer, Esq.; Jeffrey Knowles, Esq. Directors: Linda Goldstein, Mike Ackerman, Rick Cesari, Dan Danielson, Denise Dubarry Hay, Rollie Froehlig, Larry Jellen, Jack Kirby, Mark Lavin, Shigeru Ohashi, Rick Petry, Steve Pittenridgh, Richard Prochnow, Randy Ronning, Robert Rosenblatt, Bret Saxton, Mark Thornton & Reiner Weihofen.

Electronic Service Dealers Association, 4927 W. Irving Park Rd., Chicago, IL, 60641. Phone: (773) 282-9400.

Electronics Representatives Association, 444 N. Michigan Ave., Suite 1960, Chicago, IL, 60611. Phone: (312) 527-3050. Fax: (312) 527-3783. E-mail: info@era.org Web Site: www.era.org.

Mike Kunz, chmn of bd.

FCBA (Federal Communications Bar Association), 1020 19th St. N.W., Suite 325, Washington, DC, 20036-6101. Phone: (202) 293-4000. Fax: (202) 293-4317. E-mail: fcba@fcba.org Web Site: www.fcba.org.

Festival of Nouveau Cinema Montreal, 3805 Blvd. Saint-Lavrent, Montreal, PQ, H2W 7X9. Canada. Phone: (514) 282-0004. Fax: (514) 282-6664. E-mail: ngirard@nouveaucinema.ca Web Site: www.nouveaucinema.ca.

Foundation for American Communications (FACS), 85 South Grand Ave., Pasadena, CA, 91105. Phone: (626) 584-0010. Fax: (626) 584-0627. E-mail: facs@facsnet.org Web Site: www.facsnet.org.

Leesburg, VA 20176, 3 Wirt St. N.W. Phone: (703) 737-3570. (888) 739-7865.

Free TV Australia Ltd., (formerly Commercial Television Australia CTVA). 44 Avenue Rd., Mosman, N.S.W., 02088. Australia. Phone: 61 2 8968 7100. Fax: 61 2 9969 3520. E-mail: contact@freetv.com.au Web Site: www.freevaust.com.au.

Provides a forum for discussion of industry matters by its members & is the pub voice of the industry on a wide range of issues & has represented the coml free-to-air TV industry for over 40 years.

Hollywood Radio & Television Society, 13701 Riverside Dr., Suite 205, Sherman Oaks, CA, 91423. Phone: (818) 789-1182. Fax: (818) 789-1210. E-mail: info@hrts.org Web Site: www.hrts.org.

IEEE, 445 Hoes Ln., Piscataway, NJ, 08854-1331. Phone: (732) 981-0060. Fax: (732) 981-1721. Web Site: www.ieee.org.

New York, NY 10016-5997, 3 Park Ave. 17th Fl. Phone: (212) 419-7900. Fax: (212) 752-4929.

Independent Film & Television Alliance (IFTA), 10850 Wilshire Blvd., 9th Fl., Los Angeles, CA, 90024-4321. Phone: (310) 446-1000. Fax: (310) 446-1600. E-mail: info@ifta-online.org Web Site: www.ifta-online.org.

Directors: Glen Basner, Steve Bickel, Alison Thompson, Nicolas Chartier, Roger Corman, Pierre David, Peter Elson, Kimberly Ferguson, Antony Ginnane, Peter Graham, Robert Hayward, Avi Lerner, Mark Lindsay, Nicole Mackey, Nicholas Meyer, Bobby Meyers, Michael Weiser, Kevin Williams & Andrew Stevens.

Intercollegiate Broadcasting System Inc., 367 Windsor Hwy., New Windsor, NY, 12553-7900. Phone: (845) 565-0003. Fax: (845) 565-7446. E-mail: ibshq@aol.com Web Site: www.ibsradio.org.

Directors: John Murphy, Fritz Kass, Norm Prusslin & Chuck Platt.

The International Academy of Television Arts & Sciences, 888 7th St., 5th Fl., New York, NY, 10019. Phone: (212) 489-6969. Fax: (212) 489-6557. E-mail: info@emmys.tv Web Site: www.iemmys.tv.

International Advertising Association, (The global partnership of advertisers, agencies & media.). 521 Fifth Ave., Suite 1807, New York, NY, 10175. Phone: (212) 557-1133. Fax: (212) 983-0455. E-mail: iaa@iaaglobal.org Web Site: www.iaaglobal.org.

Wendy Burrell, mgng dir. Directors: Richard Corner.

International Animated Film Society, ASIFA-Hollywood, 2114 Burbank Blvd., Burbank, CA, 91506. Phone: (818) 842-8330. Fax: (818) 842-5645. E-mail: asifaalert-subscribe @yahoogroups.com Web Site: www.asifa-hollywood.org.

Directors: Jerry Beck, Stephen Worth, Bob Miller, Tom Knott, Frank Gladstone, David Derks, Margaret Kerry-Wilcox, Larry Loc & Will Ryan.

International Association of Audio Information Services, c/o NBRS, 1090 Don Mills Rd., Suite 303, Toronto, ON, M3C 3R6. Canada. Phone: (416) 422-4222, ext 224. Fax: (416) 422-1633. E-mail: hlusignan@nbrscanada.com Web Site: www.iaais.org.

International Communication Agency Network (ICOM), 1649 Lump Gulch Rd., Rollinsville, CO, 80474. Phone: (303) 258-9511. Fax: (303) 484-4087. E-mail: info@icomagencies.com Web Site: www.icomagencies.com.

Directors: Frank G. Weyforth.

International Institute of Communications, 35 Porland Pl., 3rd. Fl., Westcott, London, WIB IAE. Phone: (44) 207-323-9622. Fax: (44) 207- 323- 9623. E-mail: enquiries@iicom.org Web Site: www.iicom.org.

International Radio & Television Society Foundation Inc., 420 Lexington Ave., Suite 1601, New York, NY, 10170. Phone: (212) 867-6650. Fax: (212) 867-6653. Web Site: www.irts.org. E-mail: jim.cronin@irts.org

International Recording Media Association, 182 Nassau St., Suite 204, Princeton, NJ, 08542. Phone: (609) 279-1700. Fax: (609) 279-1999. E-mail: info@recordingmedia.org Web Site: www.recordingmedia.org.

Directors: Tony Perez.

KOBI-TV (NBC Affiliate), (An affil of CA & OR Bcstg Inc.). 125 S. Fir St., Medford, OR, 97501. Phone: (541) 779-5555. Fax: (541) 779-1151. Web Site: www.localnewscomesfirst.com. E-mail: kobi@kobi5.com

Klamath Falls, OR 97601. KOTI-TV, 222 S. 7th St.

League of Advertising Agencies, 915 Clifton Ave., Clifton, NJ, 07013. Phone: (973) 473-6643. Fax: (943) 473-0685. E-mail: info@weinrichadv.com Web Site: www.weinrichadv.com.

Library of American Broadcasting Foundation Inc., Box 2749, Alexandria, VA, 22301. Phone: (703) 548-6090. Fax: (703) 549-4349. E-mail: westsqn@aol.com Web Site: www.labfoundation.org.

Donald H. Kirkley, Jr. Directors: James L. Greenwald, Vincent Curtis, Arthur W. Carlson, Erwin Krasnow, Jerry Lee, Larry Taishoff, Jim Morley, Susan Ness, Don West, Richard Buckley, Perre Bouvard, Russ Withers, Carl Brazell, Michael Carter, Tim Cookerly, Sam Donaldson, James E. Duffy, Erica Farber, Skip Finley, Gary Fries, Marc Guild, Dr. Judy Kuriansky, David Kennedy, Dawson B. Nail, Allen Shaw, Ramsey Woodworth & Millard Younts.

Maryland, NO, University of Maryland, College Park. Phone: (301) 405-9160. Fax: (301) 314-2634. Web Site: www.lib.umd.edu/umcp/lab.

Magazine Publishers of America, 810 7th Ave, 24th Fl., New York, NY, 10019. Phone: (212) 872-3700. Fax: (212) 888-4217. E-mail: mpa@magazine.org Web Site: www.magazine.org.

Media Access Group at WGBH, One Quest St., Boston, MA, 02135. Phone: (617) 300-2000. Fax: (617) 300-1020. E-mail: access@wgbh.org Web Site: access.wgbh.org.

The Media Institute, 2300 Clarendon Blvd., Suite 503, Arlington, VA, 22201. Phone: (703) 243-5700. Fax: (703) 243-8808. E-mail: info@mediainstitute.org Web Site: www.mediainstitute.org.

Richard E. Wiley, chmn.

Minority Media and Telecommunications Council, 3636 16th St. N.W., B-366, Washington, DC, 20010. Phone: (202) 332-0500. Fax: (202) 332-0503. E-mail: dhonig@crosslink.net Web Site: www.mmtconline.org.

Mortgage Bankers Association of America (MBA), 1919 Pennsylvania Ave. N.W., Washington, DC, 20006-3404. Phone: (202) 557-2700. E-mail: info@mbaa.org Web Site: www.mbaa.org.

Motion Picture Association of America, 15503 Ventura Blvd., Encino, CA, 91436. Phone: (818) 995-6600. Fax: (818) 382-1795. Web Site: www.mpaa.org.

Encino, CA 91436, 15503 Ventura Blvd. Phone: (818) 995-6600. Fax: (818) 382-1778.

The Museum of Broadcast Communications, 400 N. State St., Suite 240, Chicago, IL, 60610-6860. Phone: (312) 245-8200. Fax: (312) 245-8207. Web Site: www.museum.tv.

NCTI, 9697 E. Mineral Ave., Centennial, CO, 80112-3408. Phone: (303) 797-9393. Fax: (303) 797-9394. Web Site: www.ncti.com.

The National Academy of Television Arts & Sciences, 111 W. 57th St., Suite 600, New York, NY, 10019. Phone: (212) 586-8424. Fax: (212) 246-8129. Web Site: www.emmyonline.org.

The National Academy of Television Journalists Inc., Box 289, Salisbury, MD, 21803-0289. Phone: (410) 251-2511. Fax: (410) 543-0658. E-mail: infi@goldenviddyawards.com Web Site: www.goldenviddyawards.com.

Directors: Dr. Catherine North & Cathy Roche.

National Association of Black Journalists (NABJ), 8701-A Adelphi Road, Adelphi, MD, 20783-1716. Phone: (301) 445-7100. Fax: (301) 445-7101. E-mail: nabj@nabj.org Web Site: www.nabj.org.

Directors: Ernie Suggs, Stephanie Jones, Elliott Lewis, Marsha J. Eaglin & Victor W. Vaughan.

National Association of Black Owned Broadcasters Inc. (NABOB), 1155 Connecticut Ave., N.W., 6th Fl., Washington, DC, 20036. Phone: (202) 463-8970. Fax: (202) 429-0657. E-mail: info@nabob.org Web Site: www.nabob.org.

Directors: Pierre M. Sutton, Bennie Turner, Sydney L Small, Lois E. Wright, Michael Carter, Alfred Liggins, James Wolfe, Michael Roberts, Karen Slade & Carol Moore Cutting.

National Association of Broadcasters,

See listing under Major National Associations, this section.

National Association of Hispanic Journalists, 1000 National Press Bldg., 529 14th St. N.W., Washington, DC, 20045-2001. Phone: (202) 662-7145. Phone: (888) 346-nahj. Fax: (202) 662-7144. E-mail: nahj@nahj.org Web Site: www.nahj.org.

National Association of Telecommunications Officers and Advisors, 1800 Diagonal Rd., Suite 495, Alexandria, VA, 22314. Phone: (703) 519-8035. Fax: (703) 519-8036. E-mail: info@natoa.org Web Site: www.natoa.org.

National Association of Television Program Executives International,

See listing under Major National Associations, this section.

National Association of Theatre Owners Inc. (NATO), 750 1st St. N.E., Suite 1130, Washington, DC, 20002. Phone: (202) 962-0054. Fax: (202) 962-0370. E-mail: nato@natodc.com Web Site: www.natoonline.org.

North Hollywood, CA 91602. NATO Communications (In Focus), 4605 Lankershim Blvd, #340. Phone: (818) 506-1778.

National Black Media Coalition, 145 Alderson Ave., Billings, MT, 59101. Phone: (406) 248-4450. Fax: (301) 593-3604. E-mail: webmaster@nbm.org Web Site: www.nbmc.org.

National Cable and Television Association Inc. (NCTA),

See listing under Major National Associations, this section.

National Cable Television Cooperative Inc.,

See listing under Major National Associations, this section.

National Captioning Institute (NCI), 1900 Gallows Rd., Suite 3000, Vienna, VA, 22182. Phone: (703) 917-7600. Fax: (703) 917-9878. Web Site: www.ncicap.org.

Directors: Karen O' Connor, Marc Okrand, Beth Nubbe & Jay Feinberg.

Burbank, CA 91502, 303 N. Glenoaks Blvd, Suite 200. Phone: (818) 238-0068. (818) 238-4255.

National Council for Families & Television, 6500 Wilshire Blvd., Suite 1950, Los Angeles, CA, 90048. Phone: (323) 866-6020. Fax: (310) 208-5984. E-mail: ncft@yahoo.com

National Education Association, 1201 16th St. N.W., Washington, DC, 20036-3290. Phone: (202) 833-4000. Fax: (202) 822-7974. Web Site: www.nea.org. E-mail: editorial@list.nea.org

National Electrical Manufacturers Association (NEMA), 1300 N. 17th St., Suite 1847, Rosslyn, VA, 22209. Phone: (703) 841-3200. Fax: (703) 841-5900. E-mail: webmaster@nema.org Web Site: www.nema.org.

National Federation of Community Broadcasters (NFCB), 1970 Broadway, Suite 1000, Oakland, CA, 94612. Phone: (510) 451-8200. Fax: (510) 451-8208. E-mail: nfcb@nfcb.org Web Site: www.nfcb.org.

National Federation of Press Women Inc.-NFPW, Box 5556, Arlington, VA, 22205. Phone: (703) 812-9487. Phone: (800) 780-2715. Fax: (703) 812-4555. E-mail: mhunt21@msn.com Web Site: www.nfpw.org.

Donna Penticuff, past pres; Kathryn Cordova, 2nd VP; Marsha Shuler, 1st VP.

Programs, svcs & contest categories for female & male professionals in all media.

National League of Cities, 1301 Pennsylvania Ave. N.W., Suite 550, Washington, DC, 20004. Phone: (202) 626-3000. Fax: (202) 626-3043. E-mail: info@nlc.org

James C. Hunt, pres; Bart Peterson, VP; Cynthia McCollum, 2nd VP. Directors: R. Michael Amyx, Tommy Baker & Vickie Barnett.

National Museum of Communications Inc., 2001 Plymouth Rock Dr., Richardson, TX, 75081. Phone: (972) 889-9872. Fax: (972) 889-2329. E-mail: bill46@yesterdayusa.com Web Site: www.yesterdayusa.com.

Directors: William J. Bragg, Betty Lewis & Kim Bragg.

National Newspaper Association, Box 7540, Columbia, MO, 65205-7540. Phone: (573) 882-5800. Fax: (573) 884-5490. E-mail: info@nna.org Web Site: www.nna.org.

Lynn Edinger, assoc dir. Directors: Tonda Rush.

Arlington, VA 22205, Box 5737. Phone: (703) 534-1278.

National Press Club, 529 14th St. N.W., Washington, DC, 20045. Phone: (202) 662-7500. Phone: (202) 662-7505. Fax: (202) 662-7512. E-mail: info@npcpress.org Web Site: www.press.org.

Joe Anselmo, chmn of bd. Directors: Tom Glad.

National Religious Broadcasters (NRB), 9510 Technology Dr., Manassas, VA, 20110. Phone: (703) 330-7000. Fax: (703) 330-7100. E-mail: fwright@nrb.org Web Site: www.nrb.org/nrb.

National Retail Federation, 325 7th St. N.W., Suite 1100, Washington, DC, 20004. Phone: (202) 783-7971. Phone: (800) nrf-how2. Fax: (202) 737-2849. Web Site: www.nrf.com.

National Telemedia Council Inc., 1922 University Ave., Madison, WI, 53726. Phone: (608) 218-1182. Fax: (608) 218-1183. E-mail: ntelemedia@aol.com Web Site: nationaltelemediacouncil.org.

Directors: Mary Moen, Mike Bergen, Dr. Martin Rayala, Greg Hoffmann, Vira Standiford, Keeiza Hyzer & Rosemary Lehman.

Newseum, 1101 Wilson Blvd., Arlington, VA, 22209. Phone: (703) 284-3544. Phone: (888) new-seum. E-mail: newseum@freedomforum.org Web Site: www.newseum.org.

Newspaper Association of America, 4401 Wilson Blvd., Suite 900, Arlington, VA, 22203-1867. Phone: (571) 366-1001. Fax: (571) 366-1201. Web Site: www.naa.org.

Directors: R. Gene Bell, William Blocl Jr., R. Bruce Bradley, J. Stewart Bryan III, Susan Clark-Johnson, Jean B. Clifton, Mark G. Contreras, W. Stacey Cowles, James C. Currow, R. Jack Fishman, Dennis J. Fitzsimons, Caroline D. Harrison, Alan M. Horton, Alberto Ibarguen, Julie Inskeep, George B. Irish & Robert C. Woodworth.

North American Broadcasters Association (NABA), Box 500, Stn A, Rm. 6C 300, Toronto, ON, M5W 1E6. Canada. Phone: (416) 598-9877. Fax: (416) 598-9774. E-mail: info@nabanet.com Web Site: www.nabanet.com.

Directors: Joseph Flaherty, Felix Arauji Ramirez, Ignacio Suarez, Andy Setos & Peter Smith.

North American Retail Dealers Association, 10 E. 22nd St., Suite 310, Lombard, IL, 60148-6191. Phone: (630) 953-8950. Phone: (800) 621-0298. Fax: (630) 953-8957. E-mail: nardahdq@narda.com Web Site: www.kwmu.com.

Pacific Pioneer Broadcasters, Box 4866, Valley Village, CA, 91617-4866. Phone: (323) 461-2121. Fax: (818) 768-8251. Web Site: www.ppbwebsite.org.

The Paley Center for Media, 25 W. 52nd St., New York, NY, 10019-6101. Phone: (212) 621-6600. Fax: (212) 621-6700. Web Site: www.paleycenter.org. E-mail: publicrelations@paleycenter.org

Beverly Hills, CA 90210, 465 N. Beverly Dr. Phone: (310) 786-1000. Barbara Dixon, VP/dir.

PROMAX&BDA, 9000 W. Sunset Blvd., Suite 900, Los Angeles, CA, 90069. Phone: (310) 788-7600. Phone: (800) 977-6629. Fax: (310) 788-7616. Web Site: www.promax.tv. E-mail: jim@promax.tv

Directors: David Snapp, George Pierson, Lisa Fengler, Leslie Celia, Jeannine Chanin, Tony Cleave, Ann Epstein-Cohen, Steve Delaney, Miguel Muelle, Karen Olcott, Jan Phillips, Abel Sanchez, Robin Skirboll, Anne White, Mark Stroman, Glynn Brailsford, Brian Blum, Judy Braune, Alan Cohen, Scott Danielson, C.J. Fredricksen, Lee Hunt, Kay Hutchison, Tony Lakin, Vince Manze, Brigitte McCray, Rob Middleton, Nick Miller, Michael Mischler, David Muscari, Billy Pittard, Sal Sardo, George Schweitzer, Curtis Symonds & Donna Weston.

Hong Kong China. Synapse Pacific LTD, 49 Hollywood Rd., 19th Fl.

London, NO SE1 0RF United Kingdom. Promax & BDA Europe, 61 Webber St.

Promotion Marketing Association Inc., 257 Park Ave. S., Suite 1102, New York, NY, 10010. Phone: (212) 420-1100. Fax: (212) 533-7622. E-mail: pma@pmalink.org Web Site: www.pmalink.org.

Public Radio in Mid-America (PRIMA), c/o KWMU-FM One University Blvd., University of Missouri, St. Louis, MO, 63121-4499. Phone: (314) 516-5968. Fax: (314) 516-5993. E-mail: pwente@kwmu.org Web Site: www.kwmu.org.

Directors: Dan Skinner & Jon Schwartz.

Public Radio News Directors Incorporated, 821 University Ave., Madison, NY, 53706. Phone: (608) 265-3378. Fax: (608) 263-5838. E-mail: walker@wpr.org

Directors: Dave Piznanelli, Martha Foley, Jonathan Ahl & Christine Paige-Diers.

Public Relations Society of America, 33 Maiden Ln., 11th Fl., New York, NY, 10038. Phone: (212) 460-1400. Fax: (212) 995-0757. E-mail: hq@prsa.org Web Site: www.prsa.org.

Radio Advertising Bureau,

See listing under Major National Associations, this section.

Radio and Television Museum, 2608 Mitchellville Rd., Bowie, MD, 20716. Phone: (301) 390-1020. Fax: (301)947-3338. E-mail: radiobelanger@comcast.net Web Site: www.radiohistory.org.

Brian Belanger, exec dir. Directors: Kenneth Mellgren, Ed Walker, Peter Eldridge, William McMahon, William Goodwin, Tony Young, Rob Huddleston, Chris Sterling, Don Ross, Gerald Schneider, Paul Courson, Charles Grant & Michael Rubin.

Radio & Television News Directors Foundation, 1600 K St. NW, Suite 700, Washington, DC, 20006. Phone: (202) 659-6510. Fax: (202) 223-4007. E-mail: rtndf@rtndf.org Web Site: www.rtndf.org.

Radio & Television Research Council, c/o MSA, Attn: R. Sharpe, 152 Madison Ave., Suite 801, New York, NY, 10016. Phone: (212) 481-3038. Fax: (212) 481-3071. E-mail: rtrcny@aol.com

Radio Marketing Bureau, 175 Bloor St. E., Suite 316 North Tower, Toronto, ON, M4W 3R8. Canada. Phone: (416) 922-5757. Phone: (800) 667-2346. Fax: (416) 922-6542. E-mail: info@rmb.ca Web Site: www.rmb.ca.

Radio-Television Correspondents' Association, S-325, U.S. Capitol, Washington, DC, 20510. Phone: (202) 224-6421. Fax: (202) 224-4882. Web Site: www.senate.gov/galleries.radiotv.com.

Directors: Jerry Bodlander, Bob Fuss, Edward O'Keefe, Dave McConnell, Richard Tillery & David Welna.

Radio Television News Directors Association (Canada), 2175 Shephard Ave. E., Suite 310, Toronto, ON, M2J 1W8. Canada. Phone: (416) 756-2126. Fax: (416) 364-8896. E-mail: tscott@broadcastnews.ca Web Site: www.rtndacanada.com.

Recording Industry Association of America Inc. (RIAA), 1330 Connecticut Ave. N.W., Suite 300, Washington, DC, 20036. Phone: (202) 775-0101. Fax: (202) 775-7253. Web Site: www.riaa.com.

Royal Television Society, North America Inc., Box 870501, Arizona State University, Tempe, AZ, 85287-0501. Phone: (480) 965-7661. Fax: (480) 965-1371. E-mail: royaltv@asu.edu

Satellite Broadcasting & Communications Assn. of America (SBCA), 1730 M St., N.W., Suite 600, Washington, DC, 20036-4557. Phone: (202) 349-3620. E-mail: info@sbca.org Web Site: www.sbca.com.

Society of Broadcast Engineers Inc., 9102 N. Meridian St., Suite 150, Indianapolis, IN, 46260-1896. Phone: (317) 846-9000. Fax: (317) 846-9120. E-mail: jporay@sbe.org Web Site: www.sbe.org.

Christopher H. Scherer. CPBE, CBNT, pres.

Society of Cable Telecommunications Engineers Inc., 140 Philips Rd., Exton, PA, 19341-1318. Phone: (610) 363-6888. Fax: (610) 363-5898. E-mail: scte@scte.org Web Site: www.scte.org.

Directors: Joel E. Welch, Thomas Russell & Joan Hagelin.

Professional membership assn offering information, professional dev resources, standards to cable telecommunications engineers & other professional.

Society of Environmental Journalists (SEJ), Box 2492, Jenkintown, PA, 19046. Phone: (215) 884-8174. Fax: (215) 884-8175. E-mail: sej@sej.org Web Site: www.sej.org.

Carolyn Whetzel, treas; Dina Cappiello, sec.

Society of Motion Picture & Television Engineers (SMPTE), 3 Barker Ave., 5th Fl., White Plains, NY, 10601. Phone: (914) 761-1100. Fax: (914) 761-3115. Web Site: www.smpte.org.

Society of Professional Journalists, 3909 N. Meridian St., Indianapolis, IN, 46208-4011. Phone: (317) 927-8000. Fax: (317) 920-4789. E-mail: webmaster@spj.org Web Site: www.spj.org.

Robert Leger, sec/treas; David Carlson, VP. Directors: Alvin Cross.

Twelve rgnl dirs, natl offs elected annually, two students reps, two dirs, at-large & two campus advisors at-large.

Society of Satellite Professionals International, 55 Broad St., 14th Fl., New York, NY, 10004. Phone: (212) 809-5199. Fax: (212) 825-0075. Web Site: www.sspi.org.

Directors: Richard Wolf, David Bioss, Carson Agnero, Blair Marshall, Ellen Hoff, Dan Stasi, Stephen Teller & D. Sacjoder.

The Songwriters Guild of America, 1560 Broadway, Suite 408, New York, NY, 10036. Phone: (212) 768-7902. Fax: (212) 768-7902. E-mail: ny@songwritersguild.com Web Site: www.songwritersguild.org.

Hollywood, CA 90028, 6430 Sunset Blvd. Phone: (213) 462-1108. Aaron Meza, rgnl dir.

Los Angeles, CA, 6430 Sunset Blvd.

Nashville, TN 37212, 1222 16th Ave. St, Suite 25. Phone: (615) 329-1782. Rondi Regan, rgnl dir.

Statenets National Association of State Radio Networks Inc., 17911 Harwood Ave., Homewood, IL, 60430. Phone: (708) 799-6676. Fax: (708) 799-6698. E-mail: tdobrez@statenets.com Web Site: www.statenets.com.

Syndicated Network Television Association, 630 Fifth Ave., Suite 2320, New York, NY, 10111. Phone: (212) 259-3740. Fax: (212) 259-3770. E-mail: mburg@snta.com Web Site: www.snta.com.

Telecommunications Industry Association, 2500 Wilson Blvd., Arlington, VA, 22201. Phone: (703) 907-7700. Fax: (703) 907-7727. E-mail: tia@tiaonline.org Web Site: www.tiaonline.org.

Directors: Grant Seiffert, Bill Belt, John Derr, Derek Khlopin, Jason Leuck, Anna Amselle, Henry Wieland, Maryann Lesso, David Smith, Dan Bart & Henry Cuschieri.

Beijing 100004 China. USITO, Rm. 332, 3/f Lido Office Tower, Lido Place, Jichang Rd., Jiang Tai Rd. Phone: (8610) 6430-1368/69/70/71/72. Fax: (8610) 6430-1367. E-mail: usito@usito.org. Web Site: www.usito.org. Anne Stevenson-Yang, mgng dir.

Telecommunications Research and Action Center (TRAC), Box 27279, Washington, DC, 20005. Phone: (202) 263-2950. Fax: (202) 263-2960. E-mail: trac@trac.org Web Site: www.trac.org.

Television Bureau of Advertising (TVB),

See listing under Major National Associations, this section.

Television Bureau of Canada, 160 Bloor St. E, Suite 1005, Toronto, ON, M4W 1B9. Canada. Phone: (416) 923-8813. Fax: (416) 413-3879. E-mail: tvb@tvb.ca Web Site: www.tvb.ca.

Television Critics Association, The Wichita Eagle, 825 E. Douglas Ave., Witchita, KS, 67202. Phone: (316) 268-6394. Fax: (316) 268-6627. E-mail: tca@tvcritics.org Web Site: www.tvcritics.org.

Television Operators Caucus, 1776 K Street NW, Washington, DC, 20006. Phone: (202) 719-7090. Fax: (202) 719-7546.

The Association for International Broadcasting, Box 141, Cranbrook, TN17 9AJ. United Kingdom. Phone: +44 (0) 20 7993 2557. Fax: +44 (0) 20 7993 8043.

Directors: Tom Walters, mktg; Tim Keeler, pub affrs.

Think LA, (formerly Los Angeles Advertising Agencies Association). 4223 Glencoe Ave., Suite C-100, Marina del Rey, CA, 90292. Phone: (310) 823-7320. Fax: (310) 823-7325. E-mail: info@thinkla.org Web Site: www.thinkla.org.

U.S. Conference of Catholic Bishops, Dept. of Communications, 3211 4th St. N.E., Washington, DC, 20017-1194. Phone: (202) 541-3000. Fax: (202) 541-3173. Web Site: www.usccb.org.

Veteran Wireless Operators Association Inc., Box 1003, Peck Slip, New York, NY, 10272-1003. E-mail: vwoa@interactive.net Web Site: www.vwoa.org.

Directors: Richard T. Kenney & D. I. Temple.

Veterans Bedside Network, (The Veterans Hospital Radio & TV Guild.). 10 Fiske Pl., Rm. 301, Mount Vernon, NY, 10550. Phone: (914) 699-6069. Fax: (914) 667-0405.

Wireless Communications Association International, Inc., 1333 H St. N.W., Suite 700, Washington, DC, 20005-4754. Phone: (202) 452-7823. Fax: (202) 452-0041. E-mail: sonu@wcai.com Web Site: www.wcai.com.

Directors: John T. von Harz III, William Andrle Jr., T. Lauriston Hardin, Chris Farnworth & Patrick J. Gossman.

Women in Cable & Telecommunications, 14555 Avion Pkwy., Suite 250, Chantilly, VA, 20151. Phone: (703) 234-9810. Fax: (703) 817-1595. E-mail: info@wict.org Web Site: www.wict.org.

Women In Film, 8857 W. Olympic Blvd., Suite 201, Beverly Hills, CA, 90211. Phone: (310) 657-5144. Fax: (310) 657-5154. E-mail: info@wif.org Web Site: www.wif.org.

World Broadcasting Unions (WBU), Box 500, Stn. A, Rm. 6C 300, Toronto, ON, M5W 1E6. Canada. Phone: (416) 598-9877. Fax: (416) 598-9774. E-mail: info@nabanet.org Web Site: www.nabanet.com/wbu.

World Teleport Association, 55 Broad St., 14th Fl., New York, NY, 10004. Phone: (212) 825-0218. Fax: (212) 825-0075. E-mail: wta@worldteleport.org Web Site: www.worldteleport.org.

Directors: Chris Russell, David Sprechman, Gary Hatch, Oliver Badard, Nick Thompson, Yoshihiro Yohoyama & tohm Tahahasin.

State and Regional Broadcast Associations

Alabama Broadcasters Association, 2180 Pkwy. Lake Dr., Hoover, AL, 35244. Phone: (205) 982-5001. Fax: (205) 982-0015. Web Site: www.al-ba.com.

Alaska Broadcasters Association, 700 W. 41st Ave., #102, Anchorage, AK, 99503. Phone: (907) 258-2424. Fax: (907) 258-2414. E-mail: akba@gci.net Web Site: www.alaskabroadcasters.org.

Directors: Dennis Egan, Dick Olson, Greg Petrowich, Chris Fry, Ric Schmidt, Cherie Curry, Gary Donovan, Nancy Jonson.

Arizona Broadcasters Association, 426 N. 44th St., Suite 310, Phoenix, AZ, 85008. Phone: (602) 252-4833. Fax: (602) 252-5265. E-mail: aba3@mindspring.com Web Site: www.azbroadcasters.org.

J. D. Freeman, chmn; Diane Frisch, vice chmn.

Arkansas Broadcasters Association, 2024 Arkansas Valley Dr., Suite 403, Little Rock, AR, 72212. Phone: (501) 227-7564. Fax: (501) 223-9798. E-mail: mail@arkbroadcasters.org Web Site: www.arkbroadcasters.org.

Directors: Jim Beard, Dina Mason, Bob Knight, Jay Bunyard, Gary Bridgman, Bob Connell, Trey Stafford, Bobby Caldwell, Sandy Sanford, Rob Roedel, Gregg Fess, Donna Stweart, Ted Fortenberry, Ron Collar, Rob Hill, Tom Arnold, Paul Parker, Scott Siler.

California Broadcasters Association, 915 L St., Suite 1150, Sacramento, CA, 95814. Phone: (916) 444-2237. Fax: (916) 444-2043. E-mail: jberry@yourcba.com Web Site: www.cabroadcasters.org.

Kathy Baker, chmn.

Colorado Broadcasters Association, Box 2369, Breckenridge, CO, 80424. Phone: (970) 547-1388. Fax: (970) 547-1384. Web Site: www.coloradobroadcasters.org. E-mail: cobroadcasters@earthlink.net

Wick Rowland, sec/treas.

Directors: Bette Bailly; Byron Grandy; Frank Hanel; Ray Quinn; Bob Richards; Justin Sasso; Dan Smith; David Whitaker; Steve Zansberg.

Connecticut Broadcasters Association, 90 South Park St., Willimantic, CT, 06226. Phone: (860) 633-5031. Fax: (860) 456-5688. E-mail: mcrice@prodigy.net Web Site: www.ctba.org.

Florida Association of Broadcasters, 201 S. Monroe St., #201, Tallahassee, FL, 32301. Phone: (850) 681-6444. Fax: (850) 222-3957. Web Site: www.fab.org.

Georgia Association of Broadcasters Inc., 8010 Roswell Rd., Suite 150, Atlanta, GA, 30350. Phone: (770) 395-7200. Fax: (770) 395-7235. E-mail: pigue@gab.org Web Site: www.gab.org.

Idaho State Broadcasters Association, 270 N. 27th St., Suite B, Boise, ID, 83702-3167. Phone: (208) 345-3072. Fax: (208) 343-8046. E-mail: connies@idacomm.net Web Site: www.isda.org.

Illinois Broadcasters Association, 300 N. Pershing St., Suite B, Energy, IL, 62933. Phone: (618) 942-2139. Fax: (618) 988-9056. E-mail: ilbrdcst@neondsl.com Web Site: www.ilba.org.

Indiana Broadcasters Association Inc., 3003 E. 98th St., Suite 161, Indianapolis, IN, 46280. Phone: (317) 573-0119. Fax: (317) 573-0895. E-mail: indba@aol.com Web Site: www.indianabroadcasters.org.

Directors: Sally Brown, Arthur Angotti III, Roger Diehm, Tasha Mann, Lundy, Phil Hoover, Leigh Ellis, James Conner, Earl Metzger, Chuck Williams, Paulette Lees. Directors at Large: Marty Pieratt, Steve Lindell, Jeff Smulyan, Dr. Joe Misiewicz, Matt Jaquint, William Van Huss, Ron Miller, Scott Uecker, Brett Beshore.

Iowa Broadcasters Association, Box 71186, Des Moines, IA, 50325. Phone: (515) 224-7237. Fax: (515) 224-6560. E-mail: iowaiba@dwx.com Web Site: www.iowabroadcasters.com.

Kansas Assn of Broadcasters, 2709 S.W. 29th St., Topeka, KS, 66614. Phone: (785) 235-1307. Fax: (785) 233-3052. E-mail: harriet@kab.net Web Site: www.kab.net.

Kentucky Broadcasters Association, 101 Enterprise Dr., Frankfort, KY, 40601. Phone: (502) 848-0426. Fax: (502) 845-5710. E-mail: kba@kba.org Web Site: www.kba.org.

Directors: At Large Seat: Mark Thomas, Keith Casebolt. Radio Directors: Jim Freeland, Henry Lackey, Tom Ulmer. Rene Bell, Paul Lyons, Alan Burton. TV Directors: Jim Carter, Tim Gilbert, Ed Groves.

Louisiana Association of Broadcasters, 660 Florida St., Baton Rouge, LA, 70801. Phone: (225) 267-4522. Fax: (225) 267-4329. E-mail: lab@broadcasters.org Web Site: www.broadcasters.org.

Directors: George Sirven; Charles Spencer, legal counsel; Mike Barras; Tom Gay, treas; Bob Holladay; Stephen Levet; Irene Robinson; Nick Simonette; Irene Robin, vice chmn-radio; Larry Delia, vice chmn-TV.

Maine Assn of Broadcasters, 69 Sewall St., Augusta, ME, 04330. Phone: (207) 623-3870. Fax: (207) 621-0585. Web Site: www.mab.org.

Maryland-District of Columbia-Delaware Broadcasters Association, 106 Old Court Rd., Suite 300, Baltimore, MD, 21208-4038. Phone: (410) 653-4122. Fax: (410) 486-7354. E-mail: info@mdcd.com Web Site: www.mdcdd.com.

Massachusetts Broadcasters Association Inc., PMB 401, 43 Riverside Ave., Medford, MA, 02155. Phone: (800) 471-1875. Fax: (800) 471-1876. E-mail: info@massbroadcasters.org Web Site: www.massbroadcasters.org.

Directors: Donna Griffin.

Michigan Association of Broadcasters, 819 N. Washington Ave., Lansing, MI, 48906. Phone: (517) 484-7444. Fax: (517) 484-5810. E-mail: mab@michmab.com Web Site: www.michmab.com.

Directors: Duane Alverson, Don Backus, Al Blinke, Julie Brinks, Sue E. Goldsen, Paul Grzebik, Michael J. King, Carol Lawrence Dobrusin, Michael Murri, Gayle Olson, Bob Peters, W. Palmer Pyle, Jeffrey J. Scarpelli, Steve Wasserman. Honorary Board Members: Ed Christian, Alan Frank, Bruce Goldsen. Legal/Legislative Counsel: Rob Elhenicky, John J. Ronayne III.

Minnesota Broadcasters Association, 3033 Excelsior Blvd., Suite 301, Minneapolis, MN, 55416. Phone: (612) 926-8123. Fax: (612) 926-9761. E-mail: jdubois@minnesotabroadcasters.com Web Site: www.minnesotabroadcasters.com.

Directors: Mike Neudecker, chmn; Rosanne Rybak; Steve Woodbury; Brett Paradis; John J. Sowada; Mike Iazzo; Dennis Wahlstrom; Ed Smith. Legal Counsels: Terry Moore; Gregg Skall.

Mississippi Association of Broadcasters, 855 S. Pear Orchard Rd., Suite 403, Ridgeland, MS, 39157. Phone: (601) 957-9121. Fax: (601) 957-9175. E-mail: jlett2@earthlink.net Web Site: msbroadcasters.org.

Missouri Broadcasters Association, Box 104445, Jefferson City, MO, 65110-4445. Phone: (573) 636-6692. Fax: (573) 634-8258. E-mail: dhicks@mbaweb.org Web Site: www.mbaweb.org.

Directors: Rick McCoy, Dave Ervin, Glen Callanan, Dave Alpert, Gary Exline, Gary Leonard, Mike Smythe.

Montana Broadcasters Association, 1914 Rainbow Bend Dr., Bonner, MT, 59823. Phone: (406) 244-4622. Fax: (406) 244-5518. E-mail: mba@mtbroadcasters.org Web Site: www.mtbroadcasters.org.

Nebraska Broadcasters Association, 12020 Shamrock Plaza, Suite 200, Omaha, NE, 68154. Phone: (402) 778-5178. Fax: (402) 778-5131. E-mail: marty@ne-ba.org Web Site: www.ne-ba.org.

Nevada Broadcasters Association, 1050 E. Flamingo Rd., Suite S-102, Las Vegas, NV, 89119. Phone: (702) 794-4994. Fax: (702) 794-4997. E-mail: rdfnba@aol.com Web Site: www.nevadabroadcasters.org.

Mary Ozer, chmn; Tony Bonnici, chmn elect.

New Hampshire Association of Broadcasters, 707 Chestnut St., Manchester, NH, 03104. Phone: (603) 627-9600. Fax: (603) 627-9603. E-mail: info@nhab.org Web Site: www.nhab.org.

New Jersey Broadcasters Association, 348 Applegarth Rd., Monroe Twp., NJ, 08831. Phone: (609) 860-0111. Fax: (609) 860-0110. E-mail: njba@njba.com Web Site: www.njba.com.

Directors: Arthur Camiolo, Josh Gertzog, Charles McCreery, Joseph M. Bilotta, Dan Spears, Richard Swetits, John F. Garziglia & Thomas R. Ray.

New Mexico Broadcasters Association, 2333 Wisconsin N.E., Albuquerque, NM, 87110. Phone: (505) 881-4444. Fax: (505) 881-5353. E-mail: info@nmba.org Web Site: www.nmba.org.

Directors: Matt Martinez, Gene Dow & Milt McConnell.

New York Market Radio Broadcasters Association (NYMRAD), 261 Madison Ave., 23rd Fl., New York, NY, 10016. Phone: (646) 254-4493. Fax: (646) 254-4498. E-mail: db@nymrad.org Web Site: www.nymrad.org.

New York State Broadcasters Association Inc., 1805 Western Ave., Albany, NY, 12203. Phone: (518) 456-8888. Fax: (518) 456-8943. E-mail: info@nysbroadcasters.org Web Site: www.nysbroadcaster.org.

North Carolina Assoc of Broadcasters, Box 627, Raleigh, NC, 27602. Phone: (919) 821-7300. Fax: (919) 839-0304. Web Site: www.ncbroadcast.com.

Directors: Don

North Dakota Broadcasters Association, Box 3178, Bismarck, ND, 58502-3178. Phone: (701) 258-1332. Fax: (701) 250-6372. E-mail: bethh@ndba.org Web Site: www.ndba.org.

Directors: Syd Stewart, Barry Schumaier, Larry Timpe, Tim Ost, Carol Anhorn, Darren Lenertz, George Smith.

Northern California Broadcasters Association, 50 Francisco St., Suite 450, San Francisco, CA, 94133. Phone: (415) 292-5700. Fax: (415) 292-5790. E-mail: tdevoto@ncradio.com Web Site: www.ncbaradio.com.

Ohio Association of Broadcasters, 88 E. Broad St., Suite 1180, Columbus, OH, 43215-3525. Phone: (614) 228-4052. Fax: (614) 228-8133. E-mail: oab@oab.org Web Site: www.oab.org.

Oklahoma Association of Broadcasters, 6520 N. Western, Suite 104, Oklahoma City, OK, 73116. Phone: (405) 848-0771. Fax: (405) 848-0772. E-mail: smith@oakok.org Web Site: www.oabok.org.

Oregon Assn of Broadcasters, 7150 S.W. Hampton St., Suite 240, Portland, OR, 97223-8366. Phone: (503) 443-2299. Fax: (503) 443-2488. E-mail: theoab@theoab.org Web Site: www.theoab.org.

John Mielke, chmn of bd; Ron Hren, vice chmn. Directors: Cary Jones, Kieran Clarke, Dan Manciu, Bill Ashenden, Joe Costello, Angela Pursel & Lee Perkins.

Directors: J. Dominic Monahan, legal counsel; Ron Hren, Kenn Brown, Kieran Clarke, Gary Jones, Dan Manciu, Angela Pursel, Robert Dove, Bill Ashenden, James Boyd, Lee Perkins, John Mielke, John Rice, Paul Steinle.

Pennsylvania Association of Broadcasters, 8501 Paxton St., Hummelstown, PA, 17036. Phone: (717) 482-4820. Fax: (717) 482-1111. E-mail: rwyckoff@pab.org Web Site: www.pab.org.

Rhode Island Broadcasters Association, 58 Cool Spring Dr., Cranston, RI, 02920. Phone: (401) 255-8200. E-mail: 1needham@ribroadcasters.com

South Carolina Broadcasters Association, One Harbison Way, Suite 112, Columbia, SC, 29212. Phone: (803) 732-1186. Fax: (803) 732-4085. E-mail: scba@scba.net Web Site: www.scba.net.

South Dakota Broadcasters Association, Box 1037, 106 W. Capital Ave., Pierre, SD, 57501. Phone: (605) 224-1034. Fax: (605) 224-7426. E-mail: info@sdba.org Web Site: www.sdba.org.

Directors: Roger Currier, Lorin Larsen, Cindy McNeill, Monte Loos, J.P. Skelly , Mark Millage, Bob Miller, assoc bd members.

Tennessee Association of Broadcasters, 50 Music Sq. W., Suite 900, Nashville, TN, 37203. Phone: (615) 321-1626. Fax: (615) 824-0054. E-mail: tabtn@bellsouth.net Web Site: www.tabtn.org.

Texas Association of Broadcasters, 502 E. 11th St., Suite 200, Austin, TX, 78701. Phone: (512) 322-9944. Fax: (512) 322-0522. E-mail: tab@tab.org Web Site: www.tab.org.

Directors: Ann Arnold, exec dir; Roger Bare; Bob Cohen; Danny Baker; Frank Carter; Pedro Gase; Jason Hightower; Brian Jones; John Kerr; Mike Lee; Mark Masepohl; Mark McKay; Tom O'Brien; Jackie Rutledge; Ted Wernn; Rodney Zent.

Utah Broadcasters Association, 1600 S. Main St., Salt Lake City, UT, 84115. Phone: (801) 486-9521. Fax: (801) 484-7294. Web Site: www.utahbroadcasters.com.

Vermont Association of Broadcasters, Box 4489, Burlington, VT, 05406. Phone: (802) 476-8789. Fax: (802) 476-8789 (Call First). Web Site: www.vab.org. E-mail: annoyes@aol.com

Judith Leech, sec; Dennis Snyder, treas.

Virginia Association of Broadcasters, 600 Peter Jefferson Pkwy, Suite 300, Charlottesville, VA, 22911. Phone: (434) 977-3716. Fax: (434) 979-2439. E-mail: doug.easter@easterassociates.com Web Site: www.vab.net.

Directors: Michael Guild.

Directors: Nick Nicholson; Harrison Pittman; Larry Saunders; Randy Smith; Jack Dempsey; Tex Meyer; Robert Scutari; Doris Newcomb; Bob Peterson; Linda Forem; Bob Willloughby; Francis Wood; John Schick; Kenneth Hill.

Virginia Public Radio Association, c/o WCVE, 23 Sesame St., Richmond, VA, 23235. Phone: (804) 320-1301. Fax: (804) 320-8729. E-mail: bmiller@ideastations.org

Washington State Association of Broadcasters, 724 Columbia St., Suite 310, Olympia, WA, 98501-1249. Phone: (360) 705-0774. Fax: (360) 705-0873. E-mail: wa-broadcasters@earthlink.net Web Site: www.wsab.org.

West Virginia Broadcasters Association, 140 7th Ave., South Charleston, WV, 25303-1452. Phone: (304) 744-2143. Fax: (304) 744-1764. E-mail: wvba@wvba.com Web Site: www.wvba.com.

Directors: Jay Phillipone, treas; Mike Smith, past pres, WOAY-TV

Wisconsin Broadcasters Association, 44 E. Mifflin St., Suite 900, Madison, WI, 53703. Phone: (608) 255-2600. Phone: (800) 236-1922. Fax: (608) 256-3986. E-mail: jlaabs@aol.com Web Site: www.wi-broadcasters.org.

Directors: Edward Allen III, Ellis Bromberg, Juli Buehler, Jim Hall, Bill Hurwitz, Tom Koser, Al Lancaster, Dean Maytag, Kira LaFond, Jeff Tyler, Bob Miller & Jill Sommers.

Wausau, WI 54403, 1908 Grand Ave. Laurin Jorstad, WAOW-TV.

Wausau, WI 54402, Box 2048. Bob Jung, Midwest Communications.

West Bend, WI 53095, Box 933. Jim Hodges, WBKV/WBWI.

Directors: Gregg Albert, Edward Allen III, Ellis Bromberg, Juli Buehler, Scott Chorski, Bill Hurwitz, Kira Lafond, Dean Maytag, Bob Miller, Jeff Robinson, Don Rosette, Jill Sommers.

Wyoming Association of Broadcasters, 7217 Hawthorne Dr., Cheyenne, WY, 82009. Phone: (307) 632-7622. Fax: (307) 638-3469. E-mail: grottski@aol.com Web Site: www.wyomingbroadcasting.org.

Directors: Larry Cross, chmn; Roger Gelder, vice chmn; Steve Core, vice chmn.

State and Regional Cable Associations

Alabama Cable Telecommunications Association, Box 230666, Montgomery, AL, 36123-0666. Phone: (334) 271-2281. Fax: (334) 271-2260. E-mail: alacable@aol.com Web Site: www.alcta.com.

Arizona-New Mexico Cable Telecommunications Association, 3875 N. 44th St., Suite 300, Phoenix, AZ, 85018. Phone: (602) 955-4122. Fax: (602) 955-4505. E-mail: info@azcable.org Web Site: www.azcable.org.

Directors: Susan Bitter Smith, exec dir; Rocky Rinehart, pres

Arkansas Cable Telecommunications Association, 411 South Victory, Suite 201A, Little Rock, AR, 72201. Phone: (501) 907-6440. E-mail: info@arcta.org Web Site: www.arcta.org.

Directors: Mike Wilson, Dennis Yocum, Garry Bowman, Harold Kimmel, Rick Smith, Harvey Oxner, Jay Butler, Doug Martin.

Broadband Cable Association of Pennsylvania, 127 State St., Harrisburg, PA, 17101. Phone: (717) 214-2000. Fax: (717) 214-2020. Web Site: www.bcata.com.

Broadband Communications Association of Washington, 216 First Ave. S., Suite 260, Seattle, WA, 98104. Phone: (206) 652-9303. Fax: (206) 652-8297. E-mail: rmain@broadbandwashington.org Web Site: www.broadbandwashington.org.

Directors: Janet Turpen, Steve Kipp, Jerry Rotondo, Bob Lam, Matt Zavala, Bruce Gladner & Carlos Gutirrez.

Directors: Janet Turpen, Steve Kipp, Steve Holmes, Bob Rubery, Jim Penney, Carlos Gutierrez, Bruce Gladner.

Cable Telecommunications Assn. of Maryland, Delaware & District of Col., 2530 Riva Rd., #316, Annapolis, MD, 21401. Phone: (410) 266-9111. Fax: (410) 266-6133. E-mail: ctaofmd-de-dc@msn.com

Cable Telecommunications Association of New York Inc., 80 State St., 10th Fl., Albany, NY, 12207. Phone: (518) 463-6676. Fax: (518) 463-0574. E-mail: cttany@ny.rr.com Web Site: www.cabletvny.com.

Directors: Lisa Rosenblum, chmn.

Cable Television & Communications Association of Illinois, 2400 E. Devon, Suite 317, Des Plaines, IL, 60018. Phone: (847) 297-4520. Fax: (847) 297-3865. E-mail: ctc2400@aol.com

Cable Television Association of Georgia, 6175 Barfield Rd., Suite 220, Atlanta, GA, 30328. Phone: (404) 252-4371. Fax: (404) 252-0215. E-mail: info@gacable.com Web Site: www.gacable.com.

Directors: Kim Gage, chmn; Michael Clemons.

California Cable & Telecommunications Assn., 360 22nd St., Suite 750, Oakland, CA, 94612. Phone: (510) 628-8043. Fax: (510) 628-8334. E-mail: sr@calcable.org Web Site: www.calcable.org.

Lee Perron, chmn; Tan Carlock, sec; Roger Keating, treas.

Sacramento, CA 95814, 1121 L St, Suite 400. Phone: (916) 446-7732. Fax: (916) 446-1605.

Directors: Leo Brennan, chmn; Don Schena, sect; Jeffrey Schwall, treas.

Colorado Cable TV Association, 1512 Larimer St., Suite 700, Denver, CO, 80202. Phone: (303) 607-0486. Fax: (303) 436-1191. E-mail: pboyle@rbwpolicy.com Web Site: www.cocabletv.com.

Hawaii Cable Television Association, 200 Akamainui St., Mililani, HI, 96789. Phone: (808) 625-8359. Fax: (808) 625-5888. E-mail: kbeuret@oceanic.com

Idaho Cable Telecommunications Association, 1015 W. Hays Street, Boise, ID, 83702. Phone: (208) 344-6633. Fax: (208) 344-0077. Web Site: www.idahocable.com.

Directors: Michelle Cameron, pres, Russ Young, past pres.

Indiana Cable Telecommunications Association Inc., 201 N. Illinois, Suite 1560, Indianapolis, IN, 46204. Phone: (317) 237-2288. Fax: (317) 237-2290. Web Site: www.incable.org. E-mail: toakes@incable.org

Directors: Rachel McKay & Nicole Roenl.

Iowa Cable & Telecommunications Association, Box 3627, Des Moines, IA, 50322. Phone: (515) 276-0006. Fax: (515) 309-3779. E-mail: tomgraves@mchsi.com

Directors: Jerry Kittelson, pres; Jeff Olson, VP.

Kansas Cable Telecommunications Association, 815 S.W. Topeka Blvd., 2nd Floor, Topeka, KS, 66612. Phone: (785) 290-0018. Fax: (785) 232-1703. E-mail: johnfed@cox.net

Directors: Gary Shorman, pres; Joe Michael, VP; Jay Allbaugh, treas; Mike Flood; Pat James; Linda Jurgensen; Patrick Knorr; Tom Krewson; Clarence Matlock; Rob Moel; Roger Ponder.

Kentucky Cable Telecommunications Association, Box 415, Burkesville, KY, 42717. Phone: (270) 864-5352. Fax: (270) 864-3110. E-mail: juddph@mchsi.com Web Site: www.kycable.com.

Jim Finch, assoc dir; Jim Hays III, assoc dir; Robert Thacker, assoc dir.

Louisiana Cable & Telecommunications Association, 763 North St., Baton Rouge, LA, 70802. Phone: (225) 387-5960. Fax: (225) 383-6705. E-mail: lcta@lacable.com Web Site: www.lacable.com.

Michigan Cable Telecommunications Association, 412 W. Ionia St., Lansing, MI, 48933. Phone: (517) 482-2622. Fax: (517) 482-1819. Web Site: www.michcable.org.

Directors: Colleen M. McNamara, exec dir; Bob McCann, sec; Rick Clark, treas; Tim Ransberge, pres, Ron Orlando, VP.

Mid-America Cable Telecommunications Association, 223 E. Capitol Ave., Jefferson City, MO, 65102. Phone: (573) 635-5588. Fax: (573) 635-1778. E-mail: info@midamericacable.tv Web Site: www.midamericacable.tv.

Directors: Charlotte McClure, chmn; Tom Krewson, first VP; Brian Thompson, second VP; Vic Davis, sect/treas; Richard Bates, LeaAnn Quist; Carol Rothwell; Chance Russell; Bill Severn, assoc dir: Debby Exon; Kim Francis; David Headley; Deirdre LaVerdiere; Tyler Leach; Blake Miller; Rick Moravec; Oscar Ordaz; Joe Scott; Larry Stiffelman; Wendy Tobias. Ex-Officio Dir: mary Campbell; Greg Harrison; Rob Marshall; Emeritus: Ron Marnell.

Minnesota Cable Communications Association, 1885 University Ave, Suite 320, St. Paul, MN, 55104. Phone: (651) 641-0268. Fax: (651) 641-0319. E-mail: mmartin@mnccn.com Web Site: www.mncca.com.

Mississippi Cable Telecommunications Association, 1501 Lakeland Dr., Suite 301, Jackson, MS, 39216. Phone: (601) 981-3646. Fax: (601) 981-5547. E-mail: mcta@bellsouth.net Web Site: www.mctaweb.com.

Missouri Cable Telecommunications Association, 223 E. Capitol Ave., Box 1895, Jefferson City, MO, 65102-1895. Phone: (573) 635-1915. Fax: (573) 635-1778. E-mail: gpharrison@mchsi.com Web Site: www.missouricabletv.com.

Nebraska Cable Communications Association, 1233 Lincoln Mall, Suite 203, Lincoln, NE, 68508. Phone: (402) 474-3242. E-mail: mary@campbellassociates.net

Directors: Mary Campbell, exec dir; Bridgit Farley, assoc dir; John Fullenkamp, legal counsel; Randy Bang, Valerie Kramer, Dick Bates, Mike Kohler, Gerald Lampe, Greg Harrison, LeaAnn Quist.

Nevada State Cable Telecommunications Association, 2210 Sugar Bowl Ct., Reno, NV, 89511-9175. Phone: (775) 852-2253. Fax: (775) 852-2403. E-mail: nscta@aol.com

Directors: Steve Schorr; Leon Brennan, VP; Scott Dockery, sec/treas; Marsha Berkbigler.

New England Cable Telecommunications Association Inc. (NECTA), 10 Forbes Rd., Suite 440W, Braintree, MA, 02184. Phone: (781) 843-3418. Fax: (781) 849-6267. E-mail: info@necta.info

New Jersey Cable Telecommunications Association, 124 W. State St., Trenton, NJ, 08608. Phone: (609) 392-3223. Fax: (609) 394-0074. Web Site: www.cablenj.org.

Directors: Adam Falk, chmn.

North Carolina Cable Telecommunications Association, Box 1347, Raleigh, NC, 27602. Phone: (919) 834-7113. Fax: (919) 839-0304. E-mail: lreynolds@nccta.com Web Site: www.nccta.com.

North Central Cable Television Association, 1885 University Ave., Suite 320, St. Paul, MN, 55104. Phone: (651) 641-0268. Fax: (651) 641-0319. E-mail: mncableassc@comcast.net

Director: Scott Geston, pres.

Ohio Cable Telecommunications Association, 50 W. Broad St., Suite 1118, Columbus, OH, 43215. Phone: (614) 461-4014. Fax: (614) 461-9326. E-mail: octa@octa.org Web Site: www.octa.org.

Directors: Tom Dawson, pres; Kevin Hayes, VP; Rhonda Fraas, sec; Kevin Flanigan, treas; Other Directors: Bob Gessner, Zakee Rashid, Virgil Reed, Dex Sedwick, Steve Trippe. Associate Director: Gayle Hanrahan.

Oklahoma Cable and Telecommunications Association, 301 N.W. 63rd, Suite 400, Oklahoma City, OK, 73116. Phone: (405) 843-8855. Fax: (405) 843-8934. E-mail: octa@coxatwork.com Web Site: www.okcable.net.

Dave Bialis, chmn; Andy Dearth, vice chmn; George Wilburn, sec/treas. Directors: Bill Drewry, Tim Tippit, Nicole Evans, Johnny Bowen, David Wall, Leon Pfeifer, Danny Thompson, Ed Perry, Holly Henderson & Tim Easley.

Oregon Cable Telecommunications Association, 1249 Commercial St. S.E., Salem, OR, 97302. Phone: (503) 362-8838. Fax: (503) 399-1029. E-mail: mdewey@oregoncable.com Web Site: www.oregoncable.com.

Tennessee Cable Telecommunications Association, 611 Commerce St., Suite 2706, Nashville, TN, 37203. Phone: (615) 256-7037. Fax: (615) 254-9710. E-mail: sb@tcta.net Web Site: www.tcta.net.

Texas Cable & Telecommunications Association, 506 W. 16th St., Austin, TX, 78701. Phone: (512) 474-2082. Fax: (512) 474-0966. Web Site: www.txcable.com.

Utah Cable Telecommunications Association, 10714 South Jordon Gateway, Ste 260, South Jordon, UT, 84905. Phone: (801) 619-6660. Fax: (801) 619-6661. E-mail: nancy@urta.org

Director: Shane Baggs, pres.

Virginia Cable Telecommunications Association, 1001 E. Broad St., Suite 210, Richmond, VA, 23219. Phone: (804) 780-1776. Fax: (804) 225-8036. E-mail: rlamura@vcta.com Web Site: www.vcta.com.

Director: Kirby Brooks, chmn; Danita Bowman, dir of govt. rel.

West Virginia Cable Telecommunications Association, 117 Summers St., Charleston, WV, 25301. Phone: (304) 345-2917. Fax: (304) 342-1285. E-mail: mpolen@arnoldagency.com Web Site: www.wvcta.com.

Directors: Mark Polen, Michael Kelemen (Capital Cablecomm, Charleston, WV), Joel Patten (Harnon Cable Communications, St. Albans, WV), Jim Underwood (TCI of West Virginia, Parkersburg, WV).

Wisconsin Cable Communications Association, 22 East Mifflin Street, Suite 1010, Madison, WI, 53703. Phone: (608) 256-1683. Fax: (608) 256-6222. E-mail: wcca@charterinternet.com

Directors: Mike Fox, Robert Ryan, John Miller, Bev Greenberg, Emmett Coleman, Lisa Washa. Director at large: Doug Nix, Dave Seyora. Associate Directors: Kenneth Mullane, Kate Schroeder, Jeff Fischer. Advisory Committee: Christy Benson, Randy Bunnell, Wendy Gross, Charlie Mullen, Randy Scott. JackHerbert, pres, Tim Vowell, treas, Bob Steichen, past pres, past, Thomas E. Moore, exec dir, Thomas S. Henson, regulatory affs, Nancy Magestro, office admin, Bev Kautzky, accounting mgr.

Wyoming CATV Association, 1113 Lucky Ct., Cheyenne, WY, 82001. Phone: (307) 637-3933. Phone: (307) 733-3081. Fax: (307) 637-5399. Web Site: wyocable.org.

Board Members: Clint Rodeman, pres; David Alexanderson, VP; Mary Johnson, sec/treas; John Nickle, past pres; Dawn Kirkwood; Marty Carollo; Marty Carollo; Wes Frost; Darlene Raymond. Assoc Board Members: Jaime Pena.

Union/Labor Groups

Actors' Equity Association (AEA), (AFL-CIO). 165 W. 46th St., New York, NY, 10036. Phone: (212) 869-8530. Fax: (212) 719-9815. Web Site: www.actorsequity.org.

Mark Zimmerman, pres; John Connolly, exec dir.

Los Angeles, CA 90036, 5757 Wilshire Blvd, Suite 1. Phone: (323) 634-1750. Fax: (323) 634-1777.

San Francisco, CA 94104, 350 Sansome St., Ste 900. Phone: (415) 391-3838. Fax: (415) 391-0102. Joel Reamer, business rep.

Orlando, FL 32821, 10319 Orangewood Blvd. Phone: (407) 345-8600. Fax: (407) 345-1522. Brian Spitler, business rep.

Chicago, IL 60603, 125 S. Clark St. Phone: (312) 641-0393. Fax: (312) 641-6365. Kathryn Lamkey, central rgnl dir.

Labor Union for Theatrical Actors & Stage mgrs.

Affiliated Property Craftsperson (IATSE Local 44), (IATSE, AFL-CIO). 12021 Riverside Dr., North Hollywood, CA, 91607. Phone: (818) 769-2500. Fax: (818) 769-3111. E-mail: ejennings@local144.org Web Site: www.local44.org.

Elliott Jennings, sec/treas.

American Federation of Labor-Congress of Industrial Organizations (AFL-CIO), 815 16th St. N.W., Washington, DC, 20006. Phone: (202) 637-5000. Fax: (202) 637-5058. E-mail: bholton@afleio.org Web Site: www.aflcio.org.

John J. Sweeney, pres; Linda Chavez-Thompson, exec VP.

American Federation of Musicians, United States & Canada, 1501 Broadway, Suite 600, New York, NY, 10036. Phone: (212) 869-1330. Fax: (212) 764-6134. Web Site: www.afm.org.

Thomas F. Lee, pres; Sam Folio, sec/treas.

Washington, DC 20036, 1717 K St. NW, Suite 500. Phone: (202) 463-0772. Fax: (202) 466-9009. Hal Ponder.

Don Mills, ON M3C 2E9 Canada, 75 The Donway W, Suite 1010. Phone: (416) 391-5161. Fax: (416) 391-5165. Bobby Herriot.

Los Angeles, CA 90010. Los Angeles, 3550 Wilshire Blvd, Suite 1900. Phone: (213) 251-4510. Fax: (213) 251-4520.

Representing 110,000 professional musicians throughout the United States & Canada.

American Federation of Television & Radio Artists (AFTRA), (AFL-CIO). 260 Madison Ave., New York, NY, 10016. Phone: (212) 532-0800. Fax: (212) 532-2242. Web Site: www.aftra.com.

Roberta Reardon, pres.

Los Angeles, CA 90036, 5757 Wilshire Blvd., 9th Fl. Phone: (323) 634-8100. Fax: (323) 634-8194. John Russum.

AFTRA represents 77,000 professional actors, singers, dancers, announcers, newspersons, sportscasters & disc jockeys throughout the country who work in TV, radio, comls, industrial & educ videos, interactive media & the recording industry.

American Guild of Musical Artists, 1430 Broadway, 14th Fl, New York, NY, 10018-3308. Phone: (212) 265-3687. Fax: (212) 262-9088. E-mail: agma@musicalartists.org Web Site: www.musicalartists.org.

Linda Mays, pres.

Branch offices in Los Angeles, CA; Chicago, IL; San Francisco, CA; New Orleans, LA; Seattle, WA; Dallas, TX; Washington, DC; Boston, MA; Philadelphia, PA.

American Guild of Variety Artists, (AFL-CIO). 363 7th Ave 17th Fl., New York, NY, 10001. Phone: (212) 675-1003. Fax: (212) 633-0097. E-mail: agva@aol.com Web Site: www.agva.com.

Rod McKuen, pres; David Cullum, VP.

Los Angeles, CA 91607, 4741 Laurel Canyon Blvd. Phone: (818) 508-9984. Fax: (818) 508-3029.

Labor Union for performers in live venues.

The Animation Guild (IATSE Local 839), 4729 Lankershim Blvd., North Hollywood, CA, 91602. Phone: (818) 766-7151. Fax: (818) 506-4805. E-mail: info@animationguild.org Web Site: www.animationguild.org.

The labor union for animation & CG artists & technicians in southern CA.

Art Directors Guild & Scenic Title and Graphic Artists (IATSE Local 800), 11969 Ventura Blvd., Suite 200, Studio City, CA, 91604. Phone: (818) 762-9995. Fax: (818) 762-9997. E-mail: lydia@artdirectors.org Web Site: www.artdirectors.org.

Scott Roth, exec dir; Tom Walsh, pres.

Broadcast-Television Recording Engineers (IBEW Local 45), 6255 Sunset Blvd., Suite 721, Hollywood, CA, 90028. Phone: (323) 851-5515. Fax: (323) 466-1793. E-mail: feedback@ibew45.org Web Site: www.ibew45.org.

Represents radio, TV, cable engirs, federal, county, city electronic tech.

Communications Workers of America (CWA), (AFL-CIO). 501 3rd St. N.W., Washington, DC, 20001-2797. Phone: (202) 434-1100. Fax: (202) 434-1279. E-mail: jmiller@cwa-union.org Web Site: www.cwa-union.org.

Larry Cohen, pres; Jeff Miller, dir communications.

Directors Guild of America Inc. (DGA), 7920 Sunset Blvd., Los Angeles, CA, 90046. Phone: (310) 289-2000. Fax: (310) 289-2029. Web Site: www.dga.org.

Michael Apted, pres; Jay Roth, exec dir.

Chicago, IL 60611, 400 N. Michigan Ave, Suite 307. Phone: (312) 644-5050. Fax: (312) 644-5775.

New York, NY 10019, 110 W. 57th St. Phone: (212) 581-0370. Fax: (212) 581-1441. Chris Lomdino, eastern exec dir.

Illustrators & Matte Artists (IATSE Local 790), 13245 Riverside Dr., Suite 300-A, Sherman Oaks, CA, 91423. Phone: (818) 784-6555. Fax: (818) 784-2004. E-mail: local790@earthlink.net

Joseph Musso, pres; Marjo Bernay, business rep.

International Alliance of Theatrical Stage Employees, Moving Picture (IATSE), 1430 Broadway, 20th Fl, New York, NY, 10018. Phone: (212) 730-1770. Fax: (212) 730-7809. Web Site: www.iatse-intl.org.

Thomas Short, intl pres.

Toronto, ON M5A 1N1 Canada, 258 Adelaide St E, Suite 403. Phone: (416) 362-3569. Fax: (416) 362-3483.

Toluca Lake, CA 91602, 10045 Riverside Dr. Phone: (818) 980-3499. Fax: (818) 980-3496. Joseph Aredas.

International Association of Machinists and Aerospace Workers (IAM), 9000 Machinists Pl, Upper Marlboro, MD, 20772-2687. Phone: (301) 967-4500. E-mail: websteward@goiam.org Web Site: www.iamaw.org.

R. Thomas Buffenbarger, pres; Warren Mart, sec/treas.

International Brotherhood of Electrical Workers (IBEW), (AFL-CIO). 900 7th St. N.W., Washington, DC, 20001. Phone: (202) 728-6026. Fax: (202) 728-6295. E-mail: broadcasting@ibew.org Web Site: www.ibew.org.

Edwin D. Hill, pres; Peter Homes, dir.

Represent worker in all areas of bcstg & cable.

International Cinematographers Guild, 7715 Sunset Blvd., Suite 300, Los Angeles, CA, 90046. Phone: (323) 876-0160. Fax: (323) 876-6383. Web Site: www.cameraguild.com.

Gary Dunham, pres; Bruce C. Doering, exec dir.

Orlando, FL 32835, 7463 Conroy-Windermere Rd, Suite A. Phone: (407) 295-5577. Fax: (407) 295-5335.

Park Ridge, IL 60068, 1411 Peterson Ave, Suite 102. Phone: (847) 692-9900. Fax: (847-692-5607.

New York, NY 10011, 80 Either Ave, 14th Fl. Phone: (212) 647-7300. Fax: (212) 647-7317.

Represents our members' contracts & activities.

International Sound Technicians (IATSE Local 695), (IATSE, MPMO). 5439 Cahuenga Blvd., North Hollywood, CA, 91601. Phone: (818) 985-9204. Phone: (323) 877-1052. Fax: (818) 760-4681. E-mail: local695@695.com Web Site: www.695.com.

Mark Ulano, pres; James Osburn, business agent.

International Union of Electronic & Communications Workers of America, 501 3rd St. N.W., Washington, DC, 20001. Phone: (202) 434-1100. Web Site: iue-cwa.org.

James Clark, pres; Bill Gray, dir info svcs.

Laboratory Film/Video Technicians & Cinetechnicians (IATSE Local 683), Box 7429, Burbank, CA, 91510-7429. Phone: (818) 252-5628. Fax: (818) 252-4962. E-mail: scottgeorge@mindsprings.com

Bill Milano, pres; Scott George, business agent.

New York, NY 10036-5741. IATSE, 1515 Broadway, Suite 601.

Serves the Labor Union.

Make-Up Artist & Hairstylists Guild (IATSE Local 706), 828 N. Hollywood Way, Burbank, CA, 91505. Phone: (818) 295-3933. Fax: (818) 295-3930. E-mail: info@ialocal706.org Web Site: www.local706.org.

Susan Cabral-Ebert, pres; Tommy Cole, business rep.

Union for make-up artists, hair stylist's in motion picture industry.

Motion Picture Costumers (IATSE Local 705), 4731 Laurel Canyon Blvd. Suite 201, Valley Village, CA, 91607. Phone: (818) 487-5655. Fax: (818) 487-5663. E-mail: mpc705@aol.com Web Site: www.motionpicturecostumers.org.

Sandra Berke Jordan, pres; Buffy Snyder, business rep.

The gathering (by rental or puchase) of costumes for film & TV. The construction of new costumes (wardrobe). The fitting & handling of costumes (wardrobe) during filming.

Motion Picture Editors Guild (IATSE Local 700), 7715 W. Sunset Blvd., Suite 200, Hollywood, CA, 90046. Phone: (323) 876-4770. Fax: (323) 876-0861. E-mail: mail@editorsguild.com Web Site: www.editorsguild.com.

Chicago, IL 60631. Chicago, 6317 N. Northwest Hwy. Phone: (773) 594-6598. Fax: (773) 594-6599.

New York, NY 10013. New York, 145 Hudson St., Suite 201. Phone: (212) 302-0700. Fax: (212) 302-1091. Paul Moore, asst exec dir.

Labor union representin post-production employees.

Motion Picture Set Painters & Sign Writers (IATSE Local 729), 1811 W. Burbank Blvd., Burbank, CA, 91506-1314. Phone: (818) 842-7729. Fax: (818) 846-3729. Web Site: www.ialocal729.com.

Kirk Hansen, pres; George Palazzo, business rep.

National Association of Broadcast Employees & Technicians, (Communications Workers of America, AFL-CIO). 501 3rd St. N.W., 5th Fl., Washington, DC, 20001. Phone: (202) 434-1254. Fax: (202) 434-1426. E-mail: nabet@cwa-union.org Web Site: www.nabetcwa.org.

John Clark, pres.

New York, NY 10106, Local 11, 888 7th Ave., Suite 4511. Phone: (212) 757-3065. John S. Clark, pres NABET-CWA.

Representing employees in the bcstg, cable TV & related industries.

The Newspaper Guild, (CWA). 501 3rd St. N.W., Washington, DC, 20001. Phone: (202) 434-7177. Fax: (202) 434-1472. Web Site: www.newsguild.org. E-mail: guild@cwa-union.org

Linda K. Foley, pres; Bernie Lunzer, sec/treas.

Ottawa, ON K2C 3P1 Canada, Baxter Centre, 1050 Baxter Rd, Unit 7B. Phone: (613) 820-9777. Arnold Amber, Canadian dir.

Represents newsroom & other media workers in U.S., Canada & Puerto Rico.

Office & Professional Employees International Union, 265 W 14th St, 6th Fl, New York, NY, 10011. Phone: (212) 675-3210.

Michael Goodwin, pres.

Professional Musicians, Local 47, AFM, 817 N. Vine St., Hollywood, CA, 90038-3779. Phone: (323) 462-2161. Fax: (323) 461-5260. Web Site: www.promusic47.com.

Hal Espinosa, pres; Vince Trombetta, VP; Serena Kay Williams, sec/treas.

Musicians, vocalists, orchestrators, copyists, composers, conductors, contractors & librarians, referral service & recording studio.

Screen Actors Guild, 5757 Wilshire Blvd., Los Angeles, CA, 90036. Phone: (323) 954-1600. Fax: (323) 549-6656. Web Site: www.sag.org.

Alan Rosenberg, pres.

Contract administration, negotiation & enforcement, residual payment processing, regulation franchising of talent agents, membership, record-keeping & communication.

Script Supervisors/Continuity & Allied Prodution Specialists Guild Local 871, IATSE, (IATSE). 11519 Chandler Blvd., North Hollywood, CA, 91601. Phone: (818) 509-7871. Fax: (818) 506-1555. E-mail: ialocal871@aol.com Web Site: www.ialocal871.org.

"We think for a living." Script Supervisors/Continuity. Teleprompter Operators, Production Office Coordinator, Art Department Coordinator, Production Accountants & Assistants.

Service Employees International Union (SEIU), 1313 L St NW, Washington, DC, 20005. Phone: (202) 898-3200. Fax: (202) 350-6614. E-mail: burgera@seiu.org Web Site: www.seiu.org.

Andrew Stern, pres; Anna Burger, sec/treas.

Set Designers & Model Makers (IATSE Local 847), 13245 Riverside Dr., Ste 300-A, Sherman Oaks, CA, 91423. Phone: (818) 784-6555. Fax: (818) 784-2004. E-mail: Local847@earthlink.net

Jim Wallace, pres; Marjo Bernay, business rep.

Studio Electrical Lighting Technicians (IATSE Local 728), 14629 Nordhoff St., Panorama City, CA, 91402. Fax: (818) 891-5288. E-mail: loc728@iatse728.org Web Site: www.iatse728.org.

Patric J. Abaravich, pres.

United Electrical, Radio & Machine Workers of America (UE), One Gateway Ctr., Suite 1400, Pittsburgh, PA, 15222-1416. Phone: (412) 471-8919. Fax: (412) 471-8999. Web Site: www.ranknfile-ue.org. E-mail: ue@ranknfile-ue.org

John H. Hovis, pres; Bruce Klipple, sec/treas.

United Scenic Artists (IATSE Local USA 829), 29 W. 38th St. 15th Fl., New York, NY, 10018. Phone: (212) 581-0300. Fax: (212) 977-2011. E-mail: administrator@usa829.org Web Site: www.usa829.org.

Michael McBride, business agent.

Los Angeles, CA 90036, 5225 Wilshire Blvd, Suite 506. Phone: (323) 965-0957. Fax: (323) 965-0958. Charles Berliner, rgnl business rep.

Miami, FL 33173, 10459 SW 78th St. Phone: (305) 596-4772. Fax: (305) 596-6095. David Goodman, rgnl business rep.

Chicago, IL 60601, 203 N. Wabash, Suite 1210. Phone: (312) 857-0829. Fax: (312) 857-0819. Chris Phillips, rgnl business rep.

Representing designers of set, costume, lighting, sound, scenic artists, computer arts and art dept coordinators in the entertainment industry.

Writers Guild of America, East Inc. (WGAE), 555 W. 57th St., Suite 1230, New York, NY, 10019. Phone: (212) 767-7800. Fax: (212) 582-1909. E-mail: info@wgaeast.org Web Site: www.wgaeast.org.

Warren Leight, pres.

Writers Guild of America, West Inc. (WGAW), 7000 W. Third St., Los Angeles, CA, 90048-4329. Phone: (323) 951-4000. Fax: (323) 782-4800. Web Site: www.wga.org. E-mail: gscott@wga.org

Patrick Verrone, pres; Gabriel Scott, dir public affrs.

WGAW represents writers primarily for the purpose of collective bargaining in the motion picture, bcst, cable & new technologies industries.

Trade Shows

Arizona Cable Telecommunication Association Annual Meeting. SHOW MANAGEMENT: Arizona Cable Telecommunication Association. ASSOCIATION ADDRESS: 3610 N. 44th Ave., Suite 240, Phoenix, AZ 85018. Contact: Susan Bitter Smith, executive dir. (602) 955-4122. FAX: (602) 955-4505. www.azcable.org. E-mail: info@azcable.org.

Audio Engineering Society Convention (Public/Trade). SHOW DATE: May 2008, Vienna, Austria. SHOW MANAGEMENT: Audio Engineering Society (AES), 60 E. 42nd St., Rm. 2520, New York, NY 10165. Contact: Roger Furness, exec dir. (212) 661-8528. FAX: (212) 682-0477. Web site: http://www.aes.org. SHOW MANAGEMENT STATEMENT: This show provides an annual marketplace for professional audio equipment engineers. PROFILE OF ATTENDEES: Audio engineers.

Broadcast Cable Financial Management Association. SHOW DATE: May 13-15 2008, The Fairmont, Dallas, TX. SHOW MANAGEMENT: BCFM, 550 Frontage Rd., Suite 3600, Northfield, IL 60093. Contact: Mary M. Collins, pres/CEO. (847) 716-7000. FAX: (847) 716-7004. Web: www.bcfm.com. SHOW MANAGEMENT STATEMENT: BCFM sponsors an annual conference with exhibits targeting financial, HR, MIS, executive mgmt from TV, radio & cable plus association in auditing data processing, credit & collections. SHOW HISTORY: Annual. First Year of Show: 1960.

Broadcasting 2008 (Public/Trade). SHOW DATE November 2008 Vancouver. SHOW MANAGEMENT: Canadian Association of Broadcasters, 350 Sparks St., Suite 306, Ottawa, ON, Canada K1R 7S8. Contact: Marye Menard-Bos, senior dir. (613) 233-4035. FAX: (613) 233-6961. E-mail: cab@cab-acr.ca. PROFILE OF EXHIBITORS: Companies who provide products or svcs to Canadian bcst (radio-TV) operators. PROFILE ATTENDEES: Senior mgmt, owners, engineers & new directors. Set Rotation Pattern: East/West rotation.

CAB Sales Management Conference. SHOW DATE: April 2008, Hilton, Chicago, IL. SHOW MANAGEMENT: Cabletelevision Advertising Bureau, Inc., 830 Third Ave., 2nd Fl., New York, NY 10022. Contact: Nancy Lagos. (212) 508-1200. FAX: (212) 832-3268.

Cable-Tec Expo. SHOW DATE: June 24-27, 2008, Philadelphia. SHOW MANAGEMENT: Society of Cable Television Engineers, 140 Phillips Rd., Exton, PA 19341. Contact: Lori Bower. (610) 363-6888. FAX: (610) 363-5898. SHOW SPONSOR: Same as show management. SHOW MANAGEMENT STATEMENT: The SCTE Cable-Tec Expo brings together engineers, technicians & technical executives from the United States who represent cable systems, MSOs & independent operators. A fulfillment to the growing need to address the technical end of the cable TV industry. PROFILE OF EXHIBITORS: Construction equipment, signal distribution, cable casting, transmission/receiving, test equipment, microware/mds, digital systems, system testing quality control & other related hardware. PROFILE OF ATTENDEES: Engineers, technicians, executives of MSOs & independent operators. SHOW HISTORY: Concurrent with rotating Annual Engineering Conference. First Year of Show: 1983.

Forum 2008. The Association of Cable Communicators. SHOW DATE: March 30-April 2, 2008, Renaissance Mayflower Hotel, Washington, DC. SHOW MANAGEMENT: CTPAA. ASSOCIATION ADDRESS: Box 75007, Washington, DC 20013. SHOW MANAGEMENT STATEMENT: Educational conference for cable pub affrs professionals. PROFILE OF ATTENDEES; Programmers, MSO, & loc system personnel. Contact: Steve Jones. (202) 222-2373. SHOW FREQUENCY: Annual. First Year of Show: 1985.

Great Lakes Broadcasting Conference & Expo (Public/Trade). SHOW DATE: March 10-11, 2008 Devosplace, Grand Rapids. SHOW MANAGEMENT: Michigan Association of Broadcasters, 819 N. Washington Ave., Lansing, MI 48906. (517) 484-7444. FAX: (517) 484-5810. E-mail: michmab@aol.com. SHOW SPONSOR: Same as show management. SHOW MANAGEMENT STATEMENT: Regional bcstg trade show which provides speakers & seminars to educate members while showcasing bcst equipment & svcs. PROFILE OF EXHIBITORS: Broadcast equipment, products & svcs. PROFILE OF ATTENDEES: General mgrs, engineers, sls mgrs, account executives, news directors & anybody involved in the audio, video or bcst industries. SHOW HISTORY: Annual.

INFOCOMM International. SHOW DATE: June 2008, SHOW MANAGEMENT: International Communications Industries Association Inc., 11242 Waples Mill Rd., Suite 200, Fairfax, VA 22030. Contact: Jason C. McGraw, senior VP. (703) 273-7200. FAX: (703) 278-8082. SHOW SPONSOR: Same as show management. SHOW MANAGEMENT STATEMENT: INFOCOMM International is the exposition of the video, computer, audiovisual, presentation & multimedia communications industries. It is sponsored by the International Communications Industries Association (ICIA). ICIA brings together more than 1,400 firms selling video, audiovisual & computer products & svcs. Members are dealers, software producers, independent representatives, rental companies & others serving users in communications, training, business, govt & education. In addition to the trade show, there are annual conventions, meetings, seminars, courses & institutes of many organizations who bring their members & interested associates to the event. PROFILE OF EXHIBITORS: Manufacturers & producers of video, computer & audiovisual-based communications & information products & producers of software programs for all technologies represented. PROFILE OF ATTENDEES: Dealers who sell video, audio, presentation systems & installations to professional users of these media; production & post-production companies, adv & PR agencies, laboratories & other users of video & related technologies. SHOW HISTORY: Annual. First Year of Show: 1983.

Interwire Trade Exposition. SHOW DATE: May 2-7, 2009, Cleveland, Ohio (I-X Center). SHOW MANAGEMENT: Wire Association International, 1570 Boston Post Rd., Box 578, Guilford, CT 06437. Contact: Steve Fetteroll, exec dir. (203) 453-2777. FAX: (203) 453-8384. SHOW SPONSOR: Same as show management. SHOW MANAGEMENT STATEMENT: Interwire provides a marketplace for the wire & cable industry. It is a trade show for wire manufacturers, fabricators, suppliers, other buyers & users with admin, engrg, tech & purchasing personnel in attendance from the United States, Europe, Asia, South & Central Americas. Product classifications include wire machinery, spring machinery, fasteners, fabricators, fiber optics, chemical coatings, accessories & other wire-related products. PROFILE OF EXHIBITORS: Machinery & accessories; fiber optics, chemicals, coatings, lubricants, dies, compounds, spools, reels, packaging, measuring, testing equipment, wire, cable, fasteners & fabricated wire products. PROFILE OF ATTENDEES: Gen & admin mgmt; engrg, opns, production; tech, rsch & dev, quality control; purchasing; sls & mktg concerned with wire industry. SHOW HISTORY: Biennial.

Kentucky Broadcasters Association Fall Convention. SHOW MANAGEMENT: Kentucky Broadcasters Association, 101 Enterprise Dr., Frankfort, KY 40601. Contact: Gary White, pres. (888) 843-5221. FAX: (502) 843-5710. SHOW MANAGEMENT STATEMENT: This convention's purpose is to further train bcstrs on both legal aspects of bcstrs as well as sls & mgmt. PROFILE OF EXHIBITORS: Electronics, computers, program syndications, bcst suppliers (radio & TV). PROFILE OF ATTENDEES: Owners & mgmt throughout the state of Kentucky. SHOW HISTORY: First Year of Show: 1953.

Louisiana Krewe of Cable Show. SHOW DATE: January 30-31, 2008, February 18-19, 2009, New Orleans, LA. SHOW MANAGEMENT: Louisiana Cable & Telecommunications Association, 763 North St., Baton Rouge, LA 70802. (225) 387-5960. FAX: (225) 383-6705. Web site: http://www.lacable.com. SHOW FREQUENCY: Annual.

Minnesota Broadcasters Association Annual Conference & Expo (Public/Trade). SHOW MANAGEMENT: Minnesota Broadcasters Association, 3033 Excelsier Blvd., Suite 301, Minneapolis, MN 55416. Contact: Jim DuBois, pres/CEO. (612) 926-8123. FAX: (612) 926-9761. E-mail: jdubois@minnesotabroadcasters.com. SHOW MANAGEMENT STATEMENT: The show includes exhibits for bcstg, cable, production, & other electronic media.

NAB 2008. SHOW DATE: April 12-17, 2008, Las Vegas Convention Center, Las Vegas, NV. SHOW MANAGEMENT: National Association of Broadcasters, 1771 N St. N.W., Washington, DC 20036. (202) 429-5300. FAX: (202) 429-7427. SHOW MANAGEMENT STATEMENT: The Convention is an annual gathering of radio, TV, video, post-production, multimedia & telecommunications professionals worldwide. Sessions cover aspects of electronic media, both tech & management-oriented. PROFILE OF EXHIBITORS: Manufacturers, distributors of products, equipment & svcs for the radio, TV, video, post-production, film multimedia & telecommunications industries. PROFILE OF ATTENDEES: Owners & managers of radio, TV, video, post-production & users of multimedia & telecommunications products & svcs. SHOW HISTORY: First Year of Show: 1922.

NATPE 2008. DATE: January 2008, Mandalay Bay Resort Convention Center, Las Vegas, NV. SHOW MANAGEMENT: National Association of Television Program Executives, 5757 Wilshire Blvd., Penthouse 10, Los Angeles, CA 90036. Contact: Nick Orfanopoulos, sr VP Conferences/Special Events. (310) 453-4440. FAX: (310) 453-5258. E-mail: orfan@aol.com. SHOW SPONSOR: Same as show management. SHOW MANAGEMENT STATEMENT: This show provides an annual marketplace for syndicated TV programs, first-run &/or off-network programs, as well as aspects of the TV business. PROFILE OF EXHIBITORS: Major studios, independent producers, marketers, cable networks, electronic retailers, new media producers, United States & intl program distributors. PROFILE OF ATTENDEES: Advertisers, cable MSOs, bcst stns & networks, producers, marketers, bankers, merchandisers, talent agents, new media, United States & overseas program distributors & TV programmers. SHOW HISTORY: First Year of Show: 1984.

National Religious Broadcasters Convention and Exposition (Public/Trade). SHOW DATE: March 7-12, 2008, Nashville, TN. SHOW MANAGEMENT: National Religious Broadcasters Association, 9510 Technology Dr., Manassas, VA 20110. Contact: David Keith. (703) 330-7000. FAX: (703) 330-7100. Website: www.nrb.org. SHOW SPONSOR: Same as show management. SHOW MANAGEMENT STATEMENT:

The NRB Convention and Exposition is an annual gathering of manufacturers & distributors of bcst equipment, computers, radio & TV programs, consultant svcs, gospel music, publishing & other related items to the religious communications industry. PROFILE OF EXHIBITORS: Manufacturers & distributors of consumer & bcst audio & video equipment, computers, radio & TV programs, publishers, & miscellaneous items relating to the religious field. PROFILE OF ATTENDEES: Radio & TV executives, ministers, denominational executives, musicians, adv executives, educators & Christian bookstore owners. SHOW HISTORY: Annual. First Year of Show: 1944.

NCTA. SHOW DATES: May 2008, Las Vegas, NV. SHOW SPONSOR: National Cable Television Association, 1724 Massachusetts Ave., N.W., Washington, DC 20036. Contact: Barbara York. (202) 775-3669. FAX: (202) 775-3692. Web site: http://www.thenationalshow.com. SPONSOR STATEMENT: NCTA's Annual Convention and International Exposition provides the industry's largest, most comprehensive showcase for equipment, progmg & enhanced svcs for cable TV, broadband & telecommunications systems. PROFILE OF EXHIBITORS: All industries related to cable TV & broadband svcs including hardware manufacturers, equipment suppliers, programming networks, enhanced svcs including VOD, PVR, software & technology providers. PROFILE OF ATTENDEES: Includes multiple system operators, independent operators, program networks, media & entertainment companies, investment & financial institutions, telecommunications providers, enhanced svc providers & the press. SHOW HISTORY: Annual. First Year of Show: 1950.

Promax & BDA 2008 International (Trade). SHOW MANAGEMENT: Promax & BDA, 9000 W. Sunset Blvd., Suite 900, Los Angeles, CA 90069. Contact: Gregg Balko. (310) 788-7600. FAX: (310) 788-7676. SHOW SPONSOR: Same as show management. SHOW MANAGEMENT STATEMENT: The purpose of the show is to bring together bcst promoters & designers in the electronic media, all incorporated in one show. PROFILE OF EXHIBITORS: Music video production; computer animation & graphics hardware; stn design & image packages; adv premiums & incentives. PROFILE OF ATTENDEES: Promax International serves those individuals responsible for the mktg, adv, promotion & publicizing of TV stns, networks, production companies, radio stns & cable systems on a national & international level. BDA consists of an international membership of art directors, designers & graphic artists.

RAB 2008 Marketing Leadership Conference. SHOW DATE: February 11-13, 2008, Hyatt Regency/Atlanta Downtown, Atlanta, GA. SHOW MANAGEMENT: Radio Advertising Bureau, 261 Madison Ave., 23rd Fl., New York, NY. Contact: Dana Honors. (972) 753-6740. FAX: (972) 753-6802. SHOW SPONSOR: Same as show management. SHOW MANAGEMENT STATEMENT: RAB is the largest gathering of sls and mgmt professionals in the radio industry. The Conference represents more than 5,000 members radio stns, radio stn sls managers, bcst groups, radio networks, stn representatives, network executives & associated industry organizations from the 50 states and 23 foreign countries. To educate, train radio sls & mktg professionals. The goal is to raise the level of professionalism in radio sls & mktg. PROFILE OF EXHIBITORS: Manufacturers, distributors or suppliers of a product or svc for any of the following categories: computer/software programs; radio & TV products; radio networks; specialty adv; program syndication; sls & mktg rsch; sls consulting. PROFILE OF ATTENDEES: Radio stn sls mgrs, gen mgrs, network executives & group heads from throughout the United States. SHOW HISTORY: First Year of Show: 1980.

SATELLITE 2008. SHOW DATE: February 25-28 2008, The Washington Convention Center, Washington, DC. SHOW MANAGEMENT: Access Intelligence, LLC. SHOW SPONSOR: Via Satellite Magazine. ASSOCIATION ADDRESS: 1201 Seven Locks Rd., Suite 300, Potomac, MD 20854. (301) 354-2000. FAX (301) 354-2315. Web site: www.satellitetoday.com. SHOW MANAGEMENT STATEMENT: This is the largest Satellite specific conference & exhibition in the world. SATELLITE 2007 provides you with an unequalled opportunity to meet key, senior satellite company executives from the United States, Europe, the Pacific Rim, South America and Africa. This is your chance to strengthen existing professional and business relationships as well as initiate new ones, before, during, and after the conference sessions and throughout our expansive exhibit hall. PROFILE OF EXHIBITORS: Satellite operators, end users, manufacturers, svc providers, launch vehicle operators teleports, consumer svc providers. PROFILE OF ATTENDEES: Distributors bcsters, programmers, VSAI network providers, satellite operators, launch vehicle svc providers & manufacturers. SHOW FREQUENCY: Annual. First Year of Show: 1976.

Society of Motion Picture and Television Engineers Annual Technical Conference and Exhibition. SHOW DATE: October 2008, Hollywood, CA (Public/Trade). SHOW MANAGEMENT: Society of Motion Picture and Television Engineers, 595 W. Hartsdale Ave., White Plains, NY 10607. Contact: Dianne Gabriele. (914) 761-1100. FAX: (914) 761-3115. SHOW SPONSOR: Same as show management. SHOW MANAGEMENT STATEMENT: The show provides an annual marketplace for equipment & supplies for production, engrg & purchasing personnel in worldwide professional motion-picture & bcst-TV industries. Product classifications include TV, cable, production, postproduction, laboratory & field-production equipment. PROFILE OF EXHIBITORS: Manufacturers, dealers & distributors of professional TV-bcst & motion-picture equipment. PROFILE OF ATTENDEES: Senior bcst & film-engineering personnel from the motion-picture & bcst-TV industries. SHOW HISTORY: Annual. Set Rotation Pattern: East Coast odd-numbered years, West Coast even-numbered years.

WCA International Symposium and Business Expo. SHOW DATE: January 29-February 1, 2008, Fairmont Hotel, San Jose, CA. SHOW MANAGEMENT: Carl Berndtson. SHOW SPONSOR: Wireless Communications Association International. ASSOCIATION ADDRESS: 1333 H St. N.W., Suite 700 W., Washington, DC 20005. (202) 452-7823. FAX (202) 452-0041. Web site: www.wcai.com. Contact: Carl Berndtson (978) 371-1792. PROFILE OF EXHIBITORS: Excellent mix of wireless broadband operators, vendors and integrators. PROFILE OF ATTENDEES: Executive and upper management; CTOs from wireless broadband industry. SHOW FREQUENCY: Annual. First Year of Show: 1983.

Vocational and Career Development Schools

The Art Institute of Pittsburgh, 420 Blvd. of the Allies, Pittsburgh, PA, 15219. Phone: (412) 263-6600. Fax: (412) 263-3715. E-mail: admissions-aip@aii.edu Web Site: www.aip.aii.edu.

Hans Westman, dir media arts & animation.

Courses offered include audio recording & production, engrg, EFP video production, bcst media, feature writing, scriptwriting, legal issues, non-linear editing, image manipulation, filmmaking & multicamera field production.

Broadcast Center, 2360 Hampton Ave., St. Louis, MO, 63139. Phone: (314) 647-8181. Fax: (314) 647-1575. E-mail: jberry@yourcba.com Web Site: www.broadcastcenterinfo.com.

Linda Hoy, VP opns.

Training in mktg & time sls, coml & program production, bcst journalism & bcst performance. Training includes voice training & dev for bcstg; announcing training including news, comls, DJ & sportscasting; news & coml copywriting.

Broadcasting Institute of Maryland, 7200 Harford Rd., Baltimore, MD, 21234. Phone: (410) 254-2770. Phone: (800) 942-9246. Fax: (410) 254-5357. E-mail: info@bim.org Web Site: www.bim.org.

John C. Jeppi Sr., pres; John I. Perry, VP; Lois Carringan, dir.

Courses offered include comprehensive course in radio & TV bcstg; majors available in radio, TV production, news & sports.

Brown College, 1440 Northland Dr., Mendota Heights, MN, 55120. Phone: (651) 905-3400. Fax: (651) 905-3550. E-mail: lwright@browncollege.edu Web Site: www.browncollege.edu.

Lisa Wright, dept chair school of bcstg radio & TV.

The Associate of Applied Science degree in Radio Broadcasting is designed to help develop an on-air presence as well as the technical-hands on, and writing skills needed for positions in this growing industry. The Associate of Applied Science degree in Television Production is designed to prepare students for an entry level position in a number of areas including: broadcast TV stations, industrial video firms, cable companies and satellite operations.

Carolina School of Broadcasting, 3435 Performance Rd., Charlotte, NC, 28214. Phone: (704) 395-9272. Fax: (704) 395-9698. E-mail: csbnc@bellsouth.net Web Site: www.csbradiotv.com.

Courses offered include a non-tech bcstg group session & an in-stn training lab course; announcing, production, copywriting, news, digital coml production, sls & administration. Day & night sessions. In-stn training, full- or part-time as determined by stn & student. TV facilities & stereo control room; resident training in studio & stn opns at coml radio & TV stns worldwide. Digital audio & non-linear editing for TV.

Cleveland Institute of Electronics, 1776 E. 17th St., Cleveland, OH, 44114. Phone: (216) 781-9400. Phone: (800) 243-6446. Fax: (216) 781-0331. Web Site: www.cie-wc.edu. E-mail: instruct@cie-wc.edu

John R. Drinko, pres; Scott D. Katzenmeyer, VP administration; Paul Valvoda, controller; Keith Conn, dean of instruction.

Virginia Beach, VA 233455. World College, 5193 Shore Dr, Suite 105. Phone: (757) 464-4600. (800) 696-7532. Fax: (757) 464-3687. E-mail: instruct@cie-wc.edu. Web Site: www.worldcollege.edu. John R. Drinko, pres.

Offers associate in applied science degree in electronics engrg technology, bsct engrg. FCC license preparation & cable technician training.

Clover Park Technical College, 4500 Steilacoom Blvd. S.W., Lakewood, WA, 98499-4098. Phone: (253) 589-5800. Fax: (253) 589-5797. Web Site: www.cptc.edu.

Bcst training since 1954. Comprehensive Associate degree program in all aspects of radio stn opn prepares students for entry-level employment. Course includes staff experience at 51-kw *KVTI(FM). A two-year state college. Assoc of Applied Technology degree programs.

Columbia College Hollywood, 18618 Oxnard St., Tarzana, CA, 91356-1411. Phone: (818) 345-8414. Fax: (818) 345-9053. E-mail: info@columbiacollege.edu Web Site: www.columbiacollege.edu.

Jan Stanley Mason, pres; Mark J. Stratton, admin dir.

Courses offered in TV/video production & cinema; degree program includes classes in directing, studio lighting, camera opns, videotape editing, scriptwriting, film editing, sound mixing, asst camera & script supervision. A.A. degree in TV/video production; B.A. degree in TV/video, cinema, and cinema/TV combination.

Columbia School of Broadcasting, (Washington, DC Metro Area). 3947 University Dr., 2nd Fl., Fairfax, VA, 22030-2506. Phone: (703) 591-6000. Fax: (703) 591-6147. E-mail: djtrain@columbiaschoolbroadcas.com Web Site: www.columbiaschoolbroadcas.com.

William Butler, pres.

Courses offered include English & Sp radio announcing (voice-over, newscaster, DJ, sportscaster, traffic/weather reporter & interviews/talk show host), TV announcing, radio play-by-play sportscasting & basic radio production. Distance educ & resident courses offer comprehensive training for entry-level bcstg positions. Founded in 1964.

Columbia School of Broadcasting, Metro Washington DC Communications Ctr., 3947 Univ. Dr. 2nd Fl, Fairfax, VA, 22030-2506. Phone: (800) 362-0660. Fax: (703) 591-6147. E-mail: bill@csbamerica.com Web Site: www.csbamerica.com.

William Butler, pres.

Courses offered include radio, TV announcing, radio, TV sls, internet bcstg, digital journalism & production.

Connecticut School of Broadcasting, Inc., 63 Bay State Rd., Boston, MA, 02215. Phone: (617) 267-2006. Fax: (617) 267-2004. E-mail: btilden@gocsb.com Web Site: www.800tvradio.com.

Brian Stone, pres/CEO; David Banner, COO; Beverly Tilden, mktg dir; Jason Muth, VP recruit & admissions; Katie MacKay, mktg asst; Scott Knight, exec VP.

Pawcatuck, CT 06379. GCP Pawcatuck LLC, 185 S. Broad St., 3rd Fl. #303. Phone: (860) 599-1108. Fax: (860) 599-5915. Mwlanie Mariano, campus coord.

Tampa, FL 33619. GCP Tampa LLC, 3901 Coconut Palm Dr., Sabal Business Center II, Suite 105. Phone: (813) 740-0990. Fax: (813) 663-0085. Kelly Crain, dir.

Westbury, LI, NY 11590. GCP Westbury LLC, 1400 Old Country Rd, Suite 211. Phone: (516) 338-1000. Fax: (516) 338-1170. Marty Herstein, dir.

Farmington, CT 06032. GCP Farmington, Media Park, 130 Birdseye Rd. Phone: (860) 677-7577. Fax: (860) 677-1141. Stacey Buba, dir.

Stratford, CT 06615. GCP Stratford LLC, 80 Ferry Blvd. Phone: (203) 378-5155. Fax: (203) 378-4330. Joe LaChance, dir., FL 33403. GCP Palm Beach Gardens LLC, 3450 Northlake Blvd., Suite 110. Phone: (561) 842-2000. Fax: (561) 775-8390. Dave Duran, rgnl dir, Skip Kelly, campus coord.

Davie, FL 33328. GCP Davie LLC, 3538 S. University Dr., University Park Plaza. Phone: (954) 474-3700. Fax: (954) 474-7404. Angie Lopez, dir. (Fort Lauderdale area).

Atlanta, GA 30338. GCP Atlanta LLC, 1117 Perimeter Ctr W., Suite N-301. Phone: (770) 522-8803. Fax: (770) 522-9876. Aarib Elya, dir, Peter Bernier, campus coord.

Needham, MA 02494. GCP Needham LLC, 73 TV Pl. Phone: (781) 235-2050. Fax: (781) 444-0406. Steve Williams, dir, Matt Sawyer, office mgr.

Cherry Hill, NJ 08002. GCP Cherry Hill LLC, One Cherry Hill, #201. Phone: (856) 755-1200. Fax: (856) 755-0865. Tom DeFranco, dir, Nicole McClintock, campus coord. (Philadelphia area).

Hasbrouck Heights, NJ 07604. GCP Hasbrouck Heights, 377 Rt. 17 S., #140. Phone: (201) 288-5800. Fax: (201) 288-7966. Janet Hutsebaut, dir, Kevin Foley, career svcs. (NYC area).

Austin, TX 78759. GCP CSB Austin LLC, 9600 Great Hills Tr, Suite 200-E. Phone: (512) 340-1420. Fax: (512) 340-1430. Randy Ahrens, dir, Jason Seale, campus coord.

Irving, TX 75038. GCP CSB Dallas LLC, 5605 N. MacArthur Blvd, Suite 220. Phone: (214) 441-9941. Fax: (214) 441-9942. Scott Powell, dir, Kristin Tran, campus coord.

Arlington, VA 22202. GCP Arlington, 2170 Crystal Plaza Arcade, #38. Phone: (703) 415-7600. Fax: (703) 415-0238. R.J. Narsavage, dir, Marcella Jones, campus coord.

Courses (day & evening): On-air performance: Radio, TV, Internet bcst. Production courses: Digital audio & video, linear & non-linear avid editing. Other communications courses: Sports, voice-overs, sls, proms, mktg, wireless & multi-media technology.

Dunwoody College of Technology, 818 Dunwoody Blvd., Minneapolis, MN, 55403. Phone: (612) 374-5800. Fax: (612) 374-4128. E-mail: info@dunwoody.edu Web Site: www.dunwoody.edu.

C. Ben Wright, pres; Brian Seviola, mktg mgr.

Courses offered include assoc in electronics tech degree, computer technician, radio-TV, industrial electronics technician, digital electronics specialists, electronics technician, TV specialists, certificate programs, aviation electronics, & info mgmt systems.

Education Direct, 925 Oak St., Scranton, PA, 18515. Phone: (570) 342-7701. Fax: (570) 961-4888. Web Site: www.educationdirect.com. E-mail: info@educationdirect.com

Dan Conrad, gen mgr.

Diploma courses include basic electronics, electronics technology, basic computer progmg, TV/VCR repair or personal computer repair, Java progmg, internet web page design, electricians, telecommunications technician. Center for Degree Studies: specialized assoc degree in electronics technology & electrical, mechanical, civil & industrial engrg technology; specialized assoc degree in business mgmt, mktg, finance, accounting or applied computer science; Internet technology in web progmg, Internet technology multimedia, Internet technology in e-commerce administration, graphic design, DC maintenance technology.

Grantham University, 2101 Wilson Blvd, Suite 110, Arlington, VA, 22201. Fax: (703) 465-1273. E-mail: info@grantham.edu Web Site: www.grantham.edu.

Roy Winter, pres; Joanna Boldt, dir of student affrs.

Courses offered include computer science, electronics engrg tech & computer engrg tech by correspondence, leading to A.S. & B.S. degrees.

The Illinois Center for Broadcasting, 55 West 22nd St., Suite 240, Lombard, IL, 60148. Phone: (630) 916-1700. Fax: (630) 916-1764. E-mail: director.chicago@beonair.com Web Site: www.beonair.com.

Patrick Johnson, school dir.

10-month, hands-on course in radio & TV bcstg procedures & techniques.

International College of Broadcasting, 6 S. Smithville Rd., Dayton, OH, 45431. Phone: (937) 258-8251. Fax: (937) 258-8714. Web Site: www.icbcollege.com. E-mail: admissions@icbcollege.com

J. Michael LeMaster, pres.

Courses offered include radio, TV, cameraman, CATV, disc jockey, news, sports & audio/recording engrg. Assoc degree in communication arts available in radio/TV & video production/recording audio engrg. Diploma programs offered in audio/recording engrg & bcstg.

Madison Media Institute-College of Media Arts, 2702 Agriculture Dr., Madison, WI, 53718. Phone: (608) 663-2000. Fax: (608) 442-0141. E-mail: swh@madisonmedia.com Web Site: www.madisonmedia.edu.

Chris Hutchings, pres; Steve Hutchings, VP & dir admissions.

Courses offered include digital media design, production, video, motion graphics, recording & music technology. Accredited by Accrediting Commission of Career Schools & Colleges of Technology.

The New England Institute of Art, 10 Brookline Pl. W., Boston, MA, 02445. Phone: (617) 739-1700. Fax: (617) 582-4500. Web Site: www.artinstitutes.edu/boston. E-mail: neiaadm@aii.edu

Debra Leahy, dir. media arts; Fran Berger, dir pub rel.

Programs offered: Bachelor's Degree in Graphic Design, Photography, Digital Film & Video Production, Interactive Media Design, Media Arts & Animation, Interior Design, Audio & Media Technology. Associate's Degrees in Audio Production, Broadcasting,

New England School of Communications, One College Cir., Bangor, ME, 04401. Phone: (207) 941-7176. Fax: (207) 947-3987. E-mail: info@nescom.edu Web Site: www.nescom.edu.

Courses offered include announcing, bcst sls, writing for bcst, TV production, sound recording, voice/diction, news/sports reporting, adv & PR, pub speaking, video graphics, desktop publishing, print journalism & web design.

New School University, 2 W. 13th St., New York, NY, 10011. Phone: (212) 229-8903. Fax: (212) 229-5357. E-mail: baronej@newschool.edu Web Site: www.newschool.edu.

Carol Wilder, chmn media & film studies; Dawnja Burris, assoc chmn media studies.

Courses offered include TV writing workshop; writing for TV, films & radio; TV production workshop; voice & speech for theater & TV; seminars on TV comls; writing TV comls. Offers certificate in film/TV studies, B.A., B.A./M.A., M.A. in media studies.

Northland Community & Technical College (KSRQ-FM), 1101 Hwy. 1 E., Thief River Falls, MN, 56701. Phone: (218) 681-0701. Phone: (800) 959-6282. Fax: (218) 681-0774. Web Site: www.northlandcollege.edu.

Julie Olson, dir public rel.

Courses offered include a diploma-earning program in radio bcstg working on a 24,000 kw educ FM stn.

The Ohio Center for Broadcasting-Cincinnati, 6703 Madison Rd., Cincinnati, OH, 45227. Phone: (513) 271-6060. Fax: (513) 271-6135. Web Site: www.beonair.com.

Eric Armstrong, opns VP.

10-month, hands-on course in Radio & TV procedures & techniques.

The Ohio Center for Broadcasting-Cleveland, 9000 Sweet Valley Drive, Valley View, OH, 44125. Phone: (216) 447-9117. Fax: (216) 642-9232. E-mail: ocb@beonair.com Web Site: www.beonair.com.

Robert Mills, pres.

Denver, CO 80214, 1310 Wadsworth Blvd. Phone: (303) 937-7070.

10-month, hands-on course on radio & TV bcstg & techniques. Fully accredited by ACCSCT. Graduates earn 36 quarter credit hrs with diploma. Instructors are professional bcstrs. Nationally accredited by ACCSCT. Full time placement assistance.

The Poynter Institute for Media Studies, 801 3rd St. S., St. Petersburg, FL, 33701. Phone: (727) 821-9494. Fax: (727) 821-0583. E-mail: oseifert@poynter.org Web Site: www.poynter.org.

Karen Brown Dunlap, pres.

Seminars & conferences for print & bcst & online journalists. Courses for TV/radio include stn leadership, new leaders in the newsroom, newsroom mgmt, ethical decision-making, anchors as newsroom leaders investigative reporting, power reporting, computer-assisted journalism, visual storytelling & ethics, & producing newscasts.

Specs Howard School of Broadcast Arts Inc., 19900 W. Nine Mile Rd., Southfield, MI, 48075-3953. Phone: (248) 358-9000. Fax: (248) 746-9772. E-mail: info@specshoward.edu Web Site: www.specshoward.edu.

Dick Kemen, VP industry rels; Jonathan Liebman, pres/CEO.

Courses offered include radio & TV bcstg & production. Accredited by ACCSCT.

Technical Career Institutes, 320 W. 31st St., New York, NY, 10001. Phone: (212) 594-4000. Fax: (212) 629-3937. Web Site: www.tcicollege.net. E-mail: admissions@tcicollege.edu

James Melville, pres; Bonnie Price, public rel.

Tech courses offered include electronics engrg, EETT & IETC office technology, computerized accounting, bldg maintenance, air conditioning, heating & refrigeration technology. Assoc degree available.

Western Technical College, 304 6th St. N., La Crosse, WI, 54601. Phone: (608) 785-9200. Fax: (608) 785-9407. E-mail: westpfahlr@westertc.edu Web Site: www.westerntc.edu.

Lee Rasch, pres; Joan Pierce, dir media arts.

Courses in the Visual Communications associates's degree include design fundamentals, audio production, media technologies, digital photography, Adobe Photoshop & illustrator, video production & web design.

Universities and Colleges with Broadcasting or Journalism Programs

Universities and Colleges Offering Degrees in Broadcasting

Alabama
Jacksonville State U. Jacksonville 36265.
Montevallo, U. of Montevallo 35115.

Arizona
Pima Community College Tucson 85709.

Arkansas
Arkansas State U. State University 72467.

California
California State U. Fullerton 92831.
California State U. Long Beach 90840.
California State U. Los Angeles 90032.
California State University-Monterey Bay Seaside 93955.
California State U. Northridge 91330.
Loyola Marymount U. Los Angeles 90045.
San Francisco State U. San Francisco 94132.
Vanguard Southern California U. of Costa Mesa 92626.

Connecticut
Manchester Community College Manchester 06045.

Delaware
Delaware State U. Dover 19901.

District of Columbia
Howard U. Washington 20059.

Florida
City College Fort Lauderdale 33309.

Georgia
Georgia Southern U. Statesboro 30460.

Idaho
Idaho U. of Moscow 83844.

Illinois
Chicago State U. Chicago 60628.
Chicago-Kennedy King College, City Colleges of Chicago 60621.
Lake Land College, Mattoon 61938.
Lewis and Clark Community College Godfrey 62035.
Parkland College Champaign 61821.

Indiana
Ball State U. Muncie 47306.
Butler U. Indianapolis 46208.
Huntington College Huntington 46750.

Iowa
Northern Iowa, U. of Cedar Falls 50614.

Kansas
Independence Community College Independence 67301.

Kentucky
Business and Tech-Lexington, National College of Lexington 40508.
Northern Kentucky U. Highland Heights 41099.

Louisiana
Bossier Parish Community College, Bossier City 71111.
Louisiana at Monroe, U. of Monroe 71209.

Maryland
Morgan State U. Baltimore 21251.

Massachusetts
Emerson College Boston 02116.
Massachusetts, U. of Amherst 01003.

Michigan
Central Michigan U. Mt. Pleasant 48859.
Grand Valley State U. Allendale 49401.
Michigan State U. East Lansing 48824.
Northern Michigan U. Marquette 49855.
Siena Heights U. Adrian 49221.
Wayne State U. Detroit 48201.

Minnesota
Lake Superior College Duluth 55811.
Northwestern College St. Paul 55113.
Southwest State U. Marshall 56258.

Mississippi
Coahoma Community College Clarksdale 38614.

Missouri
Central Missouri State U. Warrensburg 64093.

Montana
Montana, U. of Missoula 59812-6480.

Nebraska
Grace U. Omaha 68108.
Nebraska at Omaha, U. of Omaha 68182.

New Jersey
Montclair State U. Upper Montclair 07043.

New York
Cuny Brooklyn College Brooklyn 11210.
Hofstra U. Hempstead 11549. BA-BS.
Long Island Unversity-C W Post Campus Brookville 11548.
Mercy College-Main Campus Dobbs Ferry 10522.
Syracuse U. Syracuse 13244.

North Carolina
Appalachian State U. Boone 28608.
Campbell Unversity Inc. Buies Creek 27506.
North Carolina Agricultural and Technical State U. Greenboro 27401.

North Dakota
Minot State U. Minot 58707.

Ohio
Baldwin-Wallace College Berea 44017.
Cedarville U. Cedarville 45314.
Central State U. Wilberforce 45384.
Hocking College Nelsonville 45764.
International College of Broadcasting Dayton 45431.
Cincinnati-Main Campus U. of Cincinnati 45221.

Oklahoma
Oklahoma State U. Stillwater 74078.
Oral Roberts U. Tulsa 74171.

Oregon
Lane Community College Eugene 97405.

Pennsylvania
Kutztown, U. of Kutztown 19530.
Marywood U. Scranton 18509.

Puerto Rico
Puerto Rico-Ponce, Pontifical Catholic U. of Ponce 00717.

South Carolina
Bob Jones U. Greenville 29208.
North Greenville College Tigerville 29688.

Tennessee
Draughons Juniors College Inc. Nashville 37217.
Freed-Hardeman U. Henderson 38340.

Texas
Alvin Community College Alvin 77511.
Austin Community College Austin 78752.
Central Texas College Killeen 76540.
Del Mar College Corpus Christi 78404.
McLennan Community College Waco 76708.
Odessa College Odessa 79764.
San Ontonio College San Antonio 78212.
Stephen F. Austin State U. Nacogdoches 75962.

Utah
Brigham Young U. Provo 84602.

Vermont
Castleton State College Castleton 05735.

Virginia
Regent U. Virginia Beach 23464.

Washington
Eastern Washington U. Cheney 99004.

Wisconsin
Wisconsin-Oshkosh, U. of Oshkosh 54901.

Wyoming

Central Wyoming College Riverton 82501.

Two-Year Colleges Offering Programs in Broadcasting

Alabama

Spring Hill College Mobile 36608.

Arkansas

Arkansas State University-Main Campus State University 72467.

California

City College of San Francisco San Francisco 94112. .

Connecticut

Manchester Community College Manchester 06045.

District of Columbia

Howard Unversity Washington 20059.

Florida

University of Miami Coral Gables 33124.

Illinois

Lewis and Clark Community College Godfrey 62035.
Northwestern University Evanston 60208.

Indiana

Unversity of Southern Indiana Evansville 47712.

Iowa

Saint Ambrose Unversity Davenport 52803.

Kansas

Indpendence Community College Independence 67301.

Kentucky

Campbellsville Unversity Campbellsville 42718.
Eastern Kentucky Unversity Richmond 40475.
National College Business and Tech-Lexington Lexington 40508.
Northern Kentucky Uunversity Highland Heights 41099.
Western Kentucky Unversity Bowling Gree4n 42101.

Louisiana

Bossier Parish Community College,fv0 Bossier City 71111.
University of Louisiana at Monroe Monroe 74209.

Massachusetts

Newbury College-Brookline Brookline 02445.

Michigan

Michigan State Unversity East Lansing 48824.
Northern Michigan Unversity Marquette 49855.

Minnesota

Lake Superior College Duluth 55811.
Northland Community and Techical College Thief River Falls 56701.
Northwestern College St. Paul 55113.
Southwest State Unversity Marshall 56258.

Mississippi

Coahoma Community College Clarksdale 38614.
Hinds Community College, Raymond 39154.
Northeast Mississippi Community College Booneville 38829.

Missouri

Northwest Missouri State Unversity Maryville 64468.

Montana

Unversity of Montana Missoula 59812-6480.

Nebraska

Grace Unversity Omaha 68108.

New York

Ithaca College Ithaca 14850.

North Dakota

Minot State Unversity Minot 58707.

Ohio

Cedarville Unversity Cedarville 45314.
Hocking College Nelsonville 45764.
International College of Broadcasting Dayton 45431.
Kent State Unversity-Main Campus Kent 44242.
Unversity of Cincinnati-Main Campus Cincinnati 45221.
Youngstown State Unversity Youngstown 44555.

Oklahoma

Oklahoma State Unversity Stillwater 74078.

Oregon

Lane Community College Eugene 97405.
Mt. Hood Community College Gresham 97030.

Pennsylvania

Marywood U. Scranton 18509.

South Carolina

Bob Jones Unversity Greenville 29614.

Tennessee

Draughons Juniors College Inc. Nashville 37217.

Texas

Alvin Community College Alvin 77511.
Austin Community College Austin 78736.
Central Texas College Killeen 76549.
Del Mar College, Corpus Christi 78404.
McLennan Community College Waco 76708.

Utah

Dixie State College Utah 84770.
Weber State Unversity Ogden 84408.

Wyoming

Central Wyoming College Riverton 82501.

Universities and Colleges Offering Degrees in Journalism and Mass Communication

Alabama

Alabama, U. of Tuscaloosa 35487.
Auburn U. Auburn 36849-5211.
Spring Hill College Mobile 36608.
Troy State U.Main Campus, Troy 36082.

Alaska

Alaska U. of Fairbanks 99775.

Arizona

Arizona, U. of, Tucson 85721.
Arizona State U. Tempe 85287.
Northern Arizona U. Flagstaff 86011.
Prescott College Prescott 86301.

Arkansas

Arkansas, U. of Fayetteville 72701.
Arkansas State U. State University 72467.
Arkansas Tech U. Russellville 72801.
Harding U. Searcy 72149.
John Brown U. Siloam Springs 72761.

California

Biola U. La Mirada 90634.
California State Polytechnic U. San Luis Obispo 93407.
California State U. Chico, Chico 95929.
California State U. Long Beach, Long Beach 90840.
California State U. Los Angeles, Los Angeles 90032.
California State U. Northridge, Northridge 91330.
California State U. Sacramento, Sacramento 95819.
Chapman U. Orange 92866.
Humboldt State U. Arcata 95521.
La Verne, U. of La Verne 91750.
Menlo College Atherton 94027-4185.
Pepperdine U. Malibu 90263.
Point Loma Nazarene U. San Diego 92106.
San Diego State U. San Diego 92182.
San Francisco, U. of San Francisco 94117.
San Francisco State U. San Francisco 94132.
San Jose State U. San Jose 95192.
Santa Ana College Santa Ana 92706.
Southwestern College Chula Vista 91910.

Colorado

Colorado, U. of Boulder 80309.
Denver, U. of Denver 80208.
Mesa State College Grand Junction 81502.
Metropolitan State College of Denver Denver 80217.
Northern Colorado, U. of Greeley 80639.
Southern Colorado, U. of Pueblo 81001.

Connecticut

Connecticut, U. of Storrs 06269.
Quinnipiac U. Hamden 06518.
Southern Connecticut State U. New Haven 06515.

Delaware

Delaware State U. Newark 19716.

District of Columbia

American U. Washington 20016.
Catholic U. School of Law Washington 20064.
George Washington U. Washington 20052.
Howard U. Washington 20059.

Florida

Central Florida, U. of Orlando 32816.
Edward Waters College Jacksonville 32209.
Florida, U. of Gainesville 32611.
Florida Southern College Lakeland 33801.
Miami, U. of Coral Gables 33124.
Trinity International U. Miami 33169.

Georgia

Berry College Mount Berry 30149.
Brenau U. Gainesville 30501.
Georgia, U. of Athens 30602.
Georgia State U. Atlanta 30303.
Mercer U. at Macon Macon 31207.
Toccoa Falls College Toccoa Falls 30598.

Hawaii

Hawaii at Manoa, U. of Honolulu 96822.
Hawaii Pacific U. Honolulu 96813-2807.

Idaho

Boise State U. Boise 83725.
Idaho, U. of Moscow 83744.
Idaho State U. Pocatello 83709.

Illinois

Bradley U. Peoria 61625.
Columbia College Chicago, Chicago 60605.
Governors State U. University Park 60466.
Illinois College Jacksonville 62650.
Illinois, U. of Urbana 61801.
Loyola, U. of Chicago 60626.
MacMurray College Jacksonville 62650.
Northern Illinois U. DeKalb 60115.
Northwestern U. Evanston 60208.
Roosevelt U. Chicago 60605.
Southern Illinois U. Carbondale, Carbondale 629016.
Southern Illinois U. Edwardsville 62026-1775.
St Francis, U. of Joliet 60435.
Western Illinois U. Macomb 61455.

Indiana

Anderson U. Anderson 46012.
Ball State U. Muncie 47306.
Butler U. Indianapolis 46208.
Earlham College Richmond 47374.
Franklin College Franklin 46131.
Hunting College Huntington 46750.
Indiana State U. Terre Haute 47809.
Indiana U. Bloomington 47405.
Indianapolis, U. of Indianapolis 46227.
Indiana University-Purdue Unversity-Indianapolis Indianapolis 46202.
Saint Mary-of-the-Woods College Saint Mary-of-the-Woods 47876.
Southern Indiana, U. of Evansville 47712-3596.
Taylor U., Fort Wayne Campus Fort Wayne 46807.

Iowa

Buena Vista U. Storm Lake 50588.
Dordt College Sioux Center 51250.
Drake U. Des Moines 50311.
Grand View College Des Moines 50316.
Iowa, U. of Iowa City 52242.
Iowa State U. Ames 50011.
Loras College Dubuque 52004.
Morningside College Sioux City 51106.
Northern Iowa, U. of Cedar Falls 50614.
Saint Ambrose U. Davenport 52803.

Kansas

Allen County Community College Iola 66749.
Friends U. Wichita 67213.
Independence Community College Independence 67301.
Kansas, U. of Lawrence 66045.
Kansas State U. Manhattan 66506.
Tabor College Hillsboro 67063.
Washburn U. Topeka 66621.

Kentucky

Asbury College Wilmore 40390.
Campbellsville U. Campbellsville 42718.
Eastern Kentucky U. Richmond 40475-3102.
Kentucky, U. of Lexington 40506-0042.
Lindsey Wilson College Columbia 42728.
Murray State U. Murray 42071.
Northern Kentucky U. Highland Heights 41076.
Western Kentucky U. Bowling Green 42101.

Louisiana

Grambling State U. Grambling 71254.
Louisiana State U. Baton Rouge 70803-7202.
Louisiana State U. Shreveport 71115.
Louisiana Tech U. Ruston 71272-0045.
Louisiana at Lafayette, U. of Lafayette 70503.
Louisiana at Monroe, U. of Monroe 71209-0322.
McNeese State U. Lake Charles 70609.
New Orleans, U. of New Orleans 70148.
Nicholls State U. Thibodaux 70310.
Northwestern State U. of Louisiana Natchitoches 71497.
Southern U. and A&M College Baton Rouge 70813.
Southeastern Louisiana U. Hammond 70402.
Xavier, U. of Louisiana New Orleans 70125.

Maine

Maine, U. of Orono 04469.

Maryland

Bowie State U. Bowie 20715.
Columbia Union College Takoma 20912.

Massachusetts

Boston U. Boston 02216.
Emerson College Boston 02116.
Massachusetts, U. of Amherst 01003.
Newbury College Brookline 02445.
Northeastern U. Boston 02115.
Suffolk U. Boston 02114.

Michigan

Andrews University Berrier Springs 49104.
Calvin College Grand Rapids 49546.
Central Michigan U. Mt. Pleasant 48859.
Detroit Mercy, U. of Detroit 48219-0900.
Eastern Michigan U. Ypsilanti 48197-4210.
Grand Valley State U. Allendale 49401.
Madonna U. Livonia 48150.
Michigan, U. of Ann Arbor 48109-1285.
Michigan State U. East Lansing 48824.
Northern Michigan U. Marquette 49855.
Oakland U. Rochester 48309.
Spring Arbor U. Spring Arbor 49283.
Wayne State U. Detroit 48201.
Western Michigan U. Kalamazoo 49008.

Minnesota

Bemidji State U. Bemidji 56601.
Concordia College at Moorhead Moorhead 56562.
Minnesota State U. Mankato 56001.
Minnesota, U. of Minneapolis 55455.
Minnesota State U. Moorehead 56563.
Northland Community and Techinical College Thief River Falls 56701.
St. Cloud State U. St Cloud 56301.
St. Mary's U. Winona 55987.
St. Thomas, U. of St. Paul 55105.
Winona State U. Winona 55987.

Mississippi

Alcorn State U. Alcorn State 39096.
Hinds community College Raymond 39154.
Jackson State U. Jackson 39217.
Mississippi, U. of University 38677.
Mississippi State U. Mississippi State 39762.
Northeast Mississippi Community Booneville 38829.
Rust College Holly Springs 38635.
Tougaloo College Tougaloo 39174.

Missouri

Central Missouri State U. Warrensburg 64093.
College of the Ozarks Point Lookout 65726.
Crowder College Neosho 64850.
Druy U. Springfield 65802.
Evangel U. Springfield 65802.
Lincoln U. Jefferson City 65102.
Lindenwood U. St Charles 63301.
Maryville U. St. Louis 63141.
Missouri, U. of Columbia 65211.
Missouri-Kansas City, U. of Kansas City 64110.
Missouri Southern State College Joplin 64801-1595.
Missouri-St. Louis, U. of St. Louis 63121.
Missouri Western State College St. Joseph 64507.
Southwest Missouri State U. Springfield 65804.
Stephens College Columbia 65215.
Webster U. St. Louis 63119.

Montana

Montana, U. of Missoula 59812.
Montana State U. Billings 59101.

Nebraska

Creighton U. Omaha 68178.
Grace U. Omaha 68108.
Hasting College Hasting 68902.
Midland Lutheran College Fremont 68025.
Nebraska-Kearney, U. of Kearney 68849.
Nebraska-Lincoln, U. of Lincoln 68588-0443.
Nebraska at Omaha, U. of Omaha 68182.
Wayne State College Wayne 68787.

Nevada

Nevada-Reno, U. of Reno 89557.

New Hampshire

Franklin Pierce College Rindge 03461.
Keene State College Keene 03435.

New Jersey

Rutgers U. New Brunswick 08903.
Rutgers U. Newark 07102.

New Mexico

New Mexico, U. of Albuquerque 87131.
New Mexico State U. Las Cruces 88003.

New York

Columbia U. New York 10027.
Cuny Brooklyn College Buffalo 11210.
Fulton-Montgomery Community College Johnstown 12095.
Hofstra U. Hempstead 11549.
Iona College New Rochelle 10801-1890.
Ithaca College Ithaca 14850.
Long Island University-The Brooklyn Campus Brooklyn 11201.
Long Island University-C W Post Campus Brookville 11548.
Mercy College-Main Campus Dobbs Ferry 10522.
New York U. New York 10003.

Pace U. New York 10038.
St. John's Unversity-New York Jamaica 11439.
SUNDY College at New Paltz New Paltz 12561.
Syracuse U. Syracuse 13244-2100.
Utica College of Syracuse U. Utica 13502.

North Carolina

Appalachian State U. Boone 28608.
Campbell U. Buies Creek 27506.
East Carolina U. Greenville 27858.
Elon U. Elon 27244.
Gardner-Webb U. Boiling Springs 28017.
Meredith College Raleigh 27607.
North Carolina, U. of Chapel Hill 27599.
North Carolina-Asheville, U. of Asheville 28804.
North Carolina-Pembroke, U. of Pembroke 28372.
Winston-Salem State U. Winston Salem 27110.

North Dakota

North Dakota, U. of Grand Forks 58202.

Ohio

Akron U of Akron 44325.
Ashland U. Ashland 44805.
Bowling Green State U. Bowling Green 43403.
Capital U. Columbus 43209.
Cincinnati, U. of Cincinnati 45221.
Dayton U. of Dayton 45469.
Findlay The U. of 45840.
Kent State U. Kent 44242.
Marietta College Marietta 45750.
Ohio State U. Columbus 43210-1339.
Ohio U. Athens 45701.
Ohio Wesleyan U. Delaware 43015.
Otterbein College Westerville 43081.
Wright State Unversity-Main Campus Dayton 45435.
Xavier U. Cincinnati 45207-5171.
Youngstown State U. Youngstown 44555.

Oklahoma

Carl Albert State College Poteau 74953.
Central Oklahoma, U. of Edmond 73034.
East Central U. Ada 74820.
Northeastern State U. Tahlequah 74464.
Oklahoma, U. of Norman 73019.
Oklahoma Baptist U. Shawnee 74804.
Oklahoma Christian, U. Oklahoma City 73136.
Oklahoma State U. Stillwater 74078-0195.
Southern Nazarene U. Bethany 73008.
Tulsa, U. of Tulsa 74104.

Oregon

George Fox U. Newberg 97132.
Linfield College McMinnville 97128.
Oregon, U. of Eugene 97403-1275.
Portland, U. of Portland 97203.
Southern Oregon U. Ashland 97520.

Pennsylvania

Bloomsburg U. of Bloomsburg 17815.
Duquesne U. Pittsburgh 15282.
Elizabethtown College Elizabethtown 17022.
Indiana U. of Pennsylvania Indiana 15705.
Lehigh U. Bethlehem 18015.
Lock Haven U. Lock Heaven 17745.
Messiah College Grantham 17027.
Pennsylvania State U. University Park 16802.
Pittsburgh, U. of Johnstown 15904.
Point Park U. Pittsburgh 15222-1984.
Shippensburg U. Shippensburg 17257.
Susquehanna U. Sellingsgrove 17870.
Temple U. Philadelphia 19122-6080.

Puerto Rico

Sacred Heart U. of Santurce 00907.

Rhode Island

Rhode Island, U. of Kingston 02881.

South Carolina

Benedict College Columbia 29204.
Bob Jones U. Greenville 29614.
Francis Marion U. Florence 29506.
Morris College Sumter 29150.
North Greenville College Tigerville 29688.
South Carolina, U. of Columbia 29208.
Winthrop U. Rock Hill 29733.

South Dakota

Black Hills State U. Spearfish 57799.
South Dakota, U. of Vermillion 57069.
South Dakota State U. Brookings 57007.

Tennessee

Austin Peay State U. Clarksville 37044.
East Tennessee State U. Johnson City 37614.
Lee U. Cleveland 37320.
Memphis, U. of Memphis 38152.
Middle Tennessee State U. Murfreesboro 37132.
Southern Adventist U. Collegedale 37315.
Tennessee, The U. of Knoxville 37996-0332.
Tennessee Technological U. Cookeville 38505.
Trevecca Nazarene U. Nashville 37210.

Texas

Abilene Christian U. Abilene 79699.
Amarillo College Amarillo 79109.
Angelo State U. San Angelo 76909.
Baylor U. Waco 76798.
Central Texas College Killeen 76549.
Houston Baptist U. Houston 77074
Houston-Unversity Park U. of Houston 77204.
North Texas U. of Denton 76203.
Prairie View A&M U. Prairie 77446.
Sam Houston State U. Huntsville 77341.
Southern Methodist U. Dallas 75205.
Stephen F. Austin State U. Nacogdoches 75962.
Texas A&M U. College Station 77843.
Texas A&M University-Commerce Commerce 75429.
Texas Christian U. fort Worth 76129.
Texas Southern U. Houston 77004.
Texas Tech U. Lubbock 79409.
Texas Wesleyan U. Forth 76105.
Texas Women's U. Denton 76204-5828.
Trinity U. San Antonio 78212-7200.
West Texas A&M U. Canyon 79016.

Utah

Brigham Young U. Provo 84602.
Dixie State College of Utah Utah 84770.
Southern Utah U. Cedar City 84720.
Utah, U. of Salt Lake City 84112-0491.
Utah State U. Logan 84322-4605.
Weber State U. Ogden 84408.

Vermont

Castleton State College Castleton 05735.
Lyndon State College Lyndonville 05851.

Virginia

Emory and Henry College Emory 24327.
Hampton U. Hampton 23668.
Norfolk State U. Norfolk 23504.
Radford U. Radford 24141.
Regent U. Virginia Beach 23464.
Richmond, U. of Richmond 23173.
Virginia Commonwealth U. Richmond 23284.
Virginia Union U. Richmond 23220.
Washington and Lee U. Lexington 24450.

Washington

Central Washington U. Ellensburg 98926.
Eastern Washington U. Spokane 99201.
Seattle U. Seattle 98122.
Walla Walla College College Place 99324.
Washington, U. of Seattle 98195-3740.
Washington State U. Pullman 99164-2520.
Western Washington U. Bellingham 98225-9101.
Whitworth College Spokane 99251.

West Virginia

Alderson Broaddus College Philippi 26416.
Marshall U. Huntington 25755.
West Virginia U. Morgantown 26506.
West Virginia U. Parkersburg 26101.

Wisconsin

Marquette U. Milwaukee 53201.
Wisconsin-Eau Claire, U. of Eau Claire 54702.
Wisconsin-La Crosse, U. of La Crosse 54601.
Wisconsin-Madison, U. of Madison 53706-1497.
Wisconsin-Madison, U. of Madison 53706-1563.
Wisconsin-Milwaukee, U. of Milwaukee 53201.
Wisconsin-Oshkosh, U. of Oshkosh 54901.
Wisconsin-River Falls, U. of River Falls 54022.
Wisconsin-Stevens Point, U. of Stevens Point 54481.

Wyoming

Laramine County Community college Cheyenne 82007.
Wyoming, U. of Laramie 82071-3904.

Major Broadcasting and Cable Awards

AAAS Science Journalism Awards, 1200 New York Ave. N.W., Washington, DC, 20005. Phone: (202) 326-6440. Fax: (202) 789-0455. E-mail: media@aaas.org Web Site: www.aaas.org/aboutaaas/awards/sja/index.shtml.

The AAAS Science Journalism Awards represent the pinnacle of achievement for professional journalists in the science writing field. The awards recognize outstanding reporting for a gen audience & honor individuals (rather than institutions, publishers or employers) for their coverage of the sciences, engrg & mathematics. The Awards will be presented at the AAAS Annual Meeting in Boston in Feb 2008. The award for each category is $3,000. AAAS will reimburse winners for reasonable travel & hotel expenses. In cases of multiple authors, only one person's travel expenses will be covered. Eligibility Period: July 1, 2006 through June 30, 2007. Deadline for entries: Aug 1, 2007. Contact: Ginger Pinholster, AAAS Science Journalism Awards, 1200 New York Ave. N.W. Washington, DC 20005.

Academy of Television Arts and Sciences Emmy Awards, Academy of Television Arts & Sciences, 5220 Lankershim Blvd., North Hollywood, CA, 91601. Phone: (818) 754-2800. Fax: (818) 754-2836. E-mail: shore@emmys.org Web Site: www.emmys.org.

Primetime Emmy. Eligibility Period: June 1, 2007 - May 31, 2008. Deadline for entries: March 30.

Alliance for Community Media Hometown Video Festival Awards, 666 11th St. N.W., Suite 740, Washington, DC, 20001-4542. Phone: (202) 393-2650. Fax: (202) 393-2653. E-mail: dvinsel@tctv.net Web Site: www.alliancecm.org.

The Hometown Videi Awards honors & promotes community media & loc cable programs that are first distributed on Public Educ & Govt (PEG) access cable TV channels. . Eligibility Period: An annual event. Deadline for entries: Early March. Contact: Bunnie Riedel, Exec Dir, 666 11th St. N.W., Suite 740, Washington, DC 20001-4542.

The American Legion Fourth Estate Award, Attn: Public Relations, The American Legion National Hqtrs., 700 N Pennsylvania St., Indianapolis, IN, 46204. Phone: (317) 630-1253. Fax: (317) 630-1368. E-mail: pr@legion.org Web Site: www.legion.org.

Must accept award at annual National Convention. Eligibility Period: January-2007-December-2007. Deadline for entries: Jan 31 of each year. Contact: Public Relations, The American Legion National Headquarters, 700 N. Pennsylvania St., Indianapolis, IN 46204.

American Women in Radio and Television Inc. Awards, 8405 Greensboro Dr., Suite 800, McLean, VA, 22102. Phone: (703) 506-3290. Fax: (703) 506-3266. E-mail: info@awrt.org Web Site: www.awrt.org.

The Gracie Allen Awards strive to encourage the positive, realistic portrayal of women in entertainment, news, commercials, features & other progmg. Deadline for entries: See Website. Contact: Amy Lotz, Mgr, AWRT, 1595 Spring Hill Road, Ste 330, Vienna, VA 22182.

Armstrong Awards, Armstrong Foundation, Columbia Univ., 500 W. 120th St., Rm. 1312 S.W. Mudd Bldg., New York, NY, 10027. Phone: (212) 854-4718. Phone: (212) 854-3121. Fax: (212) 854-7837. E-mail: kkg1@columbia.edu Web Site: www.armstrongfoundation.org.

Awards to electrical engrg students, rsch scientists in the field of communications, AM, FM radio stns for excellence & originality in bcstg. Eligibility Period: The annual awards program covers the period from January-December 31. Deadline for entries: September 1. Contact: Ken Goldstein, pres, Armstrong Foundation, Columbia University, S.W. Mudd Hall, 500 W. 120th St., Room 1311, New York, NY 10027.

The Association for Women in Communications, 3337 Duke St., Alexandria, VA, 22314. Phone: (703) 370-7436. Fax: (703) 370-7437. E-mail: clarion@womcom.org Web Site: www.womcom.org.

The Clarion symbolize excellence in clear, concise communications, honor greatness in more than 130 categories across all communications disciplines. Eligibility Period: Jan 1, 2007-Dec 31, 2007. Deadline for entries: March 24, 2008.

Association of Cable Communicators, Beacon Awards, 25 Massachusetts Ave. N.W., Suite 100, Washington, DC, 20001. Phone: (202) 222-2370. Fax: (202) 222-2371. E-mail: services@cablecommunicators.org Web Site: www.cablecommunicators.org.

Beacon Awards honor execellence in pub affrs throughout the cable industry. Individuals from MSOs, cable systems, progmg networks & state assn are honored for their achievements in the areas of PR, community outreach, customer svc, edu & govt affrs. Finalists & winners achieve peer, industry & loc mkt recognition for their exemplary work while setting a standard for their colleagues to emulate. Eligibility Period: November 2006-November 2007. Deadline for entries: 2008.

BDA International Design Award, BDA International, 2029 Century Park E., Suite 555, Los Angeles, CA, 90067. Phone: (310) 712-0040. Fax: (310) 712-0039. Web Site: www.bda.tv. E-mail: adrienne@promax.tv

Acknowledges outstanding contributions in design for the screen & its associated promotions. Eligibility Period: January 1-December 31. Deadline for entries: January. Contact: Adrienne Alwag, BDA Design-awards cordinator, 2029 Century Park E., Suite 555, Los Angeles, CA 90067.

Batten Fellows Program, Box 6550, The Darden School, University of Virginia, Charlottesville, VA, 22906. Phone: (434) 924-7739. Fax: (434) 243-8708. E-mail: ahs4c@virginia.edu Web Site: www.darden.virginia.edu.

Batten Media Fellowships allow professionals engaged in all aspects of the media to pursue an MBA degree at the Univ. of VA Darden Graduate School of Business Admin. These Fellowships cover tuition and fees for both years of the MBA Program & include a tipend to cover a portion of living expenses. Each year up to three media professionals may be selected as Batten Fellows. Deadline for entries: No deadline. Contact: Director of Financial Aid, The Darden School, University of Virginia, Box 6550, Charlottesville, VA 22906.

The John Bayliss Broadcast Foundation Internships, Scholarships Programs, Box 51126, Pacific Grove, CA, 93950. Phone: (831) 655-5229. Fax: (831) 655-5228. E-mail: khfranke@baylissfoundation.org Web Site: www.baylissfoundation.org.

Application & information available on web site.

Broadcast Cable Financial Management Assn. "Avatar" Award, 550 W Frontage Rd., Suite 3600, Northfield, IL, 60093-1243. Phone: (847) 716-7000. Fax: (847) 716-7004. E-mail: info@bcfm.com Web Site: www.bcfm.com.

Awarded to a person who is considered the embodiment of a know model or category; an exemplar. Eligibility Period: Annually. Deadline for entries: Jan 30th. Contact: BCFM, Buz Buzogany, pres, Bcst Cable Financial Mgmt Assoc, 701 Lee St., Suite 640, Des Plaines, IL 60016.

Broadcast Education Association, 1771 N St. N.W., Washington, DC, 20036-2891. Phone: 202-429-3935. Fax: (202) 775-2981. E-mail: beainfo@beaweb.org Web Site: www.beaweb.org.

BEA is the professional dev assoc for professors, industry professionals, students involved in teaching, rsch related to radio, TV & electronic media. BEA administers fifteen scholarships annually to honor bcstrs & the bcst industry. The Two-Year Award $1,500 is for study at schools offering only freshman & sophomore instruction or for use at a four-year school by a graduate of a BEA two-year campus. All others scholarships are awarded to juniors, seniors & graduate students at BEA Member institutions. Eligibility Period: School Year. Deadline for entries: Sept 15. Contact: BEA Scholarships, 1771 N. St. N.W., Washington, DC 20036.

Heywood Broun Award, The Newspaper Guild-CWA, 501 Third St. N.W., Washington, DC, 20001. Phone: (202) 434-7177. Fax: (202) 434-1472. Web Site: www.newsguild.org. E-mail: guild@cwa-union.org

Recognizes individual journalistic achievement by members of the working media, particularly if it helps right a wrong or correct an injustice. Eligibility Period: Preceding calendar year. Deadline for entries: Last Friday in January. Contact: Andy Zipser, Editor of the Guild Reporter, The Newspaper Guild.

The CLIO Awards Ltd., 770 Broadway, 6th Fl., New York, NY, 10001. Phone: (212) 683-4300. Fax: (212) 683-4796. Web Site: www.clioawards.com.

Awards honoring advertising & design excellence worldwide. Eligibility Period: 12 month calendar year. Deadline for entries: call for details, they vary per medium. Contact: Andrew Jaffe, exec dir, 220 5th Ave., Suite 1500, New York, NY 10001.

John Chancellor Award, Columbia Univ Graduate School of Journalism, 2950 Broadway, MC 3805, New York, NY, 10027. Phone: (212) 854-5047. Fax: (212) 854-3148. E-mail: chancelloraward@jrn.columbia.edu Web Site: www.jrn.colummbia.edu.

John Chancellor Award recognizes and rewards a single journalist whose reporting over time shows courage, integrity, curiosity and intelligence and epitomizes the role of journalism in a free society. The $25,000 annual prize honors the legacy of John Chancellor, the pioneering television correspondent and longtime anchor for NBC News. This award is intended to honor the sustained achievement of a single journalist, who may not be well-known nationally, but whose cumulative accomplishments are exemplary. Eligibility Period: body of work over time. Deadline for entries: June 15, any year

Christopher Video Contest for College Students, The Christophers, 5 Hanover Sq., New York, NY, 10005. Phone: (212) 759-4050. Fax: (212) 838-5073. E-mail: youth@christophers.org Web Site: www.christophers.org/contests.html.

Eight cash prizes, from $300-$100. Enter short films of 5 minutes or less on the theme " One Person Can Make a Difference". Deadline for entries: June 6, 2008

Corporation for Public Broadcasting, 401 9th St. N.W., Washington, DC, 20004. Phone: (202) 879-9600. Fax: (202) 879-9700. Web Site: www.cpb.org.

The Edward R. Murrow Award recognizes individual whose work has fostered the growth, quality & image of public radio. Contact: Systems & Station Development, CPB, 901 E St. N.W., Washington, DC 20004-2006.

DGA Awards, Directors Guild of America Inc., 7920 Sunset Blvd., Los Angeles, CA, 90046. Phone: (310) 289-5333. Fax: (310) 289-5384. E-mail: allisonh@dga.org Web Site: www.dga.org.

Outstanding Directorial Achievement. Eligibility Period: Calendar year. Deadline for entries: Nov 2005. Contact: Directors Guild of America Inc., 7920 Sunset Blvd., Los Angeles, CA 90046.

Alfred I. duPont-Columbia University Awards, duPont Center for Broadcast Journalism. Columbia Univ Graduate School of Journalism, 2950 Broadway, Rm. 709B, New York, NY, 10027. Phone: (212) 854-5047. Fax: (212) 854-3148. E-mail: dupontawards@jrn.columbia.edu Web Site: www.dupont.org.

Honoring excellence in radio & TV news programs in the United States, the bcst equivalent of the Pulitzer Prizes. Eligibility Period: July 1, 2007- June 30, 2008. Deadline for entries: July 1st. Contact: Madiha Tahir or Jonnet Abeles. duPont Center, Columbia University Graduate School of Journalism, 2950 Broadway MC 3805, New York, NY 10027.

EDGE Awards - Entertainment Industries Council Inc., 1760 Reston Pkwy., Suite 415, Reston, VA, 20190-3303. Phone: (703) 481-1414. Fax: (703) 481-1418. Fax: (818) 333-5005 (West Coast). E-mail: eiceast@eiconline.org Web Site: www.prismawards.com.

Entertainment depiction of gun educ to promote a decrease in gun violence & promote firearm satety. Eligibility Period: Sept 2006 - Aug 2007. Deadline for entries: Sept 2007

Freedoms Foundation National Awards, Freedoms Foundation at Valley Forge, 1601 Valley Forge Rd., Valley Forge, PA, 19482-0706. Phone: (610) 933-8825. Fax: (610) 935-0522. E-mail: csantangelo@ffvf.org Web Site: www.ffvf.org.

George Washington Honor Metal presented to the most significant & dynamic TV & radio programs which increase an understanding of our American Way of Life. May be a one-time work or series of works. Eligibility Period: June 1, 2006-June I, 2007. Deadline for entries: June 1, 2007. Contact: Freedoms Foundation at Valley Forge, Awards Department, Rt. 23, 1601 Valley Forge Rd., Valley Forge, PA 19482-0706.

Gabriel Awards, Catholic Academy for Communications Arts Professionals, 1645 Brook Lynn Dr., Suite 2, Dayton, OH, 45432-1944. Phone: (937) 458-0265. Fax: (937) 458-0263. E-mail: admin@catholicacademy.org Web Site: www.catholicacademy.org.

Nine-inch silver figure mounted on base of polished wood, symbolizing the communication of God's Word to Humanity. Awarded to programs that affirm the dignity of human beings, recognize & uphold values such as community, creativity, tolerance, justice, compassion & dedication to excellence. Eligibility Period: Calendar year 2007. Deadline for entries: Mar 2008

Global Media Awards, (For Excellence in Population Reporting). The Population Institute, 107 2nd St. N.E., Washington, DC, 20002. Phone: (202) 544-3300. Fax: (202) 544-0068. E-mail: web@populationinstitute.org Web Site: www.populationinstitute.org.

Awards honors those who have contributed to creating awareness of population related issues through their journalistic endeavors. Eligibility Period: September-August. Deadline for entries: September 1. Contact: Global Media Awards Coord. The Population Institute, 107 2nd St. N.E., Washington, DC 20002.

Golden Mike Award, Broadcasters Foundation of Amercia, 7 Lincoln Ave., Greenwich, CT, 06830. Phone: (203) 862-8577. Fax: (203) 629-5739. E-mail: ghhbcast@aol.com Web Site: www.broadcastersfoundation.org.

Broadcast industry excellent svc. Contact: Gordon Hastings, pres/CEO, Broadcast Foundation, 296 Old Church Rd., Greenwich, CT 06830.

Golden Viddy Award, Box 289, National Academy of TV Journalists Inc., Salisbury, MD, 21803. Phone: (410) 548-5343. Fax: (410) 543-0658. E-mail: nbayne@shore.intercom.net Web Site: goldenviddyawards.com.

Award: Gold Statue. Eligibility Period: Year before. Deadline for entries: Third week of April. Contact: Neil Bayne, The National Academy of Television Journalists Inc., Box 31, Salisbury, MD 21803.

Hugo Awards, Chicago International TV Competition, 32 W. Randolph St.,Suite 600, Chicago, IL, 60601. Phone: (312) 683-0121. E-mail: info@chicagofilmfestival.com Web Site: www.chicagofilmfestival.com.

Gold HUGO to best overall production. Silver HUGO, Gold Plaque, Silver Plaque, certificate of Merit to best production with a specific category. Contact: Entry Coordinator, Chicago International Television Competition, 32 W. Randolph St., Suite 600, Chicago, IL 60610.

IRE Annual Awards for Investigative Reporting, 138 Neff Annex, Missouri School of Journalism, Columbia, MO, 65211. Phone: (573) 882-6668. Fax: (573) 884-8151. E-mail: beth@ire.org Web Site: www.ire.org/contest.

The annual IRE Awards recognize outstanding investigative work in print & bcst. The contest also helps to identify the techniques & resources used to complete each story. Eligibility Period: January-December 2007. Deadline for entries: First week of January 2008.

IRE Tom Renner Award for Crime Reporting, 138 Neff Annex, UMC-Journalism, Columbia, MO, 65211. Phone: (573) 882-2042. Fax: (573) 882-5431. E-mail: info@ire.org Web Site: www.ire.org.

For the best investigative crime reporting. Eligibility Period: Open. Deadline for entries: January. Contact: Len Bruzzese, Dept Dir, IRE, 138 Neff Annex, UMC-Journalism, Columbia, MO 65211.

International Broadcasting Awards, Hollywood Radio & TV Society, 13701 Riverside Dr., Suite 205, Sherman Oaks, CA, 91423. Phone: (818) 789-1182. Fax: (818) 789-1210. E-mail: info@hrts.org Web Site: www.hrts.org.

Currently on hiatus. Eligibility Period: Commercials: Must have been transmitted in previous year. Deadline for entries: Dec. 6. Contact: Dave Ferrara, exec dir, Hollywood Radio & TV Society, 13701 Riverside Drive, Suite 205, Sherman Oaks, CA 91423.

International Emmy Awards, The International Academy of Television Arts & Sciences, 888 7th Ave., Suite 506, New York, NY, 10019. Phone: (212) 489-6969. Fax: (212) 489-6557. E-mail: info@iemmys.tv Web Site: www.iemmys.tv.

The International Emmy Award recognites excellence in progmg produced outside of the US. Deadline for entries: April 1st.

International Radio & Television Society Foundation Gold Medal, 420 Lexington Ave., Suite 1601, New York, NY, 10170. Phone: (212) 867-6650. Fax: (212) 867-6653. Web Site: www.irts.org.

A gold medal is presented annually for significant career long contributions to the integrity, health & success of the electronic media industry. Eligibility Period: Nominations accepted until Sept of previous year. Contact: Joyce Tudryn, pres, International Radio & TV Society Foundation, 420 Lexington Ave., New York, NY 10170.

Robert F. Kennedy Journalism Awards, 1367 Connecticut Ave. N.W., Suite 200, Washington, DC, 20036. Phone: (202) 463-7575. Fax: (202) 463-6606. E-mail: info@rfkmemorial.org Web Site: www.rfkmemorial.org.

Awards honoring outstanding reporting of problems of the disavantaged. Recognizes authors, journalists, & human rights activists that have exposed injustice. Eligibility Period: Calendar year. Deadline for entries: January. Contact: Robert F. Kennedy Journalism Awards, 1367 Connecticut Ave. N.W., Suite 200, Washington, DC 20036.

Knight-Wallace Journalism Fellows, Wallace House, 620 Oxford Rd., Ann Arbor, MI, 48104. Phone: (734) 998-7666. Fax: (734) 998-7979. E-mail: wpalms@umich.edu Web Site: www.kwfellows.org.

To help those who have demonstrated superior ability commitment & leadership to reach peak performance in svc to leaders, viewers, & listeners of the American public. Eligibility Period: academic year (September-April). Deadline for entries: February 1. Contact: Charles R. Eisendrath, Michigan Journalism Fellows, Wallace House, 620 Oxford Rd., Ann Arbor, MI 48104.

The Livingston Awards for Young Journalists, Wallace House, 620 Oxford Rd., Ann Arbor, MI, 48104. Phone: (734) 998-7575. Fax: (734) 998-7979. Web Site: www.livawards.org. E-mail: LivingstonAwards@umich.edu

Purpose of award is to recognize & further develop the abilities of young journalists. Eligibility Period: Calendar year for entries. Deadline for entries: February 1. Contact: Charles R. Eisendrath, The Livingston Awards, Wallace House, 620 Oxford Rd., Ann Arbor, MI 48104.

Mark of Excellence, 3909 N. Meridian St., Society of Professional Journalists, Indianapolis, IN, 46208-4011. Phone: (317) 927-8000. Fax: (317) 920-4789. E-mail: awards@spj.org Web Site: www.spj.org.

Honors the best in student journalism. Open to all students enrolled in a college or university & studying for an academic degree. Eligibility Period: Entries must have been bcst in the calendar year. Deadline for entries: January 31. Contact: MOE Awards Coord, SPJ, 3909 N. Meridian St., Indianapolis, IN 46208.

Paul Miller Washington Reporting Fellowships, National Press Foundation, 1211 Connecticut Ave NW, Suite 310, Washington, DC, 20036. Phone: (202) 663-7280. Fax: (202) 530-2855. E-mail: nolan@nationalpress.org Web Site: www.nationalpress.org.

To increase journalists' knowledge of complex issues in order to improve public understanding. Recognizes & encourages excellence in journalism. Eligibility Period: Seminars from Sept-May. Deadline for entries: Applications accepted in the spring. Contact: Nolan Walters, 1211 Connecticut Ave NW, Suite 310, Washington, DC 20036.

Missouri Honor Medal, Missouri School of Journalism, 120 Neff Hall, Columbia, MO, 65211. Phone: (573) 882-6686. Fax: (573) 884-5400. E-mail: dukesb@missouri.edu Web Site: journalism.missouri.edu.

The Missouri Honor Metal for Distinguished Service in Journalism is given to recognize the highest standards of excellence. Eligibility Period: Open. Deadline for entries: March of each year. Contact: Dean, School of Journalism, 103 Neff Hall, University of Missouri-Columbia, Columbia, MO 65211.

The Mobius Advertising Awards, 713 S. Pacific Coast Hwy., Suite A, Redondo Beach, CA, 90277-4233. Phone: (310) 540-0959. Fax: (310) 316-8905. E-mail: mobiusinfo@mobiusawards.com Web Site: www.mobiusawards.com.

The world's oldest independent advertising & package design competitions. Eligibility Period: Oct 1, 2006 - Oct 1, 2007. Deadline for entries: Oct 1 yearly, late entries accepted (see website). Contact: J.W. Anderson, chmn; Patricia Meyer, exec dir, 841 N. Addison Ave., Elmhurst, IL 60126-1291.

NAB Crystal Radio Awards, NAB Radio, 1771 N St. N.W., Washington, DC, 20036-2891. Phone: (202) 775-3511. Fax: (202) 775-3523. E-mail: csuever@nab.org Web Site: www.nab.org.

Given to radio stns for their year-round commitment to community svc. Eligibility Period: January 1-December 31, 2007. Deadline for entries: Feb 1, 2008. Contact: Chris Suever, NAB Radio, 1771 N St. N.W., Washington, DC 20036-2891.

NAB Marconi Radio Awards, NAB Radio, 1771 N St. N.W., Washington, DC, 20036. Phone: (202) 775-3511. Fax: (202) 775-3523. E-mail: csuever@nab.org Web Site: www.nab.org.

Given to stns & on-air personalities for bcst excellence. Eligibility Period: January 1-December 31, 2007. Contact: NAB Radio, 1771 N St. N.W., Washington, DC 20036.

NAB National Radio Award, 1771 N St. N.W., Washington, DC, 20036-2891. Phone: (202) 775-3511. Fax: (202) 775-3523. E-mail: csuever@nab.org Web Site: www.nab.org.

Recognizes individual who has made a significant or on-going contribution to the radio in a leadership capacity. Contact: Chris Suever, NAB Radio, 1771 N St. N.W., Washington, DC 20036-2891.

NPPA Annual TV News Photography & Editing Competition, c/o Rich Murphy, WTSP-TV 10, 11450 Gandy Blvd., St. Petersburg, FL, 33702. Phone: (727) 577-8582. Fax: (727) 576-6924. E-mail: info@nppa.org Web Site: www.nppa.org.

Contest showcases the best news photography, editing in print, TV & on the website. Eligibility Period: Jan 1-Dec31 (annually). Deadline for entries: Jan 31, postmarked by midnight. Contact: Rich Murphy, contest chmn, WFLA, 200 South Parker St, Tampa, FL 33606; rmurphy@wfla.com.

National Academy of Television Arts and Sciences "Emmy" Awards, 111 W. 57th St., Suite 600, New York, NY, 10019. Phone: (212) 586-8424. Fax: (212) 246-8129. Web Site: www.emmyonline.org.

Recognizes outstanding achievements in all phases of TV, including progmg, directing, writing, performing, etc. The National Academy of TV Arts & Sciences also presents the Sports Emmy & the tech achievement awards, news & documentary awards, & community svc awards covering the entire previous year. Contact: Allan Benish, exec VP, 111 W. 57th St., Suite 1020, New York, NY 10019.

National Association of Broadcasters (NAB) Engineering Achievement Awards, 532249. 1771 N St. N.W., Washington, DC, 20036-2891. Phone: (202) 429-5300. Fax: (202) 429-5461. E-mail: nab@nab.org Web Site: www.nab.org.

Given to industry leaders for significant contributions that have advanced bcst engrg.

National Association of Broadcasters (NAB) International Bcstg Excellence Award, 1771 N St., N.W., Washington, DC, 20036-2891. Phone: (202) 429-5360. Fax: (202) 429-5461. E-mail: edorey@nab.org Web Site: www.nab.org/lag/international/award.asp.

The NAB International Broadcasting Excellence Award was established in 1995 to recognized international bcst organizations that have demonstrated exceptional leadership in advancing the bcst industry & svcs they provide to their community & audiences. The awards are presented during the International Reception held at NAB's spring convention in Las Vegas. Deadline for entries: Feb 15, 2008. Contact: Emily Dorey at the address listed above.

National Association of Broadcasters (NAB) Distinguished Service Award, 1771 N St. N.W., Washington, DC, 20036. Phone: (202) 429-5368. Fax: (202) 429-4199. E-mail: nab@nab.org Web Site: www.nab.org.

DSA Award given annually to any bcstr, whether or not actively engaged in the operations end of the bcstg industry, who has made a significant & lasting contribution to the American system of bcstg by virtue of singular achievement or continuing service for, or on behalf of, the industry. The recipient is required to be present at the convention to receive the award. Established in 1953, the award is presented at the annual convention. Deadline for entries: Dec. 2. Contact: NAB Distinguished Service Award, 1771 N St. N.W., Washington, DC 20036.

National Awards for Education Reporting, Education Writers Association, 2122 P St. N. W. # 201, Washington, DC, 20037. Phone: (202) 452-9830. Fax: (202) 452-9837. E-mail: ewa@ewa.org Web Site: www.ewa.org.

This prestigious contest awards prizes in 18 different categories & is the only independently judged educ writing competition of its kind in the United States. Eligibility Period: Published during current calendar year. Deadline for entries: Mid-Jan. Contact: Education Writers Assn, 1331 H St. N.W., Suite 307, Washington, DC 20005.

National Headliner Awards, Box 239, 226 Mt. Vernon Ave., Northfield, NJ, 08225. Phone: (609) 646-8896. Fax: (609) 646-8826. Web Site: www.nationalheadlinerawards.com. E-mail: infoheadliners@aol.com

Awards in print, photography, radio, TV & internet, 60 plus awards to be given, four grand awards, cash prize of $1,500 each. Deadline for entries: Jan 1, 2008-Dec 31, 2008. Contact: Michael Schurman, exec dir, National Headliners Club, Box 239, Northfield, NJ 08225.

The New York Festivals International Radio Programming & Promotion Awards, 260 W. 39th St., 10 th Fl., New York, NY, 10018. Phone: (212) 643-4800. Fax: (212) 643-0170. E-mail: info@newyorkfestivals.com Web Site: www.newyorkfestivals.com.

Recognizes the world's best work in radio progmg & promotion. Eligibility Period: Mar-Mar. Deadline for entries: Feb-Mar. Contact: Bilha Goldberg, festival dir, The New York Festivals, 186 5th Ave, New York, NY 10010.

The New York Festivals International TV, Cinema & Radio Advertising Awards, 260 W. 39th St., 10 th Fl., New York, NY, 10018. Phone: (212) 643-4800. Fax: (212) 643-0170. Web Site: www.newyorkfestivals.com. E-mail: info@newyorkfestivals.com

Recognizes the world's best work in TV, radio consumer advertising & pub service announcements. Eligibility Period:

June-June. Deadline for entries: June 30-USA / September 15-overseas. Contact: Bilha Goldberg, festivel dir, The New York Festivals, 186 Fifth Ave., New York, NY 10010.

The Ollie Awards, American Center for Children and Media, 5400 N. Saint Louis Ave., Chicago, IL, 60625. Phone: (773) 509-5510. Fax: (773) 509-5303. E-mail: dkleeman@atgonline.org Web Site: www.centerforchildrenandmedia.org.

Award for excellence in children's TV. Eligibility Period: To be announced. Deadline for entries: To be announced. Contact: David Kleeman, exec dir, American Center for Children and Media, 5400 N. Saint Louis Ave., Chicago, IL 60625.

Overseas Press Club Awards, 40 W. 45th St., New York, NY, 10036. Phone: (212) 626-9220. Fax: (212) 626-9210. Web Site: www.opcofamerica.org.

Overseas Press Club of America has 21 awards including: Hal Boyle Award, Bob Considine Award, Robert Capa Gold Medal Award, Olivier Rebbot Award, John Faber Award, Feature Photography Award, Edward R. Murrow Award, The Carl Spielvogel Award & The Lowell Thomas Award. Eligibility Period: Covers entries from Jan 1-Dec 31, of award year. Deadline for entries: End of Jan for previous year's work. Contact: Sonya K. Fry, exec dir, Overseas Press Club Awards, 320 E. 42nd St., New York, NY 10017.

PRISM Awards, Entertainment Industries Council Inc., PRISM Awards, Entertainment Industries Council Inc., 1760 Reston Pkwy., Suite 415, Reston, VA, 20190-3303. Phone: (703) 481-1414. Fax: (703) 481-1418. Fax: (818) 333-5005 (West Coast). E-mail: eiceast@eiconline.org Web Site: www.prismawards.com.

Recognize the accurate depiction of substance abuse, mental health in entertainment progmg, TV, film, comic books & interactive media. Eligibility Period: Jan - Dec during current calendar year. Deadline for entries: December

PROMAX & BDA Awards, 9000 W. Sunset Blvd., Los Angeles, CA, 90069-2906. Phone: (310) 788-7600. Fax: (310) 788-7616. Web Site: www.promax.tv. E-mail: awards@promax.tv

Recognizes the best in prom & mktg. Eligibility Period: Previous calender year. Deadline for entries: Jan & Mar. Contact: PROMAX & BDA, 2029 Century Park E., Suite 555, Los Angeles, CA 90067-2906.

George Foster Peabody Award, H.W. Grady College of Journalism, University of Georgia, Athens, GA, 30602-3018. Phone: (706) 542-3787. Fax: (706) 542-9273. E-mail: peabody@uga.edu Web Site: www.peabody.uga.edu.

The George Foster Peabody Awards were first awarded in1941 for radio programs bcst in 1940. The awards recognize distinguished achievement & meritorious svcs by radio & TV nets, stns, producing organizations & individuals. Eligibility Period: Jan 1 - Dec 31, 2006. Deadline for entries: Jan 15, 2007.

George Polk Awards, Long Island Univ.,The English Dept, University Plaza, Brooklyn, NY, 11201. Phone: (718) 488-1115. Fax: (718) 243-0766. Fax: (718) 246-6302. Web Site: www.liu.edu/cwis/bklyn/polk/polk.html.

A plaque. Eligibility Period: 2007. Deadline for entries: Jan 9, 2008. Contact: c/o The English Dept. Robert Spector PhD., curator, George Polk Awards, Long Island Univ., The Brooklyn Ctr., University Plaza, Brooklyn, NY 11201.

RTNDA Edward R. Murrow Awards, 1600 K St., Suite 700, Washington, DC, 20006-2838. Phone: (202) 659-6510. Phone: 800-80-RTNDA. Fax: (202) 223-4007. E-mail: rtnda@rtndf.org Web Site: www.rtnda.org.

Honors those whose work has fostered the growth, quality & positive image of public radio. Eligibility Period: Previous calendar year. Deadline for entries: Jan 31. Contact: RTNDA, 1000 Connecticut Ave. N.W., Suite 615, Washington, DC 20036.

Radio-Mercury Awards, 22 Cortland St., 17th Fl., New York, NY, 10001. Phone: (212) 681-7207. Fax: (212) 681-7223. E-mail: mercury@rab.com Web Site: www.radiomercuryaward.com.Since the first Radio-Mercury Awards competition and awards ceremony in 1992, close to $2 million has been awarded to honor the writers and producers in the finest radio commercials in America. A Mercury Award is recognized as the highest honor a radio commercial can achieve. The Radio-Mercury Awards is the richest nationwide competition devoted exclusively to honoring & rewarding excellence in Radio creative. Prizes consist of the $100,000 Grand Prize, plus cash prizes in gen station produced & spanish-lanuage categories. In addition, a charitable donation is awarded in conjuction with the Public Service Announcement Award & the Radio-Mercury Student competition awards a cash prize to the winning school. The awards are presented annually at a gala luncheon ceremony in New York City, and are governed by the Radio Creative Fund, a non-profit corporation funded by the Radio industry . Eligibility Period: Paid radio advertisements that aired on a U.S. radio stn during the calendar year are eligible. Deadline for entries: February. Contact: Wendy Frech, Radio-Mercury Awards, 261 Madison Ave., 23rd Fl., New York, NY 10016 or E-mail: mercury@rab.com.

Bart Richards Award for Media Criticism, 302 James Bldg., Pennsylvania State Univ., University Park, PA, 16801-3867. Phone: (814) 865-8801. Fax: (814) 863-6134. E-mail: sws102@psu.edu Web Site: www.comm.psu.edu/bart.

The Bart Richards Award honors work that evaluates news media coverage of significant subjects or issues. The award is intended to recognnize constructively critical articles, books, & electronic media reports; academic, other rsch; reports by media ombudsmen & journalism watchdog groups. Eligibility Period: Jan 1, 2008-Dec 31, 2008. Deadline for entries: Jan 31, 2009. Contact: The Bart Richards Award for Media Criticism, The Pennsylvania State Univ., College of Communications, 302 James Bldg., University Park, PA 16801-3867.

Scripps Howard Foundation - Jack R. Howard Awards, Scripps Howard Foundation, 312 Walnut St., Cincinnati, OH, 45202-4067. Phone: (513) 977-3030. Fax: (513) 977-3800. E-mail: porters@scripps.com Web Site: www.scripps.com/foundation.

Honors the best investigative or in-depth reporting of events covered by TV & radio stns or cable systems. Any program or series is eligible in TV/cable or radio categories. Eligibility Period: Jan 1, 2007-Dec 31, 2007. Deadline for entries: Jan 31, 2008. Contact: Patty Cottingham, exec dir, Scripps Howard Foundation, Box 5380, Cincinnati, OH 45201.

Sigma Delta Chi Distinguished Service Awards Society of Professional Journalists, 3909 N. Meridian St., Indianapolis, IN, 46208. Phone: (317) 927-8000. Fax: (317) 920-4789. E-mail: awards@spj.org Web Site: www.spj.org.

A series of annual awards recognizing excellence in many categories of professional journalism. Eligibility Period: Work published in the previous calendar year. Deadline for entries: Febr 9, SDX; July 6, Pulliam Fellowship. Contact: SDX Awards coord, SPJ, 3909 N. Meridian St., Indianapolis, In 46208.

Silver Anvil Awards, Public Relations Society of America Inc., 33 Maiden Ln., 11th Fl., New York, NY, 10003-5150. Phone: (212) 460-1456. Fax: (212) 995-0757. E-mail: awards@prsa.org Web Site: www.prsa.org.

Best public rel practices. Deadline for entries: Varies - check website. Contact: Carla Voth, Public Relations Society of America Inc., 33 Irving Pl., 3rd Fl., New York, NY 10003.

Silver Gavel Awards, American Bar Association, Div for Pub Education, 541 N. Fairbanks Ct., Chicago, IL, 60611. Phone: (312) 988-5738. Fax: (312) 988-5494. E-mail: howardkaplan@staff.abanet.org Web Site: www.abanet.org /publiced/gavel.

Annually recognizes eligible entries from communications media that have been exemplary in fostering pub understanding of the law & legal system. Eligibility Period: Eligible radio & TV programs (including network, cable, syndicated, loc & ind productions) for the competition must have been originally bcst or presented between Jan. 1 & Dec. 31 of the same year. Deadline for entries: Mid Jan annually. Contact: The Gavel Awards Program, American Bar Association, Division for Public Education, 541 N. Fairbanks Ct, Chicago, IL 60611.

Society of Motion Picture & Television Engineers Awards, 3 Barker Ave., 5th Fl., White Plains, NY, 10601. Phone: (914) 761-1100. Fax: (914) 761-3115. Web Site: www.smpte.org.Citation for Outstanding Svc to the Society recognizes individuals for dedicated svc to the society. The Presidential Proclamation recognizes individuals of established & outstanding status & reputation in the motion picture & TV industries worldwide. Eastman Kodak Medal Award recognizes outstanding contributions that lead to new or unique educ programs using motion pictures, TV, high-speed & instrumental photography or other photographic sciences. The award recognizes dev in equipment, systems or instructional applications that advance the educ process at any or all levels. The John Grierson International Gold Medal Award recognizes significant tech achievements related to the production of documentary motion picture films. The Journal Award recognizes the outstanding paper originally published in the Journal of the Society during the previous calendar year. The Technicolor/Herbert T. Kalmus Gold medal Award recognizes outstanding contributions in the dev of color films, processing, techniques or equipment useful in making color motion pictures for theater or TV use. The Fuji Gold Medal Award recognizes outstanding engrg achievements in the design & dev of new or enhanced techniques &/or equipment that have contributed significantly to the advancement of photographic or electronic image origination . Deadline for entries: January 15. Contact: Fred Motts, exec dir, Society of Motion Pictures & TV Engineers, 595 W. Hartsdale Ave., White Plains, NY, 10607.

Sunscan Award, Entertainment Industries Council, Inc., 1760 Reston Pkwy, Suite 415, Reston, VA, 20190-3303. Phone: (703) 481-1414. Fax: (703) 481-1418. Fax: (818) 333-5005 (West Coast). E-mail: eiceast@eiconline.org Web Site: www.eiconline.org.

Accurate portrayal of skin cancer awareness & sun safety. Eligibility Period: Jan 2007 - Dec 2007. Deadline for entries: Jan 5, 2008.

Voice of Democracy Scholarship Program, VFW National Hqtrs., 406 W. 34th St., Kansas City, MO, 64111. Phone: (816) 968-1117. Fax: (816) 968-1149. E-mail: kharmer@vfw.org Web Site: www.vfw.org.

College scholarships can be won by writing & recording a 3 to 5 minutes audio essay on a chosen annual theme. Go to www. vfw.org, click on programs, click on VFW scholarships. Eligibility Period: 9th, 10th, 11th & 12th grade students. Deadline for entries: November 1st. Contact: Your high school counselor, local VFW Post, or Voice of Democracy Scholarship Program, VFW National Hqtrs, 406 W. 34th St., Kansas City, MO 64111.

Ida B. Wells Award, Northwestern Univ., Medill School of Journalism, 1845 Sheridan Rd., Evanston, IL, 60208. Phone: (847) 467-2579. Fax: (847) 491-2370. E-mail: m-awards@northwestern.edu

Lawrence, KS 66647. Sam Adams (Curator), Ida B.Wells Award, 1552 Alvamar Dr.

To give tangible & highly visible recognition to an individual or group of individuals & their company. Deadline for entries: April 1st each year. Contact: Charles Whitaker (send a future correspondence to him).

Western Heritage Awards-The Wrangler, National Cowboy & Western Heritage Museum, 1700 N.E. 63rd St., Oklahoma City, OK, 73111. Phone: (405) 478-2250. Fax: (405) 478-4714. E-mail: lyndahaller@nationalcowboymuseum.org Web Site: www.nationalcowboymuseum.org.

Bronze sculpture of a cowboy on horseback. All project entries must have a Western theme. Eligibility Period: Must be aired/ between Jan1 & Dec 31 of each year. Deadline for entries: Dec 31 of each year. Contact: Lynda Haller, dir PR, National Cowboy Hall of Fame, 1700 N.E. 63rd St., Oklahoma City, OK 73111.

Section I

Government

Federal Communications Commission Executives and Staff

Headquarters: The Portals, 445 12th St. S.W., Washington, DC 20554. (888) 225-5322.

CHAIRMAN

Kevin J. Martin (202) 418-1000 8 B201; Confidential Assistant Lori Alexiou (202) 418-2102; Chief of Staff Daniel Gonzalez; Senior Legal Advisor Michelle Carey (Media Issues); Legal Advisor Ian Dillner (Wireless Issues); Legal Advisor Erika Olsen (Acting); Staff Assistant Vivette Hart (202) 418-2927; Staff Assistant Demetrice Bess (202) 418-1020; Administrative Management Specialist Thomasine Greely.

COMMISSIONERS

COMMISSIONER Robert M. McDowell (202) 418-2200 8-C302; Chief of Staff John W. Hunter (Senior Legal Advisor, Wireline Issues); Legal Advisor Angela E. Giancarlo (Wireless International Issues); Legal Advisor Cristina Chou Pauze (Media Issues); Deputy Chief of Staff Brigid Nealon Calamis.

COMMISSIONER Michael J. Copps (202) 418-2000 8 B115; Confidential Assistant Carolyn Conyers; Senior Legal Advisor Rick C. Chessen; Legal Advisor Scott M. Deutchman (Competition & Universal Service); Legal Advisor Bruce Liang Gottieb (Wireless & International Issues); Staff Assistant Renee Coles (202) 418-2000.

COMMISSIONER Jonathan S. Adelstein (202) 418-2300 8-A302; Confidential Assistant Katie Yocum (Director of Outreach); Senior Legal Advisor Barry Ohlson (Legal Advisor for Spectrum & International Issues); Legal Advisor Scott Bergmann (Wireless Issues); Legal Advisor Rudy Brioche (Media Issues); Special Assistant Tamika Jones; Staff Assistant Tajuana Dill.

OFFICE OF ADMINISTRATIVE LAW JUDGES

Richard L. Sippel/Chief Administrative Law Judge (202) 418-2280; Arthur I. Steinberg/Administrative Law Judge (202) 418-2255.

OALJ Administrative and Management Staff

Mary Gosse/Administrative officer; Patricia Ducksworth/Legal Technician

OFFICE OF COMMUNICATIONS BUSINESS OPPORTUNITIES

Carolyn Fleming Williams/Director (202) 418-0990.

OFFICE OF ENGINEERING AND TECHNOLOGY

Julius P. Knapp/Chief (202) 418-2470; Ira Keltz/Acting Deputy Chief (202) 418-2470; Ronald Repasi/Acting Deputy Chief (202) 418-2470; Alan Stillwell/Senior Associate Chief (202) 418-2470; Bruce A. Romano/Associate Chief (Legal) (202) 418-2470.

Administrative and Management Office

Xenia Hajicosti/Assistant Chief for Management (202) 418-2470.

Policy and Rules Division

Geraldine Matise/Deputy Chief (202) 418-2472; Mark Settle/Deputy Chief (202) 418-2472.

Spectrum Policy Branch

Jamison Prime/Chief (202) 418-2472.

Technical Rules Branch

Karen Rackley/Chief (202) 418-2472/A161.

Spectrum Coordination Branch

Kathryn Hosford/Chief (202) 418-2472.

Electromagnetic Compatibility Division

Ira Keltz/Chief (202) 418-2475.

Technical Analysis Branch

Ronald Chase/Branch Chief (202) 418-2475.

Experimental Licensing Branch

James R. Burtle/Chief (202) 418-2475.

Network Technology Division

Jeffery Goldthorp/Division Chief (202) 418-2478.

Laboratory (Columbia, MD)

Rashmi Doshi/Chief (301) 362-3000 LAB; Vacant/Deputy Chief (301) 362-3000 LAB.

Equipment Authorization Branch

Joe Dichoso/Chief (301) 362-3000 LAB.

Technical Research Branch

William Hurst/Chief (301) 362-3000 LAB.

Auditing and Compliance Branch

Raymond LaForge/Chief (301) 362-3000 LAB.

Customer Service Branch

Sandra Haase/Chief (301) 362-3013 LAB.

OFFICE OF GENERAL COUNSEL

Samuel Feder/General Counsel (202) 418-1700; Matthew Berry/Deputy General Counsel (202) 418-1700; P. Michele Ellison/Deputy General Counsel (202) 418-1700; Ajit Pai/Associate General Counsel (202) 418-1700; Jacob Lewis/Associate General Counsel (Counselor to the General Counsel) (202) 418-1700.

Litigation Division

Daniel M. Armstrong/Associate General Counsel & Chief (202) 418-1740; Richard K. Welch/Acting Deputy Associate General Counsel (202) 418-1740; Susan L. Launer/Deputy Chief & Deputy Associate General Counsel-Trial & Enforcement (202) 418-1740.

Administrative Law Division

Joel Kaufman/Deputy Associate General Counsel & Chief (202) 418-1720; Patrick J. Carney/Assistant General Counsel for Ethics (202) 418-1720; Maureen Duignan/Assistant General Counsel (202) 418-1720; Debra A. Weiner/Assistant General Counsel (202) 418-1720; Marilyn Sonn/Assistant General Counsel (202) 418-1720; David E. Horowitz/Assistant General Counsel (202) 418-1720; Diane Griffin Holland/Assistant General Counsel (202) 418-1720; Christopher Killion/Deputy Associate General Counsel (202) 418-1720.

OFFICE OF INSPECTOR GENERAL

Kent Nilsson/Inspector General (202) 418-0476 2C762; Jon Stover/Deputy Inspector General (202) 418-0390 2C76.

OFFICE OF LEGISLATIVE AFFAIRS

S. Kevin Washington/Director (202) 418-1900 8-C432; Gregory R. Vades/Special Counsel (202) 418-1900 8-C432; Lori Holy Maarbjerg/Attorney/Advisor (202) 418-1900 8-C432; Jim Balaguer/Legisative Analyst (202) 418-1900; Connie Chapman/Lead Congressional Liaison Specialist (202) 418-1900 8-C432; Joy Medley/Congressional Liaison Specialist (202) 418-1900 8-C432; Diane Atkinson/Congressional Liaison Specialist (202) 418-1900 8-C432; Lavenya Williams/Administrative Support Specialist (202) 418-1900 8-C432.

OFFICE OF MANAGING DIRECTOR

Anthony Dale/Agency's Managing Director (202) 418-1919 1-C144; Mindy Ginsburg/Deputy Managing Director (202) 418-1919; Joseph Hall/Deputy Managing Director (202) 418-1919; Mark Stone/Deputy Managing Director (202) 418-1919.

Office of The Secretary

Marlene Dortch/Secretary (202) 418-0300 TW-B204; William F. Caton/Deputy Secretary.

Information Resources Group

Sheryl A. Segal/Associate Secretary (202) 418-0234 TW-B204; Jacqueline Coles/Manager/Agenda & Publications Group (202) 418-0320 TW-B204. ;

Library

(202) 418-0450 TW-B505.

Commissioners

Kevin J. Martin, **Chairman**

Michael J. Copps, Jonathan S. Adelstein, Deborah Taylor Tate, Robert M. McDowell

Office of Inspector General

Office of Engineering & Technology
- Electromagnetic Compatibility Div.
- Laboratory Div.
- Policy & Rules Div.
- Administrative Staff

Office of General Counsel
- Administrative Law Div.
- Litigation Div.

Office of Managing Director
- Human Resources Management
- Information Technology Center
- Financial Operations
- Administrative Operations
- Performance Eval. & Records Mgmt
- Secretary

Office of Media Relations
- Media Services Staff
- Internet Services Staff
- Audio-Visual Services Staff

Office of Administrative Law Judges

Office of Strategic Planning & Policy Analysis

Office of Communications Business Opportunities

Office of Workplace Diversity

Office of Legislative Affairs

Public Safety & Homeland Security Bureau
- Admin. & Mgmt. Office.
- Policy Div.
- Public Communications Outreach & Operations Div.
- Communications Systems Analysis Div.

Wireline Competition Bureau
- Admin. & Mgmt. Office.
- Competition Policy Div.
- Pricing Policy Div.
- Telecommunications Access Policy Div.
- Industry Analysis & Technology Div.

Enforcement Bureau
- Office of Management & Resources
- Telecommunications Consumers Div.
- Spectrum Enforcement Div.
- Market Disputes Resolution Div.
- Investigations & Hearings Div.
- Regional & Field Offices

Wireless Telecommunications Bureau
- Management & Resources Staff
- Auctions & Spectrum Access Div.
- Spectrum Mgmt. Resources & Technologies Div.
- Spectrum & Competition Policy Div.
- Mobility Div.
- Broadband Div.

Media Bureau
- Mgmt. & Resources Staff
- Office of Com. & Industry Info.
- Policy Div.
- Industry Analysis Div.
- Engineering Div.
- Office of Broadcast License Policy
- Audio Div.
- Video Div.

Consumer & Governmental Affairs Bureau
- Admin & Mgmt. Office
- Info. & Resources Mgmt. Office
- Consumer Inquiries & Complaints Div.
- Consumer Policy Div.
- Reference Information Center
- Disability Rights Office
- Consumer Affairs & Outreach Div.
- Office of Intergovernmental Affairs

International Bureau
- Management & Administrative Staff
- Policy Div.
- Satellite Div.
- Strategic Analysis & Negotiations Div.

Assistant Managing Director-Human Resources Management

(202) 418-0100 1-A100.

Recruitment and Staffing Service Center

(202) 418-0293 Room 1-200.

Learning and Development Service Center

Jerry Liebes/Chief (202) 418-1582 1-A124.

Information Technology Center

Room 1 C361.

Associate Chief Information Officers

Dr. Greg Parhman (202) 720-2525.

Office of Personnel Management

Mark Reger/Chief Financial Officer 1-A625.

Associate Managing Director-Performance Evaluation and Records Management

Room 1-A838.

OFFICE OF MEDIA RELATIONS

David H. Fiske/Director (202) 418-0513 CY-C314, (888) 225-5322; Audrey Spivack/Associate Director; Meribeth McCarrick/Associate Director; Sharon Hurd/Releases & Digest.

Audio Visual Center

Dann Oliver/Manager (202) 418-0460 TW-A206A.

OFFICE OF STRATEGIC PLANNING AND POLICY ANALYSIS

Catherine Bohigian/Chief (202) 418-2030 7-C347; Elizabeth Andrion/Deputy Chief (202) 418-2030; Michelle Connolly/Commission's Chief Economist (202) 418-2030.

OFFICE OF WORKPLACE DIVERSITY

June Taylor/Acting Director (202) 418-1799 5-C750; Linda Miller/EEO Program Manager (202) 418-1799; Lawrence S. Schaffner/Senior Legal Advisor (202) 418-1799; Kenneth Heredia/Office Automation Clerk (202) 418-1799; Rosalind Bailey/Staff Assistant (202) 418-1799.

CONSUMER & GOVERNMENTAL AFFAIRS BUREAU

cathy Seidel/Bureau Chief (202) 418-1400 5-C754.

Office of Intergovernmental Affairs

Alice Elder/Chief (202) 418-7619 5-A660.

Office of Information Resource Management

Stephen Ebner Chief (202) 418-2147 4-A525; Roger Goldblatt/Deputy Chief (202) 418-1035 4-A523.

Consumer Inquiries and Complaints Division

Jeffrey Tignor/Acting Chief (202) 418-2516 4-C763.

Information Access and Privacy Office

Sumita Mukhoty/Director (202) 418-1110.

Consumer Policy Division

Erica McMahon/Chief (202) 418-0346 5-A803.

Disability Rights Office

Thomas E. Chandler/Chief (202) 418-1475 CY-B523; Cheryl J. King/Deputy Chief (202) 418-2284 CY-A636.

Reference Information Center

Bill Cline/Chief (202) 418-0267 CY-B533.

Consumer Affairs and Outreach Division

Louis Sigalos/Division Chief (202) 418-0614 3-A662.

Consumer Publications Branch

Stacey Reuben-Mesa/Branch Chief (202) 418-0254.

Consumer Advisory Committee

Scott Marshall/Designated Federal Officer (202) 418-2809.

ENFORCEMENT BUREAU

Kris Monteith/Bureau Chief (202) 418-7450 7-C723; Michael Carowitz/Chief of Staff/Associate Bureau (202) 418-7450 7-C723; Ellen Engleman-Conners/Senior Deputy Bureau Chief (202) 418-7450 7-C723; Robert Ratcliffe/Deputy Bureau Chief (202) 418-7450 7-C723; Susan McNeil/Deputy Bureau Chief (202) 418-7450 7-C723; Gene Fullano/Acting Deputy Bureau Chief (202) 418-7450 7-C723; George R. Dillon/Assistant Bureau Chief (202) 418-7450 7-C723; William Davenport/Assistant Bureau Chief (202) 418-7450 7-C723; Jon Minkoff/Legal Advisor (202) 418-7450 7-C723; Priya Shrinivasan/Legal Advisor (202) 418-7450 7-C723; Janice Wise/Director of Media Relations (202) 418-7450 7-C723; Debbie Smoot/Special Assistant to the Bureau Chief (202) 418-7450 7-C723; Arneatta Strange/Staff Assistant (202) 418-7450 7-C723.

Office of Homeland Security

Kenneth Moran/Acting Director (202) 418-0802.

Investigations and Hearing Division

Hillary S. De Nigro/Chief (202) 418-1420; Ben Bartolome/Deputy Chief (202) 418-1420; Trent B. Harkrader/Deputy Chief (202) 418-1420; Jennifer A. Lewis/Assistant Chief (202) 418-1420; Elizabeth Mumaw/Assistant Chief (202) 418-1420; Vickie S. Robinson/Assistant Chief (202) 418-1420; Hugh Boyle/Chief Auditor (202) 418-1420.

Market Disputes Resolution Division

Alexander Starr/Division Chief (202) 418-7330; Lisa Griffin/Deputy Chief (202) 418-7330; Rosemary McEnery/Deputy Division Chief (202) 418-7330; Lisa Saks/Assistant Chief (202) 418-7330.

Spectrum Enforcement Division

Kathryn Berthot/Chief (202) 418-1160; Ricardo Durham/Senior Deputy Chief (202) 418-1160; Tom Spavins/ Assistant Chief - Economics (202) 418-1160; Riley Hollingsworth/Special Counsel (202) 418-1160; Norman Goldstein/Special Counsel (202) 418-1160.

Equipment Development Group (Hiram, GA)

Scott Parker/Acting Director.

Telecommunications Consumers Division

Colleen Heitkamp/Division Chief (202) 418-7320; Kurt Schroeder/Deputy Chief (202) 418-7320; Marcy Greene/Deputy Chief (202) 418-7320; Sharon D. Lee/Associate Chief (202) 418-7320; Christopher Olsen/ Special Counsel (202) 418-7320; Mary Romano/Special Advisor (202) 418-7320.

North East Region

Russell (Joe) D. Monie/Regional Director (Chicago, IL) (847) 813-4670; G. Michael Moffitt/Deputy Regional Director; Sharon Webber/Regional Counsel.

District Directors

Dennis Loria, Boston, MA Office; James Roop, Chicago, IL Office (Acting); James T. Higgins, Columbia, MD Operations Center; Daniel W. Noel, New York, NY Office; Gene J. Stanbro, Philadelphia, PA Office; James Bridgewater, Detroit, MI Office.

South Central Region

Dennis (Denny) P. Carlton/Director (816) 316-1243; Loyd P. Perry/Deputy Regional Director (Houston, TX) (713) 983-6104; Diane Law-Hsu/Regional Counsel.

District Directors

Douglas G. Miller, Atlanta, GA Office; James D. Wells, Dallas, TX Office; Robert McKinney, Kansas City, MO Office; Leroy "Bud" Hall, New Orleans, LA Office; Ralph Barlow, Tampa, FL Office.

Western Region

Rebecca L. Dorch/Regional Director (Denver, CO) (925) 416-9661; Leo Cirbo/Deputy Regional Director (Denver, CO) (925) 416-9661; Margaret Egler, Regional Counsel.

District Directors

Nikki Shears, Denver, CO Office Catherine Deaton, Los Angeles, CA Office Bill Zears, San Diego, CA Office Thomas Van Stavern, San Francisco, CA Office Kristine McGowan, Seattle, WA Office

INTERNATIONAL BUREAU

Helen Domenici/Bureau Chief (202) 418-0437; John Giusti/Deputy Bureau Chief (202) 418-1407; Roderick Porter/Deputy Bureau Chief (202) 418-0437; Breck Blalock/Chief of Staff and Associate Bureau Chief (202) 418-8191; Linda L. Haller Sloan/Associate Bureau Chief (on Detail) (202) 418-1408; Narda Jones/Assistant Bureau Chief (202) 418-2489; Tom Sullivan/Assistant Bureau Chief for Management (202) 418-0411; Gardner Foster/Legal Advisor (202) 418-1990; Francis Gutierrez/Legal Advisor (202) 418-7370; Jerry Duvall/Chief Economist (202) 418-2616.

Policy Division

James L. Ball/Division Chief (202) 418-1460 7-A760; George S. Li/Deputy Division Chief (Operations) (202) 418-1460; JoAnn Ekblad/Assistant Division Chief (202) 418-1372; Howard Griboff/Acting Deputy Chief (202) 418-0657; David Krech/Assistant Division Chief (202) 418-7443; Paul Locke/Assistant Division Chief (Engineering) (202) 418-0756; David Strickland/Assistant Division Chief (202) 418-0977.

Satellite Division

Front Office Staff: Robert Nelson/Chief (202) 418-2341; Cassandra Thomas/Deputy Division Chief (202) 418-0719 6-A666; Fern Jarmulnek/Deputy Division Chief (202) 418-0719 6-A760; Karl Kensinger/Associate Division Chief (202) 418-0773 6-A663; Steven Spaeth/Assistant Division Chief (202) 418-1539 6 C407;

Policy Branch

Andrea Kelly/Chief (202) 418-7877 6-A521.

Systems Analysis Branch

Scott Kotler/Branch Chief (202) 418-0596 6 C411.

Engineering Branch

Kathyrn Medley/Acting Chief (202) 418-1211.

Strategic Analysis and Negotiations Division

Kathryn O'Brien/Division Chief (202) 418-0439; Linda Dubroof/Deputy Division Chief (202) 418-2335; Jennifer Gilsenan/Deputy Division Chief (202) 418-0757; Alexander Royblat/Assistant Division Chief (202) 418-7501; Larry Olson/Assistant Chief Engineer (202) 418-2142; Linda Armstrong/Senior Legal Advisor (202) 418-7490; Pam Gerr/Special Counsel (202) 418-0422.

Cross Border Negotiations & Treaty Compliance Branch

James Ballis/Chief (202) 418-2146.

Regional & Bilateral Affairs Branch

Brad Lerner/Acting Chief; Emily Talaga/Acting Chief (202) 418-7396.

Multilateral Negotiations & Industry Analysis Branch

Brad Lerner/Acting Chief (202) 418-7066. Emily Talaga/Acting Chief.

International Radiocommunications Branch

Dante Ibarra/Chief (202) 418-0610.

MEDIA BUREAU

Monica Desai/Bureau Chief (202) 418-7200; Roy Stewart/Senior Deputy Bureau Chief (202) 418-7200; Rosemary C. Harold/Deputy Bureau Chief (202) 418-7200; Elizabeth Andrion/Acting Deputy Bureau Chief (202) 418-7200; Thomas Horan/Chief of Staff (202) 418-7200; Sarah Whitesell/Associate Bureau Chief (202) 418-7200; Andrew Long/Associate Bureau Chief (202) 418-7200; William D. Freedman/Associate Bureau Chief (202) 418-7200; Mary Diamond/Associate Bureau Chief (Media Relations) (202) 418-7200; Christopher L. Robbins/Associate Bureau Chief (202) 418-7200; Tracy Waldon/Chief Economist (202) 418-7200; Heather Dixon/Special Counsel (202) 418-7200; Vacant/Legal Advisor.

Management and Resources Staff

Yvette Barrett/Assistant Bureau Chief for Management (202) 418-7200; Michael E. Teaney/Deputy Assistant Chief for Management (202) 418-7200.

Office of Communications and Industry Information

Michael S. Perko/Chief (202) 418-7200.

Office of Broadcast License Policy

Roy J. Stewart/Chief (202) 418-2600.

Audio Division

(202) 418-2700.

Video Division

(202) 418-1600.

Policy Division

Mary Beth Murphy/Division Chief (202) 418-2120; Steven A. Broeckaert/Deputy Division Chief; John B. Norton/Deputy Division Chief; Robert Baker/Assistant Chief; Eloise Gore/Assistant Chief; Ronald Parver/Assistant Chief; Lewis Pulley/Assistant Chief.

Industry Analysis Division

Royce D. Sherlock/Division Chief (202) 418-2330; Mania K. Baghdadi/Deputy Division Chief (202) 418-2120; Marcia A. Glauberman/Deputy Division Chief (202) 418-2330; Judith Herman/Assistant Chief (202) 418-2330; Daniel Hodes/Senior Economic Advisor (202) 418-2330.

Engineering Division

Michael L. Lance/Deputy Division Chief (202) 418-7014.

WIRELESS TELECOMMUNICATIONS BUREAU

Fred Campbell/Bureau Chief (202) 418-0600 3-C252; Jim Schlichting/Deputy Bureau Chief (202) 418-0600; Cathy Massey/Deputy Bureau Chief (202) 418-0600; Jane Jackson/Associate Bureau Chief (202) 418-0600; Mary Shultz/Assistant Bureau Chief/Senior Technical Advisor (202) 418-0600; Walter Strack/Chief Economist (202) 418-0600; Blaise Scinto/Senior Counsel for Spectrum Policy Task Force (202) 418-0600; Paul Murray/Legal Advisor (202) 418-0600; Chelsea Haga Fallon/Acting Special Counsel for Media and Public Affairs (202) 418-0600; David Hu/Legal Advisor (202) 418-0600; Brent Greenfield/Attorney Advisor (202) 418-0600.

Management and Resources Staff

Lois Jones/Assistant Bureau Chief for Management (202) 418-0600.

Auctions and Spectrum Access Division

Margaret Wiener/Division Chief (202) 418-0660.

Broadband Division

Joel Taubenblatt/Division Chief (202) 418-BITS (2487) 3-C124.

Mobility Division

Roger Noel/Division Chief (202) 418-0620 6-6411.

Public Safety & Critical Infrastructure Division

Michael Wilhelm/Division Chief (202) 418-0680.

Spectrum & Competition Policy Division

John Branscome/Division Chief (202) 418-1310; Paul D'Ari/Deputy Chief/Spectrum Policy (202) 418-1310; Jeffrey Steinberg/Deputy Chief/Infrastructure Policy (202) 418-1310; Nese Guendelsberger/Deputy Chief/Competition Policy (202) 418-1310; Rachel Kazan/Associate Chief/Industry Analysis & Chief of Staff (202) 418-1310; Ziad Sleem/Asociate Chief/Technician Policy (202) 418-1310; Dan Abeyta/Assistant Chief/NEPA Adjudications (202) 418-1310; John Borkowski/Assistant Chief/Spectrum Access (202) 418-1310; Aaron Goldschmidt/Assistant Chief/NEPA Policy (202) 418-1310; Susan Singer/Assistant Chief/Secondary Market Transactions.

Spectrum Management Resource & Technology Division

1-888-225-5322; (202) 418-0600.

WIRELINE COMPETITION BUREAU

Tom Navin/Acting Bureau Chief (202) 418-1500 5-C450; Julie Veach/Deputy Bureau Chief (202) 418-1500; Jake Jennings/Associate Bureau Chief (202) 418-1500; Marcus Maher/Associate Bureau Chief (202) 418-1500; Donald Stockdale/Associate Bureau Chief & Bureau Chief Economist (202) 418-1500; Kirk Burgee/Chief of Staff (202) 418-1500; Randy Clarke/Legal Counsel to the Bureau Chief (202) 418-1500; Mark Wigfield/Public Affairs Specialist (202) 418-1500 5-C424.

Administrative and Management Office

Val Brock/Assistant Bureau Chief (202) 418-0118.

Competition Policy Division

Christi Shewman/Division Chief (202) 418-1580 5-B125; William Dever/Deputy Chief (202) 418-1580; Tim Stelzig/Assistant Chief (202) 418-1580; Jeremy Miller/Deputy Chief (202) 418-1580; Ann Stevens/Associate Deputy Chief (202) 418-1580.

Pricing Policy Division

Albert Lewis/Division Chief (202) 418-1520 5-A225; Deena M. Shetler/Deputy Division Chief (202) 418-1520; Pam Arluk/Assistant Division Chief (202) 418-1520.

Telecommunications Access Policy Division

Jeremey Marcus/Division Chief (202) 418-7400; Thomas Buckley/Senior Deputy Division Chief (202) 418-7400; Erik Olsen/Deputy Chief (202) 418-7400; Jennifer McKee/Deputy Chief (202) 418-7400; Cheryl Callahan/Assistant Chief (202) 418-7400; Gina Spade/Assistant Chief (202) 418-7400; Mark Seifert/Assistant Chief (202) 418-7400.

Industry Analysis and Technology Division

Alan I. Feldman/Acting Division Chief (202) 418-0940; Cathy H. Zima/Acting Deputy Division Chief (202) 418-7380; Ellen Burton/Assistant Division Chief (202) 418-0958; John Adesalu/Industry Economist (202) 418-7097; Ginny Kennedy/Management Analyst (202) 418-2328; Jamal Mazuri/Technology Specialist (202) 418-0069.

U.S. Government Agencies of Interest to TV and Radio

Department of Agriculture, 1400 Independence Ave. S.W., Washington, DC, 20250. Phone: (202) 720-4623. Fax: (202) 720-5043. Web Site: www.usda.gov. E-mail: ed.lloydd@usda.gov

Terri Teuber, dir office of communications; Nichole Andrews, deputy dir office of communications; Ed Lloyd, press sec.

Wkly TV satellite newsfeed. Daily radio newsline. Wkly radio features on CD.

Department of Commerce, 1401 Constitution Ave. N.W., Suite 5040, Washington, DC, 20230. Phone: (202) 482-2000. Fax: (202) 482-2639. E-mail: rsilva@ntia.doc.gov Web Site: www.doc.gov/opa.

Ranjit DeSilva, dir public affrs.

Department of Defense, Pentagon, Washington, DC, 20301. Phone: (703) 545-6700. Fax: (703) 695-4299. Fax: (703) 695-9080. Web Site: www.defense.gov.

Department of Energy, 1000 Independence Ave. S.W., Washington, DC, 20585. Phone: (202) 586-5000. Phone: (202) 586-5806 (Press Off). Fax: (202) 586-9987 (Dir.). Fax: (202) 586-4403. E-mail: craig.stevens@hhqq.doe.gov Web Site: www.energy.gov.

Craig Stevens, deputy dir.

Five business lines encompass everything that DOE does: energy, resouces, natl security, environmental quality, science & technology, & economic productivity.

Department of Health and Human Services, 200 Independence Ave. S.W., Washington, DC, 20201. Phone: (202) 619-0257. Phone: (202) 690-6343. Fax: (202) 690-7203. Web Site: www.hhs.gov.

Suzy Francis, dir public affrs; Ellen Field, deputy asst sec.

Department of Justice, Office of Public Aff., 950 Pennsylvania Ave. N.W., Washington, DC, 20530. Phone: (202) 514-2007. E-mail: AskDOJ@usdoj.gov Web Site: www.justice.gov.

Tasia Scolinos, dir.

Department of Labor, 200 Constitution Ave. N.W., Washington, DC, 20210. Phone: (202) 693-5000. Phone: (202) 693-4650 (pub affrs). Fax: (202) 693-4674. Web Site: www.dol.gov.

Elaine L. Chao, sec.

Fosters, promotes, & develops the welfare of working people.

Department of State, 2201 C St. N.W., Washington, DC, 20520. Phone: (202) 647-4000. Web Site: www.state.gov.

Department of the Treasury, 1500 Pennsylvania Ave. N.W., 3442 MT, Washington, DC, 20020. Phone: (202) 622-2000. Phone: (202) 622-2960 (press off.). Fax: (202) 622-6415. Web Site: www.treasury.gov.

John W. Snow, sec; Christopher Smith, chief of staff.

Department of Transportation, 400 7th St. S.W., Washington, DC, 20590. Phone: (202) 366-4000. Web Site: www.dot.gov.

Norman Y. Mineta, sec; Robert Johnson, dir public affrs.

Mission: To serve the United States by ensuring a fast, safe, efficient, accessible and convenient transportation system that meets our vital natl interests and enhances the quality of life of the American people, today & into the future.

Executive Office of the President, The White House, 1600 Pennsylvania Ave. N.W., Washington, DC, 20500. Phone: (202) 456-1414. Fax: (202) 456-2461. Web Site: www.whitehouse.gov. E-mail: comments@whitehouse.gov

Tony Snow, press sec.

Federal Communications Commission, 445 12th St. S.W., Washington, DC, 20554. Phone: (888) 225-5322. Phone: (866) 418-0232. Web Site: www.fcc.gov. E-mail: david.fiske@fcc.gov

Kevin J. Martin, chmn; David Fiske, dir media rel.

(For full listing of commissioners & staff, see FCC Executives & Staff.)

Federal Emergency Management Agency, Office of Policy & Regional Operations, 500 C St. S.W., Washington, DC, 20472. Phone: (800) 621-FEMA(3362). E-mail: FEMAOPA@dhs.gov Web Site: www.fema.gov.

R. David Paulison, acting under sec.

Comprehensive info source on emergency preparedness & federal disaster response & recovery.

Federal Trade Commission, 600 Pennsylvania Ave. N.W., Washington, DC, 20580. Phone: (202) 326-2222. Fax: (202) 326-3366. Web Site: www.ftc.gov. E-mail: mkatz@ftc.gov

Deborah Platt Majoras, chmn; Mitchell Katz, office of public affrs.

House Appropriations Committee, Rm. H218, Capitol Bldg, Washington, DC, 20515. Phone: (202) 225-2771. Phone: (202) 225-3351 (commerce/justice/state subcomm). Web Site: http://appropriations.house.gov/.

Jerry Lewis, chmn.

Funds government agencies.

House Committee on Energy and Commerce, 2125 Rayburn House Office Bldg., Washington, DC, 20515-6115. Phone: (202) 225-2927. Web Site: energycommerce.house.gov.

Joe Barton, chmn.

The Committee, the oldest legislative standing committee in the U.S. House of Representatives, has served as the principal guide for the House in matters relating to the promof commerce and to the public's health & marketplace interests.

House Committee on the Judiciary, 2138 Rayburn House Office Bldg., Washington, DC, 20515-6216. Phone: (202) 225-3951. Web Site: www.house.gov/judiciary. E-mail: judiciary@mail.house.gov

F. James Sensenbrenner, Jr., chmn.

The Committee on the Judiciary has been called the lawyer for the House of Representatives because of its jurisdiction over matters relating to the administration of justice in Federal courts, administrative bodies, and law enforcement agencies. Its infrequent but important role in impeachment proceedings has also brought it much attention.

National Aeronautics & Space Administration (NASA), 300 E St. S.W., Washington, DC, 20546. Phone: (202) 358-0001. Fax: (202) 358-3469. E-mail: public-inquiries@hq.nasa.gov Web Site: www.nasa.gov.

Michael Griffin, admin.

National Labor Relations Board, 1099 14th St. N.W., Washington, DC, 20570-0001. Phone: (202) 273-1991. Fax: (202) 208-3013. E-mail: dparker@nlrb.gov Web Site: www.nlrb.gov.

Robert J. Battista, chmn; Ronald Meisburg, gen counsel.

The NLRB adjudicates unfair labor practice charges & conducts union representation elections under the Natl Labor Rel Act.

National Science Foundation, 4201 Wilson Blvd., Arlington, VA, 22230. Phone: (703) 292-8070. Fax: (703) 292-9087. E-mail: info@nsf.gov Web Site: www.nsf.gov.

Dr. Arden L. Bement, Jr., dir.

National Telecommunications and Information Administration, 14th & Constitution N.W., Washington, DC, 20230. Phone: (202) 482-7002. Fax: (202) 219-2077. Web Site: www.ntia.doc.gov.

John M. R. Kneuer, asst sec; Meredith Attwell Baker, deputy asst sec.

NTIA serves as the principal advisor to the exec branch on domestic & international communication & info issues.

Securities and Exchange Commission, 450 5th St. N.W., Washington, DC, 20549. Phone: (202) 942-8088. Phone: (202) 551-5400 (sec). Fax: (202) 942-9628. Web Site: www.sec.gov.

Christopher Cox, chmn; R. Corey Booth, CIO.

Administers federal securities laws that protect investors. These laws ensure that securities markets are fair & provide sanctions for enforcement.

Senate Appropriations Committee, Rm. SD-128, Dirksen Senate Office Bldg., Washington, DC, 20510. Phone: (202) 224-3471. Web Site: http://appropriations.senate.gov/.

Senate Committee on Commerce, Science, and Transportation, Dirksen Senate Office Bldg., Suite 508, Washington, DC, 20510-6125. Phone: (202) 224-0411 (minority). Phone: (202) 224-1251 (majority). Fax: (202) 224-1259 (majority). Fax: (202) 228-0303 (minority). Web Site: http://commerce.senate.gov/.

Ted Stevens, chmn.

Jurisdiction includes communications, aviation, consumer affairs, foreign commerce & tourism, oceans & fisheries, science, technolgy & space, surface transportion & merchant marine, manufactruring & competiveness.

Senate Judiciary Committee, 224 Dirksen Senate Office Bldg, Washington, DC, 20510-6275. Phone: (202) 224-5225. Fax: (202) 224-9102. Web Site: http://judiciary.senate.gov/.

Arlen Specter, chmn.

U.S. Department of Education, 400 Maryland Ave. S.W., Washington, DC, 20202. Phone: (202) 401-2000. Fax: (202) 401-0689. E-mail: opa@ed.gov Web Site: www.ed.gov.

Provides daily audio svc for radio news; arranges bcst interviews with sr department officials.

U.S. District Court for the District of Columbia, 333 Constitution Ave. N.W., Washington, DC, 20001. Phone: (202) 354-3000. Web Site: www.dcd.uscourts.gov.

Thomas F. Hogan, chief judge & district; Sheldon Snook, admin asst to chief judge; Elizabeth H. Paret, deputy circuit exec.

Hears civil & criminal cases that arise under federal law, cases involving the U.S. Constitution, disputes between two states, or cases in which the United States is a party.

U.S. Advisory Commission on Public Diplomacy, 301 4th St. S.W., Washington, DC, 20547. Phone: (202) 203-7880. Fax: (202) 203-7886. Web Site: http://www.state.gov/r/adcompd/.

Athena Katsoulos, exec dir; Jamice Clayton, admin off.

A bipartisan presidentially appointed panel created by Congress to oversee U.S. government activities intended to understand, inform & influence foreign publics.

U.S. Court of Appeals for the District of Columbia Circuit, 333 Constitution Ave. N.W., Washington, DC, 20001-2866. Phone: (202) 216-7000. Fax: (202) 273-0988. Fax: (202) 219-8530. Web Site: www.cadc.uscourts.gov.

Appeals from District Court cases. Appeals from federal agency decisions.

U.S. Supreme Court, One First St. N.E, Washington, DC, 20543. Phone: (202) 479-3211 (pub. info. off.). Phone: (202) 479-3030 (visitor info.). E-mail: pio@sc-us.gov Web Site: www.supremecourtus.gov/.

John G. Roberts, Jr., Chief Justice; David H. Souter, assoc justice; Clarence Thomas, assoc justice; Ruth Bader Ginsburg, assoc justice; John Paul Stevens, assoc justice; Antonin Scalia, assoc justice; Anthony M. Kennedy, assoc justice; Stephen G. Breyer, assoc justice; Samuel A. Alito, Jr., assoc justice.

U.S. State Cable Regulatory Agencies

Connecticut Department of Public Utility Control, 10 Franklin Sq., New Britain, CT, 06051. Phone: (860) 827-1553. Fax: (860) 827-2613. E-mail: dpuc.information@po.state.ct.us Web Site: www.state.ct.us/dpuc.

Delaware Public Service Commission, 861 Silver Lake Blvd. Cannon Bldg., Suite 100, Dover, DE, 19904. Phone: (302) 739-4247. Fax: (302) 739-4849. E-mail: ronnette.brown@state.de.us Web Site: www.state.de.us/delpsc. Bruce Burcat, exec dir.

Hawaii Cable Television Division, Box 541, Dept. of Commerce & Consumer Affairs, Honolulu, HI, 96809. Phone: (808) 586-2620. Fax: (808) 586-2625. E-mail: cabletv@dcca.hawaii.gov Web Site: www.hawii.gov/dcca/catv. Mark Recktenwald, dir; Clyde S. Sonobe, admin.

Massachusetts Department of Telecommunications & Energy, (Cable Television Division). One South Station, Boston, MA, 02110. Phone: (617) 305-3580. Fax: (617) 478-2590. E-mail: cable.inquiry@state.ma.us Web Site: www.state.ma.us/dpu/catv. Alicia C. Matthews, dir & cable div.

New Jersey Office of Cable Television, 2 Gateway Center, Newark, NJ, 07102. Phone: (973) 648-2670. Fax: (973) 648-3135. Web Site: http://www.state.nj.us/bpu/home/cable.shtml. Celeste M. Fasone, dir; Charles A. Russell, deputy dir.

New York State Department of Public Service, 3 Empire State Plaza, Albany, NY, 12223-1350. Phone: (518) 474-1668. Fax: (518) 474-5616. E-mail: robert_mayer@dps.state.ny.us Web Site: www.dps.state.ny.us. Robert Mayer, dir.

Regulatory Commission of Alaska, 701 W. 8th Ave., Suite 300, Anchorage, AK, 99501. Phone: (907) 276-6222. Fax: (907) 276-0160. E-mail: rca_mail@rca.state.ak.us Web Site: www.state.ak.us/rca. G. Nanette Thompson, chmn/exec dir.

Rhode Island Division of Public Utilities and Carriers, 89 Jefferson Blvd., Warwick, RI, 02888. Phone: (401) 941-4500. Fax: (401) 941-9248. E-mail: mary.kent@ripuc.org Web Site: http://www.ripuc.org/. Eric A. Palazao, assoc admin cable TV.

Vermont Public Service Board, Chittenden Bank Bldg., 4th Fl., 112 State St., Drawer 20, Montpelier, VT, 05620-2701. Phone: (802) 828-2358. Fax: (802) 828-3351. E-mail: clerk@psb.state.vt.us Web Site: www.state.vt.us/psb. James Volz, chmn.